CHILTON®

ASIAN SERVICE MANUAL 2010 EDITION VOLUME II HYUNDAI KIA LEXUS

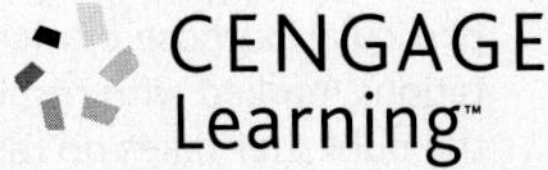

Australia • Brazil • Japan • Korea • Mexico • Singapore • Spain • United Kingdom • United States

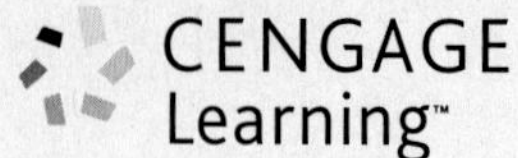

CHILTON®
Asian Service Manual
2010 Edition
Volume II
Hyundai, Kia, Lexus

Vice President, Technology Professional Business Unit:
Gregory L. Clayton

Publisher, Technology Professional Business Unit:
David Koontz

Director of Marketing:
Beth A. Lutz

Director Education Production:
Carolyn Miller

Marketing Manager:
Jennifer Barbic

Marketing Coordinator:
Rachael Torres

Chilton Content Specialist:
Paula Baillie

Graphical Designer:
Melinda Possinger

Art Director:
Benjamin Gleeksman

Sr. Content Project Manager:
Elizabeth C. Hough

Senior Editor:
Christine L. Sheeky

Editors:
Ken Burdette
Sherry Burdette
Nick D'Andrea
Celia Hanlon
John Howard
Maureen Lazarz
Kyla Nyjordet
Lance Williams

Printed in the United States of America
1 2 3 4 5 6 7 14 13 12 11 10

For product information and technology assistance, contact us at **Professional & Career Group customer Support, 1-800-648-7450.**
For permission to use material from this text or product, submit all requests online at **www.cengage.com/permissions.**
Further permissions questions can be e-mailed to **permissionrequest@cengage.com**

ISBN-13: 978-1-1110-3765-9
ISBN-10: 1-1110-3765-5
ISSN: 1939-621X

Chilton
5 Maxwell Drive
Clifton Park, NY 12065-2919
USA

Chilton products are represented in Canada by Nelson Education, Ltd.

NOTICE TO THE READER

Publisher does not warrant or guarantee any of the products described herein or perform any independent analysis in connection with any of the product information contained herein. Publisher does not assume, and expressly disclaims, any obligation to obtain and include information other than that provided to it by the manufacturer.

The reader is expressly warned to consider and adopt all safety precautions that might be indicated by the activities described herein and to avoid all potential hazards. By following the instructions contained herein, the reader willingly assumes all risks in connection with such instructions.

The publisher makes no representations or warranties of any kind, including but not limited to, the warranties of fitness for particular purpose or merchantability, nor are any such representations implied with respect to the material set forth herein, and the publisher takes no responsibility with respect to such material. The publisher shall not be liable for any special, consequential, or exemplary damages resulting, in whole or part, from the readers' use of, or reliance upon, this material.

Contents

Sections

Model Index

USING THIS INFORMATION

Organization

To find where a particular model section or procedure is located, look in the Table of Contents. Main topics are listed with the page number on which they may be found. Following the main topics is an alphabetical listing of all of the procedures within the section and their page numbers.

Manufacturer and Model Coverage

This product covers 2009–2010 Asian models that are produced in sufficient quantities to warrant coverage, and which have technical content available from the vehicle manufacturers before our publication date. Although this information is as complete as possible at the time of publication, some manufacturers may make changes which cannot be included here. While striving for total accuracy, the publisher cannot assume responsibility for any errors, changes, or omissions that may occur in the compilation of this data.

Part Numbers and Special Tools

Part numbers and special tools are recommended by the publisher and vehicle manufacturer to perform specific jobs. Before substituting any part or tool for the one recommended, you must be completely satisfied that neither your personal safety, nor the performance of the vehicle will be endangered.

ACKNOWLEDGEMENT

The publisher would like to express appreciation to the following vehicle manufacturers for their assistance in producing this manual: Hyundai Group, including Hyundai and Kia Motor. No further reproduction or distribution of the material in this manual is allowed without the expressed written permission of the vehicle manufacturers and the publisher.

PRECAUTIONS

Before servicing any vehicle, please be sure to read all of the following precautions, which deal with personal safety, prevention of component damage, and important points to take into consideration when servicing a motor vehicle:

- Always wear safety glasses or goggles when drilling, cutting, grinding or prying.
- Steel-toed work shoes should be worn when working with heavy parts. Pockets should not be used for carrying tools. A slip or fall can drive a screwdriver into your body.
- Work surfaces, including tools and the floor should be kept clean of grease, oil or other slippery material.
- When working around moving parts, don't wear loose clothing. Long hair should be tied back under a hat or cap, or in a hair net.
- Always use tools only for the purpose for which they were designed. Never pry with a screwdriver.
- Keep a fire extinguisher and first aid kit handy.
- Always properly support the vehicle with approved stands or lift.
- Always have adequate ventilation when working with chemicals or hazardous material.
- Carbon monoxide is colorless, odorless and dangerous. If it is necessary to operate the engine with vehicle in a closed area such as a garage, always use an exhaust collector to vent the exhaust gases outside the closed area.
- When draining coolant, keep in mind that small children and some pets are attracted by ethylene glycol antifreeze, and are quite likely to drink any left in an open container, or in puddles on the ground. This will prove fatal in sufficient quantity. Always drain the coolant into a sealable container.
- To avoid personal injury, do not remove the coolant pressure relief cap while the engine is operating or hot. The cooling system is under pressure; steam and hot liquid can come out forcefully when the cap is loosened slightly. Failure to follow these instructions may result in personal injury. The coolant must be recovered in a suitable, clean container for reuse. If the coolant is contaminated it must be recycled or disposed of correctly.
- When carrying out maintenance on the starting system be aware that heavy gauge leads are connected directly to the battery. Make sure the protective caps are in place when maintenance is completed. Failure to follow these instructions may result in personal injury.
- Do not remove any part of the engine emission control system. Operating the engine without the engine emission control system will reduce fuel economy and engine ventilation. This will weaken engine performance and shorten engine life. It is also a violation of Federal law.
- Due to environmental concerns, when the air conditioning system is drained, the refrigerant must be collected using refrigerant recovery/recycling equipment. Federal law requires that refrigerant be recovered into appropriate recovery equipment and the process be conducted by qualified technicians who have been certified by an approved organization, such as MACS, ASI, etc. Use of a recovery machine dedicated to the appropriate refrigerant is necessary to reduce the possibility of oil and refrigerant incompatibility concerns. Refer to the instructions provided by the equipment manufacturer when removing refrigerant from or charging the air conditioning system.
- Always disconnect the battery ground when working on or around the electrical system.
- Batteries contain sulfuric acid. Avoid contact with skin, eyes, or clothing. Also, shield your eyes when working near batteries to protect against possible splashing of the acid solution. In case of acid contact with skin or eyes, flush immediately with water for a minimum of 15 minutes and get prompt medical attention. If acid is swallowed, call a physician immediately. Failure to follow these instructions may result in personal injury.
- Batteries normally produce explosive gases. Therefore, do not allow flames, sparks or lighted substances to come near the battery. When charging or working near a battery, always shield your face and protect your eyes. Always provide ventilation. Failure to follow these instructions may result in personal injury.

• When lifting a battery, excessive pressure on the end walls could cause acid to spew through the vent caps, resulting in personal injury, damage to the vehicle or battery. Lift with a battery carrier or with your hands on opposite corners. Failure to follow these instructions may result in personal injury.

• Observe all applicable safety precautions when working around fuel. Whenever servicing the fuel system, always work in a well-ventilated area. Do not allow fuel spray or vapors to come in contact with a spark, open flame, or excessive heat (a hot drop light, for example). Keep a dry chemical fire extinguisher near the work area. Always keep fuel in a container specifically designed for fuel storage; also, always properly seal fuel containers to avoid the possibility of fire or explosion. Do not smoke or carry lighted tobacco or open flame of any type when working on or near any fuel-related components.

• Fuel injection systems often remain pressurized, even after the engine has been turned OFF. The fuel system pressure must be relieved before disconnecting any fuel lines. Failure to do so may result in fire and/or personal injury.

• The evaporative emissions system contains fuel vapor and condensed fuel vapor. Although not present in large quantities, it still presents the danger of explosion or fire. Disconnect the battery ground cable from the battery to minimize the possibility of an electrical spark occurring, possibly causing a fire or explosion if fuel vapor or liquid fuel is present in the area. Failure to follow these instructions can result in personal injury.

• The EPA warns that prolonged contact with used engine oil may cause a number of skin disorders, including cancer! You should make every effort to minimize your exposure to used engine oil. Protective gloves should be worn when changing oil. Wash your hands and any other exposed skin areas as soon as possible after exposure to used engine oil. Soap and water, or waterless hand cleaner should be used.

• Some vehicles are equipped with an air bag system, often referred to as a Supplemental Restraint System (SRS) or Supplemental Inflatable Restraint (SIR) system. The system must be disabled before performing service on or around system components, steering column, instrument panel components, wiring and sensors. Failure to follow safety and disabling procedures could result in accidental air bag deployment, possible personal injury and unnecessary system repairs.

• Always wear safety goggles when working with, or around, the air bag system. When carrying a non-deployed air bag, be sure the bag and trim cover are pointed away from your body. When placing a non-deployed air bag on a work surface, always face the bag and trim cover upward, away from the surface. This will reduce the motion of the module if it is accidentally deployed.

• Electronic modules are sensitive to electrical charges. The ABS module can be damaged if exposed to these charges.

• Brake pads and shoes may contain asbestos, which has been determined to be a cancer-causing agent. Never clean brake surfaces with compressed air. Avoid inhaling brake dust. Clean all brake surfaces with a commercially available brake cleaning fluid.

• When replacing brake pads, shoes, discs or drums, replace them as complete axle sets.

• When servicing drum brakes, disassemble and assemble one side at a time, leaving the remaining side intact for reference.

• Brake fluid often contains polyglycol ethers and polyglycols. Avoid contact with the eyes and wash your hands thoroughly after handling brake fluid. If you do get brake fluid in your eyes, flush your eyes with clean, running water for 15 minutes. If eye irritation persists, or if you have taken brake fluid internally, immediately seek medical assistance.

• Clean, high quality brake fluid from a sealed container is essential to the safe and proper operation of the brake system. You should always buy the correct type of brake fluid for your vehicle. If the brake fluid becomes contaminated, completely flush the system with new fluid. Never reuse any brake fluid. Any brake fluid that is removed from the system should be discarded. Also, do not allow any brake fluid to come in contact with a painted or plastic surface; it will damage the paint.

• Never operate the engine without the proper amount and type of engine oil; doing so will result in severe engine damage.

• Timing belt maintenance is extremely important! Many models utilize an interference- type, non freewheeling engine. If the timing belt breaks, the valves in the cylinder head may strike the pistons, causing potentially serious (also time-consuming and expensive) engine damage.

• Disconnecting the negative battery cable on some vehicles may interfere with the functions of the on-board computer system(s) and may require the computer to undergo a relearning process once the negative battery cable is reconnected.

• Steering and suspension fasteners are critical parts because they affect performance of vital components and systems and their failure can result in major service expense. They must be replaced with the same grade or part number or an equivalent part if replacement is necessary. Do not use a replacement part of lesser quality or substitute design. Torque values must be used as specified during reassembly.

HYUNDAI

Accent

1

SPECIFICATIONS AND MAINTENANCE CHARTS

ENGINE AND VEHICLE IDENTIFICATION

Engine							Model Year	
Code ①	Liters (cc)	Cu. In.	Cyl.	Fuel Sys.	Engine Type	Eng. Mfg.	Code ②	Year
C	1.6 (1599)	97.57	I4	MPFI	DOHC	Hyundai	9	2009
							A	2010

MPFI: Multi-Point Fuel Injection

DOHC: Double Overhead Camshafts

① 8th digit of VIN

② 10th digit of VIN

37655_ACCE_C0001

GENERAL ENGINE SPECIFICATIONS

All measurements are given in inches.

Year	Model	Engine Displacement Liters	Engine Series VIN	Net Horsepower @ rpm	Net Torque @ rpm (ft. lbs.)	Bore x Stroke (in.)	Com-pression Ratio	Oil Pressure @ rpm
2009	Accent	1.6	C	110@6000	106@4500	3.012 x 3.425	10.0:1	15.6@Idle
2010	Accent	1.6	C	110@6000	106@4500	3.012 x 3.425	10.0:1	15.6@Idle

① 368@6500 regular fuel. 375@6500 premium fuel

② 324@3500 regular fuel. 333@3500 premium fuel

37655_ACCE_C0002

GASOLINE ENGINE TUNE-UP SPECIFICATIONS

Year	Engine Displacement Liters	Engine VIN	Spark Plug Gap (in.)	Ignition Timing (deg.) MT	Ignition Timing (deg.) AT	Fuel Pump (psi)	Idle Speed (rpm) MT	Idle Speed (rpm) AT	Valve Clearance In.	Valve Clearance Ex.
2009	1.6	C	0.039-0.043	①	①	49.0-50.5	②	②	HYD	HYD
2010	1.6	C	0.039-0.043	①	①	49.0-50.5	②	②	HYD	HYD

Follow the figures on the label if they differ from those in this chart.

HYD: Hydraulic

① Ignition timing is computer controlled and is not adjustable

② Idle speed is maintained by the Electronic Control Module (ECM)

37655_ACCE_C0003

CAPACITIES

Year	Model	Engine Displacement Liters	Engine VIN	Engine Oil with Filter (qts.)	Transmission (pts.) Manual	Transmission (pts.) Auto. ①	Fuel Tank (gal.)	Cooling System (qts.)
2009	Accent	1.6	C	3.5	4.0	12.9	11.9	5.8-6.1
2010	Accent	1.6	C	3.5	4.0	12.9	11.9	5.8-6.1

NOTE: All capacities are approximate. Add fluid gradually and check to be sure a proper fluid level is obtained.

37655_ACCE_C0004

FLUID SPECIFICATIONS

Year	Model	Engine Displacement Liters	Engine ID/VIN	Engine Oil	Auto. Trans.	Manual Trans.	Power Steering Fluid	Brake Master Cylinder
2009	Accent	1.6	C	5W-20	①	②	PSF-3	③
2010	Accent	1.6	C	5W-20	①	②	PSF-3	③

DOT: Department Of Transportation

① DIAMOND ATF SP-III, SK ATF SP-III

② GENUINE PART MTF 75W/85 (API GL - 4)

③ DOT 3, DOT 4, or equivalent

37655_ACCE_C0005

VALVE SPECIFICATIONS

Year	Engine Displacement Liters	Engine VIN	Seat Angle (deg.)	Face Angle (deg.)	Spring Test Pressure (lbs. @ in.)	Spring Installed Height (in.)	Stem-to-Guide Clearance (in.) Intake	Stem-to-Guide Clearance (in.) Exhaust	Stem Diameter (in.) Intake	Stem Diameter (in.) Exhaust
2009	1.6	C	45	45	94.5-104.3 @1.071	NA	0.0008-0.0020	0.0014-0.0026	0.2348-0.2354	0.2343-0.2348
2010	1.6	C	45	45	94.5-104.3 @1.071	NA	0.0008-0.0020	0.0014-0.0026	0.2348-0.2354	0.2343-0.2348

NA: Not Available

37655_ACCE_C0006

CAMSHAFT AND BEARING SPECIFICATIONS CHART

All measurements are given in inches.

Year	Engine Displ. Liters	Engine ID/VIN	Journal Dia.	Brg. Oil Clearance	Shaft End-play	Runout	Journal Bore	Lobe Height Intake	Lobe Height Exhaust
2009	1.6	C	1.0616-1.0622	0.0008-0.0024	0.0039-0.0079	NA	NA	1.7224-1.7303	1.7382-1.7460
2010	1.6	C	1.0616-1.0622	0.0008-0.0024	0.0039-0.0079	NA	NA	1.7224-1.7303	1.7382-1.7460

NA: Not Available

37655_ACCE_C0007

CRANKSHAFT AND CONNECTING ROD SPECIFICATIONS

All measurements are given in inches.

Year	Engine Displacement Liters	Engine VIN	Crankshaft Main Brg. Journal Dia.	Crankshaft Main Brg. Oil Clearance	Crankshaft Shaft End-play	Crankshaft Thrust on No.	Connecting Rod Journal Diameter	Connecting Rod Oil Clearance	Connecting Rod Side Clearance
2009	1.6	C	1.9665-1.9672	①	0.0020-0.0069	3	1.8898-1.8905	0.0007-0.0014	0.0039-0.0098
2010	1.6	C	1.9665-1.9672	①	0.0020-0.0069	3	1.8898-1.8905	0.0007-0.0014	0.0039-0.0098

① No. 1, 2, 4, 5 are 0.0009-0.0016 inch
No. 3 is 0.0011-0.0018 inch

37655_ACCE_C0008

PISTON AND RING SPECIFICATIONS

All measurements are given in inches.

Year	Engine Displ. Liters	Engine VIN	Piston Clearance	Ring Gap Top Compression	Ring Gap Bottom Compression	Ring Gap Oil Control	Ring Side Clearance Top Compression	Ring Side Clearance Bottom Compression	Ring Side Clearance Oil Control
2009	1.6	C	0.0008-0.0016	0.0059-0.0118	0.0138-0.0197	0.0079-0.0276	0.0016-0.0033	0.0016-0.0033	0.0031-0.0069
2010	1.6	C	0.0008-0.0016	0.0059-0.0118	0.0138-0.0197	0.0079-0.0276	0.0016-0.0033	0.0016-0.0033	0.0031-0.0069

37655_ACCE_C0009

TORQUE SPECIFICATIONS

All readings in ft. lbs.

Year	Engine Displacement Liters	Engine VIN	Cylinder Head Bolts	Main Bearing Bolts	Rod Bearing Bolts	Crankshaft Damper Bolts	Flywheel Bolts	Manifold Intake	Manifold Exhaust	Spark Plugs	Oil Pan Drain Plug
2009	1.6	C	①	40-43	23-25	101-109	87-94	11-15	22-25	15-22	29-33
2010	1.6	C	①	40-43	23-25	101-109	87-94	11-15	22-25	15-22	29-33

① Step 1: 22 ft. lbs.
Step 2: Plus 90 degrees
Step 3: Loosen to 0 ft. lbs.
Step 4: 22 ft. lbs.
Step 5: Plus 90 degrees

37655_ACCE_C0010

WHEEL ALIGNMENT

Year	Model		Caster Range (+/-Deg.)	Caster Preferred Setting (Deg.)	Camber Range (+/-Deg.)	Camber Preferred Setting (Deg.)	Toe-in (Deg.)
2009	Accent	Front	0.50	①	0.50	0.00	0.00 +/- 0.20
		Rear	—	—	0.50	-1.00	0.20 - 0.60
2010	Accent	Front	0.50	①	0.50	0.00	0.00 +/- 0.20
		Rear	—	—	0.50	-1.00	0.20 - 0.60

① Manual Steering 0.58 degrees. Power Steering 4.0 degrees.

37655_ACCE_C0011

TIRE, WHEEL AND BALL JOINT SPECIFICATIONS

Year	Model	OEM Tires Standard	OEM Tires Optional	Tire Pressures (psi) Front	Tire Pressures (psi) Rear	Wheel Size	Ball Joint Inspection	Lug Nut Torque (ft. lbs.)
2009	Accent	P175/70R14	P185/65R14	①	①	5/5.5J x 14	②	65-79
			P195/55R15			5.5J x 15		
2010	Accent	P175/70R14	P185/65R14	①	①	5/5.5J x 14	②	65-79
			P195/55R15			5.5J x 15		

① Refer to placard on vehicle for proper inflation pressure.

② Replace if any measurable movement is found.

37655_ACCE_C0012

BRAKE SPECIFICATIONS

All measurements in inches unless noted

Year	Model		Brake Disc Original Thickness	Brake Disc Minimum Thickness	Brake Disc Maximum Runout	Brake Drum Diameter Original Inside Diameter	Brake Drum Diameter Max. Wear Limit	Brake Drum Diameter Maximum Machine Diameter	Minimum Lining Thickness	Brake Caliper Bracket Bolts (ft. lbs.)	Brake Caliper Mounting Bolts (ft. lbs.)
2009	Accent	F	0.870	0.790	0.001	—	—	—	0.079	47-54	16-23
		R	—	—	—	8.000	①	①	0.039	—	—
2010	Accent	F	0.870	0.790	0.001	—	—	—	0.079	47-54	16-23
		R	—	—	—	8.000	①	①	0.039	—	—

① Drum roundness Service Limit: 0.00236 inch

37655_ACCE_C0013

SCHEDULED MAINTENANCE INTERVALS

HYUNDAI—ACCENT

TO BE SERVICED	TYPE OF SERVICE	VEHICLE MILEAGE INTERVAL (x1000)												
		7.5	15	22.5	30	37.5	45	52.5	60	67.5	75	82.5	90	97.5
Engine oil & filter	R	✓	✓	✓	✓	✓	✓	✓	✓	✓	✓	✓	✓	✓
Tire rotation	S/I	✓	✓	✓	✓	✓	✓	✓	✓	✓	✓	✓	✓	✓
Vacuum hose	S/I	✓	✓	✓	✓	✓	✓	✓	✓	✓	✓	✓	✓	✓
Automatic transaxle fluid ①	S/I		✓		✓		✓		✓		✓		✓	
Brake pads, calipers & rotors	S/I		✓		✓		✓		✓		✓		✓	
Driveshafts & boots	S/I		✓		✓		✓		✓		✓		✓	
Air cleaner filter ②	R	✓	✓	✓	✓	✓	✓	✓	✓	✓	✓	✓	✓	✓
A/C refrigerant	S/I		✓		✓		✓		✓		✓		✓	
Brake fluid	I				✓				✓				✓	
Engine coolant ③	R				✓				✓				✓	
Fuel hose, vapor hose & fuel filter cap	S/I				✓				✓				✓	
Spark plugs	R				✓				✓				✓	
Spark plugs (Platinum coated)	R								✓					
Spark plugs (Iridium coated) 100,000 mile replacement	R													
Bolts & nuts on chassis & body	S/I		✓		✓		✓		✓		✓		✓	
Drive belts	S/I				✓				✓				✓	
Exhaust pipe connections, muffler & suspension bolts	S/I		✓		✓		✓		✓		✓		✓	
Manual transaxle oil ④	S/I				✓				✓				✓	
Brake hoses & lines	S/I		✓		✓		✓		✓		✓		✓	
Rear brake drums, linings & parking brake	S/I				✓				✓				✓	
Steering gear rack, linkage & boots	S/I				✓				✓				✓	
Suspension ball joints & dust covers	S/I		✓		✓		✓		✓		✓		✓	
Timing belt ⑤	S/I				✓				✓				✓	
Climate control air filter ⑥	R													
Fuel tank air filter ⑦	S/I		✓		✓		✓		✓		✓		✓	
Fuel filter	R							✓						
Fuel lines & connections	S/I				✓				✓				✓	
Vacuum & crankcase ventilation hoses	S/I				✓				✓				✓	

R: Replace S/I: Service or Inspect

① For Accent, replace at 105,000 miles

② For Accent, replace at 30,000 miles

③ For Accent, replace every 24 months or 30,000 miles

④ For Accent, replace at 60,000 miles

⑤ For Accent, replace at 60,000 miles. See Severe Service.

⑥ For Accent, replace at 12,000 miles, or 12 months. See Severe Service.

⑦ For Accent, replace at 30,000 miles.

FREQUENT OPERATION MAINTENANCE (SEVERE SERVICE)

If a vehicle is operated under any of the following conditions it is considered severe service:

- Extremely dusty areas.

- 50% or more of the vehicle operation is in 90°F (32°C) or higher temperatures, or constant operation in temperatures below 32°F (0°C).

- Prolonged idling (vehicle operation in stop and go traffic).

- Frequent short running periods (engine does not warm to normal operating temperatures).

- Police, taxi, delivery usage or trailer towing usage.

Oil & oil filter: change every 3,000 miles.

Brake pads, calipers & rotors: service or inspect every 7,500 miles.

Driveshaft boots: service or inspect every 7,500 miles

Steering gear rack, linkage & boots: service or inspect every 7,500 miles.

Air cleaner filter: service or inspect every 15,000 miles.

Automatic transaxle fluid & filter: replace every 30,000 miles.

Rear brake drums & linings: service or inspect every 15,000 miles.

Spark plugs: service or inspect every 24,000 miles.

Timing belt: replace every 37,500 miles, or 48 months.

Climate control air filter: replace as necessary.

37655_ACCE_C0014

BRAKES — INFORMATION AND PRECAUTIONS

ANTI-LOCK SYSTEMS

- Certain components within the ABS system are not intended to be serviced or repaired individually.
- Do not use rubber hoses or other parts not specifically specified for and ABS system. When using repair kits, replace all parts included in the kit. Partial or incorrect repair may lead to functional problems and require the replacement of components.
- Lubricate rubber parts with clean, fresh brake fluid to ease assembly. Do not use shop air to clean parts; damage to rubber components may result.
- Use only DOT 3 brake fluid from an unopened container.
- If any hydraulic component or line is removed or replaced, it may be necessary to bleed the entire system.
- A clean repair area is essential. Always clean the reservoir and cap thoroughly before removing the cap. The slightest amount of dirt in the fluid may plug an orifice and impair the system function. Perform repairs after components have been thoroughly cleaned; use only denatured alcohol to clean components. Do not allow ABS components to come into contact with any substance containing mineral oil; this includes used shop rags.
- The Anti-Lock control unit is a microprocessor similar to other computer units in the vehicle. Ensure that the ignition switch is **OFF** before removing or installing controller harnesses. Avoid static electricity discharge at or near the controller.
- If any arc welding is to be done on the vehicle, the control unit should be unplugged before welding operations begin.

DISC AND DRUM SYSTEMS

CAUTION

Dust and dirt accumulating on brake parts during normal use may contain asbestos fibers from production or aftermarket brake linings. Breathing excessive concentrations of asbestos fibers can cause serious bodily harm. Exercise care when servicing brake parts. Do not sand or grind brake lining unless equipment used is designed to contain the dust residue. Do not clean brake parts with compressed air or by dry brushing. Cleaning should be done by dampening the brake components with a fine mist of water, then wiping the brake components clean with a dampened cloth. Dispose of cloth and all residue containing asbestos fibers in an impermeable container with the appropriate label. Follow practices prescribed by the Occupational Safety and Health Administration (OSHA) and the Environmental Protection Agency (EPA) for the handling, processing, and disposing of dust or debris that may contain asbestos fibers.

BRAKES — BLEEDING THE BRAKE SYSTEM

BLEEDING PROCEDURE

BLEEDING PROCEDURE

See Figure 1.

1. Before servicing the vehicle, refer to the Precautions Section.

WARNING

The reservoir on the master cylinder must be at the MAX (upper) level mark at the start of bleeding procedure and checked after bleeding each brake caliper. Add fluid as required.

2. Make sure the brake fluid level in the reservoir is at the MAX (upper) level line.
3. Have someone slowly pump the brake pedal several times, and then apply steady pressure.
4. Loosen the right-rear brake bleed screw to allow air to escape from the system. Then tighten the bleed screw securely.
5. Repeat the procedure for each wheel in the sequence shown below until air bubbles no longer appear in the fluid.
6. Refill the master cylinder reservoir to the MAX (upper) level line.

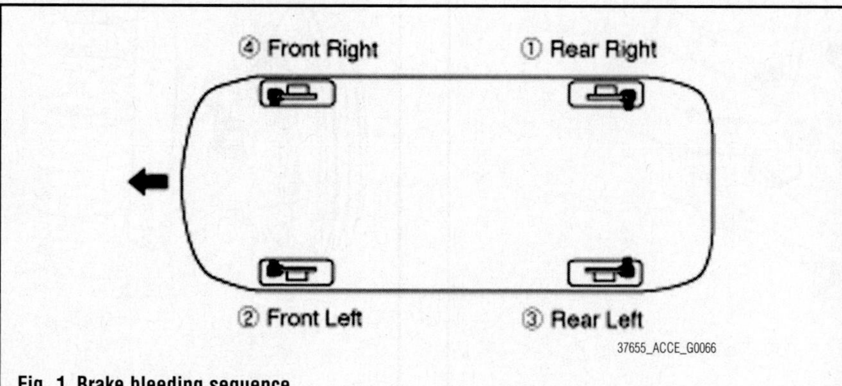

Fig. 1 Brake bleeding sequence

BLEEDING THE ABS SYSTEM

This procedure should be followed to ensure adequate bleeding of air and the filling of the ABS unit, the brake lines, and the master cylinder with brake fluid.

1. Before servicing the vehicle, refer to the Precautions Section.
2. Remove the reservoir cap and fill the brake reservoir with brake fluid.

WARNING

If there is any brake fluid on any painted surface, wash it off immediately.

➡When pressure bleeding, do not depress the brake pedal. Recommended brake fluid: DOT3 or DOT4.

3. Connect a clear plastic tube to the wheel cylinder bleeder plug and insert the other end of the tube into a clear plastic bottle that is half filled with clean brake fluid.
4. Connect the Hi-Scan Pro® to the data link connector located underneath the dash panel.
5. Select and operate according to the instructions on the Hi-Scan Pro® screen.

CAUTION

You must obey the maximum operating time of the ABS motor with the Hi-Scan Pro® to prevent the motor pump from burning.

6. Select Hyundai vehicle diagnosis.
7. Select vehicle name.
8. Select Anti-Lock Brake system.
9. Select air bleeding mode.
10. Press "YES" to operate motor pump and solenoid valve.

WARNING

Wait 60 seconds before operating the air bleeding or damage to the motor may occur.

11. Wait 60 seconds before operating the air bleeding.
12. Pump the brake pedal several times, and then loosen the bleeder screw until fluid starts to run out without bubbles. Then, close the bleeder screw.
13. Repeat until there are no more bubbles in the fluid for each wheel.

BRAKES — ANTI-LOCK BRAKE SYSTEM (ABS)

SPEED SENSORS

REMOVAL & INSTALLATION

Front

See Figures 2 through 4.

1. Before servicing the vehicle, refer to the Precautions Section.
2. Disconnect the negative battery cable.
3. Remove the front wheel speed sensor mounting bolt.
4. Remove the front wheel speed sensor bracket.
5. Remove the front wheel guard.
6. Disconnect the wheel speed sensor connector.
7. Remove the front wheel speed sensor.

To install:

8. Installation is the reverse of the removal procedure.

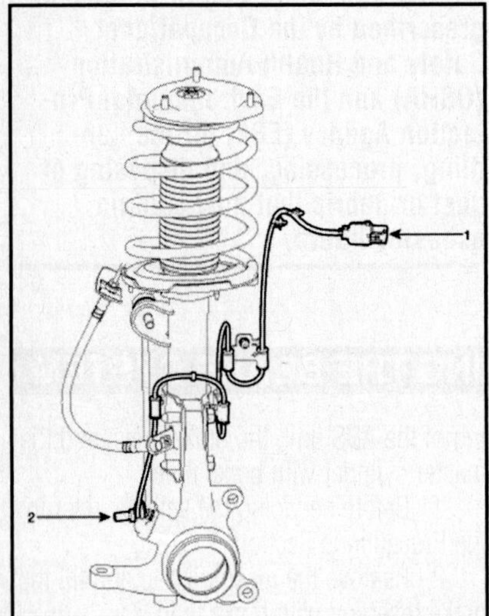

Fig. 2 Front wheel speed sensor (2) and connector (1)

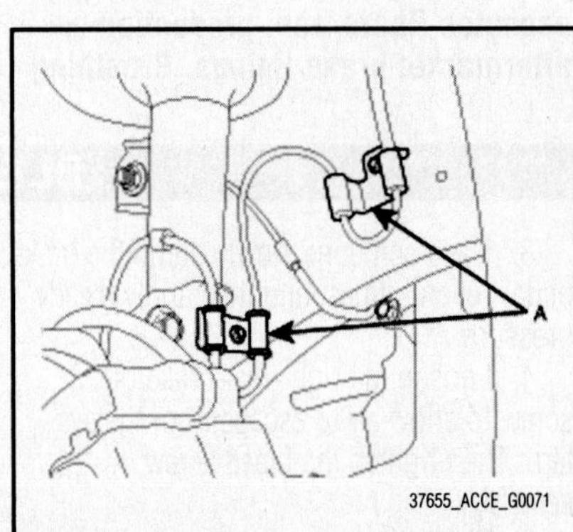

Fig. 4 Front wheel speed sensor bracket (A)

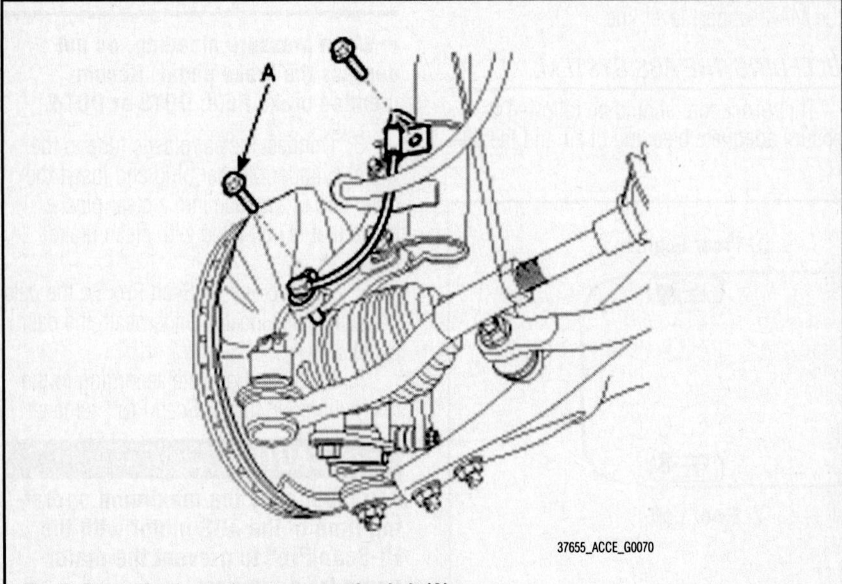

Fig. 3 Front wheel speed sensor mounting bolt (A)

Rear

See Figures 5 and 6.

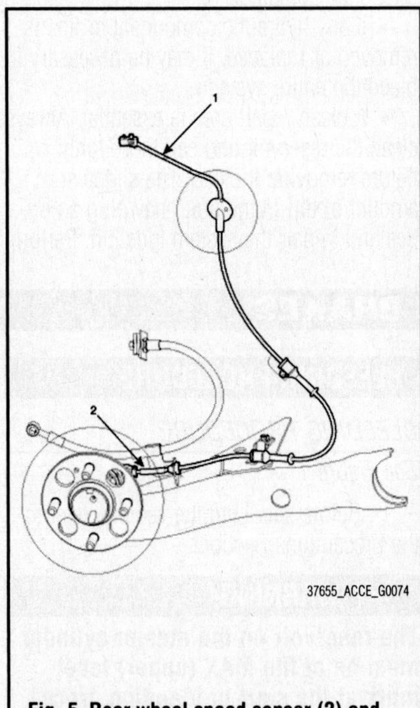

Fig. 5 Rear wheel speed sensor (2) and connector (1)

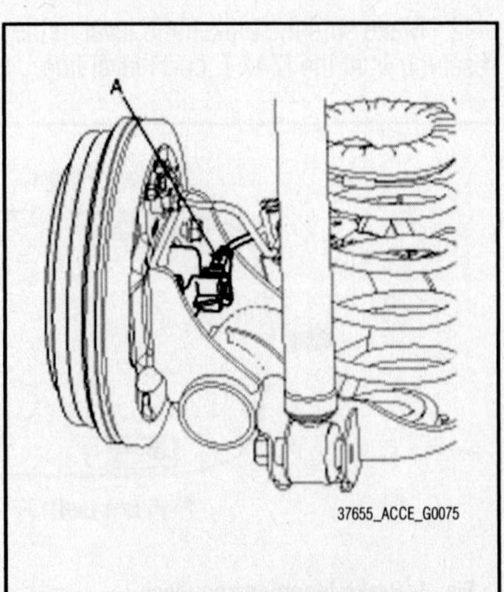

Fig. 6 Rear wheel speed retainer (A)

1. Before servicing the vehicle, refer to the Precautions Section.
2. Disconnect the negative battery cable.
3. Remove the rear wheel speed sensor retainer.
4. Remove the rear seat assembly. Refer to Seats Removal & Installation, in the Body Interior section.
5. Remove the rear wheel speed trim housing and rear pillar trim.
6. Disconnect the rear wheel speed sensor connector.
7. Remove the rear wheel speed sensor.

To install:

8. Installation is the reverse of the removal procedure.

BRAKES

FRONT DISC BRAKES

BRAKE CALIPER

REMOVAL & INSTALLATION

See Figures 7 through 9.

1. Before servicing the vehicle, refer to the Precautions Section.
2. Raise and safely support the vehicle.
3. Remove the front wheel and tire from front hub.

WARNING

Be careful not to damage the hub bolts.

4. Remove guide rod bolts and raise the caliper. Check the hoses and pin boots for damage and deterioration.
5. Remove the brake hose from the caliper.

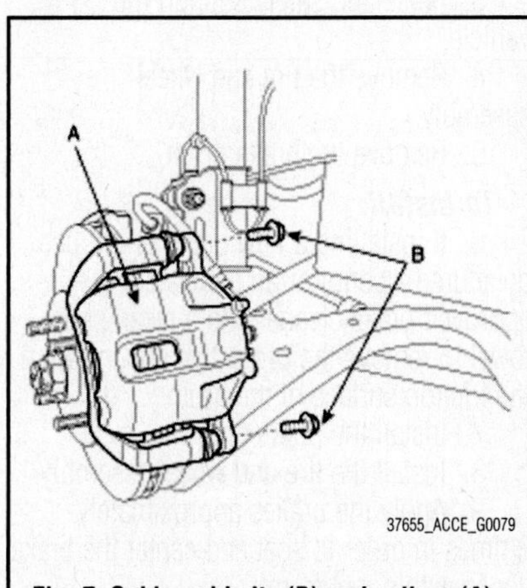

Fig. 7 Guide rod bolts (B) and caliper (A)

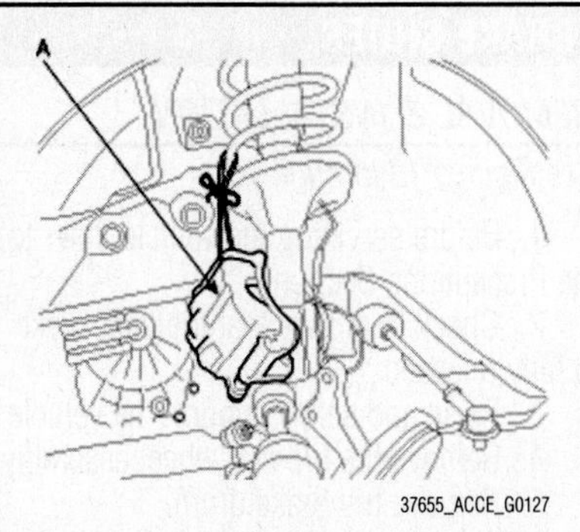

Fig. 8 Remove the caliper (A) and secure the caliper with a wire

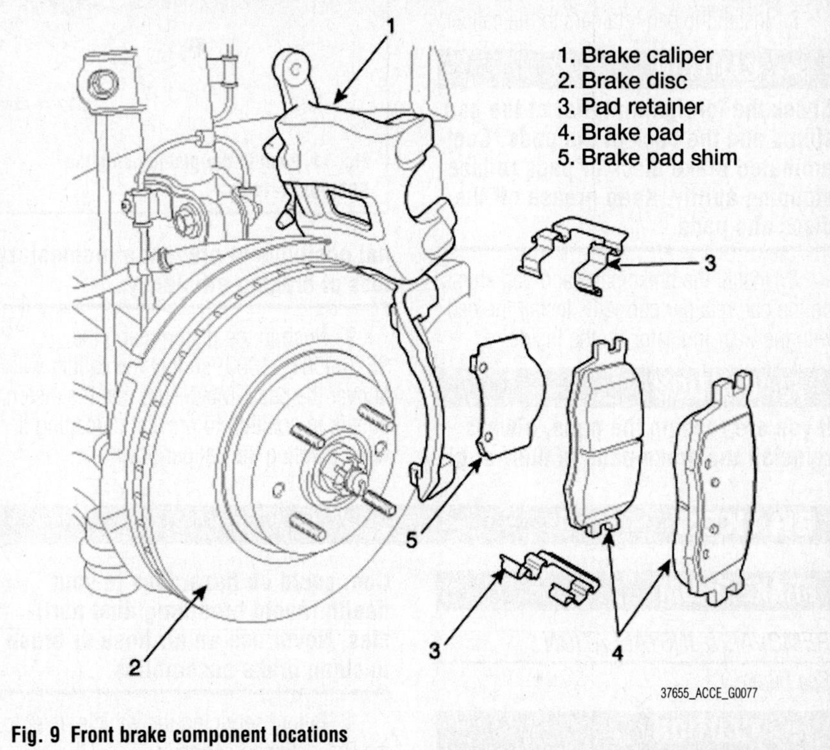

Fig. 9 Front brake component locations

6. Remove the caliper mounting bolts.
7. Remove the caliper.

To install:

8. Pivot the caliper down into position. Install the brake caliper and caliper bolts. Tighten the bolts to: 62–69 ft. lbs. (83–93 Nm)
9. Being careful not to damage the pin boot, install the guide rod bolt and tighten it to 16–24 ft. lbs. (22–32 Nm).
10. Install the brake hose onto the caliper. Tighten to 9–12 ft. lbs. (13–17 Nm).

➡Replace sealing washer whenever hose is removed.

DISC BRAKE PADS

REMOVAL & INSTALLATION

See Figures 7 through 11.

1. Before servicing the vehicle, refer to the Precautions Section.
2. Raise and safely support the vehicle.
3. Remove the front wheel and tire from front hub.

WARNING

Be careful not to damage the hub bolts.

4. Remove guide rod bolt and raise the caliper. Check the hoses and pin boots for damage and deterioration.
5. Remove the caliper mounting bolts, and hang the caliper assembly to one side.

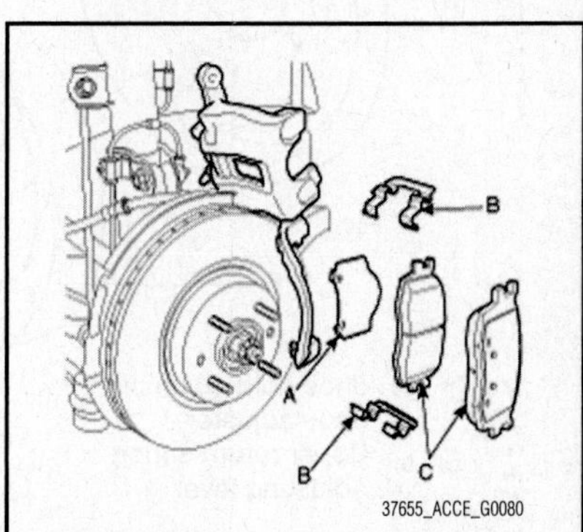

Fig. 10 Pad shims (A), pad retainers (B), and pads (C)

⁂ WARNING

To prevent damage to the caliper assembly or brake hose, use a short piece of wire to hang the caliper from the undercarriage.

6. Remove the pad shims, pad retainers and pads.

To install:

7. Install the pad retainers to the caliper.

⁂ CAUTION

Check the foreign material at the pad shims and the back of the pads. Contaminated brake discs or pads reduce stopping ability. Keep grease off the discs and pads.

8. Install the brake pads and pad shims on the pad retainer correctly. Install the pad with the wear indicator on the inside.

⁂ CAUTION

If you are reusing the pads, always reinstall the brake pads in their original positions to prevent a momentary loss of braking efficiency.

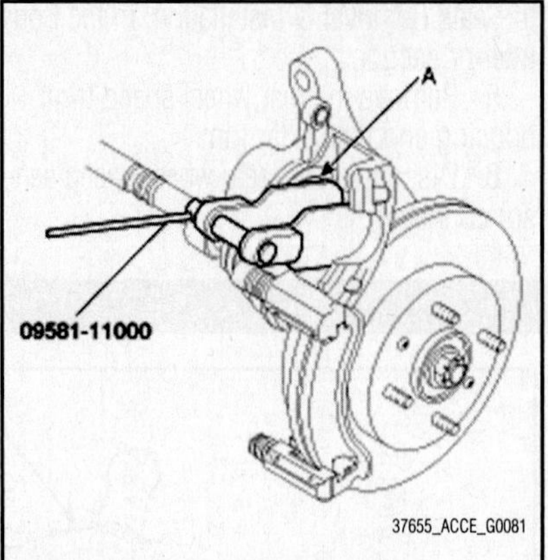

Fig. 11 Push in the piston using the SST(09581-11000)

9. Push in the piston using the SST(09581-11000) so that the caliper will fit over the pads. Make sure that the piston boot is in position to prevent damaging it when pivoting the caliper down.

10. Pivot the caliper down into position. Install the brake caliper and caliper bolts. Tighten the bolts to: 62–69 ft. lbs. (83–93 Nm)

11. Being careful not to damage the pin boot, install the guide rod bolt and tighten it to 16–24 ft. lbs. (22–32 Nm).

12. Install the front wheel and tire.

13. Refill the master cylinder reservoir to the MAX line.

14. Bleed the brake system.

15. Depress the brake pedal several times to make sure the brakes work, then test-drive.

➡Engagement of the brake may require a greater pedal stroke immediately after the brake pads have been replaced as a set. Several applications of the brake will restore the normal pedal stroke.

16. After installation, check for leaks at hose and line joints or connections, and retighten if necessary

BRAKES — REAR DRUM BRAKES

BRAKE DRUM

REMOVAL & INSTALLATION

See Figure 12.

⁂ CAUTION

Frequent inhalation of brake pad dust, regardless of material composition, could be hazardous to your health. Avoid breathing dust particles. Never use an air hose or brush to clean brake assemblies.

1. Before servicing the vehicle, refer to the Precautions Section.

2. Check to ensure that the park brake is fully released.

1. Shoe hold down pin
2. Shoe adjuster
3. Upper return spring
4. Adjusting lever
5. Shoe
6. Adjusting spring
7. Lower return spring

37655_ACCE_G0084

Fig. 12 Rear drum brake assembly

3. Raise and safely support the vehicle.

4. Remove the tire and wheel assembly.

5. Remove the brake drum.

To install:

6. If installing a new brake drum, use denatured alcohol or an equivalent approved brake cleaner and a clean shop towel to remove the protective coating from the friction surface of the drum.

7. Install the brake drum.

8. Install the tire and wheel assembly.

9. Apply the brakes approximately 3 times in order to seat and center the brake shoes within the drum.

10. Lower the vehicle.

BRAKE SHOES

REMOVAL & INSTALLATION

See Figures 13 through 16.

1. Before servicing the vehicle, refer to the Precautions Section.

2. Check to ensure that the park brake is fully released.

3. Raise and safely support the vehicle.

4. Remove the tire and wheel assembly.

5. Remove the brake drum.

6. Remove the shoe hold spring and shoe hold pin.

7. Remove the upper return spring.

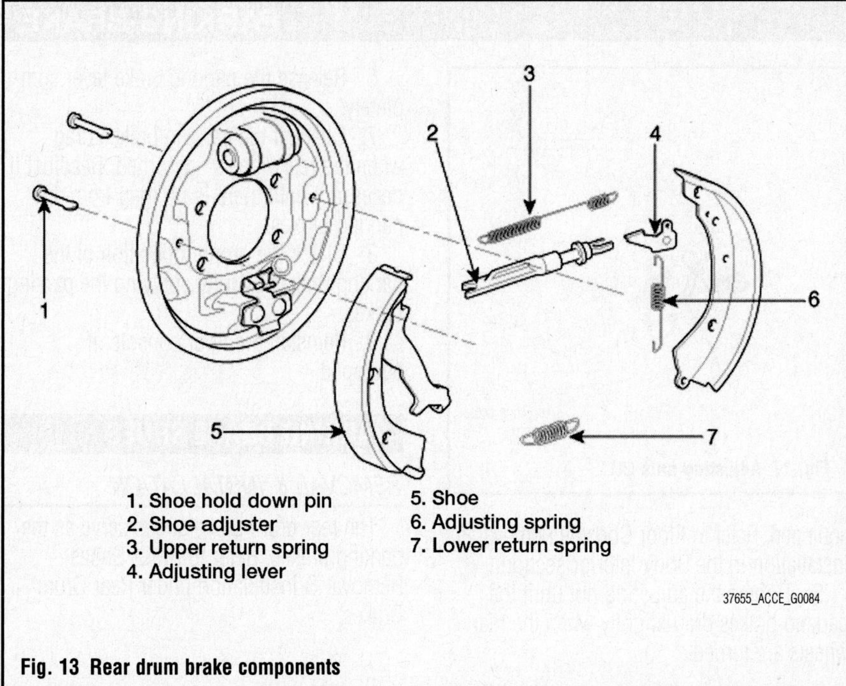

Fig. 13 Rear drum brake components

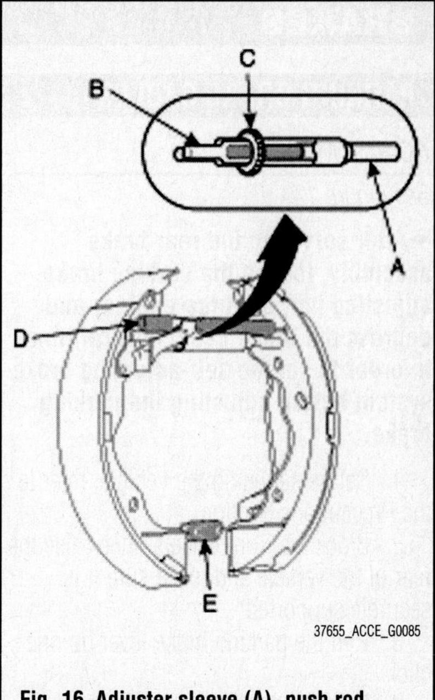

Fig. 16 Adjuster sleeve (A), push rod female (B), and adjuster bolt (C)

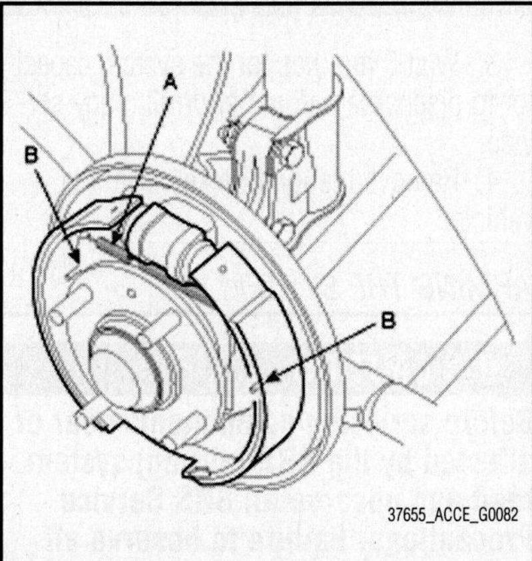

Fig. 14 Upper return spring (A) and shoe hold pin (B)

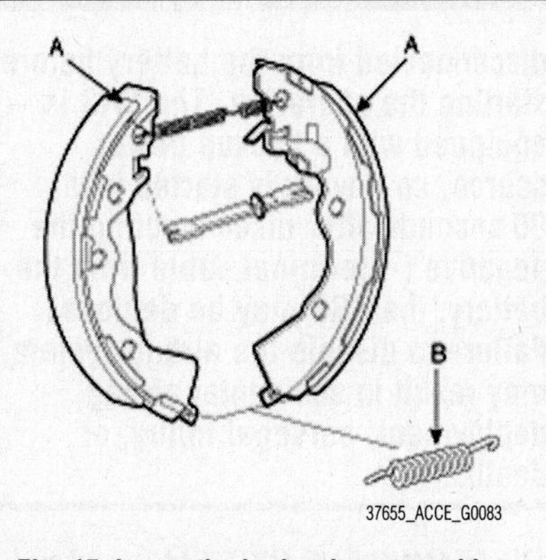

Fig. 15 Lower the brake shoe assembly (A), and remove the lower return spring (B)

8. Lower the brake shoe assembly, and remove the lower return spring. Make sure not to damage the dust cover on the wheel cylinder.
9. Disconnect the parking brake cable from the parking brake lever.
10. Remove the brake shoe assembly.

To install:

11. Connect the parking brake cable to the parking brake lever.
12. Clean the threaded portions of adjuster sleeve and push rod female. Coat the threads of the adjuster assembly with grease. To shorten the clevises, turn the adjuster bolt.
13. Hook the shoe adjuster lever, then install it to the brake shoe.
14. Install the adjuster assembly and upper return spring.

⁂ WARNING

Be careful not to damage the wheel cylinder dust covers.

15. Install the lower return spring.
16. Apply brake cylinder grease, or equivalent rubber grease, to the sliding surfaces and brake shoe ends and opposite edges of the shoes.

➡Be careful not to get grease on the brake linings.

17. Install the brake shoes onto the backing plate.
18. Install the shoe hole down pins and the shoe hole down springs.
19. Install the rear brake drum.
20. Install the tire and wheel assembly.
21. Depress the brake pedal several times to set the self-adjusting brake.
22. Adjust the parking brake, as necessary.

ADJUSTMENT

These vehicles have a self-adjusting mechanism in the rear drum brake assembly.

BRAKES — PARKING BRAKE

PARKING BRAKE CABLES

ADJUSTMENT

See Figure 17.

➡After servicing the rear brake assembly, loosen the parking brake adjusting nut, start the engine, and depress the brake pedal several times in order to set the self-adjusting brake system before adjusting the parking brake.

1. Before servicing the vehicle, refer to the Precautions Section.
2. Block the front wheels, then raise the rear of the vehicle and make sure it is securely supported.
3. Pull the parking brake lever up one click.
4. Remove the floor console, if equipped. Refer to Floor Console Removal & Installation in the Body Interior section.

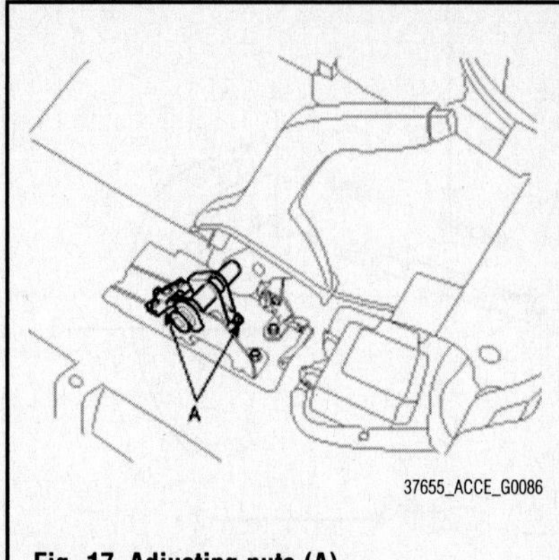

Fig. 17 Adjusting nuts (A)

5. Tighten the adjusting nut until the parking brakes drag slightly when the rear wheels are turned.
6. Release the parking brake lever completely.
7. Check if the parking brakes drag when the rear wheels are turned. Readjust if necessary until there is no drag from the parking brakes.
8. Check the proper operation of the parking brakes by fully applying the parking brakes.
9. Reinstall the floor console, if equipped.

PARKING BRAKE SHOES

REMOVAL & INSTALLATION

The rear drum brake shoes serve as the parking brakes. Refer to Brake Shoes Removal & Installation under Rear Drum Brakes.

CHASSIS ELECTRICAL — AIR BAG (SUPPLEMENTAL RESTRAINT SYSTEM)

GENERAL INFORMATION

⁂ CAUTION

These vehicles are equipped with an air bag system. The system must be disarmed before performing service on, or around, system components, the steering column, instrument panel components, wiring and sensors. Failure to follow the safety precautions and the disarming procedure could result in accidental air bag deployment, possible injury and unnecessary system repairs.

SERVICE PRECAUTIONS

⁂ CAUTION

Disconnect and isolate the battery negative cable before beginning any airbag system component diagnosis, testing, removal, or installation procedures. Wait at least 90 seconds after the ignition switch is turned off and the negative (-) terminal cable is disconnected from the battery before starting the operation. The SRS is equipped with a backup power source, so if work is started within 90 seconds after disconnecting the negative (-) terminal cable from the battery, the SRS may be deployed. Failure to disable the airbag system may result in accidental airbag deployment, personal injury, or death.

DISARMING THE SYSTEM

⁂ CAUTION

Before servicing components near or affected by the SRS (airbag) system, read and observe all SRS Service Precautions. Failure to observe all precautions may result in accidental airbag deployment, personal injury, or death.

1. Before servicing the vehicle, refer to the Precautions Section.
2. Disconnect and isolate the negative battery cable.
3. Wait 3 minutes for the system capacitor to discharge before performing any service.
4. Remove the ignition key from the vehicle.

ARMING THE SYSTEM

⁂ CAUTION

Before servicing components near or affected by the SRS (airbag) system, read and observe all SRS Service Precautions. Failure to observe all precautions may result in accidental airbag deployment, personal injury, or death.

1. Before servicing the vehicle, refer to the Precautions Section.
2. Reconnect the negative battery cable.
3. Turn the ignition switch to the **RUN** position.
4. Confirm proper system operation:
   a. Turn the ignition switch ON; the SRS indicator light should be turned on for about six seconds and then go off.

DRIVE TRAIN

DRIVEN DISC & PRESSURE PLATE

REMOVAL & INSTALLATION

See Figures 18 through 21.

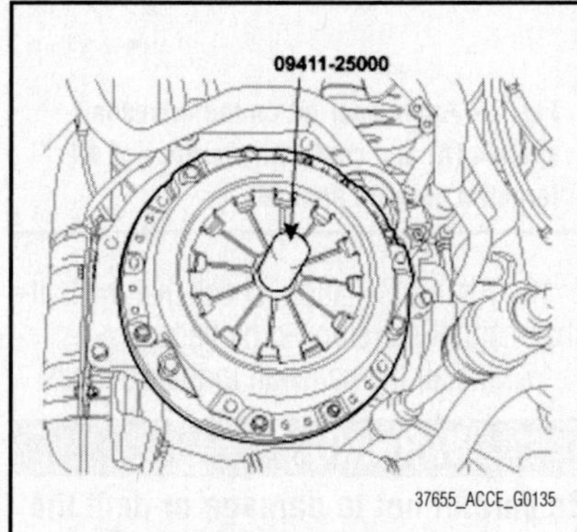

Fig. 18 Insert the special tool (09411-25000)

1. Before servicing the vehicle, refer to the Precautions Section.
2. Remove the transaxle assembly. Refer to Manual Transaxle Assembly Removal & Installation.
3. Insert the special tool (09411-25000) in the clutch disc to prevent the disc from shifting.
4. Loosen the bolts which attach the clutch cover to the flywheel in a star pattern. Loosen the bolts in succession, 1 or 2 turns at a time, to avoid bending the cover flange.

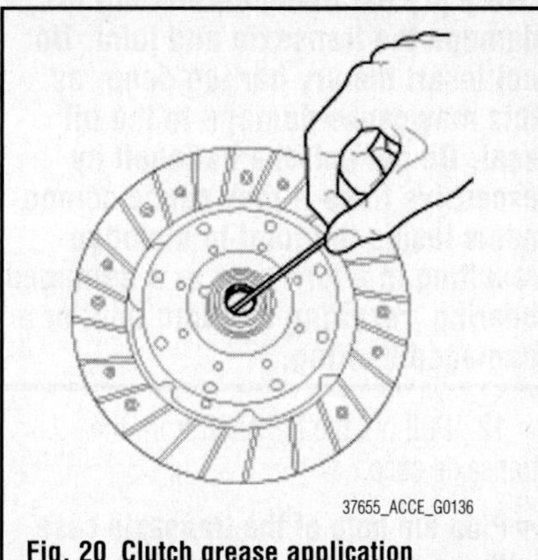

Fig. 20 Clutch grease application

➡Do not clean the clutch disc or the release bearing with cleaning solvent.

5. Remove clutch cover assembly and then driven disc and pressure plate.

To install:

6. Apply multipurpose grease to the spline of the disc.

**** WARNING**

When installing the clutch, apply grease to each part, but be careful not to apply excessive grease. It can cause clutch slippage and vibration.

7. Install the driven disc and pressure plate into the clutch cover assembly.
8. Install the clutch disc assembly to the flywheel using the special tool (09411-25000).

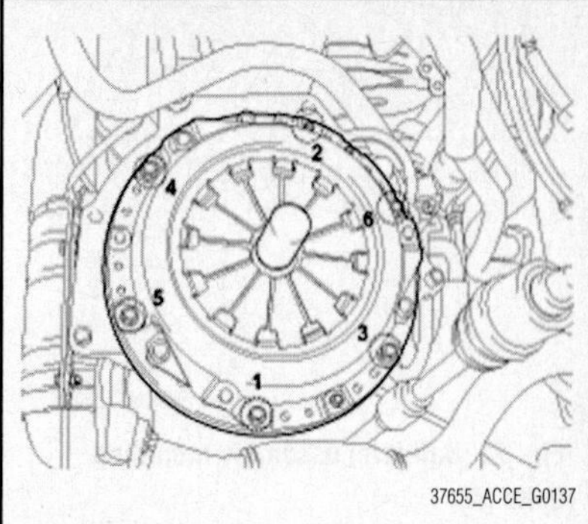

Fig. 21 Clutch cover bolt tightening sequence

9. Install the clutch cover assembly to the flywheel and temporarily tighten the bolts 1 or 2 steps at a time in a star pattern. Tightening torque for clutch cover bolt: 11–16 ft. lbs. (15–22 Nm).
10. Install the transaxle assembly to the engine.

ADJUSTMENTS

The clutch system is hydraulic and requires no adjustment.

FRONT HALFSHAFTS

REMOVAL & INSTALLATION

See Figures 22 through 26.

1. Before servicing the vehicle, refer to the Precautions Section.
2. Raise and safely support the vehicle.
3. Remove the front wheel and tire.
4. Remove the drain plug and drain the transaxle oil.

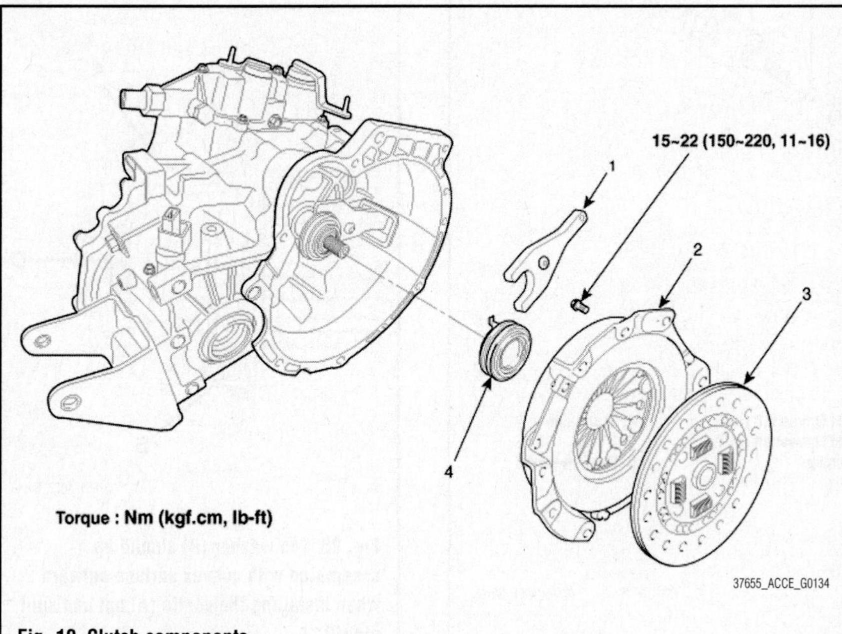

Fig. 19 Clutch components

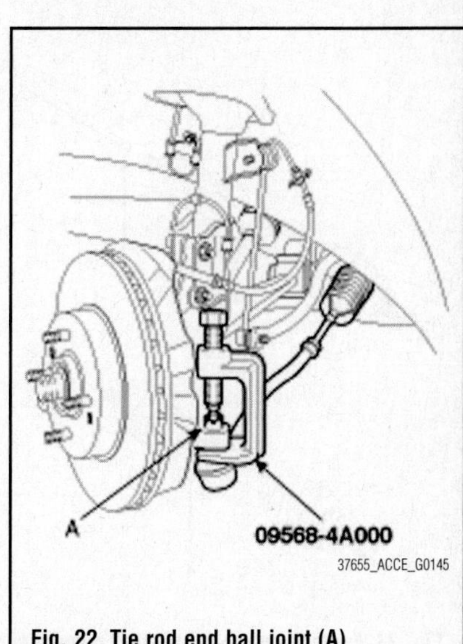

Fig. 22 Tie rod end ball joint (A)

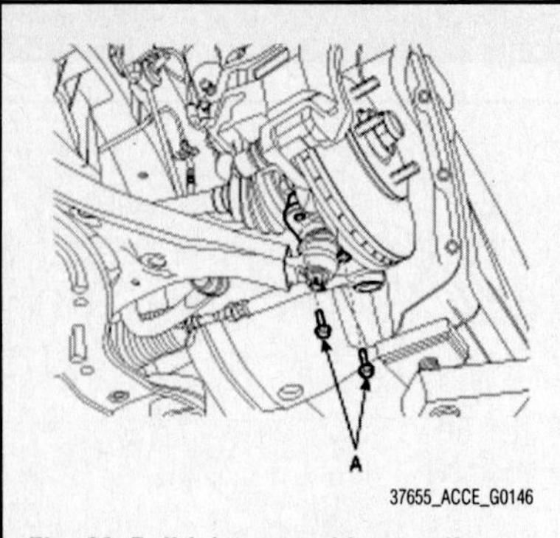

Fig. 23 Ball joint assembly mounting bolts (A)

5. Remove the split pin, castle nut and washer from the front hub assembly.
6. Disconnect the tie rod end ball joint from the knuckle using a special tool (09568-4A000).
7. Remove the wheel speed sensor from the knuckle.
8. Remove the ball joint assembly mounting bolts from the knuckle.
9. Using a plastic hammer, disconnect the halfshaft from the front hub assembly.
10. Push the front hub assembly outward and separate the driveshaft from the hub assembly.
11. Insert a pry bar between the transaxle case and joint, and separate the halfshaft from the transaxle assembly

WARNING

Use a pry bar being careful not to damage the transaxle and joint. Do not insert the pry bar too deep, as this may cause damage to the oil seal. Do not pull the halfshaft by excessive force it may cause components inside the joint to dislodge resulting in a torn boot or a damaged bearing. resulting in a torn boot or a damaged bearing.

12. Pull out the halfshaft from the transaxle case.

➡Plug the hole of the transaxle case with the oil seal cap to prevent contamination. Support the halfshaft properly. Replace the retainer ring whenever the halfshaft is removed from the transaxle case.

To install:

13. Apply gear oil on the driveshaft splines and contacting surface of differential case oil seal.
14. Replace circlip with a new one.

➡When replacing the circlip, be careful not to install a different kind of circlip.

15. Before installing the halfshaft, set the opening side of the circlip facing downward.

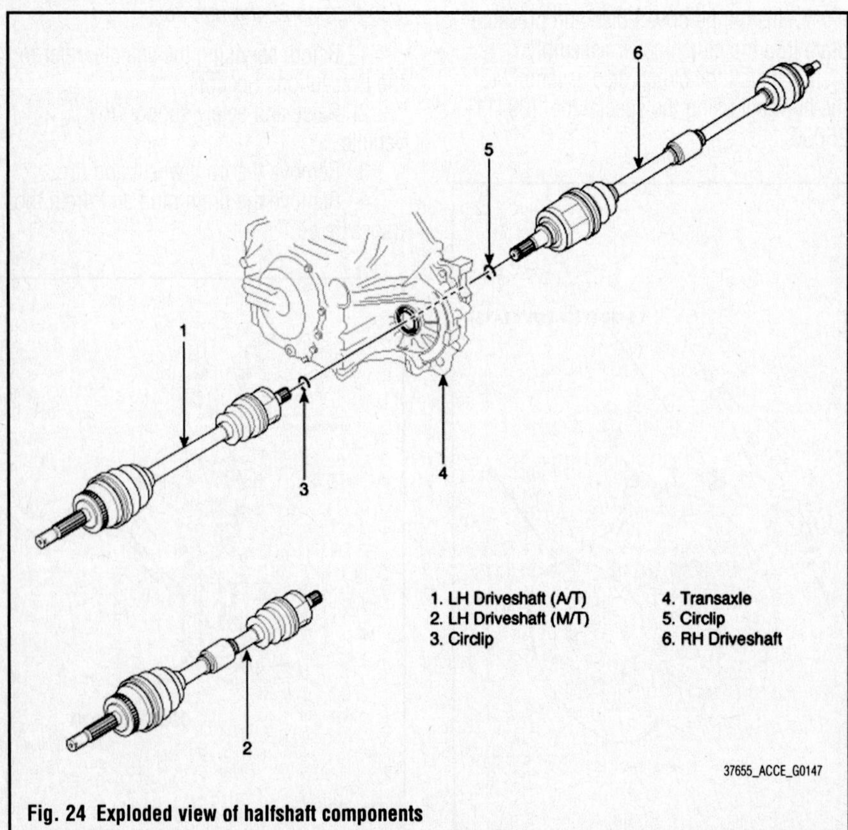

Fig. 24 Exploded view of halfshaft components

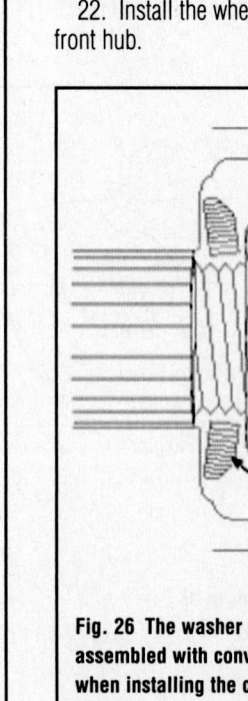

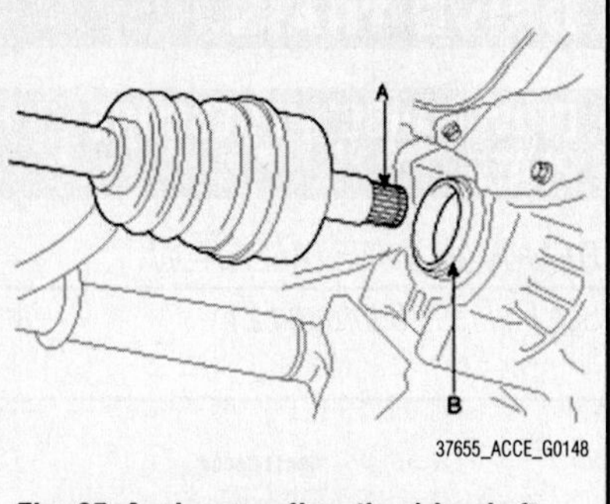

Fig. 25 Apply gear oil on the driveshaft splines (A) and contacting surface of differential case oil seal (B)

16. After installation, check that the halfshaft cannot be removed by hand.
17. Install the halfshaft to the axle hub.

WARNING

Be careful not to damage or dent the boots.

18. Install the ball joint assembly mounting bolt to the knuckle. Tightening torque: 74–89 ft. lbs. (100–120 Nm).
19. Install the front wheel speed sensor cable bracket mounting bolt.
20. Install the washer, castle nut, and split pin to the front hub assembly. Tightening torque: 148–192 ft. lbs. (200–260 Nm). The washer should be assembled with convex surface outward when installing the castle nut and split pin.
21. Replace the drain plug and refill the transaxle oil.
22. Install the wheel and the tire to the front hub.

Fig. 26 The washer (B) should be assembled with convex surface outward when installing the castle (A) nut and split pin (C)

ENGINE COOLING

ENGINE FAN

REMOVAL & INSTALLATION

See Figure 27.

1. Before servicing the vehicle, refer to the Precautions Section.
2. Disconnect the negative battery cable.
3. Drain the cooling system. Remove the radiator cap to speed draining.
4. Remove the upper and lower radiator hoses.
5. Remove the Automatic Transaxle Fluid (ATF) oil cooler hoses.
6. Disconnect the fan motor connector.
7. Remove the cooling fan mounting bolts and remove cooling fan.

To install:

8. Install the cooling fan mounting bolts, tightening to 5–8 ft. lbs. (7–11 Nm).
9. Connect the fan motor connector.
10. Install the upper radiator hose and lower radiator hose.
11. Install the ATF oil cooler hoses.
12. Connect the negative battery cable.
13. Fill with engine coolant.
14. Start engine and check for leaks.

RADIATOR

REMOVAL & INSTALLATION

See Figure 27.

1. Before servicing the vehicle, refer to the Precautions Section.
2. Disconnect the negative battery cable.
3. Drain the cooling system. Remove the radiator cap to speed draining.
4. Remove the upper and lower radiator hoses.
5. Remove the Automatic Transaxle Fluid (ATF) oil cooler hoses.
6. Disconnect the fan motor connector.
7. Remove the cooling fan mounting bolts and remove cooling fan.
8. Remove the radiator upper bracket, then pull up the radiator.

To install:

9. Install the radiator upper bracket, tightening the bolts to 5–8 ft. lbs. (7–11 Nm).
10. Install the cooling fan mounting bolts, tightening to 5–8 ft. lbs. (7–11 Nm).
11. Connect the fan motor connector.
12. Install the upper radiator hose and lower radiator hose.
13. Install the ATF oil cooler hoses.
14. Connect the negative battery cable.
15. Fill with engine coolant.
16. Start engine and check for leaks.

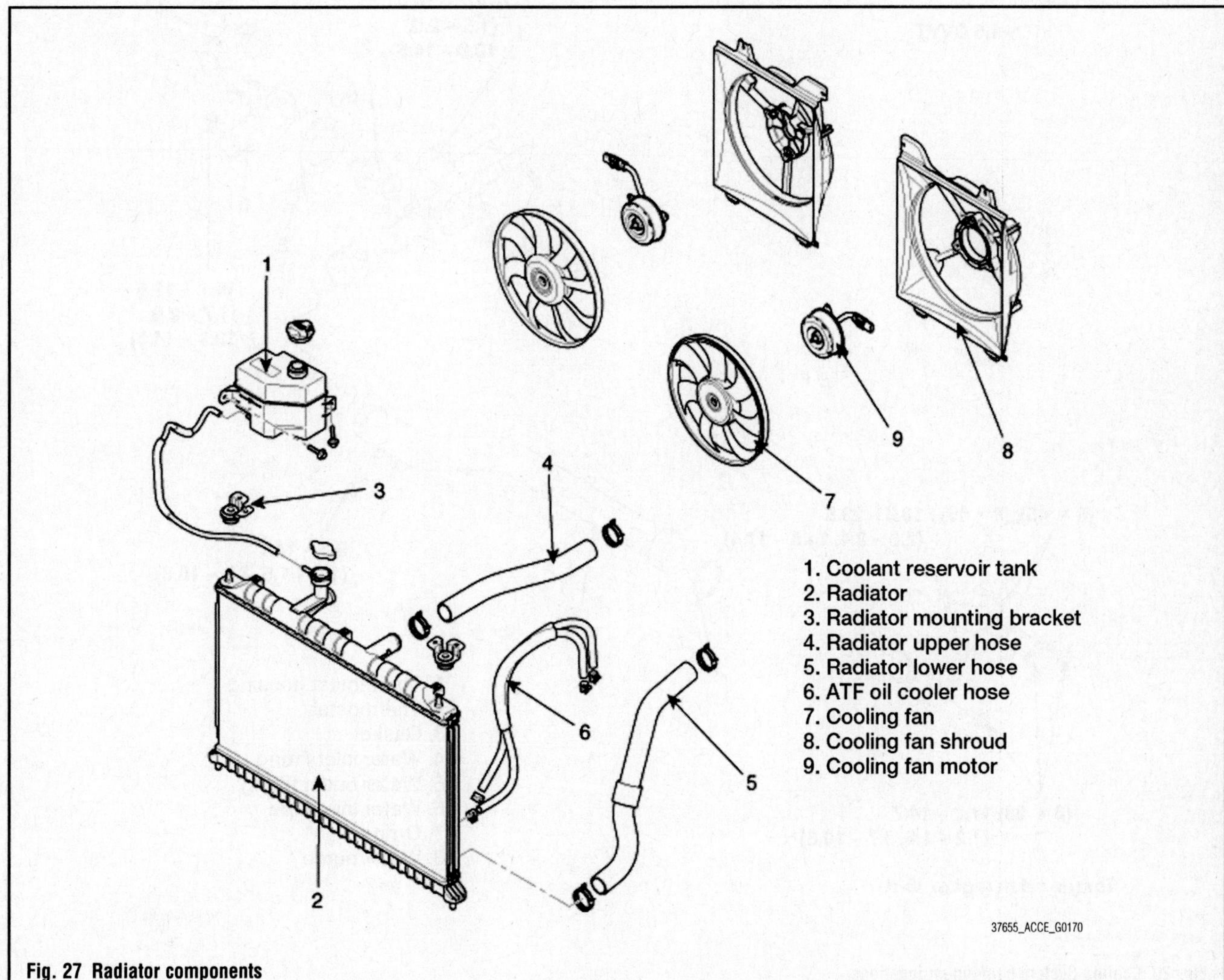

Fig. 27 Radiator components

THERMOSTAT

REMOVAL & INSTALLATION

See Figures 28 and 29.

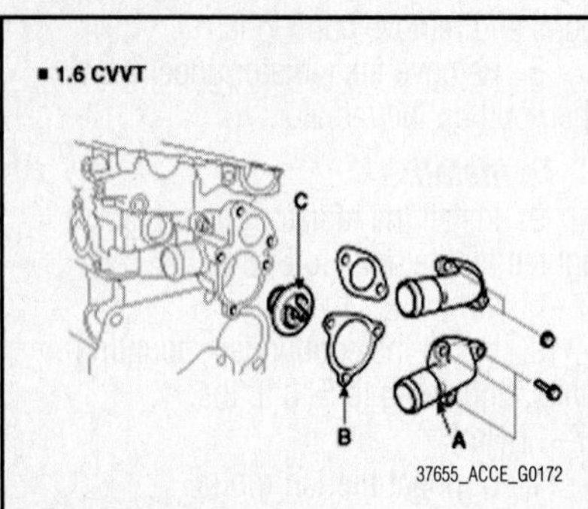

Fig. 28 Water inlet fitting (A), gasket (B), and thermostat (C)

✲✲ CAUTION

Make sure the engine and radiator are cool to the touch prior to beginning this procedure in order to prevent scalding or burns.

1. Before servicing the vehicle, refer to the Precautions Section.

✲✲ WARNING

Disassembly of the thermostat would have an adverse effect, causing a lowering of cooling efficiency.

2. Drain the engine coolant so its level is below thermostat.
3. Remove the water inlet fitting, gasket and thermostat

To install:

4. Installation is the reverse of removal.

WATER PUMP

REMOVAL & INSTALLATION

See Figures 29 and 30.

✲✲ CAUTION

The system is under high pressure when the engine is hot. To avoid danger of releasing scalding engine coolant, remove the cap only when the engine is cool.

1. Before servicing the vehicle, refer to the Precautions Section.
2. Drain the engine coolant.

■ 1.6 CVVT

14.7 ~ 19.6 (1.5 ~ 2.0 ~ 10.8 ~ 14.5)

14.7 ~ 19.6 (1.5 ~ 2.0, 10.8 ~ 14.5)

16.7 ~ 19.6 (1.7 ~ 2.0, 12.3 ~ 14.5)

(8 × 45), (8 × 65) : 19.6 ~ 23.5 (2.0 ~ 2.4, 14.5 ~ 17.4)

9.8 ~ 14.7 (1.0 ~ 1.5, 7.2 ~ 10.8)

(8 × 28) 11.8 ~ 14.7 (1.2 ~ 1.5, 8.7 ~ 10.8)

1. Thermostat housing
2. Thermostat
3. Gasket
4. Water inlet fitting
5. Water outlet fitting
6. Water inlet pipe
7. O-ring
8. Water pump

Torque : N.m (kgf.m, lb-ft)

37655_ACCE_G0171

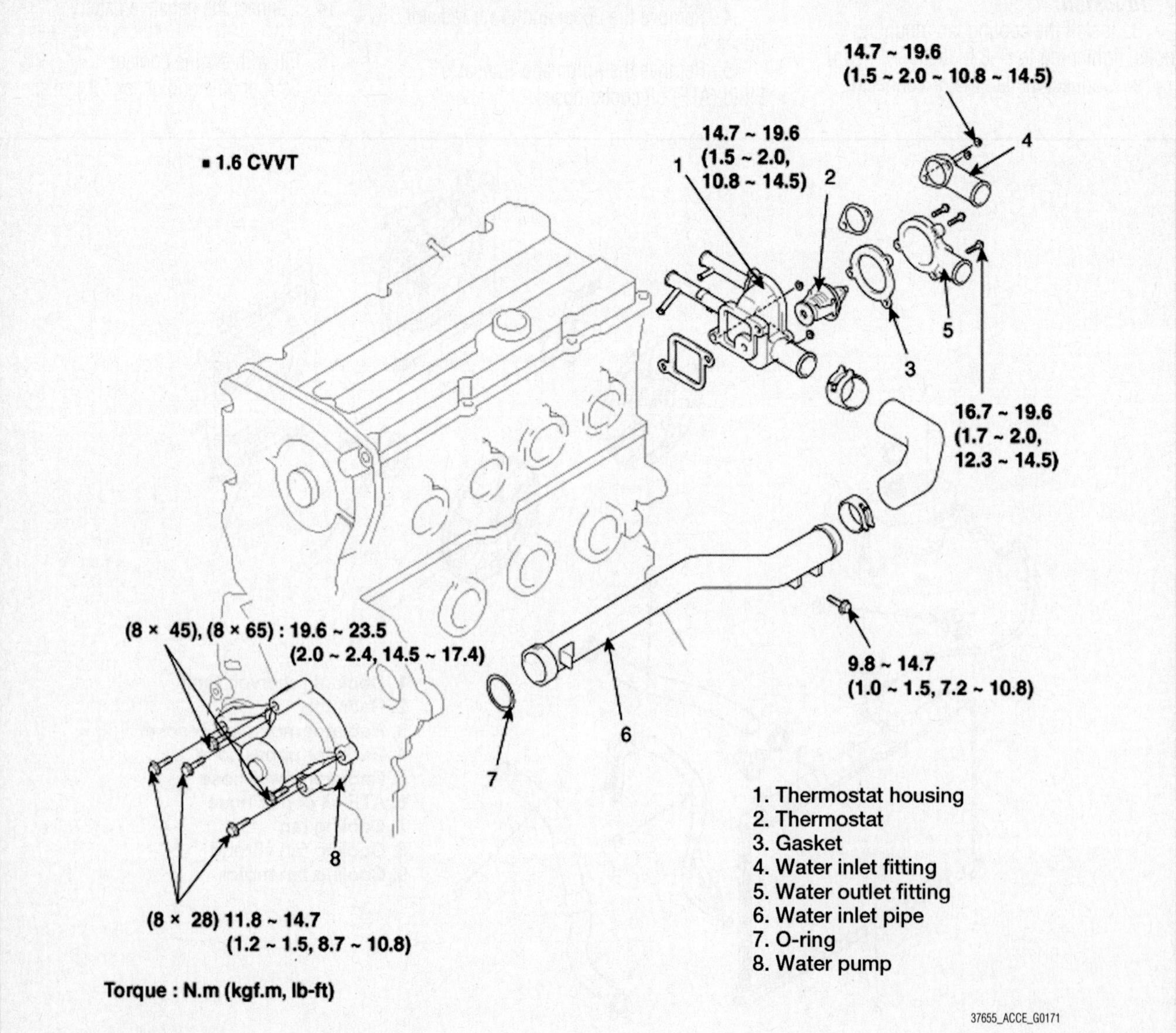

Fig. 29 Cooling System component locations

3. Loosen the water pump pulley bolts.
4. Remove the drive belts. Refer to Accessory Drive Belt Removal & Installation in the Engine Mechanical Section.
5. Remove the water pump pulley.
6. Remove the timing belt. Refer to Timing Belt Removal & Installation in the Engine Mechanical Section.
7. Remove the timing belt idler.
8. Remove the water pump:
   a. Remove the 2 bolts and the alternator brace.
   b. Remove the 3 bolts and remove the water pump and gasket.

To install:

9. Install the water pump and a new gasket with the 3 bolts, tightening the bolts to 9–11 ft. lbs. (12–15 Nm).
10. Install the alternator brace with the 2 bolts, tightening the bolts to 15–17 ft. lbs. (20–24 Nm).
11. Install the timing belt idler.
12. Install the timing belt.
13. Install the water pump pulley.
14. Install the drive belts.
15. Tighten the water pump pulley bolts to 70–86 inch lbs. (8–10 Nm).
16. Fill with engine coolant.
17. Start engine and check for leaks.
18. Recheck engine coolant level.

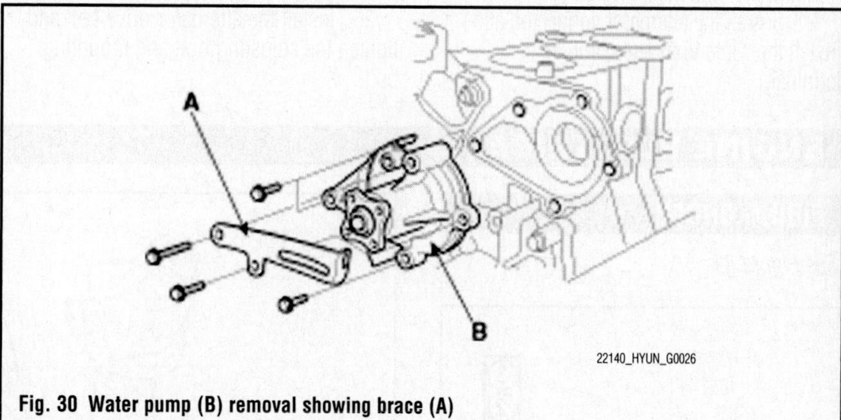

Fig. 30 Water pump (B) removal showing brace (A)

ENGINE ELECTRICAL

CHARGING SYSTEM

ALTERNATOR

REMOVAL & INSTALLATION

See Figures 31 and 32.

1. Before servicing the vehicle, refer to the Precautions Section.
2. Disconnect the battery negative terminal first, then the positive terminal.
3. Temporarily loosen the water pump pulley bolts.
4. Remove the alternator drive belt, after loosening the adjusting bolt and mounting bolt.
5. Remove the power steering pump belt. Refer to Accessory Drive Belt Removal & Installation in the Engine Mechanical section.
6. Remove the water pump pulley.
7. Remove the power steering pump.
8. Remove the power steering pump bracket.
9. Disconnect the alternator connector, and remove the cable from alternator "B" terminal.
10. Remove the adjusting bolt.
11. Remove the mounting bolt.
12. Remove the alternator brace.
13. Pull out the through bolt.
14. Remove the alternator.

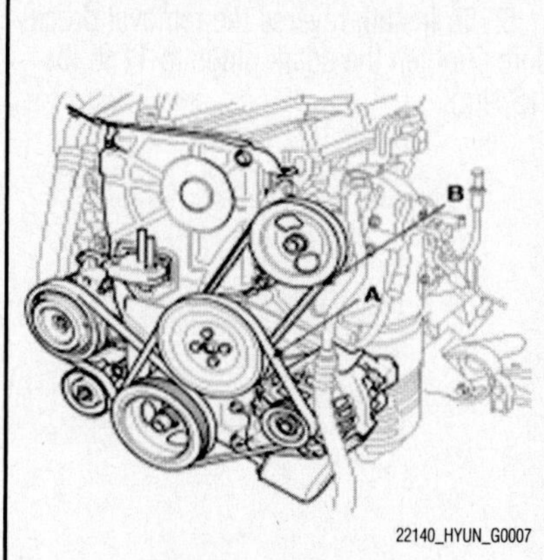

Fig. 31 Remove the alternator drive belt (A) and the power steering pump belt (B)

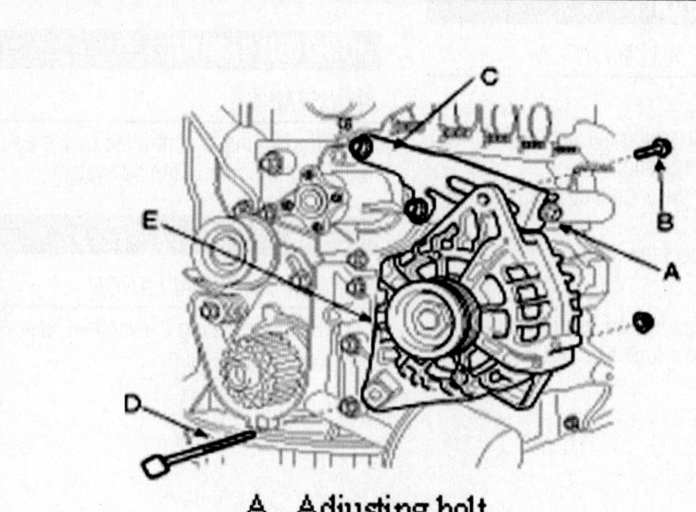

Fig. 32 Removing the bolts, brace, and alternator

To install:

15. Install or connect the following:
16. Install the alternator.
17. Install the through bolt.
18. Install the alternator brace.
19. Install the mounting bolt.
20. Install the adjusting bolt.
21. Install the alternator connector, and install the cable from alternator "B" terminal.
22. Install the power steering pump bracket.
23. Install the power steering pump.
24. Install the water pump pulley and pre-tighten the bolts.
25. Install the power steering pump belt and adjust the tension.
26. Install the alternator drive belt and tighten the adjusting bolt and mounting bolt.
27. Adjust the alternator belt tension and torque the bolts to the following specifications:
    a. Pivot bolt to 14–18 ft. lbs. (19–25 Nm).
    b. Adjustment bolt to 14–20 ft. lbs. (19–28 Nm).
28. Tighten the water pump pulley bolts to 70–86 inch lbs. (8–10 Nm).
29. Connect the battery terminals.

ENGINE ELECTRICAL

IGNITION SYSTEM

FIRING ORDER

See Figure 33.

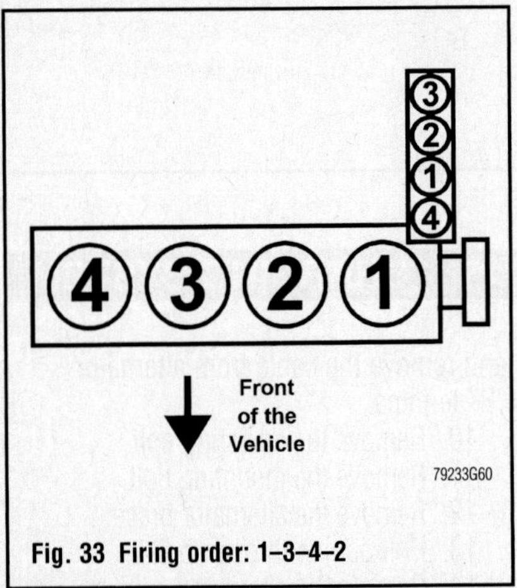

Fig. 33 Firing order: 1–3–4–2

IGNITION COILS

REMOVAL & INSTALLATION

See Figure 34.

1. Before servicing the vehicle, refer to the Precautions Section.
2. Remove the engine cover (as necessary).
3. Disconnect the ignition coil connector.

➡**When removing the ignition coil connector, pull the lock pin and push the clip.**

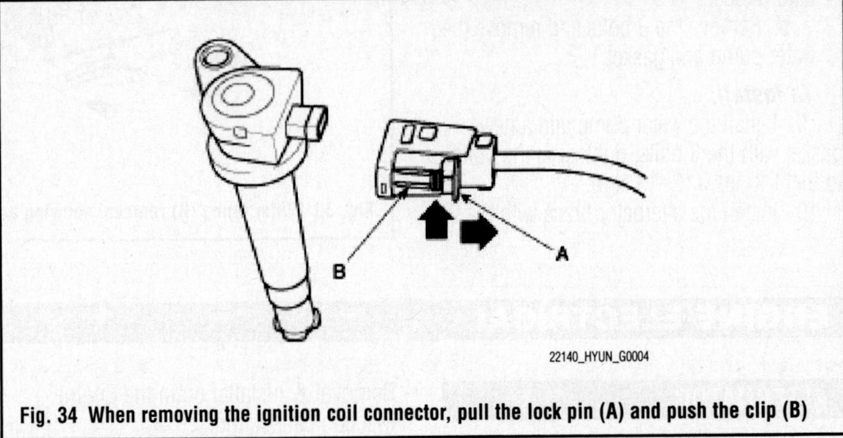

Fig. 34 When removing the ignition coil connector, pull the lock pin (A) and push the clip (B)

4. Remove the ignition coil.

To install:

5. Installation is the reverse of removal. Tighten the ignition coil bolt to 70–86 inch lbs. (8–10 Nm).

IGNITION TIMING

ADJUSTMENT

Ignition timing is controlled by the electronic control ignition timing system.

SPARK PLUGS

REMOVAL & INSTALLATION

1. Before servicing the vehicle, refer to the Precautions Section.
2. Remove the engine cover (as necessary).
3. Remove the ignition coil. Refer to Ignition Coils Removal & Installation.
4. Use a spark plug socket and wrench to remove the spark plugs.

⁂ WARNING

Be careful that no contaminates enter through the spark plug holes.

➡**Check the electrode gap on the spark plugs before installation. Specification: 0.039–0.043 in. (1.0–1.1 mm).**

To install:

5. To install, reverse the removal procedure. Tighten the spark plugs to 11 ft. lbs. (15 Nm).

STARTER

REMOVAL & INSTALLATION

See Figures 35 and 36.

The starting system includes the battery, starter, solenoid switch, ignition switch, inhibitor switch (Automatic Transaxle), ignition lock switch, connection wires and the battery cable. When the ignition key is turned to the start position, current flows and energizes the starter motor's solenoid coil. The solenoid plunger and clutch shift lever are activated, and the clutch pinion engages the ring gear. The contacts close and the starter motor cranks. In order to prevent damage caused by excessive rotation of the starter armature when the engine starts, the clutch pinion gear overruns.

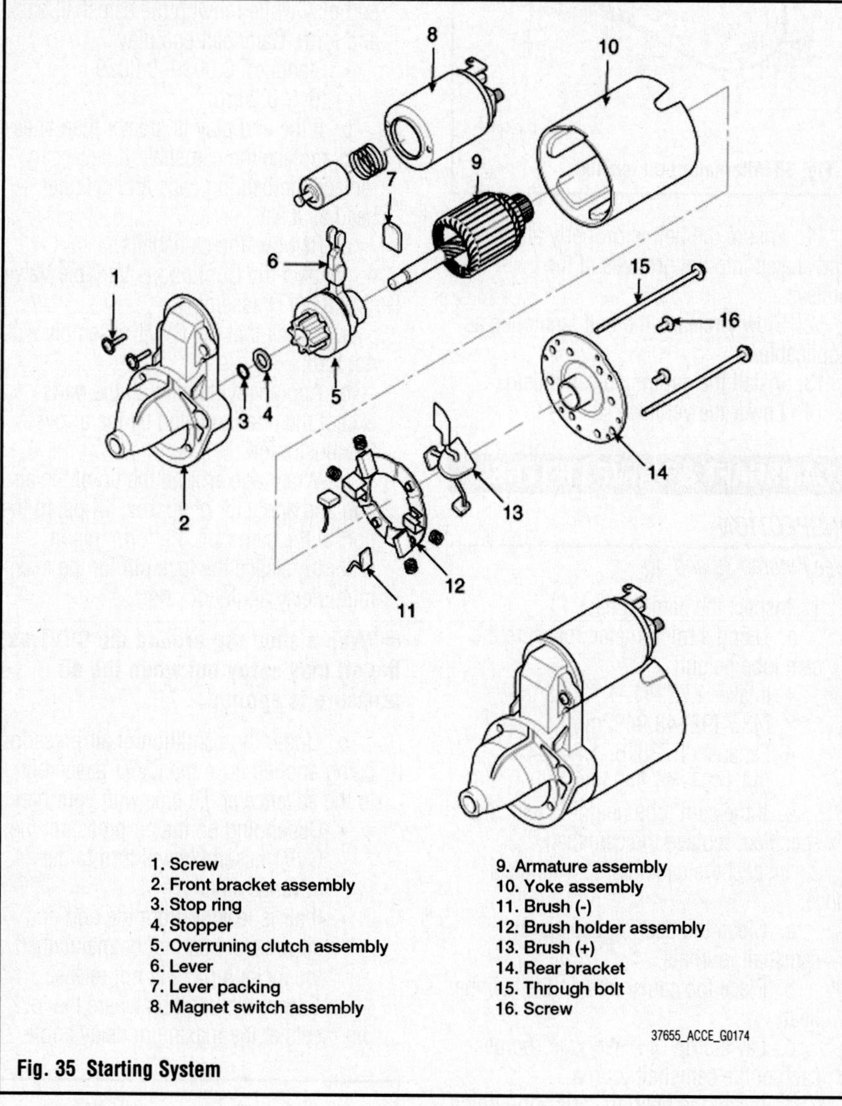

Fig. 35 Starting System

1. Before servicing the vehicle, refer to the Precautions Section.
2. Disconnect the battery negative cable.
3. Remove the air cleaner assembly.
4. For manual transaxle, remove the shift cable and bracket.
5. Disconnect the starter cable from the B terminal on the solenoid.
6. Disconnect the connector from the S terminal.
7. Remove the 2 bolts holding the starter, then remove the starter.

To install:

8. Installation is the reverse of removal. Tighten starter motor bolts to 19–24 ft. lbs. (27–34 Nm).

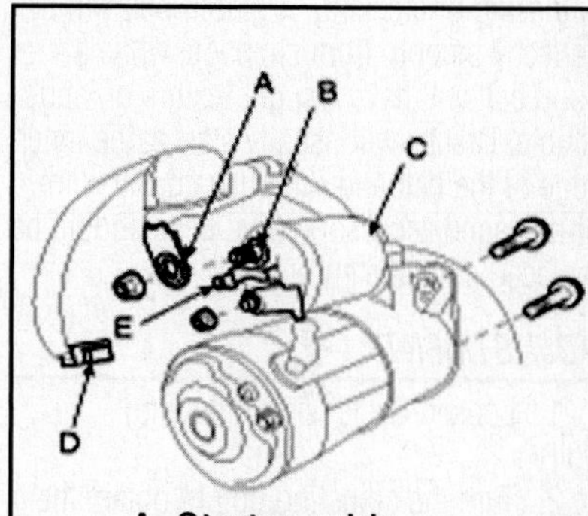

Fig. 36 Starter cable, connector, B terminal, and S terminal

ENGINE MECHANICAL

ACCESSORY DRIVE BELTS

ACCESSORY BELT ROUTING

See Figure 37.

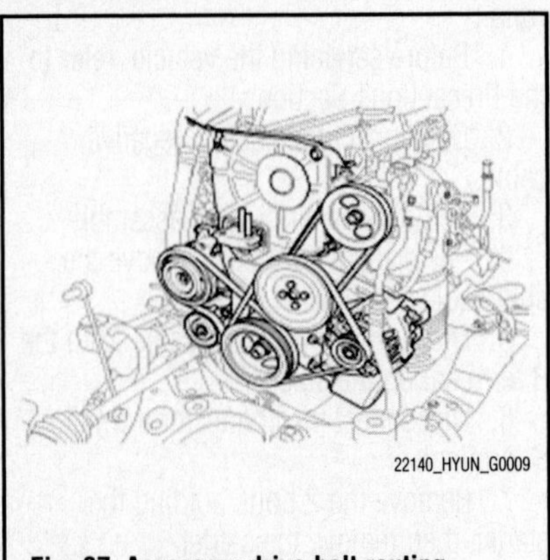

Fig. 37 Accessory drive belt routing

INSPECTION

Inspect the accessory drive belt for signs of glazing or cracking. A glazed belt will be perfectly smooth from slippage, while a good belt will have a slight texture of fabric visible. Cracks will usually start at the inner edge of the belt and run outward. All worn or damaged accessory drive belts should be replaced immediately.

ADJUSTMENT

1. Loosen the tension mounting bolt.
2. Turn the adjusting bolt to obtain the proper belt tension, then retighten the mounting bolt.
3. Recheck the deflection of the drive belt.

REMOVAL & INSTALLATION

See Figure 38.

1. Before servicing the vehicle, refer to the Precautions Section.
2. Raise and support the vehicle.
3. Remove the engine splash shield.
4. Rotate the belt tensioner clockwise to release the tension, as necessary.
5. Remove the belt from the pulley.
6. Slowly release the belt tensioner, as applicable.
7. Remove the belt.

To install:

8. Install the belt to the pulley.
9. Rotate the drive belt tensioner clockwise.
10. Install the belt.

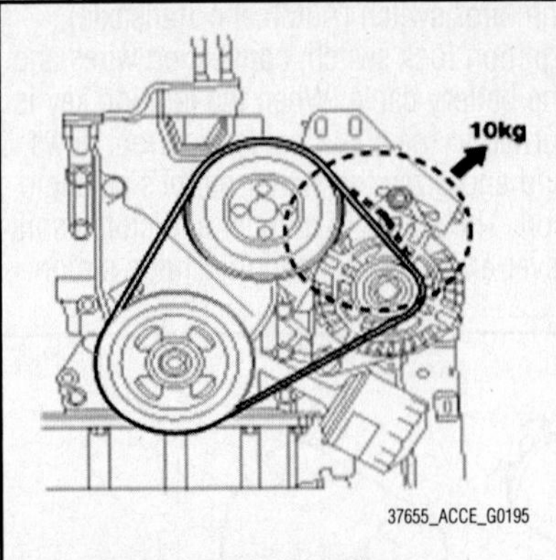

Fig. 38 Alternator belt tension

11. Ensure the belt is properly aligned and seated into the grooves of the pulleys.
12. Slowly release the belt tensioner, as applicable.
13. Install the engine splash shield.
14. Lower the vehicle.

CAMSHAFT & VALVE LIFTERS

INSPECTION

See Figures 39 and 40.

1. Inspect the cam lobes:
   a. Using a micrometer, measure the cam lobe height:
      - Intake: 1.72241–1.73028 in. (43.7492–43.9492mm)
      - Exhaust: 1.73816–1.74604 in. (44.1494–44.3494mm)

   b. If the cam lobe height is less than specified, replace the camshaft.
2. Inspect the camshaft journal clearance:
   a. Clean the bearing caps and camshaft journals.
   b. Place the camshafts on the cylinder head.
   c. Lay a strip of Plastigage® across each of the camshaft journal.
   d. Install the bearing caps and tighten the bolts with specified torque.

⁂ WARNING

Do not turn the camshaft.

   e. Remove the bearing caps.
   f. Measure the Plastigage® at its widest point. Bearing oil clearance:
      - Standard: 0.0008–0.0024 in. (0.020–0.061mm)
      - Limit: 0.0039 in. (0.1mm)

   g. If the oil clearance is greater than specified, replace the camshaft. If necessary, replace the bearing caps and cylinder head as a set.
   h. Completely remove the Plastigage®.
   i. Remove the camshafts.
3. Inspect the camshaft end play:
   a. Install the camshafts.
   b. Using a dial indicator, measure the end play while moving the camshaft back and forth. Camshaft end play:
      - Standard: 0.0039–0.0079 in. (0.1–0.2mm)

   c. If the end play is greater than specified, replace the camshaft. If necessary, replace the bearing caps and cylinder head as a set.
   d. Remove the camshafts.
4. Inspect the Continuous Variable Valve Timing (CVVT) assembly:
   a. Check that the CVVT assembly will not turn.
   b. Apply vinyl tape to all the parts except the one indicated by the arrow in the illustration.
   c. Wrap tape around the tip of the air gun and apply air of approx. 14 psi to the port of the camshaft. Perform this in order to release the lock pin for the maximum delay angle locking.

➡Wrap a shop rag around the CVVT as the oil may spray out when the air pressure is applied.

   d. Under the condition of air pressure being applied, turn the CVVT assembly to the advance angle side with your hand.
      - Depending on the air pressure, the CVVT assembly will turn to the advance side
      - If air is leaking from the port and air pressure cannot be maintained, the locking pin will not release

   e. Except the position where the lock pin meets at the maximum delay angle,

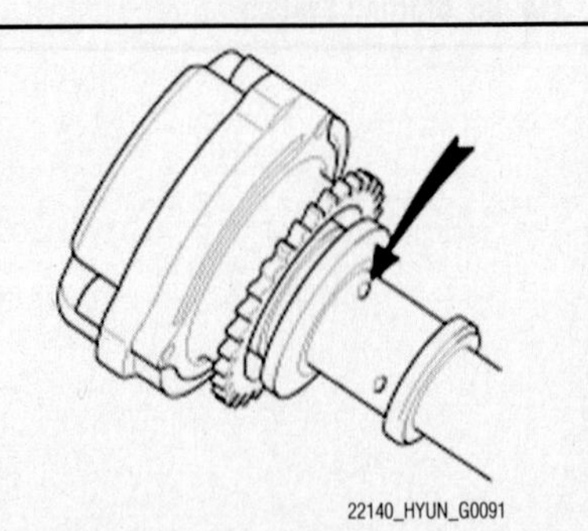

Fig. 39 Apply vinyl tape to the CVVT on all parts except the one indicated by the arrow

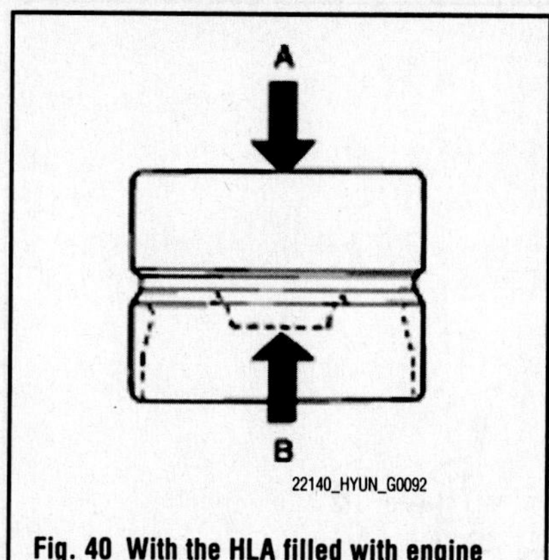

Fig. 40 With the HLA filled with engine oil, hold A and press B by hand

let the CVVT assembly turn back and forth and check the movable range and that there is no disturbance.

f. The CVVT should move smoothly in the range of about 20°.

g. Turn the CVVT assembly with your hand and lock it at the maximum delay angle position.

5. Inspect the Hydraulic Lash Adjuster (HLA):

a. With the HLA filled with engine oil, hold A and press B by hand.

b. If B moves, replace the HLA.

REMOVAL & INSTALLATION

See Figures 41 through 49.

Engine removal is not required for this procedure. Use a fender cover to avoid damaging painted surfaces. To avoid damage, unplug the wiring connectors carefully while holding the connector portion. Mark all wiring and hoses to avoid misconnection. Inspect the timing belt before removal. Turn the crankshaft pulley so that the No. 1 piston is at Top Dead Center (TDC).

1. Before servicing the vehicle, refer to the Precautions Section.
2. Disconnect the terminals from battery and remove the battery.
3. Remove the engine cover.
4. Remove the undercover.

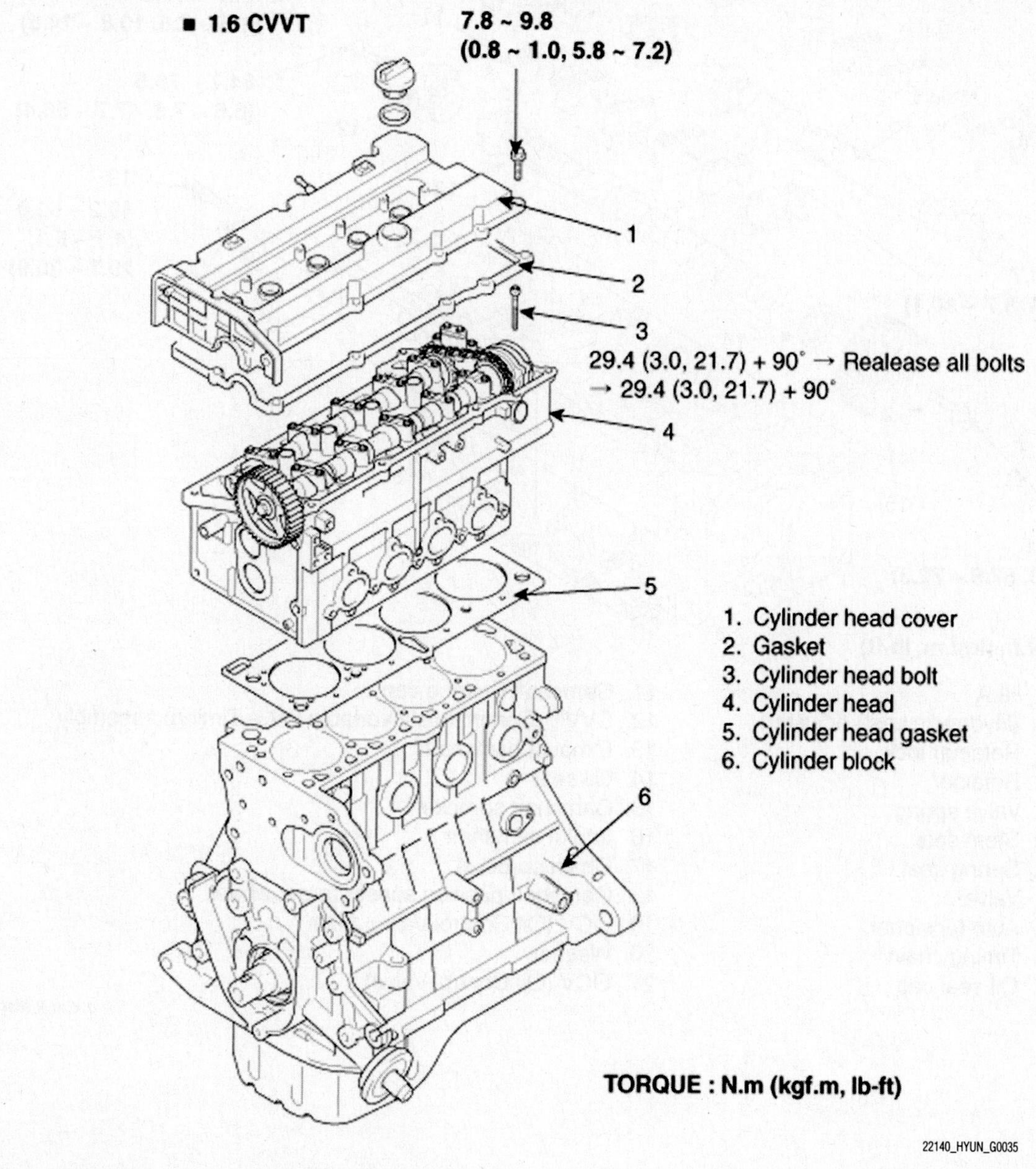

■ 1.6 CVVT

7.8 ~ 9.8
(0.8 ~ 1.0, 5.8 ~ 7.2)

29.4 (3.0, 21.7) + 90° → Realease all bolts
→ 29.4 (3.0, 21.7) + 90°

1. Cylinder head cover
2. Gasket
3. Cylinder head bolt
4. Cylinder head
5. Cylinder head gasket
6. Cylinder block

TORQUE : N.m (kgf.m, lb-ft)

22140_HYUN_G0035

Fig. 41 Exploded view of cylinder head and engine block

■ 1.6 CVVT

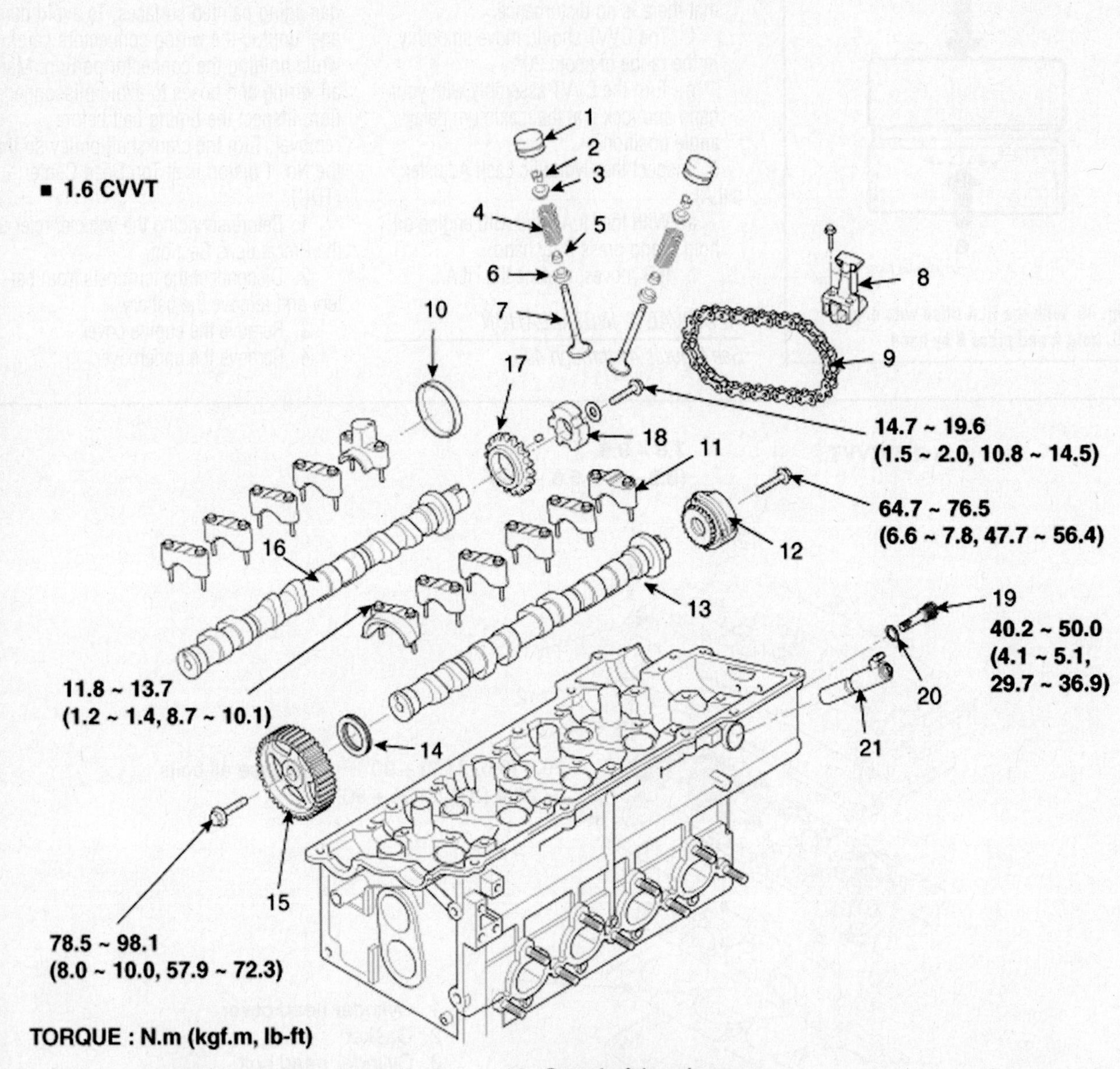

TORQUE : N.m (kgf.m, lb-ft)

1. HLA (Hydraulic Lash Adjuster)
2. Retainer lock
3. Retainer
4. Valve spring
5. Stem seal
6. Spring seat
7. Valve
8. Auto tensioner
9. Timing chain
10. Oil seal cap
11. Camshaft bearing cap
12. CVVT (Continuously Variable Valve Timing) assembly
13. Exhaust camshaft
14. Oil seal
15. Camshaft sprocket
16. Intake camshaft
17. Chain sprocket
18. Camshaft position sensor target wheel
19. OCV (Oil Control Valve) Filter
20. Washer
21. OCV (Oil Control Valve)

22140_HYUN_G0036

Fig. 42 Exploded view of cylinder head and related components

5. Drain the engine coolant. Remove the radiator cap to speed draining.
6. Remove the intake air hose and air cleaner assembly:
   a. Disconnect the breather hose from intake air hose.
   b. Remove the intake air hose and air cleaner upper cover.
   c. Disconnect the ECM connectors.
   d. Remove the air cleaner element and air cleaner lower cover.
7. Remove the battery tray.
8. Remove the upper radiator hose and lower radiator hose.
9. Remove the heater hoses.
10. Remove the fuel hose.
11. Remove the accelerator cable by loosening the lock-nut, then slip the cable end out of the throttle linkage.
12. Disconnect the Throttle Position Sensor (TPS) connector and the MAP sensor connector.
13. Remove the engine wire harness connectors and wire harness clamps from cylinder head and the intake manifold:
   a. Disconnect the rear oxygen sensor connector.
   b. Disconnect the air conditioner compressor switch connector.
   c. Disconnect the knock sensor connector.
   d. Disconnect the injector connectors.
   e. Remove the wire harness bracket.
   f. Disconnect the Idle Speed Actuator (ISA) connector.
   g. Disconnect the front oxygen sensor connector.
   h. Disconnect the Crankshaft Position Sensor (CKP) connector.
   i. Disconnect the Oil Control Valve (OCV) connector.
   j. Disconnect the ignition coil connector.
   k. Disconnect the ignition coil condenser connector.
   l. Disconnect the Camshaft Position Sensor (CMP) connector.
   m. Disconnect the ground cable.
   n. Remove the wire harness bracket.
14. Disconnect the Purge Control Solenoid Valve (PCSV) hose.
15. Remove the brake booster vacuum hose.
16. Remove the power steering pump and fix the pump to vehicle with a wire.
17. Remove the ignition coil.
18. Remove the exhaust manifold. Refer to Exhaust Manifold Removal & Installation.
19. Remove the intake manifold. Refer to Intake Manifold Removal & Installation.
20. Remove the timing belt. Refer to Timing Belt Removal & Installation.

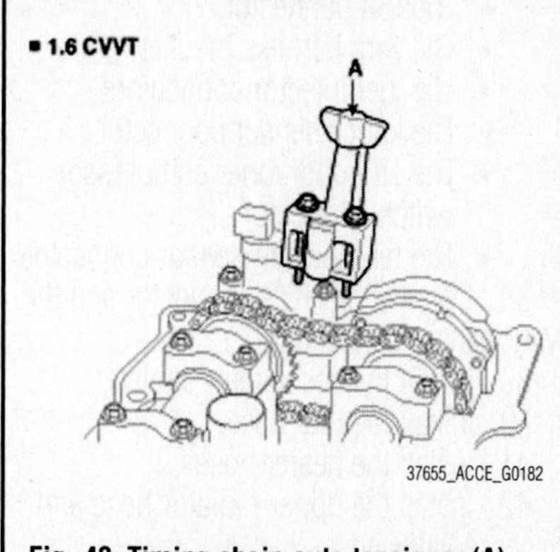

Fig. 43 Timing chain auto tensioner (A)

21. Remove the cylinder head cover. Refer to Cylinder Head Cover Removal & Installation.
22. Remove the camshaft sprocket.
23. Remove the timing chain auto tensioner.
24. Remove the camshaft bearing caps and camshafts.

To install:

Thoroughly clean all parts to be assembled. Rotate the crankshaft, set the No. 1 piston at TDC.

25. Install the camshafts:
   a. Align the camshaft timing chain with the intake timing chain sprocket and exhaust timing chain sprocket as shown.
   b. Install the camshaft and bearing caps and tighten to 108–120 inch lbs. (12–14 Nm).
   c. Install the timing chain auto tensioner and tighten to 72–84 inch lbs. (8–10 Nm).
26. Using the SST (09221 - 21000), install the camshaft bearing oil seal.
27. Install the camshaft sprocket.
28. Install the cylinder head cover.

➡Before installing the cylinder head cover gasket, thoroughly clean the cylinder head cover and the groove.

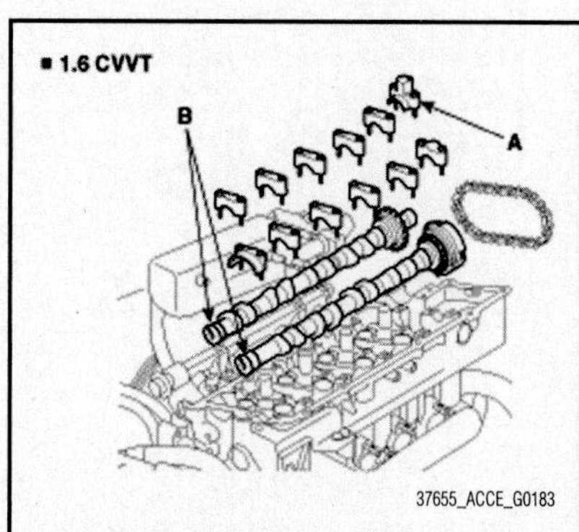

Fig. 44 Camshaft bearing caps (A) and camshafts (B)

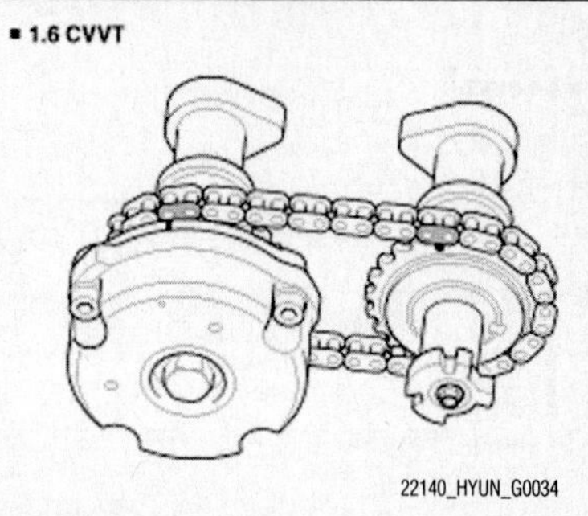

Fig. 45 Align the camshaft timing chain with the intake timing chain sprocket and exhaust timing chain sprocket

When installing, make sure the cylinder head cover gasket is seated securely in the corners of the recesses with no gap.

29. Install the cylinder head cover gasket in the groove of the cylinder head cover.
   a. Apply liquid gasket to the head cover gasket at the corners of the recess.

➡Use liquid gasket, Loctite® No. 5999. Check that the mating surfaces are clean and dry before applying liquid gasket. After assembly, wait at least 30 minutes before filling the engine with oil.

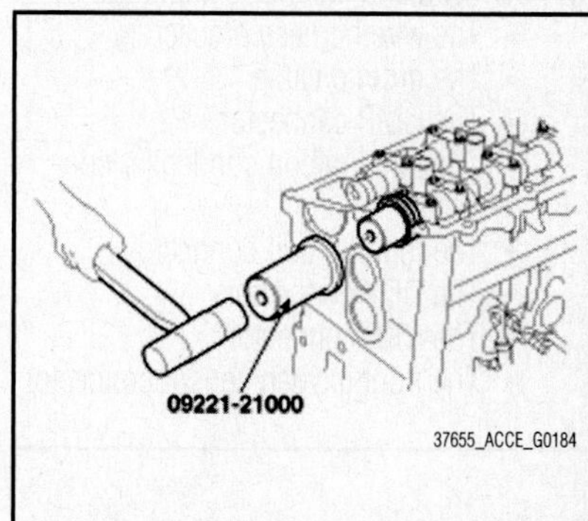

Fig. 46 Install the camshaft bearing oil seal

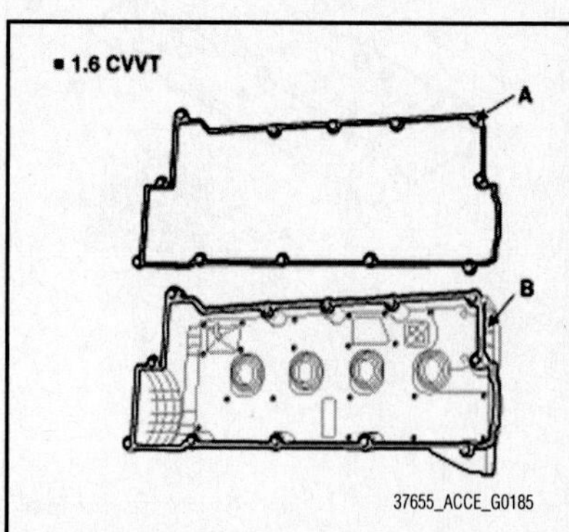

Fig. 47 Install the cylinder head cover gasket

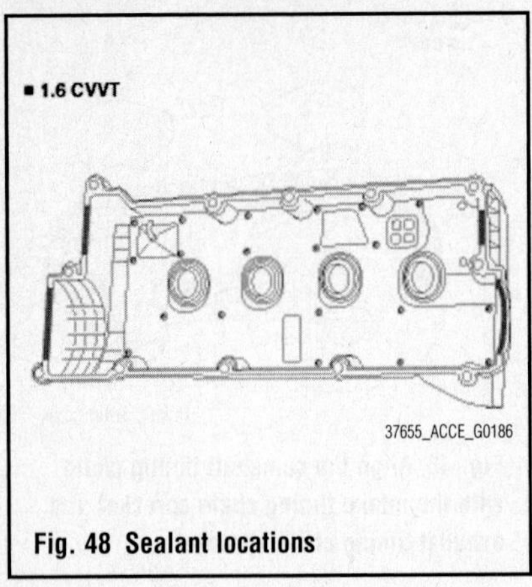

Fig. 48 Sealant locations

b. Install the cylinder head cover with bolts.

- Step 1: Pre-tighten all bolts by 36–48 inch lbs. (4–5 Nm)
- Step 2: Tighten to 72–84 inch lbs. (8–10 Nm)

30. Install the timing belt.
31. Install the intake manifold.
32. Install the exhaust manifold.
33. Install the ignition coil.
34. Install the power steering pump.
35. Install the brake booster hose.
36. Connect the hose of the PCSV side.
37. Install the engine wire harness connectors and wire harness clamps to the cylinder head and the intake manifold:

- The wire harness bracket
- The ground cable
- The CMP connector
- The ignition coil condenser connector
- The ignition coil connector
- The OCV connector
- The CKP connector
- The front oxygen sensor connector
- The ISA connector
- The wire harness bracket
- The fuel injector connectors
- The knock sensor connector
- The air conditioner compressor switch connector
- The rear oxygen sensor connector

38. Connect the TPS connector and the MAP sensor connector.
39. Install the accelerator cable.
40. Install the fuel hose.
41. Install the heater hoses.
42. Install the upper radiator hose and lower radiator hose.
43. Install the battery tray.
44. Install the intake air hose and air cleaner assembly.
45. Install the air cleaner element and air cleaner lower cover and tighten to 72–84 inch lbs. (8–10 Nm).
46. Connect the ECM connectors.
47. Install the intake air hose and air cleaner upper cover.
48. Connect the breather hose to intake air hose.
49. Install the undercover.
50. Install the engine cover and tighten to 36–48 inch lbs. (4–6 Nm).
51. Install the battery and connect the battery terminals.
52. Fill with engine coolant.
53. Start the engine and check for leaks.
54. Recheck engine coolant level and oil level.

CATALYTIC CONVERTER

REMOVAL & INSTALLATION

Refer to Combination Manifold Removal & Installation.

COMBINATION MANIFOLD

REMOVAL & INSTALLATION

See Figures 50 through 54.

1. Before servicing the vehicle, refer to the Precautions Section.
2. Disconnect the negative battery cable.
3. Remove the engine cover.
4. Remove the front oxygen sensor connector.
5. Remove the front muffler heat protector.
6. Remove the front muffler.
7. Remove the stay of the exhaust manifold and catalytic converter assembly.
8. Remove the heat protector.
9. Remove the exhaust manifold and catalytic converter assembly.

Fig. 49 Bolt tightening sequence

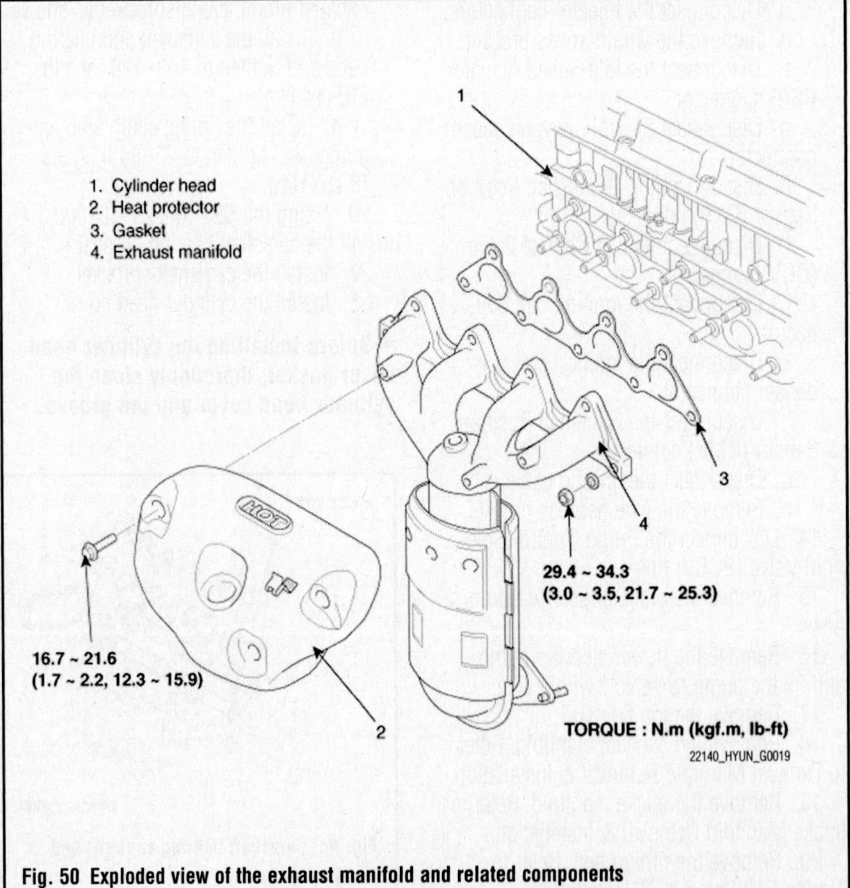

Fig. 50 Exploded view of the exhaust manifold and related components

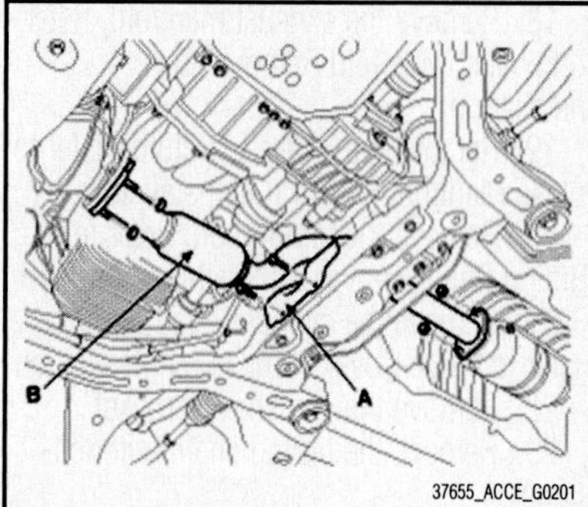

Fig. 51 Front muffler (B) and front muffler heat protector (A)

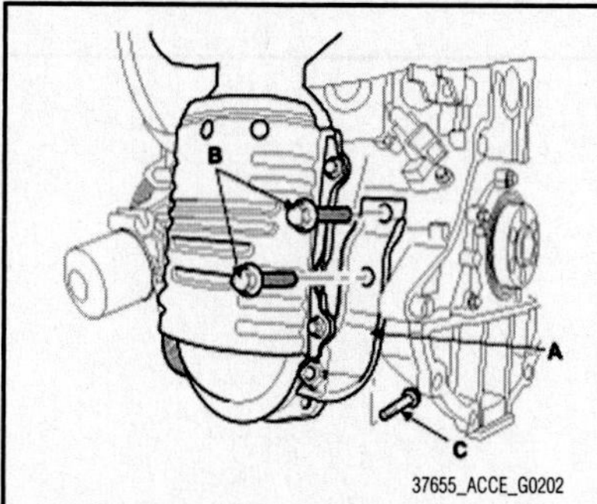

Fig. 52 Exhaust manifold and catalytic converter assembly stay (A) and bolts (B, C)

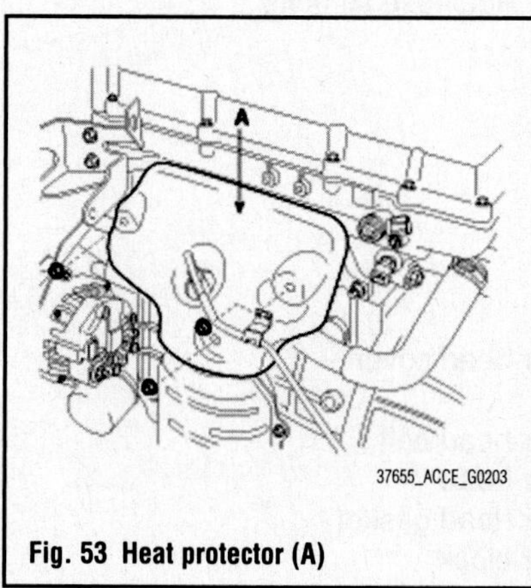

Fig. 53 Heat protector (A)

To install:

10. Install the exhaust manifold and catalytic converter assembly and tighten to 22–25 ft. lbs. (29–34 Nm).
11. Install the heat protector and tighten to 12–16 ft. lbs. (17–22 Nm).
12. Install the stay of the exhaust manifold and catalytic converter assembly. Tighten large bolts to 25–29 ft. lbs. (34–39 Nm); smaller bolts to 22–29 ft. lbs. (29–39 Nm).
13. Install the front muffler and tighten to 22–29 ft. lbs. (29–39 Nm).
14. Install the front muffler heat protector and tighten to 72–108 inch lbs. (8–12 Nm).

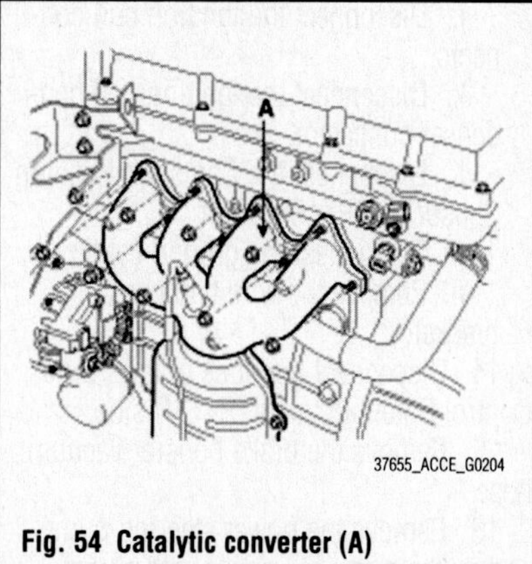

Fig. 54 Catalytic converter (A)

15. Install the front oxygen sensor connector.
16. Install the engine cover.
17. Install the negative battery cable.

CRANKSHAFT FRONT SEAL

REMOVAL & INSTALLATION

See Figures 55 and 56.

1. Before servicing the vehicle, refer to the Precautions Section.
2. Remove the front right wheel and tire.
3. Remove the accessory drive belt. Refer to Accessory Drive Belts Removal & Installation.
4. Remove the crankshaft damper pulley.
5. Remove the crankshaft damper.
6. With appropriate tool, remove the crankshaft front seal.

To install:

7. Install the crankshaft front seal.
8. Install the crankshaft damper. Torque mounting bolt to 101–109 ft. lbs. (137–148 Nm)
9. Install the accessory drive belt.
10. Install the right front tire and wheel.

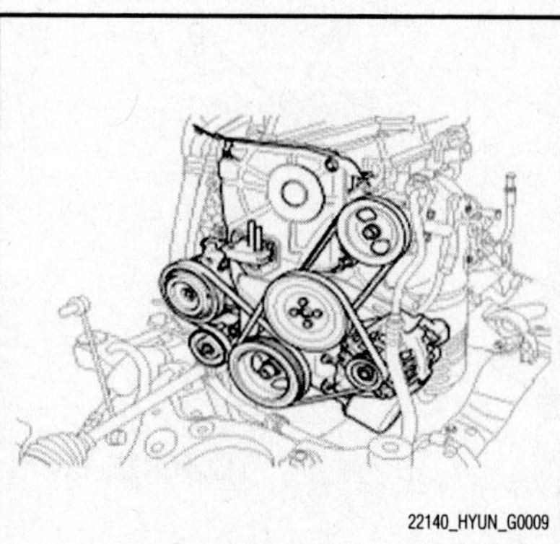

Fig. 65 Accessory belts location example

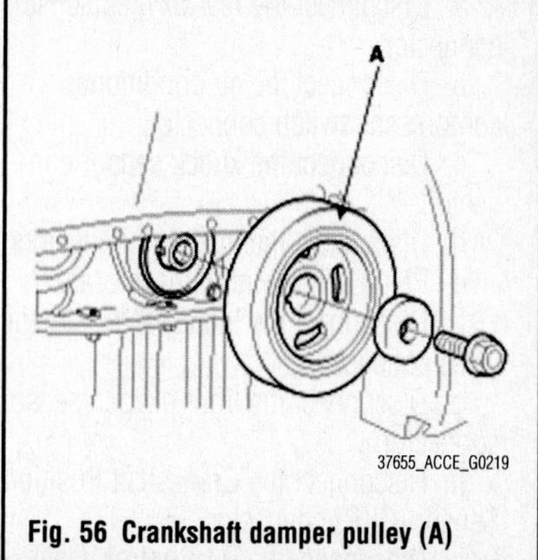

Fig. 56 Crankshaft damper pulley (A)

CYLINDER HEAD

REMOVAL & INSTALLATION

See Figures 57 through 71.

Engine removal is not required for this procedure. Use a fender cover to avoid damaging painted surfaces. To avoid damage, unplug the wiring connectors carefully while holding the connector portion. Mark all wiring and hoses to avoid misconnection. Inspect the timing belt before removal. Turn the crankshaft pulley so that the No. 1 piston is at Top Dead Center (TDC).

1. Before servicing the vehicle, refer to the Precautions Section.
2. Disconnect the terminals from battery and remove the battery.
3. Remove the engine cover.
4. Remove the undercover.
5. Drain the engine coolant. Remove the radiator cap to speed draining.
6. Remove the intake air hose and air cleaner assembly:
   a. Disconnect the breather hose from intake air hose.
   b. Remove the intake air hose and air cleaner upper cover.
   c. Disconnect the ECM connectors.
   d. Remove the air cleaner element and air cleaner lower cover.
7. Remove the battery tray.
8. Remove the upper radiator hose and lower radiator hose.
9. Remove the heater hoses.
10. Remove the fuel hose.
11. Remove the accelerator cable by loosening the lock-nut, then slip the cable end out of the throttle linkage.
12. Disconnect the Throttle Position Sensor (TPS) connector and the MAP sensor connector.
13. Remove the engine wire harness connectors and wire harness clamps from cylinder head and the intake manifold:

a. Disconnect the rear oxygen sensor connector.

b. Disconnect the air conditioner compressor switch connector.

c. Disconnect the knock sensor connector.

d. Disconnect the injector connectors.

e. Remove the wire harness bracket.

f. Disconnect the Idle Speed Actuator (ISA) connector.

g. Disconnect the front oxygen sensor connector.

h. Disconnect the Crankshaft Position Sensor (CKP) connector.

i. Disconnect the Oil Control Valve (OCV) connector.

j. Disconnect the ignition coil connector.

k. Disconnect the ignition coil condenser connector.

l. Disconnect the Camshaft Position Sensor (CMP) connector.

m. Disconnect the ground cable.

n. Remove the wire harness bracket.

14. Disconnect the hose of the Purge Control Solenoid Valve (PCSV) side.

15. Remove the brake booster vacuum hose.

16. Remove the power steering pump and fix the pump to vehicle with a wire.

17. Remove the ignition coil.

18. Remove the exhaust manifold. Refer to Exhaust Manifold Removal & Installation.

19. Remove the intake manifold. Refer to Intake Manifold Removal & Installation.

20. Remove the timing belt. Refer to Timing Belt Removal & Installation.

21. Remove the cylinder head cover. Refer to Valve Cover Removal & Installation.

22. Remove the camshaft sprocket.

23. Remove the timing chain auto tensioner.

24. Remove the camshaft bearing caps and camshafts.

25. Remove the Oil Control Valve (OCV).

26. Remove the OCV filter(A).

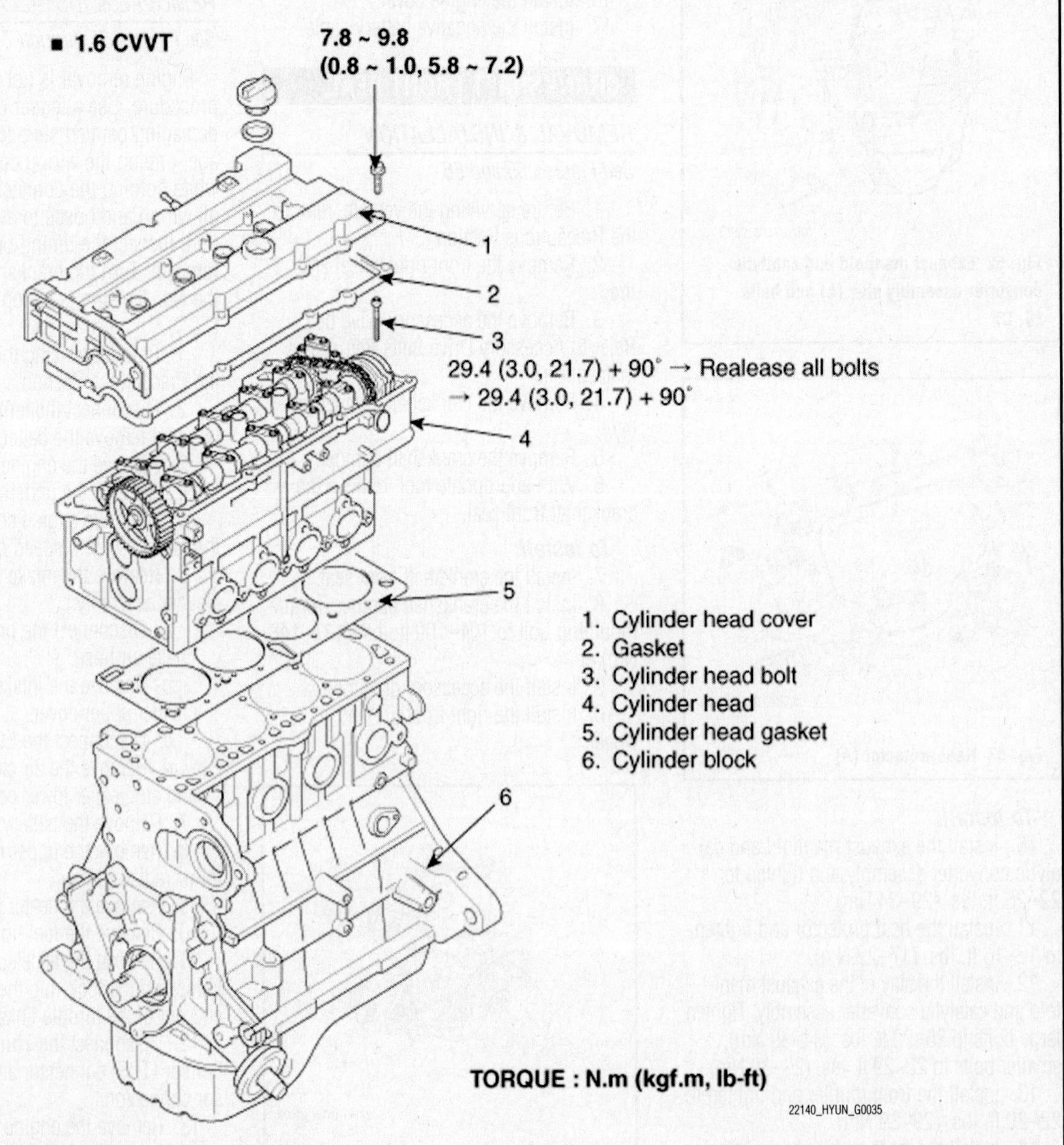

Fig. 57 Exploded view of cylinder head and engine block

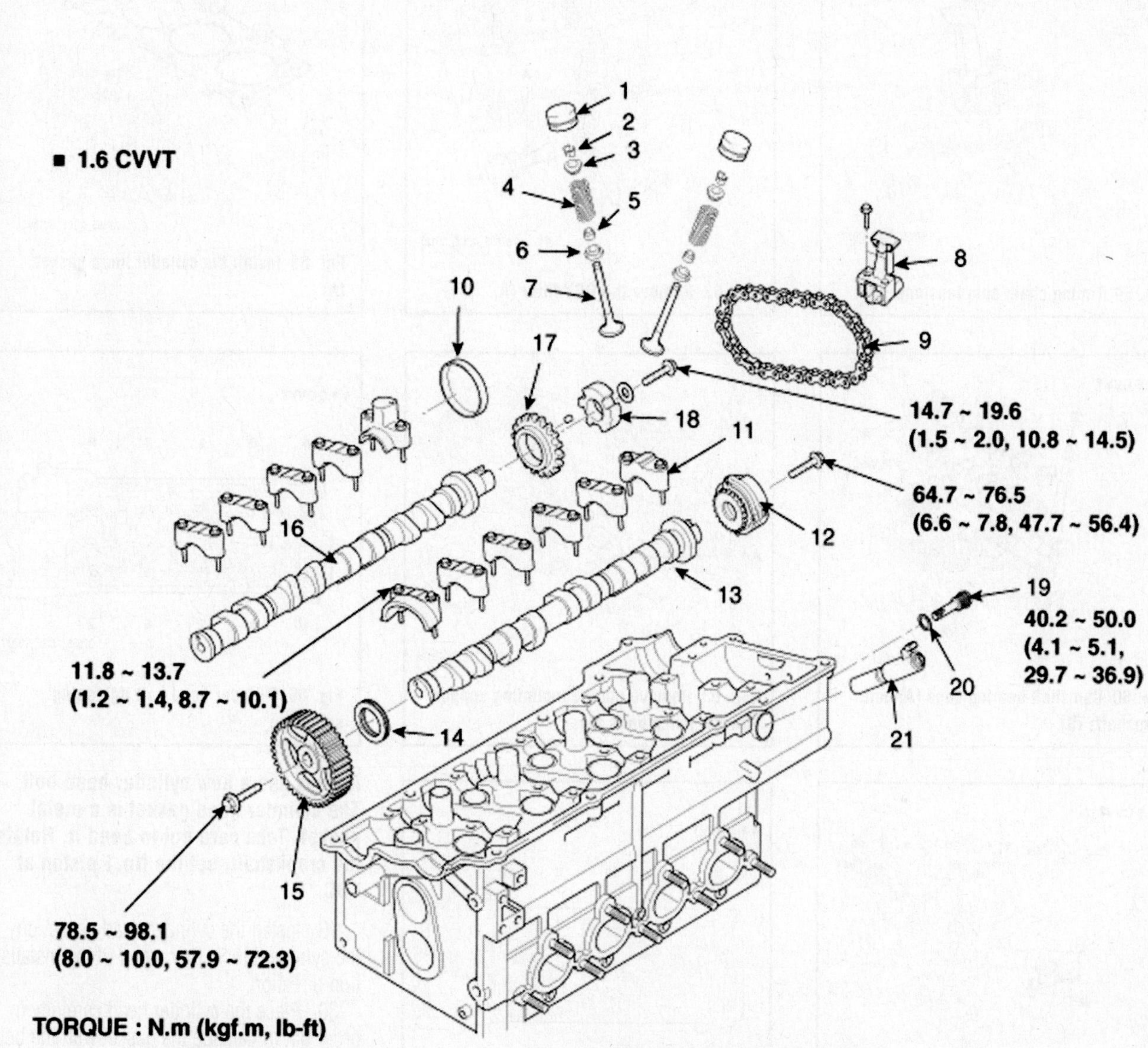

TORQUE : N.m (kgf.m, lb-ft)

1. HLA (Hydraulic Lash Adjuster)
2. Retainer lock
3. Retainer
4. Valve spring
5. Stem seal
6. Spring seat
7. Valve
8. Auto tensioner
9. Timing chain
10. Oil seal cap
11. Camshaft bearing cap
12. CVVT (Continuously Variable Valve Timing) assembly
13. Exhaust camshaft
14. Oil seal
15. Camshaft sprocket
16. Intake camshaft
17. Chain sprocket
18. Camshaft position sensor target wheel
19. OCV (Oil Control Valve) Filter
20. Washer
21. OCV (Oil Control Valve)

22140_HYUN_G0036

Fig. 58 Exploded view of cylinder head and related components

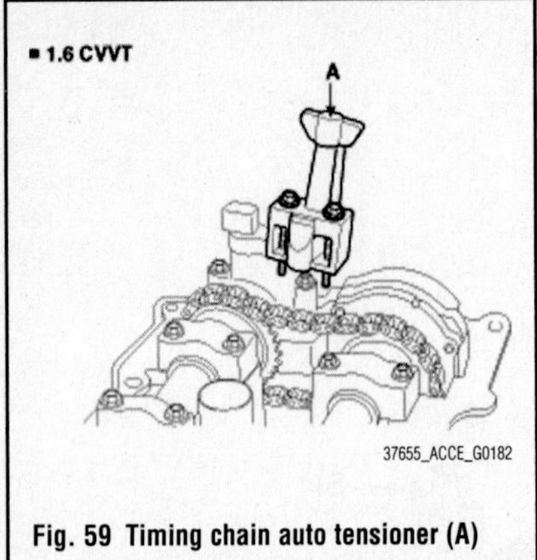

Fig. 59 Timing chain auto tensioner (A)

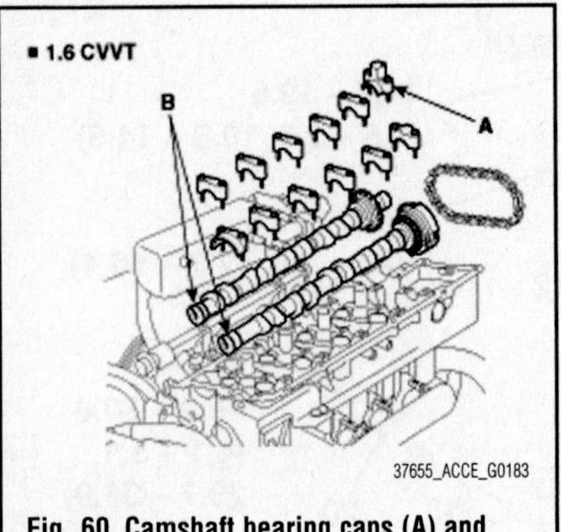

Fig. 60 Camshaft bearing caps (A) and camshafts (B)

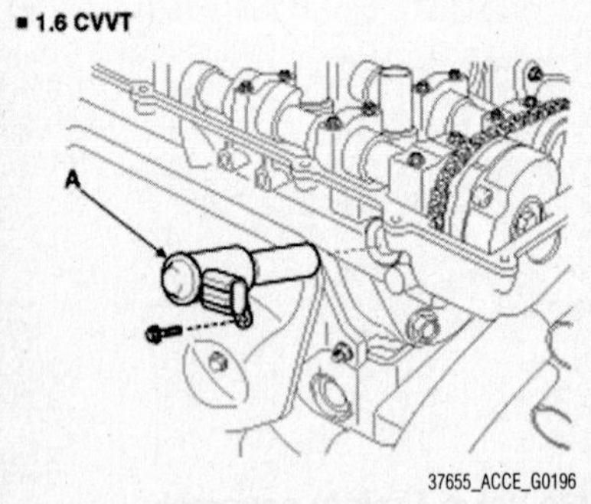

Fig. 61 Remove the Oil Control Valve (OCV) (A)

27. Remove the engine mounting support bracket fixing bolts.

28. Remove the cylinder head bolts, then remove the cylinder head:

a. Using 8mm hexagon wrench, uniformly loosen and remove the 10 cylinder head bolts, in several passes, in the sequence shown.

WARNING

Head warpage or cracking could result from removing bolts in an incorrect order.

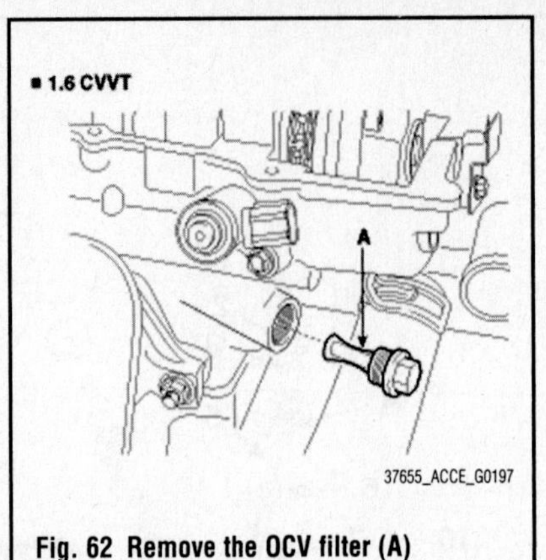

Fig. 62 Remove the OCV filter (A)

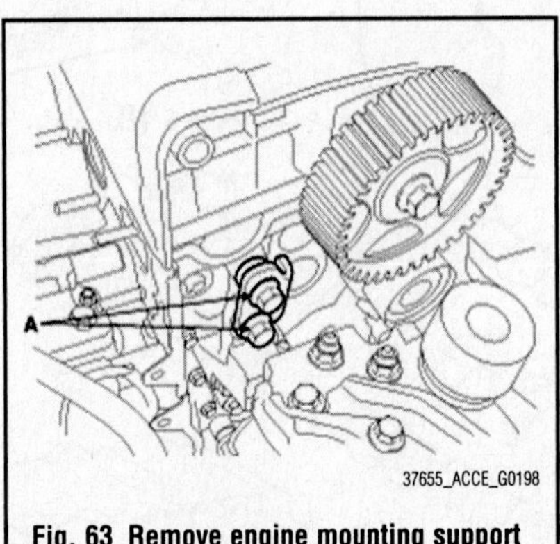

Fig. 63 Remove engine mounting support bracket fixing bolts (A)

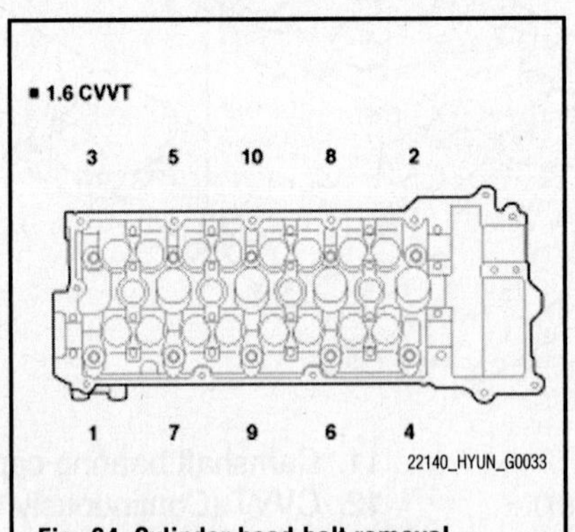

Fig. 64 Cylinder head bolt removal sequence

b. Lift the cylinder head from the dowels on the cylinder block and replace the cylinder head on wooden blocks on a bench.

WARNING

Be careful not to damage the contact surfaces of the cylinder head and cylinder block.

To install:

➡Thoroughly clean all parts to be assembled. Always use a new cylinder head and manifold gasket.

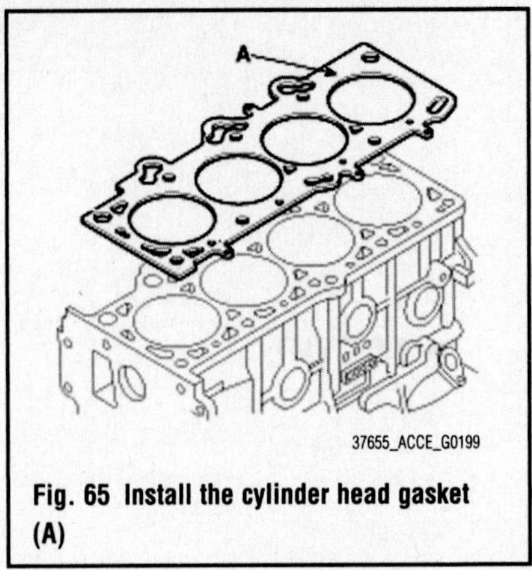

Fig. 65 Install the cylinder head gasket (A)

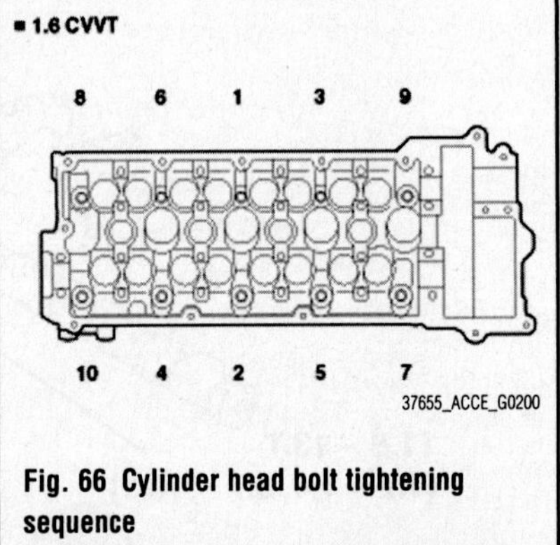

Fig. 66 Cylinder head bolt tightening sequence

Always use a new cylinder head bolt. The cylinder head gasket is a metal gasket. Take care not to bend it. Rotate the crankshaft, set the No.1 piston at TDC.

29. Install the cylinder head gasket on the cylinder block. Be careful of the installation direction.

30. Place the cylinder head carefully in order not to damage the gasket with the bottom part of the end.

31. Install the cylinder head bolts:

a. Apply a light coat if engine oil on the threads and under the heads of the cylinder head bolts.

b. Using an 8mm and 10mm hexagon wrench, install and tighten the 10 cylinder head bolts and plate washers, in several passes, in the sequence shown.

- Step 1: 22 ft. lbs. (29 Nm) plus 90°
- Step 2: Release all bolts
- Step 3: 22 ft. lbs. (29 Nm) plus 90°

32. Install the engine mounting support bracket fixing bolts.

33. Install the OCV filter and tighten to 30–37 ft. lbs. (40–50 Nm).

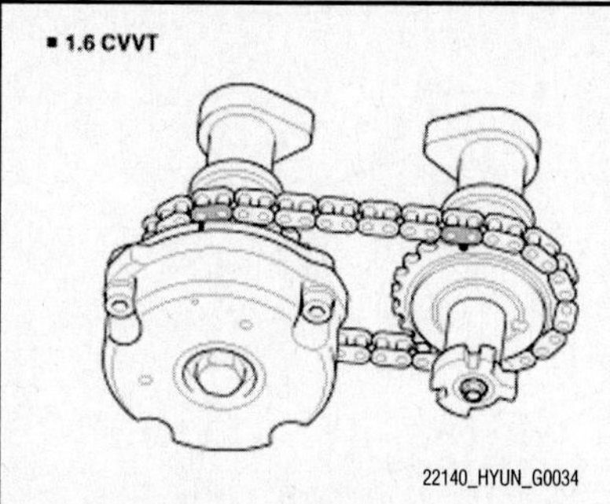

Fig. 67 Align the camshaft timing chain with the intake timing chain sprocket and exhaust timing chain sprocket

➡Always use a new OCV(Oil Control Valve) filter gasket. Keep clean the OCV(Oil Control Valve) filter.

34. Install the OCV filter and tighten to 86–104 inch lbs. (10–12 Nm).

⁂ WARNING

Do not reuse the OCV when dropped. Keep the OCV clean. Do not hold the OCV sleeve during servicing. When the OCV is installed on the engine, do not move the engine with holding the OCV yoke.

35. Install the camshafts:
   a. Align the camshaft timing chain with the intake timing chain sprocket and exhaust timing chain sprocket as shown.
   b. Install the camshaft and bearing caps and tighten to 108–120 inch lbs. (12–14 Nm).
   c. Install the timing chain auto tensioner and tighten to 72–84 inch lbs. (8–10 Nm).
36. Using the SST (09221 - 21000), install the camshaft bearing oil seal.
37. Install the camshaft sprocket.
38. Install the cylinder head cover.

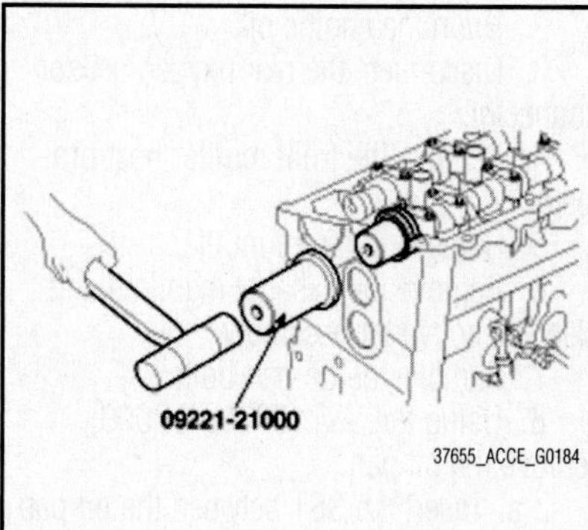

Fig. 68 Install the camshaft bearing oil seal

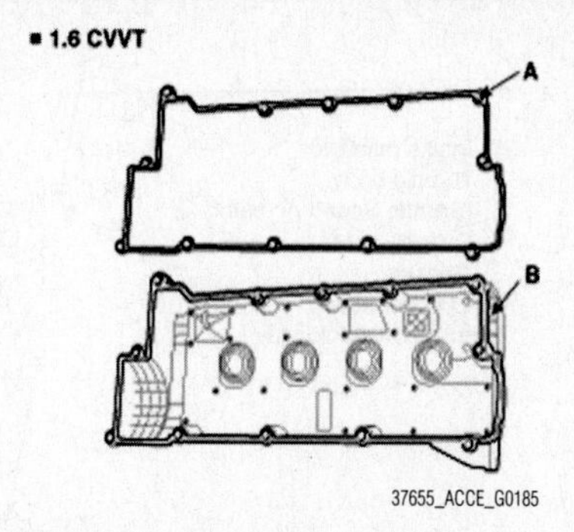

Fig. 69 Install the cylinder head cover gasket

➡Before installing the cylinder head cover gasket, thoroughly clean the cylinder head cover and the groove. When installing, make sure the cylinder head cover gasket is seated securely in the corners of the recesses with no gap.

39. Install the cylinder head cover gasket in the groove of the cylinder head cover.
   a. Apply liquid gasket to the head cover gasket at the corners of the recess.

➡Use liquid gasket, Loctite® No. 5999. Check that the mating surfaces are clean and dry before applying liquid gasket. After assembly, wait at least 30 minutes before filling the engine with oil.

   b. Install the cylinder head cover with bolts.
   - Step 1: Pre-tighten all bolts by 36–48 inch lbs. (4–5 Nm)
   - Step 2: Tighten to 72–84 inch lbs. (8–10 Nm)
40. Install the timing belt.
41. Install the intake manifold.
42. Install the exhaust manifold.
43. Install the ignition coil.
44. Install the power steering pump.

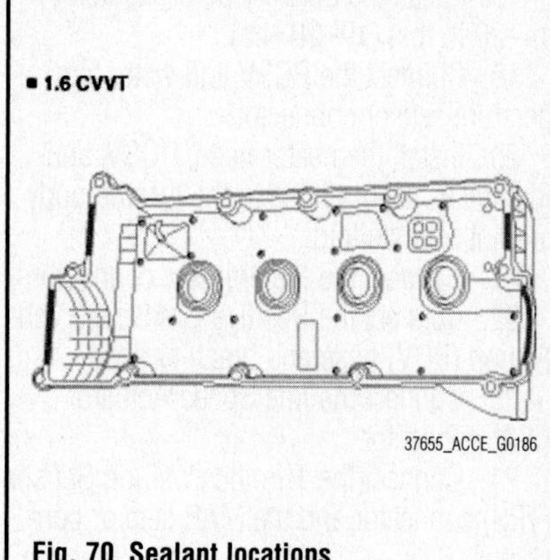

Fig. 70 Sealant locations

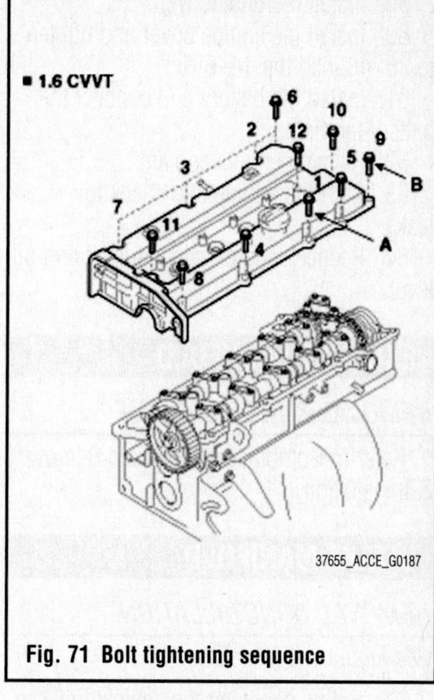

Fig. 71 Bolt tightening sequence

45. Install the brake booster hose.
46. Connect the hose of the PCSV side.
47. Install the engine wire harness connectors and wire harness clamps to the cylinder head and the intake manifold:
   - The wire harness bracket
   - The ground cable
   - The CMP connector
   - The ignition coil condenser connector
   - The ignition coil connector
   - The OCV connector
   - The CKP connector
   - The front oxygen sensor connector
   - The ISA connector
   - The wire harness bracket
   - The fuel injector connectors
   - The knock sensor connector
   - The air conditioner compressor switch connector
   - The rear oxygen sensor connector
48. Connect the TPS connector and the MAP sensor connector.
49. Install the accelerator cable.
50. Install the fuel hose.
51. Install the heater hoses.
52. Install the upper radiator hose and lower radiator hose.
53. Install the battery tray.
54. Install the intake air hose and air cleaner assembly.
55. Install the air cleaner element and air cleaner lower cover and tighten to 72–84 inch lbs. (8–10 Nm).
56. Connect the ECM connectors.
57. Install the intake air hose and air cleaner upper cover.
58. Connect the breather hose to intake air hose.

59. Install the undercover.
60. Install the engine cover and tighten to 36–48 inch lbs. (4–6 Nm).
61. Install the battery and connect the battery terminals.
62. Fill with engine coolant.
63. Start the engine and check for leaks.
64. Recheck engine coolant level and oil level.

EXHAUST MANIFOLD

REMOVAL & INSTALLATION

Refer to Combination Manifold Removal & Installation.

INTAKE MANIFOLD

REMOVAL & INSTALLATION

See Figures 72 and 73.

1. Before servicing the vehicle, refer to the Precautions Section.
2. Relieve the fuel system pressure.
3. Drain the cooling system.
4. Disconnect the negative battery cable.
5. Remove the engine cover.
6. Remove the accelerator cable.
7. Disconnect the Throttle Position Sensor (TPS) connector and the MAP sensor connector.
8. Disconnect the Idle Speed Actuator (ISA) connector.
9. Disconnect the Positive Crankcase Ventilation (PCV) hose and breather hose.
10. Disconnect the fuel injector connectors.

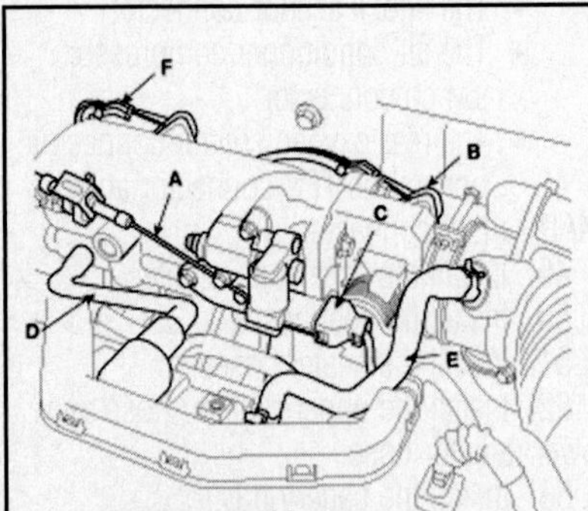

Fig. 72 Remove the accelerator cable, the TPS connector, and the MAP sensor connector. Disconnect the ISA connector, the PCV hose, and breather hose

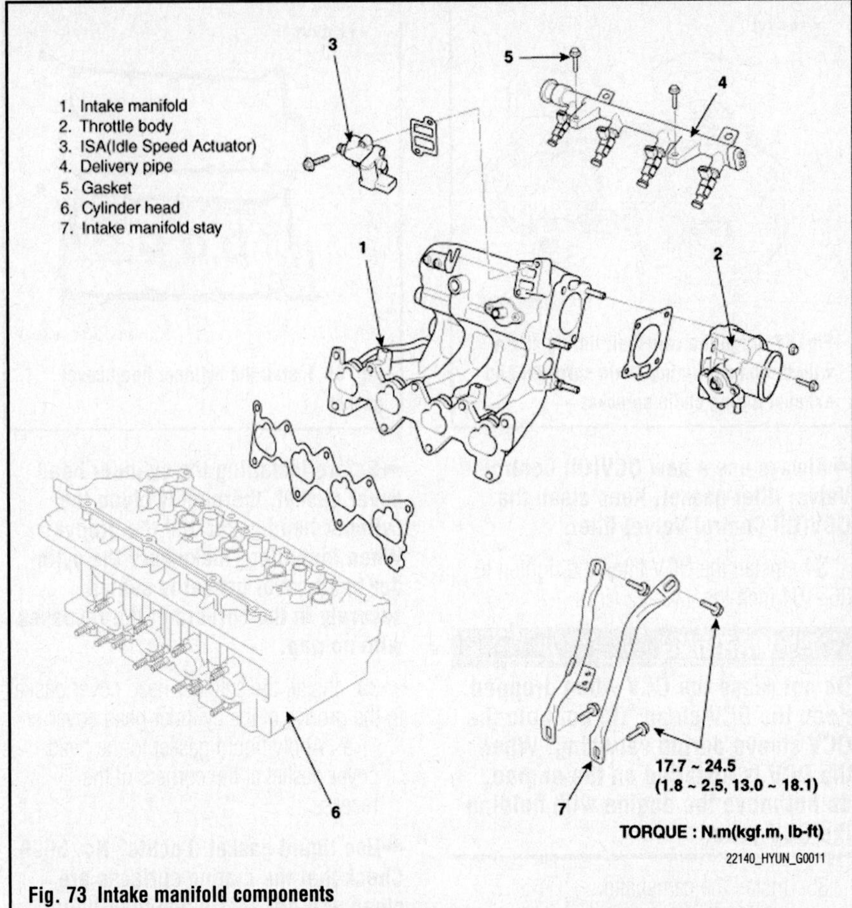

Fig. 73 Intake manifold components

11. Remove the heater hose, Purge Control Solenoid Valve (PCSV), and the brake vacuum hose from the throttle body and intake manifold.
12. Disconnect the PCSV and water temperature sensor connector.
13. Remove the delivery pipe.
14. Remove the intake manifold stay.
15. Remove the intake manifold.

To install:

16. Install the intake manifold. Tighten to 11–15 ft. lbs. (15–20 Nm).
17. Install the intake manifold stay. Tighten to 13–18 ft. lbs. (18–25 Nm).
18. Install the delivery pipe. Tighten to 14–20 ft. lbs. (19–28 Nm).
19. Connect the PCSV and water temperature sensor connector.
20. Install the heater hose, PCSV, and the brake vacuum hose to the throttle body and intake manifold.
21. Connect the fuel injector connectors.
22. Connect the Positive Crankcase Ventilation (PCV) hose and breather hose.
23. Connect the Idle Speed Actuator (ISA) connector.
24. Connect the Throttle Position Sensor (TPS) connector and the MAP sensor connector.
25. Install the accelerator cable.
26. Install the engine cover.
27. Connect the negative battery cable.
28. Fill the cooling system.
29. Start the engine and check for leaks.

OIL PAN

REMOVAL & INSTALLATION

See Figures 74.

1. Before servicing the vehicle, refer to the Precautions Section.
2. Drain the engine oil.
3. Disconnect the rear oxygen sensor connector.
4. Remove the front muffler heat protector.
5. Remove the front muffler.
6. Remove the exhaust manifold and catalytic converter assembly.
7. Remove the oil pan bolts.
8. Using the SST (09215-3C000), remove the oil pan:
    a. Insert the SST between the oil pan and the ladder frame by tapping it with a plastic hammer in the direction of arrow No. 1.

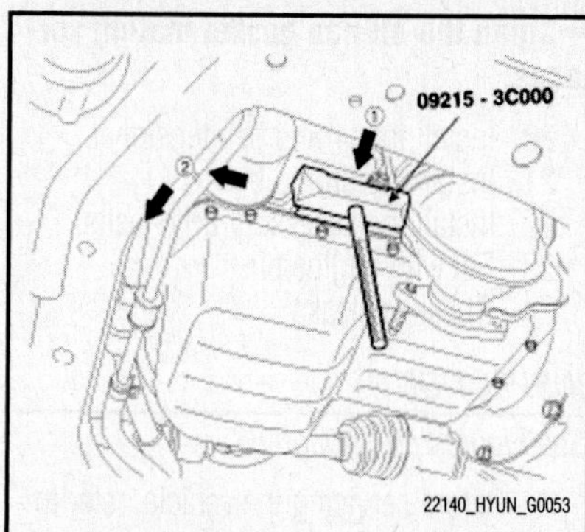

Fig. 74 Using the SST (09215-3C000) to remove the oil pan

b. After tapping the SST with a plastic hammer along the direction of arrow No. 2 around more than ⅔ of the edge of the oil pan, remove it from the ladder frame.

WARNING

Do not turn over the SST abruptly without tapping, or damage may occur to the SST or oil pan.

To install:

9. Using a razor blade and gasket scraper, carefully remove all the old packing material from the gasket surfaces.
10. Check that the mating surfaces are clean and dry before applying liquid gasket.
11. Apply liquid gasket as an even bead, centered between the edges of the mating surface. Use liquid gasket: TB1217H or equivalent. Apply a bead ⅛ inch (3mm) wide to the oil pan. To prevent leakage of oil, apply liquid gasket to the inner threads of the bolt holes.

➡Do not install the parts if 5 minutes or more have elapsed since applying the liquid gasket. Instead, reapply liquid gasket after removing the residue.

12. Install the oil pan with the bolts. Uniformly tighten the bolts in several passes to 84–108 inch lbs. (10–12 Nm).
13. Install the exhaust manifold and catalytic converter assembly.
14. Install the front muffler and tighten to 22–29 ft. lbs. (29–39 Nm).
15. Install the front muffler heat protector.
16. Connect the rear oxygen sensor connector.

➡After assembly, wait at least 30 minutes before filling the engine with oil.

17. Fill with engine oil.

OIL PUMP

REMOVAL & INSTALLATION

See Figures 75 through 79.

1. Before servicing the vehicle, refer to the Precautions Section.
2. Remove the oil pan, refer to Oil Pan Removal & Installation.
3. Remove the accessory drive belts. Refer to Accessory Drive Belt Removal & Installation.
4. Align the timing marks: Turn the crankshaft pulley, and align its groove with the timing mark "T" of the timing belt cover.
5. Remove the timing belt. Refer to Timing Belt Removal & Installation.
6. Remove the timing belt tensioner.
7. Remove the alternator. Refer to Alternator Removal & Installation in the Engine Electrical section.
8. Remove the air conditioner compressor tensioner bracket.
9. Remove the front case and oil pump:
   a. Remove the screw from the pump housing, then separate the housing and cover.
   b. Remove the inner rotor and outer rotor.

To install:

10. Place the inner and outer rotors into the front case with the marks facing the oil pump cover side.
11. Install the oil pump cover to the front case with the 7 screws and tighten to 52–61 inch lbs. (6–7 Nm).
12. Check that the oil pump turns freely.
13. Install the oil pump on the cylinder block:
   a. Place a new front case gasket on the cylinder block.
   b. Apply engine oil to the lip of the oil pump seal. Then, install the oil pump onto the crankshaft.
   c. When the pump is in place, clean any excess grease off the crankshaft and check that the oil seal lip is not distorted.

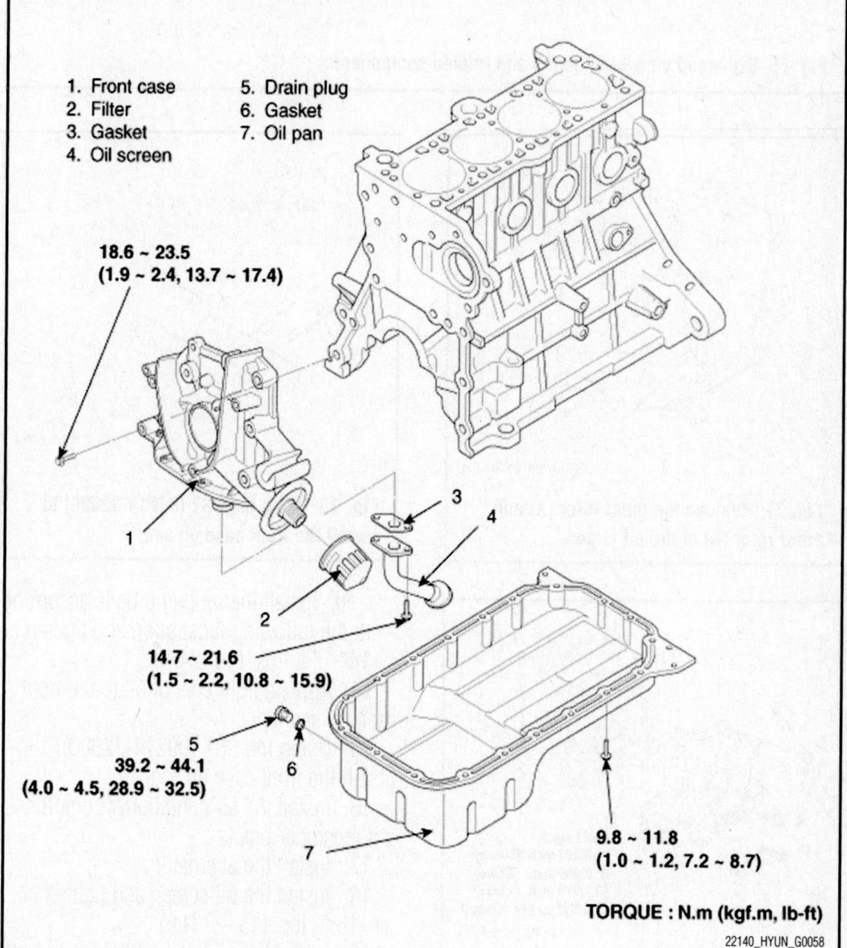

Fig. 75 Expanded view of lubrication system components

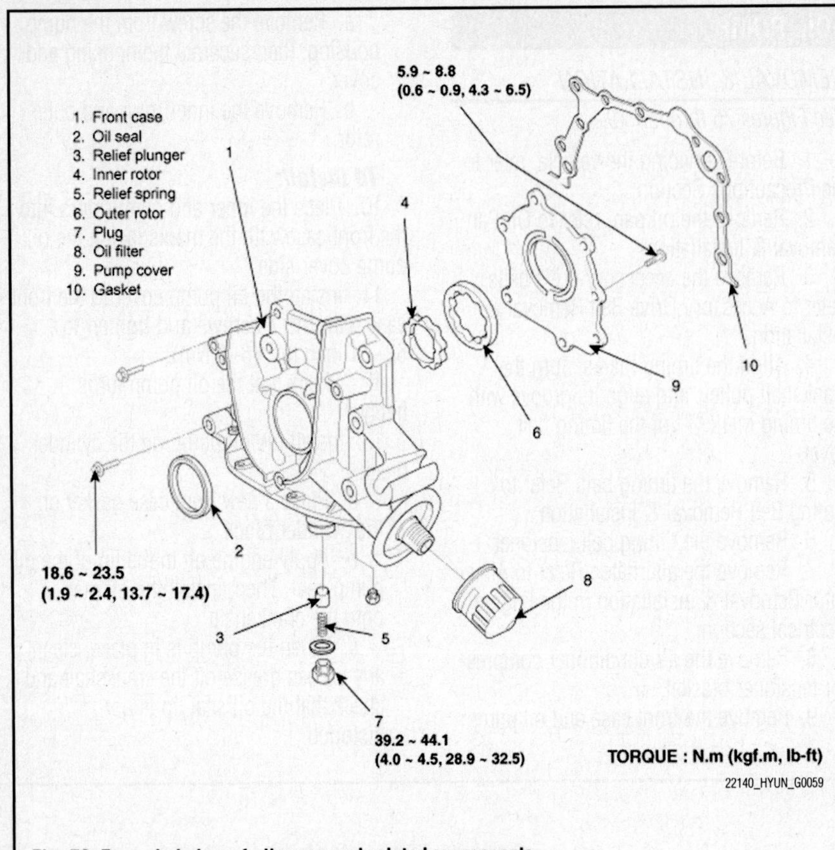

Fig. 76 Expanded view of oil pump and related components

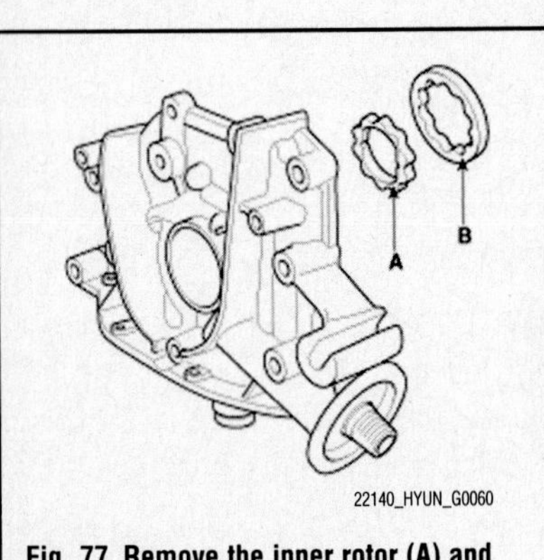

Fig. 77 Remove the inner rotor (A) and outer rotor (B) of the oil pump

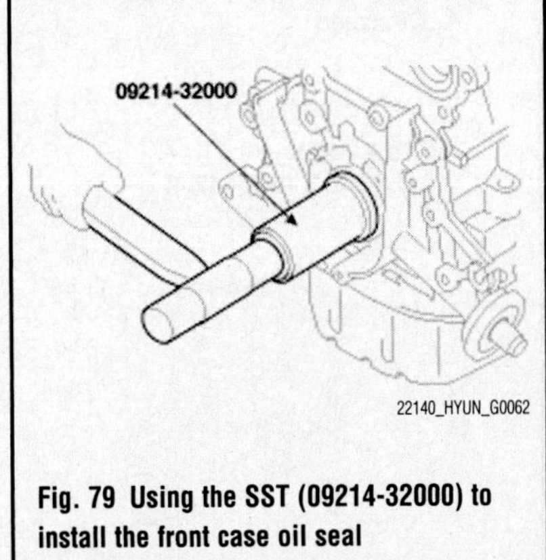

Fig. 79 Using the SST (09214-32000) to install the front case oil seal

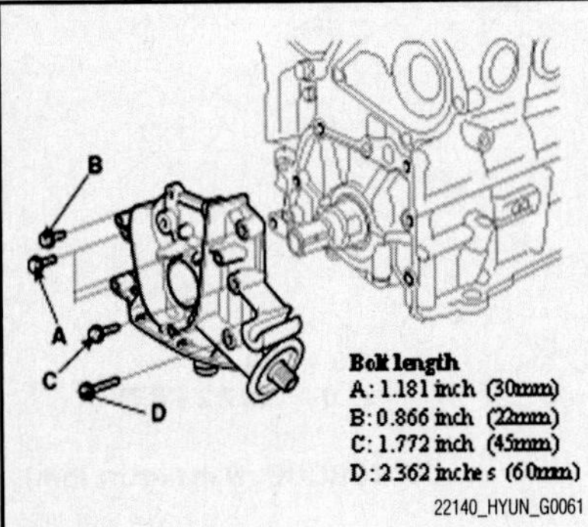

Fig. 78 Install the oil pump bolts in the correct location

d. Install the oil pump bolts according to the following illustration and tighten to 14–17 ft. lbs. (19–24 Nm).

14. Apply a light coat of oil to the front case oil seal lip.
15. Using the SST (09214-32000), install the front case oil seal.
16. Install the air conditioner compressor tensioner bracket.
17. Install the alternator.
18. Install the oil screen and tighten to 11–16 ft. lbs. (15–22 Nm).
19. Install the oil pan and tighten to 7–9 ft. lbs. (10–12 Nm).

➡Clean the oil pan gasket mating surfaces.

20. Install the timing belt tensioner.
21. Install the timing belt.
22. Install the accessory drive belts.
23. Fill with engine oil.
24. Check for leaks.

INSPECTION

See Figures 80 through 83.

1. Before servicing the vehicle, refer to the Precautions Section.
2. Remove the relief plunger: Remove the plug, spring, and relief plunger.
3. Inspect the relief plunger:
   a. Coat the plunger with engine oil.
   b. Check that it falls smoothly into the plunger hole by its own weight. If it does not, replace the relief plunger. If necessary, replace the front case.
4. Inspect the relief valve spring:
   a. Inspect for a distorted or broken relief valve spring.
   b. Standard value:
   - Free height: 1.8346 in. (46.6mm)
   - Load: 13.4 lbs. plus or minus 0.9 lbs./1.5787 in. (6.1 kg plus or minus 0.4kg/40.1mm)
5. Inspect the rotor side clearance:
   a. Using a feeler gauge and precision straight edge, measure the clearance between the rotors and precision straight edge.
   b. Side clearance:
   - Inner rotor: 0.0016–0.0033 in. (0.04–0.085mm)
   - Outer rotor: 0.0016–0.0035 in. (0.04–0.09mm)
   c. If the side clearance is greater than maximum, replace the rotors as a set. If necessary, replace the front case.
6. Inspect the rotor tip clearance:
   a. Using a feeler gauge, measure the tip clearance between the inner and outer rotor tips.

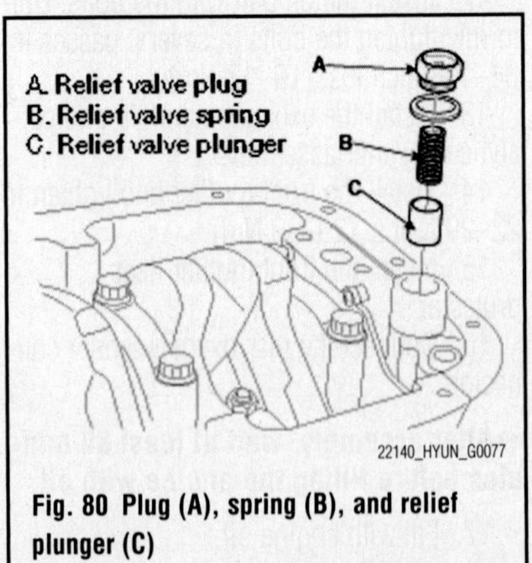

Fig. 80 Plug (A), spring (B), and relief plunger (C)

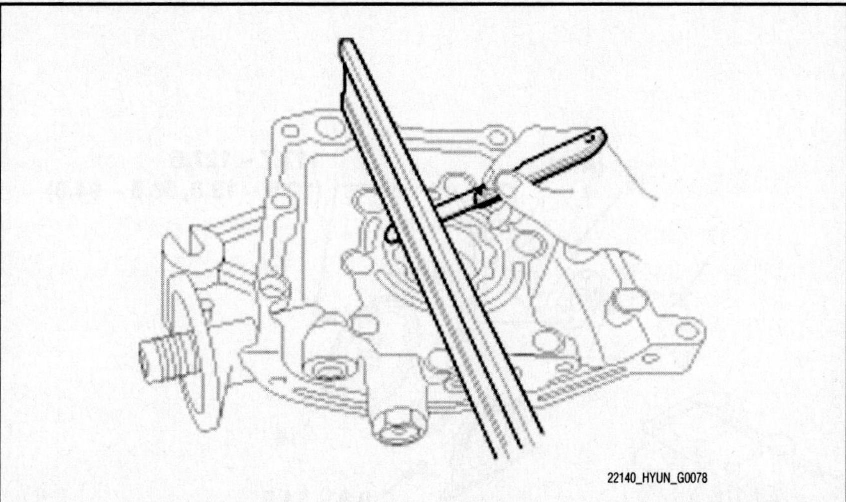

Fig. 81 Using a feeler gauge and precision straight edge to measure the side clearance of the oil pump

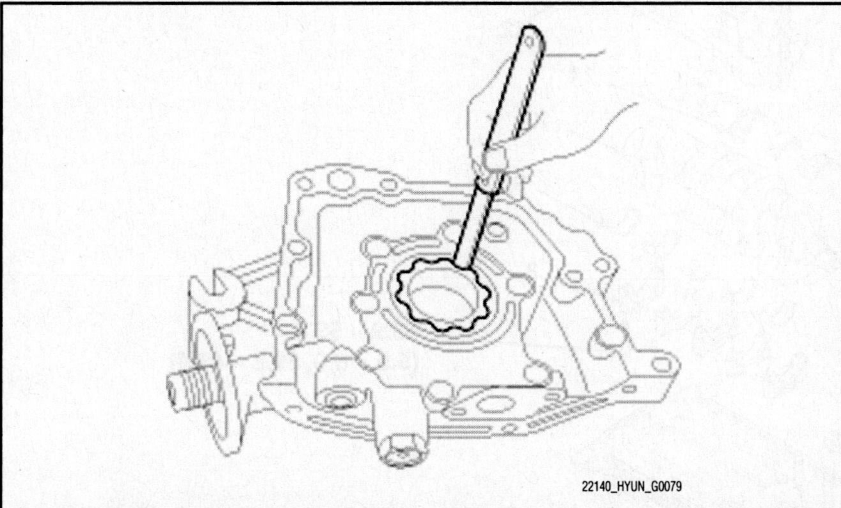

Fig. 82 Using a feeler gauge, measure the tip clearance between the inner and outer rotor tips of the oil pump

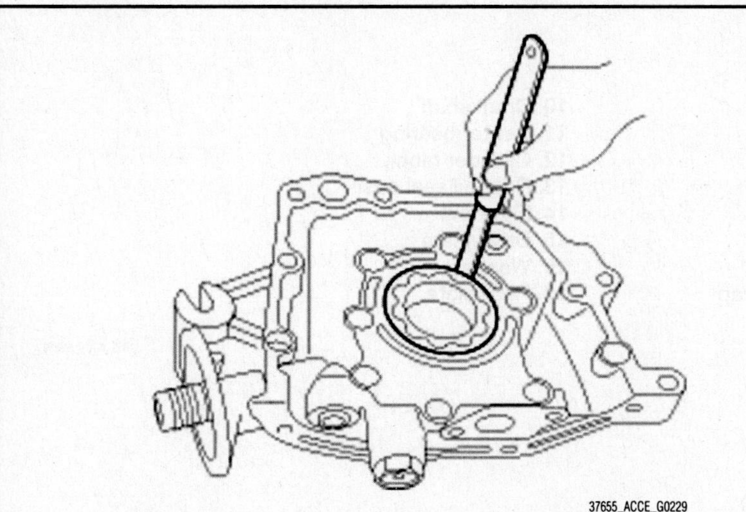

Fig. 83 Using a feeler gauge, measure the clearance between the outer rotor and body

b. Tip clearance:

- 0.0010–0.0027 in. (0.025–0.069mm)

c. If the tip clearance is greater than specified, replace the rotors as a set.

7. Inspect the rotor body clearance:

a. Using a feeler gauge, measure the clearance between the outer rotor and body.

b. Body clearance:

- 0.0024–0.0035 in. (0.060–0.090mm)

c. If the body clearance is greater than specified, replace the rotors as a set. If necessary, replace the front case.

8. Install the relief plunger and spring into the front case hole.

9. Install the plug. Tightening torque: 29–36 ft. lbs. (39–49 Nm).

PISTON & RING

POSITIONING

See Figure 84.

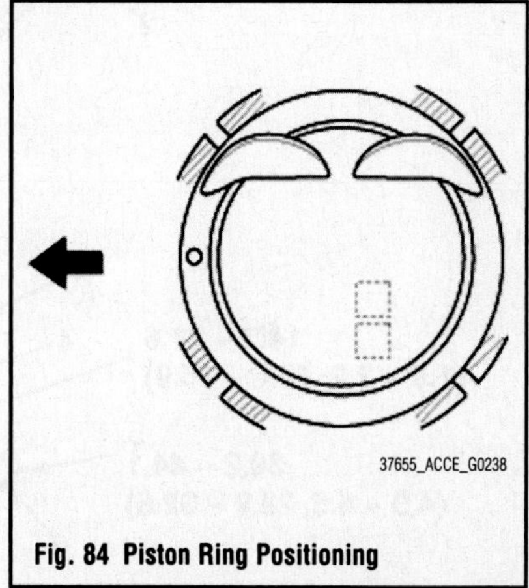

Fig. 84 Piston Ring Positioning

REAR MAIN SEAL

REMOVAL & INSTALLATION

See Figures 85 and 86.

1. Before servicing the vehicle, refer to the Precautions Section.
2. Remove the transaxle. Refer to Manual or Automatic Transaxle Removal & Installation in the Drive Train section.
3. Remove clutch cover assembly and then the clutch disc assembly. Refer to Clutch Driven Disc & Pressure Plate Removal & Installation in the Drive Train section.
4. Remove the flywheel.
5. Remove the oil seal case.
6. Remove the 5 bolts and the rear oil seal case.

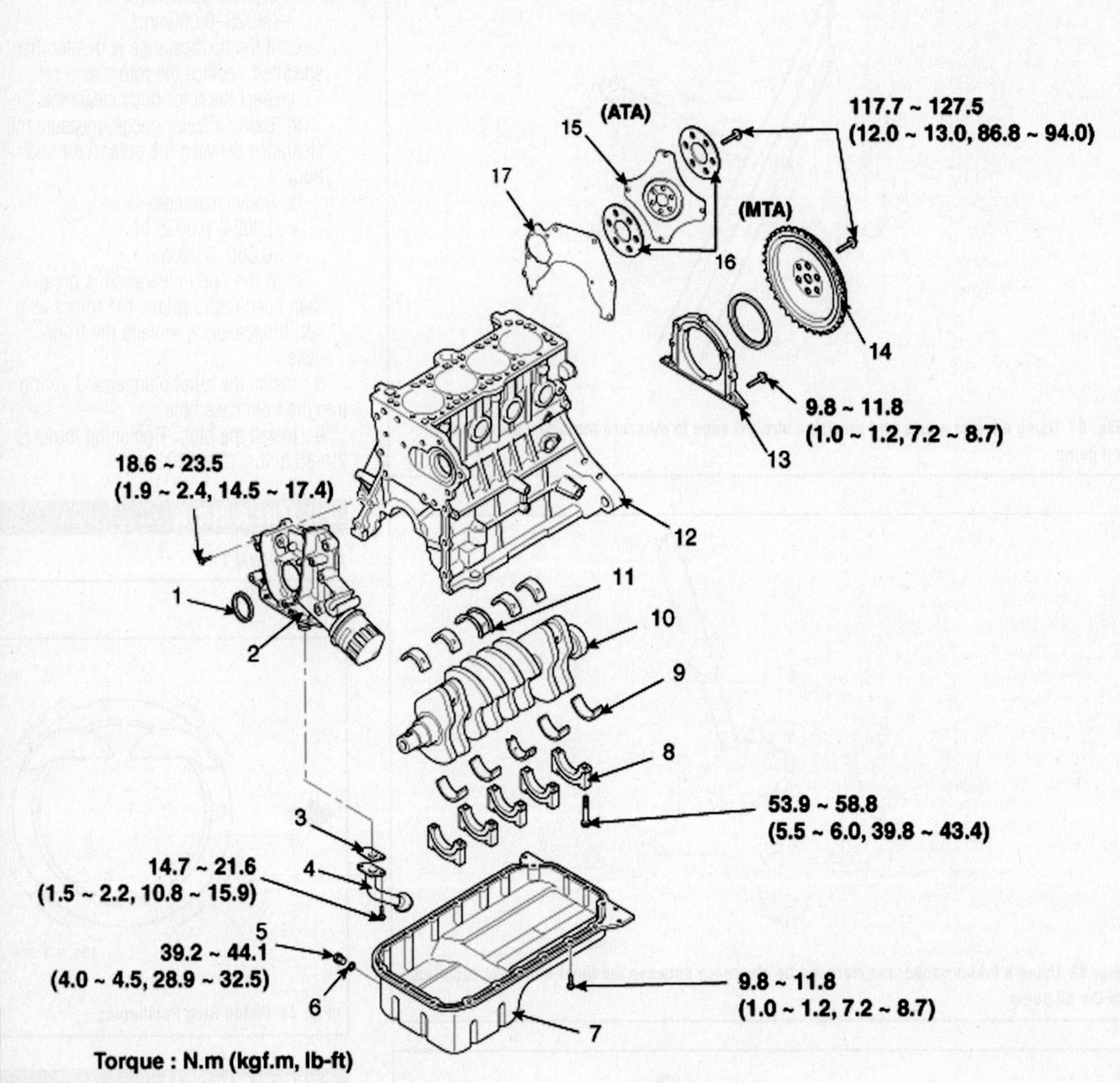

Fig. 85 Cylinder block exploded view, showing flywheel

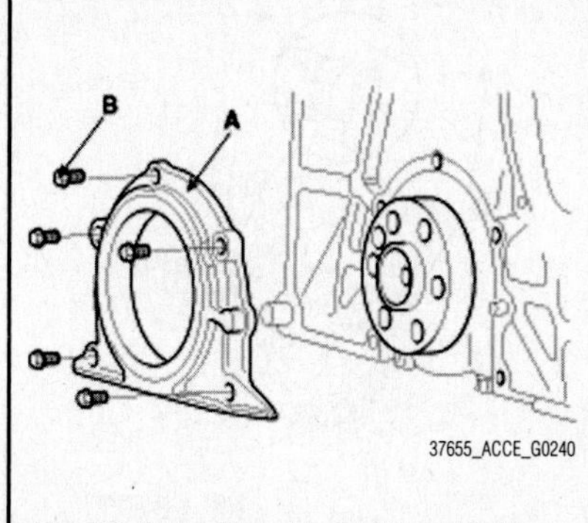

Fig. 86 Rear oil seal case (A) and bolts (B)

To install:

7. Install the rear oil seal case:

a. Using a razor blade and gasket scraper, remove all the old packing material from the gasket surfaces.

➡**Check that the mating surfaces are clean and dry before applying liquid gasket.**

b. Apply liquid gasket (LOCTITE 5900 or equivalent) as an even bead, centered between the edges of the mating surface.

c. Install the rear oil seal case with the 5 bolts and tighten the bolts to 86–104 inch lbs. (10–12 Nm).

8. Install the flywheel.

9. Install the clutch cover assembly and then the clutch disc assembly.

10. Install the transaxle.

TIMING BELT & SPROCKETS

REMOVAL & INSTALLATION

See Figures 87 through 107.

1. Before servicing the vehicle, refer to the Precautions Section.

2. Disconnect the battery cables and remove the battery.

3. Remove the engine cover.

4. Remove the right front wheel.

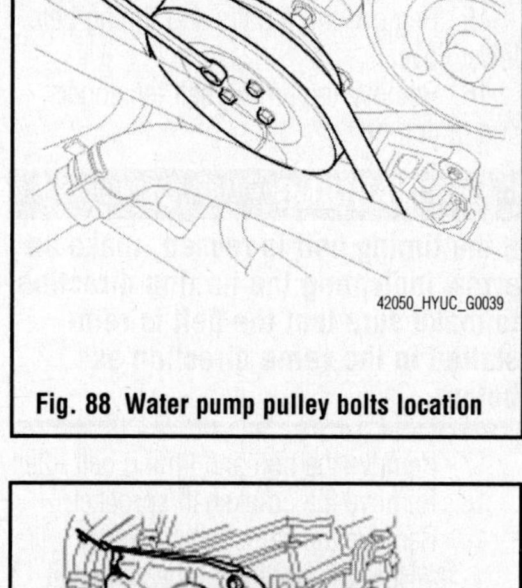

Fig. 88 Water pump pulley bolts location

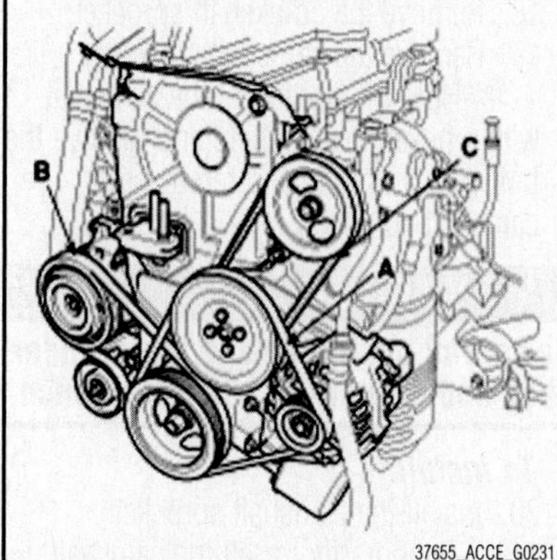

Fig. 89 Alternator drive belt (A). air conditioner belt (B) and power steering pump belt (C)

5. Remove the 2 bolts and the right-hand side cover.

6. Temporarily loosen the water pump pulley bolts.

7. Remove alternator drive belt. Refer to Accessory Drive Belt Removal & Installation.

8. Remove air conditioner compressor drive belt.

9. Remove power steering pump drive belt.

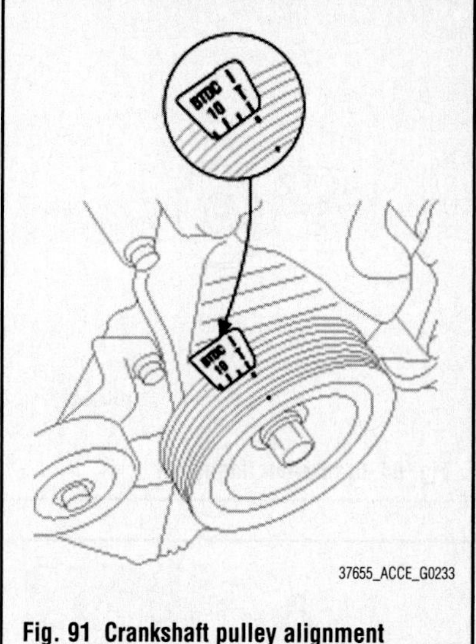

Fig. 91 Crankshaft pulley alignment

10. Remove the 4 bolts and the water pump pulley.

11. Remove the 4 bolts and the timing belt upper cover.

12. Turn the crankshaft pulley, and align its groove with timing mark "T" of the timing belt cover. Check that the timing mark of camshaft sprocket is aligned with the timing mark of cylinder head cover (No.1 cylinder compression TDC position).

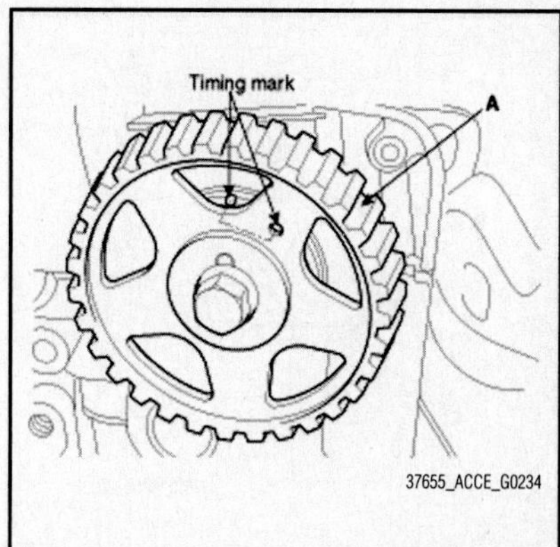

Fig. 92 Camshaft sprocket (A) alignment

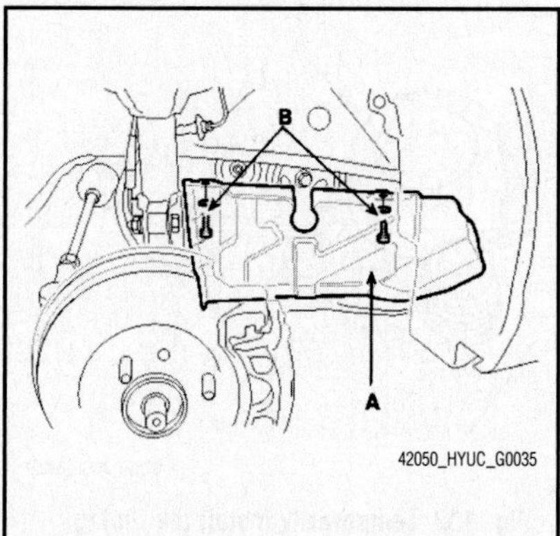

Fig. 87 Right-hand side cover location

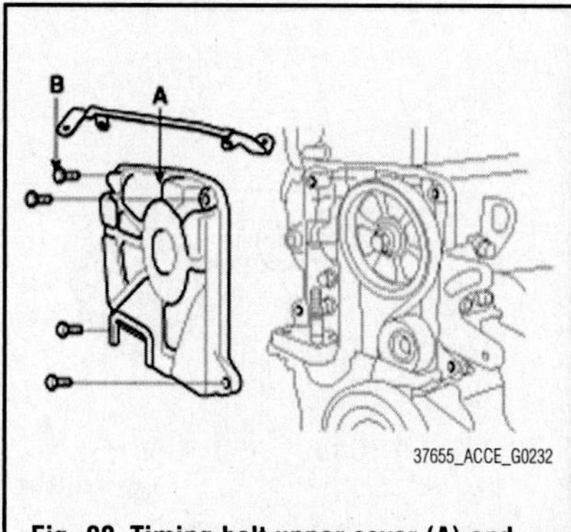

Fig. 90 Timing belt upper cover (A) and bolts (B)

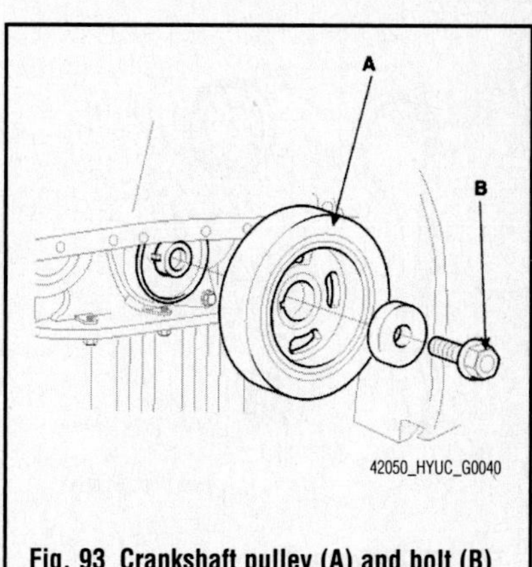

Fig. 93 Crankshaft pulley (A) and bolt (B)

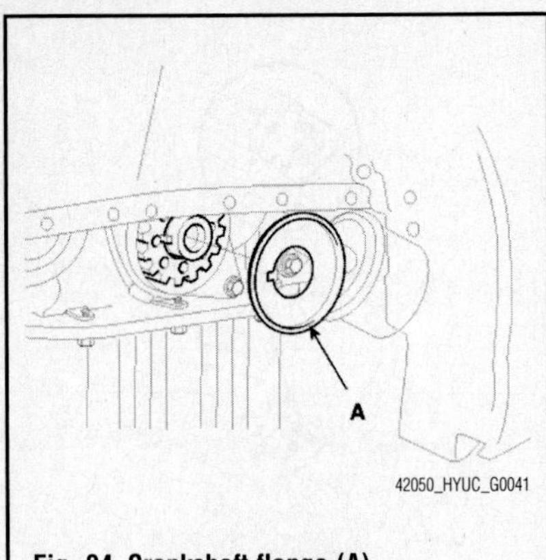

Fig. 94 Crankshaft flange (A)

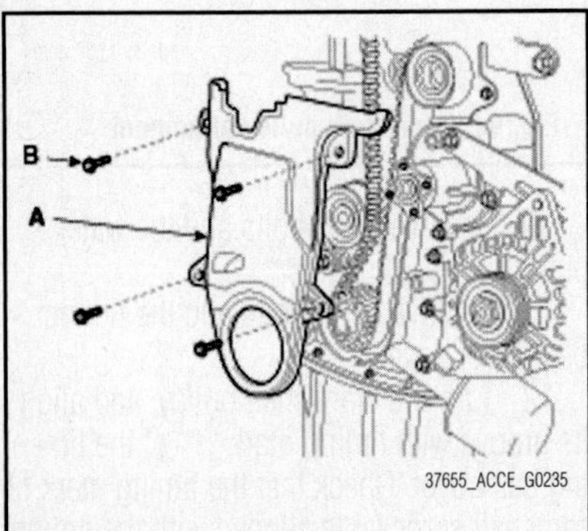

Fig. 95 Timing belt lower cover (A) and bolts (B)

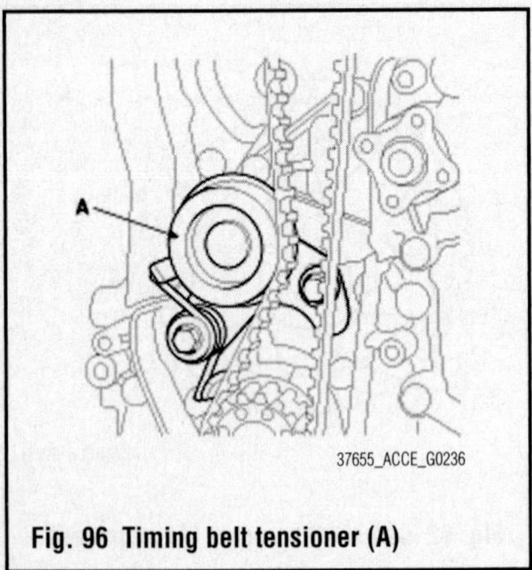

Fig. 96 Timing belt tensioner (A)

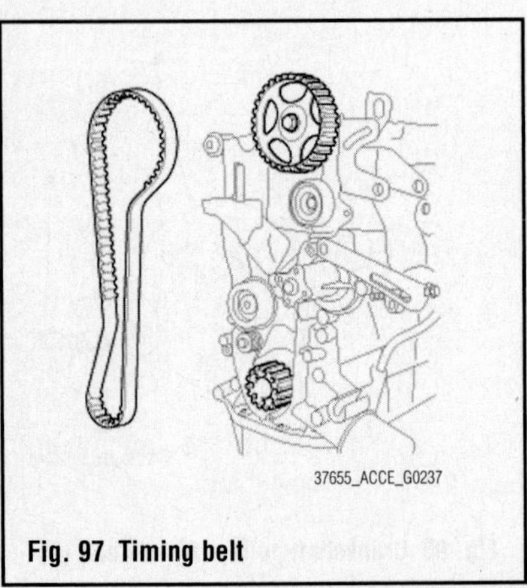

Fig. 97 Timing belt

13. Remove the crankshaft pulley bolt and crankshaft pulley.
14. Remove the crankshaft flange.
15. Remove the 4 bolts and timing belt lower cover.
16. Remove the timing belt tensioner and timing belt.

WARNING

If the timing belt is reused, make an arrow indicating the turning direction to make sure that the belt is reinstalled in the same direction as before.

17. Remove the bolt and timing belt idler.
18. Remove the crankshaft sprocket.
19. Remove the camshaft sprocket:
    a. Hold the portion of the camshaft with a hexagonal wrench, and remove the bolt with a wrench and remove the camshaft sprocket.

WARNING

Be careful not to damage the cylinder head and valve lifter with the wrench.

To install:

20. Install the camshaft sprocket:
    a. Temporarily install the camshaft sprocket bolt.

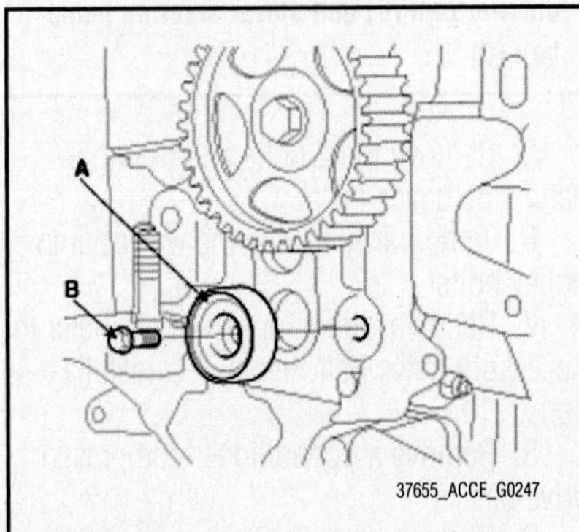

Fig. 98 Remove the bolt (B) and timing belt idler (A)

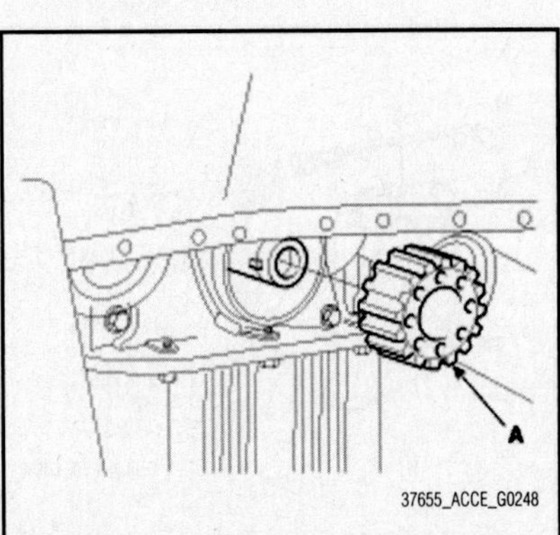

Fig. 99 Remove the crankshaft sprocket (A)

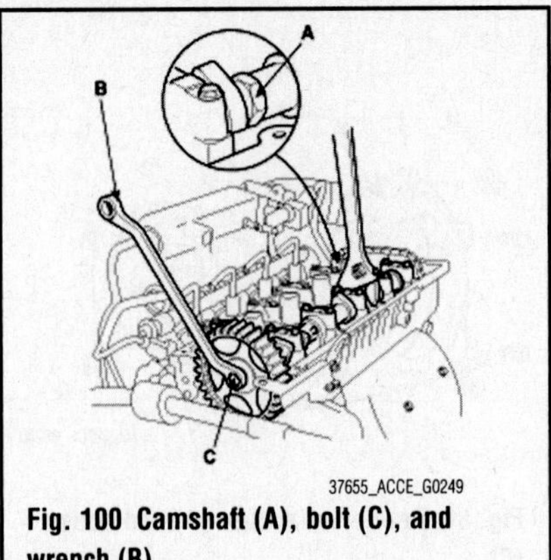

Fig. 100 Camshaft (A), bolt (C), and wrench (B)

    b. Hold the portion of the camshaft with a hexagonal wrench, and tighten the bolt with a wrench to 58–72 ft. lbs. (79–98 Nm).
21. Install the cylinder head cover.
22. Install the crankshaft sprocket.
23. Align the timing marks of the camshaft sprocket and crankshaft sprocket with the No.1 piston placed at top dead center and its compression stroke.
24. Install the idler pulley and tighten the bolt to 31–40 ft. lbs. (42–54 Nm).
25. Temporarily install the timing belt tensioner.
26. Install the belt so as not give slack at

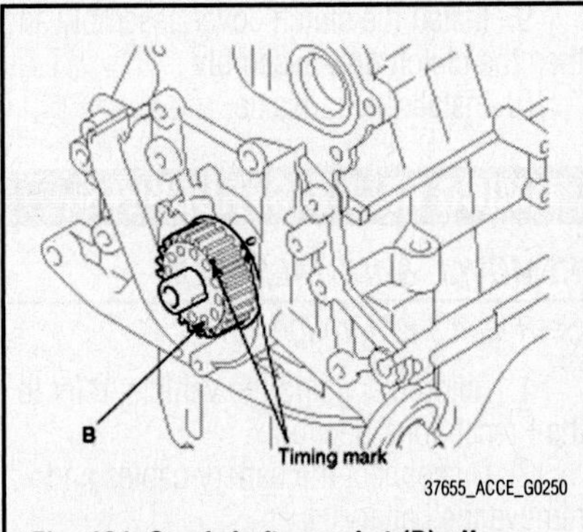

Fig. 101 Crankshaft sprocket (B) alignment

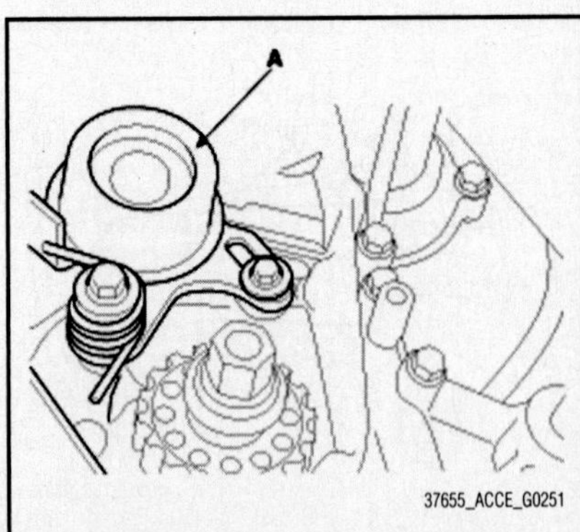

Fig. 102 Temporarily install the timing belt tensioner (A)

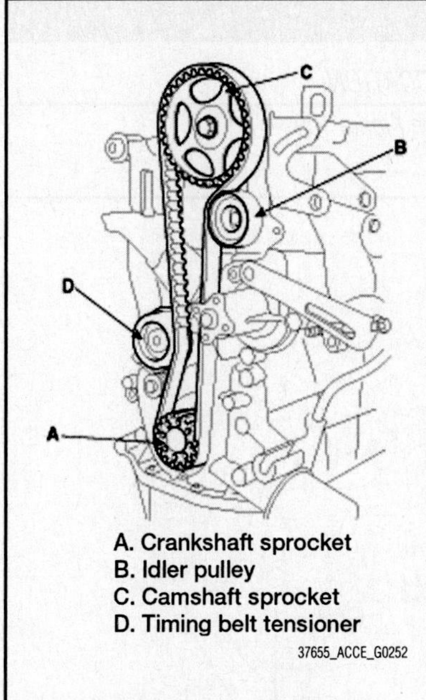

Fig. 103 Timing belt installation

each center of the shaft. Use the following order when installing timing belt:

- Crankshaft sprocket
- Idler pulley
- Camshaft sprocket
- Timing belt tensioner

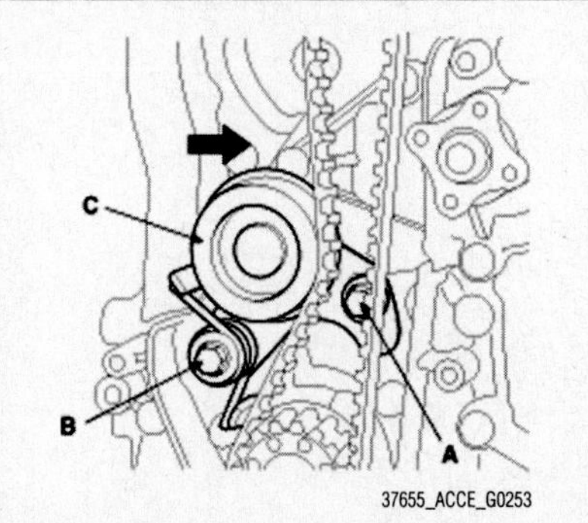

Fig. 104 Tensioner (C) and mounting bolts (A, B)

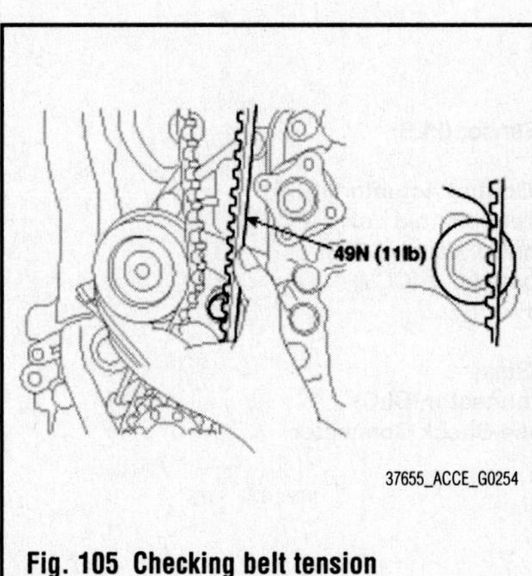

Fig. 105 Checking belt tension

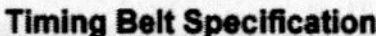

Timing Belt Specifications

Item	Specifications
S (mm)	249.6
M (gf/cm²)	0.4543
W (mm)	22
f (Hz)	70.4 ~ 87.9
T (kgf)	16 ± 3.5

37655_ACCE_G0255

Fig. 106 Timing Belt Specifications

27. Adjust the timing belt tension:
   a. Loosen the tensioner pulley mounting bolt and apply tension to the timing belt.
   b. After checking the alignment between each sprocket and each timing belt tooth, tighten the mounting bolts one by one to 15–20 ft. lbs. (20–27 Nm).
   c. Recheck the belt tension. Verify that when the tensioner and the tension side of the timing belt are pushed in horizontally with a moderate force [approx. 11 lbs. (49 N)], the timing belt cog end is approx. ½ of the tensioner mounting bolt head radius (across flats) away from the bolt head center.
   d. Timing belt tension measuring procedure (by a sonic tension gauge): Rotate crankshaft in clockwise direction to set 1st piston on top dead center (TDC) and rotate crankshaft in counterclockwise to 90° then measure the belt tension in the middle of tension side span (in arrow direction of above illustration) by free vibration method.

✲✲ WARNING

Avoid rotating the crankshaft in a counter clockwise direction. Engine damage could occur.

   e. Conversion equation of frequency into tension:
   - T = (4 / 9.8) x S_ x M x W x f_ / 100000000
   - S : Measured belt span (mm)
   - M : Unit weight of belt (gf/cm_)
   - W : Belt width (mm)
   - f : Transverse natural frequency of belt (Hz)

28. Turn the crankshaft two turns in the operating direction (clockwise) and realign the crankshaft sprocket and camshaft sprocket timing mark.
29. Install the timing belt lower cover and tighten the 5 bolts to 70–86 inch lbs. (8–10 Nm).
30. Install the flange and crankshaft pulley, and then tighten crankshaft pulley bolt. Make sure that crankshaft sprocket pin fits the small hole in the pulley. Tighten to 101–109 ft. lbs. (137–147 Nm).
31. Install the timing belt upper cover and tighten the 4 bolts to 70–86 inch lbs. (8–10 Nm).
32. Install the water pump pulley and bolts.
33. Install the power steering pump drive belt.
34. Install the air conditioner compressor drive belt.
35. Install the alternator drive belt.
36. Install the right-hand side cover and bolts.
37. Install the right front wheel.
38. Install the engine cover and tighten the bolts to 70–104 ft. lbs. (8–12 Nm).

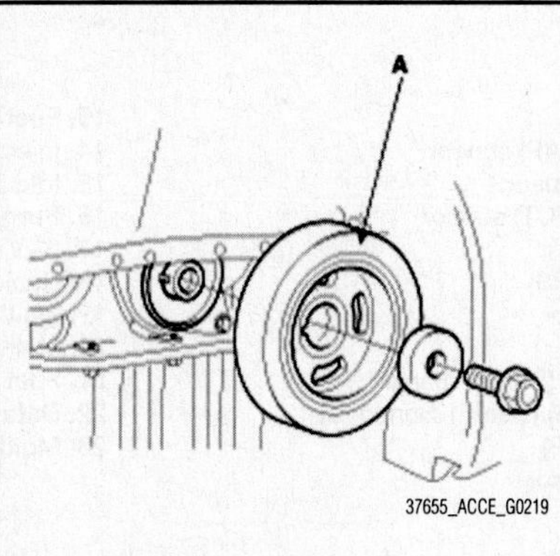

Fig. 107 Crankshaft pulley (A)

VALVE LASH

ADJUSTMENT

All engines use hydraulic valve lash adjusters. Valve lash adjustments are not necessary or possible on these engines.

ENGINE PERFORMANCE & EMISSION CONTROLS

COMPONENT LOCATIONS

See Figure 108.

CAMSHAFT POSITION (CMP) SENSOR

LOCATION

See Figure 109.

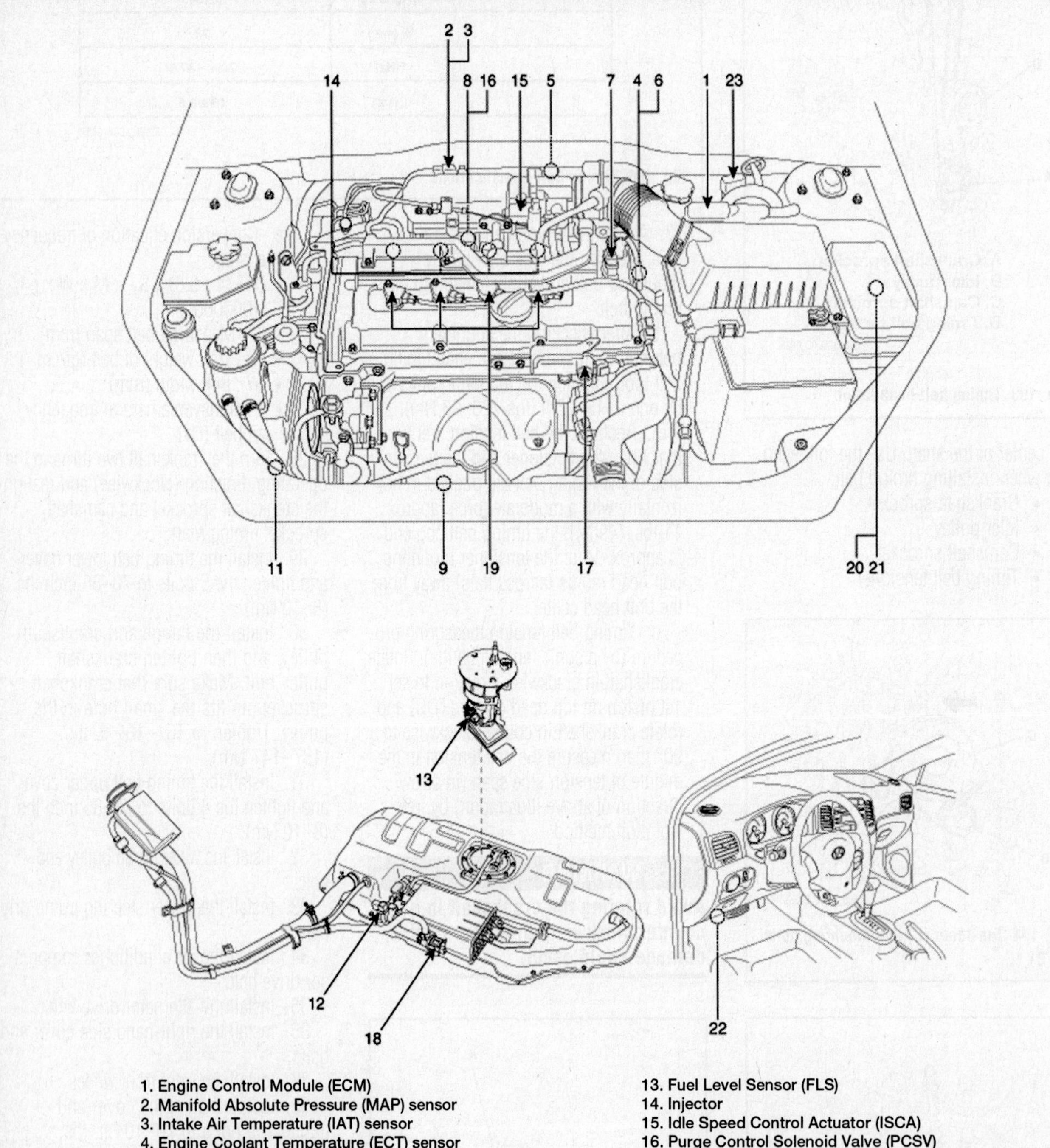

Fig. 108 Underhood sensor locations

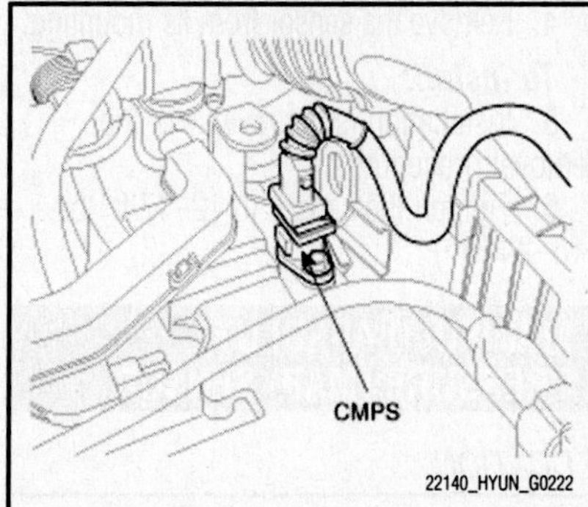

Fig. 109 Camshaft Position (CMP) sensor location—1.6L engine

Refer to the accompanying illustrations for Camshaft Position (CMP) sensor location.

REMOVAL & INSTALLATION

1. Before servicing the vehicle, refer to the Precautions Section.
2. Disconnect the negative battery cable.
3. Disconnect the connector from the CMP sensor.
4. Remove the bolt that retains the CMP sensor.
5. Remove the CMP sensor.

To install:

6. Installation is the reverse of the removal procedure.
7. Tighten the bolt that retains the CMP sensor to: 88–106 inch lbs. (10–12 Nm).

CRANKSHAFT POSITION (CKP) SENSOR

LOCATION

See Figure 110.

Refer to the accompanying illustrations for Crankshaft Position (CKP) sensor location.

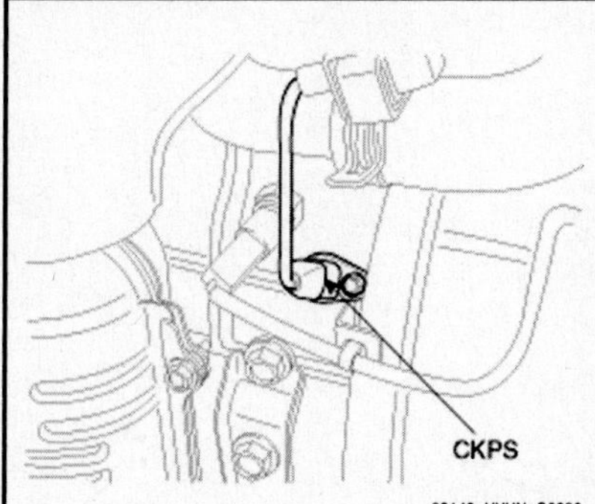

Fig. 110 Crankshaft Position (CKP) sensor location

REMOVAL & INSTALLATION

1. Before servicing the vehicle, refer to the Precautions Section.
2. Disconnect the negative battery cable.
3. Disconnect the connector from the sensor.
4. Remove the bolt that retains the sensor in place.
5. Remove the sensor from its mounting.

To install:

6. Installation is the reverse of the removal procedure.
7. Tighten the sensor retaining bolt to: 44–53 inch lbs. (5–6 Nm).

ELECTRONIC CONTROL MODULE (ECM)

LOCATION

See Figure 111.

➡For the Accent (automatic transmission), refer to Powertrain Control Module (PCM). Some manufacturers refer to the Electronic Control Module (ECM) as the Engine Control Module (ECM).

Refer to the accompanying illustrations for Electronic Control Module (ECM) location.

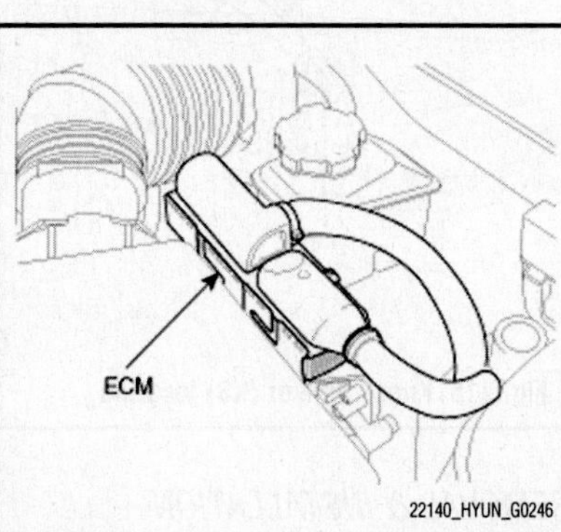

Fig. 111 Electronic Control Module (ECM) location

REMOVAL & INSTALLATION

See Figure 112.

1. Before servicing the vehicle, refer to the Precautions Section.

➡Replacing the ECM requires an appropriate scan tool to program required information.

2. Turn ignition switch off.
3. Disconnect the negative battery cable from the battery.
4. Disconnect the ECM connector(s) (A).
5. Remove the ECM mounting bolts (B) and remove the ECM from the air cleaner assembly.

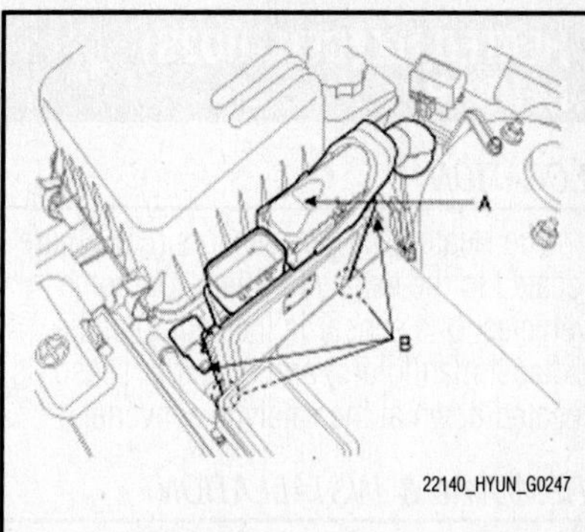

Fig. 112 Disconnect the ECM connector(s) (A) and remove the ECM mounting bolts (B)

To install:

6. Installation is the reverse of the removal.
7. Tighten the ECM mounting bolts to: 86–104 inch lbs. (10–12 Nm).

ENGINE COOLANT TEMPERATURE (ECT) SENSOR

LOCATION

See Figure 113.

The Engine Coolant Temperature (ECT) sensor is located in the engine coolant passage of the cylinder head for detecting the engine coolant temperature.

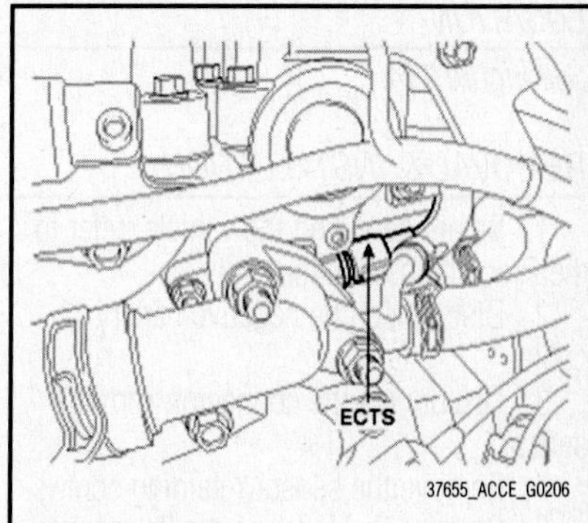

Fig. 113 Engine coolant temperature sensor location

REMOVAL & INSTALLATION

1. Before servicing the vehicle, refer to the Precautions Section.
2. Drain the coolant to a level below the bottom of the sensor.
3. Disconnect the ground cable from the battery and then remove the sensor connector.
4. Remove the coolant temperature sensor.

To install:

5. Reverse the removal procedure.
6. Tighten the sensor to 15–29 ft. lbs. (20–39 Nm).

HEATED OXYGEN (HO2S) SENSOR

LOCATION

The Heated Oxygen Sensors (HO2S) are located in the exhaust system. On some vehicles, one sensor is located up at the exhaust manifold(s) and the other sensor is located down at the catalytic converter.

REMOVAL & INSTALLATION

1. Before servicing the vehicle, refer to the Precautions Section.
2. Disconnect the electrical connector from the sensor.
3. Remove the oxygen sensor.

To install:

4. Installation is the reverse of the removal procedure.

➡Apply anti-seize compound to the threaded portion of the sensor, prior to installation. Never apply anti-seize compound to the protector of the sensor.

5. Tighten sensor to 36–44 ft. lbs. (49–59 Nm)

INTAKE AIR TEMPERATURE (IAT) SENSOR

LOCATION

See Figure 114.

REMOVAL & INSTALLATION

1. Before servicing the vehicle, refer to the Precautions Section.
2. Disconnect the negative battery cable.
3. Disconnect the connector from the sensor.
4. Remove the sensor retaining screws.
5. Remove the sensor from its mounting.

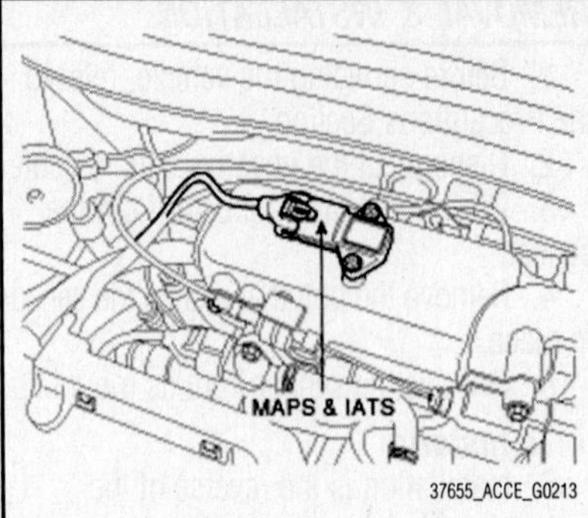

Fig. 114 Intake Air Temperature (IAT) Sensor location

To install:

6. Installation is the reverse of the removal procedure.

KNOCK SENSOR (KS)

LOCATION

See Figure 115.

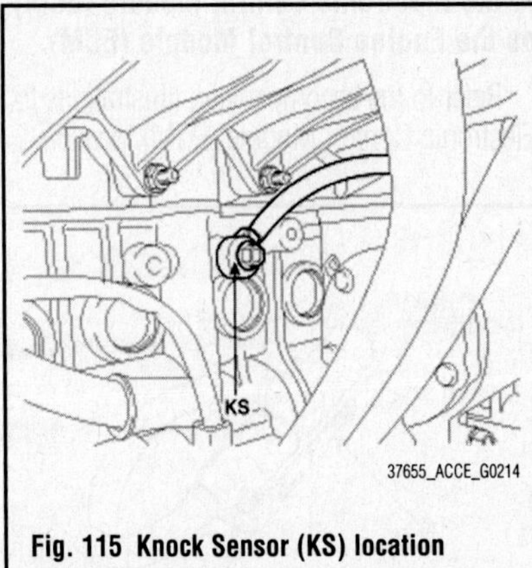

Fig. 115 Knock Sensor (KS) location

REMOVAL & INSTALLATION

1. Before servicing the vehicle, refer to the Precautions Section.
2. Disconnect the negative battery cable.
3. Disconnect the sensor connector.
4. Remove the sensor from its mounting.

To install:

5. Installation is the reverse of the removal procedure.
6. Tighten the sensor to 12–17 ft. lbs. (16–24 Nm).

MANIFOLD ABSOLUTE PRESSURE (MAP) SENSOR

LOCATION

See Figure 116.

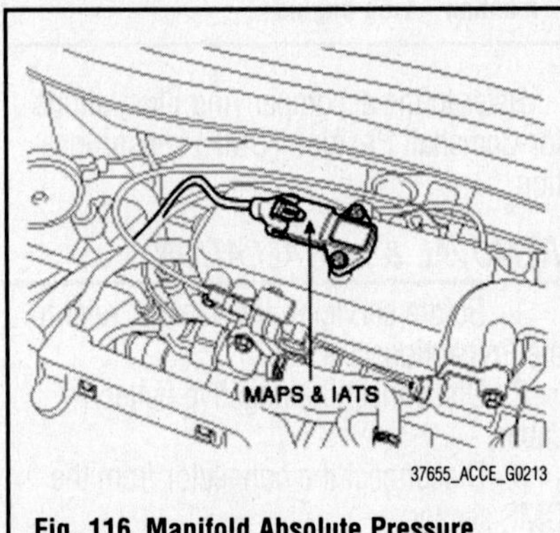

Fig. 116 Manifold Absolute Pressure (MAP) Sensor location

REMOVAL & INSTALLATION

1. Before servicing the vehicle, refer to the Precautions Section.
2. Disconnect the negative battery cable.
3. Disconnect the connector from the sensor.
4. Remove the sensor retaining screws.
5. Remove the sensor from its mounting.

To install:

6. Installation is the reverse of the removal procedure.
7. Tighten the MAP sensor installation bolt to: 62–80 inch lbs. (10–12 Nm).

FUEL SYSTEM SERVICE PRECAUTIONS

Safety is the most important factor when performing, not only fuel system maintenance, but any type of maintenance. Failure to conduct maintenance and repairs in a safe manner may result in serious personal injury or death. Maintenance and testing of the vehicle's fuel system components can be accomplished safely and effectively by adhering to the following rules and guidelines:

- To avoid the possibility of fire and personal injury, always disconnect the negative battery cable unless the repair or test procedure requires that battery voltage be applied.
- Always relieve the fuel system pressure prior to disconnecting any fuel system component (injector, fuel rail, pressure regulator, etc.), fitting, or fuel line connection. Exercise extreme caution whenever relieving fuel system pressure to avoid exposing skin, face, and eyes to fuel spray. Please be advised that fuel under pressure may penetrate the skin or any part of the body that it contacts.
- Always place a shop towel or cloth around the fitting or connection prior to loosening to absorb any excess fuel due to spillage. Ensure that all fuel spillage (should it occur) is quickly removed from engine surfaces. Ensure that all fuel soaked cloths or towels are deposited into a suitable waste container.
- Always keep a dry chemical (Class B) fire extinguisher near the work area.
- Do not allow fuel spray or fuel vapors to come into contact with a spark or an open flame.
- Always use a back-up wrench when loosening and tightening fuel line connection fittings. This will prevent unnecessary stress and torsion to fuel line piping.
- Always replace worn fuel fitting O-rings with new. Do not substitute fuel hose or equivalent where fuel pipe is installed.

RELIEVING FUEL SYSTEM PRESSURE

1. Before servicing the vehicle, refer to the Precautions Section.

CAUTION

Be sure to reduce the fuel pressure before disconnecting the fuel feed hose, otherwise fuel will spill out.

2. Disconnect the fuel pump connector.
3. Start the engine and wait until the fuel in the fuel line is exhausted.
4. After the engine stalls, turn the ignition switch to the OFF position and disconnect the negative battery cable.

FUEL FILTER

REMOVAL & INSTALLATION

The fuel delivery system integrates the fuel filter with the fuel pump. Refer to Fuel Pump Removal & Installation.

FUEL PUMP

REMOVAL & INSTALLATION

See Figures 117 through 120.

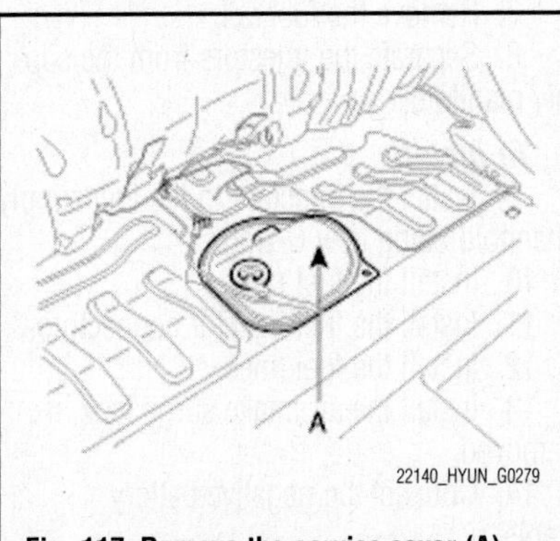

Fig. 117 Remove the service cover (A)

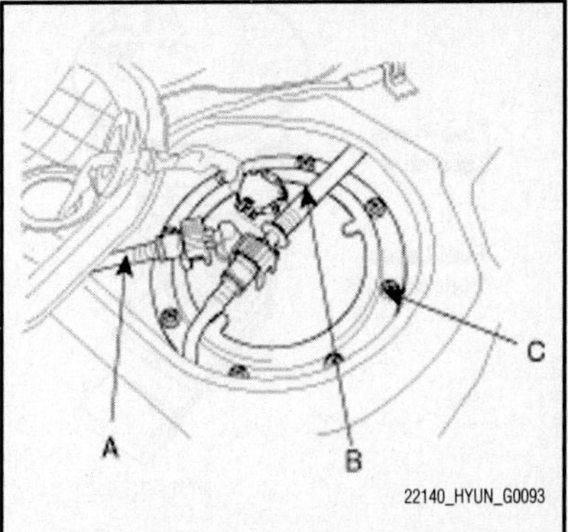

Fig. 119 Disconnect the fuel feed line (A) and canister hoses (B). Unscrew the fuel pump mounting bolts (C)

1. Before servicing the vehicle, refer to the Precautions Section.
2. Remove the rear seat cushion
3. Remove the service cover.
4. Relieve the fuel system pressure. Refer to Relieving Fuel System Pressure.
5. Disconnect the fuel feed line and canister hoses.
6. Unscrew the fuel pump mounting bolts and remove the fuel pump assembly.

To install:

7. Install the fuel pump assembly.
8. Install the fuel pump mounting bolts and tighten to 17–26 inch lbs. (2–3 Nm).

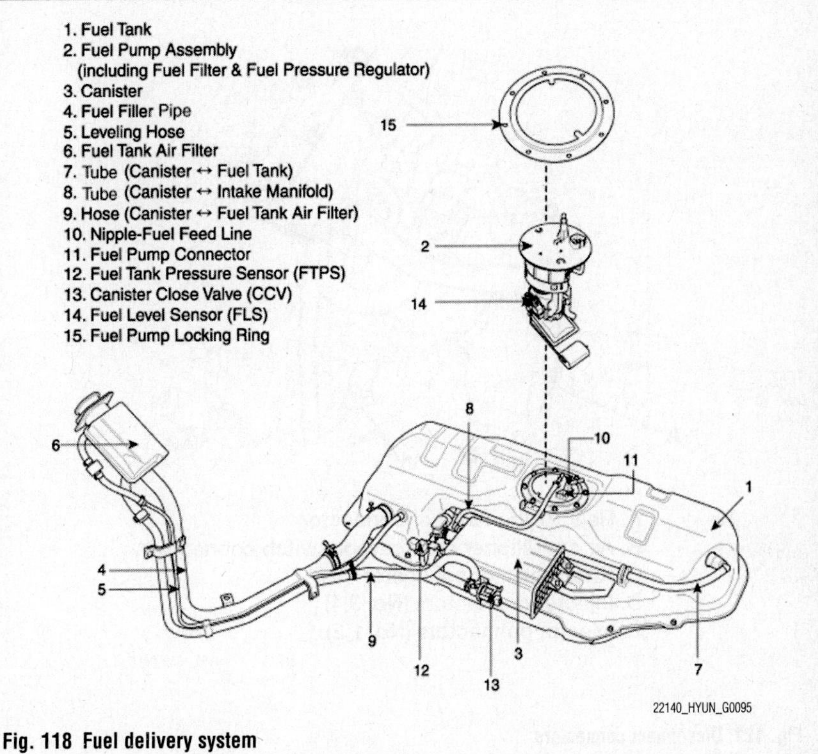

Fig. 118 Fuel delivery system

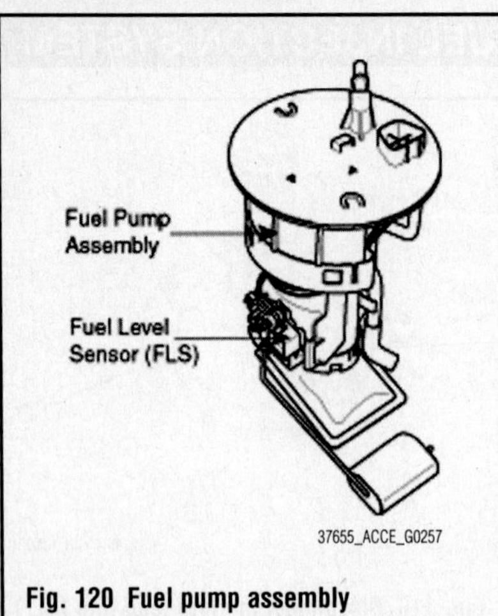

Fig. 120 Fuel pump assembly

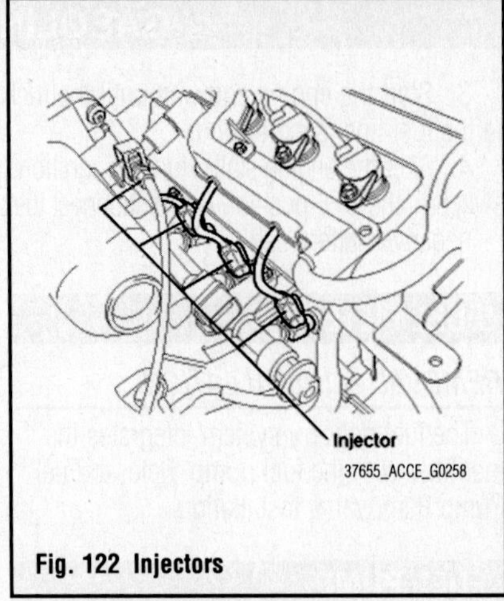

Fig. 122 Injectors

9. Connect the fuel feed line and canister hoses.

FUEL RAIL & INJECTORS

REMOVAL & INSTALLATION

See Figures 121 and 122.

1. Before servicing the vehicle, refer to the Precautions Section.
2. Disconnect the negative battery cable.
3. Relieve the fuel system pressure. Refer to Relieving Fuel System Pressure.
4. Remove the air intake surge tank, if necessary.
5. Remove the fuel lines.
6. Disconnect the fuel injector connectors.
7. Remove the fuel rail.
8. Separate the injectors from the supply manifold.

To install:

9. Install the injectors to the fuel supply manifold using new O-rings.
10. Install the fuel rail.
11. Install the fuel injector connectors.
12. Install the fuel lines.
13. Install the air intake surge tank, if removed.
14. Connect the negative battery cable.
15. Start the engine and check for leaks.

FUEL TANK

REMOVAL & INSTALLATION

See Figures 123 through 125.

1. Before servicing the vehicle, refer to the Precautions Section.
2. Remove the rear seat cushion.
3. Remove the service cover under the back seat.
4. Disconnect the fuel pump connector.
5. Start the engine and wait until the fuel in the fuel line is exhausted.
6. After the engine stalls, turn the ignition switch to the OFF position.
7. Disconnect the negative battery connection.
8. Disconnect the fuel feed line and canister hose.
9. Raise and safely support the vehicle.
10. Remove the center muffler.
11. Support the fuel tank with a jack.
12. Unscrew the brake hose mounting bolts on the left-hand and right-hand sides.
13. Disconnect the fuel filler pipe, the leveling hose, and canister hose.

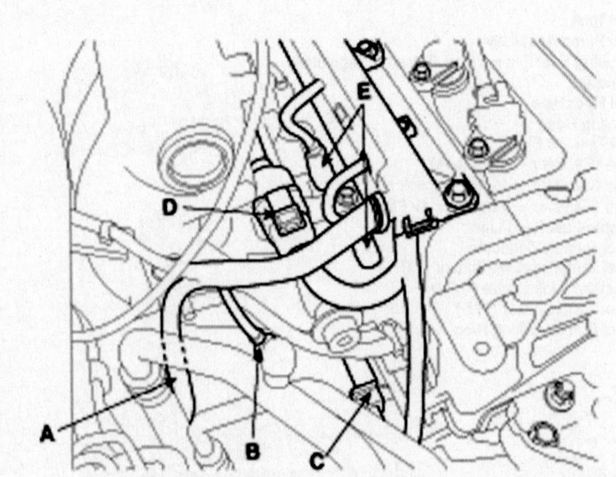

Fig. 121 Disconnect connectors

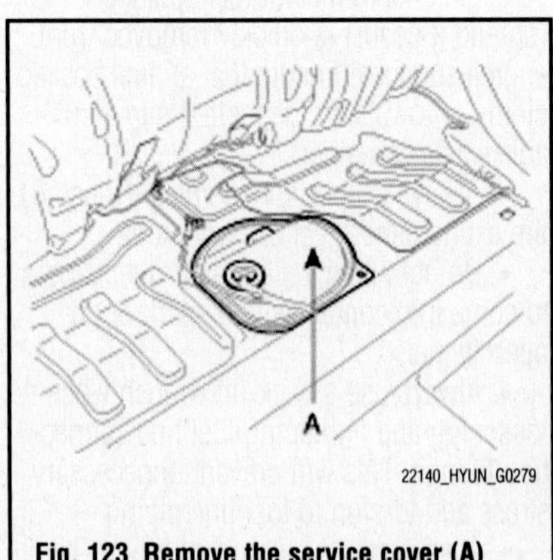

Fig. 123 Remove the service cover (A)

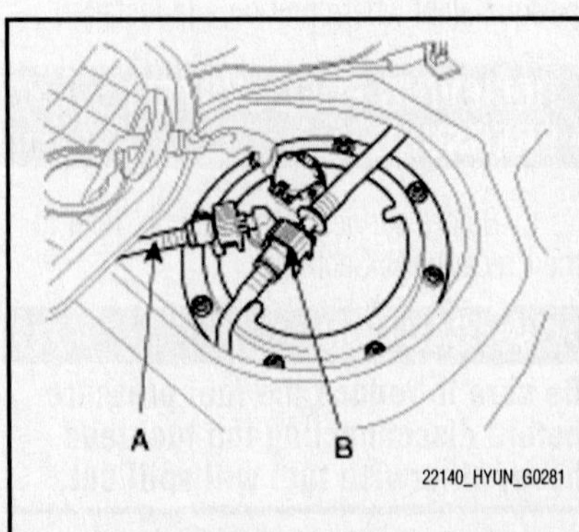

Fig. 124 Disconnect the fuel feed line (A) and canister hose (B)

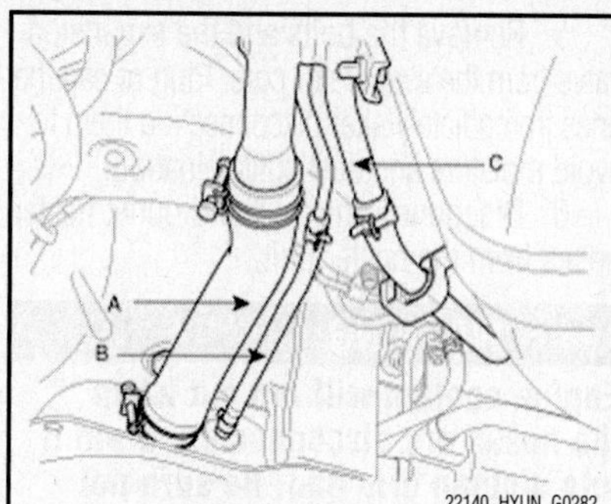

Fig. 125 Disconnect the fuel filler pipe (A), the leveling hose (B), and canister hose (C)

14. Unscrew the fuel tank mounting bolts and nuts, and then remove the fuel tank.

To install:

15. Installation is the reverse of the removal.
16. Tighten the fuel tank mounting bolts to: 29–40 ft. lbs. (39–54 Nm).

IDLE SPEED

ADJUSTMENT

Idle speed is maintained by the Powertrain Control Module (PCM). No adjustment is necessary or possible.

THROTTLE BODY

REMOVAL & INSTALLATION

See Figure 126.

1. Before servicing the vehicle, refer to the Precautions Section.
2. Turn the ignition **OFF**.
3. Remove the engine cover.
4. Remove the throttle body electrical connector.
5. Remove the throttle body bolts.
6. Remove the throttle body and gasket.

To install:

7. Clean the throttle body gasket mating surfaces.

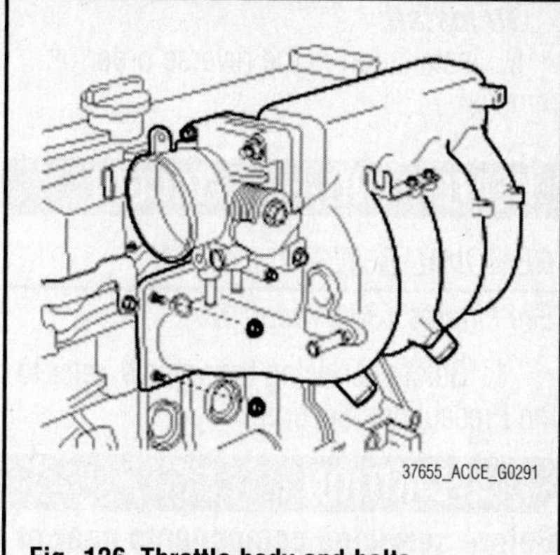

Fig. 126 Throttle body and bolts

8. Install the throttle body and NEW gasket.
9. Install the throttle body bolts and tighten to 21 ft. lbs. (28 Nm).
10. Install the throttle body electrical connector.
11. Install the engine cover.

HEATING & AIR CONDITIONING SYSTEM

BLOWER MOTOR

REMOVAL & INSTALLATION

See Figures 127 through 129.

1. Before servicing the vehicle, refer to the Precautions Section.

✲✲ CAUTION

Before servicing components near or affected by the SRS (air bag) system, read and observe all SRS Service Precautions. Refer to Supplemental Restraint System (SRS), in the Chassis Electrical section. Failure to observe all precautions may result in accidental airbag deployment, personal injury, or death. Refer to Airbag Removal & Installation.

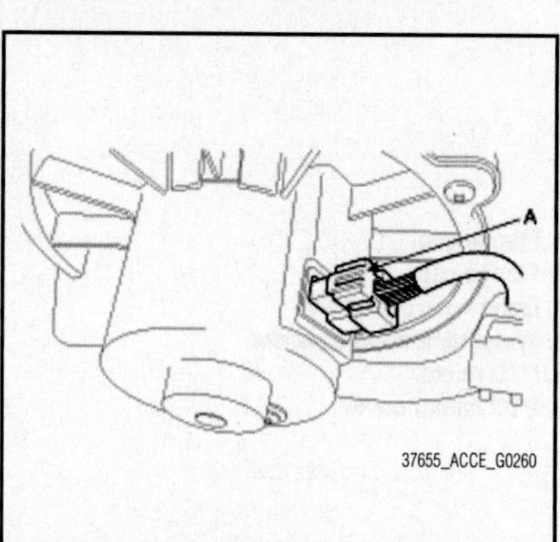

Fig. 127 Blower motor connectors

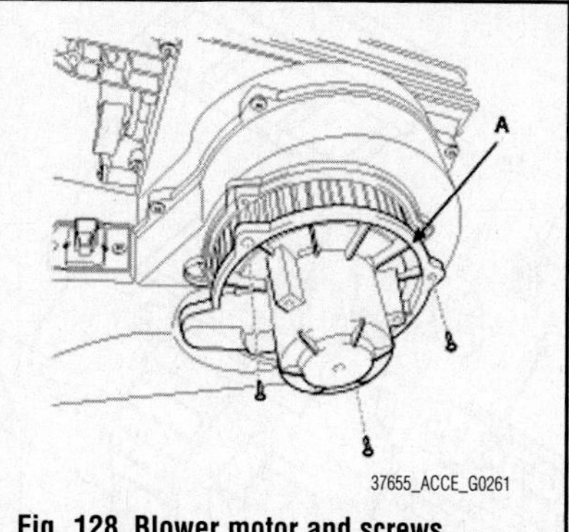

Fig. 128 Blower motor and screws

2. Disconnect the negative battery cable and wait 3 minutes for the SRS memory to drain.

✲✲ CAUTION

After disconnecting the negative battery cable, wait for at least 3 minutes for the SRS module to deplete its stored energy.

3. Disconnect the blower motor connectors.
4. Remove the mounting screws and the blower motor.

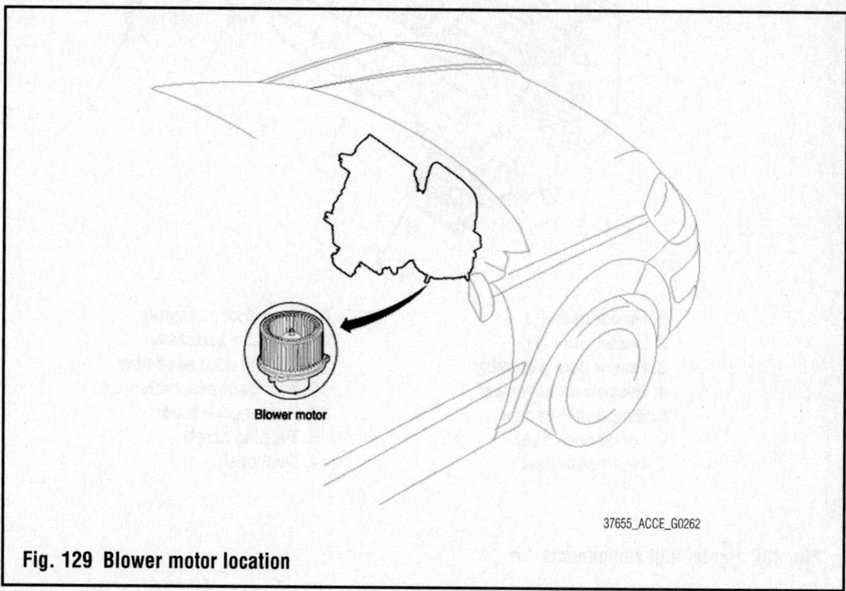

Fig. 129 Blower motor location

To install:

5. Installation is the reverse order of removal.

HEATER UNIT/HEATER CORE

REMOVAL & INSTALLATION

See Figures 130 through 137.

1. Before servicing the vehicle, refer to the Precautions Section.

CAUTION

Before servicing components near or affected by the SRS (air bag) system, read and observe all SRS Service Precautions. Refer to Supplemental Restraint System (SRS), in the Chassis Electrical section. Failure to observe all precautions may result in accidental airbag deployment, personal injury, or death. Refer to Airbag Removal & Installation.

2. Disconnect the negative battery cable and wait 3 minutes for the SRS memory to drain.

CAUTION

After disconnecting the negative battery cable, wait for at least 3 minutes for the SRS module to deplete its stored energy.

3. Recover the refrigerant with a recovery/ recycling/ charging station. Refer to Refrigerant Recovery in the Heating & Air Conditioning Maintenance section.
4. When the engine is cool, drain the engine coolant from the radiator. Refer to Coolant in the Engine Cooling section.
5. Remove the bolts and the expansion valve from the evaporator core. Plug or cap the lines immediately after disconnecting them to avoid moisture and dust contamination.
6. Disconnect the inlet and outlet heater hoses from the heater unit.

WARNING

Engine coolant will run out when the hoses are disconnected; drain it into a clean drip pan. Be sure not to let coolant spill on electrical parts or painted surfaces. If any coolant spills, rinse it off immediately.

7. Remove the instrument panel. Refer to Instrument Panel Removal & Installation in the Body Interior section.
8. Disconnect the connectors from the temperature control actuator, the mode

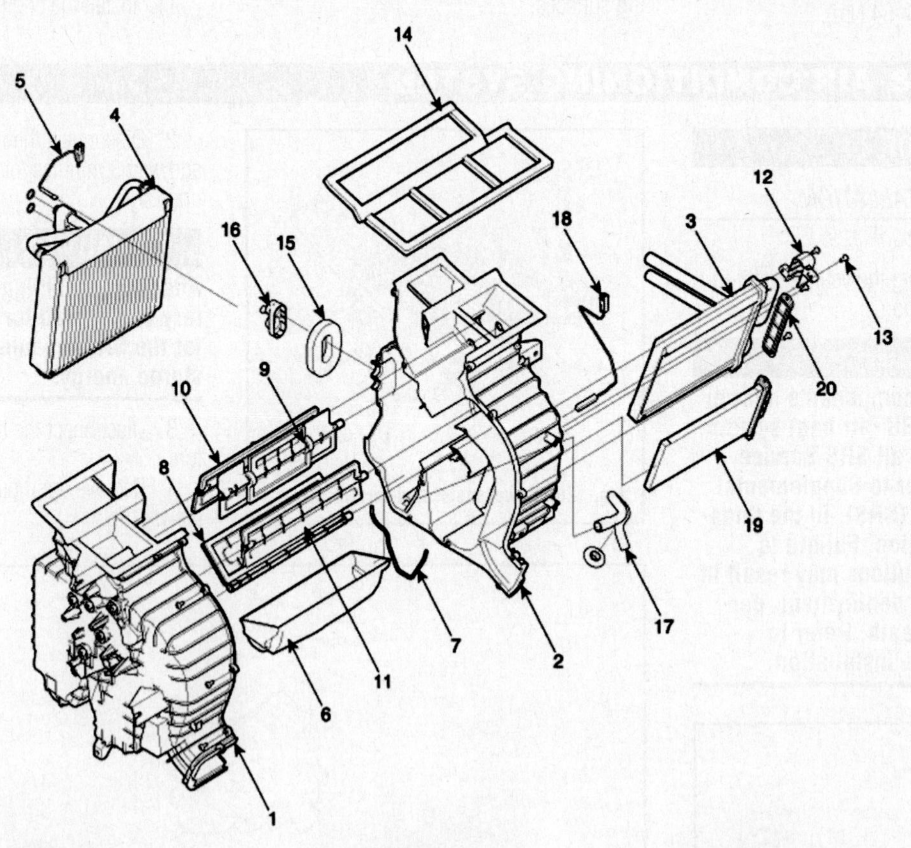

Fig. 130 Heater unit components

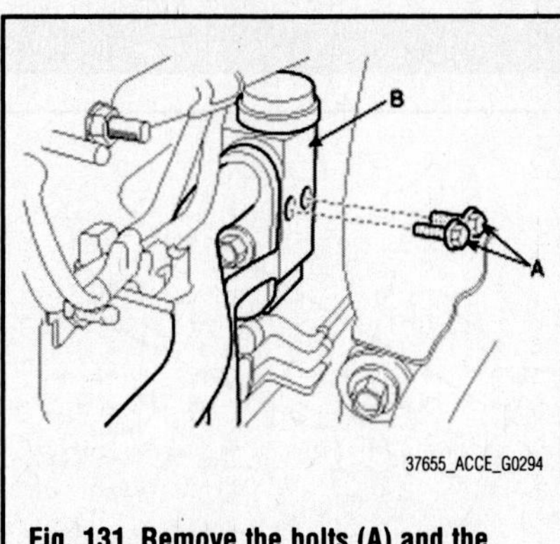

Fig. 131 Remove the bolts (A) and the expansion valve (B)

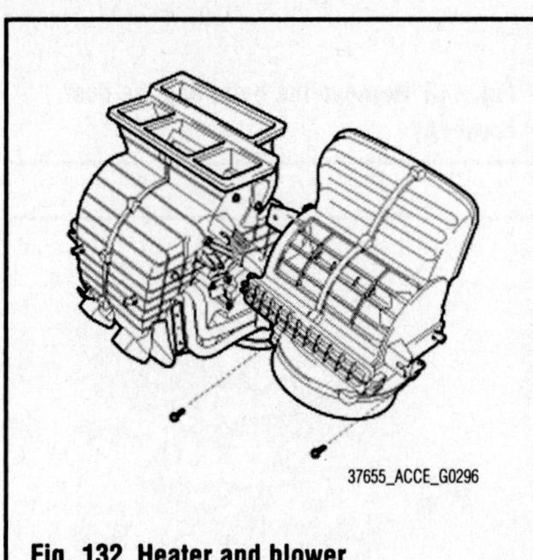

Fig. 132 Heater and blower

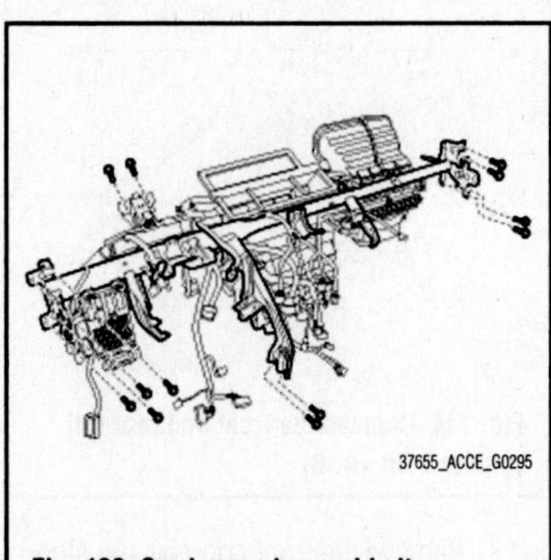

Fig. 133 Cowl cross bar and bolts

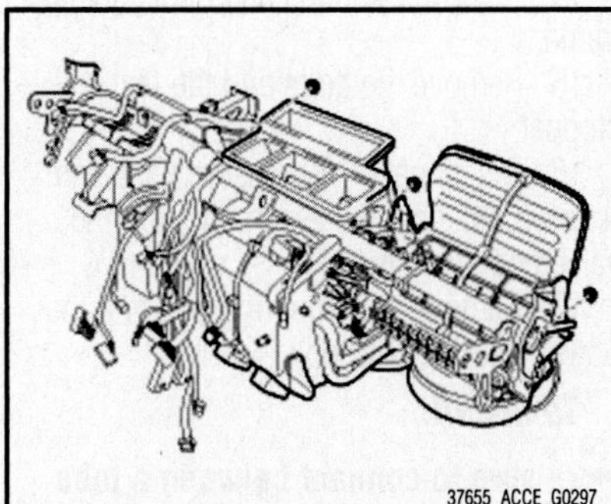

Fig. 134 Disconnect the heater and blower unit from the cowl cross bar assembly

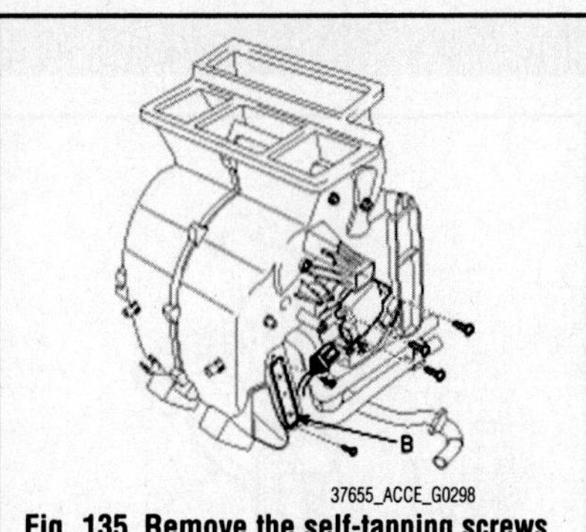

Fig. 135 Remove the self-tapping screws (A) and remove the PTC heater unit (B) or cover

control actuator and the evaporator temperature sensor.

9. Remove the heater and blower mounting bolts.

10. Unscrew 12 bolts and then remove the cowl cross bar and heater and blower unit assembly simultaneously. Refer to Instrument Panel Removal & Installation in the Body Interior section.

11. Disconnect the heater and blower unit from the cowl cross bar assembly.

12. Remove the self-tapping screws and remove the PTC heater unit or cover, as applicable.

13. Remove the heater core and cover.

14. Be careful that the inlet and outlet pipe are not bent during heater core removal, and pull out the heater core.

To install:

15. Installation is the reverse order of removal, and note these items :

a. If you're installing a new evaporator, add refrigerant oil (ND-OIL8).

b. Replace the O-rings with new ones at each fitting, and apply a thin coat of refrigerant oil before installing them. Be sure to use the right O-rings for R-134a to avoid leakage. Immediately after using the oil, replace the cap on the container, and seal it to avoid moisture absorption.

c. Apply sealant to the grommets.

d. Make sure that there is no air leakage.

e. Charge the system and test its performance.

f. Do not interchange the inlet and outlet heater hoses and install the hose clamps securely.

g. Refill the cooling system with engine coolant.

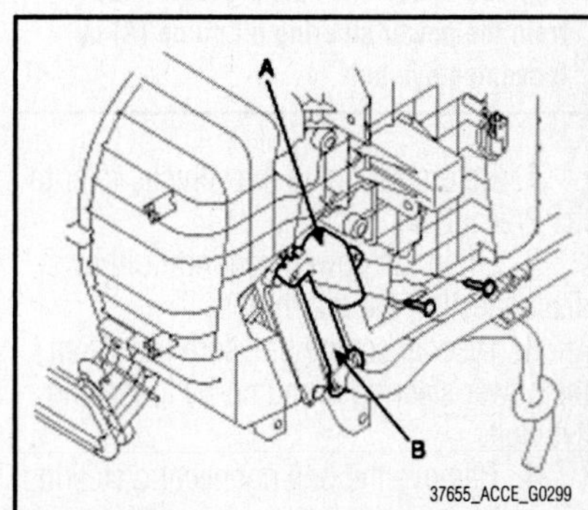

Fig. 136 Remove the heater core (B) and cover (A)

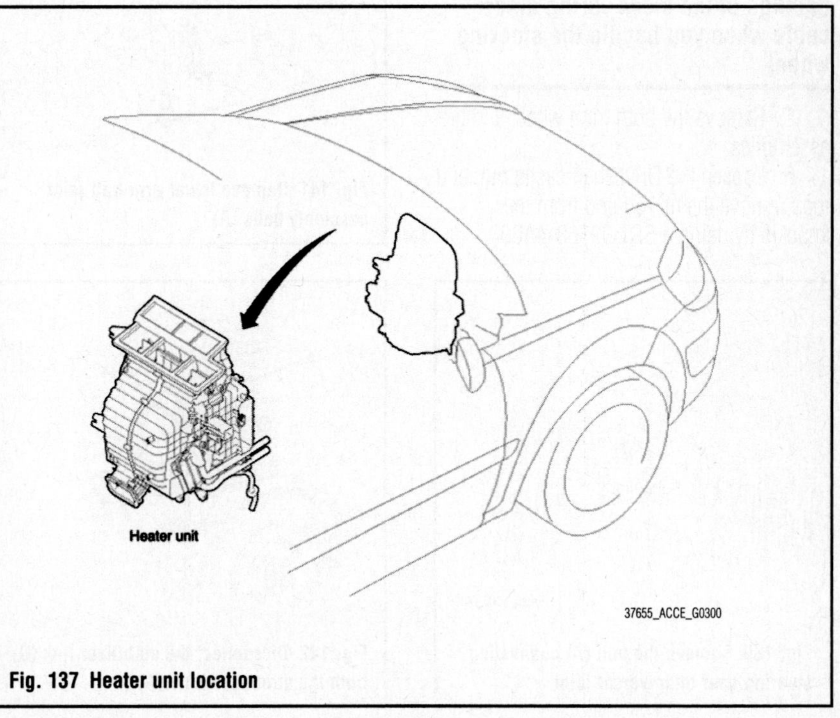

Fig. 137 Heater unit location

STEERING

POWER RACK & PINION STEERING GEAR

REMOVAL & INSTALLATION

See Figures 138 through 148.

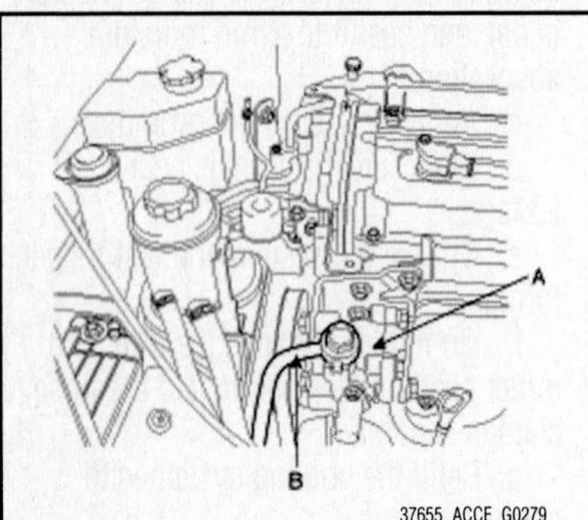

Fig. 138 Disconnect the pressure tube (B) from the power steering oil pump (A) by loosening eye bolt

1. Before servicing the vehicle, refer to the Precautions Section.
2. Drain the power steering fluid by disconnecting the return hose.
3. Disconnect the pressure tube from the power steering oil pump by loosening eye bolt.
4. Remove the bolt connecting steering gear to universal joint.

⁂ WARNING

Keep the neutral-range to prevent the damage of the clock spring inner cable when you handle the steering wheel.

5. Remove the both front wheel and tire assemblies.
6. Loosen the Driveshaft castle nut, and then remove the tie rod end from the knuckle by using a SST 09568-4A000.

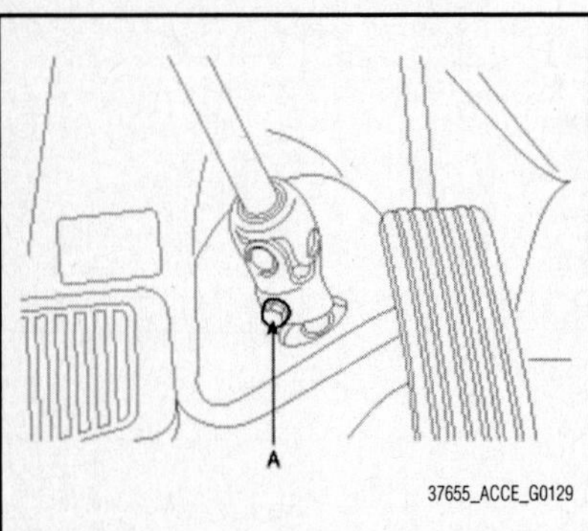

Fig. 139 Remove the bolt (A) connecting steering gear to universal joint

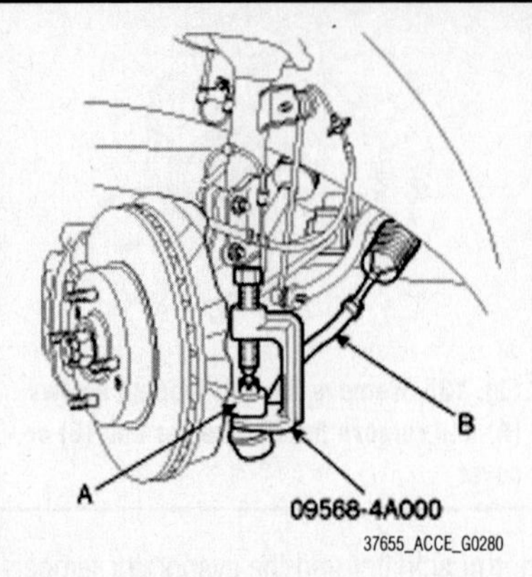

Fig. 140 Remove the tie rod end (A) from the knuckle (B) using a SST 09568-4A000

7. Remove lower arm ball joint assembly bolts.
8. Disconnect the stabilizer link from the strut assembly.
9. Repeat on the other side.
10. Remove the bolts and the dust cover.
11. Remove the bolts and the heat protector.
12. Remove the muffler mounting rubber.

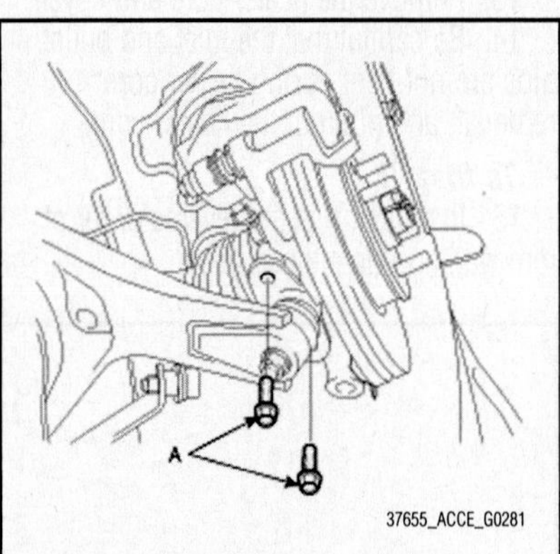

Fig. 141 Remove lower arm ball joint assembly bolts (A)

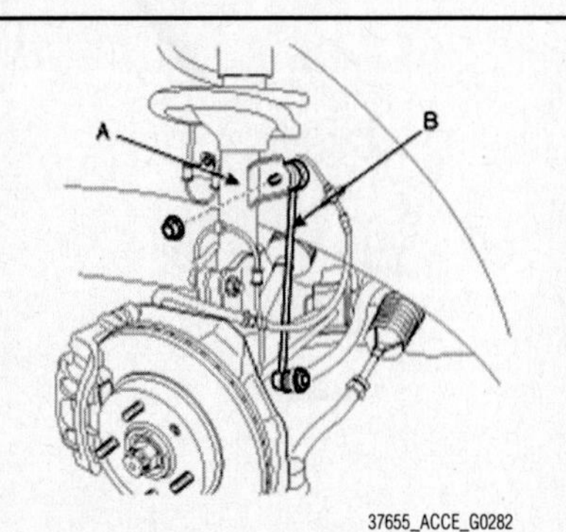

Fig. 142 Disconnect the stabilizer link (B) from the strut assembly (A)

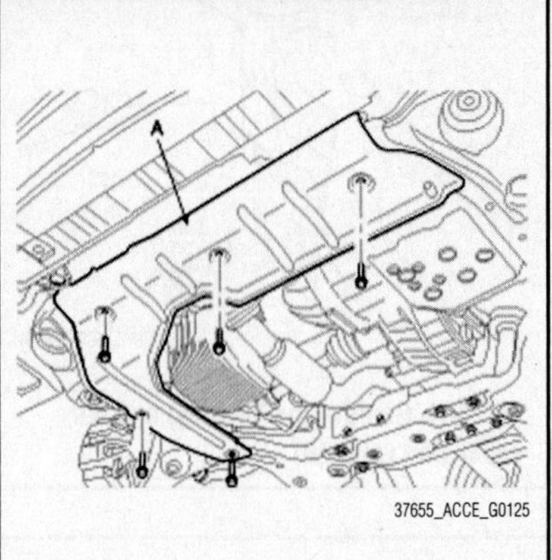

Fig. 143 Remove the bolts and the dust cover (A)

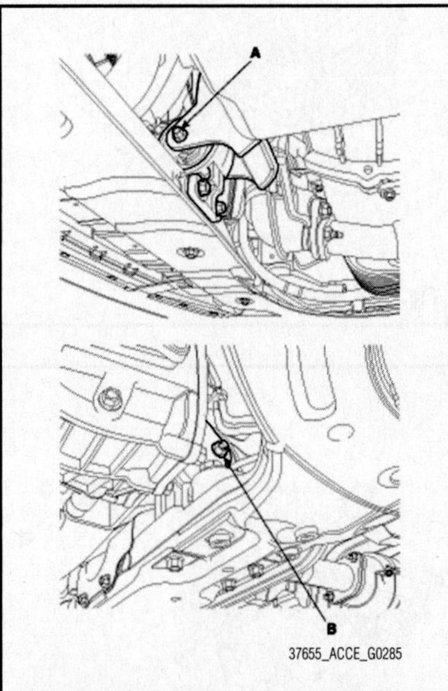

Fig. 144 Remove the front and rear roll stopper bolts (A, B)

13. Remove the front and rear roll stopper bolts.
14. Remove the bolts and nuts and subframe.
15. Remove the bolts and the rear roll stopper.
16. Disconnect the pressure hose and return tube and hose assembly from the valve body housing.
17. Remove the mounting bolt and the power steering gear box.

To install:

➡Be sure to connect between a tube and hose as shown.

18. Installation is the reverse of removal, noting the following:

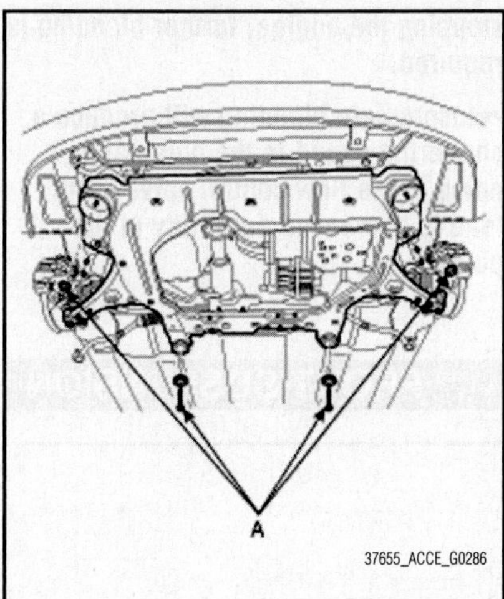

Fig. 145 Remove the bolts and nuts (A) and sub-frame

a. Tighten the power steering gear box mounting bolt to 43–58 ft. lbs. (60–80 Nm).

b. Tighten the rear roll stopper bolts to 36–43 ft. lbs. (50–60 Nm).

c. Tighten the subframe bolts and nuts to 69–87 ft. lbs. (95–120 Nm).

d. Tighten the front and rear roll stopper bolts to 36–47 ft. lbs. (50–65 Nm).

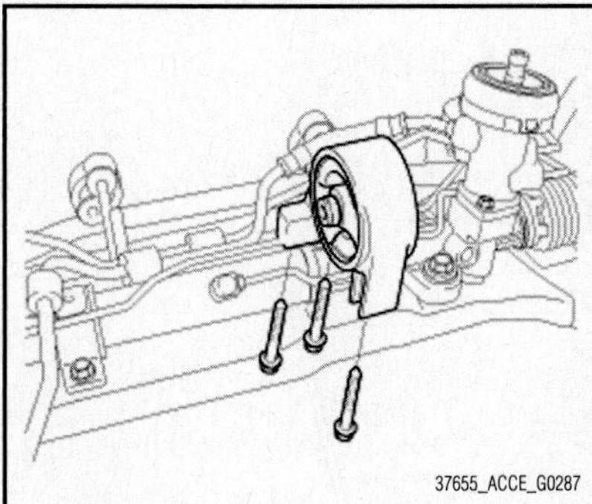

Fig. 146 Remove the bolts and the rear roll stopper

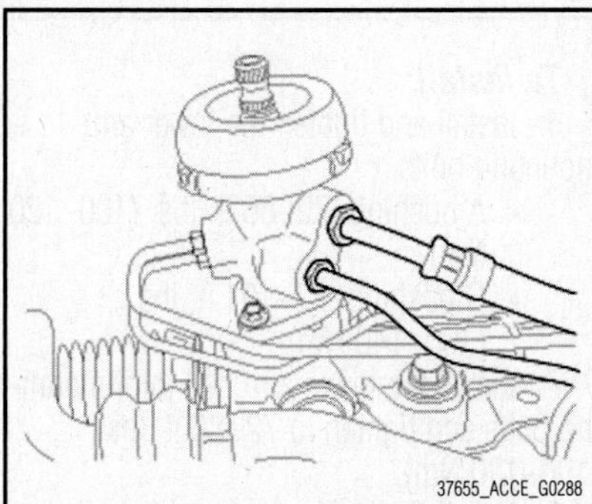

Fig. 147 Remove the mounting bolt (B) and the power steering gear box (A)

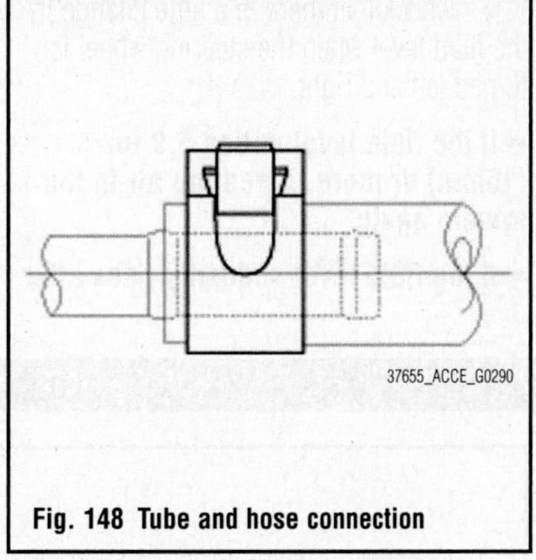

Fig. 148 Tube and hose connection

e. Tighten the stabilizer link to 25–33 ft. lbs. (35–45 Nm).

f. Tighten the lower arm ball joint assembly bolts to 72–86 ft. lbs. (100–120 Nm).

g. Tighten the tie rod end to 17–26 ft. lbs. (24–34 Nm).

h. Tighten the steering gear to universal joint bolt to 9–13 ft. lbs. (13–18 Nm).

i. Tighten the eye bolt to 40–47 ft. lbs. (55–65 Nm).

19. After installation, bleed the power steering system.

20. Adjust the wheel alignment.

POWER STEERING PUMP

REMOVAL & INSTALLATION

See Figures 149 and 150.

1. Before servicing the vehicle, refer to the Precautions Section.
2. Remove the eye bolt and disconnect the pressure tube from the oil pump.
3. Disconnect the suction hose from the suction pipe.
4. Remove the tension adjusting bolt, and then remove the oil pump drive belt.
5. Remove the power steering pump mounting bolt and nut, and then remove the power steering pump assembly from the pump bracket.

➡Be careful not to spill fluid from the power steering oil pump.

To install:

6. Installation is the reverse of the removal procedure, noting the following:

a. Tighten the power steering pump to 14–20 ft. lbs. (20–27 Nm).

b. Tighten the oil pump drive belt bolt to 18–24 ft. lbs. (35–35 Nm).

c. Tighten the eye belt bolt to 40–47 ft. lbs. (55–65 Nm).

➡The pressure tube does not twist and come in contact with other components.

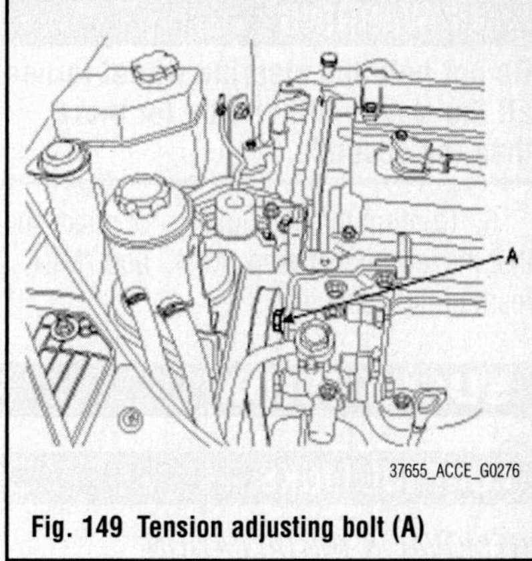

Fig. 149 Tension adjusting bolt (A)

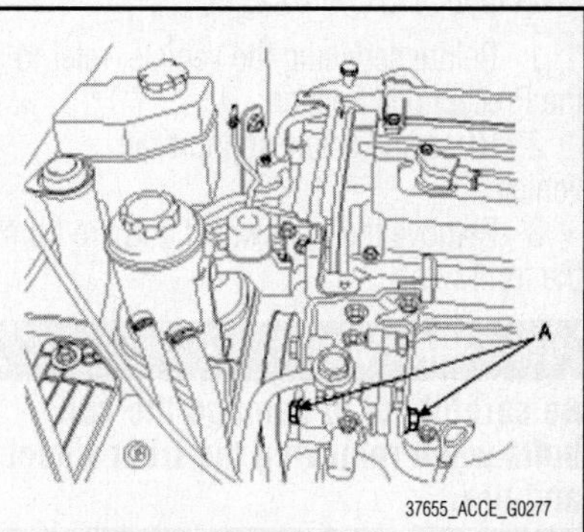

Fig. 150 Power steering pump mounting bolt and nut (A)

7. Adjust the drive belt tension.
8. Add power steering fluid.
9. Bleed the power steering system.
10. Check the oil pump pressure.

BLEEDING

1. Fill the power steering fluid reservoir up to the "MAX" position with specified fluid.
2. Jack up the front wheels.
3. Disconnect the ignition coil high tension cable, and then, while operating the starter motor intermittently (for 15 to 20 seconds), turn the steering wheel all the way to the left and then to the right five or six times.

➡When bleeding fluid, replenish with the fluid so that the level does not fall below the bottom of the filter. If air bleeding is done while the vehicle is idling, the air will be broken up and absorbed into the fluid. Be sure to do the bleeding only while cranking.

4. Connect the high tension cable, and then start the engine (idling).
5. Turn the steering wheel to the left and then to the right, until there are no air bubbles in the oil reservoir.

⁂ WARNING

Do not hold the steering wheel turned all the way to either side for more than ten seconds.

6. Confirm that the fluid is not milky and that the level is between "MAX" and "MIN" mark on the reservoir.

7. Check that there is a little change in the fluid level when the steering wheel is turned left and right.

➡If the fluid level varies 5.9 in. (15mm) or more, bleed the air in the system again.

➡If the fluid level suddenly rises after stopping the engine, further bleeding is required.

➡Incomplete bleeding will produce a chattering sound in the pump and noise in the flow control valve, and lead to decreased durability of the pump.

SUSPENSION

FRONT SUSPENSION

CONTROL LINKS

REMOVAL & INSTALLATION

See Figures 151 and 152.

1. Before servicing the vehicle, refer to the Precautions Section.
2. Raise and safely support the vehicle.
3. Remove the front wheel and tire from the front hub.

⁂ WARNING

Be careful not to damage the hub bolts when removing the front wheel and tire.

4. Remove the stabilizer bar control link by removing the upper and lower mounting nuts.
5. Remove the stabilizer bar control link.

To install:

6. Install the stabilizer bar control link.
7. Tighten the stabilizer bar control link upper and lower mounting nuts to 25–32 ft. lbs. (35–45 Nm)
8. Install the front wheels, and lower the vehicle. Tighten the wheel nuts to: 65–79 ft. lbs. (90–110 Nm).

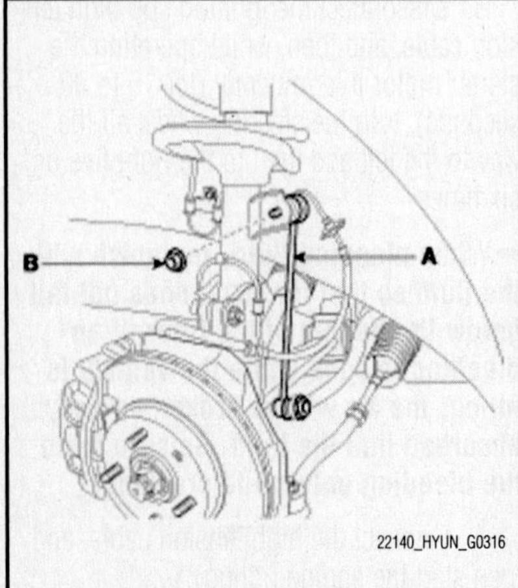

Fig. 151 Stabilizer bar control link (A) and nut (B)

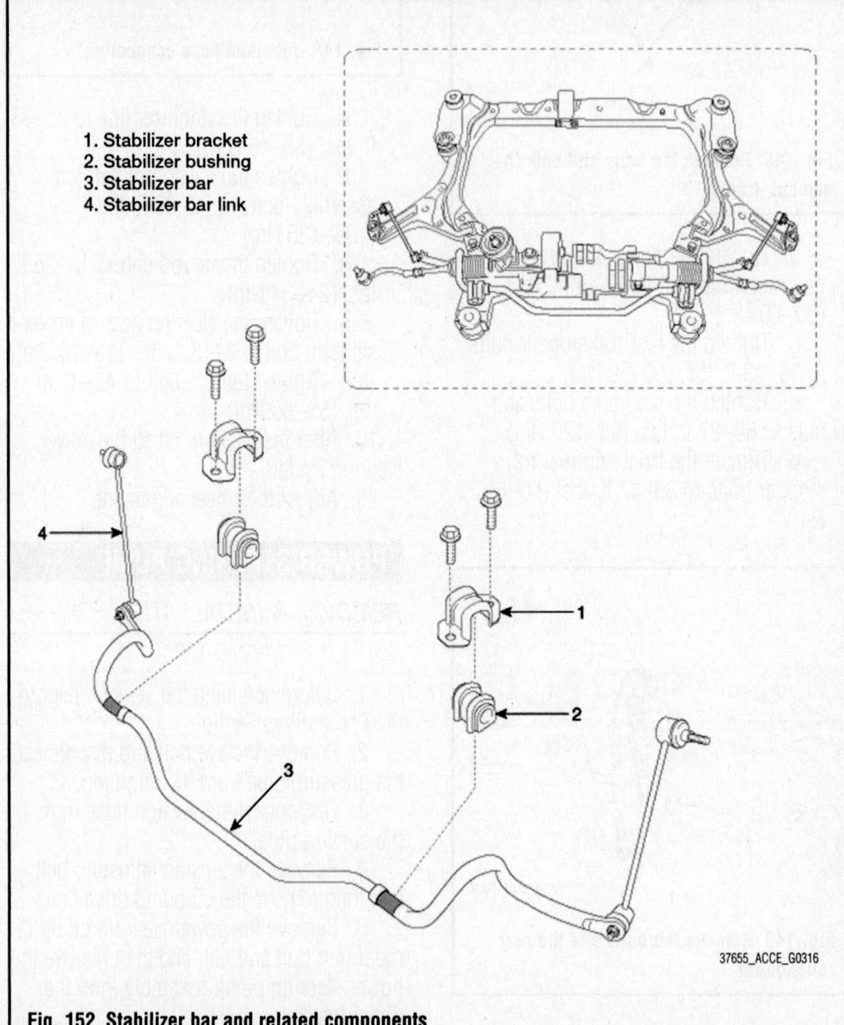

Fig. 152 Stabilizer bar and related components

LOWER CONTROL ARM

REMOVAL & INSTALLATION

See Figures 153 through 155.

1. Before servicing the vehicle, refer to the Precautions Section.
2. Raise and safely support the vehicle.
3. Remove the front wheel and tire.
4. Remove the lower arm ball joint mounting bolts.
5. Remove the lower arm mounting bolts.

To install:

6. Install and tighten the lower arm mounting bolts.
   - A bushing: 72–86 ft. lbs. (100–120 Nm)
   - G bushing: 72–101 ft. lbs. (100–140 Nm)
7. Install the lower arm ball joint mounting bolts and tighten to 72–86 ft. lbs. (100–120 Nm).
8. Install the front wheel and tire.
9. Check and/or adjust the wheel alignment.

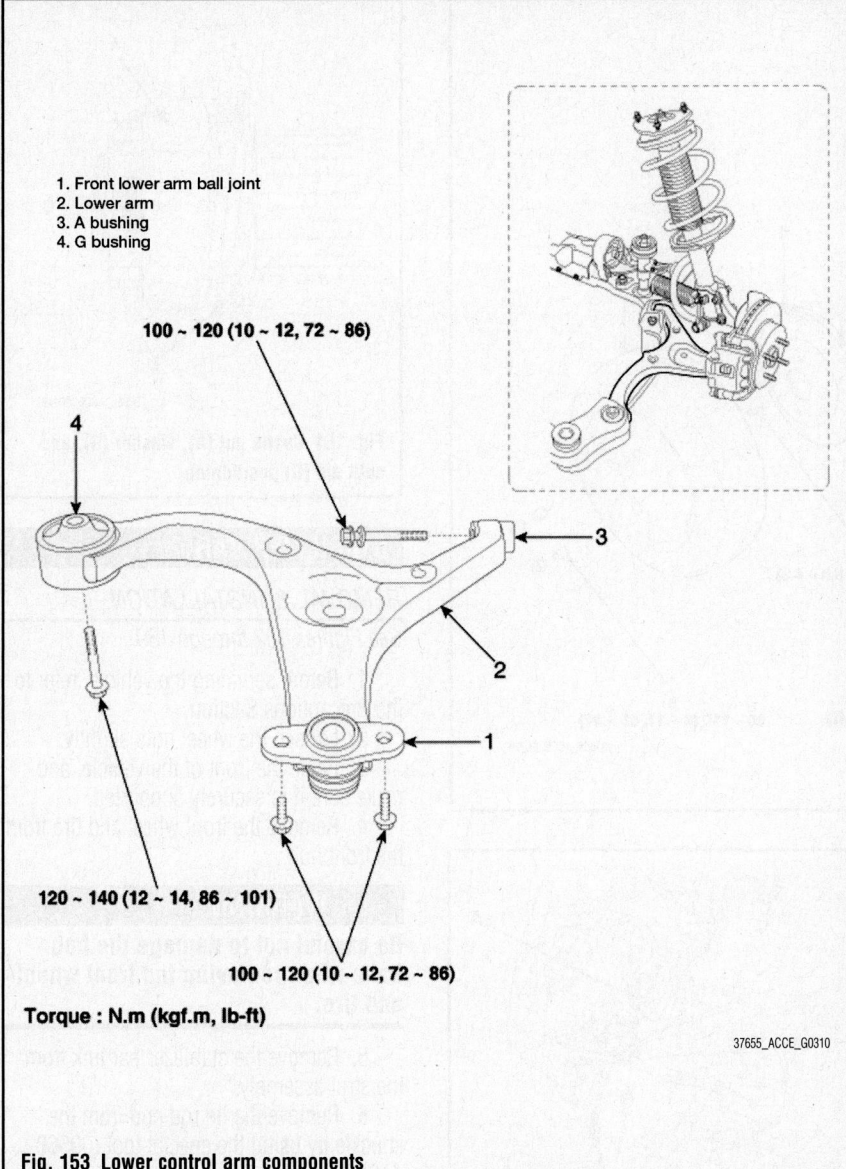

1. Front lower arm ball joint
2. Lower arm
3. A bushing
4. G bushing

100 ~ 120 (10 ~ 12, 72 ~ 86)

120 ~ 140 (12 ~ 14, 86 ~ 101)

100 ~ 120 (10 ~ 12, 72 ~ 86)

Torque : N.m (kgf.m, lb-ft)

37655_ACCE_G0310

Fig. 153 Lower control arm components

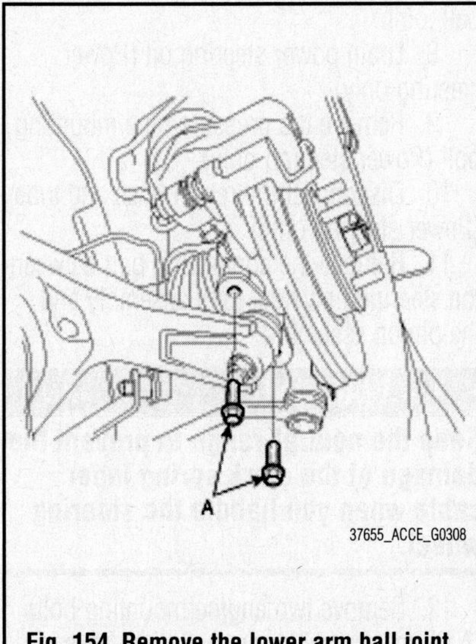

Fig. 154 Remove the lower arm ball joint mounting bolts (A)

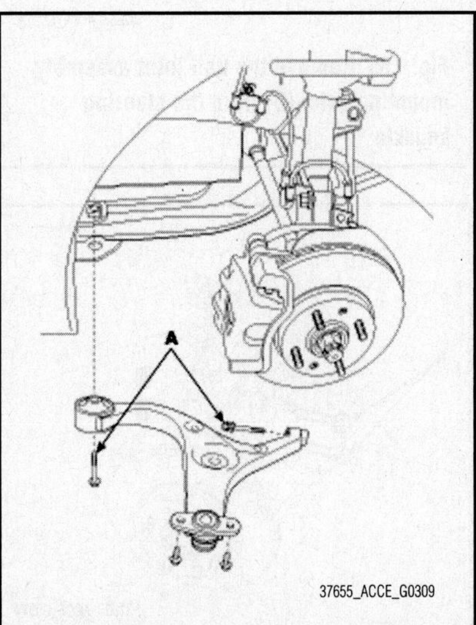

Fig. 155 Remove the lower arm mounting bolts (A)

STEERING KNUCKLE

REMOVAL & INSTALLATION

See Figures 156 through 161.

1. Before servicing the vehicle, refer to the Precautions Section.
2. Raise and safely support the vehicle.
3. Remove the front wheel and tire.

WARNING

Be careful not to damage the hub bolts when removing the front wheel and tire.

4. Remove the caliper assembly from knuckle and suspend it with wire.
5. Remove the split pin, castle nut and washer from the front hub assembly.
6. Remove the wheel speed sensor from the knuckle.
7. Disconnect the tie rod end ball joint from the knuckle using the special tool (09568-4A000).
8. Remove the ball joint assembly mounting bolts from the steering knuckle.
9. Remove the strut assembly mounting bolts.
10. Remove the hub and knuckle as an assembly.

WARNING

Be careful not to damage the boot and tone wheel.

To install:

11. Preparation for installation:
    a. Check the hub for cracks and the splines for wear.
    b. Check the brake disc for scoring and damage.
    c. Check the steering knuckle for cracks.
    d. Check the bearing for cracks or damage.
12. Installation is the reverse of removal, noting the following:
13. Tighten the strut assembly mounting bolts to 72–86 ft. lbs. (100–120 Nm).
14. Tighten the ball joint assembly mounting bolt to 72–86 ft. lbs. (100–120 Nm).
15. Install the washer, the castle nut, and split pin to the front hub assembly and tighten to 145–188 ft. lbs. (200–260 Nm).
    a. The washer should be assembled with convex surface outward when installing the castle nut and split pin.

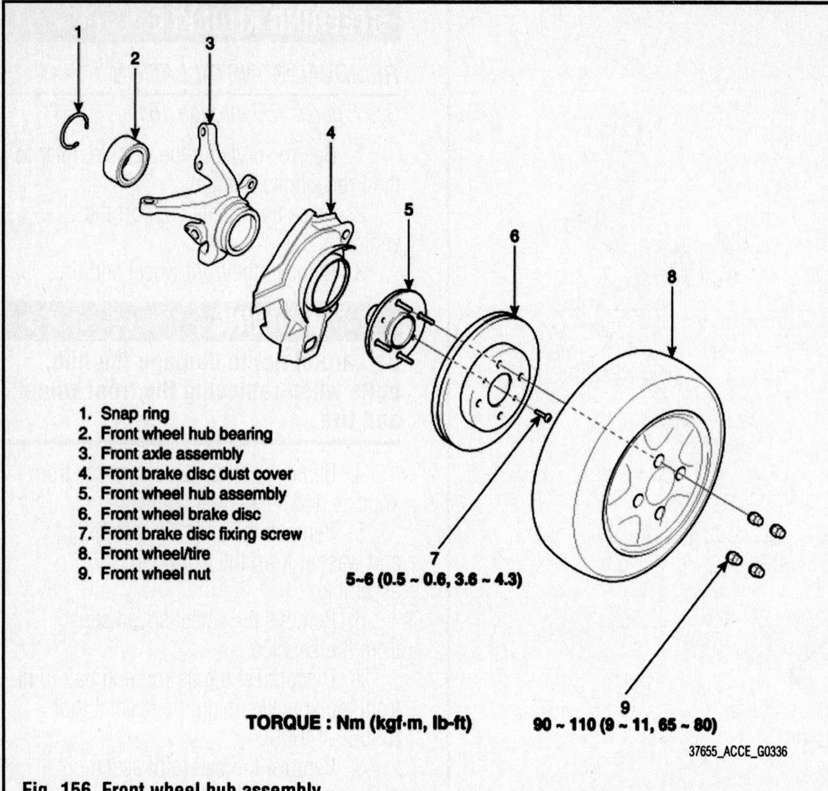

Fig. 156 Front wheel hub assembly

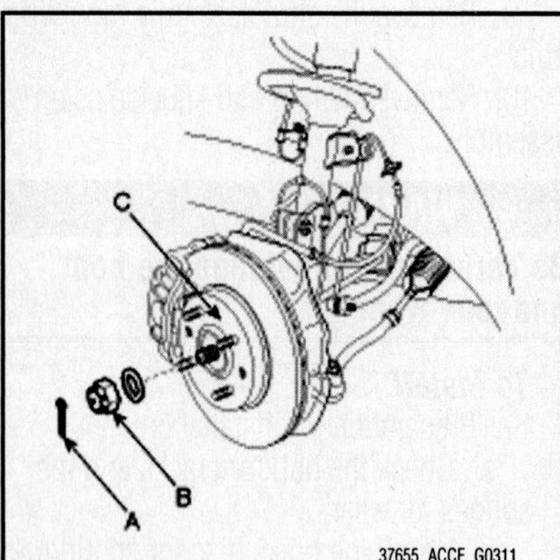

Fig. 157 Remove the split pin (A), the castle nut (B), and the washer from the front hub (C)

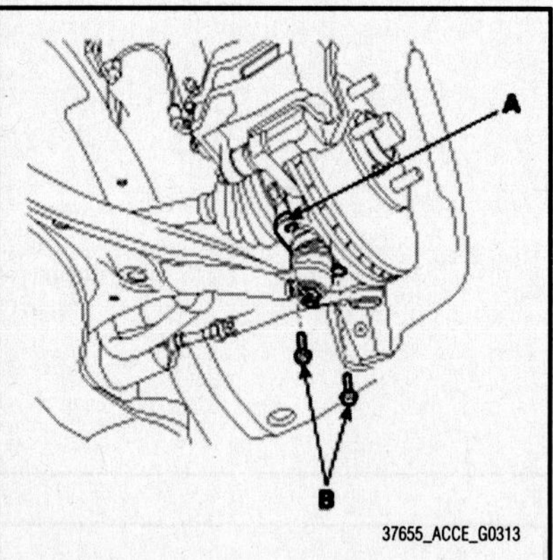

Fig. 159 Remove the ball joint assembly mounting bolts (B) from the steering knuckle

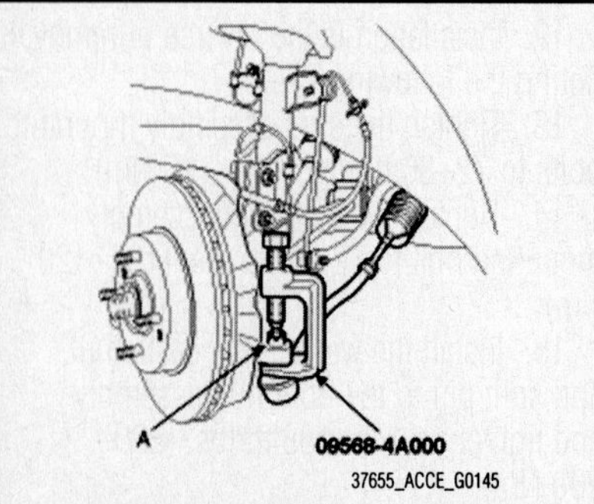

Fig. 158 Disconnect the tie rod end ball joint (A) from the knuckle using the special tool (09568-4A000)

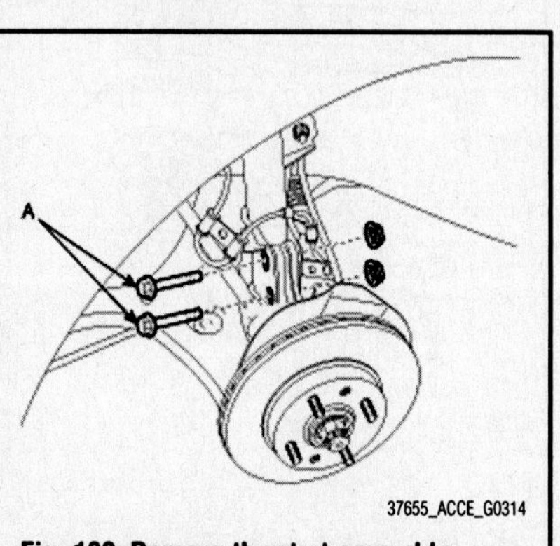

Fig. 160 Remove the strut assembly mounting bolts (A)

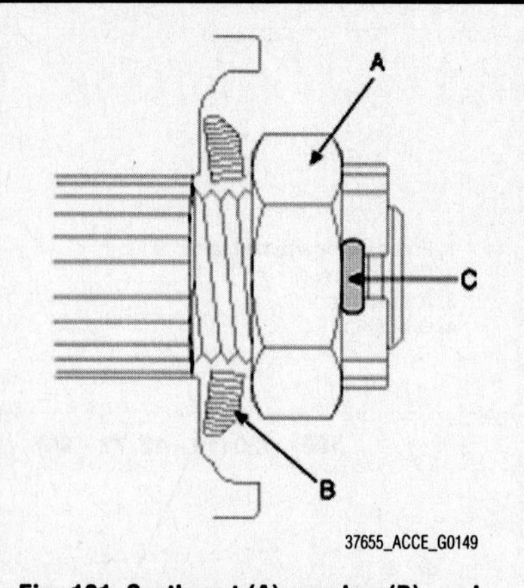

Fig. 161 Castle nut (A), washer (B), and split pin (C) positioning

STABILIZER BAR

REMOVAL & INSTALLATION

See Figures 162 through 169.

1. Before servicing the vehicle, refer to the Precautions Section.
2. Loosen the wheel nuts slightly.
3. Raise the front of the vehicle, and make sure it is securely supported.
4. Remove the front wheel and tire from the front hub.

WARNING

Be careful not to damage the hub bolts when removing the front wheel and tire.

5. Remove the stabilizer bar link from the strut assembly.
6. Remove the tie rod end from the knuckle by using the special tool (09568-4A000).
7. Remove the two bolts for lower arm ball joint.
8. Drain power steering oil (Power steering only).
9. Remove the pressure pipe mounting bolt (Power steering only).
10. Disconnect the return hose and tube (Power steering only).
11. Remove the connecting bolt between the steering universal joint assembly and the pinion assembly.

WARNING

Keep the neutral-range to prevent the damage of the clock spring inner cable when you handle the steering wheel.

12. Remove two engine mounting bolts (A,B) and six subframe mounting bolts in order to remove the subframe.

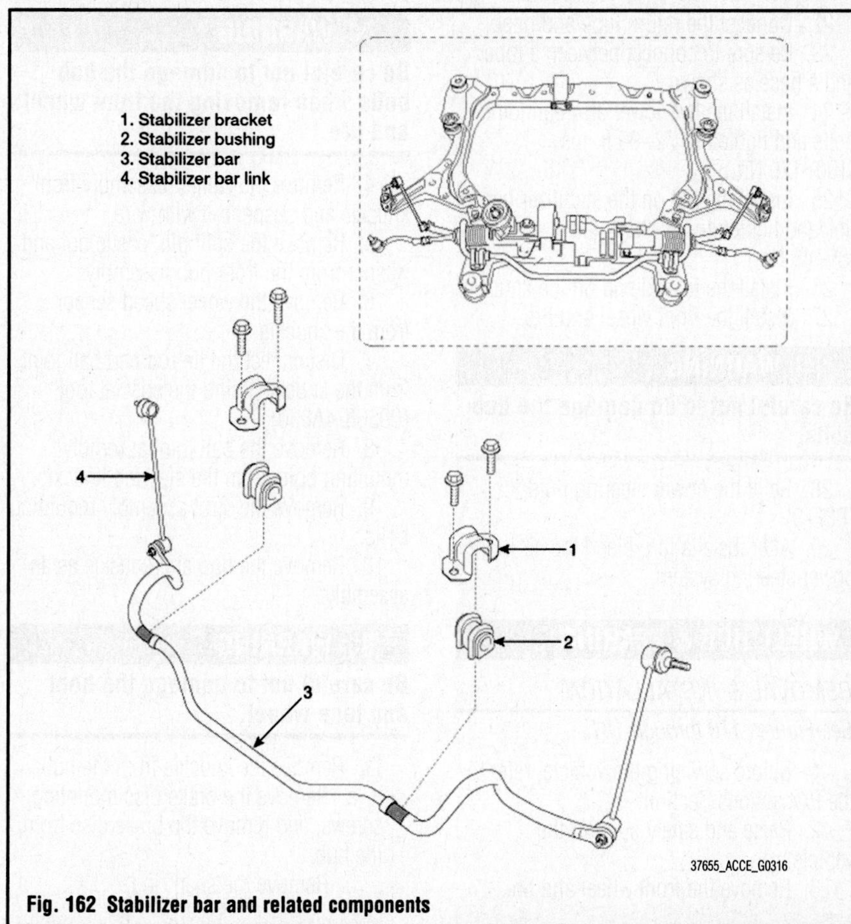

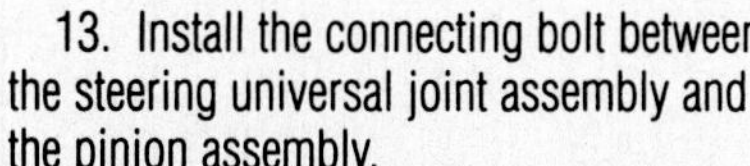
Fig. 162 Stabilizer bar and related components

13. Install the connecting bolt between the steering universal joint assembly and the pinion assembly.

14. Remove the stabilizer bracket and bushing.

15. Remove the stabilizer bracket and bushing on the opposite side in the same way.

16. Remove the stabilizer bar.

WARNING

Be careful not to do damage to pressure tubes.

To install:

17. Install the bushing on the stabilizer bar. Bring clamp of stabilizer bar into contact with bushing.

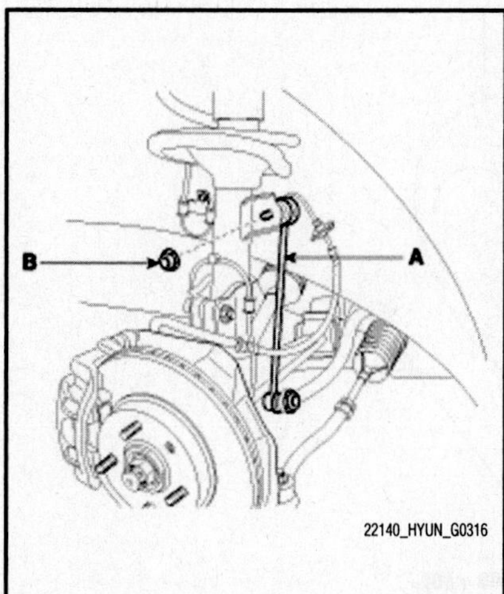

Fig. 163 Stabilizer bar control link (A) and nut (B)

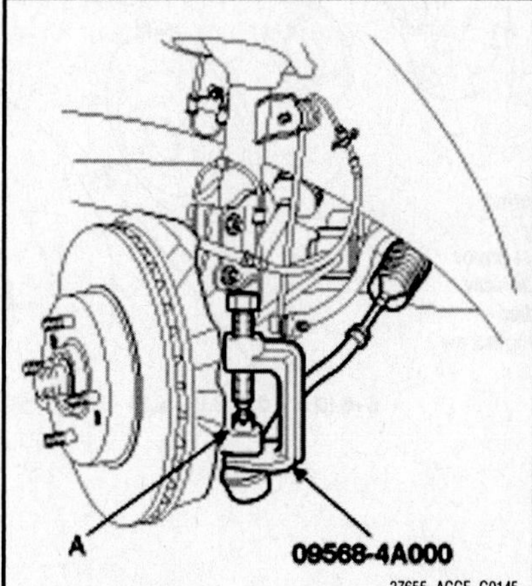

Fig. 164 Remove the tie rod end from the knuckle by using the special tool (09568-4A000)

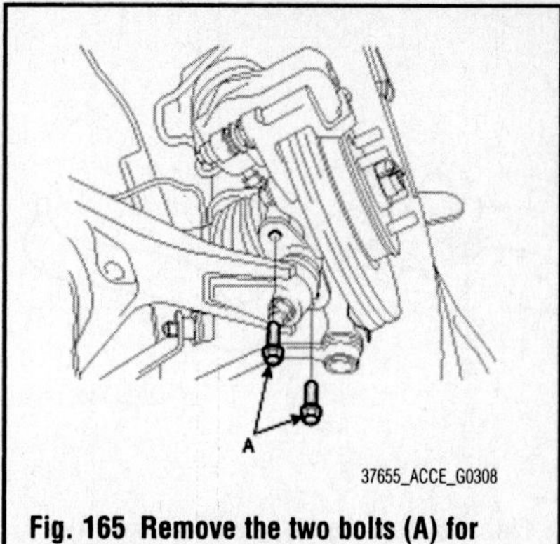

Fig. 165 Remove the two bolts (A) for lower arm ball joint

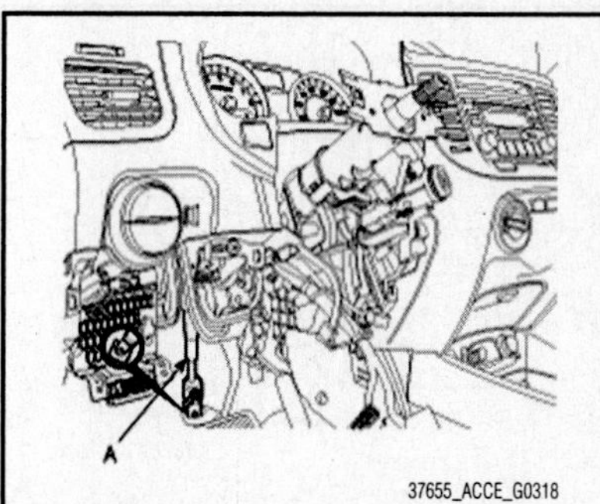

Fig. 166 Remove the connecting bolt (A) between the steering universal joint assembly and the pinion assembly

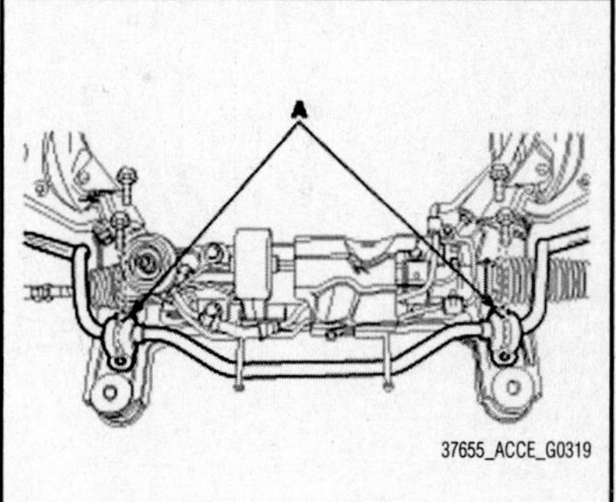

Fig. 167 Remove two stabilizer brackets and bushings

18. Install the bracket on the bushing.

19. After tightening the bolts of the bushing bracket temporarily, install the bushing bracket on the opposite side and tighten to 32–39 ft. lbs. (45–55 Nm).

20. Install the six subframe mounting bolts, then the two engine mounting bolts. Tighten the engine mounting bolts to 36–47 ft. lbs. (50–65 Nm), and the subframe mounting bolts to 68–86 ft. lbs. (95–120 Nm).

21. Install the pressure pipe mounting bolt and tighten to 39–47 ft. lbs. (55–65 Nm).

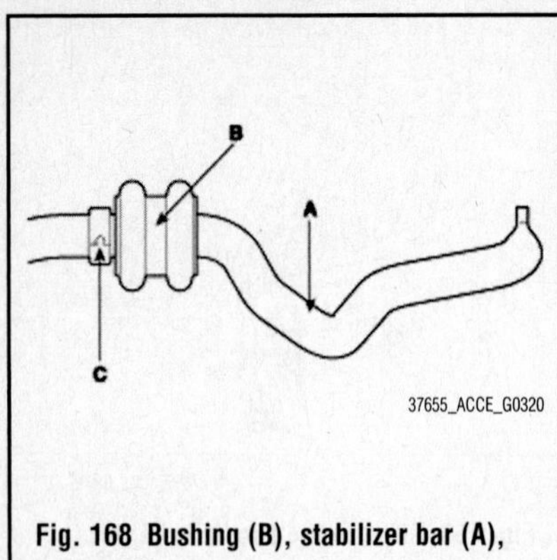

Fig. 168 Bushing (B), stabilizer bar (A), clamp (C)

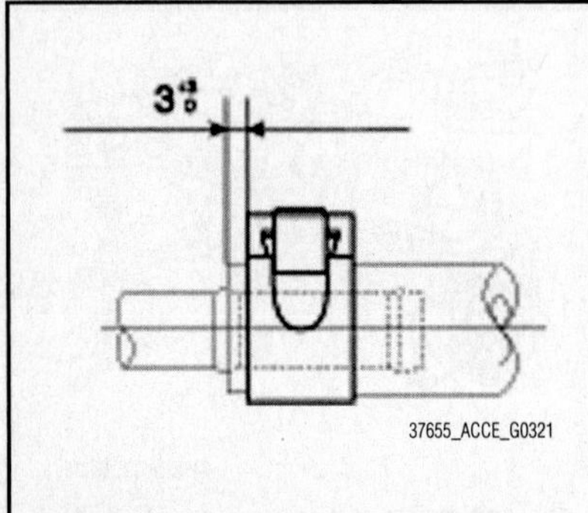

Fig. 169 Be sure to connect between a tube and a hose as shown

22. Connect the return tube and hose.
23. Be sure to connect between a tube and a hose as shown.
24. Install the two lower arm ball joint bolts and tighten to 72–86 ft. lbs. (100–120 Nm).
25. Install the nut on the stabilizer bar link and tighten to 25–32 ft. lbs. (35–45 Nm).
26. Install the tie rod end on the knuckle.
27. Install the front wheel and tire.

WARNING

Be careful not to do damage the hub bolts.

28. Refill the power steering fluid (PSF-3).
29. After installation, bleed the air in the power steering system.

WHEEL HUB & BEARING

REMOVAL & INSTALLATION

See Figures 170 through 181.

1. Before servicing the vehicle, refer to the Precautions Section.
2. Raise and safely support the vehicle.
3. Remove the front wheel and tire.

WARNING

Be careful not to damage the hub bolts when removing the front wheel and tire.

4. Remove the caliper assembly from knuckle and suspend it with wire.
5. Remove the split pin, castle nut and washer from the front hub assembly.
6. Remove the wheel speed sensor from the knuckle.
7. Disconnect the tie rod end ball joint from the knuckle using the special tool (09568-4A000).
8. Remove the ball joint assembly mounting bolts from the steering knuckle.
9. Remove the strut assembly mounting bolts.
10. Remove the hub and knuckle as an assembly.

WARNING

Be careful not to damage the boot and tone wheel.

11. Remove the knuckle from the hub:
    a. Remove the brake disc mounting screws, and remove the brake disc from the hub.
    b. Remove the snap ring.

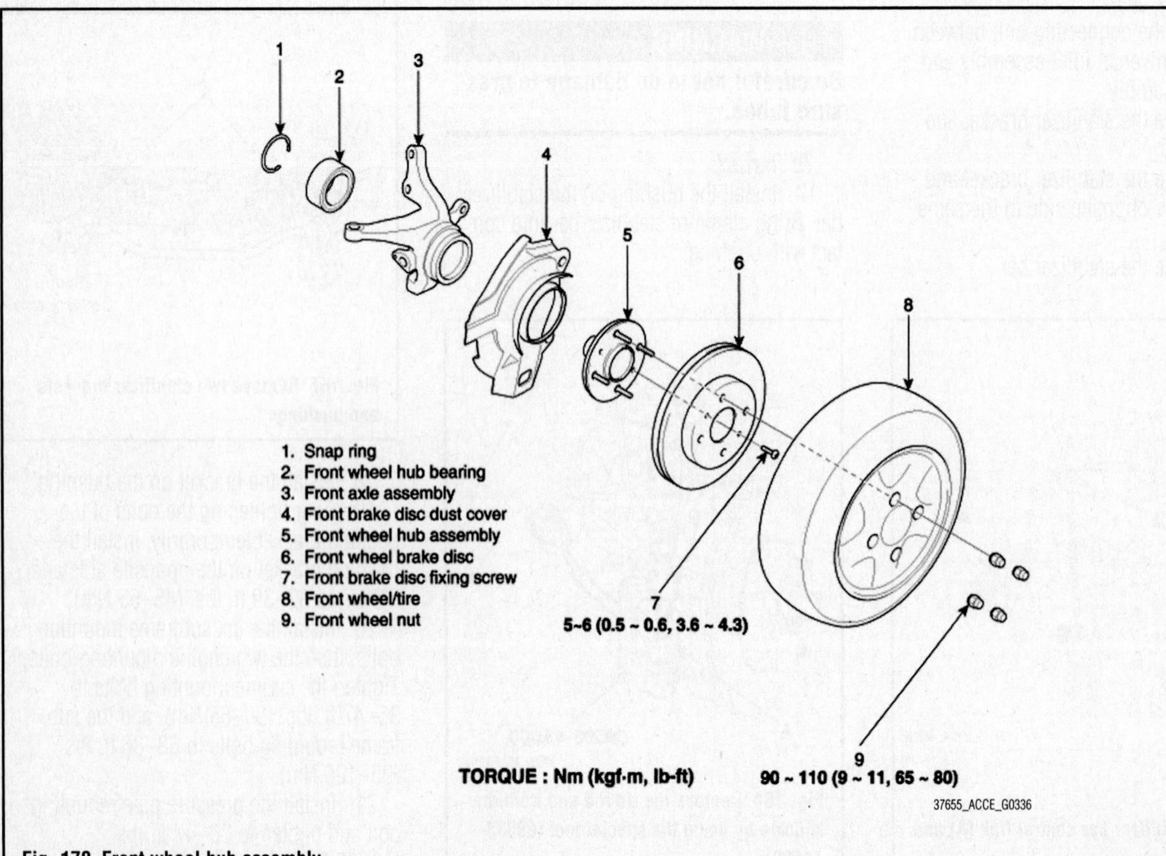

Fig. 170 Front wheel hub assembly

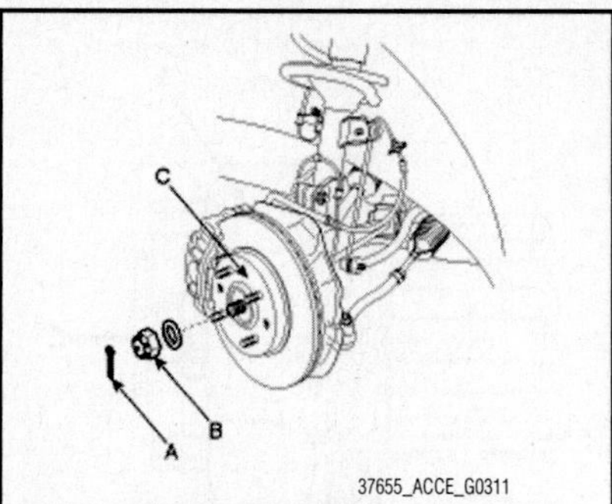

Fig. 171 Remove the split pin (A), the castle nut (B), and the washer from the front hub (C)

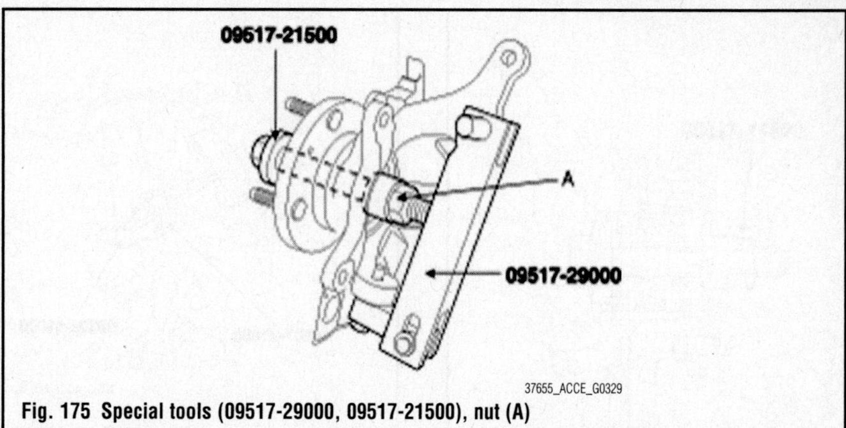

Fig. 175 Special tools (09517-29000, 09517-21500), nut (A)

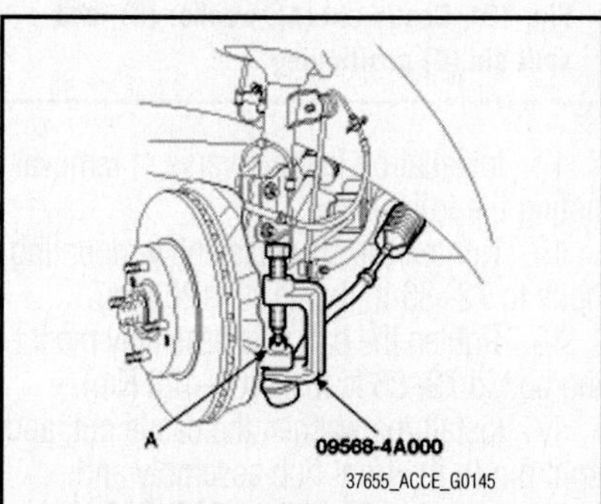

Fig. 172 Disconnect the tie rod end ball joint (A) from the knuckle using the special tool (09568-4A000)

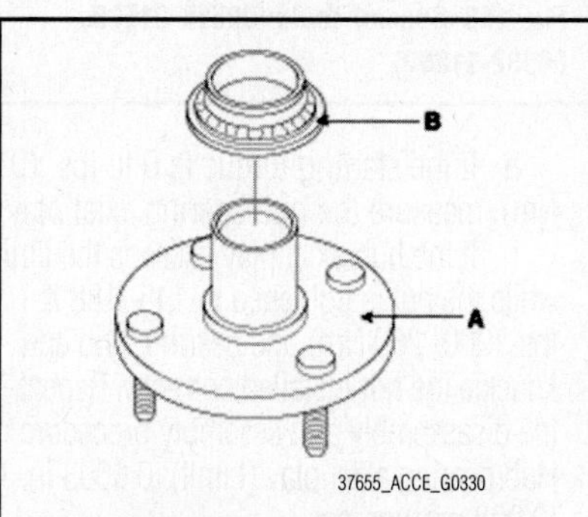

Fig. 176 Remove the bearing inner race (B) from the hub (A) using the special tool (09495-33000)

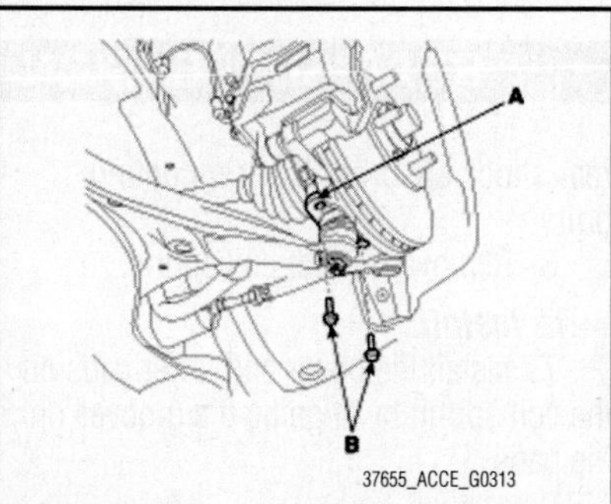

Fig. 173 Remove the ball joint assembly mounting bolts (B) from the steering knuckle

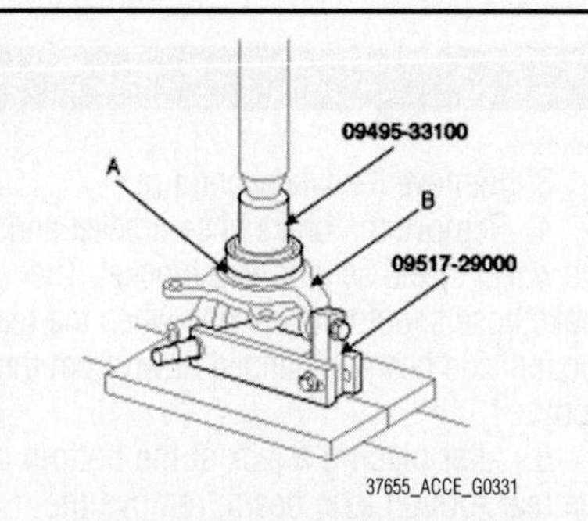

Fig. 177 Remove the outer race (A) from the knuckle (B) using the special tools (09495-33100, 09517-29000)

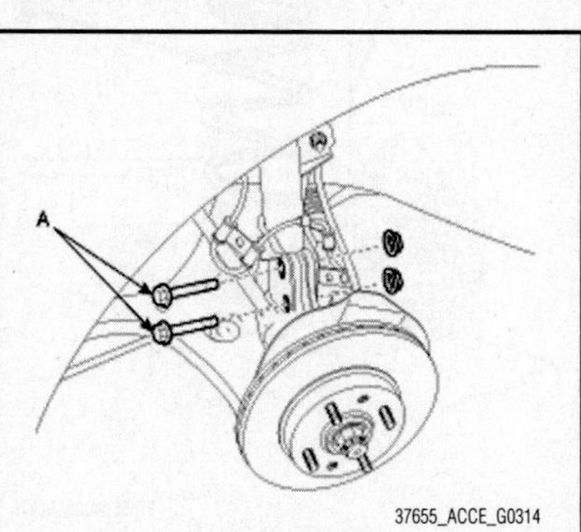

Fig. 174 Remove the strut assembly mounting bolts (A)

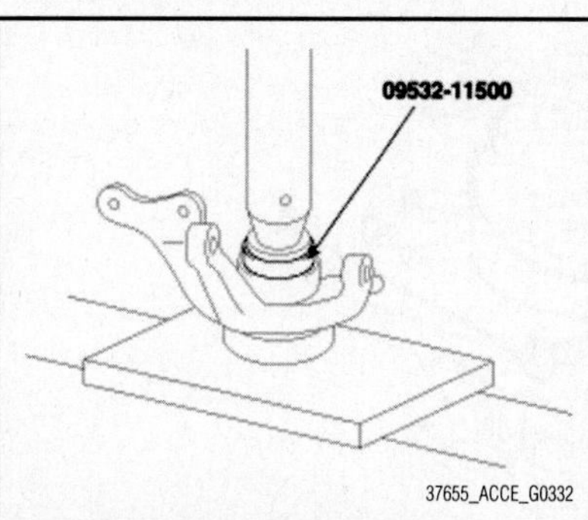

Fig. 178 Using the Special Tool (09532-11500), press-in the bearing to the knuckle

c. Install the special tools (09517-29000, 09517-21500) as shown.

d. Separate the hub from the knuckle by turning the nut of the special tool (09517-21500).

e. Using a plastic hammer, remove the dust cover from the knuckle.

f. Remove the bearing inner race from the hub using the special tool (09495-33000).

g. Remove the wheel bearing outer race from the knuckle using the special tools (09495-33100, 09517-29000).

To install:

12. Preparation for installation:

a. Check the hub for cracks and the splines for wear.

b. Check the brake disc for scoring and damage.

c. Check the steering knuckle for cracks.

d. Check the bearing for cracks or damage.

13. Install the knuckle to the hub:

a. Apply multi-purpose grease to the contacting surface of the knuckle hub and bearing thinly.

b. Using the Special Tool (09532-11500), press-in the bearing to the knuckle.

c. Install the snap ring.

WARNING

Press-in the outer race of the wheel bearing to prevent damage to the bearing assembly. When installing a bearing assembly, always use a new one. The right and the left bearings must be replaced as a matched set.

d. Using a plastic hammer, install the dust cover.

e. Press-in the hub to the knuckle. Press fit load : 20–25 KN (2000–2500 kgf)

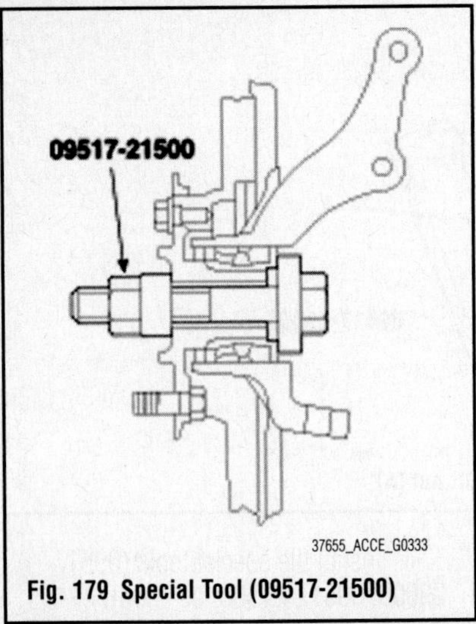

Fig. 179 Special Tool (09517-21500)

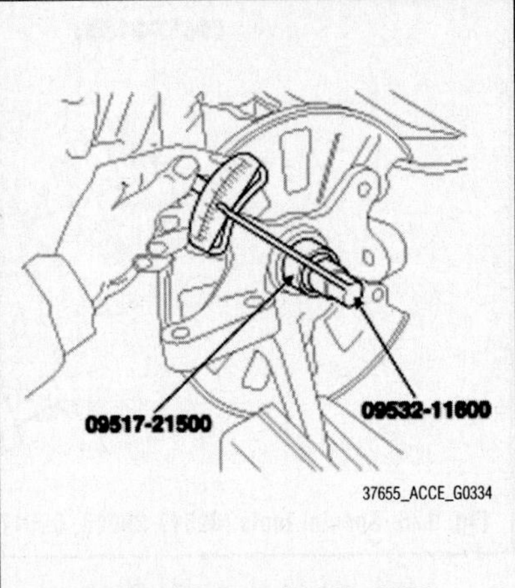

Fig. 180 Special Tools (09517-21500, 09532-11600)

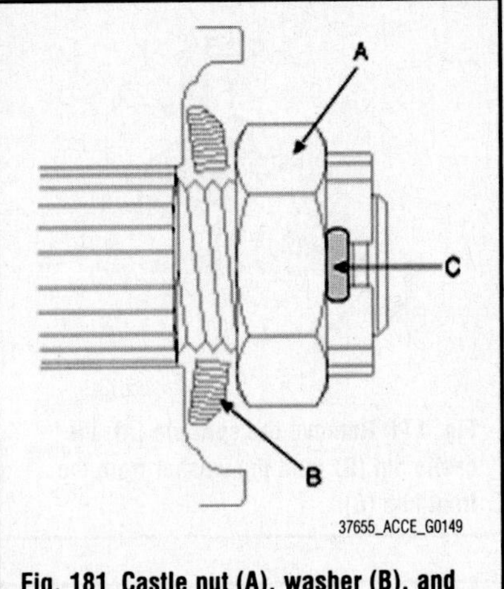

Fig. 181 Castle nut (A), washer (B), and split pin (C) positioning

⁂ WARNING

Press-in the inner race of the wheel bearing to prevent damage to the bearing assembly.

f. Tighten the hub and the knuckle to 144–188 ft. lbs. (200–260 Nm) using the Special Tool (09517-21500).

g. Measure the hub bearing starting torque. Hub bearing starting torque (Limit): 11 inch lbs. (1.3 Nm) or less.

h. If the starting torque is 0 ft. lbs. (0 Nm), measure the hub bearing axial play.

i. If the hub axial play exceeds the limit while the nut is tightened to 145–188 ft. lbs. (200–260 Nm), the bearing, hub and knuckle are not installed correctly. Repeat the disassembly and assembly procedure. Hub bearing axial play (Limit): 0.0003 in. (0.008 mm) or less.

j. Remove the Special Tool.

k. Fix the brake disc with the mounting screws.

14. Installation is the reverse of removal, noting the following:

15. Tighten the strut assembly mounting bolts to 72–86 ft. lbs. (100–120 Nm).

16. Tighten the ball joint assembly mounting bolt to 72–86 ft. lbs. (100–120 Nm).

17. Install the washer, the castle nut, and split pin to the front hub assembly and tighten to 145–188 ft. lbs. (200–260 Nm).

a. The washer should be assembled with convex surface outward when installing the castle nut and split pin.

COIL SPRING

REMOVAL & INSTALLATION

See Figures 182 through 185.

1. Before servicing the vehicle, refer to the Precautions Section.
2. Raise and safely support the vehicle.
3. Remove the wheel and tire.
4. Remove the brake hose bracket and the wheel speed sensor wire bracket. The brake hose should not expand when the rear torsion axle beam is hanged down from the body.
5. After placing a jack at the bottom of the rear torsion axle beam, remove the rear shock absorber lower mounting bolt.
6. Remove the rear coil spring.

To install:

7. Install the upper and lower pads on the coil spring by aligning the grooves on the pads.

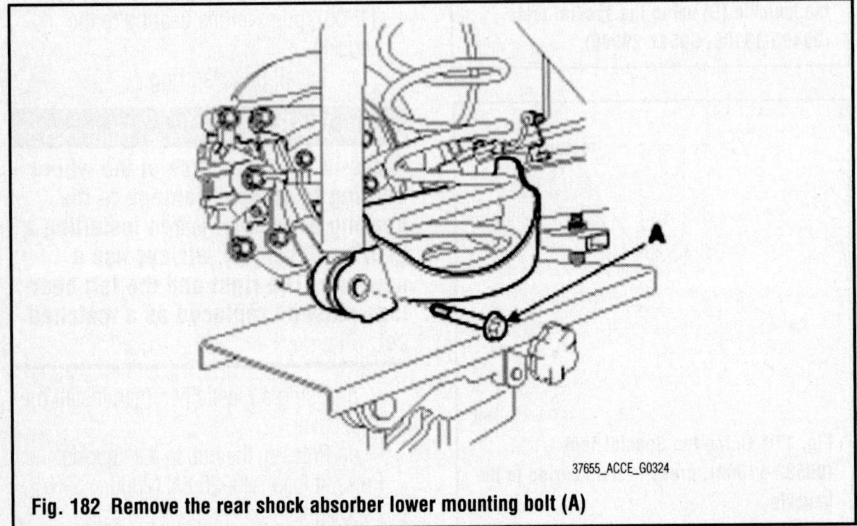

Fig. 182 Remove the rear shock absorber lower mounting bolt (A)

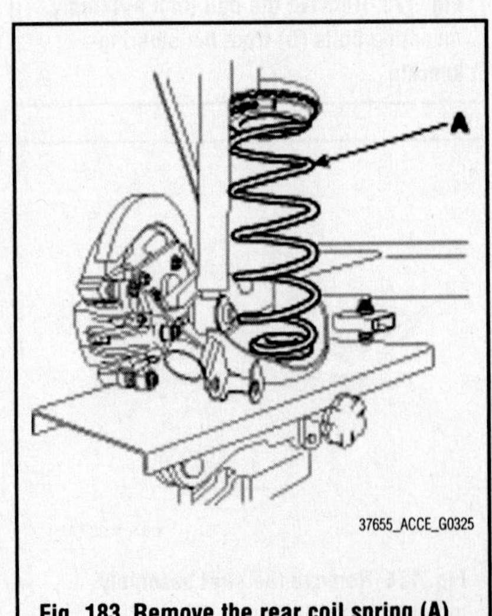

Fig. 183 Remove the rear coil spring (A)

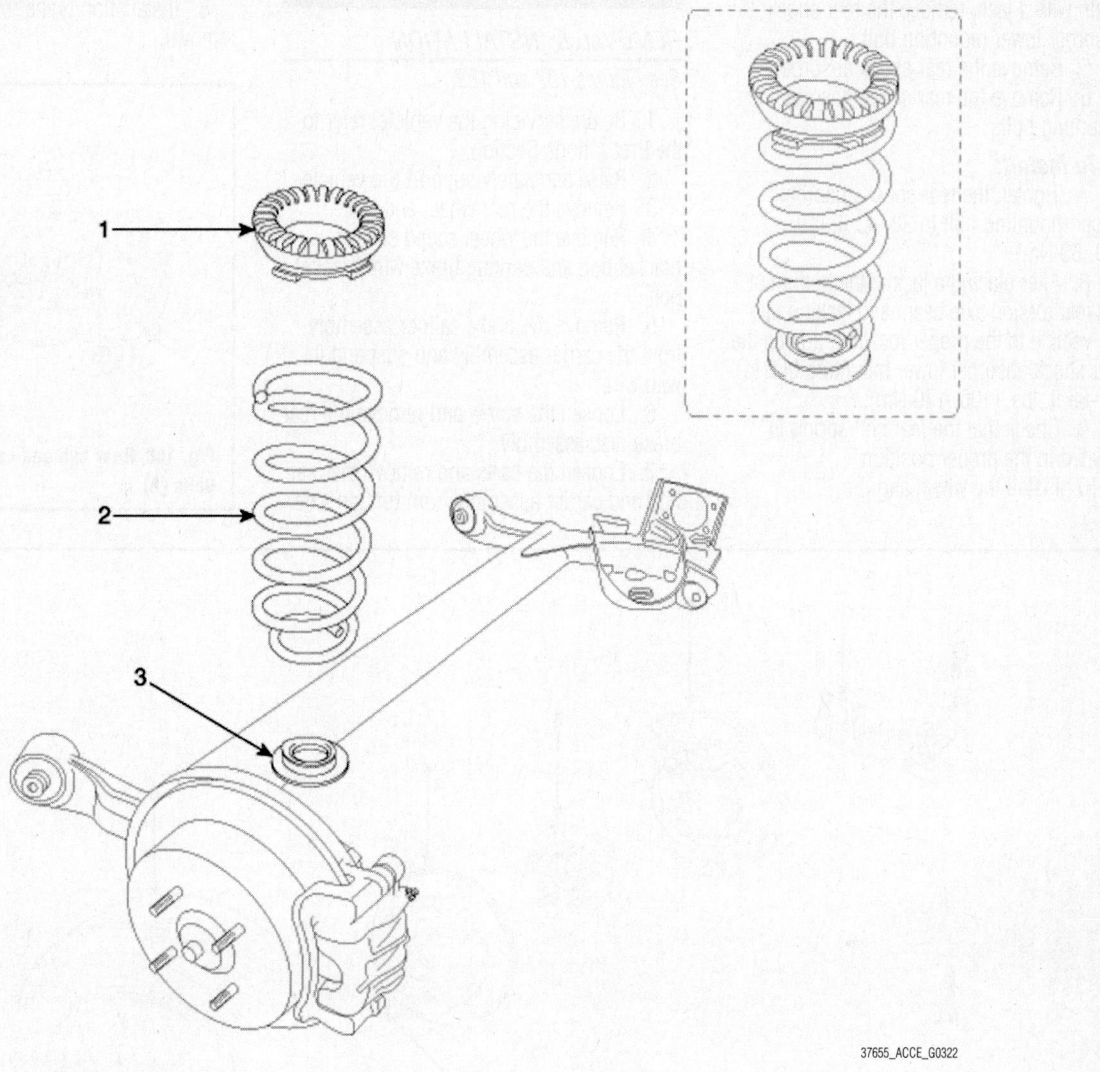

Fig. 184 Rear coil spring and components

8. Place the coil spring with the pads on the torsion axle beam and support it with a jack.

9. Install the rear shock absorber mounting bolt by lifting the rear torsion axle beam and tighten to 72–86 ft. lbs. (100–120 Nm).

10. Install the wheel speed sensor wire bracket bolt and the brake pressure hose bracket bolt.

11. Install the wheel and tire.

SHOCK ABSORBER

REMOVAL & INSTALLATION

See Figures 182 and 186.

1. Before servicing the vehicle, refer to the Precautions Section.

2. Raise and safely support the vehicle.

3. Remove the wheel and tire.

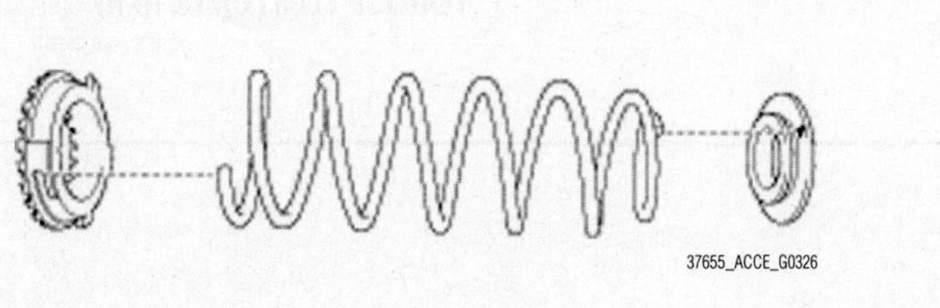

Fig. 185 Aligning the grooves on the upper and lower pads

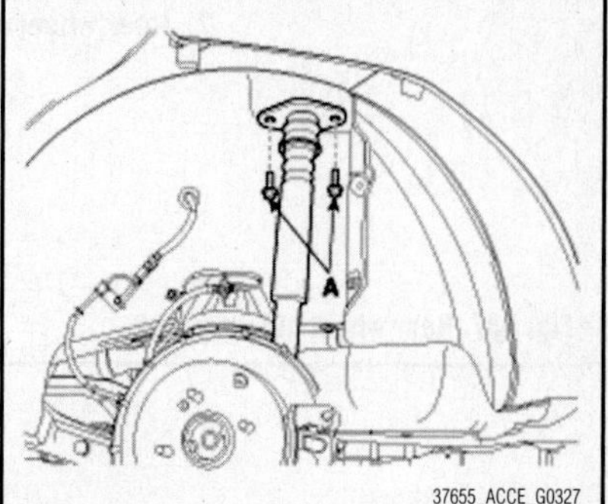

Fig. 186 Rear shock absorber mounting bolts (A)

4. After supporting the rear torsion axle beam with a jack, remove the rear shock absorber lower mounting bolt.
5. Remove the rear shock absorber.
6. Remove the rear shock absorber mounting bolts.

To install:

7. Tighten the rear shock absorber upper mounting bolt to 28–43 ft. lbs. (40–60 Nm).
8. After placing a jack at the bottom of the rear torsion axle beam and jacking up the vehicle to the proper location, tighten the rear shock absorber lower mounting bolt to 72–86 ft. lbs. (100–120 Nm).
9. Check that the rear coil spring is located in the proper position.
10. Install the wheel and tire.

WHEEL HUB & BEARING

REMOVAL & INSTALLATION

See Figures 187 and 188.

1. Before servicing the vehicle, refer to the Precautions Section.
2. Raise and safely support the vehicle.
3. Remove the rear wheel and tire.
4. Remove the wheel speed sensor bracket bolt and parking brake wire bracket bolt.
5. Remove the brake caliper assembly from the carrier assembly and suspend it with wire.
6. Loosen the screw and remove the rear brake disc assembly.
7. Loosen the bolts and remove the rear hub and carrier assembly from torsion axle.

To install:

8. Installation is the reverse of removal.

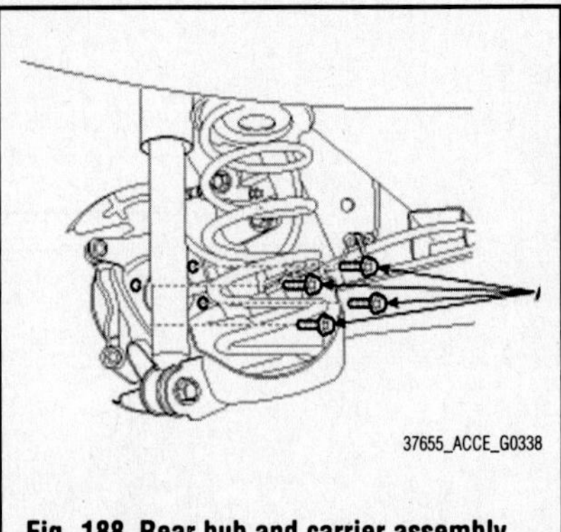

Fig. 188 Rear hub and carrier assembly bolts (A)

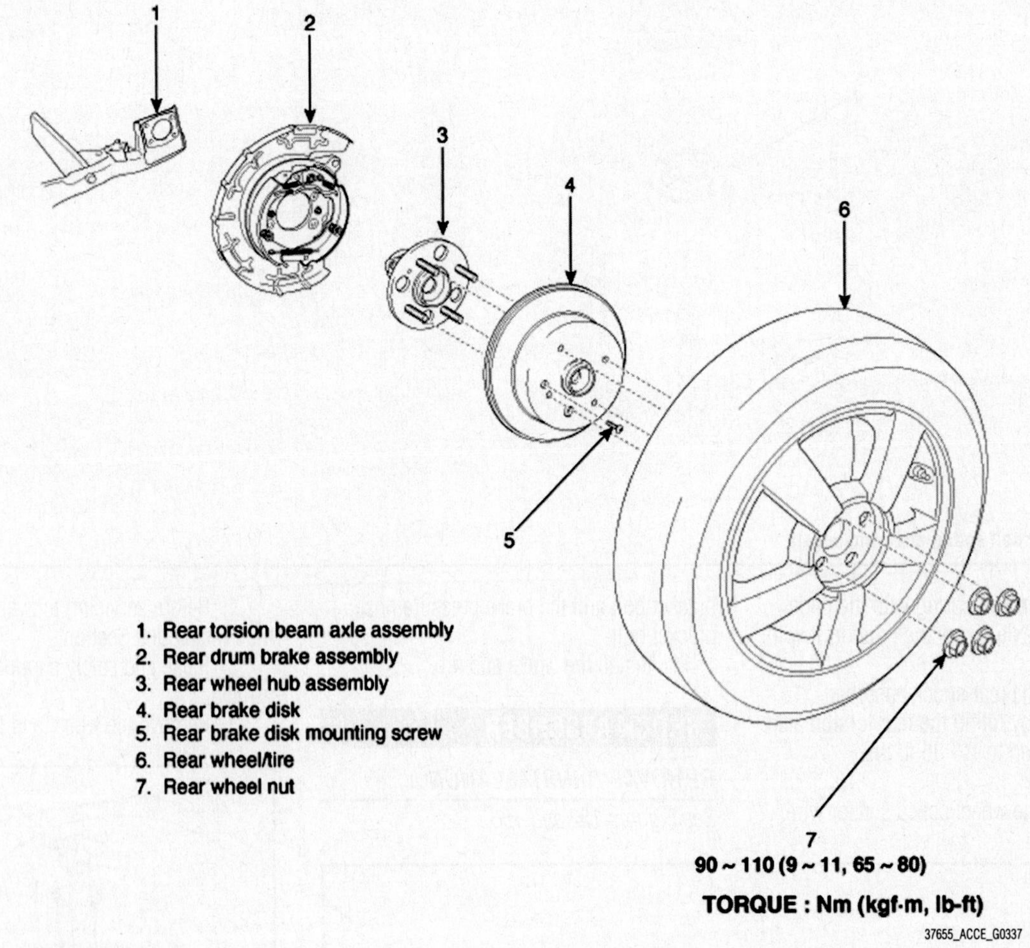

1. Rear torsion beam axle assembly
2. Rear drum brake assembly
3. Rear wheel hub assembly
4. Rear brake disk
5. Rear brake disk mounting screw
6. Rear wheel/tire
7. Rear wheel nut

7
90 ~ 110 (9 ~ 11, 65 ~ 80)

TORQUE : Nm (kgf·m, lb-ft)

37655_ACCE_G0337

Fig. 187 Rear wheel hub assembly

HYUNDAI

Azera

2

SPECIFICATIONS AND MAINTENANCE CHARTS

ENGINE AND VEHICLE IDENTIFICATION

Engine							Model Year	
Code ①	Liters (cc)	Cu. In.	Cyl.	Fuel Sys.	Engine Type	Eng. Mfg.	Code ②	Year
D	3.3 (3342)	203.86	V6	MPFI	DOHC	Hyundai	9	2009
F	3.8 (3778)	230.55	V6	MPFI	DOHC	Hyundai	A	2010

MPFI: Multi-Point Fuel Injection

DOHC: Double Overhead Camshafts

① 8th digit of VIN

② 10th digit of VIN

37655_AZER_C0001

GENERAL ENGINE SPECIFICATIONS

All measurements are given in inches.

Year	Model	Engine Displacement Liters	Engine Series VIN	Net Horsepower @ rpm	Net Torque @ rpm (ft. lbs.)	Bore x Stroke (in.)	Com-pression Ratio	Oil Pressure @ rpm
2009	Azera GLS	3.3	D	234@6000	226@3500	3.622 x 3.299	10.4:1	18.8@1000
	Azera Limited	3.8	F	263@6000	257@4500	3.780 x 3.743	10.4:1	18.8@1000
2010	Azera GLS	3.3	D	234@6000	226@3500	3.622 x 3.299	10.4:1	18.8@1000
	Azera Limited	3.8	F	263@6000	257@4500	3.780 x 3.743	10.4:1	18.8@1000

37655_AZER_C0002

GASOLINE ENGINE TUNE-UP SPECIFICATIONS

Year	Engine Displacement Liters	Engine VIN	Spark Plug Gap (in.)	Ignition Timing (deg.)		Fuel Pump (psi)	Idle Speed (rpm)		Valve Clearance	
				MT	AT		MT	AT	In.	Ex.
2009	3.3	D/F	0.039-0.043	①	①	54.3-55.8	③	③	HYD	HYD
	3.8	E/F	0.039-0.043	①	①	54.3-55.8	③	③	HYD	HYD
2010	3.3	D/F	0.039-0.043	①	①	54.3-55.8	③	③	HYD	HYD
	3.8	E/F/H	0.039-0.043	①	①	54.3-55.8	③	③	HYD	HYD

Follow the figures on the label if they differ from those in this chart.

HYD: Hydraulic

① Ignition timing is computer controlled and is not adjustable

③ Idle speed is maintained by the Electronic Control Module (ECM)

37655_AZER_C0003

CAPACITIES

Year	Model	Engine Displacement Liters	Engine VIN	Engine Oil with Filter (qts.)	Transmission (pts.) Manual	Auto. ①	Fuel Tank (gal.)	Cooling System (qts.)
2009	Azera	3.3	D	5.5	—	11.5	19.8	9.1
	Azera	3.8	F	5.5	—	11.5	19.8	9.1
2010	Azera	3.3	D	5.5	—	11.5	19.8	9.1
	Azera	3.8	F	5.5	—	11.5	19.8	9.1

NOTE: All capacities are approximate. Add fluid gradually and check to be sure a proper fluid level is obtained.

37655_AZER_C0004

FLUID SPECIFICATIONS

Year	Model	Engine Displacement Liters	Engine ID/VIN	Engine Oil	Auto. Trans.	Manual Trans.	Power Steering Fluid	Brake Master Cylinder
2009	Azera	3.3	D	5W-20	①	—	PSF-3	②
	Azera	3.8	F	5W-20	①	—	PSF-3	②
2010	Azera	3.3	D	5W-20	①	—	PSF-3	②
	Azera	3.8	F	5W-20	①	—	PSF-3	②

DOT: Department Of Transportation

① DIAMOND ATF SP-III, SK ATF SP-III

② DOT 3, DOT 4, or equivalent

37655_AZER_C0005

VALVE SPECIFICATIONS

Year	Engine Displacement Liters	Engine VIN	Seat Angle (deg.)	Face Angle (deg.)	Spring Test Pressure (lbs. @ in.)	Spring Installed Height (in.)	Stem-to-Guide Clearance (in.) Intake	Exhaust	Stem Diameter (in.) Intake	Exhaust
2009	3.3	D/F	44.75-45.20	45.25-45.75	90.4-96.2 @0.953	NA	0.0008-0.0019	0.0012-0.0021	0.2151-0.2157	0.2149-0.2153
	3.8	E/F	44.75-45.20	45.25-45.75	90.4-96.2 @0.953	NA	0.0008-0.0019	0.0012-0.0021	0.2151-0.2157	0.2149-0.2153
2010	3.3	D/F	44.75-45.20	45.25-45.75	90.4-96.2 @0.953	NA	0.0008-0.0019	0.0012-0.0021	0.2151-0.2157	0.2149-0.2153
	3.8	E/F	44.75-45.20	45.25-45.75	90.4-96.2 @0.953	NA	0.0008-0.0019	0.0012-0.0021	0.2151-0.2157	0.2149-0.2153

NA: Not Available

37655_AZER_C0006

CAMSHAFT AND BEARING SPECIFICATIONS CHART

All measurements are given in inches.

Year	Engine Displ. Liters	Engine ID/VIN	Journal Dia.	Brg. Oil Clearance	Shaft End-play	Runout	Journal Bore	Lobe Height Intake	Lobe Height Exhaust
2009	3.3	D/F	①	②	0.0008-0.0071	NA	NA	1.8228	1.8031
	3.8	E/F	①	②	0.0008-0.0071	NA	NA	1.8425	1.8031
2010	3.3	D/F	①	②	0.0008-0.0071	NA	NA	1.8228	1.8031
	3.8	E/F	①	②	0.0008-0.0071	NA	NA	1.8425	1.8031

NA: Not Available

① Intake No. 1 is 1.1009-1.1016 inch

Intake No. 2, 3, 4 are 0.9430-0.9437 inch

Exhaust No.1 is 1.1009-1.1016 inch

Exhaust No. 2, 3, 4 are 0.9430-0.9437 inch

② Intake No. 1 is 0.0008-0.0022 inch

Intake No. 2, 3, 4 are 0.0012-0.0026 inch

Exhaust No.1 is 0.0008-0.0022 inch

Exhaust No. 2, 3, 4 are 0.0012-0.0026 inch

37655_AZER_C0007

CRANKSHAFT AND CONNECTING ROD SPECIFICATIONS

All measurements are given in inches.

Year	Engine Displacement Liters	Engine VIN	Crankshaft Main Brg. Journal Dia.	Crankshaft Main Brg. Oil Clearance	Crankshaft Shaft End-play	Crankshaft Thrust on No.	Connecting Rod Journal Diameter	Connecting Rod Oil Clearance	Connecting Rod Side Clearance
2009	3.3	D/F	2.7142-2.7149	0.0008-0.0016	0.0039-0.0110	3	2.1635-2.1642	0.0015-0.0022	0.0039-0.0098
	3.8	E/F	2.7142-2.7149	0.0008-0.0016	0.0039-0.0110	3	2.1635-2.1642	0.0015-0.0022	0.0039-0.0098
2010	3.3	D	2.7142-2.7149	0.0008-0.0016	0.0039-0.0110	3	2.1635-2.1642	0.0015-0.0022	0.0039-0.0098
	3.8	E/F	2.7142-2.7149	0.0008-0.0016	0.0039-0.0110	3	2.1635-2.1642	0.0015-0.0022	0.0039-0.0098

37655_AZER_C0008

PISTON AND RING SPECIFICATIONS

All measurements are given in inches.

Year	Engine Displ. Liters	Engine VIN	Piston Clearance	Ring Gap Top Compression	Ring Gap Bottom Compression	Ring Gap Oil Control	Ring Side Clearance Top Compression	Ring Side Clearance Bottom Compression	Ring Side Clearance Oil Control
2009	3.3	D/F	0.0008-0.0016	0.0067-0.0126	0.0126-0.0185	0.0078-0.0275	0.0016-0.0031	0.0012-0.0027	0.0024-0.0059
	3.8	E/F	0.0008-0.0016	0.0067-0.0126	0.0126-0.0185	0.0078-0.0275	0.0016-0.0031	0.0012-0.0027	0.0024-0.0059
2010	3.3	D/F	0.0008-0.0016	0.0067-0.0126	0.0126-0.0185	0.0078-0.0275	0.0016-0.0031	0.0012-0.0027	0.0024-0.0059
	3.8	E/F	0.0008-0.0016	0.0067-0.0126	0.0126-0.0185	0.0078-0.0275	0.0016-0.0031	0.0012-0.0027	0.0024-0.0059

NA: Not Available

37655_AZER_C0009

TORQUE SPECIFICATIONS

All readings in ft. lbs.

Year	Engine Displacement Liters	Engine VIN	Cylinder Head Bolts	Main Bearing Bolts	Rod Bearing Bolts	Crankshaft Damper Bolts	Flexplate Bolts	Manifold Intake	Manifold Exhaust	Spark Plugs	Oil Pan Drain Plug
2009	3.3	D	①	②	③	210-224	53-56	14-17	29-33	15-22	25-33
	3.8	F	①	②	③	210-224	53-56	14-17	29-33	15-22	25-33
2010	3.3	D	①	②	③	210-224	53-56	14-17	29-33	15-22	25-33
	3.8	F	①	②	③	210-224	53-56	14-17	29-33	15-22	25-33

① Step 1: 29 ft. lbs.
Step 2: Plus 120 degrees
Step 3: Plus 90 degrees

② M11 bolts (inner) Step 1: 36 ft. lbs.
M11 bolts (inner) Step 2: Plus 90 degrees
M8 bolts (outer) Step 1: 15 ft. lbs.
M8 bolts (outer) Step 2: Plus 120 degrees
M8 bolts (side): 22-23 ft. lbs.

③ Step 1: 15 ft. lbs.
Step 2: Plus 90 degrees

37655_AZER_C0010

WHEEL ALIGNMENT

Year	Model		Caster Range (+/-Deg.)	Caster Preferred Setting (Deg.)	Camber Range (+/-Deg.)	Camber Preferred Setting (Deg.)	Toe-in (Deg.)
2009	Azera	Front	1.00	+4.83	0.50	0.00	0.00 +/- 0.20
		Rear	—	—	0.50	-0.50	0.20 +/- 0.20
2010	Azera	Front	1.00	+4.83	0.50	0.00	0.00 +/- 0.20
		Rear	—	—	0.50	-0.50	0.20 +/- 0.20

37655_AZER_C00011

TIRE, WHEEL AND BALL JOINT SPECIFICATIONS

Year	Model	OEM Tires Standard	OEM Tires Optional	Tire Pressures (psi) Front	Tire Pressures (psi) Rear	Wheel Size	Ball Joint Inspection	Lug Nut Torque (ft. lbs.)
2009	Azera	P225/60R16	P235/55R17	①	①	6.5J x 16 7.0J x 17	②	65-80
2010	Azera	P225/60R16	P235/55R17	①	①	6.5J x 16 7.0J x 17	②	65-80

① Refer to placard on vehicle for proper inflation pressure.

② Replace if any measurable movement is found.

37655_AZER_C00012

BRAKE SPECIFICATIONS

All measurements in inches unless noted

Year	Model		Brake Disc Original Thickness	Brake Disc Minimum Thickness	Brake Disc Maximum Runout	Brake Drum Diameter Original Inside Diameter	Brake Drum Diameter Max. Wear Limit	Brake Drum Diameter Maximum Machine Diameter	Minimum Lining Thickness	Brake Caliper Bracket Bolts (ft. lbs.)	Brake Caliper Mounting Bolts (ft. lbs.)
2009	Azera	F	1.100	1.040	0.002	—	—	—	0.079	58-72	16-23
		R	0.390	0.310	0.002	—	—	—	0.080	58-72	16-23
2010	Azera	F	1.100	1.040	0.002	—	—	—	0.079	58-72	16-23
		R	0.390	0.310	0.002	—	—	—	0.080	58-72	16-23

① Drum roundness Service Limit: 0.00236 inch

37655_AZER_C00013

SCHEDULED MAINTENANCE INTERVALS

HYUNDAI—AZERA

TO BE SERVICED	TYPE OF SERVICE	VEHICLE MILEAGE INTERVAL (x1000)												
		7.5	15	22.5	30	37.5	45	52.5	60	67.5	75	82.5	90	97.5
Engine oil & filter	R	✓	✓	✓	✓	✓	✓	✓	✓	✓	✓	✓	✓	✓
Automatic transaxle fluid ①	S/I		✓		✓		✓		✓		✓		✓	
Brake pads, calipers & rotors	S/I		✓		✓		✓		✓		✓		✓	
Driveshafts & boots	S/I		✓		✓		✓		✓		✓		✓	
Air cleaner filter ②	R	✓	✓	✓	✓	✓	✓	✓	✓	✓	✓	✓	✓	✓
A/C refrigerant	S/I		✓		✓		✓		✓		✓		✓	
Brake fluid	I				✓				✓				✓	
Engine coolant ③	R				✓				✓				✓	
Fuel hose, vapor hose & fuel filter cap	S/I	✓	✓	✓	✓	✓	✓	✓	✓	✓	✓	✓	✓	✓
Spark plugs (Iridium coated)	R													✓
Drive belts	S/I				✓				✓				✓	
Exhaust pipe connections, muffler & suspension bolts	S/I		✓		✓		✓		✓		✓		✓	
Valve clearance ④	S/I								✓					
Electronic throttle control	S/I		✓		✓		✓		✓		✓		✓	
Brake hoses & lines	S/I		✓		✓		✓		✓		✓		✓	
Rear brake discs, linings & parking brake	S/I				✓				✓				✓	
Steering gear rack, linkage & boots	S/I	✓	✓	✓	✓	✓	✓	✓	✓	✓	✓	✓	✓	✓
Power steering pump, belt, hoses	S/I		✓		✓		✓		✓		✓		✓	
Fuel tank air filter ⑤	S/I		✓		✓		✓		✓		✓		✓	
Climate control air filter ⑥	R													
Fuel filter ⑦	R		✓		✓		✓		✓		✓		✓	
Fuel lines & connections	S/I	✓	✓	✓	✓	✓	✓	✓	✓	✓	✓	✓	✓	✓
Vacuum & crankcase ventilation hoses	S/I				✓				✓				✓	

R: Replace S/I: Service or Inspect

① Replace at 105,000 miles
② Replace at 30,000 miles
③ Replace every 24 months or 30,000 miles
④ Replace at 60,000 miles
⑤ Replace every 30,000 miles
⑥ Replace at 10,000 miles, or 12 months. See Severe Service.
⑦ Replace at 37,500 miles.

FREQUENT OPERATION MAINTENANCE (SEVERE SERVICE)

If a vehicle is operated under any of the following conditions it is considered severe service:

- Extremely dusty areas.

- 50% or more of the vehicle operation is in 90°F (32°C) or higher temperatures, or constant operation in temperatures below 32°F (0°C).

- Prolonged idling (vehicle operation in stop and go traffic).

- Frequent short running periods (engine does not warm to normal operating temperatures).

- Police, taxi, delivery usage or trailer towing usage.

Oil & oil filter: change every 3,000 miles.

Brake pads, calipers & rotors: service or inspect more frequently

Driveshaft boots: service or inspect more frequently

Steering gear rack, linkage & boots: service or inspect more frequently

Air cleaner filter: service or inspect more frequently

Automatic transaxle fluid & filter: replace every 30,000 miles.

Climate control air filter: replace as necessary.

37655_AZER_C0014

BRAKES — INFORMATION AND PRECAUTIONS

ANTI-LOCK SYSTEMS

• Certain components within the ABS system are not intended to be serviced or repaired individually.

• Do not use rubber hoses or other parts not specifically specified for and ABS system. When using repair kits, replace all parts included in the kit. Partial or incorrect repair may lead to functional problems and require the replacement of components.

• Lubricate rubber parts with clean, fresh brake fluid to ease assembly. Do not use shop air to clean parts; damage to rubber components may result.

• Use only DOT 3 brake fluid from an unopened container.

• If any hydraulic component or line is removed or replaced, it may be necessary to bleed the entire system.

• A clean repair area is essential. Always clean the reservoir and cap thoroughly before removing the cap. The slightest amount of dirt in the fluid may plug an orifice and impair the system function. Perform repairs after components have been thoroughly cleaned; use only denatured alcohol to clean components. Do not allow ABS components to come into contact with any substance containing mineral oil; this includes used shop rags.

• The Anti-Lock control unit is a microprocessor similar to other computer units in the vehicle. Ensure that the ignition switch is **OFF** before removing or installing controller harnesses. Avoid static electricity discharge at or near the controller.

• If any arc welding is to be done on the vehicle, the control unit should be unplugged before welding operations begin.

DISC AND DRUM SYSTEMS

CAUTION

Dust and dirt accumulating on brake parts during normal use may contain asbestos fibers from production or aftermarket brake linings. Breathing excessive concentrations of asbestos fibers can cause serious bodily harm. Exercise care when servicing brake parts. Do not sand or grind brake lining unless equipment used is designed to contain the dust residue. Do not clean brake parts with compressed air or by dry brushing. Cleaning should be done by dampening the brake components with a fine mist of water, then wiping the brake components clean with a dampened cloth. Dispose of cloth and all residue containing asbestos fibers in an impermeable container with the appropriate label. Follow practices prescribed by the Occupational Safety and Health Administration (OSHA) and the Environmental Protection Agency (EPA) for the handling, processing, and disposing of dust or debris that may contain asbestos fibers.

BRAKES — BLEEDING THE BRAKE SYSTEM

BLEEDING PROCEDURE

BLEEDING PROCEDURE

See Figure 1.

1. Before servicing the vehicle, refer to the Precautions Section.

WARNING

The reservoir on the master cylinder must be at the MAX (upper) level mark at the start of bleeding procedure and checked after bleeding each brake caliper. Add fluid as required.

2. Make sure the brake fluid level in the reservoir is at the MAX (upper) level line.
3. Have someone slowly pump the brake pedal several times, and then apply steady pressure.

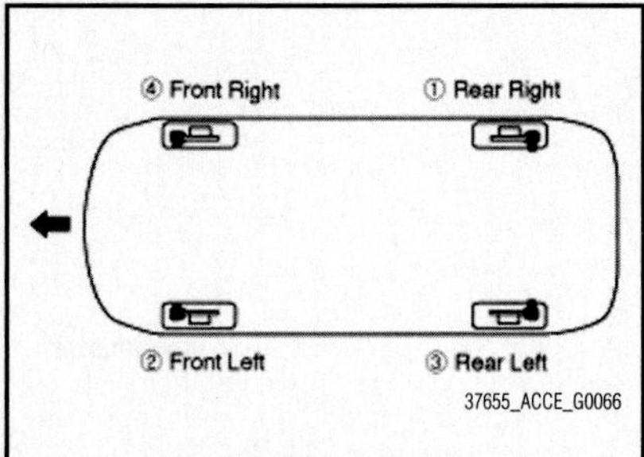

Fig. 1 Brake bleeding sequence

4. Loosen the right-rear brake bleed screw to allow air to escape from the system. Then tighten the bleed screw securely.
5. Repeat the procedure for each wheel in the sequence shown below until air bubbles no longer appear in the fluid.
6. Refill the master cylinder reservoir to the MAX (upper) level line.

BLEEDING THE ABS SYSTEM

This procedure should be followed to ensure adequate bleeding of air and the filling of the ABS unit, the brake lines, and the master cylinder with brake fluid.

1. Before servicing the vehicle, refer to the Precautions Section.
2. Remove the reservoir cap and fill the brake reservoir with brake fluid.

WARNING

If there is any brake fluid on any painted surface, wash it off immediately.

➡When pressure bleeding, do not depress the brake pedal. Recommended brake fluid: DOT3 or DOT4.

3. Connect a clear plastic tube to the wheel cylinder bleeder plug and insert the other end of the tube into a clear plastic bottle that is half filled with clean brake fluid.
4. Connect the Hi-Scan Pro® to the data link connector located underneath the dash panel.
5. Select and operate according to the instructions on the Hi-Scan Pro® screen.

CAUTION

You must obey the maximum operating time of the ABS motor with the Hi-Scan Pro® to prevent the motor pump from burning.

6. Select Hyundai vehicle diagnosis.
7. Select vehicle name.
8. Select Anti-Lock Brake system.
9. Select air bleeding mode.
10. Press "YES" to operate motor pump and solenoid valve.

WARNING

Wait 60 seconds before operating the air bleeding or damage to the motor may occur.

11. Wait 60 seconds before operating the air bleeding.
12. Pump the brake pedal several times, and then loosen the bleeder screw until fluid starts to run out without bubbles. Then, close the bleeder screw.
13. Repeat until there are no more bubbles in the fluid for each wheel.

SPEED SENSORS

REMOVAL & INSTALLATION

Front

See Figures 2 through 6.

1. Before servicing the vehicle, refer to the Precautions Section.
2. Remove the front wheel speed sensor mounting bolt.
3. Remove the two wire bracket bolts.
4. Remove the front wheel guard.
5. Disconnect the wheel speed sensor connector.
6. Remove the front wheel speed sensor.

To install:

7. Installation is the reverse of the removal procedure.

1
2
37655_AZER_G0089

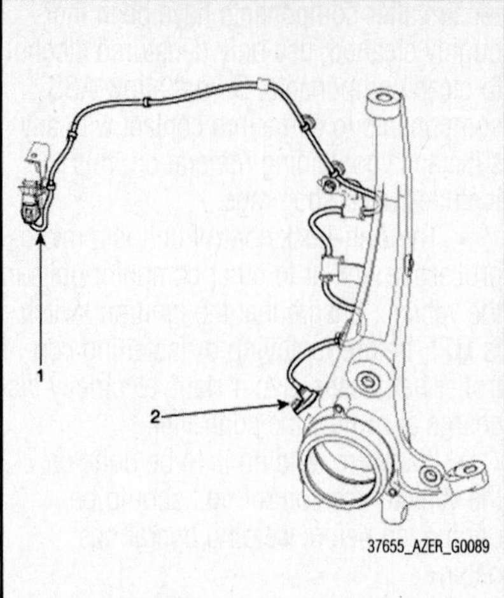

Fig. 3 Front wheel speed sensor (2) and connector (1)

A
B
37655_AZER_G0090

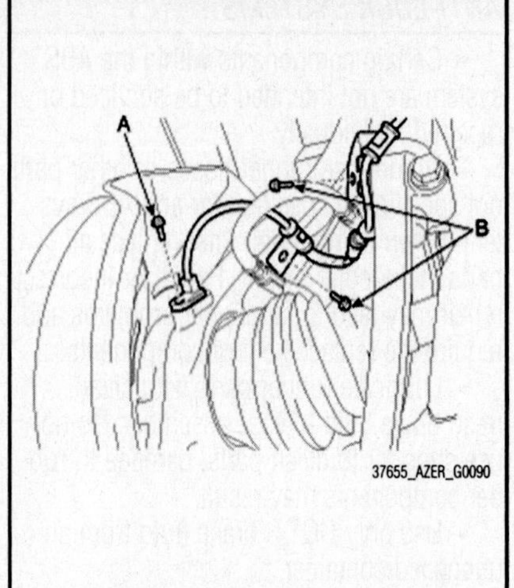

Fig. 4 Remove the front wheel speed sensor mounting bolt (A)

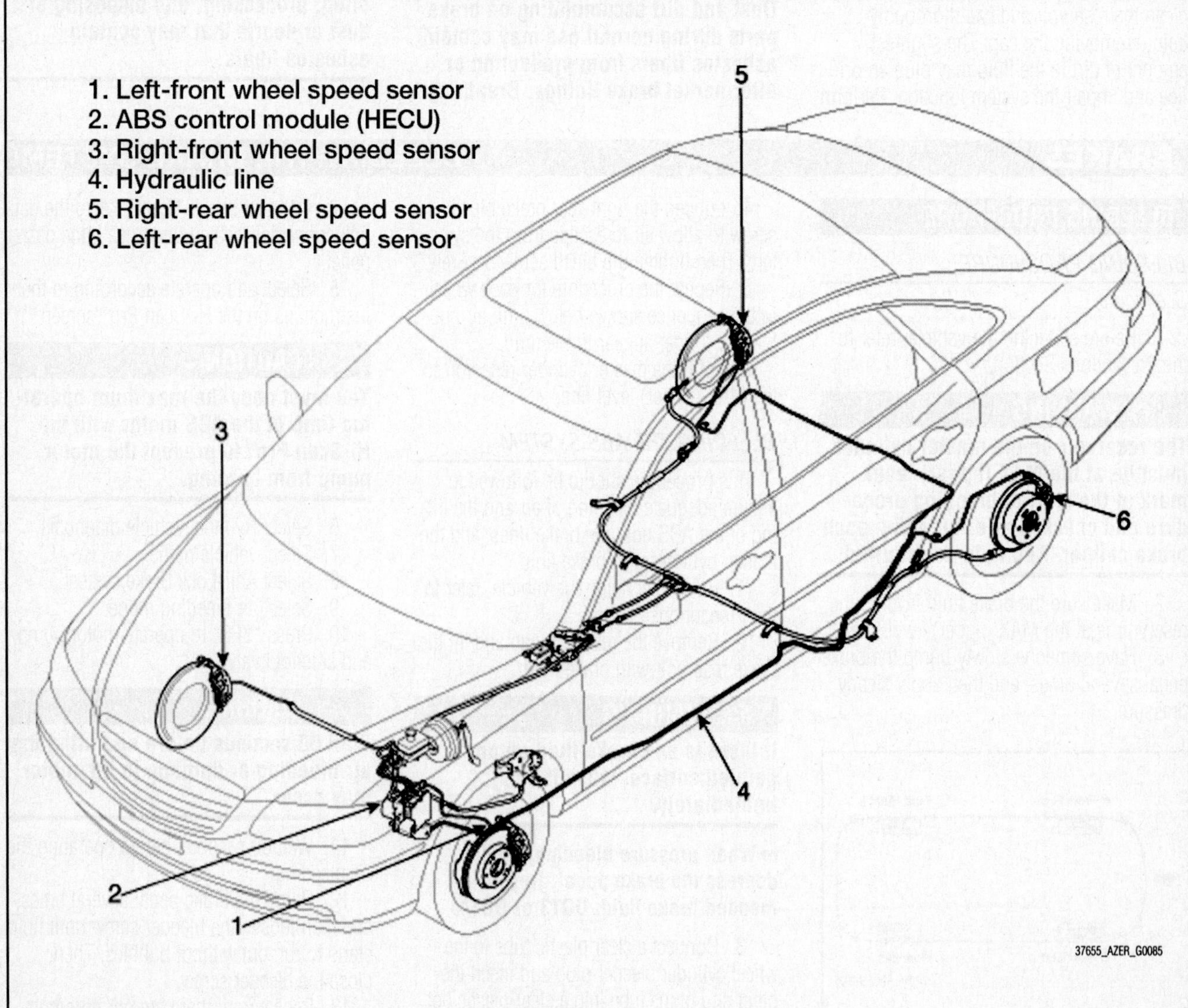

1. Left-front wheel speed sensor
2. ABS control module (HECU)
3. Right-front wheel speed sensor
4. Hydraulic line
5. Right-rear wheel speed sensor
6. Left-rear wheel speed sensor
5
3
6
4
2
1
37655_AZER_G0085

Fig. 2 ABS system components

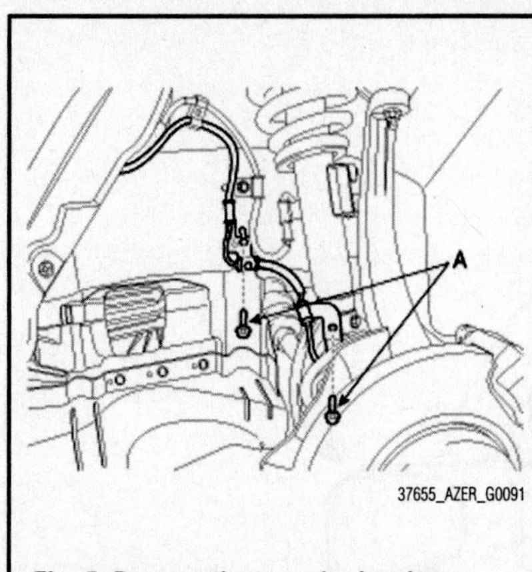

Fig. 5 Remove the two wire bracket bolts (A)

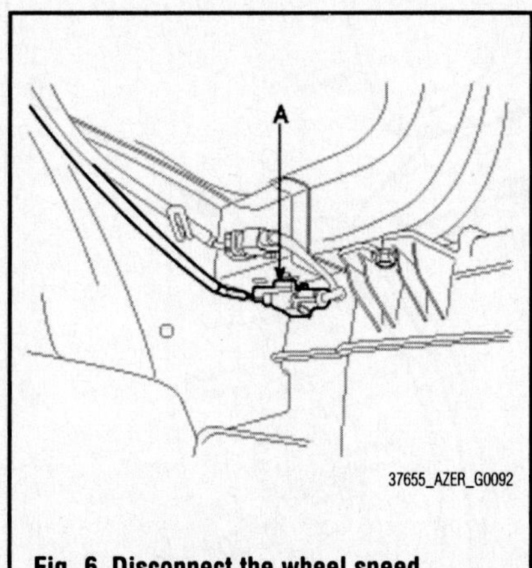

Fig. 6 Disconnect the wheel speed sensor connector (A)

Rear

See Figures 7 through 9.

1. Before servicing the vehicle, refer to the Precautions Section.
2. Remove the rear wheel speed sensor mounting bolt.
3. Remove the rear seat side pad then disconnect the rear wheel speed sensor connector.

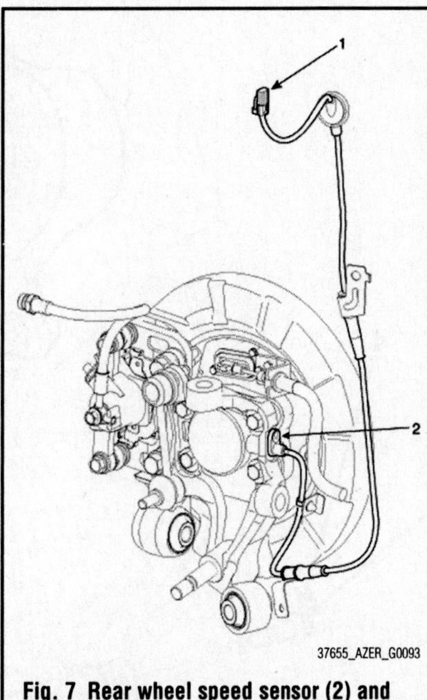

Fig. 7 Rear wheel speed sensor (2) and connector (1)

To install:

4. Installation is the reverse of the removal procedure.

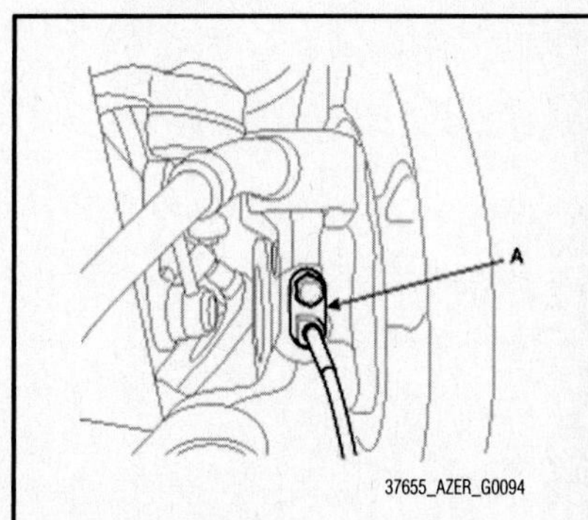

Fig. 8 Remove the rear wheel speed sensor mounting bolt (A)

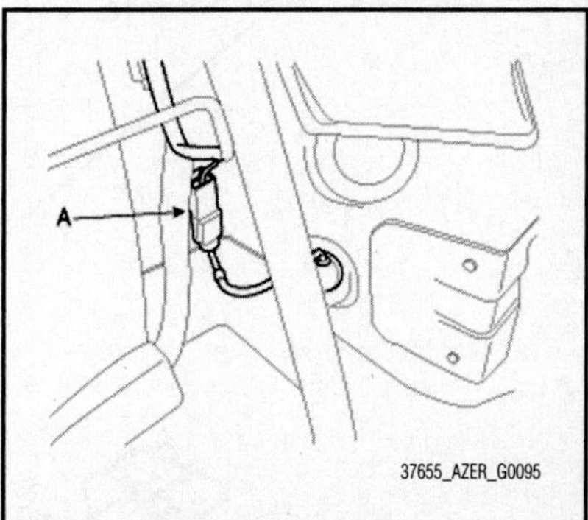

Fig. 9 Disconnect the rear wheel speed sensor connector (A)

BRAKES

FRONT DISC BRAKES

BRAKE CALIPER

REMOVAL & INSTALLATION

See Figures 10 through 12.

1. Before servicing the vehicle, refer to the Precautions Section.
2. Raise and safely support the vehicle.
3. Remove the front wheel and tire from front hub.

WARNING

Be careful not to damage the hub bolts.

4. Remove guide rod bolts and raise the caliper. Check the hoses and pin boots for damage and deterioration.
5. If replacing the caliper, remove the brake hose from the caliper.
6. Remove the caliper mounting bolts.
7. Remove the caliper.

To install:

8. Pivot the caliper down into position. Install the brake caliper and caliper bolts. Tighten the bolts to: 58–72 ft. lbs. (80–100 Nm).
9. Being careful not to damage the pin boot, install the guide rod bolt and tighten it to 16–23 ft. lbs. (22–32 Nm).
10. If replacing the caliper, install the brake hose onto the caliper. Tighten to 10–12 ft. lbs. (14–17 Nm).

➡Replace sealing washer whenever hose is removed.

11. Install wheel and tire.

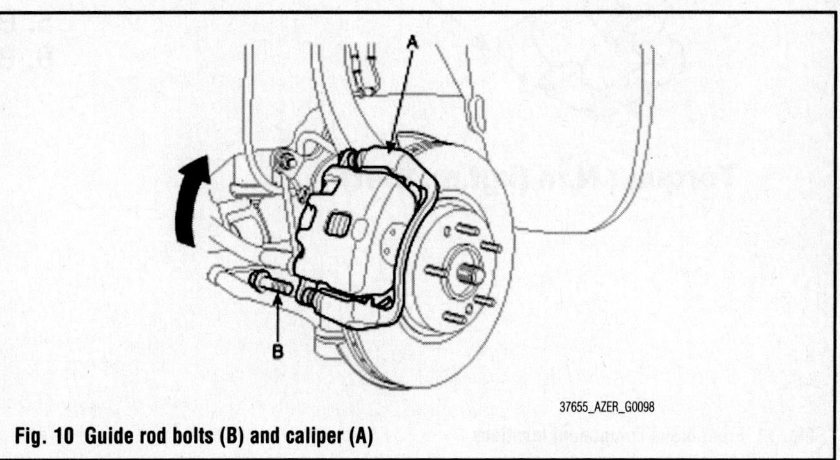

Fig. 10 Guide rod bolts (B) and caliper (A)

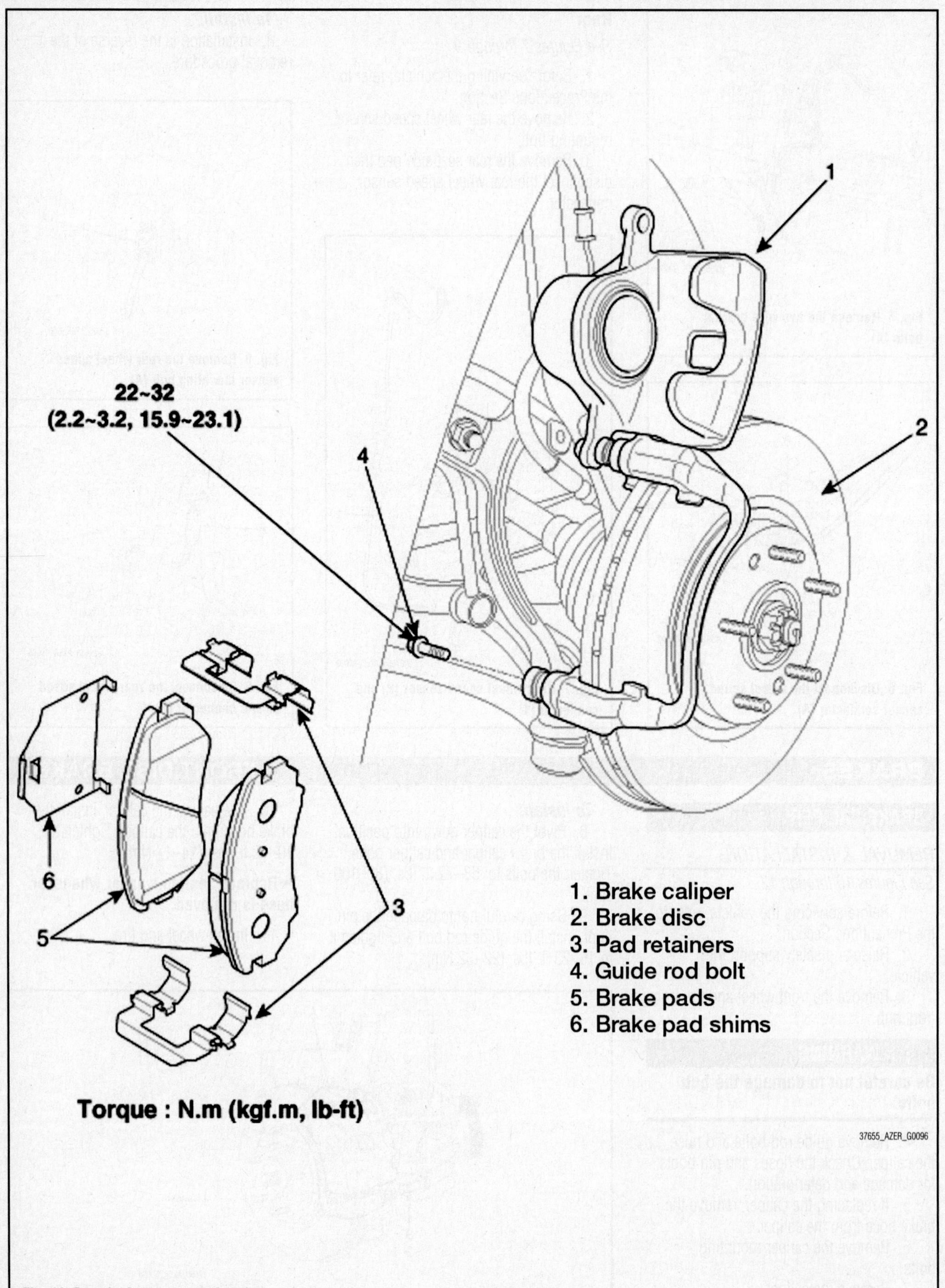

Fig. 11 Front brake component locations

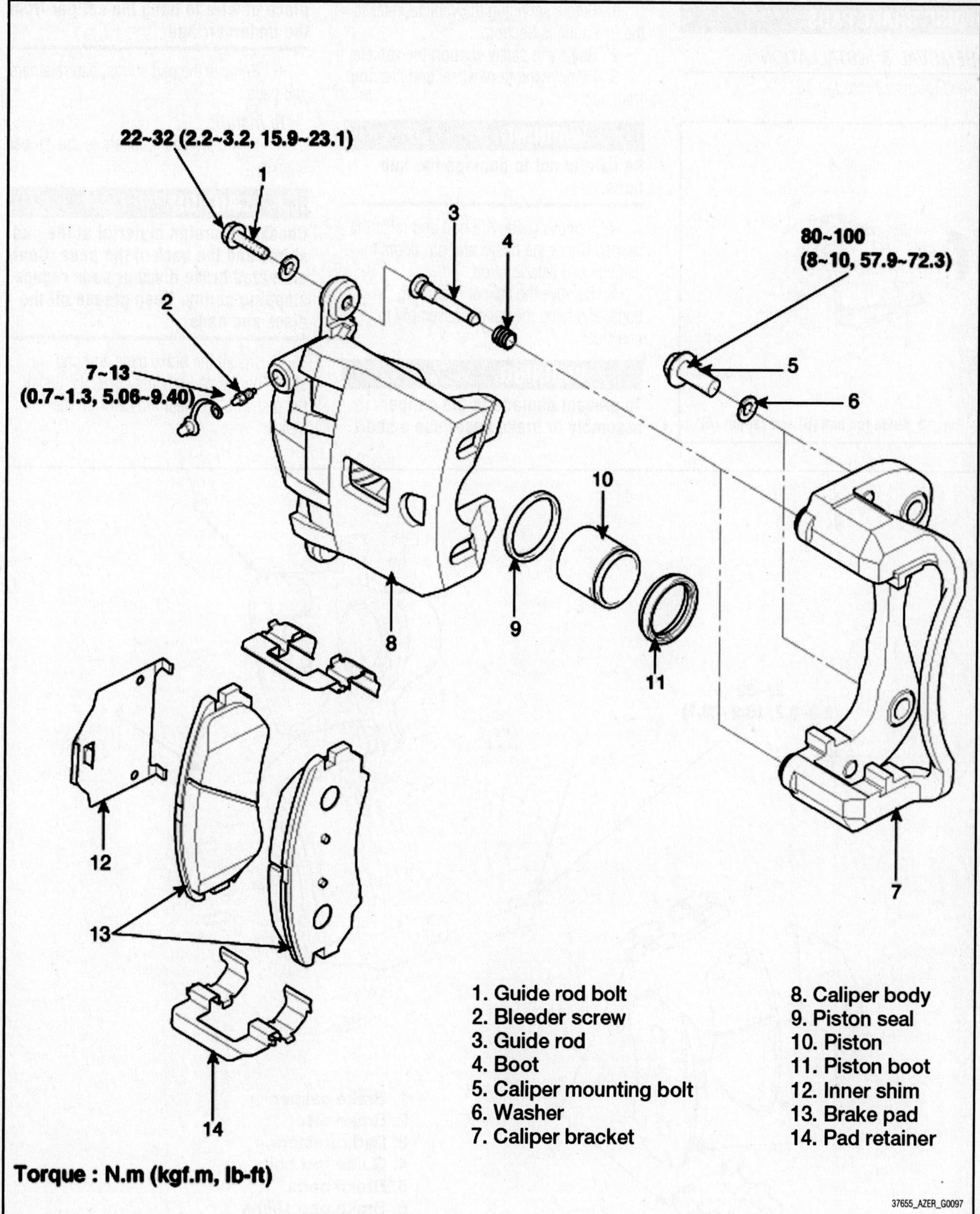

Fig. 12 Front brake components

DISC BRAKE PADS

REMOVAL & INSTALLATION

See Figures 13 through 18.

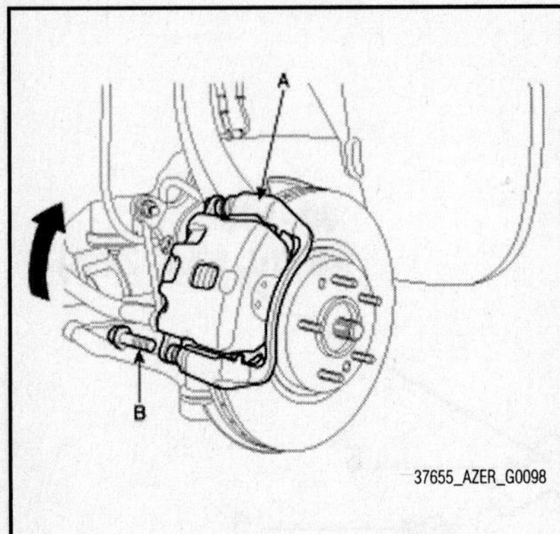

Fig. 13 Guide rod bolt (B) and caliper (A)

1. Before servicing the vehicle, refer to the Precautions Section.
2. Raise and safely support the vehicle.
3. Remove the front wheel and tire from front hub.

⁂ WARNING

Be careful not to damage the hub bolts.

4. Remove guide rod bolt and raise the caliper. Check the hoses and pin boots for damage and deterioration.
5. Remove the caliper mounting bolts, and hang the caliper assembly to one side.

⁂ WARNING

To prevent damage to the caliper assembly or brake hose, use a short piece of wire to hang the caliper from the undercarriage.

6. Remove the pad shims, pad retainers and pads.

To install:

7. Install the pad retainers to the caliper bracket.

⁂ CAUTION

Check the foreign material at the pad shims and the back of the pads. Contaminated brake discs or pads reduce stopping ability. Keep grease off the discs and pads.

8. Install the brake pads and pad shims on the pad retainer correctly. Install the pad with the wear indicator on the inside.

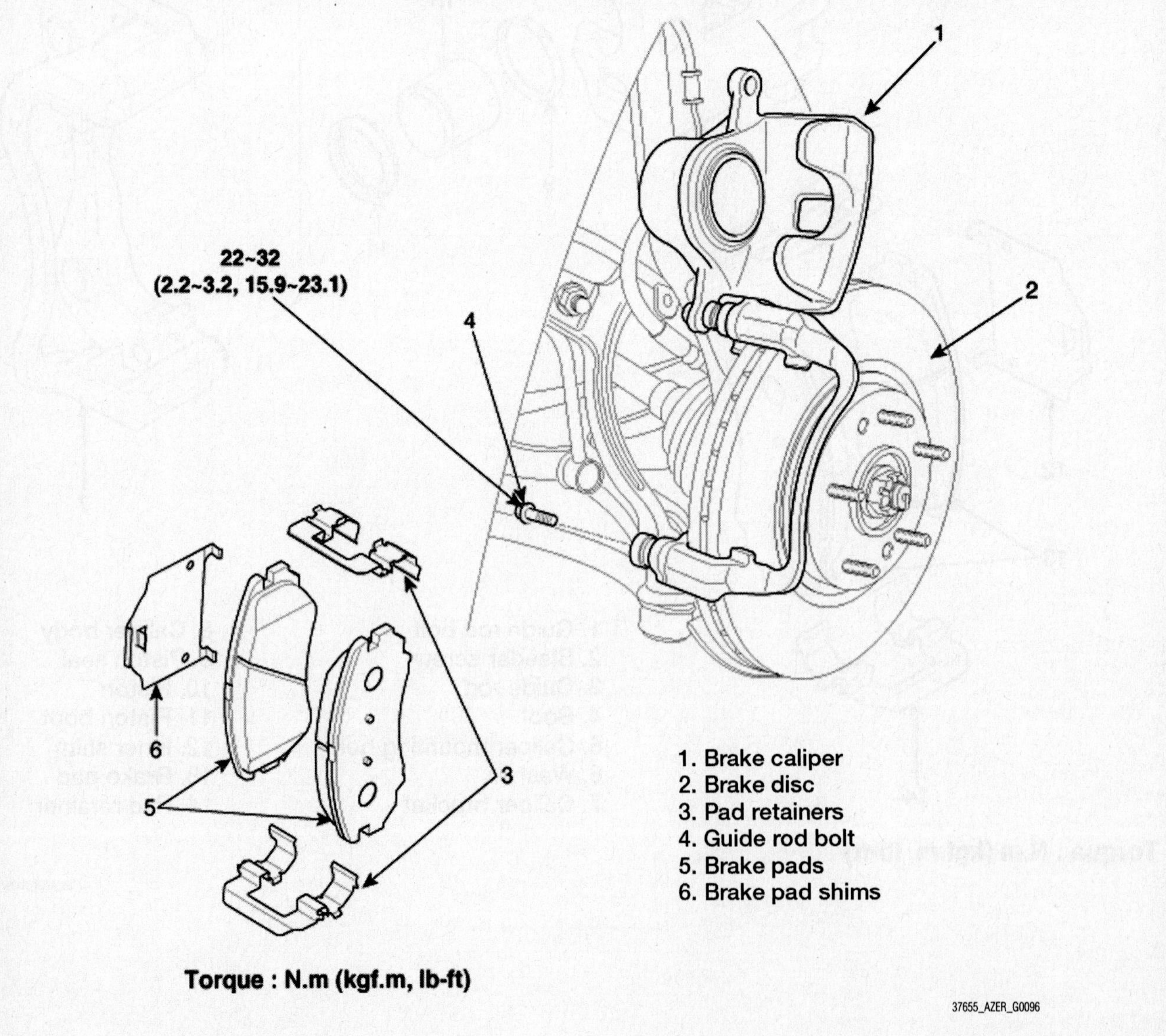

22~32 (2.2~3.2, 15.9~23.1)

1. Brake caliper
2. Brake disc
3. Pad retainers
4. Guide rod bolt
5. Brake pads
6. Brake pad shims

Torque : N.m (kgf.m, lb-ft)

37655_AZER_G0096

Fig. 14 Front brake component locations

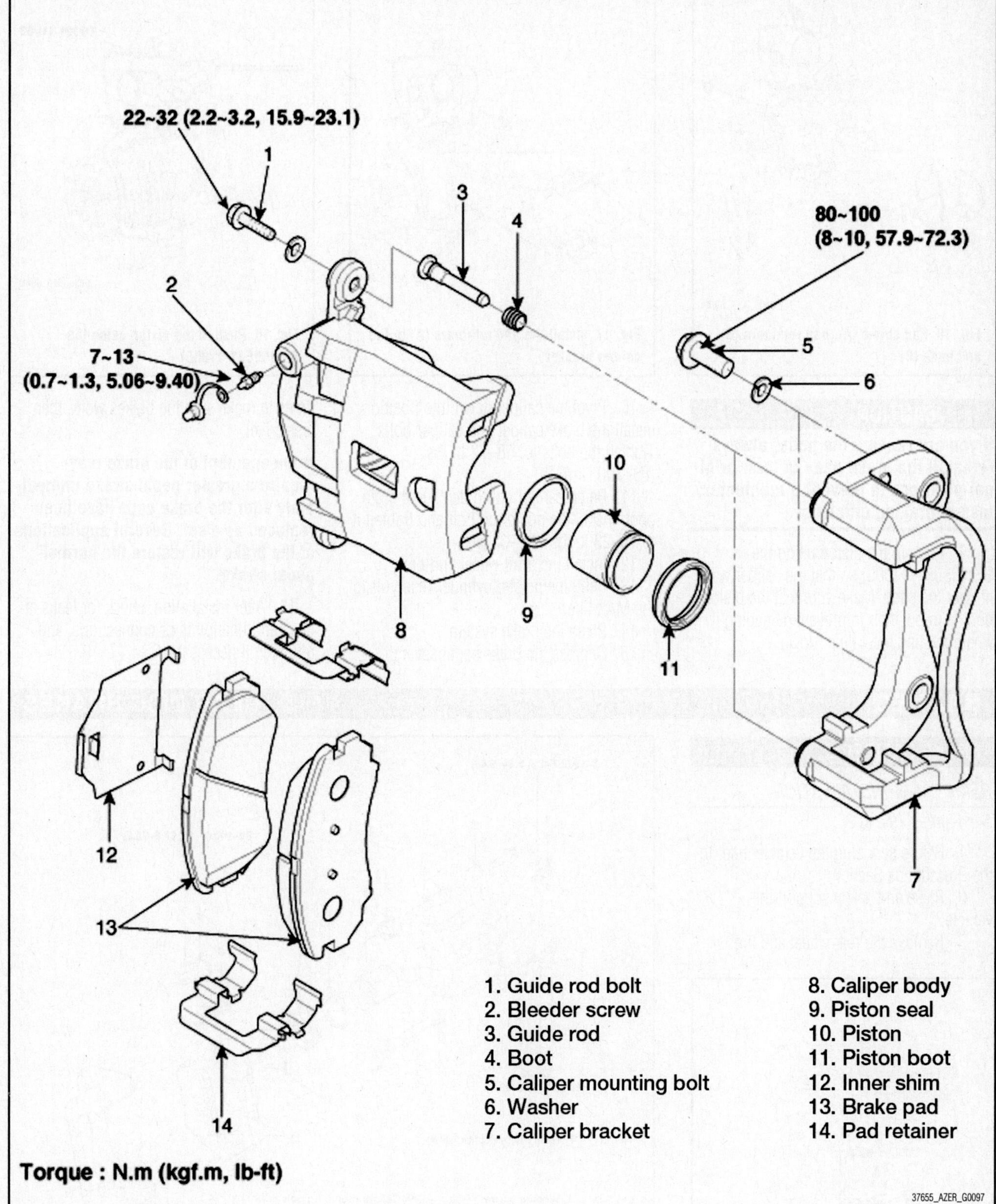

Fig. 15 Front brake components

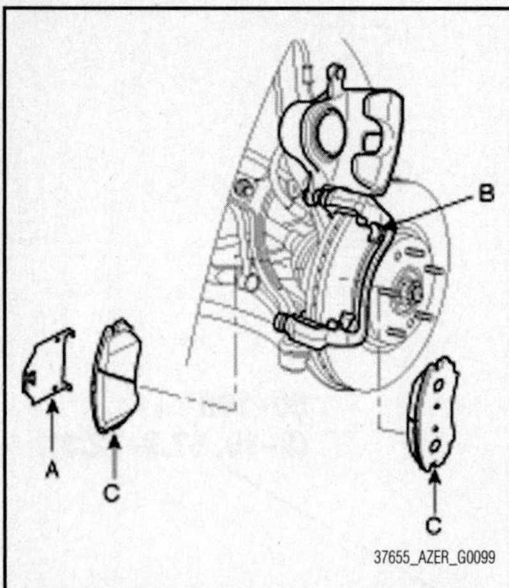

Fig. 16 Pad shims (A), pad retainers (B), and pads (C)

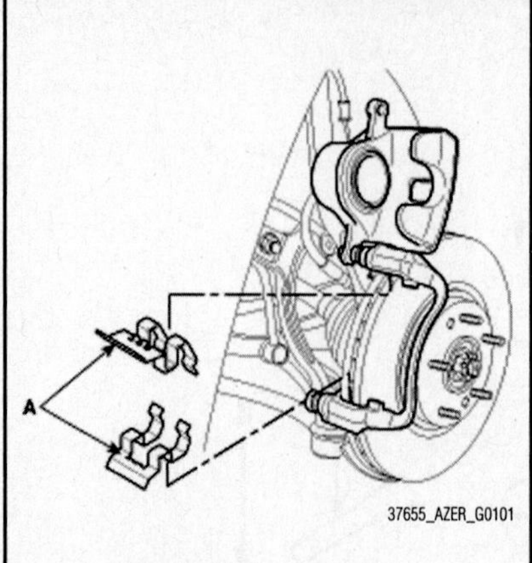

Fig. 17 Install the pad retainers (A) to the caliper bracket

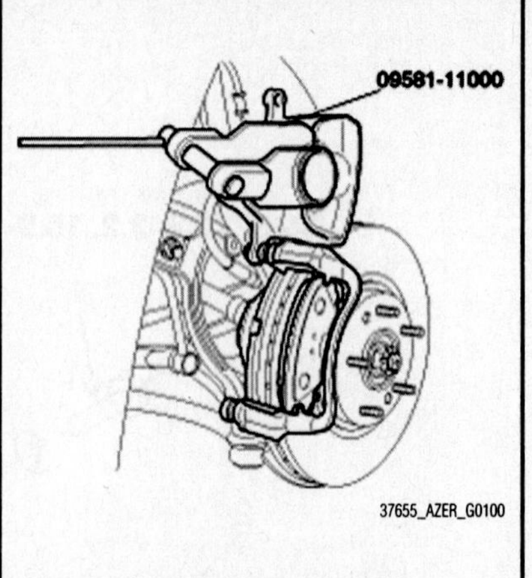

Fig. 18 Push in the piston using the SST(09581-11000)

CAUTION

If you are reusing the pads, always reinstall the brake pads in their original positions to prevent a momentary loss of braking efficiency.

9. Push in the piston using the SST(09581-11000) so that the caliper will fit over the pads. Make sure that the piston boot is in position to prevent damaging it when pivoting the caliper down.
10. Pivot the caliper down into position. Install the brake caliper and caliper bolts. Tighten the bolts to: 58–72 ft. lbs. (80–100 Nm).
11. Being careful not to damage the pin boot, install the guide rod bolt and tighten it to 16–23 ft. lbs. (22–32 Nm).
12. Install the front wheel and tire.
13. Refill the master cylinder reservoir to the MAX line.
14. Bleed the brake system.
15. Depress the brake pedal several times to make sure the brakes work, then test-drive.

➡Engagement of the brake may require a greater pedal stroke immediately after the brake pads have been replaced as a set. Several applications of the brake will restore the normal pedal stroke.

16. After installation, check for leaks at hose and line joints or connections, and retighten if necessary.

BRAKES — REAR DISC BRAKES

BRAKE CALIPER

REMOVAL & INSTALLATION

See Figures 19 and 20.

1. Before servicing the vehicle, refer to the Precautions Section.
2. Raise and safely support the vehicle.
3. Remove the rear wheel and tire.

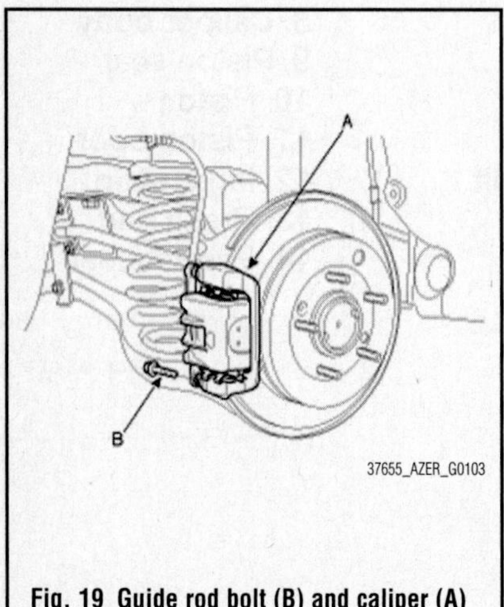

Fig. 19 Guide rod bolt (B) and caliper (A)

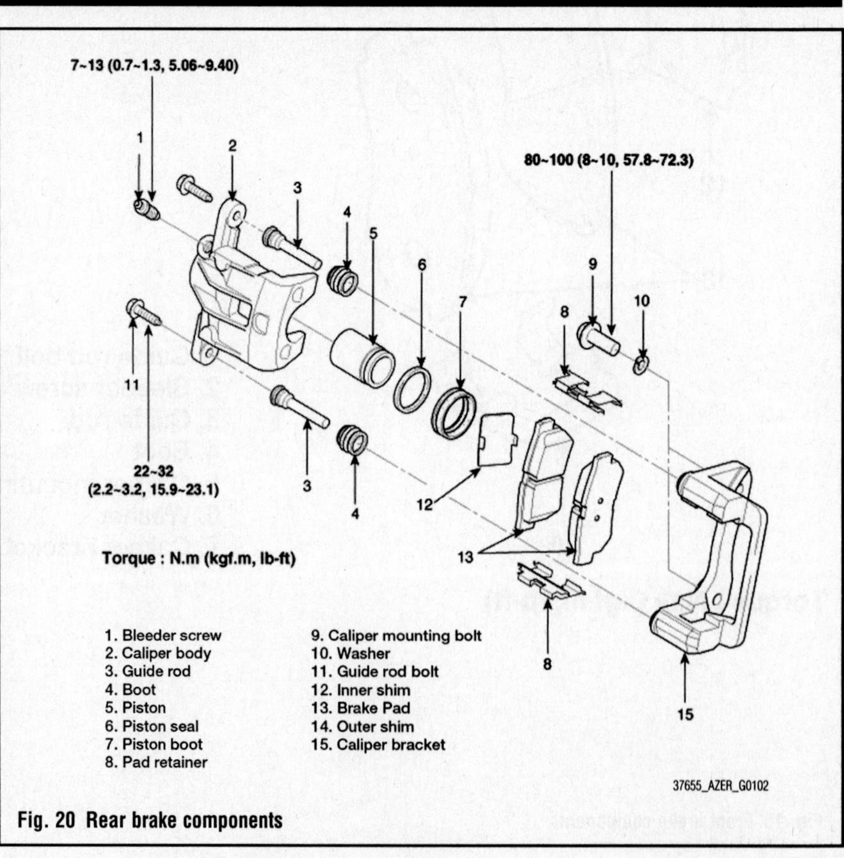

Fig. 20 Rear brake components

⁂ WARNING

Be careful not to damage the hub bolts.

4. Remove guide rod bolts and raise the caliper. Check the hoses and pin boots for damage and deterioration.
5. If replacing the caliper, remove the brake hose from the caliper.
6. Remove the caliper mounting bolts.
7. Remove the caliper.

To install:

8. Pivot the caliper down into position. Install the brake caliper and caliper bolts. Tighten the bolts to: 58–72 ft. lbs. (80–100 Nm).
9. Being careful not to damage the pin boot, install the guide rod bolt and tighten it to 16–23 ft. lbs. (22–32 Nm).
10. If replacing the caliper, install the brake hose onto the caliper. Tighten to 10–12 ft. lbs. (14–17 Nm).

➡Replace sealing washer whenever hose is removed.

11. Install wheel and tire.

DISC BRAKE PADS

REMOVAL & INSTALLATION

See Figures 19 through 22.

1. Before servicing the vehicle, refer to the Precautions Section.
2. Raise and safely support the vehicle.
3. Remove the rear wheel and tire.

⁂ WARNING

Be careful not to damage the hub bolts.

4. Remove guide rod bolt and raise the caliper. Check the hoses and pin boots for damage and deterioration.
5. Remove the caliper mounting bolts, and hang the caliper assembly to one side.

⁂ WARNING

To prevent damage to the caliper assembly or brake hose, use a short piece of wire to hang the caliper from the undercarriage.

6. Remove the pad shims, pad retainers and pads.

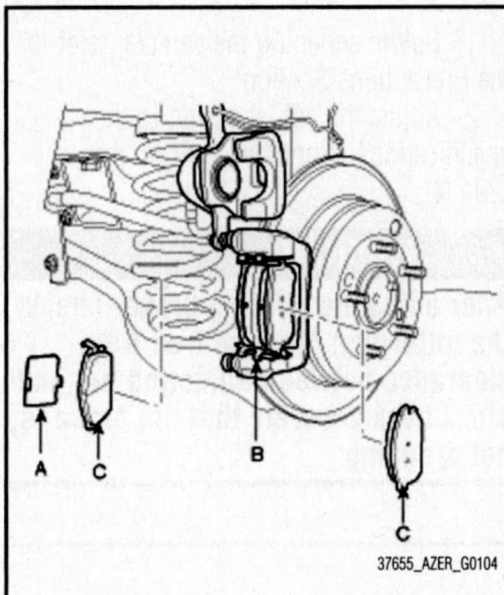

Fig. 21 Pad shims (A), pad retainers (B), and pads (C)

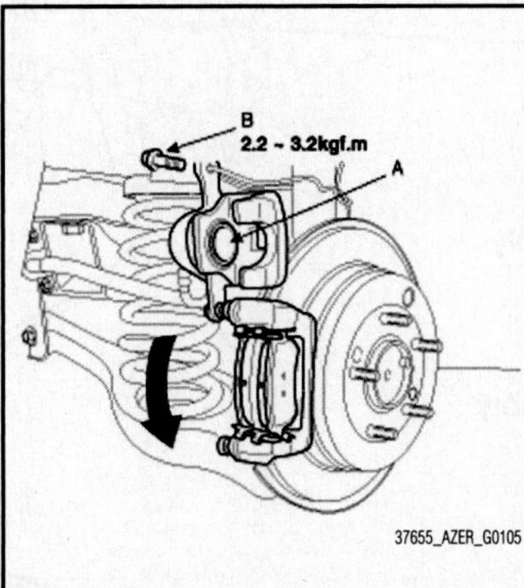

Fig. 22 Push in the piston (A) and install the guide rod bolt (B)

To install:

7. Install the pad retainers to the caliper bracket.

⁂ CAUTION

Check the foreign material at the pad shims and the back of the pads. Contaminated brake discs or pads reduce stopping ability. Keep grease off the discs and pads.

8. Install the brake pads and pad shims on the pad retainer correctly. Install the pad with the wear indicator on the inside.

⁂ CAUTION

If you are reusing the pads, always reinstall the brake pads in their original positions to prevent a momentary loss of braking efficiency.

9. Push in the piston so that the caliper will fit over the pads. Make sure that the piston boot is in position to prevent damaging it when pivoting the caliper down.
10. Pivot the caliper down into position. Install the brake caliper and caliper bolts. Tighten the bolts to: 58–72 ft. lbs. (80–100 Nm).
11. Being careful not to damage the pin boot, install the guide rod bolt and tighten it to 16–23 ft. lbs. (22–32 Nm).
12. Install the rear wheel and tire.
13. Refill the master cylinder reservoir to the MAX line.
14. Bleed the brake system.
15. Depress the brake pedal several times to make sure the brakes work, then test-drive.

➡Engagement of the brake may require a greater pedal stroke immediately after the brake pads have been replaced as a set. Several applications of the brake will restore the normal pedal stroke.

16. After installation, check for leaks at hose and line joints or connections, and retighten if necessary.

PARKING BRAKE CABLES

ADJUSTMENT

See Figure 23.

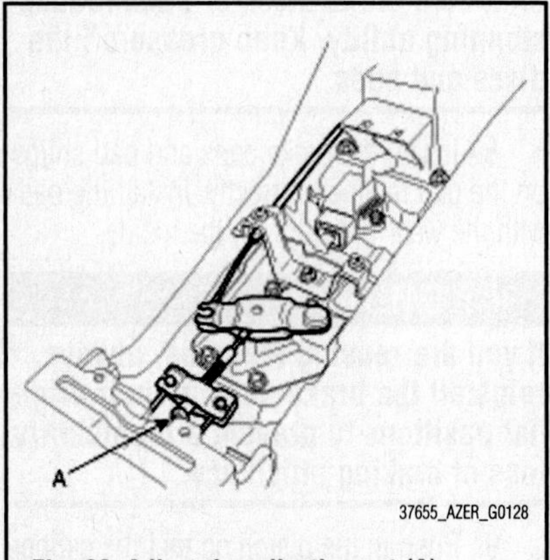

Fig. 23 Adjust the adjusting nut (A)

➡The parking brake adjustment must be carried out after adjusting the rear shoe.

➡Depress the brake pedal several times in order to set the self-adjusting brake system before adjusting the parking brake.

1. Before servicing the vehicle, refer to the Precautions Section.
2. Adjust the adjusting nut to the specifications: 7 notches at 66 ft. lbs. (294 N).

⁂ WARNING

After adjusting parking brake, check the following: Must be free from clearance between adjusting nut and pin. Check securely that the brake is not dragging

PARKING BRAKE SHOES

REMOVAL & INSTALLATION

See Figures 24 through 32.

1. Before servicing the vehicle, refer to the Precautions Section.
2. Remove the lower instrument panel, after loosening the parking lever.
3. Remove the parking brake mounting bolt and the wire.
4. Remove the floor console.
5. Remove the parking brake wire by removing the adjusting nut.
6. Raise the front of the vehicle, and make sure it is securely supported.
7. Remove the rear wheel and tire from the rear hub.
8. Remove the caliper assembly from the carrier and suspend it with wire.

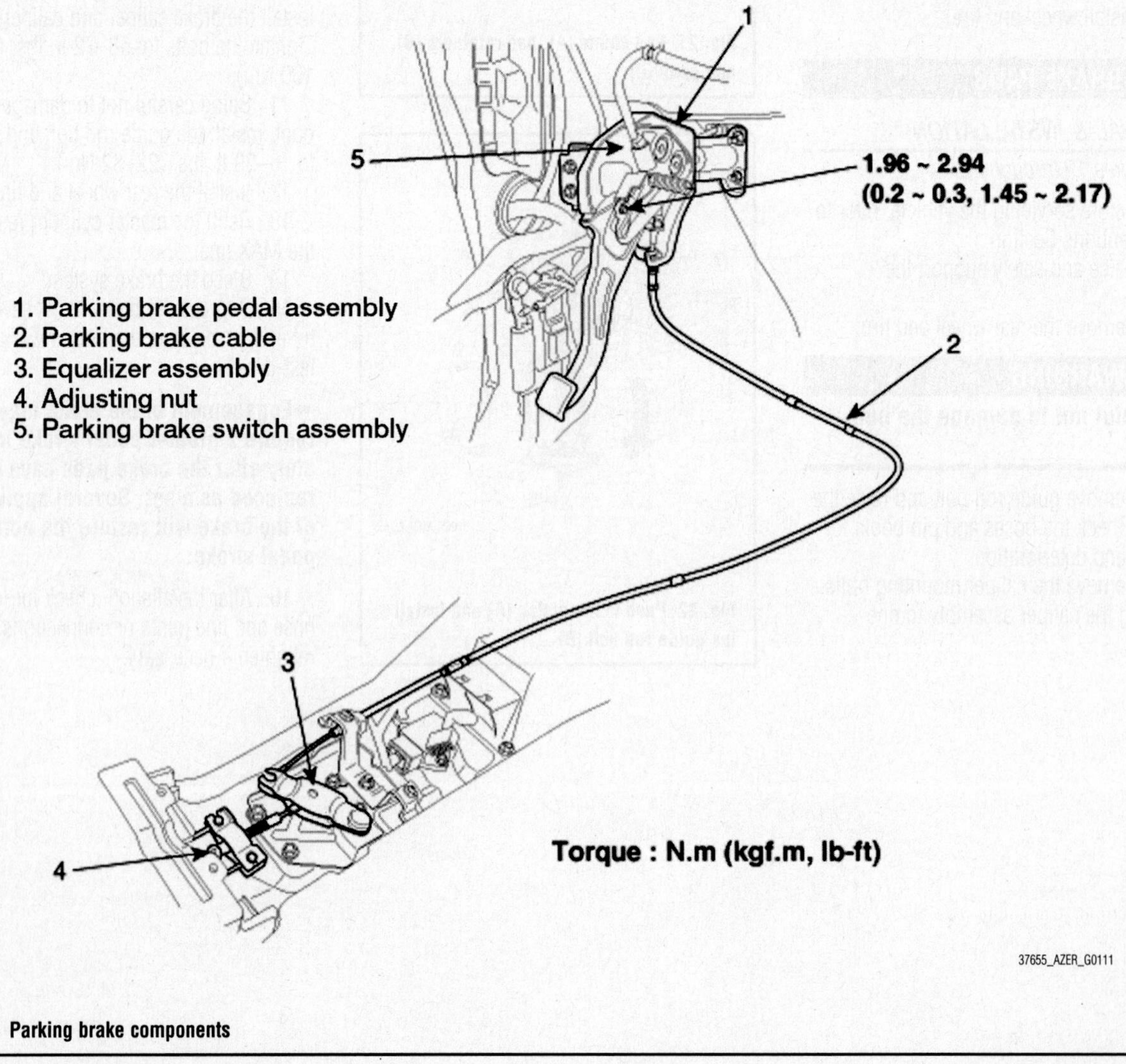

1. Parking brake pedal assembly
2. Parking brake cable
3. Equalizer assembly
4. Adjusting nut
5. Parking brake switch assembly

1.96 ~ 2.94
(0.2 ~ 0.3, 1.45 ~ 2.17)

Torque : N.m (kgf.m, lb-ft)

37655_AZER_G0111

Fig. 24 Parking brake components

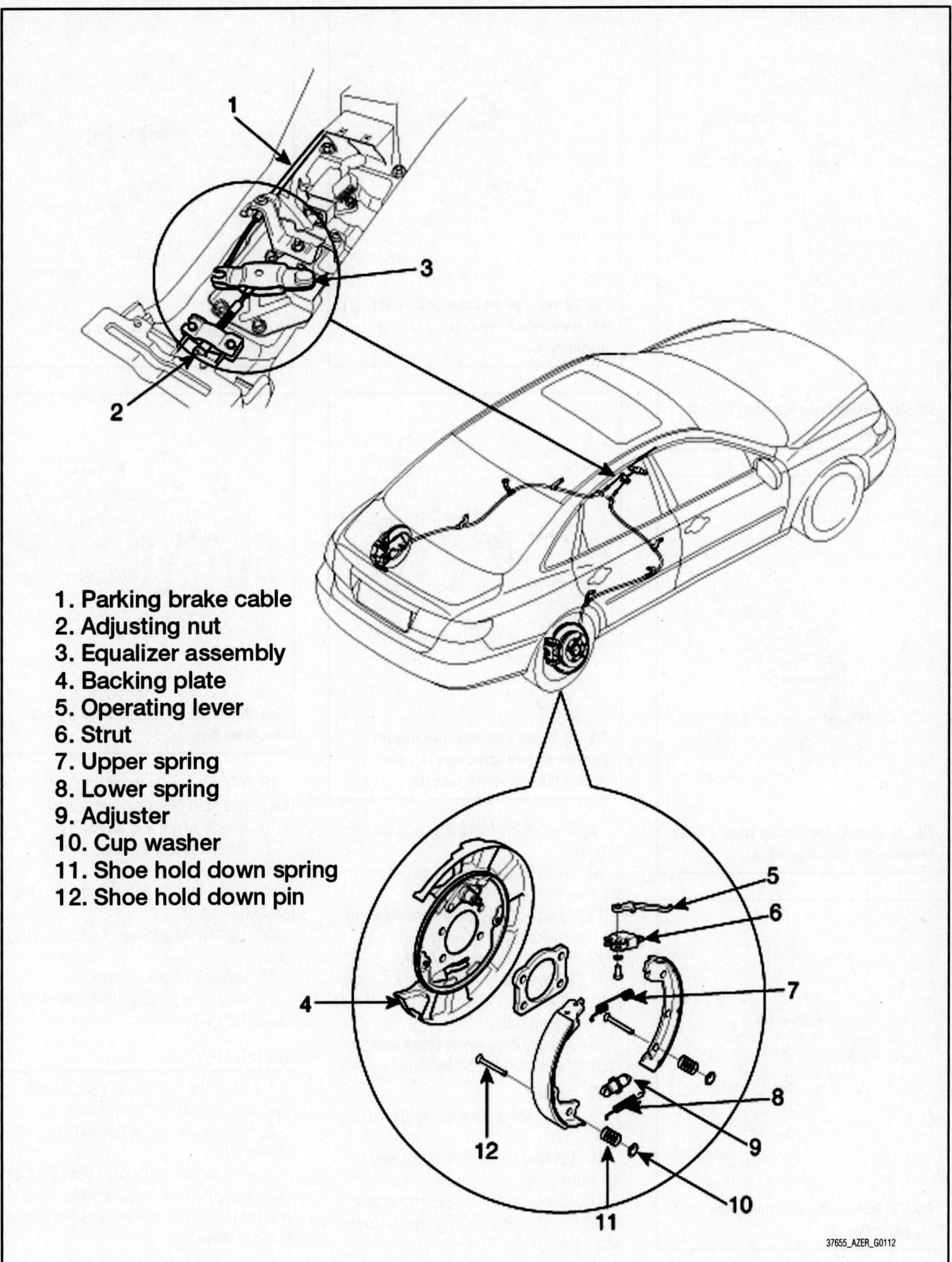

Fig. 25 Parking brake component locations

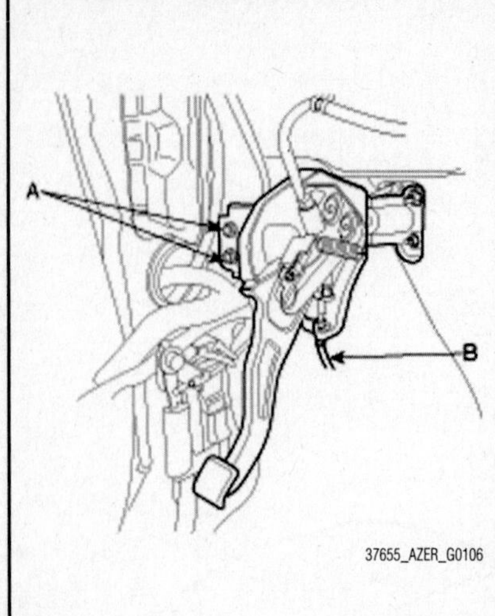

Fig. 26 Parking brake mounting bolt (A) and wire (B)

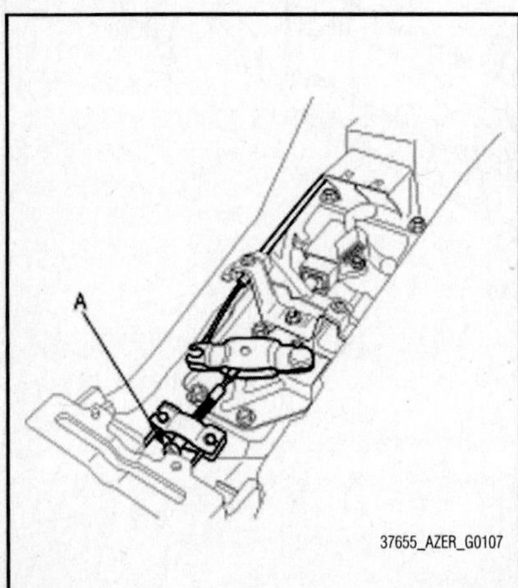

Fig. 27 Remove the parking brake wire by removing the adjusting nut (A)

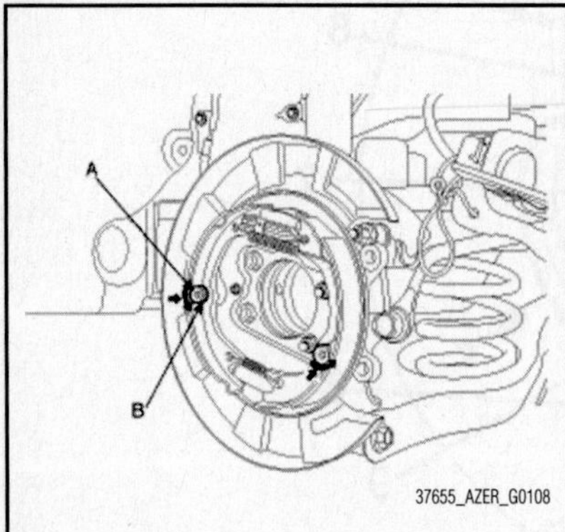

Fig. 28 Remove the shoe hold down pin (A) and spring (B)

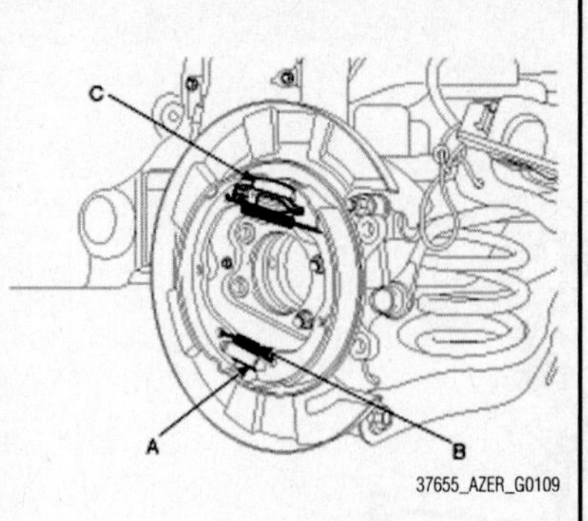

Fig. 29 Remove the adjuster assembly (A) and lower return spring (B), and strut assembly (C)

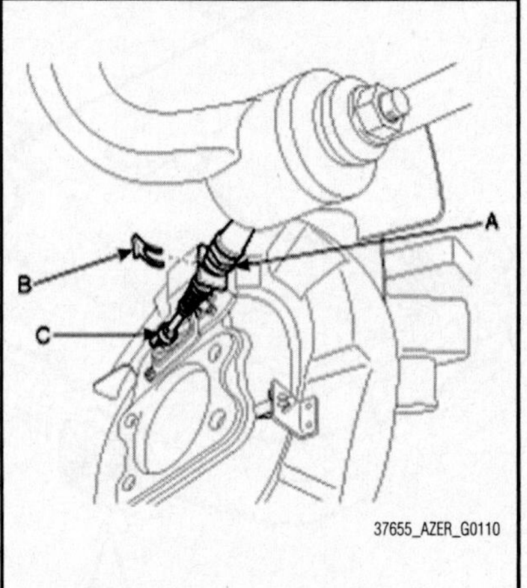

Fig. 30 Remove the retaining ring (B) from the parking brake wire (A), and remove the connecting hook (C)

9. Remove the brake disc and the rear axle hub.
10. While fastening the spring, remove the shoe hold down pin and spring.
11. Remove the adjuster assembly and the lower return spring.
12. Remove the strut assembly.
13. Remove the retaining ring from the parking brake wire, which is in the rear side of the backing plate.
14. Remove the parking brake wire connecting hook from the brake shoe.

To install:

15. Install the brake shoe to the backing plate.
16. Connect the parking brake wire to the brake shoe.
17. After installing the strut assembly, install the adjuster assembly and the lower return spring.

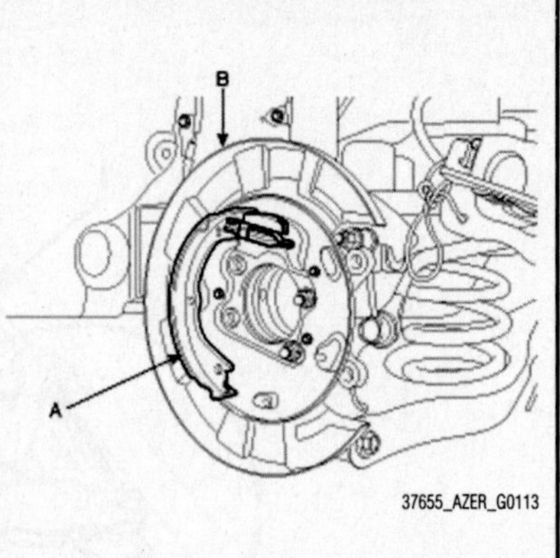

Fig. 31 Install the brake shoe (A) to the backing plate (B)

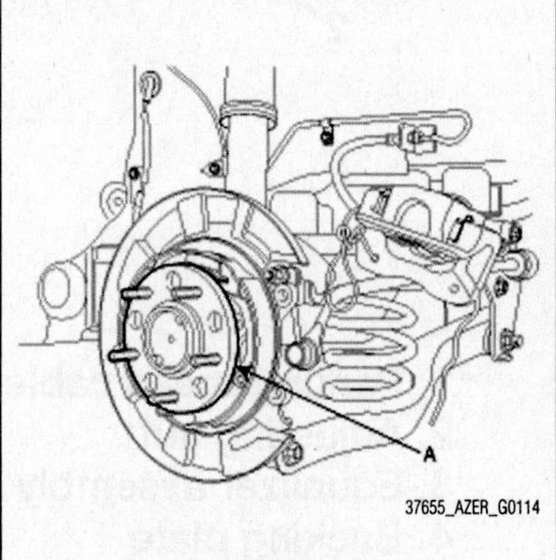

Fig. 32 Install the rear axle hub (A) and the brake disc

18. While pressing the spring, install the brake shoe hold down pin and spring.
19. Grease where it is necessary.
20. Install the rear axle hub and the brake disc.
21. Install the rear wheel and tire to the rear hub.
22. Tighten the parking brake adjusting nut.
23. Install the floor console.
24. Install the parking brake mounting bolt and the wire.

ADJUSTMENT

1. Raise the front of the vehicle, and make sure it is securely supported.
2. Remove the rear wheel and tire from the rear hub.
3. After removing the plug from the disc, rotate the toothed wheel by a screw driver until the disc is not moving, and then return it by 5 notches.

GENERAL INFORMATION

⁂ CAUTION

These vehicles are equipped with an air bag system. The system must be disarmed before performing service on, or around, system components, the steering column, instrument panel components, wiring and sensors. Failure to follow the safety precautions and the disarming procedure could result in accidental air bag deployment, possible injury and unnecessary system repairs.

SERVICE PRECAUTIONS

⁂ CAUTION

Disconnect and isolate the battery negative cable before beginning any airbag system component diagnosis, testing, removal, or installation procedures. Wait at least 90 seconds after the ignition switch is turned off and the negative (-) terminal cable is disconnected from the battery before starting the operation. The SRS is equipped with a backup power source, so if work is started within 90 seconds after disconnecting the negative (-) terminal cable from the battery, the SRS may be deployed. Failure to disable the airbag system may result in accidental airbag deployment, personal injury, or death.

DISARMING THE SYSTEM

⁂ CAUTION

Before servicing components near or affected by the SRS (airbag) system, read and observe all SRS Service Precautions. Failure to observe all precautions may result in accidental airbag deployment, personal injury, or death.

1. Before servicing the vehicle, refer to the Precautions Section.
2. Disconnect and isolate the negative battery cable.
3. Wait 3 minutes for the system capacitor to discharge before performing any service.
4. Remove the ignition key from the vehicle.

ARMING THE SYSTEM

⁂ CAUTION

Before servicing components near or affected by the SRS (airbag) system, read and observe all SRS Service Precautions. Failure to observe all precautions may result in accidental airbag deployment, personal injury, or death.

1. Before servicing the vehicle, refer to the Precautions Section.
2. Reconnect the negative battery cable.
3. Turn the ignition switch to the **RUN** position.
4. Confirm proper system operation:
   a. Turn the ignition switch ON; the SRS indicator light should be turned on for about six seconds and then go off.

DRIVE TRAIN

FRONT HALFSHAFTS

REMOVAL & INSTALLATION

See Figures 33 through 46.

1. Before servicing the vehicle, refer to the Precautions Section.
2. Remove the wheel and tire assembly.
3. Remove the split pin and halfshaft castle nut and washer from the front hub.

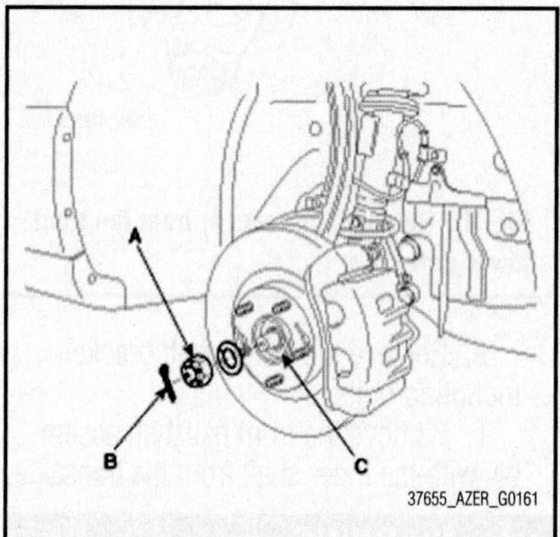

Fig. 33 Remove the split pin (B) and halfshaft castle nut (A) and washer from the front hub (C)

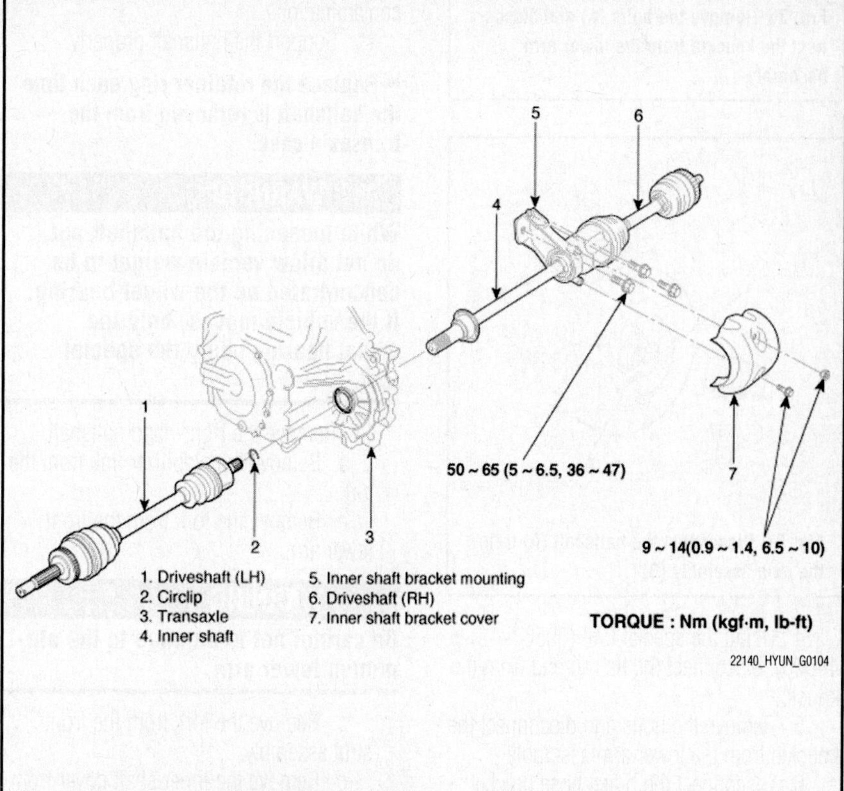

Fig. 34 Exploded view of halfshaft components

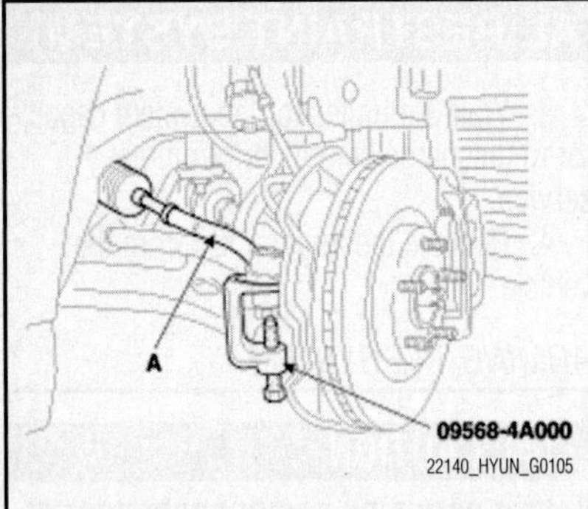

Fig. 35 Using the special tool (09568-4A000), disconnect the tie rod end from the knuckle

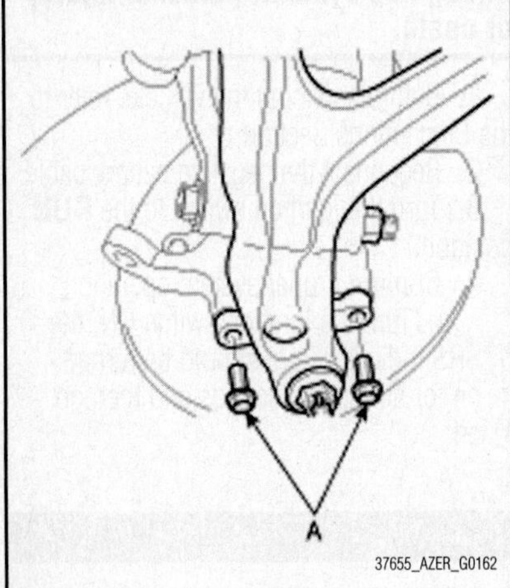

Fig. 36 Remove the bolts (A) and disconnect the knuckle from the lower arm assembly

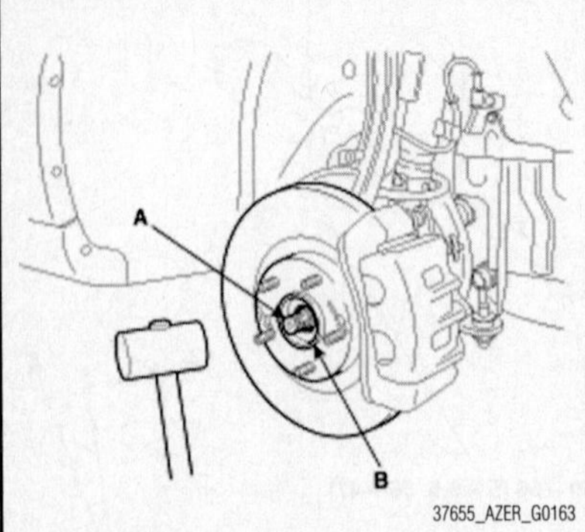

Fig. 37 Disconnect the halfshaft (A) from the axle assembly (B)

4. Using the special tool (09568-4A000), disconnect the tie rod end from the knuckle.
5. Remove the bolts and disconnect the knuckle from the lower arm assembly.
6. Disconnect the brake hose bracket from the knuckle.

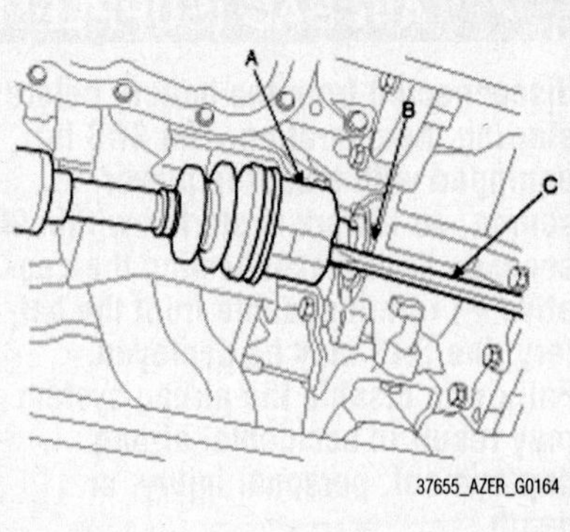

Fig. 38 Remove the left-hand halfshaft (A) from the transaxle using a pry bar (C)

7. Disconnect the wheel speed sensor bracket from the knuckle.
8. Using a plastic hammer, disconnect the halfshaft from the axle assembly.
9. Remove the left-hand halfshaft from the transaxle using a pry bar.

WARNING

Use a pry bar so you do not damage the joint. If you pull the halfshaft by excessive force, components inside the joint can be displaced causing the boot to be torn and the bearing to be damaged.

10. Plug the transaxle case opening with an oil seal cap in order to avoid contamination.
11. Support the halfshaft properly.

➡Replace the retainer ring each time the halfshaft is removed from the transaxle case.

WARNING

While loosening the halfshaft nut, do not allow vehicle weight to be concentrated on the wheel bearing. If the vehicle moves, hold the wheel bearing using the special tool.

12. Remove the right-hand halfshaft.
    a. Remove the stabilizer link from the fork.
    b. Remove the fork from the front lower arm.

WARNING

Be careful not to damage to the aluminum lower arm.

    c. Remove the fork from the front strut assembly.
    d. Remove the inner shaft cover from the inner shaft bracket.

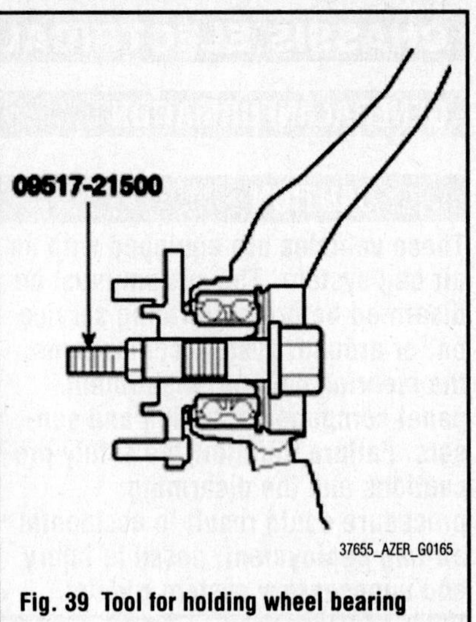

Fig. 39 Tool for holding wheel bearing

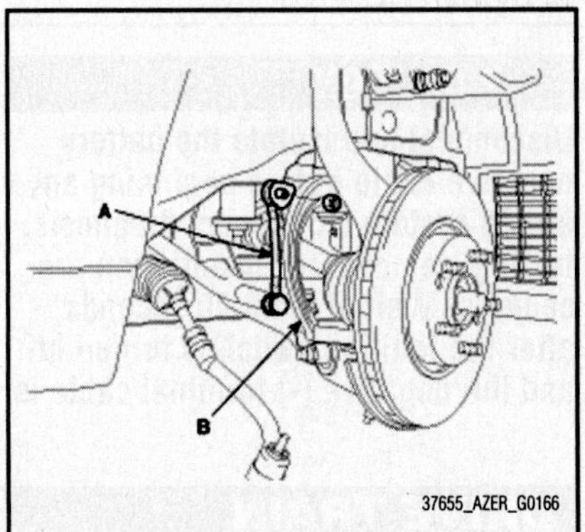

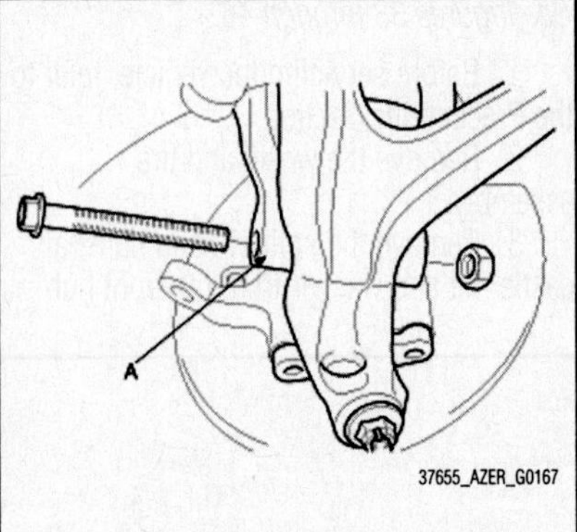

Fig. 40 Remove the stabilizer (A) link from the fork (B)—Right-hand

Fig. 41 Remove the fork (A) from the front lower arm

    e. Remove the inner shaft bracket mounting bolts.
    f. Remove the front halfshaft assembly with the inner shaft from the transaxle.

WARNING

Do not try to disconnect the inner shaft from the halfshaft. Because they cannot be disconnected once

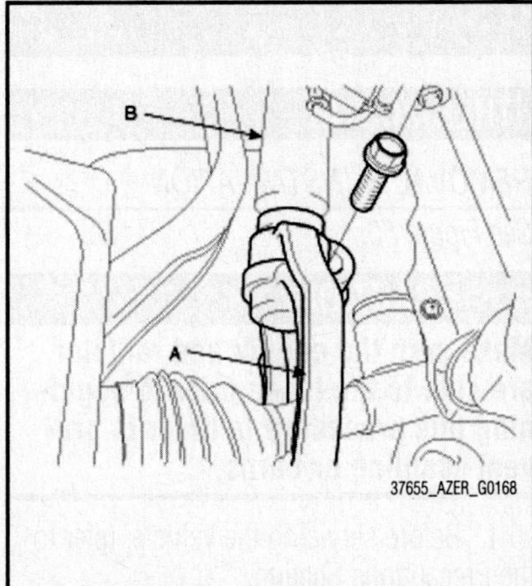

Fig. 42 Remove the fork (A) from the front strut assembly (B)

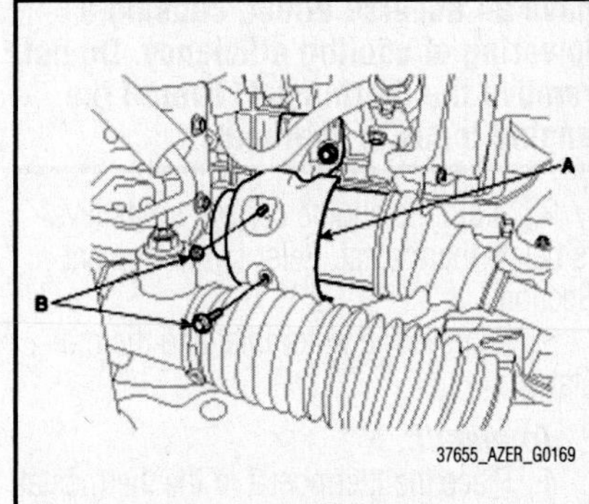

Fig. 43 Remove the inner shaft cover (A) from the inner shaft bracket (B)

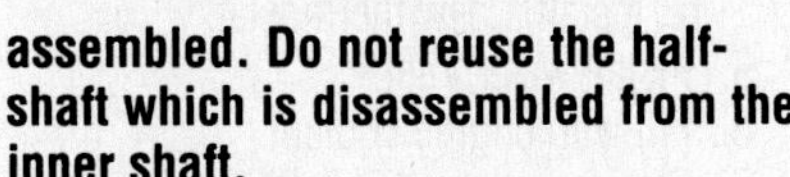

assembled. Do not reuse the halfshaft which is disassembled from the inner shaft.

To install:

13. Replace the circlips with new ones after removal.
14. Apply gear oil on the halfshaft splines and the contacting surface of differential case oil seal.
15. After installation, check that the halfshaft cannot be removed.
16. Install the right-hand halfshaft.
    a. Install the inner shaft bracket mounting bolt and tighten to 36–47 ft. lbs. (50–65 Nm).
    b. Install the inner shaft cover by

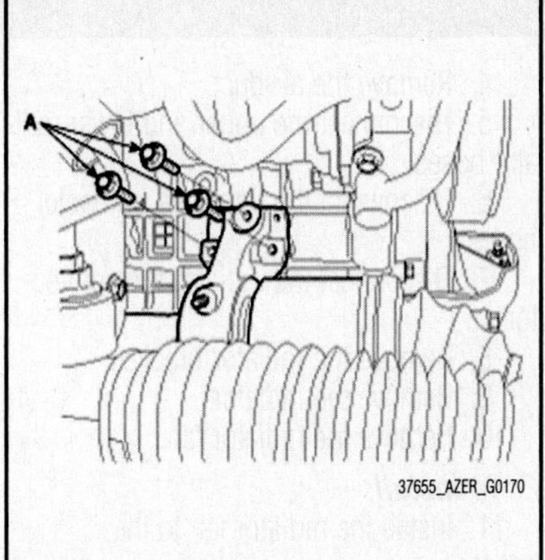

Fig. 44 Remove the inner shaft bracket mounting bolts (A)

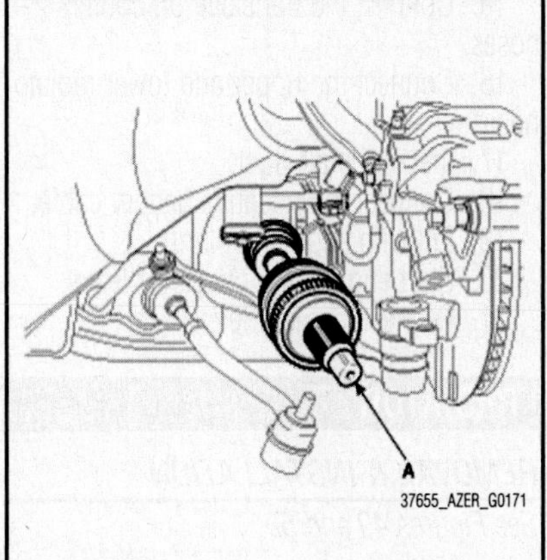

Fig. 45 Remove the front halfshaft assembly (A)

installing the cover mounting bolts and tighten to 78–120 inch lbs. (9–14 Nm).
    c. Install the fork to the front strut assembly and tighten to 44–59 ft. lbs. (60–80 Nm).
    d. Install the connecting bolt between the fork and the lower arm and tighten to 101–118 ft. lbs. (140–160 Nm).
    e. Install the stabilizer link to the fork and tighten to 74–88 ft. lbs. (100–120 Nm).
17. Install the halfshaft into the front axle assembly.

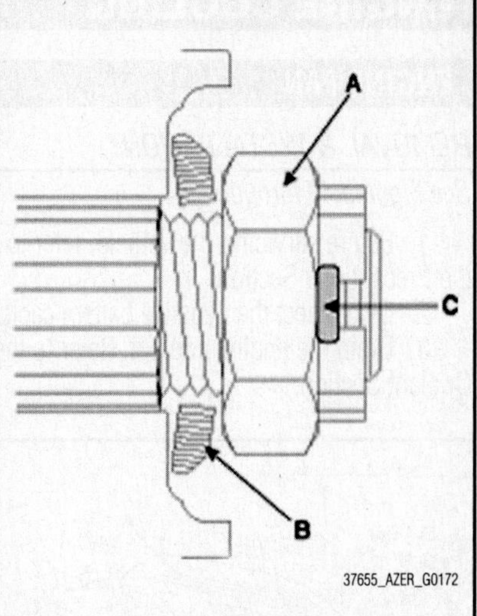

Fig. 46 Install the washer (B) with convex surface outward, and install the lock nut (A) and the split pin (C)

WARNING

Be careful not to damage the boot.

18. Install the knuckle in the lower arm assembly and tighten the bolts and tighten to 74–88 ft. lbs. (100–120 Nm).
19. Install the tie rod end in the knuckle and tighten to 18–25 ft. lbs. (24–34 Nm).
20. Install the wheel speed sensor in the knuckle.
21. Install the brake hose bracket to the front knuckle.
22. After installing the washer with convex surface outward, install the lock nut and the split pin. Tighten the lock nut to 148–207 ft. lbs. (200–280 Nm).
23. Install the wheel and tire assembly.

CV-BOOTS INSPECTION

1. Check the halfshaft boots for damage and deterioration.
2. Check splines for wear and damage.
3. Check the ball joints for wear and operating condition.
4. Check the boots for damage and deterioration.

ENGINE COOLING

ENGINE FAN

REMOVAL & INSTALLATION

See Figures 47 through 49.

1. Before servicing the vehicle, refer to the Precautions Section.
2. Disconnect the negative battery cable.
3. Drain the engine coolant. Refer to the Coolant Section.

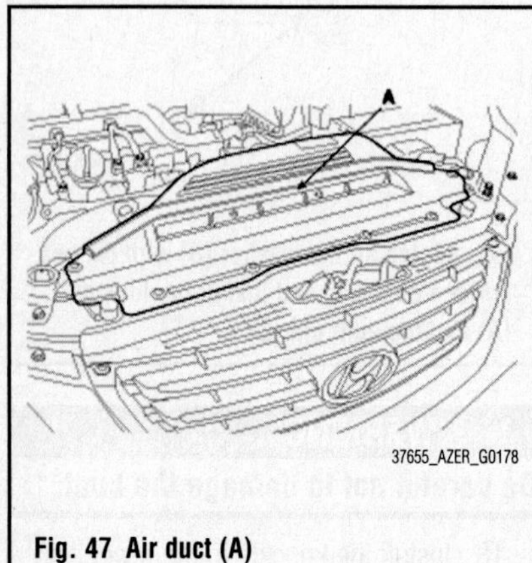

Fig. 47 Air duct (A)

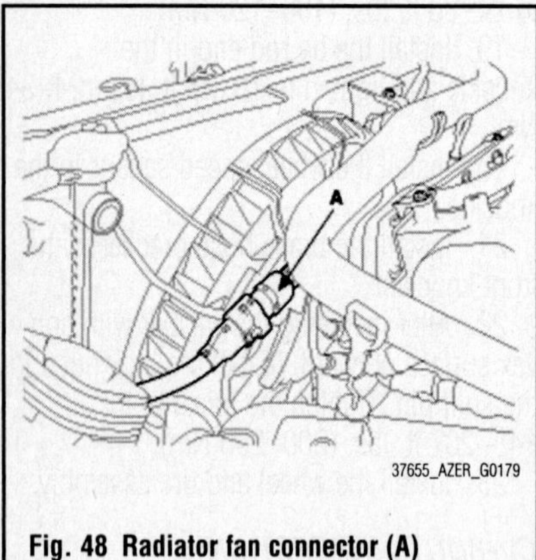

Fig. 48 Radiator fan connector (A)

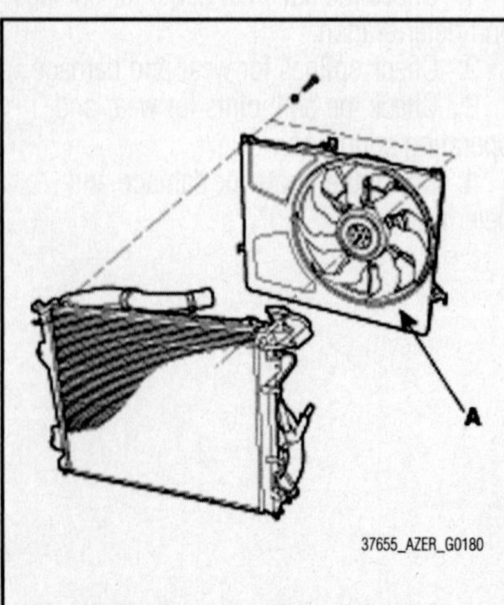

Fig. 49 Remove the radiator fan (A)

4. Remove the air duct.
5. Disconnect the upper and lower radiator hoses.
6. Disconnect the transaxle oil cooler hoses.
7. Disconnect the radiator fan connector.
8. Remove the radiator bracket.
9. Remove the radiator.
10. Remove the radiator fan.

To install:

11. Install the radiator fan to the radiator.
12. Install the radiator.
13. Install the radiator bracket.
14. Connect the radiator fan connector.
15. Connect the transaxle oil cooler hoses.
16. Connect the upper and lower radiator hoses.
17. Install the air duct.
18. Connect the negative battery cable.
19. Fill with engine coolant.
20. Start engine and check for leaks.
21. Recheck coolant level.

RADIATOR

REMOVAL & INSTALLATION

See Figures 49 and 50.

1. Before servicing the vehicle, refer to the Precautions Section.
2. Disconnect the negative battery cable.
3. Drain the engine coolant. Refer to the Coolant Section.
4. Remove the air duct.
5. Disconnect the upper and lower radiator hoses.
6. Disconnect the transaxle oil cooler hoses.
7. Disconnect the radiator fan connector.
8. Remove the radiator bracket.
9. Remove the radiator.

To install:

10. Install the radiator.
11. Install the radiator bracket.
12. Connect the radiator fan connector.
13. Connect the transaxle oil cooler hoses.
14. Connect the upper and lower radiator hoses.
15. Install the air duct.
16. Connect the negative battery cable.
17. Fill with engine coolant.
18. Start engine and check for leaks.
19. Recheck coolant level.

THERMOSTAT

REMOVAL & INSTALLATION

See Figure 50.

**** CAUTION**

Make sure the engine and radiator are cool to the touch prior to beginning this procedure in order to prevent scalding or burns.

1. Before servicing the vehicle, refer to the Precautions Section.

**** WARNING**

Removal of the thermostat would have an adverse effect, causing a lowering of cooling efficiency. Do not remove the thermostat, even if the engine tends to overheat.

2. Drain the engine coolant so its level is below thermostat. Refer to the Coolant Section.
3. Remove the water inlet and thermostat.

To install:

4. Place the thermostat in the thermostat housing:
   a. Install the thermostat with the jiggle valve upward.
   b. Install a new thermostat.
5. Install water inlet.
6. Fill with engine coolant.
7. Start engine and check for leaks.

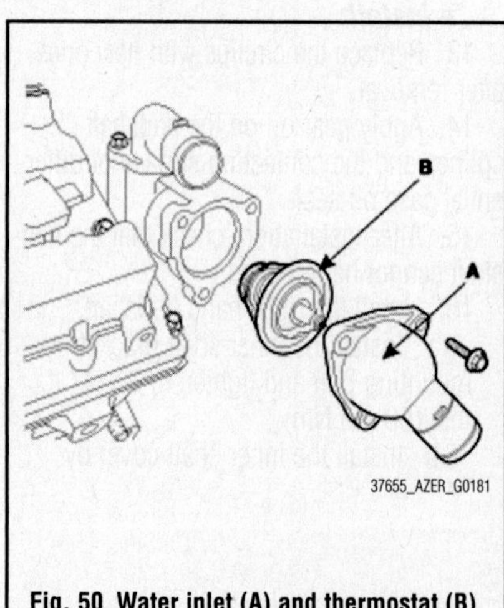

Fig. 50 Water inlet (A) and thermostat (B)

WATER PUMP

REMOVAL & INSTALLATION

See Figures 51 through 54.

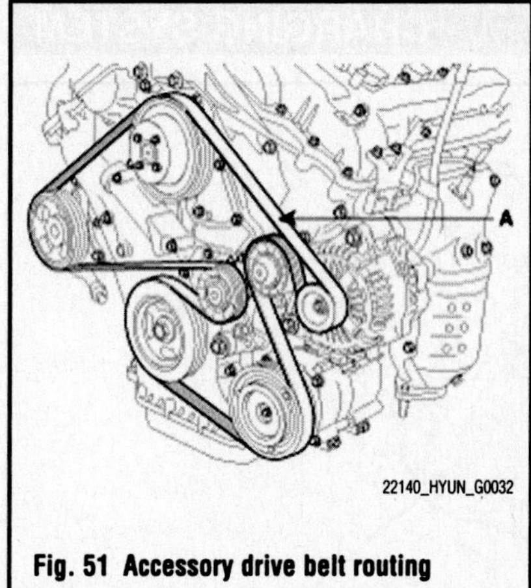

Fig. 51 Accessory drive belt routing

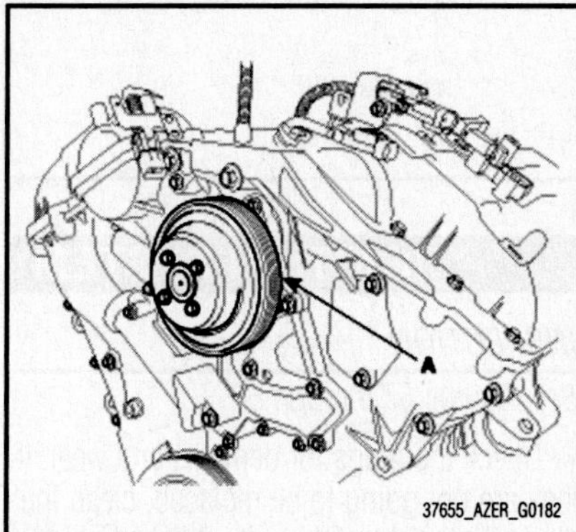

Fig. 52 Remove the 4 bolts and pump pulley (A)

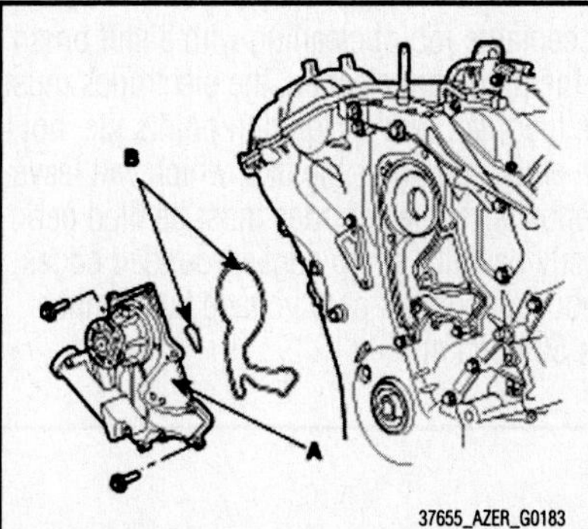

Fig. 53 Remove the water pump (A) and gasket (B)

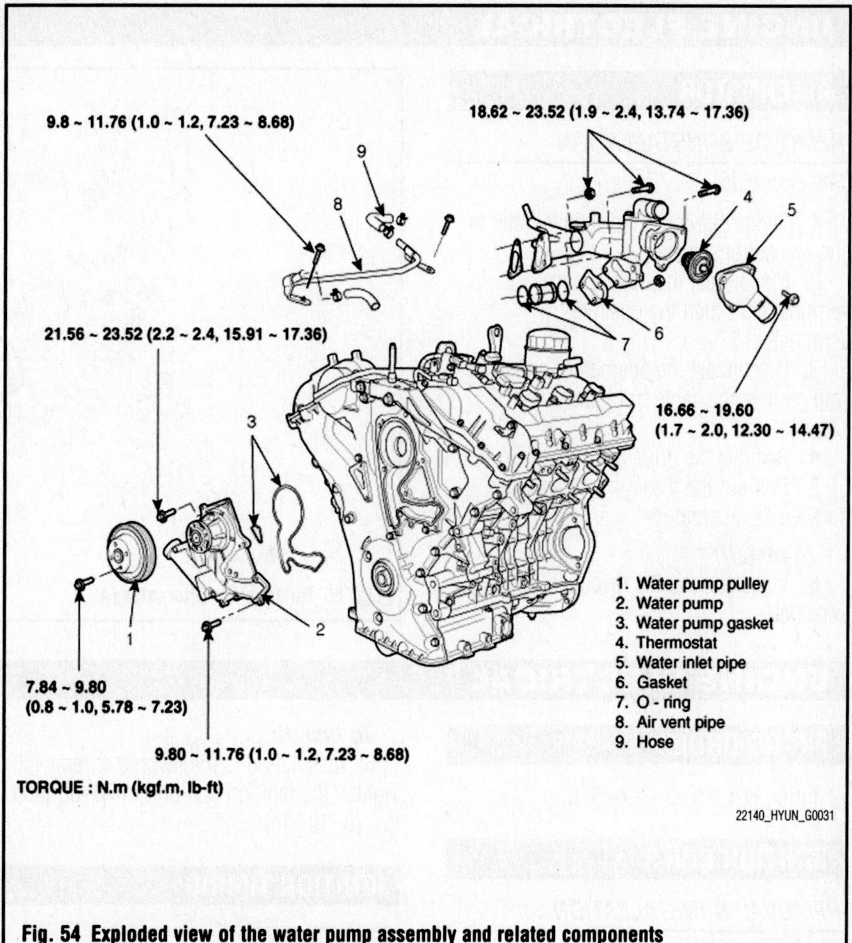

Fig. 54 Exploded view of the water pump assembly and related components

CAUTION

The system is under high pressure when the engine is hot. To avoid danger of releasing scalding engine coolant, remove the cap only when the engine is cool.

1. Before servicing the vehicle, refer to the Precautions Section.
2. Drain the engine coolant. Refer to Coolant section.
3. Remove the accessory drive belt.
4. Remove the 4 bolts and pump pulley.
5. Remove the water pump and gasket.

To install:

➡Clean the contact face before assembly.

6. Install the water pump and a new gasket with 12 bolts. Tightening torque: 16–17 ft. lbs. (22–24 Nm); 87–104 inch lbs. (10–12 Nm).
7. Install the 4 bolts and the pump pulley. Tightening torque: 69–104 inch lbs. (8–10 Nm).
8. Install the accessory drive belt.
9. Fill with engine coolant.
10. Start engine and check for leaks.
11. Recheck engine coolant level.

ENGINE ELECTRICAL — CHARGING SYSTEM

ALTERNATOR

REMOVAL & INSTALLATION

See Figure 55.

1. Before servicing the vehicle, refer to the precautions section.
2. Disconnect the battery negative terminal first, then the positive terminal.
3. Disconnect the alternator connector, and remove the cable from alternator "B" terminal.
4. Remove the drive belt.
5. Pull out the through bolt and then remove the alternator.

To install:

6. Installation is the reverse of removal.

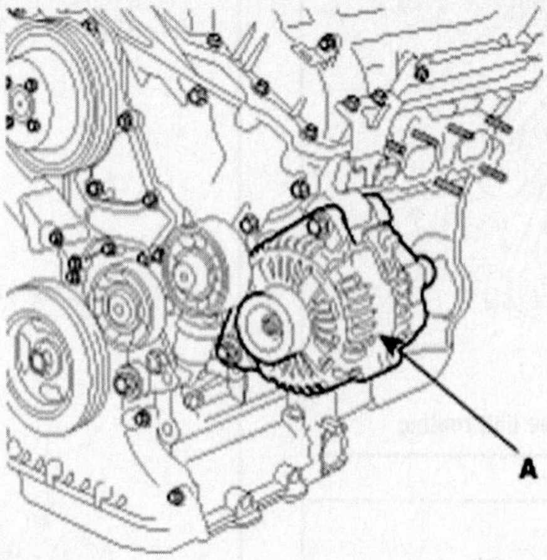

37655_AZER_G0184

Fig. 55 Remove the alternator (A)

ENGINE ELECTRICAL — IGNITION SYSTEM

FIRING ORDER

Firing order: 1–2–3–4–5–6

IGNITION COILS

REMOVAL & INSTALLATION

See Figure 56.

1. Before servicing the vehicle, refer to the Precautions Section.
2. Disconnect the negative battery cable.
3. Remove the engine cover (as necessary).
4. Disconnect the ignition coil connector.

➡When removing the ignition coil connector, pull the lock pin and push the clip.

5. Remove the ignition coil.

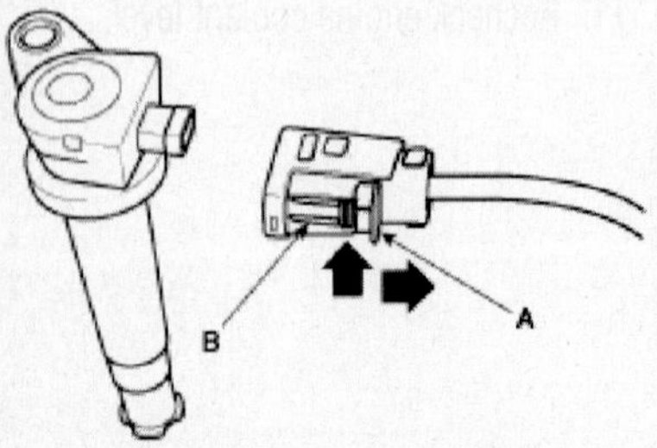

22140_HYUN_G0004

Fig. 56 When removing the ignition coil connector, pull the lock pin (A) and push the clip (B)

To install:

6. Installation is the reverse of removal. Tighten the ignition coil bolt to 70–86 inch lbs. (8–10 Nm).

IGNITION TIMING

ADJUSTMENT

Ignition timing is controlled by the electronic control ignition timing system.

SPARK PLUGS

REMOVAL & INSTALLATION

1. Before servicing the vehicle, refer to the Precautions Section.
2. Disconnect the negative battery cable.
3. Remove the engine cover (as necessary).
4. Remove the ignition coil. Refer to Ignition Coils Removal & Installation.
5. Use a spark plug socket and wrench to remove the spark plugs.

⁂ WARNING

Be careful that no contaminates enter through the spark plug holes.

➡Check the electrode gap on the spark plugs before installation. Specification: 0.039–0.043 in. (1.0–1.1 mm).

To install:

6. To install, reverse the removal procedure. Tighten the spark plugs to 11 ft. lbs. (15 Nm).

INSPECTION

See Figures 57 through 60.

Check the plugs for deposits and wear. If they are not going to be replaced, clean the plugs thoroughly. Remember that any kind of deposit will decrease the efficiency of the plug. Plugs can be cleaned on a spark plug cleaning machine, which can sometimes be found in service stations, or you can do an acceptable job of cleaning with a stiff brush. If the plugs are cleaned, the electrodes must be filed flat. Use an ignition points file, not an emery board or the like, which will leave deposits. The electrodes must be filed perfectly flat with sharp edges; rounded edges reduce the spark plug voltage by as much as 50 percent.

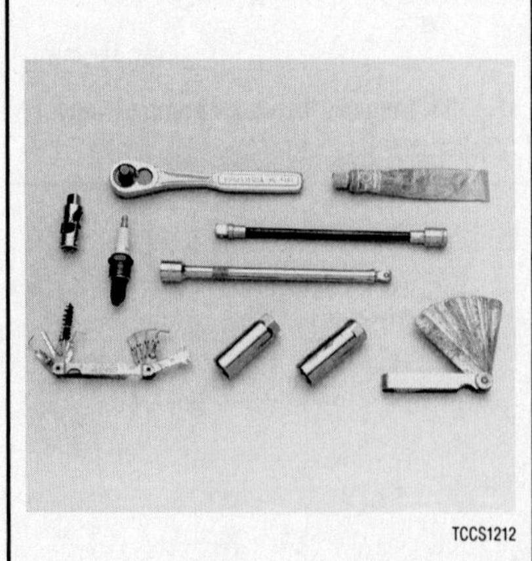

TCCS1212

Fig. 57 A variety of tools and gauges are needed for spark plug service

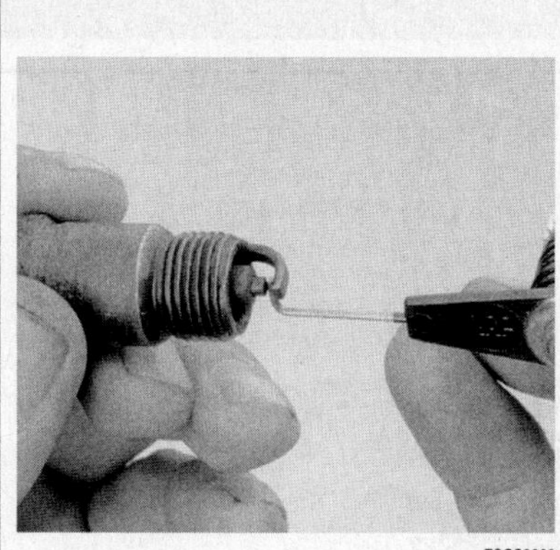

Fig. 58 Checking the spark plug gap with a feeler gauge

Check spark plug gap before installation. The ground electrode (the L-shaped one connected to the body of the plug) must be parallel to the center electrode and the specified size wire gauge (please refer to the Tune-Up Specifications chart for details) must pass between the electrodes with a slight drag.

➡**NEVER adjust the gap on a used platinum type spark plug.**

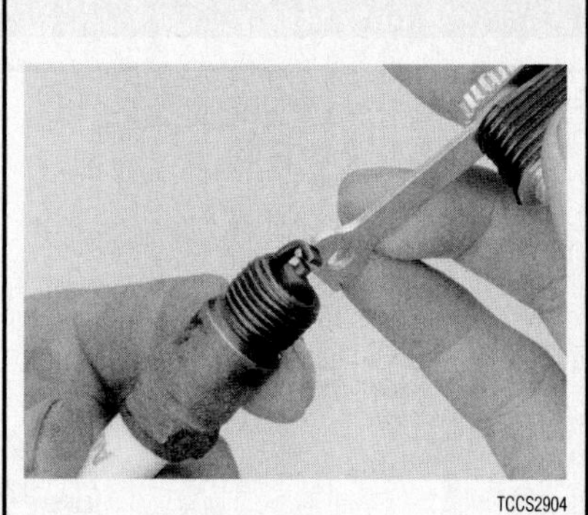

Fig. 59 Adjusting the spark plug gap

Always check the gap on new plugs as they are not always set correctly at the factory. Do not use a flat feeler gauge when measuring the gap on a used plug, because the reading may be inaccurate. A round-wire type gapping tool is the best way to check the gap. The correct gauge should pass through the electrode gap with a slight drag. If you're in doubt, try one size smaller and one larger. The smaller gauge should go through easily, while the larger one shouldn't go through at all. Wire gapping tools usually have a bending tool attached. Use

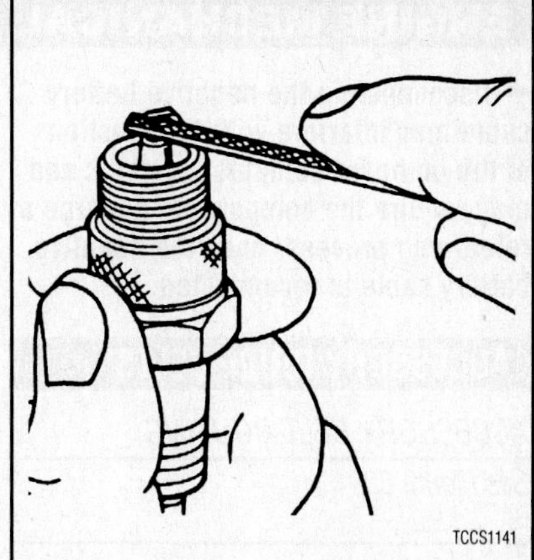

Fig. 60 If the standard plug is in good condition, the electrode may be filed flat—WARNING: do not file platinum plugs

that to adjust the side electrode until the proper distance is obtained. Absolutely never attempt to bend the center electrode. Also, be careful not to bend the side electrode too far or too often as it may weaken and break off within the engine, requiring removal of the cylinder head to retrieve it.

ENGINE ELECTRICAL

STARTING SYSTEM

STARTER

REMOVAL & INSTALLATION

See Figure 61.

1. Before servicing the vehicle, refer to the Precautions Section.
2. Disconnect the battery negative cable.
3. Disconnect the starter cable from the B terminal on the solenoid.
4. Disconnect the connector from the S terminal.
5. Remove the 2 bolts holding the starter, then remove the starter.

To install:

6. Installation is the reverse of removal. Tighten starter motor bolts to 19–24 ft. lbs. (27–34 Nm).

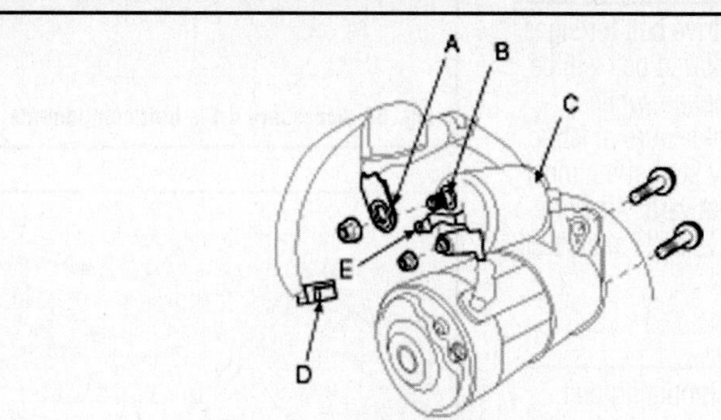

Fig. 61 Starter cable, connector, B terminal, and S terminal

ENGINE MECHANICAL

➡Disconnecting the negative battery cable may interfere with the functions of the on board computer systems and may require the computer to undergo a relearning process, once the negative battery cable is reconnected.

ACCESSORY DRIVE BELTS

ACCESSORY BELT ROUTING

See Figure 62.

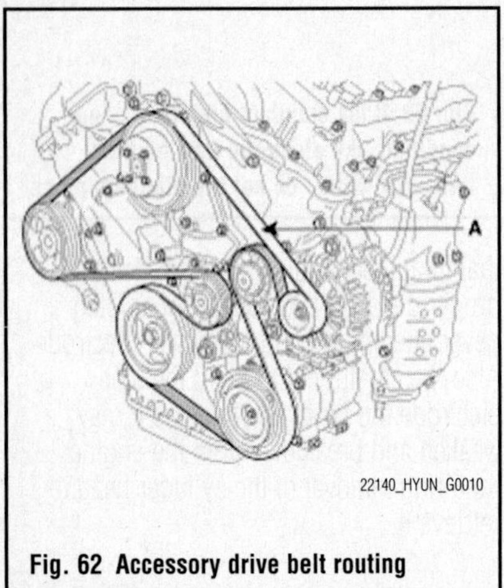

Fig. 62 Accessory drive belt routing

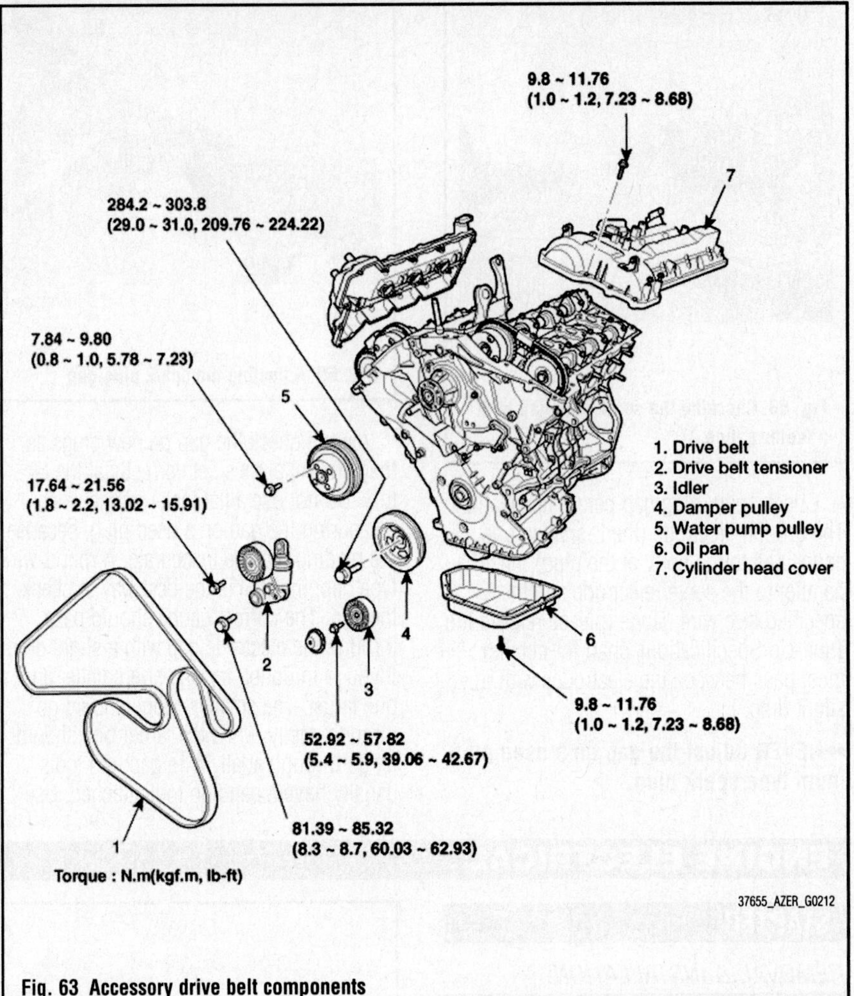

Fig. 63 Accessory drive belt components

INSPECTION

Inspect the accessory drive belt for signs of glazing or cracking. A glazed belt will be perfectly smooth from slippage, while a good belt will have a slight texture of fabric visible. Cracks will usually start at the inner edge of the belt and run outward. All worn or damaged accessory drive belts should be replaced immediately.

ADJUSTMENT

1. Loosen the tension mounting bolt.
2. Turn the adjusting bolt to obtain the proper belt tension, then retighten the mounting bolt.
3. Recheck the deflection of the drive belt.

REMOVAL & INSTALLATION

See Figures 63 and 64.

1. Before servicing the vehicle, refer to the Precautions Section.
2. Disconnect the negative battery cable.
3. Raise and safely support the vehicle.
4. Rotate the drive belt tensioner clockwise to release the drive belt tension.
5. Remove the drive belt from the alternator.

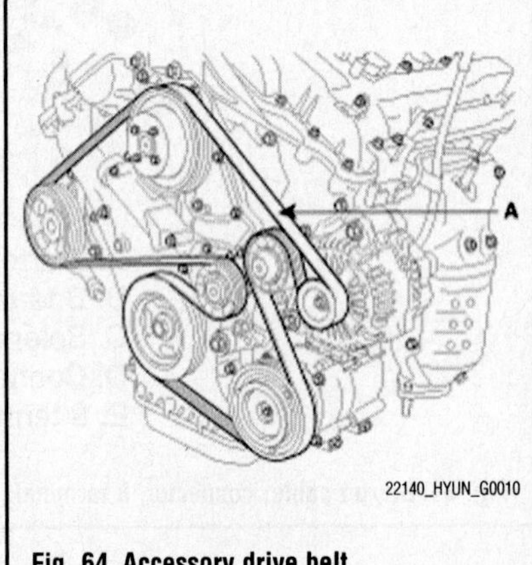

Fig. 64 Accessory drive belt

6. Slowly release the drive belt tensioner.
7. Remove the drive belt from the accessory drive pulleys.

To install:

8. Install the drive belt to the accessory drive pulley.
9. Rotate the drive belt tensioner clockwise.
10. Install the drive belt to the alternator.
11. Ensure the drive belt is properly aligned and seated into the grooves of the accessory drive pulleys.
12. Slowly release the drive belt tensioner.
13. Lower the vehicle.
14. Connect the negative battery cable.

CAMSHAFT & VALVE LIFTERS

INSPECTION

See Figure 65.

1. Inspect the cam lobes.
    a. Using a micrometer, measure the cam lobe height.
    b. If the cam lobe height is less than specified, replace the camshaft:
    - Intake: 1.8425 in. (46.8mm)
    - Exhaust: 1.8031 in. (45.8mm)

2. Inspect the camshaft journal clearance.

a. Clean the bearing caps and camshaft journals.

b. Place the camshafts on the cylinder head.

c. Lay a strip of Plastigage® across each of the camshaft journal.

d. Install the bearing caps and tighten the bolts with specified torque.

➡Do not turn the camshaft.

e. Remove the bearing caps.

f. Measure the Plastigage® at its widest point.

g. If the oil clearance is greater than specified, replace the camshaft. If necessary, replace the bearing caps and cylinder head as a set.

h. Intake:

- No. 1 Journal: 0.0008–0.0022 in. (0.020–0.057mm)
- No. 2, 3, 4 Journal: 0.0012–0.0026 in. (0.030–0.067mm)

i. Exhaust:

- No. 1 Journal: 0.0008–0.0022 in. (0.020–0.057mm)
- No. 2, 3, 4 Journal: 0.0012–0.0026 in. (0.030–0.067mm)

j. Completely remove the Plastigage®.

k. Remove the camshafts.

3. Inspect the camshaft end play.

a. Install the camshafts.

b. Using a dial indicator, measure the end play while moving the camshaft back and forth.

c. If the end play is greater than specified, replace the camshaft. If necessary, replace the bearing caps and cylinder head as a set.

- Camshaft End Play: 0.0008–0.0071 in. (0.02–0.18mm)

d. Remove the camshafts.

4. Inspect the Continuous Variable Valve Timing (CVVT) assembly.

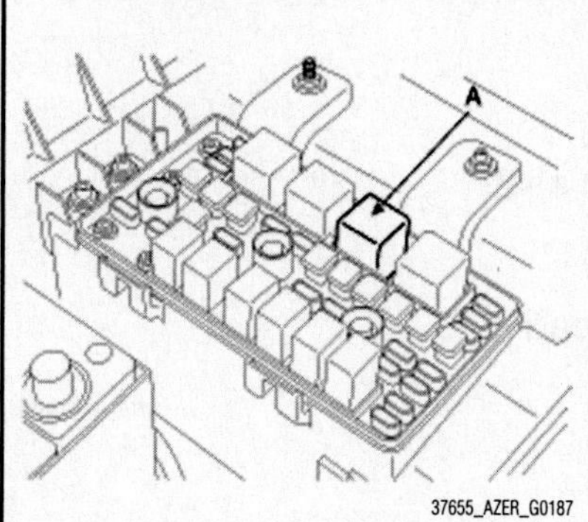

Fig. 65 Apply vinyl tape to the CVVT on all parts except the one indicated by the arrow

a. Check that the CVVT assembly will not turn.

b. Apply vinyl tape to all the parts except the one indicated by the arrow in the illustration.

c. Wrap tape around the tip of the air gun and apply air of approx. 21 psi to the port of the camshaft. Perform this in order to release the lock pin for the maximum delay angle locking.

➡Wrap a shop rag around the CVVT as the oil may spray out when the air pressure is applied.

d. Under the condition of air pressure being applied, turn the CVVT assembly to the advance angle side with your hand.

- Depending on the air pressure, the CVVT assembly will turn to the advance side
- If air is leaking from the port and air pressure cannot be maintained, the locking pin will not release

5. Except the position where the lock pin meets at the maximum delay angle, let the CVVT assembly turn back and forth and check the movable range and that there is no disturbance.

a. The CVVT should move smoothly in the range of about 22.5°.

b. Turn the CVVT assembly with your hand and lock it at the maximum delay angle position (counter-clockwise).

REMOVAL & INSTALLATION

See Figures 66 through 70.

⁂ WARNING

Use a fender cover to avoid damaging painted surfaces. To avoid damage, unplug the wiring connectors carefully while holding the connector portion. Mark all wiring and hoses to avoid misconnection. Inspect the timing belt before removing.

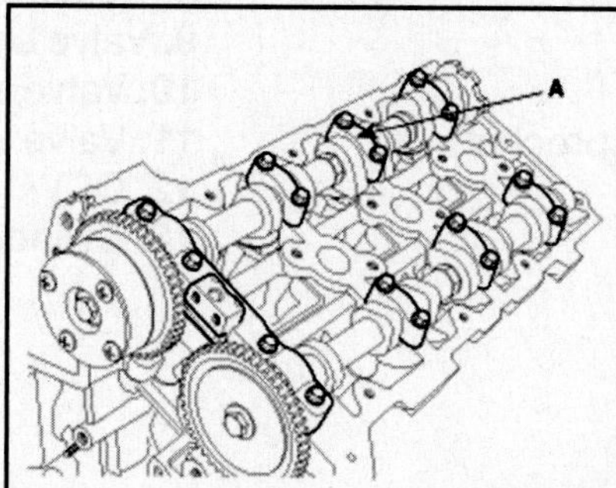

Fig. 66 Remove the camshaft bearing cap (A)

Fig. 67 Remove the camshaft assembly (A)

➡Turn the crankshaft pulley so that the No. 1 piston is at Top Dead Center (TDC).

➡Engine removal is required for this procedure.

1. Before servicing the vehicle, refer to the Precautions Section.
2. Disconnect the negative battery cable.
3. Remove the exhaust manifold. Refer to Exhaust Manifold Removal & Installation.
4. Remove the intake manifold. Refer to Intake Manifold Removal & Installation.
5. Remove the timing chain. Refer to Timing Chain & Sprockets Removal & Installation.
6. Remove the water temperature control assembly.
7. Remove the camshaft bearing cap.
8. Remove the camshaft assembly.

To install:

9. Thoroughly clean all parts to be assembled. Rotate the crankshaft, set the No. 1 piston at TDC.
10. Install the Continuously Variable Valve Timing (CVVT) and camshaft sprocket. Tightening torque: 48–56 ft. lbs. (65–76 Nm).

a. Install camshaft-inlet to dowel pin of CVVT assembly. At this time, do not install to oil hole of camshaft-inlet.

b. Hold the hexagonal head wrench portion of the camshaft with a vise, and install the bolt and CVVT assembly.

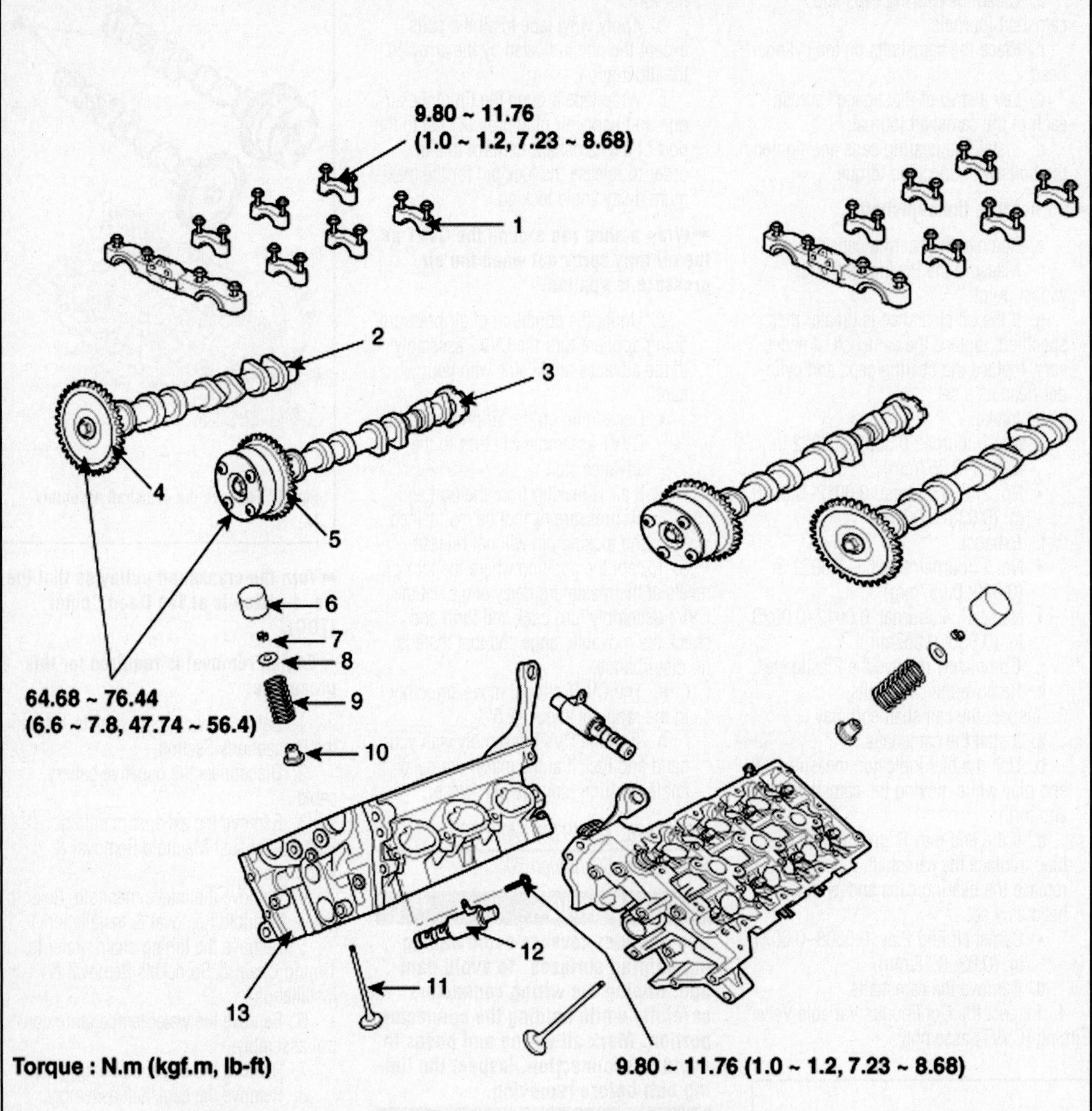

Fig. 68 Exploded view of cylinder head and related components

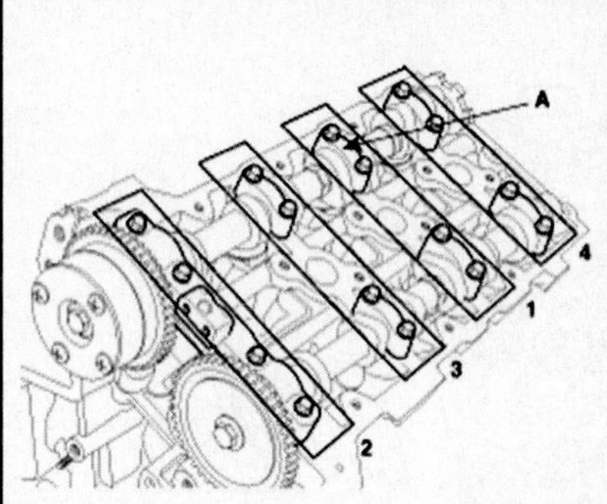

Fig. 69 Install the camshaft bearing caps in the sequence shown

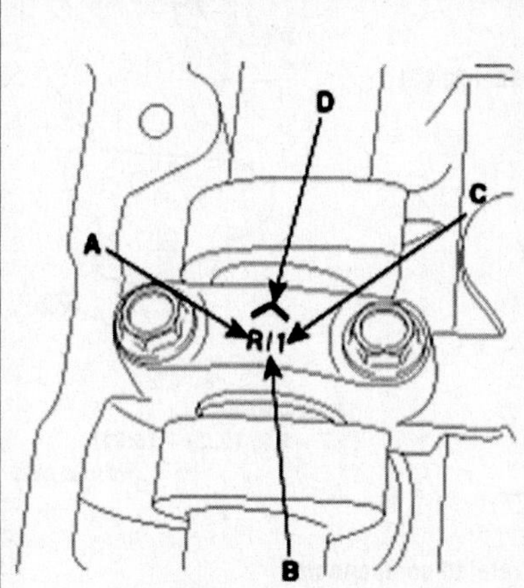

Fig. 70 Be careful to properly position the camshaft bearing caps according to its markings

c. Do not rotate the CVVT assembly when the camshaft is installed to the dowel pin of the CVVT assembly.

11. Install the camshafts.

a. Apply a light coat of engine oil on camshaft journals.

b. Assemble the key groove of camshaft rear side to the same level of head top surface.

c. Be careful to get the right bank, left bank, intake side, and exhaust side in the correct position before assembling.

12. Install the camshaft bearing caps in the sequence shown:

a. Step 1—Tightening torque: 52 inch lbs. (6 Nm).

b. Step 2—Tightening torque: 87–104 inch lbs. (10–12 Nm).

WARNING

Be careful to properly position the right bank, left bank, intake side, exhaust side, and front mark on the camshaft bearing caps while assembling.

WARNING

Rotate the crankshaft so as not to contact the valves to the pistons by positioning the pistons 0.3937 in. (10mm) below the top of the cylinder block.

13. Install the water temperature control assembly.
14. Install the timing chain.
15. Check and adjust the valve clearance, as necessary.
16. Install the exhaust manifold.
17. Install the intake manifold.
18. Connect the negative battery cable.
19. Fill with engine coolant.
20. Start the engine and check for leaks.
21. Recheck the engine coolant level and oil level.

CATALYTIC CONVERTER

REMOVAL & INSTALLATION

See Figure 71.

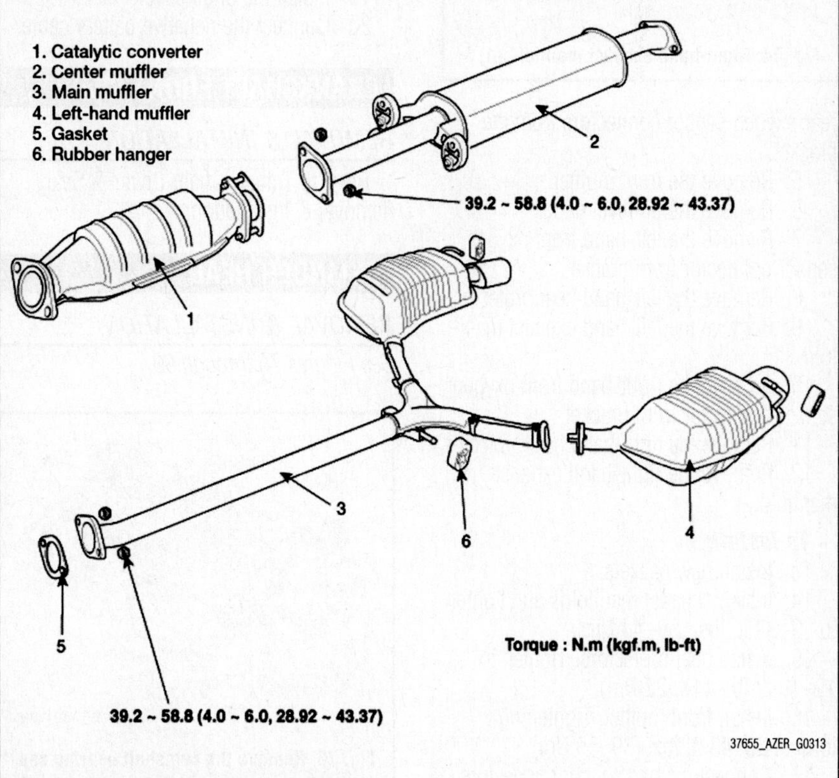

Fig. 71 Post-catalytic converter location

This system consists of pre and post catalytic converters. For pre-catalytic converter, refer to Combination Manifold Removal & Installation.

COMBINATION MANIFOLD

REMOVAL & INSTALLATION

See Figures 72 through 75.

1. Before servicing the vehicle, refer to the Precautions Section.
2. Disconnect the negative battery cable.
3. Remove the undercover.
4. Remove the left-hand and right-hand

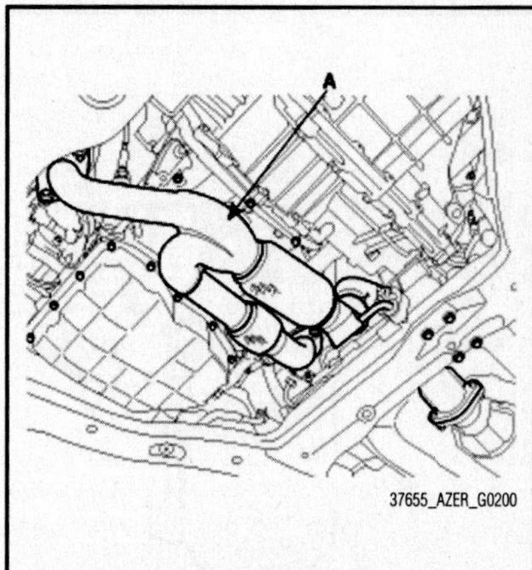

Fig. 72 Remove the front muffler (A)

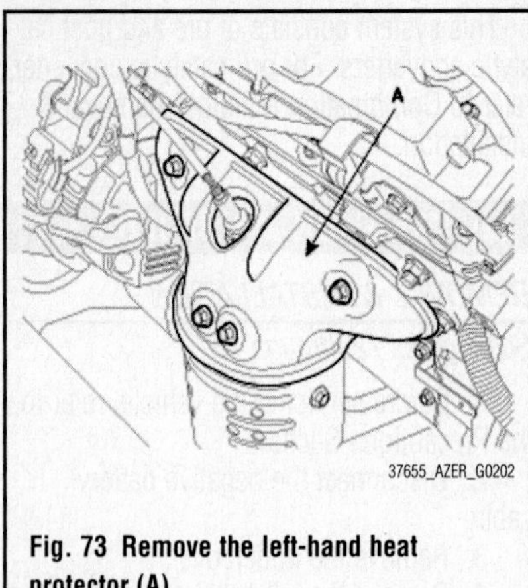

Fig. 73 Remove the left-hand heat protector (A)

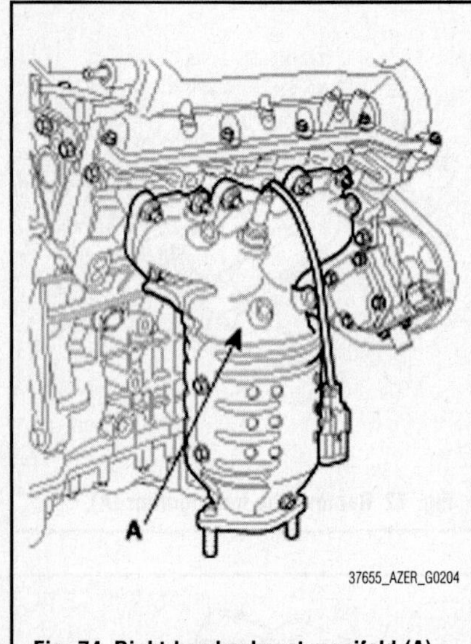

Fig. 74 Right-hand exhaust manifold (A)

rear oxygen sensor connectors from the bracket.

5. Remove the front muffler.
6. Remove the oil level gauge.
7. Remove the left-hand front oxygen sensor connector from bracket.
8. Remove the left-hand heat protector.
9. Remove the left-hand exhaust manifold.
10. Remove the right-hand front oxygen sensor connector from bracket.
11. Remove the right-hand heat protector.
12. Remove the right-hand exhaust manifold.

To install:

13. Install new gaskets.
14. Install exhaust manifolds and tighten to 29–33 ft. lbs. (39–44 Nm).
15. Install heat protectors. Tighten to 12–16 ft. lbs. (17–22 Nm).
16. Install front muffler. Tightening torque: 29–43 ft. lbs. (39–59 Nm).
17. Connect oxygen sensor connectors.
18. Install the oil level gauge.

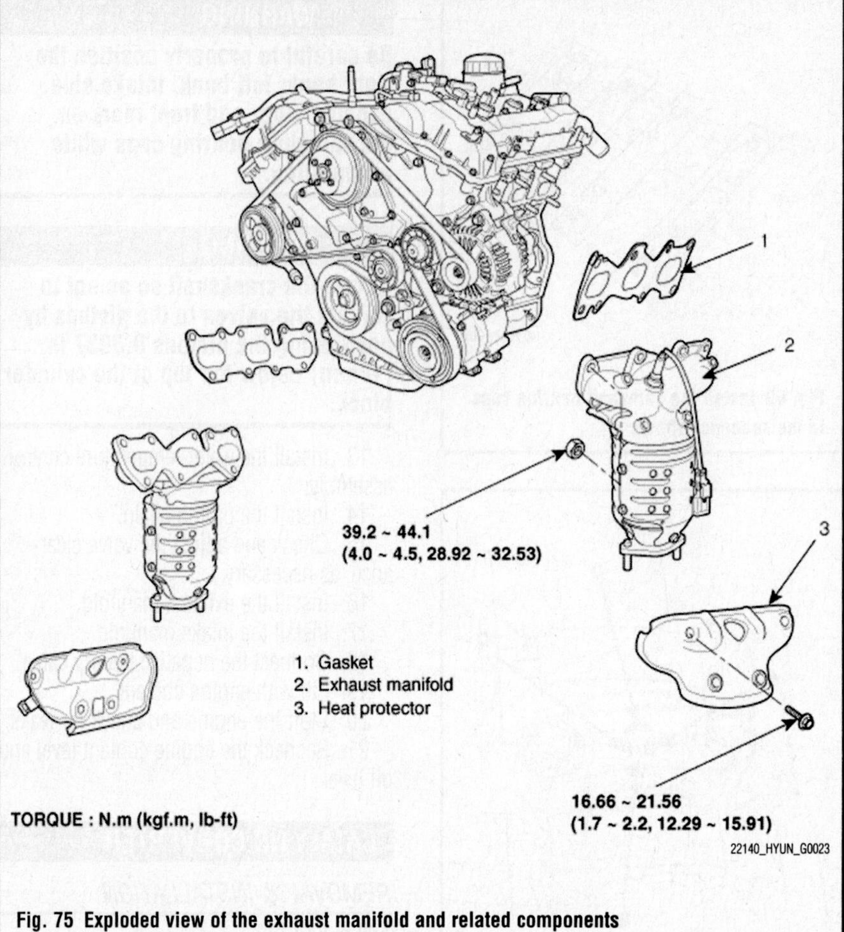

Fig. 75 Exploded view of the exhaust manifold and related components

19. Install the undercover.
20. Connect the negative battery cable.

CRANKSHAFT FRONT SEAL

REMOVAL & INSTALLATION

Refer to Timing Chain Cover & Seal Removal & Installation.

CYLINDER HEAD

REMOVAL & INSTALLATION

See Figures 76 through 90.

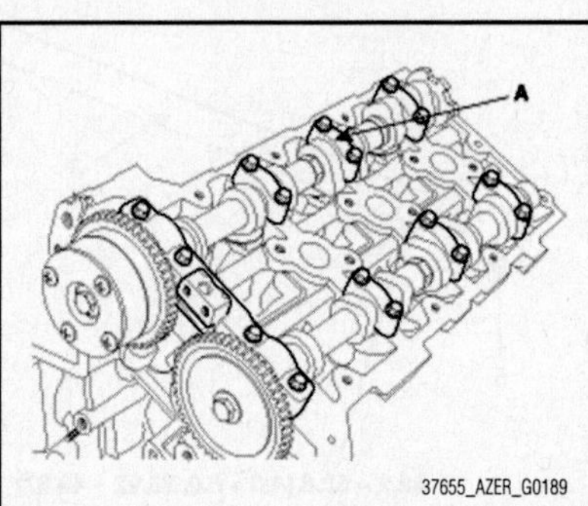

Fig. 76 Remove the camshaft bearing cap (A)

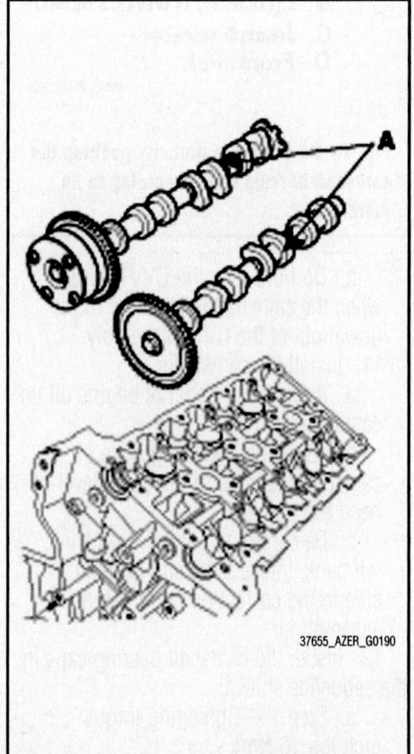

Fig. 77 Remove the camshaft assembly (A)

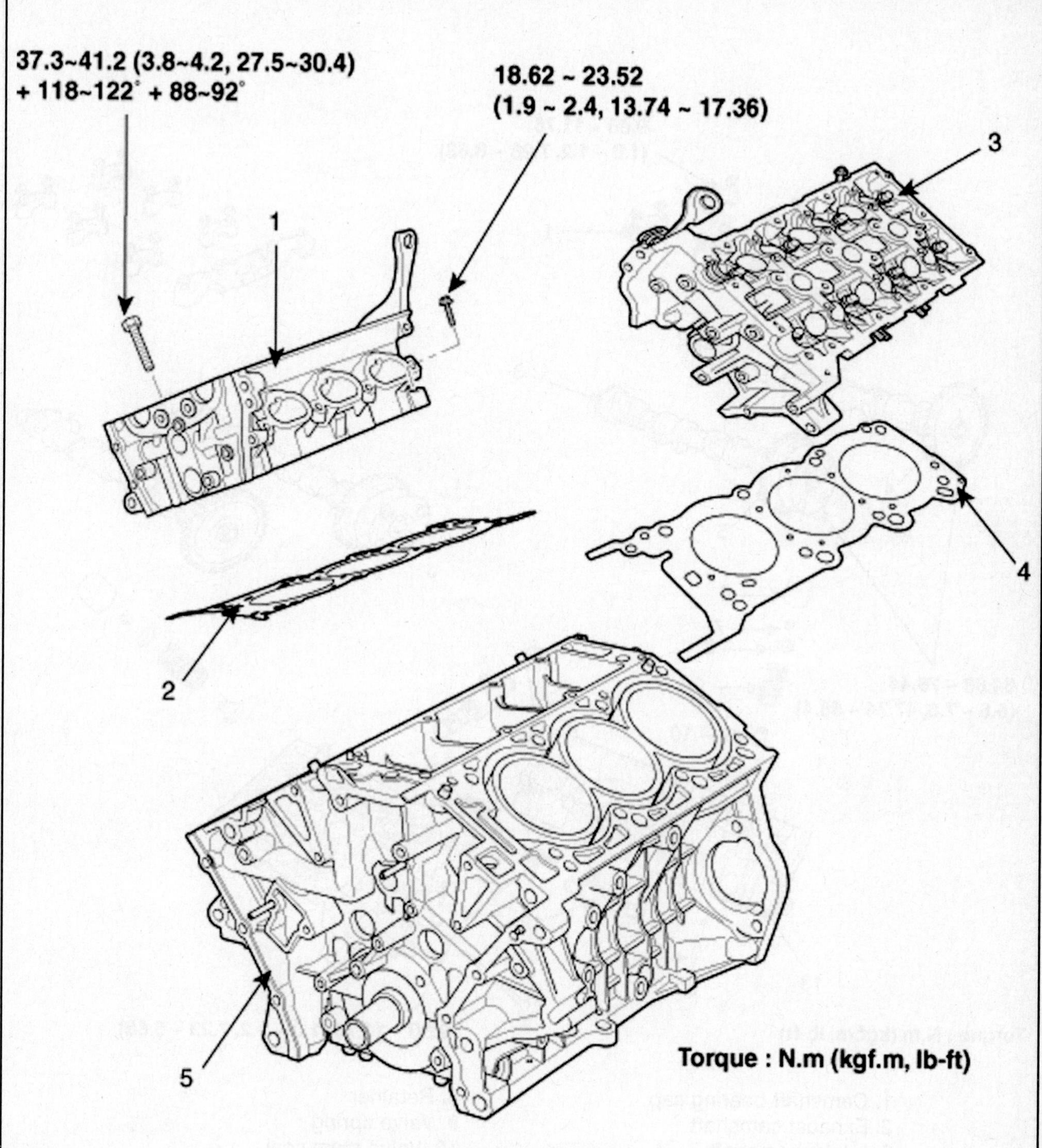

Fig. 78 Exploded view of cylinder head and engine block

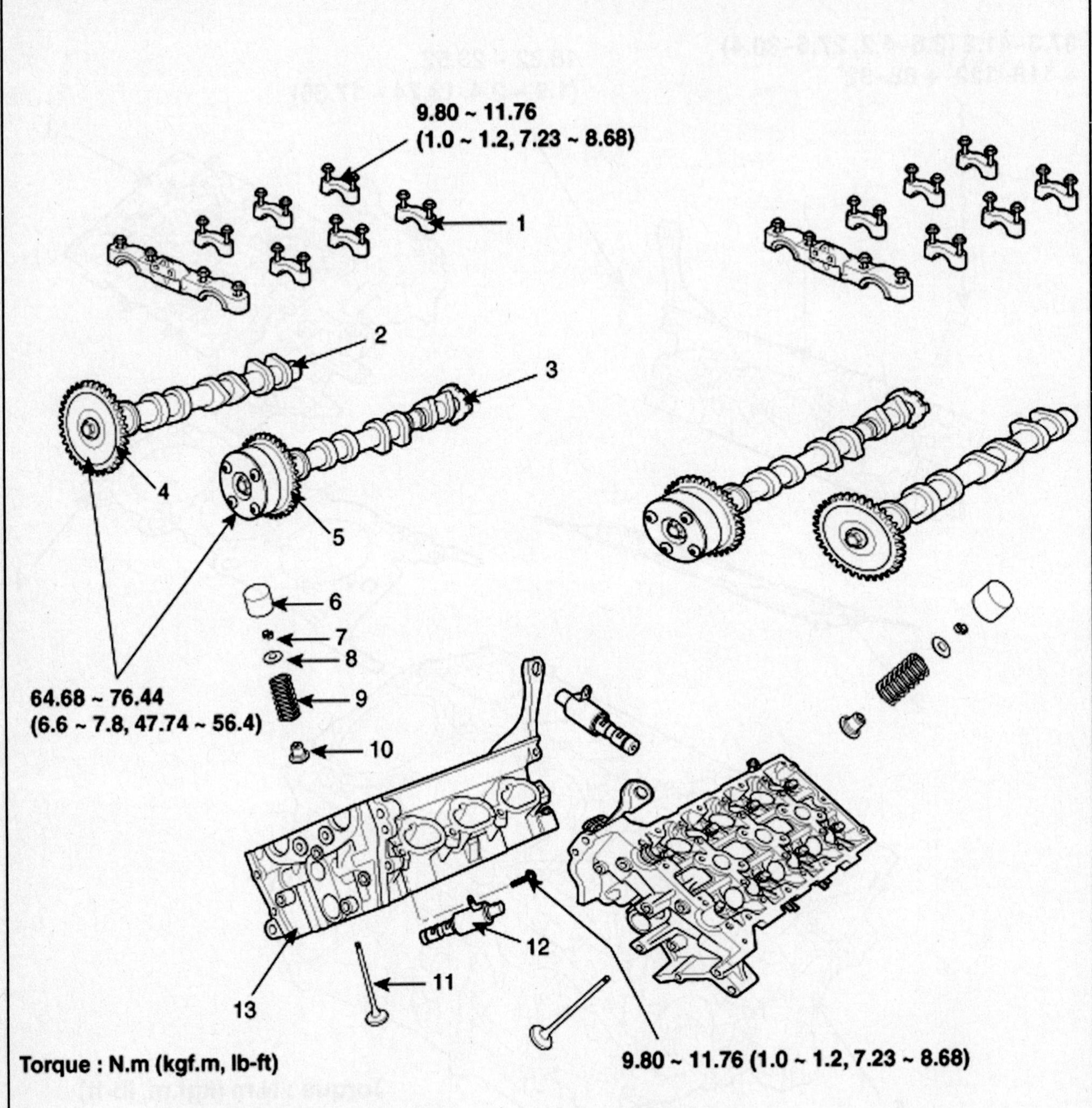

1. Camshaft bearing cap
2. Exhaust camshaft
3. Intake camshaft
4. Exhaust camshaft sprocket
5. CVVT assembly
6. MLA
7. Retainer lock
8. Retainer
9. Valve spring
10. Valve stem seal
11. Valve
12. OCV
13. Cylinder head

37655_AZER_G0192

Fig. 79 Exploded view of cylinder head and related components

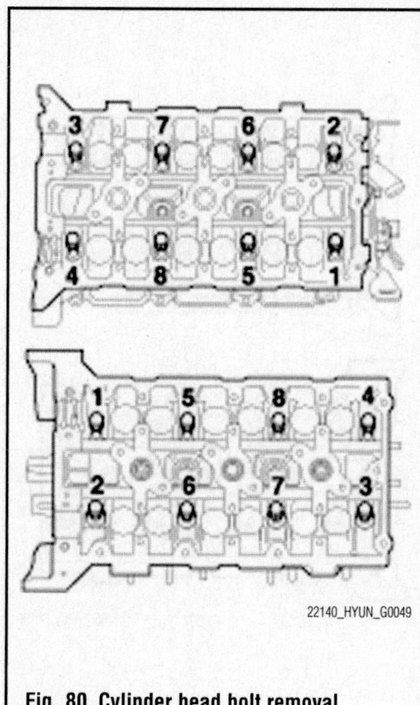

Fig. 80 Cylinder head bolt removal sequence

✱✱ WARNING

Use a fender cover to avoid damaging painted surfaces. To avoid damaging the cylinder head, wait until the engine coolant temperature drops below normal temperature before removing it. When handling a metal gasket, take care not to fold the gasket or damage the contact surface of the gasket. To avoid damage, unplug the wiring connectors carefully while holding the connector portion. Mark all wiring and hoses to avoid misconnection. Inspect the timing belt before removing.

➡Turn the crankshaft pulley so that the No. 1 piston is at Top Dead Center (TDC).

➡Engine removal is required for this procedure.

1. Before servicing the vehicle, refer to the Precautions Section.
2. Disconnect the negative battery cable.
3. Remove the exhaust manifold. Refer to Exhaust Manifold Removal & Installation.
4. Remove the intake manifold. Refer to Intake Manifold Removal & Installation.
5. Remove the timing chain. Refer to Timing Chain & Sprockets Removal & Installation.
6. Remove the water temperature control assembly.
7. Remove the camshaft bearing cap.
8. Remove the camshaft assembly.
9. Remove the cylinder head bolts, then remove cylinder head.
    a. Uniformly loosen and remove the 16 cylinder head bolts, in several passes, in the sequence shown.
    b. Remove the 16 cylinder head bolts and plate washers.

✱✱ WARNING

Head warpage or cracking could result from removing bolts in an incorrect order.

    c. Lift the cylinder head from the dowels on the cylinder block and place the cylinder head on wooden blocks on a bench.

✱✱ WARNING

Be careful not to damage the contact surfaces of the cylinder head and cylinder block.

To install:

Thoroughly clean all parts to be assembled. Always use a new head and manifold gasket. The cylinder head gasket is a metal gasket. Take care not to bend it. Rotate the crankshaft, set the No. 1 piston at TDC.

10. Ensure the sealant locations on the cylinder head and cylinder block are free of engine oil or any debris.
11. Apply sealant on the cylinder block top face before assembling cylinder head

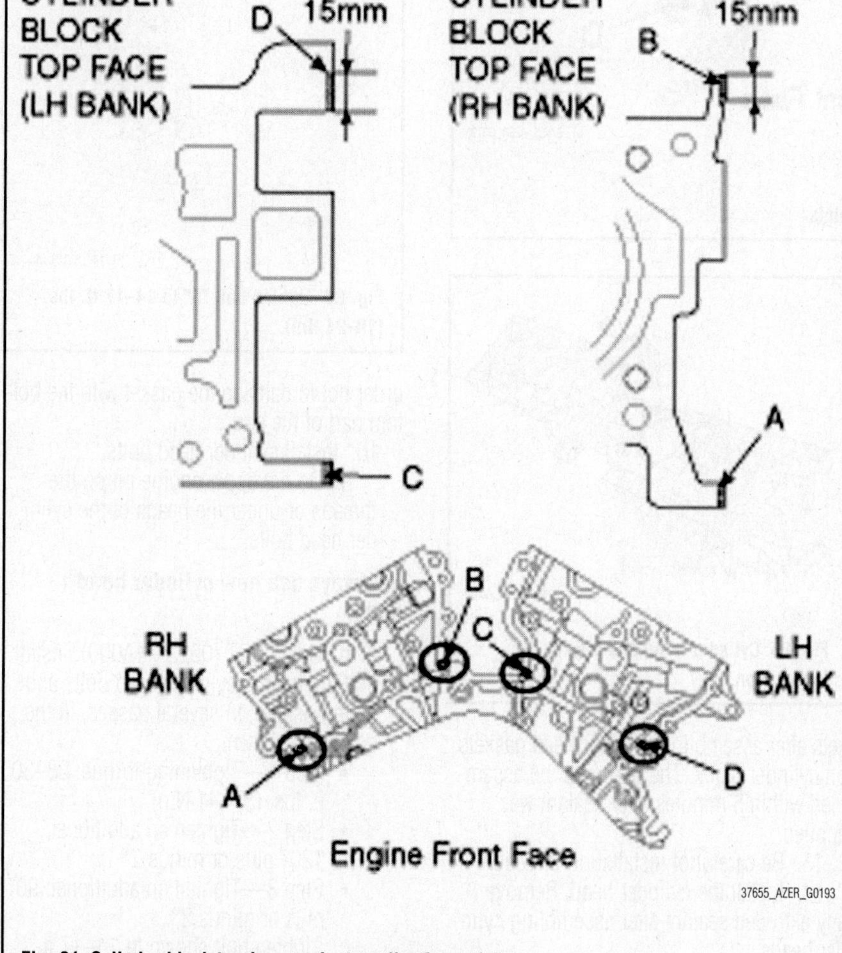

Fig. 81 Cylinder block top face sealant application points

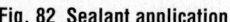

Fig. 82 Sealant application

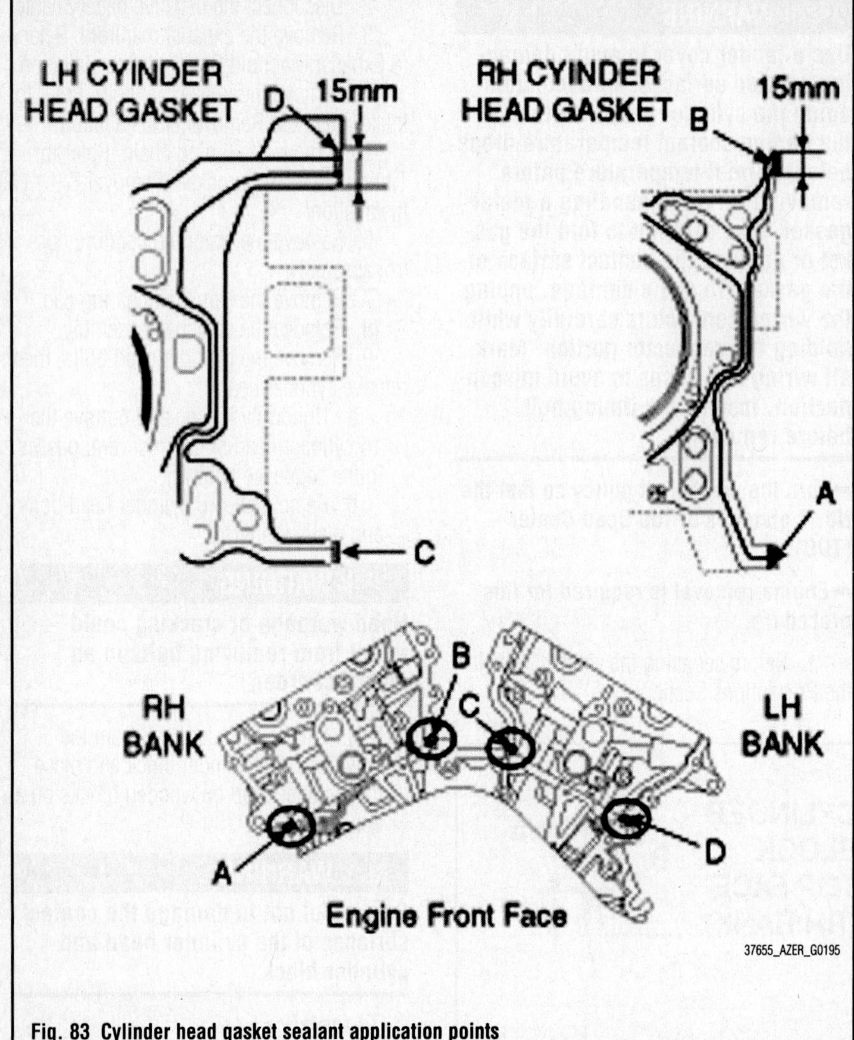

Fig. 83 Cylinder head gasket sealant application points

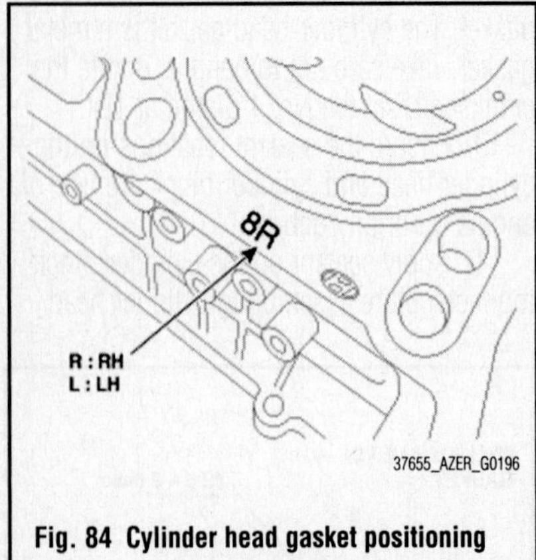

Fig. 84 Cylinder head gasket positioning

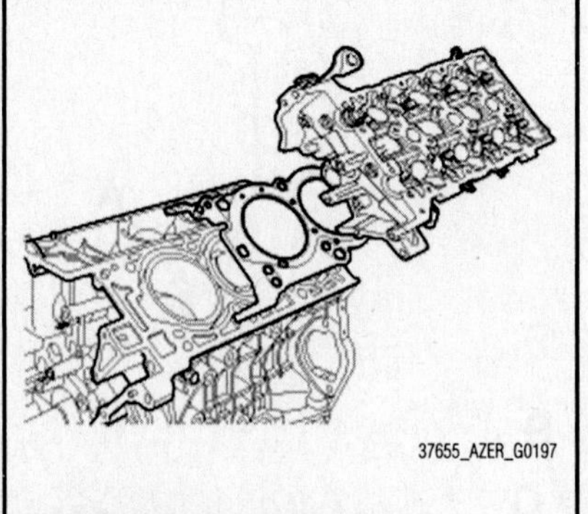

Fig. 85 Cylinder head and gasket installation

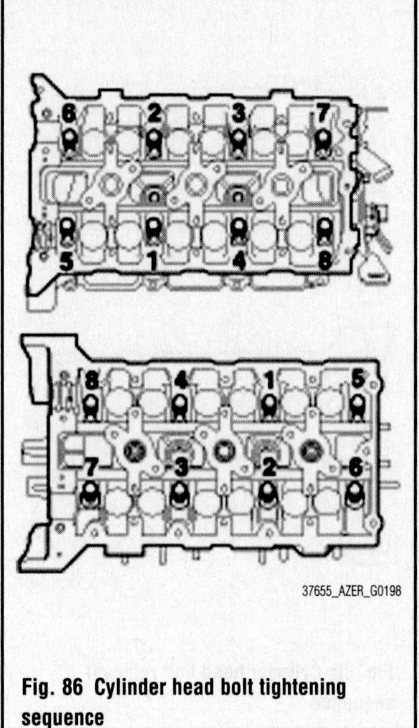

Fig. 86 Cylinder head bolt tightening sequence

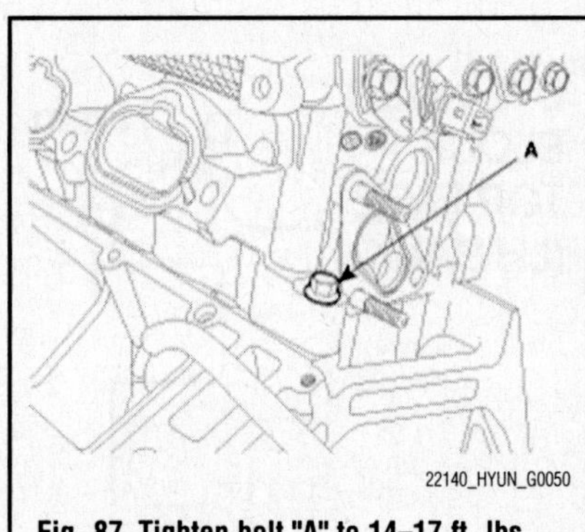

Fig. 87 Tighten bolt "A" to 14–17 ft. lbs. (19–24 Nm)

gaskets. The part must be assembled within 5 minutes after sealant is applied.

- Bead width: 0.08–0.12 in. (2–3mm)
- Sealant location: 0.04–0.06 in. (1.0–1.5mm) from block surface
- Recommended sealant: Liquid sealant TB1217H

12. Apply sealant on cylinder head gaskets after assembling cylinder head gaskets on cylinder block. The part must be assembled within 5 minutes after sealant was applied.
13. Be careful of installation direction.
14. Install the cylinder head. Remove any extruded sealant after assembling cylinder heads.
15. Place the cylinder head carefully in order not to damage the gasket with the bottom part of the end.
16. Install cylinder head bolts.
    a. Do not apply engine oil on the threads or under the heads of the cylinder head bolts.

➡Always use new cylinder head bolts.

b. Using SST (09221-4A000), install and tighten the cylinder head bolts and plate washers, in several passes, in the sequence shown.

- Step 1—Tightening torque: 28–30 ft. lbs. (37–41 Nm)
- Step 2—Tighten an additional: 120° plus or minus 2°
- Step 3—Tighten an additional: 90° plus or minus 2°

c. Tighten bolt shown to 14–17 ft. lbs. (19–24 Nm).

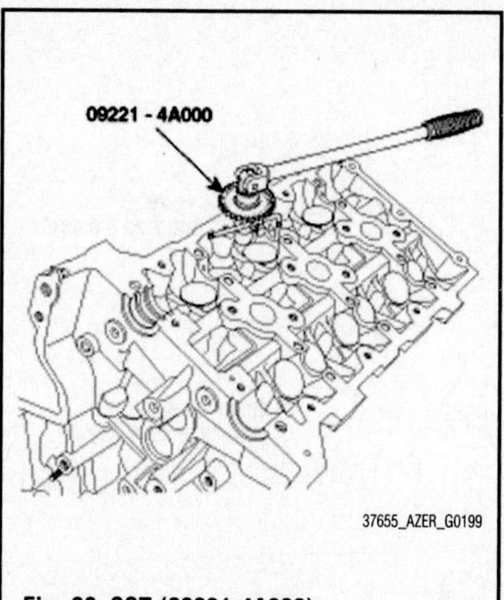

Fig. 88 SST (09221-4A000)

17. Install the Continuously Variable Valve Timing (CVVT) and camshaft sprocket. Tightening torque: 48–56 ft. lbs. (65–76 Nm).
   a. Install camshaft-inlet to dowel pin of CVVT assembly. At this time, do not install to oil hole of camshaft-inlet.
   b. Hold the hexagonal head wrench portion of the camshaft with a vise, and install the bolt and CVVT assembly.
   c. Do not rotate the CVVT assembly when the camshaft is installed to the dowel pin of the CVVT assembly.
18. Install the camshafts.
   a. Apply a light coat of engine oil on camshaft journals.
   b. Assemble the key groove of camshaft rear side to the same level of head top surface.
   c. Be careful to get the right bank, left bank, intake side, and exhaust side in the correct position before assembling.
19. Install the camshaft bearing caps in the sequence shown:
   a. Step 1—Tightening torque: 52 inch lbs. (6 Nm).

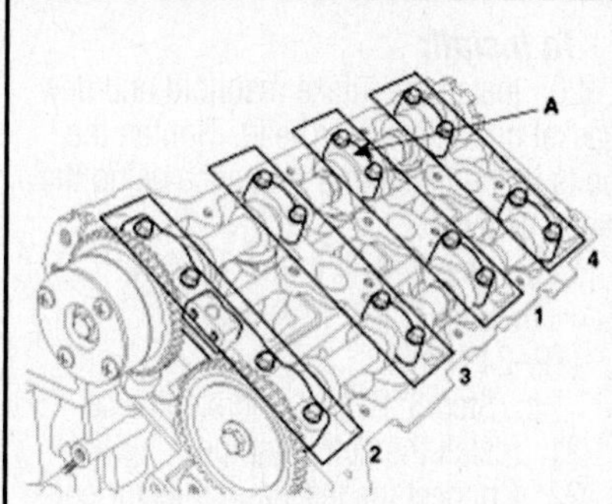

Fig. 89 Install the camshaft bearing caps in the sequence shown

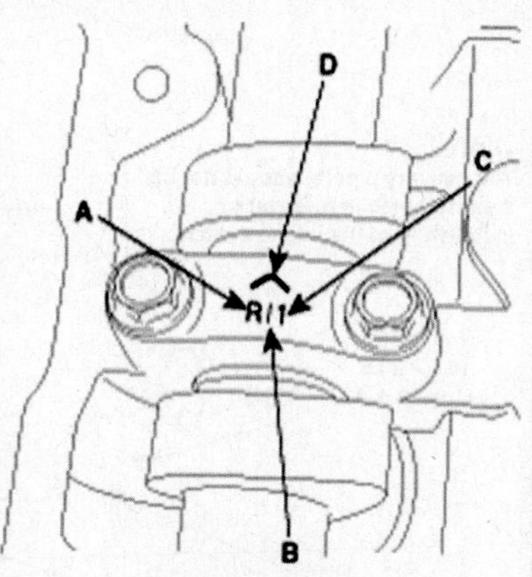

Fig. 90 Be careful to properly position the camshaft bearing caps according to its markings

   b. Step 2—Tightening torque: 87–104 inch lbs. (10–12 Nm).

WARNING

Be careful to properly position the right bank, left bank, intake side, exhaust side, and front mark on the camshaft bearing caps while assembling.

WARNING

Rotate the crankshaft so as not to contact the valves to the pistons by positioning the pistons 0.3937 in. (10mm) below the top of the cylinder block.

20. Install the water temperature control assembly.
21. Install the timing chain.
22. Check and adjust the valve clearance, as necessary.
23. Install the exhaust manifold.
24. Install the intake manifold.
25. Connect the negative battery cable.
26. Fill with engine coolant.
27. Start the engine and check for leaks.
28. Recheck the engine coolant level and oil level.

EXHAUST MANIFOLD

REMOVAL & INSTALLATION

Refer to Combination Manifold Removal & Installation.

INTAKE MANIFOLD

REMOVAL & INSTALLATION

See Figures 91 through 97.

1. Before servicing the vehicle, refer to the Precautions Section.
2. Relieve the fuel system pressure.
3. Drain the cooling system.
4. Remove the negative battery cable.

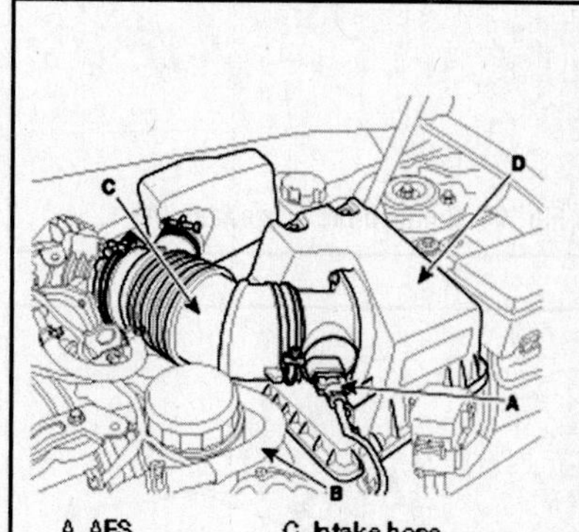

Fig. 91 Disconnect AFS, breather hose, air cleaner upper cover, and intake hose

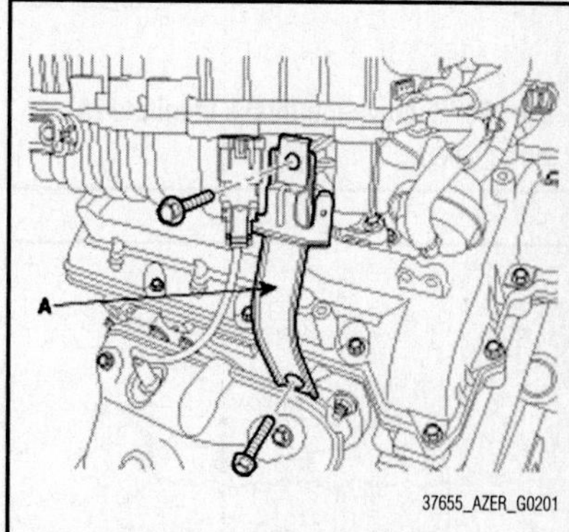

Fig. 92 Remove the surge tank stay (A)

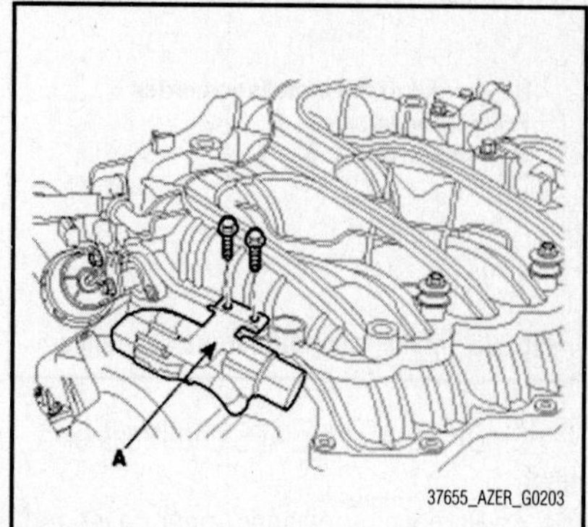

Fig. 93 Remove the connector bracket (A)

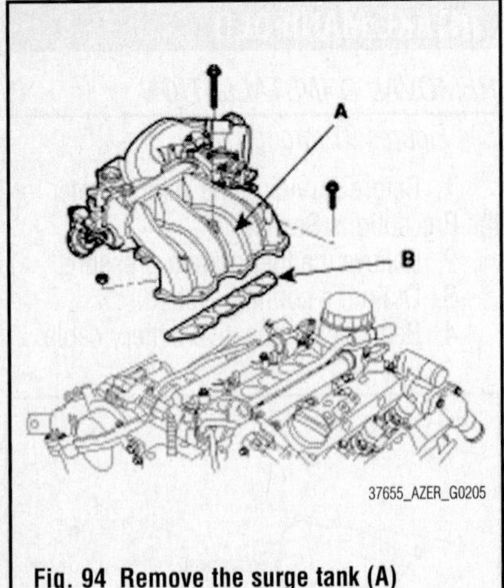

Fig. 94 Remove the surge tank (A)

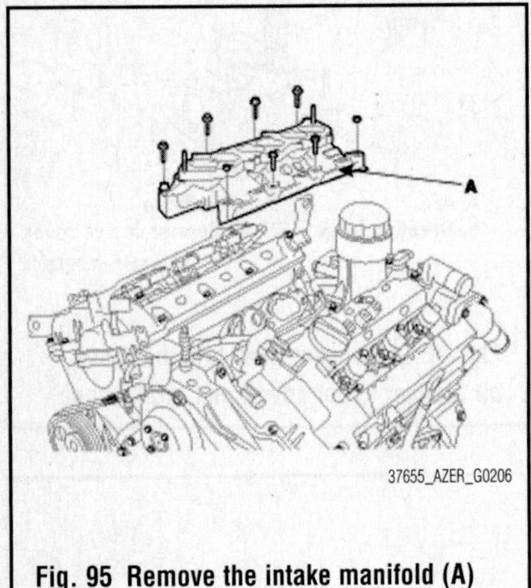

Fig. 95 Remove the intake manifold (A) and gasket.

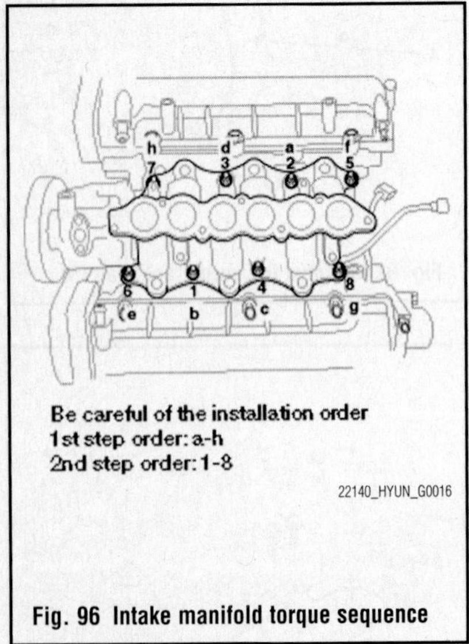

Fig. 96 Intake manifold torque sequence

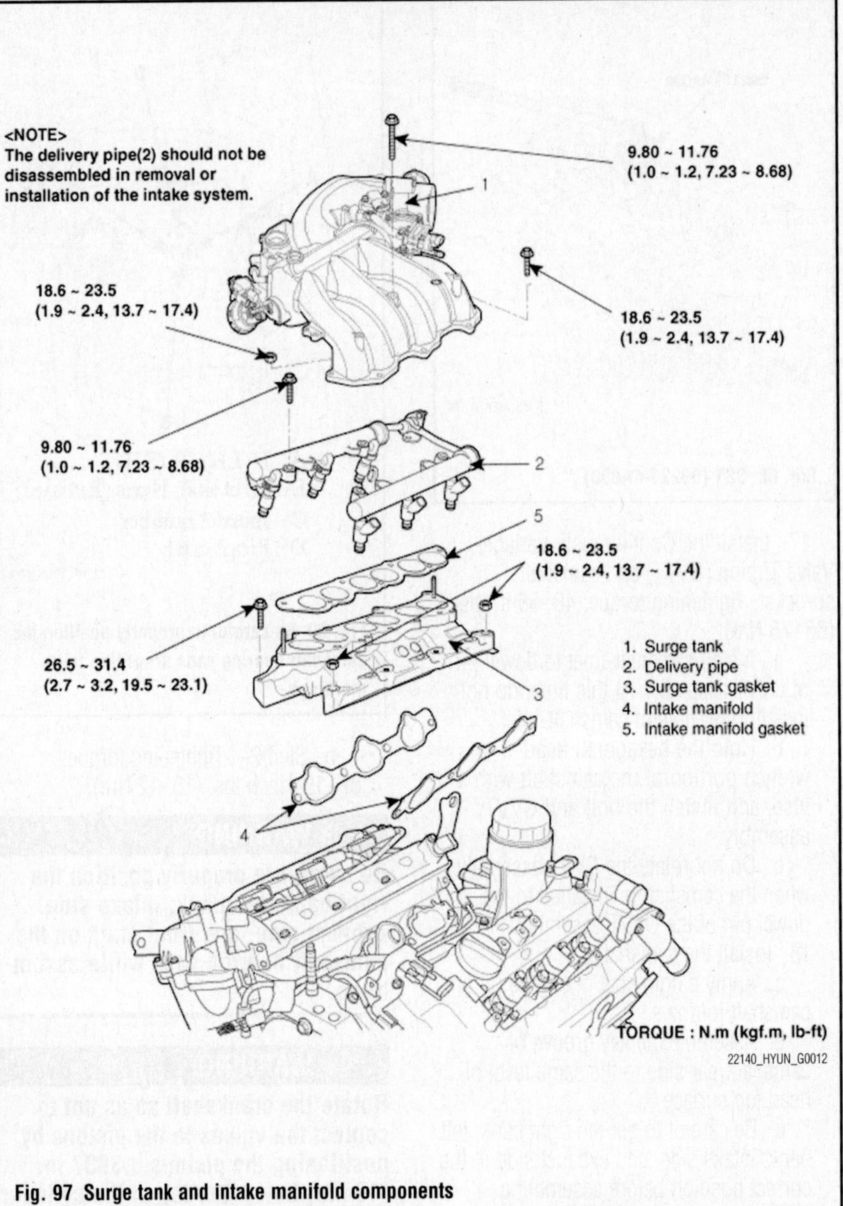

Fig. 97 Surge tank and intake manifold components

5. Disconnect the AFS and breather hose.

6. Remove air cleaner upper cover and intake hose.

7. Disconnect the right-hand oxygen sensor connector.

8. Disconnect the right-hand injector connector and ignition coil connector.

9. Disconnect the Purge Control Solenoid Valve (PCSV) connector, Manifold Absolute Pressure (MAP) sensor connector, and PCSV hose.

10. Disconnect the Electronic Throttle Control (ETC) connector and knock sensor connector.

11. Disconnect the water hoses from ETC.

12. Disconnect the PCV hose.

13. Disconnect the brake vacuum hose.

14. Remove the surge tank stay.

15. Remove the connector bracket from surge tank.

16. Remove the surge tank.

17. Disconnect the breather pipe assembly.

18. Disconnect the left-hand injector connector.

19. Remove the intake manifold and gasket.

To install:

20. Install the intake manifold and new gasket on the cylinder head. Tighten the bolts in the illustrated sequence using the steps below:

a. Step 1: 3–4 ft. lbs. (4–6 Nm).

b. Step 2: 14–17 ft. lbs. (19–24 Nm).

c. Step 3: Repeat 2nd step twice.

21. Install the delivery pipe.

22. Connect the left-hand injector connector.

23. Connect the breather pipe assembly. Tighten to 84–108 inch lbs. (10–12 Nm).

24. Install the surge tank. Tighten long bolt to 84–108 inch lbs. (10–12 Nm); short bolt and nut to 14–17 ft. lbs. (19–24 Nm).

25. Install the connector bracket on the surge tank. Tighten to 5–8 ft. lbs. (7–11 Nm).

26. Install the surge tank stay. Tighten to 20–23 ft. lbs. (27–31 Nm); 14–17 ft. lbs. (19–24 Nm).

27. Connect the brake vacuum hose.

28. Connect the PCV hose.

29. Connect the water hoses to the ETC.

30. Install the ETC bracket. Tighten to 12–19 ft. lbs. (16–26 Nm).

31. Connect the ETC connector and knock sensor connector.

32. Connect the PCSV connector, MAP sensor connector and PCSV hose.

33. Connect the right-hand injector connector and ignition coil connector.

34. Connect the right-hand oxygen sensor connector.

35. Install the air cleaner upper cover and in take hose.

36. Connect the AFS and breather hose.

OIL PAN

REMOVAL & INSTALLATION

See Figure 98.

1. Before servicing the vehicle, refer to the Precautions Section.
2. Drain the engine oil.
3. Remove the oil pan bolts.
4. Using the SST (09215-3C000), remove the oil pan.
   a. Insert the SST between the oil pan and the ladder frame by tapping it with a plastic hammer in the direction of arrow.
   b. After tapping the SST with a plastic hammer along the direction of arrow around more than ⅔ of the edge of the oil pan, remove it from the ladder frame.

WARNING

Do not turn over the SST abruptly without tapping or damage may occur to the SST or oil pan.

To install:

5. Using a razor blade and gasket scraper, carefully remove all the old packing material from the gasket surfaces.
6. Check that the mating surfaces are clean and dry before applying liquid gasket.
   a. Apply liquid gasket as an even bead, centered between the edges of the mating surface. Use liquid gasket: TB1217H or equivalent. Bead width: 0.1 inch (2.5mm).
   b. To prevent leakage of oil, apply liquid gasket to the inner threads of the bolt holes.

➡Do not install the parts if 5 minutes or more have elapsed since applying the liquid gasket. Instead, reapply liquid gasket after removing the residue.

7. Install the oil pan with the bolts. Uniformly tighten the bolts in several passes. Tightening torque: 84–108 inch lbs. (10–12 Nm).

➡After assembly, wait at least 30 minutes before filling the engine with oil.

8. Fill with engine oil.
9. Start the engine and check for leaks.
10. Recheck the engine oil level.

OIL PUMP

REMOVAL & INSTALLATION

See Figures 99 through 103.

1. Before servicing the vehicle, refer to the Precautions Section.
2. Remove the lower oil pan. Refer to Oil Pan Removal & Installation.
3. Remove the oil pump chain cover.
4. Remove the oil pump chain sprocket.
5. Remove the oil pump.

To install:

6. Install the oil pump using a new O-ring. Tighten the bolts to 15–17 ft. lbs. (21–23 Nm).

➡Always use a new O-ring.

7. Install the oil pump sprocket and oil pump chain on the oil pump. Tightening torque: 14–16 ft. lbs. (19–22 Nm).
8. Install the oil pump chain cover. Tightening torque: 87–104 inch lbs. (10–12 Nm).
9. Install the lower oil pan.
10. After assembly, wait at least 30 minutes before filling the engine with oil to allow the gasket material to cure.
11. Fill the engine with oil.
12. Start the engine and check for leaks.
13. Recheck the engine oil level.

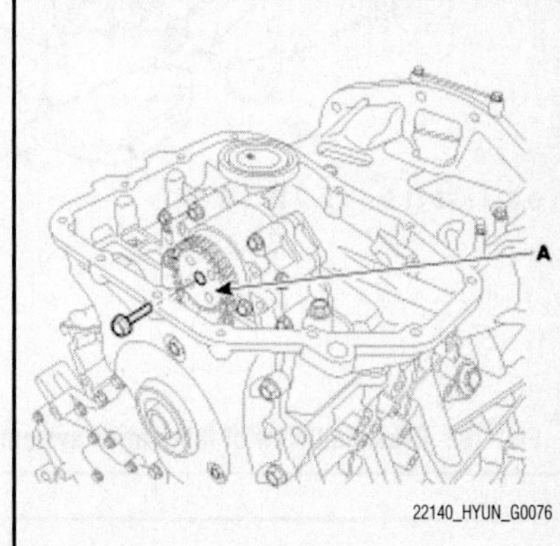

Fig. 100 Remove the oil pump chain sprocket (A)

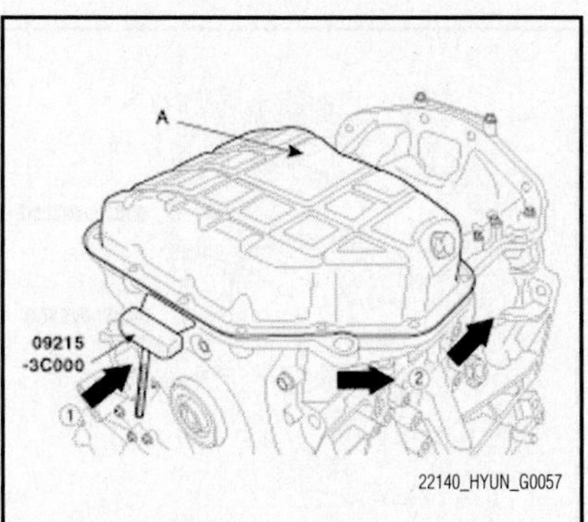

Fig. 98 Using the SST (09215-3C000) to remove the oil pan

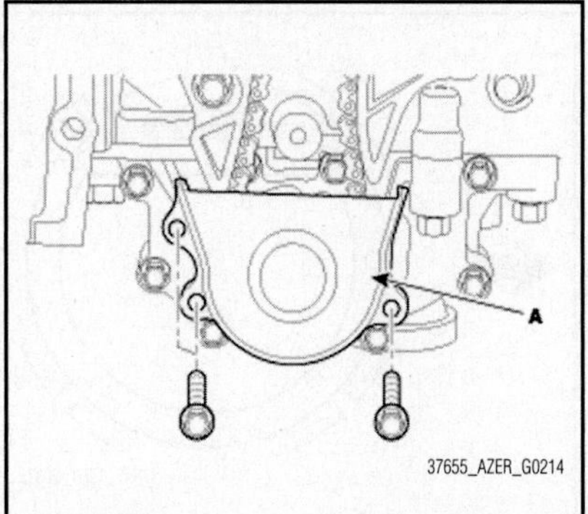

Fig. 99 Remove the oil pump chain cover (A) and bolts

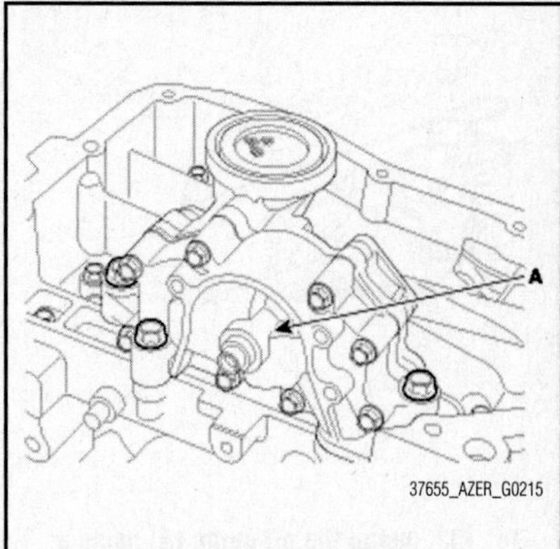

Fig. 101 Remove the oil pump (A)

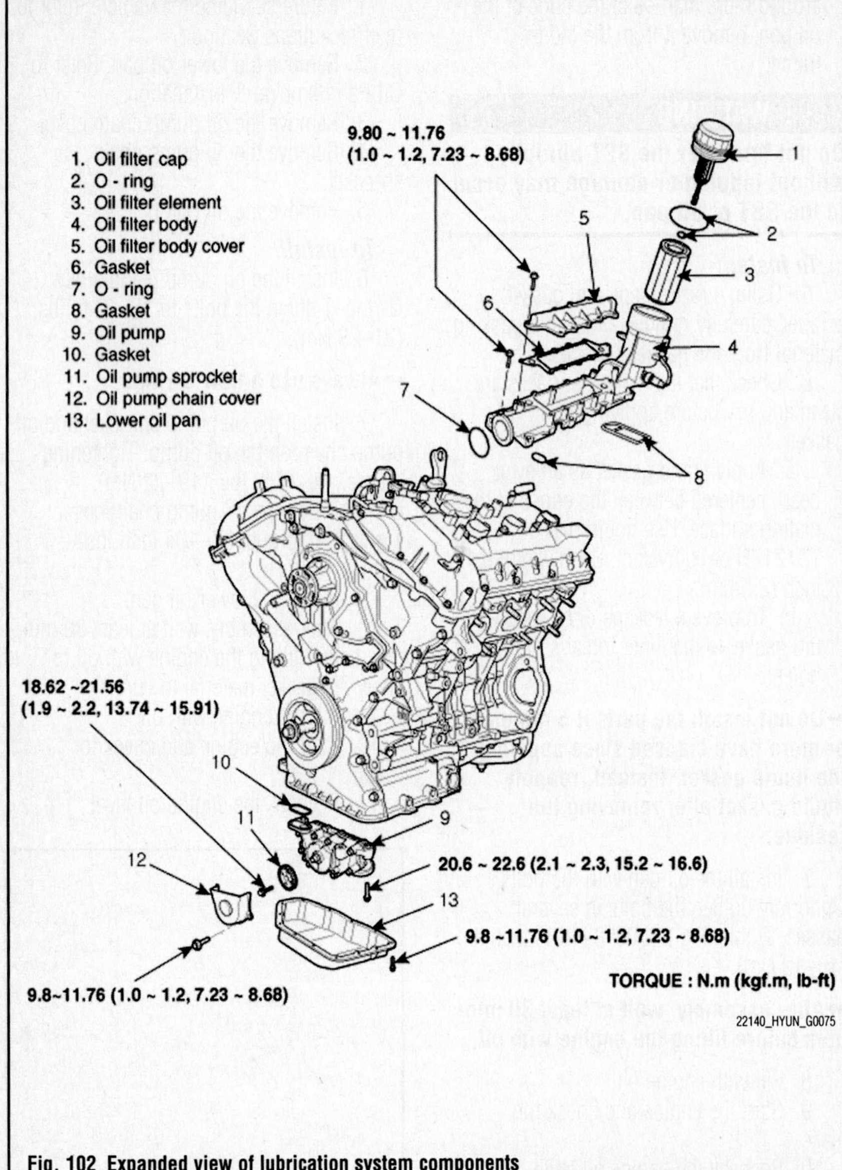

Fig. 102 Expanded view of lubrication system components

Fig. 103 Install the oil pump (A) using a new O-ring (B)

PISTON & RING

POSITIONING

See Figure 104.

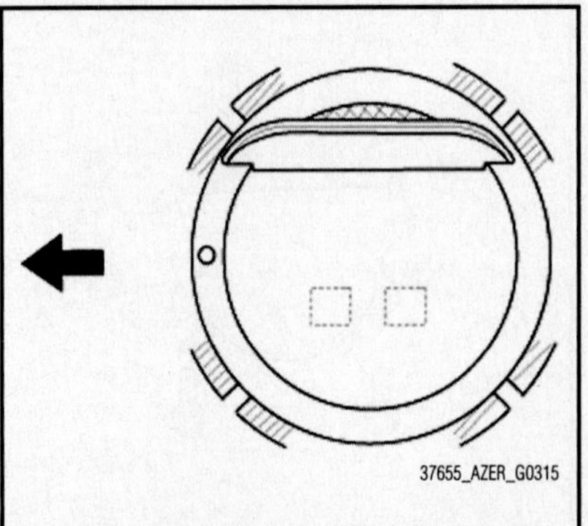

Fig. 104 Position the piston rings so that the ring ends are as shown

REAR MAIN SEAL

REMOVAL & INSTALLATION

See Figures 105 through 107.

1. Before servicing the vehicle, refer to the Precautions Section.
2. Disconnect the negative battery cable.
3. Remove the transaxle assembly.
4. Remove the flexplate.
5. Remove the rear oil seal case.

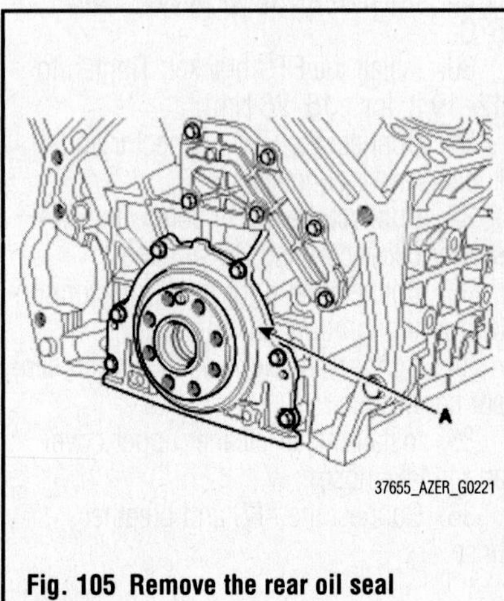

Fig. 105 Remove the rear oil seal case (A)

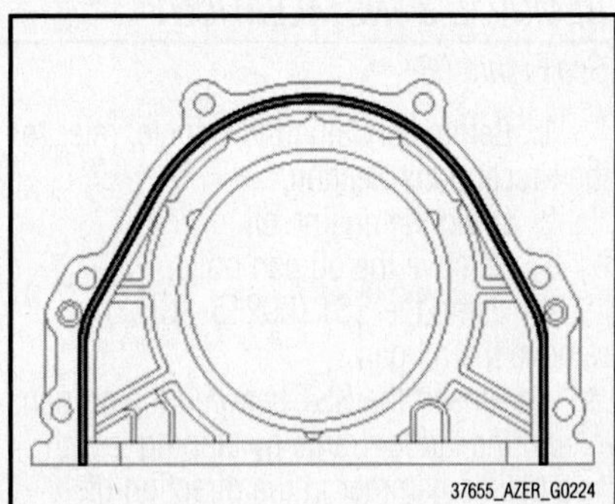

Fig. 106 Rear oil seal case sealant application

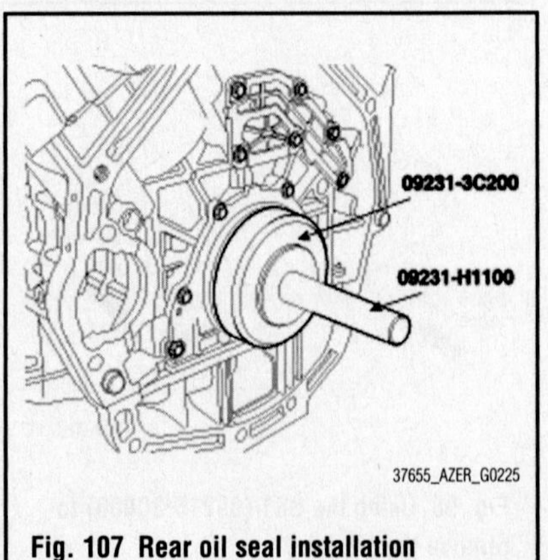

Fig. 107 Rear oil seal installation

6. Remove the rear oil seal from the case.

To install:

7. Before assembling the rear oil seal case, the liquid sealant TB1217H should be applied to the oil drain cover.

➡The part must be assembled within 5 minutes after sealant was applied. Apply sealant to the inner threads of the bolt holes.

8. Install the rear oil seal case and tighten to 84–108 inch lbs. (8–10 Nm).
9. Using SST(09231-3C200, 09231-H1100), install the rear oil seal.
10. After assembly, wait at least 30 minutes before filling the engine with oil to allow the gasket material to cure.
11. Install the flexplate.
12. Install the transaxle.
13. Connect the negative battery cable.
14. Check for oil leaks, and repair as necessary.

TIMING CHAIN COVER & SEAL

REMOVAL & INSTALLATION

See Figures 108 through 122.

1. Before servicing the vehicle, refer to the Precautions Section.

➡Use fender covers to avoid damaging painted surfaces.

➡To avoid damage, unplug the wiring connectors carefully while holding the connector portion.

➡Mark all wiring and hoses to avoid misconnection.

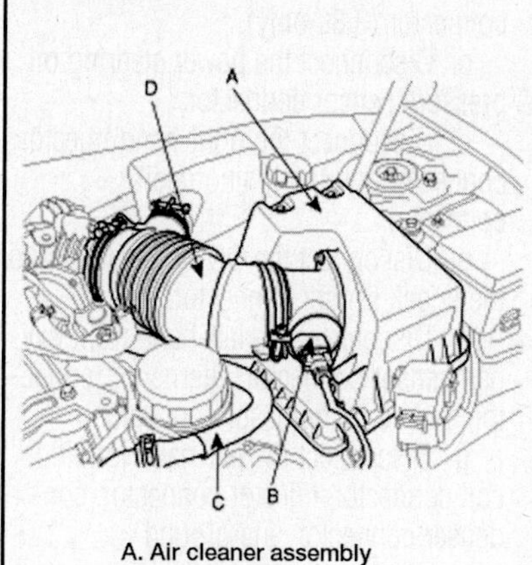

Fig. 108 Remove the intake air hose and air cleaner assembly

➡Turn the crankshaft pulley so that the No. 1 piston is at Top Dead Center (TDC).

2. Disconnect the negative terminal from the battery.

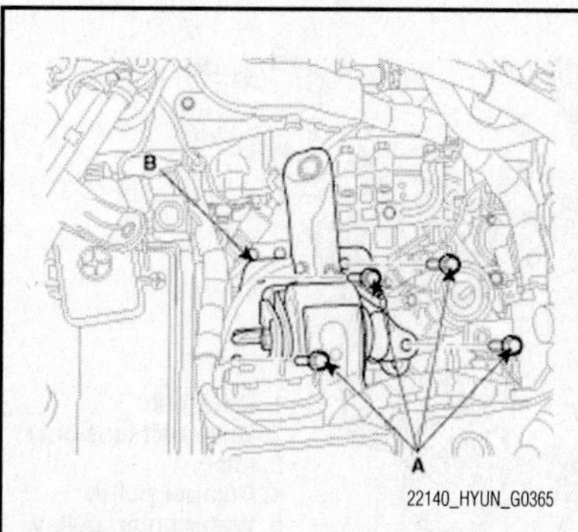

Fig. 109 Loosen the transaxle mounting bolts (A) without removing the transaxle mounting (B)

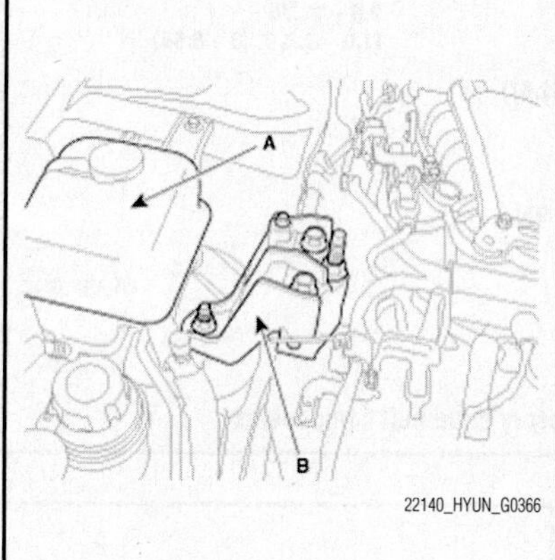

Fig. 110 Remove the engine coolant reservoir tank (A) and the engine mounting bracket (B)

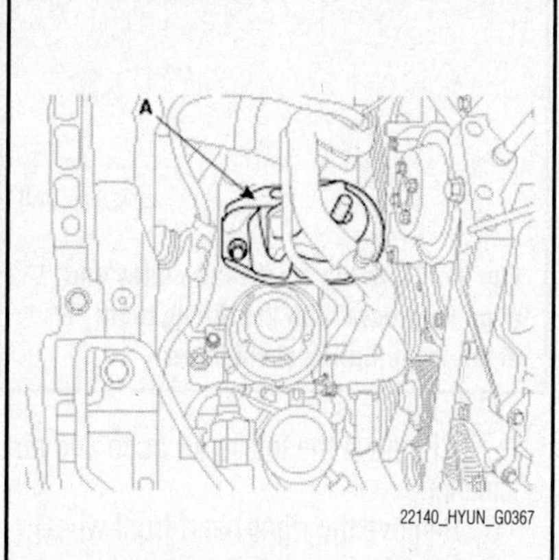

Fig. 111 Remove the No. 1 engine mounting through the lower position of the A/C pipe line

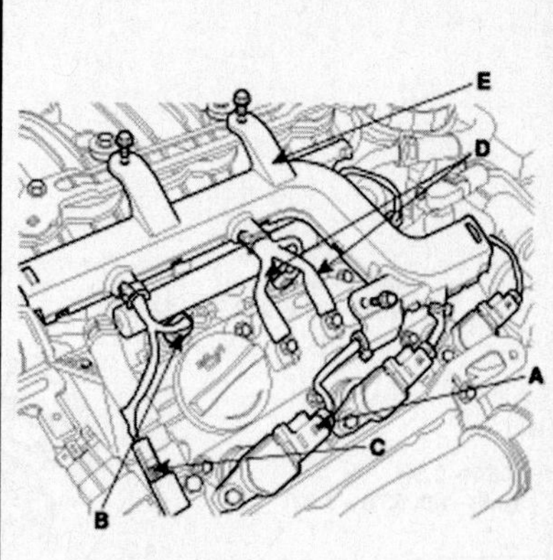

Fig. 112 Engine wiring

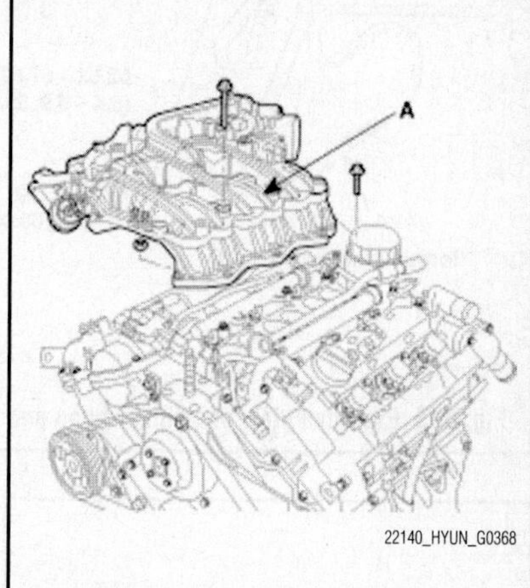

Fig. 113 Remove the surge tank (A)—3.3L engine

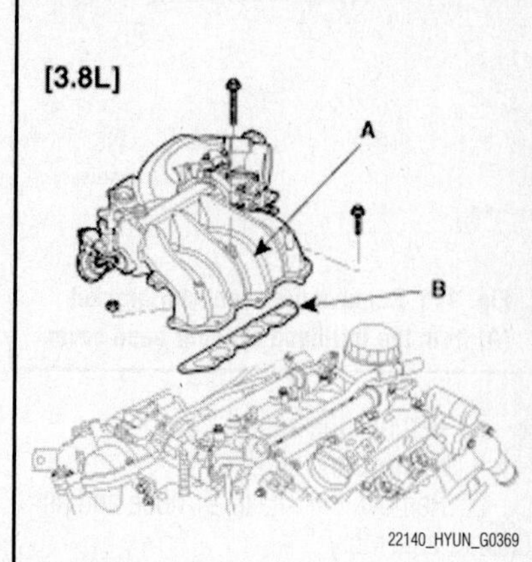

Fig. 114 Remove the surge tank (A)—3.8L engine

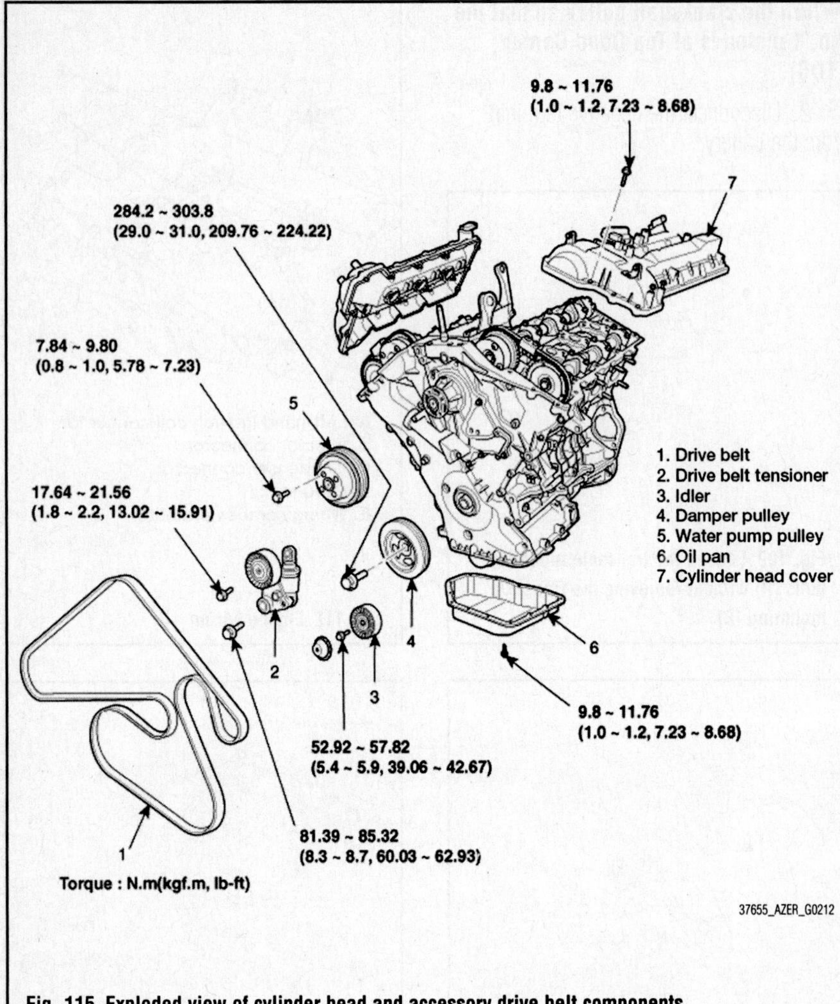

Fig. 115 Exploded view of cylinder head and accessory drive belt components

Fig. 116 Remove the connector bracket (A) from the left-hand cylinder head cover

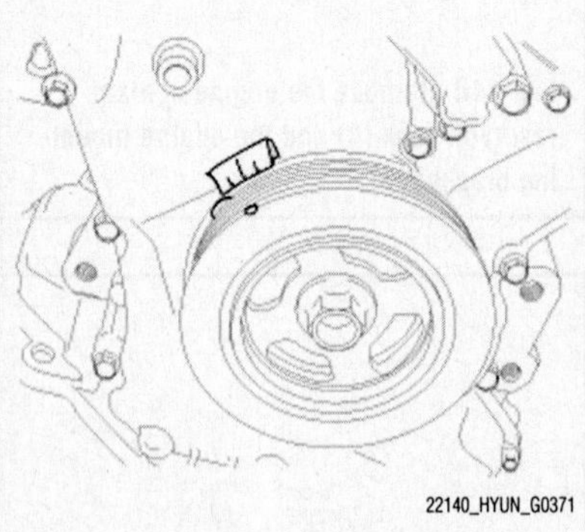

Fig. 117 Turn the crankshaft pulley and align its groove with the timing mark "T" of the lower timing chain cover

3. Remove the engine cover.
4. Remove the air duct.
5. Remove the intake air hose and air cleaner assembly.
   a. Disconnect the AFS connector.
   b. Disconnect the breather hose from air cleaner hose.
   c. Disconnect the ECM connector.
   d. Remove the intake air hose and air cleaner.
6. Remove the right-hand front wheel.
7. Remove the undercover.
8. Remove the right-hand side cover.
9. Drain the engine coolant. Remove the radiator cap to speed draining. Refer to Coolant in the Engine Cooling section.

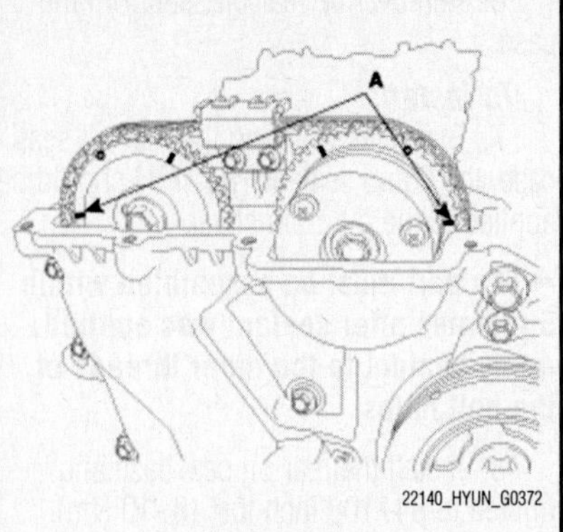

Fig. 118 Check that the mark (A) of the camshaft timing sprockets are in straight line on the cylinder head surface as shown

10. Remove the upper radiator hose.
11. Drain the engine oil.
12. Remove the lower oil pan. Refer to Oil Pan, Removal & Installation.
13. Place a jack underneath the upper oil pan.
14. Just loosen the transaxle mounting bolts without removing the transaxle mounting.
15. Remove the engine coolant reservoir tank (A).
16. Remove the engine mounting bracket (B).
17. Loosen the A/C pipe bracket mounting bolt.
18. Remove the No. 1 engine mounting (A) through the lower position of the A/C pipe line.
19. Remove the surge tank and engine wiring.
   a. Disconnect the right-hand oxygen sensor connector.
   b. Disconnect the VIS solenoid valve connector (3.8L only).
   c. Disconnect the power steering oil pressure sensor connector.
   d. Disconnect the right-hand injector connector and the ignition coil connector.
   e. Disconnect the OCV connector and the knock sensor connector.
   f. Disconnect the left-hand front oxygen sensor connector, alternator connector, and air compressor connector.
   g. Disconnect the left-hand ignition coil connector, injector connector, condenser connector, and ground
   h. Remove the wiring harness protector.
   i. Disconnect the left-hand CMPS and oil pressure switch connector.
   j. Disconnect the ETC connector and the knock sensor connector.
   k. Disconnect the PCSV connector,

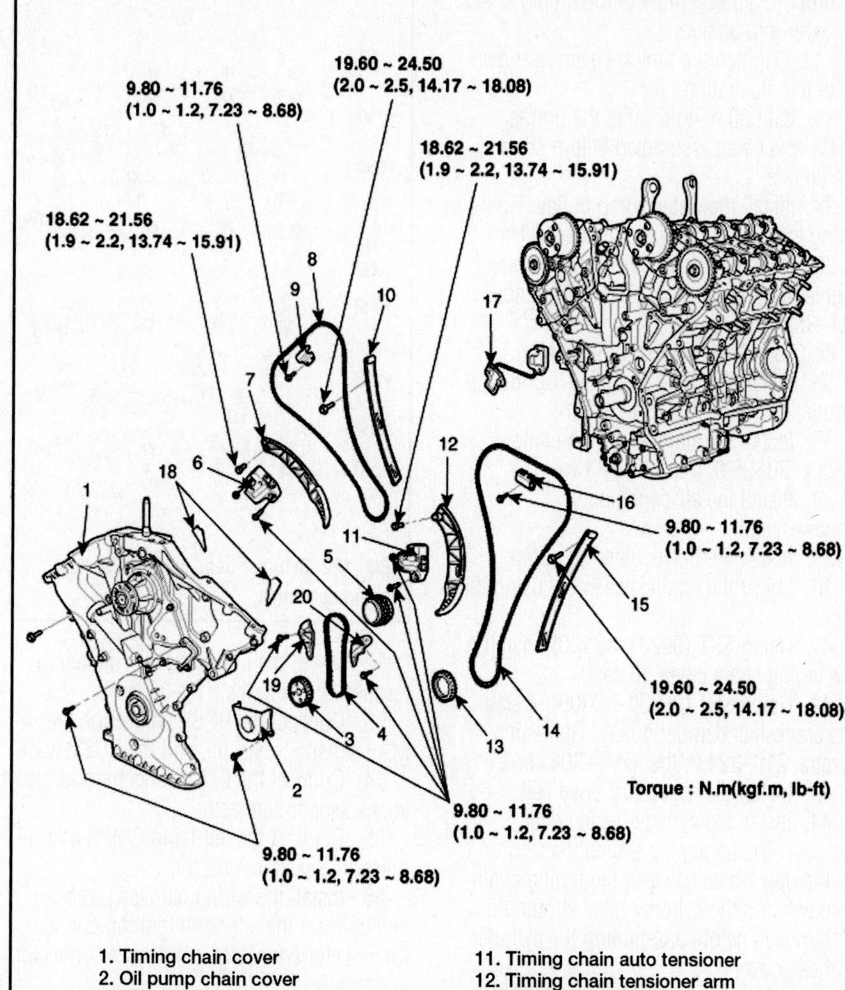

Fig. 119 Exploded view of timing chain and related components

the MAP sensor connector, and the PCSV hose.

l. Remove the Electronic Throttle Control (ETC) bracket.

m. Disconnect the water hoses from the ETC.

n. Disconnect the PCV hose.

o. Disconnect the brake vacuum hose.

p. Remove the surge tank stay.

q. Remove the connector bracket from the surge tank.

r. Remove the surge tank.

➡Cover the inlet of intake manifold with a clean woven stuff or vinyl cover to prevent foreign materials from entering.

20. Remove the cylinder head cover.

a. Remove the connector bracket from the left-hand cylinder head cover.

b. Disconnect the right-hand ignition coil connector, the condenser connector and remove the wiring bracket.

c. Remove the left-hand and right-hand ignition coils.

d. Remove the left-hand and right-hand cylinder head covers.

➡Cover the upside of engine head with a clean vinyl cover to prevent foreign materials from entering.

21. Set No. 1 cylinder to TDC/ compression.

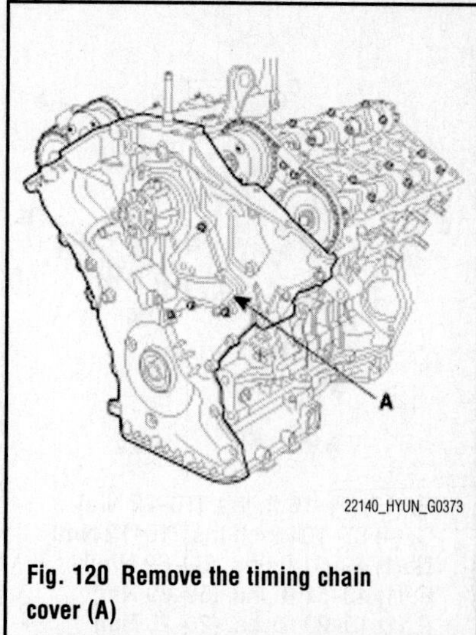

Fig. 120 Remove the timing chain cover (A)

a. Turn the crankshaft pulley and align its groove with the timing mark "T" of the lower timing chain cover.

⁂ WARNING

Do not rotate engine counterclockwise.

b. Check that the mark of the camshaft timing sprockets are in straight line on the cylinder head surface as shown in the illustration. If not, turn the crankshaft one revolution (360°). Do not rotate engine counterclockwise.

22. Remove the accessory drive belt. Refer to Accessory Drive Belts, Removal & Installation.

23. Remove the crankshaft damper pulley. Refer to Crankshaft Damper, Removal & Installation.

➡Use the SST (flywheel stopper, 09231-3C300) to remove the crankshaft pulley bolt, after removing the starter.

24. Lift up the engine assembly by using the jack.

25. Remove the power steering pump.

26. Remove the air conditioner compressor.

27. Remove the alternator.

28. Remove the drive belt idler.

29. Remove the drive belt auto tensioner.

30. Remove the water pump pulley.

31. Remove the timing chain cover.

⁂ WARNING

Be careful not to damage the contact surfaces of the cylinder block, cylinder head, or timing chain cover.

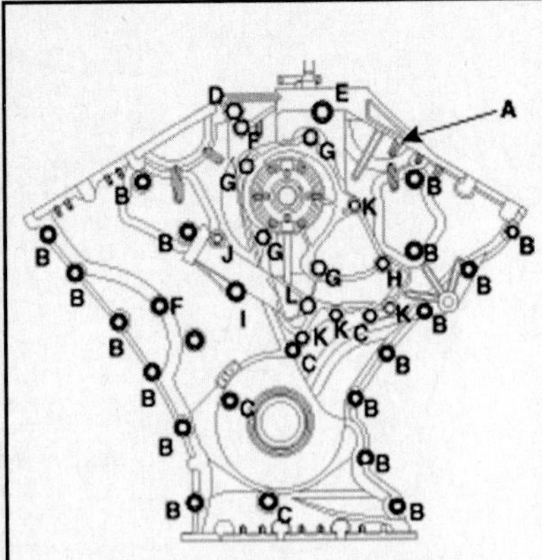

B. (17) 14–16 ft. lbs. (19–22 Nm)
C. (4) 87–104 inch lbs. (10–12 Nm)
D. (1) 43–51 ft. lbs. (59–69 Nm)
E. (1) 43–51 ft. lbs. (59–69 Nm)
F. (2) 18–20 ft. lbs. (25–26 Nm)
G. (4) 16–17 ft. lbs. (22–24 Nm)
H. (1) 87–104 inch lbs. (10–12 Nm)
I. (1) 87–104 inch lbs. (10–12 Nm)
J. (1) 87–104 inch lbs. (10–12 Nm)
K. (4) 87–104 inch lbs. (10–12 Nm)
L. (1) 16–20 ft. lbs. (22–26 Nm)

37655_AZER_G0211

Fig. 121 Timing chain cover bolt tightening specifications

➡Before removing the timing chain, mark the right-hand/left-hand timing chain with an identification based on the location of the sprocket as the identification mark on the chain for TDC can be erased accidentally.

To install:

32. Install the timing chain cover.
   a. The sealant locations on the chain cover and on the counter parts (cylinder head, cylinder block, and lower oil pan) must be free of engine oil and any debris.
   b. Before assembling the timing chain cover, the liquid sealant TB 1217H should be applied on the gap between cylinder head and cylinder block.

➡The part must be assembled within 5 minutes after the sealant is applied. Use a bead width of 0.1 inch (2.5mm).

   c. After applying liquid sealant TB1217H on the timing chain cover, the part must be assembled within 5 minutes.

➡The sealant should be applied in a continuous bead. Bead width: 0.1 inch (2.5mm).

   d. Install a new gasket to the timing chain cover.
   e. The dowel pins on the cylinder block and holes on the timing chain cover should be used as a reference in order to aid assembly of the timing chain cover into position.
   f. Tighten the timing cover as shown in the illustration.
33. Wait 30 minutes after the timing chain cover was assembled before starting the engine.
34. Install the water pump pulley. Tightening torque: 69–87 inch lbs. (8–10 Nm).
35. Install the drive belt auto tensioner. Tightening torque: Bolt (B): 60–63 ft. lbs. (81–85 Nm).
   Bolt (C): 13–16 ft. lbs. (18–22 Nm).
36. Install the drive belt idler. Tightening torque: 39–43 ft. lbs. (53–58 Nm).
37. Install the alternator. Tightening torque: 20–25 ft. lbs. (27–33 Nm).
38. Install the air conditioner compressor.
39. Install the power steering pump.
40. Lower the engine assembly using the jack.
41. Using SST (09231-3C100), install the timing chain cover oil seal.
42. Using SST (09231-3C300), install the crankshaft damper pulley. Tightening torque: 210–224 ft. lbs. (284–304 Nm).
43. Install the accessory drive belt.
44. Install the cylinder head cover.
   a. The hardening sealant located on the upper area between the timing chain cover and the cylinder head should be removed before assembling the cylinder head cover.
   b. After applying sealant (TB1217H), it should be assembled within 5 minutes. Bead width: 0.1 inch (2.5mm).
   c. Wait 30 minutes after the cylinder head cover was assembled before starting the engine.
   d. Install the cylinder head cover bolts as shown in the illustration. Tightening torque: 87–104 inch lbs. (10–12 Nm).

➡Do not reuse the cylinder head cover gasket.

45. Install the ignition coils.
46. Connect the right-hand ignition coil connector, the condenser connector, and install the wiring bracket.
47. Install the connector bracket from the left-hand cylinder head cover.
48. Install the surge tank. Tightening torque: 87–104 inch lbs. (10–12 Nm).
   a. Install the connector bracket on the surge tank. Tightening torque: 60–91 inch lbs. (7–11 Nm).
   b. Install the surge tank stay. Tightening torque: 20–23 ft. lbs. (27–31 Nm).
49. Connect the brake vacuum hose.
50. Connect the PCV hose.
51. Connect the water hoses to the ETC.

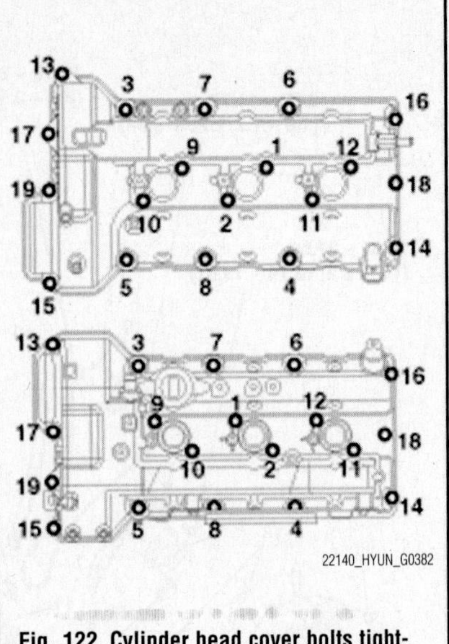

22140_HYUN_G0382

Fig. 122 Cylinder head cover bolts tightening sequence

52. Install the ETC bracket. Tightening torque: 12–19 ft. lbs. (16–25 Nm).
53. Connect the PCSV connector, the MAP sensor connector, and the PCSV hose.
54. Connect the ETC connector and the knock sensor connector.
55. Connect the left-hand CMPS and oil pressure switch connector.
56. Install the wiring harness protector, and connect the left-hand ignition connector, injector connector, condenser connector, and ground.
57. Connect the OCV connector and knock sensor connector.
58. Connect the left-hand front oxygen sensor connector, alternator connector, and air compressor connector.
59. Connect the right-hand injector connector and the ignition coil connector.
60. Connect the power steering oil pressure sensor connector.
61. Connect the right-hand oxygen sensor connector and VIS solenoid valve connector, as applicable.
62. Install the No. 1 engine mounting through the lower position of A/C pipe line. Tightening torque: 36–47 ft. lbs. (49–64 Nm).
63. Install the A/C pipe bracket mounting bolt.
64. Install the engine coolant reservoir tank.
65. Install the engine mounting bracket. Tightening torque: 47–62 ft. lbs. (64–83 Nm).
66. Install the transaxle mounting bolts. Tightening torque: 36–47 ft. lbs. (49–64 Nm).
67. Remove the jack from the upper oil pan.

68. Install the lower oil pan. Uniformly tighten the bolts in several passes. Tightening torque: 87–104 inch lbs. (10–12 Nm).
69. Install the upper radiator hose.
70. Install the side cover.
71. Install the undercover.
72. Install the front wheels.
73. Install the intake air hose and air cleaner assembly.
 a. Install the intake air hose and air cleaner.
 b. Connect the ECM connector.
 c. Connect the breather hose to air cleaner hose.
 d. Connect the AFS connector.
 e. Install the air duct.
74. Install the engine cover.
75. Connect the negative terminal to the battery.
76. Refill the engine with the proper amount and type of engine oil.
77. Refill the radiator and reservoir tank with engine coolant.
78. Bleed the air from the cooling system.
79. Run the engine and check for leaks.

TIMING CHAIN & SPROCKETS

REMOVAL & INSTALLATION

See Figures 123 through 129.

1. Before servicing the vehicle, refer to the Precautions Section.

➡Use fender covers to avoid damaging painted surfaces.

➡To avoid damage, unplug the wiring connectors carefully while holding the connector portion.

➡Mark all wiring and hoses to avoid misconnection.

➡Turn the crankshaft pulley so that the No. 1 piston is at Top Dead Center (TDC).

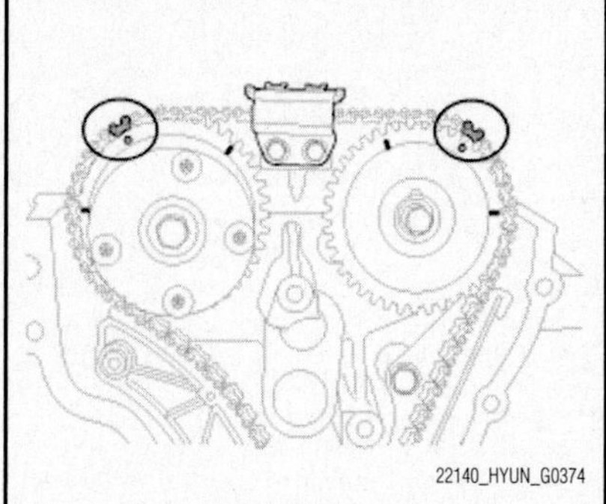

Fig. 123 Timing marks on chain and camshaft sprockets illustrated

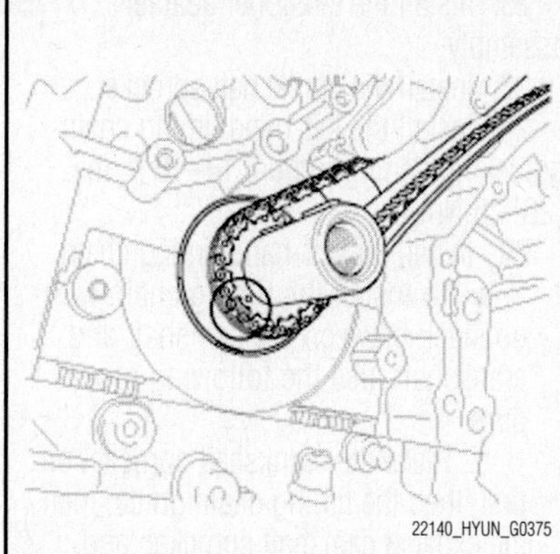

Fig. 124 Timing marks on chain and crankshaft sprocket illustrated

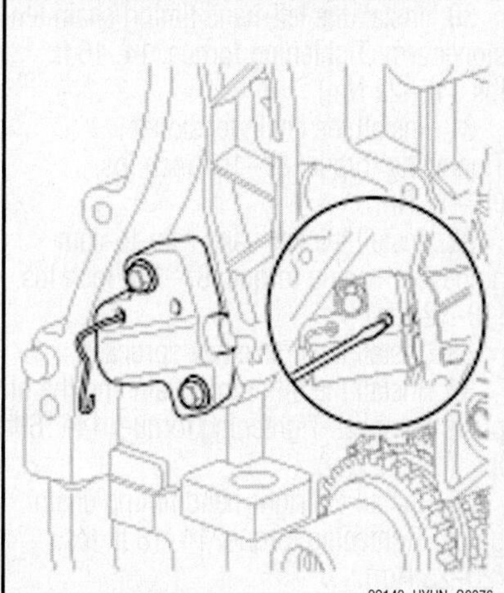

Fig. 125 Install a set pin after compressing the right-hand timing chain tensioner

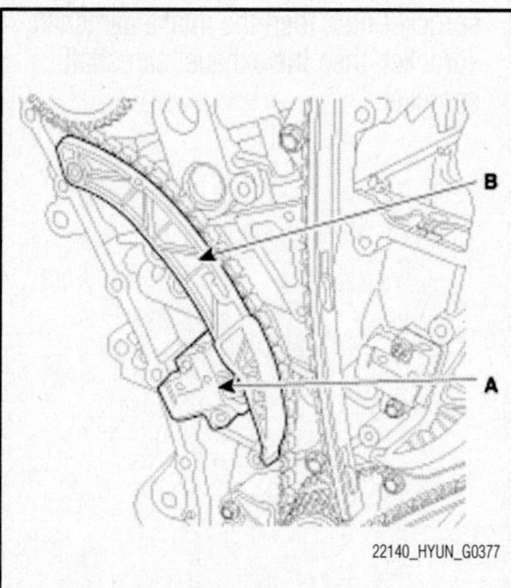

Fig. 126 Remove the right-hand timing chain auto tensioner (A) and the right-hand timing chain tensioner arm (B)—3.3L and 3.8L engines

2. Disconnect the negative terminal from the battery.
3. Remove the timing chain cover. Refer to Timing Chain Cover & Seal Removal & Installation.
4. Install a set pin after compressing the right-hand timing chain tensioner.
5. Remove the right-hand cam-to-cam guide.
6. Remove the right-hand timing chain auto tensioner and the right-hand timing chain tensioner arm.
7. Remove the right-hand timing chain.
8. Remove the right-hand timing chain guide.
9. Remove the oil pump chain cover.
10. Remove the oil pump chain tensioner assembly.
11. Remove the oil pump chain guide.
12. Remove the oil pump chain sprocket and oil pump chain.
13. Remove the crankshaft sprocket that drives the oil pump.
14. Install a set pin after compressing the left-hand timing chain tensioner.
15. Remove the left-hand cam-to-cam guide.
16. Remove the left-hand timing chain auto tensioner and left-hand timing chain tensioner arm.
17. Remove the left-hand timing chain.
18. Remove the left-hand timing chain guide.
19. Remove the crankshaft sprocket.
20. Remove the tensioner adapter assembly.

To install:

21. Check the camshaft sprocket and crankshaft sprocket for abnormal wear, cracks, or damage. Replace as necessary.
22. Inspect the tensioner arm and chain guide for abnormal wear, cracks, or damage. Replace as necessary.

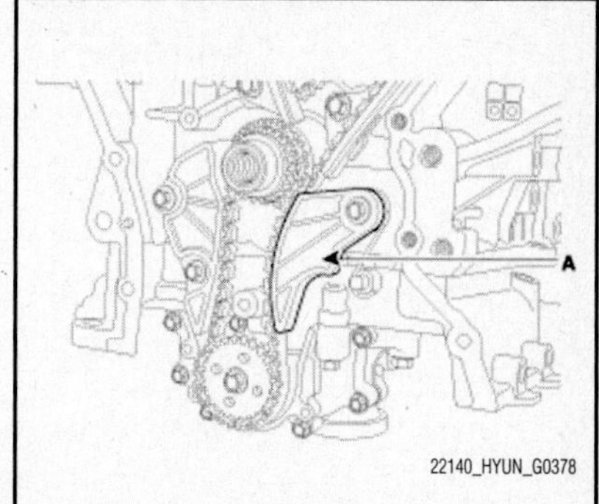

Fig. 127 Remove the oil pump chain tensioner assembly (A)

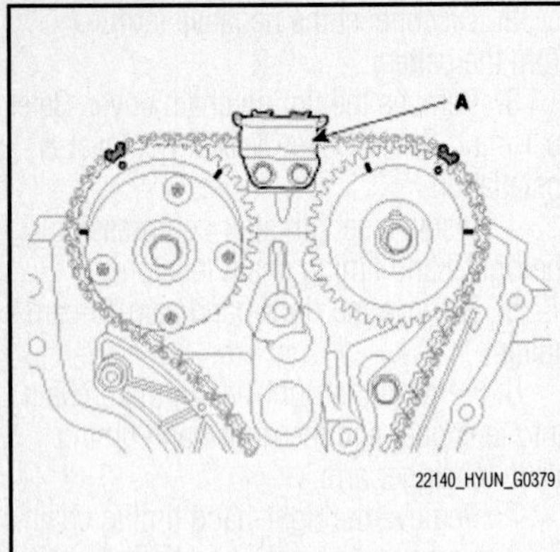

Fig. 128 Remove the left-hand cam-to-cam guide (A)

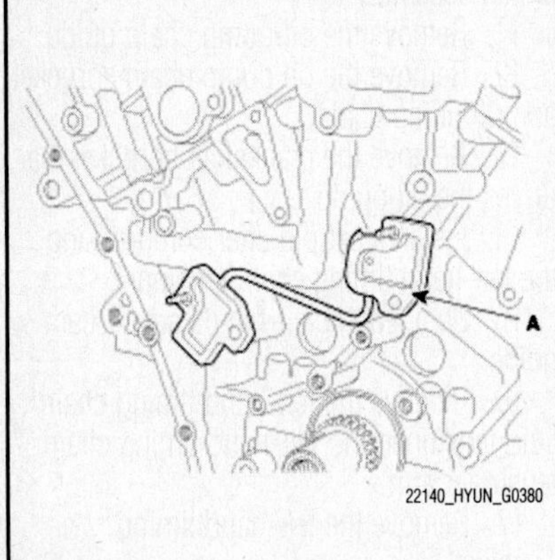

Fig. 129 Remove the tensioner adapter assembly (A)

23. Check that the tensioner piston moves smoothly when the ratchet pawl is released.
24. Install the jack to the upper oil pan.
25. The key of crankshaft must be aligned with the timing mark of timing chain cover. Then, piston of No. 1 cylinder is placed at the TDC on the compression stroke.
26. Install the tensioner adapter assembly.
27. Install the crankshaft sprocket.
28. Install the left-hand timing chain guide. Tightening torque: 14–18 ft. lbs. (20–25 Nm).
29. Install the left-hand timing chain.
    a. To install the timing chain with no slack between the camshaft and crankshaft, use the following procedure.
    b. Place the crankshaft sprocket on first, then the timing chain guide, then the exhaust camshaft sprocket, and finally the intake camshaft sprocket.
    c. The timing mark of each sprockets should be matched with the timing mark (color link) of the timing chain when installing the timing chain.
30. Install the left-hand timing chain tensioner arm. Tightening torque: 14–16 ft. lbs. (19–22 Nm).
31. Install the chain tensioner. Tightening torque: 87–104 inch lbs. (10–12 Nm).
32. Install the left-hand cam-to-cam guide. Tightening torque: 87–104 inch lbs. (10–12 Nm).
33. Install the crankshaft sprocket.
34. Install the oil pump chain and the oil pump sprocket. Tightening torque: 14–16 ft. lbs. (19–22 Nm).
35. Install the right-hand timing chain guide. Tightening torque: 14–18 ft. lbs. (20–25 Nm).
36. Install the right-hand timing chain.
    a. To install the timing chain with no slack between the camshaft and the crankshaft, use the following procedure.
    b. Place the chain on the crankshaft sprocket first, then the intake camshaft sprocket, then the exhaust camshaft sprocket.
    c. The timing mark of each of the sprockets must be matched with the timing mark (color link) of timing chain when installing the timing chain.
37. Install the right-hand timing chain tensioner arm. Tightening torque: 14–16 ft. lbs. (19–22 Nm).
38. Install the right-hand timing chain auto tensioner. Tightening torque: 87–104 inch lbs. (10–12 Nm).
39. Install the right-hand cam-to-cam guide. Tightening torque: 87–104 inch lbs. (10–12 Nm).
40. Install the oil pump chain guide. Tightening torque: 87–104 inch lbs. (10–12 Nm).
41. Install the oil pump chain tensioner assembly. Tightening torque: 87–104 inch lbs. (10–12 Nm).
42. Pull out the pins of hydraulic tensioner (left-hand & right-hand).
43. Install the oil pump chain cover. Tightening torque: 87–104 inch lbs. (10–12 Nm).
44. After rotating the crankshaft 2 revolutions in a clockwise direction (viewed from front), confirm that the timing marks are aligned.

WARNING

Always turn the crankshaft clockwise.

45. Install the timing chain cover.
46. Connect the negative terminal to the battery.
47. Refill the engine with the proper amount and type of engine oil.
48. Refill the radiator and reservoir tank with engine coolant.
49. Bleed the air from the cooling system.
50. Run the engine and check for leaks.

ENGINE PERFORMANCE & EMISSION CONTROLS

COMPONENT LOCATIONS

See Figure 130.

CAMSHAFT POSITION (CMP) SENSOR

LOCATION

See Figures 131 and 132.

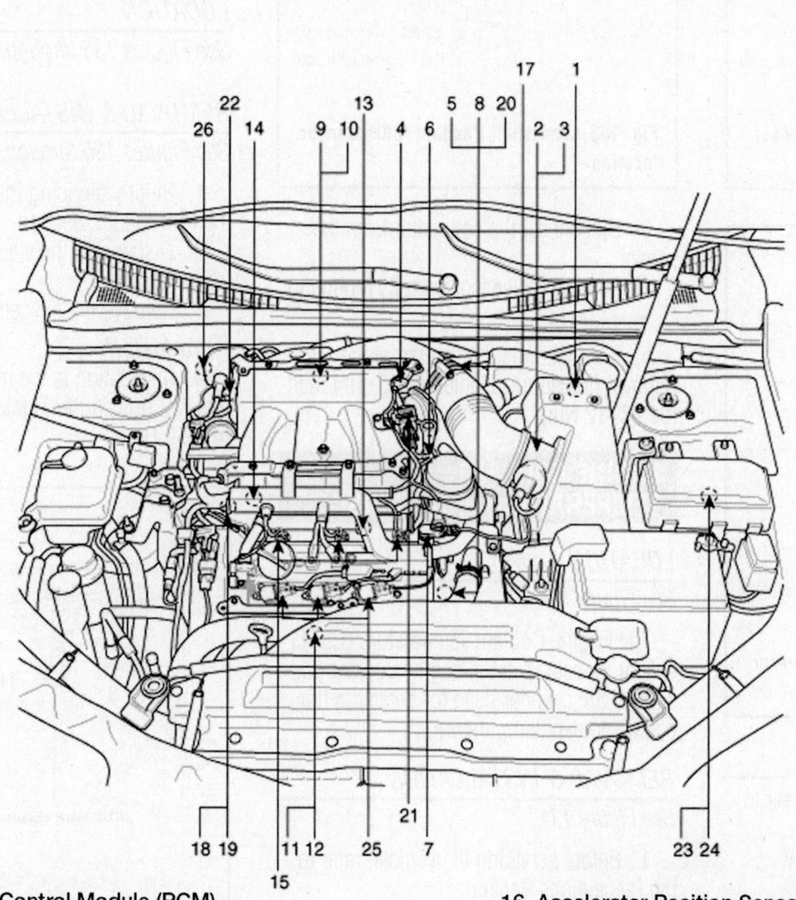

1. Powertrain Control Module (PCM)
2. Mass Air Flow Sensor (MAFS)
3. Intake Air Temperature Sensor (IATS)
4. Manifold Absolute Pressure Sensor (MAPS)
5. Engine Coolant Temperature Sensor (ECTS)
6. Camshaft Position Sensor (CMPS) [Bank 1]
7. Camshaft Position Sensor (CMPS) [Bank 2]
8. Crankshaft Position Sensor (CKPS)
9. Heated Oxygen Sensor (HO2S) [Bank 1 / Sensor 1]
10. Heated Oxygen Sensor (HO2S) [Bank 1 / Sensor 2]
11. Heated Oxygen Sensor (HO2S) [Bank 2 / Sensor 1]
12. Heated Oxygen Sensor (HO2S) [Bank 2 / Sensor 2]
13. Knock Sensor (KS) No.1
14. Knock Sensor (KS) No. 2
15. Injector
16. Accelerator Position Sensor (APS)
17. ETC Module [Throttle Position Sensor (TPS) + ETC Motor]
18. CVVT Oil Control Valve (OCV) [Bank 1]
19. CVVT Oil Control Valve (OCV) [Bank 2]
20. CVVT Oil Temperature Sensor (OTS)
21. Purge Control Solenoid Valve (PCSV)
22. Variable Intake Solenoid (VIS) Valve
23. Fuel Pump Relay
24. Main Relay
25. Ignition Coil
26. Power Steering Pressure Sensor (PSPS)
27. Data Link Connector (DLC)
28. Fuel Tank Pressure Sensor (FTPS)
29. Canister Close Valve (CCV)
30. Fuel Level Sensor (FLS)

37655_AZER_G0229

Fig. 130 Underhood sensor locations

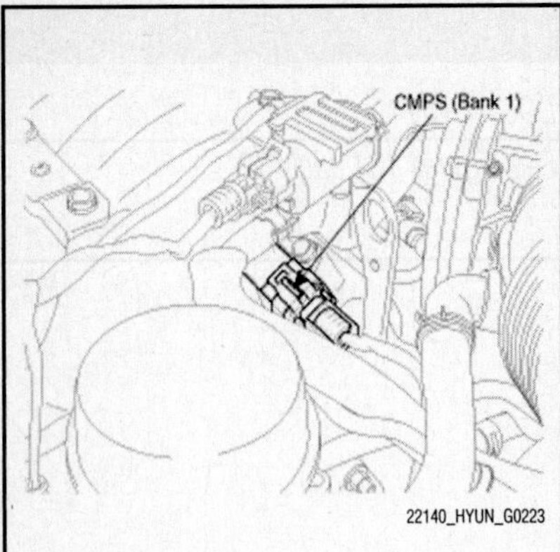

Fig. 131 Camshaft Position (CMP) sensor location (Bank 1)

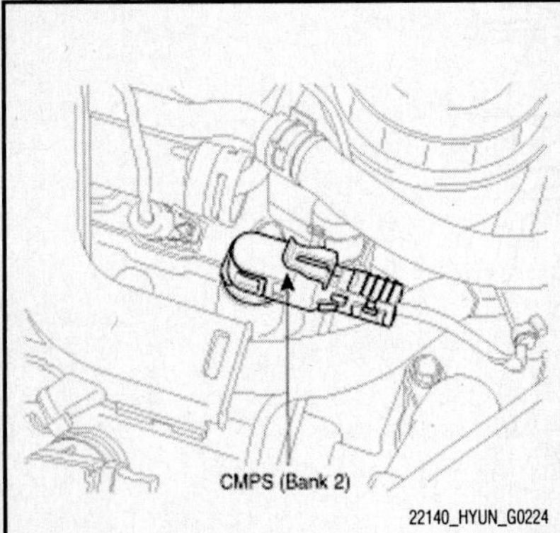

Fig. 132 Camshaft Position (CMP) sensor location (Bank 2)

REMOVAL & INSTALLATION

1. Before servicing the vehicle, refer to the Precautions Section.
2. Disconnect the negative battery cable.
3. Disconnect the connector from the CMP sensor.
4. Remove the CMP sensor.

To install:

5. Installation is the reverse of the removal procedure. Tighten the bolts to 61–86 inch lbs. (7–10 Nm).

CRANKSHAFT POSITION (CKP) SENSOR

LOCATION

See Figure 133.

REMOVAL & INSTALLATION

1. Before servicing the vehicle, refer to the Precautions Section.
2. Disconnect the negative battery cable.

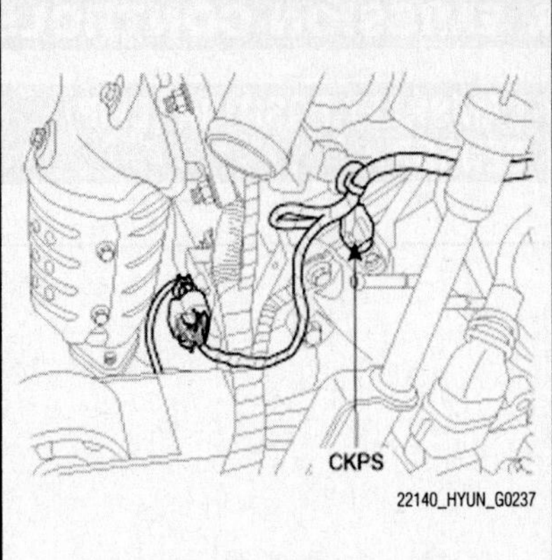

Fig. 133 Crankshaft Position (CKP) sensor location

3. Disconnect the connector from the sensor.
4. Remove the sensor from its mounting.

To install:

5. Installation is the reverse of the removal procedure. Tighten to 70–104 inch lbs. (8–12 Nm).

ENGINE COOLANT TEMPERATURE (ECT) SENSOR

LOCATION

See Figure 134.

The Engine Coolant Temperature (ECT) sensor is located in the engine coolant passage of the cylinder head for detecting the engine coolant temperature.

REMOVAL & INSTALLATION

See Figure 134.

1. Before servicing the vehicle, refer to the Precautions Section.
2. Drain the coolant to a level below the bottom of the sensor.

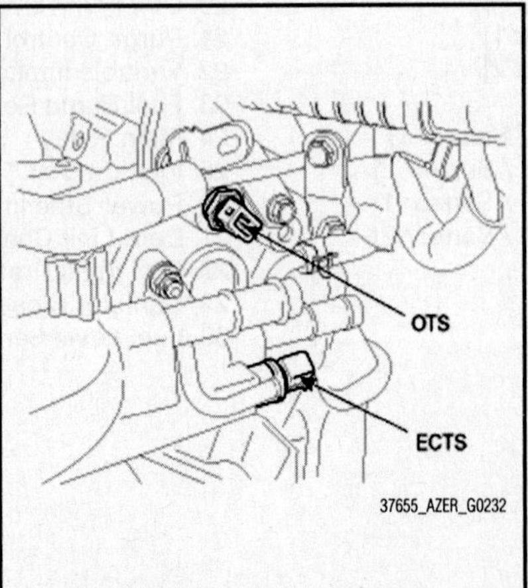

Fig. 134 Engine Coolant Temperature (ECT) sensor

3. Disconnect the ground cable from the battery and then remove the sensor connector.
4. Remove the ECT sensor.

To install:

5. Reverse the removal procedure. Tighten to 15–29 ft. lbs. (20–39 Nm).

HEATED OXYGEN SENSOR (HO2S)

LOCATION

See Figures 135 through 138.

REMOVAL & INSTALLATION

See Figures 135 through 138.

1. Before servicing the vehicle, refer to the Precautions Section.
2. Disconnect the electrical connector from the sensor.
3. Remove the oxygen sensor.

To install:

4. Installation is the reverse of the removal procedure. Tighten to 36–43 ft. lbs. (49–59 Nm).

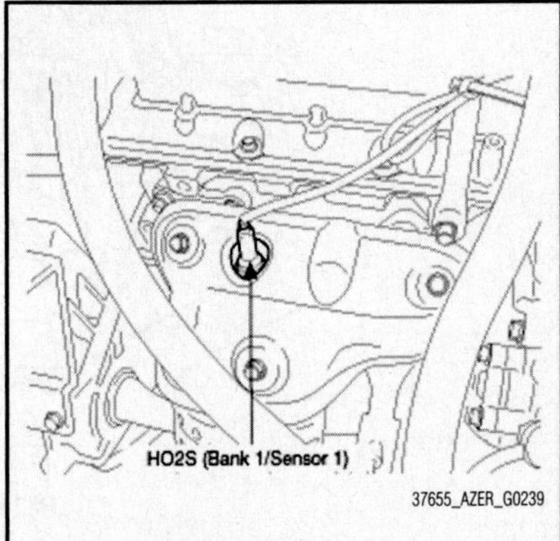

Fig. 135 Heated Oxygen Sensor (HO2S) Bank 1/Sensor 1

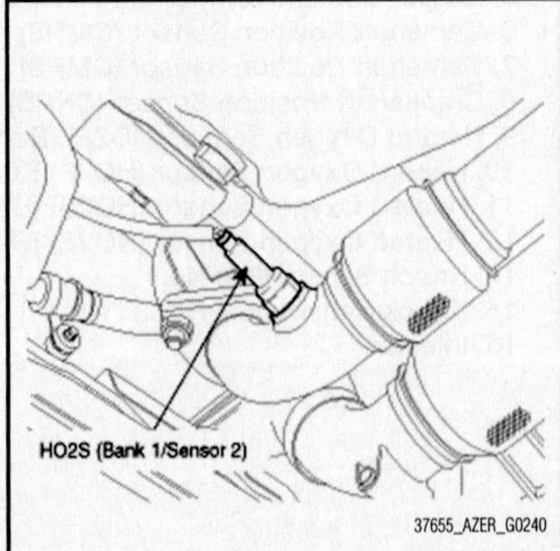

Fig. 136 Heated Oxygen Sensor (HO2S) Bank 1/Sensor 2

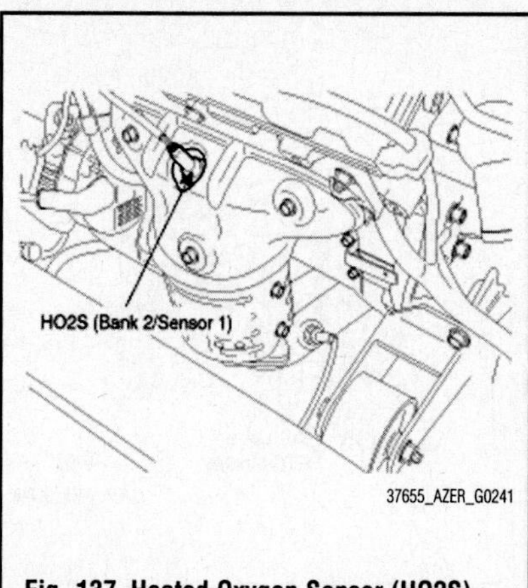

Fig. 137 Heated Oxygen Sensor (HO2S) Bank 2/Sensor 1

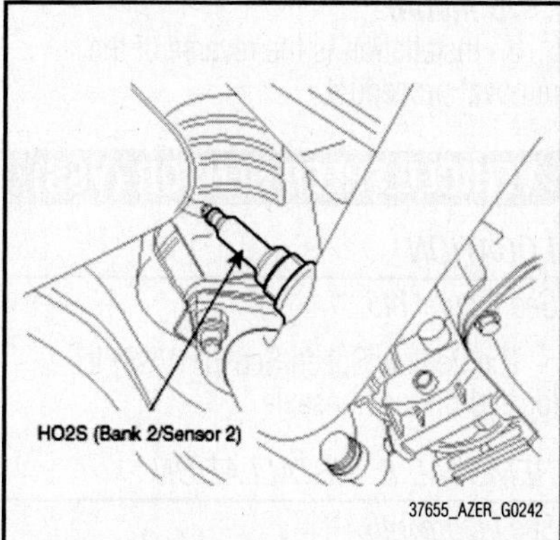

Fig. 138 Heated Oxygen Sensor (HO2S) Bank 2/Sensor 2

➡Apply anti-seize compound to the threaded portion of the sensor, prior to installation. Never apply anti-seize compound to the protector of the sensor.

INTAKE AIR TEMPERATURE (IAT) SENSOR

LOCATION

See Figure 139.

The Intake Air Temperature (IAT) sensor is mounted in the intake air hose of the air cleaner assembly. The IAT is integrated inside the Mass Air Flow (MAF) sensor.

REMOVAL & INSTALLATION

See Figure 139.

1. Before servicing the vehicle, refer to the Precautions Section.
2. Disconnect the negative battery cable.
3. Disconnect the connector from the sensor.

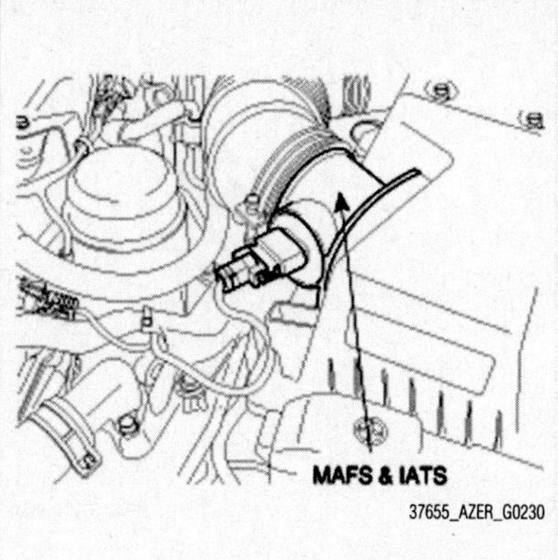

Fig. 139 Intake Air Temperature (IAT) sensor and Mass Air Flow (MAF) sensor

4. Remove the sensor retaining screws.
5. Remove the sensor from its mounting.

To install:

6. Installation is the reverse of the removal procedure.

KNOCK SENSOR (KS)

LOCATION

See Figure 140.

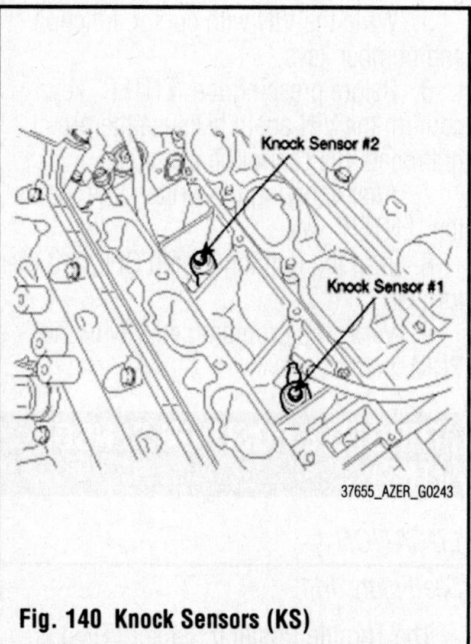

Fig. 140 Knock Sensors (KS)

REMOVAL & INSTALLATION

See Figure 140.

1. Before servicing the vehicle, refer to the Precautions Section.
2. Disconnect the negative battery cable.
3. Disconnect the sensor connector.
4. Remove the sensor from its mounting.

To install:

5. Installation is the reverse of the removal procedure.
6. Tighten the sensor to 12–17 ft. lbs. (16–24 Nm).

MASS AIR FLOW (MAF) SENSOR

LOCATION

See Figure 141.

The Mass Air Flow (MAF) sensor is mounted in the intake air hose of the air cleaner assembly. This sensor is combined with the Intake Air Temperature (IAT) sensor.

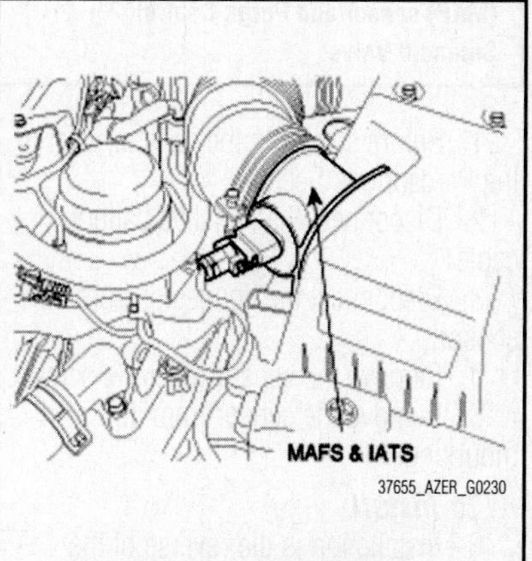

Fig. 141 Mass Air Flow (MAF) sensor and Intake Air Temperature (IAT) sensor

REMOVAL & INSTALLATION

See Figure 141.

1. Before servicing the vehicle, refer to the Precautions Section.
2. Disconnect the negative battery cable.
3. Disconnect the connector from the sensor.
4. Remove the air cleaner and air intake assembly, as required.
5. Remove the sensor from its mounting.

To install:

6. Installation is the reverse of the removal procedure.

MANIFOLD ABSOLUTE PRESSURE (MAP) SENSOR

LOCATION

See Figure 142.

REMOVAL & INSTALLATION

See Figure 142.

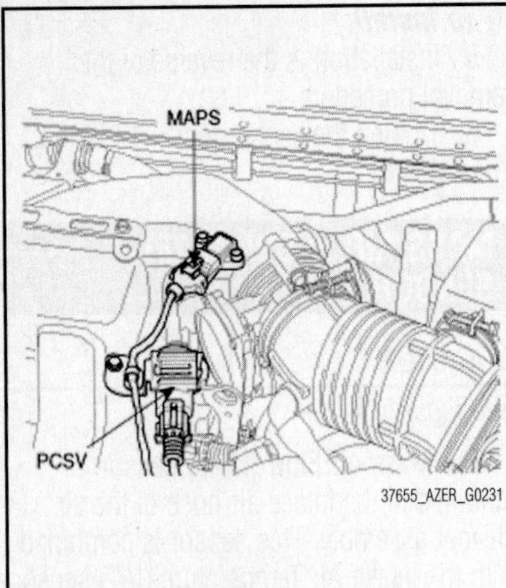

Fig. 142 Manifold Absolute Pressure (MAP) sensor and Purge Control Solenoid Valve

1. Before servicing the vehicle, refer to the Precautions Section.
2. Disconnect the negative battery cable.
3. Disconnect the connector from the sensor.
4. Remove the sensor retaining screws.
5. Remove the sensor from its mounting.

To install:

6. Installation is the reverse of the removal procedure. Tighten the bolts to 78–104 inch lbs. (9–12 Nm).

POWERTRAIN CONTROL MODULE (PCM)

LOCATION

See Figure 143.

REMOVAL & INSTALLATION

See Figure 143.

1. Before servicing the vehicle, refer to the Precautions Section.
2. Turn the ignition switch off.
3. Disconnect the negative battery cable from the battery.
4. Disconnect the PCM connector(s).
5. Remove the PCM mounting bolts and remove the PCM.

To install:

6. Installation is the reverse of the removal. Tighten the bolts to 86–104 inch lbs. (10–12 Nm).

Reset Procedure

➡When replacing a PCM, the VIN must be programmed in the PCM. If there is no VIN in PCM memory, the fault code (DTC P0630) is set.

Fig. 143 Powertrain Control Module (PCM) location

⁂ WARNING

The programmed VIN cannot be changed. When writing the VIN, confirm the VIN carefully

1. Select "Vehicle" and "Engine".
2. Select "VIN WRITING".
3. Check the PCM status:
   - VIRGIN: VIN is not programmed
   - LEARNT: VIN has been already programmed
4. Is the PCM status "VIRGIN"? If YES, go to the next step. If NO, END.
5. Write the VIN with cursor, function and number keys.
6. Before pressing the "ENTER" key, confirm the VIN again because the programmed VIN cannot be changed.
7. After verifying the written VIN, press the "ENTER" key.
8. Turn the ignition switch OFF, and then turn ON.
9. Verify the programmed VIN in the PCM memory.

THROTTLE POSITION SENSOR (TPS)

LOCATION

See Figure 144.

The Throttle Position Sensor (TPS) is mounted on the throttle body.

REMOVAL & INSTALLATION

See Figure 144.

1. Before servicing the vehicle, refer to the Precautions Section.
2. Disconnect the negative battery cable.
3. Disconnect the sensor connector.
4. Remove the sensor retaining screws.
5. Remove the sensor from its mounting.

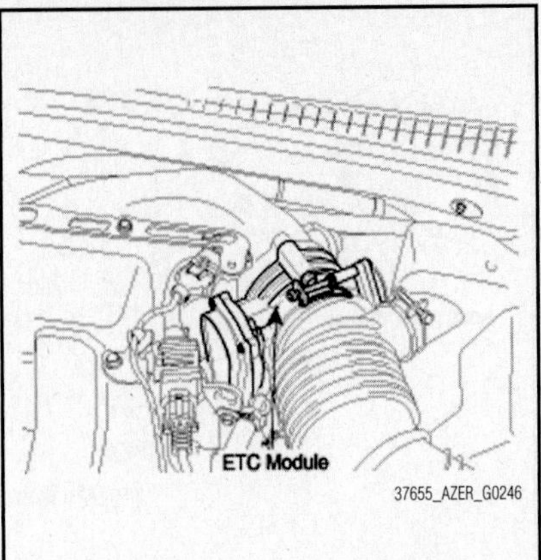

Fig. 144 Throttle Position Sensor (TPS) location

To install:

6. Installation is the reverse of the removal procedure.

VEHICLE SPEED SENSOR (VSS)

LOCATION

See Figure 145.

The Vehicle Speed Sensor (VSS) is located on the transaxle.

REMOVAL & INSTALLATION

See Figure 145.

1. Before servicing the vehicle, refer to the Precautions Section.
2. Raise and support the vehicle safely.
3. Place a drip pan below the Vehicle Speed Sensor (VSS) to catch any spilled fluid when it is removed.
4. Disconnect the VSS connector.
5. Remove the sensor from its mounting.

To install:

6. Installation is the reverse of the removal procedure.
7. Replace any lost transmission fluid.

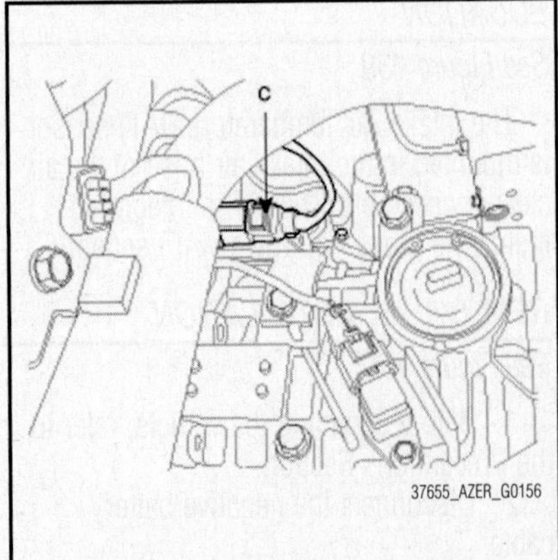

Fig. 145 Vehicle Speed Sensor (VSS) location (C)

FUEL SYSTEM SERVICE PRECAUTIONS

Safety is the most important factor when performing, not only fuel system maintenance, but any type of maintenance. Failure to conduct maintenance and repairs in a safe manner may result in serious personal injury or death. Maintenance and testing of the vehicle's fuel system components can be accomplished safely and effectively by adhering to the following rules and guidelines:

- To avoid the possibility of fire and personal injury, always disconnect the negative battery cable unless the repair or test procedure requires that battery voltage be applied.
- Always relieve the fuel system pressure prior to disconnecting any fuel system component (injector, fuel rail, pressure regulator, etc.), fitting, or fuel line connection. Exercise extreme caution whenever relieving fuel system pressure to avoid exposing skin, face, and eyes to fuel spray. Please be advised that fuel under pressure may penetrate the skin or any part of the body that it contacts.
- Always place a shop towel or cloth around the fitting or connection prior to loosening to absorb any excess fuel due to spillage. Ensure that all fuel spillage (should it occur) is quickly removed from engine surfaces. Ensure that all fuel soaked cloths or towels are deposited into a suitable waste container.
- Always keep a dry chemical (Class B) fire extinguisher near the work area.
- Do not allow fuel spray or fuel vapors to come into contact with a spark or an open flame.
- Always use a back-up wrench when loosening and tightening fuel line connection fittings. This will prevent unnecessary stress and torsion to fuel line piping.
- Always replace worn fuel fitting O-rings with new. Do not substitute fuel hose or equivalent where fuel pipe is installed.

RELIEVING FUEL SYSTEM PRESSURE

1. Before servicing the vehicle, refer to the Precautions Section.
2. Open the access panel in the trunk.
3. Disconnect the fuel pump connector.
4. Start the engine and allow it to run until it stalls.
5. Turn the ignition switch to the **OFF** position.
6. Disconnect the negative battery cable.
7. Attach the fuel pump harness connector.

FUEL FILTER

REMOVAL & INSTALLATION

The fuel delivery system integrates the fuel filter with the in-tank fuel pump. To service this filter, remove the fuel pump. Refer to Fuel Pump Removal & Installation.

FUEL PUMP

REMOVAL & INSTALLATION

See Figures 146 through 148.

1. Before servicing the vehicle, refer to the Precautions Section.
2. Relieve the fuel system pressure. Refer to Relieving Fuel System Pressure.
3. Disconnect the connectors.
4. Unfasten the fuel pump mounting bolts and remove the fuel pump assembly.

To install:

5. Install the fuel pump assembly.
6. Install the fuel pump mounting bolts. Tightening the bolts/nuts to: 17–26 inch lbs. (2–3 Nm).
7. Connect the connectors.
8. Replace the access panel.

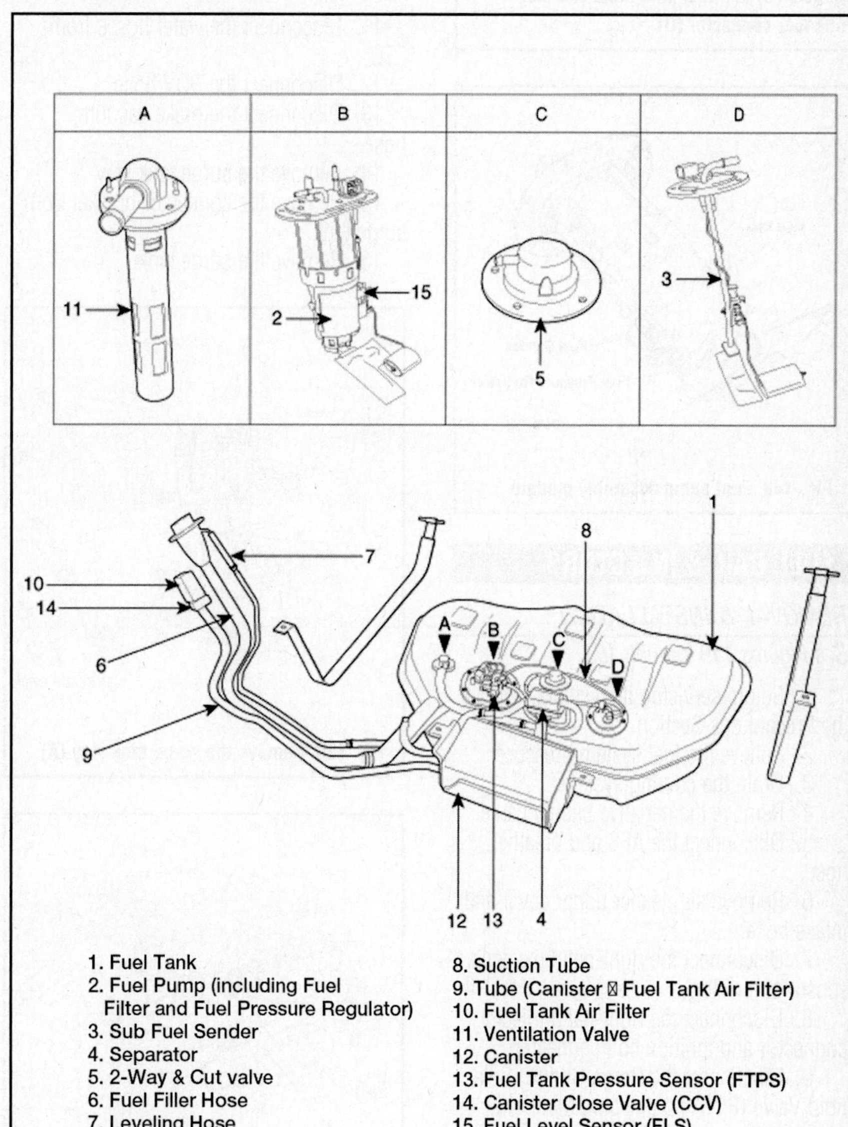

Fig. 146 Fuel delivery system

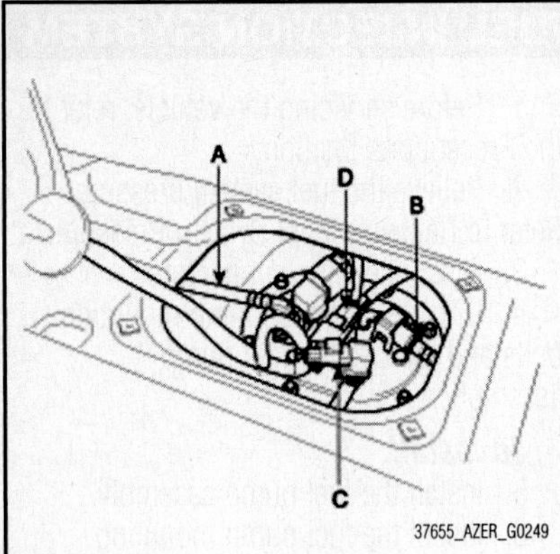

Fig. 147 Disconnect the suction tube quick-connector (A), fuel feed quick-connector (B), fuel tank pressure sensor connector(C) and fuel pump and sub fuel sender connector (D)

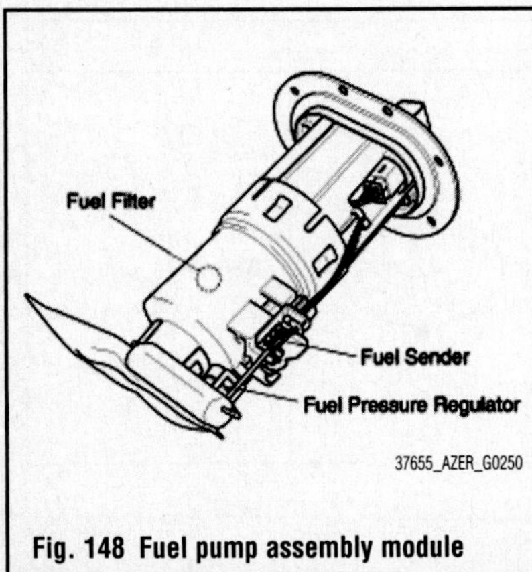

Fig. 148 Fuel pump assembly module

FUEL RAIL & INJECTORS

REMOVAL & INSTALLATION

See Figures 149 through 153.

1. Before servicing the vehicle, refer to the Precautions Section.
2. Relieve the fuel system pressure.
3. Drain the cooling system.
4. Remove the negative battery cable.
5. Disconnect the AFS and breather hose.
6. Remove air cleaner upper cover and intake hose.
7. Disconnect the right-hand oxygen sensor connector.
8. Disconnect the right-hand injector connector and ignition coil connector.
9. Disconnect the Purge Control Solenoid Valve (PCSV) connector, Manifold Absolute Pressure (MAP) sensor connector, and PCSV hose.
10. Disconnect the Electronic Throttle Control (ETC) connector and knock sensor connector.

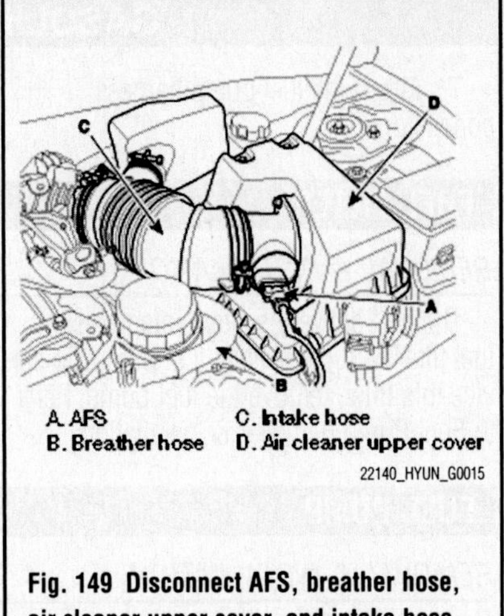

Fig. 149 Disconnect AFS, breather hose, air cleaner upper cover, and intake hose

11. Disconnect the water hoses from ETC.
12. Disconnect the PCV hose.
13. Disconnect the brake vacuum hose.
14. Remove the surge tank stay.
15. Remove the connector bracket from surge tank.
16. Remove the surge tank.

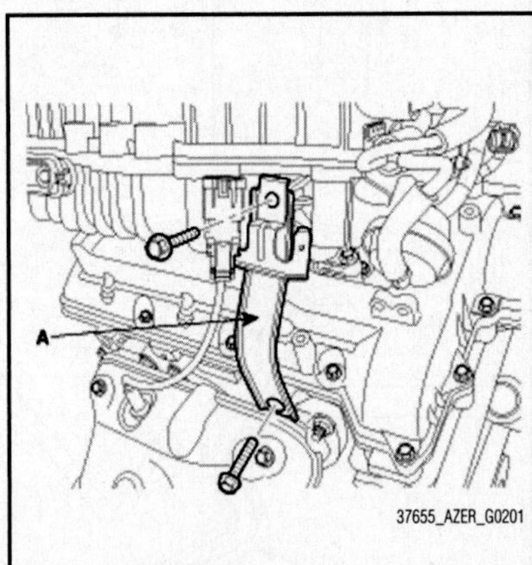

Fig. 150 Remove the surge tank stay (A)

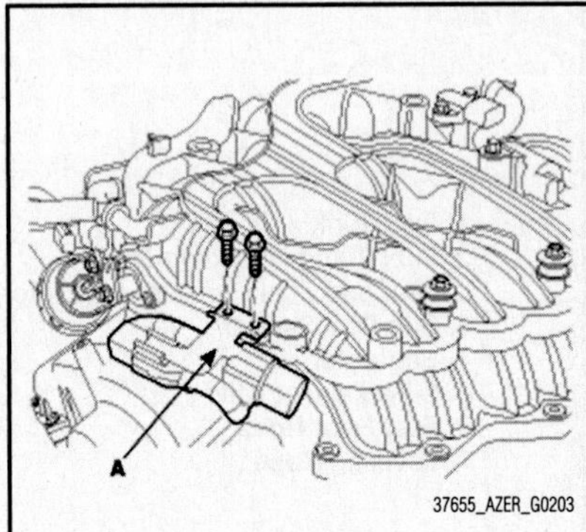

Fig. 151 Remove the connector bracket (A)

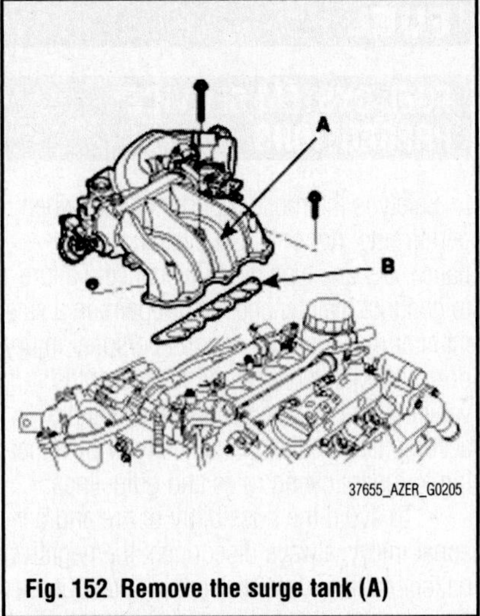

Fig. 152 Remove the surge tank (A)

17. Disconnect the breather pipe assembly.
18. Disconnect the left-hand injector connector.
19. Remove the fuel rail and injectors.

To install:

20. Install the fuel rail and injectors and tighten the bolts to 87–104 inch lbs. (10–12 Nm).
21. Connect the left-hand injector connector.
22. Connect the breather pipe assembly. Tighten to 84–108 inch lbs. (10–12 Nm).
23. Install the surge tank. Tighten long bolt to 84–108 inch lbs. (10–12 Nm); short bolt and nut to 14–17 ft. lbs. (19–24 Nm).
24. Install the connector bracket on the surge tank. Tighten to 5–8 ft. lbs. (7–11 Nm).
25. Install the surge tank stay. Tighten to 20–23 ft. lbs. (27–31 Nm); 14–17 ft. lbs. (19–24 Nm).
26. Connect the brake vacuum hose.
27. Connect the PCV hose.
28. Connect the water hoses to the ETC.
29. Install the ETC bracket. Tighten to 12–19 ft. lbs. (16–26 Nm).
30. Connect the ETC connector and knock sensor connector.
31. Connect the PCSV connector, MAP sensor connector and PCSV hose.
32. Connect the right-hand injector connector and ignition coil connector.
33. Connect the right-hand oxygen sensor connector.
34. Install the air cleaner upper cover and in take hose.
35. Connect the AFS and breather hose.
36. Start the engine and check for leaks.

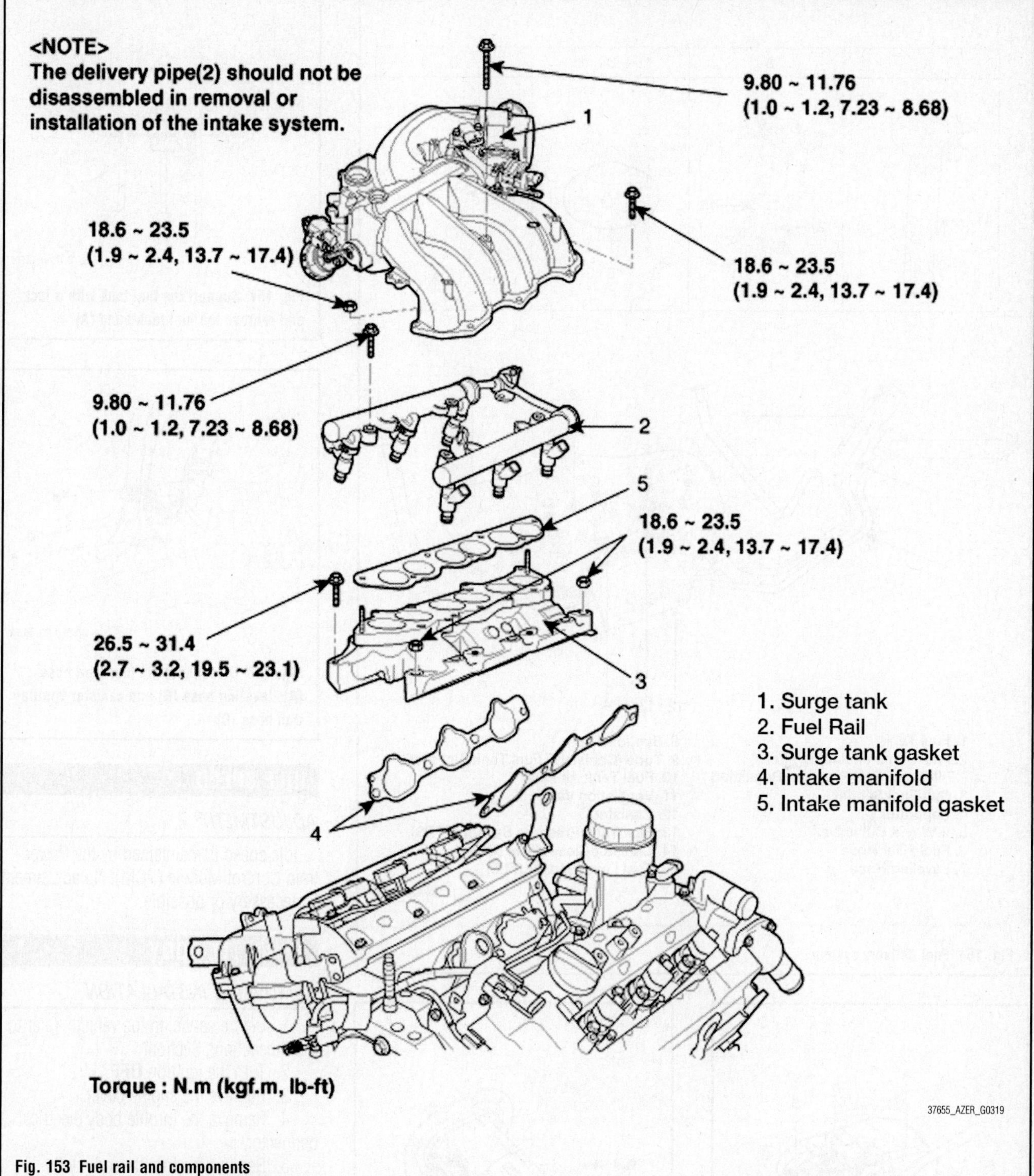

Fig. 153 Fuel rail and components

FUEL TANK

REMOVAL & INSTALLATION

See Figures 154 through 158.

1. Before servicing the vehicle, refer to the Precautions Section.
2. Relieve the fuel system pressure. Refer to Relieving Fuel System Pressure.
3. Disconnect the connectors.
4. Unfasten the fuel pump mounting bolts and remove the fuel pump assembly.
5. Raise and safely support the vehicle.
6. Remove the center muffler and main muffler, as needed.
7. Support the fuel tank with a jack.
8. Disconnect the fuel hoses.
9. Disconnect the ventilation hose connecting the separator with the canister.
10. By moving the jack down slowly, remove the fuel tank from the vehicle.

To install:

11. Installation is the reverse of the removal, noting the following:
    a. Tighten the fuel pump mounting bolts to 17–26 inch lbs. (2–3 Nm).
    b. Tighten the fuel tank mounting bolts to 29–40 ft. lbs. (39–54 Nm).

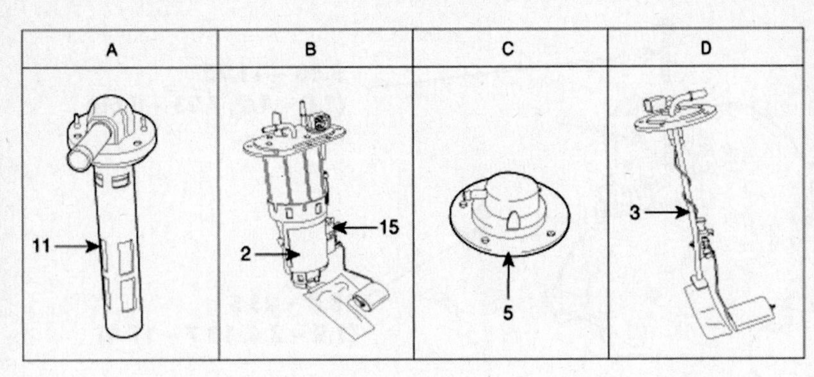

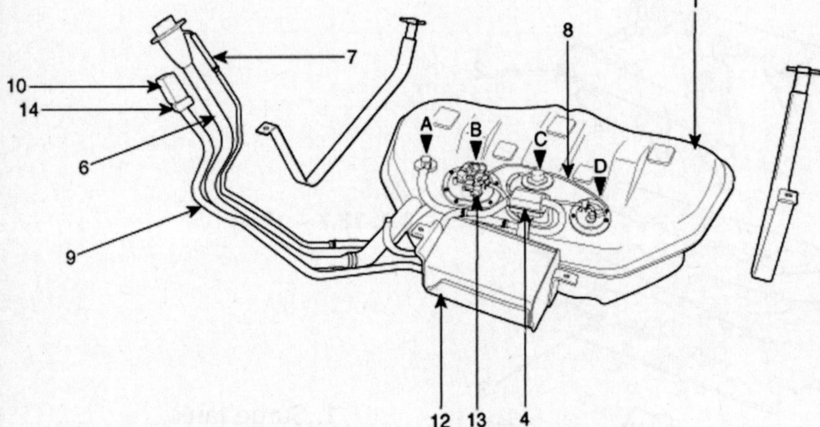

1. Fuel Tank
2. Fuel Pump (including Fuel Filter and Fuel Pressure Regulator)
3. Sub Fuel Sender
4. Separator
5. 2-Way & Cut valve
6. Fuel Filler Hose
7. Leveling Hose
8. Suction Tube
9. Tube (Canister ⊠ Fuel Tank Air Filter)
10. Fuel Tank Air Filter
11. Ventilation Valve
12. Canister
13. Fuel Tank Pressure Sensor (FTPS)
14. Canister Close Valve (CCV)
15. Fuel Level Sensor (FLS)

37655_AZER_G0251

Fig. 154 Fuel delivery system

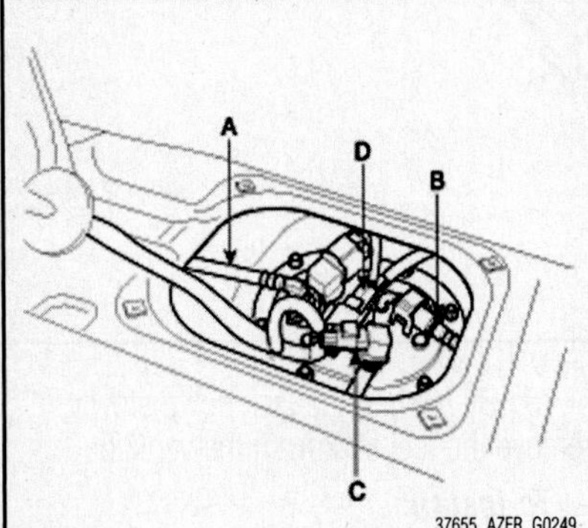

Fig. 155 Disconnect the suction tube quick-connector (A), fuel feed quick-connector (B), fuel tank pressure sensor connector(C) and fuel pump and sub fuel sender connector (D)

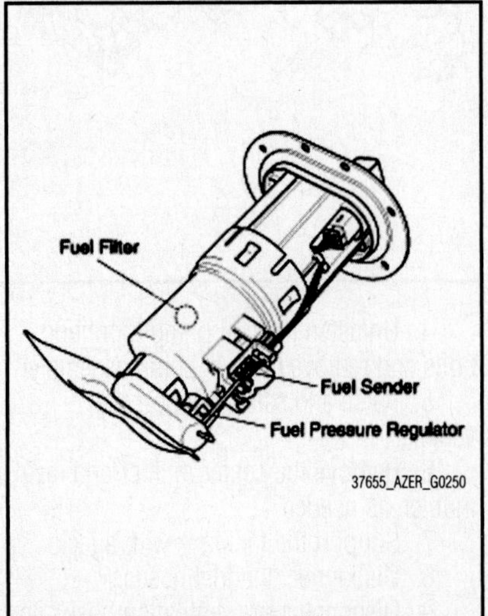

Fig. 156 Fuel pump assembly module

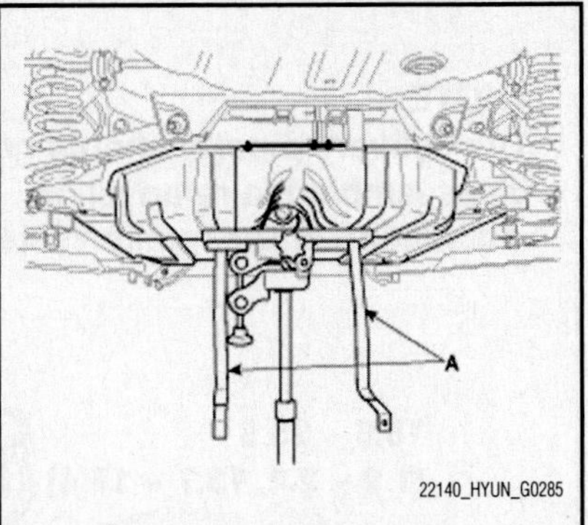

Fig. 157 Support the fuel tank with a jack and remove the fuel tank band (A)

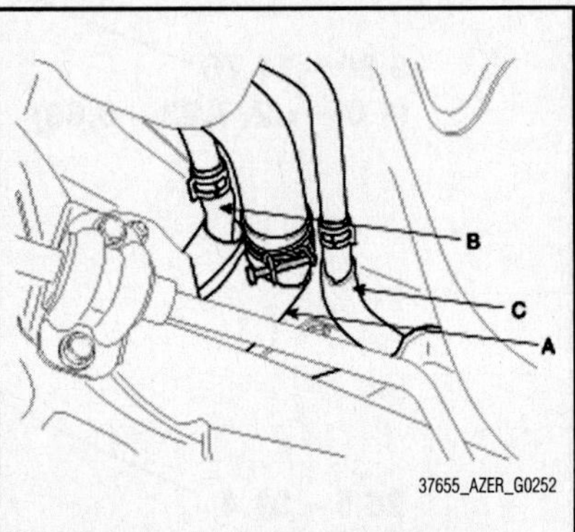

Fig. 158 Disconnect the fuel feed hose (A), leveling hose (B) and canister ventilation hose (C)

IDLE SPEED

ADJUSTMENT

Idle speed is maintained by the Powertrain Control Module (PCM). No adjustment is necessary or possible.

THROTTLE BODY

REMOVAL & INSTALLATION

1. Before servicing the vehicle, refer to the Precautions Section.
2. Turn the ignition **OFF**.
3. Remove the engine cover.
4. Remove the throttle body electrical connector.
5. Remove the throttle body bolts.
6. Remove the throttle body and gasket.

To install:

7. Clean the throttle body gasket mating surfaces.
8. Install the throttle body and NEW gasket.
9. Install the throttle body bolts and tighten to 21 ft. lbs. (28 Nm).
10. Install the throttle body electrical connector.
11. Install the engine cover.

HEATING & AIR CONDITIONING SYSTEM

BLOWER MOTOR

REMOVAL & INSTALLATION

See Figures 159 through 162.

1. Before servicing the vehicle, refer to the Precautions Section.

CAUTION

Before servicing components near or affected by the SRS (air bag) system, read and observe all SRS Service Precautions. Refer to Supplemental Restraint System (SRS), in the Chassis Electrical section. Failure to observe all precautions may result in accidental airbag deployment, personal injury, or death. Refer to Airbag Removal & Installation.

2. Disconnect the negative battery cable and wait 3 minutes for the SRS memory to drain.

CAUTION

After disconnecting the negative battery cable, wait for at least 3 minutes for the SRS module to deplete its stored energy.

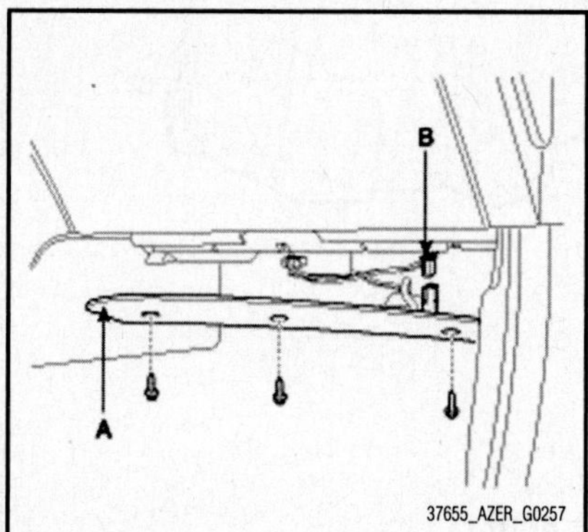

Fig. 159 Step lamp, screws, cover (A), and connector (B)

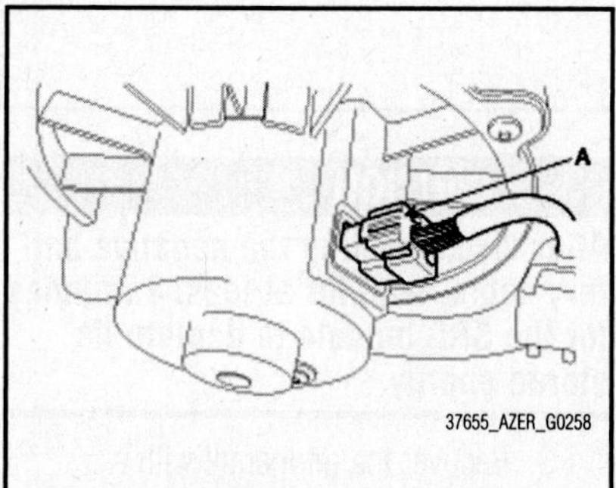

Fig. 160 Blower motor connector (A)

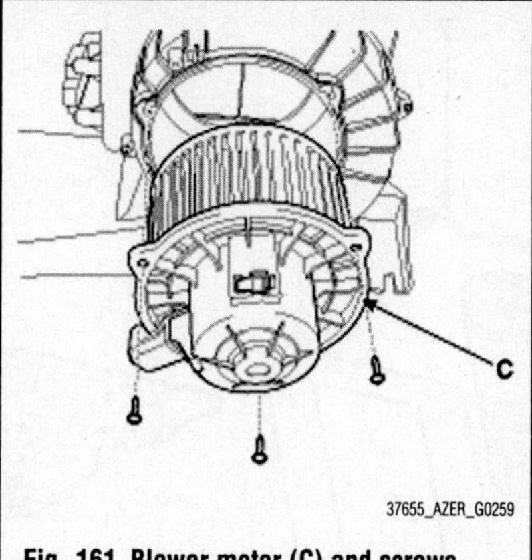

Fig. 161 Blower motor (C) and screws

3. Remove the three step lamp cover screws, remove the cover, and disconnect the connector.
4. Disconnect the blower motor connector.
5. Remove the mounting screws and the blower motor.

To install:

6. Installation is the reverse order of removal.

HEATER UNIT/CORE

REMOVAL & INSTALLATION

See Figures 163 through 169.

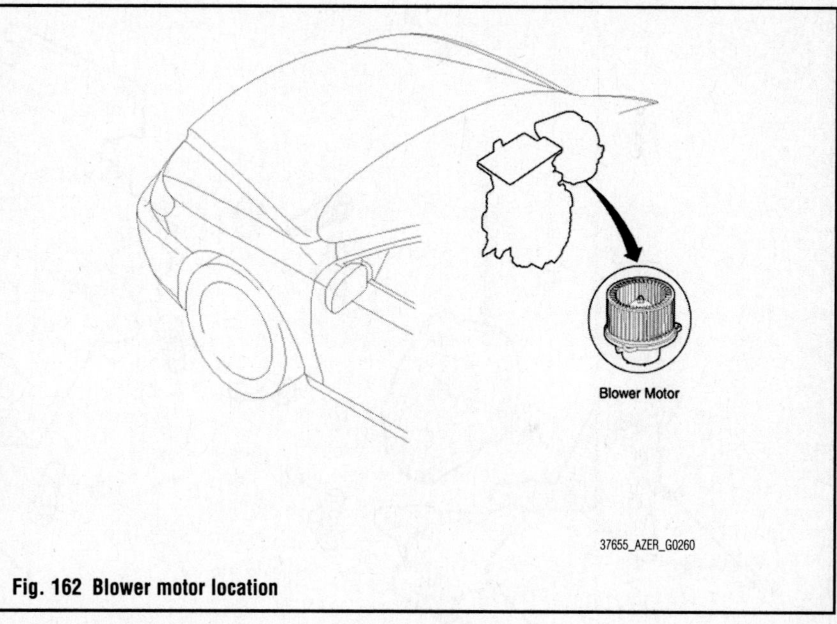

Fig. 162 Blower motor location

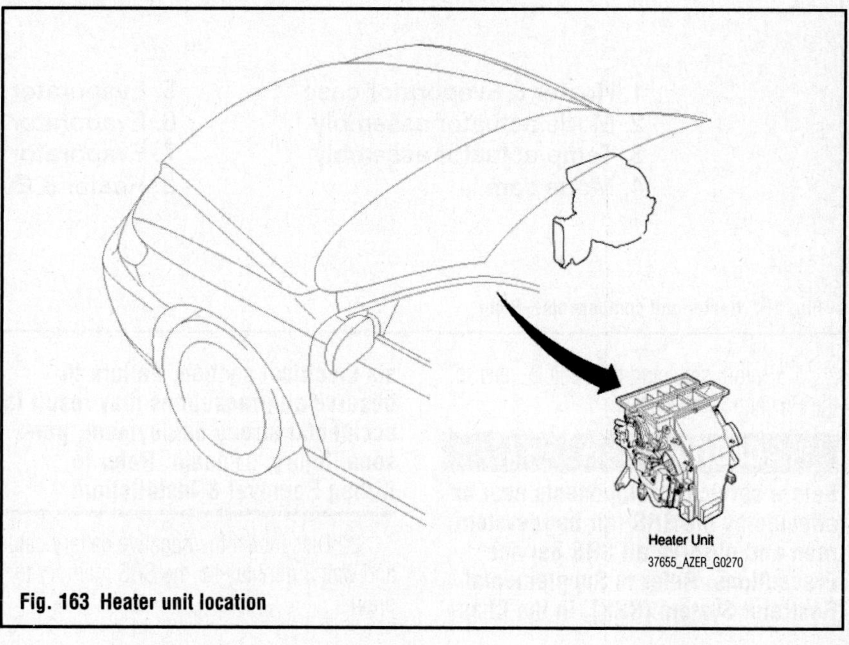

Fig. 163 Heater unit location

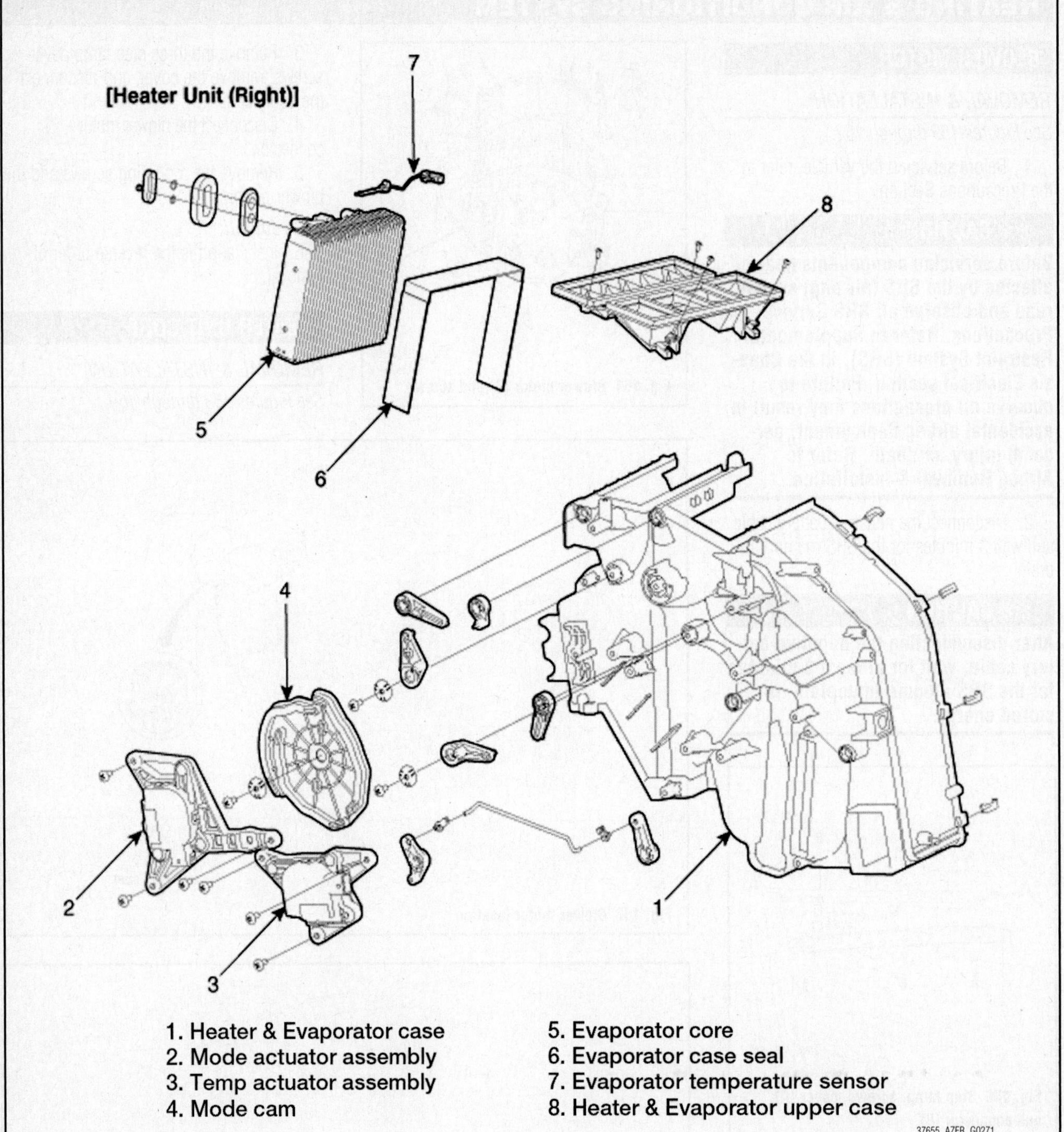

Fig. 164 Heater unit components—Right

1. Before servicing the vehicle, refer to the Precautions Section.

CAUTION

Before servicing components near or affected by the SRS (air bag) system, read and observe all SRS Service Precautions. Refer to Supplemental Restraint System (SRS), in the Chassis Electrical section. Failure to observe all precautions may result in accidental airbag deployment, personal injury, or death. Refer to Airbag Removal & Installation.

2. Disconnect the negative battery cable and wait 3 minutes for the SRS memory to drain.

CAUTION

After disconnecting the negative battery cable, wait for at least 3 minutes for the SRS module to deplete its stored energy.

3. Recover the refrigerant with a recovery/ recycling/ charging station.

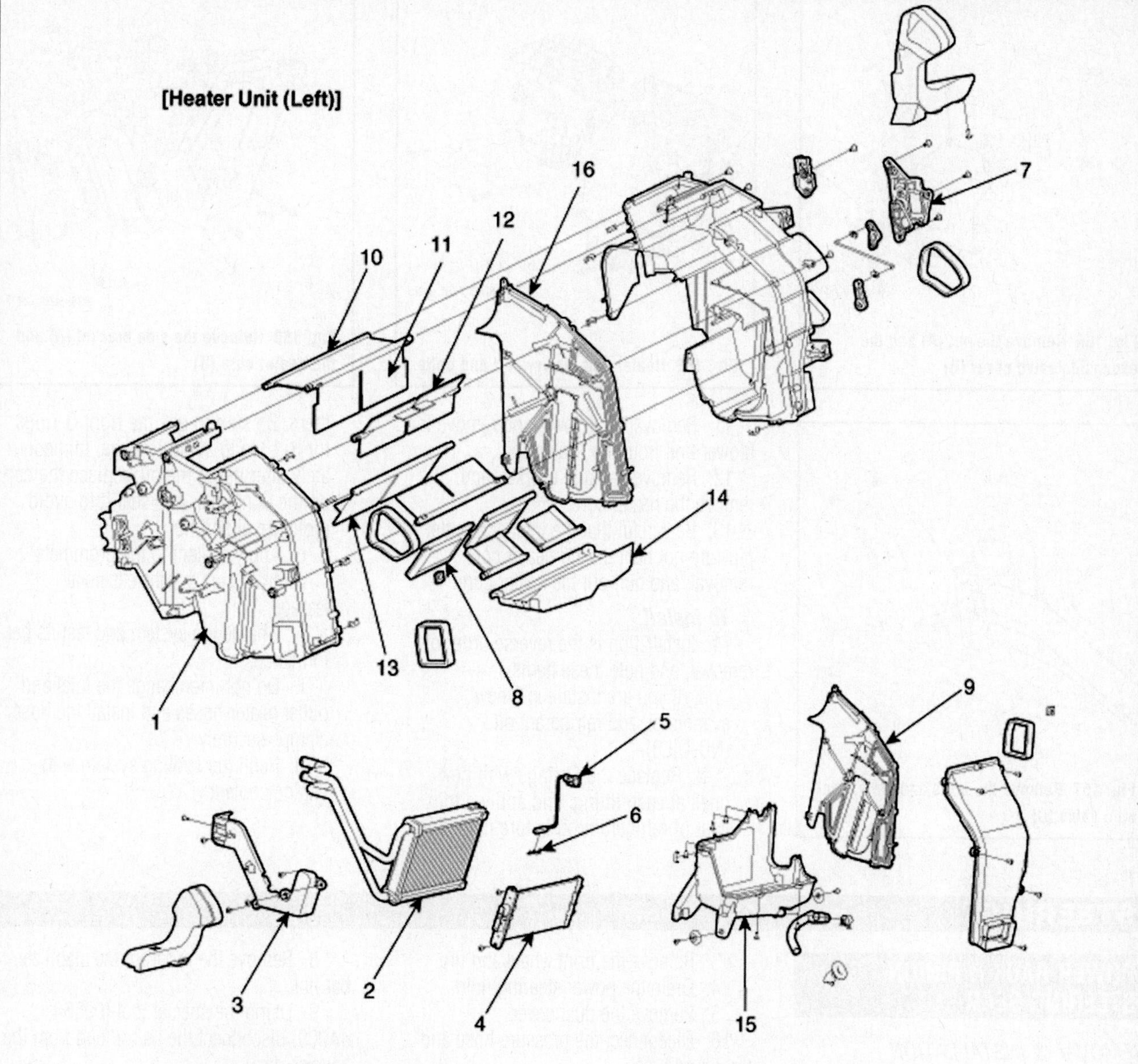

Fig. 165 Heater unit components—Left

4. When the engine is cool, drain the engine coolant from the radiator. Refer to Coolant in the Engine Cooling section.

5. Remove the nut and the expansion valve cover.

6. Remove the bolts and the expansion valve from the evaporator core. Plug or cap the lines immediately after disconnecting them to avoid moisture and dust contamination.

7. Disconnect the inlet and outlet heater hoses from the heater unit.

WARNING

Engine coolant will run out when the hoses are disconnected; drain it into a clean drip pan. Be sure not to let coolant spill on electrical parts or painted surfaces. If any coolant spills, rinse it off immediately.

8. Remove the instrument panel.

9. Remove the cowl cross bar assembly.

10. Remove the mounting bolts, and remove the heater and blower unit.

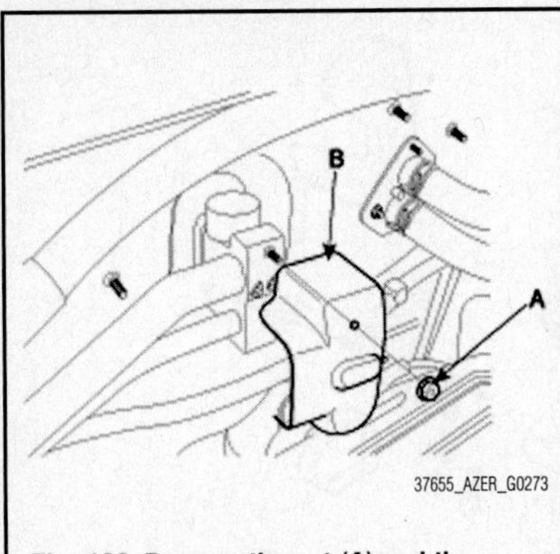

Fig. 166 Remove the nut (A) and the expansion valve cover (B)

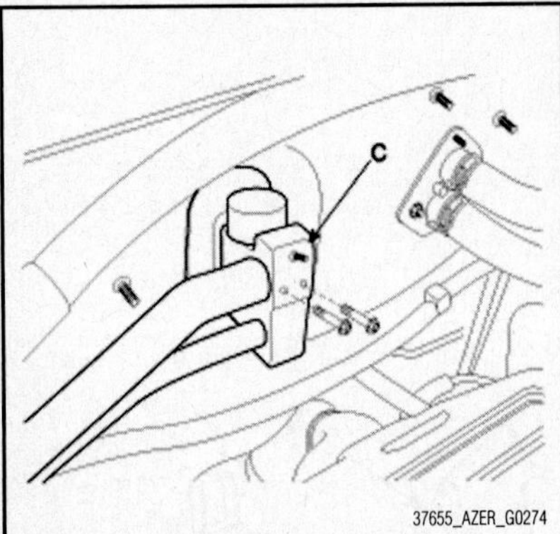

Fig. 167 Remove the bolts and the expansion valve (C)

Fig. 168 Heater and blower unit and bolts

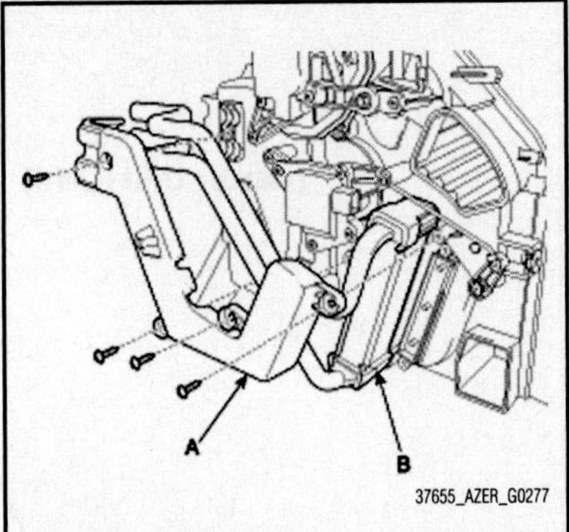

Fig. 169 Remove the side bracket (A) and the heater core (B)

11. Remove the screws, and remove the blower unit from the heater unit.
12. Remove the side bracket, and remove the heater core.
13. Be careful that the inlet and outlet pipe are not bent during heater core removal, and pull out the heater core.

To install:

14. Installation is the reverse order of removal, and note these items :
    a. If you are installing a new evaporator, add refrigerant oil (ND-OIL8).
    b. Replace the O-rings with new ones at each fitting, and apply a thin coat of refrigerant oil before installing them. Be sure to use the right O-rings for R-134a to avoid leakage. Immediately after using the oil, replace the cap on the container, and seal it to avoid moisture absorption.
    c. Apply sealant to the grommets.
    d. Make sure that there is no air leakage.
    e. Charge the system and test its performance.
    f. Do not interchange the inlet and outlet heater hoses and install the hose clamps securely.
    g. Refill the cooling system with engine coolant.

STEERING

POWER RACK & PINION STEERING GEAR

REMOVAL & INSTALLATION

See Figures 170 through 179.

1. Before servicing the vehicle, refer to the Precautions Section.
2. Raise and safely support the vehicle.
3. Remove the front wheel and tire.
4. Drain the power steering fluid.
5. Remove the dust cover.
6. Disconnect the pressure hose and the return tube.
7. Remove the pressure sensor connector from the pressure hose.
8. Remove the nut from the stabilizer bar link.
9. Using the special tool (09568-4A000) disconnect the tie rod end from the knuckle arm.
10. Remove the lower arm mounting bolts.

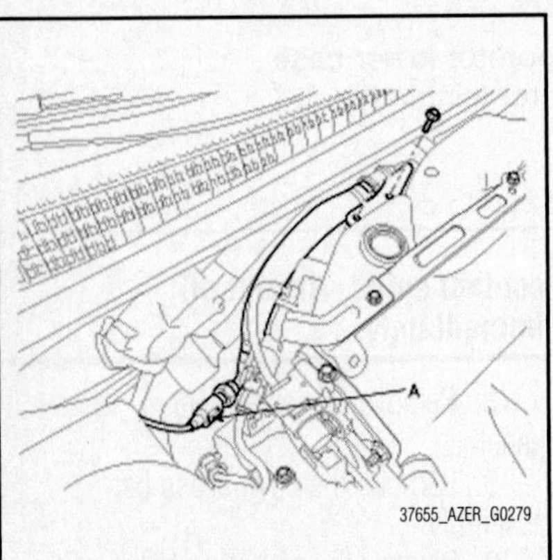

Fig. 170 Pressure sensor connector (A)

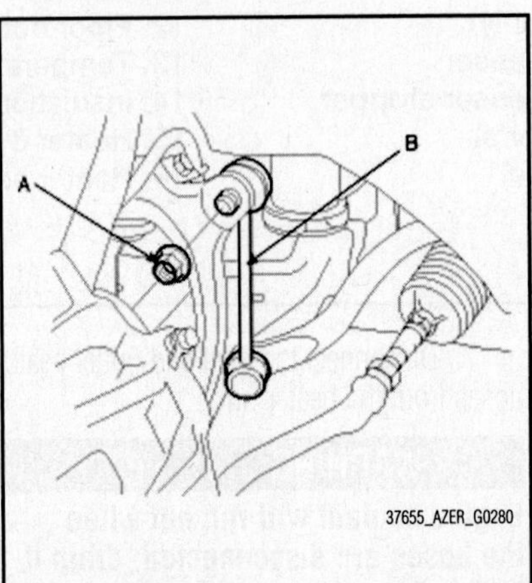

Fig. 171 Remove the nut (A) from the stabilizer bar link (B)

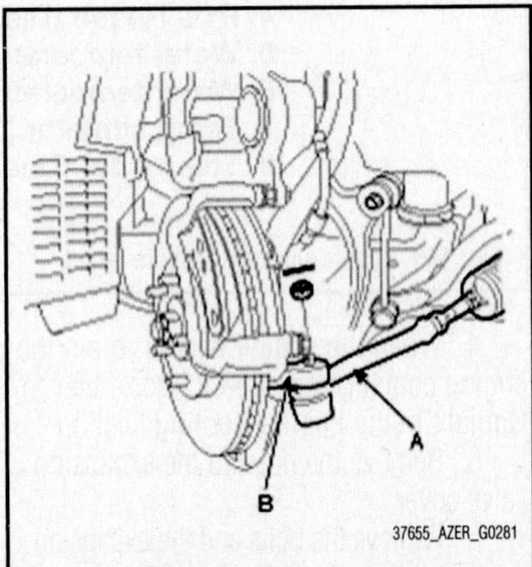

Fig. 172 Disconnect the tie rod end (A) from the knuckle arm (B)

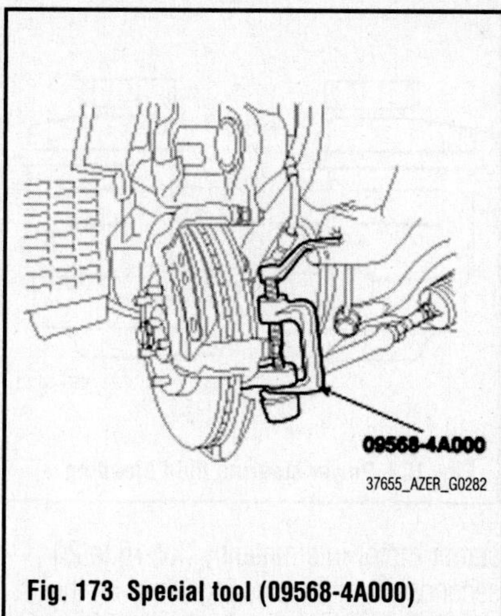

Fig. 173 Special tool (09568-4A000)

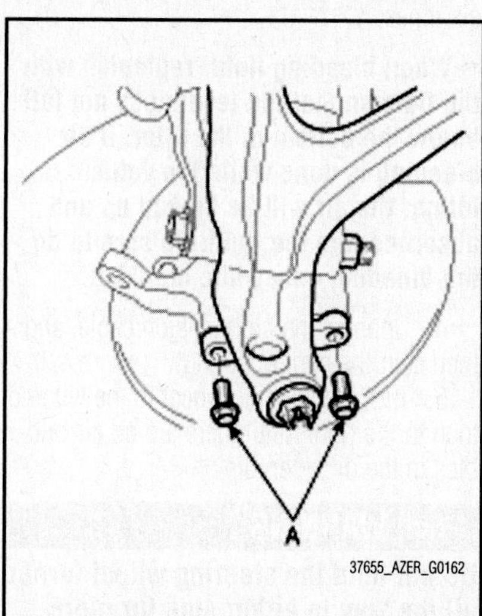

Fig. 174 Remove the lower arm mounting bolts (A)

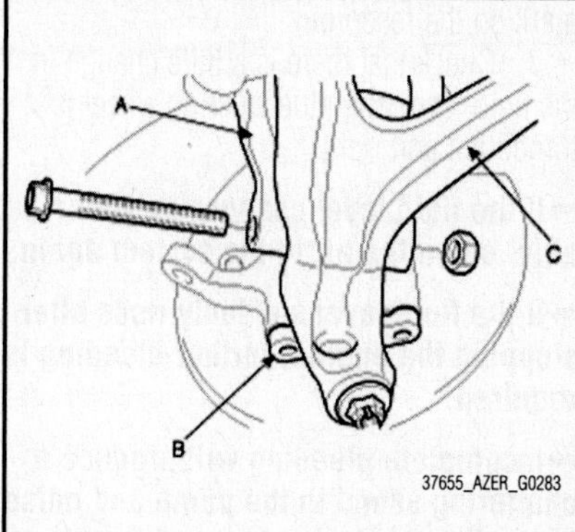

Fig. 175 Remove the front fork (A) and the knuckle ball joint (B) from the front lower arm (C)

11. Remove the front fork and the knuckle ball joint from the front lower arm. Be careful not to damage to the aluminum lower arm.
12. Remove the joint assembly connecting bolt.

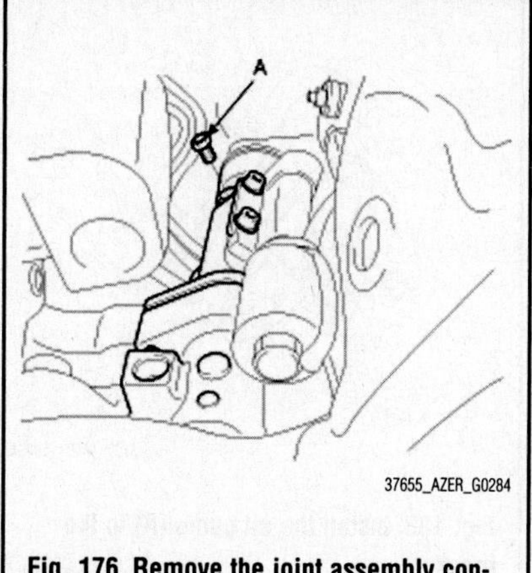

Fig. 176 Remove the joint assembly connecting bolt (A)

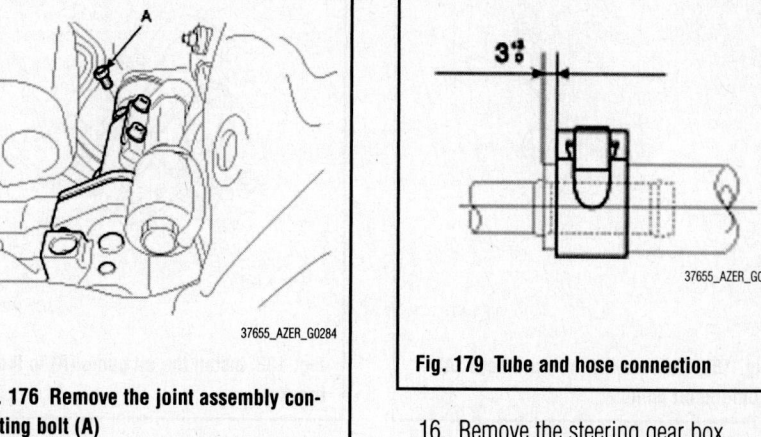

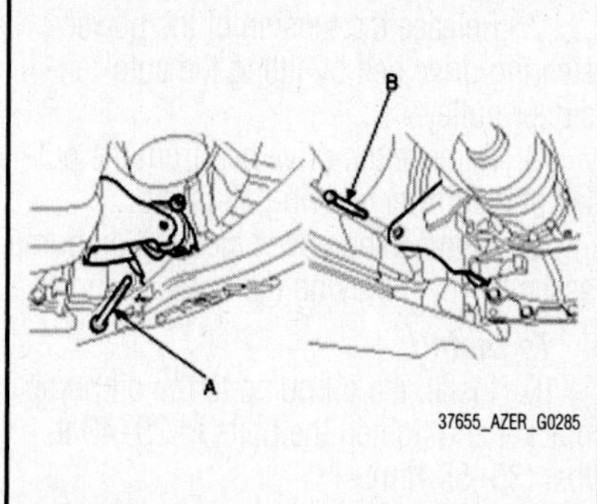

Fig. 177 Remove the connecting bolts (A, B) of front and rear roll stopper

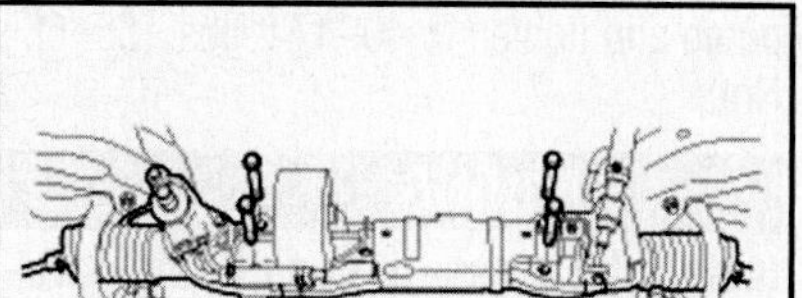

Fig. 178 Remove the steering gear box mounting bolts and remove the steering gear box assembly (A) and the mounting rubber

WARNING

Keep the neutral-range to prevent the damage of the clock spring inner cable when you handle the steering wheel.

13. Remove the connecting bolts (A, B) of front and rear roll stopper.
14. Remove the crossmember mounting bolts.
15. Remove the heat protecting cover mounting bolts.

Fig. 179 Tube and hose connection

16. Remove the steering gear box mounting bolts and remove the steering gear box assembly and the mounting rubber.

WARNING

When removing the gear box, pull it out carefully and slowly to avoid damaging the boots.

17. Disconnect the pressure hose and the return tube.

To install:

➡Be sure to connect between a tube and a hose as shown in the illustration.

18. Installation is the reverse of removal, noting the following tightening specifications:
    - Pressure hose to gear box: 9–13 ft. lbs. (12–18 Nm)
    - Return tube to gear box: 9–13 ft. lbs. (12–18 Nm)
    - Tie rod end lock nut: 36–40 ft. lbs. (50–55 Nm)
    - Pinion and valve assembly to self locking nut : 14–22 ft. lbs. (20–30 Nm)
    - Lock nut: 36–51 ft. lbs. (50–70 Nm)
    - Tie rod end self locking nut: 17–25 ft. lbs. (24–34 Nm)
    - Mounting bracket to crossmember: 43–58 ft. lbs. (60–80 Nm)
19. After installation, bleed the air in the power steering system.
20. Check front wheel alignment.

POWER STEERING PUMP

REMOVAL & INSTALLATION

See Figures 180 through 183.

1. Before servicing the vehicle, refer to the Precautions Section.
2. Raise and safely support the vehicle.

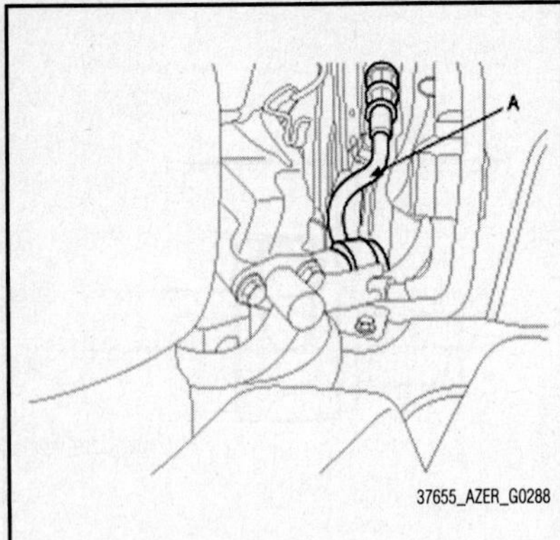

Fig. 180 Remove the pressure hose (A) from the oil pump

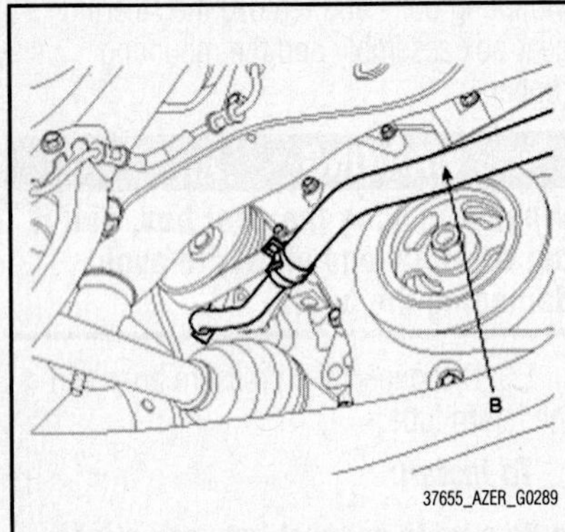

Fig. 181 Remove the suction hose (B) from the suction pipe

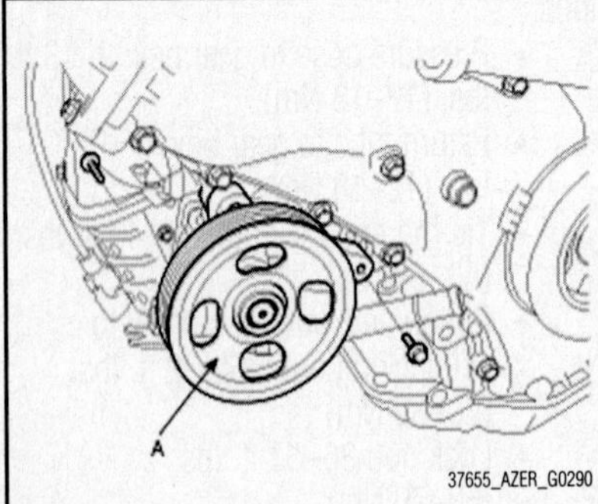

Fig. 182 Remove the power steering oil pump assembly

3. Remove the front wheel and tire.
4. Drain the power steering fluid.
5. Remove the dust cover.
6. Remove the pressure hose from the oil pump and the suction hose from the suction pipe, then drain the power steering oil.

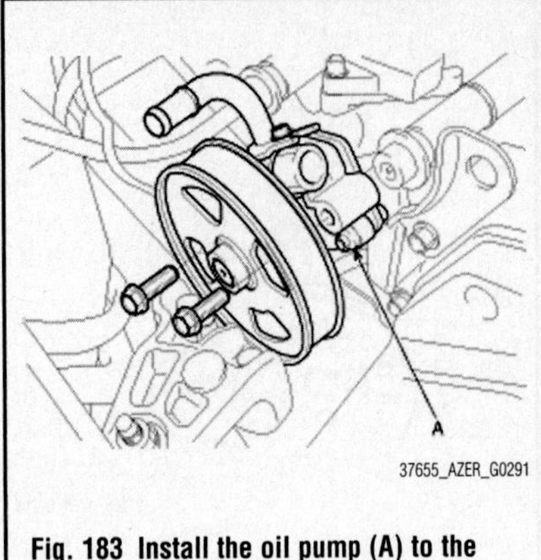

Fig. 183 Install the oil pump (A) to the bracket

7. Release the tension of the power steering drive belt by lifting the auto-tensioner pulley.
8. Remove the drive belt from the pulley of the power steering oil pump.
9. Remove the power steering oil pump assembly by removing the bolts as shown.

To install:

10. Install the oil pump to the oil pump bracket and tighten the bolts to 25–40 ft. lbs. (35–55 Nm).
11. Install the drive belt by pulling the auto tensioner.
12. Install the suction hose.
13. Install the pressure hose to the oil pump and tighten to 40–47 ft. lbs. (55–65 Nm).

⁂ WARNING

Install the pressure hose being careful so that it does not twist and come in contact with other components.

14. Add power steering fluid (PSF-4).
15. Bleed the system.
16. Check the oil pump pressure.

BLEEDING

See Figure 184.

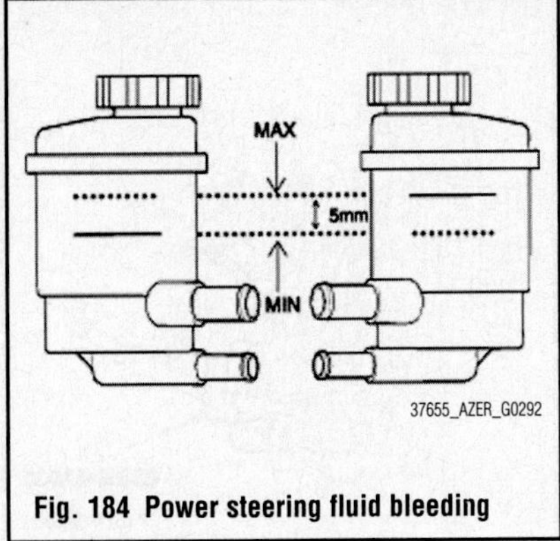

Fig. 184 Power steering fluid bleeding

1. Fill the power steering fluid reservoir up to the "MAX" position with specified fluid.
2. Jack up the front wheels.
3. Disconnect the ignition coil high tension cable, and then, while operating the starter motor intermittently (for 15 to 20 seconds), turn the steering wheel all the way to the left and then to the right five or six times.

➡When bleeding fluid, replenish with the fluid so that the level does not fall below the bottom of the filter. If air bleeding is done while the vehicle is idling, the air will be broken up and absorbed into the fluid. Be sure to do the bleeding only while cranking.

4. Connect the high tension cable, and then start the engine (idling).
5. Turn the steering wheel to the left and then to the right, until there are no air bubbles in the oil reservoir.

⁂ WARNING

Do not hold the steering wheel turned all the way to either side for more than ten seconds.

6. Confirm that the fluid is not milky and that the level is between "MAX" and "MIN" mark on the reservoir.
7. Check that there is a little change in the fluid level when the steering wheel is turned left and right.

➡If the fluid level changes considerably, bleed the air in the system again.

➡If the fluid level suddenly rises after stopping the engine, further bleeding is required.

➡Incomplete bleeding will produce a chattering sound in the pump and noise in the flow control valve, and lead to decreased durability of the pump.

CONTROL LINKS

REMOVAL & INSTALLATION

See Figures 185 and 186.

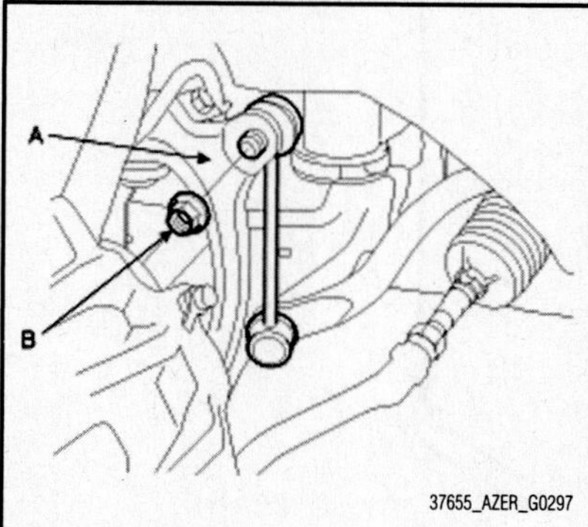

Fig. 185 Stabilizer link, nut (B) and fork (A)

1. Before servicing the vehicle, refer to the Precautions Section.
2. Raise and safely support the vehicle.
3. Remove the front wheel and tire from the front hub.

⁂ WARNING

Be careful not to damage the hub bolts when removing the front wheel and tire.

4. Remove the stabilizer link from the fork.

To install:

5. Install the stabilizer link to the fork and tighten to 72–87 ft. lbs. (100–120 Nm).
6. Install the front wheels and lower the vehicle.

LOWER CONTROL ARM

REMOVAL & INSTALLATION

See Figures 187 through 190.

1. Before servicing the vehicle, refer to the Precautions Section.
2. Raise and safely support the vehicle.
3. Remove the front wheel and tire.
4. Remove the lower arm ball joint mounting bolts.
5. Remove the front lower arm fork mounting bolt from the fork.
6. Remove the lower arm mounting bolts.

To install:

7. Install and tighten the lower arm mounting bolts and tighten to 101–116 ft. lbs. (140–160 Nm).

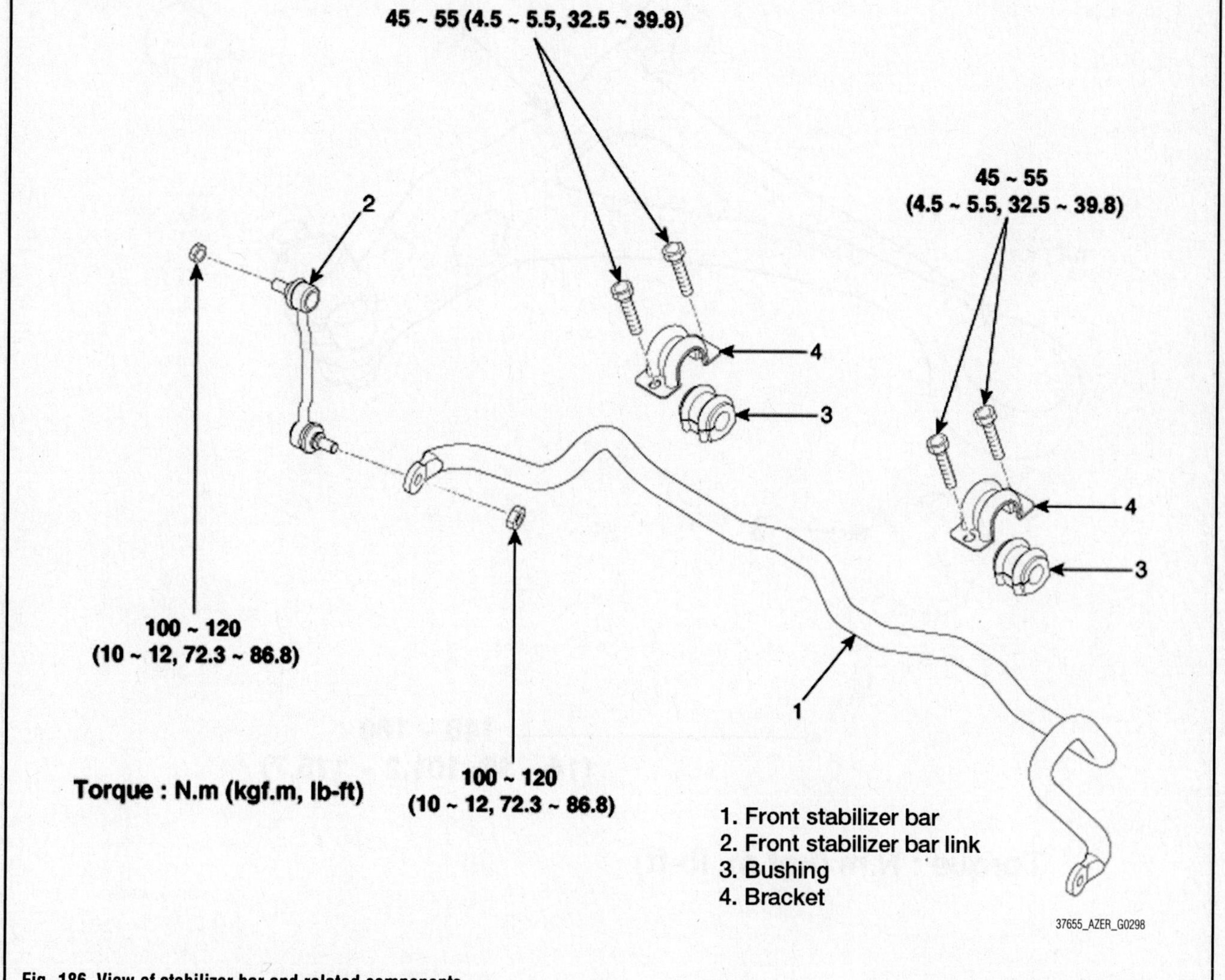

45 ~ 55 (4.5 ~ 5.5, 32.5 ~ 39.8)
45 ~ 55
(4.5 ~ 5.5, 32.5 ~ 39.8)
100 ~ 120
(10 ~ 12, 72.3 ~ 86.8)
100 ~ 120
(10 ~ 12, 72.3 ~ 86.8)
Torque : N.m (kgf.m, lb-ft)
1. Front stabilizer bar
2. Front stabilizer bar link
3. Bushing
4. Bracket
37655_AZER_G0298

Fig. 186 View of stabilizer bar and related components

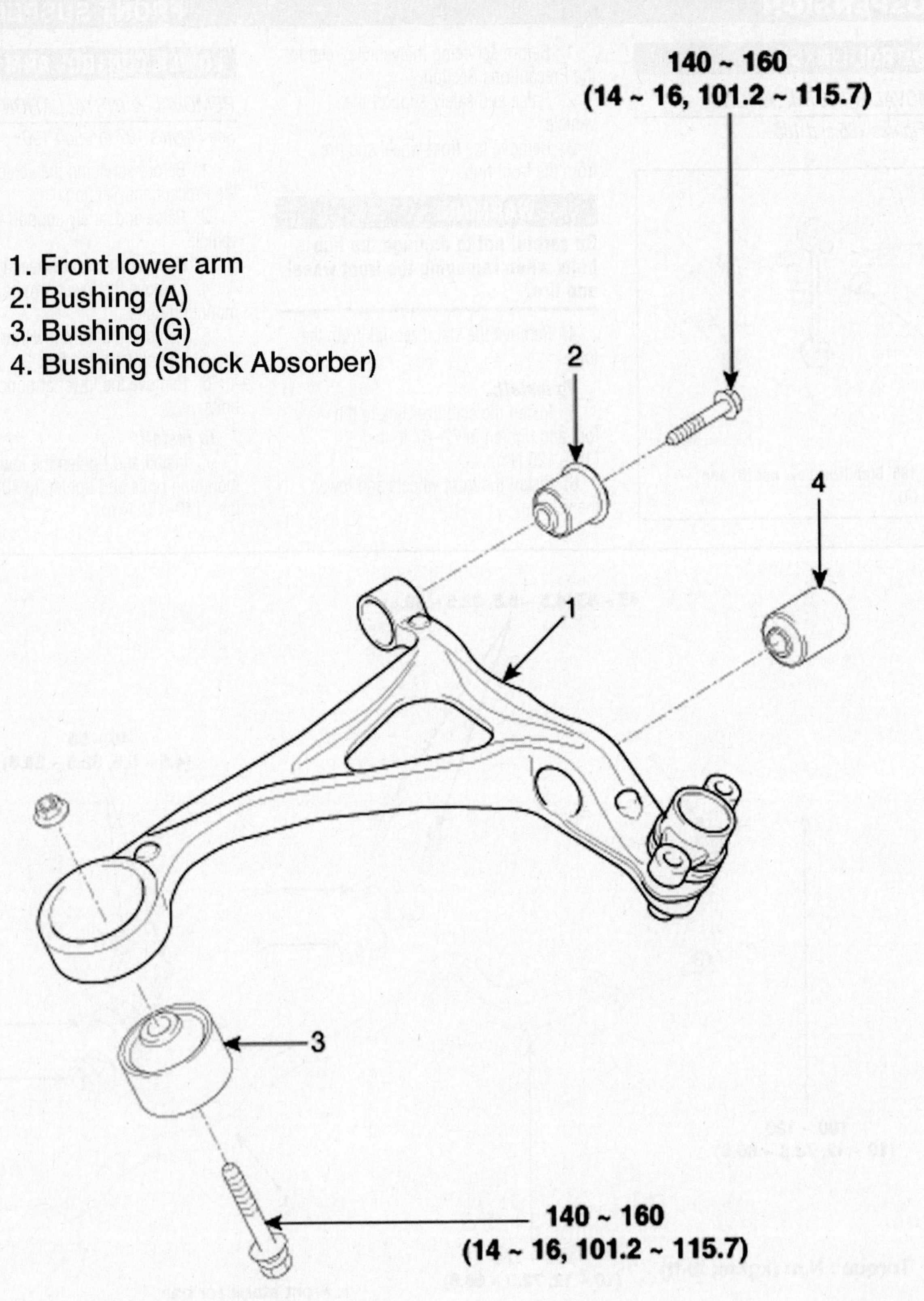

Fig. 187 Lower control arm components

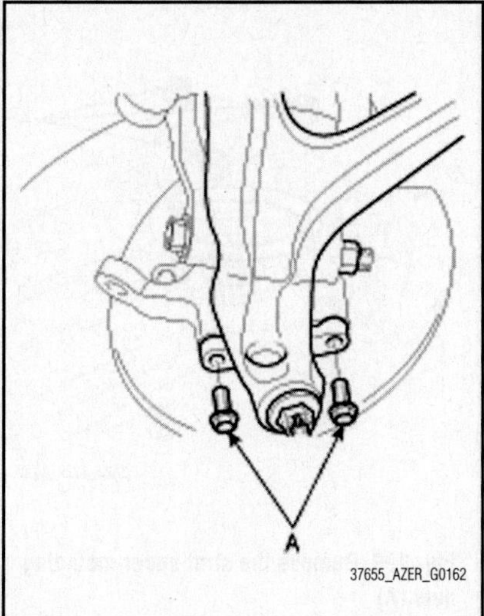

Fig. 188 Remove the lower arm ball joint mounting bolts (A)

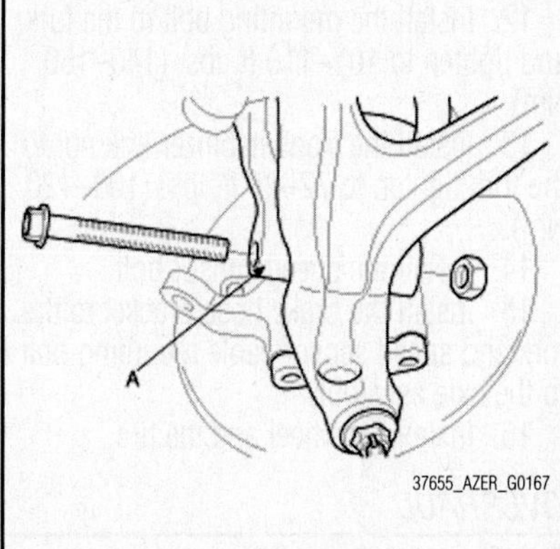

Fig. 189 Remove the front lower arm fork mounting bolt from the fork (A)

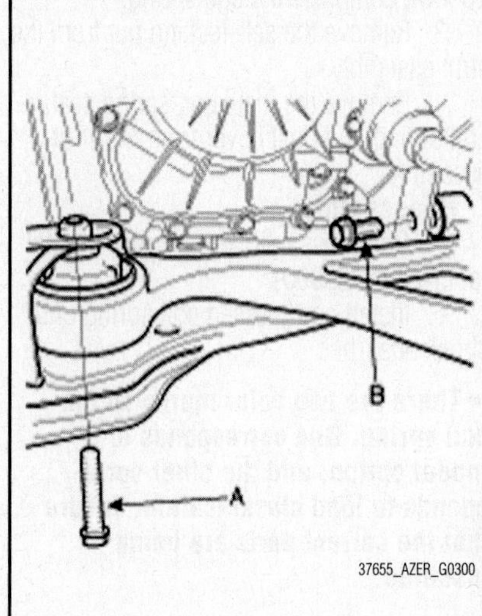

Fig. 190 Remove the lower arm mounting bolts (A, B)

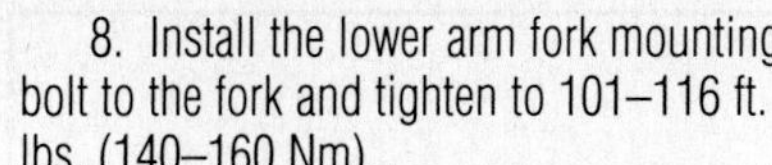

8. Install the lower arm fork mounting bolt to the fork and tighten to 101–116 ft. lbs. (140–160 Nm).
9. Install the lower arm ball joint mounting bolts and tighten to 72–87 ft. lbs. (100–120 Nm).
10. Install the front wheel and tire.
11. Check and/or adjust the wheel alignment.

STEERING KNUCKLE

REMOVAL & INSTALLATION

See Figures 188, 191 through 195.

1. Before servicing the vehicle, refer to the Precautions Section.
2. Raise and safely support the vehicle.
3. Remove the front wheel and tire.

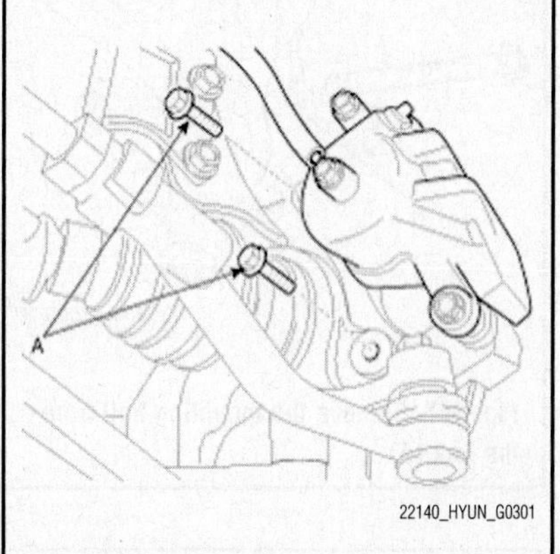

Fig. 191 Remove the brake caliper mounting bolts (A)

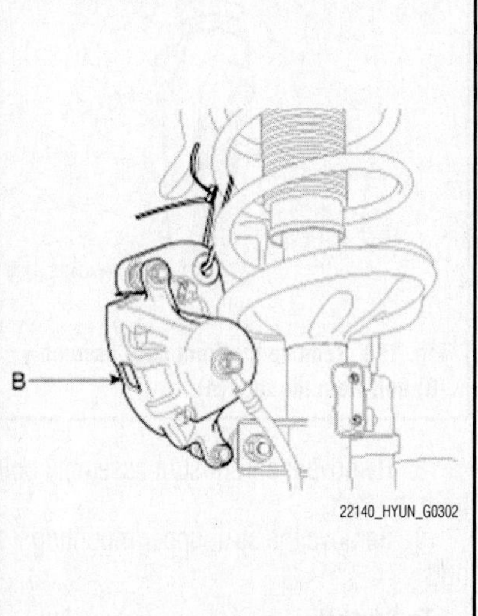

Fig. 192 Hang the brake caliper assembly (B) with a wire

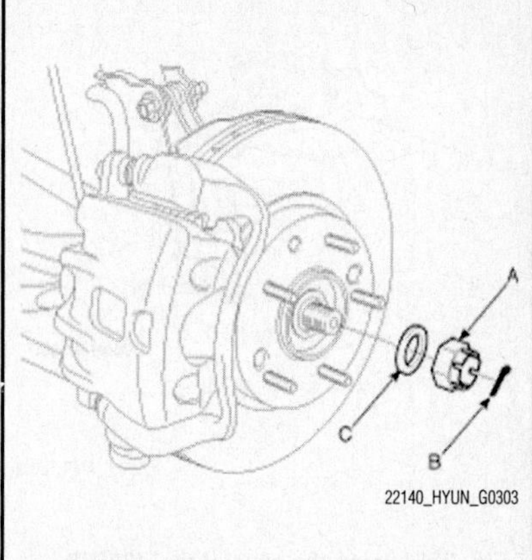

Fig. 193 Remove the split pin (B), the castle nut (A), and the washer (C) from the front hub

WARNING

Be careful not to damage the hub bolts when removing the front wheel and tire.

4. Remove the wheel speed sensor from the knuckle.
5. Disconnect the brake hose bracket from the knuckle.
6. Remove the brake caliper mounting bolts, and then hang the brake caliper assembly with a wire.

WARNING

Do not suspend the brake caliper assembly from the brake hose or damage may occur to the hose.

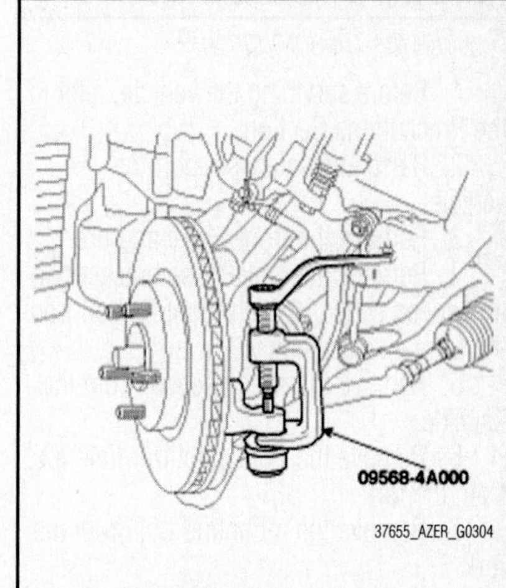

Fig. 194 Using the special tool (09568-4A000), disconnect the tie rod end from the knuckle

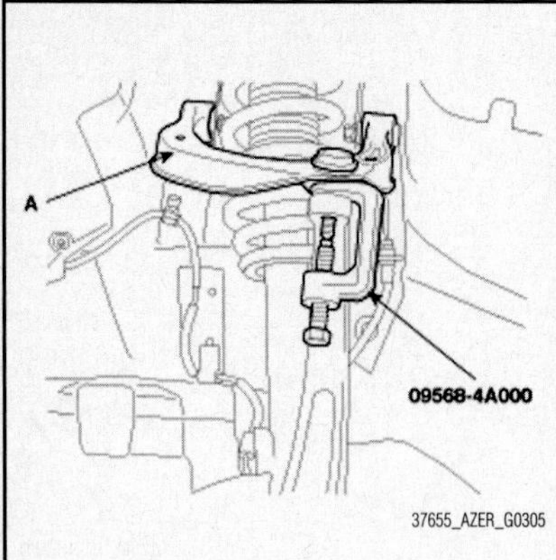

Fig. 195 Using the special tool (09568-4A000), disconnect the knuckle from the upper arm assembly (A)

7. Remove the split pin and castle nut from the front hub.
8. Remove the 2 bolts and disconnect the knuckle from the lower arm assembly.
9. Using the special tool (09568-4A000), disconnect the tie rod end from the knuckle.
10. Using a plastic hammer, disconnect the driveshaft from the axle hub.
11. Loosen the upper arm mounting nut but do not remove it.
12. Using the special tool (09568-4A000), disconnect the knuckle from the upper arm assembly.
13. Remove the steering knuckle.

To install:

14. Installation is the reverse of removal.

STRUT

REMOVAL & INSTALLATION

See Figures 196 through 199.

1. Before servicing the vehicle, refer to the Precautions Section.
2. Raise and safely support the vehicle.
3. Remove the front wheel and tire.
4. Remove the brake hose bracket bolt and speed sensor cable mounting bolt from the front axle assembly.
5. Remove the speed sensor from the knuckle.
6. Remove the front stabilizer link nut from the fork.
7. Remove the mounting bolt from the fork.

⁂ WARNING

Be careful not to damage the aluminum lower arm.

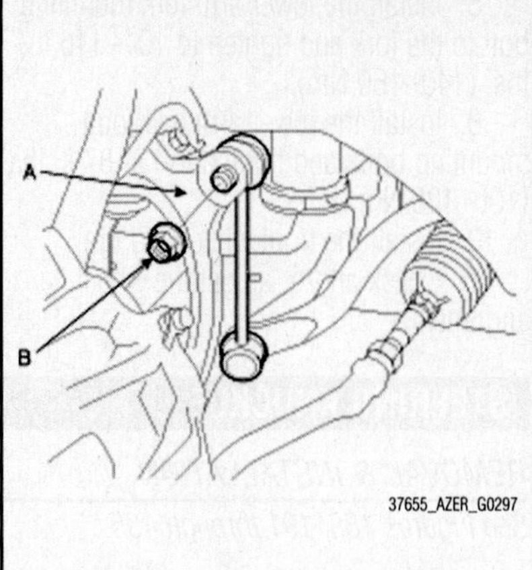

Fig. 196 Remove the front stabilizer link nut (B) from the fork (A)

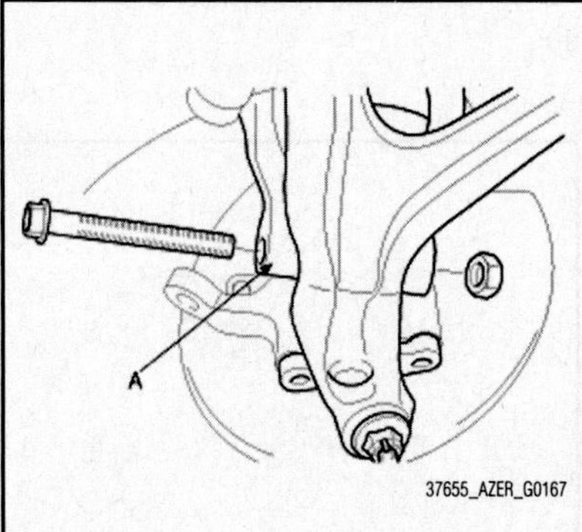

Fig. 197 Remove the mounting bolt from the fork (A)

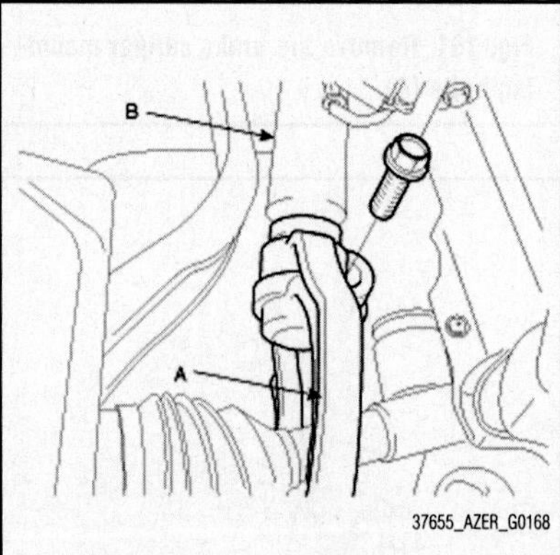

Fig. 198 Remove the front strut assembly (B) bolt from the fork (A)

8. Remove the front strut assembly bolt from the fork.
9. Remove the strut upper mounting nuts.

To install:

10. Install the strut upper mounting nuts and tighten to 33–43 ft. lbs. (45–60 Nm).

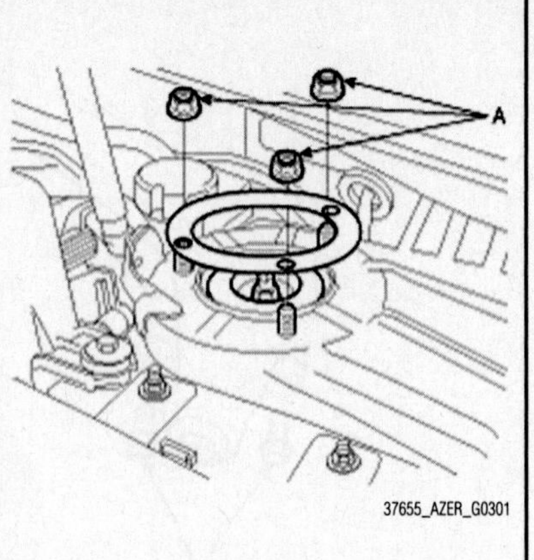

Fig. 199 Remove the strut upper mounting nuts (A)

11. Install the fork mounting bolt to the strut assembly with the I.D. mark facing outward and tighten to 43–58 ft. lbs. (60–80 Nm).
12. Install the mounting bolt to the fork and tighten to 101–116 ft. lbs. (140–160 Nm).
13. Install the front stabilizer link nut to the fork tighten to 72–87 ft. lbs. (100–120 Nm).
14. Install the speed sensor bolt.
15. Install the brake hose bracket to the fork and speed sensor cable mounting bolt to the axle assembly.
16. Install the wheel and the tire.

OVERHAUL

See Figures 200 and 201.

1. Before servicing the vehicle, refer to the Precautions Section.
2. Using the special tool (09546-26000), compress the coil spring.
3. Remove the self-locking nut from the strut assembly.
4. Remove the insulator, spring seat, coil spring and dust cover from the strut assembly.

Reassembly:

5. Compress coil spring using special tool (09546-26000).
6. Install compressed coil spring onto shock absorber.

➡There are two color marks on the coil spring. One corresponds to model option, and the other corresponds to load classification. Ensure that the correct parts are being installed.

➡Install the coil spring with the identification mark directed toward the knuckle.

7. After fully extending the piston rod, install the spring upper seat and insulator assembly.

8. After seating the upper and lower ends of the coil spring in the upper and lower spring seat grooves correctly, tighten new self-locking nut temporarily.

9. Remove the special tool (09546-26000).

10. Tighten the self-locking nut to 15–18 ft lbs. (20–25 Nm).

WARNING

Do not reuse the self-locking nut.

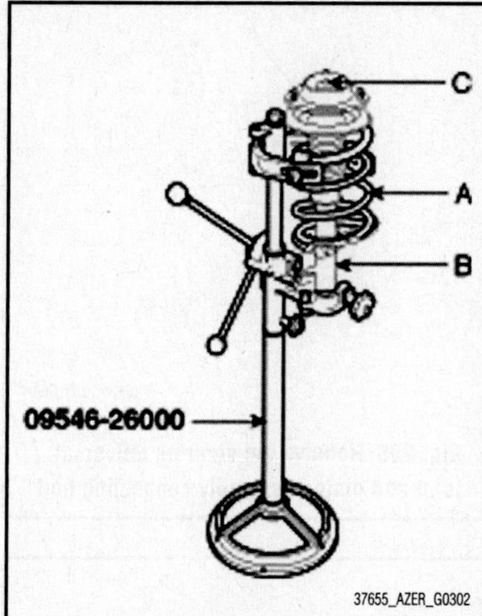

Fig. 200 Using the special tool (09546-26000), compress the coil spring (A), remove the self-locking nut (C) from the strut (B)

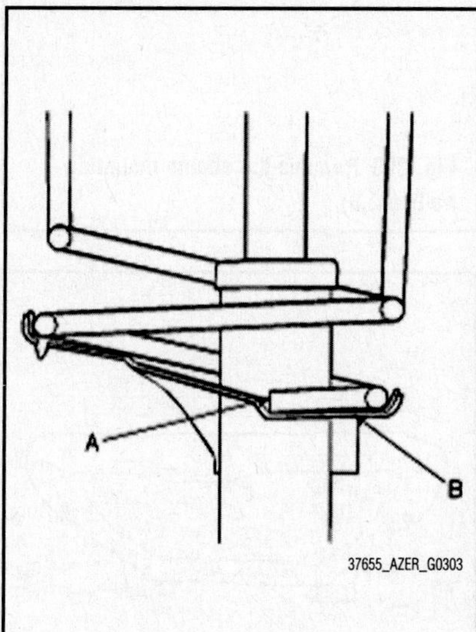

Fig. 201 Seat the upper and lower ends of the coil spring (A) in the upper and lower spring seat grooves (B)

STABILIZER BAR

REMOVAL & INSTALLATION

See Figures 202 through 213.

1. Before servicing the vehicle, refer to the Precautions Section.
2. Raise and safely support the vehicle.
3. Remove the front wheel and tire from the front hub.

WARNING

Be careful not to damage the hub bolts when removing the front wheel and tire.

4. Remove the stabilizer link from the fork.
5. After removing both sides of the tie rod end self-locking nuts, remove the ball joint by using the special tool (09568-4A000).

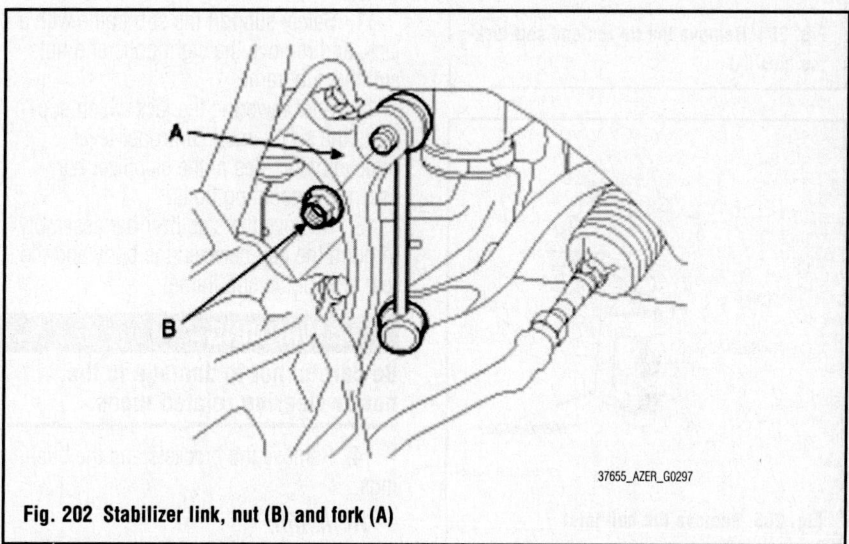

Fig. 202 Stabilizer link, nut (B) and fork (A)

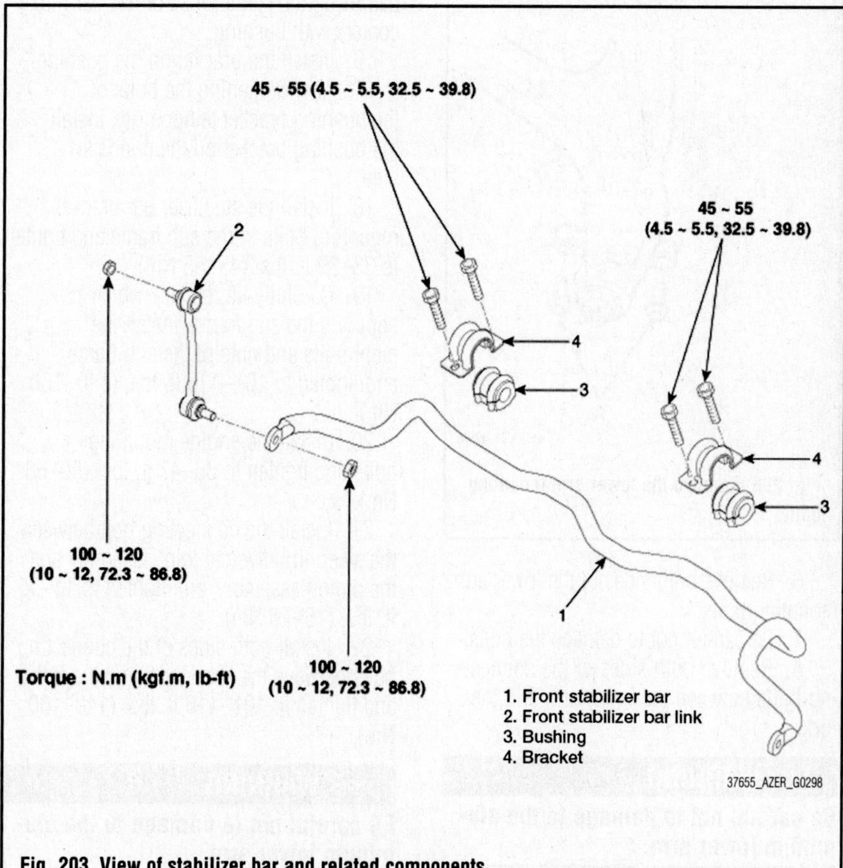

Fig. 203 View of stabilizer bar and related components

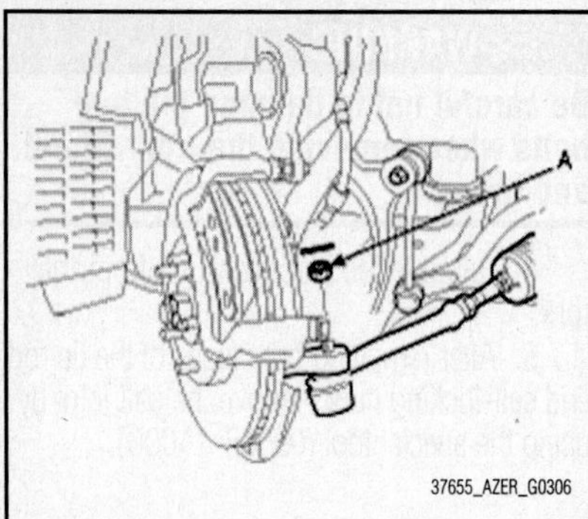

Fig. 204 Remove the tie rod end self-locking nuts (A)

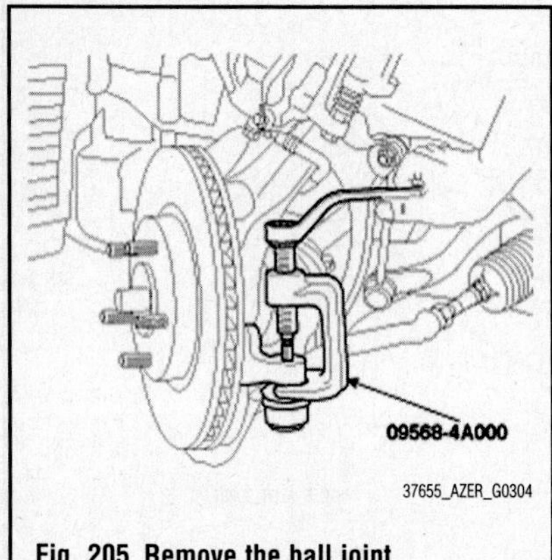

Fig. 205 Remove the ball joint

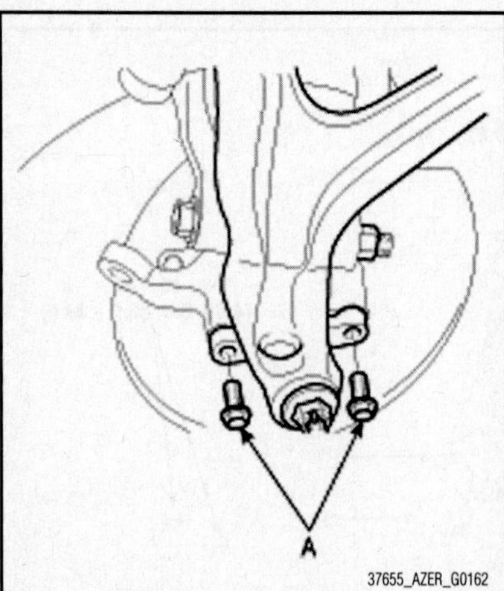

Fig. 206 Remove the lower arm mounting bolts

6. Remove both sides of the lower arm mounting bolts.
7. Be careful not to damage the bolts.
8. Remove both sides of the connecting bolts between the lower arm and the fork.

WARNING

Be careful not to damage to the aluminum lower arm.

9. Remove the connecting bolt between the steering universal joint assembly and the pinion assembly.

WARNING

Keep the neutral-range to prevent the damage of the clock spring inner cable when you handle the steering wheel.

10. Remove the engine mounting bolts.
11. Safely support the sub frame with a jack, and remove the eight bolts and nuts and the sub frame.
12. After lowering the jack which supports the sub frame in a proper level, remove both sides of the stabilizer bar assembly mounting bolts.
13. Remove the stabilizer bar assembly through the gap between the body and the rear side of the sub frame.

WARNING

Be careful not to damage to the power steering related tubes.

14. Remove the brackets and the bushings.

To install:

15. Install the bushing on the stabilizer bar. Bring the clamp of stabilizer bar into contact with bushing.
16. Install the bracket on the bushing.
17. After tightening the bolts of the bushing bracket temporarily, install the bushing bracket on the opposite side.
18. Install the stabilizer bar bracket mounting bolts to the sub frame and tighten to 33–39 ft. lbs. (45–55 Nm).
19. Carefully lift the jack which supports the sub frame, install the eight bolts and nuts of the sub frame, and tighten to 101–116 ft. lbs. (140–160 Nm).
20. Install the engine mounting bolts and tighten to 36–47 ft. lbs. (50–65 Nm).
21. Install the connecting bolt between the steering universal joint assembly and the pinion assembly, and tighten to 11–15 ft. lbs. (15–20 Nm).
22. Install both sides of the connecting bolts between the lower arm and the fork and tighten to 101–116 ft. lbs. (140–160 Nm).

WARNING

Be careful not to damage to the aluminum lower arm.

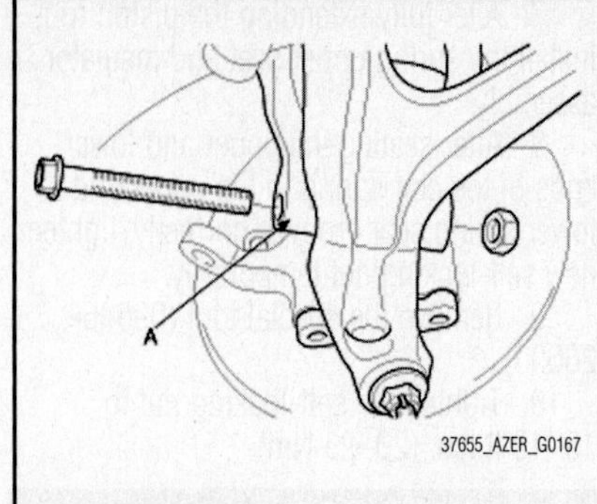

Fig. 207 Remove the connecting bolts between the lower arm and the fork

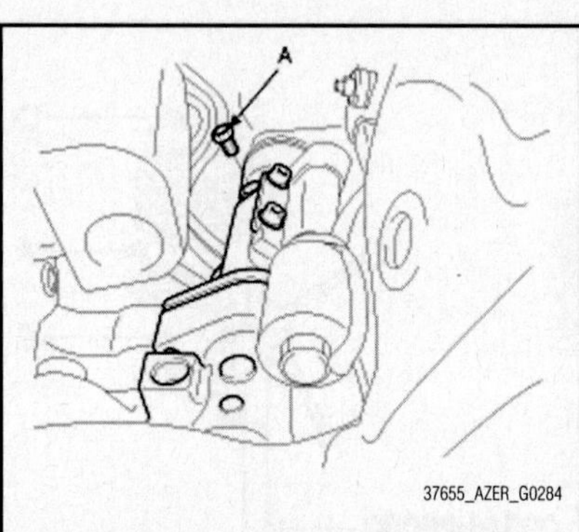

Fig. 208 Remove the steering universal joint and pinion assembly connecting bolt

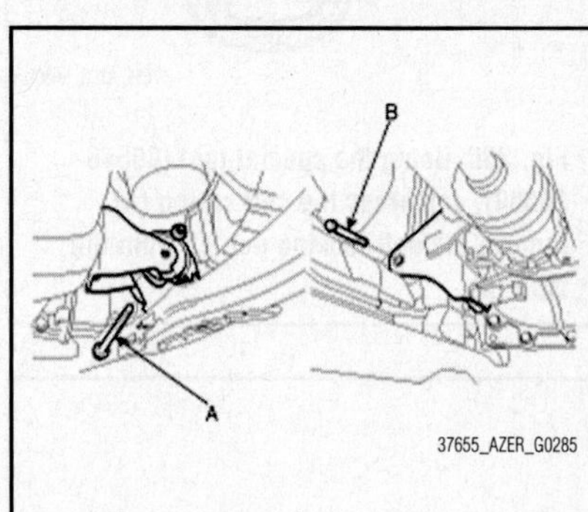

Fig. 209 Remove the engine mounting bolts (A, B)

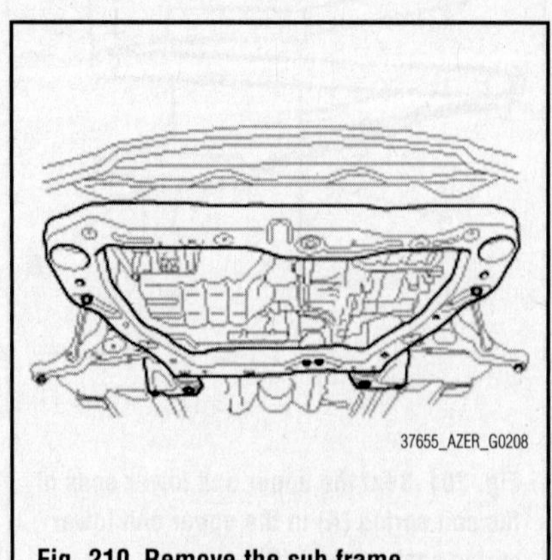

Fig. 210 Remove the sub frame

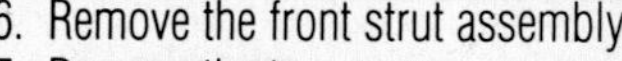

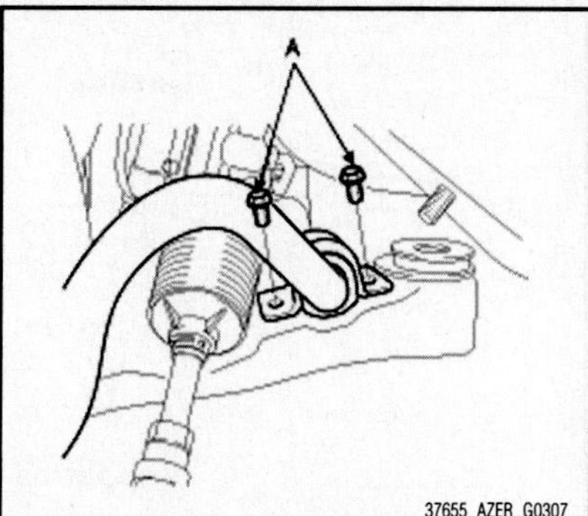

Fig. 211 Remove the stabilizer bar assembly mounting bolts (A)

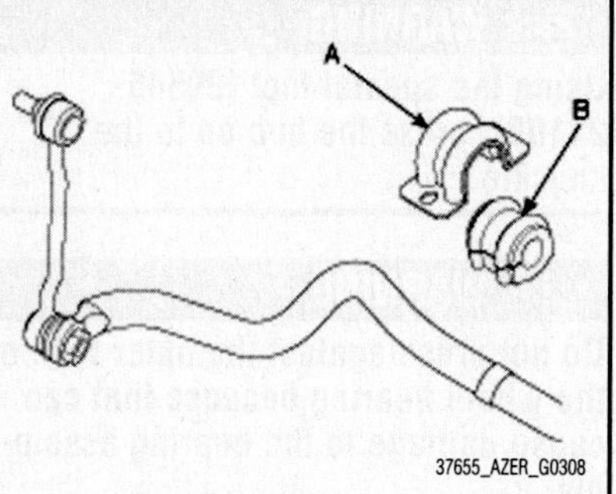

Fig. 212 Remove the brackets (A) and the bushings (B)

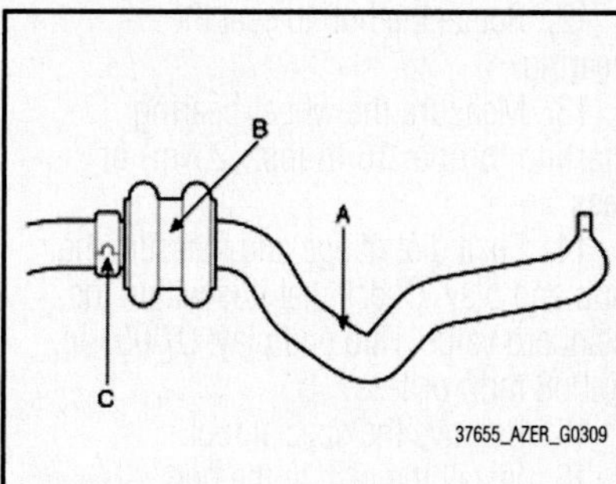

Fig. 213 Install the bushing (B) on the stabilizer bar (A), bringing the clamp (C) into contact with the bushing

23. Install both sides of the lower arm mounting bolts and tighten to 72–87 ft. lbs. (100–120 Nm).
24. Install both sides of the tie rod end self-locking nuts and tighten to 36–40 ft. lbs. (50–55 Nm).
25. Install the stabilizer link to the fork and tighten to 72–87 ft. lbs. (100–120 Nm).
26. Install the wheels and tires.

UPPER BALL JOINT

REMOVAL & INSTALLATION

The upper ball joints are replaced with the upper control arms as an assembly.

UPPER CONTROL ARM

REMOVAL & INSTALLATION

See Figures 214 through 216.

1. Before servicing the vehicle, refer to the Precautions Section.
2. Raise and safely support the vehicle.
3. Remove the front wheel and tire from the front hub.
4. Remove the upper arm ball joint self-locking nut and the snap pin.
5. Using the special tool (09568-4A000), disconnect the upper arm ball joint from the knuckle.

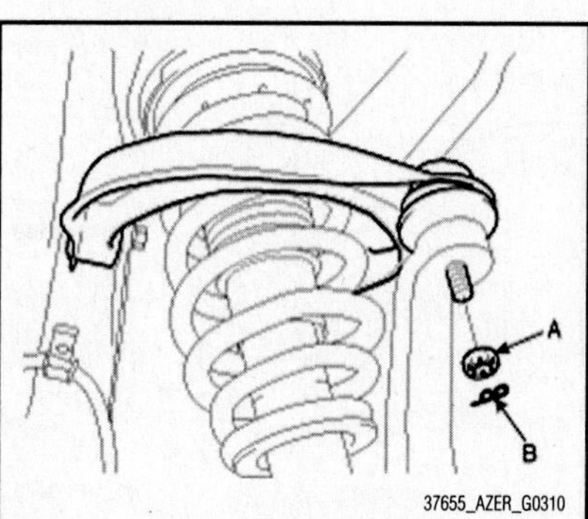

Fig. 214 Remove the upper arm ball joint self-locking nut (A) and the snap pin (B)

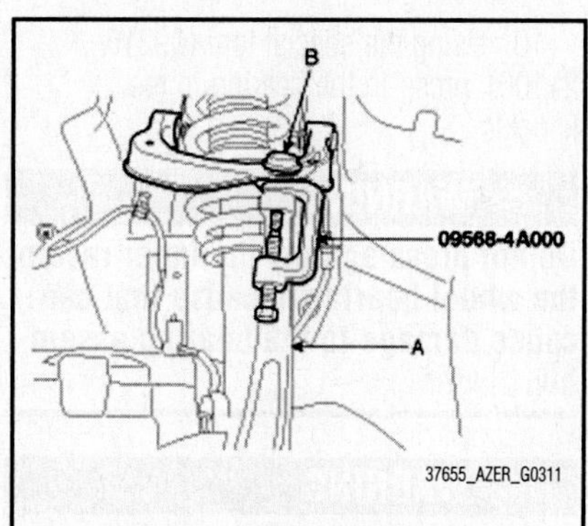
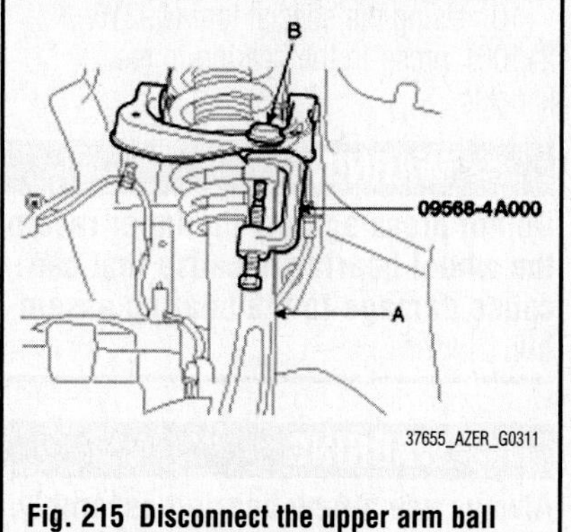

Fig. 215 Disconnect the upper arm ball joint (B) from the knuckle (A)

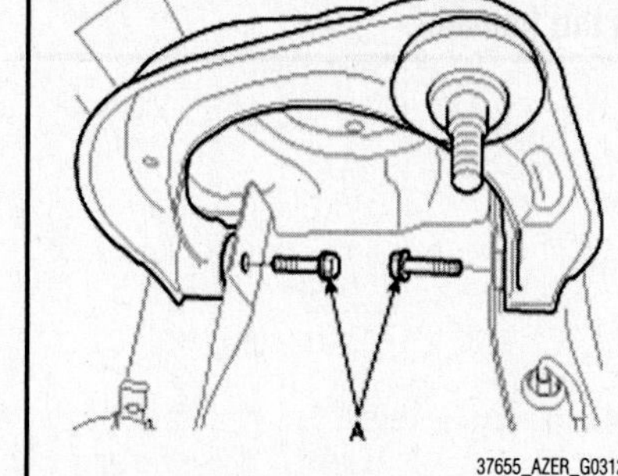

Fig. 216 Remove the two upper arm mounting bolts (A) from the body

6. Remove the front strut assembly.
7. Remove the two upper arm mounting bolts from the body.

To install:

8. Install the two upper arm mounting bolts to the body and tighten to 40–47 ft. lbs. (55–65 Nm).
9. Install the front strut assembly.
10. Install the upper arm ball joint self-locking nut and the snap pin and tighten to 25–33 ft. lbs. (35–45 Nm).
11. Install the wheel and the tire.

WHEEL HUB & BEARING

REMOVAL & INSTALLATION

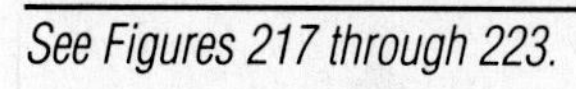

See Figures 217 through 223.

1. Before servicing the vehicle, refer to the Precautions Section.
2. Remove the steering knuckle. Refer to Steering Knuckle Removal & Installation.
3. Remove the brake disc from the front hub.
4. Remove the snap ring.
5. Using the special tool (09545-34100), disconnect the hub from the knuckle.
6. Using the special tools (09432-11000, 09545-34100), remove the wheel bearing inner race from the hub.

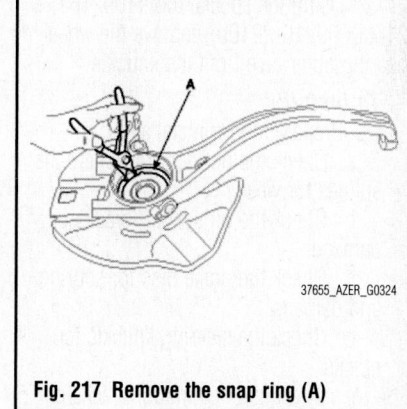

Fig. 217 Remove the snap ring (A)

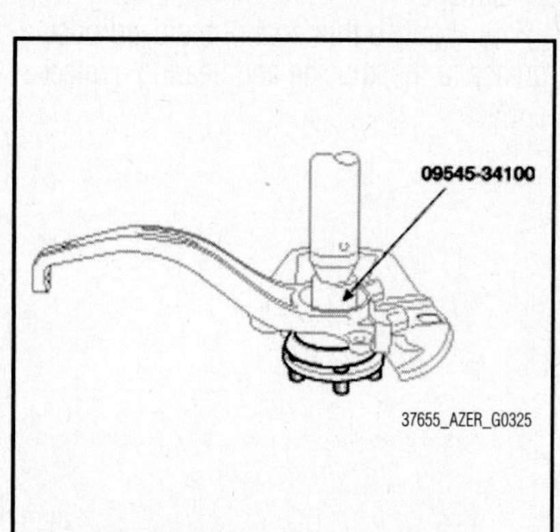

Fig. 218 Disconnect the hub from the knuckle

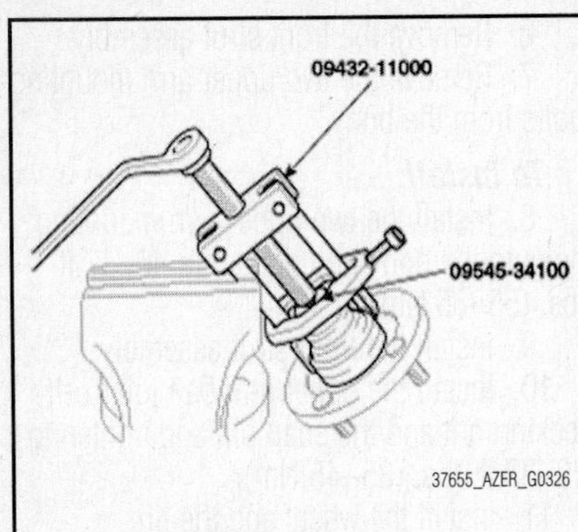

Fig. 219 Remove the wheel bearing inner race from the hub

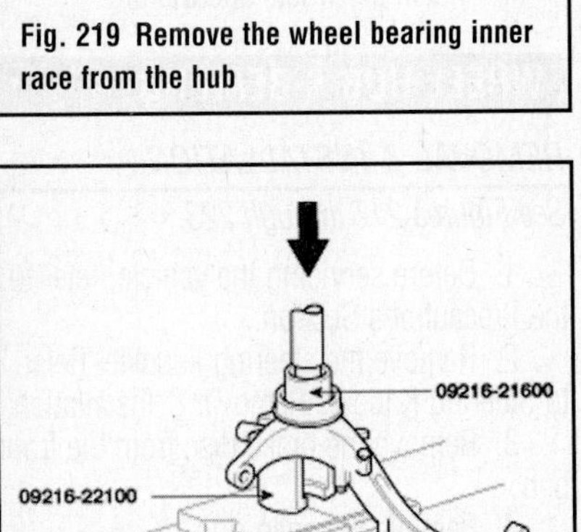

Fig. 220 Remove the wheel bearing outer race from the knuckle

7. Using the special tools (09216-21600, 09216-22100), remove the wheel bearing outer race from the knuckle.

To install:

8. Preparation for installation:
   a. Check the hub for cracks and the splines for wear.
   b. Check the oil seal for cracks or damage.
   c. Check the brake disc for scoring and damage.
   d. Check the steering knuckle for cracks.
   e. Check the bearing for cracks or damage.
9. Apply a thin coat of multi-purpose grease to the knuckle and bearing contact surface.

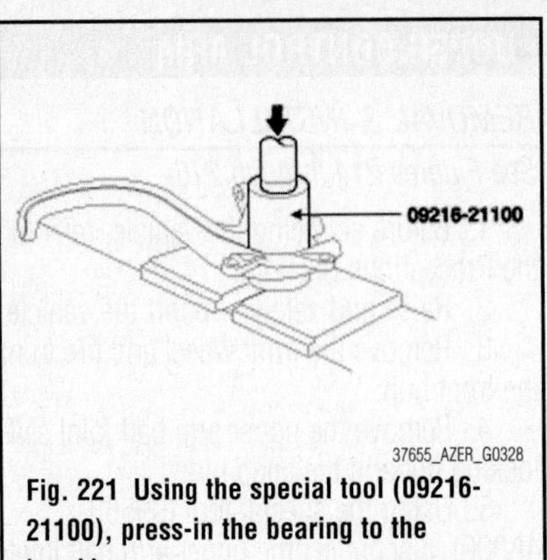

Fig. 221 Using the special tool (09216-21100), press-in the bearing to the knuckle

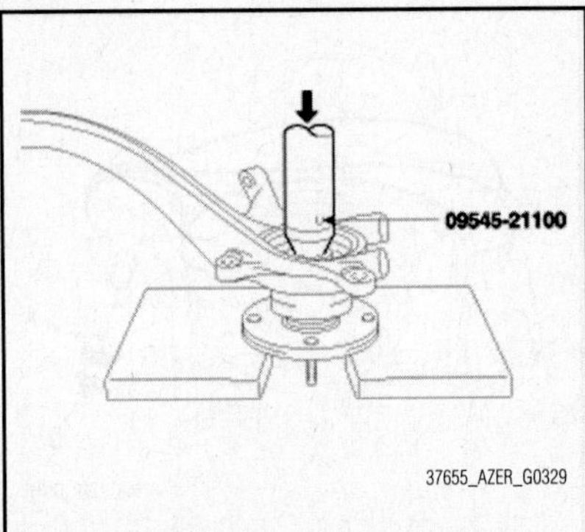

Fig. 222 Using the special tool (09545-21100), press the hub on to the knuckle

10. Using the special tool (09216-21100), press-in the bearing to the knuckle.

WARNING

Do not press against the inner race of the wheel bearing because that can cause damage to the bearing assembly.

WARNING

Always use a new bearing assembly.

WARNING

Install the snap ring into the groove of the knuckle.

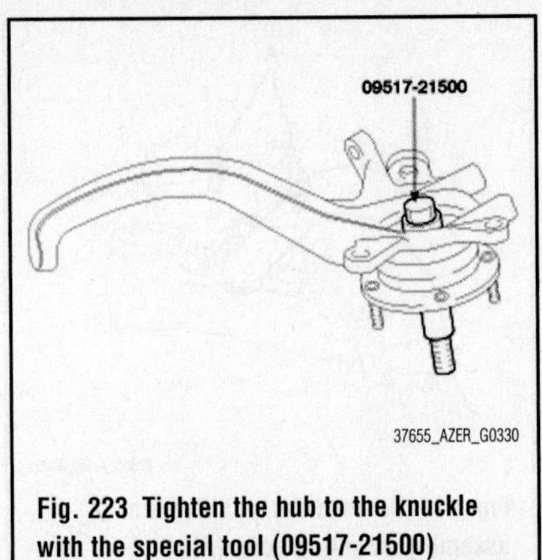

Fig. 223 Tighten the hub to the knuckle with the special tool (09517-21500)

WARNING

Using the special tool (09545-21100), press the hub on to the knuckle.

WARNING

Do not press against the outer race of the wheel bearing because that can cause damage to the bearing assembly.

11. Tighten the hub to the knuckle to 148 ft. lbs. (200 Nm) with the special tool (09517-21500).
12. Rotate the hub to seat the bearing.
13. Measure the wheel bearing starting torque: 16 ft. lbs. (2 Nm) or less.
14. Fix a dial gauge and measure the hub end play. Check that it is within the standard value. Hub end play: 0.0003 in. (0.008 mm) or less
15. Remove the special tool.
16. Install the disc to the hub.
17. Install the steering knuckle.
18. Check and/or adjust the wheel alignment.

ADJUSTMENT

The front wheel bearing is a sealed unit and is not adjustable.

COIL SPRING

REMOVAL & INSTALLATION

Refer to Lower Control Arm Removal and Installation

CONTROL ARMS

REMOVAL & INSTALLATION

Rear Assist Arm

See Figure 224.

1. Before servicing the vehicle, refer to the Precautions Section.
2. Raise and safely support the vehicle.
3. Remove the rear wheel and tire.

CAUTION

Be careful not to damage the hub bolts when removing the rear wheel and tire.

4. Remove the assist arm mounting bolt from the rear knuckle.
5. Remove the assist arm mounting bolt from the cross member.

To install:

6. Install the assist arm mounting bolt to the cross member and tighten to 80–87 ft. lbs. (110–120 Nm)
7. Install the assist arm mounting bolt from the rear knuckle and tighten to 101–116 ft. lbs. (140–160 Nm).
8. Install the wheel and the tire to the rear hub.

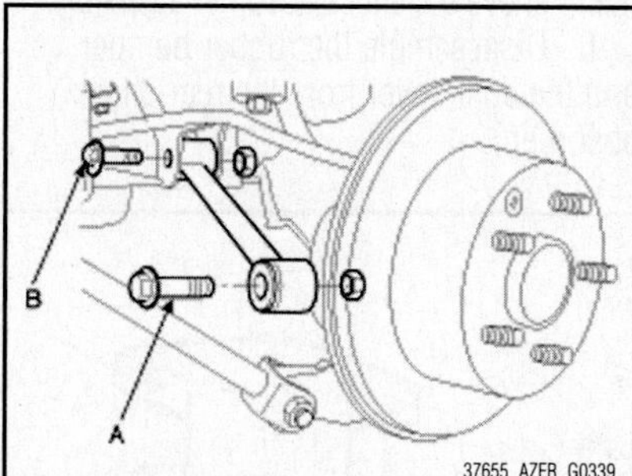

Fig. 224 Remove the assist arm mounting bolts (A, B) from the rear knuckle and the cross member

Lower Control Arm

See Figure 225.

1. Before servicing the vehicle, refer to the Precautions Section.
2. Raise and safely support the vehicle.
3. Remove the rear wheel and tire.

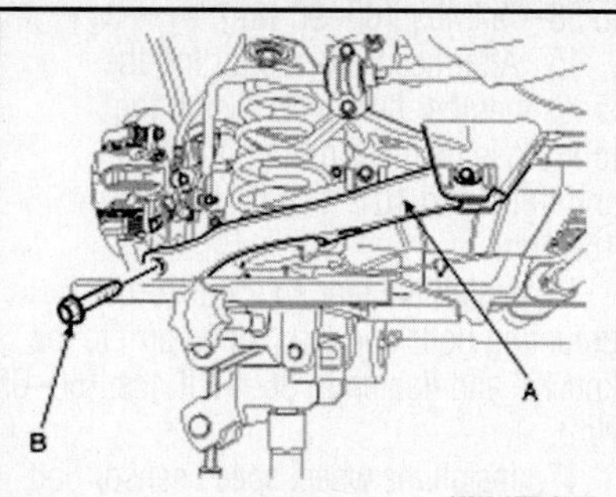

Fig. 225 Remove the lower arm bolt (A) from the rear knuckle, while supporting the lower arm (B)

4. Be careful not to damage the hub bolts when removing the rear wheel and tire.
5. Remove the lower arm bolt from the rear knuckle, while supporting the lower arm with a jack.
6. Remove the spring, the lower seat, and the upper pad.
7. Remove the lower arm mounting bolts from the cross member.

To install:

8. Install the lower arm mounting bolts to the cross member and tighten to 80–87 ft. lbs. (110–120 Nm).
9. Install the spring, the lower seat, and the upper pad.
10. Install the lower arm bolt to the rear knuckle and tighten to 101–116 ft. lbs. (140–160 Nm), while supporting the lower arm with a jack.
11. Install the wheel and the tire to the rear hub.

Upper Control Arm

See Figures 226 through 231.

1. Before servicing the vehicle, refer to the Precautions Section.
2. Raise and safely support the vehicle.
3. Remove the rear wheel and tire.
4. Be careful not to damage the hub bolts when removing the rear wheel and tire.
5. Remove the wheel speed sensor bolt and the parking brake cable.
6. Remove the brake caliper mounting bolts, and then place the brake caliper assembly with wire.
7. Remove the rear shock absorber mounting bolts and the trailing arm from the knuckle.
8. Remove the four rear muffler mounting rubbers and the connecting bolts.

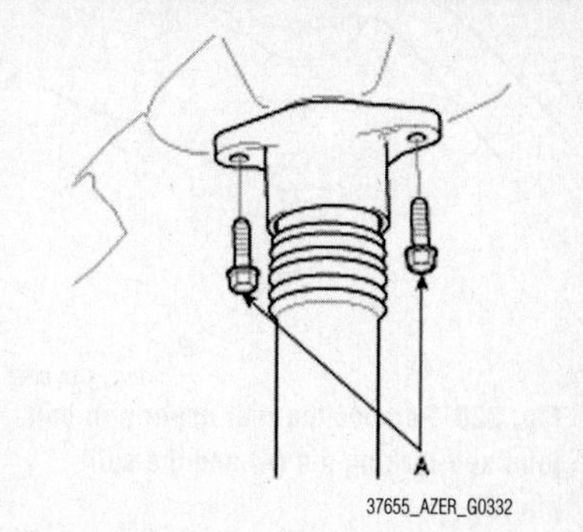

Fig. 226 Remove the two rear shock absorber assembly mounting bolts (A)

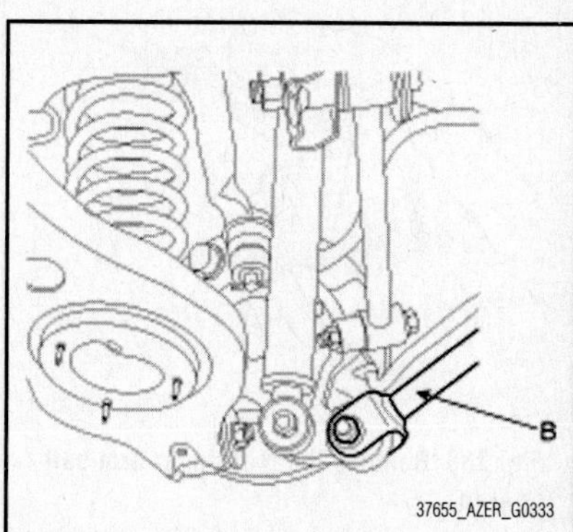

Fig. 227 Remove the trailing arm from the knuckle (B)

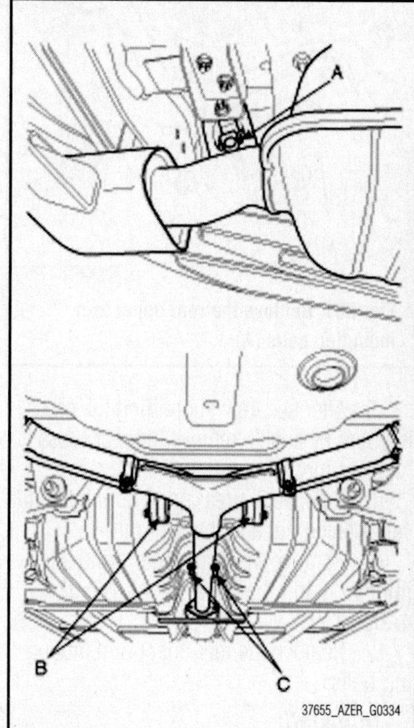

Fig. 228 Remove the four rear muffler mounting rubbers (A, B) and the connecting bolts (C)

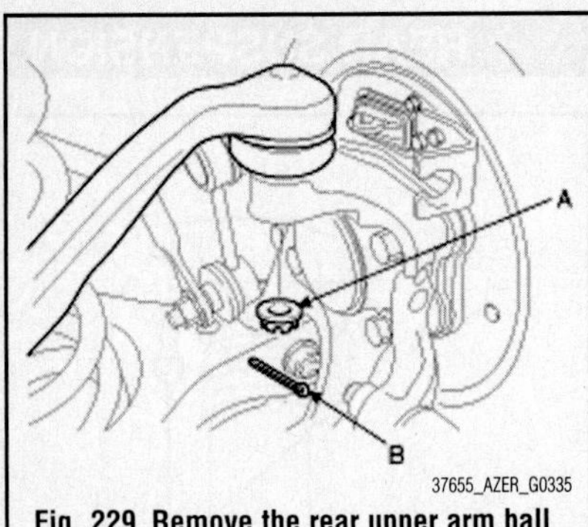

Fig. 229 Remove the rear upper arm ball joint self-locking nut (A) and the split pin (B)

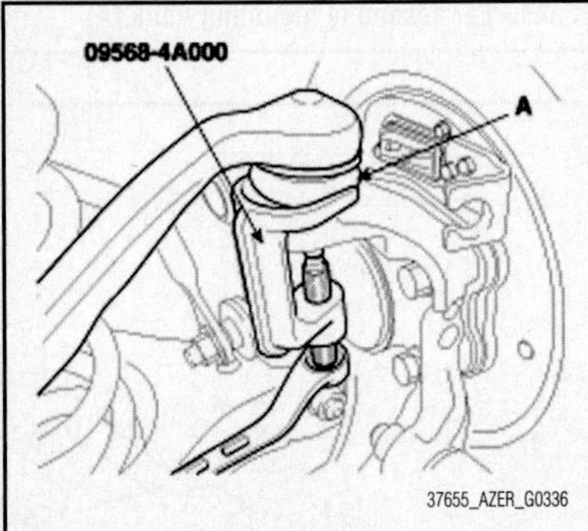

Fig. 230 Remove the rear upper arm ball joint (A)

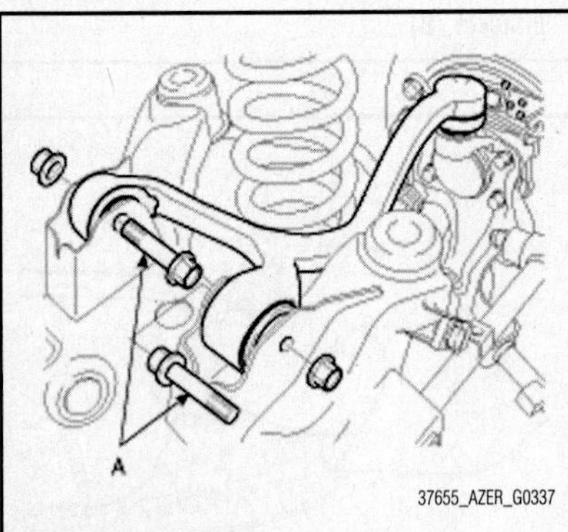

Fig. 231 Remove the rear upper arm mounting bolts (A)

9. After securely supporting the cross member by a jack, remove the four cross member mounting bolts.

10. Remove the rear upper arm ball joint self-locking nut and the split pin.

11. Remove the rear upper arm ball joint by using the special tool (09568-4A000).

12. Remove the rear upper arm mounting bolts.

To install:

13. Install the rear upper arm mounting bolts and tighten to 72–87 ft. lbs. (100–120 Nm).

14. Install the rear upper arm ball joint self-locking nut and the split pin and tighten to 58–65 ft lbs. (80–90 Nm).

15. After securely supporting the cross member by a jack, Install the four cross member mounting bolts and tighten to 101–116 ft. lbs. (140–160 Nm).

16. Install the rear shock absorber mounting bolts and the trailing arm to the knuckle and tighten to 36–47 ft. lbs. (50–65 Nm).

17. Install the wheel speed sensor bolt and the parking brake cable.

18. Install the brake caliper mounting bolts and tighten to 58–72 ft. lbs. (80–100 Nm).

19. Install the four rear muffler mounting rubbers and the connecting bolts and tighten to 29–43 ft. lbs. (40–60 Nm).

20. Install the wheel and the tire to the rear hub.

Trailing Arm

See Figures 232 and 233.

1. Before servicing the vehicle, refer to the Precautions Section.

2. Raise and safely support the vehicle.

3. Remove the rear wheel and tire.

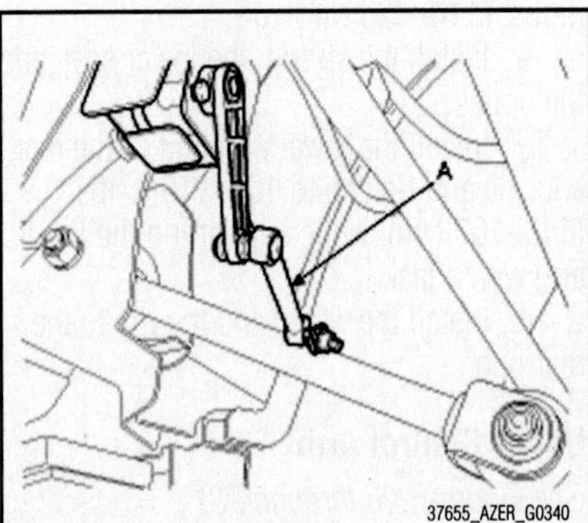

Fig. 232 Remove the head lamp leveling link (A)

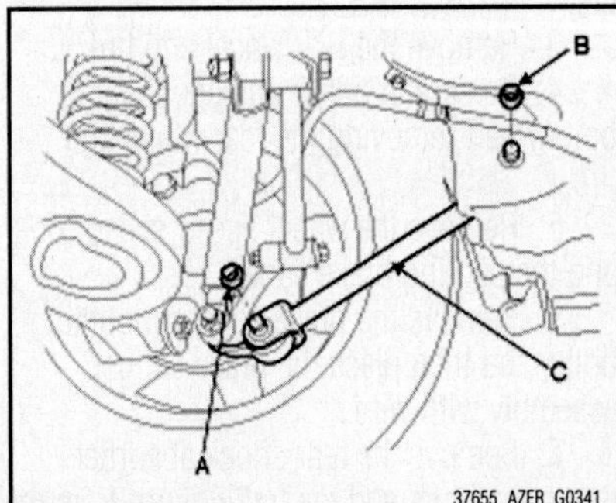

Fig. 233 Remove the trailing arm mounting nuts (A, B) from the knuckle and the body and remove the trailing arm (C)

4. Be careful not to damage the hub bolts when removing the rear wheel and tire.

5. Remove the trailing arm mounting nut from the rear knuckle.

6. Remove the head lamp leveling link, and then remove the right trailing arm.

7. Remove the trailing arm mounting nut from the body.

8. Remove the trailing arm.

To install:

9. Replace the trailing arm.

10. Install the trailing arm nuts:

➡Fully tighten the trailing arm mounting nuts with the vehicle on the ground in unloaded condition.

a. Install the trailing arm mounting nut and tighten to 101–116 ft. lbs. (140–160 Nm).

b. Install the trailing arm bracket mounting nut and tighten to 101–116 ft. lbs. (140–160 Nm).

11. Install the right trailing arm, and then install the head lamp leveling link.

SHOCK ABSORBER

REMOVAL & INSTALLATION

See Figures 234 and 235.

1. Before servicing the vehicle, refer to the Precautions Section.

2. Support the vehicle under the lower control arm.

3. Remove the rear wheel.

4. Remove the two rear shock absorber assembly mounting bolts.

5. Remove the rear shock absorber assembly nut from the rear knuckle, then remove the shock absorber assembly.

6. Disassemble the rubber bumper and the dust cover from the rear shock absorber.

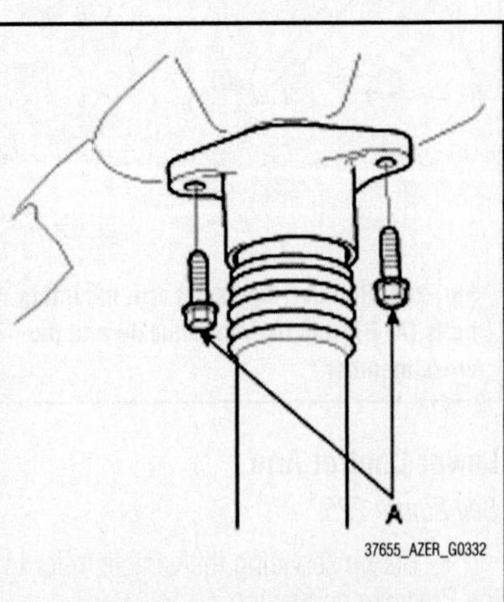

Fig. 234 Remove the two rear shock absorber assembly mounting bolts (A)

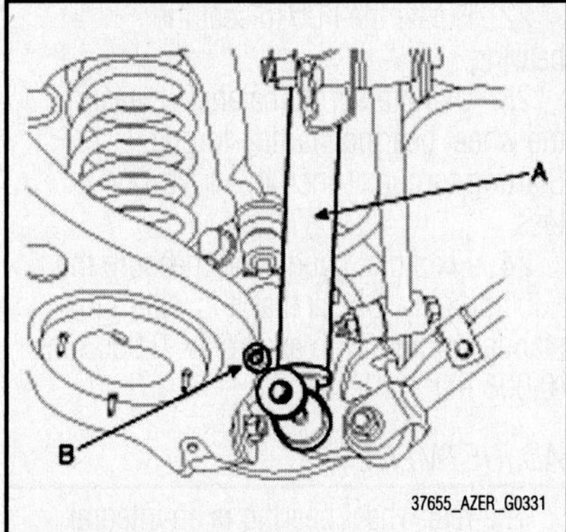

Fig. 235 Remove the rear shock absorber assembly nut (B) from the rear knuckle, then remove the shock absorber assembly (A)

To install:

7. Install the shock absorber. Tighten the upper bolts to 36–47 ft. lbs. (50–65 Nm). Tighten the lower nut to 101–116 ft. lbs. (140–160 Nm)
8. Install the rear wheel.

TESTING

1. Check the rubber parts for damage or deterioration.
2. Check for correct height and proper return of shock absorber to original height.
3. Check the shock absorber for abnormal resistance or unusual sounds.
4. Check for oil leakage around seals.
5. Replace if necessary.

WHEEL HUB & BEARING

REMOVAL & INSTALLATION

See Figures 236 through 242.

1. Before servicing the vehicle, refer to the Precautions Section.
2. Raise and safely support the vehicle.

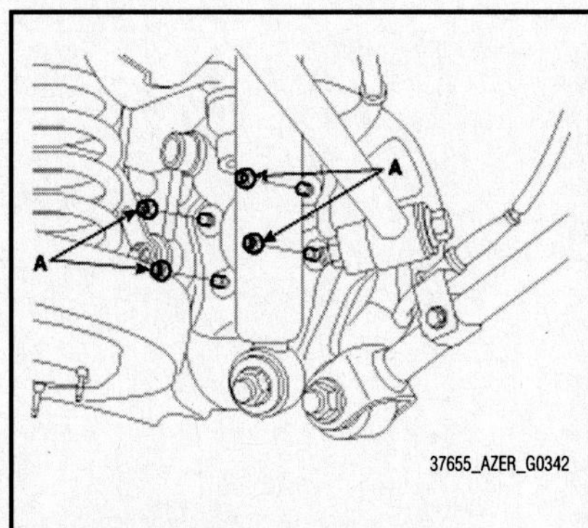

Fig. 236 Remove the rear axle hub mounting bolts (A)

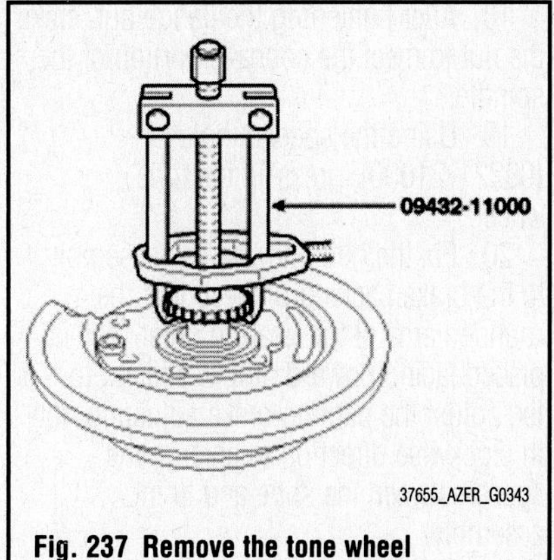

Fig. 237 Remove the tone wheel

Fig. 238 After unstaking the flange nut, remove the nut

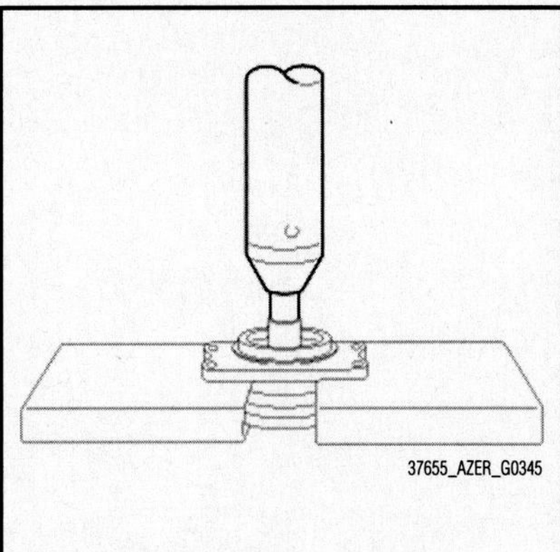

Fig. 239 Press out the rear axle hub

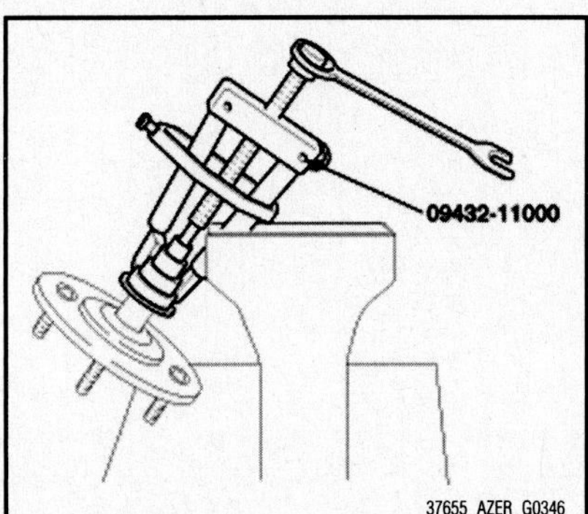

Fig. 240 Remove the bearing inner race from the axle hub

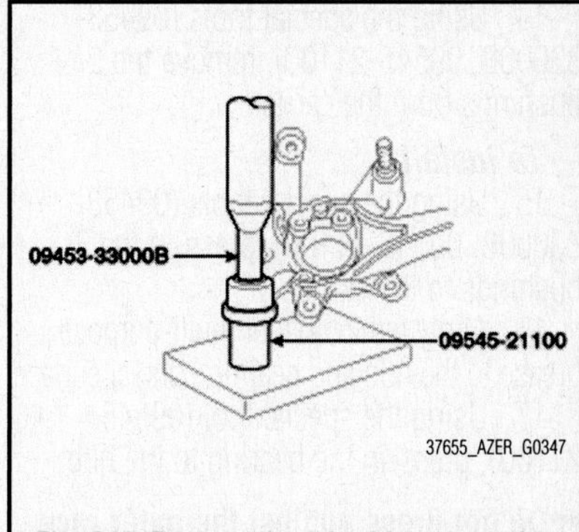

Fig. 241 Remove the 2 bushings from the carrier

3. Remove the rear wheel and tire.
4. Be careful not to damage the hub bolts when removing the rear wheel and tire.
5. Remove the wheel speed sensor from the carrier.
6. Remove the bolts, and then remove the caliper assembly from the carrier assembly and suspend it with wire.
7. Remove the brake disc.
8. Remove the rear axle hub mounting bolts.
9. Using the special tool (09432-11000), remove the tone wheel.
10. Remove the carrier assembly.
11. After unstaking the flange nut, remove the nut.
12. While supporting the flange area of the bearing outer race, press out the rear axle hub.
13. Using the special tool (09432-11000), remove the bearing inner race from the axle hub.

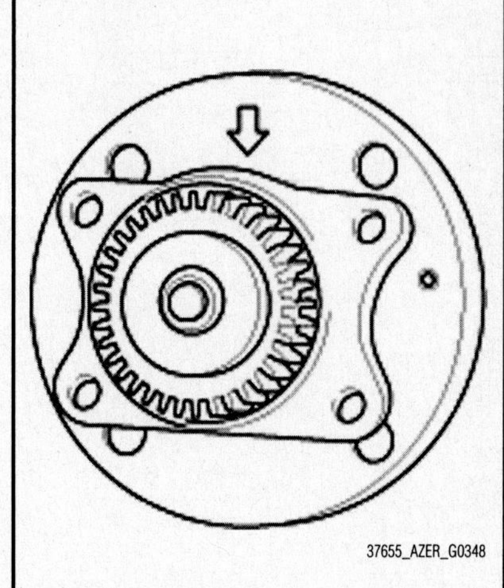

Fig. 242 Fix the hub and bearing assembly to the brake backing plate so that the rounded area of the bearing outer race is placed facing upward

14. Using the special tools (09453-33000B, 09545-21100), remove the 2 bushings from the carrier.

To install:

15. Using the special tools (09453-33000B, 09545-21100), press-in the 2 bushings to the carrier.
16. Apply a thin coat of multi-purpose grease to the hub and bearing contact surface.
17. Using the special tool (09545-21100), press-in the bearing to the hub.

➡Do not press against the outer race of the bearing because that can cause damage to the bearing assembly. Always use a new bearing assembly.

18. After tightening the flange nut, stake the nut to meet the concave portion of the spindle.
19. Using the special tool (09221-21000), press-in the tone wheel.
20. Fix the hub and bearing assembly to the brake backing plate so that the rounded area of the bearing outer race is placed facing upward. If it is difficult to fix, adjust the parking brake adjusting nut in clockwise direction to enlarge the space between the shoe and lining assembly.
21. Tighten the 4 bolts to 44–52 ft. lbs. (60–70 Nm).
22. Rotate the hub to seat the bearing.
23. Using a spring balance, measure the wheel bearing starting torque. Starting torque: 1 inch lb. (1 Nm) or less.
24. Fix a dial gauge and measure the hub end play. Check that it is within the standard value. Hub axial play: 0.0003 in, (0.008 mm) or less

ADJUSTMENT

The rear wheel bearing is an integral part of the rear hub. No adjustment is possible.

HYUNDAI

Elantra

3

SPECIFICATIONS AND MAINTENANCE CHARTS

ENGINE AND VEHICLE IDENTIFICATION

Engine							Model Year	
Code ①	Liters (cc)	Cu. In.	Cyl.	Fuel Sys.	Engine Type	Eng. Mfg.	Code ②	Year
③	2.0 (1975)	120.52	I4	MPFI	DOHC	Hyundai	9	2009
							A	2010

MPFI: Multi-Point Fuel Injection

DOHC: Double Overhead Camshafts

① 8th digit of VIN

② 10th digit of VIN

③ Elantra Sedan code D, Elantra Touring code E

37655_ELAN_C0001

GENERAL ENGINE SPECIFICATIONS

All measurements are given in inches.

Year	Model	Engine Displacement Liters	Engine Series VIN	Net Horsepower @ rpm	Net Torque @ rpm (ft. lbs.)	Bore x Stroke (in.)	Com-pression Ratio	Oil Pressure @ rpm
2009	Elantra	2.0	D/E	138@6000	136@4500	3.228 x 3.681	10.1:1	35.5@1500
2010	Elantra	2.0	D/E	138@6000	136@4500	3.228 x 3.681	10.1:1	35.5@1500

37655_ELAN_C0002

GASOLINE ENGINE TUNE-UP SPECIFICATIONS

Year	Engine Displacement Liters	Engine VIN	Spark Plug Gap (in.)	Ignition Timing (deg.)		Fuel Pump (psi)	Idle Speed (rpm)		Valve Clearance	
				MT	AT		MT	AT	In.	Ex.
2009	2.0	D/E	0.039-0.043	①	①	49.0-50.5	②	②	HYD	HYD
2010	2.0	D/E	0.039-0.043	①	①	49.0-50.5	②	②	HYD	HYD

NOTE: The Vehicle Emission Control Information label reflects specification changes made during production. Follow the figures on the label if they differ from those in this chart.

HYD: Hydraulic

① Ignition timing is preset and cannot be adjusted

② Idle speed is maintained by the Electronic Control Module (ECM)

37655_ELAN_C0003

CAPACITIES

Year	Model	Engine Displacement Liters	Engine VIN	Engine Oil with Filter (qts.)	Transmission (pts.) Manual	Transmission (pts.) Auto. ①	Fuel Tank (gal.)	Cooling System (qts.)
2009	Elantra	2.0	D/E	4.0	4.0	13.8	14.0	6.9-7.0
2010	Elantra	2.0	D/E	4.0	4.0	13.8	14.0	6.9-7.0

NOTE: All capacities are approximate. Add fluid gradually and check to be sure a proper fluid level is obtained.

① Drain and refill

37655_ELAN_C0004

FLUID SPECIFICATIONS

Year	Model	Engine Displacement Liters	Engine ID/VIN	Engine Oil	Auto. Trans.	Manual Trans.	Power Steering Fluid	Brake Master Cylinder
2009	Elantra	2.0	D/E	5W-20	①	②	NA	③
2010	Elantra	2.0	D/E	5W-20	①	②	NA	③

NA: Does not apply

DOT: Department Of Transportation

① DIAMOND ATF SP-III, SK ATF SP-III

② GENUINE PART MTF 75W/85 (API GL - 4)

③ DOT 3, DOT 4, or equivalent

37655_ELAN_C0005

VALVE SPECIFICATIONS

Year	Engine Displacement Liters	Engine VIN	Seat Angle (deg.)	Face Angle (deg.)	Spring Test Pressure (lbs. @ in.)	Spring Installed Height (in.)	Stem-to-Guide Clearance (in.) Intake	Stem-to-Guide Clearance (in.) Exhaust	Stem Diameter (in.) Intake	Stem Diameter (in.) Exhaust
2009	2.0	D/E	45	NS	87.1-93.7 @1.201	NS	0.0008-0.0019	0.0014-0.0026	0.2348-0.2354	0.2343-0.2348
2010	2.0	D/E	45	NS	87.1-93.7 @1.201	NS	0.0008-0.0019	0.0014-0.0026	0.2348-0.2354	0.2343-0.2348

NA: Not Supplied

37655_ELAN_C0006

CAMSHAFT AND BEARING SPECIFICATIONS CHART

All measurements are given in inches.

Year	Engine Displ. Liters	Engine ID/VIN	Journal Dia.	Brg. Oil Clearance	Shaft End-play	Runout	Journal Bore	Lobe Height Intake	Lobe Height Exhaust
2009	2.0	D/E	1.1023	0.0008-0.0024	0.0039-0.0059	NS	NS	1.7527-1.7605	1.7487-1.7566
2010	2.0	D/E	1.1023	0.0008-0.0024	0.0039-0.0059	NS	NS	1.7527-1.7605	1.7487-1.7566

NA: Not Supplied

37655_ELAN_C0007

CRANKSHAFT AND CONNECTING ROD SPECIFICATIONS

All measurements are given in inches.

	Engine		Crankshaft				Connecting Rod		
Year	Displacement Liters	Engine VIN	Main Brg. Journal Dia.	Main Brg. Oil Clearance	Shaft End-play	Thrust on No.	Journal Diameter	Oil Clearance	Side Clearance
2009	2.0	D/E	2.2418-2.2426	0.0011-0.0018	0.0023-0.0100	3	1.7695-1.7703	0.0009-0.0017	0.0039-0.0100
2010	2.0	D/E	2.2418-2.2426	0.0011-0.0018	0.0023-0.0100	3	1.7695-1.7703	0.0009-0.0017	0.0039-0.0100

37655_ELAN_C0008

PISTON AND RING SPECIFICATIONS

All measurements are given in inches.

	Engine			Ring Gap			Ring Side Clearance		
Year	Displ. Liters	Engine VIN	Piston Clearance	Top Compression	Bottom Compression	Oil Control	Top Compression	Bottom Compression	Oil Control
2009	2.0	D/E	0.0008-0.0016	0.0079-0.0138	0.0146-0.0205	0.0078-0.0236	0.0015-0.0031	0.0012-0.0027	NS
2010	2.0	D/E	0.0008-0.0016	0.0090-0.0149	0.0177-0.0236	0.0078-0.0236	0.0015-0.0031	0.0012-0.0027	NS

NA: Not Supplied

37655_ELAN_C0009

TORQUE SPECIFICATIONS

All readings in ft. lbs.

Year	Engine Displacement Liters	Engine VIN	Cylinder Head Bolts	Main Bearing Bolts	Rod Bearing Bolts	Crankshaft Damper Bolts	Flywheel Bolts	Manifold Intake	Manifold Exhaust	Spark Plugs	Oil Pan Drain Plug
2009	2.0	D/E	①	②	36-38	116-123	87-94	13-17	31-40	15-21	29-33
2010	2.0	D/E	①	②	36-38	116-123	87-94	13-17	31-40	15-21	29-33

① M10: 17-20 ft. lbs. Plus 60-65 degrees
M12: 20-33 ft. lbs. Plus 60-65 degrees

② Tighten: 20-23 ft. lbs. Plus 60-64 degrees

37655_ELAN_C0010

WHEEL ALIGNMENT

Year	Model		Caster Range (+/-Deg.)	Caster Preferred Setting (Deg.)	Camber Range (+/-Deg.)	Camber Preferred Setting (Deg.)	Toe-in (Deg.)
2009	Elantra	Front	0.50	+4.32	0.50	-0.60	0 +/- 0.20
		Rear	—	—	0.50	-1.00	0.22 +/- 0.20
2010	Elantra	Front	0.50	+4.32	0.50	-0.60	0 +/- 0.20
		Rear	—	—	0.50	-1.00	0.22 +/- 0.20

37655_ELAN_C0011

TIRE, WHEEL AND BALL JOINT SPECIFICATIONS

Year	Model	OEM Tires		Tire Pressures (psi)		Wheel	Ball Joint	Lug Nut
		Standard	Optional	Front	Rear	Size	Inspection	Torque (ft. lbs.)
2009	Elantra	P195/65R15	P205/55R16	①	①	6.0J x 15 6.5J x 16/17	②	65-80
2010	Elantra	P195/65R15	P205/55R16	①	①	6.0J x 15 6.5J x 16/17	②	65-80

① Refer to placard on vehicle for proper inflation pressure.
② Replace if any measurable movement is found.

37655_ELAN_C0012

BRAKE SPECIFICATIONS

All measurements in inches unless noted

Year	Model		Brake Disc			Brake Drum Diameter			Minimum	Brake Caliper	
			Original Thickness	Minimum Thickness	Maximum Runout	Original Inside Diameter	Max. Wear Limit	Maximum Machine Diameter	Lining Thickness	Bracket Bolts (ft. lbs.)	Mounting Bolts (ft. lbs.)
2009	Elantra	F	1.020	0.940	0.002	—	—	—	0.079	58-72	16-23
		R	0.390	0.330	0.002	—	—	—	0.079	36-43	16-23
	Elantra	F	1.020	0.940	0.002	—	—	—	0.079	58-72	16-23
		R	—	—	—	8.000	①	①	0.039	—	—
2010	Elantra	F	1.020	0.940	0.002	—	—	—	0.079	58-72	16-23
		R	0.390	0.330	0.002	—	—	—	0.079	36-43	16-23
	Elantra	F	1.020	0.940	0.002	—	—	—	0.079	58-72	16-23
		R	—	—	—	8.000	①	①	0.039	—	—

① Drum roundness Service Limit: 0.00236 inch

37655_ELAN_C0013

SCHEDULED MAINTENANCE INTERVALS

HYUNDAI—Elantra Sedan, Elantra Touring

TO BE SERVICED	TYPE OF SERVICE	VEHICLE MILEAGE INTERVAL (x1000)												
		7.5	15	22.5	30	37.5	45	52.5	60	67.5	75	82.5	90	97.5
Engine oil & filter	R	✓	✓	✓	✓	✓	✓	✓	✓	✓	✓	✓	✓	✓
Brake pads, calipers & rotors	S/I	✓	✓	✓	✓	✓	✓	✓	✓	✓	✓	✓	✓	✓
Driveshafts & boots	S/I	✓	✓	✓	✓	✓	✓	✓	✓	✓	✓	✓	✓	✓
Air cleaner filter ①	S/I	✓	✓	✓	✓	✓	✓	✓	✓	✓	✓	✓	✓	✓
A/C refrigerant	S/I	✓	✓	✓	✓	✓	✓	✓	✓	✓	✓	✓	✓	✓
Brake fluid	S/I	✓	✓	✓	✓	✓	✓	✓	✓	✓	✓	✓	✓	✓
Engine coolant ②	R								✓				✓	
Fuel hose, vapor hose & fuel filter cap	S/I				✓				✓				✓	
Spark plugs: Platinum (Iridium) 100,000 mile replacement	R								✓					
Suspension mounting bolts and nuts, upper and lower ball joint.	S/I	✓	✓	✓	✓	✓	✓	✓	✓	✓	✓	✓	✓	✓
Drive belts	S/I	✓	✓	✓	✓	✓	✓	✓	✓	✓	✓	✓	✓	✓
Exhaust pipe connections, muffler & suspension bolts	S/I	✓	✓	✓	✓	✓	✓	✓	✓	✓	✓	✓	✓	✓
Valve clearance ③	S/I								✓					
Electronic throttle control	S/I		✓		✓		✓		✓		✓		✓	
Brake hoses & lines	S/I	✓	✓	✓	✓	✓	✓	✓	✓	✓	✓	✓	✓	✓
Rear brake discs, drums, linings & parking brake	S/I	✓	✓	✓	✓	✓	✓	✓	✓	✓	✓	✓	✓	✓
Steering gear rack, linkage & boots	S/I	✓	✓	✓	✓	✓	✓	✓	✓	✓	✓	✓	✓	✓
Fuel tank air filter ④	S/I		✓		✓		✓		✓		✓		✓	
Climate control air filter	R	✓	✓	✓	✓	✓	✓	✓	✓	✓	✓	✓	✓	✓
Fuel filter	S/I				✓				✓				✓	
Fuel lines & connections	S/I				✓				✓				✓	
Vacuum hose	S/I	✓	✓	✓	✓	✓	✓	✓	✓	✓	✓	✓	✓	✓
Timing belt ⑤	S/I								✓					
Timing belt tensioner	S/I								✓				✓	
Manual transaxle fluid ⑥	S/I													
Automatic transaxle fluid ⑥	S/I													
Crankcase ventilation hoses	S/I				✓				✓				✓	

R: Replace S/I: Service or Inspect

① Replace every 30,000 miles

② Replace every 24 months or 30,000 miles after 60,000 miles

③ Adjust every 60,000 miles

④ Replace every 30,000 miles

⑤ Replace at 90,000 miles

⑥ Inspect every 40,000 miles.

FREQUENT OPERATION MAINTENANCE (SEVERE SERVICE)

If a vehicle is operated under any of the following conditions it is considered severe service:

- **Repeatedly driving short distance of less than 5miles (8km) in normal temperature or less than 10miles (16km) in freezing temperature**
- **Extensive engine idling or low speed driving for long distance**
- **Driving on rough, dusty, muddy, unpaved, graveled or salt- spread roads**
- **Driving in areas using salt or other corrosive materials or in very cold weather**
- **Driving in heavy traffic area over 90°F (32°C)**
- **Driving on uphill, downhill, or mountain road**
- **Towing a Trailer, or using a camper, or roof rack**
- **Driving as a patrol car, taxi, other commercial use or vehicle towing**
- **Driving over 100 MPH (170 Km/h)**
- **Frequently driving in stop-and-go conditions**

Oil & oil filter: change every 3,750 miles.

Brake pads, calipers & rotors: service or inspect more frequently

Brake drums, linings, parking brake: service or inspect more frequently

Driveshaft boots: service or inspect every 7500 miles

Steering gear box, linkage & boots, lower and upper ball joints: service or inspect more frequently

Air cleaner filter: replace more frequently

Automatic transaxle fluid: replace every 60,000 miles.

Manual transaxle fluid: replace every 80,000 miles.

Climate control air filter: replace more frequently

Spark plugs: replace more frequently

Timing belt idler/tensioner: replace every 60,000 miles or 48 months

37655_ELAN_C0014

BRAKES — INFORMATION AND PRECAUTIONS

ANTI-LOCK SYSTEMS

- Certain components within the ABS system are not intended to be serviced or repaired individually.
- Do not use rubber hoses or other parts not specifically specified for and ABS system. When using repair kits, replace all parts included in the kit. Partial or incorrect repair may lead to functional problems and require the replacement of components.
- Lubricate rubber parts with clean, fresh brake fluid to ease assembly. Do not use shop air to clean parts; damage to rubber components may result.
- Use only DOT 3 brake fluid from an unopened container.
- If any hydraulic component or line is removed or replaced, it may be necessary to bleed the entire system.
- A clean repair area is essential. Always clean the reservoir and cap thoroughly before removing the cap. The slightest amount of dirt in the fluid may plug an orifice and impair the system function. Perform repairs after components have been thoroughly cleaned; use only denatured alcohol to clean components. Do not allow ABS components to come into contact with any substance containing mineral oil; this includes used shop rags.
- The Anti-Lock control unit is a microprocessor similar to other computer units in the vehicle. Ensure that the ignition switch is **OFF** before removing or installing controller harnesses. Avoid static electricity discharge at or near the controller.
- If any arc welding is to be done on the vehicle, the control unit should be unplugged before welding operations begin.

DISC AND DRUM SYSTEMS

⁂ CAUTION

Dust and dirt accumulating on brake parts during normal use may contain asbestos fibers from production or aftermarket brake linings. Breathing excessive concentrations of asbestos fibers can cause serious bodily harm. Exercise care when servicing brake parts. Do not sand or grind brake lining unless equipment used is designed to contain the dust residue. Do not clean brake parts with compressed air or by dry brushing. Cleaning should be done by dampening the brake components with a fine mist of water, then wiping the brake components clean with a dampened cloth. Dispose of cloth and all residue containing asbestos fibers in an impermeable container with the appropriate label. Follow practices prescribed by the Occupational Safety and Health Administration (OSHA) and the Environmental Protection Agency (EPA) for the handling, processing, and disposing of dust or debris that may contain asbestos fibers.

BRAKES — BLEEDING THE BRAKE SYSTEM

BLEEDING PROCEDURE

BLEEDING PROCEDURE

See Figures 1 and 2.

1. Before servicing the vehicle, refer to the Precautions Section.

⁂ WARNING

The reservoir on the master cylinder must be at the MAX (upper) level mark at the start of bleeding procedure and checked after bleeding each brake caliper. Add fluid as required.

2. Make sure the brake fluid level in the reservoir is at the MAX (upper) level line.
3. Have someone slowly pump the brake pedal several times, and then apply steady pressure.
4. Loosen the right-rear brake bleed screw to allow air to escape from the system. Then tighten the bleed screw securely.

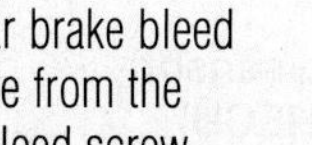

5. Repeat the procedure for each wheel in the sequence shown below until air bubbles no longer appear in the fluid.
6. Refill the master cylinder reservoir to the MAX (upper) level line.

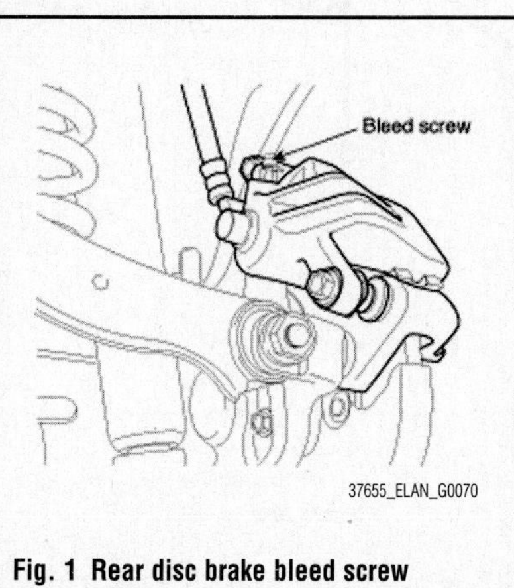

Fig. 1 Rear disc brake bleed screw

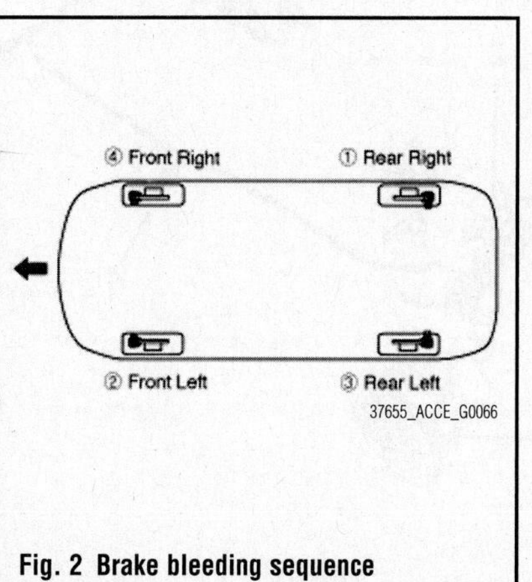

Fig. 2 Brake bleeding sequence

BLEEDING THE ABS SYSTEM

This procedure should be followed to ensure adequate bleeding of air and the filling of the ABS unit, the brake lines, and the master cylinder with brake fluid.

1. Before servicing the vehicle, refer to the Precautions Section.
2. Remove the reservoir cap and fill the brake reservoir with brake fluid.

⁂ WARNING

If there is any brake fluid on any painted surface, wash it off immediately.

➡When pressure bleeding, do not depress the brake pedal. Recommended brake fluid: DOT3 or DOT4.

3. Connect a clear plastic tube to the wheel cylinder bleeder plug and insert the other end of the tube into a clear plastic bottle that is half filled with clean brake fluid.
4. Connect the Hi-Scan Pro® to the data link connector located underneath the dash panel.
5. Select and operate according to the instructions on the Hi-Scan Pro® screen.

⁂ CAUTION

You must obey the maximum operating time of the ABS motor with the Hi-Scan Pro® to prevent the motor pump from burning.

6. Select Hyundai vehicle diagnosis.
7. Select vehicle name.
8. Select Anti-Lock Brake system.
9. Select air bleeding mode.
10. Press "YES" to operate motor pump and solenoid valve.

⁂ WARNING

Wait 60 seconds before operating the air bleeding or damage to the motor may occur.

11. Wait 60 seconds before operating the air bleeding.
12. Pump the brake pedal several times, and then loosen the bleeder screw until fluid starts to run out without bubbles. Then, close the bleeder screw.
13. Repeat until there are no more bubbles in the fluid for each wheel.

BRAKES — ANTI-LOCK BRAKE SYSTEM (ABS)

SPEED SENSORS

REMOVAL & INSTALLATION

Front

See Figures 3 and 4.

1. Before servicing the vehicle, refer to the Precautions Section.
2. Disconnect the negative battery cable.
3. Remove the front wheel speed sensor mounting bolt.
4. Remove the wire bracket bolts, if applicable.
5. Remove the front wheel guard.
6. Disconnect the wheel speed sensor connector.
7. Remove the front wheel speed sensor.

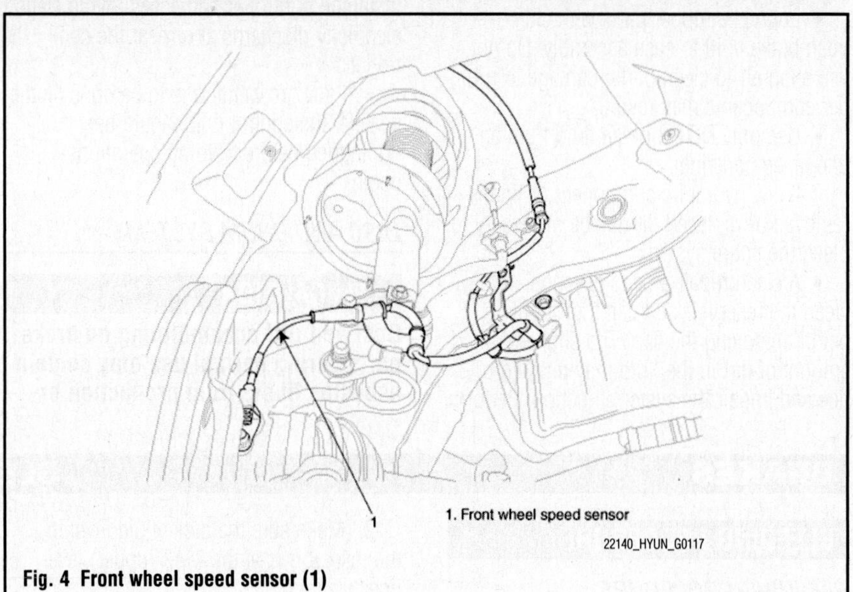

Fig. 4 Front wheel speed sensor (1)

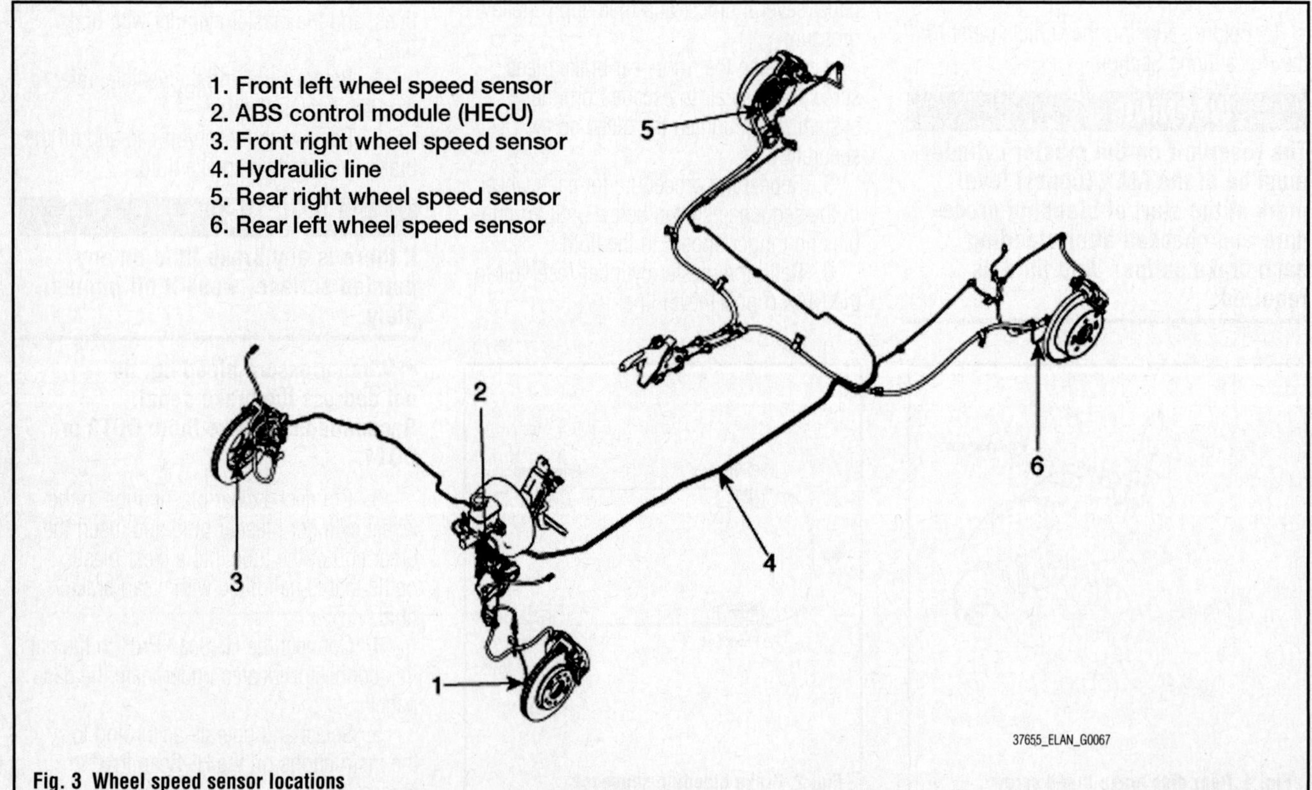

Fig. 3 Wheel speed sensor locations

To install:

8. Installation is the reverse of the removal procedure.

Rear

See Figures 3 and 5.

1. Before servicing the vehicle, refer to the Precautions Section.
2. Disconnect the negative battery cable.
3. Remove the rear wheel speed sensor mounting bolt.
4. Remove the rear seat side pad then disconnect the rear wheel speed sensor connector.
5. Remove the rear wheel speed sensor.

To install:

6. Installation is the reverse of the removal procedure.

1. Rear wheel speed sensor
1
22140_HYUN_G0118
Fig. 5 Rear wheel speed sensor (1)

BRAKES — FRONT DISC BRAKES

BRAKE CALIPER

REMOVAL & INSTALLATION

See Figures 6 through 9.

1. Before servicing the vehicle, refer to the Precautions Section.
2. Disconnect the negative battery cable.
3. Raise and safely support the vehicle.
4. Remove the front wheel and tire from front hub.

⁂ WARNING

Be careful not to damage the hub bolts.

5. Remove the brake hose bolt and guide rod bolts from the caliper assembly.

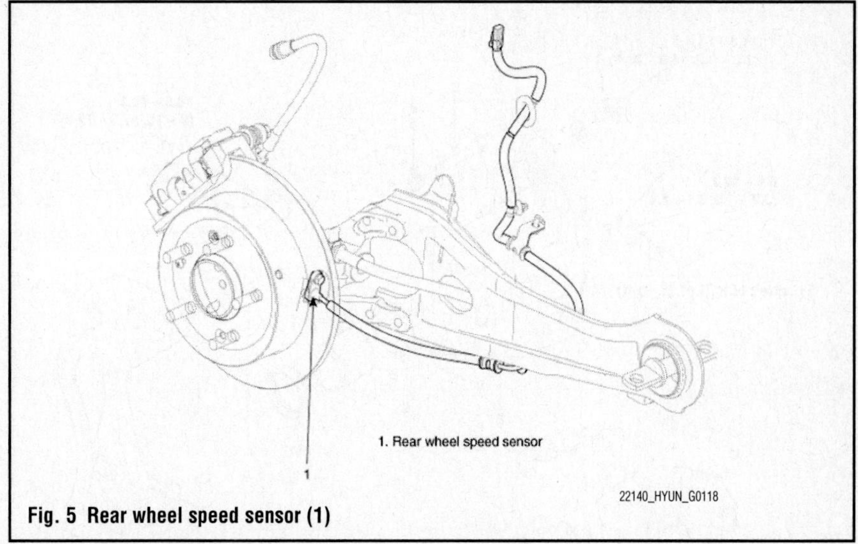

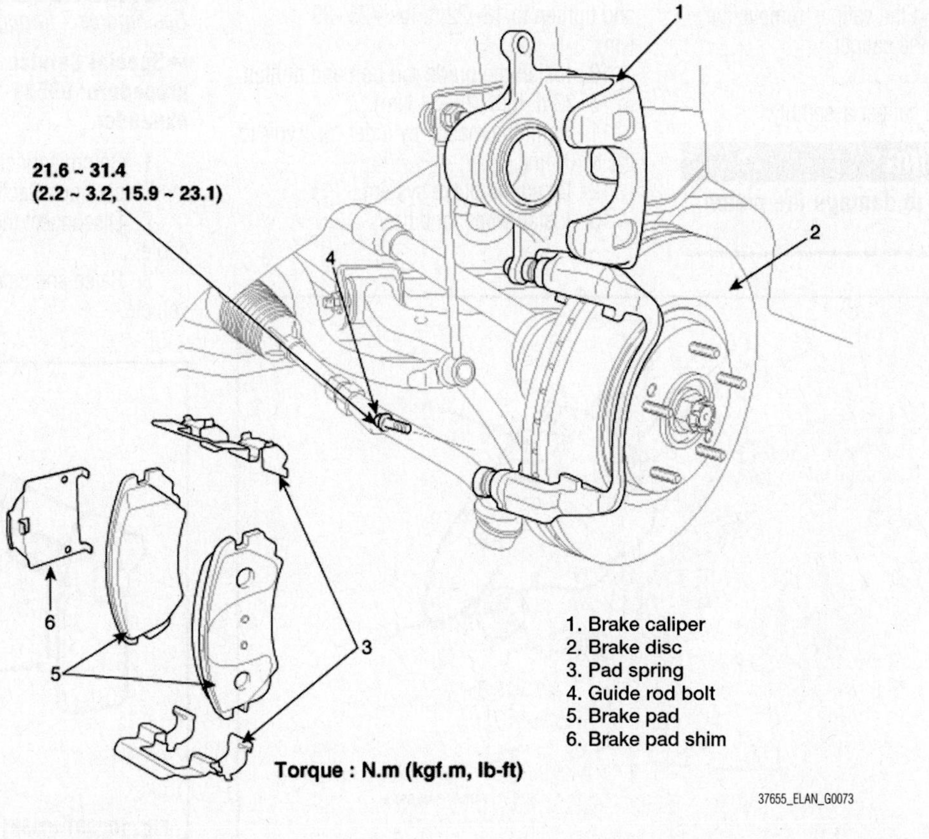

21.6 ~ 31.4
(2.2 ~ 3.2, 15.9 ~ 23.1)
1
2
3
4
5
6
1. Brake caliper
2. Brake disc
3. Pad spring
4. Guide rod bolt
5. Brake pad
6. Brake pad shim
Torque : N.m (kgf.m, lb-ft)
37655_ELAN_G0073
Fig. 6 Front disc brake components

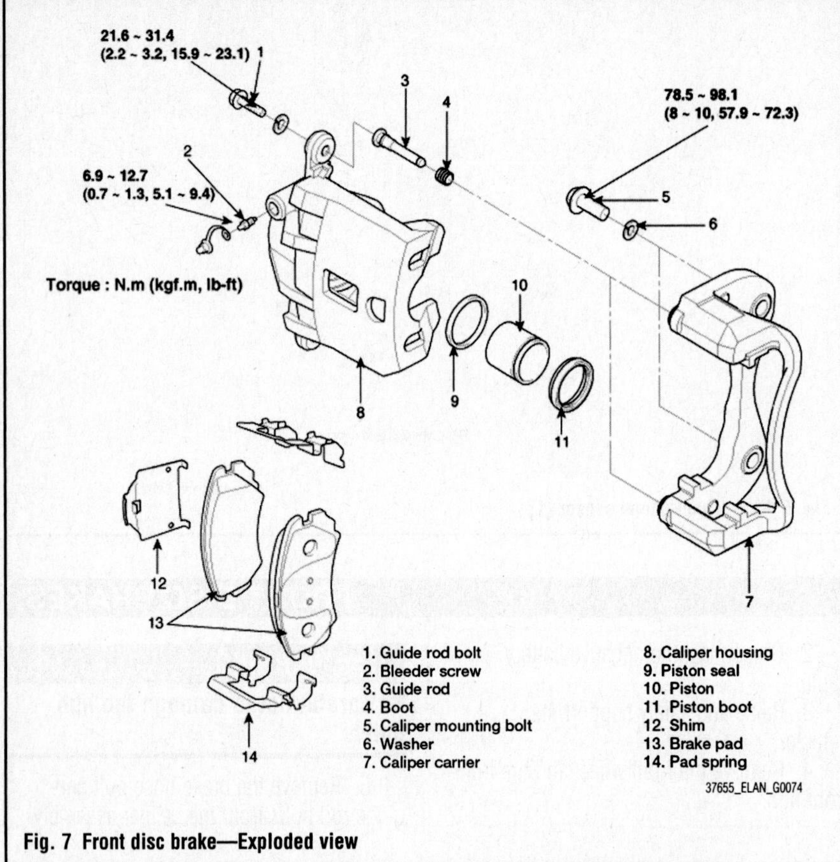

Fig. 7 Front disc brake—Exploded view

6. Remove the caliper assembly.

7. If replacing the caliper, remove the brake hose from the caliper.

To install:

8. Install the caliper assembly.

⁂ WARNING

Be careful not to damage the piston pin boot.

9. Install the brake hose bolt and tighten to 18–22 ft. lbs. (25–29 Nm),

10. Install the guide rod bolt and tighten to 16–23 ft. lbs. (22–31 Nm).

11. Refill the master cylinder reservoir to the MAX line.

12. Bleed the brake system.

13. Install wheel and tire.

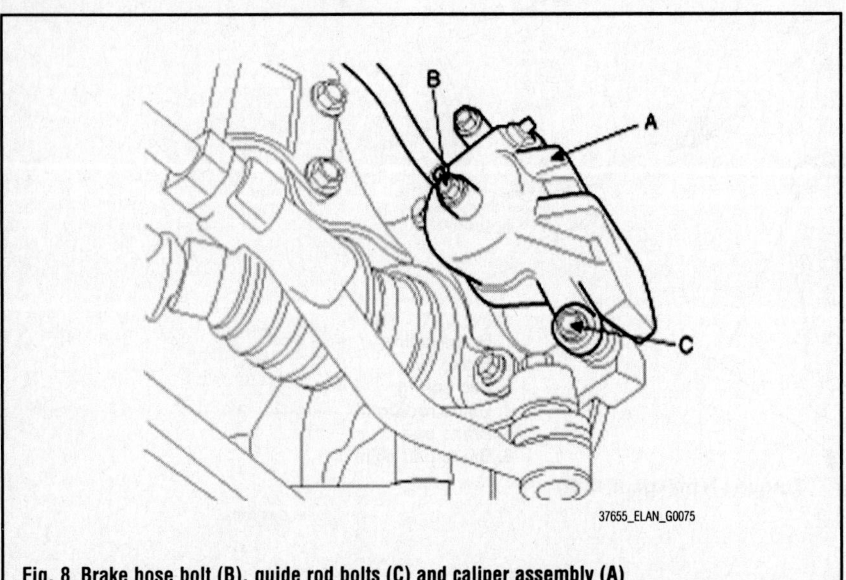

Fig. 8 Brake hose bolt (B), guide rod bolts (C) and caliper assembly (A)

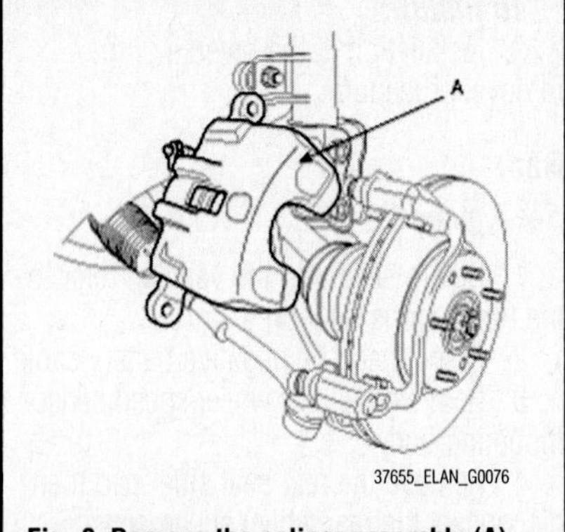

Fig. 9 Remove the caliper assembly (A)

➡Engagement of the brake may require a greater pedal stroke immediately after the brake pads have been replaced as a set. Several applications of the brake will restore the normal pedal stroke.

14. After installation, check for leaks at hose and line joints or connections, and retighten if necessary.

15. Depress the brake pedal several times to make sure the brakes work, and then test-drive.

DISC BRAKE PADS

REMOVAL & INSTALLATION

See Figures 7 through 13.

➡Special Service Tool used in this procedure: 09581-11000, piston expander.

1. Before servicing the vehicle, refer to the Precautions Section.

2. Disconnect the negative battery cable.

3. Raise and safely support the vehicle.

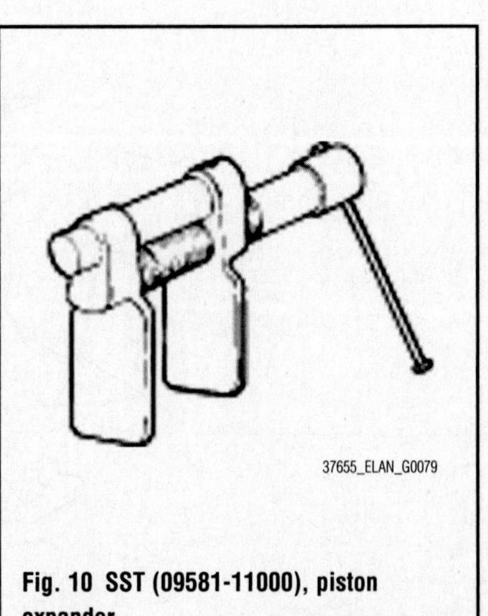

Fig. 10 SST (09581-11000), piston expander

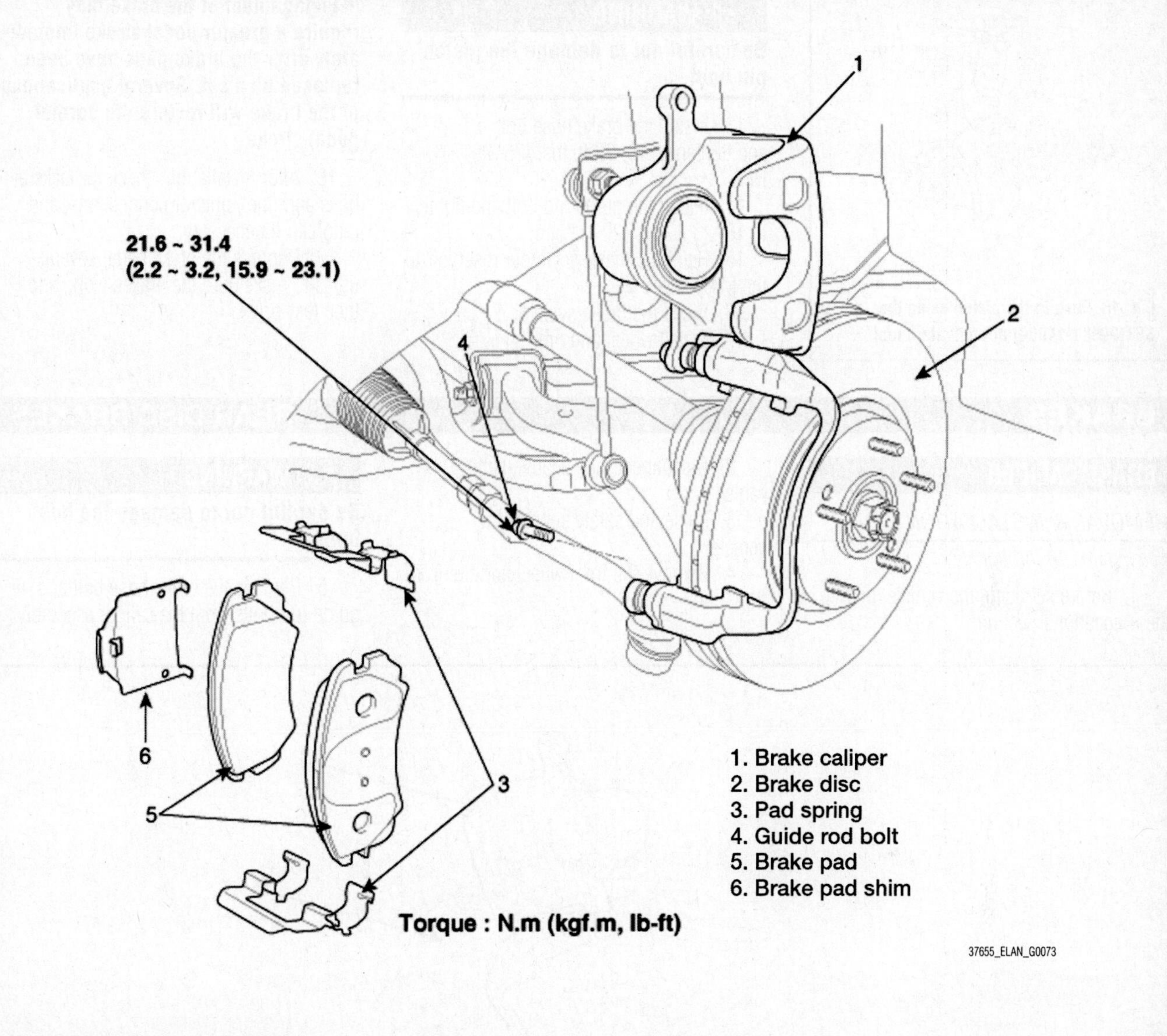

21.6 ~ 31.4
(2.2 ~ 3.2, 15.9 ~ 23.1)

1. Brake caliper
2. Brake disc
3. Pad spring
4. Guide rod bolt
5. Brake pad
6. Brake pad shim

Torque : N.m (kgf.m, lb-ft)

37655_ELAN_G0073

Fig. 11 Front disc brake components

4. Remove the front wheel and tire from front hub.

WARNING

Be careful not to damage the hub bolts.

5. Remove the brake hose bolt and guide rod bolts from the caliper assembly.
6. Remove the caliper assembly.
7. If replacing the caliper, remove the brake hose from the caliper.
8. Remove the pads, the pad shims and the pad springs from the caliper bracket.

To install:

9. Install the pad springs to the caliper bracket.
10. Install the brake pads and pad shims on the pad retainer correctly. Install the pad with the wear indicator on the inside. If you are reusing the pads, always reinstall the brake pads in their original positions to prevent a momentary loss of braking efficiency.

CAUTION

Check the foreign material at the pad shims and the back of the pads. Contaminated brake discs or pads reduce stopping ability. Keep grease off the discs and pads.

11. Push in the piston using the SST (09581-11000) or equivalent tool, so that the caliper will fit over the pads. Make sure that the piston boot is in position to prevent damaging it when installing the caliper.
12. Install the caliper assembly.

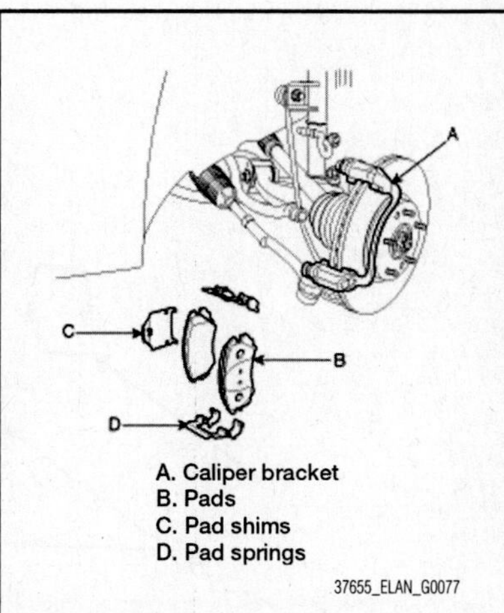

Fig. 12 Remove the pads, the pad shims and the pad springs from the caliper bracket

09581-11000
37655_ELAN_G0078

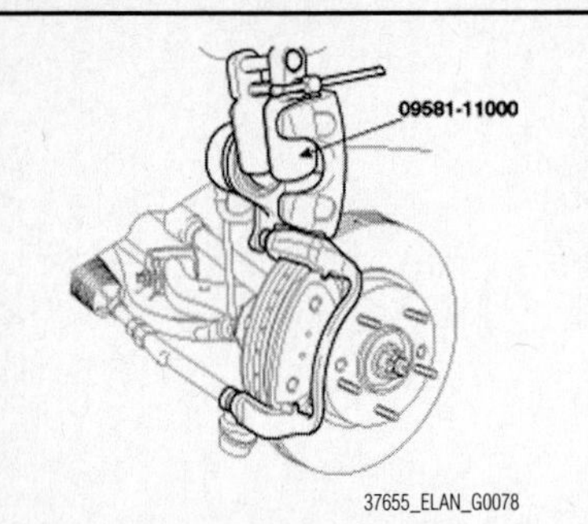

Fig. 13 Push in the piston using the SST(09581-11000) or equivalent tool

⁂ WARNING

Be careful not to damage the piston pin boot.

13. Install the brake hose bolt and tighten to 18–22 ft. lbs. (25–29 Nm),
14. Install the guide rod bolt and tighten to 16–23 ft. lbs. (22–31 Nm).
15. Refill the master cylinder reservoir to the MAX line.
16. Bleed the brake system.
17. Install wheel and tire.

➡Engagement of the brake may require a greater pedal stroke immediately after the brake pads have been replaced as a set. Several applications of the brake will restore the normal pedal stroke.

18. After installation, check for leaks at hose and line joints or connections, and retighten if necessary.
19. Depress the brake pedal several times to make sure the brakes work, and then test-drive.

BRAKES — REAR DISC BRAKES

BRAKE CALIPER

REMOVAL & INSTALLATION

See Figures 14 through 17.

1. Before servicing the vehicle, refer to the Precautions Section.
2. Disconnect the negative battery cable.
3. Raise and safely support the vehicle.
4. Remove the front wheel and tire from front hub.

⁂ WARNING

Be careful not to damage the hub bolts.

5. Remove the brake hose bolt and guide rod bolts from the caliper assembly.

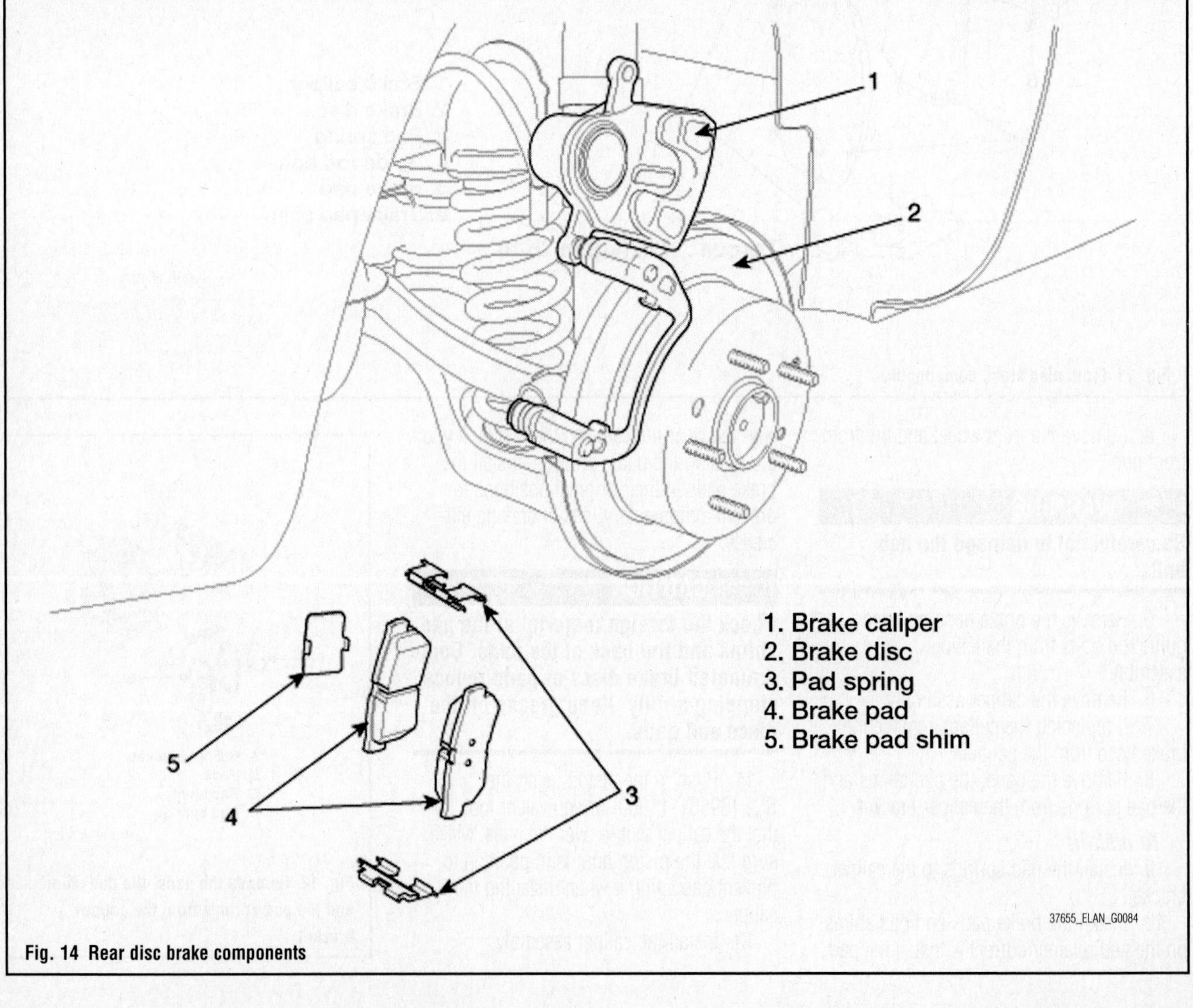

1
2
3
4
5
1. Brake caliper
2. Brake disc
3. Pad spring
4. Brake pad
5. Brake pad shim
37655_ELAN_G0084

Fig. 14 Rear disc brake components

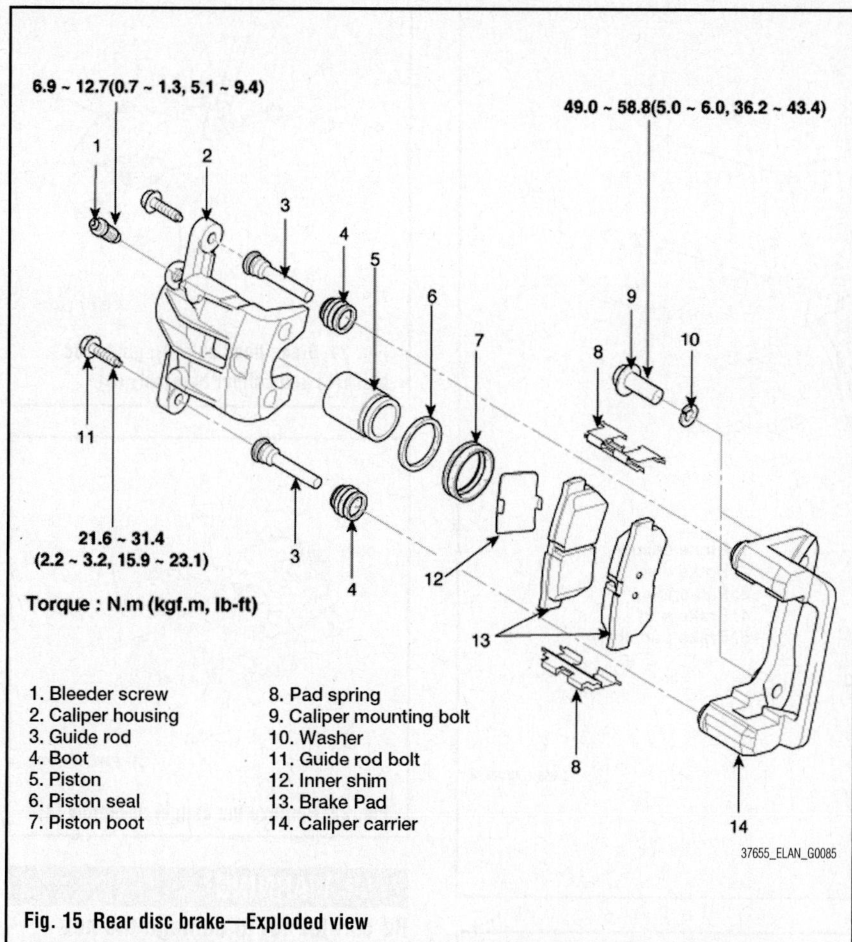

Fig. 15 Rear disc brake—Exploded view

6. Remove the caliper assembly.
7. If replacing the caliper, remove the brake hose from the caliper.

To install:

8. Install the caliper assembly.

WARNING

Be careful not to damage the piston pin boot.

9. Install the brake hose bolt and tighten to 18–22 ft. lbs. (25–29 Nm),
10. Install the guide rod bolt and tighten to 16–23 ft. lbs. (22–31 Nm).
11. Refill the master cylinder reservoir to the MAX line.
12. Bleed the brake system.
13. Install the wheel and tire.

➡Engagement of the brake may require a greater pedal stroke immediately after the brake pads have been replaced as a set. Several applications of the brake will restore the normal pedal stroke.

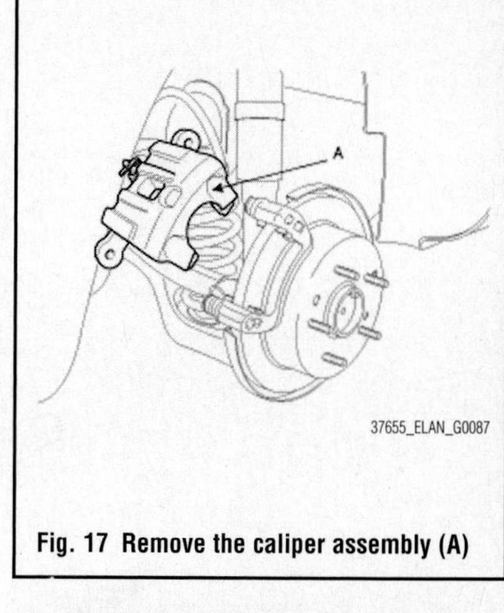

Fig. 17 Remove the caliper assembly (A)

14. After installation, check for leaks at hose and line joints or connections, and retighten if necessary.
15. Depress the brake pedal several times to make sure the brakes work, and then test-drive.

DISC BRAKE PADS

REMOVAL & INSTALLATION

See Figures 18 through 24.

➡Special Service Tool used in this procedure: 09581-11000, piston expander

1. Before servicing the vehicle, refer to the Precautions Section.
2. Disconnect the negative battery cable.
3. Raise and safely support the vehicle.
4. Remove the front wheel and tire from front hub.

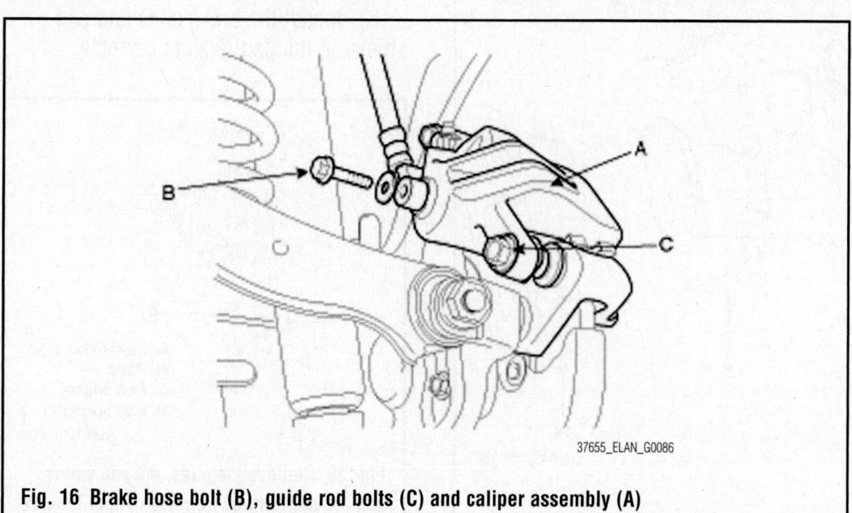

Fig. 16 Brake hose bolt (B), guide rod bolts (C) and caliper assembly (A)

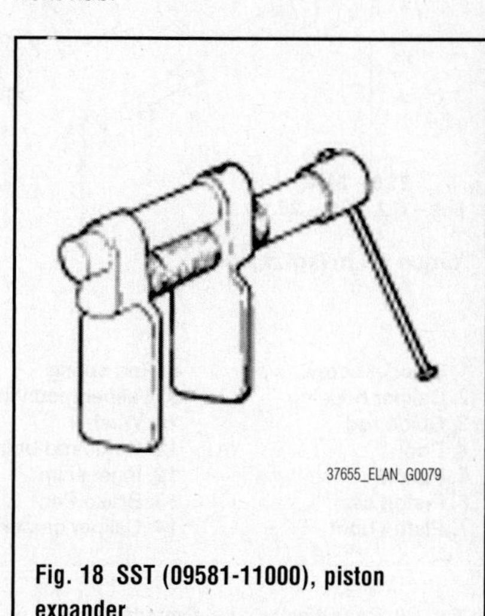

Fig. 18 SST (09581-11000), piston expander

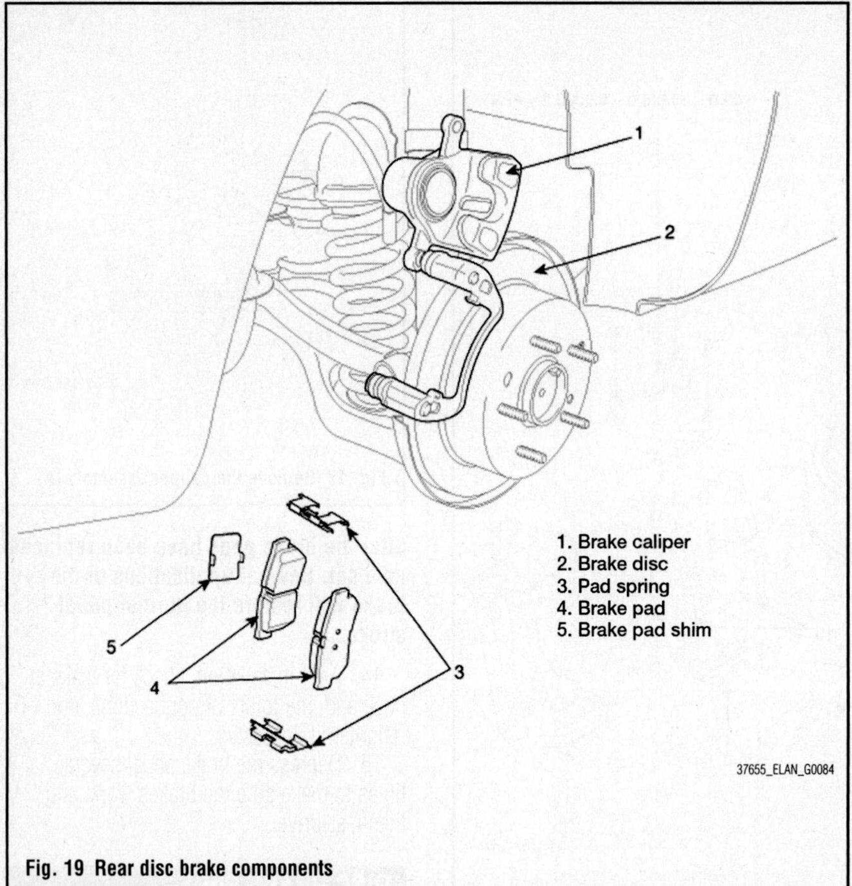

Fig. 19 Rear disc brake components

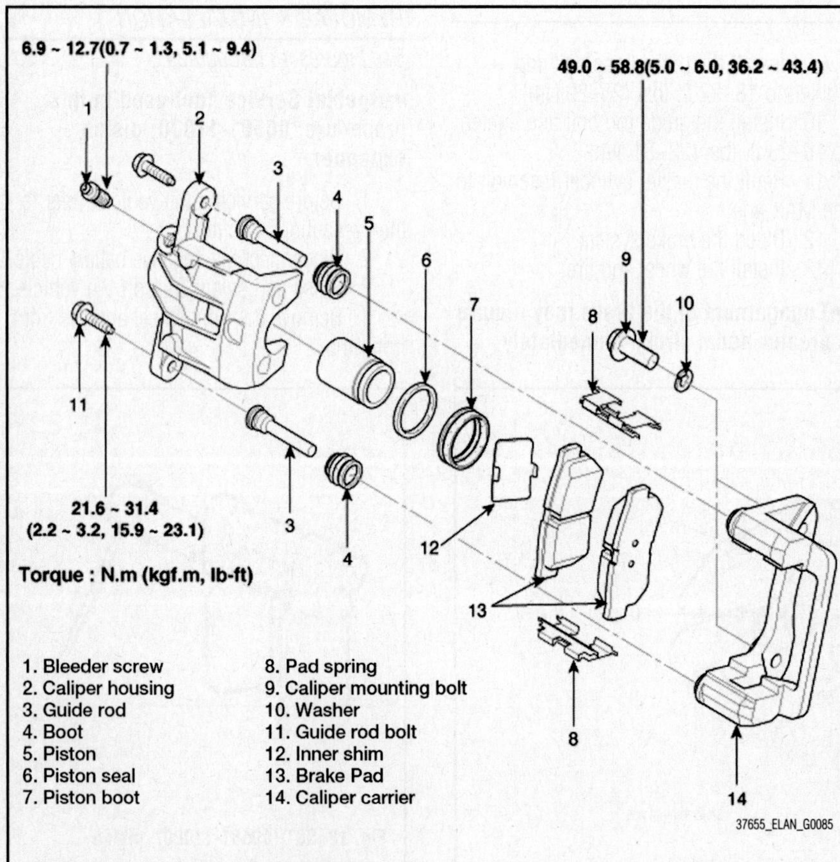

Fig. 20 Rear disc brake—Exploded view

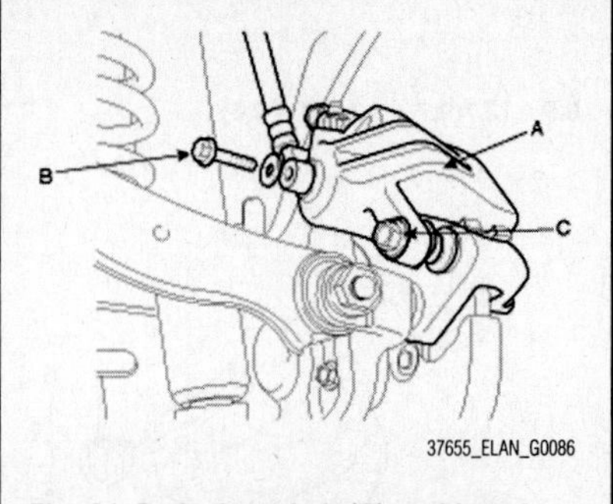

Fig. 21 Brake hose bolt (B), guide rod bolts (C) and caliper assembly (A)

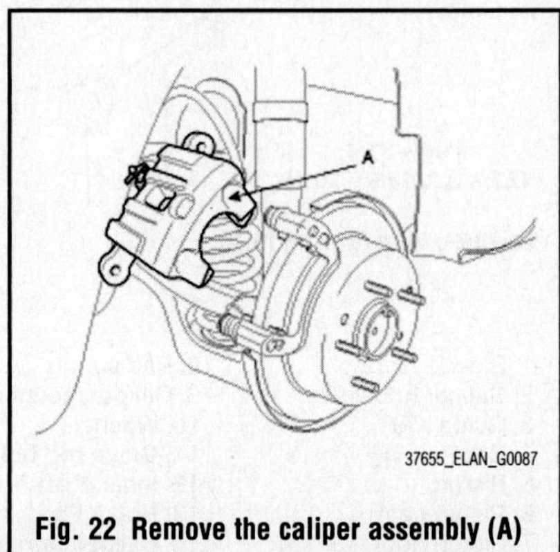

Fig. 22 Remove the caliper assembly (A)

WARNING

Be careful not to damage the hub bolts.

5. Remove the brake hose bolt and guide rod bolts from the caliper assembly.
6. Remove the caliper assembly.
7. If replacing the caliper, remove the brake hose from the caliper.
8. Remove the pads, the pad shims, and the pad springs from the caliper bracket.

To install:

9. Install the pad springs to the caliper bracket.
10. Install the brake pads and pad shims on the pad retainer correctly.

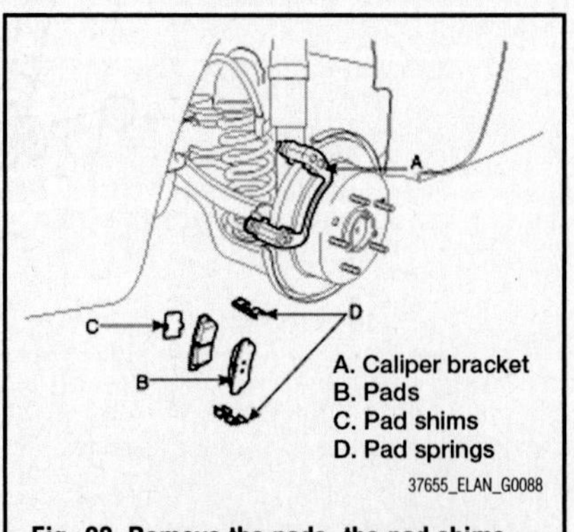

Fig. 23 Remove the pads, the pad shims and the pad springs from the caliper bracket

Install the pad with the wear indicator on the inside. If you are reusing the pads, always reinstall the brake pads in their original positions to prevent a momentary loss of braking efficiency.

✲✲ CAUTION

Check the foreign material at the pad shims and the back of the pads. Contaminated brake discs or pads reduce stopping ability. Keep grease off the discs and pads.

11. Push in the piston using the SST (09581-11000) or equivalent tool, so that the caliper will fit over the pads. Make sure that the piston boot is in position to prevent damaging it when installing the caliper.
12. Install the caliper assembly.

✲✲ WARNING

Be careful not to damage the piston pin boot.

13. Install the brake hose bolt and tighten to 18–22 ft. lbs. (25–29 Nm),
14. Install the guide rod bolt and tighten to 16–23 ft. lbs. (22–31 Nm).

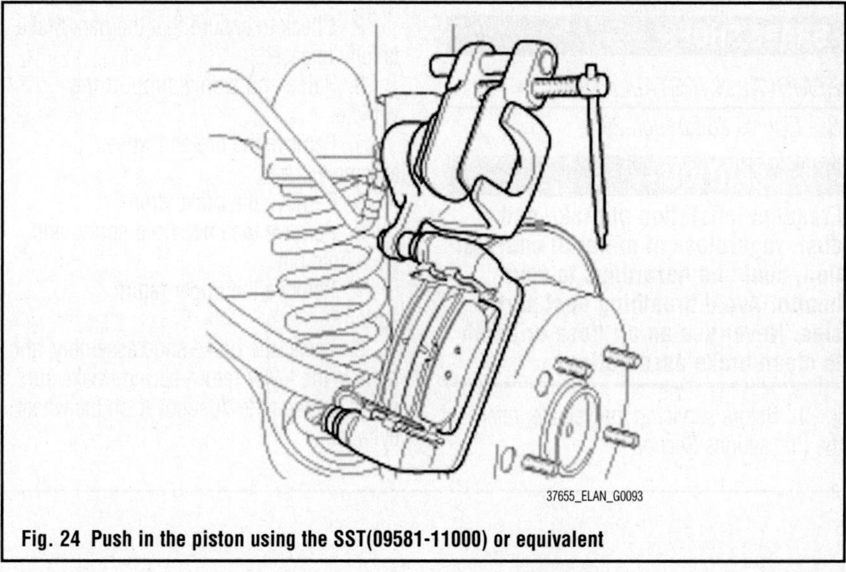

Fig. 24 Push in the piston using the SST(09581-11000) or equivalent

15. Refill the master cylinder reservoir to the MAX line.
16. Bleed the brake system.
17. Install wheel and tire.

➡**Engagement of the brake may require a greater pedal stroke immediately after the brake pads have been replaced as a set. Several applications of the brake will restore the normal pedal stroke.**

18. After installation, check for leaks at hose and line joints or connections, and retighten if necessary.
19. Depress the brake pedal several times to make sure the brakes work, and then test-drive.

BRAKES — REAR DRUM BRAKES

BRAKE DRUM

REMOVAL & INSTALLATION

See Figure 25.

✲✲ CAUTION

Frequent inhalation of brake pad dust, regardless of material composition, could be hazardous to your health. Avoid breathing dust particles. Never use an air hose or brush to clean brake assemblies.

1. Before servicing the vehicle, refer to the Precautions Section.
2. Check to ensure that the park brake is fully released.
3. Raise and safely support the vehicle.
4. Remove the tire and wheel assembly.
5. Remove the brake drum.

To install:

6. If installing a new brake drum, use denatured alcohol or an equivalent approved brake cleaner and a clean shop towel to remove the protective coating from the friction surface of the drum.
7. Install the brake drum.
8. Install the tire and wheel assembly.
9. Apply the brakes approximately 3 times in order to seat and center the brake shoes within the drum.
10. Lower the vehicle.

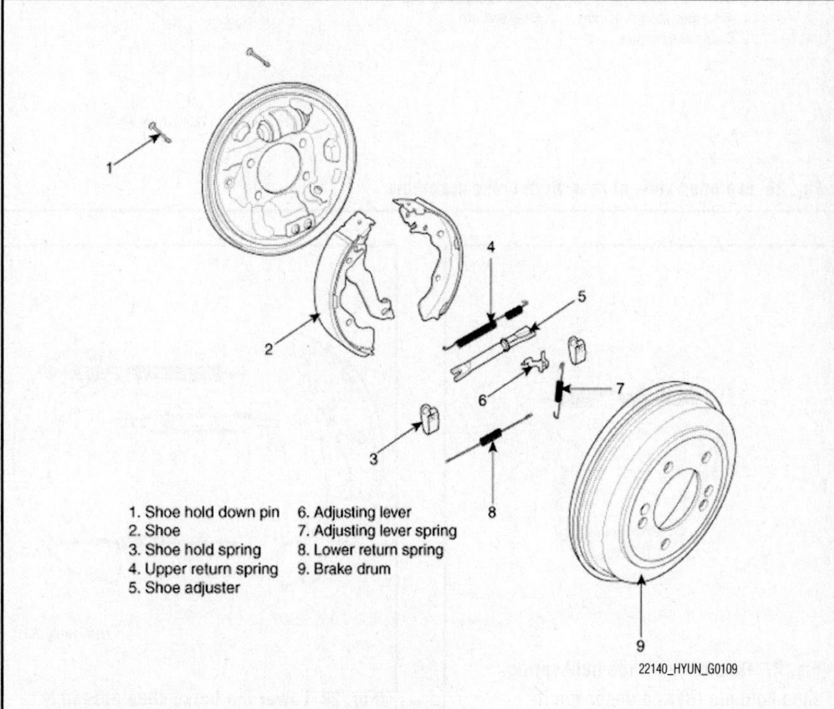

Fig. 25 Exploded view of rear drum brake assembly

BRAKE SHOES

REMOVAL & INSTALLATION

See Figures 26 through 29.

CAUTION

Frequent inhalation of brake pad dust, regardless of material composition, could be hazardous to your health. Avoid breathing dust particles. Never use an air hose or brush to clean brake assemblies.

1. Before servicing the vehicle, refer to the Precautions Section.
2. Check to ensure that the park brake is fully released.
3. Raise and safely support the vehicle.
4. Remove the tire and wheel assembly.
5. Remove the brake drum.
6. Remove the shoe hold spring and shoe hold pin.
7. Remove the upper return spring.
8. Lower the brake shoe assembly, and remove the lower return spring. Make sure not to damage the dust cover on the wheel cylinder.

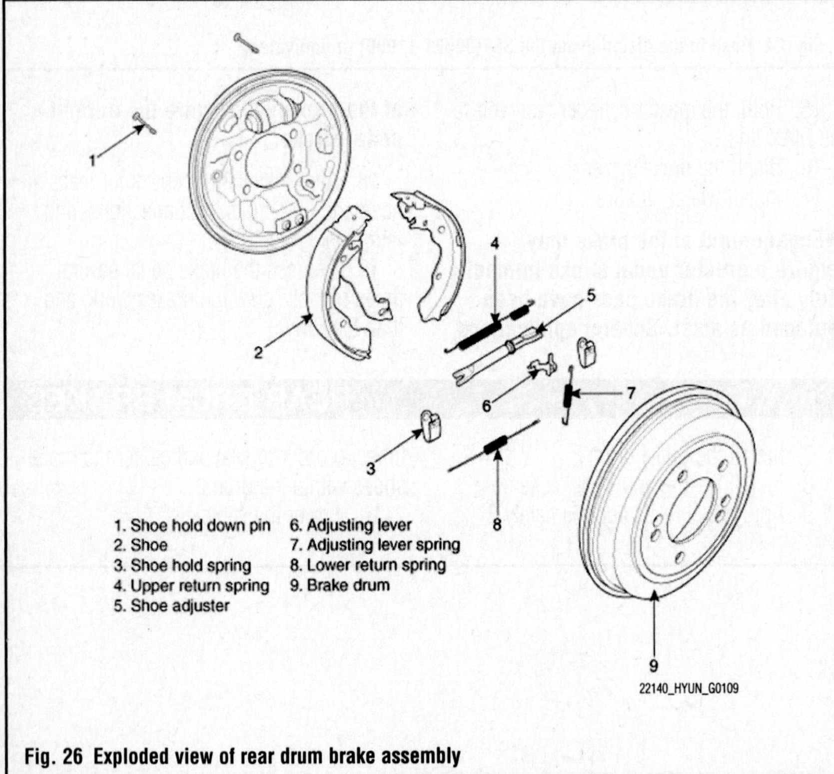

Fig. 26 Exploded view of rear drum brake assembly

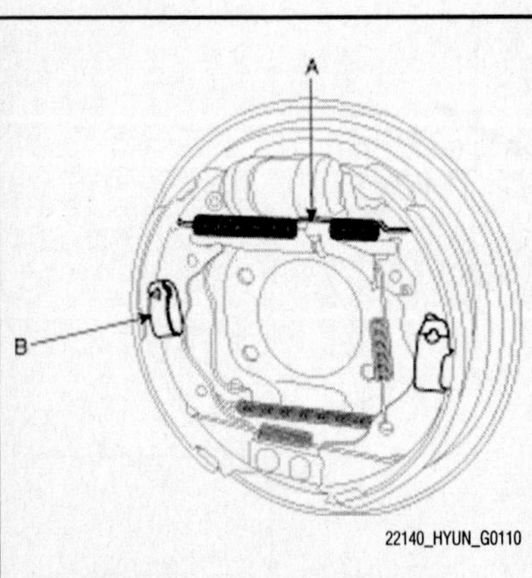

Fig. 27 Remove the shoe hold spring, shoe hold pin (B) and upper return spring (A)

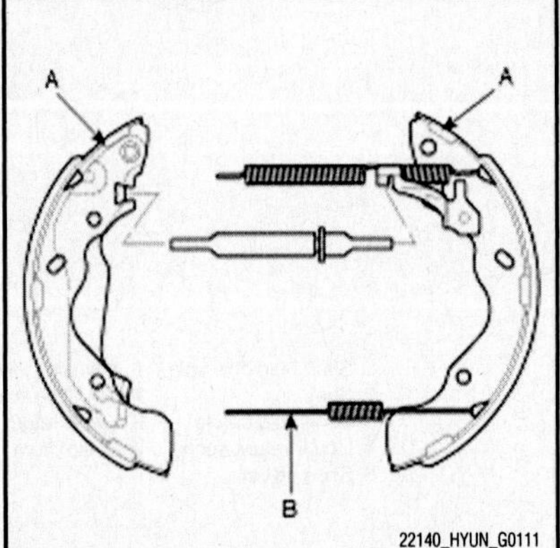

Fig. 28 Lower the brake shoe assembly (A), and remove the lower return spring (B)

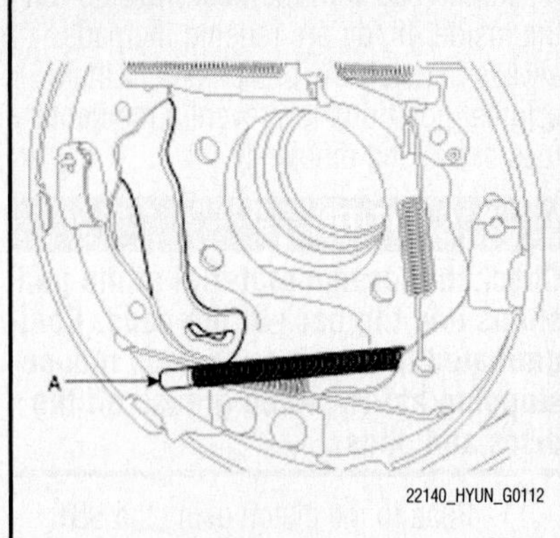

Fig. 29 Remove the parking brake cable (A) from the brake assembly

9. Remove the parking brake cable from the brake assembly.
10. Remove the brake shoe assembly.

To install:

11. Connect the parking brake cable (A) to the brake assembly.
12. Clean the threaded portions of adjuster sleeve and push rod female. Coat the threads of the adjuster assembly with grease. To shorten the clevises, turn the adjuster bolt.
13. Hook the shoe adjuster lever, then install it to the brake shoe.
14. Install the adjuster assembly and upper return spring.

WARNING

Be careful not to damage the wheel cylinder dust covers.

15. Install the lower return spring.
16. Apply brake cylinder grease, or equivalent rubber grease, to the sliding surfaces and brake shoe ends and opposite edges of the shoes.

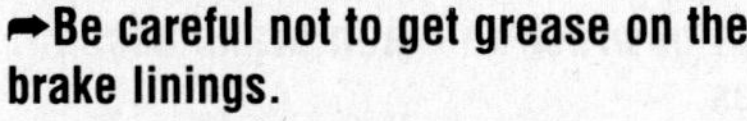

➡Be careful not to get grease on the brake linings.

17. Install the brake shoes onto the backing plate.
18. Install the shoe hole down pins and the shoe hole down springs.
19. Install the rear brake drum.
20. Install the tire and wheel assembly.
21. Depress the brake pedal several times to set the self-adjusting brake.
22. Adjust the parking brake, as necessary.

ADJUSTMENT

These vehicles have a self-adjusting mechanism in the rear drum brake assembly.

BRAKES — PARKING BRAKE

PARKING BRAKE CABLES

ADJUSTMENT

See Figure 30.

1. Before servicing the vehicle, refer to the Precautions Section.

➡**After repairing the parking brake shoe, adjust the brake shoe clearance, and then adjust the parking brake lever stroke.**

2. Raise and securely support the vehicle.
3. Pull the parking brake lever up one click.
4. Remove the floor console, if equipped.

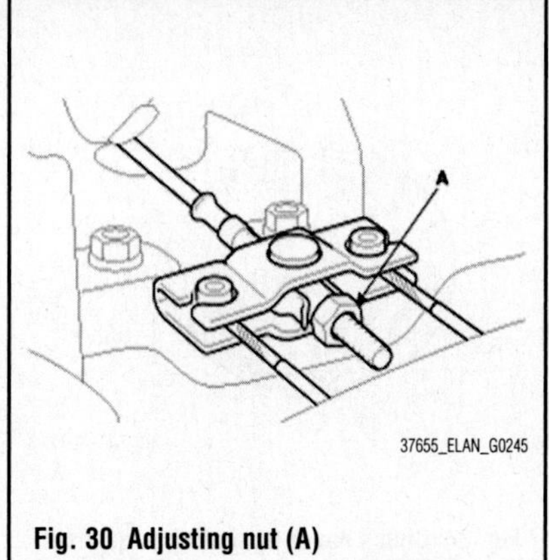

Fig. 30 Adjusting nut (A)

5. Adjust the parking brake lever stroke by turning adjusting nut. Stroke: 7 clicks.
6. Release the parking brake lever completely.
7. Check if the parking brakes drag when the rear wheels are turned. Readjust if necessary until there is no drag from the parking brakes.
8. Check the proper operation of the parking brakes by fully applying the parking brakes.
9. Reinstall the floor console, if equipped.

PARKING BRAKE SHOES

REMOVAL & INSTALLATION

The rear drum brake shoes serve as the parking brakes. Refer to the procedures under Rear Drum Brakes, Removal & Installation.

CHASSIS ELECTRICAL — AIR BAG (SUPPLEMENTAL RESTRAINT SYSTEM)

GENERAL INFORMATION

CAUTION

These vehicles are equipped with an air bag system. The system must be disarmed before performing service on, or around, system components, the steering column, instrument panel components, wiring and sensors. Failure to follow the safety precautions and the disarming procedure could result in accidental air bag deployment, possible injury and unnecessary system repairs.

SERVICE PRECAUTIONS

CAUTION

Disconnect and isolate the battery negative cable before beginning any airbag system component diagnosis, testing, removal, or installation procedures. Wait at least 90 seconds after the ignition switch is turned off and the negative (-) terminal cable is disconnected from the battery before starting the operation. The SRS is equipped with a backup power source, so if work is started within 90 seconds after disconnecting the negative (-) terminal cable from the battery, the SRS may be deployed. Failure to disable the airbag system may result in accidental airbag deployment, personal injury, or death.

DISARMING THE SYSTEM

CAUTION

Before servicing components near or affected by the SRS (airbag) system, read and observe all SRS Service Precautions. Failure to observe all precautions may result in accidental airbag deployment, personal injury, or death.

1. Before servicing the vehicle, refer to the Precautions Section.
2. Disconnect and isolate the negative battery cable.
3. Wait 3 minutes for the system capacitor to discharge before performing any service.
4. Remove the ignition key from the vehicle.

ARMING THE SYSTEM

CAUTION

Before servicing components near or affected by the SRS (airbag) system, read and observe all SRS Service Precautions. Failure to observe all precautions may result in accidental airbag deployment, personal injury, or death.

1. Before servicing the vehicle, refer to the Precautions Section.
2. Reconnect the negative battery cable.
3. Turn the ignition switch to the **RUN** position.
4. Confirm proper system operation:
   a. Turn the ignition switch ON; the SRS indicator light should be turned on for about six seconds and then go off.

CLOCKSPRING CENTERING

See Figure 31.

1. Prior to installing the clockspring, confirm that the front wheels are pointed straight ahead.
2. Turn clockspring fully clockwise.
3. Turn clockspring counter-clockwise 3 revolutions.
4. Matchmark the clockspring housing.
5. Confirm clockspring centering in each direction from matchmark.

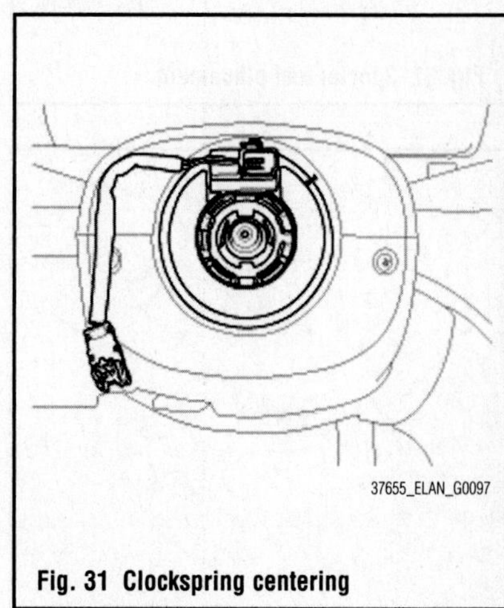

Fig. 31 Clockspring centering

DRIVE TRAIN

CLUTCH

REMOVAL & INSTALLATION

See Figures 32 through 34.

1. Before servicing the vehicle, refer to the Precautions Section.
2. Remove the transaxle assembly. Refer to Manual Transaxle Assembly, Removal & Installation.
3. Insert the special tool (09411-11000) in the clutch disc to prevent the disc from falling.
4. Loosen the bolts which attach the clutch cover to the flywheel in a star pattern. Loosen the bolts in succession, 1 or 2 turns at a time, to avoid bending the cover flange.

➡**Do not clean the clutch disc or the release bearing with cleaning solvent.**

5. Remove clutch cover assembly and then driven disc and pressure plate.

To install:

⁂ WARNING

When installing the clutch, apply grease to each part, but be careful not to apply excessive grease. It can cause clutch slippage and shudder.

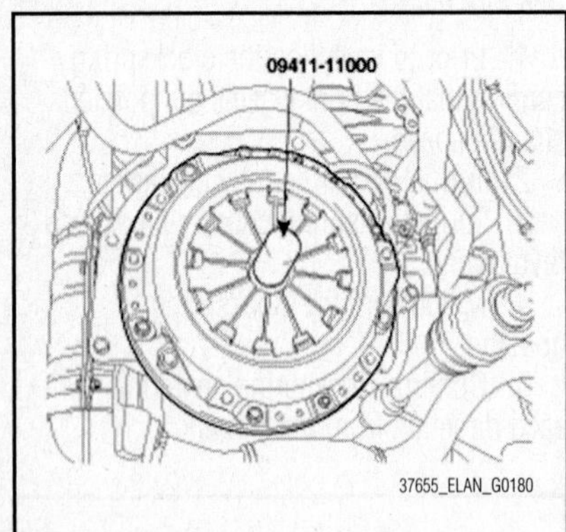

Fig. 32 Special tool placement

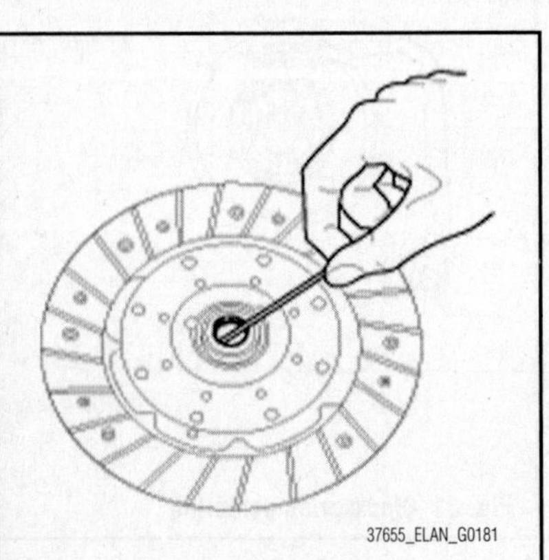

Fig. 33 Clutch grease application

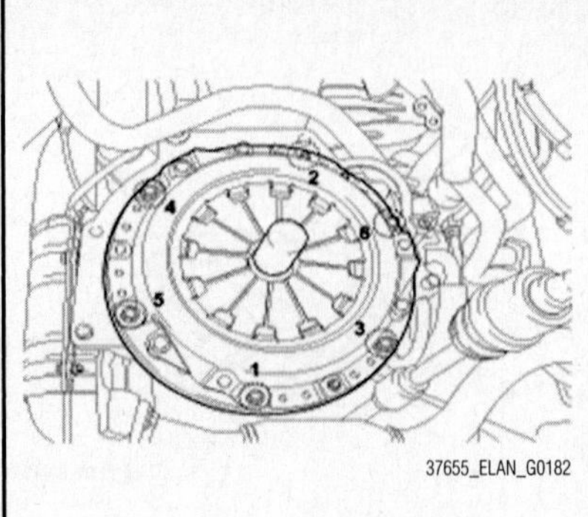

Fig. 34 Clutch cover tightening sequence

6. Install the driven disc and pressure plate in to the clutch cover assembly.
7. Install the clutch disc assembly to the flywheel using the special tool (09411-11000).
8. Install the clutch cover assembly to the flywheel and temporarily tighten the bolts 1 or 2 steps at a time in a star pattern to 11–16 ft. lbs. (15–22 Nm).
9. Install the transaxle assembly to the engine.

CV-JOINT

OVERHAUL

See Figures 35 through 49.

➡**Special Service Tools (SST) needed for this procedure: Bearing Remover, No. 09495-33000, and Band Installer, No. 09495-3K000.**

➡**Do not disassemble the BJ assembly. Special grease must be applied to the driveshaft joint. Do not substitute with another type of grease. The boot band should be replaced with a new one.**

1. Before servicing the vehicle, refer to the Precautions Section.

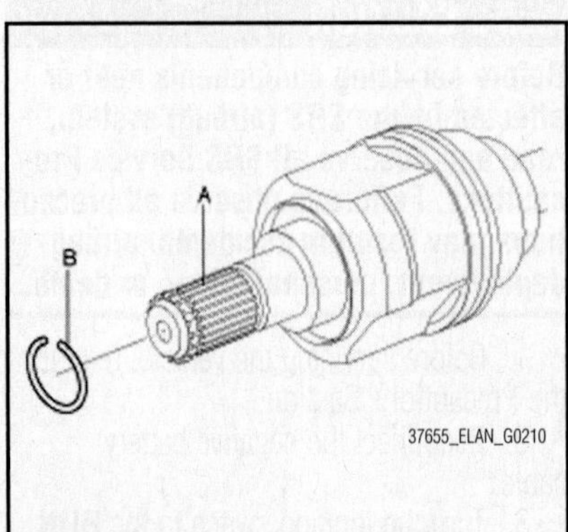

Fig. 35 Remove the circlip (B) from halfshaft splines (A)

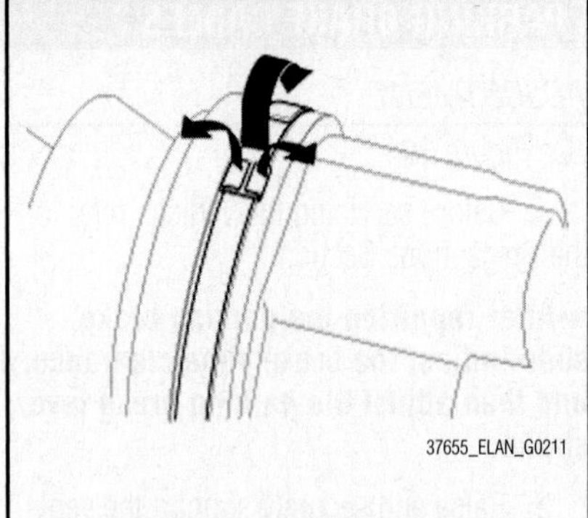

Fig. 36 Remove the boot clamps from the transaxle side TJ case

2. Remove the circlip from halfshaft splines of the transaxle side TJ case.
3. Remove both boot clamps from the transaxle side TJ case.
   a. Using a flat-tipped screwdriver, remove both clamps of the transaxle side.
4. Pull out the boot from the transaxle side joint (TJ).
5. After removing the boot, wipe the grease in the TJ case.
6. Matchmark the spider roller assembly, TJ case, and shaft splines, as shown.

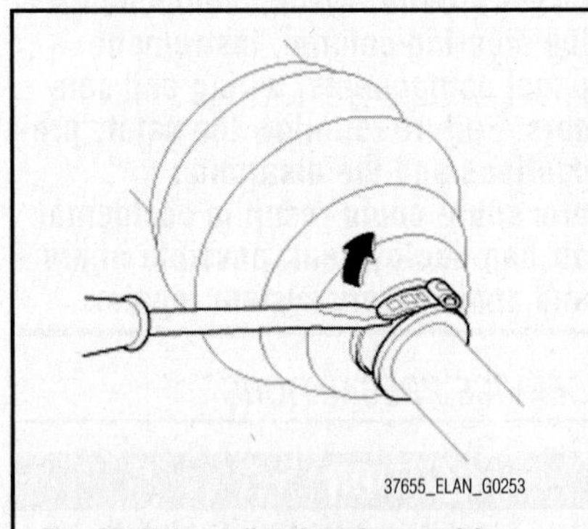

Fig. 37 Using a flat-tipped screwdriver, remove both clamps of the transaxle side

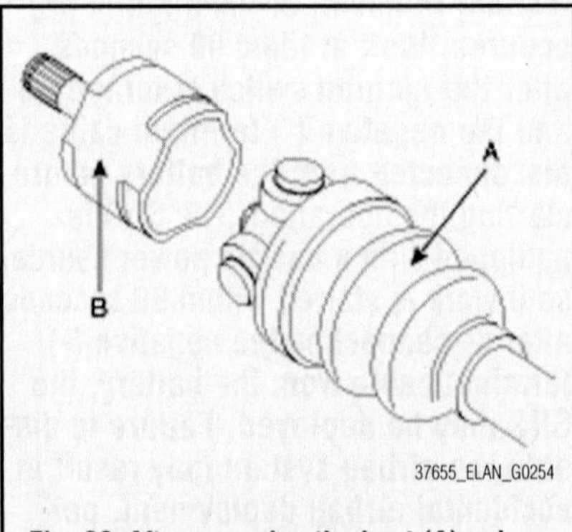

Fig. 38 After removing the boot (A), wipe the grease in the TJ case (B)

7. Using snap ring pliers or a flat-tipped screwdriver, remove the circlip.

8. Remove the spider assembly from the driveshaft using the special tool (09495-33000).

9. Clean the spider assembly.

10. Remove the boot of the transaxle side joint (TJ).

➡If reusing the boot, wrap tape around the driveshaft splines to protect the boot.

11. Using pliers or a flat-tipped screwdriver, remove the clamps on both sides of the dynamic damper.

12. Attach a vise to the driveshaft as shown.

13. Apply soap powder on the shaft to prevent damage to the shaft spline and the dynamic damper when the dynamic damper is removed.

14. Carefully separate the dynamic damper from the shaft.

To install:

15. Wrap tape around the driveshaft splines (TJ. side) to prevent damage to the boots.

16. Reassemble the dynamic damper by keeping the shaft straight and tightening the dynamic damper with a dynamic damper band.

17. Install the TJ boot bands and TJ boot.

18. Install the spider assembly and the circlip to the spline on the halfshaft, aligning the matchmarks.

19. Install grease into the TJ to equal the amount removed upon cleaning.

20. Install the TJ boots.

21. To control the air in the TJ boot, keep the specified distance between the boot bands when they are tightened.

22. Using the SST (09495-3K000), secure the TJ boot bands. Clearance: 0.079 in. (2.0 mm) or less.

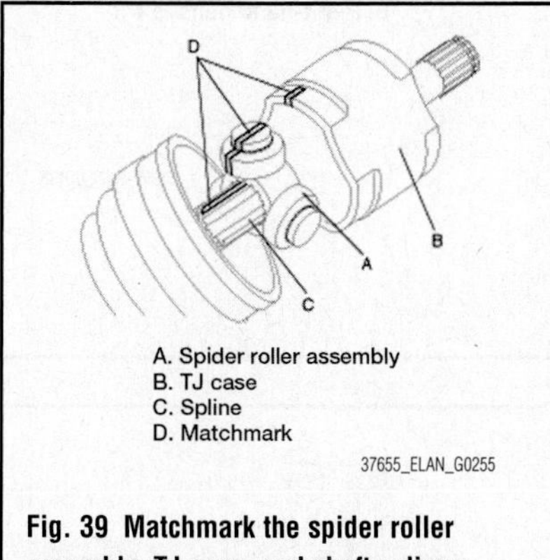

Fig. 39 Matchmark the spider roller assembly, TJ case, and shaft splines

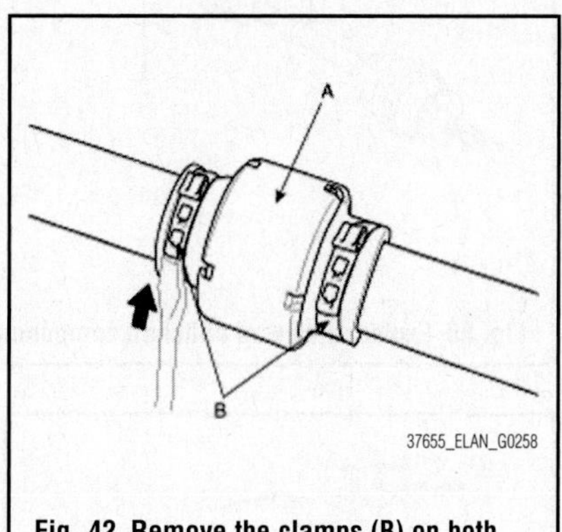

Fig. 42 Remove the clamps (B) on both sides of the dynamic damper (A)

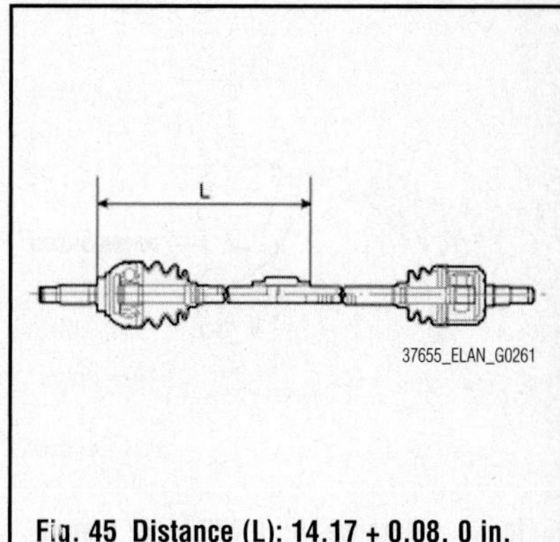

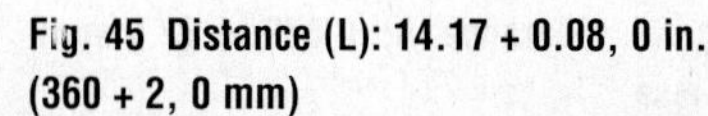

Fig. 45 Distance (L): 14.17 + 0.08, 0 in. (360 + 2, 0 mm)

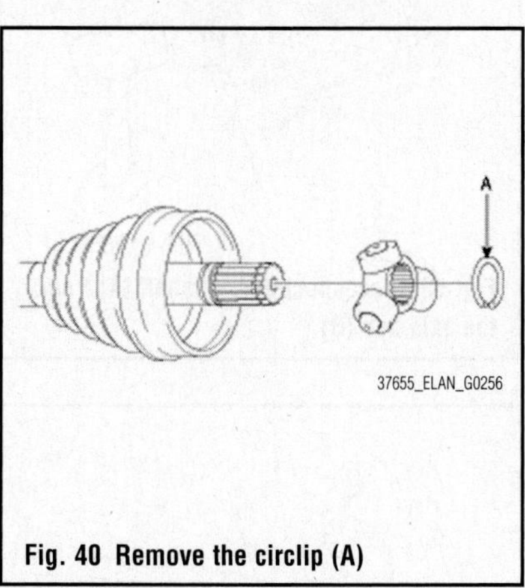

Fig. 40 Remove the circlip (A)

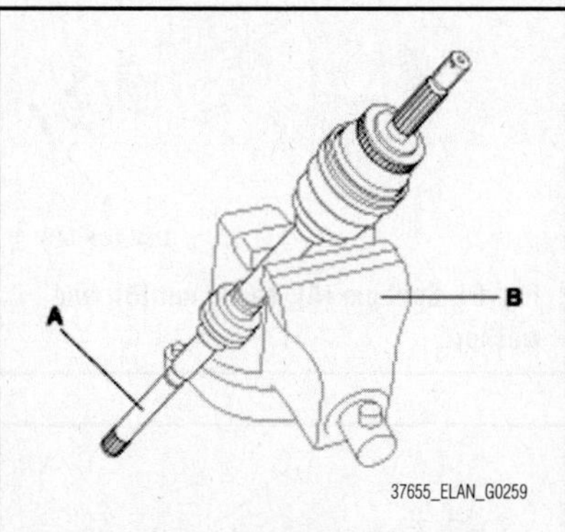

Fig. 43 Attach a vise (B) to the halfshaft (A)

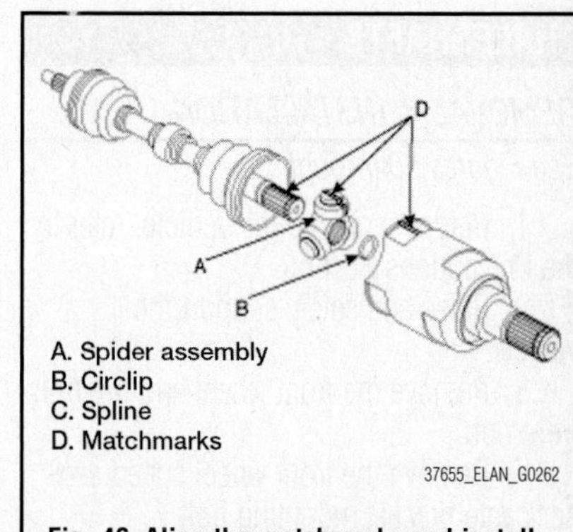

Fig. 46 Align the matchmarks and install the spider assembly and circlip

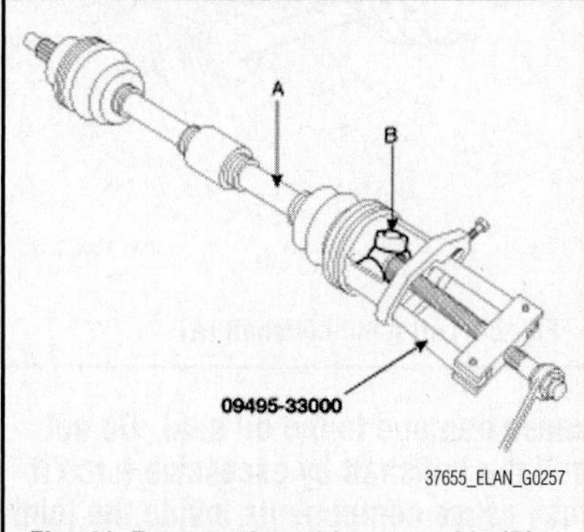

Fig. 41 Remove the spider assembly (B) from the driveshaft (A)

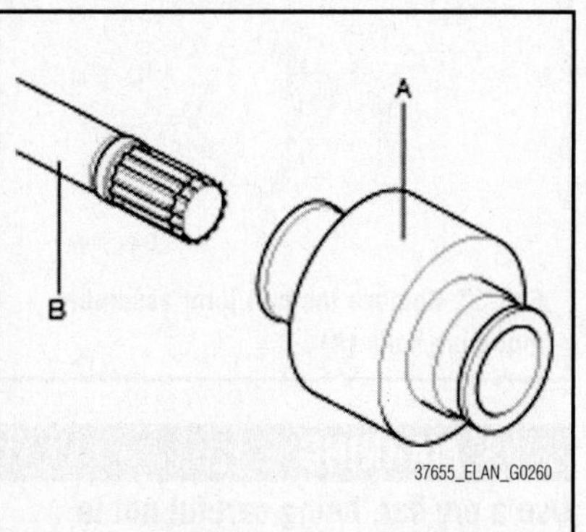

Fig. 44 Separate the dynamic damper (A) from the shaft (B)

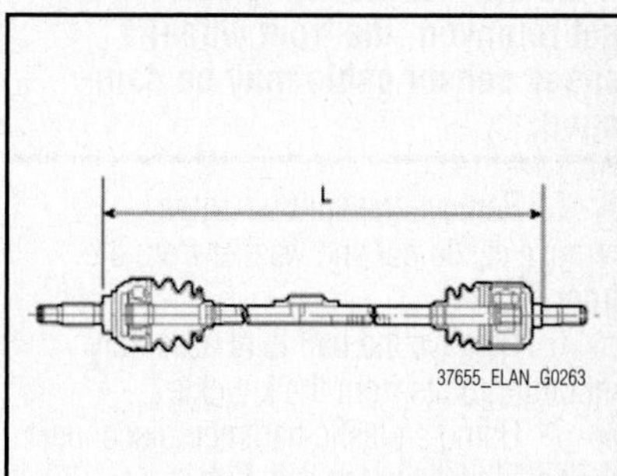

Fig. 47 Left-hand side distance (L): 21.11 in. (536.2mm), Right-hand side distance (L): 31.75 in. (806.5mm)

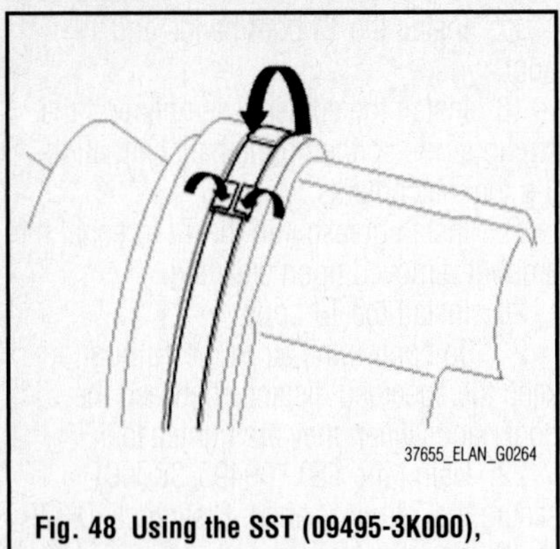
Fig. 48 Using the SST (09495-3K000), secure the TJ boot bands

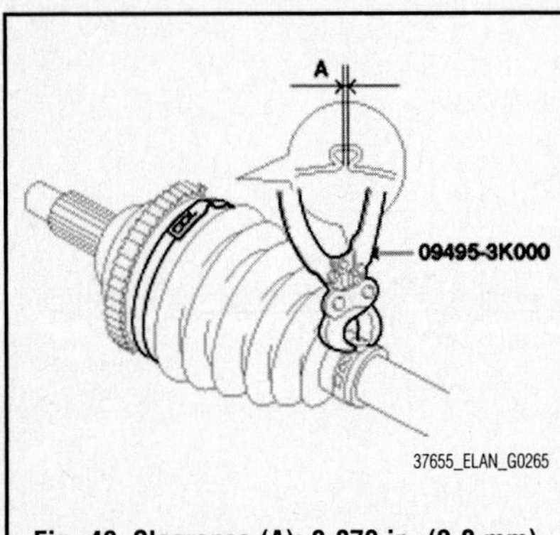

Fig. 49 Clearance (A): 0.079 in. (2.0 mm) or less

HALFSHAFTS

REMOVAL & INSTALLATION

See Figures 50 through 59.

1. Before servicing the vehicle, refer to the Precautions Section.
2. Raise and safely support the vehicle.
3. Remove the front wheel and tire from front hub.
4. Remove the front wheel speed sensor cable bracket mounting bolt.

WARNING

If the bracket mounting bolt is not removed, the front wheel speed sensor cable may be damaged.

5. Remove the split pin, then remove castle nut and washer from the front hub.
6. Remove the ball joint assembly mounting bolts from the knuckle.
7. Using a plastic hammer, disconnect the halfshaft from the axle hub.
8. Insert a pry bar between the transaxle case and joint case, and separate the halfshaft from the transaxle case.

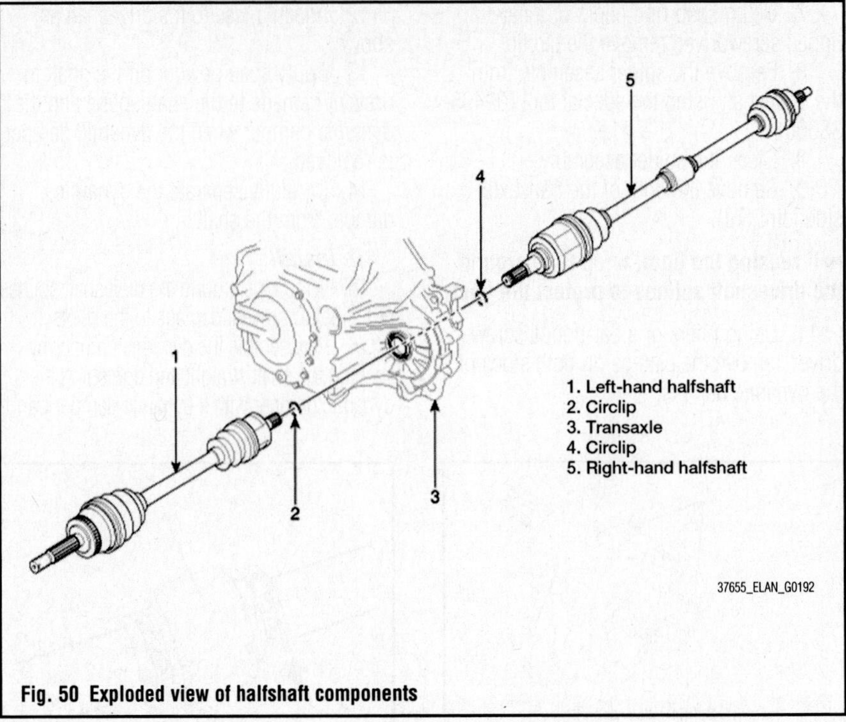

Fig. 50 Exploded view of halfshaft components

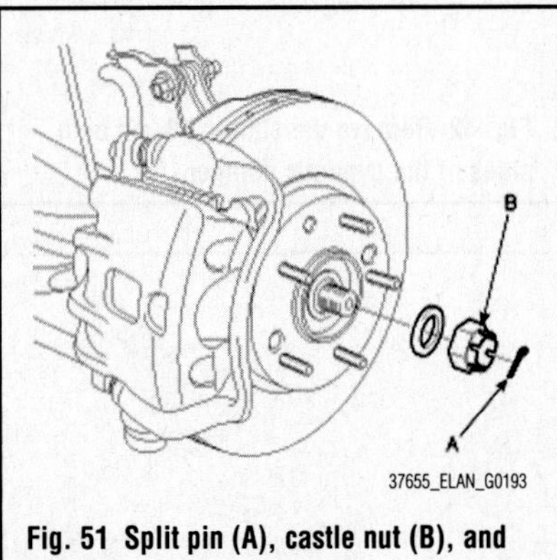

Fig. 51 Split pin (A), castle nut (B), and washer

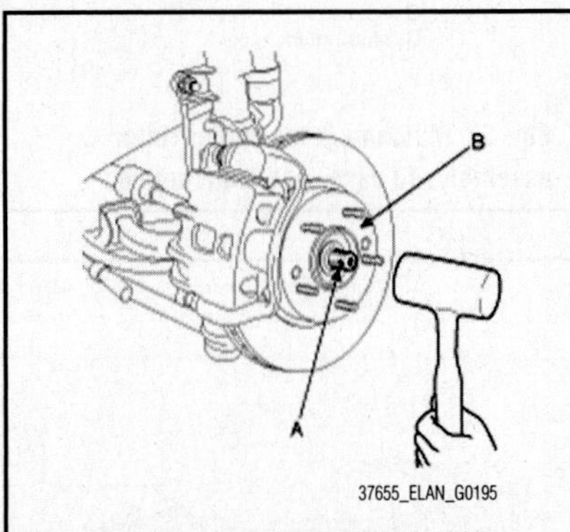

Fig. 53 Disconnect the halfshaft (A) from the axle hub (B)

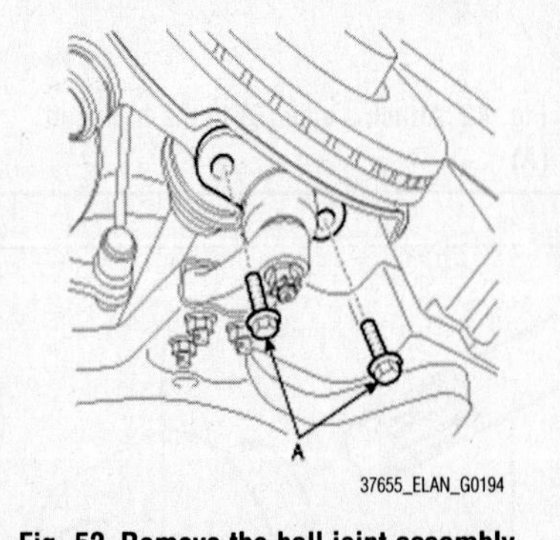

Fig. 52 Remove the ball joint assembly mounting bolts (A)

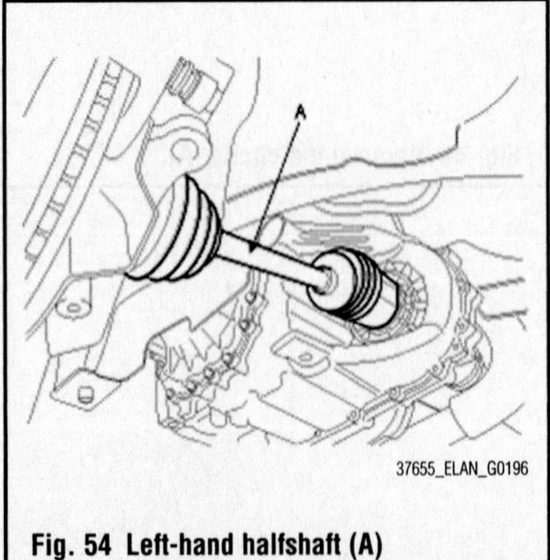

Fig. 54 Left-hand halfshaft (A)

WARNING

Use a pry bar, being careful not to damage the transaxle and joint. Do not insert the pry bar too deep, as this may cause damage to the oil seal. Do not pull the halfshaft by excessive force it may cause components inside the joint to dislodge resulting in a torn boot or a damaged bearing.

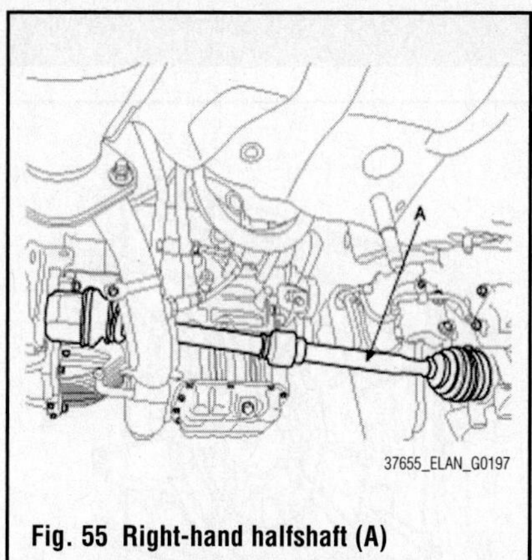

Fig. 55 Right-hand halfshaft (A)

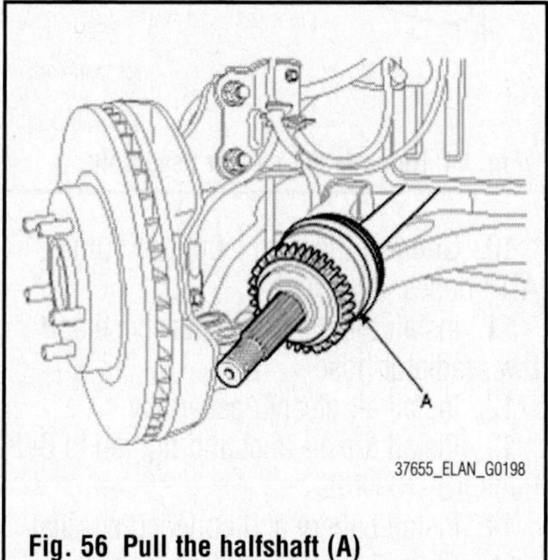

Fig. 56 Pull the halfshaft (A)

9. Pull the halfshaft from the transaxle case.

➡**Plug the hole of the transaxle case with the oil seal cap to prevent contamination. Support the halfshaft properly. Replace the retainer ring whenever the halfshaft is removed from the transaxle case.**

To install:

10. Apply gear oil on the oil seal contacting surface of transaxle case and the halfshaft splines.

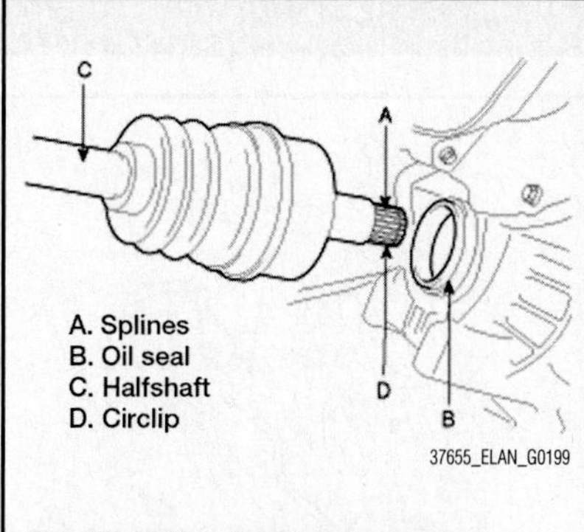

Fig. 57 Halfshaft and components installation preparation and positioning

11. Replace circlip with a new one.

➡**Be careful not to install a different kind of circlip, when replacing the circlip.**

12. Before installing the halfshaft, set the opening side of the circlip facing downward.
13. After installation, check that the halfshaft cannot be removed by hand.
14. Install the halfshaft to the axle hub.

⁂ WARNING

Be careful not to damage or dent the boots.

15. Install the ball joint assembly mounting bolt to the knuckle. Tightening torque: 72–87 ft. lbs. (98–118 Nm).
16. Install the washer, castle nut, and split pin to the front hub assembly. Tightening torque: 145–203 ft. lbs. (196–275 Nm).

➡**The washer should be assembled with convex surface outward when installing the castle nut and split pin.**

17. Install the front wheel speed sensor cable bracket mounting bolt. Tightening torque: 61–96 inch lbs. (7–11 Nm).

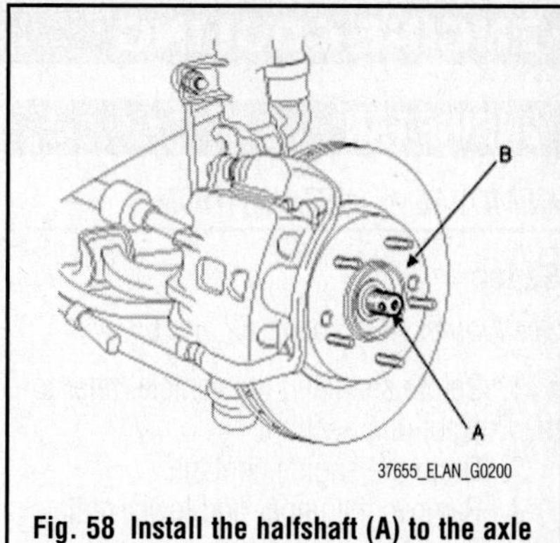

Fig. 58 Install the halfshaft (A) to the axle hub (B)

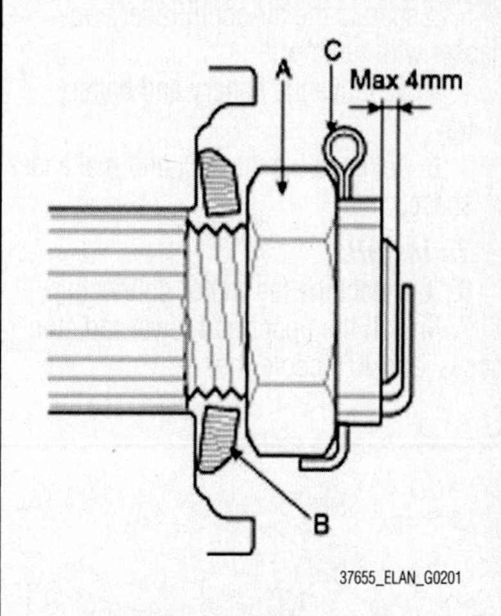

Fig. 59 The washer (B) should be assembled with convex surface outward when installing the castle nut (A)

18. Install the wheel and the tire to the front hub. Tightening torque: 65–80 ft. lbs. (88–108 Nm).

CV-BOOTS INSPECTION

1. Check the boots for leaks, damage, or deterioration.

ENGINE COOLING

ENGINE FAN

REMOVAL & INSTALLATION

Sedan

See Figures 60 through 62 and 65.

1. Before servicing the vehicle, refer to the Precautions Section.
2. Drain the engine coolant.
3. Remove the upper and lower radiator hoses, and ATF cooler hoses.
4. Disconnect the fan motor connector.
5. Separate the air conditioner condenser with radiator.
   a. Remove the battery and battery tray.
   b. Remove the air duct and make task space.

To install:

6. Connect the fan motor connector.
7. Install the upper and lower radiator hoses, and ATF cooler hoses.

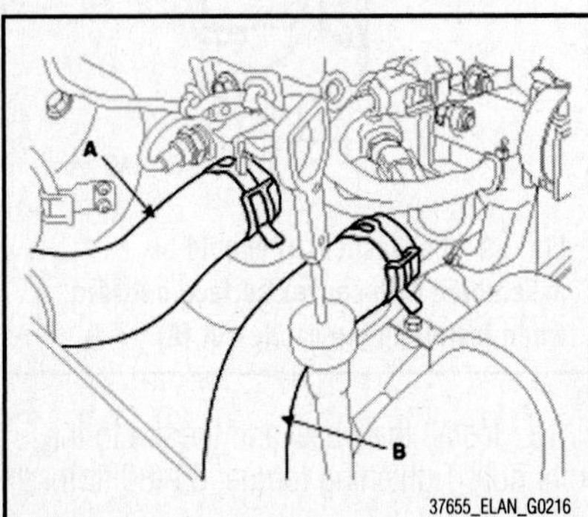

Fig. 60 Remove the upper and lower radiator hoses (A, B), and ATF cooler hoses

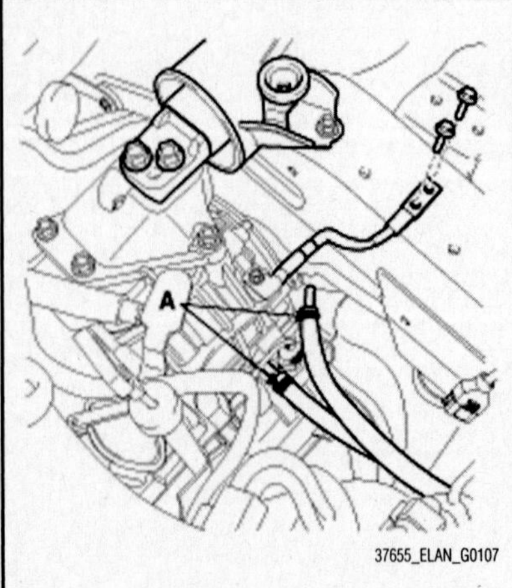

Fig. 61 Automatic Transaxle Fluid (ATF) cooler hose (A)

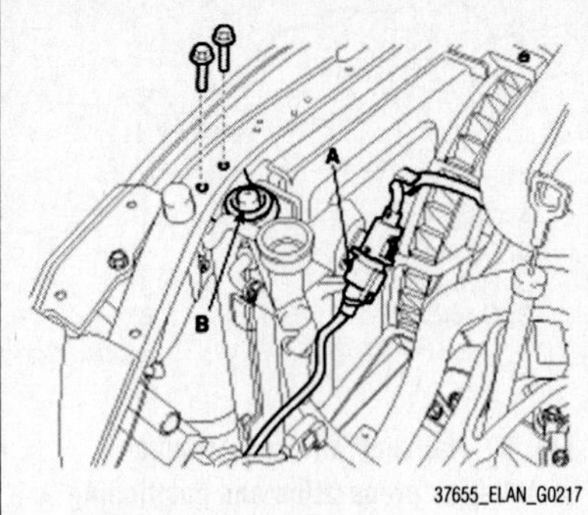

Fig. 62 Disconnect the fan motor connector (A)

8. Fill with engine coolant.
9. Start engine and check for leaks.

Touring Model

See Figures 61, 63 and 64.

1. Remove the radiator cap to speed draining.
2. Loosen the radiator drain plug and drain engine coolant.
3. Disconnect the terminals and remove the battery.
4. Remove the air duct.
5. Remove the air cleaner assembly.
   a. Disconnect the power train module (PCM) connector.
   b. Disconnect the intake hose.
   c. Remove the air cleaner assembly.
6. Disconnect the auto transaxle fluid (ATF) hose.
7. Disconnect the fan motor connector and remove the radiator mounting bracket.
8. Remove the blower assembly.

To install:

9. Connect the fan motor connector and install the radiator mounting bracket and torque to 7–8 ft. lbs. (9–11 Nm).

Fig. 63 Remove air duct

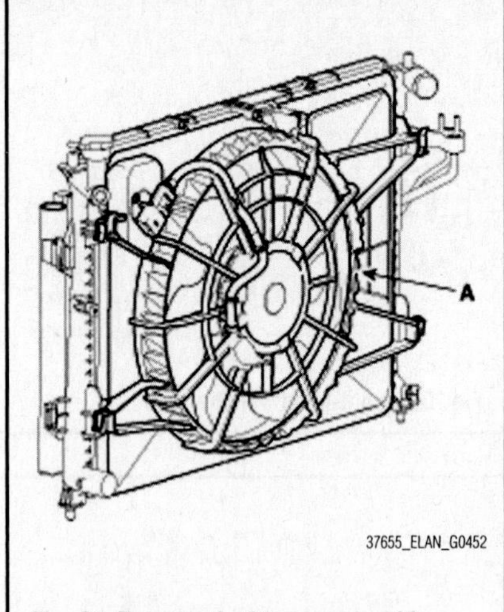

Fig. 64 Remove the blower assembly

10. Connect the auto transaxle fluid (ATF) hoses.
11. Install the upper radiator hose and lower radiator hose.
12. Install air cleaner assembly.
13. Install the air duct and tighten to 6–8 ft. lbs. (8–10 Nm).
14. Install battery and connect terminal.
15. Fill with engine coolant.
16. Start engine and check for leaks.

RADIATOR

REMOVAL & INSTALLATION

Sedan

See Figures 60 through 62 and 65.

1. Before servicing the vehicle, refer to the Precautions Section.
2. Drain the engine coolant.
3. Remove the upper and lower radiator hoses, and ATF cooler hoses.
4. Disconnect the fan motor connector.

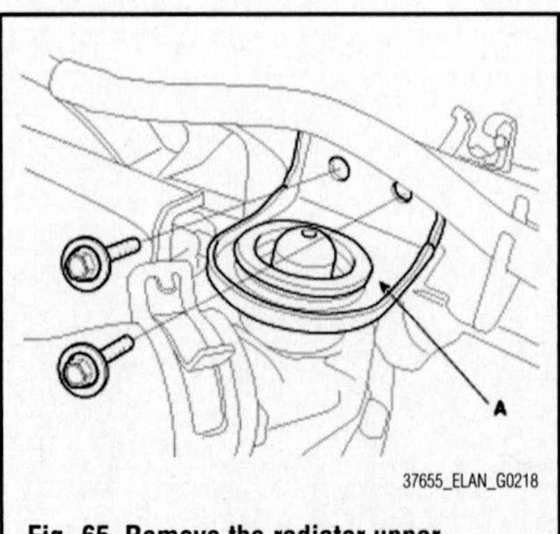

Fig. 65 Remove the radiator upper bracket (A), and then pull up the radiator

5. Separate the air conditioner condenser with radiator.
   a. Remove the battery and battery tray.
   b. Remove the air duct and make task space.
6. Remove the radiator upper bracket, and then pull up the radiator.

To install:

7. Install the cooling fan to the radiator.
8. Install the radiator at the air conditioner condenser, in the reverse order of removal.
9. Connect the fan motor connector.
10. Install the upper and lower radiator hoses, and ATF cooler hoses.
11. Fill with engine coolant.
12. Start engine and check for leaks.

Touring Model

See Figures 61, 63 and 64.

1. Remove the radiator cap to speed draining.
2. Loosen the radiator drain plug and drain engine coolant.
3. Disconnect the terminals and remove the battery.
4. Remove the air duct.
5. Remove the air cleaner assembly.
   a. Disconnect the power train module (PCM) connector.
   b. Disconnect the intake hose.
   c. Remove the air cleaner assembly.
6. Disconnect the auto transaxle fluid (ATF) hose.
7. Disconnect the fan motor connector and remove the radiator mounting bracket.
8. Remove the blower assembly.
9. After pulling back the condenser attaching bracket and remove the radiator assembly.

To install:

10. Install the radiator.
11. Connect the fan motor connector and install the radiator mounting bracket and torque to 7–8 ft. lbs. (9–11 Nm).
12. Connect the auto transaxle fluid (ATF) hoses.
13. Install the upper radiator hose and lower radiator hose.
14. Install air cleaner assembly.
15. Install the air duct and tighten to 6–8 ft. lbs. (8–10 Nm).
16. Install battery and connect terminal.
17. Fill with engine coolant.
18. Start engine and check for leaks.

THERMOSTAT

REMOVAL & INSTALLATION

See Figures 66 and 67.

CAUTION

Make sure the engine and radiator are cool to the touch prior to beginning this procedure in order to prevent scalding or burns.

1. Before servicing the vehicle, refer to the Precautions Section.
2. Drain the coolant down to thermostat level or below.

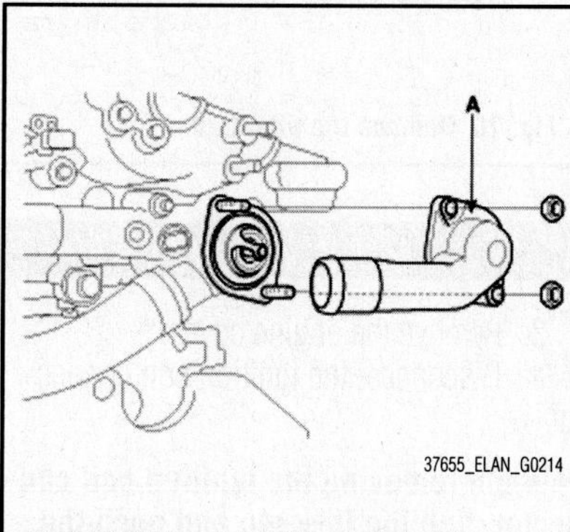

Fig. 66 Water inlet (A), gasket and thermostat

3. Remove the water inlet and gasket.
4. Remove the thermostat.

To install:

5. Place thermostat in thermostat housing:

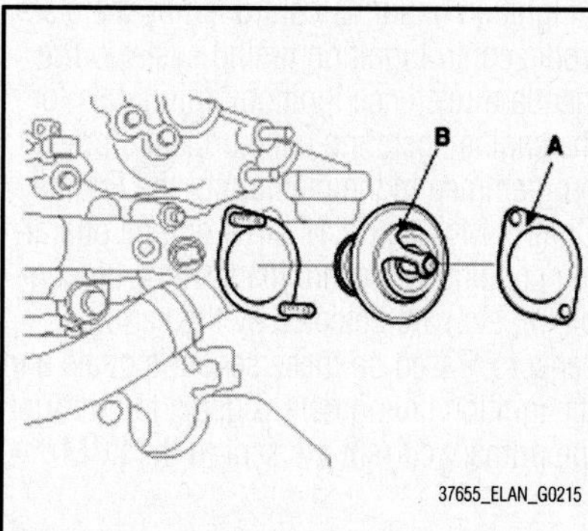

Fig. 67 Install a new gasket (A) to the thermostat (B)

   a. Install the thermostat with the jiggle valve upward.
   b. Install a new gasket to the thermostat.
6. Install water inlet and tighten to 11–15 ft. lbs. (15–20 Nm).
7. Fill with engine coolant.
8. Start engine and check for leaks.

WATER PUMP

REMOVAL & INSTALLATION

See Figure 68.

CAUTION

The system is under high pressure when the engine is hot. To avoid danger of releasing scalding engine coolant, remove the cap only when the engine is cool.

1. Before servicing the vehicle, refer to the Precautions Section.
2. Drain the engine coolant.
3. Remove the drive belts. Refer to Accessory Drive Belt Removal & Installation, in the Engine Mechanical section.
4. Remove the timing belt and timing belt idler. Refer to Timing Belt Removal & Installation, in the Engine Mechanical section.

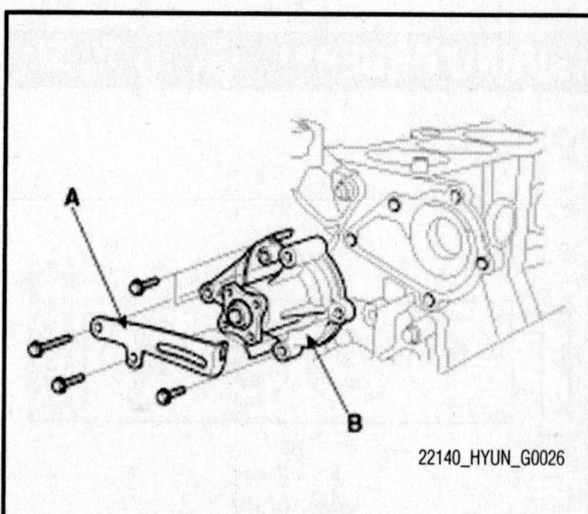

Fig. 68 Water pump (B) removal showing brace (A) and bolts (C)

5. Remove the water pump:
   a. Remove the bolts and pump pulley.
   b. Remove the bolts and alternator brace.
   c. Remove the water pump and gasket.

To install:

6. Install the water pump and a new gasket with the 3 bolts. Tightening torque: 15–20 ft. lbs. (20–27 Nm).
7. Install the alternator brace with the 2 bolts. Tightening torque: 15–20 ft. lbs. (20–27 Nm).
8. Install the pump pulley and bolts. Tightening torque: 72–84 inch lbs. (8–10 Nm).
9. Install the timing belt idler.
10. Install the timing belt.
11. Install the drive belts.
12. Fill with engine coolant.
13. Start engine and check for leaks.
14. Recheck engine coolant level.

ENGINE ELECTRICAL — CHARGING SYSTEM

ALTERNATOR

REMOVAL & INSTALLATION

See Figures 69 and 70.

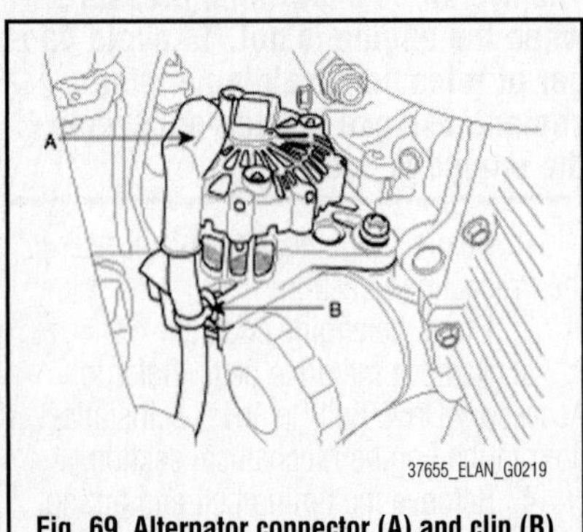

Fig. 69 Alternator connector (A) and clip (B)

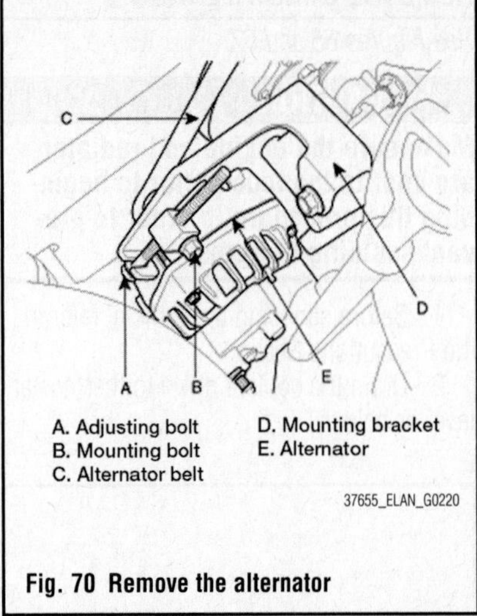

Fig. 70 Remove the alternator

1. Before servicing the vehicle, refer to the precautions section.
2. Disconnect the battery negative terminal first, then disconnect the positive terminal.
3. Disconnect the alternator connector and "B" terminal cable from the alternator. Loosen the clip.
4. Remove the adjusting bolt, mounting bolt, the alternator belt and the alternator mounting bracket.
5. Pull out the through bolt, then remove the alternator.

To install:

6. Installation is the reverse of removal.
7. Adjust the alternator belt tension after installation.

ENGINE ELECTRICAL — IGNITION SYSTEM

FIRING ORDER

See Figure 71.

Fig. 71 2.0L engine Firing order: 1–3–4–2 Distributorless ignition system

IGNITION COIL PACK

REMOVAL & INSTALLATION

1. Before servicing the vehicle, refer to the Precautions Section.
2. Disconnect the negative battery cable.
3. Remove the engine cover.
4. Disconnect the ignition coil connector.

➡**When removing the ignition coil connector, pull the lock pin and push the clip.**

5. Remove the ignition coil.

To install:

6. Installation is the reverse of removal.

IGNITION TIMING

ADJUSTMENT

Ignition timing is controlled by the electronic control ignition timing system. The standard reference ignition timing data for the engine operating conditions are pre-programmed in the memory of the Engine Control Module (ECM). The engine operating conditions (speed, load, warm-up condition, etc.) are detected by the various sensors. Based on these sensor signals and the ignition timing data, signals to interrupt the primary current are sent to the ECM. The ignition coil is activated, and timing is controlled.

SPARK PLUGS

REMOVAL & INSTALLATION

1. Before servicing the vehicle, refer to the Precautions Section.
2. Remove the engine cover.
3. Disconnect the spark plug cables from the spark plugs.
4. Use a spark plug socket and wrench to remove the spark plugs.

WARNING

Be careful that no contaminates enter through the spark plug holes.

➡**Check the electrode gap on the spark plugs before installation. Specification: 0.039–0.043 inch (1.0–1.1 mm).**

To install:

5. To install, reverse the removal procedure. Tighten the spark plugs to 15–21 ft. lbs. (20–30 Nm).

ENGINE ELECTRICAL

STARTER

REMOVAL & INSTALLATION

1. Before servicing the vehicle, refer to the Precautions Section.
2. Disconnect the negative battery cable.
3. Remove the speedometer cable and shift cable from the transaxle.
4. Remove the starter motor assembly.

To install:

5. Installation is the reverse of removal.

ENGINE MECHANICAL

➡**Disconnecting the negative battery cable may interfere with the functions of the on board computer systems and may require the computer to undergo a relearning process, once the negative battery cable is reconnected.**

ACCESSORY DRIVE BELTS

ACCESSORY BELT ROUTING

See Figure 72.

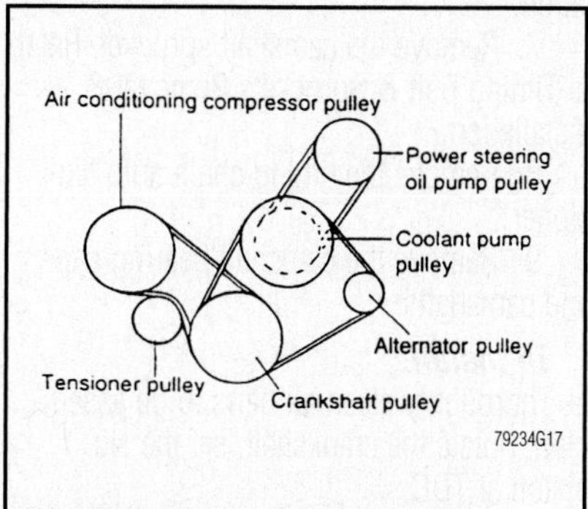

Fig. 72 Accessory drive belt routing

INSPECTION

Inspect the accessory drive belt for signs of glazing or cracking. A glazed belt will be perfectly smooth from slippage, while a good belt will have a slight texture of fabric visible. Cracks will usually start at the inner edge of the belt and run outward. All worn or damaged accessory drive belts should be replaced immediately.

ADJUSTMENT

1. Loosen the tension mounting bolt.
2. Turn the adjusting bolt to obtain the proper belt tension, then retighten the mounting bolt.
3. Recheck the deflection of the drive belt.

REMOVAL & INSTALLATION

1. Before servicing the vehicle, refer to the Precautions Section.
2. Raise and support the vehicle.
3. Remove the engine splash shield.
4. Rotate the drive belt tensioner clockwise to release the drive belt tension.
5. Remove the drive belt from the alternator.
6. Slowly release the drive belt tensioner.
7. Remove the drive belt from the accessory drive pulleys.

To install:

8. Install the drive belt to the accessory drive pulley.
9. Rotate the drive belt tensioner clockwise.
10. Install the drive belt to the alternator.
11. Ensure the drive belt is properly aligned and seated into the grooves of the accessory drive pulleys.
12. Slowly release the drive belt tensioner.
13. Install the engine splash shield.
14. Lower the vehicle.

CAMSHAFT AND VALVE LIFTERS

INSPECTION

See Figures 73 and 74.

1. Inspect the cam lobes.
   a. Using a micrometer, measure the cam lobe height.
   b. If the cam lobe height is less than the specifications, replace the camshaft.
2. Inspect the camshaft journal clearance.
   a. Clean the bearing caps and camshaft journals.
   b. Place the camshafts on the cylinder head.
   c. Lay a strip of Plastigage across each of the camshaft journal.
   d. Install the bearing caps.

➡**Do not turn the camshaft.**

   e. Remove the bearing caps.
   f. Measure the Plastigage at its widest point.
   g. If the oil clearance is greater than the specifications, replace the camshaft. If necessary, replace the bearing caps and cylinder head as a set.
   h. Completely remove the Plastigage.
   i. Remove the camshafts.

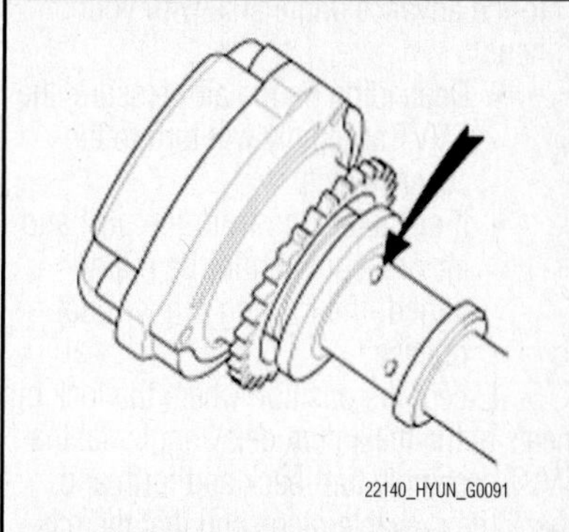

Fig. 73 Apply vinyl tape to the CVVT on all parts except the one indicated by the arrow

3. Inspect the camshaft end play.
   a. Install the camshafts.
   b. Using a dial indicator, measure the end play while moving the camshaft back and forth.
   c. If the end play is greater than specified, replace the camshaft. If necessary, replace the bearing caps and cylinder head as a set.
   d. Remove the camshafts.
4. Inspect the Continuous Variable Valve Timing (CVVT) assembly.
   a. Check that the CVVT assembly will not turn.
   b. Apply vinyl tape to all the parts except the one indicated by the arrow in the illustration.

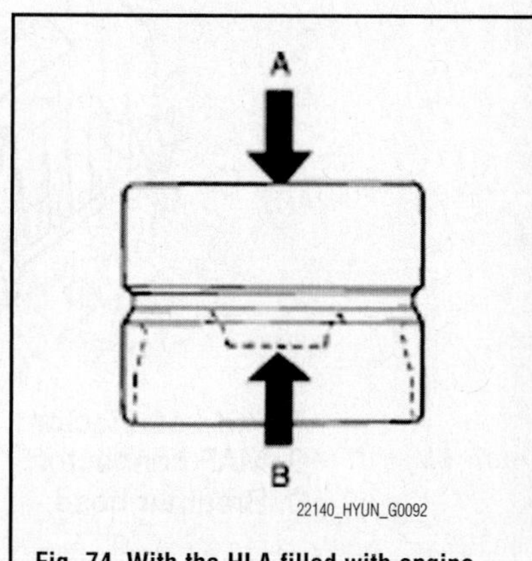

Fig. 74 With the HLA filled with engine oil, hold A and press B by hand

c. Wrap tape around the tip of the air gun and apply air of approx. 14 psi to the port of the camshaft. Perform this in order to release the lock pin for the maximum delay angle locking.

➡Wrap a shop rag around the CVVT as the oil may spray out when the air pressure is applied.

d. Under the condition of air pressure being applied, turn the CVVT assembly to the advance angle side with your hand.

- Depending on the air pressure, the CVVT assembly will turn to the advance side
- If air is leaking from the port and air pressure cannot be maintained, the locking pin will not release

5. Except the position where the lock pin meets at the maximum delay angle, let the CVVT assembly turn back and forth and check the movable range and that there is no disturbance.

a. The CVVT should move smoothly in the range of about 20°.

b. Turn the CVVT assembly with your hand and lock it at the maximum delay angle position.

6. Inspect the Hydraulic Lash Adjuster (HLA).

a. With the HLA filled with engine oil, hold A and press B by hand.

b. If B moves, replace the HLA.

REMOVAL & INSTALLATION

See Figures 75 through 81.

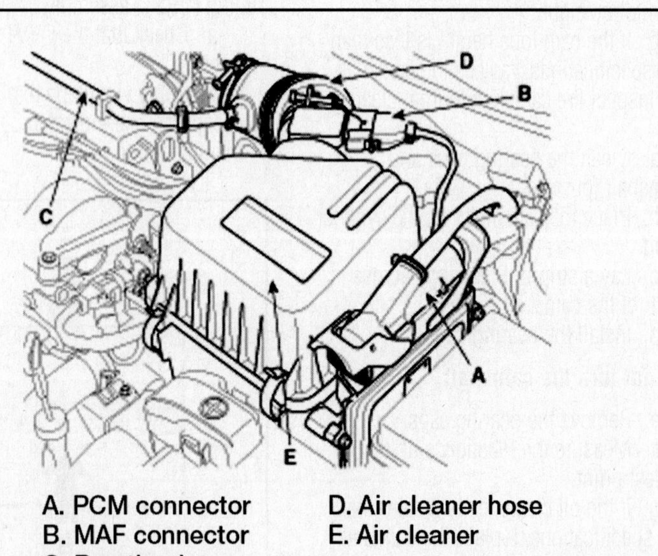

Fig. 75 Air cleaner and components

Fig. 76 Timing chain auto tensioner (A) and stopper pin (B)

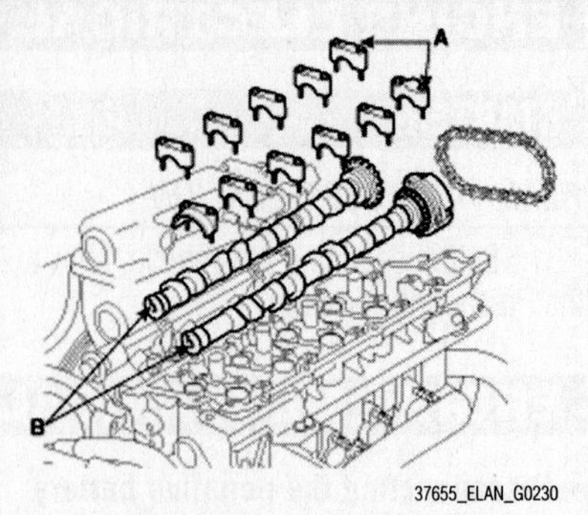

Fig. 77 Remove the camshaft bearing caps (A) and camshafts (B)

➡Special Service Tool (SST) required for this service: 09221-21000, Camshaft Oil Seal Installer.

Engine removal is not required for this procedure. Use a fender cover to avoid damaging painted surfaces. To avoid damage, unplug the wiring connectors carefully while holding the connector portion. Mark all wiring and hoses to avoid misconnection. Inspect the timing belt before removal. Turn the crankshaft pulley so that the No. 1 piston is at Top Dead Center (TDC).

1. Before servicing the vehicle, refer to the Precautions Section.
2. Disconnect the terminals from battery and remove the battery.
3. Remove the engine cover.
4. Remove the intake air hose and air cleaner assembly:

a. Disconnect the PCM connectors.

b. Disconnect the MAF connector.

c. Disconnect the breather hose from cleaner air hose.

d. Remove the intake air hose and air cleaner assembly.

5. Remove the cylinder head cover.
6. Remove the timing belt. Refer to Timing Belt & Sprockets Removal & Installation.
7. Remove the camshaft sprocket. Refer to Timing Belt & Sprockets Removal & Installation.
8. Remove the timing chain auto tensioner.
9. Remove the camshaft bearing caps and camshafts.

To install:

Thoroughly clean all parts to be assembled. Rotate the crankshaft, set the No. 1 piston at TDC.

10. Install the camshafts:

a. Align the camshaft timing chain with the intake timing chain sprocket and exhaust timing chain sprocket as shown.

b. Install the camshafts and bearing caps. Tightening torque: 10–11 ft. lbs. (14–15 Nm).

c. Install the timing chain auto tensioner. Tightening torque: 72–84 inch lbs. (8–10 Nm).

d. Remove the auto tensioner stopper pin.

e. Check and adjust valve clearance.

11. Using a seal installer, install the camshaft bearing oil seal.
12. Install the camshaft sprocket.
13. Install the timing belt.
14. Install the cylinder head cover.
15. Install the breather hose, intake air hose and air cleaner assembly, and connect the connectors.
16. Install the engine cover.
17. Connect the negative battery cable.
18. Start the engine and check for leaks.

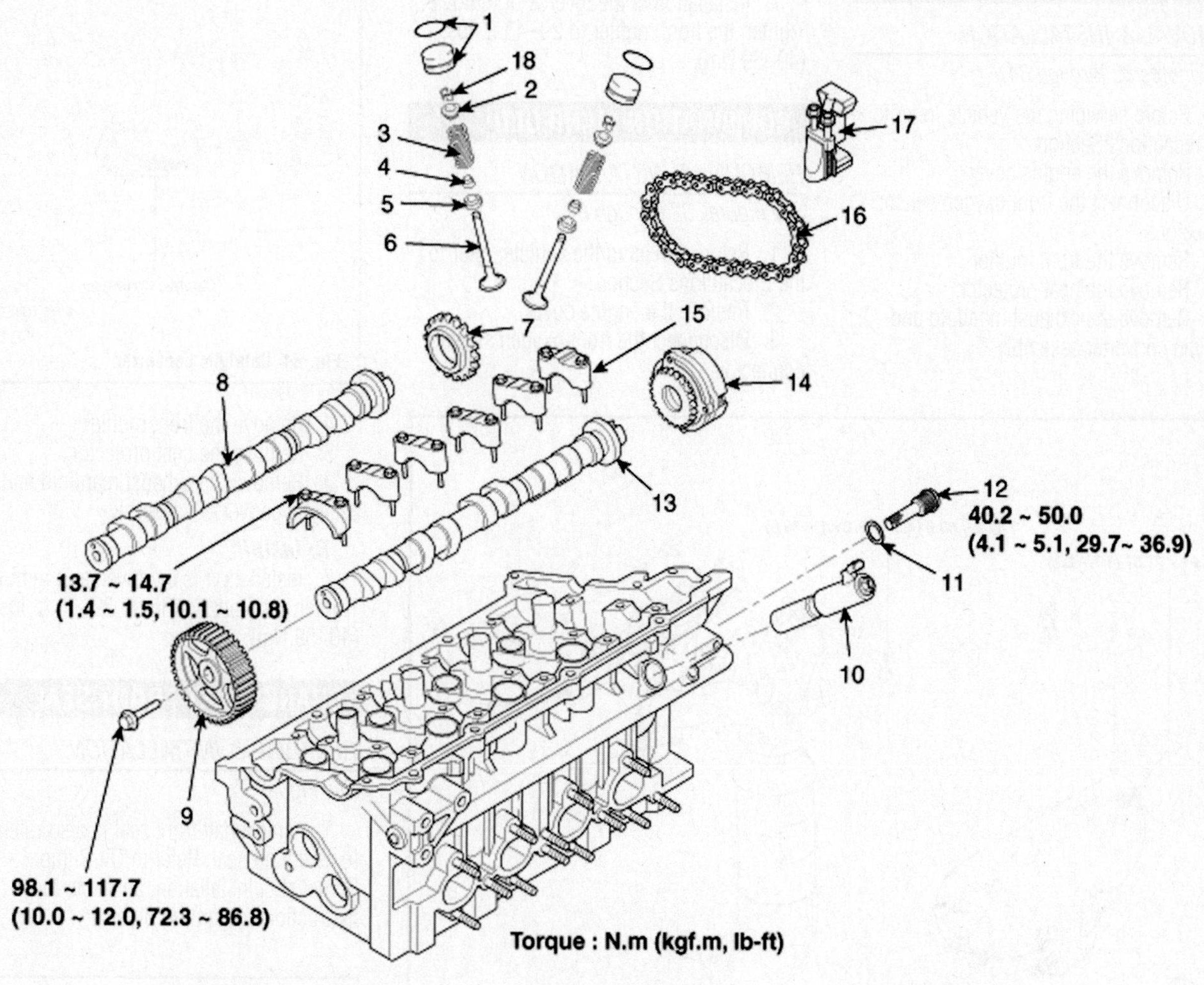

1
18
2
3
4
5
6
17
16
7
15
8
14
13
12
40.2 ~ 50.0
(4.1 ~ 5.1, 29.7~ 36.9)
11
13.7 ~ 14.7
(1.4 ~ 1.5, 10.1 ~ 10.8)
10
9
98.1 ~ 117.7
(10.0 ~ 12.0, 72.3 ~ 86.8)
Torque : N.m (kgf.m, lb-ft)

1. Mechanical Lash Adjuster (MLA)
2. Retainer
3. Valve spring
4. Stem seal
5. Spring seat
6. Valve
7. Chain sprocket
8. Intake camshaft
9. Camshaft sprocket
10. Oil Control Valve (OCV)
11. Washer
12. OCV filter
13. Exhaust camshaft
14. CVVT assembly
15. Camshaft bearing cap
16. Timing chain
17. Auto Tensioner
18. Retainer lock

37655_ELAN_G0231

Fig. 78 Exploded view of cylinder head and related components

37655_ELAN_G0225

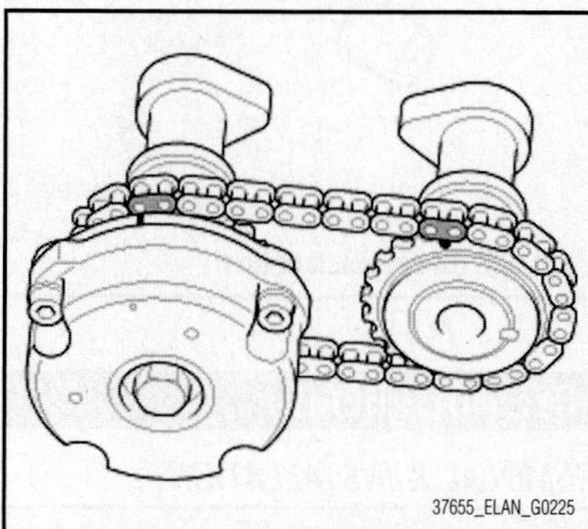

Fig. 79 Align the camshaft timing chain with the intake timing chain sprocket and exhaust timing chain sprocket

B
A

37655_ELAN_G0232

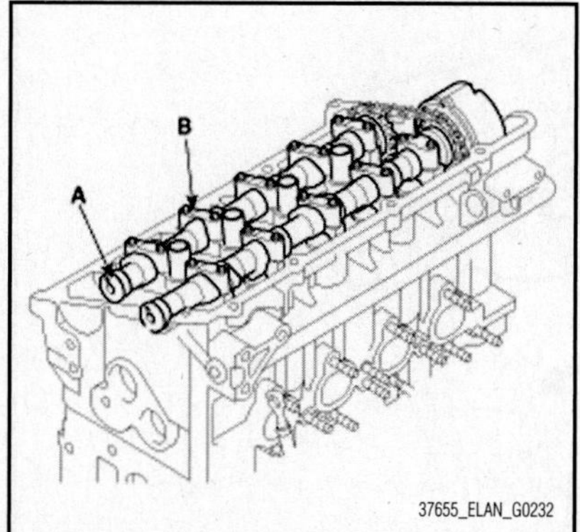

Fig. 80 Install the camshafts (A) and bearing caps (B)

09221-21000

37655_ELAN_G0233

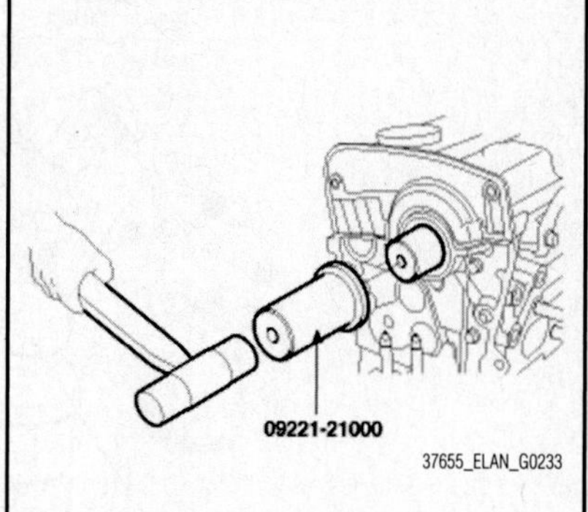

Fig. 81 Using the seal installer, install the camshaft bearing oil seal

CATALYTIC CONVERTER

REMOVAL & INSTALLATION

See Figures 82 through 84.

1. Before servicing the vehicle, refer to the Precautions Section.
2. Remove the engine cover.
3. Disconnect the front oxygen sensor connector.
4. Remove the front muffler.
5. Remove the heat protector.
6. Remove the exhaust manifold and catalytic converter assembly.

To install:

7. Installation is the reverse of removal. Tighten the front muffler to 29–43 ft. lbs. (40–59 Nm).

COMBINATION MANIFOLD

REMOVAL & INSTALLATION

See Figures 82 through 84.

1. Before servicing the vehicle, refer to the Precautions Section.
2. Remove the engine cover.
3. Disconnect the front oxygen sensor connector.

42.2 ~ 53.9 (4.3 ~5.5, 31.1 ~ 39.8)
16.67 ~ 21.57 (1.7 ~ 2.2)
49.0 ~ 58.8
(5.0 ~ 6.0, 36.2 ~ 43.4)
Torque : N.m (kgf.m, lb-ft)

1. Cylinder head
2. Gasket
3. Exhaust manifold
4. Front oxygen sensor
5. Heat protector

37655_ELAN_G0268

Fig. 82 Exploded view of the exhaust manifold and related components

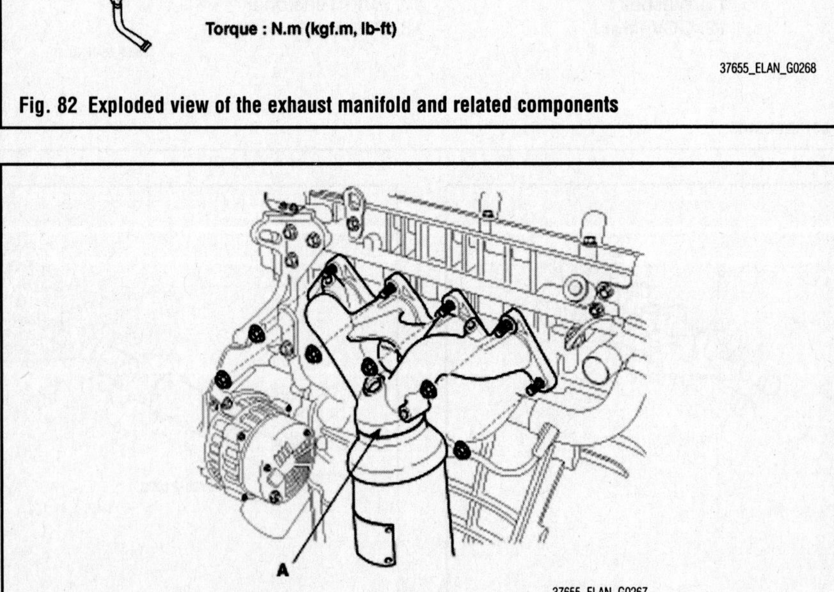

Fig. 83 Remove the exhaust manifold and catalytic converter assembly (A)

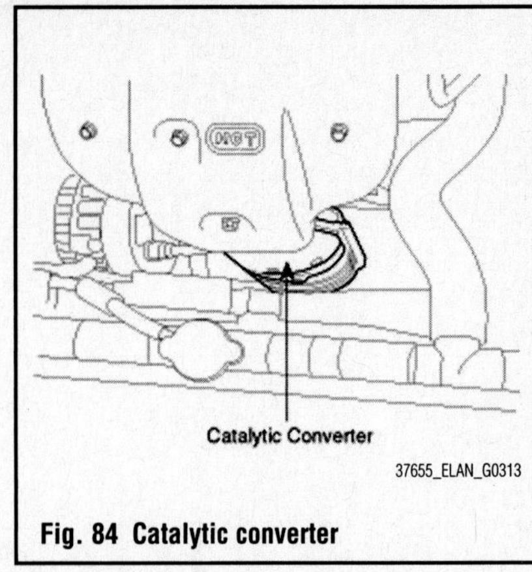

Fig. 84 Catalytic converter

4. Remove the front muffler.
5. Remove the heat protector.
6. Remove the exhaust manifold and catalytic converter assembly.

To install:

7. Installation is the reverse of removal. Tighten the front muffler to 29–43 ft. lbs. (40–59 Nm).

CRANKSHAFT FRONT SEAL

REMOVAL & INSTALLATION

See Figure 85.

The crankshaft front seal is also referred to as the oil seal. Refer to Oil Pump Removal & Installation, and note the instructions regarding oil seal.

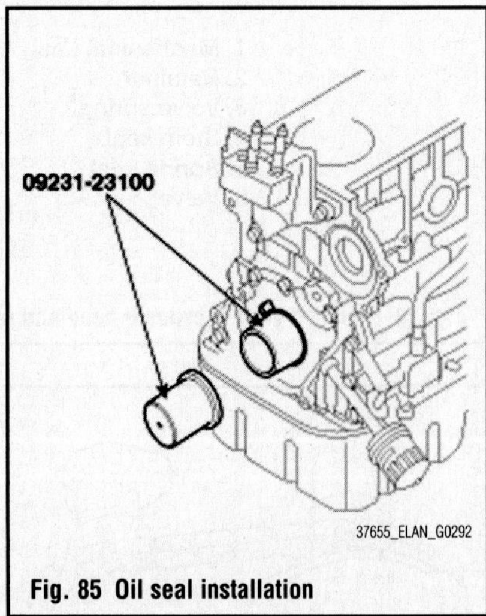

Fig. 85 Oil seal installation

CYLINDER HEAD

REMOVAL & INSTALLATION

See Figures 86 through 97.

Engine removal is not required for this procedure. Use a fender cover to avoid damaging painted surfaces. To avoid damage, unplug the wiring connectors carefully while holding the connector portion. Mark all wiring and hoses to avoid misconnection. Inspect the timing belt before removal. Turn the crankshaft pulley so that the No. 1 piston is at Top Dead Center (TDC).

1. Before servicing the vehicle, refer to the Precautions Section.
2. Disconnect the terminals from battery and remove the battery.
3. Remove the engine cover.
4. Drain the engine coolant. Remove the radiator cap to speed draining.
5. Remove the intake air hose and air cleaner assembly:
    a. Disconnect the PCM connectors.
    b. Disconnect the MAF connector.
    c. Disconnect the breather hose from cleaner air hose.
    d. Remove the intake air hose and air cleaner assembly.
6. Remove the upper radiator hose and lower radiator hose.
7. Remove the heater hoses.
8. Remove the engine wire harness connectors and wire harness clamps from the cylinder head and the intake manifold:
    a. Disconnect the Oil Control Valve (OCV) connector.
    b. Disconnect the Oil Temperature Sensor (OTS) connector.
    c. Disconnect the Engine Coolant Temperature (ECT) sensor connector.
    d. Disconnect the ignition coil connector.
    e. Disconnect the Throttle Position Sensor (TPS) connector.
    f. Disconnect the Idle Speed Actuator (ISA) connector.
    g. Disconnect the Camshaft Position Sensor (CMP) connector.
    h. Disconnect the injector connectors.
    i. Disconnect the knock sensor connector.
    j. Disconnect the Purge Control Solenoid Valve (PCSV) connector.
    k. Disconnect the front oxygen sensor connector.
9. Remove the fuel inlet hose from the delivery pipe.
10. Remove the PCSV hose.
11. Remove the brake booster vacuum hose.
12. Remove the accelerator cable and the auto-cruise cable by loosening the locknut, then slip the cable end out of the throttle linkage.
13. Remove the ignition coil.
14. Remove the PCV hose.
15. Remove the cylinder head cover.
16. Remove the timing belt. Refer to Timing Belt & Sprockets Removal & Installation.

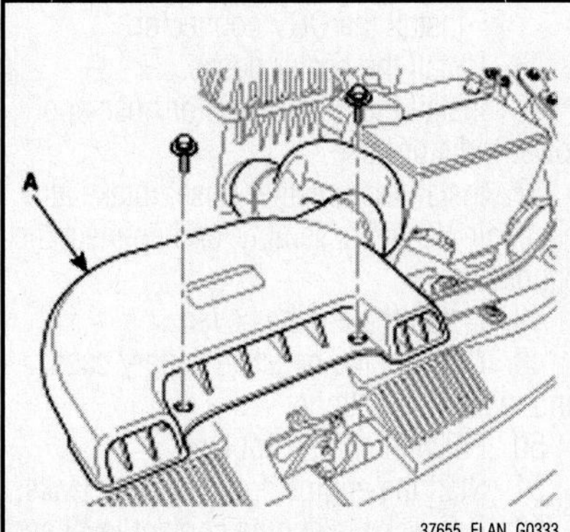

Fig. 86 Air duct (A)—Touring model

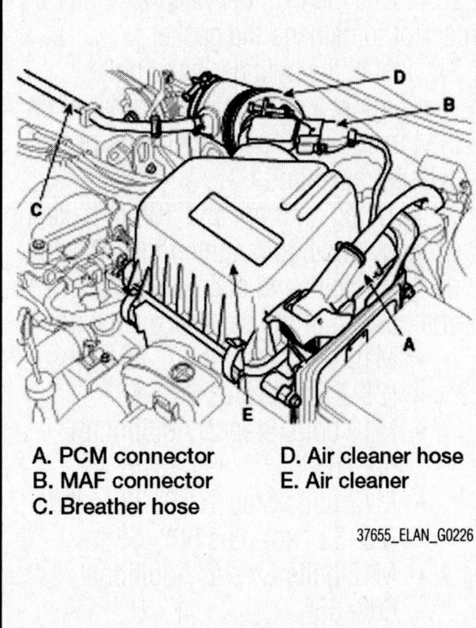

Fig. 87 Air cleaner and components

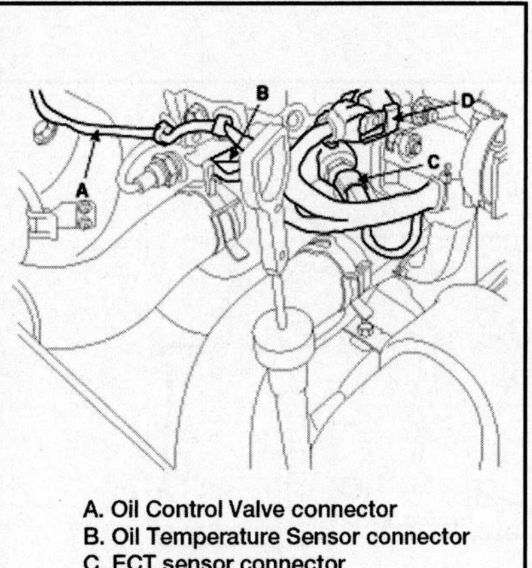

Fig. 88 Remove the engine wire harness connectors and wire harness clamps

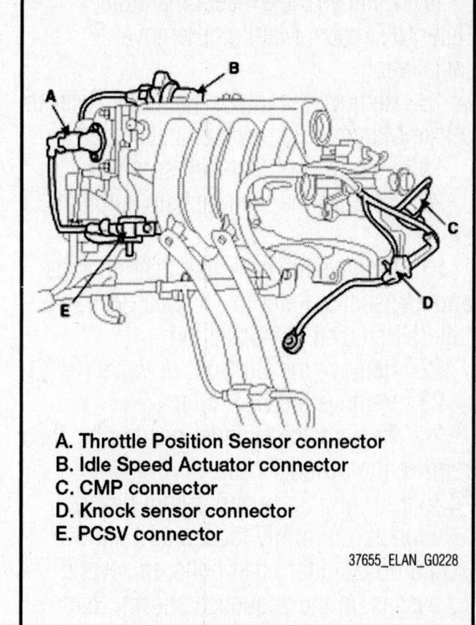

Fig. 89 Intake manifold connectors

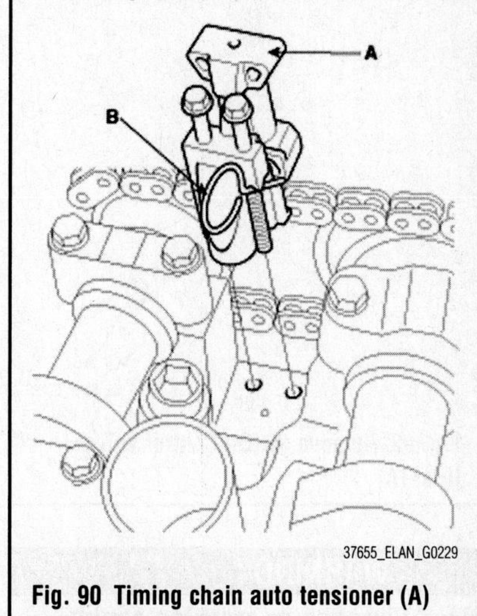

Fig. 90 Timing chain auto tensioner (A) and stopper pin (B)

Fig. 91 Remove the Oil Control Valve (A)

17. Remove the exhaust manifold. Refer to Exhaust Manifold Removal & Installation
18. Remove the intake manifold. Refer to Intake Manifold Removal & Installation.
19. Remove the camshaft sprocket.
20. Remove the timing chain auto tensioner.
21. Remove the camshaft bearing caps and camshafts. Refer to Camshaft & Valve Lifters Removal & Installation.
22. Remove the Oil Control Valve (OCV).
23. Remove the OCV filter.
24. Remove the cylinder head bolts, then remove the cylinder head:
  a. Using 8mm and 10mm hexagon wrench, uniformly loosen and remove the 10 cylinder head bolts, in several passes, in the sequence shown. Remove the 10 cylinder head bolts and plate washers.

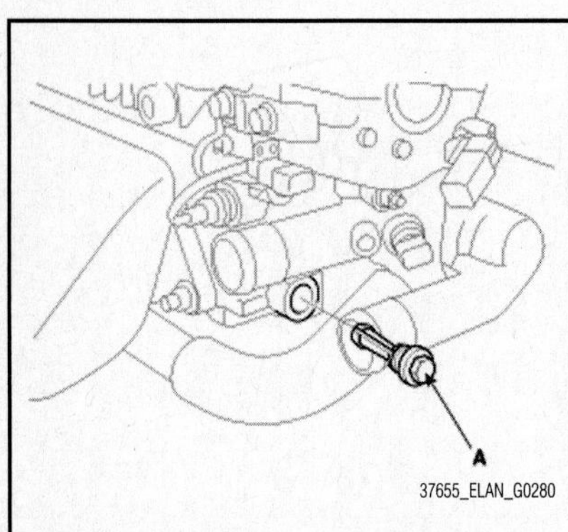

Fig. 92 Remove the Oil Control Valve filter (A)

WARNING

Head warpage or cracking could result from removing bolts in an incorrect order.

  b. Lift the cylinder head from the dowels on the cylinder block and replace the cylinder head on wooden blocks on a bench.

WARNING

Be careful not to damage the contact surfaces of the cylinder head and cylinder block.

To install:

➡Thoroughly clean all parts to be assembled. Always use a new cylinder head and manifold gasket. Always use a new cylinder head bolt. The cylinder head gasket is a metal gasket. Take care not to bend it. Rotate the crankshaft, set the No. 1 piston at TDC.

25. Install the cylinder head gasket on the cylinder block. Be careful of the installation direction.
26. Place the cylinder head carefully in order not to damage the gasket.
27. Install the cylinder head bolts:
  a. Apply a light coat if engine oil on the threads and under the heads of the cylinder head bolts.
  b. Using an 8mm and 10mm hexagon wrench, install and tighten the 10 cylinder head bolts and plate washers, in several passes, in the sequence shown.
   - M10 bolts Step 1: 17–20 ft. lbs. (23–27 Nm) plus 60°–65°
   - M10 bolts Step 2: Additional 60°–65°
   - M12 bolts Step 1: 20–23 ft. lbs. (28–31 Nm) plus 60°–65°
   - M12 bolts Step 2: Additional 60°–65°
28. Install the OCV filter and tighten to 30–37 ft. lbs. (40–50 Nm).

➡Always use a new OCV filter gasket and keep the OCV filter clean.

29. Install the OCV. Tightening torque: 84–108 inch lbs. (10–12 Nm).

➡Do not reuse the OCV when dropped. Keep the OCV clean. Do not hold the OCV sleeve during servicing. When the OCV is installed on the engine, do not move the engine while holding the OCV yoke.

30. Install the camshafts. Refer to Camshaft & Valve Lifters Removal & Installation.
31. Check and adjust the valve clearance.
32. Using the SST (09221-21000), or equivalent, install the camshaft bearing oil seal.
33. Install the camshaft sprocket and tighten bolt to 73–87 ft. lbs. (98–118 Nm).
34. Install the timing belt.
35. Install the cylinder head cover.
36. Install the intake manifold.
37. Install the exhaust manifold.
38. Install the PCV.
39. Install the ignition coil.
40. Install the accelerator and the auto-cruise cables.
41. Install the brake booster hose.
42. Install the PCSV hose.
43. Install the fuel inlet hose.
44. Install the engine wire harness connectors and wire harness clamps to the cylinder head and the intake manifold:
  a. Install the front oxygen sensor connector.
  b. Install the knock sensor connector.
  c. Install the fuel injector connectors.
  d. Install the CMP connector.
  e. Install the PCSV connector.
  f. Install the ISA connector.
  g. Install the TPS connector.
  h. Install the ignition coil connector.
  i. Install the ECT sensor connector.
  j. Install the oil temperature sensor connector.
  k. Install the OCV connector.
45. Install the heater hoses.
46. Install the upper radiator hose and lower radiator hose.
47. Install the breather hose, intake air hose, air cleaner assembly, and connect the connectors.
48. Install the engine cover.
49. Connect the negative battery cable and install the battery.
50. Fill with engine coolant.
51. Start the engine and check for leaks.
52. Recheck the engine coolant level and oil level.

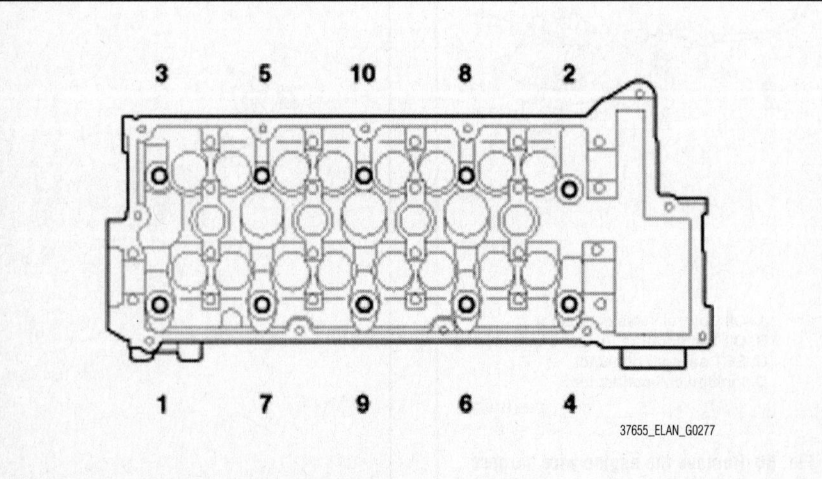

Fig. 93 Cylinder head bolt removal sequence

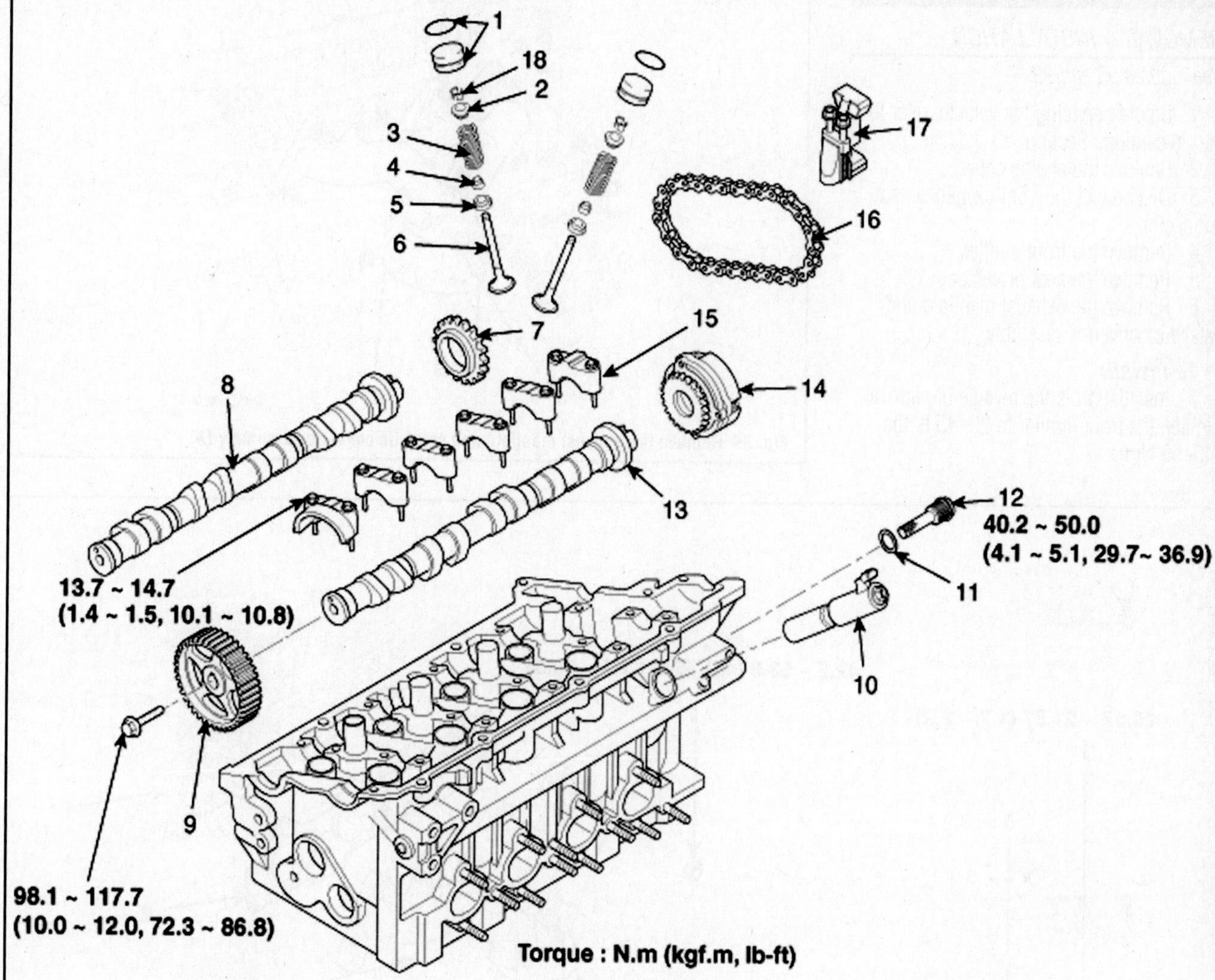

1
18
2
3
4
5
6
17
16
15
7
14
8
13
12
40.2 ~ 50.0
(4.1 ~ 5.1, 29.7~ 36.9)
11
13.7 ~ 14.7
(1.4 ~ 1.5, 10.1 ~ 10.8)
10
9
98.1 ~ 117.7
(10.0 ~ 12.0, 72.3 ~ 86.8)
Torque : N.m (kgf.m, lb-ft)

1. Mechanical Lash Adjuster (MLA)
2. Retainer
3. Valve spring
4. Stem seal
5. Spring seat
6. Valve
7. Chain sprocket
8. Intake camshaft
9. Camshaft sprocket
10. Oil Control Valve (OCV)
11. Washer
12. OCV filter
13. Exhaust camshaft
14. CVVT assembly
15. Camshaft bearing cap
16. Timing chain
17. Auto Tensioner
18. Retainer lock

37655_ELAN_G0231

Fig. 94 Exploded view of cylinder head and related components

A

37655_ELAN_G0276

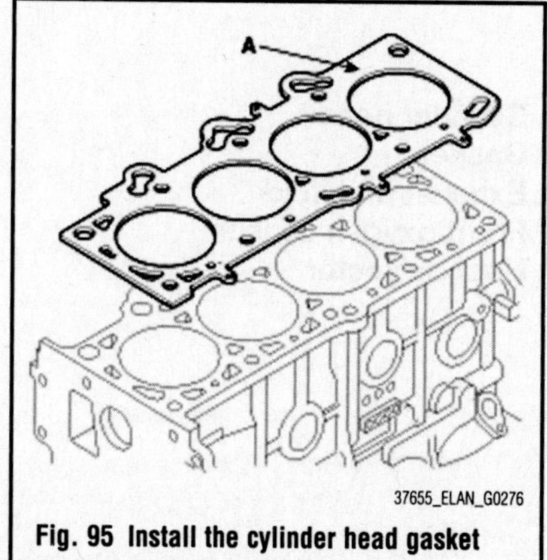

Fig. 95 Install the cylinder head gasket

8
6
1
3
9
M12
10
4
2
5
7

37655_ELAN_G0278

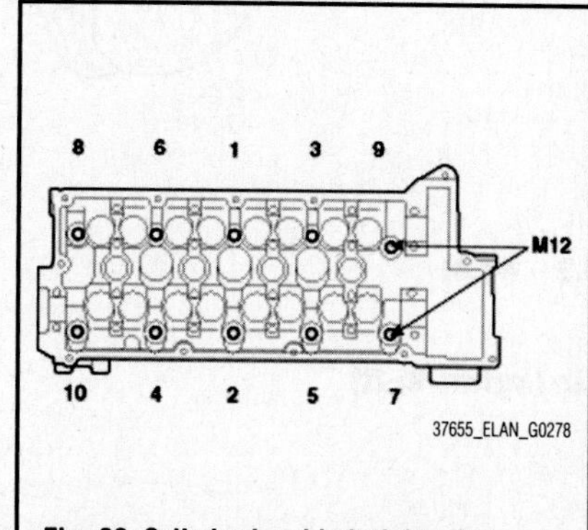

Fig. 96 Cylinder head bolt tightening sequence

09221-21000

37655_ELAN_G0233

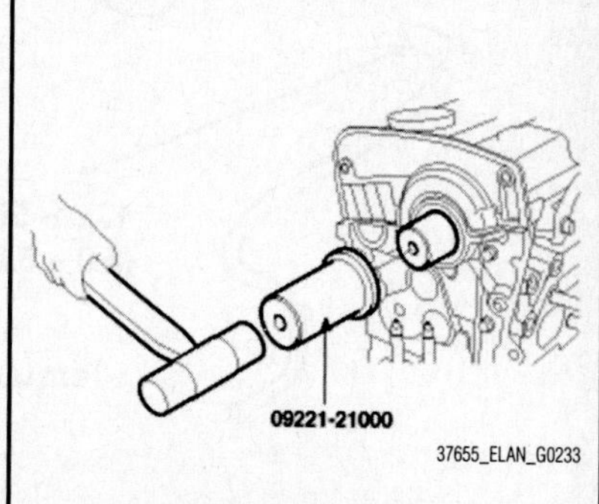

Fig. 97 Using the SST (09221-21000), install the camshaft bearing oil seal

EXHAUST MANIFOLD

REMOVAL & INSTALLATION

See Figures 98 and 99.

1. Before servicing the vehicle, refer to the Precautions Section.
2. Remove the engine cover.
3. Disconnect the front oxygen sensor connector.
4. Remove the front muffler.
5. Remove the heat protector.
6. Remove the exhaust manifold and catalytic converter assembly.

To install:

7. Installation is the reverse of removal. Tighten the front muffler to 29–43 ft. lbs. (40–59 Nm).

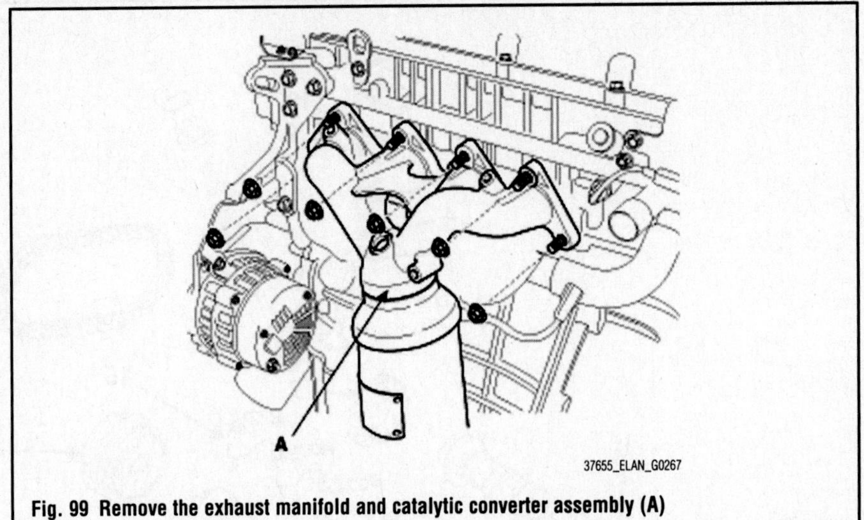

Fig. 99 Remove the exhaust manifold and catalytic converter assembly (A)

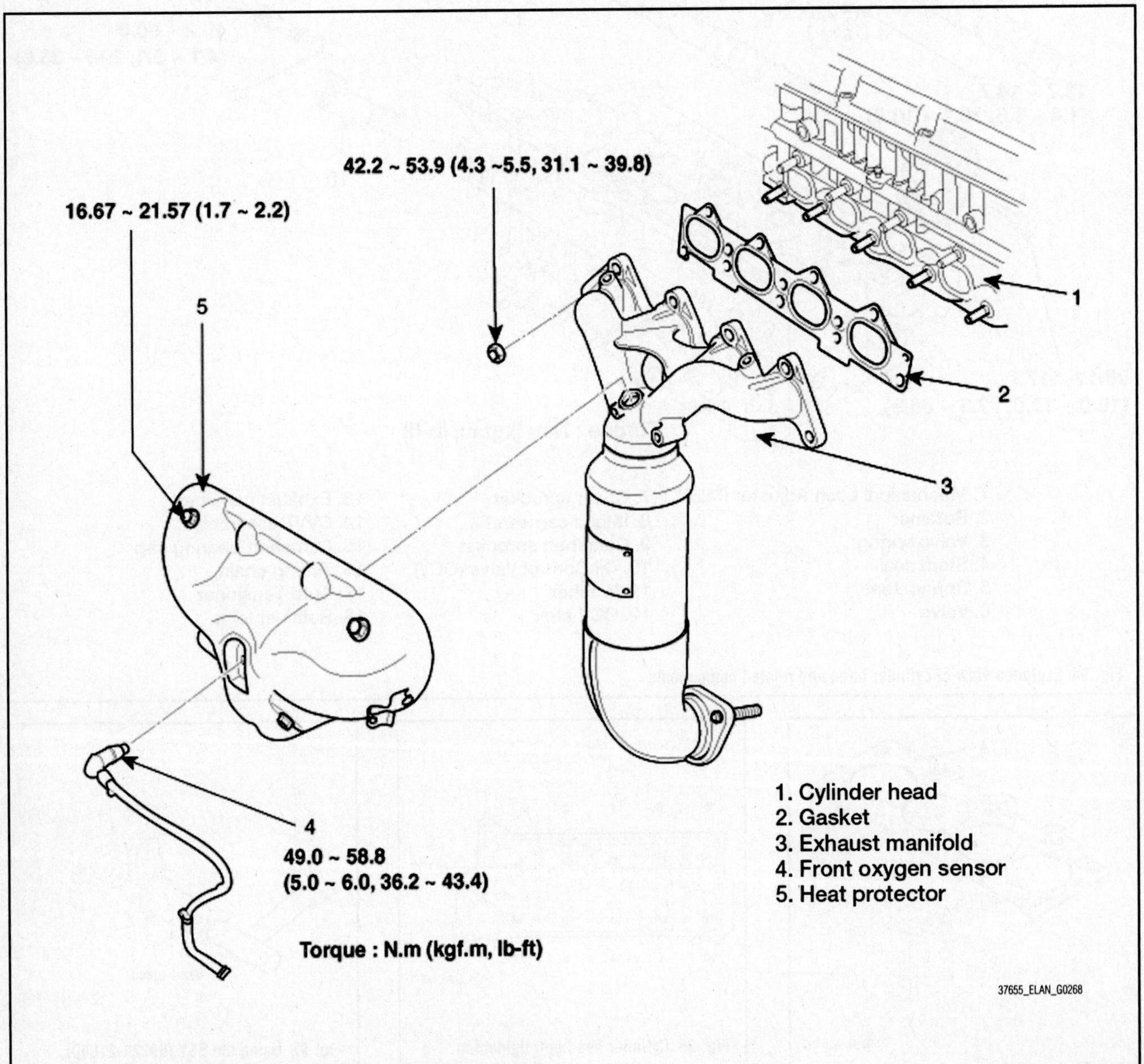

42.2 ~ 53.9 (4.3 ~5.5, 31.1 ~ 39.8)

16.67 ~ 21.57 (1.7 ~ 2.2)

5

1

2

3

4

49.0 ~ 58.8
(5.0 ~ 6.0, 36.2 ~ 43.4)

1. Cylinder head
2. Gasket
3. Exhaust manifold
4. Front oxygen sensor
5. Heat protector

Torque : N.m (kgf.m, lb-ft)

37655_ELAN_G0268

Fig. 98 Exploded view of the exhaust manifold and related components

INTAKE MANIFOLD

REMOVAL & INSTALLATION

See Figures 100 through 103.

1. Before servicing the vehicle, refer to the Precautions Section.
2. Relieve the fuel system pressure.
3. Drain the cooling system.
4. Disconnect the negative battery cable.
5. Remove the engine cover.
6. Remove the Throttle Position Sensor (TPS) connector and the Idle Speed Actuator (ISA) connectors.
7. Disconnect the Positive Crankcase Ventilation (PCV) hose and the breather hose.
8. Disconnect the accelerator cable.
9. Remove the delivery pipe.
10. Disconnect the Purge Control Solenoid Valve (PCSV) hose and the brake booster hose from the intake manifold and throttle body assembly.
11. Remove the intake manifold stay.
12. Remove the intake manifold assembly.

To install:

13. Installation is the reverse of removal, noting the following:

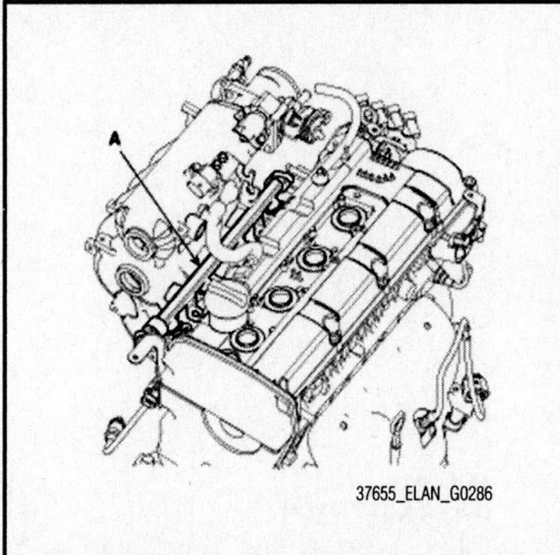

Fig. 101 Remove the delivery pipe (A)

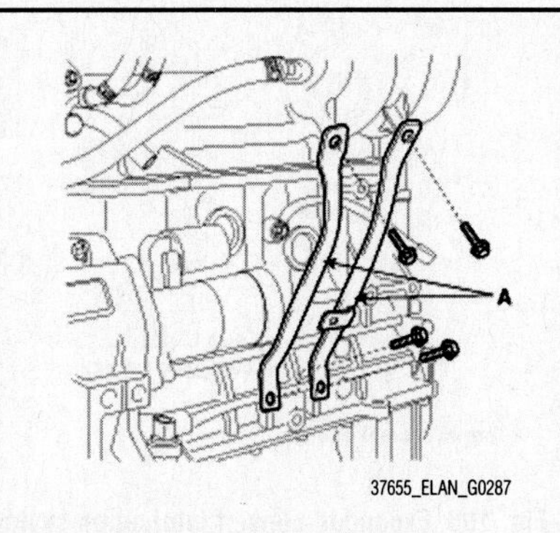

Fig. 102 Intake manifold stay (A)

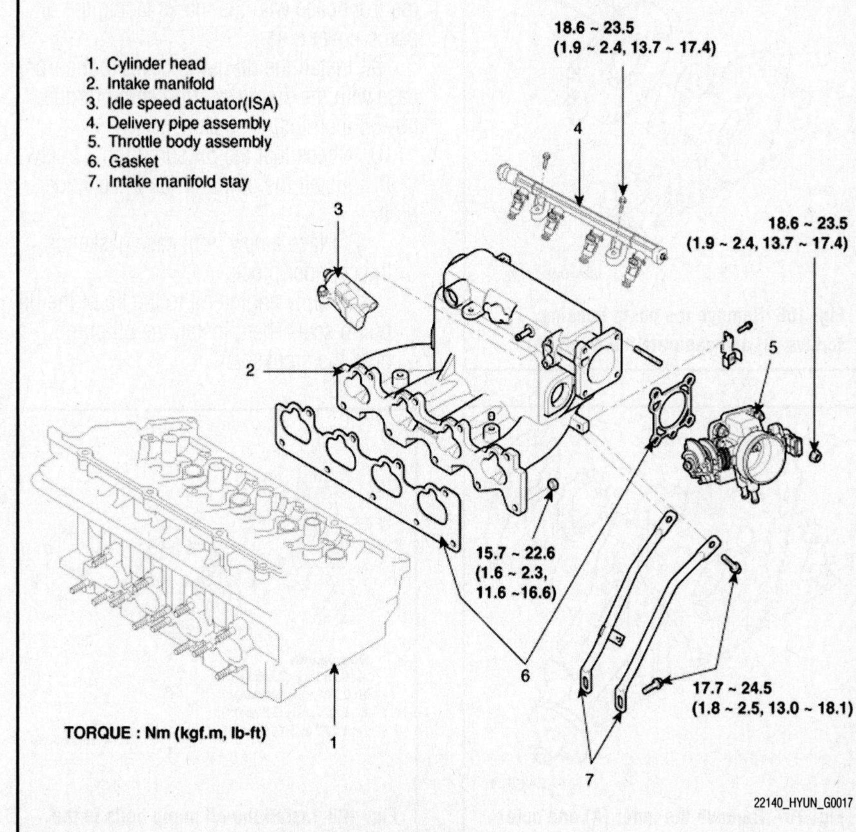

Fig. 100 Surge tank and intake manifold components

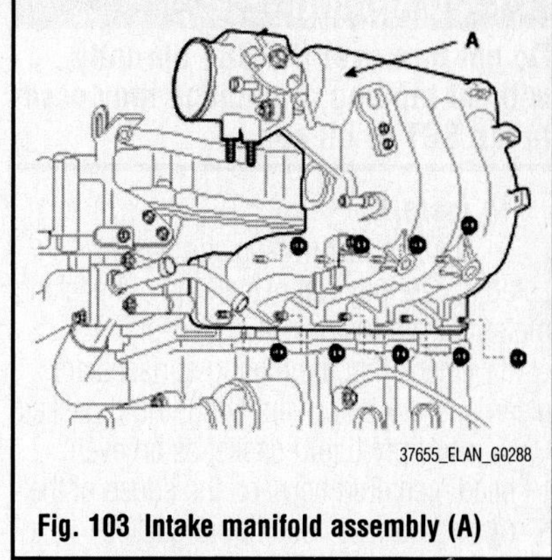

Fig. 103 Intake manifold assembly (A)

a. Use new gaskets
b. Tighten the intake manifold to 12–17 ft. lbs. (16–23 Nm)
c. Tighten the intake manifold stay to 13–18 ft. lbs. (18–25 Nm)

14. Fill the cooling system.
15. Start the engine and check for leaks.

OIL PAN

REMOVAL & INSTALLATION

See Figure 104

Special Service Tool (SST) used for this procedure: 09215-3C000, Oil Pan Removal Tool.

1. Before servicing the vehicle, refer to the Precautions Section.
2. Drain the engine oil.
3. Remove the oil pan bolts.
4. Using the SST (09215-3C000), remove the oil pan.

a. Insert the SST between the oil pan and the ladder frame by tapping it with a plastic hammer in the direction of arrow.
b. After tapping the SST with a plastic hammer along the direction of arrow around more than ⅔ of the edge of the oil pan, remove it from the ladder frame.

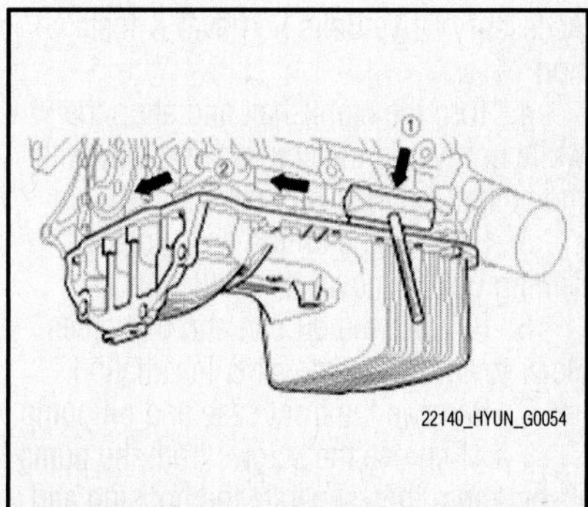

Fig. 104 Using the SST (09215-3C000) to remove the oil pan

⁂ WARNING

Do not turn over the SST abruptly without tapping or damage may occur to the SST or oil pan.

To install:

5. Using a razor blade and gasket scraper, carefully remove all the old packing material from the gasket surfaces.

6. Check that the mating surfaces are clean and dry before applying liquid gasket.

 a. Apply liquid gasket as an even bead, centered between the edges of the mating surface. Use liquid gasket: TB1217H or equivalent. Apply a bead ⅛ inch (3mm) wide to the oil pan.

 b. To prevent leakage of oil, apply liquid gasket to the inner threads of the bolt holes.

⁂ WARNING

Do not install the parts if 5 minutes or more have elapsed since applying the liquid gasket. Instead, reapply liquid gasket after removing the residue.

7. Install the oil pan with the bolts. Uniformly tighten the bolts in several passes. Tightening torque: 84–108 inch lbs. (10–12 Nm).

➡After assembly, wait at least 30 minutes before filling the engine with oil.

8. Fill with engine oil.

OIL PUMP

REMOVAL & INSTALLATION

See Figures 105 through 109.

Special Service Tool (SST) used for this procedure: 09231-23100, Crankshaft Front Oil Seal Installer.

1. Before servicing the vehicle, refer to the Precautions Section.
2. Drain the engine oil.
3. Remove the drive belts. Refer to Accessory Drive Belts Removal & Installation.
4. Turn the crankshaft and align the white groove on the crankshaft pulley with the pointer on the lower cover.
5. Remove the timing belt. Refer to Timing Belt Removal & Installation.
6. Remove the oil pan and oil screen. Refer to Oil Pan Removal & Installation.
7. Remove the front case and oil pump.

 a. Remove the screws from the pump housing, then separate the housing and cover.

 b. Remove the inner and outer rotors.

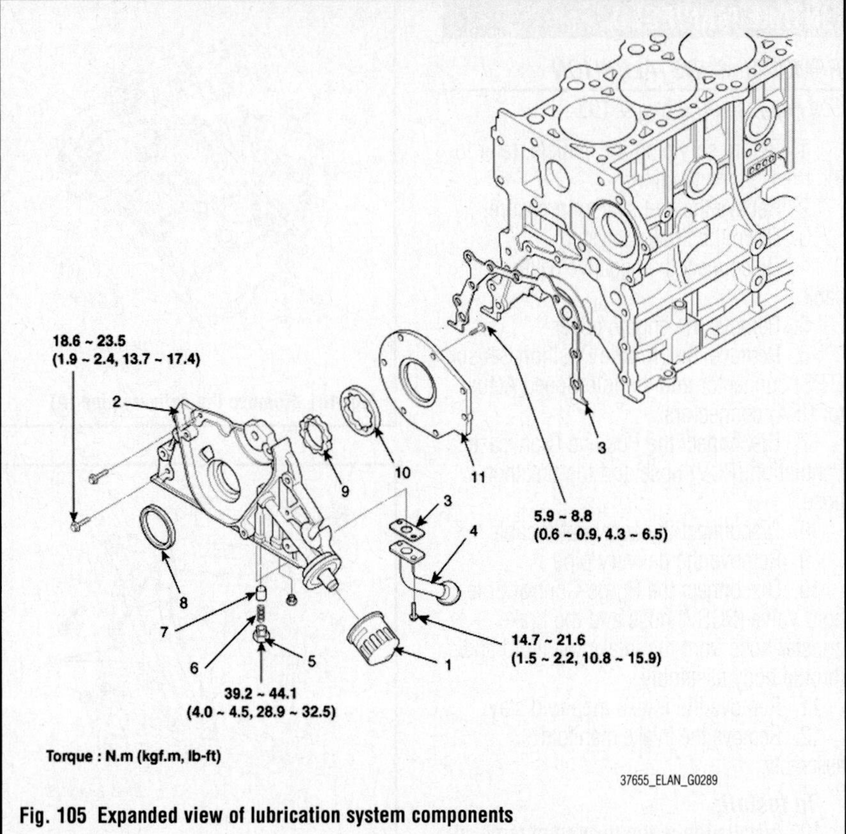

Fig. 105 Expanded view of lubrication system components

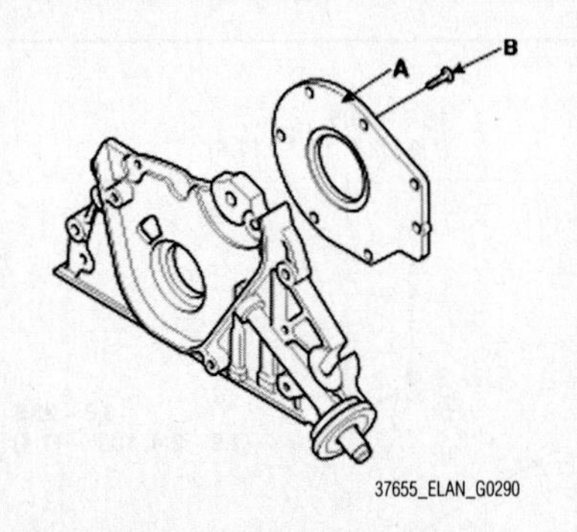

Fig. 106 Remove the pump housing screws (B) and separate the cover (A)

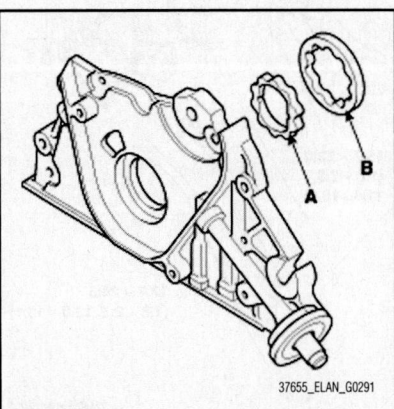

Fig. 107 Remove the inner (A) and outer (B) rotors

To install:

8. Place the inner and outer rotors into the front case with the marks facing the oil pump cover side.

9. Install the oil pump cover to the front case with the 7 screws. Tightening torque: 48–84 inch lbs. (6–9 Nm).

10. Check that the oil pump turns freely.

11. Install the oil pump on the cylinder block.

 a. Place a new front case gasket on the cylinder block.

 b. Apply engine oil to the lip of the oil pump seal. Then, install the oil pump onto the crankshaft.

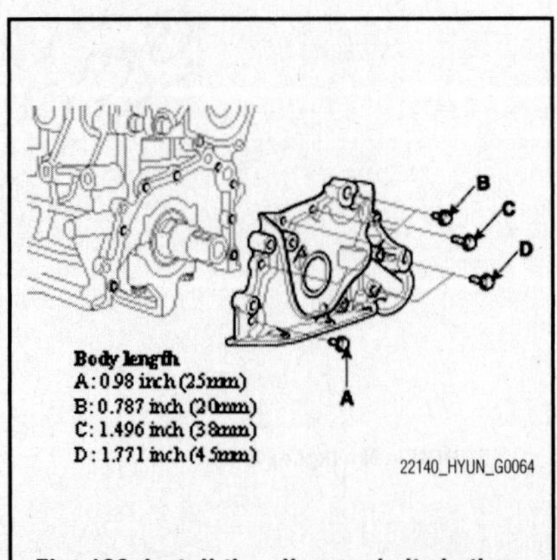

Fig. 108 Install the oil pump bolts in the correct position

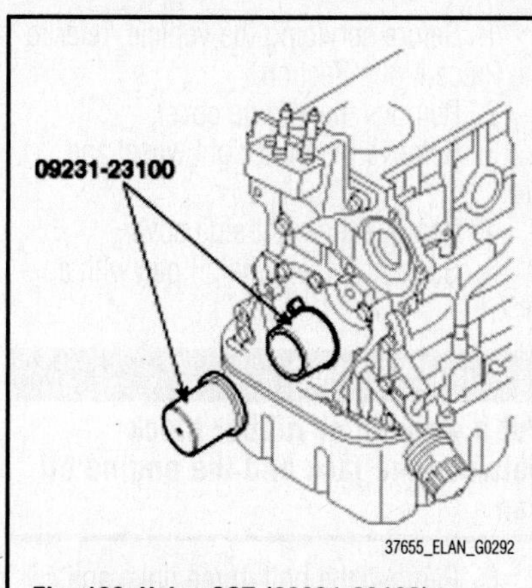

Fig. 109 Using SST (09231-23100) to install the oil seal

c. When the pump is in place, clean any excess grease off the crankshaft and check that the oil seal lip is not distorted.

d. Install the oil pump bolts in the correct position as illustrated. Tightening torque: 14–17 ft. lbs. (19–24 Nm).

12. Apply a light coat of oil to the seal lip.
13. Using the SST (09231-23100), install the oil seal.
14. Install the oil screen.
15. Install the oil pan.
16. Ensure that the crankshaft aligns with the white groove on the crankshaft pulley and the pointer on the lower cover.
17. Install the timing belt.
18. Install the accessory drive belts.
19. Fill with engine oil.
20. Check for leaks.

INSPECTION

See Figures 110 through 112.

1. Before servicing the vehicle, refer to the Precautions Section.
2. Remove the relief plunger: Remove the plug, spring, and relief plunger.
3. Inspect the relief plunger:
   a. Coat the plunger with engine oil.
   b. Check that it falls smoothly into the plunger hole by its own weight.
   c. If it does not, replace the relief plunger.
   d. If necessary, replace the front case.
4. Inspect the relief valve spring:
   a. Inspect for a distorted or broken relief valve spring.
   b. Standard value:
   - Free height: 1.724 in. (43.8mm)
   - Load: 8.14 lbs./1.579 in. (3.7kg/40.1mm)
5. Inspect the rotor side clearance:
   a. Using a feeler gauge and precision straight edge, measure the clearance between the rotors and precision straight edge.
   b. Side clearance:
   - Outer gear: 0.0016–0.0035 in. (0.04–0.09mm)
   - Inner gear: 0.0016–0.0033 in. (0.04–0.085mm)

   c. If the side clearance is greater than maximum, replace the rotors as a set. If necessary, replace the front case.
6. Inspect the rotor tip clearance:
   a. Using a feeler gauge, measure the tip clearance between the inner and outer rotor tips.
   b. Tip clearance: 0.0010–0.0027 in. (0.025–0.069mm)
   c. If the tip clearance is greater than specified, replace the rotors as a set.
7. Inspect the rotor body clearance:
   a. Using a feeler gauge, measure the clearance between the outer rotor and body.
   b. Body clearance: 0.0047–0.0073 in. (0.12–0.185mm)
   c. If the body clearance is greater than specified, replace the rotors as a set. If necessary, replace the front case.
8. Install the relief plunger and spring into the front case hole.

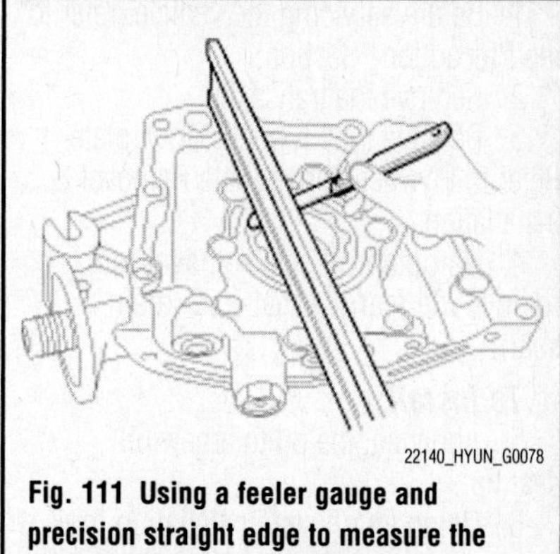

Fig. 111 Using a feeler gauge and precision straight edge to measure the side clearance of the oil pump

9. Install the plug. Tightening torque: 29–36 ft. lbs. (39–49 Nm).

PISTON AND RING

POSITIONING

See Figure 113.

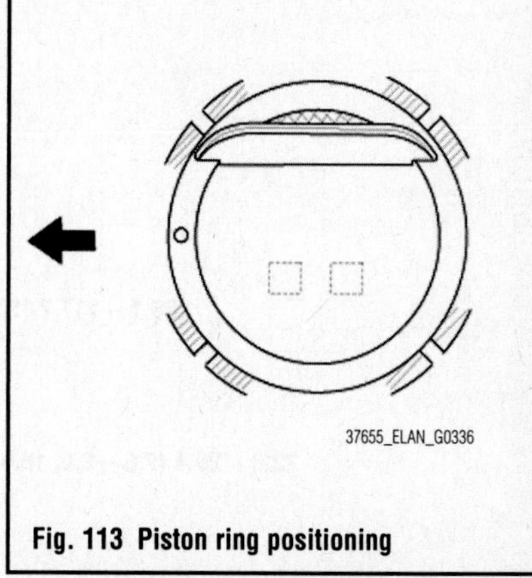

Fig. 113 Piston ring positioning

REAR MAIN SEAL

REMOVAL & INSTALLATION

See Figure 114.

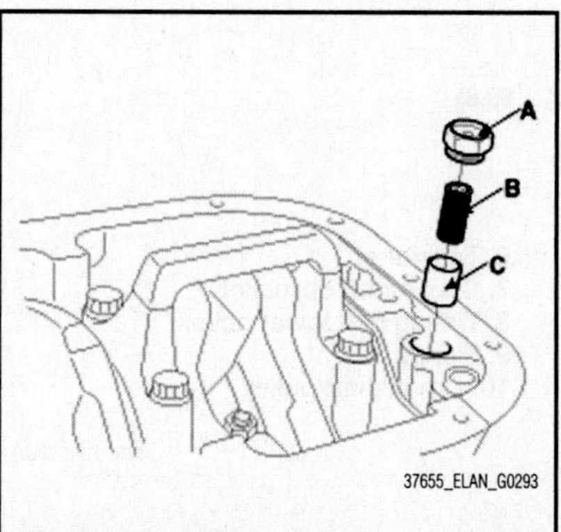

Fig. 110 Remove the plug (A), spring (B), and relief plunger (C)

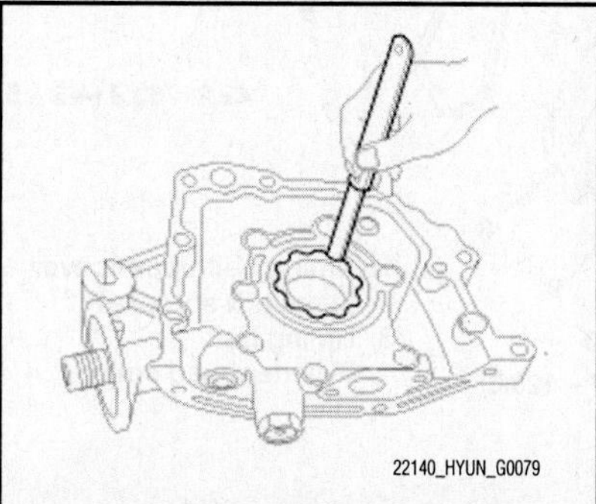

Fig. 112 Using a feeler gauge, measure the tip clearance between the inner and outer rotor tips of the oil pump

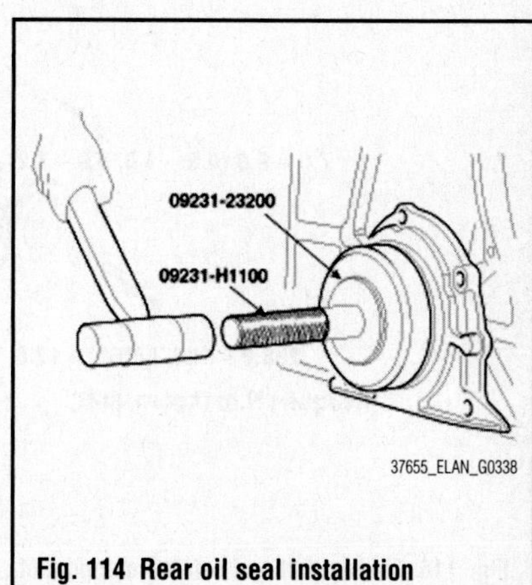

Fig. 114 Rear oil seal installation

1. Before servicing the vehicle, refer to the Precautions Section.
2. Remove the transaxle.
3. Remove the Flywheel/Drive plate. Refer to Flywheel/Drive Plate Removal & Installation.
4. Using an oil seal removal tool, remove the rear oil seal (rear main seal).

To install:

5. Apply engine oil to a new oil seal lip.
6. Using an oil seal installation tool and a hammer, tap in the oil seal until its surface is flush with the rear oil seal retainer edge.
7. Install the Flywheel/Drive plate.
8. Install the transaxle.

TIMING BELT & SPROCKETS

REMOVAL & INSTALLATION

See Figures 115 through 134.

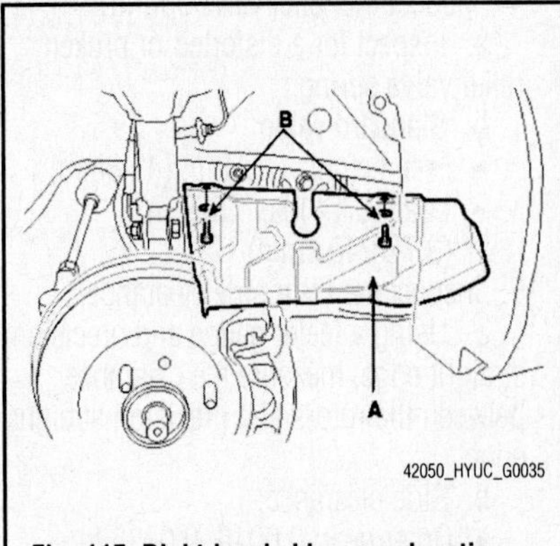

Fig. 115 Right-hand side cover location

1. Before servicing the vehicle, refer to the Precautions Section.
2. Remove the engine cover.
3. Remove the front right wheel and tire.
4. Remove the right side cover.
5. Support the engine oil pan with a jack.

WARNING

Put a wooden or rubber block between the jack and the engine oil pan.

6. Remove the bolt, three nuts, and engine mounting bracket.
7. Remove the bolt and stay plate.
8. Temporarily loosen the water pump pulley bolts.
9. Remove the alternator belt.

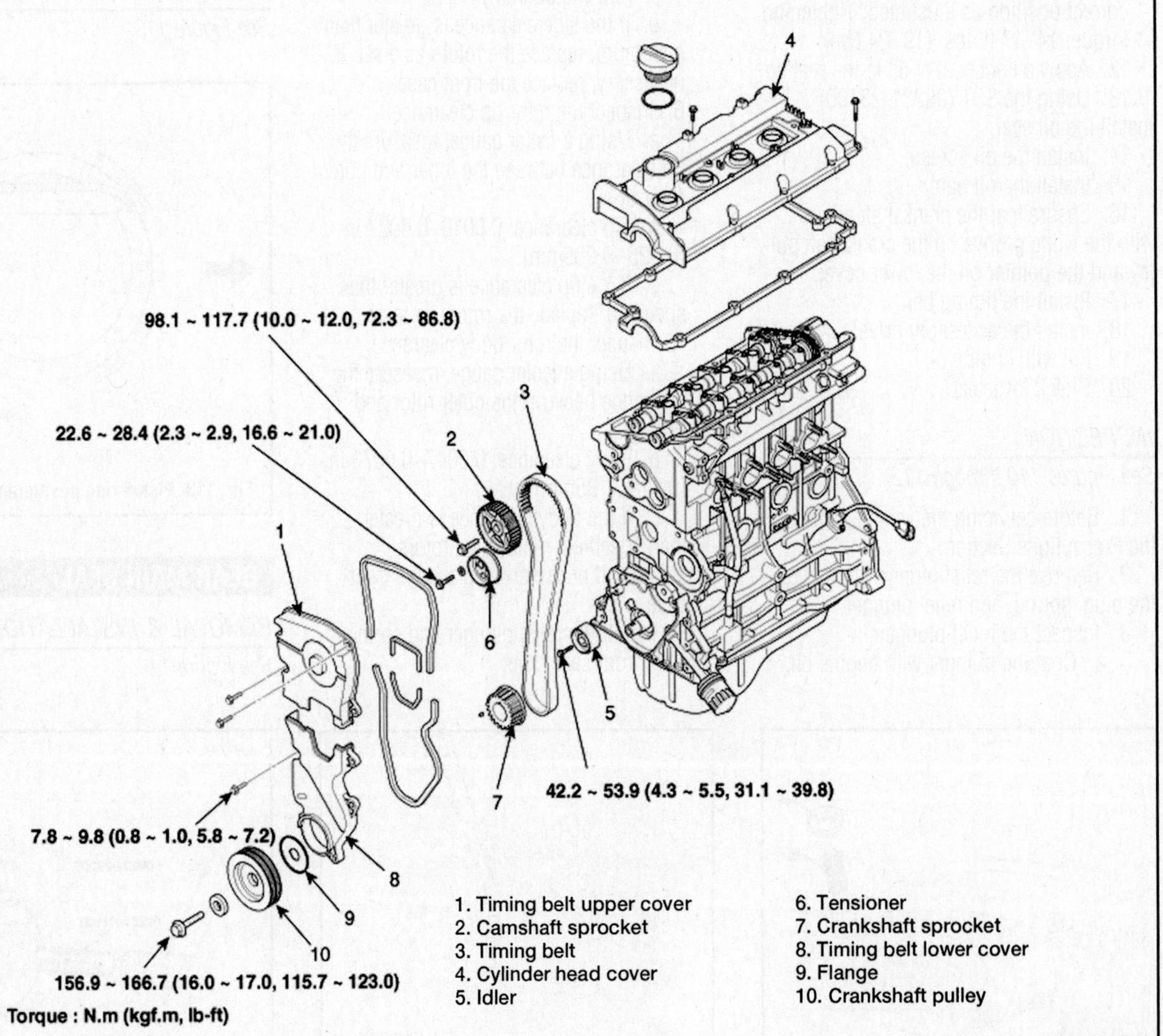

98.1 ~ 117.7 (10.0 ~ 12.0, 72.3 ~ 86.8)
22.6 ~ 28.4 (2.3 ~ 2.9, 16.6 ~ 21.0)
42.2 ~ 53.9 (4.3 ~ 5.5, 31.1 ~ 39.8)
7.8 ~ 9.8 (0.8 ~ 1.0, 5.8 ~ 7.2)
156.9 ~ 166.7 (16.0 ~ 17.0, 115.7 ~ 123.0)
Torque : N.m (kgf.m, lb-ft)

1. Timing belt upper cover
2. Camshaft sprocket
3. Timing belt
4. Cylinder head cover
5. Idler
6. Tensioner
7. Crankshaft sprocket
8. Timing belt lower cover
9. Flange
10. Crankshaft pulley

37655_ELAN_G0269

Fig. 116 Timing belt and related components

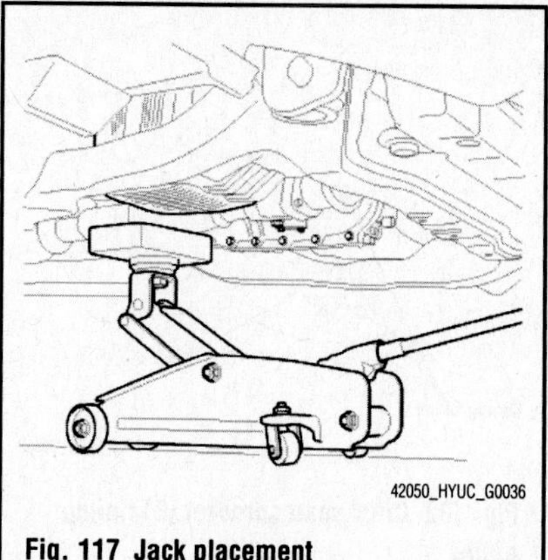

Fig. 117 Jack placement

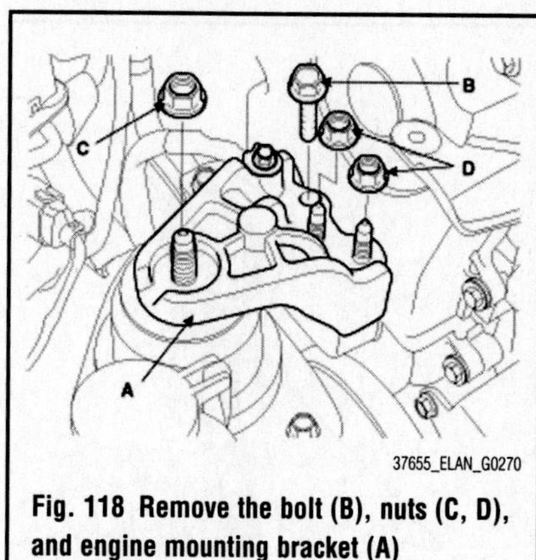

Fig. 118 Remove the bolt (B), nuts (C, D), and engine mounting bracket (A)

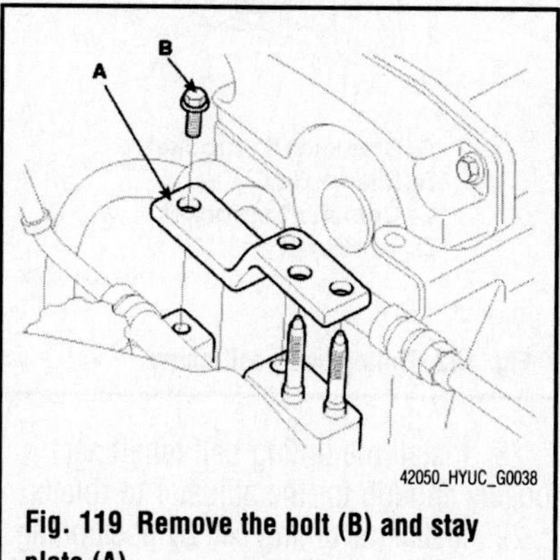

Fig. 119 Remove the bolt (B) and stay plate (A)

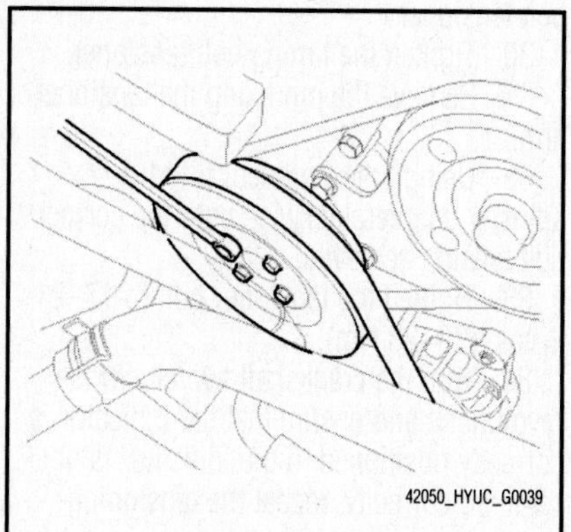

42050_HYUC_G0039

Fig. 120 Water pump pulley bolts

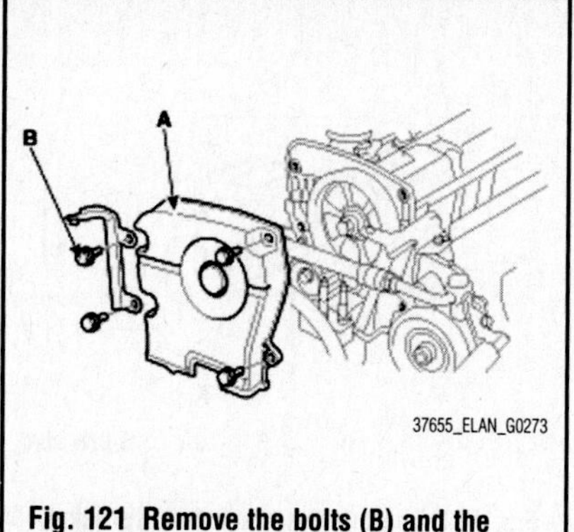

Fig. 121 Remove the bolts (B) and the timing belt upper cover (A)

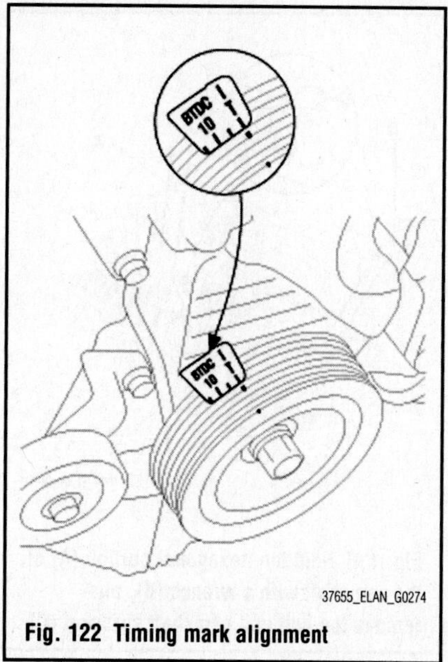

Fig. 122 Timing mark alignment

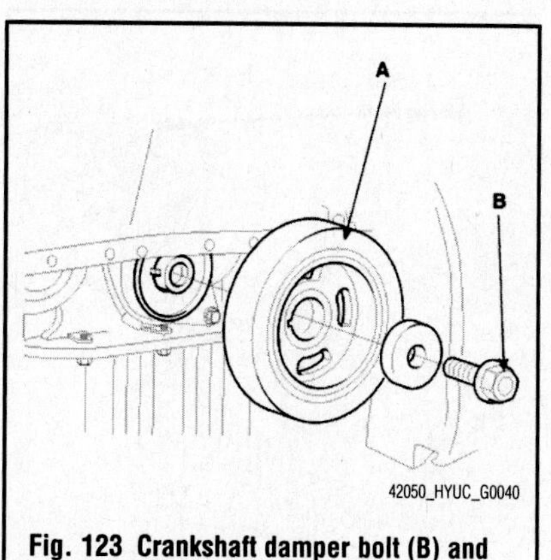
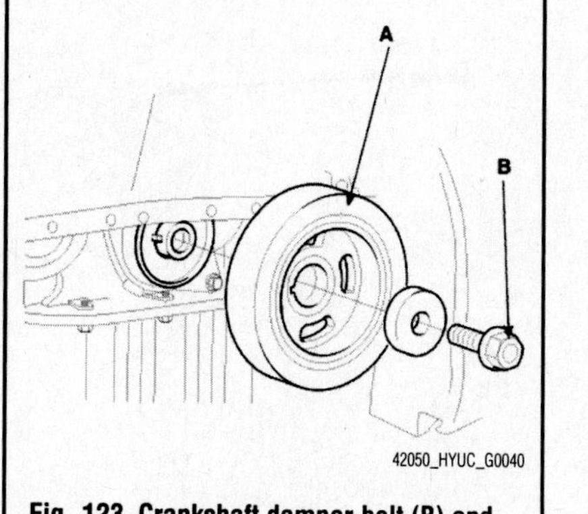

Fig. 123 Crankshaft damper bolt (B) and the crankshaft damper (A)

10. Remove the air compressor belt.

11. Remove the power steering belt.

12. Remove the four bolts and water pump pulley.

13. Remove the four bolts and the timing belt upper cover.

A

42050_HYUC_G0041

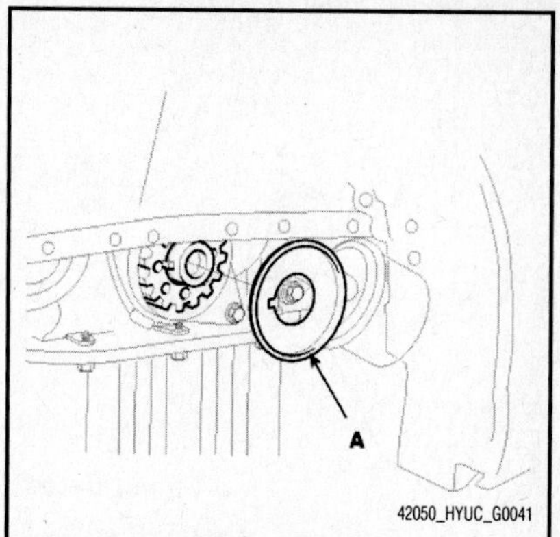

Fig. 124 Crankshaft flange (A)

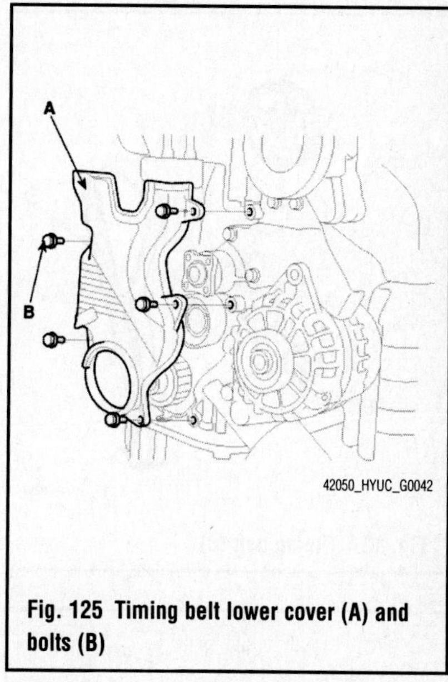

Fig. 125 Timing belt lower cover (A) and bolts (B)

14. Turn the crankshaft damper (pulley), aligning the groove with the timing mark "T" of the timing belt cover.

15. Remove the crankshaft damper bolt and the crankshaft damper.

16. Remove the crankshaft flange.

17. Remove the five bolts and timing belt lower cover.

18. Remove the timing belt tensioner and timing belt.

➡If the timing belt is reused, make an arrow indicating the turning direction to make sure that the belt is reinstalled in the same direction as before.

19. Remove the bolt and timing belt idler

20. Remove the crankshaft sprocket.

21. Remove the cylinder head cover.

22. Remove camshaft sprocket.

a. Hold the hexagonal portion of the camshaft with a wrench, and remove the bolt and camshaft sprocket.

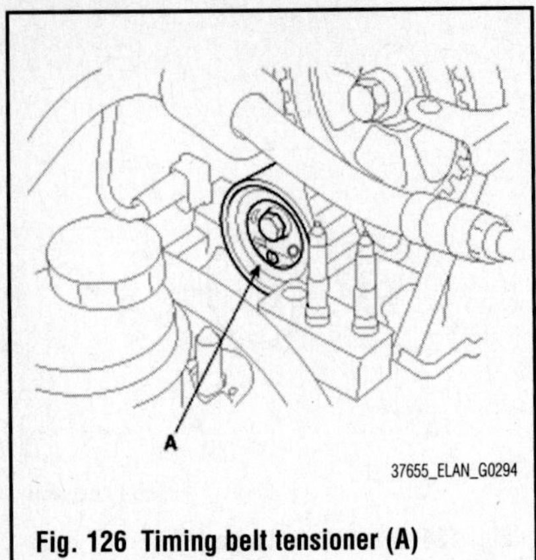

Fig. 126 Timing belt tensioner (A)

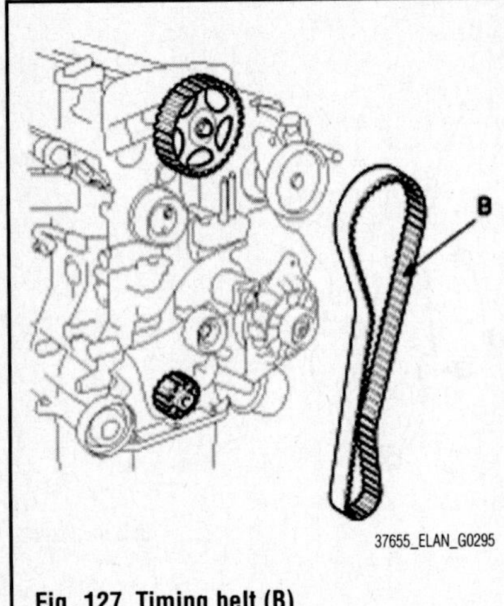

Fig. 127 Timing belt (B)

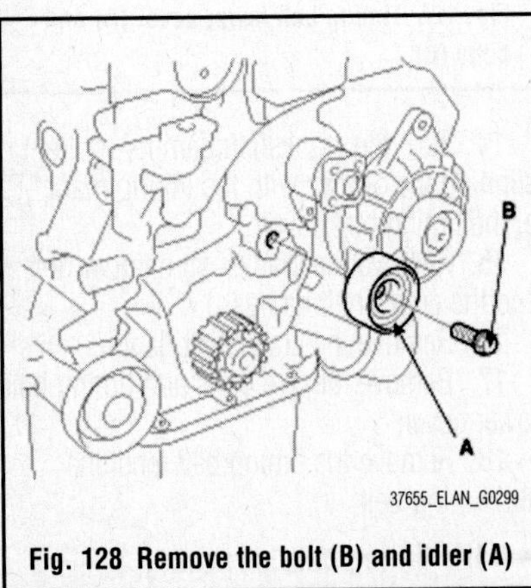

Fig. 128 Remove the bolt (B) and idler (A)

WARNING

Be careful not to damage the cylinder head and valve lifter with the wrench.

To install:

23. Install the camshaft sprocket and tighten the bolt to 72–87 ft. lbs. (98–118 Nm).
24. Install cylinder head cover.
25. Install the crankshaft sprocket.
26. Align the timing marks of the camshaft sprocket and crankshaft sprocket with the No. 1 piston placed at top dead center and its compression stroke.
27. Install the idler pulley and tighten the bolt to 31–40 ft. lbs. (42–54 Nm).

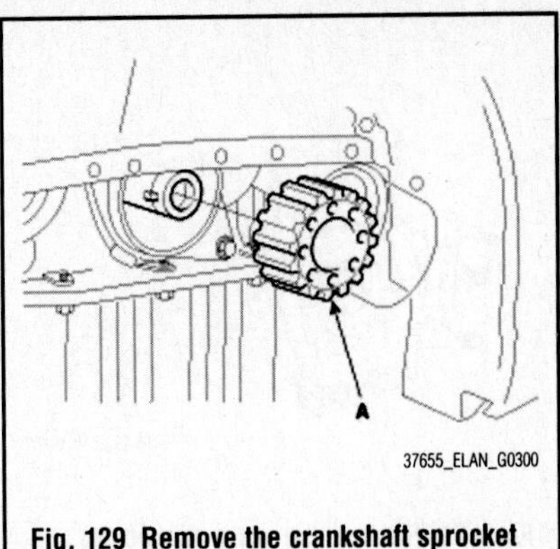

Fig. 129 Remove the crankshaft sprocket (A)

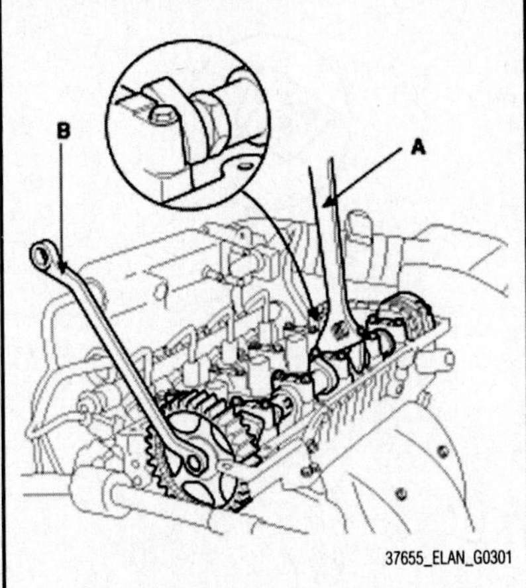

Fig. 130 Hold the hexagonal portion (A) of the camshaft with a wrench (B), and remove the bolt and camshaft sprocket (C)

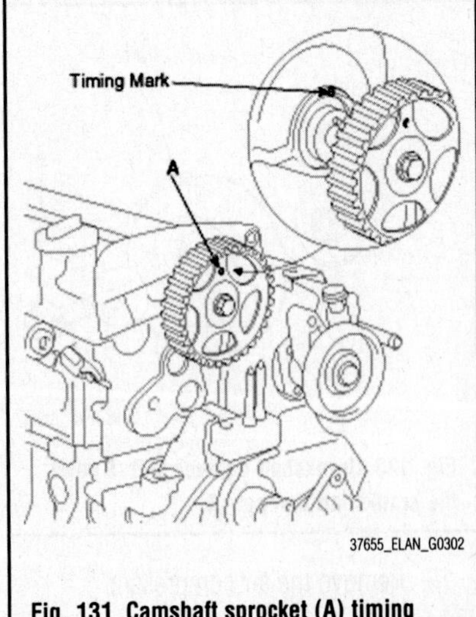

Fig. 131 Camshaft sprocket (A) timing marks

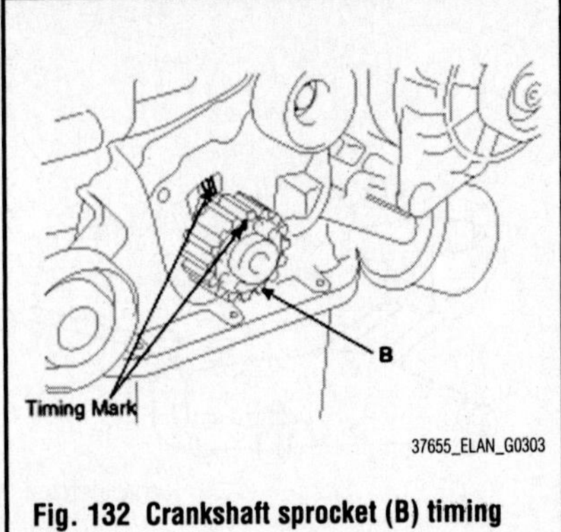

Fig. 132 Crankshaft sprocket (B) timing marks

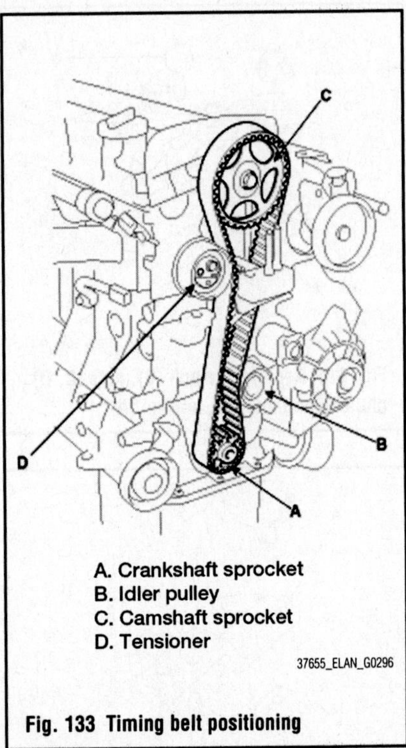

Fig. 133 Timing belt positioning

28. Install the timing belt tensioner loosely enough for the adjuster to rotate.
29. Install the timing belt by positioning it over the crankshaft sprocket, the idler pulley, the camshaft sprocket, and the timing belt tensioner.
30. Tighten the timing belt tensioner.
31. Remove the pin fixing the tensioner arm.
32. Using a hex wrench, turn the adjuster counterclockwise until the correct belt tension achieved.
33. Tighten the tensioner bolt to 17–21 ft. lbs. (23–28 Nm).
34. Turn the crankshaft two clockwise revolutions and ensure that the indicator is correctly positioned. If the indicator is not centered correctly, repeat the tensioning procedure.

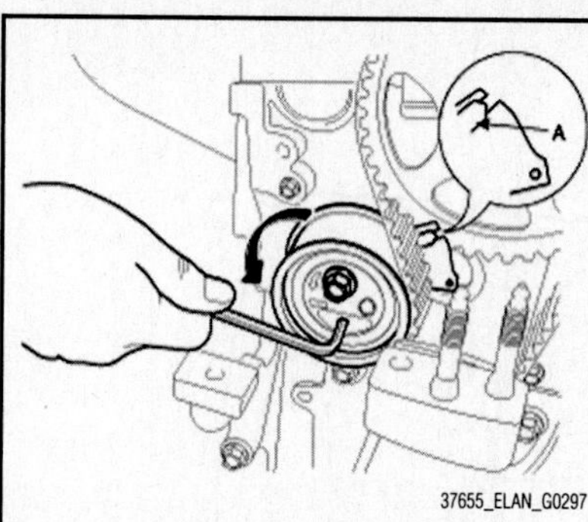

Fig. 134 Timing belt tension adjustment, showing tensioner arm indicator (A)

WARNING

Do not rotate the adjuster clockwise.

35. Install the timing belt lower cover and tighten to 70–86 inch lbs. (8–10 Nm).
36. Install the crankshaft damper flange.
37. Install the crankshaft damper and tighten the mounting bolt to 116–123 ft. lbs. (157–167 Nm). Make sure that crankshaft sprocket pin fits the small hole in the pulley.
38. Install the timing belt upper cover and tighten the bolts to 70–86 inch lbs. (8–10 Nm).
39. Install the coolant pump pulley.
40. Install the power steering belt.
41. Install the air compressor bolt.
42. Install the alternator belt.
43. Install the engine mount bracket.
    a. Install the stay plate and tighten the bolt to 31–40 ft. lbs. (42–54 Nm).
    b. Install engine mount bracket. Tighten the 17mm nut to 51–69 ft. lbs. (69–93 Nm), and the 14mm nuts to 36–47 ft. lbs. (49–64 Nm).
44. Lower the jack and remove from under the vehicle.
45. Install the right side cover.
46. Install the right front wheel and tire.
47. Install the engine cover.

ENGINE PERFORMANCE & EMISSION CONTROLS

COMPONENT LOCATIONS

See Figures 135 through 140.

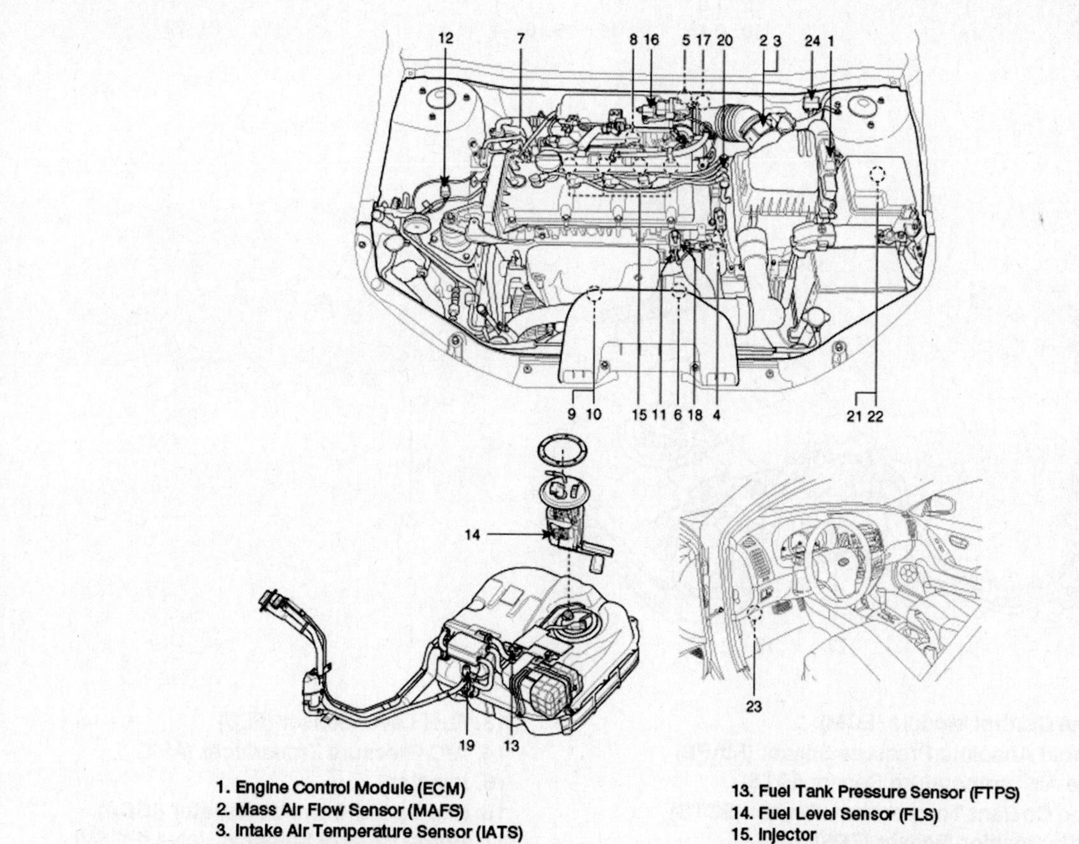

Fig. 135 Engine control system component locations—Sedan

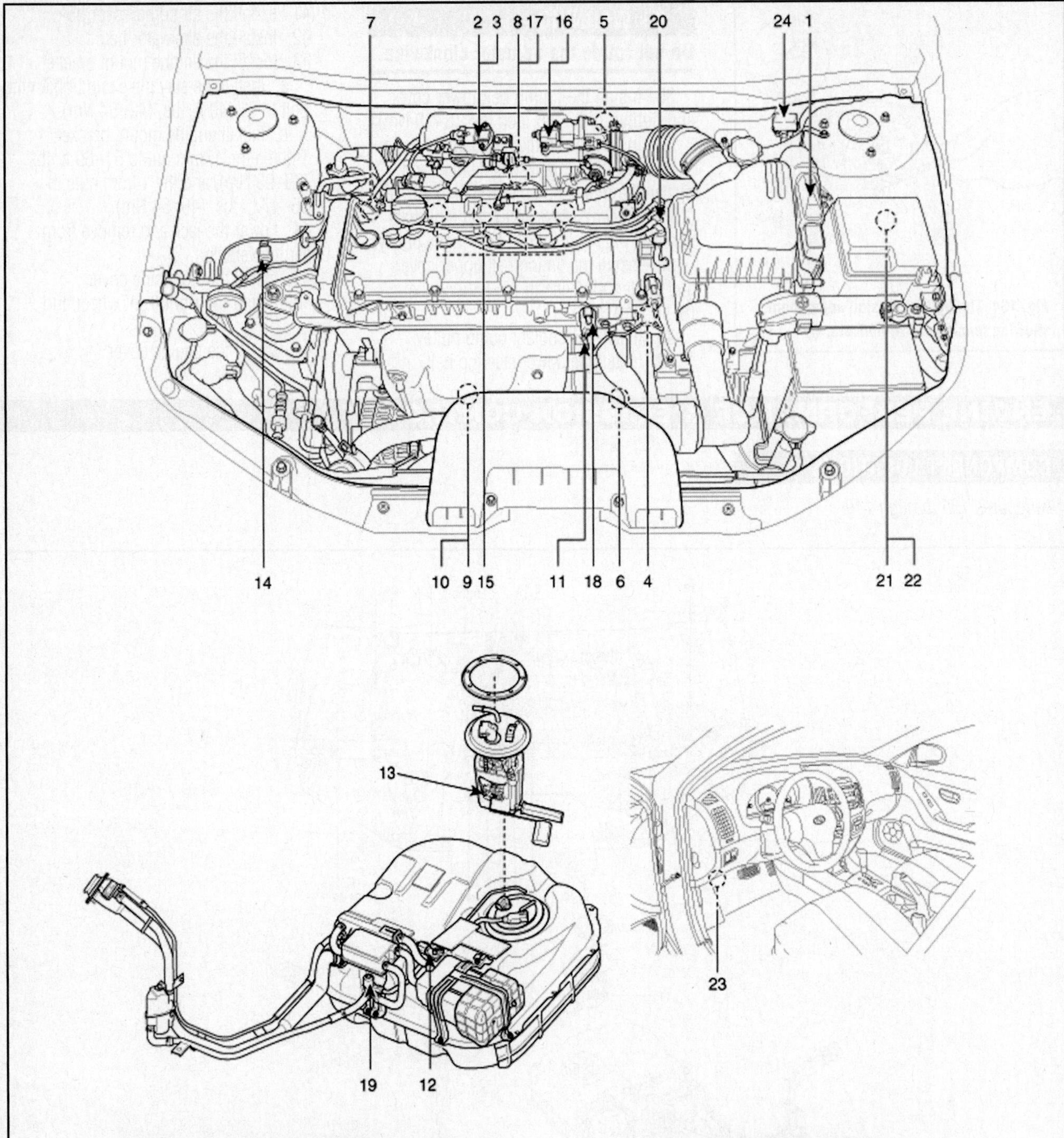

1. Engine Control Module (ECM)
2. Manifold Absolute Pressure Sensor (MAPS)
3. Intake Air Temperature Sensor (IATS)
4. Engine Coolant Temperature Sensor (ECTS)
5. Throttle Position Sensor (TPS)
6. Crankshaft Position Sensor (CKPS)
7. Camshaft Position Sensor (CMPS)
8. Knock Sensor (KS)
9. Heated Oxygen Sensor (HO2S) [Bank 1/Sensor 1]
10. Heated Oxygen Sensor (HO2S) [Bank 1/Sensor 2]
11. CVVT Oil Temperature Sensor (OTS)
12. Fuel Tank Pressure Sensor (FTPS)
13. Fuel Level Sensor (FLS)
14. A/C Pressure Transducer (APT)
15. Injector
16. Idle Speed Control Actuator (ISCA)
17. Purge Control Solenoid Valve (PCSV)
18. CVVT Oil Control Valve (OCV)
19. Canister Close Valve (CCV)
20. Ignition Coil
21. Main Relay
22. Fuel Pump Relay
23. Data Link Connector (DLC)
24. Multi-Purpose Check Connector

37655_ELAN_G0474

Fig. 136 Engine control system component locations—Touring model

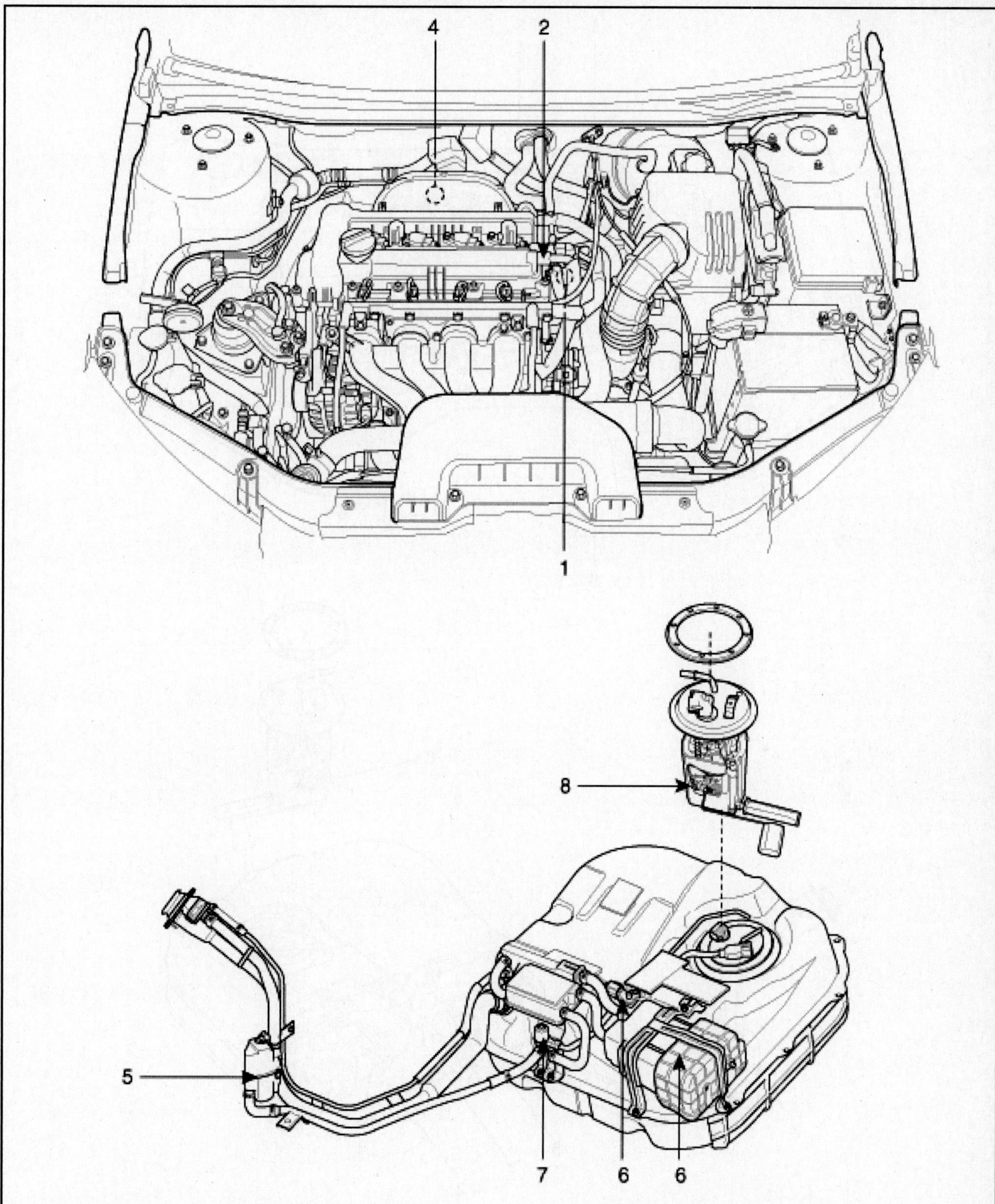

Fig. 137 Emission control system component locations—Sedan

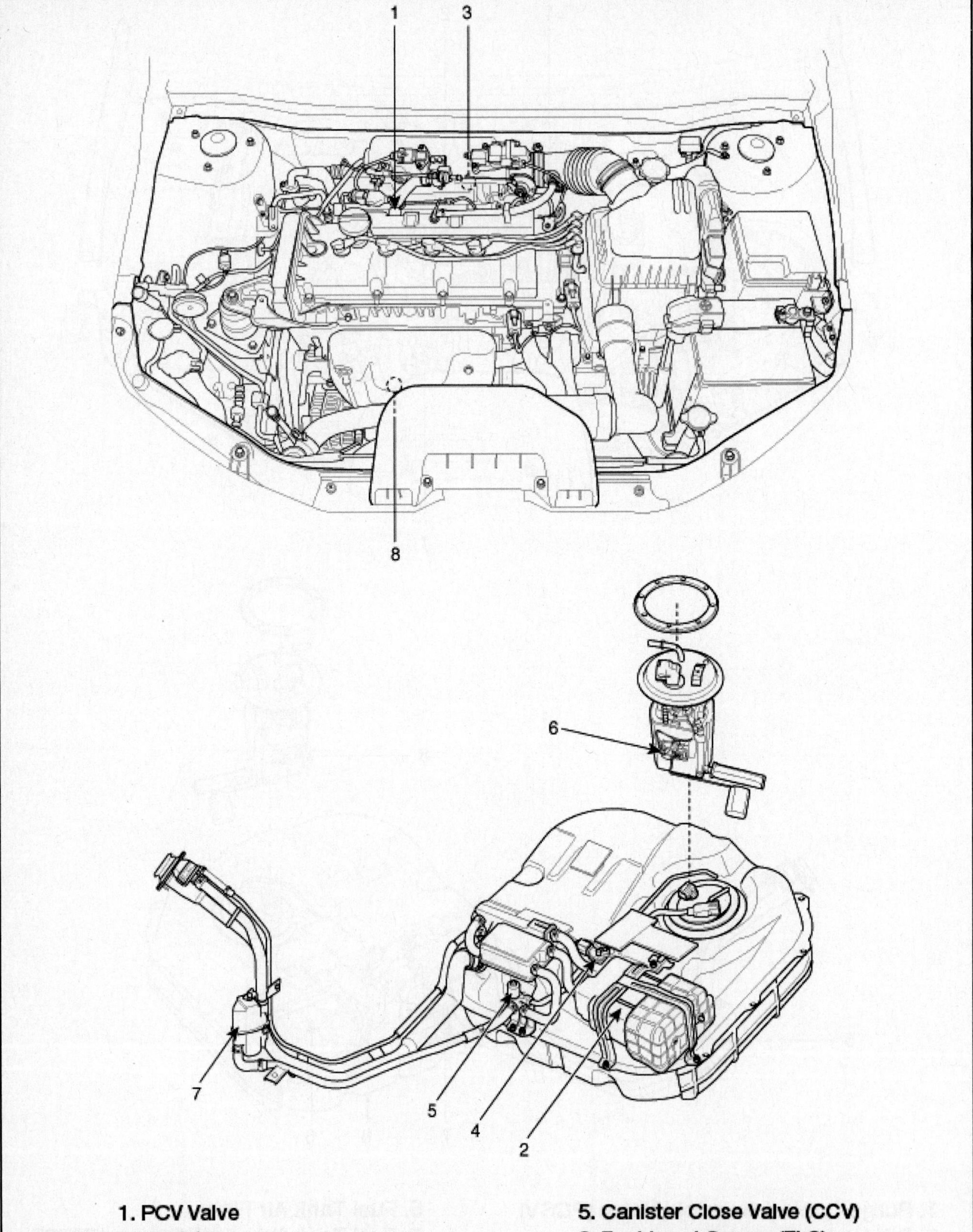

1. PCV Valve
2. Canister
3. Purge Control Solenoid Valve (PCSV)
4. Fuel Tank Pressure Sensor (FTPS)
5. Canister Close Valve (CCV)
6. Fuel Level Sensor (FLS)
7. Fuel Tank Air Filter
8. Catalytic Converter (MCC)

37655_ELAN_G0475

Fig. 138 Emission control system component locations—Touring model

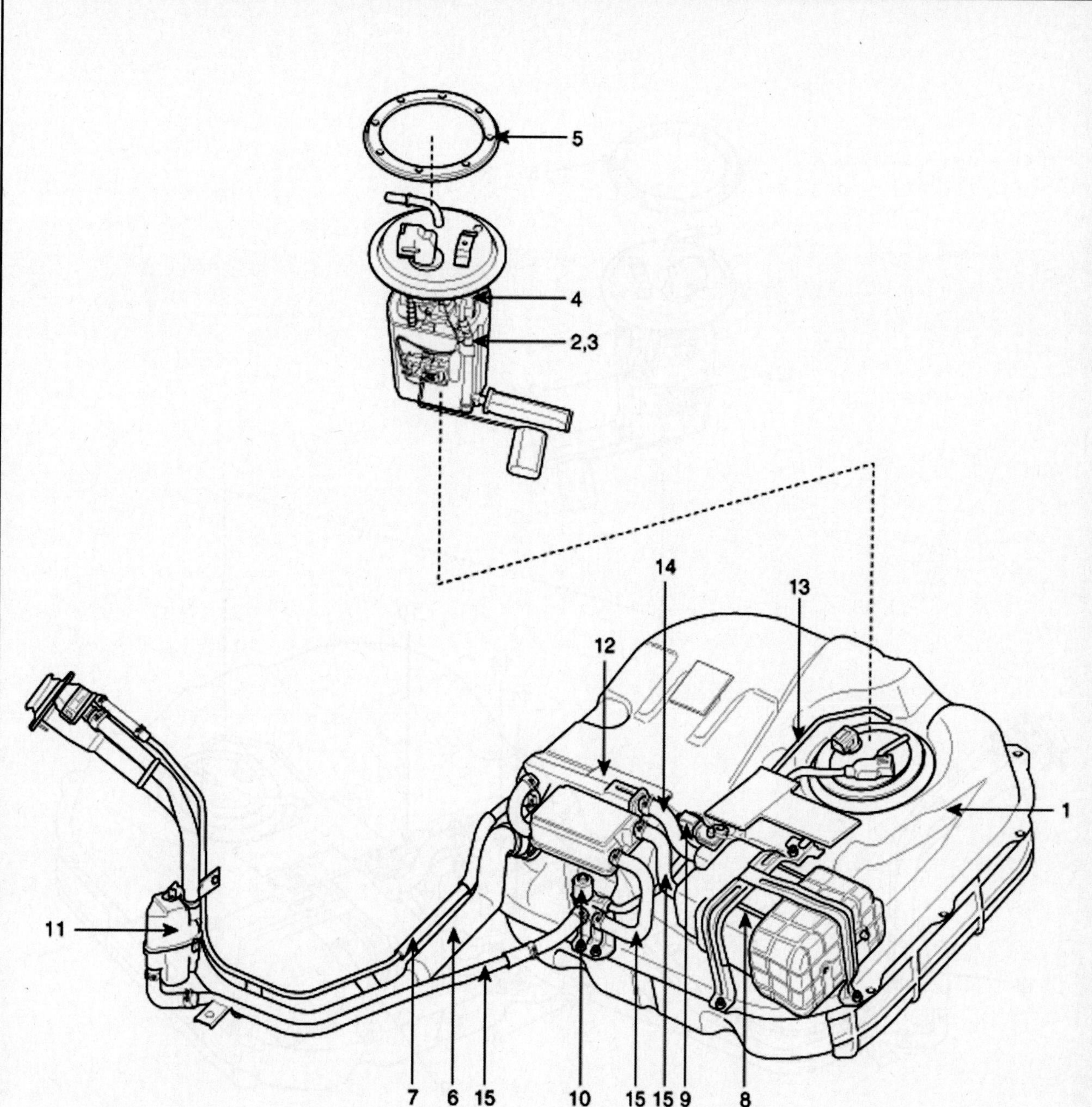

1. Fuel Tank
2. Fuel Pump
3. Fuel Filter (Included Fuel Pump)
4. Fuel Pressure Regulator
5. Fuel Pump Plate Cover
6. Fuel Filler Hose
7. Leveling Hose
8. Canister
9. Fuel Tank Pressure Sensor (FTPS)
10. Canister Close Valve (CCV)
11. Fuel Tank Air Filter
12. Separator
13. Tube (Canister to Intake Manifold)
14. Hose (Canister to Fuel Tank)
15. Hose (Canister to Fuel Tank Air Filter)

37655_ELAN_G0477

Fig. 139 Fuel system component locations—Sedan

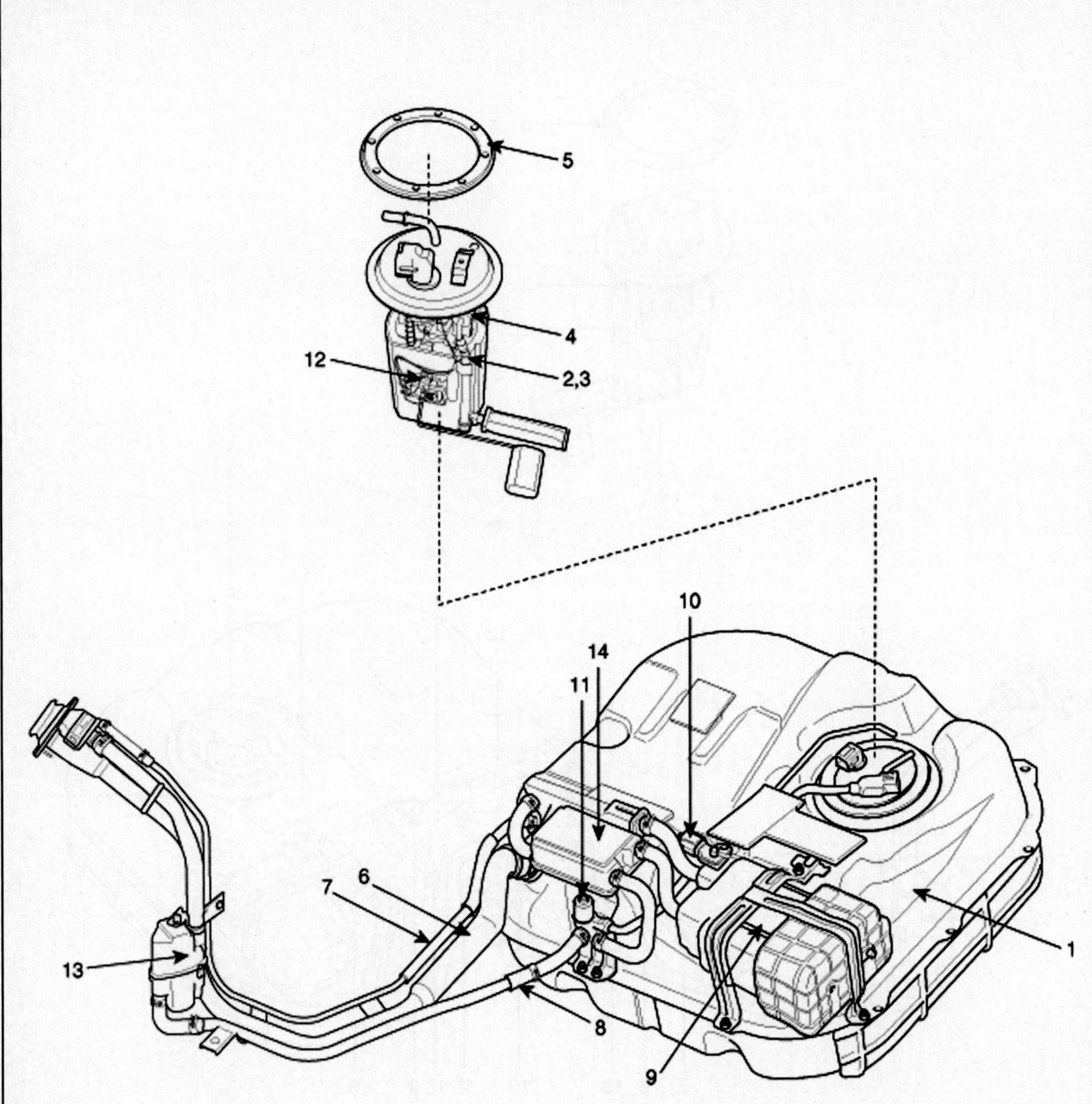

1. Fuel Tank
2. Fuel Pump
3. Fuel Filter (Included Fuel Pump)
4. Fuel Pressure Regulator
5. Fuel Pump Plate Cover
6. Fuel Filler Hose
7. Leveling Hose
8. Ventilation Hose
9. Canister
10. Fuel Tank Pressure Sensor (FTPS)
11. Canister Close Valve (CCV)
12. Fuel Level Sensor (FLS)
13. Fuel Tank Air Filter
14. Separator

37655_ELAN_G0476

Fig. 140 Fuel system component locations—Touring model

CAMSHAFT POSITION (CMP) SENSOR

LOCATION

See Figure 141.

REMOVAL & INSTALLATION

See Figure 141.

1. Before servicing the vehicle, refer to the Precautions Section.
2. Disconnect the negative battery cable.
3. Disconnect the connector from the CMP sensor.
4. Remove the CMP sensor.

To install:

5. Installation is the reverse of the removal procedure.

CRANKSHAFT POSITION (CKP) SENSOR

LOCATION

See Figure 142.

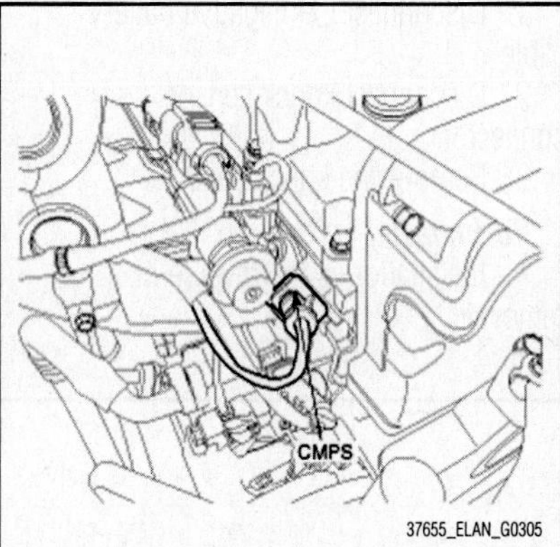

Fig. 141 Camshaft Position (CMP) sensor location

REMOVAL & INSTALLATION

See Figure 142.

1. Before servicing the vehicle, refer to the Precautions Section.
2. Disconnect the negative battery cable.
3. Disconnect the connector from the sensor.
4. Remove the sensor from its mounting.

To install:

5. Installation is the reverse of the removal procedure. Tighten to 70–104 inch lbs. (8–12 Nm).

ELECTRONIC CONTROL MODULE (ECM)

LOCATION

See Figure 143.

REMOVAL & INSTALLATION

See Figure 143.

1. Turn ignition switch off.
2. Disconnect the negative battery cable.
3. Disconnect the ECM connector.
4. Unscrew the ECM mounting bolts and remove the ECM from the air cleaner assembly.

To install:

5. Install a new ECM and tighten the mounting bolts to 86–104 inch lbs. (10–12 Nm).

RESET

The VIN (Vehicle Identification Number) is a number that has the vehicle's information. When replacing an PCM, the VIN must be programmed in the ECM/PCM. If there is no VIN in ECM/PCM memory, the fault code (DTC P0630) is set.

➡The programmed VIN cannot be changed. When writing the VIN, confirm the VIN carefully

1. Select "Vehicle" and "Engine".
2. Select "VIN WRITING".
3. Check the ECM/PCM status. (VIRGIN: VIN is not programmed, LEARNT: VIN has already been programmed) Is the PCM status "VIRGIN"? If yes, go to the next step. If no, the VIN has already been programmed.
4. Write the VIN with cursor, function and number keys.

WARNING

Before pressing the "ENTER" key, confirm the VIN again because the programmed VIN cannot be changed.

5. After verifying the written VIN, press the "ENTER" key.
6. Turn the ignition switch OFF, and then turn ON.
7. Verify the programmed VIN in the PCM memory.

ENGINE COOLANT TEMPERATURE (ECT) SENSOR

LOCATION

See Figure 144.

REMOVAL & INSTALLATION

See Figure 144.

1. Before servicing the vehicle, refer to the Precautions Section.
2. Drain the coolant to a level below the bottom of the sensor.

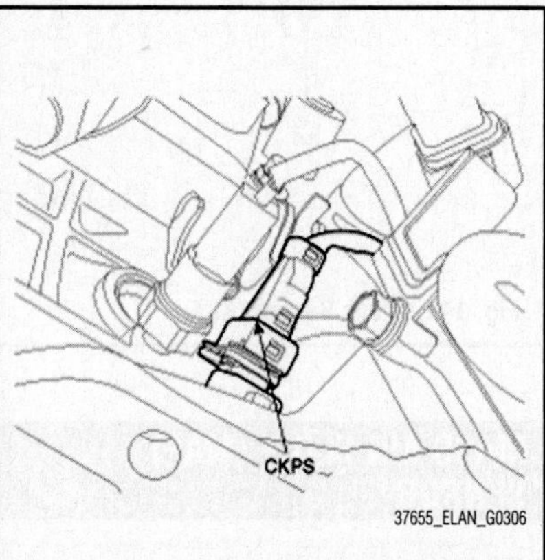

Fig. 142 Crankshaft Position (CKP) sensor location

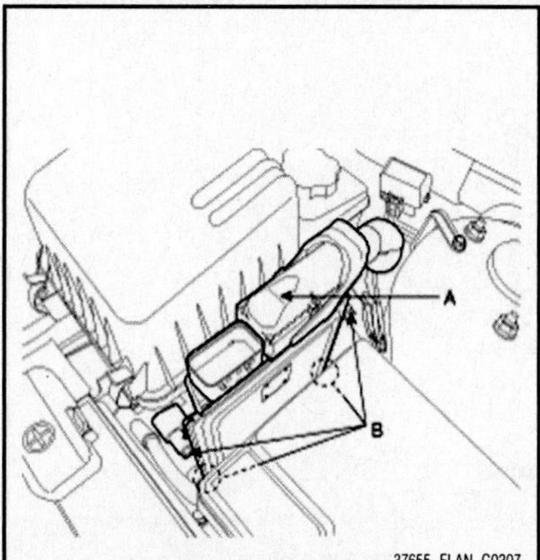

Fig. 143 ECM, connector (A), and mounting bolts (B)

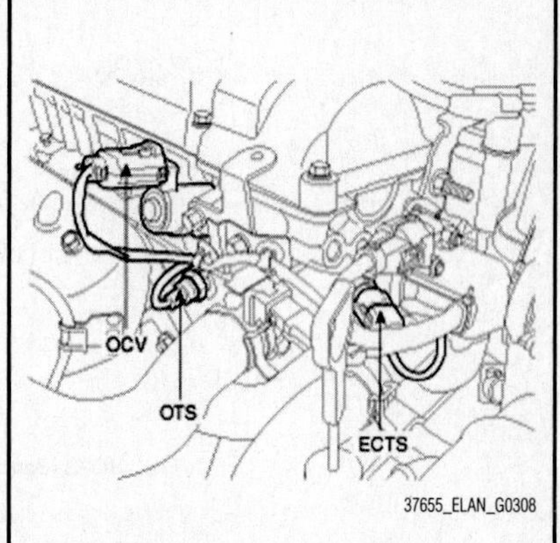

Fig. 144 Engine Coolant Temperature Sensor (ECTS)

3. Disconnect the negative battery cable.
4. Remove the sensor connector.
5. Remove the coolant temperature sensor.

To install:

6. Reverse the removal procedure.

HEATED OXYGEN SENSOR (HO2S)

LOCATION

See Figures 145 and 146.

REMOVAL & INSTALLATION

See Figures 145 and 146.

1. Before servicing the vehicle, refer to the Precautions Section.
2. Disconnect the negative battery cable.
3. Disconnect the oxygen sensor connector.
4. Remove the oxygen sensor.

To install:

5. Installation is the reverse of removal.

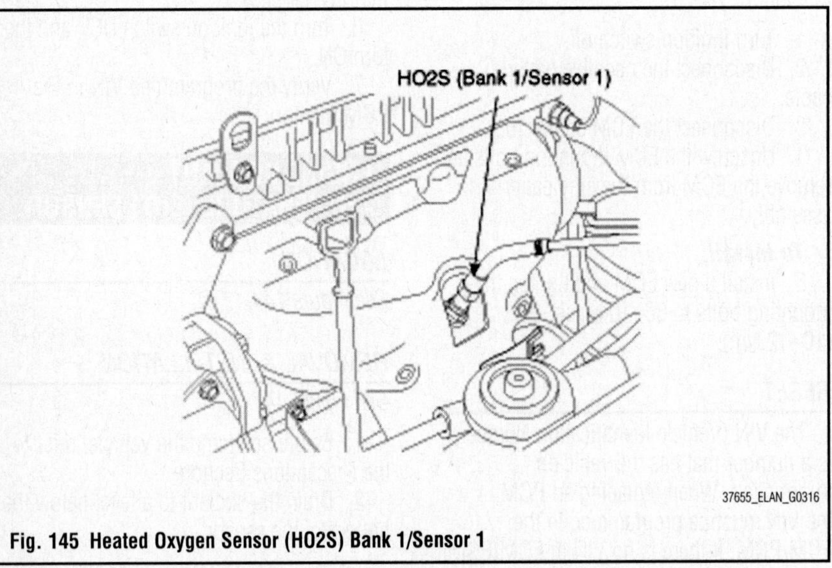

Fig. 145 Heated Oxygen Sensor (HO2S) Bank 1/Sensor 1

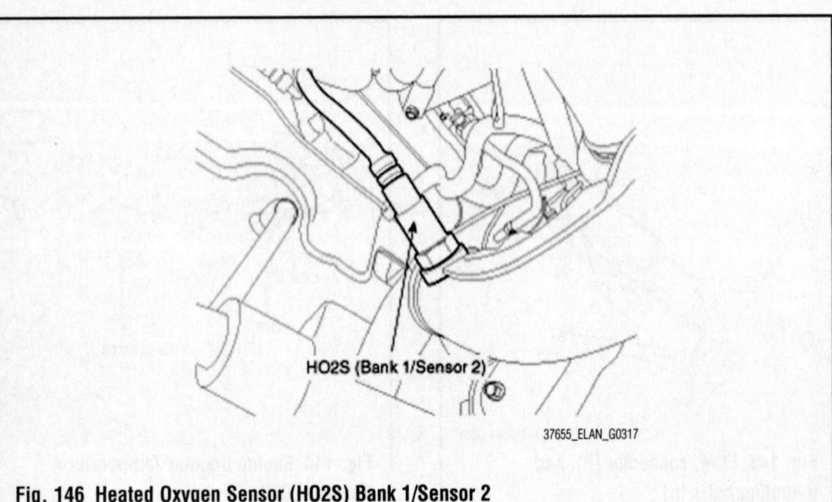

Fig. 146 Heated Oxygen Sensor (HO2S) Bank 1/Sensor 2

INTAKE AIR TEMPERATURE (IAT) SENSOR

LOCATION

See Figure 147.

REMOVAL & INSTALLATION

See Figure 147.

1. Disconnect the negative battery cable.
2. Disconnect Intake Air Temperature (IAT) sensor connector.
3. Remove the IAT sensor.

To install:

4. Installation is the reverse of removal.

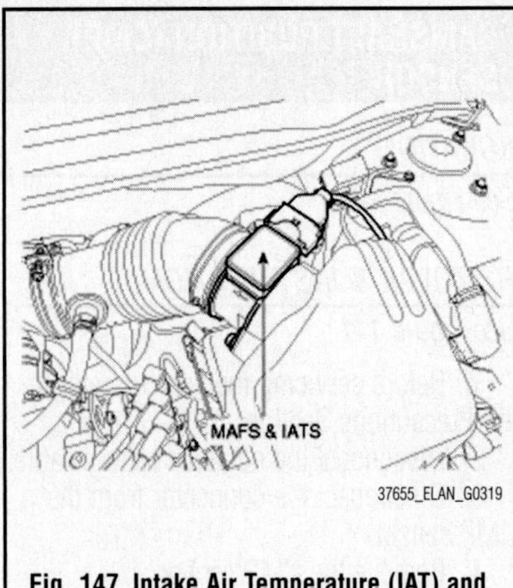

Fig. 147 Intake Air Temperature (IAT) and MAF sensor

KNOCK SENSOR (KS)

LOCATION

See Figure 148.

REMOVAL & INSTALLATION

See Figure 148.

1. Disconnect the negative battery cable.
2. Disconnect knock sensor connector.
3. Remove the knock sensor.

To install:

4. Installation is the reverse of removal.

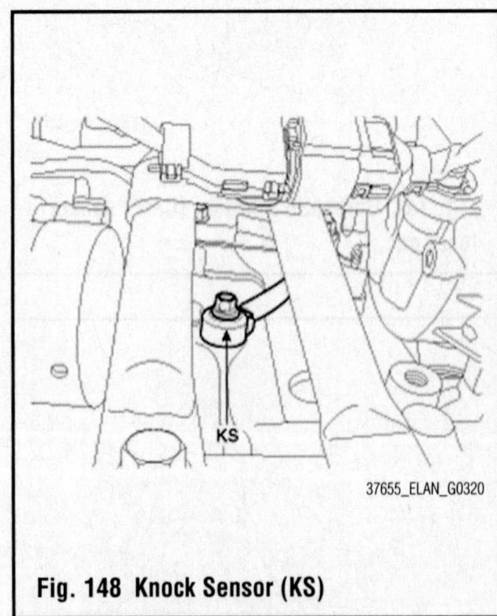

Fig. 148 Knock Sensor (KS)

MASS AIR FLOW (MAF) SENSOR

LOCATION

See Figure 147.

REMOVAL & INSTALLATION

See Figure 147.

1. Disconnect the negative battery cable.
2. Disconnect the Mass Air Flow (MAF) sensor connector.
3. Remove the MAF sensor.

To install:

4. Installation is the reverse of removal.

THROTTLE POSITION SENSOR (TPS)

LOCATION

See Figure 149.

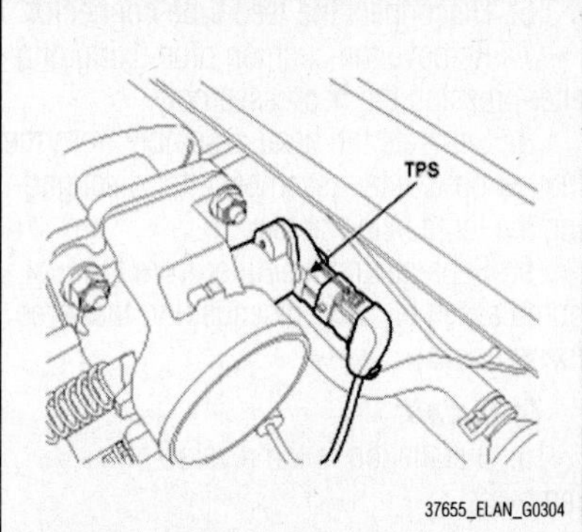

Fig. 149 Throttle Position Sensor (TPS) location

REMOVAL & INSTALLATION

See Figure 149.

1. Before servicing the vehicle, refer to the Precautions Section.
2. Disconnect the negative battery cable.
3. Disconnect the sensor connector.
4. Remove the sensor from its mounting.

To install:

5. Installation is the reverse of the removal procedure.

FUEL — GASOLINE FUEL INJECTION SYSTEM

FUEL SYSTEM SERVICE PRECAUTIONS

Safety is the most important factor when performing not only fuel system maintenance, but any type of maintenance. Failure to conduct maintenance and repairs in a safe manner may result in serious personal injury or death. Work on a vehicle's fuel system components can be accomplished safely and effectively by adhering to the following rules and guidelines.

- To avoid the possibility of fire and personal injury, always disconnect the negative battery cable unless the repair or test procedure requires that battery voltage be applied.
- Always relieve the fuel system pressure prior to disconnecting any fuel system component (injector, fuel rail, pressure regulator, etc.) fitting or fuel line connection. Exercise extreme caution whenever relieving fuel system pressure to avoid exposing skin, face and eyes to fuel spray. Please be advised that fuel under pressure may penetrate the skin or any part of the body that it contacts.
- Always place a shop towel or cloth around the fitting or connection prior to loosening to absorb any excess fuel due to spillage. Ensure that all fuel spillage is quickly removed from engine surfaces. Ensure that all fuel-soaked cloths or towels are deposited into a flame-proof waste container with a lid.
- Always keep a dry chemical (Class B) fire extinguisher near the work area.
- Do not allow fuel spray or fuel vapors to come into contact with a spark or open flame.
- Always use a second wrench when loosening or tightening fuel line connection fittings. This will prevent unnecessary stress and torsion on fuel piping. Always follow the proper torque specifications.
- Always replace worn fuel fitting O-rings with new ones. Do not substitute fuel hose where rigid pipe is installed.

FUEL SYSTEM PRESSURE

RELIEVING

See Figure 150.

1. Before servicing the vehicle, refer to the Precautions Section.
2. Remove the rear seat cushion.
3. Open the service cover.
4. Disconnect the fuel pump connector.
5. Start the engine and allow it to run until the fuel in fuel line is exhausted and it stalls.
6. Turn the ignition switch to the **OFF** position.
7. Disconnect the negative battery cable.

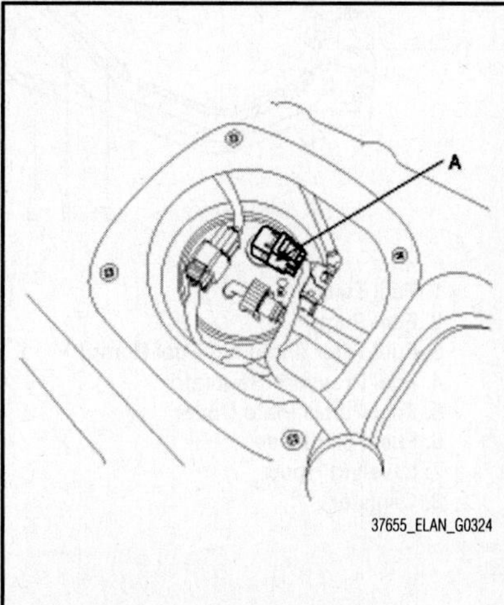

Fig. 150 Disconnect the fuel pump connector (A)

➡Be sure to reduce the fuel pressure before disconnecting the fuel feed hose, otherwise fuel will spill out.

FUEL FILTER

REMOVAL & INSTALLATION

See Figures 151 through 154.

1. Before servicing the vehicle, refer to the Precautions Section.
2. Relieve the fuel system pressure. Refer to Relieving Fuel System Pressure.
3. The fuel delivery system integrates the fuel filter with the in-tank fuel pump. To service this filter, remove the fuel pump. Refer to Fuel Pump, Removal & Installation.
4. Disconnect the electric pump & sender wiring connector and remove the regulator cap.
5. Disconnect the electric pump wiring connector from the pump.

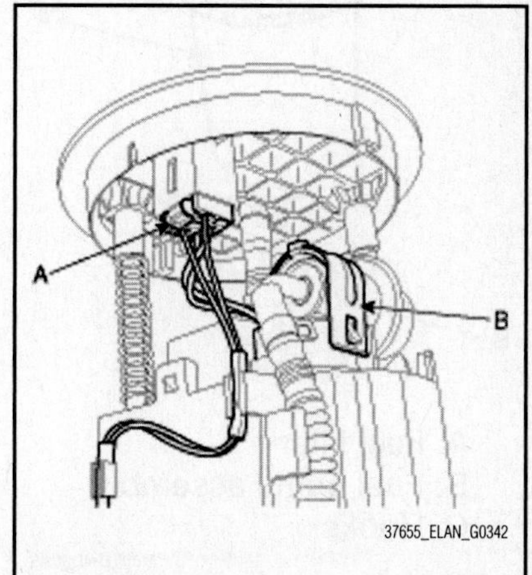

Fig. 151 Disconnect the electric pump & sender wiring connector (A) and remove the regulator cap (B)

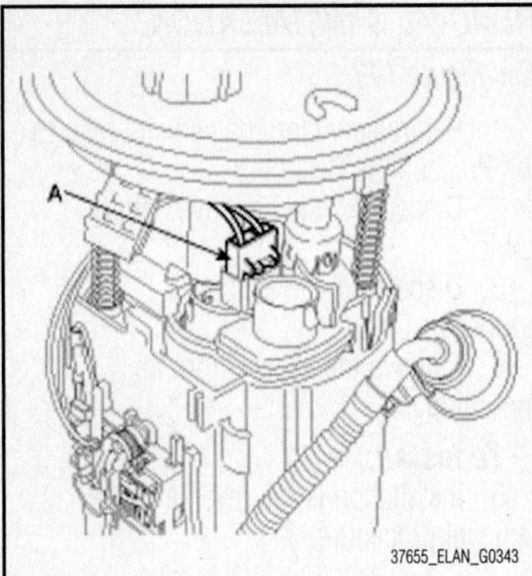

Fig. 152 Disconnect the electric pump wiring connector (A) from the pump

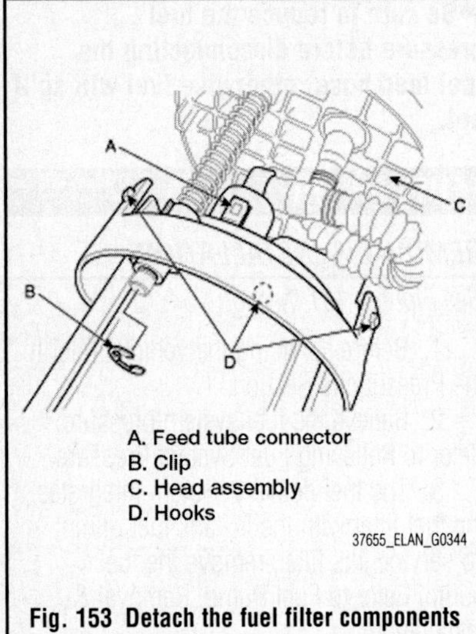

Fig. 153 Detach the fuel filter components

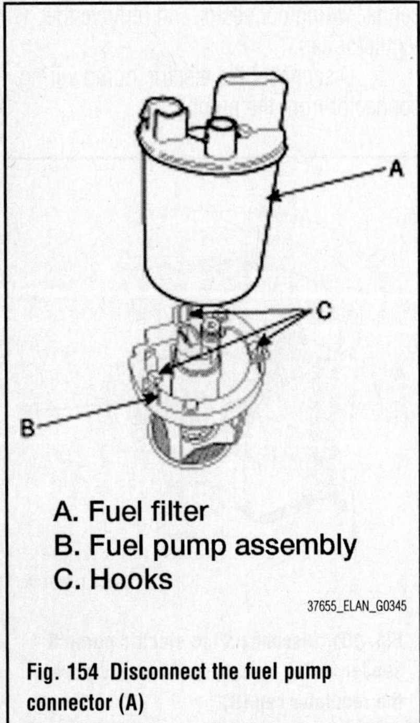

Fig. 154 Disconnect the fuel pump connector (A)

6. Disconnect the feed tube connector.
7. Remove the cushion pipe fixing clip after pressing the head assembly.
8. Separate the head assembly from the fuel pump & filter assembly after disengaging the three fixing hooks.
9. Separate the fuel filter from the fuel pump assembly after disengaging the three fixing hooks.

To install:

10. Installation is the reverse of removal.

FUEL PUMP

REMOVAL & INSTALLATION

See Figures 155 through 159.

1. Before servicing the vehicle, refer to the Precautions Section.

➡Special service tool used for this procedure: SST: 09310-2B200, Fuel Pump Plate Cover Wrench.

2. Relieve the fuel system pressure. Refer to Relieving Fuel System Pressure.
3. Disconnect the fuel feed tube quick-connector, the vacuum tube quick-connector and canister close valve connector.
4. Remove the rubber cover.
5. Remove the fuel pump plate cover with the special service tool (SST: 09310-2B200) and remove the fuel pump assembly.

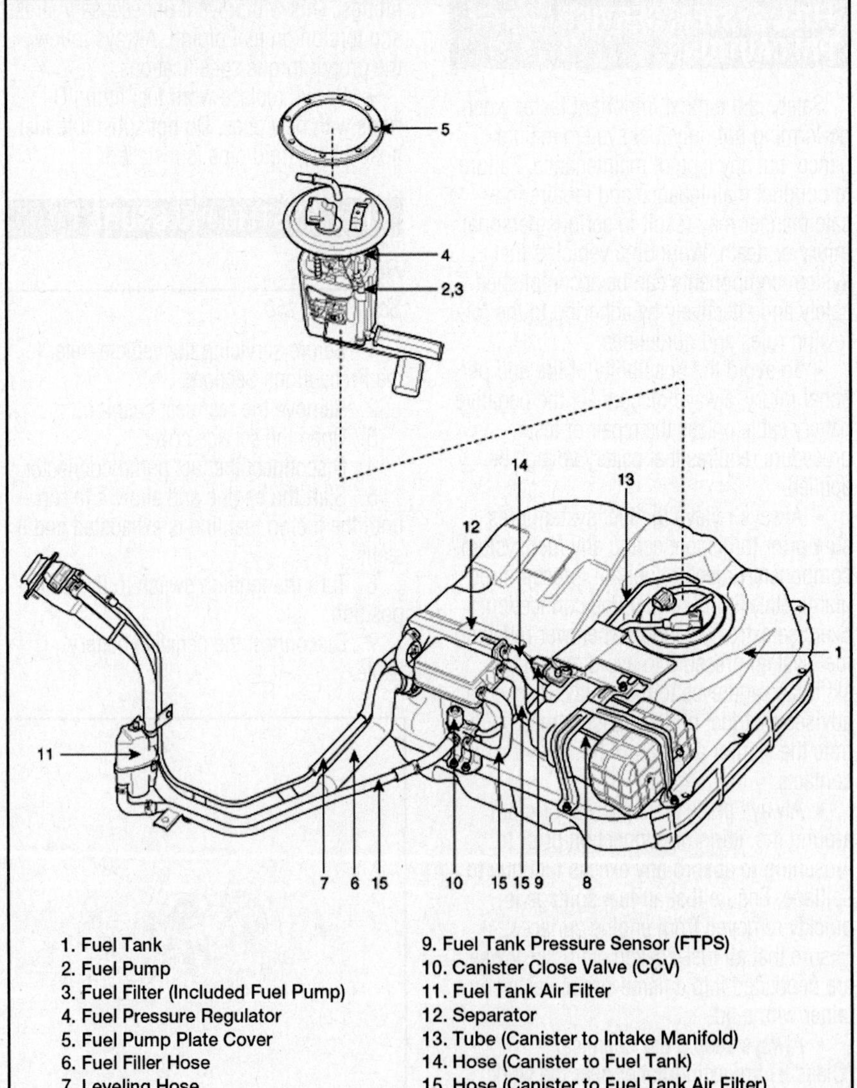

Fig. 155 Fuel delivery system—Sedan

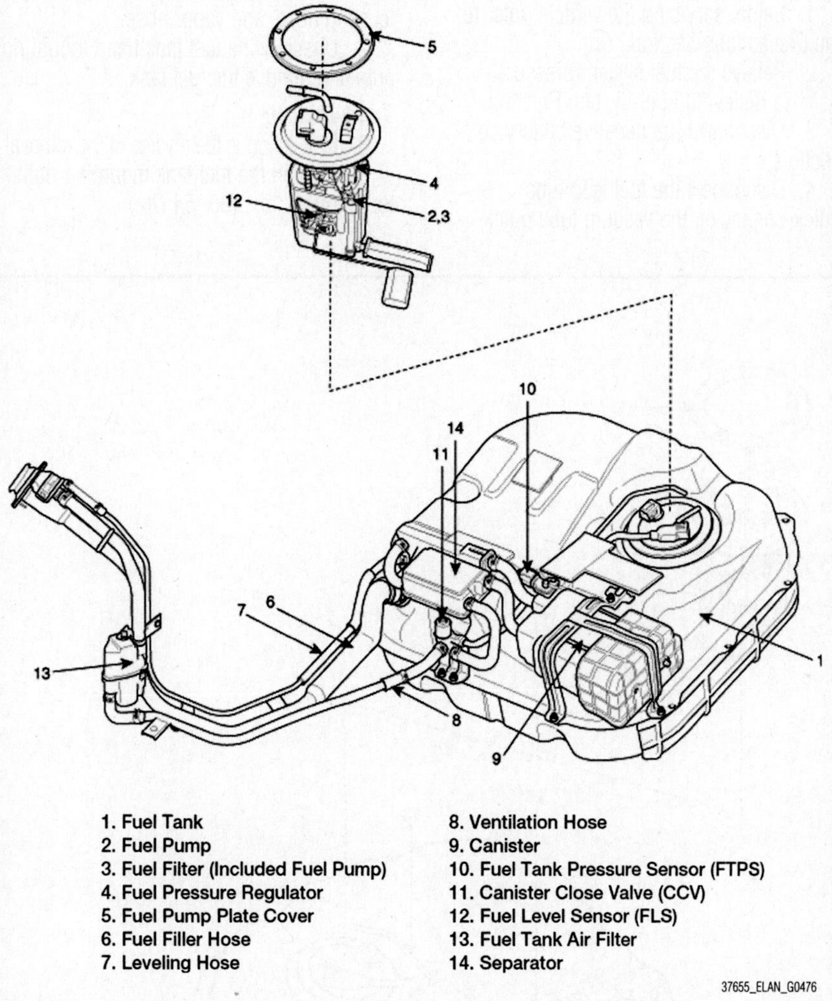

1. Fuel Tank
2. Fuel Pump
3. Fuel Filter (Included Fuel Pump)
4. Fuel Pressure Regulator
5. Fuel Pump Plate Cover
6. Fuel Filler Hose
7. Leveling Hose
8. Ventilation Hose
9. Canister
10. Fuel Tank Pressure Sensor (FTPS)
11. Canister Close Valve (CCV)
12. Fuel Level Sensor (FLS)
13. Fuel Tank Air Filter
14. Separator

37655_ELAN_G0476

Fig. 156 Fuel delivery system—Touring model

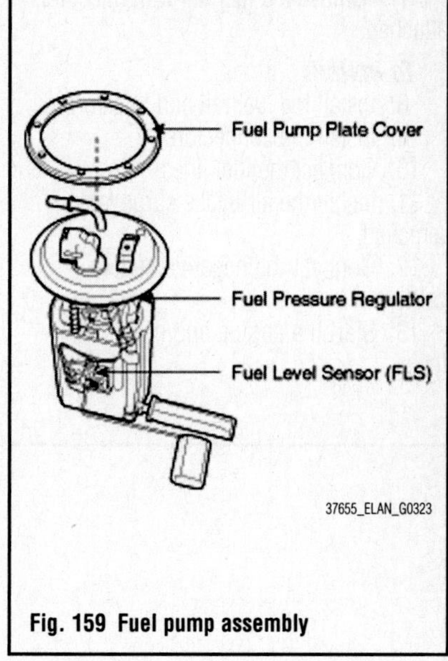

Fig. 159 Fuel pump assembly

To install:

6. Installation is the reverse of the removal procedure. Tighten the fuel pump plate cover to 58–72 ft. lbs. (79–98 Nm).

FUEL RAIL AND INJECTOR

REMOVAL & INSTALLATION

See Figure 160.

1. Before servicing the vehicle, refer to the Precautions Section.
2. Relieve the fuel system pressure. Refer to Relieving Fuel System Pressure.
3. Disconnect the negative battery cable.
4. Remove the air intake surge tank, if necessary.
5. Disconnect the fuel lines.
6. Disconnect the fuel injector connectors.

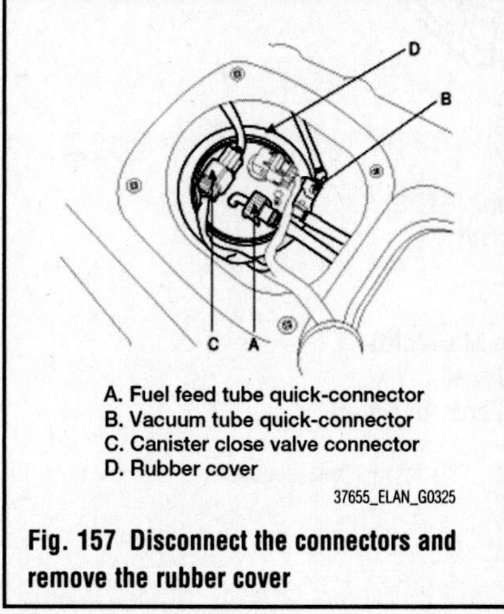

Fig. 157 Disconnect the connectors and remove the rubber cover

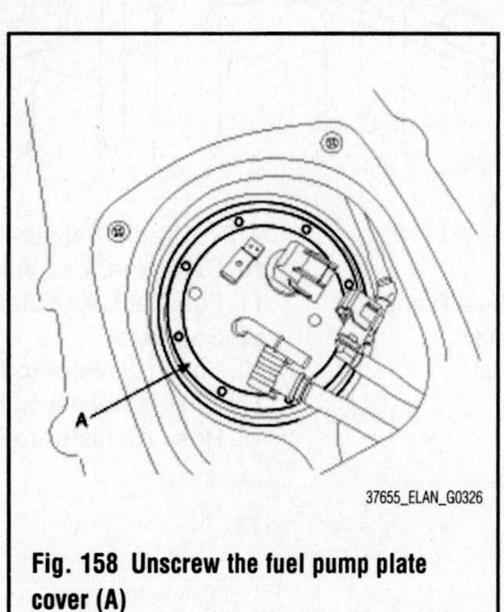

Fig. 158 Unscrew the fuel pump plate cover (A)

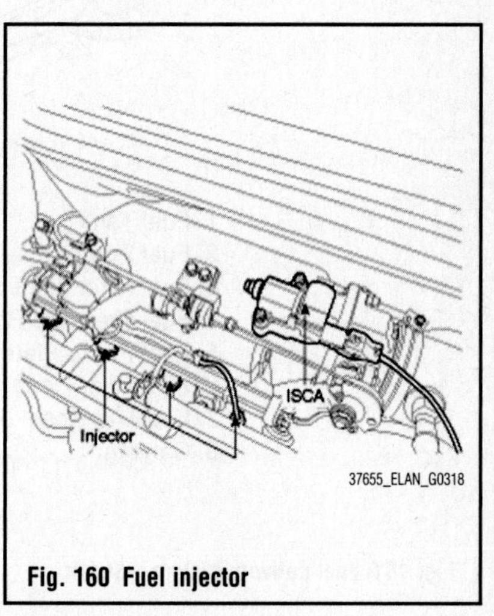

Fig. 160 Fuel injector

7. Remove the fuel rail with injectors attached.

To install:

8. Install the fuel rail and injectors.
9. Install the connectors.
10. Connect the fuel lines.
11. Install the air intake surge tank, if removed.
12. Connect the negative battery cable.
13. Start the engine and check for leaks.

FUEL TANK

REMOVAL & INSTALLATION

See Figures 161 through 165.

1. Before servicing the vehicle, refer to the Precautions Section.
2. Relieve the fuel system pressure. Refer to Relieving Fuel System Pressure.
3. Disconnect the negative battery connection.
4. Disconnect the fuel feed tube quick-connector, the vacuum tube quick-connector and canister close valve connector.
5. Raise and safely support the vehicle and support the fuel tank with a jack.
6. Disconnect the fuel filler hose, the leveling hose, and vapor hose.
7. Unscrew the fuel tank band mounting nuts and remove the fuel tank.

To install:

8. Installation is the reverse of the removal.
9. Tighten the fuel tank mounting bolts to: 29–40 ft. lbs. (39–54 Nm).

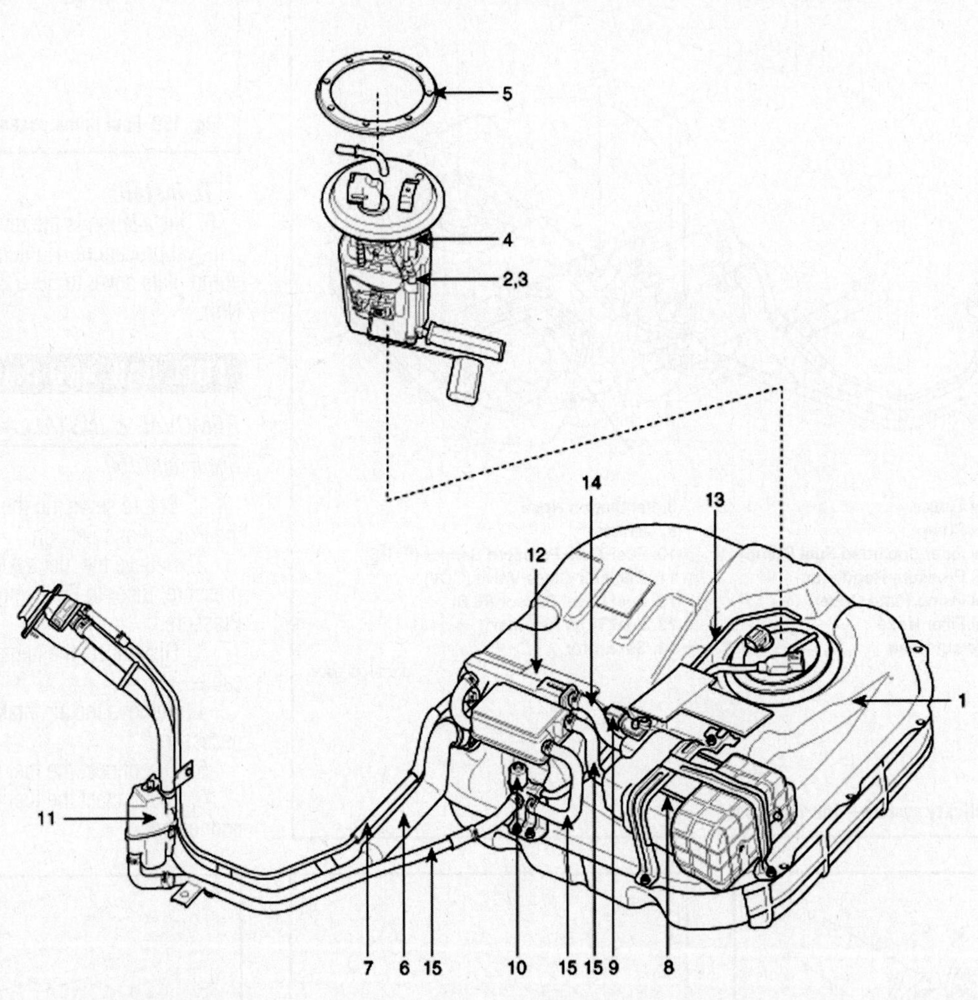

1. Fuel Tank
2. Fuel Pump
3. Fuel Filter (Included Fuel Pump)
4. Fuel Pressure Regulator
5. Fuel Pump Plate Cover
6. Fuel Filler Hose
7. Leveling Hose
8. Canister
9. Fuel Tank Pressure Sensor (FTPS)
10. Canister Close Valve (CCV)
11. Fuel Tank Air Filter
12. Separator
13. Tube (Canister to Intake Manifold)
14. Hose (Canister to Fuel Tank)
15. Hose (Canister to Fuel Tank Air Filter)

37655_ELAN_G0477

Fig. 161 Fuel delivery system—Sedan

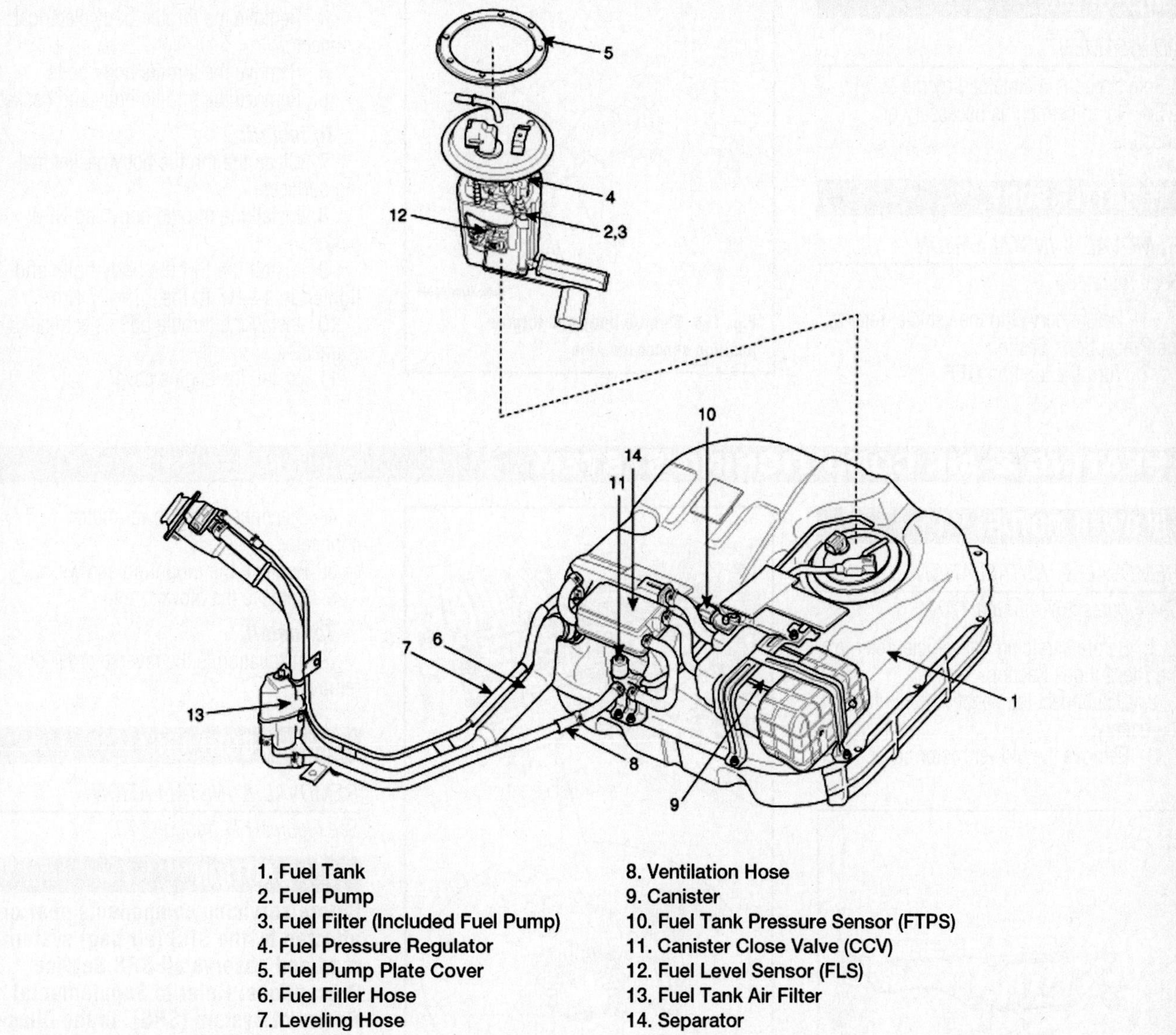

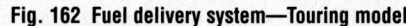

5
4
12
2,3
10
14
11
6
7
13
8
9
1
1. Fuel Tank
2. Fuel Pump
3. Fuel Filter (Included Fuel Pump)
4. Fuel Pressure Regulator
5. Fuel Pump Plate Cover
6. Fuel Filler Hose
7. Leveling Hose
8. Ventilation Hose
9. Canister
10. Fuel Tank Pressure Sensor (FTPS)
11. Canister Close Valve (CCV)
12. Fuel Level Sensor (FLS)
13. Fuel Tank Air Filter
14. Separator
37655_ELAN_G0476

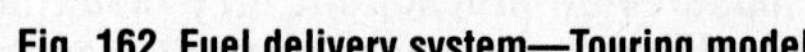

Fig. 162 Fuel delivery system—Touring model

B
C
A
37655_ELAN_G0339

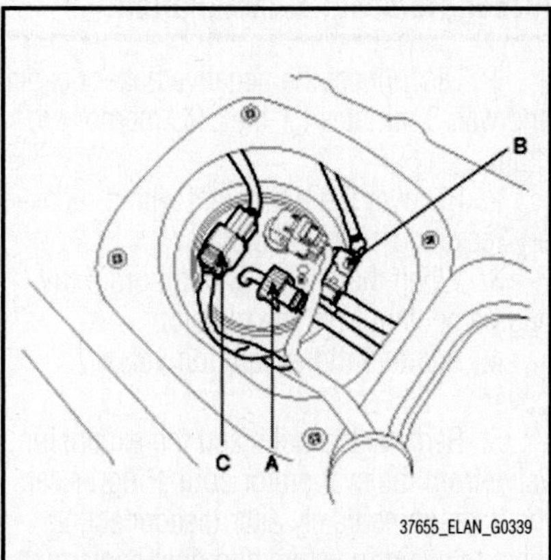

Fig. 163 Disconnect the fuel feed tube quick-connector (A), the vacuum tube quick-connector (B) and canister close valve connector (C)

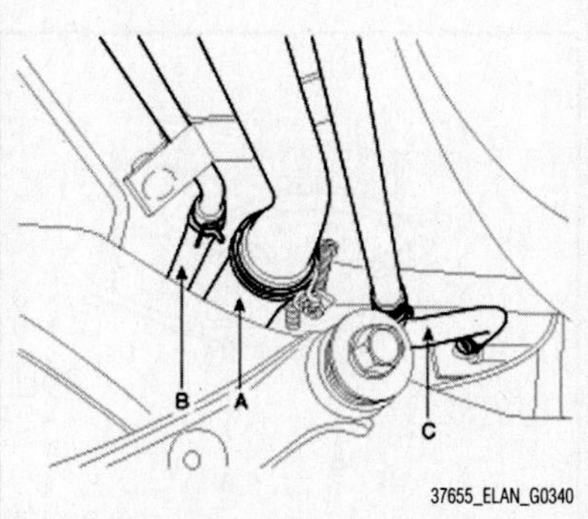

B
A
C
37655_ELAN_G0340

Fig. 164 Disconnect the fuel filler hose (A), the leveling hose (B), and vapor hose (C)

A
B
37655_ELAN_G0341

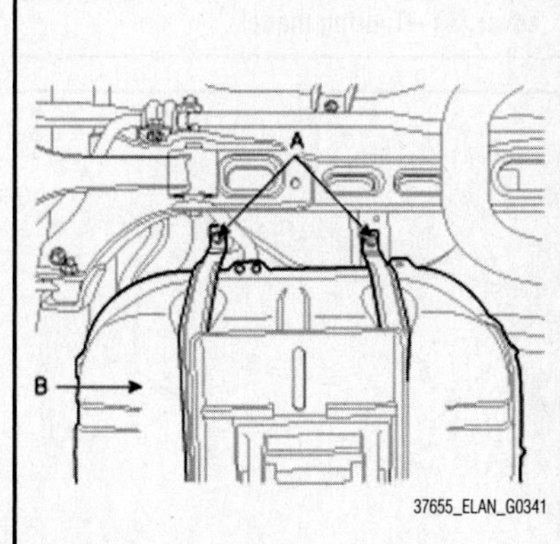

Fig. 165 Unscrew the fuel tank band mounting nuts (A) and remove the fuel tank (B)

IDLE SPEED

ADJUSTMENT

Idle speed is maintained by the ECM. No adjustment is necessary or possible.

THROTTLE BODY

REMOVAL & INSTALLATION

See Figure 166.

1. Before servicing the vehicle, refer to the Precautions Section.
2. Turn the ignition **OFF**.

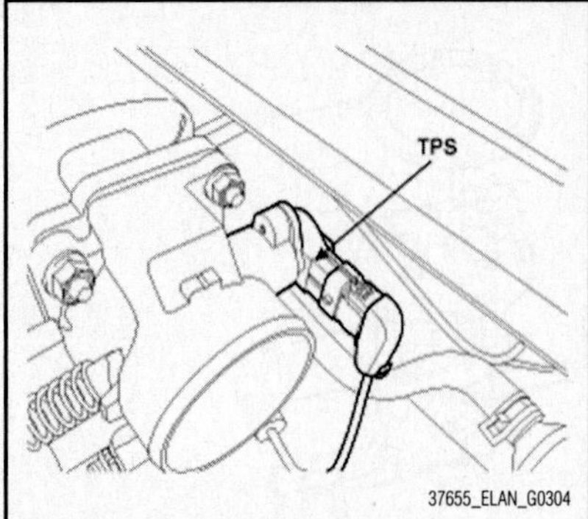

Fig. 166 Throttle body and throttle position sensor location

3. Remove the engine cover.
4. Remove the throttle body electrical connector.
5. Remove the throttle body bolts.
6. Remove the throttle body and gasket.

To install:

7. Clean the throttle body gasket mating surfaces.
8. Install the throttle body and NEW gasket.
9. Install the throttle body bolts and tighten to 14–17 ft. lbs. (19–27 Nm).
10. Install the throttle body electrical connector.
11. Install the engine cover.

HEATING & AIR CONDITIONING SYSTEM

BLOWER MOTOR

REMOVAL & INSTALLATION

See Figures 167 through 170.

1. Before servicing the vehicle, refer to the Precautions Section.
2. Disconnect the negative cable from the battery.
3. Remove the blower motor cover.

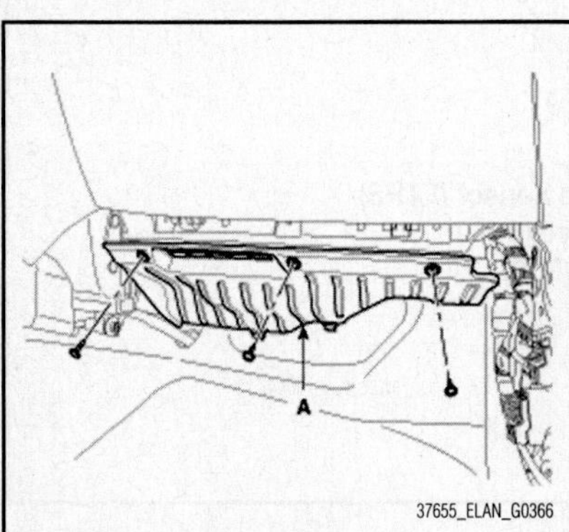

Fig. 167 Primary blower motor cover (A)—Touring model

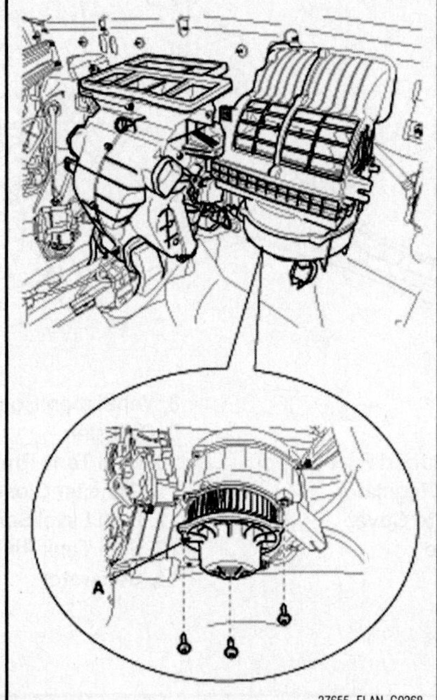

Fig. 169 Blower motor (A) and screws—Sedan

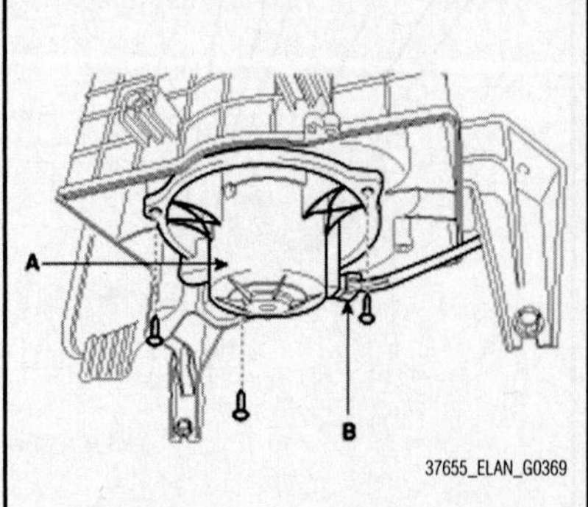

Fig. 170 Blower motor (A) and screws—Touring model

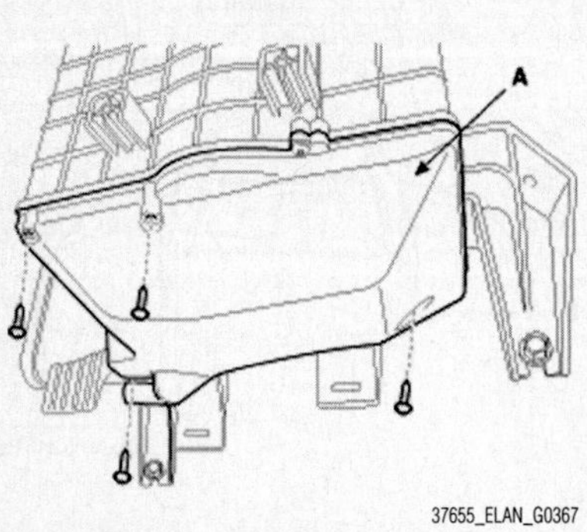

Fig. 168 Secondary blower motor cover (A)—Touring model

4. Disconnect the blower motor connector.
5. Remove the mounting screws.
6. Remove the blower motor.

To install:

7. Installation is the reverse order of removal.

HEATER CORE/UNIT

REMOVAL & INSTALLATION

See Figures 171 through 178.

⁂ CAUTION

Before servicing components near or affected by the SRS (air bag) system, read and observe all SRS Service Precautions. Refer to Supplemental Restraint System (SRS), in the Chassis Electrical section. Failure to observe all precautions may result in accidental airbag deployment, personal injury, or death. Refer to Airbag Removal & Installation.

1. Disconnect the negative battery cable and wait 3 minutes for the SRS memory to drain.
2. Recover the refrigerant with a recovery/recycling/charging station.
3. When the engine is cool, drain the engine coolant from the radiator.
4. Remove the expansion valve cover.
5. Remove the bolts and the expansion valve from the evaporator core. Plug or cap the lines immediately after disconnecting them to avoid moisture and dust contamination.
6. Disconnect the inlet and outlet heater hoses from the heater unit.

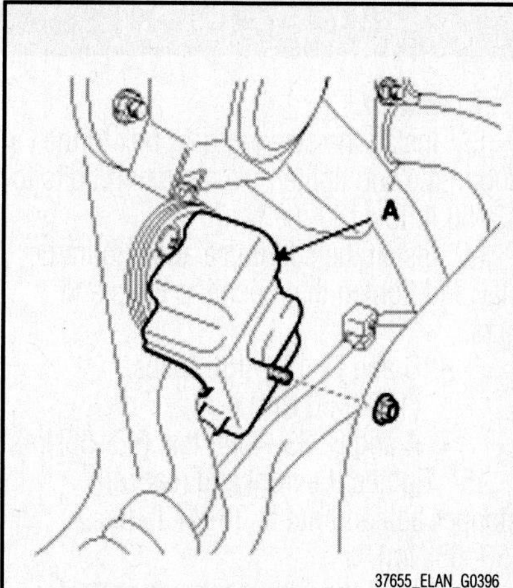

Fig. 171 Remove the expansion valve cover (A)

➡**Engine coolant will run out when the hoses are disconnected; drain it into a clean drip pan. Be sure not to let coolant spill on electrical parts or painted surfaces. If any coolant spills, rinse it off immediately.**

7. Remove the instrument panel.
8. Remove the cowl cross bar assembly.

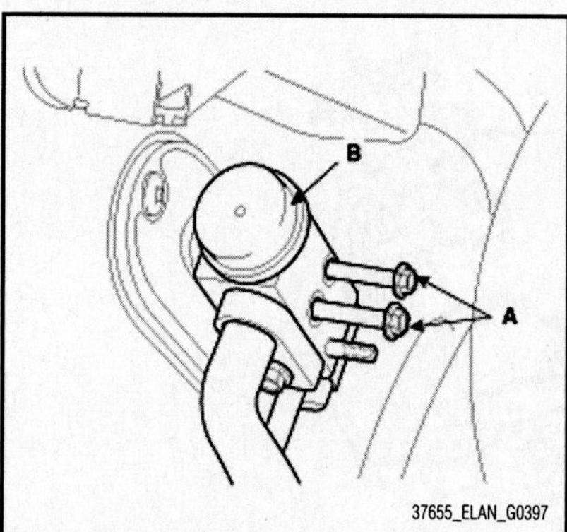

Fig. 172 Remove the bolts (A) and the expansion valve (B) from the evaporator core

9. Disconnect the connectors from the temperature control actuator, the mode control actuator and the evaporator temperature sensor.
10. Remove the mounting nuts and the heater and blower unit.
11. Remove the screws and blower unit from the heater unit.
12. Remove the cover and pull out the heater core. Be careful not to bend the inlet and outlet pipes.
13. Remove the heater unit lower case.

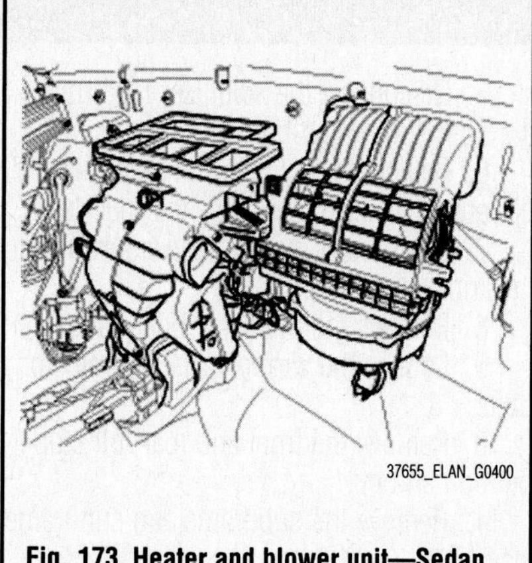

Fig. 173 Heater and blower unit—Sedan

14. Remove the evaporator core.

To install:

15. Install in the reverse order of removal, noting the following:
   a. If installing a new evaporator, add refrigerant oil ND-OIL8.
   b. Replace the O-rings with new ones at each fitting, and apply a thin coat of refrigerant oil before installing them. Be sure to use the right O-rings for R-134a to avoid leakage. Immediately after using the oil, replace the cap on the container, and seal it to avoid moisture absorption.
   c. Do not spill the refrigerant oil on the vehicle; it may damage the paint; if the refrigerant oil contacts the paint, wash it off immediately.
   d. Apply sealant to the grommets.
   e. Make sure that there is no air leakage.
16. Charge the A/C system and test the performance.

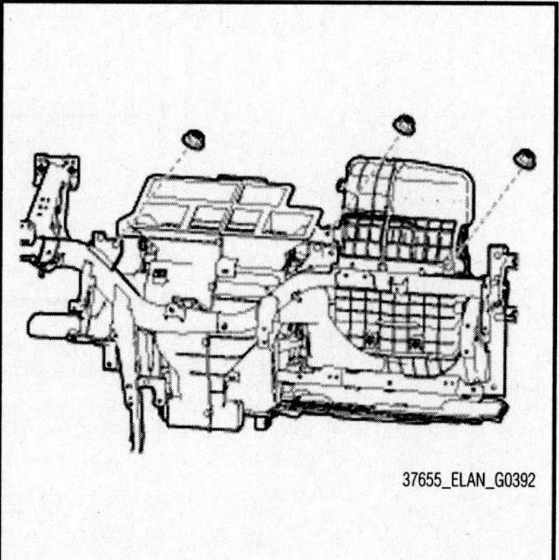

Fig. 174 Heater and blower unit—Touring model

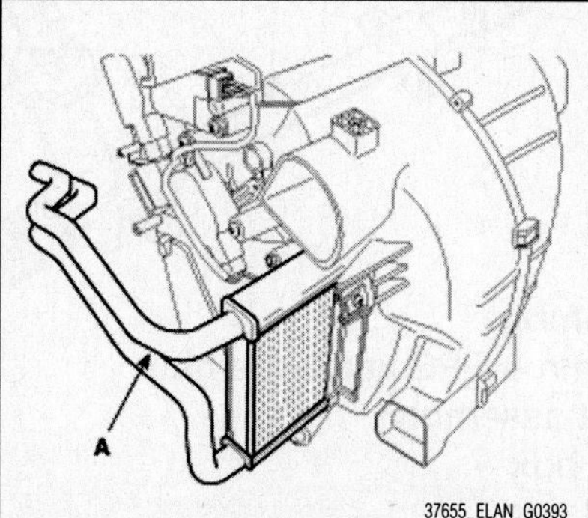

Fig. 175 Pull out the heater core without bending the inlet and outlet pipes (A)

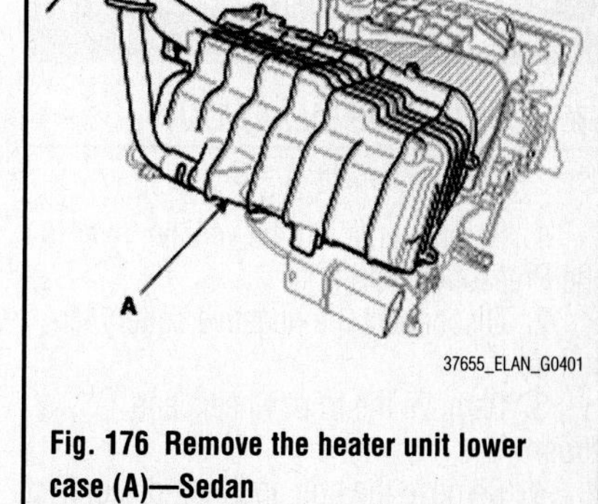

Fig. 176 Remove the heater unit lower case (A)—Sedan

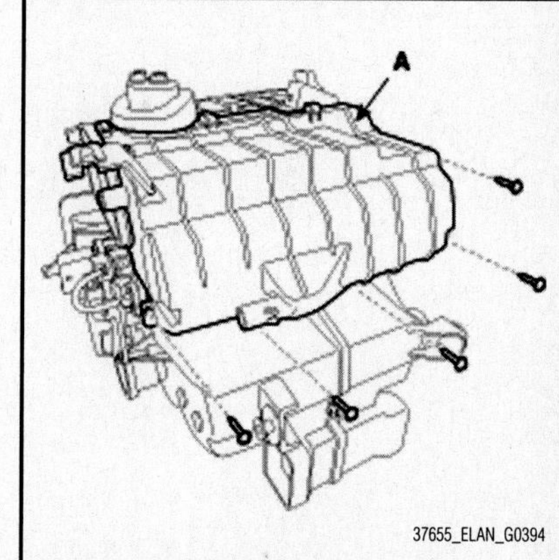

Fig. 177 Remove the heater unit lower case (A)—Touring model

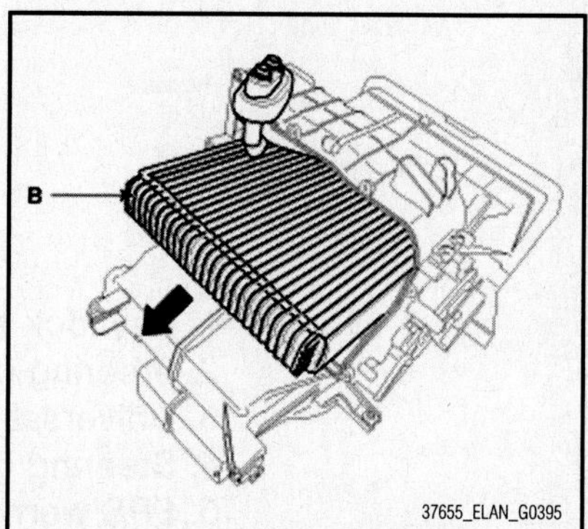

Fig. 178 Remove the evaporator core (B)—Sedan, Touring model similar

STEERING

POWER RACK & PINION STEERING GEAR

REMOVAL & INSTALLATION

See Figures 179 through 187.

1. Before servicing the vehicle, refer to the Precautions Section.
2. Disconnect the negative battery terminal.
3. Remove the front wheels and tires.
4. Remove the bolt and then disconnect the universal joint assembly with the pinion of the steering gear box.
5. Disconnect the stabilizer link from the front strut assembly.
6. Remove the split pin and castle nut and disconnect the tie rod end with the knuckle.
7. Disconnect the front lower arm from the knuckle.
8. Remove the muffler rubber hanger.
9. Loosen the wiring harness protector bolts.
10. Remove the front and rear roll stopper bolt and nut.
11. Remove the sub frame and sub frame stay.
12. Remove the steering gear box from the sub frame.

To install:

13. Install the steering gear box to the sub frame and tighten the mounting bolts to 43–58 ft. lbs. (60–80 Nm).
14. Install the sub frame and sub frame stay and tighten the mounting bolts and nuts:
    - C and D: 116–130 ft. lbs. (160–180 Nm)
    - A and B: 33–40 ft. lbs. (45–55Nm)
15. Tighten the front and rear roll stopper bolt and nut to 36–47 ft. lbs. (50–65 Nm).
16. Install the wiring harness protector to the sub-frame.

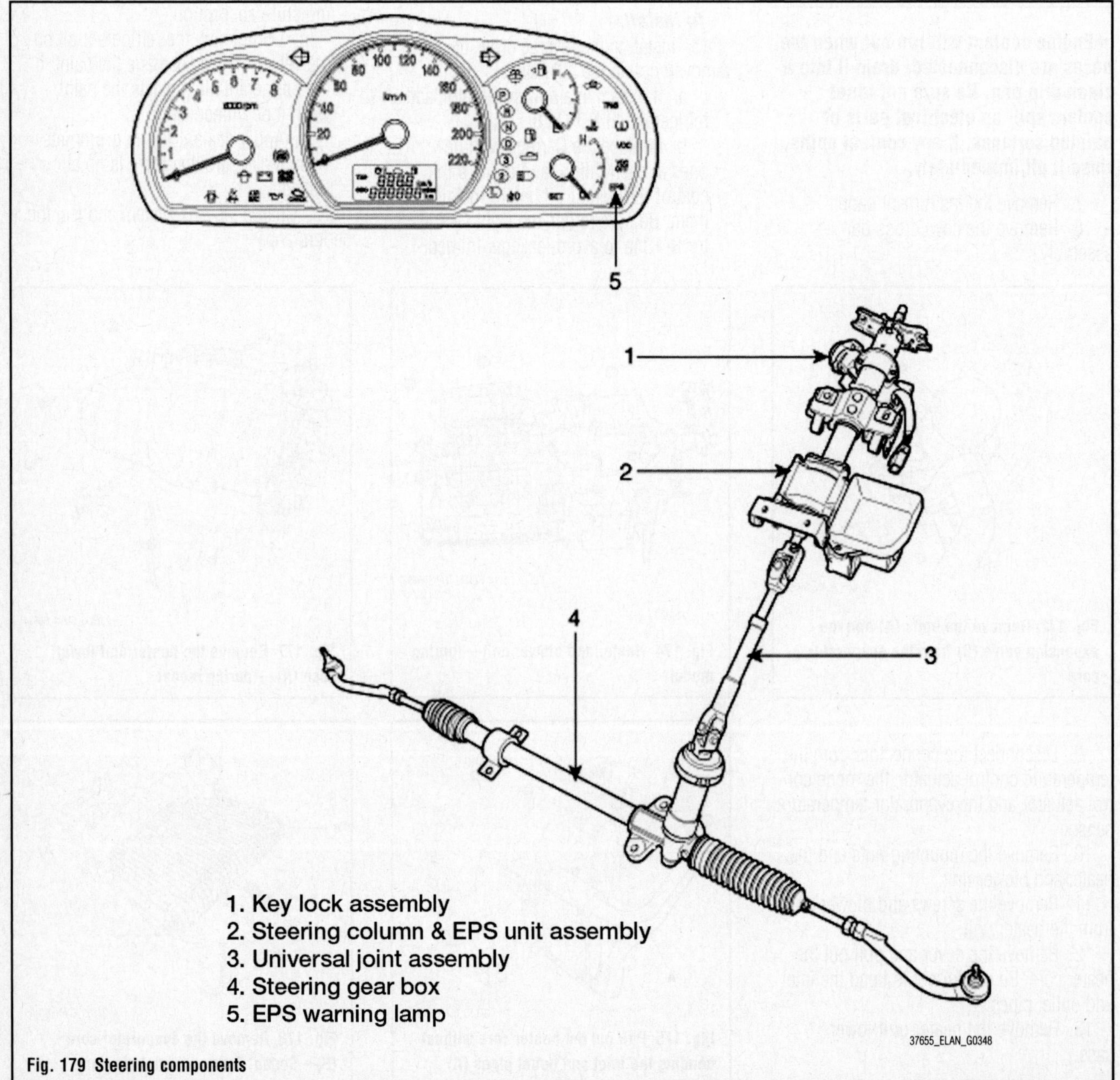

Fig. 179 Steering components

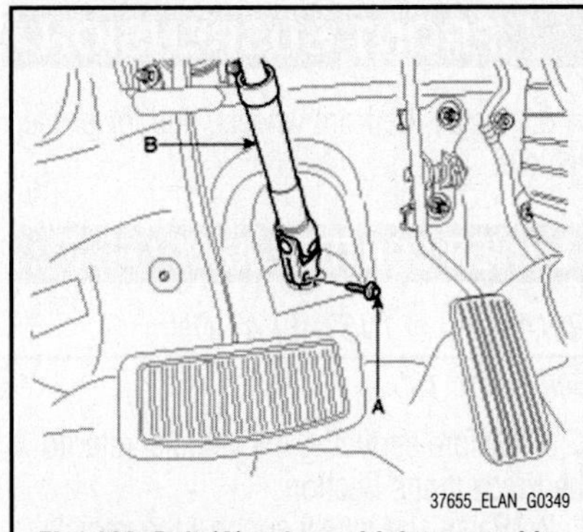

Fig. 180 Bolt (A) universal joint assembly (B) location

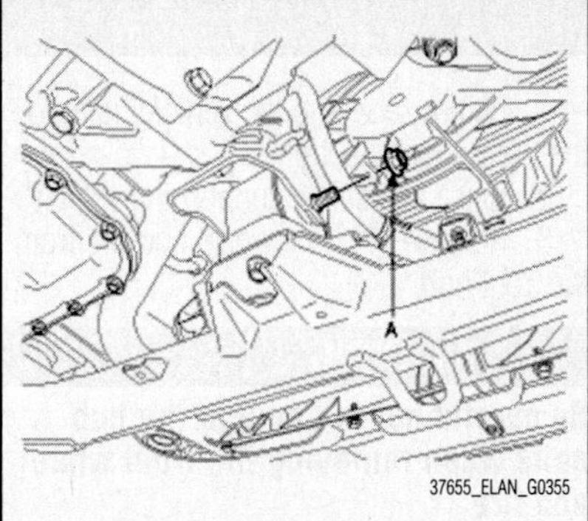

Fig. 183 Front roll stopper bolt and nut location (A)

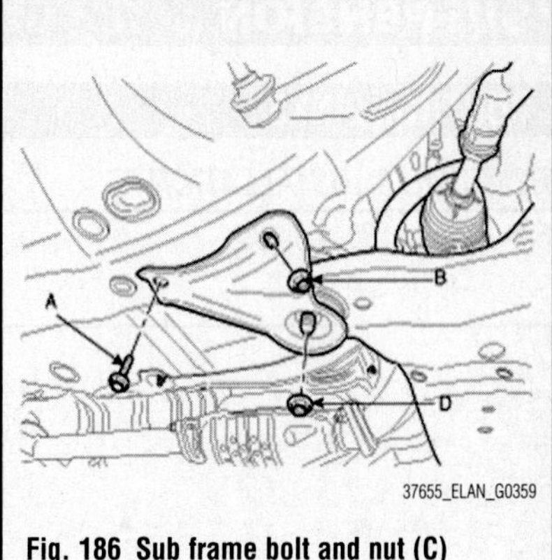

Fig. 186 Sub frame bolt and nut (C)

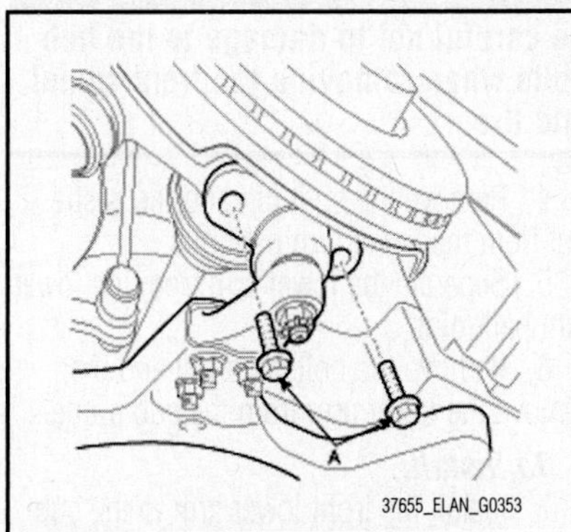

Fig. 181 Remove the bolts (A) and the front lower arm from the knuckle

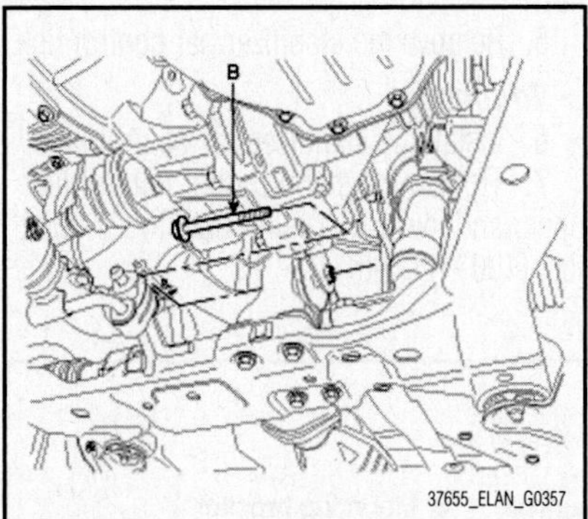

Fig. 184 Rear roll stopper bolt and nut location (B)

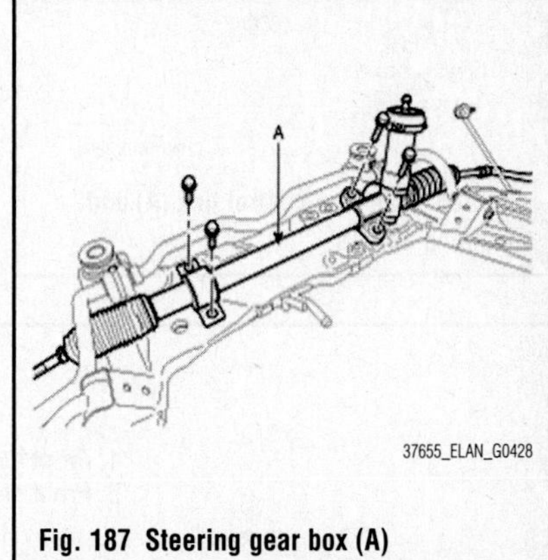

Fig. 187 Steering gear box (A)

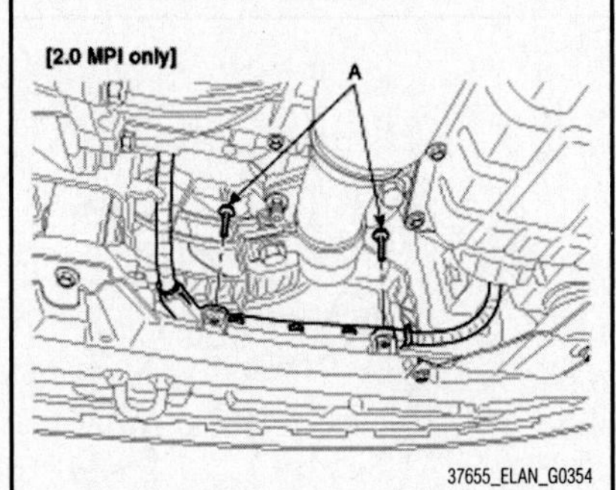

Fig. 182 Wiring harness protector bolt (A) location

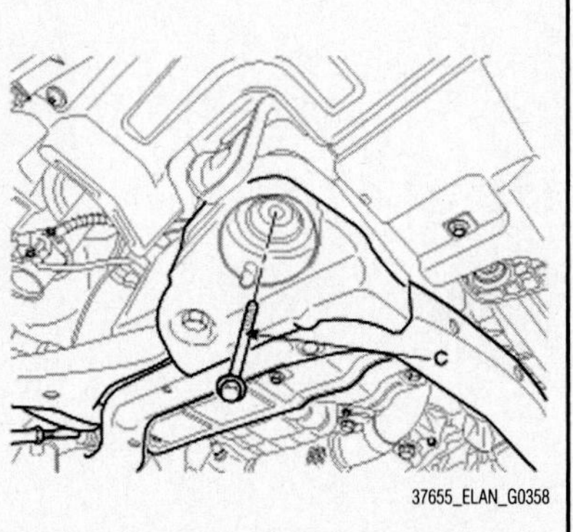

Fig. 185 Sub frame bolt and nut (A, B, D)

17. Install the muffler rubber hanger.
18. Connect the lower arm with the knuckle by tightening the bolts to 72–87 ft. lbs. (100–120 Nm).
19. Connect the tie rod end with the knuckle and install the castle nut and split pin. Tighten castle nut to 17–25 ft. lbs. (24–34 Nm).
20. Connect the stabilizer link with the front strut assembly and tighten the nut to 72–87 ft. lbs. (100–120 Nm).
21. Connect the universal joint assembly with the pinion of the steering gear box and tighten the bolt to 22–25 ft. lbs. (30–35 Nm).
22. Install the front wheels and tires.
23. Check and adjust the front wheel alignment.
24. Following the instructions in a compatible scan tool, perform the Absolute Steering Position (ASP) calibration procedure.

WARNING

If the ASP calibration procedure is not performed, EPS warning lamp will be turned on or flicker and vehicle may pull to the left or the right.

CONTROL LINKS

REMOVAL & INSTALLATION

See Figures 188 and 189.

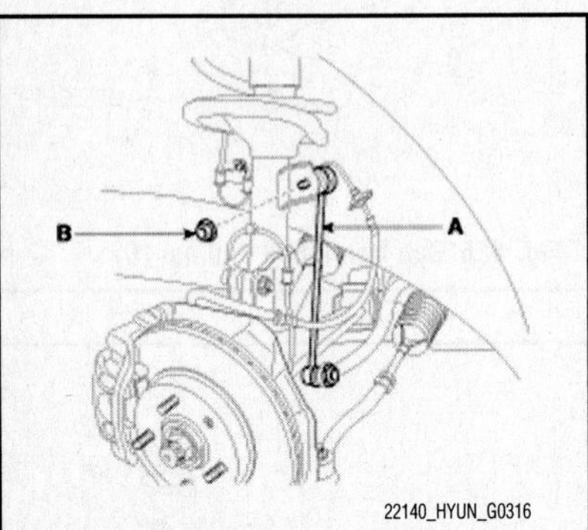

Fig. 188 Stabilizer control link (A) and mounting nuts (B)

1. Before servicing the vehicle, refer to the Precautions Section.
2. Raise and safely support the vehicle.
3. Remove the front wheel and tire from the front hub.

⁂ WARNING

Be careful not to damage the hub bolts when removing the front wheel and tire.

4. Remove the stabilizer bar control link by removing the upper and lower mounting nuts.
5. Remove the stabilizer bar control link.

To install:

6. Install the stabilizer bar control link.
7. Tighten the stabilizer bar control link upper and lower mounting nuts to 72–87 ft. lbs. (100–120 Nm).

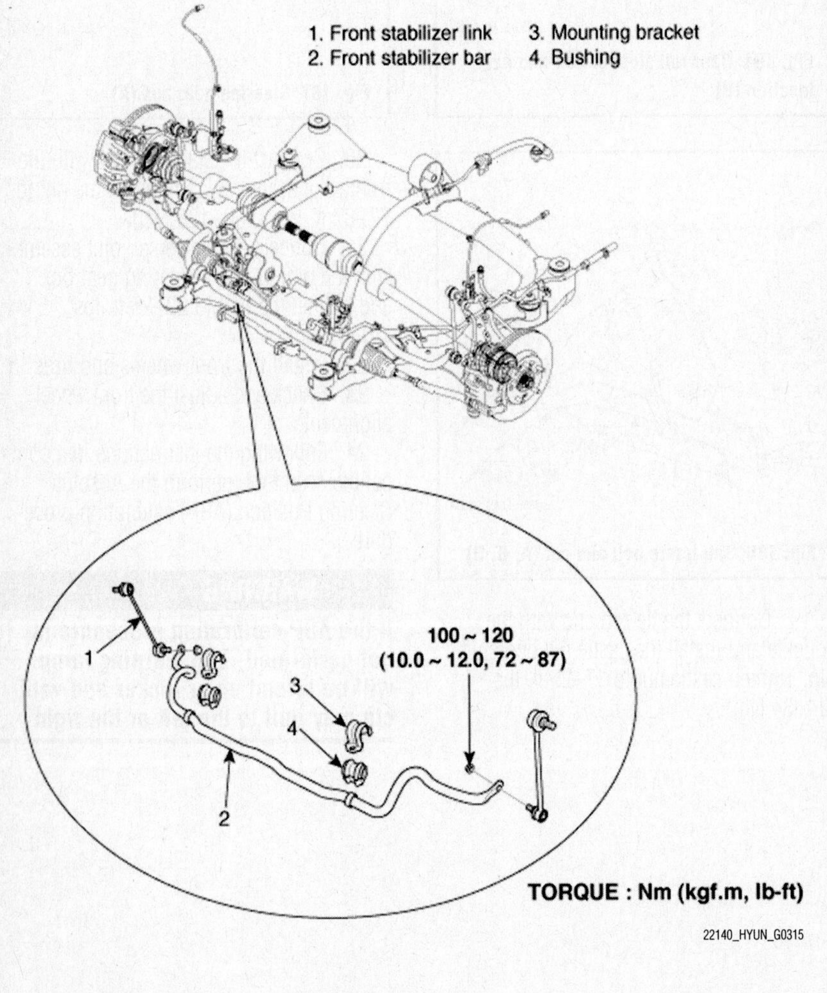

Fig. 189 Stabilizer bar and related components

8. Install the front wheels, and lower the vehicle.

LOWER CONTROL ARM

REMOVAL & INSTALLATION

See Figure 190.

1. Before servicing the vehicle, refer to the Precautions Section.
2. Raise and safely support the vehicle.
3. Remove the front wheel and tire.

⁂ WARNING

Be careful not to damage to the hub bolts when removing the front wheel and tire.

4. Remove the split pin and the castle nut from the lower arm ball joint.
5. Separate the lower arm from the lower arm ball joint.
6. Remove the bolts and nut and then remove the lower arm from the sub frame.

To install:

7. Install the front lower arm to the sub frame and tighten the bolts and nuts:
   - A: 101–116 ft. lbs. (140–160 Nm)
   - B: 72–87 ft. lbs. (100–120 Nm).

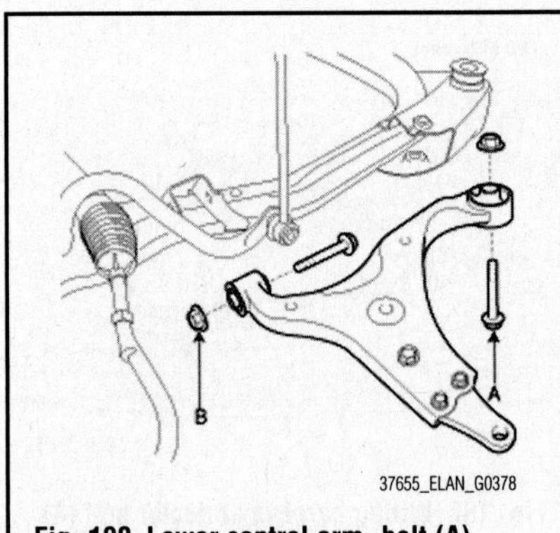

Fig. 190 Lower control arm, bolt (A), and nut (B)

8. Connect the lower arm with the ball joint and then install the castle nut and the split pin. Tighten castle nut to 58–65 ft. lbs. (80–90 Nm).
9. Install the front wheel and tire, and lower the vehicle.

STEERING KNUCKLE

REMOVAL & INSTALLATION

See Figures 191 through 199.

1. Before servicing the vehicle, refer to the Precautions Section.

2. Raise and safely support the vehicle.

3. Remove the front wheel and tire.

⁂ WARNING

Be careful not to damage the hub bolts when removing the front wheel and tire.

4. Remove the wheel speed sensor from the knuckle.

5. Remove the brake caliper mounting bolts, and then hang the brake caliper assembly with a wire.

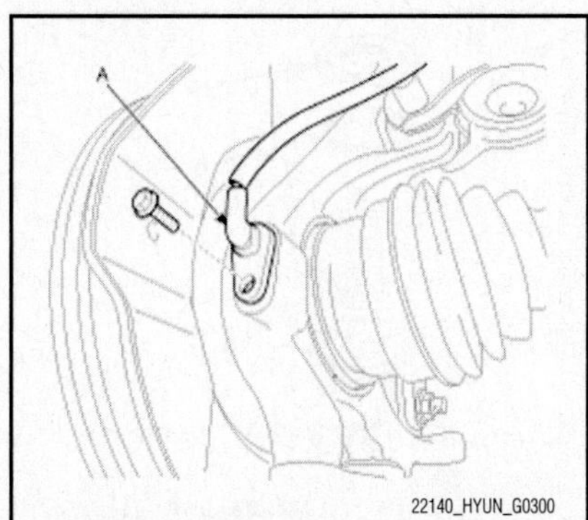

Fig. 191 Remove the wheel speed sensor (A) from the knuckle

⁂ WARNING

Do not suspend the brake caliper assembly from the brake hose or damage may occur to the hose.

6. Remove the split pin, then remove the castle nut and washer from the front hub.

7. Remove the ball joint assembly mounting bolts from the steering knuckle.

8. Remove the tie rod end ball joint from the knuckle.

   a. Remove the split pin and castle nut and disconnect the ball joint from the steering knuckle.

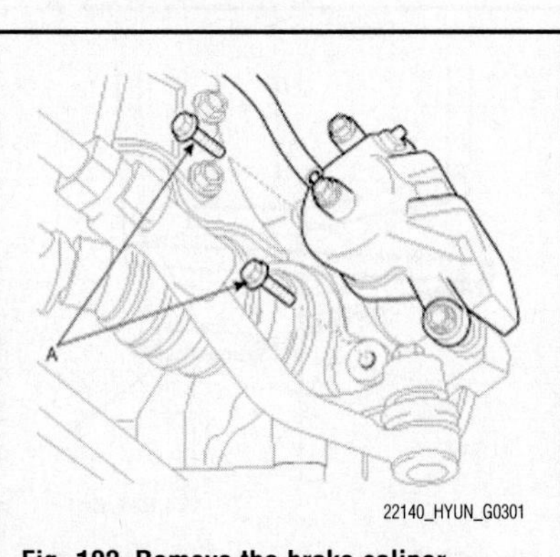

Fig. 192 Remove the brake caliper mounting bolts (A)

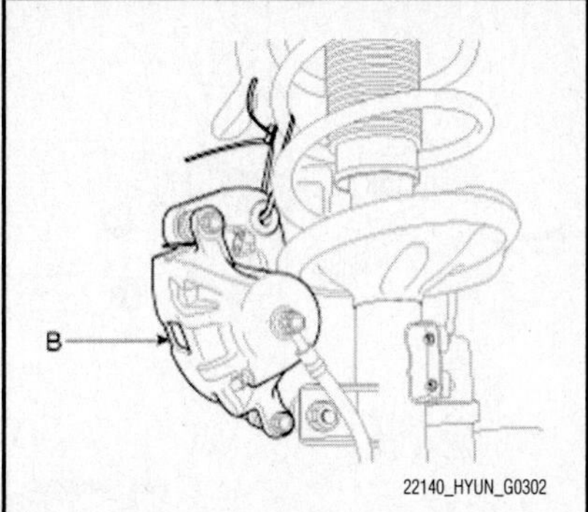

Fig. 193 Hang the brake caliper assembly (B) with a wire

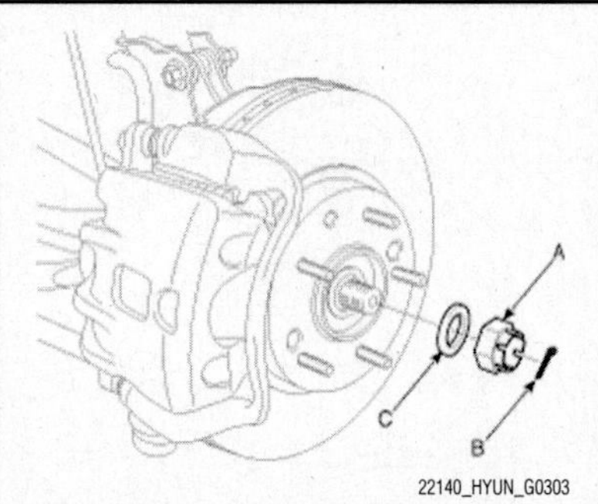

Fig. 194 Remove the split pin (B), the castle nut (A), and the washer (C) from the front hub

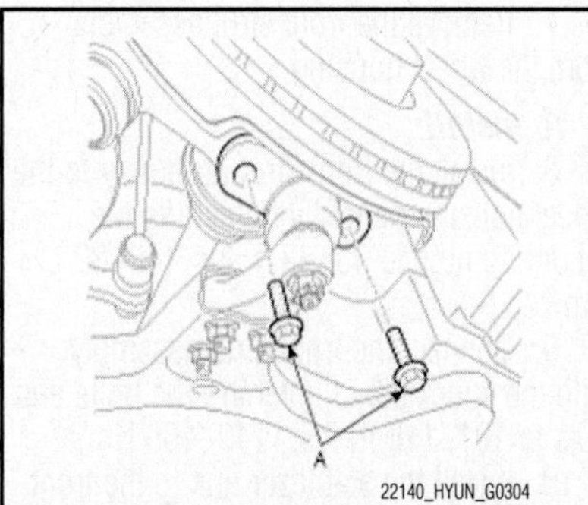
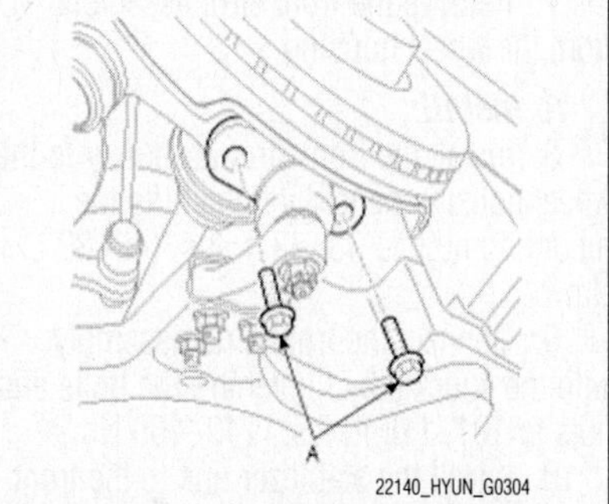

Fig. 195 Remove the ball joint assembly mounting bolts (A) from the steering knuckle

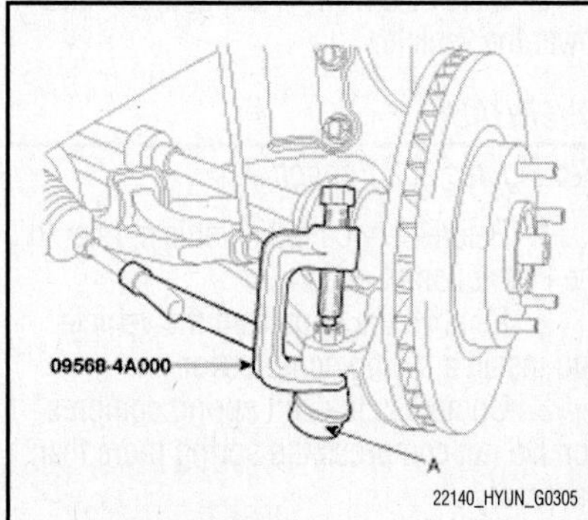

Fig. 196 Disconnect the ball joint (A) from the steering knuckle

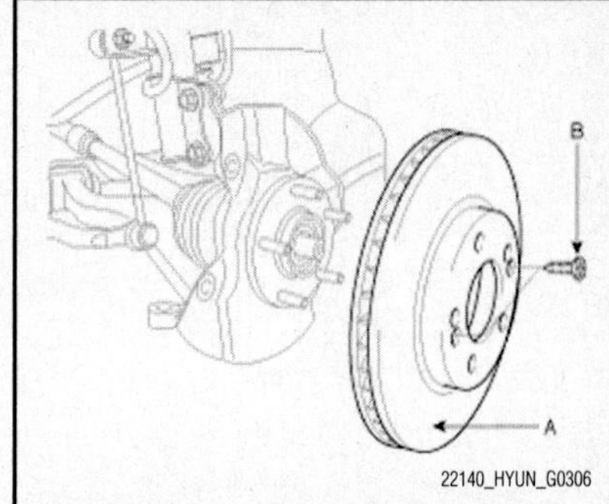

Fig. 197 Remove the brake disc (A) from the front hub assembly after removing the screws (B)

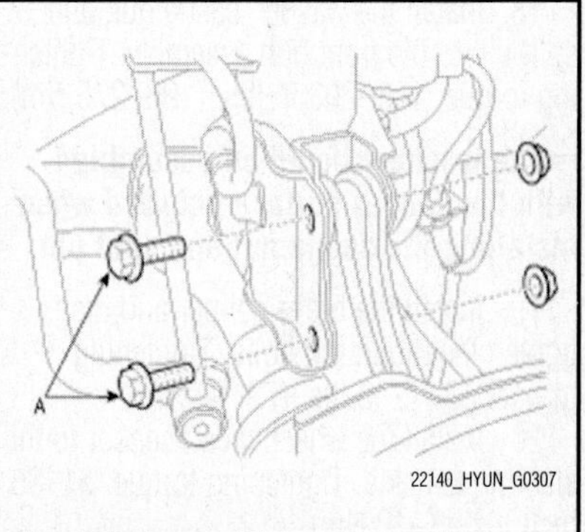

Fig. 198 Remove the strut assembly mounting bolts (A) and nuts

9. Remove the brake disc from the front hub assembly after removing the screws.

10. Remove the strut assembly mounting bolts and nuts.

11. Remove the hub and knuckle assembly.

⁂ WARNING

Be careful not to damage the boot and rotor teeth.

To install:

12. Install the hub and knuckle assembly to the halfshaft.

13. Install the knuckle to the strut assembly and tighten the mounting bolts and nuts. Tightening torque: 101–116 ft. lbs. (137–157 Nm).

14. Install the brake disc to the front hub assembly and tighten the screws.

15. Install the tie rod end ball joint to the knuckle.

16. Install the nut and split pin. Tightening torque: 17–25 ft. lbs. (24–33 Nm).

17. Install the ball joint assembly mounting bolt to the steering knuckle. Tightening torque: 72–87 ft. lbs. (98–118 Nm).

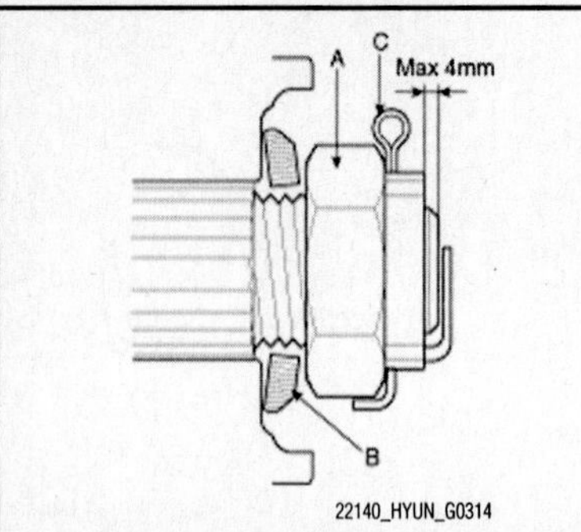

Fig. 199 Place the washer with the convex surface outward when installing the castle nut and split pin

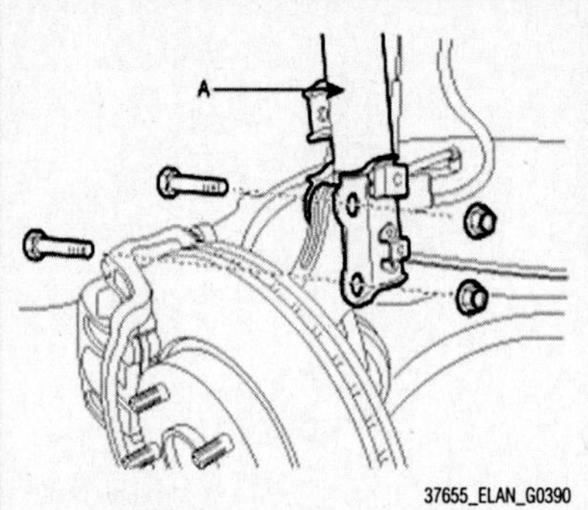

Fig. 200 Remove the front strut (A) and bolts from the knuckle

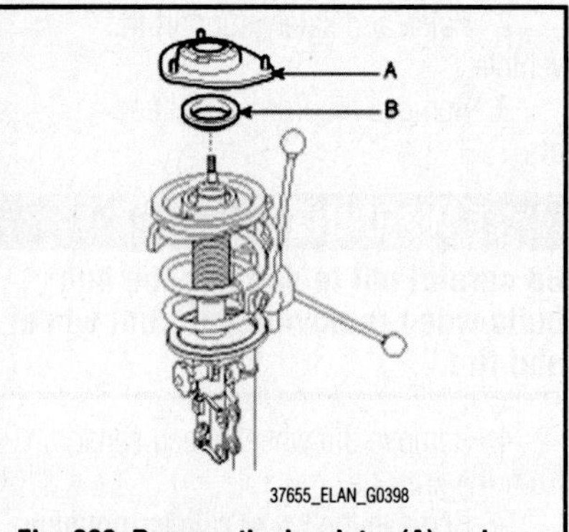

Fig. 202 Remove the insulator (A) and strut bearing (B)

18. Install the washer, castle nut, and split pin to the front hub assembly. Tightening torque: 145–203 ft. lbs. (196–275 Nm).

➡The washer should be assembled with the convex surface outward when installing the castle nut and split pin.

19. Install the brake caliper and then tighten the mounting bolts. Tightening torque: 58–72 ft. lbs. (79–98 Nm).

20. Install the wheel speed sensor to the steering knuckle. Tightening torque: 61–86 inch lbs. (7–10 Nm).

21. Install the wheel and the tire to the front hub. Tightening torque: 65–80 ft. lbs. (88–108 Nm).

WARNING

Be careful not to damage the hub bolts when installing the front wheel and tire.

STRUT

REMOVAL & INSTALLATION

See Figures 200 and 201.

1. Before servicing the vehicle, refer to the Precautions Section.
2. Raise and safely support the vehicle.
3. Remove the front wheel and tire.

WARNING

Be careful not to damage to the hub bolts when removing the front wheel and tire.

4. Remove the brake hose and the wheel speed sensor bracket from the front strut assembly by loosening the mounting bolts.
5. Disconnect the stabilizer link with the front strut assembly after loosening the nut. Refer to Control Links Removal & Installation.
6. Disconnect the front strut assembly from knuckle.

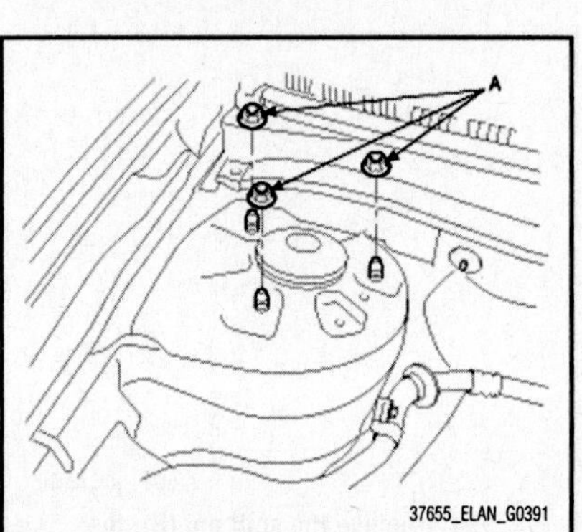

Fig. 201 Remove the front strut and nuts (A) from the wheel housing

7. Remove the front strut assembly from the wheel housing.

To install:

8. Install the front strut assembly to the wheel housing panel by tightening the mounting nuts to 33–43 ft. lbs. (45–60 Nm).
9. Connect the front strut assembly with the knuckle by tightening the bolts and nuts to 101–116 ft. lbs. (140–160 Nm).
10. Install the stabilizer link to the front strut assembly and then tighten the nut to 72–87 ft. lbs. (100–120 Nm).
11. Install the brake hose and wheel speed sensor bracket to front strut assembly.
12. Install the front wheel and tire, and lower the vehicle.

OVERHAUL

See Figures 202 through 204.

1. Before servicing the vehicle, refer to the Precautions Section.
2. Remove the strut from the vehicle and install a spring compressor.
3. Compress the coil spring compressor. Do not compress the spring more than necessary.
4. Loosen the self locking nut.
5. Remove the insulator assembly and the strut bearing.

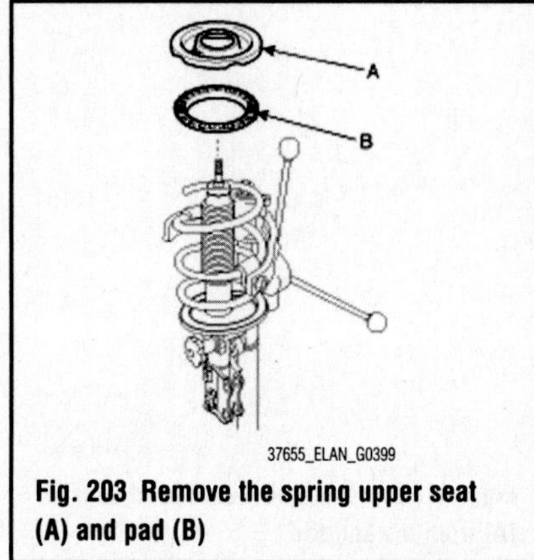

Fig. 203 Remove the spring upper seat (A) and pad (B)

6. Remove the spring upper seat and pad.
7. Remove the dust cover and the bumper rubber.
8. Remove the coil spring and the spring lower pad.
9. Remove the shock absorber from the spring compressor.

To install:

10. Install the front shock absorber to spring compressor.
11. Install the spring lower pad so that the protrusions fit in the holes in the spring lower seat.

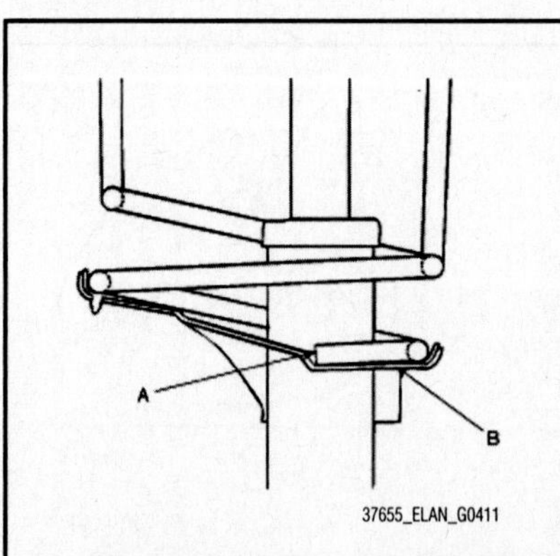

Fig. 204 Positioning coil spring (A) fit the lower pad (B)

12. Put the coil spring on the spring lower pad.

13. Clean the piston rod and then install the bumper rubber and the dust cover to piston rod.

14. Install the spring upper pad and seat.

WARNING

When installing the coil spring, align the spring end with the grooves of spring pad and seat correctly.

15. Compress the coil spring.

16. Fully extend the piston rod and then install the strut bearing and the insulator.

17. Tighten a new self locking nut to 36–51 ft. lbs. (50–70 Nm).

18. Remove the strut from spring compressor.

STABILIZER BAR

REMOVAL & INSTALLATION

See Figures 205 through 213.

1. Before servicing the vehicle, refer to the Precautions Section.

2. Raise and safely support the vehicle.

3. Remove the front wheel and tire.

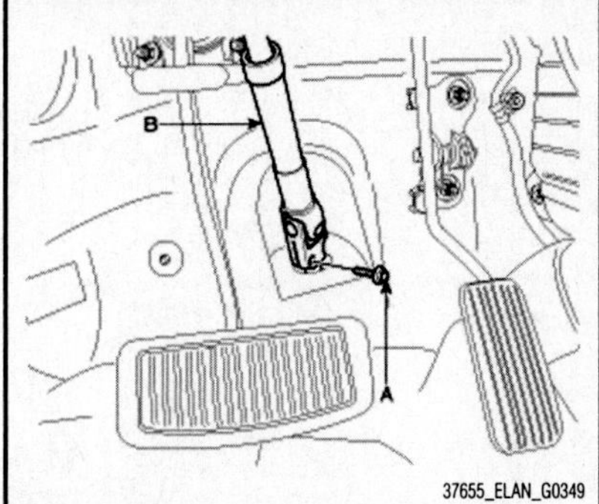

Fig. 206 Bolt (A) universal joint assembly (B) location

4. Remove the bolt and then disconnect the universal joint assembly with the pinion of the steering gear box.

5. Disconnect the stabilizer link from the front strut assembly.

6. Remove the split pin and castle nut and disconnect the tie rod end with the knuckle.

7. Disconnect the front lower arm from the knuckle.

8. Remove the muffler rubber hanger.

9. Loosen the wiring harness protector bolts.

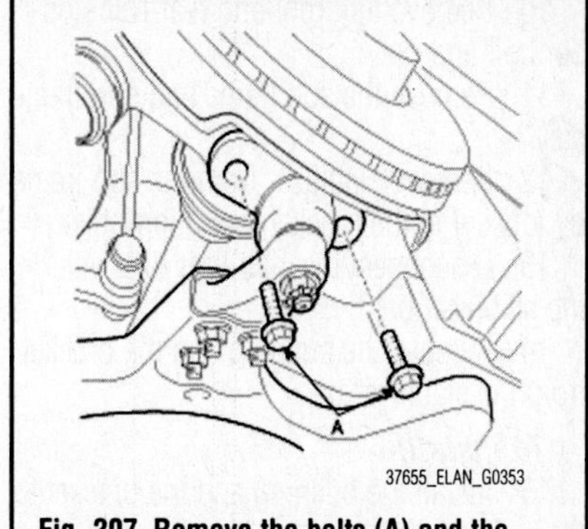

Fig. 207 Remove the bolts (A) and the front lower arm from the knuckle

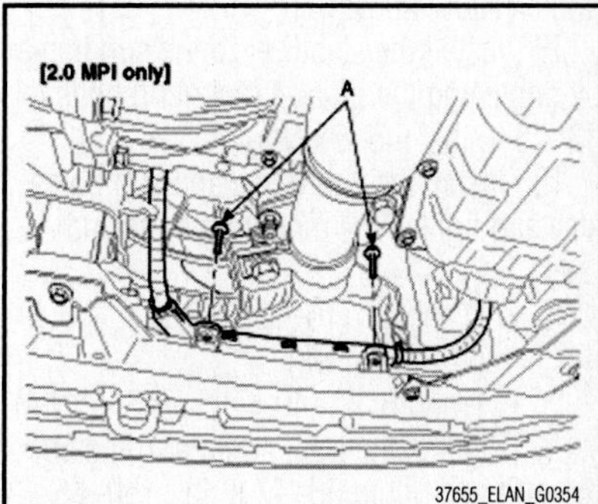

Fig. 208 Wiring harness protector bolt (A) location

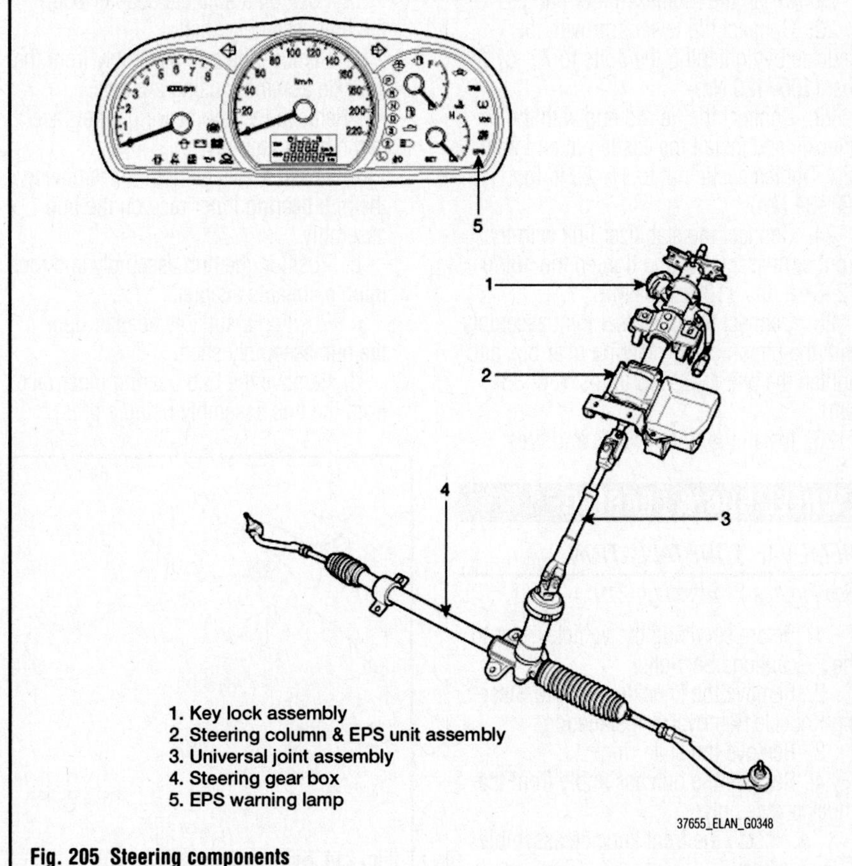

Fig. 205 Steering components

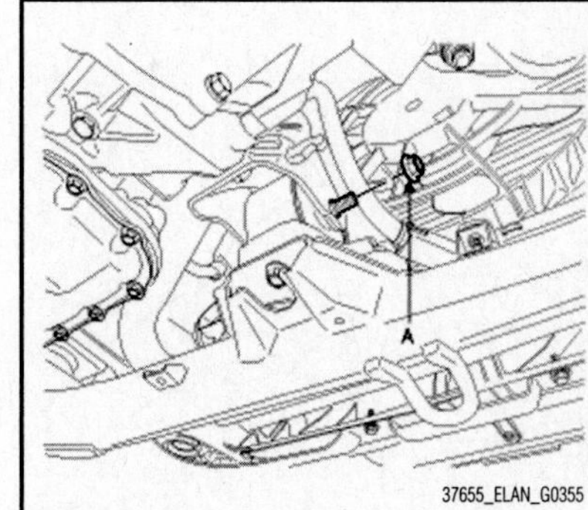

Fig. 209 Front roll stopper bolt and nut location (A)

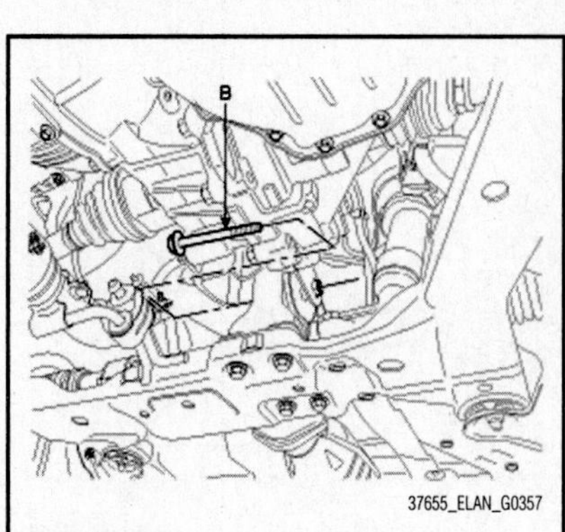

Fig. 210 Rear roll stopper bolt and nut location (B)

10. Remove the front and rear roll stopper bolt and nut.

11. Remove the sub frame and sub frame stay.

12. Remove stabilizer from the sub frame by loosening the bracket mounting bolts.

13. Disconnect the stabilizer link with the stabilizer bar.

14. Remove the bushing and the bracket from the stabilizer bar.

To install:

15. Install the bushing and the bracket to the stabilizer bar.

16. Connect the stabilizer link with the stabilizer bar by tightening the nut to 72–87 ft lbs. (100–120 Nm).

17. Install the stabilizer to the sub frame by tightening the bracket mounting bolts to 33–40 ft. lbs. (45–55 Nm).

18. Install the sub frame and sub frame stay and tighten the mounting bolts and nuts:

- C and D: 116–130 ft. lbs. (160–180 Nm)
- A and B: 33–40 ft. lbs. (45–55Nm)

19. Tighten the front and rear roll stopper bolt and nut to 36–47 ft. lbs. (50–65 Nm).

20. Install the wiring harness protector to the sub-frame.

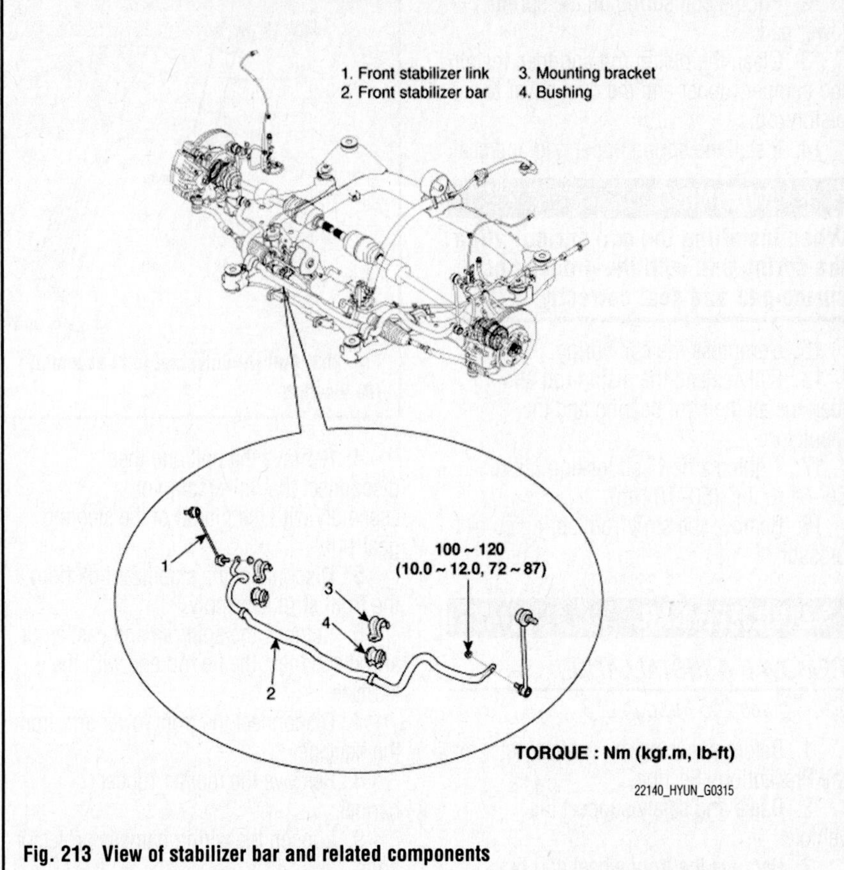

Fig. 213 View of stabilizer bar and related components

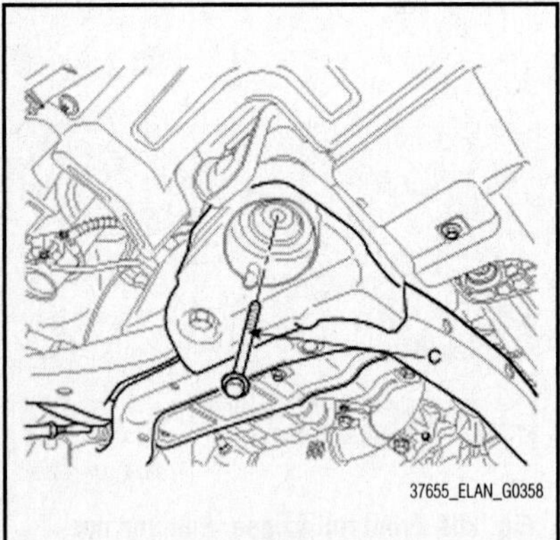

Fig. 211 Sub frame bolt and nut (A, B, D)

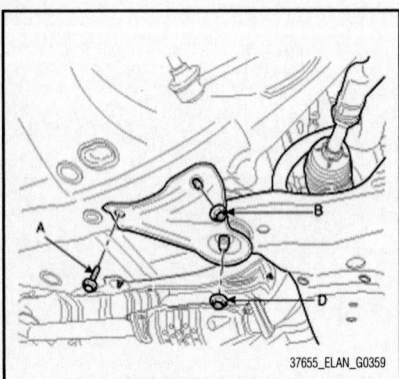

Fig. 212 Sub frame bolt and nut (C)

21. Install the muffler rubber hanger.

22. Connect the lower arm with the knuckle by tightening the bolts to 72–87 ft. lbs. (100–120 Nm).

23. Connect the tie rod end with the knuckle and install the castle nut and split pin. Tighten castle nut to 17–25 ft. lbs. (24–34 Nm).

24. Connect the stabilizer link with the front strut assembly and tighten the nut to 72–87 ft. lbs. (100–120 Nm).

25. Connect the universal joint assembly with the pinion of the steering gear box and tighten the bolt to 22–25 ft. lbs. (30–35 Nm).

26. Install the front wheels and tires.

WHEEL HUB & BEARING

REMOVAL & INSTALLATION

See Figures 214 through 219.

1. Before servicing the vehicle, refer to the Precautions Section.

2. Remove the knuckle. Refer to Steering Knuckle Removal & Installation.

3. Remove the snap ring.

4. Remove the hub assembly from the knuckle assembly:

a. Install the front knuckle assembly on a press.

b. Position a suitable adapter upon the hub assembly shaft.

c. Remove the hub assembly from the knuckle assembly using a press.

5. Remove the hub bearing inner race from the hub assembly:

a. Install a suitable tool for removing the hub bearing inner race on the hub assembly.

b. Position the hub assembly and tool upon a suitable adapter.

c. Position a suitable adapter upon the hub assembly shaft.

d. Remove the hub bearing inner race from the hub assembly using a press.

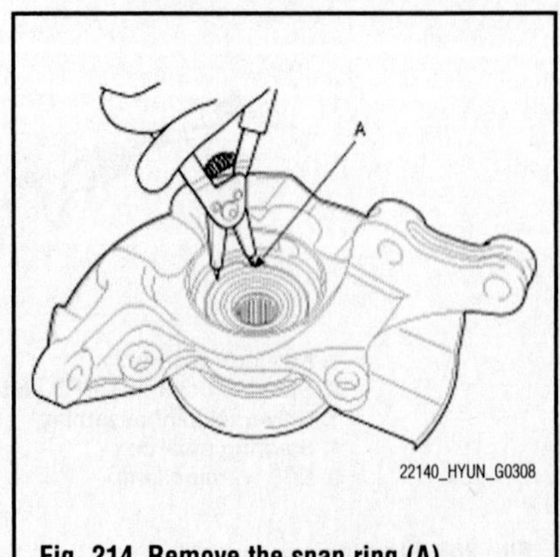

Fig. 214 Remove the snap ring (A)

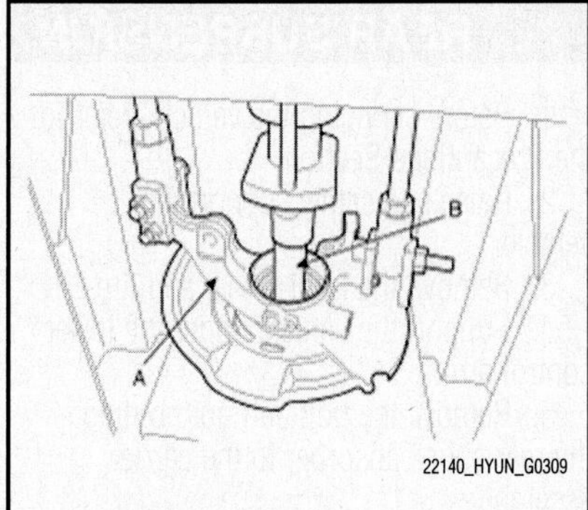

Fig. 215 Install the front knuckle assembly (A) on a press with an adapter (B) upon the hub assembly shaft

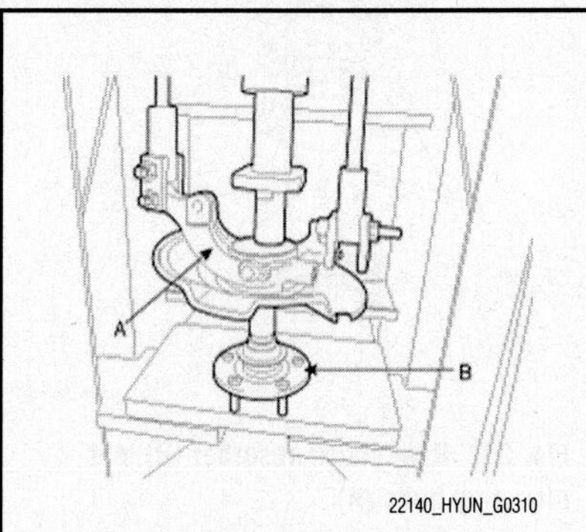

Fig. 216 Use a press to remove the hub assembly (B) from the knuckle assembly (A)

6. Remove the hub bearing outer race from the knuckle assembly:
   a. Position the hub assembly upon a suitable adapter.
   b. Position a suitable adapter upon the hub bearing outer race.
   c. Remove the hub bearing outer race from the steering knuckle assembly by using a press.

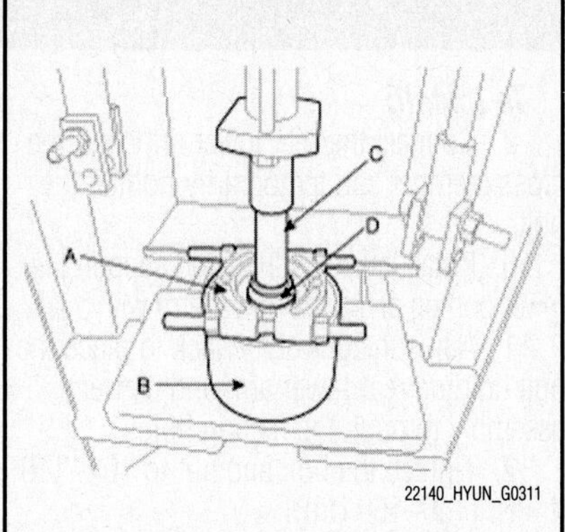

Fig. 217 Remove the hub bearing inner race

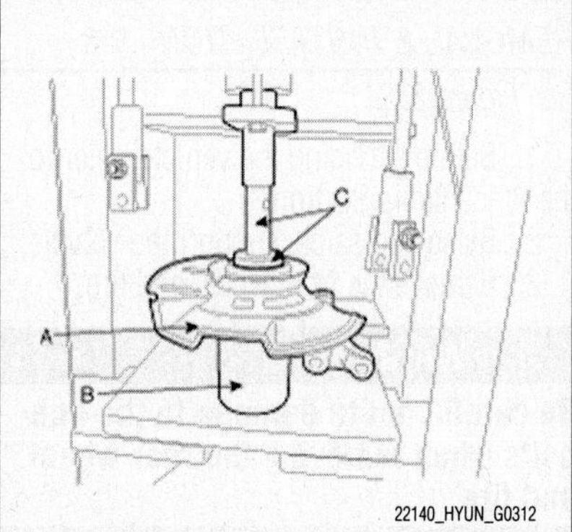

Fig. 218 Remove the hub bearing outer race

To install:

7. Install the hub bearing to the knuckle assembly:
   a. Position the knuckle assembly on a press.
   b. Position a new hub bearing upon the steering knuckle assembly.
   c. Position a suitable adapter upon the hub bearing.

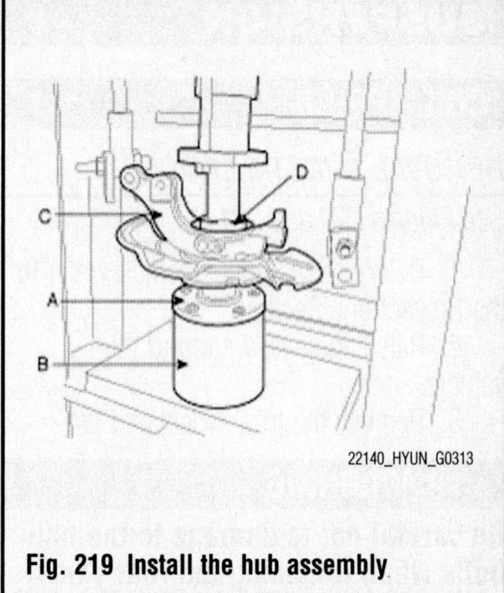

Fig. 219 Install the hub assembly

   d. Install the hub bearing to the steering knuckle assembly by using a press.

WARNING

Do not press against the inner race of the hub bearing or damage may occur to the bearing assembly.

8. Install the hub assembly to the knuckle assembly:
   a. Position the hub assembly upon a suitable adapter.
   b. Position the knuckle assembly upon the hub assembly.
   c. Position a suitable adapter upon the hub bearing.
   d. Install the hub assembly to the knuckle assembly by using a press.
9. Install the snap ring.
10. Install the knuckle.

ADJUSTMENT

The front wheel bearing is a sealed unit and is not adjustable.

COIL SPRING

REMOVAL & INSTALLATION

See Figures 220 and 221.

1. Before servicing the vehicle, refer to the Precautions Section.
2. Raise and safely support the vehicle.
3. Remove the front wheel and tire.

WARNING

Be careful not to damage to the hub bolts when removing the rear wheel and tire.

4. Support the lower portion of the rear lower arm with a jack.
5. Temporarily loosen the bolt holding the cross member to the rear lower arm, do not remove.
6. Remove the bolt and nut holding the rear lower arm to the carrier assembly.
7. Lower the jack and remove the coil spring and the spring pad.
8. Remove the rear lower arm from the cross member by removing the bolt completely.

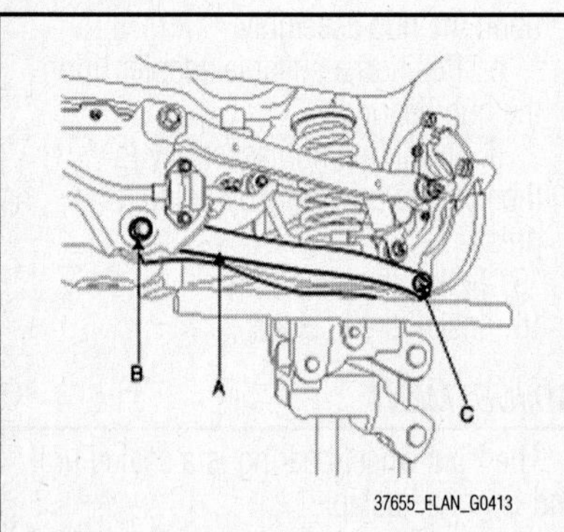

Fig. 220 Rear lower arm (A), bolt (B), and nut (C)

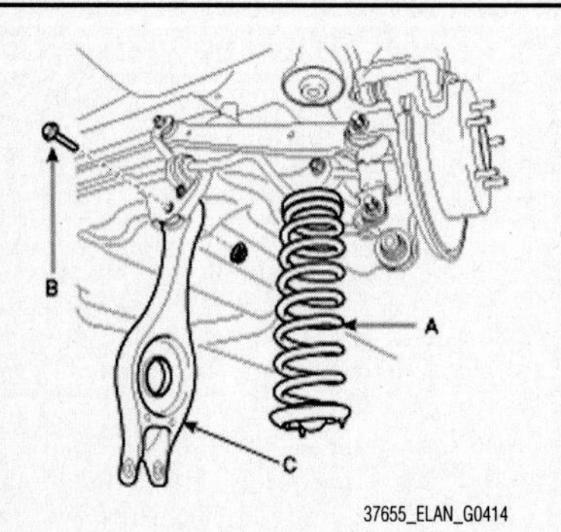

Fig. 221 Coil spring (A), rear lower arm (C), and bolt (B)

To install:

9. Connect the rear lower arm with the cross member and temporarily tighten the bolt.
10. Install the coil spring and support the lower portion of the rear lower arm with a jack.
11. Adjust height of the jack to place the bolt holding rear lower arm and carrier assembly through the mating holes.
12. Tighten the bolt and nut to 101–116 ft. lbs. (140–160 Nm).
13. Install the rear wheel and tire, and lower the vehicle

CONTROL ARMS/LINKS

REMOVAL & INSTALLATION

See Figure 222.

1. Before servicing the vehicle, refer to the Precautions Section.
2. Raise and safely support the vehicle.
3. Remove the front wheel and tire.

WARNING

Be careful not to damage to the hub bolts when removing the rear wheel and tire.

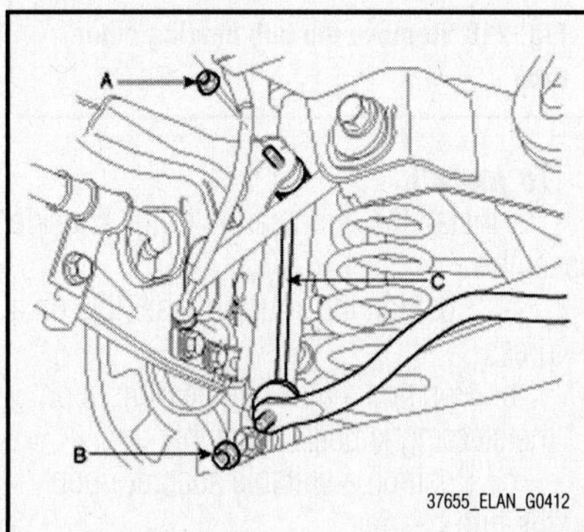

Fig. 222 Remove the stabilizer link (C) by loosening the nuts (A, B)

4. Remove the stabilizer link by loosening the nuts.

To install:

5. Connect the rear stabilizer link between the rear stabilizer bar and the trailing arm and then tighten the nuts to 33–40 ft. lbs. (44–55 Nm).
6. Install the rear wheel and tire, and lower the vehicle

SHOCK ABSORBER

REMOVAL & INSTALLATION

See Figure 223.

1. Before servicing the vehicle, refer to the Precautions Section.
2. Raise and safely support the vehicle.
3. Remove the front wheel and tire.
4. Support the vehicle under the lower control arm.
5. Remove the bolt and nut holding the rear shock absorber to the carrier assembly.
6. Remove the rear shock absorber from the wheel housing.

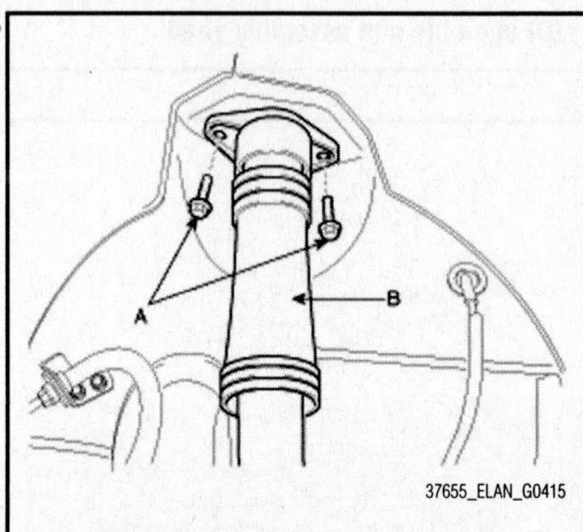

Fig. 223 Rear shock absorber (B) and mounting bolts (A)

To install:

7. Install the rear shock absorber to the wheel housing panel by tightening the bracket mounting bolts to 36–47 ft. lbs. (50–65 Nm).
8. Connect the rear shock absorber with the carrier assembly and tighten the bolt and nut to 101–116 ft. lbs. (140–160 Nm).
9. Install the rear wheel and tire, and lower the vehicle

TESTING

1. Check the rubber parts for damage or deterioration.
2. Check for correct height and proper return of shock absorber to original height.
3. Check the shock absorber for abnormal resistance or unusual sounds.
4. Check for oil leakage around seals.
5. Replace if necessary.

STABILIZER BAR

REMOVAL & INSTALLATION

See Figure 224.

1. Before servicing the vehicle, refer to the Precautions Section.
2. Raise and safely support the vehicle.
3. Remove the front wheel and tire.

WARNING

Be careful not to damage to the hub bolts when removing the rear wheel and tire.

4. Remove the stabilizer link by loosening the nuts.
5. Remove the rear stabilizer bar from the cross member by loosening the bracket mounting bolts.
6. Remove the mounting bracket and bushing from the rear stabilizer bar.

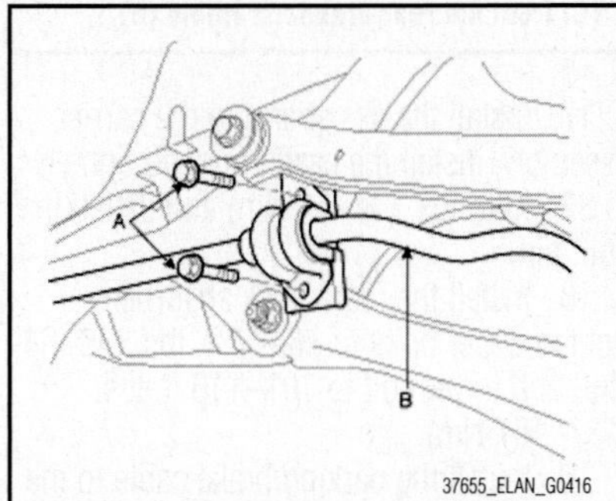

Fig. 224 Remove the rear stabilizer bar (B) from the crossmember by loosening the bracket mounting bolts (A)

To install:

7. Install the mounting bracket and bushing to the rear stabilizer bar.
8. Install the rear stabilizer bar to the cross member by tightening the bracket mounting bolts to 33–40 ft. lbs. (45–55 Nm).
9. Connect the rear stabilizer link between the rear stabilizer bar and the trailing arm and then tighten the nuts to 33–40 ft. lbs. (45–55 Nm).
10. Install the rear wheel and tire, and lower the vehicle

WHEEL HUB & BEARING

REMOVAL & INSTALLATION

With Rear Disc Brakes

See Figures 225 through 228.

1. Release the parking brake.
2. Loosen the wheel nuts slightly.
3. Raise the vehicle, and make sure it is securely supported.
4. Remove the rear wheel and tire from rear hub.

WARNING

Be careful not to damage to the hub bolts when removing the rear wheel and tire.

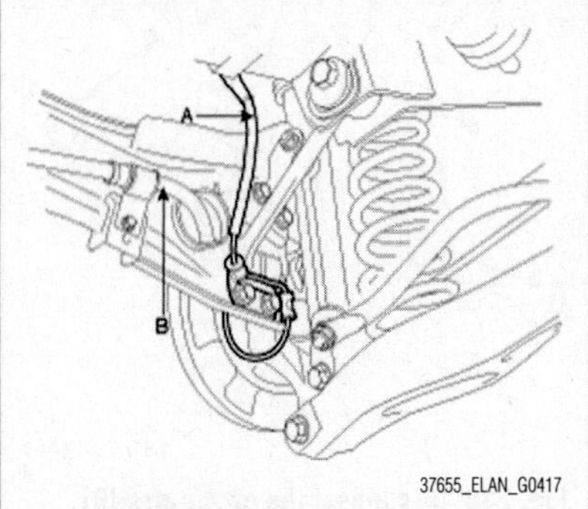

Fig. 225 Remove the wheel speed sensor (A) and parking brake cable (B)

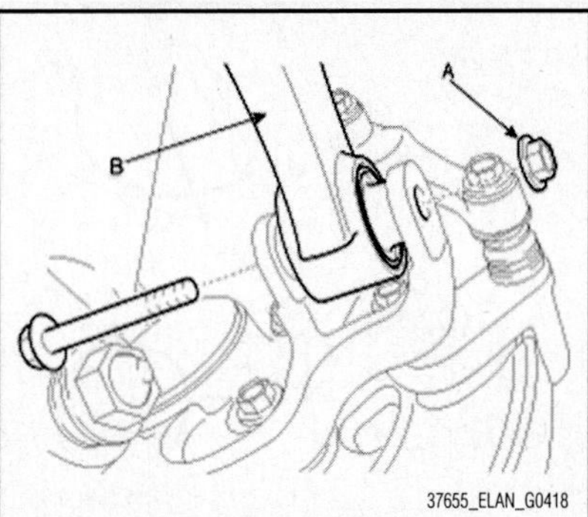

Fig. 226 Disconnect the upper arm (B) from the carrier assembly after loosen the nut (A)

5. While supporting the lower arm with a jack, remove the mounting bolt from the rear lower arm to the rear carrier.
6. Loosen the mounting nut of the cross member and the rear lower arm, then remove the coil spring by lowering the jack.
7. Remove the wheel speed sensor and parking brake cable.
8. Disconnect the upper arm from the carrier assembly after loosening the nut.
9. Remove the brake caliper mounting bolts, and support the caliper assembly with mechanics wire.
10. Remove the rear brake disc assembly after removing the retaining screw.

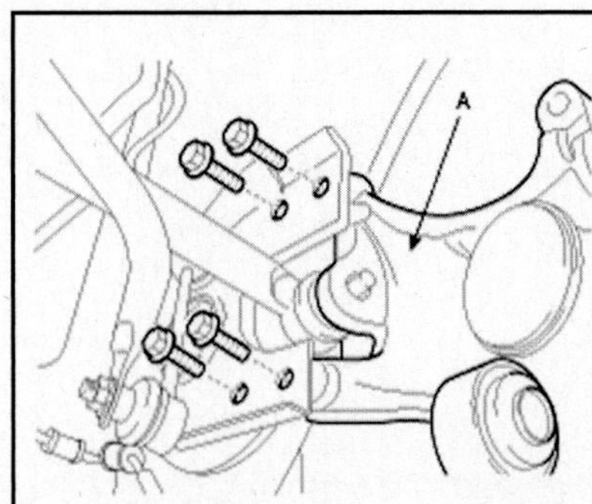

Fig. 227 Remove the carrier assembly (A) from the trailing arm

11. Remove the rear shock absorber. Refer to Shock Absorber in this section.
12. Remove the split pin and castle nut from the assist arm.
13. Disconnect the assist arm from the carrier assembly.
14. Remove the carrier assembly from the trailing arm.
15. Remove the rear hub assembly from the rear brake assembly.

To install:

16. Install the rear hub assembly and rear brake assembly to the rear carrier, tighten mounting bolts to 43–55 ft. lbs. (59–69 Nm).

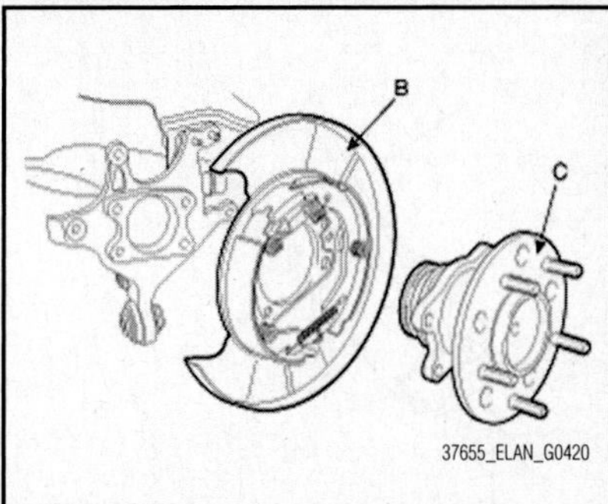

Fig. 228 Remove the rear hub bolts (A) from hub assembly (C) and from the rear brake assembly (B)

17. Install the carrier assembly to the trailing arm, tighten the mounting bolts to 25–40 ft. lbs. (34–54 Nm).
18. Install the assist arm to the carrier assembly, install the castle nut and tighten to 32–40 ft. lbs. (44–54 Nm) and install the split pin.
19. Install the rear shock absorbers, tighten upper bolts to 36–47 ft. lbs. (49–64 Nm) and lower bolt to 101–116 ft. lbs. (137–157 Nm).
20. Install the rear brake disc assembly and tighten retaining screw.
21. Install the brake caliper, tighten the brake caliper mounting bolts to 36–43 ft. lbs. (49–59 Nm).
22. Install the upper arm from the carrier assembly, tighten the nut to72–87 ft. lbs. (91–118 Nm).
23. Install the wheel speed sensor and parking brake cable.
24. Install the coil spring on the rear lower arm, slowly jack-up the rear lower arm using jack.
25. While supporting the lower arm with a jack, install the rear lower arm mounting bolt to the rear carrier and tighten to 101–116 ft. lbs. (137–157 Nm).
26. Install the wheel and the tire to the rear hub.

With Rear Drum Brakes

See Figures 229 through 232.

1. Release the parking brake.
2. Loosen the wheel nuts slightly.
3. Raise the vehicle, and make sure it is securely supported.
4. Remove the rear wheel and tire from rear hub.

WARNING

Be careful not to damage to the hub bolts when removing the rear wheel and tire.

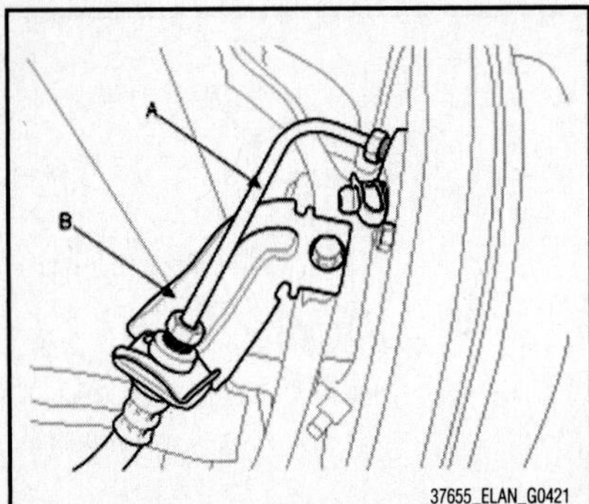

Fig. 229 Remove the brake line (A) and bracket (B)

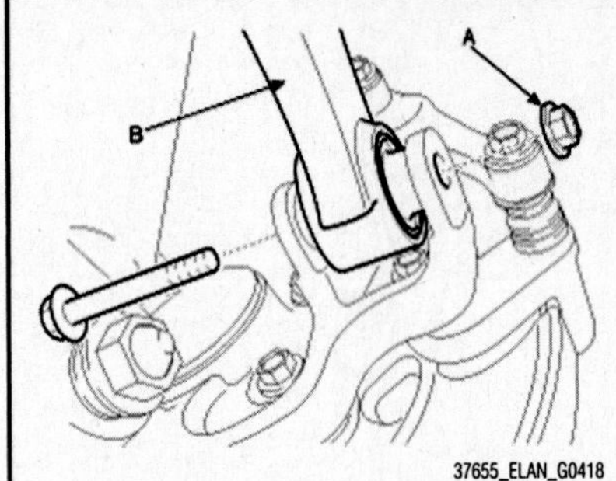

Fig. 230 Disconnect the upper arm (B) from the carrier assembly after loosen the nut (A)

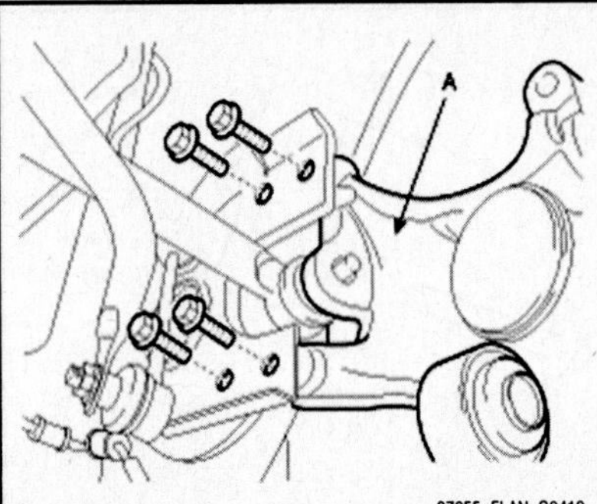

Fig. 231 Remove the carrier assembly (A) from the trailing arm

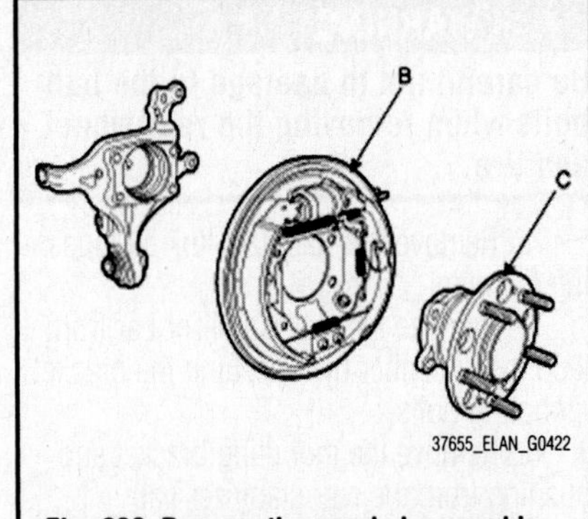

Fig. 232 Remove the rear hub assembly (C) from the rear brake assembly (B)

5. While supporting the lower arm with a jack, remove the mounting bolt from the rear lower arm to the rear carrier.
6. Loosen the mounting nut of the cross member and the rear lower arm, then remove the coil spring by lowering the jack.
7. Remove the brake line and bracket.
8. Disconnect the upper arm from the carrier assembly after loosening the nut.
9. Remove the rear brake drum assembly after removing retaining screw.
10. Remove the rear shock absorber. Refer to Shock Absorber Removal & Installation.
11. Remove the split pin and castle nut from the assist arm.
12. Disconnect the assist arm from the carrier assembly.
13. Remove the carrier assembly from the trailing arm.
14. Remove the rear hub assembly from the rear brake assembly.

To install:

15. Install the rear hub assembly and rear brake assembly to the rear carrier, tighten mounting bolts to 43–55 ft. lbs. (59–69 Nm).
16. Install the carrier assembly to the trailing arm, tighten the mounting bolts to 25–40 ft. lbs. (34–54 Nm).
17. Install the assist arm to the carrier assembly, install the castle nut and tighten to 32–40 ft. lbs. (44–54 Nm) and install the split pin.
18. Install the rear shock absorbers, tighten upper bolts to 36–47 ft. lbs. (49–64 Nm) and lower bolt to 101–116 ft. lbs. (137–157 Nm).
19. Install the parking brake cable to the brake assembly.
20. Install the rear brake drum assembly and tighten retaining screw.
21. Install the upper arm from the carrier assembly, tighten the nut to72–87 ft. lbs. (91–118 Nm).
22. Install the brake line and bracket.
23. Install the coil spring on the rear lower arm, slowly jack-up the rear lower arm using jack.
24. While supporting the lower arm with a jack, install the rear lower arm mounting bolt to the rear carrier and tighten to 101–116 ft. lbs. (137–157 Nm).
25. Install the wheel and the tire to the rear hub.

ADJUSTMENT

The rear wheel bearing is an integral part of the rear hub. No adjustment is possible.

HYUNDAI

Genesis Coupe

4

SPECIFICATIONS AND MAINTENANCE CHARTS

ENGINE AND VEHICLE IDENTIFICATION

Engine							Model Year	
Code ①	Liters (cc)	Cu. In.	Cyl.	Fuel Sys.	Engine Type	Eng. Mfg.	Code ②	Year
D	2.0 (1998)	121.92	I4	MPFI	DOHC	Hyundai	A	2010
H	3.8 (3778)	230.55	V6	MPFI	DOHC	Hyundai		

MPFI: Multi-Point Fuel Injection

DOHC: Double Overhead Camshafts

① 8th digit of VIN

② 10th digit of VIN

37655_GENC_C0001

GENERAL ENGINE SPECIFICATIONS

All measurements are given in inches.

Year	Model	Engine Displacement Liters	Engine Series VIN	Net Horsepower @ rpm	Net Torque @ rpm (ft. lbs.)	Bore x Stroke (in.)	Com-pression Ratio	Oil Pressure @ rpm
2010	Genesis	2.0	D	210@6000	223@2000	3.385 x 3.385	9.4:1	35.5@1000
	Genesis	3.8	H	306@6300	266@4700	3.780 x 3.743	10.4:1	18.8@1000

37655_GENC_C0002

GASOLINE ENGINE TUNE-UP SPECIFICATIONS

Year	Engine Displacement Liters	Engine VIN	Spark Plug Gap (in.)	Ignition Timing (deg.)		Fuel Pump (psi)	Idle Speed (rpm)		Valve Clearance	
				MT	AT		MT	AT	In.	Ex.
2010	2.0	D	0.039-0.043	①	①	54.3-55.8	②	②	HYD	HYD
	3.8	H	0.039-0.043	①	①	54.3-55.8	②	②	HYD	HYD

Follow the figures on the label if they differ from those in this chart.

HYD: Hydraulic

NA: Not Applicable

① Ignition timing is computer controlled and is not adjustable

② Idle speed is maintained by the Electronic Control Module (ECM)

37655_GENC_C0003

CAPACITIES

Year	Model	Engine Displacement Liters	Engine VIN	Engine Oil with Filter (qts.)	Transmission (pts.)		Fuel Tank (gal.)	Cooling System (qts.)
					Manual	Auto.		
2010	Genesis	2.0	D	5.8	4.2	NS	17.2	5.8
	Genesis	3.8	H	5.8	4.6	NS	17.2	9.5

NOTE: All capacities are approximate. Add fluid gradually and check to be sure a proper fluid level is obtained.
NS: Not Supplied

37655_GENC_C0004

FLUID SPECIFICATIONS

Year	Model	Engine Displacement Liters	Engine ID/VIN	Engine Oil	Auto. Trans.	Manual Trans.	Power Steering Fluid	Brake Master Cylinder
2010	Genesis	2.0	D	5W-20	①	②	PSF-3	③
	Genesis	3.8	H	5W-20	④	②	PSF-3	③

DOT: Department Of Transportation
NA: Not Applicable
① Shell 1375.4 ATF
② MS721-40
② DOT 3 or DOT 4 hydraulic brake fluid
③ APOLLOIL

37655_GENC_C0005

VALVE SPECIFICATIONS

Year	Engine Displacement Liters	Engine VIN	Seat Angle (deg.)	Face Angle (deg.)	Spring Test Pressure (lbs. @ in.)	Spring Installed Height (in.)	Stem-to-Guide Clearance (in.)		Stem Diameter (in.)	
							Intake	Exhaust	Intake	Exhaust
2010	2.0	D	44.75-45.20	45.25-45.75	40.5-43.2 @1.3779	NS	0.0008-0.0019	0.0012-0.0021	0.2151-0.2157	0.2149-0.2153
	3.8	H	44.75-45.20	45.25-45.75	90.4-96.2 @0.953	NS	0.0008-0.0019	0.0012-0.0021	0.2151-0.2157	0.2149-0.2153

NS: Not Supplied

37655_GENC_C0006

CAMSHAFT AND BEARING SPECIFICATIONS CHART

All measurements are given in inches.

Year	Engine Displ. Liters	Engine ID/VIN	Journal Dia.	Brg. Oil Clearance	Shaft End-play	Runout	Journal Bore	Lobe Height Intake	Lobe Height Exhaust
2010	2.0	D	①	②	0.0008-0.0071	NS	NS	1.8582	1.8031
	3.8	H	③	④	0.0008-0.0071	NS	NS	1.8425	1.8031

NA: Not Supplied

① Intake No. 1 is 1.1811 inch
Intake No. 2, 3, 4 are 0.9449 inch
Exhaust No.1 is 1.4173 inch
Exhaust No. 2, 3, 4 are 0.9449 inch

② Intake No. 1 is 0.0008-0.0022 inch
Intake No. 2, 3, 4 are 0.0012-0.0026 inch
Exhaust No.1 is 0.0008-0.0022 inch
Exhaust No. 2, 3, 4 are 0.0012-0.0026 inch

③ Intake No. 1 is 1.1009-1.1015 inch
Intake No. 2, 3, 4 are 0.9430-0.9437 inch
Exhaust No.1 is 1.1009-1.1016 inch
Exhaust No. 2, 3, 4 are 0.9430-0.9437 inch

④ Intake No. 1 is 0.0011-0.0022 inch
Intake No. 2, 3, 4 are 0.0012-0.0026 inch
Exhaust No.1 is 0.0011-0.0022 inch
Exhaust No. 2, 3, 4 are 0.0012-0.0026 inch

37655_GENC_C0007

CRANKSHAFT AND CONNECTING ROD SPECIFICATIONS

All measurements are given in inches.

Year	Engine Displacement Liters	Engine VIN	Crankshaft Main Brg. Journal Dia.	Crankshaft Main Brg. Oil Clearance	Crankshaft Shaft End-play	Crankshaft Thrust on No.	Connecting Rod Journal Diameter	Connecting Rod Oil Clearance	Connecting Rod Side Clearance
2010	2.0	D	2.0449-2.0456	0.0007-0.0014	0.0027-0.0098	3	1.8879-1.8886	0.0009-0.0016	0.0039-0.0100
	3.8	H	2.7142-2.7149	0.0008-0.0016	0.0039-0.0110	3	2.1635-2.1642	0.0015-0.0022	0.0039-0.0098

NS: Not Supplied

37655_GENC_C0008

PISTON AND RING SPECIFICATIONS

All measurements are given in inches.

Year	Engine Displ. Liters	Engine VIN	Piston Clearance	Ring Gap Top Compression	Ring Gap Bottom Compression	Ring Gap Oil Control	Ring Side Clearance Top Compression	Ring Side Clearance Bottom Compression	Ring Side Clearance Oil Control
2010	2.0	D	0.0005-0.0013	0.0059-0.0118	0.0145-0.0204	0.0078-0.0275	0.0019-0.0031	0.0015-0.0031	0.0023-0.0051
	3.8	H	0.0008-0.0016	0.0067-0.0126	0.0126-0.0185	0.0078-0.0275	0.0016-0.0031	0.0012-0.0027	0.0024-0.0059

37655_GENC_C0009

TORQUE SPECIFICATIONS

All readings in ft. lbs.

Year	Engine Displacement Liters	Engine VIN	Cylinder Head Bolts	Main Bearing Bolts	Rod Bearing Bolts	Crankshaft Damper Bolts	Flexplate Bolts	Manifold Intake	Manifold Exhaust	Spark Plugs	Oil Pan Drain Plug
2010	2.0	D	①	②	③	123-130	87-94	14-20	36-40	N/S	N/S
	3.8	H	④	⑤	⑥	210-224	53-56	14-17	29-33	15-22	25-33

① Step 1: 24 - 27 ft. lbs.
Step 2: Plus 90 degrees
Step 3: Plus 90 degrees

② Step 1: 11 ft. lbs.
Step 2: Plus 22 degrees
Step 2: Plus 120 degrees

③ Step 1: 15 ft. lbs.
Step 2: Plus 90 degrees

④ Step 1: 29 ft. lbs.
Step 2: Plus 120 degrees
Step 3: Plus 90 degrees

⑤ M11 bolts (inner) Step 1: 36 ft. lbs.
M11 bolts (inner) Step 2: Plus 90 degrees
M8 bolts (outer) Step 1: 15 ft. lbs.
M8 bolts (outer) Step 2: Plus 120 degrees
M8 bolts (side): 22-23 ft. lbs.

⑥ Step 1: 15 ft. lbs.
Step 2: Plus 90 degrees

37655_GENC_C0010

WHEEL ALIGNMENT

Year	Model		Caster Range (+/-Deg.)	Caster Preferred Setting (Deg.)	Camber Range (+/-Deg.)	Camber Preferred Setting (Deg.)	Toe-in (Deg.)
2010	Genesis	Front	0.50	+7.45	0.50	-0.50	0.28 +/- 0.16
		Rear	—	—	0.50	-1.50	0.16 +/- 0.20

37655_GENC_C0011

TIRE, WHEEL AND BALL JOINT SPECIFICATIONS

Year	Model	OEM Tires		Tire Pressures (psi)		Wheel	Ball Joint	Lug Nut
		Standard	Optional	Front	Rear	Size	Inspection	Torque (ft. lbs.)
2010	Front	P225/45VR18	P225/40YR19	①	①	6.5J x 18	②	65-80
	Rear	P245/45VR18	P245/40YR19			7.5J x 19		

① Refer to placard on vehicle for proper inflation pressure.

② Replace if any measurable movement is found.

37655_GENC_C0012

BRAKE SPECIFICATIONS

All measurements in inches unless noted

Year	Model		Brake Disc			Brake Drum Diameter			Minimum	Brake Caliper	
			Original Thickness	Minimum Thickness	Maximum Runout	Original Inside Diameter	Max. Wear Limit	Maximum Machine Diameter	Lining Thickness	Bracket Bolts (ft. lbs.)	Mounting Bolts (ft. lbs.)
2010	Genesis	F	1.100	1.040	0.002	—	—	—	0.079	58-72	16-23
	General	R	0.510	0.450	0.002	—	—	—	0.079	58-72	16-23
	Genesis	F	1.100	1.020	0.002	—	—	—	0.079	65-76	—
	Brembo	R	0.790	0.710	0.002	—	—	—	0.079	58-72	—

37655_GENC_C0013

SCHEDULED MAINTENANCE INTERVALS

HYUNDAI—Genesis Coupe

TO BE SERVICED	TYPE OF SERVICE	VEHICLE MILEAGE INTERVAL (x1000)												
		7.5	15	22.5	30	37.5	45	52.5	60	67.5	75	82.5	90	97.5
Engine oil & filter ①	R	✓	✓	✓	✓	✓	✓	✓	✓	✓	✓	✓	✓	✓
Automatic transaxle fluid ②	S/I					✓					✓			
Manual transmission	S/I				✓				✓				✓	
Brake pads, calipers/rotors	S/I		✓		✓		✓		✓		✓		✓	
Driveshafts & boots	S/I		✓		✓		✓		✓		✓		✓	
Air cleaner filter ③	R	✓	✓	✓	✓	✓	✓	✓	✓	✓	✓	✓	✓	✓
A/C refrigerant	S/I		✓		✓		✓		✓		✓		✓	
Brake fluid	I				✓				✓				✓	
Engine coolant ④	R								✓					
Fuel hose, vapor hose & fuel filter cap	S/I				✓				✓				✓	
Spark plugs (Iridium coated) 100,000 mile replacement	R													
Bolts & nuts on chassis & body	S/I		✓		✓		✓		✓		✓		✓	
Drive belts	S/I								✓		✓		✓	
Exhaust pipe connections, muffler & suspension bolts	S/I		✓		✓		✓		✓		✓		✓	
Valve clearance ⑤	S/I								✓					
Electronic throttle control	S/I		✓		✓		✓		✓		✓		✓	
Brake hoses & lines	S/I		✓		✓		✓		✓		✓		✓	
Rear brake discs, linings & parking brake	S/I				✓				✓				✓	
Steering gear rack, linkage & boots	S/I		✓		✓		✓		✓		✓		✓	
Power steering pump, hoses	S/I		✓		✓		✓		✓		✓		✓	
Power Steering fluid	S/I	✓	✓	✓	✓	✓	✓	✓	✓	✓	✓	✓	✓	✓
Fuel tank air filter	S/I				✓				✓				✓	
Propeller shaft	S/I		✓		✓		✓		✓		✓		✓	
Rear axle oil	S/I				✓				✓				✓	
Climate control air filter ⑥	R		✓		✓		✓		✓		✓		✓	
Fuel filter	R				✓				✓				✓	
Vacuum & crankcase ventilation hoses	S/I	✓	✓	✓	✓	✓	✓	✓	✓	✓	✓	✓	✓	✓

R: Replace S/I: Service or Inspect

① For 2.0 Liter engine, initial at 3000 miles and every 4800 miles or 6 months thereafter

② Add only specified fluid.

③ Replace at 30,000 miles

④ Replace every 24 months or 30,000 miles

⑤ Inspect for excessive tappet noise and/or engine vibration and adjust if necessary.

⑥ Replace at 15,000 miles, or 12 months. See Severe Service.

37655_GENC_C0014

SCHEDULED MAINTENANCE INTERVALS

HYUNDAI—Genesis Coupe

FREQUENT OPERATION MAINTENANCE (SEVERE SERVICE)

If a vehicle is operated under any of the following conditions it is considered severe service:

Repeatedly driving short distance of less than 5 miles (8 km) in normal temperature or less than 10 miles (16 km) in freezing temperature

Driving on rough, dusty, muddy, unpaved, graveled or salt- spread roads

Driving in areas using salt or other corrosive materials or in very cold weather

Extensive engine idling or low speed driving for long distances

Driving in sandy areas

Driving in heavy traffic area over 90°F (32°C)

Driving on uphill, downhill, or mountain road

Towing a Trailer, or using a camper, or roof rack

Driving as a patrol car, taxi, other commercial use or vehicle towing

Driving over 106 mph (170 km/h)

Frequently driving in stop-and-go conditions

Oil & oil filter: change 2.0 Liter engine every 3000 miles, for 3.8 Liter engine every 3,750 miles.

Brake pads, calipers & rotors: service or inspect more frequently

Driveshaft boots: service or inspect every 7500 miles.

Steering gear rack, linkage & boots: service or inspect more frequently

Air cleaner filter: service or inspect more frequently

Manual transaxle: replace every 60,000 miles.

Climate control air filter: replace as necessary.

Spark plugs replace more frequently

Rear axle oil replace every 60,000 miles.

Propeller shaft inspect every 7,500 miles.

37655_GENC_C0015

BRAKES — INFORMATION AND PRECAUTIONS

ANTI-LOCK SYSTEMS

• Certain components within the ABS system are not intended to be serviced or repaired individually.

• Do not use rubber hoses or other parts not specifically specified for and ABS system. When using repair kits, replace all parts included in the kit. Partial or incorrect repair may lead to functional problems and require the replacement of components.

• Lubricate rubber parts with clean, fresh brake fluid to ease assembly. Do not use shop air to clean parts; damage to rubber components may result.

• Use only DOT 3 brake fluid from an unopened container.

• If any hydraulic component or line is removed or replaced, it may be necessary to bleed the entire system.

• A clean repair area is essential. Always clean the reservoir and cap thoroughly before removing the cap. The slightest amount of dirt in the fluid may plug an orifice and impair the system function. Perform repairs after components have been thoroughly cleaned; use only denatured alcohol to clean components. Do not allow ABS components to come into contact with any substance containing mineral oil; this includes used shop rags.

• The Anti-Lock control unit is a microprocessor similar to other computer units in the vehicle. Ensure that the ignition switch is **OFF** before removing or installing controller harnesses. Avoid static electricity discharge at or near the controller.

• If any arc welding is to be done on the vehicle, the control unit should be unplugged before welding operations begin.

DISC AND DRUM SYSTEMS

CAUTION

Dust and dirt accumulating on brake parts during normal use may contain asbestos fibers from production or aftermarket brake linings. Breathing excessive concentrations of asbestos fibers can cause serious bodily harm. Exercise care when servicing brake parts. Do not sand or grind brake lining unless equipment used is designed to contain the dust residue. Do not clean brake parts with compressed air or by dry brushing. Cleaning should be done by dampening the brake components with a fine mist of water, then wiping the brake components clean with a dampened cloth. Dispose of cloth and all residue containing asbestos fibers in an impermeable container with the appropriate label. Follow practices prescribed by the Occupational Safety and Health Administration (OSHA) and the Environmental Protection Agency (EPA) for the handling, processing, and disposing of dust or debris that may contain asbestos fibers.

BRAKES — BLEEDING THE BRAKE SYSTEM

BLEEDING PROCEDURE

BLEEDING PROCEDURE

See Figures 1 through 4.

➡Do not reuse the drained fluid.

➡Always use genuine DOT3/DOT4 brake Fluid. Using a non-genuine DOT3/DOT4 brake fluid can cause corrosion and decrease the life of the system.

➡Make sure no dirt or other foreign matter is allowed to contaminate the brake fluid.

➡The reservoir on the master cylinder must be at the MAX (upper) level mark at the start of bleeding procedure and checked after bleeding each brake caliper. Add fluid as required.

1. Before servicing the vehicle, refer to the Precautions Section.
2. Make sure the brake fluid in the reservoir is at the MAX(upper) level line.
3. Have someone slowly pump the brake pedal several times, and then apply pressure.
4. Loosen the right-rear brake bleed screw to allow air to escape from the system. Then tighten the bleed screw securely.
5. Repeat the procedure for each wheel in the sequence shown below until air bubbles no longer appear in the fluid.
6. Refill the master cylinder reservoir to the MAX (upper) level line.

BLEEDING THE ABS SYSTEM

See Figures 5 through 7.

This procedure should be followed to ensure adequate bleeding of air and the filling of the ABS unit, the brake lines, and the master cylinder with brake fluid.

1. Before servicing the vehicle, refer to the Precautions Section.
2. Remove the reservoir cap and fill the brake reservoir with brake fluid.

WARNING

If there is any brake fluid on any painted surface, wash it off immediately.

➡When pressure bleeding, do not depress the brake pedal. Recommended brake fluid: DOT3 or DOT4.

3. For 2.0 AT and ESC Only: Disconnect the vacuum switch connector.
4. Connect a clear plastic tube to the wheel cylinder bleeder plug and insert the other end of the tube into a clear plastic bottle that is half filled with clean brake fluid.
5. Connect the scan tool to the data link connector located underneath the dash panel.
6. Select and operate according to the instructions on the scan tool screen.

CAUTION

You must obey the maximum operating time of the ABS motor with the scan tool to prevent the motor pump from burning.

7. Select vehicle name.
8. Select Anti-Lock Brake system.
9. Select air bleeding mode.
10. Press "OK" to operate motor pump and solenoid valve.

WARNING

Wait 60 seconds before operating the air bleeding or damage to the motor may occur.

11. Wait 600 seconds before operating the air bleeding.
12. Pump the brake pedal several times, and then loosen the bleeder screw until fluid starts to run out without bubbles. Then, close the bleeder screw.
13. Repeat until there are no more bubbles in the fluid for each wheel.
14. Tighten the bleeder screw to 61–113 inch lbs. (9–13 Nm).
15. For 2.0 AT and ESC Only: Connect the vacuum switch connector.

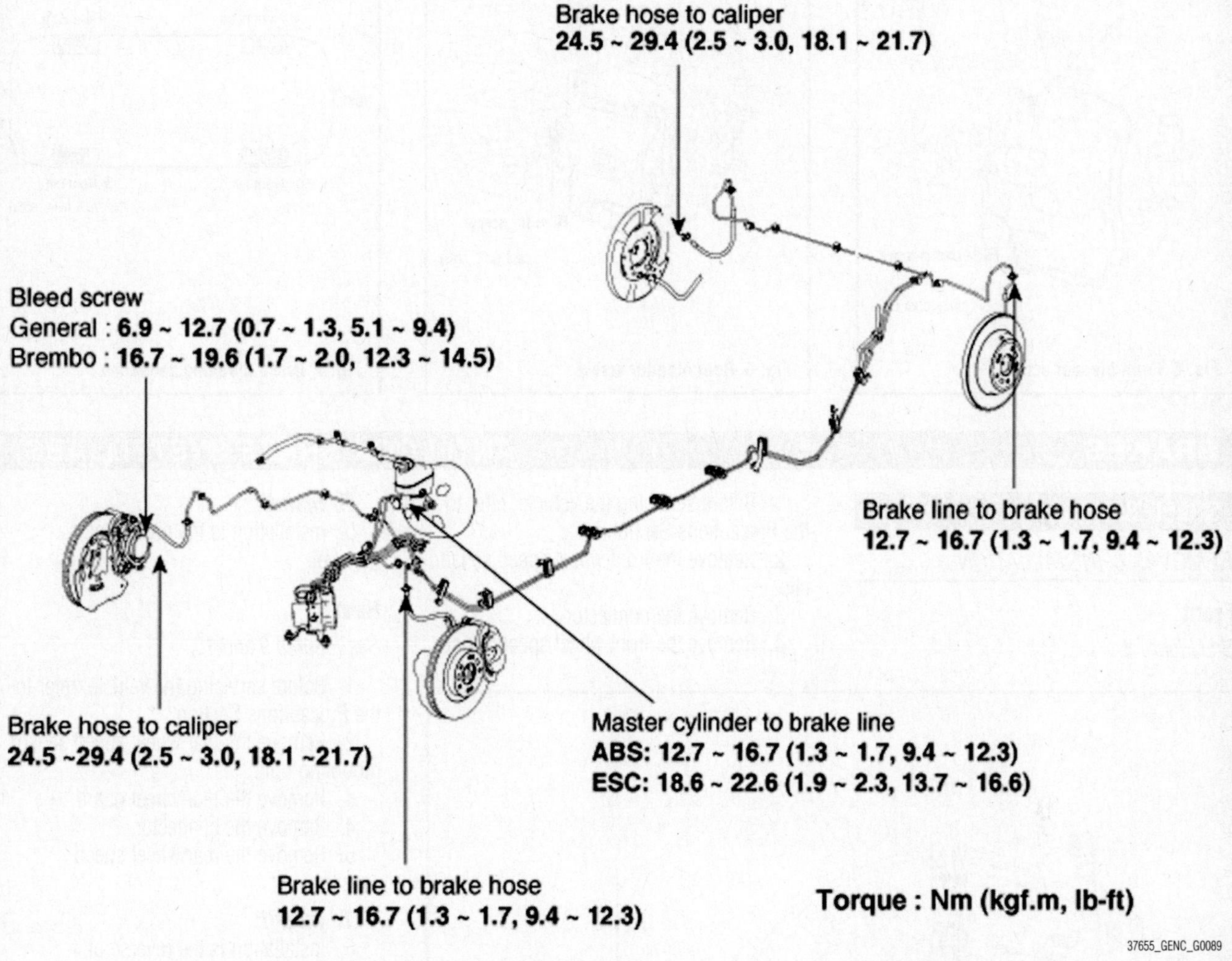

Brake hose to caliper
24.5 ~ 29.4 (2.5 ~ 3.0, 18.1 ~ 21.7)
Bleed screw
General : 6.9 ~ 12.7 (0.7 ~ 1.3, 5.1 ~ 9.4)
Brembo : 16.7 ~ 19.6 (1.7 ~ 2.0, 12.3 ~ 14.5)
Brake line to brake hose
12.7 ~ 16.7 (1.3 ~ 1.7, 9.4 ~ 12.3)
Brake hose to caliper
24.5 ~29.4 (2.5 ~ 3.0, 18.1 ~21.7)
Master cylinder to brake line
ABS: 12.7 ~ 16.7 (1.3 ~ 1.7, 9.4 ~ 12.3)
ESC: 18.6 ~ 22.6 (1.9 ~ 2.3, 13.7 ~ 16.6)
Brake line to brake hose
12.7 ~ 16.7 (1.3 ~ 1.7, 9.4 ~ 12.3)
Torque : Nm (kgf.m, lb-ft)
37655_GENC_G0089

Fig. 1 Brake lines

A
A
(Brembo)
(General)
37655_GENC_G0090

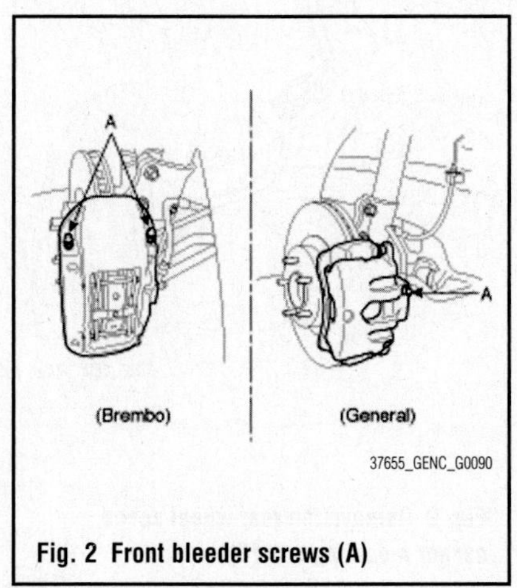

Fig. 2 Front bleeder screws (A)

A
A
(Brembo)
(General)
37655_GENC_G0091

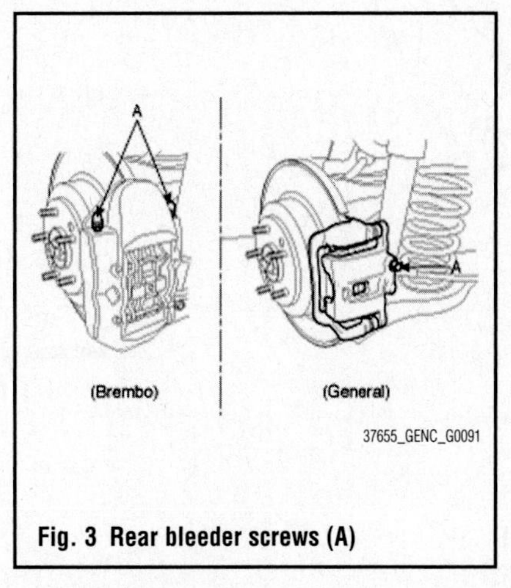

Fig. 3 Rear bleeder screws (A)

④ Front right
① Rear right
② Front left
③ Rear left
37655_GENC_G0085

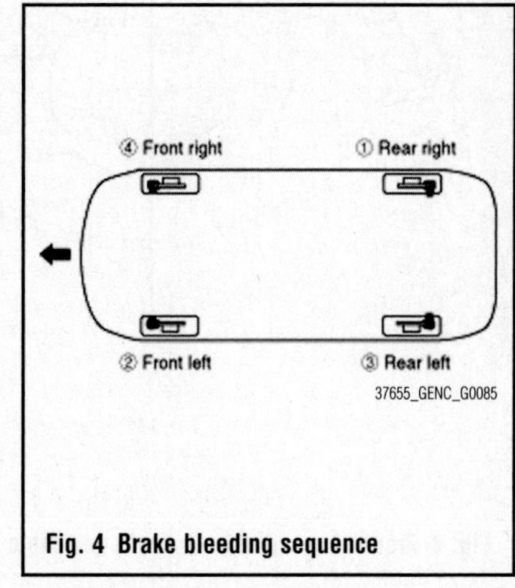

Fig. 4 Brake bleeding sequence

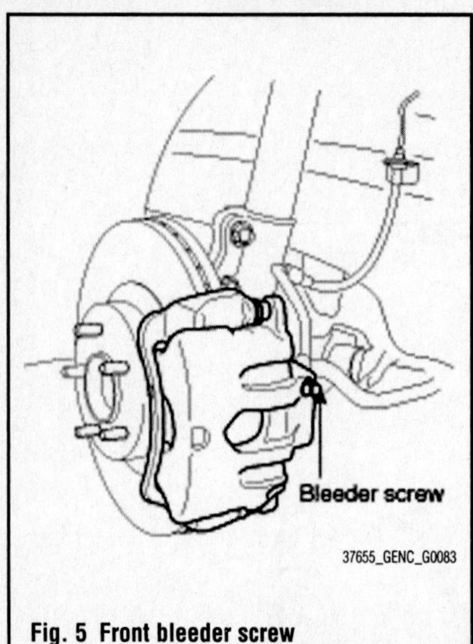

Fig. 5 Front bleeder screw

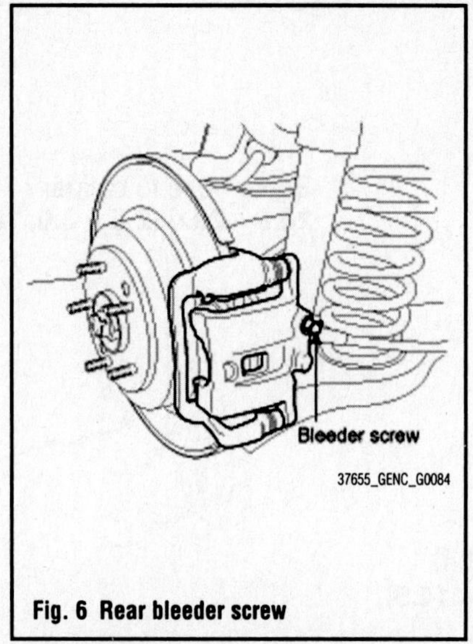

Fig. 6 Rear bleeder screw

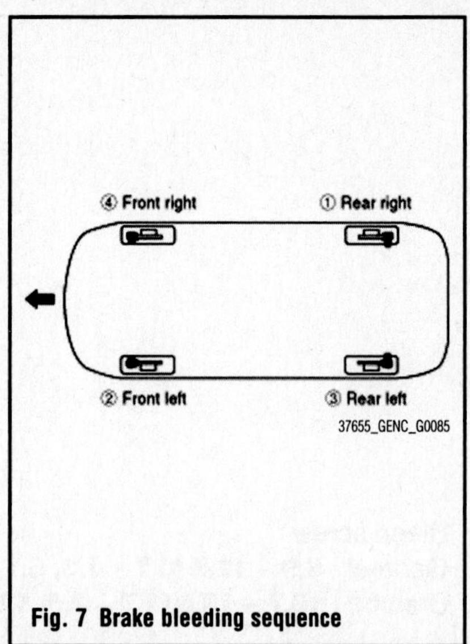

Fig. 7 Brake bleeding sequence

BRAKES — ANTI-LOCK BRAKE SYSTEM (ABS)

SPEED SENSORS

REMOVAL & INSTALLATION

Front

See Figure 8.

1. Before servicing the vehicle, refer to the Precautions Section.
2. Remove the front wheel speed sensor clip.
3. Remove the connector.
4. Remove the front wheel speed sensor.

To install:

5. Installation is the reverse of removal.

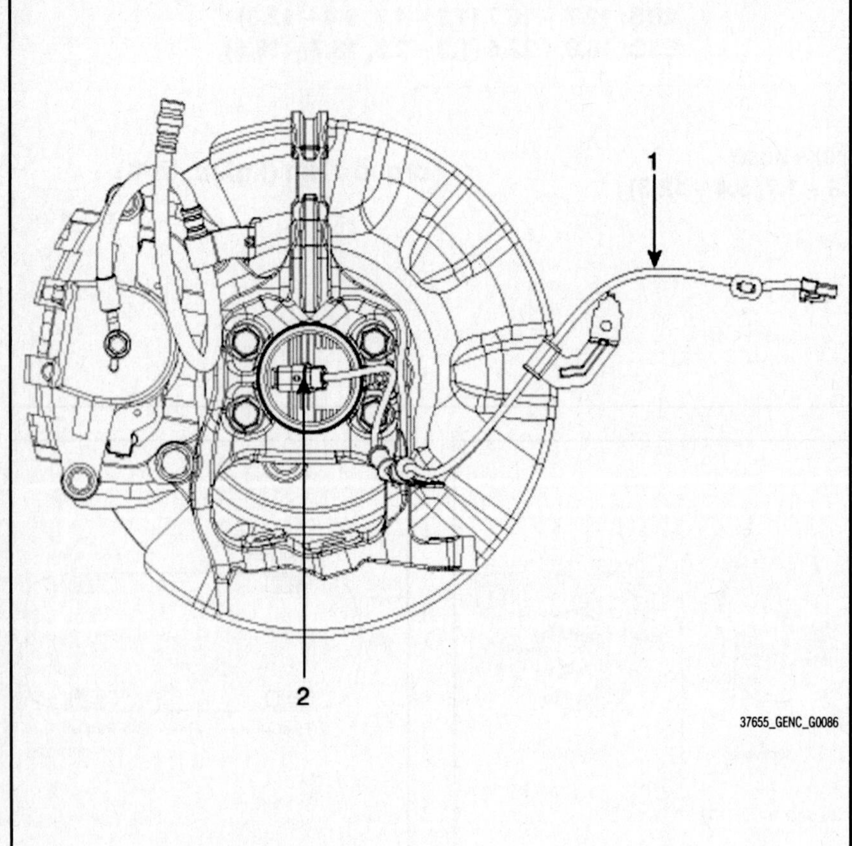

Fig. 8 Front wheel speed sensor (2) and cable (1)

Rear

See Figures 9 and 10.

1. Before servicing the vehicle, refer to the Precautions Section.
2. Remove the rear wheel speed sensor mounting bolt.
3. Remove the rear wheel guard.
4. Remove the connector.
5. Remove the rear wheel speed sensor.

To install:

6. Installation is the reverse of removal.

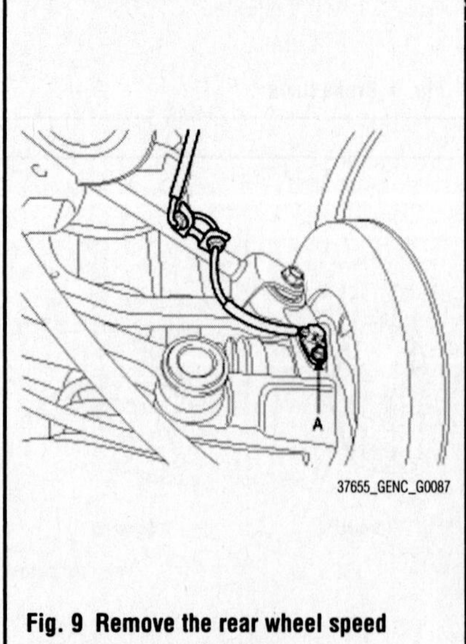

Fig. 9 Remove the rear wheel speed sensor mounting bolt (A)

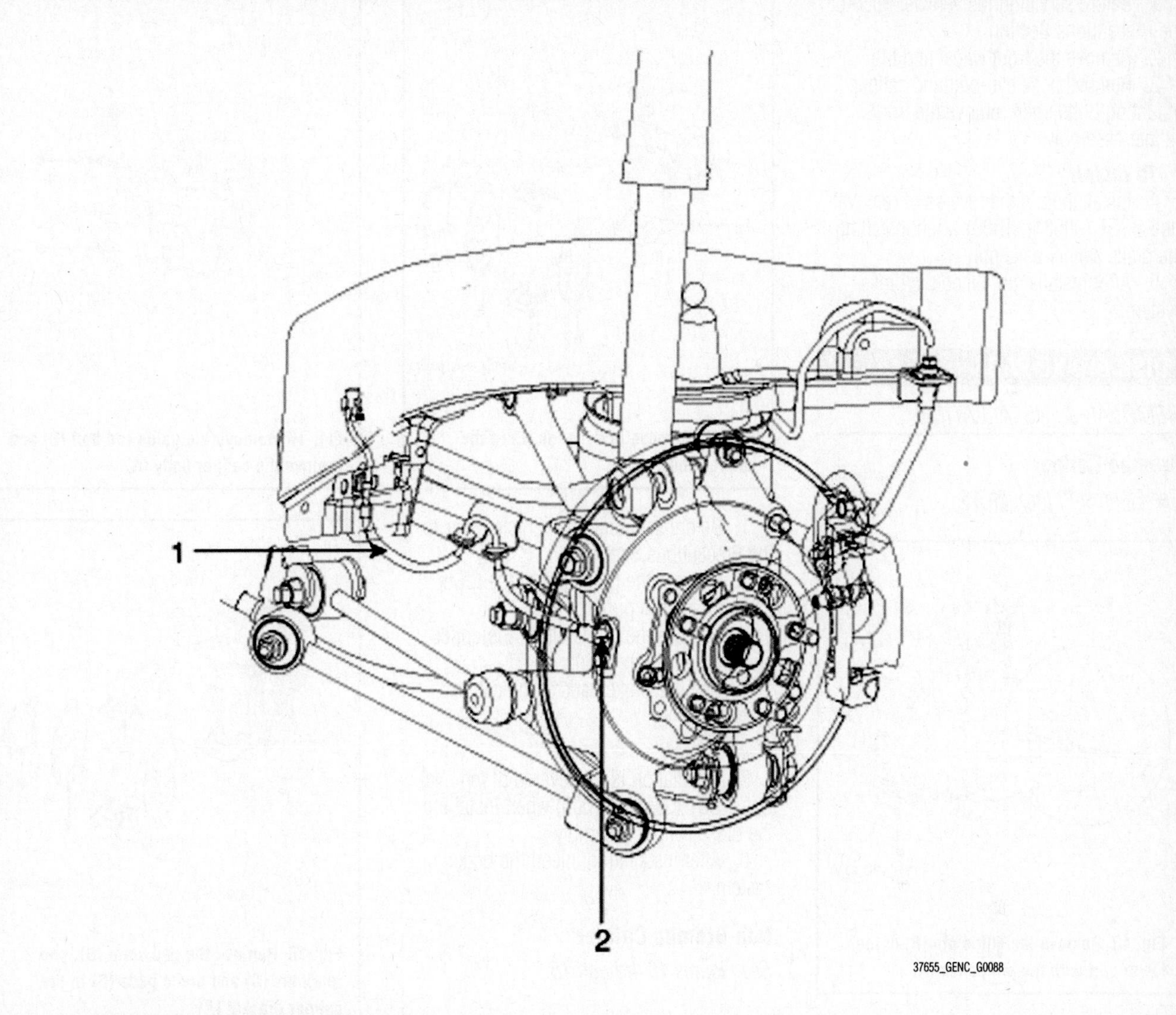

Fig. 10 Rear wheel speed sensor (2) and cable (1)

BRAKES

FRONT DISC BRAKES

BRAKE CALIPER

REMOVAL & INSTALLATION

Brembo Caliper

See Figure 11.

1. Before servicing the vehicle, refer to the Precautions Section.
2. Remove the front wheel and tire.
3. Remove the hose eye-bolt and caliper mounting bolts, then remove the front caliper assembly.

To install:

4. Installation is the reverse of removal. Use a SST (09581-11000) when installing the brake caliper assembly.
5. After installation, bleed the brake system.

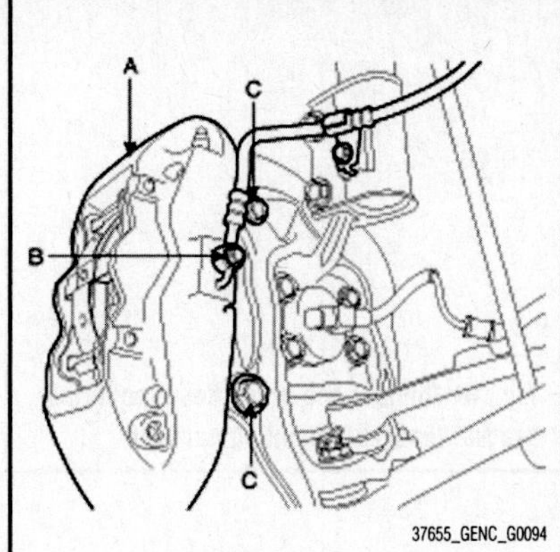

Fig. 11 Remove the hose eye-bolt (B) and caliper mounting bolts (C), then remove the front caliper assembly (A)

Non-Brembo Caliper

See Figure 12.

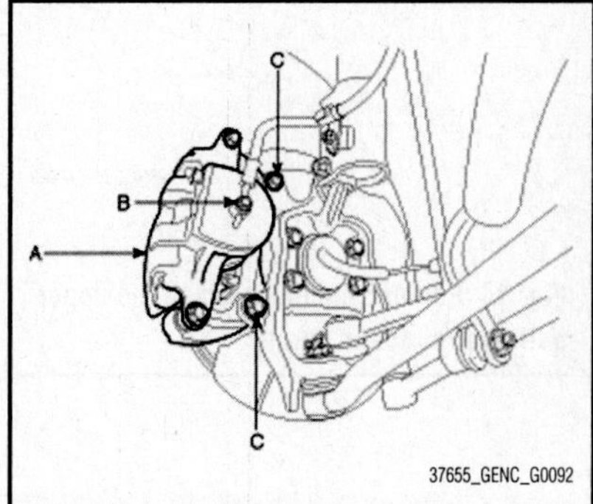

Fig. 12 Remove hose eye-bolt (B) and caliper mounting bolts (C), then remove the front caliper assembly (A)

1. Before servicing the vehicle, refer to the Precautions Section.
2. Remove the front wheel and tire.
3. Remove hose eye-bolt and caliper mounting bolts, then remove the front caliper assembly.

To install:

4. Installation is the reverse of removal. Use a SST (09581-11000) when installing the brake caliper assembly.
5. After installation, bleed the brake system.

DISC BRAKE PADS

REMOVAL & INSTALLATION

Brembo Caliper

See Figures 13 through 15.

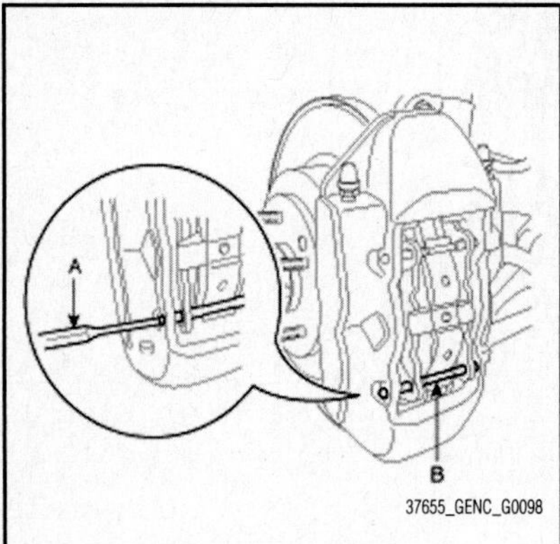

Fig. 13 Remove the guide pin (B) of the lower part with the pin punch (A)

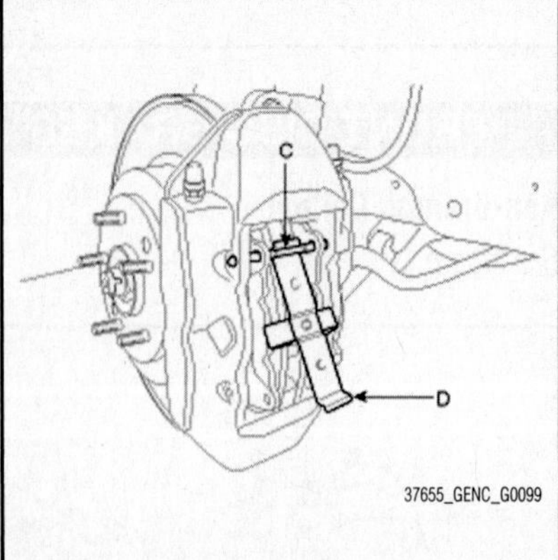

Fig. 14 Remove the guide pin of the upper part and retraction spring

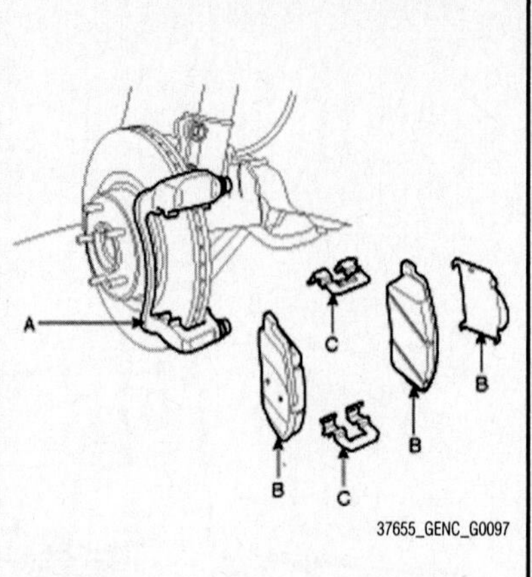

Fig. 15 Remove brake pads (E) at the caliper body

1. Before servicing the vehicle, refer to the Precautions Section.
2. Remove the guide pin of the lower part with the pin punch.
3. Remove the guide pin of the upper part and retraction spring.
4. Remove brake pads at the caliper body.

To install:

5. Installation is the reverse of removal. Use a SST (09581-11000) when installing the brake caliper assembly.
6. After installation, bleed the brake system.

Non-Brembo Caliper

See Figures 16 through 18.

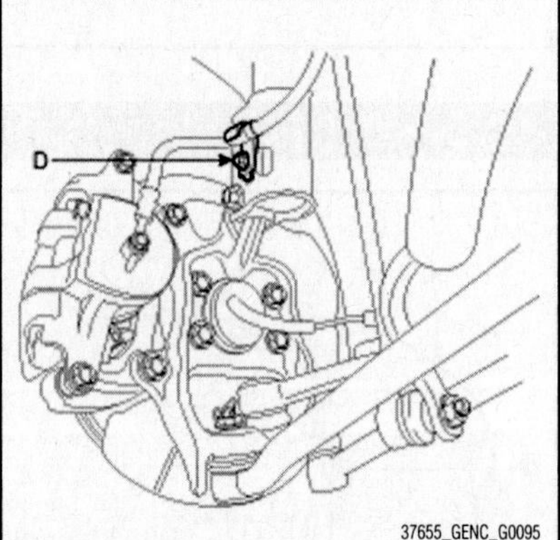

Fig. 16 Remove the brake hose mounting bracket (knuckle mounting part: D)

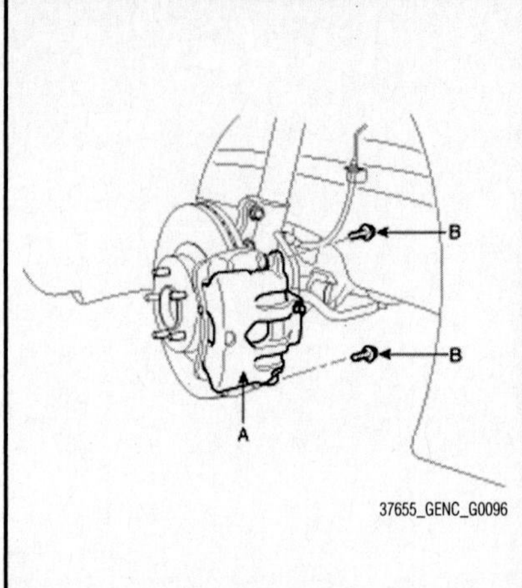

Fig. 17 Remove the guide rod bolt (B) and remove the caliper body (A)

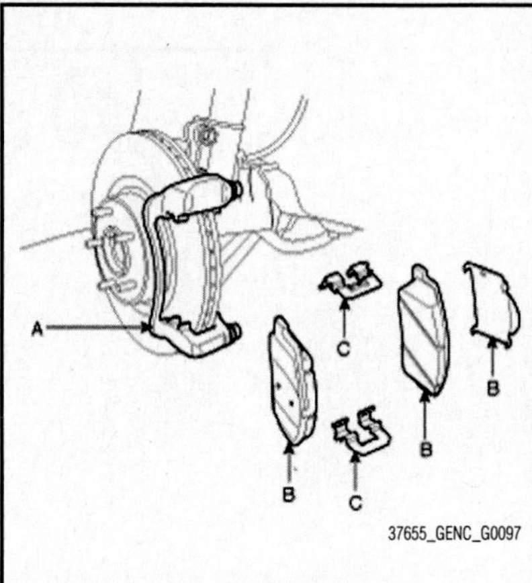

Fig. 18 Remove the pad shim (B), pad retainers (C) and brake pads (B) in the caliper bracket (A)

1. Before servicing the vehicle, refer to the Precautions Section.
2. Remove the brake hose mounting bracket.
3. Remove the guide rod bolt and remove the caliper body.
4. Remove the pad shim, pad retainers and brake pads in the caliper bracket.

To install:

5. Installation is the reverse of removal. Use a SST (09581-11000) when installing the brake caliper assembly. Tighten the guide rod bolt to 16–23 ft. lbs. (22–31 Nm).
6. After installation, bleed the brake system.

BRAKE CALIPER

REMOVAL & INSTALLATION

Brembo Caliper

See Figure 19.

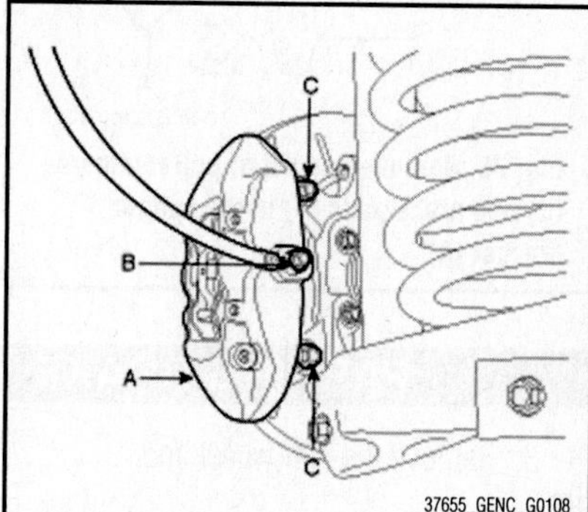

Fig. 19 Remove the hose eye-bolt (B) and caliper mounting bolts (C), then remove the rear caliper assembly (A)

1. Before servicing the vehicle, refer to the Precautions Section.
2. Remove the rear wheel and tire.
3. Remove the hose eye-bolt and caliper mounting bolts, then remove the rear caliper assembly.

To install:

4. Installation is the reverse of removal. Use a SST (09581-11000) when installing the brake caliper assembly.
5. After installation, bleed the brake system.

Non-Brembo Caliper

See Figure 20.

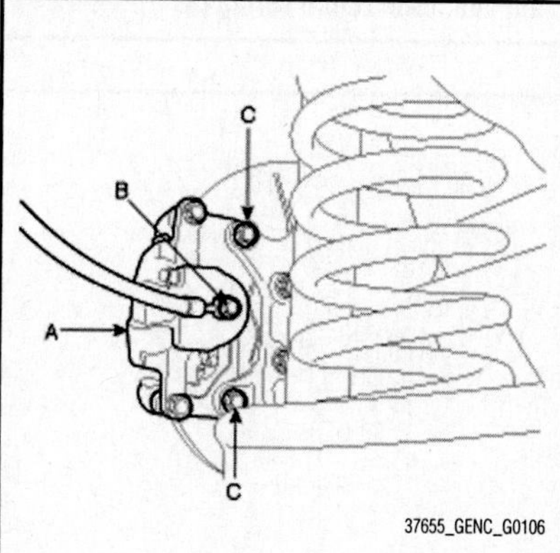

Fig. 20 Remove the hose eye-bolt (B) and caliper mounting bolts (C), then remove the rear caliper assembly (A)

1. Before servicing the vehicle, refer to the Precautions Section.
2. Remove the rear wheel and tire.
3. Remove the hose eye-bolt and caliper mounting bolts, then remove the rear caliper assembly.

To install:

4. Installation is the reverse of removal. Use a SST (09581-11000) when installing the brake caliper assembly.
5. After installation, bleed the brake system.

DISC BRAKE PADS

REMOVAL & INSTALLATION

Brembo Caliper

See Figures 21 through 23.

1. Before servicing the vehicle, refer to the Precautions Section.

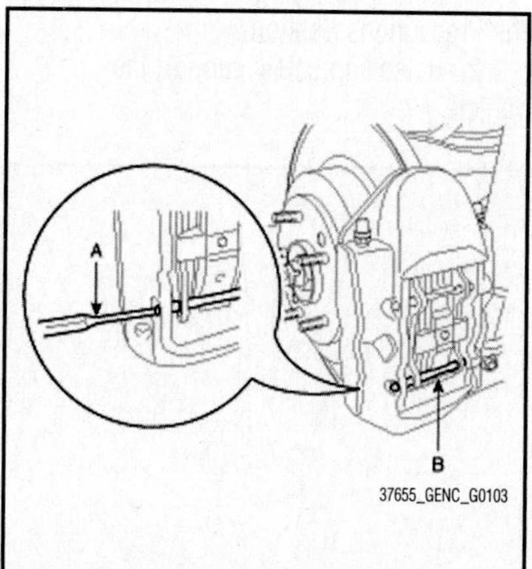

Fig. 21 Remove the guide pin (B) of the lower part with the pin punch (A)

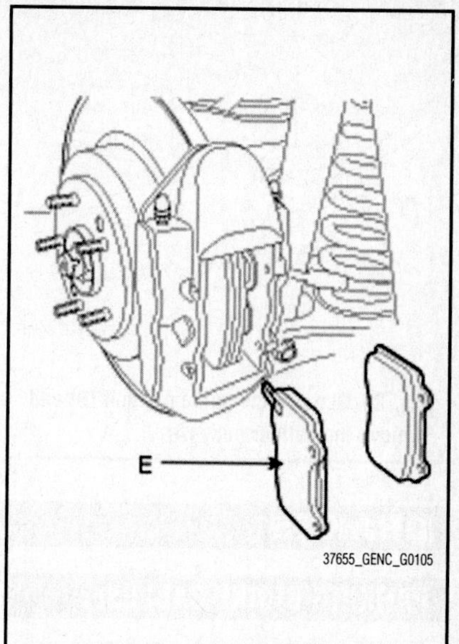

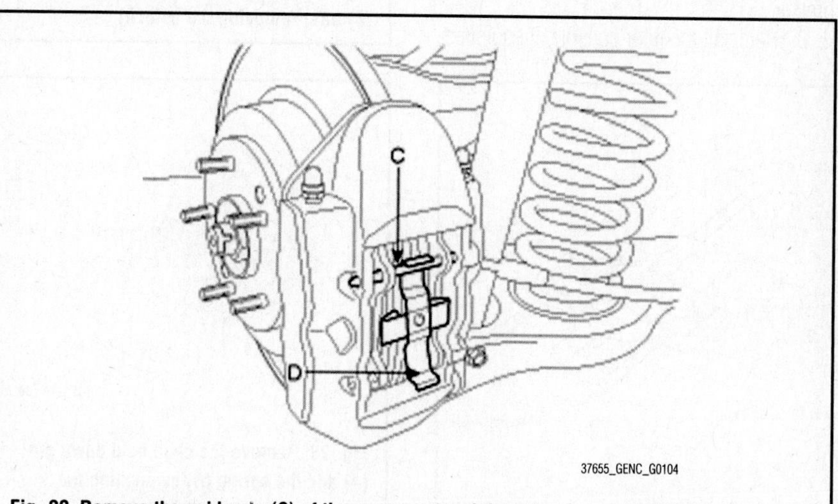

Fig. 22 Remove the guide pin (C) of the upper part and the retraction spring (D)

Fig. 23 Remove brake pads (E) at the caliper body

2. Remove the guide pin of the lower part with the pin punch.
3. Remove the guide pin of the upper part and the retraction spring.
4. Remove brake pads at the caliper body.

To install:

5. Installation is the reverse of removal. Use a SST (09581-11000) when installing the brake caliper assembly. Tighten the guide rod bolt to 16–23 ft. lbs. (22–31 Nm).
6. After installation, bleed the brake system.

Non-Brembo Caliper

See Figures 24 and 25.

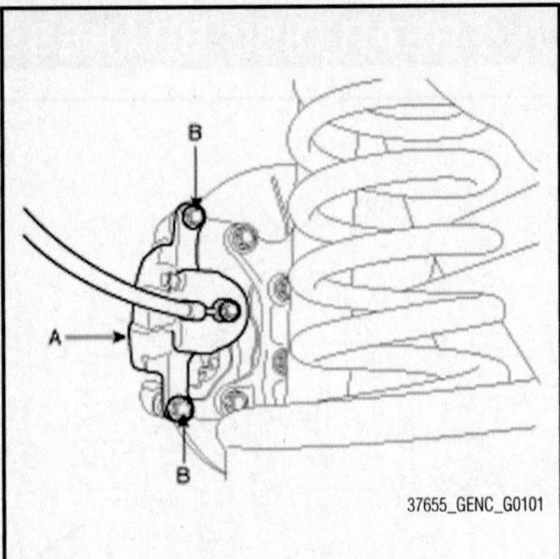

Fig. 24 Remove the guide rod bolt (B) and remove the caliper body (A)

1. Before servicing the vehicle, refer to the Precautions Section.
2. Remove the guide rod bolt and remove the caliper body.
3. Before servicing the vehicle, refer to the Precautions Section.
4. Remove pad shim, pad retainers and brake pads in the caliper bracket.

To install:

5. Installation is the reverse of removal. Use a SST (09581-11000) when installing the brake caliper assembly. Tighten the guide rod bolt to 16–23 ft. lbs. (22–31 Nm).
6. After installation, bleed the brake system.

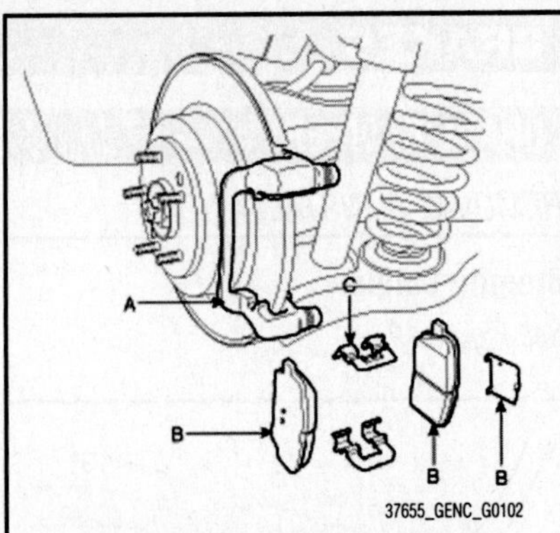

Fig. 25 Remove pad shim, pad retainers (C) and brake pads (B) in the caliper bracket (A)

BRAKES

PARKING BRAKE

PARKING BRAKE CABLES

ADJUSTMENT

See Figure 26.

1. Before servicing the vehicle, refer to the Precautions Section.

➡After repairing the parking brake shoe, adjust the brake shoe clearance, and then adjust the parking brake lever stroke.

2. Raise and safely support the vehicle.
3. Remove the floor console, if equipped.
4. Adjust the parking brake lever stroke by turning the adjusting nut. Stroke: 5 clicks.
5. Release the parking brake lever completely.
6. Check if the parking brakes drag when the rear wheels are turned. Readjust if necessary, until there is no drag from the parking brakes.
7. Check the proper operation of the parking brakes by fully applying the parking brakes.
8. Reinstall the floor console, if equipped.

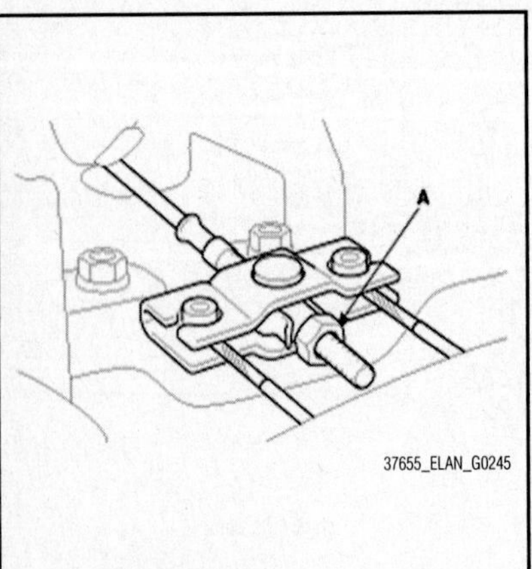

Fig. 26 Adjusting nut (A)

PARKING BRAKE SHOES

REMOVAL & INSTALLATION

See Figures 27 through 30.

1. Before servicing the vehicle, refer to the Precautions Section.
2. Raise and safely support the vehicle.

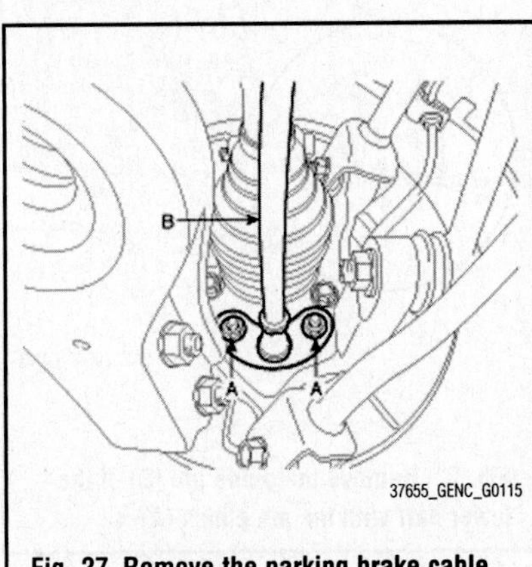

Fig. 27 Remove the parking brake cable (B) after removing the bolt (A)

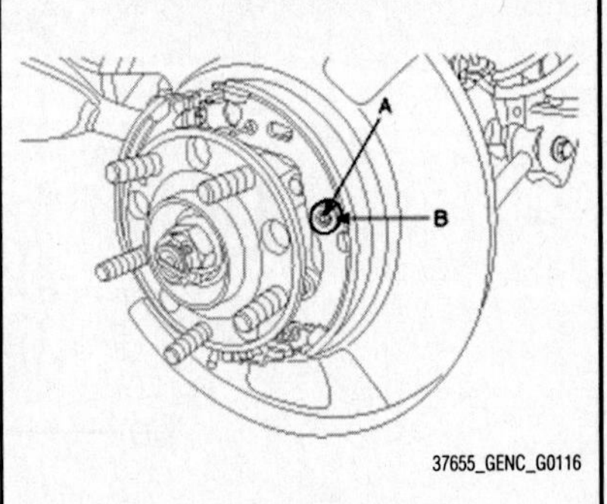

Fig. 28 Remove the shoe hold down pin (A) and the spring (B) by pushing the retainer spring and turning the pin

3. Remove the rear wheel and tire.
4. Remove the brake caliper and rear disc brake.
5. Remove the parking brake cable after removing the bolt.

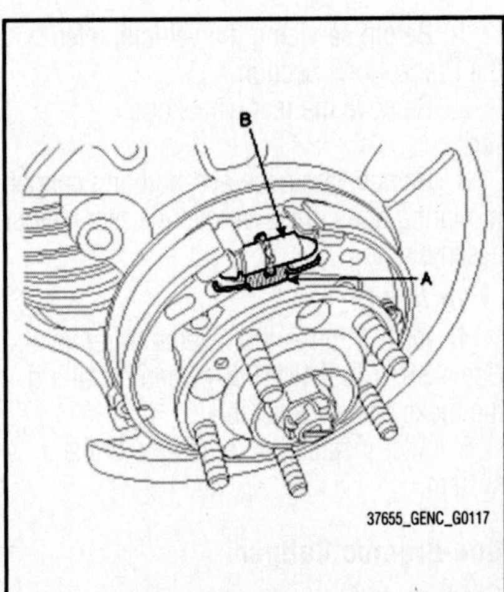

Fig. 29 Remove the adjuster assembly (B) and the lower return spring (A)

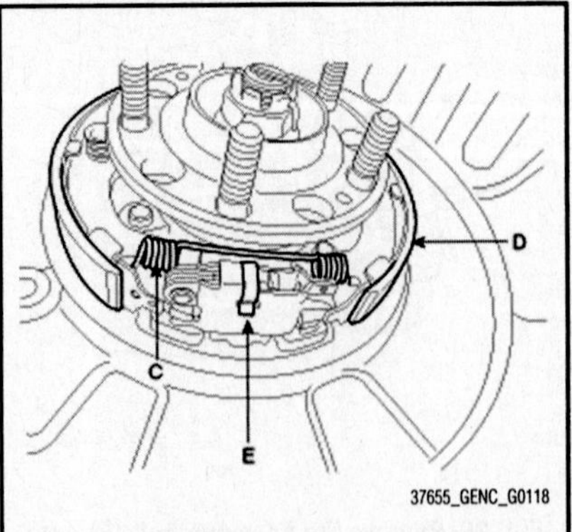

Fig. 30 Remove the upper return spring (C) and the brake shoes (D)

6. Remove the shoe hold down pin and the spring by pushing the retainer spring and turning the pin.

7. Remove the adjuster assembly and the lower return spring.

8. Remove the upper return spring and the brake shoes.

To install:

9. Install the operating lever assembly.

10. Install the upper return spring and the brake shoes.

11. Install the adjuster assembly and the lower return spring.

12. While pressing the spring, install the brake shoe hold down pin and spring.

13. Install the parking brake cable and tighten the bolt to 40–47 ft. lbs. (54–64 Nm).

14. How to install the DIH cable (Quick Fit type):

a. Put the inner cable into the knuckle hole in DIH lever operating direction when installing the cable.

b. Confirm by pulling the cable that cable is fixed certainly before installing the bolt.

15. Install the rear brake disc, then adjust the rear brake shoe clearance.

16. Install the brake caliper assembly.

17. Install the tire and wheel.

18. If the parking brake shoe or the brake disc are replaced by a new one, perform the brake shoe bed-in procedure:

a. While operating the parking brake pedal with 15 ft. lbs. (69 N) effort, drive the vehicle 0.31 miles (500 meters) at 37 mph (60 kph).

b. Repeat the above procedure more than two times.

c. Must be held on at 30% uphill.

19. After adjusting parking brake, note the following:

a. The parking pedal operates at 154 lbs. (686 N).

b. Check that all parts move smoothly.

c. The parking brake indicator lamp must be on after the parking pedal is worked and must be off after the pedal is released.

ADJUSTMENT

1. Before servicing the vehicle, refer to the Precautions Section.

2. Remove the rear wheel and tire.

3. Remove the plug from the disc.

4. Rotate the toothed wheel of adjuster by a screw driver until the disc is not moving, and then return it by 5 notches in the opposite direction.

5. Install the plug.

6. Install wheel and tire.

CHASSIS ELECTRICAL

AIR BAG (SUPPLEMENTAL RESTRAINT SYSTEM)

GENERAL INFORMATION

CAUTION

These vehicles are equipped with an air bag system. The system must be disarmed before performing service on, or around, system components, the steering column, instrument panel components, wiring and sensors. Failure to follow the safety precautions and the disarming procedure could result in accidental air bag deployment, possible injury and unnecessary system repairs.

SERVICE PRECAUTIONS

CAUTION

Disconnect and isolate the battery negative cable before beginning any airbag system component diagnosis, testing, removal, or installation procedures. Wait at least 90 seconds after the ignition switch is turned off and the negative (-) terminal cable is disconnected from the battery before starting the operation. The SRS is equipped with a backup power source, so if work is started within 90 seconds after disconnecting the negative (-) terminal cable from the battery, the SRS may be deployed. Failure to disable the airbag system may result in accidental airbag deployment, personal injury, or death.

DISARMING THE SYSTEM

CAUTION

Before servicing components near or affected by the SRS (airbag) system, read and observe all SRS Service Precautions. Failure to observe all precautions may result in accidental airbag deployment, personal injury, or death.

1. Disconnect and isolate the negative battery cable.

2. Wait 3 minutes for the system capacitor to discharge before performing any service.

3. Remove the ignition key from the vehicle.

ARMING THE SYSTEM

CAUTION

Before servicing components near or affected by the SRS (airbag) system, read and observe all SRS Service Precautions. Failure to observe all precautions may result in accidental airbag deployment, personal injury, or death.

1. Reconnect the negative battery cable.

2. Turn the ignition switch to the **RUN** position.

3. Confirm proper system operation:

a. Turn the ignition switch ON; the SRS indicator light should be turned on for about six seconds and then go off.

CLOCKSPRING CENTERING

1. Connect the clock spring harness connector and horn harness connector to the clock spring.

Set the clockspring center position by aligning the marks between the clock spring and the cover. Matchmark and turn the clockspring clockwise to the stop and then 3 revolutions counterclockwise.

DRIVE TRAIN

DIFFERENTIAL CARRIER ASSEMBLY

REMOVAL & INSTALLATION

See Figures 31 and 32.

1. Before servicing the vehicle, refer to the Precautions Section.
2. Disconnect the negative battery cable.
3. Raise and safely support the vehicle.
4. Remove the rear wheels and tires.
5. Drain the differential gear oil.
6. Remove the rear halfshafts. Refer to Halfshaft Removal & Installation.
7. Remove the propeller shaft assembly. Refer to Propeller Shaft Removal & Installation.
8. Remove the differential carrier assembly mounting bolts.
9. Remove the differential assembly.

To install:

10. Installation is the reverse of removal, noting the following:
    a. Tighten the differential carrier assembly mounting bolts to 58–72 ft. lbs. (80–100 Nm).

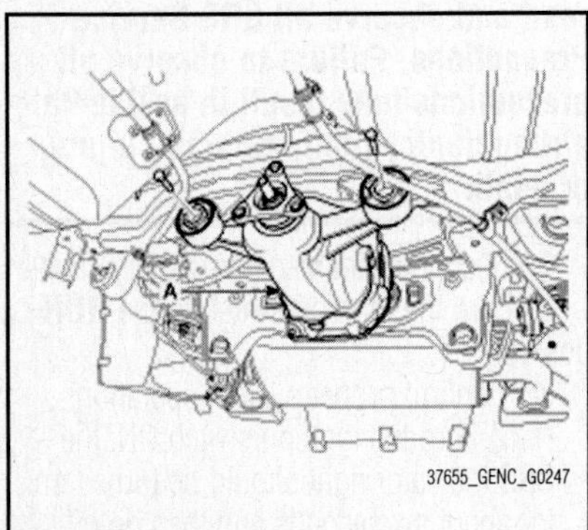

Fig. 31 Remove the differential assembly (A)

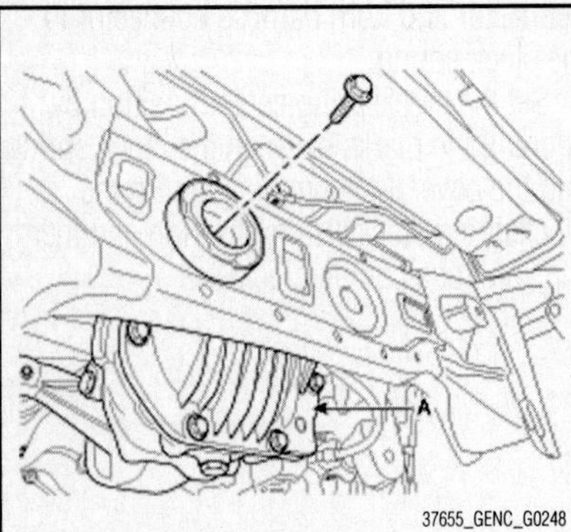

Fig. 32 Remove the differential carrier assembly mounting bolts

DRIVEN DISC & PRESSURE PLATE

REMOVAL & INSTALLATION

See Figures 33 through 35.

➡Special Service Tool used for this procedure: SST 09411-43000, Clutch Disc Guide.

1. Before servicing the vehicle, refer to the Precautions Section.
2. Remove the transmission assembly.
3. Insert the special tool (09411-43000) in the clutch disc to prevent the disc from shifting.
4. Remove the bolts which attach the clutch cover to the flywheel in a star pattern. Remove the bolts in succession, one or two turns at a time, to avoid bending the cover.

✲✲ WARNING

Do not clean the clutch disc or the release bearing with cleaning solvent.

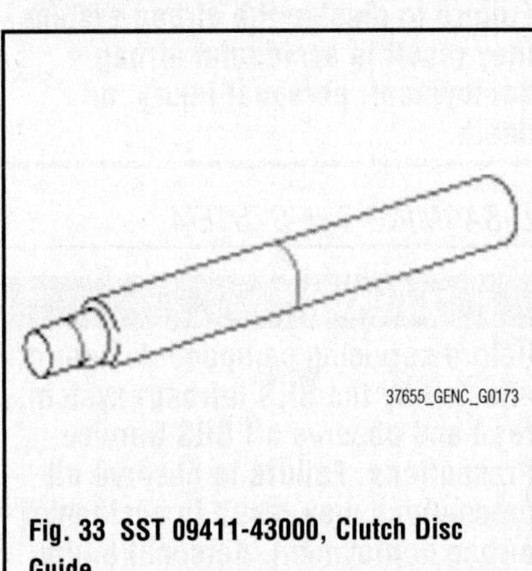

Fig. 33 SST 09411-43000, Clutch Disc Guide

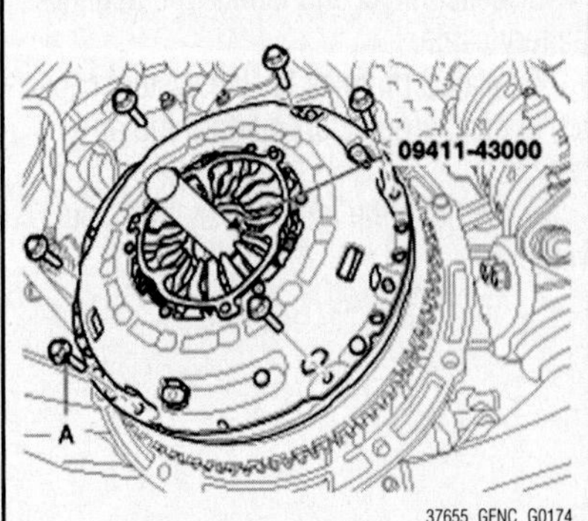

Fig. 34 Insert the special tool (09411-43000), and remove the clutch cover bolts

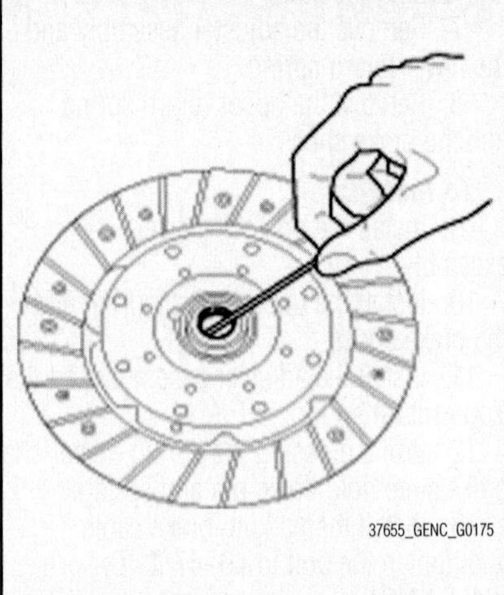

Fig. 35 Apply multipurpose grease to the spline of the disc

To install:

5. Apply multipurpose grease to the spline of the disc. Grease: CASMOLY L 9508

➡When installing the clutch, apply grease to each part, but be careful not to apply excessive grease. It can cause clutch slippage and vibration.

6. Install the clutch disc assembly to the flywheel using the special tool (09411-43000).

✲✲ WARNING

Be sure to place the face of the disc labeled 'T/M SIDE' toward the transmission.

7. Install the clutch cover assembly to the flywheel and temporarily tighten the bolts one or two steps at a time in a star pattern. Tighten the clutch cover bolts to 18–26 ft. lbs. (25–36 Nm).
8. Remove the clutch disc guide (09411-43000).
9. Install the transmission assembly to the engine.

HALFSHAFTS

REMOVAL & INSTALLATION

See Figures 36 through 46.

1. Before servicing the vehicle, refer to the Precautions Section.
2. Disconnect the negative battery cable.
3. Raise and safely support the vehicle.
4. Remove the rear wheels and tires.
5. Remove the brake caliper mounting bolts, and then place the brake caliper assembly with wire.

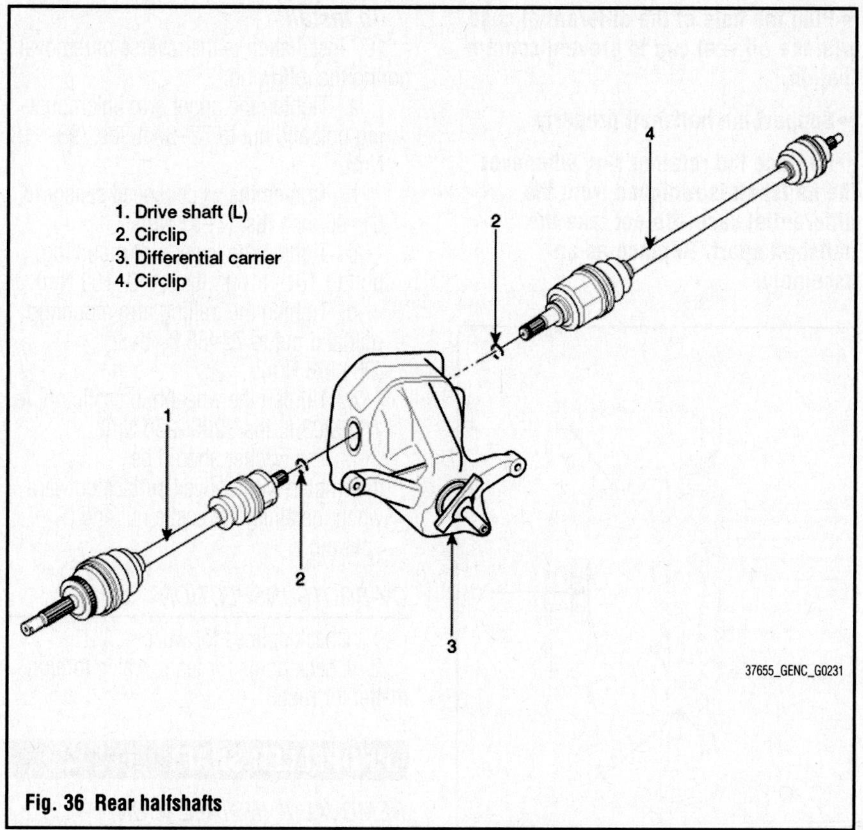

Fig. 36 Rear halfshafts

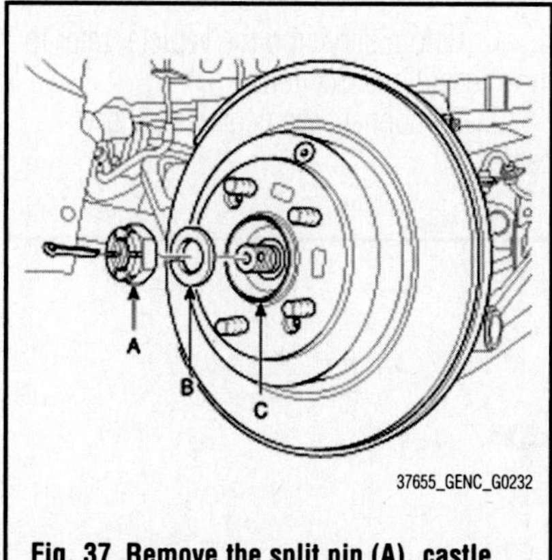

Fig. 37 Remove the split pin (A), castle nut (B) and washer (C)

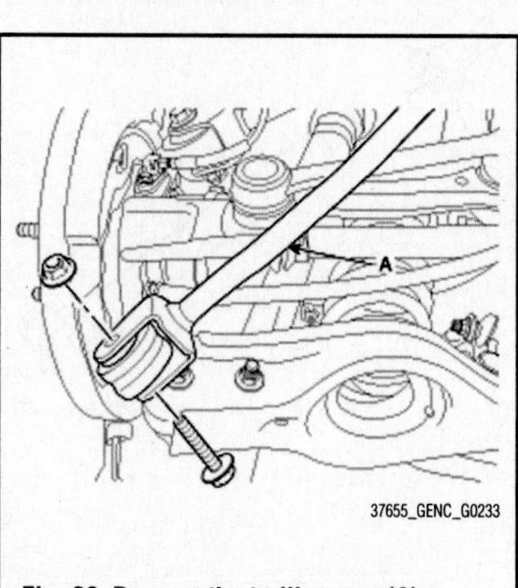

Fig. 38 Remove the trailing arm (A)

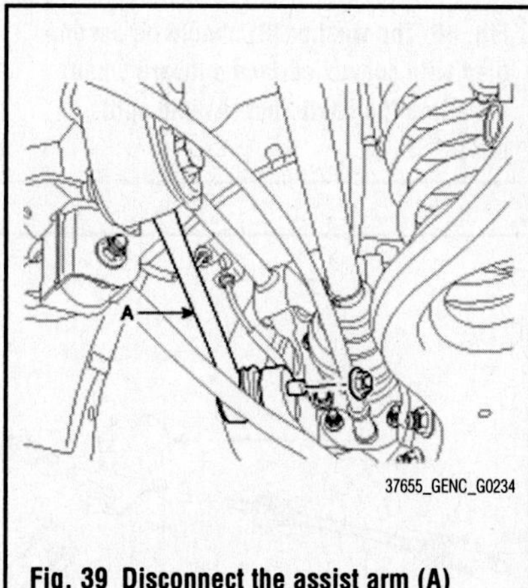

Fig. 39 Disconnect the assist arm (A)

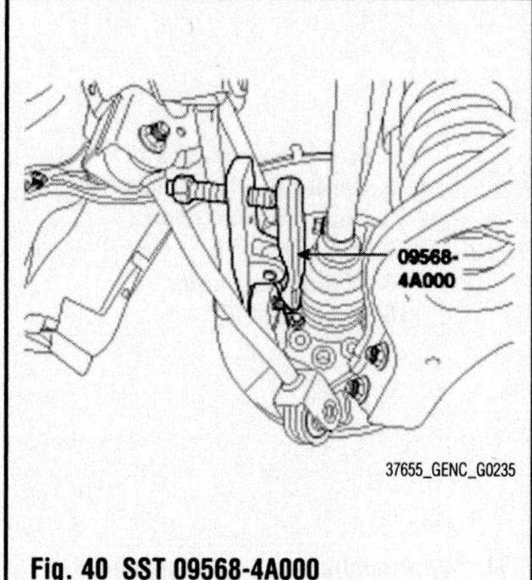

Fig. 40 SST 09568-4A000

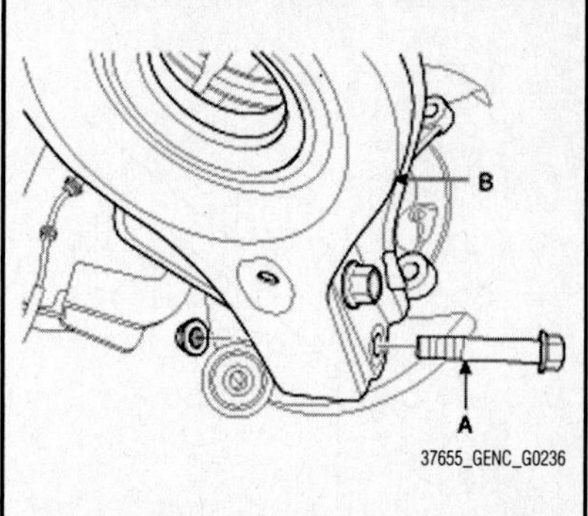

Fig. 41 Remove lower arm mounting bolt (A) and then remove the lower arm (B)

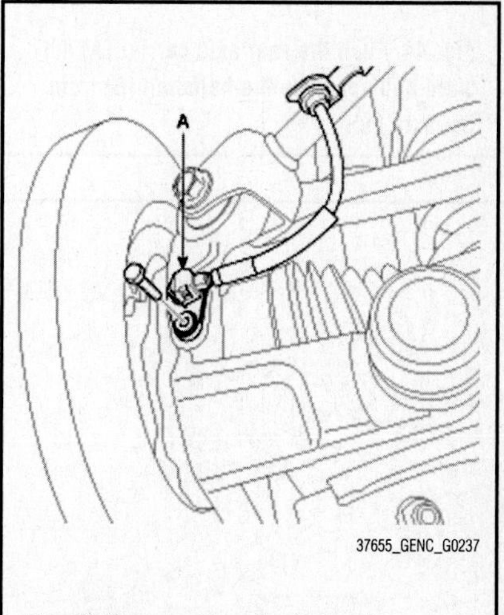

Fig. 42 Remove the wheel speed sensor (A)

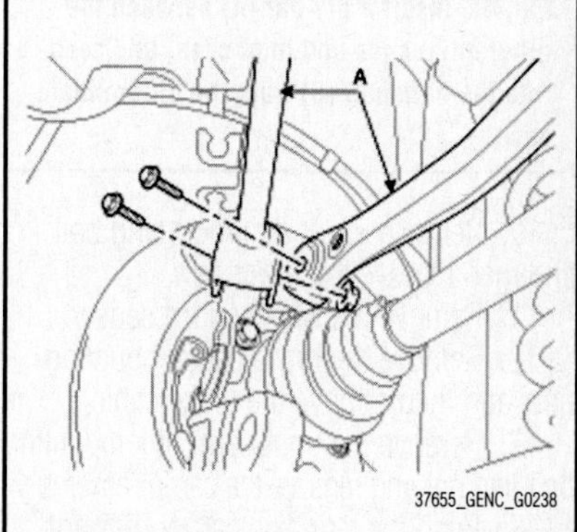

Fig. 43 Remove the upper arm (A) link mounting bolt and nut and remove the carrier assembly (B)

6. Remove the split pin, and then the remove castle nut and washer from the wheel hub.
7. Remove the rear brake lining.
8. Remove the trailing arm mounting bolt and nut and remove the trailing arm.
9. Remove the assist arm mounting nut and disconnect the assist arm.

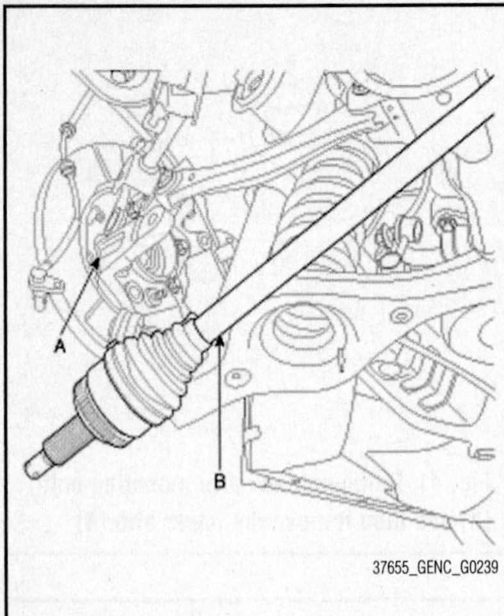

Fig. 44 Push the rear axle carrier (A) outward and separate the halfshaft (B) from the axle hub

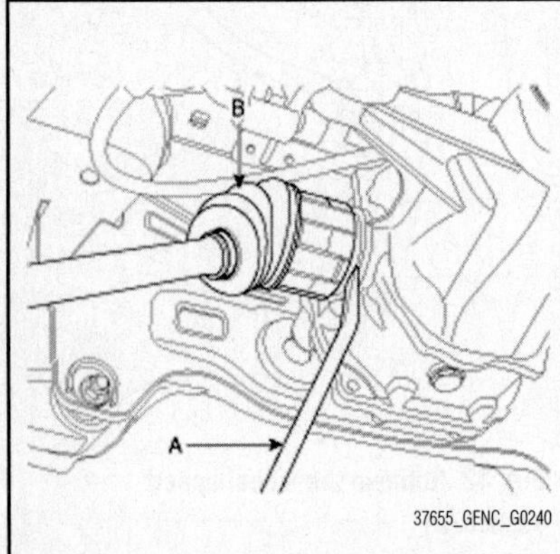

Fig. 45 Insert a pry bar (A) between the differential case and joint case, and separate the halfshaft (B) from the differential case

10. Remove lower arm mounting bolt and then remove the lower arm.
11. Remove the wheel speed sensor.
12. Remove the brake cable mounting nuts and then remove the brake cable.
13. Remove the upper arm link mounting bolt and nut and remove the carrier assembly.
14. Push the rear axle carrier outward and separate the halfshaft from the axle hub.
15. Insert a pry bar between the differential case and joint case, and separate the halfshaft from the differential case.

➡Be careful not to damage the differential and joint. Do not insert the pry bar too deep, as this may cause damage to the oil seal. Do not pull the halfshaft by excessive force it may cause components inside the joint kit to dislodge resulting in a torn boot or a damaged bearing.

➡Plug the hole of the differential case with the oil seal cap to prevent contamination.

➡Support the halfshaft properly.

➡Replace the retainer ring whenever the halfshaft is removed from the differential case. Do not take the halfshaft apart. Replace as an assembly.

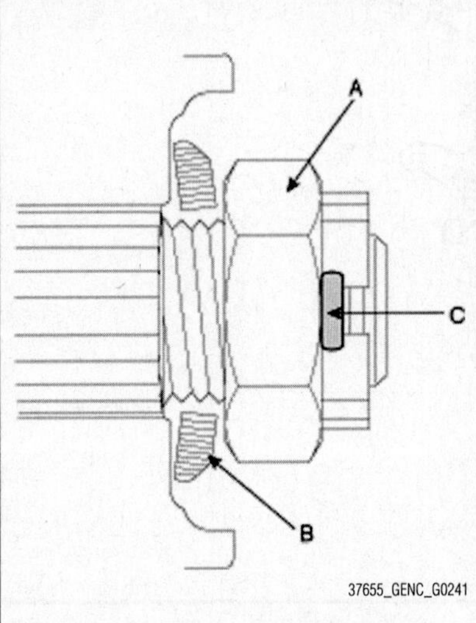

Fig. 46 The washer (B) should be assembled with convex surface outward when installing the castle nut (A) and split pin (C)

To install:

16. Installation is the reverse of removal, noting the following:
  a. Tighten the upper arm link mounting bolt and nut to 72–87 ft. lbs. (98–118 Nm).
  b. Tighten the wheel speed sensor to 61–96 inch lbs. (7–11 Nm).
  c. Tighten the lower arm mounting bolt to 101–116 ft. lbs. (140–160 Nm).
  d. Tighten the trailing arm mounting bolt and nut to 72–88 ft. lbs. (98–118 Nm).
  e. Tighten the wheel hub castle nut to 145–203 ft. lbs. (200–280 Nm).
  f. The washer should be assembled with convex surface outward when installing the castle nut and split pin.

CV-BOOTS INSPECTION

1. Check splines for wear.
2. Check boots for tears, water, foreign matter, or rust.

PROPELLER SHAFT

REMOVAL & INSTALLATION

See Figures 47 through 51.

1. Before servicing the vehicle, refer to the Precautions Section.
2. Disconnect the negative battery cable.

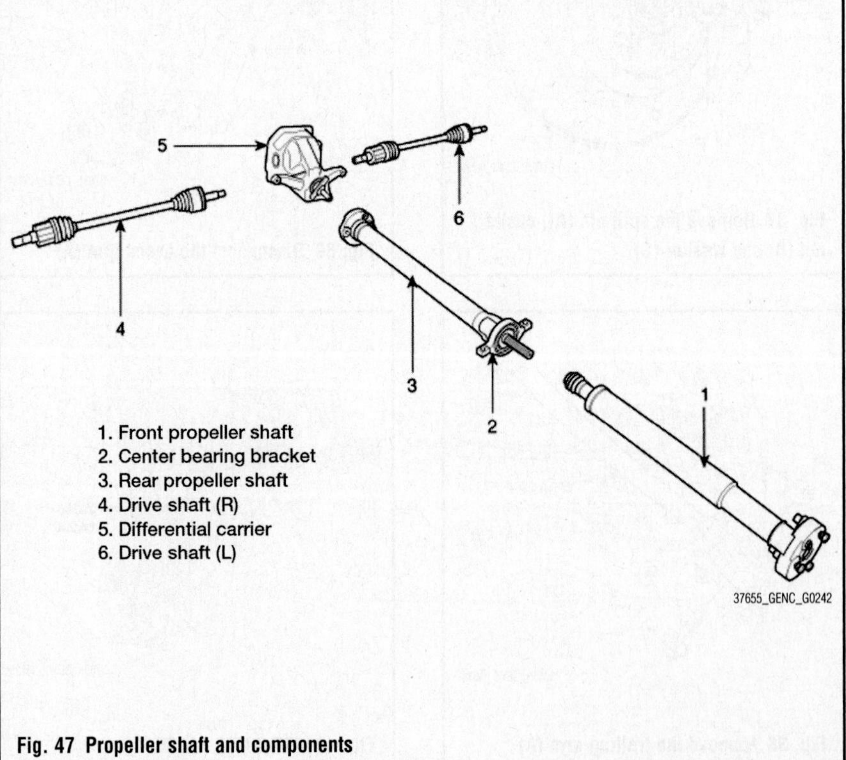

Fig. 47 Propeller shaft and components

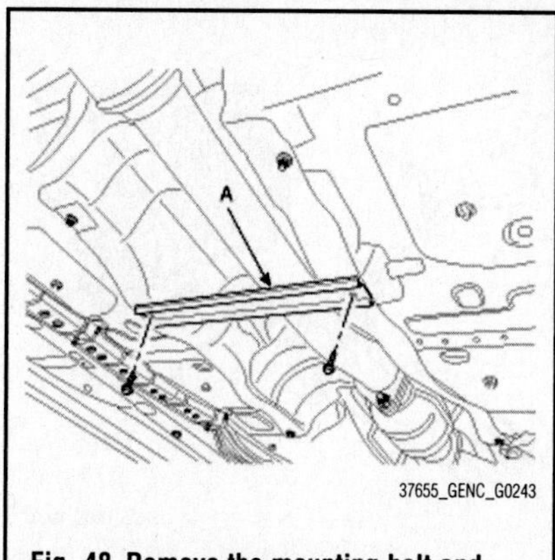

Fig. 48 Remove the mounting bolt and then remove the bracket (A)

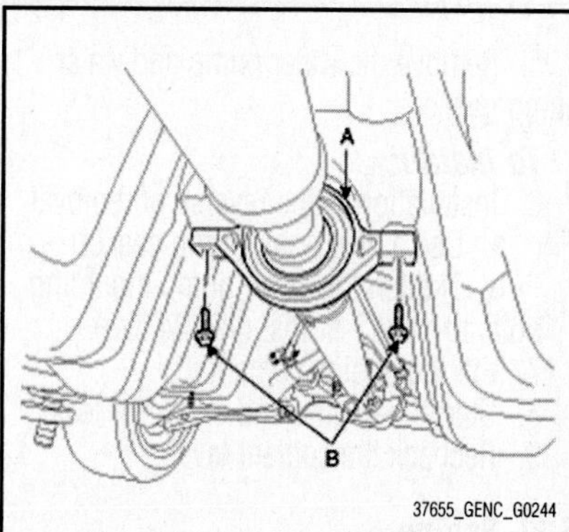

Fig. 49 Remove the center bearing bracket (A) mounting bolts (B)

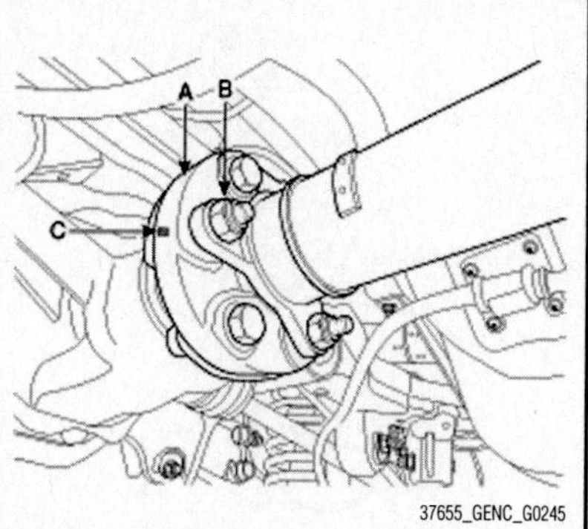

Fig. 50 After making a match mark (C) on the rubber coupling (A) and rear differential companion (B), remove the propeller shaft mounting bolts

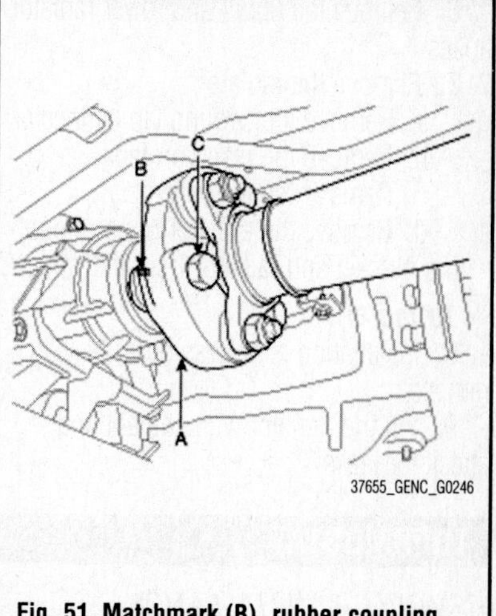

Fig. 51 Matchmark (B), rubber coupling (A), mounting bolts (C)

3. Raise and safely support the vehicle.
4. Remove the rear wheels and tires.
5. Remove the mounting bolt and then remove the bracket.
6. Remove the rear muffler.
7. Remove the mounting bolts, and then remove the heating bracket.
8. Remove the center bearing bracket mounting bolts.
9. After making a match mark on the rubber coupling and rear differential companion, remove the propeller shaft mounting bolts.

➡Use the hexagonal wrench to prevent damage of bolt head when removing bolts.

To install:

10. Installation is the reverse of removal, noting the following:
    a. Tighten the propeller shaft mounting bolts to 65–80 ft. lbs. (90–110 Nm).

➡Be careful to position the mounting bolts in their original positions. Balance the propeller shaft if necessary.

ENGINE COOLING

RADIATOR

REMOVAL & INSTALLATION

2.0L Engine

See Figure 52.

1. Before servicing the vehicle, refer to the Precautions Section.
2. Disconnect the battery negative cable.
3. Drain the engine coolant.
4. Disconnect the breather hose, the vacuum hose and remove the air duct and the air cleaner assembly.
5. Remove the radiator upper hose.
6. Disconnect the intercooler inlet hose and the radiator lower hose.
7. Disconnect the BPS connector and intercooler outlet hose.
8. Remove the radiator.
    a. Remove the cooling fan connector.
    b. Remove the reservoir tank.
    c. Remove the fan assembly.
    d. Remove the radiator upper mounting bracket and radiator from the vehicle.

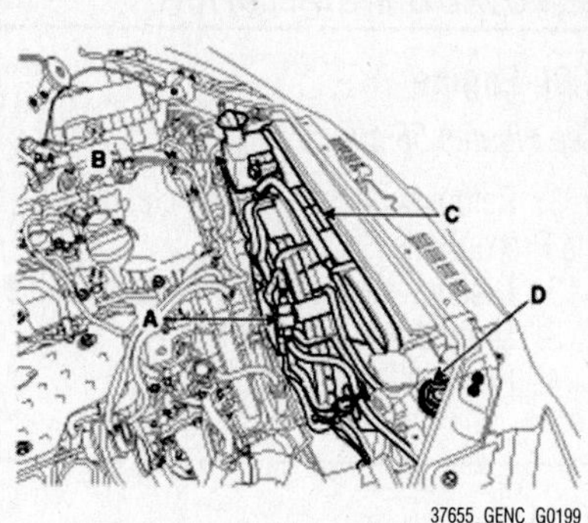

Fig. 52 Remove the cooling fan connector (A), reservoir tank (B), fan assembly (C), and radiator brackets (D)

To install:

9. Installation is reverse order of removal.
10. Fill the radiator with coolant and check for leaks.

3.8L Engine

See Figure 53.

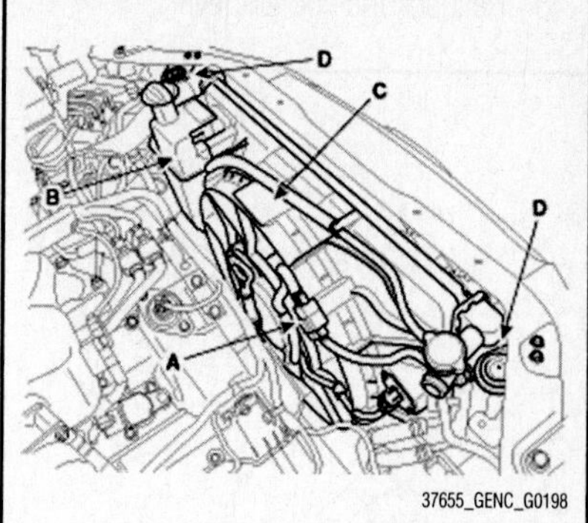

Fig. 53 Remove the cooling fan connector (A), reservoir tank (B), fan assembly (C), and radiator brackets (D)

1. Before servicing the vehicle, refer to the Precautions Section.
2. Disconnect the battery negative cable.
3. Drain the engine coolant.
4. Remove the air duct.
5. Remove the air cleaner assembly after removing the AFS connector.

6. Remove the upper and lower radiator hoses.
7. Remove the radiator.
   a. Remove the cooling fan connector.
   b. Remove the reservoir tank.
   c. Remove the fan assembly.
   d. Remove the radiator upper mounting bracket and radiator from the vehicle.

To install:

8. Installation is reverse order of removal.
9. Fill the radiator with coolant and check for leaks.

THERMOSTAT

REMOVAL & INSTALLATION

2.0L Engine

See Figure 54.

1. Before servicing the vehicle, refer to the Precautions Section.
2. Disconnect the battery negative cable.
3. Drain the engine coolant so its level is below the thermostat.
4. Remove the water inlet fitting and thermostat.

To install:

5. Installation is the reverse of removal.
   a. Install the thermostat with jiggle valve upward.
   b. Tighten the water inlet fitting to 14–17 ft. lbs. (19–24 Nm)
6. Fill the engine coolant.
7. Start the engine and check for leaks.
8. Recheck the coolant level.

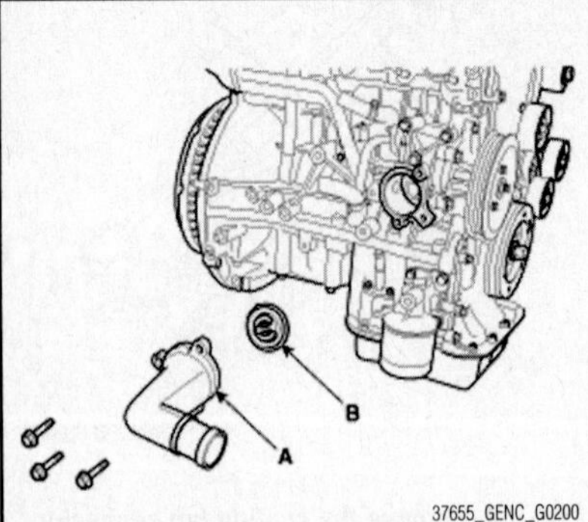

Fig. 54 Remove the water inlet fitting (A) and thermostat (B)

3.8L Engine

See Figure 55.

1. Before servicing the vehicle, refer to the Precautions Section.
2. Disconnect the battery negative cable.

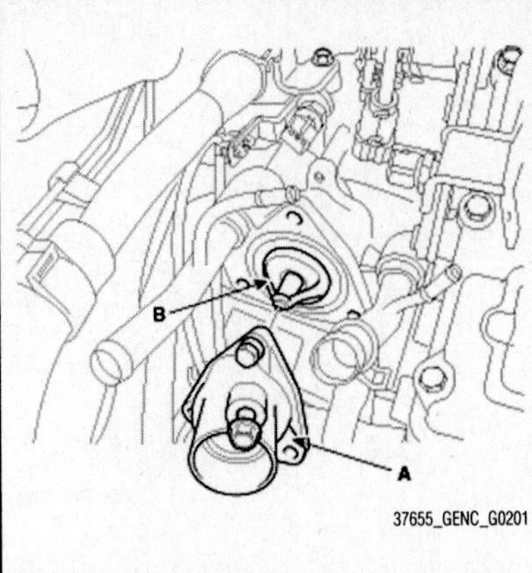

Fig. 55 Remove the water inlet fitting (A) and thermostat (B)

3. Drain the engine coolant so its level is below the thermostat.
4. Remove the water inlet fitting and thermostat.

To install:

5. Place the thermostat in thermostat housing.
   a. Install the thermostat with jiggle valve upward.
   b. Install a new thermostat.
6. Tighten the water inlet fitting to 12–14 ft. lbs. (17–20 Nm)
7. Fill the engine coolant.
8. Start the engine and check for leaks.
9. Recheck the coolant level.

WATER PUMP

REMOVAL & INSTALLATION

2.0L Engine

See Figures 56 and 57.

1. Before servicing the vehicle, refer to the Precautions Section.
2. Disconnect the battery negative cable.
3. Drain the engine coolant.
4. Remove the drive belt.

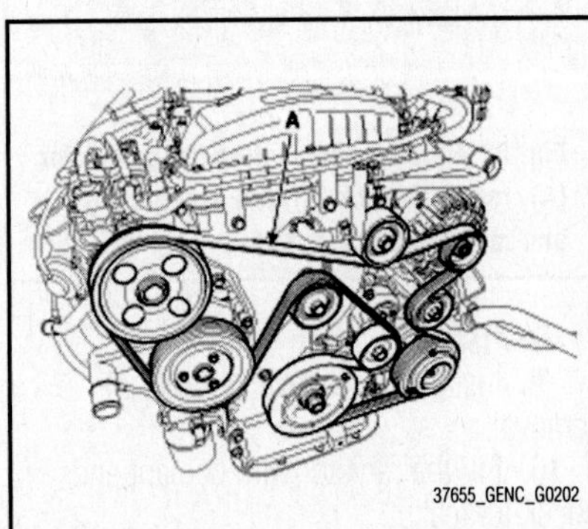

Fig. 56 Remove the drive belt (A)–2.0L engine

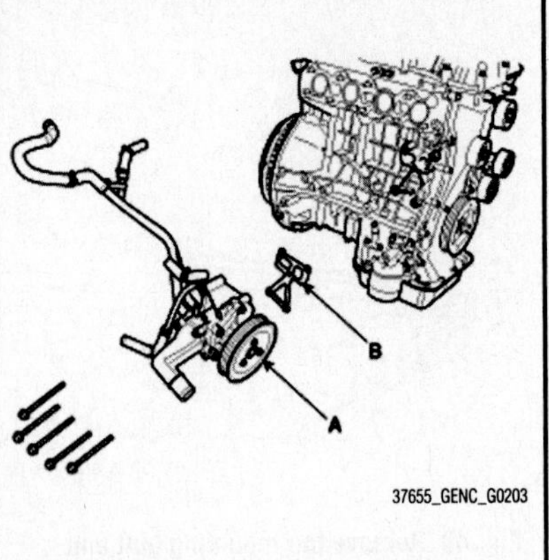

Fig. 57 Remove the water pump (A) and gasket (B)

5. Remove the water pump and water pump gasket.

To install:

6. Installation is the reverse of removal.
   a. Use a new water pump gasket.
   b. Tighten the water pump mounting bolts to 14–17 ft. lbs. (19–24 Nm).
7. Fill the engine coolant.
8. Start the engine and check for leaks.
9. Recheck the coolant level.

3.8L Engine

See Figures 58 through 60.

1. Before servicing the vehicle, refer to the Precautions Section.
2. Disconnect the battery negative cable.
3. Remove the drive belt.
4. Remove the water pump pulley.
5. Remove the water pump pulley.

To install:

6. Installation is the reverse of removal.
7. Fill the engine coolant.
8. Start the engine and check for leaks.
9. Recheck the coolant level.

Fig. 58 Remove the drive belt (A)—3.8L engine

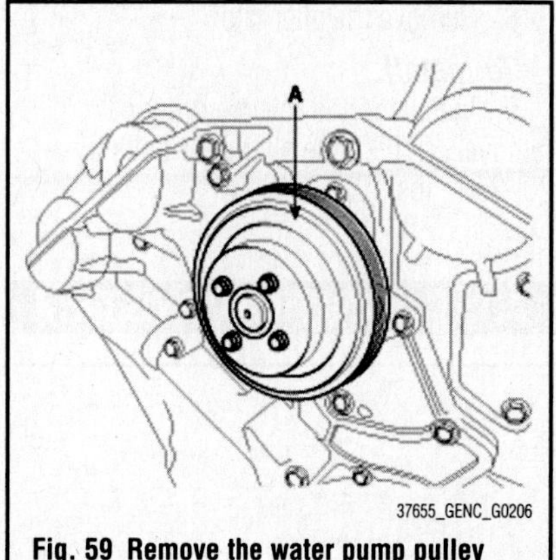

Fig. 59 Remove the water pump pulley (A)

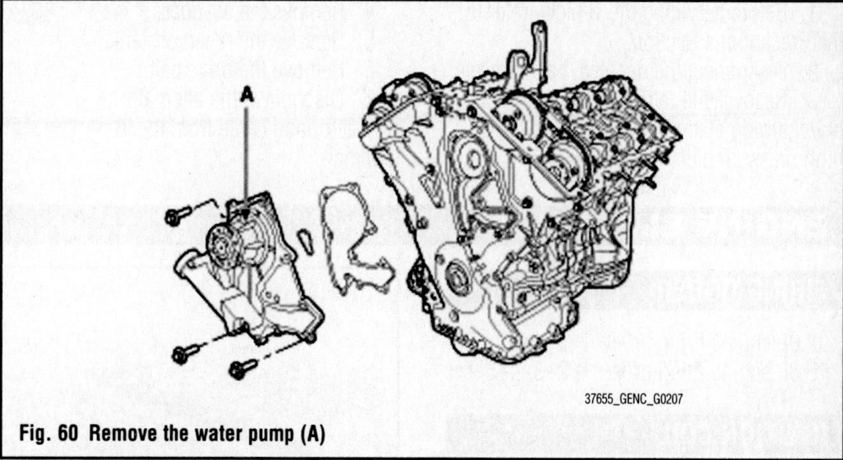

Fig. 60 Remove the water pump (A)

ENGINE ELECTRICAL

CHARGING SYSTEM

ALTERNATOR

REMOVAL & INSTALLATION

2.0L Engine

See Figures 61 and 62.

1. Before servicing the vehicle, refer to the Precautions Section.
2. Disconnect the negative battery cable.
3. Disconnect the alternator connector and cable.
4. Remove the drive belt and alternator.

To install:

5. Installation is the reverse order of removal. Tighten the alternator bolts to 33–40 ft. lbs. (44–54 Nm).

3.8L Engine

See Figures 63 through 66.

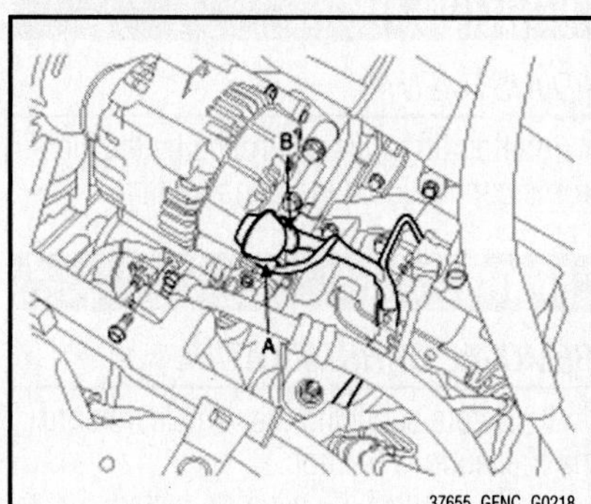

Fig. 61 Disconnect the alternator connector (B) and cable (A)

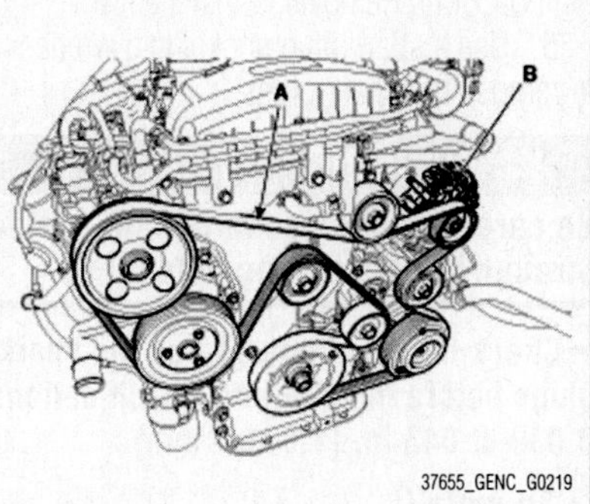

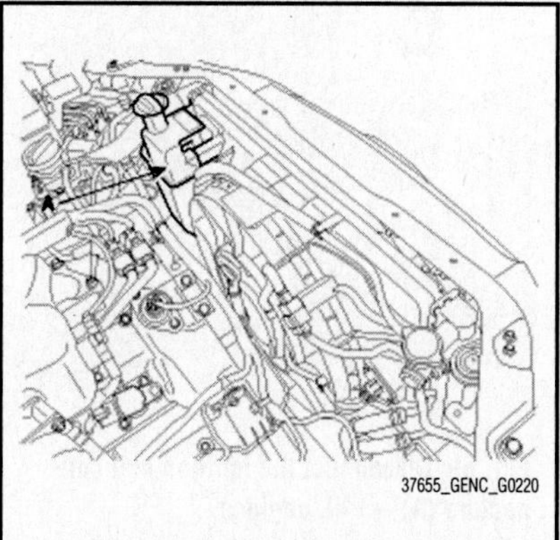

Fig. 63 Remove the reservoir tank (B)

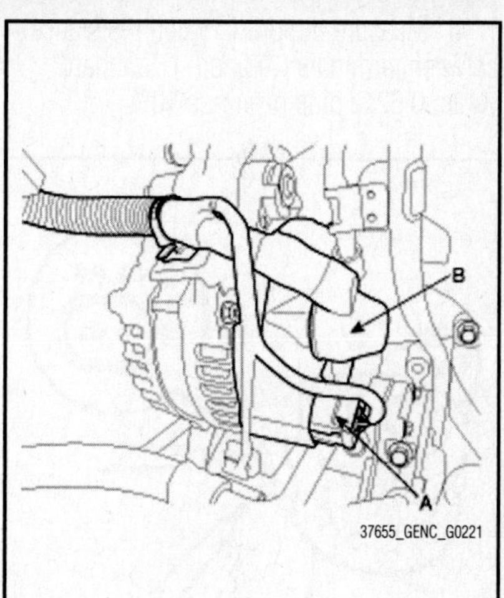

Fig. 65 Disconnect the alternator connector (A) and cable (B) from the 'B' terminal

37655_GENC_G0219
Fig. 62 Remove the drive belt (A) and alternator (B)

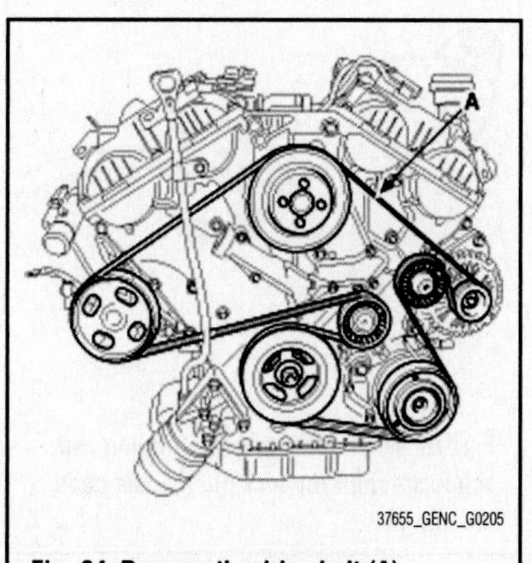

Fig. 64 Remove the drive belt (A)

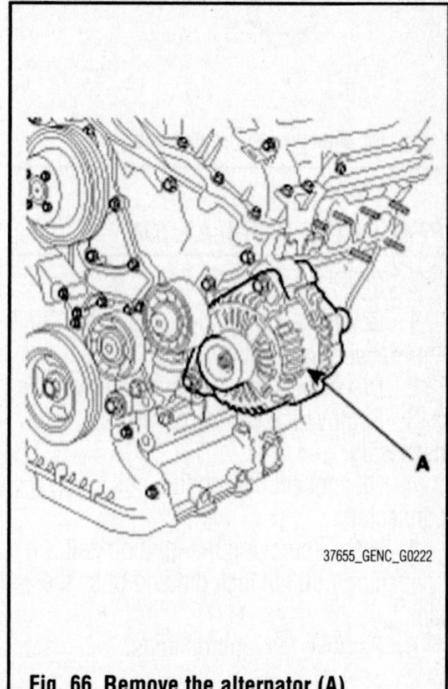

Fig. 66 Remove the alternator (A)

1. Before servicing the vehicle, refer to the Precautions Section.
2. Disconnect the negative battery cable.
3. Recover the refrigerant with a recovery/charging station and remove the A/C high pressure pipe.
4. Remove the air duct.
5. Remove the reservoir tank.
6. Remove the drive belt.
7. Disconnect the alternator connector and cable from the 'B' terminal.
8. Remove the alternator.

To install:

9. Installation is reverse order of removal. Tighten the alternator bolts to 20–25 ft. lbs. (27–33 Nm).

ENGINE ELECTRICAL — IGNITION SYSTEM

FIRING ORDER

2.0L engine firing order: 1–3–4–2
3.8L engine firing order: 1–2–3–4–5–6

IGNITION COIL

TESTING

See Figure 67.

1. Measure the primary coil resistance between terminals (+) and (-). Standard value: 0.62Ω plus or minus 10%

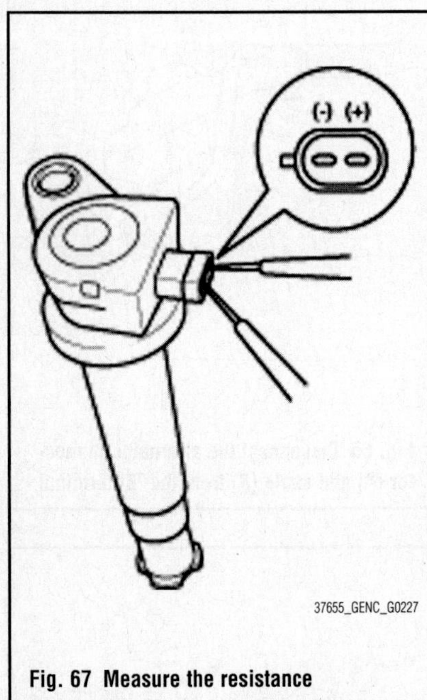

Fig. 67 Measure the resistance

REMOVAL & INSTALLATION

See Figures 68 through 71.

1. Before servicing the vehicle, refer to the Precautions Section.
2. Disconnect the negative battery cable.
3. Remove the engine cover, as necessary.
4. Disconnect the ignition coil connectors.
5. When removing the ignition coil connector, pull the lock pin and push the clip.
6. Remove the ignition coils.

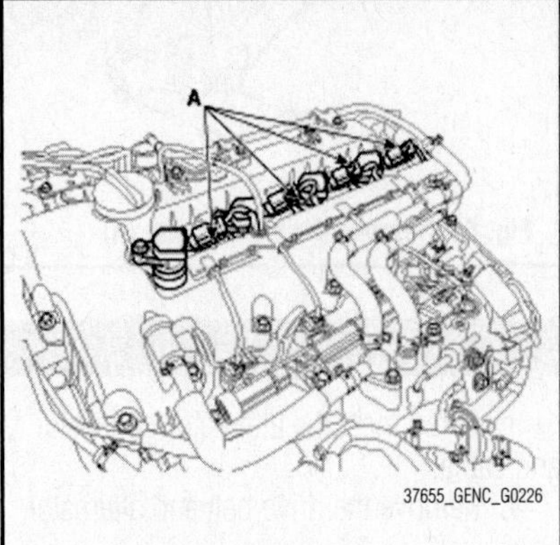

Fig. 68 Disconnect the ignition coil connectors (A)—2.0L engines

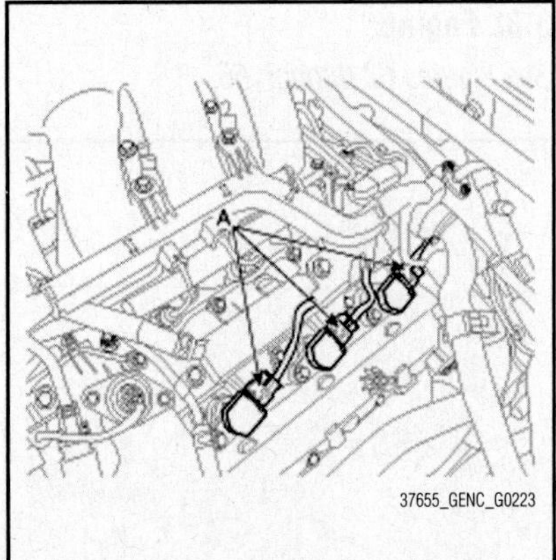

Fig. 69 Disconnect the ignition coil connectors (A)—3.8L engines

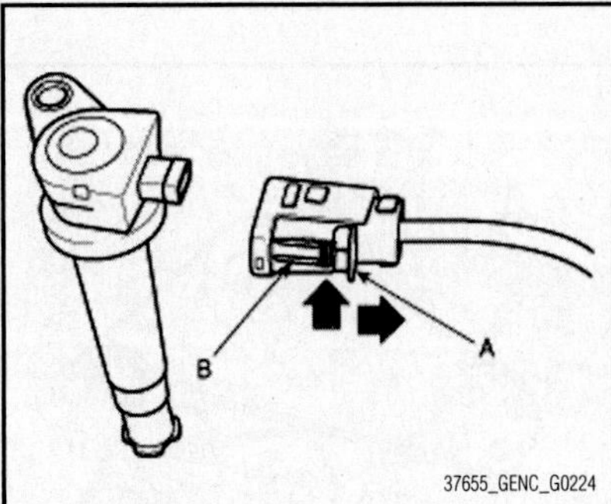

Fig. 70 When removing the ignition coil connector, pull the lock pin (A) and push the clip (B)

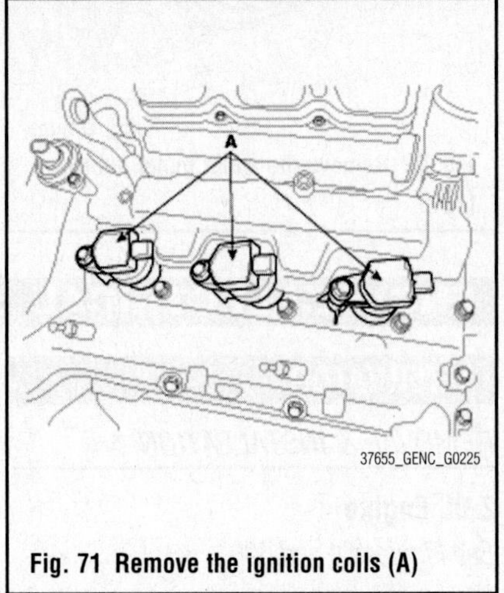

Fig. 71 Remove the ignition coils (A)

To install:

7. Installation is the reverse of removal.

IGNITION TIMING

ADJUSTMENT

Ignition timing is controlled by the electronic control ignition timing system.

SPARK PLUGS

REMOVAL & INSTALLATION

1. Before servicing the vehicle, refer to the Precautions Section.
2. Disconnect the negative battery cable.
3. Remove the engine cover (as necessary).
4. Remove the ignition coil. Refer to Ignition Coils Removal & Installation.
5. Use a spark plug socket and wrench to remove the spark plugs.

**** WARNING**

Be careful that no contaminates enter through the spark plug holes.

➡Check the electrode gap on the spark plugs before installation. Specification: 0.039–0.043 in. (1.0–1.1 mm).

To install:

6. To install, reverse the removal procedure.

STARTER

REMOVAL & INSTALLATION

2.0L Engine

See Figures 72 through 74.

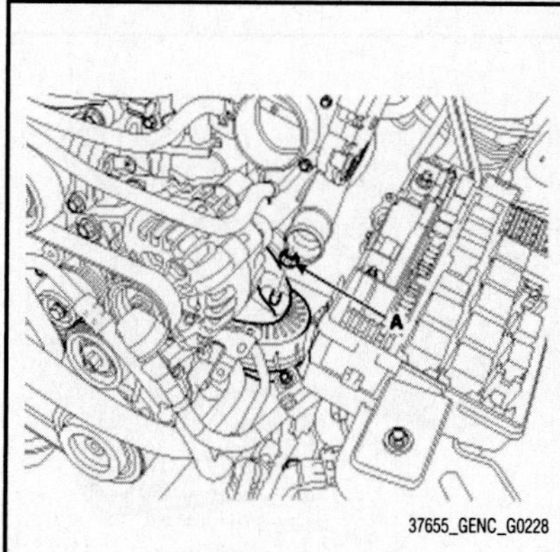

Fig. 72 Remove the left-hand engine mounting bracket nut (A)

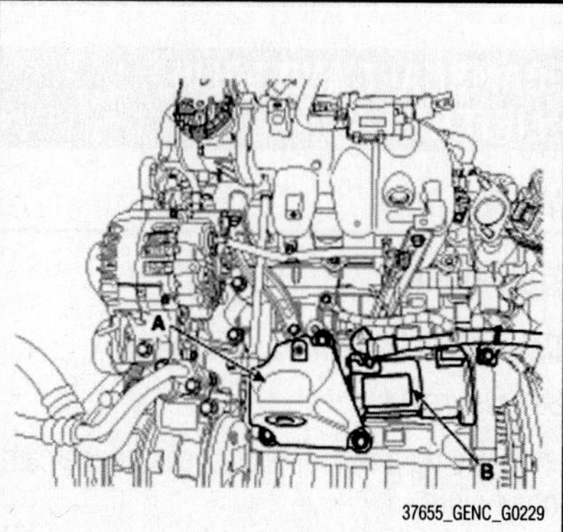

Fig. 73 Remove the engine support bracket (A), and then remove cable and connector from starter (B)

1. Before servicing the vehicle, refer to the Precautions Section.
2. Disconnect the negative battery cable.
3. Install a jack under the engine oil pan. Insert the rubber block between engine oil pan and jack to prevent the damage of oil pan.
4. Remove the left-hand engine mounting bracket nut.
5. Lift up the engine assembly slightly by using a jack to get access to the side of engine.
6. Remove the engine support bracket, and then remove cable and connector from starter.
7. Remove the bolts and starter.

To install:

8. Installation is the reverse of removal.
    a. Tighten the mounting bolts to 31–40 ft. lbs. (42–54 Nm).
    b. Tighten the engine support bracket bolts to 36–47 ft. lbs. (49–64 Nm).
    c. Tighten the left-hand engine mounting bracket nut to 47–62 ft. lbs. (64–83 Nm).

3.8L Engines

See Figures 75 through 78.

1. Before servicing the vehicle, refer to the Precautions Section.
2. Disconnect the negative battery cable.
3. Before servicing the vehicle, refer to the Precautions Section.
4. Disconnect the negative battery cable.
5. Remove the engine cover, as applicable.
6. Remove the air duct.
7. Remove the AFS connector and air cleaner assembly.
8. Remove the starter cover.
9. Disconnect the starter connector and cable from the 'B' terminal.
10. Install a jack under the engine oil pan. Insert the rubber block between engine oil pan and jack to prevent the damage of oil pan.

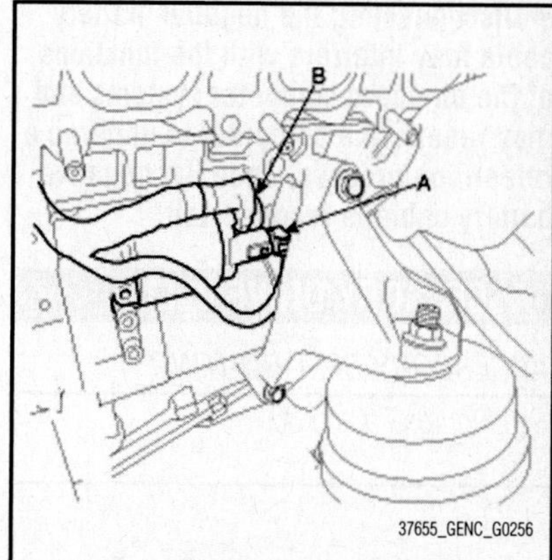

Fig. 76 Disconnect the starter connector (A) and cable (B) from the 'B' terminal

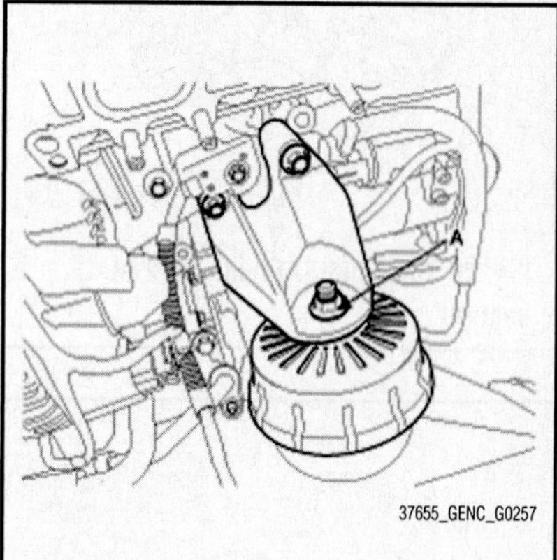

Fig. 77 Remove the engine mounting nut (A)

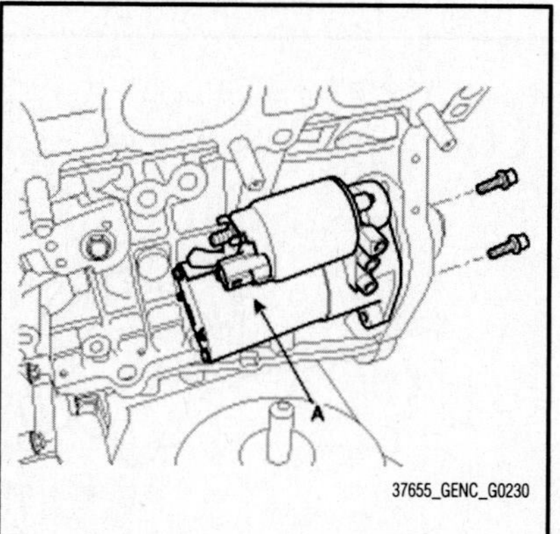

Fig. 74 Remove the starter (A)

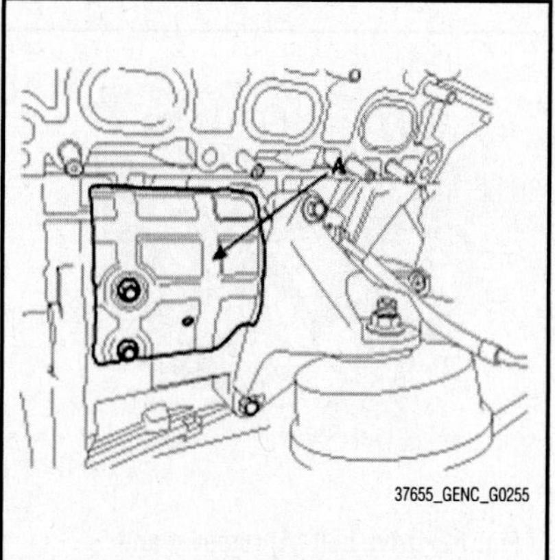

Fig. 75 Remove the starter cover (A)

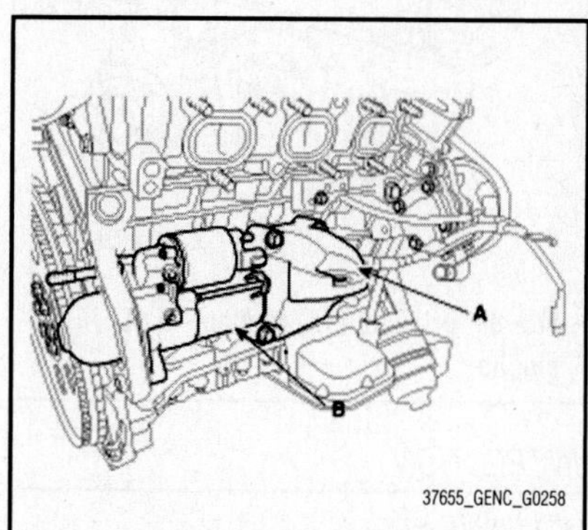

Fig. 78 Remove the right-hand engine support bracket (A) and the starter (B)

11. Remove the engine mounting nut.
12. Lift up the engine assembly slightly by using a jack to access the side of the engine.
13. Remove the right-hand engine support bracket.
14. Remove the starter.

To install:

15. Installation is the reverse of removal, noting the following:
    a. Tighten the support bracket bolts to 47–62 ft. lbs. (64–83 Nm).
    b. Tighten the starter mounting bolts to 36–47 ft. lbs. (49–64 Nm).
    c. Tighten the starter nuts to 31–40 ft. lbs. (42–54 Nm).
    d. Tighten the engine mounting nut to 47–62 ft. lbs. (64–83 Nm).
    e. Tighten the starter cover bolts to 78–121 inch lbs. (9–14 Nm).

ENGINE MECHANICAL

➡Disconnecting the negative battery cable may interfere with the functions of the on board computer systems and may require the computer to undergo a relearning process, once the negative battery cable is reconnected.

ACCESSORY DRIVE BELTS

ACCESSORY BELT ROUTING

See Figures 79 and 80.

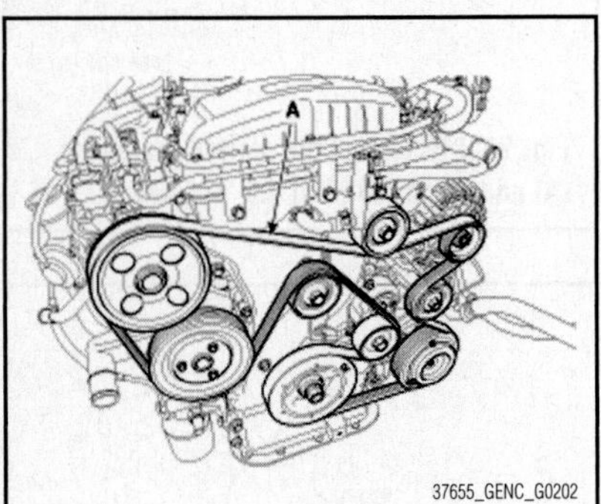
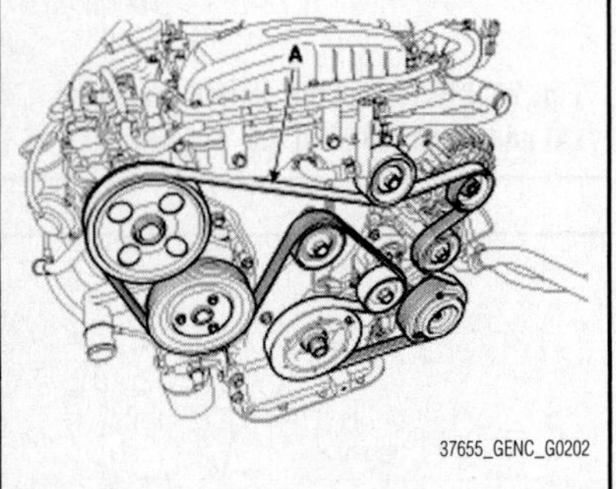

Fig. 79 Drive belt (A) routing—2.0L engine

37655_GENC_G0205

Fig. 80 Drive belt (A) routing—3.8L engine

INSPECTION

See Figure 81.

1. Visually check the belt for excessive wear, frayed cords etc. Cracks on the rib side of a belt are considered acceptable. If the belt has chunks missing from the ribs, it should be replaced.
2. If any defect has been found, replace the drive belt.

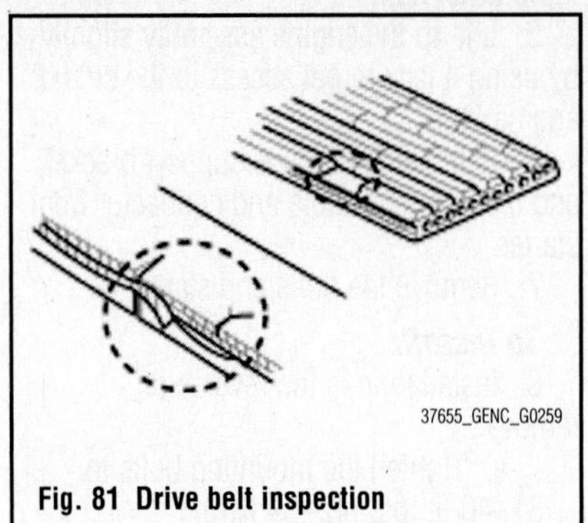

Fig. 81 Drive belt inspection

ADJUSTMENT

1. No adjustment is possible or necessary.

REMOVAL & INSTALLATION

See Figures 82 and 83.

1. Before servicing the vehicle, refer to the Precautions Section.
2. Disconnect the negative battery cable.
3. Turn the tensioner clockwise and loosen, and then remove the drive belt.

To install:

4. Installation is the reverse of removal.

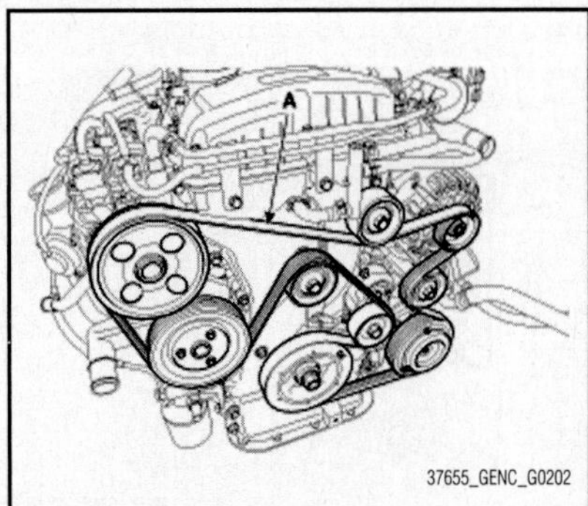

Fig. 82 Drive belt (A) removal and installation—2.0L engine

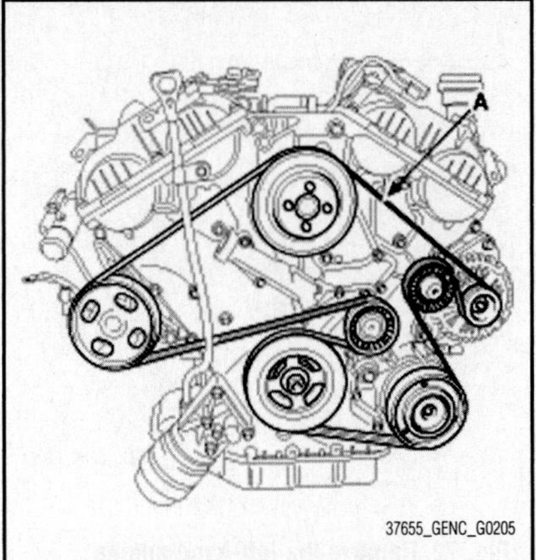

Fig. 83 Drive belt (A) removal and installation—3.8L engine

CAMSHAFT AND VALVE LIFTERS

INSPECTION

2.0L Engine

Camshaft Lobe Measurement

See Figure 84.

1. Using a micrometer, measure the cam lobe height.
    a. Intake height specification: 1.720–1.78 in. (43.70–43.90mm)
    b. Exhaust height specification: 1.768–1.776 in. (44.90–45.10mm)
2. If the cam lobe height is less than standard, replace the camshaft.

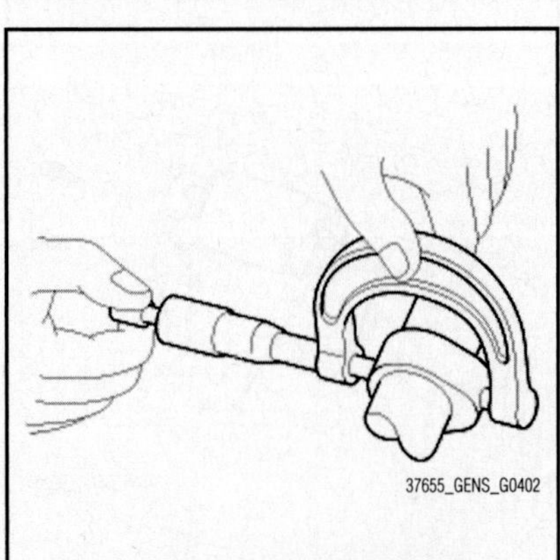

Fig. 84 Camshaft lobe measurement

Camshaft End Play

See Figure 85.

1. Install the camshafts.
2. Install the bearing cap and thrust bearing cap with specified torque.
3. Using a dial indicator, measure the end play while moving the camshaft back and forth.
    a. End play specification: 0.0015—0.0062 in. (0.04—0.16 mm)
4. If the end play is greater than maximum, replace the camshaft. If necessary, replace cylinder head.

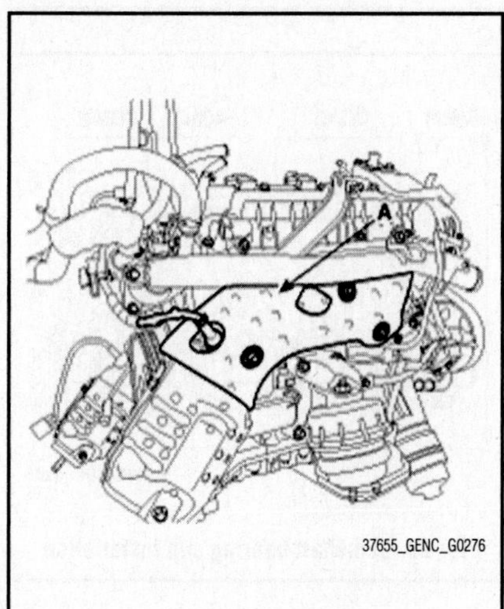

Fig. 85 Camshaft end play measurement

3.8L Engine

CAMSHAFT LOBE MEASUREMENT

See Figure 84.

1. Using a micrometer, measure the cam lobe height.
    a. Intake height specification: 1.8425 in. (46.8mm)
    b. Exhaust height specification: 1.8031 in. (45.8mm)
2. If the cam lobe height is less than standard, replace the camshaft.

CAMSHAFT END PLAY

See Figure 86.

1. Install the camshafts.
2. Install the bearing cap and thrust bearing cap with specified torque.
3. Using a dial indicator, measure the end play while moving the camshaft back and forth.
    a. End play specification: 0.0008—0.0071 in. (0.02—0.18 mm)
4. If the end play is greater than maximum, replace the camshaft. If necessary, replace cylinder head.

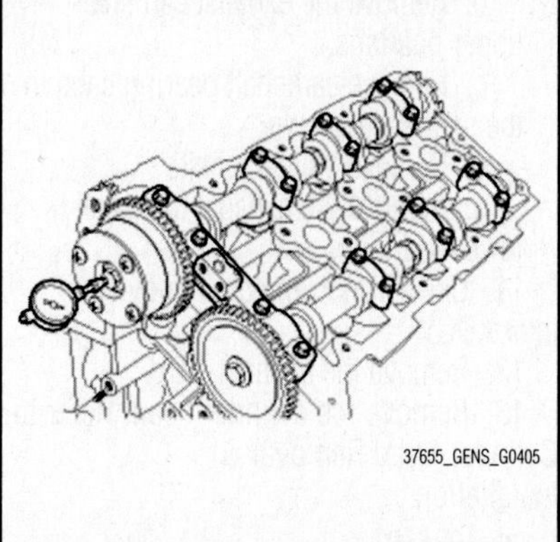

Fig. 86 Camshaft end play measurement

REMOVAL & INSTALLATION

2.0L Engine

See Figures 87 through 95.

WARNING

Use fender covers to avoid damaging painted surfaces. To avoid damaging the cylinder head, wait until the engine coolant temperature drops below normal temperature (68°F [20°C]) before removing it. When handling a metal gasket, take care not to fold the gasket or damage the contact surface of the gasket. To avoid damage, unplug the wiring connectors carefully while holding the connector portion. Mark all wiring and hoses to avoid misconnection. Turn the crankshaft pulley so that the No. 1 piston is at top dead center.

1. Before servicing the vehicle, refer to the Precautions Section.
2. Disconnect the negative battery cable.

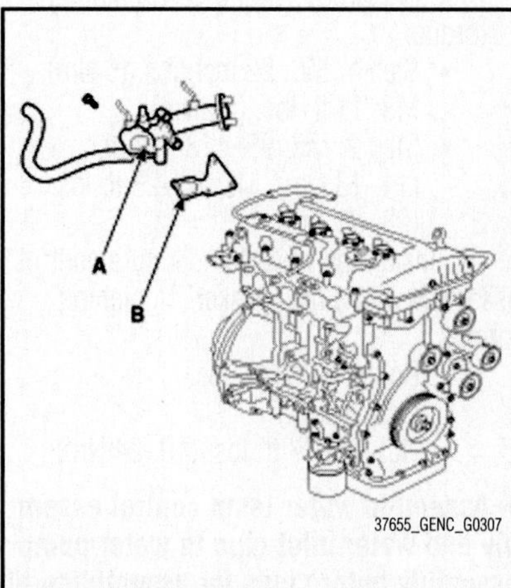

Fig. 87 Remove the water temperature control assembly (A) and gasket (B)

Fig. 88 Remove the front camshaft bearing cap (A)

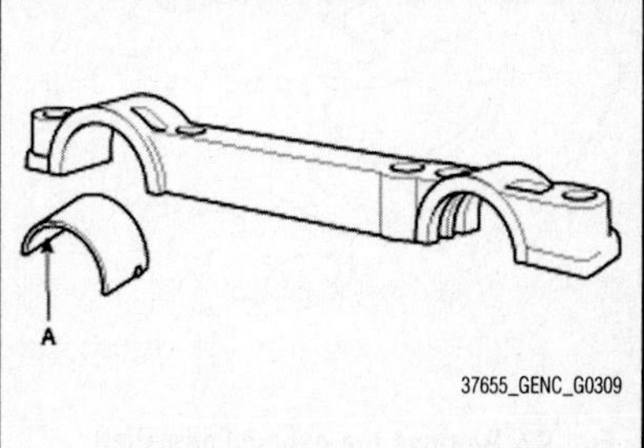

Fig. 89 Remove the exhaust camshaft upper bearing (A)

3. Remove the heater hoses.
4. Remove the intake and exhaust manifolds. Refer to Intake Manifold Removal & Installation and Exhaust Manifold Removal & Installation.
5. Drain the engine coolant.
6. Remove the water pump and gasket. Refer to Water Pump Removal & Installation in the Engine Cooling section.

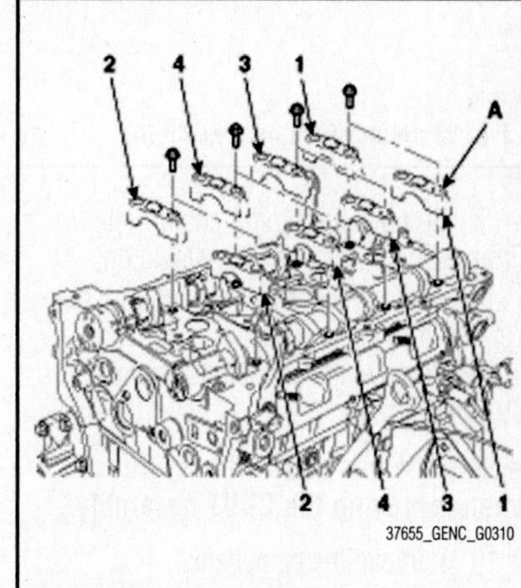

Fig. 90 Camshaft bearing cap (A) removal sequence

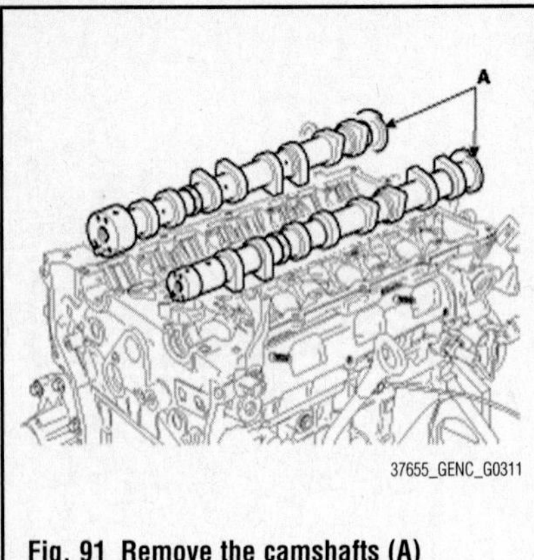

Fig. 91 Remove the camshafts (A)

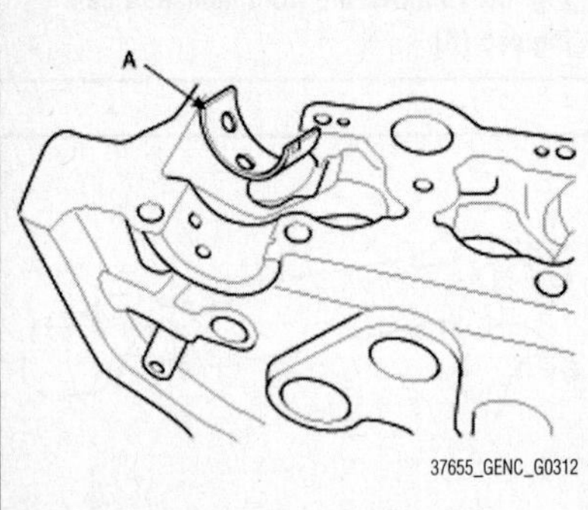

Fig. 92 Remove the exhaust camshaft lower bearings (A)

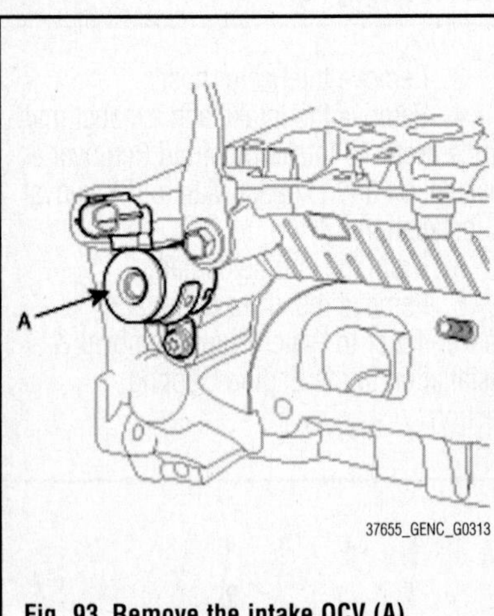

Fig. 93 Remove the intake OCV (A)

7. Remove the timing chain. Refer to Timing Chain Removal & Installation.
8. Remove the water temperature control assembly and gasket.
9. Remove the intake and exhaust CVVT assembly.

➡Hold the camshaft with a wrench when removing the CVVT assembly.

10. Remove the camshafts:
    a. Remove the front camshaft bearing caps.
    b. Remove the exhaust camshaft upper bearings.
    c. Remove camshaft bearing caps in the sequence shown.
    d. Remove the camshafts.
    e. Remove the exhaust camshaft lower bearings.
11. Use a Torx® wrench, remove the intake OCV.
12. Remove the exhaust OCV.
13. Remove the cylinder head. Refer to Cylinder Head Removal & Installation.

To install:

➡Thoroughly clean all parts to be assembled. Turn the crankshaft pulley so that the No. 1 piston is at top dead center. Always use a new head and manifold gasket.

14. Install the OCV filter. Keep the OCV filter clean.
15. Install the cylinder head and gasket.
16. Install the OCV and tighten to 86–104 inch lbs. (10–12 Nm).

➡Do not reuse the OCV when dropped. Keep the OCV filter clean. Do not hold the OCV sleeve during servicing. When the OCV is installed on the engine, do not move the engine while holding the OCV yoke.

17. Install the camshafts:
    a. Apply a light coat of engine oil on camshaft journals.
    b. Install the exhaust camshaft lower bearings.
    c. Install the camshafts.
    d. Install the exhaust camshaft upper bearings.
    e. Install the camshaft bearing caps in their proper locations. Tightening order: Group A, Group, Group C. Tightening torque:
        - Step 1: M6: 52 inch lbs. (6 Nm), M8: 11 ft. lbs. (15 Nm)
        - Step 2: M6: 95–113 inch lbs. (11–13 Nm), M8: 20–23 ft. lbs. (28–31 Nm)
18. Install the water temperature control assembly with a new gasket. Tightening torque:
    - Bolts: 11–14 ft. lbs. (15–20 Nm)
    - Nut: 14–17 ft. lbs. (19–24 Nm)

➡Assemble water temp control assembly and water inlet pipe to water pump assembly before nuts for assembling of water inlet pipe to be tightened. Always use a new O-ring.

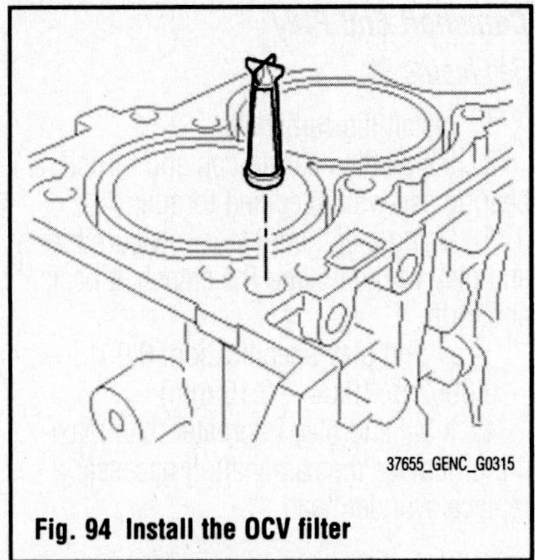

Fig. 94 Install the OCV filter

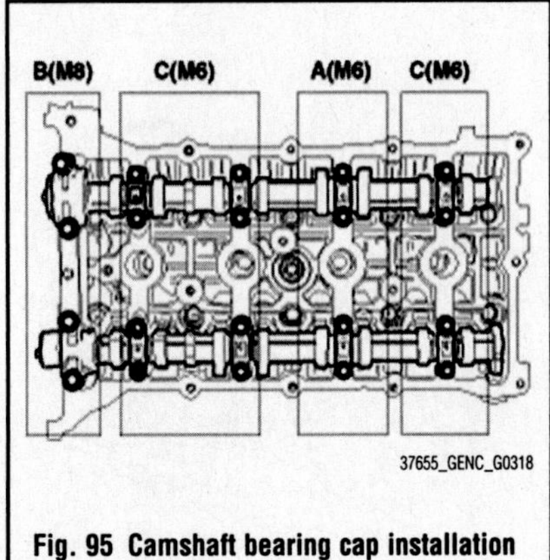

Fig. 95 Camshaft bearing cap installation

19. Install the timing chain.
20. Check and adjust valve clearance.
21. Install the cylinder head cover.
22. Install the intake and exhaust manifolds.
23. Install the heater hoses.
24. Connect the negative battery terminal

3.8L Engine

See Figures 96 through 112.

1. Before servicing the vehicle, refer to the Precautions Section.
2. Disconnect the negative battery cable.

⁂ WARNING

Use fender covers to avoid damaging painted surfaces. To avoid damage, unplug the wiring connectors carefully while holding the connector portion. Mark all wiring and hoses to avoid misconnection.

➡Turn the crankshaft pulley so that the No. 1 piston is at top dead center.

3. Remove the timing chain. Refer to Timing Chain & Sprockets Removal & Installation.

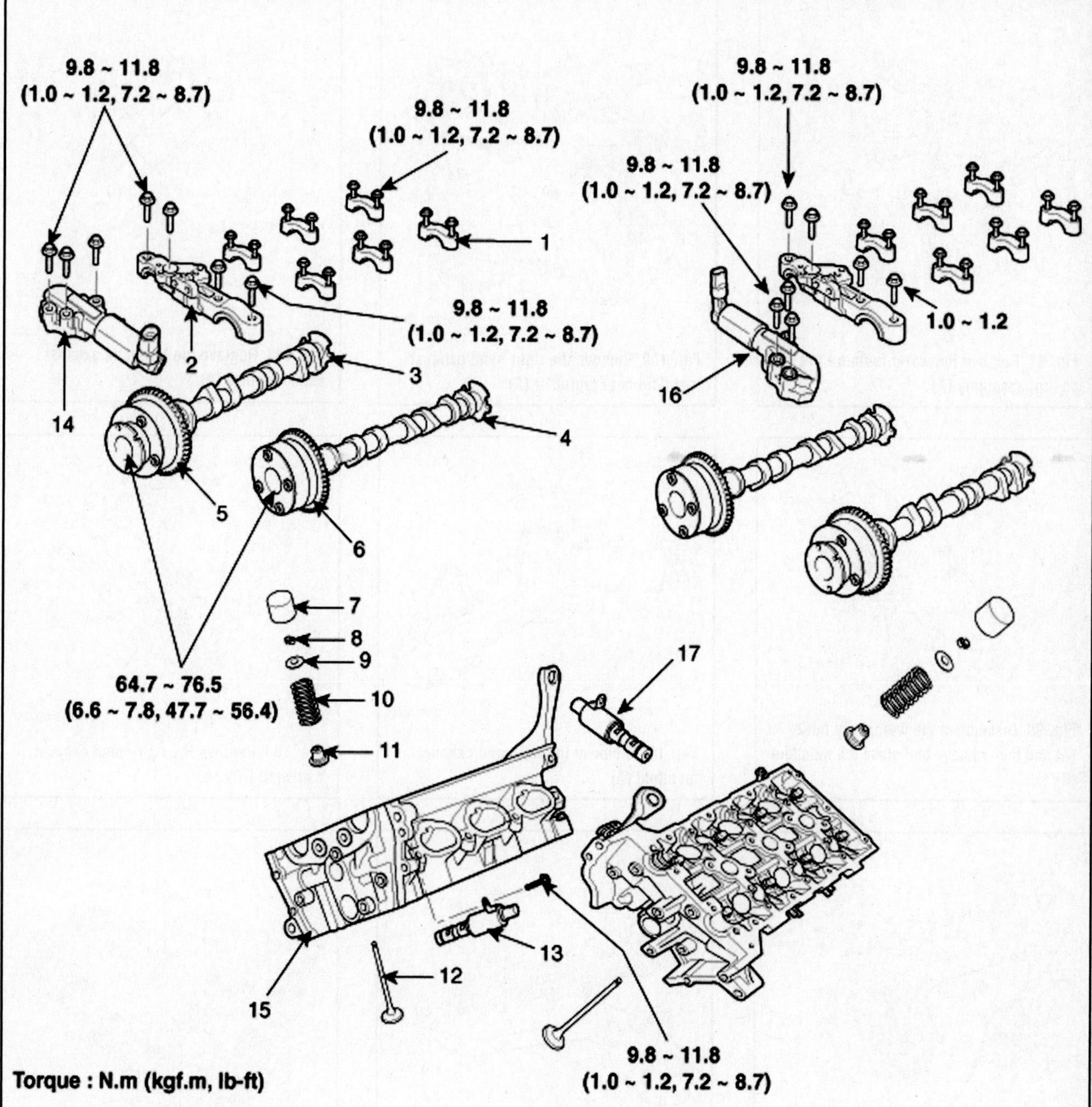

1. Camshaft bearing cap
2. Camshaft thrust bearing cap
3. Exhaust camshaft
4. Intake camshaft
5. Exhaust CVVT assembly
6. Intake CVVT assembly
7. Mechanical lash adjuster
8. Retainer lock
9. Retainer
10. Valve spring
11. Valve stem seal
12. Valve
13. Intake camshaft OCV (Right-hand)
14. Exhaust camshaft OCV (Right-hand)
15. Cylinder head
16. Exhaust camshaft OCV (Left-hand)
17. Intake camshaft OCV (Left-hand)

37655_GENC_G0273

Fig. 96 Camshafts and related components

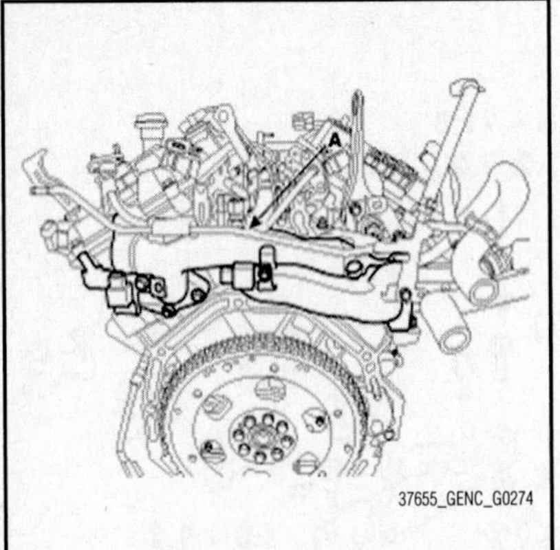

Fig. 97 Remove the water temperature control assembly (A)

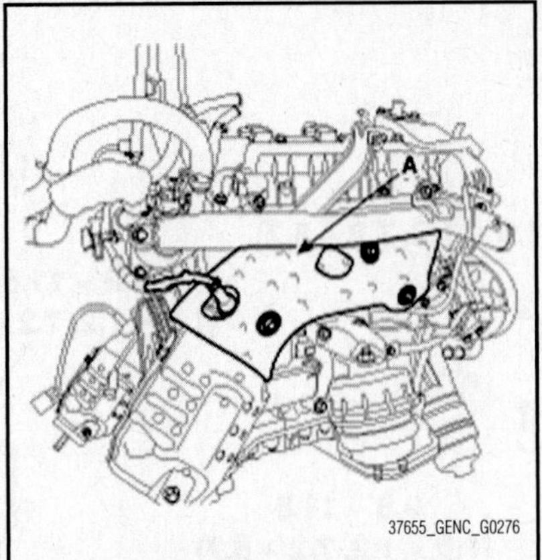

Fig. 100 Remove the right-hand exhaust manifold heat protector (A)

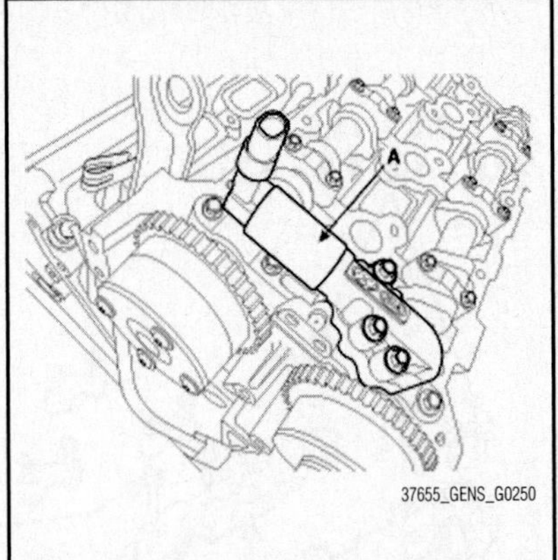

Fig. 103 Remove the left-hand exhaust camshaft OCV (A)

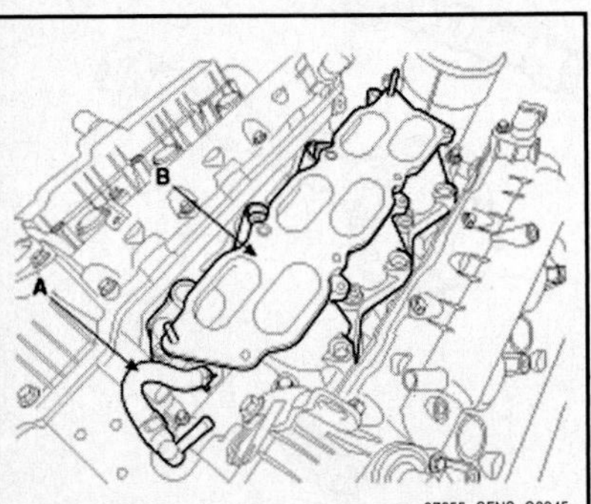

Fig. 98 Disconnect the water vent hose (A) and then remove the intake the manifold (B)

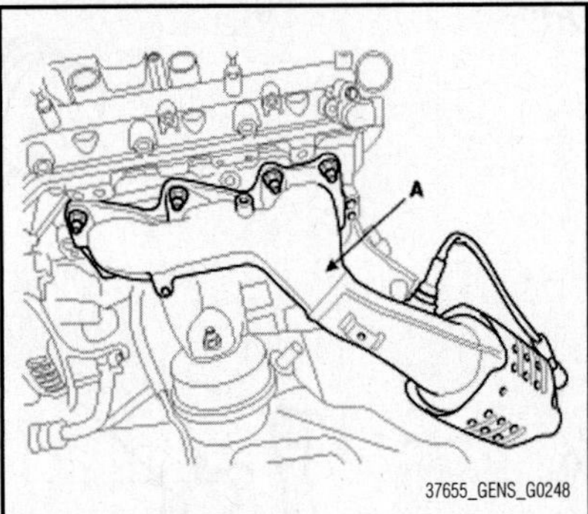

Fig. 101 Remove the left-hand exhaust manifold (A)

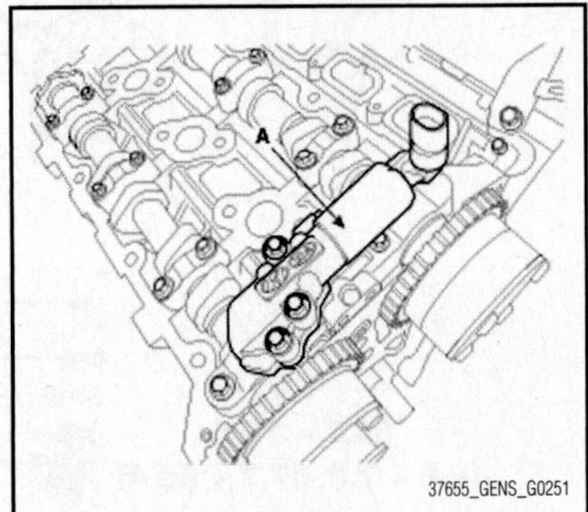

Fig. 104 Remove the right-hand exhaust camshaft OCV (A)

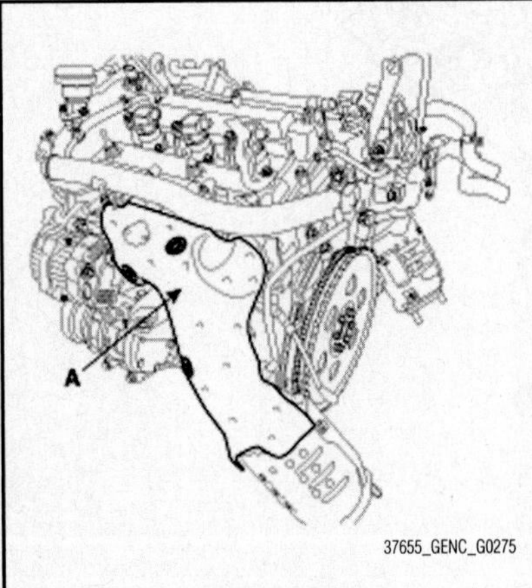
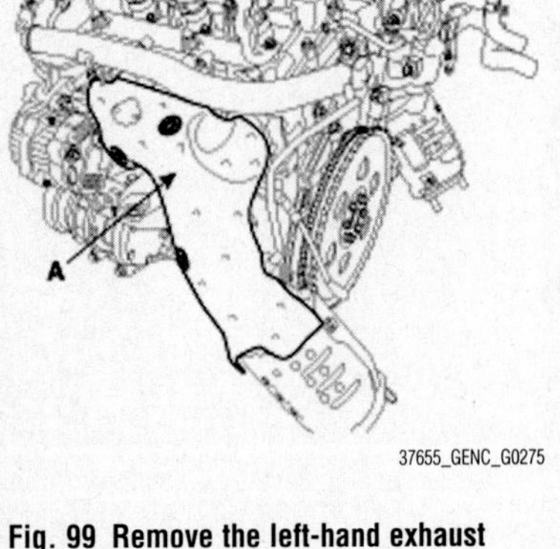

Fig. 99 Remove the left-hand exhaust manifold heat protector (A)

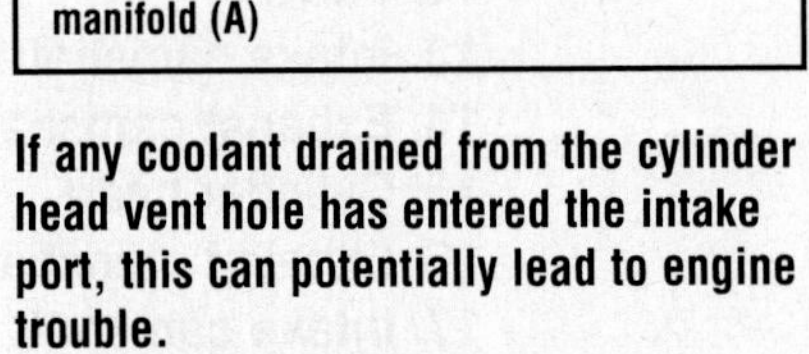

Fig. 102 Remove the right-hand exhaust manifold (A)

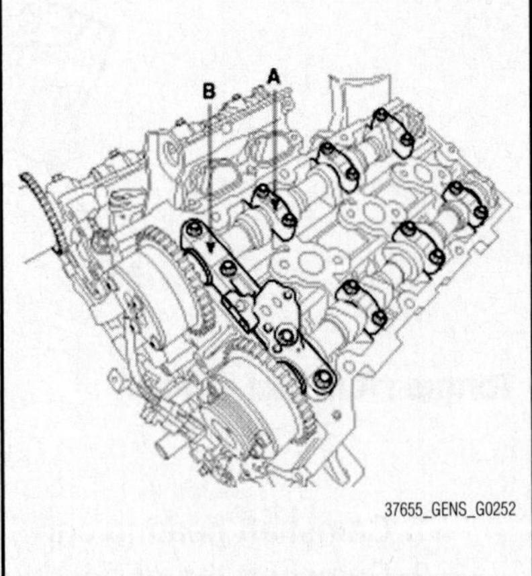

Fig. 105 Remove the left-hand camshaft bearing cap (A) and thrust bearing cap (B)

4. Drain the engine coolant.
5. Disconnect the heater hoses.
6. Remove the water temperature control assembly.
7. Disconnect the water vent hose and then remove the intake the manifold. Refer to Intake Manifold Removal & Installation.

➡**Be sure to drain the engine coolant before removing the intake manifold. If any coolant drained from the cylinder head vent hole has entered the intake port, this can potentially lead to engine trouble.**

8. Remove the left-hand and right-hand exhaust manifold heat protectors. Refer to Exhaust Manifold Removal & Installation.
9. Remove the left-hand and right-hand exhaust manifolds. Refer to Exhaust Manifold Removal & Installation.
10. Remove the left-hand and right-hand exhaust camshaft OCV.
11. Remove the left-hand and right-hand camshaft bearing caps and thrust bearing caps.
12. Remove the left-hand and right-hand camshaft assemblies.

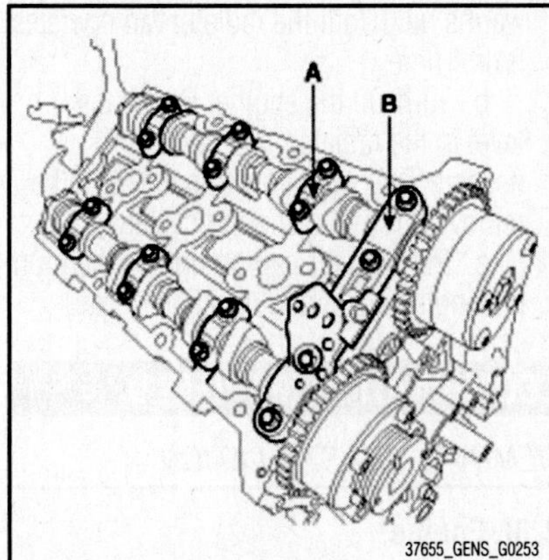

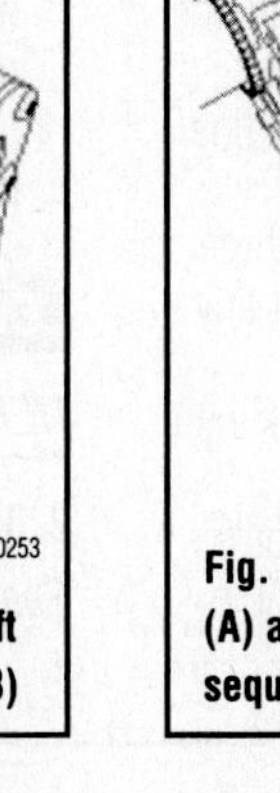

37655_GENS_G0253

Fig. 106 Remove the right-hand camshaft bearing cap (A) and thrust bearing cap (B)

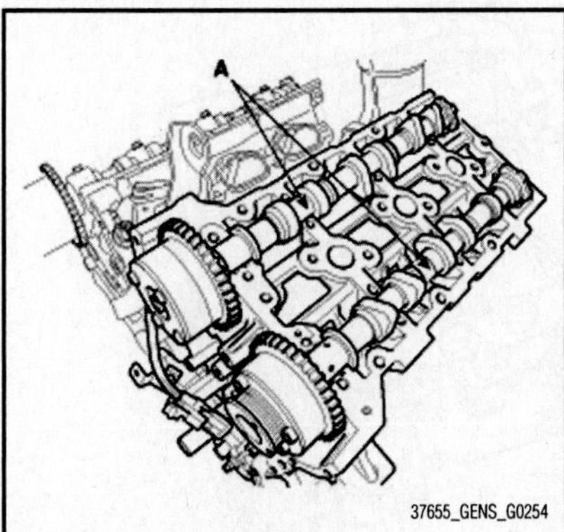

37655_GENS_G0254

Fig. 107 Remove the left-hand camshaft assembly (A)

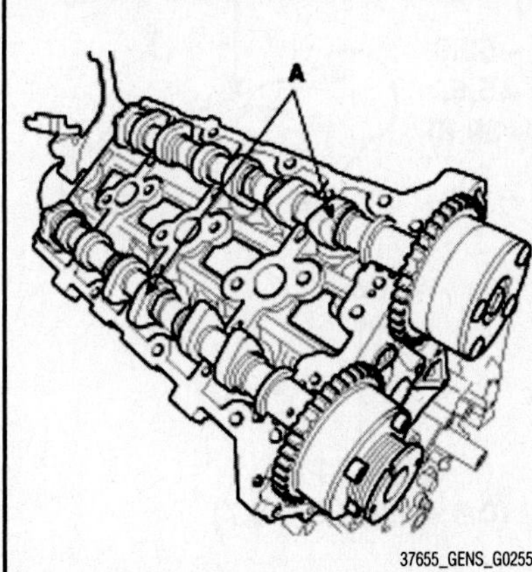

37655_GENS_G0255

Fig. 108 Remove the right-hand camshaft assembly (A)

To install:

➡Thoroughly clean all parts to be assembled. Turn the crankshaft pulley so that the No. 1 piston is at top dead center.

13. Install the left-hand and right-hand camshaft assemblies.

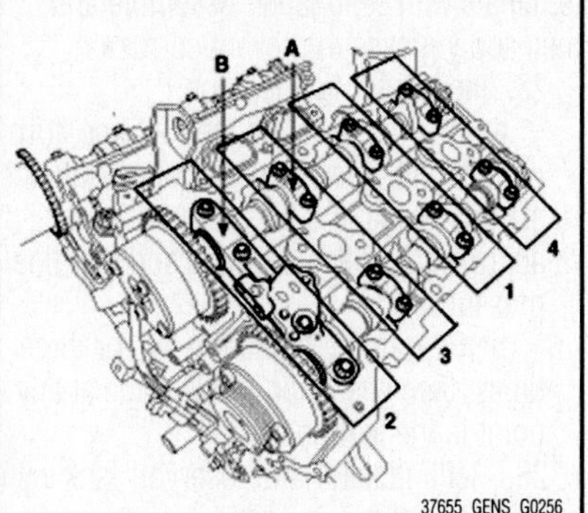

37655_GENS_G0256

Fig. 109 Left-hand camshaft bearing cap (A) and thrust bearing cap (B) installation sequence

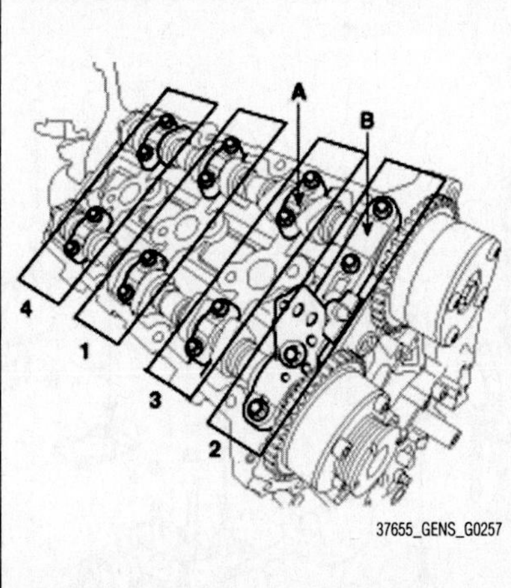

37655_GENS_G0257

Fig. 110 Right-hand camshaft bearing cap (A) and thrust bearing cap (B) installation sequence

➡Apply a light coat of engine oil on camshaft journals. Assemble the key groove of camshaft rear side to the same level of head top surface. Be careful the right, left bank, intake, exhaust side before assembling.

14. Install the left-hand and right-hand camshaft bearing caps and thrust bearing caps. Tightening torque:

- First Step: 52 inch lbs. (6 Nm)
- Second Step: 86–104 inch lbs. (10–12 Nm)

WARNING

Be sure to install the thrust bearing cap bolts and the bearing cap bolts in the correct place.

WARNING

Be careful of the right, left bank, intake, exhaust side before assembling.

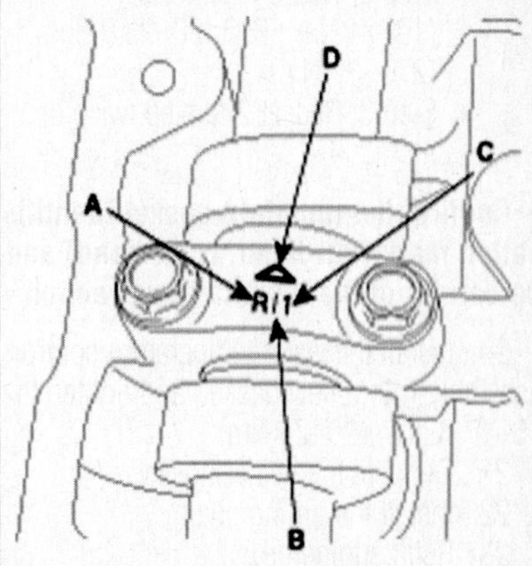

A: L (Left-hand), R(Right-hand)
B: I (Intake), None or E (Exhaust)
C: Journal number
D: Front mark

37655_GENS_G0258

Fig. 111 Position the camshaft bearing cap according to its markings

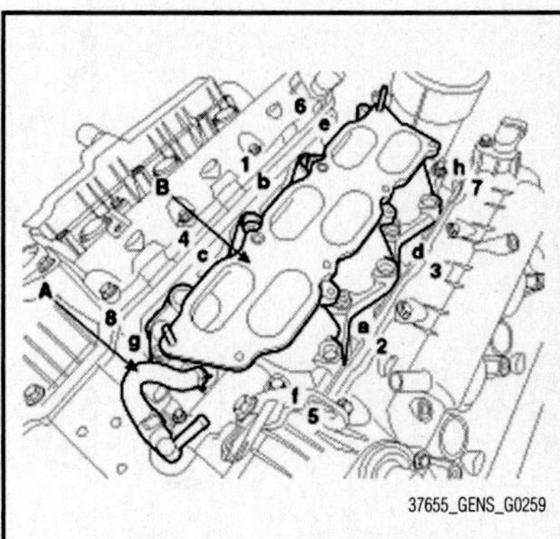

37655_GENS_G0259

Fig. 112 a–h: Step 1 order, 1–8: Step 2 order

WARNING

Rotate the crankshaft so as not to contact the valves with the pistons by making the pistons below 0.3937 in. (10mm) from the top of cylinder block.

15. Install the left-hand and right-hand exhaust camshaft OCV and tighten to 86–104 inch lbs. (10–12 Nm).

16. Install the cylinder head covers.

17. Install the left-hand and right-hand exhaust manifolds with new gaskets, and tighten to 29–33 ft. lbs. (39–44 Nm).

18. Install the left-hand and right-hand exhaust manifold heat protectors and tighten to 86–104 inch lbs. (10–12 Nm).

19. Install the intake the manifold with a new gasket, and connect the water vent hose. Tightening torque:

- Step 1: 35–52 inch lbs. (4–6 Nm)

- Step 2: Nut: 14–17 ft. lbs. (19–24 Nm), Bolt: 20–23 ft. lbs. (27—31 Nm)
- Step 3: Repeat 2nd step twice or more.

➡**Confirm the manifold gasket identification mark (left-hand, right-hand) and be careful of the installation direction.**

20. Install the water temperature control assembly with a new gasket, and tighten to 15–17 ft. lbs. (20–24 Nm).
21. Connect the heater hoses.
22. Install the timing chain.
23. Refill engine oil.
24. Clean the battery posts and cable terminals with sandpaper. Assemble and then apply grease to prevent corrosion.
25. Inspect for fuel leakage:
    a. After assembling the fuel line, turn on the ignition switch (do not operate the starter) so that the fuel pump runs for approximately two seconds and fuel line pressurizes.
    b. Repeat this operation two or three times, then check for fuel leakage at any point in the fuel lines.
26. Refill radiator and reservoir tank with engine coolant.
27. Bleed air from the cooling system:
    a. Start engine and let it run until it warms up. (Until the radiator fan operates 3 or 4 times.)
    b. Turn Off the engine. Check the level in the radiator, add coolant if needed. This will allow trapped air to be removed from the cooling system.
    c. Put radiator cap on tightly, then run the engine again and check for leaks.

COMBINATION MANIFOLD

REMOVAL & INSTALLATION

2.0L Engine

See Figures 113 through 121.

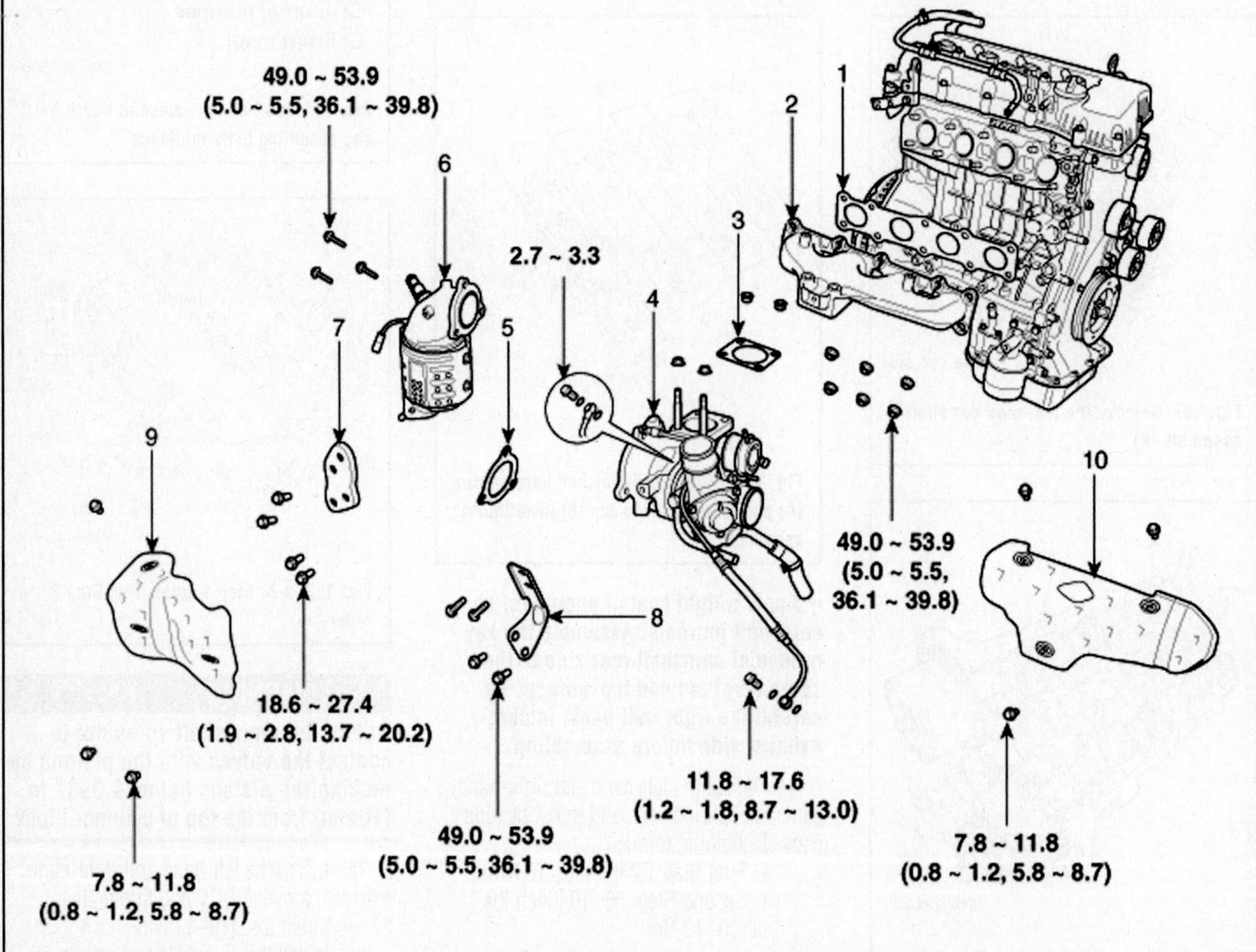

Fig. 113 Combination manifold and turbocharger and components

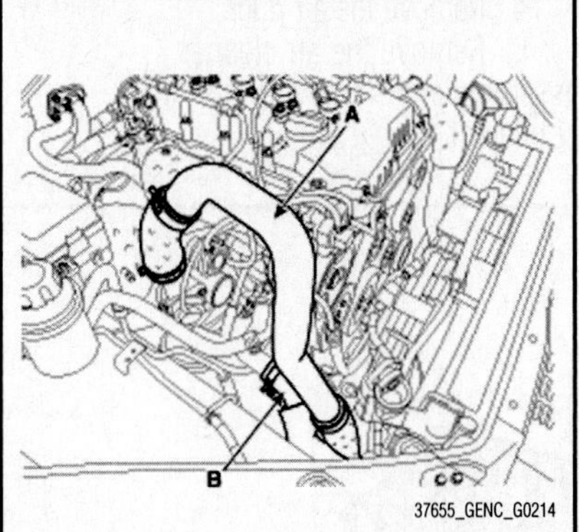

Fig. 114 Disconnect the intercooler inlet hose (A)

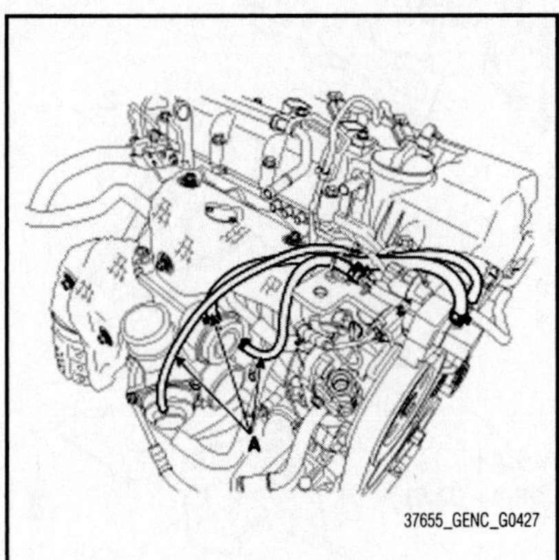

Fig. 115 Remove the turbo control hoses (A)

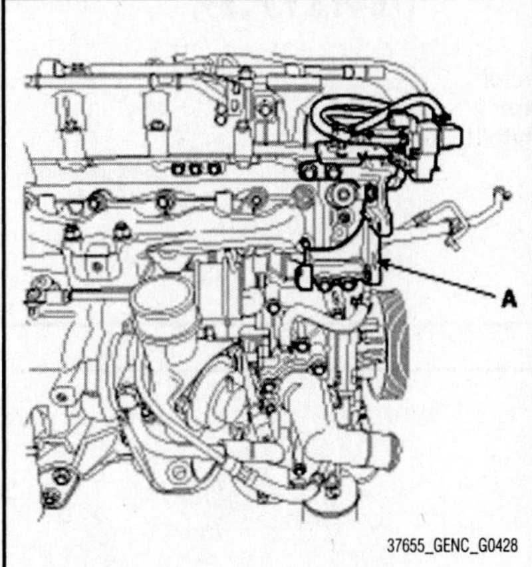

Fig. 116 Remove the steering pump bracket (A)

1. Before servicing the vehicle, refer to the Precautions Section.

➡Check that engine is cool enough to work.

2. Disconnect the negative battery cable.

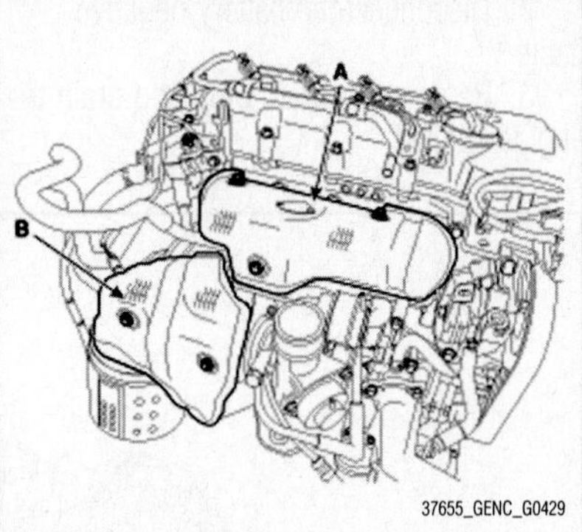

Fig. 117 Remove the exhaust manifold heat protector (A) and turbocharger heat protector (B)

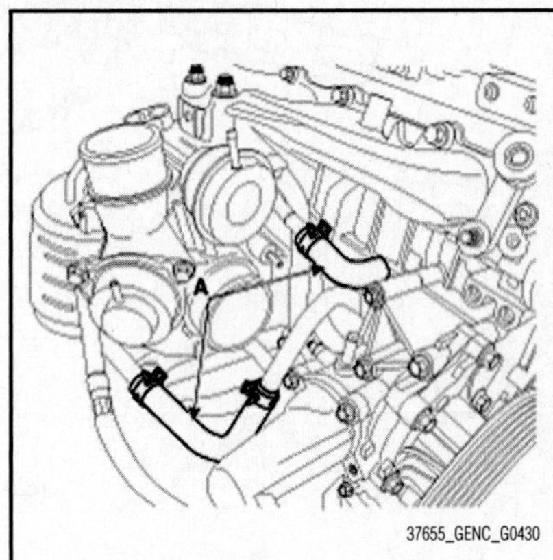

Fig. 118 Remove the coolant hoses (A) from turbocharger

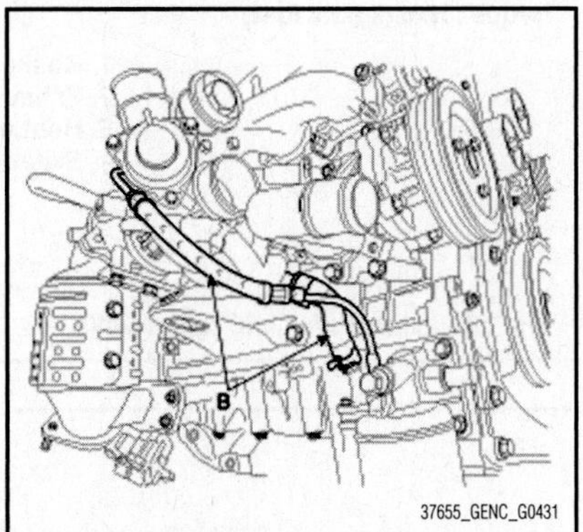

Fig. 119 Remove the oil hoses (B) from turbocharger

3. Drain the engine coolant.
4. Remove the front muffler.
5. Disconnect the breather hose and vacuum hose.
6. Remove the air duct and air cleaner assembly.
7. Disconnect the intercooler inlet hose.
8. Remove the drive belt. Refer to Accessory Drive Belt Removal & Installation.

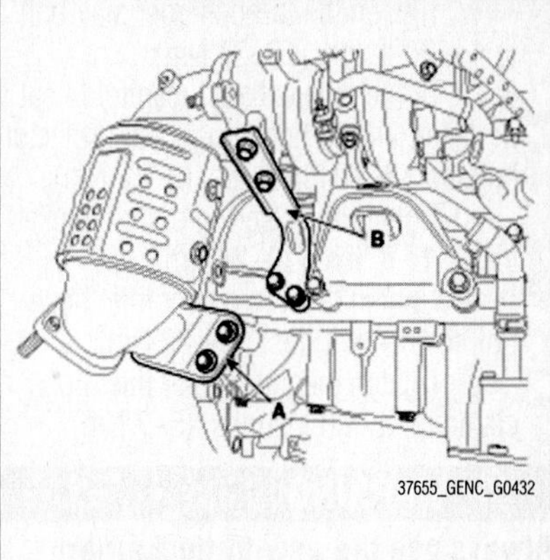

Fig. 120 Remove the turbocharger stay bolt (A) and WCC stay bolt (B)

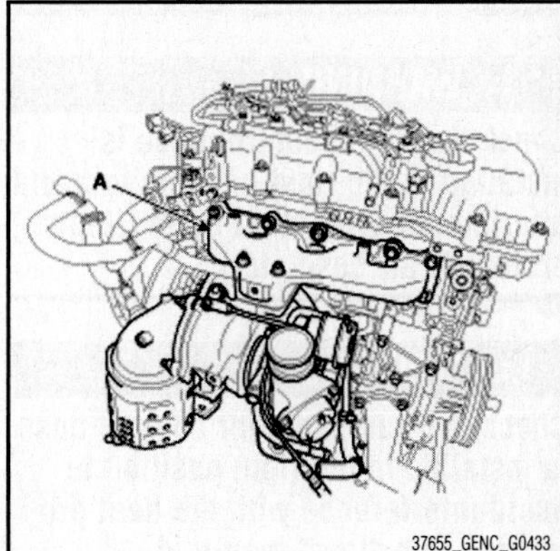

Fig. 121 Remove the exhaust manifold and turbocharger assembly (A) and gasket

9. Remove the turbo control hoses.
10. Remove the power steering pump.
11. Remove the steering pump bracket.
12. Remove the exhaust manifold heat protector and turbocharger heat protector.

➡Check that the engine is cool enough to work.

13. Remove the oxygen sensor connectors.
14. Remove the coolant hoses and oil hoses from turbocharger.
15. Remove the turbocharger stay bolt and WCC stay bolt.
16. Remove the exhaust manifold and turbocharger assembly and gasket.

To install:

17. Installation is the reverse of removal, noting the following:
    a. Tighten the exhaust manifold and turbocharger assembly bolts to 36–40 ft. lbs. (49–54 Nm).
    b. Tighten the WCC stay bolt to 36–40 ft. lbs. (49–54 Nm).

c. Tighten the turbocharger stay bolt to 14–20 ft. lbs. (19–27 Nm).

d. Tighten the exhaust manifold heat protector and turbocharger heat protector bolts to 70–104 inch lbs. (8–12 Nm).

e. Tighten the steering pump bracket to 13–17 ft. lbs. (20–24 Nm).

f. Tighten the intercooler inlet hose bolt to 11–14 ft. lbs. (15–20 Nm).

g. Tighten the intercooler inlet hose clamp to 43–60 inch lbs. (5–7 Nm).

✲✲ WARNING

Always use the new turbocharger, exhaust manifold and WCC gaskets when replacing. Always use a new turbocharger nuts and, exhaust manifold nuts when it is removed.

✲✲ WARNING

Check that turbo control hose is installed in the right position to avoid interference with other parts (Heat protector, air hose etc.).

✲✲ WARNING

Check that turbocharger coolant hose is installed in the right position to avoid interference with the heat protector and exhaust manifold.

3.8L Engine

See Figures 122 through 129.

1. Before servicing the vehicle, refer to the Precautions Section.

✲✲ WARNING

To avoid damage, unplug the wiring connectors carefully while holding the connector portion. Mark all wiring and hoses to avoid misconnection.

2. Disconnect the battery negative cable.

3. Remove the drain plug and drain the engine coolant.

4. Remove the air duct.

5. Remove the air cleaner assembly after removing the AFS connector.

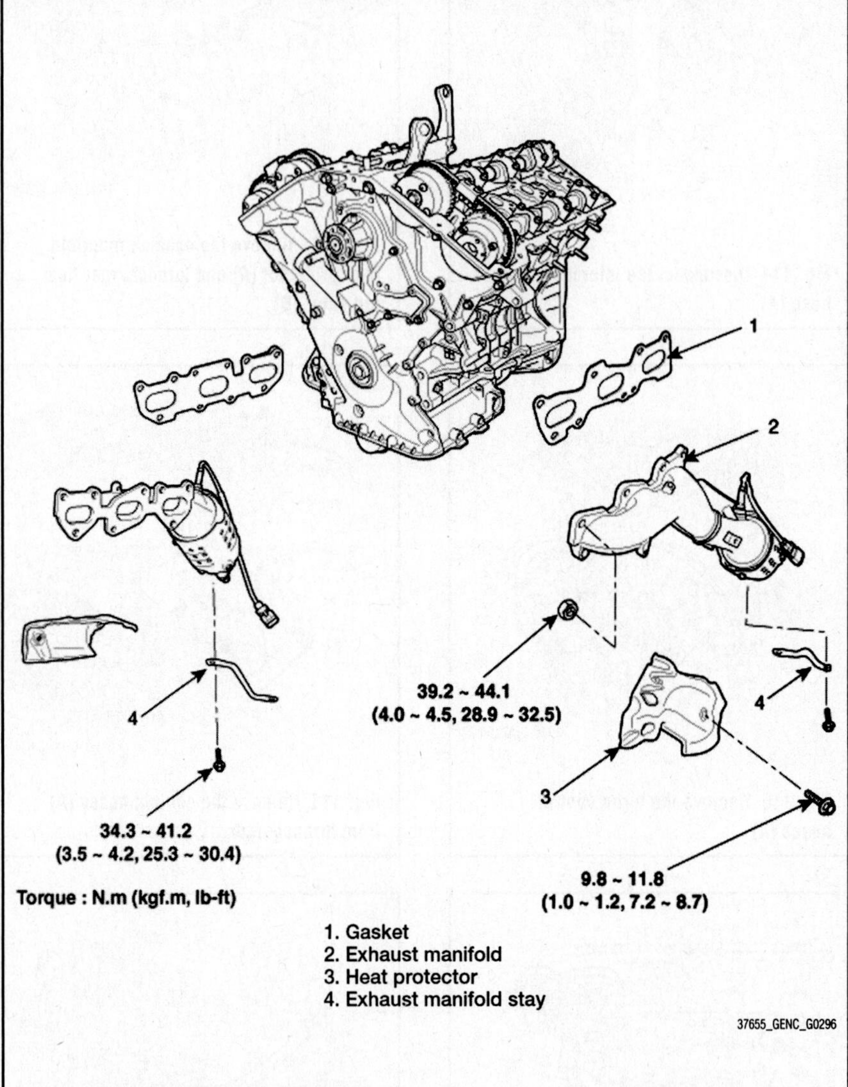

Fig. 123 Exhaust manifold components

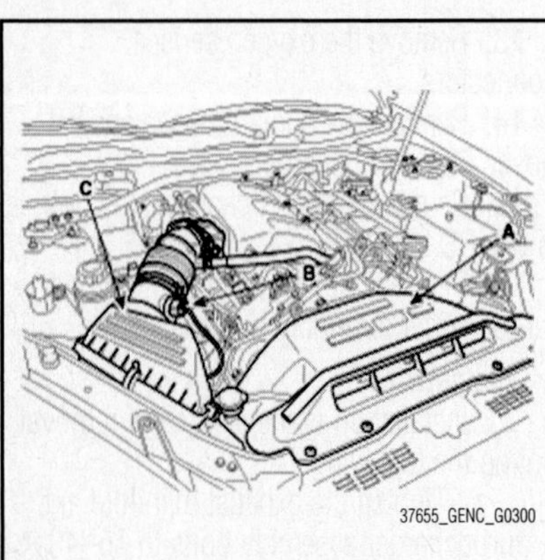

Fig. 122 Remove the AFS connector (B) and the air cleaner assembly (C)

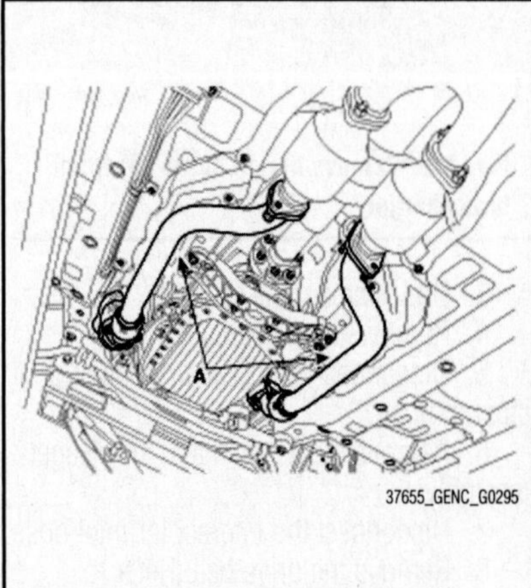

Fig. 124 Remove the front muffler (A)

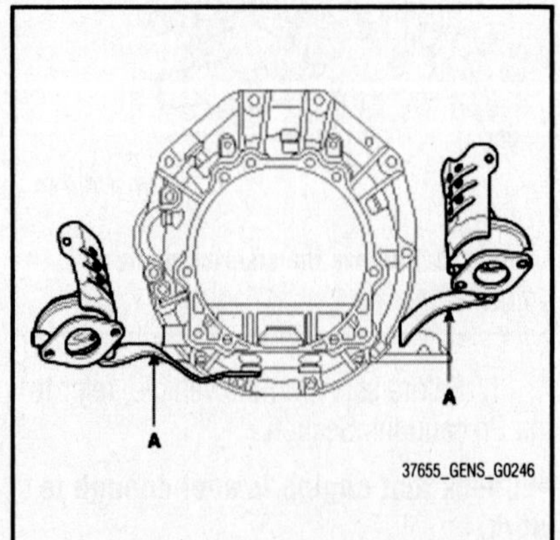

Fig. 125 Remove the exhaust manifold stays (A)

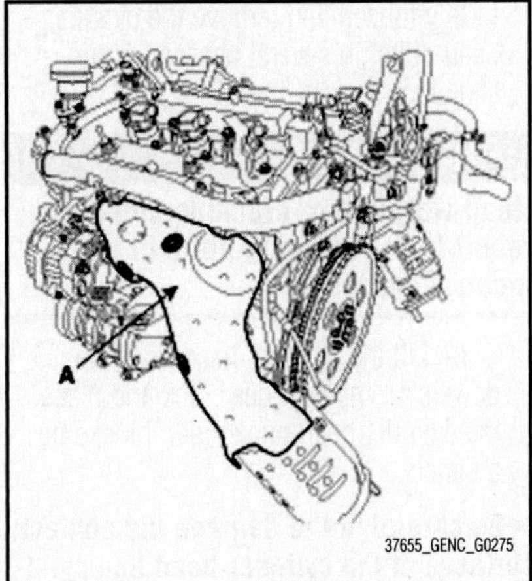

Fig. 126 Remove the left-hand exhaust manifold heat protector (A)

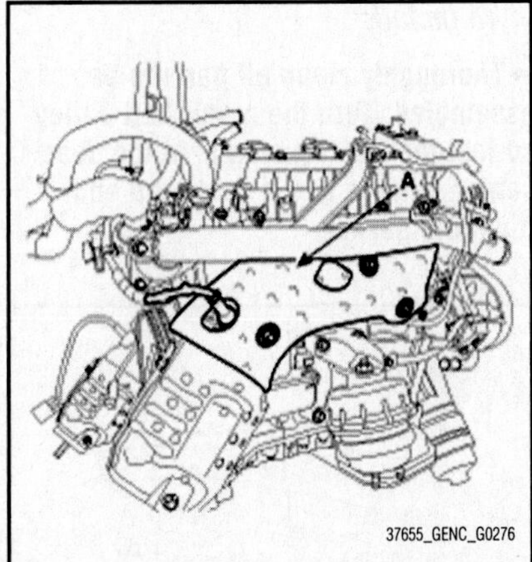

Fig. 127 Remove the right-hand exhaust manifold heat protector (A)

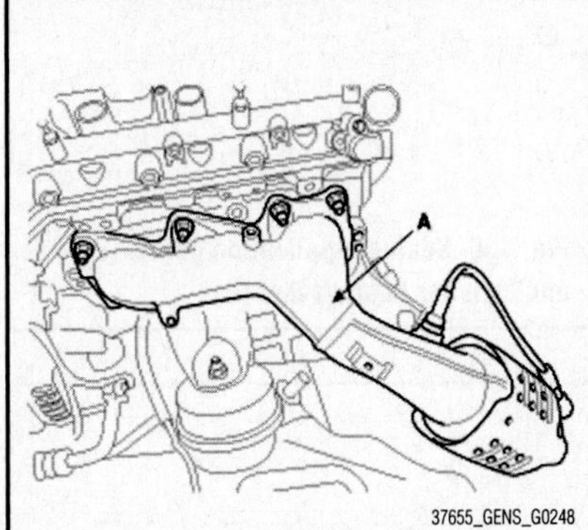

Fig. 128 Remove the left-hand exhaust manifold (A)

6. Remove the upper and lower radiator hoses.
7. Remove the front muffler.
8. Remove the exhaust manifold stays.
9. Remove the left-hand side coolant pipe and hose and then remove the exhaust manifold heat protector.

Fig. 129 Remove the right-hand exhaust manifold (A)

10. Remove the right-hand side coolant pipe and hose and then remove the exhaust manifold heat protector.
11. Remove the left-hand/right-hand exhaust manifolds.

To install:

12. Install the left-hand/right-hand exhaust manifolds and tighten to 29–33 ft. lbs. (39–44 Nm).
13. Install the right-hand exhaust manifold heat protector and tighten to 86–104 inch lbs. (10–12 Nm)
14. Install the right-hand coolant pipe and hose and tighten to 15–17 ft. lbs. (20–24 Nm).
15. Install the left-hand exhaust manifold heat protector and tighten to 86–104 inch lbs. (10–12 Nm)
16. Install the left-hand coolant pipe and hose and tighten to 15–17 ft. lbs. (20–24 Nm).
17. Install the front muffler and tighten to 29–43 ft. lbs. (39–59 Nm).
18. Install the exhaust manifold stays and tighten to 25–30 ft. lbs. (34–41 Nm).
19. Install the radiator hoses.
20. Install the air cleaner assembly and then connect the AFS connector.
21. Install the air duct.
22. Connect the negative battery cable.
23. Refill engine coolant.

CRANKSHAFT FRONT SEAL

REMOVAL & INSTALLATION

3.8L Engine

See Figures 130 and 131.

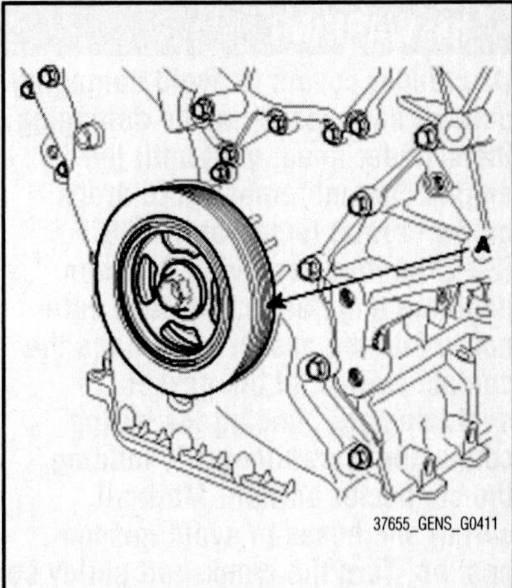

Fig. 130 Crankshaft pulley removal—3.8L engine

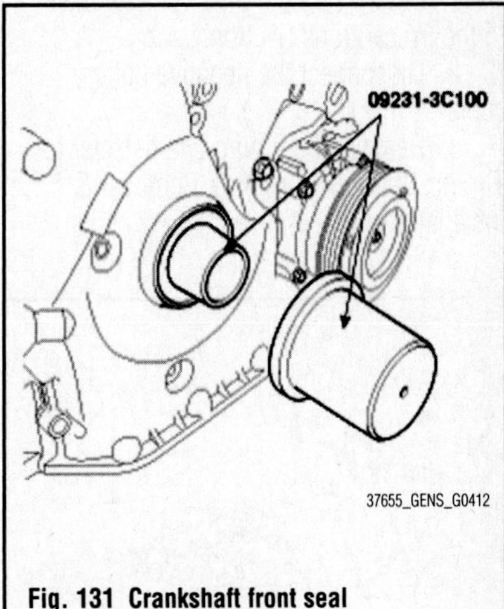

Fig. 131 Crankshaft front seal installation—3.8L engine

1. Before servicing the vehicle, refer to the Precautions Section.
2. Remove the accessory drive belts. Refer to Accessory Drive Belt Removal & Installation.
3. Remove the crankshaft pulley bolt and pulley.
4. Using seal removal tool, remove the front crankshaft seal.

To install:

5. Using seal installation tool, install the front crankshaft seal.
6. Installation is reverse of removal.
   a. Tighten crankshaft bolt to 210—224 ft. lbs. (284—304 Nm).

CYLINDER HEAD

REMOVAL & INSTALLATION

2.0L Engine

See Figures 132 through 137.

WARNING

Use fender covers to avoid damaging painted surfaces. To avoid damaging the cylinder head, wait until the engine coolant temperature drops below normal temperature (68°F [20°C]) before removing it. When handling a metal gasket, take care not to fold the gasket or damage the contact surface of the gasket. To avoid damage, unplug the wiring connectors carefully while holding the connector portion. Mark all wiring and hoses to avoid misconnection. Turn the crankshaft pulley so that the No. 1 piston is at top dead center.

1. Before servicing the vehicle, refer to the Precautions Section.
2. Disconnect the negative battery cable.
3. Remove the timing chain. Refer to Timing Chain & Sprockets Removal & Installation.
4. Remove the heater hoses.
5. Remove the intake and exhaust manifolds. Refer to Intake Manifold Removal & Installation and Exhaust Manifold Removal & Installation.
6. Drain the engine coolant.
7. Remove the water pump and gasket. Refer to Water Pump Removal & Installation in the Engine Cooling section.
8. Remove the water temperature control assembly and gasket.
9. Remove the intake and exhaust CVVT assembly.

➡Hold the camshaft with a wrench when removing the CVVT assembly.

10. Remove the camshafts. Refer to Camshafts & Valve Lifters Removal & Installation.
11. Use a Torx® wrench, remove the intake OCV.
12. Remove the exhaust OCV.
13. Remove the cylinder head:
    a. Using triple square wrench, uniformly loosen and remove the cylinder head bolts, in several passes, in the sequence shown.

WARNING

Head warpage or cracking could result from removing bolts in an incorrect order.

    b. Lift the cylinder head from the dowels on the cylinder block and place the cylinder head on wooden blocks on a bench.

➡Be careful not to damage the contact surfaces of the cylinder head and cylinder block.

14. Remove the cylinder head gasket.

To install:

➡Thoroughly clean all parts to be assembled. Turn the crankshaft pulley so that the No. 1 piston is at top dead center. Always use a new head and manifold gasket.

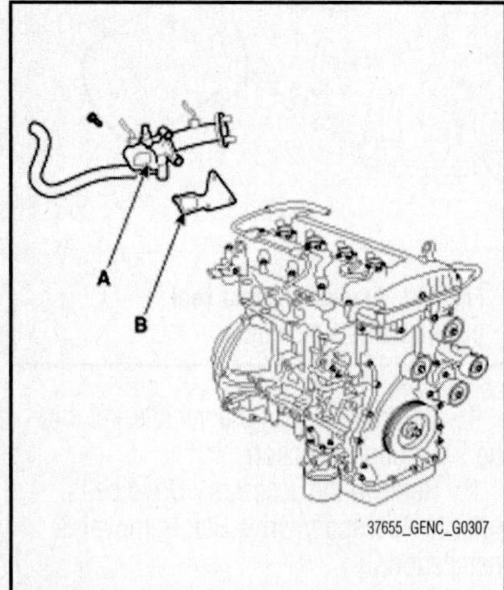

Fig. 132 Remove the water temperature control assembly (A) and gasket (B)

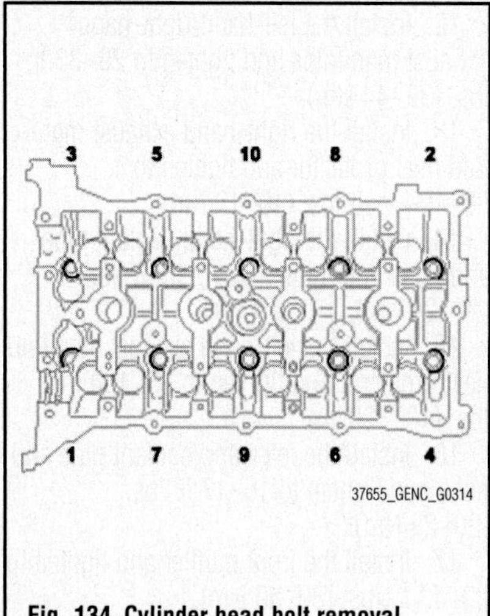

Fig. 134 Cylinder head bolt removal sequence

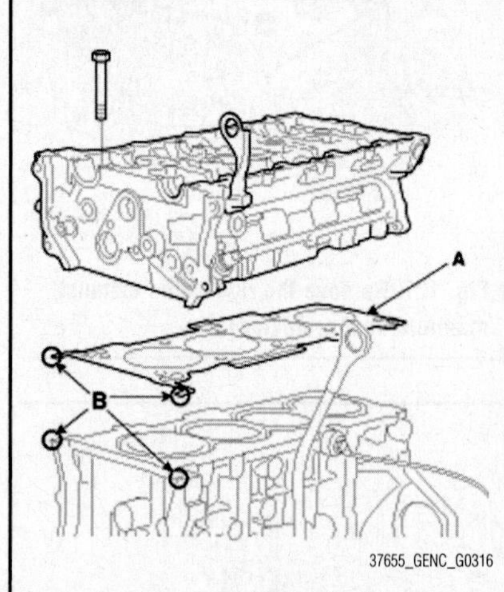

Fig. 136 Sealant application points (B) and cylinder head gasket (A)

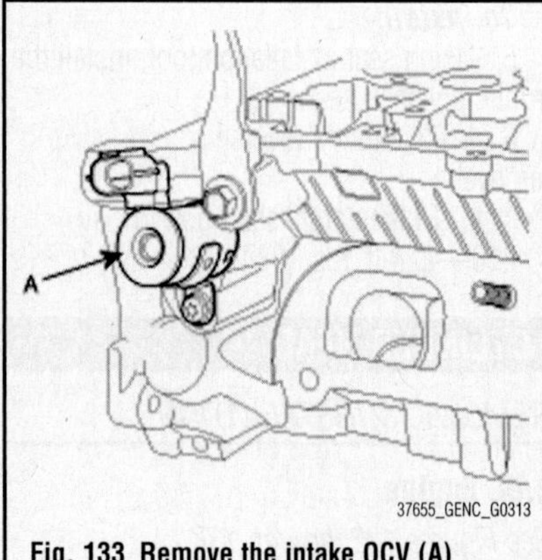

Fig. 133 Remove the intake OCV (A)

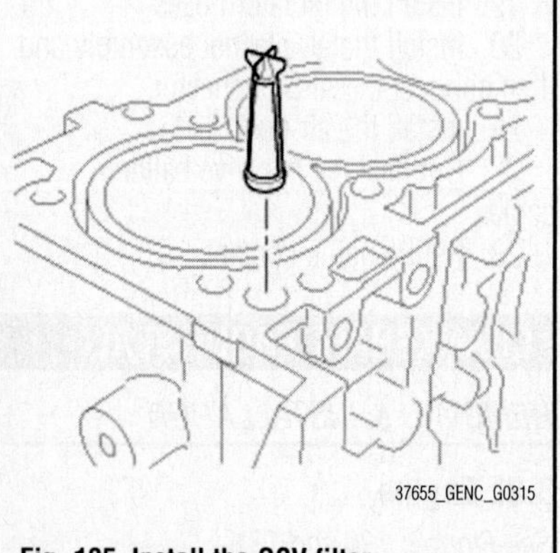

Fig. 135 Install the OCV filter

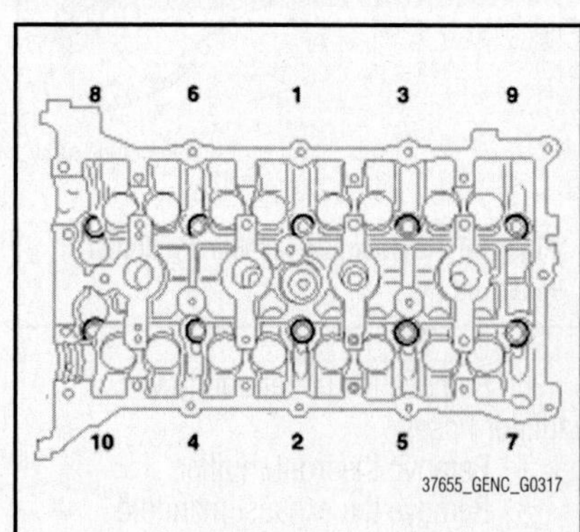

Fig. 137 Cylinder head bolt tightening sequence

15. Install the OCV filter. Keep the OCV filter clean.

16. Install the cylinder head gasket (A) on the cylinder block.

➡Be careful of the installation direction. Apply liquid gasket (Loctite 5900H) on the edge of cylinder head gasket upside and downside. (At the position 'B') After applying sealant, assemble the cylinder head in five minutes.

17. Place the cylinder head carefully in order not to damage the gasket with the bottom part of the end.

18. Install cylinder head bolts:

a. Apply a light coat if engine oil on the threads and under the heads of the cylinder head bolts.

b. Using hexagon wrench, install and tighten the 10 cylinder head bolts and plate washers, in several passes, in the sequence shown. Tightening torque: 24–27 ft. lbs. (32–36 Nm), plus 90–95°, plus 90–95°.

➡Always use new cylinder head bolts.

19. Install the OCV. Tightening torque: 86–104 inch lbs. (10–12 Nm).

➡Do not reuse the OCV if dropped. Keep the OCV filter clean. Do not hold the OCV sleeve during servicing. When the OCV is installed on the engine, do not move the engine with holding the OCV yoke.

20. Install the camshafts.

21. Install the water temperature control assembly with a new gasket, and tighten the bolts to 11–14 ft. lbs. (15–20 Nm) and nuts to 14–17 ft. lbs. (19–24 Nm).

➡Assemble water temp control assembly and water inlet pipe to water pump assembly before nuts for assembling of water inlet pipe to be tightened. Always use a new O-ring.

22. Install the timing chain.

23. Check and adjust valve clearance.

24. Install the cylinder head cover.

25. Install the intake and exhaust manifolds.

26. Install the heater hoses.

27. Connect the negative battery cable.

3.8L Engine

See Figures 138 through 155.

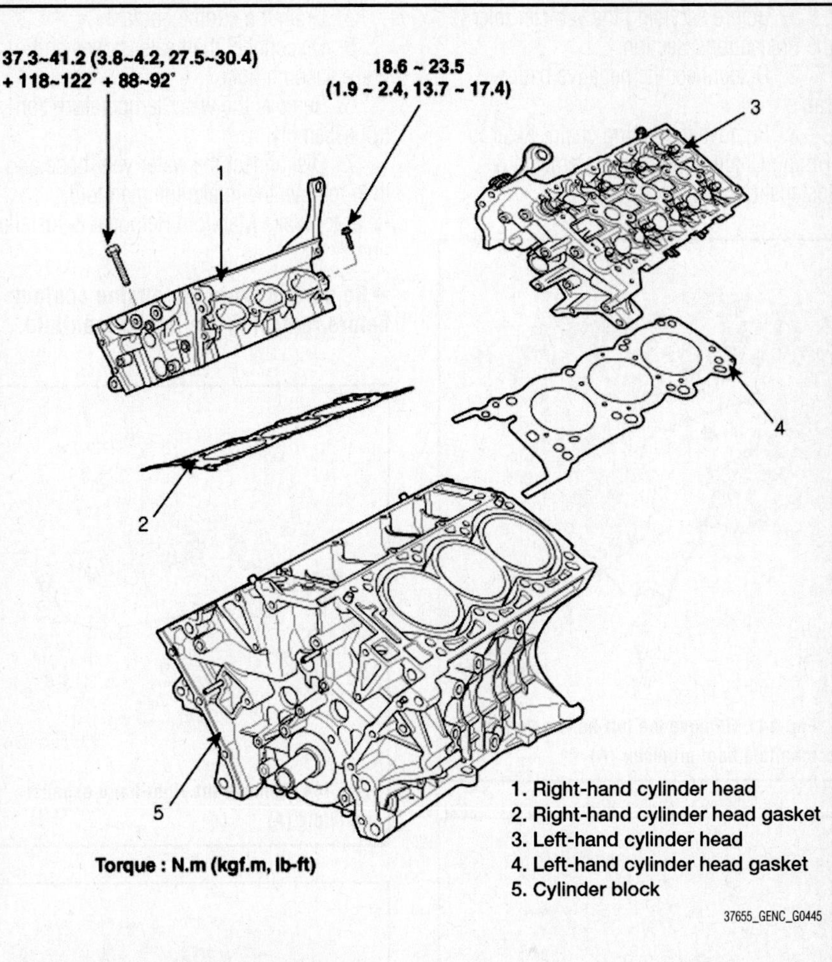

Fig. 138 Cylinder head and components

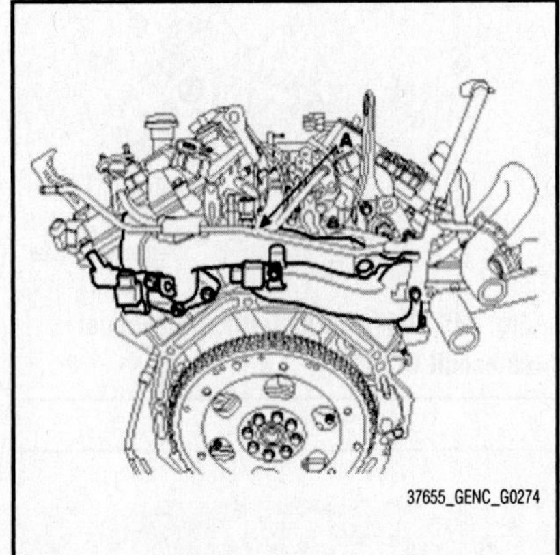

Fig. 139 Remove the water temperature control assembly (A)

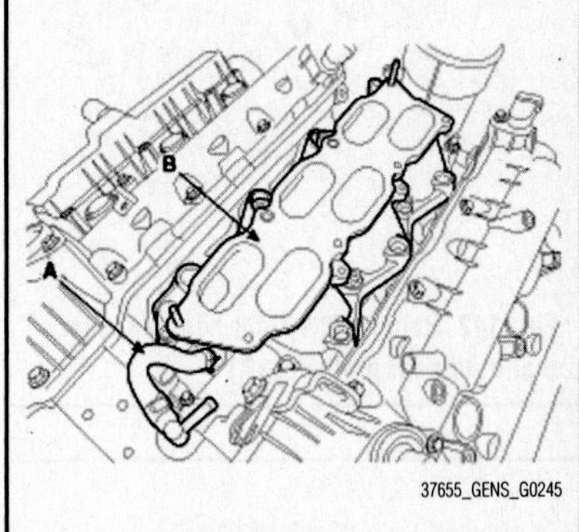

Fig. 140 Disconnect the water vent hose (A) and then remove the intake the manifold (B)

⁂ WARNING

Use fender covers to avoid damaging painted surfaces. To avoid damaging the cylinder head, wait until the engine coolant temperature drops below normal temperature (68°F [20°C]) before removing it. When handling a metal gasket, take care not to fold the gasket or damage the contact surface of the gasket. To avoid damage, unplug the wiring connectors carefully while holding the connector portion. Mark all wiring and hoses to avoid misconnection. Turn the crankshaft pulley so that the No. 1 piston is at top dead center.

1. Before servicing the vehicle, refer to the Precautions Section.

2. Disconnect the negative battery cable.

3. Remove the timing chain. Refer to Timing Chain & Sprockets Removal & Installation.

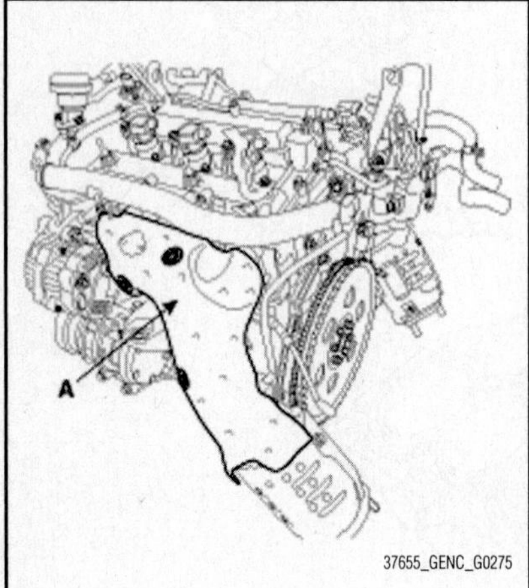

Fig. 141 Remove the left-hand exhaust manifold heat protector (A)

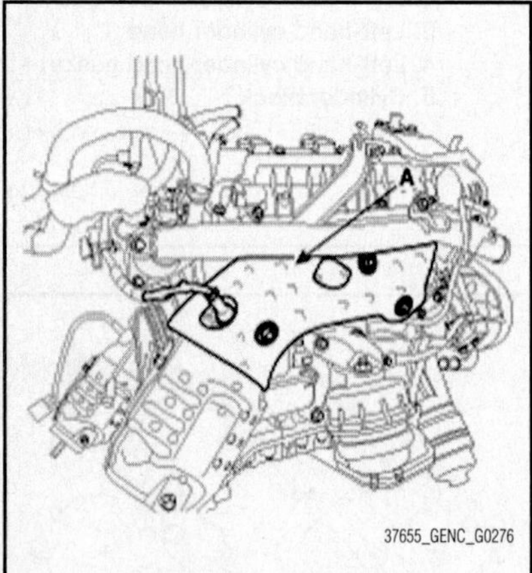

Fig. 142 Remove the right-hand exhaust manifold heat protector (A)

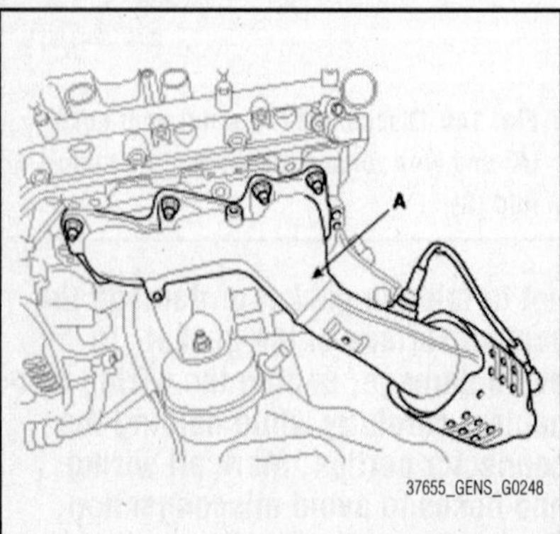

Fig. 143 Remove the left-hand exhaust manifold (A)

4. Drain the engine coolant.

5. Disconnect the heater hoses and brake vacuum hose.

6. Remove the water temperature control assembly.

7. Disconnect the water vent hose and then remove the intake the manifold. Refer to Intake Manifold Removal & Installation.

➡**Be sure to drain the engine coolant before removing the intake manifold.**

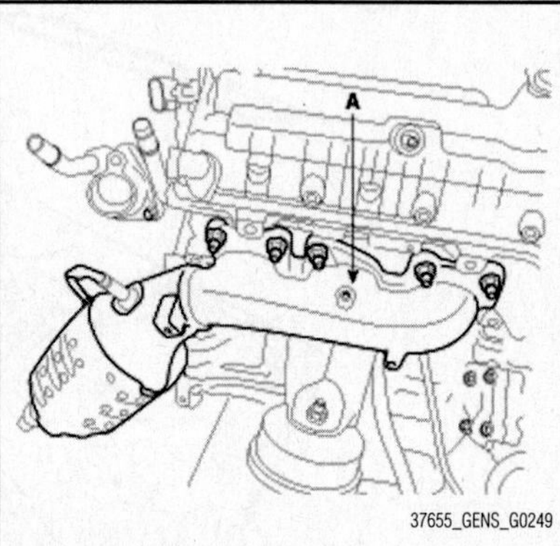

Fig. 144 Remove the right-hand exhaust manifold (A)

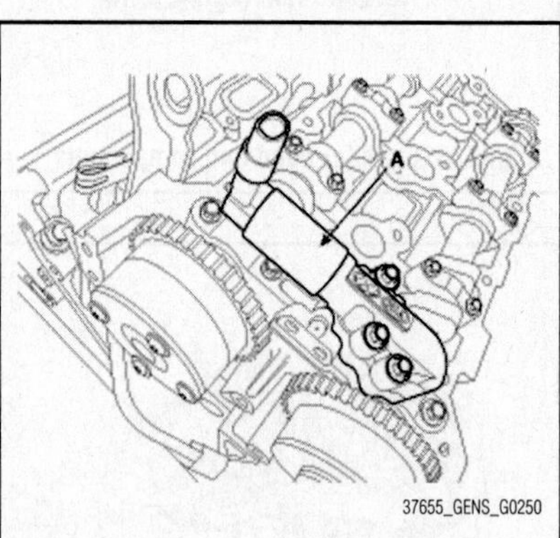

Fig. 145 Remove the left-hand exhaust camshaft OCV (A)

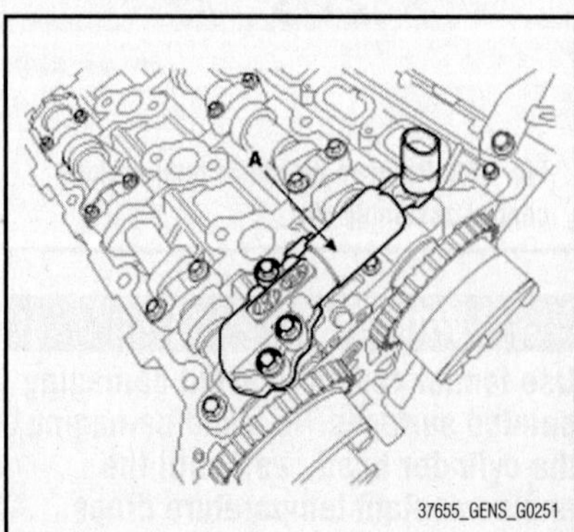

Fig. 146 Remove the right-hand exhaust camshaft OCV (A)

If any coolant drained from the cylinder head vent hole has entered the intake port, this can potentially lead to engine trouble.

8. Remove the left-hand and right-hand exhaust manifold heat protectors. Refer to Exhaust Manifold Removal & Installation.

9. Remove the left-hand and right-hand exhaust manifolds. Refer to Exhaust Manifold Removal & Installation.

10. Remove the left-hand and right-hand exhaust camshaft OCV. Refer to Camshaft Removal & Installation.

11. Remove the left-hand and right-hand camshaft bearing caps and thrust bearing caps. Refer to Camshaft Removal & Installation.

12. Remove the left-hand and right-hand camshaft assemblies. Refer to Camshaft Removal & Installation.

13. Remove the cylinder head.

a. Uniformly loosen and remove the cylinder head bolts, in several passes, in the sequence shown.

WARNING

Head warpage or cracking could result from removing bolts in an incorrect order.

b. Lift the cylinder head from the dowels on the cylinder block and place the cylinder head on wooden blocks on a bench.

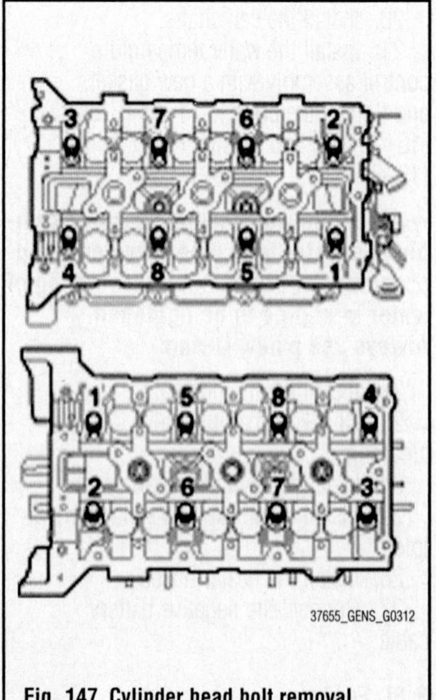

Fig. 147 Cylinder head bolt removal sequence

➡**Be careful not to damage the contact surfaces of the cylinder head and cylinder block.**

To install:

➡**Thoroughly clean all parts to be assembled. Turn the crankshaft pulley so that the No. 1 piston is at top dead center.**

14. Install the cylinder head.

➡**The sealant locations on cylinder head and cylinder block must be free of engine oil, etc. The part must be assembled within 5 minutes after sealant was applied.**

a. Apply sealant on cylinder block top face before assembling cylinder head gaskets. Bead width: 0.078–0.118 in. (2.0–3.0mm). Sealant locations: 0.039–0.059 in. (1.0–1.5mm). Recommended sealant: Liquid sealant TB1217H.

b. Apply sealant on cylinder head gaskets after assembling cylinder head gaskets on cylinder block.

c. Install the cylinder head.

d. Remove the extruded sealant after assembling cylinder heads.

15. Install cylinder head bolts.

a. Do not apply engine oil on the threads and under the heads of the cylinder head bolts.

b. Using SST (09221-4A000), install and tighten the cylinder head bolts and plate washers, in several passes, in the sequence shown. Tightening torque:

- Head bolt: 28–30 ft. lbs. (37–41 Nm) plus 118–122° plus 88–92°
- Bolt "A": 14–17 ft. lbs. (19–24 Nm)

16. Install the left-hand and right-hand camshaft assemblies, camshaft bearing caps and thrust bearing caps. Refer to Camshafts & Valve Lifters Removal & Installation.

17. Install the left-hand and right-hand exhaust camshaft OCV and tighten to 86–104 inch lbs. (10–12 Nm).

18. Install the cylinder head cover.

19. Install the left-hand and right-hand exhaust manifolds with new gaskets, and tighten to 29–33 ft. lbs. (39–44 Nm).

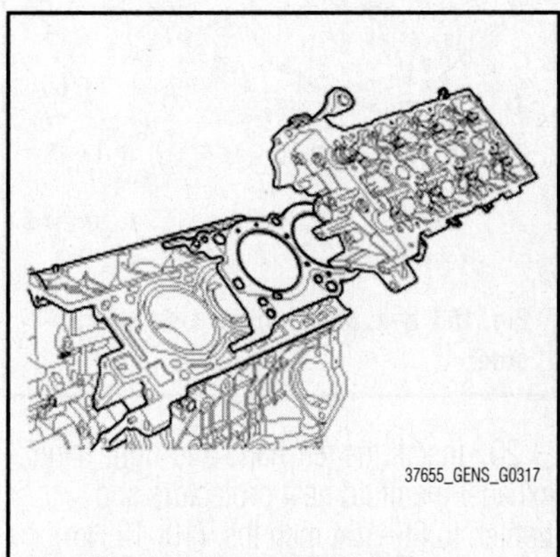

Fig. 152 Cylinder head installation

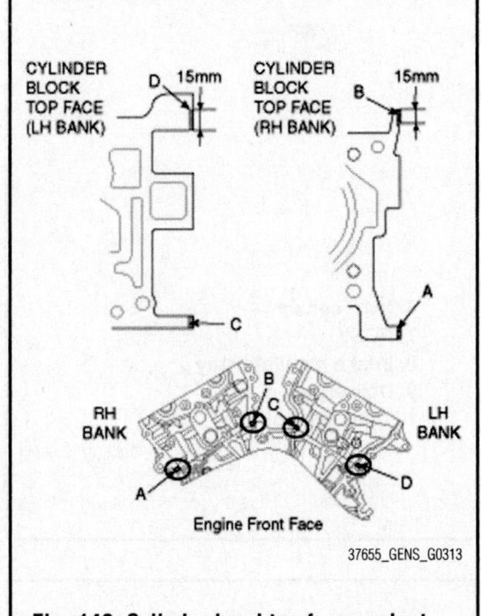

Fig. 148 Cylinder head top face sealant locations

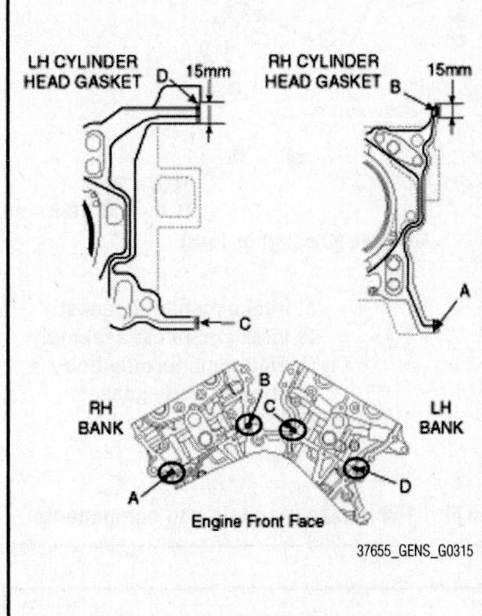

Fig. 150 Cylinder head gasket sealant locations

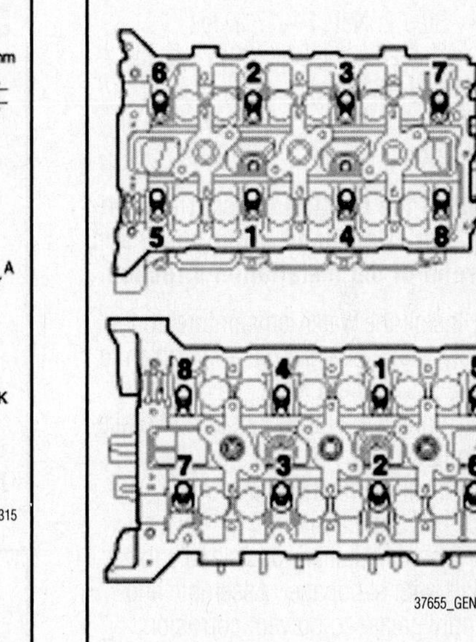

Fig. 153 Cylinder head bolt tightening sequence

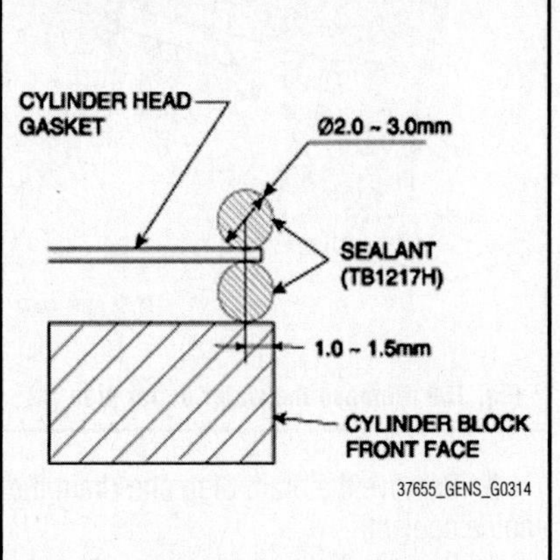

Fig. 149 Cylinder head top face sealant application

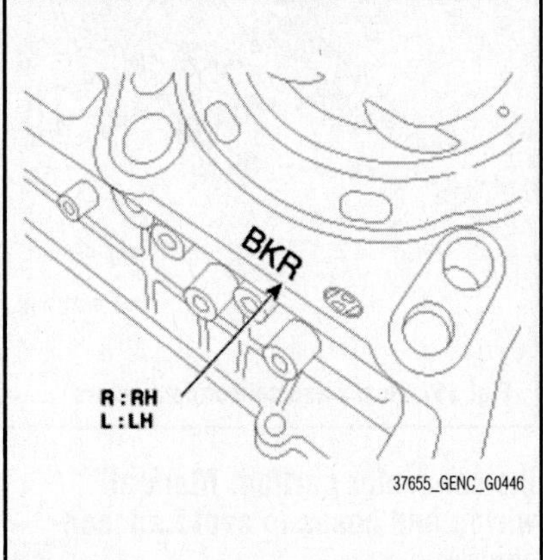

Fig. 151 Be careful of the installation direction

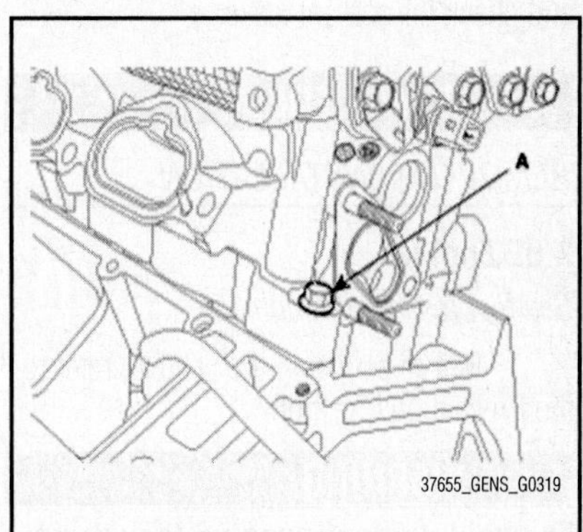

Fig. 154 Cylinder head bolt "A"

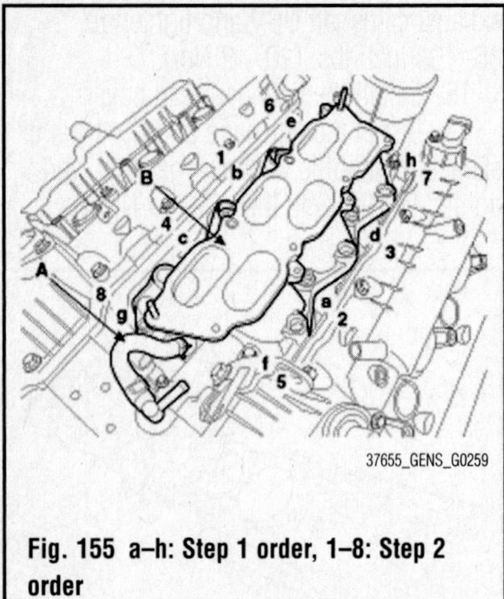

Fig. 155 a–h: Step 1 order, 1–8: Step 2 order

20. Install the left-hand and right-hand exhaust manifold heat protectors and tighten to 86–104 inch lbs. (10–12 Nm).
21. Install the intake the manifold with a new gasket, and connect the water vent hose. Tightening torque:
    - Step 1: 35–52 inch lbs. (4–6 Nm)
    - Step 2: Nut: 14–17 ft. lbs. (19–24 Nm), Bolt: 20–23 ft. lbs. (27—31 Nm)
    - Step 3: Repeat 2nd step twice or more.

➡Confirm the manifold gasket identification mark (left-hand, right-hand) and be careful of the installation direction.

22. Install the water temperature control assembly with a new gasket, and tighten to 15–17 ft. lbs. (20–24 Nm).
23. Connect the heater hoses and brake vacuum hose.
24. Install the timing chain.
25. Refill engine oil.
26. Clean the battery posts and cable terminals with sandpaper. Assemble and then apply grease to prevent corrosion.
27. Inspect for fuel leakage.
28. Refill radiator and reservoir tank with engine coolant.
29. Bleed air from the cooling system and check for coolant leaks.

INTAKE MANIFOLD

REMOVAL & INSTALLATION

2.0L Engine

See Figures 156 through 160.

1. Before servicing the vehicle, refer to the Precautions Section.

⁂ WARNING

To avoid damage, unplug the wiring connectors carefully while holding the connector portion. Mark all wiring and hoses to avoid misconnection.

2. Disconnect the battery negative cable.
3. Remove the drain plug and drain the engine coolant.
4. Remove the air duct.
5. Remove the ETC connector and intercooler outlet hose.

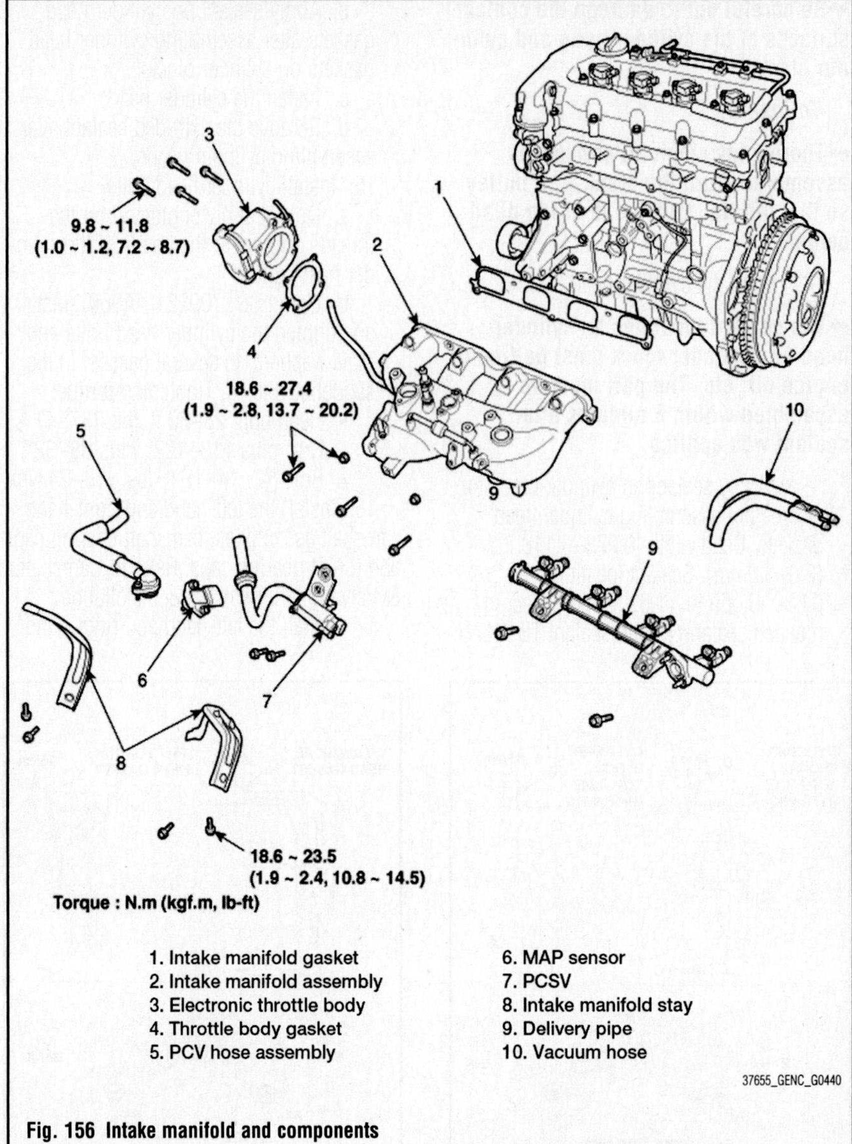

Fig. 156 Intake manifold and components

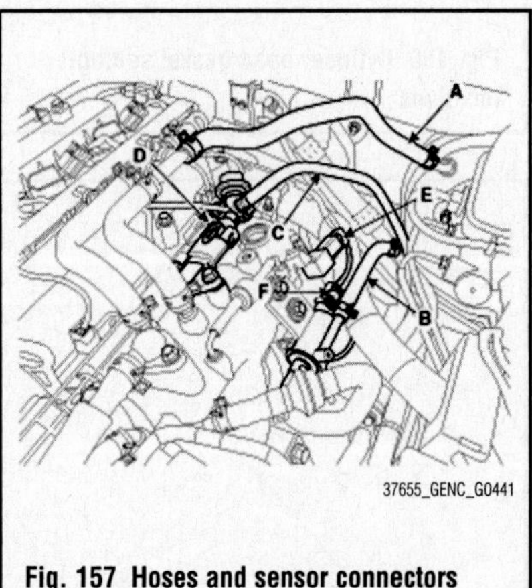
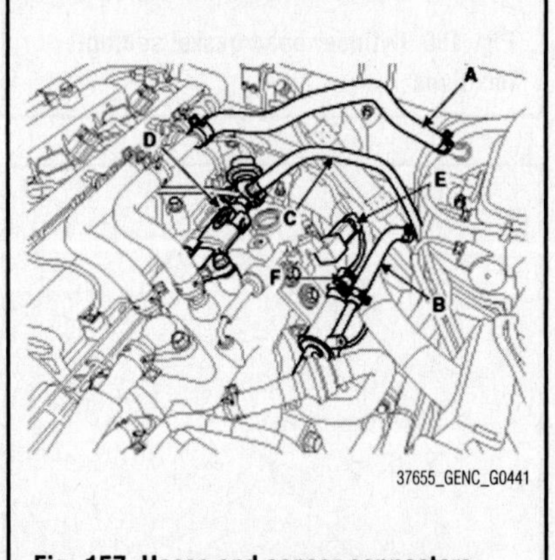

Fig. 157 Hoses and sensor connectors

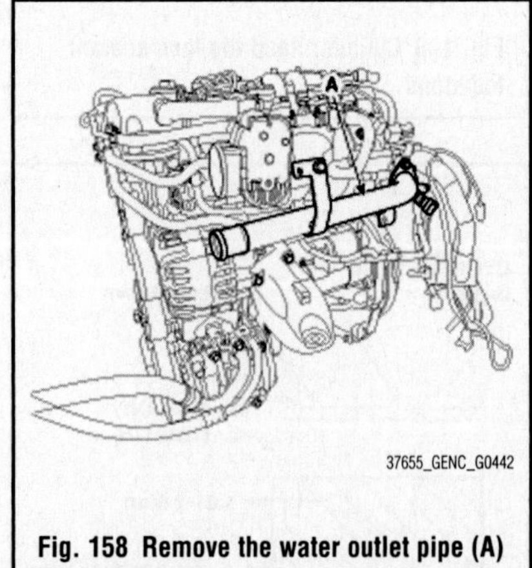

Fig. 158 Remove the water outlet pipe (A)

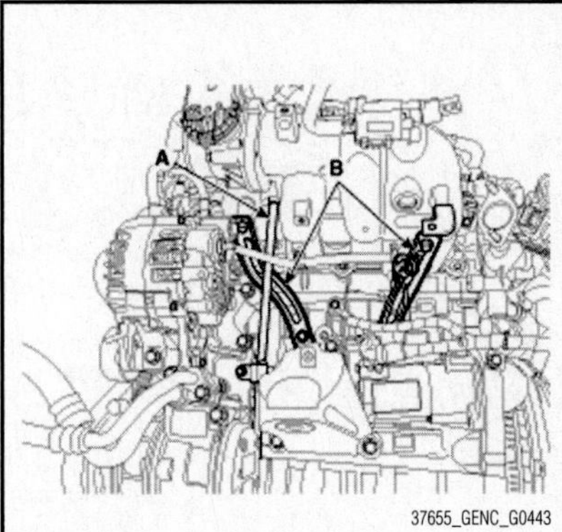

Fig. 159 Remove the oil level gauge (A) and intake manifold stay (B) bolt

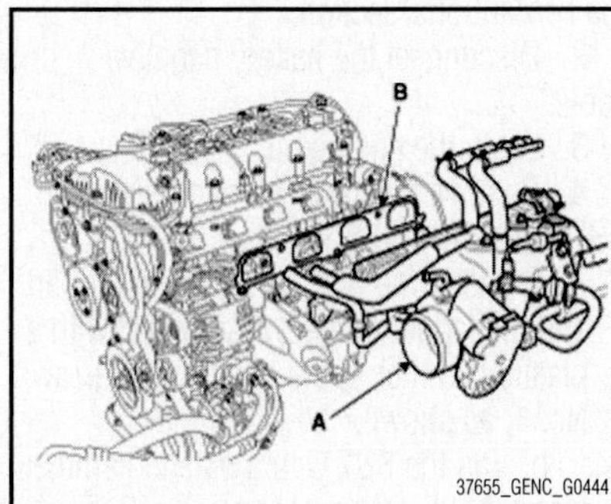

Fig. 160 Remove the intake manifold (A) and gasket (B)

6. Disconnect the vacuum hose, PCSV hose, fuel hose, MAP sensor connector, condenser connector, and PCSV connector.
7. Remove the PCV hose, coolant hoses and vacuum hoses.
8. Remove the injector connectors.
9. Remove the water outlet pipe.
10. Remove the oil level gauge and intake manifold stay bolt.
11. Remove the intake manifold (A) and gasket (B).

To install:

12. Installation is the reverse of removal, noting the following:
   a. Tighten the intake manifold to 14–20 ft. lbs. (19–27 Nm).
   b. Tighten the oil level gauge to 70–104 inch lbs. (8–12 Nm).
   c. Tighten the intake manifold stay bolt to 14–17 ft. lbs. (19–23 Nm).
   d. Tighten the water outlet pipe to 14–17 ft. lbs. (19–24 Nm).

3.8L Engine

See Figures 161 through 166.

1. Before servicing the vehicle, refer to the Precautions Section.

✲✲ WARNING

To avoid damage, unplug the wiring connectors carefully while holding the connector portion. Mark all wiring and hoses to avoid misconnection.

2. Disconnect the battery negative cable.
3. Remove the drain plug and drain the engine coolant.

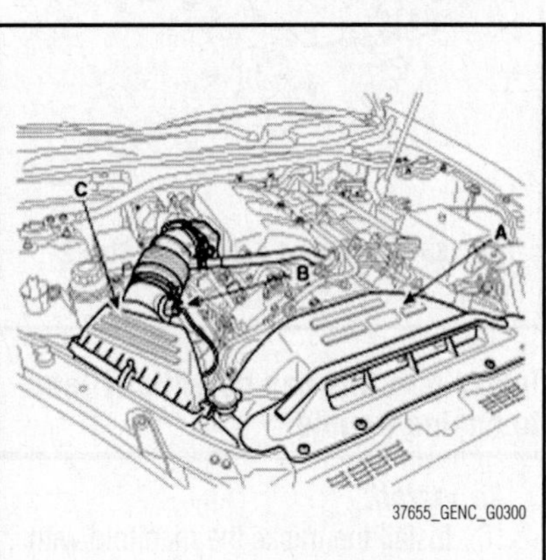

Fig. 161 Remove the AFS connector (B) and the air cleaner assembly (C)

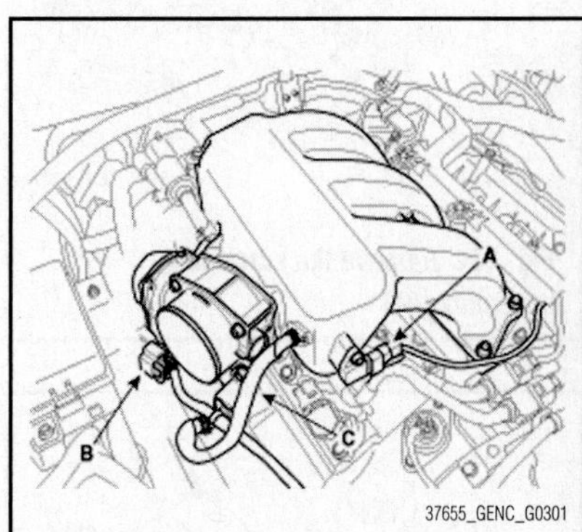

Fig. 163 Disconnect the MAP sensor connector (A), ETC connector (B), and PCV hose (C)

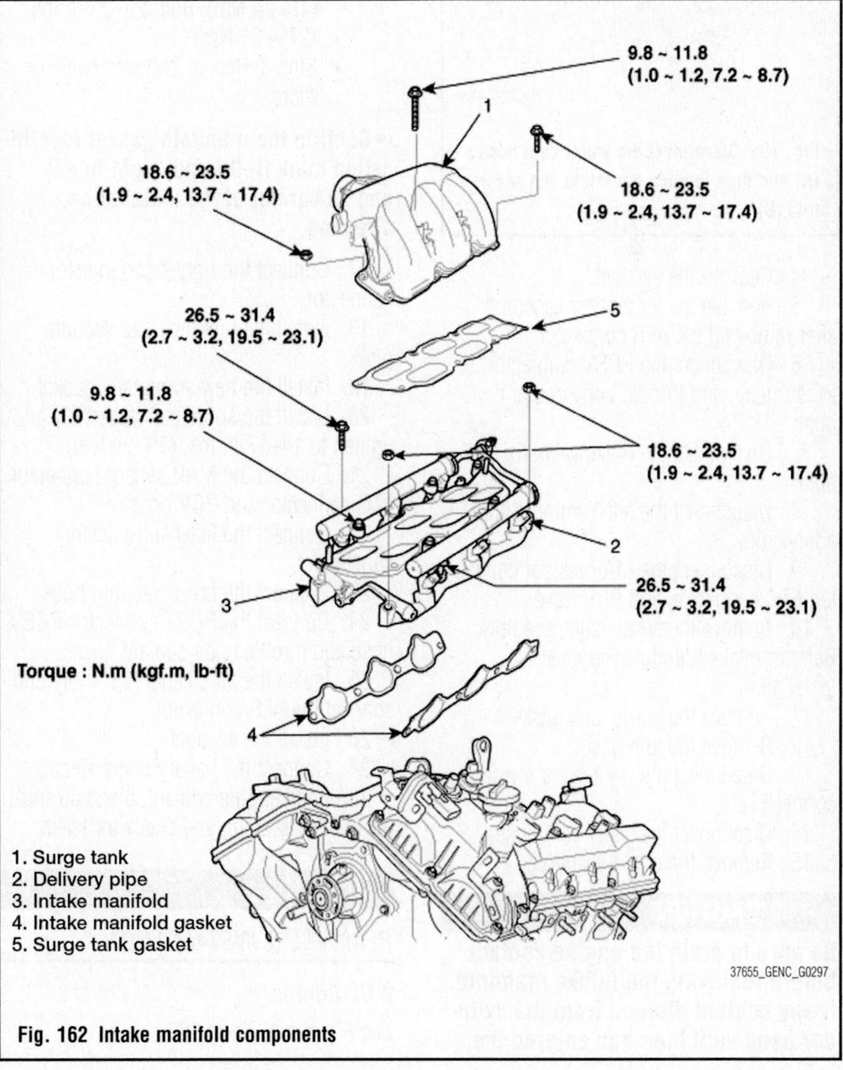

Fig. 162 Intake manifold components

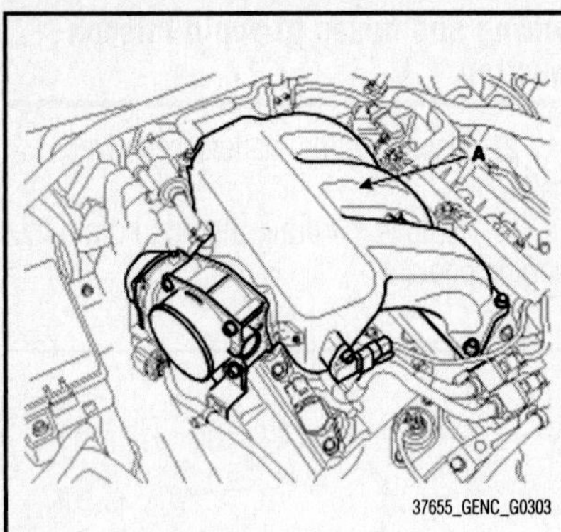

Fig. 164 Remove the surge tank assembly (A)

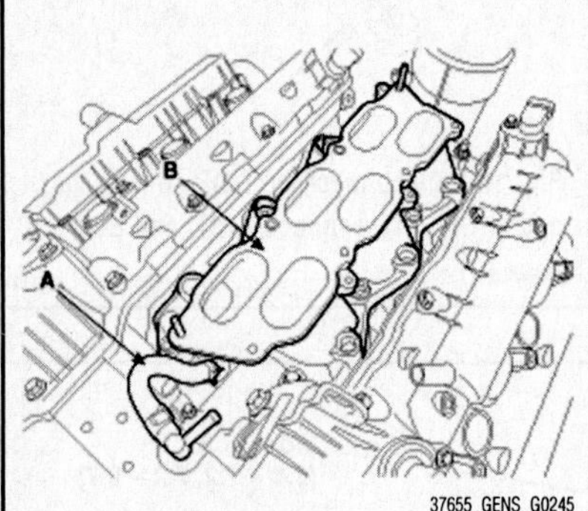

Fig. 165 Disconnect the water vent hose (A) and then remove the intake the manifold (B)

4. Remove the air duct.
5. Remove the air cleaner assembly after removing the AFS connector.
6. Disconnect the PCSV connector, PCSV hose, and throttle body coolant hose.
7. Disconnect the brake vacuum hose.
8. Disconnect the left-hand injector connectors.
9. Disconnect the MAP sensor connector, ETC connector, and PCV hose.
10. Remove the surge tank assembly. Refer to Intake Manifold Removal & Installation.
11. Remove the surge tank gasket.
12. Remove the fuel hose.
13. Disconnect the right-hand injector connectors.
14. Disconnect the water vent hose.
15. Remove the intake manifold.

CAUTION

Be sure to drain the engine coolant before removing the intake manifold. If any coolant drained from the cylinder head vent hole has entered the intake port this can potentially lead to engine trouble.

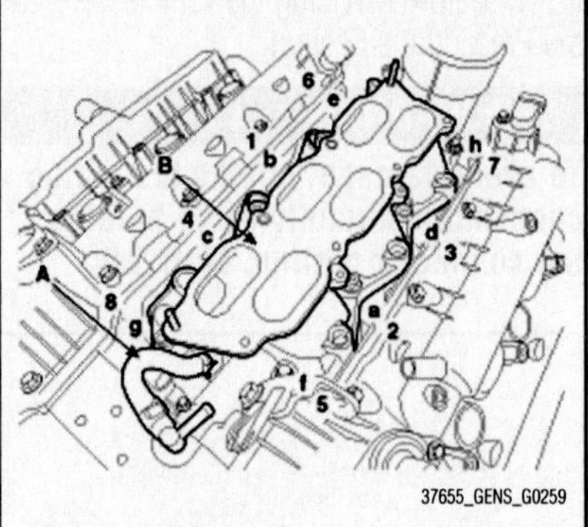

Fig. 166 a–h: Step 1 order, 1–8: Step 2 order

To install:

16. Install the intake the manifold with a new gasket, and connect the water vent hose. Tightening torque:
   - Step 1: 35–52 inch lbs. (4–6 Nm)
   - Step 2: Nut: 14–17 ft. lbs. (19–24 Nm), Bolt: 20–23 ft. lbs. (27—31 Nm)
   - Step 3: Repeat 2nd step twice or more.

➡Confirm the manifold gasket identification mark (left-hand, right-hand) and be careful of the installation direction.

17. Connect the right-hand injector connector.
18. Install the fuel pipe and vacuum pipe.
19. Install the new surge tank gasket.
20. Install the surge tank assembly and tighten to 14–17 ft. lbs. (19–24 Nm).
21. Connect the MAP sensor connector, ETC connector and PCV hose.
22. Connect the left-hand injector connectors.
23. Connect the brake vacuum hose.
24. Connect the PCSV connector PCSV hose and throttle body coolant hoses.
25. Install the air cleaner assembly and connect the AFS connector.
26. Install the air duct.
27. Connect the battery negative cable.
28. Refill engine coolant, bleed air from the coolant system, and check for leaks.

OIL PAN

REMOVAL & INSTALLATION

2.0L Engine

See Figures 167 and 168.

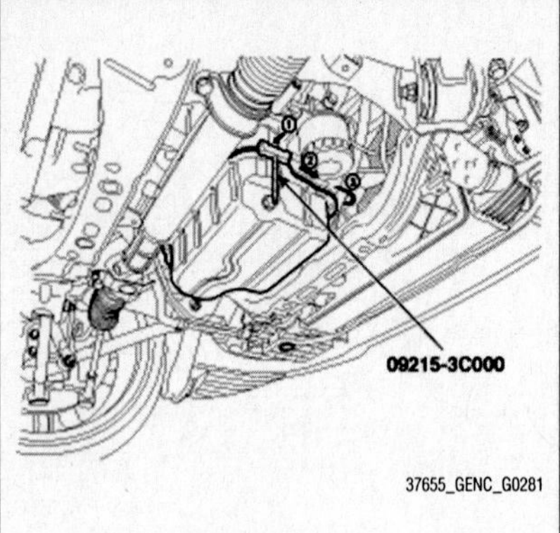

Fig. 167 Tap the SST in the direction of the No. 1 and 2 arrows

1. Before servicing the vehicle, refer to the Precautions Section.
2. Disconnect the battery negative cable.
3. Drain the engine oil.
4. Remove the lower oil pan using the SST (09215-3C000):
   a. Insert the SST between the oil pan and the ladder frame by tapping it with a plastic hammer in the direction of arrow No. 1, as shown.
   b. Tap the SST with a plastic hammer along the direction of arrow No. 2, as shown, around more than ⅔ of the edge of the oil pan.
   c. Remove the oil pan from the ladder frame.

WARNING

Do not turn over the SST abruptly without tapping, as this will damage the tool. Be careful not to damage the oil pan contact surfaces.

To install:

5. Using a gasket scraper, remove all the old packing material from the gasket surfaces.

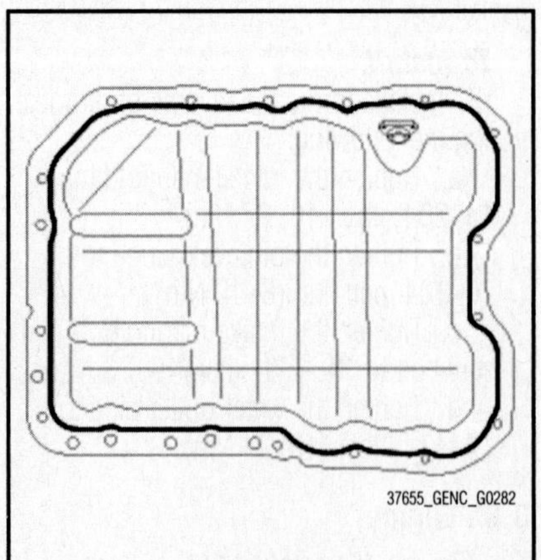

Fig. 168 Sealant application

6. Before assembling the oil pan, the liquid sealant Loctite® 5900H or ThreeBond®1217H should be applied on oil pan. The part must be assembled within 5 minutes after applying the sealant.

WARNING

When applying sealant gasket, sealant must not protrude into the inside of oil pan. To prevent leakage of oil, apply sealant gasket to the inner threads of the bolt holes.

7. Install the oil pan. Uniformly tighten the bolts in several passes. Tightening torque: 86–104 inch lbs. (10–12 Nm).

8. After assembly, wait at least 30 minutes before filling the engine with oil.

3.8L Engine

See Figures 169 and 170.

1. Before servicing the vehicle, refer to the Precautions Section.
2. Drain the engine oil.
3. Remove the lower oil pan. Insert the blade of SST (09215-3C000) between the upper oil pan and lower oil pan. Cut off applied sealer and remove the lower oil pan.

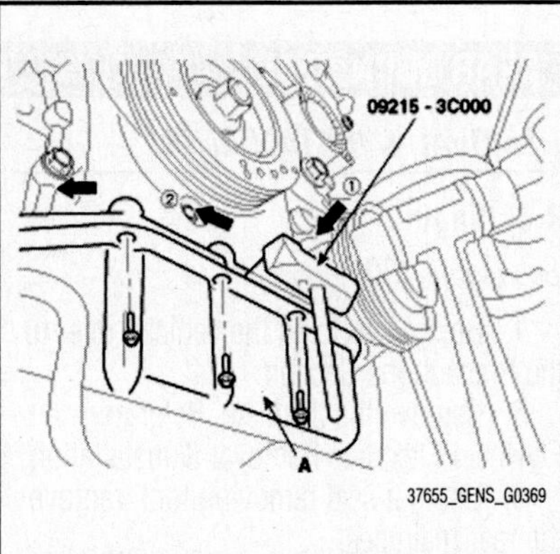

Fig. 169 Remove the lower oil pan (A)

Fig. 170 Oil pan sealant application

➡Insert the SST between the oil pan and the ladder frame by tapping it with a plastic hammer in the direction of arrow. After tapping the SST with a plastic hammer along the direction of arrow around more than ⅔ edge of the oil pan, remove it from the ladder frame. Do not turn over the SST abruptly without tapping. It be result in damage of the SST. Be careful not to damage the contact surfaces of the upper oil pan and lower oil pan.

To install:

4. Install the lower oil pan.
   a. Using a gasket scraper, remove all the old packing material from the gasket surfaces.
   b. Before assembling the oil pan, the liquid sealant TB 1217H should be applied on oil pan. The part must be assembled within 5 minutes after the sealant was applied. Bead width: 0.1 in. (2.5mm)

➡Clean the sealing face before assembling two parts. Remove harmful foreign matters on the sealing face before applying sealant. When applying sealant gasket, sealant must not be protruded into the inside of oil pan. To prevent leakage of oil, apply sealant gasket to the inner threads of the bolt holes.

   c. Install the oil pan. Uniformly tighten the bolts in several passes. Tightening torque: 86–104 inch lbs. (10–12 Nm).
5. After assembly, wait at least 30 minutes before filling the engine with oil.

OIL PUMP

REMOVAL & INSTALLATION

2.0L Engine

See Figures 171 and 172.

1. Before servicing the vehicle, refer to the Precautions Section.
2. Drain the engine oil.
3. Remove the oil pan.
4. Remove the timing chain.
5. Remove the oil pump, oil pump chain and sprocket.

To install:

6. Align the crankshaft key with the mating surface of main bearing cap. This will position the piston of the No. 1 cylinder at the top dead center on the compression stroke.
7. Assemble the crankshaft sprocket on the crankshaft with the front mark facing outward.
8. Tighten the oil pump tensioner bolt after placing the tensioner spring on the dowel pin located in the ladder frame, and then insert the stopper pin to secure the oil pump chain tensioner. Tightening torque: 86–104 inch lbs. (10–12 Nm).
9. Install the oil pump chain on the crankshaft sprocket.
10. Install the oil pump assembly, as follows:
    a. Tighten the oil pump mounting bolts to 19 ft. lbs. (26 Nm).
    b. Loosen the oil pump mounting bolts.
    c. Tighten the oil pump mounting bolts to 70 inch lbs. (8 Nm).
    d. Tighten the oil pump mounting bolts to 12 ft. lbs. (17 Nm).
    e. Tighten the oil pump mounting bolts to 19 ft. lbs. (26 Nm).

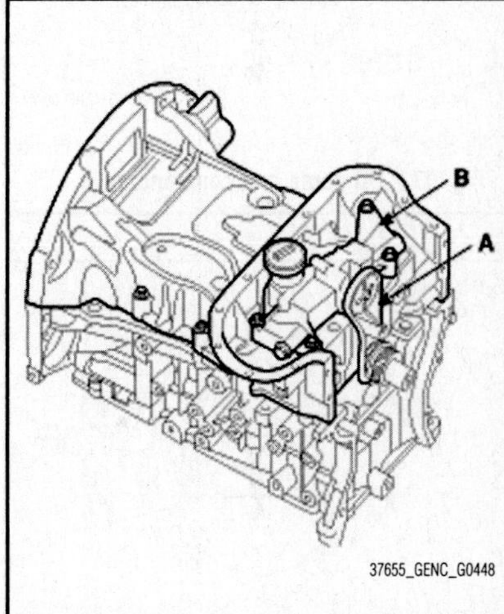

Fig. 171 Remove the oil pump (B), oil pump chain (A) and sprocket

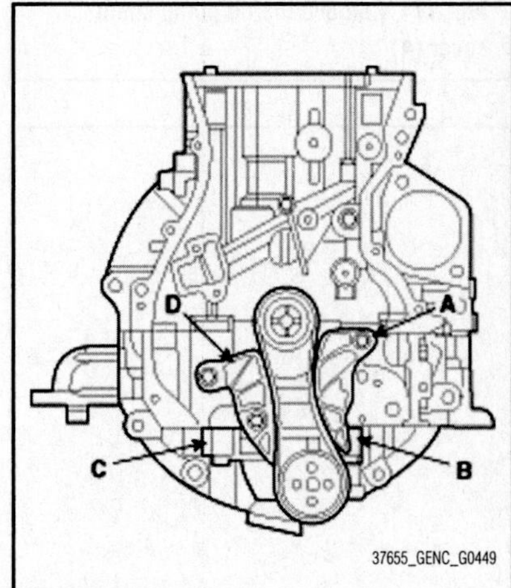

Fig. 172 Oil pump tensioner bolt (A), Oil pump chain tensioner (B), and oil pump chain guide (D)

11. Remove the tensioner pin after installing the oil pump chain guide. Tighten the oil pump chain guide to 86–104 inch lbs. (10–12 Nm).

12. Install the timing chain.

3.8L Engine

See Figures 173 through 177.

1. Before servicing the vehicle, refer to the Precautions Section.

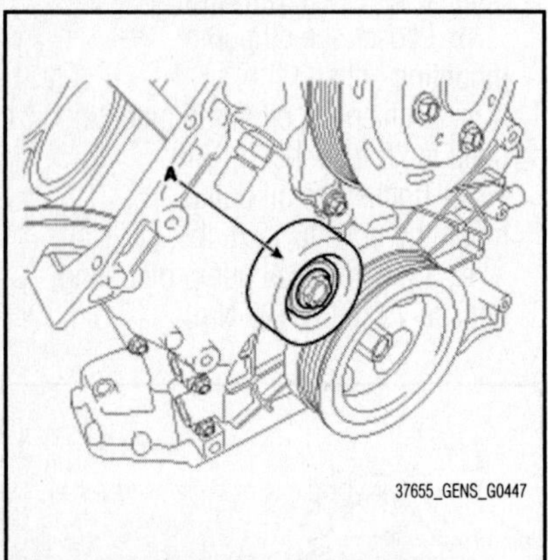

Fig. 173 Oil pump and components

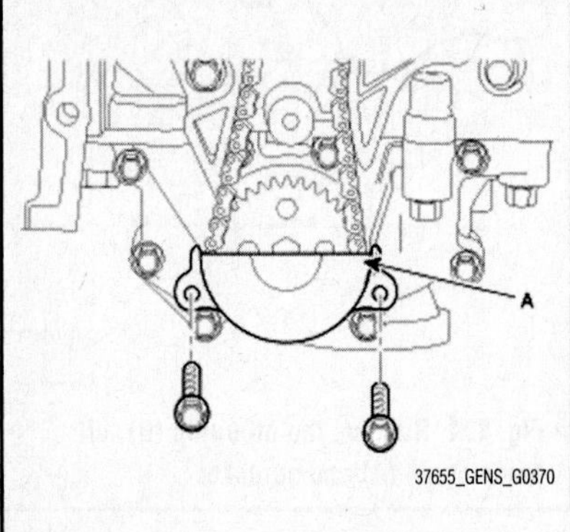

Fig. 174 Remove the oil pump chain cover (A)

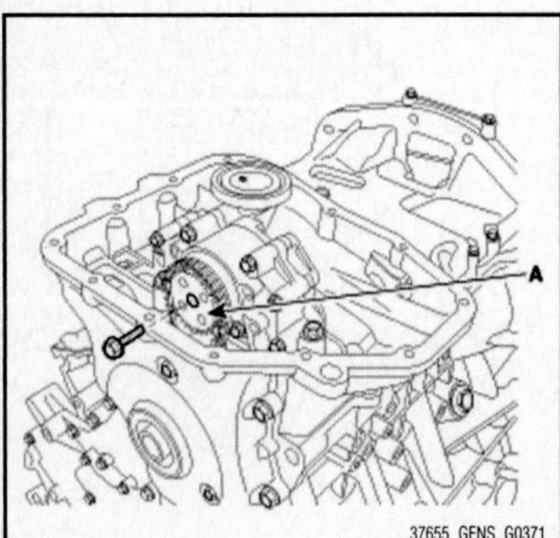

Fig. 175 Remove the oil pump chain sprocket (A)

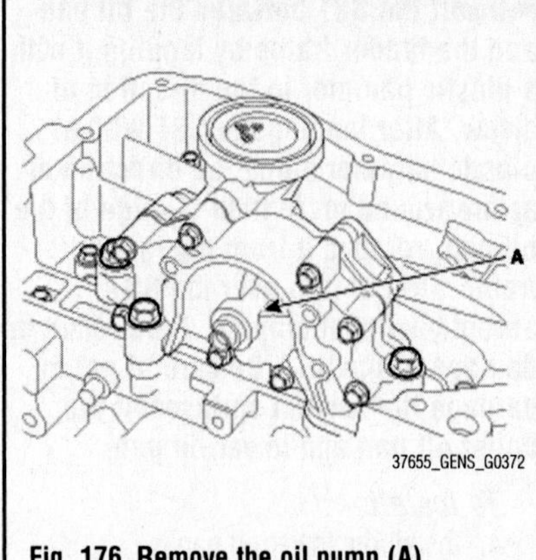

Fig. 176 Remove the oil pump (A)

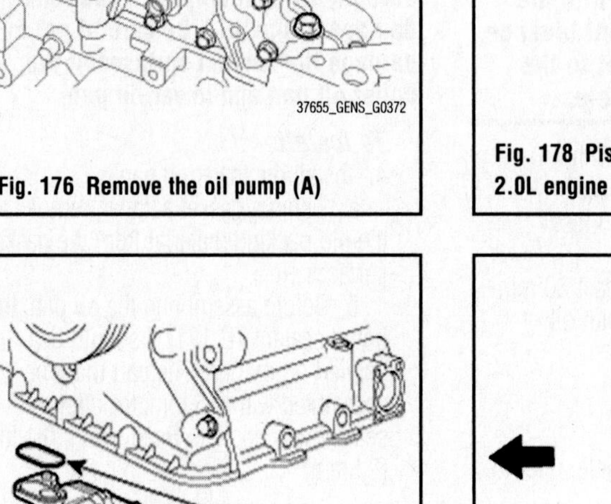

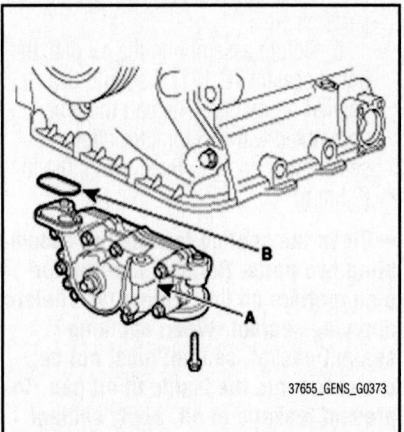

Fig. 177 Install the oil pump (A) and a new O-ring (B)

2. Drain the engine oil.
3. Remove the lower oil pan. Refer to Oil Pan Removal & Installation.
4. Remove the oil pump chain cover.
5. Remove the oil pump chain sprocket.
6. Remove the oil pump.

To install:

7. Install the oil pump (A). Always use a new O-ring (B). Tightening torque: 15–17 ft. lbs. (21–23 Nm).
8. Install the oil pump sprocket and the oil pump chain on the oil pump. Tightening torque: 14–16 ft. lbs. (19–22 Nm).
9. Install the oil pump chain cover. Tightening torque: 86–104 inch lbs. (10–12 Nm).
10. Install the lower oil pan.
11. After assembly, wait at least 30 minutes before filling the engine with oil.

PISTON AND RING

POSITIONING

See Figures 178 and 179.

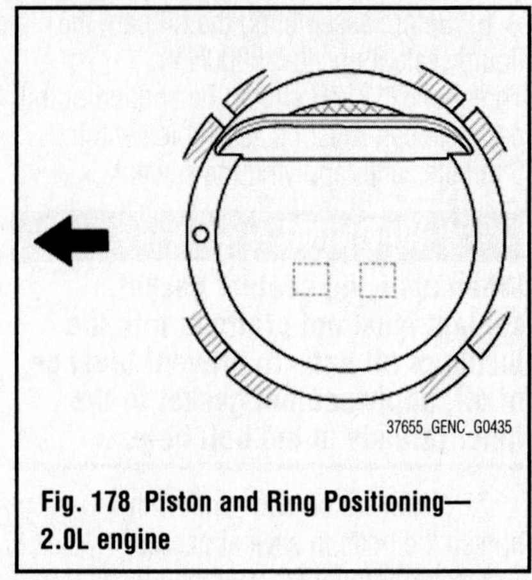

Fig. 178 Piston and Ring Positioning—2.0L engine

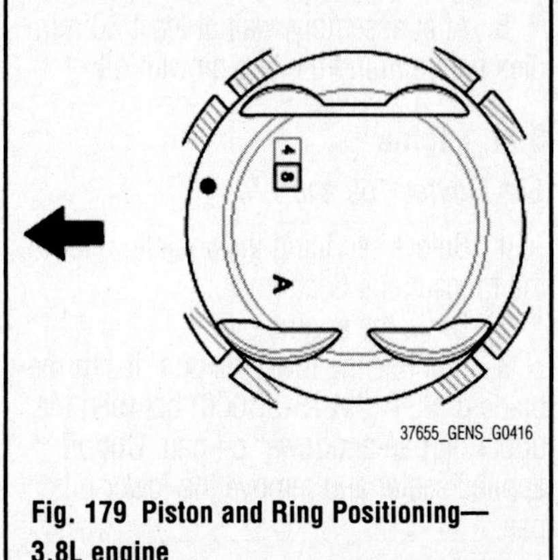

Fig. 179 Piston and Ring Positioning—3.8L engine

REAR MAIN SEAL

REMOVAL & INSTALLATION

3.8L Engine

See Figure 180.

1. Before servicing the vehicle, refer to the Precautions Section.
2. Remove the flexplate. Refer to Flywheel/Flexplate Removal & Installation.
3. Using a seal removing tool, remove the rear main seal.

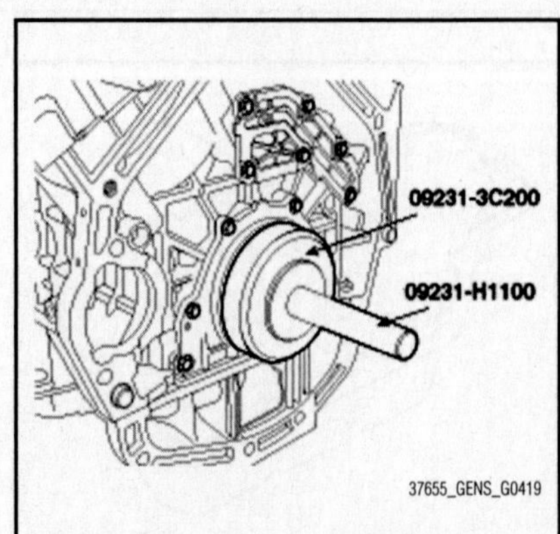

Fig. 180 Rear main seal installation—3.8L engine

To install:

4. Using the illustrated service tool, install the oil seal.

TIMING CHAIN & SPROCKETS

REMOVAL & INSTALLATION

2.0L Engine

See Figures 181 through 194.

➡Special Service Tools used in this procedure: SST 09231-2M100,09231-2J210 (Crankshaft pulley adapter and holder,), SST 09215-3C000 (Oil pan removal tool)

➡Use fender covers to avoid damaging painted surfaces. To avoid damage, unplug the wiring connectors carefully while holding the connector portion. Mark all wiring and hoses to avoid mis-connection.

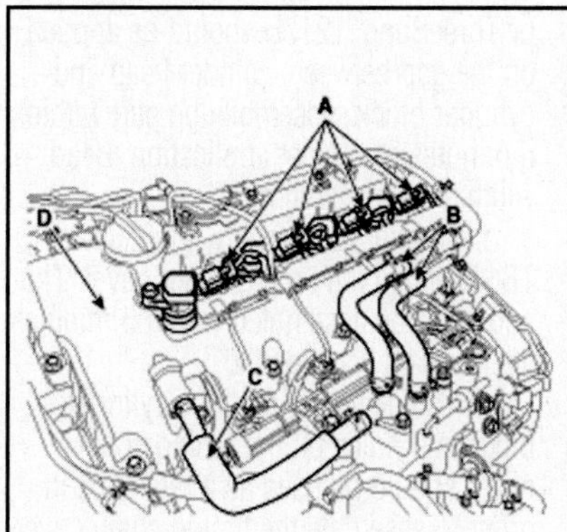

A. Ignition coil connectors
B. Vacuum hoses
C. PCV hose
D. Cylinder head cover

37655_GENC_G0278

Fig. 181 Ignition coil connectors (A), cylinder head cover (D), vacuum hoses (B) and PCV hose (C)

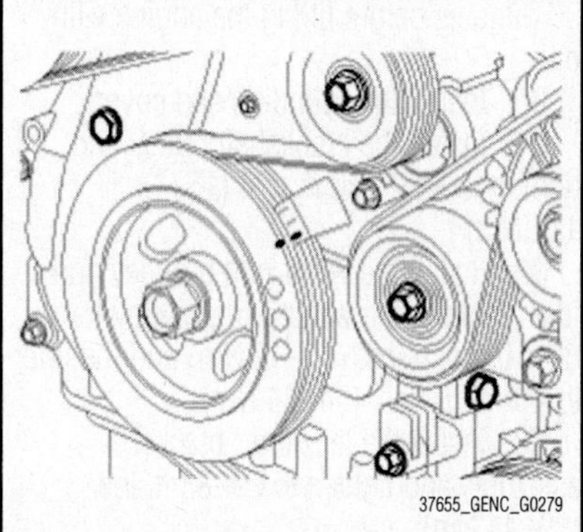

Fig. 182 Set No. 1 cylinder to TDC/ compression

1. Before servicing the vehicle, refer to the Precautions Section.
2. Disconnect the battery negative cable.
3. Drain the engine coolant.

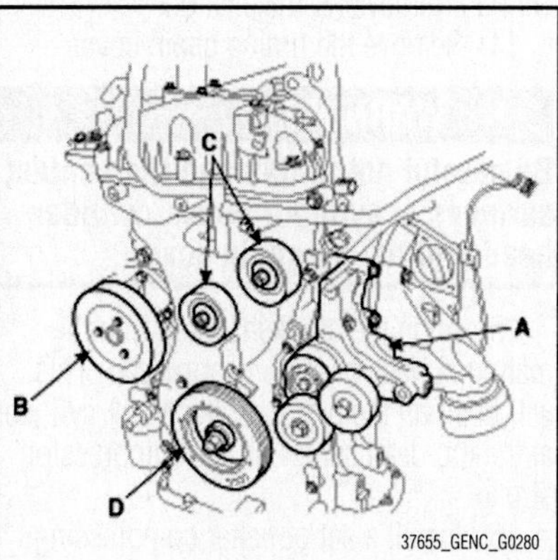

Fig. 183 Remove the tensioner assembly (A), water pump assembly (B), idler (C) and crankshaft pulley (D)

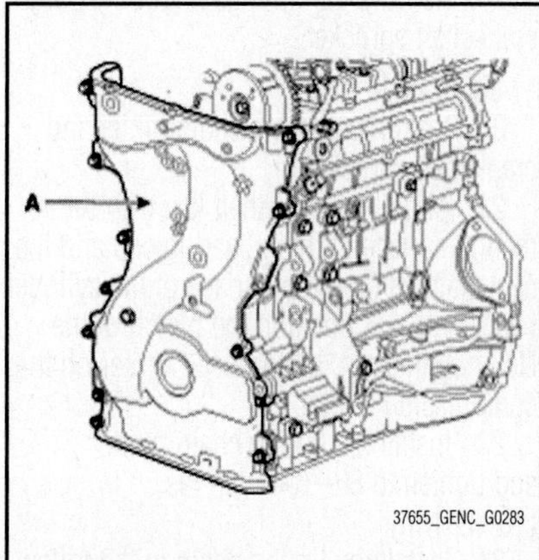

Fig. 184 Remove the timing chain cover (A)

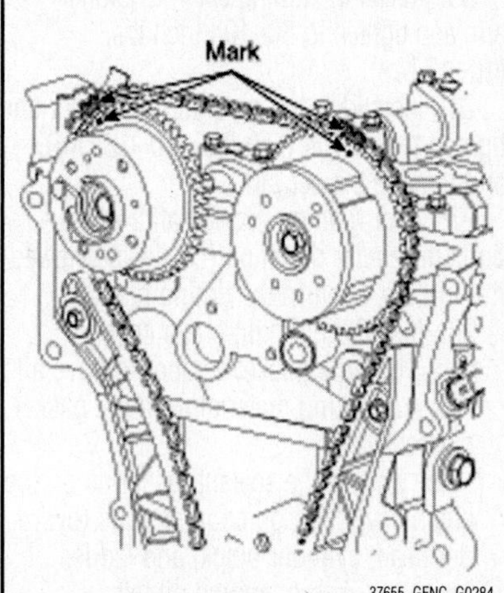

Fig. 185 Camshaft timing mark positioning

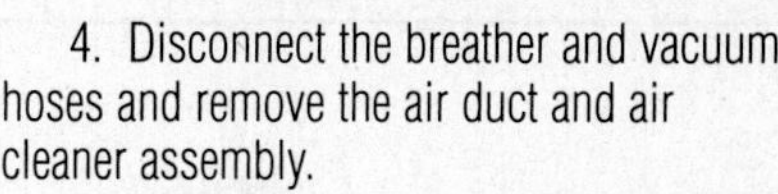

4. Disconnect the breather and vacuum hoses and remove the air duct and air cleaner assembly.
5. Remove the upper and lower radiator hoses and the intercooler inlet hose.

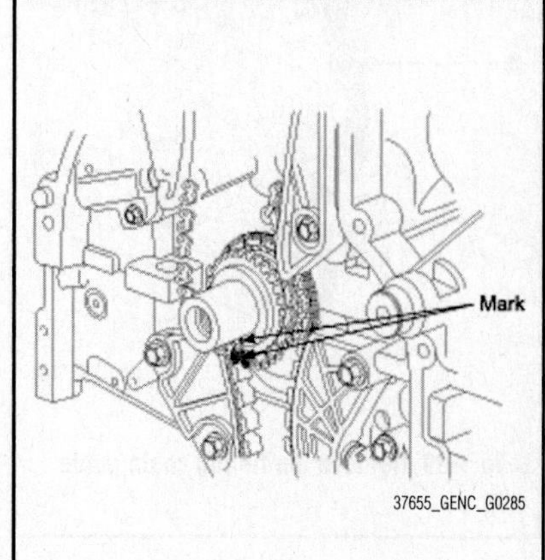

Fig. 186 Crankshaft timing mark positioning

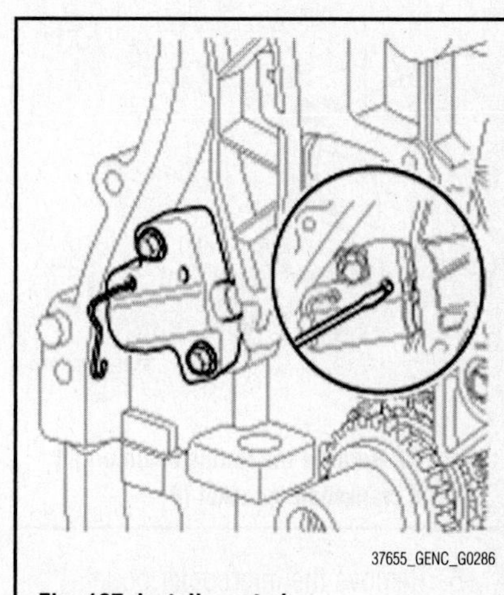

Fig. 187 Install a set pin

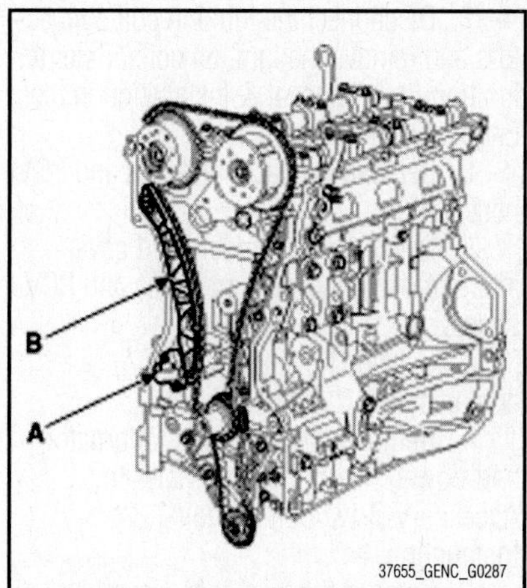

Fig. 188 Remove the timing chain tensioner (A) and timing chain tensioner arm (B)

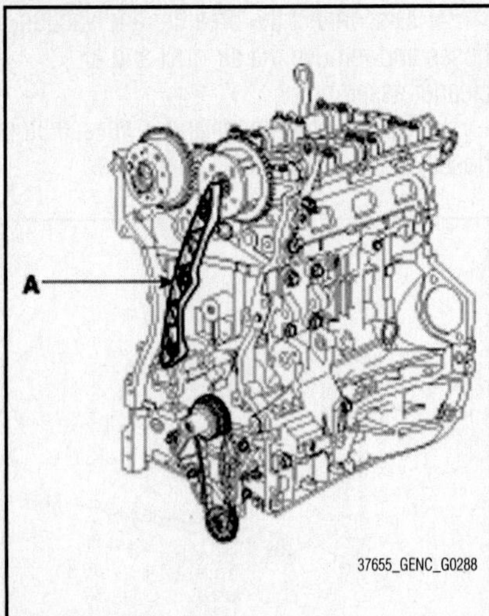

Fig. 189 Remove the timing chain guide (A)

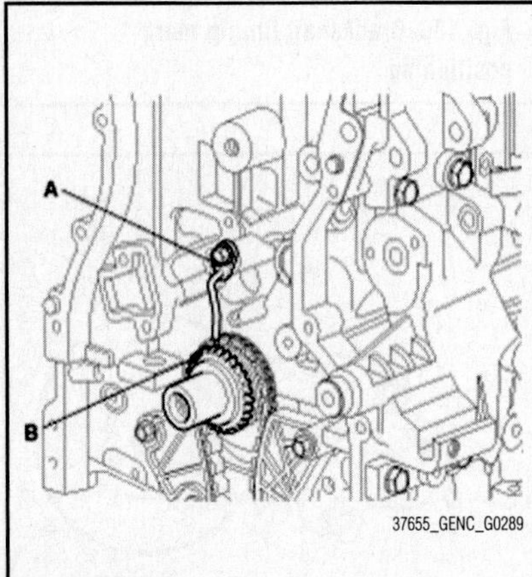

Fig. 190 Remove the timing chain oil jet (A) and crankshaft sprocket (B)

6. Remove the intercooler outlet hose after disconnecting the BPS connector.

7. Disconnect the ignition coil connectors and remove the ignition coils. Refer to Ignition Coil Removal & Installation in the Engine Electrical section.

8. Remove the vacuum hoses and PCV hose.

9. Remove the cylinder head cover after removing the vacuum hoses and PCV hose.

10. Set No. 1 cylinder to TDC/compression.

11. Remove the drive belt, alternator and power steering pump. Refer to Accessory Drive Belt Removal & Installation.

12. Remove the tensioner assembly, water pump assembly, idler and crankshaft pulley. Refer to Crankshaft Balancer Removal & Installation.

➡Use the SST (Crankshaft pulley adapter and holder, 09231-2M100,09231-2J210) to remove the crankshaft pulley bolt.

13. Remove the lower oil pan. Refer to Oil Pan Removal & Installation.

14. Remove the timing chain cover.

WARNING

Be careful not to damage the contact surfaces of cylinder block, cylinder head and timing chain cover.

15. Align the crankshaft key with the mating surface of main bearing cap. This will position the piston of the No. 1 cylinder at the top dead center on the compression stroke.

16. Install a set pin after compressing the timing chain tensioner.

17. Remove the timing chain tensioner and timing chain tensioner arm.

18. Remove the timing chain.

19. Remove the timing chain guide.

20. Remove the timing chain oil jet and crankshaft sprocket.

To install:

21. Install the timing chain oil jet and crankshaft sprocket.

22. Align the crankshaft key with the mating surface of main bearing cap and the camshaft with the top surface of the cylinder head. This will position the piston of the No. 1 cylinder at the top dead center on the compression stroke.

23. Install the timing chain guide and tighten to 86–104 inch lbs. (10–12 Nm).

24. Install the timing chain in the following sequence: crankshaft sprocket, timing chain guide, intake CVVT assembly, exhaust CVVT assembly.

25. Install the timing chain tensioner arm and tighten to 86–104 inch lbs. (10–12 Nm).

26. Install the timing chain tensioner and tighten to 86–104 inch lbs. (10–12 Nm), and remove the set pin.

27. After rotating crankshaft 2 revolutions in regular direction (clockwise viewed from front), confirm the timing mark.

28. Install the timing chain cover:

a. Using a gasket scraper remove all the old packing material from the gasket surfaces.

b. Ensure the sealant locations on the chain cover and on counter parts (cylinder head, cylinder block, and ladder frame) are free of engine oil, etc.

c. Before assembling the timing chain cover, the liquid sealant Loctite® 5900H or ThreeBond®1217H should be applied on the gap between cylinder head and cylinder block. Assemble the part within 5 minutes of sealant application. Bead width: 0.1 in. (2.5mm)

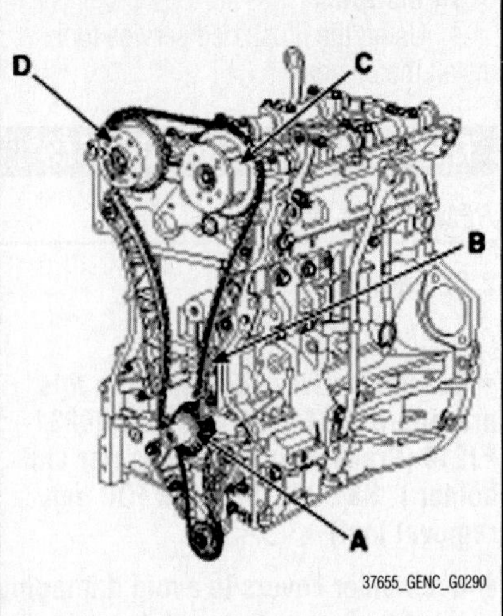

Fig. 191 Timing chain installation sequence: crankshaft sprocket (A), timing chain guide (B), intake CVVT assembly (C), exhaust CVVT assembly (D)

d. Continuously apply liquid sealant TB1217H on the timing chain cover. The part must be assembled within 5 minutes after sealant was applied.

e. The dowel pins on the cylinder block and holes on the timing chain cover should be used as a reference in order to assemble the timing chain cover to be in exact position. Tighten the timing chain bolts as follows:

- M6: 70–86 inch lbs. (8–10 Nm)
- M8: 14–17 ft. lbs. (19–23 Nm)

f. The firing and/or blow out test should not be performed within 30 minutes after the timing chain cover was assembled.

29. Install the oil pan. Wait at least 30 minutes before filling the engine with oil.

30. Install the cylinder head cover.

31. Install the crankshaft pulley and tighten to 123–130 ft. lbs. (167–177 Nm).

32. Install the water pump pulley and idler. Tighten the water pump pulley to 86–104 inch lbs. (10–12 Nm) and the idler to 40–47 ft. lbs. (54–34 Nm).

33. Install the tensioner bracket assembly and tighten to 29–33 ft. lbs. (39–44 Nm).

34. Install the alternator and tighten to 36–47 ft. lbs. (49–64 Nm).

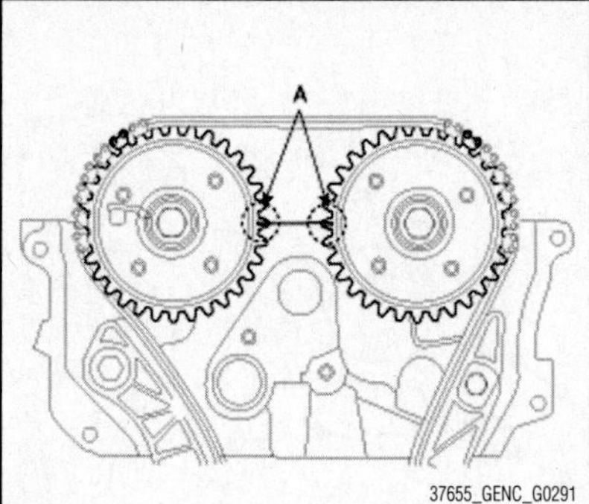

Fig. 192 Crankshaft timing mark positioning

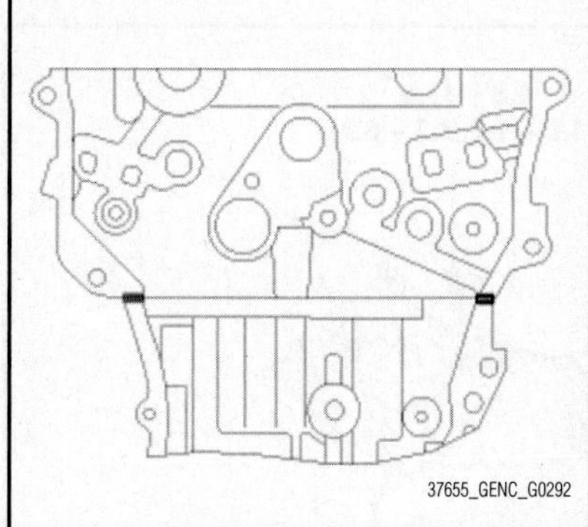

Fig. 193 Apply sealant to the gap between the cylinder head and the cylinder block

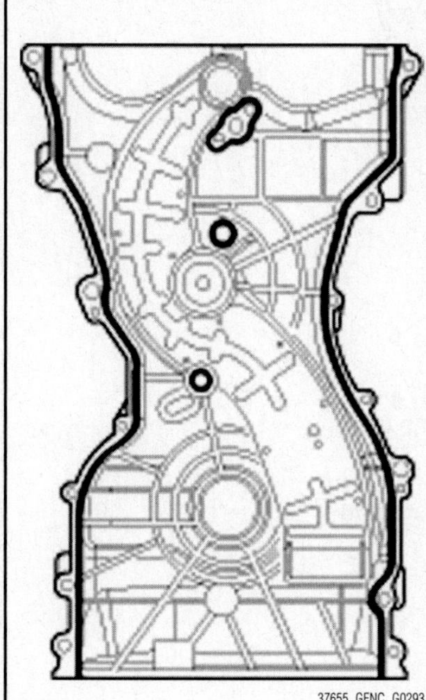

Fig. 194 Timing chain cover sealant application

35. Install the power steering pump and tighten to 12–15 ft. lbs. (17–20 Nm).
36. Install the ignition coil and connect the ignition coil connector and tighten to 86–104 inch lbs. (10–12 Nm).
37. Install the vacuum hose and PCV hose.
38. Install the intercooler inlet hose.
39. Install the intercooler outlet hose and connect the BPS connector.
40. Install the radiator upper hose.
41. Install the air cleaner assembly and air duct.
42. Connect the breather hose and vacuum hose.
43. Connect the negative battery cable.
44. Refill engine oil.
45. Clean the battery posts and cable terminals with sandpaper. Assemble and then apply grease to prevent corrosion.
46. Inspect for fuel leakage.
47. Refill radiator and reservoir tank with engine coolant.
48. Bleed air from the cooling system.

3.8L Engine

See Figures 195 through 234.

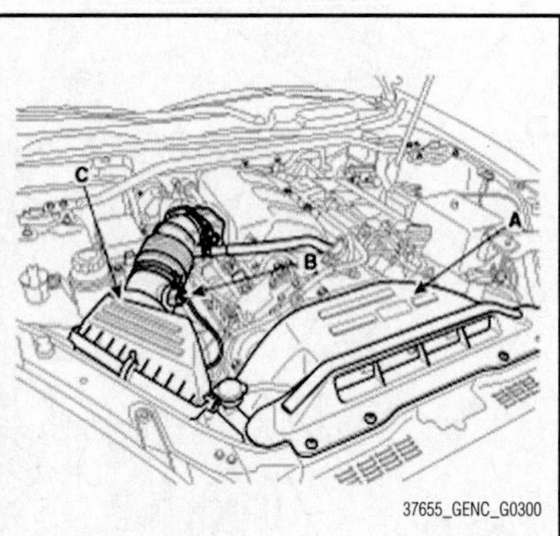

Fig. 195 Remove the AFS connector (B) and the air cleaner assembly (C)

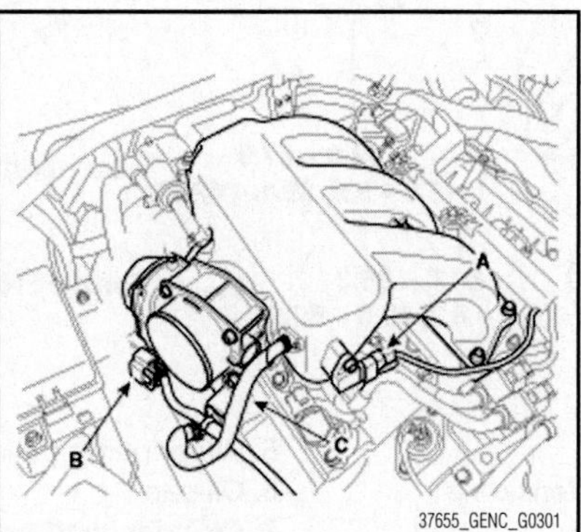

Fig. 196 Disconnect the MAP sensor connector (A), ETC connector (B), and PCV hose (C)

➡Special Service tools used for this procedure: SST 09231-2J210, 09231-2J200 (Crankshaft pulley adapter and holder), and SST 09215-3C000 (Oil pan removal tool)

➡Use fender covers to avoid damaging painted surfaces. To avoid damage, unplug the wiring connectors carefully while holding the connector portion. Mark all wiring and hoses to avoid misconnection. Turn the crankshaft pulley so that the No. 1 piston is at top dead center.

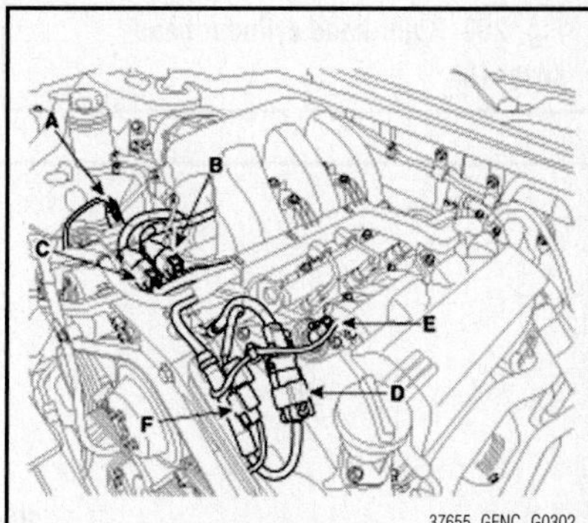

Fig. 197 Disconnect the engine connectors

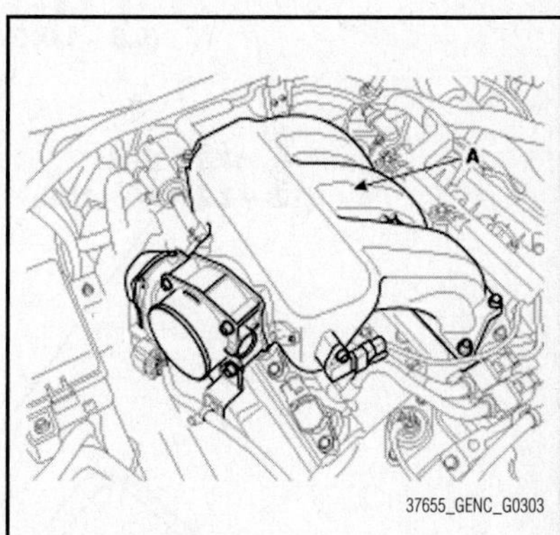

Fig. 198 Remove the surge tank assembly (A)

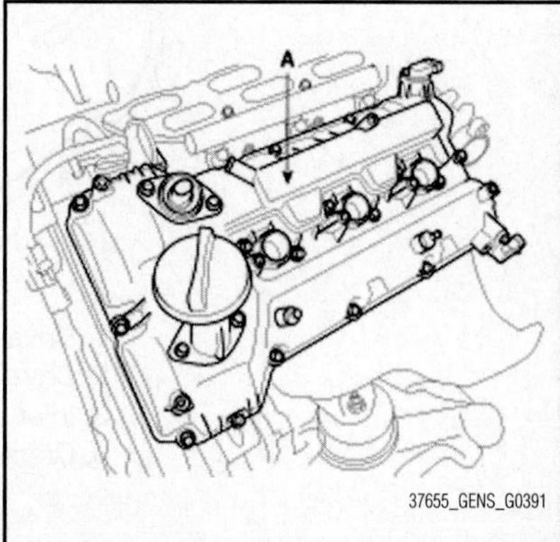

Fig. 199 Left-hand cylinder head cover (A)

A
37655_GENS_G0392

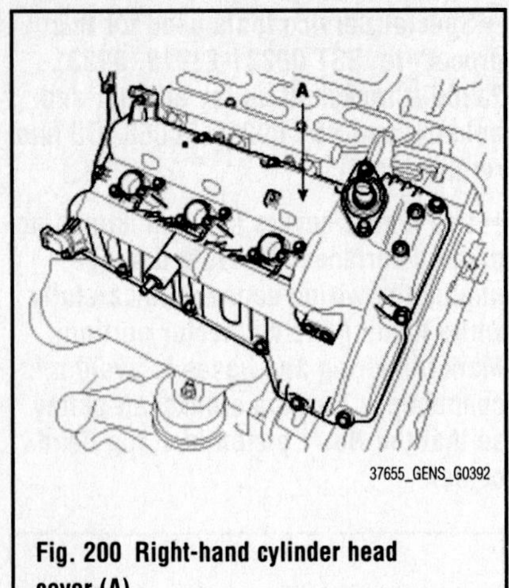

Fig. 200 Right-hand cylinder head cover (A)

A
37655_GENS_G0394

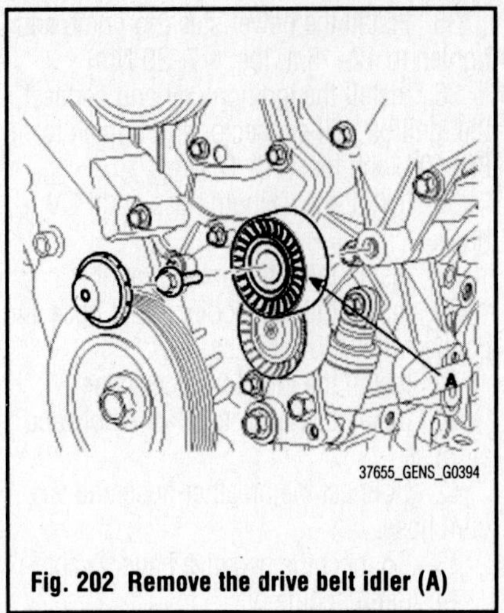

Fig. 202 Remove the drive belt idler (A)

C
B
A
37655_GENS_G0395

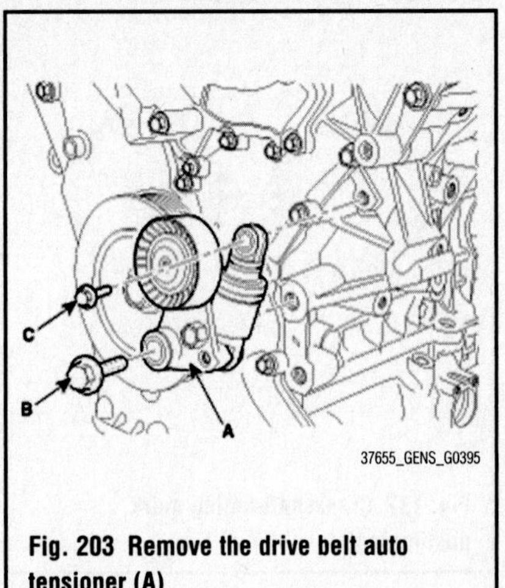

Fig. 203 Remove the drive belt auto tensioner (A)

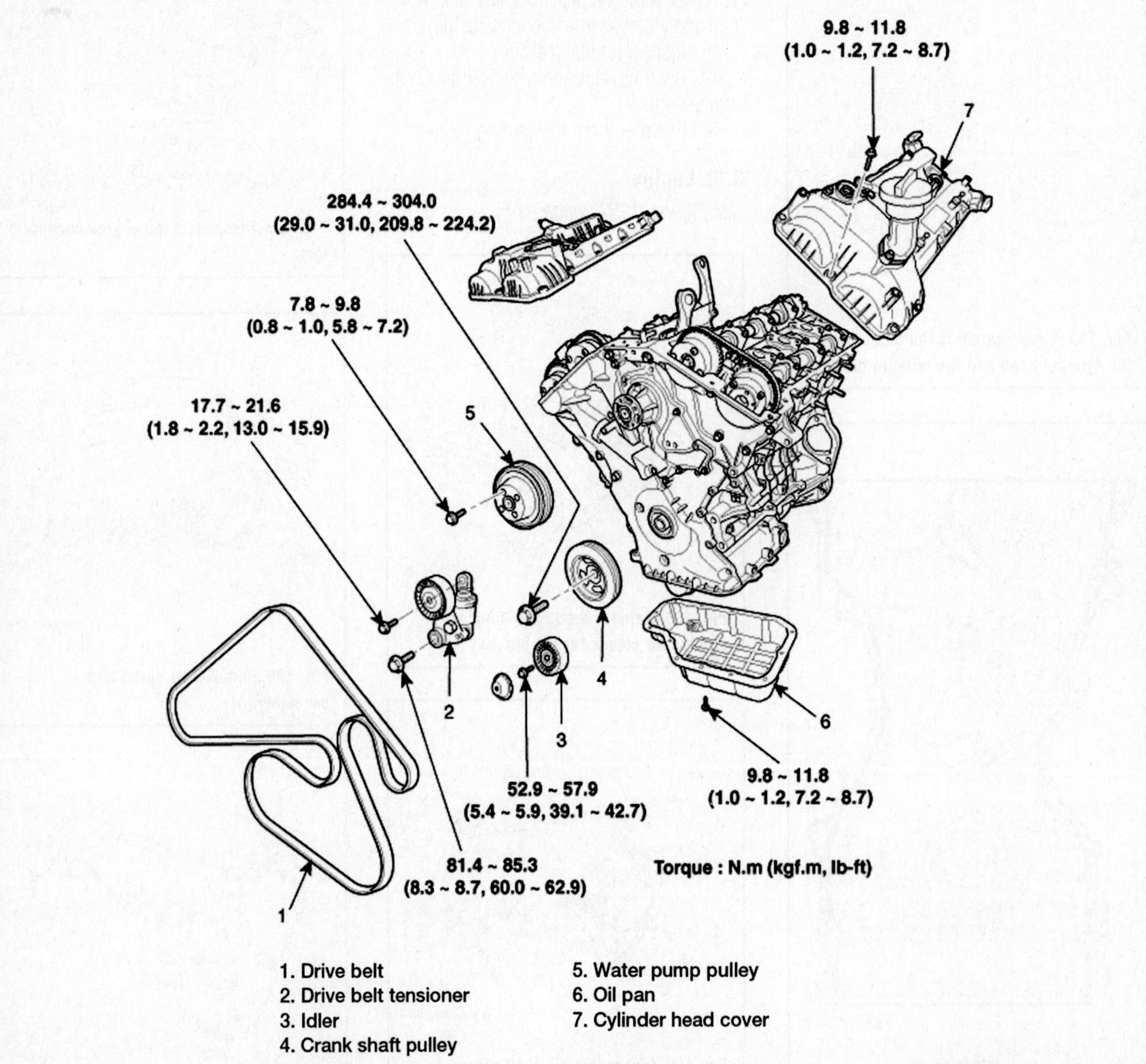

9.8 ~ 11.8
(1.0 ~ 1.2, 7.2 ~ 8.7)
7
284.4 ~ 304.0
(29.0 ~ 31.0, 209.8 ~ 224.2)
7.8 ~ 9.8
(0.8 ~ 1.0, 5.8 ~ 7.2)
17.7 ~ 21.6
(1.8 ~ 2.2, 13.0 ~ 15.9)
5
2
4
3
6
9.8 ~ 11.8
(1.0 ~ 1.2, 7.2 ~ 8.7)
52.9 ~ 57.9
(5.4 ~ 5.9, 39.1 ~ 42.7)
81.4 ~ 85.3
(8.3 ~ 8.7, 60.0 ~ 62.9)
Torque : N.m (kgf.m, lb-ft)
1
1. Drive belt
2. Drive belt tensioner
3. Idler
4. Crank shaft pulley
5. Water pump pulley
6. Oil pan
7. Cylinder head cover
37655_GENS_G0393

Fig. 201 Drive belt and components

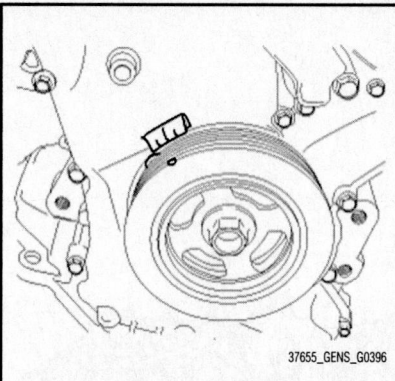

37655_GENS_G0396

Fig. 204 Turn the crankshaft pulley clockwise and align its groove with the timing mark "T" of the lower timing chain cover

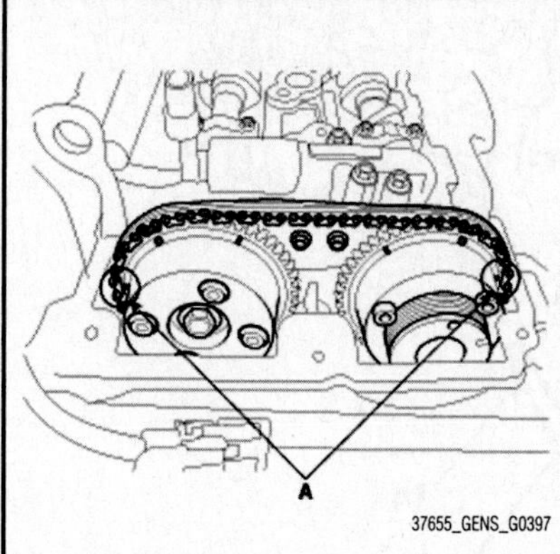

Fig. 205 Check that the mark (A) of the camshaft timing sprockets are in straight line on the cylinder head surface

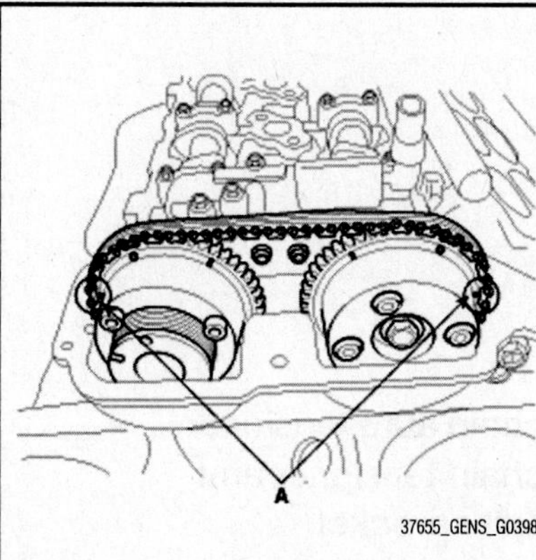

Fig. 206 Check that the mark (A) of the camshaft timing sprockets are in straight line on the cylinder head surface

1. Before servicing the vehicle, refer to the Precautions Section.
2. Disconnect the battery negative cable.
3. Remove the drain plug and drain the engine coolant.

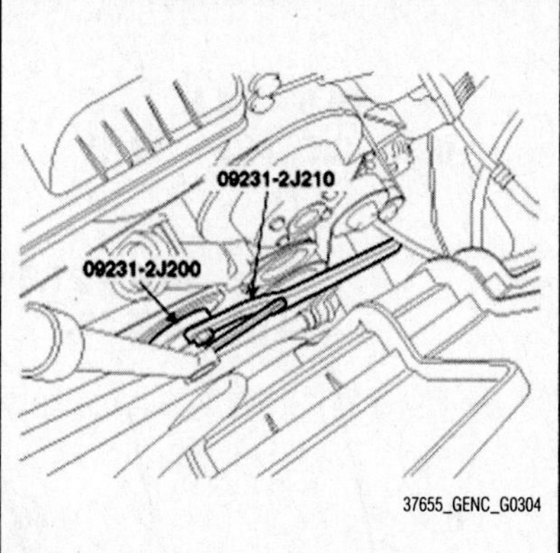

Fig. 207 Use the SST (09231-2J210, 09231-2J200) to fix the crankshaft pulley

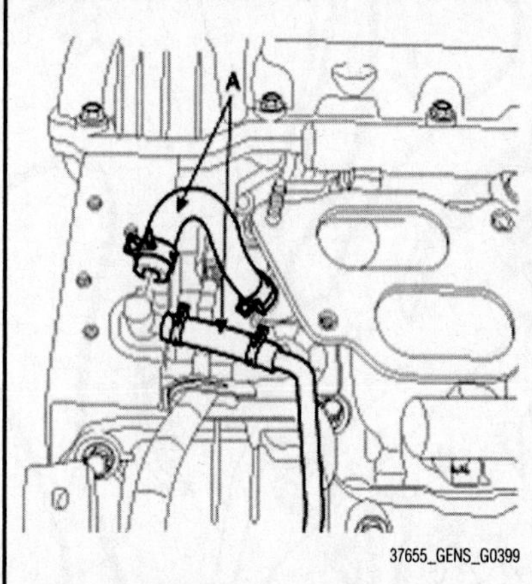

Fig. 208 Remove the water vent hose (A) from the timing chain cover

4. Remove the engine cover.
5. After recovering refrigerant, remove the high and low pressure pipe.
6. Remove the air duct.
7. Remove the air cleaner assembly after removing the AFS connector.
8. Remove the upper and lower radiator hoses.
9. Disconnect the engine wiring connectors.
   a. Disconnect the power steering oil pressure switch connector and right-hand knock sensor connector.
   b. Disconnect the MAP sensor connector, ETC connector, and PCV hose.
   c. Disconnect the right-hand exhaust OCV connector, right-hand injector connector, right-hand ignition coil connector, left-hand/right-hand intake OCV connector, left-hand exhaust OCV connector, and the oil pressure switch connector.
   d. Disconnect the left-hand injector connectors and left-hand ignition coil connectors.
   e. Disconnect the PCSV connector, PCSV hose, and throttle body coolant hose.
   f. Disconnect the left-hand exhaust CMP sensor connector.
   g. Disconnect the left-hand intake CMP sensor connector, and the water temperature sensor connector.
   h. Disconnect the right-hand intake CMP sensor connector, and the oil temperature sensor connector.
   i. Disconnect the right-hand exhaust CMP sensor connector.
10. Remove the left-hand side coolant pipe and hose.
11. Remove the right-hand side coolant pipe.
12. Remove the surge tank assembly. Refer to Intake Manifold Removal & Installation.
13. Disconnect the right-hand ignition coil connector and the injector connector.
14. Remove the left-hand/right-hand ignition coils.
15. Remove the left-hand/right-hand cylinder head covers.
16. Remove the drive belt after removing the oil pressure gauge. Refer to Accessory Drive Belt Removal & Installation.
17. Remove the power steering pump. Refer to Power Steering Pump Removal & Installation in the Steering section.
18. Remove the air conditioner compressor.
19. Remove the alternator. Refer to Alternator Removal & Installation in the Engine Electrical section.
20. Remove the drive belt idler.
21. Remove the drive belt auto tensioner.
22. Remove the water pump pulley.
23. Remove the oil filter body.
24. Set No. 1 cylinder to TDC/ compression:
   a. Turn the crankshaft pulley clockwise and align its groove with the timing mark "T" of the lower timing chain cover.
   b. Check that the mark of the camshaft timing sprockets are in straight line on the cylinder head surface as shown. If not, turn the crankshaft clockwise one revolution (360°).

➡Do not rotate engine counterclockwise.

25. Remove the lower oil pan. Refer to Oil Pan Removal & Installation
26. Remove the crankshaft pulley. Refer to Crankshaft Damper Removal & Installation.

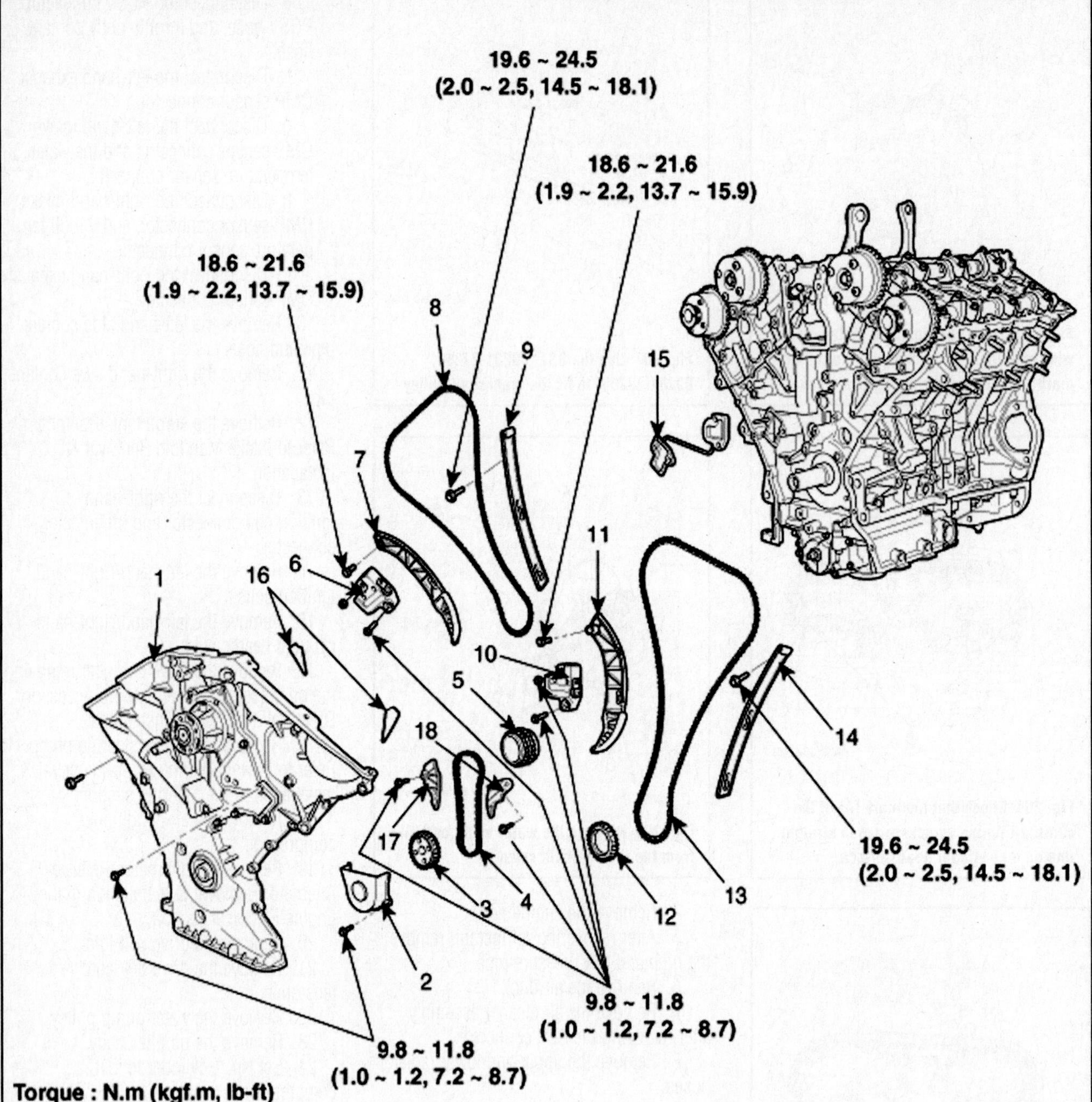

1. Timing chain cover
2. Oil pump chain cover
3. Oil pump sprocket
4. Oil pump chain
5. Crankshaft sprocket
6. Timing chain auto tensioner
7. Timing chain tensioner arm
8. Timing chain
9. Timing chain guide
10. Timing chain auto tensioner
11. Timing chain tensioner arm
12. Crankshaft sprocket
13. Timing chain
14. Timing chain guide
15. Tensioner adapter
16. Gasket
17. Oil pump chain guide
18. Oil pump tensioner assembly

37655_GENS_G0421

Fig. 209 Timing chain and components

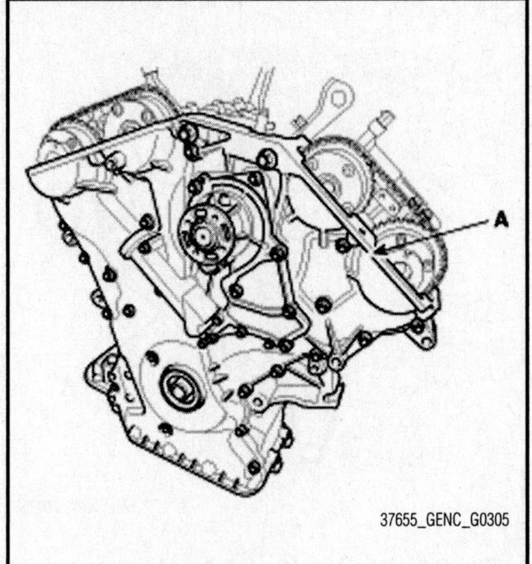

Fig. 210 Remove the timing chain cover (A)

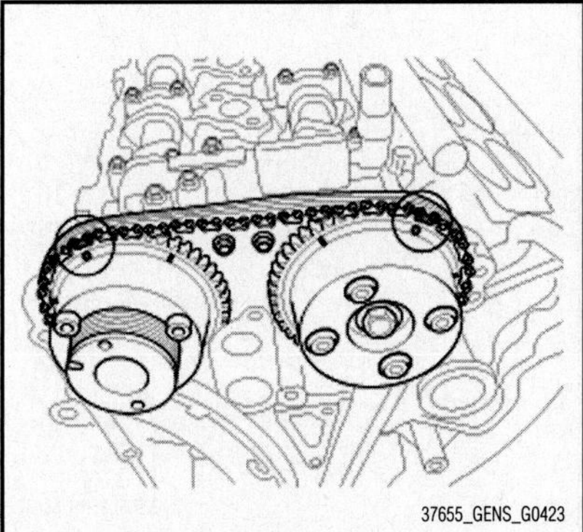

Fig. 211 Timing chain to camshaft position location—left-hand side

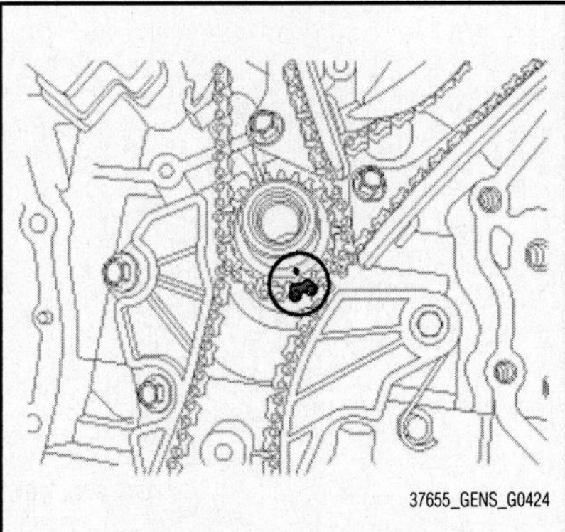

Fig. 212 Timing chain to crankshaft position location

➡Use the SST (09231-2J210, 09231-2J200) to hold the crankshaft pulley.

27. Remove the water vent hose from the timing chain cover.
28. Remove the timing chain cover.

➡Be careful not to damage the contact surfaces of cylinder block, cylinder head and timing chain cover.

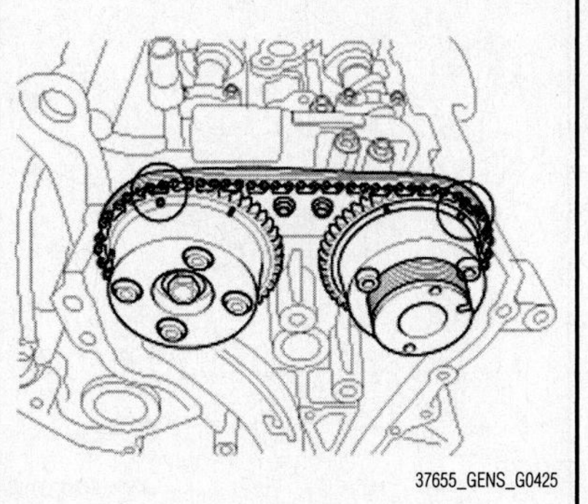

Fig. 213 Timing chain to camshaft position location—right-hand side

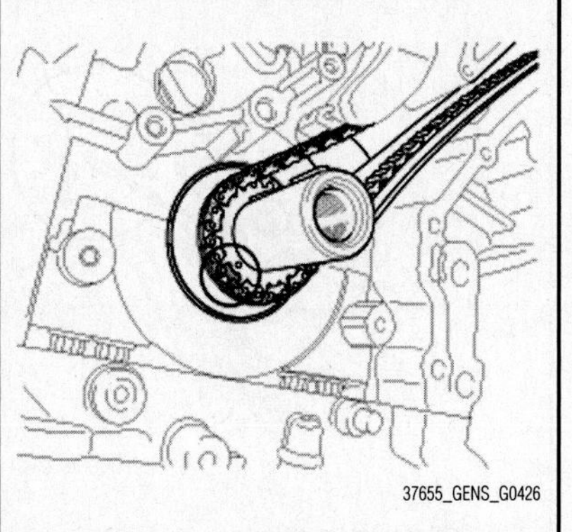

Fig. 214 Timing chain to crankshaft sprocket location

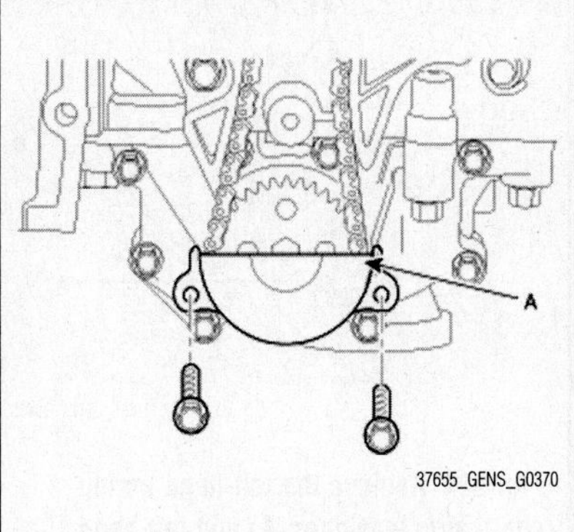

Fig. 215 Remove the oil pump chain cover (A)

a. Before removing the timing chain, matchmark the right-hand/left-hand timing chain based on the location of the sprocket.

29. Remove the oil pump chain cover.
30. Remove the oil pump chain tensioner assembly.
31. Remove the oil pump chain guide.
32. Install a set pin after compressing the right-hand timing chain tensioner.
33. Remove the right-hand timing chain auto tensioner and the right-hand timing chain tensioner arm.

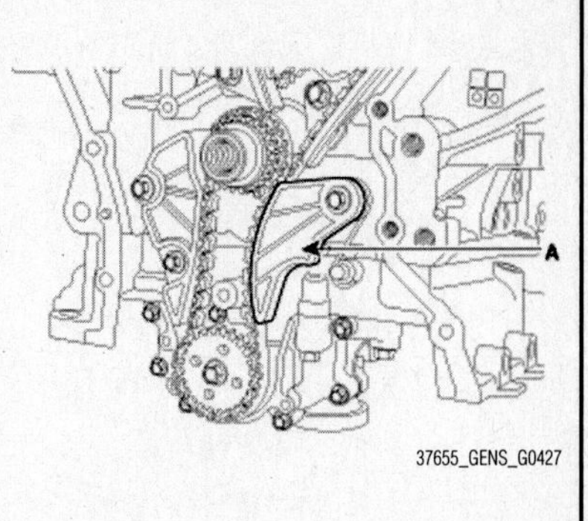

Fig. 216 Remove the oil pump chain tensioner assembly (A)

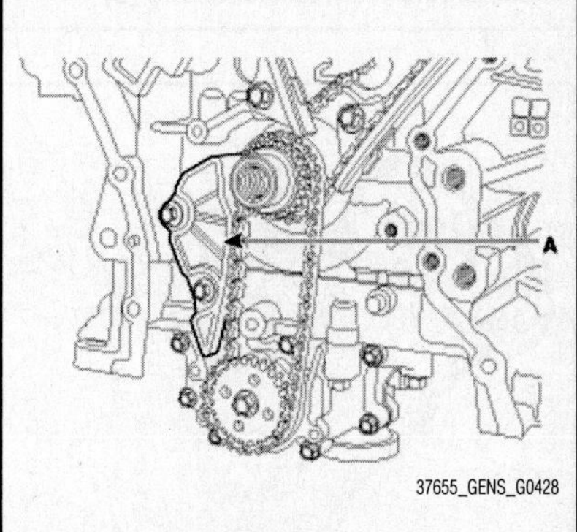

Fig. 217 Remove the oil pump chain guide (A)

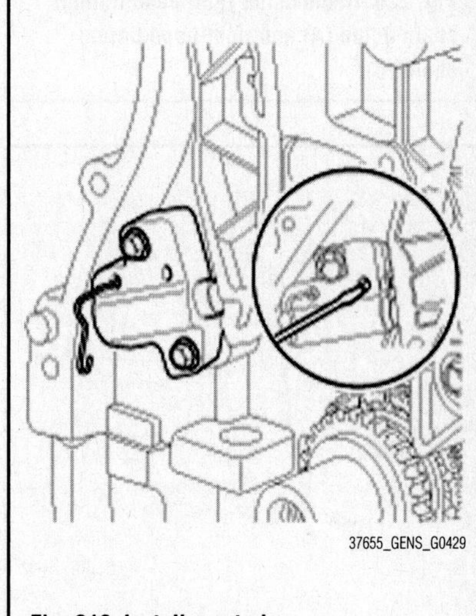

Fig. 218 Install a set pin

34. Remove the right-hand timing chain guide and right-hand timing chain.
35. Remove the oil pump chain sprocket and oil pump chain.
36. Remove the crankshaft sprocket. (O/P and right-hand camshaft drive).

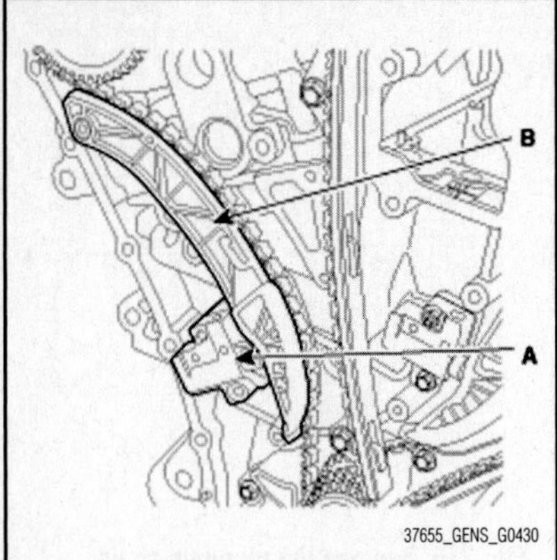

Fig. 219 Remove the right-hand timing chain auto tensioner (A) and the right-hand timing chain tensioner arm (B)

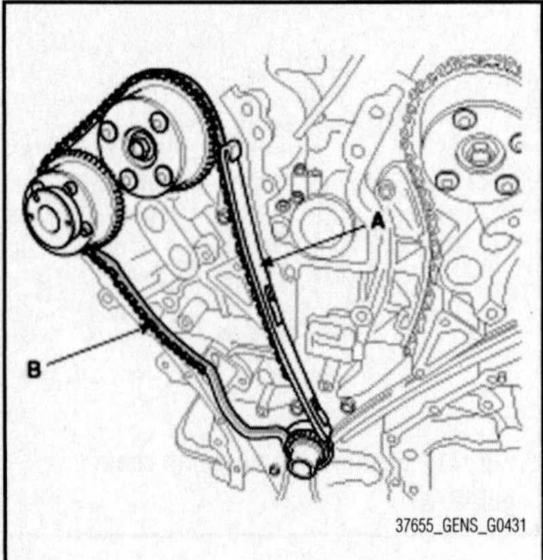

Fig. 220 Remove the right-hand timing chain guide (A) and right-hand timing chain (B)

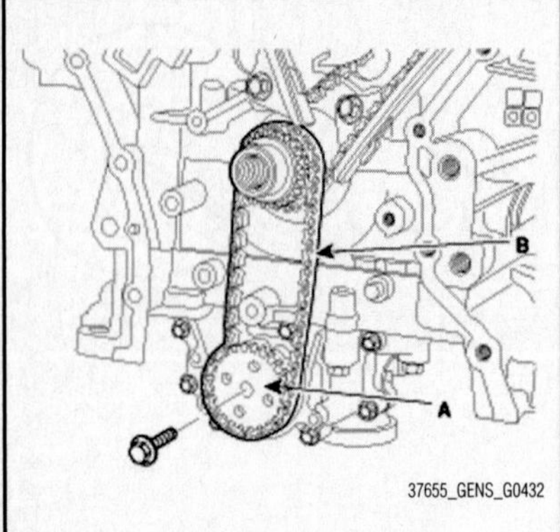

Fig. 221 Remove the oil pump chain sprocket (A) and oil pump chain (B)

37. Install a set pin after compressing the left-hand timing chain tensioner.

38. Remove the left-hand timing chain auto tensioner and left-hand timing chain tensioner arm.

39. Remove the left-hand timing chain guide and left-hand timing chain.

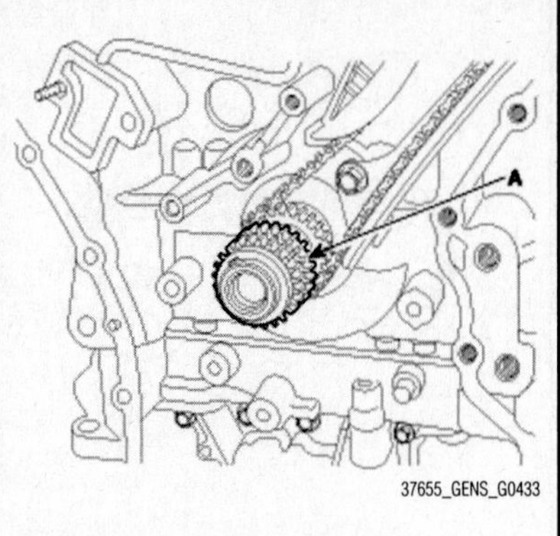

Fig. 222 Remove the crankshaft sprocket (A)

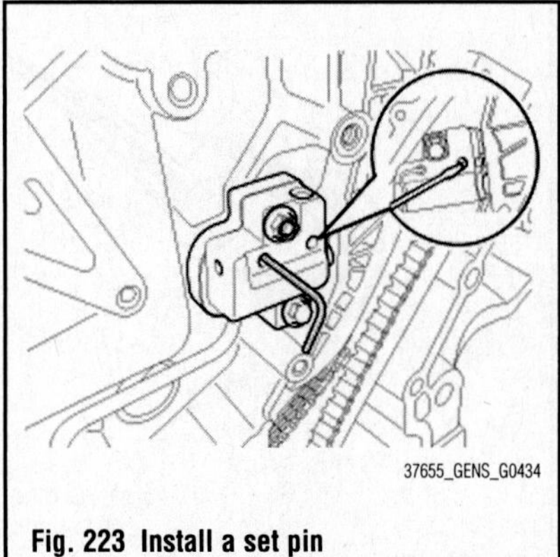

Fig. 223 Install a set pin

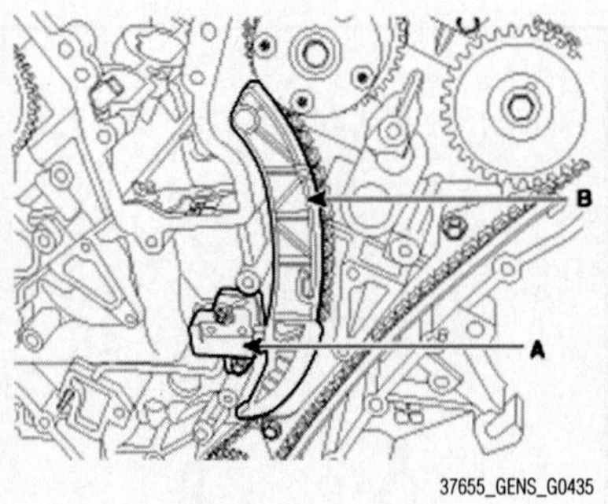

Fig. 224 Remove the left-hand timing chain auto tensioner (A) and left-hand timing chain tensioner arm (B)

40. Remove the crankshaft sprocket. (Left-hand camshaft drive)

41. Remove the tensioner adapter assembly.

To install:

42. Align the crankshaft key with the timing chain cover timing mark. This will position the No. 1 cylinder piston at the top dead center on compression stroke.

43. Install the tensioner adapter assembly.

44. Install the crankshaft sprocket. (Left-hand camshaft drive)

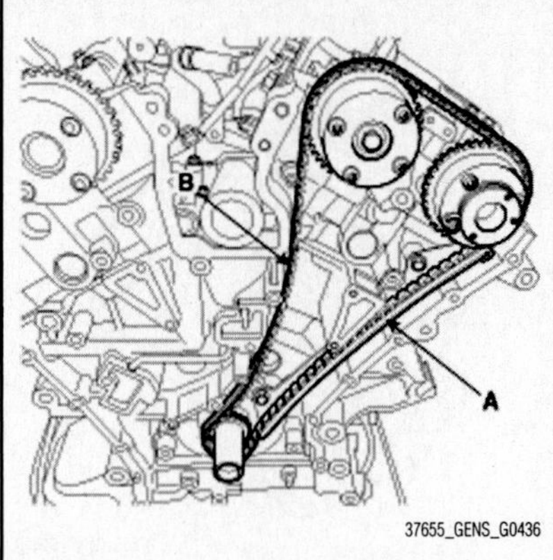

Fig. 225 Remove the left-hand timing chain guide (A) and left-hand timing chain (B)

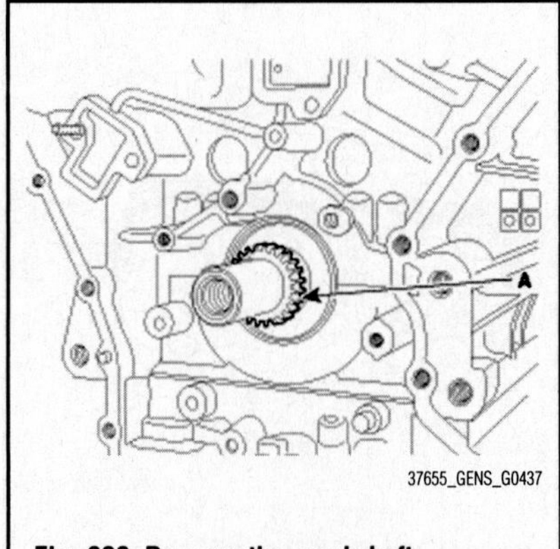

Fig. 226 Remove the crankshaft sprocket (A)

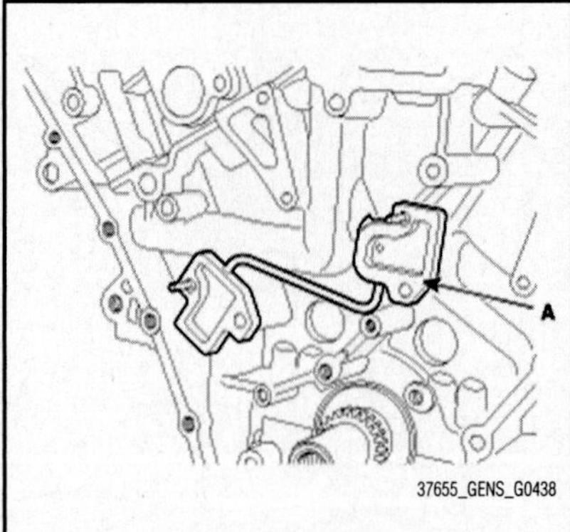

Fig. 227 Remove the tensioner adapter assembly (A)

45. Install the left-hand timing chain guide and left-hand timing chain and tighten to 15–18 ft. lbs. (20–25 Nm).

➡To install the timing chain with no slack between each shaft (cam, crank), install in the following sequence: Crankshaft sprocket, Timing chain guide, Exhaust camshaft sprocket, Intake camshaft sprocket.

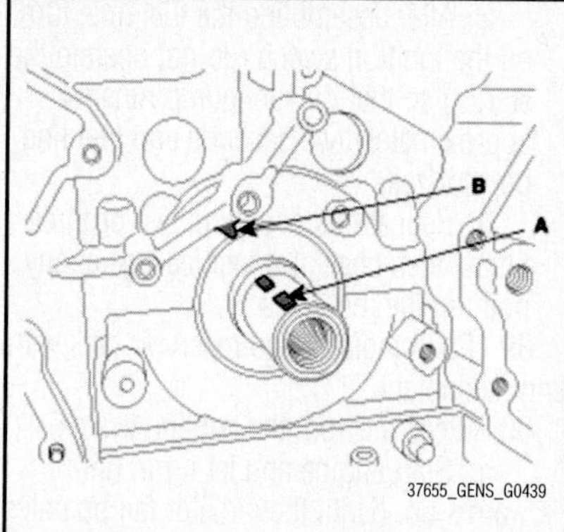

Fig. 228 Align the crankshaft key with the timing chain cover timing mark

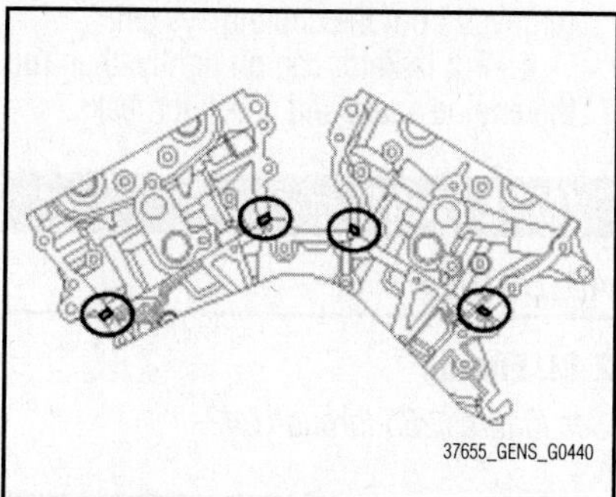

Fig. 229 Apply sealant to the gap between the cylinder head and the cylinder block

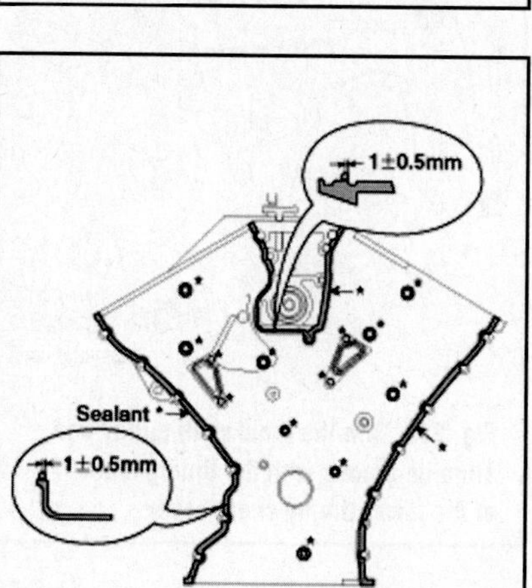

Fig. 230 Timing chain cover sealant application

The timing mark of each sprockets should be matched with timing mark (color link) of timing chain at installing timing chain.

46. Install the left-hand timing chain tensioner arm and tighten to 14–16 ft. lbs. (19–22 Nm).

47. Install the left-hand timing chain auto tensioner and tighten to 86–104 inch lbs. (10–12 Nm).

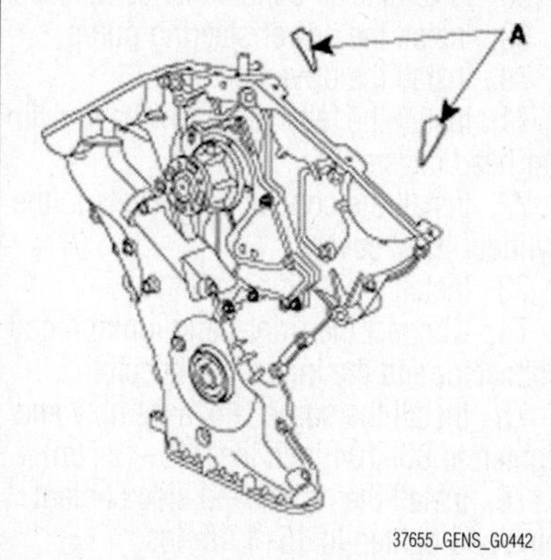

Fig. 231 Install new gaskets to the timing chain cover

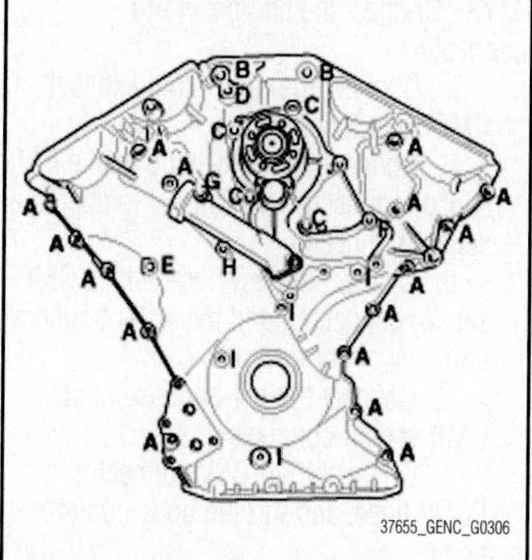

Fig. 232 Timing chain cover bolt tightening sequence

48. Install the crankshaft sprocket (O/P and right-hand camshaft drive)

49. Install the oil pump chain sprocket and oil pump chain and tighten to 14–16 ft. lbs. (19–22 Nm).

50. Install the right-hand timing chain guide and right-hand timing chain and tighten to 15–18 ft. lbs. (20–25 Nm).

➡To install the timing chain with no slack between each shaft (cam, crank), install in the following sequence: Crankshaft sprocket, Timing chain guide, Intake camshaft sprocket, Exhaust camshaft sprocket. The timing mark of each sprockets should be matched with timing mark (color link) of timing chain at installing timing chain.

51. Install the right-hand timing chain tensioner arm and tighten to 14–16 ft. lbs. (19–22 Nm).

52. Install the right-hand timing chain auto tensioner and tighten to 86–104 inch lbs. (10–12 Nm).

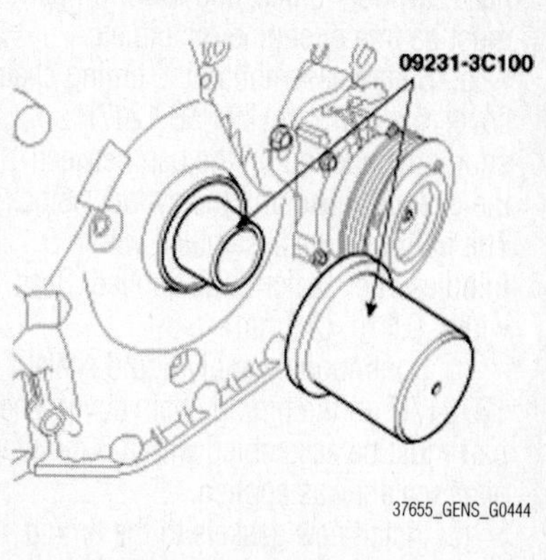

Fig. 233 Using SST (09231-3C100), install the timing chain cover oil seal

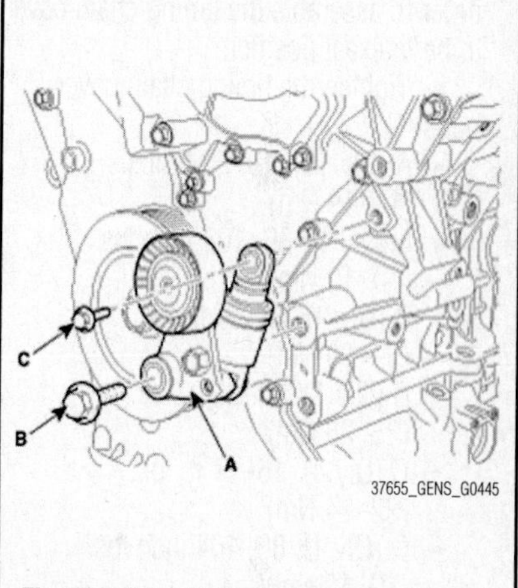

Fig. 234 Drive belt auto tensioner (A) and bolts (B, C)

53. Install the oil pump chain guide and tighten to 86–104 inch lbs. (10–12 Nm).

54. Install the oil pump chain tensioner assembly and tighten to 86–104 inch lbs. (10–12 Nm).

55. Remove the pins from the hydraulic tensioner (left-hand and right-hand).

56. Install the oil pump chain cover and tighten to 86–104 inch lbs. (10–12 Nm).

57. After rotating the crankshaft 2 revolutions in regular direction (clockwise viewed from front), confirm the timing mark.

WARNING

Always turn the crankshaft clockwise. Turning the crankshaft counter clockwise before building up oil pressure in the hydraulic timing chain tensioner may result in the chain disengaging from the sprocket teeth.

58. Install the timing chain cover.

a. The sealant locations on chain cover and on counter parts (cylinder head, cylinder block, and lower oil pan) must be free of engine oil and etc.

b. Before assembling the timing chain cover, the liquid sealant TB 1217H should be applied on the gap between the cylinder head and the cylinder block. The part must be assembled within 5 minutes after sealant was applied. Bead width: 0.1 in. (2.5mm)

c. Continuously apply liquid sealant TB1217H on the timing chain cover. The part must be assembled within 5 minutes after sealant was applied.

d. Install new gaskets to the timing chain cover.

e. The dowel pins on the cylinder block and holes on the timing chain cover should be used as a reference in order to assemble the timing chain cover to be in exact position.

f. Tighten the timing chain cover bolts:

- A (Qty 16): 19–25 ft. lbs. (14–19 Nm)
- F (Qty 1): 86–104 inch lbs. (10–12 Nm)
- B (Qty 2): 43–51 ft. lbs. (59–69 Nm)
- D (Qty 1): 18–20 ft. lbs. (25–26 Nm)
- C (Qty 4): 16–17 ft. lbs. (22–24 Nm)
- G (Qty 1): 86–104 inch lbs. (10–12 Nm)
- H (Qty 1): 86–104 inch lbs. (10–12 Nm)
- I (Qty 4): 86–104 inch lbs. (10–12 Nm)
- E (Qty 1): 18–20 ft. lbs. (25–26 Nm)

g. The firing and/or blow out test should not be perfórmed within 30 minutes after the timing chain cover was assembled.

59. Install the water vent hose to the timing chain cover.

60. Using SST (09231-3C100), install the timing chain cover oil seal.

61. Install the lower oil pan.

62. Install the crankshaft pulley and tighten to 210–224 ft. lbs. (284–304 Nm).

63. Install the water pump pulley and tighten to 70–86 inch lbs. (8–10 Nm).

64. Install the oil filter assembly.

65. Install the drive belt auto tensioner. Tightening torque:

- Bolt (B): 60–63 ft. lbs. (81–85 Nm)
- Bolt (C): 13–16 ft. lbs. (18–22 Nm)

66. Install the drive belt idler and tighten to 39–43 ft. lbs. (53–58 Nm).

67. Install the alternator.

68. Install the air conditioner compressor.

69. Install the power steering pump.

70. Install the drive belt.

71. Install the left-hand/right-hand cylinder head covers.

72. Install the connector brackets to the cylinder head covers.

73. Install the ignition coils.

74. Connect the right-hand ignition coil connector and the injector connector.

75. Install the surge tank assembly and tighten to 86–104 inch lbs. (10–12 Nm).

76. Install the right-hand side coolant pipe and tighten to 15–17 ft. lbs. (20–24 Nm).

77. Install the left-hand side coolant pipe and hose. Tighten to15–17 ft. lbs. (20–24 Nm).

78. Connect the engine wiring connectors.

a. Connect the right-hand exhaust CMP sensor connector.

b. Connect the right-hand intake CMP sensor connector, and the oil temperature sensor connector.

c. Connect the left-hand intake CMP sensor connector, and the water temperature sensor connector.

d. Connect the left-hand exhaust CMP sensor connector.

e. Connect the PCSV connector, PCSV hose, and throttle body coolant hose.

f. Connect the left-hand injector connector and left-hand ignition coil connectors.

g. Connect the right-hand exhaust OCV connector, right-hand injector connector, right-hand ignition coil connector, left-hand/right-hand intake OCV connector, left-hand exhaust OCV connector, and the oil pressure switch connector.

h. Connect the MAP sensor connector, ETC connector, and PCV hose.

i. Connect the power steering oil pressure switch connector and right-hand knock sensor connector.

79. Install the cooling fan.

80. Install the fuel hose and brake vacuum hose.

81. Install the radiator hoses.

82. Install the AFS connector and air cleaner assembly.

83. Install the air duct.

84. Install the engine cover.

85. Connect the battery negative cable.

86. Refill engine oil.

87. Clean the battery posts and cable terminals with sandpaper. Assemble and then apply grease to prevent corrosion.

88. Inspect for fuel leakage.

a. After assembling the fuel line, turn on the ignition switch (do not operate the starter) so that the fuel pump runs for approximately two seconds and fuel line pressurizes.

b. Repeat this operation two or three times, then check for fuel leakage at any point in the fuel lines.

89. Refill radiator and reservoir tank with engine coolant.

90. Bleed air from the cooling system.

a. Start engine and let it run until it warms up. (Until the radiator fan operates 3 or 4 times.)

b. Turn Off the engine. Check the level in the radiator, add coolant if needed. This will allow trapped air to be removed from the cooling system.

c. Put radiator cap on tightly, then run the engine again and check for leaks.

VALVE LASH

ADJUSTMENT

2.0L Engine

See Figures 235 through 242.

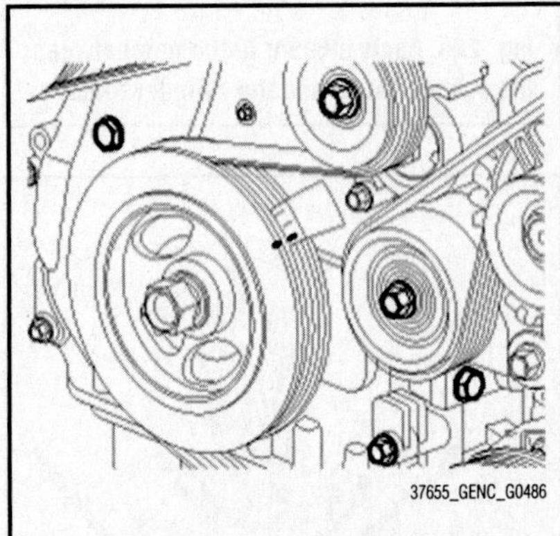

Fig. 235 Turn the crankshaft pulley and align its groove with the timing mark "T" of the lower timing chain cover

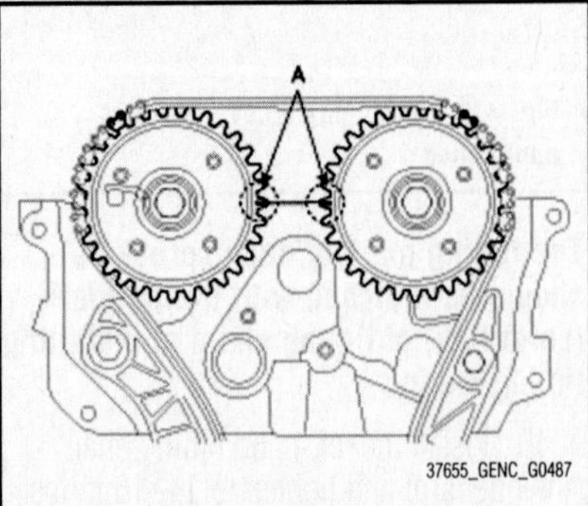

Fig. 236 Check that the mark (A) of the camshaft timing sprockets are in straight line on the cylinder head

➡**Special Service Tool used for this procedure: SST 09240-2G000, Auto tensioner locking tool**

➡**Inspect and adjust the valve clearance when the engine is cold [(Engine coolant temperature: 68°F (20°C)] and cylinder head is installed on the cylinder block.**

1. Remove the cylinder head cover.
2. Set the No. 1 cylinder to TDC/compression:
   a. Turn the crankshaft pulley and align its groove with the timing mark "T" of the lower timing chain cover.
   b. Check that the mark of the camshaft timing sprockets are in straight line on the cylinder head surface as shown in the illustration. If not, turn the crankshaft one revolution (360°).
3. Inspect the valve clearance:
   a. Check only the valve indicated as shown. (No. 1 cylinder: TDC/Compression) measure the valve clearance. Using a thickness gauge, measure the clearance between the tappet and the base circle of camshaft. Record the out-of-specification valve clearance measurements. They will be used later to determine the required replacement adjusting tappet. Specification: Engine coolant temperature: (68°F (20°C). Valve Clearance Limit:
      - Intake: 0.0039–0.0118 in. (0.10–0.30mm)
      - Exhaust: 0.0079–0.0157 in. (0.20–0.40mm)

   b. Turn the crankshaft pulley one revolution (360°) and align the groove with timing mark "T" of the lower timing chain cover.
   c. Check only valves indicated as shown. (No. 4 cylinder: TDC/compression). Measure the valve clearance.
4. Adjust the intake and exhaust valve clearance.
   a. Set the No. 1 cylinder to the TDC/compression.
   b. Matchmark the timing chain and camshaft timing sprockets.
   c. Remove the service hole bolt of the timing chain cover.

➡**The bolt must NOT be reused once it has been assembled.**

   d. Insert the SST (09240-2G000) in the service hole of the timing chain cover and release the ratchet.
   e. Remove the front camshaft bearing cap.
   f. Remove the exhaust camshaft bearing cap and exhaust camshaft.
   g. Remove the intake camshaft bearing cap and intake camshaft.

➡**When disconnecting the timing chain from the camshaft timing sprocket, hold the timing chain.**

   h. Tie down timing chain so that it doesn't move.

➡**Be careful not to drop anything inside timing chain cover.**

   i. Measure the thickness of the removed tappet using a micrometer.
   j. Calculate the thickness of a new tappet so that the valve clearance comes within the specified value. Valve clearance (Engine coolant temperature: 20°C), (T=Thickness of removed tappet,

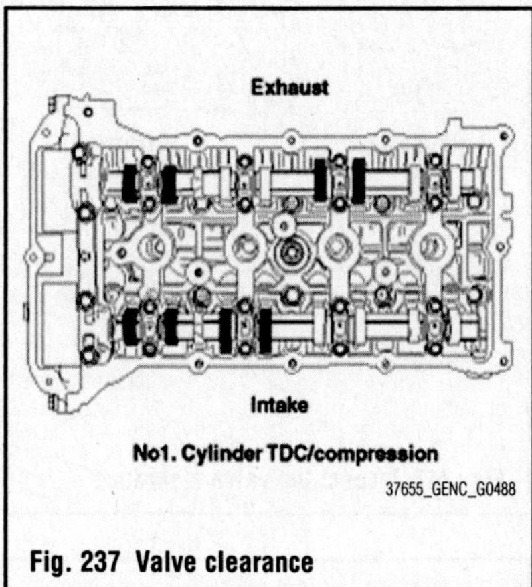

Fig. 237 Valve clearance

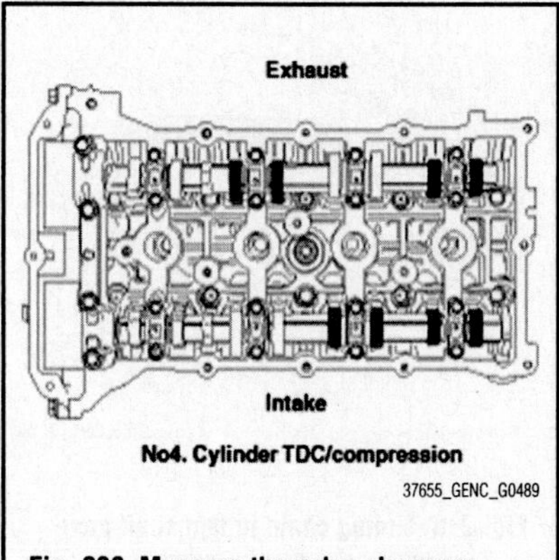

Fig. 238 Measure the valve clearance

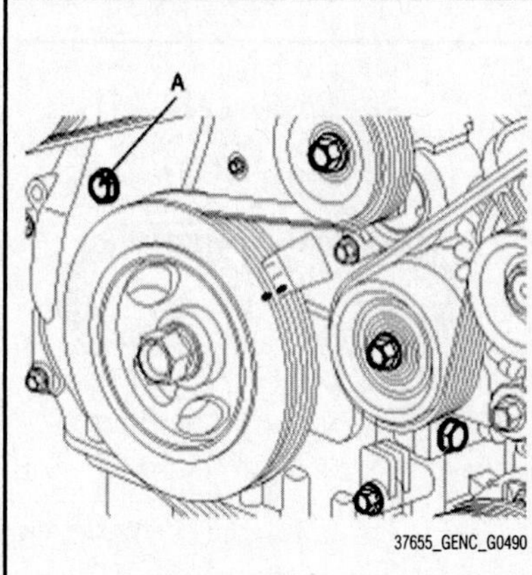

Fig. 239 Remove the service hole bolt (A) of the timing chain cover

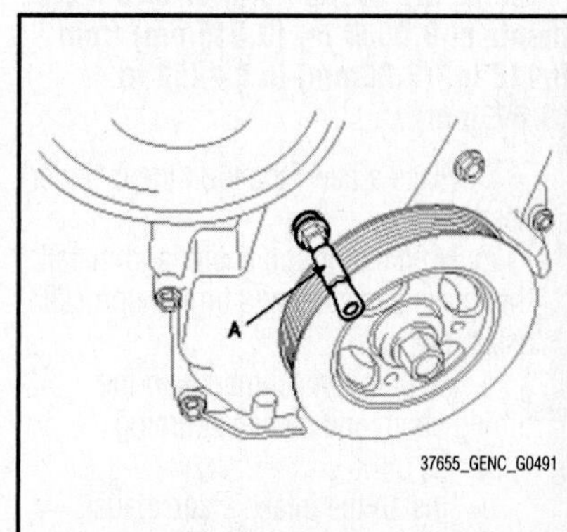

Fig. 240 Insert the SST (A) (09240-2G000) in the service hole of the timing chain cover and release the ratchet

Fig. 241 Remove the front camshaft bearing cap (A)

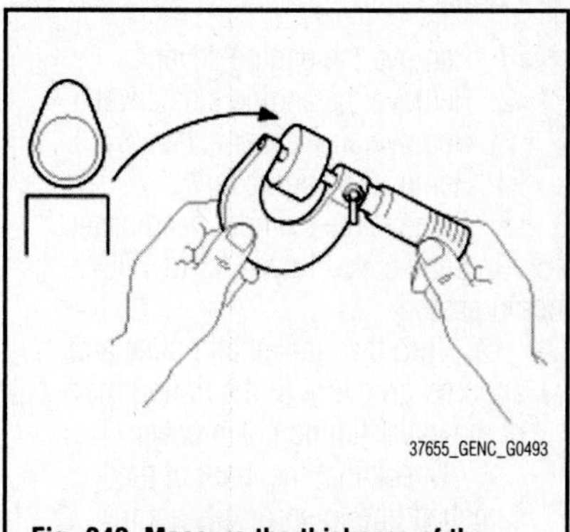

Fig. 242 Measure the thickness of the removed tappet using a micrometer

A=Measured valve clearance, N=Thickness of new tappet):

- Intake: N = T plus [A minus 0.0079 in. (0.20mm)]
- Exhaust: N = T plus [A minus 0.0118 in. (0.30mm)]

k. Select a new tappet with a thickness as close as possible to the calculated value.

➡Shims are available in 47 size increments of 0.0006 in. (0.015mm) from 0.118 in. (3.00mm) to 0.1452 in. (3.690mm).

l. Place a new tappet on the cylinder head.

m. Hold the timing chain, and install the intake camshaft and timing sprocket assembly.

n. Align the matchmarks on the timing chain and camshaft timing sprocket.

o. Install the intake and exhaust camshaft.

p. Install the front bearing cap.

q. Install the service hole bolt. Tightening torque:9–11 ft. lbs. (12–15 Nm)

r. Turn the crankshaft two turns in the operating direction (clockwise) and realign crankshaft sprocket and camshaft sprocket timing marks.

s. Recheck the valve clearance. Valve clearance (Engine coolant temperature : 20°C) Specification:

- Intake: 0.0067–0.0090 in. (0.17–0.23mm)
- Exhaust: 0.0106–0.0129 in. (0.27–0.33mm)

3.8L Engine

See Figures 243 through 252.

➡Inspect and adjust the valve clearance when the engine is cold [(Engine coolant temperature: 68°F (20°C)] and cylinder head is installed on the cylinder block.

1. Remove the engine cover.
2. Remove the engine side cover.
3. Remove air cleaner assembly.
4. Remove the surge tank.
5. Remove the cylinder head cover.
6. Set the No. 1 cylinder to TDC/ compression:

a. Turn the crankshaft pulley and align its groove with the timing mark "T" of the lower timing chain cover.

b. Check that the mark of the camshaft timing sprockets are in straight line on the cylinder head surface as shown in the illustration. If not, turn the crankshaft one revolution (360°).

⁂ WARNING

Do not rotate engine counterclockwise.

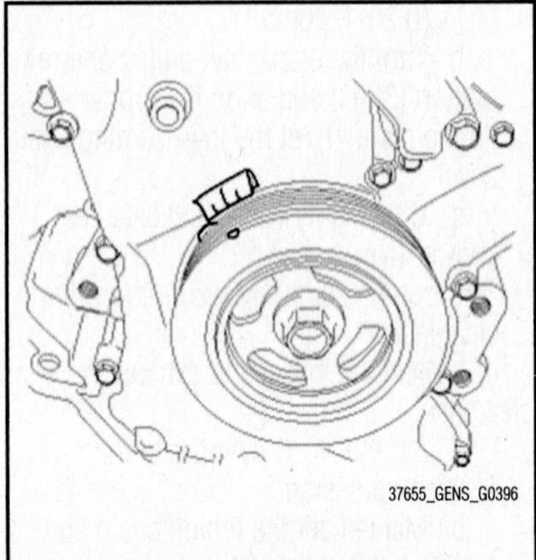

Fig. 243 Turn the crankshaft pulley and align its groove with the timing mark "T" of the lower timing chain cover

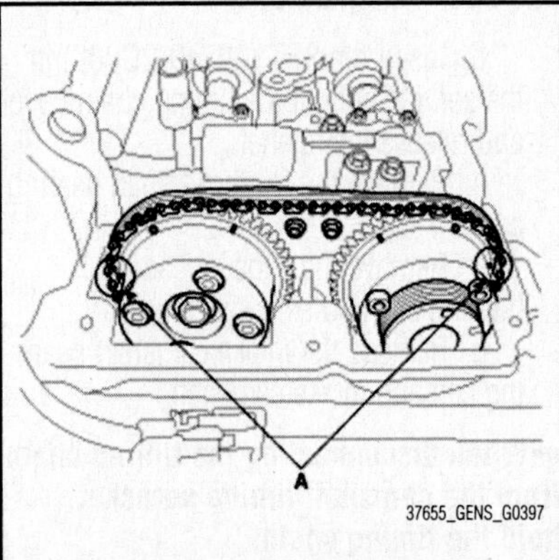

Fig. 244 Check that the mark (A) of the camshaft timing sprockets are in straight line on the cylinder head surface

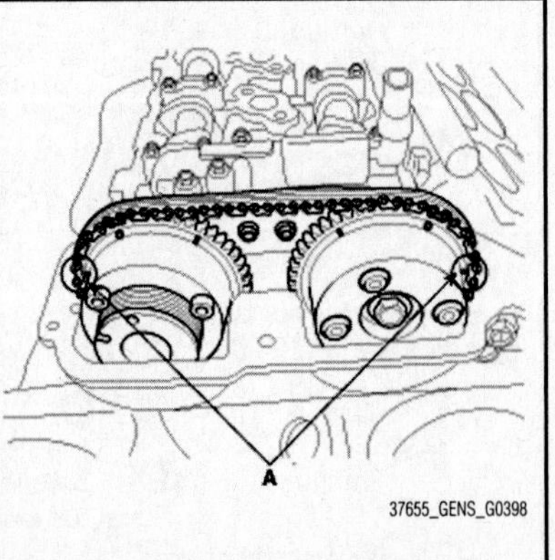

Fig. 245 Check that the mark (A) of the camshaft timing sprockets are in straight line on the cylinder head surface

7. Inspect the valve clearance. With No. 1 cylinder at TDC, inspect clearances only on the valves shown in diagram below. Measurement method:

a. Using a thickness gauge, measure the clearance between the tappet and the base circle of camshaft.

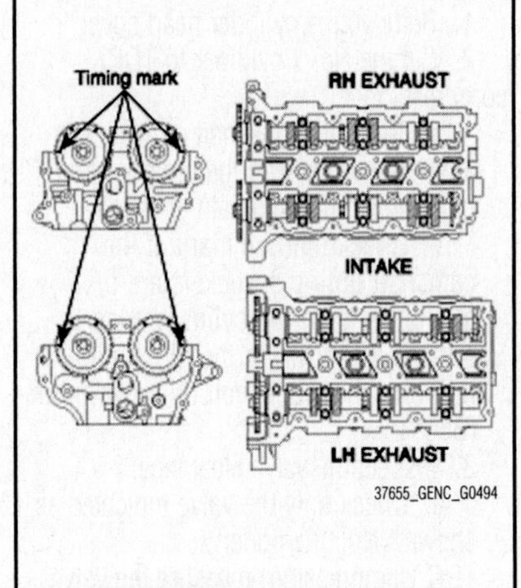

Fig. 246 Measure the valve clearance

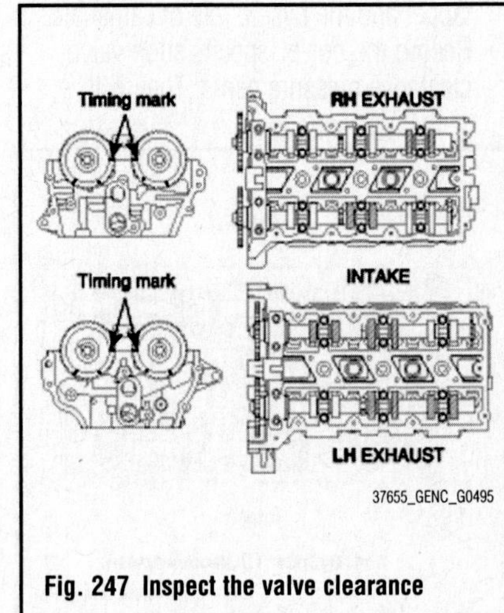

Fig. 247 Inspect the valve clearance

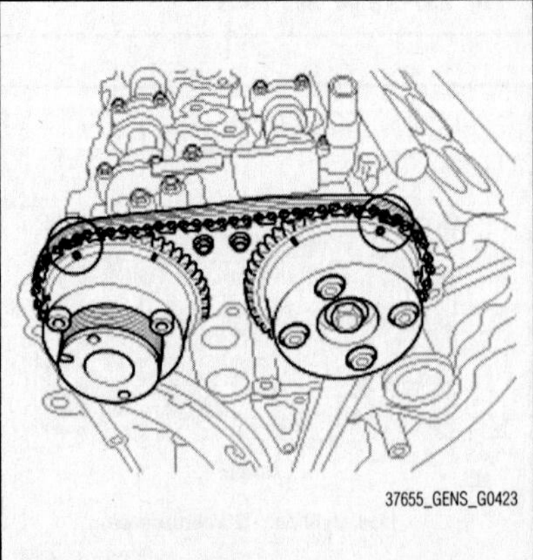

Fig. 248 Timing chain to camshaft position location—left-hand side

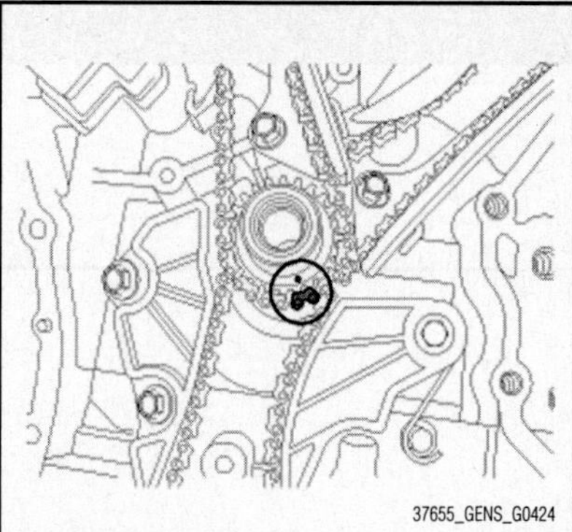

Fig. 249 Timing chain to crankshaft position location

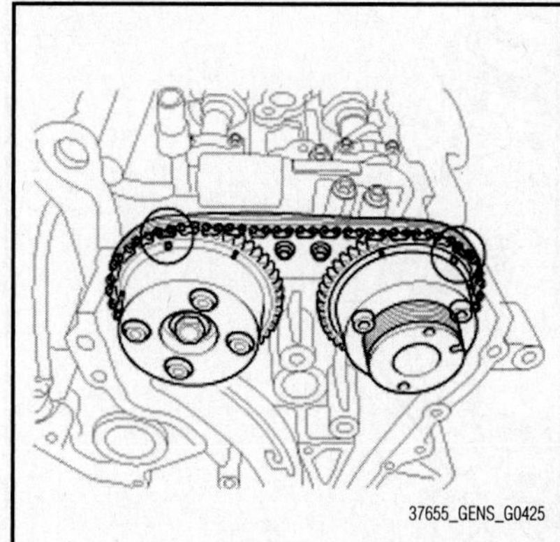

Fig. 250 Timing chain to camshaft position location—right-hand side

b. Record the out-of-specification valve clearance measurements. They will be used later to determine the required replacement adjusting tappet. Valve Clearance Specification (Engine coolant temperature: 68°F (20°C) Limit:

- Intake: 0.0039–0.0118 in. (0.10–0.30mm)
- Exhaust: 0.0078–0.0157 in. (0.20–0.40mm)

c. Turn the crankshaft pulley clockwise one revolution (360°) and align the groove with timing mark "T" of the lower timing chain cover.

d. With No. 4 cylinder at TDC inspect clearances only the valves shown in diagram below. (Refer to the procedure above.)

8. Adjust the intake and exhaust valve clearance.

a. Set the No. 1 cylinder to the TDC/compression.

b. Remove the timing chain. Refer to Timing Chain & Sprockets Removal & Installation.

➡Before removing the timing chain, mark the left-hand/right-hand timing chain with an identification based on the location of the sprocket because the identification mark on the chain for TDC (Top Dead Center) can be erased.

9. Remove the left-hand/right-hand camshaft bearing caps, thrust bearing caps, and camshaft assemblies. Refer to Camshaft & Valve Lifter Removal & Installation.
10. Remove the MLA.
11. Measure the thickness of the removed tappet using a micrometer.
12. Calculate the thickness of a new tappet so that the valve clearance comes within the specified value. T=Thickness of removed tappet, A=Measured valve clearance, N=Thickness of new tappet

- Intake: N = T plus [A minus 0.0079 in. (0.20mm)]
- Exhaust: N = T plus [A minus 0.0118 in. (0.30mm)]

13. Select a new tappet with a thickness as close as possible to the calculated value.

➡Shims are available in 41 size increments of 0.0006 in. (0.015mm) from 0.118 in. (3.00mm) to 0.1417 in. (3.600mm).

14. Place a new tappet on the cylinder head. Apply engine oil at the selected tappet on the periphery and top surface.

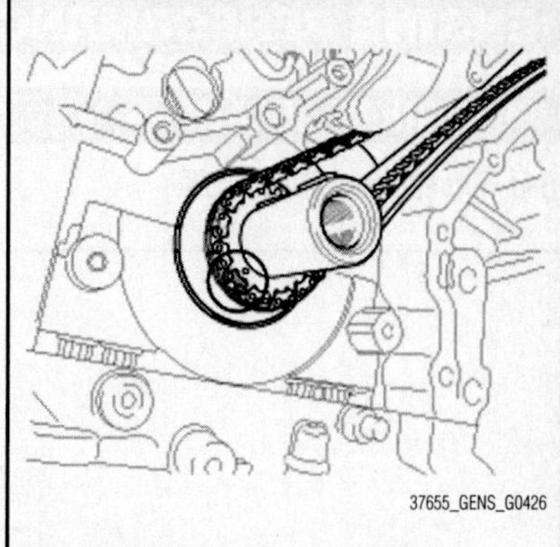

Fig. 251 Timing chain to crankshaft sprocket location

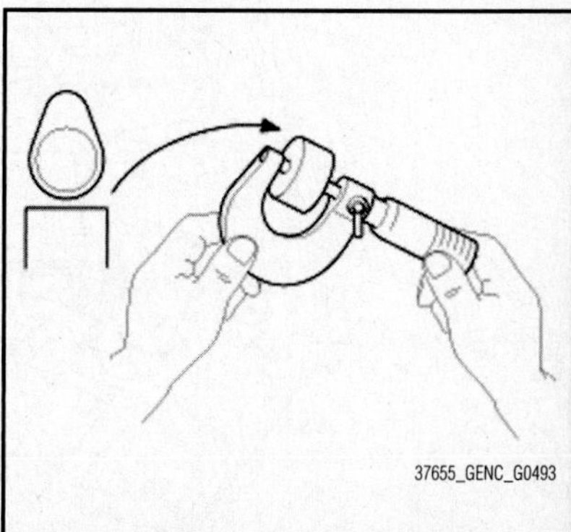

Fig. 252 Measure the thickness of the removed tappet using a micrometer

15. Install the intake and exhaust camshaft.
16. Install the bearing caps.
17. Install the timing chain.
18. Turn the crankshaft two turns in the operating direction (clockwise) and realign crankshaft sprocket and camshaft sprocket timing marks.
19. Recheck the valve clearance. Valve Clearance (Engine coolant temperature: 68°F (20°C)] Specification:

- Intake: 0.0067–0.0090 in. (0.17–0.23mm)
- Exhaust: 0.0106–0.0129 in. (0.27–0.33mm)

ENGINE PERFORMANCE & EMISSION CONTROLS

COMPONENT LOCATIONS

See Figures 253 through 258.

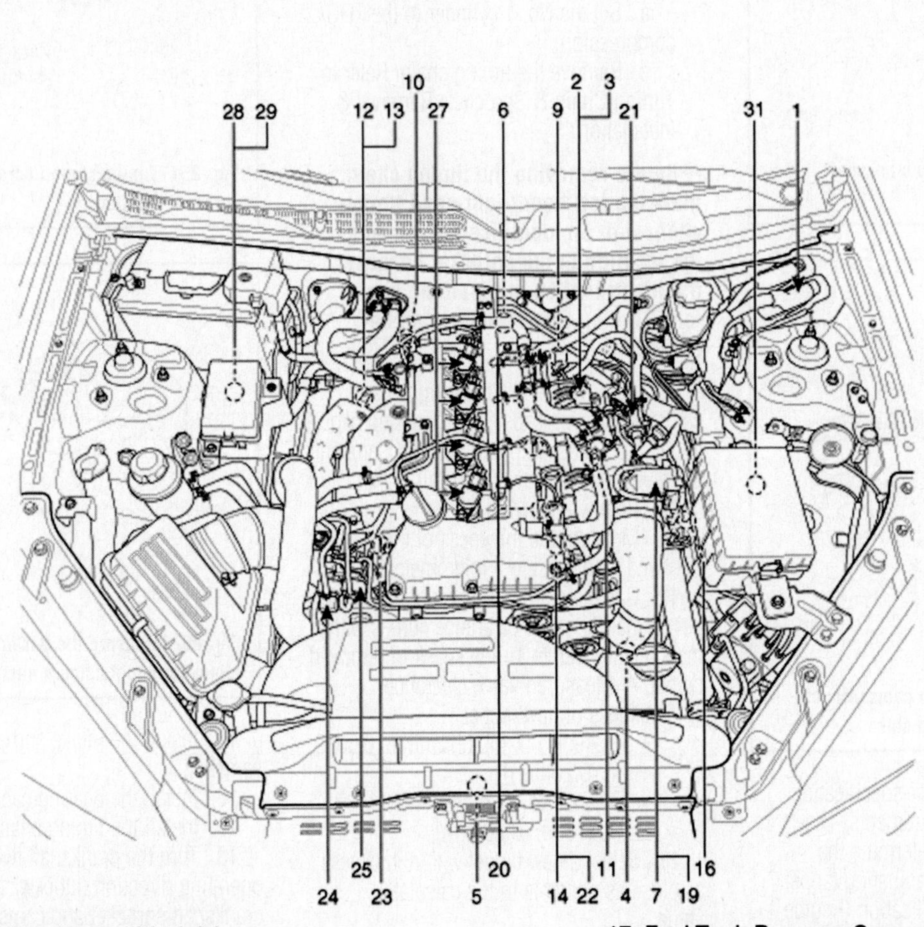

1. ECM (Engine Control Module)
2. Manifold Absolute Pressure Sensor (MAP) #1
3. Intake Air Temperature Sensor (IAT)
4. Manifold Absolute Pressure Sensor (MAP) #2
5. Ambient Temperature Sensor (ATS)
6. Engine Coolant Temperature Sensor (ECT)
7. Throttle Position Sensor (TPS) [integrated into ETC Module]
8. Crankshaft Position Sensor (CKP)
9. Camshaft Position Sensor (CMP) [Bank 1/Intake]
10. Camshaft Position Sensor (CMP) [Bank 1/Exhaust]
11. Knock Sensor (KS)
12. Heated Oxygen Sensor (HO2S) [Bank 1/Sensor 1]
13. Heated Oxygen Sensor (HO2S) [Bank 1/Sensor 2]
14. CVVT Oil Temperature Sensor (OTS)
15. Accelerator Position Sensor (APP)
16. A/C Pressure Transducer (APT)
17. Fuel Tank Pressure Sensor (FTP)
18. Fuel Level Sensor (FLS)
19. ETC Motor [integrated into ETC Module]
20. Injector
21. Purge Control Solenoid Valve (PCSV)
22. CVVT Oil Control Valve (OCV) [Bank 1/Intake]
23. CVVT Oil Control Valve (OCV) [Bank 1/Exhaust]
24. WGT Control Solenoid Valve
25. RCV Control Solenoid Valve
26. Canister Close Valve (CCV)
27. Ignition Coil
28. Main Relay
29. Fuel Pump Relay
30. Data Link Connector (DLC) [16 Pin]
31. Multi-Purpose Check Connector [20 Pin]

37655_GENC_G0451

Fig. 253 Engine Control System Components—2.0L engines

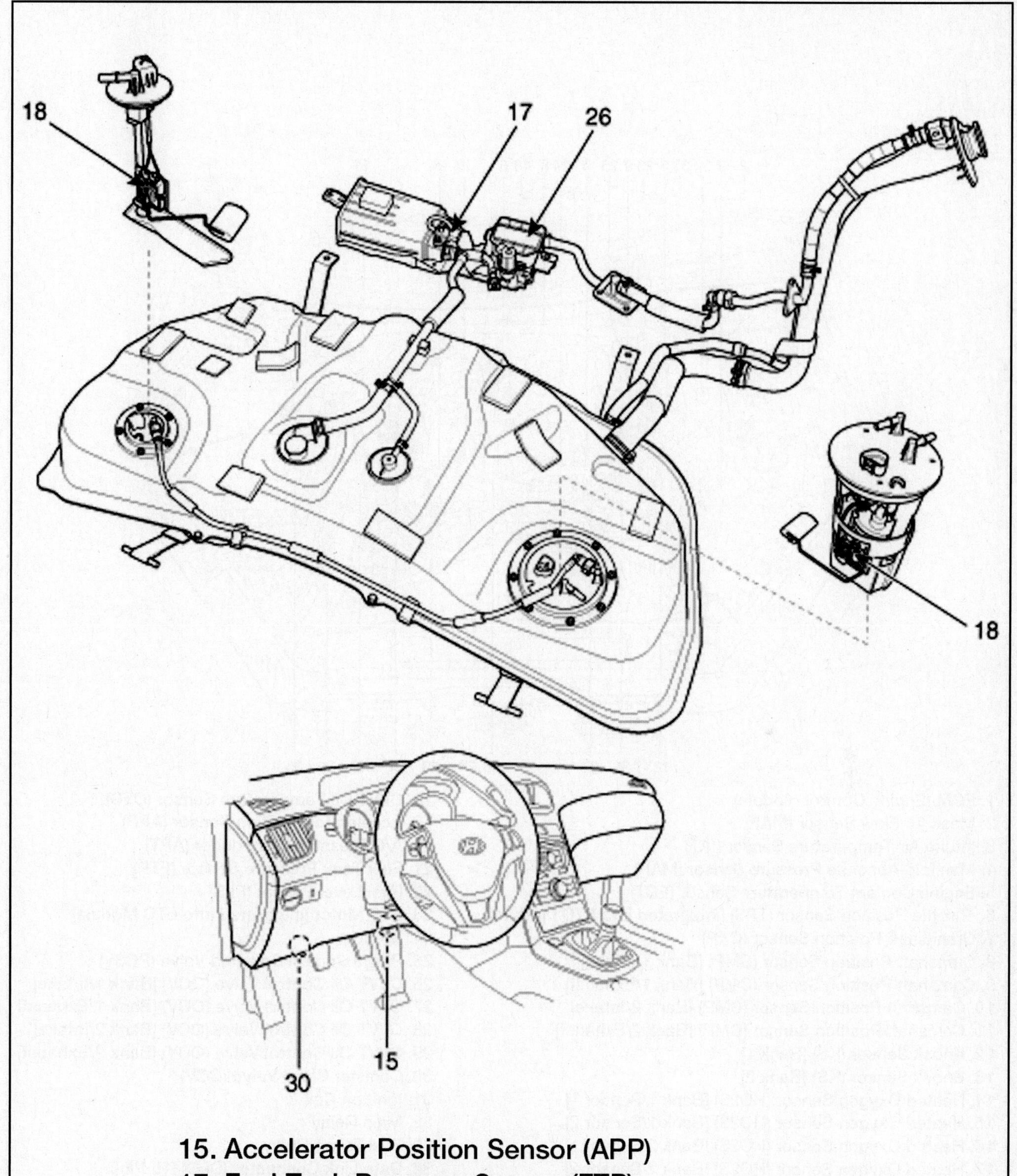

15. Accelerator Position Sensor (APP)
17. Fuel Tank Pressure Sensor (FTP)
18. Fuel Level Sensor (FLS)
26. Canister Close Valve (CCV)
30. Data Link Connector (DLC) [16 Pin]

37655_GENC_G0454

Fig. 254 Sensors and connectors—2.0L engines

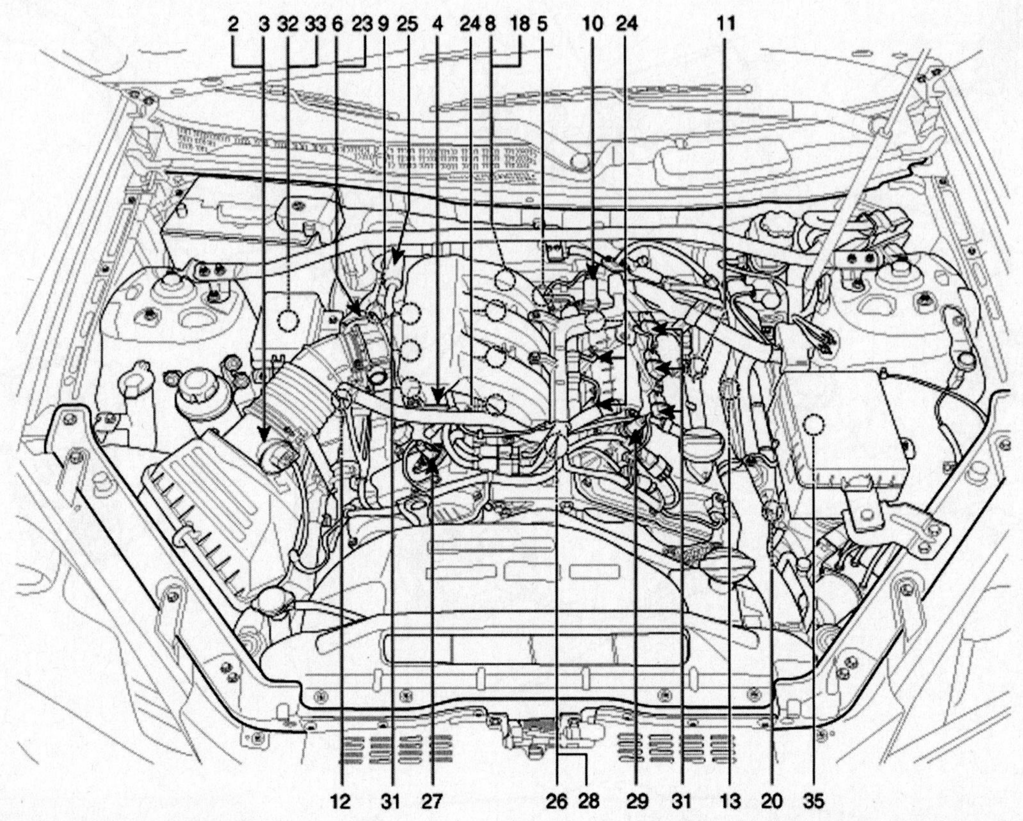

1. ECM (Engine Control Module)
2. Mass Air Flow Sensor (MAF)
3. Intake Air Temperature Sensor (IAT)
4. Manifold Absolute Pressure Sensor (MAP)
5. Engine Coolant Temperature Sensor (ECT)
6. Throttle Position Sensor (TPS) [integrated into ETC Module]
7. Crankshaft Position Sensor (CKP)
8. Camshaft Position Sensor (CMP) [Bank 1/Intake]
9. Camshaft Position Sensor (CMP) [Bank 1/Exhaust]
10. Camshaft Position Sensor (CMP) [Bank 2/Intake]
11. Camshaft Position Sensor (CMP) [Bank 2/Exhaust]
12. Knock Sensor (KS) [Bank 1]
13. Knock Sensor (KS) [Bank 2]
14. Heated Oxygen Sensor (HO2S) [Bank 1/Sensor 1]
15. Heated Oxygen Sensor (HO2S) [Bank 1/Sensor 2]
16. Heated Oxygen Sensor (HO2S) [Bank 2/Sensor 1]
17. Heated Oxygen Sensor (HO2S) [Bank 2/Sensor 2]
18. CVVT Oil Temperature Sensor (OTS)
19. Accelerator Position Sensor (APP)
20. A/C Pressure Transducer (APT)
21. Fuel Tank Pressure Sensor (FTP)
22. Fuel Level Sensor (FLS)
23. ETC Motor [integrated into ETC Module]
24. Injector
25. Purge Control Solenoid Valve (PCSV)
26. CVVT Oil Control Valve (OCV) [Bank 1/Intake]
27. CVVT Oil Control Valve (OCV) [Bank 1/Exhaust]
28. CVVT Oil Control Valve (OCV) [Bank 2/Intake]
29. CVVT Oil Control Valve (OCV) [Bank 2/Exhaust]
30. Canister Close Valve (CCV)
31. Ignition Coil
32. Main Relay
33. Fuel Pump Relay
34. Data Link Connector (DLC) [16 Pin]
35. Multi-Purpose Check Connector [20 Pin]

37655_GENC_G0452

Fig. 255 Engine Control System Components—3.8L engines

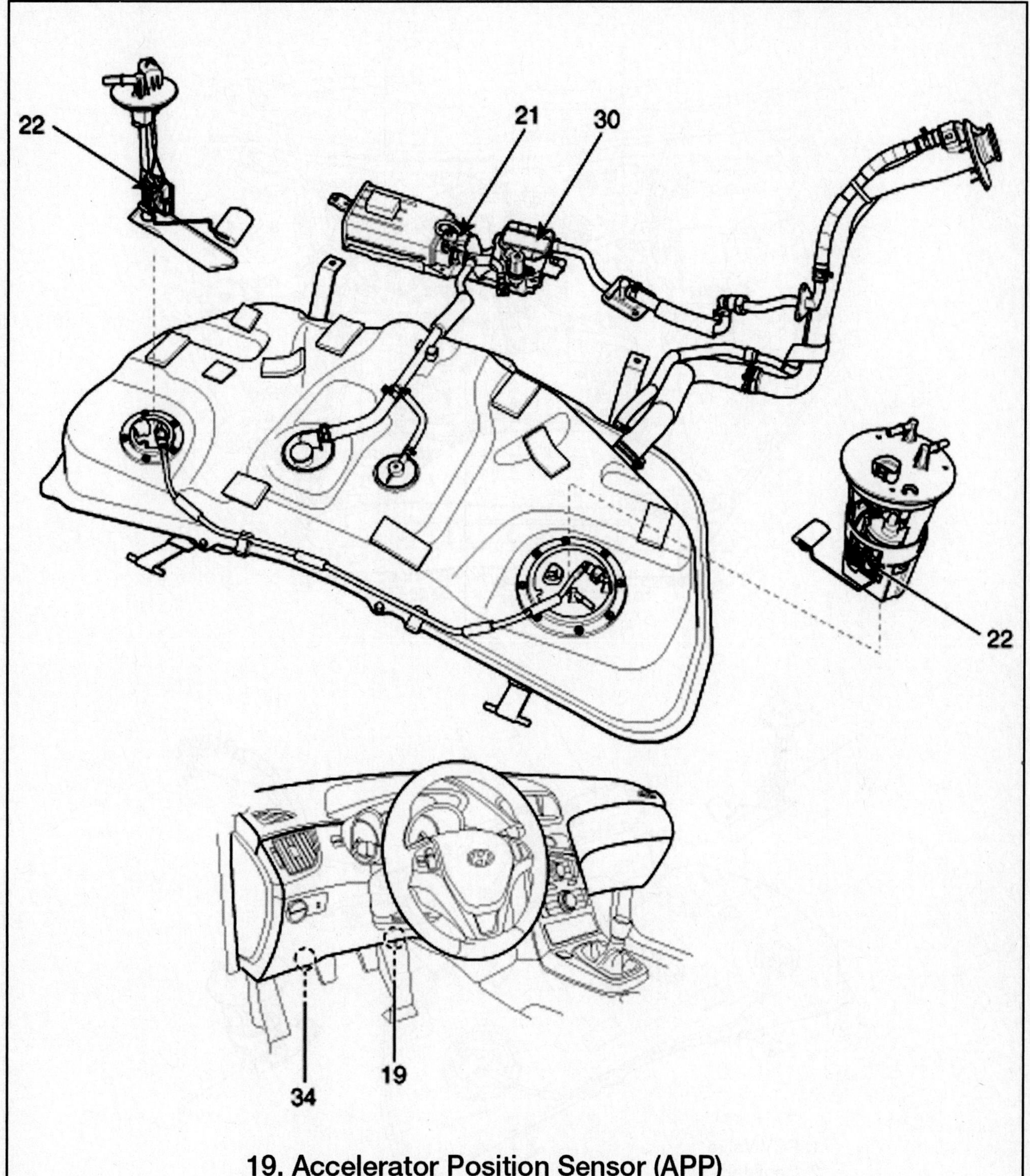

19. Accelerator Position Sensor (APP)
21. Fuel Tank Pressure Sensor (FTPS)
22. Fuel Level Sensor (FLS)
30. Canister Close Valve (CCV)
34. Data Link Connector (DLC) [16 Pin]

37655_GENC_G0453

Fig. 256 Sensors and connectors—3.8L engines

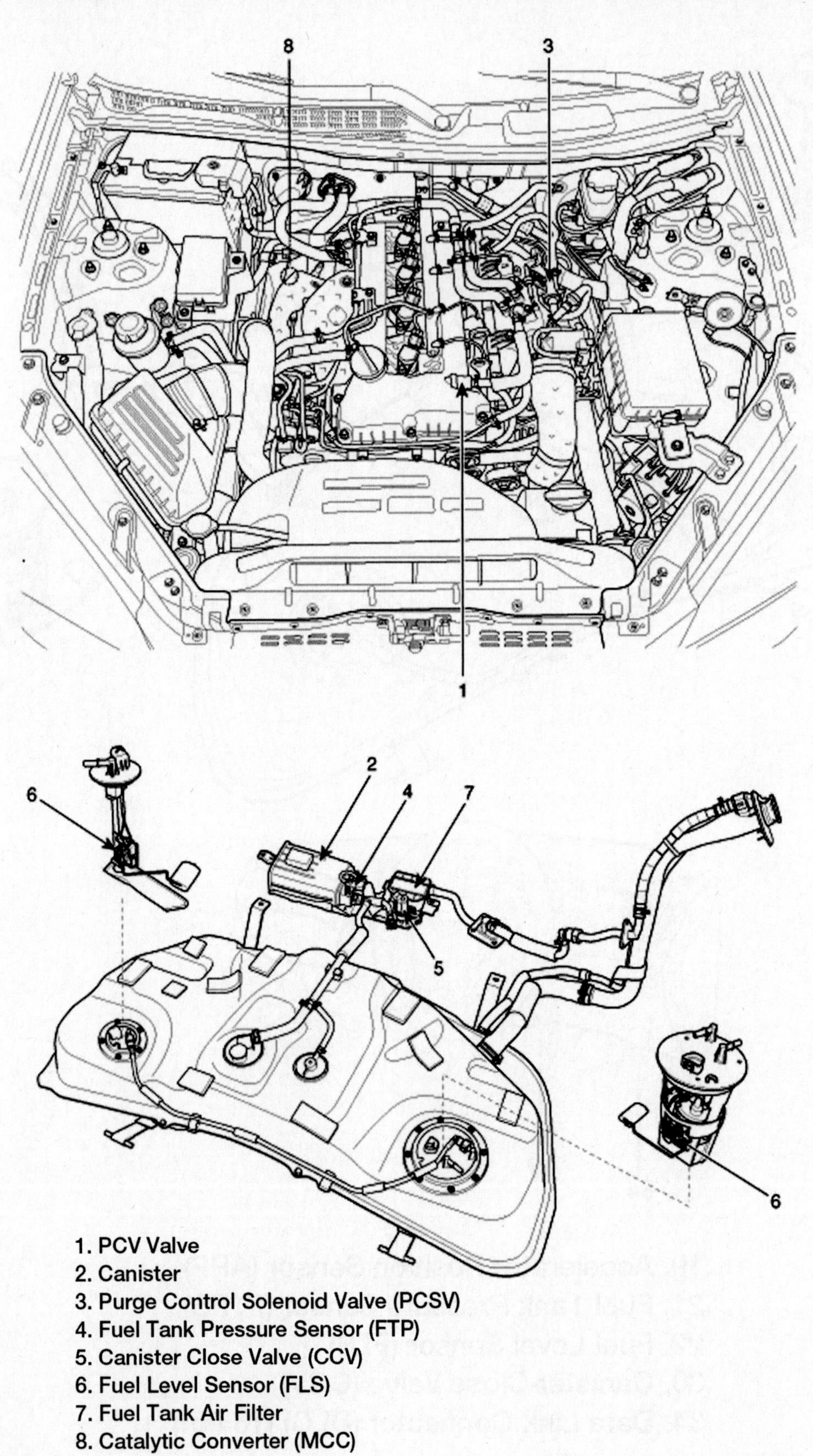

Fig. 257 Emission control components—2.0L engine

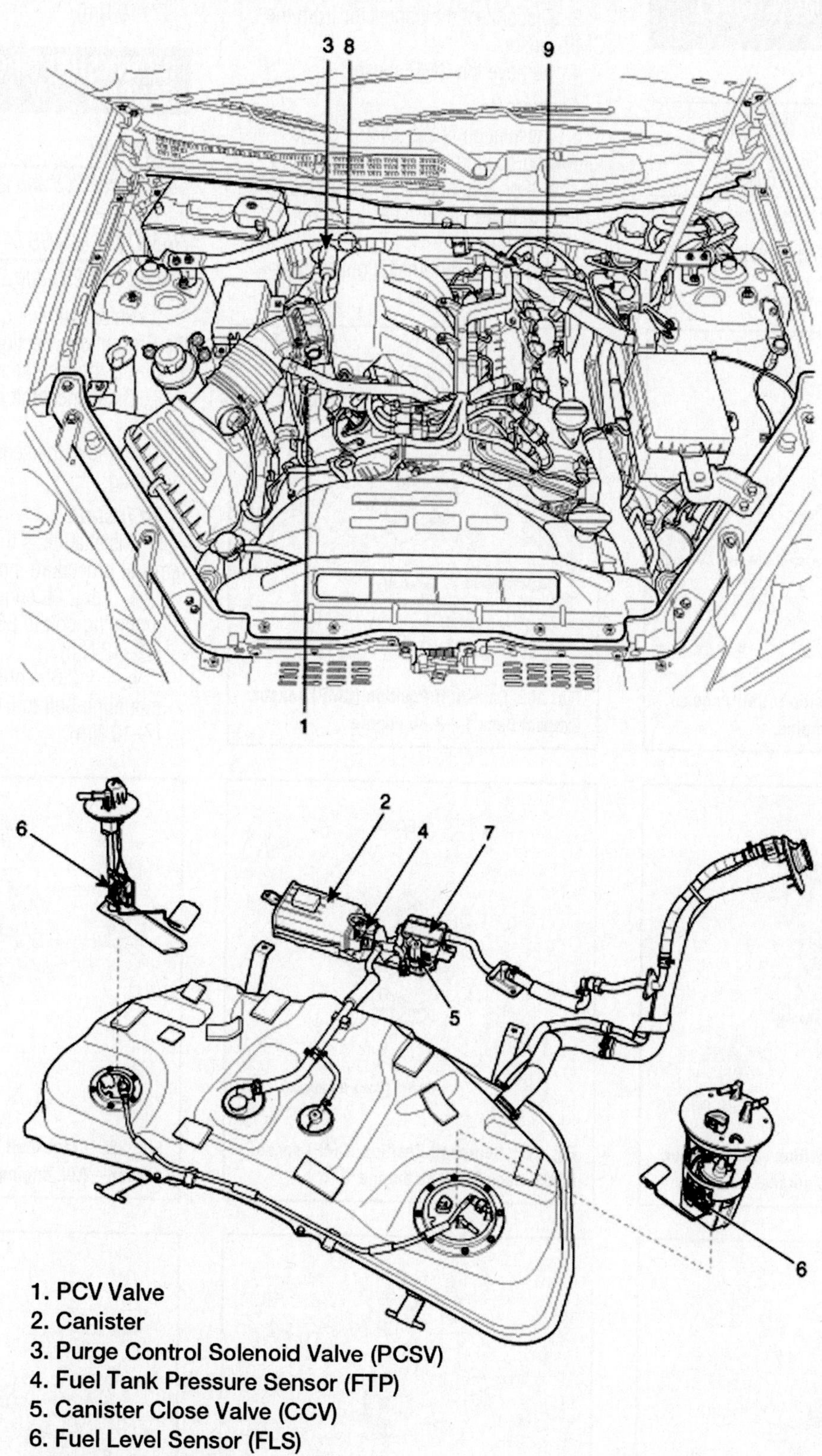

Fig. 258 Emission control components—3.8L engine

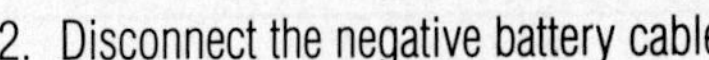

CAMSHAFT POSITION (CMP) SENSOR

LOCATION

See Figures 259 through 264.

REMOVAL & INSTALLATION

See Figures 259 through 264.

1. Before servicing the vehicle, refer to the Precautions Section.
2. Disconnect the negative battery cable.
3. Disconnect the connector from the CMP sensor.
4. Remove the CMP sensor.

To install:

5. Installation is the reverse of the removal procedure, noting the following:
   a. For 2.0L engines, tighten the mounting bolt to 86–104 inch lbs. (10–12 Nm).
   b. For 3.8L engines, tighten the mounting bolt to 61–86 inch lbs. (7–10 Nm).

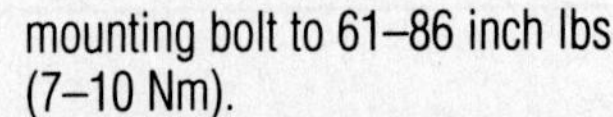

CRANKSHAFT POSITION (CKP) SENSOR

LOCATION

See Figures 265 and 266.

REMOVAL & INSTALLATION

See Figures 265 and 266.

1. Before servicing the vehicle, refer to the Precautions Section.
2. Disconnect the negative battery cable.
3. Disconnect the connector from the sensor.
4. Remove the sensor from its mounting.

To install:

5. Installation is the reverse of the removal procedure, noting the following:
   a. For 2.0L engines, tighten the mounting bolt to 86–104 inch lbs. (10–12 Nm).
   b. For 3.8L engines, tighten the mounting bolt to 61–86 inch lbs. (7–10 Nm).

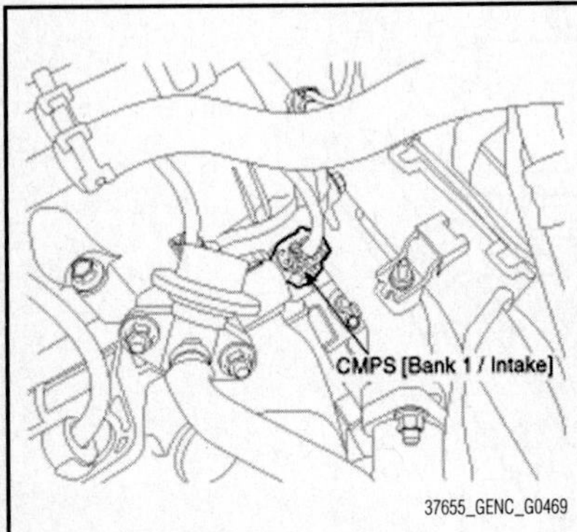

Fig. 259 Camshaft Position (CMP) sensor, Intake Bank 1—2.0L engine

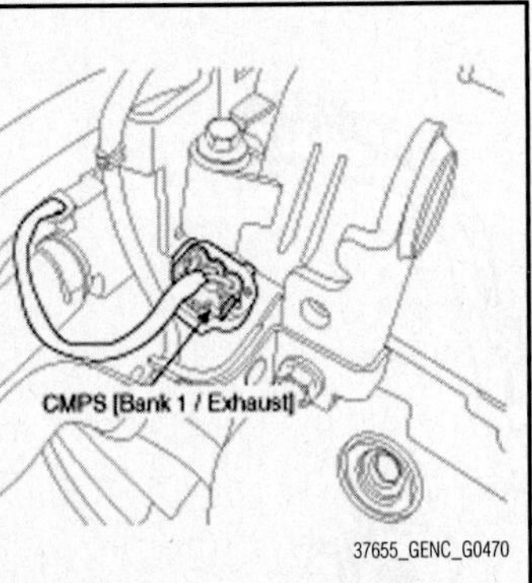

Fig. 260 Camshaft Position (CMP) sensor, Exhaust Bank 1—2.0L engine

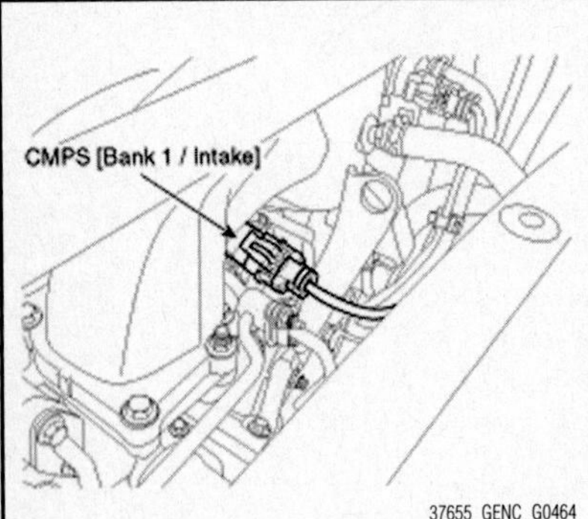

Fig. 261 Camshaft Position (CMP) sensor, Intake Bank 1—3.8L engine

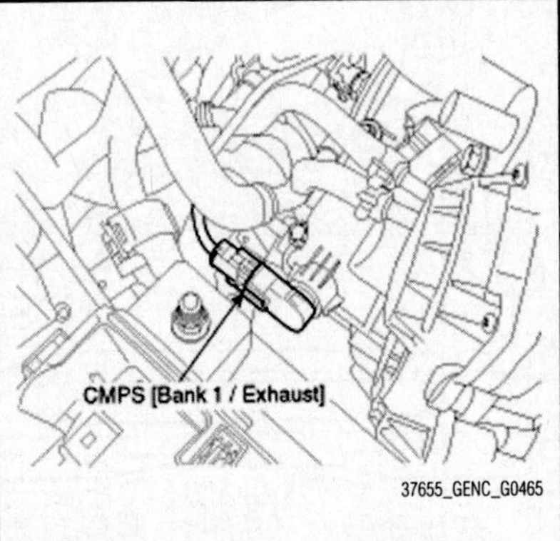

Fig. 262 Camshaft Position (CMP) sensor, Exhaust Bank 1—3.8L engine

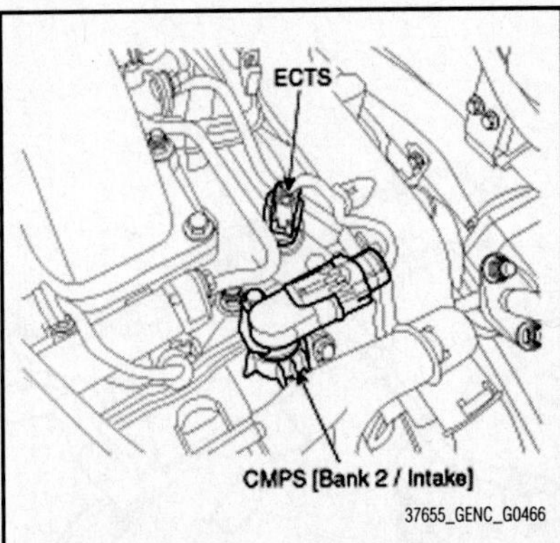

Fig. 263 Camshaft Position (CMP) sensor, Intake Bank 2—3.8L engine

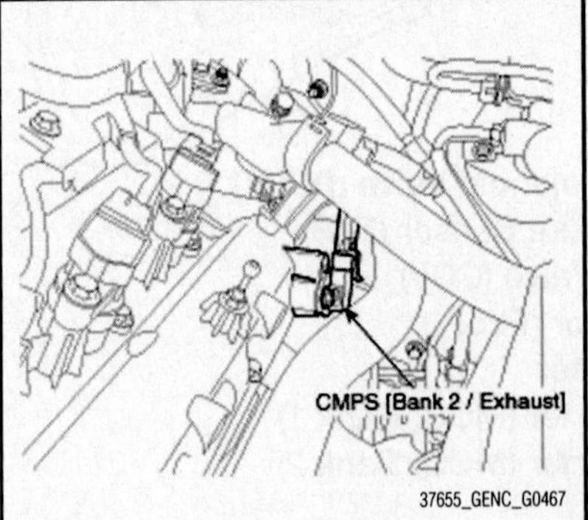

Fig. 264 Camshaft Position (CMP) sensor, Exhaust Bank 2—3.8L engine

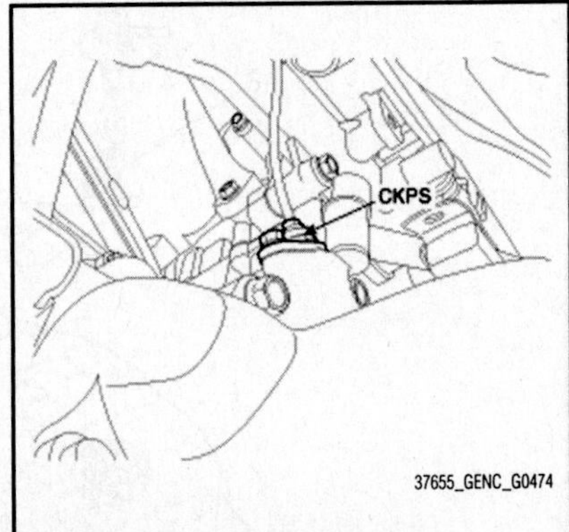

Fig. 265 Crankshaft Position (CKP) sensor—2.0L engine

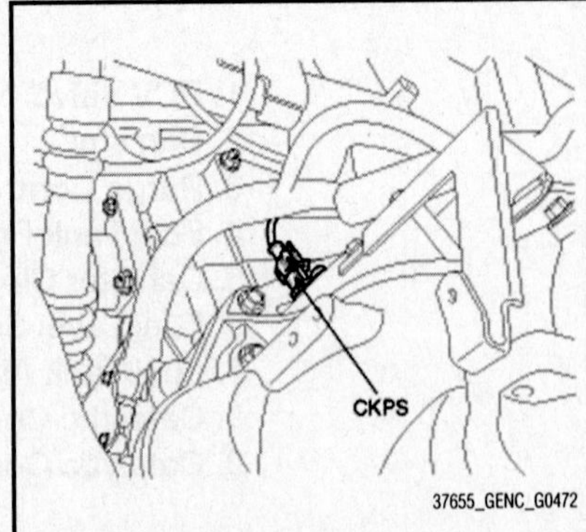

Fig. 266 Crankshaft Position (CKP) sensor—3.8L engine

ELECTRONIC CONTROL MODULE (ECM)

LOCATION

See Figures 267 and 268.

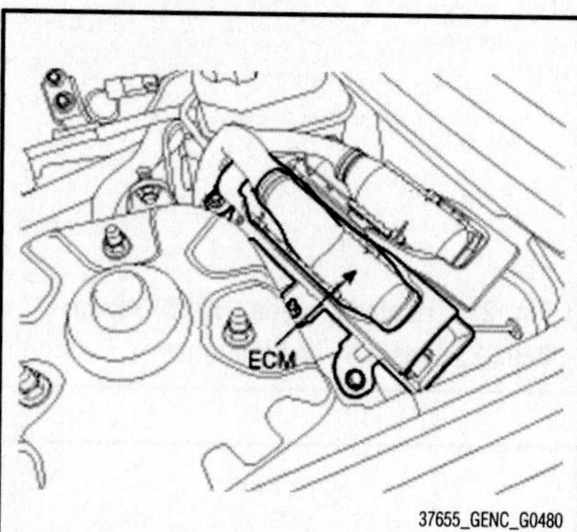

Fig. 267 Electronic Control Module (ECM)—2.0L engine

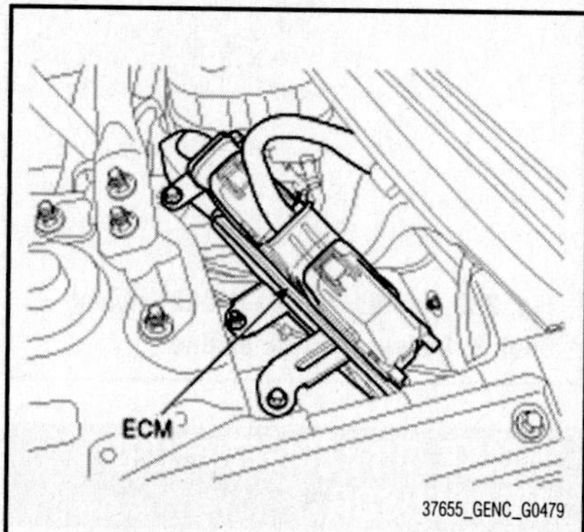

Fig. 268 Electronic Control Module (ECM)—3.8L engine

REMOVAL & INSTALLATION

See Figures 269 and 270.

1. Before servicing the vehicle, refer to the Precautions Section.

➡Replacement of the ECM will require a scan tool.

➡If the vehicle is equipped with immobilizer, replacement of the ECM will require proprietary equipment in order to perform the necessary "Key Teaching" procedure.

2. Follow scan tool instructions prior to ECM removal.
3. Turn ignition switch OFF and disconnect the negative battery cable.
4. Remove the cover, as necessary.
5. For 2.0L engines, remove the front strut assembly. Refer to Struts Removal & Installation in the Front Suspension section.

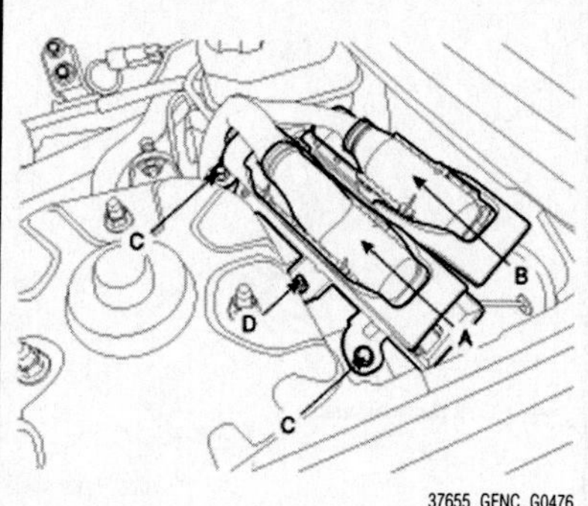

Fig. 269 ECM connector (A) and TCM connector (B); ECM bracket installation bolts (C) and nut (D)—2.0L engine

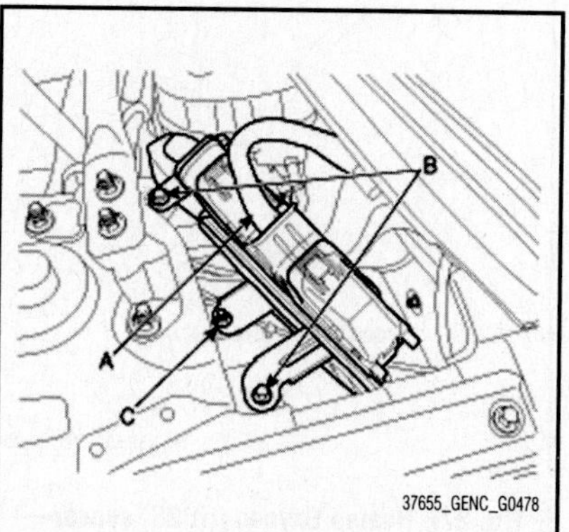

Fig. 270 Disconnect the ECM connector (A) and remove bracket installation bolts (C) and nut (D)—3.8L engine

6. Disconnect the ECM connector.
7. Remove the ECM and the bracket installation bolts and nut.
8. After removing the installation bolts, remove the ECM from the bracket.

To install:

9. If the vehicle is equipped with immobilizer, perform the "Key Teaching" procedure together.
10. Installation is reverse of removal. Tighten the ECM installation bolts to 86–104 inch lbs. (10–12 Nm).

VIN REPROGRAMMING

➡Reprogramming of the VIN will require a scan tool.

➡When replacing an ECM, the VIN must be programmed in the ECM. If there is no VIN in ECM memory, the fault code (DTC P0630) is set.

WARNING

The programmed VIN cannot be changed. When writing the VIN, confirm the VIN carefully.

1. Select "VIN Writing" function in "Vehicle S/W Management".
2. Select "Write VIN" in "ID Resister".
3. Input the VIN.
4. Turn the ignition switch OFF, then back ON.

ENGINE COOLANT TEMPERATURE (ECT) SENSOR

LOCATION

See Figures 271 and 272.

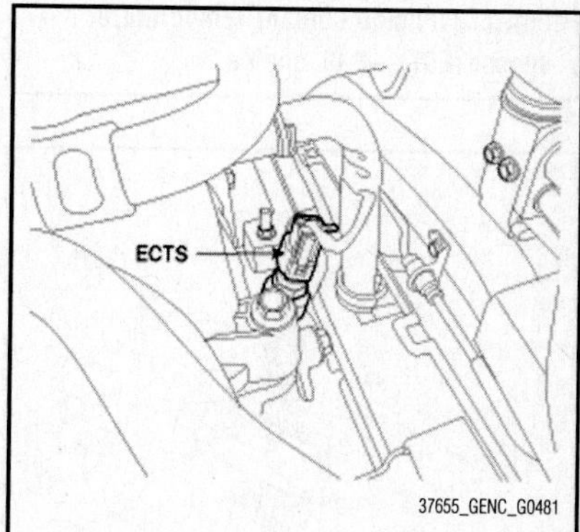

Fig. 271 Engine Coolant Temperature sensor (ECT)—2.0L engine

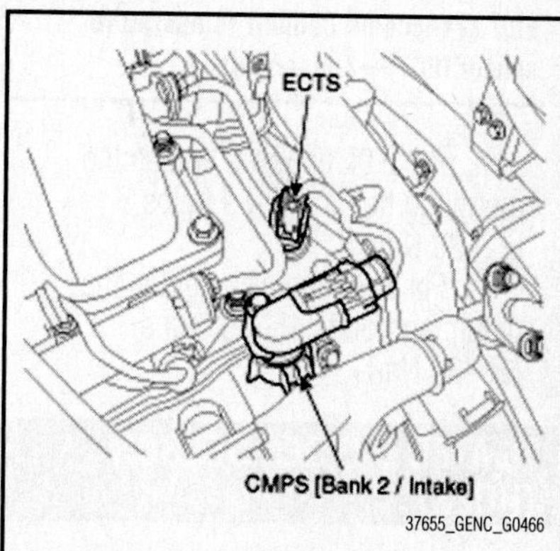

Fig. 272 Engine Coolant Temperature sensor (ECT)—3.8L engine

REMOVAL & INSTALLATION

See Figures 273 and 274.

1. Before servicing the vehicle, refer to the Precautions Section.
2. Turn ignition switch OFF and disconnect the negative battery cable.
3. Turn the ignition switch OFF.
4. Disconnect the ECT sensor connector.
5. Remove the ECT sensor.

To install:

6. Installation is the reverse of removal.

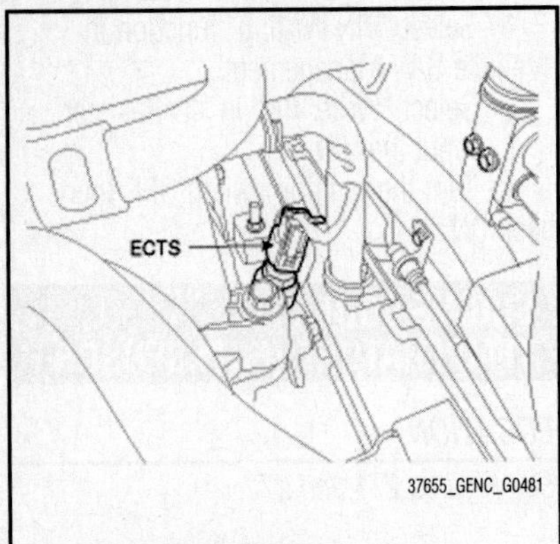

Fig. 273 Engine Coolant Temperature sensor (ECT)—2.0L engine

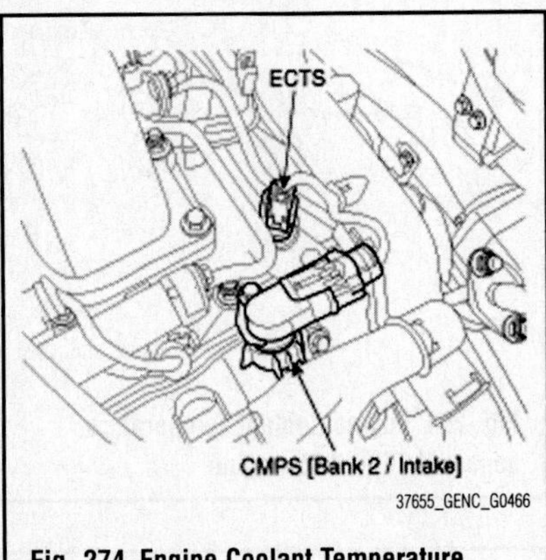

Fig. 274 Engine Coolant Temperature sensor (ECT)—3.8L engine

a. For 2.0L engines, tighten the mounting bolt to 29–39 ft. lbs. (22–29 Nm).

b. For 3.8L engines, tighten the mounting bolt to 15–29 ft. lbs. (20–39 Nm).

HEATED OXYGEN SENSOR (HO2S)

LOCATION

See Figures 275 through 280.

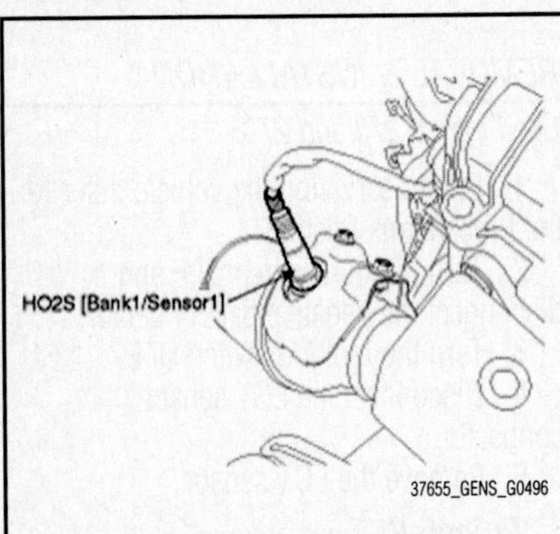

Fig. 275 Heated Oxygen (HO2S) sensor—Bank 1, Sensor 1—2.0L engine

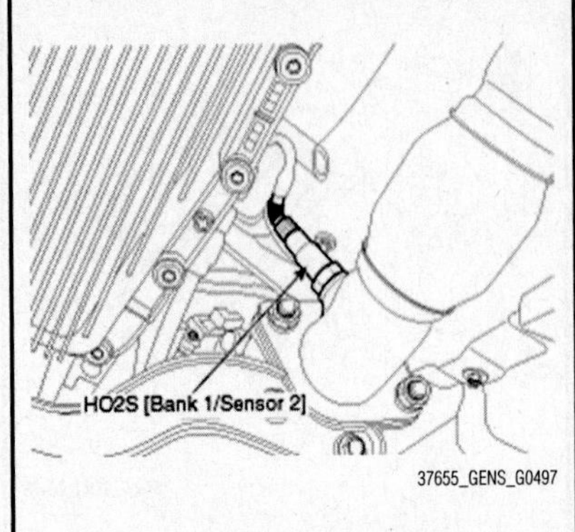

Fig. 276 Heated Oxygen (HO2S) sensor—Bank 1, Sensor 2—2.0L engine

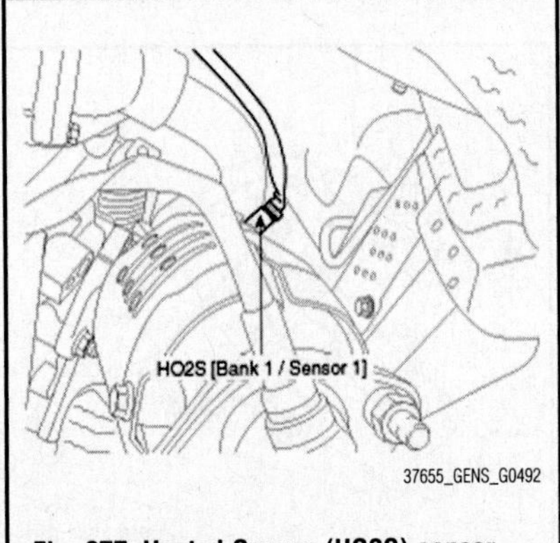
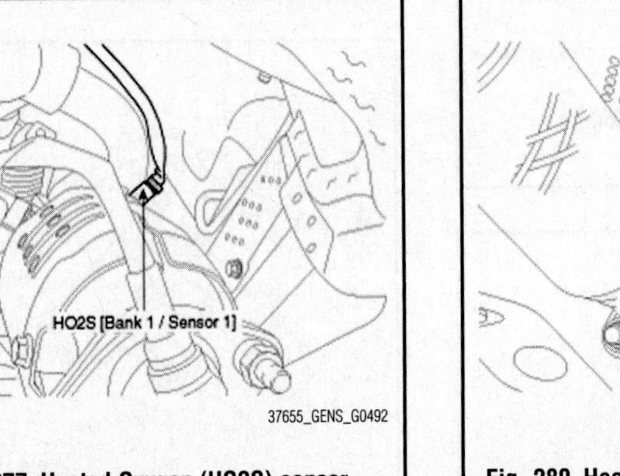

Fig. 277 Heated Oxygen (HO2S) sensor—Bank 1, Sensor 1—3.8L engine

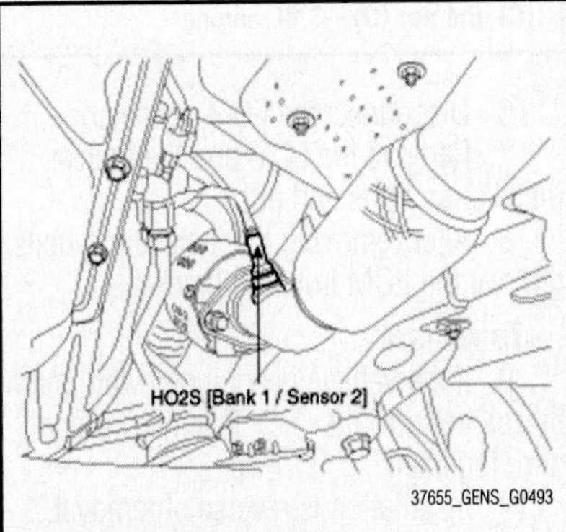

Fig. 278 Heated Oxygen (HO2S) sensor—Bank 1, Sensor 2—3.8L engine

REMOVAL & INSTALLATION

1. Before servicing the vehicle, refer to the Precautions Section.
2. Disconnect the negative battery cable.
3. Disconnect the oxygen sensor connector.
4. Remove the oxygen sensor.

To install:

5. Installation is the reverse of removal. Tighten the HO2S to 29–36 ft. lbs. (39–49 Nm).

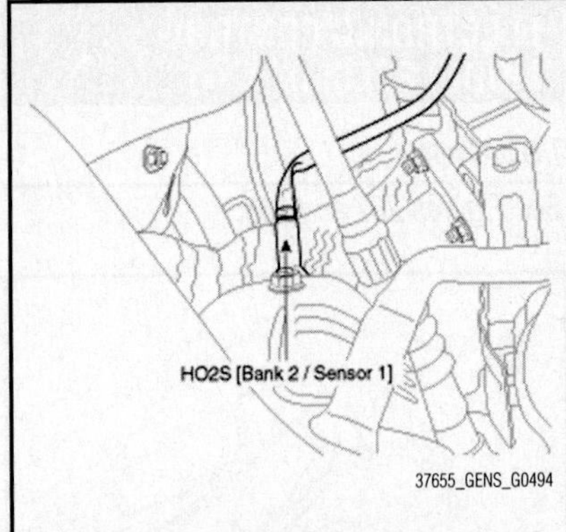

Fig. 279 Heated Oxygen (HO2S) sensor—Bank 2, Sensor 1—3.8L engine

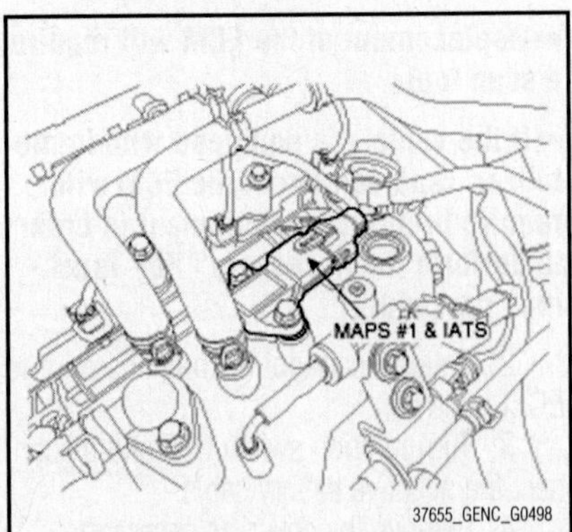

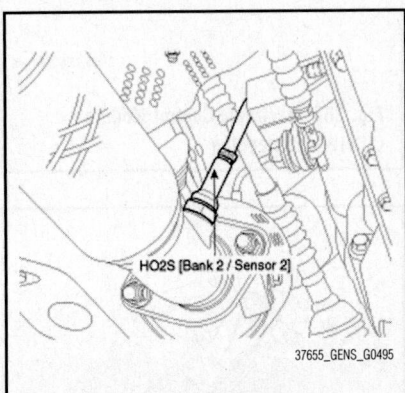

Fig. 280 Heated Oxygen (HO2S) sensor—Bank 2, Sensor 2—3.8L engine

INTAKE AIR TEMPERATURE (IAT) SENSOR

LOCATION

See Figures 281 and 282.

REMOVAL & INSTALLATION

See Figures 281 and 282.

1. Before servicing the vehicle, refer to the Precautions Section.

Fig. 281 Intake Air Temperature (IAT) and No. 1 MAP sensor—2.0L engine

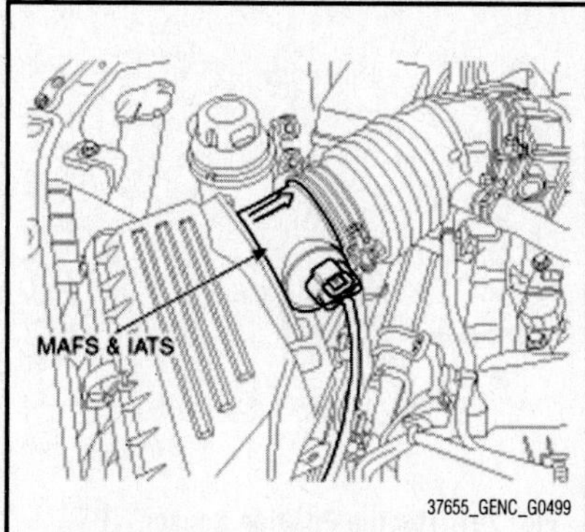

Fig. 282 Intake Air Temperature (IAT) and MAF sensor—3.8L engine

2. Disconnect the negative battery cable.
3. Disconnect Intake Air Temperature (IAT) sensor connector.
4. Remove the IAT sensor.

To install:

5. Installation is the reverse of removal.

KNOCK SENSOR (KS)

LOCATION

See Figures 283 through 285.

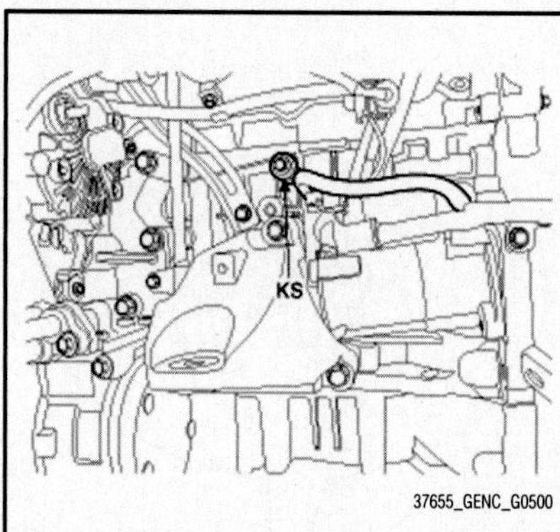

Fig. 283 Knock Sensor (KS)—2.0L engine

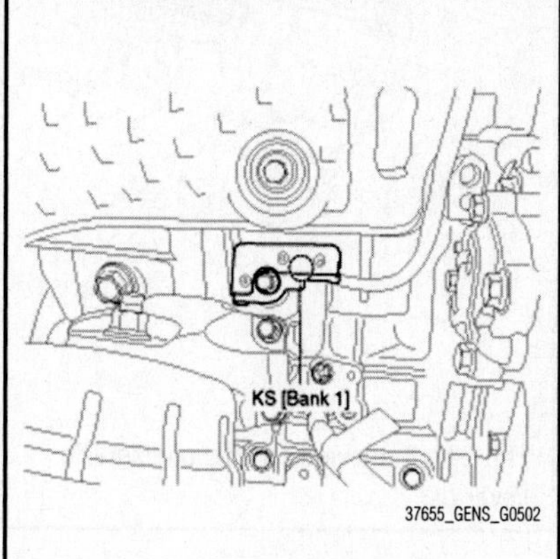

Fig. 284 Knock Sensor (KS), Bank 1—3.8L engine

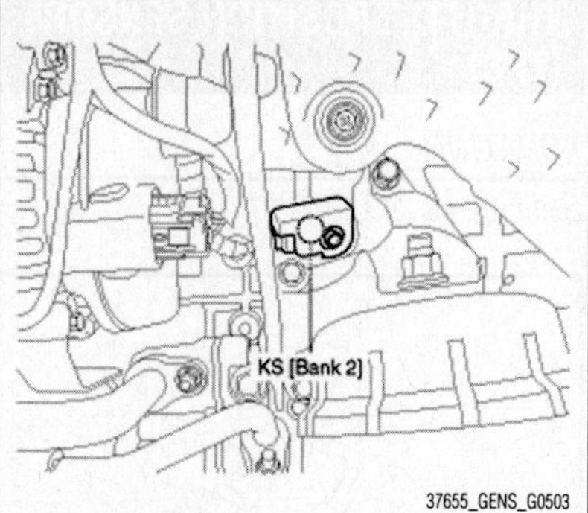

Fig. 285 Knock Sensor (KS), Bank 2—3.8L engine

REMOVAL & INSTALLATION

See Figures 282 through 284.

1. Before servicing the vehicle, refer to the Precautions Section.
2. Disconnect the negative battery cable.
3. Disconnect knock sensor connector.
4. Remove the knock sensor.

To install:

5. Installation is the reverse of removal.

MASS AIR FLOW (MAF) SENSOR

LOCATION

3.8L Engine

See Figure 286.

REMOVAL & INSTALLATION

3.8L Engine

See Figure 286.

1. Before servicing the vehicle, refer to the Precautions Section.
2. Disconnect the negative battery cable.

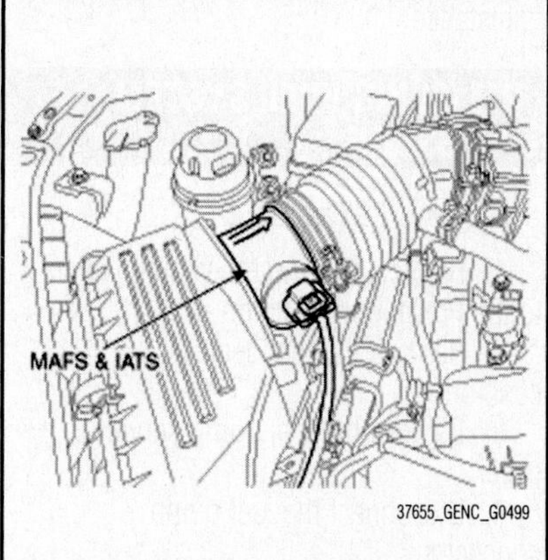

Fig. 286 MAF sensor and Intake Air Temperature (IAT)—3.8L engine

3. Disconnect the Mass Air Flow (MAF) sensor connector.
4. Remove the MAF sensor.

To install:

5. Installation is the reverse of removal. Tighten the MAF sensor installation bolt to 35–52 inch lbs. (4–6 Nm).

MANIFOLD ABSOLUTE PRESSURE (MAP) SENSOR

LOCATION

See Figures 287 through 289.

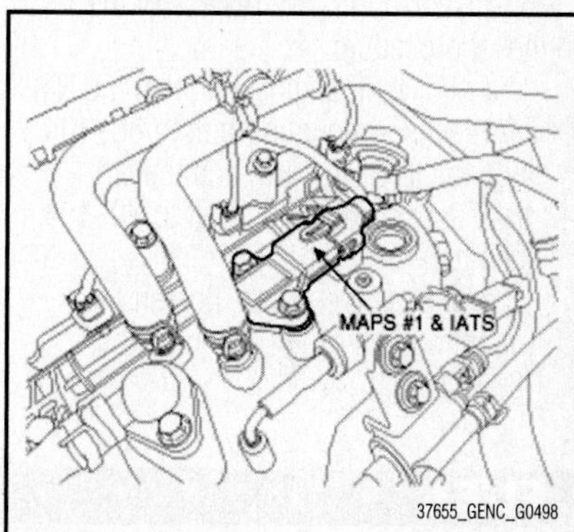

Fig. 287 No. 1 MAP sensor and Intake Air Temperature (IAT)—2.0L engine

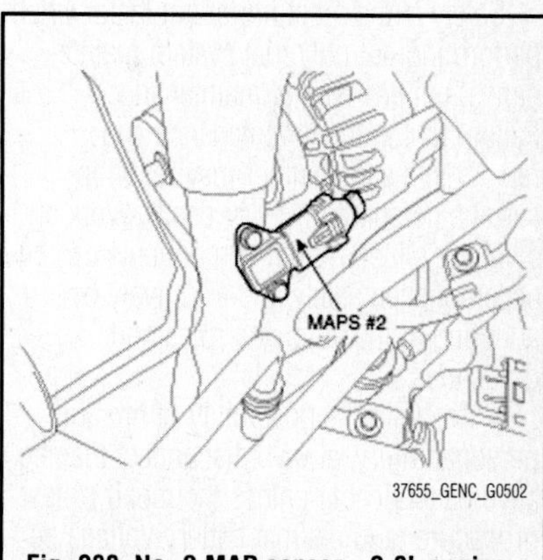

Fig. 288 No. 2 MAP sensor—2.0L engine

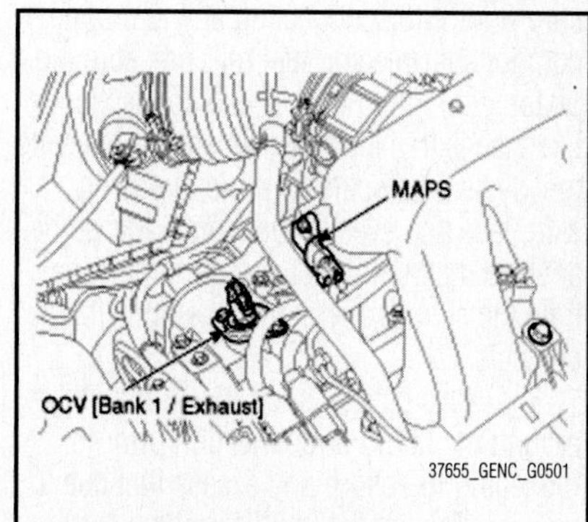

Fig. 289 Manifold Absolute Pressure (MAP) sensor—3.8L engine

REMOVAL & INSTALLATION

See Figures 287 through 289.

1. Before servicing the vehicle, refer to the Precautions Section.
2. Disconnect the negative battery cable.
3. Disconnect the connector from the sensor.
4. Remove the sensor retaining screws.
5. Remove the sensor from its mounting.

To install:

6. Installation is the reverse of the removal procedure.
   a. For 2.0L engines, tighten the No. 1 MAP sensor mounting bolt to 86–104 inch lbs. (10–12 Nm), and the No. 2 MAP sensor mounting bolt to 70–104 inch lbs. (8–9 Nm).
   b. For 3.8L engines, tighten the mounting bolt to 78–104 inch lbs. (9–12 Nm).

THROTTLE POSITION SENSOR (TPS)

LOCATION

See Figures 290 and 291.

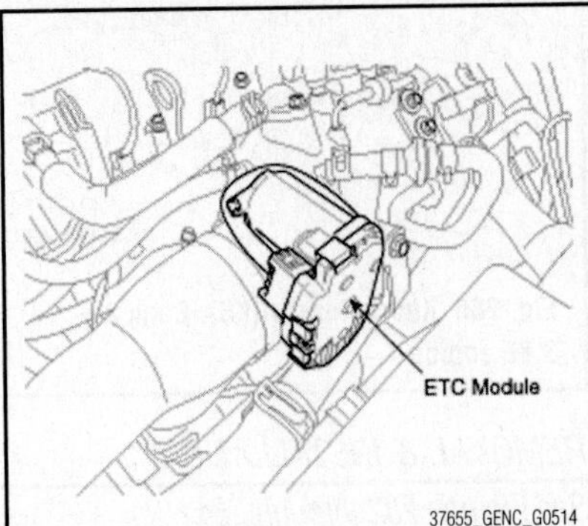

Fig. 290 Throttle Position Sensor (TPS) location, integrated into the Electronic Throttle Control (ETC) module—2.0L engine

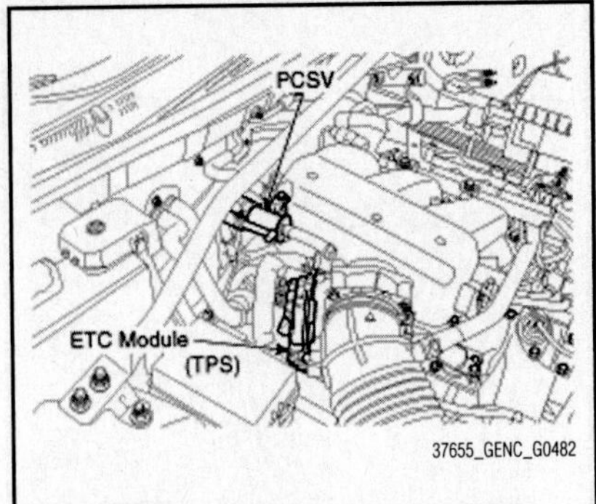

Fig. 291 Throttle Position Sensor (TPS) location—3.8L engine

The Throttle Position Sensor (TPS) is located on the throttle body/Electronic Throttle Control module.

REMOVAL & INSTALLATION

The Throttle Position Sensor (TPS) is located on the throttle body/Electronic Throttle Control module. See Throttle Body Removal & Installation.

FUEL — GASOLINE FUEL INJECTION SYSTEM

FUEL SYSTEM SERVICE PRECAUTIONS

Safety is the most important factor when performing not only fuel system maintenance, but any type of maintenance. Failure to conduct maintenance and repairs in a safe manner may result in serious personal injury or death. Work on a vehicle's fuel system components can be accomplished safely and effectively by adhering to the following rules and guidelines.

- To avoid the possibility of fire and personal injury, always disconnect the negative battery cable unless the repair or test procedure requires that battery voltage be applied.
- Always relieve the fuel system pressure prior to disconnecting any fuel system component (injector, fuel rail, pressure regulator, etc.) fitting or fuel line connection. Exercise extreme caution whenever relieving fuel system pressure to avoid exposing skin, face and eyes to fuel spray. Please be advised that fuel under pressure may penetrate the skin or any part of the body that it contacts.
- Always place a shop towel or cloth around the fitting or connection prior to loosening to absorb any excess fuel due to spillage. Ensure that all fuel spillage is quickly removed from engine surfaces. Ensure that all fuel-soaked cloths or towels are deposited into a flame-proof waste container with a lid.
- Always keep a dry chemical (Class B) fire extinguisher near the work area.
- Do not allow fuel spray or fuel vapors to come into contact with a spark or open flame.
- Always use a second wrench when loosening or tightening fuel line connection fittings. This will prevent unnecessary stress and torsion on fuel piping. Always follow the proper torque specifications.
- Always replace worn fuel fitting O-rings with new ones. Do not substitute fuel hose where rigid pipe is installed.

RELIEVING FUEL SYSTEM PRESSURE

See Figures 292 and 293.

1. Before servicing the vehicle, refer to the Precautions Section.
2. Remove the rear seat cushion.
3. Remove the fuel pump service cover.
4. Disconnect the fuel pump connector.
5. Start the engine and run the engine until the fuel in fuel line is exhausted.

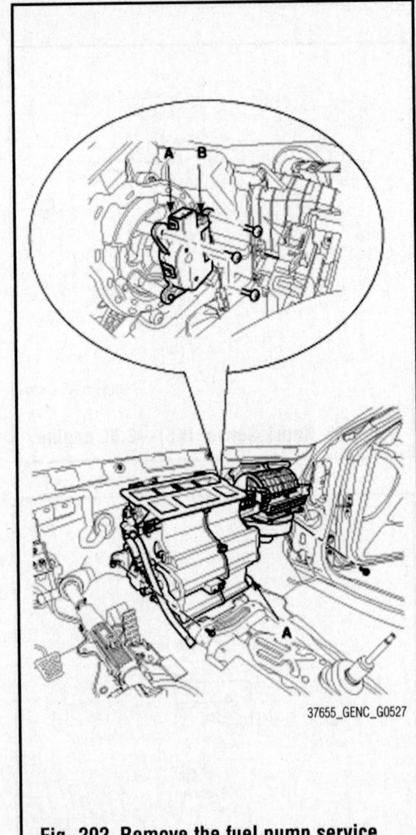

Fig. 292 Remove the fuel pump service cover (A)

6. After engine stops, turn the ignition switch OFF, and disconnect the negative battery terminal.

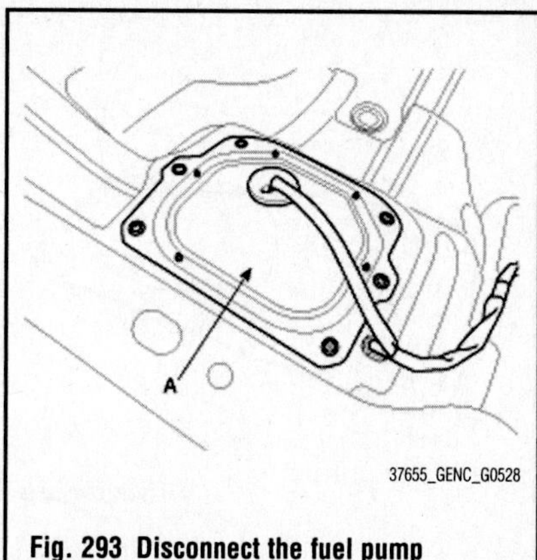

Fig. 293 Disconnect the fuel pump connector (A)

FUEL FILTER

REMOVAL & INSTALLATION

See Figures 294 through 297.

1. Before servicing the vehicle, refer to the Precautions Section.
2. Remove the fuel pump. Refer to Fuel Pump Removal & Installation.

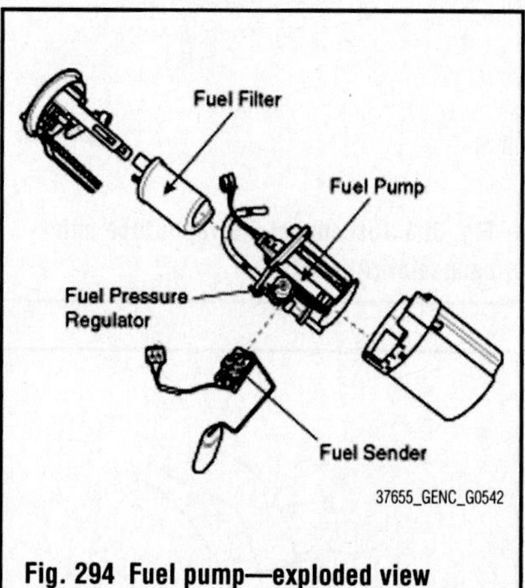

Fig. 294 Fuel pump—exploded view

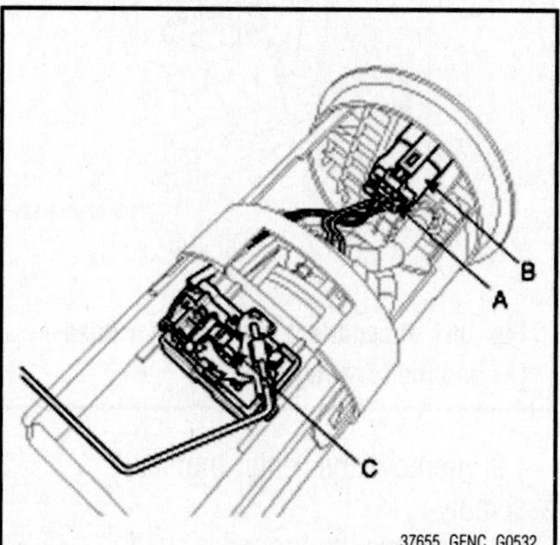

Fig. 295 Disconnect the electric pump wiring connector (A) and the fuel sender connector (B), remove the fuel sender (C)

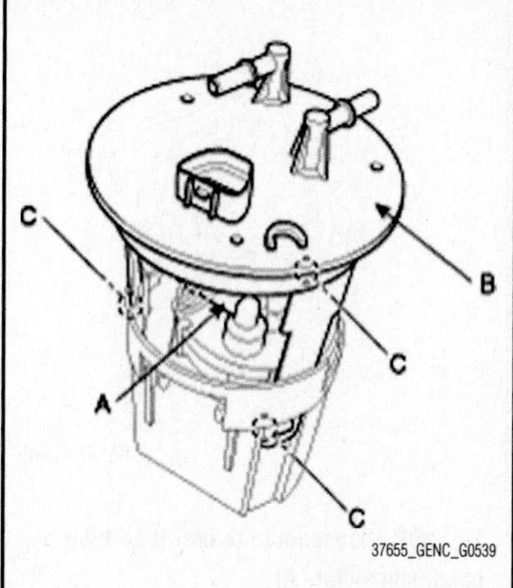

Fig. 296 Disconnect the fuel feed line (A), and separate the head assembly (B) with the hooks (C) released

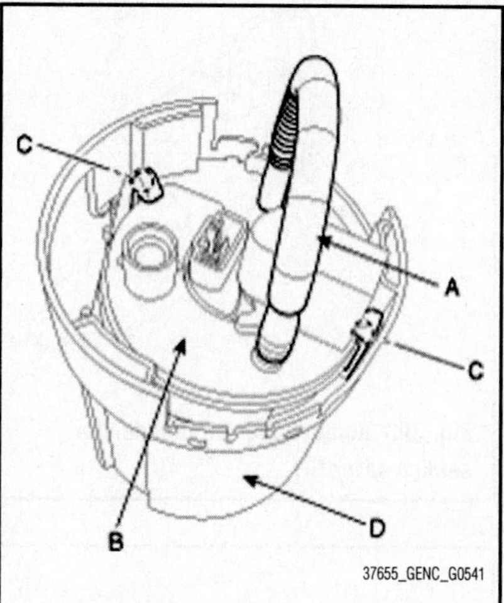

Fig. 297 Disconnect the regulator hose (A) from the fuel filter (B) and separate the fuel filter from the reservoir (D) with the hooks (C) released

3. Disconnect the electric pump wiring connector and the fuel sender connector.
4. Remove the fuel sender.
5. Disconnect the fuel feed line from the fuel filter.
6. Separate the head assembly with the hooks released.
7. Disconnect the regulator hose from the fuel filter.
8. Separate the fuel filter from the reservoir with the hooks released.

To install:

9. Installation is the reverse of removal.

FUEL PUMP

REMOVAL & INSTALLATION

See Figures 298 and 299.

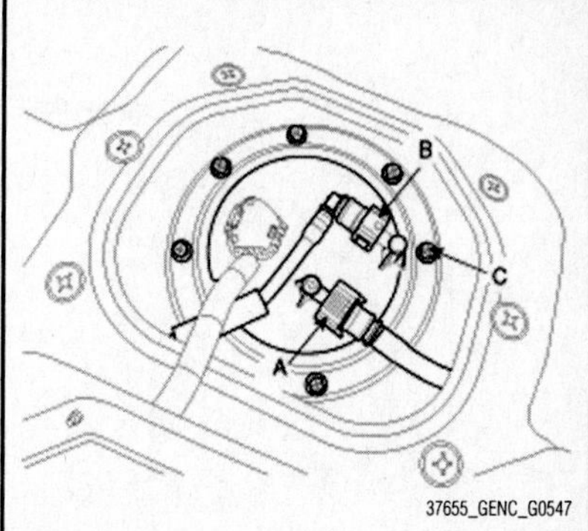

Fig. 298 Disconnect the fuel feed tube quick-connector (A) and the suction tube quick-connector (B) and remove the installation bolts (C)

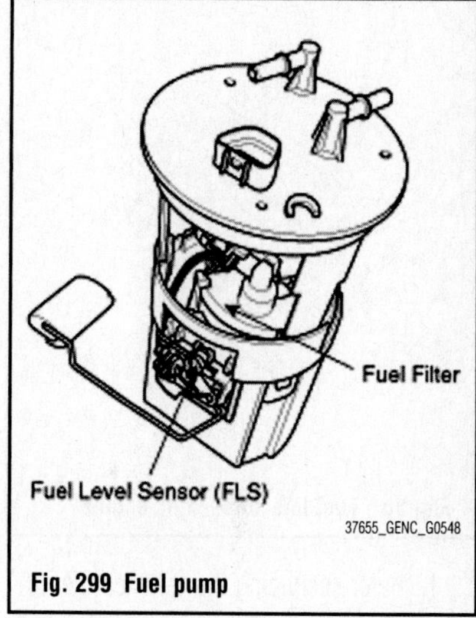

Fig. 299 Fuel pump

1. Before servicing the vehicle, refer to the Precautions Section.
2. Relieve the fuel system pressure. Refer to the Fuel System Pressure Relieving procedure.
3. Disconnect the fuel feed tube quick-connector and the suction tube quick-connector.
4. Remove the installation bolts.
5. Remove the fuel pump from the fuel tank.

To install:

6. Installation is the reverse of removal. Tighten the installation bolts to 17–26 inch lbs. (2–3 Nm).

➡When installing the fuel pump module, be careful not to get the seal-ring entangled.

FUEL RAIL AND INJECTOR

REMOVAL & INSTALLATION

See Figures 300 and 301.

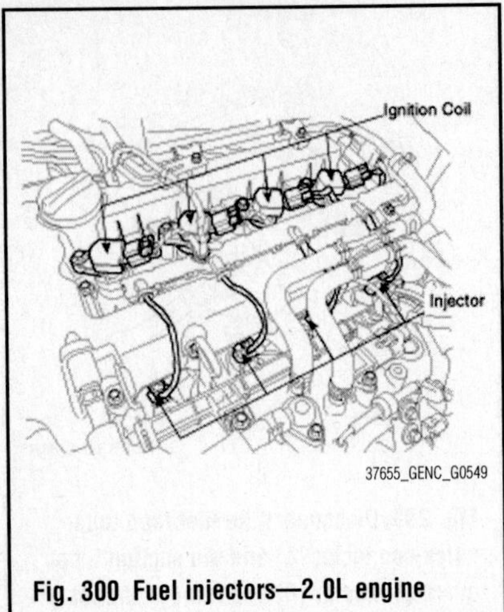

Fig. 300 Fuel injectors—2.0L engine

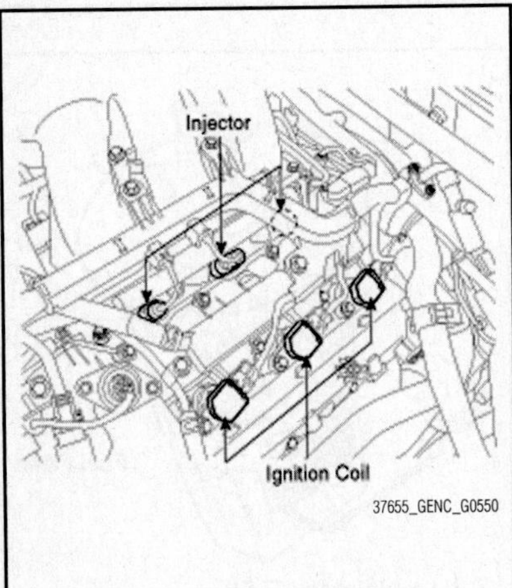

Fig. 301 Fuel injectors—3.8L engine

1. Before servicing the vehicle, refer to the Precautions Section.
2. Disconnect the negative battery cable.
3. Remove the engine cover, as necessary.
4. Disconnect the injector connector.
5. Remove the fuel injectors.

To install:

6. Installation is the reverse of removal.

FUEL TANK

REMOVAL & INSTALLATION

See Figures 302 through 309.

1. Before servicing the vehicle, refer to the Precautions Section.
2. Relieve the fuel system pressure. Refer to the Fuel System Pressure Relieving procedure.
3. Disconnect the fuel feed tube quick-connector.
4. Remove the sub fuel sender service cover.

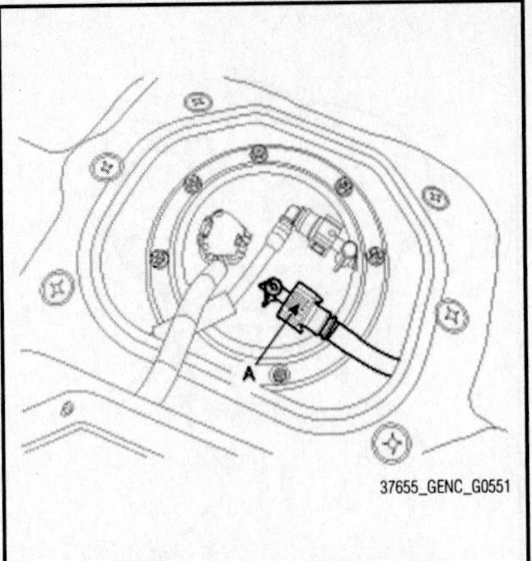

Fig. 302 Disconnect the fuel feed tube quick-connector (A)

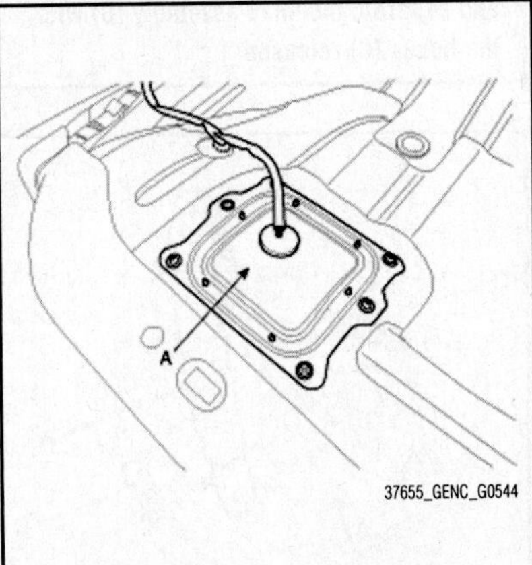

Fig. 303 Remove the sub fuel sender service cover (A)

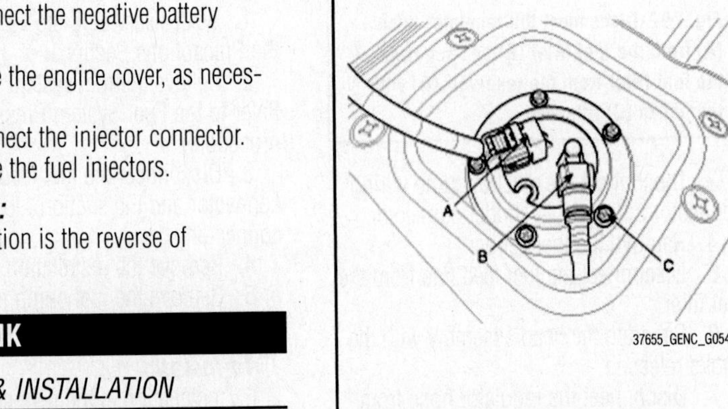

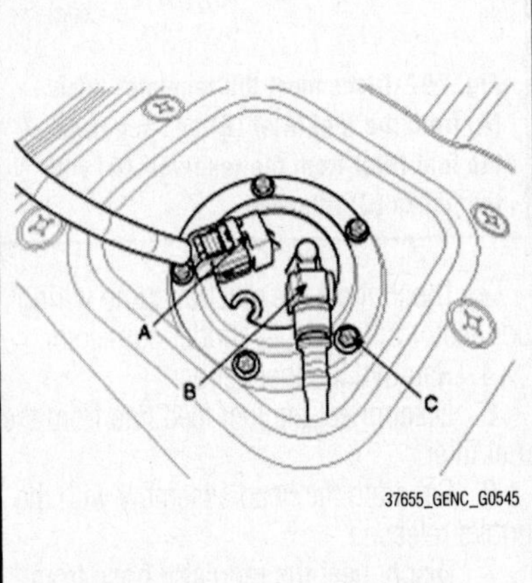

Fig. 304 Remove the sub fuel sender service cover (A)

5. Disconnect the sub fuel sender connector.
6. Remove the canister service cover in the trunk.
7. Disconnect the vapor tube quick-connector.
8. Lift the vehicle.

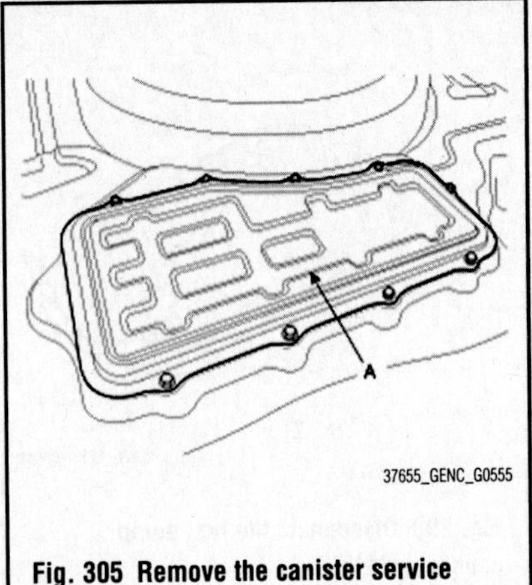

Fig. 305 Remove the canister service cover (A)

Fig. 306 Disconnect the vapor tube quick-connector (A)

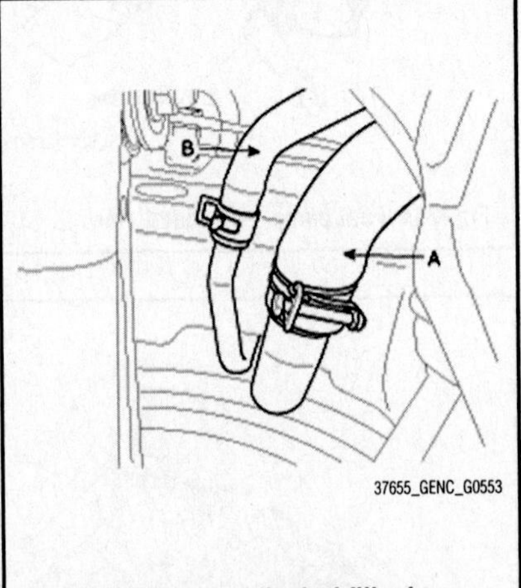

Fig. 307 Disconnect the fuel filler hose (A) and the leveling hose (B)

9. Remove the center muffler assembly.
10. Remove the propeller shaft.
11. Disconnect the fuel filler hose and the leveling hose.
12. Remove the brake line bracket installation bolts.

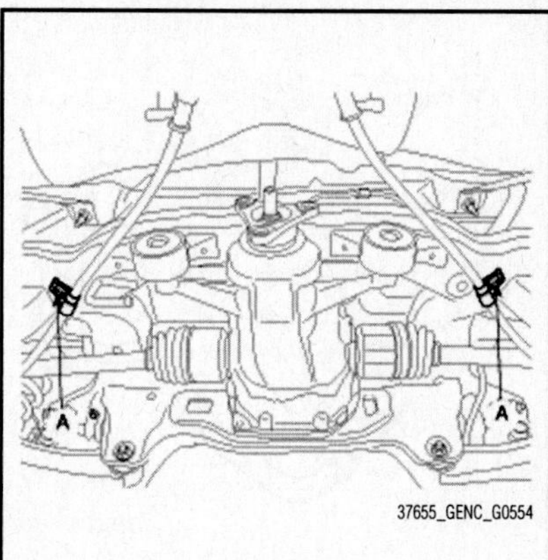

Fig. 308 Remove the brake line bracket installation bolts (A)

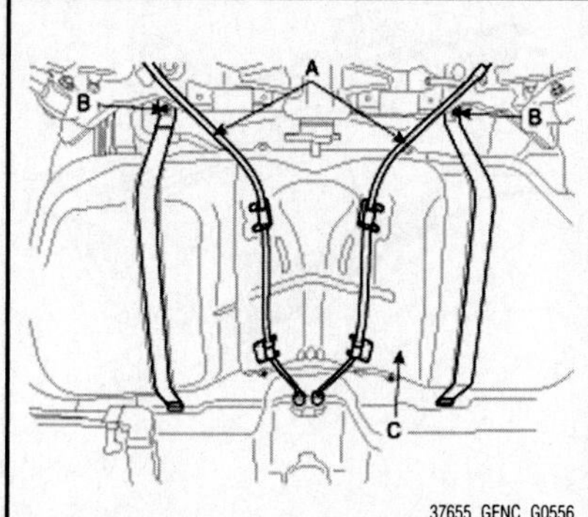

Fig. 309 Detach the parking brake cable (A), remove the mounting nuts (B) and fuel tank (C)

13. Detach the parking brake cable from the fuel tank.
14. Remove the mounting nuts and the fuel tank from the vehicle.

To install:

15. Installation is the reverse of removal. Tighten the fuel tank band installation nut to 29–40 ft. lbs. (39–54 Nm).

THROTTLE BODY

REMOVAL & INSTALLATION

See Figures 310 and 311.

The throttle body is part of the Electronic Throttle Control (ETC) system.

1. Before servicing the vehicle, refer to the Precautions Section.
2. Disconnect the negative battery cable.
3. Remove the air cleaner assembly.
4. Disconnect the ETC connector.
5. Remove the vacuum hoses, necessary.
6. Drain engine coolant and disconnect throttle body coolant hoses, as necessary.
7. Remove the ETC mounting bolts.
8. Remove the ETC.

To install:

9. Installation is the reverse of removal.

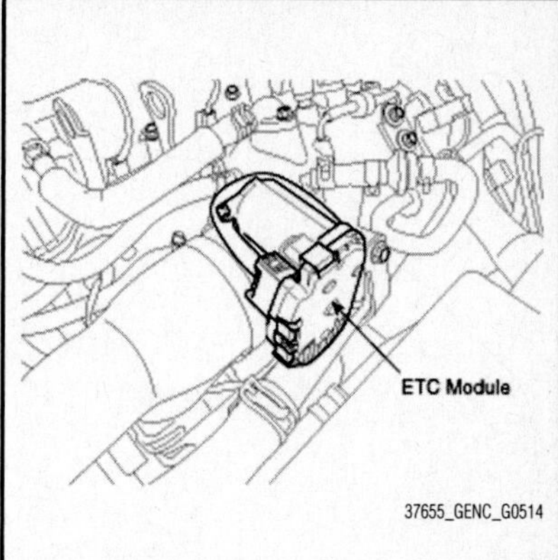

Fig. 310 Throttle body and ETC module—2.0L engine

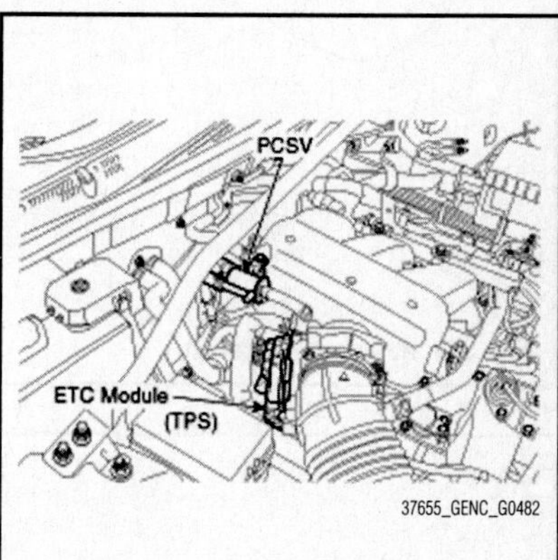

Fig. 311 Throttle body and ETC module—3.8L engine

HEATING & AIR CONDITIONING SYSTEM

BLOWER MOTOR

REMOVAL & INSTALLATION

See Figure 312.

1. Before servicing the vehicle, refer to the Precautions Section.

CAUTION

Before servicing components near or affected by the SRS (airbag) system, read and observe all SRS Service Precautions. Failure to observe all precautions may result in accidental airbag deployment, personal injury, or death.

2. Disconnect the battery negative cable and wait for at least three minutes before beginning work.
3. Remove the driver side instrument panel under cover.
4. Disconnect the blower motor connector.

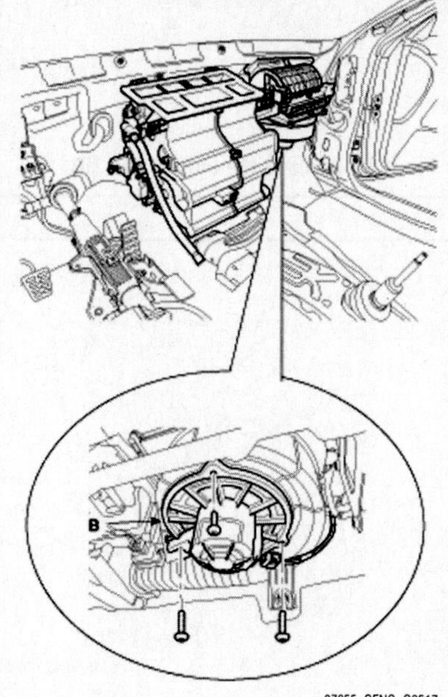

Fig. 312 Blower unit location (B)

5. Remove the blower motor mounting screws.
6. Remove the blower motor.

To install:

7. Installation is the reverse of removal.

HEATER CORE/UNIT

REMOVAL AND INSTALLATION

See Figures 313 through 320.

1. Before servicing the vehicle, refer to the Precautions Section.

CAUTION

Before servicing components near or affected by the SRS (airbag) system, read and observe all SRS Service Precautions. Failure to observe all precautions may result in accidental airbag deployment, personal injury, or death.

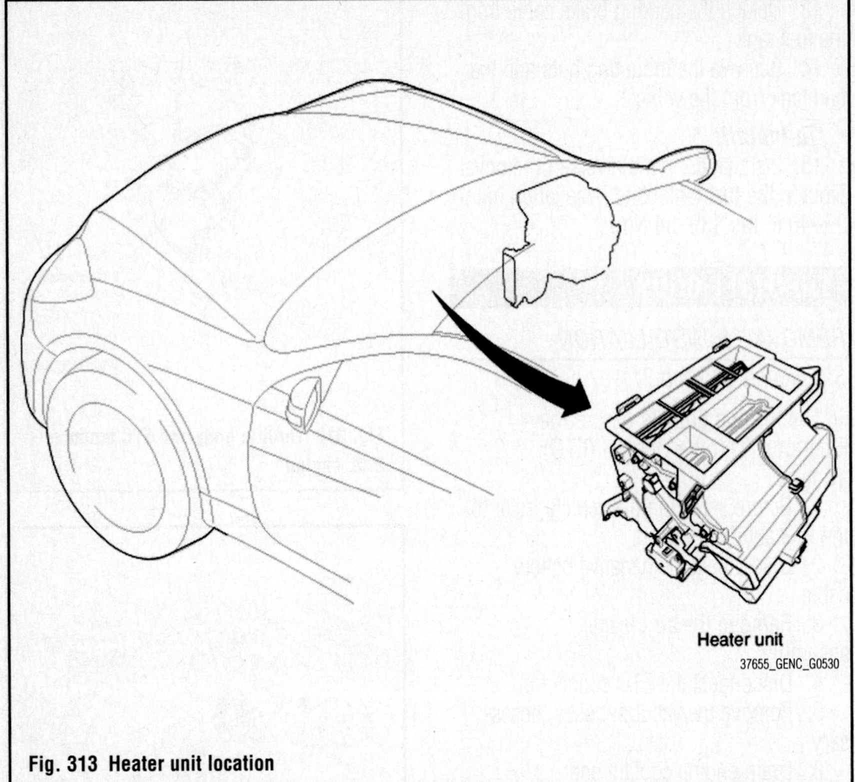

Fig. 313 Heater unit location

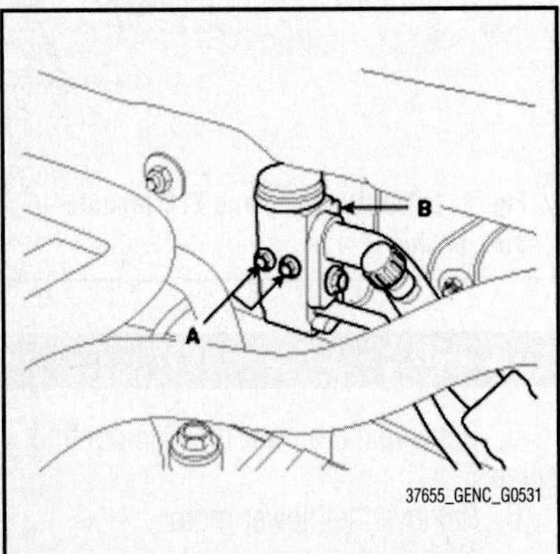

Fig. 314 Remove the bolts (A) and the expansion valve (B) from the evaporator core

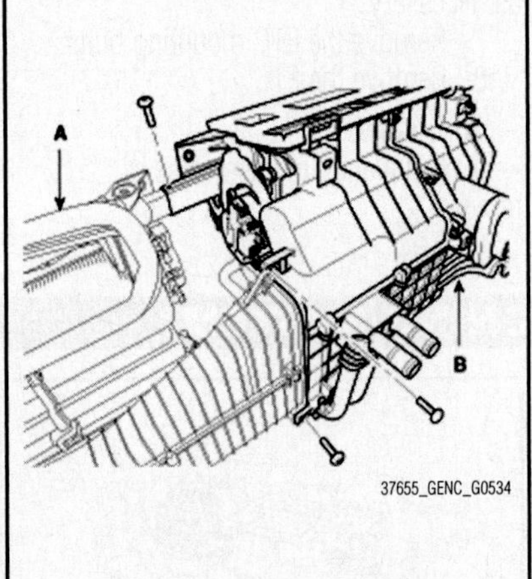

Fig. 316 Remove the blower unit (B)

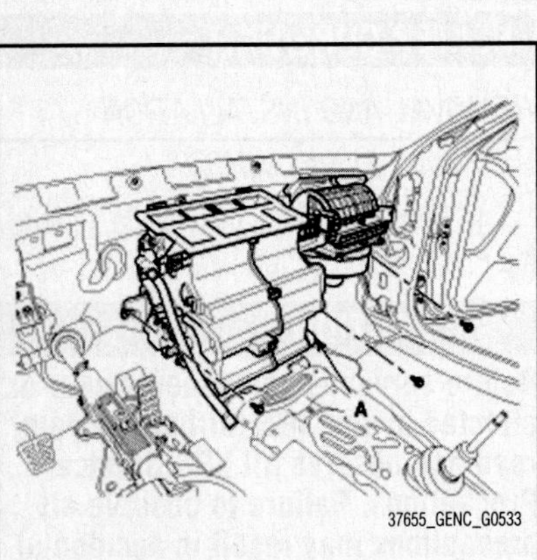

Fig. 315 Remove the heater and blower unit

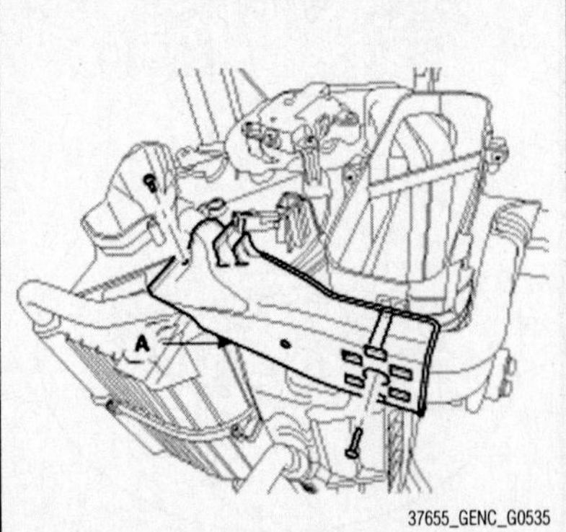

Fig. 317 Remove the heater core cover (A)

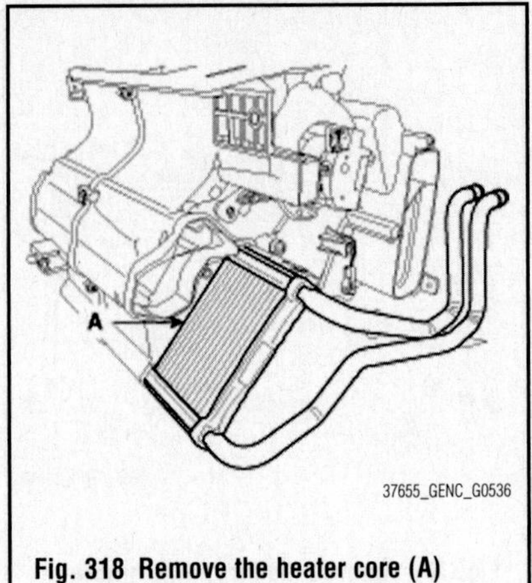

Fig. 318 Remove the heater core (A)

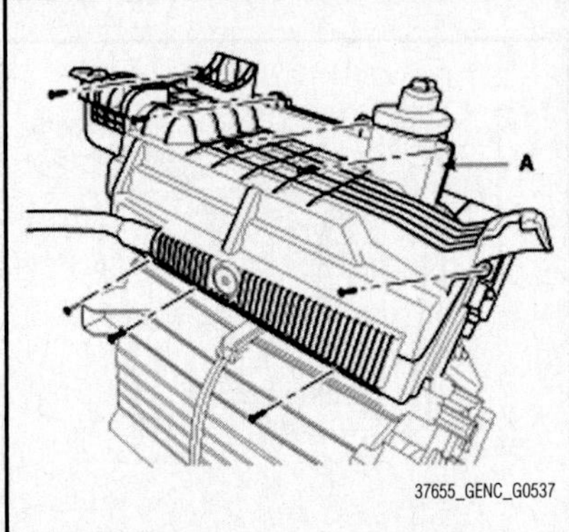

Fig. 319 Remove the heater unit lower case (A)

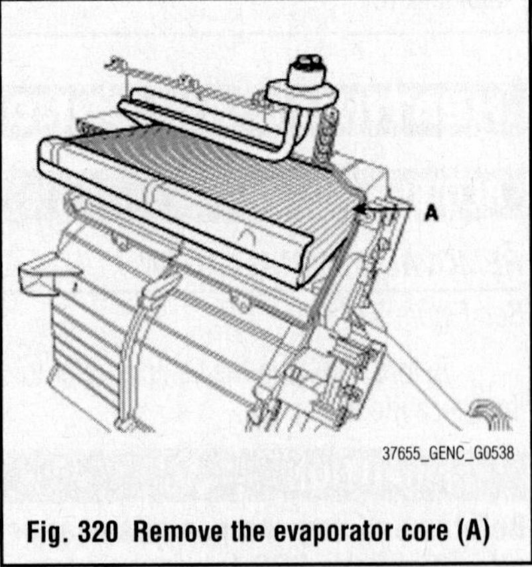

Fig. 320 Remove the evaporator core (A)

2. Disconnect the battery negative cable and wait for at least three minutes before beginning work.

3. Recover the refrigerant with a recovery/recycling/charging station.

4. When the engine is cool, drain the engine coolant from the radiator.

5. Remove the bolts and the expansion valve from the evaporator core. Plug or cap the lines immediately after disconnecting them to avoid moisture and dust contamination.

6. Disconnect the inlet and outlet heater hoses from the heater unit.

WARNING

Engine coolant will spill when the hoses are disconnected; drain it into a clean drip pan. Be sure not to let coolant spill on electrical parts or painted surfaces. If any coolant spills, rinse it off immediately.

7. Remove the instrument panel.
8. Remove the cowl cross bar assembly.
9. Remove the heater and blower bolts, and then remove the heater and blower unit.
10. Remove the 2 blower unit screws, and then remove the blower unit from the heater unit.
11. Remove the heater core cover.
12. Be careful that the inlet and outlet pipe are not bent during heater core removal, and pull out the heater core.
13. Remove the heater unit lower case.
14. Remove the evaporator core.
15. Be careful that the inlet and outlet pipe are not bent during heater core removal, and pull out the heater core.

To install:

16. Installation is the reverse order of removal, and note these items:
    a. If you're installing a new evaporator, add refrigerant oil (ND-OIL8).
    b. Replace the O-rings with new ones at each fitting, and apply a thin coat of refrigerant oil before installing. Be sure to use the right O-rings for R-134a to avoid leakage.
    c. Immediately after using the oil, replace the cap on the container, and seal it to avoid moisture absorption.
    d. Do not spill the refrigerant oil on the vehicle; it may damage paint; if the refrigerant oil contacts the paint, wash off immediately.
    e. Apply sealant to the grommets.
    f. Make sure that there is no air leakage.
    g. Charge the system and test its performance.
    h. Do not interchange the inlet and outlet heater hoses and install the hose clamps securely.
    i. Refill the cooling system with engine coolant.

STEERING

POWER RACK & PINION STEERING GEAR

REMOVAL & INSTALLATION

See Figure 321.

1. Before servicing the vehicle, refer to the Precautions Section.
2. Drain the power steering fluid.
3. Remove the front wheel and tire.
4. Remove the bolt and then disconnect the universal joint assembly from the pinion of the steering gear box.
5. Remove the tie rod end castle nut.
6. Remove the tie rod end from the front axle.
7. Remove steering gearbox from the cross member by removing the bracket mounting bolts.

To install:

8. Installation is the reverse of removal, noting the following:
    a. Tighten castle nut to 43—58 ft. lbs. (60—80 Nm).

POWER STEERING PUMP

REMOVAL & INSTALLATION

See Figures 322 through 325.

1. Before servicing the vehicle, refer to the Precautions Section.
2. Drain the power steering fluid.
3. Remove the drive belt.
4. Disconnect the pressure tube and return hose from the power steering pump.

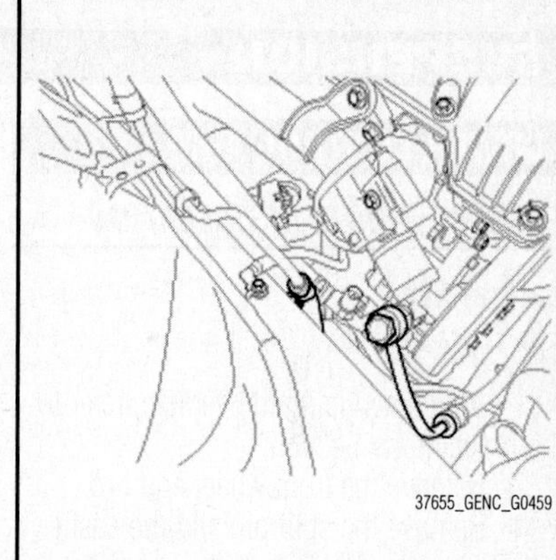

Fig. 323 Disconnect the pressure tube and return hose from the power steering pump—3.8L engine

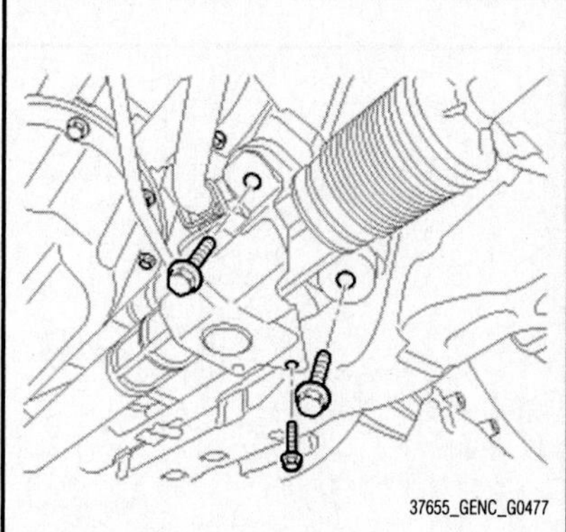

Fig. 321 Remove steering gearbox from the cross member by loosening the bracket mounting bolts

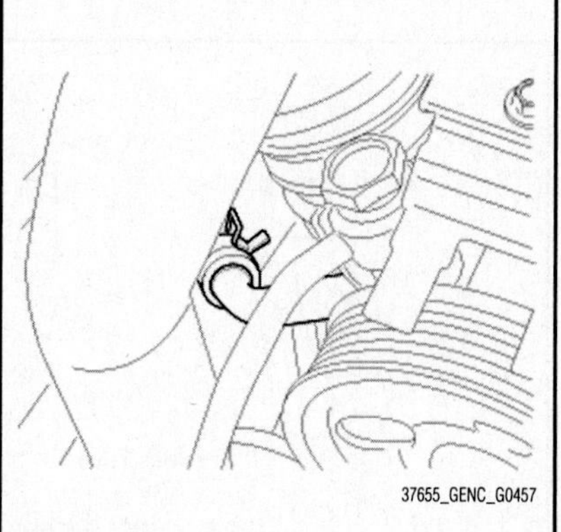

Fig. 322 Disconnect the pressure tube and return hose from the power steering pump—2.0L engine

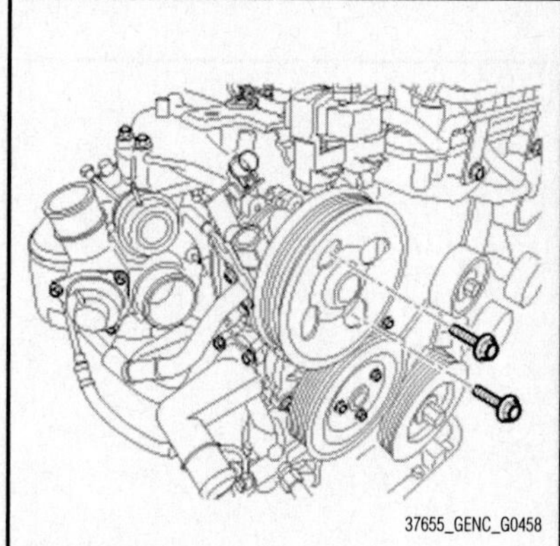

Fig. 324 Remove the mounting bolts and then remove the power steering pump—2.0L engine

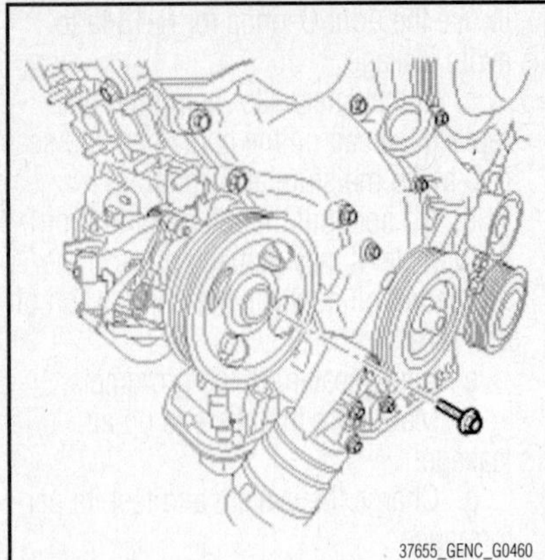

Fig. 325 Remove the mounting bolts and then remove the power steering pump—3.8L engine

5. Remove the mounting bolts and then remove the power steering pump.

To install:

6. Installation is the reverse of removal.

BLEEDING

➡Always use genuine power steering fluid. Using other type of power steering fluid or ATF can cause increased wear and poor steering in cold weather.

1. Fill the reservoir with the power steering fluid up to the level of 'COLD MAX' marked on the reservoir.

➡While conducting the following operations, keep replenishing the reservoir so that the fluid level can be always between the 'COLD MAX' and the 'COLD MIN' marked on the reservoir.

2. Jack up the front wheels.
3. Crank the engine 1–2 times by turning the ignition key very quickly from the 'On' position to the 'Start' position, but do not start the engine.

CAUTION

Be careful not to start the engine. If starting the engine before performing the steps 3 through 4, it may cause an abnormal noise during power steering pump operation.

4. Turn the steering wheel from lock to lock 5–6 times for 15– seconds.
5. Start the engine and keep turning the steering wheel from lock to lock until air bubbles stop appearing in the reservoir with the engine idle.
6. Check the color and level of the power steering fluid in the reservoir and then replenish the reservoir up to the 'COLD MAX' level as required.

➡If the fluid level moves up and down when turning the steering wheel, the fluid overflows out of the reservoir when the turning off the engine or the fluid has white color, it indicates that air bubbles have not been removed sufficiently from the power steering system. Therefore, repeat the steps 5 through 6 as required.

SUSPENSION

FRONT SUSPENSION

LOWER CONTROL ARM

REMOVAL AND & INSTALLATION

Lateral Arm

See Figure 326.

1. Before servicing the vehicle, refer to the Precautions Section.
2. Remove the front wheel and tire.
3. Remove the split pin and the castle nut.
4. Separate the lateral arm from the front axle ball joint by using SST (09568-2J100).
5. Remove the bolts and nuts and then remove the lateral arm from the sub frame.

To install:

6. Installation is the reverse of removal, noting the following:
   a. Tighten the castle nuts to 65–80 ft. lbs. (90–110 Nm).
   b. Tighten the lateral arm to the sub frame bolt and nut to 101–116 ft. lbs. (140–160 Nm).

Tension Arm

See Figures 327 through 329.

1. Before servicing the vehicle, refer to the Precautions Section.
2. Remove the front wheel and tire.
3. Remove the tension arm castle nut.

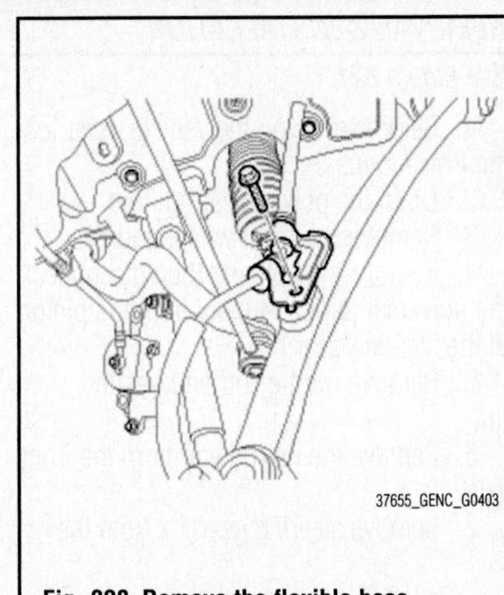

Fig. 328 Remove the flexible hose

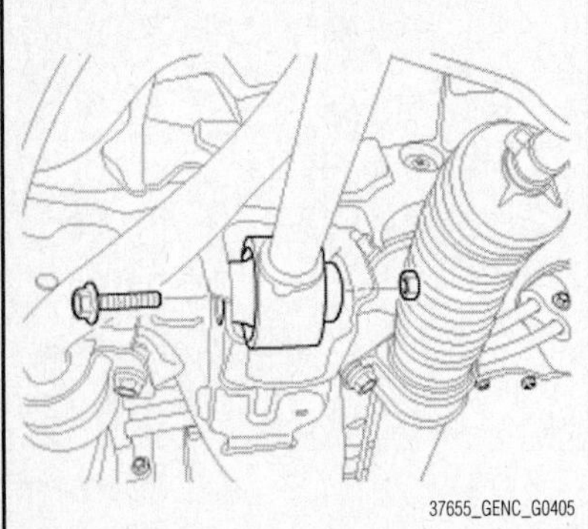

Fig. 326 Remove the bolts and nuts and then remove the lateral arm from the sub frame

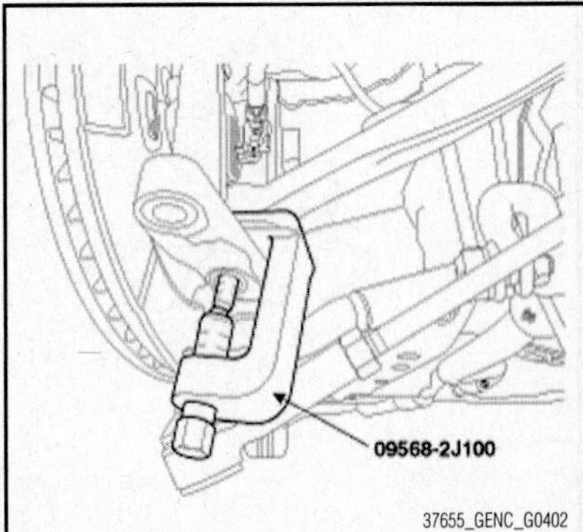

Fig. 327 Separate the tension arm from the front axle ball joint by using SST (09568-2J100)

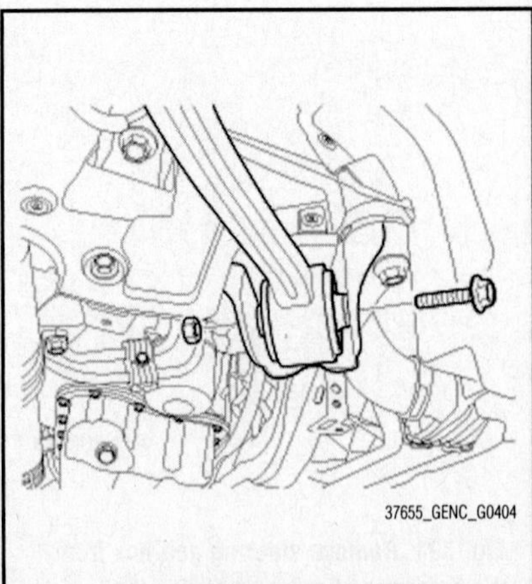

Fig. 329 Remove the flexible hose

4. Separate the tension arm from the front axle ball joint by using SST (09568-2J100).

5. Remove the flexible hose.

6. Remove the bolts and nuts, and then remove the tension arm from the sub frame.

To install:

7. Installation is the reverse of removal, noting the following:

a. Tighten the tension arm to sub frame bolts and lock nuts to 101–116 ft. lbs. (140–160 Nm).

b. Tighten the castle nuts to 58–65 ft. lbs. (80–90 Nm).

c. Tighten flex hose bracket to 5–8 ft. lbs. (7–11 Nm).

STRUT

REMOVAL & INSTALLATION

See Figures 330 through 332.

1. Before servicing the vehicle, refer to the Precautions Section.

2. Remove the front wheel and tire.

3. Disconnect the stabilizer link from the front strut assembly by removing the nuts.

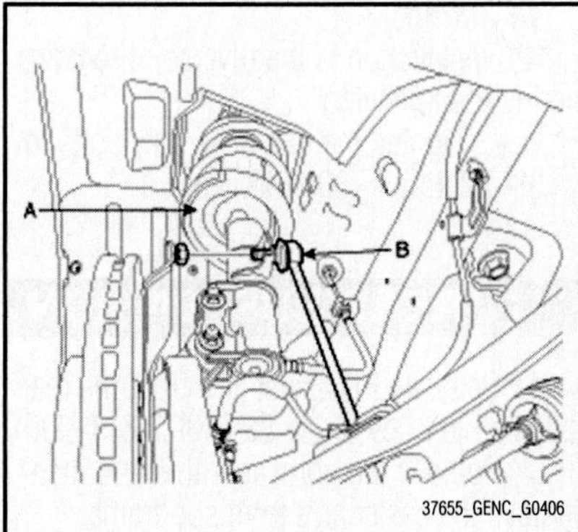

Fig. 330 Disconnect the stabilizer link (B) from the front strut assembly (A)

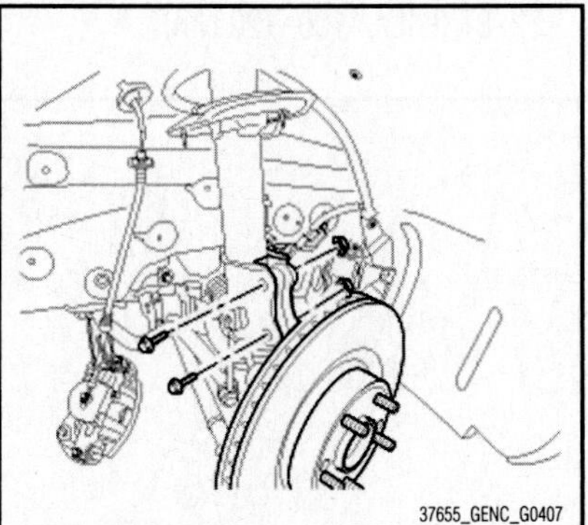

Fig. 331 Disconnect the front strut assembly with the knuckle by removing the bolt and nut

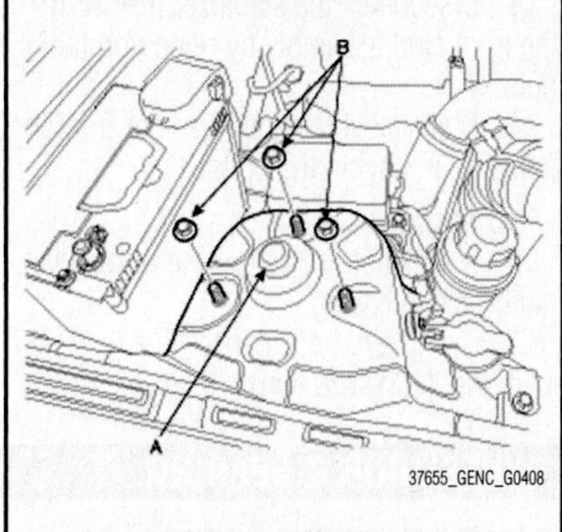

Fig. 332 Remove the strut cap (A) and remove the strut mounting nuts (B)

4. Disconnect the front strut assembly with the knuckle by removing the bolt and nut.

5. Remove the strut cap.

6. Remove the strut mounting nuts.

To install:

7. Installation is the reverse of removal, noting the following:

a. Tighten the strut assembly to knuckle mounting bolt to 101–116 ft. lbs. (140–160 Nm).

b. Tighten the strut mounting nuts to 32–47 ft. lbs. (45–65 Nm).

c. Tighten the stabilizer link nuts to 72–87 ft. lbs. (100–120 Nm).

OVERHAUL

See Figures 333 and 334.

1. Using the special tool (09546-26000), compress the coil spring.

2. Remove the self-locking nut.

3. Remove the insulator, spring seat, coil spring and dust cover from the strut assembly.

4. Reassembly is the reverse of removal, noting the following:

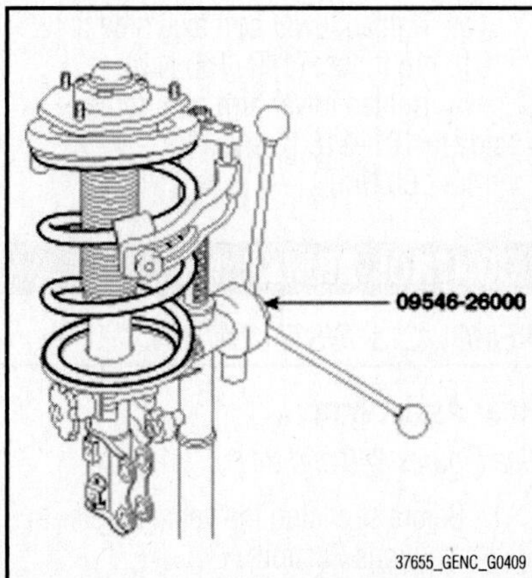

Fig. 333 Compress the coil spring

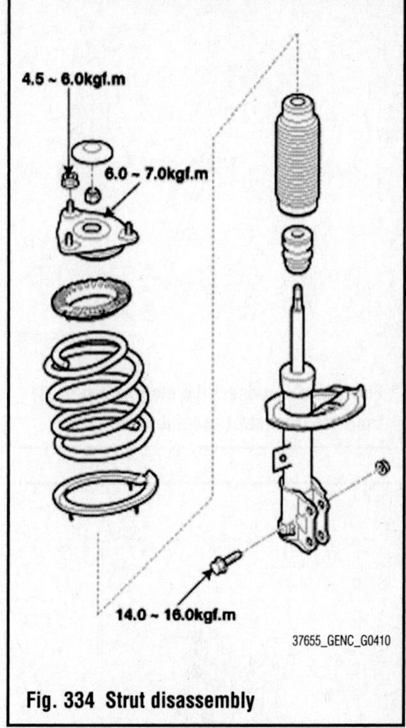

Fig. 334 Strut disassembly

a. Tighten the strut self locking nut to 44–51 ft. lbs. (60–69 Nm).

STRUT BAR

REMOVAL & INSTALLATION

See Figure 335.

1. Remove the strut bar nuts.

To install:

2. Installation is the reverse of removal.

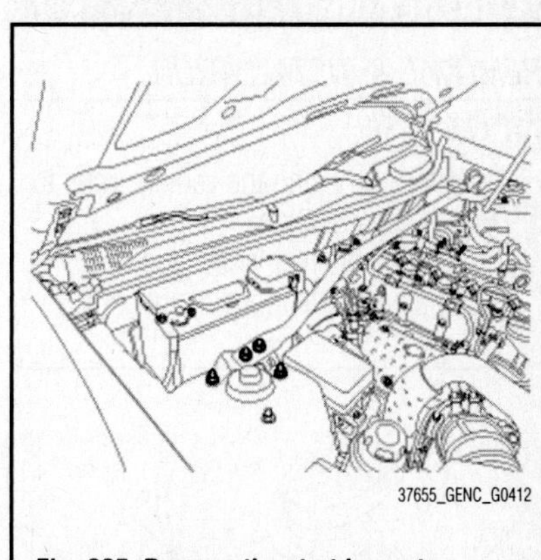

Fig. 335 Remove the strut bar nuts

STABILIZER BAR

REMOVAL & INSTALLATION

See Figures 336 and 337.

1. Before servicing the vehicle, refer to the Precautions Section.

2. Remove the front wheel and tire.

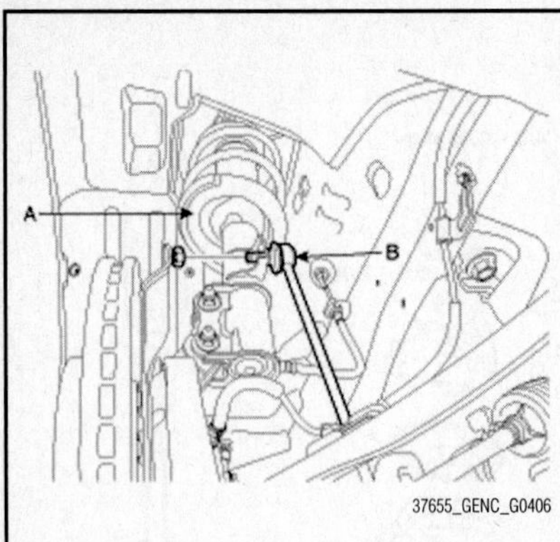

Fig. 336 Disconnect the stabilizer link (B) from the front strut assembly (A)

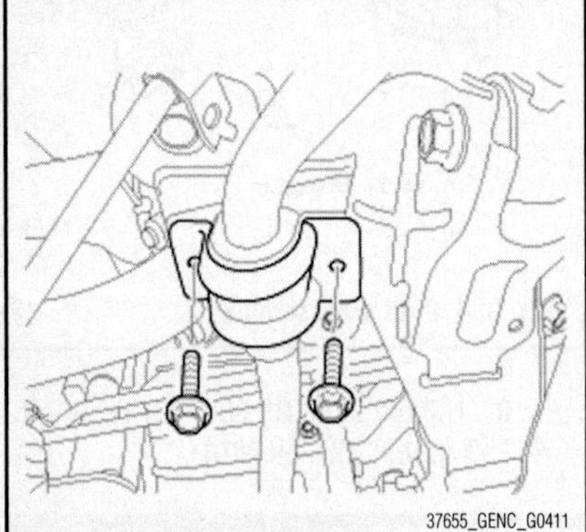

Fig. 337 Disconnect the stabilizer bar from the frame by removing the bolts (A)

3. Disconnect the stabilizer link from the front strut assembly by removing the nuts.
4. Disconnect the stabilizer bar from the frame by removing the bolts.

To install:

5. Installation is the reverse of removal, noting the following:
   a. Tighten the stabilizer link to 72–87 ft. lbs. (100–120 Nm).

WHEEL HUB & BEARING

REMOVAL & INSTALLATION

See Figure 338.

1. Before servicing the vehicle, refer to the Precautions Section.
2. Remove the front wheel and tire.
3. Remove the mounting screw and the brake disc.
4. Remove the caliper mounting bolts, and then hang the brake caliper on a wire. Refer to Front Disc Brake Removal & Installation in the Brake section.
5. Remove the tie rod end from the knuckle. Refer to Steering Linkage Removal & Installation in the Steering section.
6. Remove the tension arm mounting bolt, and then remove the tension arm from knuckle. Refer to Lower Control Arm Removal & Installation.

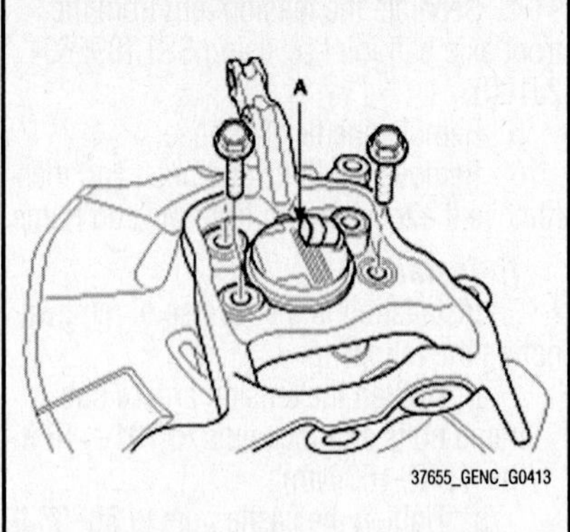

Fig. 338 Remove the hub assembly (A) from knuckle assembly

7. Remove the lateral arm mounting bolt, and then remove the lateral arm from knuckle. Refer to Lower Control Arm Removal & Installation.
8. Remove the strut mounting bolts.
9. Remove the hub and knuckle assembly.
10. Remove the hub assembly from knuckle assembly.
11. Remove the dust cover mounting bolts, and then remove the dust cover.

To install:

12. Installation is the reverse of removal, noting the following:
    a. Tighten the hub assembly bolts to 58–72 ft. lbs. (80–100 Nm).

SUSPENSION

REAR SUSPENSION

COIL SPRING

REMOVAL & INSTALLATION

See Figure 339.

1. Before servicing the vehicle, refer to the Precautions Section.
2. Remove the rear wheel and tire.
3. Remove the rear shock absorber.

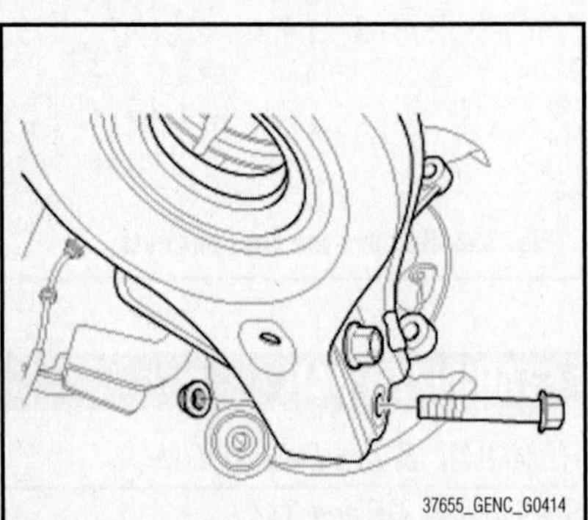

Fig. 339 Remove the bolt, and nuts remove the lower arm from rear axle

4. Remove the bolts and nuts and then remove the lower arm from rear axle.
5. Remove the bolts and nuts and then remove the lower arm from sub frame.
6. Remove the coil spring.

To install:

7. Installation is the reverse of removal, noting the following:
   a. Tighten lower arm axle side to 101–116 ft. lbs. (140–160 Nm).
   b. Tighten lower arm sub frame side to 101–116 ft. lbs. (140–160 Nm).

CONTROL ARMS/LINKS

REMOVAL & INSTALLATION

Rear Assist Arm

See Figures 340 and 341.

1. Before servicing the vehicle, refer to the Precautions Section.
2. Remove the rear wheel and tire.
3. Remove the bolts and nuts.
4. Separate the assist arm from the rear axle ball joint by using SST (09568-34000).
5. Remove the bolts and nuts and then remove the assist arm from sub frame.

To install:

6. Installation is the reverse of removal, noting the following:
   a. Tighten bolt and lock nut to 72–87 ft. lbs. (100–120 Nm).

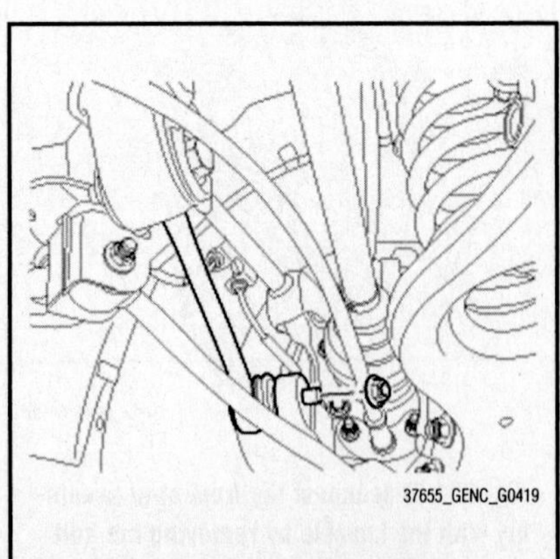

Fig. 340 Remove the bolts and nuts

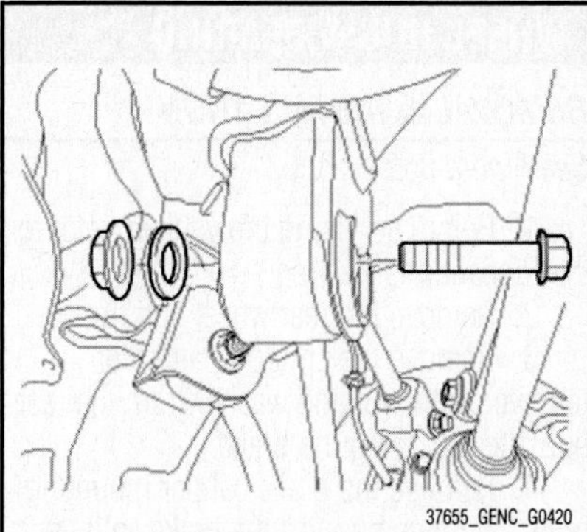

Fig. 341 Remove the bolts and nuts and then remove the assist arm from sub frame

b. Tighten bolt and lock nut to 101–116 ft. lbs. (140–160 Nm).

Rear Lower Arm

See Figure 339.

1. Before servicing the vehicle, refer to the Precautions Section.
2. Remove the rear wheel and tire.
3. Remove the rear shock absorber.
4. Remove the bolts and nuts and then remove the lower arm from rear axle.
5. Remove the bolts and nuts and then remove the lower arm from sub frame.
6. Remove the coil spring.

To install:

7. Installation is the reverse of removal, noting the following:
   a. Tighten lower arm axle side to 101–116 ft. lbs. (140–160 Nm).
   b. Tighten lower arm sub frame side to 101–116 ft. lbs. (140–160 Nm).

Rear Upper Arm

See Figures 342 and 343.

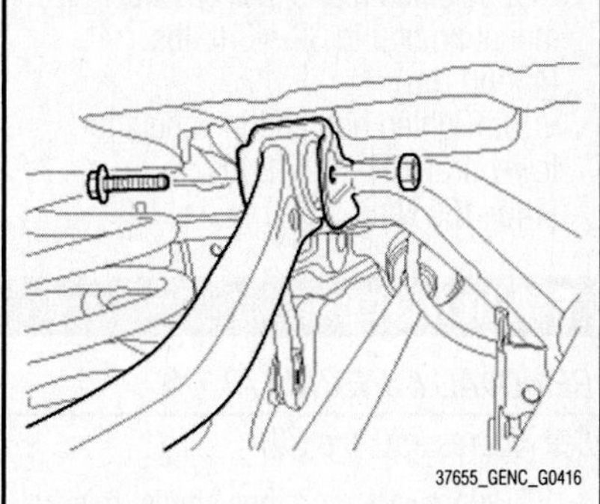

Fig. 343 Remove the bolts and nuts and then remove the rear upper arm from sub frame

1. Before servicing the vehicle, refer to the Precautions Section.
2. Remove the rear wheel and tire.
3. Remove the bolts and nuts and then remove the rear upper arm from rear axle.
4. Remove the bolts and nuts and then remove the rear upper arm from sub frame.

To install:

5. Installation is the reverse of removal, noting the following:
   a. Tighten upper arm axle side to 101–116 ft. lbs. (140–160 Nm).
   b. Tighten upper arm sub frame side to 72–87 ft. lbs. (100–120 Nm).

Front Upper Arm

See Figures 344 and 345.

1. Before servicing the vehicle, refer to the Precautions Section.
2. Remove the rear wheel and tire.
3. Remove the brake hose bracket.
4. Remove the bolts and nuts and then remove the rear upper arm from rear axle.
5. Remove the bolts and nuts and then remove the front upper arm from sub frame.

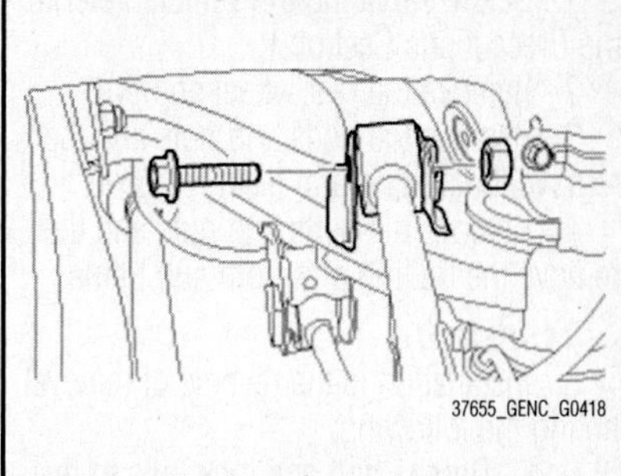

Fig. 345 Remove the bolts and nuts and then remove the front upper arm from sub frame

To install:

6. Installation is the reverse of removal, noting the following:
   a. Tighten upper arm axle side to 72–87 ft. lbs. (100–120 Nm).
   b. Tighten upper arm sub frame side to 72–87 ft. lbs. (100–120 Nm).

Trailing Arm

See Figures 346 and 347.

Fig. 346 Remove the trailing arm bolt and lock nuts from rear axle

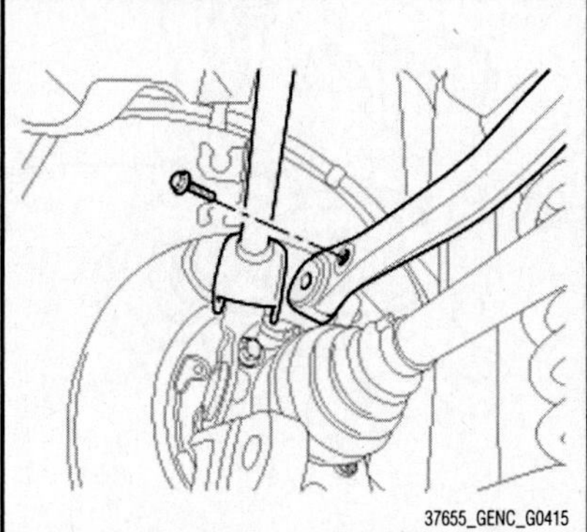

Fig. 343 Remove the bolts and nuts and then remove the rear upper arm from rear axle

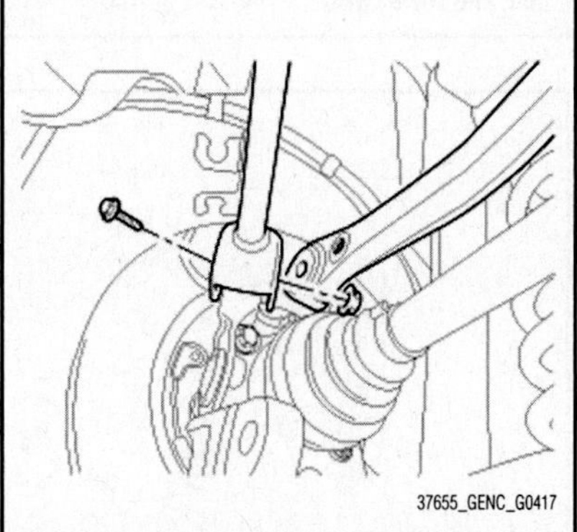

Fig. 344 Remove the bolts and nuts and then remove the front upper arm from rear axle

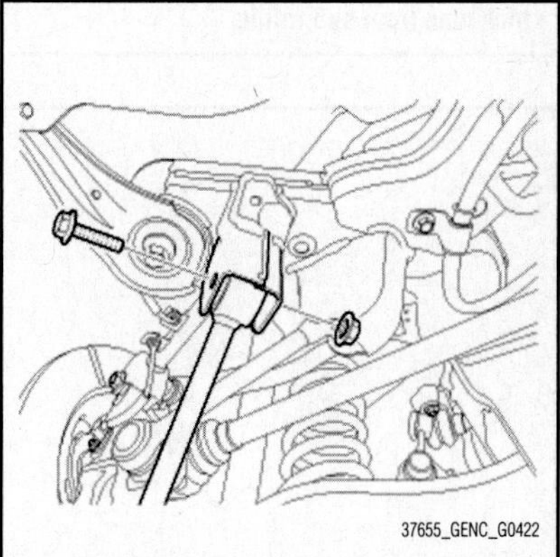

Fig. 347 Remove the trailing arm bolt and lock nuts from sub frame

1. Before servicing the vehicle, refer to the Precautions Section.
2. Remove the rear wheel and tire.
3. Remove the bolts and nuts and then remove the trailing arm from rear axle.
4. Remove the bolts and nuts and then remove the trailing arm from sub frame.

To install:

5. Installation is the reverse of removal, noting the following:
   a. Tighten bolt and lock nuts to rear axle 72–87 ft. lbs. (100–120 Nm).
   b. Tighten bolt and lock nuts to sub frame 72–87 ft. lbs. (100–120 Nm).

SHOCK ABSORBER

REMOVAL & INSTALLATION

See Figures 348 and 349.

1. Before servicing the vehicle, refer to the Precautions Section.
2. Remove the rear wheel and tire.
3. Remove the bolts and nuts and then remove the rear shock absorber from the lower arm.
4. Remove the mounting bolts.

To install:

5. Installation is the reverse of removal, noting the following:
   a. Tighten rear shock absorber top mounting bolt to 32–43 ft. lbs. (45–60 Nm).
   b. Tighten bolt and lock nuts to lower arm 101–116 ft. lbs. (140–160 Nm).

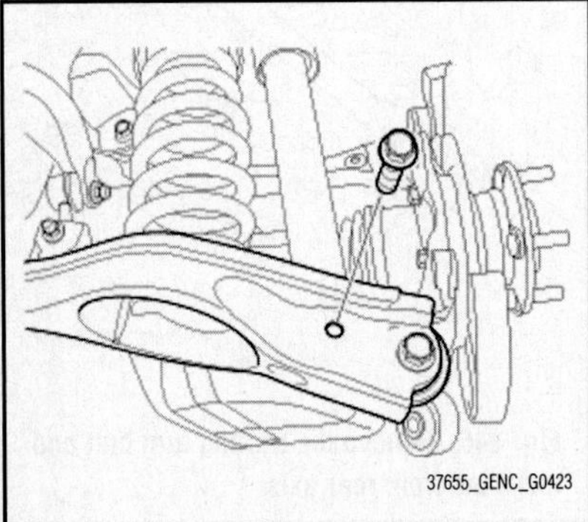

Fig. 348 Remove the trailing arm bolt and lock nuts from sub frame

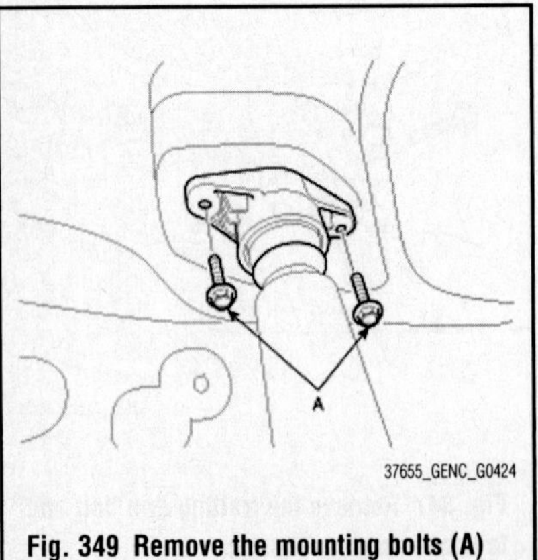

Fig. 349 Remove the mounting bolts (A)

STABILIZER BAR

REMOVAL & INSTALLATION

See Figures 350 and 351.

1. Before servicing the vehicle, refer to the Precautions Section.
2. Remove the rear wheel and tire.
3. Remove the nuts and then remove the stabilizer link from stabilizer bar and lower arm.
4. Remove the mounting bolts.

To install:

5. Installation is the reverse of removal, noting the following:
   a. Tighten the link nuts to 72–87 ft. lbs. (100–120 Nm).
   b. Tighten the mounting bolts to 36–47 ft. lbs. (50–65 Nm).

Fig. 350 Remove the nuts and then remove the stabilizer link from stabilizer bar and lower arm

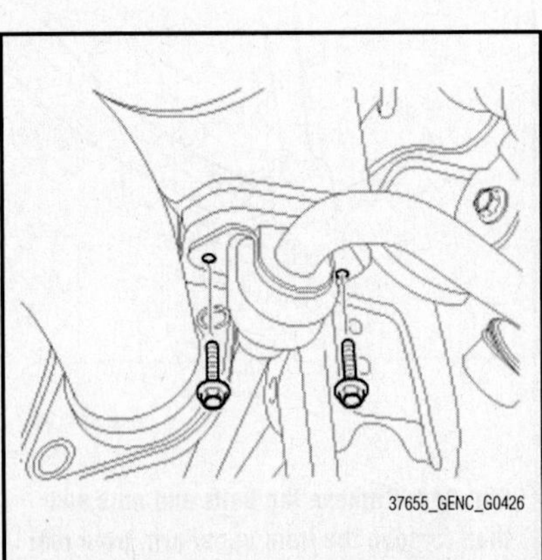

Fig. 351 Remove the mounting bolts

WHEEL HUB & BEARING

REMOVAL & INSTALLATION

See Figure 352.

1. Before servicing the vehicle, refer to the Precautions Section.
2. Remove the rear wheel and tire.
3. Remove the split pin, and then remove castle nut and washer from the rear hub after applying the brake.
4. Remove the brake caliper mounting bolts, and then support the brake caliper assembly with wire.
5. Remove the rear brake disc.
6. Remove the rear break lining. Refer to Rear Disc Brake Removal & Installation in the Brake section.
7. Remove the assist arm and the trailing arm from the rear axle carrier. Refer to Rear Assist Arm and Trailing Arm, in Control Arms Removal & Installation.
8. Remove lower arm mount bolt from rear axle carrier. Refer to Rear Lower Arm, in Control Arms Removal & Installation.
9. Remove the upper arm link mounting bolt and then remove the upper arm link. Refer to Rear Upper Arm, in Control Arms Removal & Installation.
10. Loosen the brake cable mount nuts and then remove the brake cable.
11. Remove the rear hub assembly.

To install:

12. Installation is the reverse of removal.
    a. Tighten the castle nut to 145–202 ft. lbs. (200–280 Nm).
    b. Tighten hub bolts to 58–73 ft. lbs. (79–98 Nm).

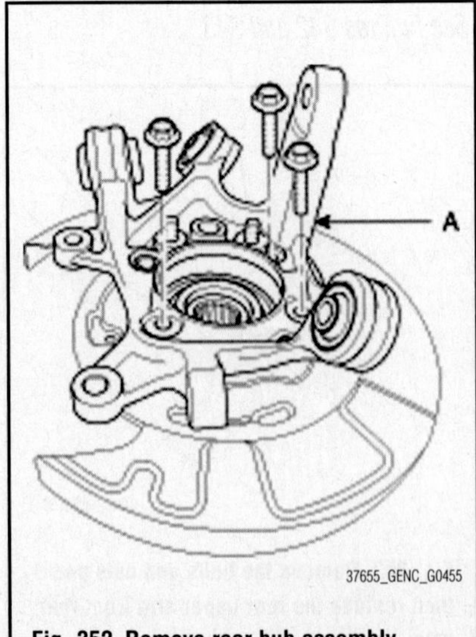

Fig. 352 Remove rear hub assembly

HYUNDAI

Genesis Sedan

5

SPECIFICATIONS AND MAINTENANCE CHARTS

ENGINE AND VEHICLE IDENTIFICATION

Engine							Model Year	
Code ①	Liters (cc)	Cu. In.	Cyl.	Fuel Sys.	Engine Type	Eng. Mfg.	Code ②	Year
E	3.8 (3778)	230.55	V6	MPFI	DOHC	Hyundai	9	2009
F	4.6 (4627)	282.40	V8	MPFI	DOHC	Hyundai	A	2010

MPFI: Multi-Point Fuel Injection

DOHC: Double Overhead Camshafts

① 8th digit of VIN

② 10th digit of VIN

37655_GENS_C0001

GENERAL ENGINE SPECIFICATIONS

All measurements are given in inches.

Year	Model	Engine Displacement Liters	Engine Series VIN	Net Horsepower @ rpm	Net Torque @ rpm (ft. lbs.)	Bore x Stroke (in.)	Com-pression Ratio	Oil Pressure @ rpm
2009	Genesis	3.8	E	290@6200	264@4500	3.780 x 3.743	10.4:1	18.8@1000
	Genesis	4.6	F	375@6500	333@3500	3.413 x 2.953	10.4:1	22.8@1000
2010	Genesis	3.8	E	290@6200	264@4500	3.780 x 3.743	10.4:1	18.8@1000
	Genesis	4.6	F	375@6500	333@3500	3.413 x 2.953	10.4:1	22.8@1000

37655_GENS_C0002

GASOLINE ENGINE TUNE-UP SPECIFICATIONS

Year	Engine Displacement Liters	Engine VIN	Spark Plug Gap (in.)	Ignition Timing (deg.)		Fuel Pump (psi)	Idle Speed (rpm)		Valve Clearance	
				MT	AT		MT	AT	In.	Ex.
2009	3.8	E	0.039-0.043	NA	①	54.3-55.8	NA	②	HYD	HYD
	4.6	F	0.039-0.043	NA	①	54.3-55.8	NA	②	HYD	HYD
2010	3.8	E	0.039-0.043	NA	①	54.3-55.8	NA	②	HYD	HYD
	4.6	F	0.039-0.043	NA	①	54.3-55.8	NA	②	HYD	HYD

Follow the figures on the label if they differ from those in this chart.

HYD: Hydraulic

NA: Not Applicable

① Ignition timing is computer controlled and is not adjustable

② Idle speed is maintained by the Electronic Control Module (ECM)

37655_GENS_C0003

CAPACITIES

Year	Model	Engine Displacement Liters	Engine VIN	Engine Oil with Filter (qts.)	Transmission (pts.)		Fuel Tank (gal.)	Cooling System (qts.)
					Manual	Auto.		
2009	Genesis	3.8	E	5.8	NA	NS	19.3	5.8
	Genesis	4.6	F	6.9	NA	NS	20.3	8.9
2010	Genesis	3.8	E	5.8	NA	NS	19.3	5.8
	Genesis	4.6	F	6.9	NA	NS	20.3	8.9

NOTE: All capacities are approximate. Add fluid gradually and check to be sure a proper fluid level is obtained.

NA: Not Applicable

NS: Not Supplied

37655_GENS_C0004

FLUID SPECIFICATIONS

Year	Model	Engine Displacement Liters	Engine ID/VIN	Engine Oil	Auto. Trans.	Manual Trans.	Power Steering Fluid	Brake Master Cylinder
2009	Genesis	3.8	E	5W-20	①	NA	PSF-3	②
	Genesis	4.6	F	5W-20	③	NA	NA	②
2010	Genesis	3.8	E	5W-20	①	NA	PSF-3	②
	Genesis	4.6	F	5W-20	③	NA	NA	②

DOT: Department Of Transportation

NA: Not Applicable

① NWS 9638 ATF

② DOT 3 or DOT 4 hydraulic brake fluid

② Shell 1375.4 ATF

37655_GENS_C0005

VALVE SPECIFICATIONS

Year	Engine Displacement Liters	Engine VIN	Seat Angle (deg.)	Face Angle (deg.)	Spring Test Pressure (lbs. @ in.)	Spring Installed Height (in.)	Stem-to-Guide Clearance (in.)		Stem Diameter (in.)	
							Intake	Exhaust	Intake	Exhaust
2009	3.8	E	44.75-45.20	45.25-45.75	90.4-96.2 @0.953	NS	0.0008-0.0019	0.0012-0.0021	0.2151-0.2157	0.2149-0.2153
	4.6	F	44.75-45.20	45.25-45.75	56.6-62.6 @1.496	NS	0.0008-0.0018	0.0012-0.0021	0.2348-0.2354	0.2346-0.2350
2010	3.8	E	44.75-45.20	45.25-45.75	90.4-96.2 @0.953	NS	0.0008-0.0019	0.0012-0.0021	0.2151-0.2157	0.2149-0.2153
	4.6	F	44.75-45.20	45.25-45.75	56.6-62.6 @1.496	NS	0.0008-0.0018	0.0012-0.0021	0.2348-0.2354	0.2346-0.2350

NS: Not Supplied

37655_GENS_C0006

CAMSHAFT AND BEARING SPECIFICATIONS CHART

All measurements are given in inches.

Year	Engine Displ. Liters	Engine ID/VIN	Journal Dia.	Brg. Oil Clearance	Shaft End-play	Runout	Journal Bore	Lobe Height Intake	Lobe Height Exhaust
2009	3.8	E	①	②	0.0008-0.0071	NS	NS	1.8425	1.8031
	4.6	F	③	④	0.0047-0.0086	NS	NS	1.64	1.63
2010	3.8	E	①	②	0.0008-0.0071	NS	NS	1.8425	1.8031
	4.6	F	③	④	0.0047-0.0086	NS	NS	1.64	1.63

NA: Not Supplied

① Intake No. 1 is 1.1009-1.1016 inch
Intake No. 2, 3, 4 are 0.9430-0.9437 inch
Exhaust No.1 is 1.1009-1.1016 inch
Exhaust No. 2, 3, 4 are 0.9430-0.9437 inch

② Intake No. 1 is 0.0008-0.0022 inch
Intake No. 2, 3, 4 are 0.0012-0.0026 inch
Exhaust No.1 is 0.0008-0.0022 inch
Exhaust No. 2, 3, 4 are 0.0012-0.0026 inch

③ Intake/Exhaust Outer No. 1 is 1.4159-1.4165 inch
Intake/Exhaust Outer No. 2, 3, 4 are 1.0222-1.0228 inch
Intake/Exhaust Inner No.1 is 1.4175-1.4181 inch
Intake/Exhaust inner No. 2, 3, 4 are 1.0236-1.0244 inch

④ No. 1 is 0.0010-0.0022 inch
No. 2, 3, 4 are 0.0008-0.0022 inch

37655_GENS_C0007

CRANKSHAFT AND CONNECTING ROD SPECIFICATIONS

All measurements are given in inches.

Year	Engine Displacement Liters	Engine VIN	Crankshaft Main Brg. Journal Dia.	Crankshaft Main Brg. Oil Clearance	Crankshaft Shaft End-play	Crankshaft Thrust on No.	Connecting Rod Journal Diameter	Connecting Rod Oil Clearance	Connecting Rod Side Clearance
2009	3.8	E	2.7142-2.7149	0.0008-0.0016	0.0039-0.0110	3	2.1635-2.1642	0.0015-0.0022	0.0039-0.0098
	4.6	F	2.5583-2.5591	0.0002-0.0009	0.0039-0.0110	NS	2.0465-2.0472	0.0007-0.0014	0.0039-0.0118
2010	3.8	E	2.7142-2.7149	0.0008-0.0016	0.0039-0.0110	3	2.1635-2.1642	0.0015-0.0022	0.0039-0.0098
	4.6	F	2.5583-2.5591	0.0002-0.0009	0.0039-0.0110	NS	2.0465-2.0472	0.0007-0.0014	0.0039-0.0118

NS: Not Supplied

37655_GENS_C0008

PISTON AND RING SPECIFICATIONS

All measurements are given in inches.

Year	Engine Displ. Liters	Engine VIN	Piston Clearance	Ring Gap Top Compression	Ring Gap Bottom Compression	Ring Gap Oil Control	Ring Side Clearance Top Compression	Ring Side Clearance Bottom Compression	Ring Side Clearance Oil Control
2009	3.8	E	0.0008-0.0016	0.0067-0.0126	0.0126-0.0185	0.0078-0.0275	0.0016-0.0031	0.0012-0.0027	0.0024-0.0059
	4.6	F	0.0018-0.0026	0.0067-0.0126	0.0146-0.0205	0.0079-0.0276	0.0016-0.0031	0.0016-0.0031	0.0024-0.0059
2010	3.8	E	0.0008-0.0016	0.0067-0.0126	0.0126-0.0185	0.0078-0.0275	0.0016-0.0031	0.0012-0.0027	0.0024-0.0059
	4.6	F	0.0018-0.0026	0.0067-0.0126	0.0146-0.0205	0.0079-0.0276	0.0016-0.0031	0.0016-0.0031	0.0024-0.0059

37655_GENS_C0009

TORQUE SPECIFICATIONS

All readings in ft. lbs.

Year	Engine Displacement Liters	Engine VIN	Cylinder Head Bolts	Main Bearing Bolts	Rod Bearing Bolts	Crankshaft Damper Bolts	Flexplate Bolts	Manifold Intake	Manifold Exhaust	Spark Plugs	Oil Pan Drain Plug
2009	3.8	E	①	②	③	210-224	53-56	14-17	29-33	15-22	25-33
	4.6	F	④	⑤	⑥	290-297	72-80	14-19	36-40	18-21	25-33
2010	3.8	E	①	②	③	210-224	53-56	14-17	29-33	15-22	25-33
	4.6	F	④	⑤	⑥	290-297	72-80	14-19	36-40	18-21	25-33

① Step 1: 29 ft. lbs.
Step 2: Plus 120 degrees
Step 3: Plus 90 degrees

② M11 bolts (inner) Step 1: 36 ft. lbs.
M11 bolts (inner) Step 2: Plus 90 degrees
M8 bolts (outer) Step 1: 15 ft. lbs.
M8 bolts (outer) Step 2: Plus 120 degrees
M8 bolts (side): 22-23 ft. lbs.

③ Step 1: 15 ft. lbs.
Step 2: Plus 90 degrees

④ Lomg bolt Step 1: 26 ft. lbs.
Step 2: Plus 90 degrees
Step 3: Plus 120 degrees
Flange bolt: 23.9-26.8 ft. lbs.

⑤ Step 1: 27.5-30.4 ft. lbs.
Step 2: Plus 120 degrees
Flange bolt:15.9-18.8 ft. lbs.

⑥ Step 1: 16.6-19.5 ft. lbs.
Step 2: Plus 100 degrees

37655_GENS_C0010

WHEEL ALIGNMENT

Year	Model		Caster Range (+/-Deg.)	Caster Preferred Setting (Deg.)	Camber Range (+/-Deg.)	Camber Preferred Setting (Deg.)	Toe-in (Deg.)
2009	Genesis	Front	0.75	+7.63	0.50	-0.45	0.10 +/- 0.20
		Rear	—	—	0.50	-1.37	0.40 +/- 0.20
2010	Genesis	Front	0.75	+7.63	0.50	-0.45	0.10 +/- 0.20
		Rear	—	—	0.50	-1.37	0.40 +/- 0.20

37655_GENS_C0011

TIRE, WHEEL AND BALL JOINT SPECIFICATIONS

Year	Model	OEM Tires Standard	OEM Tires Optional	Tire Pressures (psi) Front	Tire Pressures (psi) Rear	Wheel Size	Ball Joint Inspection	Lug Nut Torque (ft. lbs.)
2009	Genesis	P225/55R17	P235/505R18	①	①	6.5J x 17 7.5J x 18	②	65-80
2010	Genesis	P225/55R17	P235/505R18	①	①	6.5J x 17 7.5J x 18	②	65-80

① Refer to placard on vehicle for proper inflation pressure.

② Replace if any measurable movement is found.

37655_GENS_C0012

BRAKE SPECIFICATIONS

All measurements in inches unless noted

Year	Model		Brake Disc Original Thickness	Brake Disc Minimum Thickness	Brake Disc Maximum Runout	Brake Drum Diameter Original Inside Diameter	Brake Drum Diameter Max. Wear Limit	Brake Drum Diameter Maximum Machine Diameter	Minimum Lining Thickness	Brake Caliper Bracket Bolts (ft. lbs.)	Brake Caliper Mounting Bolts (ft. lbs.)
2009	Genesis V6	F	1.100	1.040	0.002	—	—	—	0.079	58-72	16-23
		R	0.510	0.450	0.002	—	—	—	0.079	58-72	16-23
	Genesis V8	F	1.181	1.180	0.002	—	—	—	0.079	58-72	16-23
		R	0.510	0.450	0.002	—	—	—	0.079	58-72	16-23
2010	Genesis V6	F	1.100	1.040	0.002	—	—	—	0.079	58-72	16-23
		R	0.510	0.450	0.002	—	—	—	0.079	58-72	16-23
	Genesis V8	F	1.181	1.180	0.002	—	—	—	0.079	58-72	16-23
		R	0.510	0.450	0.002	—	—	—	0.079	58-72	16-23

37655_GENS_C0013

SCHEDULED MAINTENANCE INTERVALS

HYUNDAI—Genesis Sedan

TO BE SERVICED	TYPE OF SERVICE	VEHICLE MILEAGE INTERVAL (x1000)												
		7.5	15	22.5	30	37.5	45	52.5	60	67.5	75	82.5	90	97.5
Engine oil & filter	R	✓	✓	✓	✓	✓	✓	✓	✓	✓	✓	✓	✓	✓
Automatic transaxle fluid ①	S/I					✓					✓			
Frt brake pads, calipers/rotors	S/I		✓		✓		✓		✓		✓		✓	
Driveshafts & boots	S/I		✓		✓		✓		✓		✓		✓	
Air cleaner filter ②	R	✓	✓	✓	✓	✓	✓	✓	✓	✓	✓	✓	✓	✓
A/C refrigerant	S/I		✓		✓		✓		✓		✓		✓	
Brake fluid	I				✓				✓				✓	
Engine coolant ③	R								✓					
Fuel hose, vapor hose & fuel filter cap	S/I	✓	✓	✓	✓	✓	✓	✓	✓	✓	✓	✓	✓	✓
Spark plugs (Iridium coated) 100,000 mile replacement	R													
Bolts & nuts on chassis & body	S/I		✓		✓		✓		✓		✓		✓	
Drive belts	S/I				✓				✓				✓	
Exhaust pipe connections, muffler & suspension bolts	S/I		✓		✓		✓		✓		✓		✓	
Valve clearance ④	S/I								✓					
Electronic throttle control	S/I		✓		✓		✓		✓		✓		✓	
Brake hoses & lines	S/I		✓		✓		✓		✓		✓		✓	
Rear brake discs, linings & parking brake	S/I				✓				✓				✓	
Steering gear rack, linkage & boots	S/I		✓		✓		✓		✓		✓		✓	
Power steering pump, belt, hoses	S/I		✓		✓		✓		✓		✓		✓	
Fuel tank air filter ⑤	S/I		✓		✓		✓		✓		✓		✓	
Propeller shaft	S/I		✓		✓		✓		✓		✓		✓	
Rear axle oil	S/I				✓				✓				✓	
Rear differential oil	S/I					✓					✓			
Climate control air filter ⑥	R		✓		✓		✓		✓		✓		✓	
Fuel filter	R					✓					✓			
Fuel lines & connections	S/I	✓	✓	✓	✓	✓	✓	✓	✓	✓	✓	✓	✓	✓
Vacuum & crankcase ventilation hoses	S/I				✓				✓				✓	

R: Replace S/I: Service or Inspect

① Add only specified fluid.

② Replace at 30,000 miles

③ Replace every 24 months or 30,000 miles

④ 3.8 liter engine only

⑤ Replace every 30,000 miles

⑥ Replace at 15,000 miles, or 12 months. See Severe Service.

37655_GENS_C0014

SCHEDULED MAINTENANCE INTERVALS

HYUNDAI—Genesis Sedan

FREQUENT OPERATION MAINTENANCE (SEVERE SERVICE)

If a vehicle is operated under any of the following conditions it is considered severe service:

Repeatedly driving short distance of less than 5 miles (8 km) in normal temperature or less than 10 miles (16 km) in freezing temperature

Driving on rough, dusty, muddy, unpaved, graveled or salt- spread roads

Driving in areas using salt or other corrosive materials or in very cold weather

Extensive engine idling or low speed driving for long distances

Driving in sandy areas

Driving in heavy traffic area over 90°F (32°C)

Driving on uphill, downhill, or mountain road

Towing a Trailer, or using a camper, or roof rack

Driving as a patrol car, taxi, other commercial use or vehicle towing

Driving over 106 mph (170 km/h)

Frequently driving in stop-and-go conditions

Oil & oil filter: change every 3,750 miles.

Brake pads, calipers & rotors: service or inspect more frequently

Driveshaft boots: service or inspect every 7500 miles.

Steering gear rack, linkage & boots: service or inspect more frequently

Air cleaner filter: service or inspect more frequently

Automatic transaxle fluid & filter: replace every 60,000 miles.

Climate control air filter: replace as necessary.

Spark plugs replace more frequently

Rear axle oil replace every 60,000 miles.

Rear differential oil inspect every 80,000 miles.

Propeller shaft inspect every 7,500 miles.

37655_GENS_C0015

BRAKES — INFORMATION AND PRECAUTIONS

ANTI-LOCK SYSTEMS

• Certain components within the ABS system are not intended to be serviced or repaired individually.

• Do not use rubber hoses or other parts not specifically specified for and ABS system. When using repair kits, replace all parts included in the kit. Partial or incorrect repair may lead to functional problems and require the replacement of components.

• Lubricate rubber parts with clean, fresh brake fluid to ease assembly. Do not use shop air to clean parts; damage to rubber components may result.

• Use only DOT 3 brake fluid from an unopened container.

• If any hydraulic component or line is removed or replaced, it may be necessary to bleed the entire system.

• A clean repair area is essential. Always clean the reservoir and cap thoroughly before removing the cap. The slightest amount of dirt in the fluid may plug an orifice and impair the system function. Perform repairs after components have been thoroughly cleaned; use only denatured alcohol to clean components. Do not allow ABS components to come into contact with any substance containing mineral oil; this includes used shop rags.

• The Anti-Lock control unit is a microprocessor similar to other computer units in the vehicle. Ensure that the ignition switch is **OFF** before removing or installing controller harnesses. Avoid static electricity discharge at or near the controller.

• If any arc welding is to be done on the vehicle, the control unit should be unplugged before welding operations begin.

DISC AND DRUM SYSTEMS

**** CAUTION**

Dust and dirt accumulating on brake parts during normal use may contain asbestos fibers from production or aftermarket brake linings. Breathing excessive concentrations of asbestos fibers can cause serious bodily harm. Exercise care when servicing brake parts. Do not sand or grind brake lining unless equipment used is designed to contain the dust residue. Do not clean brake parts with compressed air or by dry brushing. Cleaning should be done by dampening the brake components with a fine mist of water, then wiping the brake components clean with a dampened cloth. Dispose of cloth and all residue containing asbestos fibers in an impermeable container with the appropriate label. Follow practices prescribed by the Occupational Safety and Health Administration (OSHA) and the Environmental Protection Agency (EPA) for the handling, processing, and disposing of dust or debris that may contain asbestos fibers.

BRAKES — BLEEDING THE BRAKE SYSTEM

BLEEDING PROCEDURE

BLEEDING PROCEDURE

See Figures 1 through 3.

1. Before servicing the vehicle, refer to the Precautions Section.

**** WARNING**

The reservoir on the master cylinder must be at the MAX (upper) level mark at the start of bleeding procedure and checked after bleeding each brake caliper. Add fluid as required.

2. Make sure the brake fluid level in the reservoir is at the MAX (upper) level line.
3. Have someone slowly pump the brake pedal several times, and then apply steady pressure.
4. Loosen the right-rear brake bleed screw to allow air to escape from the system. Then tighten the bleed screw securely.
5. Repeat the procedure for each wheel in the sequence shown below until air bubbles no longer appear in the fluid.
6. Refill the master cylinder reservoir to the MAX (upper) level line.

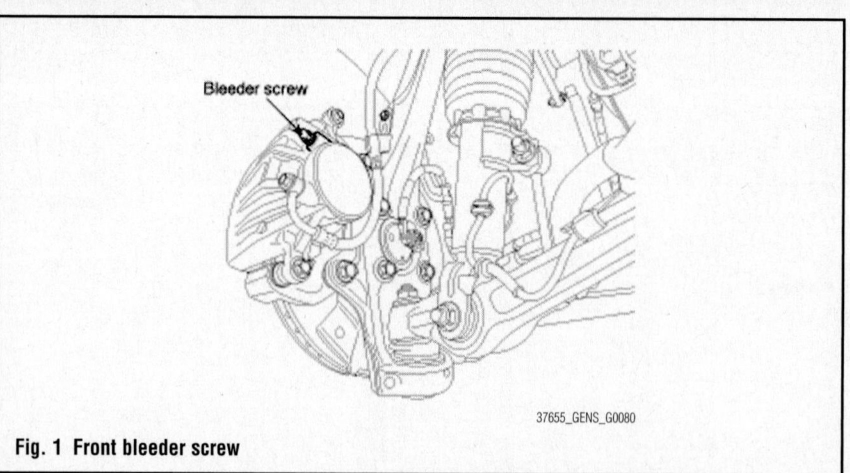

Fig. 1 Front bleeder screw

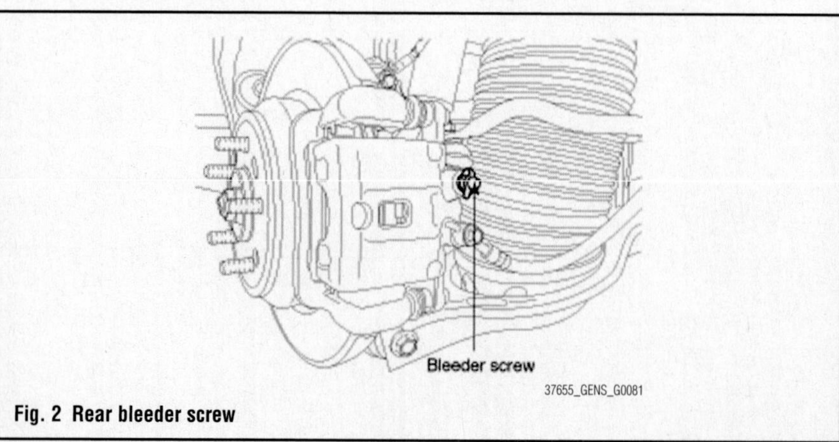

Fig. 2 Rear bleeder screw

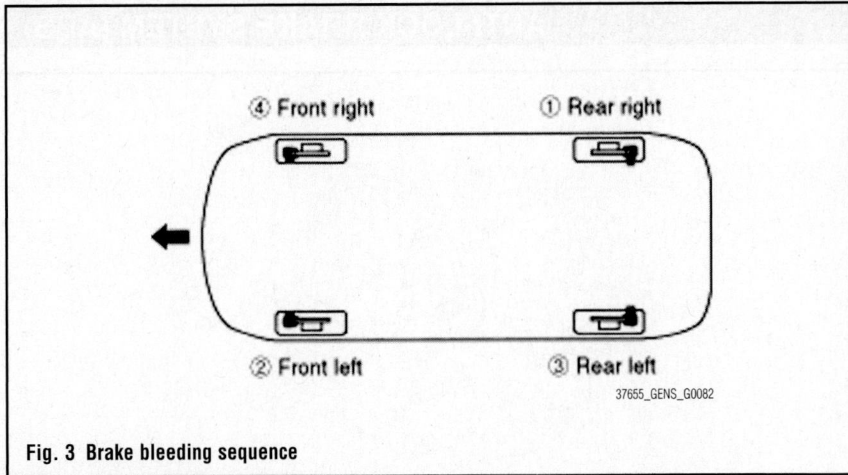

Fig. 3 Brake bleeding sequence

BLEEDING THE ABS SYSTEM

See Figures 1 through 4.

This procedure should be followed to ensure adequate bleeding of air and the filling of the ABS unit, the brake lines, and the master cylinder with brake fluid.

1. Before servicing the vehicle, refer to the Precautions Section.
2. Remove the reservoir cap and fill the brake reservoir with brake fluid.

WARNING

If there is any brake fluid on any painted surface, wash it off immediately.

➡When pressure bleeding, do not depress the brake pedal. Recommended brake fluid: DOT3 or DOT4.

3. Connect a clear plastic tube to the wheel cylinder bleeder plug and insert the other end of the tube into a clear plastic bottle that is half filled with clean brake fluid.
4. Connect the scan tool to the data link connector located underneath the dash panel.
5. Select and operate according to the instructions on the scan tool screen.

CAUTION

You must obey the maximum operating time of the ABS motor with the scan tool to prevent the motor pump from burning.

6. Select vehicle name.
7. Select Anti-Lock Brake system.
8. Select air bleeding mode.
9. Press "YES" to operate motor pump and solenoid valve.

WARNING

Wait 120 seconds before operating the air bleeding or damage to the motor may occur.

10. Wait 120 seconds before operating the air bleeding.
11. Pump the brake pedal several times, and then loosen the bleeder screw until fluid starts to run out without bubbles. Then, close the bleeder screw.
12. Repeat until there are no more bubbles in the fluid for each wheel.
13. Tighten the bleeder screw to 61–113 inch lbs. (9–13 Nm).

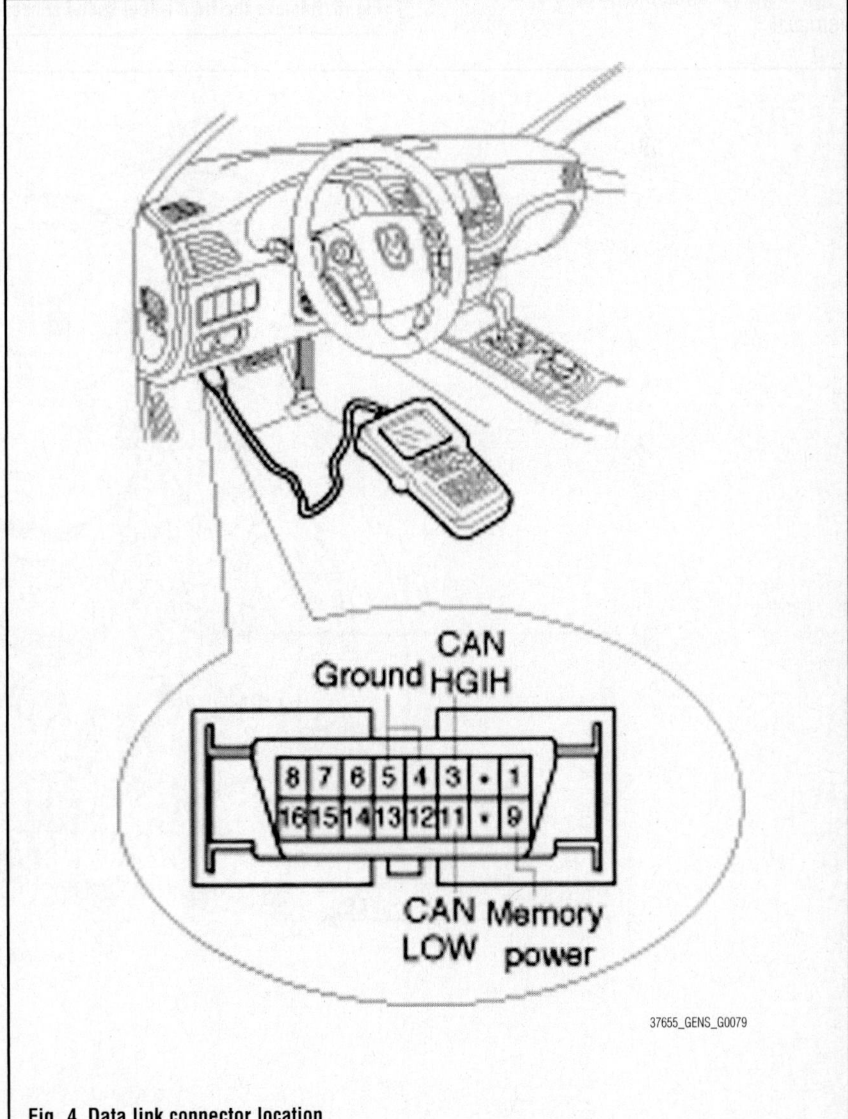

Fig. 4 Data link connector location

SPEED SENSORS

REMOVAL & INSTALLATION

Front

See Figures 5 and 6.

1. Before servicing the vehicle, refer to the Precautions Section.
2. Remove the front wheel speed sensor clip.
3. Disconnect the connector.
4. Remove the front wheel speed sensor.

To install:

5. Installation is the reverse of removal.

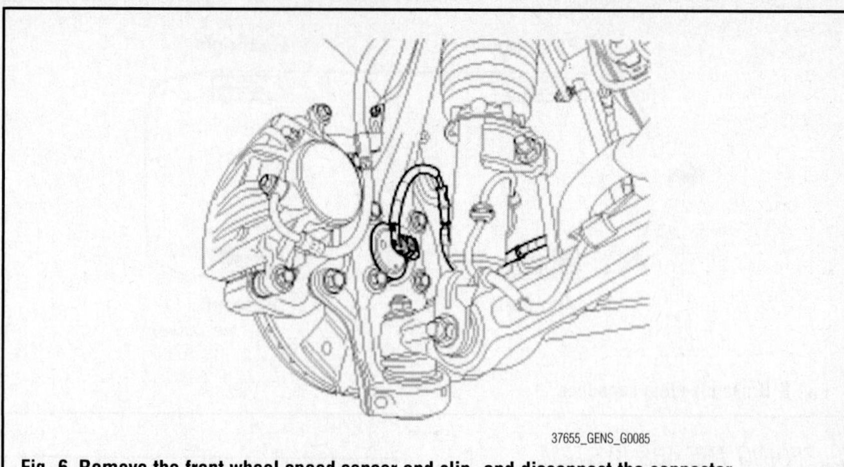

Fig. 6 Remove the front wheel speed sensor and clip, and disconnect the connector

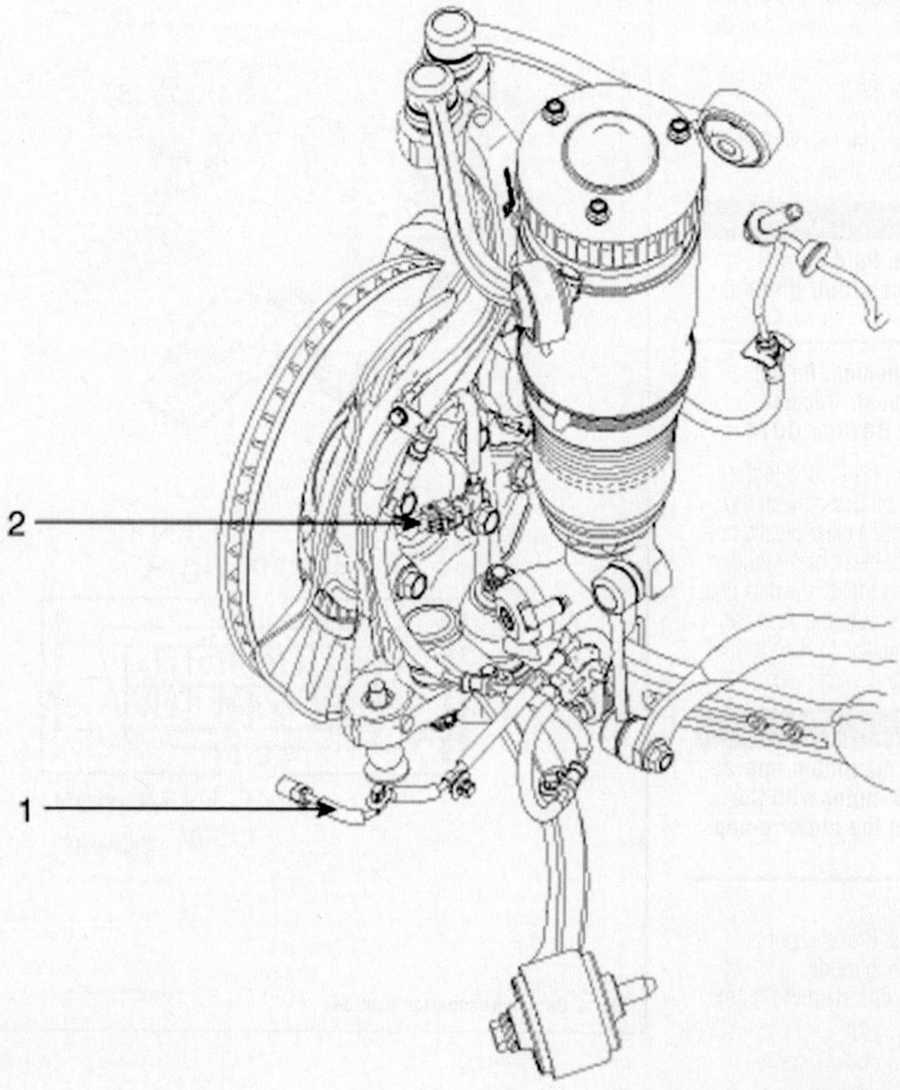

Fig. 5 Front wheel speed sensor (2) and cable (1)

Rear

See Figures 7 through 9.

1. Before servicing the vehicle, refer to the Precautions Section.
2. Remove the rear wheel speed sensor mounting bolt.
3. Remove the rear wheel guard.
4. Disconnect the connector.
5. Remove the rear wheel speed sensor.

To install:

6. Installation is the reverse of removal.

A
37655_GENS_G0086

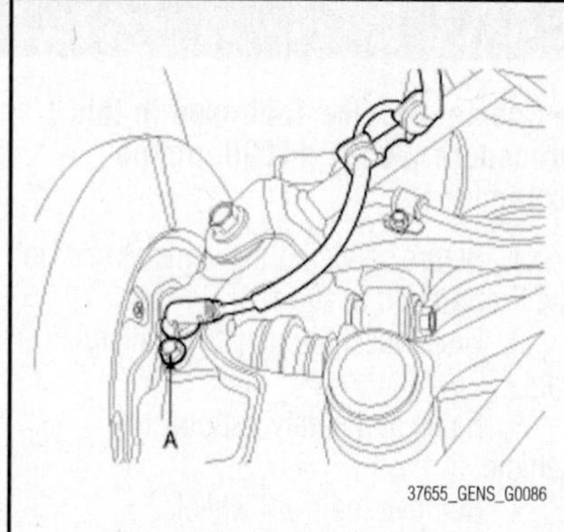

Fig. 8 Remove the rear wheel speed sensor mounting bolt (A)

A
37655_GENS_G0087

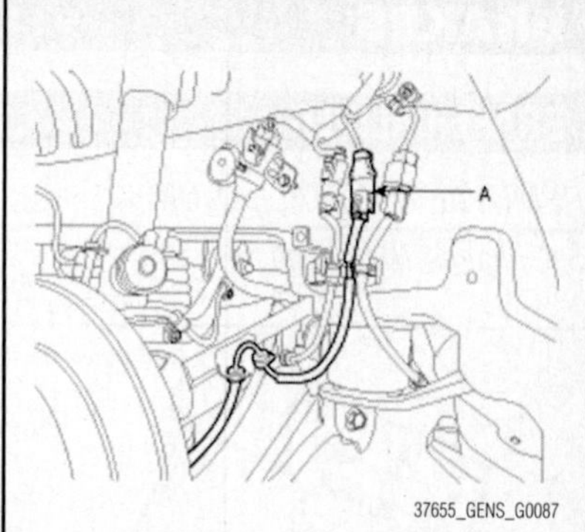

Fig. 9 Disconnect the connector (A) and remove rear wheel speed sensor

1
6.9 ~ 10.8
(0.7 ~ 1.1, 5.1 ~ 8.0)
2
Torque : N.m (kgf.m, lb-ft)
37655_GENS_G0084

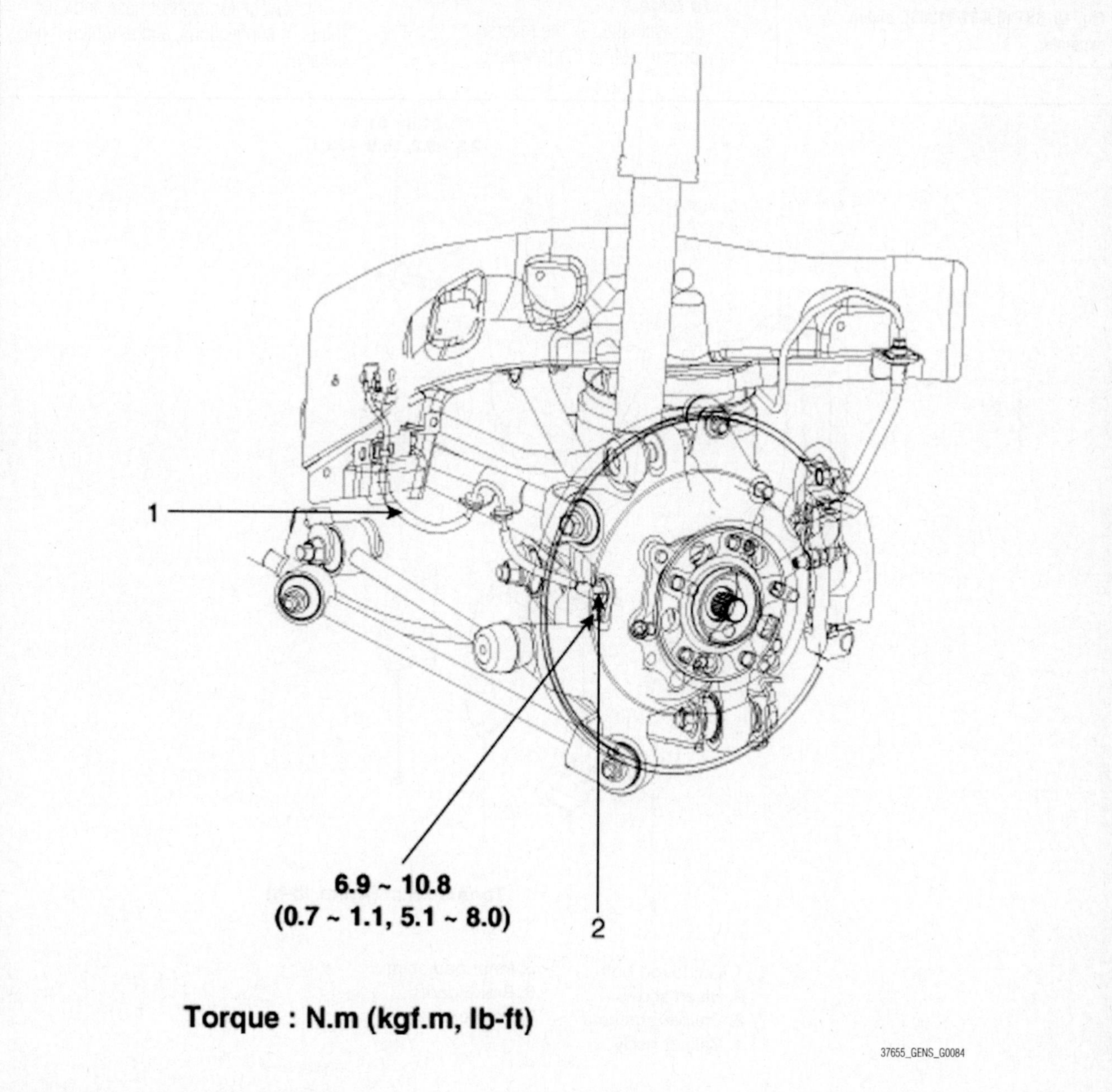

Fig. 7 Rear wheel speed sensor (2) and cable (1)

BRAKE CALIPER

REMOVAL & INSTALLATION

See Figures 10 through 14.

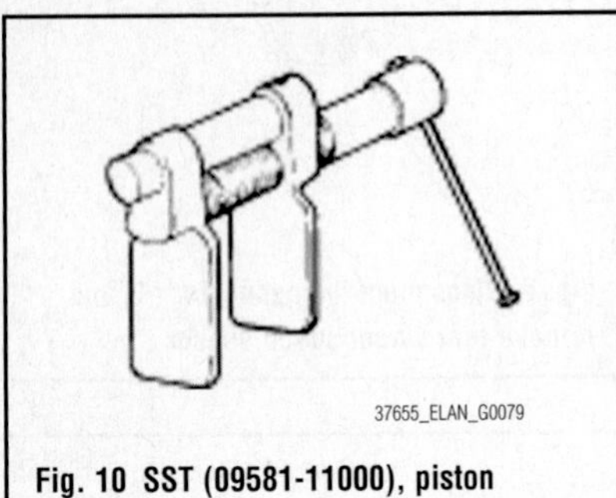

Fig. 10 SST (09581-11000), piston expander

➡Special Service Tool used in this procedure: 09581-11000, piston expander

1. Before servicing the vehicle, refer to the Precautions Section.
2. Disconnect the negative battery cable.
3. Raise and safely support the vehicle.
4. Remove the front wheel and tire.
5. Remove the hose eyebolt and caliper mounting bolts, and then remove the front caliper assembly.

To install:

6. Installation is the reverse of removal, noting the following:
7. Use the SST (09581-11000) or equivalent tool, to properly install the caliper assembly.
   a. Tighten the caliper assembly to knuckle bolts to 58–72 ft. lbs. (79–98 Nm).
   b. Tighten the brake hose to caliper assembly bolts to 18–22 ft. lbs.(25–29 Nm).
8. Refill the master cylinder reservoir to the MAX line.
9. Bleed the brake system.
10. Depress the brake pedal several times to make sure the brakes work, and then test-drive.
11. Check for leaks at hose and line joints or connections, and retighten if necessary.

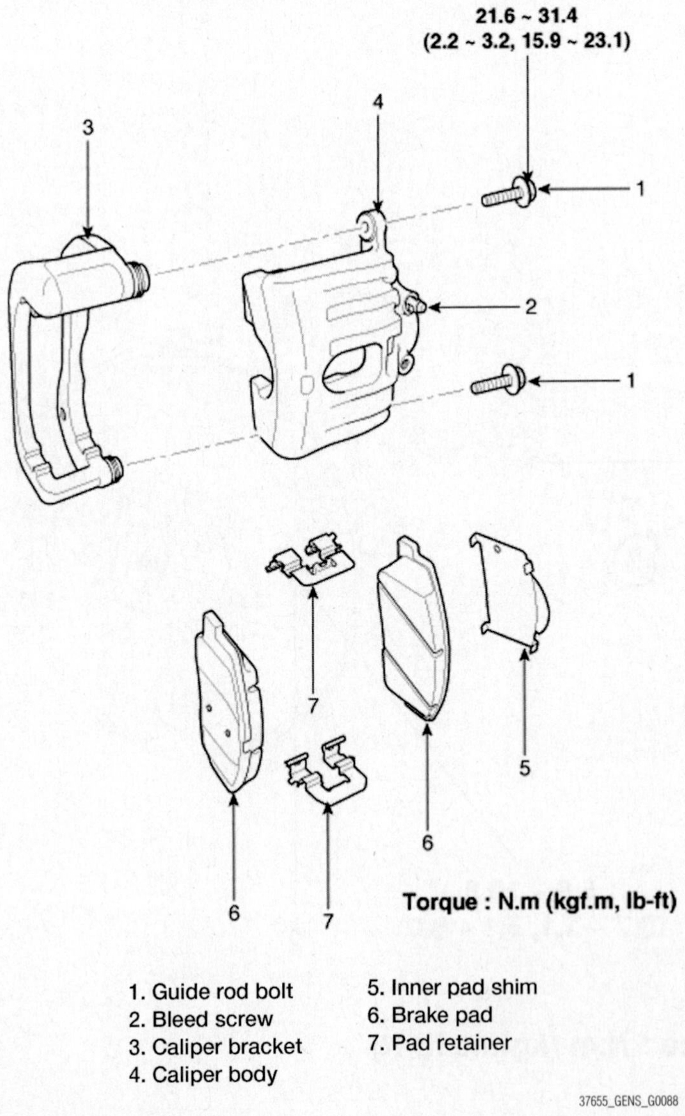

21.6 ~ 31.4
(2.2 ~ 3.2, 15.9 ~ 23.1)

Torque : N.m (kgf.m, lb-ft)

1. Guide rod bolt
2. Bleed screw
3. Caliper bracket
4. Caliper body
5. Inner pad shim
6. Brake pad
7. Pad retainer

37655_GENS_G0088

Fig. 11 Front disc brake components—3.8L engines

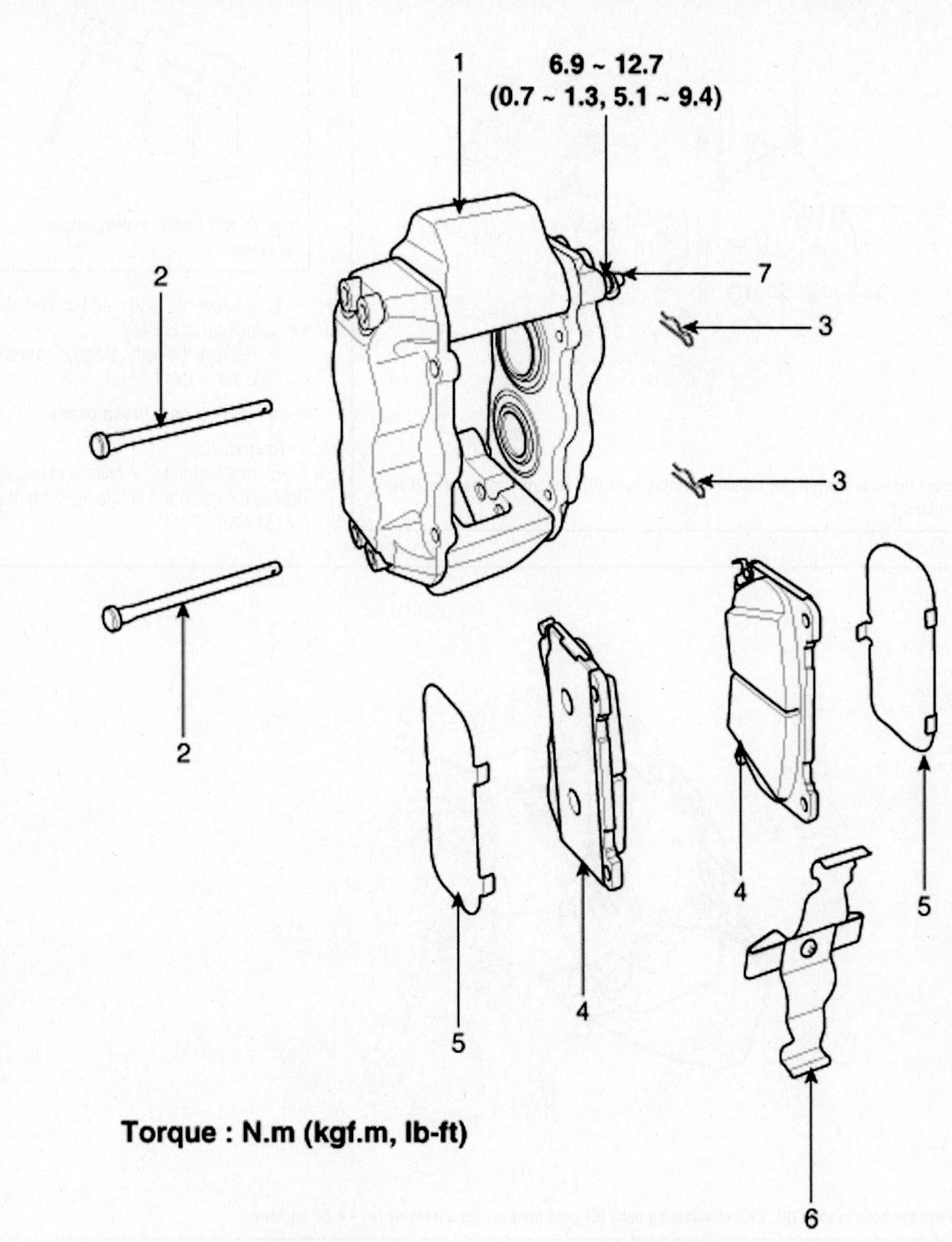

1. Caliper body
2. Guide pin
3. Locking pin
4. Brake pad
5. Pad shim
6. Retraction spring
7. Bleed screw

37655_GENS_G0089

Fig. 12 Front disc brake components—4.6L engines

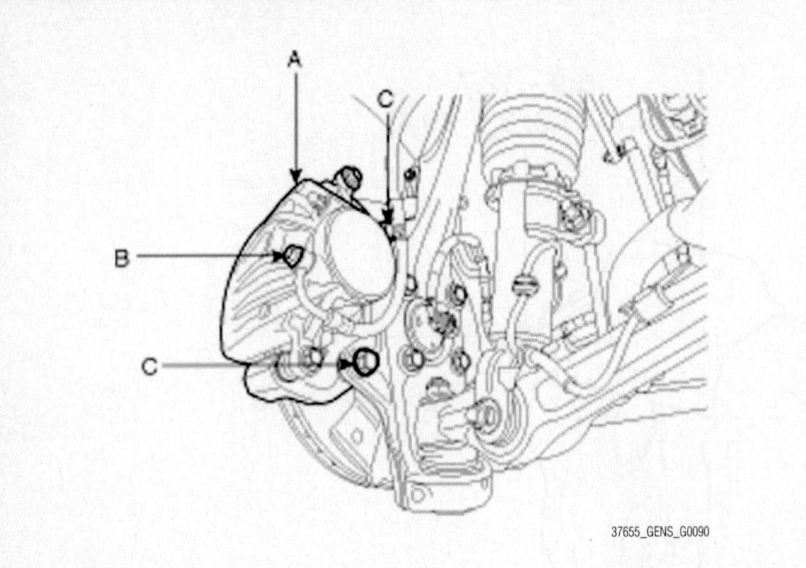

Fig. 13 Remove the hose eyebolt (B), caliper mounting bolts (C), and front caliper assembly (A)—3.8L engines

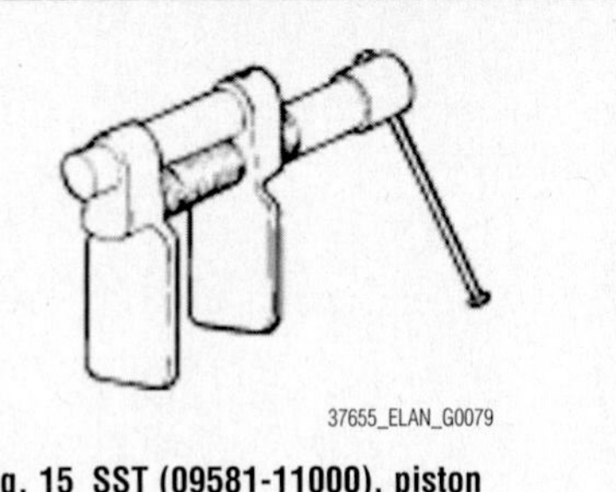

Fig. 15 SST (09581-11000), piston expander

6. Remove the guide rod bolt and pivot the caliper out of the way.

7. Remove the pads, shims, and retainers from the caliper bracket.

➡Do not step on a brake pedal.

To install:

8. Installation is the reverse of removal. Tighten the guide rod bolt to 16–23 ft. lbs. (22–31 Nm).

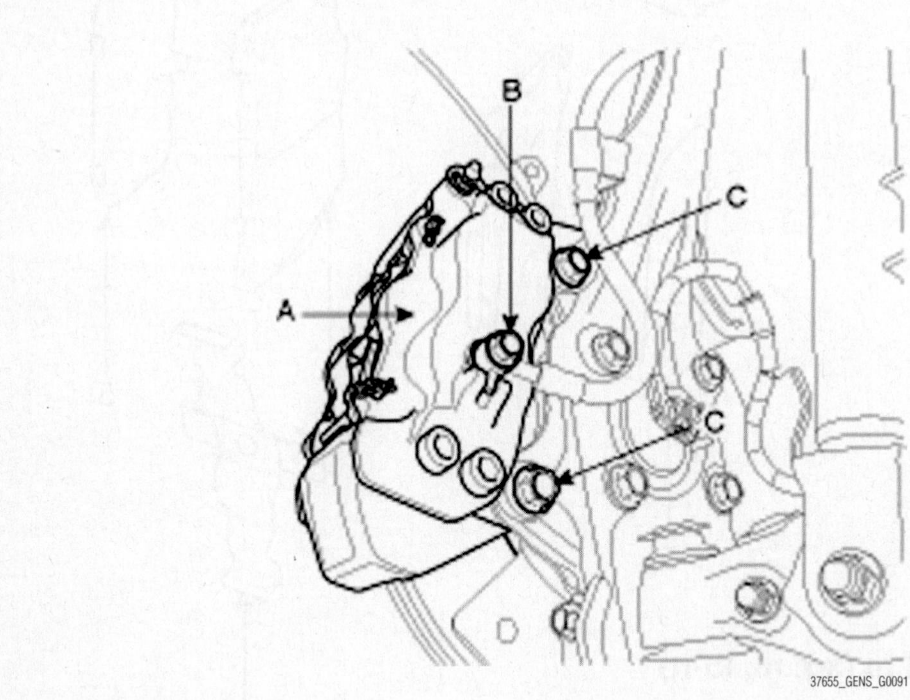

Fig. 14 Remove the hose eyebolt (B), caliper mounting bolts (C), and front caliper assembly (A)—4.6L engines

DISC BRAKE PADS

REMOVAL & INSTALLATION

3.8L Engines

See Figures 15 through 19.

➡Special Service Tool used in this procedure: 09581-11000, piston expander

1. Before servicing the vehicle, refer to the Precautions Section.

2. Disconnect the negative battery cable.

3. Raise and safely support the vehicle.

4. Remove the front wheel and tire.

5. Remove the brake hose mounting bracket.

9. Use the SST (09581-11000) or equivalent tool, to properly install the caliper assembly.

10. Refill the master cylinder reservoir to the MAX line.

11. Bleed the brake system.

12. Depress the brake pedal several times to make sure the brakes work, and then test-drive.

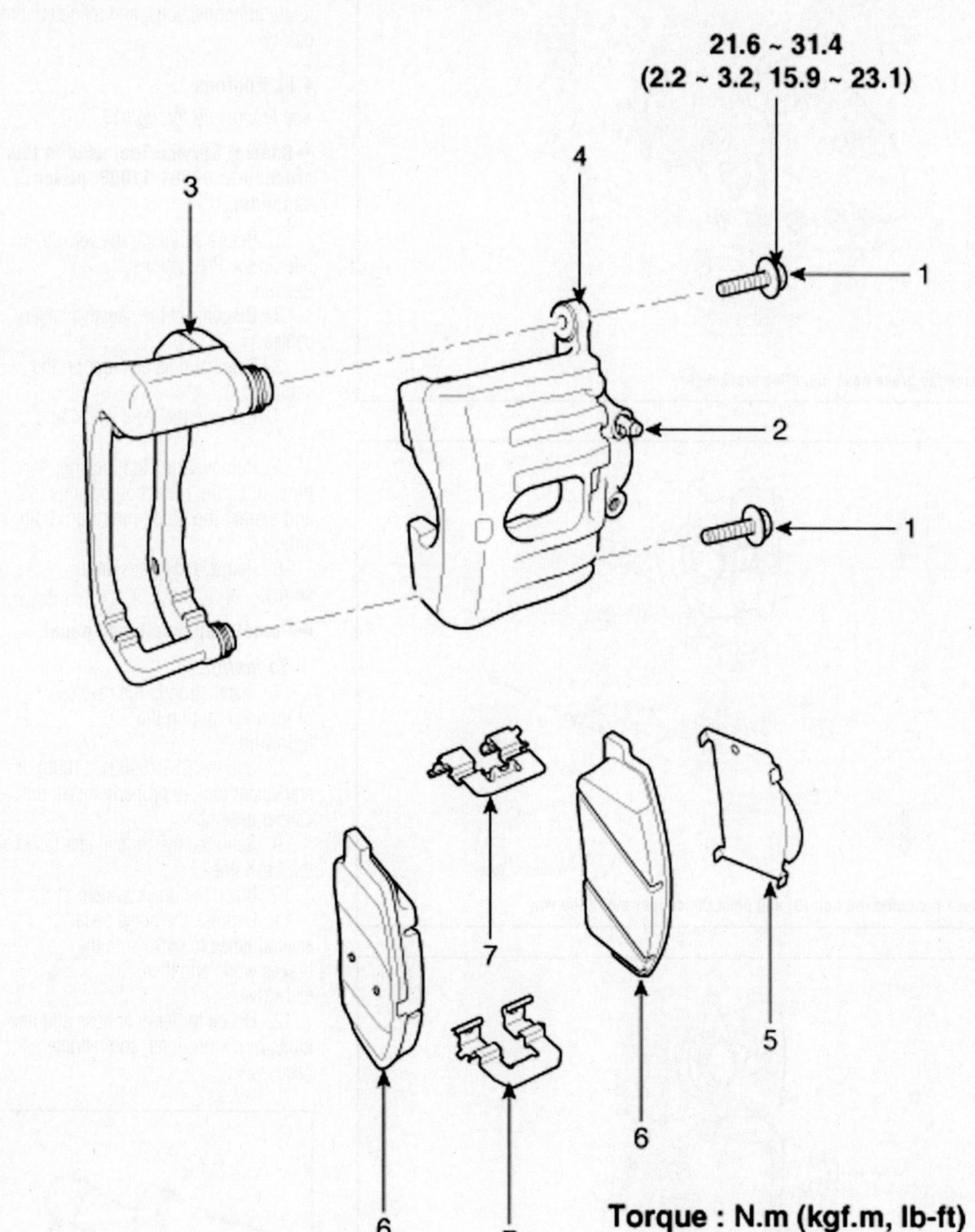

Fig. 16 Front disc brake components—3.8L engines

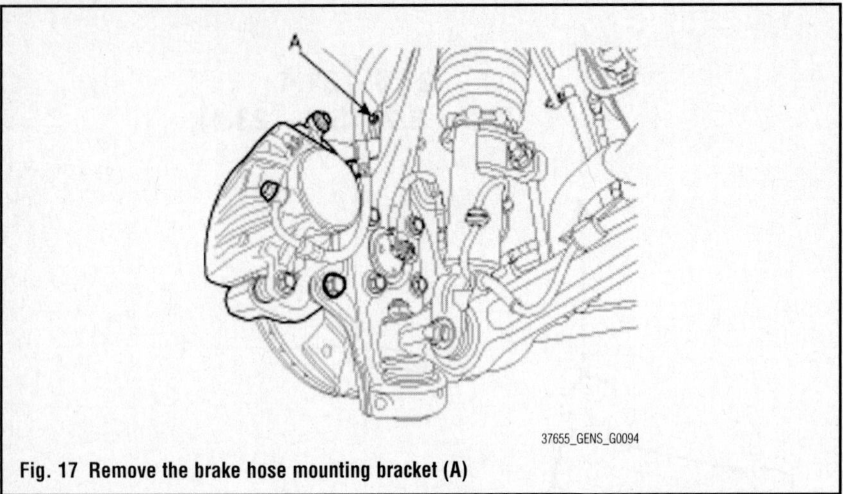

Fig. 17 Remove the brake hose mounting bracket (A)

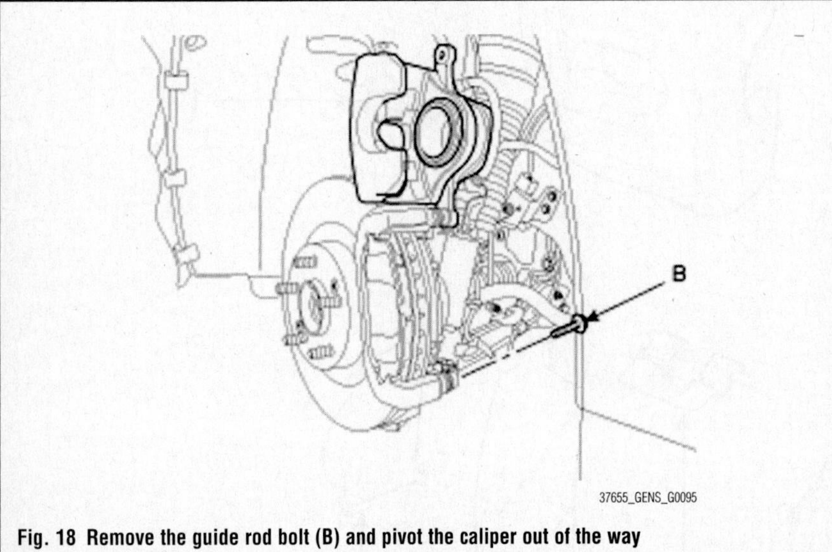

Fig. 18 Remove the guide rod bolt (B) and pivot the caliper out of the way

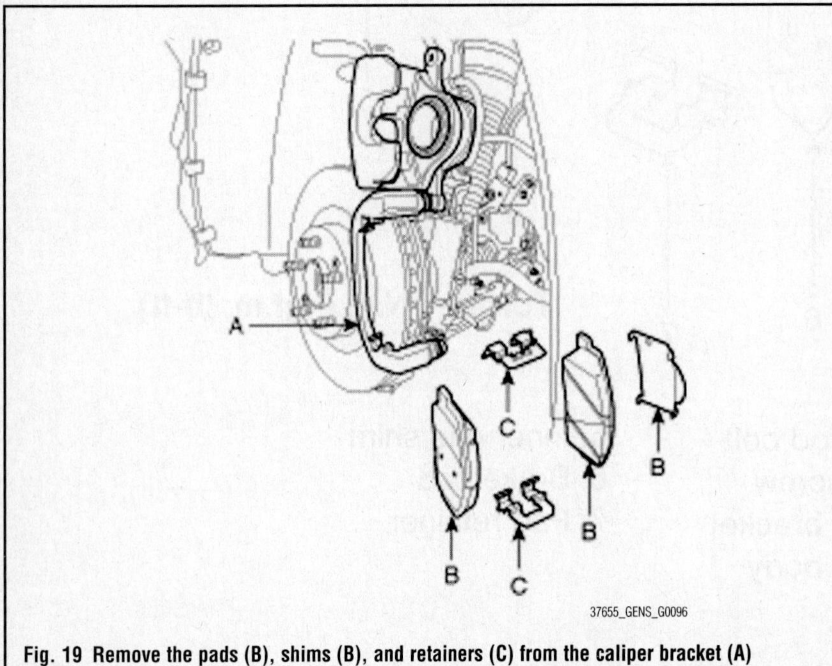

Fig. 19 Remove the pads (B), shims (B), and retainers (C) from the caliper bracket (A)

13. Check for leaks at hose and line joints or connections, and retighten if necessary.

4.6L Engines

See Figures 20 through 23.

➡Special Service Tool used in this procedure: 09581-11000, piston expander

1. Before servicing the vehicle, refer to the Precautions Section.
2. Disconnect the negative battery cable.
3. Raise and safely support the vehicle.
4. Remove the front wheel and tire.
5. Remove the side locking pins, push the retraction springs, and extract the guide pins from both sides.
6. Remove the pads and shims.

➡Do not step on a brake pedal.

To install:

7. Installation is the reverse of removal, noting the following:
8. Use the SST (09581-11000) or equivalent tool, to properly install the caliper assembly.
9. Refill the master cylinder reservoir to the MAX line.
10. Bleed the brake system.
11. Depress the brake pedal several times to make sure the brakes work, and then test-drive.
12. Check for leaks at hose and line joints or connections, and retighten if necessary.

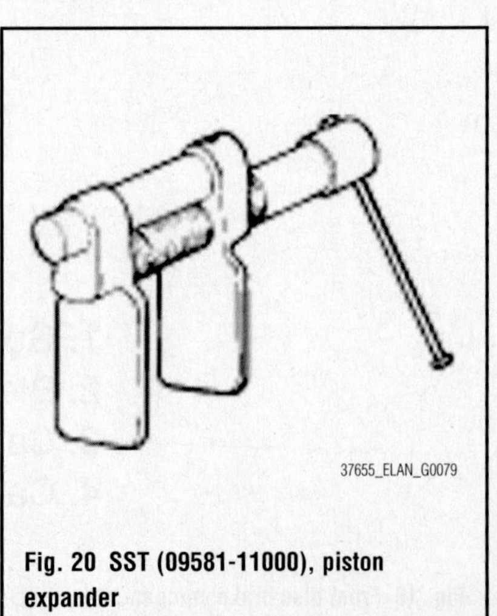

Fig. 20 SST (09581-11000), piston expander

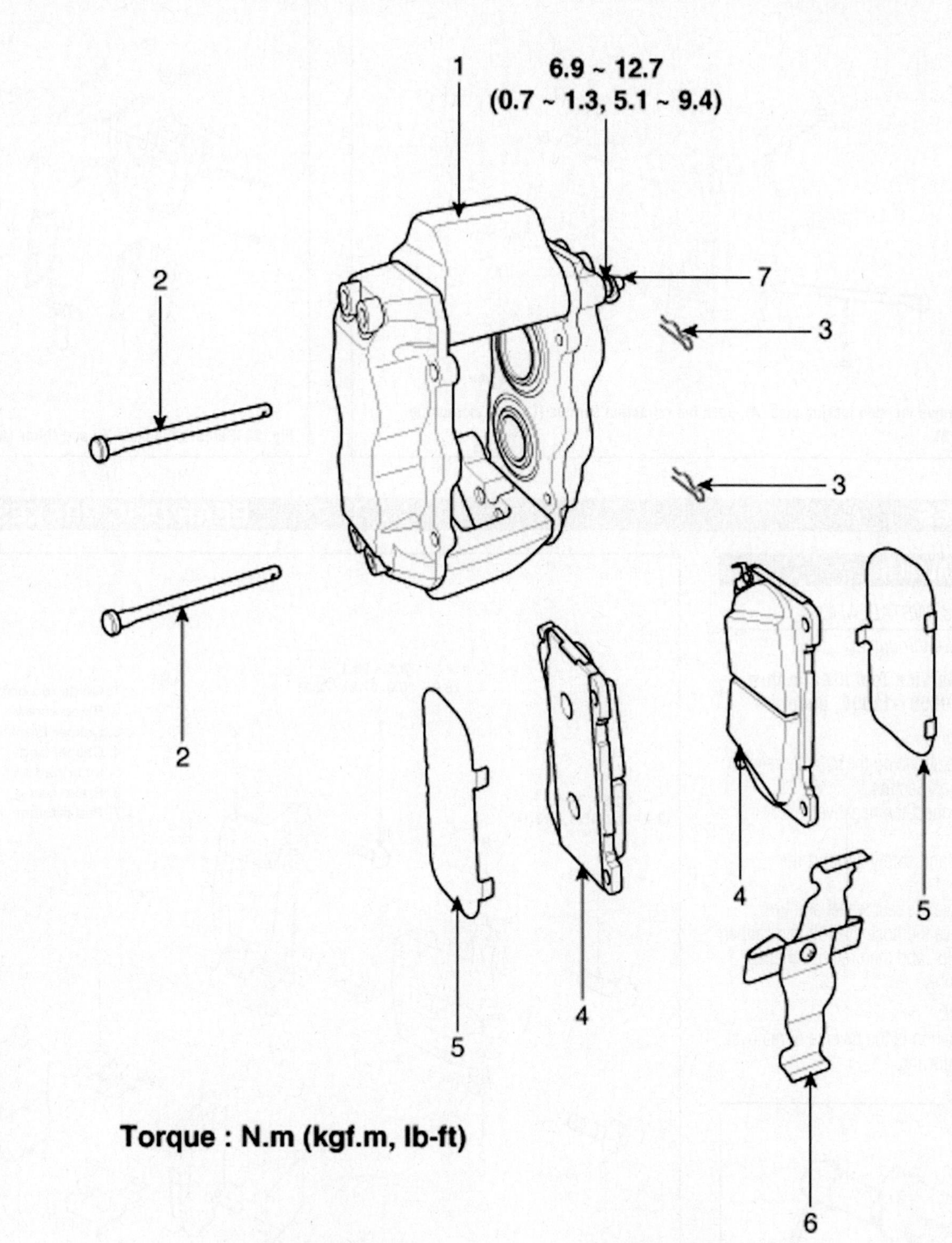

Fig. 21 Front disc brake components—4.6L engines

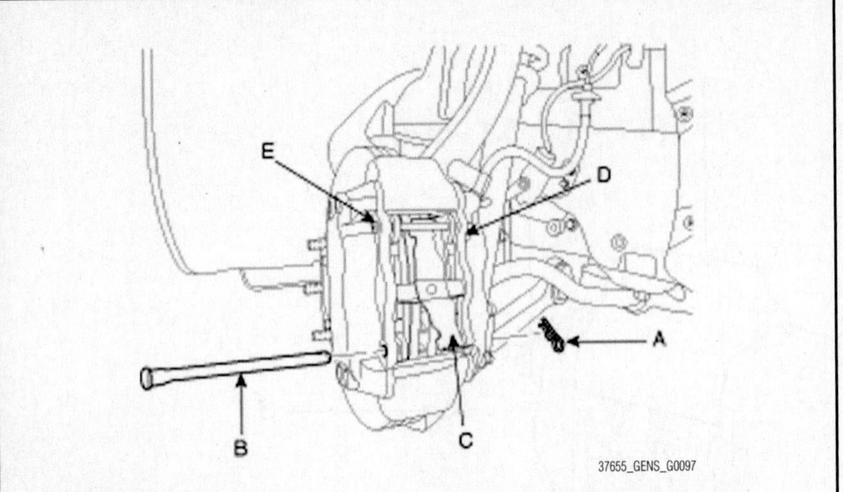

Fig. 22 Remove the side locking pins (A), push the retraction springs (C), and extract the guide pins (B)

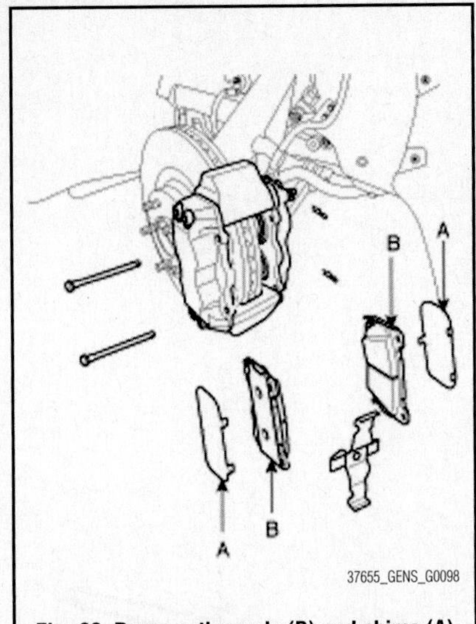

Fig. 23 Remove the pads (B) and shims (A)

BRAKE CALIPER

REMOVAL & INSTALLATION

See Figures 24 through 26.

➡Special Service Tool used in this procedure: 09581-11000, piston expander

1. Before servicing the vehicle, refer to the Precautions Section.
2. Disconnect the negative battery cable.
3. Raise and safely support the vehicle.
4. Remove the rear wheel and tire.
5. Remove the hose eyebolt and caliper mounting bolts, and then remove the rear caliper assembly.

To install:

6. Installation is the reverse of removal, noting the following:

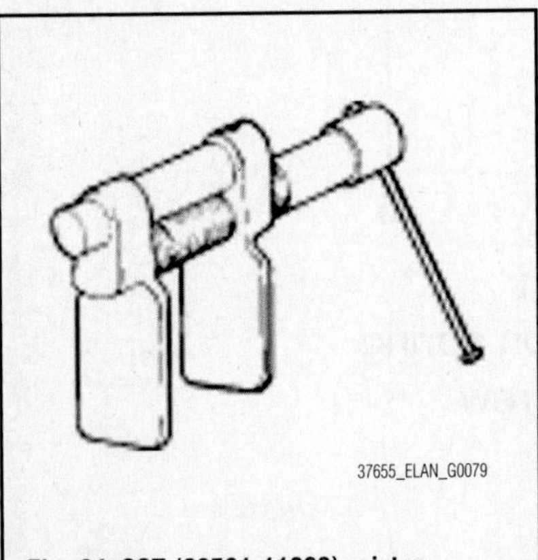

Fig. 24 SST (09581-11000), piston expander

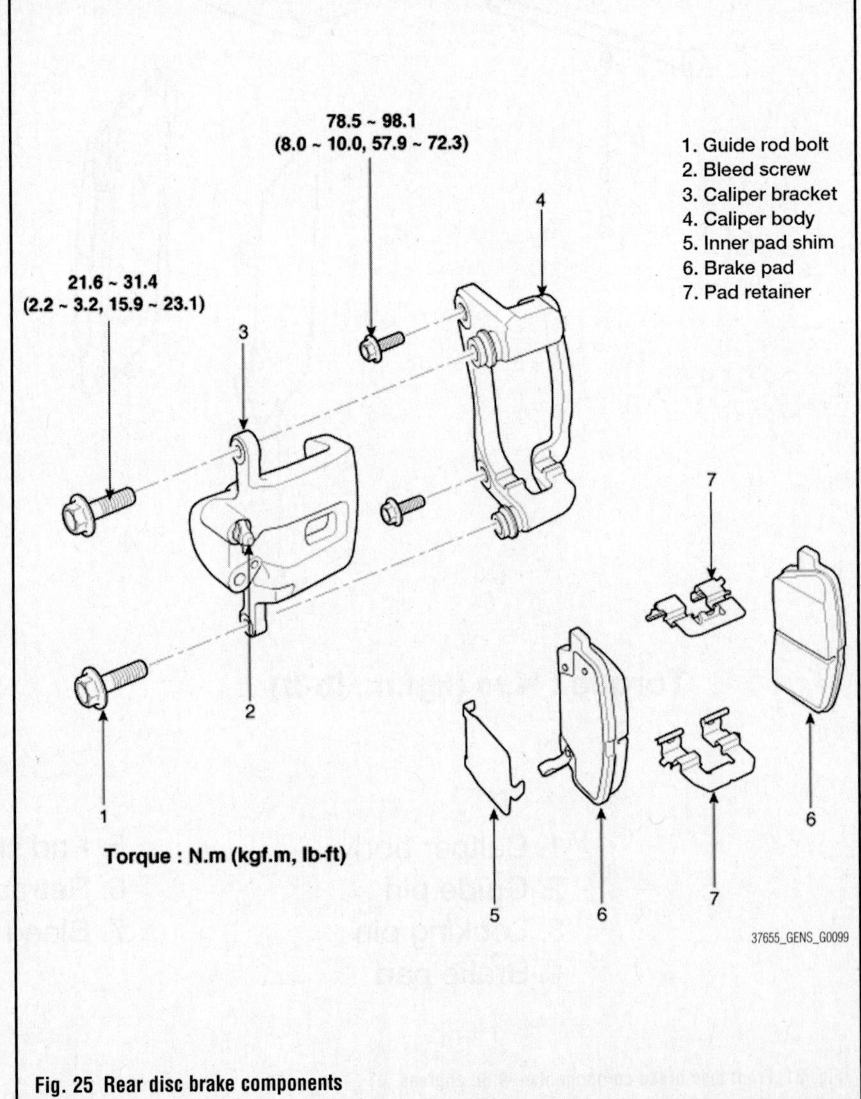

Fig. 25 Rear disc brake components

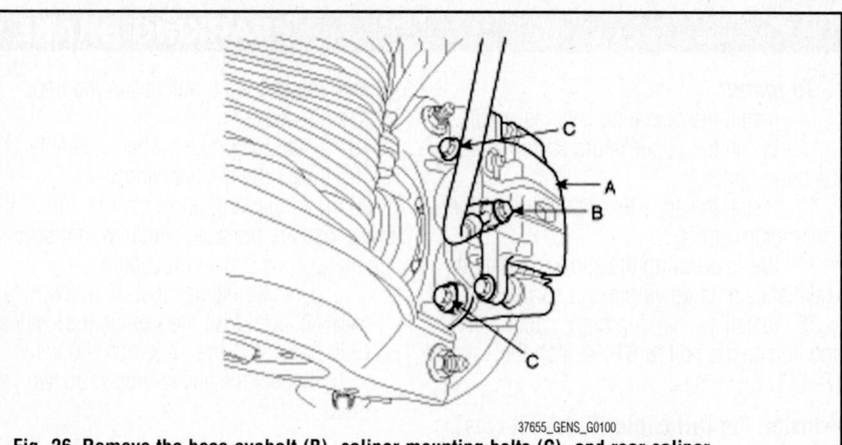

Fig. 26 Remove the hose eyebolt (B), caliper mounting bolts (C), and rear caliper assembly (A)

7. Use the SST (09581-11000) or equivalent tool, to properly install the caliper assembly.
   a. Tighten the caliper assembly bolts to 58–72 ft. lbs.(79–98 Nm).
   b. Tighten the brake hose to caliper assembly bolts to 18–22 ft. lbs.(25–29 Nm).
8. Refill the master cylinder reservoir to the MAX line.
9. Bleed the brake system.
10. Depress the brake pedal several times to make sure the brakes work, and then test-drive.
11. Check for leaks at hose and line joints or connections, and retighten if necessary.

DISC BRAKE PADS

REMOVAL & INSTALLATION

See Figures 24, 25, 27 and 28.

➡Special Service Tool used in this procedure: 09581-11000, piston expander

1. Before servicing the vehicle, refer to the Precautions Section.
2. Disconnect the negative battery cable.
3. Raise and safely support the vehicle.
4. Remove the rear wheel and tire.
5. Remove the guide rod bolt and pivot the caliper out of the way.
6. Remove the pads, shims, and retainers from the caliper bracket.

To install:

7. Installation is the reverse of removal, noting the following:
8. Use the SST (09581-11000) or equivalent tool, to properly install the caliper assembly.
9. Refill the master cylinder reservoir to the MAX line.
10. Bleed the brake system.
11. Depress the brake pedal several times to make sure the brakes work, and then test-drive.
12. Check for leaks at hose and line joints or connections, and retighten if necessary.

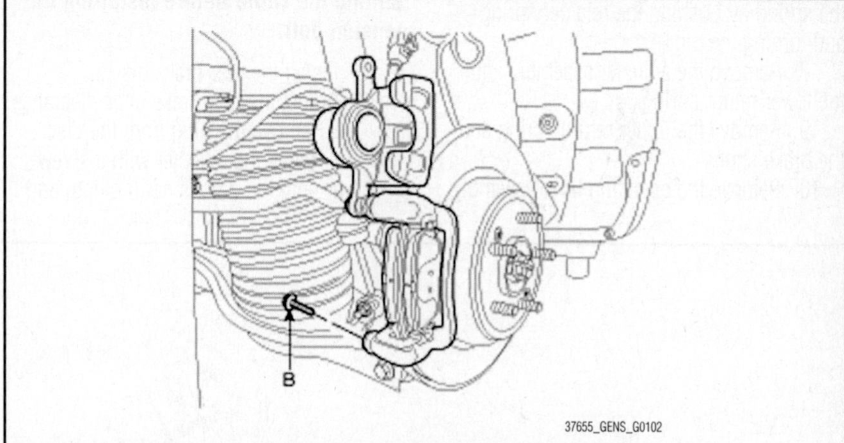

Fig. 27 Remove the guide rod bolt (B) and pivot the caliper out of the way

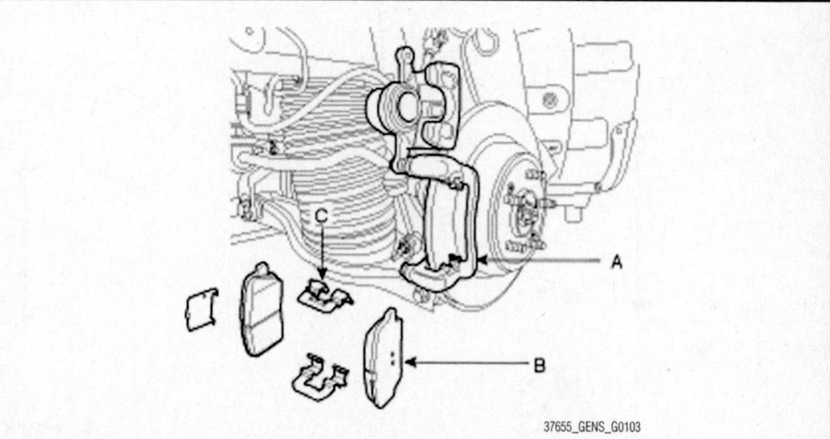

Fig. 28 Remove the pads (B), shims, and retainers (C) from the caliper bracket (A)

PARKING BRAKE SHOES

REMOVAL & INSTALLATION

See Figures 29 through 35.

1. Before servicing the vehicle, refer to the Precautions Section.
2. Disconnect the negative battery cable.
3. Raise and safely support the vehicle.
4. Remove the rear wheel and tire.
5. Remove the brake caliper and rear disc brake.
6. Remove the parking brake cable and bolt.
7. Remove the shoe hold down pin and the spring by pushing the retainer spring and turning the pin.
8. Remove the adjuster assembly and the lower return spring.
9. Remove the upper return spring and the brake shoes.
10. Remove the operating lever assembly.

To install:

11. Install the operating lever assembly.
12. Install the upper return spring and the brake shoes.
13. Install the adjuster assembly and the lower return spring.
14. While pressing the spring, install the brake shoe hold down pin and spring.
15. Install the parking brake cable and tighten the bolt to 61–96 inch lbs. (7–11 Nm).

➡Install the DIH cable (Quick Fit type) by installing the inner cable through the knuckle hole and attaching to the DIH lever. Confirm attachment by pulling the cable before installing the tension bolt.

16. Install the rear brake disc.
17. Adjust the rear brake shoe clearance.
    a. Remove the plug from the disc.
    b. Rotate the adjuster with a screwdriver until the disc is not moving, and then return it by 5 notches in the opposite direction.
18. Install the brake caliper assembly.
19. Install the tire and wheel.
20. If the parking brake shoe or the brake disc are replaced with new, perform the brake shoe bed-in procedure.
    a. Operate the pedal at 33 ft. lbs. (147 N), and drive the vehicle 0.31 miles (500 meters) at the 37.3 mph (60 kph).
    b. Repeat the above procedure two times.

ADJUSTMENT

1. Raise and safely support the vehicle.
2. Remove the rear tire and wheel.
3. Remove the plug from the disc.
4. Rotate the toothed wheel of the adjuster with a screwdriver until the disc is not moving, and then return it by 5 notches in the opposite direction.
5. Install the plug on the disc
6. Install the rear tire and wheel.

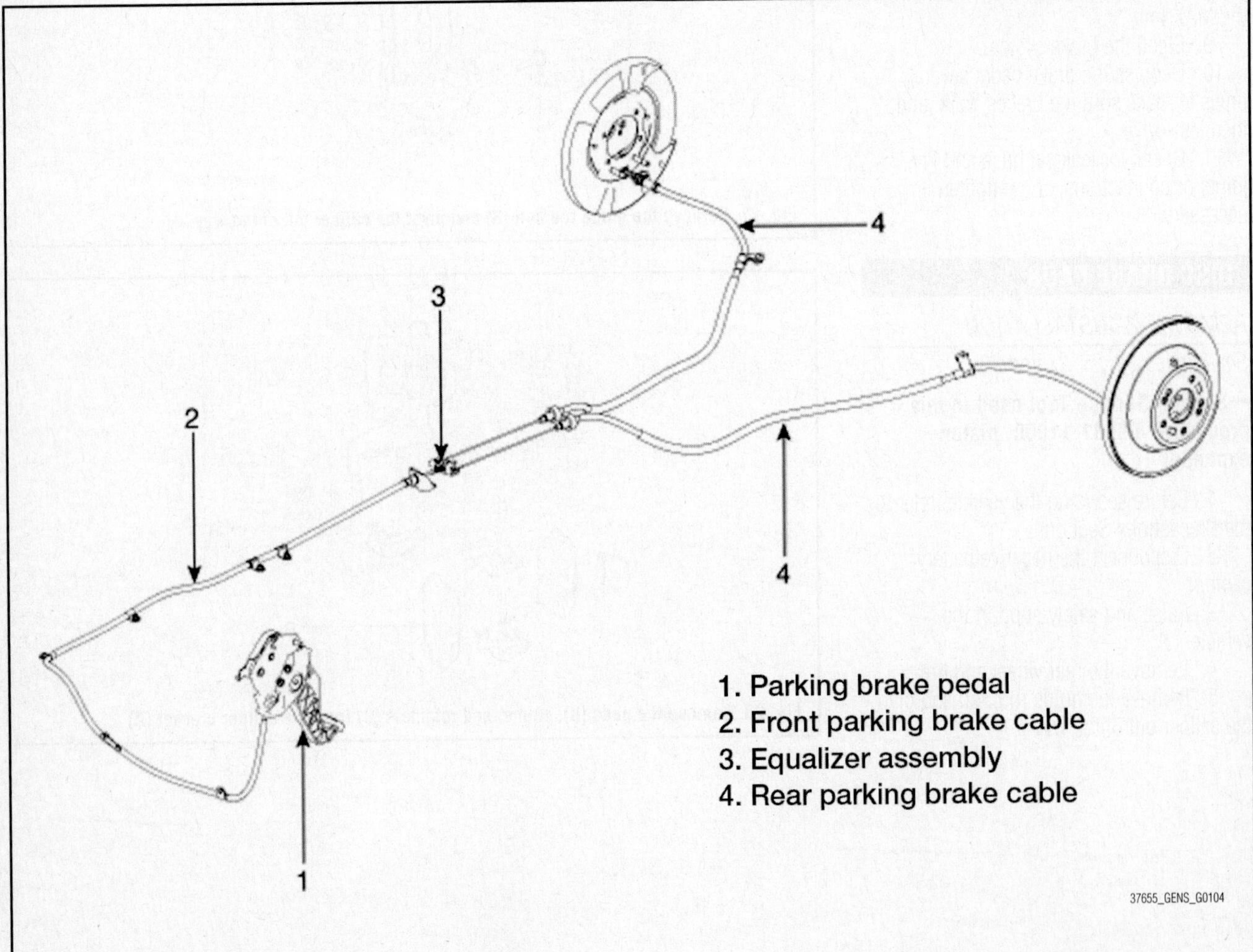

Fig. 29 Parking brake component locations

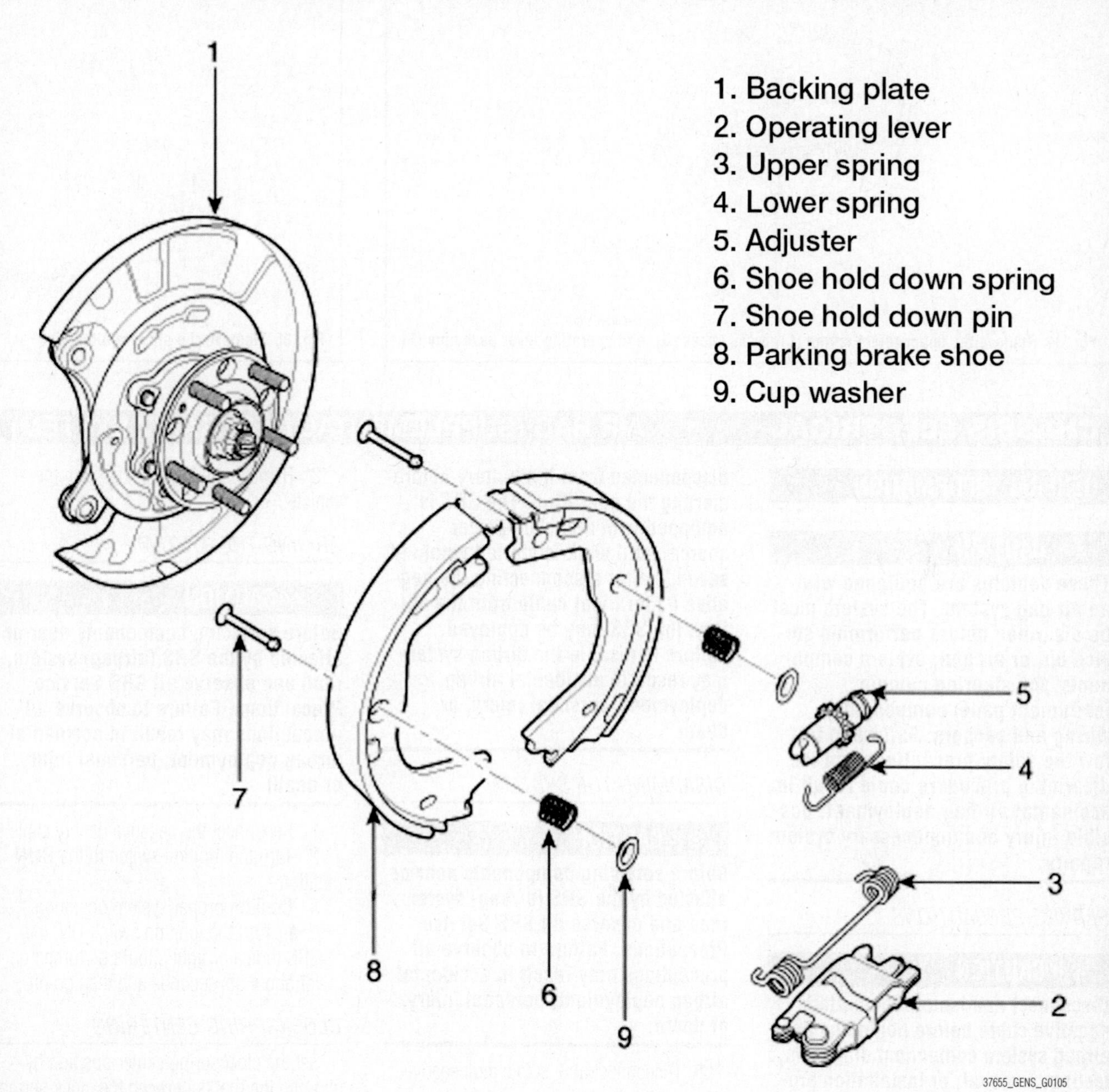

1. Backing plate
2. Operating lever
3. Upper spring
4. Lower spring
5. Adjuster
6. Shoe hold down spring
7. Shoe hold down pin
8. Parking brake shoe
9. Cup washer
1
2
3
4
5
6
7
8
9
37655_GENS_G0105

Fig. 30 Parking brake components

B
A
37655_GENS_G0106

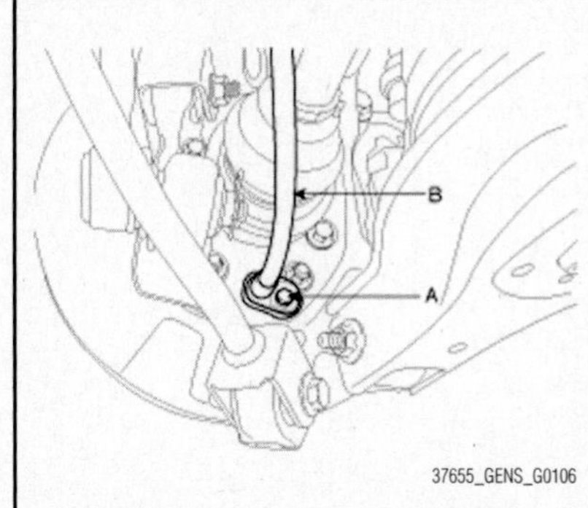

Fig. 31 Remove the parking brake cable (B) and bolt (A)

A
B
37655_GENS_G0107

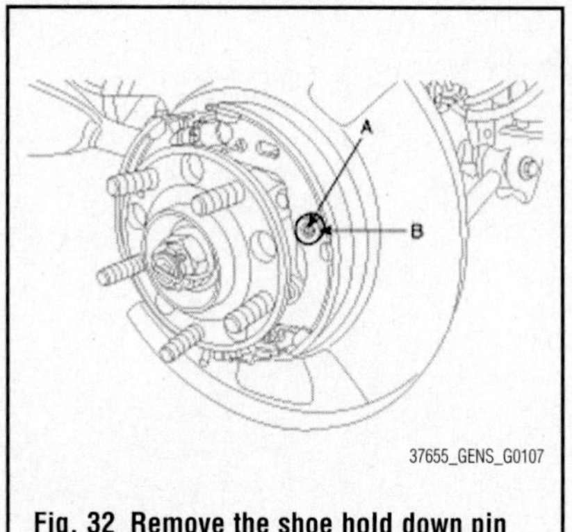

Fig. 32 Remove the shoe hold down pin (A) and spring (B)

B
A
37655_GENS_G0108

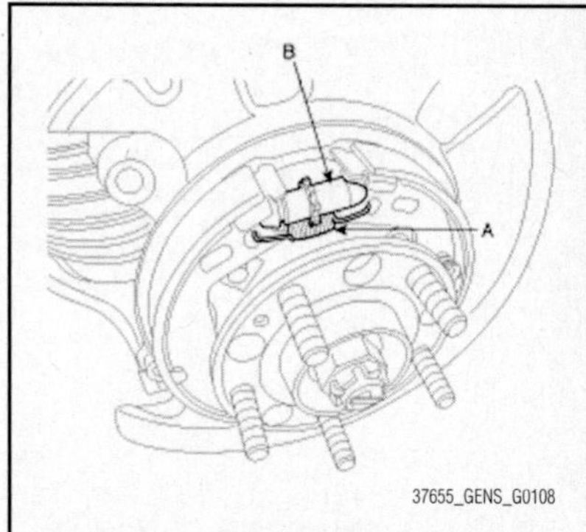

Fig. 33 Remove the adjuster assembly (B) and the lower return spring (A)

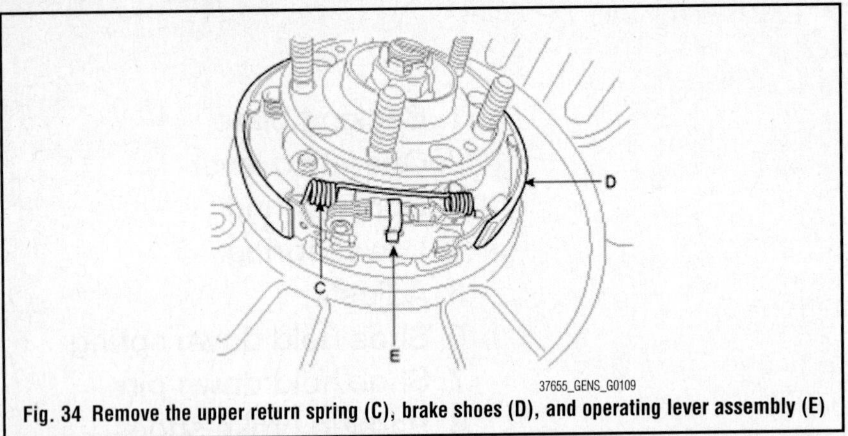

Fig. 34 Remove the upper return spring (C), brake shoes (D), and operating lever assembly (E)

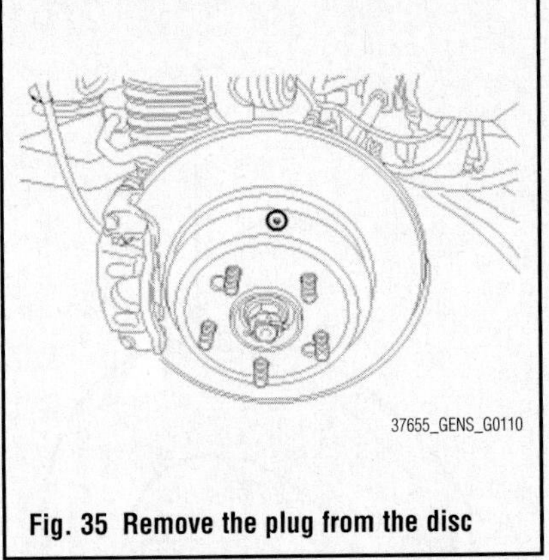

Fig. 35 Remove the plug from the disc

CHASSIS ELECTRICAL

AIR BAG (SUPPLEMENTAL RESTRAINT SYSTEM)

GENERAL INFORMATION

**** CAUTION**

These vehicles are equipped with an air bag system. The system must be disarmed before performing service on, or around, system components, the steering column, instrument panel components, wiring and sensors. Failure to follow the safety precautions and the disarming procedure could result in accidental air bag deployment, possible injury and unnecessary system repairs.

SERVICE PRECAUTIONS

**** CAUTION**

Disconnect and isolate the battery negative cable before beginning any airbag system component diagnosis, testing, removal, or installation procedures. Wait at least 90 seconds after the ignition switch is turned off and the negative (-) terminal cable is disconnected from the battery before starting the operation. The SRS is equipped with a backup power source, so if work is started within 90 seconds after disconnecting the negative (-) terminal cable from the battery, the SRS may be deployed. Failure to disable the airbag system may result in accidental airbag deployment, personal injury, or death.

DISARMING THE SYSTEM

**** CAUTION**

Before servicing components near or affected by the SRS (airbag) system, read and observe all SRS Service Precautions. Failure to observe all precautions may result in accidental airbag deployment, personal injury, or death.

1. Disconnect and isolate the negative battery cable.
2. Wait 3 minutes for the system capacitor to discharge before performing any service.
3. Remove the ignition key from the vehicle.

ARMING THE SYSTEM

**** CAUTION**

Before servicing components near or affected by the SRS (airbag) system, read and observe all SRS Service Precautions. Failure to observe all precautions may result in accidental airbag deployment, personal injury, or death.

1. Reconnect the negative battery cable.
2. Turn the ignition switch to the **RUN** position.
3. Confirm proper system operation:
   a. Turn the ignition switch ON; the SRS indicator light should be turned on for about six seconds and then go off.

CLOCKSPRING CENTERING

Set the clockspring center position by aligning the marks between the clock spring and the cover. Matchmark and turn the clockspring clockwise to the stop and then 3 revolutions counterclockwise.

DRIVE TRAIN

HALFSHAFTS

REMOVAL & INSTALLATION

See Figures 36 through 49.

1. Before servicing the vehicle, refer to the Precautions Section.
2. Disconnect the negative battery cable.
3. Raise and safely support the vehicle.
4. Remove the rear wheels and tires.
5. Remove the split pin, then the remove castle nut and washer from the front hub.
6. Remove the mounting bolts, and then remove the bracket.
7. Remove the muffler mounting nuts, and then remove the muffler.
8. Remove the rear muffler.
9. Remove the mounting bolts, and then remove the aluminum cover.
10. After making a match mark on the rubber coupling and rear differential companion, remove the propeller shaft mounting bolts.
11. Remove the bracket.
12. Remove the differential carrier assembly mounting bolts.
13. Remove the rear halfshaft assembly (B).
14. Insert a pry bar between the differential case and joint case, and separate the halfshaft from the differential case.

➡Use a pry bar, being careful not to damage the differential and joint. Do not insert the pry bar too deep, as this may cause damage to the oil seal. Do not pull the halfshaft by excessive force it may cause components inside the joint kit to dislodge resulting in a torn boot or a damaged bearing. Plug the hole of the differential case with the oil seal cap to prevent contamination. Support the halfshaft properly. Replace the retainer ring whenever the halfshaft is removed from the differential case. Do not take the halfshaft apart.

To install:

15. Installation is the reverse of removal, noting the following:
    a. Tighten the differential carrier assembly mounting bolts to 58–72 ft. lbs. (80–100 Nm).

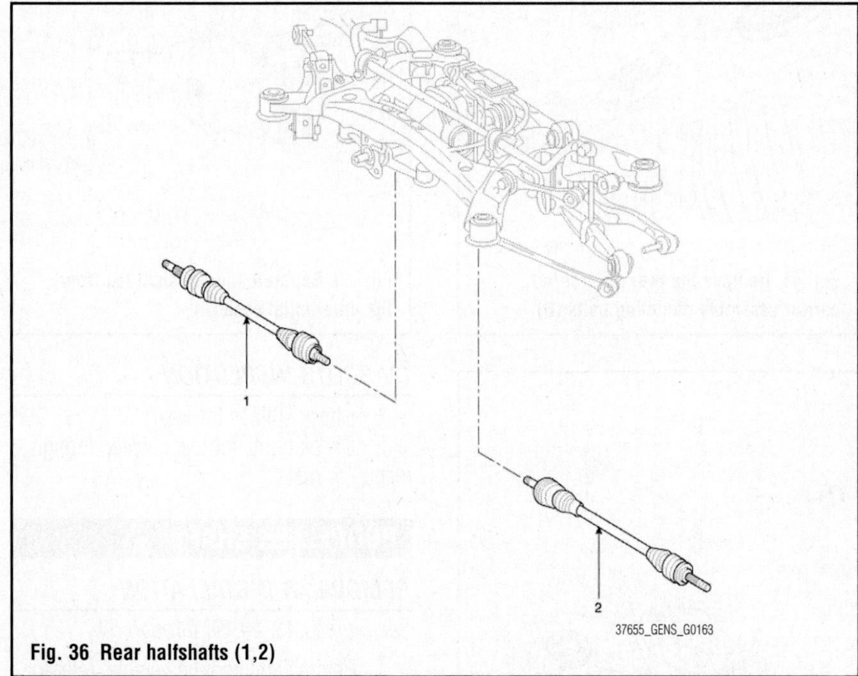

Fig. 36 Rear halfshafts (1,2)

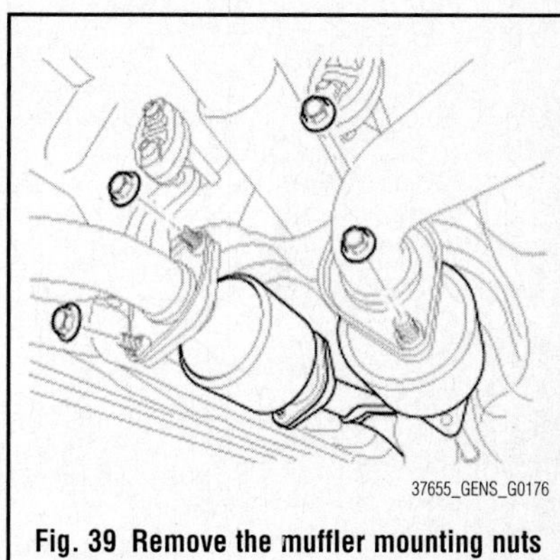

Fig. 39 Remove the muffler mounting nuts

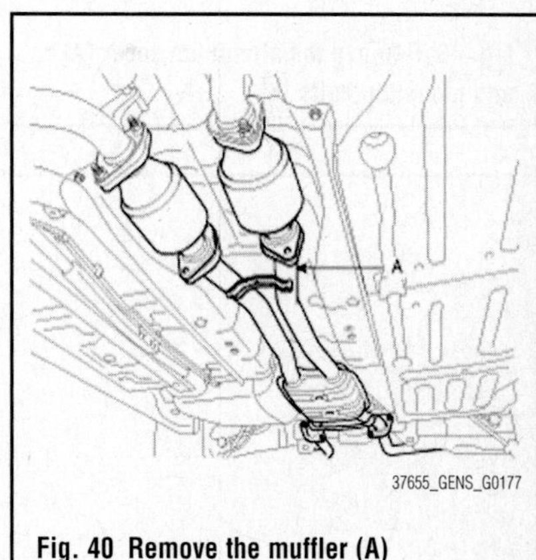

Fig. 40 Remove the muffler (A)

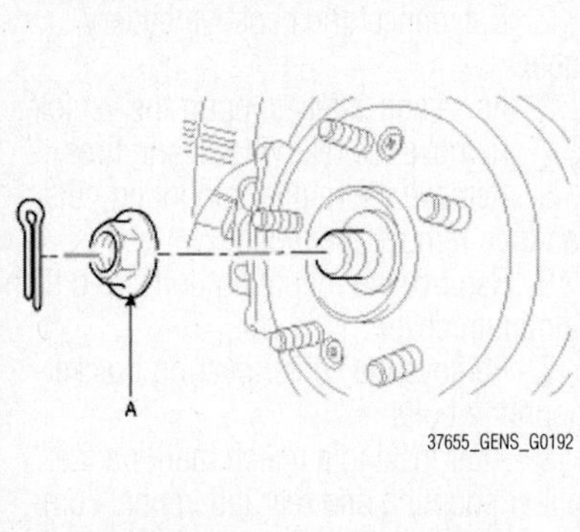

Fig. 37 Remove the split pin, castle nut (A) and washer

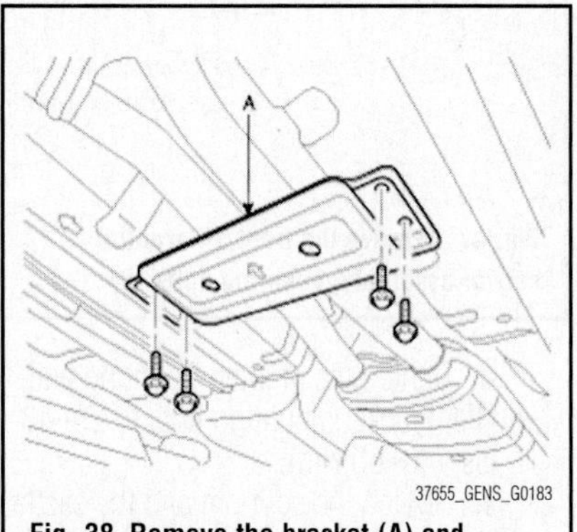

Fig. 38 Remove the bracket (A) and mounting bolts

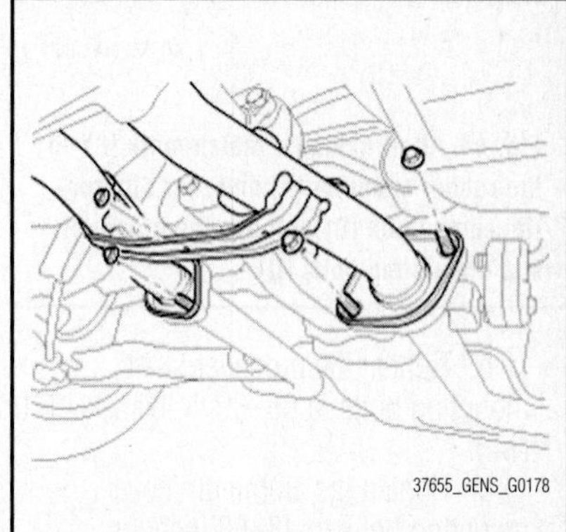

Fig. 41 Remove the rear header pipe nuts

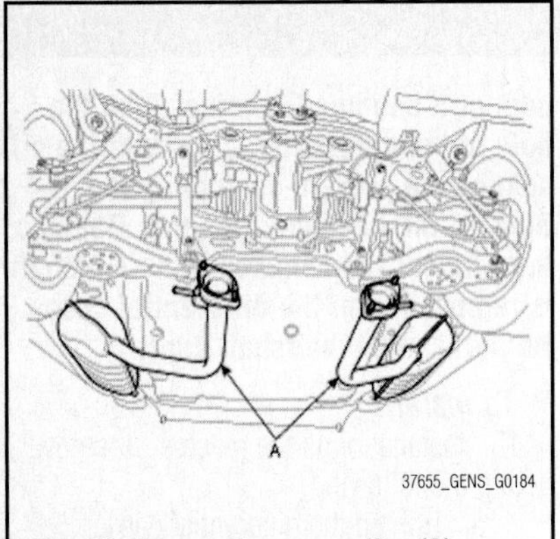

Fig. 42 Remove the rear muffler (A)

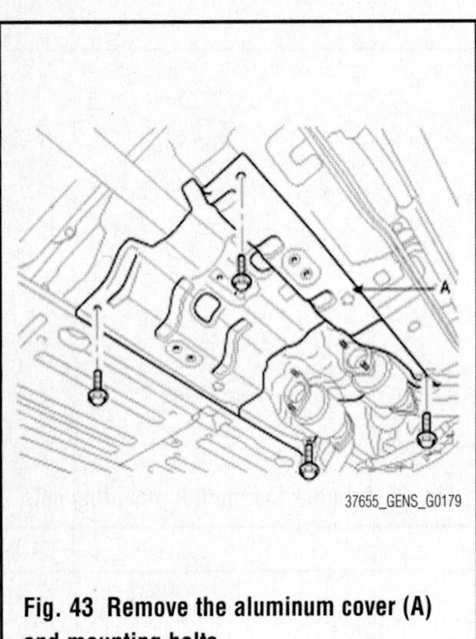

Fig. 43 Remove the aluminum cover (A) and mounting bolts

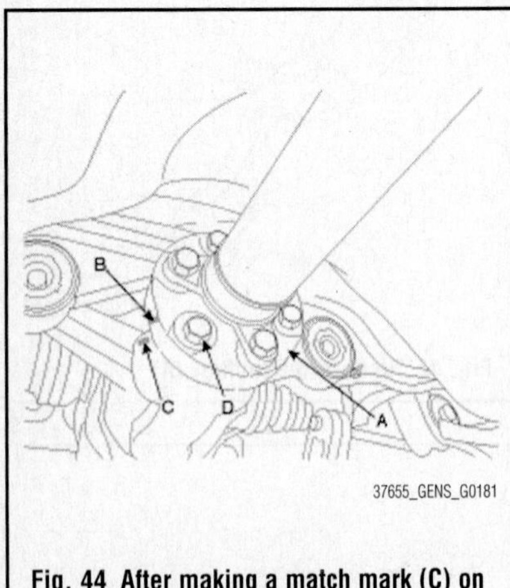

Fig. 44 After making a match mark (C) on the rubber coupling (A) and rear differential companion (B), remove the propeller shaft mounting bolts (D)

b. Tighten the propeller shaft mounting bolts to 65–80 ft. lbs. (90–110 Nm).

c. Tighten the aluminum cover mounting bolts to 43–68 inch lbs. (5–8 Nm).

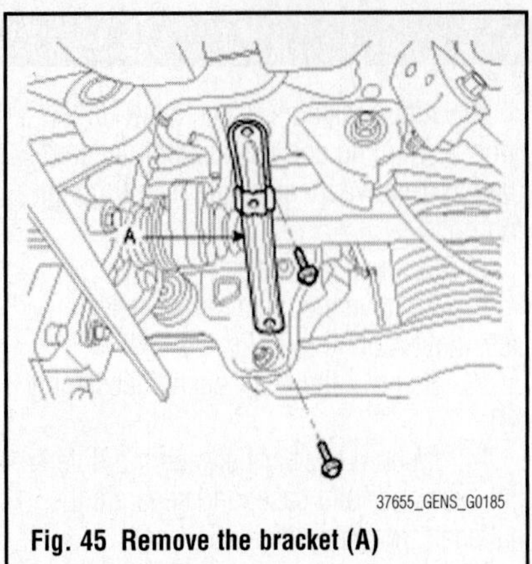

Fig. 45 Remove the bracket (A)

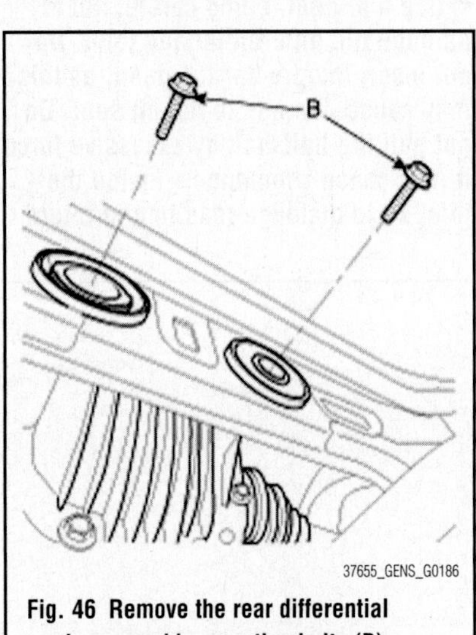

Fig. 46 Remove the rear differential carrier assembly mounting bolts (B)

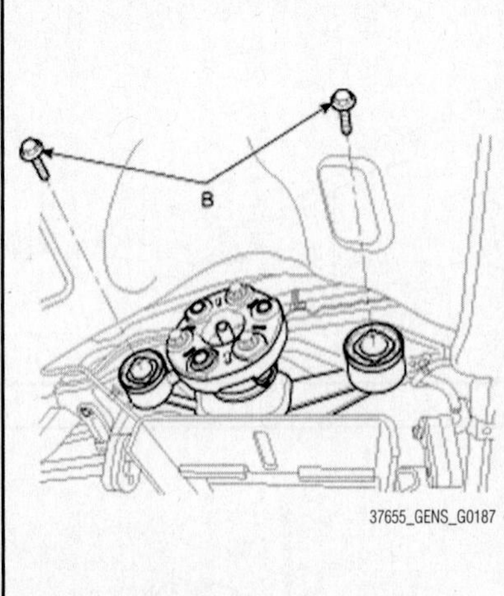

Fig. 47 Remove the front differential carrier assembly mounting bolts (B)

d. Tighten the muffler mounting and muffler bracket mounting nuts to 29–43 ft. lbs. (40–60 Nm).

e. Tighten the split pin and the castle nut and washer to 145–203 ft. lbs. (200–280 Nm).

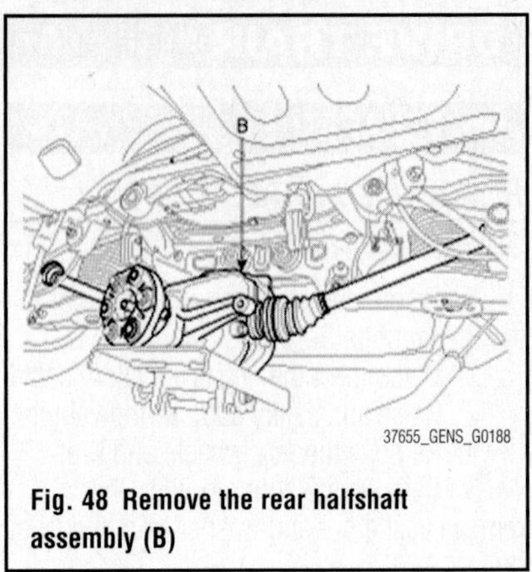

Fig. 48 Remove the rear halfshaft assembly (B)

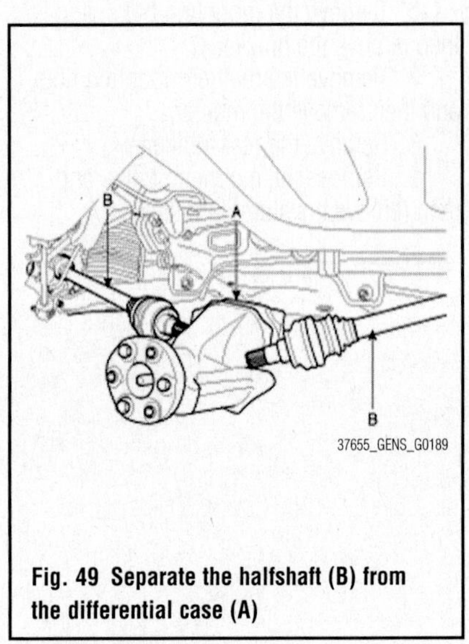

Fig. 49 Separate the halfshaft (B) from the differential case (A)

CV-BOOTS INSPECTION

1. Check splines for wear.
2. Check boots for tears, water, foreign matter, or rust.

PROPELLER SHAFT

REMOVAL & INSTALLATION

See Figures 43, 44, 50 through 54.

1. Before servicing the vehicle, refer to the Precautions Section.
2. Disconnect the negative battery cable.
3. Raise and safely support the vehicle.
4. Remove the rear wheels and tires.
5. Remove the muffler mounting nuts, and then remove the muffler.
6. Remove the mounting bolts, and then aluminum cover.
7. Remove the center bearing bracket mounting bolts.
8. After making a match mark on the rubber coupling and rear differential companion, remove the propeller shaft mounting bolts.

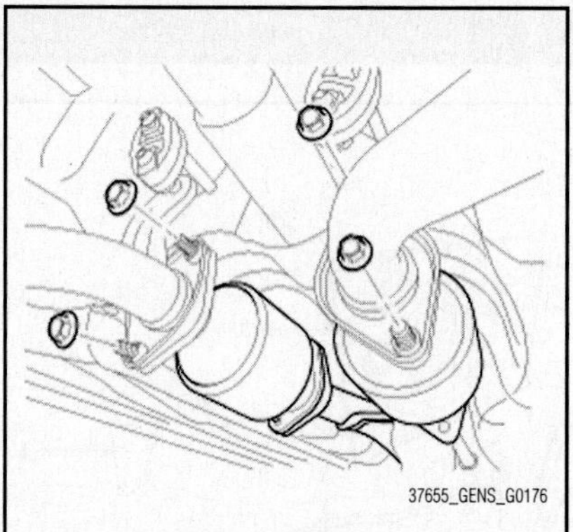

Fig. 50 Remove the muffler mounting nuts

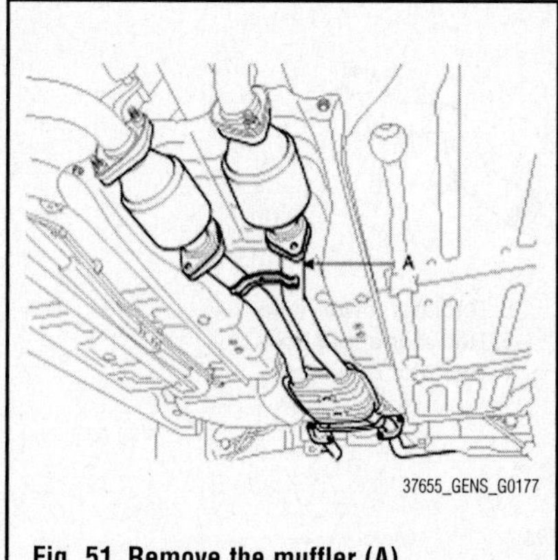

Fig. 51 Remove the muffler (A)

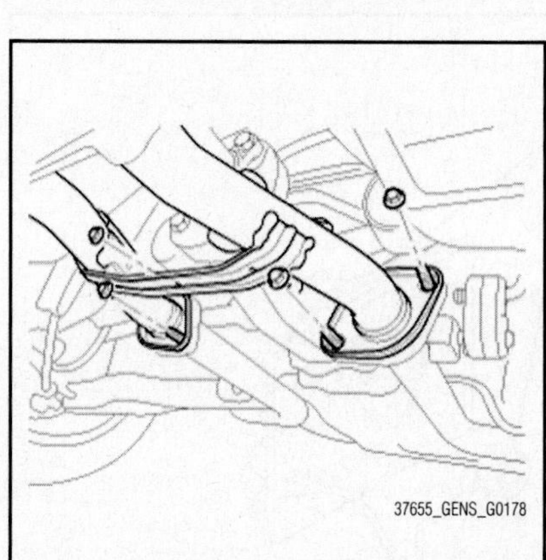

Fig. 52 Remove the rear header pipe nuts

➡Use the hexagonal wrench to prevent damage of bolt head when removing bolts.

To install:

9. Installation is the reverse of removal, noting the following:
   a. Tighten the differential carrier assembly mounting bolts to 58–72 ft. lbs. (80–100 Nm).
   b. Tighten the propeller shaft mounting bolts to 65–80 ft. lbs. (90–110 Nm).
   c. Tighten the aluminum cover mounting bolts to 43–68 inch lbs. (5–8 Nm).
   d. Tighten the center bearing bracket mounting bolts to 43–68 inch lbs. (5–8 Nm).
   e. Tighten the muffler mounting and muffler bracket mounting nuts to 29–43 ft. lbs. (40–60 Nm).

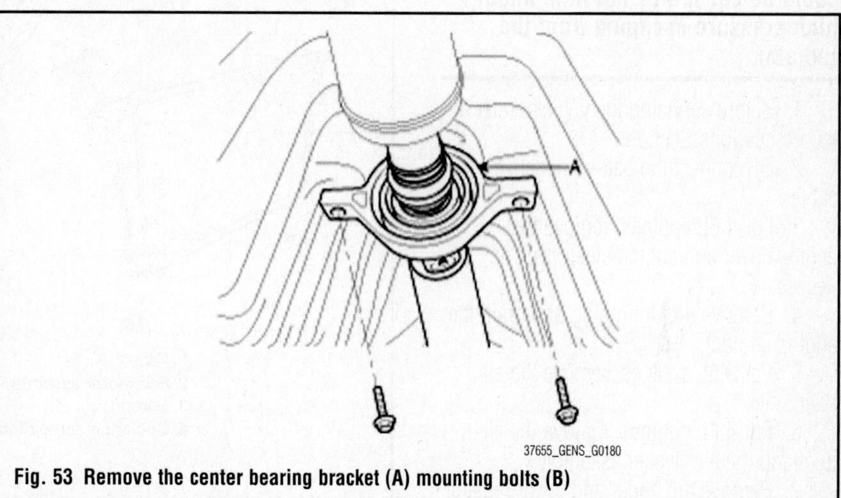

Fig. 53 Remove the center bearing bracket (A) mounting bolts (B)

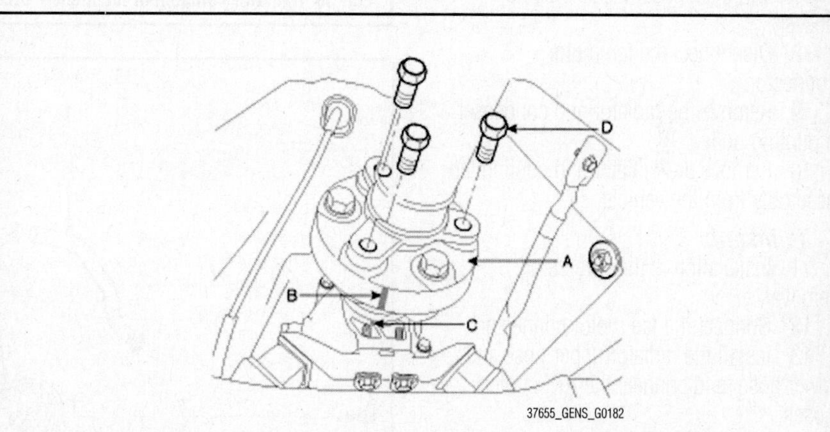

Fig. 54 Matchmark (C), rubber coupling (A), rear differential companion (B), and propeller shaft mounting bolts (D)

ENGINE COOLING

RADIATOR

REMOVAL & INSTALLATION

See Figures 55 through 60.

CAUTION

Never remove the radiator cap when the engine is hot. Serious scalding could be caused by hot fluid under high pressure escaping from the radiator.

1. Before servicing the vehicle, refer to the Precautions Section.
2. Disconnect the battery negative cable.
3. For 4.6L engines, remove the engine cover and engine side cover, as necessary.
4. Remove the drain plug and drain the engine coolant.
5. For 3.8L engines, remove the air duct.
6. For 4.6L engines, remove the air duct and the air cleaner assembly.
7. Remove the upper and lower radiator hoses, and disconnect the AFT cooler hoses.
8. Disconnect the fan motor connector.
9. Remove the radiator and condenser mounting bolts.
10. Remove the radiator and cooling fan assembly from the vehicle.

To install:

11. Installation is the reverse of removal.
12. Connect the fan motor connector.
13. Install the radiator upper hose and lower hose, and connect the ATF cooler hoses.
14. Fill the radiator with coolant and check for leaks.

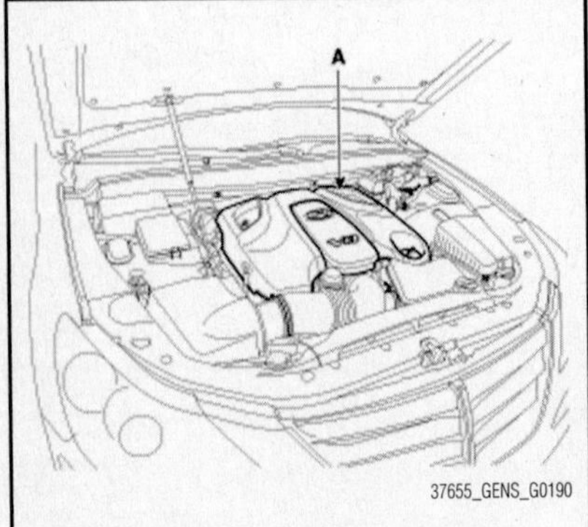

Fig. 55 Remove the engine cover (A)—4.6L engine

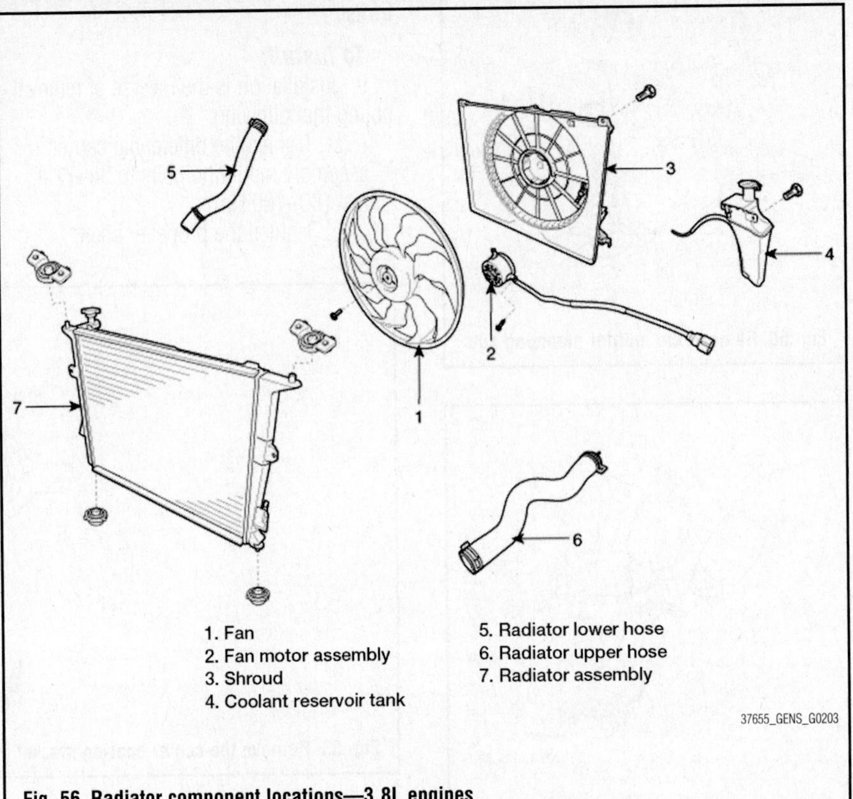

Fig. 56 Radiator component locations—3.8L engines

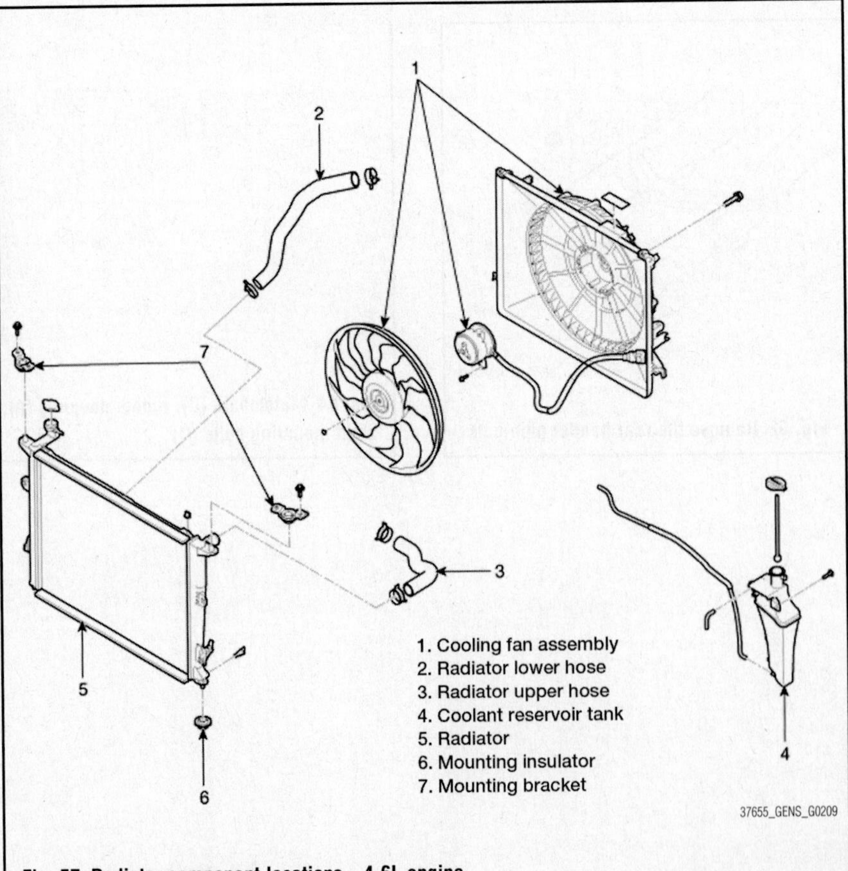

Fig. 57 Radiator component locations—4.6L engine

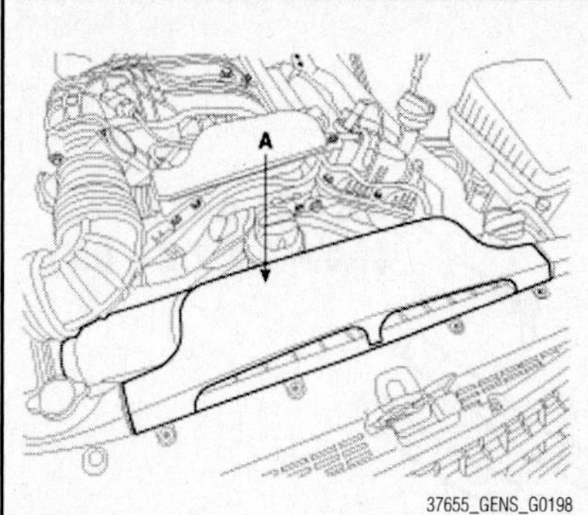

Fig. 58 Remove the air duct (A)—3.8L engines

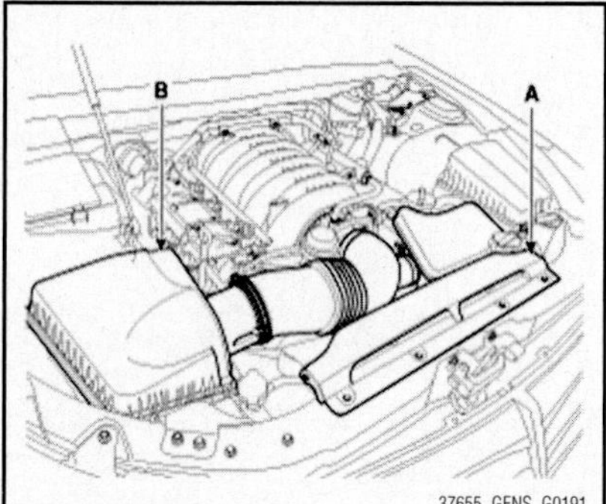

Fig. 59 Remove the air duct (A) and air cleaner assembly (B)—4.6L engine

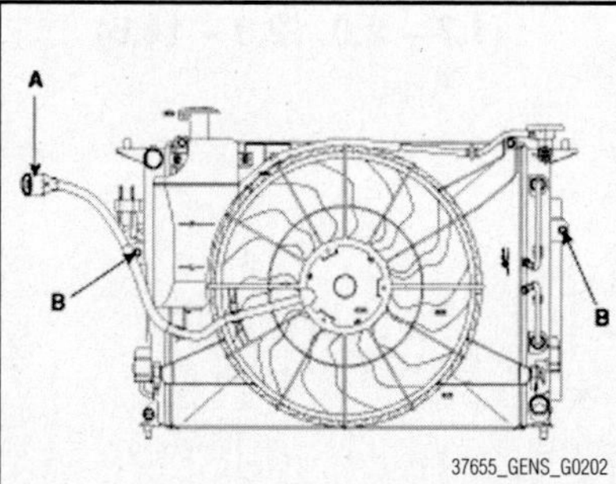

Fig. 60 Remove the fan motor connector (A) and the radiator and condenser mounting bolts (B)

THERMOSTAT

REMOVAL & INSTALLATION

3.8L Engines

See Figure 61.

1. Before servicing the vehicle, refer to the Precautions Section.
2. Disconnect the battery negative cable.
3. Drain engine coolant so its level is below thermostat.
4. Remove the water inlet fitting and the thermostat.

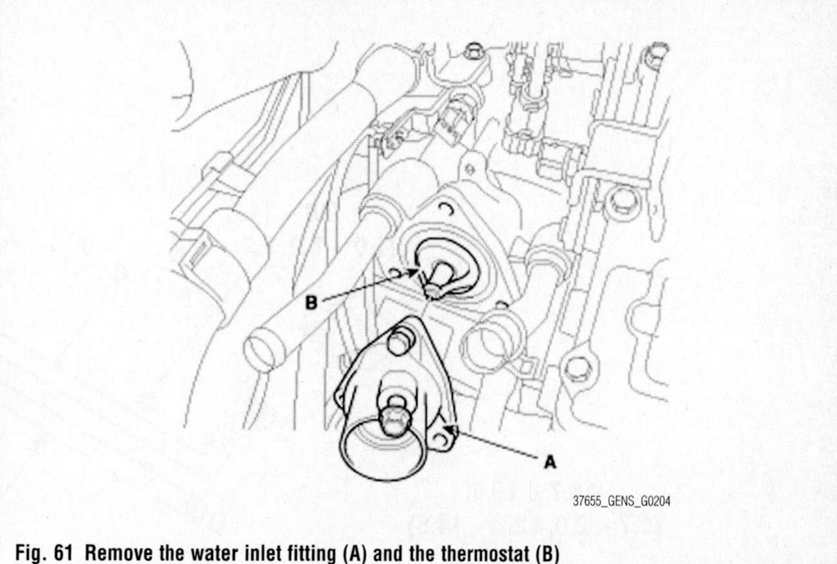

Fig. 61 Remove the water inlet fitting (A) and the thermostat (B)

To install:

5. Place the thermostat in thermostat housing.
   a. Install the thermostat with the jiggle valve upward.
   b. Install a new thermostat.
6. Install the water inlet fitting and tighten to 12–15 ft. lbs. (17–20 Nm).
7. Fill with engine coolant.
8. Start engine and check for leaks.

4.6L Engine

See Figures 62 through 68.

1. Before servicing the vehicle, refer to the Precautions Section.
2. Disconnect the battery negative cable.
3. Remove the engine cover.
4. Remove the drain plug and drain the engine coolant.

CAUTION

Never remove the radiator cap when the engine is hot. Serious scalding could be caused by hot fluid under high pressure escaping from the radiator.

5. Remove the air duct and the air cleaner assembly.
6. Disconnect the ETC module connector.
7. Disconnect the PCV hose, the PCSV hose and the water hoses from the intake manifold module and the throttle body.
8. Remove the throttle body.
9. Remove the water temperature control assembly.
10. Remove the O-ring.

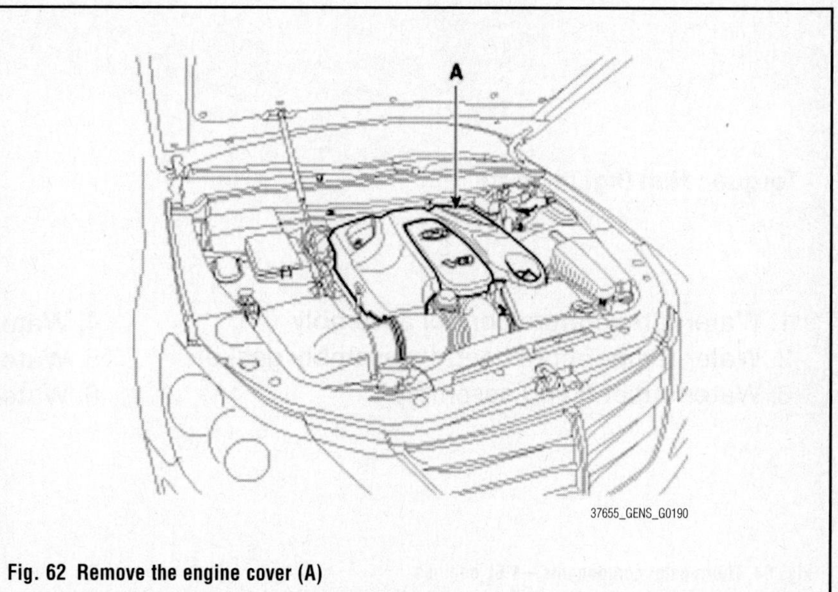

Fig. 62 Remove the engine cover (A)

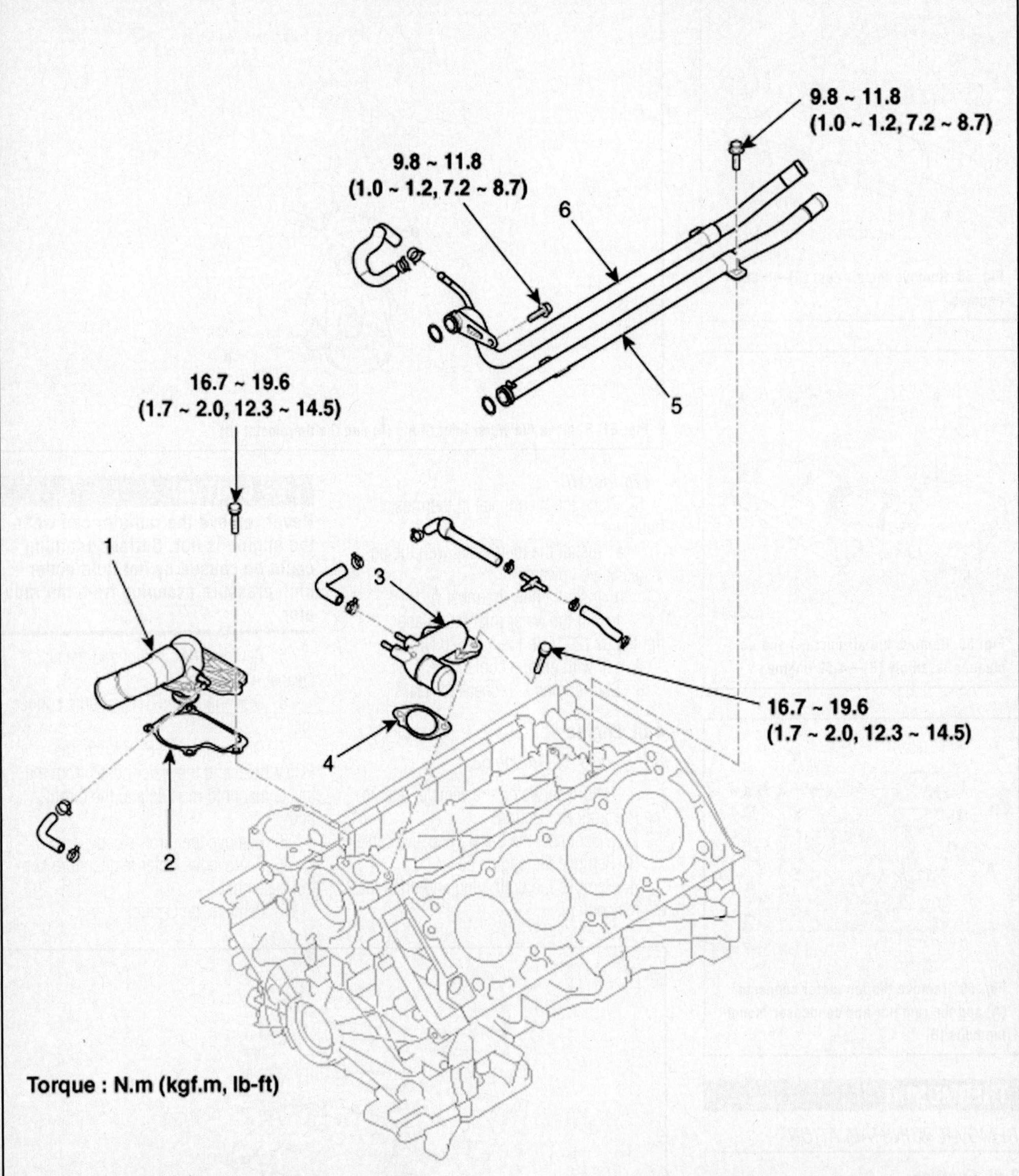

1. Water temperature control assembly
2. Water temperature control assembly gasket
3. Water outlet fitting assembly
4. Water outlet fitting assembly gasket
5. Water inlet pipe
6. Water outlet pipe

37655_GENS_G0214

Fig. 63 Thermostat components—4.6L engines

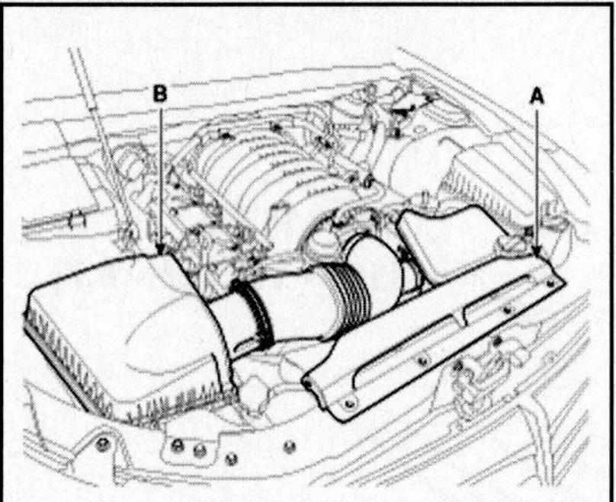

Fig. 64 Remove the air duct (A) and air cleaner assembly (B)

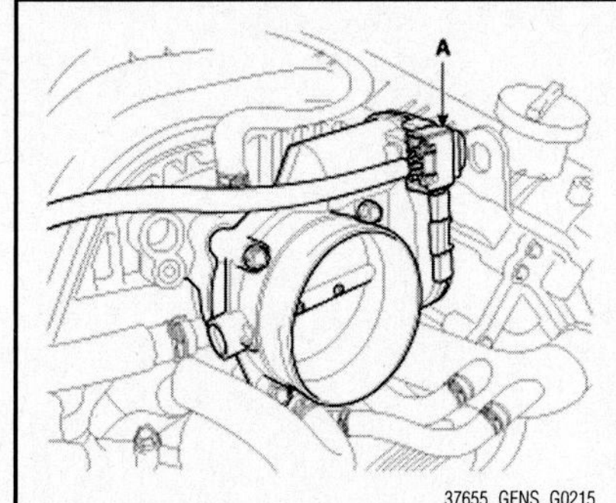

Fig. 65 Disconnect the ETC module connector (A)

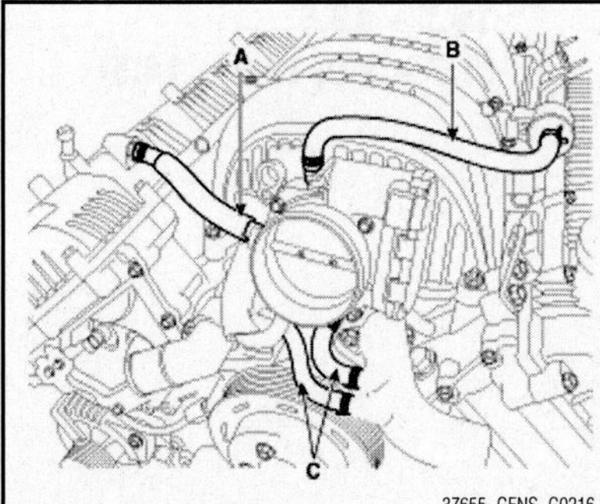

Fig. 66 Disconnect the PCV hose (A), the PCSV hose (B) and the water hoses (C)

To install:

11. Install the O-ring.
12. Install the water temperature control assembly and tighten to 12–15 ft. lbs. (17–20 Nm).

➡The protrusion of the gasket must be face the front of the engine.

13. Install the throttle body.
14. Reconnect the PCV hose, the PCSV hose and the water hoses to the intake manifold module and the throttle body.
15. Reconnect the ETC module connector.
16. Install the air duct and the air cleaner assembly and tighten the bolts to 86–104 inch lbs. (10–12 Nm).

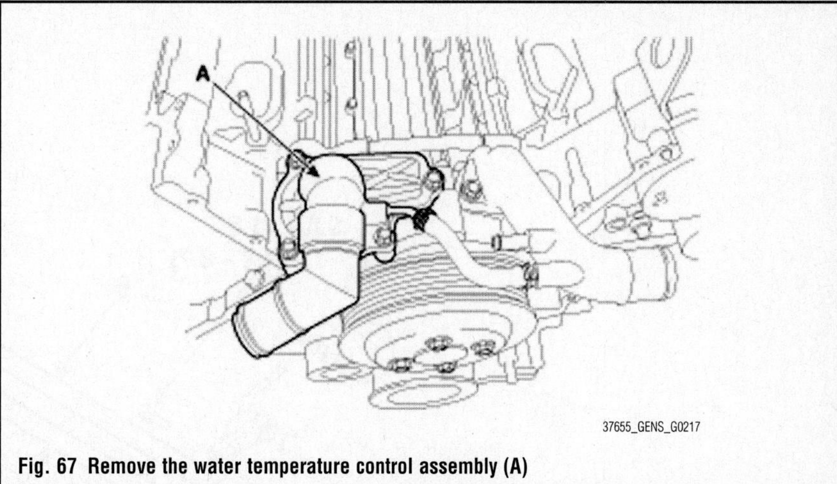

Fig. 67 Remove the water temperature control assembly (A)

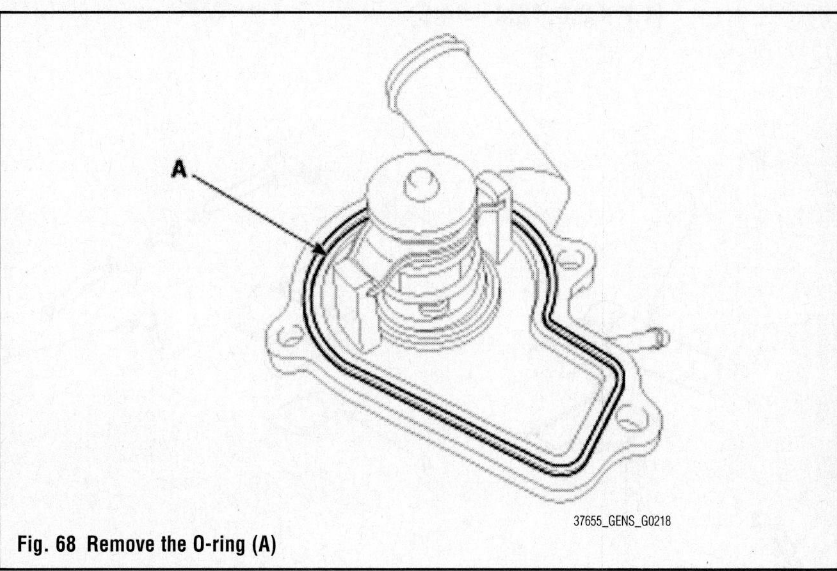

Fig. 68 Remove the O-ring (A)

17. Install the engine cover.
18. Connect the negative battery cable.
19. Fill the radiator with coolant and check for leaks.

WATER PUMP

REMOVAL & INSTALLATION

3.8L Engines

See Figures 69 through 75.

1. Before servicing the vehicle, refer to the Precautions Section.
2. Disconnect the battery negative cable.
3. Remove the drain plug and drain the engine coolant.

⁂ CAUTION

Never remove the radiator cap when the engine is hot. Serious scalding could be caused by hot fluid under high pressure escaping from the radiator.

4. Remove the drain plug and drain the engine coolant.

⁂ CAUTION

Never remove the radiator cap when the engine is hot. Serious scalding could be caused by hot fluid under high pressure escaping from the radiator.

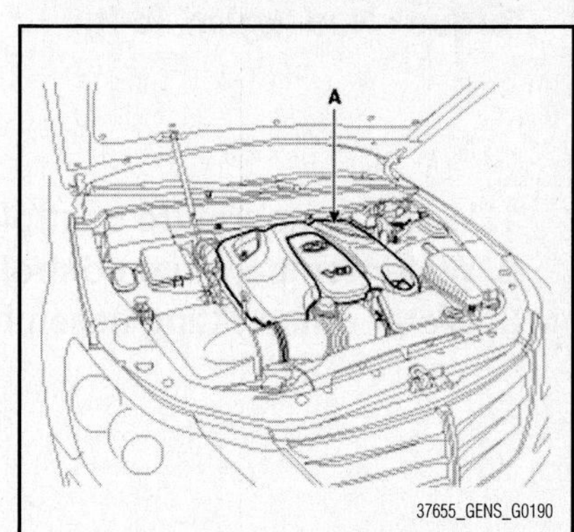

Fig. 69 Remove the engine cover (A)

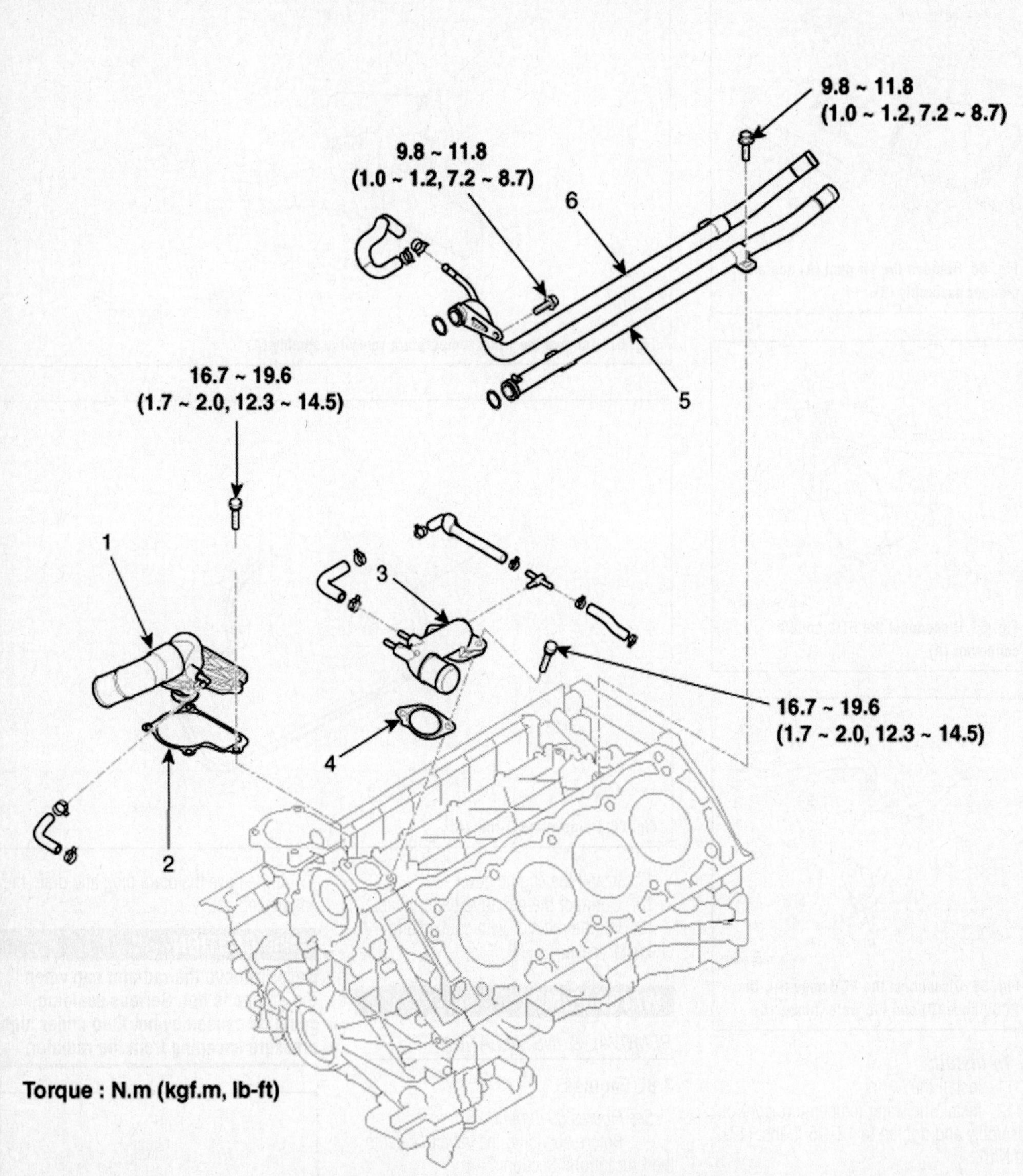

Fig. 70 Thermostat components—4.6L engines

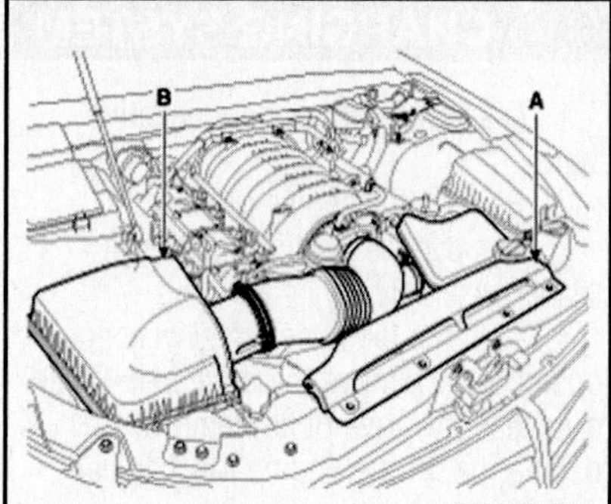

Fig. 71 Remove the air duct (A) and air cleaner assembly (B)

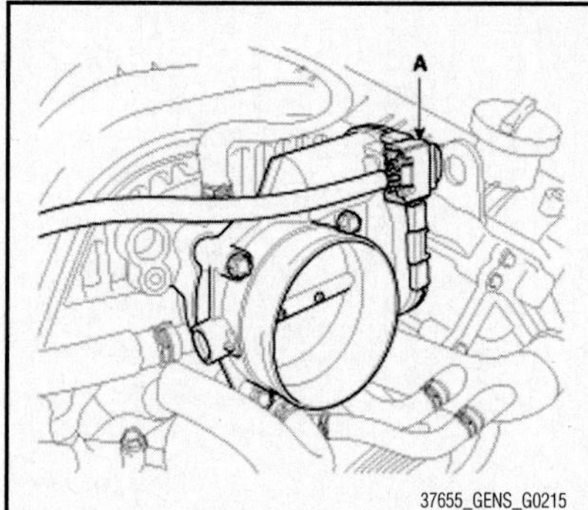

Fig. 72 Disconnect the ETC module connector (A)

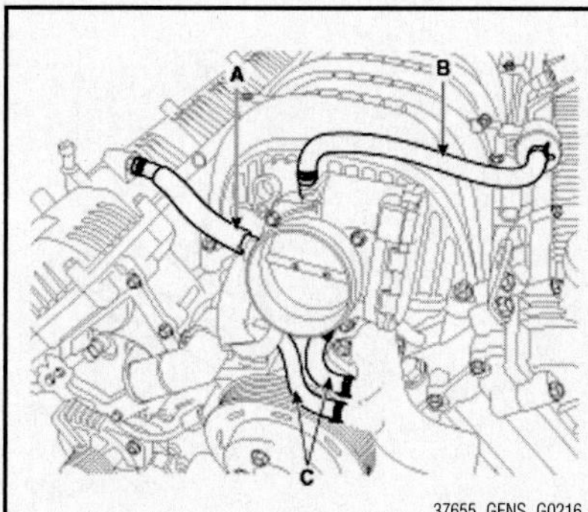

Fig. 73 Disconnect the PCV hose (A), the PCSV hose (B) and the water hoses (C)

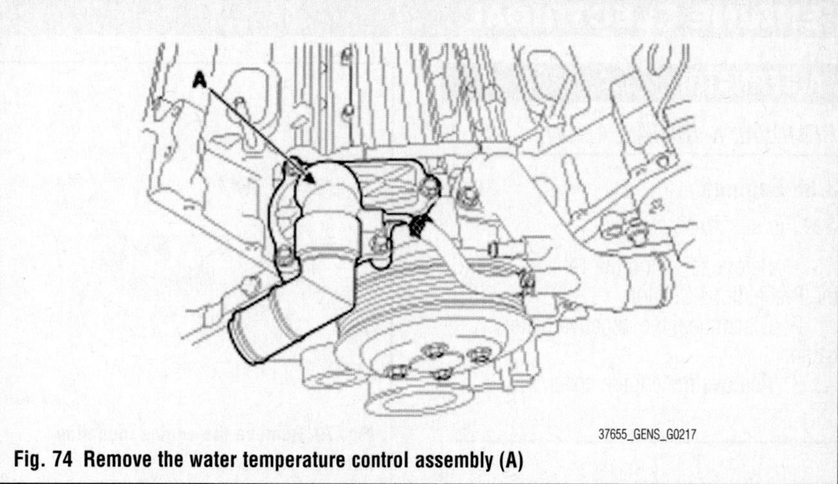

Fig. 74 Remove the water temperature control assembly (A)

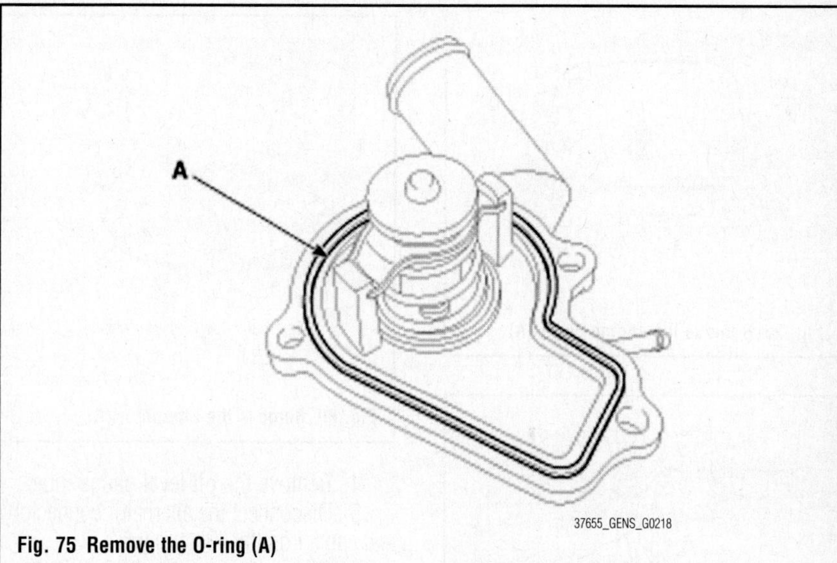

Fig. 75 Remove the O-ring (A)

5. Remove the air duct and the air cleaner assembly.

6. Disconnect the ETC module connector.

7. Disconnect the PCV hose, the PCSV hose and the water hoses from the intake manifold module and the throttle body.

8. Remove the throttle body.

9. Remove the water temperature control assembly.

10. Remove the O-ring.

To install:

11. Install the water pump and tighten the bolts to 15–18 ft. lbs. (20–25 Nm).

12. Install the water pump pulley and tighten every bolt diagonally, to 14–17 ft. lbs. (19–24 Nm).

13. Install the drive belt.

14. Install the upper and lower radiator hoses.

15. Install the air duct and the air cleaner assembly and tighten the bolts to 86–104 inch lbs. (10–12 Nm).

16. Install the engine cover.

17. Connect the negative battery cable.

18. Fill the radiator with coolant and check for leaks.

ALTERNATOR

REMOVAL & INSTALLATION

3.8L Engines

See Figures 76 through 80.

1. Before servicing the vehicle, refer to the Precautions Section.
2. Disconnect the negative battery cable.
3. Remove the engine cover.

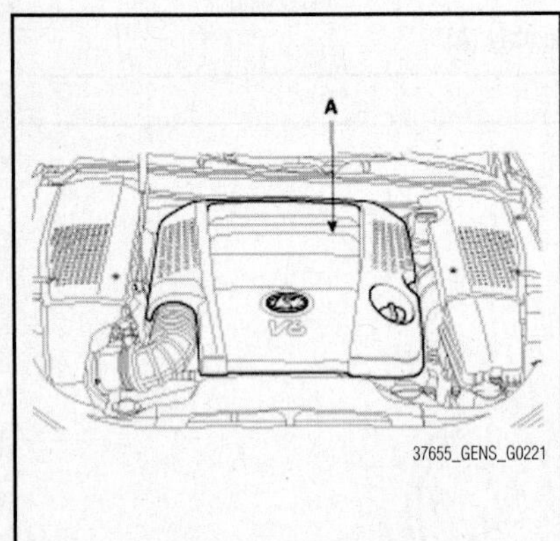

Fig. 76 Remove the engine cover (A)

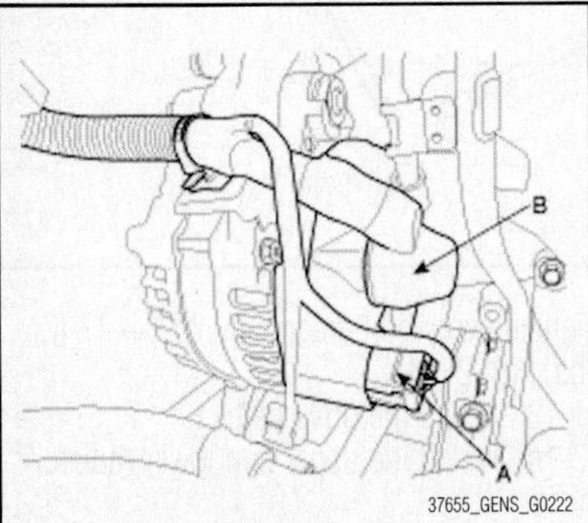

Fig. 77 Disconnect the alternator connector (A) and cable (B)

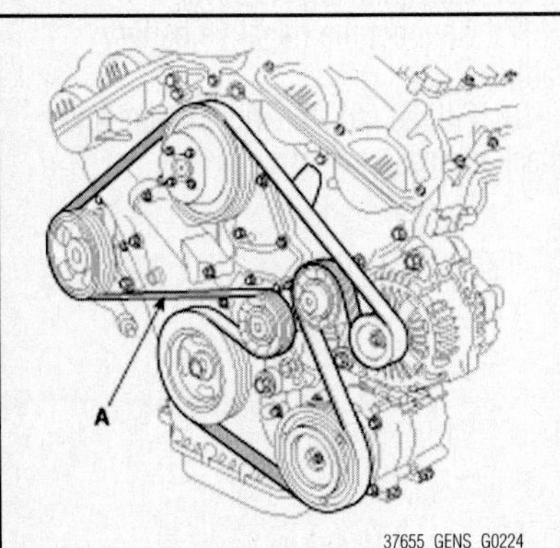

Fig. 78 Remove the drive belt (A)

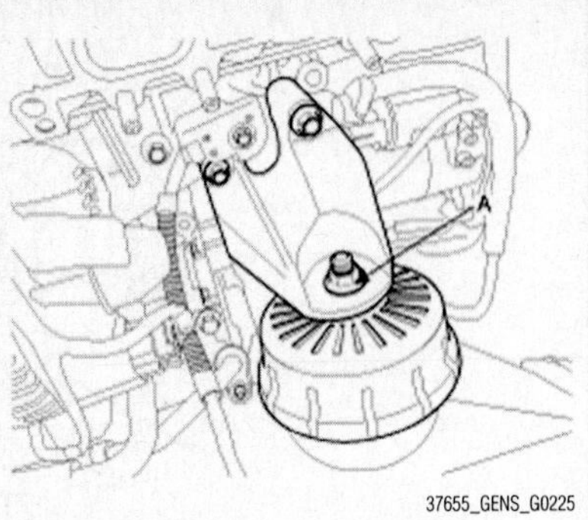

37655_GENS_G0225

Fig. 79 Remove the engine mounting nut (A)

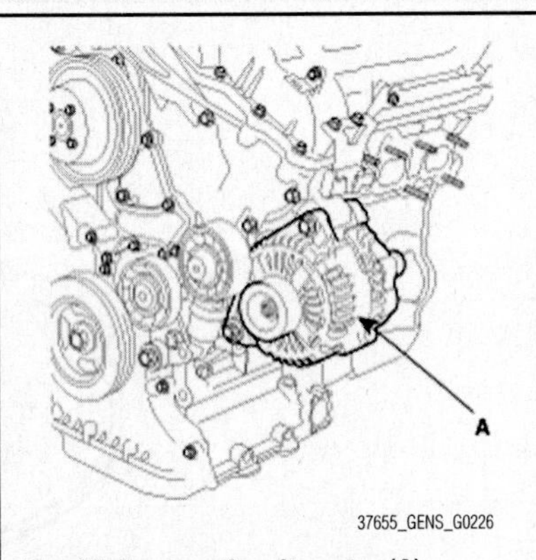

Fig. 80 Remove the alternator (A)

4. Remove the oil level gauge tube.
5. Disconnect the alternator connector and cable from the 'B' terminal.
6. Remove the drive belt. Refer to Accessory Drive Belts Removal & Installation, in the Engine Mechanical section.
7. Remove the engine undercover.
8. Position a jack under the oil pan.
9. Remove the engine mounting nut.
10. Lift up the engine assembly slightly by using a jack to gain access to the side of the engine.
11. Remove the alternator.

To install:

12. Installation is the reverse of removal, noting the following:
    a. Install the engine mounting nut and tighten to 47–62 ft. lbs. (64–83 Nm).
    b. Install the alternator and tighten to 20–25 ft. lbs. (27–33 Nm).

4.6L Engine

See Figures 81 through 85.

1. Disconnect the battery negative cable.
2. Remove the engine cover.
3. Remove the air duct and the air cleaner assembly.
4. Turn the tensioner clockwise and loosen, then remove the drive belt. Refer to Accessory Drive Belts Removal & Installation, in the Engine Mechanical section.
5. Remove the oil level gauge.
6. Remove the alternator.

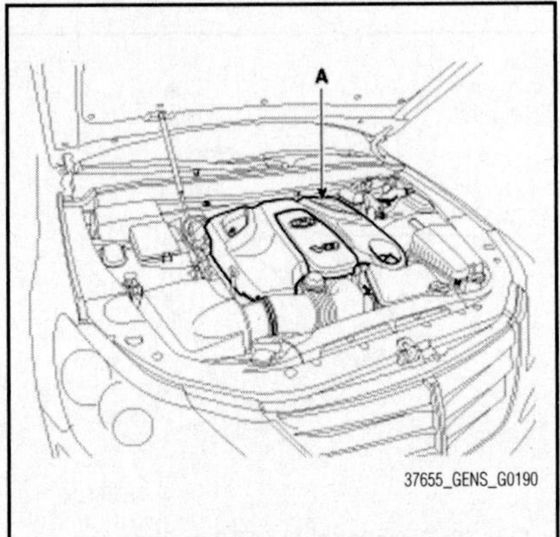

Fig. 81 Remove the engine cover (A)

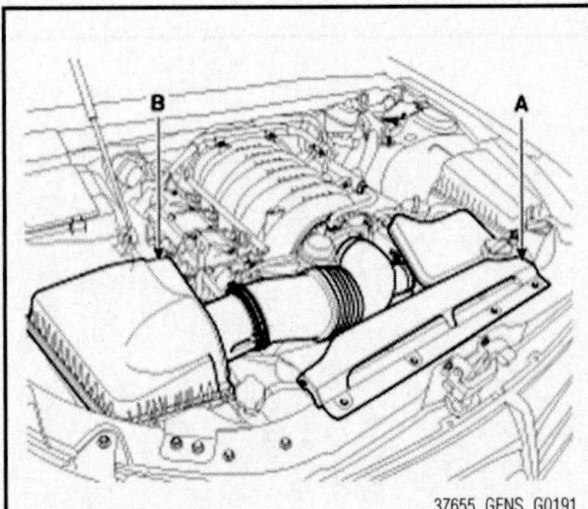

Fig. 82 Remove the air duct (A) and air cleaner assembly (B)

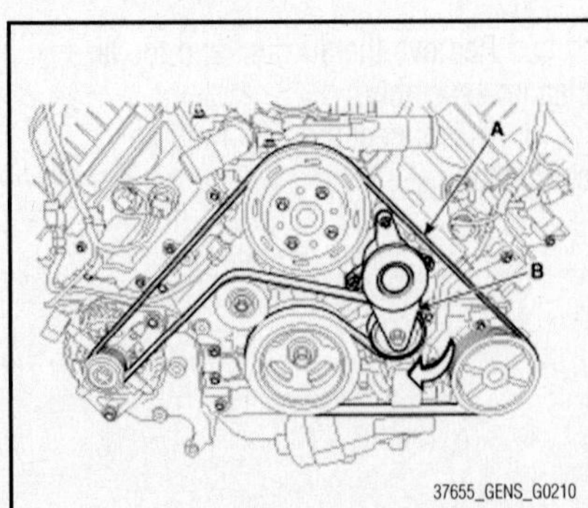

Fig. 83 Turn the tensioner (B) clockwise and loosen, then remove the drive belt (A)

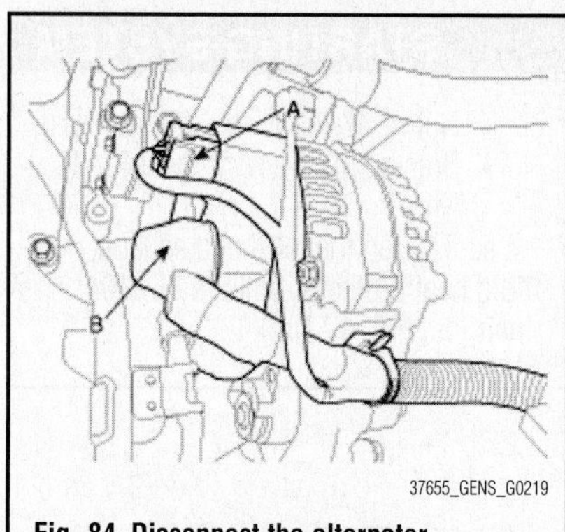

Fig. 84 Disconnect the alternator connector (A) and cable (B)

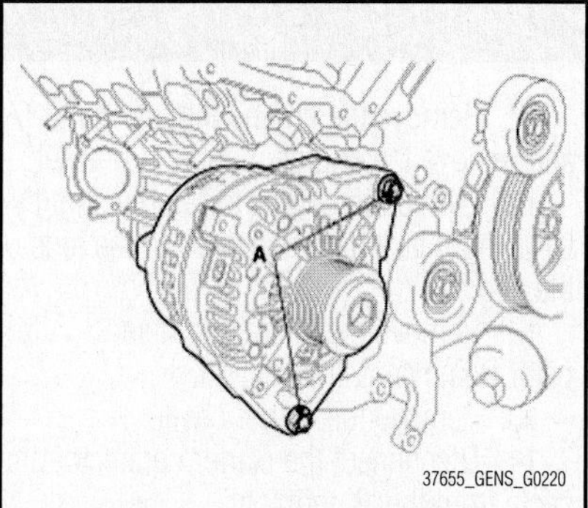

Fig. 85 Remove the alternator and mounting bolts (A)

a. Disconnect the alternator connector and cable from the 'B' terminal.

b. Remove the alternator mounting bolts, and then remove alternator from the vehicle.

To install:

7. Installation is the reverse of removal, noting the following:

a. Tighten the alternator mounting bolts to 22–30 ft. lbs. (29–41 Nm).

b. Tighten the oil level gauge to 86–104 inch lbs. (10–12 Nm).

ENGINE ELECTRICAL — IGNITION SYSTEM

FIRING ORDER

3.8L engines firing order: 1–2–3–4–5–6

4.6L Engine firing order: 1–2–7–8–4–5–6–3

IGNITION COIL

REMOVAL & INSTALLATION

See Figures 86 through 88.

1. Before servicing the vehicle, refer to the Precautions Section.
2. Disconnect the negative battery cable.
3. Remove the engine cover (as necessary).
4. Disconnect the ignition coil connectors.
5. Remove the ignition coils.

➡When removing the ignition coil connector, pull the lock pin and push the clip.

To install:

6. Installation is the reverse of removal.

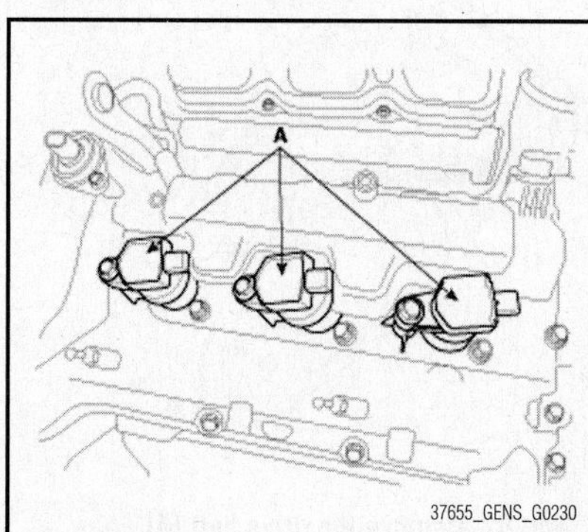

Fig. 86 Remove the ignition coils (A)—3.8L engines

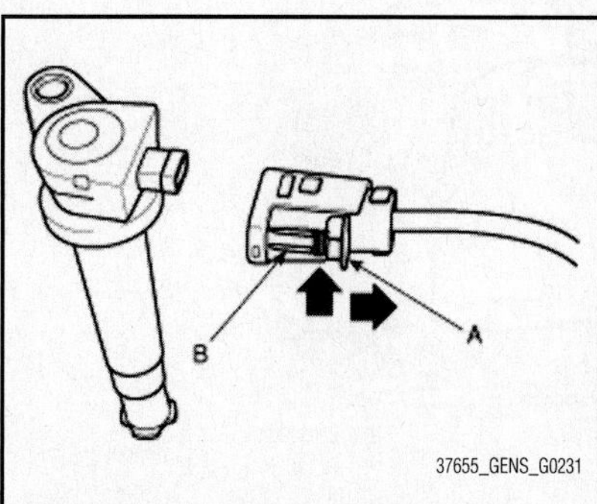

Fig. 88 When removing the ignition coil connector, pull the lock pin (A) and push the clip (B)

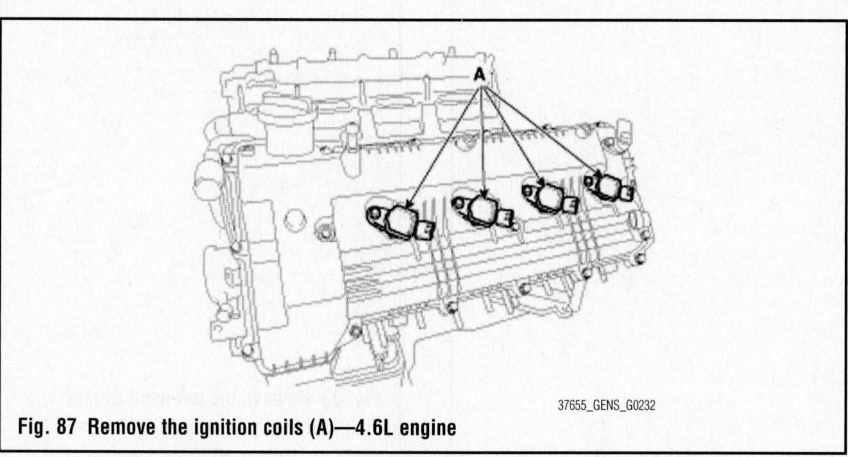

Fig. 87 Remove the ignition coils (A)—4.6L engine

IGNITION TIMING

ADJUSTMENT

Ignition timing is controlled by the electronic control ignition timing system.

SPARK PLUGS

REMOVAL & INSTALLATION

See Figure 88.

1. Before servicing the vehicle, refer to the Precautions Section.
2. Disconnect the negative battery cable.
3. Remove the engine cover (as necessary).
4. Disconnect the ignition coil connector.

➡When removing the ignition coil connector, pull the lock pin (A) and push the clip (B).

5. Remove the ignition coil.
6. Use a spark plug socket and wrench to remove the spark plugs.

✲✲ WARNING

Be careful that no contaminates enter through the spark plug holes.

To install:

➡Check the electrode gap on the spark plugs before installation. Specification: 0.039–0.043 in. (1.0–1.1 mm).

7. To install, reverse the removal procedure.

STARTER

REMOVAL & INSTALLATION

3.8L Engines

See Figures 89 through 98.

1. Before servicing the vehicle, refer to the Precautions Section.
2. Disconnect the negative battery cable.
3. Remove the engine cover.
4. Remove the oil level gauge tube.
5. Disconnect the alternator connector and cable from the 'B' terminal.
6. Remove the drive belt. Refer to Accessory Drive Belts Removal & Installation, in the Engine Mechanical section.
7. Remove the engine undercover.
8. Position a jack under the oil pan.
9. Remove the left-hand exhaust manifold heat protector. Refer to Exhaust Manifold Removal & Installation, in the Engine Mechanical section.
10. Remove the engine mounting nut (A) and remove the engine mounting bracket.
11. Lift up the engine assembly slightly by using a jack to gain access to the side of the engine.
12. Remove the alternator. Refer to Alternator Removal & Installation
13. Remove the starter cover.
14. Disconnect the starter connector and cable from the 'B' terminal.
15. Remove the starter and mounting bolts.

To install:

16. Installation is the reverse of removal, noting the following:
    a. Tighten the starter mounting bolts to 36–47 ft. lbs. (49–64 Nm).
    b. Tighten the starter cover bolts to 78–121 inch lbs. (9–14 Nm).
    c. Tighten the alternator bolts to 20–25 ft. lbs. (27–33 Nm).
    d. Tighten the engine mounting nut and bracket to:
        - Nut A: 47–62 ft. lbs. (64–83 Nm)
        - Bracket B: 36–47 ft. lbs. (49–64 Nm)
    e. Tighten the left-hand exhaust manifold heat protector bolts to 86–104 inch lbs. (10–12 Nm).

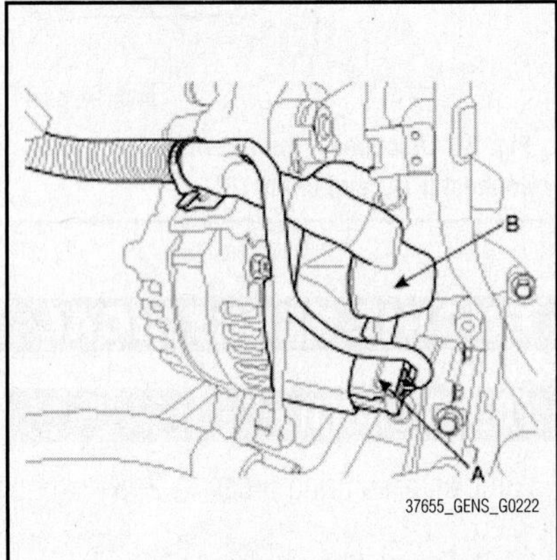

Fig. 91 Disconnect the alternator connector (A) and cable (B)

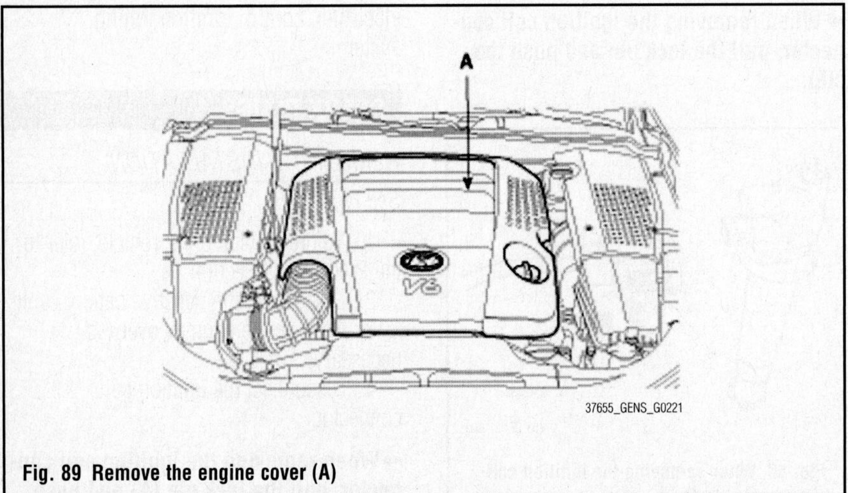

Fig. 89 Remove the engine cover (A)

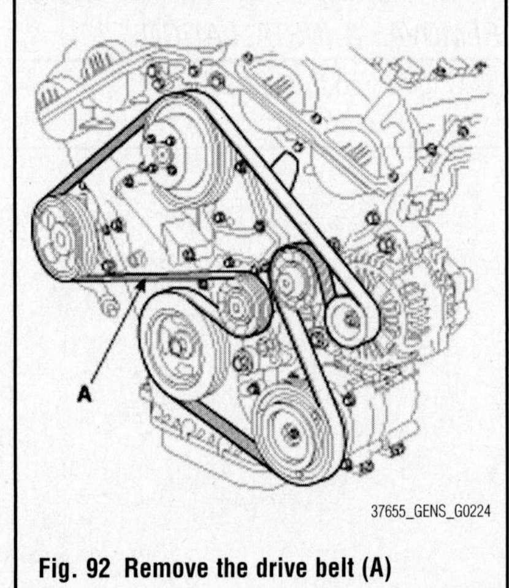

Fig. 92 Remove the drive belt (A)

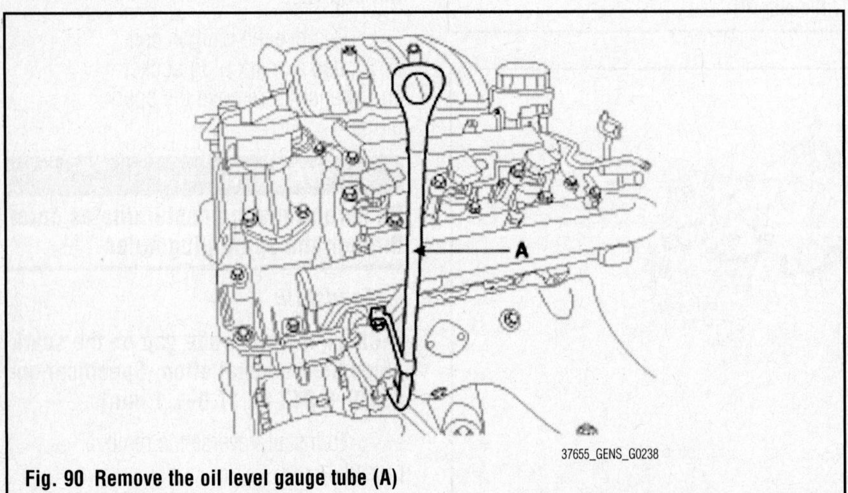

Fig. 90 Remove the oil level gauge tube (A)

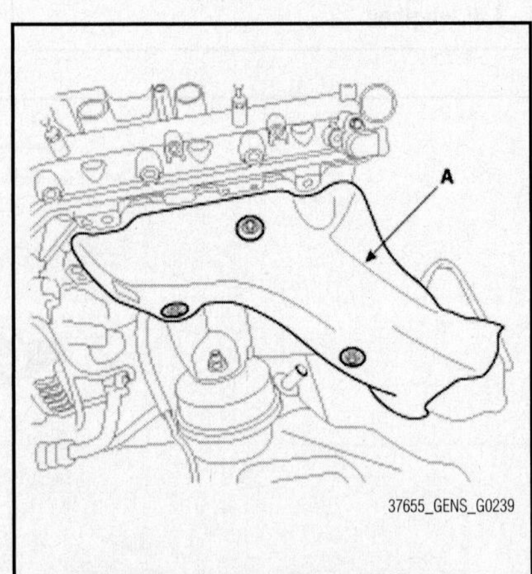

Fig. 93 Remove the left-hand exhaust manifold heat protector (A)

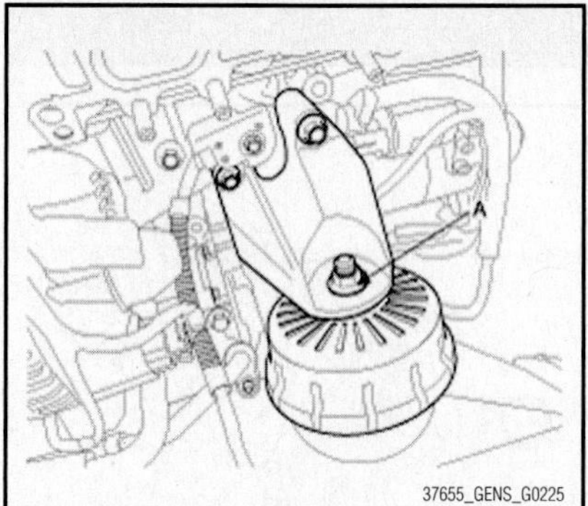

Fig. 94 Remove the engine mounting nut (A) and engine mounting bracket

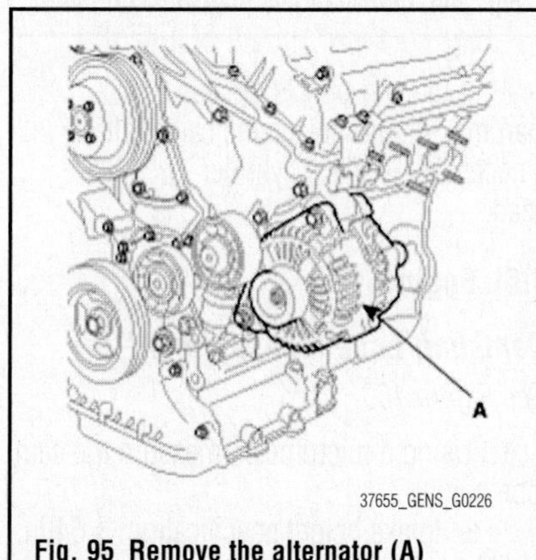

Fig. 95 Remove the alternator (A)

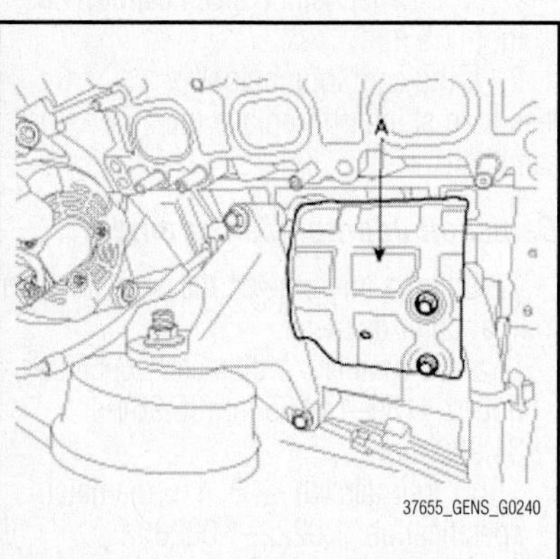

Fig. 96 Remove the starter cover (A)

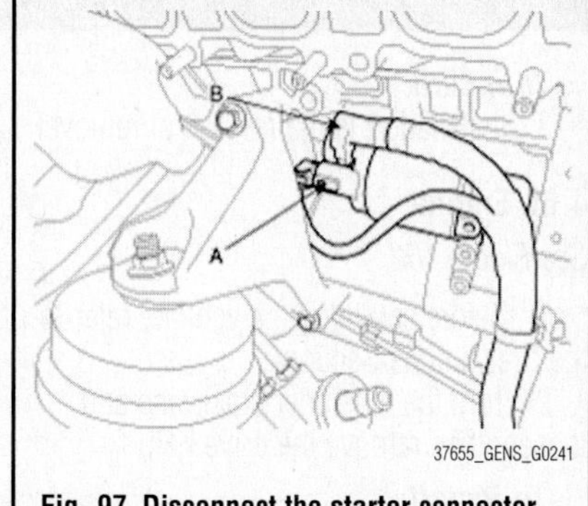

Fig. 97 Disconnect the starter connector (A) and cable (B)

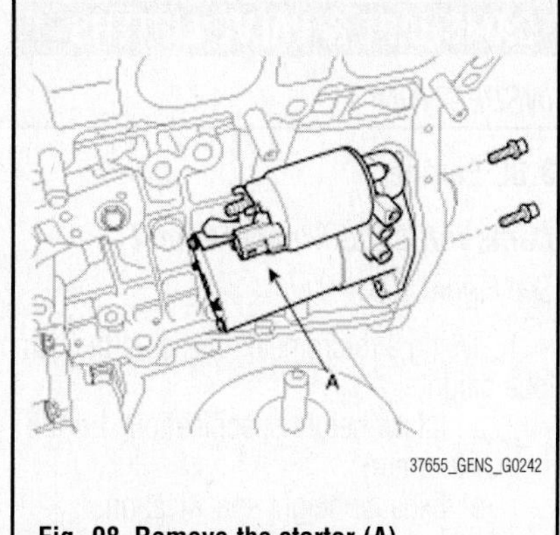

Fig. 98 Remove the starter (A)

4.6L Engine

See Figure 99.

1. Before servicing the vehicle, refer to the Precautions Section.
2. Disconnect the negative battery cable.
3. Disconnect the starter cable from the B terminal on the solenoid, and the connector from the S terminal.
4. Remove the starter mounting bolts and then remove the starter.

To install:

5. Installation is the reverse of removal. Tighten the starter motor mounting bolts to 36–47 ft. lbs. (49–64 Nm).

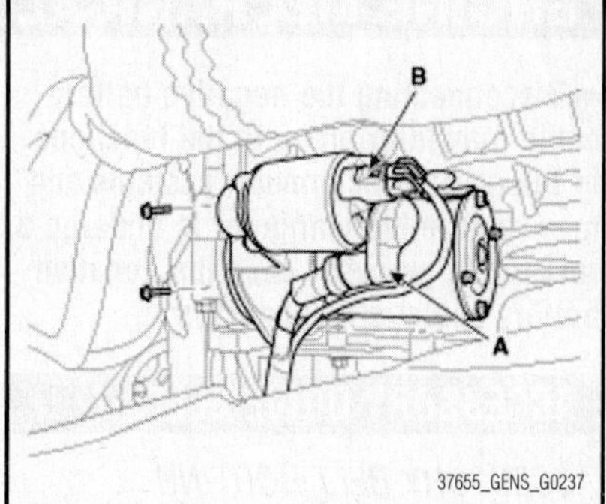

Fig. 99 Disconnect the starter cable (A) from the B terminal on the solenoid, and the connector (B) from the S terminal, and remove the mounting bolts and starter

SOLENOID OR RELAY REPLACEMENT

See Figure 100.

1. Before servicing the vehicle, refer to the Precautions Section.
2. Disconnect the negative battery cable.
3. Remove the fuse box cover.
4. Remove the starter relay.

To install:

5. Installation is the reverse of removal.

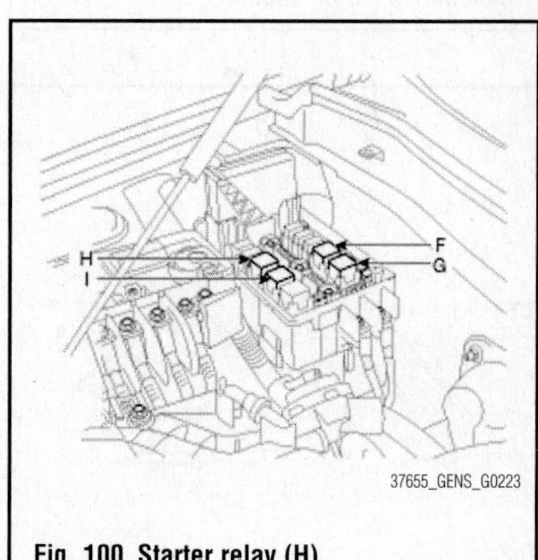

Fig. 100 Starter relay (H)

ENGINE MECHANICAL

➡Disconnecting the negative battery cable may interfere with the functions of the on board computer systems and may require the computer to undergo a relearning process, once the negative battery cable is reconnected.

ACCESSORY DRIVE BELTS

ACCESSORY BELT ROUTING

See Figures 101 and 102.

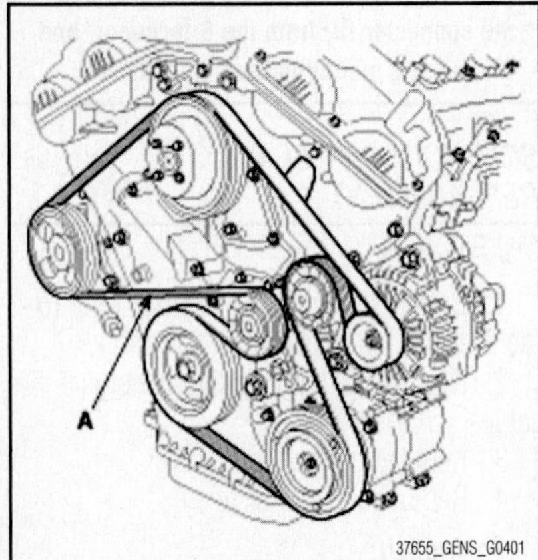

Fig. 101 Drive belt (A) removal and installation—3.8L engine

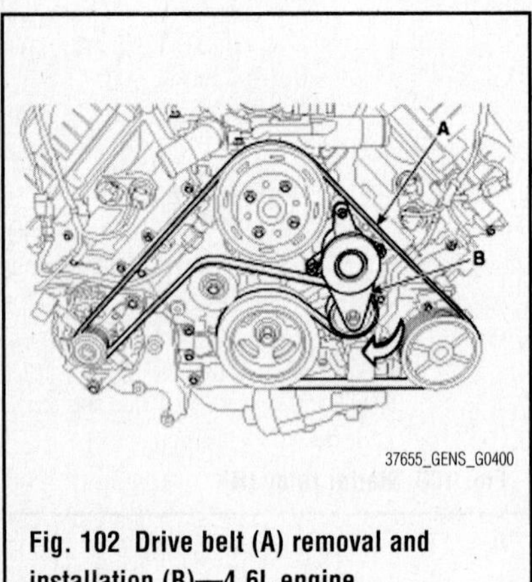

Fig. 102 Drive belt (A) removal and installation (B)—4.6L engine

ADJUSTMENT

1. No adjustment is possible or necessary.

REMOVAL & INSTALLATION

3.8L Engine

See Figure 101.

1. Before servicing the vehicle, refer to the Precautions Section.
2. Turn the tensioner clockwise and loosen, then remove the drive belt.

To install:

3. Installation is the reverse of removal.

4.6L Engine

See Figure 102.

1. Before servicing the vehicle, refer to the Precautions Section.
2. Turn the tensioner clockwise and loosen, then remove the drive belt.

To install:

3. Installation is the reverse of removal.

CAMSHAFT & VALVE LIFTERS

INSPECTION

3.8L Engine

Camshaft Lobe Measurement

See Figure 103.

1. Using a micrometer, measure the cam lobe height.
   a. Intake height specification: 1.8425 in. (46.8mm)
   b. Exhaust height specification: 1.8031 in. (45.8mm)
2. If the cam lobe height is less than standard, replace the camshaft.

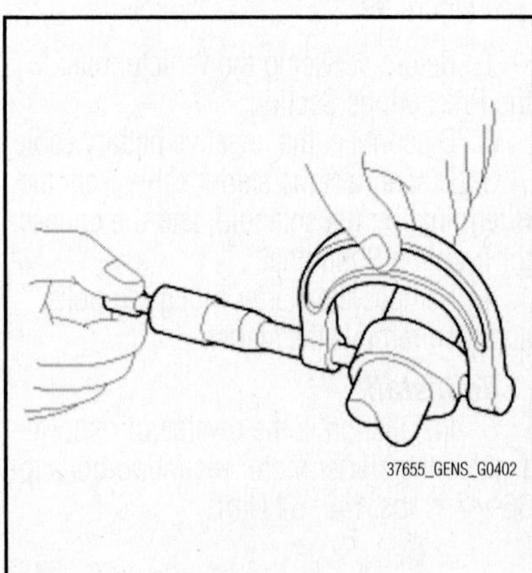

Fig. 103 Camshaft lobe measurement

Camshaft End Play

See Figure 104.

1. Install the camshafts.
2. Install the bearing cap and thrust bearing cap with specified torque.
3. Using a dial indicator, measure the end play while moving the camshaft back and forth.
   a. End play specification: 0.0008–0.0071 in. (0.02–0.18 mm)

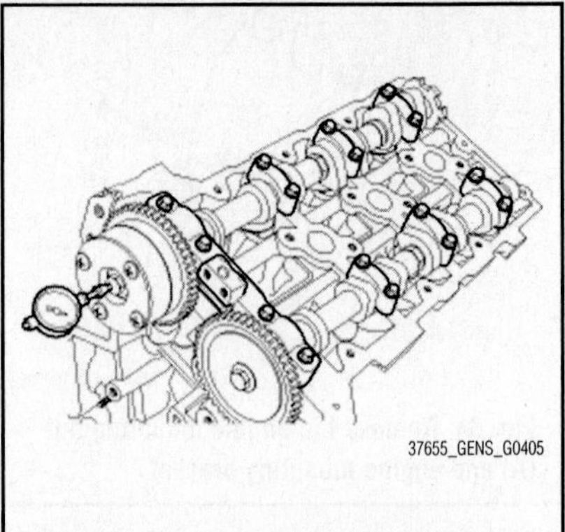

Fig. 104 Camshaft end play measurement

4. If the end play is greater than maximum, replace the camshaft. If necessary, replace cylinder head.

4.6L Engine

Camshaft Lobe Measurement

See Figure 103.

1. Using a micrometer, measure the cam lobe height.
   a. Intake height specification: 1.64 in. (41.6 mm)
   b. Exhaust height specification: 1.63 in. (41.5 mm)
2. If the cam lobe height is less than standard, replace the camshaft.

Camshaft Journal Measurement

1. Using a micrometer, measure the cam journal outer diameter.
   a. Journal No. 1 diameter specification: 1.4159–1.4165 in. (35.964–31.980 mm)
   b. Journals No. 2, 3, 4, 5 diameter specification: 1.0222–1.0228 in. (25.964–25.980 mm)

Camshaft End Play

See Figure 105.

1. Install the camshafts.
2. Install the bearing cap and thrust bearing cap with specified torque.
3. Using a dial indicator, measure the end play while moving the camshaft back and forth.
   a. End play specification: 0.0047–0.0086 in. (0.12–0.22 mm)
4. If the end play is greater than maximum, replace the camshaft. If necessary, replace cylinder head.

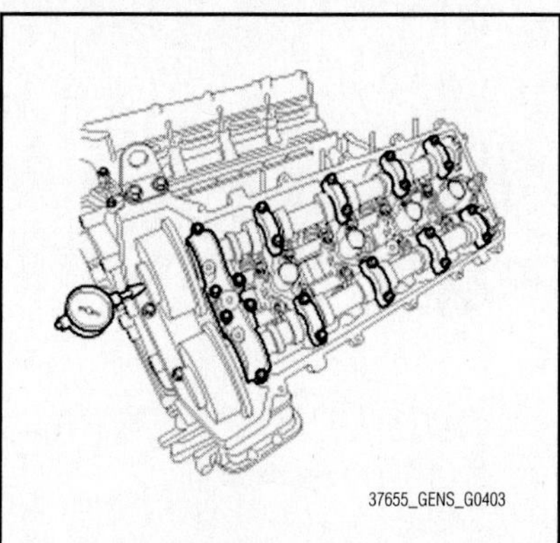

37655_GENS_G0403
Fig. 105 Camshaft end play measurement

Hydraulic Lash Adjuster

See Figure 106.

1. With the HLA filled with engine oil, hold top and press bottom by hand. If bottom moves, replace the HLA.

REMOVAL & INSTALLATION

3.8L Engine

See Figures 107 through 125.

1. Before servicing the vehicle, refer to the Precautions Section.
2. Disconnect the negative battery cable.

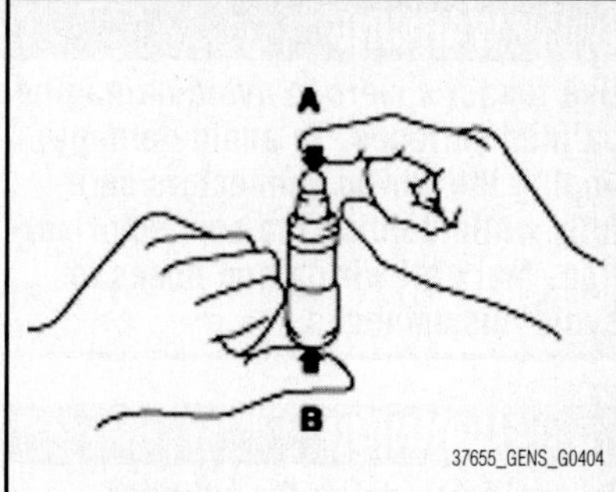

A
B
37655_GENS_G0404
Fig. 106 Hydraulic lash adjuster top (A) and bottom (B) testing

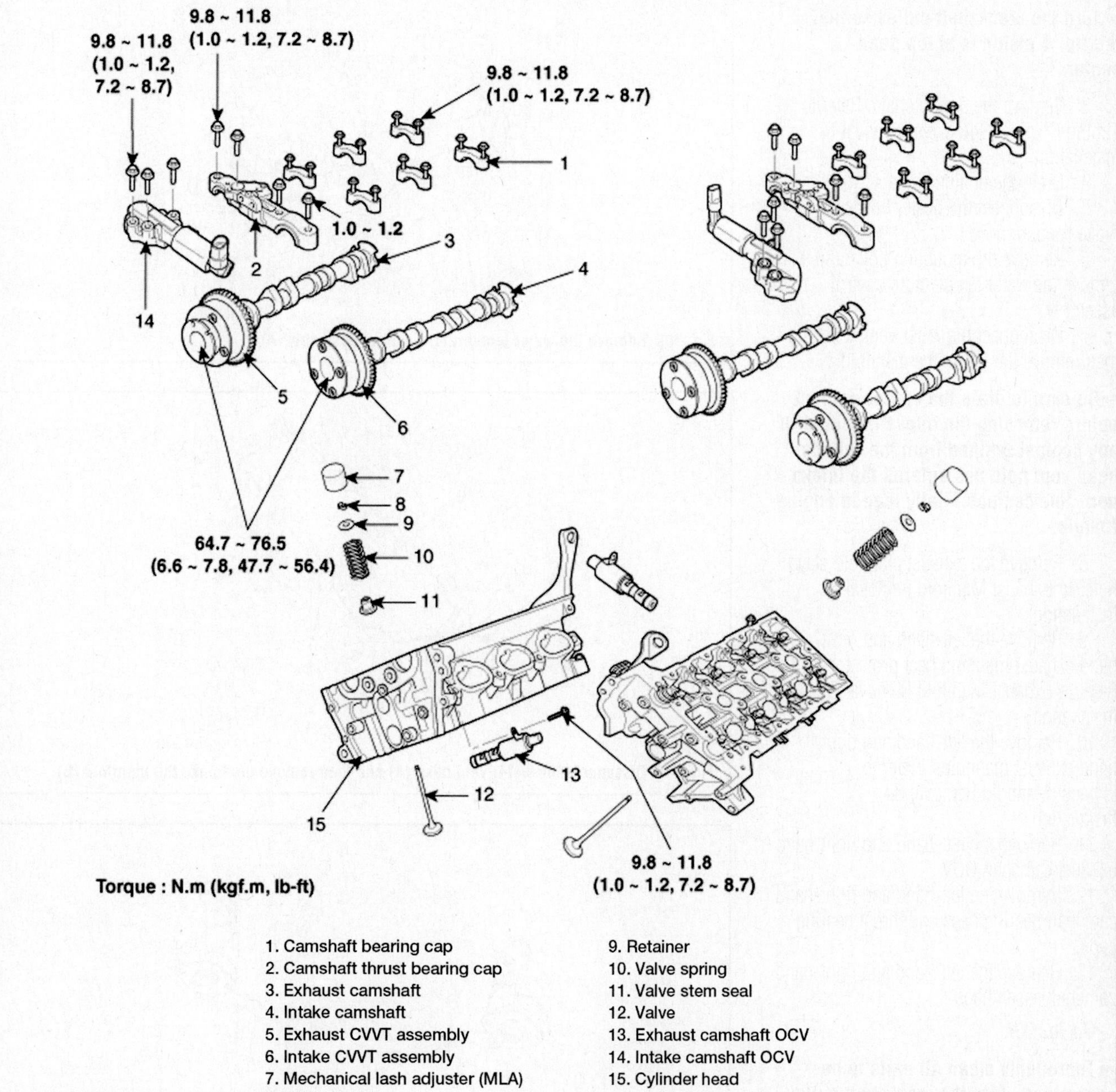

9.8 ~ 11.8
(1.0 ~ 1.2, 7.2 ~ 8.7)
9.8 ~ 11.8
(1.0 ~ 1.2,
7.2 ~ 8.7)
9.8 ~ 11.8
(1.0 ~ 1.2, 7.2 ~ 8.7)
1
1.0 ~ 1.2
2
3
4
14
5
6
7
8
9
64.7 ~ 76.5
(6.6 ~ 7.8, 47.7 ~ 56.4)
10
11
13
12
15
9.8 ~ 11.8
(1.0 ~ 1.2, 7.2 ~ 8.7)
Torque : N.m (kgf.m, lb-ft)
1. Camshaft bearing cap
2. Camshaft thrust bearing cap
3. Exhaust camshaft
4. Intake camshaft
5. Exhaust CVVT assembly
6. Intake CVVT assembly
7. Mechanical lash adjuster (MLA)
8. Retainer lock
9. Retainer
10. Valve spring
11. Valve stem seal
12. Valve
13. Exhaust camshaft OCV
14. Intake camshaft OCV
15. Cylinder head
37655_GENS_G0330
Fig. 107 Camshafts and related components

⁂ WARNING

Use fender covers to avoid damaging painted surfaces. To avoid damage, unplug the wiring connectors carefully while holding the connector portion. Mark all wiring and hoses to avoid misconnection.

⁂ WARNING

To avoid damaging the cylinder head, wait until the engine coolant temperature drops below normal temperature (68° F [20° C]) before removing it.

➡Turn the crankshaft pulley so that the No. 1 piston is at top dead center.

3. Remove the timing chain. Refer to Timing Chain & Sprockets Removal & Installation.
4. Drain the engine coolant.
5. Disconnect the heater hoses and brake vacuum hose.
6. Remove the mounting bolts and then remove the water temperature control assembly.
7. Disconnect the water vent hose and then remove the intake the manifold.

➡Be sure to drain the engine coolant before removing the intake manifold. If any coolant drained from the cylinder head vent hole has entered the intake port, this can potentially lead to engine trouble.

8. Remove the exhaust manifold stays. Refer to Exhaust Manifold Removal & Installation.
9. Remove the left-hand and right-hand exhaust manifold heat protectors. Refer to Exhaust Manifold Removal & Installation.
10. Remove the left-hand and right-hand exhaust manifolds. Refer to Exhaust Manifold Removal & Installation.
11. Remove the left-hand and right-hand exhaust camshaft OCV.
12. Remove the left-hand and right-hand camshaft bearing caps and thrust bearing caps.
13. Remove the left-hand and right-hand camshaft assemblies.

To install:

➡Thoroughly clean all parts to be assembled. Turn the crankshaft pulley so that the No. 1 piston is at top dead center.

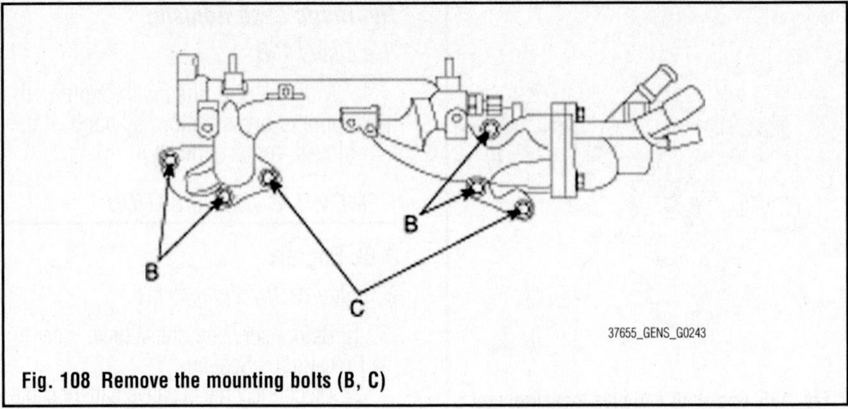

Fig. 108 Remove the mounting bolts (B, C)

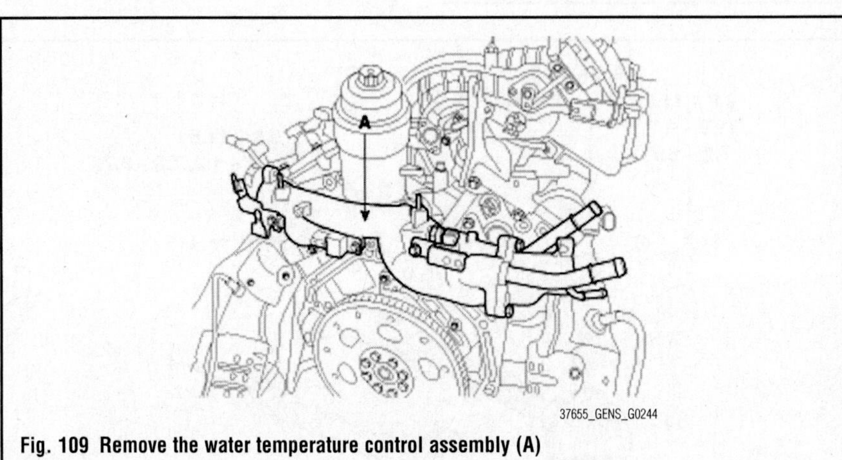

Fig. 109 Remove the water temperature control assembly (A)

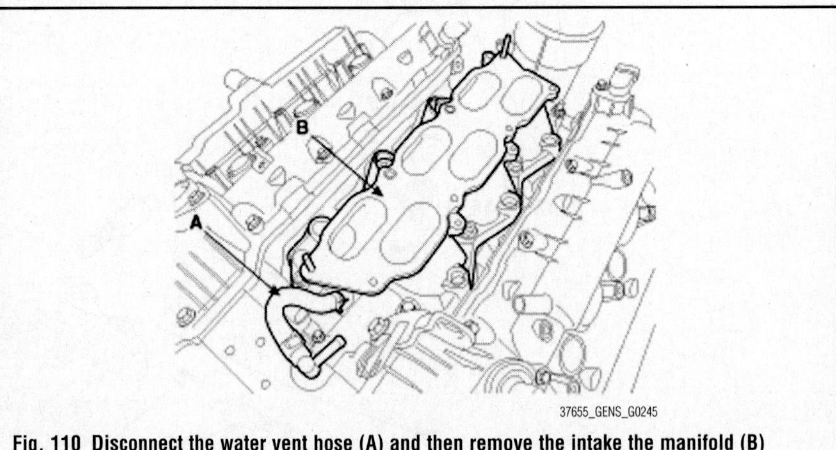

Fig. 110 Disconnect the water vent hose (A) and then remove the intake the manifold (B)

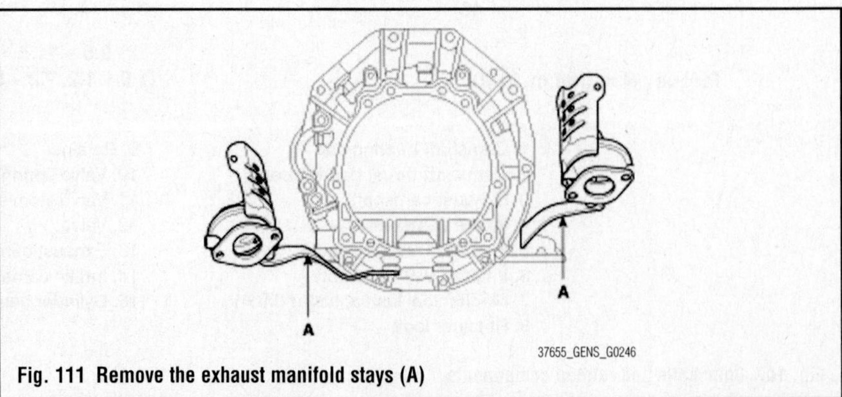

Fig. 111 Remove the exhaust manifold stays (A)

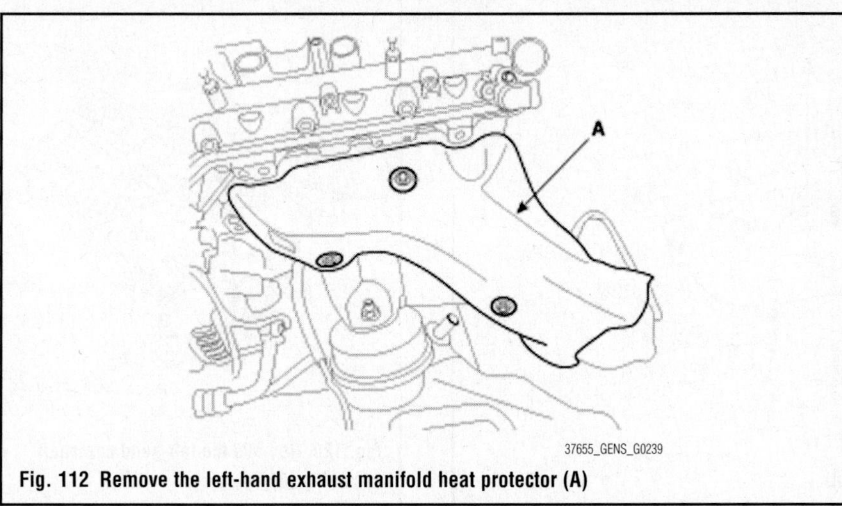

Fig. 112 Remove the left-hand exhaust manifold heat protector (A)

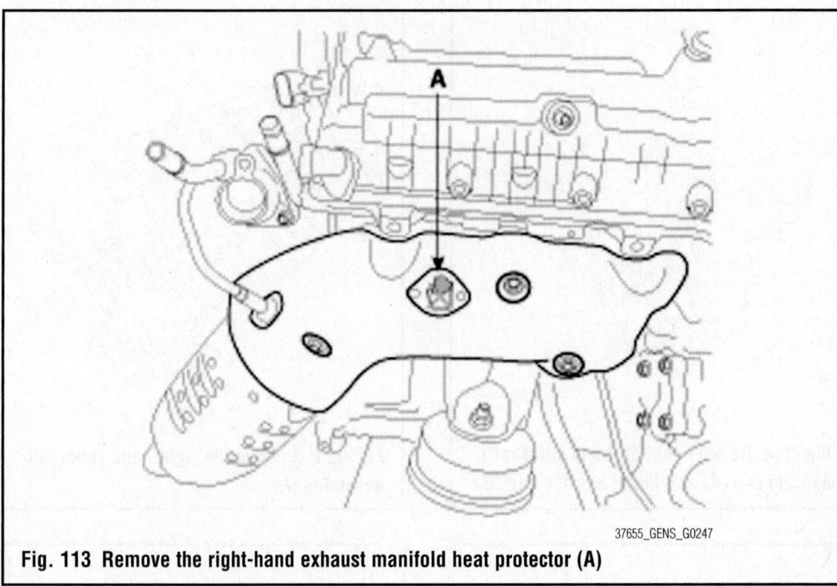

Fig. 113 Remove the right-hand exhaust manifold heat protector (A)

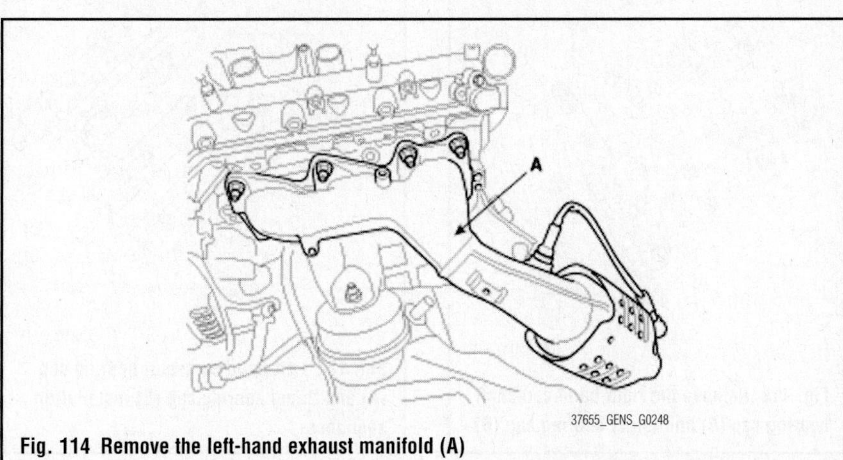

Fig. 114 Remove the left-hand exhaust manifold (A)

14. Install the left-hand and right-hand camshaft assemblies.

➡Apply a light coat of engine oil on camshaft journals. Assemble the key groove of camshaft rear side to the same level of head top surface.

Be careful the right, left bank, intake, exhaust side before assembling.

15. Install the left-hand and right-hand camshaft bearing caps and thrust bearing caps. Tightening torque:

- First Step: 52 inch lbs. (6 Nm)
- Second Step: 86–104 inch lbs. (10–12 Nm)

⁂ WARNING

Be sure to install the thrust bearing cap bolts and the bearing cap bolts in the correct place.

⁂ WARNING

Be careful of the right, left bank, intake, exhaust side before assembling.

⁂ WARNING

Rotate the crankshaft so as not to contact the valves with the pistons by making the pistons below 0.3937in. (10mm) from the top of cylinder block.

16. Install the left-hand and right-hand exhaust camshaft OCV and tighten to 86–104 inch lbs. (10–12 Nm).
17. Install the left-hand and right-hand exhaust manifolds with new gaskets, and tighten to 29–33 ft. lbs. (39–44 Nm).
18. Install the left-hand and right-hand exhaust manifold heat protectors and tighten to 86–104 inch lbs. (10–12 Nm).
19. Install the exhaust manifold stays and tighten to 25–30 ft. lbs. (34–41 Nm).
20. Install the intake the manifold with a new gasket, and connect the water vent hose. Tightening torque:
    - Step 1: 35–52 (4–6 Nm)
    - Step 2: Nut: 14–17 ft. lbs. (19–24 Nm), Bolt: 20–23 ft. lbs. (27–31 Nm)
    - Step 3: Repeat 2nd step twice or more.

➡Confirm the manifold gasket identification mark (left-hand, RH) and be careful of the installation direction.

21. Install the water temperature control assembly with a new gasket, and tighten to 15–17 ft. lbs. (20–24 Nm).
22. Connect the heater hoses and brake vacuum hose.
23. Install the timing chain.
24. Refill engine oil.
25. Clean the battery posts and cable terminals with sandpaper. Assemble and then apply grease to prevent corrosion.
26. Inspect for fuel leakage:

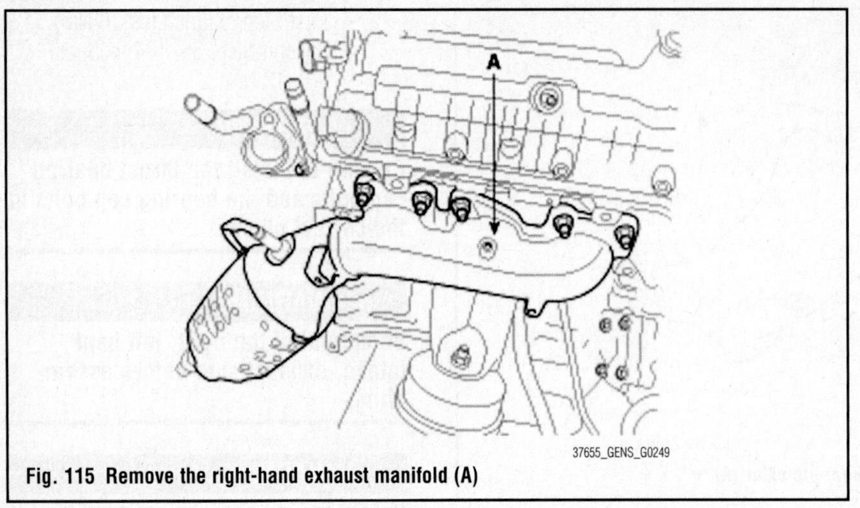

Fig. 115 Remove the right-hand exhaust manifold (A)

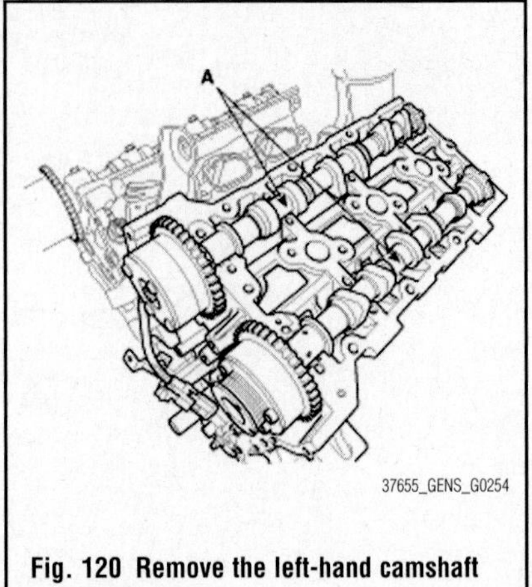

Fig. 120 Remove the left-hand camshaft assembly (A)

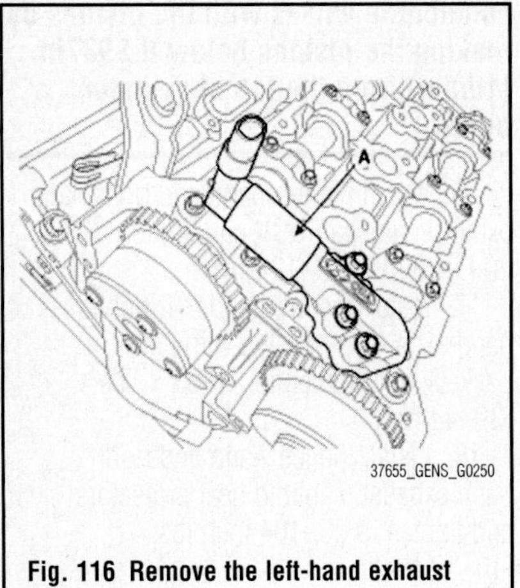

Fig. 116 Remove the left-hand exhaust camshaft OCV (A)

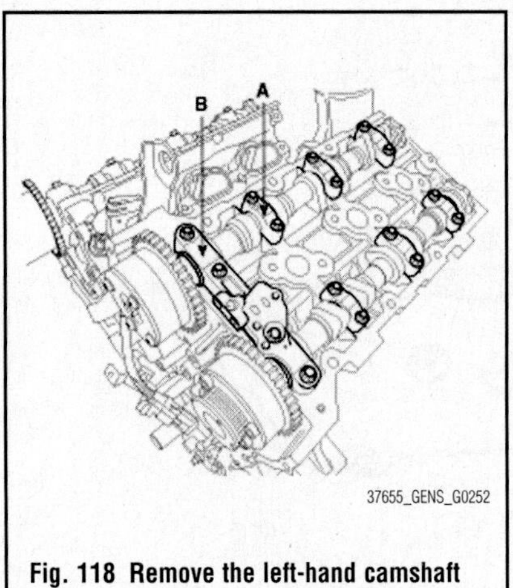

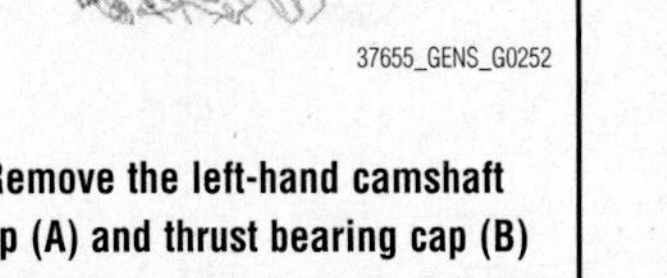

Fig. 118 Remove the left-hand camshaft bearing cap (A) and thrust bearing cap (B)

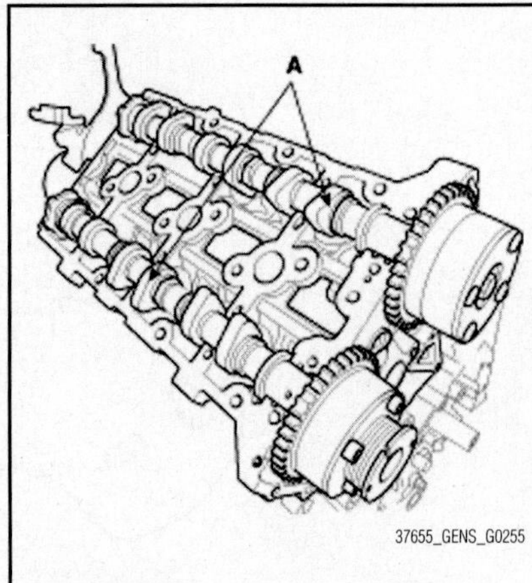

Fig. 121 Remove the right-hand camshaft assembly (A)

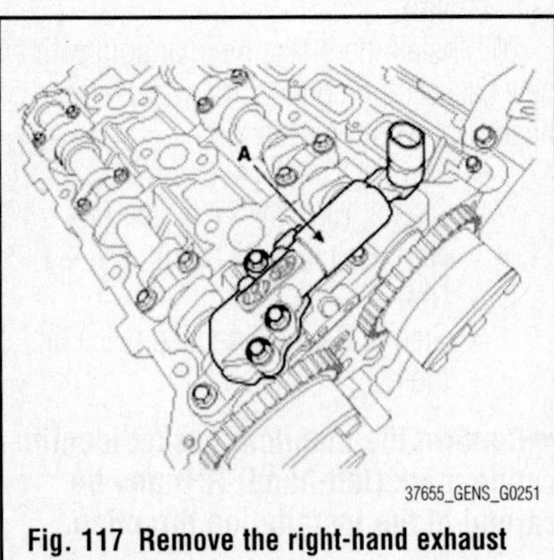

Fig. 117 Remove the right-hand exhaust camshaft OCV (A)

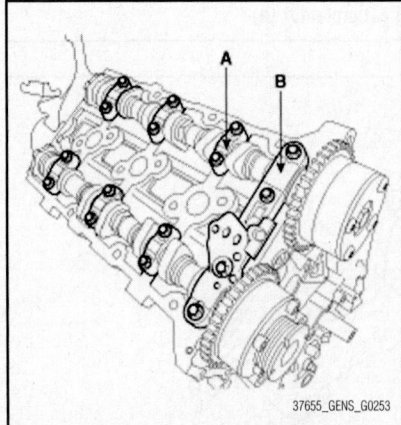

A
B
37655_GENS_G0253

Fig. 119 Remove the right-hand camshaft bearing cap (A) and thrust bearing cap (B)

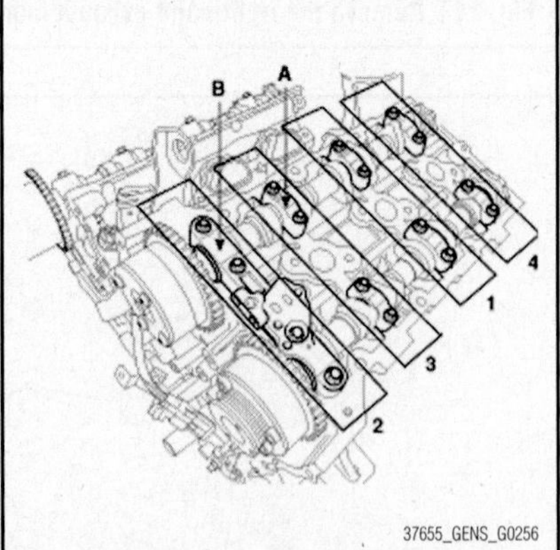

Fig. 122 Left-hand camshaft bearing cap (A) and thrust bearing cap (B) installation sequence

a. After assembling the fuel line, turn on the ignition switch (do not operate the starter) so that the fuel pump runs for approximately two seconds and fuel line pressurizes.

b. Repeat this operation two or three times, then check for fuel leakage at any point in the fuel lines.

27. Refill radiator and reservoir tank with engine coolant.

28. Bleed air from the cooling system:

a. Start engine and let it run until it warms up. (Until the radiator fan operates 3 or 4 times.)

b. Turn Off the engine. Check the level in the radiator, add coolant if needed. This will allow trapped air to be removed from the cooling system.

c. Put radiator cap on tightly, then run the engine again and check for leaks.

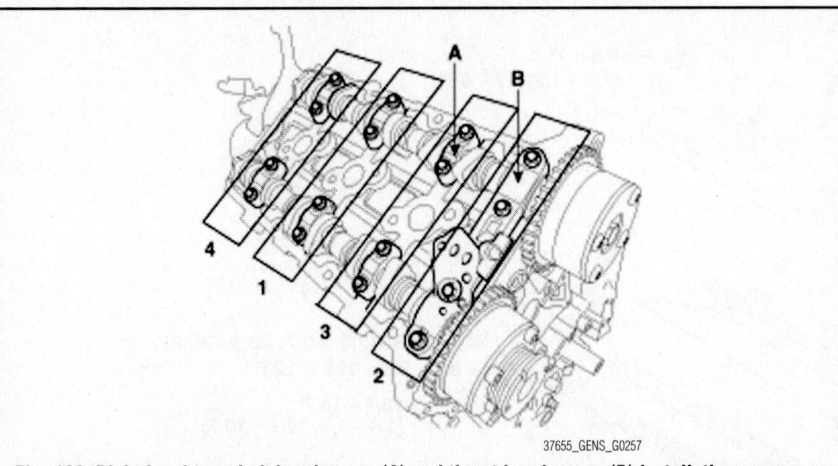

Fig. 123 Right-hand camshaft bearing cap (A) and thrust bearing cap (B) installation sequence

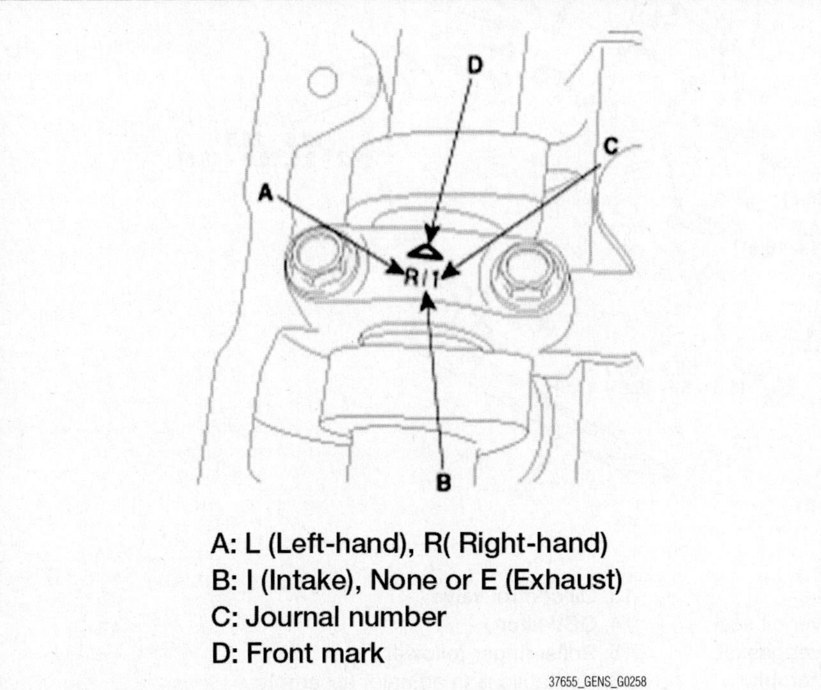

Fig. 124 Position the camshaft bearing cap according to its markings

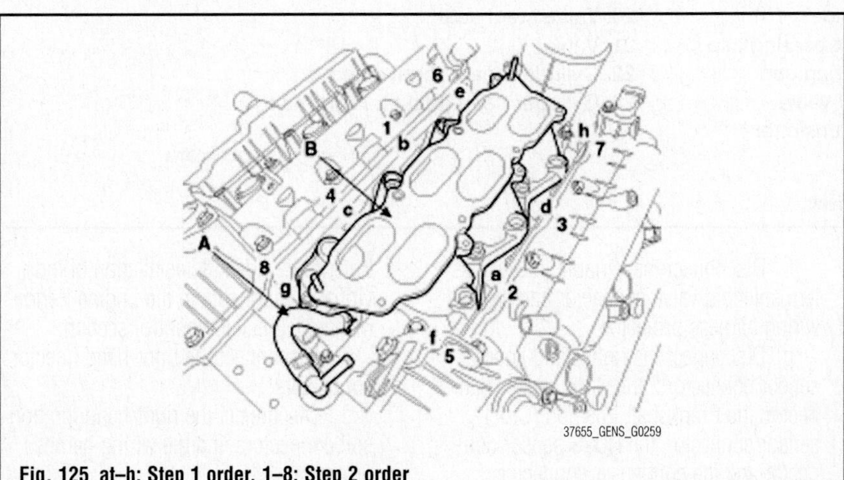

Fig. 125 at–h: Step 1 order, 1–8: Step 2 order

4.6L Engines

See Figures 126 through 160.

1. Before servicing the vehicle, refer to the Precautions Section.
2. Disconnect the negative battery cable.

⁂ WARNING

Use fender covers to avoid damaging painted surfaces. To avoid damage, unplug the wiring connectors carefully while holding the connector portion. Mark all wiring and hoses to avoid misconnection.

⁂ WARNING

To avoid damaging the cylinder head, wait until the engine coolant temperature drops below normal temperature (68° F [20° C]) before removing it.

➡Turn the crankshaft pulley so that the No. 1 piston is at top dead center.

3. Remove the engine cover.
4. Remove the engine side cover.
5. Remove the drain plug and drain the engine coolant.
6. Remove the air duct, and the air cleaner assembly.
7. Disconnect the radiator upper hose and lower hose.
8. Remove the fuel hose, the Purge Control Solenoid Valve (PCSV) hose and the brake booster vacuum hose.
9. Disconnect the heater hoses.
10. Remove the radiator and cooling fan assembly. Refer to Radiator Removal & Installation in the Engine Cooling section.
11. Turn the tensioner clockwise and loosen, then remove the drive belt. Refer to Accessory Drive Belt Removal & Installation.
12. Remove the drive belt tensioner.
13. Remove the engine wiring harness from the cylinder head.
    a. Disconnect the water temperature sensor connector and the CVVT oil control valve connectors.
    b. Disconnect the ETC module connector.
    c. Disconnect the PCSV connector and the left-hand injector connectors.
    d. Disconnect the left-hand ignition coil connectors and the wiring harness protector.

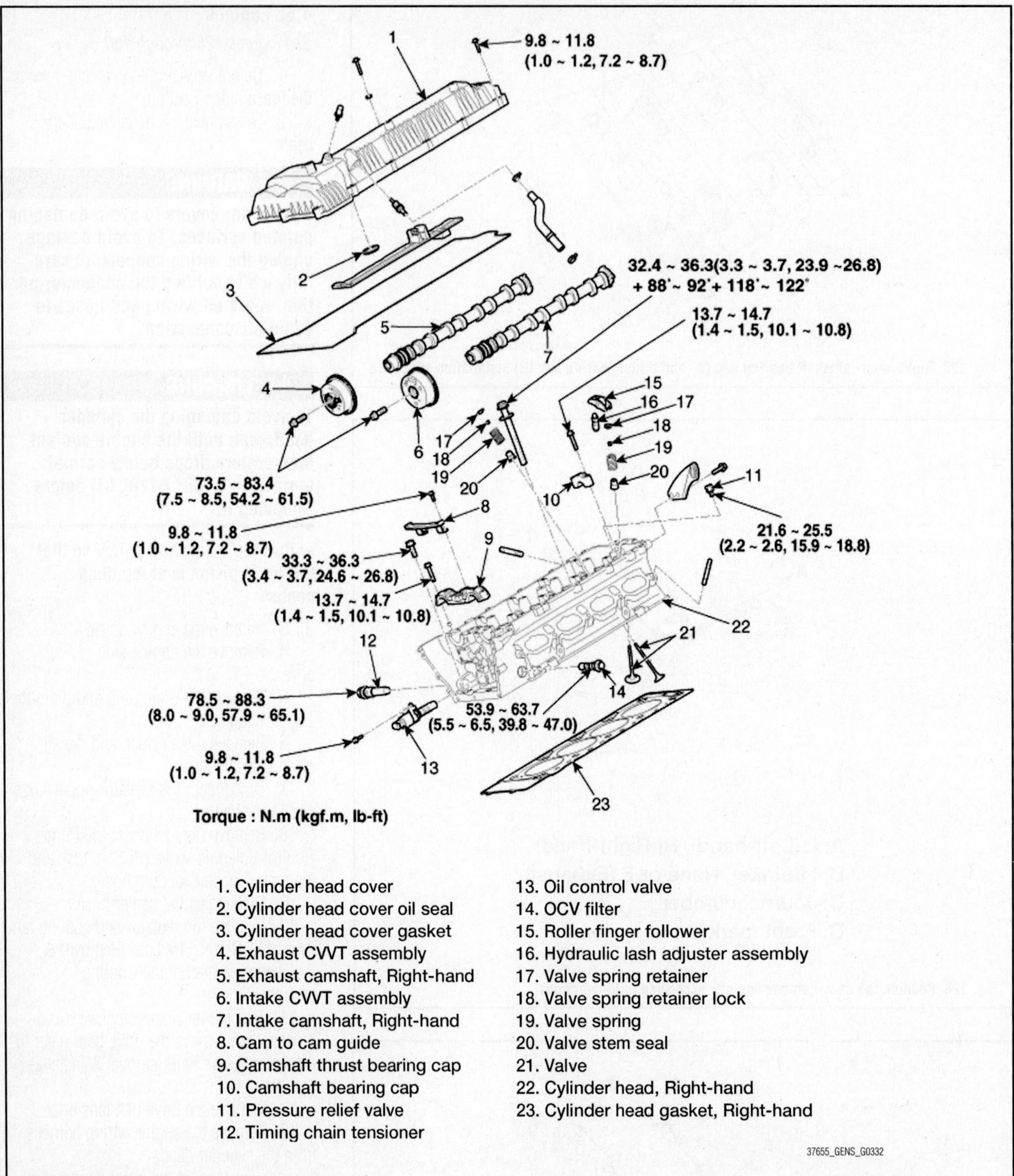

Fig. 126 Right-hand camshafts and related components

e. Disconnect the left-hand knock sensor connectors, the Camshaft Position (CMP) sensor connectors, the Oxygen sensor (HO2S) connector, and the condenser connector. Refer to Removal & Installation of individual components in the Engine Performance & Emission control section.

f. Disconnect the variable intake system solenoid valve connector and the wiring harness protector.

g. Disconnect the right-hand knock sensor connectors, the CMP sensor connector, the Crankshaft Position (CKP) sensor connector, the HO2S sensor connector and the condenser connector. Refer to Removal & Installation of individual components in the Engine Performance & Emission control section.

h. Disconnect the right-hand injector connectors.

i. Disconnect the right-hand ignition coil connectors and the wiring harness protector.

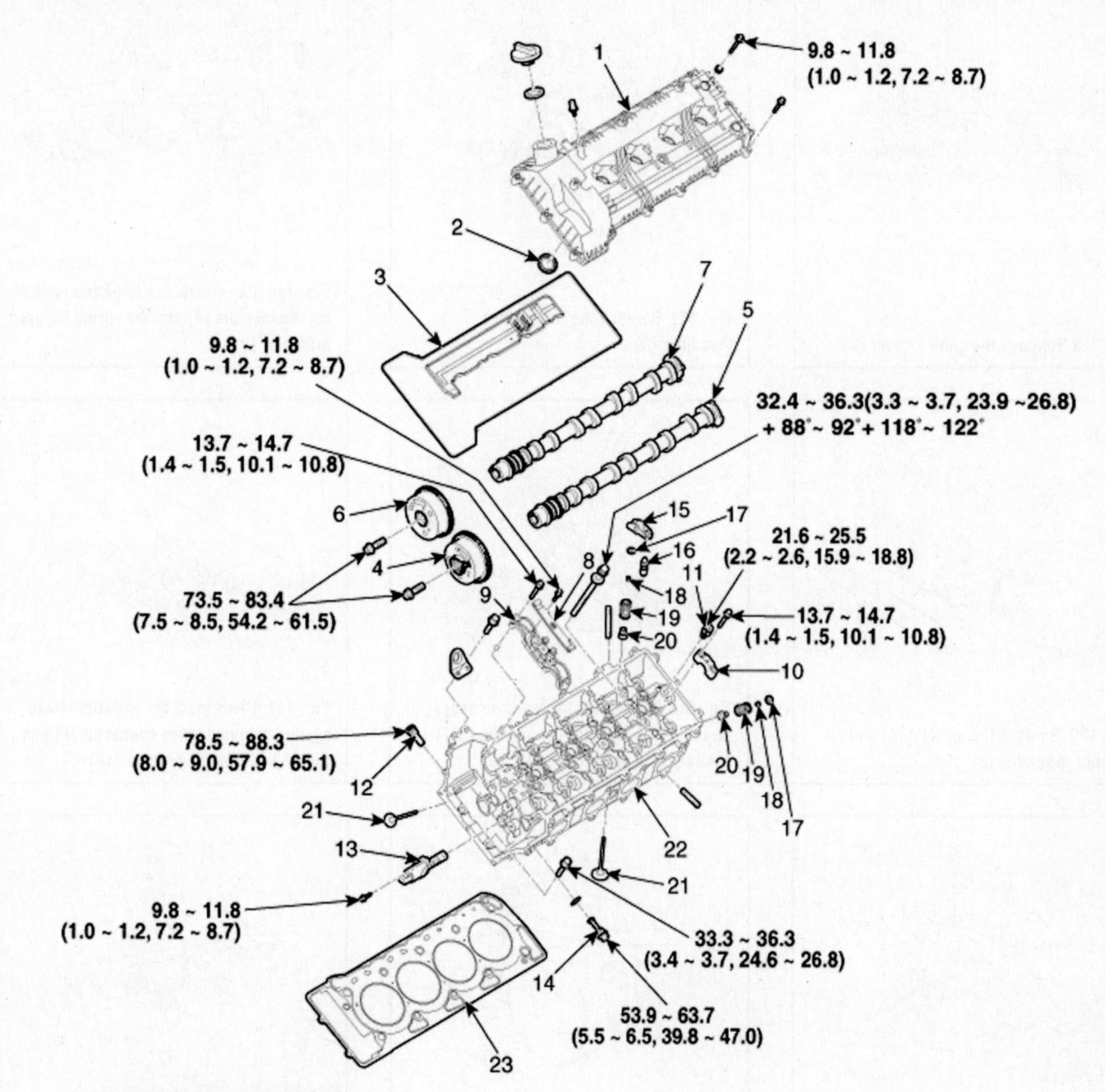

Fig. 127 Left-hand camshafts and related components

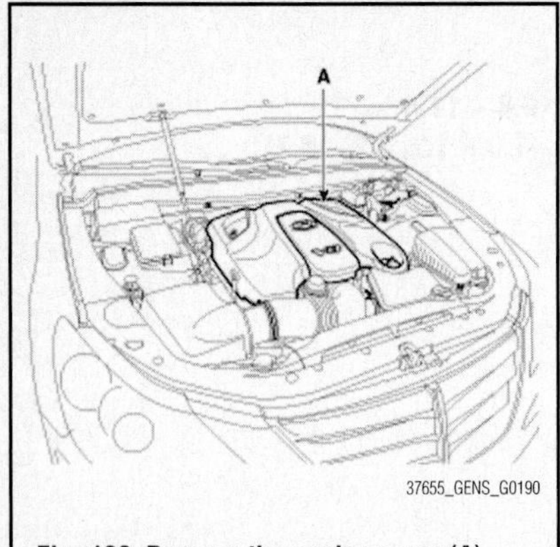

Fig. 128 Remove the engine cover (A)

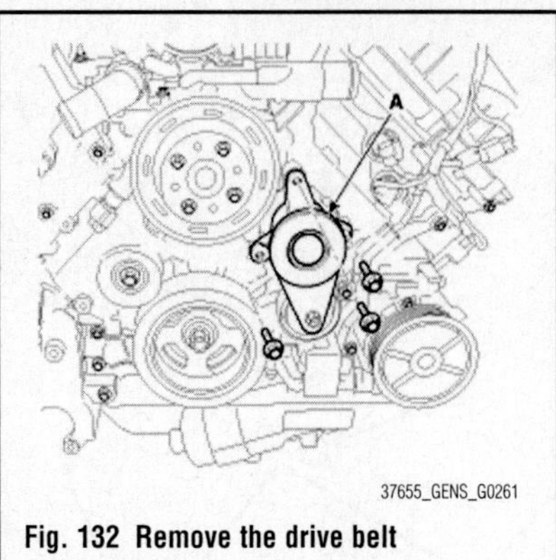

Fig. 132 Remove the drive belt tensioner (A)

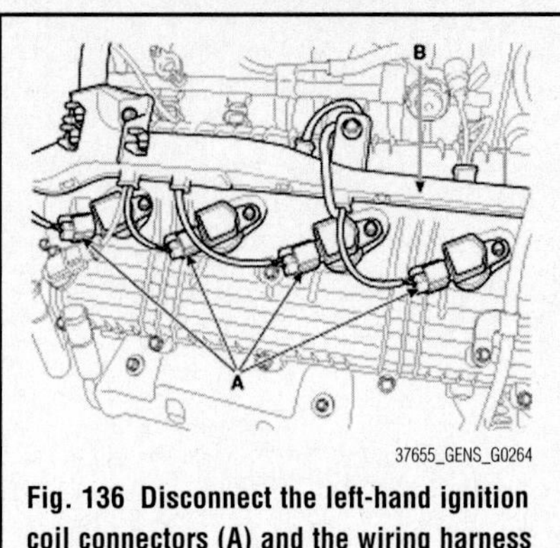

Fig. 136 Disconnect the left-hand ignition coil connectors (A) and the wiring harness protector (B)

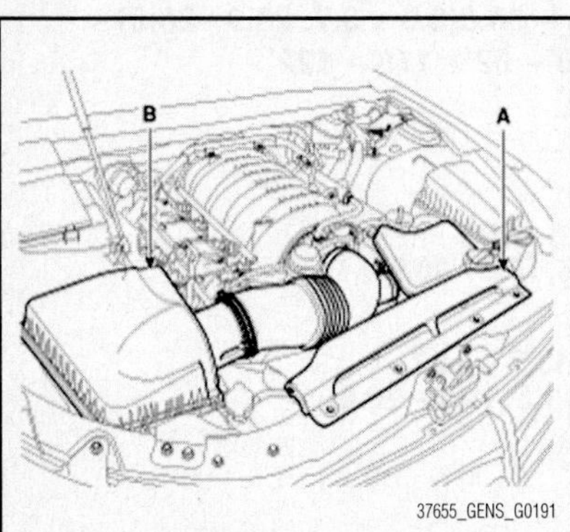

Fig. 129 Remove the air duct (A) and air cleaner assembly (B)

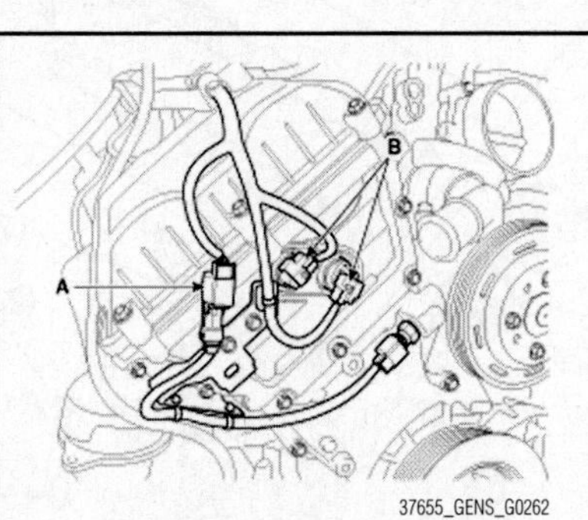

Fig. 133 Disconnect the water temperature sensor connector (A) and the CVVT oil control valve connectors (B)

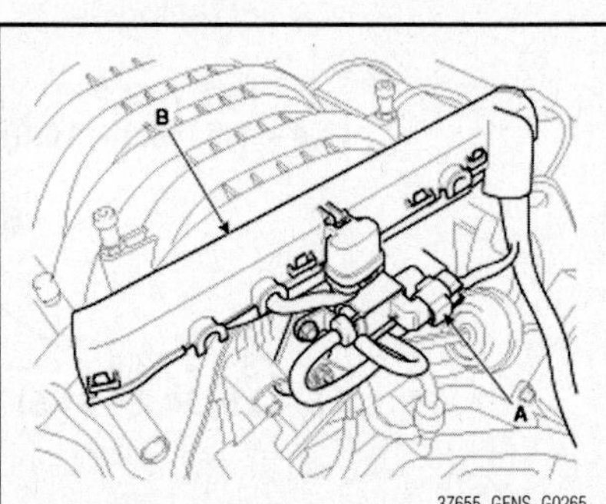

Fig. 137 Disconnect the variable intake system solenoid valve connector (A) and the wiring harness protector (B)

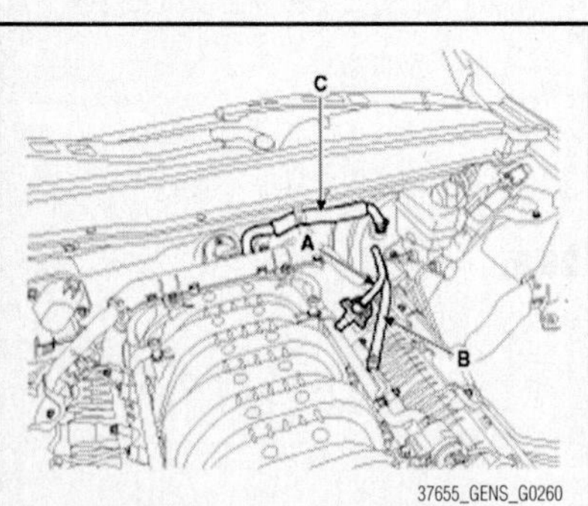

Fig. 130 Remove the fuel hose (A), the PCSV hose (B), and the brake booster vacuum hose (C)

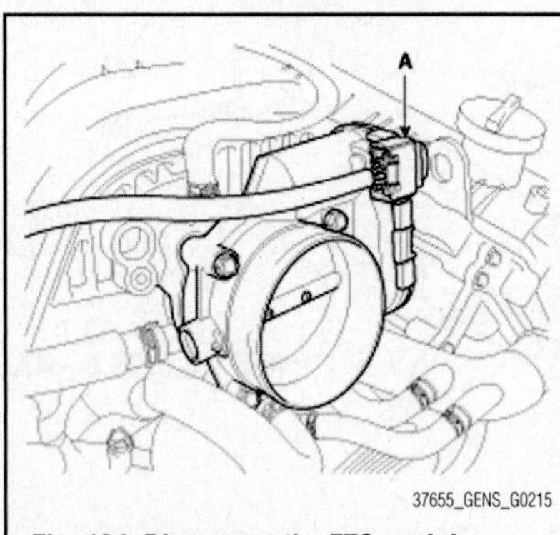

Fig. 134 Disconnect the ETC module connector (A)

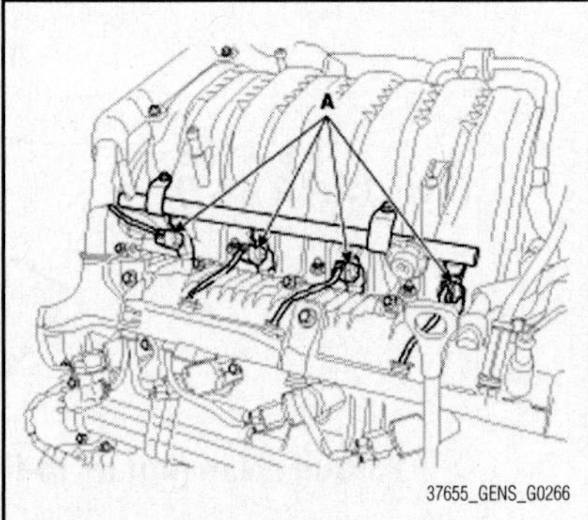

Fig. 138 Disconnect the right-hand injector connectors (A)

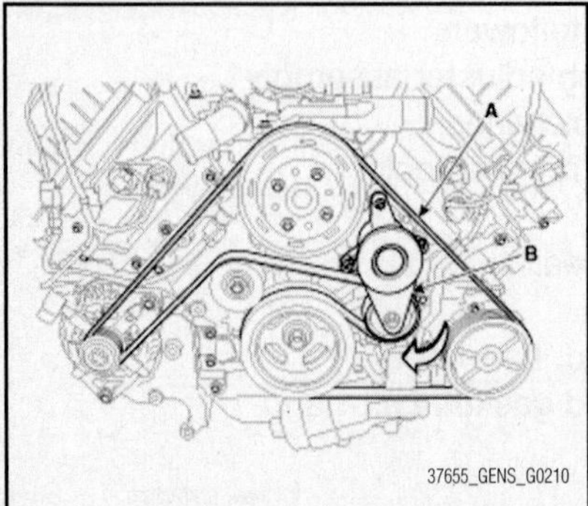

Fig. 131 Turn the tensioner (B) clockwise and loosen, then remove the drive belt (A)

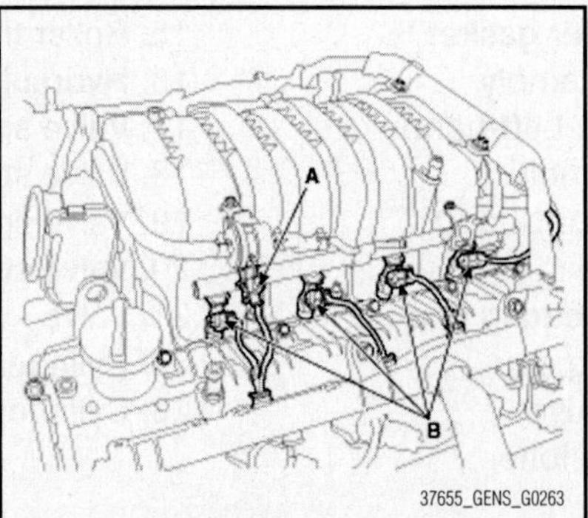

Fig. 135 Disconnect the PCSV connector (A) and the left-hand injector connectors (B)

j. Disconnect the oil pressure switch connector, the CVVT oil control valve connectors and the air conditioner compressor switch connector.

14. Remove the oil level gauge.

15. Disconnect the HO2S connector and remove the oxygen sensor connector bracket. Refer to Heated Oxygen Sensor Removal & Installation in the Engine Performance & Emission control section.

16. Remove the exhaust manifold heat protector. Refer to Exhaust Manifold Removal & Installation.

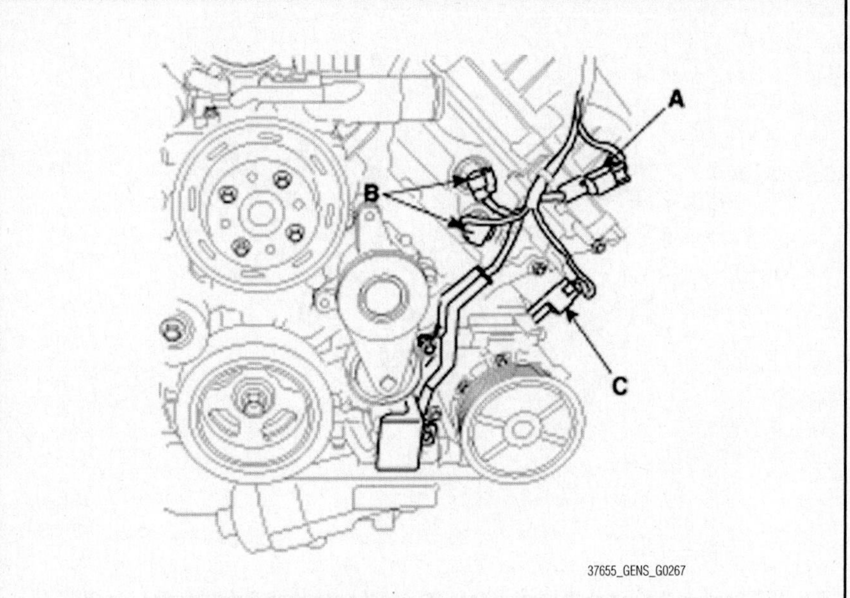

Fig. 139 Disconnect the oil pressure switch connector (A) the CVVT oil control valve connectors (B), and the air conditioner compressor switch connector (C)

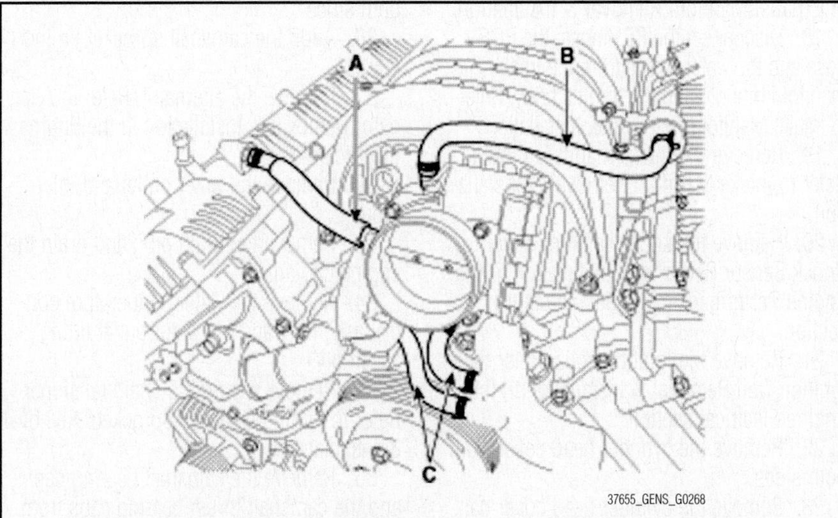

Fig. 140 Disconnect the PCV hose (A) the PCSV hose (B), and the water hoses (C) from the intake manifold module and the throttle body

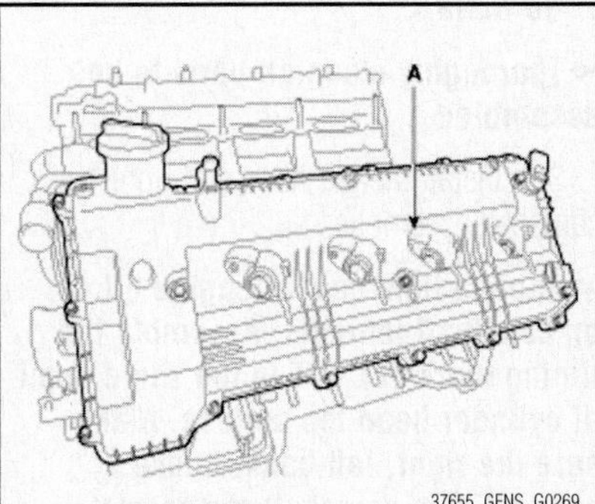

Fig. 141 Remove the cylinder head cover (A)—Left-hand side shown, right-hand side similar

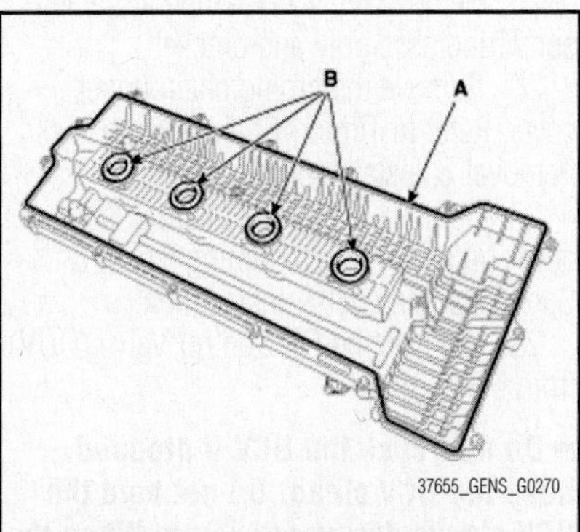

Fig. 142 Remove the cylinder head cover gasket (A) and oil seal (B))—Left-hand side shown, right-hand side similar

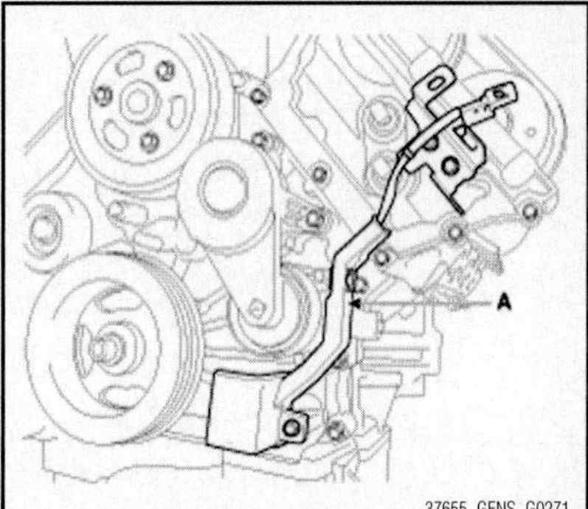

Fig. 143 Remove the oil pressure switch wiring assembly and bracket (A)

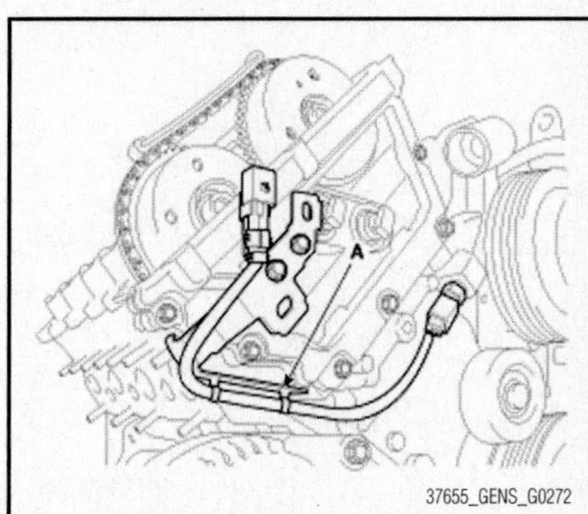

Fig. 144 Remove the water temperature sensor wiring assembly and bracket (A)

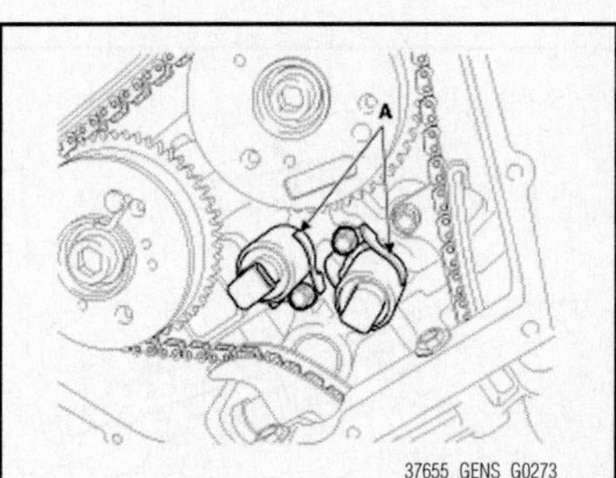

Fig. 145 Remove the oil control valve (A)—Right-hand side shown, left-hand side similar

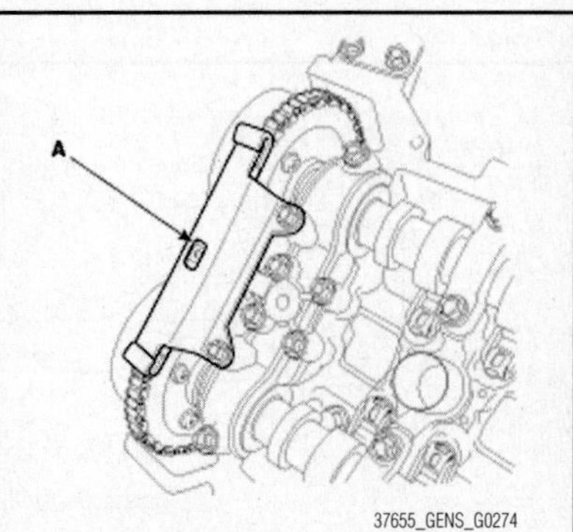

Fig. 146 Remove the cam to cam guide (A)—Left-hand side shown, right-hand side similar

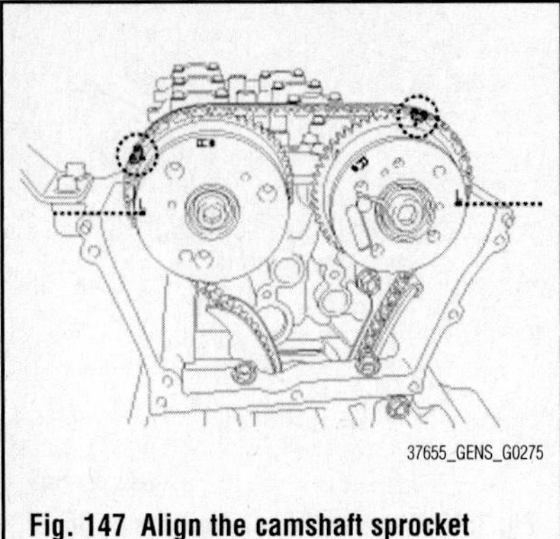

Fig. 147 Align the camshaft sprocket timing marks—Left-hand side

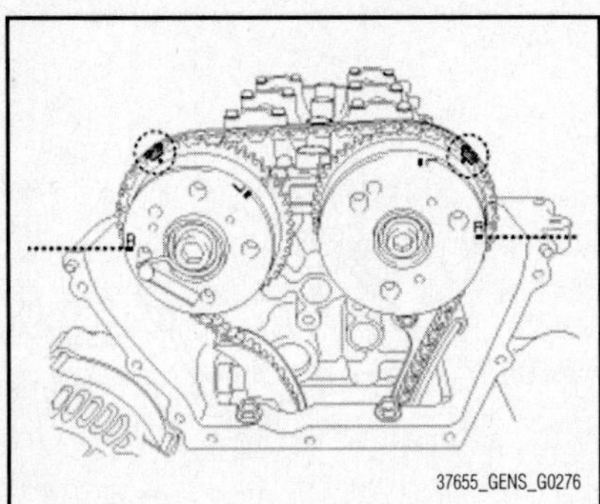

Fig. 148 Align the camshaft sprocket timing marks—Right-hand side

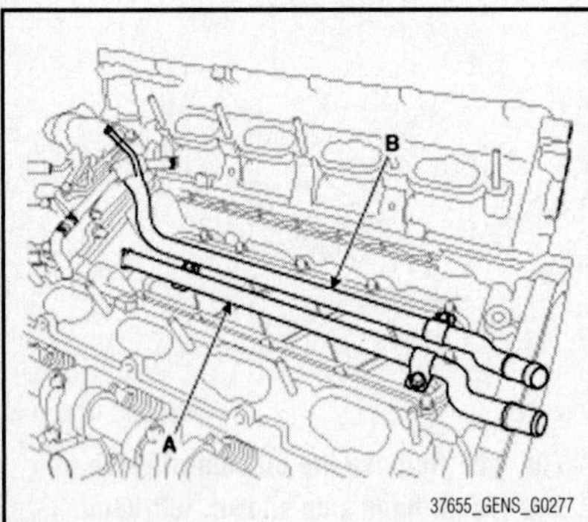

Fig. 149 Remove the water outlet (A) and inlet (B) pipes

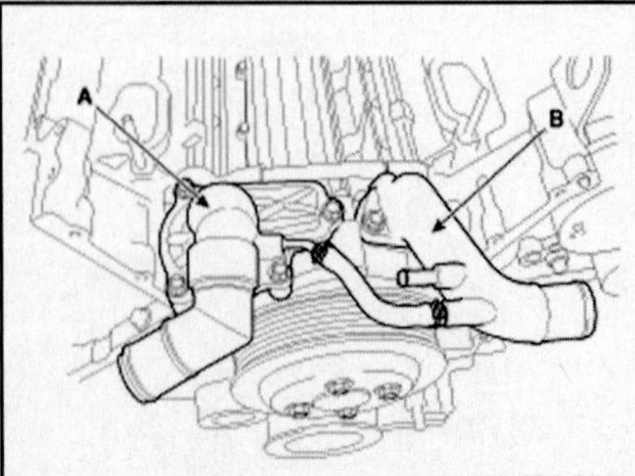

Fig. 150 Remove the water temperature control assembly (A) and the water outlet fitting assembly (B)

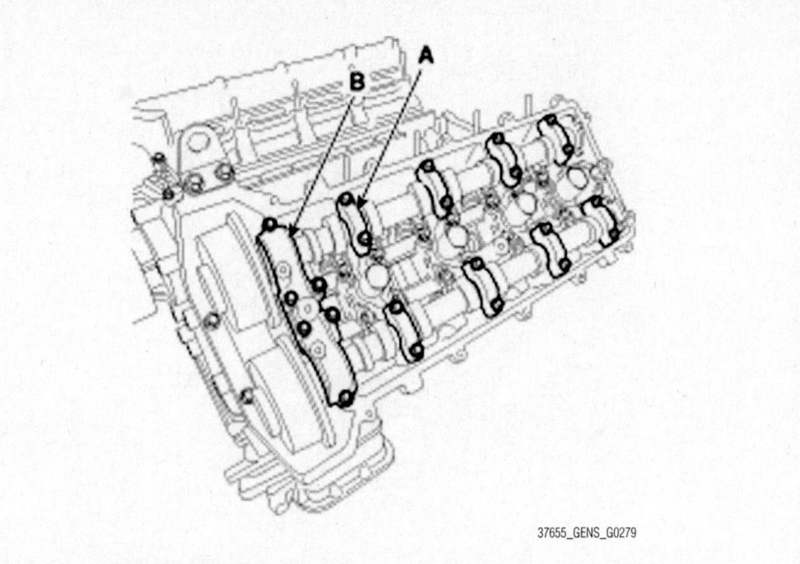

Fig. 151 Remove the camshaft bearing caps (A) and the camshaft thrust bearing cap (B)—Left-hand side shown, right-hand side similar

17. Remove the exhaust manifold. Refer to Exhaust Manifold Removal & Installation.
18. Disconnect the PCV hose, the PCSV hose and the water hoses from the intake manifold module and the throttle body. Refer to Intake Manifold Removal & Installation.
19. Remove the intake manifold module. Refer to Intake Manifold Removal & Installation.
20. Remove the knock sensors. Refer to Knock Sensor Removal & Installation in the Engine Performance & Emission control section.
21. Remove the ignition coils. Refer to Ignition Coil Removal & Installation in the Engine Electrical section.
22. Remove the cylinder head cover from both sides.
23. Remove the cylinder head cover gasket and oil seal from both sides.
24. Remove the Oil Pressure Switch (OPS) wiring assembly and bracket.
25. Remove the water temperature sensor wiring assembly and bracket.
26. Remove the timing chain upper cover. Refer to Timing Chain & Sprockets Removal & Installation.
27. Remove the timing chain upper cover oil seal. Refer to Timing Chain & Sprockets Removal & Installation.
28. Remove the Oil Control Valve (OCV) from both sides.

➡Do not reuse the OCV if dropped. Keep the OCV clean. Do not hold the OCV sleeve during servicing. When the OCV is installed on the engine, do not move the engine while holding the OCV yoke.

29. Remove the cam to cam guide from both sides.
30. Align the camshaft sprocket timing marks.
31. Remove the alternator. Refer to Alternator Removal & Installation in the Engine Electrical section.
32. Remove the water outlet and inlet pipes.
33. Remove the drain plug and drain the engine coolant.
34. Remove the water temperature control assembly and the water outlet fitting assembly.
35. Remove the timing chain tensioner. Refer to Timing Chain & Sprockets Removal & Installation.
36. Remove the camshaft bearing caps and the camshaft thrust bearing caps from both sides.
37. Remove the camshaft from both sides.

To install:

➡Thoroughly clean all parts to be assembled.

38. Install the left-hand and right-hand camshaft assemblies.

➡Apply a light coat of engine oil on camshaft journals. Assemble the timing mark of CVVT to the same level of cylinder head top surface. Make sure the right, left bank, intake, exhaust side camshaft are correct before assembling. Turn the crankshaft to set the No.1 piston at TDC position.

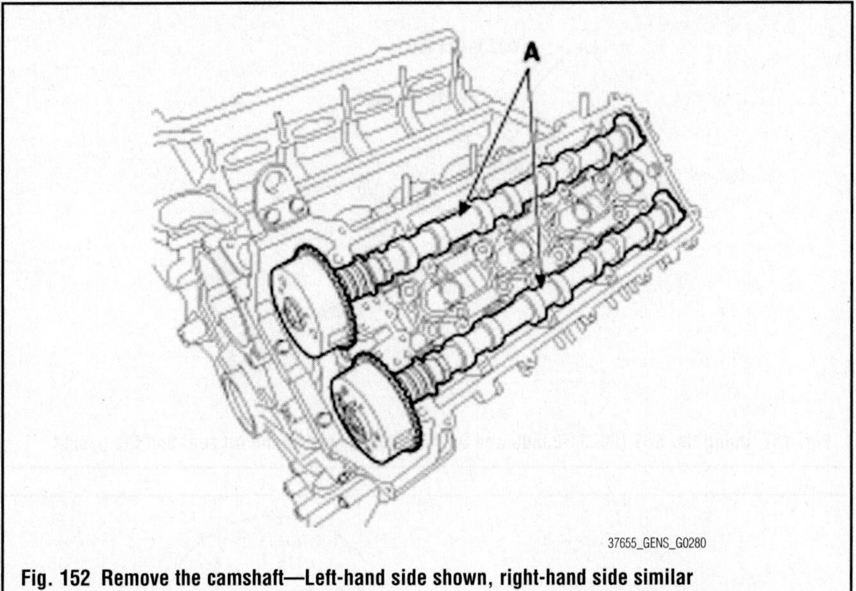

Fig. 152 Remove the camshaft—Left-hand side shown, right-hand side similar

39. Install the timing chain to the CVVT sprocket.

➡**Make sure the crankshaft is at TDC position.**

a. Align the chain installation mark of the crankshaft sprocket and the timing mark of the chain link, and then install the chain.

➡**Align the chain installation mark of the sprocket and the timing marked chain links for reassembly.**

b. To install the timing chain with no slack between each shaft (cam, crank), follow the below procedure:

- Crankshaft sprocket, then timing chain guide, then exhaust CVVT sprocket, then intake CVVT sprocket. (Left-hand Bank)
- Crankshaft sprocket, then timing chain guide, then intake CVVT sprocket, then exhaust CVVT sprocket. (Right-hand Bank)
- When reassembling the timing chain, make sure all of the chain installation marks of the sprocket and the timing marked chain links are aligned.

40. Install the camshaft bearing caps and the camshaft thrust bearing cap in the sequence shown. Tightening torque:

- 1st step: 61–70 inch lbs. (7–8 Nm)
- 2nd step: 10–11 ft. lbs. (14–15 Nm)

➡**Be careful of the right, left bank, intake, exhaust side before assembling.**

➡**Be sure to install the thrust bearing cap and bearing cap bolts in the correct places.**

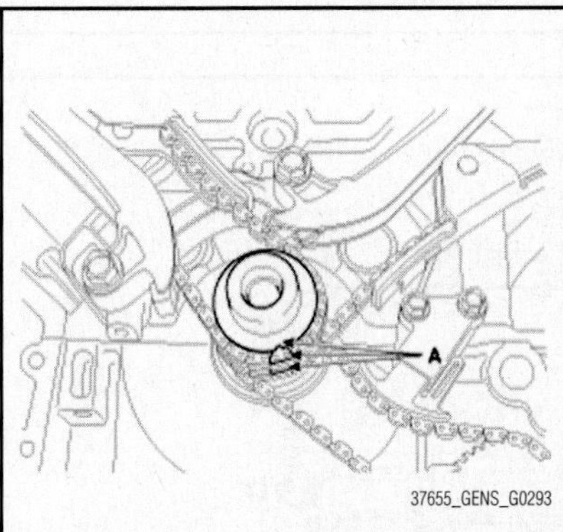

Fig. 153 Align the chain installation mark (A) of the crankshaft sprocket and the timing mark (A) of the chain link

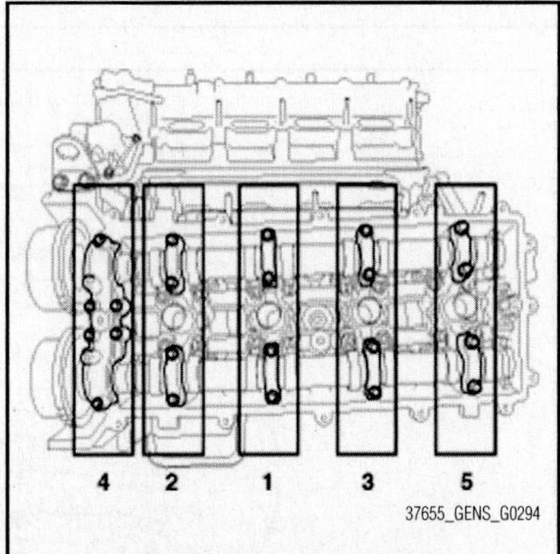

Fig. 154 Camshaft bearing caps and camshaft thrust bearing cap tightening sequence—Left-hand side

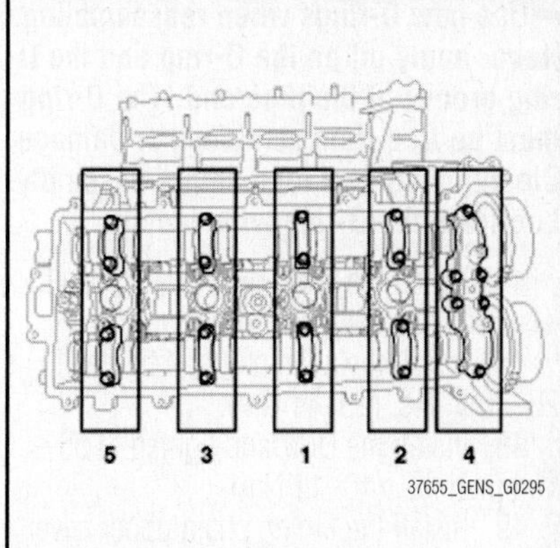

Fig. 155 Camshaft bearing caps and camshaft thrust bearing cap tightening sequence—Right-hand side

41. Install the cam-to-cam guide to both sides, and tighten to 86–104 inch lbs. (10–12 Nm).

42. Install the timing chain guide bolts and tensioners. Refer to Timing Chain & Sprockets Removal & Installation.

43. Set the timing chain tension. Refer to Timing Chain & Sprockets Removal & Installation.

44. Install the water temperature control assembly and the water outlet fitting assembly with new O-ring and new gasket, and tighten to 12–15 ft. lbs. (17–20 Nm).

45. Install the water outlet and inlet pipes and tighten to 86–104 inch lbs. (10–12 Nm).

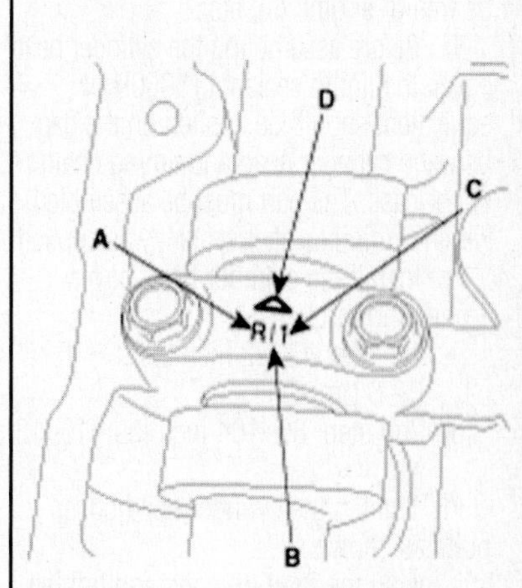

Fig. 156 Position the camshaft bearing cap according to its markings

➡Use new O-rings when reassembling. Never apply oil on the O-ring and the O-ring groove of the pipe end. The O-ring must be free from scratches or damage. Clean the contact face before assembly. Lubricate the O-ring with coolant.

46. Install the drain plug and tighten to 71–74 ft. lbs. (96–100 Nm).
47. Install the alternator and tighten to 22–30 ft. lbs. (29–41 Nm).
48. Install the OCV and tighten to 86–104 inch lbs. (10–12 Nm).
49. Install the timing chain upper cover and seal. Refer to Timing Chain & Sprockets Removal & Installation.
50. Install the exhaust manifold, gasket, and heat protector. Refer to Exhaust Manifold Removal & Installation.
51. Install the Engine Coolant Temperature Sensor (ECTS) wiring assembly and bracket. Tightening torque:
    - Wiring bracket bolts: 86–104 inch lbs. (10–12 Nm)
    - Engine coolant temperature sensor: 15–29 ft. lbs. (20–39 Nm)
52. Install the oil pressure switch wiring assembly and bracket. Tightening torque:
    - Wiring bracket bolts: 86–104 inch lbs. (10–12 Nm)
    - Oil pressure switch: 11–16 ft. lbs. (15–22 Nm)
53. Using the SST (09231-2J500 and 09231-H1100), install the oil seal and the gasket to both sides.
54. Install the cylinder head cover.
    a. The sealant locations on cylinder head and timing chain upper case must be free of engine oil, etc.
    b. Before assembling the cylinder head cover, the liquid sealant LT5900H or equivalent should be applied on the gap between cylinder head and timing chain upper case. The part must be assembled within 5 minutes after sealant was applied.
    c. Install the cylinder head cover.
    Tightening torque:
    - 1st step: 43–52 inch lbs. (5–6 Nm)
    - 2nd step: 86–104 inch lbs. (10–12 Nm)

    d. Tighten the cylinder head cover bolts as shown.
55. Install the ignition coils and tighten to 86–104 inch lbs. (10–12 Nm).
56. Reconnect the water hose.
57. Install the knock sensors and wiring assembly bracket and tighten to 48–56 ft. lbs. (65–77 Nm).
58. Install the intake manifold module. Refer to Intake Manifold Removal & Installation.

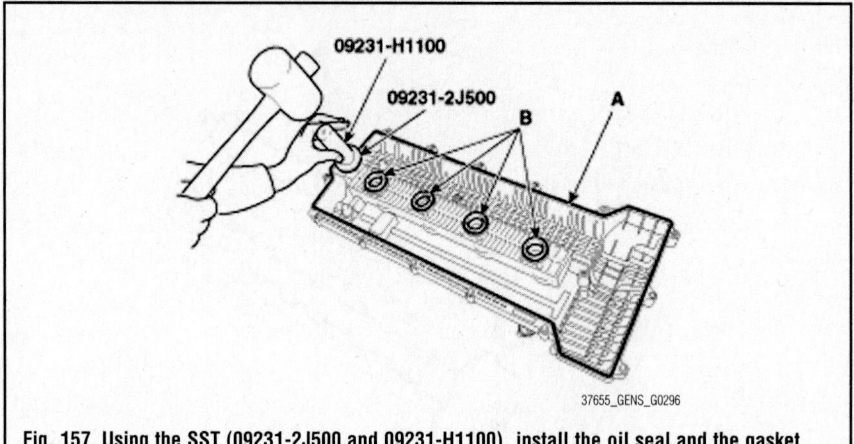

Fig. 157 Using the SST (09231-2J500 and 09231-H1100), install the oil seal and the gasket

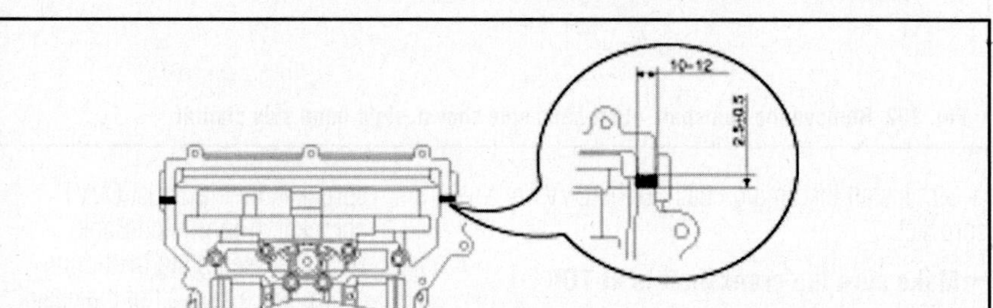

Fig. 158 Cylinder head cover sealant application points

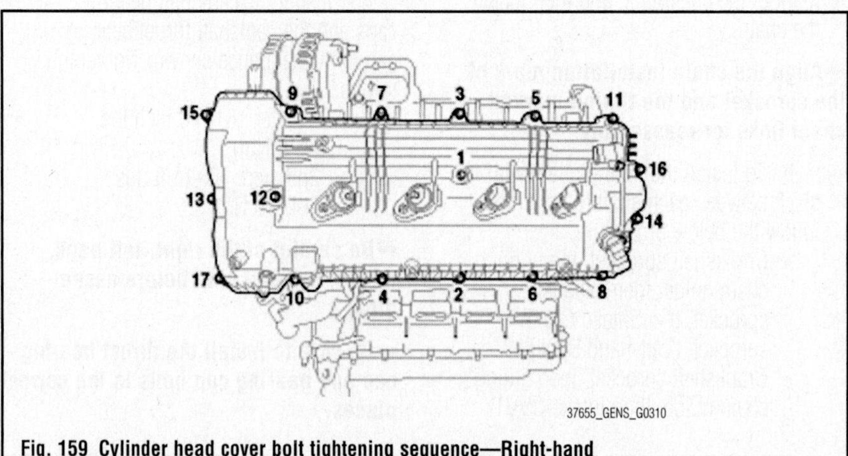

Fig. 159 Cylinder head cover bolt tightening sequence—Right-hand

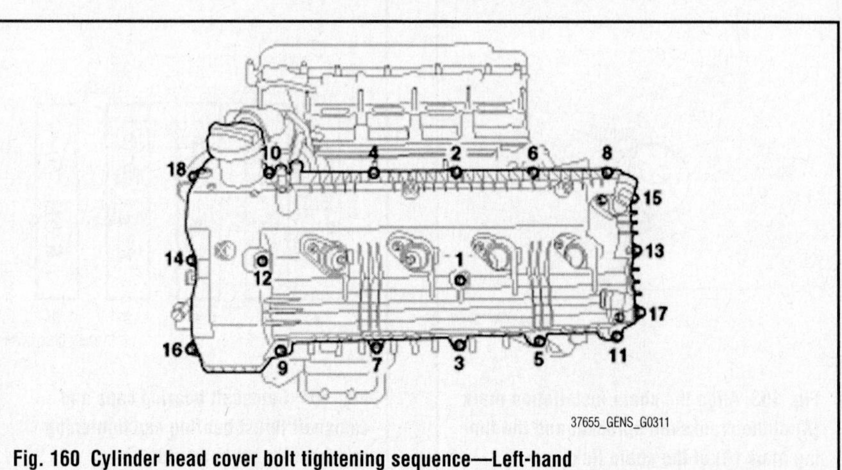

Fig. 160 Cylinder head cover bolt tightening sequence—Left-hand

59. Reconnect the PCV hose to the intake manifold module.
60. Install the HO2S connector bracket.
61. Install the oil level gauge and tighten to 86–104 inch lbs. (10–12 Nm).
62. Install the engine wiring harness to the cylinder head.
   a. Reconnect the Engine Coolant Temperature (ECT) sensor connector and the CVVT oil control valve (OCV) connectors.
   b. Reconnect the alternator connector.
   c. Reconnect the ETC module connector.
   d. Reconnect the PCSV connector and the injector connectors.
   e. Reconnect the ignition coil connectors and the wiring harness protector.
   f. Reconnect the knock sensor connectors, the CMP sensor connectors, the CKP sensor connector, the HO2S connector, the condenser connector and the ground line.
   g. Reconnect the Variable Intake System (VIS) solenoid valve connector and the wiring harness protector.
   h. Reconnect the injector connectors.
   i. Reconnect the Oil Pressure Switch (OPS) connector and the air conditioner compressor switch connector.
63. Install the drive belt tensioner and drive belt.
64. Install the radiator and cooling fan assembly.
65. Connect the heater hoses.
66. Connect the fuel hose, the PCSV hose and the brake booster vacuum hose.
67. Connect the radiator upper and lower hoses.
68. Install the air cleaner assembly.
69. Install the engine cover.
70. Connect the battery negative terminal.
71. Refill the engine oil.
72. Clean the battery posts and cable terminals with sandpaper, assemble them, and then apply grease to prevent corrosion.
73. Inspect for fuel leakage:
   a. After assembling the fuel line, turn on the ignition switch (do not operate the starter) so that the fuel pump runs for approximately two seconds and fuel line pressurizes.
   b. Repeat this operation two or three times, and then check for fuel leakage at any point in the fuel lines.
74. Refill the radiator and reservoir tank with engine coolant.
75. Bleed air from the cooling system:
   a. Start the engine and let it run until it warms up. (Until the radiator fan operates 3 or 4 times.)
   b. Turn Off the engine. Check the level in the radiator, add coolant if needed. This will allow trapped air to be removed from the cooling system.
   c. Put the radiator cap on tightly, then run the engine again and check for leaks.

COMBINATION MANIFOLD

REMOVAL & INSTALLATION

3.8L Engine

See Figure 161.

1. Before servicing the vehicle, refer to the Precautions Section.
2. Remove the left-hand/right-hand exhaust manifolds.

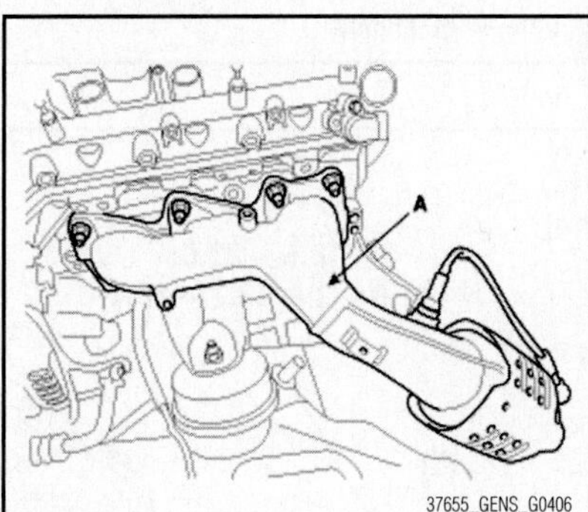

Fig. 161 Left-hand/right-hand exhaust manifolds

To install:

3. Installation is reverse of removal.

4.6L engine

See Figure 162.

1. Before servicing the vehicle, refer to the Precautions Section.
2. Remove the left-hand/right-hand exhaust manifolds.

To install:

3. Installation is reverse of removal.

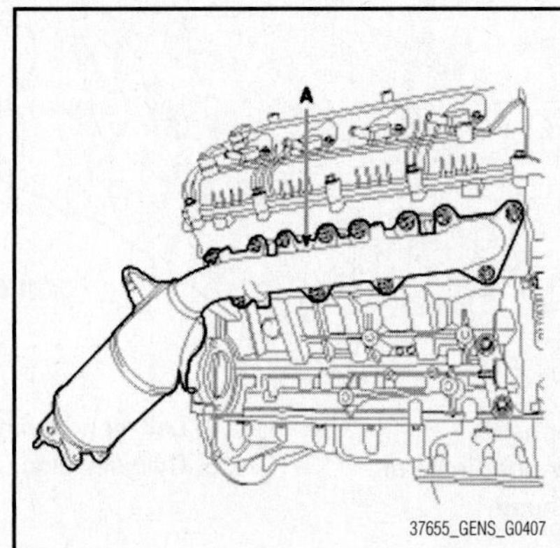

Fig. 162 Left-hand/right-hand exhaust manifolds

CRANKSHAFT FRONT SEAL

REMOVAL & INSTALLATION

3.8L Engine

See Figures 163 and 164.

1. Before servicing the vehicle, refer to the Precautions Section.
2. Remove the accessory drive belts. Refer to Accessory Drive Belt Removal & Installation.
3. Remove the crankshaft pulley bolt and pulley.
4. Using seal removal tool, remove the front crankshaft seal.

To install:

5. Using seal installation tool, install the front crankshaft seal.
6. Installation is reverse of removal.
   a. Tighten crankshaft bolt to 210–224 ft. lbs. (284–304 Nm).

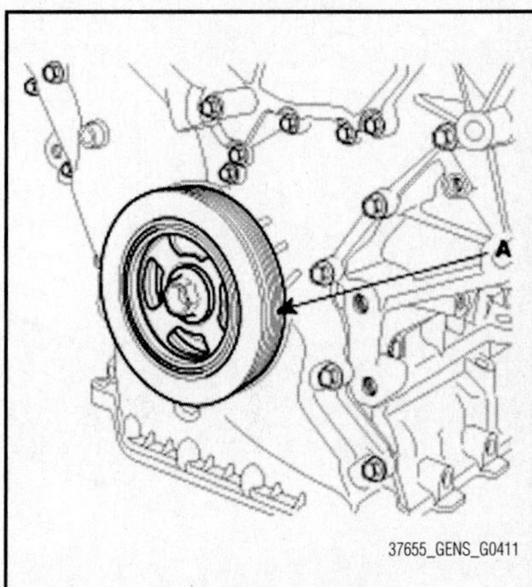

Fig. 163 Crankshaft pulley removal—3.8L engine

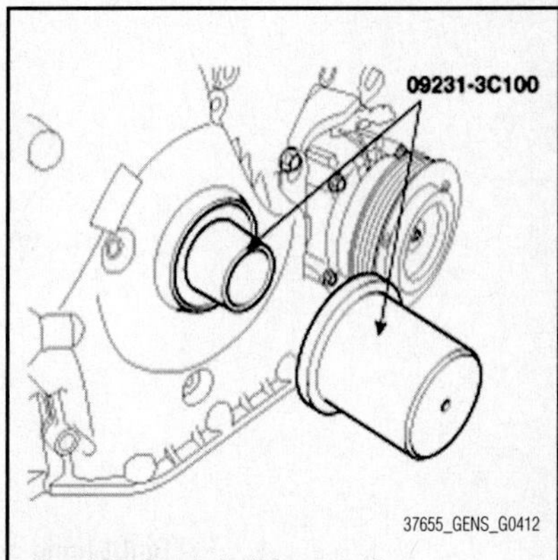

Fig. 164 Crankshaft front seal installation—3.8L engine

4.6L Engine

See Figures 165 and 166.

1. Before servicing the vehicle, refer to the Precautions Section.

A
37655_GENS_G0409

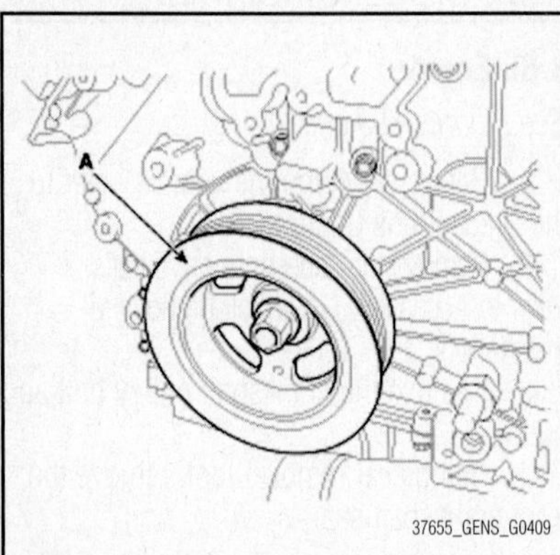

Fig. 165 Crankshaft pulley removal—4.6L engine

2. Remove the accessory drive belts. Refer to Accessory Drive Belt Removal & Installation.

A
09231-2J300
09231-H1100
37655_GENS_G0410

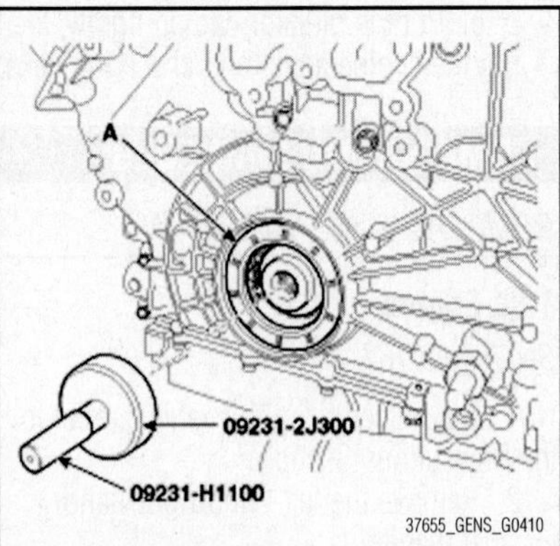

Fig. 166 Crankshaft front seal installation—4.6L engine

3. Remove the crankshaft pulley bolt and pulley.

4. Using seal removal tool, remove the front crankshaft seal.

To install:

5. Using seal installation tool, install the front crankshaft seal.

6. Installation of remaining components is reverse of removal.

a. Tighten crankshaft bolt to 289–296 ft. lbs. (392–402 Nm).

CYLINDER HEAD

REMOVAL & INSTALLATION

3.8L Engine

See Figures 167 through 193.

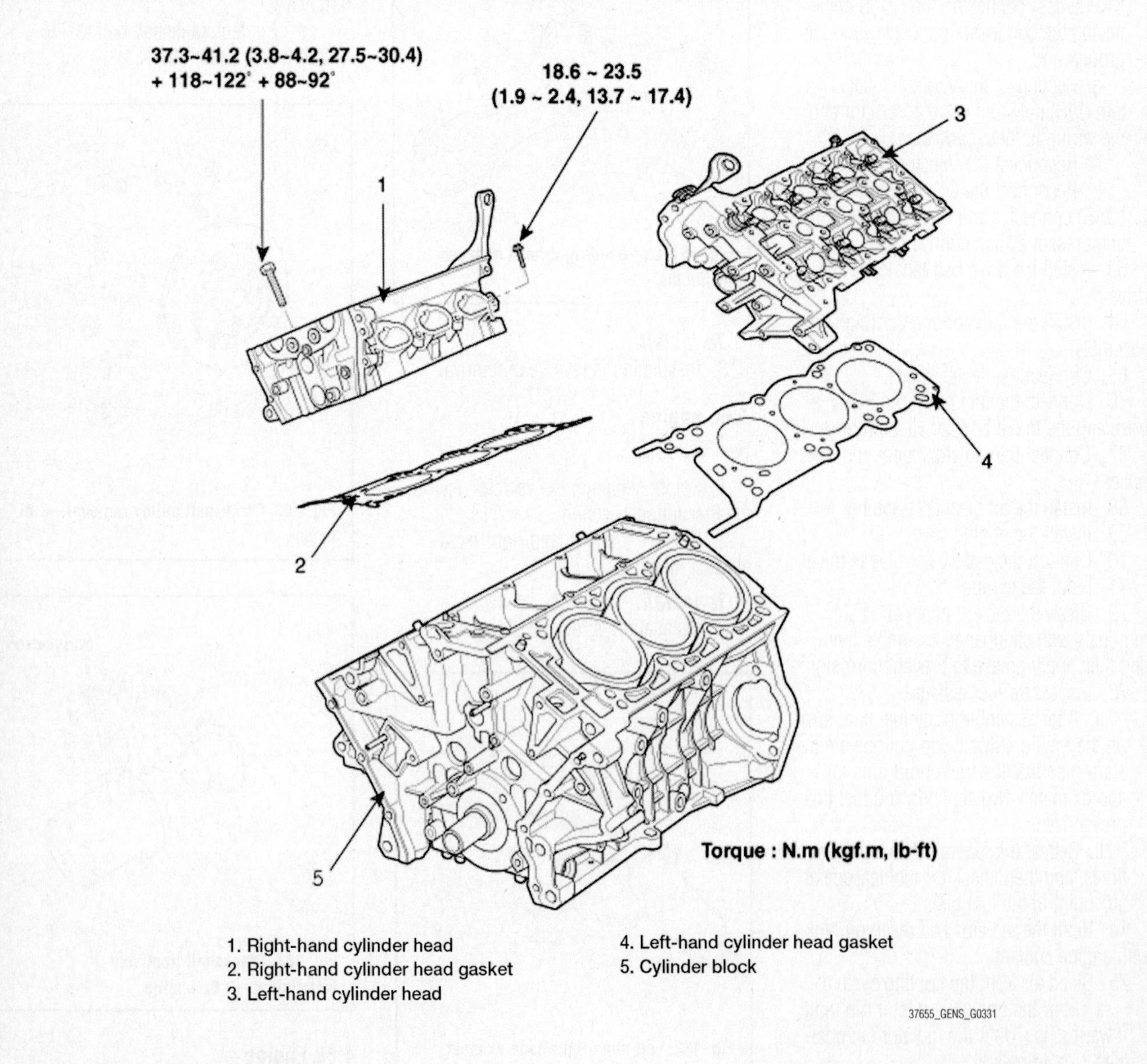

37.3~41.2 (3.8~4.2, 27.5~30.4)
+ 118~122° + 88~92°
18.6 ~ 23.5
(1.9 ~ 2.4, 13.7 ~ 17.4)
1
2
3
4
5
Torque : N.m (kgf.m, lb-ft)
1. Right-hand cylinder head
2. Right-hand cylinder head gasket
3. Left-hand cylinder head
4. Left-hand cylinder head gasket
5. Cylinder block
37655_GENS_G0331

Fig. 167 Cylinder head and components

⁂ WARNING

Use fender covers to avoid damaging painted surfaces. To avoid damaging the cylinder head, wait until the engine coolant temperature drops below normal temperature (68°F [20°C]) before removing it. When handling a metal gasket, take care not to fold the gasket or damage the contact surface of the gasket. To avoid damage, unplug the wiring connectors carefully while holding the connector portion. Mark all wiring and hoses to avoid misconnection. Turn the crankshaft pulley so that the No. 1 piston is at top dead center.

1. Before servicing the vehicle, refer to the Precautions Section.
2. Disconnect the negative battery cable.
3. Remove the timing chain. Refer to Timing Chain & Sprockets Removal & Installation.
4. Drain the engine coolant.
5. Disconnect the heater hoses and brake vacuum hose.
6. Remove the mounting bolts and then remove the water temperature control assembly.
7. Disconnect the water vent hose and then remove the intake the manifold.

➡Be sure to drain the engine coolant before removing the intake manifold. If any coolant drained from the cylinder head vent hole has entered the intake port, this can potentially lead to engine trouble.

8. Remove the exhaust manifold stays. Refer to Exhaust Manifold Removal & Installation.
9. Remove the left-hand and right-hand exhaust manifold heat protectors. Refer to Exhaust Manifold Removal & Installation.
10. Remove the left-hand and right-hand exhaust manifolds. Refer to Exhaust Manifold Removal & Installation.
11. Remove the left-hand and right-hand exhaust camshaft OCV. Refer to Camshaft Removal & Installation.
12. Remove the left-hand and right-hand camshaft bearing caps and thrust bearing caps. Refer to Camshaft Removal & Installation.
13. Remove the left-hand and right-hand camshaft assemblies. Refer to Camshaft Removal & Installation.

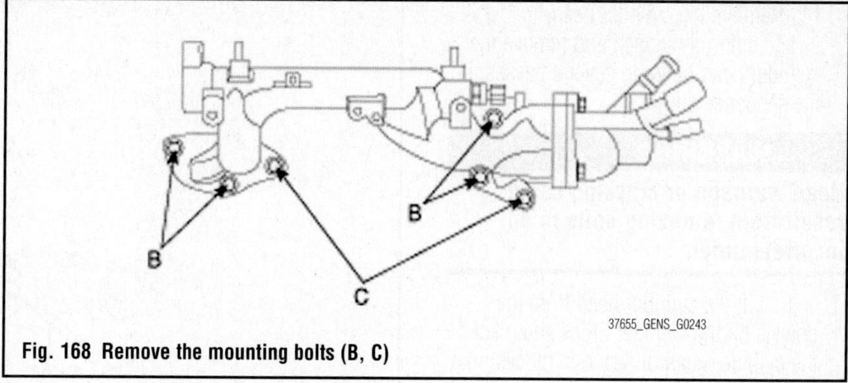

Fig. 168 Remove the mounting bolts (B, C)

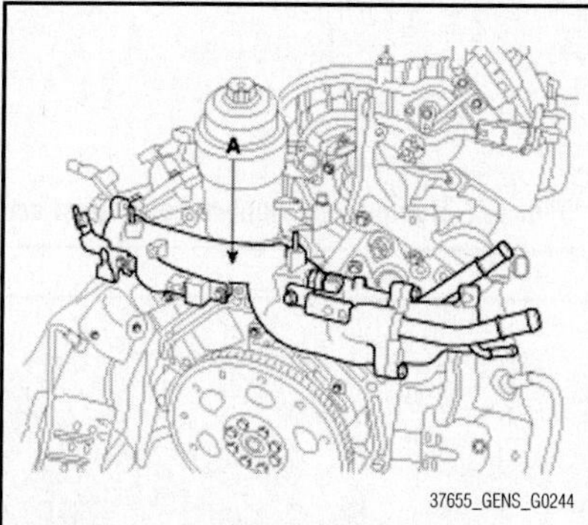

Fig. 169 Remove the water temperature control assembly (A)

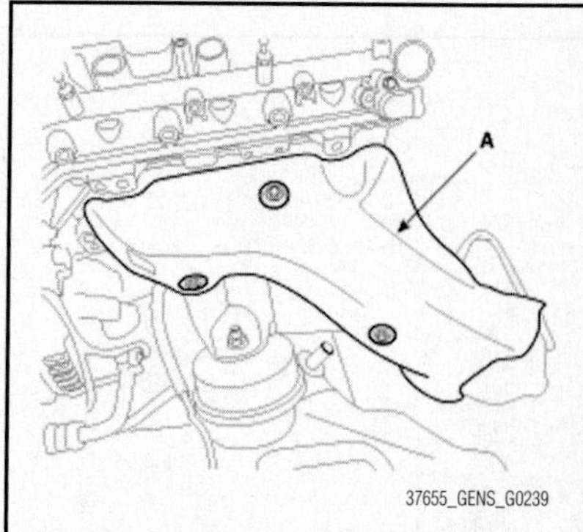

Fig. 172 Remove the left-hand exhaust manifold heat protector (A)

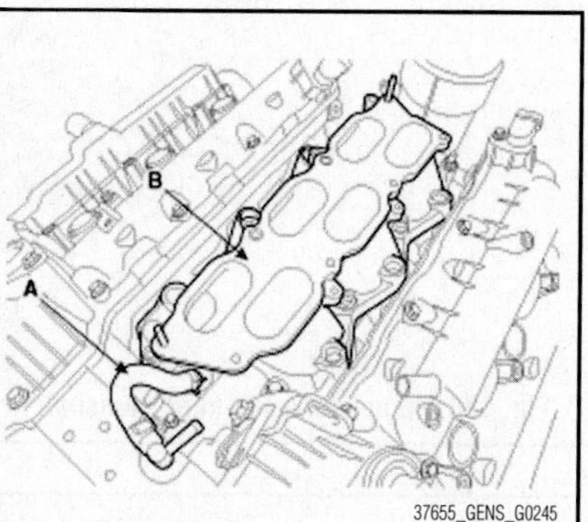

Fig. 170 Disconnect the water vent hose (A) and then remove the intake the manifold (B)

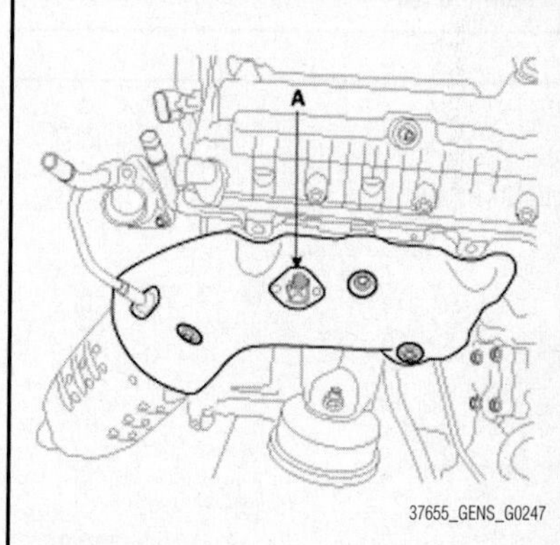

Fig. 173 Remove the right-hand exhaust manifold heat protector (A)

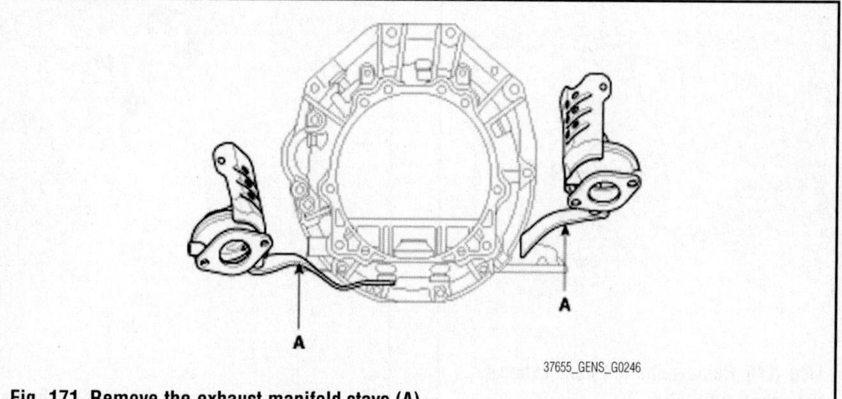

Fig. 171 Remove the exhaust manifold stays (A)

14. Remove the cylinder head.
    a. Uniformly loosen and remove the cylinder head bolts, in several passes, in the sequence shown.

⁂ WARNING

Head warpage or cracking could result from removing bolts in an incorrect order.

    b. Lift the cylinder head from the dowels on the cylinder block and place the cylinder head on wooden blocks on a bench.

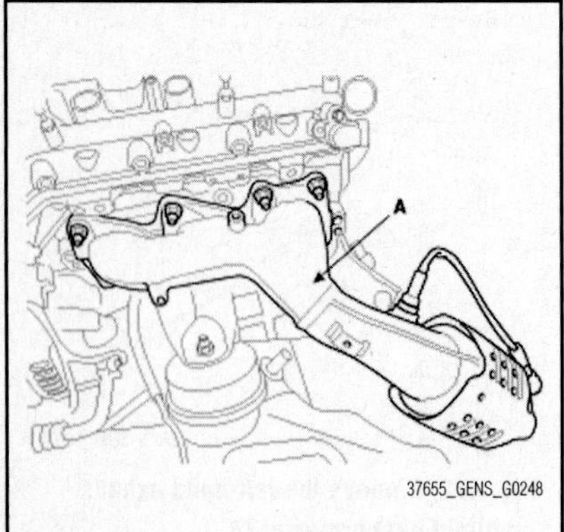

Fig. 174 Remove the left-hand exhaust manifold (A)

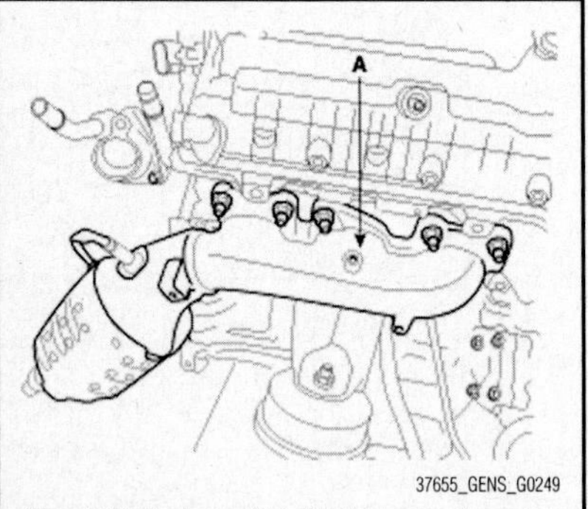

Fig. 175 Remove the right-hand exhaust manifold (A)

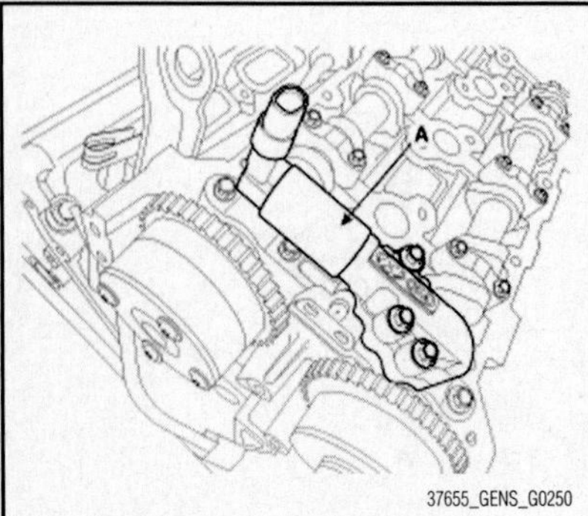

Fig. 176 Remove the left-hand exhaust camshaft OCV (A)

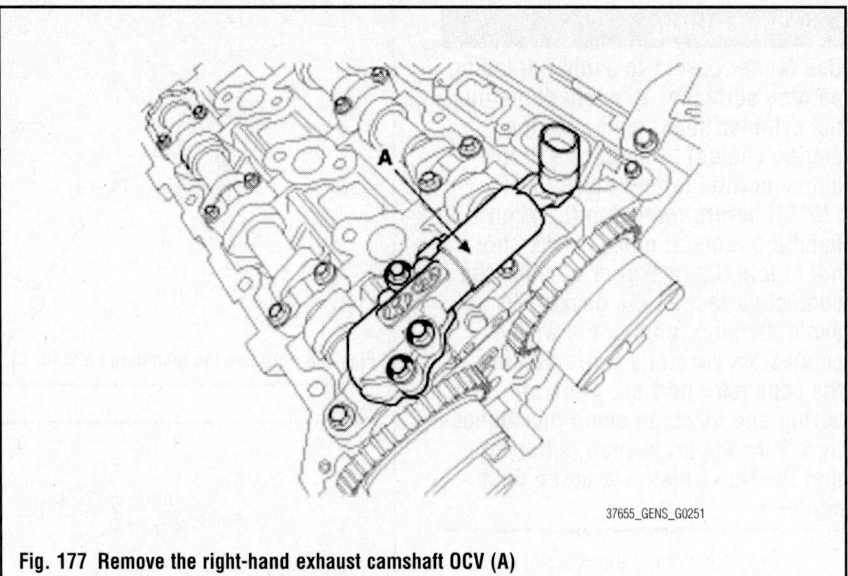

Fig. 177 Remove the right-hand exhaust camshaft OCV (A)

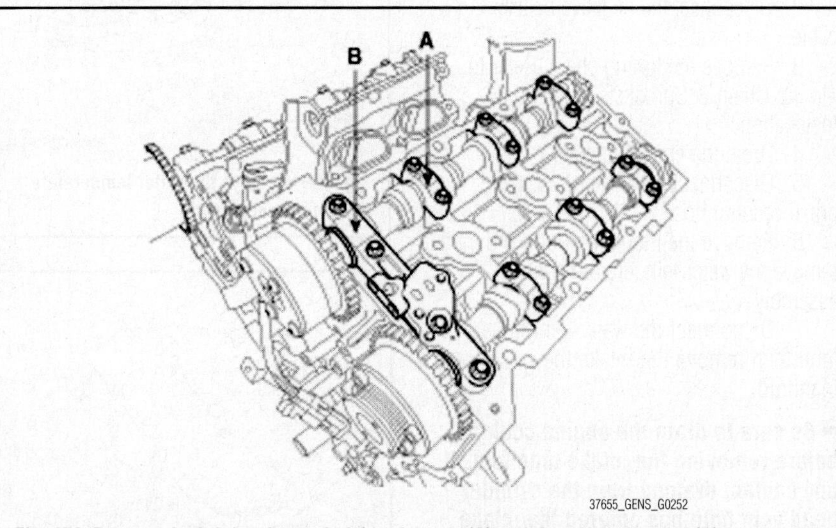

Fig. 178 Remove the left-hand camshaft bearing cap (A) and thrust bearing cap (B)

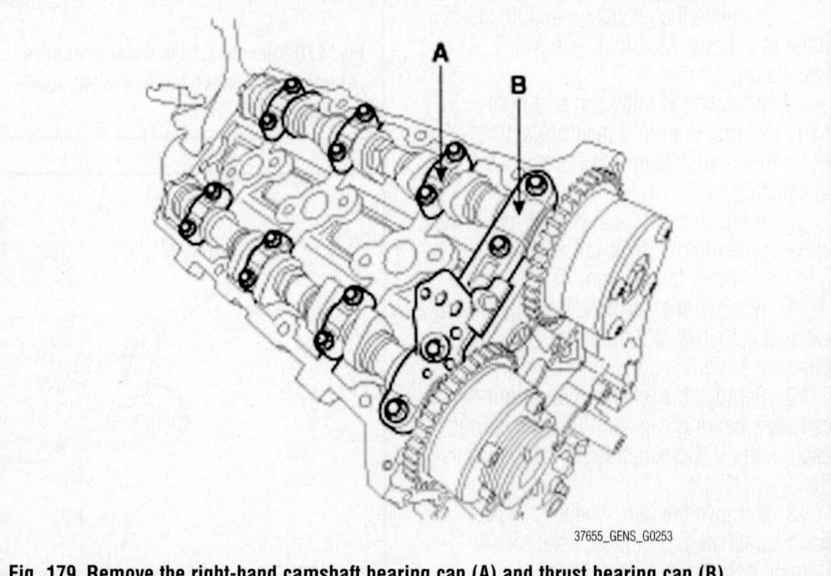

Fig. 179 Remove the right-hand camshaft bearing cap (A) and thrust bearing cap (B)

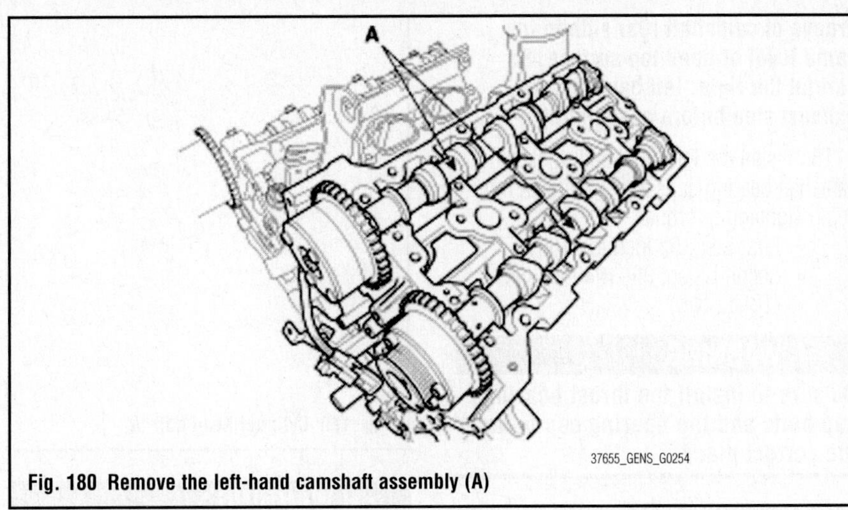

Fig. 180 Remove the left-hand camshaft assembly (A)

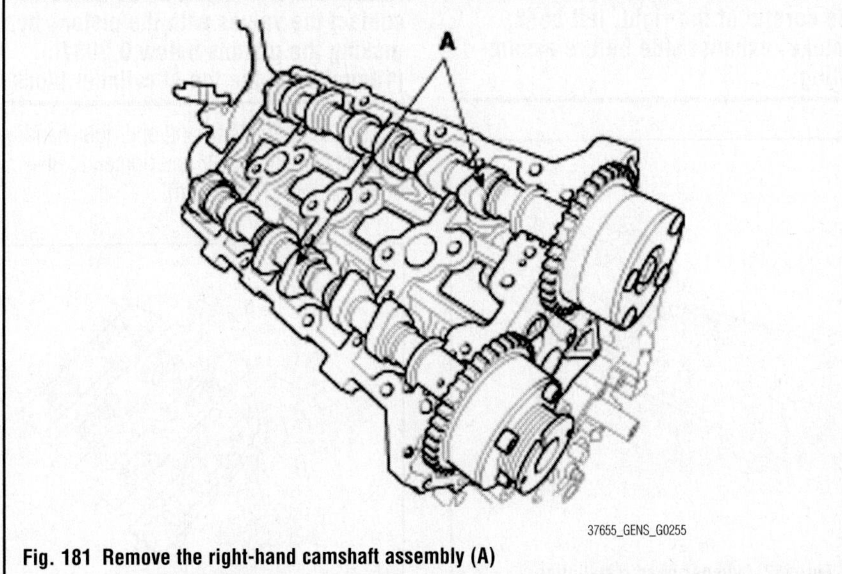

Fig. 181 Remove the right-hand camshaft assembly (A)

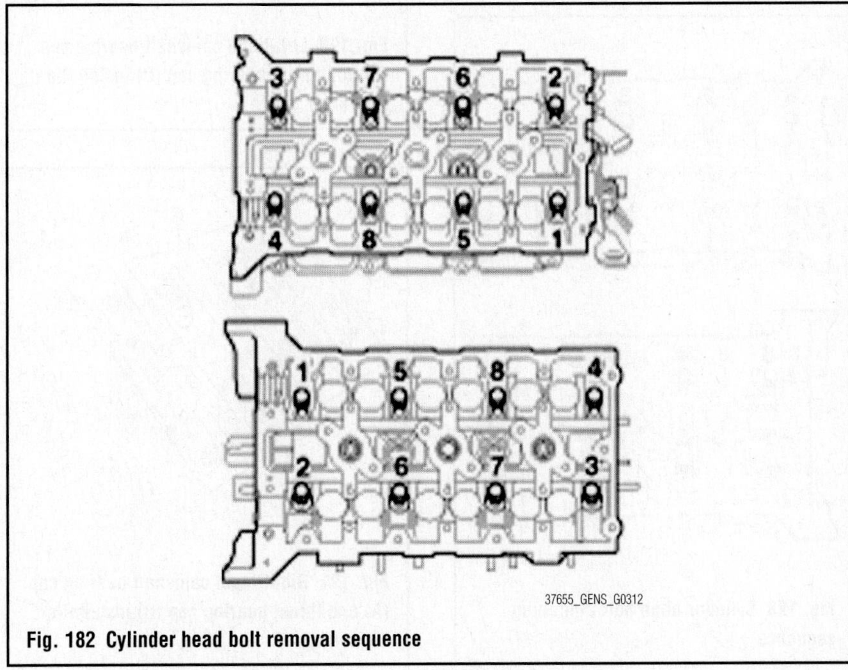

Fig. 182 Cylinder head bolt removal sequence

➡Be careful not to damage the contact surfaces of the cylinder head and cylinder block.

To install:

➡Thoroughly clean all parts to be assembled. Turn the crankshaft pulley so that the No. 1 piston is at top dead center.

15. Install the cylinder head.

➡The sealant locations on cylinder head and cylinder block must be free of engine oil and ETC. The part must be assembled within 5 minutes after sealant was applied.

a. Apply sealant on cylinder block top face before assembling cylinder head gaskets. Bead width: 0.078–0.118 in. (2.0–3.0mm). Sealant locations: 0.039–0.059 in. (1.0–1.5mm). Recommended sealant: Liquid sealant TB1217H.

b. Apply sealant on cylinder head gaskets after assembling cylinder head gaskets on cylinder block.

c. Install the cylinder head.

d. Remove the extruded sealant after assembling cylinder heads.

16. Install cylinder head bolts.

a. Do not apply engine oil on the threads and under the heads of the cylinder head bolts.

b. Using SST (09221-4A000), install and tighten the cylinder head bolts and plate washers, in several passes, in the sequence shown. Tightening torque:

- Head bolt: 28–30 ft. lbs. (37–41 Nm) plus 118–122° plus 88–92°
- Bolt "A": 14–17 ft. lbs. (19–24 Nm)

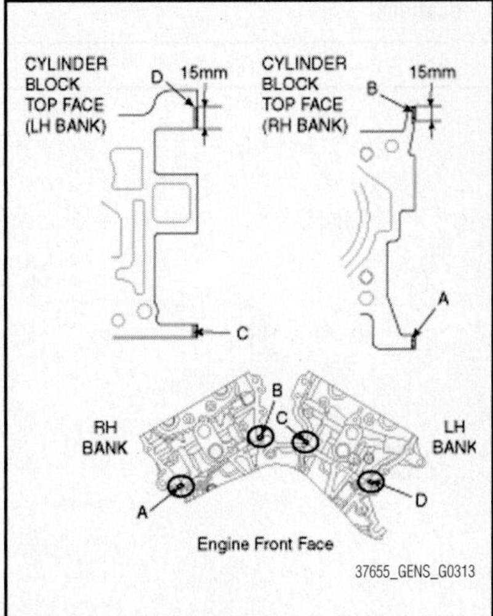

Fig. 183 Cylinder head top face sealant locations

17. Install the left-hand and right-hand camshaft assemblies.

➡Apply a light coat of engine oil on camshaft journals. Assemble the key groove of camshaft rear side to the same level of head top surface. Be careful the right, left bank, intake, exhaust side before assembling.

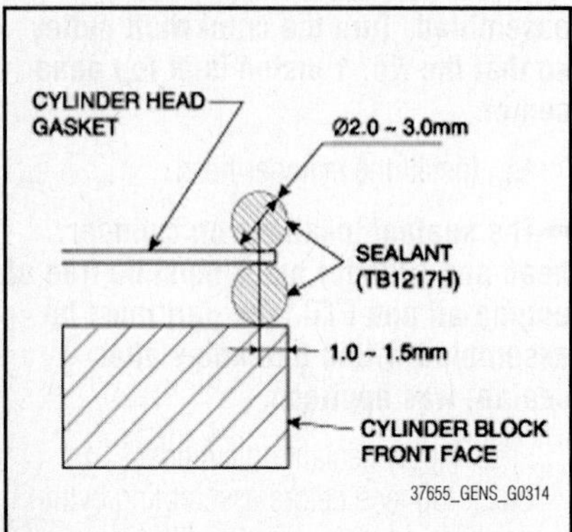

Fig. 184 Cylinder head top face sealant application

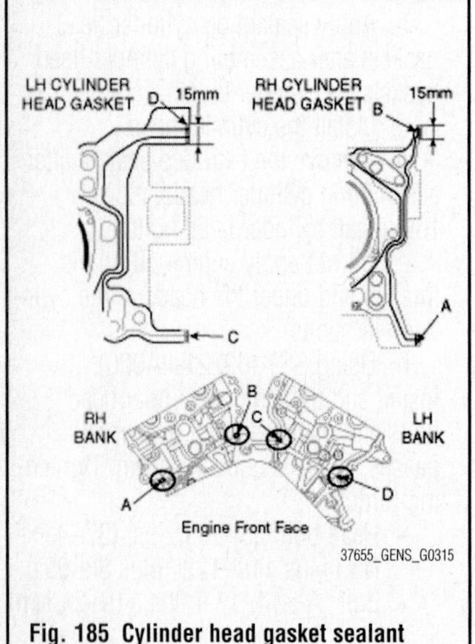

Fig. 185 Cylinder head gasket sealant locations

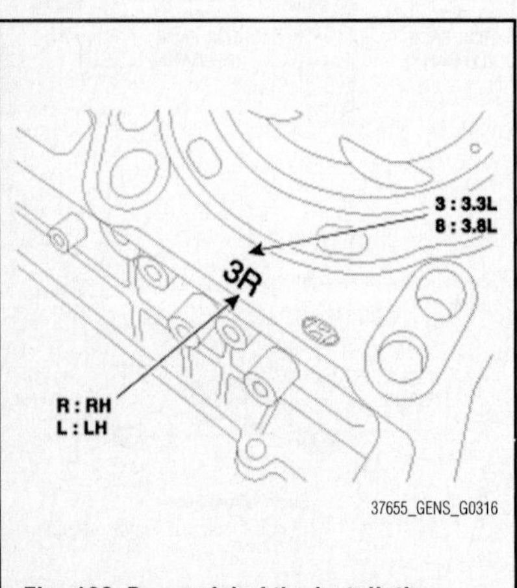

Fig. 186 Be careful of the installation direction

18. Install the left-hand and right-hand camshaft bearing caps and thrust bearing caps. Tightening torque:

- First Step: 52 inch lbs. (6 Nm)
- Second Step: 86–104 inch lbs. (10–12 Nm)

WARNING

Be sure to install the thrust bearing cap bolts and the bearing cap bolts in the correct place.

WARNING

Be careful of the right, left bank, intake, exhaust side before assembling.

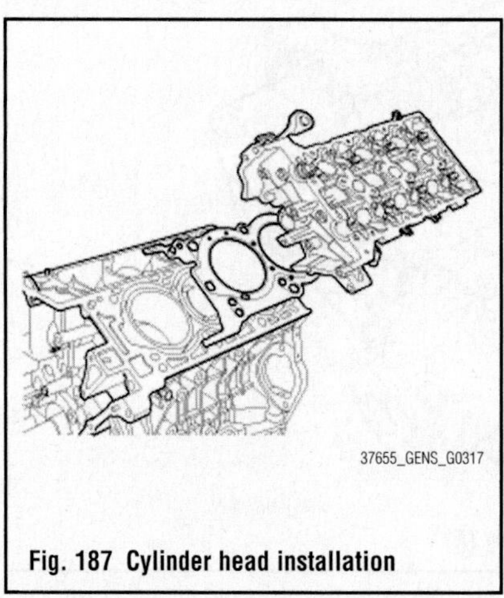

Fig. 187 Cylinder head installation

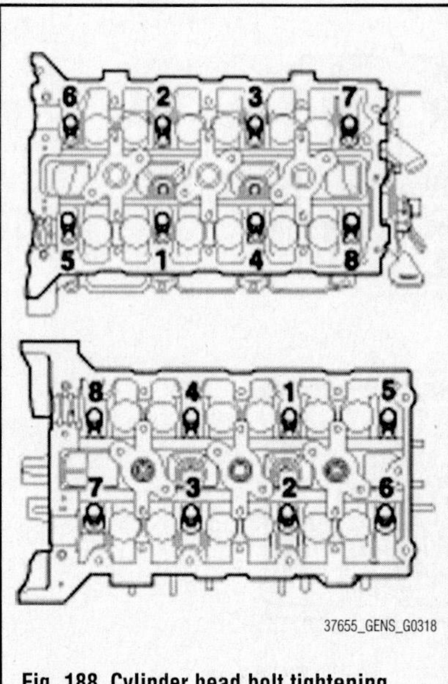

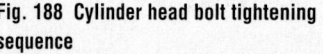

Fig. 188 Cylinder head bolt tightening sequence

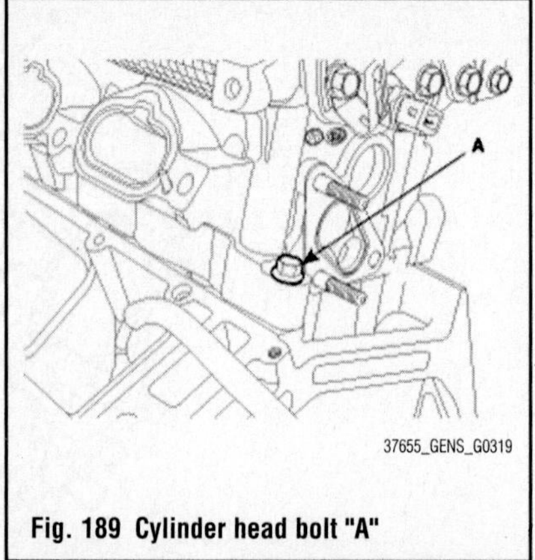

Fig. 189 Cylinder head bolt "A"

WARNING

Rotate the crankshaft so as not to contact the valves with the pistons by making the pistons below 0.3937in. (10mm) from the top of cylinder block.

19. Install the left-hand and right-hand exhaust camshaft OCV and tighten to 86–104 inch lbs. (10–12 Nm).

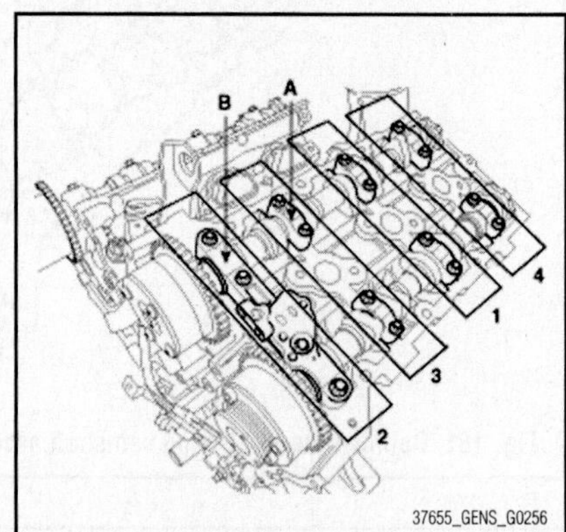

Fig. 190 Left-hand camshaft bearing cap (A) and thrust bearing cap (B) installation sequence

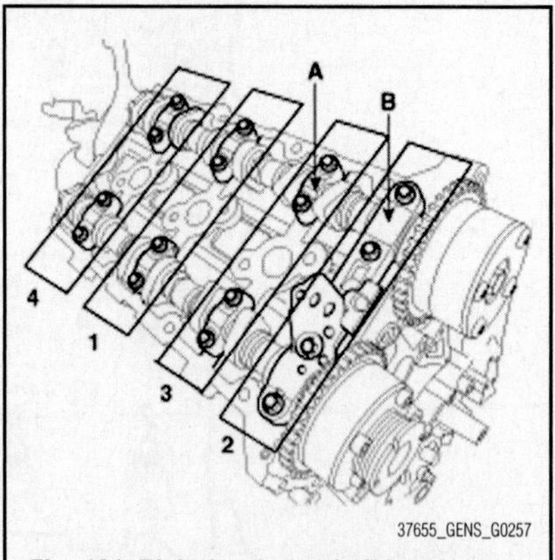

Fig. 191 Right-hand camshaft bearing cap (A) and thrust bearing cap (B) installation sequence

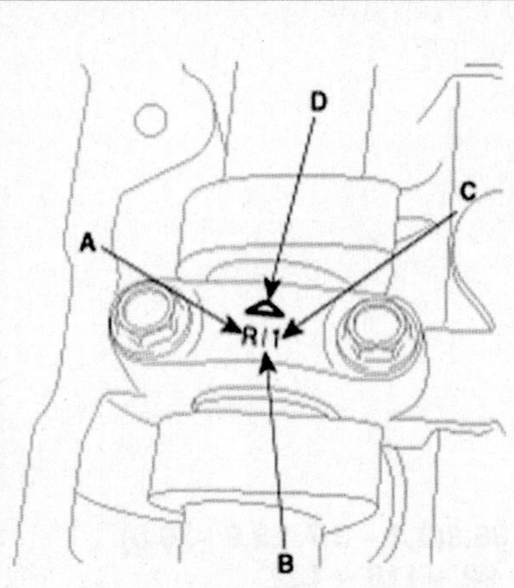

Fig. 192 Position the camshaft bearing cap according to its markings

20. Install the left-hand and right-hand exhaust manifolds with new gaskets, and tighten to 29–33 ft. lbs. (39–44 Nm).

21. Install the left-hand and right-hand exhaust manifold heat protectors and tighten to 86–104 inch lbs. (10–12 Nm).

22. Install the exhaust manifold stays and tighten to 25–30 ft. lbs. (34–41 Nm).

23. Install the intake the manifold with a new gasket, and connect the water vent hose. Tightening torque:

- Step 1: 35–52 (4–6 Nm)
- Step 2: Nut: 14–17 ft. lbs. (19–24 Nm), Bolt: 20–23 ft. lbs. (27—31 Nm)
- Step 3: Repeat 2nd step twice or more.

➡Confirm the manifold gasket identification mark (left-hand, RH) and be careful of the installation direction.

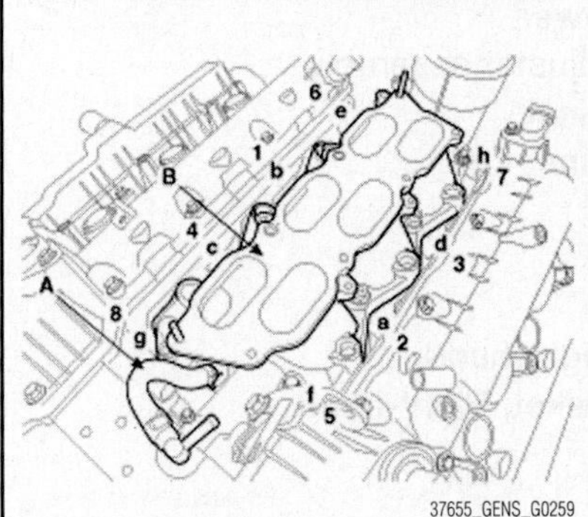

Fig. 193 a–h: Step 1 order, 1–8: Step 2 order

24. Install the water temperature control assembly with a new gasket, and tighten to 15–17 ft. lbs. (20–24 Nm).

25. Connect the heater hoses and brake vacuum hose.

26. Install the timing chain.

27. Refill engine oil.

28. Clean the battery posts and cable terminals with sandpaper. Assemble and then apply grease to prevent corrosion.

29. Inspect for fuel leakage:

a. After assembling the fuel line, turn on the ignition switch (do not operate the starter) so that the fuel pump runs for approximately two seconds and fuel line pressurizes.

b. Repeat this operation two or three times, then check for fuel leakage at any point in the fuel lines.

30. Refill radiator and reservoir tank with engine coolant.

31. Bleed air from the cooling system:

a. Start engine and let it run until it warms up. (Until the radiator fan operates 3 or 4 times.)

b. Turn Off the engine. Check the level in the radiator, add coolant if needed. This will allow trapped air to be removed from the cooling system.

c. Put radiator cap on tightly, then run the engine again and check for leaks.

4.6L Engine

See Figures 194 through 238.

1. Before servicing the vehicle, refer to the Precautions Section.

2. Disconnect the negative battery cable.

⁂ WARNING

Use fender covers to avoid damaging painted surfaces. To avoid damage, unplug the wiring connectors carefully while holding the connector portion. Mark all wiring and hoses to avoid misconnection.

⁂ WARNING

To avoid damaging the cylinder head, wait until the engine coolant temperature drops below normal temperature (68° F [20° C]) before removing it.

➡Turn the crankshaft pulley so that the No. 1 piston is at top dead center.

3. Remove the engine cover.

4. Remove the engine side cover.

5. Remove the drain plug and drain the engine coolant.

6. Remove the air duct, and the air cleaner assembly.

7. Disconnect the radiator upper hose and lower hose.

8. Remove the fuel hose, the Purge Control Solenoid Valve (PCSV) hose and the brake booster vacuum hose.

9. Disconnect the heater hoses.

10. Remove the radiator and cooling fan assembly. Refer to Radiator Removal & Installation in the Engine Cooling section.

11. Turn the tensioner clockwise and loosen, then remove the drive belt. Refer to Accessory Drive Belt Removal & Installation.

12. Remove the drive belt tensioner.

13. Remove the engine wiring harness from the cylinder head.

a. Disconnect the water temperature sensor connector and the CVVT oil control valve connectors.

b. Disconnect the ETC module connector.

c. Disconnect the PCSV connector and the left-hand injector connectors.

d. Disconnect the left-hand ignition coil connectors and the wiring harness protector.

e. Disconnect the left-hand knock sensor connectors, the Camshaft Position (CMP) sensor connectors, the Oxygen sensor (HO2S) connector, and the condenser connector. Refer to Removal & Installation of individual components in the Engine Performance & Emission control section.

f. Disconnect the variable intake system solenoid valve connector and the wiring harness protector.

g. Disconnect the right-hand knock sensor connectors, the CMP sensor connector, the Crankshaft Position (CKP) sensor connector, the HO2S sensor connector and the condenser connector. Refer to Removal & Installation of individual components in the Engine Performance & Emission control section.

h. Disconnect the right-hand injector connectors.

i. Disconnect the right-hand ignition coil connectors and the wiring harness protector.

j. Disconnect the oil pressure switch connector, the CVVT oil control valve connectors and the air conditioner compressor switch connector.

14. Remove the oil level gauge.

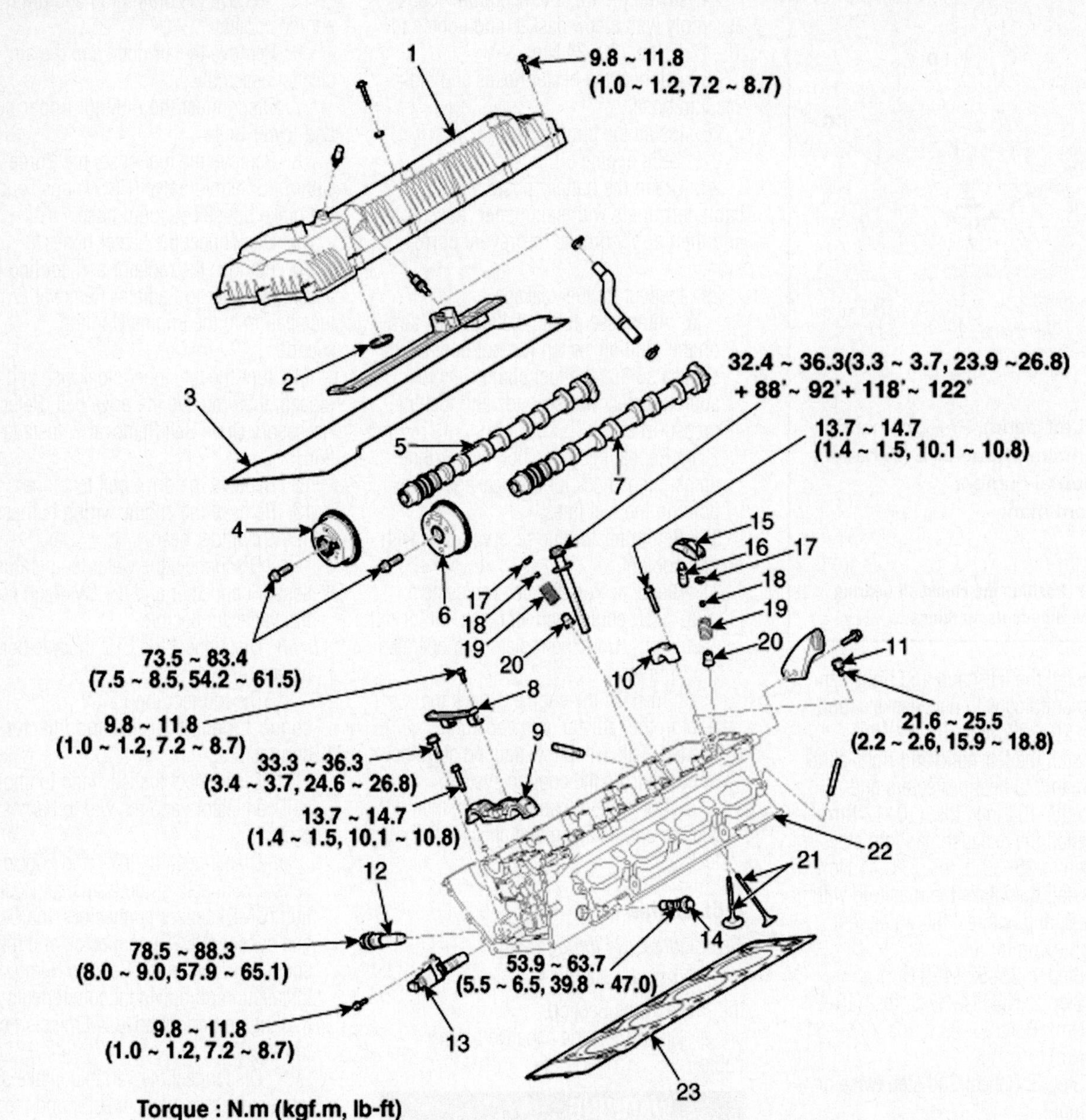

1. Cylinder head cover
2. Cylinder head cover oil seal
3. Cylinder head cover gasket
4. Exhaust CVVT assembly
5. Exhaust camshaft, Right-hand
6. Intake CVVT assembly
7. Intake camshaft, Right-hand
8. Cam to cam guide
9. Camshaft thrust bearing cap
10. Camshaft bearing cap
11. Pressure relief valve
12. Timing chain tensioner
13. Oil control valve
14. OCV filter
15. Roller finger follower
16. Hydraulic lash adjuster assembly
17. Valve spring retainer
18. Valve spring retainer lock
19. Valve spring
20. Valve stem seal
21. Valve
22. Cylinder head, Right-hand
23. Cylinder head gasket, Right-hand

37655_GENS_G0332

Fig. 194 Right-hand camshafts and related components

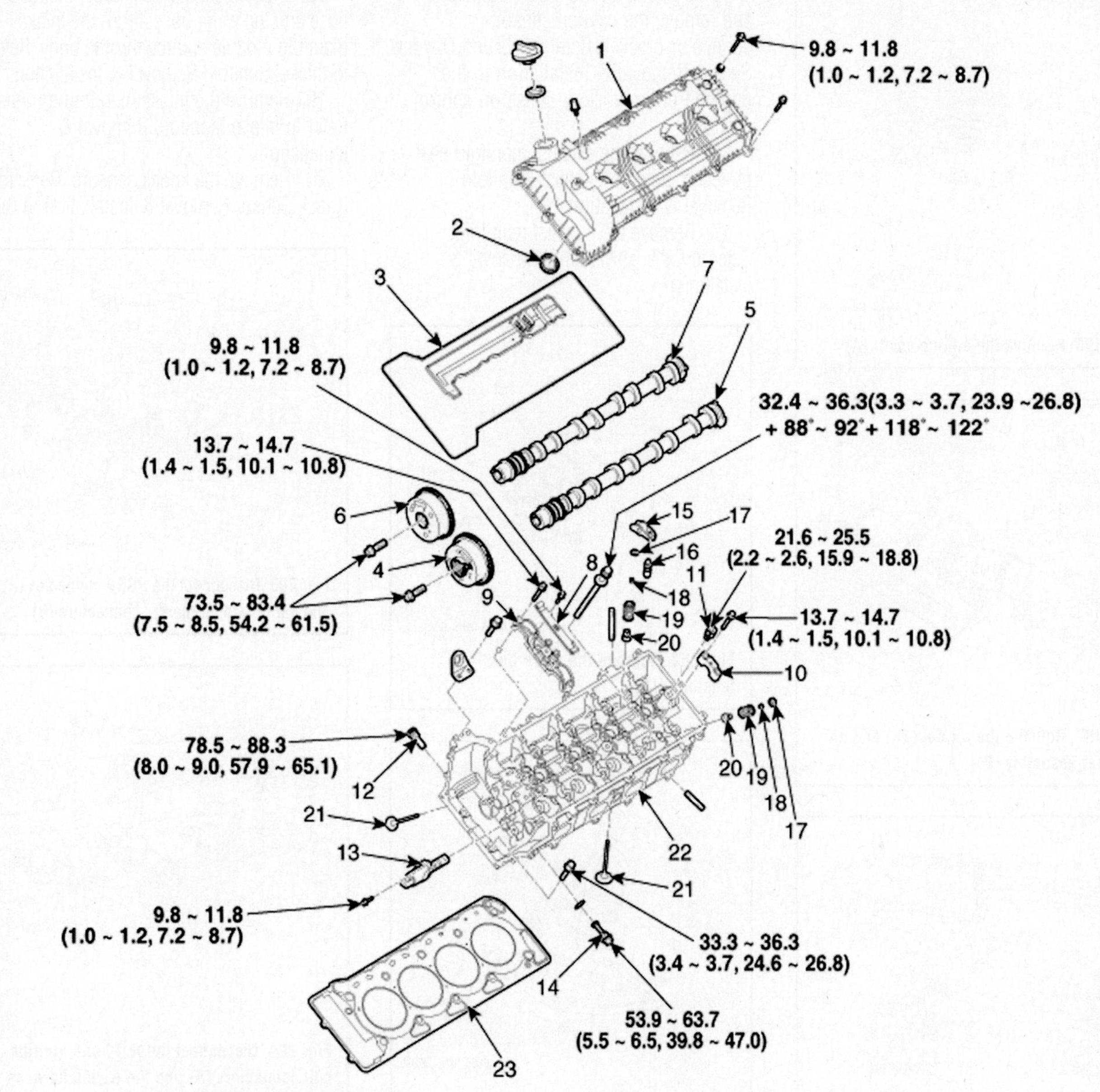

1. Cylinder head cover
2. Cylinder head cover oil seal
3. Cylinder head cover gasket
4. Exhaust CVVT assembly
5. Exhaust camshaft, Left-hand
6. Intake CVVT assembly
7. Intake camshaft, Left-hand
8. Cam to cam guide
9. Camshaft thrust bearing cap
10. Camshaft bearing cap
11. Pressure relief valve
12. Timing chain tensioner
13. Oil control valve
14. OCV filter
15. Roller finger follower
16. Hydraulic lash adjuster assembly
17. Valve spring retainer
18. Valve spring retainer lock
19. Valve spring
20. Valve stem seal
21. Valve
22. Cylinder head, Left-hand
23. Cylinder head gasket, Left-hand

37655_GENS_G0333

Fig. 195 Left-hand camshafts and related components

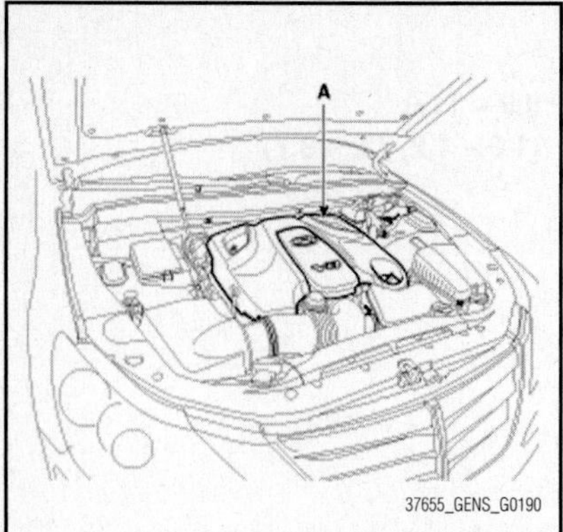

Fig. 196 Remove the engine cover (A)

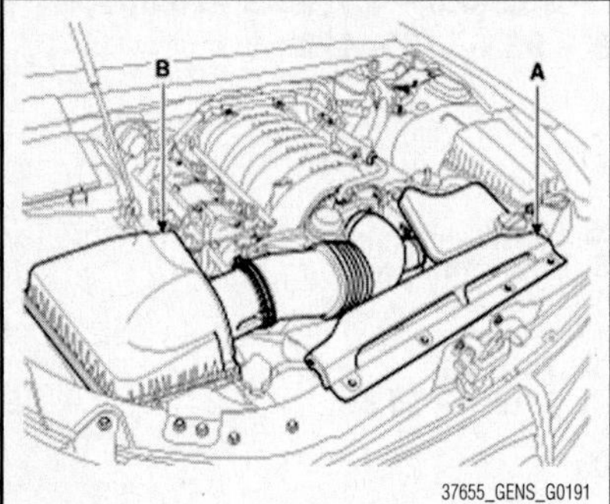

Fig. 197 Remove the air duct (A) and air cleaner assembly (B)

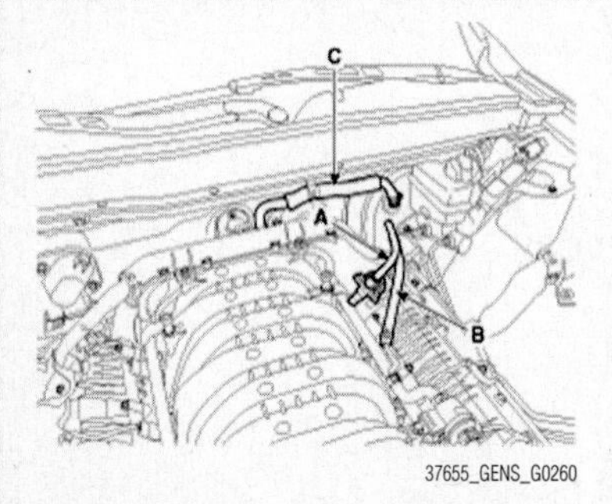

Fig. 198 Remove the fuel hose (A), the PCSV hose (B), and the brake booster vacuum hose (C)

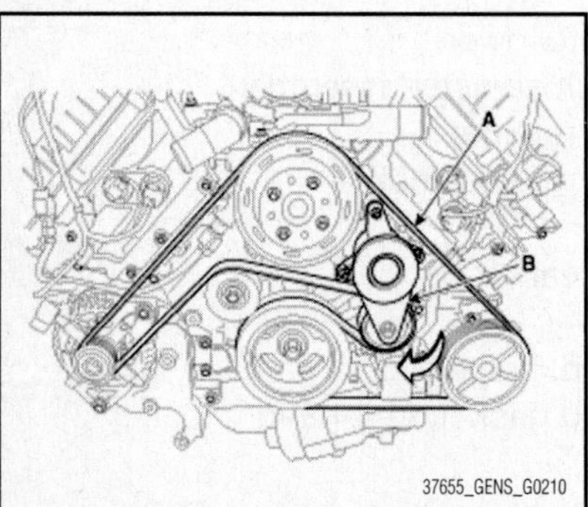

Fig. 199 Turn the tensioner (B) clockwise and loosen, then remove the drive belt (A)

15. Disconnect the HO2S connector and remove the oxygen sensor connector bracket. Refer to Heated Oxygen Sensor Removal & Installation in the Engine Performance & Emission control section.

16. Remove the exhaust manifold heat protector. Refer to Exhaust Manifold Removal & Installation.

17. Remove the exhaust manifold. Refer to Exhaust Manifold Removal & Installation.

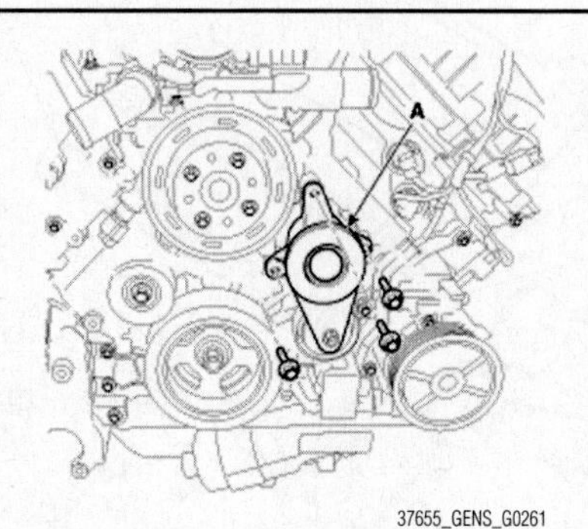

Fig. 200 Remove the drive belt tensioner (A)

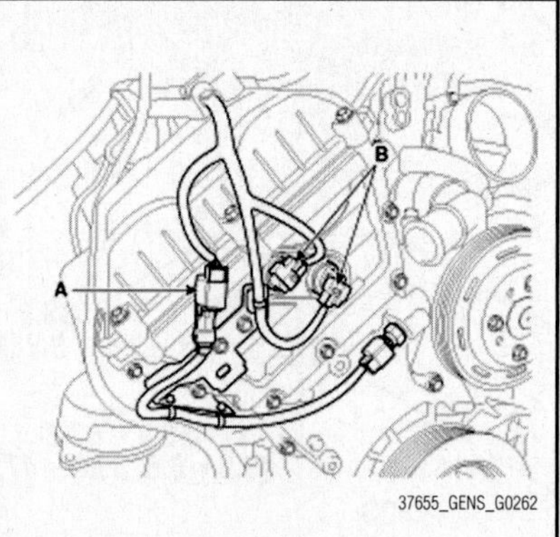

Fig. 201 Disconnect the water temperature sensor connector (A) and the CVVT oil control valve connectors (B)

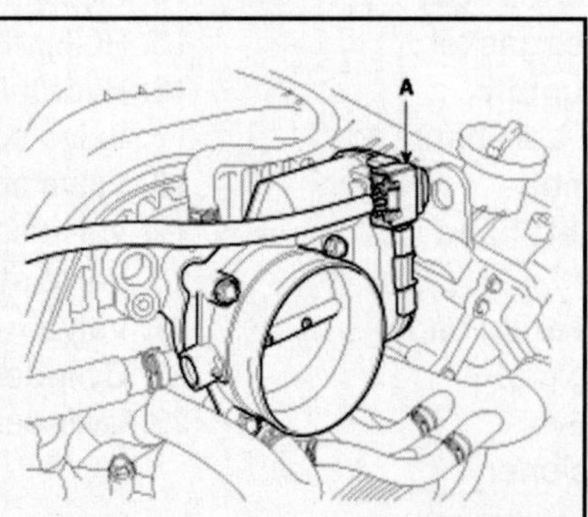

Fig. 202 Disconnect the ETC module connector (A)

18. Disconnect the PCV hose, the PCSV hose and the water hoses from the intake manifold module and the throttle body. Refer to Intake Manifold Removal & Installation.

19. Remove the intake manifold module. Refer to Intake Manifold Removal & Installation.

20. Remove the knock sensors. Refer to Knock Sensor Removal & Installation in the

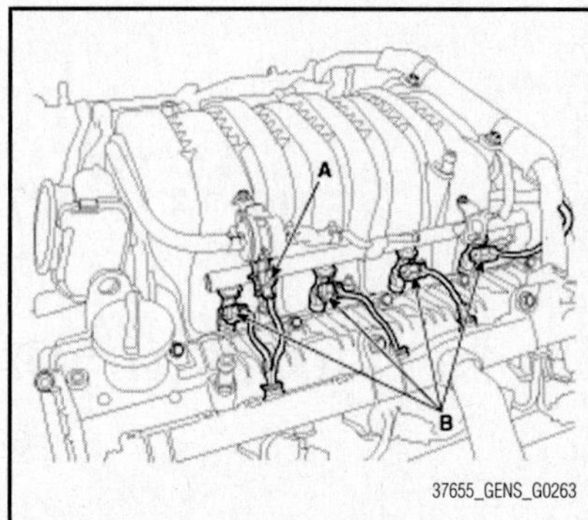

Fig. 203 Disconnect the PCSV connector (A) and the left-hand injector connectors (B)

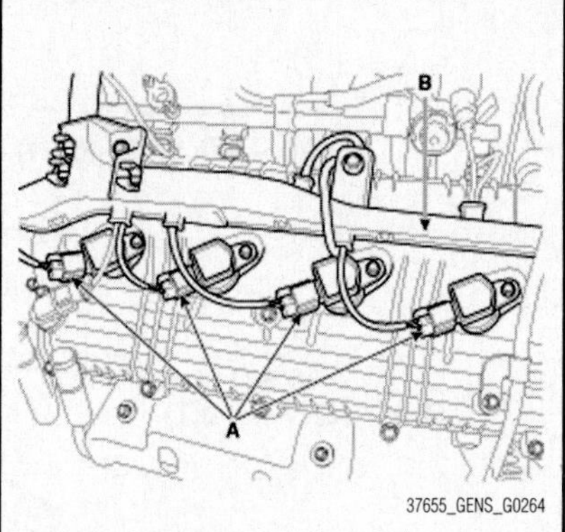

Fig. 204 Disconnect the left-hand ignition coil connectors (A) and the wiring harness protector (B)

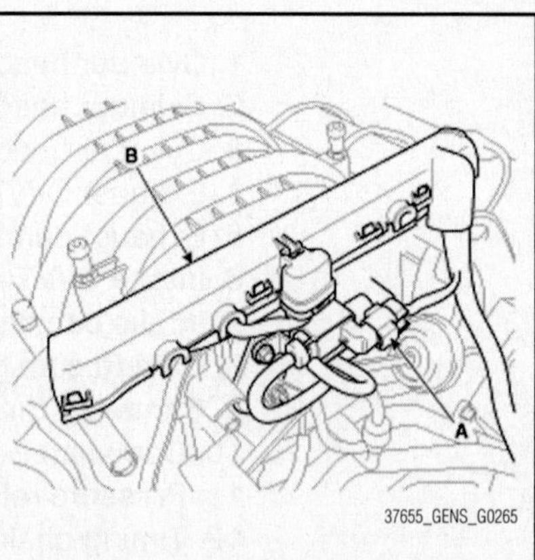

Fig. 205 Disconnect the variable intake system solenoid valve connector (A) and the wiring harness protector (B)

Engine Performance & Emission control section.

21. Remove the ignition coils. Refer to Ignition Coil Removal & Installation in the Engine Electrical section.

22. Remove the cylinder head cover from both sides.

23. Remove the cylinder head cover gasket and oil seal from both sides.

24. Remove the Oil Pressure Switch (OPS) wiring assembly and bracket.

25. Remove the water temperature sensor wiring assembly and bracket.

26. Remove the timing chain upper cover. Refer to Timing Chain & Sprockets Removal & Installation.

27. Remove the timing chain upper cover oil seal. Refer to Timing Chain & Sprockets Removal & Installation.

28. Remove the Oil Control Valve (OCV) from both sides.

➡Do not reuse the OCV if dropped. Keep the OCV clean. Do not hold the OCV sleeve during servicing. When the OCV is installed on the engine, do not move the engine while holding the OCV yoke.

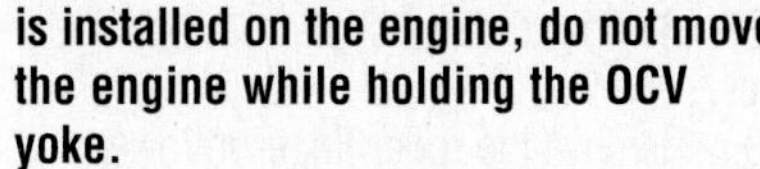

29. Remove the cam to cam guide from both sides. Refer to Camshaft Removal & Installation.

30. Align the camshaft sprocket timing marks. Refer to Camshaft Removal & Installation.

31. Remove the alternator. Refer to Alternator Removal & Installation in the Engine Electrical section.

32. Remove the water outlet and inlet pipes.

33. Remove the drain plug and drain the engine coolant.

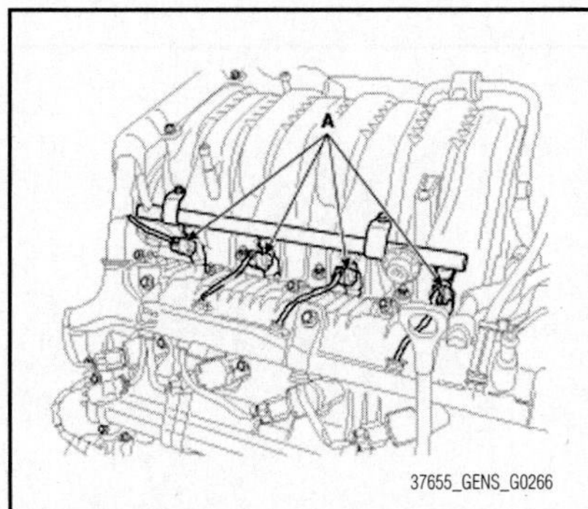

Fig. 206 Disconnect the right-hand injector connectors (A)

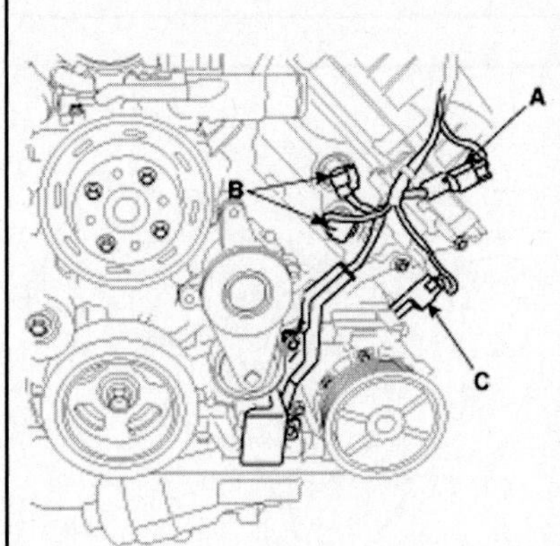

Fig. 207 Disconnect the oil pressure switch connector (A) the CVVT oil control valve connectors (B), and the air conditioner compressor switch connector (C)

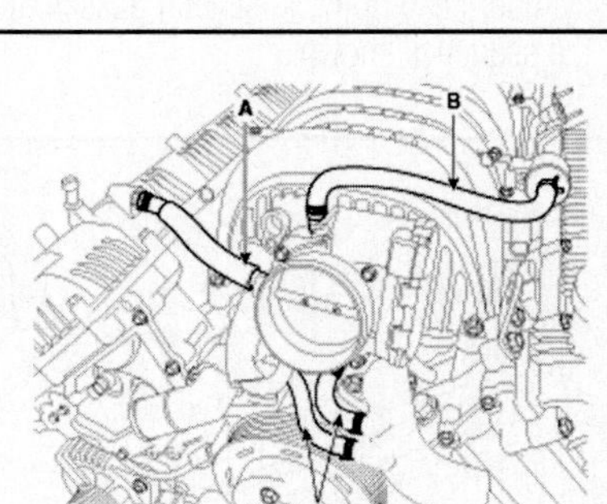

Fig. 208 Disconnect the PCV hose (A) the PCSV hose (B), and the water hoses (C) from the intake manifold module and the throttle body

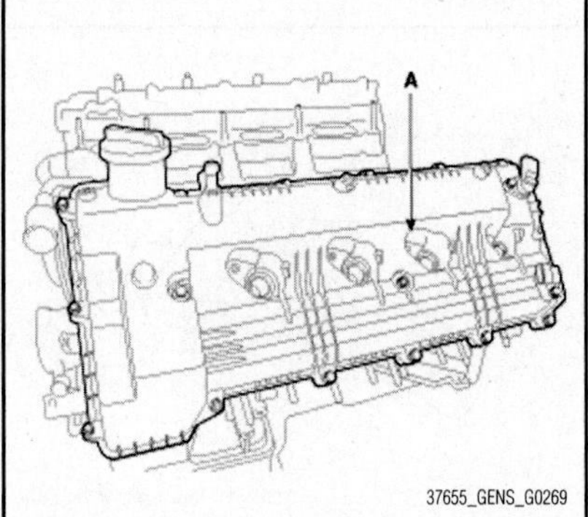

Fig. 209 Remove the cylinder head cover (A)—Left-hand side shown, right-hand side similar

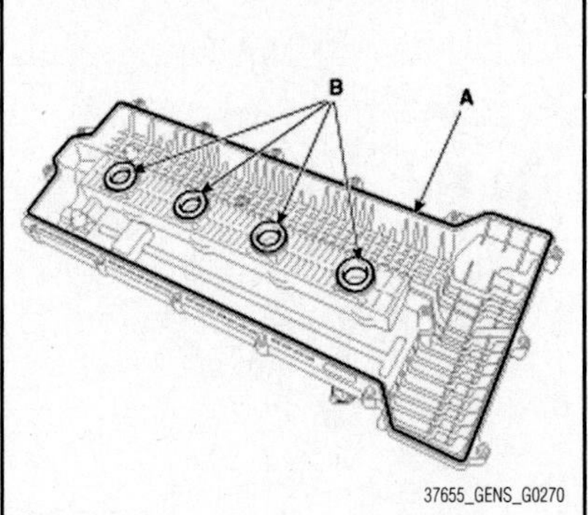

Fig. 210 Remove the cylinder head cover gasket (A) and oil seal (B))—Left-hand side shown, right-hand side similar

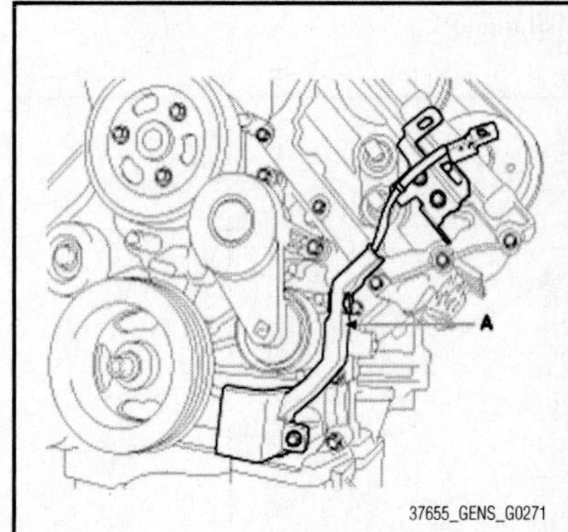

Fig. 211 Remove the oil pressure switch wiring assembly and bracket (A)

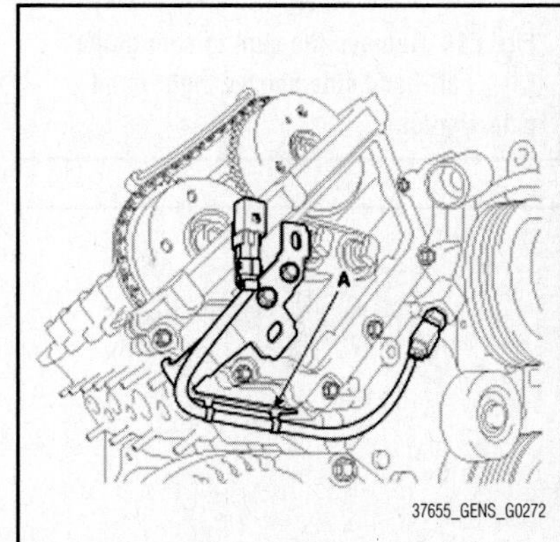

Fig. 212 Remove the water temperature sensor wiring assembly and bracket (A)

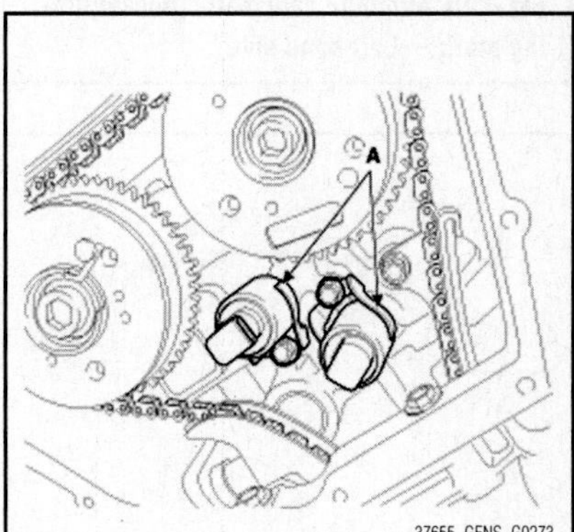

Fig. 213 Remove the oil control valve (A)—Right-hand side shown, left-hand side similar

34. Remove the water temperature control assembly and the water outlet fitting assembly.

35. Remove the timing chain tensioner. Refer to Timing Chain & Sprockets Removal & Installation.

36. Remove the camshaft bearing caps and the camshaft thrust bearing caps from both sides. Refer to Camshaft Removal & Installation.

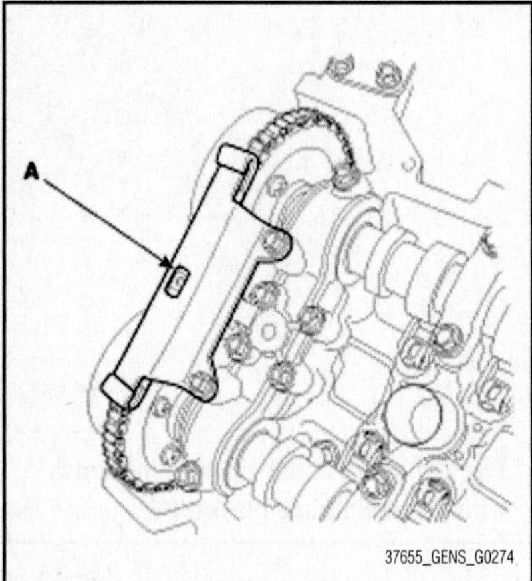

Fig. 214 Remove the cam to cam guide (A)—Left-hand side shown, right-hand side similar

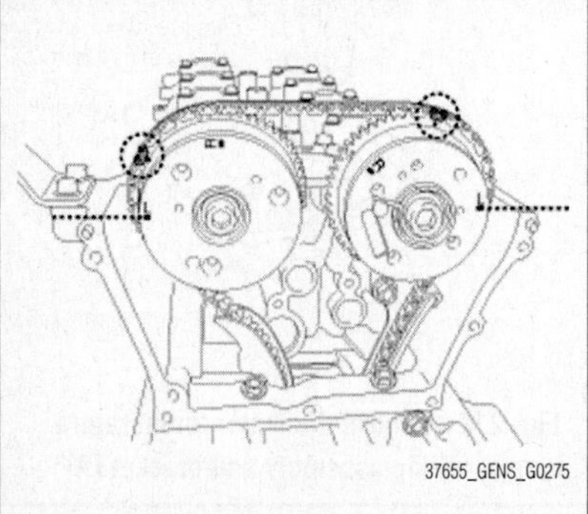

Fig. 215 Align the camshaft sprocket timing marks—Left-hand side

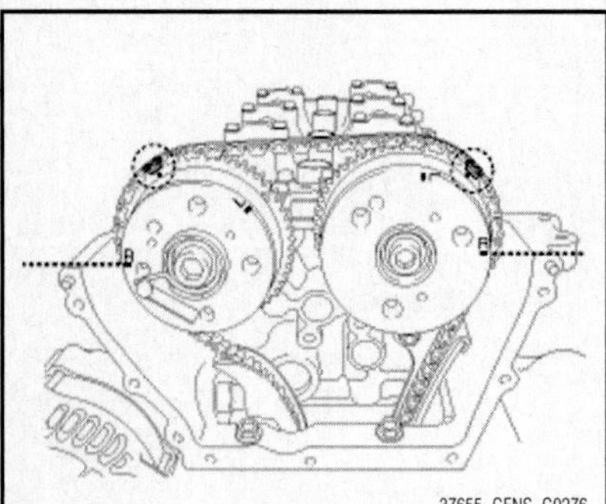

Fig. 216 Align the camshaft sprocket timing marks—Right-hand side

37. Remove the camshaft from both sides.

38. Remove the roller finger follower and the hydraulic lash adjuster assembly. Refer to Rocker Arm Removal & Installation.

39. Remove the cylinder head.

a. Uniformly loosen and remove the cylinder head bolts, in several passes, in the sequence shown.

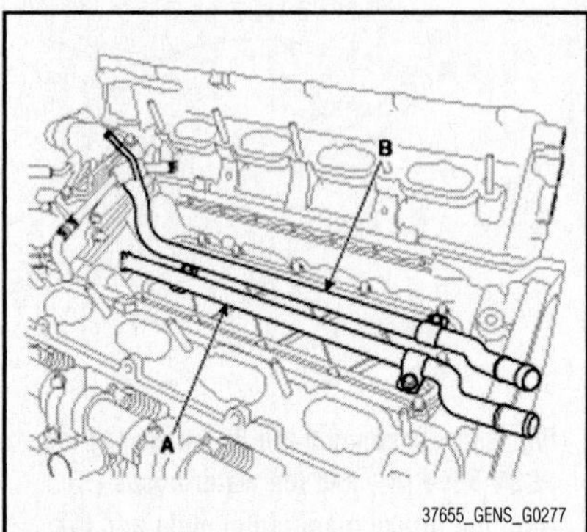

Fig. 217 Remove the water outlet (A) and inlet (B) pipes

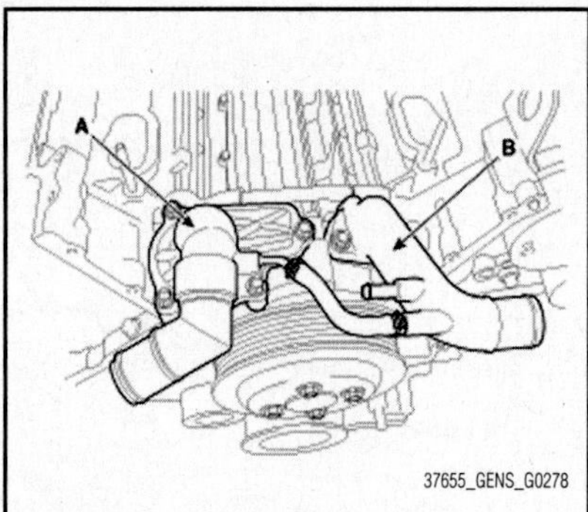

Fig. 218 Remove the water temperature control assembly (A) and the water outlet fitting assembly (B)

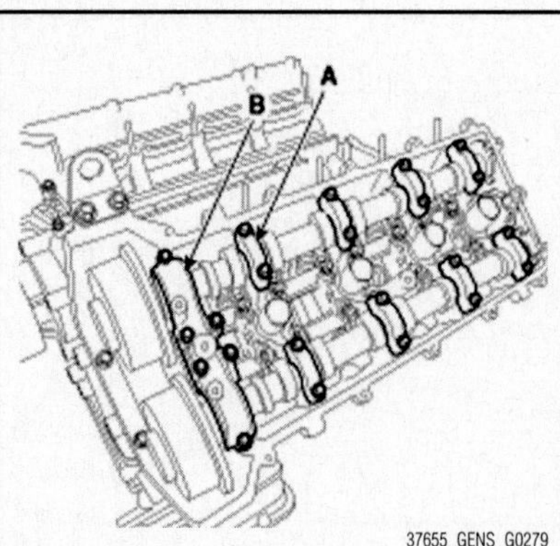

Fig. 219 Remove the camshaft bearing caps (A) and the camshaft thrust bearing cap (B)—Left-hand side shown, right-hand side similar

WARNING

Head warpage or cracking could result from removing bolts in an incorrect order.

b. Lift the cylinder head from the dowels on the cylinder block and place

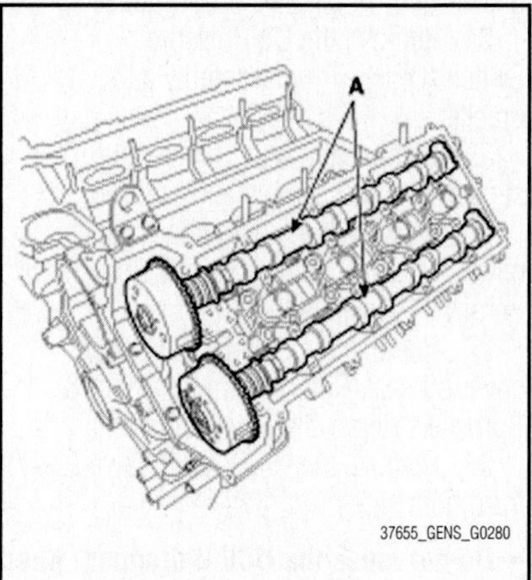

Fig. 220 Remove the camshaft—Left-hand side shown, right-hand side similar

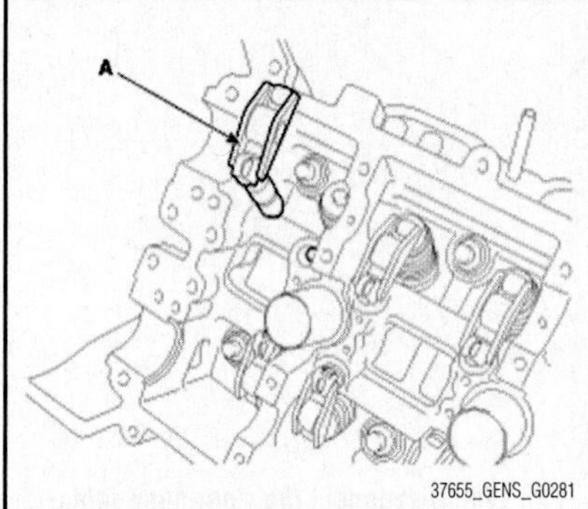

Fig. 221 Remove the roller finger follower and the hydraulic lash adjuster assembly (A)

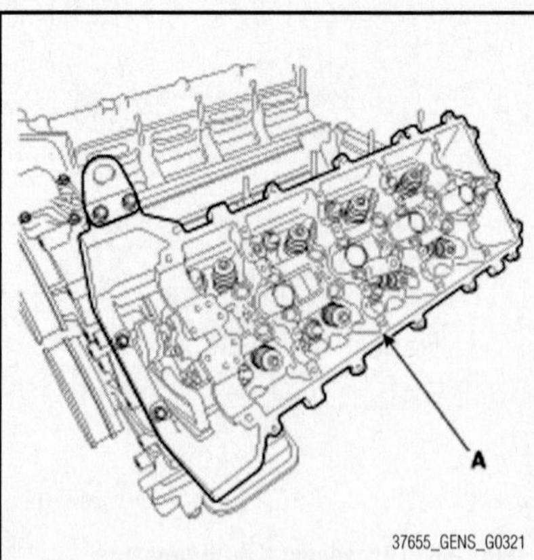

Fig. 222 Remove the cylinder head (A)—Left-hand side shown, right-hand side similar

the cylinder head on wooden blocks on a bench.

WARNING

Be careful not to damage the contact surfaces of the cylinder head and cylinder block.

40. Remove the cylinder head gasket.

To install:

➡Thoroughly clean all parts to be assembled. Always use a new head and manifold gasket. The cylinder head gasket is a metal gasket. Take care not to bend it. Rotate the crankshaft; set the No.1 piston at TDC.

41. Install the cylinder head gasket.

➡The sealant locations on the cylinder head gasket, cylinder block and timing chain lower case must be free of engine oil and etc. The part must be assembled within 5 minutes after sealant was applied.

a. Before assembling the cylinder head gasket, the liquid sealant LT5900H or equivalent should be applied on the gap between cylinder block and timing chain lower case. Bead width: 0.1378–0.1772 in. (3.5–4.5mm). Sealant locations: 0.0591–0.0984 in. (1.5–2.5mm) from timing chain lower case inner surface. Recommended sealant: Liquid sealant LT5900H or equivalent.

b. Apply sealant on the cylinder head gaskets after assembling cylinder head gaskets on cylinder block.

➡Be careful of the installation direction.

42. Install the cylinder head. Remove the extruded sealant after assembling cylinder heads.

a. Install the cylinder head bolts.

b. Do not apply engine oil on the cylinder head bolts.

c. Using SST (09221-4A000), install and tighten the cylinder head bolts and plate washers, in several passes, in the sequence shown. Tightening torque:

- Cylinder head bolts (1–10): 24–27 ft. lbs. (32–36 Nm) plus 88°92° plus 118–122°
- Flange bolts (11–12): 25–27 ft. lbs. (33–36 Nm)

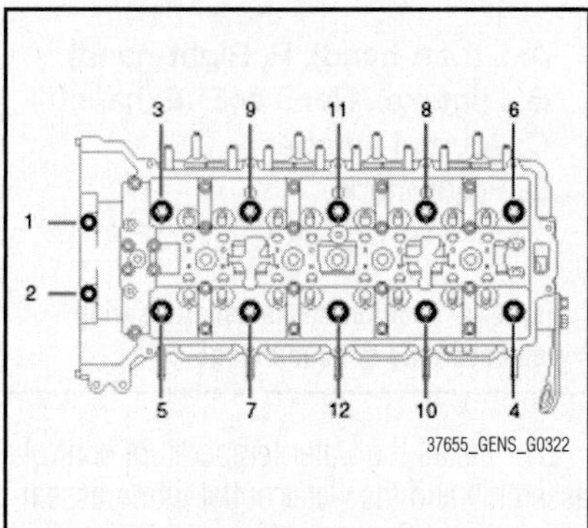

Fig. 223 Cylinder head bolt removal sequence—Right-hand

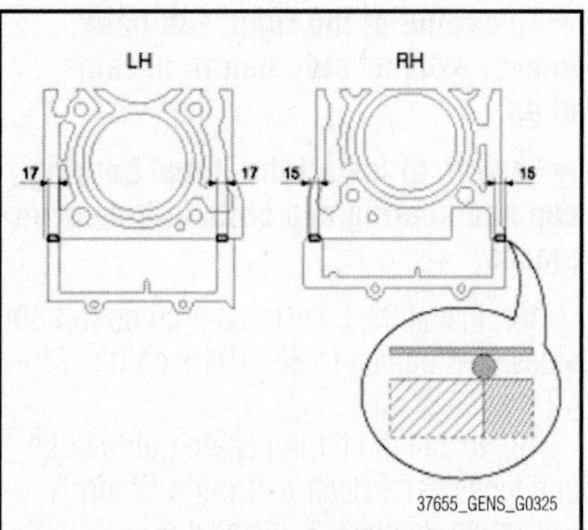

Fig. 226 Cylinder block and timing chain lower case sealant locations

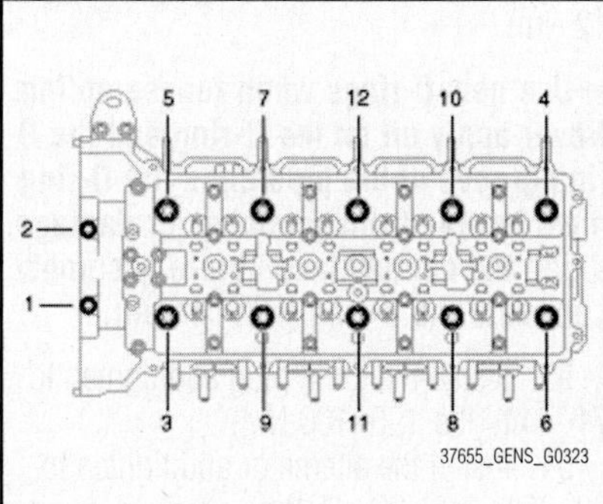

Fig. 224 Cylinder head bolt removal sequence—Left-hand

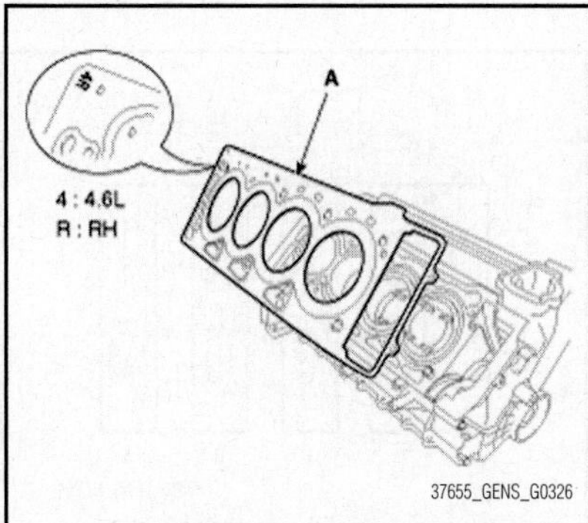

Fig. 227 Cylinder head gasket positioning—Right-hand

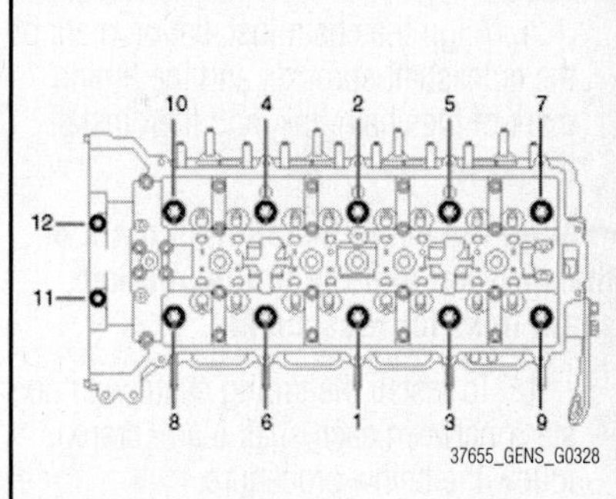

Fig. 229 Cylinder head bolt tightening sequence—Right-hand

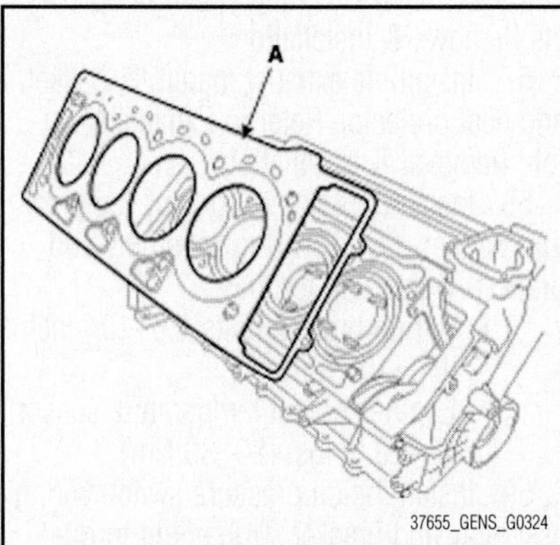

Fig. 225 Cylinder head gasket (A)—Right-hand shown, left-hand side similar

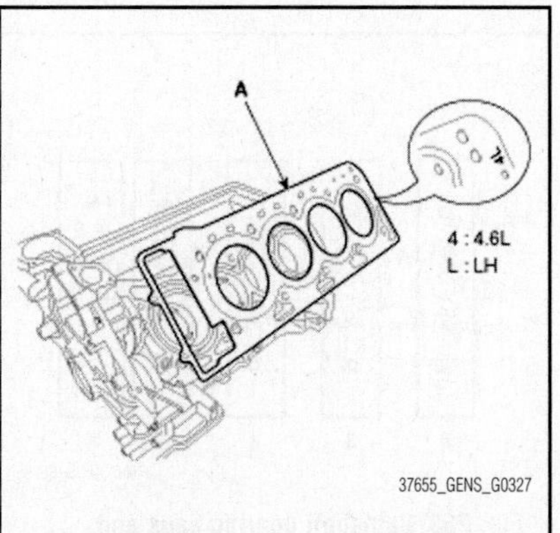

Fig. 228 Cylinder head gasket positioning—Left-hand

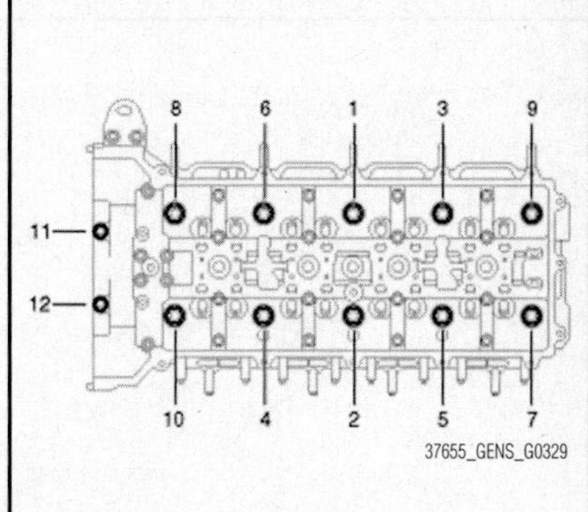

Fig. 230 Cylinder head bolt tightening sequence—Left-hand

43. Remove the roller finger follower and the hydraulic lash adjuster assembly.

44. Assemble the CVVT and camshaft assembly and tighten to 54–62 ft. lbs. (74–83 Nm).

a. Assemble the camshaft to the dowel pin of CVVT assembly. At this time, attend not to be assembled to oil hole of camshaft.

b. Hold the hexagonal head wrench portion of the camshaft with a vise, and install the bolt to the CVVT assembly.

c. Do not rotate the CVVT assembly when the camshaft is assembled to the dowel pin of CVVT assembly.

45. Install the left-hand and right-hand CVVT and camshaft assemblies.

➡Apply a light coat of engine oil on camshaft journals. Assemble the timing mark of CVVT to the same level of cylinder head top surface. Make sure the right, left bank, intake, exhaust side camshaft are correct before assembling. Turn the crankshaft to set the No.1 piston at TDC position.

46. Install the timing chain to the CVVT sprocket.

➡Make sure the crankshaft is at TDC position.

a. Align the chain installation mark of the crankshaft sprocket and the timing mark of the chain link, and then install the chain.

➡Align the chain installation mark of the sprocket and the timing marked chain links for reassembly.

b. To install the timing chain with no slack between each shaft (cam, crank), follow the below procedure:

- Crankshaft sprocket, then timing chain guide, then exhaust CVVT sprocket, then intake CVVT sprocket. (Left-hand Bank)

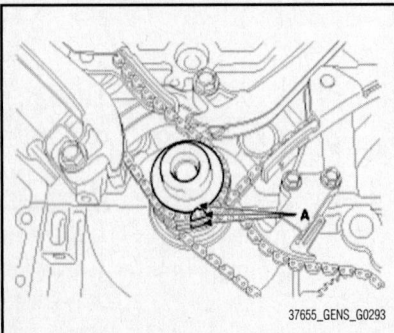

Fig. 231 Align the chain installation mark (A) of the crankshaft sprocket and the timing mark (A) of the chain link

- Crankshaft sprocket, then timing chain guide, then intake CVVT sprocket, then exhaust CVVT sprocket. (Right-hand Bank)
- When reassembling the timing chain, make sure all of the chain installation marks of the sprocket and the timing marked chain links are aligned.

47. Install the camshaft bearing caps and the camshaft thrust bearing cap in the sequence shown. Tightening torque:

- 1st step: 61–70 inch lbs. (7–8 Nm)
- 2nd step: 10–11 ft. lbs. (14–15 Nm)

➡Be careful of the right, left bank, intake, exhaust side before assembling.

➡Be sure to install the thrust bearing cap and bearing cap bolts in the correct places.

48. Install the cam-to-cam guide to both sides, and tighten to 86–104 inch lbs. (10–12 Nm).

49. Install the timing chain guide bolts and tensioners. Refer to Timing Chain & Sprockets Removal & Installation.

50. Set the timing chain tension. Refer to Timing Chain & Sprockets Removal & Installation.

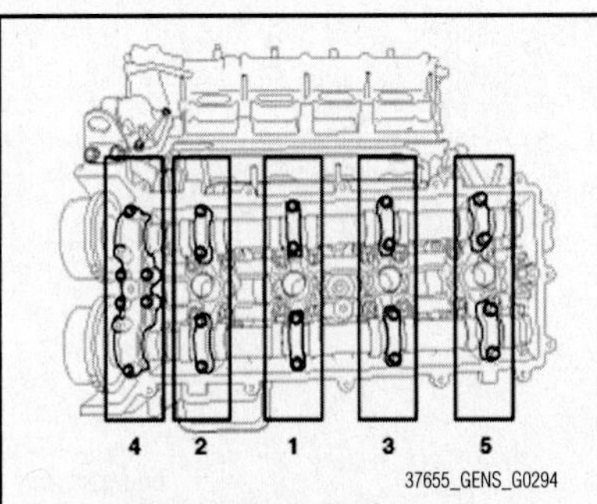

Fig. 232 Camshaft bearing caps and camshaft thrust bearing cap tightening sequence—Left-hand side

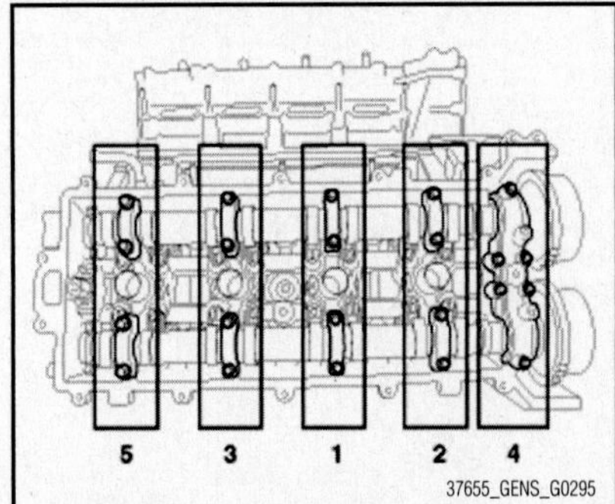

Fig. 233 Camshaft bearing caps and camshaft thrust bearing cap tightening sequence—Right-hand side

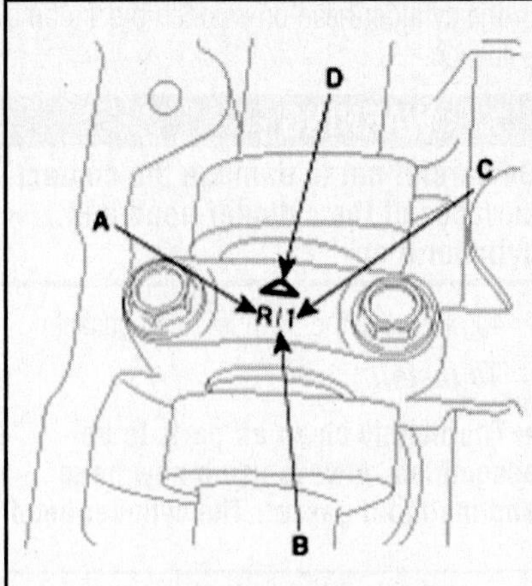

A: L (Left-hand), R(Right-hand)
B: I (Intake), None or E (Exhaust)
C: Journal number
D: Front mark

37655_GENS_G0258

Fig. 234 Position the camshaft bearing cap according to its markings

51. Install the water temperature control assembly and the water outlet fitting assembly with new O-ring and new gasket, and tighten to 12–15 ft. lbs. (17–20 Nm).

52. Install the water outlet and inlet pipes and tighten to 86–104 inch lbs. (10–12 Nm).

➡Use new O-rings when reassembling. Never apply oil on the O-ring and the O-ring groove of the pipe end. The O-ring must be free from scratches or damage. Clean the contact face before assembly. Lubricate the O-ring with coolant.

53. Install the drain plug and tighten to 71–74 ft. lbs. (96–100 Nm).

54. Install the alternator and tighten to 22–30 ft. lbs. (29–41 Nm).

55. Install the OCV and tighten to 86–104 inch lbs. (10–12 Nm).

56. Install the timing chain upper cover and seal. Refer to Timing Chain & Sprockets Removal & Installation.

57. Install the exhaust manifold, gasket, and heat protector. Refer to Exhaust Manifold Removal & Installation.

58. Install the Engine Coolant Temperature Sensor (ECTS) wiring assembly and bracket. Tightening torque:

- Wiring bracket bolts: 86–104 inch lbs. (10–12 Nm)
- Engine coolant temperature sensor: 15–29 ft. lbs. (20–39 Nm)

59. Install the oil pressure switch wiring assembly and bracket. Tightening torque:

- Wiring bracket bolts: 86–104 inch lbs. (10–12 Nm)

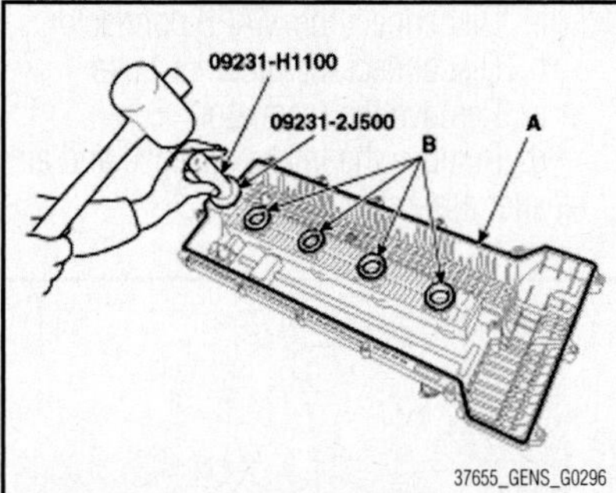

Fig. 235 Using the SST (09231-2J500 and 09231-H1100), install the oil seal and the gasket

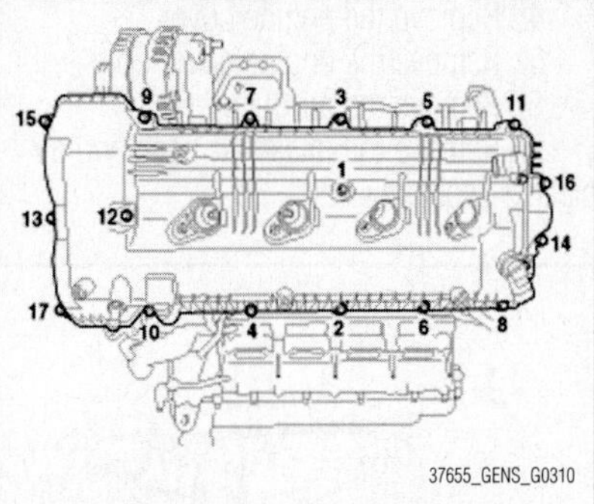

Fig. 237 Cylinder head cover bolt tightening sequence—Right-hand

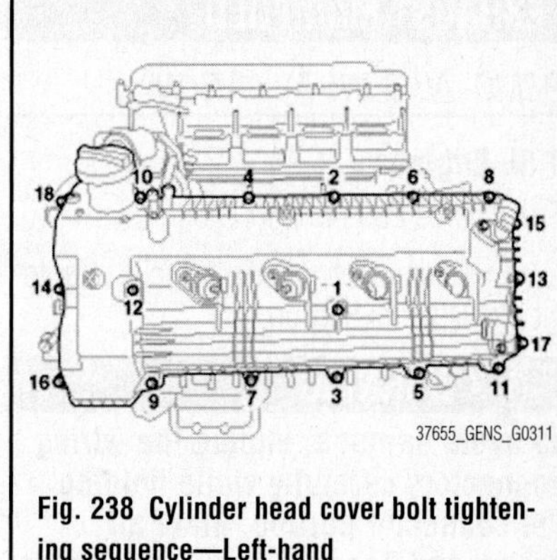

Fig. 238 Cylinder head cover bolt tightening sequence—Left-hand

- Oil pressure switch: 11–16 ft. lbs. (15–22 Nm)

60. Using the SST (09231-2J500 and 09231-H1100), install the oil seal and the gasket to both sides.

61. Install the cylinder head cover.

a. The sealant locations on cylinder head and timing chain upper case must be free of engine oil, etc.

b. Before assembling the cylinder head cover, the liquid sealant LT5900H or equivalent should be applied on the gap between cylinder head and timing chain upper case. The part must be assembled within 5 minutes after sealant was applied.

c. Install the cylinder head cover. Tightening torque:

- 1st step: 43–52 inch lbs. (5–6 Nm)
- 2nd step: 86–104 inch lbs. (10–12 Nm)

d. Tighten the cylinder head cover bolts as shown.

62. Install the ignition coils and tighten to 86–104 inch lbs. (10–12 Nm).

63. Reconnect the water hose.

64. Install the knock sensors and wiring assembly bracket and tighten to 48–56 ft. lbs. (65–77 Nm).

65. Install the intake manifold module. Refer to Intake Manifold Removal & Installation.

66. Reconnect the PCV hose to the intake manifold module.

67. Install the HO2S connector bracket.

68. Install the oil level gauge and tighten to 86–104 inch lbs. (10–12 Nm).

69. Install the engine wiring harness to the cylinder head.

a. Reconnect the Engine Coolant Temperature (ECT) sensor connector and the CVVT oil control valve (OCV) connectors.

b. Reconnect the alternator connector.

c. Reconnect the ETC module connector.

d. Reconnect the PCSV connector and the injector connectors.

e. Reconnect the ignition coil connectors and the wiring harness protector.

f. Reconnect the knock sensor connectors, the CMP sensor connectors, the CKP sensor connector, the HO2S connector, the condenser connector and the ground line.

g. Reconnect the Variable Intake System (VIS) solenoid valve connector and the wiring harness protector.

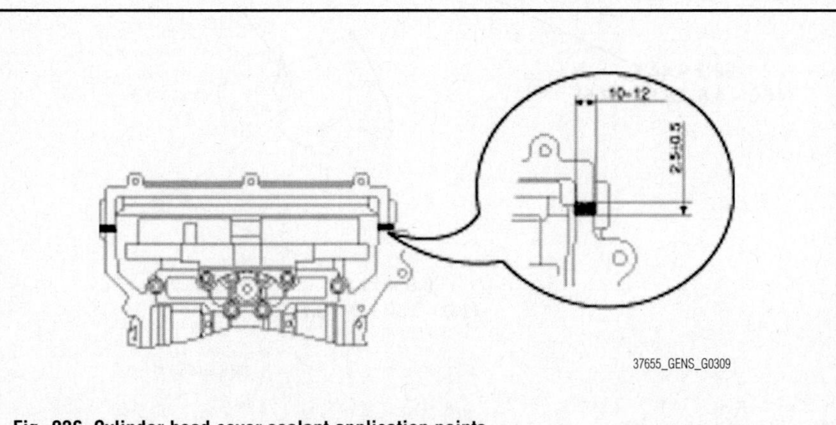

Fig. 236 Cylinder head cover sealant application points

h. Reconnect the injector connectors.

i. Reconnect the Oil Pressure Switch (OPS) connector and the air conditioner compressor switch connector.

70. Install the drive belt tensioner and drive belt.

71. Install the radiator and cooling fan assembly.

72. Connect the heater hoses.

73. Connect the fuel hose, the PCSV hose and the brake booster vacuum hose.

74. Connect the radiator upper and lower hoses.

75. Install the air cleaner assembly.

76. Install the engine cover.

77. Connect the battery negative terminal.

78. Refill the engine oil.

79. Clean the battery posts and cable terminals with sandpaper, assemble them, and then apply grease to prevent corrosion.

80. Inspect for fuel leakage:

a. After assembling the fuel line, turn on the ignition switch (do not operate the starter) so that the fuel pump runs for approximately two seconds and fuel line pressurizes.

b. Repeat this operation two or three times, and then check for fuel leakage at any point in the fuel lines.

81. Refill the radiator and reservoir tank with engine coolant.

82. Bleed air from the cooling system:

a. Start the engine and let it run until it warms up. (Until the radiator fan operates 3 or 4 times.)

b. Turn Off the engine. Check the level in the radiator, add coolant if needed. This will allow trapped air to be removed from the cooling system.

c. Put the radiator cap on tightly, then run the engine again and check for leaks.

EXHAUST MANIFOLD

REMOVAL & INSTALLATION

3.8L Engines

See Figures 239 through 246.

1. Before servicing the vehicle, refer to the Precautions Section.

⁂ WARNING

To avoid damage, unplug the wiring connectors carefully while holding the connector portion. Mark all wiring and hoses to avoid misconnection.

2. Disconnect the battery negative cable.
3. Remove the drain plug and drain the engine coolant.
4. Remove the engine cover.
5. Remove the engine cover bolts.
6. Remove the air duct.
7. Remove the intake air hose and air cleaner assembly.

A
1
2
3
4
37655_GENS_G0334

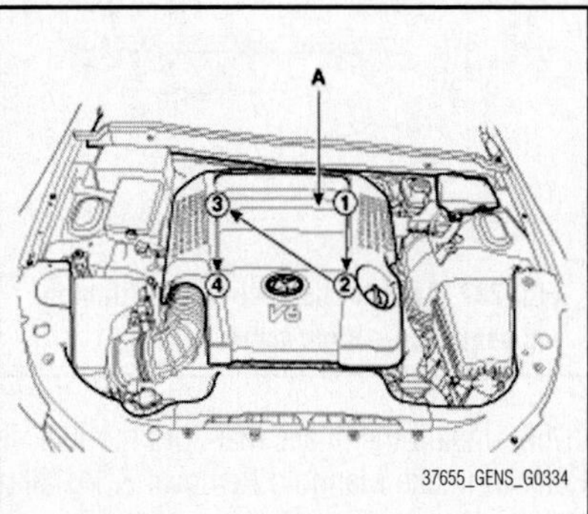

Fig. 239 Engine cover bolt removal sequence (A)

a. Disconnect the MAFS connector.
b. Disconnect the breather hose.
c. Remove the resonator.
d. Remove the intake air hose and air cleaner assembly.

A
37655_GENS_G0238

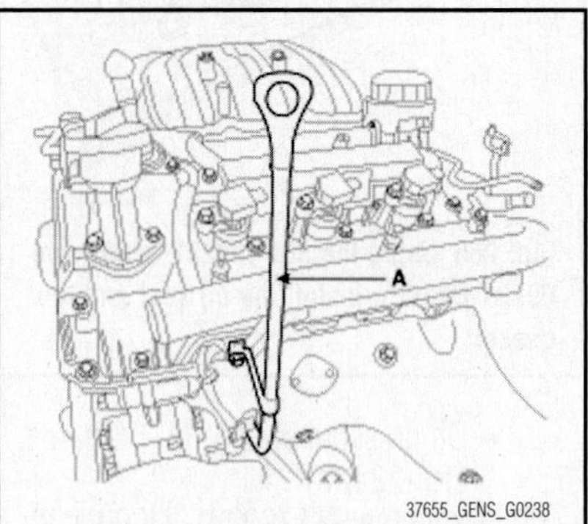

Fig. 240 Remove the oil level gauge tube (A)

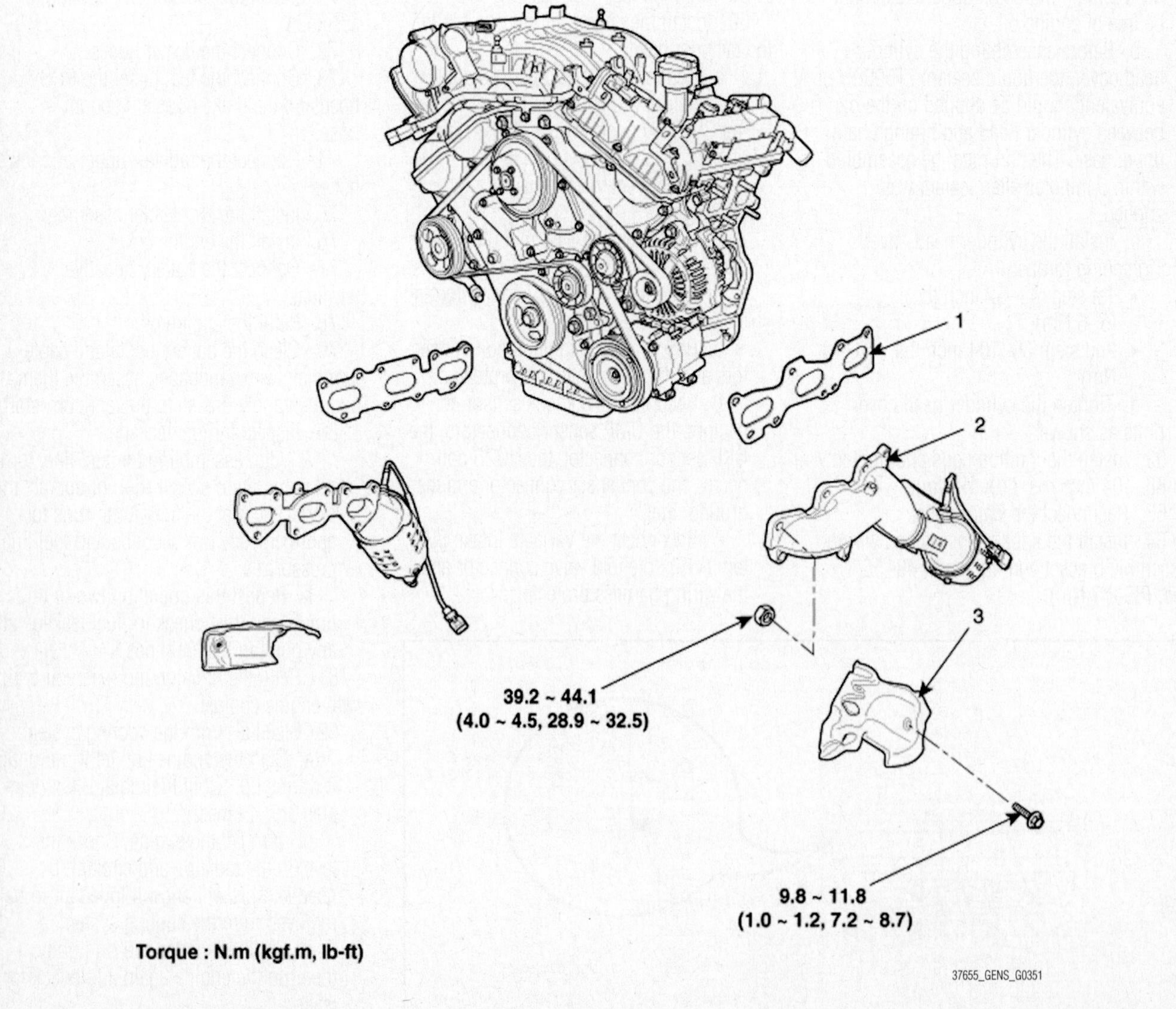

1
2
3
39.2 ~ 44.1
(4.0 ~ 4.5, 28.9 ~ 32.5)
9.8 ~ 11.8
(1.0 ~ 1.2, 7.2 ~ 8.7)
Torque : N.m (kgf.m, lb-ft)
37655_GENS_G0351

Fig. 241 Exhaust manifold (2) components, gasket (1), and heat protector (3)

8. Remove the radiator upper and lower hoses.

9. Disconnect the left-hand and right-hand oxygen sensor connectors.

10. Remove the oil level gauge tube.

11. Remove the left-hand side coolant pipe and hose.

12. Remove the throttle body coolant hose and pipe.

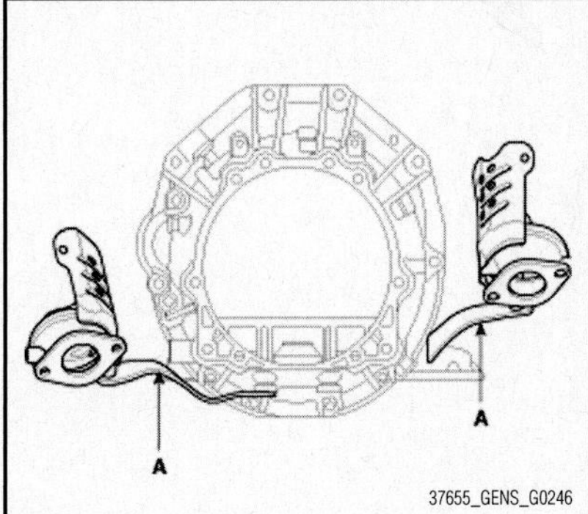

Fig. 242 Remove the exhaust manifold stays (A)

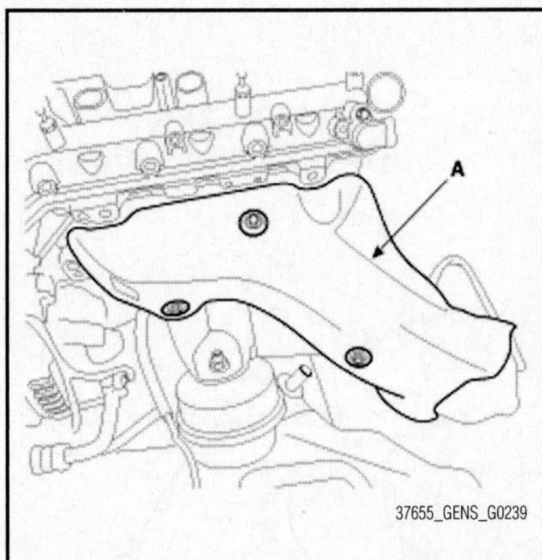

Fig. 243 Remove the left-hand exhaust manifold heat protector (A)

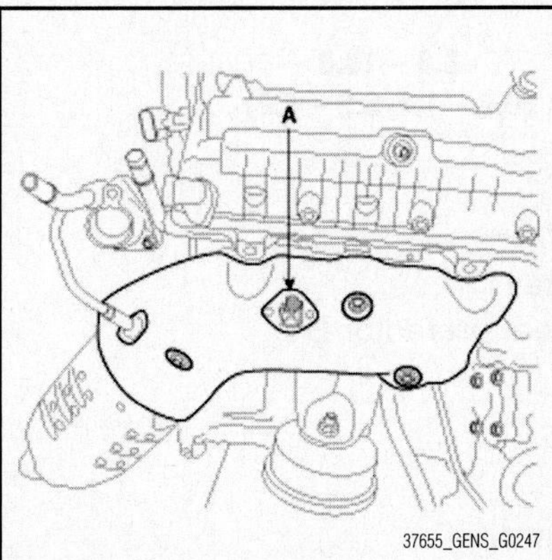

Fig. 244 Remove the right-hand exhaust manifold heat protector (A)

13. Remove the right-hand side coolant pipe.

14. Remove the exhaust manifold stays.

15. Remove the left-hand/right-hand exhaust manifold heat protectors.

16. Remove the left-hand/right-hand exhaust manifolds.

To install:

17. Install the exhaust manifolds and tighten to 29–33 ft. lbs. (39–44 Nm).

18. Install the heat protectors and tighten to 86–104 inch lbs. (10–12 Nm).

19. Install the exhaust manifold stays and tighten to 25–30 ft. lbs. (34–42 Nm).

20. Install the right-hand side coolant pipe and tighten to 15–17 ft. lbs. (20–24 Nm).

21. Install the throttle body coolant hose and pipe and tighten to 86–104 inch lbs. (10–12 Nm).

22. Install the left-hand side coolant pipe and hose and tighten:

- Bolt (B): 15–17 ft. lbs. (20–24 Nm)
- Bolt (C): 86–104 inch lbs. (10–12 Nm)

23. Install the oil level gauge tube and tighten to 14–17 ft. lbs. (19–23 Nm).

24. Connect the oxygen sensor connectors.

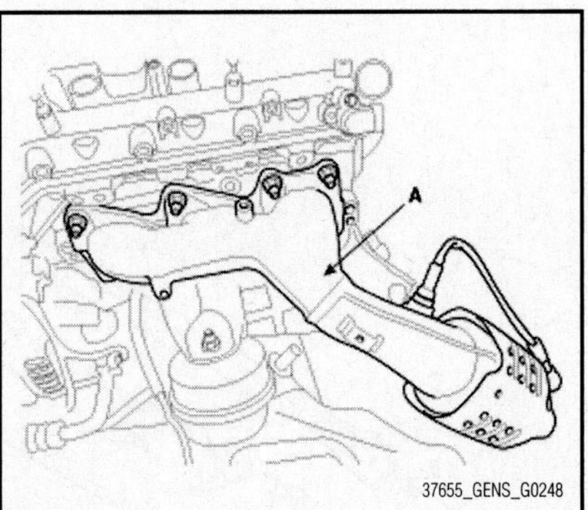

Fig. 245 Remove the left-hand exhaust manifold (A)

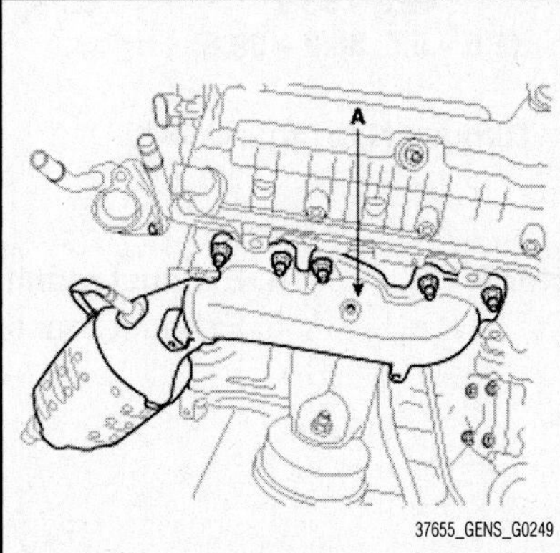

Fig. 246 Remove the right-hand exhaust manifold (A)

25. Install the radiator hoses.

26. Install the intake air hose and air cleaner assembly.

27. Install the air duct.

28. Install the engine cover.

29. Install the engine cover bolts.

30. Connect the battery negative cable.

31. Refill engine coolant.

32. Refill radiator and reservoir tank with engine coolant.

33. Bleed air from the cooling system.

a. Start engine and let it run until it warms up. (Until the radiator fan operates 3 or 4 times.)

b. Turn Off the engine. Check the level in the radiator, add coolant if needed. This will allow trapped air to be removed from the cooling system.

c. Put radiator cap on tightly, then run the engine again and check for leaks.

34. Clean the battery posts and cable terminals with sandpaper. Assemble and, then apply grease to prevent corrosion.

4.6L Engines

See Figures 247 through 254.

1. Before servicing the vehicle, refer to the Precautions Section.

WARNING

To avoid damage, unplug the wiring connectors carefully while holding the connector portion. Mark all wiring and hoses to avoid misconnection.

2. Disconnect the battery negative cable.

3. Remove the engine cover.

4. Remove the air duct and the air cleaner assembly.

5. Remove the oil level gauge.

6. Disconnect the oxygen sensor (HO2S) connector and remove the wiring bracket.

7. Remove the exhaust manifold stay and then disconnect the front muffler on both sides.

8. Remove the exhaust manifold heat protectors.

➡To remove the exhaust manifold and heat protector on left-hand side, remove the cylinder head cover on left-hand side first. Be careful not to put any harmful materials in the cylinder head during work.

9. Remove the exhaust manifold.

➡Use a universal joint wrench to loosen the upper and lower nuts on the right-hand exhaust manifold.

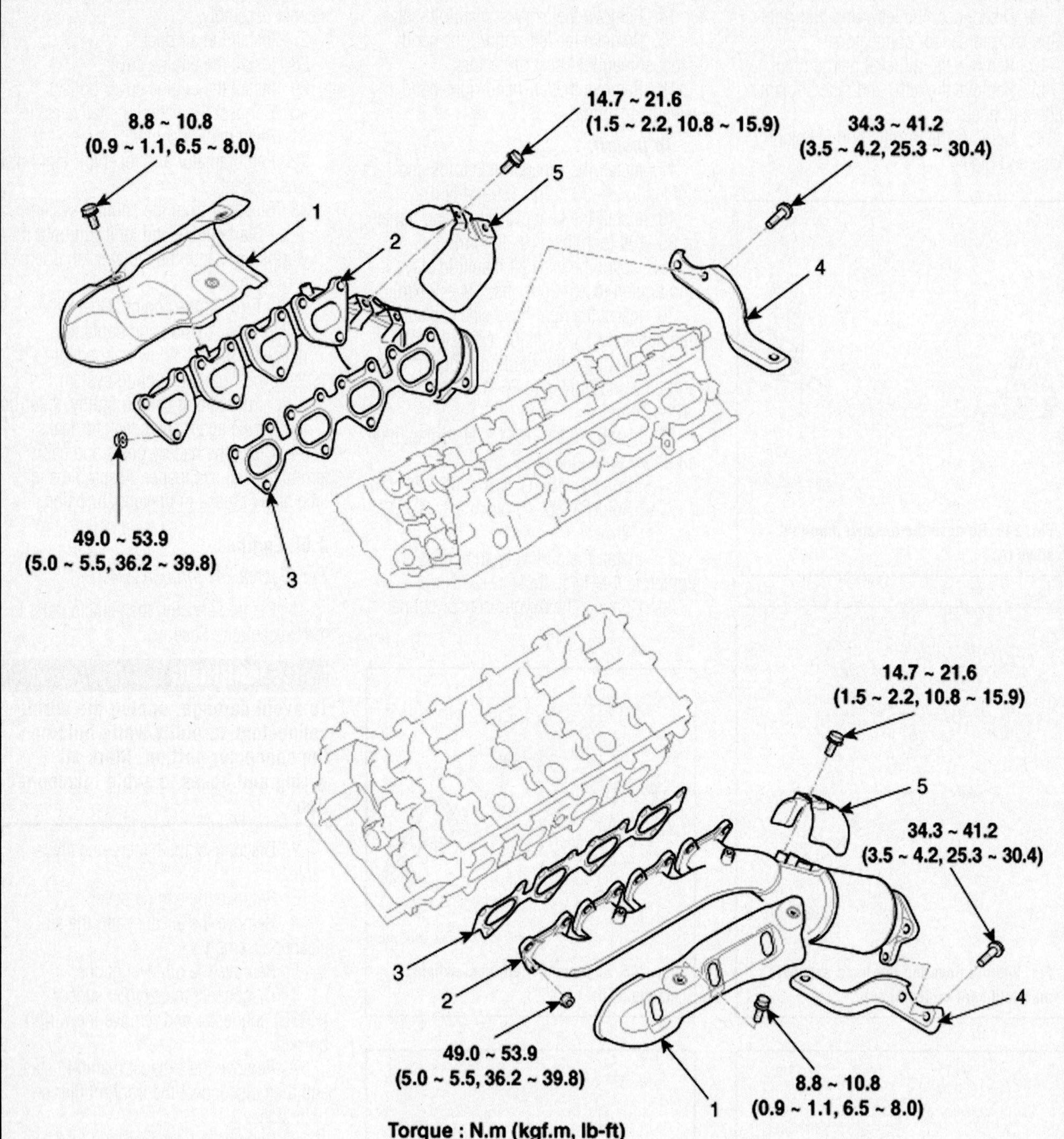

Fig. 247 Exhaust manifold and components

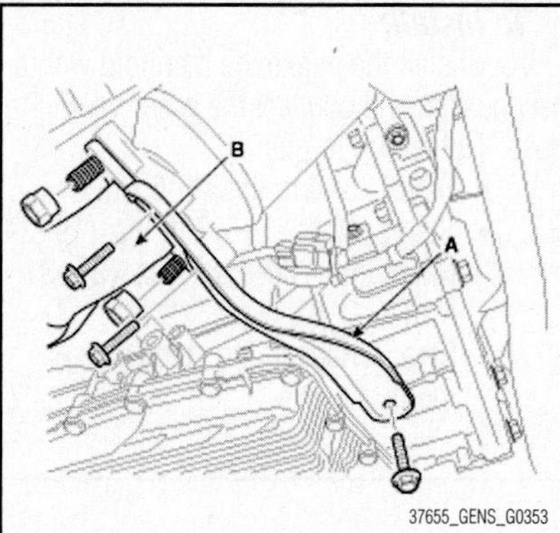

Fig. 248 Remove the exhaust manifold stay (A) and then disconnect the front muffler (B) on both sides

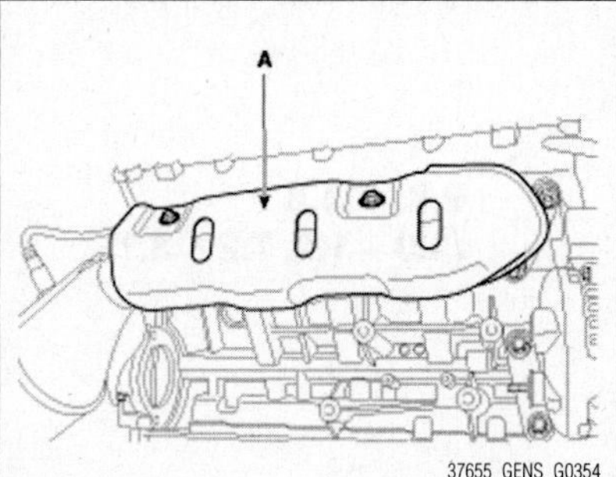

Fig. 249 Remove the exhaust manifold heat protectors (A)—Left-hand shown, right-hand similar

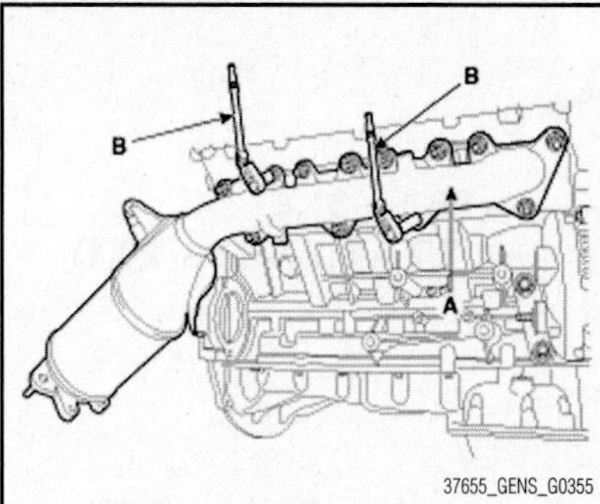

Fig. 250 Exhaust manifold (A) and universal joint wrench (B)—Left-hand

➡Use a universal joint wrench to loosen the upper nuts on the left-hand exhaust manifold. To loosen the lower nuts on the left-hand exhaust manifold, use an extension bar and an universal joint after removing the front wheel and side cover on left-hand side.

To install:

10. Install the exhaust manifold. Tightening torque: 36–40 ft. lbs. (49–54 Nm)

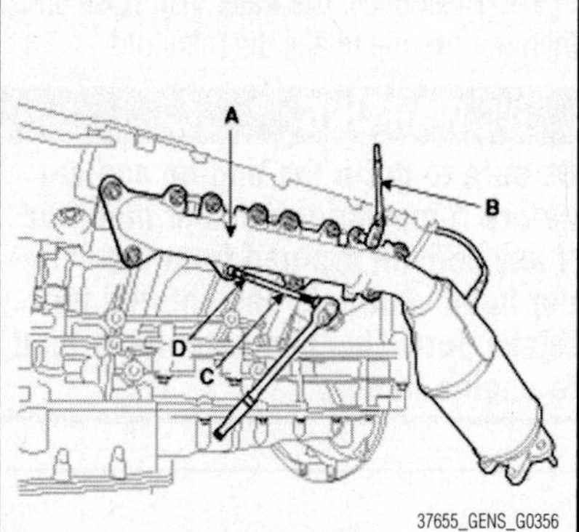

Fig. 251 Exhaust manifold (A), universal joint wrench (B), extension bar (C), and universal joint (D)—Right-hand

➡The "TOP" mark of the gasket must face the exhaust manifold.

11. Install the exhaust manifold heat protector. Tightening torque: 78–96 inch lbs. (9–11 Nm)

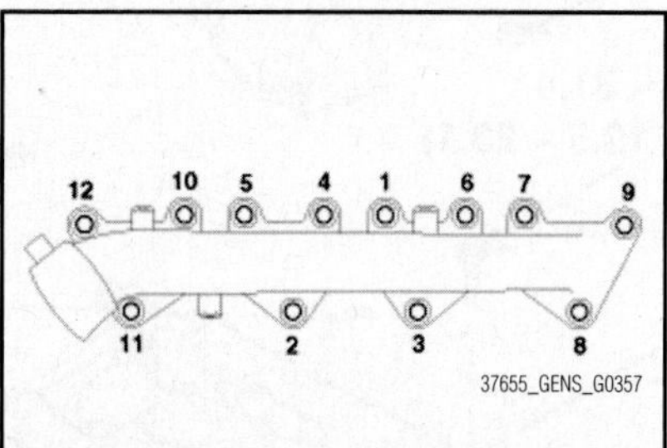

Fig. 252 Exhaust manifold bolt tightening sequence—Right-hand

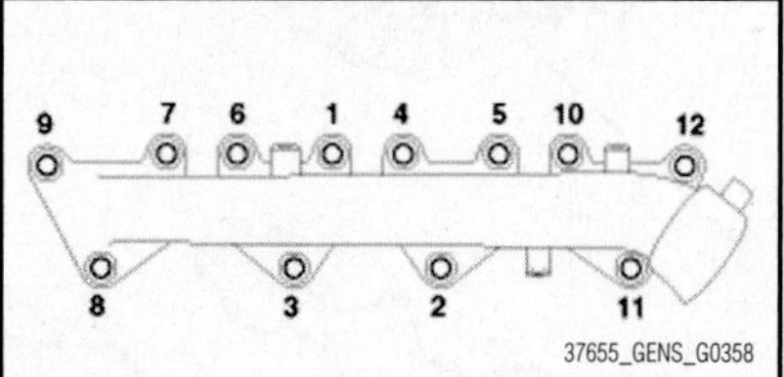

Fig. 253 Exhaust manifold bolt tightening sequence—Left-hand with engine assembly removal

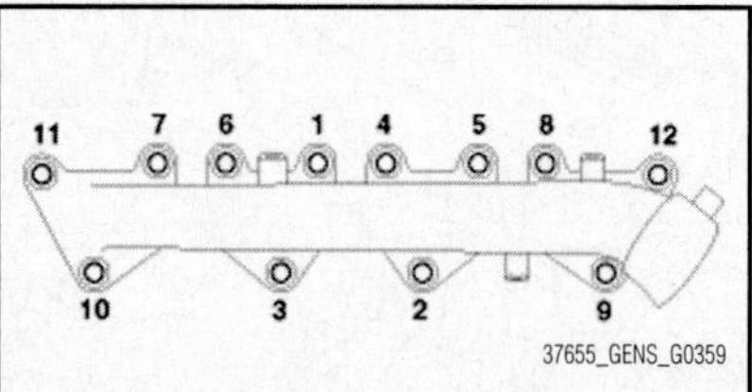

Fig. 254 Exhaust manifold bolt tightening sequence—Left-hand without engine assembly removal

➡After installing the exhaust manifold and heat protector on left-hand side, install the cylinder head.

12. Install the exhaust manifold stay and reconnect the front muffler on both sides. Tightening torque:
- Bolts (C): 25–30 ft. lbs. (34–41 Nm)
- Nuts (D): 36–40 ft. lbs. (49–54 Nm)

13. Reconnect the HO2S connector and install the wiring connector bracket.
14. Install the oil level gauge. Tightening torque: 86–104 inch lbs. (10–12 Nm).
15. Install the air cleaner assembly and air duct.
16. Install the engine cover.
17. Connect the negative battery cable.

INTAKE MANIFOLD

REMOVAL & INSTALLATION

3.8L Engines

See Figures 255 through 260.

1. Before servicing the vehicle, refer to the Precautions Section.

➡To avoid damage, unplug the wiring connectors carefully while holding the connector portion. Mark all wiring and hoses to avoid misconnection.

2. Disconnect the battery negative cable.
3. Remove the drain plug and drain the engine coolant.
4. Remove the engine cover.
5. Remove the air duct.
6. Remove the intake air hose and air cleaner assembly.
7. Remove the fuel hose.
8. Disconnect the brake vacuum hose.
9. Disconnect the engine wiring connectors.

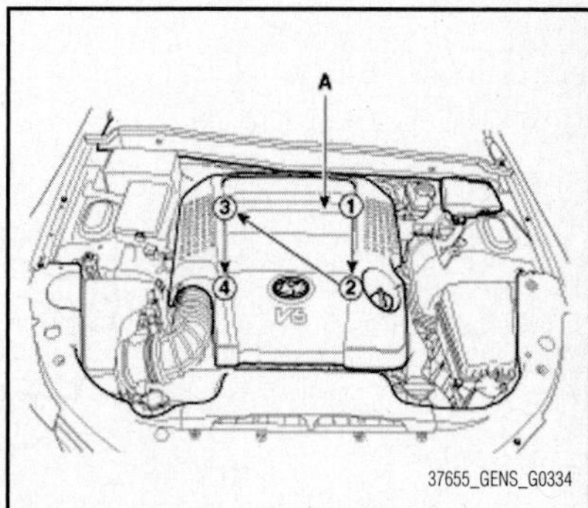

Fig. 255 Engine cover bolt removal sequence (A)

10. Remove the fuel pipe and vacuum pipe.
11. Remove the throttle body coolant hoses and PCV hose.
12. Remove the surge tank stay.
13. Remove the surge tank assembly.
14. Disconnect the right-hand injector connector.
15. Disconnect the water vent hose and then remove the intake the manifold.

⁂ WARNING

Be sure to drain the engine coolant before removing the intake manifold. If any coolant drained from the cylinder head vent hole has entered the intake port. This can potentially lead to engine trouble.

To install:

16. Install the intake the manifold with a new gasket, and connect the water vent hose. Tightening torque:
    - Step 1: 3–4 ft. lbs. (4–6 Nm)
    - Step 2: Nut—14–17 ft. lbs. (19–24 Nm), Bolt—20–23 ft. lbs. (27–31 Nm)
    - Step 3: Repeat 2nd step twice or more.

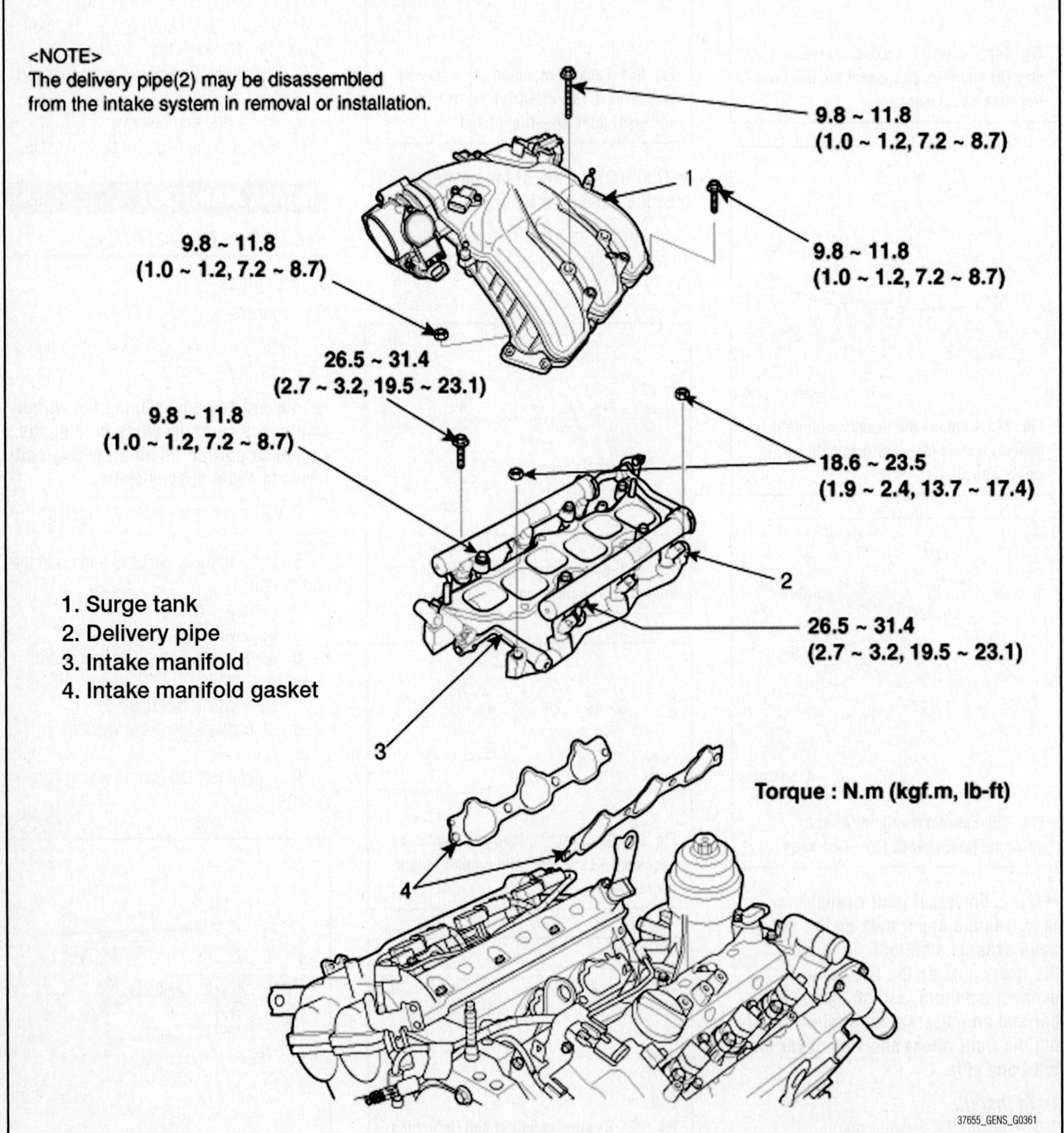

Fig. 256 Intake manifold and components

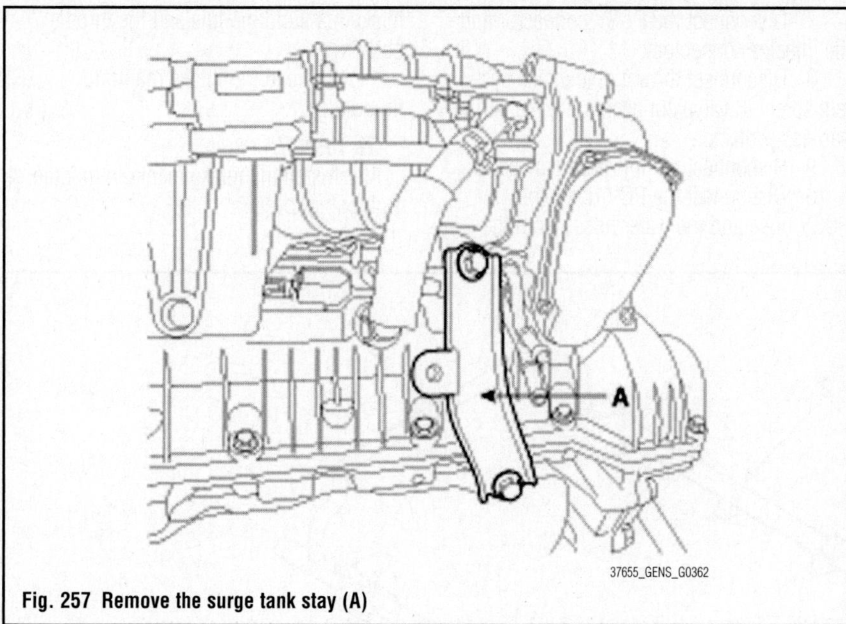

Fig. 257 Remove the surge tank stay (A)

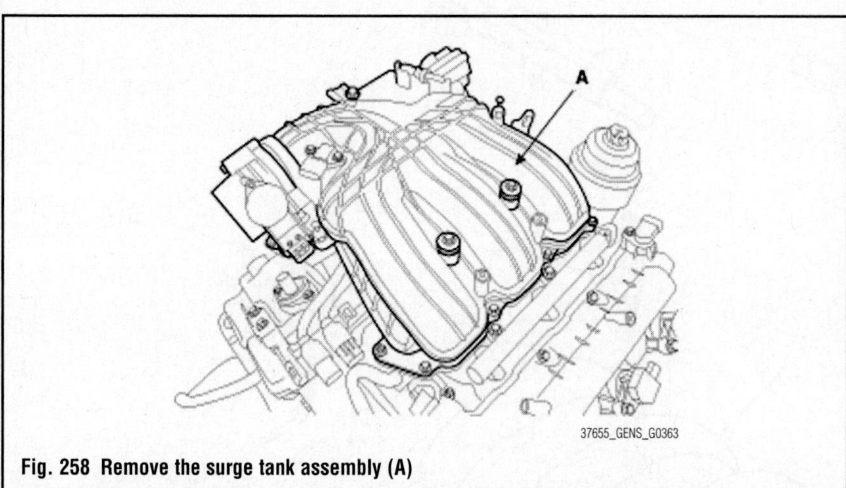

Fig. 258 Remove the surge tank assembly (A)

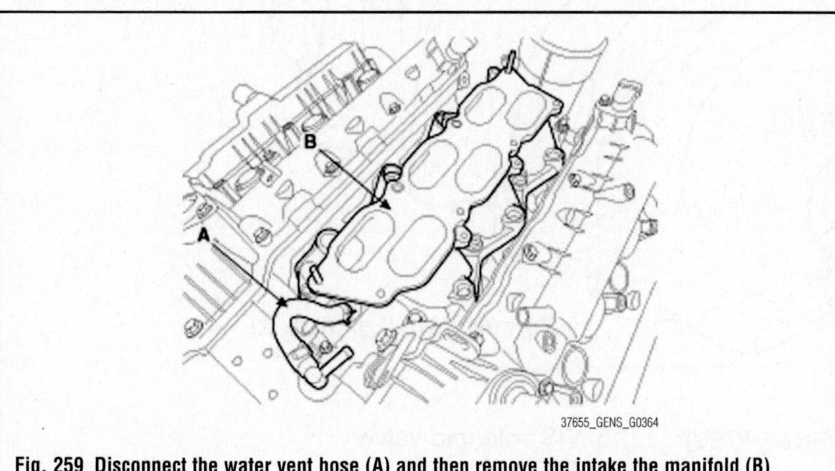

Fig. 259 Disconnect the water vent hose (A) and then remove the intake the manifold (B)

- a–h: 1st step order
- 1–8: 2nd step order

➡Confirm the manifold gasket identification mark (left-hand, RH) and be careful of the installation direction.

17. Connect the right-hand injector connector.

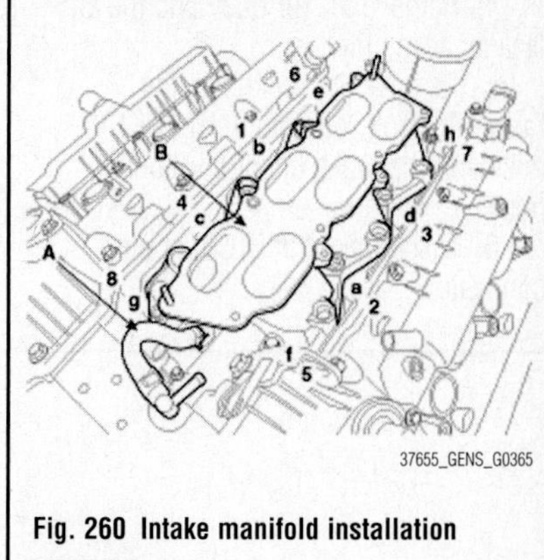

Fig. 260 Intake manifold installation sequence

18. Install the surge tank assembly. Tightening torque: 86–104 inch lbs. (10–12 Nm).
19. Install the surge tank stay. Tightening torque: 20–23 ft. lbs. (28–31 Nm).
20. Install the throttle body coolant hoses and PCV hose.
21. Install the fuel pipe and vacuum pipe. Tightening torque: 86–104 inch lbs. (10–12 Nm).
22. Connect the engine wiring connectors.
23. Connect the brake vacuum hose.
24. Install the fuel hose. Tightening torque: 86–104 inch lbs. (10–12 Nm).
25. Install the intake air hose and air cleaner assembly.
26. Install the air duct.
27. Install the engine cover.
28. Install the engine cover bolts.
29. Connect the battery negative cable.
    a. Refill engine coolant.
    b. Refill radiator and reservoir tank with engine coolant.
    c. Bleed air from the cooling system. Start engine and let it run until it warms up. (Until the radiator fan operates 3 or 4 times.) Turn Off the engine. Check the level in the radiator, add coolant if needed. This will allow trapped air to be removed from the cooling system. Put radiator cap on tightly, then run the engine again and check for leaks.
    d. Clean the battery posts and cable terminals with sandpaper. Assemble and, then apply grease to prevent corrosion.

4.6L Engines

See Figures 261 through 269.

1. Before servicing the vehicle, refer to the Precautions Section.
2. Disconnect the negative battery cable.
3. Remove the engine cover.

4. Remove the air duct, and the air cleaner assembly.

5. Remove the fuel hose, the Purge Control Solenoid Valve (PCSV) hose and the brake booster vacuum hose.

6. Disconnect the ETC module connector.

7. Disconnect the PCSV connector and the injector connectors.

8. Disconnect the variable intake system solenoid valve connector and the wiring harness protector.

9. Disconnect the injector connectors.

10. Disconnect the PCV hose, the PCSV hose and the water hoses from the intake manifold module and the throttle body.

11. Remove the intake manifold module.

To install:

12. Install the intake manifold module. Tightening torque:

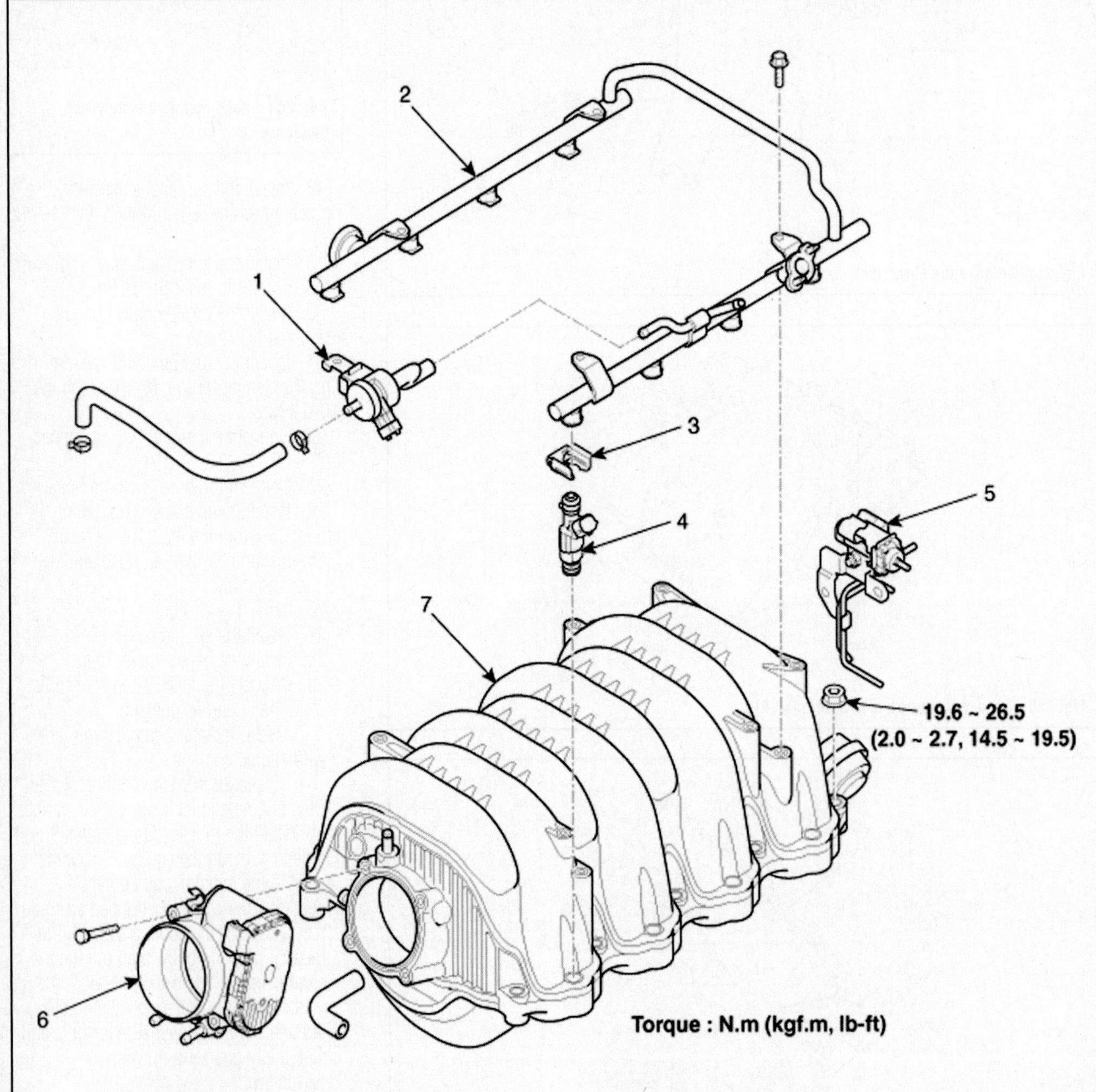

Fig. 261 Intake manifold and components

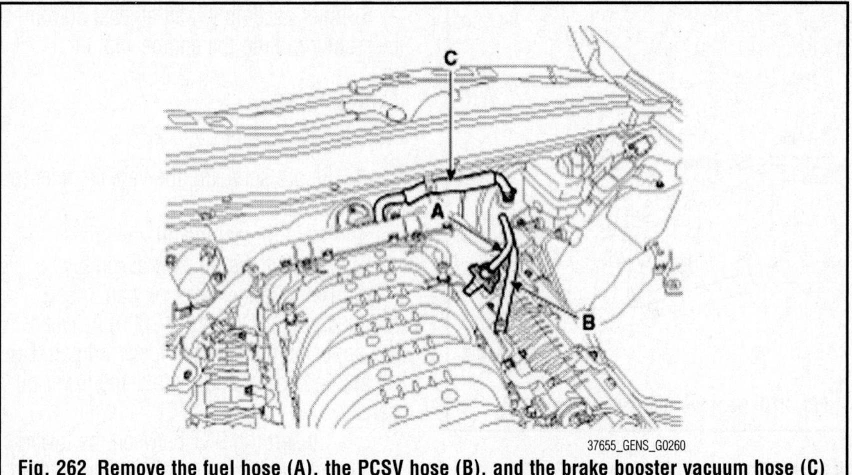

Fig. 262 Remove the fuel hose (A), the PCSV hose (B), and the brake booster vacuum hose (C)

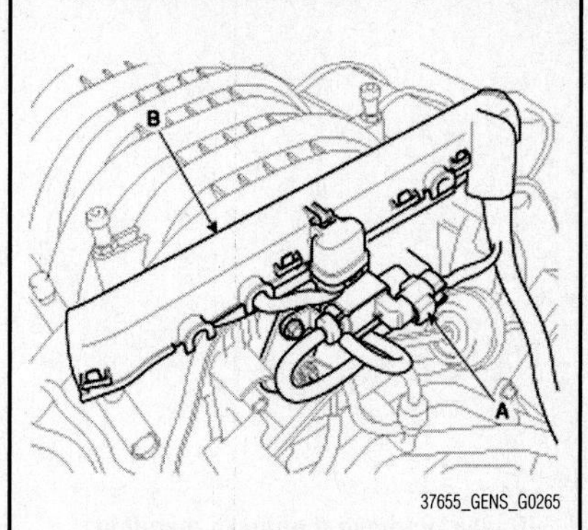

Fig. 265 Disconnect the variable intake system solenoid valve connector (A) and the wiring harness protector (B)

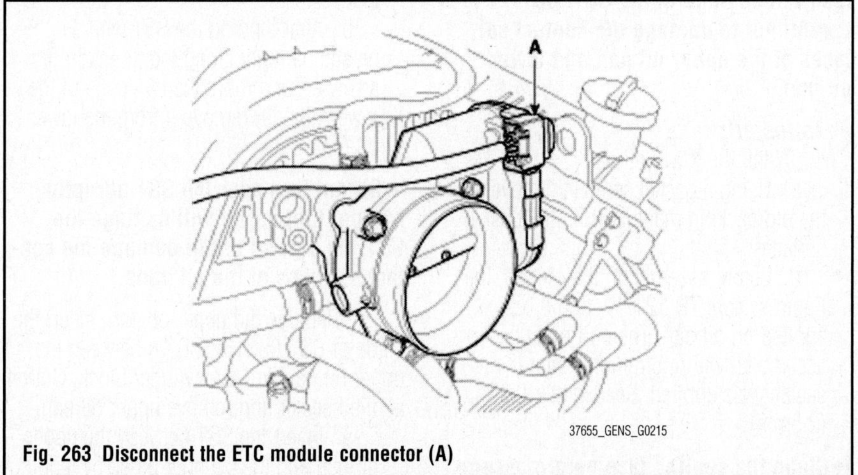

Fig. 263 Disconnect the ETC module connector (A)

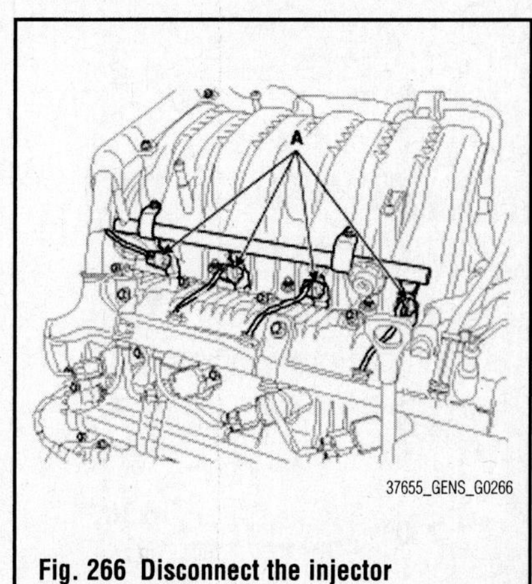

Fig. 266 Disconnect the injector connectors (A)

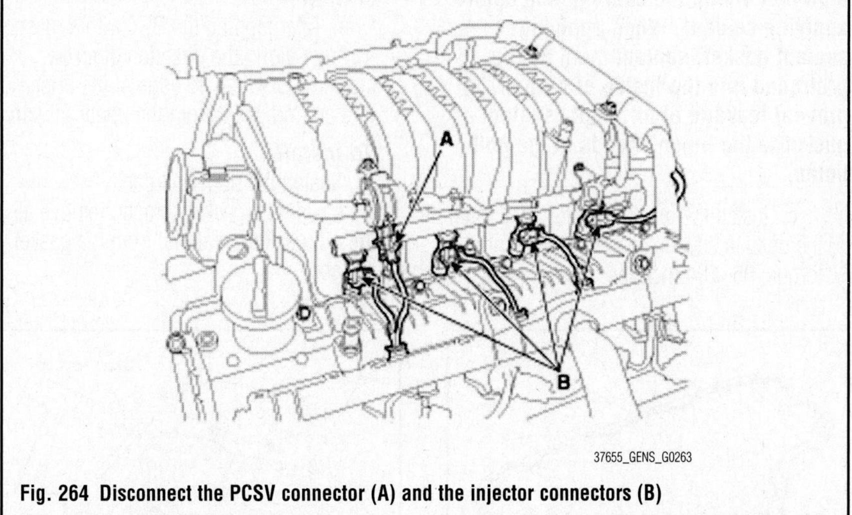

Fig. 264 Disconnect the PCSV connector (A) and the injector connectors (B)

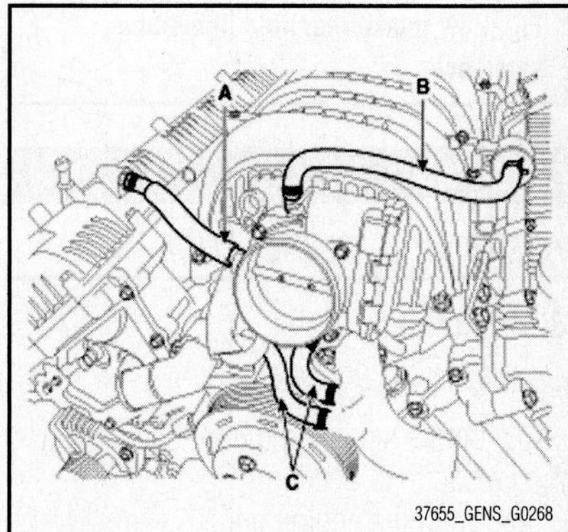

Fig. 267 Disconnect the PCV hose (A), the PCSV hose (B) and the water hoses (C) from the intake manifold module and the throttle body

- 1st step: 43–70 inch lbs. (5–8 Nm)
- 2nd step: 15–20 ft. lbs. (20–27 Nm).

13. Connect the PCV hose, the PCSV hose and the water hoses to the intake manifold module and the throttle body.
14. Reconnect the injector connectors.
15. Connect the variable intake system solenoid valve connector and the wiring harness protector.
16. Connect the PCSV connector and the injector connectors.
17. Connect the ETC module connector.
18. Connect the fuel hose, the PCSV hose and the brake booster vacuum hose.
19. Install the air cleaner assembly and air duct.
20. Install the engine cover.
21. Connect the battery negative terminal.

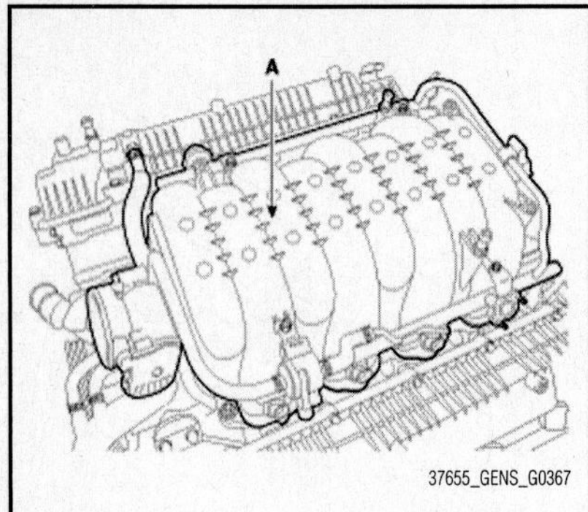

Fig. 268 Remove the intake manifold module (A)

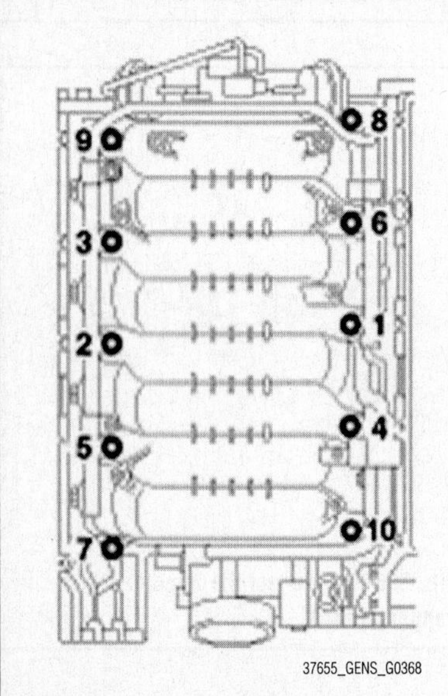

Fig. 269 Intake manifold tightening sequence

OIL PAN

REMOVAL & INSTALLATION

3.8L Engines

See Figures 270 and 271.

1. Before servicing the vehicle, refer to the Precautions Section.
2. Drain the engine oil.
3. Remove the lower oil pan. Insert the blade of SST (09215-3C000) between the upper oil pan and lower oil pan. Cut off applied sealer and remove the lower oil pan.

➡Insert the SST between the oil pan and the ladder frame by tapping it with a plastic hammer in the direction of arrow. After tapping the SST with a plastic hammer along the direction of arrow around more than ⅔ edge of the oil pan, remove it from the ladder

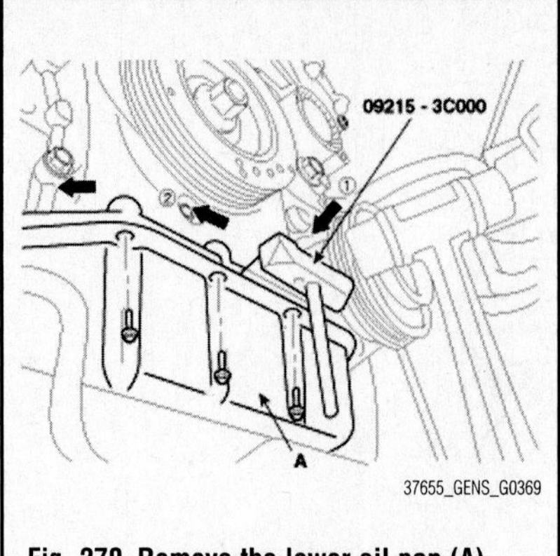

Fig. 270 Remove the lower oil pan (A)

frame. Do not turn over the SST abruptly without tapping. It be result in damage of the SST. Be careful not to damage the contact surfaces of the upper oil pan and lower oil pan.

To install:

4. Install the lower oil pan.
    a. Using a gasket scraper, remove all the old packing material from the gasket surfaces.
    b. Before assembling the oil pan, the liquid sealant TB 1217H should be applied on oil pan. The part must be assembled within 5 minutes after the sealant was applied. Bead width: 0.1 in. (2.5mm)

➡Clean the sealing face before assembling two parts. Remove harmful foreign matters on the sealing face before applying sealant. When applying sealant gasket, sealant must not be protruded into the inside of oil pan. To prevent leakage of oil, apply sealant gasket to the inner threads of the bolt holes.

    c. Install the oil pan. Uniformly tighten the bolts in several passes. Tightening torque: 86–104 inch lbs. (10–12 Nm).

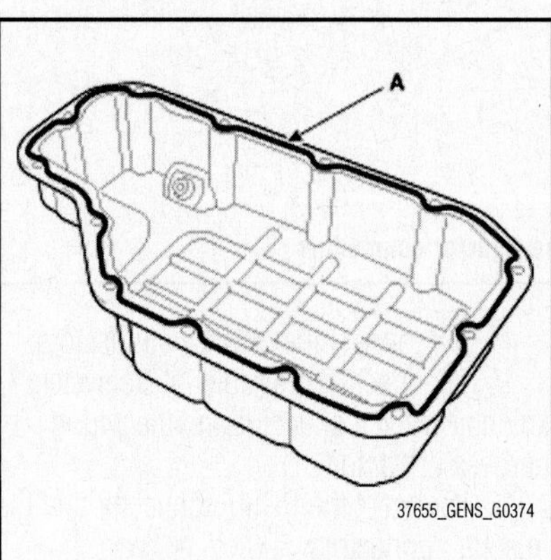

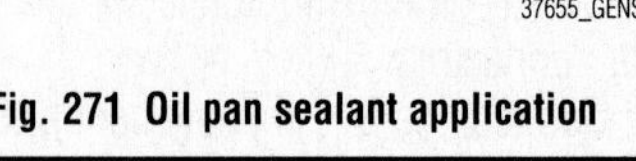

Fig. 271 Oil pan sealant application

5. After assembly, wait at least 30 minutes before filling the engine with oil.

4.6L Engine

See Figures 272 through 277.

1. Before servicing the vehicle, refer to the Precautions Section.
2. Drain the engine oil.
3. Remove the oil filter assembly.
4. Remove the lower oil pan. Insert the blade of SST(09215-3C000) between the upper oil pan and the lower oil pan. Cut off applied sealer and remove the lower oil pan.
    a. Insert the SST between the lower oil pan and the upper oil pan by tapping it with a plastic hammer in the direction of arrow.
    b. After tapping the SST with a plastic hammer along the direction of arrow around more than ⅔ edge of the lower oil pan, remove it from the lower oil pan.

➡Do not turn over the SST abruptly without tapping. It will damage the SST. Be careful not to damage the contact surfaces of the oil pans.

5. Remove the upper oil pan. Insert the blade of SST(09215-3C000) between the upper oil pan and the cylinder block. Cut off applied sealer and remove upper oil pan.
    a. Insert the SST between the upper oil pan and the cylinder block by tapping it with a plastic hammer in the direction of arrow.
    b. After tapping the SST with a plastic hammer along the direction of arrow around more than ⅔ edge of the upper oil pan, remove it from the upper oil pan.

To install:

6. Install the upper oil pan.
    a. Using a gasket scraper, remove all the old packing material from the gasket surfaces.

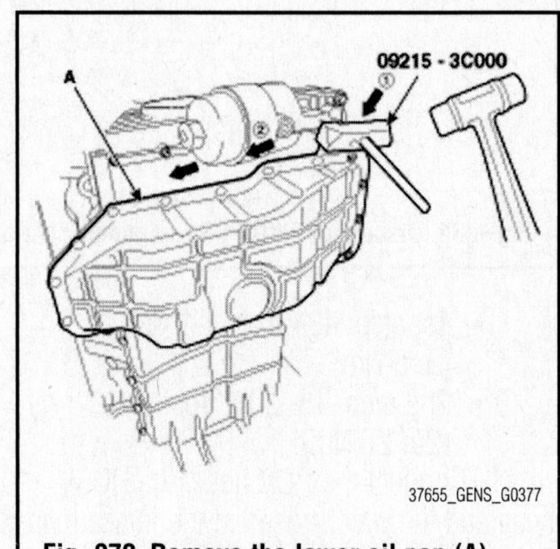

Fig. 272 Remove the lower oil pan (A)

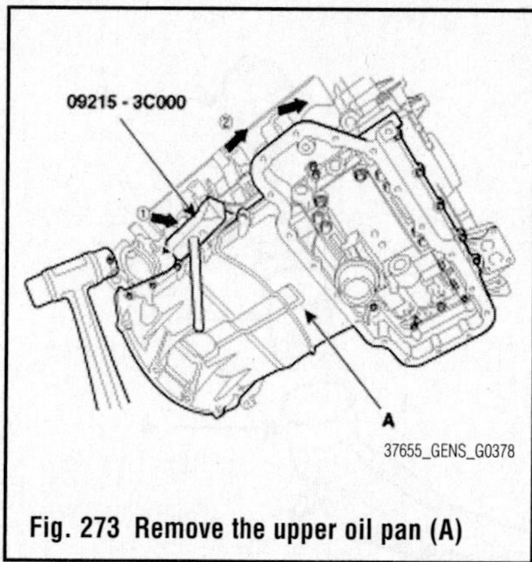

Fig. 273 Remove the upper oil pan (A)

b. Before assembling the oil pan, the liquid sealant TB1217H or LT5900H should be applied on upper oil pan. The part must be assembled within 5 minutes after the sealant was applied. Bead width: 0.1 in. (2.5mm)

➡Clean the sealing face before assembling two parts. Remove harmful foreign materials on the sealing face before applying sealant. When applying sealant gasket, sealant must not protrude into the inside of oil pan. To prevent leakage of oil, apply sealant gasket to the inner threads of the bolt holes.

c. Install the upper oil pan. Uniformly tighten the bolts in several passes. Tightening torque: 86–104 inch lbs. (10–12 Nm).

7. Install the lower oil pan.
   a. Using a gasket scraper, remove all the old packing material from the gasket surfaces.
   b. Before assembling the oil pan, the liquid sealant TB1217H or LT5900H should be applied on lower oil pan. The part must be assembled within 5 minutes after the sealant was applied. Bead width: 0.1 in. (2.5mm)

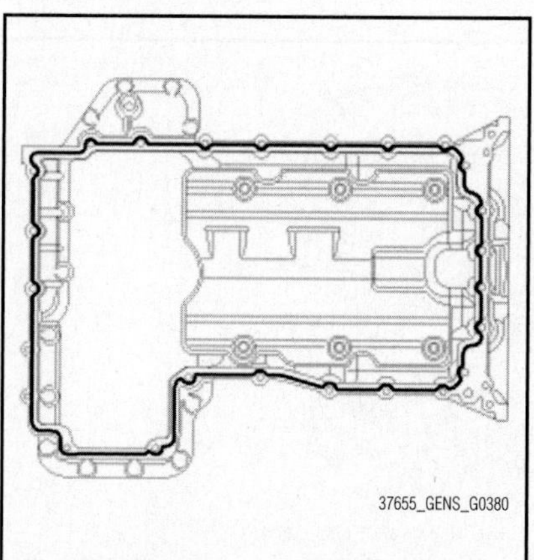

Fig. 274 Upper oil pan sealant application

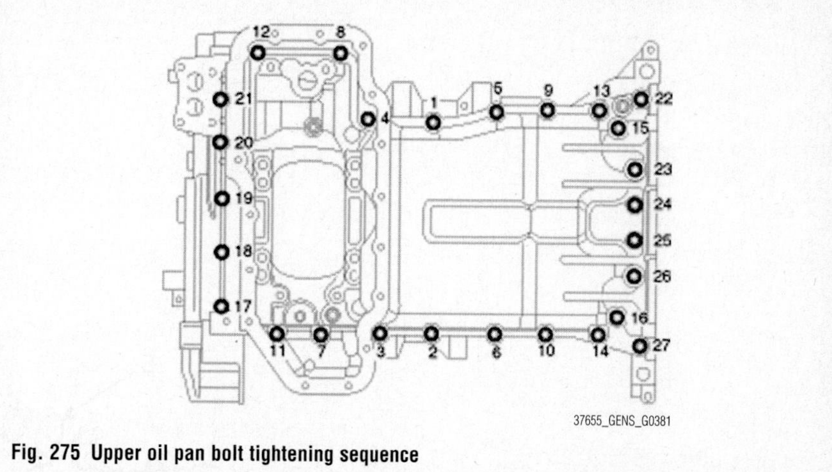

Fig. 275 Upper oil pan bolt tightening sequence

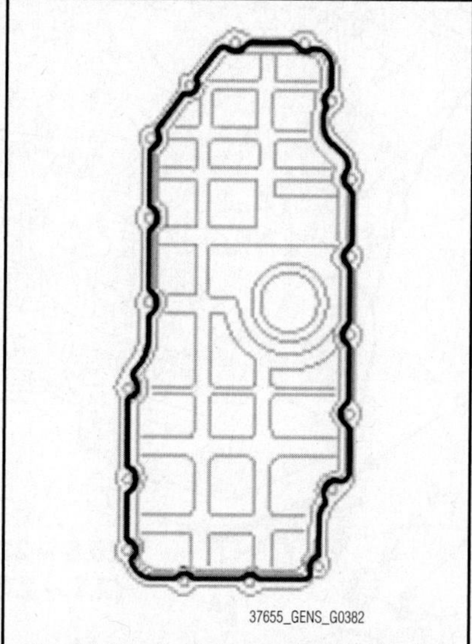

Fig. 276 Lower oil pan sealant application

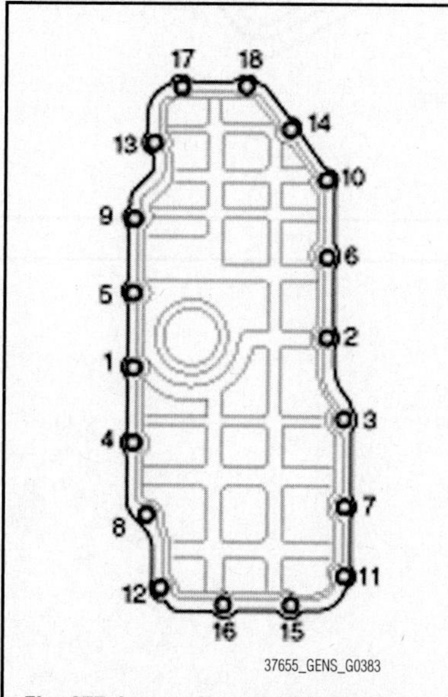

Fig. 277 Lower oil pan bolt tightening sequence

➡Clean the sealing face before assembling two parts. Remove harmful foreign materials on the sealing face before applying sealant. When applying sealant gasket, sealant must not protrude into the inside of oil pan. To prevent leakage of oil, apply sealant gasket to the inner threads of the bolt holes.

8. Install the lower oil pan. Uniformly tighten the bolts in several passes. Tightening torque: 86–104 inch lbs. (10–12 Nm).
9. Install the oil filter assembly. Tightening torque: 15–17 ft. lbs. (20–24 Nm)
10. Refill the engine oil.

OIL PUMP

REMOVAL & INSTALLATION

3.8L Engines

See Figures 278 through 282.

1. Before servicing the vehicle, refer to the Precautions Section.
2. Drain the engine oil.
3. Remove the lower oil pan. Refer to Oil Pan Removal & Installation.
4. Remove the oil pump chain cover.
5. Remove the oil pump chain sprocket.
6. Remove the oil pump.

To install:

7. Install the oil pump (A). Always use a new O-ring (B). Tightening torque: 15–17 ft. lbs. (21–23 Nm).
8. Install the oil pump sprocket and the oil pump chain on the oil pump. Tightening torque: 14–16 ft. lbs. (19–22 Nm).
9. Install the oil pump chain cover. Tightening torque: 86–104 inch lbs. (10–12 Nm).
10. Install the lower oil pan.
11. After assembly, wait at least 30 minutes before filling the engine with oil.

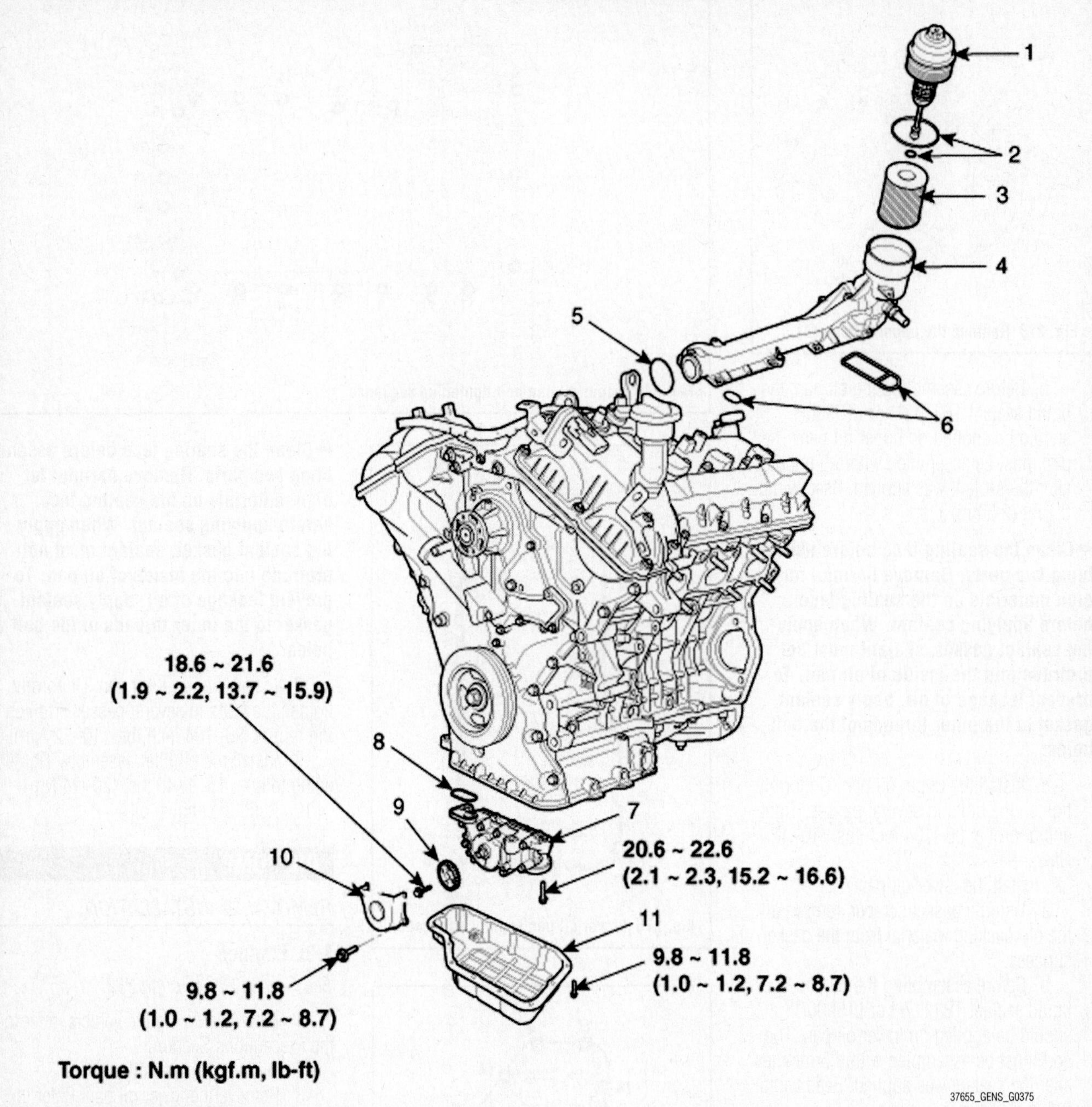

1
2
3
4
5
6
7
8
9
10
11
18.6 ~ 21.6
(1.9 ~ 2.2, 13.7 ~ 15.9)
20.6 ~ 22.6
(2.1 ~ 2.3, 15.2 ~ 16.6)
9.8 ~ 11.8
(1.0 ~ 1.2, 7.2 ~ 8.7)
9.8 ~ 11.8
(1.0 ~ 1.2, 7.2 ~ 8.7)
Torque : N.m (kgf.m, lb-ft)
37655_GENS_G0375

Fig. 278 Oil pump and components

A
37655_GENS_G0370

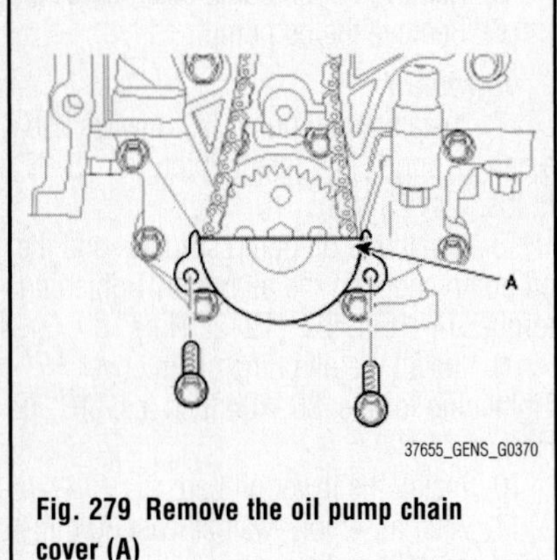

Fig. 279 Remove the oil pump chain cover (A)

A
37655_GENS_G0371

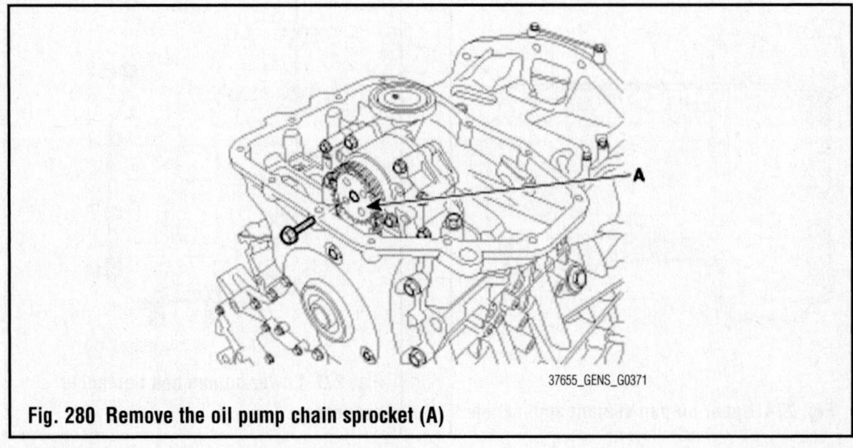

Fig. 280 Remove the oil pump chain sprocket (A)

A
37655_GENS_G0372

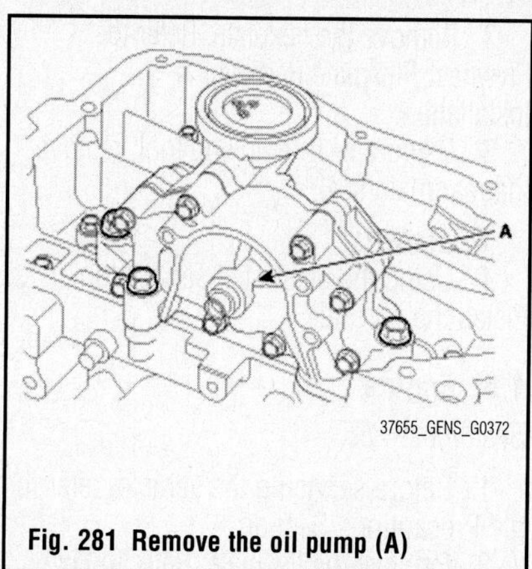

Fig. 281 Remove the oil pump (A)

B
A
37655_GENS_G0373

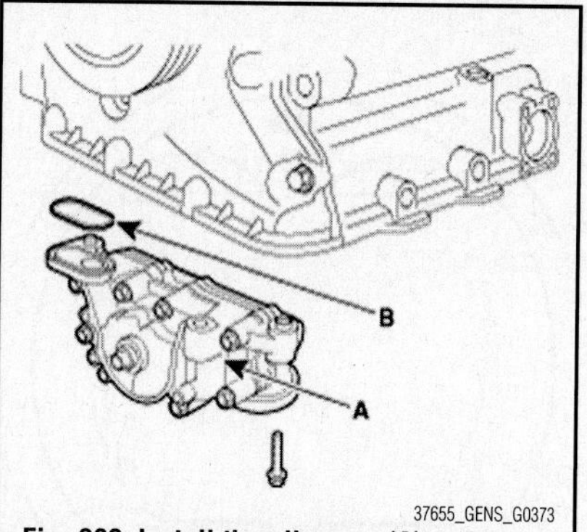

Fig. 282 Install the oil pump (A) and a new O-ring (B)

4.6L Engine

See Figures 283 and 284.

1. Before servicing the vehicle, refer to the Precautions Section.
2. Drain the engine oil.
3. Remove the oil filter assembly.
4. Remove the lower and upper oil pans. Refer to Oil Pan Removal & Installation.
5. Remove the oil pump assembly after remove the oil pump sprocket.

To install:

6. Install the oil pump assembly and then oil pump sprocket. Tightening torque:

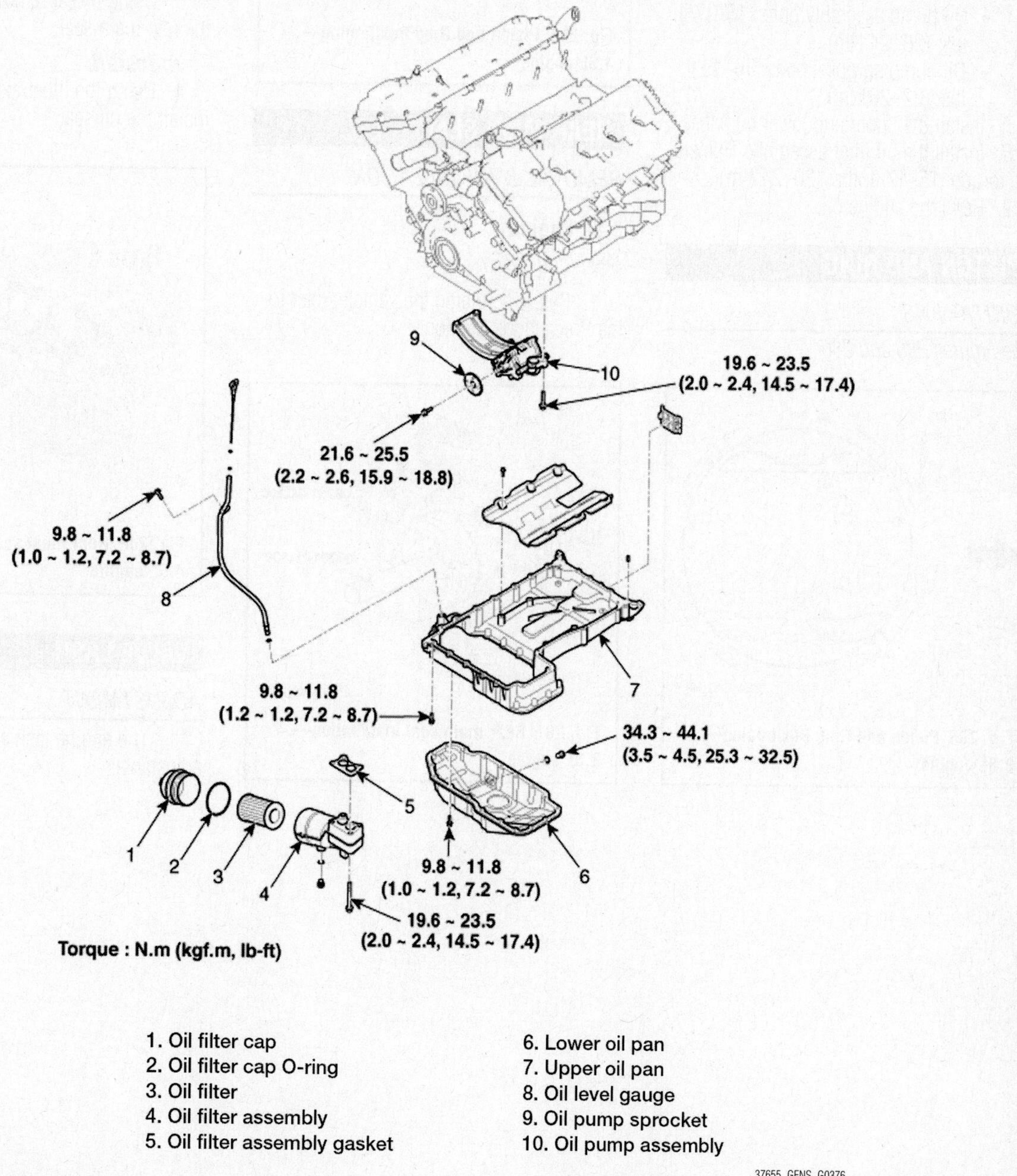

9
10
19.6 ~ 23.5
(2.0 ~ 2.4, 14.5 ~ 17.4)
21.6 ~ 25.5
(2.2 ~ 2.6, 15.9 ~ 18.8)
9.8 ~ 11.8
(1.0 ~ 1.2, 7.2 ~ 8.7)
8
9.8 ~ 11.8
(1.2 ~ 1.2, 7.2 ~ 8.7)
7
34.3 ~ 44.1
(3.5 ~ 4.5, 25.3 ~ 32.5)
5
1
2
3
4
6
9.8 ~ 11.8
(1.0 ~ 1.2, 7.2 ~ 8.7)
19.6 ~ 23.5
(2.0 ~ 2.4, 14.5 ~ 17.4)
Torque : N.m (kgf.m, lb-ft)
1. Oil filter cap
2. Oil filter cap O-ring
3. Oil filter
4. Oil filter assembly
5. Oil filter assembly gasket
6. Lower oil pan
7. Upper oil pan
8. Oil level gauge
9. Oil pump sprocket
10. Oil pump assembly
37655_GENS_G0376

Fig. 283 Oil pump and components

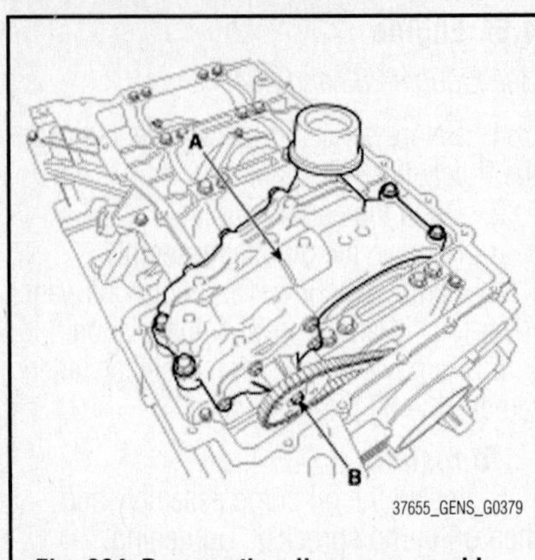

Fig. 284 Remove the oil pump assembly (A) after remove the oil pump sprocket (B)

- Oil pump assembly bolts: 15–17 ft. lbs. (20–24 Nm)
- Oil pump sprocket bolts: 16–19 ft. lbs. (22–26 Nm)

7. Install the upper and lower oil pans.
8. Install the oil filter assembly. Tightening torque: 15–17 ft. lbs. (20–24 Nm)
9. Refill the engine oil.

PISTON AND RING

POSITIONING

See Figures 285 and 286.

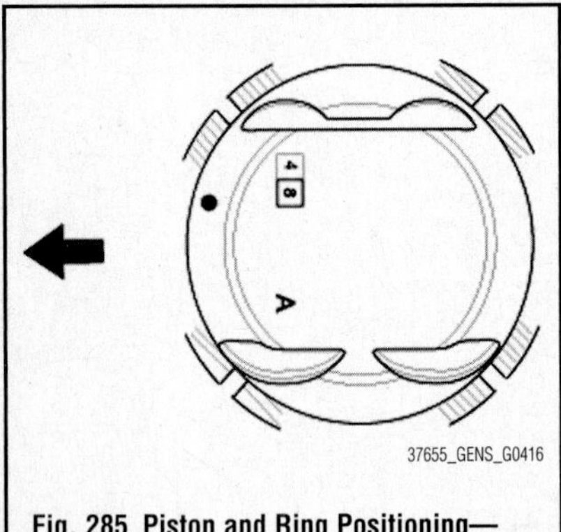

Fig. 285 Piston and Ring Positioning—3.8L engine

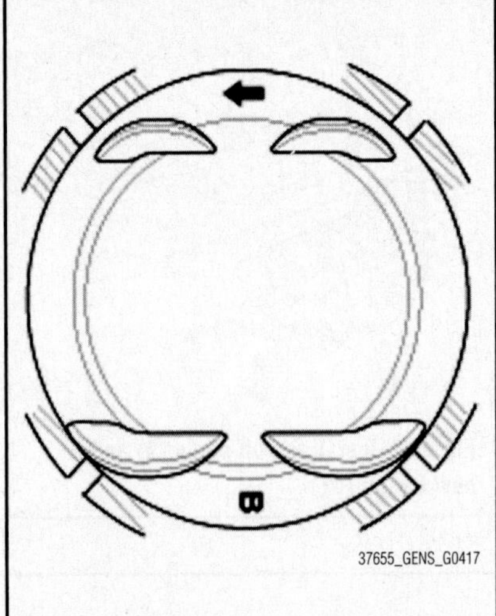

Fig. 286 Piston and Ring Positioning—4.6L engine

REAR MAIN SEAL

REMOVAL & INSTALLATION

3.8L Engine

See Figure 287.

1. Before servicing the vehicle, refer to the Precautions Section.

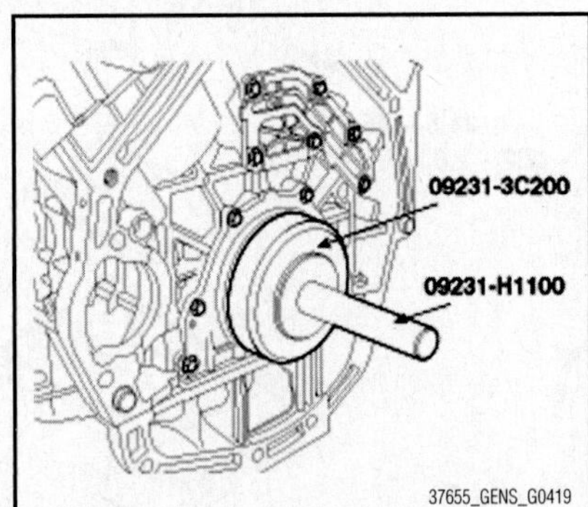

Fig. 287 Rear main seal installation—3.8L engine

2. Remove the flexplate. Refer to Flywheel/Flexplate Removal & Installation.
3. Using a seal removing tool, remove the rear main seal.

To install:

4. Using the illustrated service tool, install the oil seal.

4.6L Engine

See Figure 288.

1. Before servicing the vehicle, refer to the Precautions Section.
2. Remove the flexplate. Refer to Flywheel/Flexplate Removal & Installation.
3. Using a seal removing tool, remove the rear main seal.

To install:

4. Using the illustrated service tool, install the oil seal.

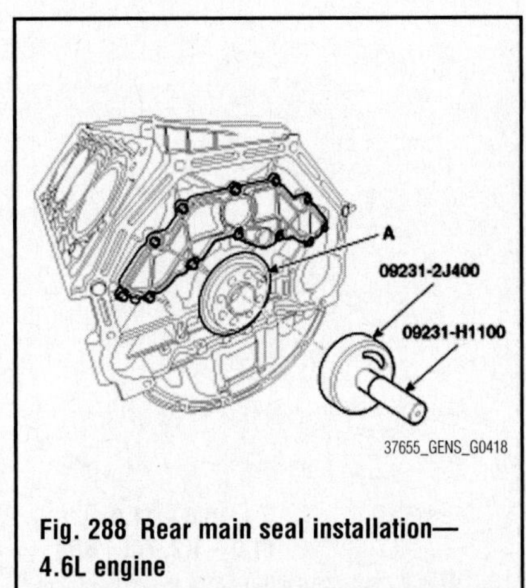

Fig. 288 Rear main seal installation—4.6L engine

VALVE LASH

ADJUSTMENT

1. The engine lash adjusters are not adjustable.

ENGINE PERFORMANCE & EMISSION CONTROLS

COMPONENT LOCATIONS

See Figures 289 through 292.

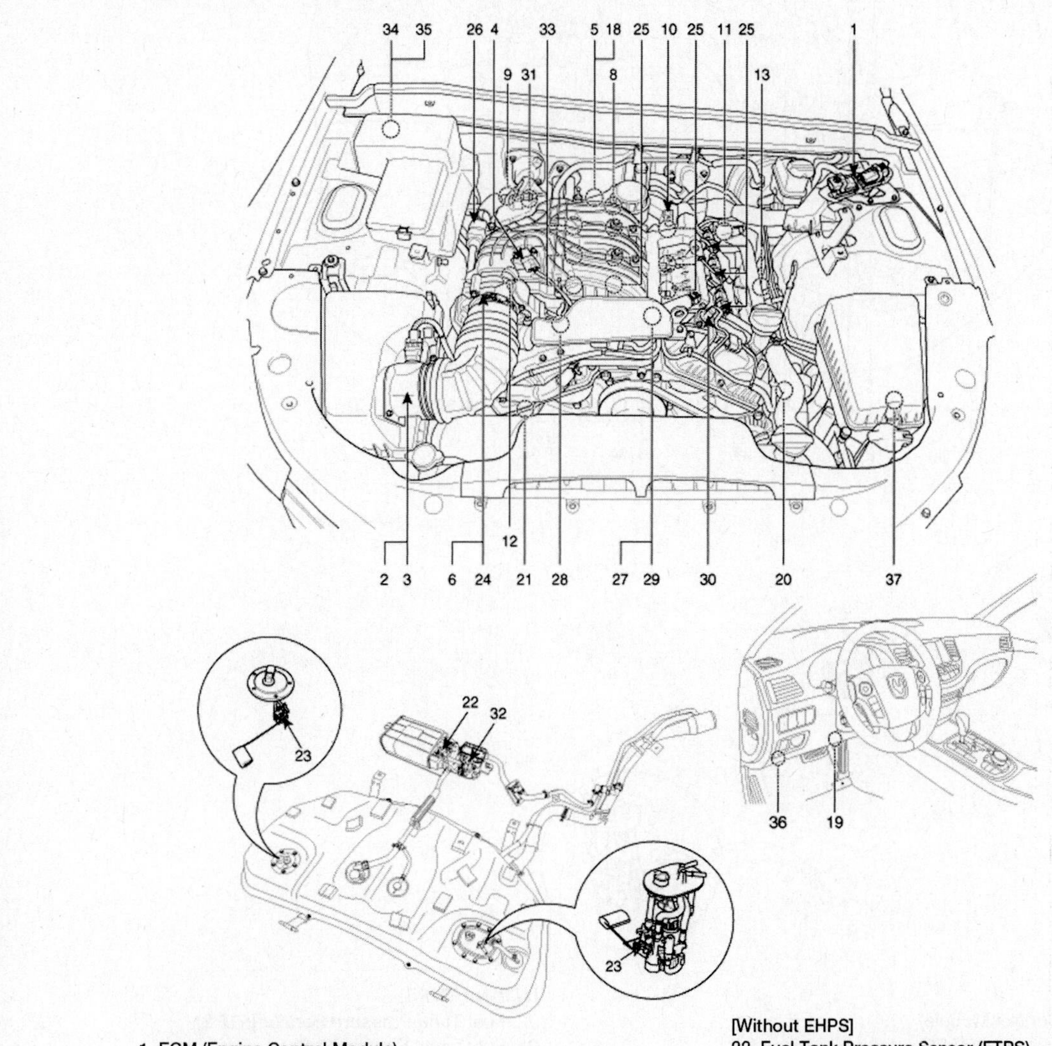

1. ECM (Engine Control Module)
2. Mass Air Flow Sensor (MAFS)
3. Intake Air Temperature Sensor (IATS)
4. Manifold Absolute Pressure Sensor (MAPS)
5. Engine Coolant Temperature Sensor (ECTS)
6. Throttle Position Sensor (TPS) [integrated into ETC Module]
7. Crankshaft Position Sensor (CKPS)
8. Camshaft Position Sensor (CMPS) [Bank 1 / Intake]
9. Camshaft Position Sensor (CMPS) [Bank 1 / Exhaust]
10. Camshaft Position Sensor (CMPS) [Bank 2 / Intake]
11. Camshaft Position Sensor (CMPS) [Bank 2 / Exhaust]
12. Knock Sensor (KS) [Bank 1]
13. Knock Sensor (KS) [Bank 2]
14. Heated Oxygen Sensor (HO2S) [Bank 1 / Sensor 1]
15. Heated Oxygen Sensor (HO2S) [Bank 1 / Sensor 2]
16. Heated Oxygen Sensor (HO2S) [Bank 2 / Sensor 1]
17. Heated Oxygen Sensor (HO2S) [Bank 2 / Sensor 2]
18. CVVT Oil Temperature Sensor (OTS)
19. Accelerator Position Sensor (APS)
20. A/C Pressure Transducer (APT)
21. Power Steering Pressure Sensor (PSPS)

[Without EHPS]

22. Fuel Tank Pressure Sensor (FTPS)
23. Fuel Level Sensor (FLS)
24. ETC Motor [integrated into ETC Module]
25. Injector
26. Purge Control Solenoid Valve (PCSV)
27. CVVT Oil Control Valve (OCV) [Bank 1 / Intake]
28. CVVT Oil Control Valve (OCV) [Bank 1 / Exhaust]
29. CVVT Oil Control Valve (OCV) [Bank 2 / Intake]
30. CVVT Oil Control Valve (OCV) [Bank 2 / Exhaust]
31. Variable Intake Solenoid (VIS) Valve
32. Canister Close Valve (CCV)
33. Ignition Coil
34. Main Relay
35. Fuel Pump Relay
36. Data Link Connector (DLC) [16 Pin]
37. Multi-Purpose Check Connector [20 Pin]

37655_GENS_G0467

Fig. 289 Engine Control System Components—3.8L engines

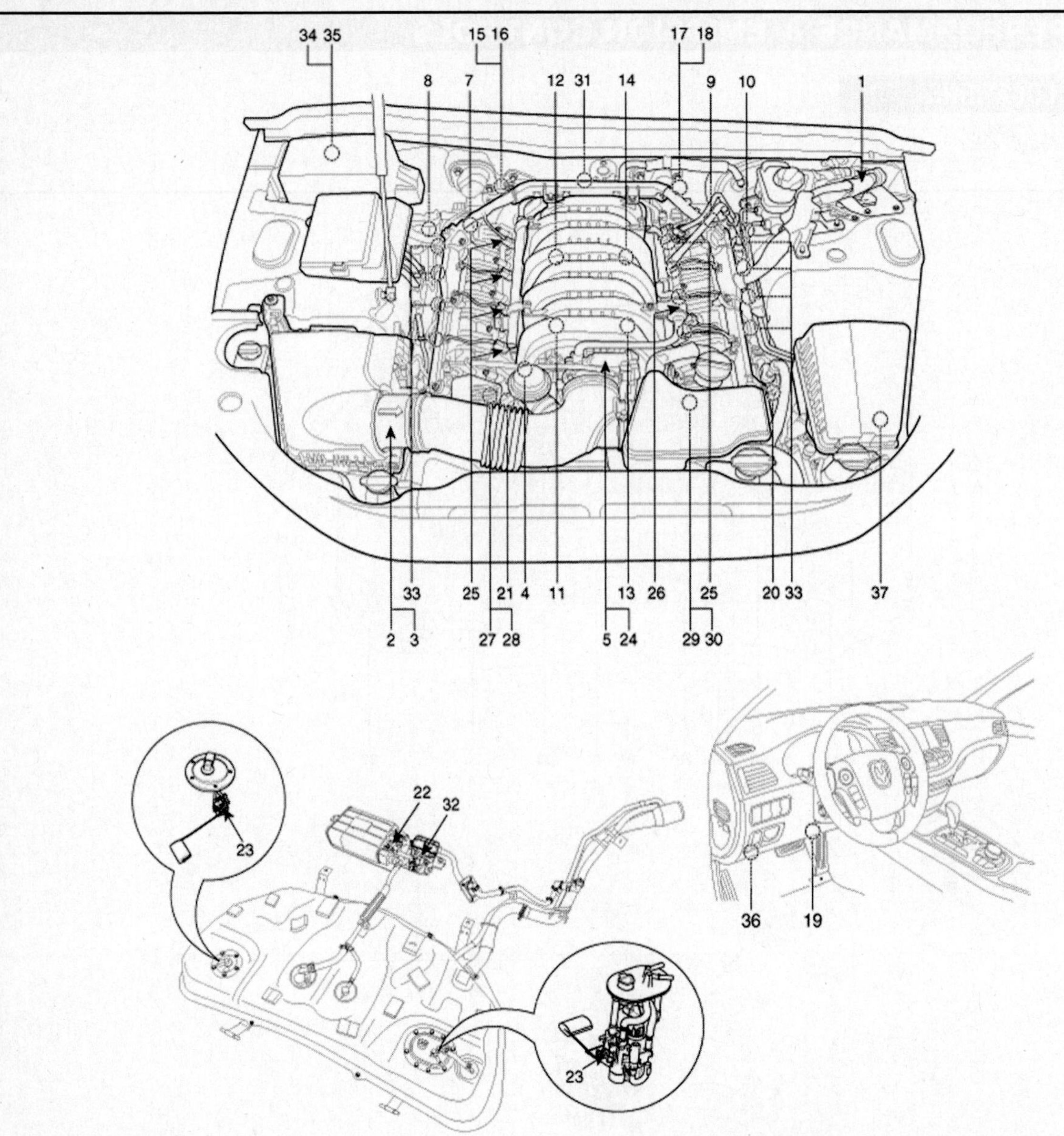

[Without EHPS}

1. ECM (Engine Control Module)
2. Mass Air Flow Sensor (MAFS)
3. Intake Air Temperature Sensor (IATS)
4. Engine Coolant Temperature Sensor (ECTS)
5. Throttle Position Sensor (TPS) [integrated into ETC Module]
6. Crankshaft Position Sensor (CKPS)
7. Camshaft Position Sensor (CMPS) [Bank 1 / Intake]
8. Camshaft Position Sensor (CMPS) [Bank 1 / Exhaust]
9. Camshaft Position Sensor (CMPS) [Bank 2 / Intake]
10. Camshaft Position Sensor (CMPS) [Bank 2 / Exhaust]
11. Knock Sensor (KS) [Bank 1/ Front]
12. Knock Sensor (KS) [Bank 1/ Rear]
13. Knock Sensor (KS) [Bank 2/ Front]
14. Knock Sensor (KS) [Bank 2/ Rear]
15. Heated Oxygen Sensor (HO2S) [Bank 1 / Sensor 1]
16. Heated Oxygen Sensor (HO2S) [Bank 1 / Sensor 2]
17. Heated Oxygen Sensor (HO2S) [Bank 2 / Sensor 1]
18. Heated Oxygen Sensor (HO2S) [Bank 2 / Sensor 2]
19. Accelerator Position Sensor (APS)
20. A/C Pressure Transducer (APT)
21. Power Steering Pressure Sensor (PSPS)
22. Fuel Tank Pressure Sensor (FTPS)
23. Fuel Level Sensor (FLS)
24. ETC Motor [integrated into ETC Module]
25. Injector
26. Purge Control Solenoid Valve (PCSV)
27. CVVT Oil Control Valve (OCV) [Bank 1 / Intake]
28. CVVT Oil Control Valve (OCV) [Bank 1 / Exhaust]
29. CVVT Oil Control Valve (OCV) [Bank 2 / Intake]
30. CVVT Oil Control Valve (OCV) [Bank 2 / Exhaust]
31. Variable Intake Solenoid (VIS) Valve
32. Canister Close Valve (CCV)
33. Ignition Coil
34. Main Relay
35. Fuel Pump Relay
36. Data Link Connector (DLC) [16 Pin]
37. Multi-Purpose Check Connector [20 Pin]

37655_GENS_G0468

Fig. 290 Engine Control System Components—4.6L engines

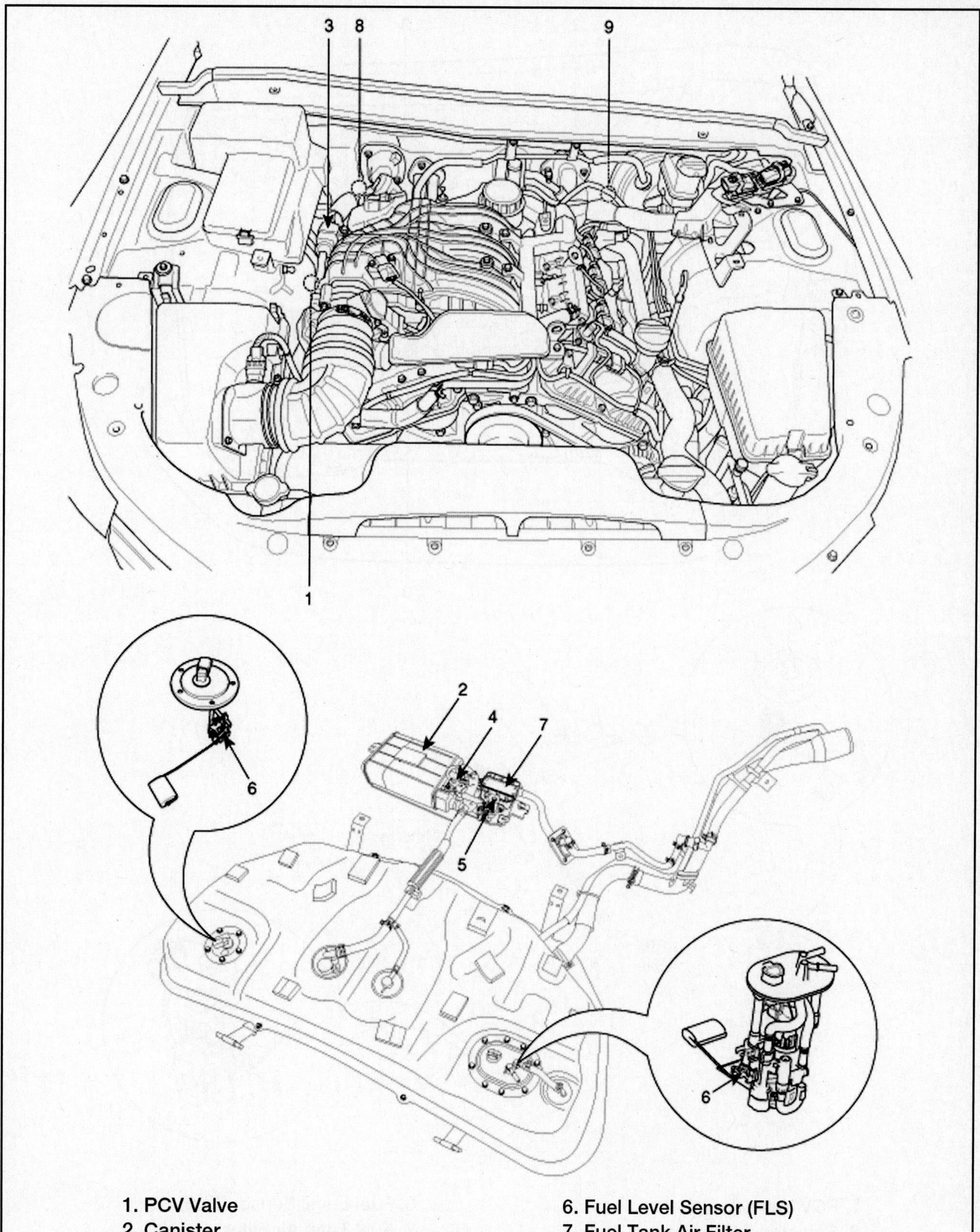

1. PCV Valve
2. Canister
3. Purge Control Solenoid Valve (PCSV)
4. Fuel Tank Pressure Sensor (FTPS)
5. Canister Close Valve (CCV)
6. Fuel Level Sensor (FLS)
7. Fuel Tank Air Filter
8. Catalytic Converter (MCC, Bank 1)
9. Catalytic Converter (MCC, Bank 2)

37655_GENS_G0485

Fig. 291 Emission Control System Components—3.8L engines

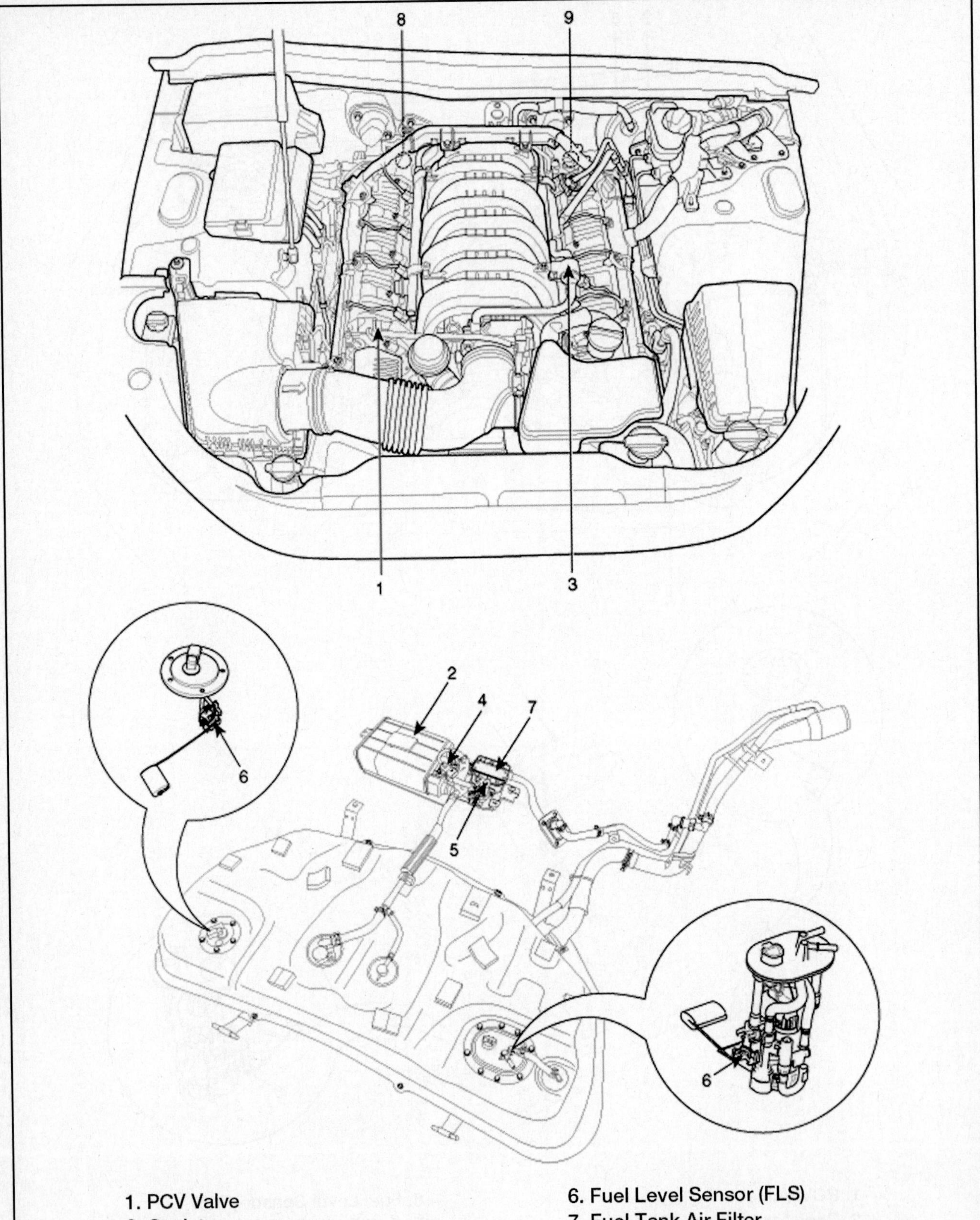

Fig. 292 Emission Control System Components—4.6L engines

CAMSHAFT POSITION (CMP) SENSOR

LOCATION

See Figures 293 through 298.

Refer to the accompanying illustrations.

REMOVAL & INSTALLATION

See Figures 293 through 298.

1. Before servicing the vehicle, refer to the Precautions Section.
2. Disconnect the negative battery cable.

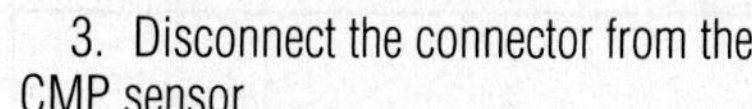

3. Disconnect the connector from the CMP sensor.
4. Remove the CMP sensor.

To install:

5. Installation is the reverse of the removal procedure, noting the following:
    a. For 3.8L engines, tighten the mounting bolt to 61–86 inch lbs. (7–10 Nm).
    b. For 4.6L engines, tighten the mounting bolt to 70–104 inch lbs. (8–12 Nm).

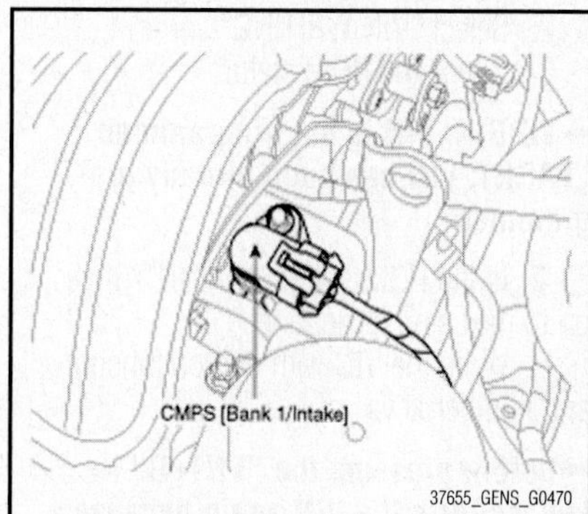

Fig. 293 Camshaft Position (CMP) sensor, Intake Bank 1—3.8L engine

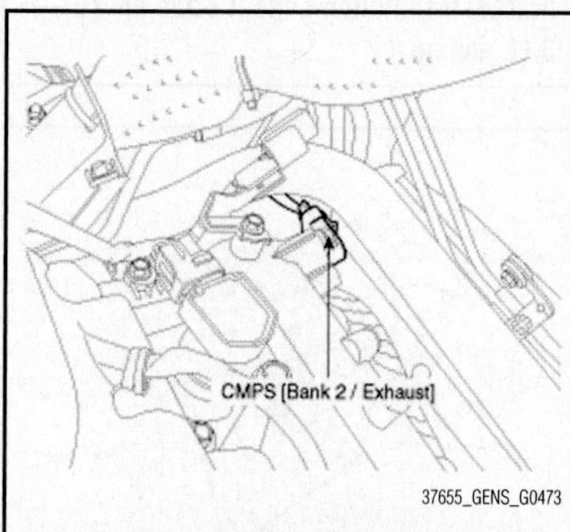

Fig. 296 Camshaft Position (CMP) sensor, Exhaust Bank 2—3.8L engine

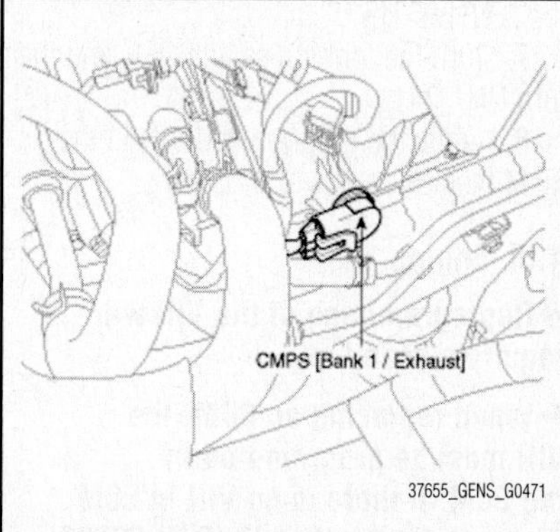

Fig. 294 Camshaft Position (CMP) sensor, Exhaust Bank 1—3.8L engine

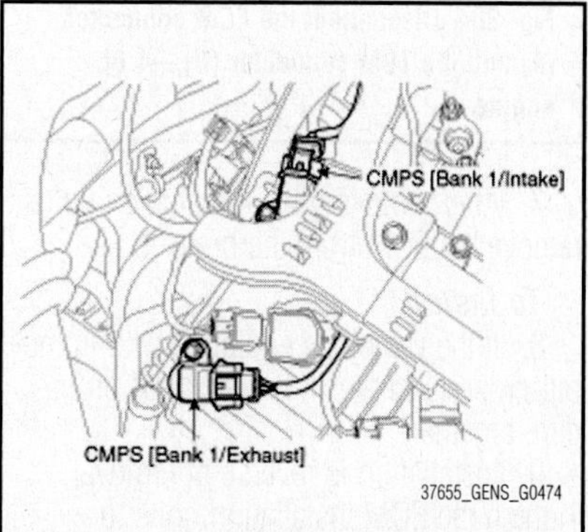

Fig. 297 Camshaft Position (CMP) sensor, Bank 1—4.6L engine

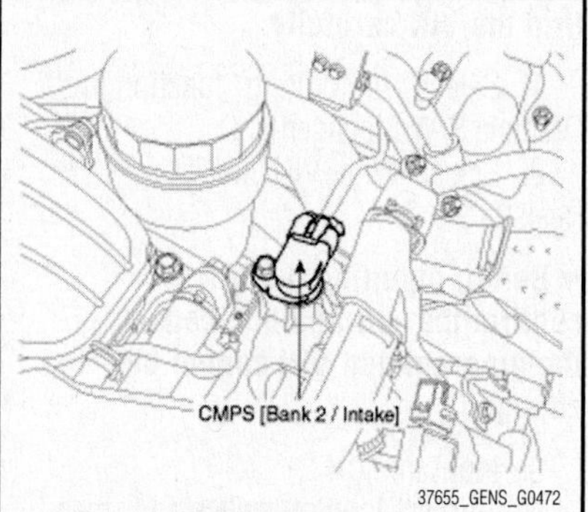

Fig. 295 Camshaft Position (CMP) sensor, Intake Bank 2—3.8L engine

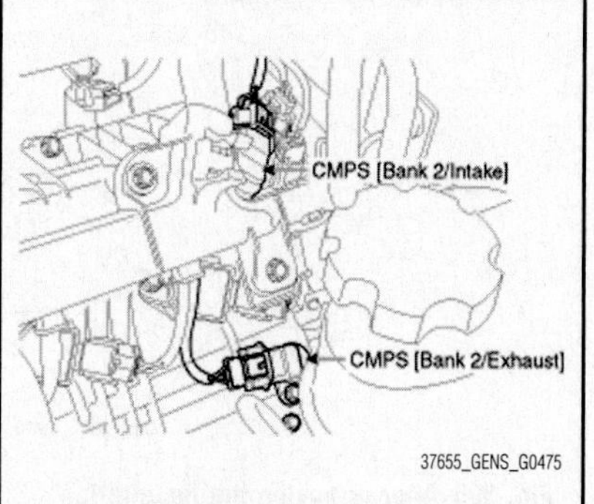

Fig. 298 Camshaft Position (CMP) sensor, Bank 2—4.6L engine

CRANKSHAFT POSITION (CKP) SENSOR

LOCATION

See Figures 299 and 300.

Refer to the accompanying illustrations.

REMOVAL & INSTALLATION

See Figures 299 and 300.

1. Before servicing the vehicle, refer to the Precautions Section.
2. Disconnect the negative battery cable.
3. Disconnect the connector from the sensor.
4. Remove the sensor from its mounting.

To install:

5. Installation is the reverse of the removal procedure, noting the following:
    a. For 3.8L engines, tighten the mounting bolt to 61–86 inch lbs. (7–10 Nm).
    b. For 4.6L engines, tighten the mounting bolt to 86–104 inch lbs. (10–12 Nm).

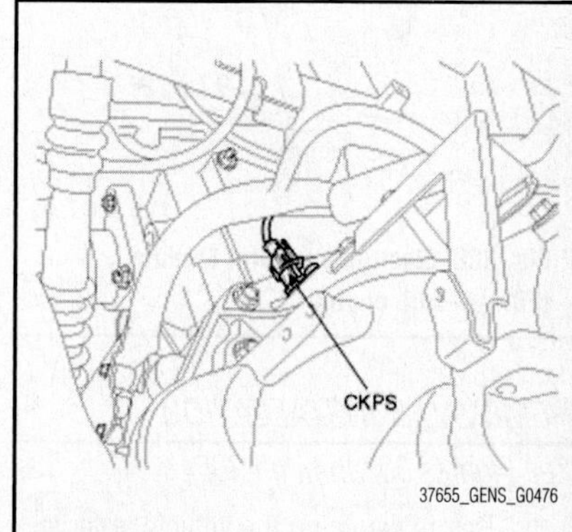

Fig. 299 Crankshaft Position (CKP) sensor—3.8L engine

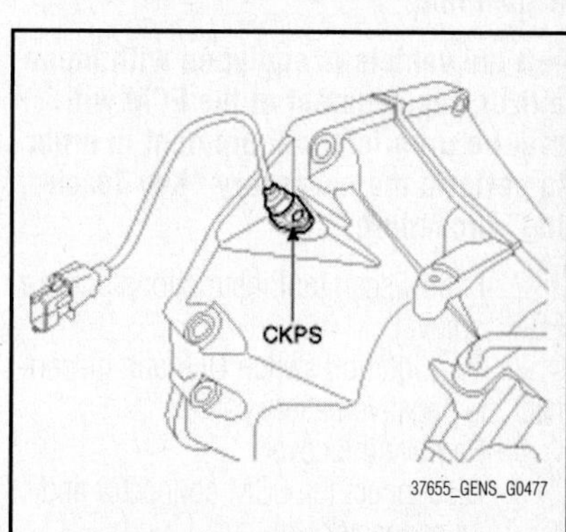

Fig. 300 Crankshaft Position (CKP) sensor—4.6L engine

ELECTRONIC CONTROL MODULE (ECM)

LOCATION

See Figures 301 and 302.

Refer to the accompanying illustrations.

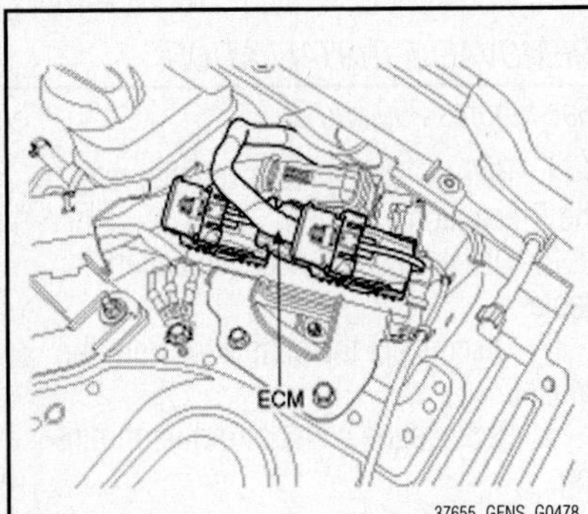

Fig. 301 Electronic Control Module (ECM)—3.8L engine

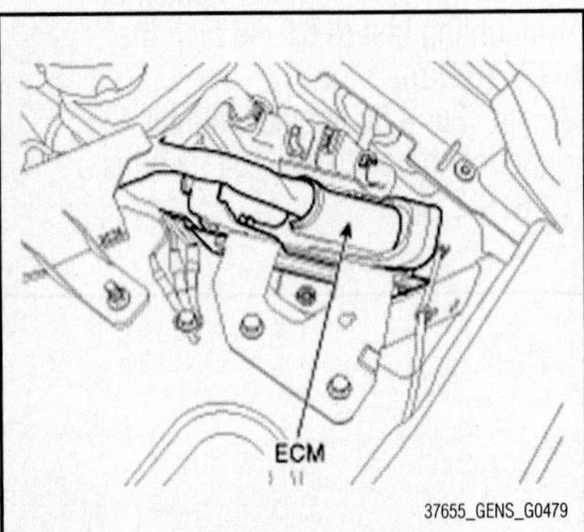

Fig. 302 Electronic Control Module (ECM)—4.6L engine

REMOVAL & INSTALLATION

See Figures 303 through 305.

1. Before servicing the vehicle, refer to the Precautions Section.

➡Replacement of the ECM will require a scan tool.

➡If the vehicle is equipped with immobilizer, replacement of the ECM will require proprietary equipment in order to perform the necessary "Key Teaching" procedure.

2. Follow scan tool instructions prior to ECM removal.
3. Turn ignition switch OFF and disconnect the negative battery cable.
4. Remove the cover.
5. Disconnect the ECM connector and the TCM connector.
6. Remove the ECM and TCM bracket installation bolts and nut.

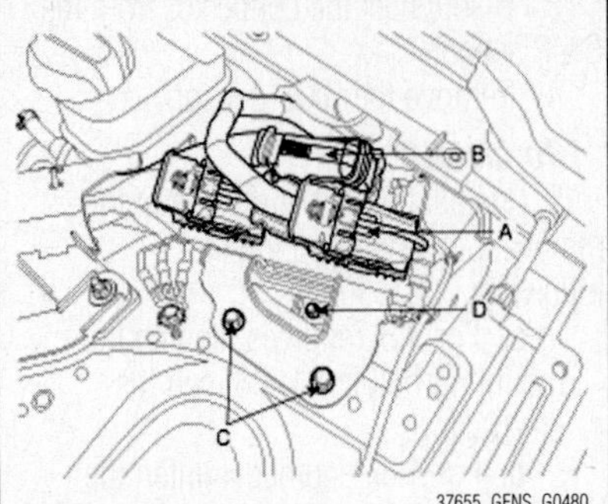

Fig. 303 Disconnect the ECM connector (A) and the TCM connector (B) and remove bracket installation bolts (C) and nut (D)—3.8L engine

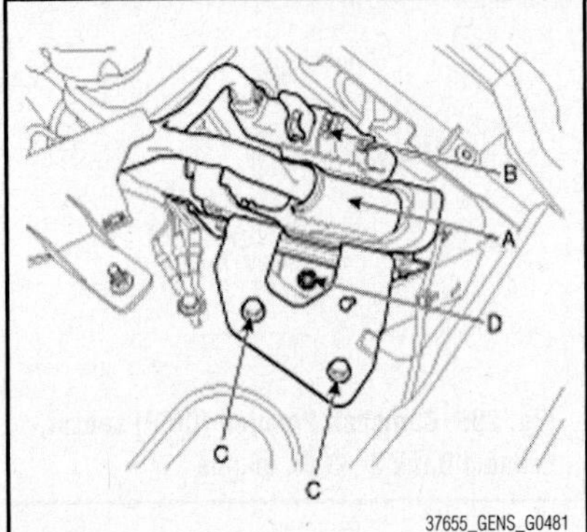

Fig. 304 Disconnect the ECM connector (A) and the TCM connector (B)—4.6L engine

7. After removing the installation bolts, remove the ECM from the bracket.

To install:

8. If the vehicle is equipped with immobilizer, perform the "Key Teaching" procedure together.
9. Installation is reverse of removal. Tighten the ECM installation bolts to 86–104 inch lbs. (10–12 Nm).

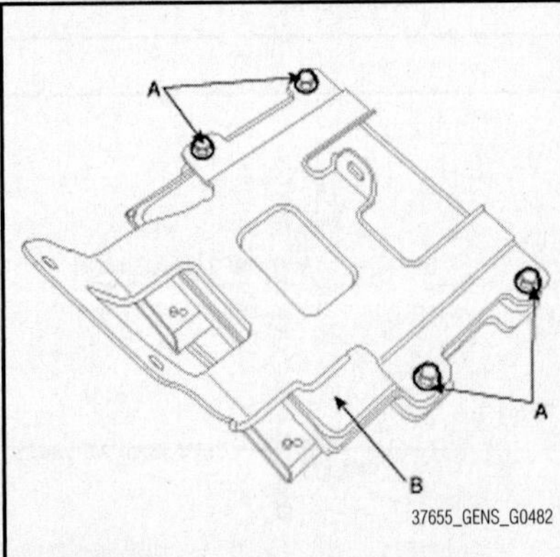

Fig. 305 After removing the installation bolts (A), remove the ECM (B) from the bracket—3.8L engine

VIN REPROGRAMMING

3.8L Engine

➡Reprogramming of the VIN will require a scan tool.

➡When replacing an ECM, the VIN must be programmed in the ECM. If there is no VIN in ECM memory, the fault code (DTC P0630) is set.

➡The programmed VIN cannot be changed. When writing the VIN, confirm the VIN carefully.

1. Select "Vehicle" and "Engine".
2. Select "VIN WRITING".
3. Check the PCM status.

➡VIRGIN: VIN is not programmed. LEARNT: VIN has been already programmed

4. Is the PCM status "VIRGIN"? If yes, go to next step. If no, end.
5. Write the VIN with cursor, function and number keys.

➡Before pressing the "ENTER" key, confirm the VIN again because the programmed VIN cannot be changed.

6. After verifying the written VIN, press the "ENTER" key.
7. Turn the ignition switch OFF, and then turn ON.
8. Verify the programmed VIN in the ECM memory.

4.6L Engine

➡Reprogramming of the VIN will require a scan tool.

➡When replacing an ECM, the VIN must be programmed in the ECM. If there is no VIN in ECM memory, the fault code (DTC P0630) is set.

➡The programmed VIN cannot be changed. When writing the VIN, confirm the VIN carefully.

1. Select "VIN Writing" function in "Vehicle S/W Management".
2. Select "Write VIN" in "ID Resister".

➡Before inputting the VIN, confirm the VIN again because the programmed VIN cannot be changed.

3. Input the VIN.
4. Turn the ignition switch OFF, then back ON.

ENGINE COOLANT TEMPERATURE (ECT) SENSOR

LOCATION

See Figures 306 and 307.

Refer to the accompanying illustrations.

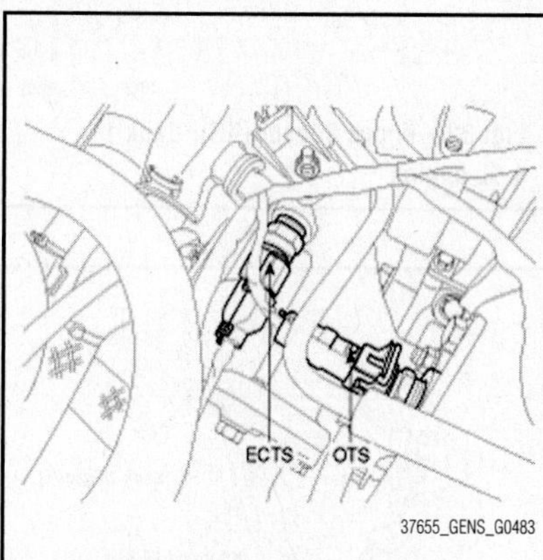

Fig. 306 Engine Coolant Temperature sensor (ECTS)—3.8L engine

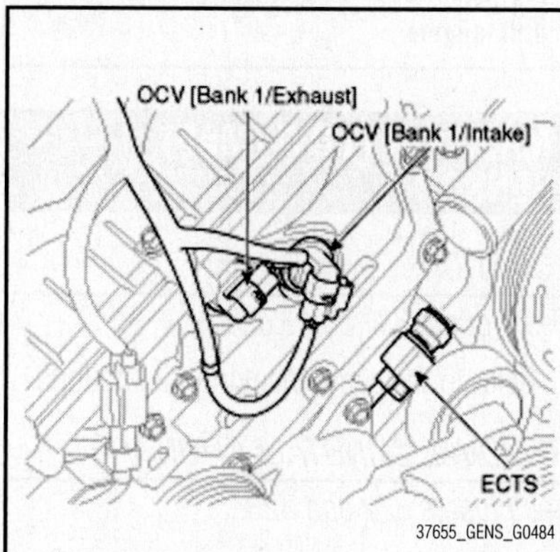

Fig. 307 Engine Coolant Temperature sensor (ECTS)—4.6L engine

REMOVAL & INSTALLATION

See Figures 306 and 307.

1. Before servicing the vehicle, refer to the Precautions Section.
2. Turn ignition switch OFF and disconnect the negative battery cable.
3. Turn the ignition switch OFF.
4. Disconnect the ECT sensor connector.
5. Remove the ECT sensor.

To install:

6. Installation is the reverse of removal. Tighten the mounting bolts to 15–29 ft. lbs. (20–39 Nm).

HEATED OXYGEN (HO2S) SENSOR

LOCATION

See Figures 308 through 315.

Refer to the accompanying illustrations.

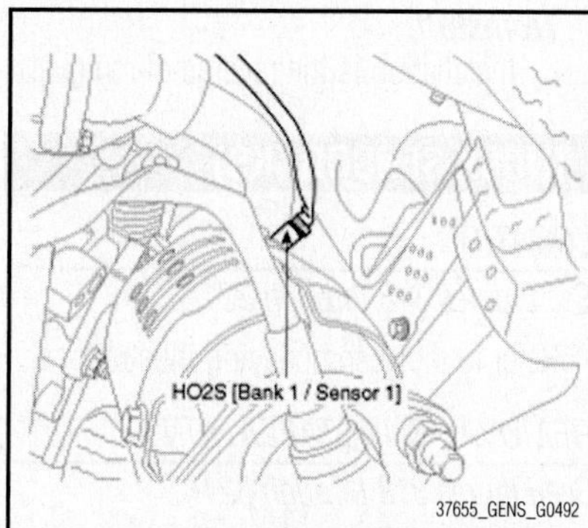

Fig. 308 Heated Oxygen (HO2S) sensor—Bank 1, Sensor 1—3.8L engine

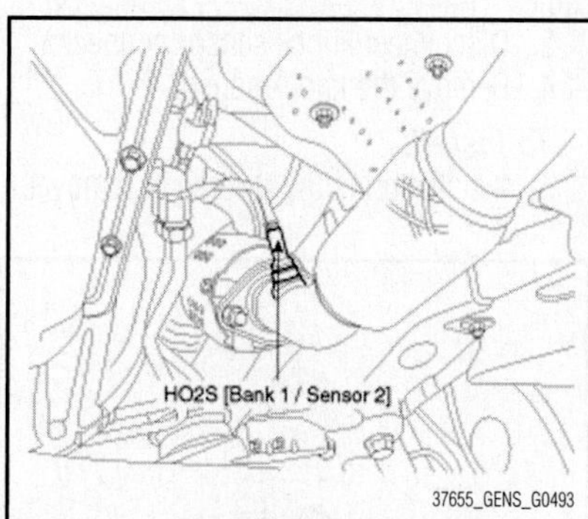

Fig. 309 Heated Oxygen (HO2S) sensor—Bank 1, Sensor 2—3.8L engine

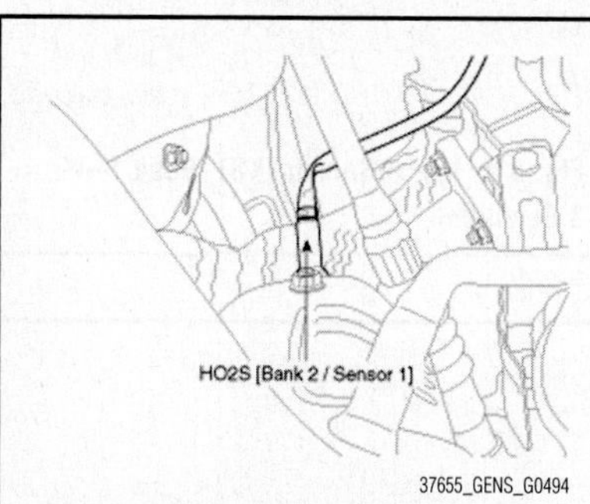

Fig. 310 Heated Oxygen (HO2S) sensor—Bank 2, Sensor 1—3.8L engine

REMOVAL & INSTALLATION

1. Before servicing the vehicle, refer to the Precautions Section.
2. Disconnect the negative battery cable.
3. Disconnect the oxygen sensor connector.

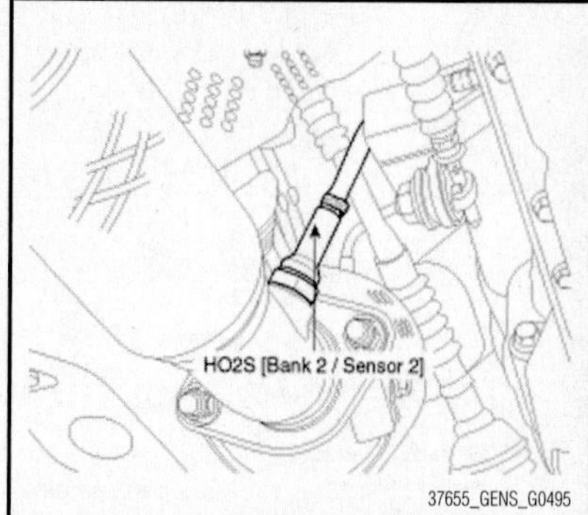

Fig. 311 Heated Oxygen (HO2S) sensor—Bank 2, Sensor 2—3.8L engine

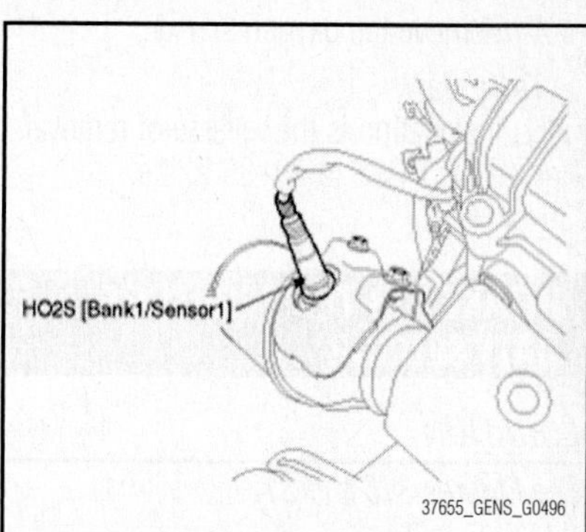

Fig. 312 Heated Oxygen (HO2S) sensor—Bank 1, Sensor 1—4.6L engine

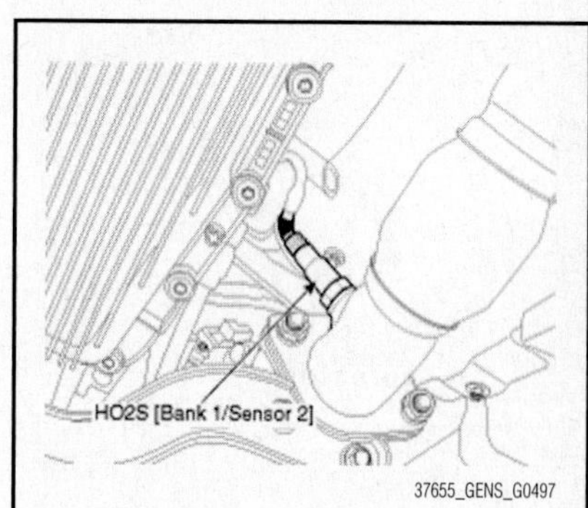

Fig. 313 Heated Oxygen (HO2S) sensor—Bank 1, Sensor 2—4.6L engine

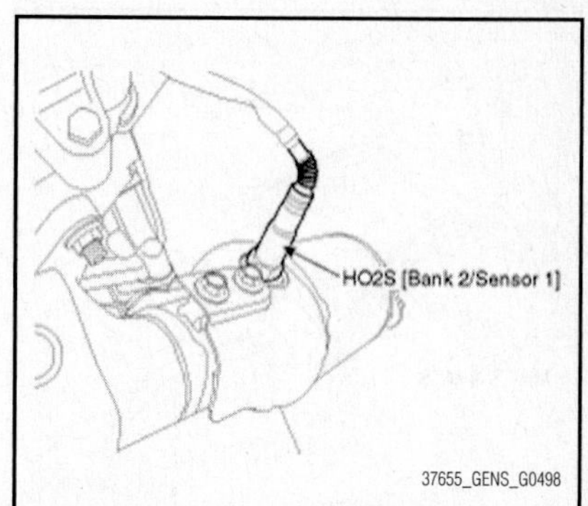

Fig. 314 Heated Oxygen (HO2S) sensor—Bank 2, Sensor 1—4.6L engine

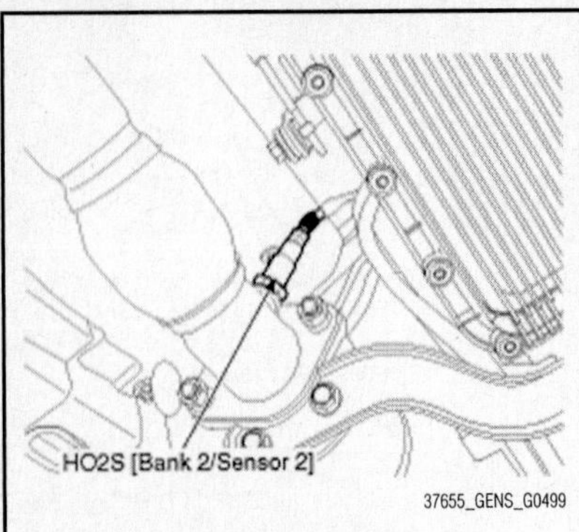

Fig. 315 Heated Oxygen (HO2S) sensor—Bank 2, Sensor 2—4.6L engine

4. Remove the oxygen sensor.

To install:

5. Installation is the reverse of removal. Tighten the HO2S to 29–36 ft. lbs. (39–49 Nm).

INTAKE AIR TEMPERATURE (IAT) SENSOR

LOCATION

See Figures 316 and 317.

Refer to the accompanying illustrations.

REMOVAL & INSTALLATION

See Figures 316 and 317.

1. Before servicing the vehicle, refer to the Precautions Section.
2. Disconnect the negative battery cable.
3. Disconnect Intake Air Temperature (IAT) sensor connector.
4. Remove the IAT sensor.

To install:

5. Installation is the reverse of removal.

KNOCK SENSOR (KS)

LOCATION

See Figures 318 through 321.

Refer to the accompanying illustrations.

REMOVAL & INSTALLATION

See Figures 318 through 321.

1. Before servicing the vehicle, refer to the Precautions Section.
2. Disconnect the negative battery cable.
3. Disconnect knock sensor connector.
4. Remove the knock sensor.

To install:

5. Installation is the reverse of removal.

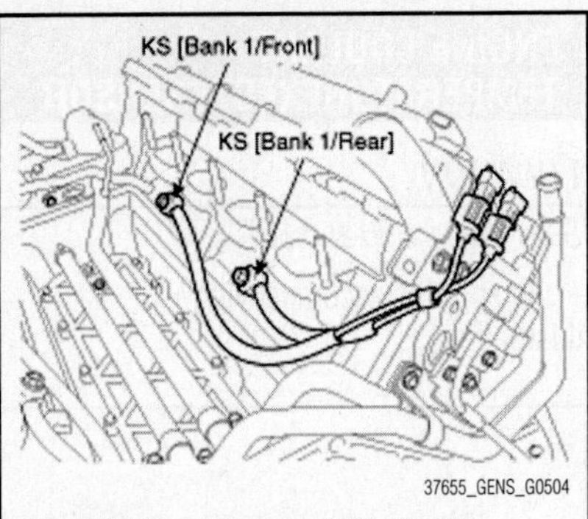

Fig. 320 Knock Sensor (KS), Bank 1—4.6L engine

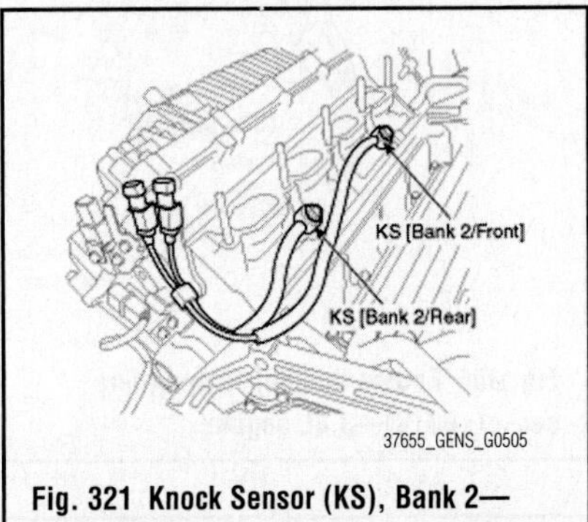

Fig. 321 Knock Sensor (KS), Bank 2—4.6L engine

MASS AIR FLOW (MAF) SENSOR

LOCATION

See Figures 322 and 323.

Refer to the accompanying illustrations.

REMOVAL & INSTALLATION

See Figures 322 and 323.

1. Before servicing the vehicle, refer to the Precautions Section.
2. Disconnect the negative battery cable.
3. Disconnect the Mass Air Flow (MAF) sensor connector.
4. Remove the MAF sensor.

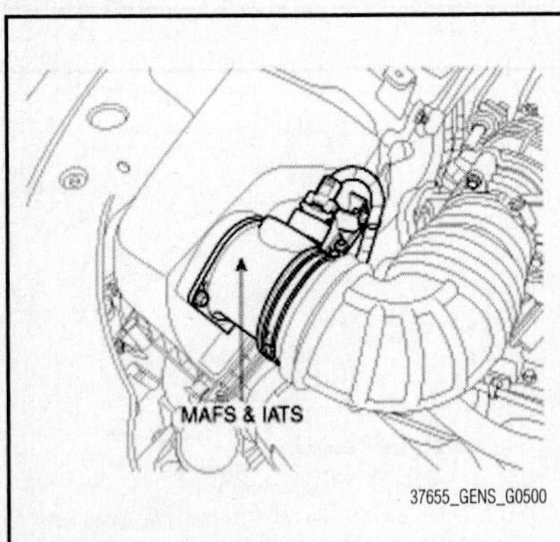

Fig. 316 Intake Air Temperature (IAT) and MAF sensor—3.8L engine

Fig. 318 Knock Sensor (KS), Bank 1—3.8L engine

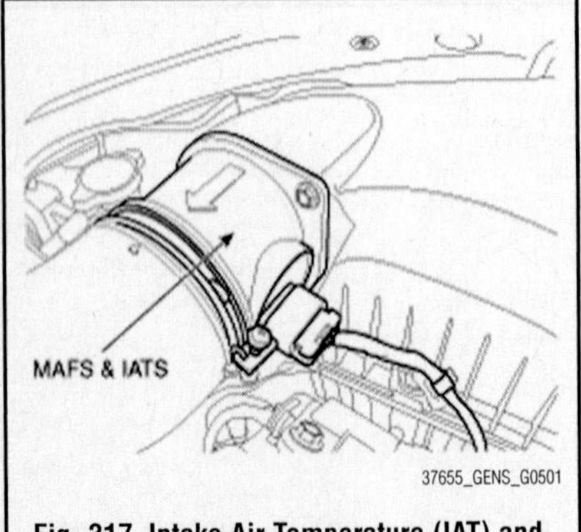

Fig. 317 Intake Air Temperature (IAT) and MAF sensor—4.6L engine

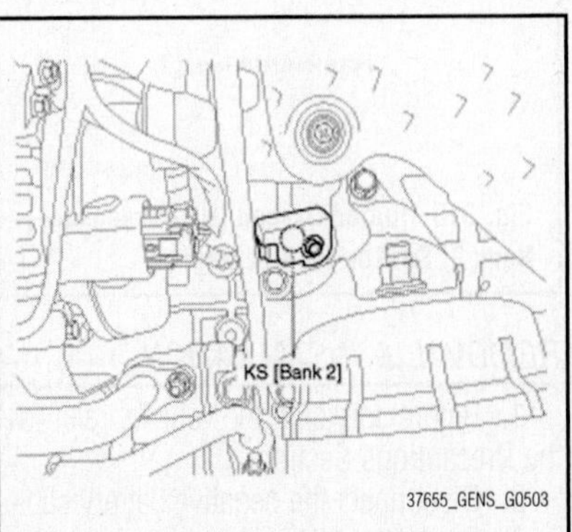

Fig. 319 Knock Sensor (KS), Bank 2—3.8L engine

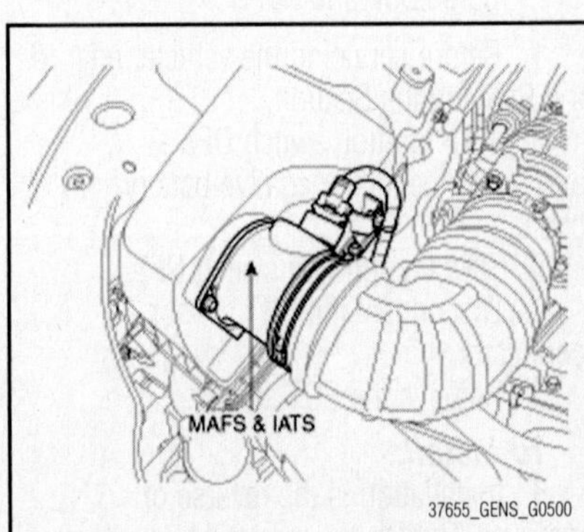

Fig. 322 MAF sensor and Intake Air Temperature (IAT)—3.8L engine

MAFS & IATS
37655_GENS_G0501

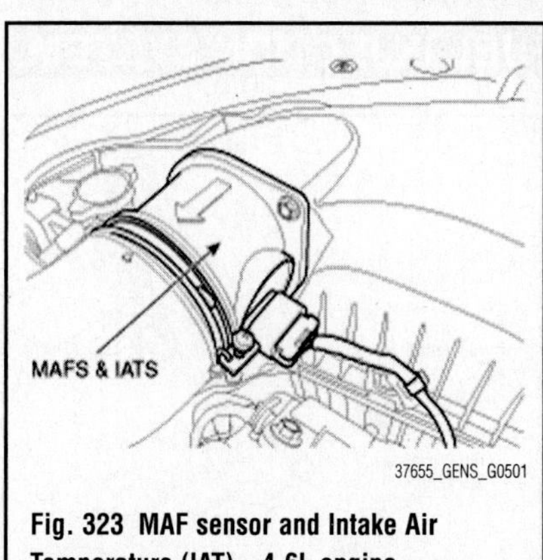

Fig. 323 MAF sensor and Intake Air Temperature (IAT)—4.6L engine

To install:

5. Installation is the reverse of removal. Tighten the MAF sensor installation bolt to 35–52 inch lbs. (4–6 Nm).

MANIFOLD ABSOLUTE PRESSURE (MAP) SENSOR

LOCATION

3.8L Engine

See Figure 324.

ETC Module
MAPS
37655_GENS_G0630

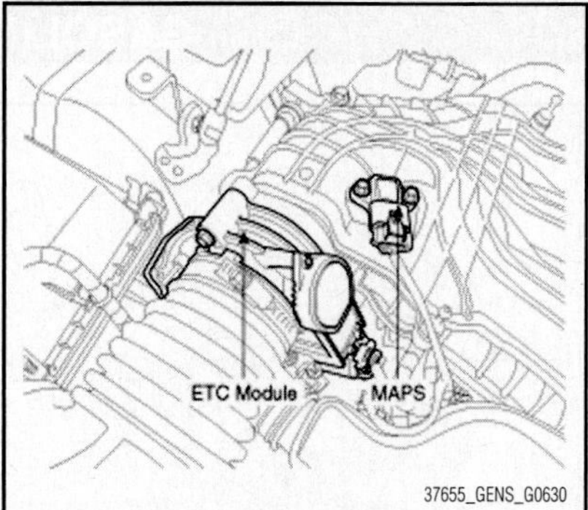

Fig. 324 Manifold Absolute Pressure (MAP) sensor—3.8L engine

REMOVAL & INSTALLATION

3.8L Engine

See Figure 324.

1. Before servicing the vehicle, refer to the Precautions Section.
2. Disconnect the negative battery cable.
3. Disconnect the connector from the sensor.
4. Remove the sensor retaining screws.
5. Remove the sensor from its mounting.

To install:

6. Installation is the reverse of the removal procedure. Tighten the bolts to 78–104 inch lbs. (9–12 Nm).

THROTTLE POSITION SENSOR (TPS)

LOCATION

See Figure 325.

The Throttle Position Sensor (TPS) is located on the throttle body/Electronic Throttle Control module.

REMOVAL & INSTALLATION

The Throttle Position Sensor (TPS) is located on the throttle body/Electronic Throttle Control module. See Throttle Body Removal & Installation.

Throttle Position Sensor (TPS)
Throttle Valve
Gear Assembly
ETC Motor
ETC Module Assembly
Gear (Idler)
37655_GENS_G0510

Fig. 325 Throttle Position Sensor (TPS) location—3.8L engine shown, 4.6L similar

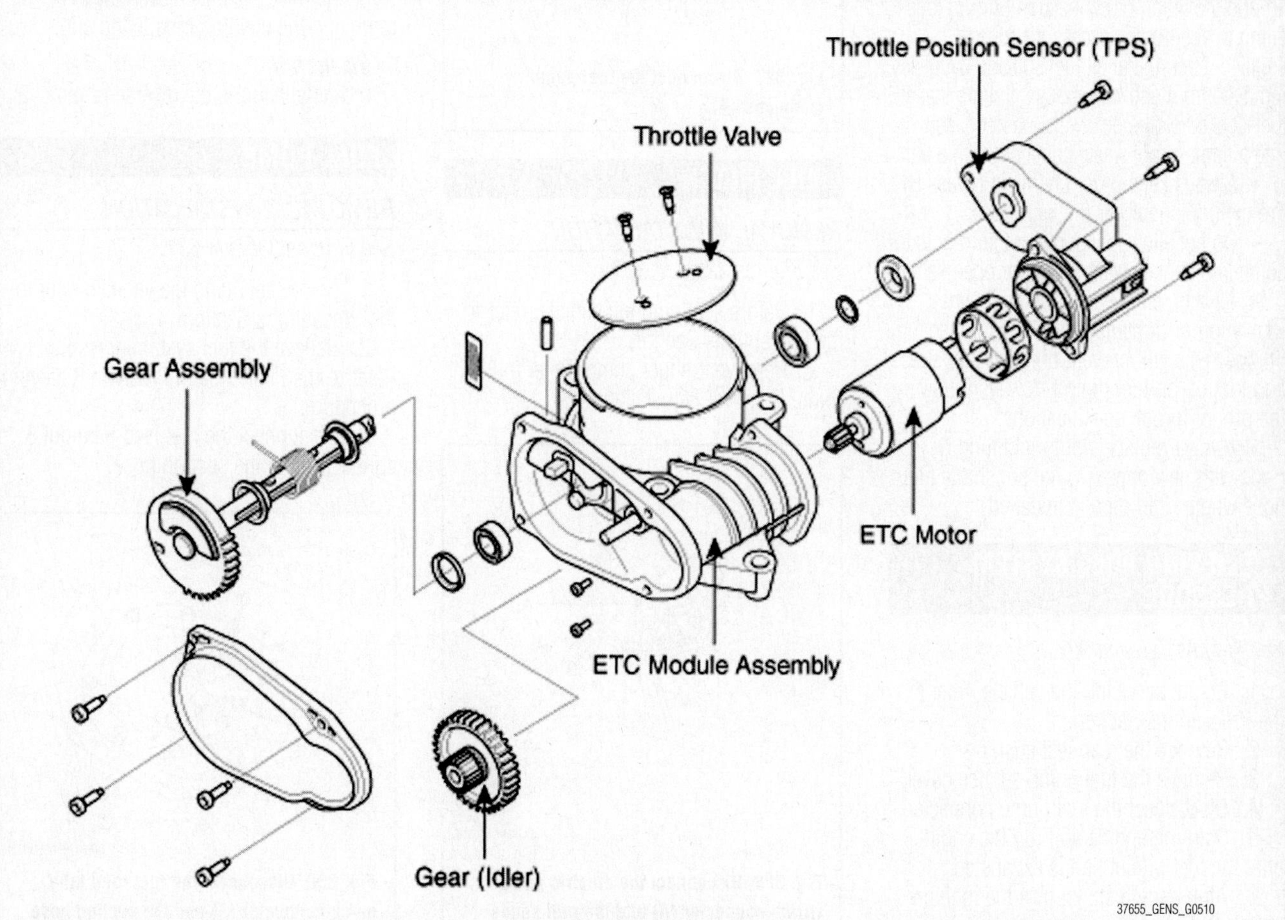

FUEL SYSTEM SERVICE PRECAUTIONS

Safety is the most important factor when performing not only fuel system maintenance, but any type of maintenance. Failure to conduct maintenance and repairs in a safe manner may result in serious personal injury or death. Work on a vehicle's fuel system components can be accomplished safely and effectively by adhering to the following rules and guidelines.

- To avoid the possibility of fire and personal injury, always disconnect the negative battery cable unless the repair or test procedure requires that battery voltage be applied.
- Always relieve the fuel system pressure prior to disconnecting any fuel system component (injector, fuel rail, pressure regulator, etc.) fitting or fuel line connection. Exercise extreme caution whenever relieving fuel system pressure to avoid exposing skin, face and eyes to fuel spray. Please be advised that fuel under pressure may penetrate the skin or any part of the body that it contacts.
- Always place a shop towel or cloth around the fitting or connection prior to loosening to absorb any excess fuel due to spillage. Ensure that all fuel spillage is quickly removed from engine surfaces. Ensure that all fuel-soaked cloths or towels are deposited into a flame-proof waste container with a lid.
- Always keep a dry chemical (Class B) fire extinguisher near the work area.
- Do not allow fuel spray or fuel vapors to come into contact with a spark or open flame.
- Always use a second wrench when loosening or tightening fuel line connection fittings. This will prevent unnecessary stress and torsion on fuel piping. Always follow the proper torque specifications.
- Always replace worn fuel fitting O-rings with new ones. Do not substitute fuel hose where rigid pipe is installed.

RELIEVING FUEL SYSTEM PRESSURE

See Figures 326 and 327.

1. Before servicing the vehicle, refer to the Precautions Section.
2. Remove the rear seat cushion.
3. Remove the fuel pump service cover.
4. Disconnect the fuel pump connector.
5. Start the engine and run the engine until the fuel in fuel line is exhausted.
6. After engine stops, turn the ignition switch OFF, and disconnect the negative battery terminal.

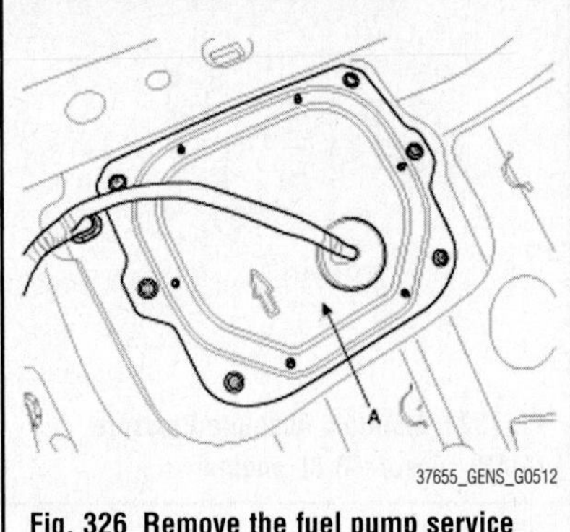

Fig. 326 Remove the fuel pump service cover (A)

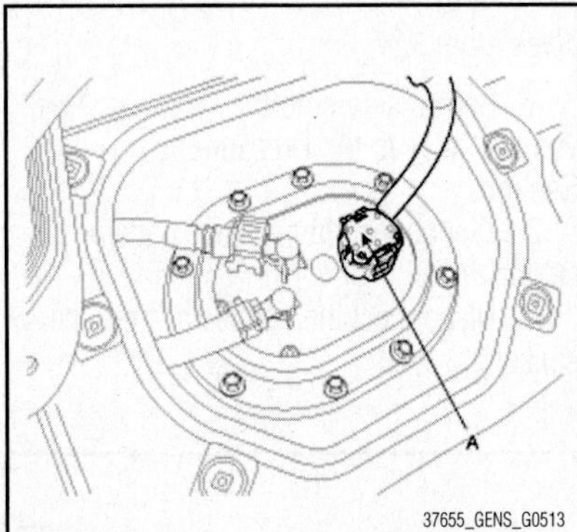

Fig. 327 Disconnect the fuel pump connector (A)

FUEL FILTER

REMOVAL & INSTALLATION

See Figures 328 and 329.

1. Before servicing the vehicle, refer to the Precautions Section.
2. Remove the fuel pump. Refer to Fuel Pump Removal & Installation.

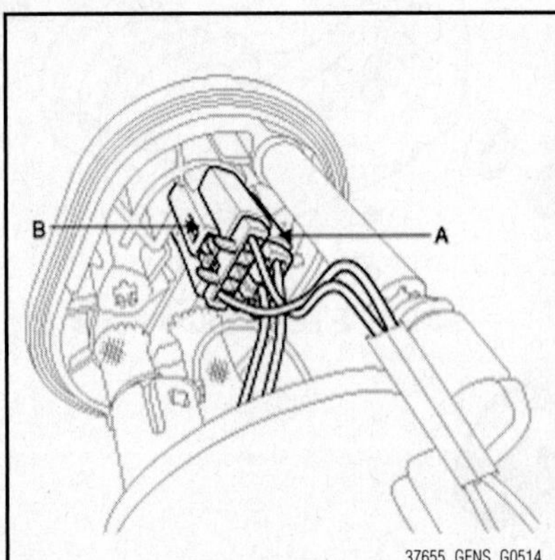

Fig. 328 Disconnect the electric pump wiring connector (A) and the fuel sender connector (B)

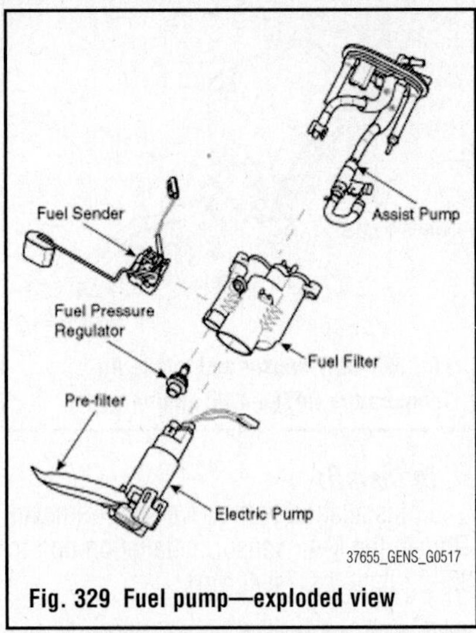

Fig. 329 Fuel pump—exploded view

3. Disconnect the electric pump wiring connector and the fuel sender connector.
4. Remove the fuel sender.
5. Remove the assist pump.
6. Disconnect the fuel feed tube.
7. Remove the electric pump and pre-filter assembly and the fuel pressure regulator.
8. Remove the plate assembly after removing the cushion pipe fixing clip.

To install:

9. Installation is the reverse of removal.

FUEL PUMP

REMOVAL & INSTALLATION

See Figures 330 and 331.

1. Before servicing the vehicle, refer to the Precautions Section.
2. Relieve the fuel system pressure. Refer to the Fuel System Pressure Relieving procedure.
3. Disconnect the fuel feed tube quick-connector and the suction hose.

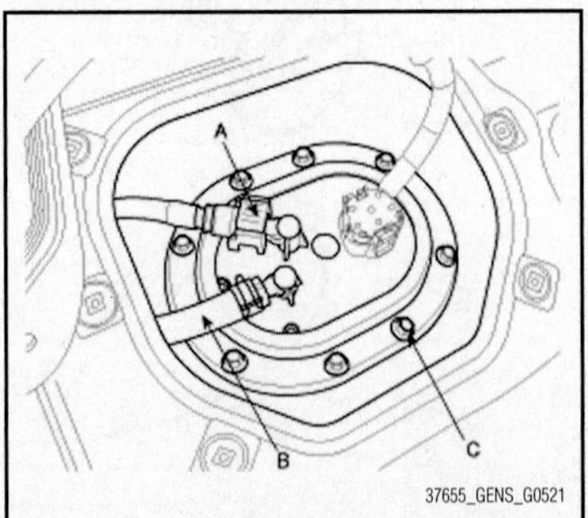

Fig. 330 Disconnect the fuel feed tube quick-connector (A) and the suction hose (B) and remove the installation bolts (C)

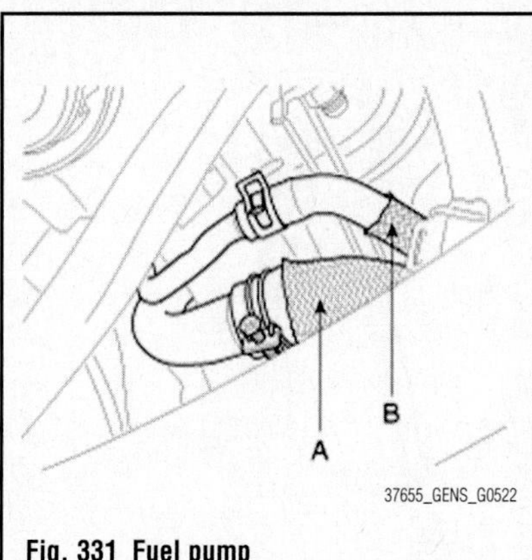

Fig. 331 Fuel pump

4. Remove the installation bolts.
5. Remove the fuel pump from the fuel tank.

To install:

6. Installation is the reverse of removal. Tighten the installation bolts to 17–26 inch lbs. (2–3 Nm).

➡When installing the fuel pump module, be careful not to get the seal-ring entangled.

FUEL RAIL AND INJECTOR

REMOVAL & INSTALLATION

3.8L Engine

1. Before servicing the vehicle, refer to the Precautions Section.
2. Remove the intake manifold. Refer to Intake Manifold Removal & Installation in the Engine Mechanical section.
3. Remove the fuel rail and injectors.

To install:

4. Installation is the reverse of removal.

4.6L Engine

See Figures 332 through 334.

1. Before servicing the vehicle, refer to the Precautions Section.
2. Disconnect the negative battery cable.
3. Remove the engine cover.
4. Remove the air duct, and the air cleaner assembly.
5. Remove the fuel hose, the Purge Control Solenoid Valve (PCSV) hose and the brake booster vacuum hose.
6. Disconnect the ETC module connector.
7. Disconnect the PCSV connector and the injector connectors.

To install:

8. Reconnect the injector connectors.

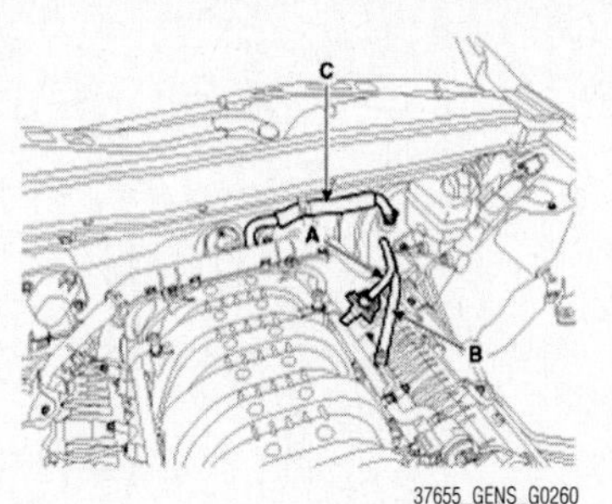

Fig. 332 Remove the fuel hose (A), the PCSV hose (B), and the brake booster vacuum hose (C)

9. Connect the variable intake system solenoid valve connector and the wiring harness protector.
10. Connect the PCSV connector and the injector connectors.
11. Connect the ETC module connector.
12. Connect the fuel hose, the PCSV hose and the brake booster vacuum hose.
13. Install the air cleaner assembly and air duct.
14. Install the engine cover.
15. Connect the battery negative terminal.

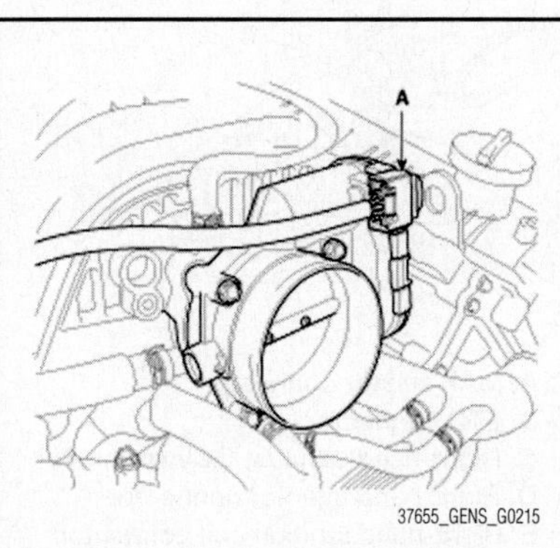

Fig. 333 Disconnect the ETC module connector (A)

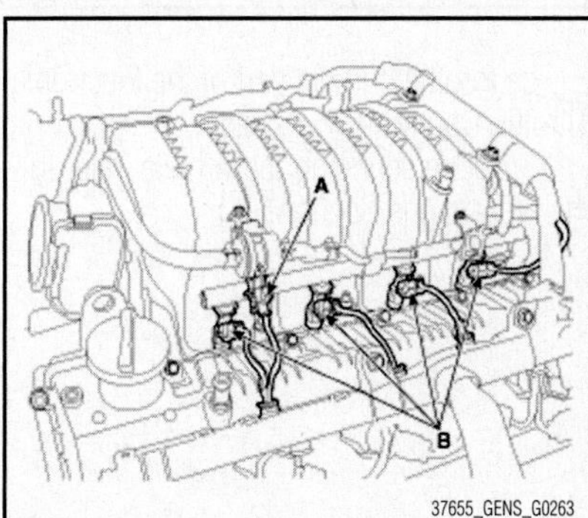

Fig. 334 Disconnect the PCSV connector (A) and the injector connectors (B)

FUEL TANK

REMOVAL & INSTALLATION

See Figures 335 through 339.

1. Before servicing the vehicle, refer to the Precautions Section.
2. Relieve the fuel system pressure. Refer to the Fuel System Pressure Relieving procedure.
3. Disconnect the fuel feed tube quick-connector.
4. Remove the sub fuel sender service cover.
5. Disconnect the sub fuel sender connector.
6. Lift the vehicle.
7. Remove the center muffler assembly.
8. Remove the propeller shaft.
9. Remove the undercover.
10. Support the fuel tank with a jack.
11. Disconnect the fuel filler hose and the leveling hose.

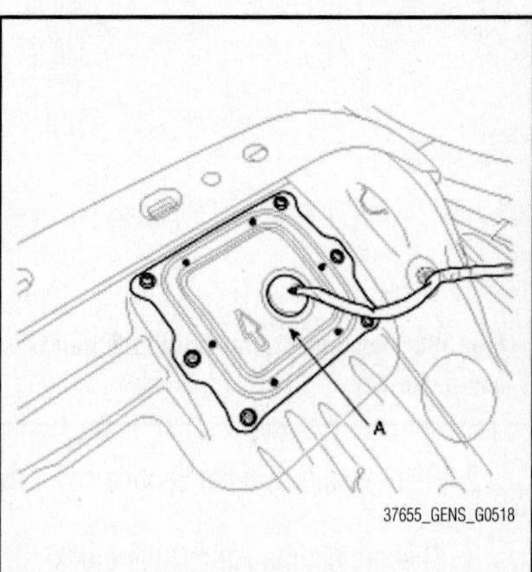

Fig. 335 Remove the sub fuel sender service cover (A)

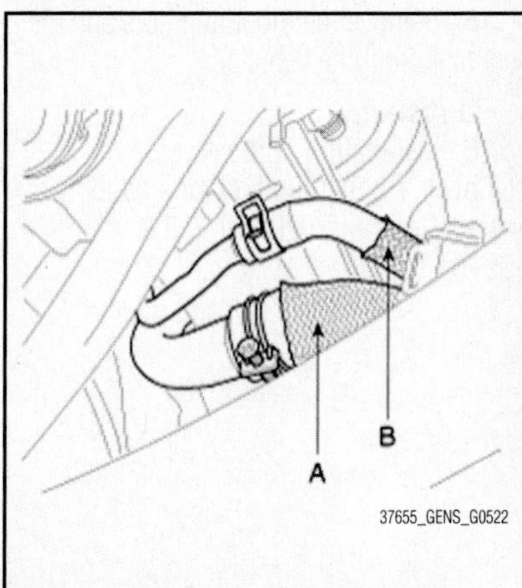

Fig. 336 Disconnect the fuel filler hose (A) and the leveling hose (B)

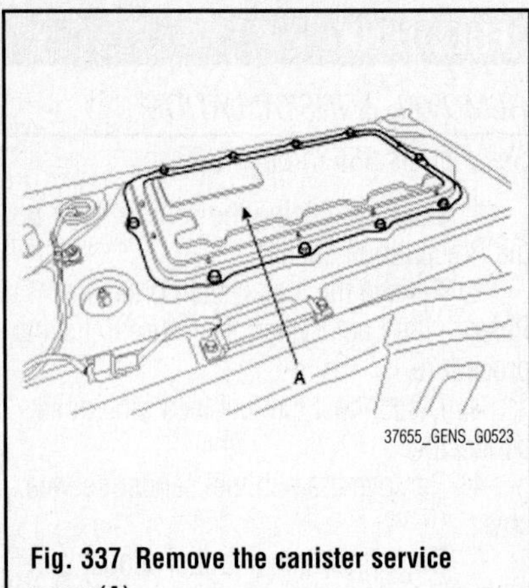

Fig. 337 Remove the canister service cover (A)

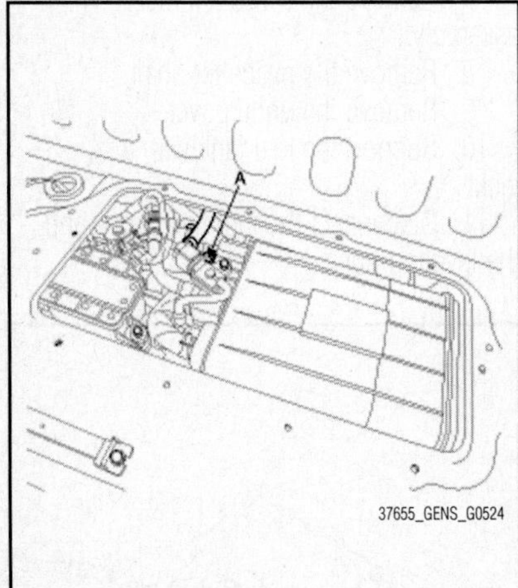

Fig. 338 Disconnect the vapor tube quick-connector (A)

12. Remove the canister service cover in the trunk.
13. Disconnect the vapor tube quick-connector.
14. Detach the parking brake cable from the fuel tank (Except Electric Parking Brake system).
15. Remove the mounting nuts and the fuel tank from the vehicle.

To install:

16. Installation is the reverse of removal. Tighten the fuel tank band installation nut to 29–40 ft. lbs. (39–54 Nm).

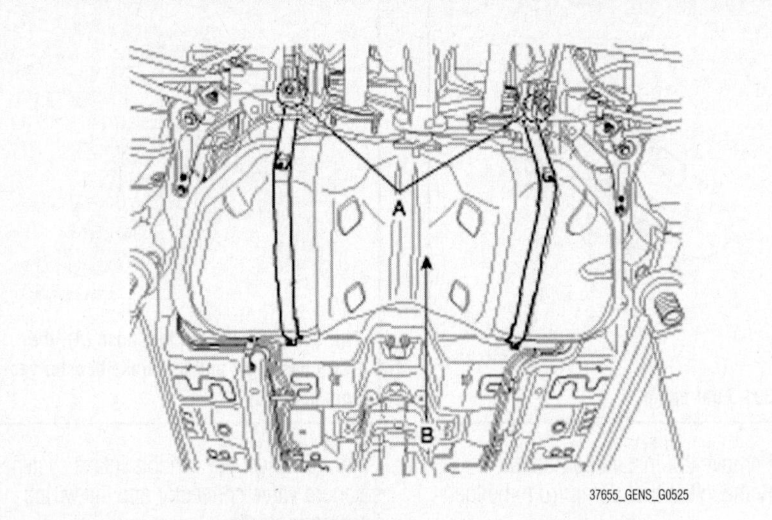

Fig. 339 Remove the mounting nuts (A) and the fuel tank (B) from the vehicle

THROTTLE BODY

REMOVAL & INSTALLATION

See Figures 340 and 341.

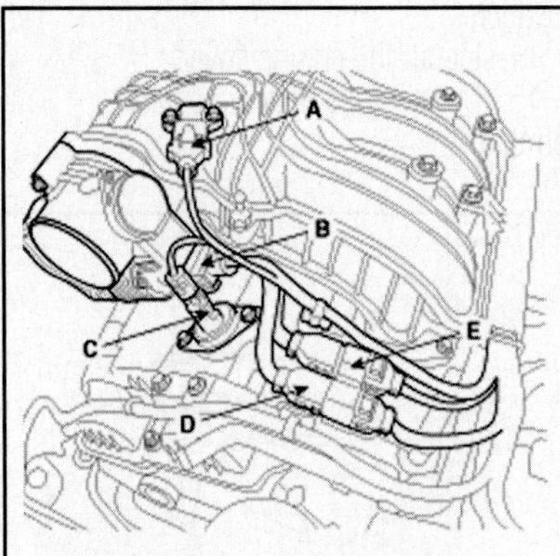

A. MAP sensor connector
B. ETC connector
C. Right-hand exhaust OCV connector
D. Right-hand injector connector
E. Right-hand ignition coil connector
37655_GENS_G0384

Fig. 340 Throttle body and ETC module connector (B)—3.8L engine

The throttle body is part of the Electronic Throttle Control (ETC) system.

1. Before servicing the vehicle, refer to the Precautions Section.

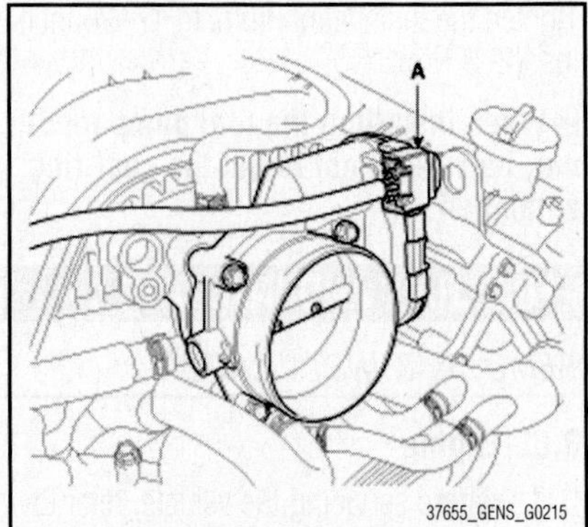

Fig. 341 Throttle body, ETC module and connector (A)—4.6L engine

2. Disconnect the negative battery cable.
3. Remove the air cleaner assembly.
4. Disconnect the ETC connector.
5. Remove the vacuum hoses, necessary.
6. Drain engine coolant and disconnect throttle body coolant hoses, as necessary.
7. Remove the ETC mounting bolts.
8. Remove the ETC.

To install:

9. Installation is the reverse of removal.

HEATING & AIR CONDITIONING SYSTEM

BLOWER MOTOR

REMOVAL & INSTALLATION

See Figures 342 and 343.

1. Before servicing the vehicle, refer to the Precautions Section.
2. Disconnect the negative battery terminal.
3. Remove the driver side instrument panel undercover.
4. Disconnect the blower motor connector.
5. Remove the blower motor mounting screws.
6. Remove the blower motor.

To install:

7. Installation is the reverse of removal.

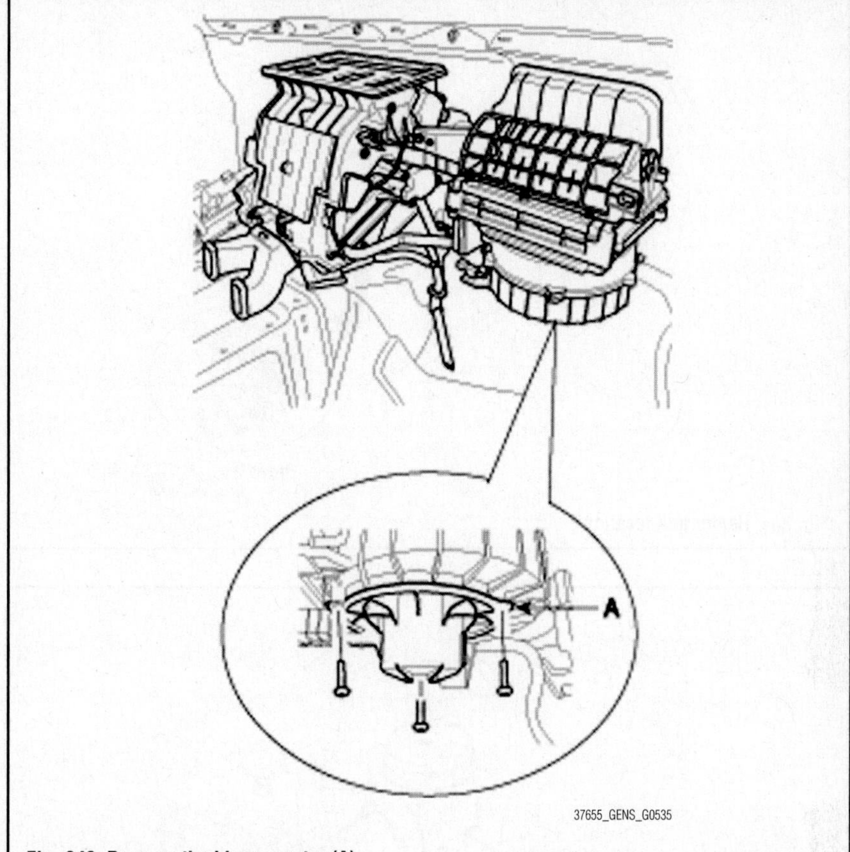

Fig. 343 Remove the blower motor (A)

HEATER UNIT/CORE

REMOVAL & INSTALLATION

See Figures 344 through 351.

1. Before servicing the vehicle, refer to the Precautions Section.
2. Disconnect the negative battery terminal.
3. Recover the refrigerant with a recovery/ recycling/ charging station.

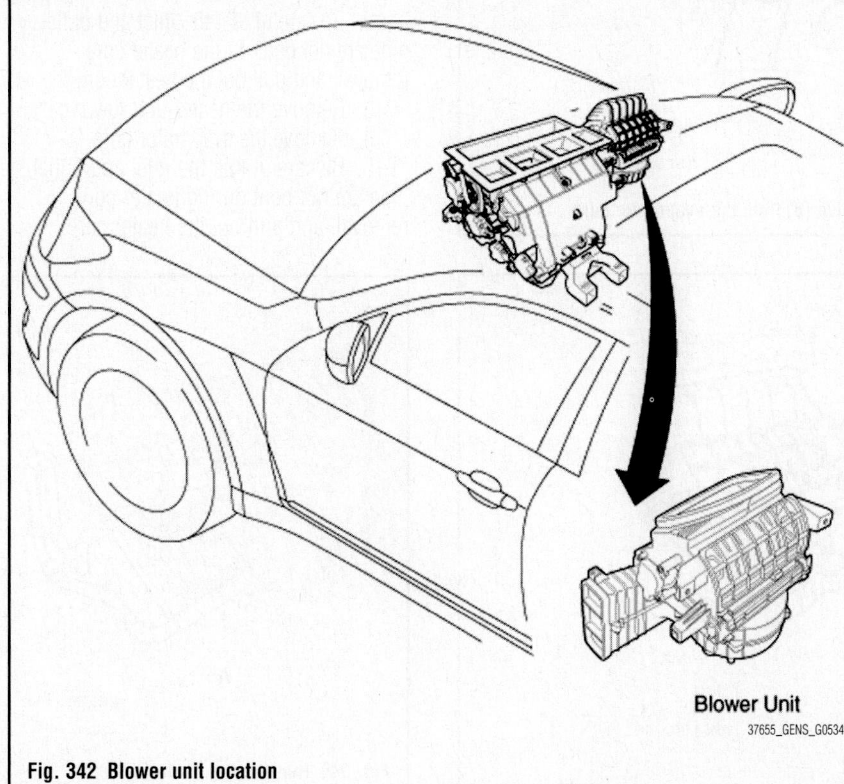

Fig. 342 Blower unit location

4. When the engine is cool, drain the engine coolant from the radiator.
5. Remove the bolts and the expansion valve from the evaporator core. Plug or cap the lines immediately after disconnecting them to avoid moisture and dust contamination.
6. Disconnect the inlet and outlet heater hoses from the heater unit.

⁂ WARNING

Engine coolant will spill when the hoses are disconnected; drain it into a clean drip pan. Be sure not to let coolant spill on electrical parts or painted surfaces. If any coolant spills, rinse it off immediately.

7. Remove the instrument panel.
8. Remove the cowl cross bar assembly.
9. Remove the heater and blower bolts, and then remove the heater and blower unit.
10. Remove the 2 blower unit screws, and then remove the blower unit from the heater unit.
11. Remove the heater core cover.

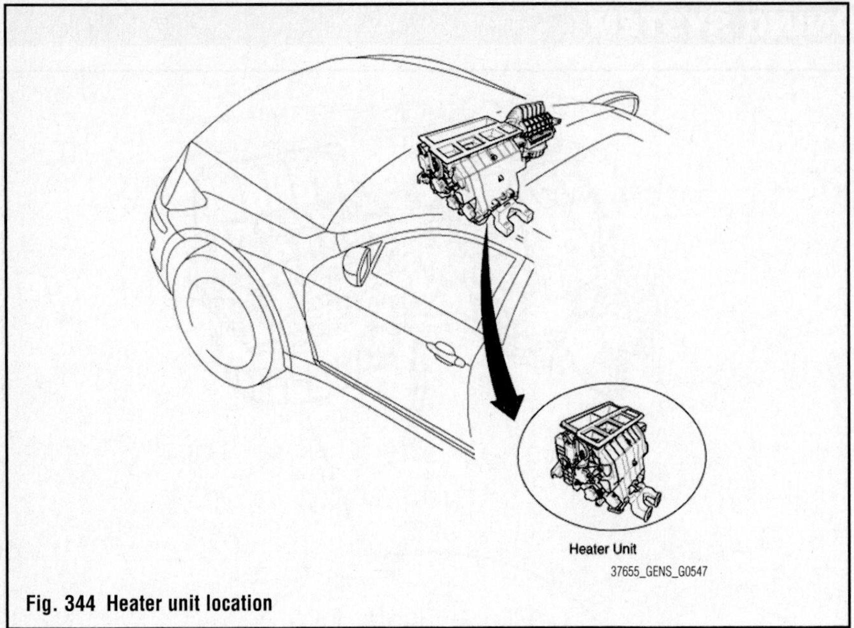

Fig. 344 Heater unit location

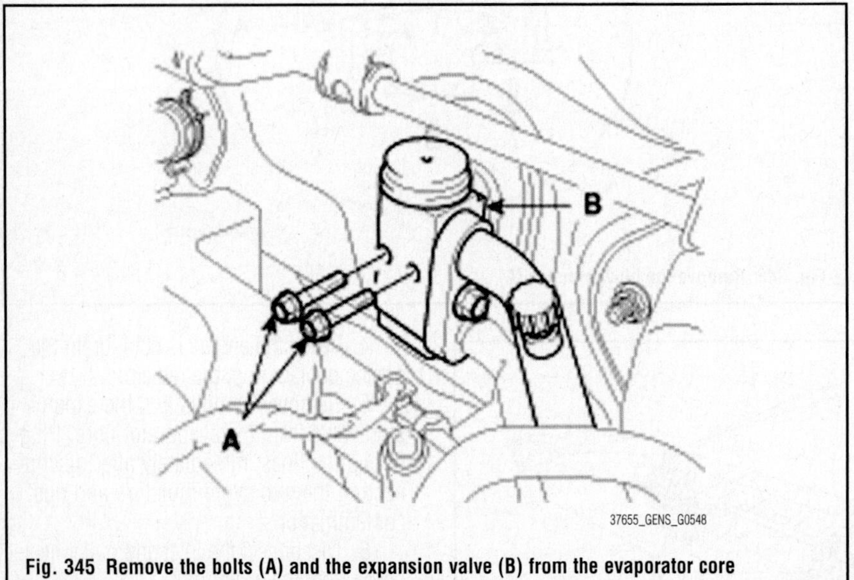

Fig. 345 Remove the bolts (A) and the expansion valve (B) from the evaporator core

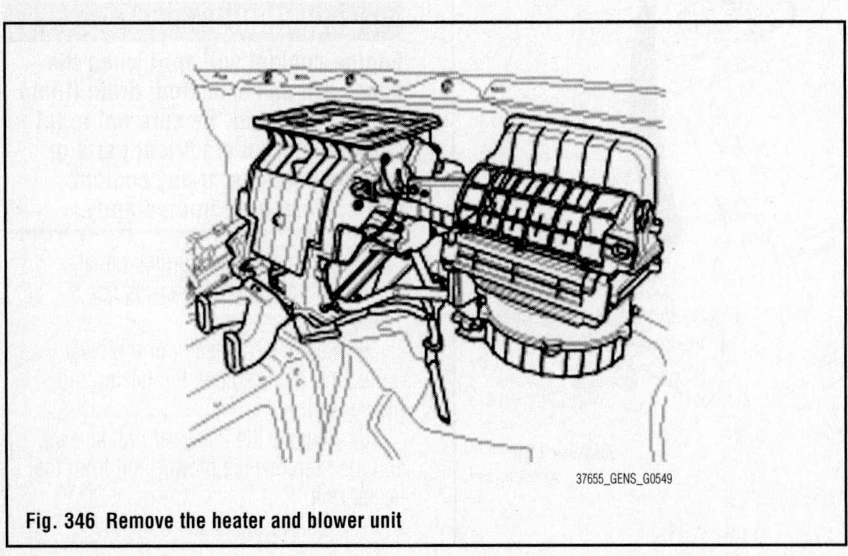

Fig. 346 Remove the heater and blower unit

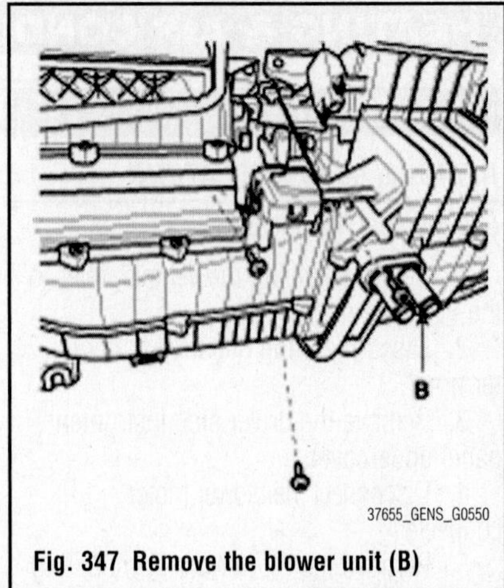

Fig. 347 Remove the blower unit (B)

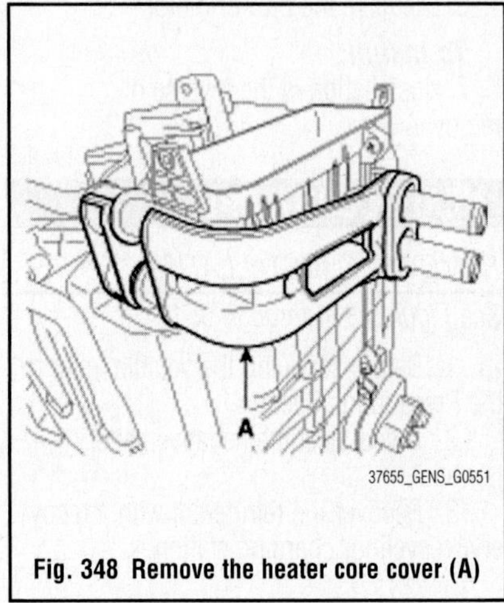

Fig. 348 Remove the heater core cover (A)

12. Be careful that the inlet and outlet pipe are not bent during heater core removal, and pull out the heater core.
13. Remove the heater unit lower case.
14. Remove the evaporator core.
15. Be careful that the inlet and outlet pipe are not bent during heater core removal, and pull out the heater core.

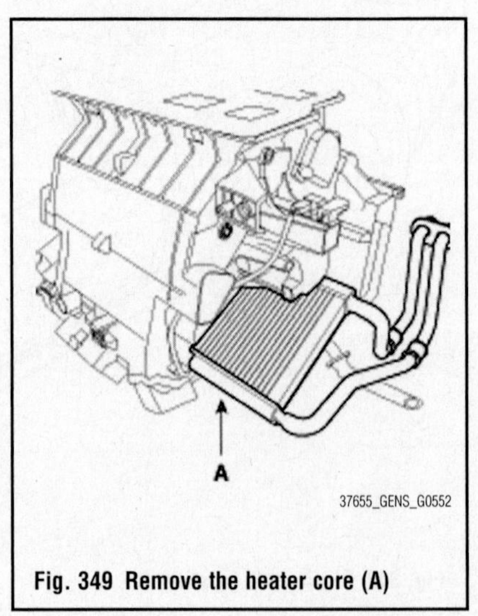

Fig. 349 Remove the heater core (A)

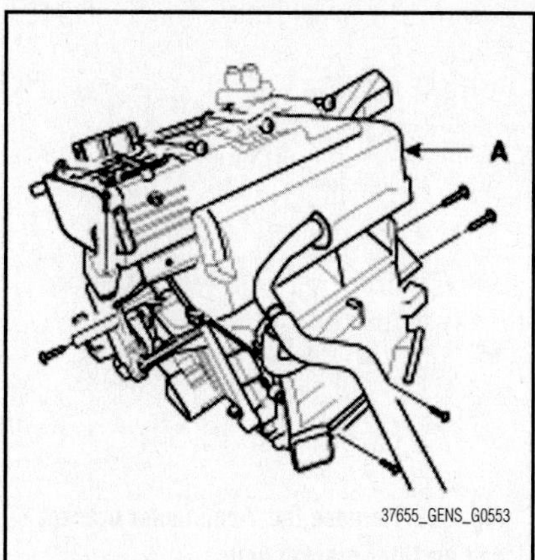

Fig. 350 Remove the heater unit lower case (A)

To install:

16. Installation is the reverse order of removal, and note these items :

a. If you're installing a new evaporator, add refrigerant oil (ND-OIL8).

b. Replace the O-rings with new ones at each fitting, and apply a thin coat of refrigerant oil before installing. Be sure to use the right O-rings for R-134a to avoid leakage.

c. Immediately after using the oil, replace the cap on the container, and seal it to avoid moisture absorption.

d. Do not spill the refrigerant oil on the vehicle; it may damage paint; if the refrigerant oil contacts the paint, wash off immediately.

e. Apply sealant to the grommets.

f. Make sure that there is no air leakage.

g. Charge the system and test its performance.

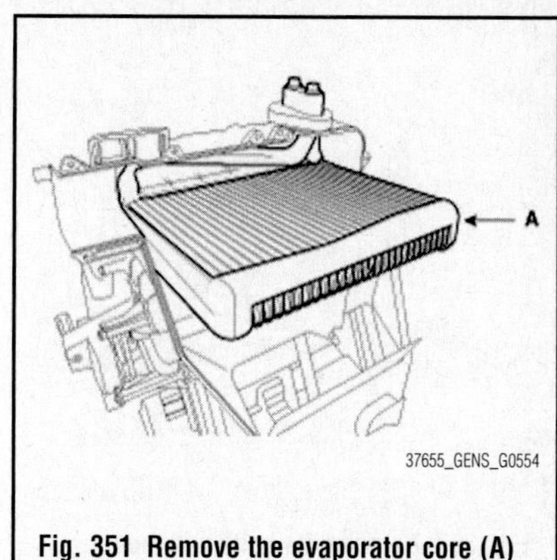

Fig. 351 Remove the evaporator core (A)

h. Do not interchange the inlet and outlet heater hoses and install the hose clamps securely.

i. Refill the cooling system with engine coolant.

STEERING

STEERING COMPONENTS

ELECTRO HYDRAULIC RACK & PINION STEERING GEAR

REMOVAL & INSTALLATION

See Figures 352 through 357.

1. Before servicing the vehicle, refer to the Precautions Section.
2. Disconnect the negative battery terminal.
3. Drain the power steering fluid.
4. Remove both front wheels and tires.
5. Remove the split pin and castle nut.
6. Disconnect the tie-rod end with the knuckle using a SST (09568-2J100).
7. Remove the bolt.

➡Keep the neutral-range to prevent the damage of the clock spring inner cable when you handle the steering wheel.

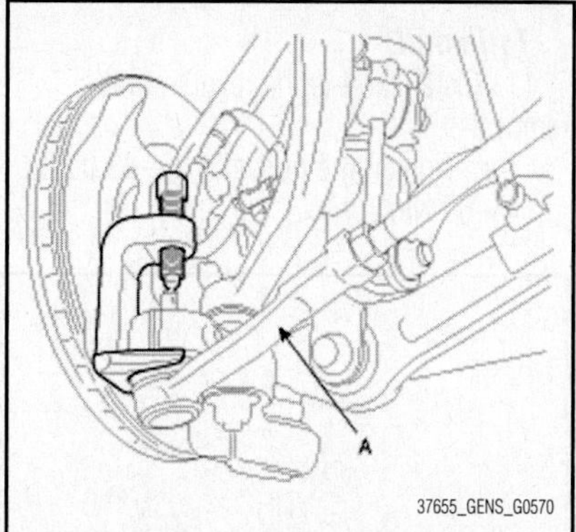

Fig. 353 Disconnect the tie-rod end (A) with the knuckle using a SST (09568-2J100)

8. Remove the tube bolts and then disconnect the pressure and return tube.
9. Remove the bolts and remove steering gear box.

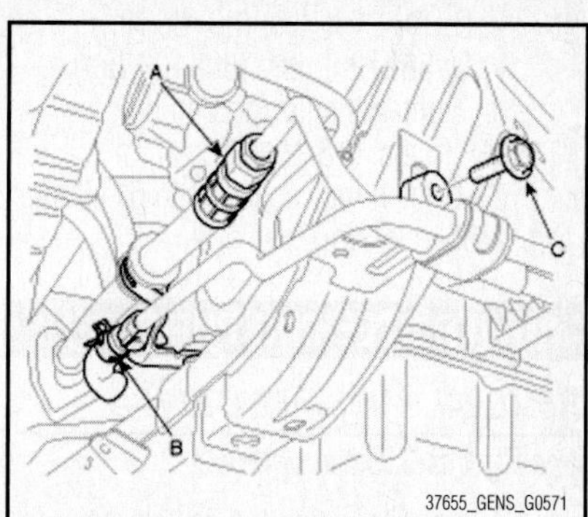

Fig. 355 Remove the tube bolts (C) and then disconnect the pressure (A) and return tube (B)

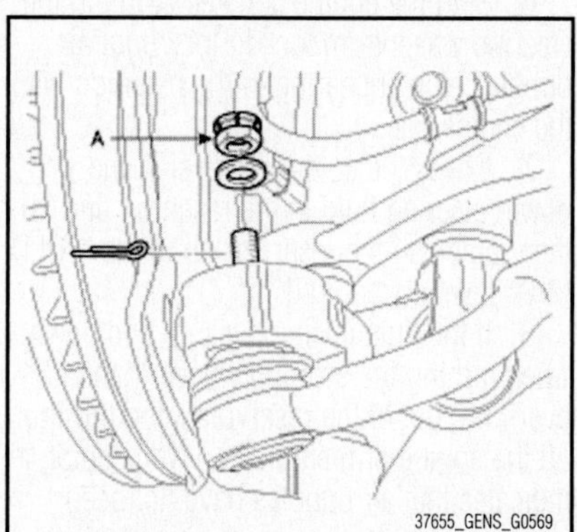

Fig. 352 Remove the split pin and castle nut (A)

Fig. 354 Remove the bolt (A)

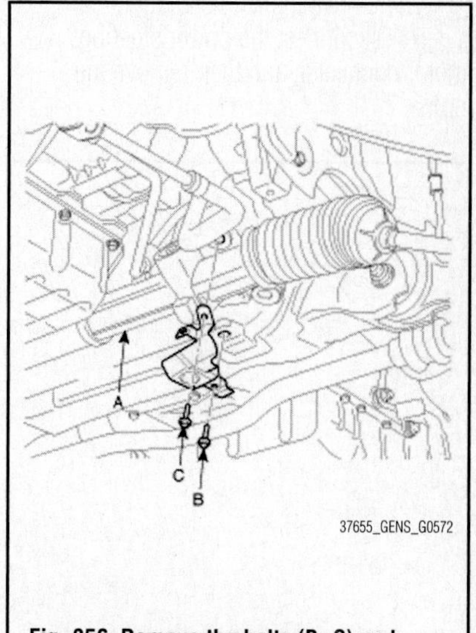

Fig. 356 Remove the bolts (B, C) and remove steering gear box (A)

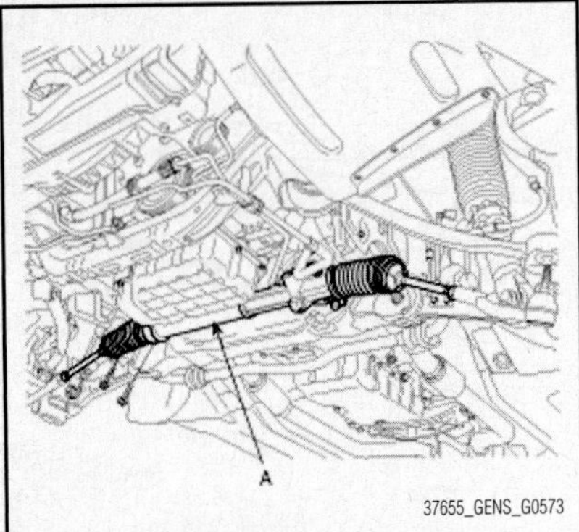

Fig. 357 Remove the bolts and then disconnect the steering gear box (A)

10. Remove the bolts and then disconnect the steering gear box.

To install:

11. Installation is the reverse of the removal, noting the following:
   a. Tighten the steering gear box bolts to:
      - A: 58–72 ft. lbs. (80–100 Nm)
      - B: 14–21 ft. lbs. (20–30 Nm)
      - C: 58–72 ft. lbs. (80–100 Nm)
   b. Tighten the tie-rod end bolt to 13–18 ft. lbs. (18–25 Nm)
   c. Tighten the castle nut to 61–80 ft. lbs. (85–110 Nm)

POWER STEERING PUMP

REMOVAL & INSTALLATION

See Figures 358 through 362.

1. Before servicing the vehicle, refer to the Precautions Section.
2. Disconnect the negative battery terminal.
3. Remove the front bumper.
4. Remove the left head light.
5. Disconnect the power steering motor connecter, and then remove the bolts.

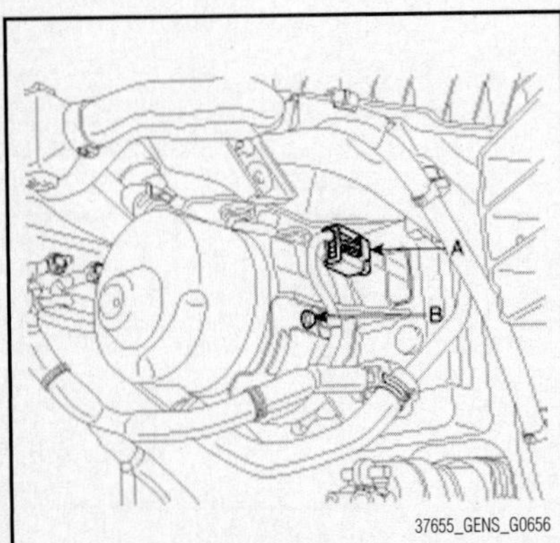

Fig. 358 Disconnect the motor connecter (A) and remove the bolts (B)

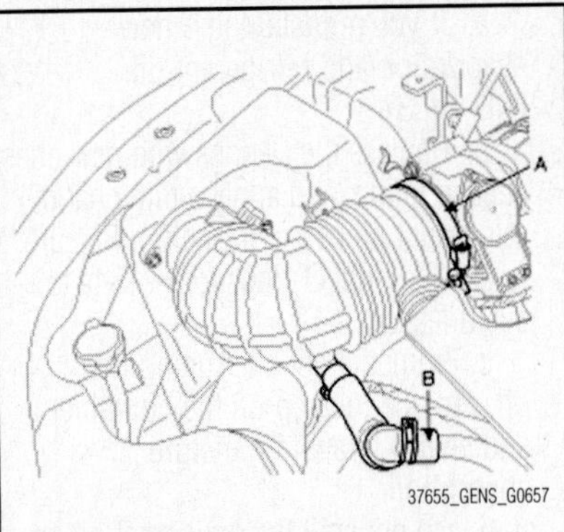

Fig. 359 Remove the air cleaner by disconnecting the seal ring (A) and hose (B)

6. Remove the air cleaner by disconnecting the seal ring and hose.
7. Remove the hose bracket bolts.
8. Remove the Y-connecter bracket and air filter bracket bolts.
9. Remove the Y-connecter bracket and air filter brackets.
10. Remove the mounting bolt, pressure hose, return hose.

To install:

11. Installation is the reverse of the removal.
   a. Tighten the hoses to 33–43 ft. lbs. (45–60 Nm).

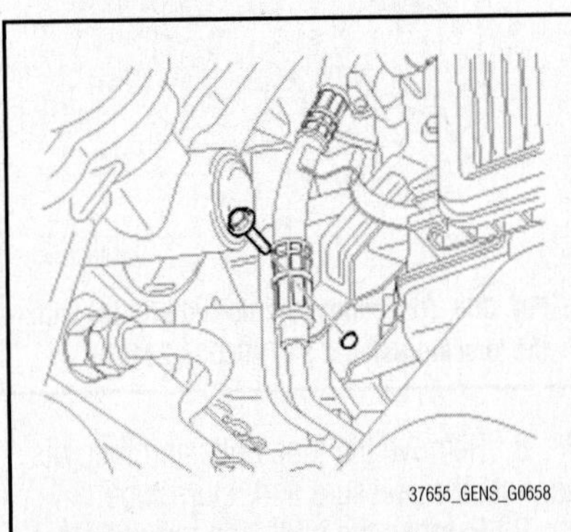

Fig. 360 Remove the hose bracket bolts

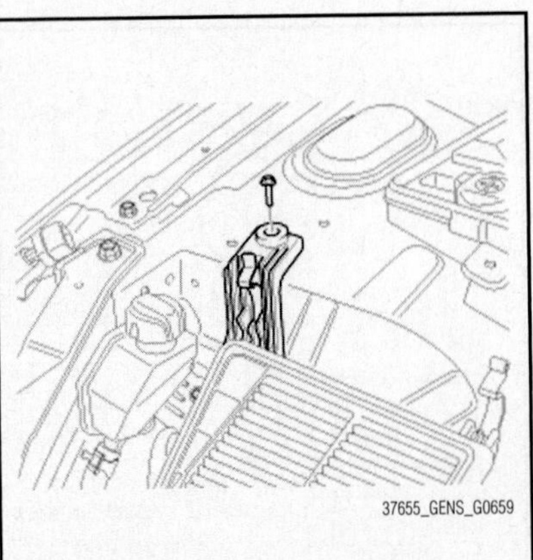

Fig. 361 Remove the bracket bolts

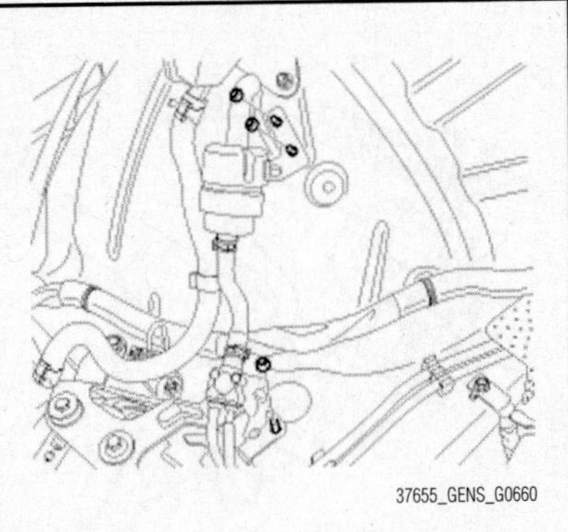

Fig. 362 Remove the Y-connecter bracket and air filter bracket bolts

BLEEDING

Electro Hydraulic Power Steering (EHPS)

➡Always use genuine Pentosin CHF202. Using other type of power steering fluid or ATF can cause increased wear and poor steering in cold weather.

1. Jack up the front wheels.
2. Fill the reservoir with the power steering fluid up to the level of 'COLD MAX' marked on the reservoir.

⁂ WARNING

Be careful not to start the engine. If starting the engine before performing the steps 3 through 4, it may cause an abnormal noise during power steering pump operation.

3. Turn the steering wheel from lock to lock 5–6 times for 15–20 seconds.
4. Crank the engine 1–2 times by turning the ignition key very quickly from the 'On' position to the 'Start' position, but do not start the engine.
5. Turn the steering wheel from lock to lock 5–6 times for 15–20 seconds.
6. Start the engine and keep turning the steering wheel from lock to lock until air bubbles stop appearing in the reservoir with the engine idle.
7. Check the color and level of the power steering fluid in the reservoir and then replenish the reservoir up to the 'COLD MAX' level as required.
8. If the fluid level moves up and down when turning the steering wheel, the fluid overflows out of the reservoir when turning off the engine or the fluid is a white color, it indicates that air bubbles have not been removed sufficiently from the power steering system. Therefore, repeat the steps 5 through 6 as required.

LOWER CONTROL ARM

REMOVAL & INSTALLATION

Lateral Arm

See Figures 363 through 366.

1. Before servicing the vehicle, refer to the Precautions Section.
2. Remove the front wheel and tire.
3. Remove the flange bolt and lock nuts.

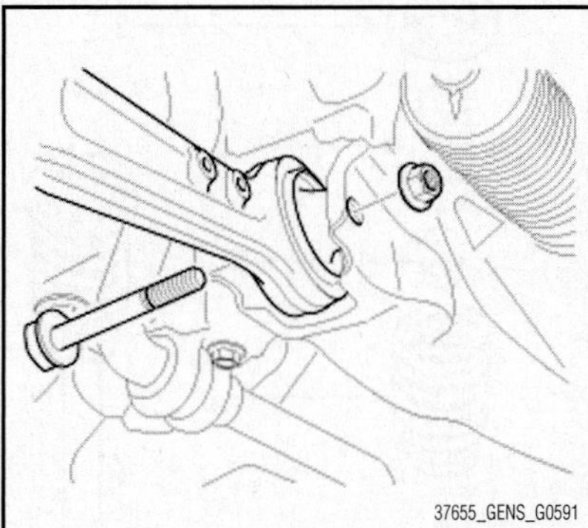

Fig. 363 Remove the flange bolt and lock nuts

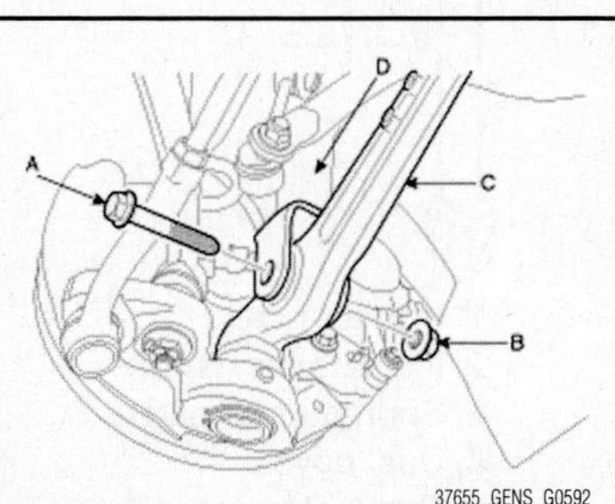

Fig. 364 Disconnect the lateral arm (C) from the front strut assembly (D) by removing the flange bolts (A) and lock nuts (B)

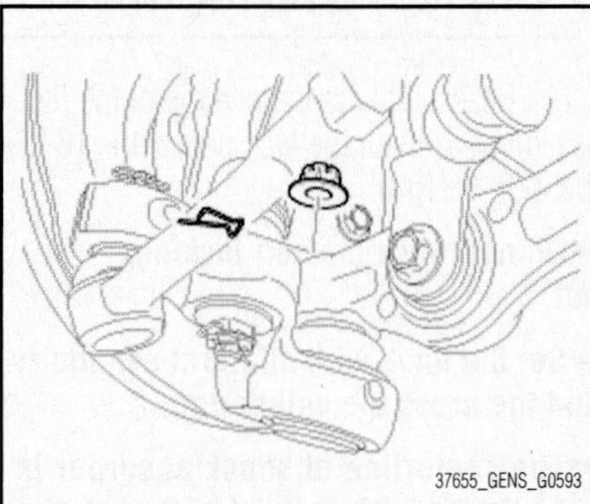

Fig. 365 Remove the split pin and castle nuts

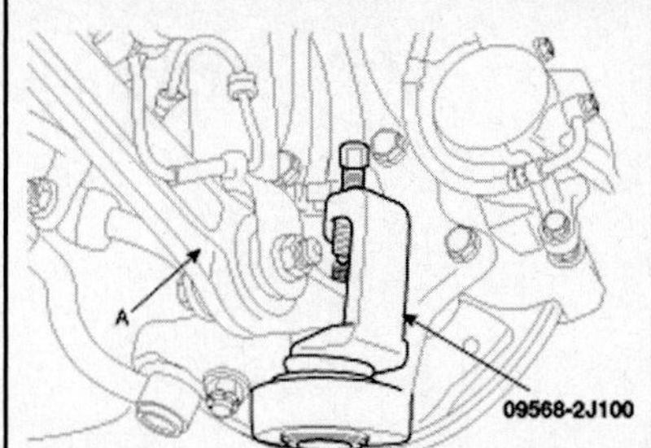

Fig. 366 Disconnect the lateral arm (A) from the front knuckle using a SST (09568-2J100)

4. Disconnect the lateral arm from the front strut assembly by removing the flange bolts and lock nuts.

➡Do not use the lock nuts again.

5. Remove the split pin and castle nuts.
6. Disconnect the lateral arm from the front knuckle using a SST (09568-2J100).

To install:

7. Installation is the reverse of removal, noting the following:
   a. Tighten the castle nuts to 65–80 ft. lbs. (90–110 Nm).
   b. Tighten the flange bolts to 101–116 ft. lbs. (140–160 Nm).

Tension Arm

See Figures 367 through 369.

1. Before servicing the vehicle, refer to the Precautions Section.
2. Remove the front wheel and tire.
3. Remove the split pin and castle nuts.
4. Disconnect the tension arm from the front knuckle using the SST (09568-2J100).

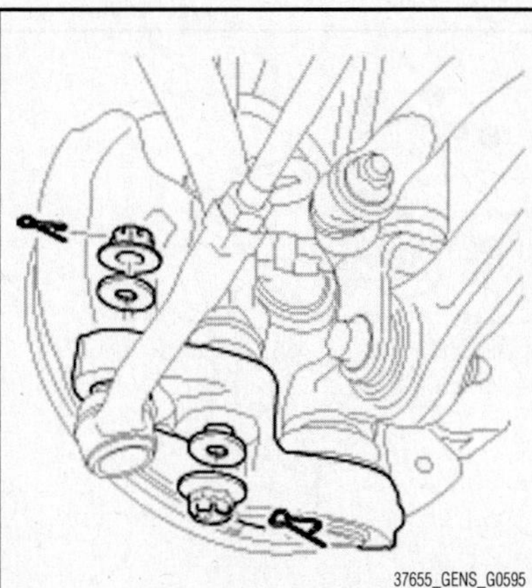

Fig. 367 Remove the split pin and castle nuts

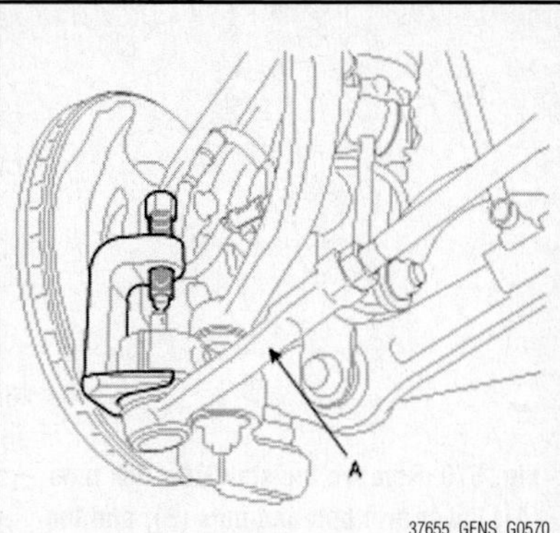

Fig. 368 Disconnect the tension arm (A) from the front knuckle using the SST (09568-2J100)

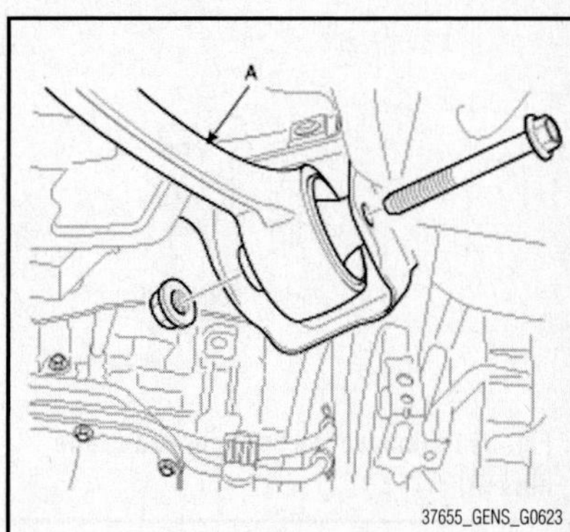

Fig. 369 Disconnect the tension arm (A) from the frame by removing the flange bolts and lock nuts

5. Disconnect the tension arm from the frame by removing the flange bolts and lock nuts.

To install:

6. Installation is the reverse of removal, noting the following:
   a. Tighten the flange bolts and lock nuts to 101–116 ft. lbs. (140–160 Nm).
   b. Tighten the castle nuts to 65–80 ft. lbs. (90–110 Nm).

STRUT

REMOVAL & INSTALLATION

See Figures 370 through 373.

1. Before servicing the vehicle, refer to the Precautions Section.
2. Remove the front wheel and tire.
3. Disconnect the stabilizer link with the front strut assembly by removing the nuts.
4. Disconnect the front strut assembly

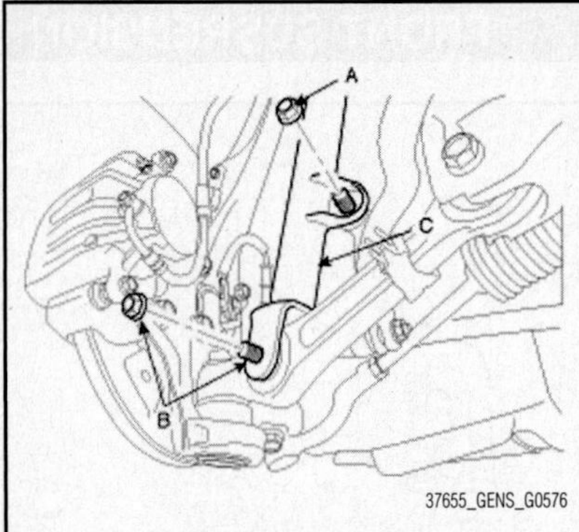

Fig. 370 Remove the stabilizer link nuts (A), lower arm bolt and nuts (B), and the front strut assembly (C)

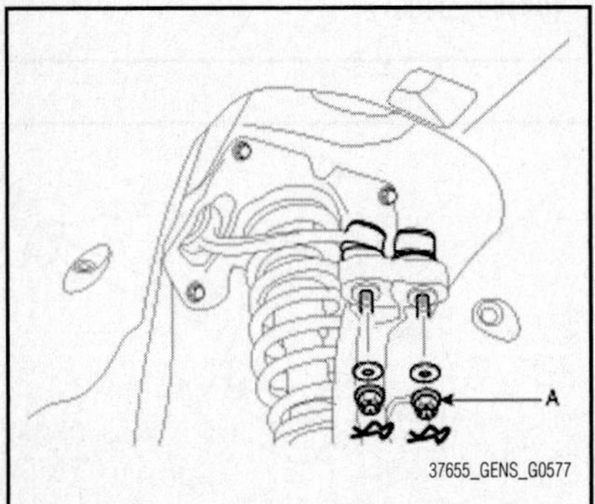

Fig. 371 Remove the split pin and castle nuts (A)

with the front lower arm by removing the bolt and nuts.

5. Remove the split pin and castle nuts.

6. Disconnect the front upper arm with the knuckle using a SST (09568-2J100).

7. Disconnect the front strut assembly with the frame by removing the mounting bolt.

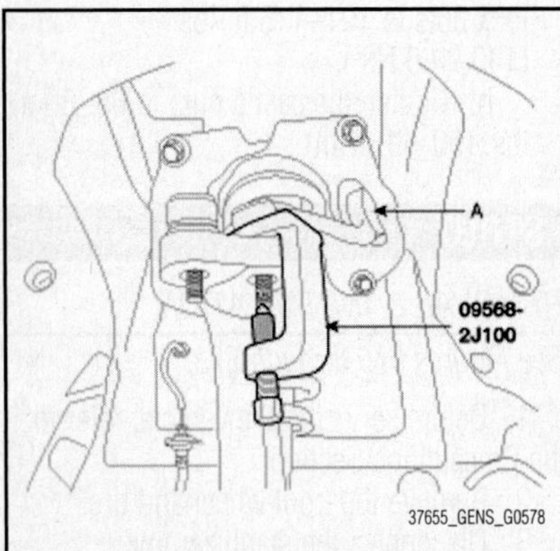

Fig. 372 Disconnect the front upper arm (A) with the knuckle using a SST (09568-2J100)

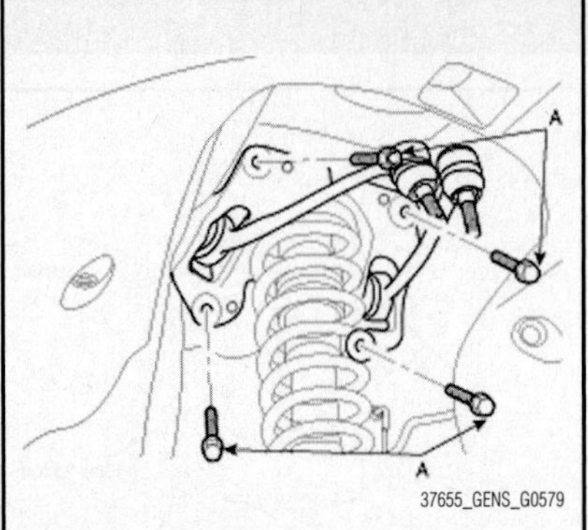

Fig. 373 Disconnect the front strut assembly with the frame by removing the mounting bolt (A)

To install:

8. Installation is the reverse of removal, noting the following:

a. Tighten the strut assembly mounting bolt to 40–47 ft. lbs. (55–65 Nm).

b. Tighten the castle nuts to 58–65 ft. lbs. (80–90 Nm).

c. Tighten the strut assembly to lower arm bolt and nuts to 101–116 ft. lbs. (140–160 Nm).

d. Tighten the stabilizer link nuts to 72–87 ft. lbs. (100–120 Nm).

OVERHAUL

See Figures 374 through 377.

1. Disconnect the bracket from the front strut assembly.

2. Compress the coil spring with a strut spring compressor. Do not compress the spring more than necessary.

3. Loosen the lock nut.

4. Disassemble the components of front strut assembly in sequence.

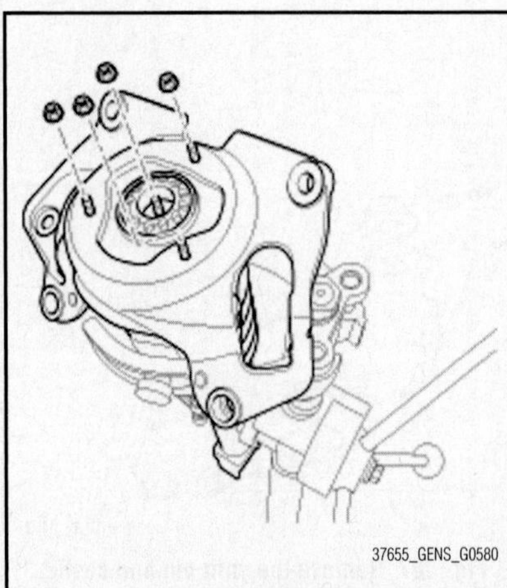

Fig. 374 Remove the retaining nuts

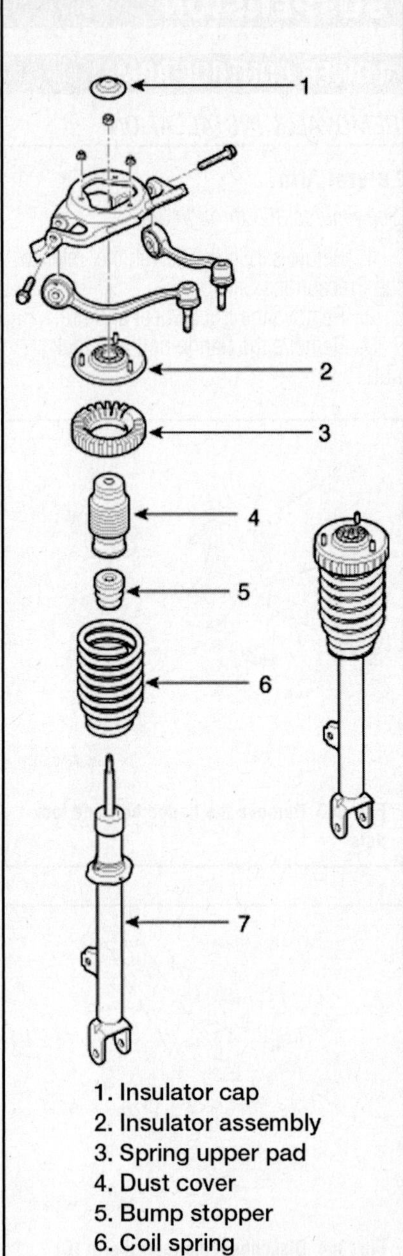

Fig. 375 Front strut assembly components

5. Reassembly is the reverse of the disassembly. Tighten the lock nut to 14–18 ft. lbs. (20–25 Nm)

➡Do not reuse the self locking nut.

➡Set the hook with bilateral symmetry and the press the coil spring.

➡The centerline of shock absorber is not aligned with coil spring, push the shock absorber to align the centerline.

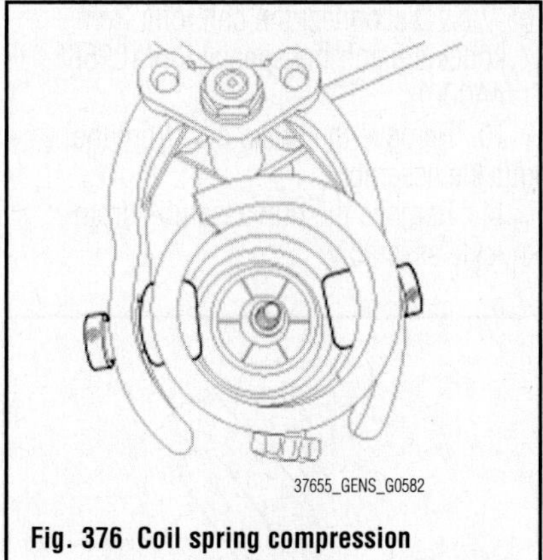

Fig. 376 Coil spring compression

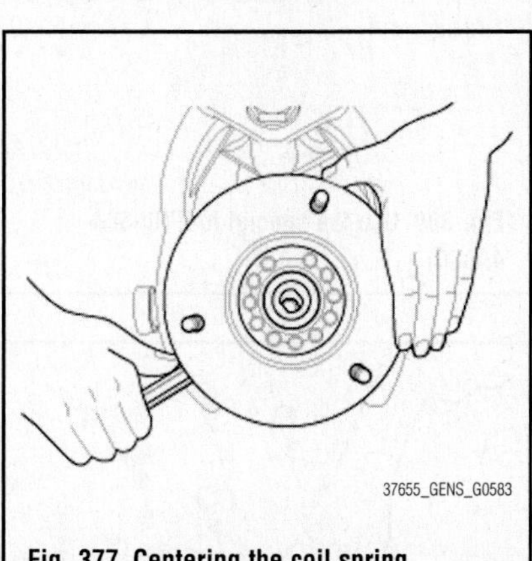

Fig. 377 Centering the coil spring

STABILIZER BAR

REMOVAL & INSTALLATION

See Figures 378 through 380.

1. Before servicing the vehicle, refer to the Precautions Section.
2. Remove the front wheel and tire.
3. Disconnect the stabilizer link from the front strut assembly by removing the nuts.
4. Disconnect the stabilizer bar from the frame by removing the bolts.

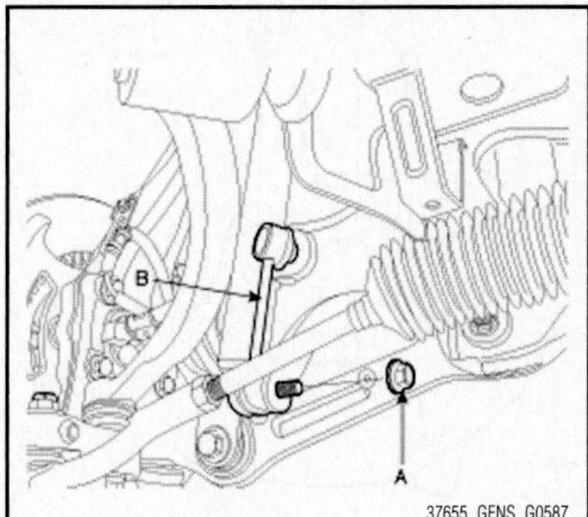

Fig. 378 Disconnect the stabilizer link (B) from the front strut assembly by removing the nuts (A)

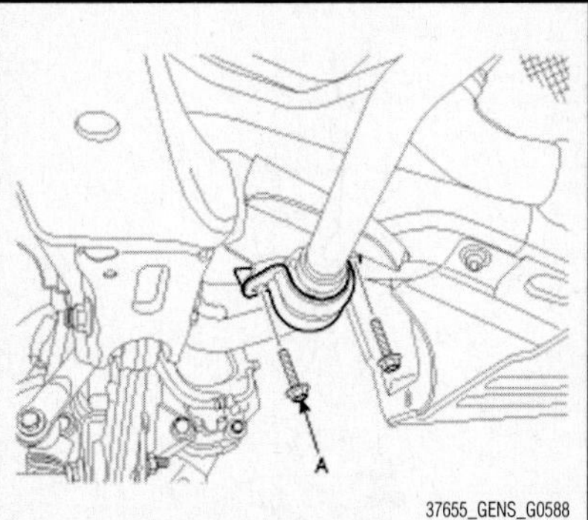

Fig. 379 Disconnect the stabilizer bar from the frame by removing the bolts (A)

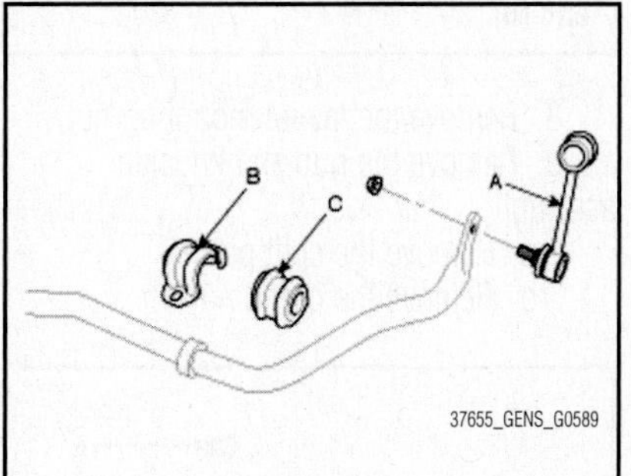

Fig. 380 Remove the stabilizer link (A), clamp (B), and bushing (C)

5. Remove the stabilizer link, clamp, and bushing.

To install:

6. Installation is the reverse of removal, noting the following:
   a. Tighten the stabilizer link to 72–87 ft. lbs. (100–120 Nm).
   b. Tighten the bolts to 45–55 ft. lbs. (33–40 Nm).
   c. Tighten the nuts to 72–87 ft. lbs. (100–120 Nm).

➡Do not use the lock nuts again.

UPPER CONTROL ARM

REMOVAL & INSTALLATION

See Figure 381.

1. Before servicing the vehicle, refer to the Precautions Section.
2. Remove the front wheel and tire.
3. Remove the front strut assembly. Refer to Strut Removal & Installation.
4. Disconnect the front upper arm from the front strut assembly bracket by removing the bolt.

To install:

5. Installation is the reverse of removal. Tighten the bolt to 72–87 ft. lbs. (100–120 Nm)

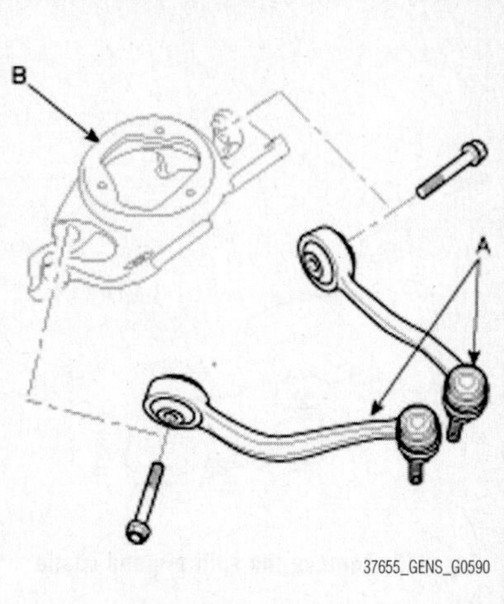

Fig. 381 Disconnect the front upper arm (A) from the front strut assembly bracket (B) by removing the bolt

WHEEL HUB & BEARING

REMOVAL & INSTALLATION

See Figures 382 through 391.

1. Before servicing the vehicle, refer to the Precautions Section.
2. Remove the front wheel and tire.
3. Remove the mounting screw and the brake disc.
4. Remove the caliper mounting bolts, and then hang the brake caliper on a wire. Refer to Front Disc Brake Removal & Installation in the Brake section.
5. Remove the tie rod end ball joint from the knuckle.
   a. Remove the split pin.
   b. Remove the castle nut.
   c. Disconnect the ball joint from knuckle using the special tool

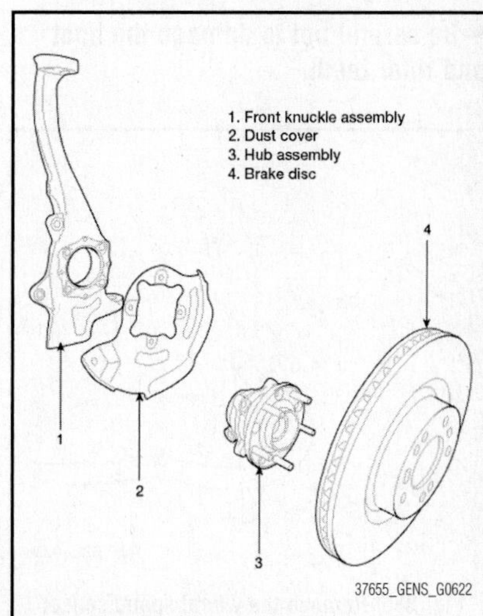

Fig. 382 Wheel hub and bearing components

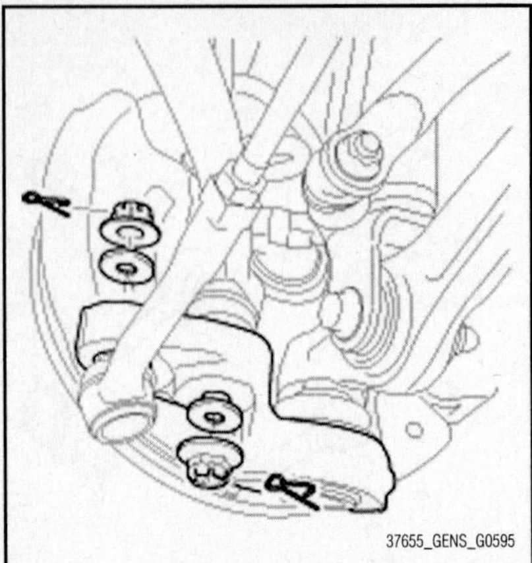

Fig. 383 Remove the split pin and castle nut

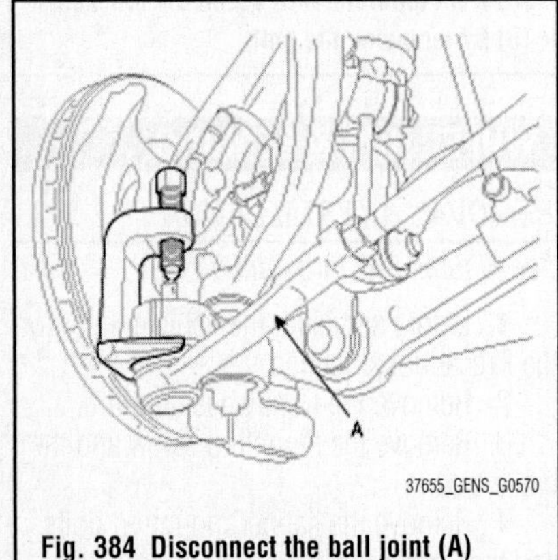

Fig. 384 Disconnect the ball joint (A)

(09568-2J100). Apply a few drops of oil to the special tool (Boot contact part).

6. Remove the wheel speed sensor, the strut lower mounting bolt, and the lower arm mounting bolt from the knuckle.

7. Remove the tension arm mounting bolt, and then remove the tension arm.

➡**Be careful not to damage the boot and rotor teeth.**

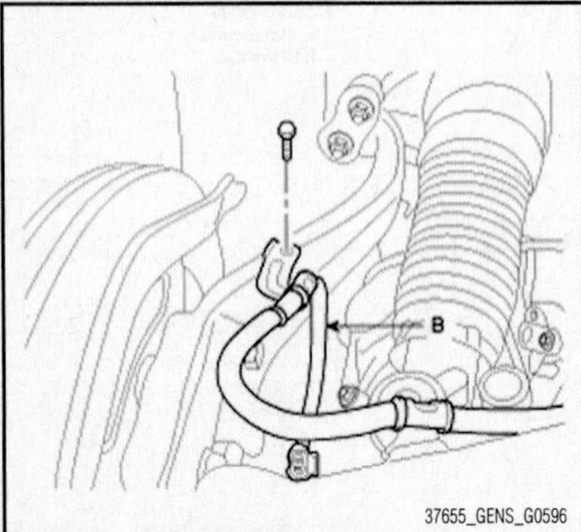

Fig. 385 Remove the wheel speed sensor (B), the strut lower mounting bolt, and the lower arm mounting bolt from the knuckle

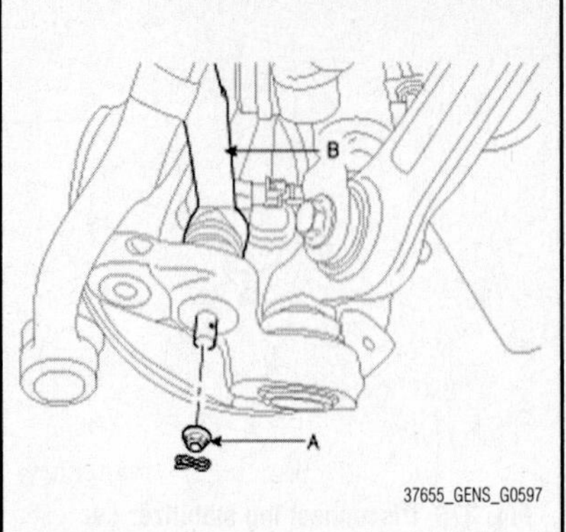

Fig. 386 Remove the tension arm mounting bolt (A), and then remove the tension arm (B)

8. Remove the lateral mounting nut.

9. Remove the hub and knuckle assembly.

a. Remove the split pin.

b. Remove the castle nut.

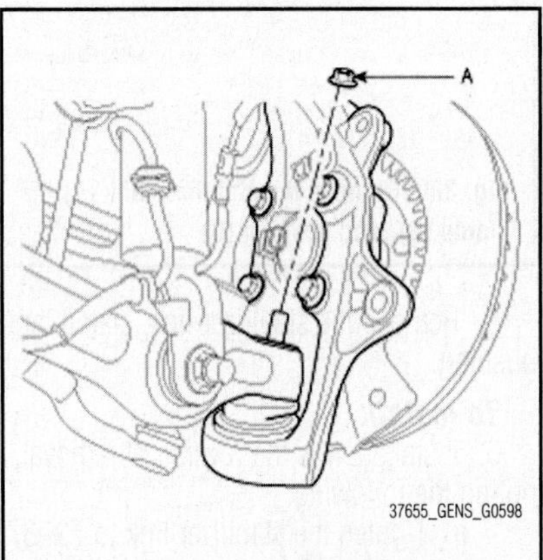

Fig. 387 Remove the lateral mounting nut (A)

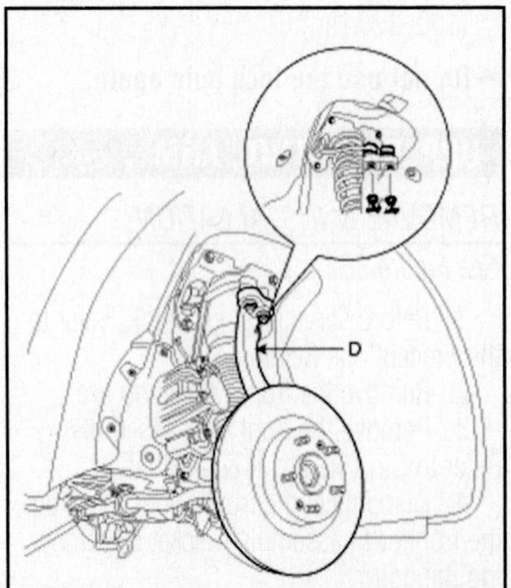

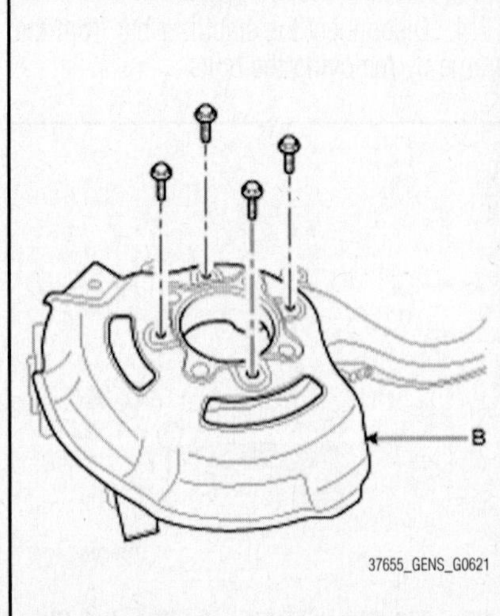

Fig. 388 Disconnect the ball joint from knuckle (D) using the special tool (09568-4A000)

c. Disconnect the ball joint from knuckle using the special tool (09568-4A000).

10. Remove the brake disc from the knuckle assembly.

11. Remove the hub assembly from knuckle assembly.

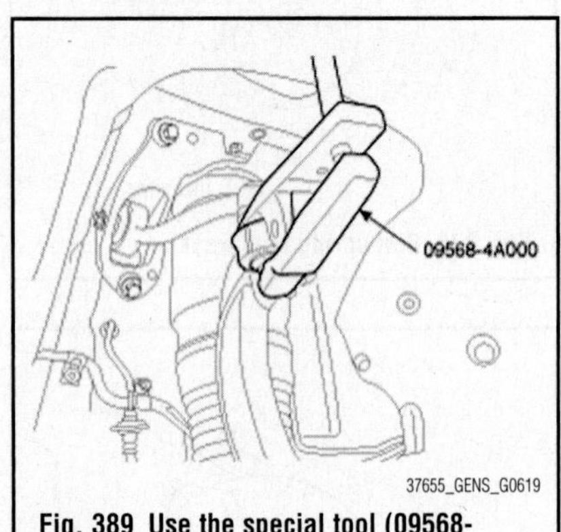

Fig. 389 Use the special tool (09568-4A000)

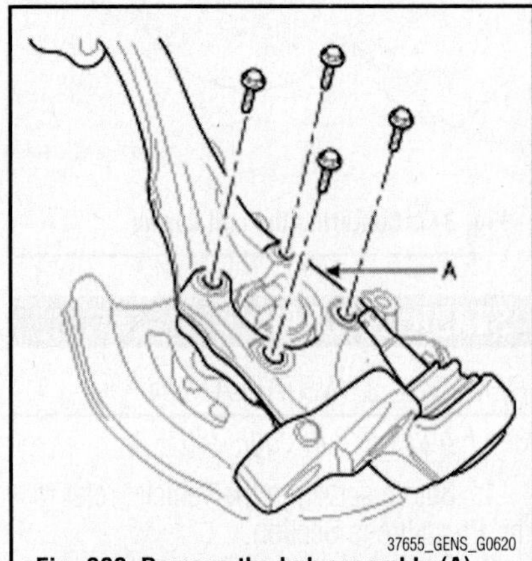

Fig. 390 Remove the hub assembly (A) from knuckle assembly

Fig. 391 Remove the dust cover mounting bolts and dust cover (B)

12. Remove the dust cover mounting bolts, and then remove the dust cover.

To install:

13. Installation is the reverse of removal, noting the following:

a. Tighten the dust cover mounting bolts to 60–95 inch lbs. (7–11 Nm).

b. Tighten the hub assembly bolts to 58–72 ft. lbs. (80–100 Nm).

c. Tighten the hub and knuckle castle nut to 25–33 ft. lbs. (35–45 Nm).

d. Tighten the lateral mounting nut to 65–80 ft. lbs. (90–110 Nm).

e. Tighten the tension arm mounting bolt to 101–116 ft. lbs. (140–160 Nm).

f. Tighten the tie rod end castle nut to 62–80 ft. lbs. (85–110 Nm).

COIL SPRING

REMOVAL & INSTALLATION

See Figures 392 and 393.

1. Before servicing the vehicle, refer to the Precautions Section.
2. Remove the rear wheel and tire.
3. Disconnect the rear height sensor and stabilizer link from the lower arm.
4. Remove the lower arm bolt, nuts, flange bolt, lock nuts.
5. Remove the coil spring.

To install:

6. Installation is the reverse of removal.
7. Tighten lower arm axle side to 101–116 ft. lbs. (140–160 Nm).
8. Tighten lower arm damper side to 72–87 ft. lbs. (100–120 Nm).
9. Tighten height sensor to 36–48 inch lbs. (4–6 Nm).

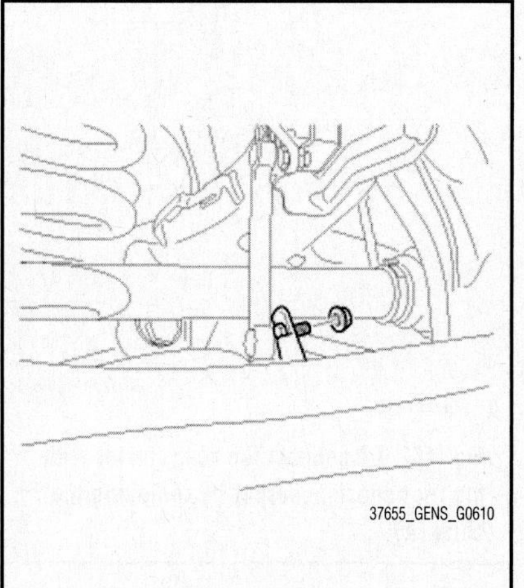

Fig. 392 Disconnect the rear height sensor and stabilizer link from the lower arm

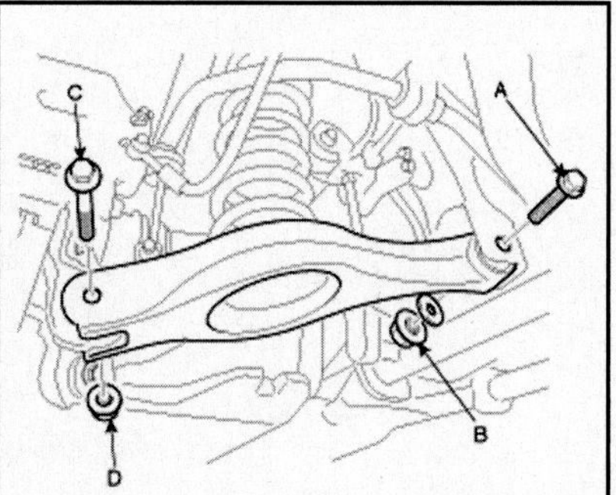

Fig. 393 Remove the bolt (A), nuts (B), flange bolt (C), lock nuts (D)

CONTROL ARMS/LINKS

REMOVAL & INSTALLATION

Rear Assist Arm

See Figures 394 through 396.

1. Before servicing the vehicle, refer to the Precautions Section.
2. Remove the rear wheel and tire.

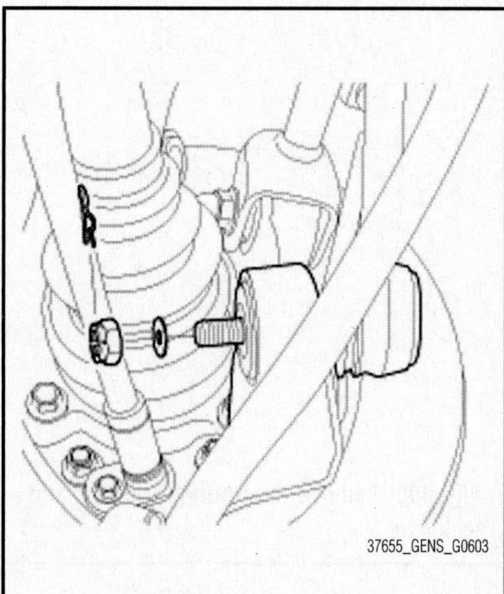

Fig. 394 Remove the rear assist arm split pin and castle nuts

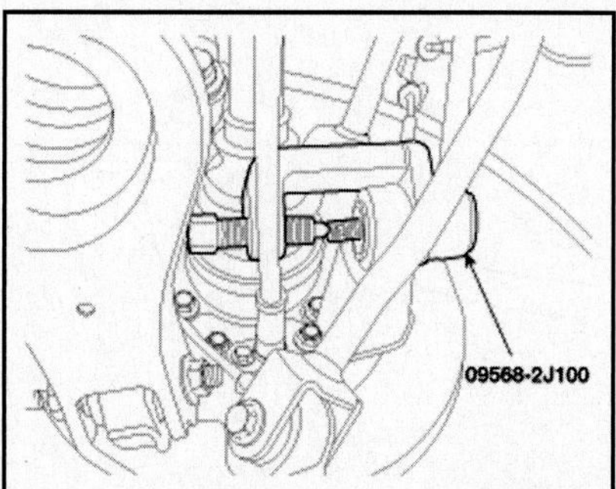

Fig. 395 Disconnect the rear assist arm with the rear carrier using the SST (09568-2J100)

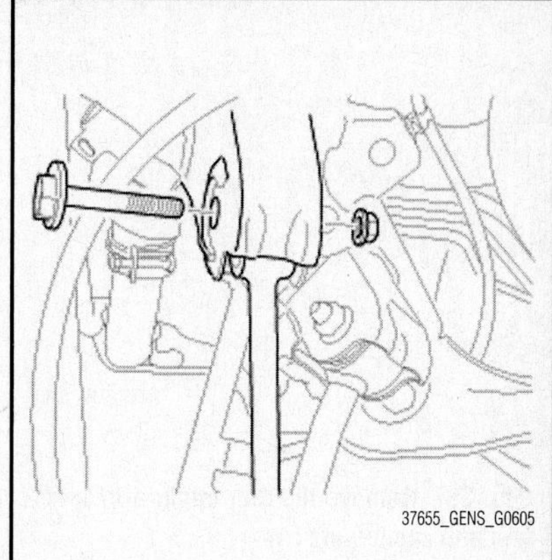

Fig. 396 Remove the rear assist arm bolt and lock nuts

3. Remove the rear assist arm split pin and castle nuts.
4. Disconnect the rear assist arm with the rear carrier using the SST (09568-2J100).
5. Remove the rear assist arm bolt and lock nuts.

To install:

6. Installation is the reverse of removal, noting the following:

a. Tighten castle nut to 58–65 ft. lbs. (80–90 Nm).

b. Tighten bolt and lock nut to 72–87 ft. lbs. (100–120 Nm).

Rear Lower Arm

See Figures 392 and 393.

1. Before servicing the vehicle, refer to the Precautions Section.
2. Remove the rear wheel and tire.
3. Disconnect the rear height sensor and stabilizer link from the lower arm.
4. Remove the lower arm bolt, nuts, flange bolt, lock nuts.

To install:

5. Installation is the reverse of removal, noting the following:
6. Tighten lower arm axle side to 101–116 ft. lbs. (140–160 Nm).
7. Tighten lower arm damper side to 72–87 ft. lbs. (100–120 Nm).
8. Tighten height sensor to 36–48 inch lbs. (4–6 Nm).

Rear Upper Arm

See Figures 397 through 399.

1. Before servicing the vehicle, refer to the Precautions Section.
2. Remove the rear wheel and tire.

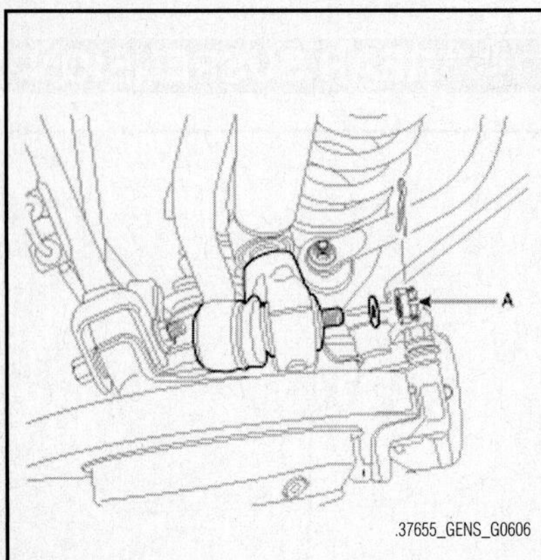

Fig. 397 Remove the rear upper arm split pin and castle nuts (A)

3. Support the lower portion of the rear axle with a jack securely.
4. Remove the split pin and castle nuts.
5. Disconnect the rear upper arm from the rear carrier using the SST (09568-2J100).
6. Remove the flange bolt and lock nuts.

To install:

7. Installation is reverse of removal. Tighten castle nut to 58–65 ft. lbs. (80–90 Nm).

Trailing Arm

See Figures 400 and 401.

1. Before servicing the vehicle, refer to the Precautions Section.
2. Remove the rear wheel and tire.
3. Remove the bolt and lock nuts.
4. Remove the trailing arm flange bolt and lock nuts.

To install:

5. Installation is the reverse of removal, noting the following:
   a. Tighten bolt and lock nuts to 72–87 ft. lbs. (100–120 Nm).
   b. Tighten flange bolt and lock nuts to 72–87 ft. lbs. (100–120 Nm).

SHOCK ABSORBER

REMOVAL & INSTALLATION

See Figures 402 and 403.

1. Before servicing the vehicle, refer to the Precautions Section.
2. Remove the rear wheel and tire.
3. Support the lower portion of the rear axle with a jack securely.
4. Disconnect the rear carrier from the rear shock absorber by removing the bolts.
5. Disconnect the rear shock absorber from the wheel housing panel by removing the mounting bolts.

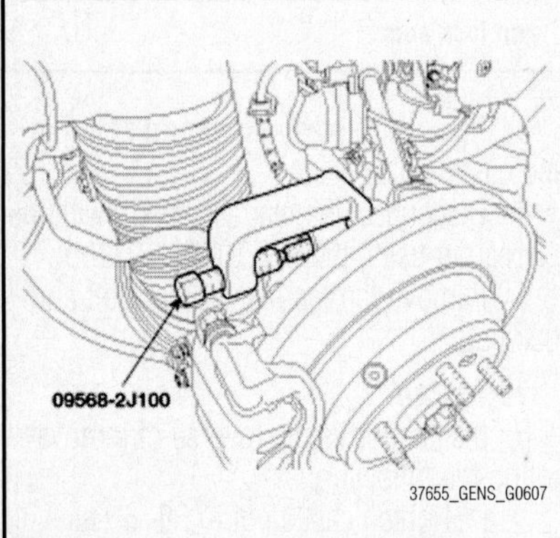

Fig. 398 Disconnect the rear upper arm (A) with the rear carrier using the SST (09568-2J100)

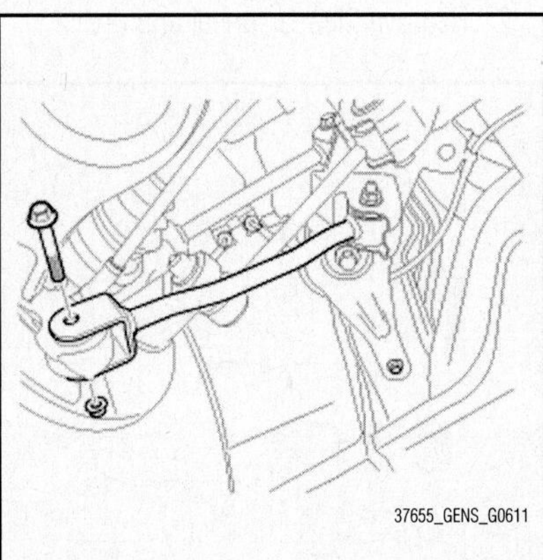

Fig. 400 Remove the trailing arm bolt and lock nuts

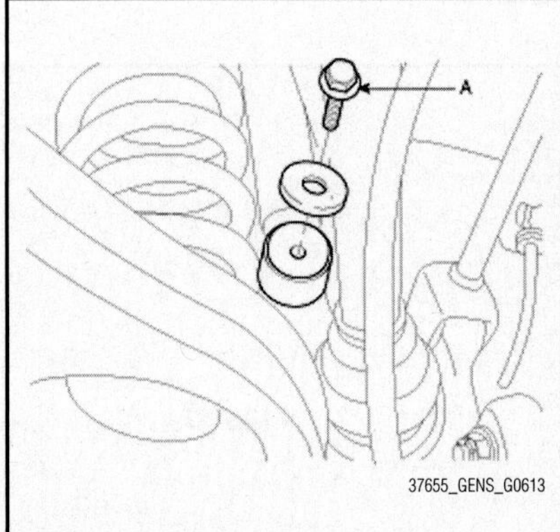

Fig. 402 Disconnect the rear carrier with the rear shock absorber by removing the bolts (A)

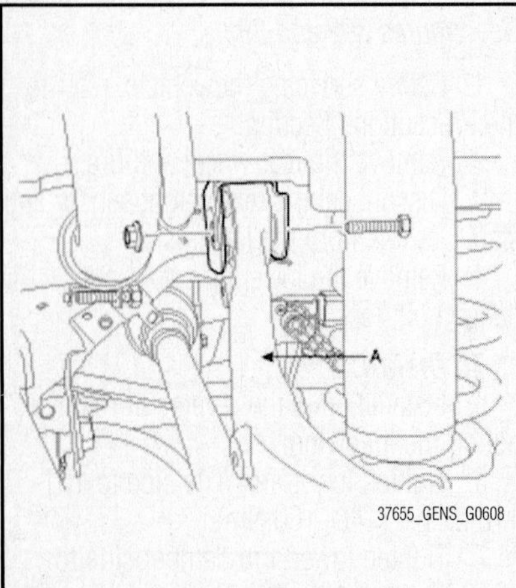

Fig. 399 Remove the rear upper arm flange bolt and lock nuts

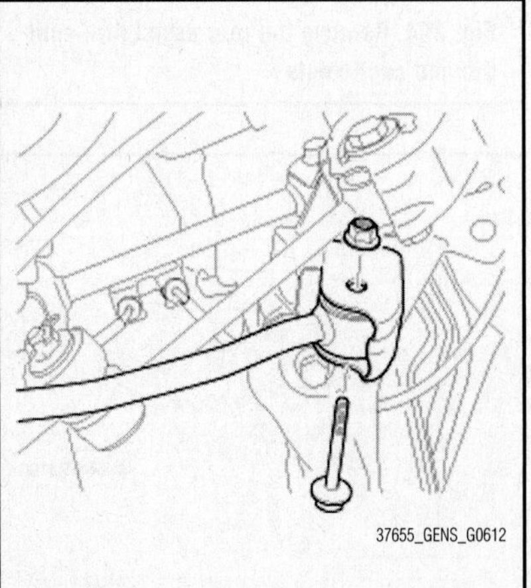

Fig. 401 Remove the trailing arm flange bolt and lock nuts

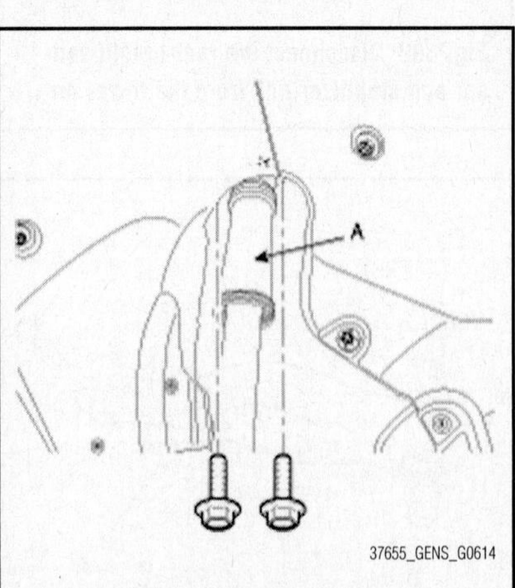

Fig. 403 Disconnect the rear shock absorber (A) with the wheel housing panel by removing the mounting bolts

To install:

6. Installation is the reverse of removal, noting the following:

a. Tighten rear shock absorber top bolt to 101–116 ft. lbs. (140–160 Nm).

b. Tighten bolt and lock nuts to 36–47 ft. lbs. (50–65 Nm).

STABILIZER BAR

REMOVAL & INSTALLATION

See Figures 404 through 406.

1. Before servicing the vehicle, refer to the Precautions Section.
2. Remove the rear wheel and tire.
3. Remove the nuts.
4. Remove the bolts.
5. Disconnect the rear stabilizer link from the stabilizer bar.

To install:

6. Installation is the reverse of removal, noting the following:

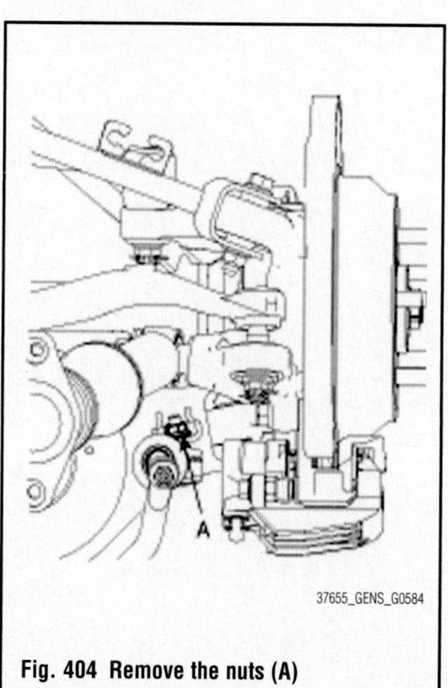

Fig. 404 Remove the nuts (A)

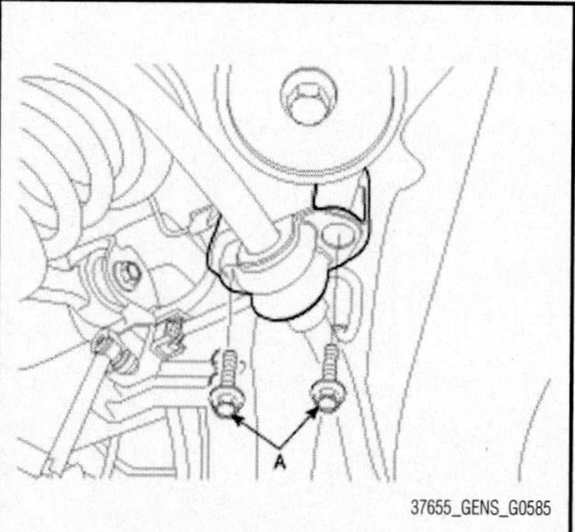

Fig. 405 Remove the bolts (A)

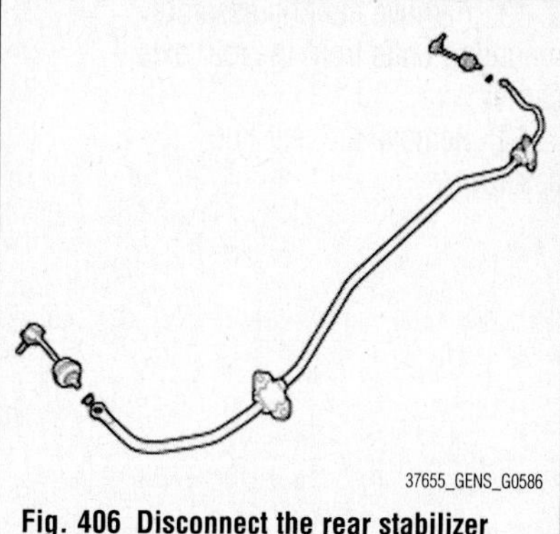

Fig. 406 Disconnect the rear stabilizer link with the stabilizer bar

a. Tighten the nuts to 36–47 ft. lbs. (50–65 Nm).

b. Tighten the bolts to 33–40 ft. lbs. (45–55 Nm).

c. Tighten the stabilizer link to 36–47 ft. lbs. (50–65 Nm).

WHEEL HUB & BEARING

REMOVAL & INSTALLATION

See Figures 407 through 410.

1. Before servicing the vehicle, refer to the Precautions Section.
2. Remove the rear wheel and tire.
3. Remove the split pin, and then remove castle nut and washer from the rear hub after applying the brake.
4. Remove the brake caliper mounting bolts, and then support the brake caliper assembly with wire.
5. Remove the rear brake disc.
6. Disconnect the parking brake cable from the operating lever.
7. Remove the rear brake lining.
8. Remove the wheel speed sensor and the parking brake cable from the rear axle carrier.
9. Remove the assist arm and the trailing arm from the rear axle carrier. Refer to Rear Assist Arm and Trailing Arm, in Control Arms Removal and Installation.
10. Remove lower arm mount bolt from rear axle carrier. Refer to Rear Lower Arm, in Control Arms Removal & Installation.
11. Remove the upper arm link mounting bolt and then remove the upper arm link.

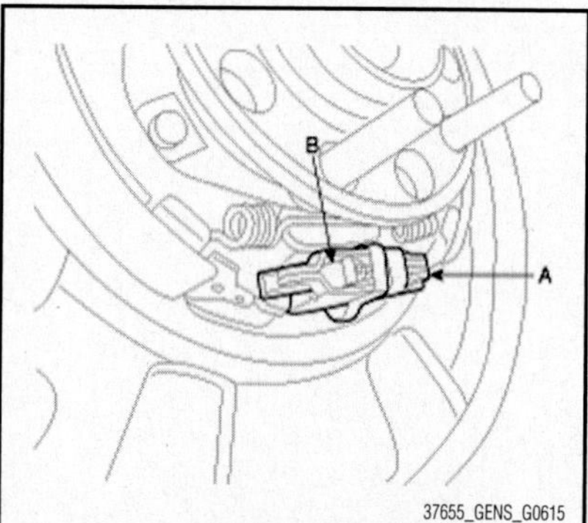

Fig. 407 Disconnect the parking brake cable (A) from operating lever (B)

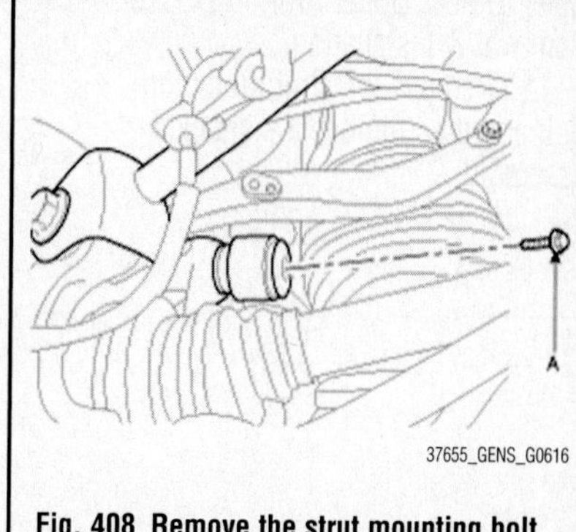

Fig. 408 Remove the strut mounting bolt (A) and then remove the knuckle assembly

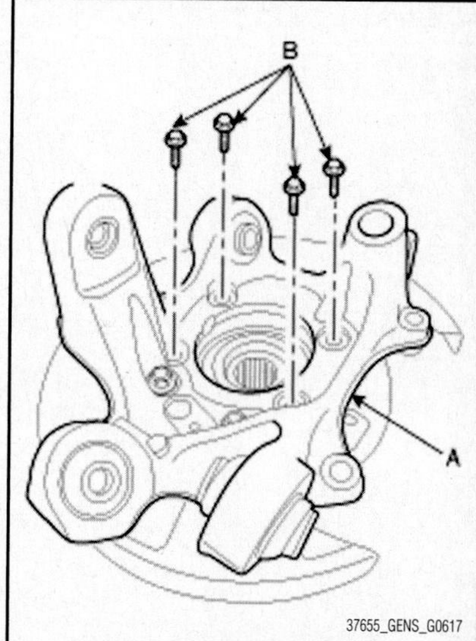

Fig. 409 Remove the hub assembly mount bolts (B) from the rear axle carrier (A)

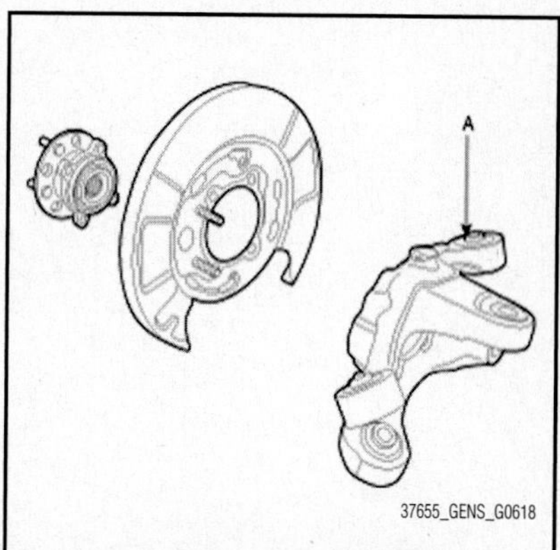

Fig. 410 Remove rear hub assembly

Refer to Rear Upper Arm, in Control Arms Removal & Installation.

12. Remove the strut mounting bolt and then remove the knuckle assembly.

13. Remove the hub assembly mounting bolts from the rear axle carrier.

14. Remove the rear hub assembly.

To install:

15. Installation is the reverse of removal. Tighten the castle nut to 145–202 ft. lbs. (200–280 Nm).

HYUNDAI

Santa Fe

6

SPECIFICATIONS AND MAINTENANCE CHARTS

ENGINE AND VEHICLE IDENTIFICATION

Engine							Model Year	
Code ①	Liters (cc)	Cu. In.	Cyl.	Fuel Sys.	Engine Type	Eng. Mfg.	Code ②	Year
B	2.4 (2359)	143.90	I4	MPFI	DOHC	Hyundai	9	2009
D	2.7 (2656)	164.80	V6	MPFI	DOHC	Hyundai	A	2010
E	3.3 (3342)	203.86	V6	MPFI	DOHC	Hyundai		
G	3.5 (3470)	213.60	V6	MPFI	DOHC	Hyundai		

MPFI: Multi-Point Fuel Injection

DOHC: Double Overhead Camshafts

① 8th digit of VIN

② 10th digit of VIN

37655_SANT_C0001

GENERAL ENGINE SPECIFICATIONS

All measurements are given in inches.

Year	Model	Engine Displacement Liters	Engine Series VIN	Net Horsepower @ rpm	Net Torque @ rpm (ft. lbs.)	Bore x Stroke (in.)	Com-pression Ratio	Oil Pressure @ rpm
2009	Santa Fe	2.7	D	185@6000	183@4000	3.413 x 2.952	10.4:1	18.8@1000
		3.3	E	242@6000	226@4500	3.622 x 3.299	10.4:1	18.8@1000
2010	Santa Fe	2.4	B	175@6000	169@3750	3.464 x 3.819	10.5:1	21.3@1000
		3.5	G	276@6300	248@5000	3.622 x 3.425	10.6:1	18.8@1000

37655_SANT_C0002

GASOLINE ENGINE TUNE-UP SPECIFICATIONS

Year	Engine Displacement Liters	Engine VIN	Spark Plug Gap (in.)	Ignition Timing (deg.) MT	Ignition Timing (deg.) AT	Fuel Pump (psi)	Idle Speed (rpm) MT	Idle Speed (rpm) AT	Valve Clearance In.	Valve Clearance Ex.
2009	2.7	D	0.039-0.043	①	①	54.3-55.8	②	②	HYD	HYD
	3.3	E	0.039-0.043	①	①	54.3-55.8	②	②	HYD	HYD
2010	2.4	B	0.039-0.043	①	①	46.9-52.6	②	②	HYD	HYD
	3.5	G	0.039-0.043	①	①	55	②	②	HYD	HYD

Follow the figures on the label if they differ from those in this chart.

HYD: Hydraulic

① Ignition timing is computer controlled and is not adjustable

② Idle speed is maintained by the Electronic Control Module (ECM)

37655_SANT_C0003

CAPACITIES

Year	Model	Engine Displacement Liters	Engine VIN	Engine Oil with Filter (qts.)	Transmission (pts.) Manual	Auto. ①	Fuel Tank (gal.)	Cooling System (qts.)
2009	Santa Fe	2.7	D	4.5	4.1	17.8	19.8	8.7
		3.3	E	4.0	NA	23.0	21.1	9.4
2010	Santa Fe	2.4	B	4.0	3.8	15.0	18.0	7.2
		3.5	G	4.5	NA	16.4	18.0	8.8

NA: Not Applicable

NOTE: All capacities are approximate. Add fluid gradually and check to be sure a proper fluid level is obtained.

① Drain and refill

37655_SANT_C0004

FLUID SPECIFICATIONS

Year	Model	Engine Displacement Liters	Engine ID/VIN	Engine Oil	Auto. Trans.	Manual Trans.	Power Steering Fluid	Brake Master Cylinder
2009	Santa Fe	2.7	D	①	②	③	PSF-4	④
		3.3	E	①	②	NA	PSF-4	④
2010	Santa Fe	2.4	B	5W-20	⑤	③	PSF-4	④
		3.5	G	5W-20	⑤	NA	PSF-4	④

NA: Not Applicable

DOT: Department Of Transportation

① Refer to owners manual

② DIAMOND ATF SP-III, SK ATF SP-III

③ GENUINE PART MTF 75W/85 (API GL - 4)

④ DOT 3, DOT 4, or equivalent

⑤ Hyundai ATF SP IV

37655_SANT_C0005

VALVE SPECIFICATIONS

Year	Engine Displacement Liters	Engine VIN	Seat Angle (deg.)	Face Angle (deg.)	Spring Test Pressure (lbs. @ in.)	Spring Installed Height (in.)	Stem-to-Guide Clearance (in.) Intake	Exhaust	Stem Diameter (in.) Intake	Exhaust
2009	2.7	D	NS	45 45.50	76.9-85.1 @1.024	NS	0.0008-0.0020	0.0014-0.0026	0.2348-0.2354	0.2343 0.2348
	3.3	E	NS	45 45.50	90.4-96.2 @0.953	NS	0.00078-0.0019	0.0012-0.0021	0.2151-0.2157	0.2149-0.2153
2010	2.4	B	44.75-45.10	45.25-45.75	85.1-90.4 @1.024	NS	0.0008-0.0019	0.0012-0.0021	0.2151-0.2157	0.2149-0.2153
	3.5	G	44.75-45.20	45.25-45.75	90.4-96.2 @0.937	NS	0.0008-0.0019	0.0012-0.0021	0.2151-0.2157	0.2149-0.2153

NS: Not Supplied

37655_SANT_C0006

CAMSHAFT AND BEARING SPECIFICATIONS CHART

All measurements are given in inches.

Year	Engine Displ. Liters	Engine ID/VIN	Journal Dia.	Brg. Oil Clearance	Shaft End-play	Runout	Journal Bore	Lobe Height Intake	Lobe Height Exhaust
2009	2.7	D	1.1009-1.1016	0.0012-0.0022	0.0039-0.0086	NS	NS	1.7520	1.7520
	3.3	E	①	②	0.0008-0.0071	NS	NS	1.8228	1.8031
2010	2.4	B	③	④	0.0039-0.0086	NS	NS	1.7401	1.7716
	3.5	G	①	②	0.0008-0.0071	NS	NS	1.8582	1.8031

NS: Not Supplied

① Intake No. 1 is 1.1009-1.1016 inch
Intake No. 2, 3, 4 are 0.9430-0.9437 inch
Exhaust No.1 is 1.1009-1.1016 inch
Exhaust No. 2, 3, 4 are 0.9430-0.9437 inch

② Intake No. 1 is 0.0011-0.0022 inch
Intake No. 2, 3, 4 are 0.0012-0.0026 inch
Exhaust No.1 is 0.0011-0.0022 inch
Exhaust No. 2, 3, 4 are 0.0012-0.0026 inch

③ Intake No. 1 is 1.1811 inch
Intake No. 2, 3, 4, 5 are 0.9449 inch
Exhaust No.1 is 1.4173 inch
Exhaust No. 2, 3, 4, 5 are 0.9449 inch

④ Intake No. 1 is 0.0008-0.0022 inch
Intake No. 2, 3, 4, 5 are 0.0017-0.0032 inch
Exhaust No. 1 is 0.0000-0.0012 inch
Exhaust No. 2, 3, 4, 5 are 0.0017-0.0032 inch

37655_SANT_C0007

CRANKSHAFT AND CONNECTING ROD SPECIFICATIONS

All measurements are given in inches.

Year	Engine Displacement Liters	Engine VIN	Crankshaft Main Brg. Journal Dia.	Crankshaft Main Brg. Oil Clearance	Crankshaft Shaft End-play	Crankshaft Thrust on No.	Connecting Rod Journal Diameter	Connecting Rod Oil Clearance	Connecting Rod Side Clearance
2009	2.7	D	2.4402-2.4409	0.0007-0.0014	0.0039-0.0098	3	1.8891-1.8898	0.0007-0.0014	0.0039-0.0098
	3.3	E	2.7142-2.7149	0.0008-0.0016	0.0039-0.0110	3	2.1635-2.1642	0.0015-0.0022	0.0039-0.0098
2010	2.4	B	2.0449-2.0456	0.0007-0.0014	0.0027-0.0098	3	1.8879-1.8886	0.0012-0.0017	0.0039-0.0100
	3.5	G	2.7142-2.7149	0.0008-0.0016	0.0039-0.0110	3	2.1635-2.1642	0.0014-0.0022	0.0039-0.0098

37655_SANT_C0008

PISTON AND RING SPECIFICATIONS

All measurements are given in inches.

Year	Engine Displ. Liters	Engine VIN	Piston Clearance	Ring Gap Top Compression	Ring Gap Bottom Compression	Ring Gap Oil Control	Ring Side Clearance Top Compression	Ring Side Clearance Bottom Compression	Ring Side Clearance Oil Control
2009	2.7	D	0.0008-0.0020	0.0059-0.0118	0.0118-0.0177	0.0078-0.0275	0.0016-0.0031	0.0012-0.0027	0.0024-0.0059
	3.3	E	0.0008-0.0016	0.0067-0.0126	0.0126-0.0185	0.0078-0.0275	0.0016-0.0031	0.0012-0.0027	0.0024-0.0059
2010	2.4	B	0.0005-0.0013	0.0059-0.0118	0.0145-0.0204	0.0078-0.0275	0.0019-0.0031	0.0015-0.0031	0.0023-0.0059
	3.5	G	0.0012-0.0020	0.0067-0.0126	0.0145-0.0204	0.0078-0.0196	0.0015-0.0031	0.0015-0.0031	0.0024-0.0059

37655_SANT_C0009

TORQUE SPECIFICATIONS

All readings in ft. lbs.

Year	Engine Displacement Liters	Engine VIN	Cylinder Head Bolts	Main Bearing Bolts	Rod Bearing Bolts	Crankshaft Damper Bolts	Flywheel Bolts	Manifold Intake	Manifold Exhaust	Spark Plugs	Oil Pan Drain Plug
2009	2.7	D	①	②	③	123-130	53-56	14-17	22-25	15-22	25-32
	3.3	E	④	⑤	③	210-224	53-56	14-17	29-33	15-22	25-33
2010	2.4	B	⑥	⑦	③	123-130	87-94	14-17	⑧	15-22	25-33
	3.5	G	④	⑤	③	210-224	53-56	⑨	⑩	15-22	25-33

① Step 1: 18 ft. lbs.
Step 2: Plus 60 degrees
Step 3: Plus 45 degrees

② M10 Step 1: 22 ft. lbs.
M10 Step 2: Plus 90 degrees
M8 Step 1: 12 ft. lbs.
M8 Step 2: Plus 90 degrees

③ Step 1: 15 ft. lbs.
Step 2: Plus 90 degrees

④ Step 1: 29 ft. lbs.
Step 2: Plus 120 degrees
Step 3: Plus 90 degrees

⑤ M11 bolts (inner) Step 1: 36 ft. lbs.
M11 bolts (inner) Step 2: Plus 90 degrees
M8 bolts (outer) Step 1: 15 ft. lbs.
M8 bolts (outer) Step 2: Plus 120 degrees
M8 bolts (side): 22-23 ft. lbs.

⑥ Step 1: 25 ft. lbs.
Step 2: Plus 90 degrees
Step 3: Plus 90 degrees

⑦ Step 1: 22 ft. lbs.
Step 2: Plus 120 degrees

⑧ Exhaust manifold heat protector bolt and stay bolt (M8): 6-9 ft. lbs.
Exhaust manifold nut: 36-40 ft. lbs.
Exhaust manifold stay bolt (M10): 31-40 ft. lbs.

⑨ Bolt: 21 ft. lbs.
Nut: 16 ft. lbs.

⑩ Stay bolt: 22 ft. lbs.
Nut: 31 ft. lbs.

37655_SANT_C0010

WHEEL ALIGNMENT

Year	Model		Caster Range (+/-Deg.)	Caster Preferred Setting (Deg.)	Camber Range (+/-Deg.)	Camber Preferred Setting (Deg.)	Toe-in (Deg.)
2009	Santa Fe	Front	0.50	+4.33	0.50	0.50	0.00 +/- 0.20
		Rear	—	—	1.00	-0.50	0.20 +/- 0.20
2010	Santa Fe	Front	0.50	+4.50	0.50	-0.50	0.00 +/- 0.20
		Rear	—	—	1.00	-0.50	0.20 +/- 0.20

37655_SANT_C0011

TIRE, WHEEL AND BALL JOINT SPECIFICATIONS

Year	Model	OEM Tires Standard	OEM Tires Optional	Tire Pressures (psi) Front	Tire Pressures (psi) Rear	Wheel Size	Ball Joint Inspection	Lug Nut Torque (ft. lbs.)
2009	Santa fe	P235/70R16	P235/60R18	32	32	7.0J×16 7.0J×18	①	65-80
2010	Santa fe	P235/70R16	P235/60R18	32	32	7.0J×16 7.0J×18	①	65-80

① Replace if any measurable movement is found.

37655_SANT_C0012

BRAKE SPECIFICATIONS

All measurements in inches unless noted

Year	Model		Brake Disc Original Thickness	Brake Disc Minimum Thickness	Brake Disc Maximum Runout	Brake Drum Diameter Original Inside Diameter	Brake Drum Diameter Max. Wear Limit	Brake Drum Diameter Maximum Machine Diameter	Minimum Lining Thickness	Brake Caliper Bracket Bolts (ft. lbs.)	Brake Caliper Mounting Bolts (ft. lbs.)
2009	Santa Fe	F	1.100	1.020	0.003	—	—	—	NS	59-74	16-23
		R	0.430	0.370	0.003	—	—	—	NS	59-74	16-23
2010	Santa Fe	F	1.100	1.020	0.003	—	—	—	NS	59-74	16-23
		R	0.430	0.370	0.003	—	—	—	NS	59-74	16-23

NS: Not Supplied

37655_SANT_C0013

SCHEDULED MAINTENANCE INTERVALS

HYUNDAI—Santa Fe

TO BE SERVICED	TYPE OF SERVICE	VEHICLE MILEAGE INTERVAL (x1000)												
		7.5	15	22.5	30	37.5	45	52.5	60	67.5	75	82.5	90	97.5
Engine oil & filter	R	✓	✓	✓	✓	✓	✓	✓	✓	✓	✓	✓	✓	✓
Driveshafts & boots	S/I		✓		✓		✓		✓		✓		✓	
Air cleaner filter ①	S/I	✓	✓	✓	✓	✓	✓	✓	✓	✓	✓	✓	✓	✓
A/C refrigerant	S/I		✓		✓		✓		✓		✓		✓	
Brake & clutch fluid	S/I				✓				✓				✓	
Engine coolant ②	R								✓				✓	
Fuel hose, vapor hose & fuel filter cap	S/I				✓				✓				✓	
Spark plugs: Platinum (Iridium) 100,000 mile replacement	R													
Suspension mounting bolts and nuts, upper and lower ball joint.	S/I		✓		✓		✓		✓		✓		✓	
Drive belts	S/I				✓				✓				✓	
Exhaust pipe connections, muffler & suspension bolts	S/I		✓		✓		✓		✓		✓		✓	
Valve clearance ③	S/I								✓					
Brake hoses & lines	S/I		✓		✓		✓		✓		✓		✓	
Rear brake discs, linings & parking brake	S/I				✓				✓				✓	
Front brake discs, linings & caliper & rotor	S/I		✓		✓		✓		✓		✓		✓	
Steering gear rack, linkage & boots	S/I		✓		✓		✓		✓		✓		✓	
Fuel tank air filter	S/I				✓				✓				✓	
Power steering belt and hoses	S/I		✓		✓		✓		✓		✓		✓	
Fuel filter	S/I				✓				✓				✓	
Fuel lines & connections	S/I				✓				✓				✓	✓
Manual transaxle fluid	S/I					✓					✓			
Automatic transaxle fluid	NS													
Transfer case ④	S/I													
Rear axle oil ④	S/I													
Climate control air filter	R		✓		✓		✓		✓		✓		✓	
Power steering fluid	S/I	✓	✓	✓	✓	✓	✓	✓	✓	✓	✓	✓	✓	✓
Vacuum hoses	S/I	✓	✓	✓	✓	✓	✓	✓	✓	✓	✓	✓	✓	✓
Crankcase ventilation hoses	S/I				✓				✓				✓	

R: Replace S/I: Service or Inspect NS: No service required

① Replace every 30,000 miles

② Replace every 24 months or 30,000 miles after 60,000 mile replacement.

③ Inspect for excessive tappet noise and/or engine vibration and adjust if necessary.

④ Inspect every 40,000 miles. Replace every time submerged in water.

37655_SANT_C0014

SCHEDULED MAINTENANCE INTERVALS

HYUNDAI—Santa Fe

FREQUENT OPERATION MAINTENANCE (SEVERE SERVICE)

If a vehicle is operated under any of the following conditions it is considered severe service:

- Repeatedly driving short distance of less than 5miles (8km) in normal temperature or less than 10miles (16km) in freezing temperature

- Extensive engine idling or low speed driving for long distance

- Driving on rough, dusty, muddy, unpaved, graveled or salt- spread roads

- Driving in areas using salt or other corrosive materials or in very cold weather

- Driving in heavy traffic area over 90°F (32°C)

- Driving on uphill, downhill, or mountain road

- Towing a Trailer, or using a camper, or roof rack

- Driving as a patrol car, taxi, other commercial use or vehicle towing

- Driving over 100 MPH (170 Km/h)

- Frequently driving in stop-and-go conditions

Oil & oil filter: change every 3,750 miles.

Brake pads, calipers & rotors: service or inspect more frequently

Brake linings, parking brake: service or inspect more frequently

Driveshaft boots: service or inspect every 7500 miles

Steering gear box, linkage & boots, lower and upper ball joints: service or inspect more frequently

Air cleaner filter: replace more frequently

Automatic transaxle fluid: replace every 60,000 miles.

Manual transaxle fluid: replace every 80,000 miles.

Climate control air filter: replace more frequently

Spark plugs: replace more frequently

Transfer case oil : replace every 80,000 miles.

Rear axle oil : replace every 80,000 miles.

Propeller shaft : service or inspect every 7500 miles

37655_SANT_C0015

BRAKES — INFORMATION AND PRECAUTIONS

ANTI-LOCK SYSTEMS

• Certain components within the ABS system are not intended to be serviced or repaired individually.

• Do not use rubber hoses or other parts not specifically specified for and ABS system. When using repair kits, replace all parts included in the kit. Partial or incorrect repair may lead to functional problems and require the replacement of components.

• Lubricate rubber parts with clean, fresh brake fluid to ease assembly. Do not use shop air to clean parts; damage to rubber components may result.

• Use only DOT 3 brake fluid from an unopened container.

• If any hydraulic component or line is removed or replaced, it may be necessary to bleed the entire system.

• A clean repair area is essential. Always clean the reservoir and cap thoroughly before removing the cap. The slightest amount of dirt in the fluid may plug an orifice and impair the system function. Perform repairs after components have been thoroughly cleaned; use only denatured alcohol to clean components. Do not allow ABS components to come into contact with any substance containing mineral oil; this includes used shop rags.

• The Anti-Lock control unit is a microprocessor similar to other computer units in the vehicle. Ensure that the ignition switch is **OFF** before removing or installing controller harnesses. Avoid static electricity discharge at or near the controller.

• If any arc welding is to be done on the vehicle, the control unit should be unplugged before welding operations begin.

DISC AND DRUM SYSTEMS

**** CAUTION**

Dust and dirt accumulating on brake parts during normal use may contain asbestos fibers from production or aftermarket brake linings. Breathing excessive concentrations of asbestos fibers can cause serious bodily harm. Exercise care when servicing brake parts. Do not sand or grind brake lining unless equipment used is designed to contain the dust residue. Do not clean brake parts with compressed air or by dry brushing. Cleaning should be done by dampening the brake components with a fine mist of water, then wiping the brake components clean with a dampened cloth. Dispose of cloth and all residue containing asbestos fibers in an impermeable container with the appropriate label. Follow practices prescribed by the Occupational Safety and Health Administration (OSHA) and the Environmental Protection Agency (EPA) for the handling, processing, and disposing of dust or debris that may contain asbestos fibers.

BRAKES — BLEEDING THE BRAKE SYSTEM

BLEEDING PROCEDURE

BLEEDING THE ABS SYSTEM

See Figures 1 and 2.

This procedure should be followed to ensure adequate bleeding of air and the filling of the ABS unit, the brake lines, and the master cylinder with brake fluid.

1. Before servicing the vehicle, refer to the Precautions Section.
2. Remove the reservoir cap and fill the brake reservoir with brake fluid.

**** WARNING**

If there is any brake fluid on any painted surface, wash it off immediately.

➡When pressure bleeding, do not depress the brake pedal. Recommended brake fluid: DOT3 or DOT4.

3. Connect a clear plastic tube to the wheel cylinder bleeder plug and insert the other end of the tube into a clear plastic bottle that is half filled with clean brake fluid.
4. Connect the scan tool to the data link connector located underneath the dash panel.
5. Select and operate according to the instructions on the scan tool screen.

**** CAUTION**

You must obey the maximum operating time of the ABS motor with the scan tool to prevent the motor pump from burning.

6. Select vehicle name.
7. Select Anti-Lock Brake system.
8. Select air bleeding mode.
9. Press "OK" to operate THE motor pump and solenoid valve.

Fig. 1 Close the bleeder screw (A)

**** WARNING**

Wait 60 seconds before operating the air bleeding or damage to the motor may occur.

10. Wait 60 seconds before operating the air bleeding.
11. Pump the brake pedal several times, and then loosen the bleeder screw until fluid starts to run out without bubbles. Then, close the bleeder screw.
12. Repeat until there are no more bubbles in the fluid for each wheel.
13. Tighten the bleeder screw to 61–113 inch. lbs. (7–13 Nm).

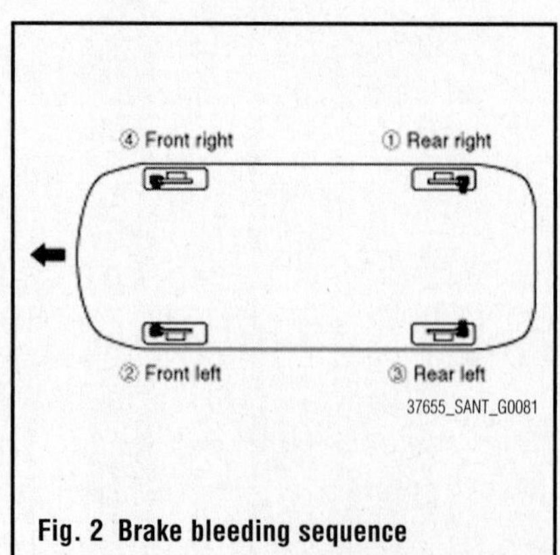

Fig. 2 Brake bleeding sequence

SPEED SENSORS

REMOVAL & INSTALLATION

Front

See Figure 3.

1. Before servicing the vehicle, refer to the Precautions Section.
2. Remove the front wheel speed sensor mounting bolt.
3. Remove the front wheel speed sensor bracket.
4. Remove the front wheel guard, as necessary.
5. Remove the front wheel speed sensor connector.
6. Remove the front wheel speed sensor.

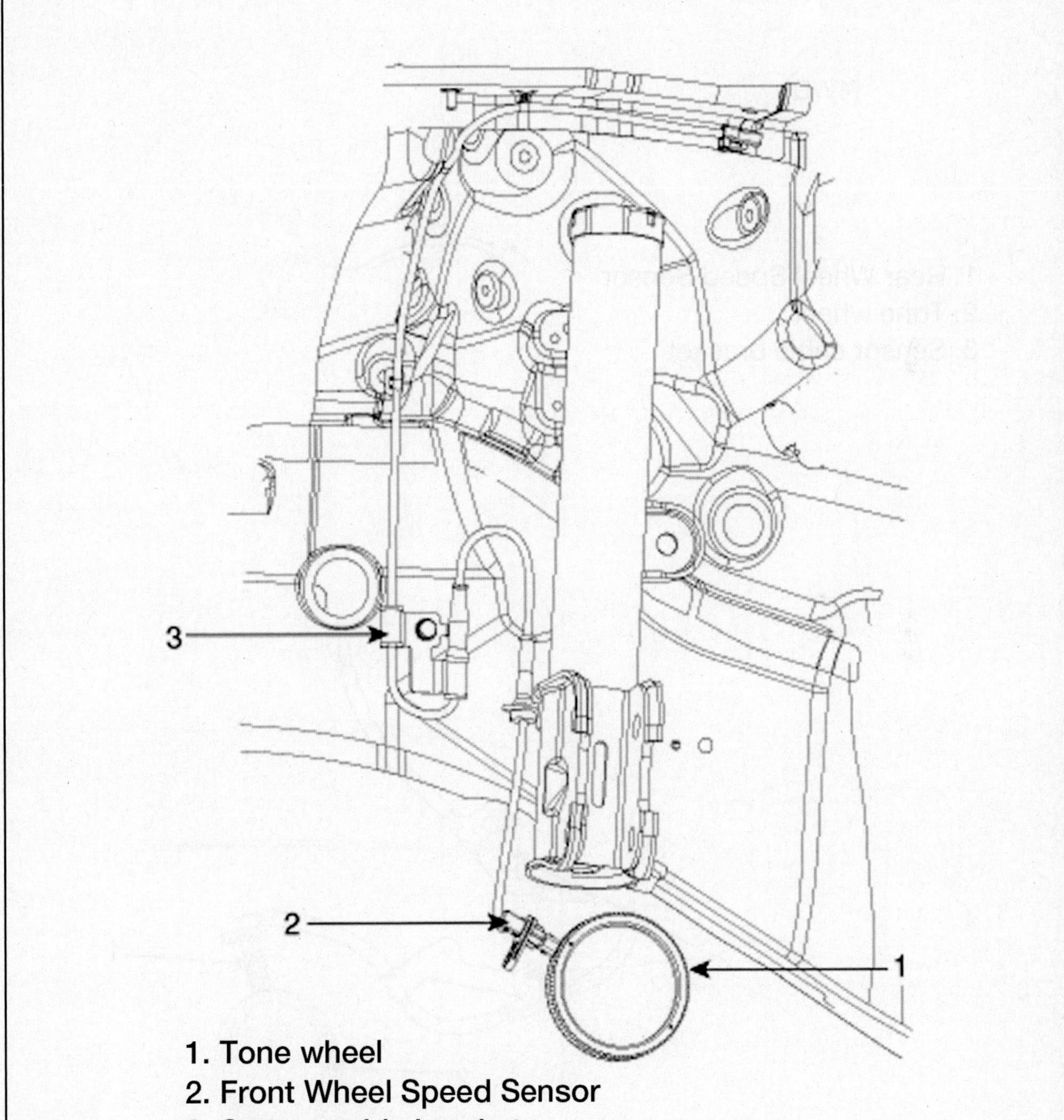

Fig. 3 Components of the front wheel speed sensor

To install:

7. Installation is the reverse of the removal procedure. Tighten the mounting bolt 60–96 inch lbs. (7–11 Nm).

Rear

See Figure 4.

1. Before servicing the vehicle, refer to the Precautions Section.
2. Remove the rear wheel speed sensor mounting bolt.
3. Remove the rear wheel speed sensor bracket.
4. Remove the rear wheel guard, as necessary.
5. Remove the rear wheel speed sensor connector.
6. Remove the rear wheel speed sensor.

To install:

7. Installation is the reverse of the removal procedure. Tighten the mounting bolt 60–96 inch lbs. (7–11 Nm).

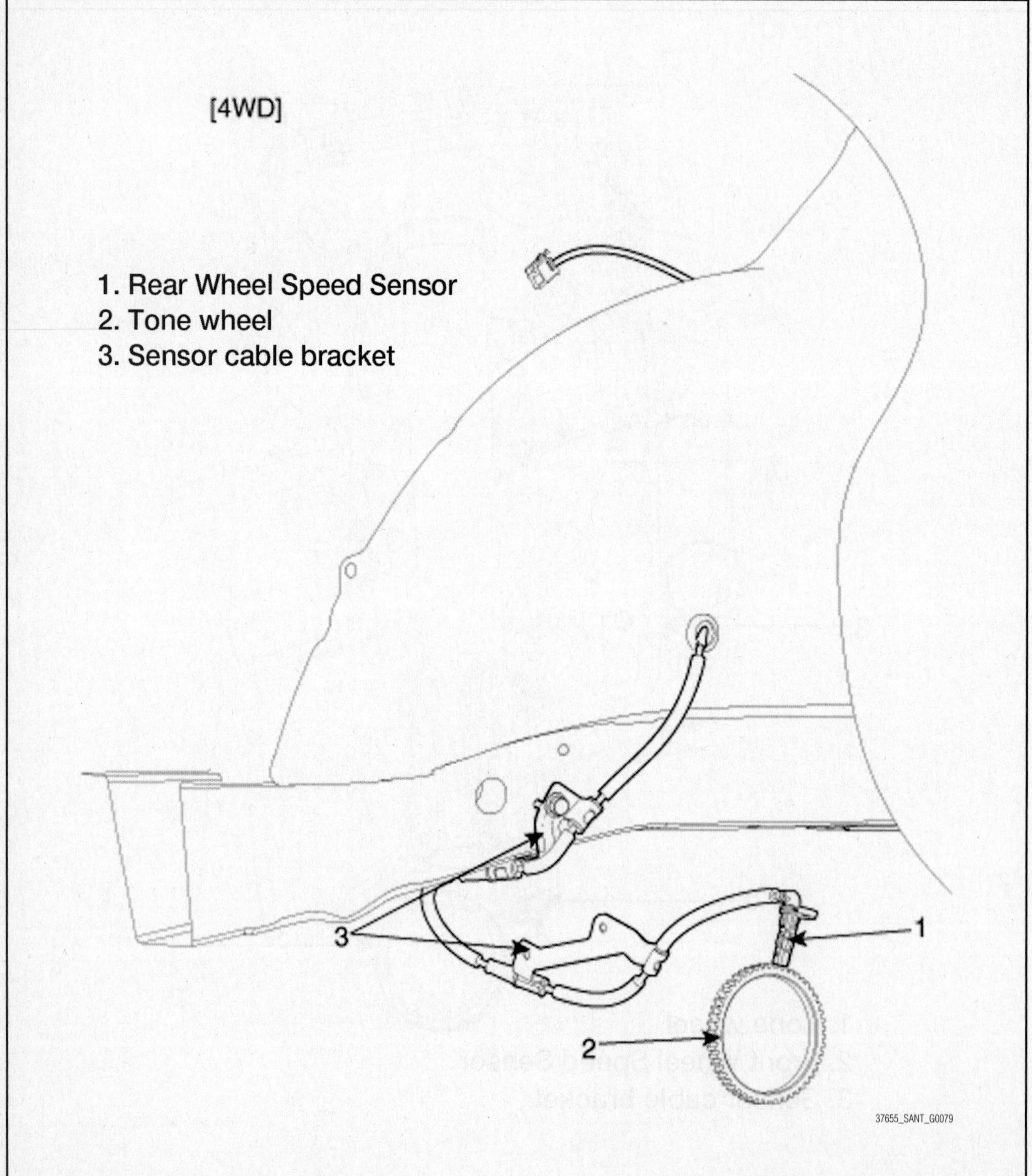

Fig. 4 Components of the rear wheel speed sensor

BRAKE CALIPER

REMOVAL & INSTALLATION

See Figures 5 through 7.

➡Special Service Tool used for this procedure: SST 09581-11000

1. Before servicing the vehicle, refer to the Precautions Section.
2. Remove the wheels and tires.
3. Remove the hose eye-bolt and the brake hose from the caliper.
4. Remove the caliper mounting bolts.
5. Remove the caliper assembly.

To install:

6. Installation is the reverse of the removal procedure, noting the following:
   a. Use the SST 09581-11000 or a C-clamp to install the brake caliper assembly.

B
C
A
C
37655_SANT_G0082

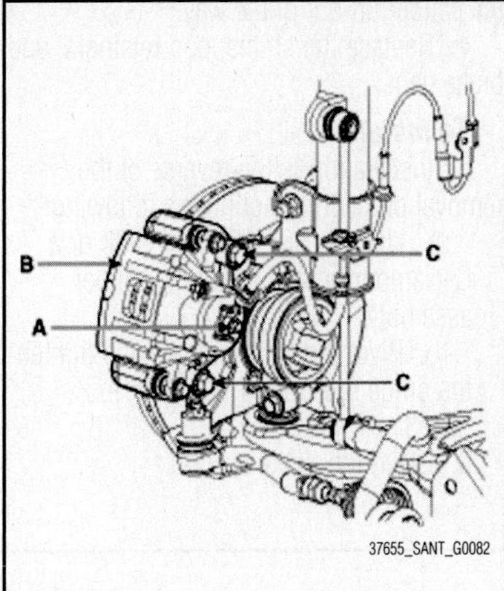

Fig. 6 Remove the hose eye-bolt (A), caliper mounting bolts (C), and front caliper (B)

09581-
11000
37655_SANT_G0090

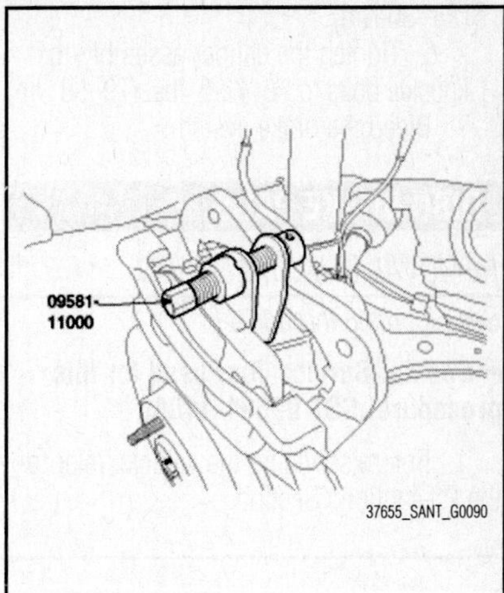

Fig. 7 Use the SST 09581-11000 or a C-clamp to install the brake caliper assembly

21.6 ~ 31.4
(2.2 ~ 3.2, 15.9 ~ 23.1)
4
1
2
3
5
6
1. Brake caliper assembly
2. Brake disc
3. Pad retainer
4. Guide rod bolt
5. Brake pad
6. Pad shim
Torque : Nm (Kgf.m, lb-ft)
37655_SANT_G0092

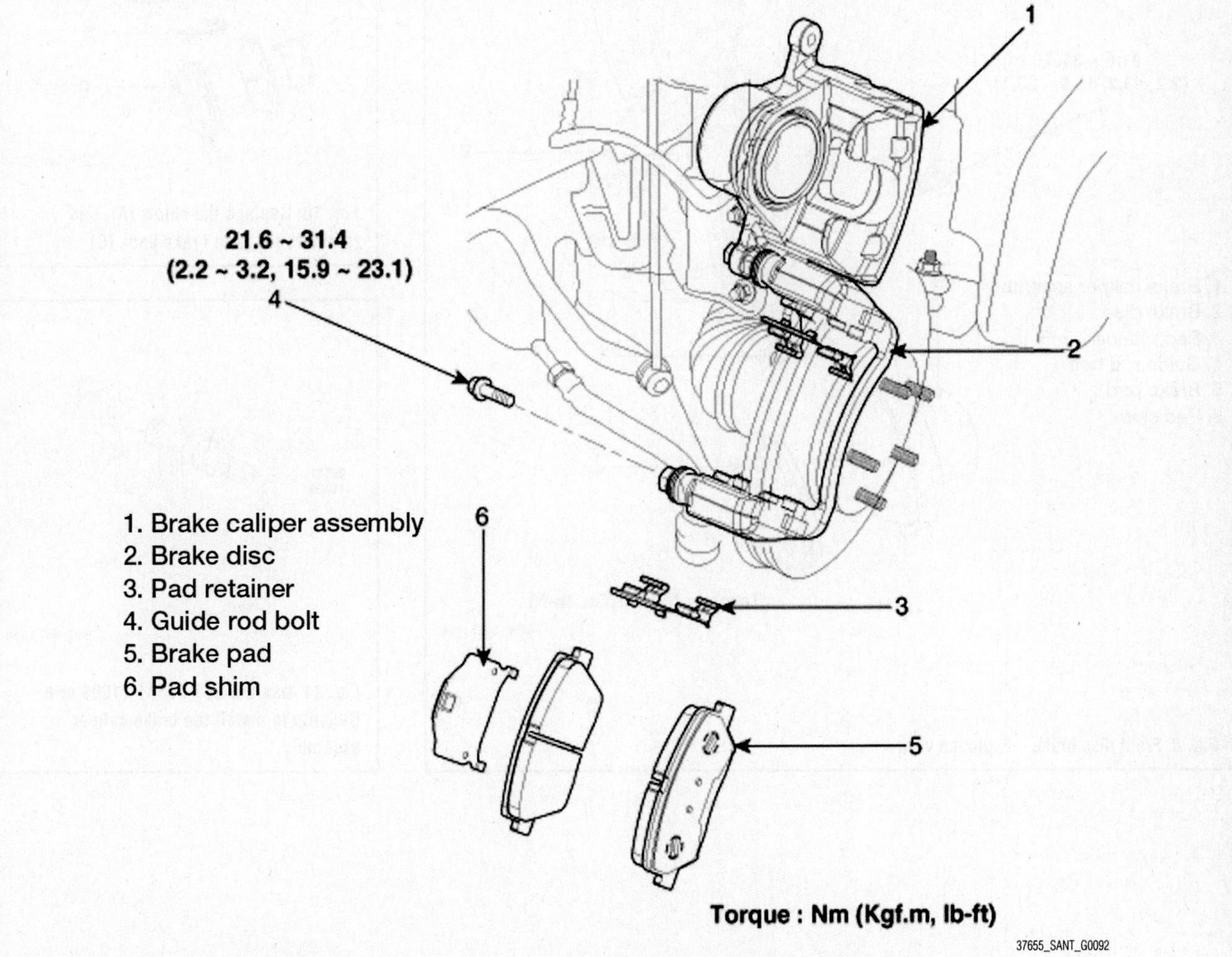

Fig. 5 Front disc brake—Exploded view

b. Tighten the brake hose-to-caliper mounting bolts to 18–22 ft. lbs. (25–30 Nm)

c. Tighten the caliper assembly-to-knuckle bolts to 58–72 ft. lbs. (79–98 Nm)

7. Bleed the brake system.

DISC BRAKE PADS

REMOVAL & INSTALLATION

See Figures 8 through 11.

➡Special Service Tool used for this procedure: SST 09581-11000

1. Before servicing the vehicle, refer to the Precautions Section.
2. Remove the wheels and tires.
3. Remove the guide rod bolt and pivot the caliper up out of the way.
4. Replace the shims, pad retainers, and brake pads.

To install:

5. Installation is the reverse of the removal procedure, noting the following:
   a. Use the SST 09581-11000 or a C-clamp to install the brake caliper assembly.
   b. Pivot the caliper down and tighten the guide rod bolt to 16–23 ft. lbs. (22–31 Nm).
6. Bleed the brake system.

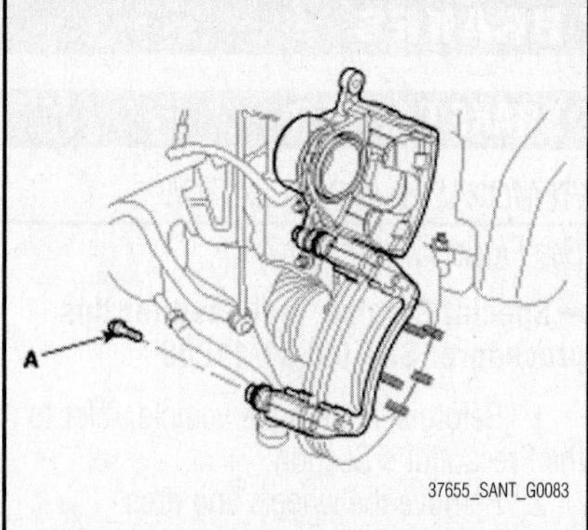

Fig. 9 Remove the guide rod bolt (A) and pivot the caliper up out of the way

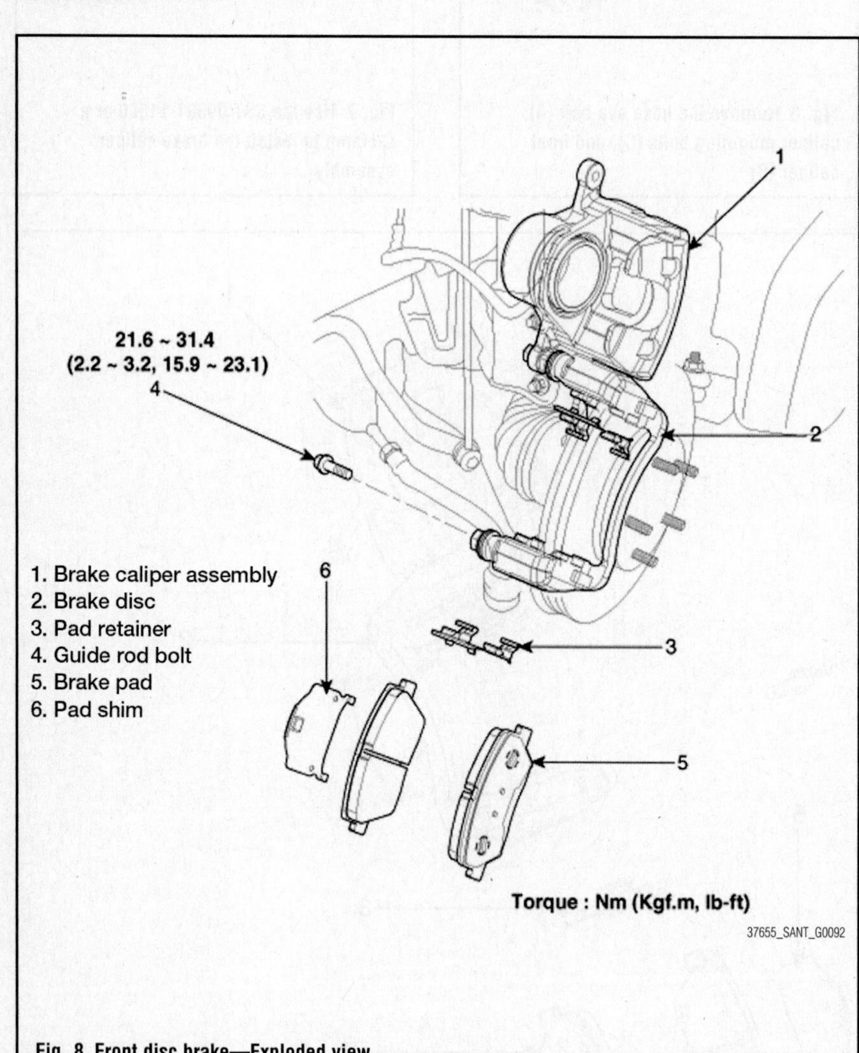

Fig. 8 Front disc brake—Exploded view

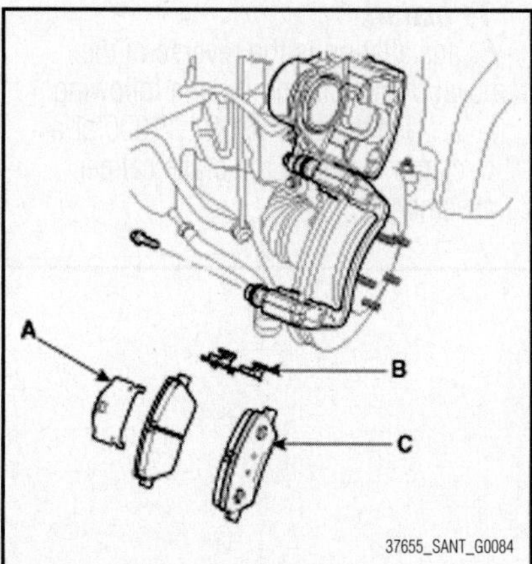

Fig. 10 Replace the shims (A), pad retainers (B), and brake pads (C)

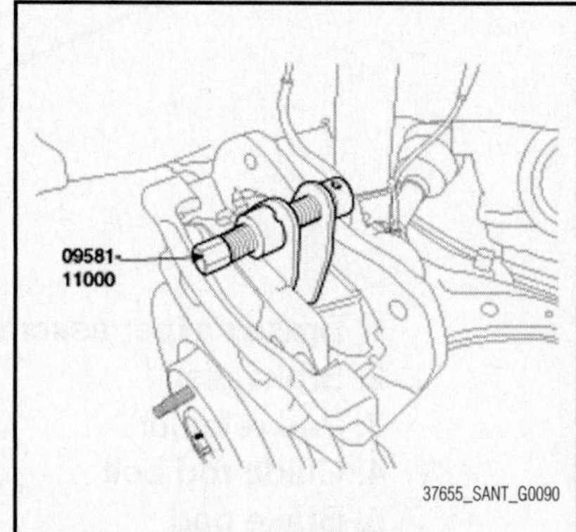

Fig. 11 Use the SST 09581-11000 or a C-clamp to install the brake caliper assembly

BRAKE CALIPER

REMOVAL & INSTALLATION

See Figures 12 through 14.

➡Special Service Tool used for this procedure: SST 09581-11000

1. Before servicing the vehicle, refer to the Precautions Section.
2. Remove the wheels and tires.
3. Remove the hose eye-bolt and the brake hose from the caliper.
4. Remove the caliper mounting bolts.
5. Remove the caliper assembly.

To install:

6. Installation is the reverse of the removal procedure, noting the following:
   a. Use the SST 09581-11000 or a C-clamp to install the brake caliper assembly.

A
B
37655_SANT_G0086

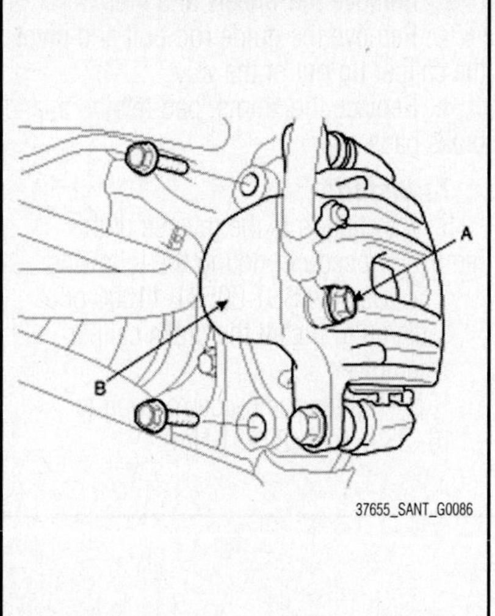

Fig. 13 Remove the hose eye-bolt (A) and rear caliper (B)

09581-11000
37655_SANT_G0091

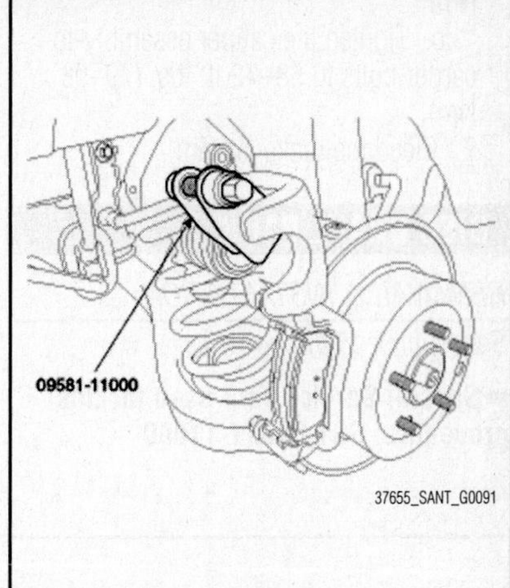

Fig. 14 Use the SST 09581-11000 or a C-clamp to install the brake caliper assembly

10
9
8
7
6
5
4
3
2
1
21.6 ~ 31.4
(2.2 ~ 3.2, 15.9 ~ 23.1)
1. Guide rod bolt
2. Caliper body
3. Piston
4. Piston seal
5. Piston boot
6. Pad shim
7. Brake pad
8. Pad retainer
9. Guide rod
10. Boot
Torque : Nm (Kgf.m, lb-ft)
37655_SANT_G0093

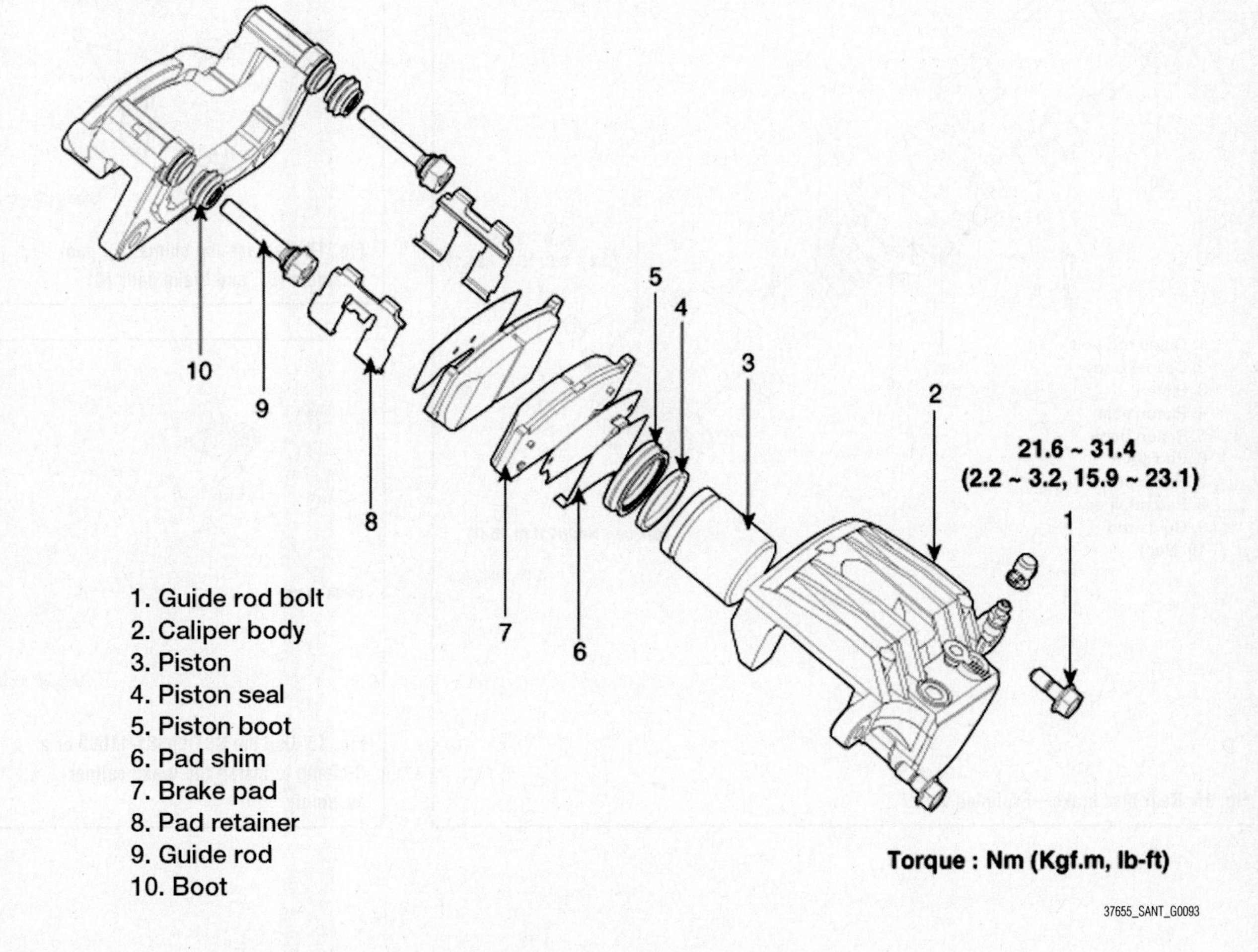

Fig. 12 Rear disc brake—Exploded view

b. Tighten the brake hose-to-caliper mounting bolts to 18–22 ft. lbs. (25–30 Nm)

c. Tighten the caliper assembly-to-carrier bolts to 58–72 ft. lbs. (79–98 Nm)

7. Bleed the brake system.

DISC BRAKE PADS

REMOVAL & INSTALLATION

See Figures 15 through 18.

➡Special Service Tool used for this procedure: SST 09581-11000

1. Before servicing the vehicle, refer to the Precautions Section.
2. Remove the wheels and tires.
3. Remove the guide rod bolt and pivot the caliper up out of the way.
4. Replace the shims, pad retainers, and brake pads.

To install:

5. Installation is the reverse of the removal procedure, noting the following:
   a. Use the SST 09581-11000 or a C-clamp to install the brake caliper assembly.
   b. Tighten the guide rod bolt to 16–23 ft. lbs. (22–31 Nm).

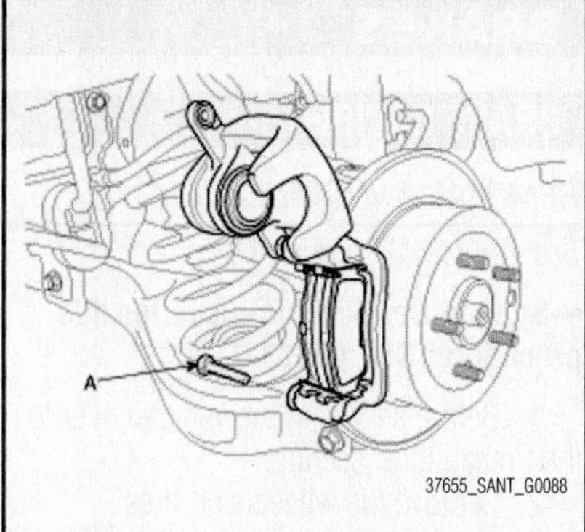

Fig. 16 Remove the guide rod bolt (A) and pivot the caliper up out of the way

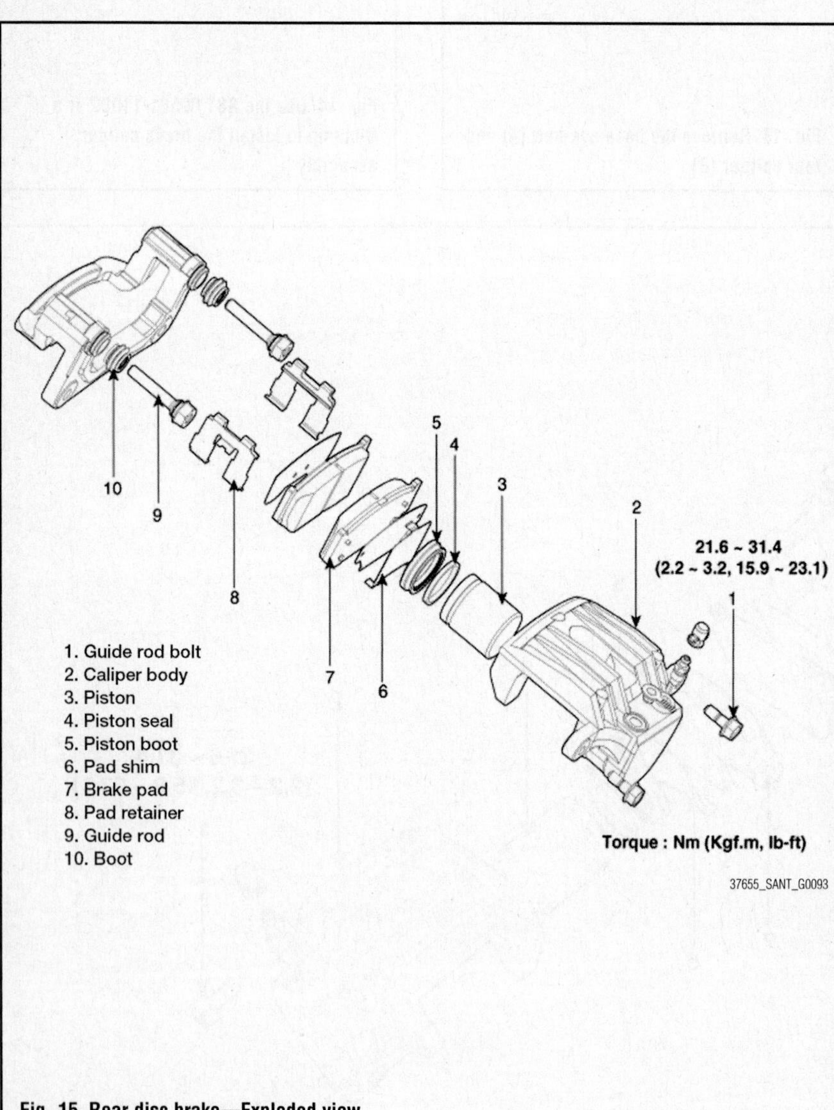

Fig. 15 Rear disc brake—Exploded view

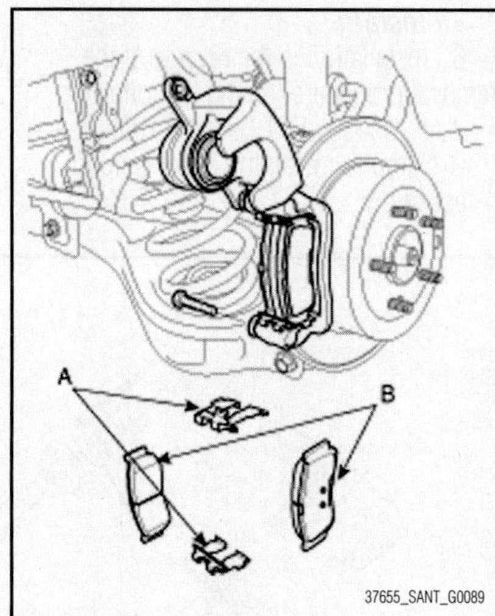

Fig. 17 Replace the shims (A), pad retainers (B), and brake pads (C)

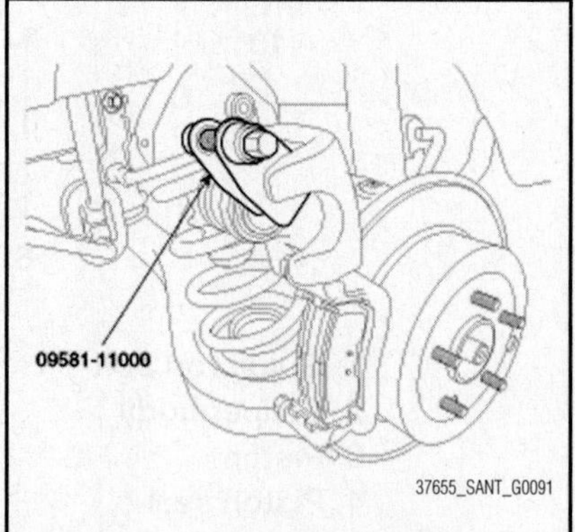

Fig. 18 Use the SST 09581-11000 or a C-clamp to install the brake caliper assembly

PARKING BRAKE CABLES

ADJUSTMENT

1. Before servicing the vehicle, refer to the Precautions Section.
2. Operate the parking brake lever through a full stoke 3 times to position the cables.
3. For hand brake, the travel must be between 6–7 notches when applying a force of approx. 44 ft. lbs. at 1.57 inches from the bottom of lever assembly.
4. For foot brake, the travel must be between 8–9 notches when applying a force of approx. 66 ft. lbs.
5. To adjust, turn equalizer adjusting nut.

➡The parking brake indicator lamp must be OFF when lever assembly is released, and ON when operating to 1 notch.

PARKING BRAKE SHOES

REMOVAL & INSTALLATION

See Figures 19 and 20.

1. Before servicing the vehicle, refer to the Precautions Section.
2. Raise and safely support the vehicle.
3. Remove the rear wheel and tire assembly.
4. Remove the brake caliper. Refer to Rear Disc Brakes, Brake Caliper Removal & Installation.
5. Remove the brake rotor.
6. Remove the clip retainer and the parking brake cable.
7. Remove the shoe hold down pin and spring by pressing and rotating the spring.
8. Remove the adjuster assembly and the lower return spring.
9. Remove the upper return spring and the brake shoes.

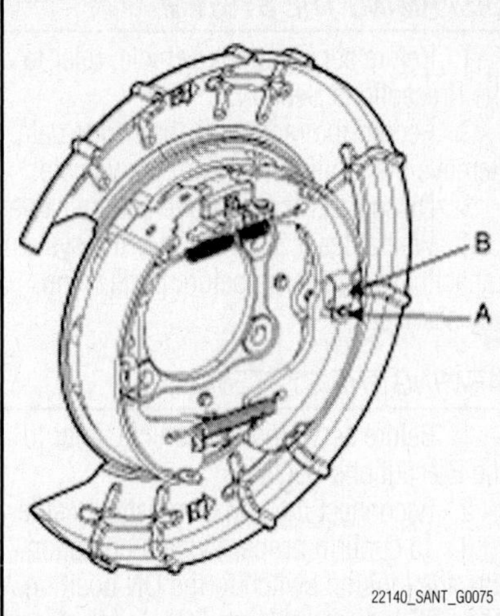

Fig. 19 Remove the shoe hold down pin (A) and spring (B) by pressing and rotating the spring

10. Remove the operating lever assembly.

To install:

11. Install the operating lever assembly.
12. Install the upper return spring and the brake shoes.
13. Install the adjuster assembly and the lower return spring.
14. Install the shoe hold down pin and spring by pressing and rotating the spring.
15. Install the parking brake cable and then install the clip.
16. Install the rear brake disc rotor.
17. Adjust the rear brake shoe clearance:
    a. Remove the plug from the disc.
    b. Rotate the toothed wheel of adjuster with a screw driver until the disc is not moving.
    c. Back off 5 notches in the opposite direction.
18. Install the brake caliper. Refer to Rear Disc Brakes, Brake Caliper, removal & installation.

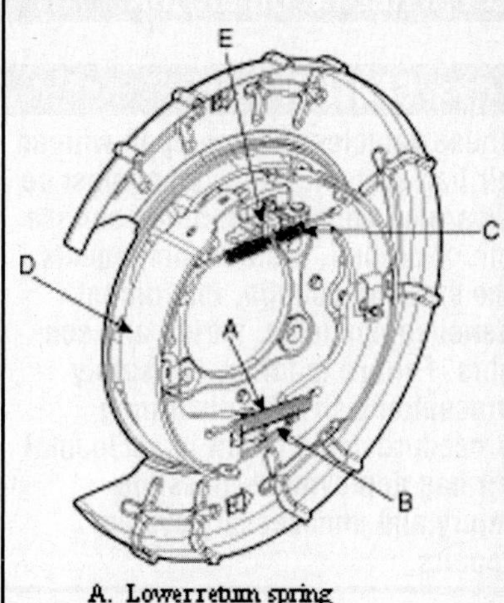

Fig. 20 Remove the adjuster assembly, the lower return spring, the upper return spring, the brake shoes, and the operating lever assembly

19. Install the tire and wheel.
20. If the parking brake shoe or the brake disc are replaced, perform the brake shoe brake-in procedure:
    a. While operating the parking brake engaged with a 15 lb. (69 N) effort, drive the vehicle 0.31 miles (500 meters) at the speed of about 37 mph (60 kph).
    b. Repeat the above procedure at least 2 times.
21. The parking brake should hold on a 30 percent grade.
22. Ensure that all parts move smoothly.
23. The parking brake indicator lamp must turn ON when the parking brake is applied and turn OFF when the parking brake is released.

CHASSIS ELECTRICAL

AIR BAG (SUPPLEMENTAL RESTRAINT SYSTEM)

GENERAL INFORMATION

CAUTION

These vehicles are equipped with an air bag system. The system must be disarmed before performing service on, or around, system components, the steering column, instrument panel components, wiring and sensors. Failure to follow the safety precautions and the disarming procedure could result in accidental air bag deployment, possible injury and unnecessary system repairs.

SERVICE PRECAUTIONS

CAUTION

Disconnect and isolate the battery negative cable before beginning any airbag system component diagnosis, testing, removal, or installation procedures. Wait at least 90 seconds after the ignition switch is turned off and the negative (-) terminal cable is disconnected from the battery before starting the operation. The SRS is equipped with a backup power source, so if work is started within 90 seconds after disconnecting the negative (-) terminal cable from the battery, the SRS may be deployed. Failure to disable the airbag system may result in accidental airbag deployment, personal injury, or death.

DISARMING THE SYSTEM

1. Before servicing the vehicle, refer to the Precautions Section.
2. Record the radio anti-theft code data. Remove the ignition key from the vehicle.
3. Disconnect the negative battery cable.
4. Wait at least 3 minutes for the system capacitor to discharge before performing any service.

ARMING THE SYSTEM

1. Before servicing the vehicle, refer to the Precautions Section.
2. Reconnect the negative battery cable.
3. To confirm proper system operation, turn the ignition switch to the ON position. The SRS indicator light will be lit for at least 6 seconds and then go off.

CLOCKSPRING CENTERING

See Figure 21.

1. Disconnect and isolate the battery negative cable. Allow the system capacitor to discharge for 3 minutes before beginning any component service. This will disable the airbag system.

CAUTION

Failure to disable the airbag system may result in accidental airbag deployment, personal injury, or death.

2. Remove the ignition key from the vehicle.
3. Connect the clockspring harness connector and horn harness connector to the clockspring.
4. Set the clockspring on neutral position and, after turning the front wheels to the straight-ahead position, install the clockspring.
   a. Check connectors and protective tube for damage, and terminals for deformities.
   b. If even one abnormal point is discovered, replace the clockspring with a new one.
5. Connect the clockspring harness connector and the steering switch harness connector to the clockspring.
6. Set the center position by getting the marks between the clockspring and the cover into line.
7. Set the center position by setting the marks between the clock spring and the cover into line. Make an array the mark by turning the clock spring clockwise to the stop and then 3 revolutions counterclockwise.

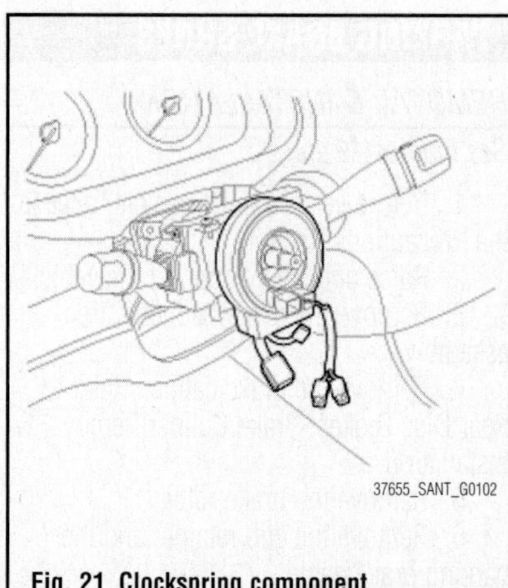

Fig. 21 Clockspring component

DRIVE TRAIN

DRIVEN DISC & PRESSURE PLATE

REMOVAL & INSTALLATION

2009 Models

2.7L Engine

See Figures 22 through 26.

1. Before servicing the vehicle, refer to the Precautions Section.
2. Remove the transaxle assembly.
3. Insert the SST (09411-25000) in the clutch disc to prevent the disc from falling.
4. Remove the bolts which attach the clutch cover to the flywheel. Loosen the bolts in succession, 1–2 turns at a time, in a star pattern.
5. Remove clutch cover assembly and the driven disc and pressure plate.

➡Do not clean the clutch disc or the release bearing with cleaning solvent.

➡Replace a clutch cover and disc as a set.

To install:

6. Install the driven disc and pressure plate in to the clutch cover assembly.
7. Using the SST (09411-43000), install a clutch disc and cover.
8. Apply multipurpose grease to the spline of the disc. Grease: CASMOLY L 9508

WARNING

Be careful not to apply excessive grease. It can cause clutch slippage and shudder.

9. Temporarily install the clutch disc assembly to the flywheel using the SST (09411-25000).
10. Tighten the bolts one or two steps at a time in a star pattern to 18–26 ft. lbs. (45–35 Nm).
11. Install the transaxle assembly.
12. Test the clutch for proper operation.

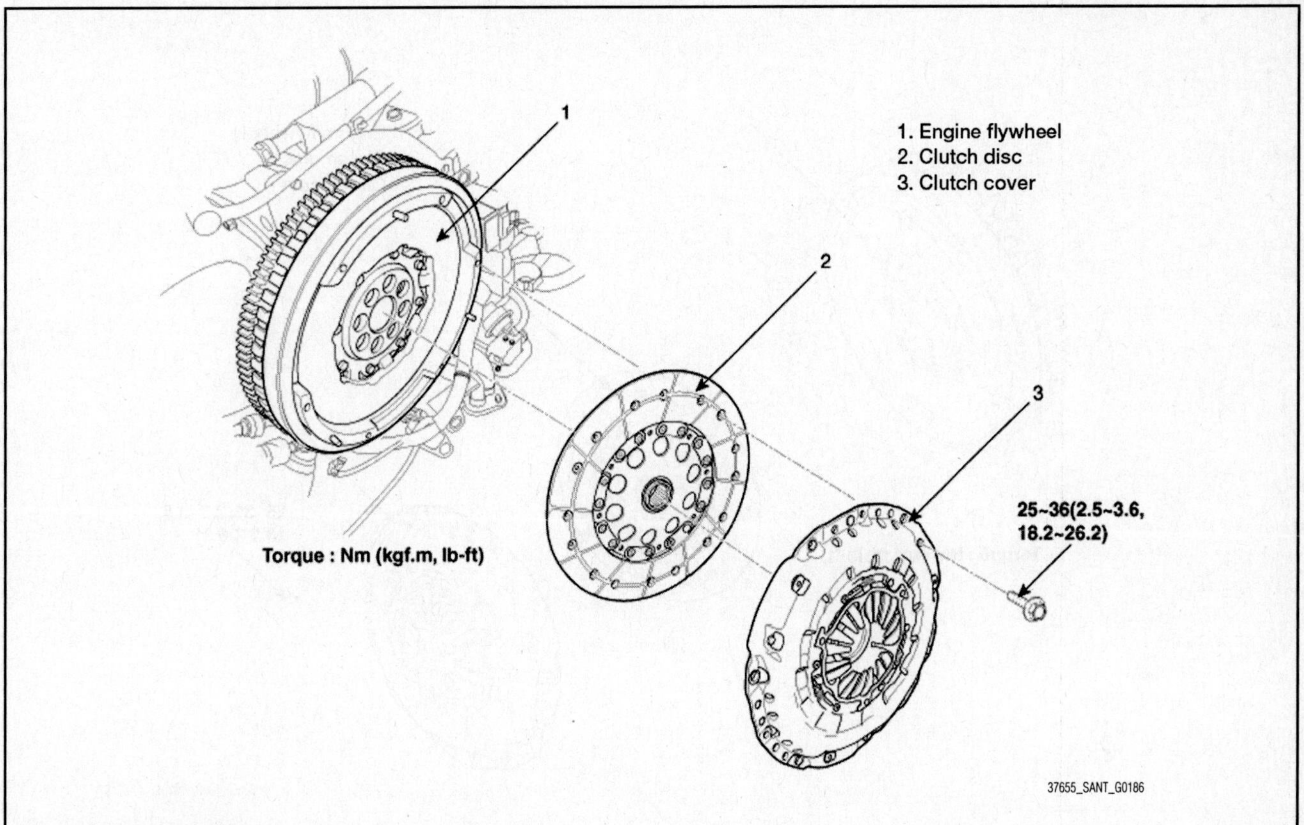

Fig. 22 Clutch components

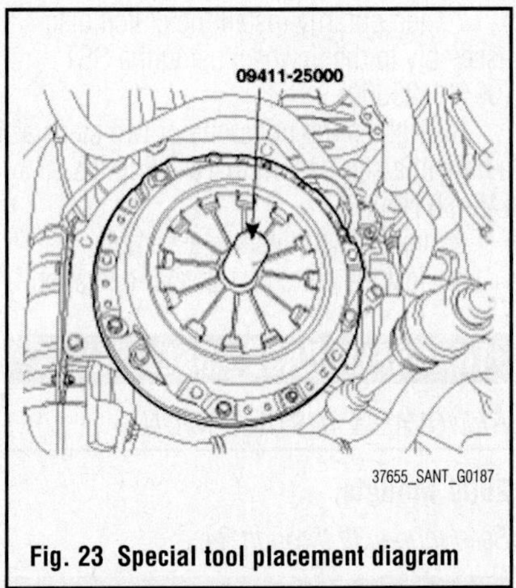

Fig. 23 Special tool placement diagram

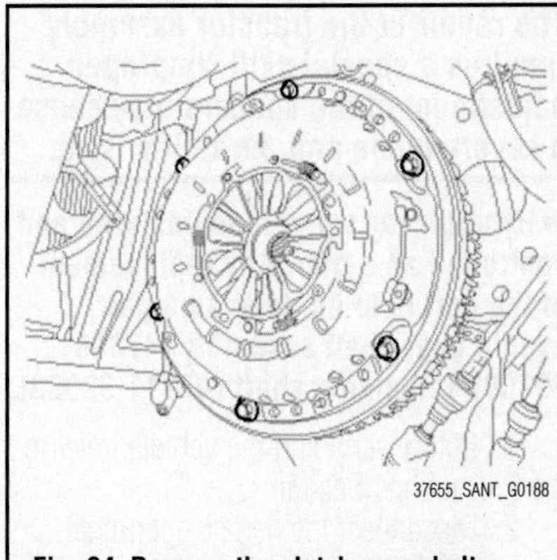

Fig. 24 Remove the clutch cover bolts

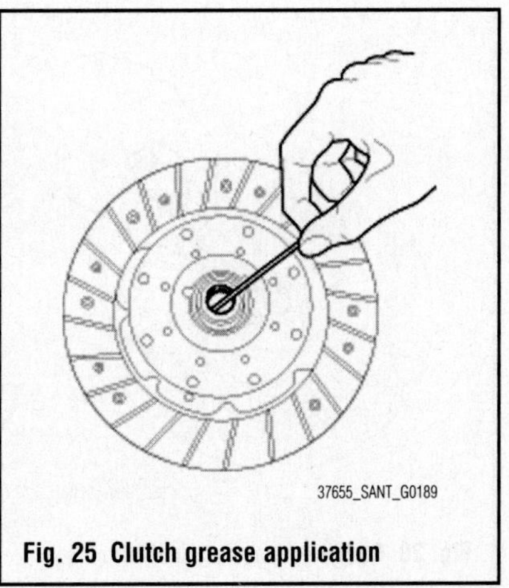

Fig. 25 Clutch grease application

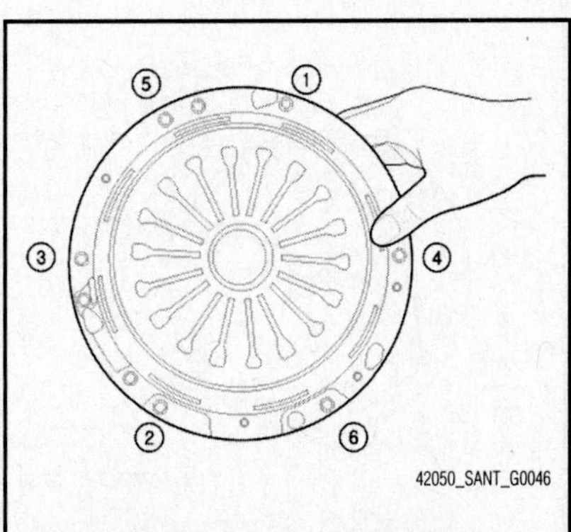

Fig. 26 Clutch cover bolt tightening sequence

2010 Models

2.4L Engine

See Figures 27 through 31.

1. Before servicing the vehicle, refer to the Precautions Section.
2. Remove the transaxle assembly.
3. Insert the SST (09411-25000) in the clutch disc to prevent the disc from falling.
4. Remove the bolts which attach the clutch cover to the flywheel. Loosen the bolts in succession, 1–2 turns at a time, in a star pattern.
5. Remove clutch cover assembly and the driven disc and pressure plate.

➡Do not clean the clutch disc or the release bearing with cleaning solvent.

➡Replace a clutch cover and disc as a set.

To install:

6. Apply multipurpose grease to the spline of the disc. Grease: CASMOLY L 9508

WARNING

Be careful not to apply excessive grease. It can cause clutch slippage and shudder.

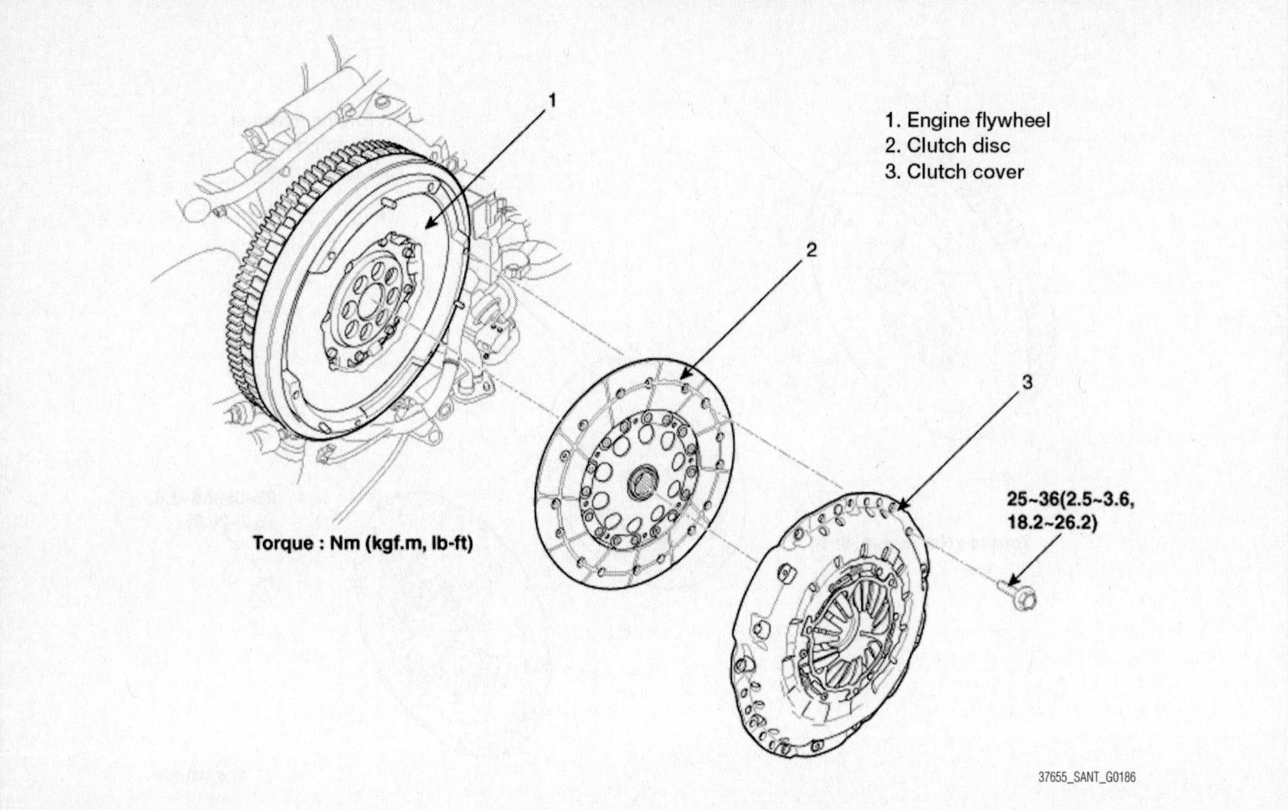

1
1. Engine flywheel
2. Clutch disc
3. Clutch cover
2
3
25~36(2.5~3.6,
18.2~26.2)
Torque : Nm (kgf.m, lb-ft)
37655_SANT_G0186

Fig. 27 Clutch components

09411-25000
37655_SANT_G0187

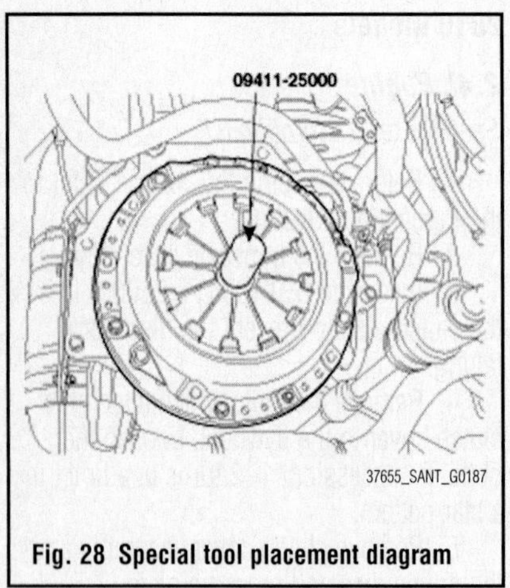

Fig. 28 Special tool placement diagram

37655_SANT_G0189

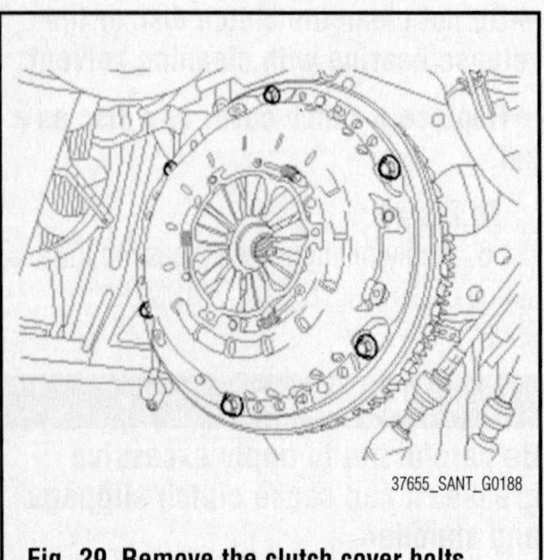

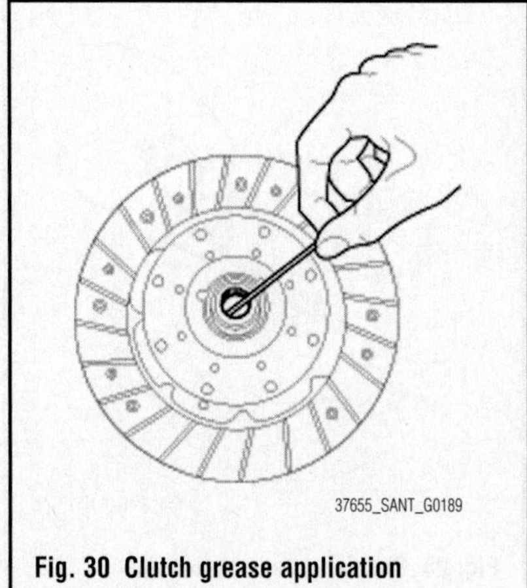

Fig. 30 Clutch grease application

37655_SANT_G0188

Fig. 29 Remove the clutch cover bolts

1
2
3
4
5
6
37655_SANT_G0190

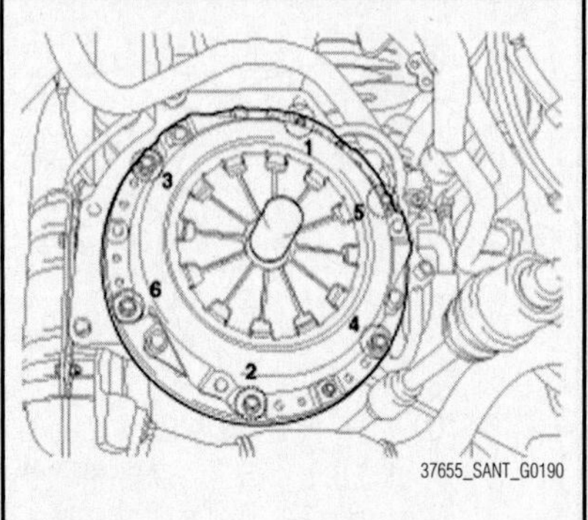

Fig. 31 Clutch cover bolt tightening sequence

7. Temporarily install the clutch disc assembly to the flywheel using the SST (09411-25000).
8. Tighten the bolts one or two steps at a time in a star pattern to 18–26 ft. lbs. (45–35 Nm).
9. Install the transaxle assembly.
10. Test the clutch for proper operation.

TRANSFER CASE ASSEMBLY

REMOVAL & INSTALLATION

2009 Models

See Figures 32 through 34.

⁂ WARNING

The repair of the transfer assembly requires a special skill. Improper adjustment of the spacers may cause a severe noise and durability issue.

➡Hypoid gear set is manufactured and controlled as a pair. Any replacement of the part is to be done as a pair, hypoid gear shaft assembly (47308-39200) and pinion shaft (47311-39000).

1. Before servicing the vehicle, refer to the Precautions Section.
2. Disconnect the negative terminal from the battery.

3. Disconnect the oxygen sensor connectors and remove the oxygen sensor connector wire clamps from the bracket.
4. Raise and safely support the vehicle.
5. Remove the right front wheel and tire.
6. Remove the right-hand engine side cover.
7. Remove the propeller shaft from the transfer case.
8. Remove the front exhaust muffler assembly.
9. Drain the transfer oil through drain plug hole.
10. Remove the pinion case mounting bolts.
11. Remove the heat protector using a hexagonal socket. Apply WD-40® if necessary to aid bolt removal.
12. Remove the exhaust manifold. Refer to Exhaust Manifold Removal & Installation in the Engine Mechanical section.
13. Remove the transfer mounting bracket.
14. Remove the transfer mounting bolts.
15. Using a flat head screw driver, remove the transfer case from the transaxle.

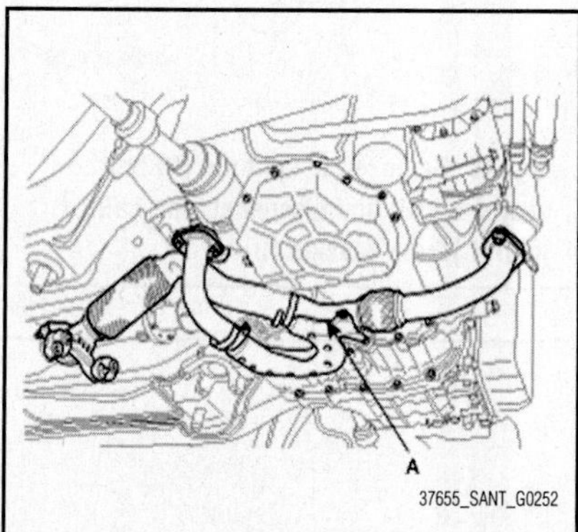

Fig. 32 Remove the front exhaust muffler assembly (A)

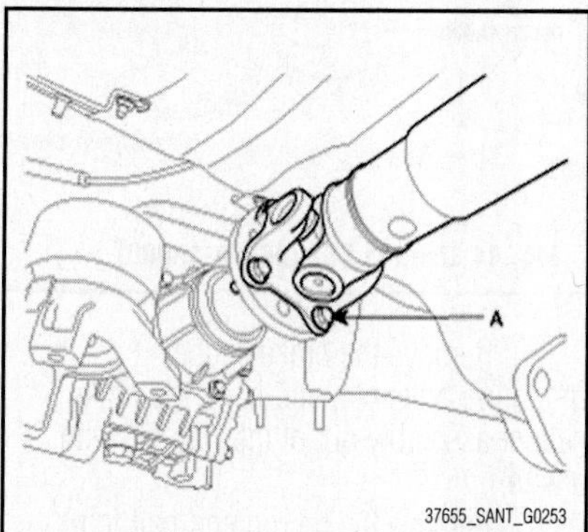

Fig. 33 Remove the pinion case mounting bolts (A)

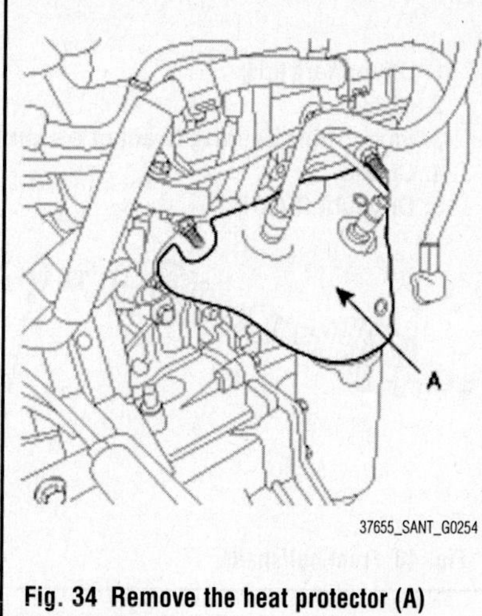

Fig. 34 Remove the heat protector (A)

To install:

16. Temporarily install the transfer assembly to the transaxle. To install the transfer easily, install it by moving the pinion left and right, and slightly rotating the inner drive shaft of transfer assembly.
17. Install the transfer case mounting bolts and tighten to 45–49 ft. lbs. (62–67 Nm).
18. Install the transfer mounting bracket. Tighten 2 bolts to 34–37 ft. lbs. (47–51 Nm), and 2 bolts to 17–20 ft. lbs. (24–28 Nm).
19. Install the exhaust manifold and tighten the mounting nuts to 22–25 ft. lbs. (30–35 Nm).
20. Install the heat protector and tighten the mounting nuts to 104–127 inch lbs. (12–15 Nm).
21. Install the pinion case and tighten the mounting bolts to 15–22 ft. lbs. (37–40 Nm).
22. Refill the transfer oil through the filler plug.
23. Install the front exhaust muffler assembly.
24. Install the propeller shaft to the transfer case assembly, lower arm ball joint mounting and steering bar tie rod ball joint.
25. Lower the vehicle.
26. Install the right-hand engine side cover and the wheel and tire.
27. Install the oxygen sensor connector brackets and connect the oxygen sensor connectors.
28. Install the air intake hose and air cleaner cover and then connect the oxygen sensor connectors.
29. Connect the battery negative cable.

2010 Models

See Figures 35 through 39.

1. Before servicing the vehicle, refer to the Precautions Section.
2. Disconnect the negative terminal from the battery.
3. Raise and safely support the vehicle.
4. Remove the front muffler.
5. Matchmark the propeller shaft at both ends and remove the propeller shaft.
6. Disconnect the right halfshaft from the transfer case.

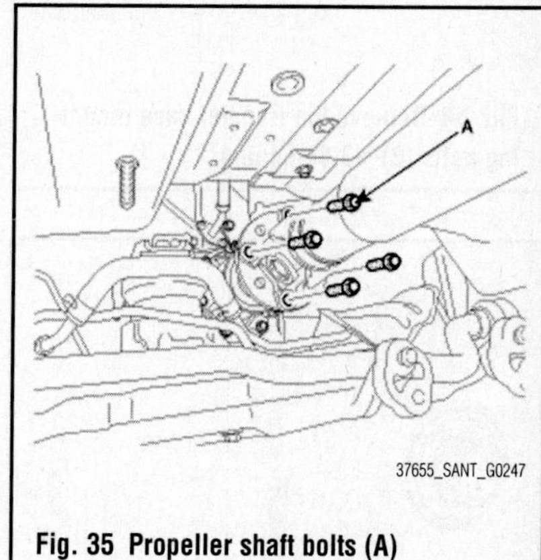

Fig. 35 Propeller shaft bolts (A)

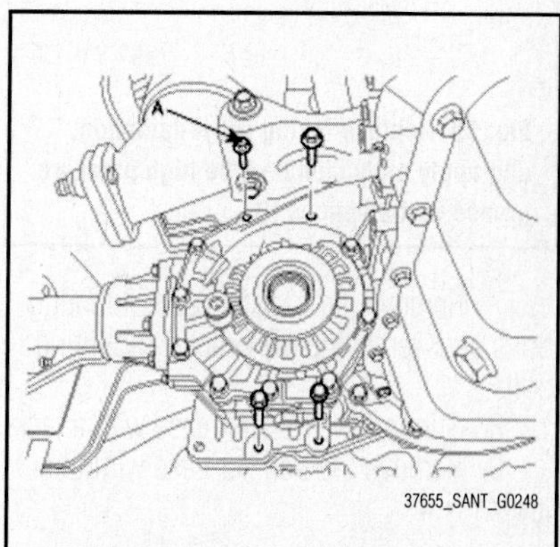

Fig. 36 Remove the transfer case mounting bolts (A)—2.4L engine

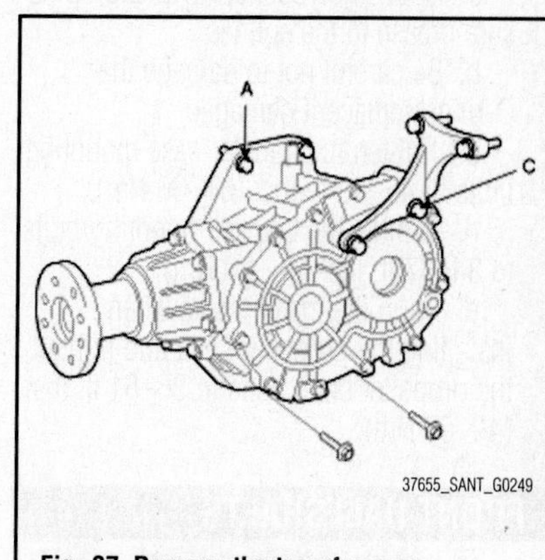

Fig. 37 Remove the transfer case mounting bolts (A) and bracket mounting bolts (C)—3.5L engine

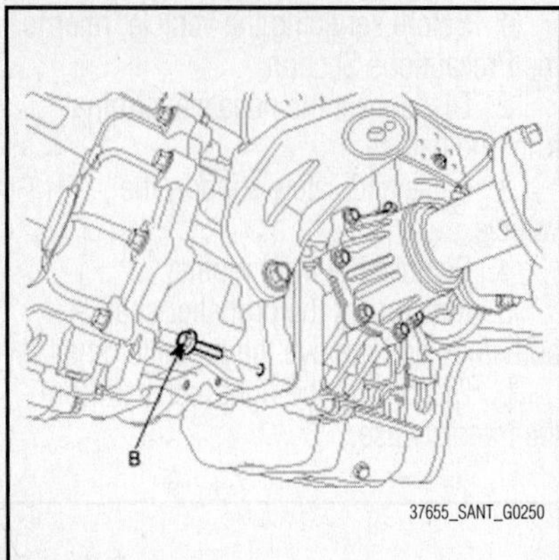

Fig. 38 Remove the transfer case mounting bolts (B)—3.5L engine

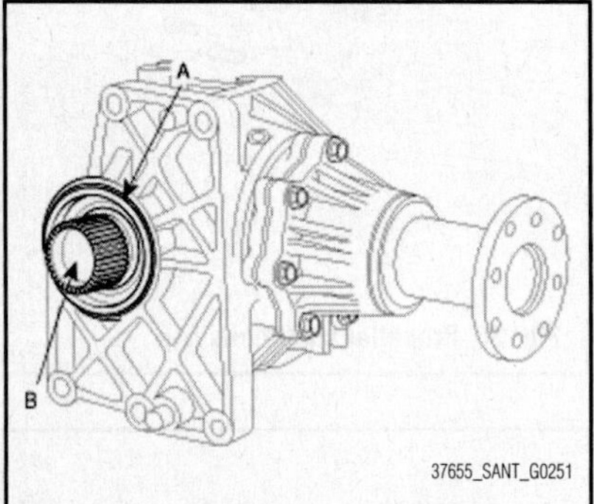

Fig. 39 Replace O-ring (A) if damaged, and apply molybdenum type high pressure grease to the splines (B)

7. Remove the transfer case mounting bolts after removing the bracket mounting bolts.
8. Support the transfer case with a jack.
9. Remove the transfer case with the lever.

To install:

10. Installation is the reverse of removal, noting the following:
    a. Apply molybdenum type high pressure grease to the splines.
    b. Be careful not to damage the O-ring. Replace if damaged.
    c. Tighten the transfer case mounting bolts to 45–49 ft. lbs. (61–66 Nm).
    d. Tighten the bracket mounting bolts to 34–37 ft. lbs. (46–50 Nm).
    e. Align the matchmarks when installing the propeller shaft and tighten the propeller shaft bolts to 36–51 ft. lbs. (49–67 Nm).

FRONT HALFSHAFTS

REMOVAL & INSTALLATION

See Figures 40 through 53.

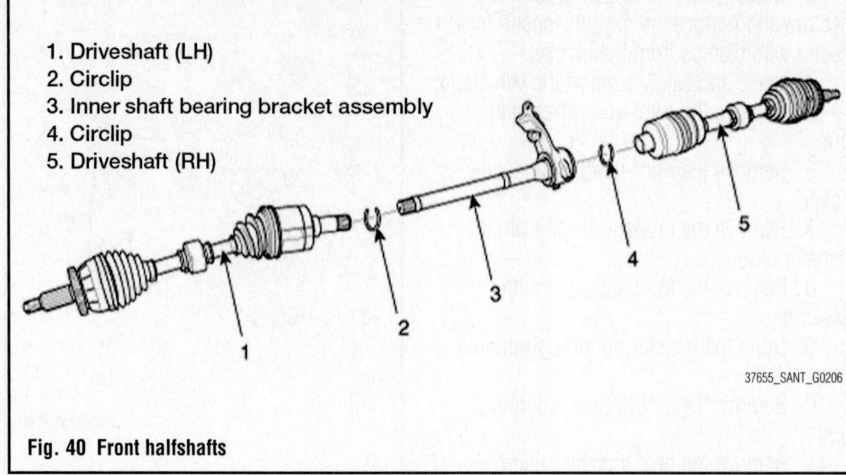

Fig. 40 Front halfshafts

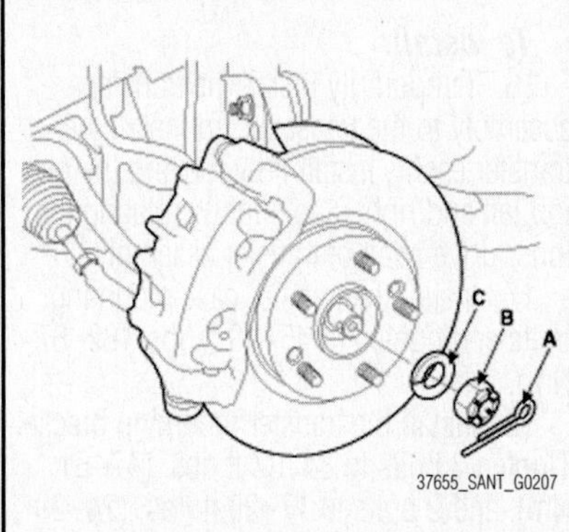

Fig. 41 Remove the split pin (A), castle nut (B) and washer (C)

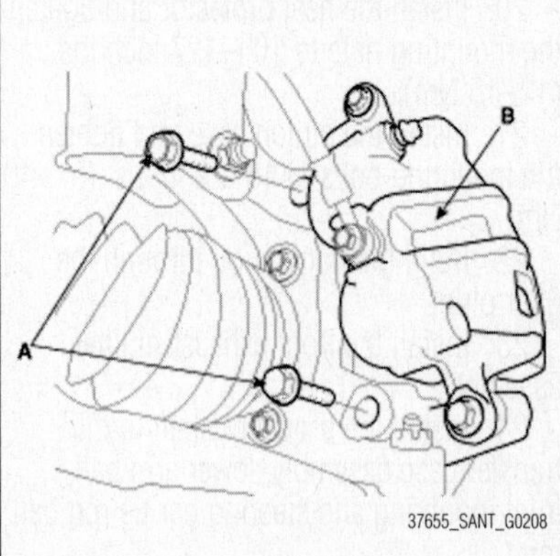

Fig. 42 Remove the brake caliper mounting bolts (A), and hang the brake caliper assembly (B)

1. Before servicing the vehicle, refer to the Precautions Section.
2. Raise and safely support the vehicle.
3. Remove the front wheels and tires.
4. Remove the split pin (A), and then remove the castle nut (B) and washer (C) from the front hub.

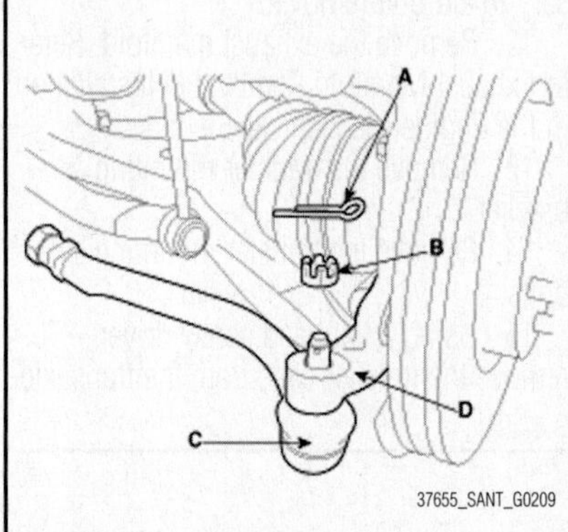

Fig. 43 Remove the split pin (A) and castle nut (B) and disconnect the ball joint (C) from knuckle (D)

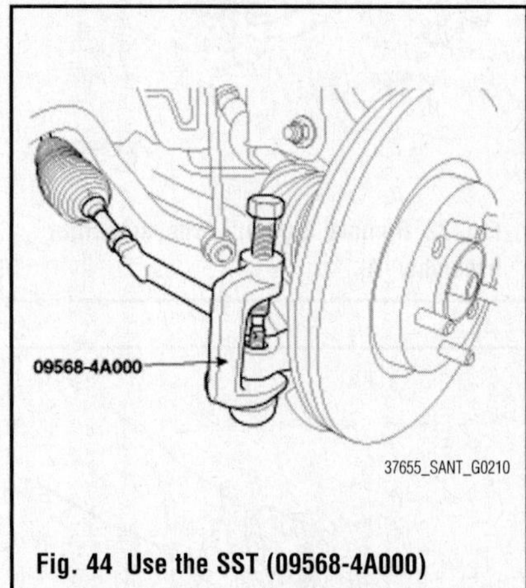

Fig. 44 Use the SST (09568-4A000)

5. Remove the brake caliper mounting bolts (A), and hang the brake caliper assembly out of the way (B) with a wire.
6. Remove the tie rod end ball joint from the knuckle:
    a. Remove the split pin.
    b. Remove the castle nut.

c. Disconnect the ball joint from knuckle using the SST (09568-4A000). Apply a few drops of oil to the boot contact part of the SST.

7. Remove the split pin and the lower arm mounting bolt from the knuckle.

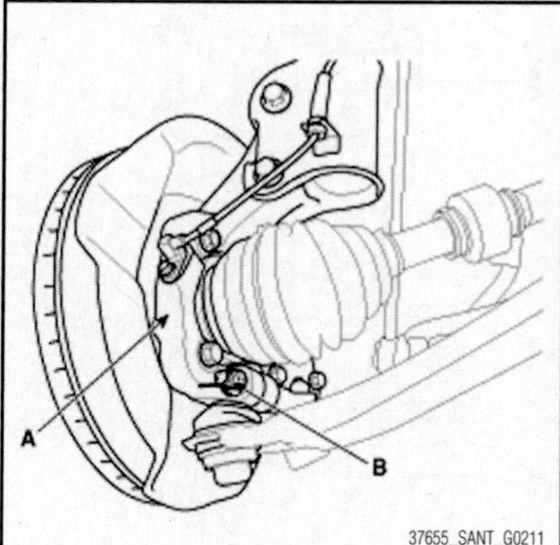

Fig. 45 Remove the split pin and the lower arm mounting bolt (B) from the knuckle (A)

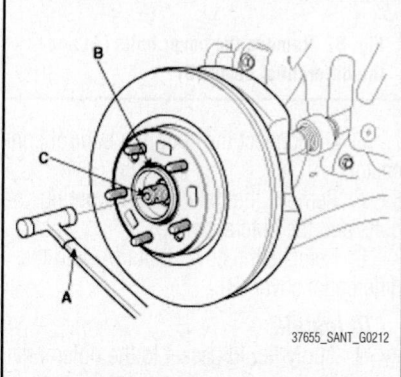

Fig. 46 Using a plastic hammer (A), disconnect halfshaft (C) from the axle hub (B)

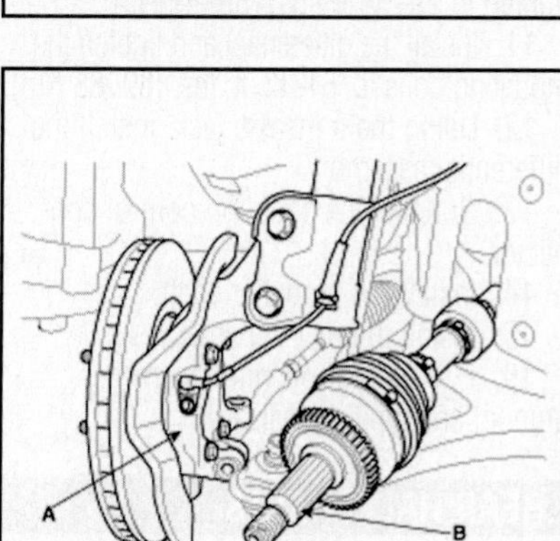

Fig. 47 Push the axle hub (A) outward and separate the halfshaft (B) from the axle hub (A)

8. Using a plastic hammer, disconnect the halfshaft from the axle hub.

9. Push the axle hub outward and separate the halfshaft from the axle hub.

10. Remove the dust cover (A).

11. Remove the mounting bolts of inner shaft bearing bracket assembly (A).

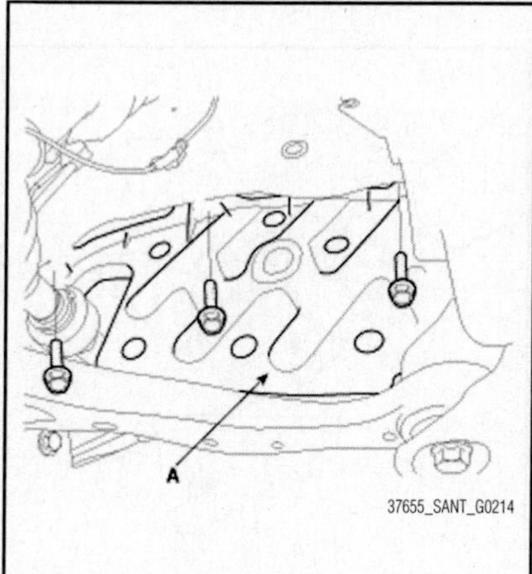

Fig. 48 Remove the dust cover (A)—Right-hand side

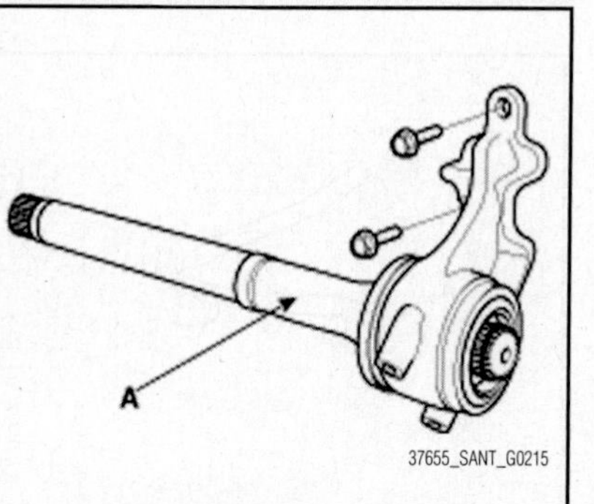

Fig. 49 Remove the inner shaft bearing bracket assembly (A) mounting bolts—2.4L and 2.7L engines

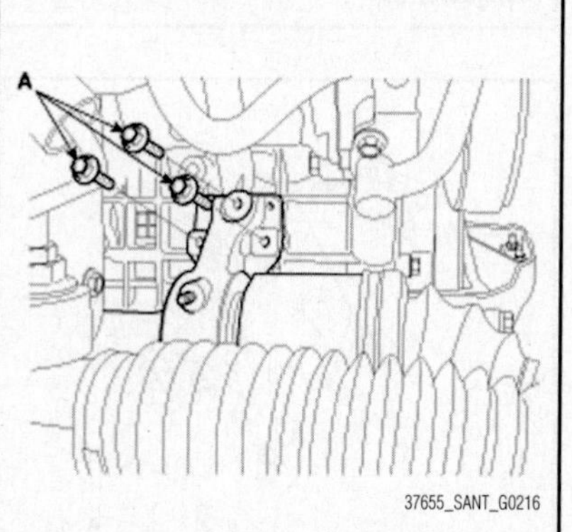

Fig. 50 Remove the inner shaft bearing bracket assembly (A) mounting bolts—3.3L and 3.5L engines

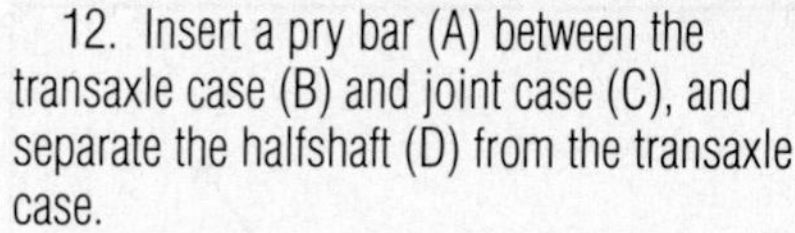

12. Insert a pry bar (A) between the transaxle case (B) and joint case (C), and separate the halfshaft (D) from the transaxle case.

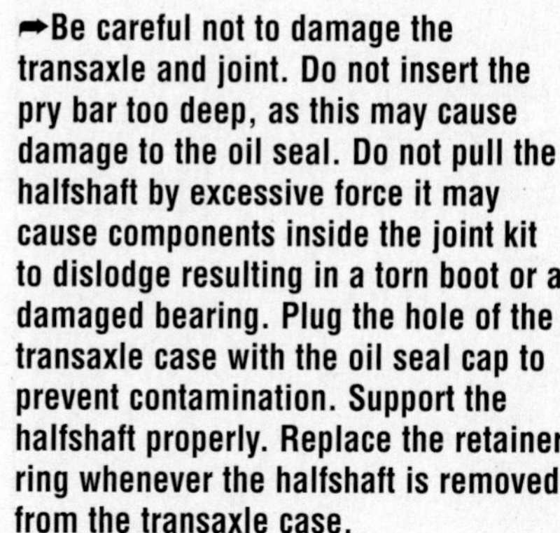

➡Be careful not to damage the transaxle and joint. Do not insert the pry bar too deep, as this may cause damage to the oil seal. Do not pull the halfshaft by excessive force it may cause components inside the joint kit to dislodge resulting in a torn boot or a damaged bearing. Plug the hole of the transaxle case with the oil seal cap to prevent contamination. Support the halfshaft properly. Replace the retainer ring whenever the halfshaft is removed from the transaxle case.

To install:

13. Apply gear oil on the oil seal contacting surface of transaxle case and the halfshaft splines.

14. Before installing the halfshaft, set the opening side of the circlip facing downward.

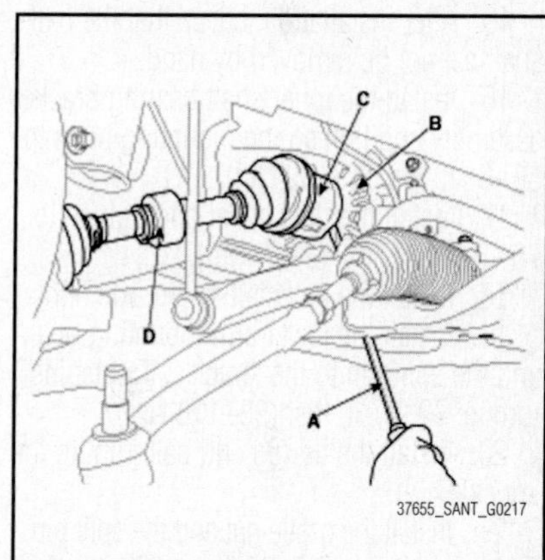

Fig. 51 Insert a pry bar (A) between the transaxle case (B) and joint case (C), and separate the halfshaft (D)

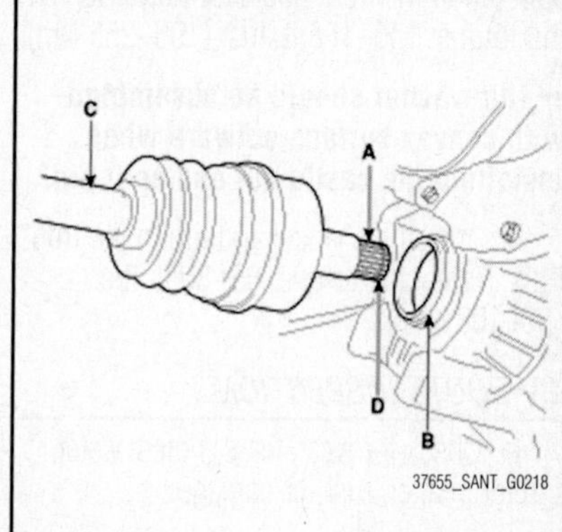

Fig. 52 Apply gear oil on the oil seal contacting surface (B) of transaxle case and the halfshaft splines (A)

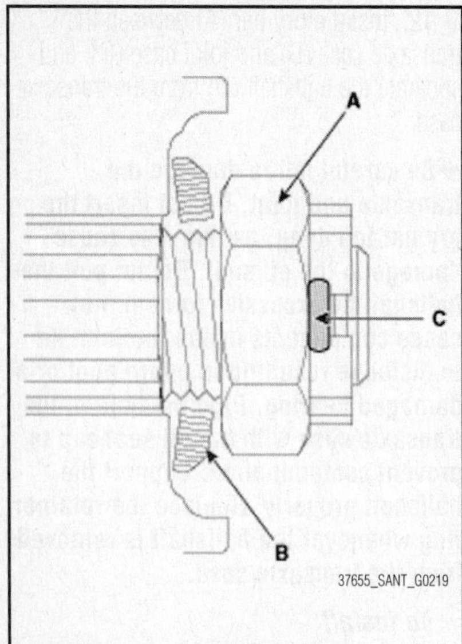

Fig. 53 The washer (B) should be assembled with convex surface outward when installing the castle nut (A) and split pin (C)

15. After installation, check that the halfshaft cannot be removed by hand.
16. Install the inner shaft bearing bracket assembly and tighten the mounting bolts to 36–51 ft. lbs. (49–69 Nm).
17. Install the dust cover and tighten to 70–104 inch lbs. (8–12 Nm).
18. Install the halfshaft to the axle hub.
19. Install the lower arm mounting bolt and the split pin to the knuckle. Tightening torque: 72–87 ft. lbs. (98–118 Nm).
20. Install the tie rod end ball joint to the knuckle.
21. Install the castle nut and the split pin. Tightening torque: 17–25 ft. lbs. (24–33 Nm).
22. Install the brake caliper and tighten the mounting bolts to 58–72 ft. lbs. (79–98 Nm).
23. Install the washer, castle nut and split pin to the front hub assembly. Tightening torque: 145–188 ft. lbs. (196–255 Nm).

➡The washer should be assembled with convex surface outward when installing the castle nut and split pin.

24. Install the wheel and tire to the front hub. Tightening torque: 65–80 ft. lbs. (88–108 Nm).

CV-BOOTS INSPECTION

1. Check for wear, tears, holes, water, foreign matter, rust, or damage.

REAR DIFFERENTIAL

REMOVAL & INSTALLATION

See Figures 54 through 57.

1. Before servicing the vehicle, refer to the Precautions Section.
2. Drain the differential gear oil.
3. Remove the rear halfshaft.
4. Remove the propeller shaft. Refer to Propeller Shaft Removal & Installation.
5. Support the differential assembly with the jack.

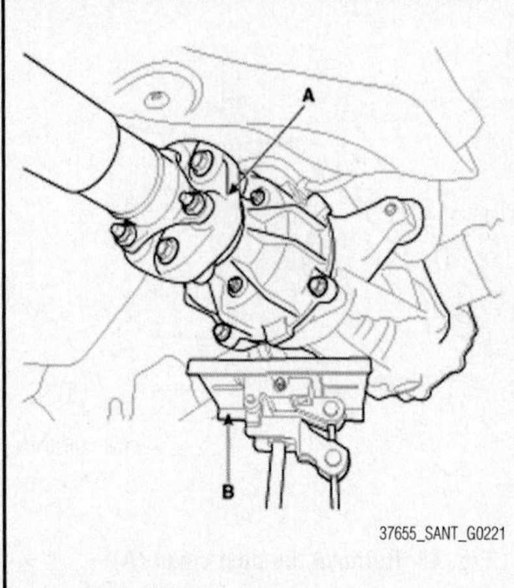

Fig. 54 Support the differential assembly (A) with the jack (B)

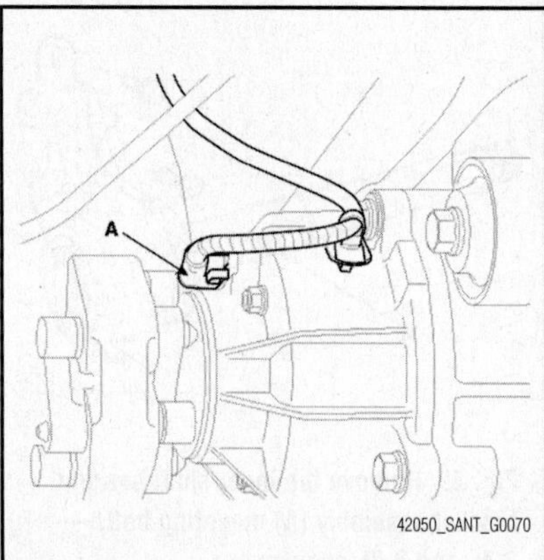

Fig. 55 Disconnect the coupling control connector (A)

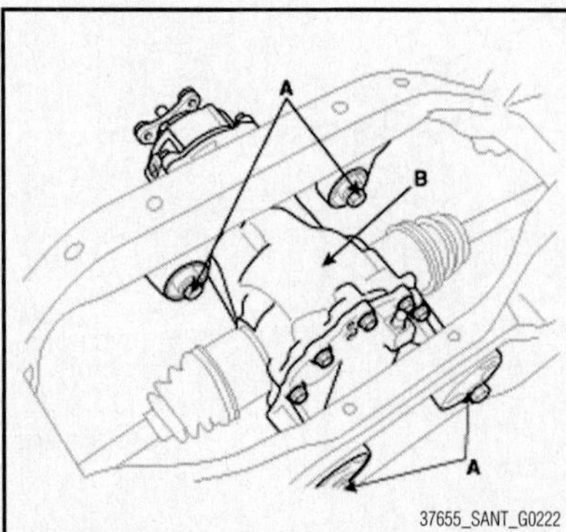

Fig. 56 Remove the differential mounting bolts (A) and the differential (B)

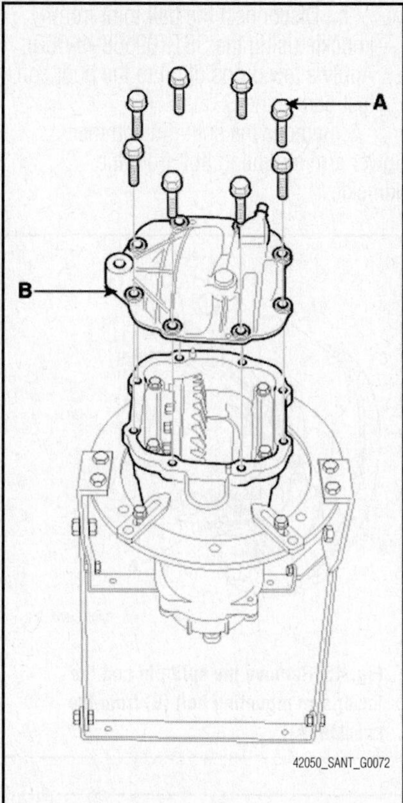

Fig. 57 Remove the cover bolts (A) and the differential cover (B)

6. Disconnect the coupling control connector.
7. Remove the differential mounting bolts and the differential.
8. Remove the cover bolts (A) and the differential cover (B).

To install:

9. Apply liquid gasket to the differential cover.
10. Install the mounting bolts and tighten to 29–36 ft. lbs. (39–49 Nm).
11. Install the differential and tighten the mounting bolts to: 51–65 ft. lbs. (69–88 Nm).
12. Using the transaxle jack, install the differential assembly.
13. Connect the coupling control connector.
14. Install the propeller shaft.
15. Install the rear halfshaft.
16. Fill the gear oil with the proper amount and type of gear oil.

REAR HALFSHAFTS

REMOVAL & INSTALLATION

See Figures 58 through 69.

1. Before servicing the vehicle, refer to the Precautions Section.
2. Raise and safely support the vehicle.

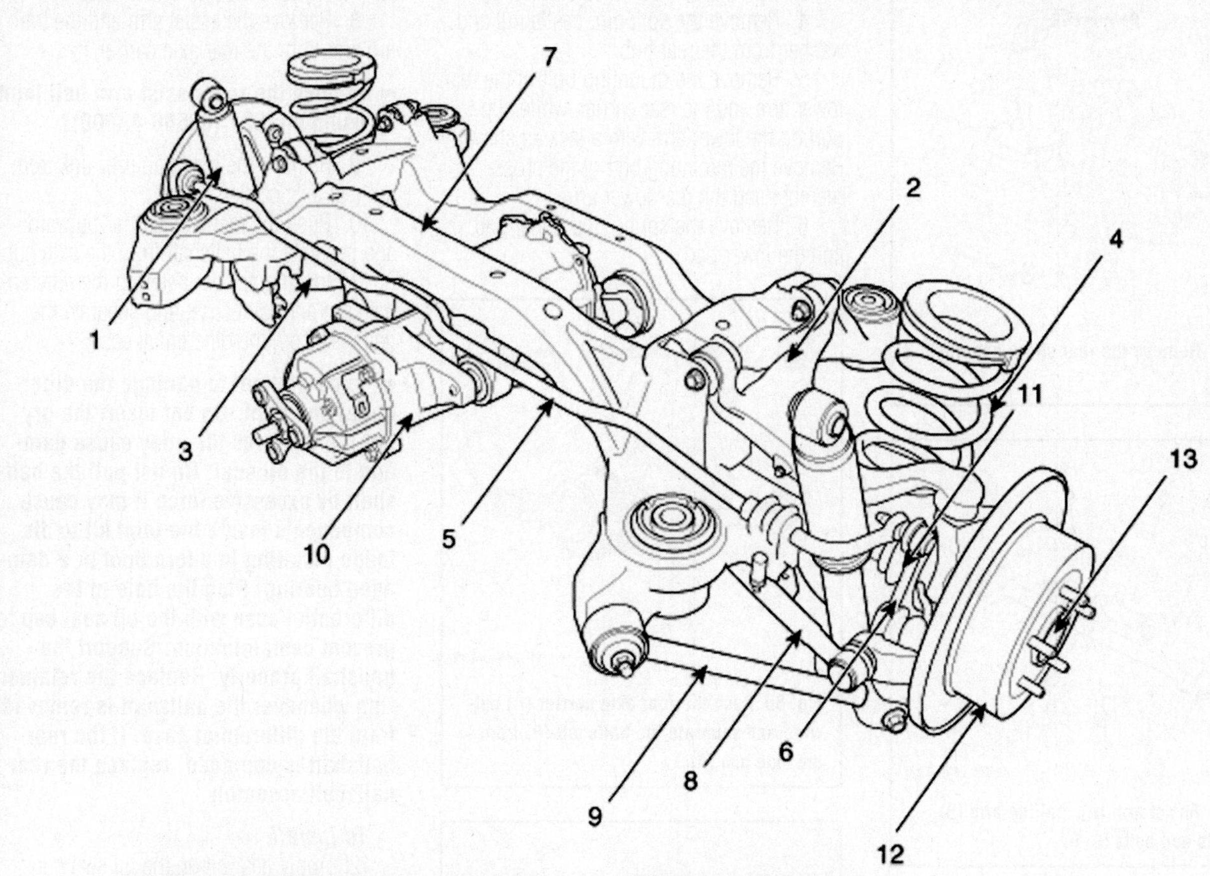

Fig. 58 Rear halfshaft components

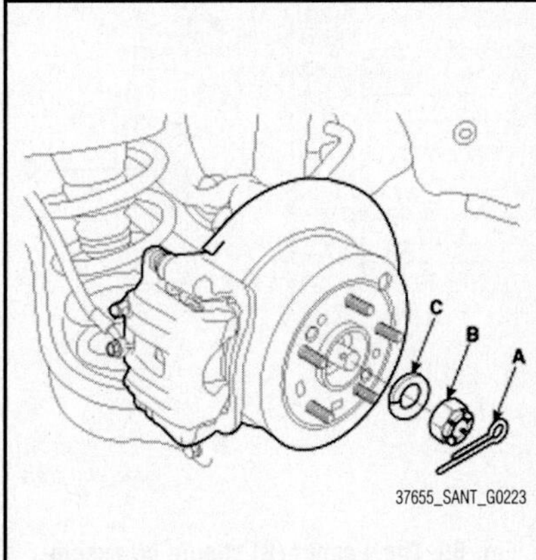

Fig. 59 Remove the split pin (A), castle nut (B) and washer (C)

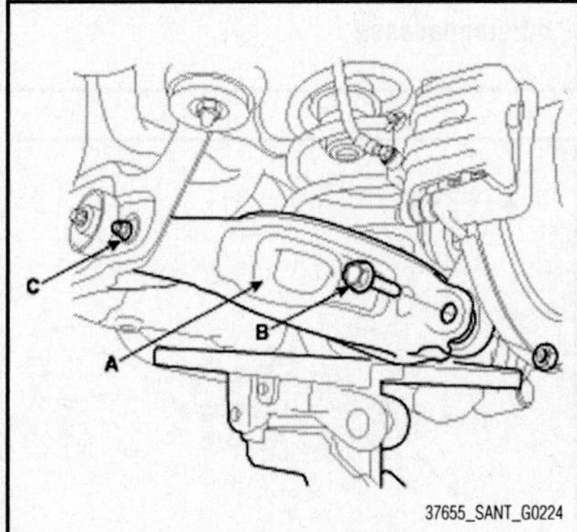

Fig. 60 Remove the rear lower arm (A) and the rear carrier mounting bolt (B), and the crossmember and the rear lower arm mounting bolt (C)

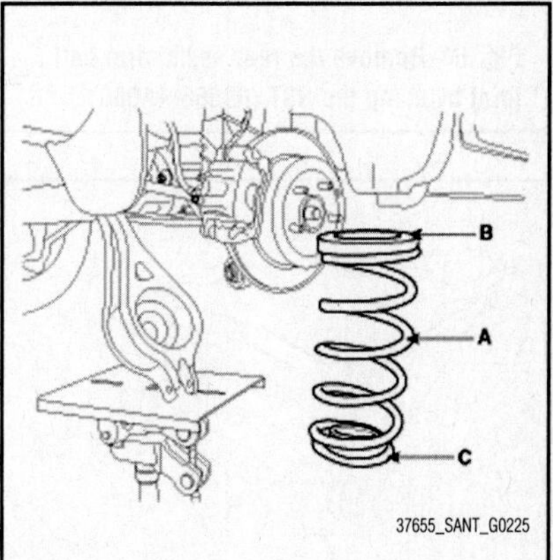

Fig. 61 Remove the spring (A), the upper pad (B) and the lower pad (C)

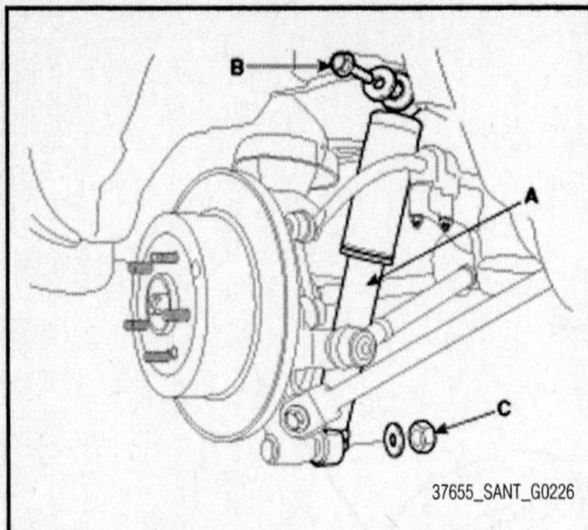

Fig. 62 Remove the rear shock absorber (A)

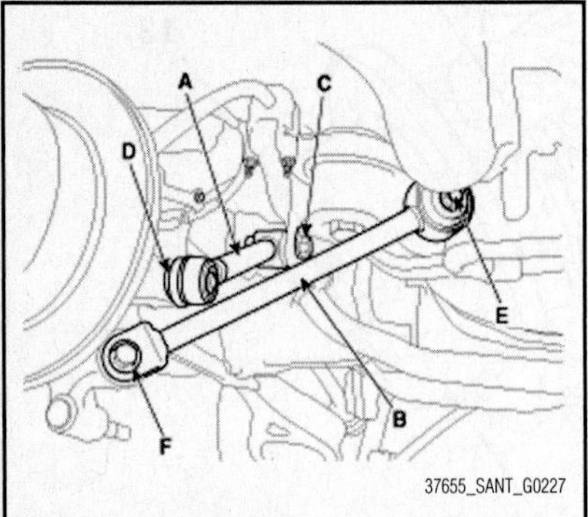

Fig. 63 Assist arm (A), trailing arm (B), and nuts and bolts (C–F)

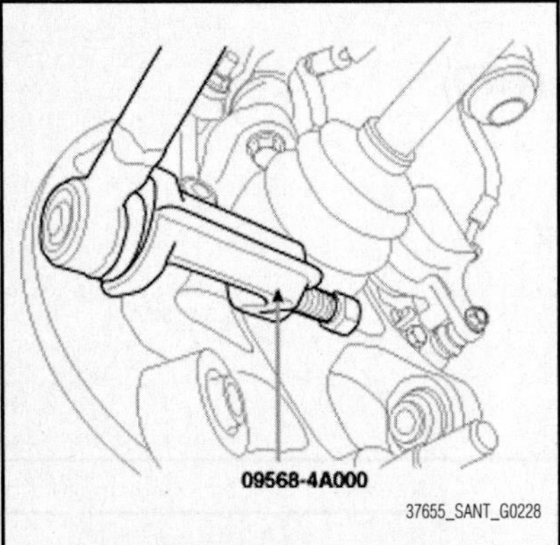

Fig. 64 Remove the rear assist arm ball joint by using the SST (09568-4A000)

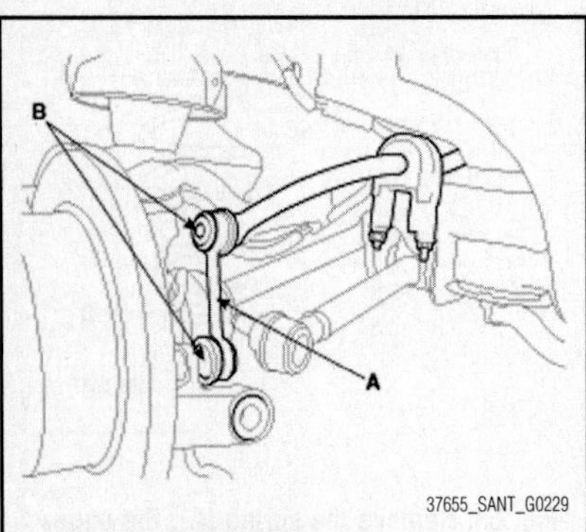

Fig. 65 Remove the rear stabilizer link (A)

3. Remove the rear wheels and tires.

4. Remove the split pin, castle nut and washer from the rear hub.

5. Remove the mounting bolt of the rear lower arm and the rear carrier, while supporting the lower arm with a jack as shown. Remove the mounting bolt of the crossmember and the rear lower arm.

6. Remove the spring, the upper pad and the lower pad.

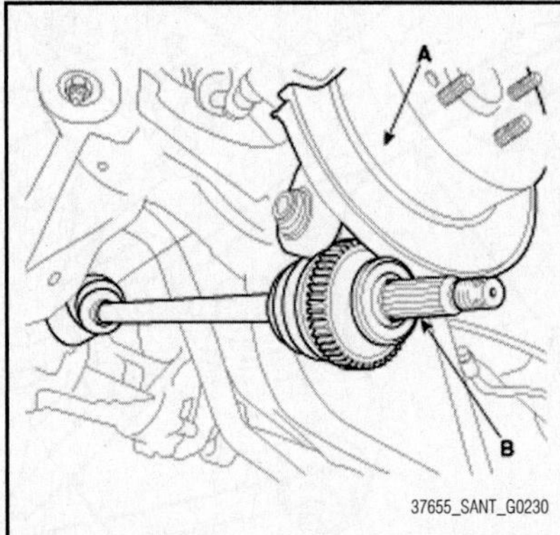

Fig. 66 Push the rear axle carrier (A) outward and separate the halfshaft (B) from the axle hub (A)

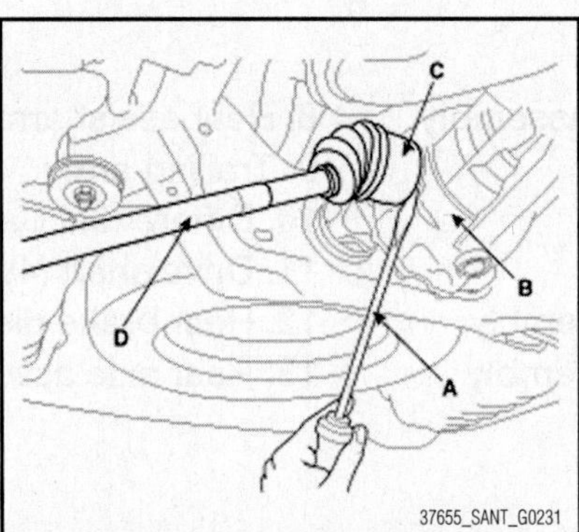

Fig. 67 Insert a pry bar (A) between the differential case (B) and joint case (C), and separate the halfshaft (D) from the differential case

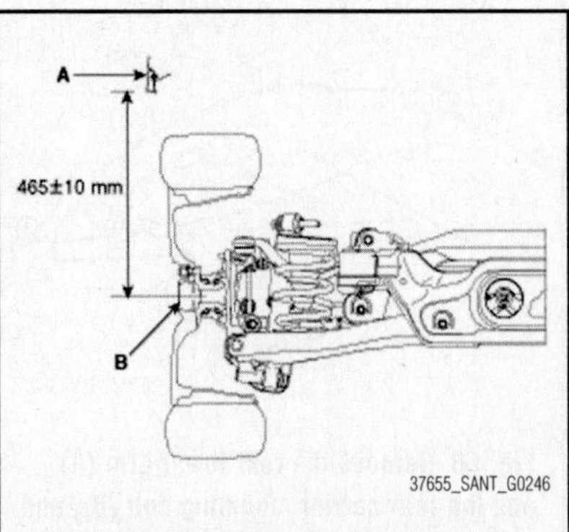

Fig. 68 Distance between the wheel housing molding (A) and hub assembly (B)

7. Remove the rear shock absorber.

8. Remove the assist arm and the trailing arm from the rear axle carrier.

➡Remove the rear assist arm ball joint by using the SST (09568-4A000).

9. Remove the rear stabilizer link from the rear axle carrier.

10. Push the rear axle carrier outward and separate the halfshaft from the axle hub.

11. Insert a pry bar between the differential case and joint case, and separate the halfshaft from the differential case.

➡Be careful not to damage the differential and joint. Do not insert the pry bar too deep, as this may cause damage to the oil seal. Do not pull the halfshaft by excessive force it may cause components inside the joint kit to dislodge resulting in a torn boot or a damaged bearing. Plug the hole of the differential case with the oil seal cap to prevent contamination. Support the halfshaft properly. Replace the retainer ring whenever the halfshaft is removed from the differential case. If the rear halfshaft is damaged, replace the rear halfshaft assembly.

To install:

12. Apply gear oil on the oil seal contacting surface of differential case and the halfshaft splines.

13. Before installing the halfshaft, set the opening side of the circlip facing downward.

14. After installation, check that the halfshaft cannot be removed by hand.

15. Install the halfshaft to the rear axle carrier assembly.

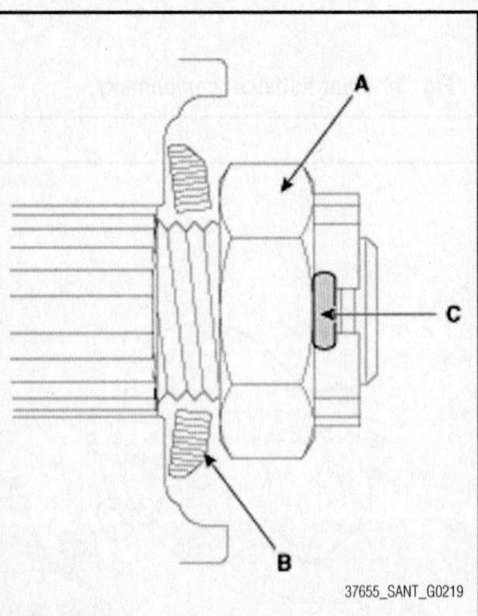

Fig. 69 The washer (B) should be assembled with convex surface outward when installing the castle nut (A) and split pin (C)

16. Install the rear stabilizer link to the rear axle carrier and tighten the nut to 43–58 ft. lbs. (59–79 Nm).

17. Install the assist arm and the trailing arm to the rear axle carrier. Tightening torque:

- Bolt (C): 101–116 ft. lbs. (137–157 Nm)
- Nut (D): 72–87 ft. lbs. (98–118 Nm)
- Nut (E): 101–116 ft. lbs. (137–157 Nm)
- Bolt (F): 101–116 ft. lbs. (137–157 Nm)

➡Before tightening the chassis components, measure the distance between the wheel housing molding and the hub assembly, as shown. Specified distance: 18.31 inches, plus or minus 0.39 in. (465mm, plus or minus 10 mm)

18. Install the rear shock absorber. Tightening torque:

- Bolt (B): 101–116 ft. lbs. (137–157 Nm)
- Nut (C): 72–87 ft. lbs. (98–118 Nm)

19. Install the spring, the upper pad and the lower pad.

20. Install the mounting bolt of the rear lower arm and the rear carrier and tighten to 101–116 ft. lbs. (137–157 Nm), while supporting the lower arm with a jack. Tighten the mounting bolt of the cross member and the rear lower arm to 101–116 ft. lbs. (137–157 Nm).

21. Install the washer, castle nut and split pin to the rear hub assembly. Tightening torque: 145–188 ft. lbs. (196–255 Nm)

22. The washer should be assembled with convex surface outward when installing the castle nut and split pin.

23. Install the wheel and tire to the rear hub. Tightening torque: 65–80 ft. lbs. (88–108 Nm).

CV-BOOTS INSPECTION

1. Check for wear, tears, holes, water, foreign matter, rust, or damage.

ENGINE COOLING

ENGINE FAN

REMOVAL & INSTALLATION

2009 Models

See Figures 70 through 72.

1. Before servicing the vehicle, refer to the Precautions Section.

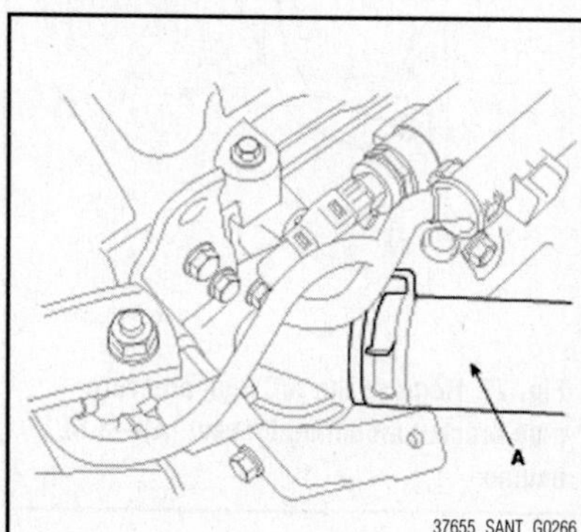

Fig. 70 Radiator upper hose (A)—2.7L engine

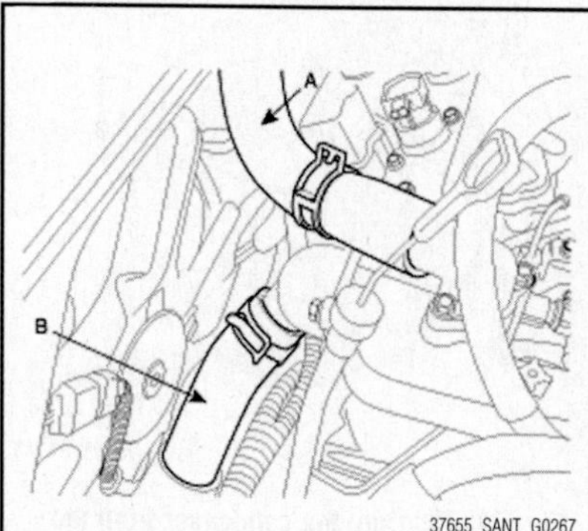

Fig. 71 Radiator upper (A) and lower hose (B)—3.3L engine

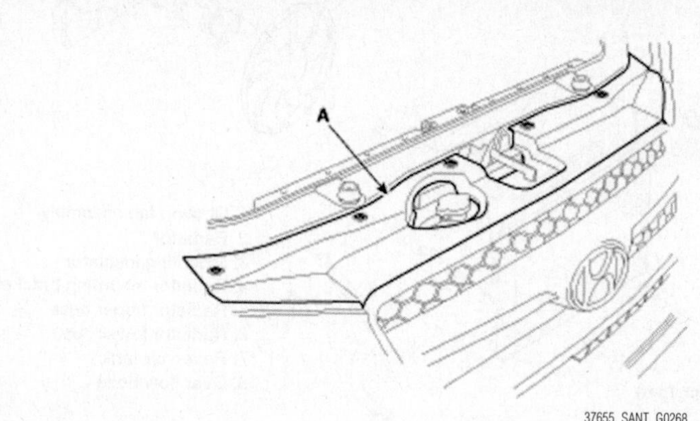

Fig. 72 Radiator grill upper cover (A)

2. Disconnect the negative battery cable.
3. Remove the air duct.
4. Drain the coolant. Remove the radiator cap to speed draining.
5. Remove the radiator upper and lower hoses.
6. Disconnect the radiator fan connectors.
7. Remove the radiator grill upper cover.
8. Remove the head lamp washer nozzle cover and front bumper.
9. Remove the radiator cap hose.
10. Remove the cooling fan.

To install:

11. Installation is in the reverse of removal.
12. Fill the radiator with coolant, bleed, and check for leaks.

2010 Models

See Figures 73 through 79.

1. Before servicing the vehicle, refer to the Precautions Section.
2. Disconnect the negative battery cable.
3. Remove the air cleaner assembly.
4. Remove the undercover.
5. Drain the coolant. Remove the radiator cap to speed draining.
6. Remove the radiator upper and lower hoses.
7. Disconnect the fan motor connector.
8. Disconnect the ATF cooler hoses (A/T only).
9. Remove the front bumper.
10. Disconnect the overflow hose from the radiator.
11. Remove the upper cover.
12. For 2.4L engines, remove the ATF cooler.
13. Remove the guide and radiator mounting bracket.
14. Remove the A/C high pressure pipe bracket mounting bracket.

15. Separate the condenser from the radiator and then remove the upper cover and cooling fan assembly.

To install:

16. Installation is in the reverse of removal, noting the following:

a. Tighten the upper cover and cooling fan assembly to 78–95 inch lbs. (9–11 Nm).

b. Tighten the A/C high pressure pipe bracket mounting bracket to 86–104 inch lbs. (10–12 Nm).

c. Tighten the radiator mounting bracket to 78–95 inch lbs. (9–11 Nm).

17. Fill the radiator with coolant, bleed, and check for leaks.

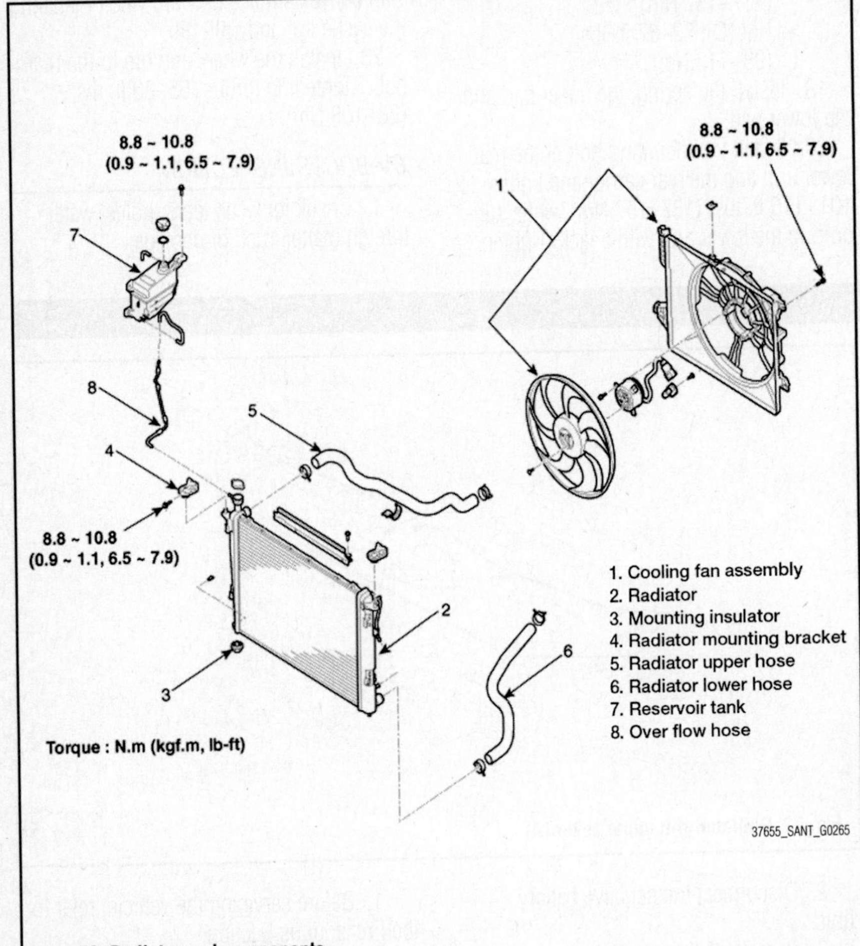

Fig. 73 Radiator and components

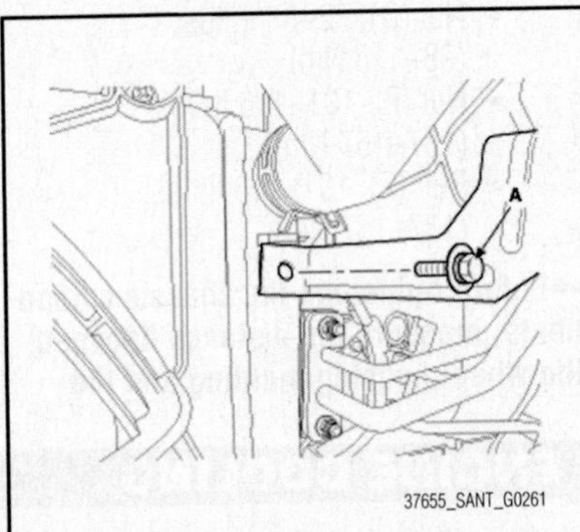

Fig. 76 Remove the A/C high pressure pipe bracket mounting bracket (A)—2.4L engine

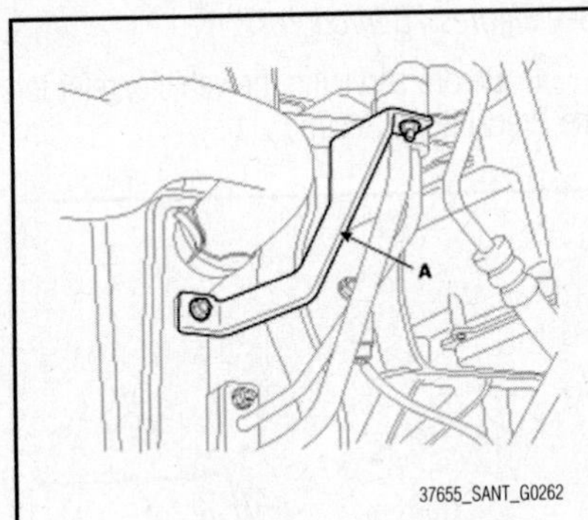

Fig. 77 Remove the A/C high pressure pipe bracket mounting bracket (A)—3.5L engine

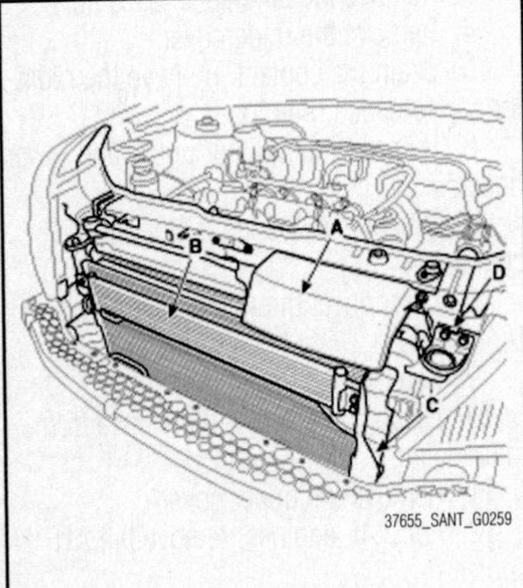

Fig. 74 Remove the upper cover (A), ATF cooler (B), guide (C), and radiator mounting bracket (D)—2.4L engine

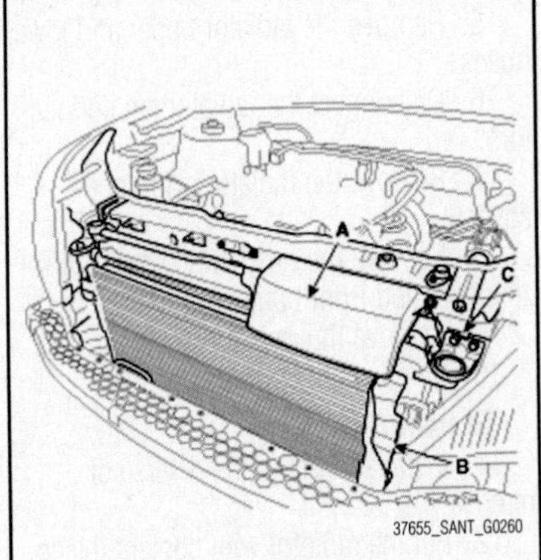

Fig. 75 Remove the upper cover (A), guide (B), and radiator mounting bracket (C)—3.5L engine

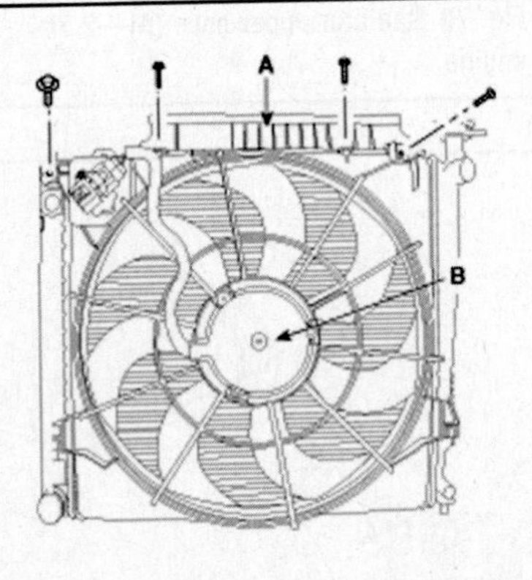

Fig. 78 Separate the condenser from the radiator and then remove the upper cover (A) and cooling fan assembly (B)—2.4L engine

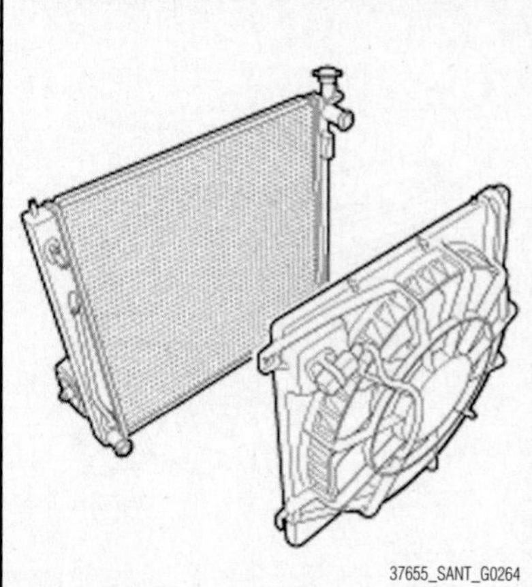
Fig. 79 Separate the condenser from the radiator and then remove the upper cover and cooling fan assembly—3.5L engine

RADIATOR

REMOVAL & INSTALLATION

2009 Models

See Figures 80 through 85.

1. Before servicing the vehicle, refer to the Precautions Section.
2. Disconnect the negative battery cable.

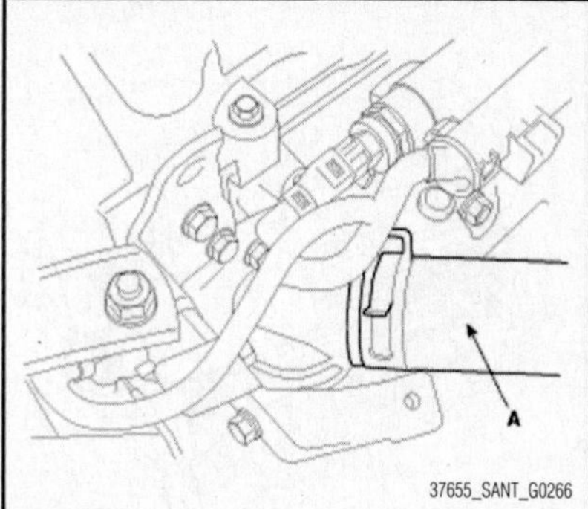

Fig. 81 Radiator upper hose (A)—2.7L engine

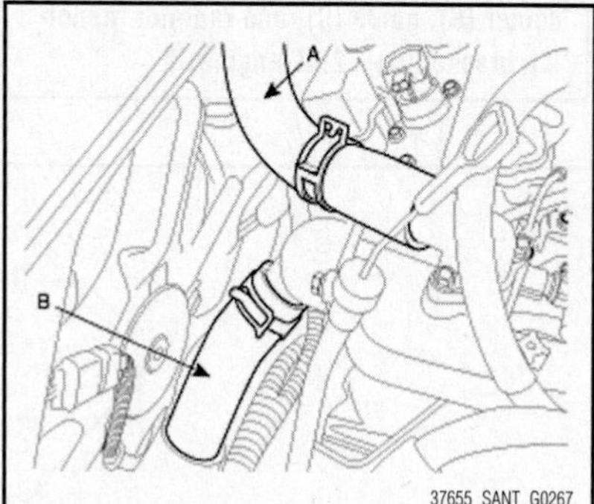

Fig. 82 Radiator upper (A) and lower hose (B)—3.3L engine

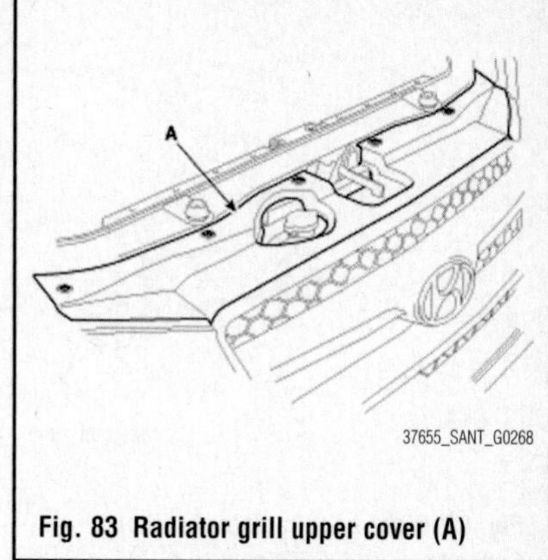

Fig. 83 Radiator grill upper cover (A)

3. Remove the air duct.
4. Drain the coolant. Remove the radiator cap to speed draining.
5. Remove the radiator upper and lower hoses.
6. Disconnect the radiator fan connectors.
7. Remove the radiator grill upper cover.
8. Remove the head lamp washer nozzle cover and front bumper.
9. Remove the radiator cap hose.
10. Remove the cooling fan.

➡Remove the bracket bolt of radiator lower hose.

11. Recover the refrigerant.
12. Remove the air conditioner condenser.
13. Remove the upper radiator bracket bolts.
14. Remove the radiator assembly from the engine.

To install:

15. Installation is in the reverse of removal.
16. Fill the radiator with coolant, bleed, and check for leaks.

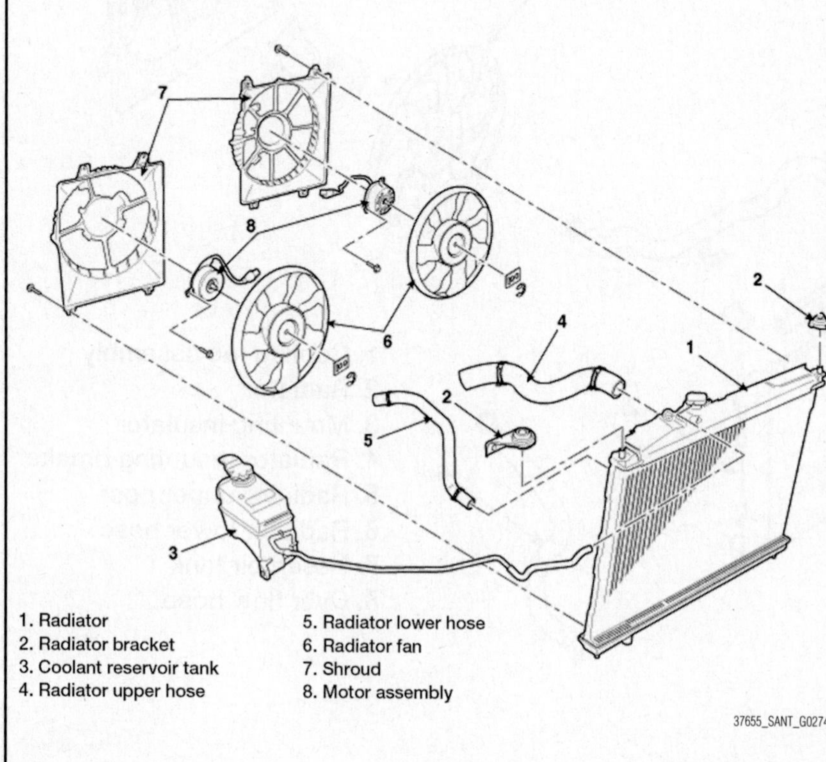

Fig. 80 Radiator and components—2.7L engine

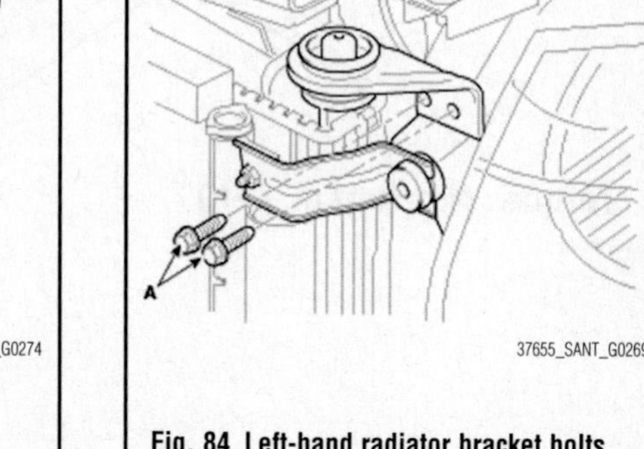

Fig. 84 Left-hand radiator bracket bolts (A)

A
37655_SANT_G0270

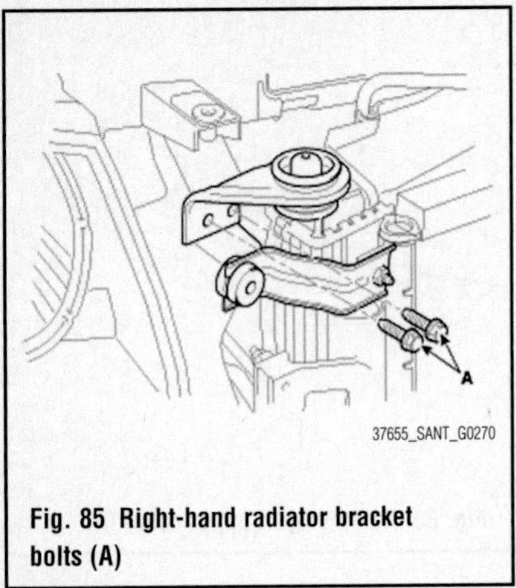

Fig. 85 Right-hand radiator bracket bolts (A)

A
B
D
C
37655_SANT_G0259

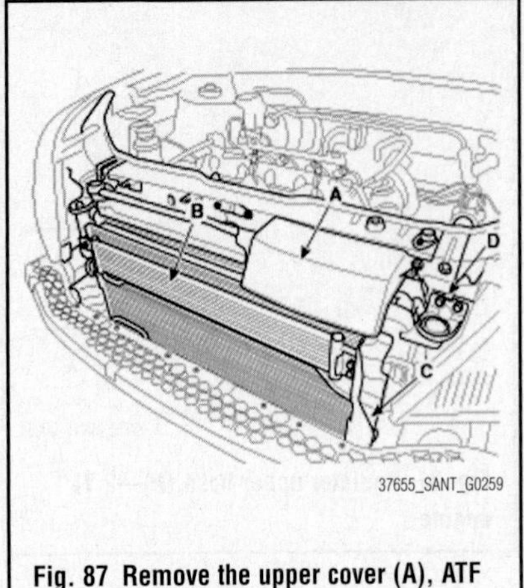

Fig. 87 Remove the upper cover (A), ATF cooler (B), guide (C), and radiator mounting bracket (D)—2.4L engine

A
C
B
37655_SANT_G0260

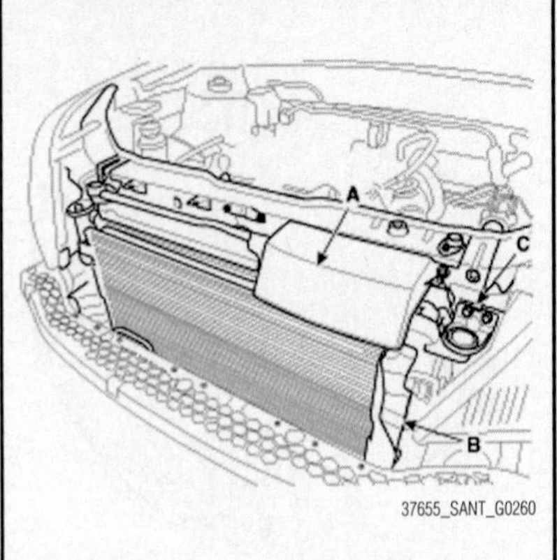

Fig. 88 Remove the upper cover (A), guide (B), and radiator mounting bracket (C)—3.5L engine

2010 Models

See Figures 86 through 92.

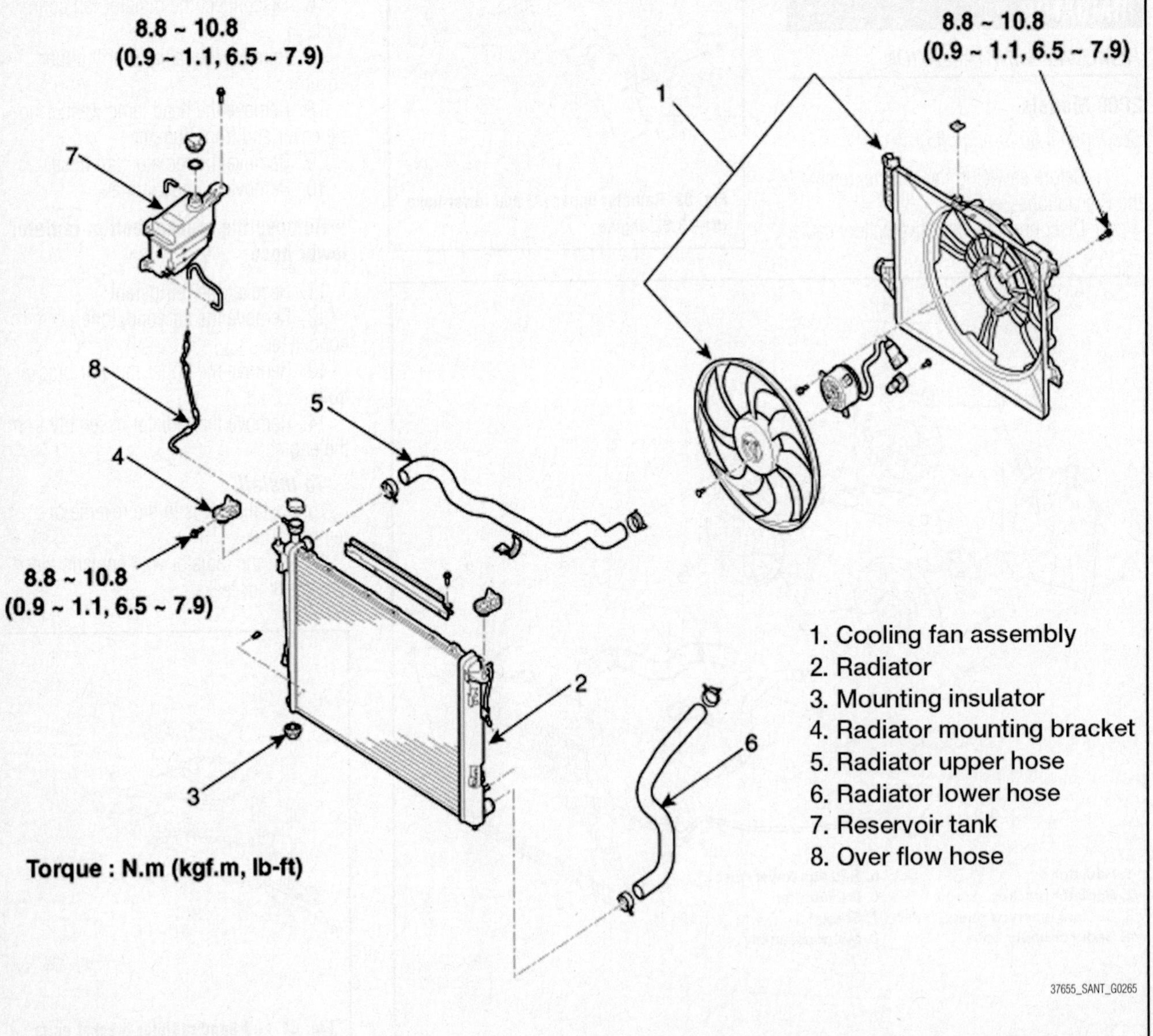

8.8 ~ 10.8
(0.9 ~ 1.1, 6.5 ~ 7.9)
8.8 ~ 10.8
(0.9 ~ 1.1, 6.5 ~ 7.9)
1
7
8
5
4
8.8 ~ 10.8
(0.9 ~ 1.1, 6.5 ~ 7.9)
2
6
3
Torque : N.m (kgf.m, lb-ft)
1. Cooling fan assembly
2. Radiator
3. Mounting insulator
4. Radiator mounting bracket
5. Radiator upper hose
6. Radiator lower hose
7. Reservoir tank
8. Over flow hose
37655_SANT_G0265

Fig. 86 Radiator and components

1. Before servicing the vehicle, refer to the Precautions Section.
2. Disconnect the negative battery cable.
3. Remove the air cleaner assembly.
4. Remove the undercover.
5. Drain the coolant.
6. Remove the radiator upper and lower hoses.
7. Disconnect the fan motor connector.
8. Disconnect the ATF cooler hoses (A/T only).
9. Remove the front bumper.
10. Disconnect the overflow hose from the radiator.
11. Remove the upper cover.
12. For 2.4L engines, remove the ATF cooler.
13. Remove the guide and radiator mounting bracket.

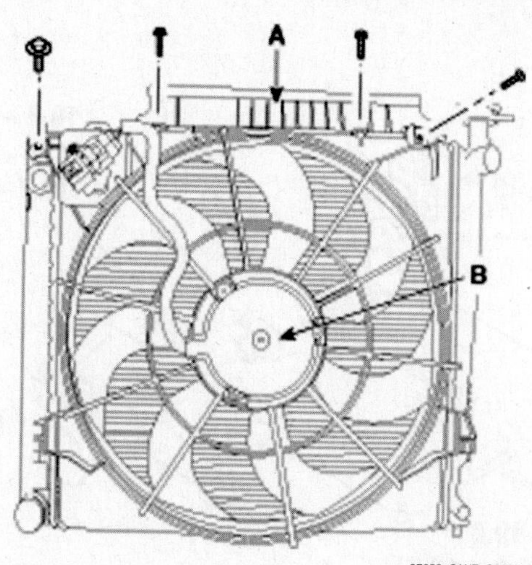

Fig. 91 Separate the condenser from the radiator and then remove the upper cover (A) and cooling fan assembly (B)—2.4L engine

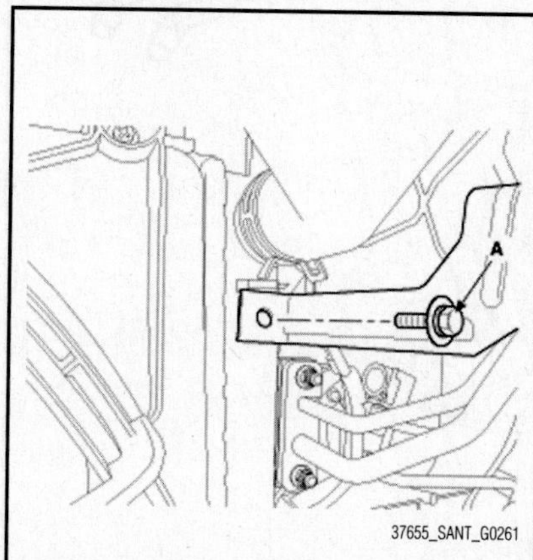

Fig. 89 Remove the A/C high pressure pipe bracket mounting bracket (A)—2.4L engine

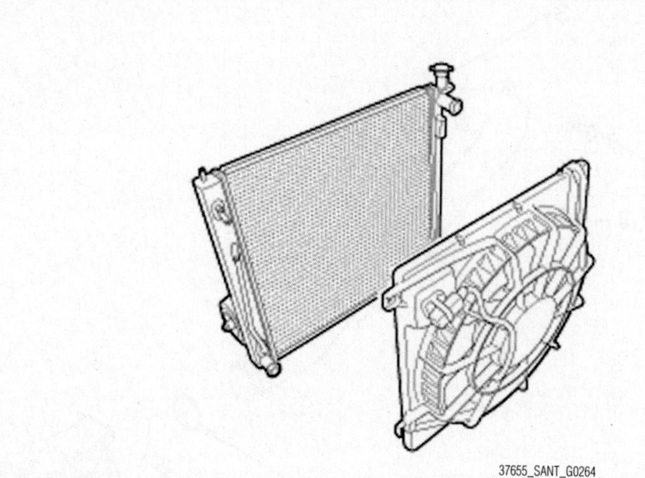

Fig. 92 Separate the condenser from the radiator and then remove the upper cover and cooling fan assembly—3.5L engine

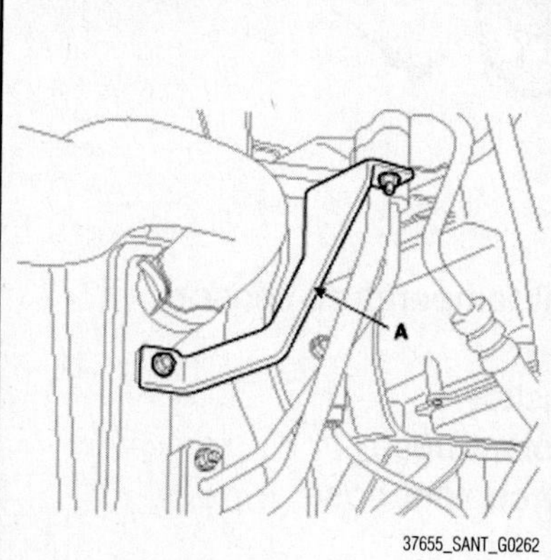

Fig. 90 Remove the A/C high pressure pipe bracket mounting bracket (A)—3.5L engine

14. Remove the A/C high pressure pipe bracket mounting bracket.
15. Separate the condenser from the radiator and then remove the upper cover and cooling fan assembly.
16. Pull radiator upper from engine room.

To install:

17. Installation is in the reverse of removal, noting the following:
   a. Tighten the upper cover and cooling fan assembly to 78–95 inch lbs. (9–11 Nm).
   b. Tighten the A/C high pressure pipe bracket mounting bracket to 86–104 inch lbs. (10–12 Nm).
   c. Tighten the radiator mounting bracket to 78–95 inch lbs. (9–11 Nm).
18. Fill the radiator with coolant, bleed, and check for leaks.

THERMOSTAT

REMOVAL & INSTALLATION

See Figures 93 through 99.

**** CAUTION**

Make sure the engine and radiator are cool to the touch prior to beginning this procedure in order to prevent scalding or burns.

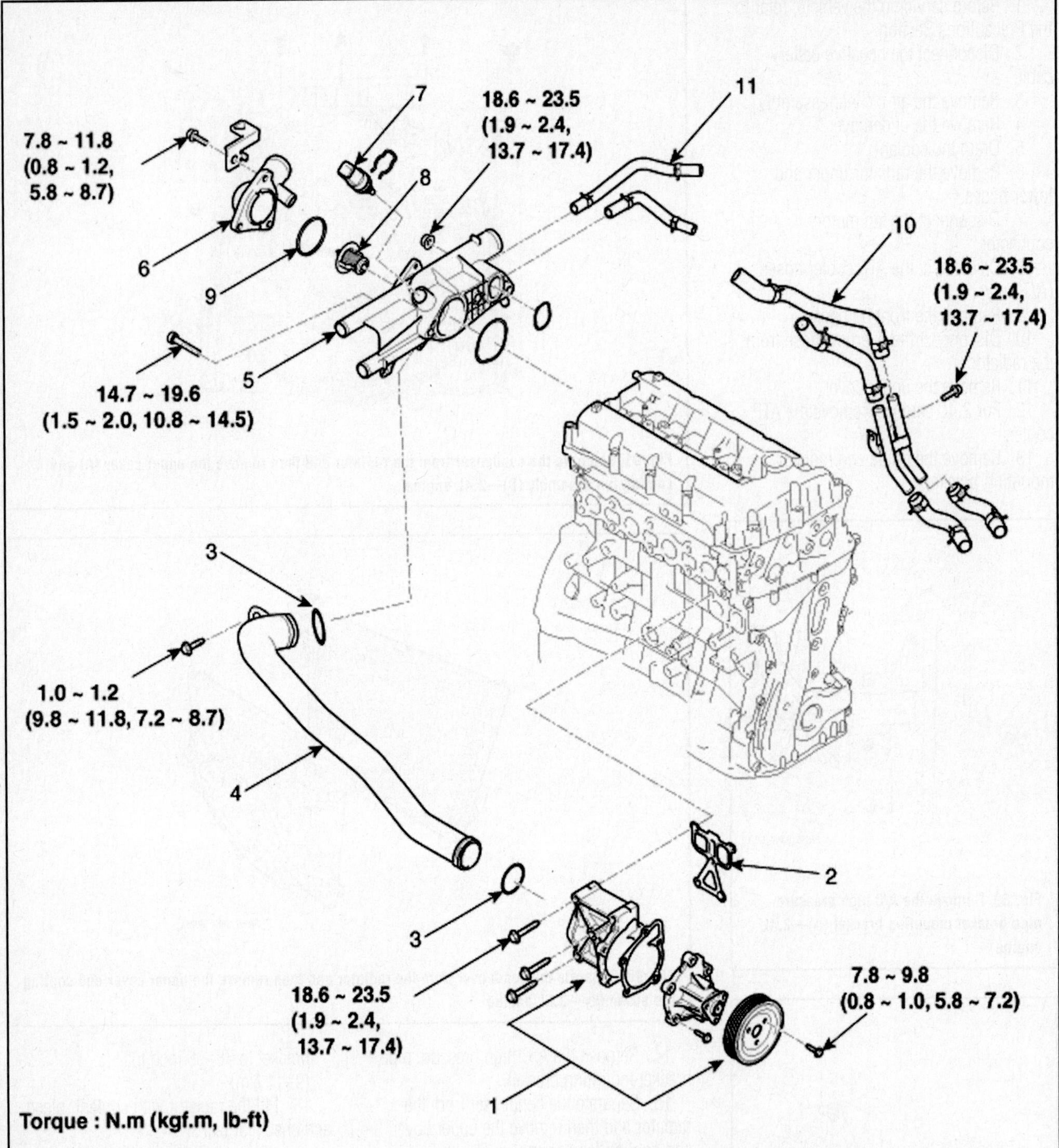

Fig. 93 Cooling system components—2.4L engine

1. Water pump
2. Water pump gasket
3. Water pipe O-ring
4. Water inlet pipe
5. Water outlet fitting
6. Thermostat
7. Water inlet fitting
8. Water temperature control assembly
9. Thermostat Housing gasket
16.7 ~ 19.6
(1.7 ~ 2.0, 12.3 ~ 14.5)
16.7 ~ 19.6
(1.7 ~ 2.0, 12.3 ~ 14.5)
29.4 ~ 41.2
(3.0 ~ 4.2, 21.7 ~ 30.4)
Torque : N.m (kgf.m, lb-ft)
14.7 ~ 21.6
(1.5 ~ 2.2, 10.8 ~ 15.9)
37655_SANT_G0275

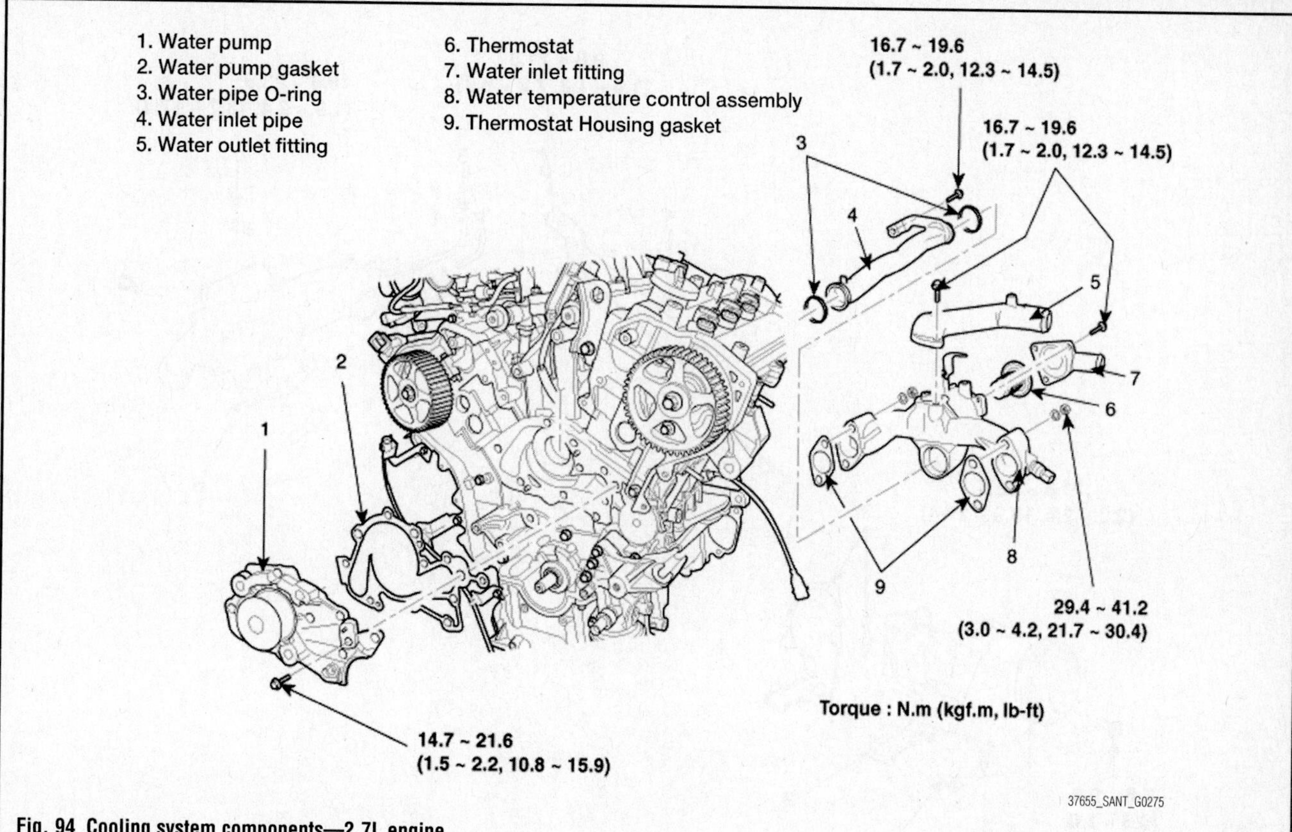

Fig. 94 Cooling system components—2.7L engine

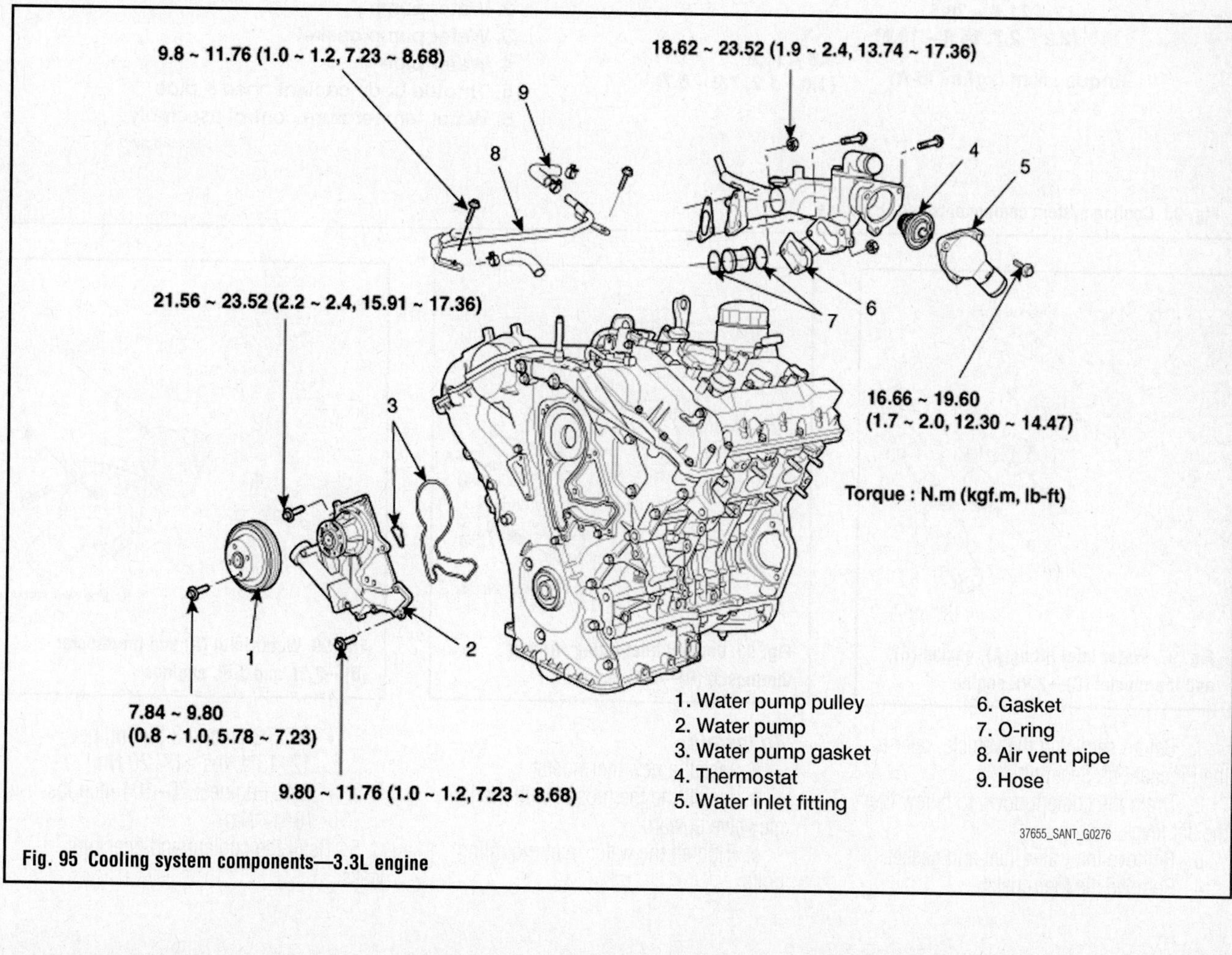

9.8 ~ 11.76 (1.0 ~ 1.2, 7.23 ~ 8.68)
18.62 ~ 23.52 (1.9 ~ 2.4, 13.74 ~ 17.36)
21.56 ~ 23.52 (2.2 ~ 2.4, 15.91 ~ 17.36)
16.66 ~ 19.60
(1.7 ~ 2.0, 12.30 ~ 14.47)
Torque : N.m (kgf.m, lb-ft)
7.84 ~ 9.80
(0.8 ~ 1.0, 5.78 ~ 7.23)
9.80 ~ 11.76 (1.0 ~ 1.2, 7.23 ~ 8.68)
1. Water pump pulley
2. Water pump
3. Water pump gasket
4. Thermostat
5. Water inlet fitting
6. Gasket
7. O-ring
8. Air vent pipe
9. Hose
37655_SANT_G0276

Fig. 95 Cooling system components—3.3L engine

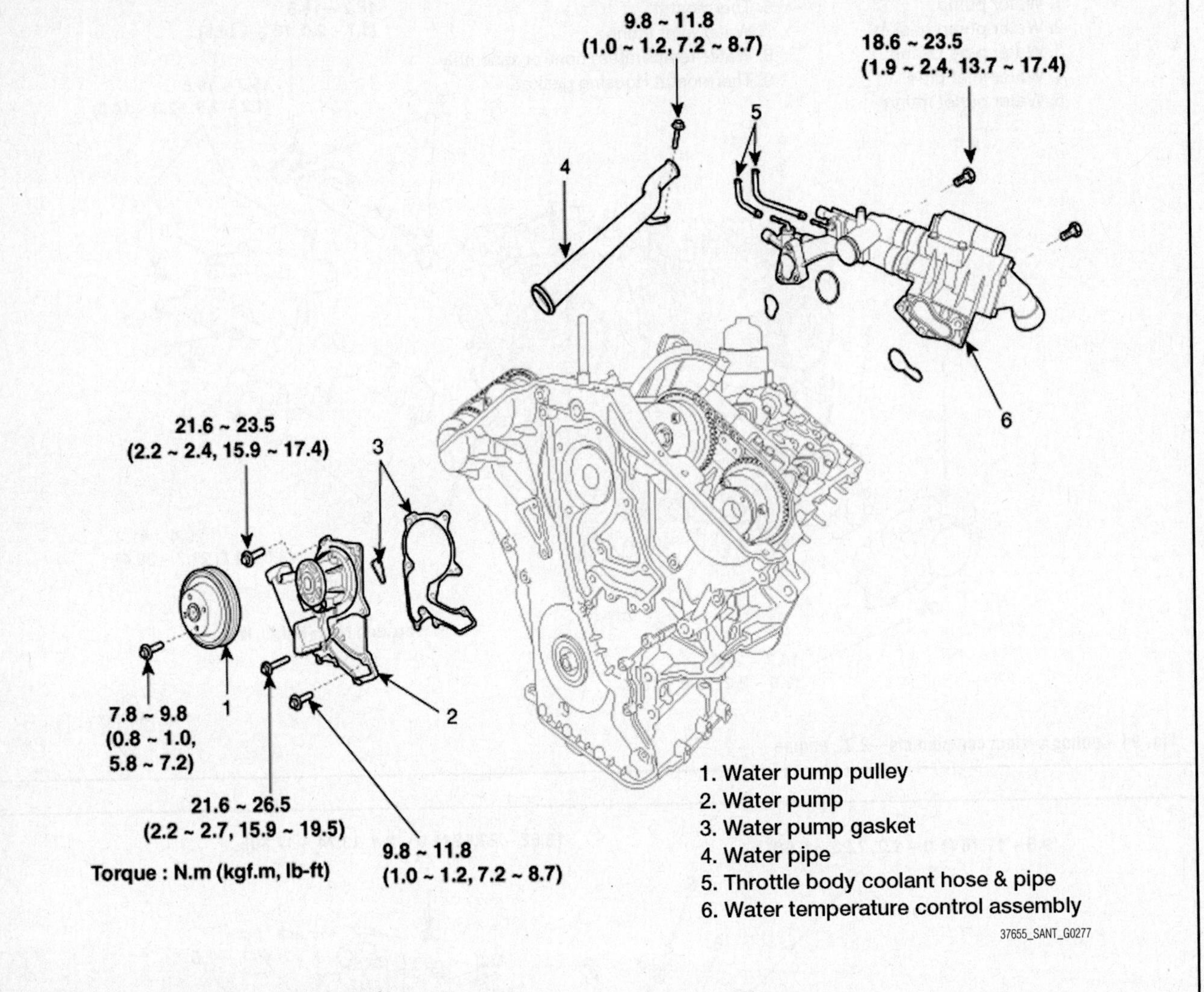

9.8 ~ 11.8
(1.0 ~ 1.2, 7.2 ~ 8.7)
18.6 ~ 23.5
(1.9 ~ 2.4, 13.7 ~ 17.4)
21.6 ~ 23.5
(2.2 ~ 2.4, 15.9 ~ 17.4)
7.8 ~ 9.8
(0.8 ~ 1.0,
5.8 ~ 7.2)
21.6 ~ 26.5
(2.2 ~ 2.7, 15.9 ~ 19.5)
9.8 ~ 11.8
(1.0 ~ 1.2, 7.2 ~ 8.7)
Torque : N.m (kgf.m, lb-ft)
1. Water pump pulley
2. Water pump
3. Water pump gasket
4. Water pipe
5. Throttle body coolant hose & pipe
6. Water temperature control assembly
37655_SANT_G0277

Fig. 96 Cooling system components—3.5L engine

37655_SANT_G0271

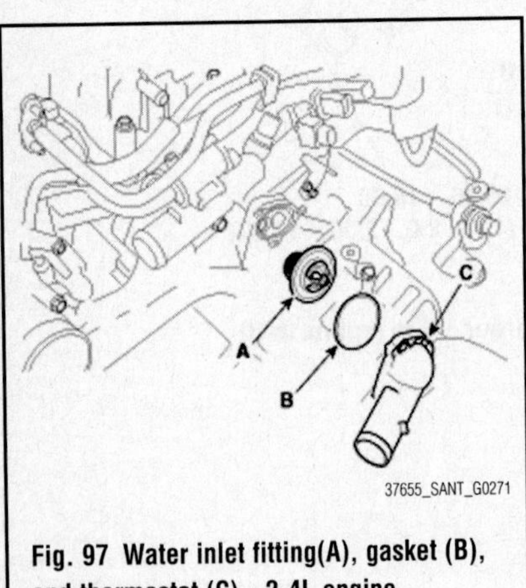

Fig. 97 Water inlet fitting(A), gasket (B), and thermostat (C)—2.4L engine

37655_SANT_G0272

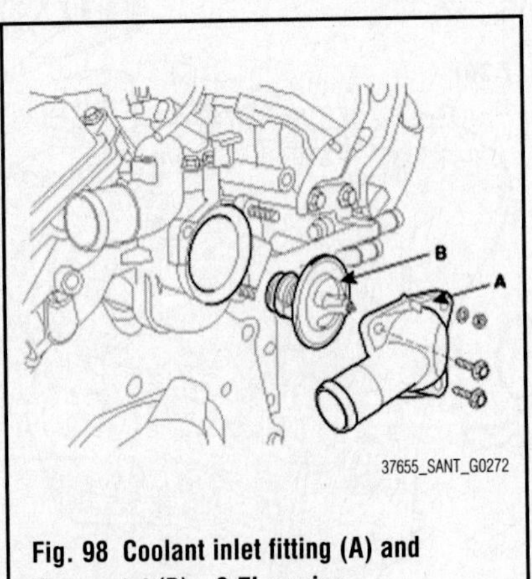

Fig. 98 Coolant inlet fitting (A) and thermostat (B)—2.7L engine

37655_SANT_G0273

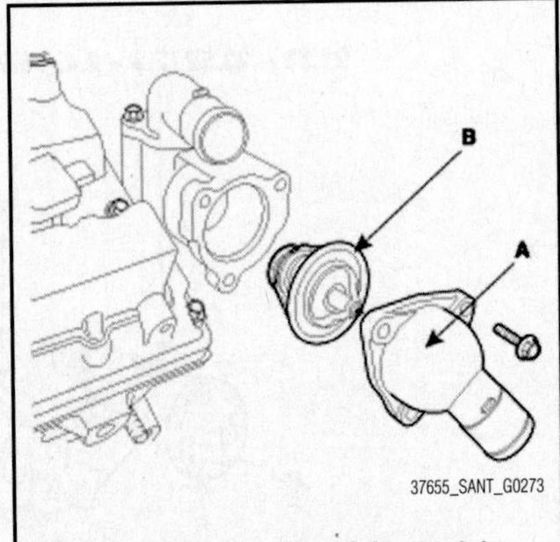

Fig. 99 Water inlet (A) and thermostat (B)—3.3L and 3.5L engines

1. Before servicing the vehicle, refer to the Precautions Section.
2. Drain the coolant down to below thermostat level.
3. Remove the water inlet and gasket.
4. Remove the thermostat.

To install:

a. Install a new thermostat.

b. Install the thermostat with the jiggle valve upward.

c. Tighten the water inlet mounting bolts:

- 2.7L, 3.3L and 3.5L Engine: 12–15 ft. lbs. (17–20 Nm)
- 2.4L Engines: 70–104 inch lbs. (8–12 Nm)

5. Refill the coolant and check for leaks.

WATER PUMP

REMOVAL & INSTALLATION

2009 Models

2.7L Engine

See Figures 100 and 101.

✲✲ CAUTION

The system is under high pressure when the engine is hot. To avoid danger of releasing scalding engine coolant, remove the cap only when the engine is cool.

1. Before servicing the vehicle, refer to the Precautions Section.
2. Drain the engine coolant.
3. Remove the accessory drive belt. Refer to Accessory Drive Belt Removal & Installation in the Engine Mechanical section.
4. Remove the timing belt.
5. Remove the water pump and gasket.

To install:

➡Clean the contact face before assembly.

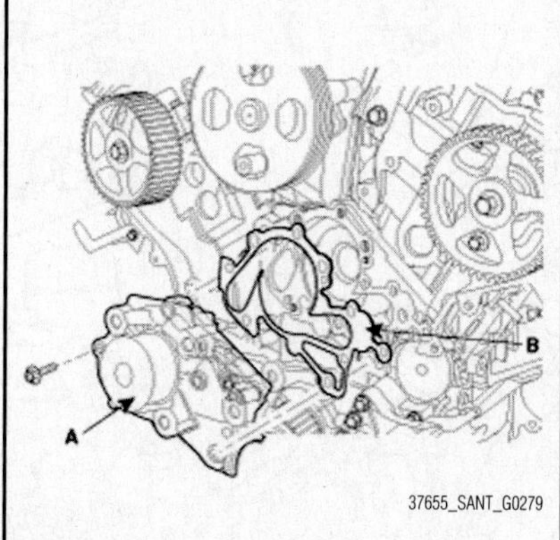

Fig. 101 Remove the water pump (A) and gasket (B)—2.7L engine

6. Install the water pump and a new gasket with the bolts. Tighten to 11–16 ft. lbs. (15–22 Nm).
7. Install the timing belt.
8. Install the accessory drive belt.
9. Fill with engine coolant.
10. Start engine and check for leaks.
11. Recheck engine coolant level.

3.3L Engine

See Figures 102 through 104.

✲✲ CAUTION

The system is under high pressure when the engine is hot. To avoid danger of releasing scalding engine coolant, remove the cap only when the engine is cool.

1. Before servicing the vehicle, refer to the Precautions Section.
2. Drain the engine coolant.
3. Remove the accessory drive belt.

1. Water pump
2. Water pump gasket
3. Water pipe O-ring
4. Water inlet pipe
5. Water outlet fitting
6. Thermostat
7. Water inlet fitting
8. Water temperature control assembly
9. Thermostat Housing gasket

16.7 ~ 19.6
(1.7 ~ 2.0, 12.3 ~ 14.5)
16.7 ~ 19.6
(1.7 ~ 2.0, 12.3 ~ 14.5)
29.4 ~ 41.2
(3.0 ~ 4.2, 21.7 ~ 30.4)
Torque : N.m (kgf.m, lb-ft)
14.7 ~ 21.6
(1.5 ~ 2.2, 10.8 ~ 15.9)
37655_SANT_G0275

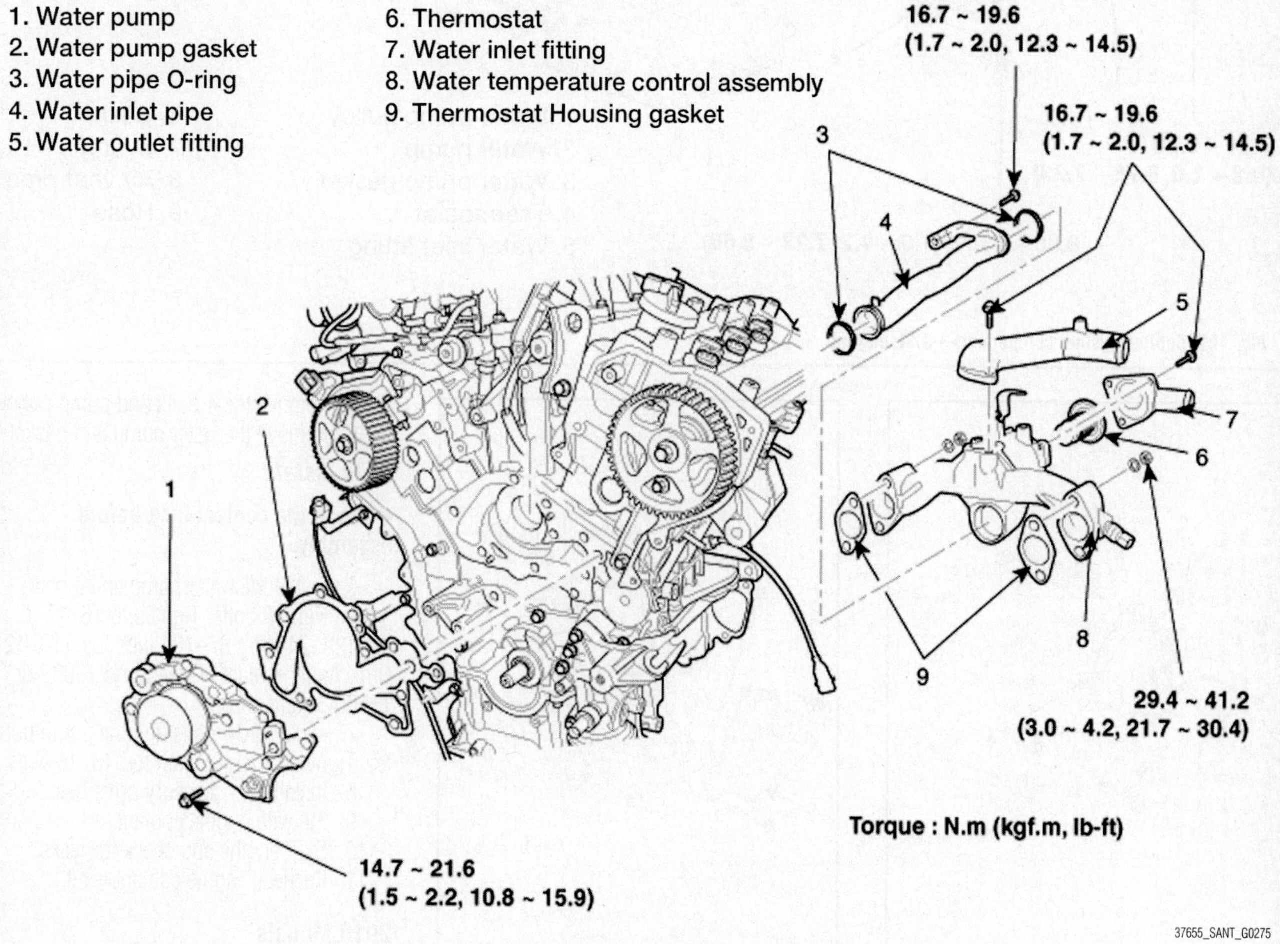

Fig. 100 Cooling system components—2.7L engine

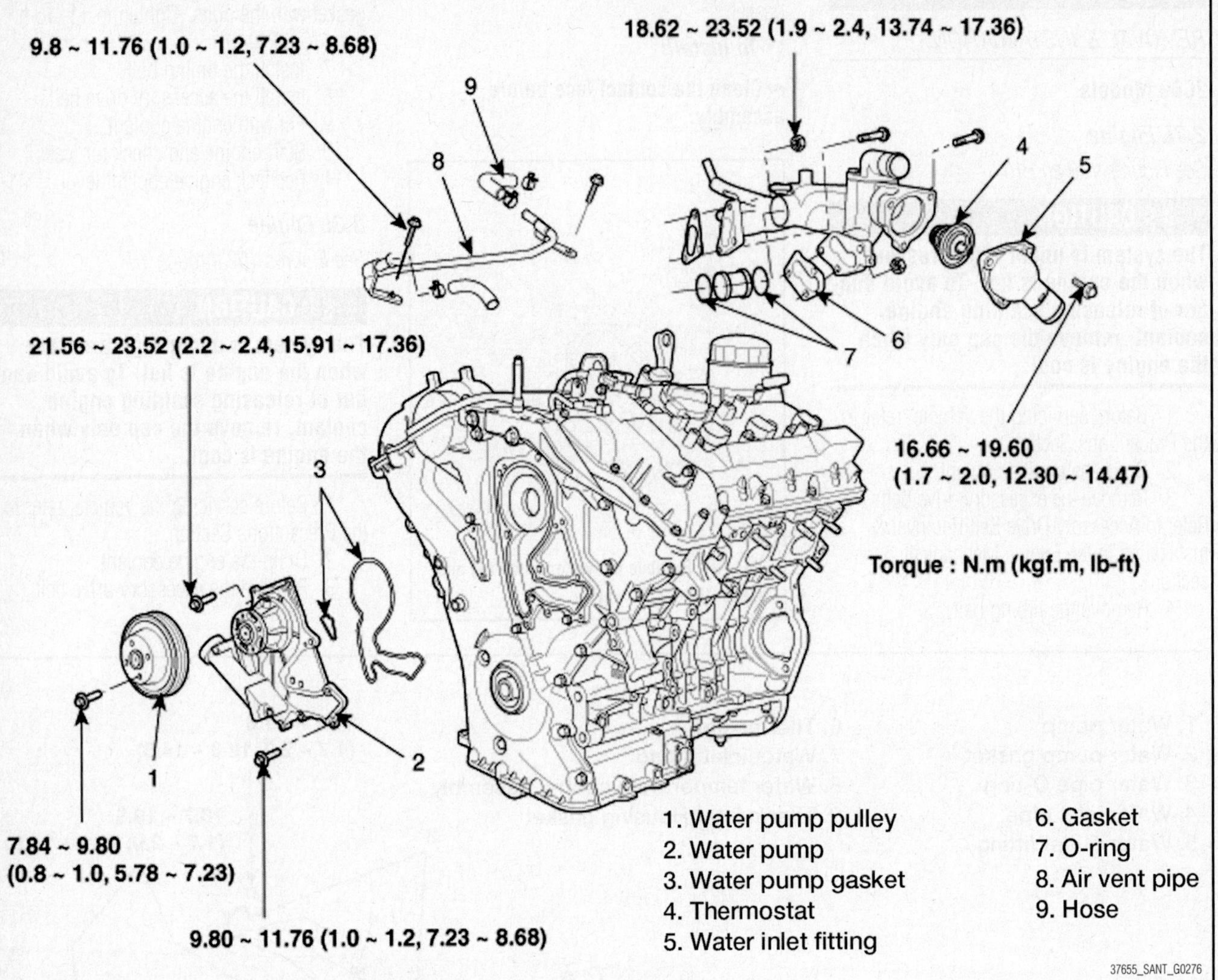

9.8 ~ 11.76 (1.0 ~ 1.2, 7.23 ~ 8.68)
18.62 ~ 23.52 (1.9 ~ 2.4, 13.74 ~ 17.36)
9
8
4
5
6
7
21.56 ~ 23.52 (2.2 ~ 2.4, 15.91 ~ 17.36)
3
16.66 ~ 19.60
(1.7 ~ 2.0, 12.30 ~ 14.47)
Torque : N.m (kgf.m, lb-ft)
1
2
7.84 ~ 9.80
(0.8 ~ 1.0, 5.78 ~ 7.23)
9.80 ~ 11.76 (1.0 ~ 1.2, 7.23 ~ 8.68)
1. Water pump pulley
2. Water pump
3. Water pump gasket
4. Thermostat
5. Water inlet fitting
6. Gasket
7. O-ring
8. Air vent pipe
9. Hose
37655_SANT_G0276

Fig. 102 Cooling system components—3.3L engine

A
37655_SANT_G0280

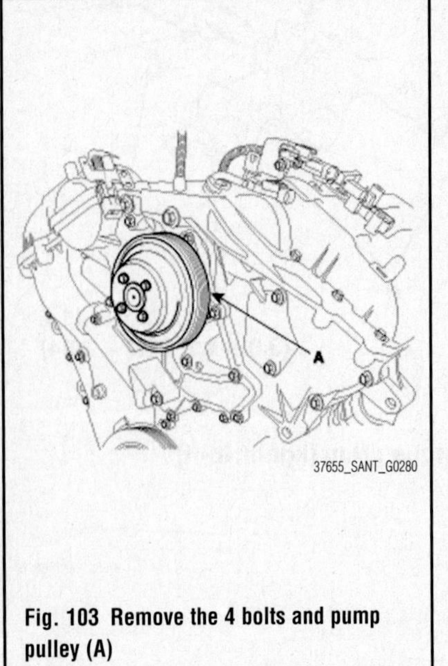

Fig. 103 Remove the 4 bolts and pump pulley (A)

B
C
A
37655_SANT_G0281

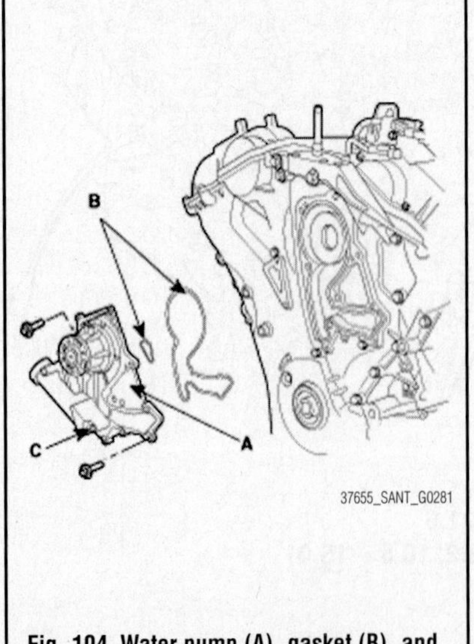

Fig. 104 Water pump (A), gasket (B), and sealant application location (C)

4. Remove the 4 bolts and pump pulley.
5. Remove the water pump and gasket.

To install:

➡Clean the contact face before assembly.

6. Install the water pump and a new gasket with 12 bolts. Tighten to 16–17 ft. lbs. (22–24 Nm); 87–104 inch lbs. (10–12 Nm). Apply sealant (Three Bond 13868 or 13860) where shown.
7. Install the 4 bolts and the pump pulley. Tighten to 70–86 inch lbs. (8–10 Nm).
8. Install the accessory drive belt.
9. Fill with engine coolant.
10. Start engine and check for leaks.
11. Recheck engine coolant level.

2010 Models

2.4L Engine

See Figures 105 and 106.

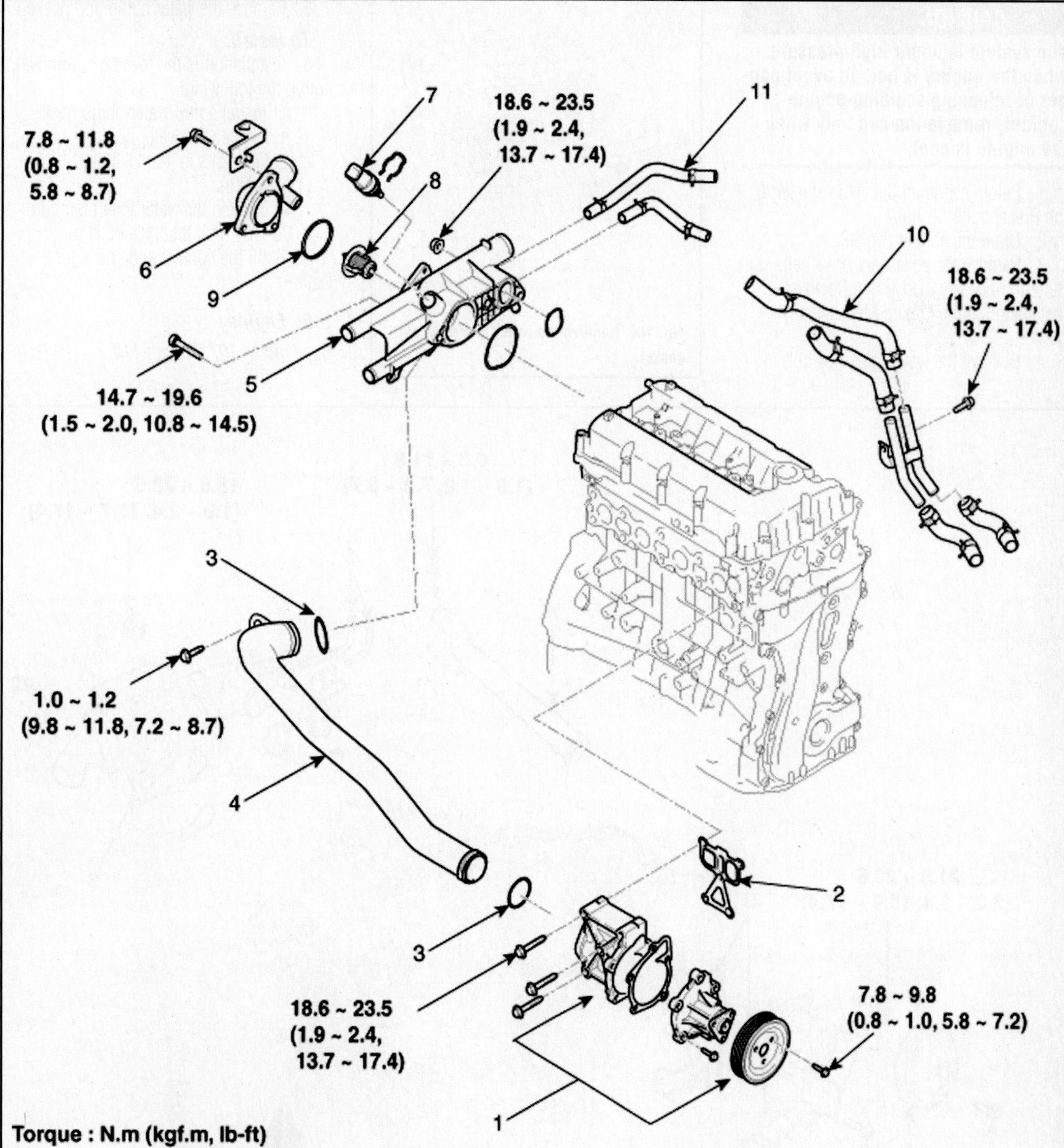

Fig. 105 Cooling system components—2.4L engine

⁂ CAUTION

The system is under high pressure when the engine is hot. To avoid danger of releasing scalding engine coolant, remove the cap only when the engine is cool.

1. Before servicing the vehicle, refer to the Precautions Section.
2. Drain the engine coolant.
3. Remove the accessory drive belt. Refer to Accessory Drive Belt Removal & Installation in the Engine Mechanical section.
4. Remove the water inlet pipe bolt.

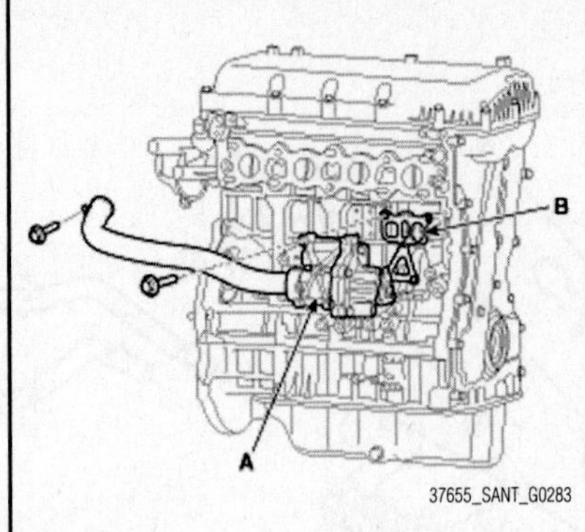

Fig. 106 Remove the water pump (A) and gasket

5. Remove the water pump and gasket.

To install:

6. Installation is the reverse of removal, noting the following:
   a. Install a new water pump gasket.
   b. Tighten the water pump mounting bolts to 14–17 ft. lbs. (19–24 Nm).
   c. Tighten the water inlet pipe bolt to 86–104 inch lbs. (10–12 Nm).
7. Refill the coolant and check for leaks.

3.5L Engine

See Figures 107 through 110.

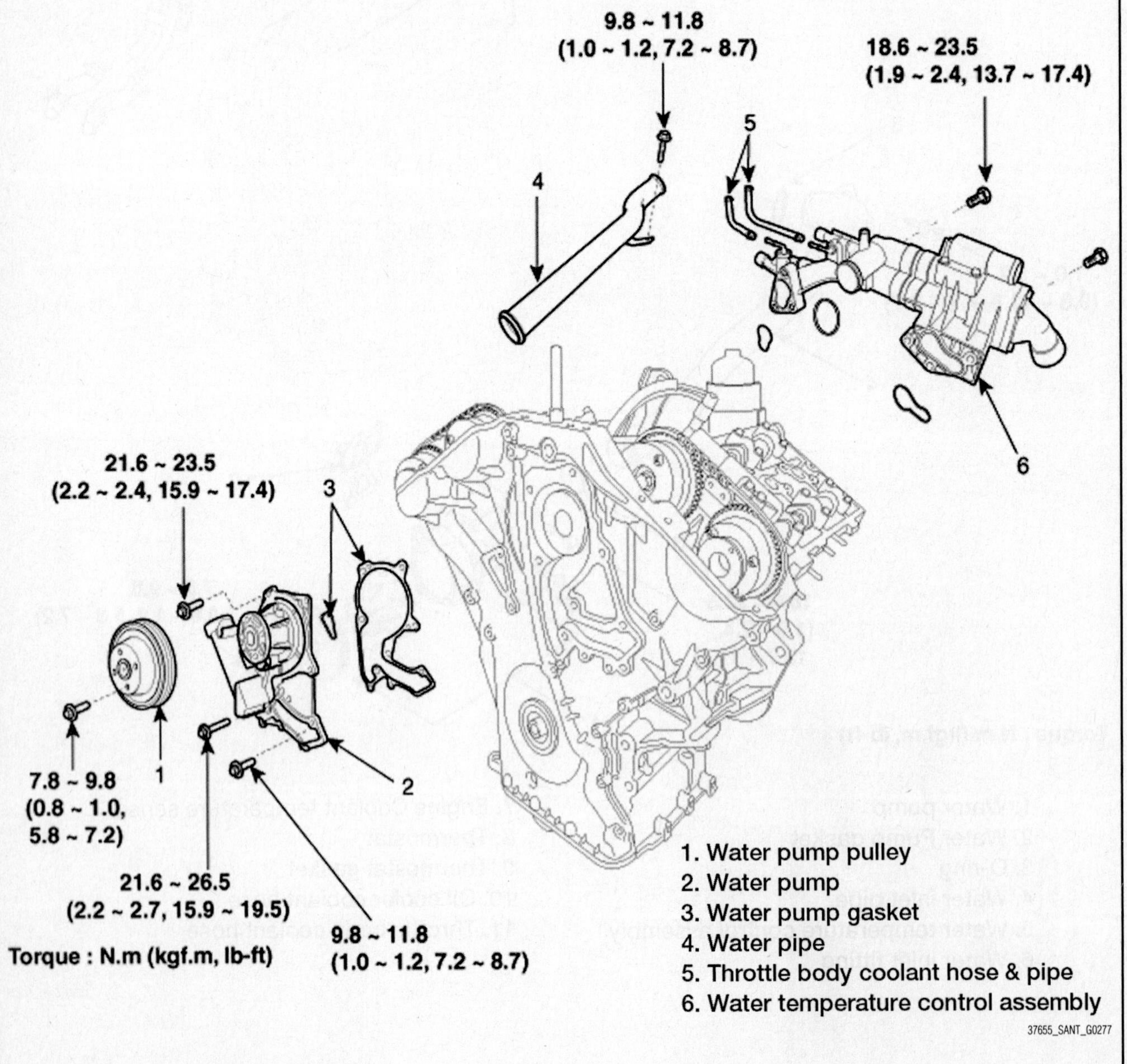

9.8 ~ 11.8
(1.0 ~ 1.2, 7.2 ~ 8.7)
18.6 ~ 23.5
(1.9 ~ 2.4, 13.7 ~ 17.4)
5
4
6
21.6 ~ 23.5
(2.2 ~ 2.4, 15.9 ~ 17.4)
3
7.8 ~ 9.8
(0.8 ~ 1.0,
5.8 ~ 7.2)
1
2
21.6 ~ 26.5
(2.2 ~ 2.7, 15.9 ~ 19.5)
9.8 ~ 11.8
(1.0 ~ 1.2, 7.2 ~ 8.7)
Torque : N.m (kgf.m, lb-ft)

1. Water pump pulley
2. Water pump
3. Water pump gasket
4. Water pipe
5. Throttle body coolant hose & pipe
6. Water temperature control assembly

37655_SANT_G0277

Fig. 107 Cooling system components—3.5L engine

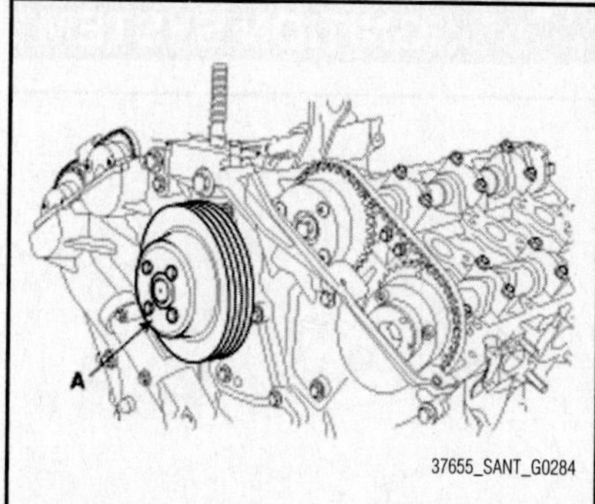

Fig. 108 Remove the water pump pulley (A)

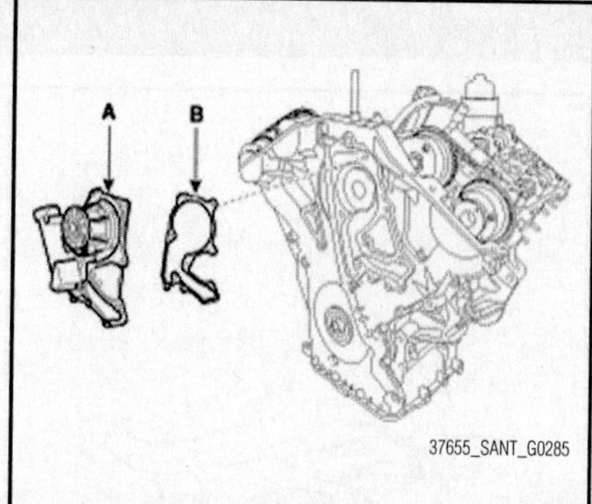

Fig. 109 Remove the water pump (A) and gasket (B)

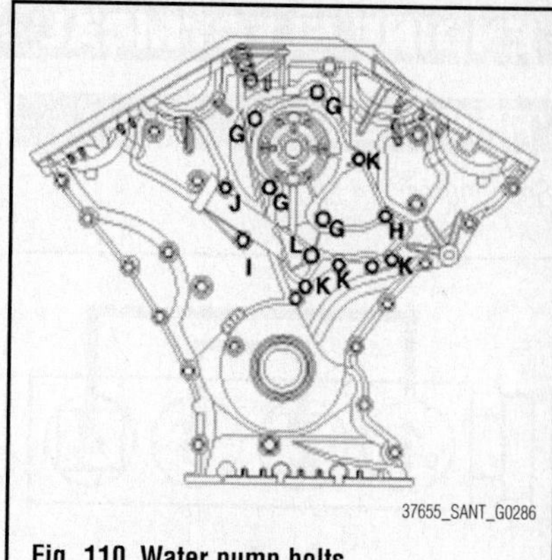

Fig. 110 Water pump bolts

CAUTION

The system is under high pressure when the engine is hot. To avoid danger of releasing scalding engine coolant, remove the cap only when the engine is cool.

1. Before servicing the vehicle, refer to the Precautions Section.
2. Drain the engine coolant.
3. Remove the accessory drive belt. Refer to Accessory Drive Belt Removal & Installation in the Engine Mechanical section.
4. Remove the water pump pulley.
5. Remove the water pump and gasket.

To install:

➡Clean the contact face before assembly.

6. Install the water pump with a new gasket. Tighten torques:
   - G (Qty 4): 16–17 ft. lbs. (22–24 Nm)
   - H (Qty 1): 86–104 inch lbs. (10–12 Nm)
   - I (Qty 1): 86–104 inch lbs. (10–12 Nm)
   - J (Qty 1): 86–104 inch lbs. (10–12 Nm)
   - K (Qty 4): 86–104 inch lbs. (10–12 Nm)
   - L (Qty 1): 16–20 ft. lbs. (22–27 Nm), New bolt
7. Tighten the water pump pulley bolts to 70–86 inch lbs. (8–10 Nm).
8. Install the drive belt.
9. Refill the coolant and check for leaks.

ENGINE ELECTRICAL

CHARGING SYSTEM

ALTERNATOR

REMOVAL & INSTALLATION

See Figures 111 through 113.

1. Before servicing the vehicle, refer to the Precautions Section.
2. Disconnect the negative battery terminal, then the positive terminal.
3. Disconnect the electrical connectors from the alternator.
4. Remove the accessory drive belt. Refer to Accessory Drive Belts, Removal & Installation, in the Engine Mechanical section.
5. Remove the alternator mounting bolts.
6. Remove the alternator.

To install:

7. Installation is the reverse of removal. Tighten the alternator mounting bolts:
   - 2.4L engine: 36–47 ft. lbs. (49–64 Nm)
   - 3.3L and 3.5L engines: 20–25 ft. lbs. (27–33 Nm)

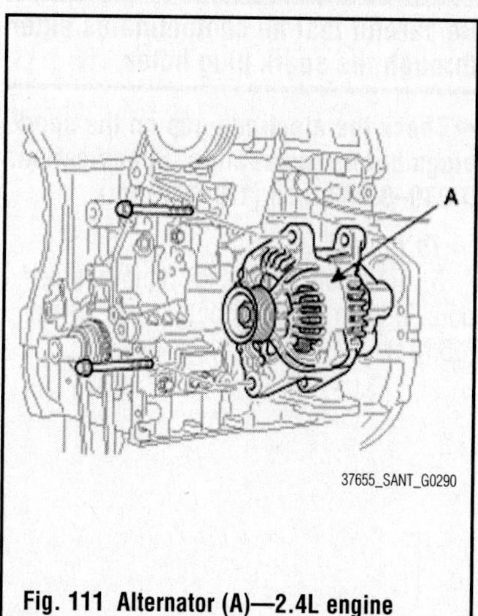

Fig. 111 Alternator (A)—2.4L engine

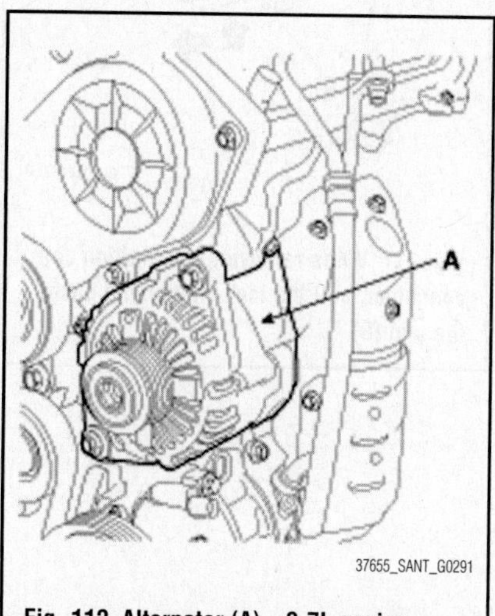

Fig. 112 Alternator (A)—2.7L engine

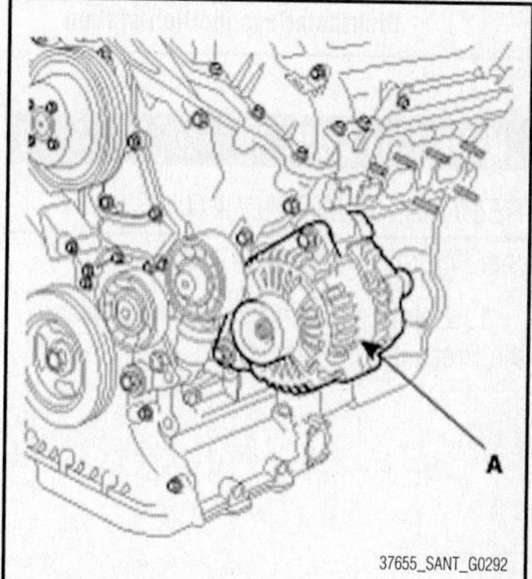

Fig. 113 Alternator (A)—3.3L and 3.5L engines

FIRING ORDER

See Figures 114 and 115.

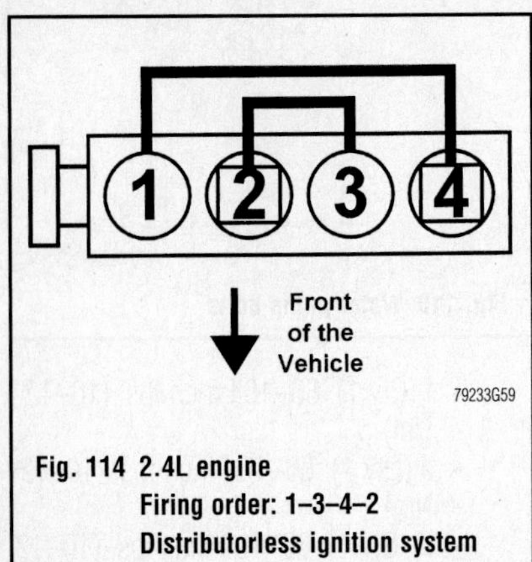

Fig. 114 2.4L engine Firing order: 1–3–4–2 Distributorless ignition system

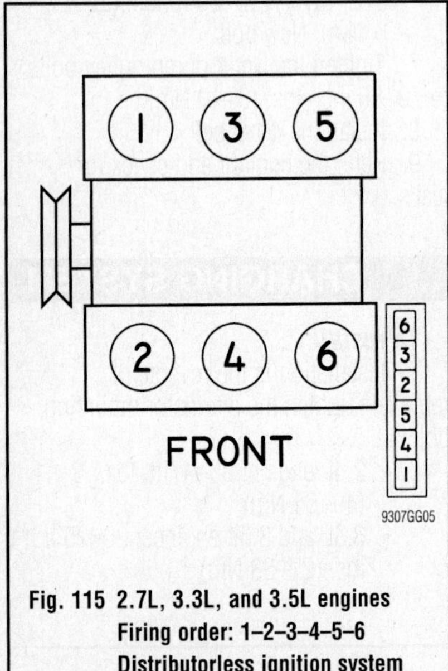

Fig. 115 2.7L, 3.3L, and 3.5L engines Firing order: 1–2–3–4–5–6 Distributorless ignition system

IGNITION COIL

REMOVAL & INSTALLATION

See Figures 116 through 118.

1. Before servicing the vehicle, refer to the Precautions Section.

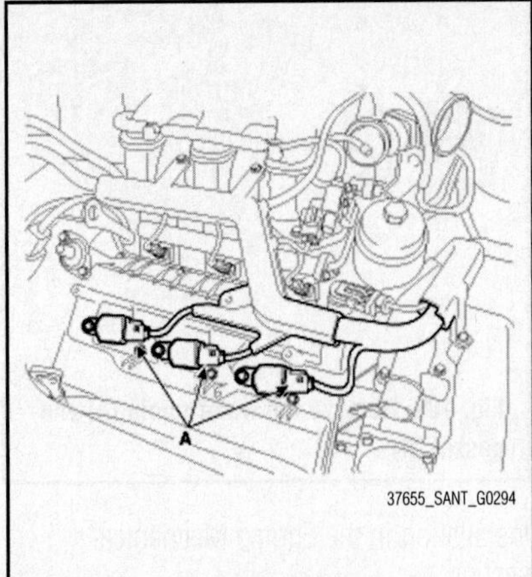

Fig. 116 Disconnect the ignition coil connector (A)—3.5L engine shown, others similar

2. Remove the engine cover (as necessary).
3. Disconnect the ignition coil connector.

➡**When removing the ignition coil connector, pull the lock pin and push the clip.**

4. Remove the ignition coil.

To install:

5. Installation is the reverse of the removal procedure.

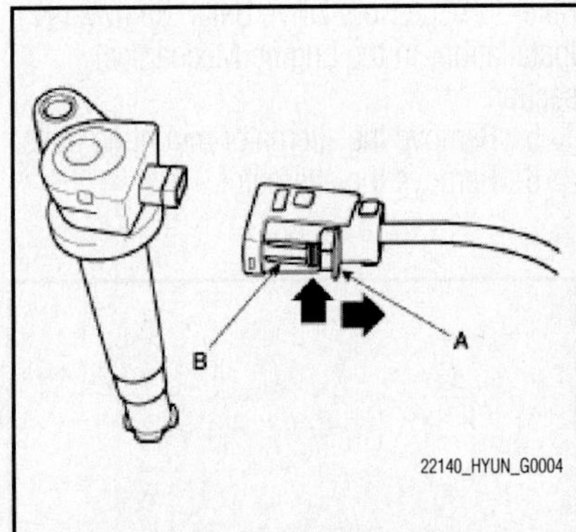

Fig. 117 When removing the ignition coil connector, pull the lock pin (A) and push the clip (B)

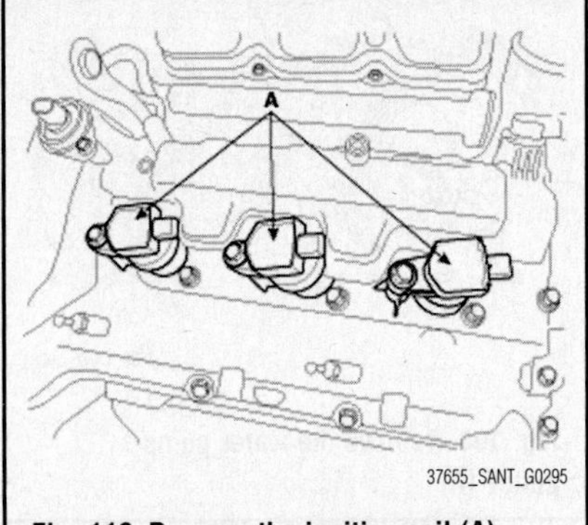

Fig. 118 Remove the ignition coil (A)—3.5L engine shown, others similar

IGNITION TIMING

ADJUSTMENT

Ignition timing is controlled by the electronic control ignition timing system. The standard reference ignition timing data for the engine operating conditions are pre-programmed in the memory of the Engine Control Module (ECM). No adjustment is necessary.

SPARK PLUGS

REMOVAL & INSTALLATION

1. Before servicing the vehicle, refer to the Precautions Section.
2. Remove the engine cover (as necessary).
3. Remove the ignition coil.
4. Use a spark plug socket and wrench to remove the spark plugs.

⁂ WARNING

Be careful that no contaminates enter through the spark plug holes.

➡**Check the electrode gap on the spark plugs before installation. Specification: 0.039–0.043 inch (1.0–1.1 mm).**

To install:

5. To install, reverse the removal procedure. Tighten the spark plugs to 11 ft. lbs. (15 Nm).

ENGINE ELECTRICAL

STARTER

REMOVAL & INSTALLATION

2009 Models

See Figure 119.

1. Before servicing the vehicle, refer to the Precautions Section.
2. Disconnect the negative battery cable.
3. Remove the starter cover.
4. Disconnect the starter cable from the B terminal on the solenoid, then disconnect the connector from the S terminal.
5. Remove the starter mounting bolts.
6. Remove the starter.

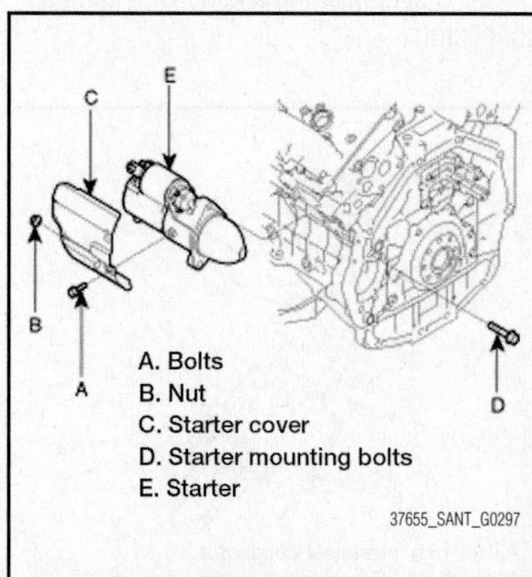

Fig. 119 Remove the starter cover and starter—2.7L and 3.3L engines

To install:

7. Installation is the reverse of removal.

2010 Models

See Figure 120.

1. Before servicing the vehicle, refer to the Precautions Section.
2. Disconnect the negative battery cable.

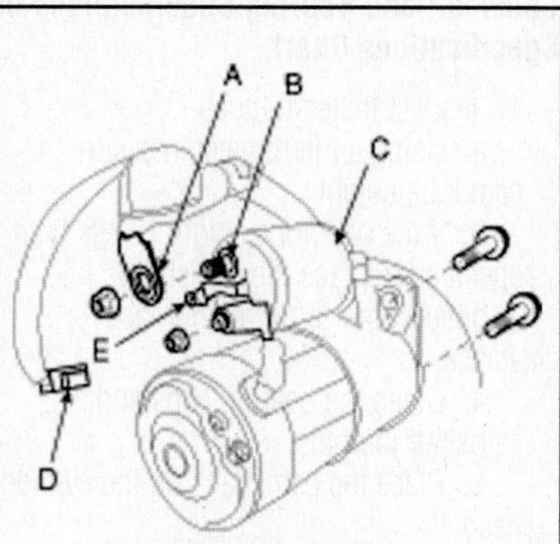

Fig. 120 Remove the starter wiring—2.4L and 3.5L engines

3. Disconnect the starter cable from the B terminal on the solenoid, then disconnect the connector from the S terminal.
4. Remove the starter mounting bolts.
5. Remove the starter.

To install:

6. Installation is the reverse of removal.

SOLENOID OR RELAY REPLACEMENT

See Figure 121.

1. Remove the fuse box cover.
2. Remove the starter relay (A).
3. Install a new starter relay.
4. Replace the fuse box cover.

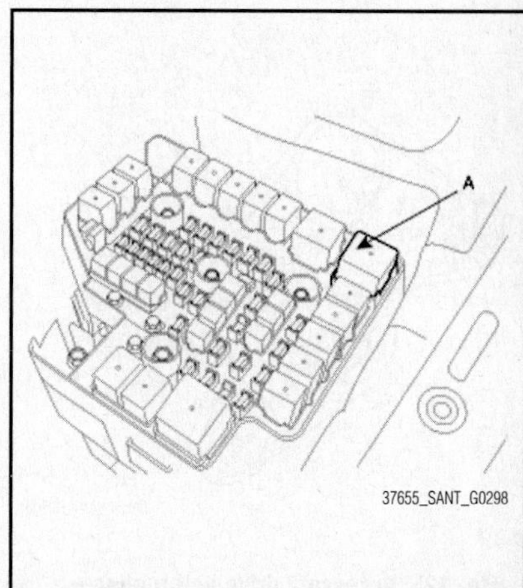

Fig. 121 Starter relay position in the fuse box.

ENGINE MECHANICAL

➡Disconnecting the negative battery cable may interfere with the functions of the on board computer systems and may require the computer to undergo a relearning process, once the negative battery cable is reconnected.

ACCESSORY DRIVE BELTS

ACCESSORY BELT ROUTING

See Figures 122 through 125.

INSPECTION

Inspect the accessory drive belt for signs of glazing or cracking. A glazed belt will be perfectly smooth from slippage, while a good belt will have a slight texture of fabric visible. Cracks will usually start at the inner edge of the belt and run outward. All worn or damaged accessory drive belts should be replaced immediately.

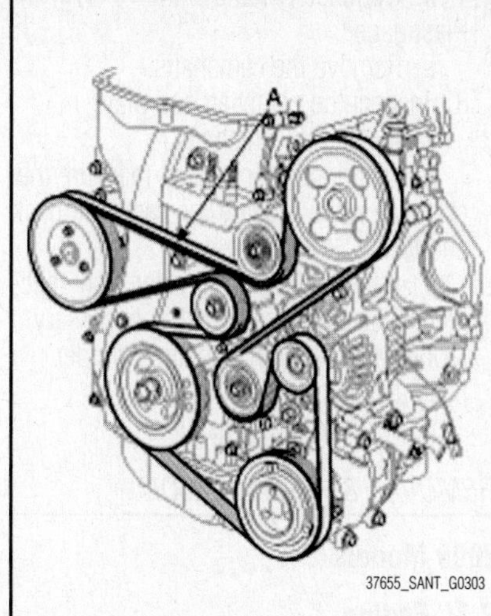

Fig. 122 Accessory drive belt routing—2.4L engine

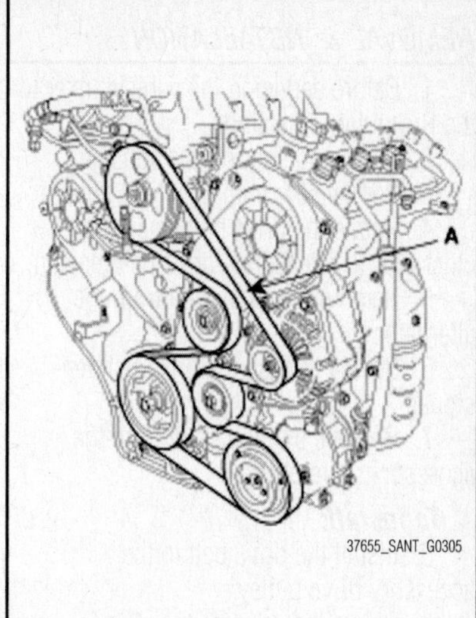

Fig. 123 Accessory drive belt routing—2.7L engine

37655_SANT_G0304

Fig. 124 Accessory drive belt routing—3.3L engine

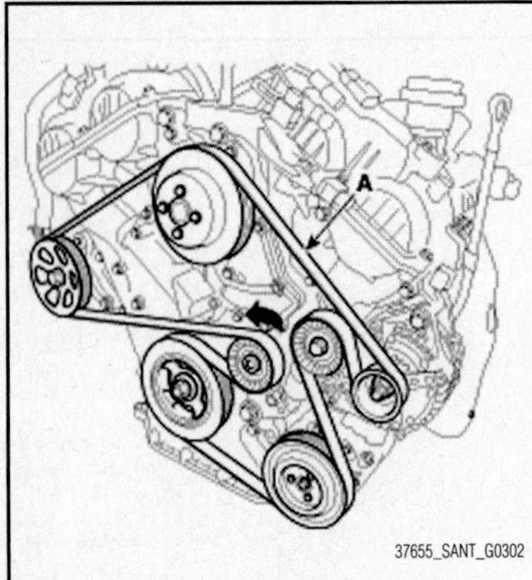

Fig. 125 Accessory drive belt routing—3.5L engine

ADJUSTMENT

The drive belt tension is provided through a drive belt tensioner.

REMOVAL & INSTALLATION

1. Before servicing the vehicle, refer to the Precautions Section.
2. Raise and support the vehicle.
3. Remove the engine splash shield.
4. Rotate the drive belt tensioner clockwise to release the drive belt tension.
5. Remove the drive belt from the alternator.
6. Slowly release the drive belt tensioner.
7. Remove the drive belt from the accessory drive pulleys.

To install:

8. Install the drive belt to the accessory drive pulley.
9. Rotate the drive belt tensioner clockwise.
10. Install the drive belt to the alternator.
11. Ensure the drive belt is properly aligned and seated into the grooves of the accessory drive pulleys.
12. Slowly release the drive belt tensioner.
13. Install the engine splash shield.
14. Lower the vehicle.

CAMSHAFT AND VALVE LIFTERS

INSPECTION

➡For camshaft specifications, refer to Camshaft and Bearing Specifications in Specifications Chart.

1. Inspect the cam lobes.
   a. Using a micrometer, measure the cam lobe height.
   b. If the cam lobe height is less than specified, replace the camshaft.
2. Inspect the camshaft journal clearance.
   a. Clean the bearing caps and camshaft journals.
   b. Place the camshafts on the cylinder head.
   c. Lay a strip of Plastigage®across each of the camshaft journal.
   d. Install the bearing caps and tighten the bolts with specified torque.

➡Do not turn the camshaft.

   e. Remove the bearing caps.
   f. Measure the Plastigage® at its widest point.
   g. If the oil clearance is greater than specified, replace the camshaft. If necessary, replace the bearing caps and cylinder head as a set.
   h. Completely remove the Plastigage®.
   i. Remove the camshafts.
3. Inspect the camshaft end play.
   a. Install the camshafts.
   b. Using a dial indicator, measure the end play while moving the camshaft back and forth.
   c. If the end play is greater than specified, replace the camshaft. If necessary, replace the bearing caps and cylinder head as a set.
   d. Remove the camshafts.

REMOVAL & INSTALLATION

2009 Models

2.7L Engine

See Figures 126 through 137.

Use fender covers to avoid damaging painted surfaces. To avoid damage, unplug the wiring connectors carefully while holding the connector portion. Mark all wiring and hoses to avoid misconnection. Inspect the timing belt before removing. Turn the crankshaft pulley so that the No. 1 piston is at Top Dead Center (TDC).

1. Before servicing the vehicle, refer to the Precautions Section.
2. Remove the air duct.
3. Disconnect the battery terminals and remove the battery.
4. Drain engine coolant. Remove the radiator cap to speed draining.
5. Remove the air cleaner assembly.
6. Remove the upper and lower radiator hoses.
7. Remove the fuel inlet hose from the delivery pipe.
8. Disconnect the engine wire harness connectors:

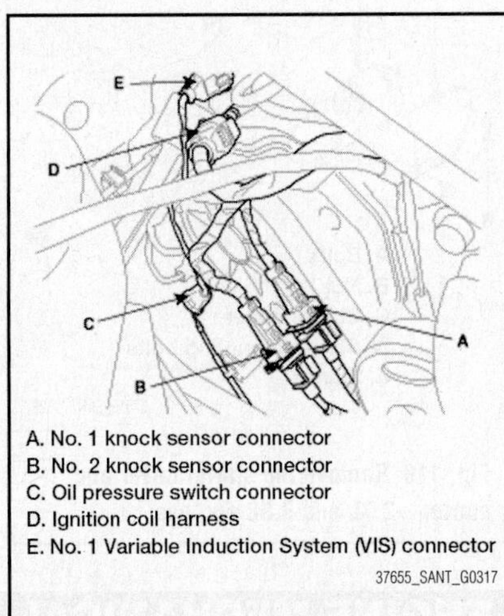

Fig. 126 Disconnect the knock sensor connectors, oil pressure switch connector, ignition coil harness, and the No. 1 VIS connector

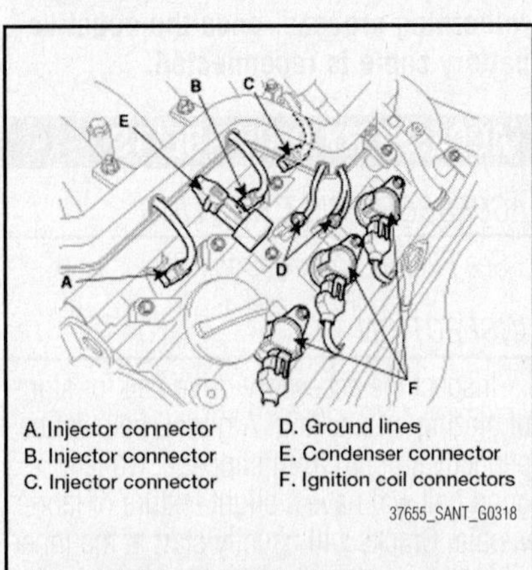

Fig. 127 Disconnect the injector connectors, ground lines, condenser connector, and the ignition coil connectors

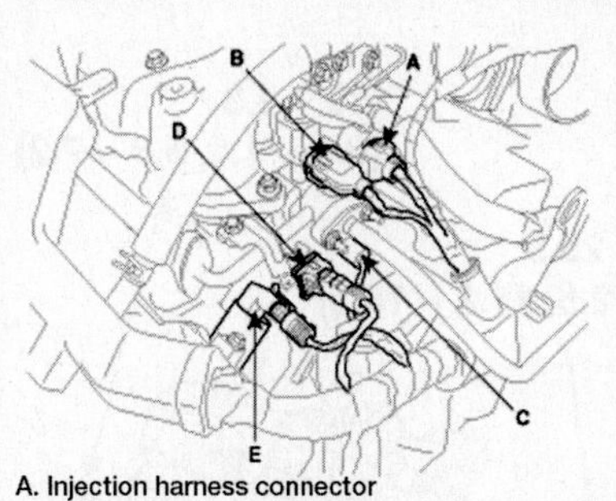

Fig. 128 Disconnect the injection harness connector, No. 2 Variable Induction System (VIS) connector, No. 1/No. 2 Oil Control Valve (OCV) connectors, and the Oil Temperature Sensor (OTS) connector

a. Disconnect the knock sensor connectors, the oil pressure switch connector, the ignition coil harness, and the No. 1 Variable Induction System (VIS) connector.

b. Disconnect the Bank 1 Front/Rear O2 sensor connectors.

c. Disconnect the injector connectors, the ground lines, the condenser connector and the ignition coil connectors.

d. Disconnect the injection harness connector, the No. 2 Variable Induction System (VIS) connector, the No. 1/No. 2 Oil Control Valve (OCV) connectors, and the Oil Temperature Sensor (OTS) connector.

e. Disconnect the Manifold Absolute Pressure (MAP) sensor connector, the Electronic Throttle Control (ETC) connector, and the Purge Control Solenoid Valve (PCSV) connector.

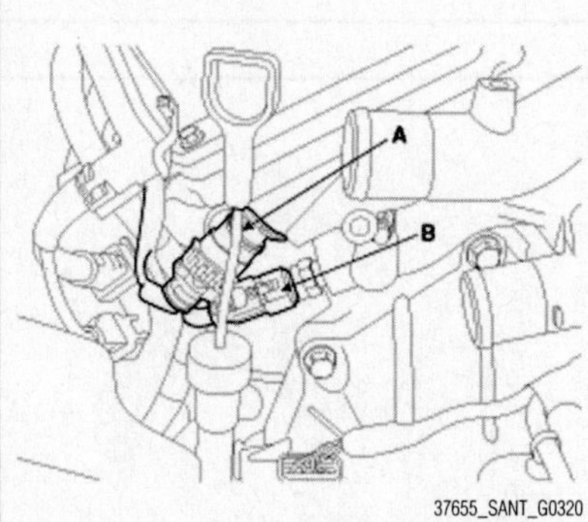

Fig. 129 Disconnect the Bank 2 CMP sensor connector (A) and the Engine Coolant Temperature (ECT) sensor connector (B)

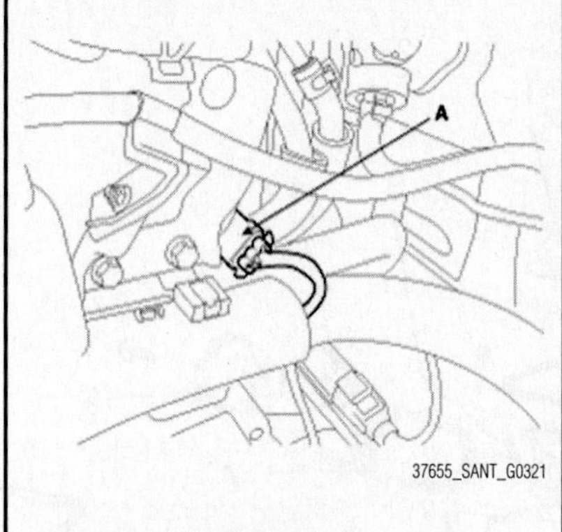

Fig. 130 Disconnect the Bank 1 CMP sensor connector (A)

f. Disconnect the alternator connector and the air conditioning compressor connector.

g. Disconnect the Bank 2 CMP sensor connector and the Engine Coolant Temperature (ECT) sensor connector.

h. Disconnect the Bank 2 Front/Rear O2 sensor connectors and the CKP sensor connector.

i. Disconnect the Bank 1 CMP sensor connector.

9. Remove the Purge Control Valve (PCV) hose.
10. Disconnect the brake vacuum hose.
11. Remove the heater hoses.
12. Remove the drive belt. Refer to Accessory Drive Belt Removal & Installation.
13. Remove the power steering pump.
14. Remove the exhaust manifold. Refer to Exhaust Manifold Removal & Installation.
15. Remove the intake manifold. Refer to Intake Manifold Removal & Installation.
16. Remove the timing belt. Refer to Timing Belt & Sprockets, Removal & Installation.
17. Remove the ignition coils.
18. Remove the water temperature control assembly.
19. Remove the cylinder head cover.
20. Remove the camshaft bearing caps.
21. Remove the camshaft timing chain tensioner.
22. Remove the camshafts.

To install:

Thoroughly clean all parts to be assembled. Rotate the crankshaft; set the No. 1 piston at TDC.

23. Install the camshafts:

a. Align the timing marks of the intake and exhaust camshaft chain sprockets and the timing chain.

b. Install the intake and exhaust camshafts on the cylinder head with the timing marks aligned.

24. Install the timing chain tensioner.

a. Insert the set pin by pressing the timing chain tensioner.

b. Install the chain tensioner in the cylinder head assembly.

c. Install the camshaft bearing caps.

- Bolt "A": 96–113 inch lbs. (11–13 Nm)
- Bolt "B": 15–19 ft. lbs. (21–23 Nm)

➡When installing the bearing caps, check the marks and be sure they are positioned properly.

➡When installing the bearing caps, turn the crankshaft to place a piston in the middle of the block because interference between valves and pistons can occur.

25. Using the SST (09214-21000), install the camshaft oil seal.

➡Before installing, apply engine oil. The camshaft cap surface should adhere to the cylinder head assembly.

26. Install the camshaft sprocket.

a. Hold the hexagonal head wrench portion of the camshaft with a wrench, and tighten the camshaft sprocket bolts to 65–80 ft. lbs. (88–108 Nm).

27. If the camshaft is replaced with new one, inspect the valve clearances and then install appropriate MLA tappet.
28. Install the cylinder head cover.
29. Install the water temperature control assembly.
30. Install the ignition coils.
31. Install the timing belt.
32. Install the intake manifold.
33. Install the exhaust manifold.
34. Install the power steering pump.
35. Install the drive belt.
36. Install the heater hoses.
37. Install the brake vacuum hose.
38. Install the Purge Control Valve (PCV) hose.
39. Install the engine wire harness connectors.
40. Install the fuel inlet hose.
41. Install the radiator hoses.
42. Install the air cleaner assembly.
43. Connect the battery terminals.
44. Install the battery.
45. Fill with engine coolant.
46. Install the side cover and the engine cover.
47. Start the engine and check for leaks.
48. Recheck the engine coolant level and oil level.

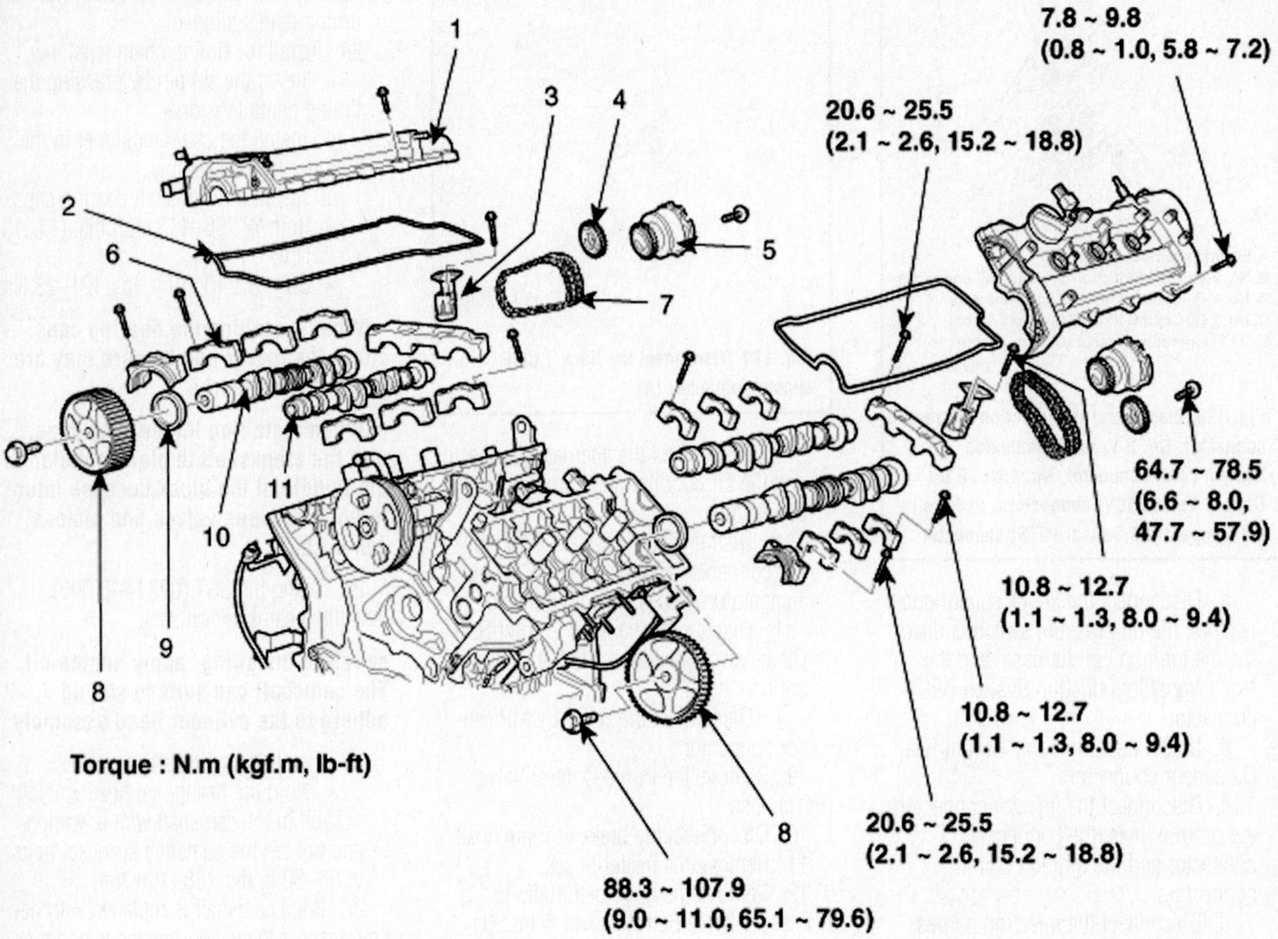

1
2
3
4
5
6
7
7.8 ~ 9.8
(0.8 ~ 1.0, 5.8 ~ 7.2)
20.6 ~ 25.5
(2.1 ~ 2.6, 15.2 ~ 18.8)
64.7 ~ 78.5
(6.6 ~ 8.0, 47.7 ~ 57.9)
10.8 ~ 12.7
(1.1 ~ 1.3, 8.0 ~ 9.4)
10.8 ~ 12.7
(1.1 ~ 1.3, 8.0 ~ 9.4)
20.6 ~ 25.5
(2.1 ~ 2.6, 15.2 ~ 18.8)
88.3 ~ 107.9
(9.0 ~ 11.0, 65.1 ~ 79.6)
8
9
10
11
8

Torque : N.m (kgf.m, lb-ft)

1. Cylinder head cover
2. Cylinder head cover gasket
3. Timing chain auto tensioner
4. Exhaust camshaft chain sprocket
5. CVVT assembly
6. Camshaft bearing cap
7. Timing chain
8. Exhaust camshaft sprocket
9. Camshaft oil seal
10. Exhaust camshaft
11. Intake camshaft

37655_SANT_G0322

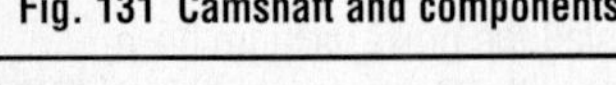

Fig. 131 Camshaft and components

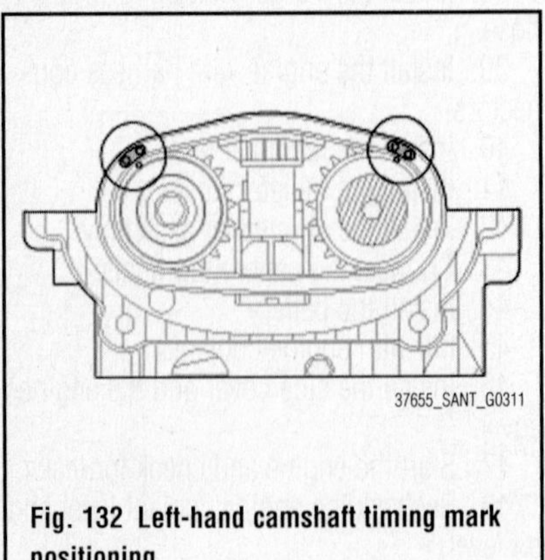

Fig. 132 Left-hand camshaft timing mark positioning

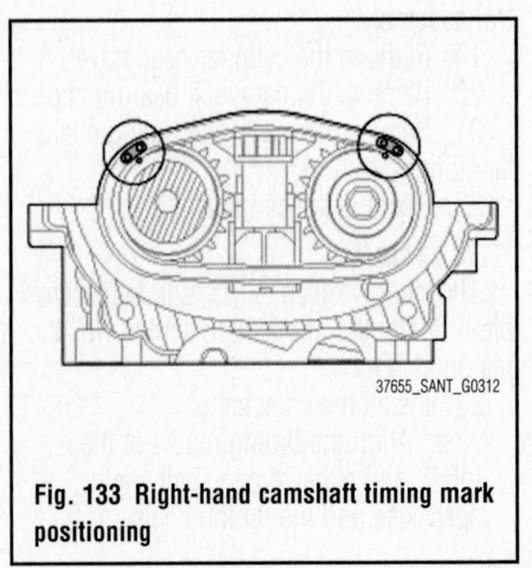

Fig. 133 Right-hand camshaft timing mark positioning

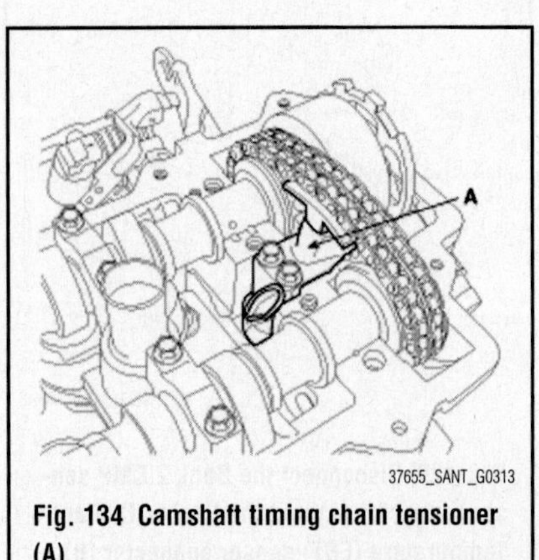

Fig. 134 Camshaft timing chain tensioner (A)

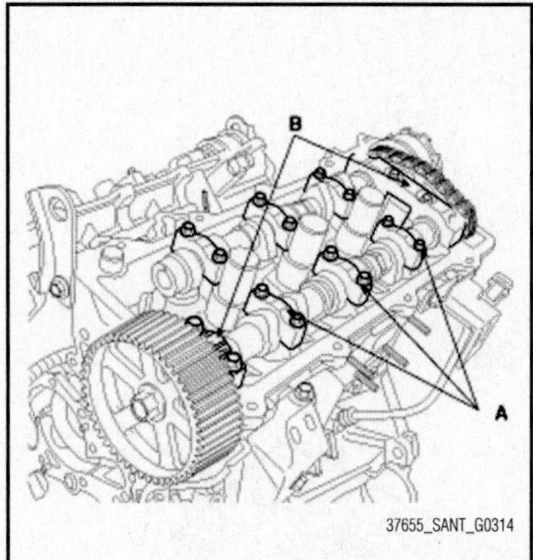

Fig. 135 Install the camshaft bearing caps (A, B)

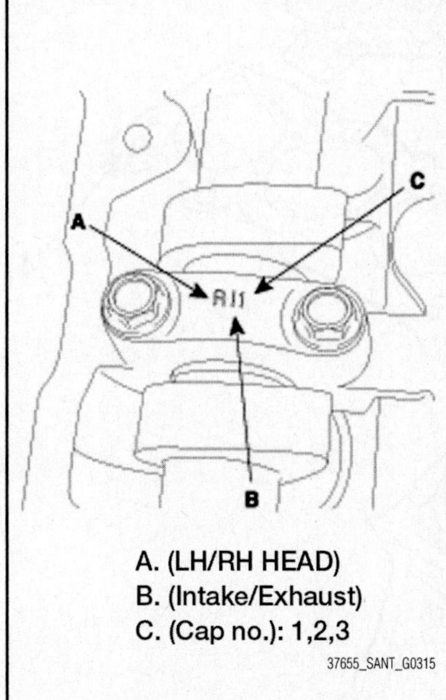

Fig. 136 Camshaft bearing cap positioning

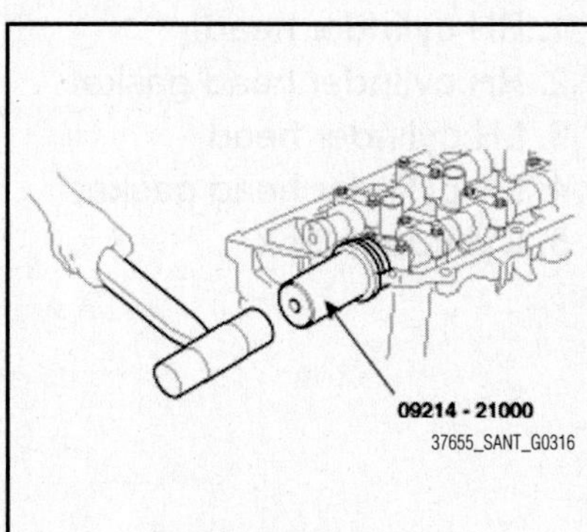

Fig. 137 Using the SST (09214-21000), install the camshaft oil seal

3.3L Engine

See Figures 138 through 143.

Engine removal is required for this procedure. Use fender covers to avoid damaging painted surfaces. To avoid damage, unplug the wiring connectors carefully while holding the connector portion. Mark all wiring and hoses to avoid misconnection. Inspect the timing belt before removing. Turn the crankshaft pulley so that the No. 1 piston is at Top Dead Center (TDC).

1. Before servicing the vehicle, refer to the Precautions Section.
2. Disconnect the negative battery cable.

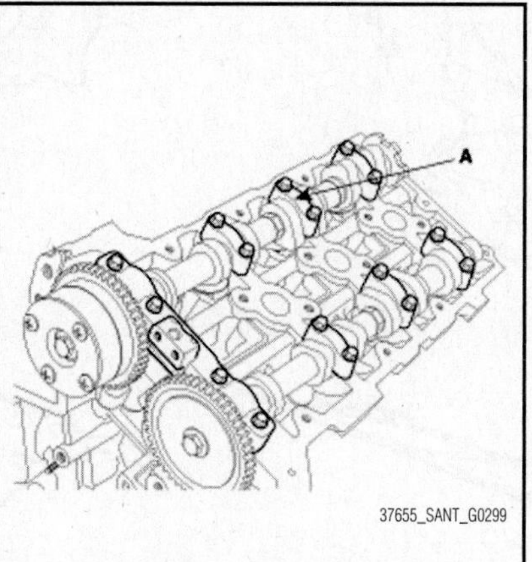

Fig. 138 Remove the camshaft bearing caps (A)

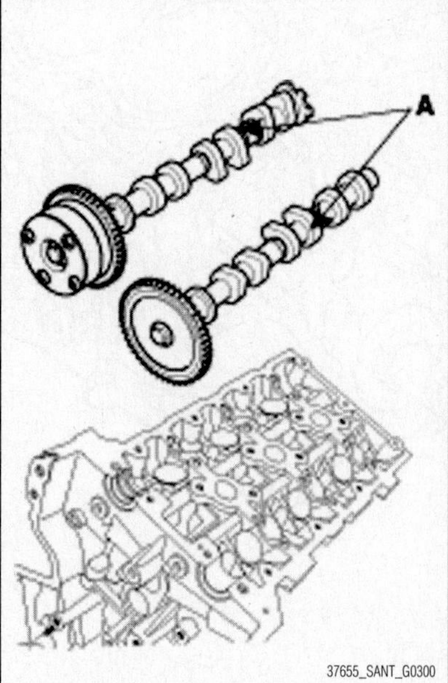

Fig. 139 Remove the camshaft assembly (A)

3. Remove the exhaust manifold. Refer to Exhaust Manifold Removal & Installation.
4. Remove the intake manifold. Refer to Intake Manifold Removal & Installation.
5. Remove the timing chain. Refer to Timing Chain & Sprockets Removal & Installation.
6. Remove the water temperature control assembly.
7. Remove the camshaft bearing caps.
8. Remove the camshaft assembly.

To install:

Thoroughly clean all parts to be assembled. Rotate the crankshaft, set the No. 1 piston at TDC.

9. Install the camshafts.
   a. Apply a light coat of engine oil on camshaft journals.
   b. Assemble the key groove of camshaft rear side to the same level of head top surface.
   c. Be careful to get the right bank, left bank, intake side, and exhaust side in the correct position before assembling.
10. Install the camshaft bearing caps in the sequence shown:
    a. Step 1—Tighten to 48 inch lbs. (6 Nm).
    b. Step 2—Tighten to 84–108 inch lbs. (10–12 Nm).

**** WARNING**

Be sure to properly position the right bank, left bank, intake side, exhaust side, and front mark on the camshaft bearing caps while assembling.

**** WARNING**

Rotate the crankshaft so as not to contact the valves to the pistons by positioning the pistons 0.3937 in. (10mm) from the top of the cylinder block.

11. Install the water temperature control assembly.
12. Install the timing chain.
13. Check and adjust the valve clearance, as necessary.
14. Install the intake manifold.
15. Install the exhaust manifold.
16. Connect the negative battery cable.
17. Fill with engine coolant.
18. Start the engine and check for leaks.
19. Recheck the engine coolant level and oil level.

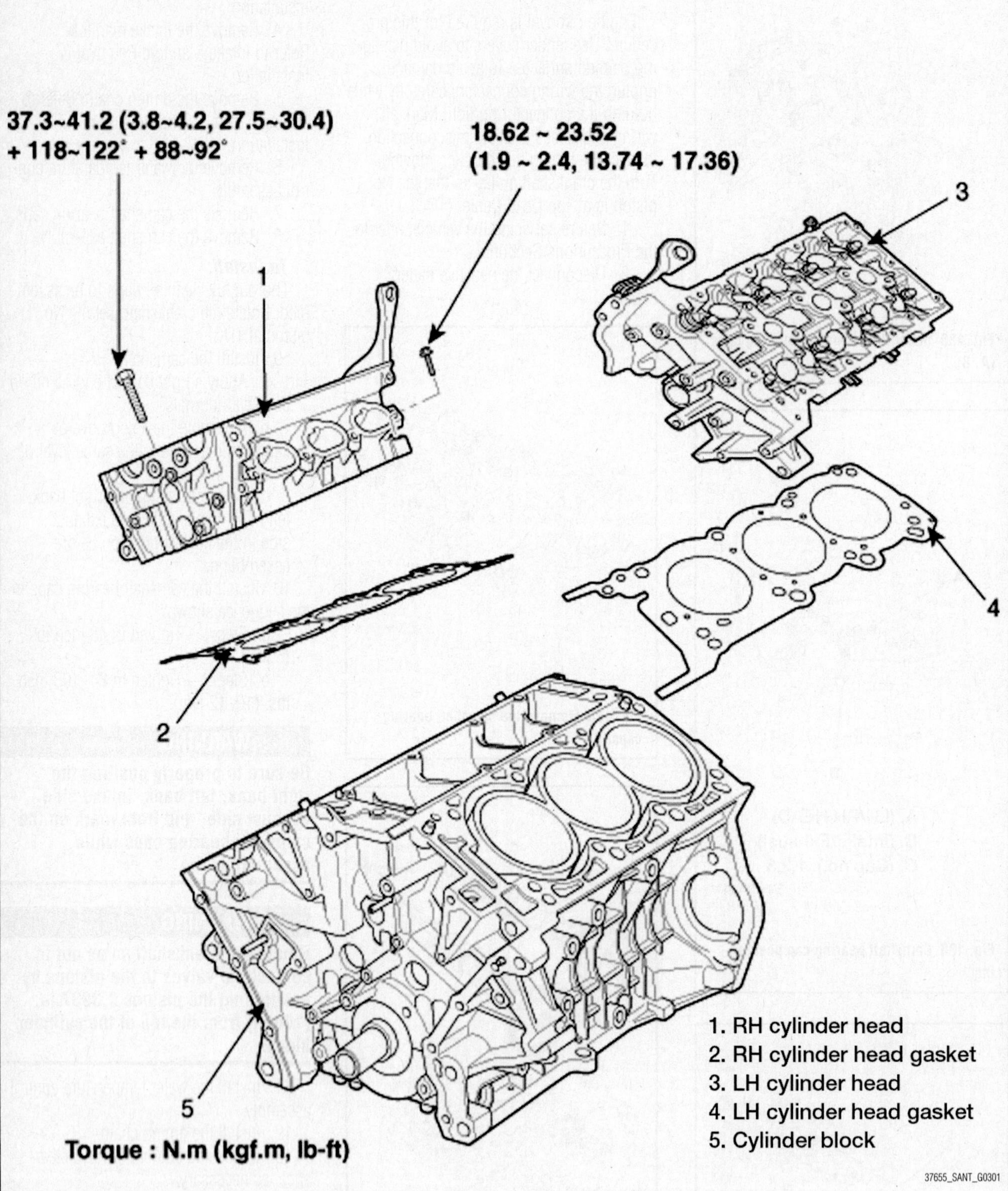

Fig. 140 Exploded view of cylinder head and engine block—3.3L engine

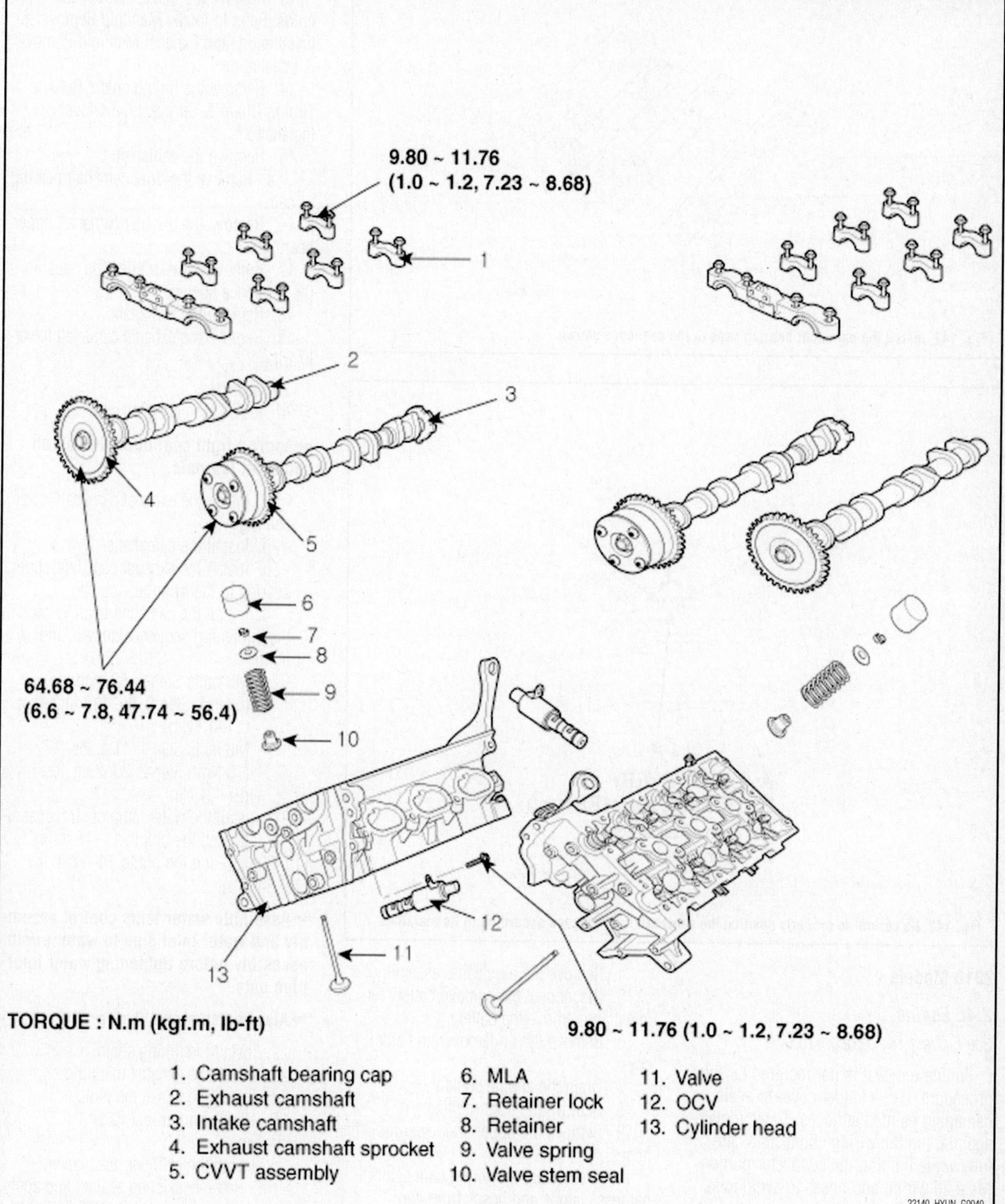

Fig. 141 Exploded view of cylinder head and related components—3.3L engine

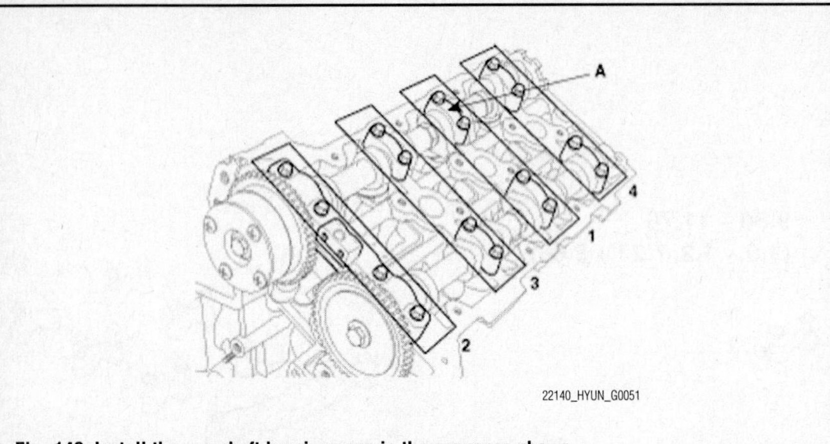

Fig. 142 Install the camshaft bearing caps in the sequence shown

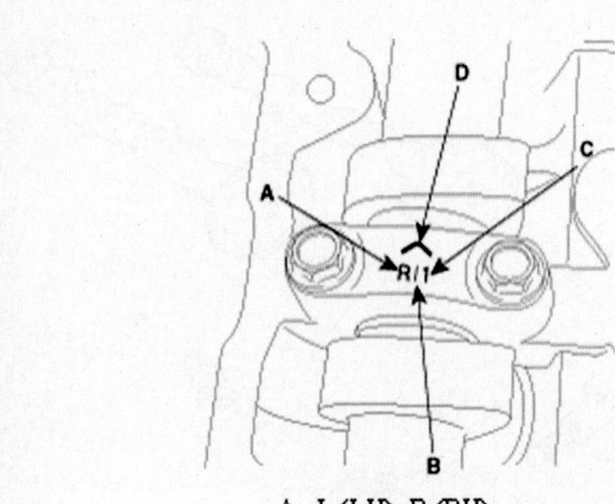

Fig. 143 Be careful to properly position the camshaft bearing caps according to its markings

2010 Models

2.4L Engine

See Figures 144 through 153.

Engine removal is not required for this procedure. Use a fender cover to avoid damaging painted surfaces. To avoid damage, unplug the wiring connectors carefully while holding the connector portion. Mark all wiring and hoses to avoid misconnection. Turn the crankshaft pulley so that the No. 1 piston is at Top Dead Center (TDC).

1. Before servicing the vehicle, refer to the Precautions Section.
2. Disconnect the negative battery cable.
3. Remove the engine cover.
4. Remove the air cleaner assembly.
5. Disconnect the positive battery terminal and remove the battery.
6. Remove the undercover and side cover.
7. Drain the engine coolant.
8. Remove the front muffler.
9. Remove the upper and lower radiator hoses.
10. Disconnect the wiring connectors, harness clamps and hoses from the cylinder head, intake manifold and exhaust manifold.
11. Disconnect the fuel hose and the heater hoses.
12. Disconnect the throttle body coolant hoses, oil cooler coolant hoses and then remove the water temperature control assembly.
13. Remove the intake and exhaust manifolds. Refer to Intake Manifold Removal & Installation, and Exhaust Manifold Removal & Installation.
14. Remove the timing chain. Refer to Timing Chain & Sprockets, Removal & Installation.
15. Remove the camshaft:
    a. Remove the front camshaft bearing cap.
16. Remove the exhaust camshaft upper bearing.
17. Remove camshaft bearing caps, in the sequence shown.
18. Remove the camshafts.
19. Remove the exhaust camshaft lower bearing.

To install:

20. Install the camshafts:

➡Apply a light coat of engine oil on camshaft journals.

    a. Install the exhaust camshaft lower bearing.
    b. Install the camshafts.
    c. Install the exhaust camshaft upper bearing to the front bearing cap.
    d. Install the camshaft bearing caps in the tightening sequence shown, and tighten:
        - M6 bolts Step 1: 52 inch lbs. (6 Nm), Step 2: 95–112 inch lbs. (11–13 Nm)
        - M8 bolts Step 1: 11 ft. lbs. (15 Nm), Step 2: 20–23 ft. lbs. (27–31 Nm)
21. Install the water temp control assembly, and tighten the bolt to 11–15 ft. lbs. (15–20 Nm) and the nut to 14–17 ft. lbs. (19–24 Nm).

➡Assemble water temp control assembly and water inlet pipe to water pump assembly before tightening water inlet pipe nuts.

➡Always use a new O-ring.

22. Install the timing chain.
23. Install the exhaust manifold.
24. Install the intake manifold.
25. Check and adjust valve clearance.
26. Install the cylinder head cover:
    a. Remove existing sealant and apply new sealant Sealant: LOCTITE® 5900H, Bead width: 0.1 in. (2.5mm)] as shown. After applying sealant, assemble within 5 minutes.

➡The firing and/or blow out test should not be performed within 30 minutes after the cylinder head cover was assembled.

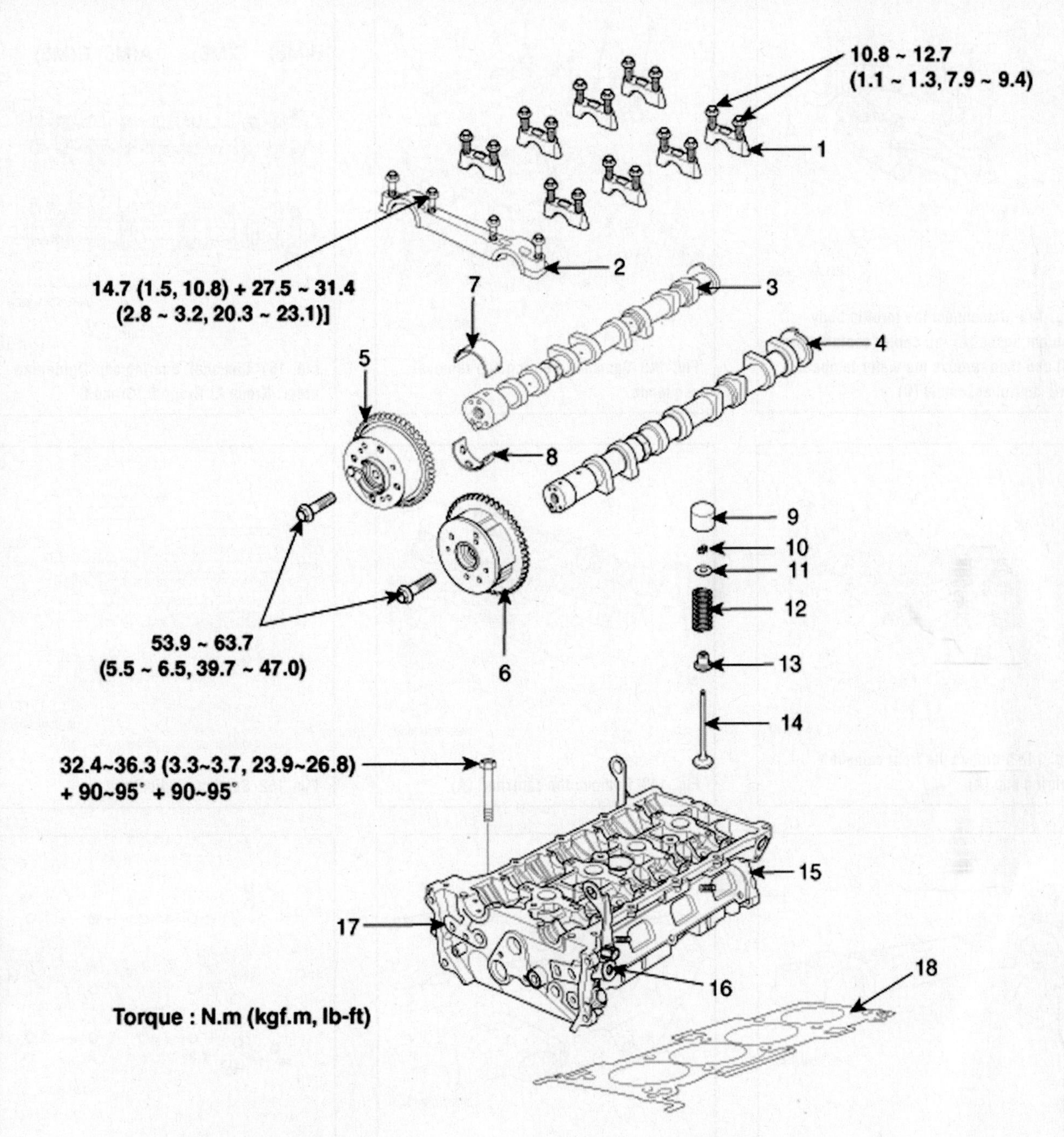

1. Camshaft bearing cap
2. Camshaft front bearing cap
3. Exhaust camshaft
4. Intake camshaft
5. Exhaust CVVT assembly
6. Intake CVVT assembly
7. Exhaust camshaft upper bearing
8. Exhaust camshaft lower bearing
9. MLA
10. Retainer lock
11. Retainer
12. Valve spring
13. Valve stem seal
14. Valve
15. Cylinder head
16. Intake OCV
17. Exhaust OCV
18. Cylinder head gasket

37655_SANT_G0324

Fig. 144 Camshaft components

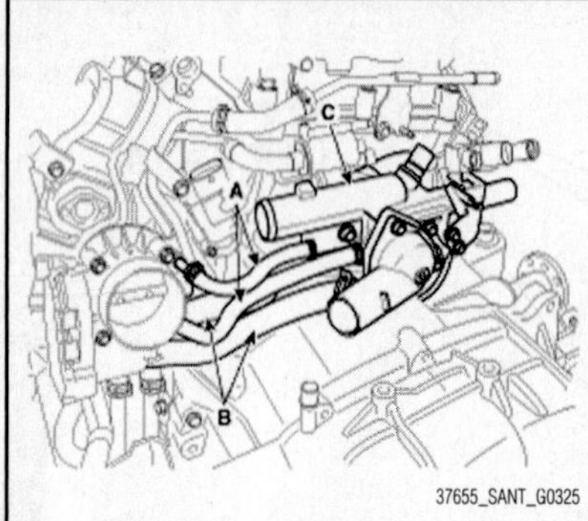

Fig. 145 Disconnect the throttle body coolant hoses (A), oil cooler coolant hoses (B) and then remove the water temperature control assembly (C)

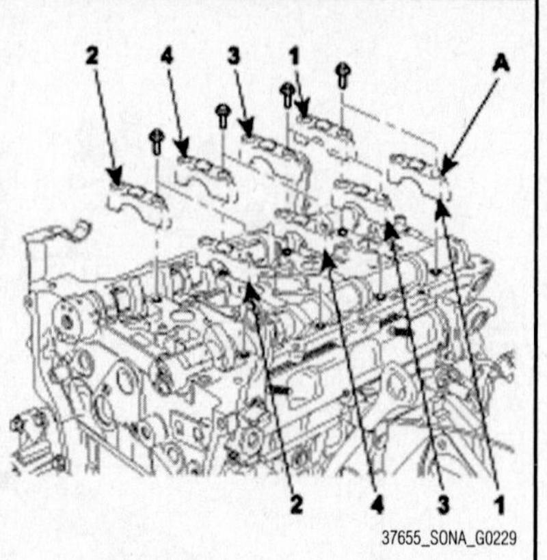

Fig. 148 Camshaft bearing cap removal sequence

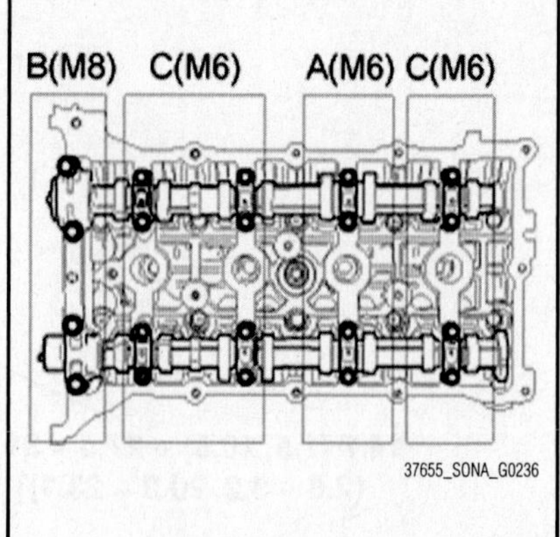

Fig. 151 Camshaft bearing cap Tightening order: Group A, Group B, Group C

Fig. 146 Remove the front camshaft bearing cap (A)

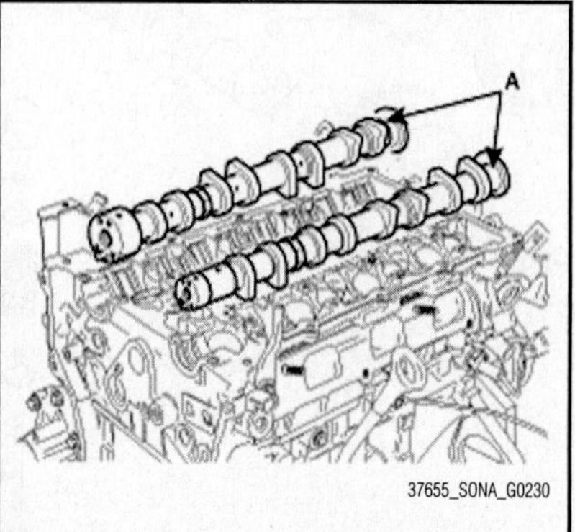

Fig. 149 Remove the camshaft (A)

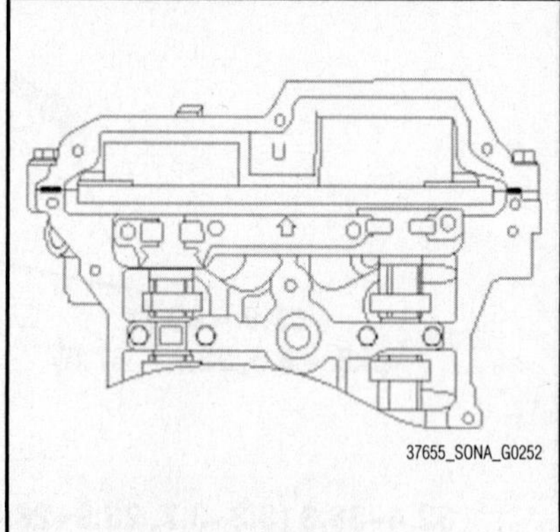

Fig. 152 Sealant application

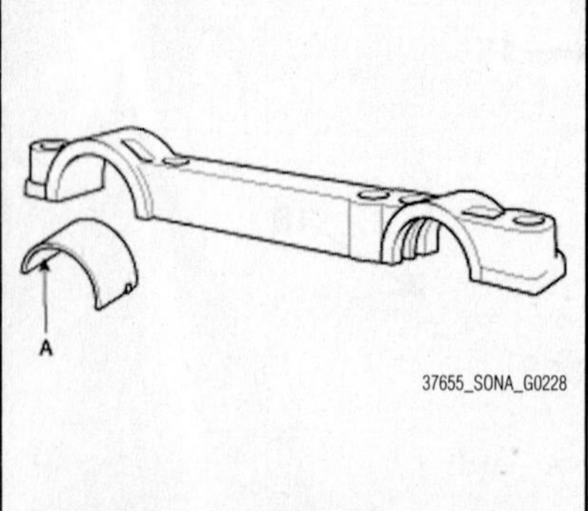

Fig. 147 Remove the exhaust camshaft upper bearing (A)

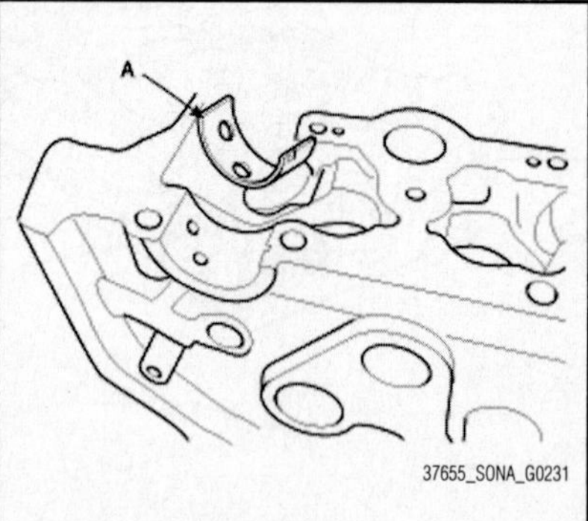

Fig. 150 Remove the exhaust camshaft lower bearing (A)

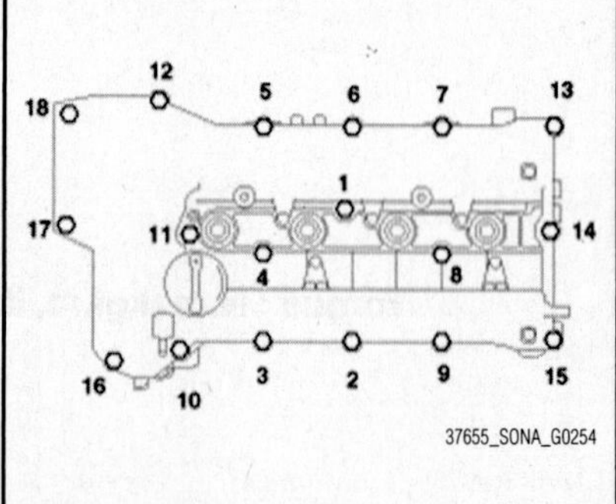

Fig. 153 Cylinder head bolt tightening sequence

b. Install the cylinder head cover bolts in the sequence shown, tightening to:

- Step 1: 35–52 inch lbs. (4–6 Nm)
- Step 2: 70–86 inch lbs. (8–10 Nm)

➡**Do not reuse cylinder head cover gasket.**

27. Connect the throttle body coolant hoses, oil cooler coolant hoses and then remove the water temperature control assembly.

28. Connect the fuel hose and the heater hoses.

29. Connect the wiring connectors, harness clamps and hoses from the cylinder head, intake manifold and exhaust manifold.

30. Install the upper and lower radiator hoses.

31. Install the front muffler.
32. Install the undercover and side cover.
33. Install the air cleaner assembly.
34. Install the engine cover.
35. Connect the battery cables.
36. Install the battery.
37. Fill with engine coolant.
38. Start the engine and check for leaks.
39. Recheck the engine coolant level and oil level.

3.5L Engine

See Figures 154 through 160.

Engine removal is not required for this procedure. Use a fender cover to avoid damaging painted surfaces. To avoid damage, unplug the wiring connectors carefully

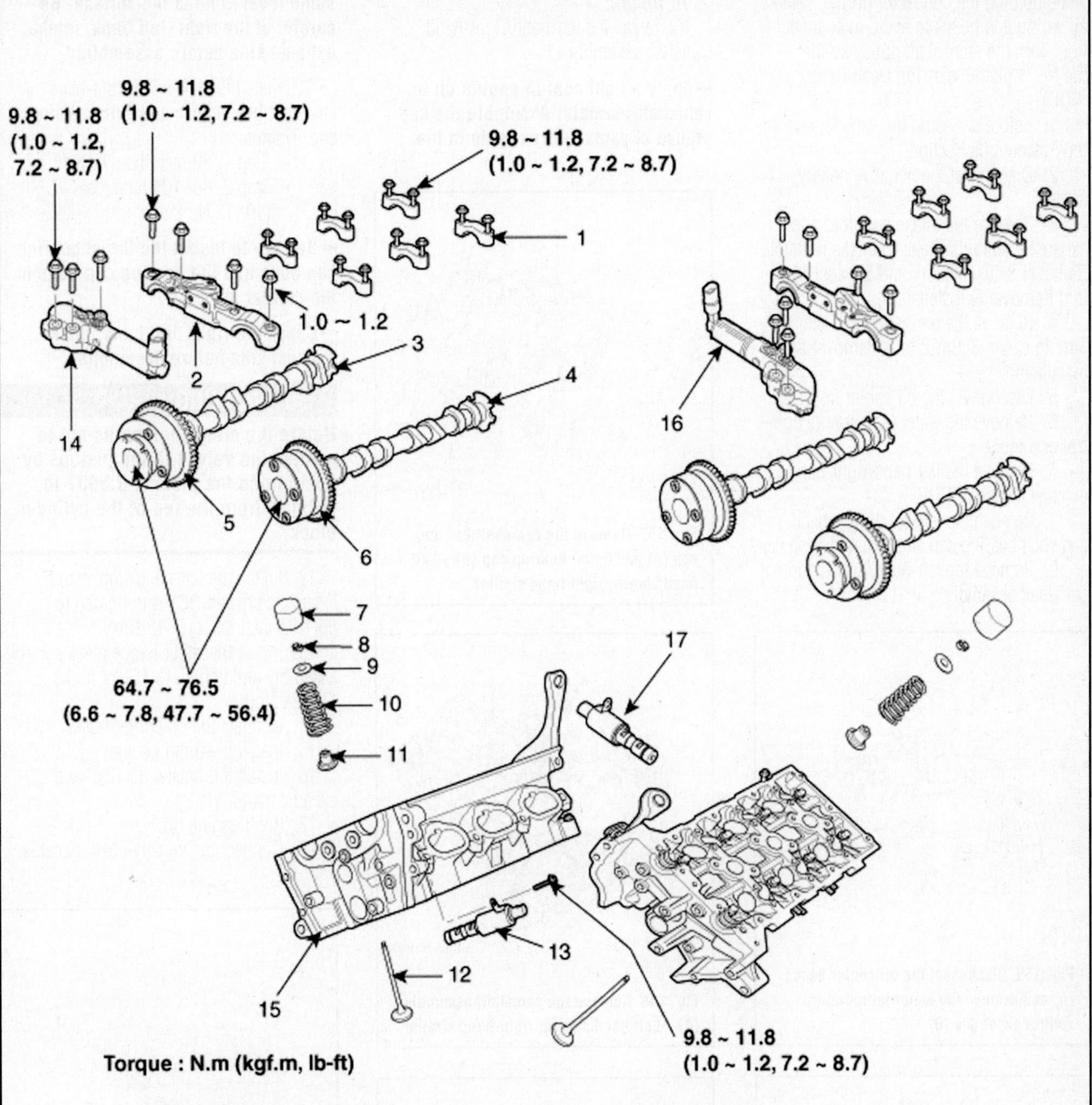

1. Camshaft bearing cap
2. Camshaft thrust bearing cap
3. Exhaust camshaft
4. Intake camshaft
5. Exhaust CVVT assembly
6. Intake CVVT assembly
7. Mechanical lash adjuster (MLA)
8. Retainer lock
9. Retainer
10. Valve spring
11. Valve stem seal
12. Valve
13. RH Exhaust camshaft OCV
14. RH Intake camshaft OCV
15. Cylinder head
16. LH Exhaust camshaft OCV
17. LH Intake camshaft OCV

37655_SANT_G0326

Fig. 154 Camshaft components

while holding the connector portion. Mark all wiring and hoses to avoid misconnection. Turn the crankshaft pulley so that the No. 1 piston is at Top Dead Center (TDC).

1. Before servicing the vehicle, refer to the Precautions Section.
2. Disconnect the negative battery cable.
3. Remove the intake manifold and exhaust manifolds. Refer to Intake Manifold Removal & Installation, and Exhaust Manifold Removal & Installation.
4. Remove the timing chain. Refer to Timing Chain & Sprockets, Removal & Installation.
5. Disconnect the oil cooler hoses.
6. Remove the water temperature control assembly.
7. Remove the left-hand/right-hand exhaust camshaft OCV.
8. Remove the left-hand/right-hand camshaft bearing cap and thrust bearing cap.
9. Remove the left-hand/right-hand camshaft assembly.

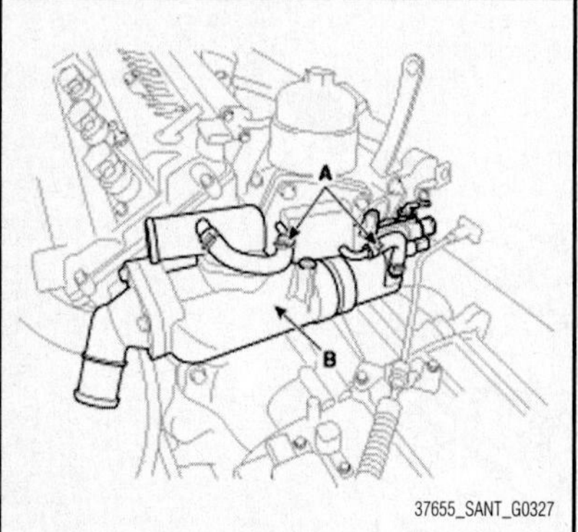

Fig. 155 Disconnect the oil cooler hoses (A) and remove the water temperature control assembly (B)

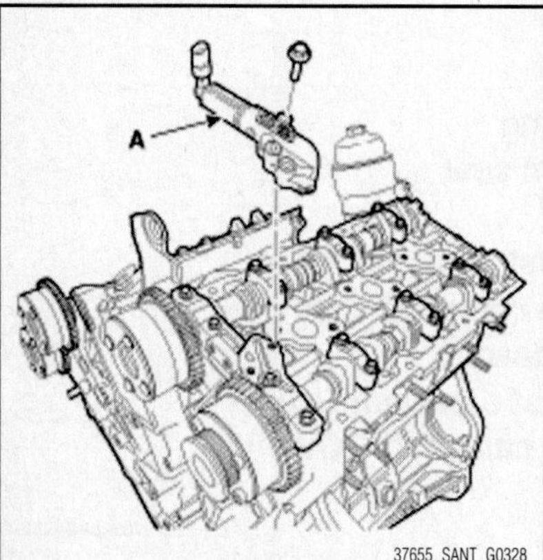

Fig. 156 Remove the exhaust camshaft OCV (A)—Left-hand shown, right-hand similar

To install:

10. Install the left-hand/right-hand camshaft assembly.

➡**Apply a light coat of engine oil on camshaft journals. Assemble the key groove of camshaft rear side to the same level of head top surface. Be careful of the right, left bank, intake, exhaust side before assembling.**

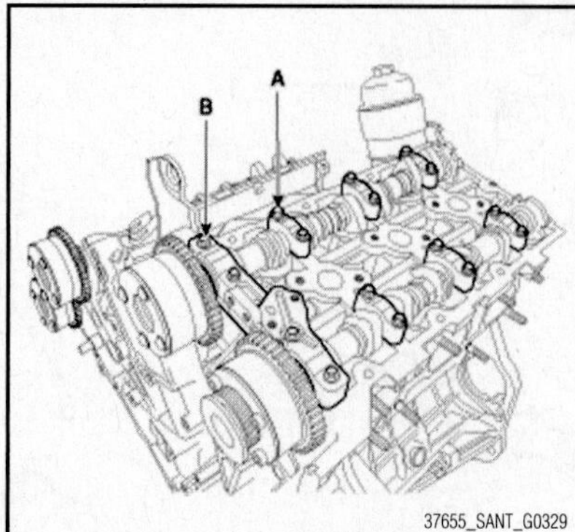

Fig. 157 Remove the camshaft bearing cap (A) and thrust bearing cap (B)—Left-hand shown, right-hand similar

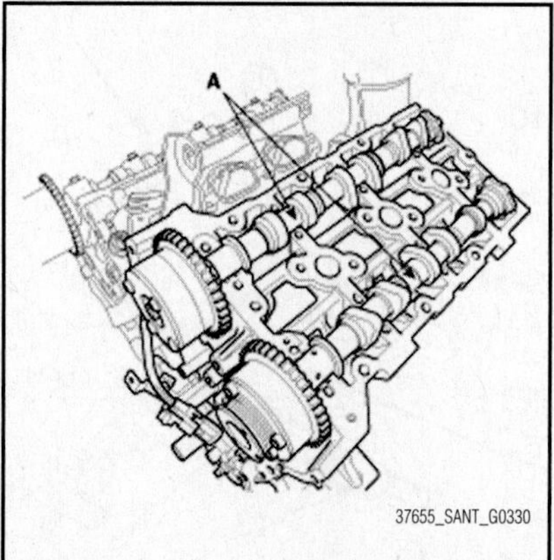

Fig. 158 Remove the camshaft assembly (A)—Left-hand shown, right-hand similar

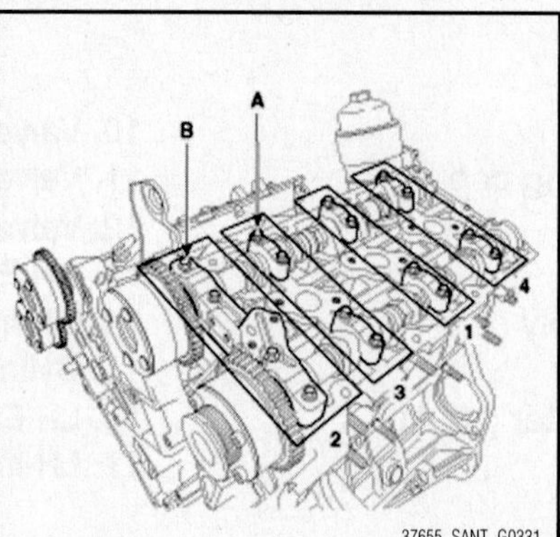

Fig. 159 Install the camshaft bearing cap (A) and thrust bearing cap (B)—Left-hand shown, right-hand similar

11. Install the left-hand/right-hand camshaft bearing cap and thrust bearing cap. Tighten:
    - Step 1: 52 inch lbs. (70 Nm)
    - Step 2: 86–104 inch lbs. (10–12 Nm)

➡**Be sure to install the thrust bearing cap bolts and the bearing cap bolts in the correct place.**

➡**Check the right, left bank, intake, exhaust side before assembling.**

WARNING

Rotate the crankshaft so as not to contact the valves to the pistons by positioning the pistons 0.3937 in. (10mm) from the top of the cylinder block.

12. Install the left-hand/right-hand exhaust camshaft OCV and tighten to 86–104 inch lbs. (10–12 Nm).
13. Install the water temperature control assembly and tighten to 15–17 ft. lbs. (20–24 Nm).
14. Connect the oil cooler hoses.
15. Install the timing chain.
16. Install the intake manifold and exhaust manifolds.
17. Refill engine oil.
18. Connect the negative battery cable.

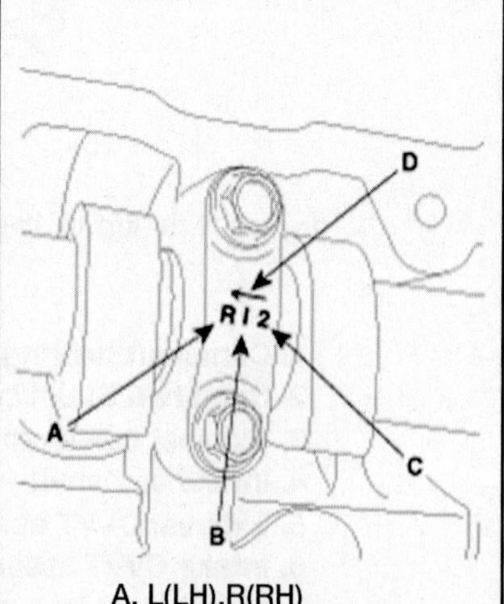

Fig. 160 Check the right, left bank, intake, exhaust side before assembling

19. Fill with engine coolant and bleed the cooling system.

20. Start the engine and check for leaks.

21. Recheck the engine coolant level and oil level.

CRANKSHAFT FRONT SEAL

REMOVAL & INSTALLATION

1. Before servicing the vehicle, refer to the Precautions Section.
2. Remove the crankshaft pulley. Refer to Crankshaft Damper Removal & Installation.
3. Using a seal removal tool, remove, the crankshaft front seal.

To install:

4. Using a seal installation tool, install the crankshaft front seal.
5. Install the crankshaft pulley.
6. Start the engine and check for leaks.

CYLINDER HEAD

REMOVAL & INSTALLATION

2009 Models

2.7L Engine

See Figures 161 through 173.

Use fender covers to avoid damaging painted surfaces. To avoid damage, unplug the wiring connectors carefully while holding the connector portion. Mark all wiring and hoses to avoid misconnection. Inspect the timing belt before removing. Turn the crankshaft pulley so that the No. 1 piston is at Top Dead Center (TDC).

1. Before servicing the vehicle, refer to the Precautions Section.
2. Remove the air duct.
3. Disconnect the battery terminals and remove the battery.
4. Drain engine coolant. Remove the radiator cap to speed draining.
5. Remove the air cleaner assembly.
6. Remove the upper and lower radiator hoses.
7. Remove the fuel inlet hose from the delivery pipe.
8. Disconnect the engine wire harness connectors:
   a. Disconnect the knock sensor connectors, the oil pressure switch connector, the ignition coil harness, and the No. 1 Variable Induction System (VIS) connector.
   b. Disconnect the Bank 1 Front/Rear O2 sensor connectors.
   c. Disconnect the injector connectors, the ground lines, the condenser connector and the ignition coil connectors.
   d. Disconnect the injection harness connector, the No. 2 Variable Induction System (VIS) connector, the No. 1/No. 2 Oil Control Valve (OCV) connectors, and the Oil Temperature Sensor (OTS) connector.
   e. Disconnect the Manifold Absolute Pressure (MAP) sensor connector, the Electronic Throttle Control (ETC) connector, and the Purge Control Solenoid Valve (PCSV) connector.
   f. Disconnect the alternator connector and the air conditioning compressor connector.
   g. Disconnect the Bank 2 CMP sensor connector and the Engine Coolant Temperature (ECT) sensor connector.
   h. Disconnect the Bank 2 Front/Rear O2 sensor connectors and the CKP sensor connector.
   i. Disconnect the Bank 1 CMP sensor connector.
9. Remove the Purge Control Valve (PCV) hose.
10. Disconnect the brake vacuum hose.

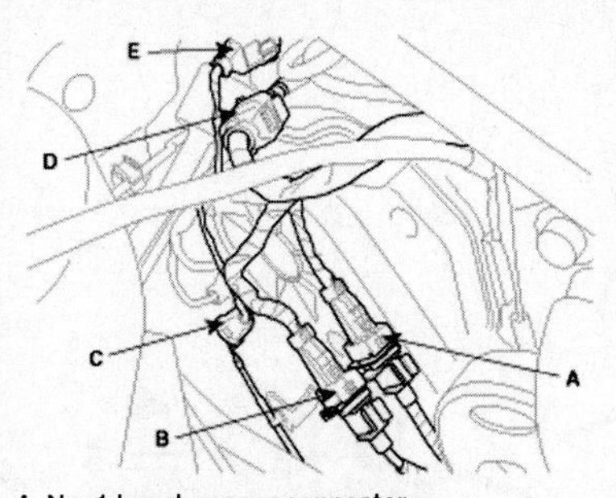

A. No. 1 knock sensor connector
B. No. 2 knock sensor connector
C. Oil pressure switch connector
D. Ignition coil harness
E. No. 1 Variable Induction System (VIS) connector

37655_SANT_G0317

Fig. 161 Disconnect the knock sensor connectors, oil pressure switch connector, ignition coil harness, and the No. 1 VIS connector

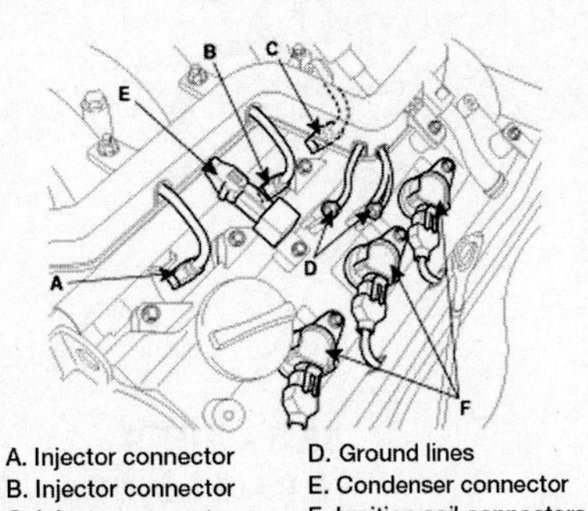

A. Injector connector
B. Injector connector
C. Injector connector
D. Ground lines
E. Condenser connector
F. Ignition coil connectors

37655_SANT_G0318

Fig. 162 Disconnect the injector connectors, ground lines, condenser connector, and the ignition coil connectors

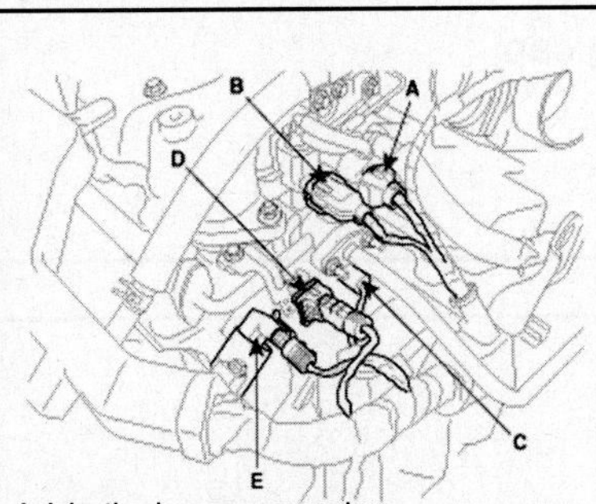

A. Injection harness connector
B. No. 2 Variable Induction System (VIS) connector
C. No. 1 Oil Control Valve (OCV) connectors
D. No. 2 Oil Control Valve (OCV) connectors
E. Oil Temperature Sensor (OTS) connector

37655_SANT_G0319

Fig. 163 Disconnect the injection harness connector, No. 2 Variable Induction System (VIS) connector, No. 1/No. 2 Oil Control Valve (OCV) connectors, and the Oil Temperature Sensor (OTS) connector

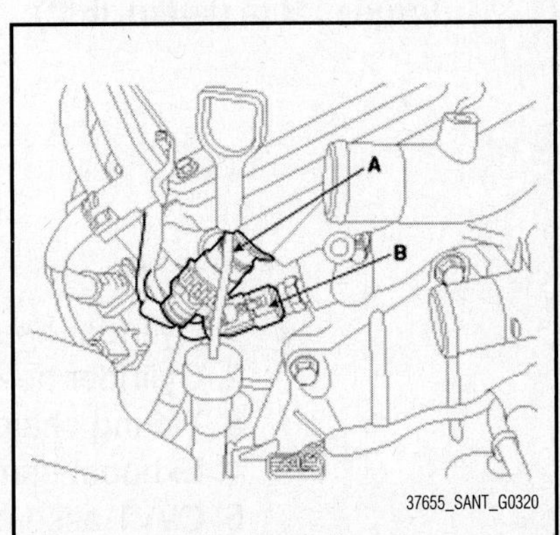

37655_SANT_G0320

Fig. 164 Disconnect the Bank 2 CMP sensor connector (A) and the Engine Coolant Temperature (ECT) sensor connector (B)

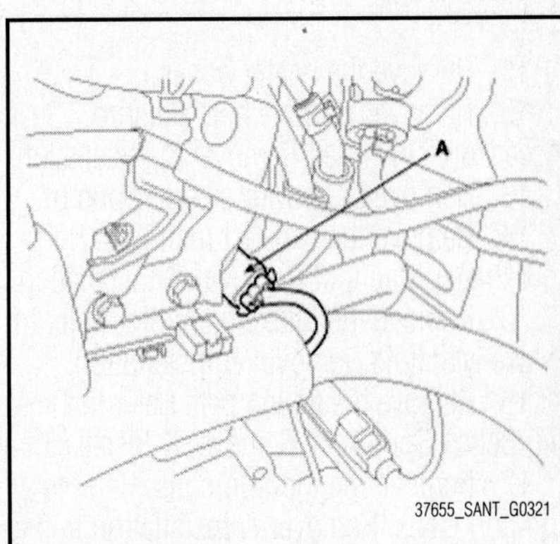

37655_SANT_G0321

Fig. 165 Disconnect the Bank 1 CMP sensor connector (A)

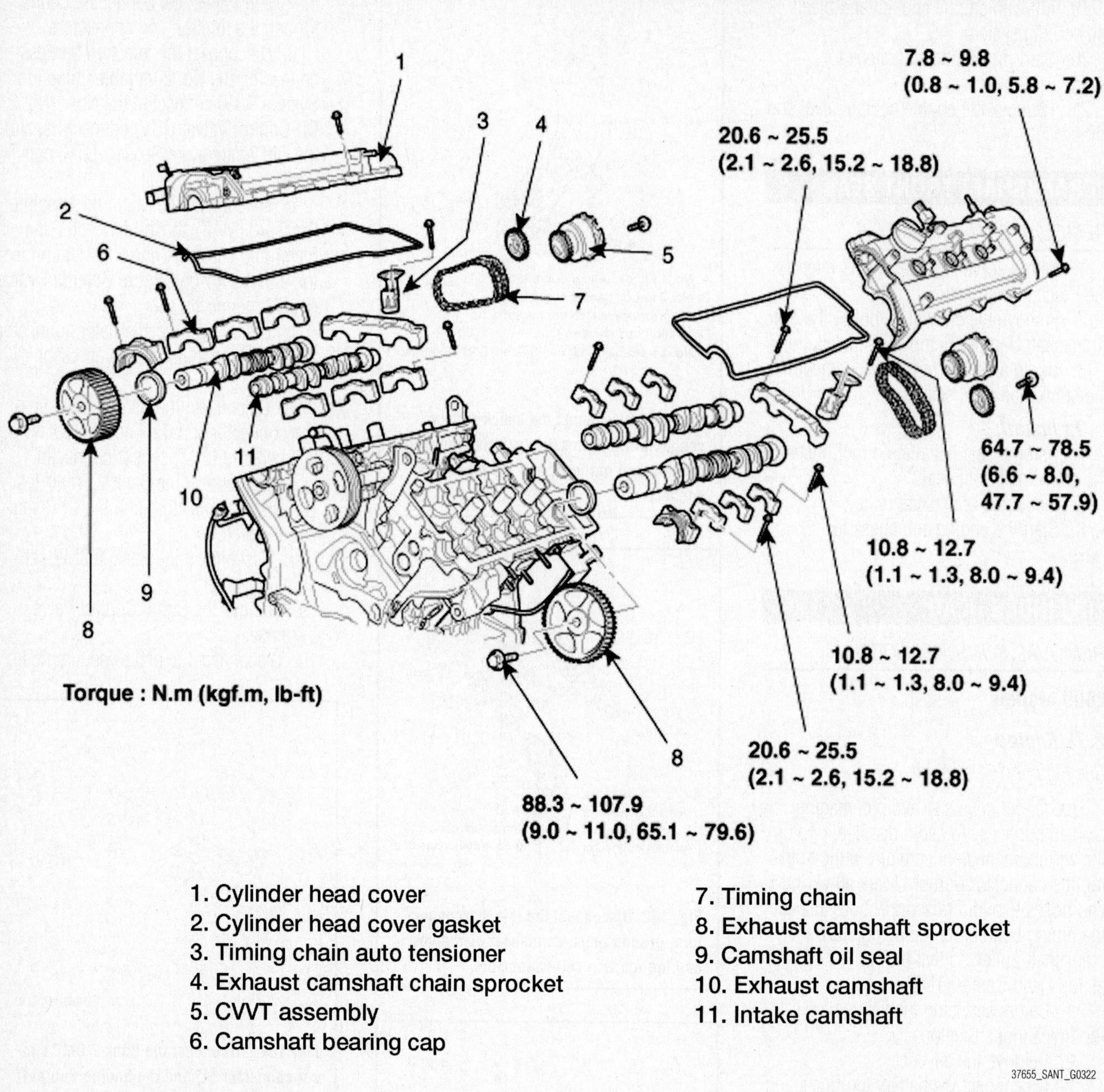

1
2
3
4
5
6
7
8
9
10
11
7.8 ~ 9.8
(0.8 ~ 1.0, 5.8 ~ 7.2)
20.6 ~ 25.5
(2.1 ~ 2.6, 15.2 ~ 18.8)
64.7 ~ 78.5
(6.6 ~ 8.0,
47.7 ~ 57.9)
10.8 ~ 12.7
(1.1 ~ 1.3, 8.0 ~ 9.4)
10.8 ~ 12.7
(1.1 ~ 1.3, 8.0 ~ 9.4)
20.6 ~ 25.5
(2.1 ~ 2.6, 15.2 ~ 18.8)
88.3 ~ 107.9
(9.0 ~ 11.0, 65.1 ~ 79.6)
Torque : N.m (kgf.m, lb-ft)
1. Cylinder head cover
2. Cylinder head cover gasket
3. Timing chain auto tensioner
4. Exhaust camshaft chain sprocket
5. CVVT assembly
6. Camshaft bearing cap
7. Timing chain
8. Exhaust camshaft sprocket
9. Camshaft oil seal
10. Exhaust camshaft
11. Intake camshaft
37655_SANT_G0322

Fig. 166 Camshaft and components

11. Remove the heater hoses.
12. Remove the drive belt. Refer to Accessory Drive Belt Removal & Installation.
13. Remove the power steering pump.
14. Remove the exhaust manifold. Refer to Exhaust Manifold Removal & Installation.
15. Remove the intake manifold. Refer to Intake Manifold Removal & Installation.
16. Remove the timing belt. Refer to Timing Belt & Sprockets, Removal & Installation.
17. Remove the ignition coils. Refer to Ignition Coils, Removal & Installation in the Engine Electrical section.
18. Remove the water temperature control assembly.

A
37655_SANT_G0419

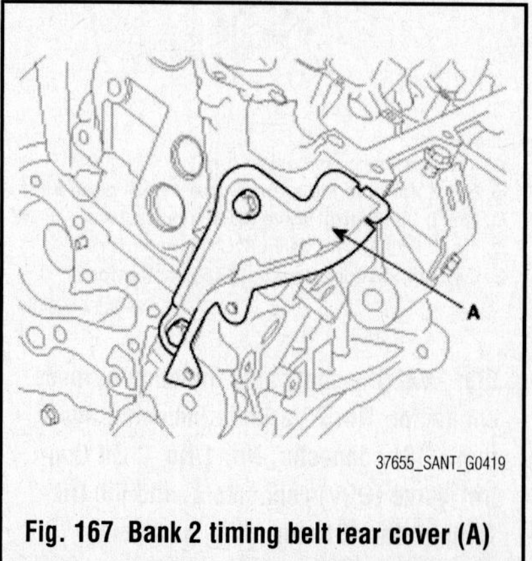

Fig. 167 Bank 2 timing belt rear cover (A)

A
37655_SANT_G0420

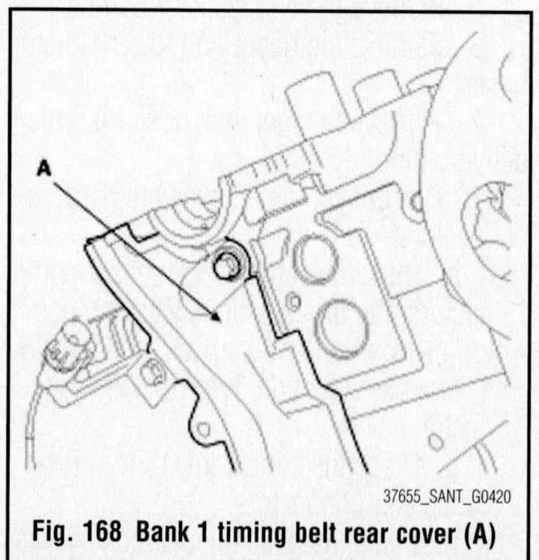

Fig. 168 Bank 1 timing belt rear cover (A)

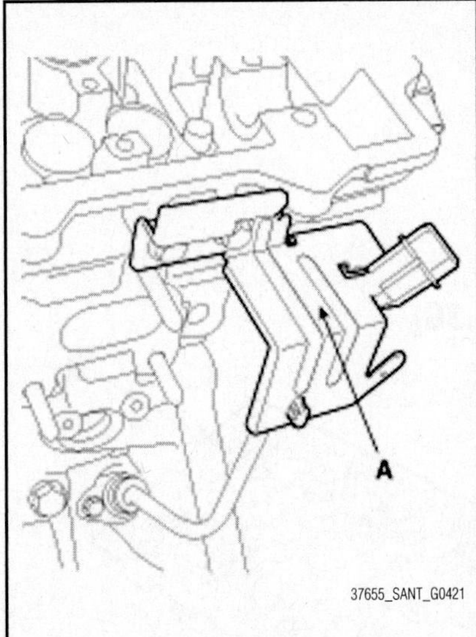

Fig. 169 Remove the CKP sensor connector bracket (A)

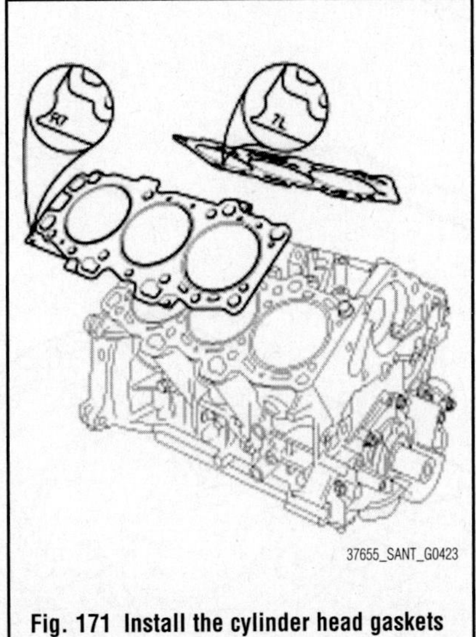

Fig. 170 Cylinder head bolt removal sequence

19. Remove the cylinder head cover.
20. Remove the camshaft bearing caps, camshaft timing chain tensioner, and the camshafts. Refer to Camshaft Removal & Installation.
21. Remove the Bank 2 timing belt rear cover.
22. Remove the Bank 1 timing belt rear cover.
23. Remove the CKP sensor connector bracket.
24. Remove the cylinder head bolts as shown.

WARNING

Head warpage or cracking could result from removing bolts in an incorrect order.

25. Remove the cylinder head assembly.

➡Ensure that the surface between the cylinder head and the block is not damaged.

To install:

Thoroughly clean all parts to be assembled. Always use a new head and manifold gasket. The cylinder head gasket is a metal gasket. Take care not to bend it. Rotate the crankshaft; set the No. 1 piston at TDC.

26. Install the cylinder head gaskets. Confirm the right-hand/left-hand positioning of the gaskets.
27. Install the cylinder head.
28. Tighten the cylinder head bolts with the in several steps as following order.

➡In assembling washers, the marked surface should face upward. In installing the cylinder head bolts, apply engine oil on the thread of the bolts and the surface of the washers.

➡Always use new cylinder head bolts.

➡Using the SST (09221-4A000), tighten the bolts.

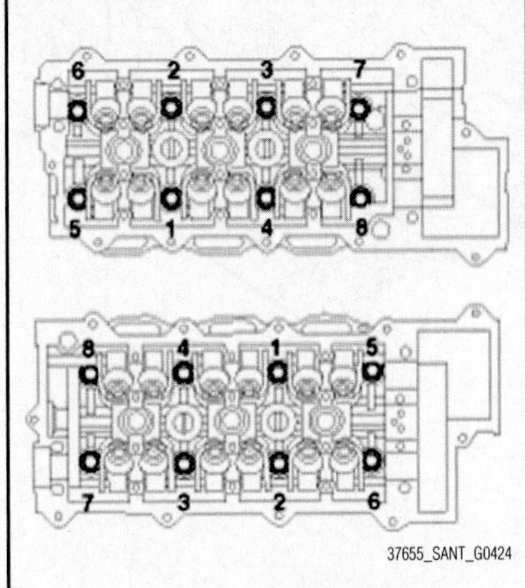

Fig. 171 Install the cylinder head gaskets

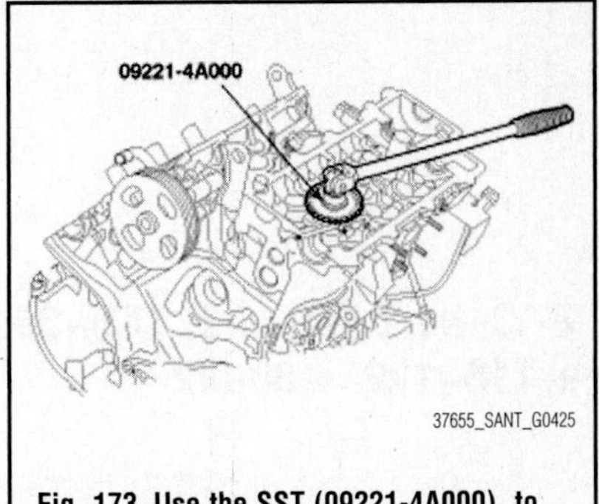

Fig. 173 Use the SST (09221-4A000), to tighten the bolts

Fig. 172 Cylinder head bolt tightening sequence

29. Install the camshafts, camshaft timing chain tensioner, camshaft bearing caps, camshaft oil seal, and camshaft sprockets.
30. Install the cylinder head cover.
31. Install the water temperature control assembly.
32. Install the ignition coils.
33. Install the timing belt.
34. Install the intake manifold.
35. Install the exhaust manifold.
36. Install the power steering pump.
37. Install the drive belt.
38. Install the heater hoses.
39. Install the brake vacuum hose.
40. Install the Purge Control Valve (PCV) hose.
41. Install the engine wire harness connectors.
42. Install the fuel inlet hose.
43. Install the radiator hoses.
44. Install the air cleaner assembly.
45. Connect the battery terminals.
46. Install the battery.
47. Fill with engine coolant.
48. Install the side cover and the engine cover.
49. Start the engine and check for leaks.
50. Recheck the engine coolant level and oil level.

3.3L Engine

See Figures 174 through 183.

Engine removal is required for this procedure. Use fender covers to avoid damaging painted surfaces. To avoid damage, unplug the wiring connectors carefully while holding the connector portion. Mark all wiring and hoses to avoid misconnection. Inspect the timing belt before removing. Turn the crankshaft pulley so that the No. 1 piston is at Top Dead Center (TDC).

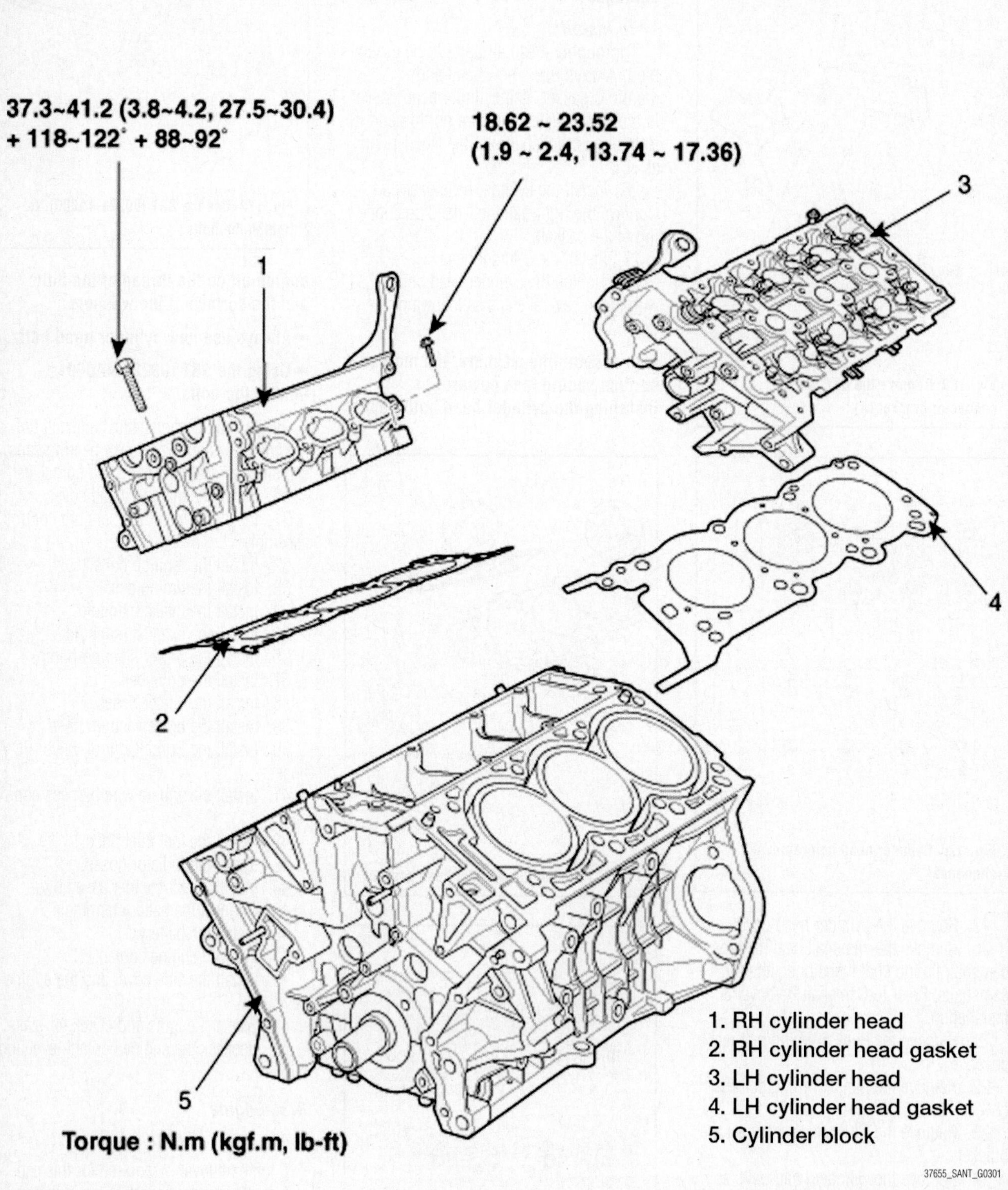

Fig. 174 Exploded view of cylinder head and engine block

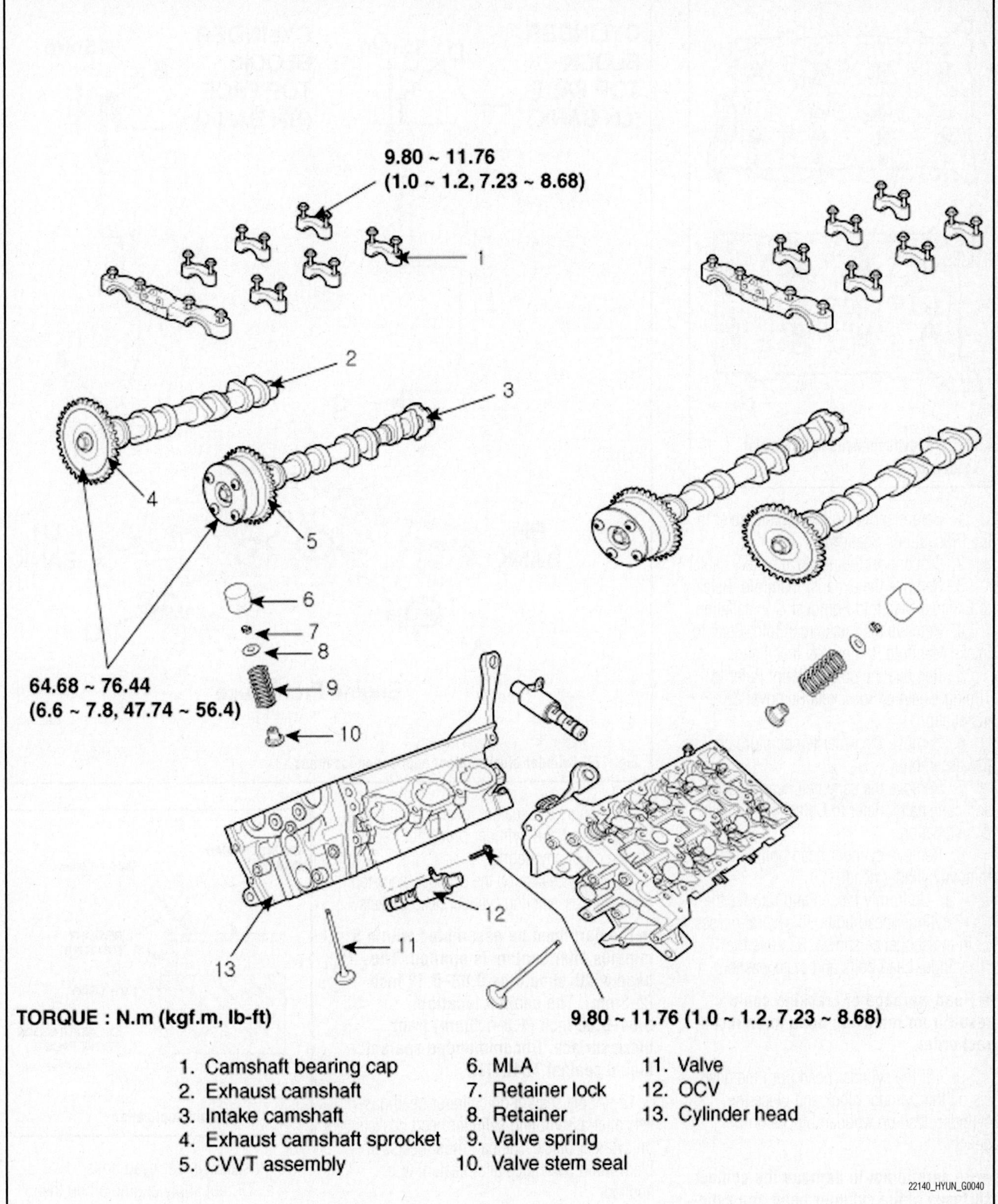

Fig. 175 Exploded view of cylinder head and related components

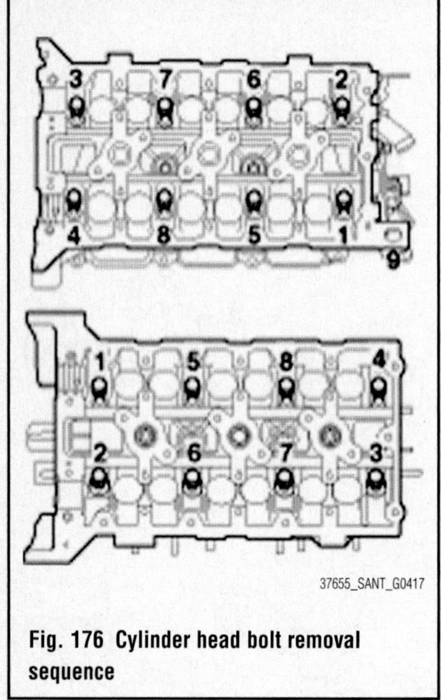

Fig. 176 Cylinder head bolt removal sequence

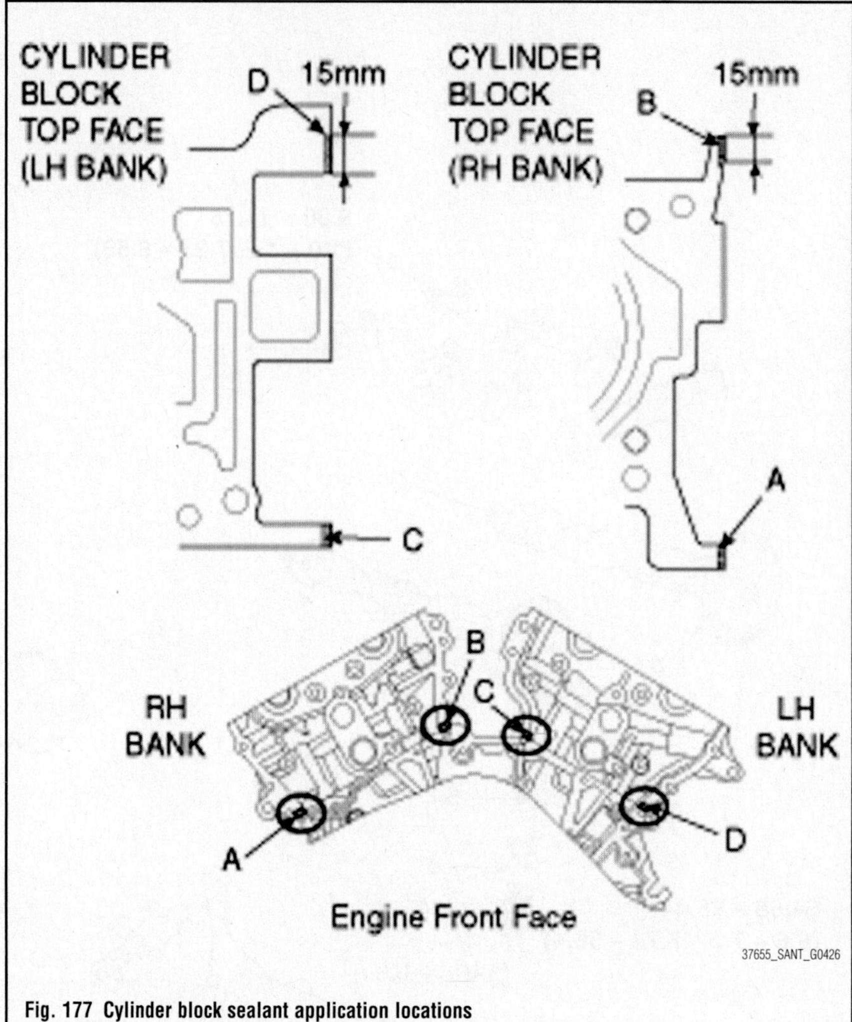

Fig. 177 Cylinder block sealant application locations

1. Before servicing the vehicle, refer to the Precautions Section.
2. Disconnect the negative battery cable.
3. Remove the exhaust manifold. Refer to Exhaust Manifold Removal & Installation.
4. Remove the intake manifold. Refer to Intake Manifold Removal & Installation.
5. Remove the timing chain. Refer to Timing Chain & Sprockets Removal & Installation.
6. Remove the water temperature control assembly.
7. Remove the camshaft bearing caps and camshafts. Refer to Camshaft Removal & Installation.
8. Remove cylinder head bolts, then remove cylinder head.
   a. Uniformly loosen and remove the 17 cylinder head bolts, in several passes, in the sequence shown. Remove the 17 cylinder head bolts and plate washers.

➡Head warpage or cracking could result from removing bolts in an incorrect order.

9. Lift the cylinder head from the dowels on the cylinder block and place the cylinder head on wooden blocks on a bench.

➡Be careful not to damage the contact surfaces of the cylinder head and cylinder block.

To install:

Thoroughly clean all parts to be assembled. Always use a new head and manifold gasket. The cylinder head gasket is a metal gasket. Take care not to bend it. Rotate the crankshaft, set the No. 1 piston at TDC.

10. Ensure the sealant locations on the cylinder head and cylinder block are free of engine oil or any debris.
11. Apply sealant on the cylinder block top face before assembling cylinder head gaskets.

➡The part must be assembled within 5 minutes after sealant is applied. The bead width should be 0.08–0.12 inch (2–3mm). The sealant location: 0.04–0.06 inch (1.0–1.5mm) from block surface. Recommended sealant: Liquid sealant TB1217H.

12. Apply sealant on cylinder head gaskets after assembling cylinder head gaskets on cylinder block. The part must be assembled within 5 minutes after sealant was applied.

➡Be careful of the installation direction.

13. Install the cylinder head. Remove any extruded sealant after assembling cylinder heads.
14. Place the cylinder head carefully in order not to damage the gasket with the bottom part of the end.

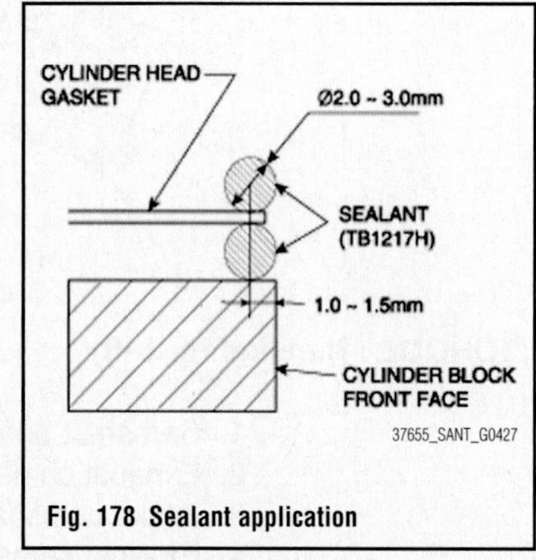

Fig. 178 Sealant application

15. Install cylinder head bolts.
   a. Do not apply engine oil on the threads or under the heads of the cylinder head bolts.
   b. Using SST (09221-4A000), install and tighten the cylinder head bolts and plate washers, in several passes, in the sequence shown.
      - Step 1—Tighten to 28–30 ft. lbs. (37–41 Nm)

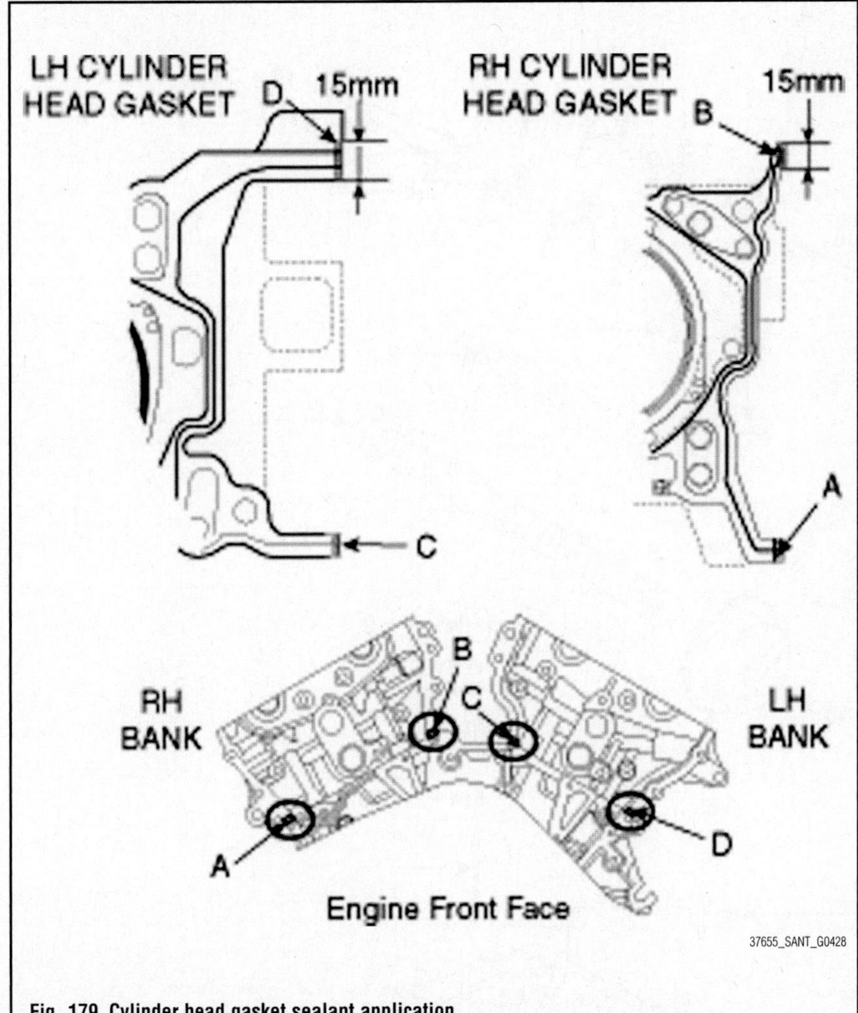

Fig. 179 Cylinder head gasket sealant application

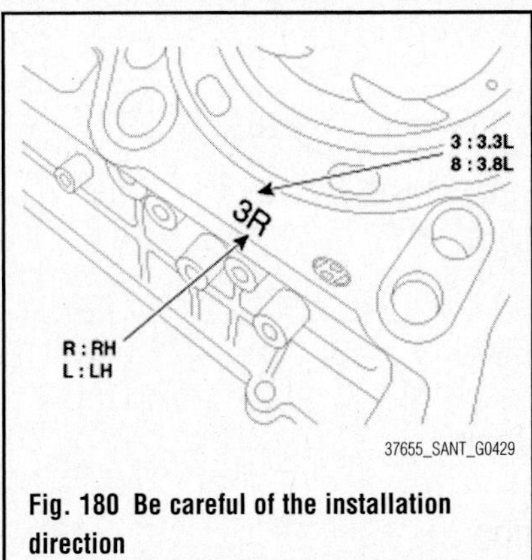

Fig. 180 Be careful of the installation direction

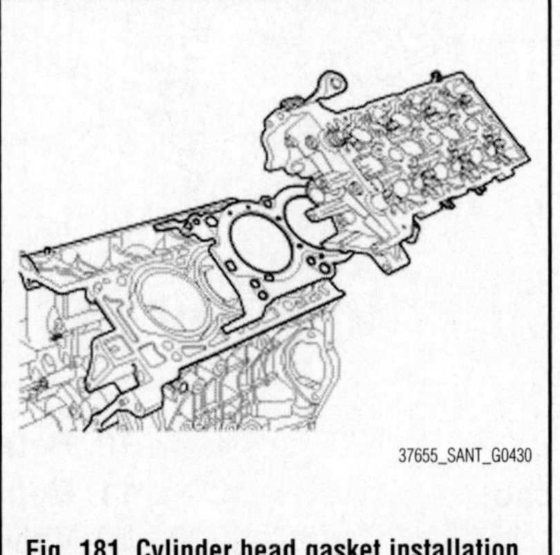

Fig. 181 Cylinder head gasket installation

- Step 2—Tighten an additional: 120° plus or minus 2°
- Step 3—Tighten an additional: 90° plus or minus 2°

c. Tighten bolt "A" to: 14–17 ft. lbs. (19–24 Nm).

➡Always use new cylinder head bolts.

⁂ WARNING

Rotate the crankshaft so as not to contact the valves to the pistons by positioning the pistons 0.3937 in. (10mm) from the top of the cylinder block.

16. Install the water temperature control assembly.

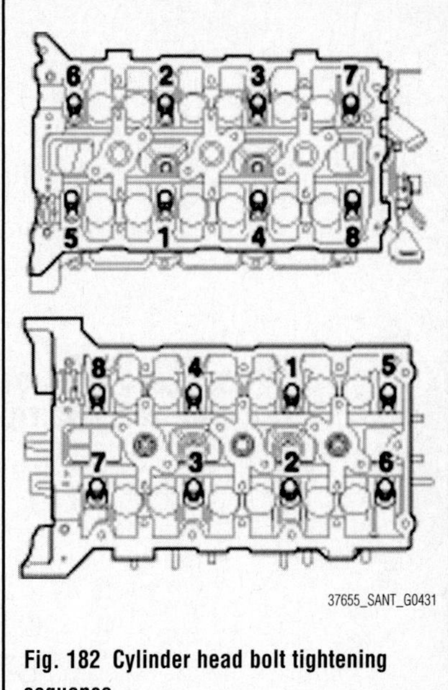

Fig. 182 Cylinder head bolt tightening sequence

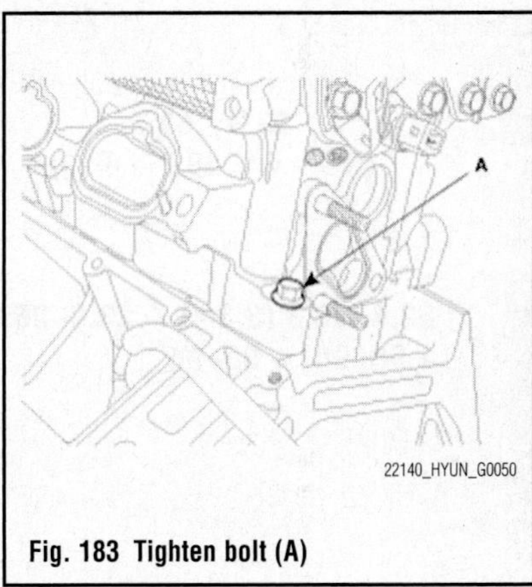

Fig. 183 Tighten bolt (A)

17. Install the timing chain.
18. Check and adjust the valve clearance, as necessary.
19. Install the intake manifold.
20. Install the exhaust manifold.
21. Connect the negative battery cable.
22. Fill with engine coolant.
23. Start the engine and check for leaks.
24. Recheck the engine coolant level and oil level.

2010 Models

2.4L Engine

See Figures 146 through 151 and 184 through 194.

Use a fender cover to avoid damaging painted surfaces. To avoid damaging the cylinder head, wait until the engine coolant temperature drops below normal temperature before removing it. When handling a metal gasket, take care not to fold the gasket or damage the contact surface of the gasket.

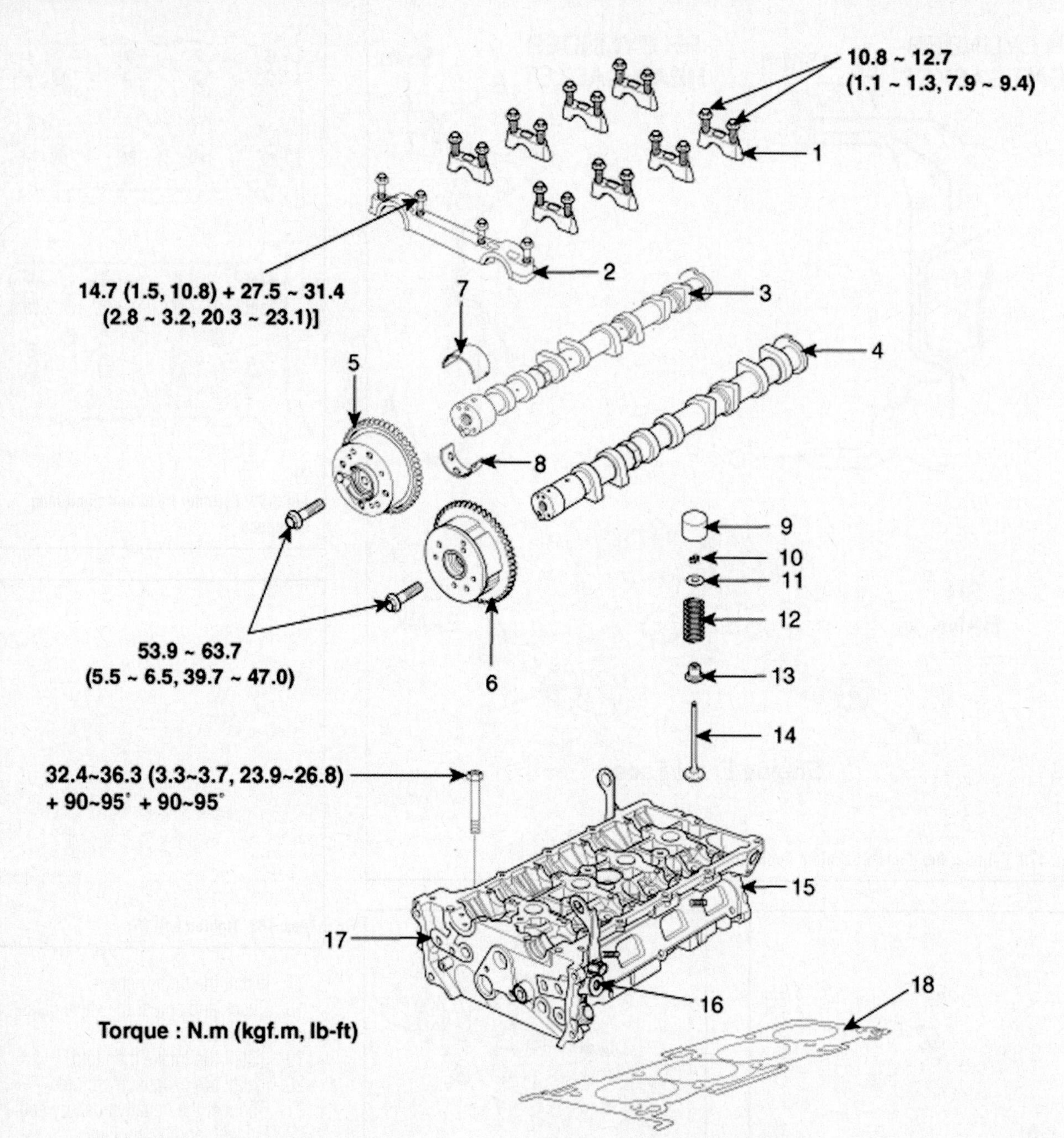

1. Camshaft bearing cap
2. Camshaft front bearing cap
3. Exhaust camshaft
4. Intake camshaft
5. Exhaust CVVT assembly
6. Intake CVVT assembly
7. Exhaust camshaft upper bearing
8. Exhaust camshaft lower bearing
9. MLA
10. Retainer lock
11. Retainer
12. Valve spring
13. Valve stem seal
14. Valve
15. Cylinder head
16. Intake OCV
17. Exhaust OCV
18. Cylinder head gasket

37655_SANT_G0324

Fig. 184 Camshaft and cylinder head components

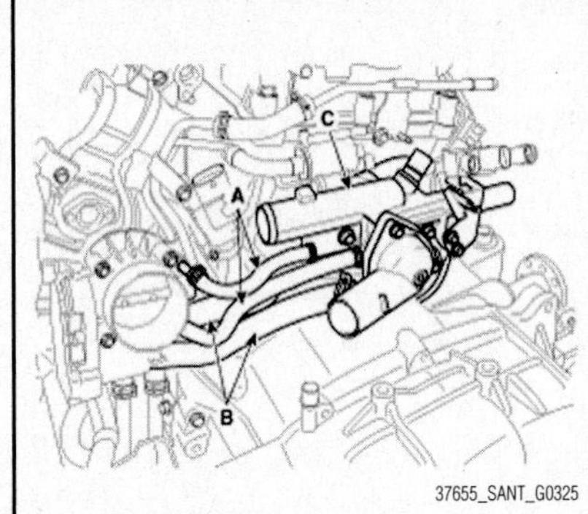

Fig. 185 Disconnect the throttle body coolant hoses (A), oil cooler coolant hoses (B) and then remove the water temperature control assembly (C)

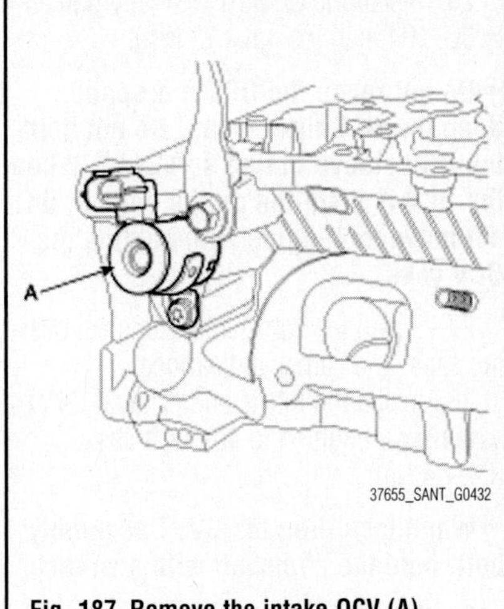

Fig. 187 Remove the intake OCV (A)

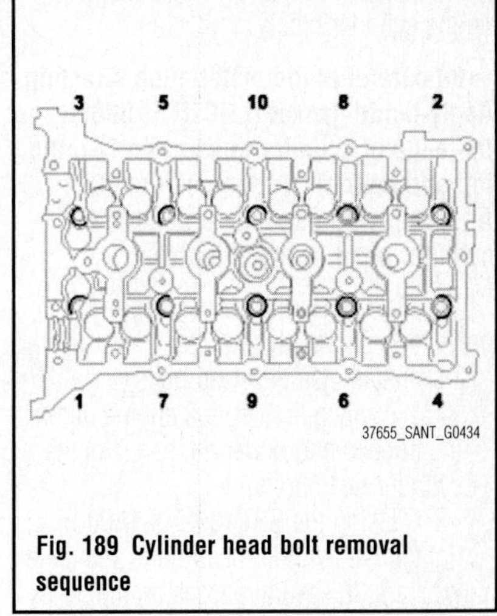

Fig. 189 Cylinder head bolt removal sequence

To avoid damage, unplug the wiring connectors carefully while holding the connector portion. Mark all wiring and hoses to avoid misconnection.

1. Before servicing the vehicle, refer to the Precautions Section.
2. Disconnect the negative battery cable.
3. Remove the engine cover.
4. Remove the air cleaner assembly.
5. Disconnect the positive battery terminal and remove the battery.
6. Remove the undercover and side cover.
7. Drain the engine coolant.
8. Remove the front muffler.
9. Remove the upper and lower radiator hoses.
10. Disconnect the wiring connectors, harness clamps and hoses from the cylinder head, intake manifold and exhaust manifold.
11. Disconnect the fuel hose and the heater hoses.
12. Disconnect the throttle body coolant hoses, oil cooler coolant hoses and then remove the water temperature control assembly.
13. Remove the intake and exhaust manifolds. Refer to Intake Manifold Removal & Installation, and Exhaust Manifold Removal & Installation.
14. Remove the timing chain. Refer to Timing Chain & Sprockets, Removal & Installation.
15. Remove the intake and exhaust CVVT assembly.

➡When removing the CVVT assembly bolt, hold the camshaft with a wrench.

16. Remove the camshafts, camshaft bearing caps, and camshaft bearings. Refer to Camshaft Removal & Installation.
17. Using a Torx® wrench, remove the intake OCV.
18. Using a Torx® wrench, remove the exhaust OCV.
19. Remove the cylinder head bolts and then remove the cylinder head.
    a. Using a triple square wrench, uniformly loosen and remove the 10 cylinder head bolts, in several passes, in the sequence shown. Remove the 10 cylinder head bolts and plate washers.

❊❊ WARNING

Head warpage or cracking could result from removing bolts in an incorrect order.

    b. Lift the cylinder head from the dowels on the cylinder block and place the cylinder head on wooden blocks on a bench.

➡Be careful not to damage the contact surfaces of the cylinder head and cylinder block.

20. Remove the cylinder head gasket.

To install:

➡Thoroughly clean all parts to be assembled. Always use a new head and manifold gasket. The cylinder head gasket is a metal gasket. Take care not to bend it. Rotate the crankshaft, set the No.1 piston at TDC.

21. Install the OCV filter. Keep the OCV filter clean.

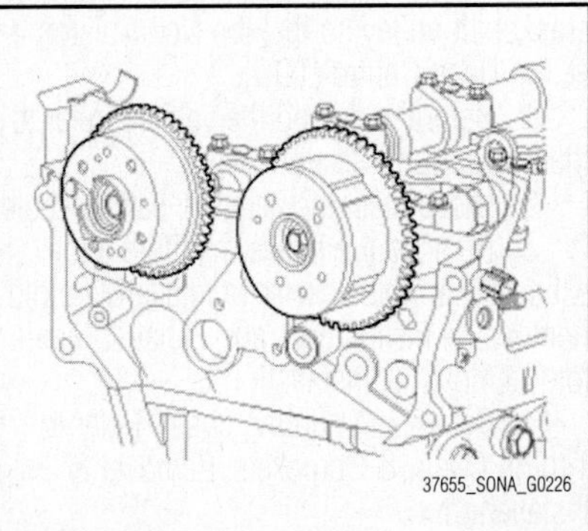

Fig. 186 Remove the intake and exhaust CVVT assembly

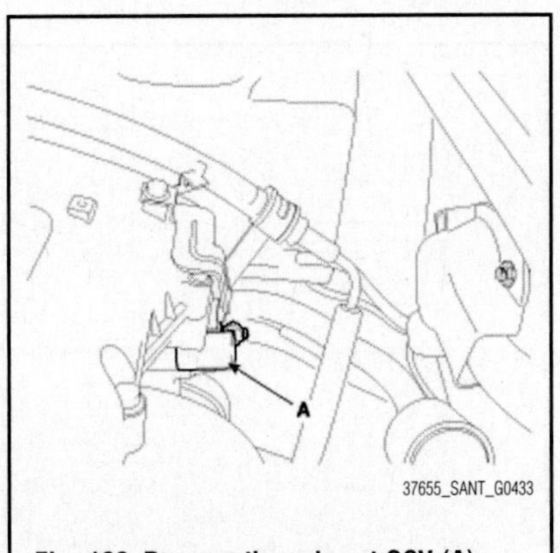

Fig. 188 Remove the exhaust OCV (A)

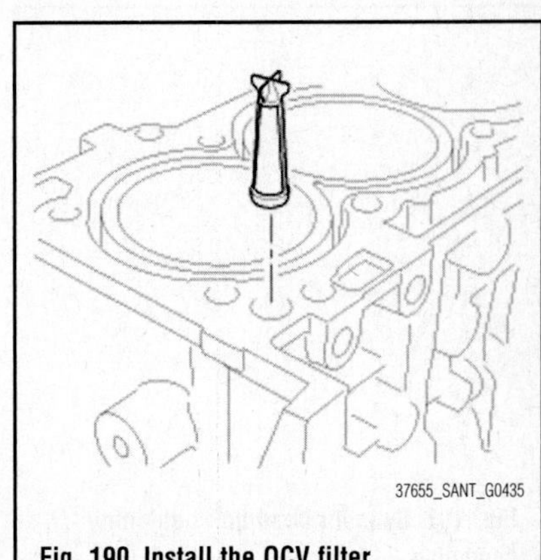

Fig. 190 Install the OCV filter

22. Install the cylinder head gasket (A) on the cylinder block.

➡Be careful of the installation direction. Apply liquid gasket (LOCTITE® 5900H) on the edge of cylinder head gasket, at the top and bottom. (At the position 'B'). After applying sealant, assemble the cylinder head in five minutes.

23. Place the cylinder head carefully to prevent damaging the gasket.

24. Install cylinder head bolts.

a. Apply a light coat if engine oil on the threads and under the heads of the cylinder head bolts.

b. Using the SST (09221-4A000), tighten the cylinder head bolts and plate washers, in several passes, in the sequence shown. Tightening torque:

- Step 1: 24–27 ft. lbs. (32–36 Nm) plus 90–95° plus an additional 90–95°

➡Always use new cylinder head bolts.

25. Install the intake OCV and tighten to 86–104 inch lbs. (10–12 Nm).

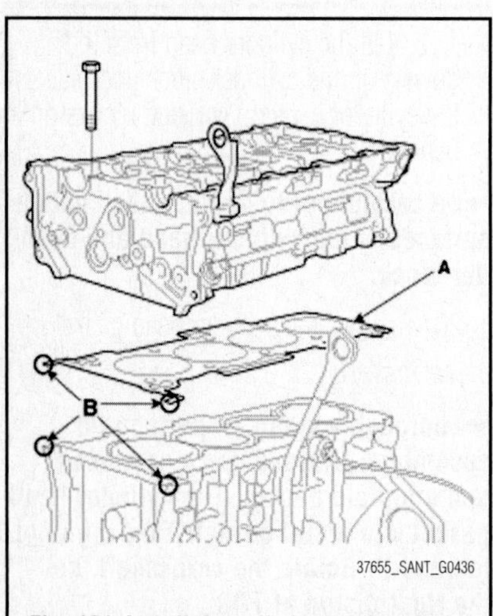

Fig. 191 Install the cylinder head gasket (A)

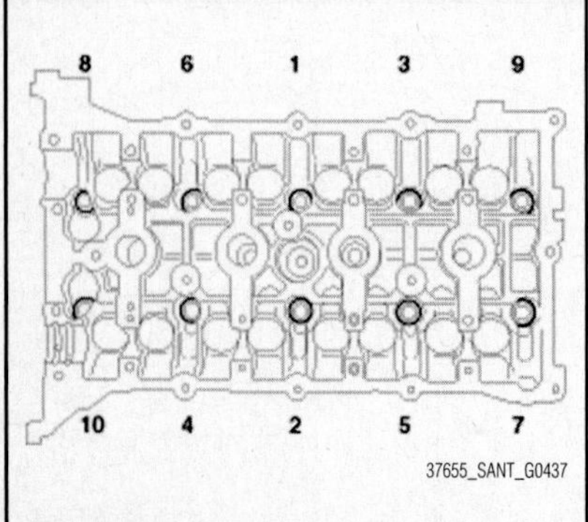

Fig. 192 Cylinder head bolt tightening sequence

26. Install the exhaust OCV and tighten to 86–104 inch lbs. (10–12 Nm).

➡Do not reuse the OCV if dropped. Keep the OCV filter clean. Do not hold the OCV sleeve during servicing. When the OCV is installed on the engine, do not move the engine with holding the OCV yoke.

27. Install the camshafts, camshaft bearing caps, and camshaft bearings.

28. Install the intake and exhaust CVVT assembly an tighten to 40–47 ft. lbs. (54–64 Nm).

➡When installing the CVVT assembly bolt, hold the camshaft with a wrench.

29. Install the water temp control assembly, and tighten the bolt to 11–15 ft. lbs. (15–20 Nm) and the nut to 14–17 ft. lbs. (19–24 Nm).

➡Assemble the water temp control assembly and water inlet pipe to water pump assembly before tightening the water inlet pipe nuts.

➡Always use a new O-ring.

30. Install the timing chain.

31. Install the exhaust manifold.

32. Install the intake manifold.

33. Check and adjust valve clearance.

34. Install the cylinder head cover:

a. Remove existing sealant and apply new sealant Sealant: LOCTITE® 5900H, Bead width: 0.1 in. (2.5mm)] as shown. After applying sealant, assemble within 5 minutes.

➡The firing and/or blow out test should not be performed within 30 minutes after the cylinder head cover was assembled.

b. Install the cylinder head cover bolts in the sequence shown, tightening to:

- Step 1: 35–52 inch lbs. (4–6 Nm)
- Step 2: 70–86 inch lbs. (8–10 Nm)

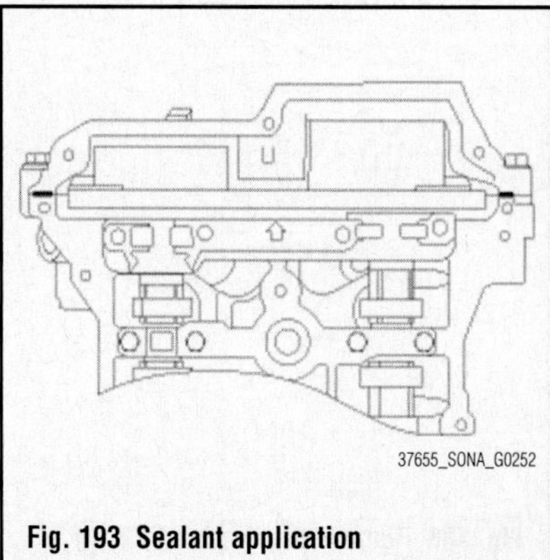

Fig. 193 Sealant application

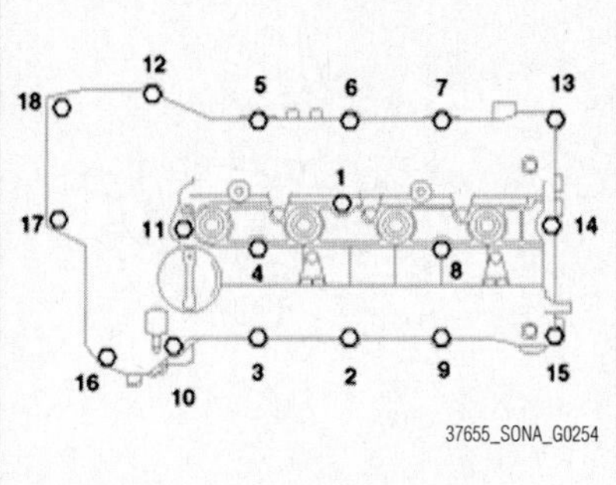

Fig. 194 Cylinder head bolt tightening sequence

➡Do not reuse cylinder head cover gasket.

35. Connect the throttle body coolant hoses, oil cooler coolant hoses and then remove the water temperature control assembly.

36. Connect the fuel hose and the heater hoses.

37. Connect the wiring connectors, harness clamps and hoses from the cylinder head, intake manifold and exhaust manifold.

38. Install the upper and lower radiator hoses.

39. Install the front muffler.

40. Install the undercover and side cover.

41. Install the air cleaner assembly.

42. Install the engine cover.

43. Connect the battery cables.

44. Install the battery.

45. Fill with engine coolant.

46. Start the engine and check for leaks.

47. Recheck the engine coolant level and oil level.

3.5L Engine

See Figures 195 through 207.

Use a fender cover to avoid damaging painted surfaces. To avoid damage, unplug the wiring connectors carefully while holding the connector portion. Mark all wiring and hoses to avoid misconnection. Turn the crankshaft pulley so that the No. 1 piston is at Top Dead Center (TDC).

1. Before servicing the vehicle, refer to the Precautions Section.

2. Disconnect the negative battery cable.

3. Remove the intake manifold and exhaust manifolds. Refer to Intake Manifold Removal & Installation, and Exhaust Manifold Removal & Installation.

4. Remove the timing chain. Refer to Timing Chain & Sprockets, Removal & Installation.

5. Disconnect the oil cooler hoses.

6. Remove the water temperature control assembly.

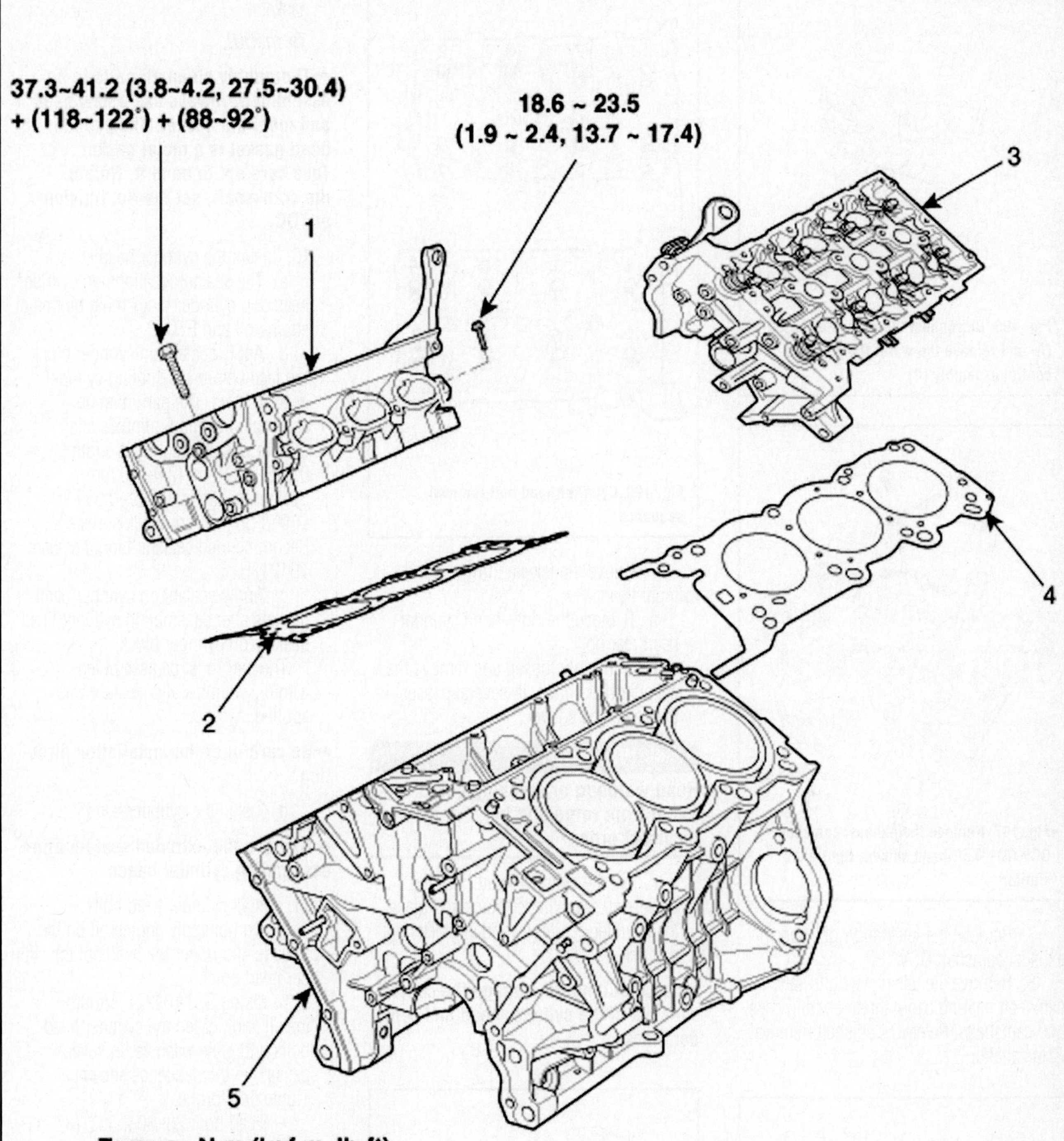

Fig. 195 Cylinder head components

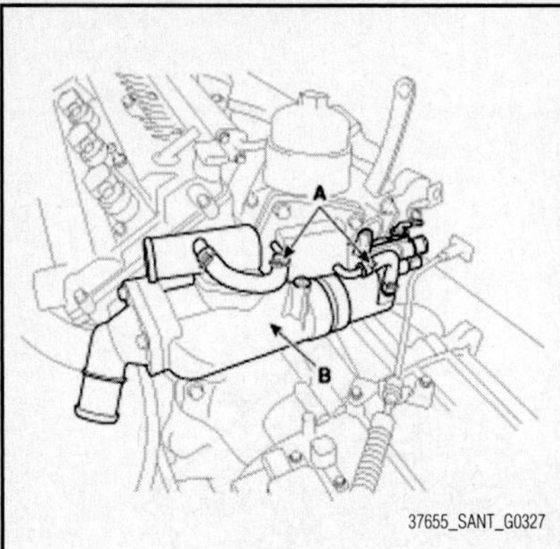

Fig. 196 Disconnect the oil cooler hoses (A) and remove the water temperature control assembly (B)

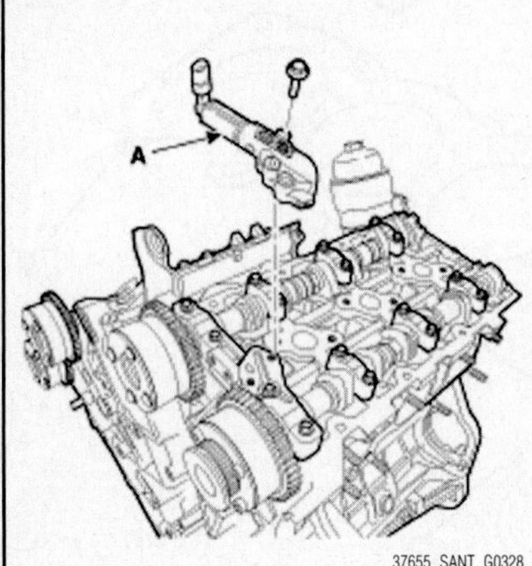

Fig. 197 Remove the exhaust camshaft OCV (A)—Left-hand shown, right-hand similar

7. Remove the left-hand/right-hand exhaust camshaft OCV.
8. Remove the left-hand/right-hand camshaft bearing caps, thrust bearing caps, and camshafts. Refer to Camshaft Removal & Installation.

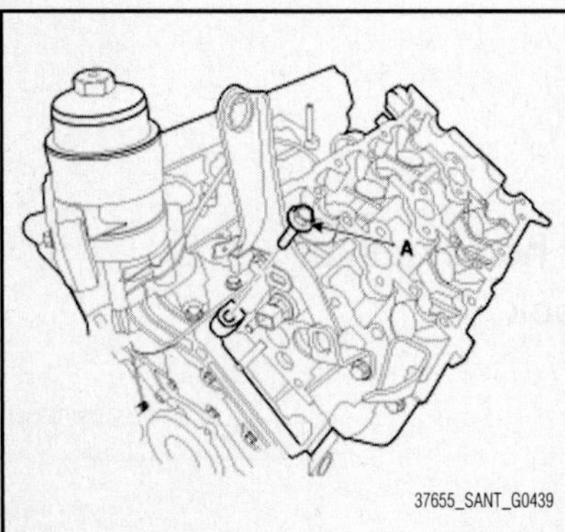

Fig. 198 Remove the right-hand cylinder head rear bolt (A)

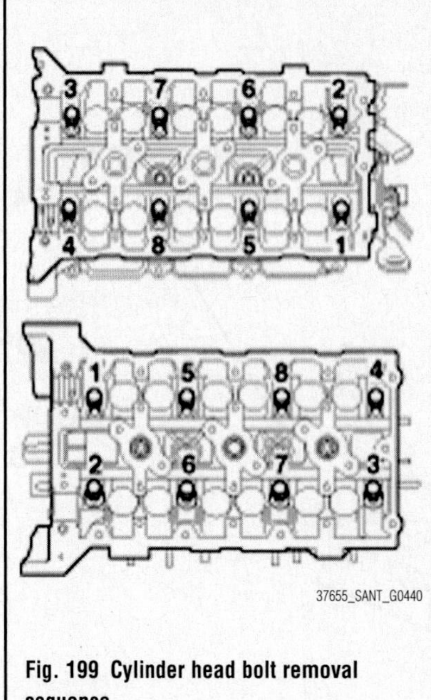

Fig. 199 Cylinder head bolt removal sequence

9. Remove the left-hand/right-hand cylinder heads.
   a. Remove the right-hand cylinder head rear bolt.
   b. Uniformly loosen and remove the cylinder head bolts, in several passes, in the sequence shown.

⁂ WARNING

Head warpage or cracking could result from removing bolts in an incorrect order.

   c. Lift the cylinder head from the dowels on the cylinder block and place the cylinder head on wooden blocks on a bench.

➡Be careful not to damage the contact surfaces of the cylinder head and cylinder block.

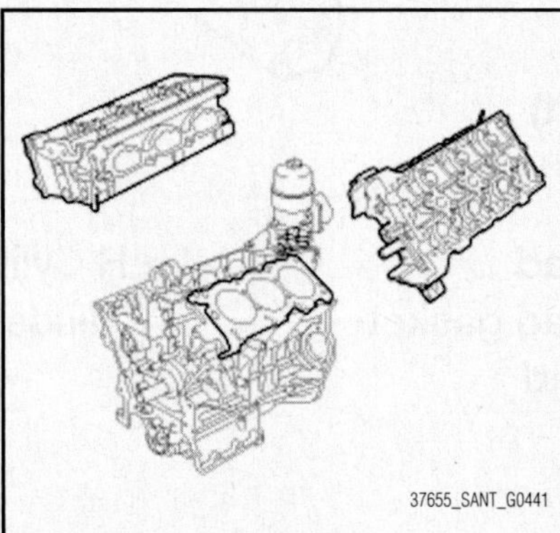

Fig. 200 Remove the cylinder head gaskets and cylinder head

   d. Remove the cylinder head gaskets.

To install:

➡Thoroughly clean all parts to be assembled. Always use a new head and manifold gasket. The cylinder head gasket is a metal gasket. Take care not to bend it. Rotate the crankshaft, set the No.1 piston at TDC.

10. Install the cylinder head.
   a. The sealant locations on cylinder head and cylinder block must be free of engine oil and ETC.
   b. Apply sealant on cylinder block top face before assembling cylinder head gaskets. The part must be assembled within 5 minutes after sealant was applied. Bead width: 0.078–0.118 in. (2.0–3.0 mm), Sealant locations: 0.039–0.059 in. (1.0–1.5mm) from block surface, Recommended sealant: Liquid sealant TB1217H
   c. Apply sealant on cylinder head gaskets after assembling cylinder head gaskets on cylinder block.
   The part must be assembled within 5 minutes after sealant was applied.

➡Be careful of the installation direction.

   d. Install the cylinder head.

➡Remove the extruded sealant after assembling cylinder heads.

11. Install cylinder head bolts.
   a. Do not apply engine oil on the threads and under the heads of the cylinder head bolts.
   b. Using SST (09221-4A000), install and tighten the cylinder head bolts and plate washers, in several passes, in the sequence shown. Tightening torque:
   - Head bolt: 28–30 ft. lbs. (37–41 Nm), plus 118–122°, plus 88–92°
   - Bolt "A": 14–17 ft. lbs. (19–24 Nm)

➡Always use a new cylinder head bolts.

12. Install the CVVT assembly and tighten to 48–56 ft. lbs. (65—77 Nm).

➡Install camshaft-inlet to dowel pin of CVVT assembly. Hold the hexagonal head wrench portion of the camshaft with a vise, and install the bolt and CVVT assembly. Do not

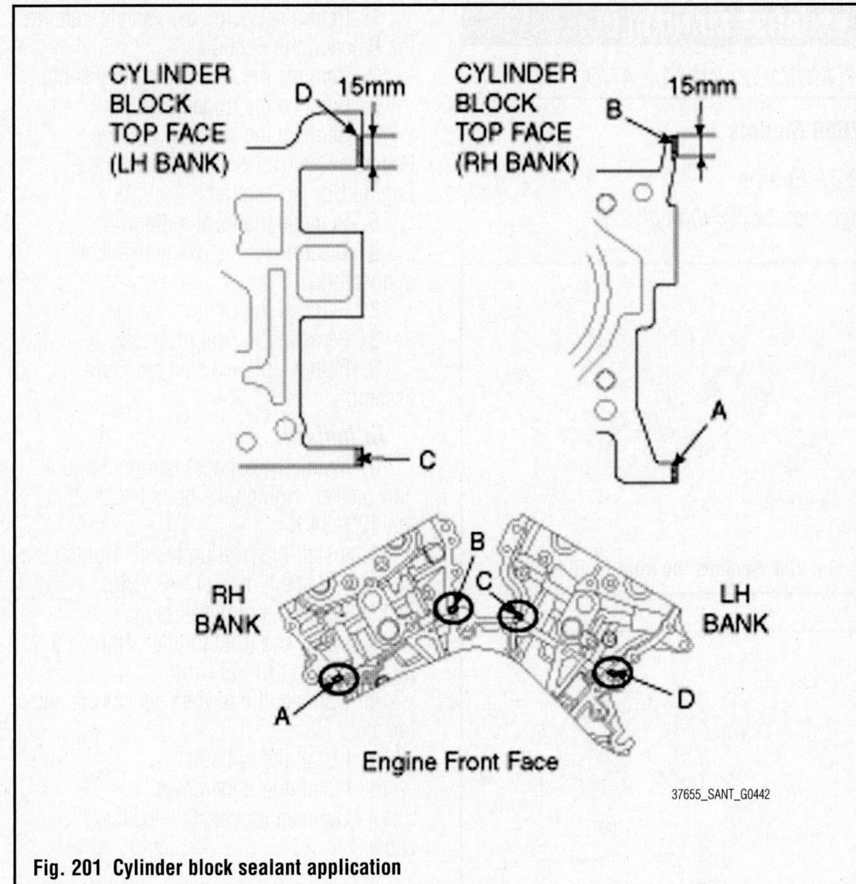

Fig. 201 Cylinder block sealant application

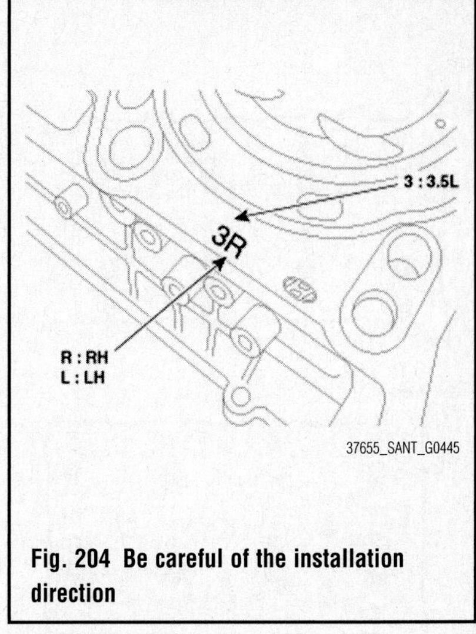

Fig. 204 Be careful of the installation direction

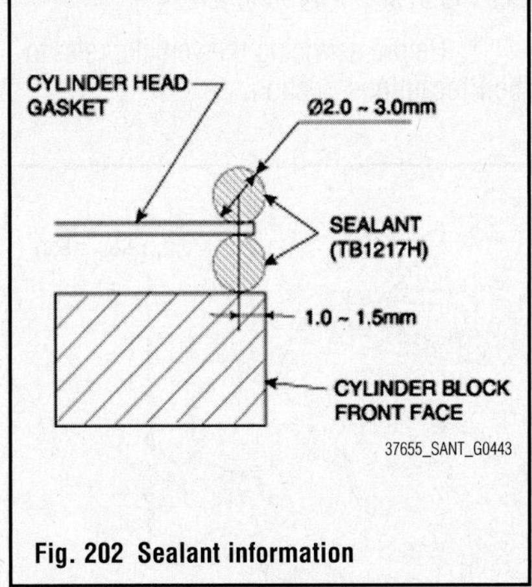

Fig. 202 Sealant information

rotate CVVT assembly when camshaft is installed to dowel pin of CVVT assembly.

13. Install the camshafts, camshaft bearing caps and thrust bearing caps.
14. Install the left-hand/right-hand exhaust camshaft OCV and tighten to 86–104 inch lbs. (10–12 Nm).
15. Install the water temperature control assembly and tighten to 15–17 ft. lbs. (20–24 Nm).
16. Connect the oil cooler hoses.
17. Install the timing chain.
18. Install the intake manifold and exhaust manifolds.
19. Refill engine oil.
20. Connect the negative battery cable.
21. Fill with engine coolant and bleed the cooling system.
22. Start the engine and check for leaks.
23. Recheck the engine coolant level and oil level.

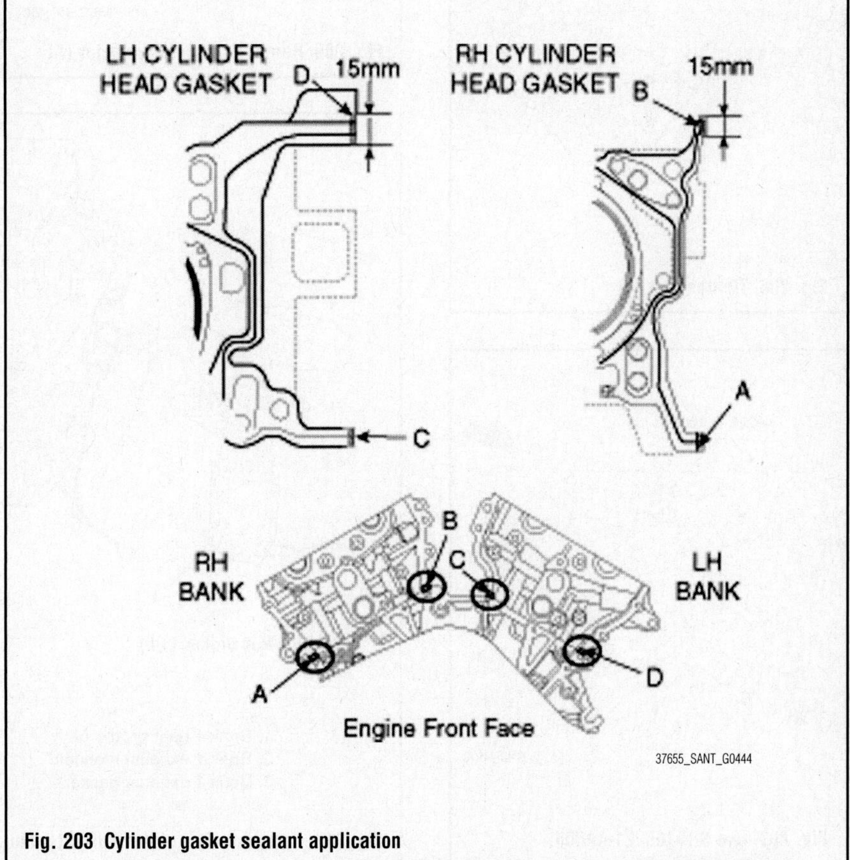

Fig. 203 Cylinder gasket sealant application

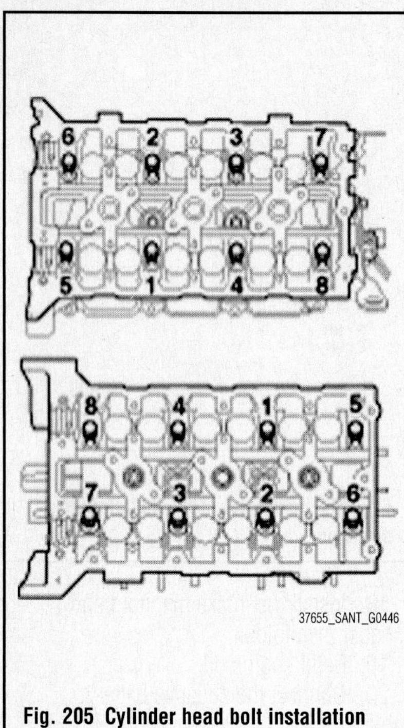

Fig. 205 Cylinder head bolt installation sequence

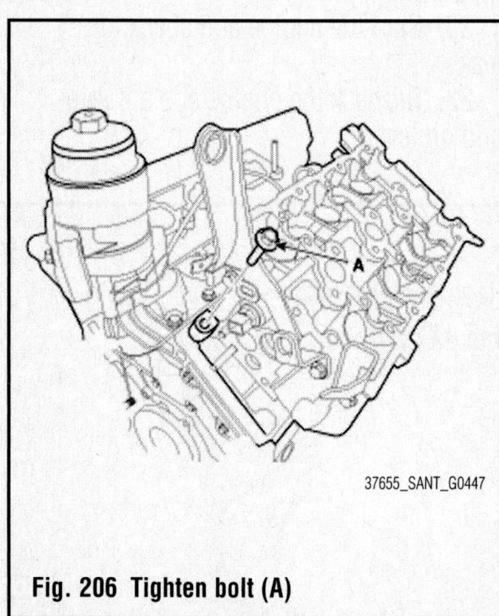

Fig. 206 Tighten bolt (A)

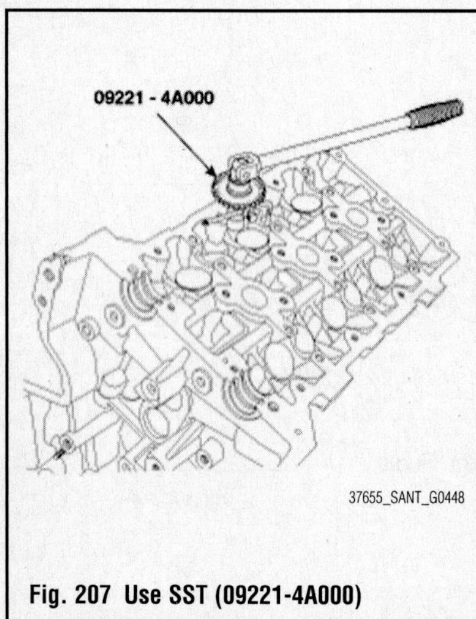

Fig. 207 Use SST (09221-4A000)

EXHAUST MANIFOLD

REMOVAL & INSTALLATION

2009 Models

2.7L Engine

See Figures 208 through 210.

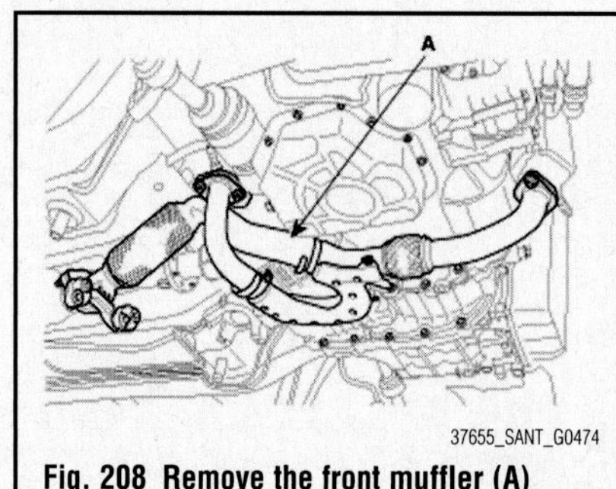

Fig. 208 Remove the front muffler (A)

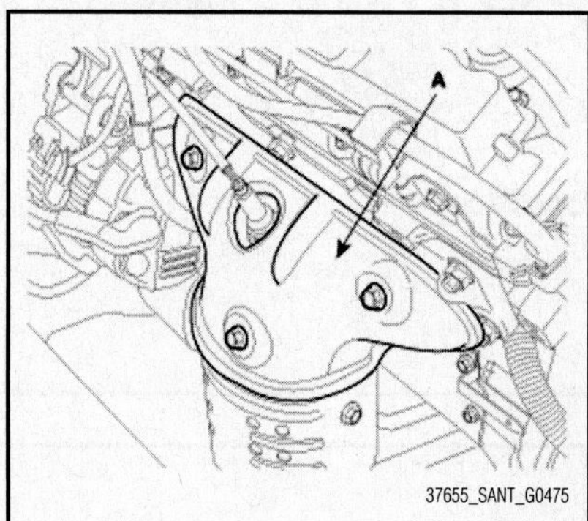

Fig. 209 Remove the heat protector (A)

1. Before servicing the vehicle, refer to the Precautions Section.
2. Remove the negative battery cable.
3. Remove the undercover.
4. Remove the radiator. Refer to Radiator Removal & Installation in the Engine Cooing section.
5. Remove the front muffler.
6. Disconnect the oxygen sensor connectors.
7. Remove the oil level gauge.
8. Remove the heat protector.
9. Remove the exhaust manifold assembly.

To install:

10. Install the exhaust manifold and a new gasket. Tighten the bolts to 22–25 ft. lbs. (29–34 Nm).
11. Install the heat protector. Tighten the bolts to 12–16 ft. lbs. (17–22 Nm).
12. Install the oil level gauge.
13. Install the front muffler. Tighten to 29–43 ft. lbs. (39–59 Nm).
14. Connect the oxygen sensor connector.
15. Install the radiator.
16. Install the undercover.
17. Connect the negative battery cable.

3.3L Engine

See Figures 211 through 213.

1. Before servicing the vehicle, refer to the Precautions Section.

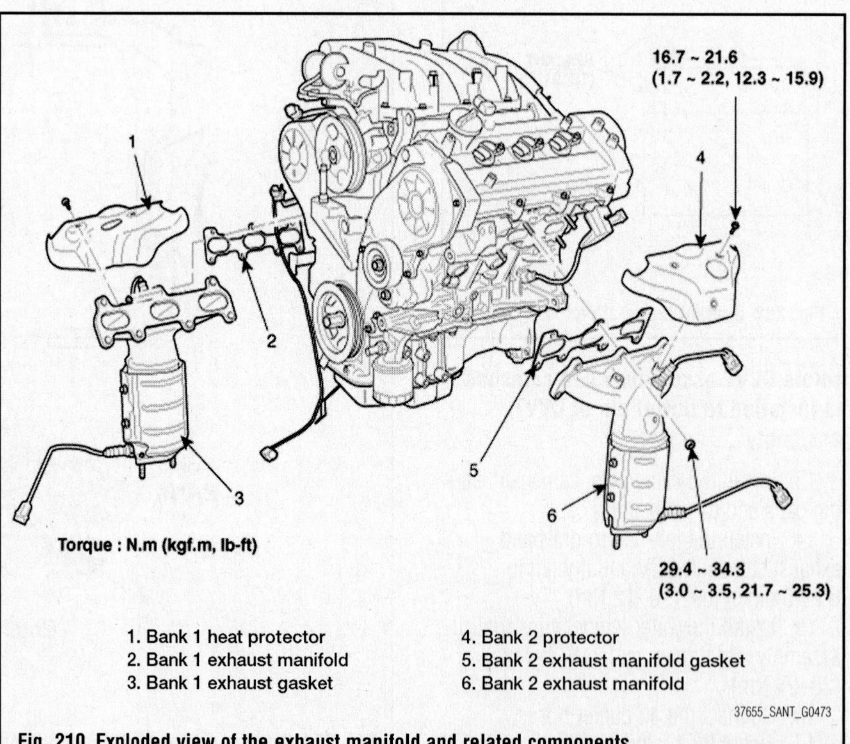

Fig. 210 Exploded view of the exhaust manifold and related components

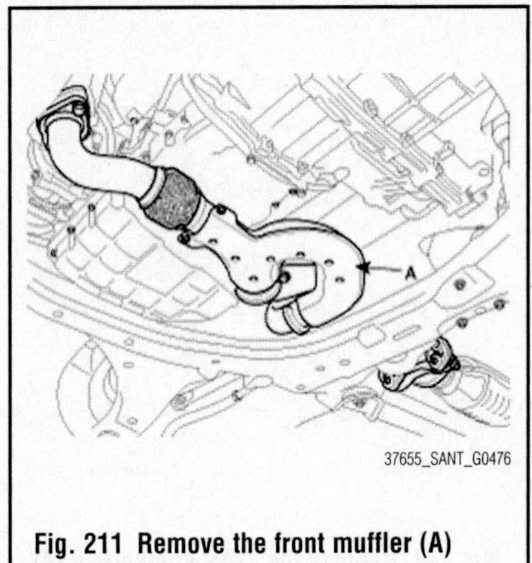

Fig. 211 Remove the front muffler (A)

A

37655_SANT_G0477

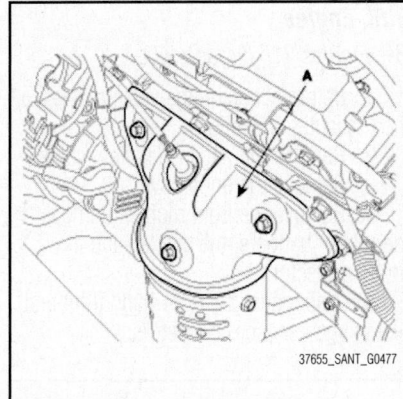

Fig. 212 Remove the left-hand heat protector (A)

2. Remove the negative battery cable.
3. Remove the undercover.
4. Remove the radiator. Refer to Radiator Removal & Installation in the Engine Cooing section.
5. Disconnect the left-hand, right-hand rear oxygen sensor connectors from the bracket.
6. Remove the front muffler.
7. Remove the oil level gauge.
8. Disconnect the left-hand front oxygen sensor connector from bracket.
9. Remove the left-hand heat protector.
10. Remove the left-hand exhaust manifold.
11. Disconnect the right-hand front oxygen sensor connector from the bracket.
12. Remove the right-hand heat protector.
13. Remove the right-hand exhaust manifold.

To install:

14. Install new gaskets and the exhaust manifolds. Tighten to 29–33 ft. lbs. (39–44 Nm).

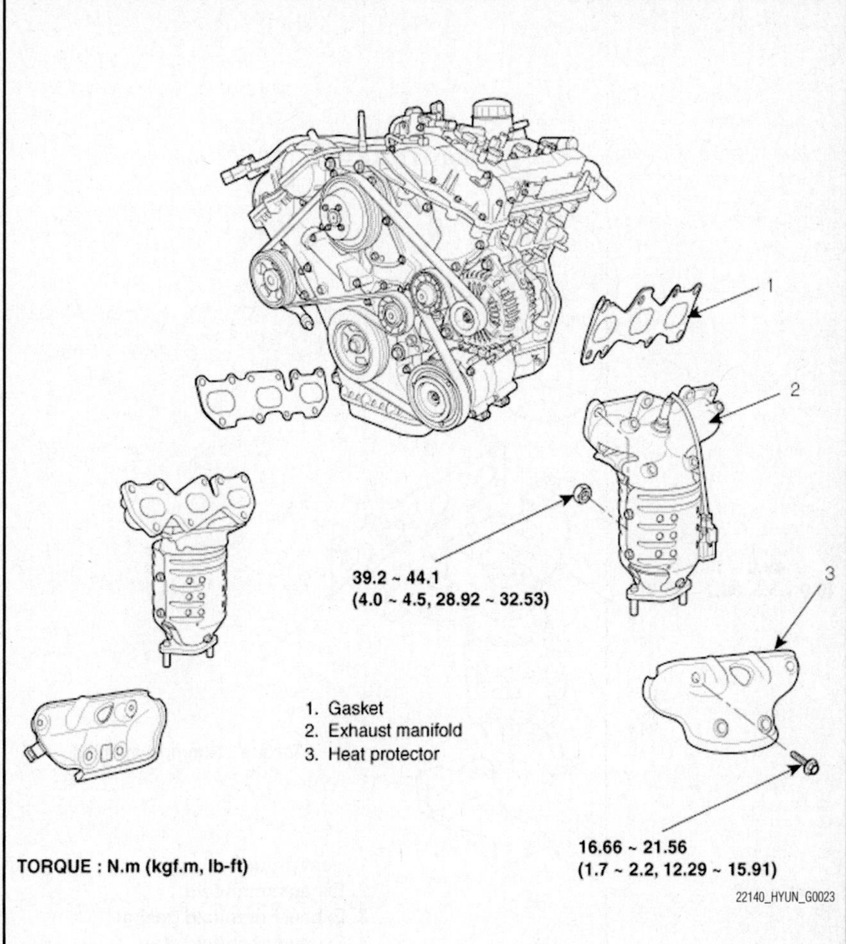

Fig. 213 Exploded view of the exhaust manifold and related components—3.3L and 3.8L engines

15. Install the heat protectors. Tighten to 12–16 ft. lbs. (17–22 Nm).
16. Install the front muffler. Tighten to 29–43 ft. lbs. (39–59 Nm).
17. Connect the oxygen sensor connectors.
18. Install the oil level gauge.
19. Install the undercover.
20. Install the radiator.
21. Connect the negative battery cable.

2010 Models

2.4L Engine

See Figures 214 through 217.

1. Before servicing the vehicle, refer to the Precautions Section.
2. Remove the negative battery cable.
3. Remove the engine cover.
4. Remove the air cleaner assembly.
5. Disconnect the oxygen sensor connectors.
6. Using the SST (Oxygen sensor socket wrench: 09392-2H100), remove the oxygen sensor (A) after disconnecting the connectors from the bracket and wiring clips, and then remove the heat protector (B).
7. Remove the front muffler.
8. Remove the exhaust manifold stay bolts, and then remove the exhaust manifold.

To install:

9. Installation is reverse order of removal, noting the following:
    a. Install a new manifold gasket.
    b. Tighten the exhaust manifold stay bolts to 38–43 ft. lbs. (52–58 Nm).
    c. Tighten the exhaust manifold nuts to 36–40 ft. lbs. (49–54 Nm).
    d. Tighten the muffler mounting bolts to 29–43 ft. lbs. (39–59 Nm).
    e. Tighten the oxygen sensor to 29–36 ft. lbs. (39–49 Nm).
    f. Tighten the heat protector bolts to 70–86 inch lbs. (8–10 Nm).

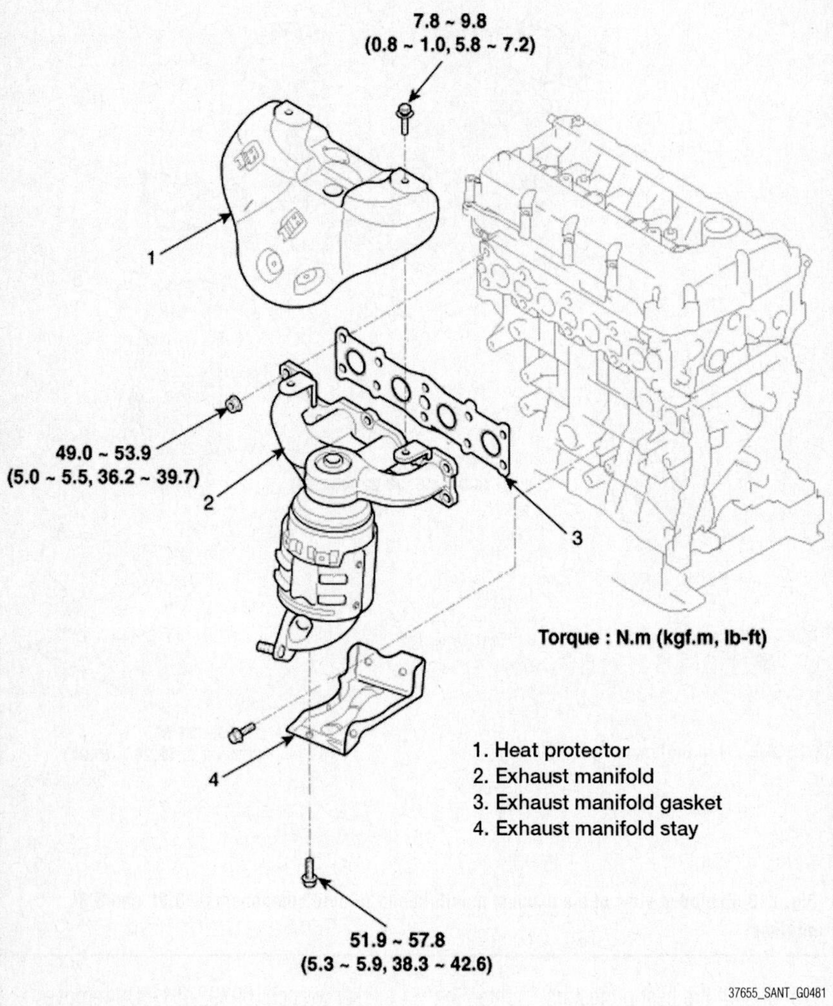

7.8 ~ 9.8
(0.8 ~ 1.0, 5.8 ~ 7.2)
1
49.0 ~ 53.9
(5.0 ~ 5.5, 36.2 ~ 39.7)
2
3
Torque : N.m (kgf.m, lb-ft)
4
1. Heat protector
2. Exhaust manifold
3. Exhaust manifold gasket
4. Exhaust manifold stay
51.9 ~ 57.8
(5.3 ~ 5.9, 38.3 ~ 42.6)
37655_SANT_G0481

Fig. 214 Exhaust manifold and components

SST:09392-2H100
B
A
37655_SANT_G0479

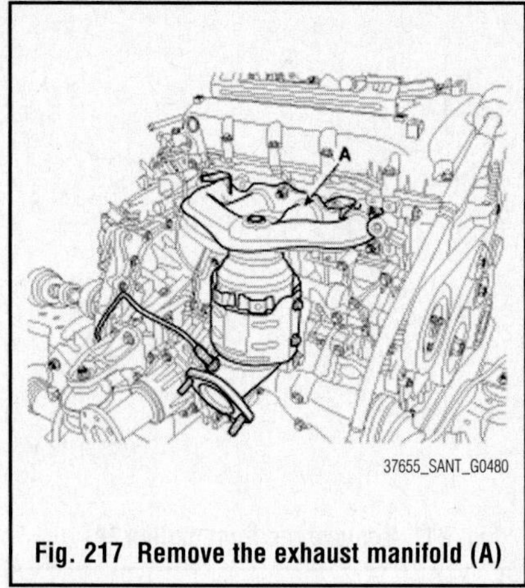

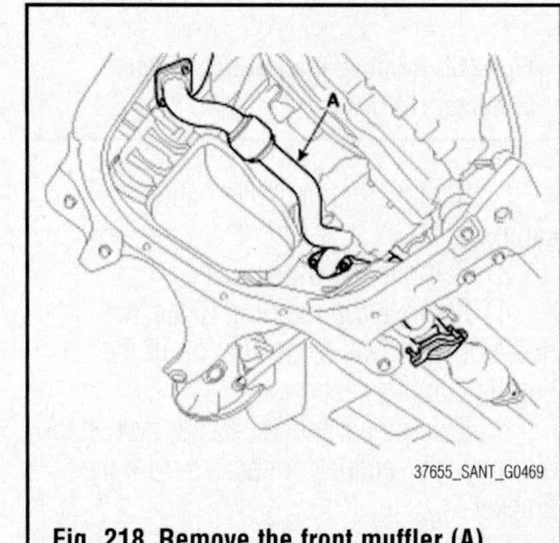

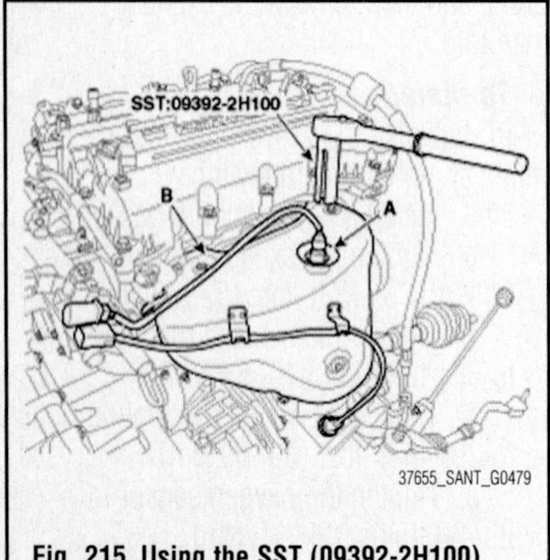

Fig. 215 Using the SST (09392-2H100), remove the oxygen sensor (A) and the heat protector (B)

A
37655_SANT_G0462

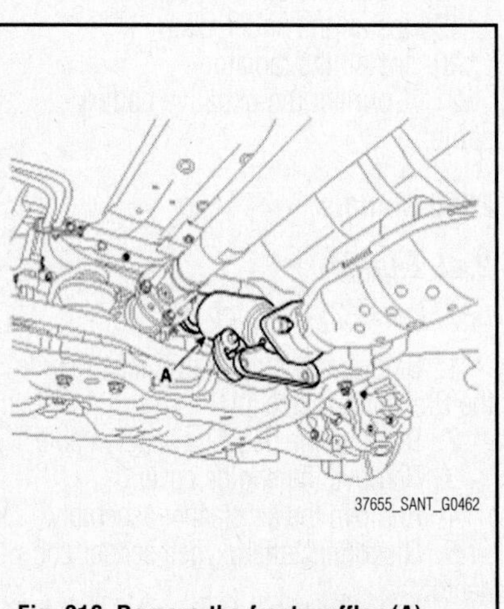

Fig. 216 Remove the front muffler (A)

A
37655_SANT_G0480

Fig. 217 Remove the exhaust manifold (A)

3.5L Engine

See Figures 218 through 222.

1. Before servicing the vehicle, refer to the Precautions Section.
2. Remove the negative battery cable.
3. Remove the undercover.
4. Remove the front muffler. Remove the rubber hangers, the bracket and the under protector.
5. Disconnect the right-hand front and rear oxygen sensor connectors.

A
37655_SANT_G0469

Fig. 218 Remove the front muffler (A)

37655_SANT_G0470

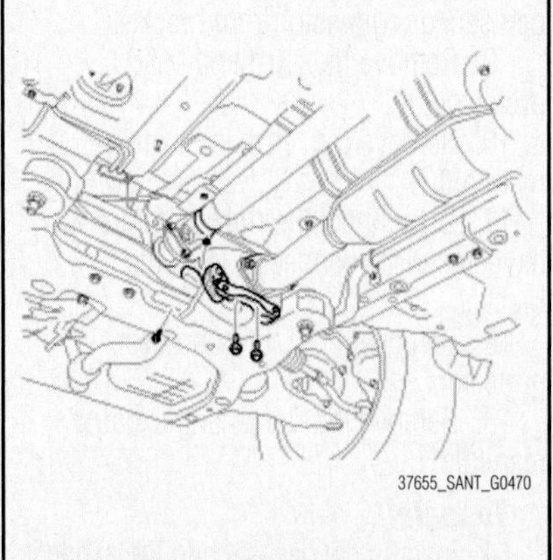

Fig. 219 Remove the rubber hangers, the bracket and the under protector

6. Disconnect left-hand front oxygen sensor connector.

7. Disconnect the left-hand rear oxygen sensor connector.

8. Remove the left-hand/right-hand exhaust manifold heat protectors.

9. Remove the manifold stays and remove the exhaust manifolds.

To install:

10. Installation is reverse order of removal, noting the following:

a. Tighten the exhaust manifold stay bolts to 18–26 ft. lbs. (25–35 Nm).

b. Tighten the exhaust manifold nuts to 29–33 ft. lbs. (39–44 Nm).

c. Tighten the heat protector bolts to 86–104 inch lbs. (10–12 Nm).

d. Tighten the muffler mounting bolts to 29–40 ft. lbs. (39–54 Nm)

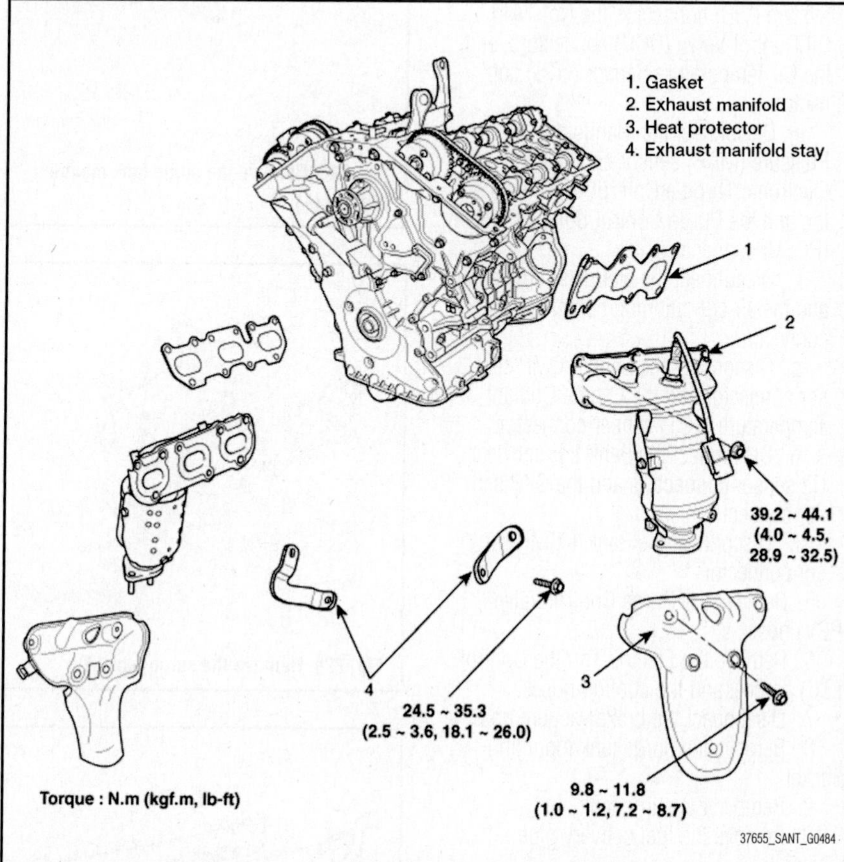

Fig. 220 Remove the manifold stays (A) and remove the exhaust manifolds (B)—one side shown, other side similar

INTAKE MANIFOLD

REMOVAL & INSTALLATION

2009 Models

2.7L Engine

See Figures 223 through 233.

1. Before servicing the vehicle, refer to the Precautions Section.

2. Disconnect the negative battery cable.

3. Remove the engine cover.

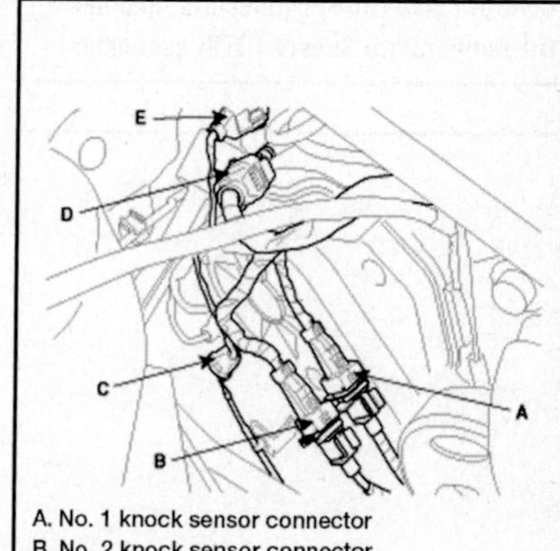

Fig. 223 Disconnect the knock sensor connectors, oil pressure switch connector, ignition coil harness, and the No. 1 VIS connector

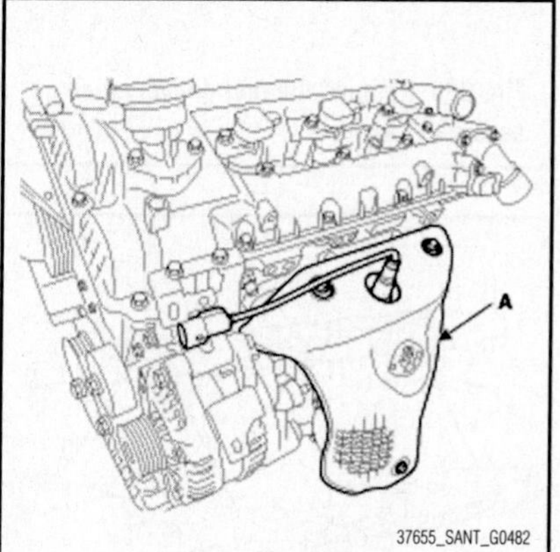

Fig. 221 Remove the left-hand/right-hand exhaust manifold heat protector (A)—one side shown, other side similar

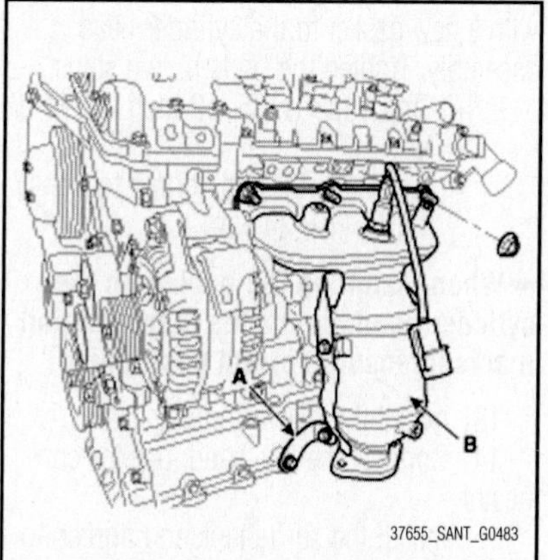

Fig. 222 Remove the manifold stays (A) and remove the exhaust manifolds (B)—one side shown, other side similar

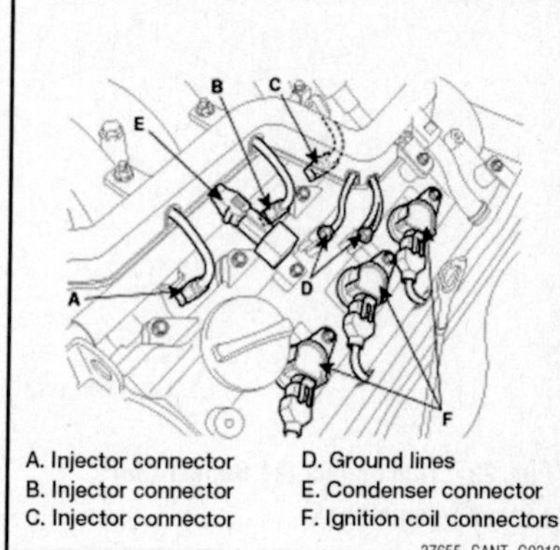

Fig. 224 Disconnect the injector connectors, ground lines, condenser connector, and the ignition coil connectors

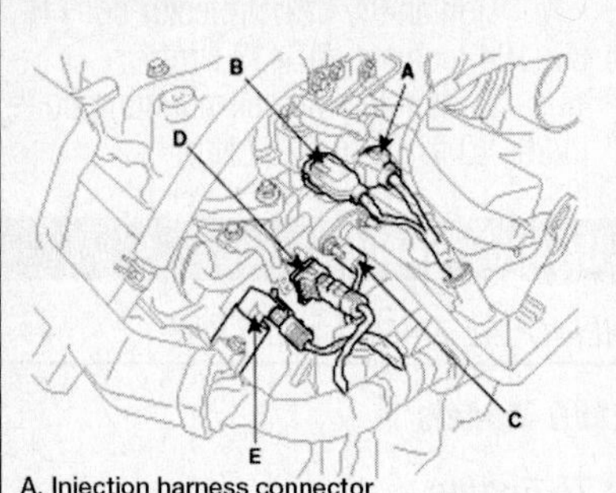

A. Injection harness connector
B. No. 2 Variable Induction System (VIS) connector
C. No. 1 Oil Control Valve (OCV) connectors
D. No. 2 Oil Control Valve (OCV) connectors
E. Oil Temperature Sensor (OTS) connector

37655_SANT_G0319

Fig. 225 Disconnect the injection harness connector, No. 2 Variable Induction System (VIS) connector, No. 1/No. 2 Oil Control Valve (OCV) connectors, and the Oil Temperature Sensor (OTS) connector

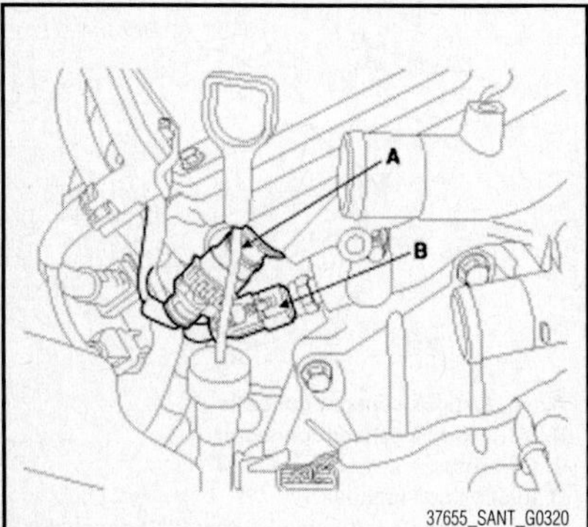

37655_SANT_G0320

Fig. 226 Disconnect the Bank 2 CMP sensor connector (A) and the Engine Coolant Temperature (ECT) sensor connector (B)

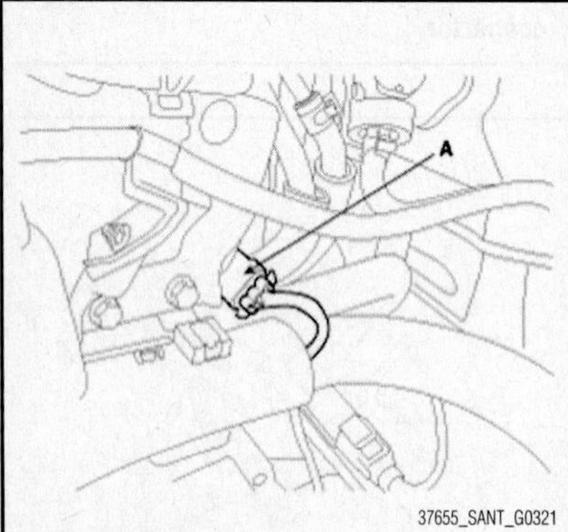

37655_SANT_G0321

Fig. 227 Disconnect the Bank 1 CMP sensor connector (A)

4. Remove the air cleaner.

a. Disconnect the knock sensor connectors, the oil pressure switch connector, the ignition coil harness, and the No. 1 Variable Induction System (VIS) connector.

b. Disconnect the Bank 1 Front/Rear O2 sensor connectors.

c. Disconnect the injector connectors, the ground lines, the condenser connector and the ignition coil connectors.

d. Disconnect the injection harness connector, the No. 2 Variable Induction System (VIS) connector, the No. 1/No. 2 Oil Control Valve (OCV) connectors, and the Oil Temperature Sensor (OTS) connector.

e. Disconnect the Manifold Absolute Pressure (MAP) sensor connector, the Electronic Throttle Control (ETC) connector, and the Purge Control Solenoid Valve (PCSV) connector.

f. Disconnect the alternator connector and the air conditioning compressor connector.

g. Disconnect the Bank 2 CMP sensor connector and the Engine Coolant Temperature (ECT) sensor connector.

h. Disconnect the Bank 2 Front/Rear O2 sensor connectors and the CKP sensor connector.

i. Disconnect the Bank 1 CMP sensor connector.

5. Remove the Purge Control Valve (PCV) hose.

6. Remove the Electric Throttle Control (ETC) bracket and the cooling hoses.

7. Disconnect the brake vacuum hose.

8. Remove the surge tank mounting bracket.

9. Remove the surge tank.

10. Remove the fuel delivery pipe assembly.

11. Remove the intake manifold assembly.

To install:

12. Install the intake manifold assembly with a new gasket to the cylinder head assembly. Tighten the bolts in two steps:

- Step 1 (a–h): 35–52 inch lbs. (4–6 Nm)
- Step 2 (1–8): and tighten to 14–17 ft, lbs. (19–24 Nm)

➡When installing the gasket on the cylinder head, check the identification marks to ensure correct positioning.

13. Install the delivery pipe.

14. Connect the left-hand injector connector.

15. Install the surge tank and tighten to 14–17 ft, lbs. (19–24 Nm).

16. Install the surge tank mounting bracket and tighten to 14–17 ft, lbs. (19–24 Nm).

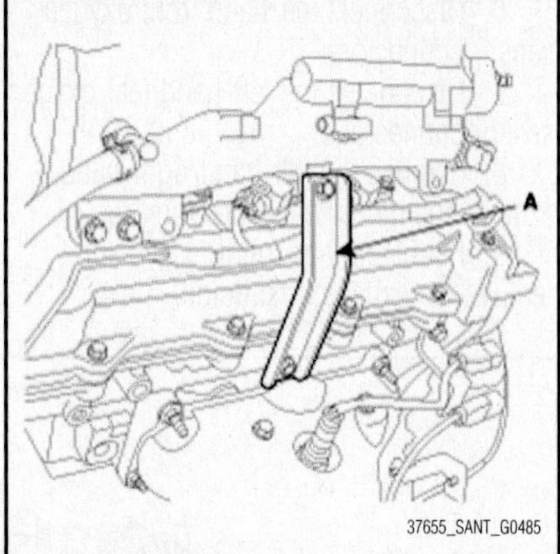

37655_SANT_G0485

Fig. 228 Remove the surge tank mounting bracket (A)

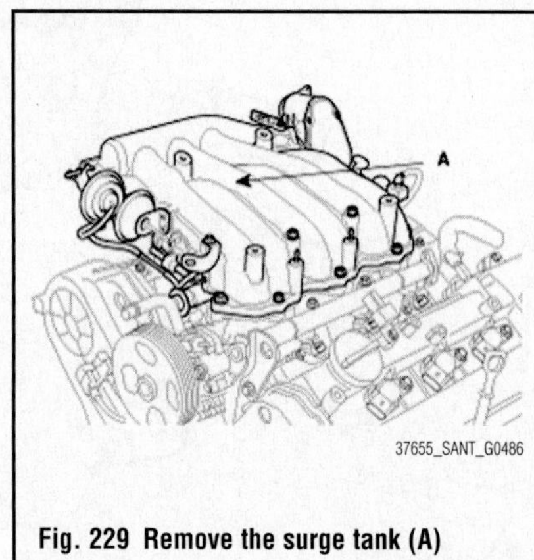

37655_SANT_G0486

Fig. 229 Remove the surge tank (A)

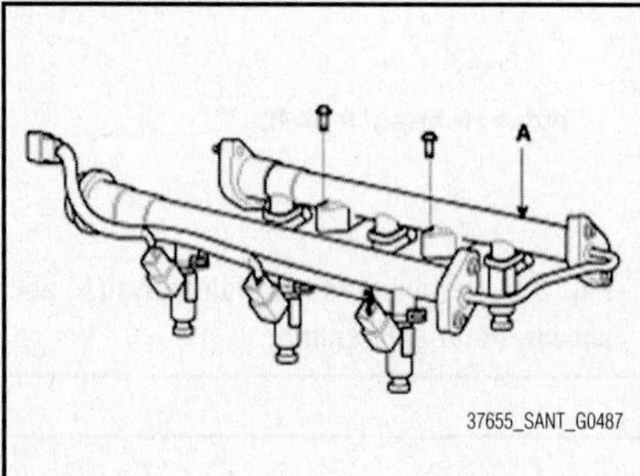

37655_SANT_G0487

Fig. 230 Remove the fuel delivery pipe assembly (A)

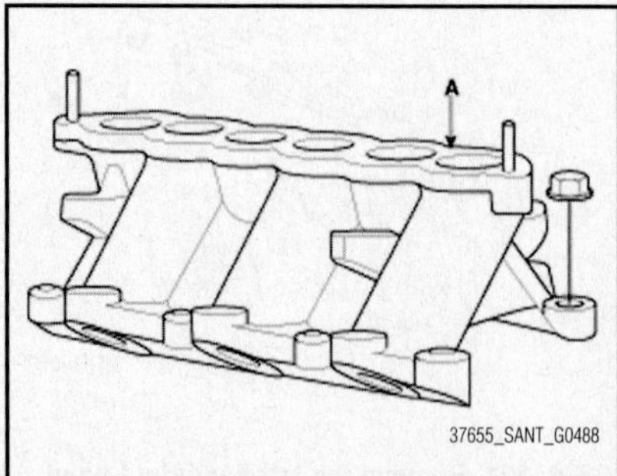

37655_SANT_G0488

Fig. 231 Remove the intake manifold assembly (A)

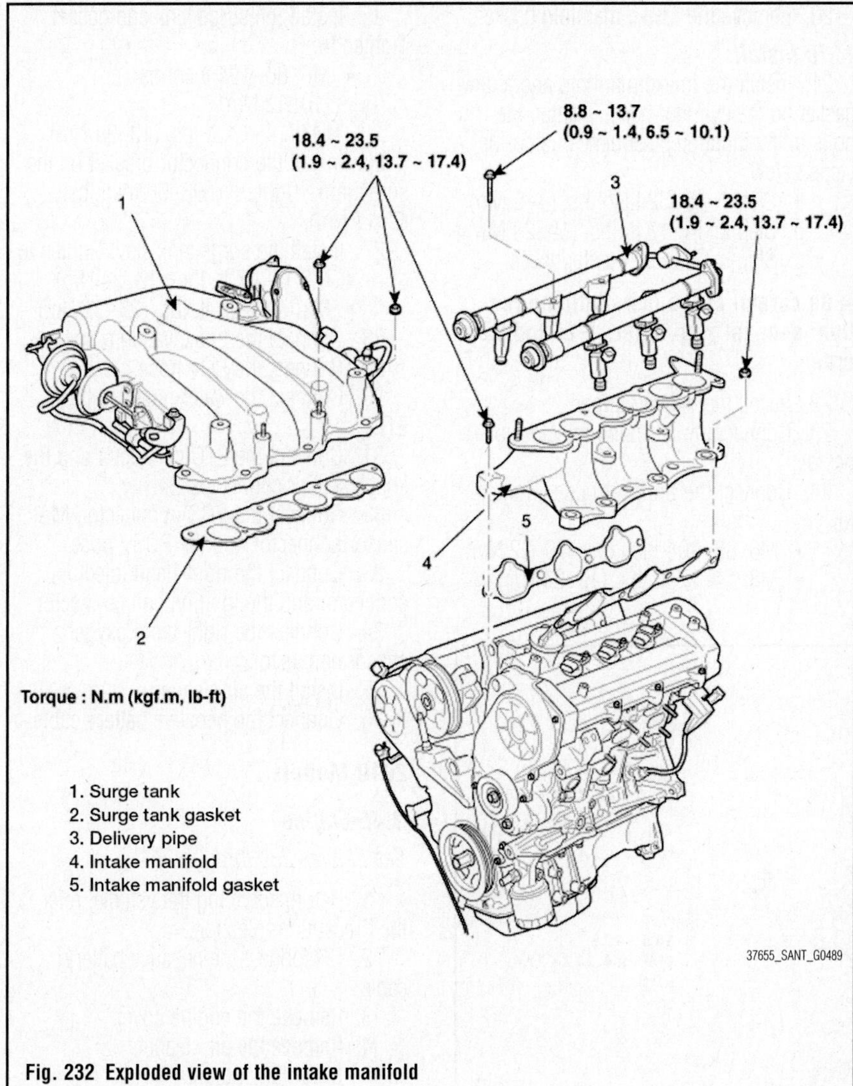

Fig. 232 Exploded view of the intake manifold

17. Install the Electronic Throttle Control (ETC) system fixing bracket.
18. Connect the hoses and connectors.
19. Install the air cleaner assembly.
20. Connect the negative battery cable.
21. Install the engine cover.
22. Start the engine and check for proper operation.

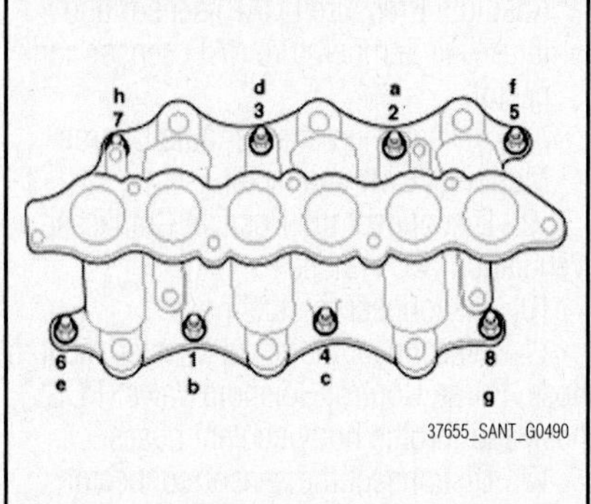

Fig. 233 Intake manifold bolt tightening sequence

3.3L Engine

See Figures 234 through 238.

1. Before servicing the vehicle, refer to the Precautions Section.

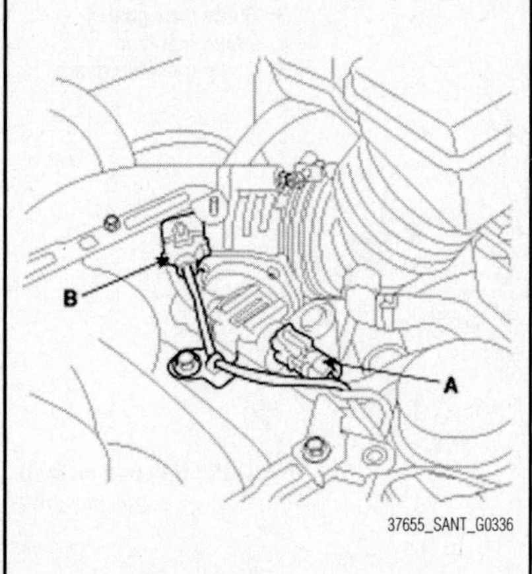

Fig. 234 PCSV connector (A) and MAP sensor connector (B)

2. Disconnect the negative battery cable.
3. Relieve the fuel system pressure.
4. Drain the cooling system.
5. Remove air cleaner and components.
6. Disconnect the right-hand oxygen sensor connector.
7. Disconnect the right-hand injector connector, ignition coil connector, and the VIS solenoid valve connector.
8. Disconnect ETC connector and knock sensor connector.
9. Disconnect the Purge Control Solenoid Valve (PCSV) connector, MAP (Manifold Absolute Pressure) sensor connector, PCSV hose, and the ECM vacuum hose.
10. Disconnect the water hoses from the ETC.
11. Disconnect the PCV hose.
12. Disconnect the brake vacuum hose.
13. Remove the surge tank stay.
14. Remove the connector bracket from the surge tank.
15. Remove the surge tank.
16. Disconnect the breather hose.
17. Disconnect the left-hand injector connector.
18. Remove the surge tank gasket.

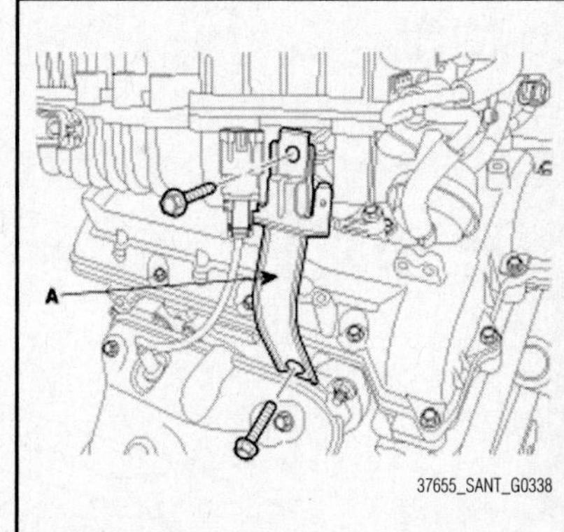

Fig. 235 Remove the surge tank stay (A)

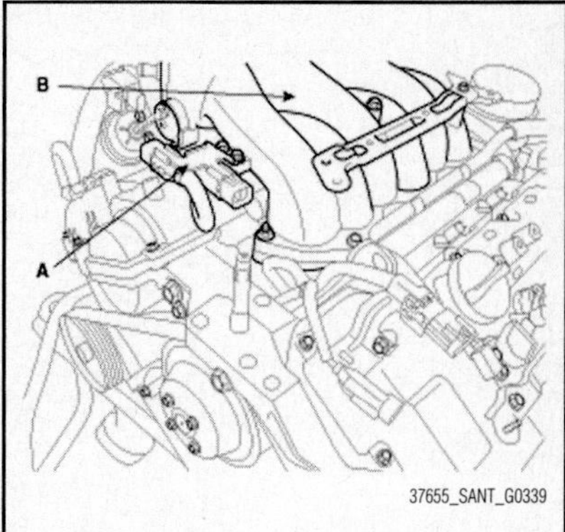

Fig. 236 Remove the connector bracket (A) and the surge tank (B)

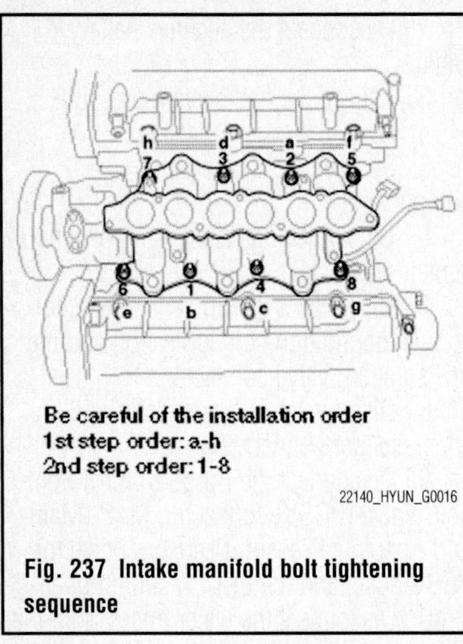

Be careful of the installation order
1st step order: a-h
2nd step order: 1-8
22140_HYUN_G0016

Fig. 237 Intake manifold bolt tightening sequence

19. Remove the intake manifold assembly.

20. Remove the intake manifold gasket.

To install:

21. Install the intake manifold and a new gasket on the cylinder head. Tighten the bolts in the illustrated sequence using the steps below:
    a. Step 1: 35–52 inch lbs. (4–6 Nm).
    b. Step 2: 14–17 ft. lbs. (19–24 Nm).
    c. Step 3: Repeat 2nd step twice.

➡Be careful of the installation direction. a–h: 1st step order, 1–8: 2nd step order

22. Install the delivery pipe.
23. Connect the left-hand injector connector.
24. Connect the breather hose. Tighten to:
    - M6: 86–104 inch lbs. (10–12 Nm)
    - M8: 14–17 ft. lbs. (19–24 Nm)

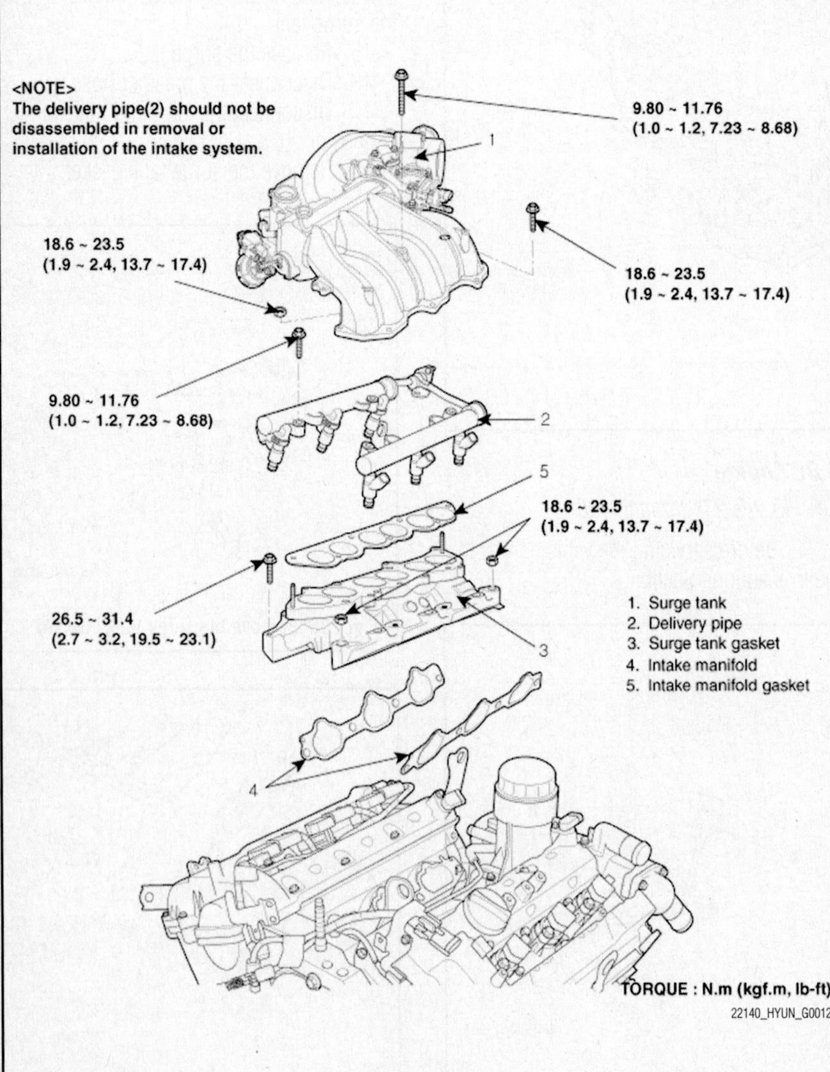

Fig. 238 Surge tank and intake manifold components

25. Install the surge tank and gasket. Tighten to:
    - M6: 86–104 inch lbs. (10–12 Nm)
    - M8: 14–17 ft. lbs. (19–24 Nm)
26. Install the connector bracket on the surge tank. Tighten to 60–96 inch lbs. (7–11 Nm).
27. Install the surge tank stay. Tighten to:
    - M8:14–17 ft. lbs. (19–24 Nm)
    - M10: 20–23 ft. lbs. (27–31 Nm)
28. Connect the brake vacuum hose.
29. Connect the PCV hose.
30. Connect the water hoses to the ETC.
31. Connect the ETC connector and the knock sensor connector.
32. Connect the PCSV connector, MAP sensor connector and the PCSV hose.
33. Connect the right-hand injector connector and the ignition coil connector.
34. Connect the right-hand oxygen sensor connector.
35. Install the air cleaner.
36. Connect the negative battery cable.

2010 Models

2.4L Engine

See Figures 239 through 244.

1. Before servicing the vehicle, refer to the Precautions Section.
2. Disconnect the negative battery cable.
3. Remove the engine cover.
4. Remove the air cleaner.
5. Drain the coolant.
6. Remove the radiator upper hose.
7. Disconnect the intake OCV (Oil control valve) connector.
    a. Disconnect the Variable Intake System (VIS) connector, oil pressure switch connector, knock sensor connector and the air compressor connector.
8. Disconnect the injector connectors.
    a. Disconnect the Electronic Throttle Control (ETC) connector and Manifold Absolute Pressure (MAP) sensor and Intake Air Temperature (IAT) sensor connector.
    b. Disconnect the Variable Charge Motion (VCM) connector.
9. Disconnect the Positive Crankcase Ventilation (PCV) hose.
10. Disconnect the fuel hose.
11. Disconnect the brake booster vacuum hose, Purge Control Solenoid Valve (PCSV) hose and throttle body coolant hoses.
12. Disconnect the sensor connector from the bracket and remove the intake manifold stay.

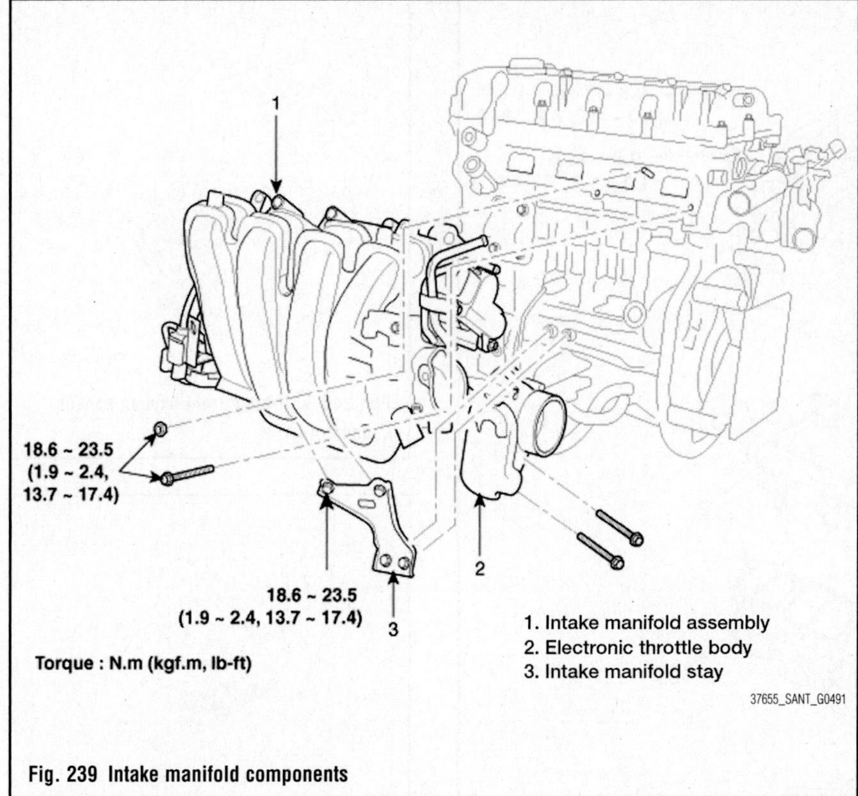

Fig. 239 Intake manifold components

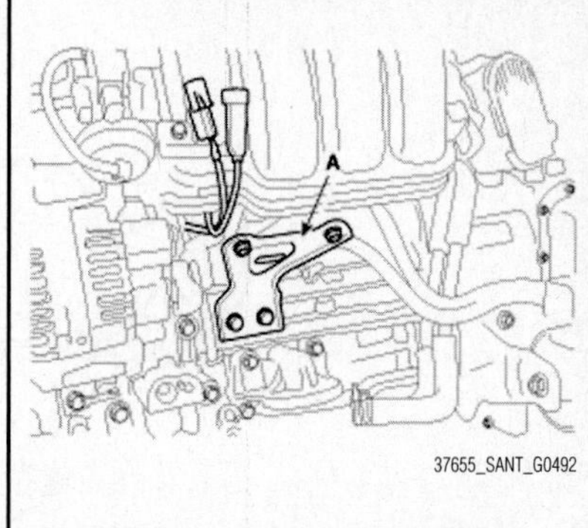

Fig. 243 Remove the intake manifold stay (A)

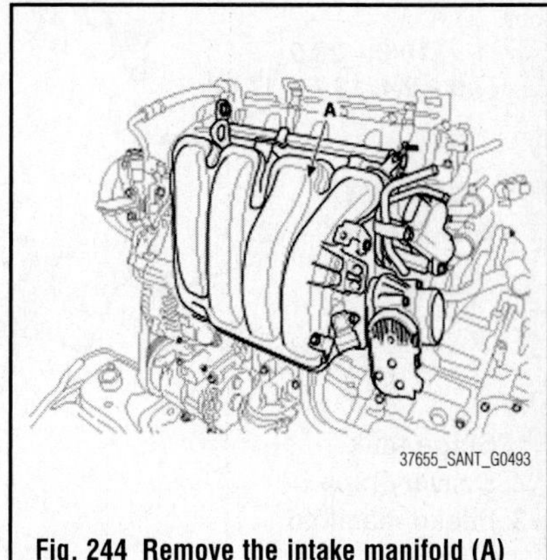

Fig. 244 Remove the intake manifold (A)

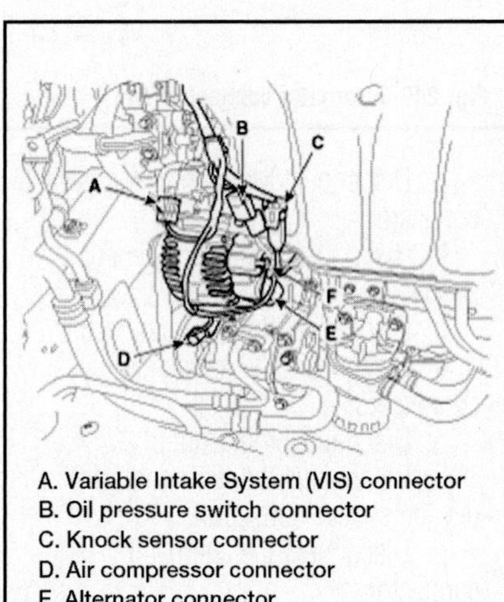

Fig. 240 Disconnect the connectors 1

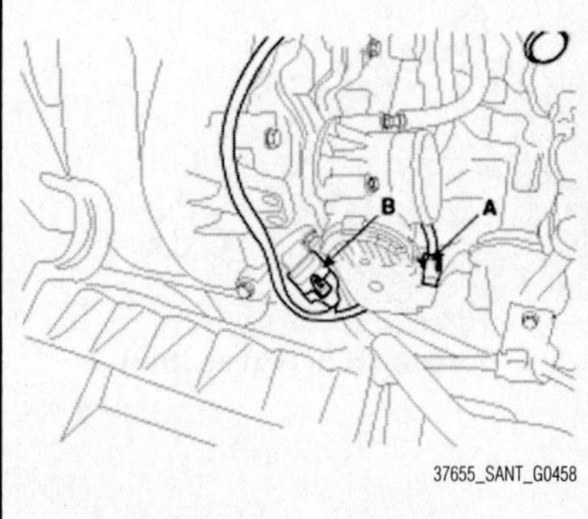

Fig. 241 Disconnect the Electronic Throttle Control (ETC) connector (A) and Manifold Absolute Pressure (MAP) sensor and Intake Air Temperature (IAT) sensor connector (B)

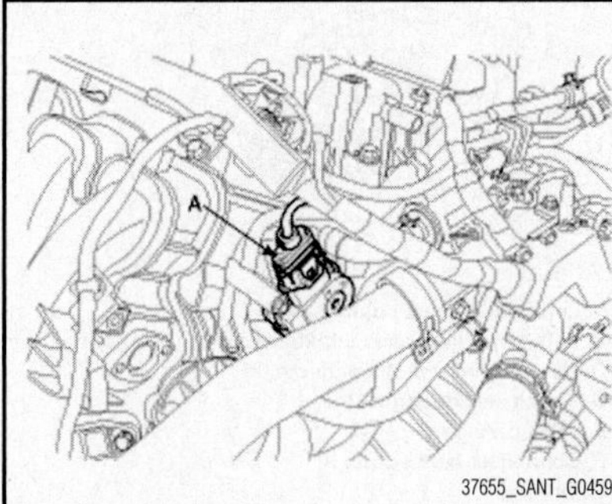

Fig. 242 Disconnect the Variable Charge Motion (VCM) connector (A)

13. Remove the intake manifold.

To install:

14. Installation is the reverse of removal, noting the following:
    a. Tighten the intake manifold stay mounting bolts to 14–17 ft. lbs. (19–24 Nm)
    b. Tighten the intake manifold mounting bolts and nuts to 14–17 ft. lbs. (19–24 Nm)

3.5L Engine

See Figures 245 through 256.

1. Before servicing the vehicle, refer to the Precautions Section.
2. Disconnect the negative battery cable.
3. Remove the engine cover.
4. Remove the air duct.
5. Remove the air cleaner.
6. Remove the undercover.
7. Drain the coolant.
8. Disconnect the wiring connectors, harness clamps and hoses from the engine:
    a. Disconnect the right-hand front and rear oxygen sensor connectors.
    b. Disconnect the VIS connector.
    c. Disconnect the right-hand exhaust camshaft OCV connector.
    d. Disconnect the right-hand ignition coil connector.
    e. Disconnect the right-hand injector connector.
    f. Disconnect the VCM connector.
    g. Disconnect the power steering switch connector.
    h. Disconnect the left-hand exhaust camshaft OCV connector.
    i. Disconnect the left-hand/right-hand intake camshaft OCV connector.

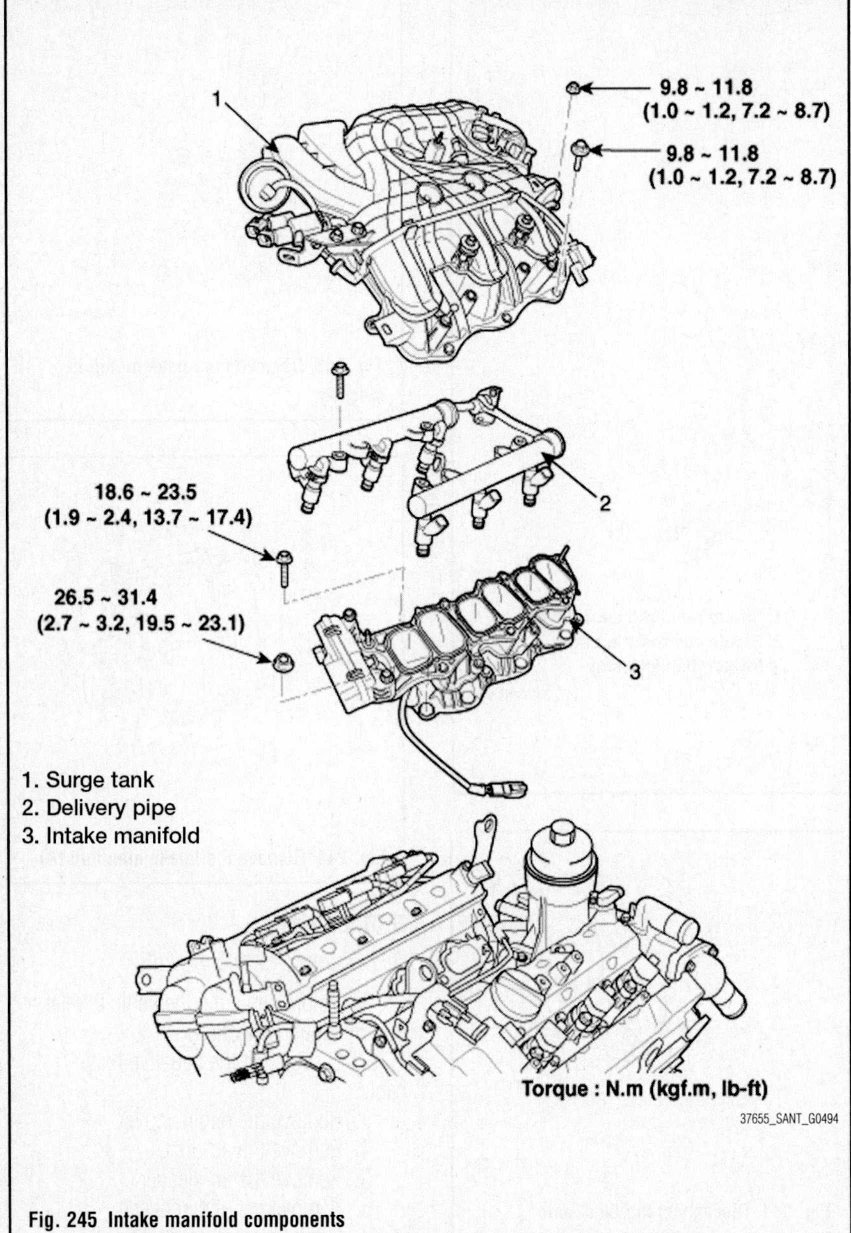

Fig. 245 Intake manifold components

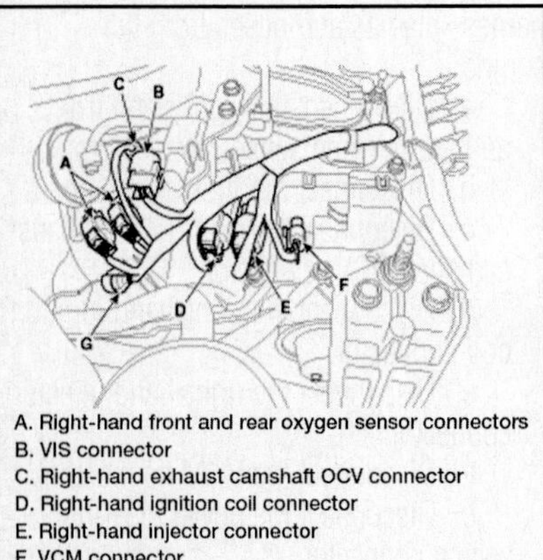

Fig. 246 Connectors 1

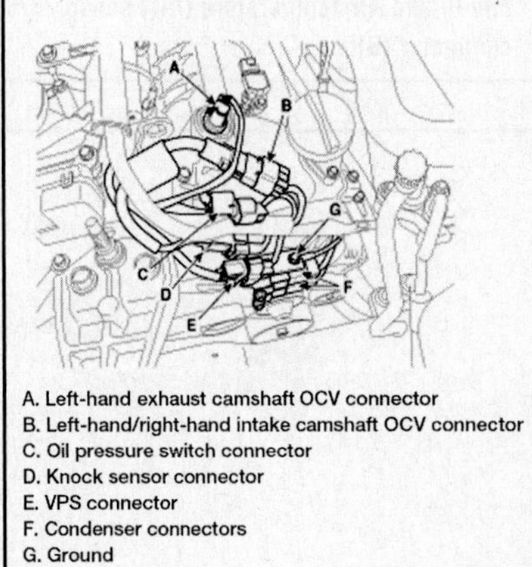

Fig. 247 Connectors 2

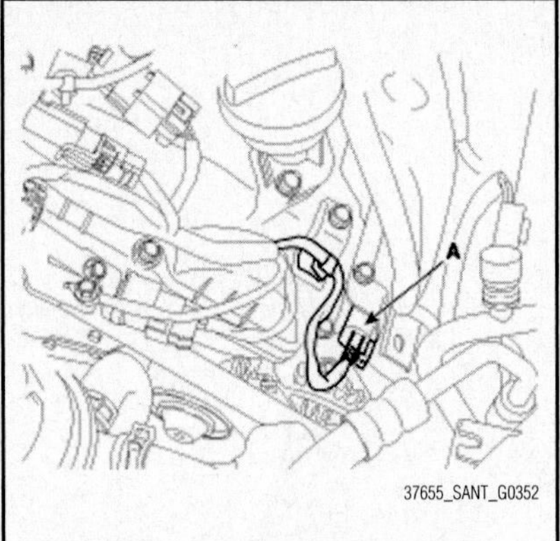

Fig. 248 Left-hand front oxygen sensor connector (A)

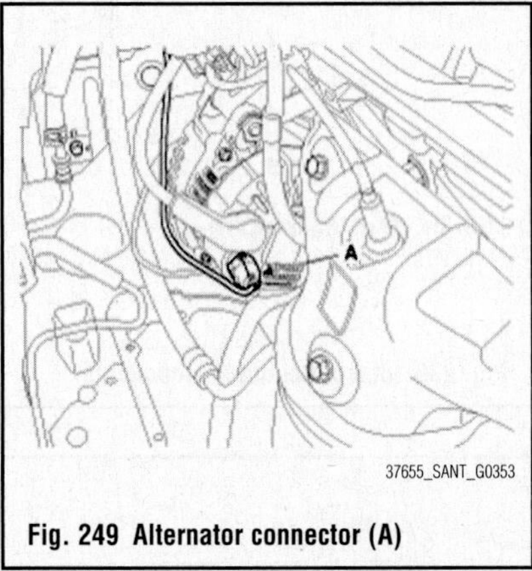

Fig. 249 Alternator connector (A)

j. Disconnect the oil pressure switch connector.
k. Disconnect the knock sensor connector.
l. Disconnect the VPS connector.
m. Disconnect the condenser connectors.
n. Disconnect the ground.
o. Disconnect the left-hand front oxygen sensor connector.
p. Disconnect the alternator connector.
q. Disconnect the left-hand ignition coil connectors.
r. Disconnect the left-hand injector connectors.
s. Disconnect VIS connector.
t. Disconnect the left-hand intake CMPS connector.
u. Disconnect MAP sensor connector.
v. Disconnect PCSV connector.
w. Disconnect right-hand intake CMPS connector.
x. Disconnect knock sensor connector.
y. Disconnect OTS connector.
z. Disconnect ETC connector.

aa. Disconnect right-hand exhaust CMPS connector.

9. Disconnect the PCSV hose, fuel hose and throttle body coolant hoses.

10. Remove the throttle body mounting bolt.

11. Disconnect the PCV hose and remove the surge tank stay.

12. Remove the surge tank.

13. Remove the delivery pipe and injector assembly.

14. Remove the intake the manifold.

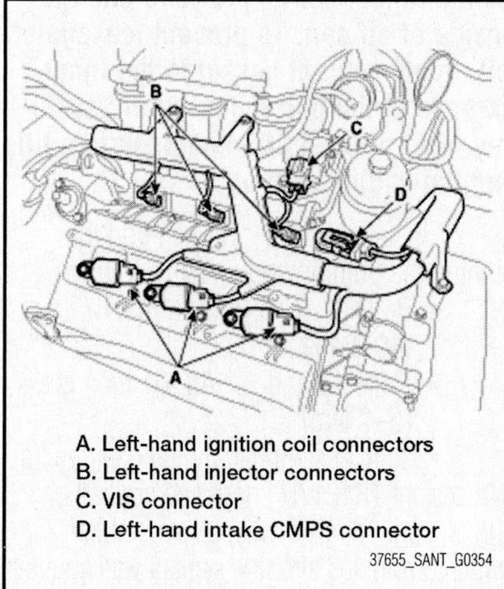

Fig. 250 Connectors 3

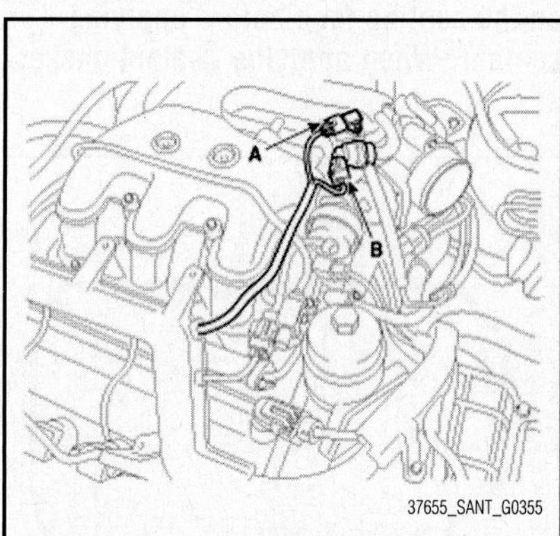

Fig. 251 MAP sensor connector (A) and PCSV connector (B)

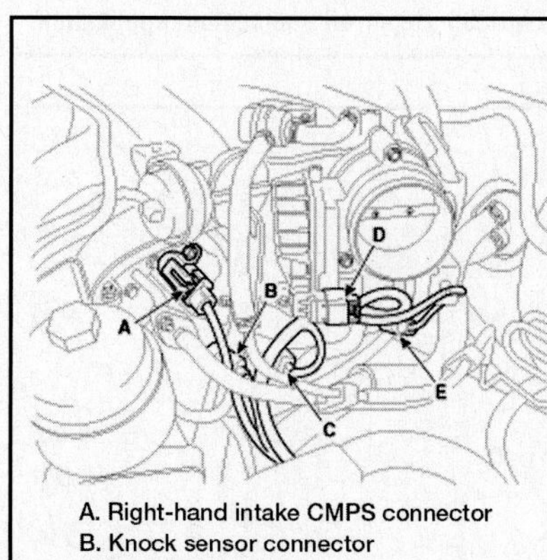

Fig. 252 Connectors 4

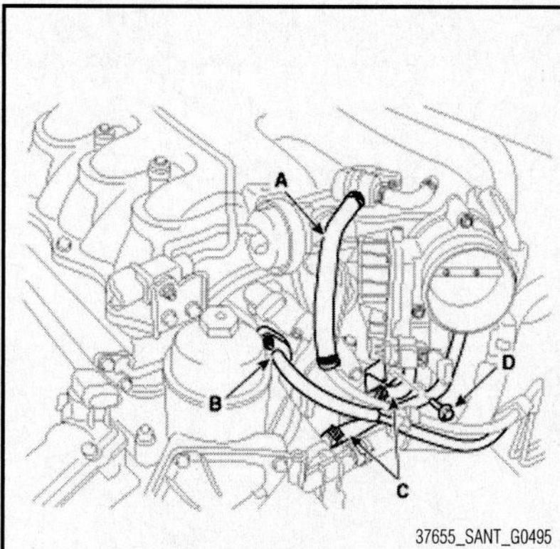

Fig. 253 Disconnect the PCSV hose (A), fuel hose (B), throttle body coolant hoses (C), and throttle body mounting bolt (D)

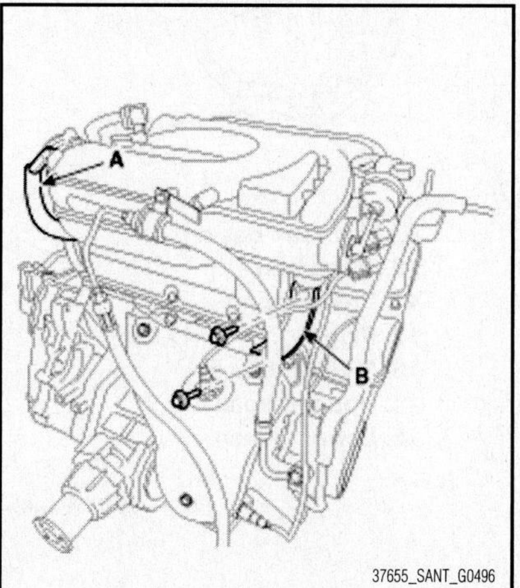

Fig. 254 Disconnect the PCV hose (A) and remove the surge tank stay (B)

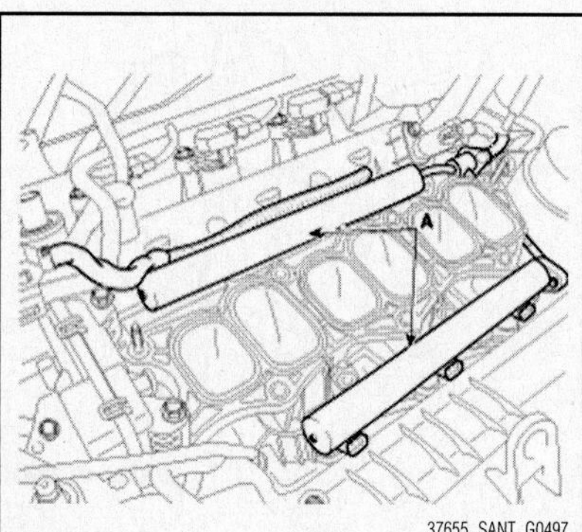

Fig. 255 Remove the delivery pipe and injector assembly (A)

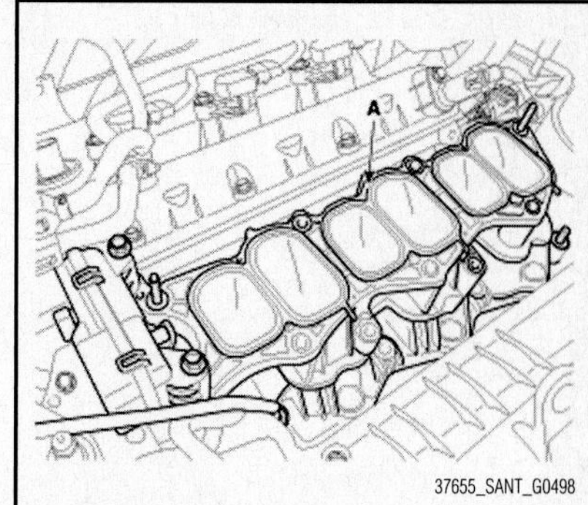

Fig. 256 Remove the intake the manifold (A)

➡Cover the inlet of cylinder head to prevent foreign materials from entering.

To install:

15. Installation is reverse order of removal, noting the following:

a. Tighten the intake manifold bolts and nuts, in several passes as shown. Tightening torque:

- Step 1: 35–52 inch lbs. (4–6 Nm)
- Step 2: Nut—14–17 ft. lbs. (19–24 Nm), Bolt—20–23 ft. lbs. (27–31 Nm)

b. Step 3: Repeat 2nd step twice.

➡Be careful of the installation direction. a–h: 1st step order, 1–8: 2nd step order

c. Tighten the throttle body mounting bolt to 14–17 ft. lbs. (19–24 Nm).

d. Tighten the surge tank stay to 20–23 ft. lbs. (28–31 Nm).

e. Tighten the surge tank to 86–104 inch lbs. (10–12 Nm).

OIL PAN

REMOVAL & INSTALLATION

2009 Models

2.7L Engine

See Figures 257 through 263.

1. Before servicing the vehicle, refer to the Precautions Section.

2. Disconnect the negative battery cable.

3. Drain the engine oil.

4. Using SST (09215-3C000), remove the lower oil pan.

5. Using SST (09215-3C000), remove the upper oil pan.

➡Be careful not to damage the contact surfaces of upper oil pan and lower oil pan.

To install:

6. Before assembling the oil pan, the liquid sealant TB1217H should be applied on the oil pan. The part must be assembled

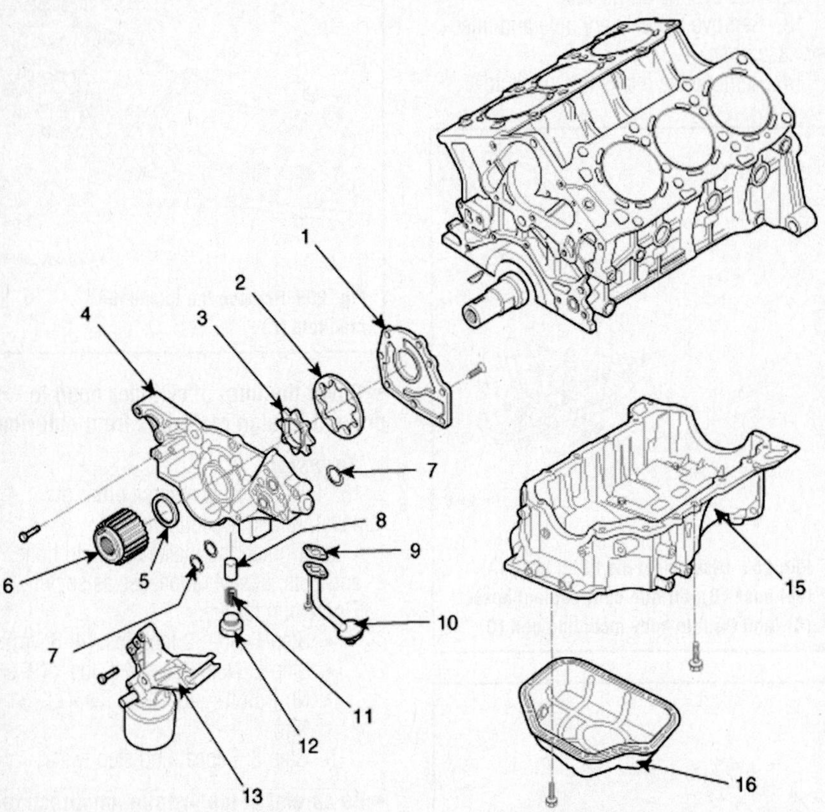

1. Oil pump cover
2. Oil pump outer rotor
3. Oil pump inner rotor
4. Oil pump case
5. Oil seal
6. Crankshaft sprocket
7. O-ring
8. Relief plunger
9. Oil screen gasket
10. Oil screen
11. Relief spring
12. Plug
13. Oil filter bracket
14. Upper oil pan
15. Lower oil pan

37655_SANT_G0499

Fig. 257 Oil pump, oil pan and related components

within 5 minutes after the sealant was applied.

➡Clean the sealing face before assembly. Remove harmful foreign materials on the sealing face before applying sealant. When applying sealant gasket, sealant must not be protrude into the inside of oil pan. To prevent leakage of oil, apply sealant gasket to the inner threads of the bolt holes. After assembly, wait at least 30 minutes before filling the engine with oil.

7. Install the upper oil pan and the tighten the bolts in sequence as follows:
   - Bolts 1–15: 14–17 ft. lbs. (19–24 Nm)
   - Bolts 16 and 17: 43–61 inch lbs. (5–7 Nm)
8. Before assembling the oil pan, the liquid sealant TB1217H should be applied on the oil pan. The part must be assembled within 5 minutes after the sealant was applied.

➡Clean the sealing face before assembly. Remove harmful foreign materials on the sealing face before applying sealant. When applying sealant gasket,

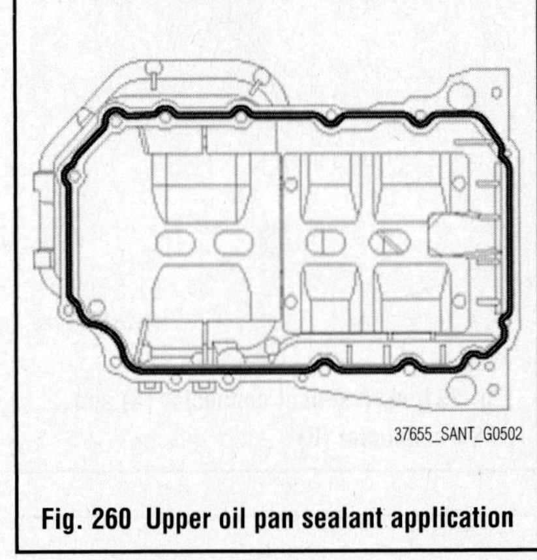

Fig. 260 Upper oil pan sealant application

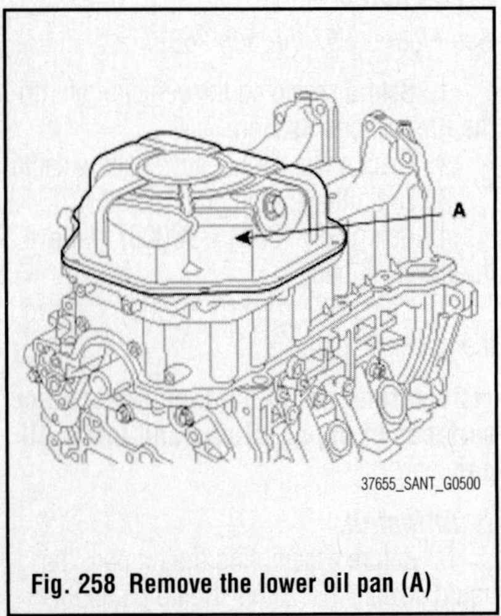

Fig. 258 Remove the lower oil pan (A)

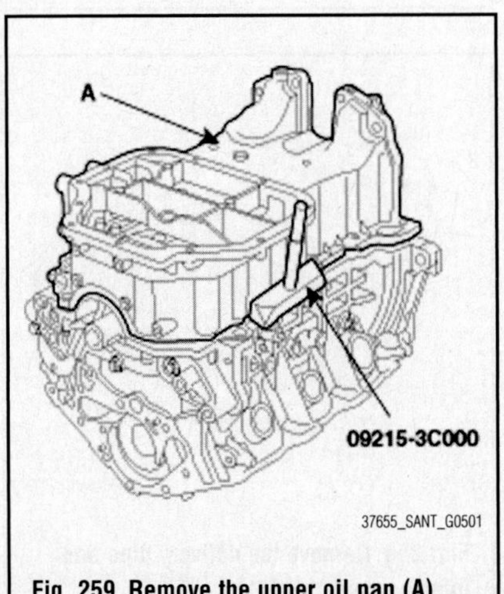

Fig. 259 Remove the upper oil pan (A)

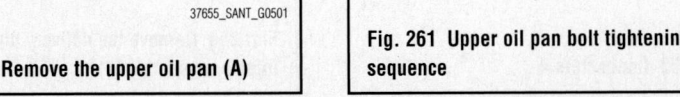

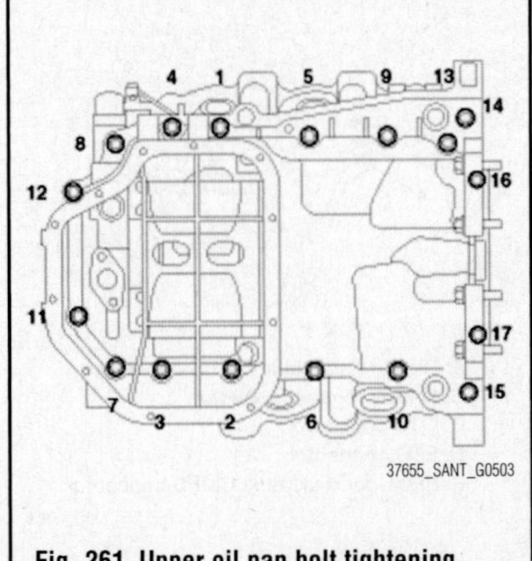

Fig. 261 Upper oil pan bolt tightening sequence

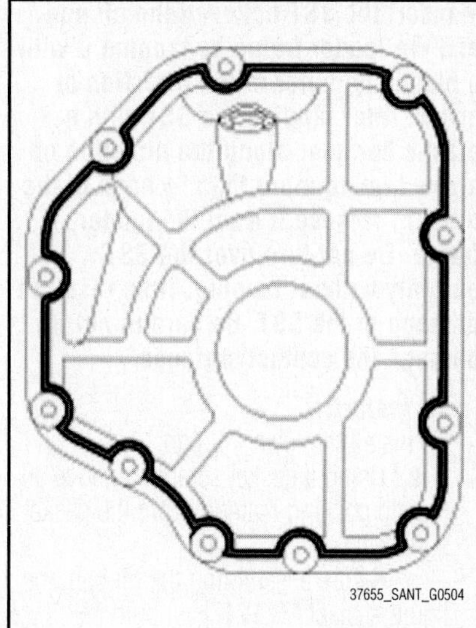

Fig. 262 Lower oil pan sealant application

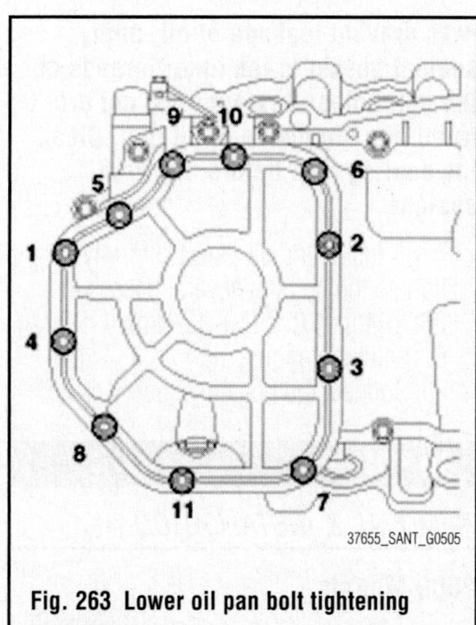

Fig. 263 Lower oil pan bolt tightening sequence

sealant must not be protrude into the inside of oil pan. To prevent leakage of oil, apply sealant gasket to the inner threads of the bolt holes. After assembly, wait at least 30 minutes before filling the engine with oil.

9. Install the lower oil pan and tighten the bolts to 86–104 inch lbs. (10–12 Nm).
10. Connect the negative battery cable.
11. Refill the crankcase with oil.

3.3L Engine

See Figures 264 through 266.

1. Before servicing the vehicle, refer to the Precautions Section.
2. Disconnect the negative battery cable.
3. Drain the engine oil.
4. Remove the oil pan bolts.

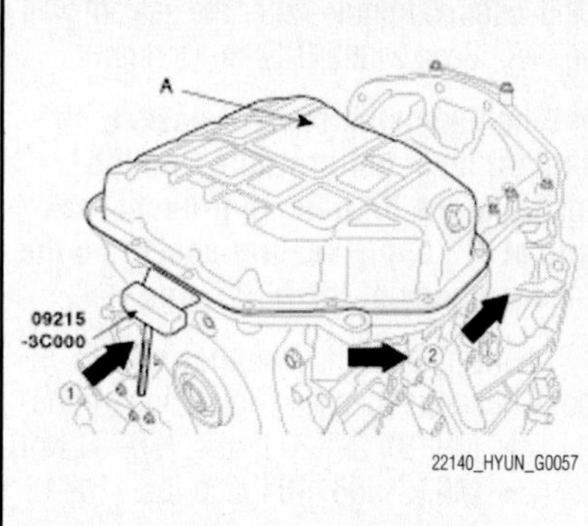

Fig. 264 Use the SST (09215-3C000) to remove the oil pan

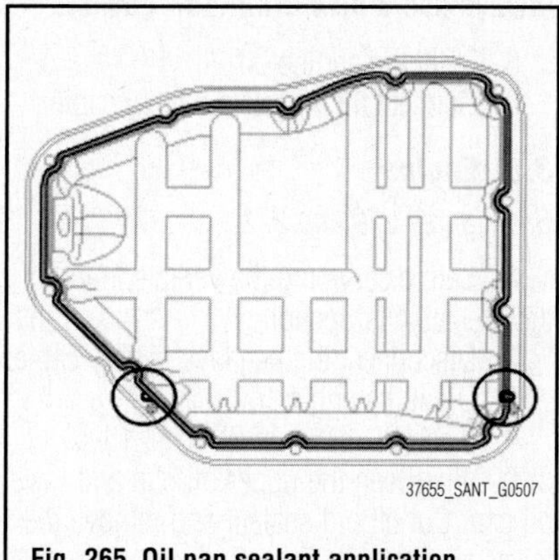

Fig. 265 Oil pan sealant application

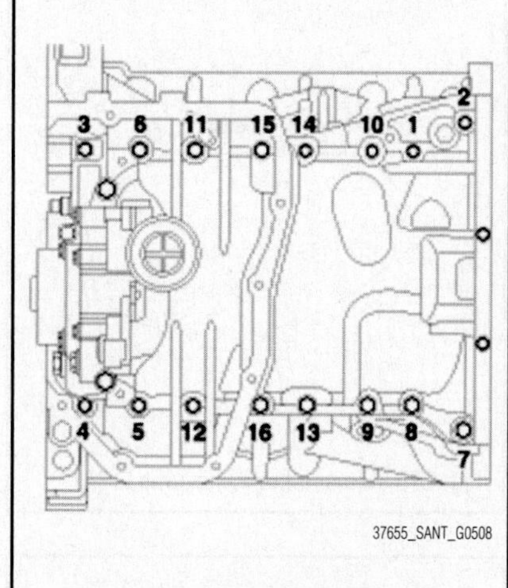

Fig. 266 Oil pan bolt tightening sequence

5. Using the SST (09215-3C000), remove the oil pan.
   a. Insert the SST between the oil pan and the ladder frame by tapping it with a plastic hammer in the direction of arrow.
   b. After tapping the SST with a plastic hammer along the direction of arrow around more than ⅔ of the edge of the oil pan, remove it from the ladder frame.

WARNING

Do not turn over the SST abruptly without tapping. Damage may occur to the SST or the oil pan.

➡Be careful not to damage the contact surfaces of upper oil pan and lower oil pan.

To install:

6. Using a gasket scraper, carefully remove all the old packing material from the gasket surfaces.
7. Before assembling the oil pan, the liquid sealant TB1217H should be applied on oil pan. The part must be assembled within 5 minutes after the sealant was applied. Bead width: 0.1 in. (2.5mm)

➡Clean the sealing face before assembly. Remove harmful foreign materials on the sealing face before applying sealant. When applying sealant gasket, sealant must not be protrude into the inside of oil pan. To prevent leakage of oil, apply sealant gasket to the inner threads of the bolt holes. After assembly, wait at least 30 minutes before filling the engine with oil.

8. Install the oil pan with the bolts. Uniformly tighten the bolts in several passes. Tighten to 86–104 inch lbs. (10–12 Nm).

➡After assembly, wait at least 30 minutes before filling the engine with oil.

9. Fill the engine with the proper type and amount of engine oil.

2010 Models

2.4L Engine

See Figures 267 through 269.

1. Before servicing the vehicle, refer to the Precautions Section.
2. Disconnect the negative battery cable.
3. Drain the engine oil.
4. Insert the blade of SST (09215-3C000) between the ladder frame and oil pan. Cut off applied sealer and remove the lower oil pan.

➡Insert the SST between the oil pan and the ladder frame by tapping it with a plastic hammer in the direction of arrow. After tapping the SST with a plastic hammer along the direction of arrow around more than ⅔ edge of the oil pan, remove it from the ladder frame. Do not use the SST as a prybar. Hold the tool in position (on gasket

line) and tap in with a light hammer. Be careful not to damage the contact surfaces of ladder frame and lower oil pan.

To install:

5. Using a gasket scraper, remove all the old packing material from the gasket surfaces.
6. Before assembling the oil pan, the liquid sealant LOCTITE® 5900H should be applied on oil pan. The part must be assembled within 5 minutes after the sealant was applied. Bead width: 0.12 in. (3.0mm)

➡When applying sealant gasket, sealant must not protrude into the inside of the oil pan. To prevent leakage of oil, apply sealant gasket on the inner threads of the bolt holes.

7. Install oil pan. Uniformly tighten the bolts in several passes. Tightening torque:
   - M9 (B): 22–25 ft. lbs. (30–34 Nm)
   - M6 (C): 86–104 inch lbs. (10–12 Nm)

➡After assembly, wait at least 30 minutes before filling the engine with oil. Always use a new drain bolt gasket.

8. Refill the engine oil.
9. Connect the negative battery cable.

3.5L Engine

See Figures 270 and 271.

1. Before servicing the vehicle, refer to the Precautions Section.
2. Disconnect the negative battery cable.
3. Drain the engine oil.
4. Insert the blade of SST(09215-3C000) between the upper oil pan and lower oil pan. Cut off old sealant and remove the lower oil pan.

➡Insert the SST between the oil pan and the ladder frame by tapping it with a plastic hammer in the direction of arrow. After tapping the SST with a plastic hammer along the direction of arrow around more than ⅔ edge of the oil pan, remove it from the ladder frame. Do not turn over the SST abruptly without tapping. It be result in damage of the SST. Be careful not to damage the contact surfaces.

To install:

5. Install the lower oil pan:
   a. Using a gasket scraper, remove all the old packing material from the gasket surfaces.
   b. Before assembling the oil pan, the liquid sealant TB 1217H should be applied. The part must be assembled within 5 minutes after the sealant was applied. Bead width: 0.1 in. (2.5mm)

➡To prevent leakage of oil, apply sealant gasket to the inner threads of the bolt holes. Sealant must not protrude into the inside of oil pan. Clean the sealing face before applying sealant.

   c. Install the oil pan. Uniformly tighten the bolts in several passes to 86–104 inch lbs. (10–12 Nm)
6. Refill the engine oil.
7. Connect the negative battery cable.

OIL PUMP

REMOVAL & INSTALLATION

2009 Models

2.7L Engine

See Figures 272 through 278.

1. Before servicing the vehicle, refer to the Precautions Section.
2. Drain engine oil.

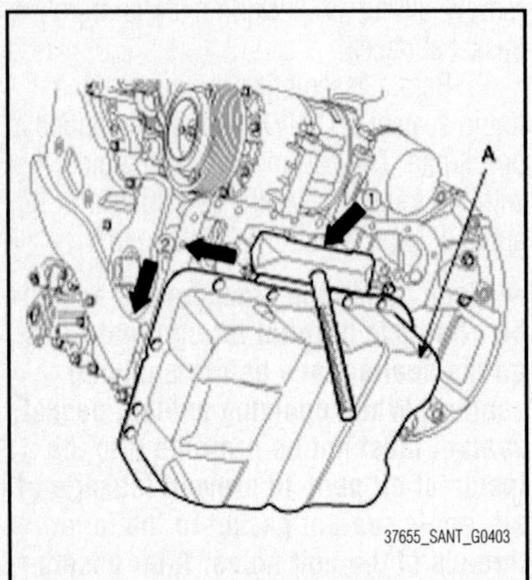

Fig. 267 Remove the oil pan (A)

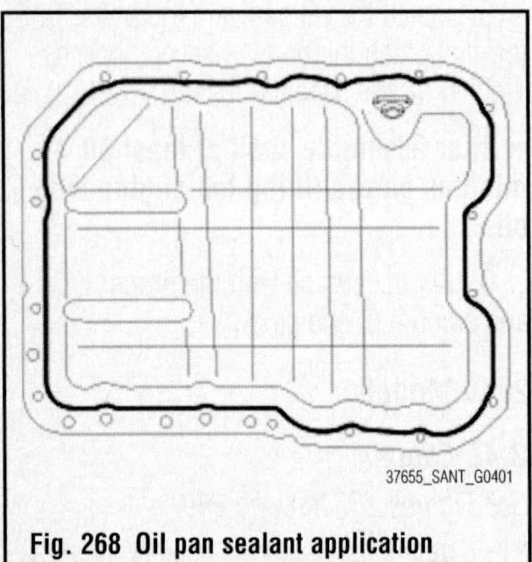

Fig. 268 Oil pan sealant application

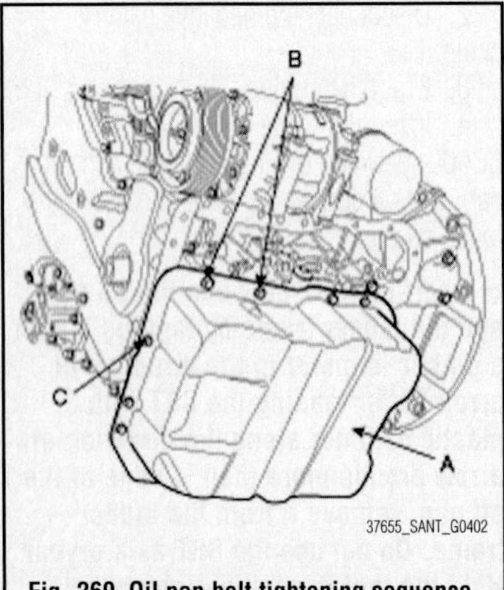

Fig. 269 Oil pan bolt tightening sequence

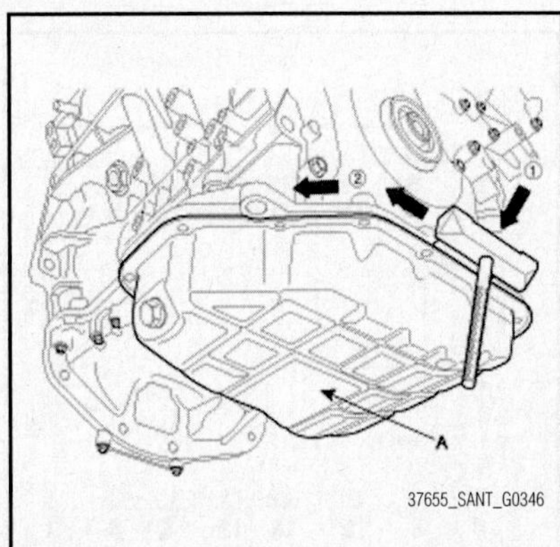

Fig. 270 Remove the lower oil pan (A)

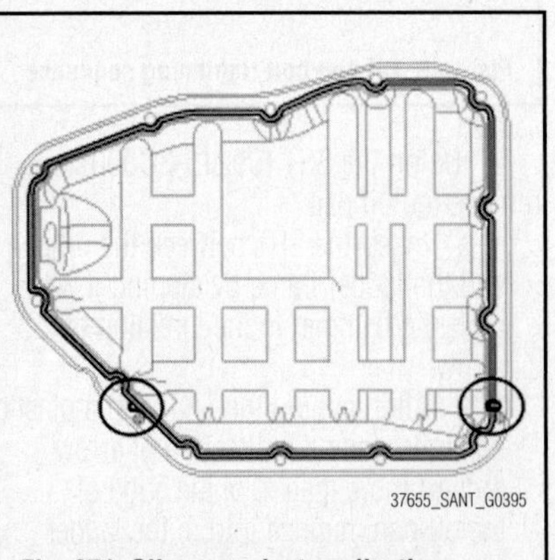

Fig. 271 Oil pan sealant application

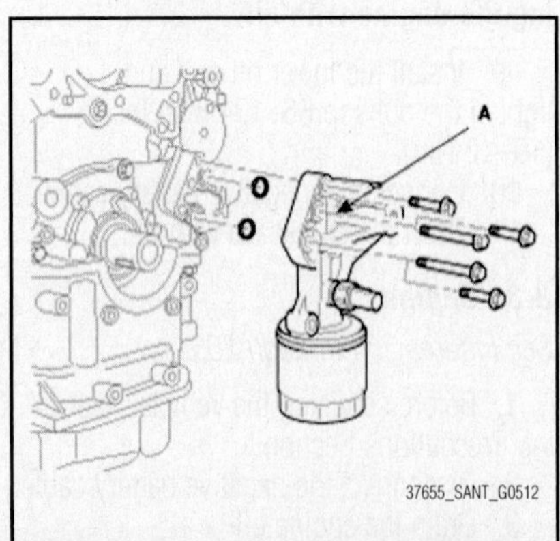

Fig. 272 Remove the oil filter bracket (A)

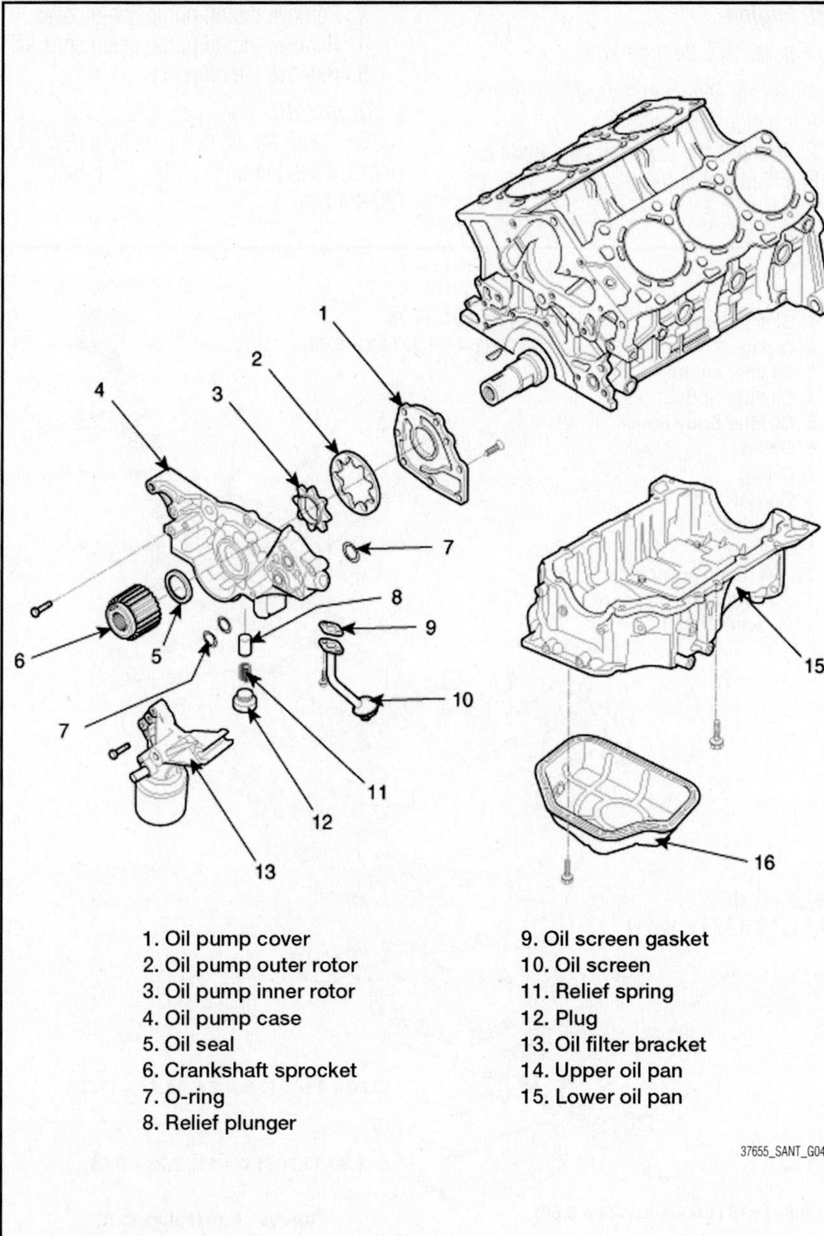

Fig. 273 Oil pump, oil pan and related components

3. Remove the right front wheel.
4. Remove the right front side cover.
5. Remove the front muffler.
6. Remove the alternator from the engine. Refer to Alternator Removal & Installation in the Engine Electrical section.
7. Remove the timing belt. Refer to Timing Belt Removal & Installation.
8. Remove the oil filter bracket.
9. Remove the lower oil pan. Refer to Oil Pan Removal & Installation.
10. Remove the oil screen.
11. Remove the upper oil pan. Refer to Oil Pan Removal & Installation.
12. Remove the oil pump case.
13. After removing the plug, remove the relief spring and the relief plunger.

To install:

14. Install the relief spring, relief plunger and tighten the plug. Tighten to 29–36 ft. lbs. (39–49 Nm).
15. Install oil pump case:
    a. Using a gasket scraper, remove all the old packing material from the gasket surfaces.
    b. Before assembling the oil pan, the liquid sealant TB1217H should be applied on the oil pan. The part must be assembled within 5 minutes after the sealant was applied. Bead width: 0.0984 in. (2.mm)

➡Clean the sealing face before assembling two parts. Remove harmful foreign materials on the sealing face before applying sealant. When applying sealant gasket, sealant must not be protrude into the inside of oil pan. To

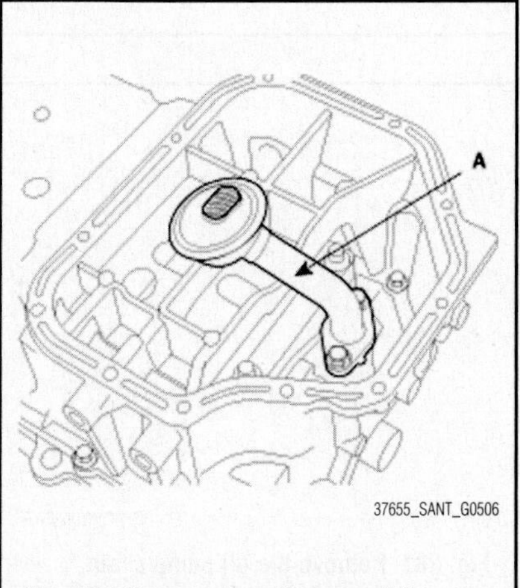

Fig. 274 Remove the oil screen (A)

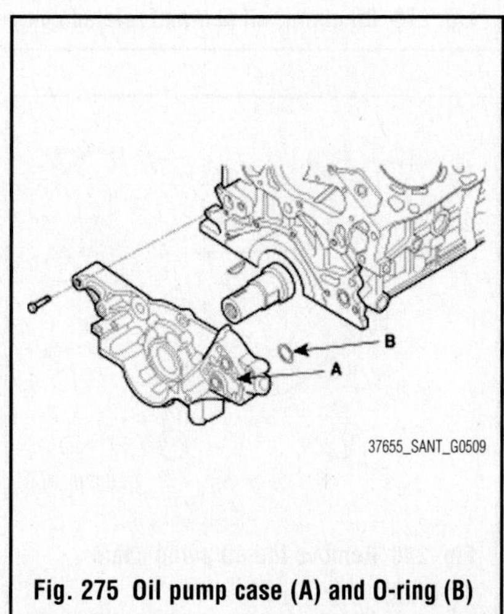

Fig. 275 Oil pump case (A) and O-ring (B)

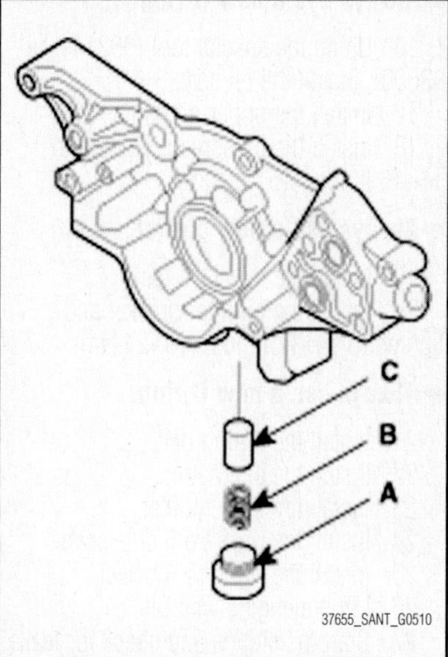

Fig. 276 Remove the plug (A), remove the relief spring (B) and the relief plunger (C)

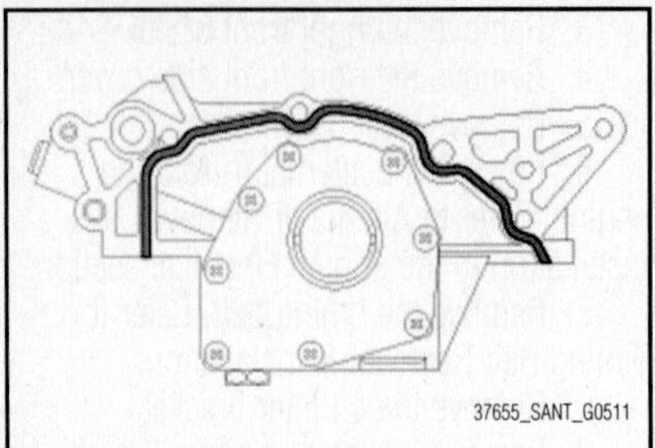

Fig. 277 Oil pump case sealant application

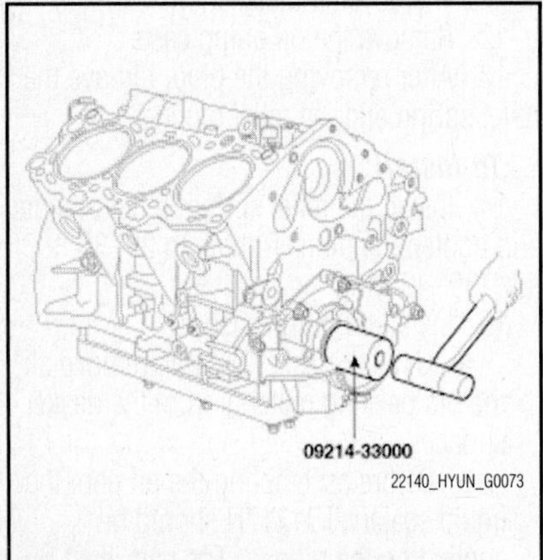

Fig. 278 Using the special tool (09214-33000), install the oil seal

prevent leakage of oil, apply sealant gasket to the inner threads of the bolt holes. After assembly, wait at least 30 minutes before filling the engine with oil.

c. Install the oil pump case and tighten to 14–17 ft. lbs. (19–24 Nm).

➡Always use a new O-ring.

16. Using the special tool (09214-33000), install the oil seal.
17. Install the upper oil pan.
18. Install the oil screen. Tighten to 11–16 ft. lbs. (15–22 Nm).

➡Always use a new gasket.

19. Install the lower oil pan.
20. Install the oil filter bracket and tighten to 14–17 ft. lbs. (19–24 Nm).

➡Always use a new O-ring.

21. Install the timing belt.
22. Install the alternator.
23. Install the front muffler.
24. Install the right front side cover.
25. Install the right front wheel.
26. Fill the engine with oil.
27. Start the engine and check for leaks.
28. Recheck the engine oil level.

3.3L Engine

See Figures 279 through 282.

1. Before servicing the vehicle, refer to the Precautions Section.
2. Remove the lower oil pan. Refer to Oil Pan Removal & Installation.
3. Remove the oil pump chain cover.
4. Remove the oil pump chain sprocket.
5. Remove the oil pump.

To install:

6. Install the oil pump using a new O-ring. Tighten the bolts to 14–17 ft. lbs. (20–24 Nm).

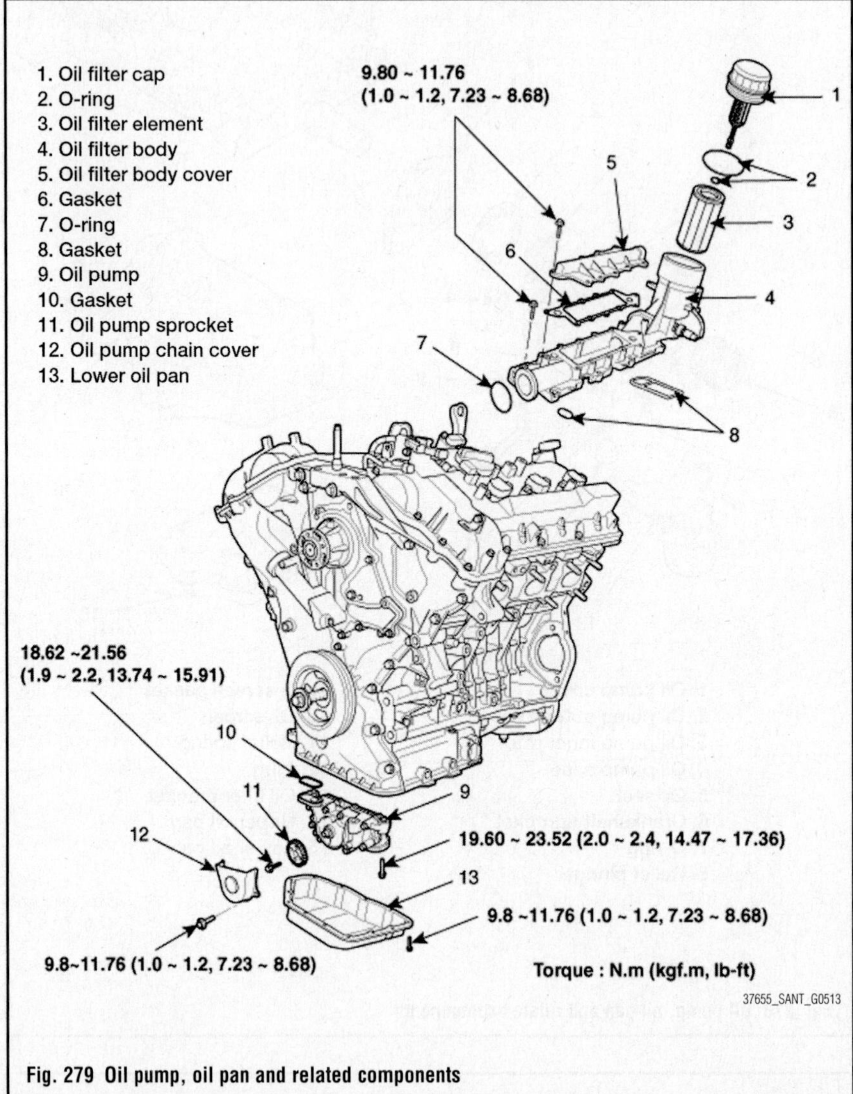

Fig. 279 Oil pump, oil pan and related components

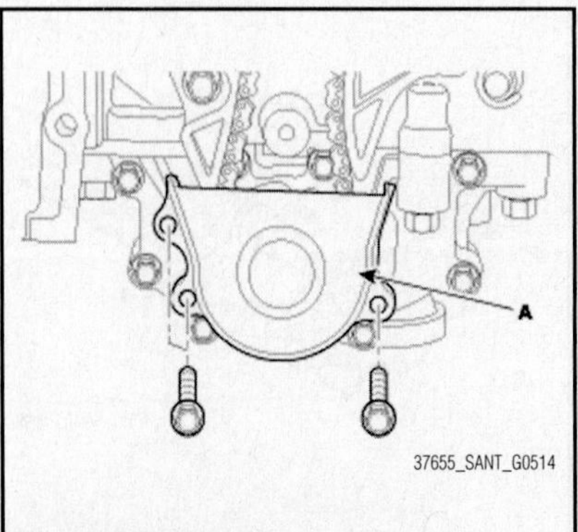

Fig. 280 Remove the oil pump chain cover (A)

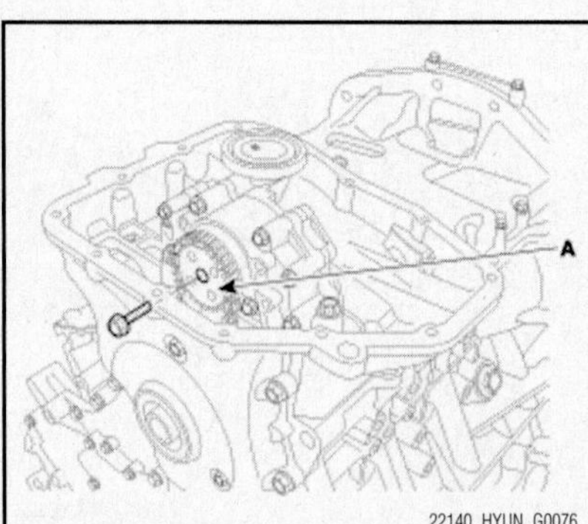

Fig. 281 Remove the oil pump chain sprocket (A)

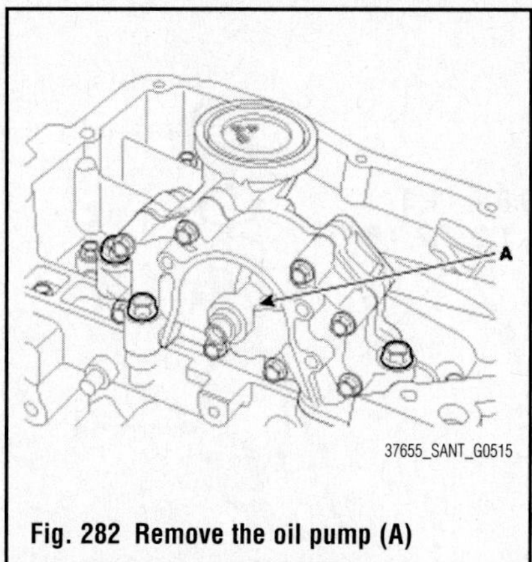

Fig. 282 Remove the oil pump (A)

➡Always use a new O-ring.

7. Install the oil pump sprocket and oil pump chain on the oil pump. Tighten to 14–16 ft. lbs. (19–22 Nm).
8. Install the oil pump chain cover. Tighten to 86–104 inch lbs. (10–12 Nm).
9. Install the lower oil pan.
10. After assembly, wait at least 30 minutes before filling the engine with oil to allow the gasket material to cure.
11. Fill the engine with oil.
12. Start the engine and check for leaks.
13. Recheck the engine oil level.

2010 Models

2.4L Engine

See Figures 283 through 287.

1. Before servicing the vehicle, refer to the Precautions Section.
2. Remove the timing chain. Refer to Timing Chain Removal & Installation.
3. Install a set pin after compressing the balance shaft chain tensioner.
4. Remove the balance shaft chain hydraulic tensioner.
5. Remove the balance shaft chain tensioner arm.
6. Remove the balance shaft chain guide.
7. Remove the oil pump and balance shaft module and balance shaft chain.

➡Do not disassemble the oil pump and balance shaft module.

To install:

8. The key of crankshaft should be aligned with the mating face of main bearing cap. This will position the piston of No. 1 cylinder at the top dead center on compression stroke.
9. Confirm the balance shaft module timing mark. Timing marks to be visually aligned with centers of adjacent cast timing notches.
10. Install balance shaft module with the timing mark of balance shaft module sprocket matched with the timing mark (color link) of balance shaft chain. Tightening torque: 17–20 ft. lbs. (23–27 Nm), plus 103–107°.
11. Bolting order:
    a. Assemble the bolts in order number as shown with seating torque of 17–20 ft. lbs. (23–27 Nm).
    b. Unfasten the bolts in reverse bolting order (4-3-2-1).

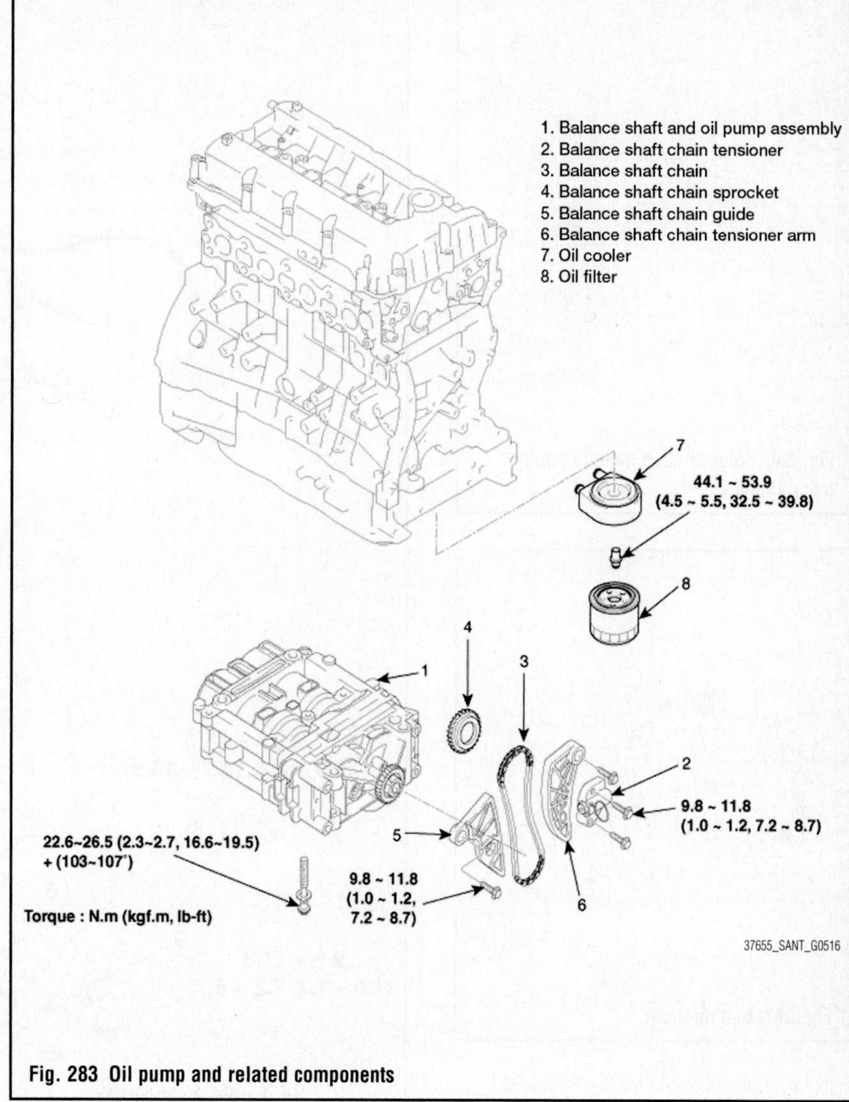

Fig. 283 Oil pump and related components

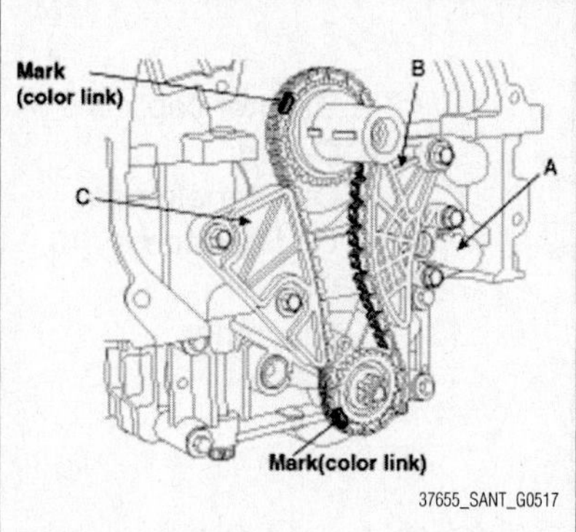

Fig. 284 Remove the balance shaft chain hydraulic tensioner (A), tensioner arm (B), and chain guide (C)

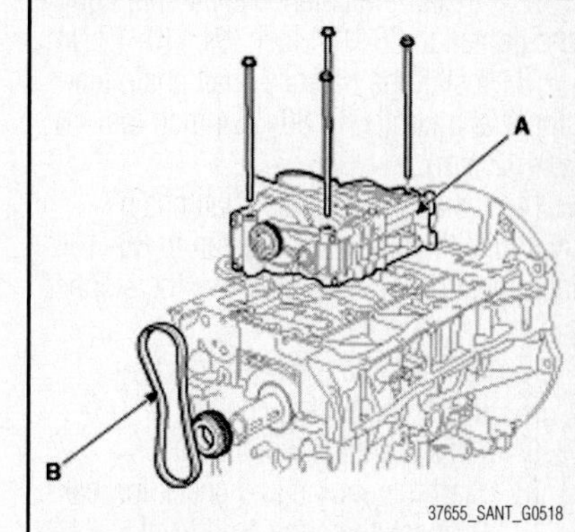

Fig. 285 Remove the oil pump and balance shaft module (A) and balance shaft chain (B)

mark
mark
37655_SANT_G0519

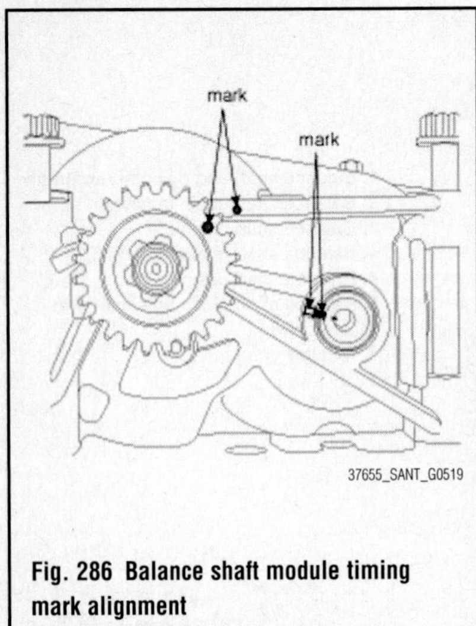

Fig. 286 Balance shaft module timing mark alignment

2
3
4
1
37655_SANT_G0520

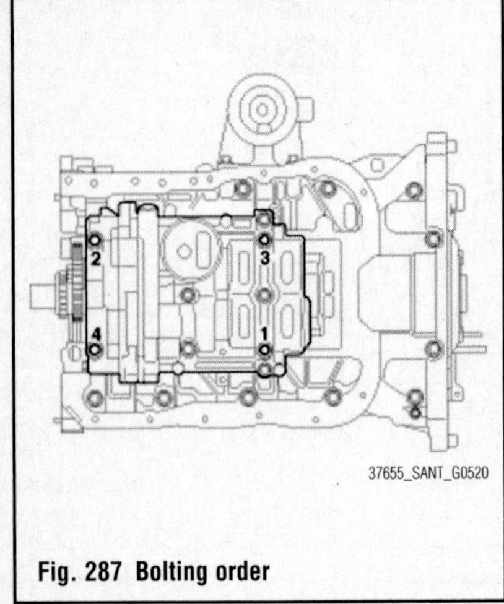

Fig. 287 Bolting order

c. Assemble the bolts in specified bolting order in the same increments.

12. Install the balance shaft chain guide and tighten to 86–104 inch lbs. (10–12 Nm).

13. Install the balance shaft chain tensioner and tighten to 86–104 inch lbs. (10–12 Nm).

14. Install the balance shaft chain hydraulic tensioner and tighten to 86–104 inch lbs. (10–12 Nm). Remove the stopper pin.

15. Confirm the timing marks.

16. Install the timing chain.

17. Fill the engine with oil.

18. Start the engine and check for leaks.

19. Recheck the engine oil level.

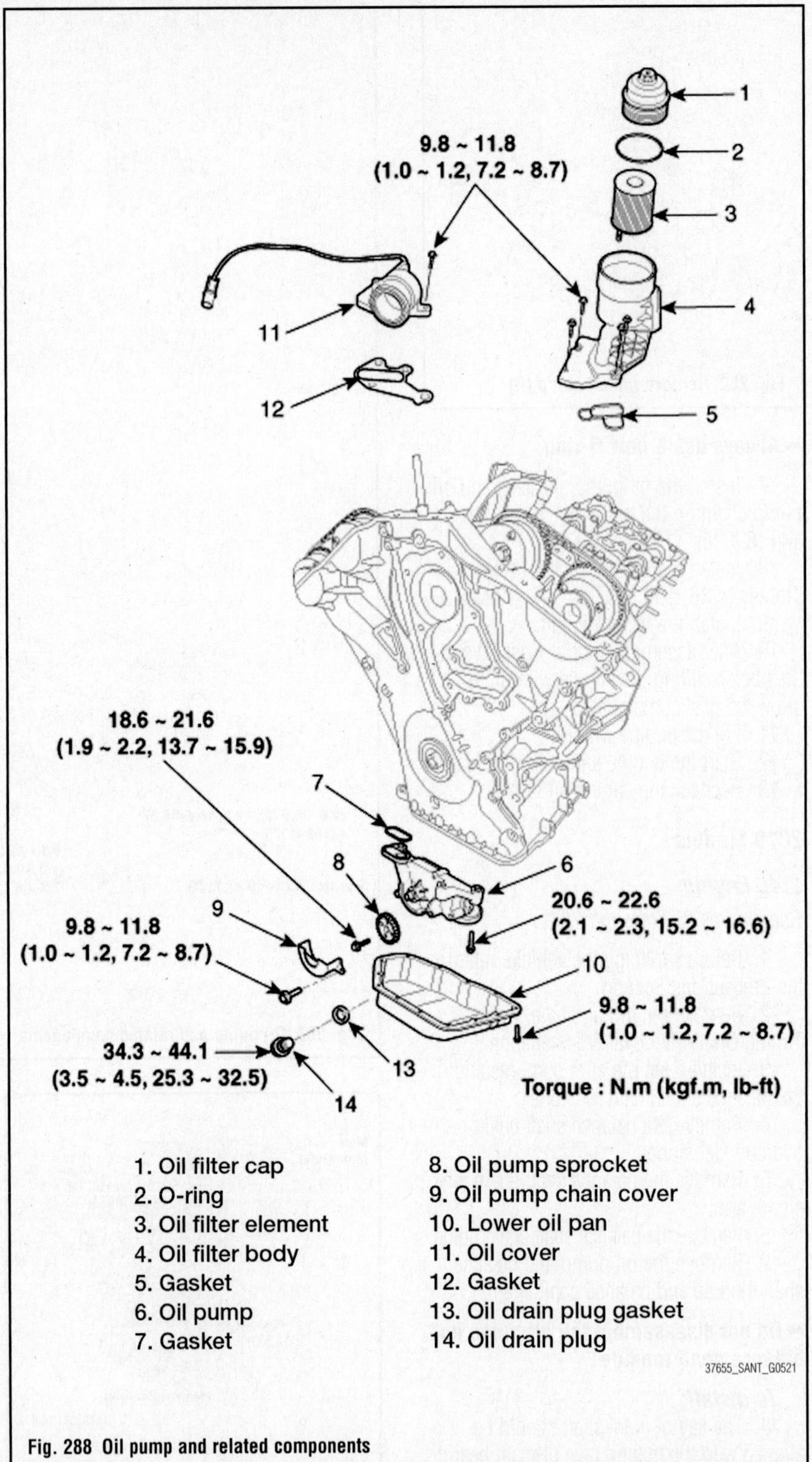

9.8 ~ 11.8
(1.0 ~ 1.2, 7.2 ~ 8.7)
1
2
3
4
5
11
12
18.6 ~ 21.6
(1.9 ~ 2.2, 13.7 ~ 15.9)
7
8
6
9
9.8 ~ 11.8
(1.0 ~ 1.2, 7.2 ~ 8.7)
20.6 ~ 22.6
(2.1 ~ 2.3, 15.2 ~ 16.6)
10
9.8 ~ 11.8
(1.0 ~ 1.2, 7.2 ~ 8.7)
34.3 ~ 44.1
(3.5 ~ 4.5, 25.3 ~ 32.5)
13
14
Torque : N.m (kgf.m, lb-ft)
1. Oil filter cap
2. O-ring
3. Oil filter element
4. Oil filter body
5. Gasket
6. Oil pump
7. Gasket
8. Oil pump sprocket
9. Oil pump chain cover
10. Lower oil pan
11. Oil cover
12. Gasket
13. Oil drain plug gasket
14. Oil drain plug
37655_SANT_G0521

Fig. 288 Oil pump and related components

3.5L Engine

See Figures 288 through 291.

1. Before servicing the vehicle, refer to the Precautions Section.

2. Drain the engine oil.

3. Remove the lower oil pan. Refer to Oil Pan Removal & Installation.

4. Remove the oil pump chain cover.

5. Remove the oil pump chain sprocket.

6. Remove the oil pump.

To install:

7. Install the oil pump and tighten to 15–17 ft. lbs. (21–23 Nm).

➡**Always use a new O-ring.**

8. Install the oil pump sprocket (A) and the oil pump chain on the oil pump and tighten to 14–16 ft. lbs. (19–22 Nm).

9. Install the oil pump chain cover and tighten to 86–104 inch lbs. (10–12 Nm).

10. Install the lower oil pan.

11. After assembly, wait at least 30 minutes before filling the engine with oil.

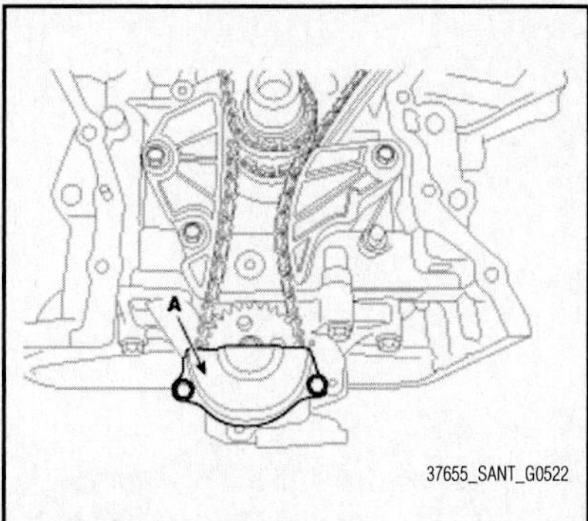

Fig. 289 Remove the oil pump chain cover (A)

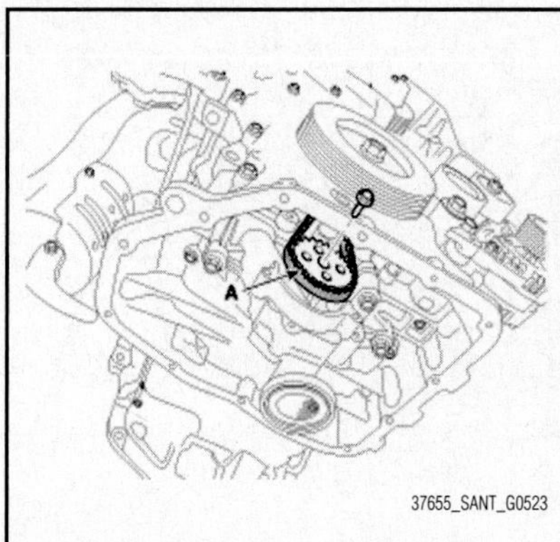

Fig. 290 Remove the oil pump chain sprocket (A)

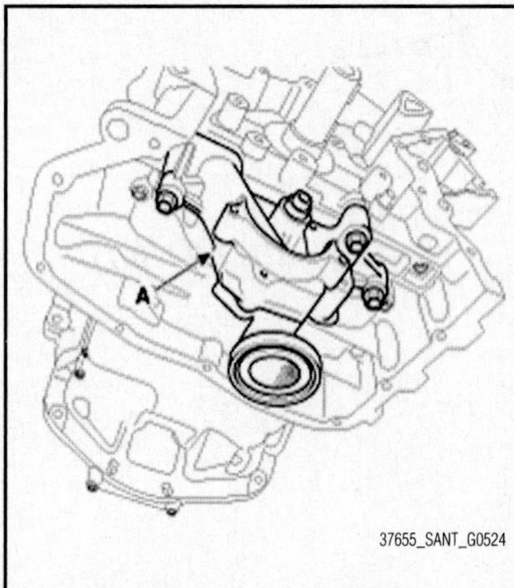

Fig. 291 Remove the oil pump (A)

12. Fill the engine with oil.
13. Start the engine and check for leaks.
14. Recheck the engine oil level.

INSPECTION

2.7L Engine

1. Check the relief plunger.
2. Apply engine oil on the plunger and check that it moves smoothly in the hole. If it does not, replace the plunger or the front case only in necessary cases.
3. Check the relief valve spring for deformation or damage.
   a. Free length: 1.7244 inches (43.8 mm)
   b. Load: 1.5787 inches at 8.21 lbs.

PISTON AND RING

POSITIONING

See Figures 292 and 293.

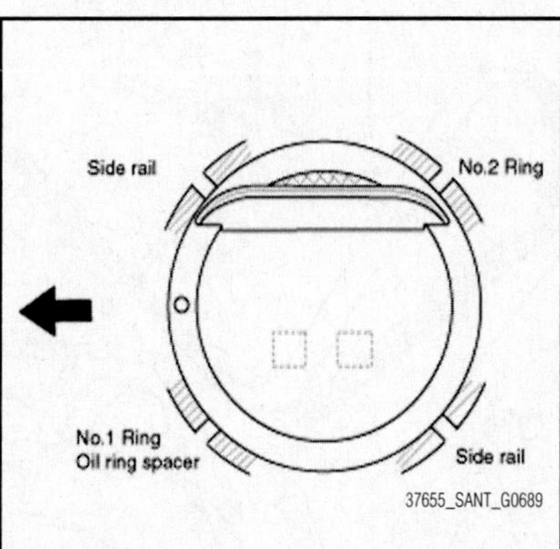

Fig. 292 Piston ring position—2.4L, 3.3L and 3.5L engines

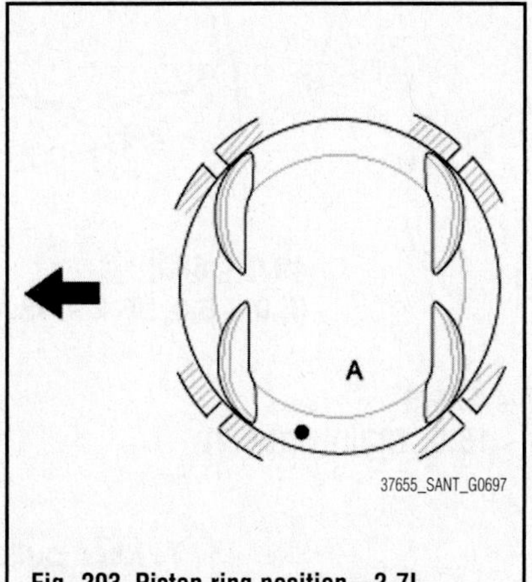

Fig. 293 Piston ring position—2.7L engine

REAR MAIN SEAL

REMOVAL & INSTALLATION

See Figure 294.

1. Install rear oil seal.
   a. Apply engine oil to a new oil seal lip.
   b. Using appropriate installation tool, tap in the oil seal until its surface is flush with the rear oil seal retainer edge.

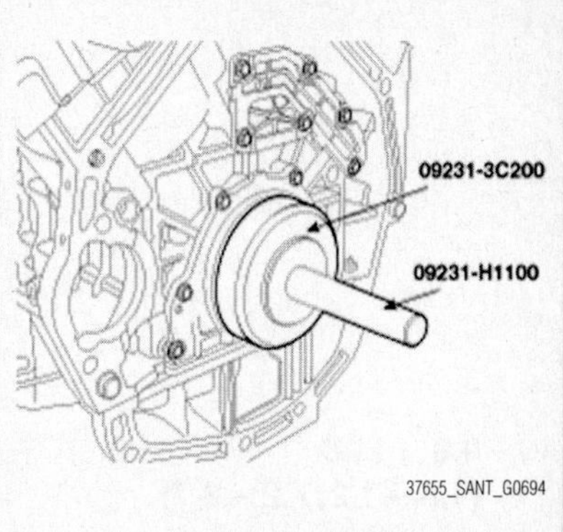

Fig. 294 Rear main seal retainer case position—3.3L and 3.5L engines

TIMING BELT FRONT COVER

REMOVAL & INSTALLATION

2.7L Engine

See Figures 295 through 301.

1. Before servicing the vehicle, refer to the Precautions Section.
2. Remove the engine cover.
3. Remove the front right wheel and tire.
4. Remove the right side cover.
5. Remove the accessory drive belt, the idler, and the tensioner.

➡When removing the accessory drive belt, hold the auto tensioner pulley bolt and turn the bolt counter clockwise.

6. Remove the timing belt upper cover.
7. Align the groove of the pulley with the timing mark of the timing belt cover by turning the crankshaft pulley clockwise.
8. Ensure that the timing mark of the camshaft sprocket is aligned with that of the cylinder head cover with No. 1 cylinder piston at Top Dead Center (TDC).
9. Support the engine oil pan with a jack.

⁂ WARNING

Put a wooden or rubber block between the jack and the engine oil pan.

10. Remove the engine mounting bracket.
11. Remove the crankshaft damper pulley.
12. Remove the timing belt lower cover.

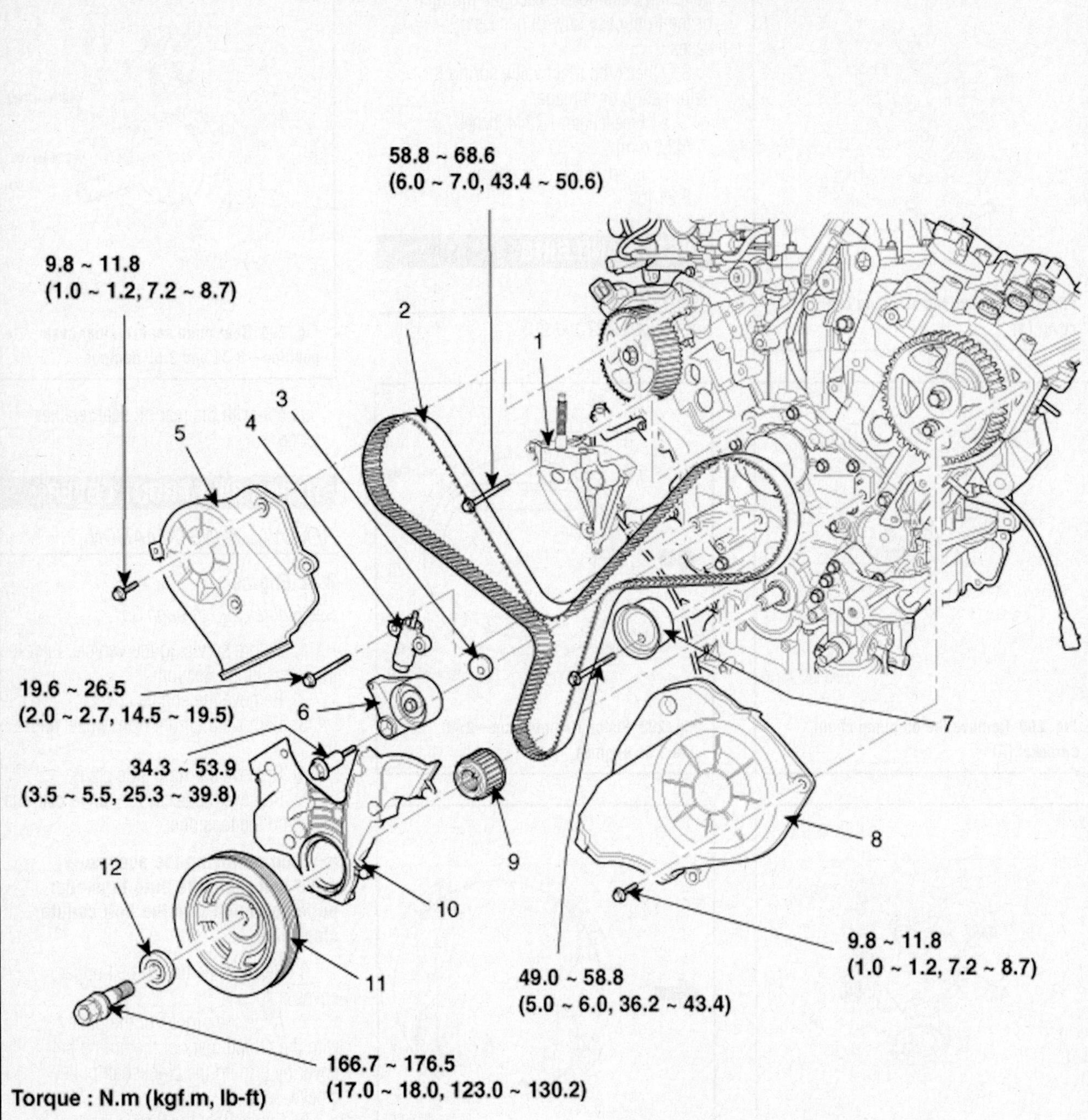

1. Engine support bracket
2. Timing belt
3. Tensioner arm assembly washer
4. Timing belt auto tensioner
5. Bank 1 timing belt upper cover
6. Timing belt tensioner arm assembly
7. Idler pulley
8. Bank 2 timing belt upper cover
9. Crankshaft sprocket
10. Timing belt lower cover
11. Damper pulley
12. Special washer

37655_SANT_G0525

Fig. 295 Timing belt components

To install:

13. Install the timing belt lower cover and tighten to 94–104 inch lbs. (10–12 Nm).
14. Install the crankshaft damper and tighten the mounting bolt to 170–177 ft. lbs. (231–240 Nm).
15. Install the engine mounting bracket and tighten to 47–62 ft. lbs. (64–83 Nm).
16. Lower the jack and remove from the vehicle.
17. Install the upper timing belt cover and tighten to 86–104 inch lbs. (10–12 Nm).
18. Making sure all timing and alignment marks match, install the drive belt, the idler, and the tensioner and tighten to 25–40 ft. (34–54 Nm).
19. Install the right side cover.
20. Install the right front tire and wheel.
21. Install the engine cover.

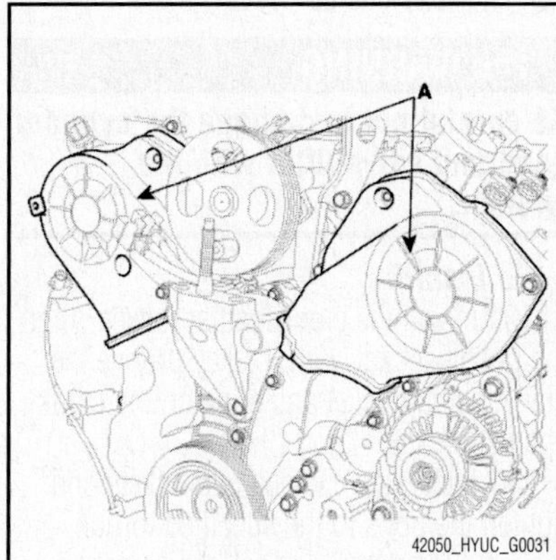

Fig. 296 Remove the timing belt upper cover (A)

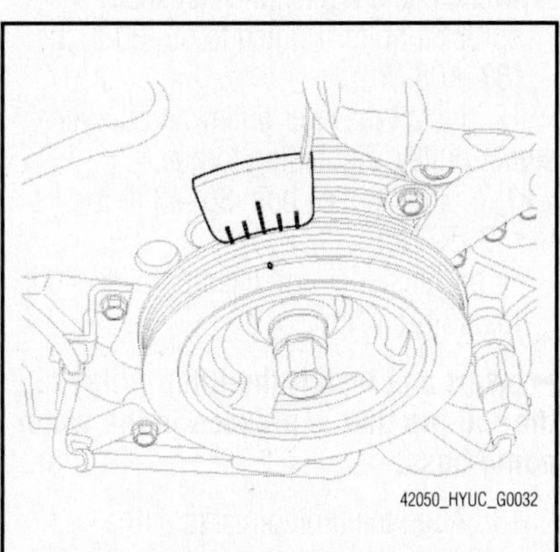

Fig. 297 Timing mark location

Fig. 298 Jack placement

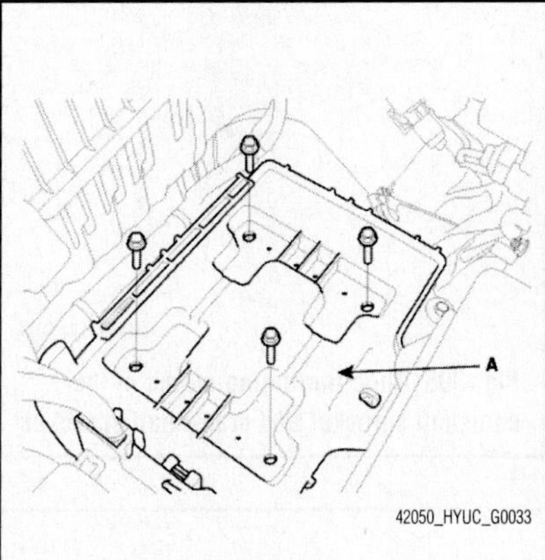

Fig. 299 Remove the engine mounting bracket (A)

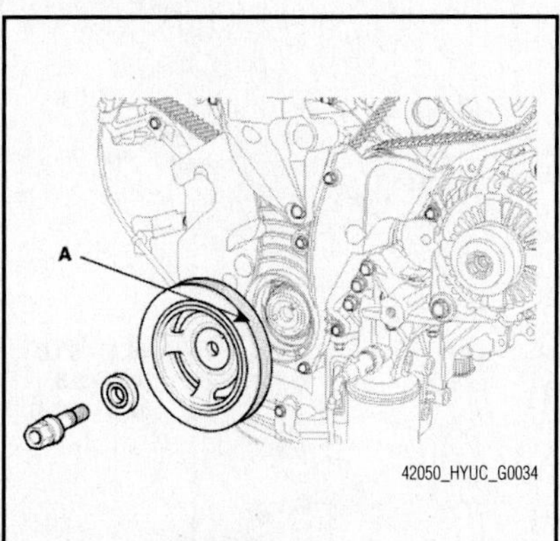

Fig. 300 Crankshaft damper pulley

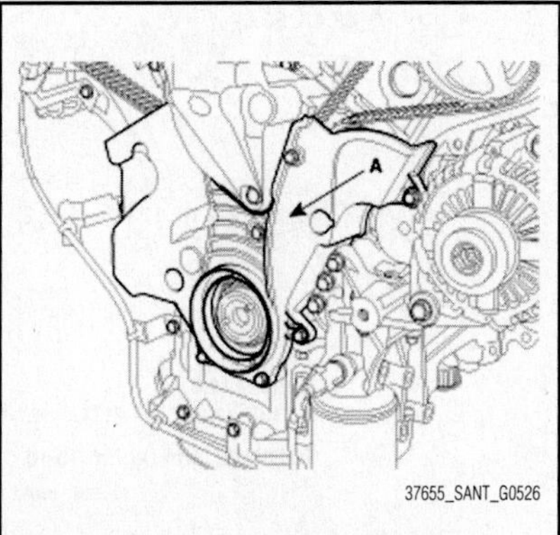

Fig. 301 Remove the timing belt lower cover (A)

TIMING BELT & SPROCKETS

REMOVAL & INSTALLATION

2.7L Engine

See Figures 302 through 309.

1. Before servicing the vehicle, refer to the Precautions Section.
2. Remove the timing belt front covers (upper and lower). Refer to Timing Belt Front Cover, Removal & Installation.
3. Remove the engine support bracket.
4. Check that timing marks of the camshaft timing pulleys and cylinder head covers are aligned. If not, turn the crankshaft 1 revolution (360°).
5. Remove the timing belt tensioner. Alternately loosen the 2 bolts and remove the tensioner.
6. Remove the timing belt.

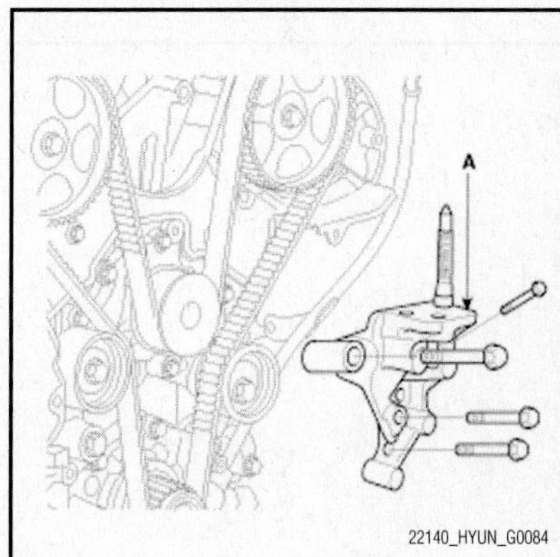

Fig. 302 Remove the engine support bracket (A)

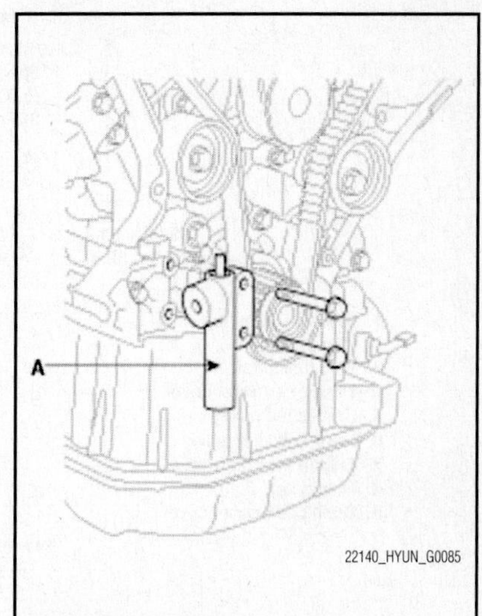

Fig. 303 Alternately loosen the 2 bolts and remove the tensioner (A)

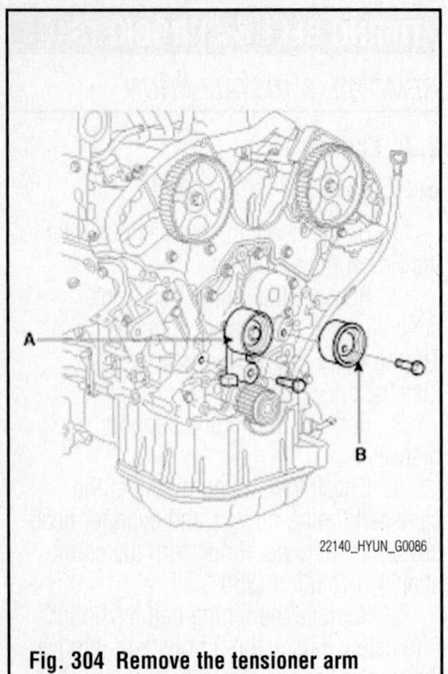

Fig. 304 Remove the tensioner arm assembly (A) and timing belt idler pulley (B)

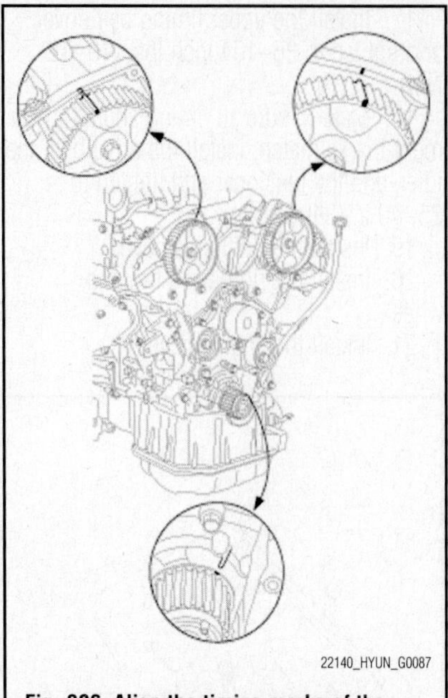

Fig. 306 Align the timing marks of the camshaft sprocket and crankshaft sprocket

1. Tensioner arm assembly
2. Idler pulley
3. Auto tensioner
4. Timing belt lower cover
5. Crankshaft pulley
6. Drive belt idler pulley
7. Engine support bracket
8. Timing belt
9. Timing belt upper cover

TORQUE : Nm (kgf.m, lb-ft)

22140_HYUN_G0080

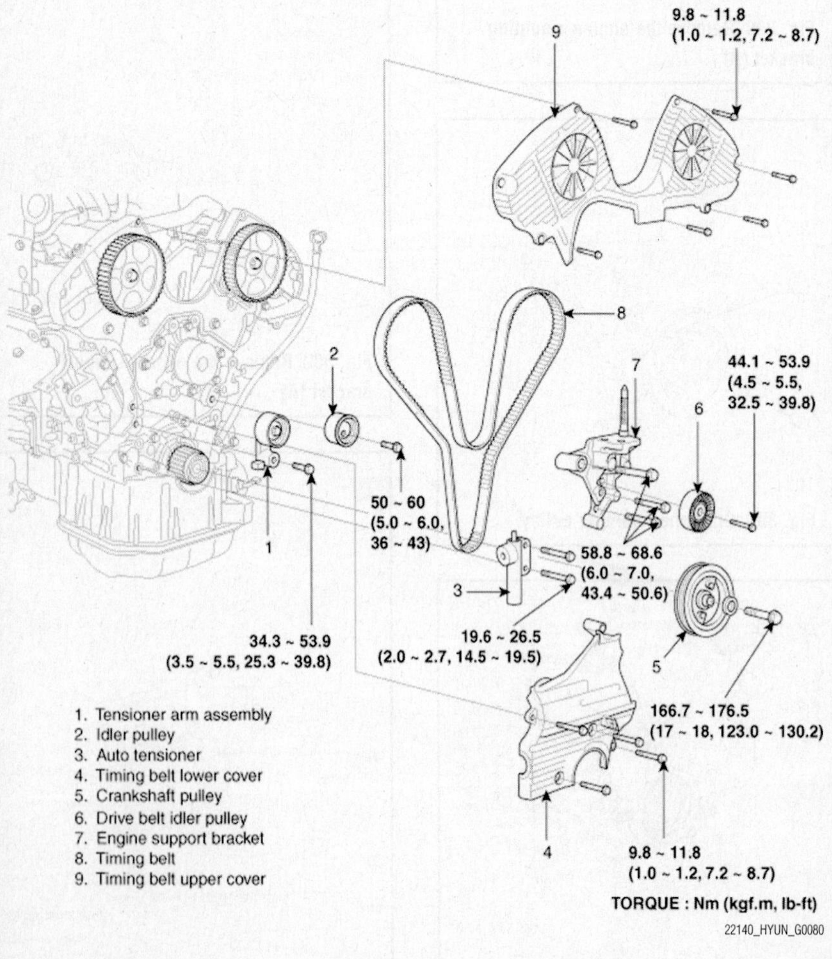

Fig. 305 Timing belt system and related components

➡If the timing belt is to be reused, make an arrow indicating the turning direction to make sure that the belt is reinstalled in the same direction as before.

7. Remove the tensioner arm assembly and timing belt idler pulley.
8. Remove the crankshaft sprocket.
9. Remove the camshaft sprockets. Hold the hexagonal head wrench portion of the camshaft with a wrench and remove the bolt and camshaft sprocket.

⁂ WARNING

Be careful not to damage the cylinder head and valve lifter with the wrench.

To install:

10. Install the crankshaft sprocket: Align the pulley set key with the key groove the crankshaft sprocket and slide on the crankshaft sprocket.
11. Install the camshaft sprockets and tighten the bolts to the specified torque.
    a. Temporarily install the camshaft sprocket bolts.
    b. Hold the hexagonal head wrench portion of the camshaft with a wrench, and tighten the camshaft sprocket bolts. Tighten to 65–80 ft. lbs. (88–108 Nm).
12. Install the idler pulley and the tensioner pulley. Tightening torque:
    a. Idler pulley bolt: 36–43 ft. lbs. (49–59 Nm).
    b. Tensioner arm fixed bolt: 25–40 ft. lbs. (34–54 Nm).

➡Insert and install the idler pulley to the roll pin that is pressed in the water pump boss.

13. Align the timing marks of the camshaft sprocket and crankshaft sprocket with the No. 1 piston placed at Top Dead Center (TDC) of its compression stroke.
14. Set the timing belt tensioner:
    a. Using a press, slowly press in the push rod.
    b. Align the holes of the push rod and housing.
    c. Pass a set pin through the holes to keep the setting position of the push rod.
    d. Release the press.
15. Install the timing belt tensioner:
    a. Temporarily install the tensioner with the 2 bolts.
    b. Alternately tighten the 2 bolts. Tighten to 15–20 ft. lbs. (20–27 Nm).
16. Install the timing belt.

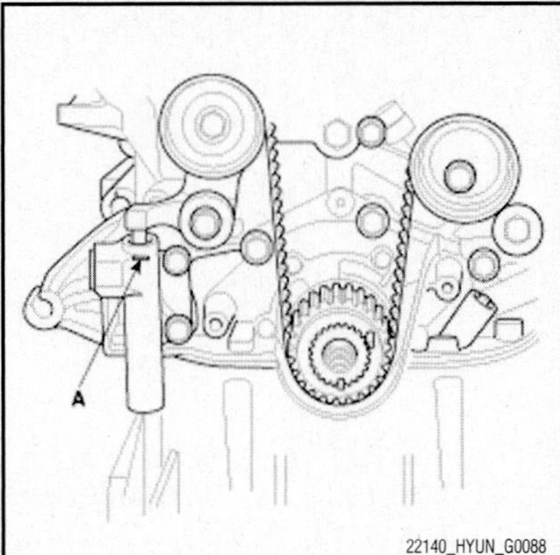

Fig. 307 Remove the set pin (A) from the tensioner

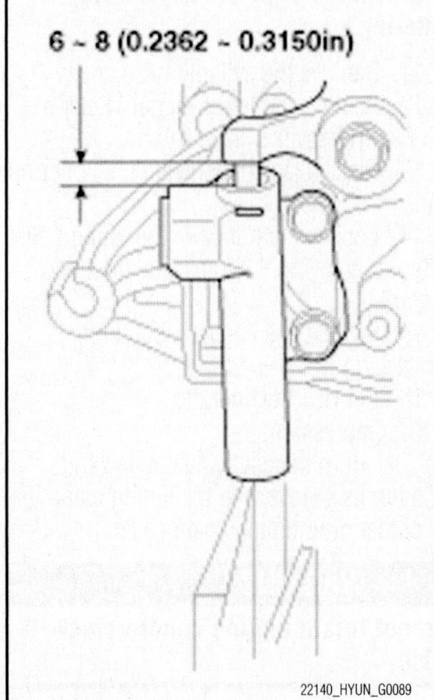

Fig. 308 The projected length of the timing belt tensioner should be 0.27–0.31 inch (7–9mm)

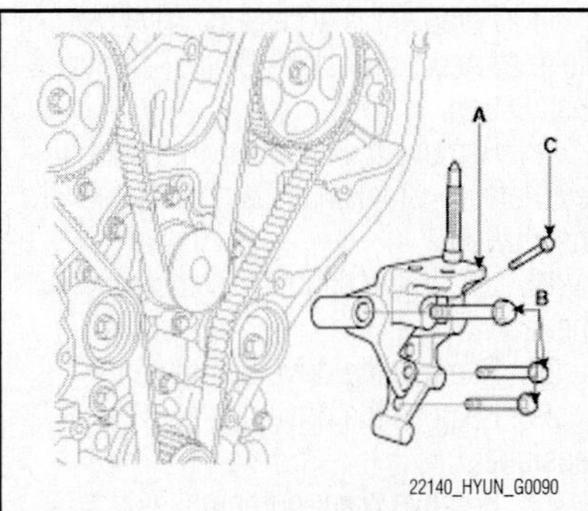

Fig. 309 Install the engine support bracket (A)

a. Remove any oil or water on the sprockets, and keep them clean.

b. Install the timing belt in this order:

- Crankshaft sprocket
- Idler pulley
- Camshaft sprocket left-hand side
- Water pump pulley
- Camshaft sprocket right-hand side
- Tensioner pulley

17. Remove the set pin from the tensioner.

18. Check the timing belt tensioner.

a. Rotate the crankshaft 2 turns clockwise and measure the projected length of the auto tensioner at TDC (No. 1 compression stroke) after 5 minutes.

b. As illustrated in the example, the projected length of the timing belt tensioner should be 0.27–0.31 in. (7–9mm).

19. Install the engine support bracket. Tightening torque of the following bolts (see illustration):

- B: 43–51 ft. lbs. (59–67 Nm)
- C: 11–16 ft. lbs. (15–22 Nm)

20. Install the timing belt front covers (upper and lower).

TIMING CHAIN & SPROCKETS

REMOVAL & INSTALLATION

2009 Models

3.3L Engine

See Figures 310 through 331.

1. Before servicing the vehicle, refer to the Precautions Section.

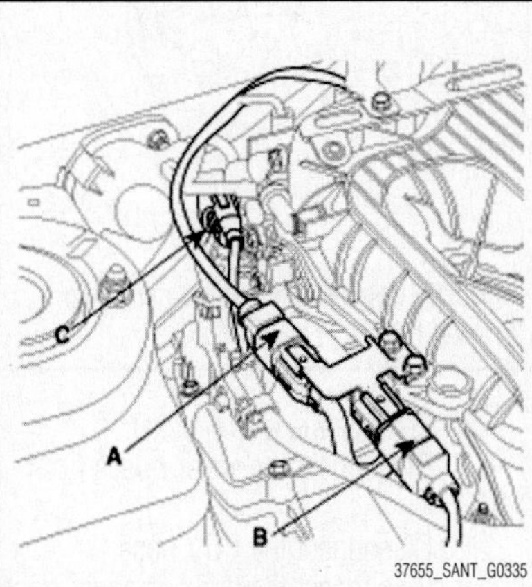

Fig. 310 Disconnect the right-hand injector connector (A), ignition coil connector (B) and VIS connector (C)

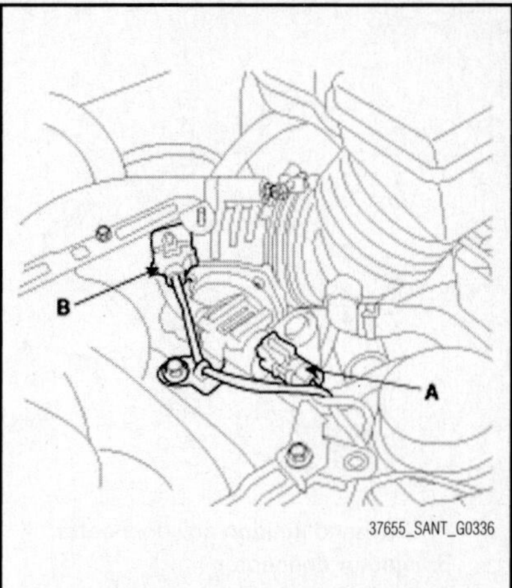

Fig. 311 Disconnect the PCSV connector (A), MAP sensor connector (B), PCSV hose and the ECM vacuum hose

➡Use fender covers to avoid damaging painted surfaces. To avoid damage, unplug the wiring connectors carefully while holding the connector portion. Mark all wiring and hoses to avoid misconnection. Turn the crankshaft pulley so that the No. 1 piston is at Top Dead Center (TDC).

2. Disconnect the negative terminal from the battery.

3. Remove the engine cover.

4. Remove the intake air hose and air cleaner assembly.

5. Remove the right-hand front wheel.

6. Remove the undercover.

7. Remove the right-hand side cover.

8. Drain the engine coolant. Remove the radiator cap to speed draining.

9. Drain the engine oil.

10. Remove the surge tank:

a. Disconnect the right-hand oxygen sensor connector.

b. Disconnect the right-hand injector connector, ignition coil connector and VIS connector.

c. Disconnect the PCSV connector, MAP sensor connector, PCSV hose and the ECM vacuum hose.

d. Disconnect the ETC connector and knock sensor connector.

e. Disconnect the OCV connector and knock sensor connector.

f. Disconnect the left-hand front oxygen sensor connector.

g. Disconnect the left-hand ignition coil connector, injector connector, condenser connector and ground, and remove the wiring harness protector.

h. Disconnect the left-hand CMPS and oil pressure switch connector.

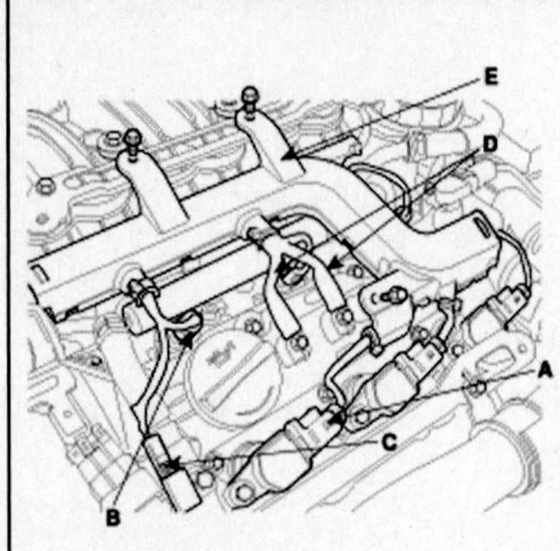

A. Left-hand ignition coil connector
B. Injector connector
C. Condenser connector
D. Ground
E. Wiring harness protector

37655_SANT_G0337

Fig. 312 Disconnect the left-hand ignition coil connector, injector connector, condenser connector and ground, and remove the wiring harness protector

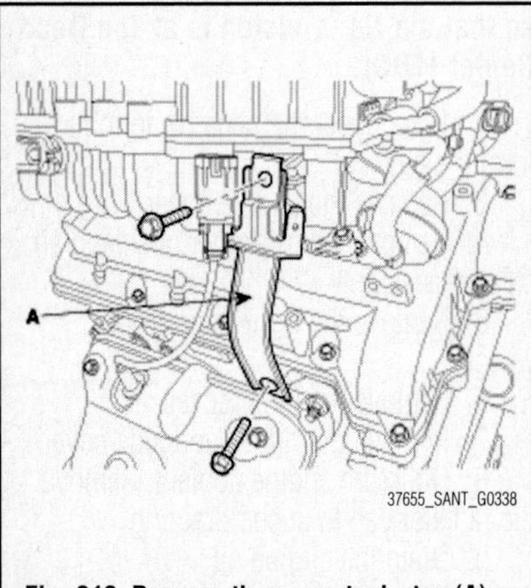

Fig. 313 Remove the surge tank stay (A)

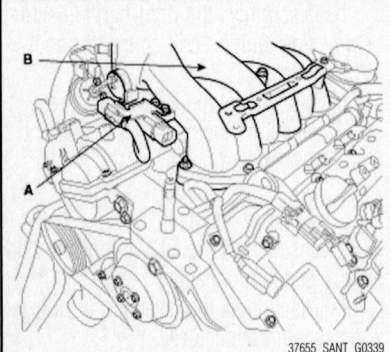

37655_SANT_G0339

Fig. 314 Remove the surge tank connector bracket (A)

i. Disconnect the right-hand CMPS connector.

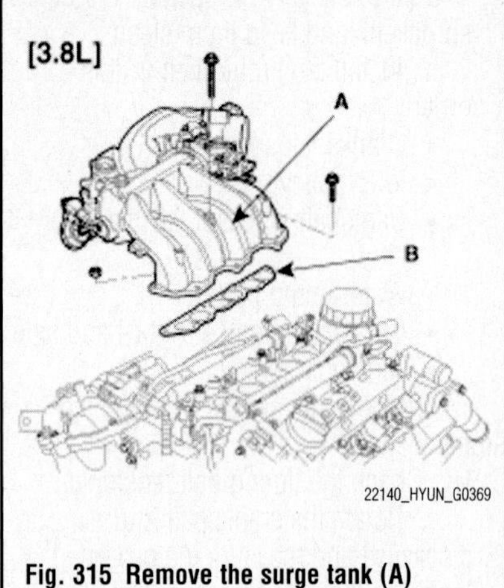

Fig. 315 Remove the surge tank (A)

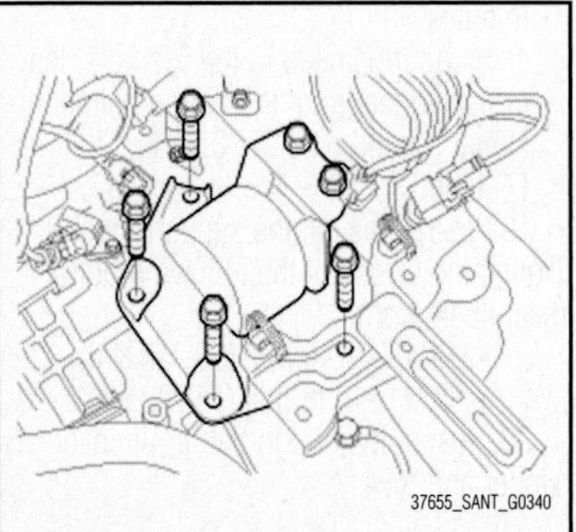

Fig. 316 Loosen the transaxle mounting bolts without removing the transaxle mounting bracket

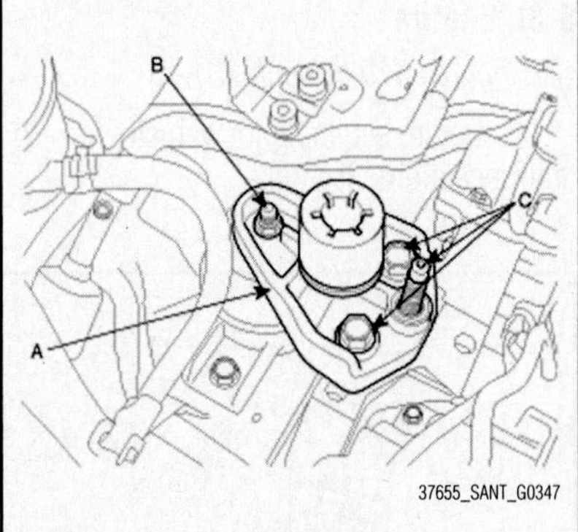

Fig. 317 Remove the engine mounting bracket (A)

j. Remove the ETC bracket.

k. Disconnect the water hoses from ETC.

l. Disconnect the PCV hose.

m. Disconnect the brake vacuum hose.

n. Remove the surge tank stay.

o. Remove the connector bracket from surge tank.

p. Remove the surge tank.

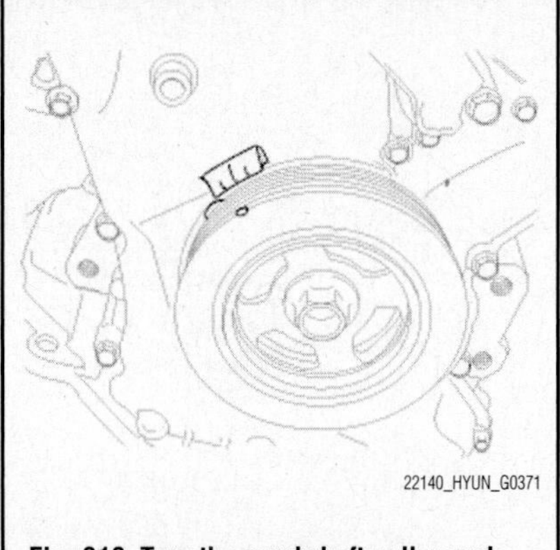

Fig. 318 Turn the crankshaft pulley and align its groove with the timing mark "T" of the lower timing chain cover

➡Cover the inlet of intake manifold to prevent foreign materials from entering.

11. Remove the cylinder head cover.
12. Remove the lower oil pan. Refer to Oil Pan, Removal & Installation.
13. Place a jack underneath the upper oil pan.
14. Loosen the transaxle mounting bolts without removing the transaxle mounting bracket.
15. Remove the engine mounting bracket.
16. Set No. 1 cylinder to TDC/compression.
    a. Turn the crankshaft pulley and align its groove with the timing mark "T" of the lower timing chain cover.

WARNING

Do not rotate engine counterclockwise.

    b. Check that the mark of the camshaft timing sprockets are in straight line on the cylinder head surface as shown. If not, turn the crankshaft one revolution (360°). Do not rotate engine counterclockwise.
17. Remove the accessory drive belt. Refer to Accessory Drive Belts, Removal & Installation.
18. Remove the crankshaft damper pulley. Refer to Crankshaft Damper, Removal & Installation.
19. Lift up the engine assembly using the jack.
20. Remove the drive belt idler.
21. Remove the drive belt auto tensioner.
22. Remove water pump pulley.
23. Remove the timing chain cover. If necessary remove the water pump first.

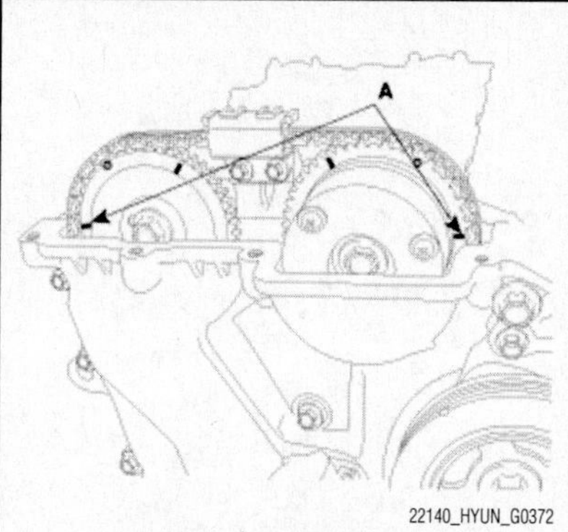

Fig. 319 Check that the mark (A) of the camshaft timing sprockets are in straight line on the cylinder head surface

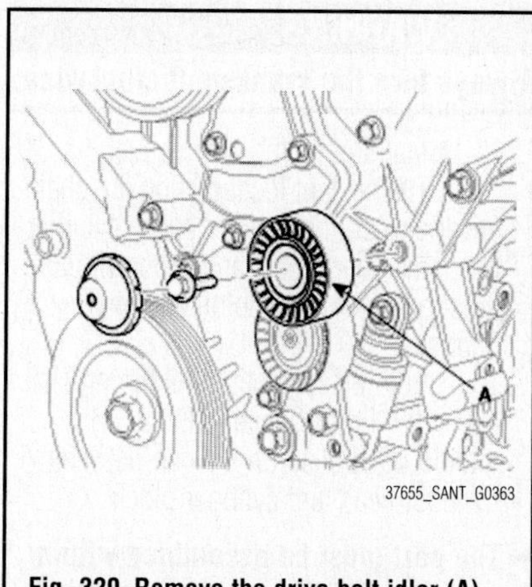

Fig. 320 Remove the drive belt idler (A)

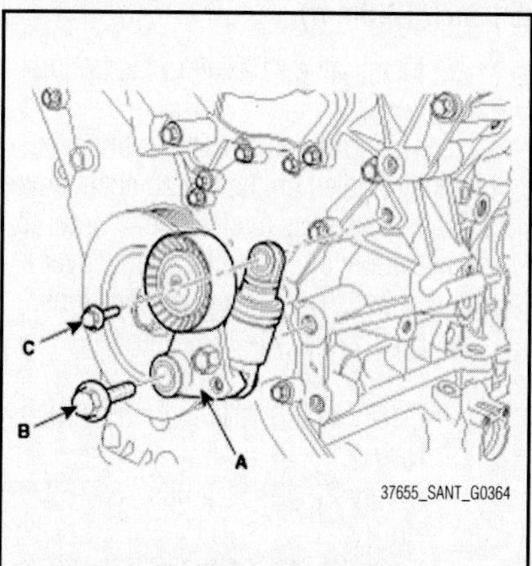

Fig. 321 Remove the drive belt auto tensioner (A) and bolts (B, C)

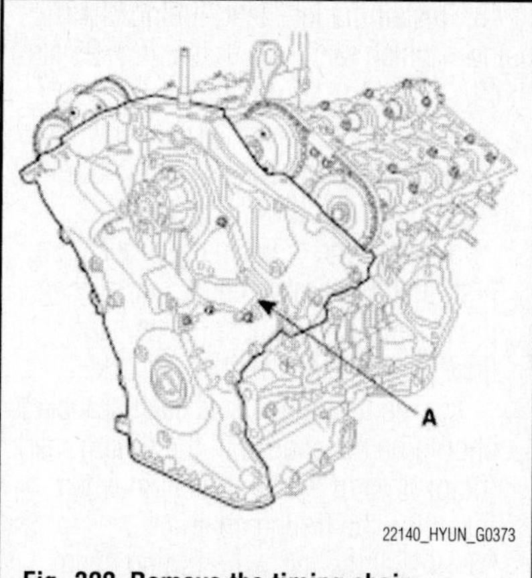

Fig. 322 Remove the timing chain cover (A)

22140_HYUN_G0374

Fig. 323 Timing marks on chain and camshaft sprockets illustrated

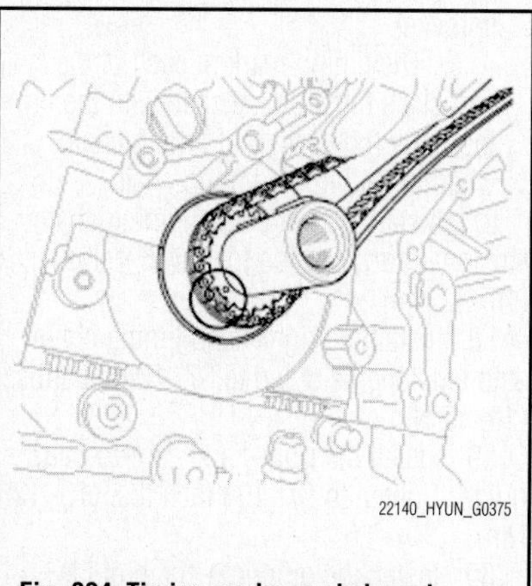

Fig. 324 Timing marks on chain and crankshaft sprocket illustrated

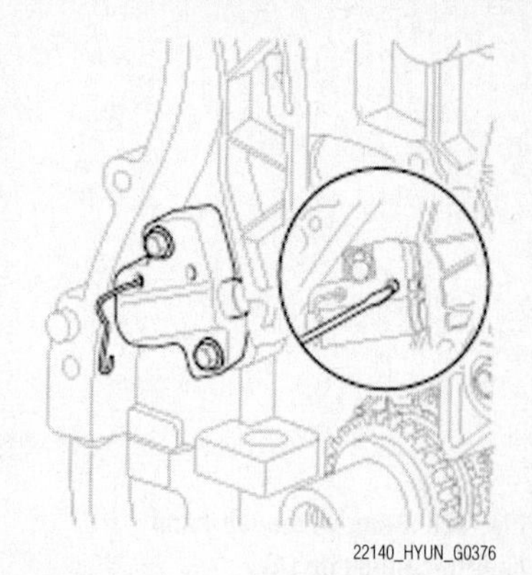

Fig. 325 Install a set pin after compressing the right-hand timing chain tensioner

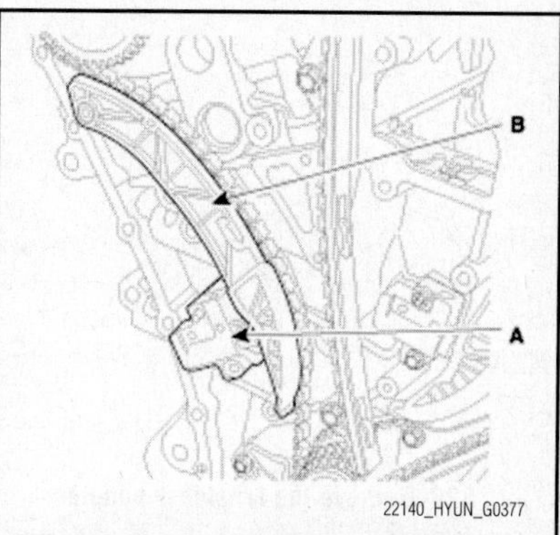

Fig. 326 Remove the right-hand timing chain auto tensioner (A) and the right-hand timing chain tensioner arm (B)

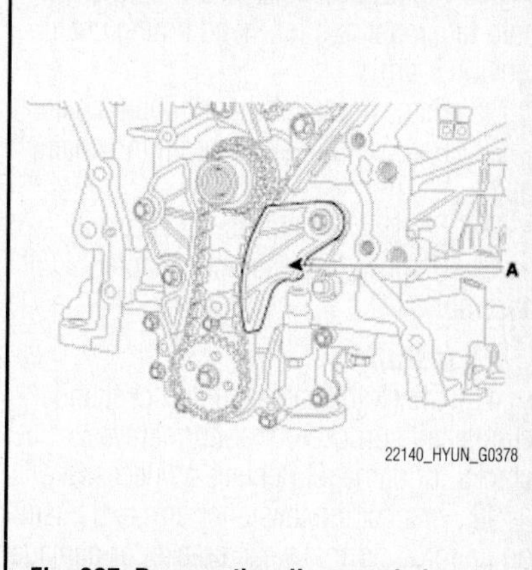

Fig. 327 Remove the oil pump chain tensioner assembly (A)

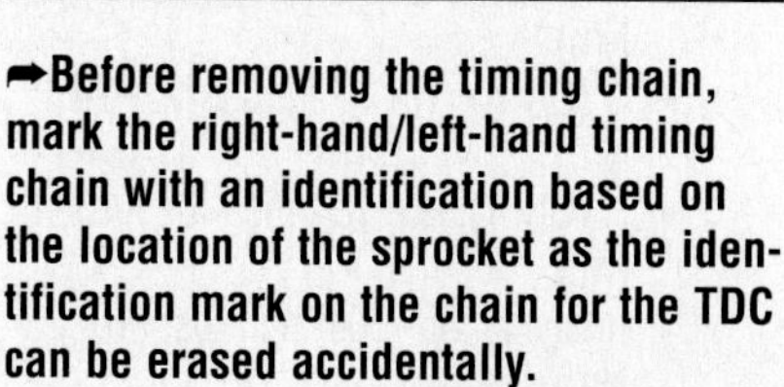
➡Before removing the timing chain, mark the right-hand/left-hand timing chain with an identification based on the location of the sprocket as the identification mark on the chain for the TDC can be erased accidentally.

24. Install a set pin after compressing the right-hand timing chain tensioner.
25. Remove the right-hand cam-to-cam guide.
26. Remove the right-hand timing chain auto tensioner and the right-hand timing chain tensioner arm.
27. Remove the right-hand timing chain.
28. Remove the right-hand timing chain guide.
29. Remove the oil pump chain cover.
30. Remove the oil pump chain tensioner assembly.
31. Remove the oil pump chain guide.
32. Remove the oil pump chain sprocket and oil pump chain.
33. Remove the crankshaft sprocket that drives the oil pump.
34. Install a set pin after compressing the left-hand timing chain tensioner.

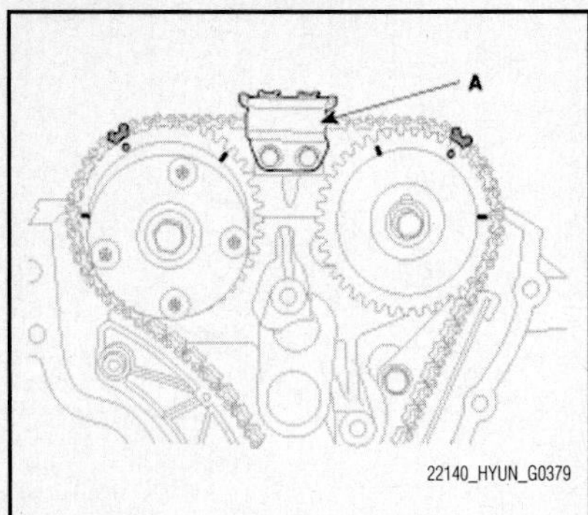

Fig. 328 Remove the left-hand cam-to-cam guide (A)

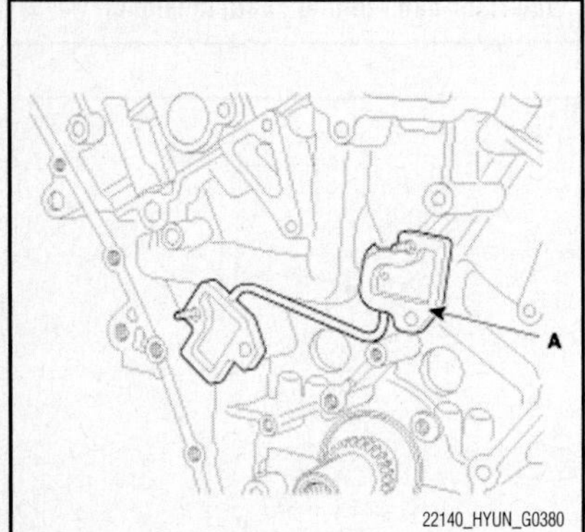

Fig. 329 Remove the tensioner adapter assembly (A)

35. Remove the left-hand cam-to-cam guide.
36. Remove the left-hand timing chain auto tensioner and left-hand timing chain tensioner arm.
37. Remove the left-hand timing chain.
38. Remove the left-hand timing chain guide.
39. Remove the crankshaft sprocket.
40. Remove the tensioner adapter assembly.

To install:

41. Check the camshaft sprocket and crankshaft sprocket for abnormal wear, cracks, or damage. Replace as necessary.
42. Inspect the tensioner arm and chain guide for abnormal wear, cracks, or damage. Replace as necessary.
43. Check that the tensioner piston moves smoothly when the ratchet pawl is released.
44. Install the jack to the upper oil pan.
45. The key of crankshaft must be aligned with the timing mark of timing chain cover. Then, piston of No. 1 cylinder is placed at the TDC on the compression stroke.
46. Install the tensioner adapter assembly.
47. Install the crankshaft sprocket.
48. Install the left-hand timing chain guide. Tighten to 14–18 ft. lbs. (20–25 Nm).
49. Install the left-hand timing chain.
   a. To install the timing chain with no slack between the camshaft and crankshaft, use the following procedure.
   b. Place the crankshaft sprocket on first, then the timing chain guide, then the exhaust camshaft sprocket, and finally the intake camshaft sprocket.
   c. The timing mark of each sprockets should be matched with the timing mark (color link) of the timing chain when installing the timing chain.
50. Install the left-hand timing chain ensioner arm. Tighten to 14–16 ft. lbs. (19–22 Nm).
51. Install the chain tensioner. Tighten to 87–104 inch lbs. (10–12 Nm).
52. Install the left-hand cam-to-cam guide. Tighten to 87–104 inch lbs. (10–12 Nm).
53. Install the crankshaft sprocket.
54. Install the oil pump chain and the oil pump sprocket. Tighten to 14–16 ft. lbs. (19–22 Nm).
55. Install the right-hand timing chain guide. Tighten to 14–18 ft. lbs. (20–25 Nm).
56. Install the right-hand timing chain.
   a. To install the timing chain with no slack between the camshaft and the crankshaft, use the following procedure.
   b. Place the chain on the crankshaft sprocket first, then the intake camshaft sprocket, then the exhaust camshaft sprocket.
   c. The timing mark of each of the sprockets must be matched with the timing mark (color link) of timing chain when installing the timing chain.
57. Install the right-hand timing chain tensioner arm. Tighten to 14–16 ft. lbs. (19–22 Nm).
58. Install the right-hand timing chain auto tensioner. Tighten to 87–104 inch lbs. (10–12 Nm).
59. Install the right-hand cam-to-cam guide. Tighten to 87–104 inch lbs. (10–12 Nm).
60. Install the oil pump chain guide. Tighten to 87–104 inch lbs. (10–12 Nm).
61. Install the oil pump chain tensioner assembly. Tighten to 87–104 inch lbs. (10–12 Nm).
62. Pull out the pins of hydraulic tensioner (left-hand and right-hand).
63. Install the oil pump chain cover. Tighten to 87–104 inch lbs. (10–12 Nm).
64. After rotating the crankshaft 2 revolutions in a clockwise direction (viewed from front), confirm that the timing marks are aligned.

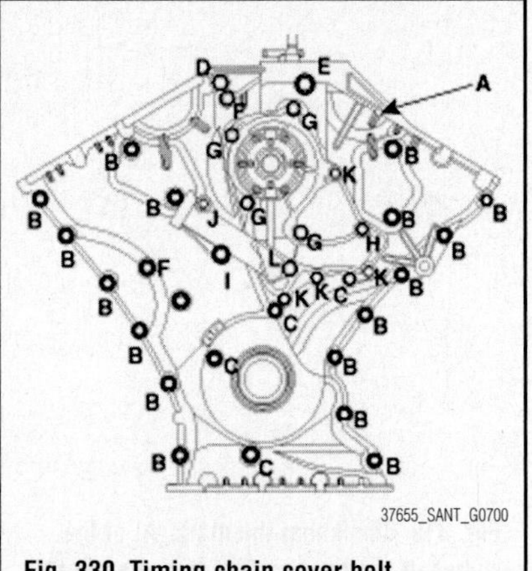

Fig. 330 Timing chain cover bolt tightening sequence

WARNING

Always turn the crankshaft clockwise.

65. Install the timing chain cover:
   a. The sealant locations on the chain cover and on the counter parts (cylinder head, cylinder block, and lower oil pan) must be free of engine oil and any debris.
   b. Before assembling the timing chain cover, the liquid sealant TB 1217H should be applied on the gap between cylinder head and cylinder block.

➡The part must be assembled within 5 minutes after the sealant is applied. Use a continuous bead width of 0.1 inch (2.5mm).

   c. Install a new gasket to the timing chain cover.
   d. The dowel pins on the cylinder block and holes on the timing chain cover should be used as a reference in order to aid assembly of the timing chain cover into position. Tightening specifications:
   - B (Qty 17): 14–16 ft. lbs. (19–22 Nm)
   - C (Qty 4): 86–104 inch lbs. (10–12 Nm)
   - D (Qty 2): 43–51 ft. lbs. (59–69 Nm)
   - E (Qty 1): 43–51 ft. lbs. (59–69 Nm)
   - F (Qty 2): 18–20 ft. lbs. (25–27 Nm)
   - G (Qty 4): 16–17 ft. lbs. (22–24 Nm)
   - H (Qty 1): 86–104 inch lbs. (10–12 Nm)
   - I (Qty 1): 86–104 inch lbs. (10–12 Nm)
   - J (Qty 1): 86–104 inch lbs. (10–12 Nm)

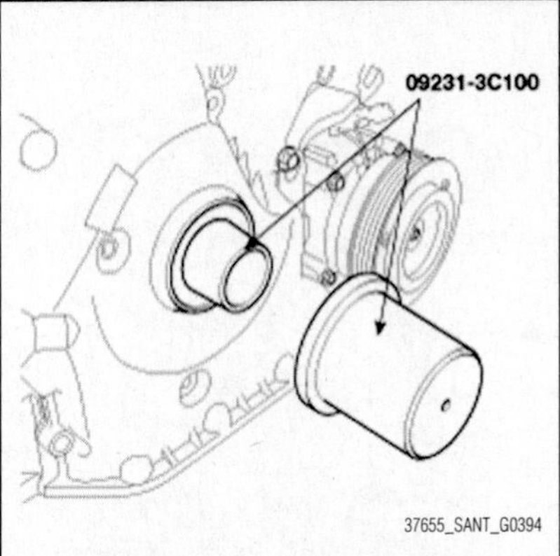

Fig. 331 Using SST (09231-3C100), install timing chain cover oil seal

- K (Qty 4): 86–104 inch lbs. (10–12 Nm)
- L (Qty 1): 16–20 ft. lbs. (22–27 Nm), New bolt

66. Wait 30 minutes after the timing chain cover was assembled before starting the engine.
67. Install the water pump pulley. Tighten to 69–87 inch lbs. (8–10 Nm).
68. Install the drive belt auto tensioner. Tighten to Bolt (B): 60–63 ft. lbs. (81–85 Nm). Bolt (C): 13–16 ft. lbs. (18–22 Nm).
69. Install the drive belt idler. Tighten to 39–43 ft. lbs. (53–58 Nm).
70. Lower the engine assembly using the jack.
71. Using SST (09231-3C100), install the timing chain cover oil seal.
72. Using SST (09231-3C300), install the crankshaft damper pulley. Tighten to 210–224 ft. lbs. (284–304 Nm).
73. Install the accessory drive belt.
74. Install the cylinder head cover.
75. Install the ignition coils.
76. Connect the right-hand ignition coil connector, the condenser connector, and install the wiring bracket.
77. Install the connector bracket from the left-hand cylinder head cover.
78. Install the surge tank. Tighten to 87–104 inch lbs. (10–12 Nm).
    a. Install the connector bracket on the surge tank. Tighten to 60–95 inch lbs. (7–11 Nm).
    b. Install the surge tank stay. Tighten to 20–23 ft. lbs. (27–31 Nm).
79. Connect the brake vacuum hose.
80. Connect the PCV hose.
81. Connect the water hoses to the ETC.
82. Install the ETC bracket.
83. Connect the right-hand CMPS connector.
84. Connect the left-hand CMPS and oil pressure switch connector.
85. Install the wiring harness protector, and connect the left-hand ignition connector, injector connector, condenser connector, and ground.
86. Connect the left-hand front oxygen sensor connector.
87. Connect the OCV connector and knock sensor connector.
88. Connect the ETC connector and knock sensor connector.
89. Connect the PCSV connector, MAP sensor connector, PCSV hose and ECM vacuum hose.
90. Connect the right-hand injector connector, ignition coil connector, and VIS connector.
91. Connect the right-hand oxygen sensor connector.
92. Install the engine mounting bracket. Tighten bolt (B) to 58–72 ft. lbs. (79–98 Nm) and bolt (C) to 47–62 ft. lbs. (64–83 Nm).
93. Install the transaxle mounting bracket bolts. Tighten to 47–62 ft. lbs. (64–83 Nm).
94. Remove the jack from the upper oil pan.
95. Install the lower oil pan.
96. Install the side cover.
97. Install the undercover.
98. Install the front wheels.
99. Install the air cleaner assembly.
100. Install the engine cover.
101. Connect the negative terminal to the battery.
102. Refill the engine with the proper amount and type of engine oil.
103. Refill the radiator and reservoir tank with engine coolant.
104. Bleed the air from the cooling system.
105. Run the engine and check for leaks.

2010 Models

2.4L Engine

See Figures 332 through 345.

➡Use fender covers to avoid damaging painted surfaces. To avoid damage, unplug the wiring connectors carefully while holding the connector portion. Mark all wiring and hoses to avoid misconnection.

1. Before servicing the vehicle, refer to the Precautions Section.
2. Disconnect the battery negative terminal.
3. Remove the engine cover.

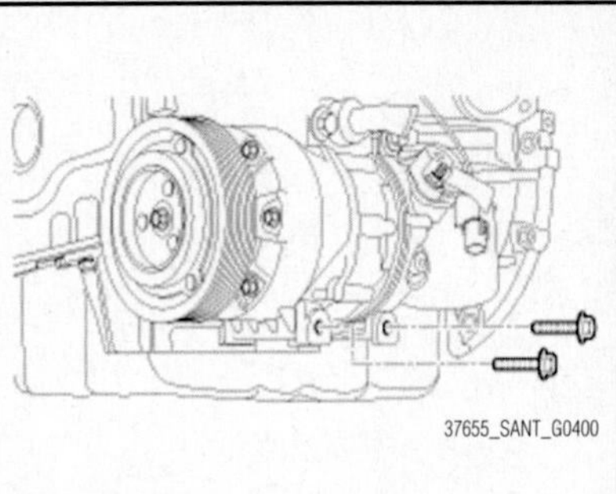

Fig. 332 Remove the air compressor lower bolts

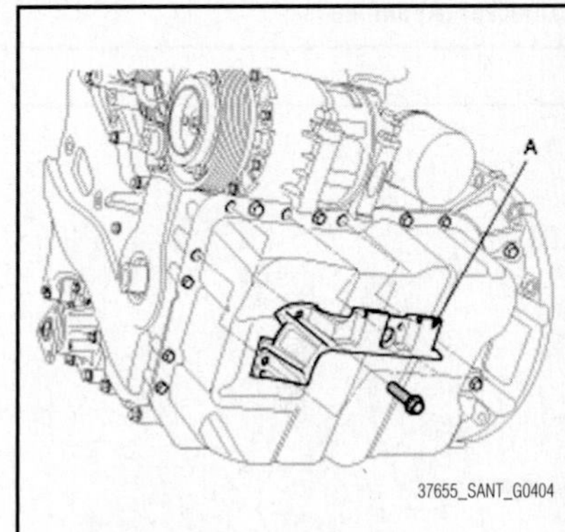

Fig. 333 Remove the air compressor bracket (A)

4. Remove the air cleaner assembly.
5. Remove the undercover and side cover.
6. Disconnect the ignition coil connectors. Refer to Ignition Coils, Removal & Installation in the Engine Electrical section.
7. Disconnect the exhaust Camshaft Position (CMP) sensor connector. Refer to CMP Sensor Removal & Installation in the Engine Performance & Emission Control section.
8. Disconnect the Positive Crankcase Ventilation (PCV) hose.
9. Remove the ignition coils. Refer to Ignition Coils, Removal & Installation in the Engine Electrical section.
10. Remove the cylinder head cover.
11. Remove the air compressor lower bolts.
12. Remove the air compressor bracket.
13. Drain the engine oil.
14. Remove the oil pan. Refer to Oil Pan Removal & Installation.
15. Install the jack to the edge of ladder frame. Insert the wooden block between ladder frame and jack.
16. Disconnect the ground cable, and then remove the engine mounting bracket.

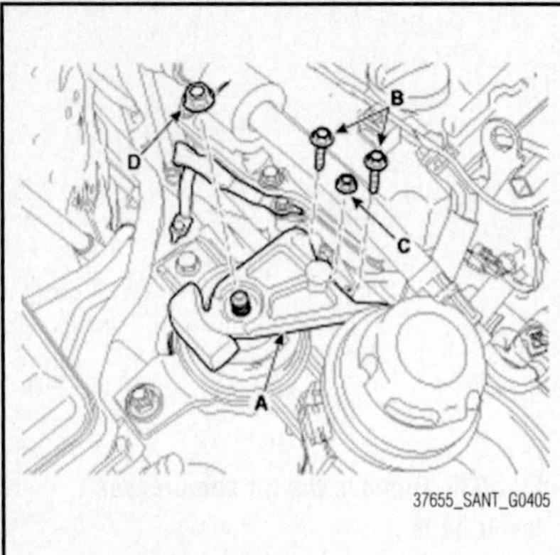

Fig. 334 Remove the engine mounting bracket (A) and bolts

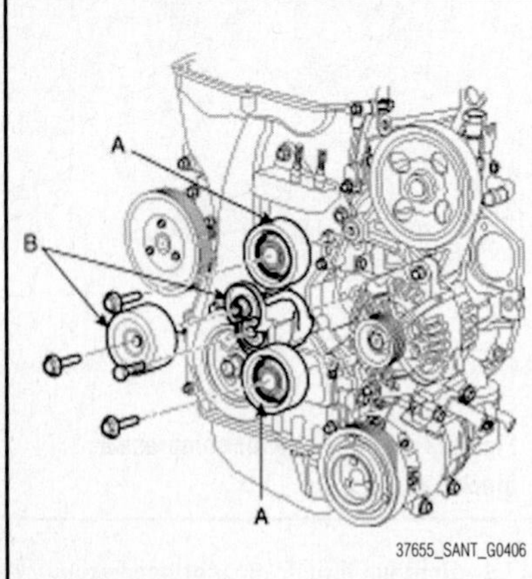

Fig. 335 Remove the idler (A) and drive belt tensioner (B)

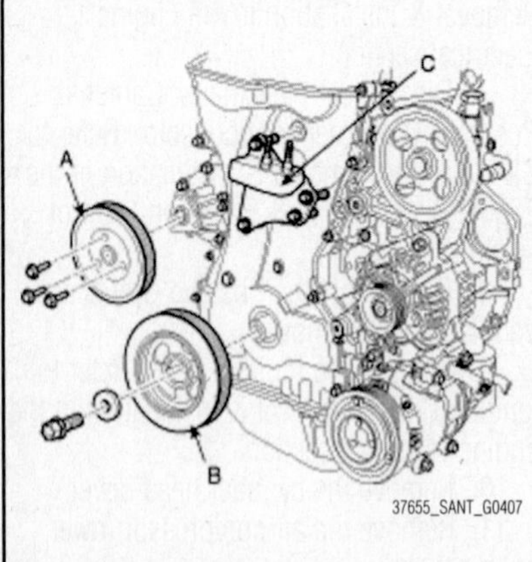

Fig. 336 Remove the water pump pulley (A), crankshaft pulley (B) and engine support bracket (C)

17. Turn the drive belt tensioner counterclockwise then remove the drive belt. Refer to Accessory Drive Belt Removal & Installation.

18. Remove the idler and drive belt tensioner. The tensioner pulley bolt has left handed threads.

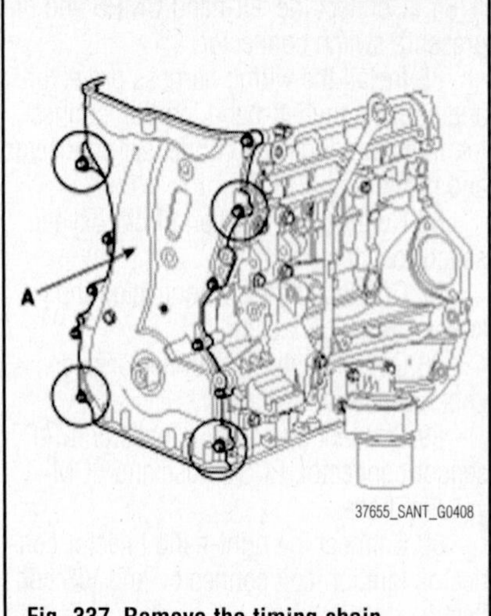

Fig. 337 Remove the timing chain cover (A)

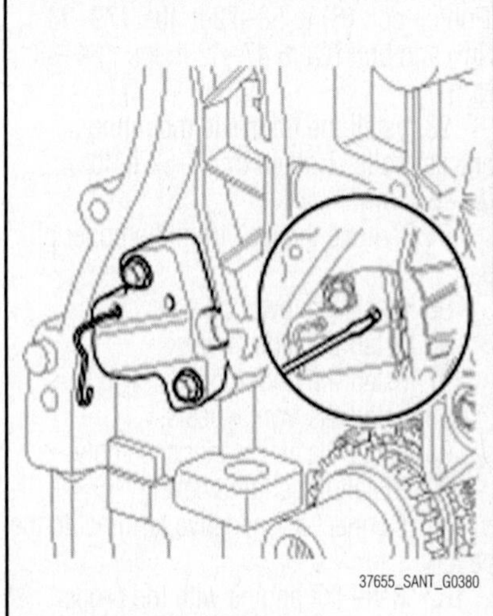

Fig. 338 Install a set pin

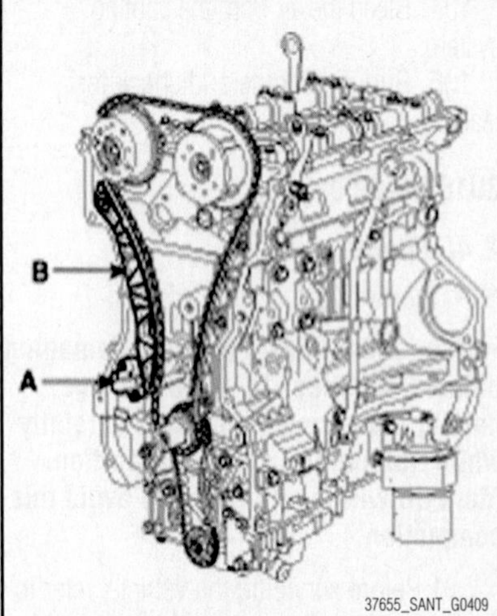

Fig. 339 Remove the timing chain tensioner (A) and timing chain tensioner arm (B)

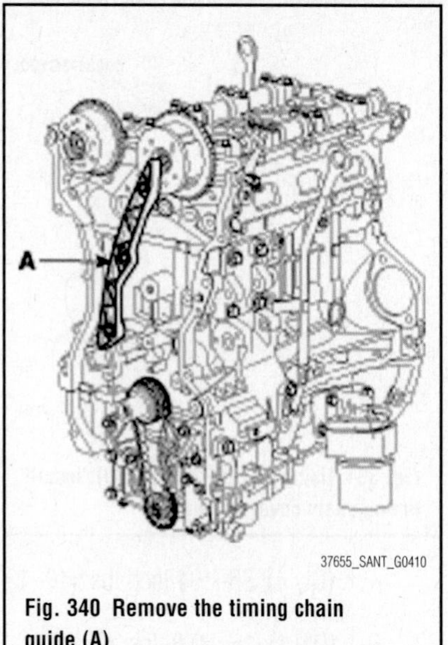

Fig. 340 Remove the timing chain guide (A)

19. Remove the water pump pulley, crankshaft pulley and engine support bracket.

➡Use the SST (Flywheel stopper, 09231-3K000) to remove the crankshaft pulley bolt.

20. Remove the timing chain cover by gently prying between the timing chain cover and the cylinder block.

➡Be careful not to damage the contact surfaces of cylinder block, cylinder head and timing chain cover.

21. The key of crankshaft should be aligned with the mating face of main bearing cap. This will position the piston of No. 1 cylinder at the top dead center on compression stroke.

22. Before removing the timing chain, matchmark the timing chain with an identification based on the location of the sprocket.

23. Install a set pin after compressing the timing chain tensioner.

24. Remove the timing chain tensioner (A) and timing chain tensioner arm (B).

25. Remove the timing chain.

26. Remove the timing chain guide.

To install:

27. Set the crankshaft so the timing marks are aligned on the crankshaft gear. Assemble the camshafts with the TDC mark of the intake CVVT assembly and exhaust CVVT assembly aligned with the top surface of cylinder head. This should position the engine at No. 1 cylinder top dead center on compression stroke.

28. Install the timing chain guide and tighten to 86–104 inch lbs. (10–12 Nm).

29. Install the timing chain. To install the timing chain with no slack between each

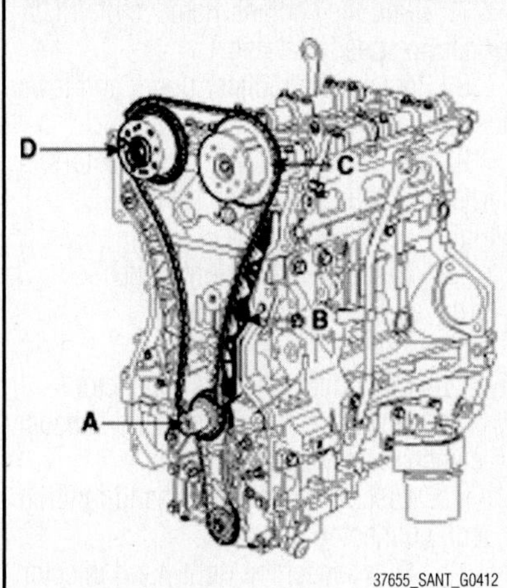

Fig. 341 Install the timing chain: Crankshaft sprocket (A), Timing chain guide (B), Intake CVVT sprocket (C), Exhaust CVVT sprocket (D)

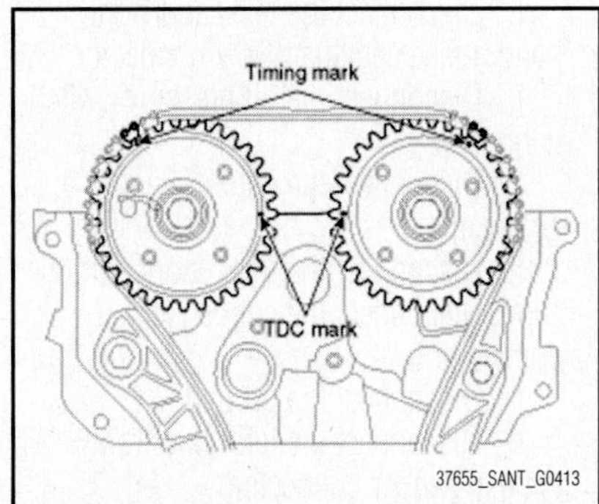

Fig. 342 Timing chain and sprocket positioning

shaft (cam, crank), install as follows: Crankshaft sprocket, Timing chain guide, Intake CVVT sprocket, Exhaust CVVT sprocket.

➡The timing mark of each sprocket should be matched with timing mark (color link) of timing chain.

30. Install timing chain tensioner arm and tighten to 86–104 inch lbs. (10–12 Nm).
31. Install timing chain auto tensioner and remove the set pin and tighten to 86–104 inch lbs. (10–12 Nm).
32. After rotating crankshaft 2 revolutions in regular direction (clockwise viewed from front), confirm the timing mark.
33. Install timing chain cover:
    a. Using a gasket scraper, remove all the old gasket material from the gasket surfaces.
    b. The sealant locations on chain cover and on counter parts (cylinder head, cylinder block, and ladder frame) must be free of engine oil, etc.
    c. Before assembling the timing chain cover, the liquid sealant LOCTITE® 5900H or THREEBOND® 1217H should be applied on the gap between cylinder head and cylinder block. The part must be assembled within 5 minutes after sealant was applied. Bead width: 0.12 in. (3.0mm)
    d. Apply liquid sealant LOCTITE® 5900H on timing chain cover. The part must be assembled within 5 minutes after sealant was applied. Bead width: 0.12 in. (3.0mm)
    e. The dowel pins on the cylinder block and holes on the timing chain cover should be used as a reference in order to assemble the timing chain cover to be in exact position. Tightening torque:
    - M6: 70–86 inch lbs. (8–10 Nm)
    - M8: 14–17 ft. lbs. (19–23 Nm)

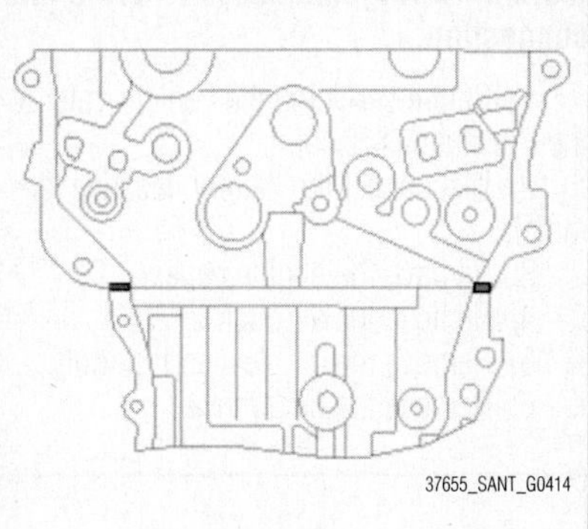

Fig. 343 Apply sealant between cylinder head and cylinder block

➡Running the engine or performing a pressure test should not be performed within 30 minutes of assembly.

34. Install the engine support bracket. Tightening torque:
    - M10: 29–33 ft. lbs. (39–44 Nm)
    - M8: 115–18 ft. lbs. (10–25 Nm)
35. Install the crankshaft pulley and tighten to 123–130 ft. lbs. (167–176 Nm).

➡Use the SST(flywheel stopper, 09231-3K000) to install the crankshaft pulley bolt.

36. Install the water pump pulley and tighten to 70–86 inch lbs. (8–10 Nm).
37. Install the drive belt tensioner and tensioner pulley and tighten to 40–47 ft. lbs. (54–64 Nm).
38. Install the idler pulley and tighten to 40–47 ft. lbs. (54–64 Nm).
39. Install the air compressor bracket and tighten to 14–15 ft. lbs. (20–24 Nm).
40. Install the compressor lower bolts and tighten to 14–18 ft. lbs. (20–25 Nm).

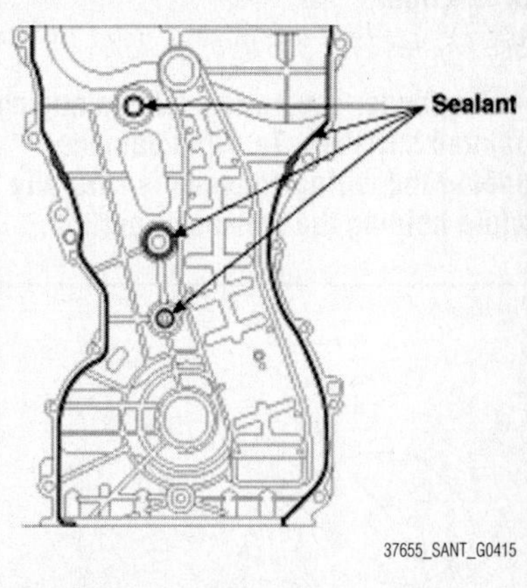

Fig. 344 Timing chain cover sealant application

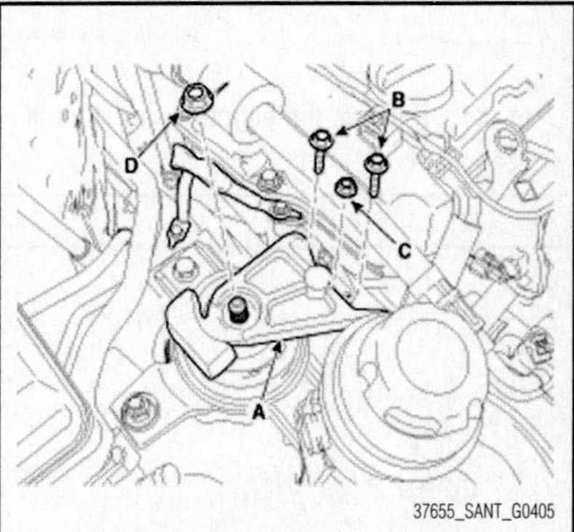

Fig. 345 Install the engine mounting bracket (A), bolts (B), and nuts (C, D)

41. Install the drive belt.
42. Install the engine mounting bracket and connect the ground cable. Tightening torque:
    - Bolts (B) and Nut (C): 47–62 ft. lbs. (64–83 Nm)
    - Nut (D): 58–80 ft. lbs. (79–108 Nm)
43. Install the oil pan.
44. Install cylinder head cover.
45. Install the ignition coils and tighten to 35–53 inch lbs. (4–6 Nm).
46. Connect the PCV hose.
47. Connect the ignition coil connectors and the exhaust CMP connector.
48. Install the undercover and side cover and tighten to 86–104 inch lbs. (10–12 Nm).
49. Install the air cleaner assembly.
50. Install the engine cover.
51. Connect the battery negative terminal.
52. Refill engine oil.
53. Inspect for fuel leakage.
54. Refill radiator and reservoir tank with engine coolant and bleed air from the cooling system.
55. Run the engine and check for leaks.

3.5L Engine

See Figures 271, 346 through 387.

➡**Use fender covers to avoid damaging painted surfaces. To avoid damage, unplug the wiring connectors carefully while holding the connector portion.**

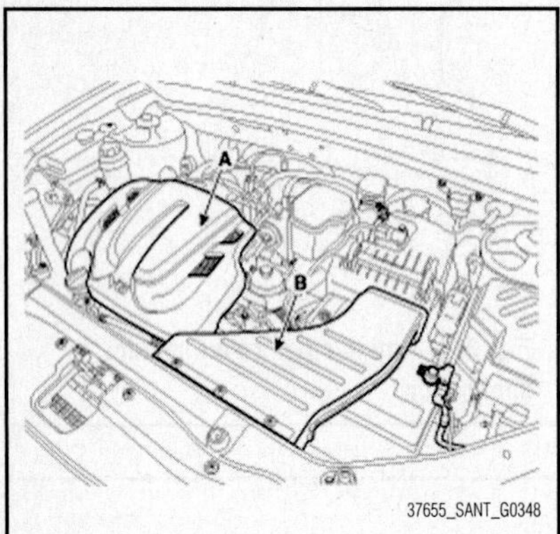

37655_SANT_G0348

Fig. 346 Remove the engine cover (A) and air duct (B)

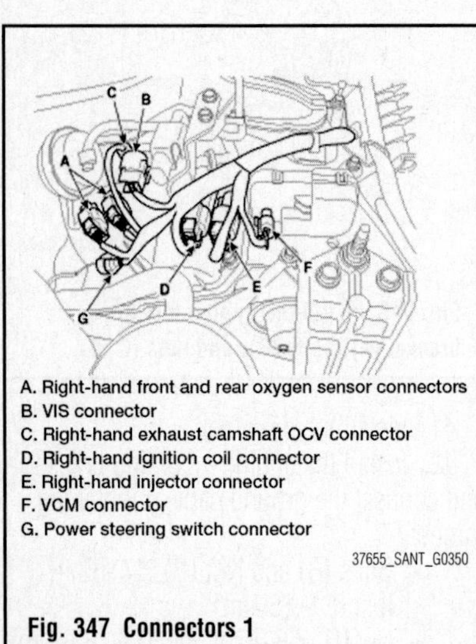

A. Right-hand front and rear oxygen sensor connectors
B. VIS connector
C. Right-hand exhaust camshaft OCV connector
D. Right-hand ignition coil connector
E. Right-hand injector connector
F. VCM connector
G. Power steering switch connector

37655_SANT_G0350

Fig. 347 Connectors 1

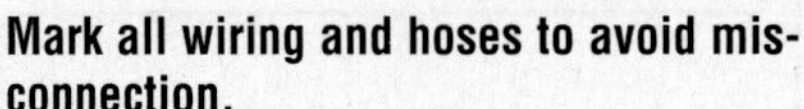

Mark all wiring and hoses to avoid misconnection.

1. Before servicing the vehicle, refer to the Precautions Section.
2. Disconnect the battery negative terminal.
3. Remove the engine cover.
4. Remove the air duct.
5. Remove the air cleaner assembly.
6. Remove the undercover.

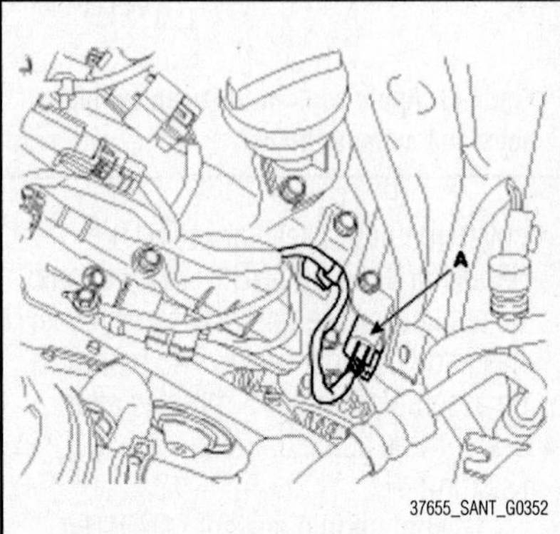

37655_SANT_G0352

Fig. 349 Left-hand front oxygen sensor connector (A)

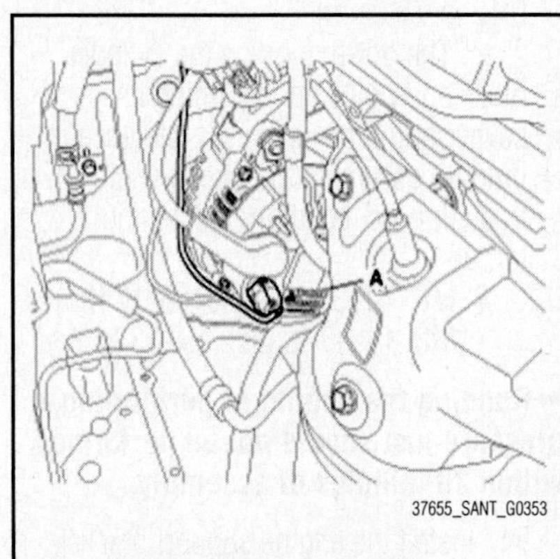

37655_SANT_G0353

Fig. 350 Alternator connector (A)

7. Drain the coolant. Remove the radiator cap to speed draining.
8. Remove the radiator upper and lower hoses.
9. Disconnect the wiring connectors, harness clamps and hoses from the engine:
    a. Disconnect the right-hand front and rear oxygen sensor connectors.
    b. Disconnect the VIS connector.
    c. Disconnect the right-hand exhaust camshaft OCV connector.
    d. Disconnect the right-hand ignition coil connector.
    e. Disconnect the right-hand injector connector.
    f. Disconnect the VCM connector.
    g. Disconnect the power steering switch connector.
    h. Disconnect the left-hand exhaust camshaft OCV connector.
    i. Disconnect the left-hand/right-hand intake camshaft OCV connector.
    j. Disconnect the oil pressure switch connector.
    k. Disconnect the knock sensor connector.
    l. Disconnect the VPS connector.
    m. Disconnect the condenser connectors.
    n. Disconnect the ground.
    o. Disconnect the left-hand front oxygen sensor connector.
    p. Disconnect the alternator connector.
    q. Disconnect the left-hand ignition coil connectors.
    r. Disconnect the left-hand injector connectors.
    s. Disconnect the VIS connector.
    t. Disconnect the left-hand intake CMPS connector.

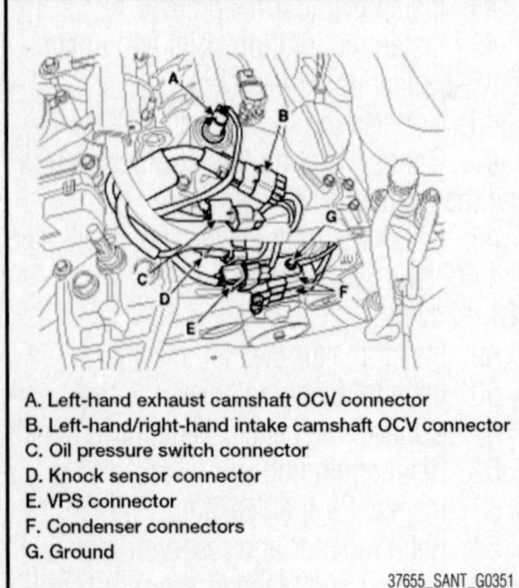

A. Left-hand exhaust camshaft OCV connector
B. Left-hand/right-hand intake camshaft OCV connector
C. Oil pressure switch connector
D. Knock sensor connector
E. VPS connector
F. Condenser connectors
G. Ground

37655_SANT_G0351

Fig. 348 Connectors 2

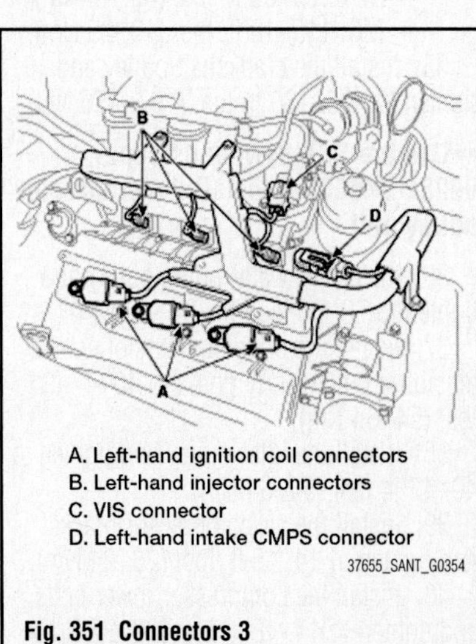

A. Left-hand ignition coil connectors
B. Left-hand injector connectors
C. VIS connector
D. Left-hand intake CMPS connector

37655_SANT_G0354

Fig. 351 Connectors 3

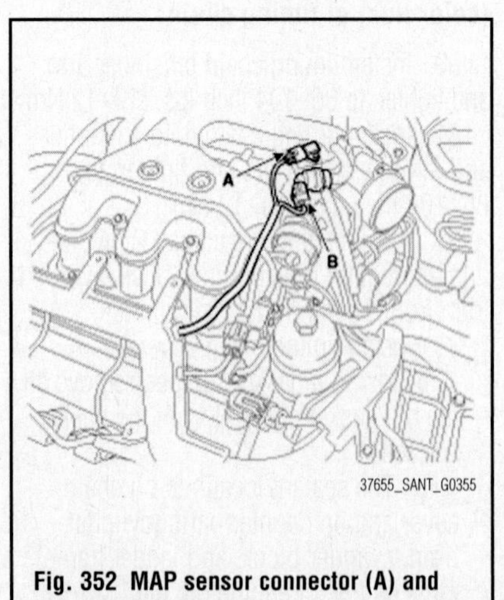

37655_SANT_G0355

Fig. 352 MAP sensor connector (A) and PCSV connector (B)

A
B
C
D
E

A. Right-hand intake CMPS connector
B. Knock sensor connector
C. OTS connector
D. ETC connector
E. Right-hand exhaust CMPS connector

37655_SANT_G0356

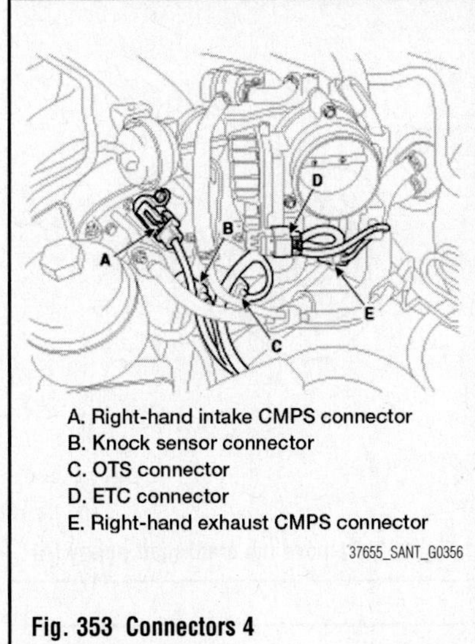

Fig. 353 Connectors 4

A

37655_SANT_G0357

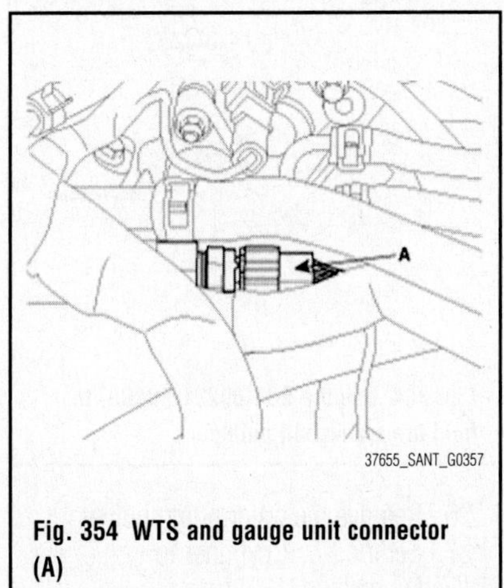

Fig. 354 WTS and gauge unit connector (A)

A
B
C

37655_SANT_G0358

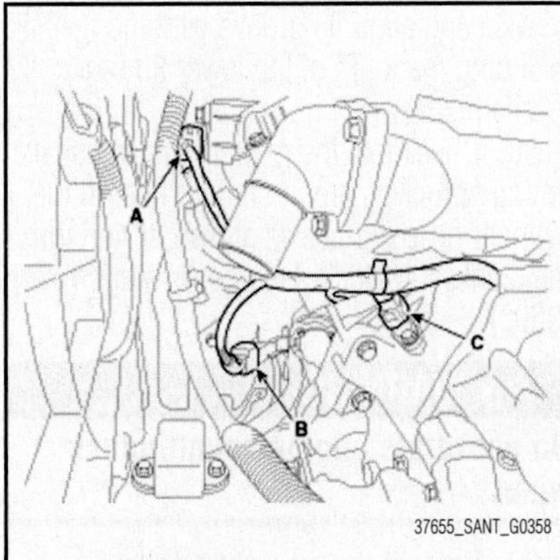

Fig. 355 Left-hand exhaust CMPS connector (A), rear oxygen sensor connector (B), and CKPS connector (C)

u. Disconnect the MAP sensor connector.

v. Disconnect the PCSV connector.

w. Disconnect the right-hand intake CMPS connector.

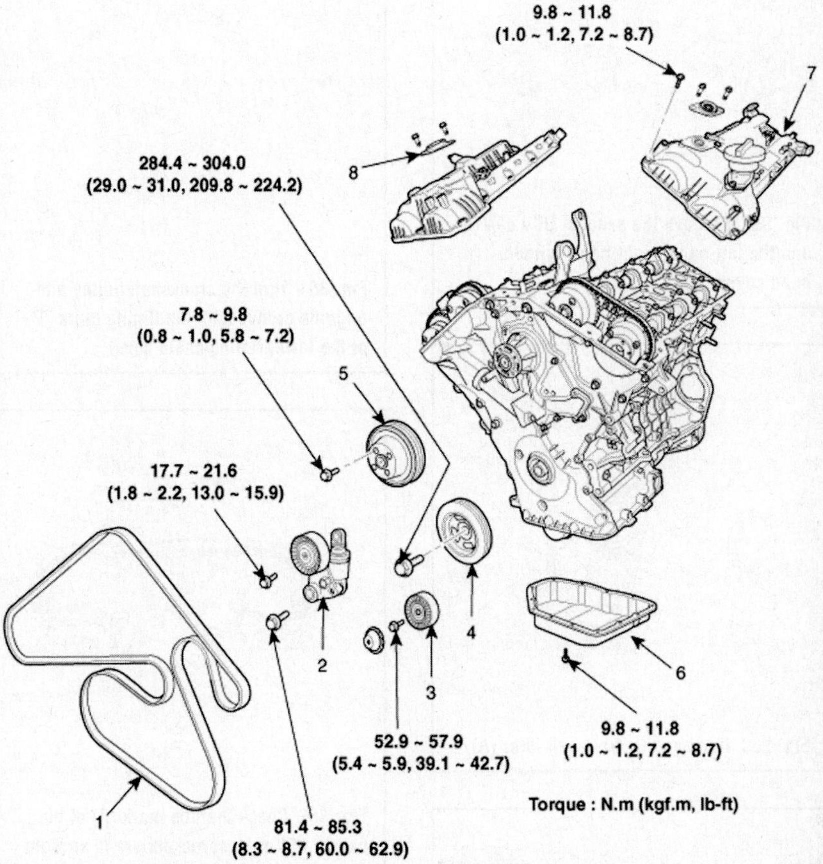

9.8 ~ 11.8
(1.0 ~ 1.2, 7.2 ~ 8.7)
7
8
284.4 ~ 304.0
(29.0 ~ 31.0, 209.8 ~ 224.2)
7.8 ~ 9.8
(0.8 ~ 1.0, 5.8 ~ 7.2)
5
17.7 ~ 21.6
(1.8 ~ 2.2, 13.0 ~ 15.9)
2
4
3
6
52.9 ~ 57.9
(5.4 ~ 5.9, 39.1 ~ 42.7)
9.8 ~ 11.8
(1.0 ~ 1.2, 7.2 ~ 8.7)
Torque : N.m (kgf.m, lb-ft)
1
81.4 ~ 85.3
(8.3 ~ 8.7, 60.0 ~ 62.9)

1. Drive belt
2. Drive belt tensioner
3. Idler
4. Crankshaft pulley
5. Water pump pulley
6. Oil pan
7. Cylinder head cover
8. OCV cap

37655_SANT_G0397

Fig. 356 Crankshaft pulley and components

x. Disconnect the knock sensor connector.

y. Disconnect the OTS connector.

z. Disconnect the ETC connector.

aa. Disconnect the right-hand exhaust CMPS connector.

bb. Disconnect the WTS and gauge unit connector.

cc. Disconnect the left-hand exhaust CMPS connector.

dd. Disconnect the left-hand rear oxygen sensor connector.

ee. Disconnect the CKPS connector.

10. Disconnect the PCSV hose, fuel hose.

11. Drain the engine oil.

12. Remove the lower oil pan. Refer to Oil Pan Removal & Installation.

37655_SANT_G0359

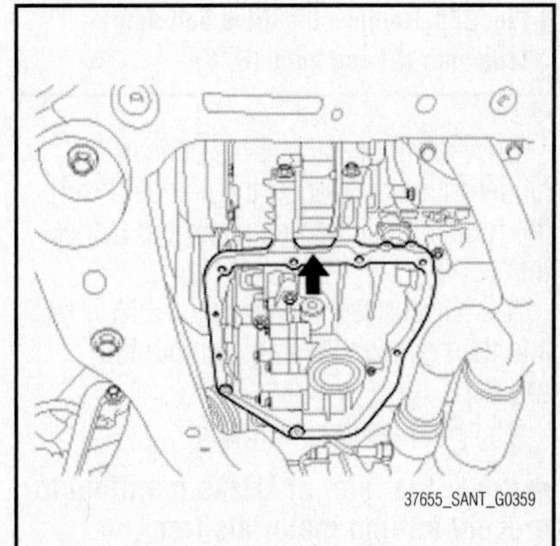

Fig. 357 Install the jack to the edge of ladder frame (A) to support the engine

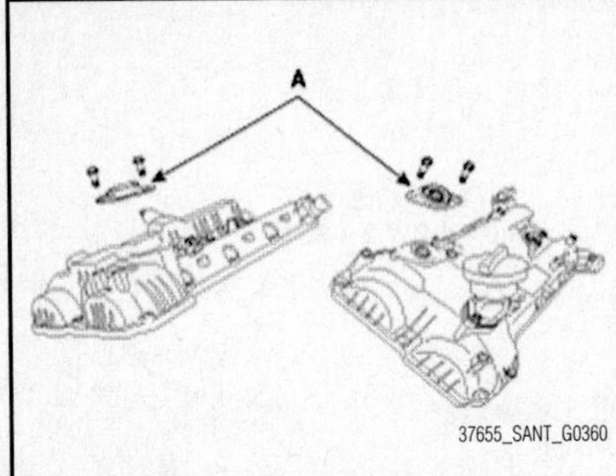

Fig. 358 Remove the exhaust OCV cap (A) and the left-hand, right-hand cylinder head covers

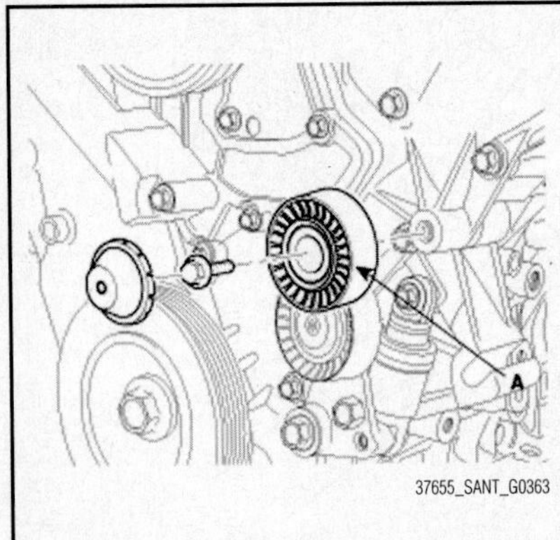

Fig. 359 Remove the drive belt idler (A)

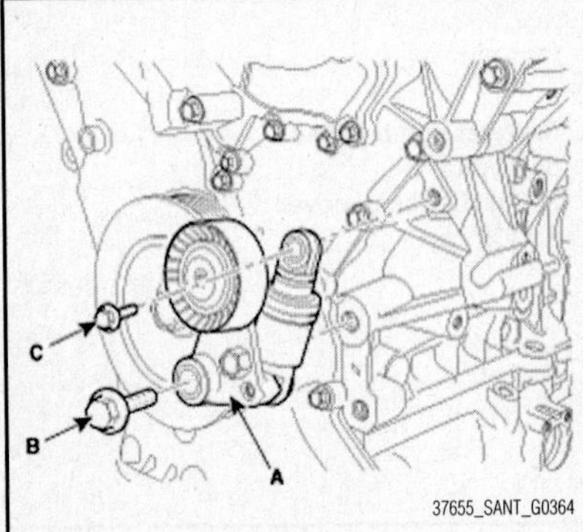

Fig. 360 Remove the drive belt auto tensioner (A) and bolts (B, C)

13. Install the jack to the edge of ladder frame to support the engine. Insert the rubber block between jack and ladder frame.

14. Disconnect the ground cable, and then remove the engine mounting bracket.

15. Remove the surge tank.

➡Cover the inlet of intake manifold to prevent foreign materials from entering.

16. Disconnect the right-hand ignition coil connectors and the right-hand injector connectors. Refer to Ignition Coils, Removal & Installation in the Engine Electrical section.

17. Remove the left-hand/right-hand ignition coils.

18. Remove the exhaust OCV cap, and then remove the left-hand, right-hand cylinder head covers.

19. Remove the drive belt. Refer to Accessory Drive Belt Removal & Installation.

20. Lift up the engine assembly with the jack.

21. Remove the power steering pump. Refer to Power Steering Pump Removal & Installation in the Steering section.

22. Remove the air conditioner compressor.

23. Remove the alternator. Refer to Alternator Removal & Installation in the Engine Electrical section.

24. Remove the drive belt idler.

25. Remove the drive belt auto tensioner.

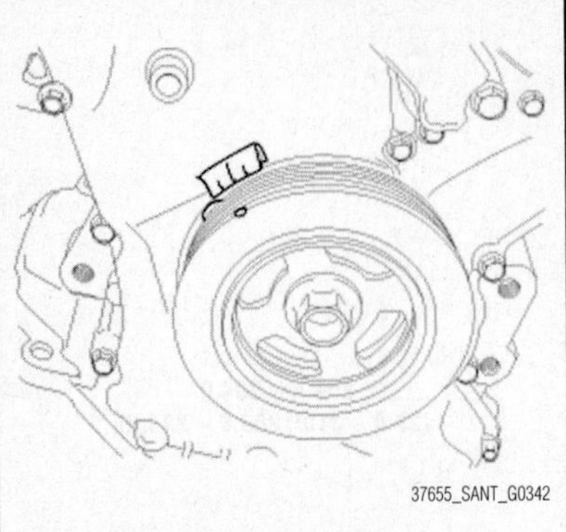

Fig. 361 Turn the crankshaft pulley and align its groove with the timing mark "T" of the lower timing chain cover

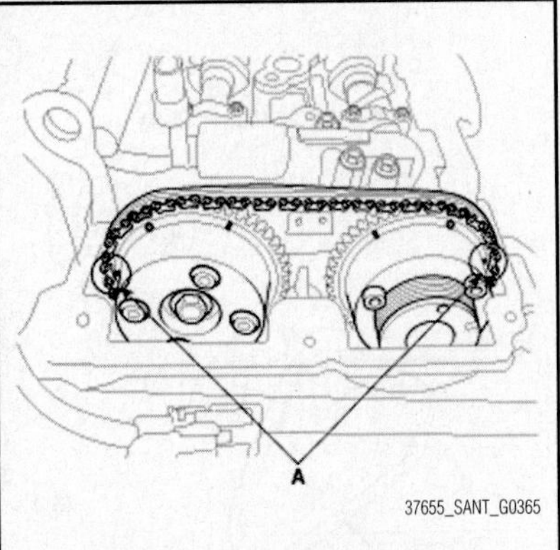

Fig. 362 Check that the mark (A) of the camshaft timing sprockets are in straight line on the cylinder head surface

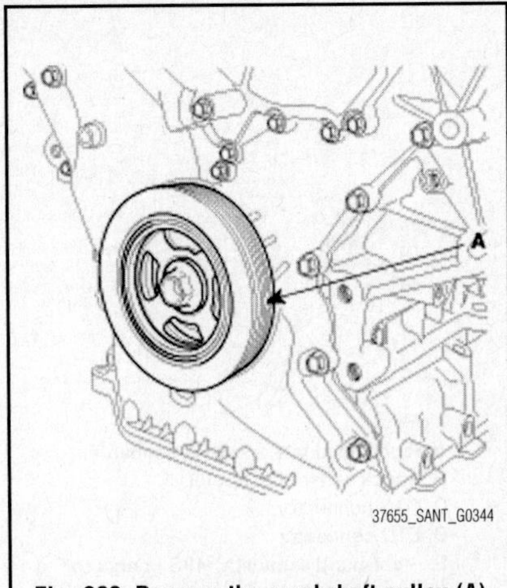

Fig. 363 Remove the crankshaft pulley (A)

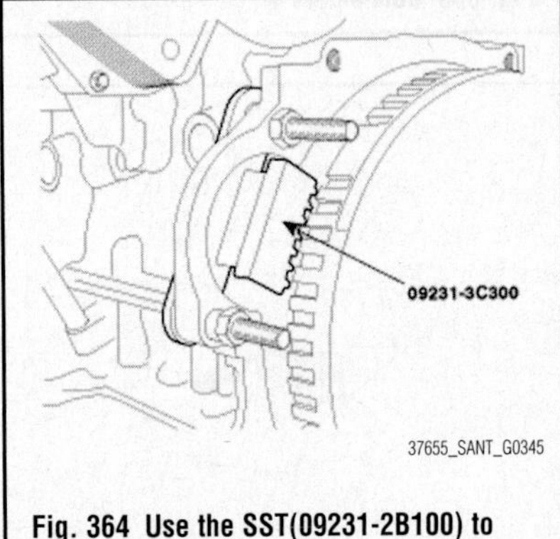

Fig. 364 Use the SST(09231-2B100) to hold the crankshaft pulley

26. Remove the water pump pulley.

27. Set the No. 1 cylinder to TDC/compression.

a. Turn the crankshaft pulley clockwise and align its groove with the timing mark "T" of the lower timing chain cover.

28. Check that the mark of the camshaft timing sprockets are in straight line on the cylinder head surface as shown. If not, turn the crankshaft clockwise one revolution (360°).

WARNING

Do not rotate engine counterclockwise.

29. Remove the crankshaft pulley.

➡Use the SST(09231-2B100) to hold the crankshaft pulley.

30. Remove the water pump and gasket. Refer to Water Pump Removal & Installation in the Engine Cooling section.

31. Remove the timing chain cover. Refer to Timing Chain Front Cover Removal & Installation.

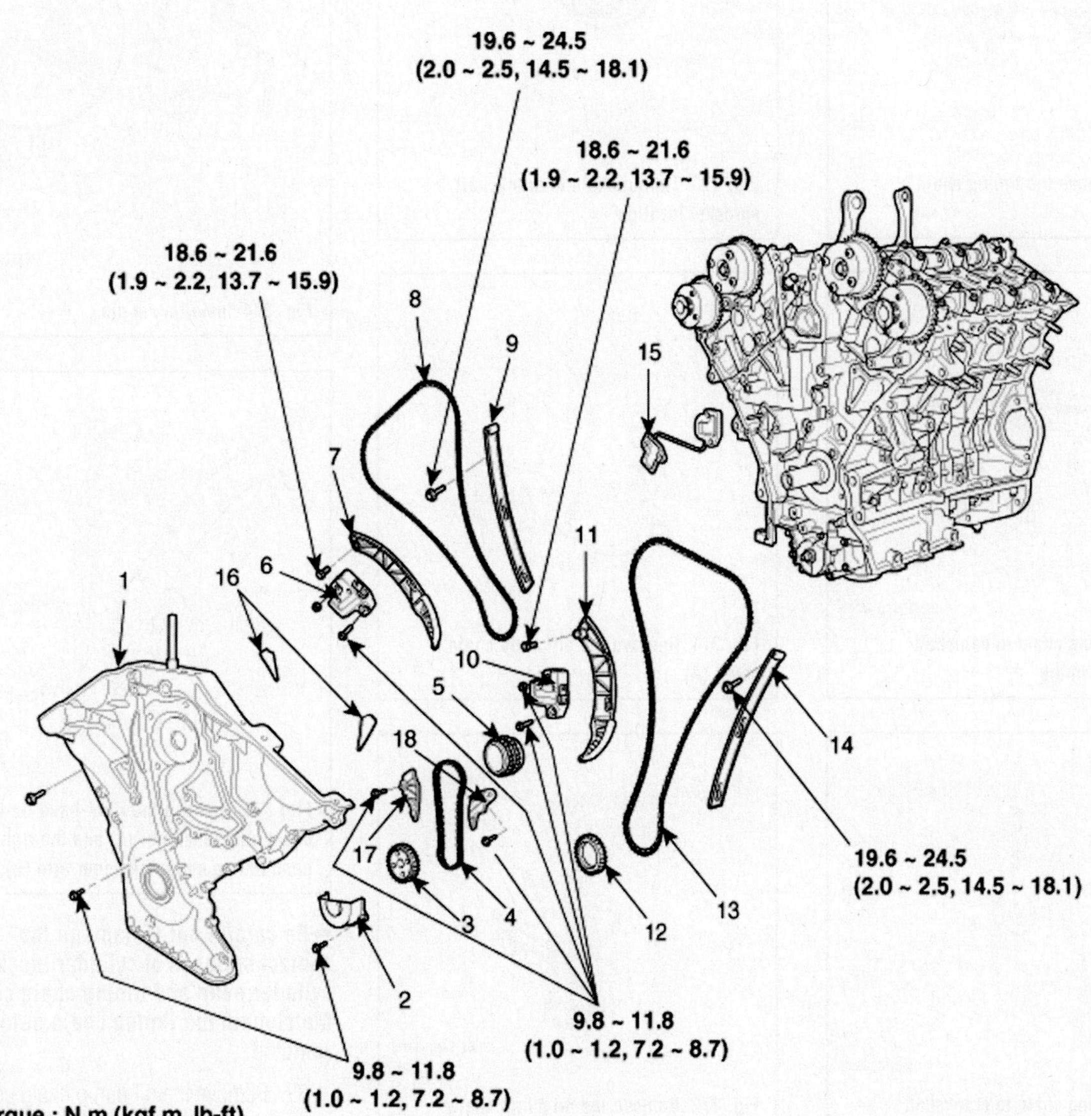

1. Timing chain cover
2. Oil pump chain cover
3. Oil pump sprocket
4. Oil pump chain
5. Crankshaft sprocket
6. Timing chain auto tensioner
7. Timing chain tensioner arm
8. Timing chain
9. Timing chain guide
10. Timing chain auto tensioner
11. Timing chain tensioner arm
12. Crankshaft sprocket
13. Timing chain
14. Timing chain guide
15. Tensioner adapter
16. Gasket
17. Oil pump chain guide
18. Oil pump tensioner assembly

37655_SANT_G0398

Fig. 365 Timing chain and components

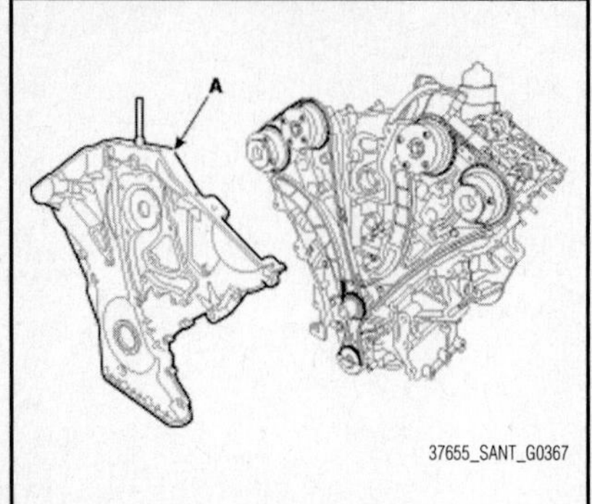

Fig. 366 Remove the timing chain cover (A)

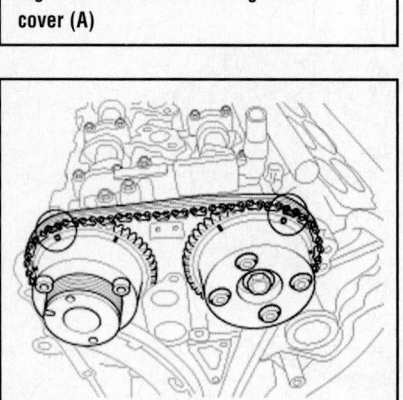

Fig. 367 Timing chain to camshaft position matchmark

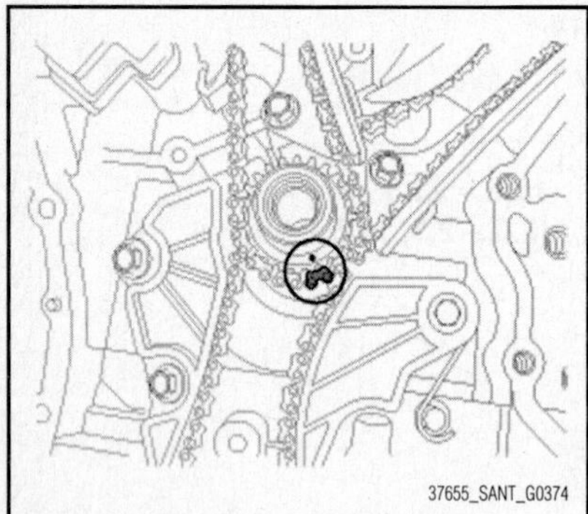

Fig. 368 Timing chain to crankshaft position location

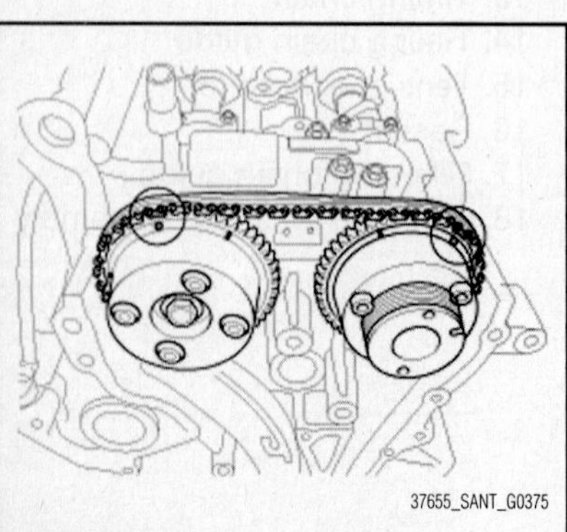

Fig. 369 Timing chain to camshaft matchmark

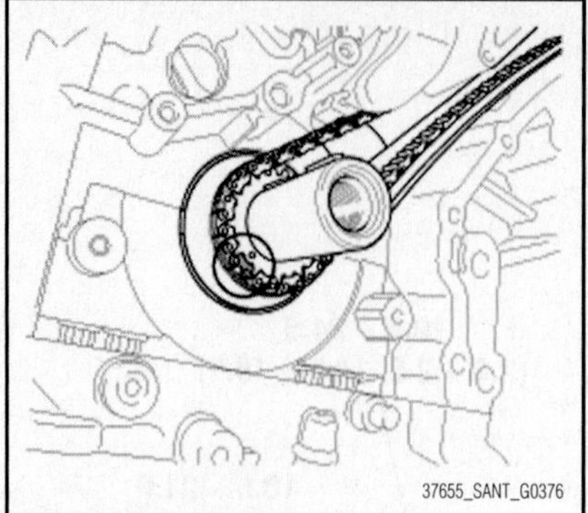
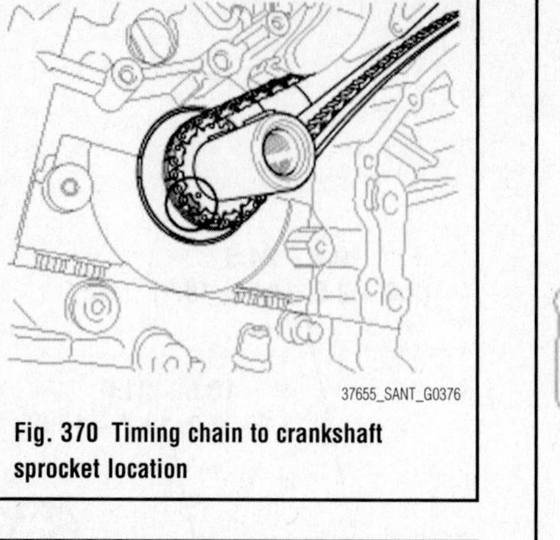

Fig. 370 Timing chain to crankshaft sprocket location

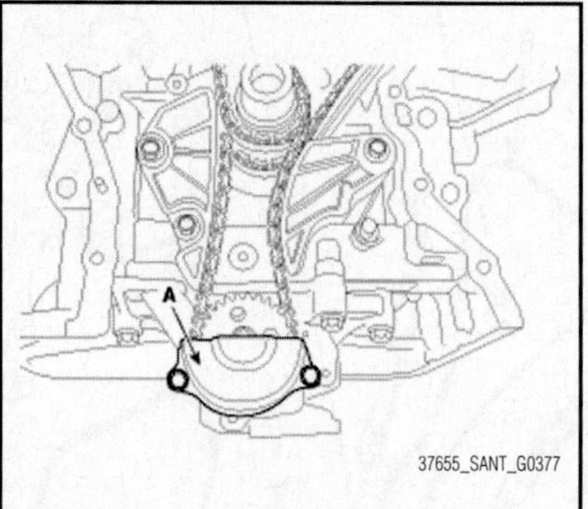

Fig. 371 Remove the oil pump chain cover (A)

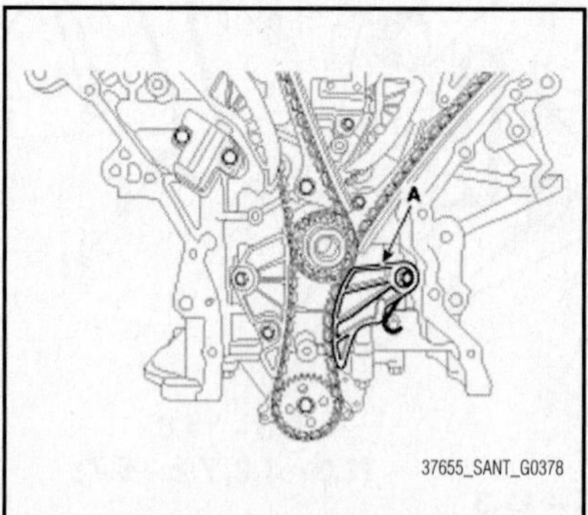
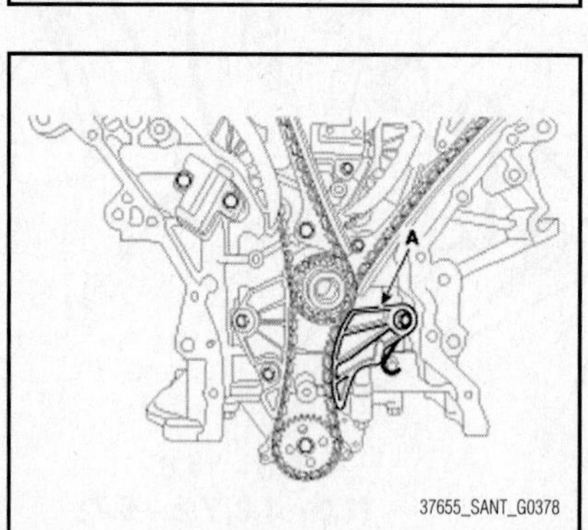

Fig. 372 Remove the oil pump chain tensioner assembly (A)

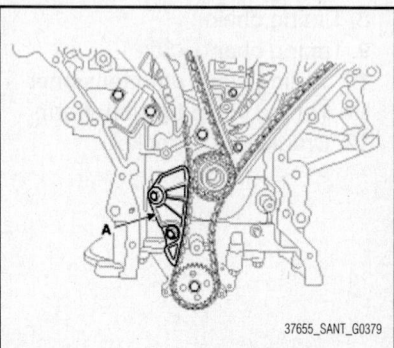

Fig. 373 Remove the oil pump chain guide (A)

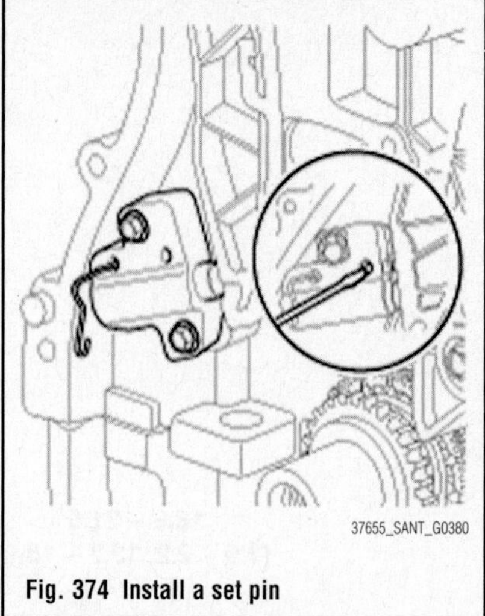

Fig. 374 Install a set pin

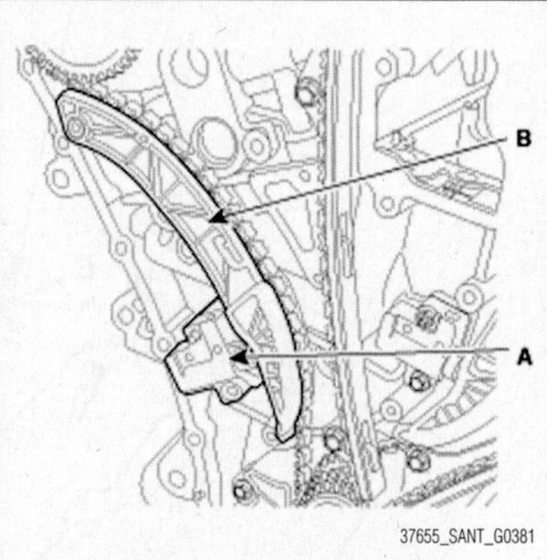

Fig. 375 Remove the right-hand timing chain auto tensioner (A) and the right-hand timing chain tensioner arm (B)

➡Be careful not to damage the contact surfaces of cylinder block, cylinder head and timing chain cover. Matchmark the timing chain before removal.

32. Remove the oil pump chain cover.
33. Remove the oil pump chain tensioner assembly.
34. Remove the oil pump chain guide.
35. Install a set pin after compressing the right-hand timing chain tensioner.
36. Remove the right-hand timing chain auto tensioner and the right-hand timing chain tensioner arm.
37. Remove the right-hand timing chain guide and right-hand timing chain.
38. Remove the oil pump chain sprocket and oil pump chain.
39. Remove the crankshaft sprocket (O/P and right-hand camshaft drive).
40. Install a set pin after compressing the left-hand timing chain tensioner.

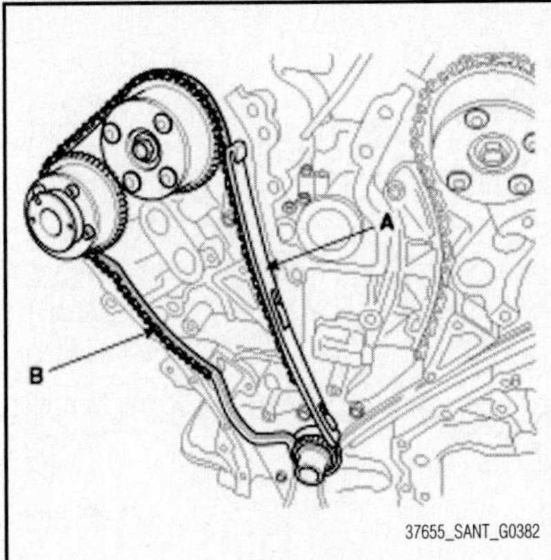

Fig. 376 Remove the right-hand timing chain guide and right-hand timing chain

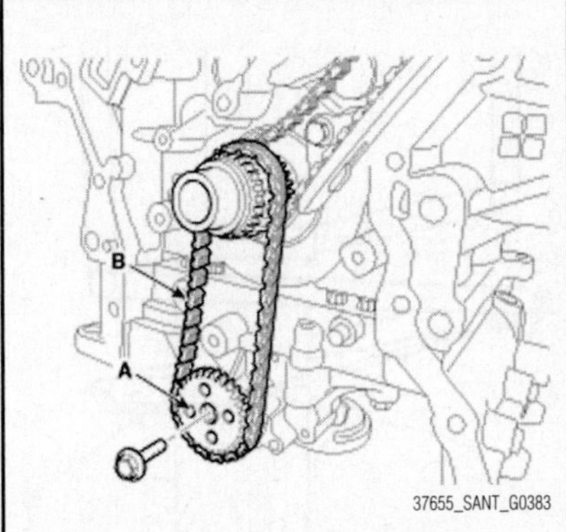

Fig. 377 Remove the oil pump chain sprocket (A) and oil pump chain (B)

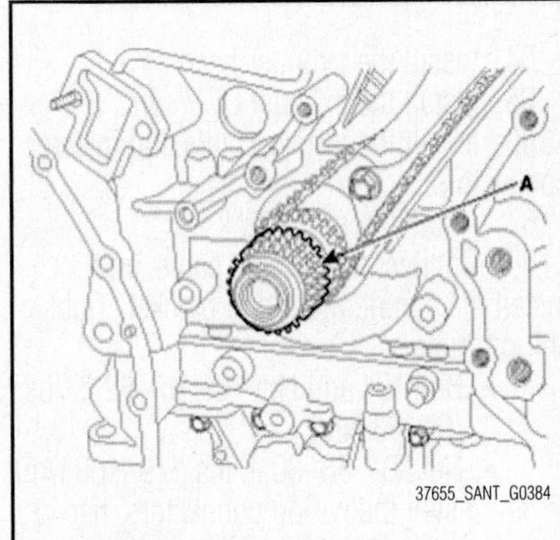

Fig. 378 Remove the crankshaft sprocket (A)

41. Remove the left-hand timing chain auto tensioner and left-hand timing chain tensioner arm.

42. Remove the left-hand timing chain guide and left-hand timing chain.

To install:

43. Install the left-hand timing chain guide and left-hand timing chain and tighten to 15–18 ft. lbs. (20–25 Nm).

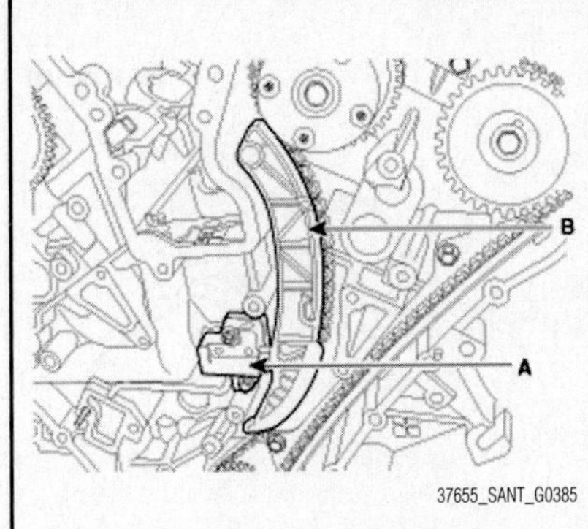
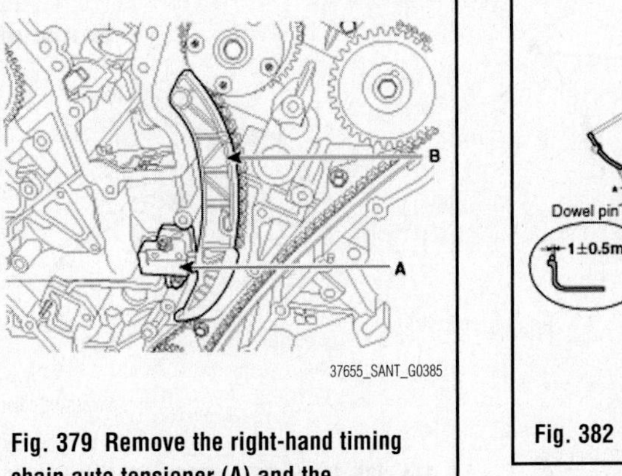

Fig. 379 Remove the right-hand timing chain auto tensioner (A) and the right-hand timing chain tensioner arm (B)

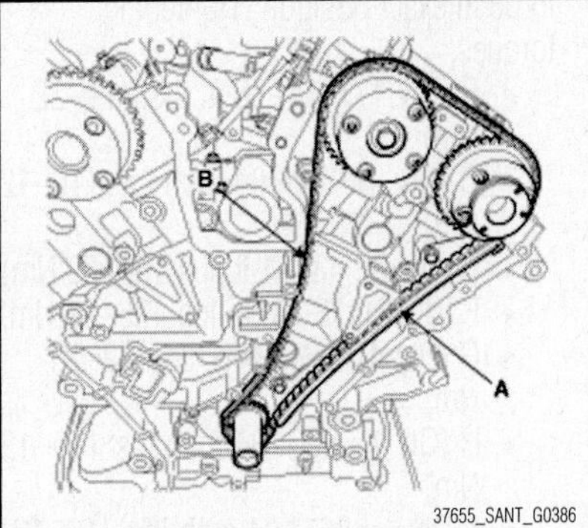

Fig. 380 Remove the left-hand timing chain guide (A) and left-hand timing chain (B)

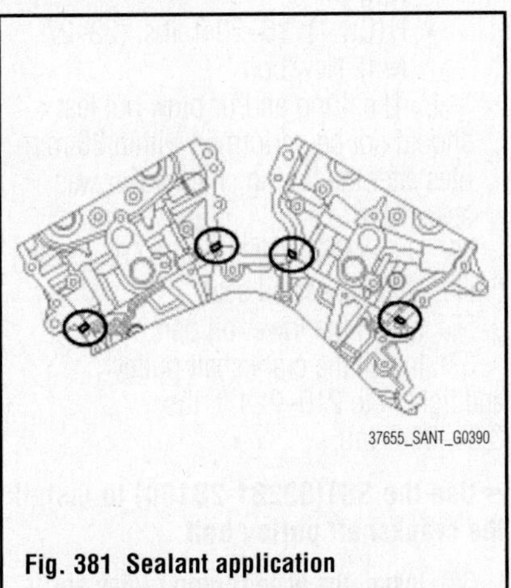

Fig. 381 Sealant application

➡Install the timing chain with no slack between each shaft (cam, crank) as follows: Crankshaft sprocket, Timing chain guide, Exhaust camshaft sprocket, Intake camshaft sprocket. The timing mark of each sprockets should be matched with timing mark (color link) of timing chain.

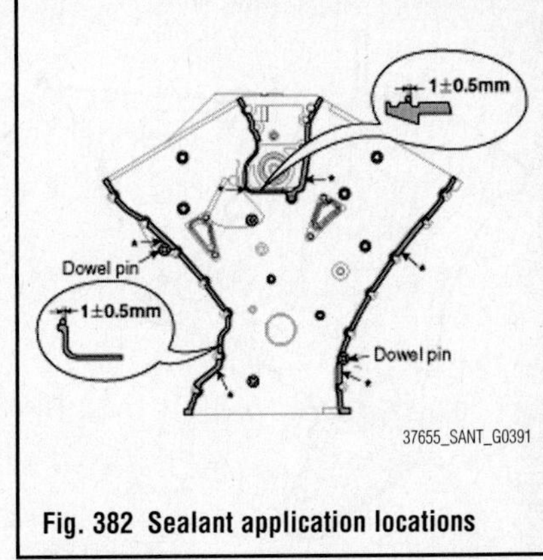

Fig. 382 Sealant application locations

44. Install the left-hand timing chain auto tensioner and left-hand timing chain tensioner arm. Tightening torque:
    - Bolt "A": 86–104 inch lbs. (10–12 Nm)
    - Bolt "B": 14–16 ft. lbs. (19–22 Nm)

45. Install the crankshaft sprocket (O/P and right-hand camshaft drive).

46. Install the oil pump chain sprocket and oil pump chain. Tighten to 14–16 ft. lbs. (19–22 Nm).

47. Install the right-hand timing chain guide and the right-hand timing chain. Tighten to 15–18 ft. lbs. (20–25 Nm).

➡Install the timing chain with no slack between each shaft (cam, crank) as follows: Crankshaft sprocket, Timing chain guide, Intake camshaft sprocket, Exhaust camshaft sprocket. The timing mark of each sprockets should be matched with timing mark (color link) of timing chain at installing timing chain.

48. Install the right-hand timing chain auto tensioner and the right-hand timing chain tensioner arm. Tightening torque:
    - Bolt "A": 86–104 inch lbs. (10–12 Nm)
    - Bolt "B": 14–16 ft. lbs. (19–22 Nm)

49. Install the oil pump chain guide. Tighten to 86–104 inch lbs. (10–12 Nm).

50. Install the oil pump chain tensioner assembly. Tighten to 86–104 inch lbs. (10–12 Nm).

51. Remove the hydraulic tensioner pins. (Left-hand and right-hand).

52. Install the oil pump chain cover. Tighten to 86–104 inch lbs. (10–12 Nm).

53. After rotating the crankshaft 2 revolutions in regular direction (clockwise viewed from front), confirm the timing mark.

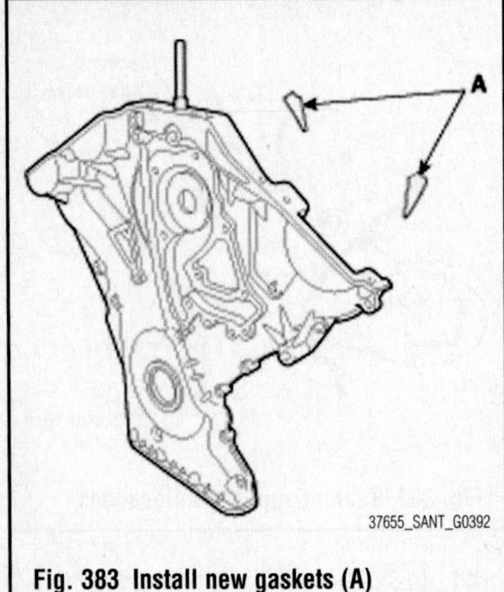

Fig. 383 Install new gaskets (A)

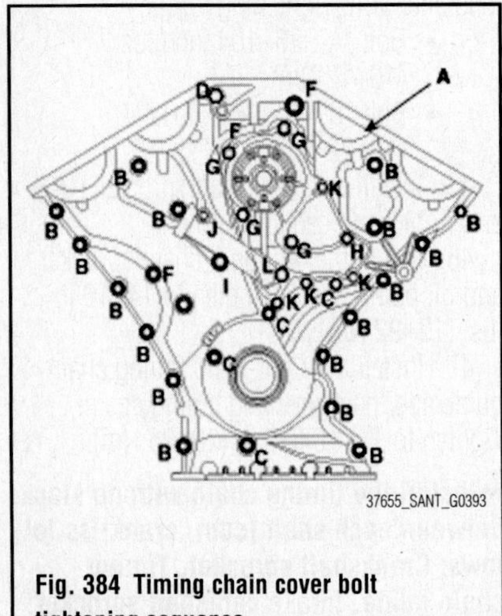

Fig. 384 Timing chain cover bolt tightening sequence

⁂ WARNING

Always turn the crankshaft clockwise.

54. Install the timing chain cover.
  a. The sealant locations on chain cover and on counter parts (cylinder head, cylinder block, and lower oil pan) must be free of engine oil and etc.
  b. Before assembling the timing chain cover, the liquid sealant TB 1217H should be applied on the gap between cylinder head and cylinder block. The part must be assembled within 5 minutes after sealant was applied.
  c. After applying liquid sealant TB1217H on timing chain cover, the part must be assembled within 5 minutes. Fill the T-joint areas with sealant. Apply sealant all around the dowel pin holes.
  d. Install new gaskets to the timing chain cover.
  e. The dowel pins on the cylinder block and holes on the timing chain cover should be used as a reference in order to assemble the timing chain cover

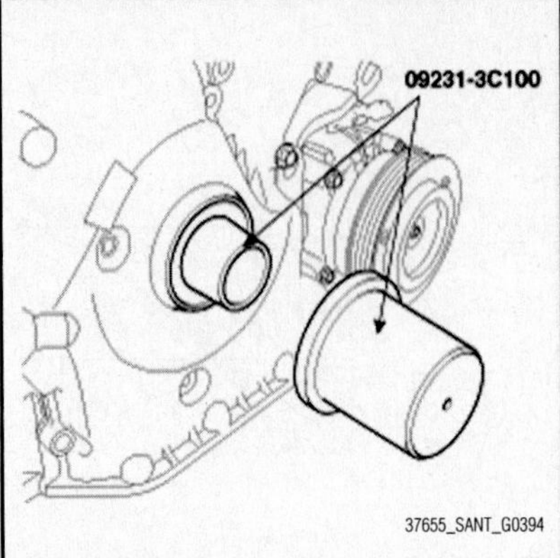

Fig. 385 Using SST (09231-3C100), install timing chain cover oil seal

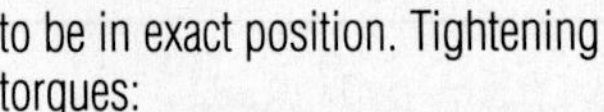
to be in exact position. Tightening torques:
- B (Qty 17): 14–16 ft. lbs. (19–22 Nm)
- C (Qty 4): 86–104 inch lbs. (10–12 Nm)
- D (Qty 2): 43–51 ft. lbs. (59–69 Nm)
- F (Qty 2): 18–20 ft. lbs. (25–27 Nm)
- G (Qty 4): 16–17 ft. lbs. (22–24 Nm)
- H (Qty 1): 86–104 inch lbs. (10–12 Nm)
- I (Qty 1): 86–104 inch lbs. (10–12 Nm)
- J (Qty 1): 86–104 inch lbs. (10–12 Nm)
- K (Qty 4): 86–104 inch lbs. (10–12 Nm)
- L (Qty 1): 16–20 ft. lbs. (22–27 Nm), New bolt

  f. The firing and/or blow out test should not be performed within 30 minutes after the timing chain cover was assembled.

55. Using SST (09231-3C100), install timing chain cover oil seal.
56. Install the lower oil pan.
57. Install the crankshaft pulley and tighten to 210–224 ft. lbs. (284–304 Nm).

➡Use the SST(09231-2B100) to install the crankshaft pulley bolt.

58. Install the water pump pulley and tighten to 70–86 inch lbs. (8–10 Nm).
59. Install the drive belt auto tensioner and idler. Tighten:
  - Bolt "B": 60–63 ft. lbs. (81–85 Nm)
  - Bolt "C": 13–16 ft. lbs. (18–22 Nm)
60. Install the alternator.
61. Install the air conditioner compressor.
62. Install the power steering pump.
63. Install the drive belt.

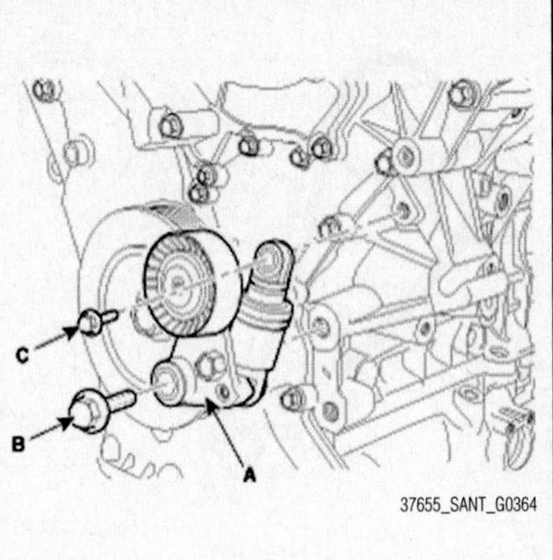

Fig. 386 Drive belt auto tensioner (A) and bolts (B, C)

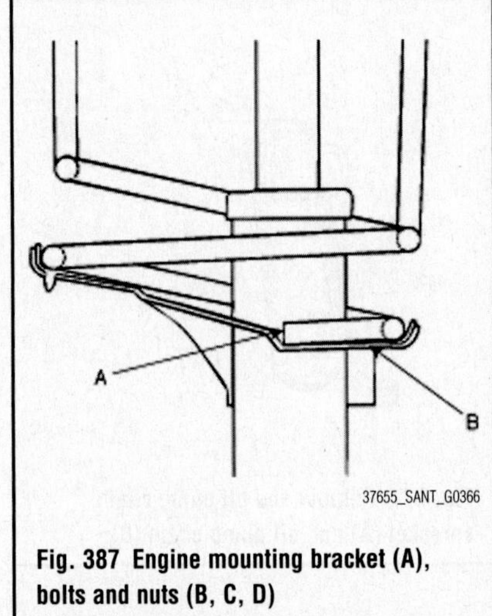

Fig. 387 Engine mounting bracket (A), bolts and nuts (B, C, D)

64. Install the cylinder head covers.
65. Install the exhaust OCV cap.
66. Install the ignition coils and injector connectors.
67. Install the surge tank.
68. Connect the ground cable, and install the engine mounting bracket. Tightening torque:
  - Bolt "B" and Nut "C": 47–62 ft. lbs. (64–83 Nm)
  - Nut "D": 58–80 ft. lbs. (79–108 Nm)
69. Install the wiring connectors, harness clamps and hoses to the engine.
70. Install the radiator hoses.
71. Install the undercover.
72. Install the air cleaner assembly and air duct.
73. Install the engine cover.
74. Refill engine oil.
75. Connect the negative battery cable.
76. Inspect for fuel leakage.
77. Refill radiator and reservoir tank with engine coolant and bleed air from the cooling system.
78. Run the engine and check for leaks.

ENGINE PERFORMANCE & EMISSION CONTROLS

COMPONENT LOCATIONS

See Figures 388 through 395.

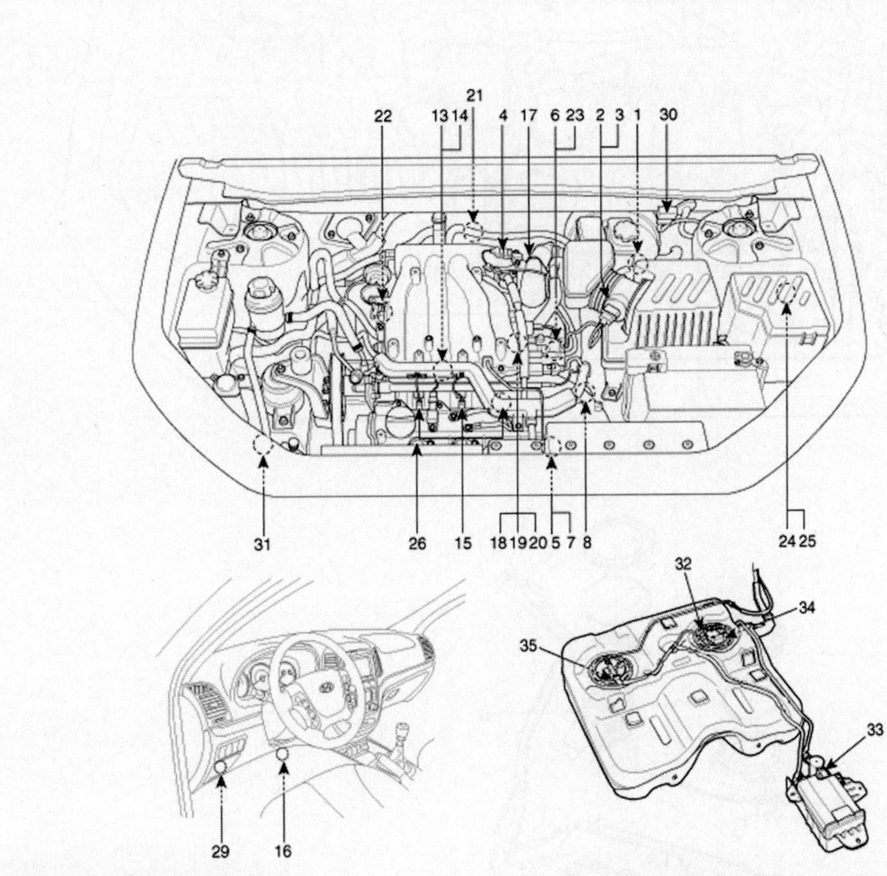

1. Engine Control Module (ECM)
2. Mass Air Flow (MAF) Sensor
3. Intake Air Temperature (IAT) Sensor
4. Manifold Absolute Pressure (MAP) Sensor
5. Engine Coolant Temperature (ECT) Sensor
6. Camshaft Position (CMP) Sensor [Bank 1]
7. Camshaft Position (CMP) Sensor [Bank 2]
8. Crankshaft Position (CKP) Sensor
9. Heated Oxygen Sensor (HO2S) [Bank 1 / Sensor 1]
10. Heated Oxygen Sensor (HO2S) [Bank 1 / Sensor 2]
11. Heated Oxygen Sensor (HO2S) [Bank 2 / Sensor 1]
12. Heated Oxygen Sensor (HO2S) [Bank 2 / Sensor 2]
13. Knock Sensor (KS) [Bank 1]
14. Knock Sensor (KS) [Bank 2]
15. Injector
16. Accelerator Position (APP) Sensor
17. ETC Module [Throttle Position Sensor (TPS) + ETC Motor]
18. CVVT Oil Control Valve (OCV) [Bank 1]
19. CVVT Oil Control Valve (OCV) [Bank 2]
20. CVVT Oil Temperature Sensor (OTS)
21. Purge Control Solenoid (PCV) Valve
22. Variable Intake Solenoid (VIS) Valve No. 1 (Surge Tank Side)
23. Variable Intake Solenoid (VIS) Valve No. 2 (Intake Manifold Side)
24. Fuel Pump Relay
25. Main Relay
26. Ignition Coil
27. Wheel Speed Sensor (WSS)
28. Vehicle Speed Sensor (VSS)
29. Data Link Connector (DLC)
30. Multi-Purpose Connector
31. A/C Pressure Transducer (APT)
32. Fuel Tank Pressure Sensor
33. Canister Close Valve (CCV)
34. Fuel Level Sensor (FLS) 1
35. Fuel Level Sensor (FLS) 2

37655_SANT_G0543

Fig. 388 Underhood and related sensor locations—2.7L engine

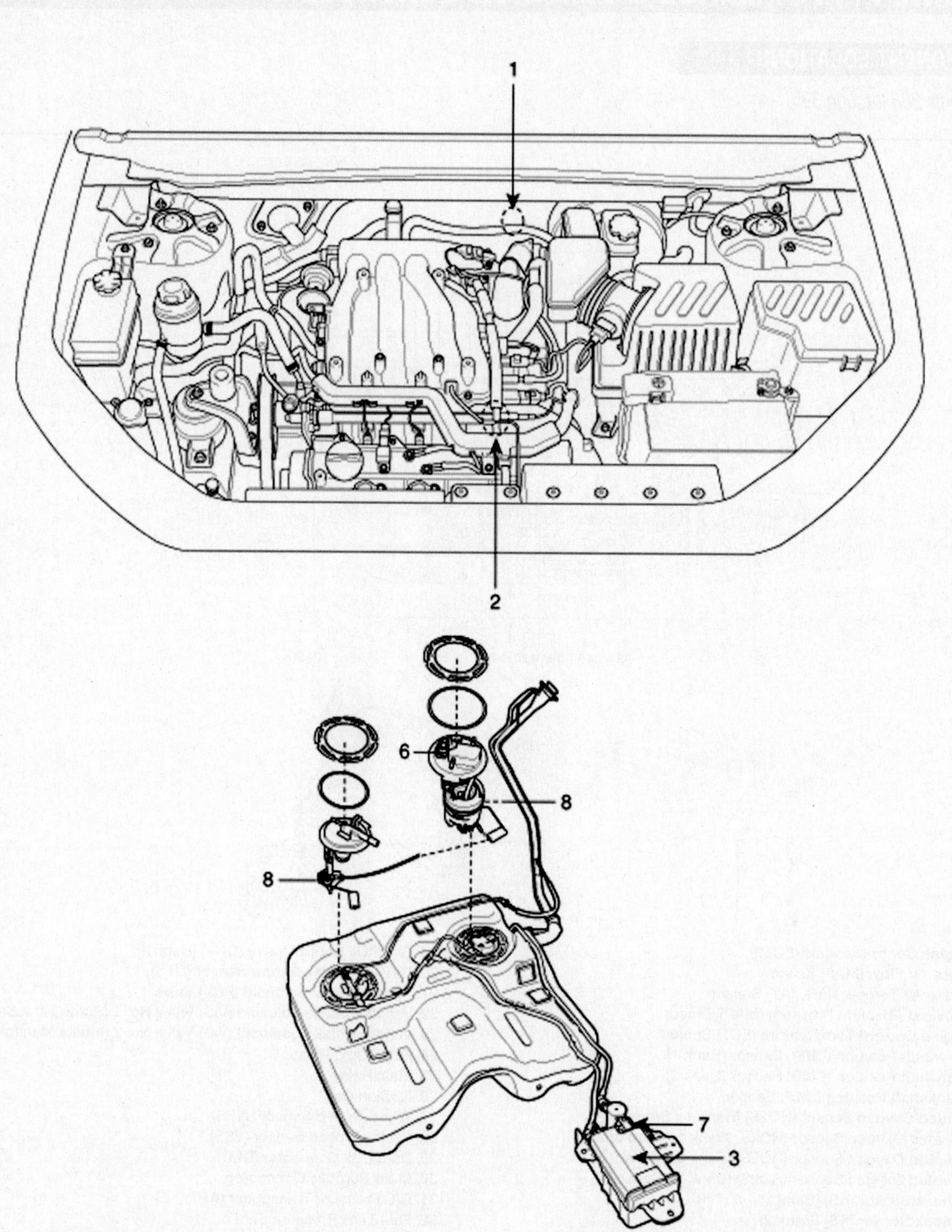

Fig. 389 Emission Control System component locations—2.7L engine

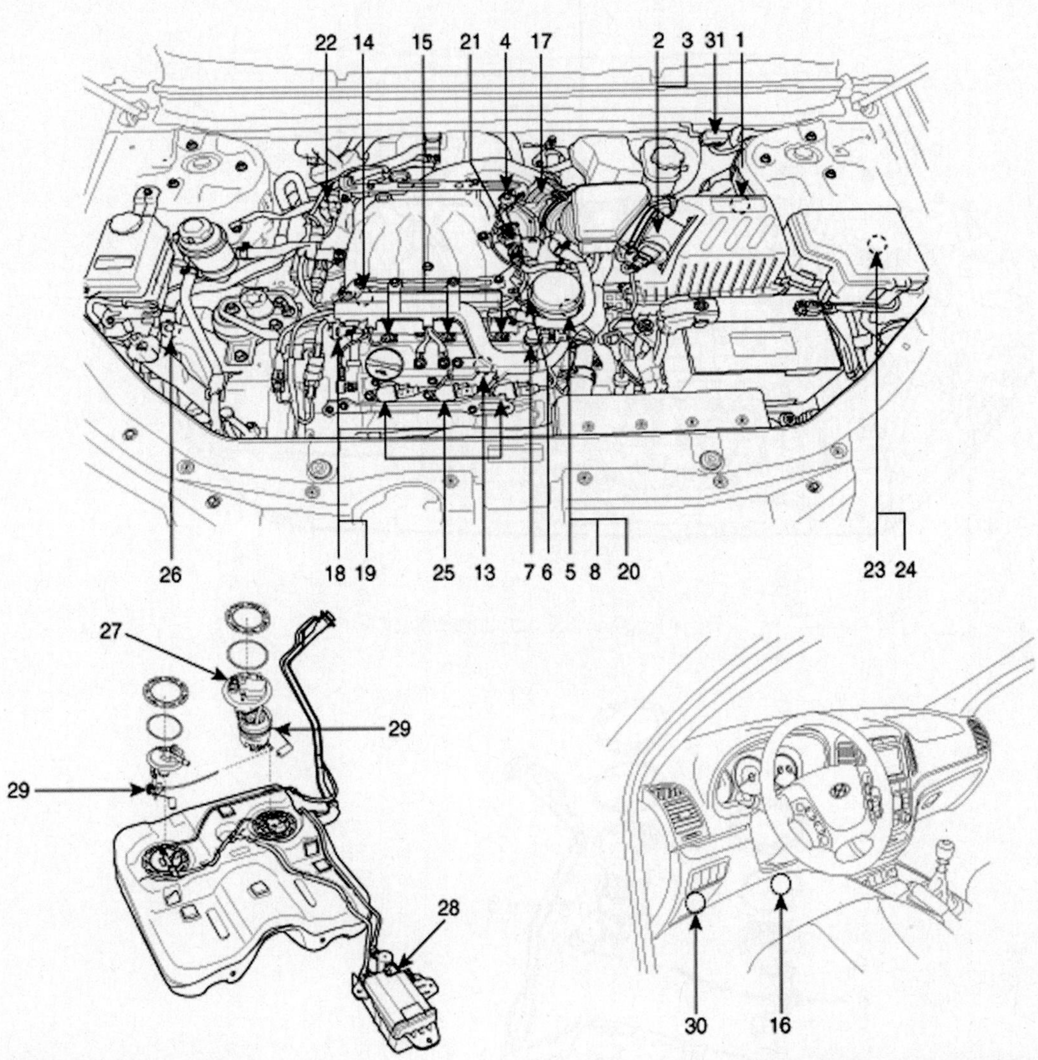

1. Powertrain Control Module (PCM)
2. Mass Air Flow (MAF) Sensor
3. Intake Air Temperature (IAT) Sensor
4. Manifold Absolute Pressure (MAP) Sensor
5. Engine Coolant Temperature (ECT) Sensor
6. Camshaft Position (CMP) Sensor [Bank 1]
7. Camshaft Position (CMP) Sensor [Bank 2]
8. Crankshaft Position (CKP) Sensor
9. Heated Oxygen Sensor (HO2S) [Bank 1 / Sensor 1]
10. Heated Oxygen Sensor (HO2S) [Bank 1 / Sensor 2]
11. Heated Oxygen Sensor (HO2S) [Bank 2 / Sensor 1]
12. Heated Oxygen Sensor (HO2S) [Bank 2 / Sensor 2]
13. Knock Sensor (KS) No. 1
14. Knock Sensor (KS) No. 2
15. Injector
16. Accelerator Position (APP) Sensor
17. ETC Module [Throttle Position Sensor (TPS) + ETC Motor]
18. CVVT Oil Control Valve (OCV) [Bank 1]
19. CVVT Oil Control Valve (OCV) [Bank 2]
20. CVVT Oil Temperature Sensor (OTS)
21. Purge Control Solenoid (PCV) Valve
22. Variable Intake Solenoid (VIS) Valve
23. Fuel Pump Relay
24. Main Relay
25. Ignition Coil
26. A/C Pressure Transducer (APT)
27. Fuel Tank Pressure (FTP) Sensor
28. Canister Close Valve (CCV)
29. Fuel Level Sensor
30. Wheel Speed Sensor (WSS)
31. Data Link Connector (DLC)
32. Multi-Purpose Check Connector

37655_SANT_G0544

Fig. 390 Underhood and related sensor locations—3.3L engine

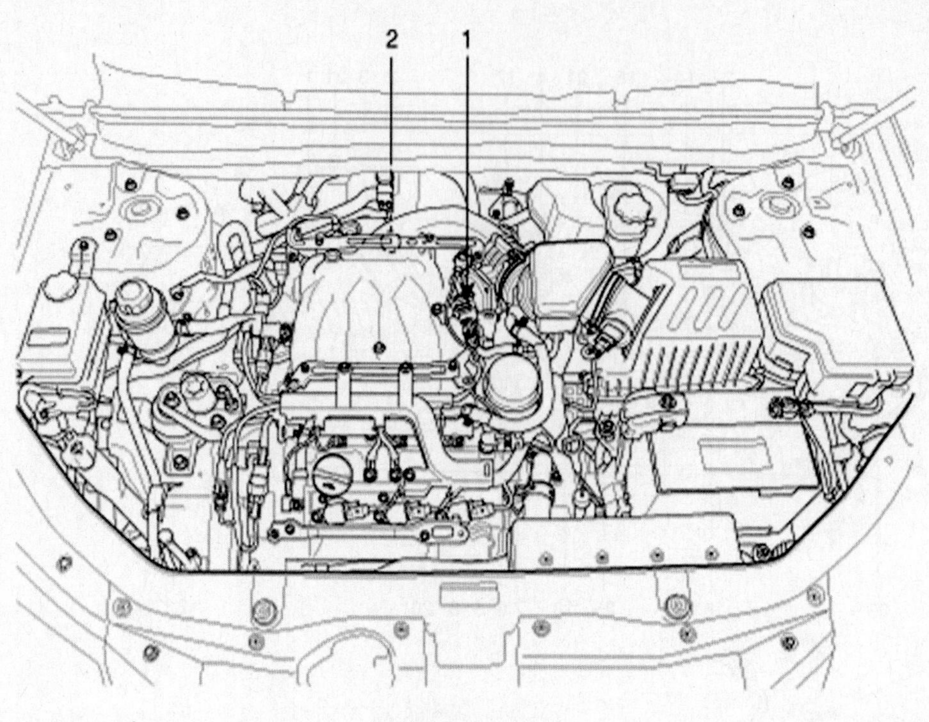

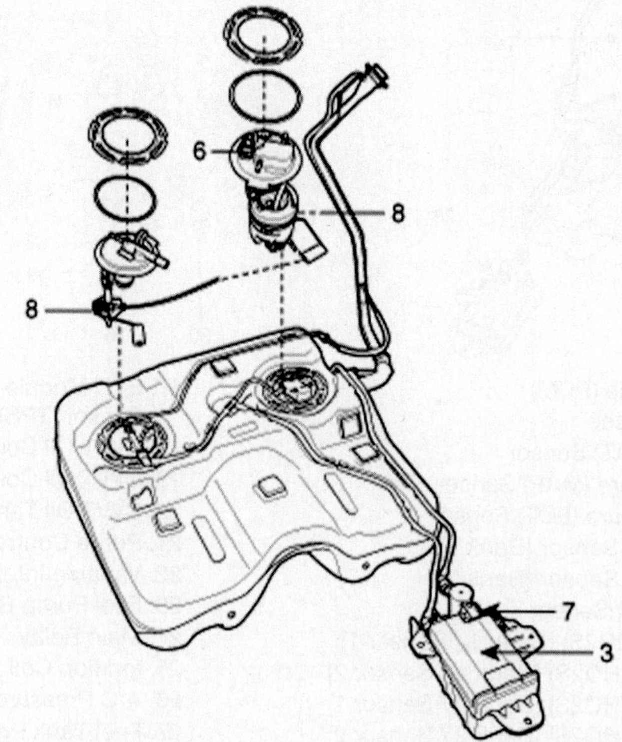

1. Purge Control Solenoid Valve (PCSV)
2. PCV Valve
3. Canister
4. Catalytic Converter (Bank1)
5. Catalytic Converter (Bank2)
6. Fuel Tank Pressure (FTP) Sensor
7. Canister Close Valve (CCV)
8. Fuel Level Sensor (FLS)

37655_SANT_G0559

Fig. 391 Emission Control System component locations—3.3L engine

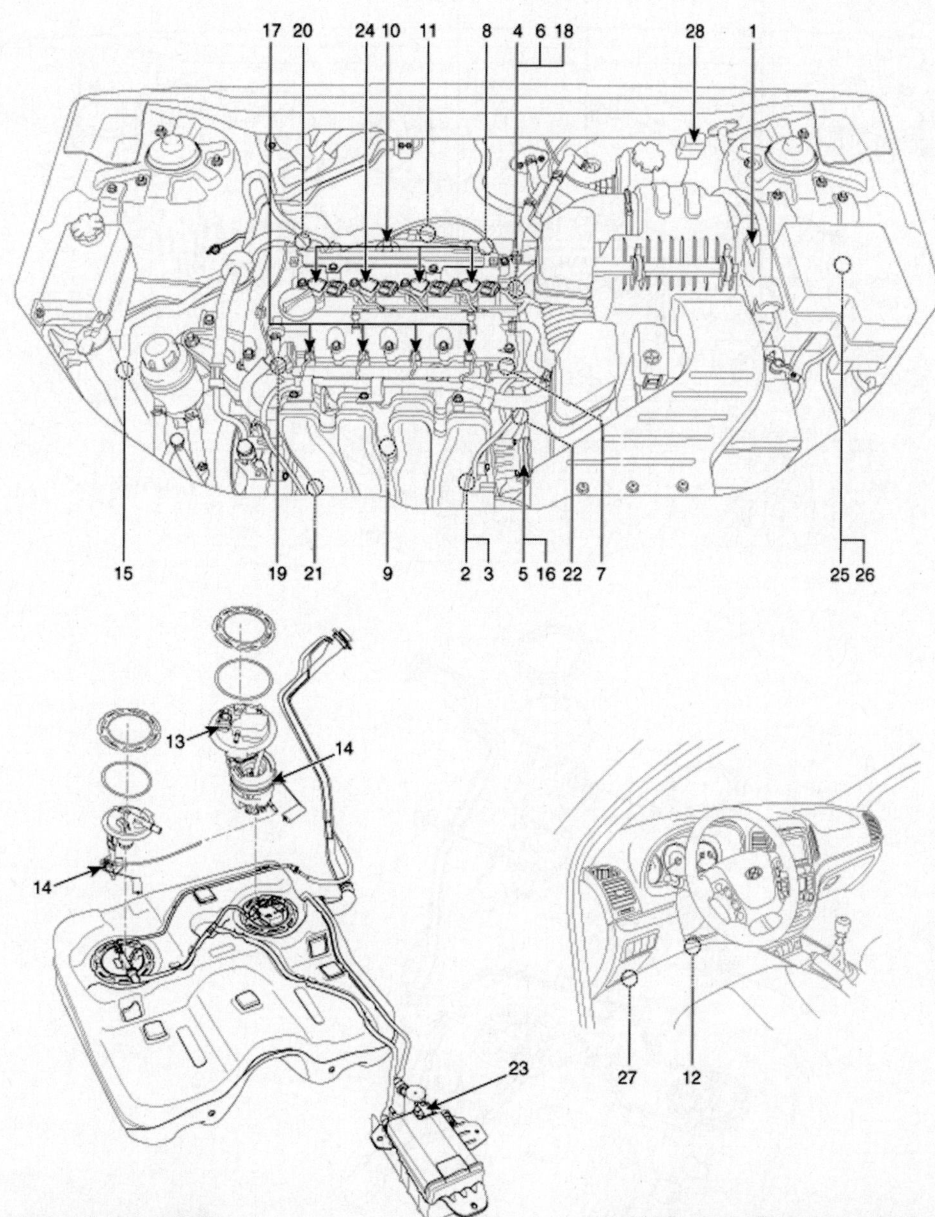

1. Engine Control Module (ECM)
2. Manifold Absolute Pressure (MAP) Sensor
3. Intake Air Temperature (IAT) Sensor
4. Engine Coolant Temperature (ECT) Sensor
5. Throttle Position Sensor (TPS) [integrated into ETC Module]
6. Crankshaft Position (CKP) Sensor
7. Camshaft Position (CMP) Sensor [Bank 1 / Intake]
8. Camshaft Position (CMP) Sensor [Bank 1 / Exhaust]
9. Knock Sensor (KS)
10. Heated Oxygen Sensor (HO2S) [Bank 1 / Sensor 1]
11. Heated Oxygen Sensor (HO2S) [Bank 1 / Sensor 2]
12. Accelerator Position (APP) Sensor
13. Fuel Tank Pressure (FTP) Sensor
14. Fuel Level Sensor (FLS)
15. A/C Pressure Transducer (APT)
16. ETC Motor [integrated into ETC Module]
17. Injector
18. Purge Control Solenoid Valve (PCV)
19. CVVT Oil Control Valve (OCV) [Bank 1 / Intake]
20. CVVT Oil Control Valve (OCV) [Bank 1 / Exhaust]
21. Variable Intake Solenoid (VIS) Valve
22. Variable Charge Motion Actuator (VCMA)
23. Canister Close Valve (CCV)
24. Ignition Coil
25. Main Relay
26. Fuel Pump Relay
27. Data Link Connector (DLC) [16 Pin]
28. Multi-Purpose Check Connector [20 Pin]

37655_SANT_G0545

Fig. 392 Underhood and related sensor locations—2.4L engine

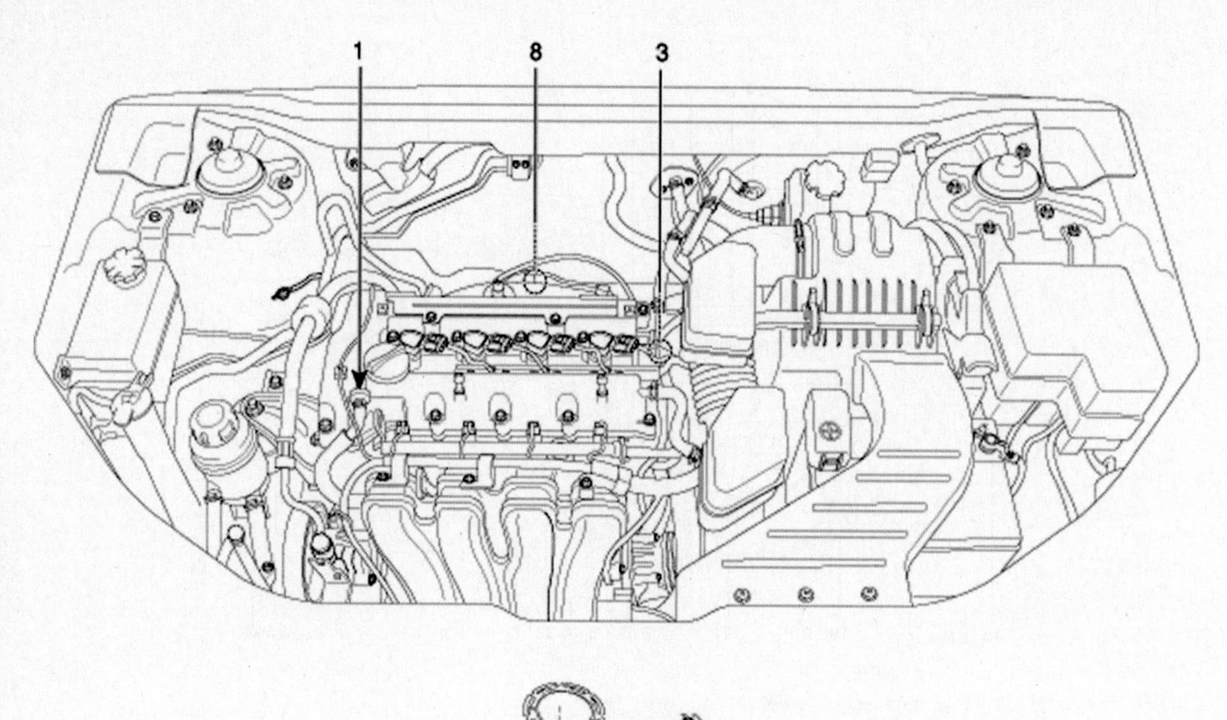

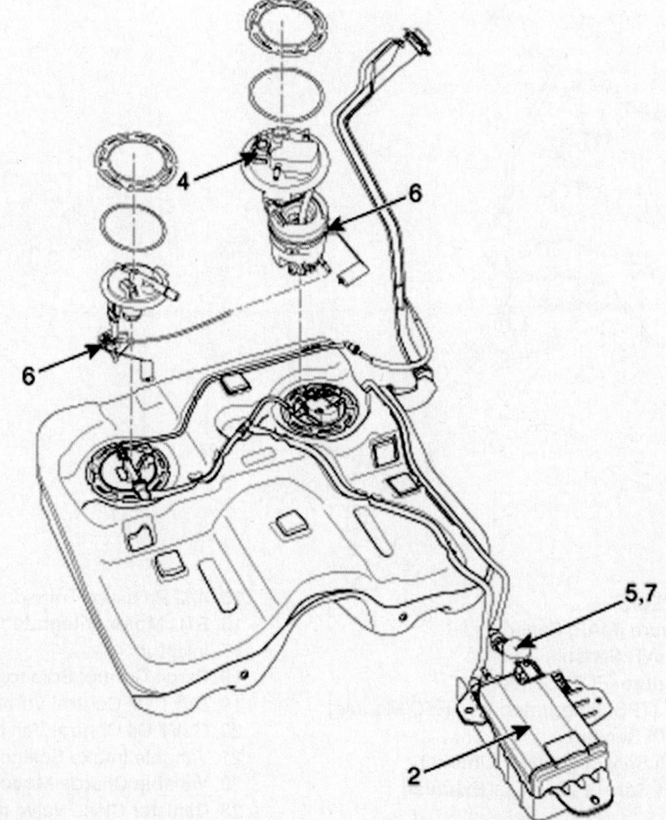

1. PCV Valve
2. Canister
3. Purge Control Solenoid Valve (PCSV)
4. Fuel Tank Pressure (FTP) Sensor
5. Canister Close Valve (CCV)
6. Fuel Level Sensor (FLS)
7. Fuel Tank Air Filter
8. Catalytic Converter

37655_SANT_G0560

Fig. 393 Emission Control System component locations—2.4L engine

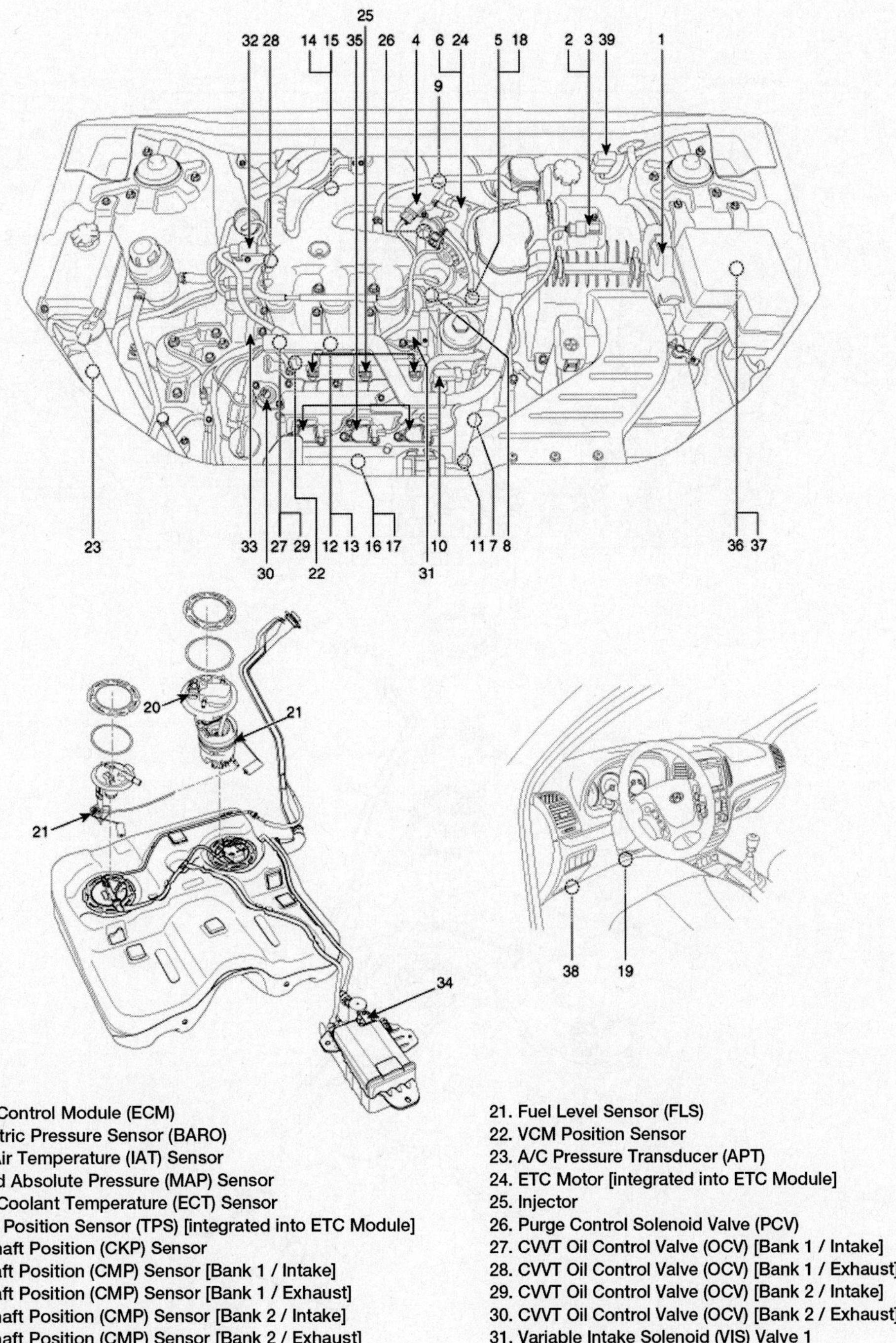

1. Engine Control Module (ECM)
2. Barometric Pressure Sensor (BARO)
3. Intake Air Temperature (IAT) Sensor
4. Manifold Absolute Pressure (MAP) Sensor
5. Engine Coolant Temperature (ECT) Sensor
6. Throttle Position Sensor (TPS) [integrated into ETC Module]
7. Crankshaft Position (CKP) Sensor
8. Camshaft Position (CMP) Sensor [Bank 1 / Intake]
9. Camshaft Position (CMP) Sensor [Bank 1 / Exhaust]
10. Camshaft Position (CMP) Sensor [Bank 2 / Intake]
11. Camshaft Position (CMP) Sensor [Bank 2 / Exhaust]
12. Knock Sensor (KS) [Bank 1]
13. Knock Sensor (KS) [Bank 2]
14. Heated Oxygen Sensor (HO2S) [Bank 1 / Sensor 1]
15. Heated Oxygen Sensor (HO2S) [Bank 1 / Sensor 2]
16. Heated Oxygen Sensor (HO2S) [Bank 2 / Sensor 1]
17. Heated Oxygen Sensor (HO2S) [Bank 2 / Sensor 2]
18. CVVT Oil Temperature Sensor (OTS)
19. Accelerator Position (APP) Sensor
20. Fuel Tank Pressure (FTP) Sensor
21. Fuel Level Sensor (FLS)
22. VCM Position Sensor
23. A/C Pressure Transducer (APT)
24. ETC Motor [integrated into ETC Module]
25. Injector
26. Purge Control Solenoid Valve (PCV)
27. CVVT Oil Control Valve (OCV) [Bank 1 / Intake]
28. CVVT Oil Control Valve (OCV) [Bank 1 / Exhaust]
29. CVVT Oil Control Valve (OCV) [Bank 2 / Intake]
30. CVVT Oil Control Valve (OCV) [Bank 2 / Exhaust]
31. Variable Intake Solenoid (VIS) Valve 1
32. Variable Intake Solenoid (VIS) Valve 2
33. Variable Charge Motion Actuator (VCMA)
34. Canister Close Valve (CCV)
35. Ignition Coil
36. Main Relay
37. Fuel Pump Relay
38. Data Link Connector (DLC) [16 Pin]
39. Multi-Purpose Check Connector [20 Pin]

37655_SANT_G0546

Fig. 394 Underhood and related sensor locations—3.5L engine

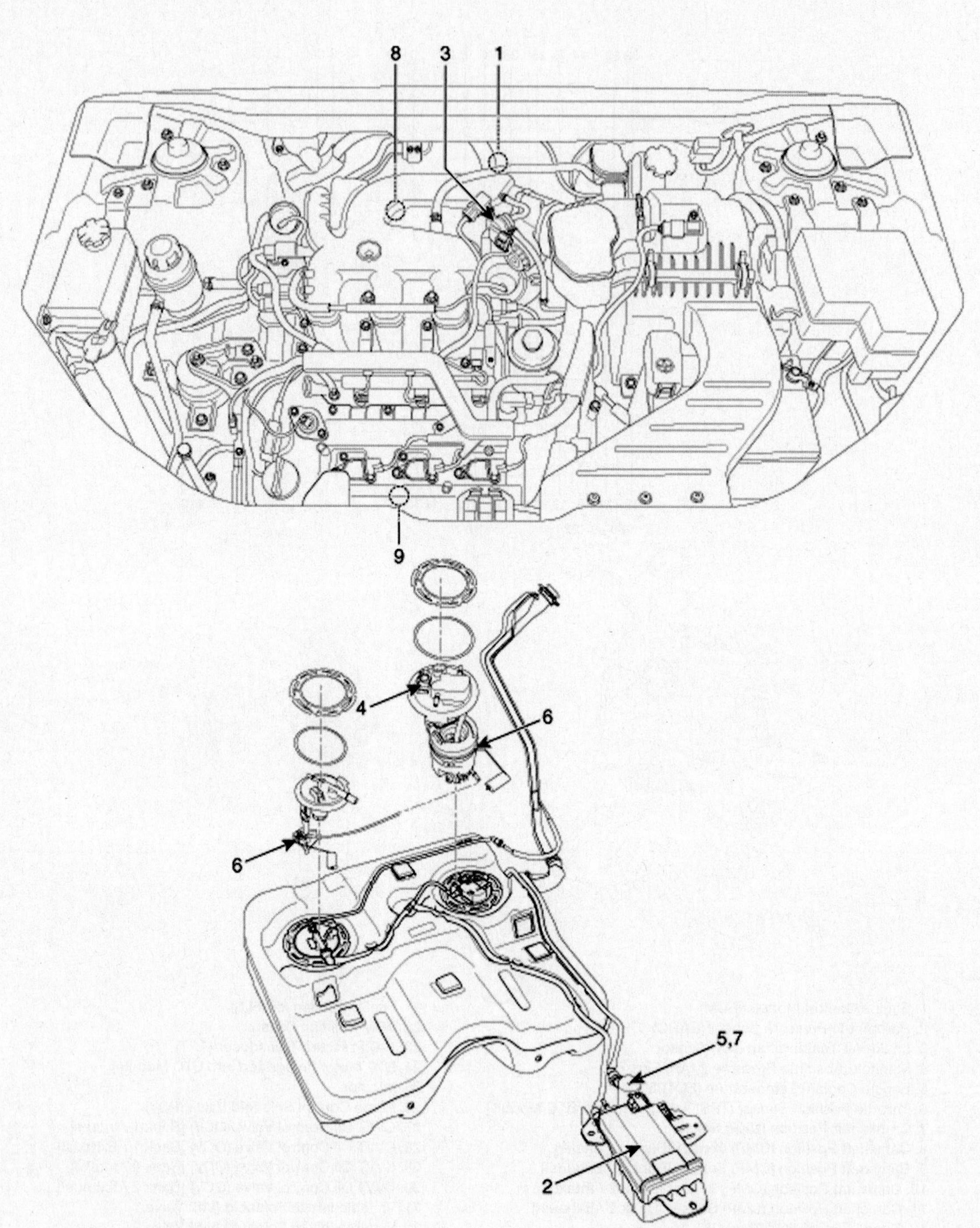

Fig. 395 Emission Control System component locations—3.5L engine

BAROMETRIC PRESSURE (BARO) SENSOR

LOCATION

See Figure 396.

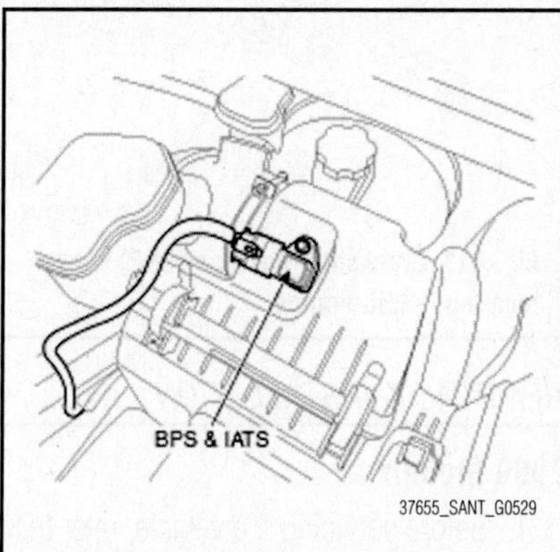

Fig. 396 Barometric Pressure sensor (A) and Intake Air Temperature (B) sensor–3.5L engine

REMOVAL & INSTALLATION

3.5L Engine

See Figure 397.

1. Before servicing the vehicle, refer to the Precautions Section.
2. Turn the ignition switch OFF and disconnect the battery negative cable.
3. Disconnect the barometric pressure sensor connector.
4. Remove the installation bolt, and then remove the sensor from the air cleaner assembly.

To install:

5. Installation is reverse of removal. Tighten the installation bolt to 35–52 inch lbs. (4–6 Nm).

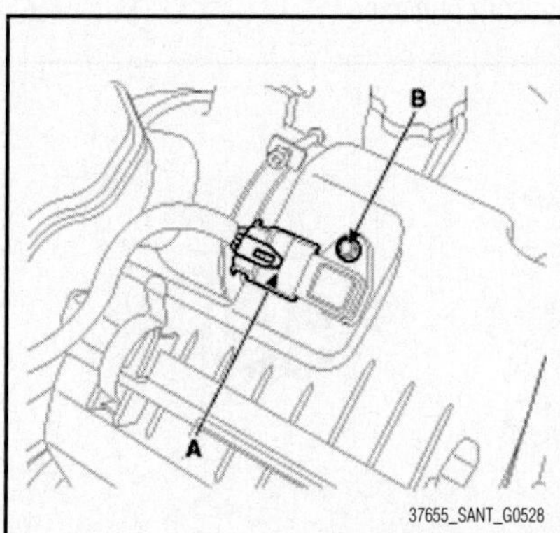

Fig. 397 Disconnect the barometric pressure sensor connector (A) and remove the installation bolt (B)

CAMSHAFT POSITION (CMP) SENSOR

LOCATION

See Figures 398 through 407.

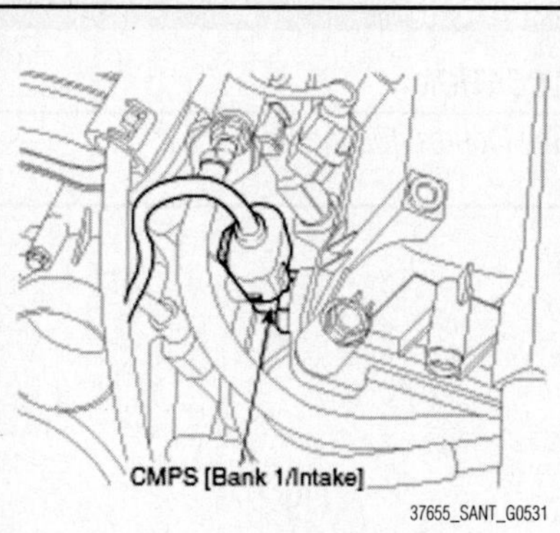

Fig. 398 Camshaft Position (CMP) sensor location (Bank 1, Intake)—2.4L engine

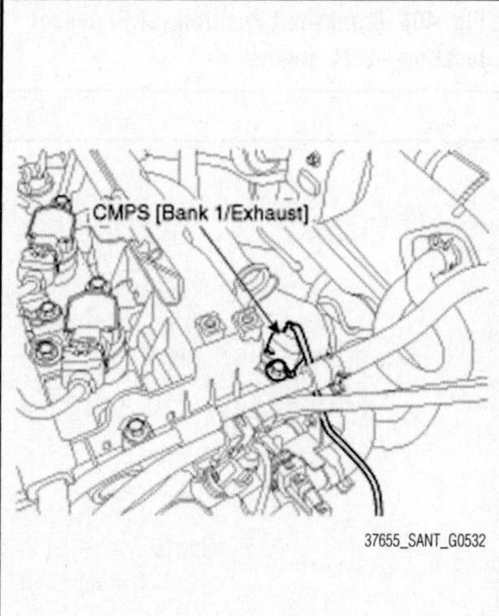

Fig. 399 Camshaft Position (CMP) sensor location (Bank 1, Exhaust)—2.4L engine

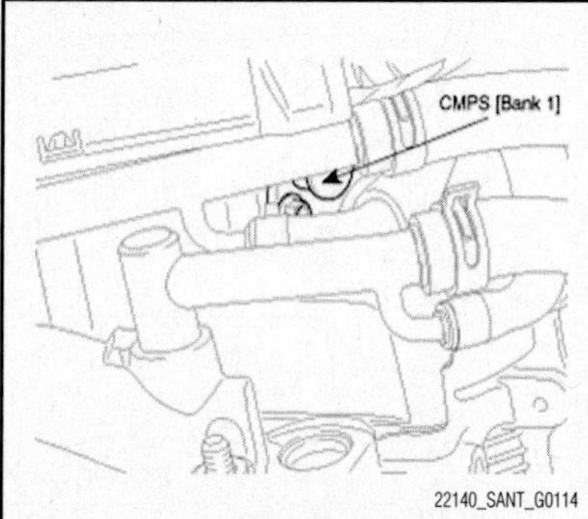

Fig. 400 Camshaft Position (CMP) sensor location (Bank 1)—2.7L engine

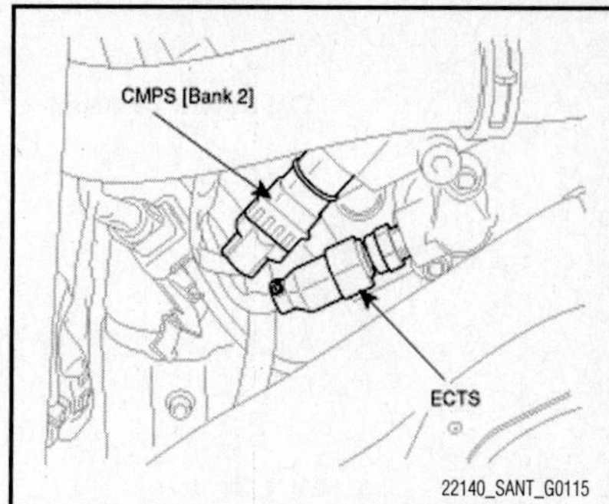

Fig. 401 Camshaft Position (CMP) sensor location (Bank 2) and Engine Coolant Temperature (ECT) sensor—2.7L engine

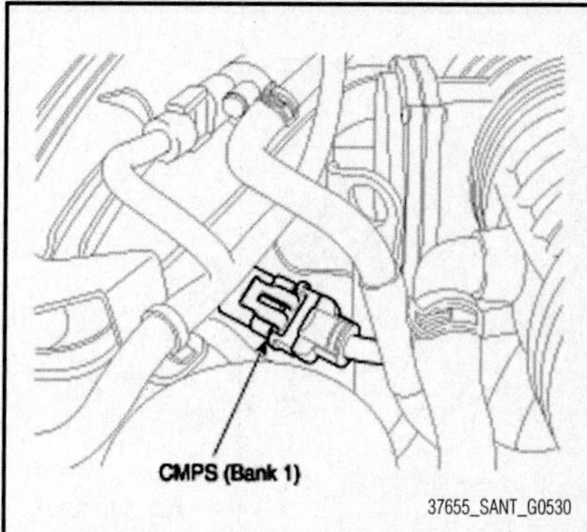

Fig. 402 Camshaft Position (CMP) sensor location (Bank 1)—3.3L engine

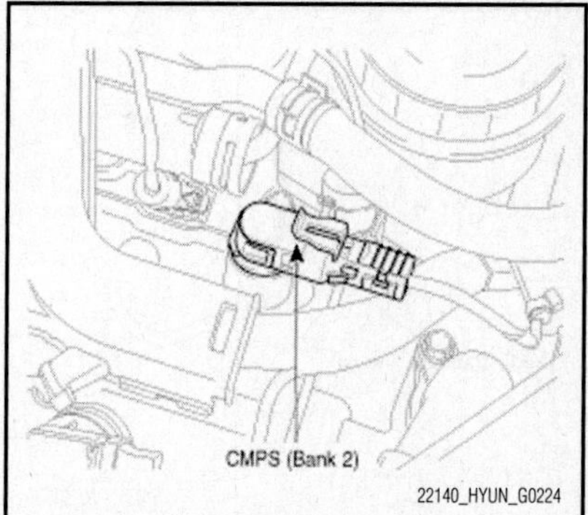

Fig. 403 Camshaft Position (CMP) sensor location (Bank 2)—3.3L engine

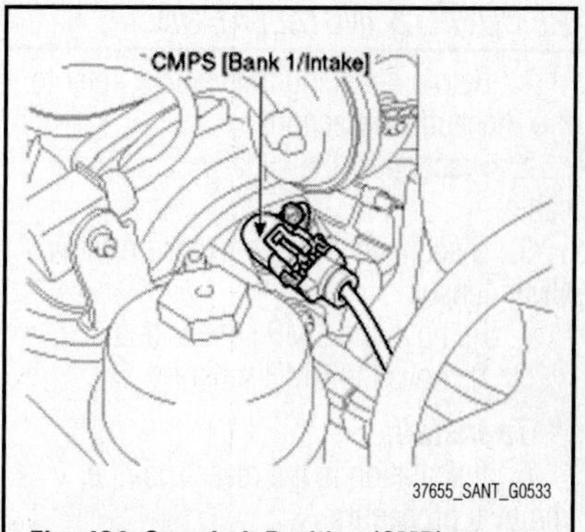

Fig. 404 Camshaft Position (CMP) sensor location (Bank 1, Intake)—3.5L engine

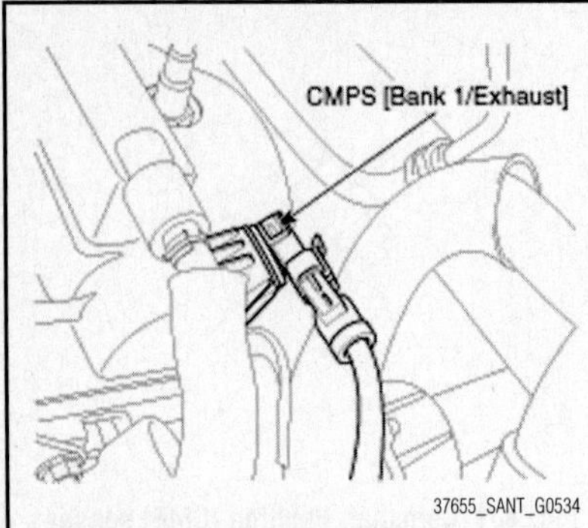

Fig. 405 Camshaft Position (CMP) sensor location (Bank 1, Exhaust)—3.5L engine

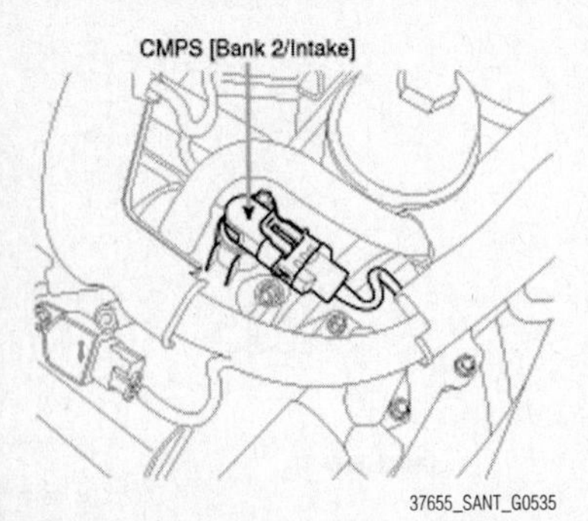

Fig. 406 Camshaft Position (CMP) sensor location (Bank 2, Intake)—3.5L engine

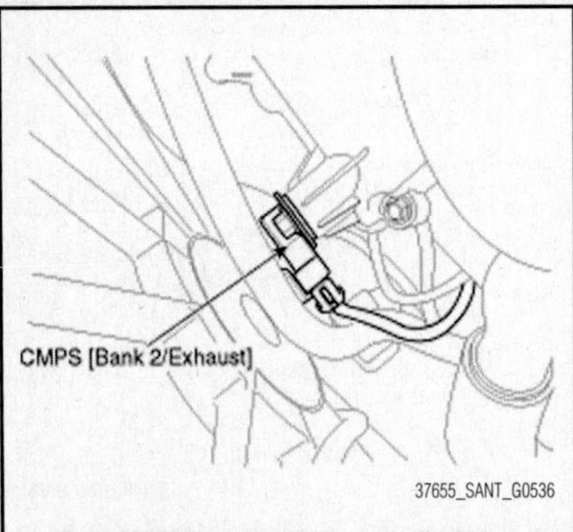

Fig. 407 Camshaft Position (CMP) sensor location (Bank 2, Exhaust)—3.5L engine

REMOVAL & INSTALLATION

1. Before servicing the vehicle, refer to the Precautions Section.
2. Disconnect the negative battery cable.
3. Disconnect the connector from the CMP sensor.
4. Remove the CMP sensor bolt.
5. Remove the CMP sensor.

To install:

6. Installation is the reverse of the removal procedure.
7. Tighten the bolt that retains the CMP sensor to:
   - 2.7L, 3.3L, and 3.5L engines: 61–86 inch lbs. (7–10 Nm)
   - 2.4L engines: 86–104 inch lbs. (9–12 Nm)

CRANKSHAFT POSITION (CKP) SENSOR

LOCATION

See Figures 408 through 411.

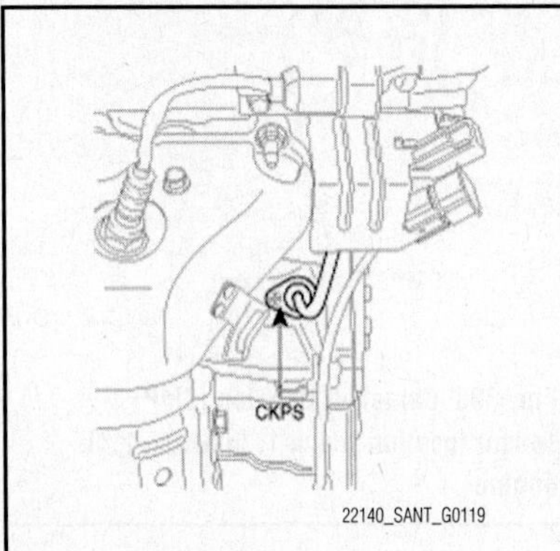

Fig. 408 Crankshaft Position (CKP) sensor location—2.7L engine

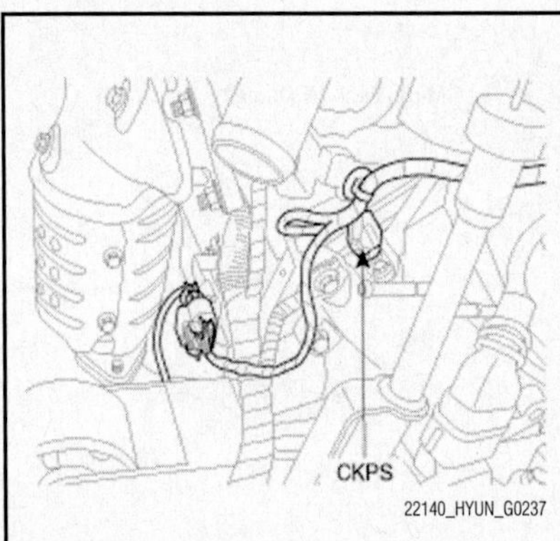

Fig. 409 Crankshaft Position (CKP) sensor location—3.3L engine

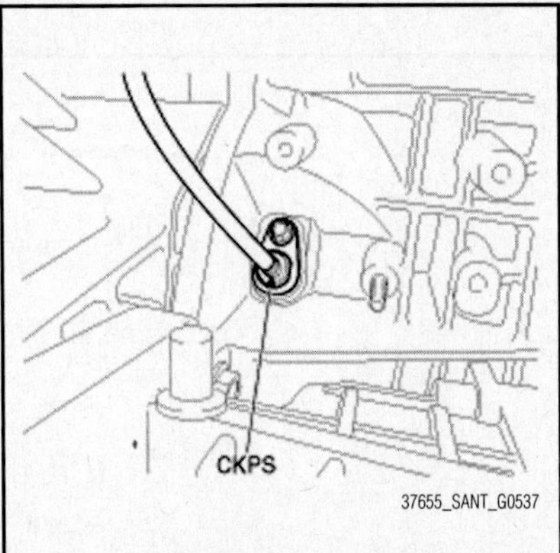

Fig. 410 Crankshaft Position (CKP) sensor location—2.4L engine

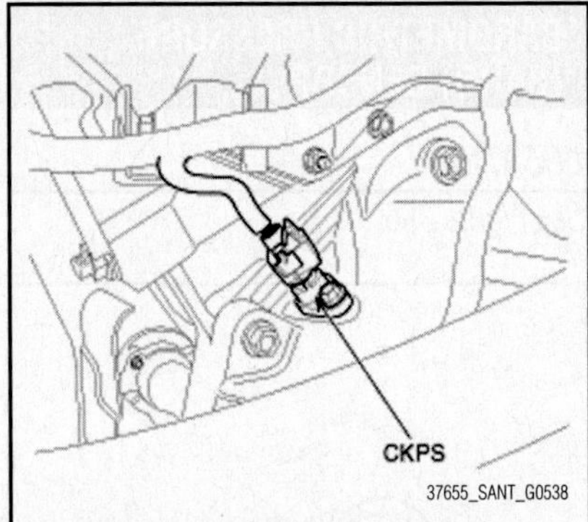

Fig. 411 Crankshaft Position (CKP) sensor location—3.5L engine

REMOVAL & INSTALLATION

2009 Models

1. Before servicing the vehicle, refer to the Precautions Section.
2. Disconnect the negative battery cable.
3. Disconnect the connector from the sensor.
4. Remove the bolt that retains the sensor in place.
5. Remove the sensor from its mounting.

To install:

6. Installation is the reverse of the removal procedure.
7. Tighten the sensor retaining bolt to:
   - 2.4L engine: 61–86 inch lbs. (7–10 Nm)
   - 3.3L engine: 70–104 inch lbs. (8–12 Nm).

2010 Models

2.4L Engine

See Figures 412 through 414.

1. Before servicing the vehicle, refer to the Precautions Section.
2. Disconnect the negative battery cable.
3. Disconnect the crankshaft position sensor connector.

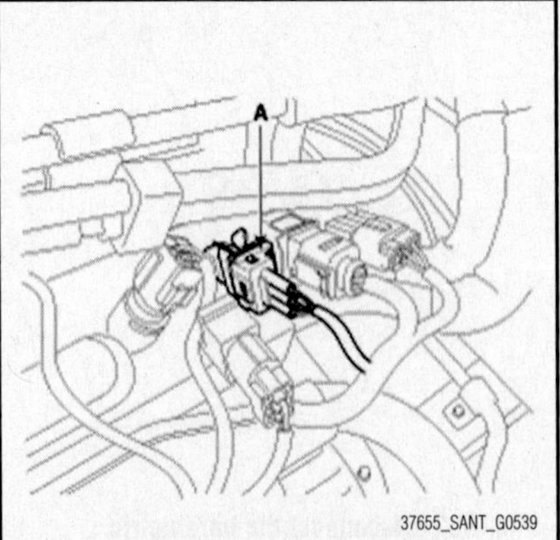

Fig. 412 Disconnect the crankshaft position sensor connector (A)

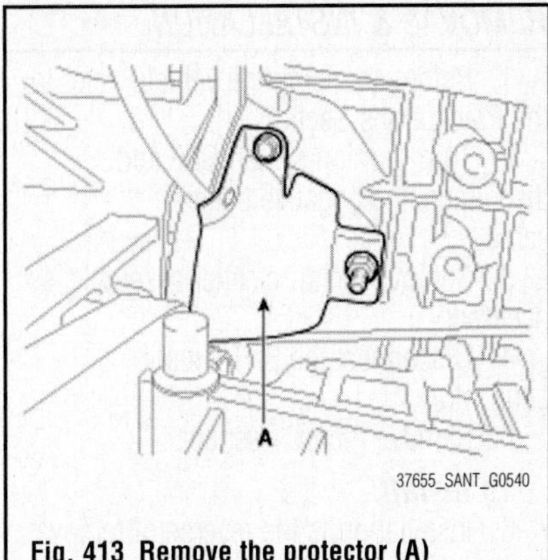

Fig. 413 Remove the protector (A)

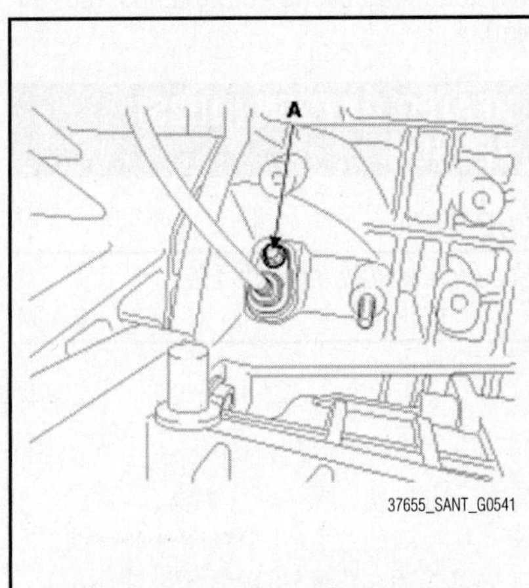

Fig. 414 Remove the installation bolt (A)

4. Remove the protector.
5. Remove the installation bolt.
6. Remove the sensor.

To install:

7. Installation is the reverse of the removal procedure, noting the following:
    a. Tighten the sensor installation bolt to 86–104 inch lbs. (10–12 Nm)
    b. Tighten the sensor protector installation bolts:
        - M8 bolts: 14–17 ft. lbs. (19–24 Nm)
        - M6 bolts: 86–104 inch lbs. (10–12 Nm)

3.5L Engine

See Figure 415.

1. Before servicing the vehicle, refer to the Precautions Section.
2. Disconnect the negative battery cable.
3. Remove the air cleaner assembly.
4. Disconnect the crankshaft position sensor connector.
5. Remove the installation bolt (B), and

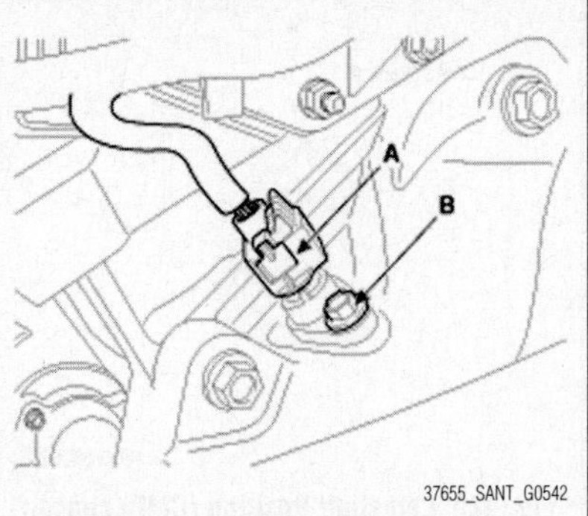

Fig. 415 Disconnect the sensor connector (A) and remove the bolt (B)

then vertically remove the sensor from the transaxle housing.

To install:

6. Installation is the reverse of the removal procedure. Tighten the sensor installation bolt to 61–86 inch lbs. (7–10 Nm)

ELECTRONIC CONTROL MODULE (ECM)

LOCATION

See Figures 416 through 419.

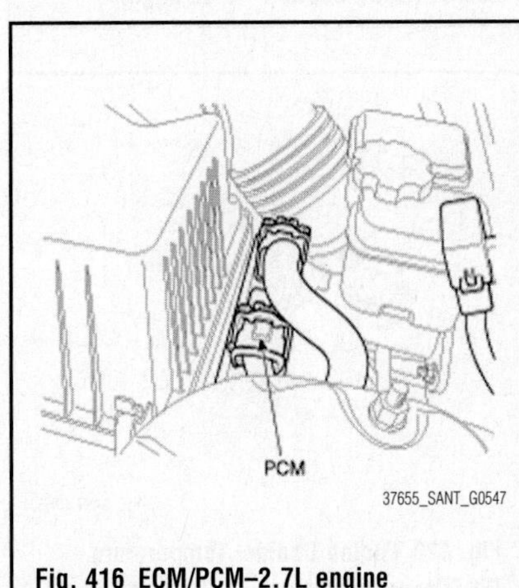

Fig. 416 ECM/PCM–2.7L engine

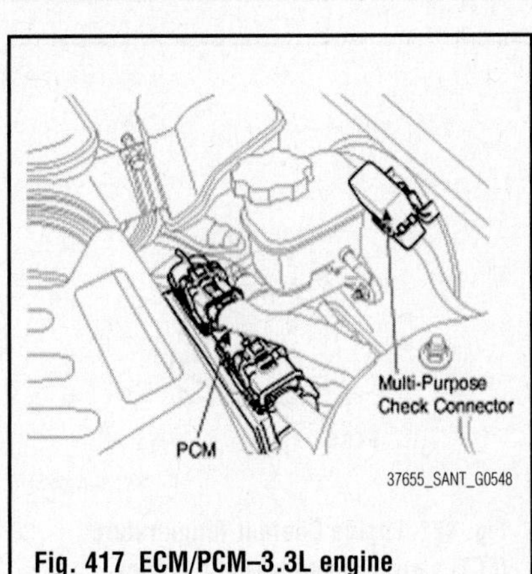

Fig. 417 ECM/PCM–3.3L engine

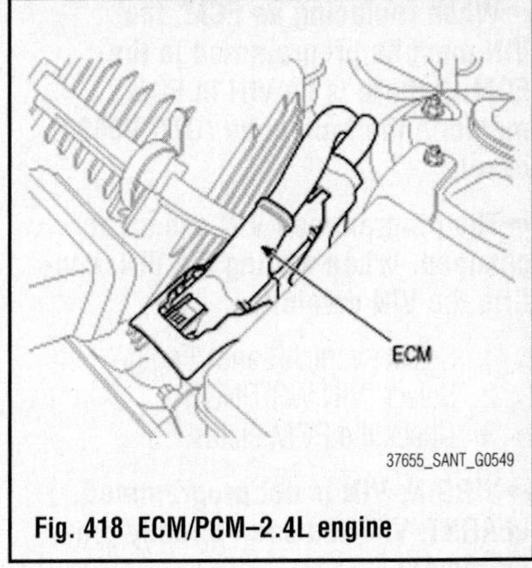

Fig. 418 ECM/PCM–2.4L engine

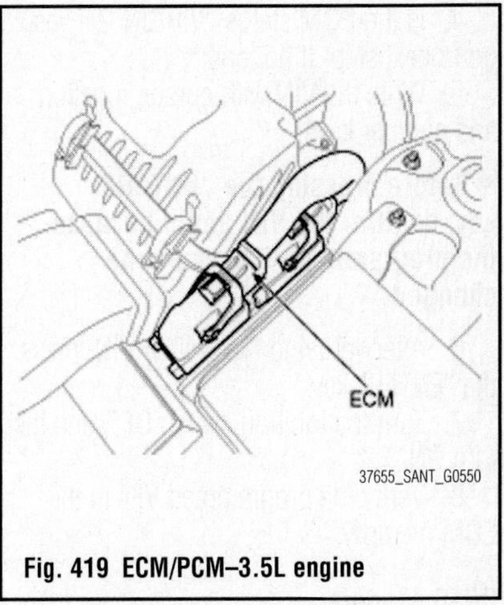

Fig. 419 ECM/PCM–3.5L engine

REMOVAL & INSTALLATION

1. Before servicing the vehicle, refer to the Precautions Section.
2. Turn ignition switch off.
3. Disconnect the negative battery cable from the battery.
4. Remove the Air Cleaner assembly, as necessary.
5. Disconnect the ECM connectors.
6. Remove the ECM mounting bolts and remove the ECM from the bracket.

To install:

7. Installation is the reverse of the removal.
8. Tighten the ECM mounting bolts to 86–104 inch lbs. (10–12 Nm).

**** WARNING**

When replacing the ECM, be careful to use the right part number, as damage to the injection system could occur.

RESET

2009 Models

➡Reprogramming of the VIN will require a scan tool.

➡When replacing an ECM, the VIN must be programmed in the ECM. If there is no VIN in ECM memory, the fault code (DTC P0630) is set.

➡The programmed VIN cannot be changed. When writing the VIN, confirm the VIN carefully.

1. Select "Vehicle" and "Engine".
2. Select "VIN WRITING".
3. Check the PCM status.

➡VIRGIN: VIN is not programmed. LEARNT: VIN has been already programmed

4. Is the PCM status "VIRGIN"? If yes, go to next step. If no, end.
5. Write the VIN with cursor, function and number keys.

➡Before pressing the "ENTER" key, confirm the VIN again because the programmed VIN cannot be changed.

6. After verifying the written VIN, press the "ENTER" key.
7. Turn the ignition switch OFF, and then turn ON.
8. Verify the programmed VIN in the ECM memory.

2010 Models

➡Reprogramming of the VIN will require a scan tool.

➡When replacing an ECM, the VIN must be programmed in the ECM. If there is no VIN in ECM memory, the fault code (DTC P0630) is set.

➡The programmed VIN cannot be changed. When writing the VIN, confirm the VIN carefully.

1. Select "VIN Writing" function in "Vehicle S/W Management".
2. Select "Write VIN" in "ID Register".

➡Before inputting the VIN, confirm the VIN again because the programmed VIN cannot be changed.

3. Input the VIN.
4. Turn the ignition switch OFF, then back ON.

ENGINE COOLANT TEMPERATURE (ECT) SENSOR

LOCATION

See Figures 420 through 423.

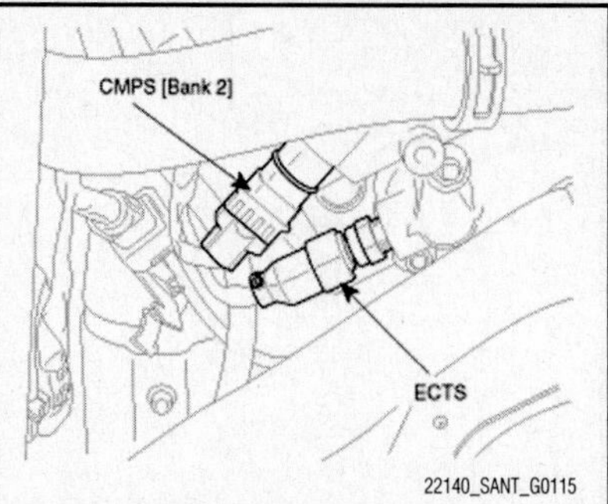

Fig. 420 Camshaft Position (CMP) sensor location (Bank 2) and Engine Coolant Temperature (ECT) sensor—2.7L engine

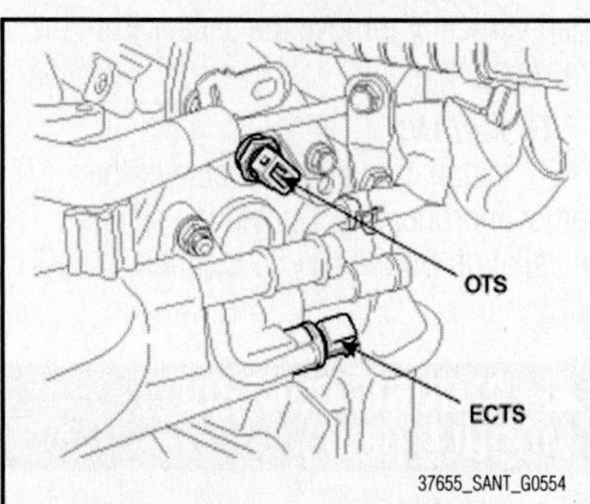

Fig. 421 Engine Coolant Temperature (ECT) sensor and CVVT Oil Temperature sensor (OTS) location—3.3L engine

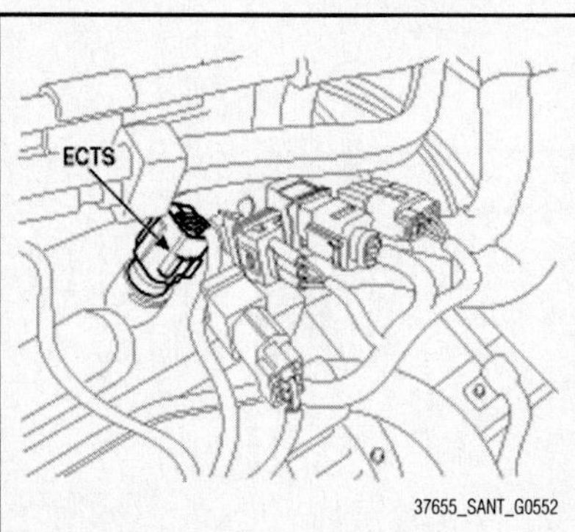

Fig. 422 Engine Coolant Temperature (ECT) sensor location—2.4L engine

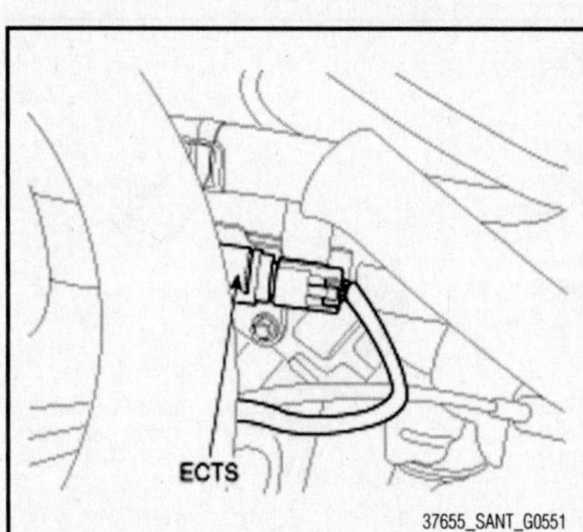

Fig. 423 Engine Coolant Temperature (ECT) sensor location—3.5L engine

REMOVAL & INSTALLATION

1. Before servicing the vehicle, refer to the Precautions Section.
2. Turn ignition switch OFF and disconnect the negative battery cable.
3. Remove the air cleaner assembly, as necessary.
4. Disconnect the ECT sensor connector.
5. Remove the ECT sensor.

To install:

6. Installation is the reverse of removal. Apply engine coolant to the O-ring. Tighten the mounting bolts to 15–29 ft. lbs. (20–39 Nm).

HEATED OXYGEN (HO2S) SENSOR

LOCATION

See Figures 424 through 436.

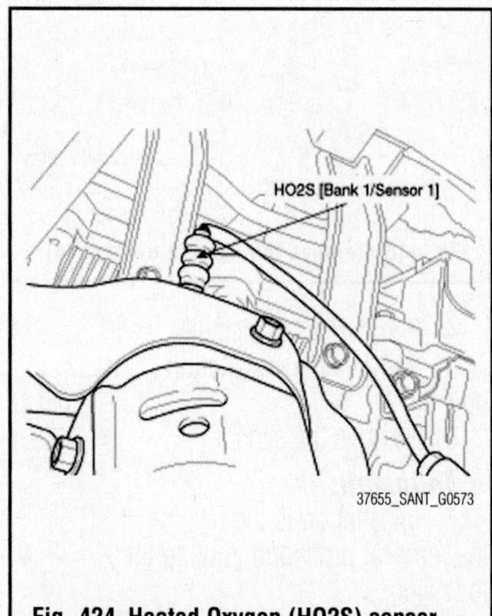

Fig. 424 Heated Oxygen (HO2S) sensor—Bank 1, Sensor 1—2.7L engine

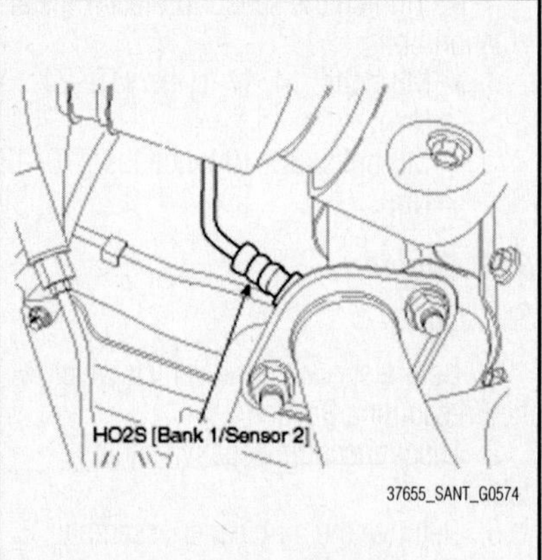

Fig. 425 Heated Oxygen (HO2S) sensor—Bank 1, Sensor 2—2.7L engine

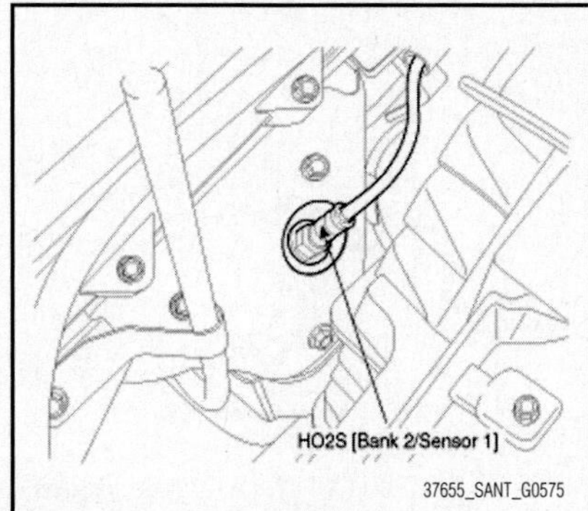

Fig. 426 Heated Oxygen (HO2S) sensor—Bank 2, Sensor 1—2.7L engine

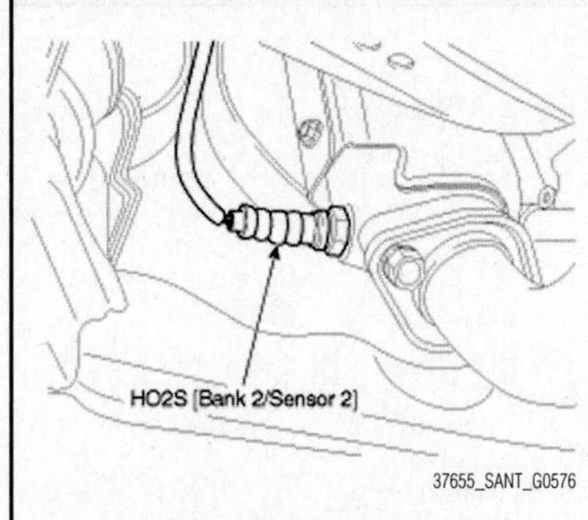

Fig. 427 Heated Oxygen (HO2S) sensor—Bank 2, Sensor 2—2.7L engine

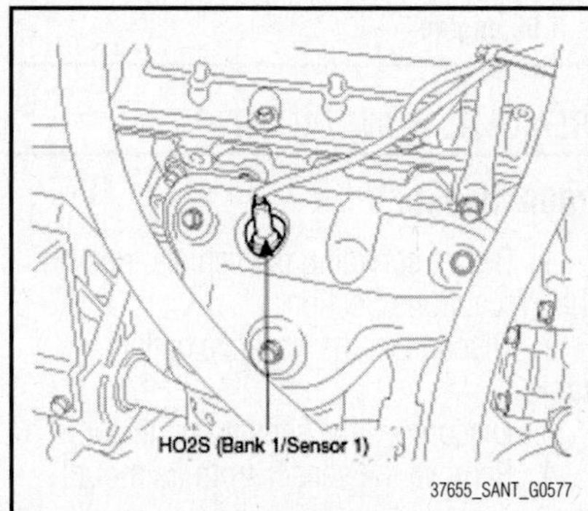

Fig. 428 Heated Oxygen (HO2S) sensor—Bank 1, Sensor 1—3.3L engine

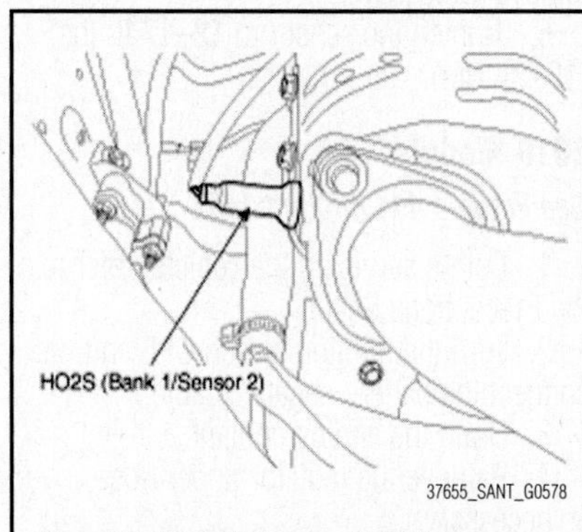

Fig. 429 Heated Oxygen (HO2S) sensor—Bank 1, Sensor 2—3.3L engine

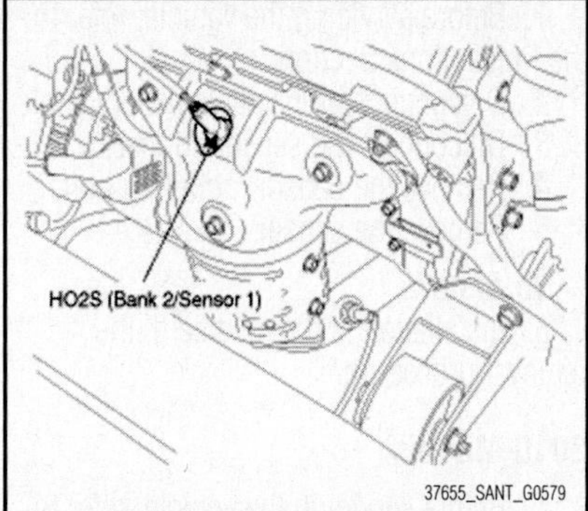

Fig. 430 Heated Oxygen (HO2S) sensor—Bank 2, Sensor 1—3.3L engine

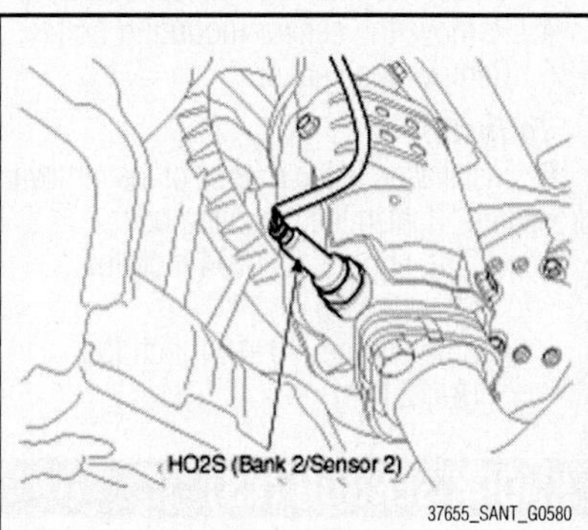

Fig. 431 Heated Oxygen (HO2S) sensor—Bank 2, Sensor 2—3.3L engine

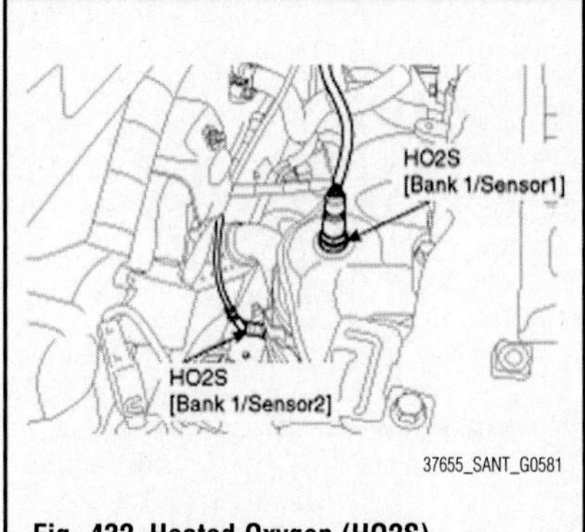

Fig. 432 Heated Oxygen (HO2S) sensors—2.4L engine

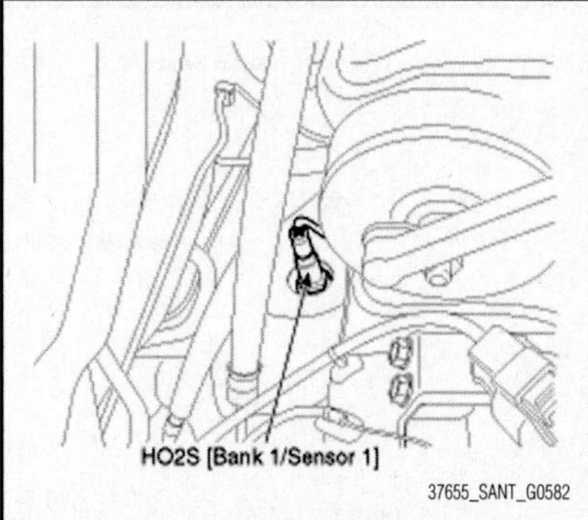

Fig. 433 Heated Oxygen (HO2S) sensor—Bank 1, Sensor 1—3.5L engine

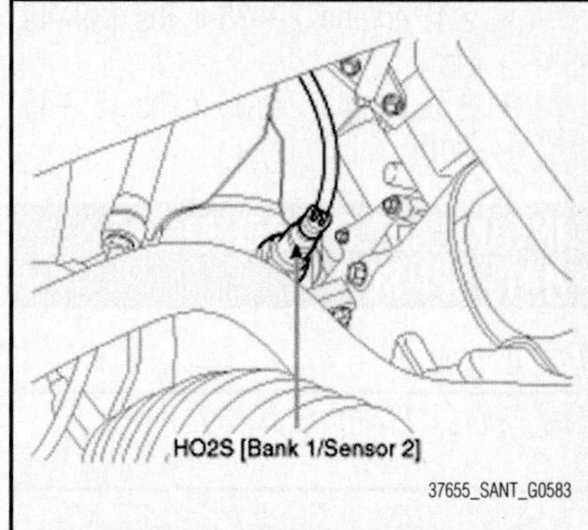

Fig. 434 Heated Oxygen (HO2S) sensor—Bank 1, Sensor 2—3.5L engine

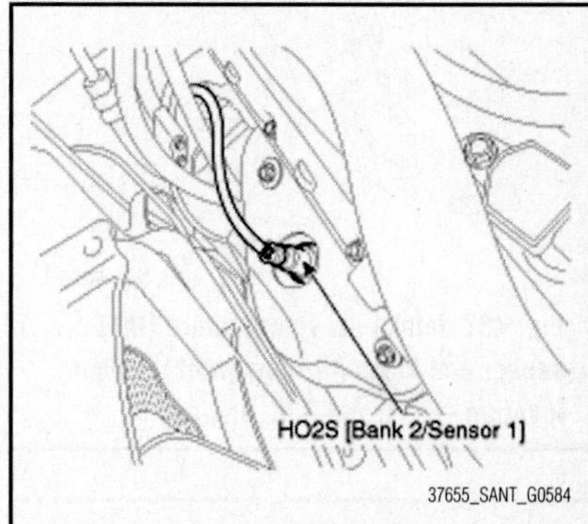

Fig. 435 Heated Oxygen (HO2S) sensor—Bank 2, Sensor 1—3.5L engine

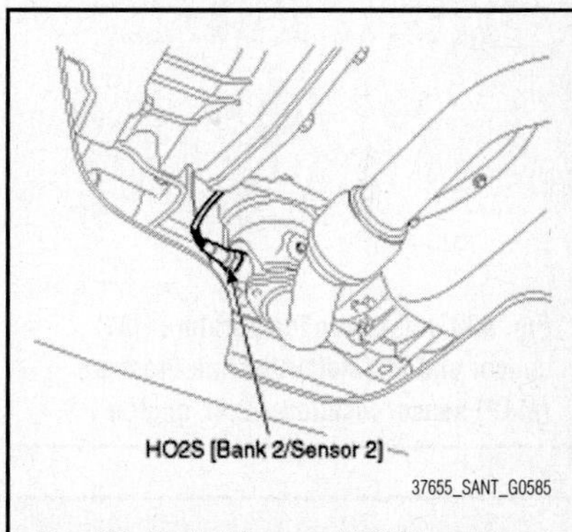

Fig. 436 Heated Oxygen (HO2S) sensor—Bank 2, Sensor 2—3.5L engine

REMOVAL & INSTALLATION

1. Before servicing the vehicle, refer to the Precautions Section.
2. Disconnect the negative battery cable.
3. Disconnect the oxygen sensor connector.
4. Remove the oxygen sensor.

To install:

5. Installation is the reverse of removal. Tighten the HO2S:
   - 2.7L engine: 25–33 ft. lbs. (34–44 Nm)
   - 3.3L engine: 36–43 ft. lbs. (49–59 Nm)

- 2.4L engine: 29–36 ft. lbs. (39–49 Nm)
- 3.5L engine: 26–33 ft. lbs. (35–45 Nm)

INTAKE AIR TEMPERATURE (IAT) SENSOR

LOCATION

See Figures 437 through 439.

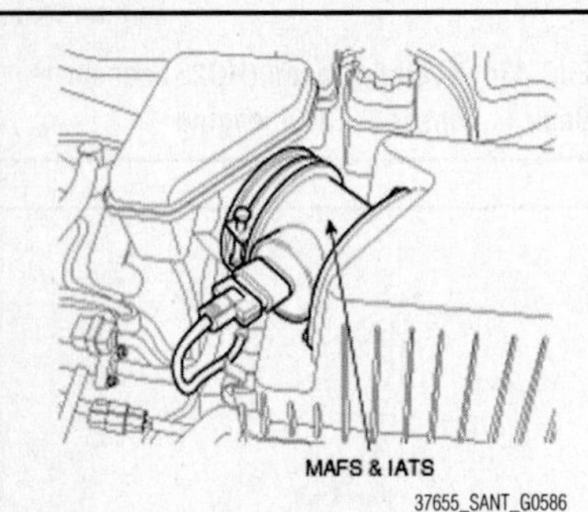

Fig. 437 Intake Air Temperature (IAT) sensor and Mass Air Flow (MAF) sensor location—2.7L and 3.3L engines

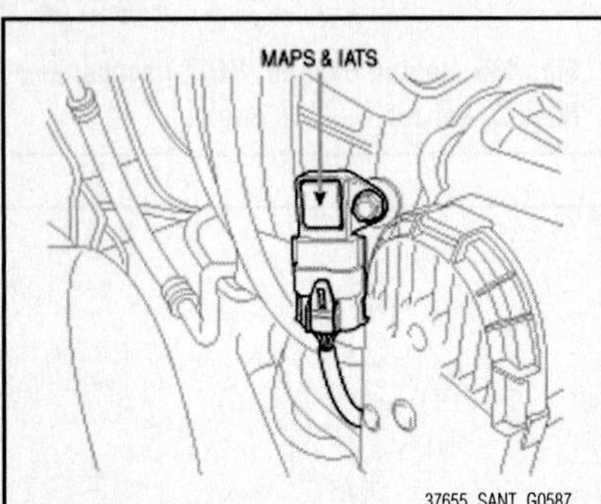

Fig. 438 Intake Air Temperature (IAT) sensor and Manifold Absolute Pressure (MAP) sensor location—2.4L engine

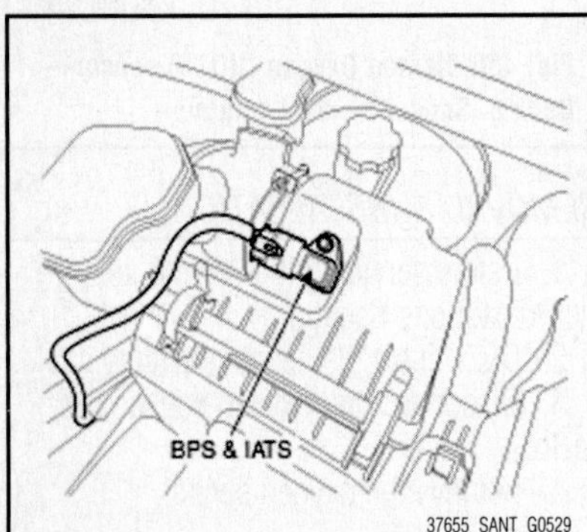

Fig. 439 Intake Air Temperature (B) sensor and Barometric Pressure sensor (A)

REMOVAL & INSTALLATION

2009 Models

1. Before servicing the vehicle, refer to the Precautions Section.
2. Disconnect the negative battery cable.
3. Disconnect the sensor connector.
4. Remove the sensor mounting bolt.
5. Remove the sensor.

To install:

6. Installation is the reverse of the removal procedure.

2010 Models

1. Before servicing the vehicle, refer to the Precautions Section.
2. Disconnect the negative battery cable.
3. Disconnect the MAP sensor connector.
4. Remove the sensor mounting bolt.
5. Remove the sensor.

To install:

6. Installation is the reverse of the removal procedure. Tighten the mounting bolt:
   - 2.4L engine: 86–104 inch lbs. (10–12 Nm)
   - 3.5L engine: 70–104 inch lbs. (8–12 Nm)

KNOCK SENSOR (KS)

LOCATION

See Figures 440 through 443.

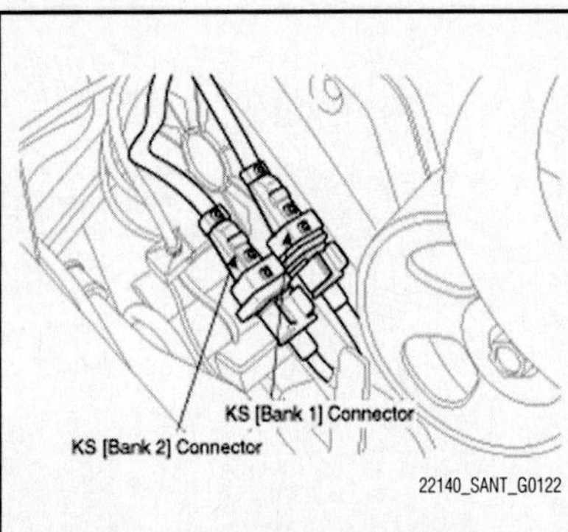

Fig. 440 Knock Sensor (KS) location—2.7L engine

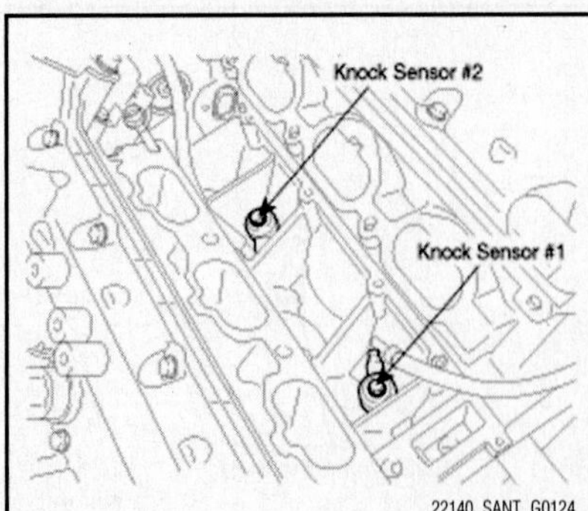

Fig. 441 Knock Sensor (KS) No. 1 and No. 2 locations—3.3L engine

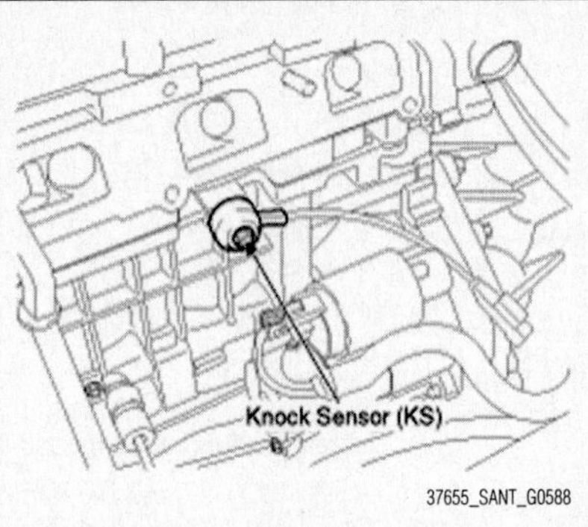

Fig. 442 Knock Sensor (KS) location—2.4L engine

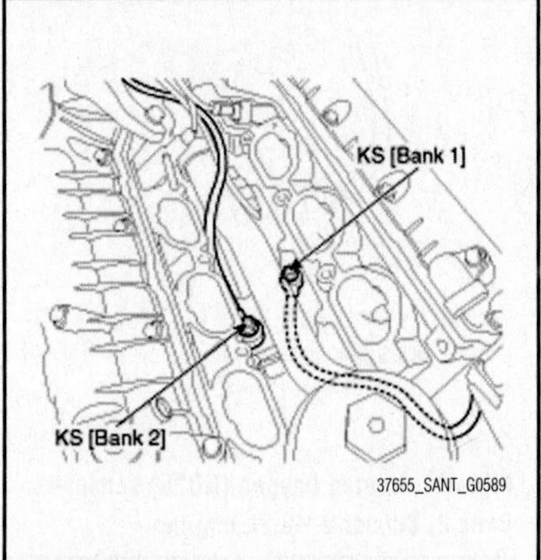

Fig. 443 Knock Sensor (KS) locations—3.5L engine

REMOVAL & INSTALLATION

2009 Models

1. Before servicing the vehicle, refer to the Precautions Section.
2. Disconnect the negative battery cable.
3. Disconnect the sensor connector.
4. Remove the sensor from its mounting.

To install:

5. Installation is the reverse of the removal procedure.
6. Tighten the sensor to 12–17 ft. lbs. (16–24 Nm).

2010 Models

See Figures 444 through 446.

1. Before servicing the vehicle, refer to the Precautions Section.
2. Turn the ignition switch OFF and disconnect the battery negative cable.
3. Drain the engine coolant.
4. Remove the radiator upper hose, as necessary.
5. Disconnect the knock sensor connector.

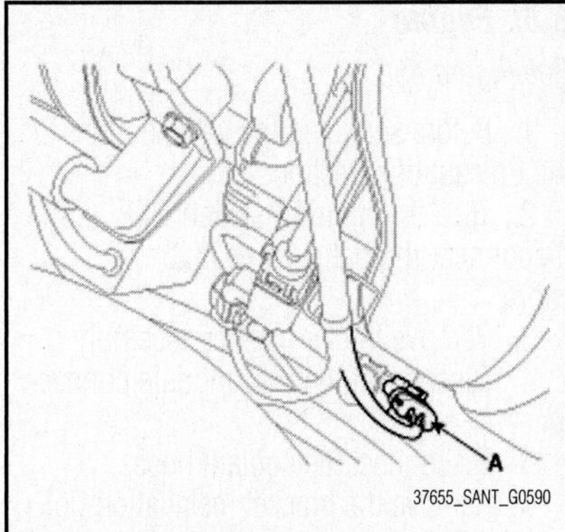

Fig. 444 Disconnect the knock sensor connector (A)—2.4L engine

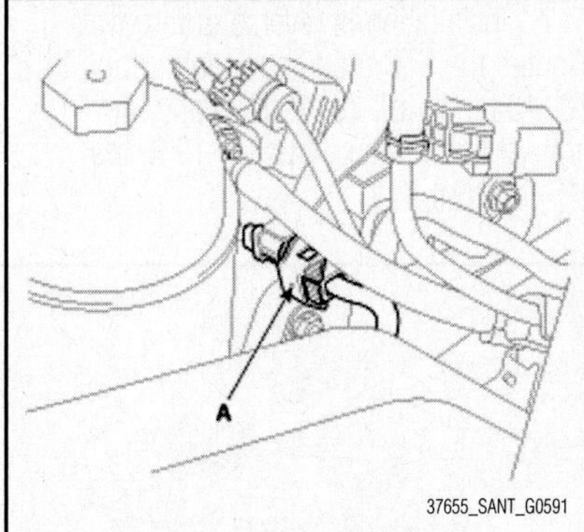

Fig. 445 Disconnect the Bank 1 knock sensor connector (A)—3.5L engine

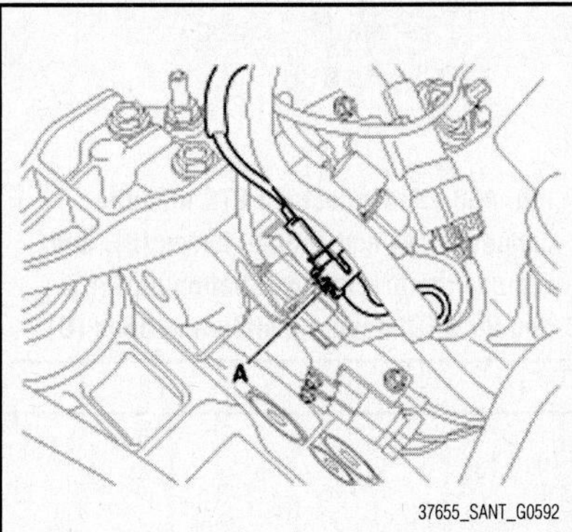

Fig. 446 Disconnect the Bank 2 knock sensor connector (A)—3.5L engine

6. Remove the intake manifold. Refer to Intake Manifold Removal & Installation in the Engine Mechanical section.
7. Remove the mounting bolt and then remove the sensor from the cylinder block.

To install:

8. Installation is the reverse of removal. Tighten the mounting bolt to 12–17 ft. lbs. (16–24 Nm).

MASS AIR FLOW (MAF) SENSOR

LOCATION

2009 Models

See Figure 447.

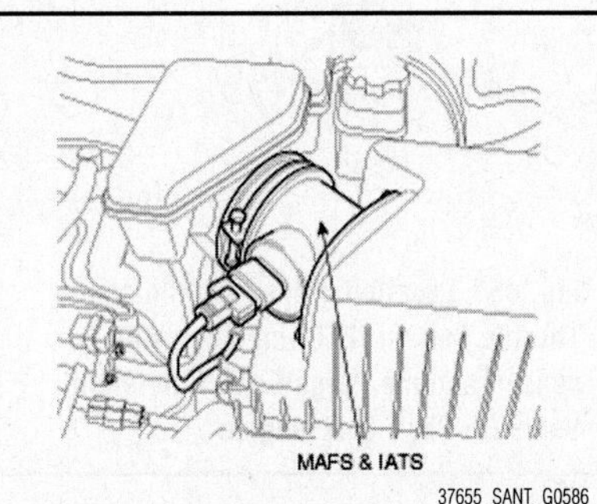

Fig. 447 Mass Air Flow (MAF) sensor and Intake Air Temperature (IAT) sensor location—2.7L and 3.3L engines

REMOVAL & INSTALLATION

1. Before servicing the vehicle, refer to the Precautions Section.
2. Disconnect the negative battery cable.
3. Disconnect the Mass Air Flow (MAF) sensor connector.
4. Remove the MAF sensor.

To install:

5. Installation is the reverse of removal. Tighten the MAF sensor installation bolt to 35–52 inch lbs. (4–6 Nm).

MANIFOLD ABSOLUTE PRESSURE (MAP) SENSOR

LOCATION

See Figures 438, 448 through 450.

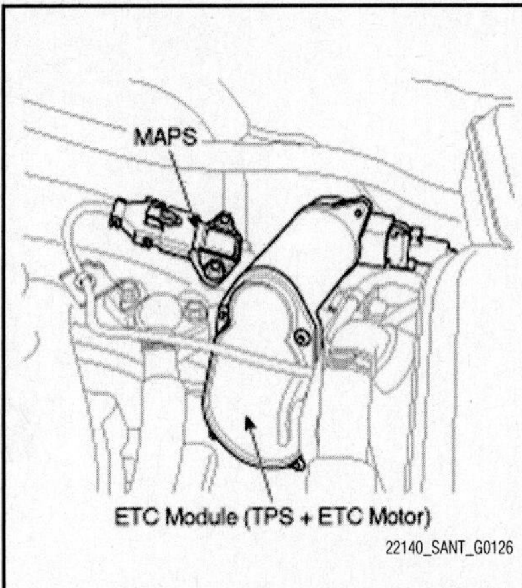

Fig. 448 Location of the MAP sensor—2.7L engine

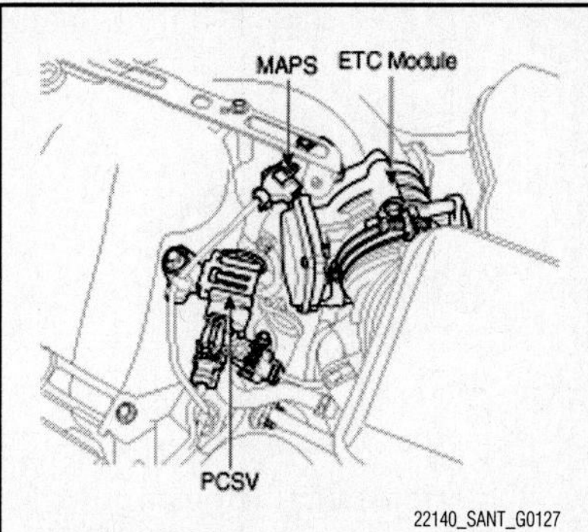

Fig. 449 Location of the MAP sensor, Purge Control Solenoid Valve (PCSV), and the Electronic Throttle Control (ETC) module—3.3L engine

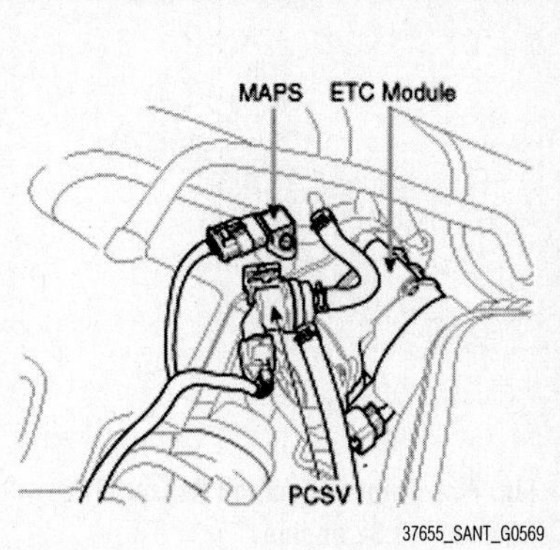

Fig. 450 MAP sensor, Purge Control Solenoid Valve (PCSV), and Electronic Throttle Control (ETC) module locations—3.5L engine

REMOVAL & INSTALLATION

1. Before servicing the vehicle, refer to the Precautions Section.
2. Disconnect the negative battery cable.
3. Disconnect the connector from the Manifold Absolute Pressure (MAP) sensor.
4. Remove the MAP sensor retaining screws.
5. Remove the MAP sensor from its mounting.

To install:

6. Installation is the reverse of the removal procedure. Tighten the MAP sensor installation bolt to: 70–104 inch lbs. (9–12 Nm).

THROTTLE POSITION SENSOR (TPS)

LOCATION

See Figures 451 through 454.

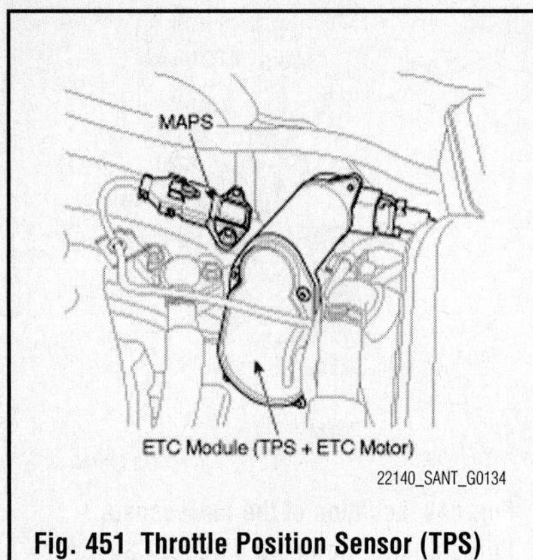

Fig. 451 Throttle Position Sensor (TPS) location—2.7L engine

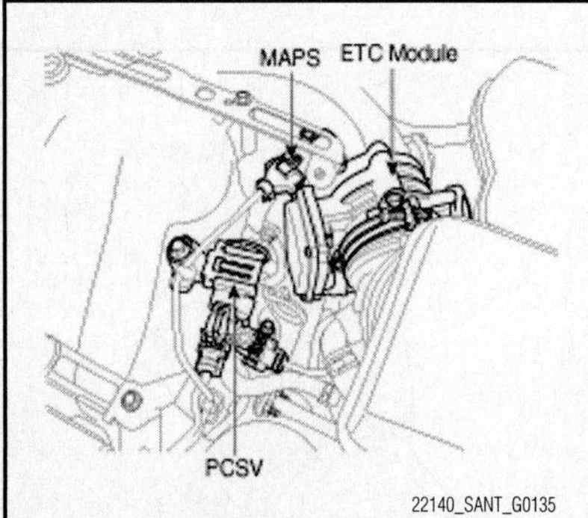

Fig. 452 Throttle Position Sensor (TPS) location—3.3L engine

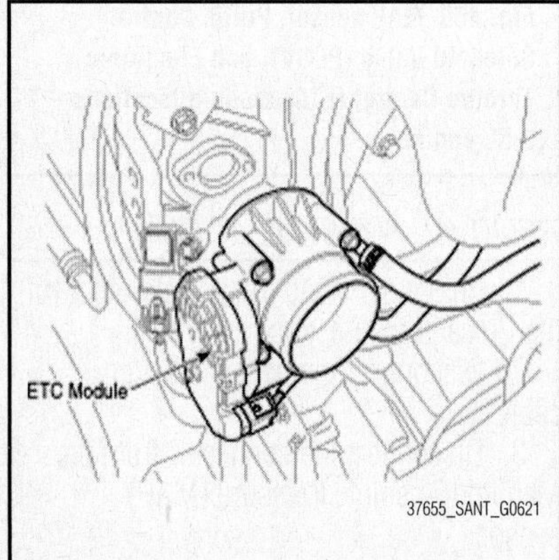

Fig. 453 Throttle Position Sensor (TPS) and ETC Module location—2.4L engine

REMOVAL & INSTALLATION

2009 Models

1. Before servicing the vehicle, refer to the Precautions Section.
2. Disconnect the negative battery cable.
3. Disconnect the Electronic Throttle Control (ETC) module connector.
4. Remove the retaining screws.
5. Remove the module.

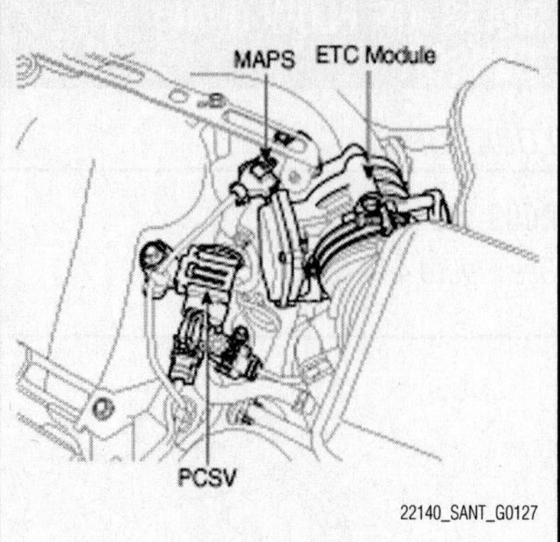

Fig. 454 Location of the Electronic Throttle Control (ETC) module, MAP sensor, and the Purge Control Solenoid Valve (PCSV)—3.3L engine

To install:

6. Installation is the reverse of the removal procedure.

2010 Models

2.4L Engine

See Figure 455.

1. Before servicing the vehicle, refer to the Precautions Section.
2. Turn the ignition switch OFF and disconnect the battery negative cable.
3. Remove the resonator and the air intake hose.
4. Disconnect the ETC module connector.
5. Disconnect the coolant hoses.
6. Remove the installation bolts, and then remove the ETC module from the engine.

To install:

7. Installation is reverse of removal. Tighten the ETC module mounting bolt to 70–104 inch lbs. (8–12 Nm)

3.5L Engine

See Figure 456.

1. Before servicing the vehicle, refer to the Precautions Section.
2. Turn the ignition switch OFF and disconnect the battery negative cable.
3. Remove the air cleaner assembly.
4. Disconnect the ETC module connector.
5. Disconnect the coolant hoses.
6. Remove the bracket installation bolt and the ETC module installation bolts, and then remove the ETC module from the engine.

To install:

7. Installation is reverse of removal. Tighten the ETC module mounting bolt to 70–104 inch lbs. (8–12 Nm) and the bracket mounting bolt to 12–19 ft. lbs. (16–26 Nm).

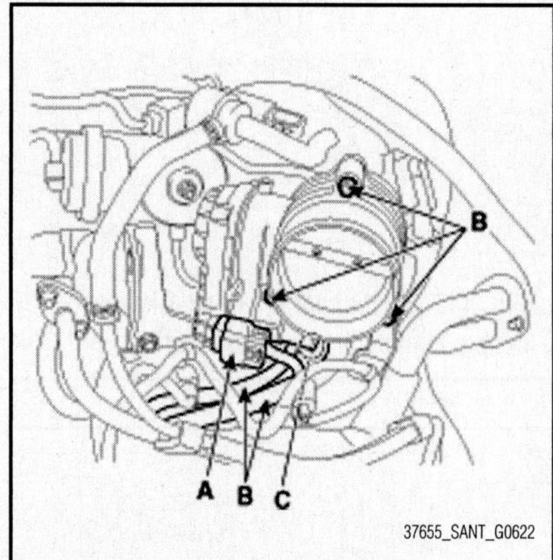

Fig. 456 Disconnect the ETC module connector (A) and coolant hoses (B), and remove the bracket installation bolt (C) and the ETC module installation bolts (D)

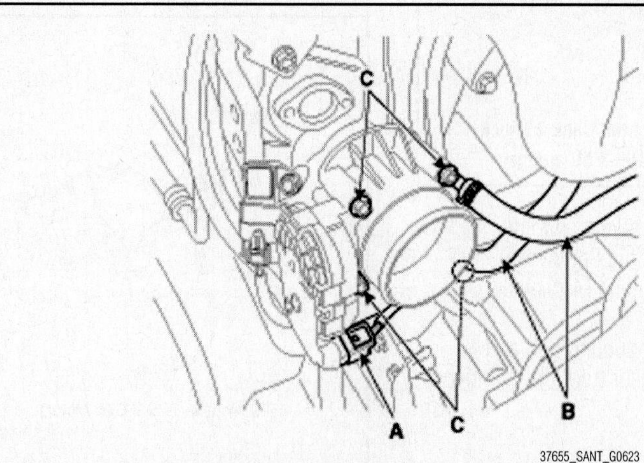

Fig. 455 Disconnect the ETC module connector (A) and coolant hoses (B), and remove the bracket installation bolt (C) and the ETC module installation bolts (D)

FUEL SYSTEM SERVICE PRECAUTIONS

Safety is the most important factor when performing not only fuel system maintenance but any type of maintenance. Failure to conduct maintenance and repairs in a safe manner may result in serious personal injury or death. Maintenance and testing of the vehicle's fuel system components can be accomplished safely and effectively by adhering to the following rules and guidelines.

- To avoid the possibility of fire and personal injury, always disconnect the negative battery cable unless the repair or test procedure requires that battery voltage be applied.
- Always relieve the fuel system pressure prior to disconnecting any fuel system component (injector, fuel rail, pressure regulator, etc.), fitting or fuel line connection. Exercise extreme caution whenever relieving fuel system pressure to avoid exposing skin, face and eyes to fuel spray. Please be advised that fuel under pressure may penetrate the skin or any part of the body that it contacts.
- Always place a shop towel or cloth around the fitting or connection prior to loosening to absorb any excess fuel due to spillage. Ensure that all fuel spillage (should it occur) is quickly removed from engine surfaces. Ensure that all fuel soaked cloths or towels are deposited into a suitable waste container.
- Always keep a dry chemical (Class B) fire extinguisher near the work area.
- Do not allow fuel spray or fuel vapors to come into contact with a spark or open flame.
- Always use a back-up wrench when loosening and tightening fuel line connection fittings. This will prevent unnecessary stress and torsion to fuel line piping.
- Always replace worn fuel fitting O-rings with new Do not substitute fuel hose or equivalent where fuel pipe is installed.

Before servicing the vehicle, make sure to also refer to the precautions in the beginning of this section as well.

RELIEVING FUEL SYSTEM PRESSURE

2009 Models

See Figures 457 through 459.

1. Before servicing the vehicle, refer to the Precautions Section.

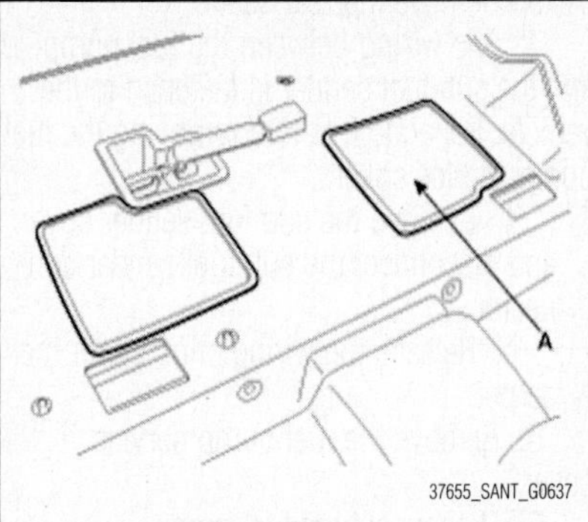

Fig. 457 Lift the carpet to access the fuel pump (A)

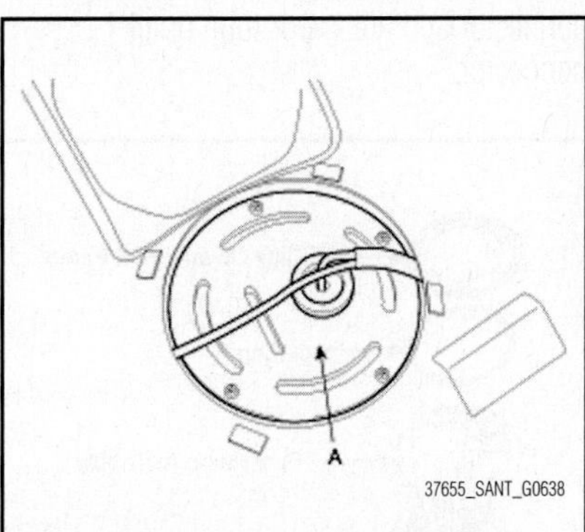

Fig. 458 Remove the fuel pump service cover (A)

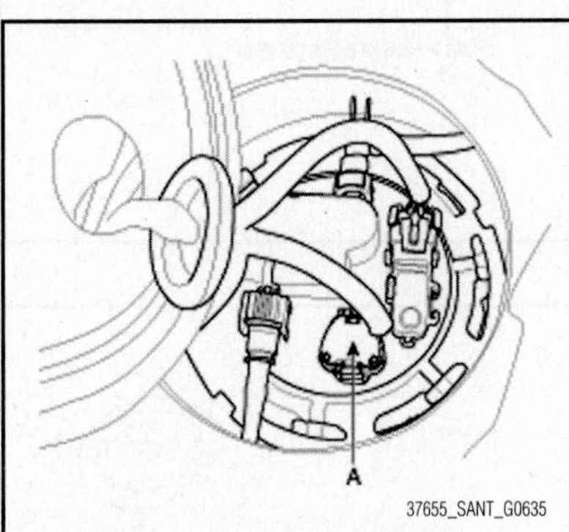

Fig. 459 Disconnect the fuel pump connector (A)

2. Remove the second seat.
3. Lift the carpet to access the fuel pump.
4. Remove the fuel pump service cover.
5. Disconnect the fuel pump connector.
6. Start the engine and wait until fuel in fuel line is exhausted.
7. After engine stops, turn the ignition switch off and disconnect the negative battery cable.

2010 Models

See Figure 460.

1. Before servicing the vehicle, refer to the Precautions Section.

➡Cover the hose connection with a shop towel to prevent residual fuel from spilling out before disconnecting any fuel connection.

2. Turn the ignition switch OFF and disconnect the negative battery cable.
3. Remove the fuel pump relay.

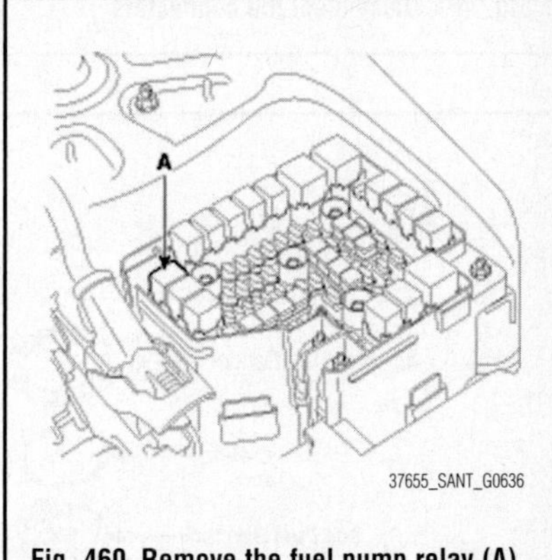

Fig. 460 Remove the fuel pump relay (A)

➡When removing the fuel pump relay, a Diagnostic Trouble Code (DTC) may occur. Delete the code with the GDS.

4. Connect the negative battery cable.
5. Start the engine and let idle, and then turn the ignition switch OFF after the engine has stopped on its own.
6. Disconnect the negative battery cable.

FUEL FILTER

REMOVAL & INSTALLATION

The fuel delivery system integrates the fuel filter with the in-tank fuel pump. To service this filter, remove the fuel pump. Refer to Fuel Pump, Removal & Installation.

FUEL PUMP

REMOVAL & INSTALLATION

2009 Models

See Figures 461 and 462.

1. Before servicing the vehicle, refer to the Precautions Section.

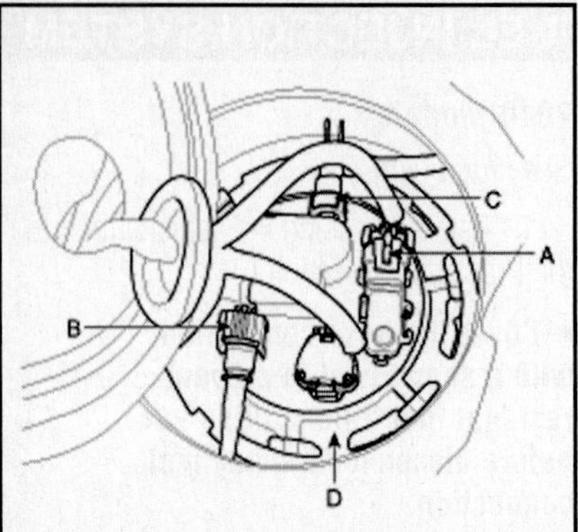

A. Fuel tank pressure sensor connector
B. Fuel feed quick-connector
C. Vacuum tube quick-connector
D. Fuel pump plate cover

Fig. 461 Disconnect the connectors

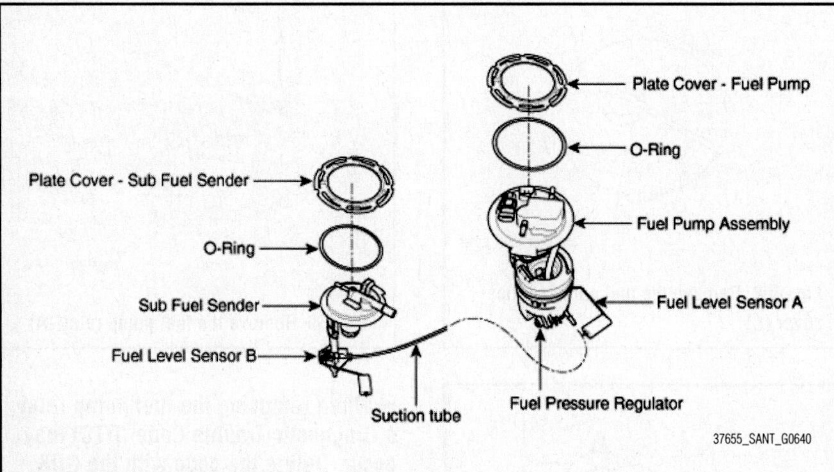

Fig. 462 Fuel pump assembly

2. Relieve the fuel system pressure. Refer to Relieving Fuel System Pressure.
3. Disconnect the fuel tank pressure sensor connector.
4. Disconnect the fuel feed quick-connector and the vacuum tube quick-connector.
5. Unfasten the fuel pump plate cover and disconnect the suction tube at the bottom of the fuel pump.
6. Remove the fuel pump from the fuel tank.

To install:

7. Installation is the reverse of removal.

2010 Models

See Figures 463 through 467.

1. Before servicing the vehicle, refer to the Precautions Section.
2. Relieve the fuel system pressure. Refer to Relieving Fuel System Pressure.
3. Remove the rear seat.
4. The wiring between the fuel pump and the sub fuel sender is fastened to the vehicle. Separate it before removing the fuel pump service cover:
   a. Remove the sub fuel sender cover and disconnect the sub fuel sender connector.
   b. Release the wiring clips under the carpet.
5. Remove the fuel pump service cover.
6. Disconnect the fuel pump connector and the fuel tank pressure sensor connector.
7. Disconnect the fuel feed tube quick connector and the vapor tube quick-connector.

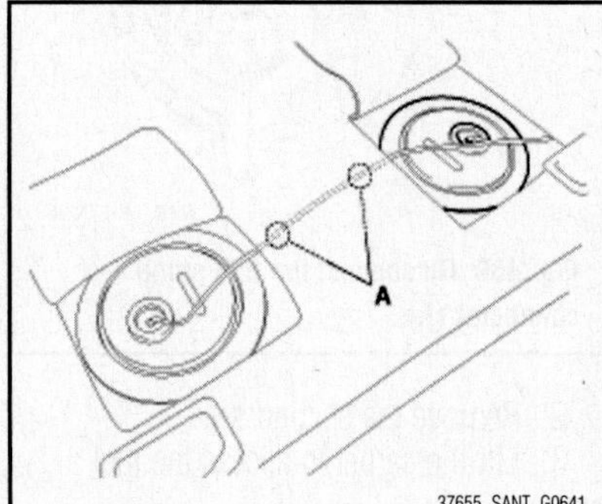

Fig. 463 Release the wiring clips (A) under the carpet

8. Remove locking ring.
9. Disconnect the suction tube at the bottom of the fuel pump, and then remove the fuel pump from the fuel tank.

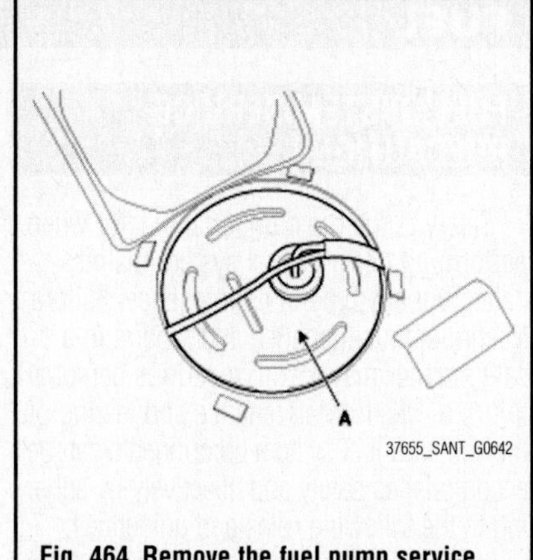

Fig. 464 Remove the fuel pump service cover (A)

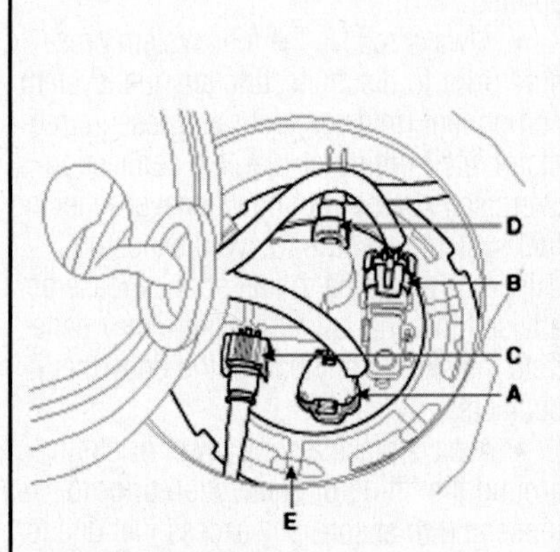

A. Fuel pump connector
B. Fuel tank pressure sensor connector
C. Fuel feed tube quick connector
D. Vapor tube quick-connector
E. Locking ring

Fig. 465 Disconnect the connectors

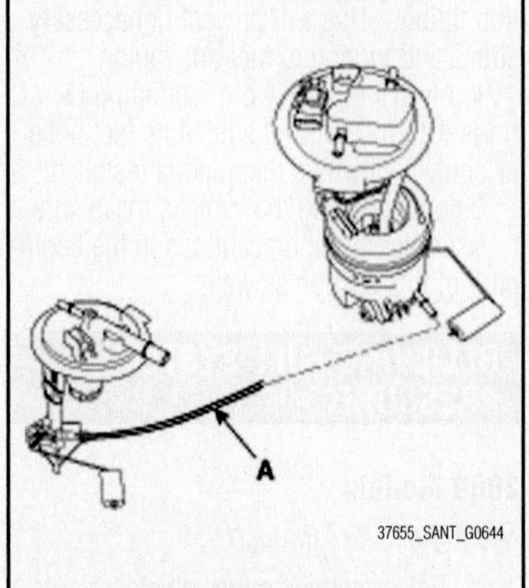

Fig. 466 Disconnect the suction tube (A)

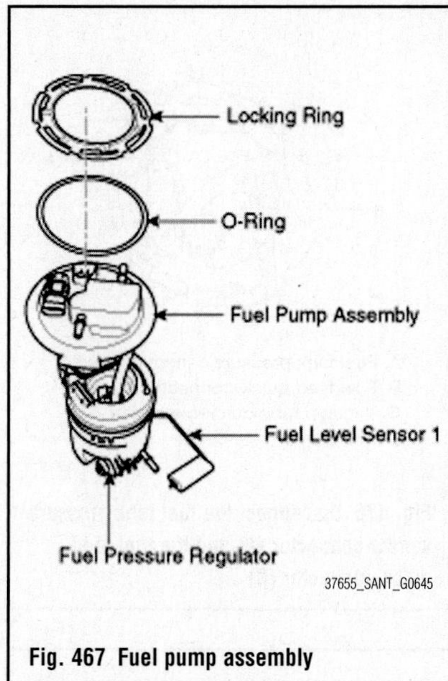

Fig. 467 Fuel pump assembly

To install:

10. Installation is the reverse of removal.

FUEL RAIL AND INJECTOR

REMOVAL & INSTALLATION

2009 Models

2.7L Engine

See Figures 468 through 470.

1. Before servicing the vehicle, refer to the Precautions Section.
2. Turn the ignition switch OFF and disconnect the battery negative cable.
3. Release the fuel pressure.

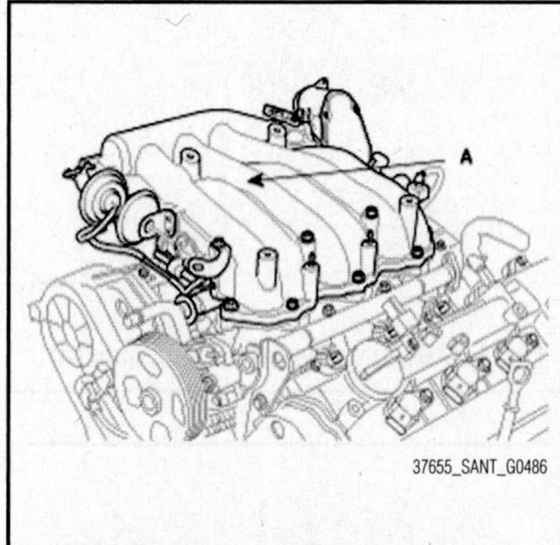

Fig. 469 Remove the surge tank (A)

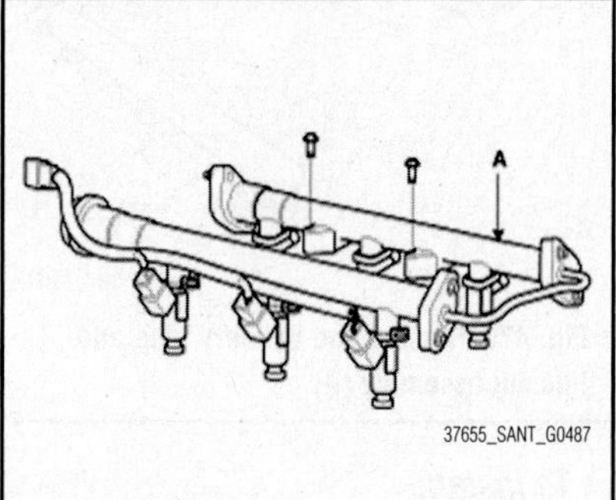

Fig. 470 Remove the fuel delivery pipe assembly (A)

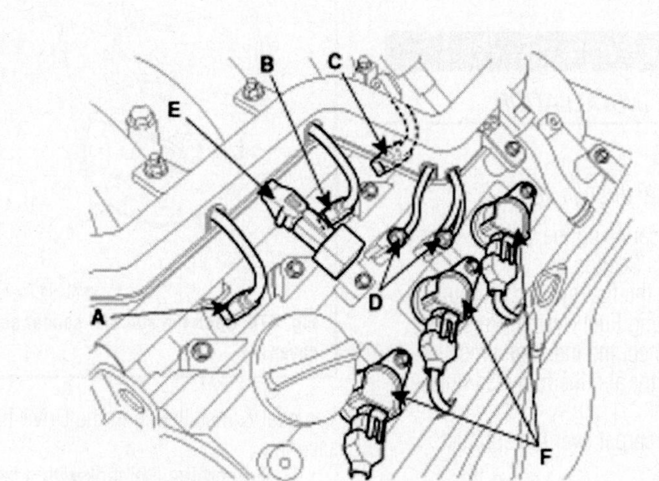

Fig. 468 Disconnect the injector connectors (A, B, C)

4. Disconnect the injector connectors, the ground lines, the condenser connector and the ignition coil connectors.
5. Remove the surge tank.
6. Remove the fuel delivery pipe assembly.

To install:

7. Installation is the reverse of removal.

3.3L Engine

See Figure 471.

1. Before servicing the vehicle, refer to the Precautions Section.
2. Turn the ignition switch OFF and disconnect the battery negative cable.
3. Release the fuel pressure.
4. Disconnect the injector connectors.
5. Remove the surge tank.

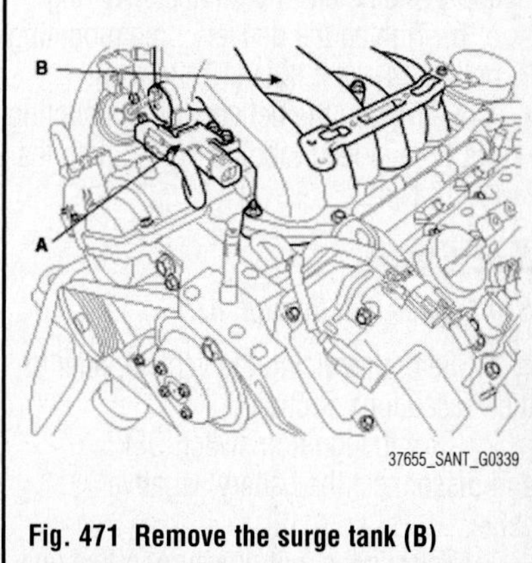

Fig. 471 Remove the surge tank (B)

6. Remove the fuel delivery pipe assembly.

To install:

7. Installation is the reverse of removal.

2010 Models

2.4L Engine

See Figure 472.

1. Before servicing the vehicle, refer to the Precautions Section.
2. Turn the ignition switch OFF and disconnect the battery negative cable.
3. Release the fuel pressure.
4. Disconnect the injector connector.
5. Remove the wiring harness bracket installation bolt.
6. Remove the installation nut, and then disconnect the fuel feed tube.
7. Remove the installation bolt, and then remove the delivery pipe and injector assembly from the engine.

To install:

8. Installation is the reverse of removal, noting the following:

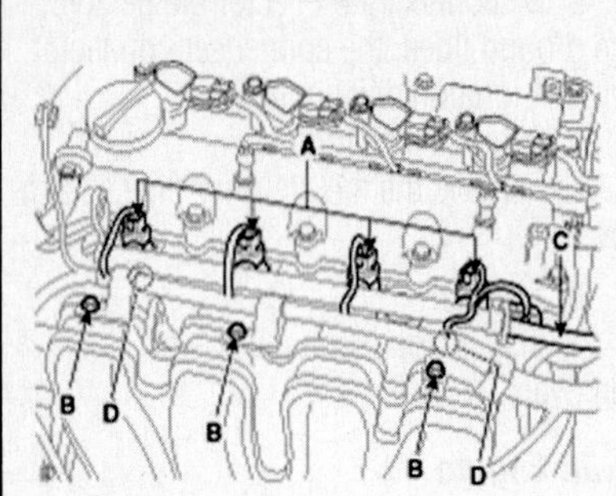

Fig. 472 Fuel rail and injector assembly, connectors, and bolts

a. Inspect the injector O-ring and apply engine oil to the injector O-ring.

b. Tighten the delivery pipe mounting bolt to 14–17 ft. lbs. (19–24 Nm).

c. Tighten the delivery pipe mounting nut (to fuel feed tube) to 70–86 inch lbs. (8–10 Nm).

3.5L Engine

See Figures 473 through 475.

1. Before servicing the vehicle, refer to the Precautions Section.
2. Turn the ignition switch OFF and disconnect the battery negative cable.
3. Release the fuel pressure in fuel line.
4. Remove the surge tank.
5. Disconnect the Bank 1 injector connector. The Bank 2 connector was disconnected when removing the surge tank.
6. Remove the delivery pipe mounting bolts.
7. Remove the installation nuts.
8. Remove the delivery pipe and injector assembly from the engine.

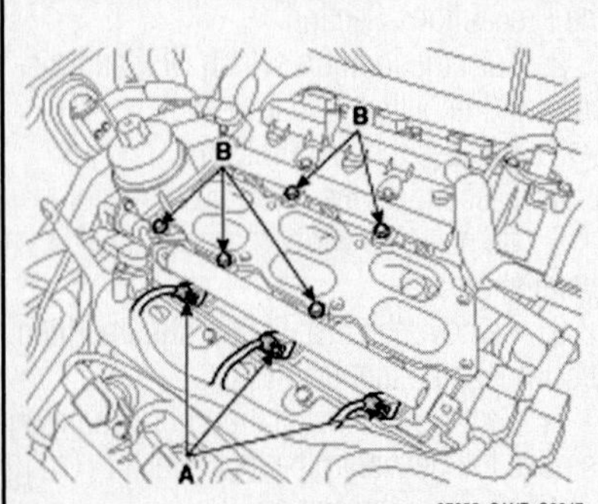

Fig. 473 Disconnect the Bank 1 injector connector (A) and remove the delivery pipe mounting bolts (B)

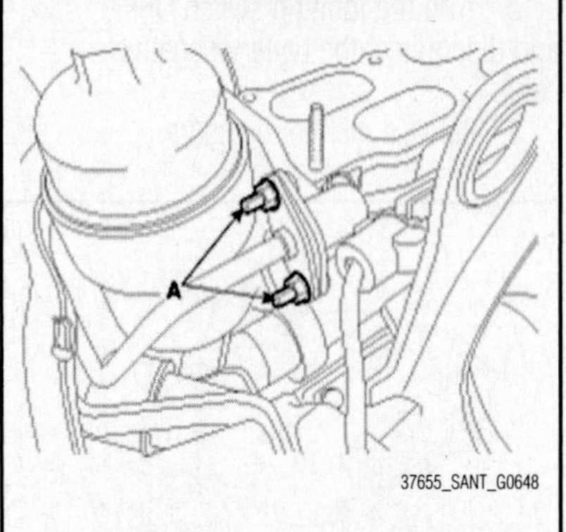

Fig. 474 Remove the installation nuts (A)

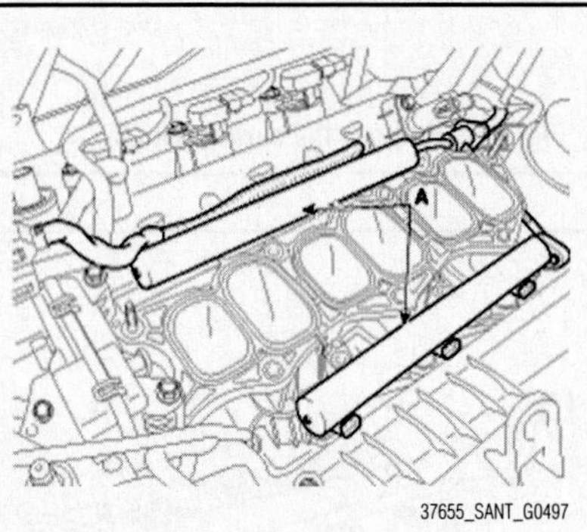

Fig. 475 Remove the delivery pipe and injector assembly (A)

To install:

9. Installation is the reverse of removal. Inspect the injector O-ring and apply engine oil to the injector O-ring. Tighten the delivery pipe mounting bolt to 78–104 inch lbs. (9–12 Nm).

FUEL TANK

REMOVAL & INSTALLATION

2009 Models

See Figures 476 through 480.

1. Before servicing the vehicle, refer to the Precautions Section.
2. Relieve the fuel system pressure. Refer to Relieving Fuel System Pressure.
3. Disconnect the fuel tank pressure sensor connector and the fuel feed quick-connector.
4. Lift the carpet over the sub fuel sender.
5. Remove the sub fuel sender service cover.
6. Disconnect the sub fuel sender connector.
7. Lift the vehicle and remove the muffler assembly.
8. For 4WD vehicles, remove the propeller shaft. Refer to Propeller Shaft Removal & Installation in the Drive Train section.

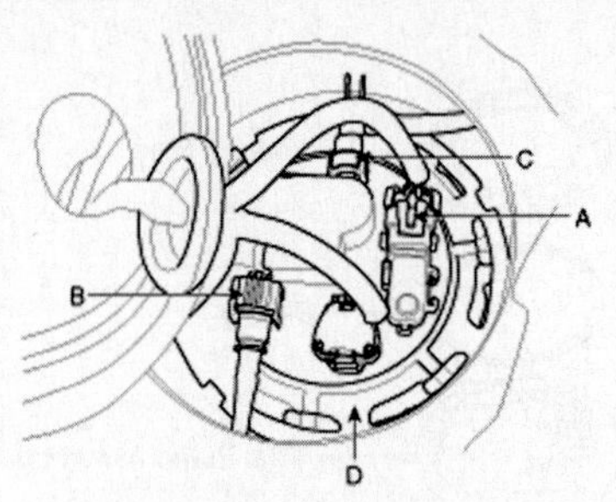

Fig. 476 Disconnect the fuel tank pressure sensor connector (A) and the fuel feed quick-connector (B)

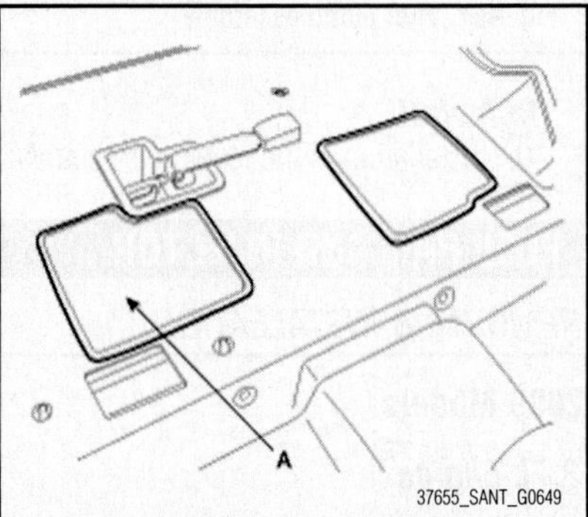

Fig. 477 Lift the carpet over the sub fuel sender (A)

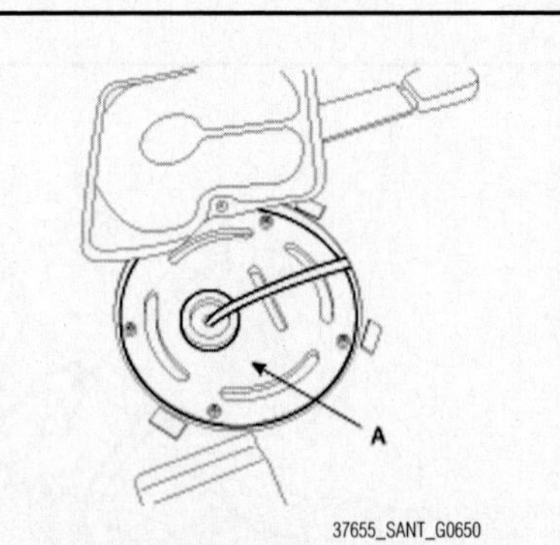

Fig. 478 Open the sub fuel sender service cover (A)

9. Support the fuel tank with a jack.
10. Disconnect the fuel filler hose, the leveling tube quick-connector and the vacuum tube quick-connector.
11. Disconnect the vacuum tube quick-connectors.
12. Unscrew the fuel tank bank mounting nuts and remove the fuel tank from the vehicle.

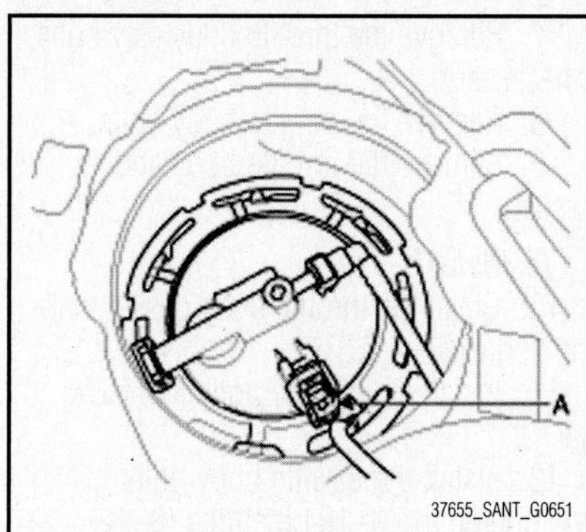

Fig. 479 Disconnect the sub fuel sender connector (A)

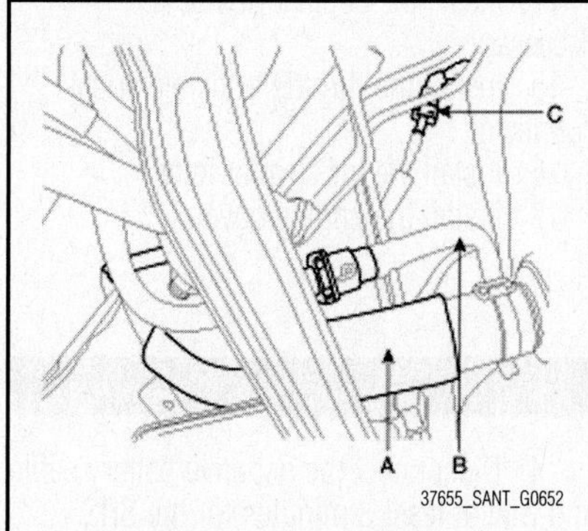

Fig. 480 Disconnect the fuel filler hose (A), the leveling tube quick-connector (B) and the vacuum tube quick-connector (C)

To install:

13. Installation is the reverse of the removal procedure. Tighten the fuel tank band mounting nuts to: 29–40 ft. lbs. (39–54 Nm).

2010 Models

See Figures 481 through 485.

1. Before servicing the vehicle, refer to the Precautions Section.
2. Relieve the fuel system pressure. Refer to Relieving Fuel System Pressure.
3. Remove the rear seat.
4. The wiring between the fuel pump and the sub fuel sender is fastened to the vehicle. Separate it before removing the fuel pump service cover:
   a. Remove the sub fuel sender cover and disconnect the sub fuel sender connector.

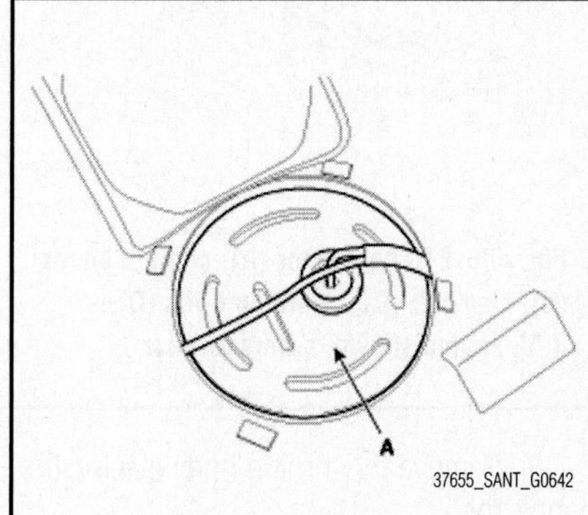

Fig. 482 Remove the fuel pump service cover (A)

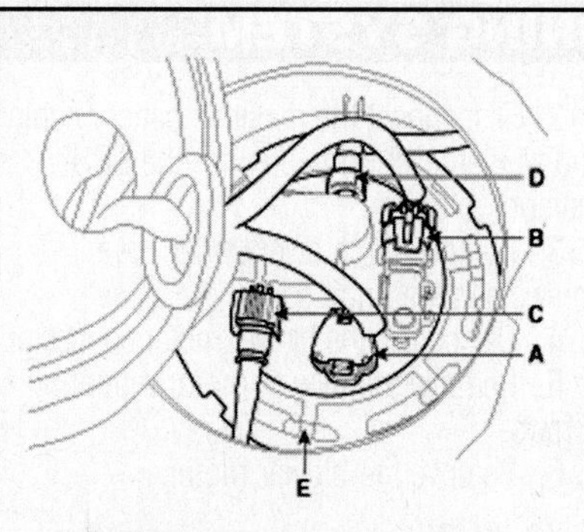

Fig. 483 Disconnect the fuel pump connector (A), fuel tank pressure sensor connector (B), and the fuel feed tube quick connector (C)

   b. Release the wiring clips under the carpet.
5. Remove the fuel pump service cover.
6. Disconnect the fuel pump connector and the fuel tank pressure sensor connector.
7. Disconnect the fuel feed tube quick connector.
8. Remove the sub fuel sender service cover.
9. Disconnect the sub fuel sender connector.
10. Lift the vehicle and support the fuel tank with a jack.
11. Remove the center muffler assembly.
12. For 4WD vehicles, remove the propeller shaft. Refer to Propeller Shaft Removal & Installation in the Drive Train section.
13. Remove the protector.
14. Disconnect the fuel filler hose.
15. Disconnect the ventilation hose quick-connector.
16. Disconnect the vapor tube quick-connector and the ventilation tube quick-connector.
17. Remove the brake line bracket.
18. Remove the fuel tank from the vehicle after removing the fuel tank bands.

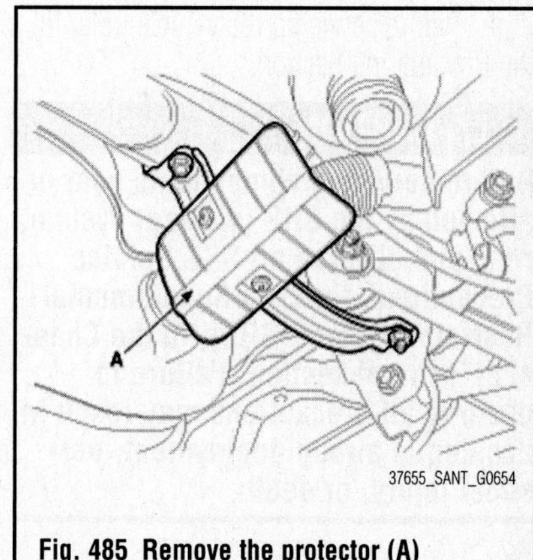

Fig. 485 Remove the protector (A)

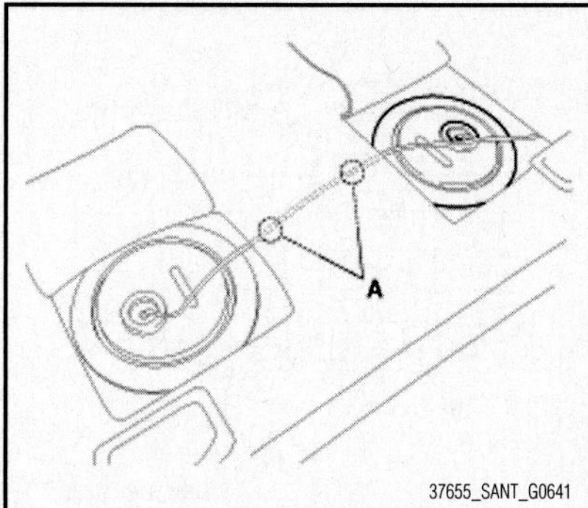

Fig. 481 Release the wiring clips (A) under the carpet

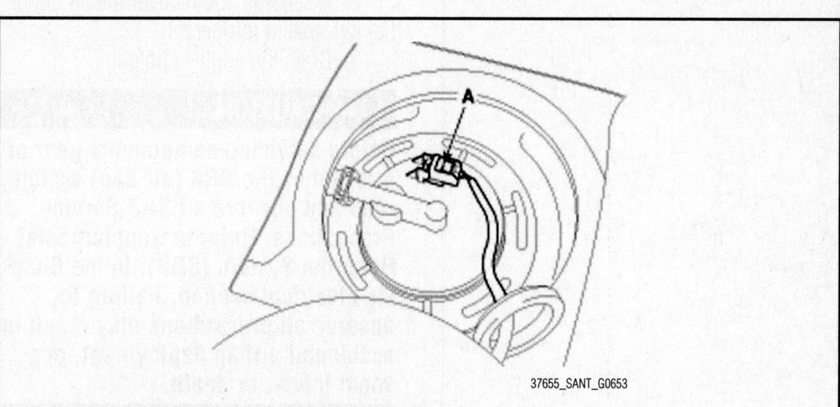

Fig. 484 Disconnect the sub fuel sender connector (A)

To install:

19. Installation is the reverse of the removal procedure. Tighten the fuel tank band mounting nuts to: 29–40 ft. lbs. (39–54 Nm).

IDLE SPEED

ADJUSTMENT

Idle speed is maintained by the Powertrain Control Module (PCM). No adjustment is necessary or possible.

THROTTLE BODY

REMOVAL & INSTALLATION

See Figure 486.

1. Before servicing the vehicle, refer to the Precautions Section.
2. Turn the ignition **OFF**.
3. Remove the engine cover.
4. Remove the air cleaner inlet.

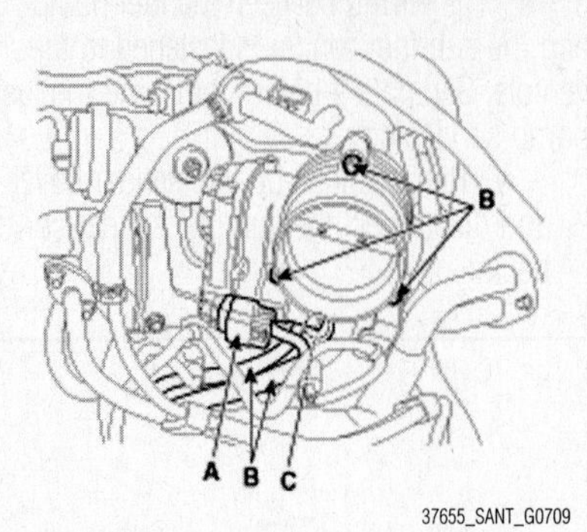

Fig. 486 ETC connector (A), coolant hoses (B), stay bolt (C), retaining bolts (D)—3.5L engine shown, others similar

5. Remove the throttle body electrical connector.
6. Remove the coolant hoses, if necessary.
7. Remove the throttle body stay bolts, if necessary.
8. Remove the throttle body bolts.
9. Remove the throttle body and gasket.

To install:

10. Clean the throttle body gasket mating surfaces.
11. Install the throttle body and NEW gasket.
12. Install the throttle body bolts and tighten to 70–104 inch lbs (8–12 Nm).
13. Install the throttle body stay bolts, if necessary.
14. Install the coolant hoses, if necessary.
15. Install the throttle body electrical connector.
16. Install the air cleaner inlet.
17. Install the engine cover.

HEATING & AIR CONDITIONING SYSTEM

BLOWER MOTOR

REMOVAL & INSTALLATION

See Figure 487.

1. Before servicing the vehicle, refer to the Precautions Section.

⁂ CAUTION

Before servicing components near or affected by the SRS (air bag) system, read and observe all SRS Service Precautions. Refer to Supplemental Restraint System (SRS), in the Chassis Electrical section. Failure to observe all precautions may result in accidental airbag deployment, personal injury, or death.

A

22140_SANT_G0147

Fig. 487 Remove the blower motor (A)

2. Disconnect the negative battery cable and wait at least 3 minutes for the SRS memory to drain.
3. Remove the blower under cover screws and the undercover.
4. Disconnect the blower motor connector.
5. Remove the blower motor mounting screws.
6. Remove the blower motor.

To install:

7. Installation is the reverse of the removal procedure.

HEATER UNIT/CORE

REMOVAL & INSTALLATION

See Figures 488 through 491.

1. Before servicing the vehicle, refer to the Precautions Section.
2. Discharge and recover the air conditioning system refrigerant.
3. Drain the engine coolant.

⁂ CAUTION

Before servicing components near or affected by the SRS (air bag) system, read and observe all SRS Service Precautions. Refer to Supplemental Restraint System (SRS), in the Chassis Electrical section. Failure to observe all precautions may result in accidental airbag deployment, personal injury, or death.

4. Disconnect the negative battery cable and wait at least 3 minutes for the SRS memory to drain.
5. Remove the bolts and expansion valve from the evaporate core and plug the lines.

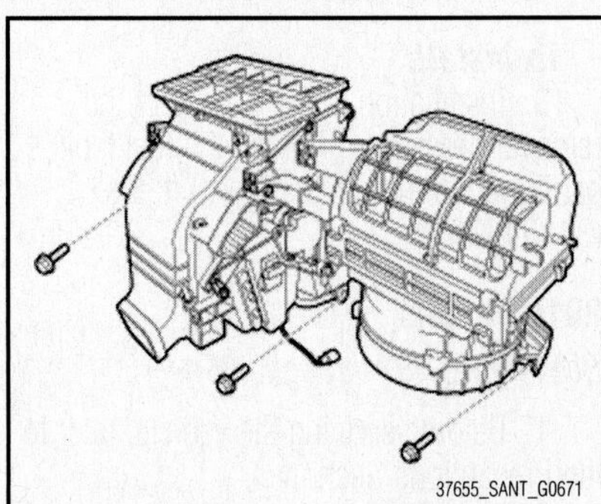

Fig. 488 Remove the heater and blower unit after removing the 3 mounting bolts

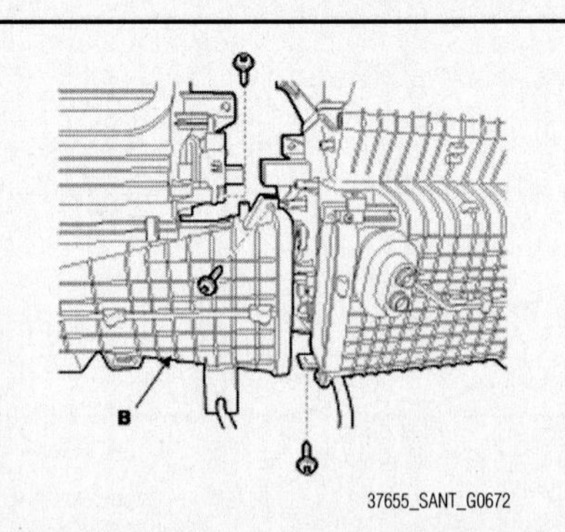

Fig. 489 Remove the blower unit (B) from heater unit after removing the 3 screws

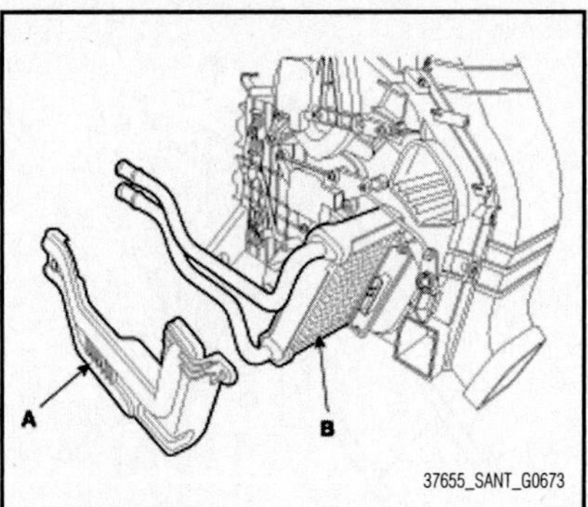

Fig. 490 Remove the heater core (B) after removing the cover (A)

6. Disconnect the heater hoses from the heater unit.
7. Remove the instrument panel.
8. Remove the cowl cross bar assembly.
9. Disconnect the connectors from the temperature control actuator, the mode control actuator and the evaporator temperature sensor and blower motor.
10. Remove the heater and blower unit after removing the mounting bolts.
11. Remove the blower unit from heater unit after removing the screws.

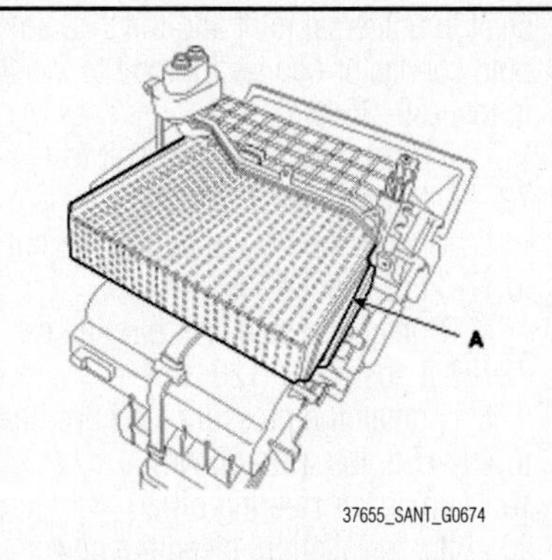

Fig. 491 Remove the evaporator core (A)

12. Remove the water temperature sensor.
13. Remove the heater core after removing the cover.
14. Remove the screws and remove the heater unit cover.
15. Remove the evaporator core.
16. Be careful that the inlet and outlet pipe are not bent during heater core removal, and pull out the heater core.

To install:

17. Installation is the reverse of the removal procedure.
18. Observe the following:
    - If you're installing a new evaporator, add refrigerant oil (ND-OIL8).
    - Replace the O-rings with new ones at each fitting, and apply a thin coat of refrigerant oil before installing them. Be sure to use the right O-rings for R-134a to avoid leakage.
    - Immediately after using the oil, replace the cap on the container, and seal it to avoid moisture absorption.
    - Apply sealant to the grommets.
    - Make sure that there is no air leakage.
    - Charge the system and test its performance.
    - Do not interchange the inlet and outlet heater hoses and install the hose clamps securely.

WARNING

Do not spill the refrigerant oil on the vehicle ; it may damage the paint ; if the refrigerant oil contacts the paint, wash it off immediately.

STEERING

POWER RACK & PINION STEERING GEAR

REMOVAL & INSTALLATION

See Figures 492 through 496.

1. Before servicing the vehicle, refer to the Precautions Section.
2. Remove the both front wheels.
3. Drain the power steering fluid by disconnecting the return hose.
4. Disconnect the pressure tube from the power steering pump by loosening the eye bolt.
5. Disconnect the stabilizer bar link from the front strut assembly.
6. Remove the tie-rod end split pin and castle nut.
7. Disconnect the tie rod end from the knuckle by using a SST (09568-4A000 or 09568-34000).
8. Remove the bolt pin and nut and disconnect the lower arm from the knuckle.
9. Repeat on the other side.
10. Remove the bolt connecting steering gear pinion shaft to universal joint and the EPS solenoid connector.

WARNING

Keep the vehicle in neutral to prevent the damage of the clock spring inner cable when you handle the steering wheel.

11. Remove the front and rear roll stopper bolts.
12. Remove the muffler hanger.
13. Remove the sub-frame s tay mounting bolts and sub-frame mounting bolts and nut and then make the space between sub-frame and frame.

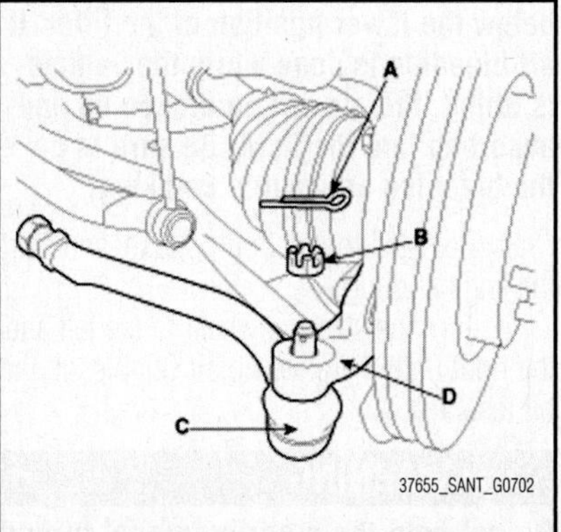

Fig. 492 Remove the split pin (A) and castle nut (B)

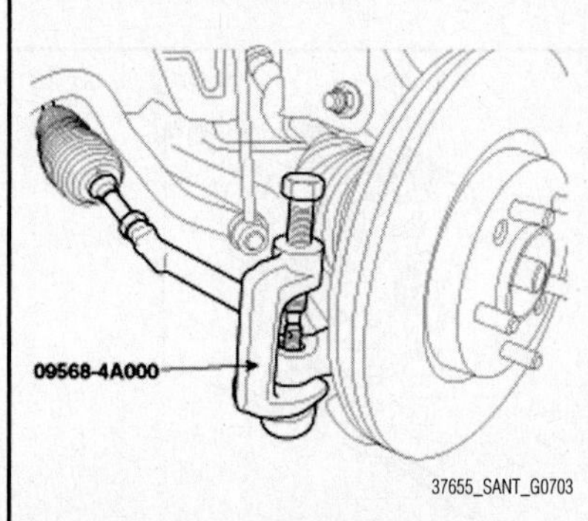

Fig. 493 Disconnect the tie rod end (A) from the knuckle by using a SST (09568-4A000 or 09568-34000)

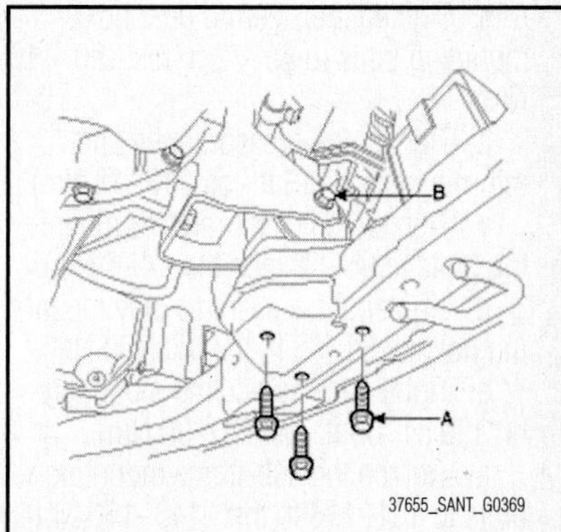

Fig. 494 Remove the front roll stopper mounting bolts (A)

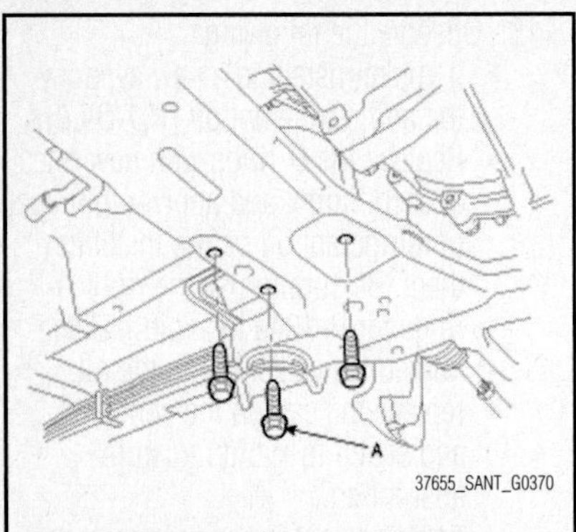

Fig. 495 Remove the rear roll stopper mounting bolts (A)

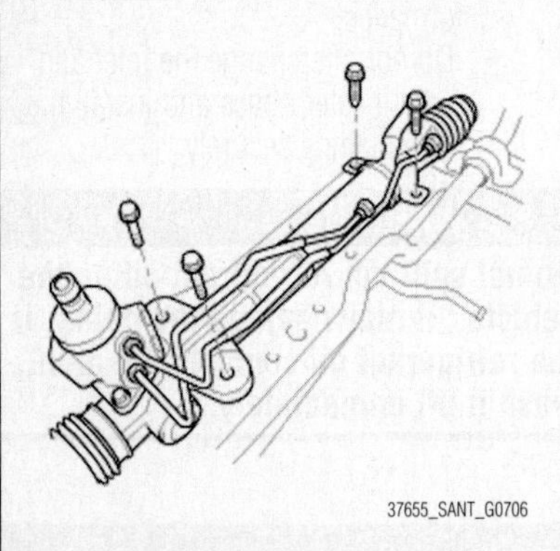

Fig. 496 Remove the steering gear box mount bolts and then pull the steering gear box to the left side

14. Remove the heat protector.
15. Disconnect the pressure tube and return tube from the valve body housing.
16. Remove the tube bracket bolts and nut.
17. Remove the steering gear box mounting bolts and then pull the steering gear box to the left side.

To install:

18. Installation is the reverse of removal noting the following:
    a. Tighten the steering gear box mounting bolts to 65–72 ft. lbs. (90–110 Nm).
    b. Tighten the pressure tube and return tube to 9–13 ft. lbs. (12–18 Nm).
    c. Tighten the heat protector mounting bolts to 35–52 inch lbs. (4–6 Nm).
    d. Tighten the sub-frame stay mounting bolts to 51–65 ft. lbs. (70–90 Nm).
    e. Tighten the sub-frame mounting nuts to 51–65 ft. lbs. (70–90 Nm).
    f. Tighten the sub-frame mounting bolts to 101–116 ft. lbs. (140–160 Nm).
    g. Tighten the roll stopper bolts to 36–47 ft. lbs. (50–65 Nm).
    h. Tighten the steering gear pinion shaft to universal joint and the EPS solenoid connector connecting bolt to 22–25 ft. lbs. (30–35 Nm).
    i. Tighten the lower arm bolt to 72–87 ft. lbs. (100–120 Nm).
    j. Tighten the tie-rod end castle nut to 17–25 ft. lbs. (24–34 Nm).
    k. Tighten the stabilizer bar link to 72–87 ft. lbs. (100–120 Nm).
    l. Tighten the pressure tube eye bolt to 40–47 ft. lbs. (55–65 Nm).
19. Add power steering oil.
20. After installation, bleed the power steering system.
21. Adjust the wheel alignment.

POWER STEERING PUMP

REMOVAL & INSTALLATION

See Figures 497 through 499.

1. Before servicing the vehicle, refer to the Precautions Section.
2. Remove the wheel guide and side cover.
3. Remove the drive belt.
4. Disconnect the oil pressure switch.
5. Disconnect the pressure tube and suction hose from the power steering pump assembly.

Fig. 497 Power steering pump–3.3L engine

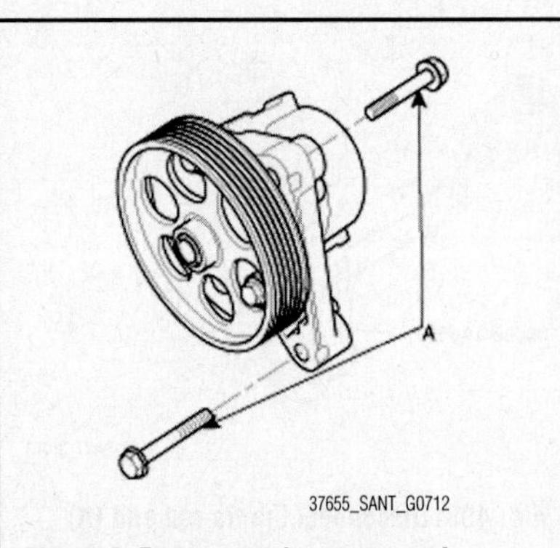

Fig. 498 Power steering pump and mounting bolts (A)–3.3L engine

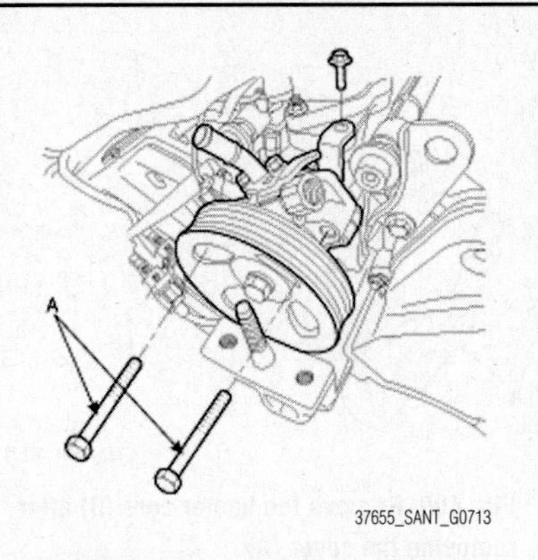

Fig. 499 Power steering pump and mounting bolts (A)–2.7L engine

6. Loosen the power steering pump mounting bolts, and then remove the power steering pump assembly from the pump bracket.

To install:

7. Installation is the reverse of removal noting the following:
    a. Tighten the power steering pump mounting bolts to 25–36 ft. lbs. (35–50 Nm).
    b. Tighten the pressure tube and suction hose to 40–47 ft. lbs. (55–65 Nm).
8. Add the power steering oil.
9. After installation, bleed the power steering system.
10. Check the oil pump pressure.

BLEEDING

1. Before servicing the vehicle, refer to the Precautions Section.
2. Remove the fuel pump fuse, then start the engine and wait for the engine to stall. Next, while operating the starting motor intermittently (for 15–20 seconds), turn the steering wheel all the way to the left and then to the right five or six times.

➡During air bleeding, replenish the fluid supply so that the level never falls below the lower position of the filter. If air bleeding is done while the vehicle is idling, the air will be broken up and absorbed into the fluid. Be sure to do the bleeding only while cranking.

3. Reinstall the fuel pump fuse, and start the engine(idling).
4. Turn the steering wheel to the left and the right until there are no air bubbles in the oil reservoir.

✲✲ WARNING

Do not hold the steering wheel turned all the way to either side for more than ten seconds.

5. Confirm that the fluid is not milky, and that the level is up to the position specified on the level gauge.

6. Confirm that there is little change in the surface of the fluid when the steering wheel is turned left and right.

➡If the surface of the fluid changes considerably, air bleeding should be done again. If the fluid level rises suddenly when the engine is stopped, it indicates that there is still air in the system. If there is air in the system, a jingling noise may be heard from the pump and the control valve may also produce unusual noises. Air in the system will shorten the life of the pump and other parts.

SUSPENSION

FRONT SUSPENSION

CONTROL LINKS

REMOVAL & INSTALLATION

See Figure 500.

1. Before servicing the vehicle, refer to the Precautions Section.
2. Raise and safely support the vehicle.
3. Remove the front wheel and tire from the front hub.

⁂ WARNING

Be careful not to damage the hub bolts when removing the front wheel and tire.

4. Remove the stabilizer bar control link by removing the upper and lower mounting nuts.
5. Remove the stabilizer bar control link.

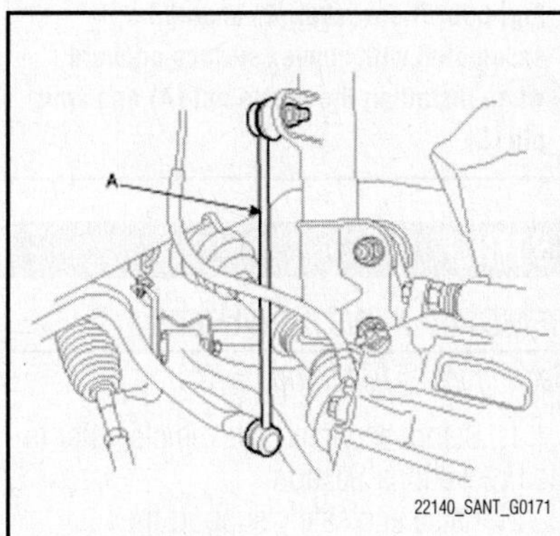

Fig. 500 Stabilizer control link (A) and mounting nuts

To install:

6. Install the stabilizer bar control link.
7. Tighten the stabilizer bar control link upper and lower mounting nuts to 72–87 ft. lbs. (100–120 Nm).
8. Install the front wheels, and lower the vehicle. Tighten the wheel lug nuts to 65–80 ft. lbs. (90–110 Nm).

LOWER CONTROL ARM

REMOVAL & INSTALLATION

See Figures 501 and 502.

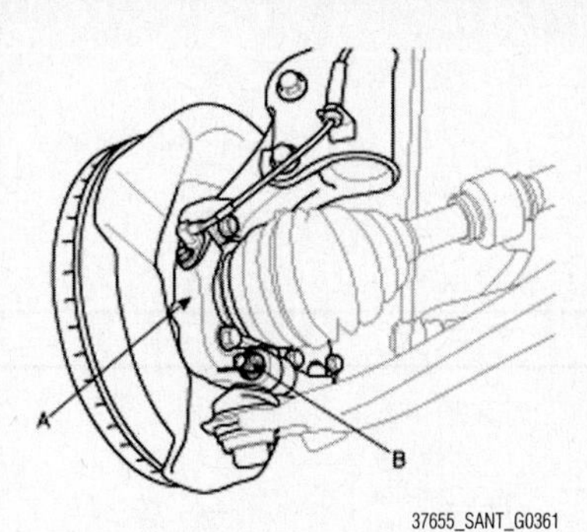

Fig. 501 Remove the split pin and front lower arm mounting bolt (B) from the knuckle (A)

1. Loosen the wheel nuts slightly.
2. Raise the vehicle, and make sure it is securely supported.
3. Remove the front wheel and tire from the front hub.

➡Be careful not to damage to the hub bolts when removing the front wheel and tire.

4. Remove the split pin and front lower arm mounting bolt from the knuckle.
5. Remove the lower arm mounting bolts and then remove the lower arm.

To install:

6. Installation is the reverse of removal. Tighten the fasteners as follows:

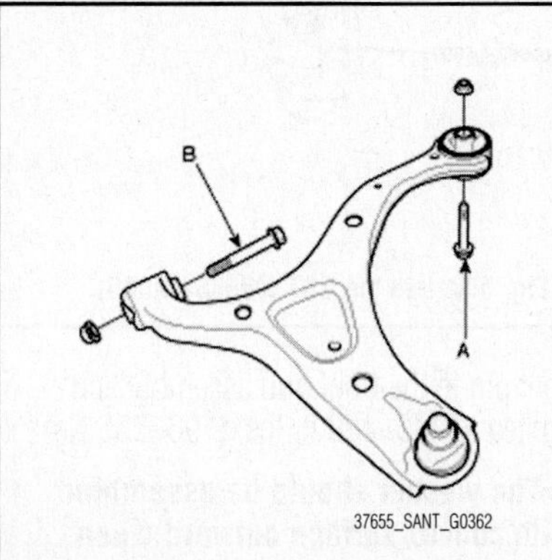

Fig. 502 Remove the lower arm mounting bolts (A, B)

- Lower control arm bolt (A): 101–116 ft. lbs. (137–157 Nm)
- Bolt (B): 101–116 ft. lbs. (137–157 Nm)
- Ball joint-to-knuckle bolt: 73–87 ft. lbs. (98–118 Nm)

7. Check the wheel alignment and adjust as necessary.

STEERING KNUCKLE

REMOVAL & INSTALLATION

See Figures 503 through 509.

1. Before servicing the vehicle, refer to the Precautions Section.
2. Raise and safely support the vehicle.
3. Remove the front wheels and tires.
4. Remove the split pin, and then remove the castle nut and washer from the front hub.
5. Remove the brake caliper mounting bolts, and hang the brake caliper assembly out of the way with a wire.
6. Remove the tie rod end ball joint from the knuckle:
   a. Remove the split pin.
   b. Remove the castle nut.
   c. Disconnect the ball joint from knuckle using the SST (09568-4A000). Apply a few drops of oil to the boot contact part of the SST.
7. Remove the wheel speed sensor, the strut lower mounting bolt and the lower arm mounting bolt from the knuckle.
8. Remove the hub and knuckle assembly.

➡Be careful not to damage the boot and rotor teeth.

To install:

9. Install the hub and knuckle assembly.
10. Install the wheel speed sensor, the strut lower mounting bolt and the lower arm mounting bolt from the knuckle).Tightening torque:
   - Wheel speed sensor (B): 61–96 inch lbs. (7–11 Nm)
   - Bolts (C): 112–127 ft. lbs. (152–172 Nm)
   - Bolt (D) : 72–87 ft. lbs. (98–118 Nm)

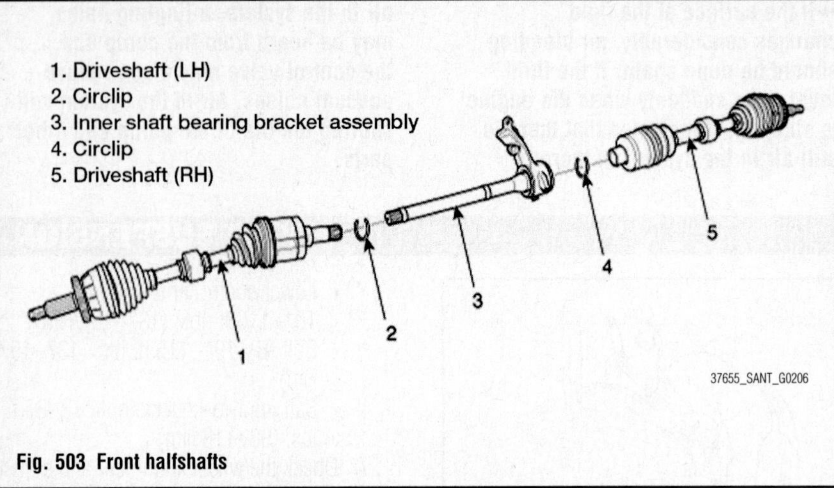

Fig. 503 Front halfshafts

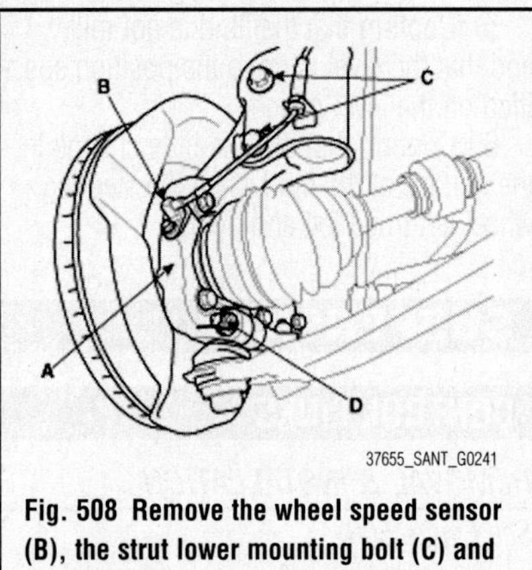

Fig. 508 Remove the wheel speed sensor (B), the strut lower mounting bolt (C) and the lower arm mounting bolt (D) from the knuckle (A)

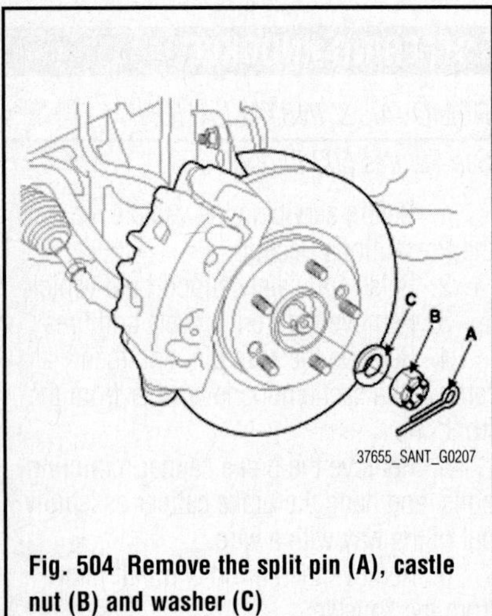

Fig. 504 Remove the split pin (A), castle nut (B) and washer (C)

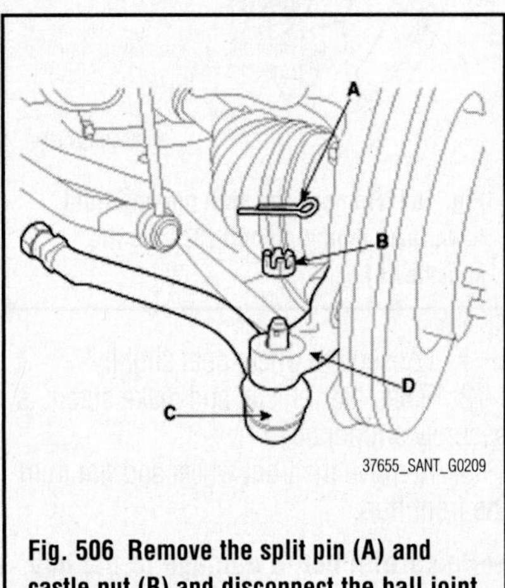

Fig. 506 Remove the split pin (A) and castle nut (B) and disconnect the ball joint (C) from knuckle (D)

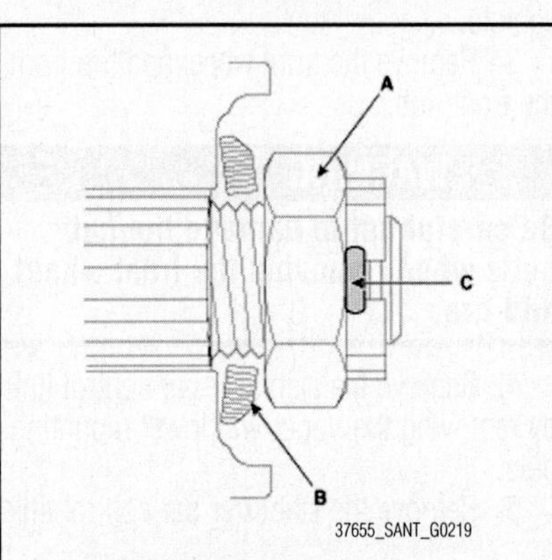

Fig. 509 The washer (B) should be assembled with convex surface outward when installing the castle nut (A) and split pin (C)

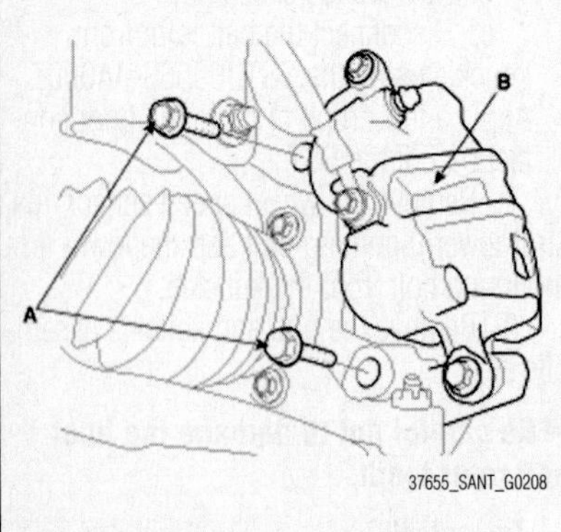

Fig. 505 Remove the brake caliper mounting bolts (A), and hang the brake caliper assembly (B)

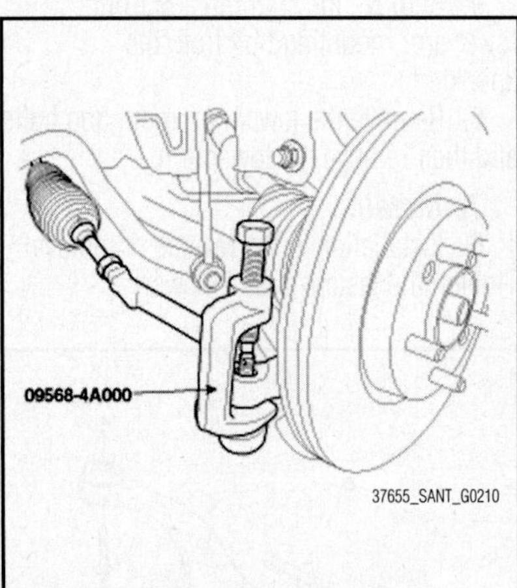

Fig. 507 Use the SST (09568-4A000)

11. Install the tie rod end ball joint to the knuckle.
12. Install the castle nut and the split pin and tighten to 17–25 ft. lbs. (24–33 Nm).
13. Install the brake caliper and tighten the mounting bolts to 58–72 ft. lbs. (79–98 Nm).
14. Install the washer, castle nut and split pin to the front hub assembly and tighten to 145–188 ft. lbs. (196–255 Nm).

➡The washer should be assembled with convex surface outward when installing the castle nut and split pin.

15. Install the wheel and tire to the front hub and tighten to 65–80 ft. lbs. (88–108 Nm).

STRUT

REMOVAL & INSTALLATION

See Figures 510 through 513.

1. Before servicing the vehicle, refer to the Precautions Section.
2. Raise and safely support the vehicle.
3. Remove the front wheels and tires.

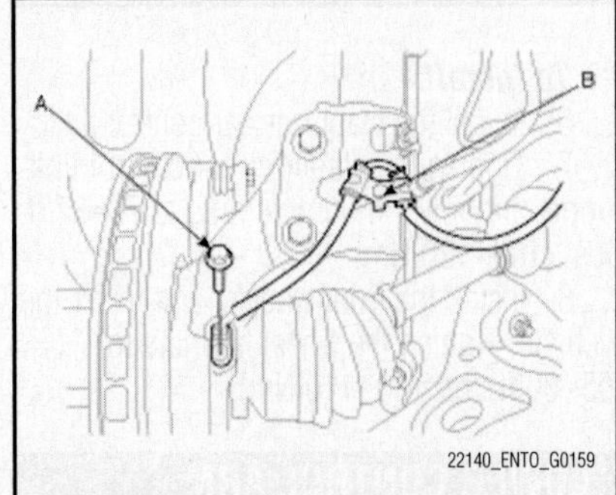

Fig. 510 Remove the speed sensor (A) and wire bracket bolts (B) from the front steering knuckle

4. Remove the brake hose bracket bolts from the strut assembly.

5. Remove the speed sensor and wire bracket bolts from the front steering knuckle.

6. Make an alignment marking on the camber adjusting bolt and strut for installation alignment approximation later.

7. Remove the front stabilizer control link and nut from the strut.

8. Remove the upper strut mounting nuts.

9. Remove the front strut mounting bolts from the knuckle.

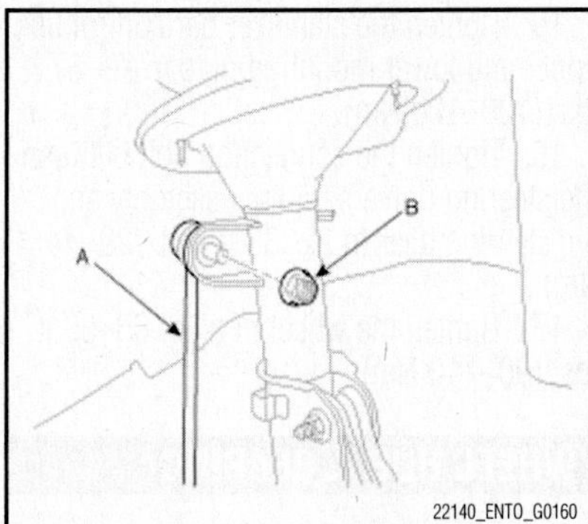

Fig. 511 Remove the front stabilizer control link (A) and nut (B) from the strut

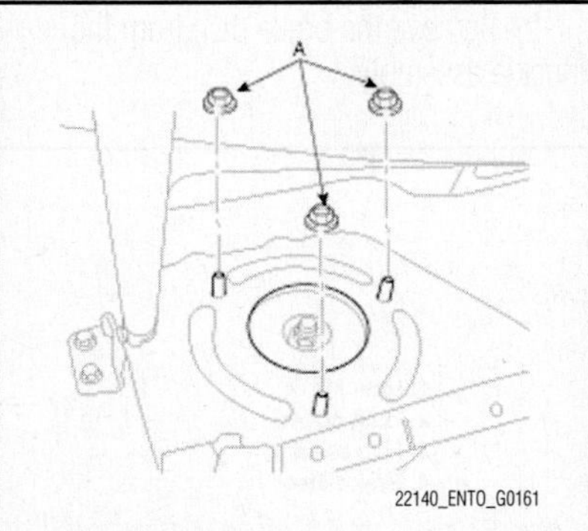

Fig. 512 Remove the upper strut mounting nuts (A)

10. Remove the strut assembly and bolts from its mounting in the steering knuckle.

To install:

11. Install the strut upper mounting nuts and tighten to 33–43 ft. lbs. (45–60 Nm).

12. Match the alignment marks made during removal and install the front strut assembly bolts to the front knuckle. Tighten to 112–127 ft. lbs. (152–172 Nm).

13. Install the front stabilizer link nut to the strut assembly and tighten to 72–87 ft. lbs. (100–120 Nm).

14. Install the speed sensor and wire bracket bolts and tighten to 60–96 inch lbs. (7–11 Nm).

15. Install the brake hose bracket bolt to the axle assembly.

16. Install the wheel and the tire to the front hub. Tighten to 65–80 ft. lbs. (90–110 Nm).

WARNING

Be careful not to damage the hub bolts when installing the front wheel and tire.

17. Check the wheel alignment and adjust as necessary.

OVERHAUL

See Figure 514

1. Before servicing the vehicle, refer to the Precautions Section.

2. Remove the strut from the vehicle and attach Service Tool 09546-2600 or another suitable spring compressor.

3. Compress the coil spring.

4. Remove the self-locking nut from the strut.

5. Remove the insulator, spring seat, coil spring, and dust cover.

6. Inspect the insulator for wear and damage, replace as required.

7. Check the rubber parts for damage or deterioration, replace as required.

8. Install the spring lower pad so that the protrusions fit the holes in the spring lower seat.

9. Compress the spring, using the spring compressor tool.

10. Install the compressed spring over the shock absorber.

➡There are 2 color identification marks on the coil spring, one indicates model option and the other indicates load classification. Install the coil spring with the identification mark directed toward the steering knuckle.

11. After fully extending the piston rod, install the spring upper seat and insulator assembly.

12. After correctly seating the upper and lower ends of the coil spring in the upper and lower spring grooves, tighten the NEW self-locking nut temporarily.

13. Remove the spring compression tool. Tighten the self-locking nut to 43–51 ft. lbs. (60–70 Nm).

14. Install the strut to the vehicle.

15. Check the wheel alignment and adjust as necessary.

STABILIZER BAR

REMOVAL & INSTALLATION

See Figures 515 through 519.

1. Before servicing the vehicle, refer to the Precautions Section.

2. Raise the front of the vehicle, and make sure it is securely supported.

3. Remove the front wheel and tire from the front hub.

WARNING

Be careful not to damage the hub bolts when removing the front wheel and tire.

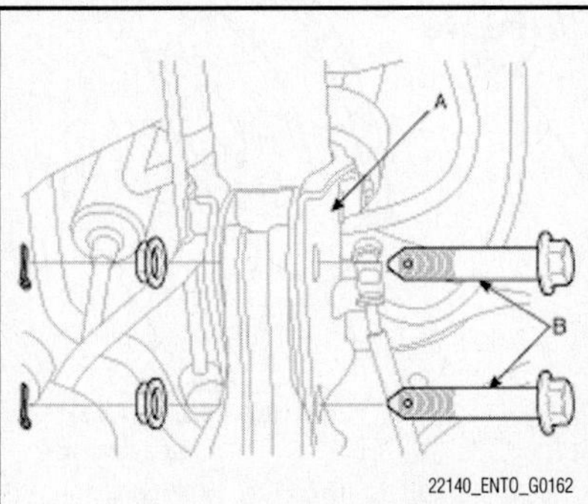

Fig. 513 Remove the strut assembly (A) and bolts (B) from its mounting

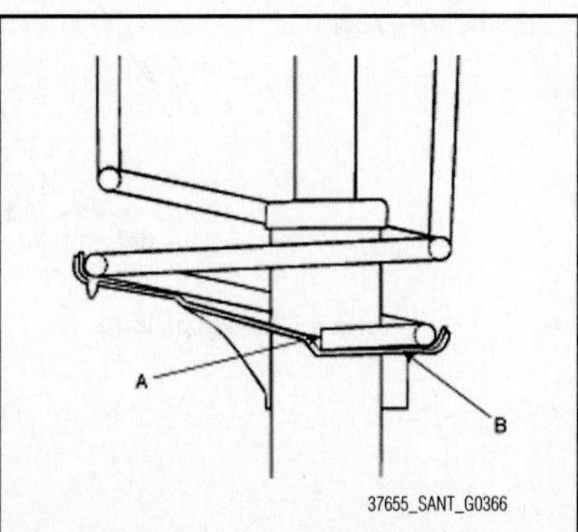

Fig. 514 Install the compressed spring (A) over the shock (B) absorber

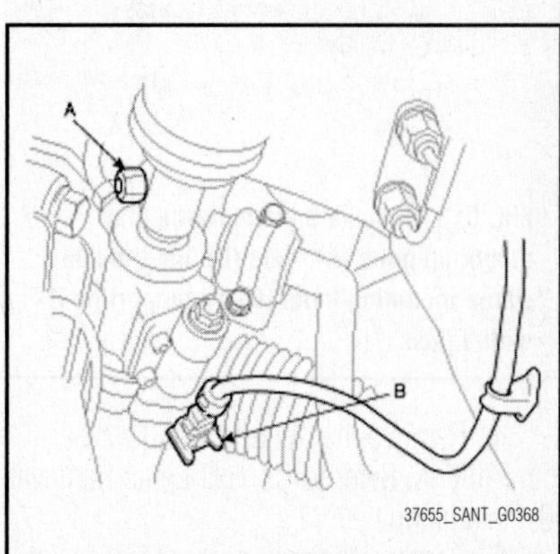

Fig. 515 Disconnect the EPS connector (B)

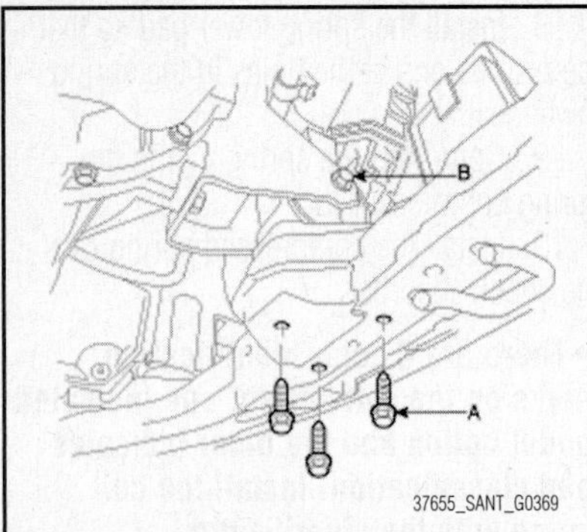

Fig. 516 Remove the front roll stopper mounting bolts (A)

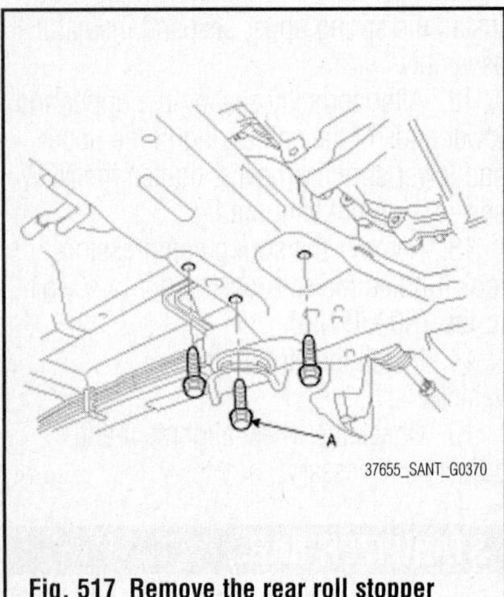

Fig. 517 Remove the rear roll stopper mounting bolts (A)

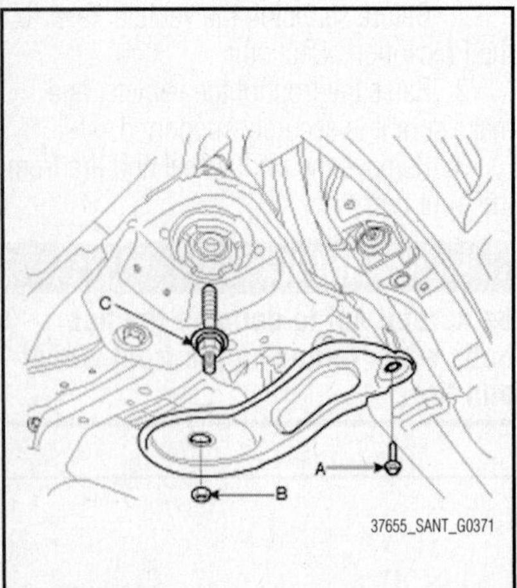

Fig. 518 Remove the sub frame stay mounting bolts (A), nuts (B) and the sub frame mounting bolts (C) by supporting it with a jack

4. Remove the nut and stabilizer bar control link. Refer to Control Links Removal & Installation.

5. Remove the split pin and front lower arm mounting bolt from the knuckle.

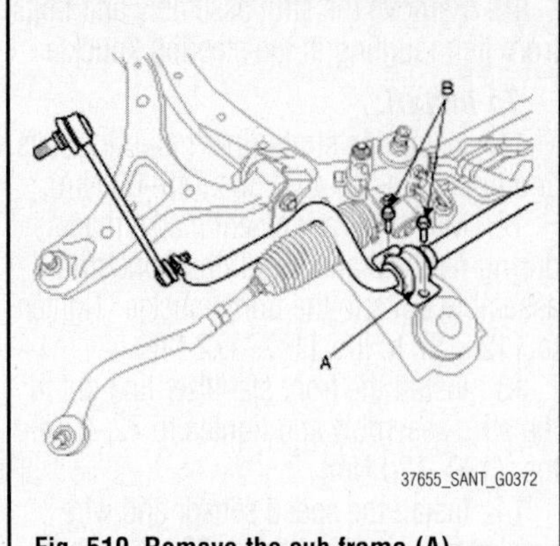

Fig. 519 Remove the sub frame (A)

6. Remove the undercover.

7. Remove the connecting bolt between the steering universal joint assembly and the steering gear.

➡Keep the neutral-range to prevent the damage of the clock spring inner cable when you handle the steering wheel.

8. Disconnect the EPS connector.

9. Remove the front muffler.

10. Remove the front roll stopper mounting bolts.

11. Remove the rear roll stopper mounting bolts.

12. Remove the sub frame stay mounting bolts, nuts and the sub frame mounting bolts by supporting it with a jack.

13. Remove the sub frame.

To install:

14. Installation is the reverse of the removal, noting the following:

a. Tighten the stabilizer bracket bolts to 36–47 ft. lbs. (49–64 Nm).

b. Tighten the sub-frame mounting bolts (A) and nuts (B) to 51–65 ft. lbs. (69–88 Nm).

c. Tighten the sub-frame mounting bolts (C) to 101–116 ft. lbs. (137–157 Nm).

d. Tighten the roll stopper bolts (A) to 36–47 ft. lbs. (49–64 Nm).

e. Tighten the roll stopper bolts (B) to 58–73 ft. lbs. (79–98 Nm).

15. Tighten the stabilizer bar control link upper and lower mounting nuts to 72–87 ft. lbs. (100–120 Nm).

16. Tighten the connecting bolt between the steering universal joint assembly and the steering gear to 22–33 ft. lbs. (29–44 Nm).

17. Tighten the wheel nuts to: 65–80 ft. lbs. (90–110 Nm).

WHEEL HUB & BEARING

REMOVAL & INSTALLATION

See Figures 520 through 524.

1. Before servicing the vehicle, refer to the Precautions Section.

2. Remove the brake disc from the knuckle assembly.

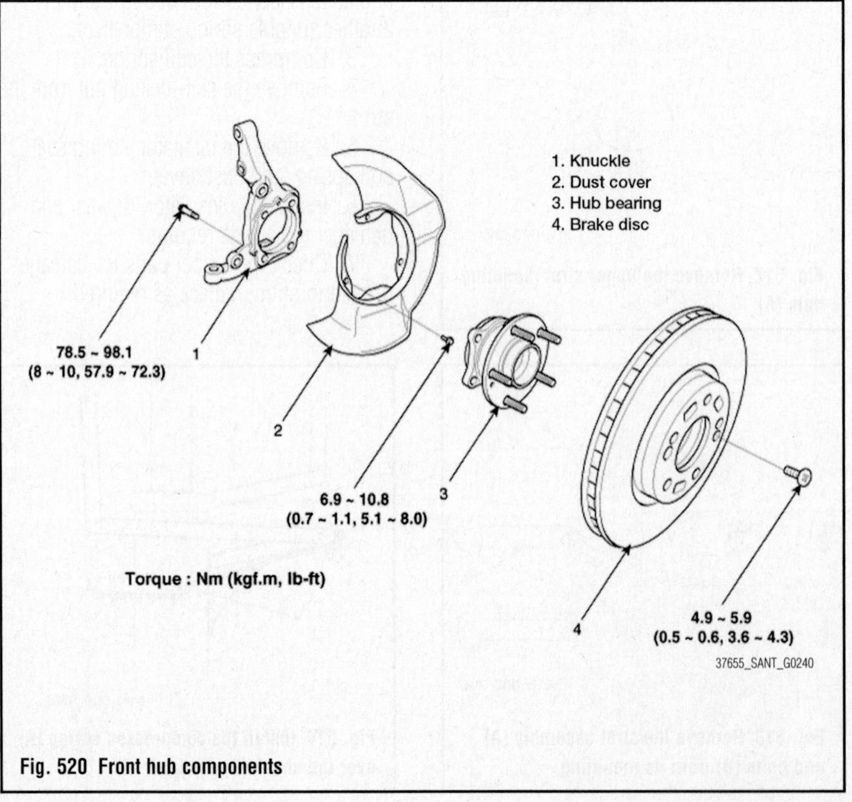

Fig. 520 Front hub components

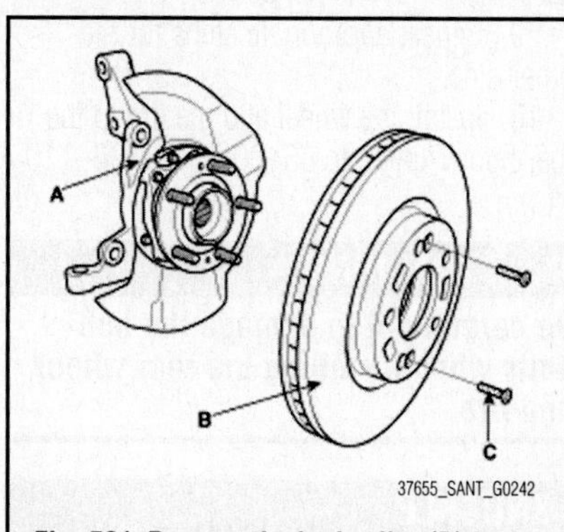

Fig. 521 Remove the brake disc (B) from the knuckle assembly (A)

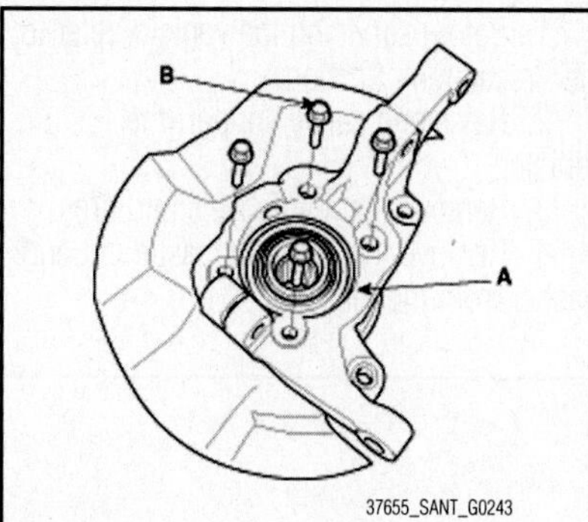

Fig. 522 Remove the hub bearing mounting bolts (B) from the knuckle (A)

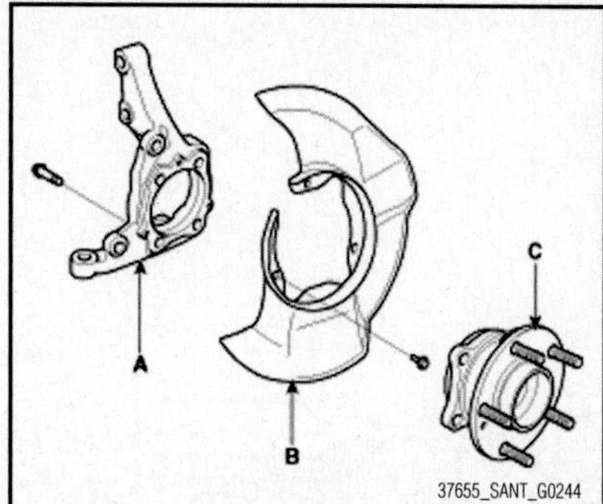
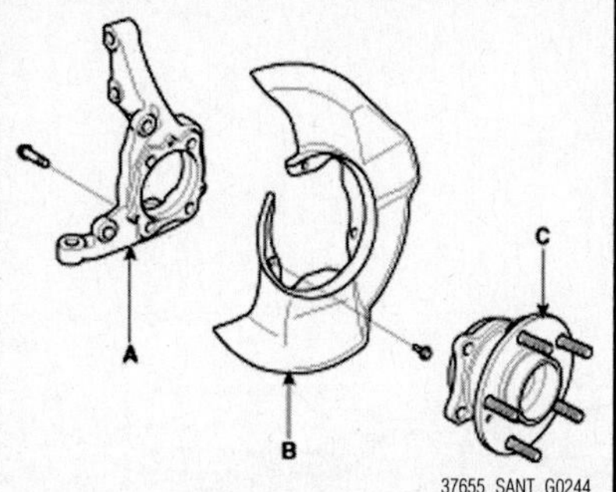

Fig. 523 Remove the hub bearing (C) and the dust cover (B) from the knuckle (A)

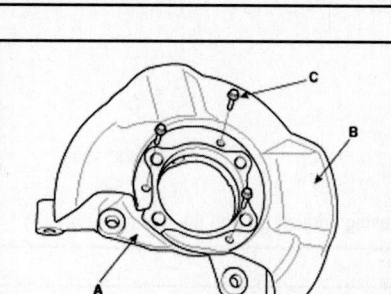

Fig. 524 Install the dust cover (B) to the knuckle (A) and tighten the mounting bolt (C)

3. Remove the hub bearing mounting bolts from the knuckle.
4. Remove the hub bearing and the dust cover from the knuckle.

➡Do not disassemble the hub bearing.

To install:

5. Check the hub for cracks and the splines for wear.
6. Check the brake disc for scoring and damage.
7. Check the knuckle for cracks.
8. Check the bearing for cracks or damage.
9. Install the dust cover to the knuckle and tighten the mounting bolt to 61–96 inch lbs. (7–11 Nm).
10. Install the hub bearing to the knuckle and then tighten the mounting bolt to 58–72 ft. lbs. (79–98 Nm).
11. Install the brake disc to the knuckle assembly and tighten the screws to 43–52 inch lbs. (5–6 Nm).

ADJUSTMENT

The wheel bearings are sealed units and are not adjustable.

SUSPENSION — REAR SUSPENSION

ASSIST ARM

REMOVAL & INSTALLATION

See Figures 525 through 527.

1. Before servicing the vehicle, refer to the Precautions Section.
2. Raise and safely support the vehicle.
3. Remove the rear wheels and tires.

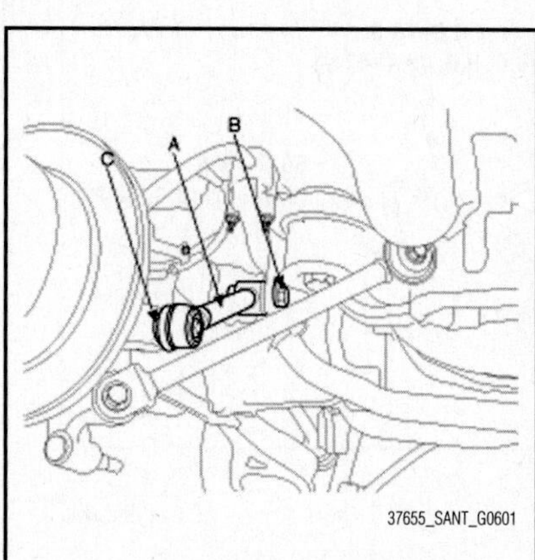

Fig. 525 Assist arm (A) to rear carrier

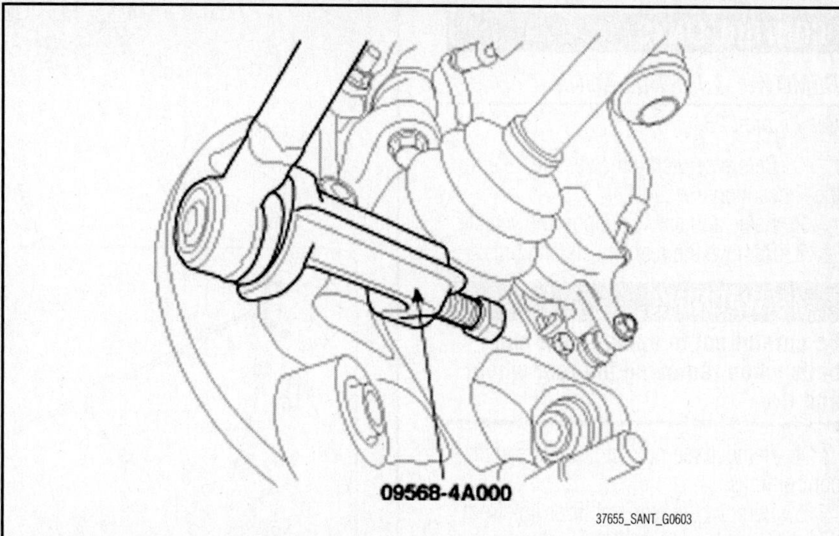

Fig. 526 Remove the rear assist arm ball joint (A) by using the special tool (09568-4A000)

⁂ WARNING

Be careful not to damage the hub bolts when removing the rear wheel and tire.

4. Remove the assist arm from the rear carrier.
5. Remove the rear assist arm ball joint by using the special tool (09568-4A000).

To install:

6. Install the assist arm to the rear carrier. Tighten the bolt to 101–116 ft. lbs. (137–157 Nm). Tighten the nut to 98–118 ft. lbs. (72–87 Nm).

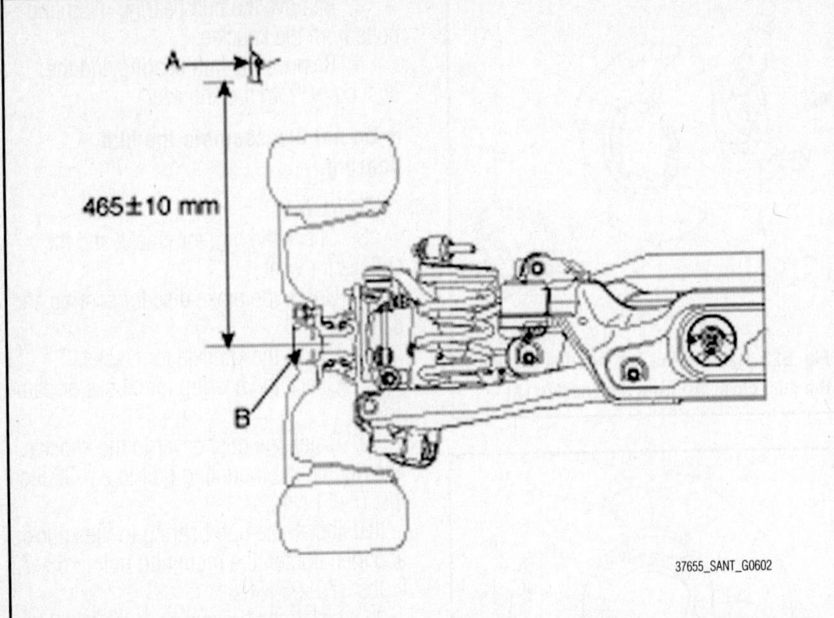

Fig. 527 Check the distance between the wheel housing garnish (A) and the hub assembly (B)

➡After checking the distance [(18.31 in., plus or minus 0.39 in. (465mm, plus or minus 10mm)] between the wheel housing garnish and the hub assembly as shown in the illustration, tighten the mounting bolts and nuts of rear chassis part to the specified torque.

7. Install the rear wheels and tires.

CONTROL LINKS

REMOVAL & INSTALLATION

See Figure 528.

1. Before servicing the vehicle, refer to the Precautions Section.
2. Raise and safely support the vehicle.
3. Remove the rear wheels and tires.

⁂ WARNING

Be careful not to damage the hub bolts when removing the rear wheel and tire.

4. Remove the nuts of the rear stabilizer control links.
5. Remove the rear stabilizer link from the stabilizer bar assembly.
6. Remove the control links from the vehicle.

To install:

7. Install the rear stabilizer control link and nut to the stabilizer bar assembly. Tighten to 36–47 ft. lbs. (50–65 Nm).
8. Install the stabilizer control link nut to the trailing arm. Tighten to 36–47 ft. lbs. (50–65 Nm).

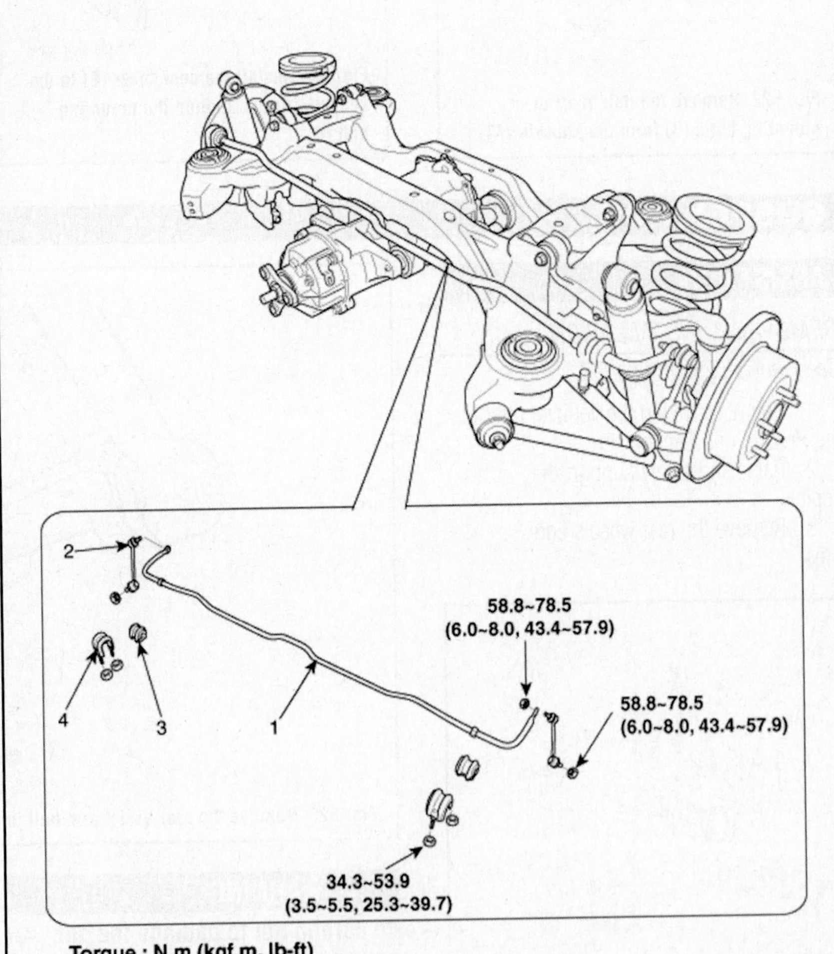

Fig. 528 Rear suspension system with stabilizer bar and control links shown

9. Repeat appropriate steps for the other side.
10. Install the wheel and the tire to the rear hub. Tighten to 65–80 ft. lbs. (90–110 Nm).

⁂ WARNING

Be careful not to damage the hub bolts when installing the rear wheel and tire.

KNUCKLE

REMOVAL & INSTALLATION

See Figures 529 through 544.

1. Before servicing the vehicle, refer to the Precautions Section.
2. Raise and safely support the vehicle.
3. Remove the rear wheels and tires.
4. Remove the split pin, castle nut and washer from the rear hub.

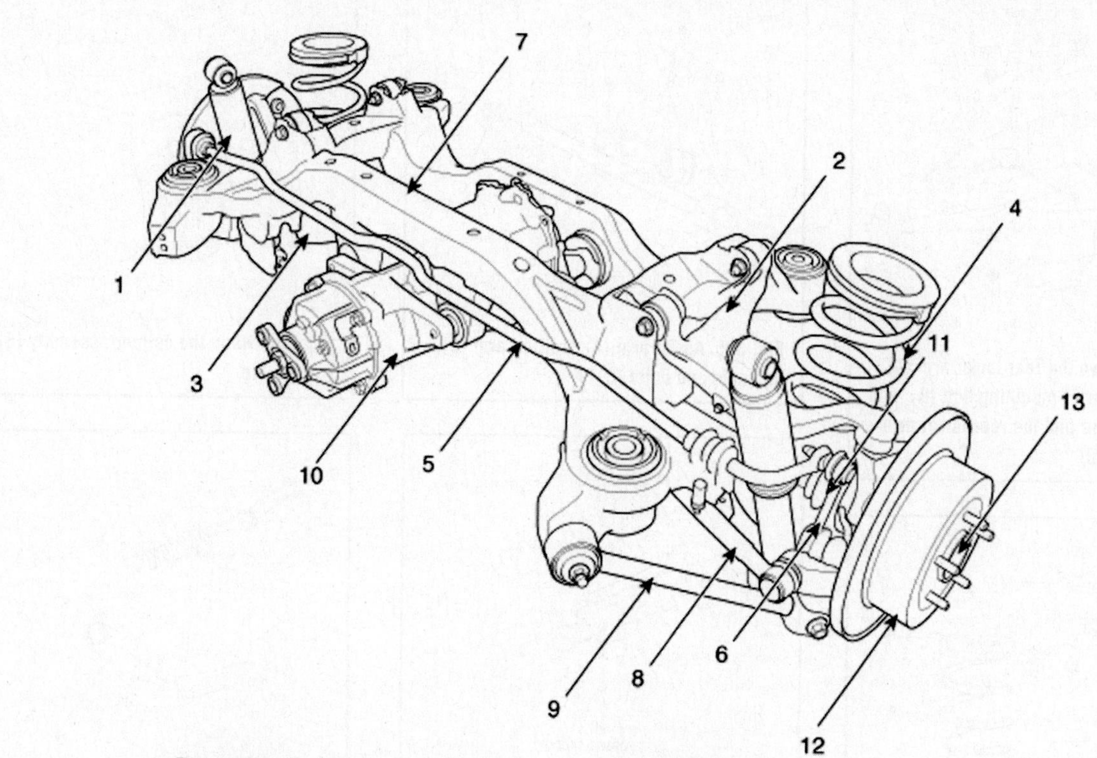

Fig. 529 Rear halfshaft components

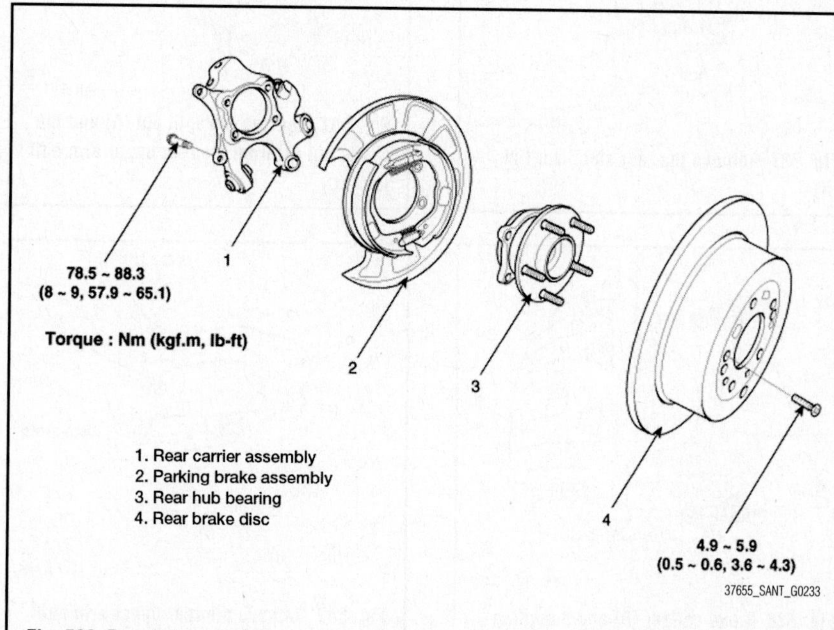

Fig. 530 Rear hub assembly

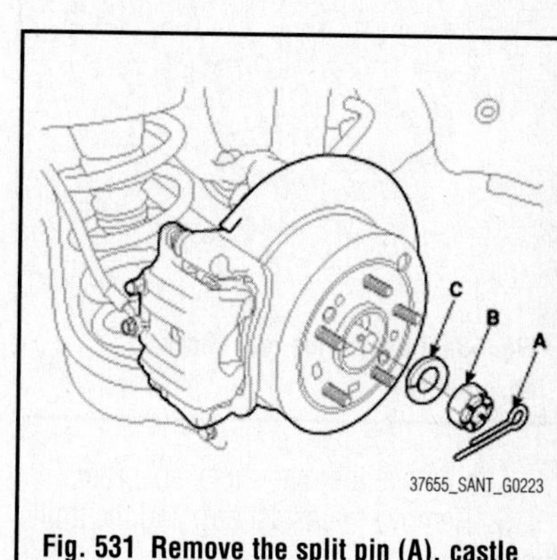

Fig. 531 Remove the split pin (A), castle nut (B) and washer (C)

5. Remove the mounting bolt of the rear lower arm and the rear carrier, while supporting the lower arm with a jack as shown. Remove the mounting bolt of the cross-member and the rear lower arm.
6. Remove the spring, the upper pad and the lower pad.

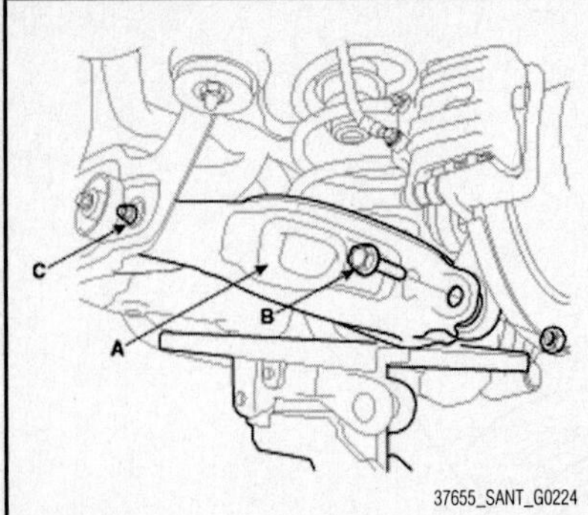

Fig. 532 Remove the rear lower arm (A) and the rear carrier mounting bolt (B), and the crossmember and the rear lower arm mounting bolt (C)

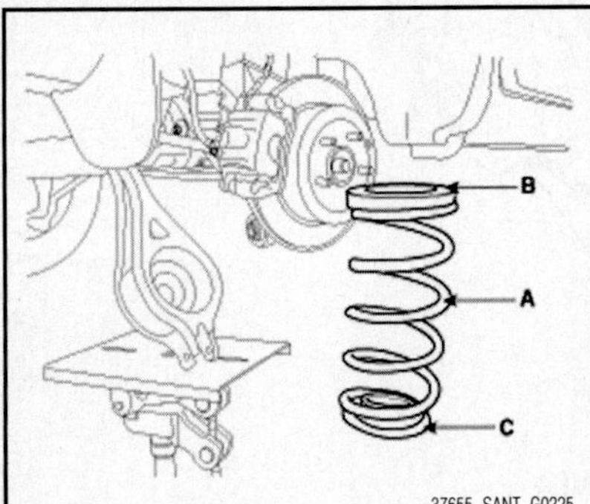

Fig. 533 Remove the spring (A), the upper pad (B) and the lower pad (C)

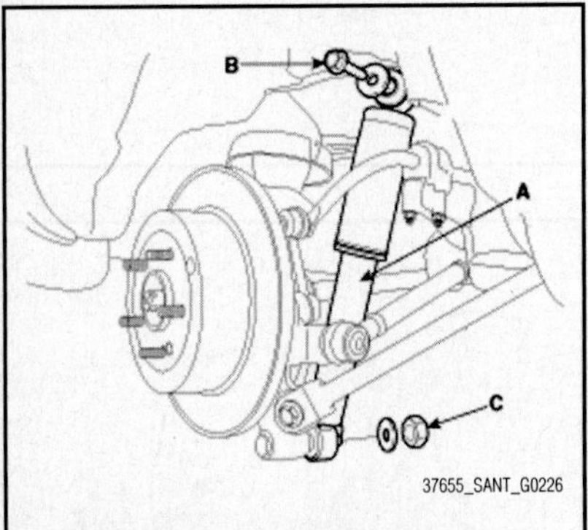

Fig. 534 Remove the rear shock absorber (A)

7. Remove the rear shock absorber.

8. Remove the assist arm and the trailing arm from the rear axle carrier.

➡Remove the rear assist arm ball joint by using the SST (09568-4A000).

9. Remove the rear stabilizer link from the rear axle carrier.

10. Remove the brake caliper mounting bolts, and then hang the brake caliper assembly out of the way with a wire.

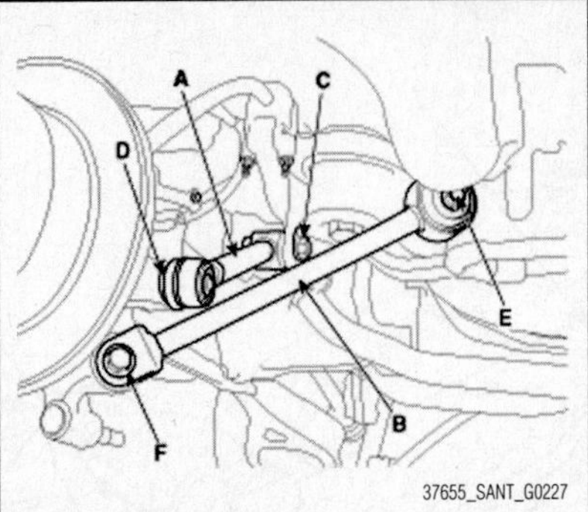

Fig. 535 Assist arm (A), trailing arm (B), and nuts and bolts (C–F)

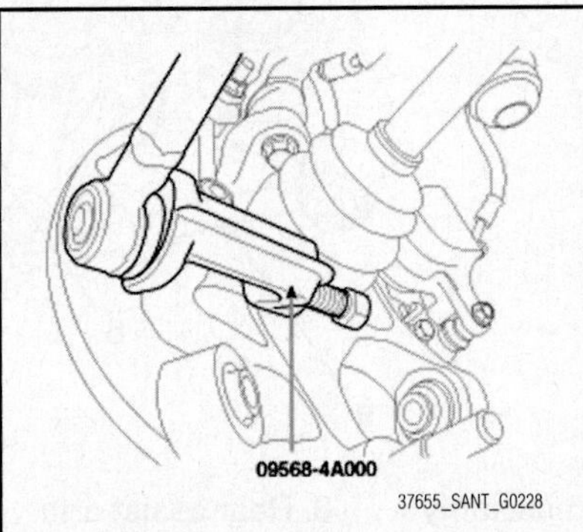

Fig. 536 Remove the rear assist arm ball joint by using the SST (09568-4A000)

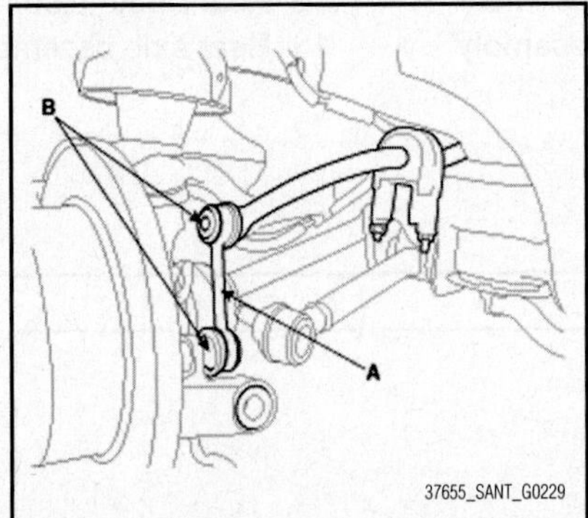

Fig. 537 Remove the rear stabilizer link (A)

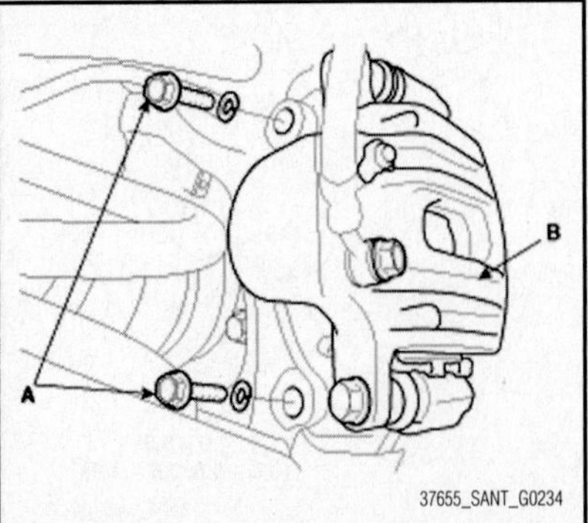

Fig. 538 Brake caliper (B) and mounting bolts (A)

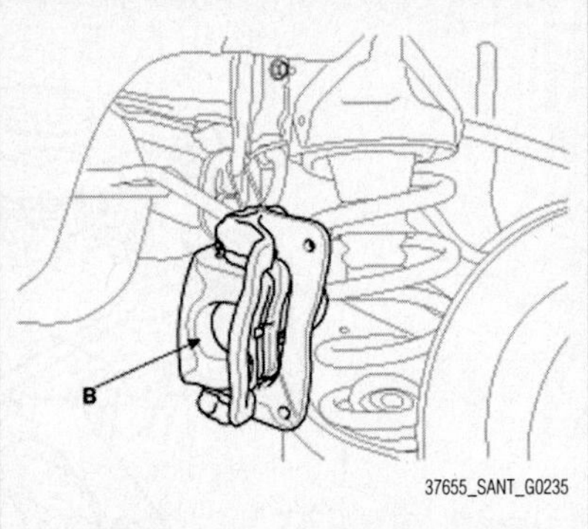

Fig. 539 Hang the caliper assembly (B) with a wire

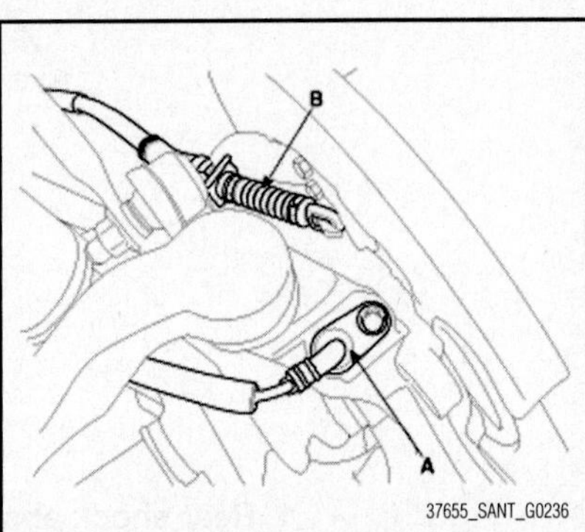

Fig. 540 Remove the wheel speed sensor (A) and the parking brake cable (B)

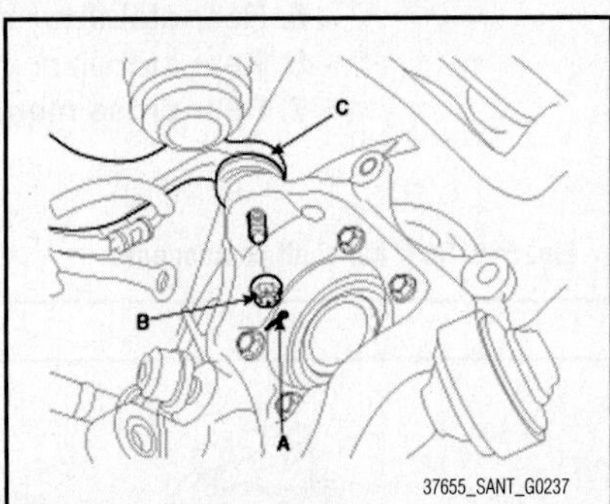

Fig. 541 Remove the split pin (A) and the castle nut (B) from the rear upper arm ball joint (C)

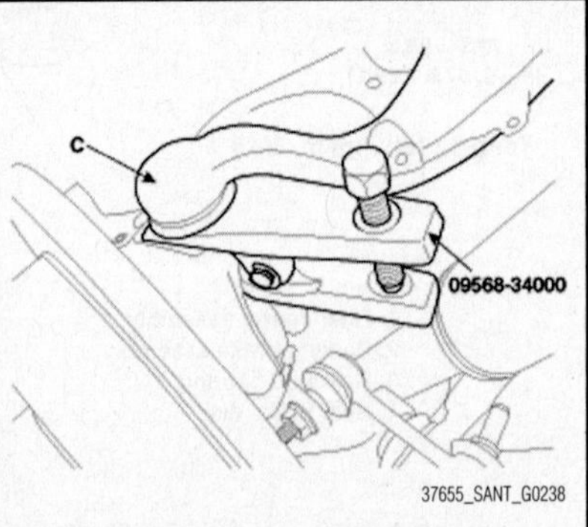

Fig. 542 Remove the rear upper arm ball joint (C), using the SST (09568-34000)

11. Remove the wheel speed sensor and the parking brake cable from the rear axle carrier.

12. Remove the split pin and the castle nut from the rear upper arm ball joint, and then remove the rear upper arm ball joint, using the SST (09568-34000).

13. Remove the rear axle carrier assembly.

To install:

14. Install the rear axle carrier assembly.

15. Install the split pin and the castle nut to the rear upper arm ball joint.

16. Install the wheel speed sensor and the parking brake cable to the rear axle carrier and tighten to 61–96 inch lbs. (7–11 Nm).

17. Install the brake caliper, then tighten the brake caliper mounting bolts to 47–54 ft. lbs. (64–74 Nm).

18. Install the rear stabilizer link to the rear axle carrier and tighten the nut to 43–58 ft. lbs. (59–79 Nm).

19. Install the assist arm and the trailing arm to the rear axle carrier. Tightening torque:

- Bolt (C): 101–116 ft. lbs. (137–157 Nm)
- Nut (D): 72–87 ft. lbs. (98–118 Nm)
- Nut (E): 101–116 ft. lbs. (137–157 Nm)
- Bolt (F): 101–116 ft. lbs. (137–157 Nm)

➡Before tightening the chassis components, measure the distance between the wheel housing molding and the hub assembly, as shown. Specified distance: 18.31 inches, plus or minus 0.39 in. (465mm, plus or minus 10 mm)

20. Install the rear shock absorber. Tightening torque:

- Bolt (B): 101–116 ft. lbs. (137–157 Nm)
- Nut (C): 72–87 ft. lbs. (98–118 Nm)

21. Install the spring, the upper pad, and the lower pad.

22. Install the mounting bolt of the rear lower arm and the rear carrier and tighten to 101–116 ft. lbs. (137–157 Nm), while supporting the lower arm with a jack. Tighten the mounting bolt of the cross member and the rear lower arm to 101–116 ft. lbs. (137–157 Nm).

23. Install the washer, castle nut and split pin to the rear hub assembly. Tightening torque: 145–188 ft. lbs. (196–255 Nm).

24. The washer should be assembled with convex surface outward when installing the castle nut and split pin.

25. Install the wheel and tire to the rear hub. Tightening torque: 65–80 ft. lbs. (88–108 Nm).

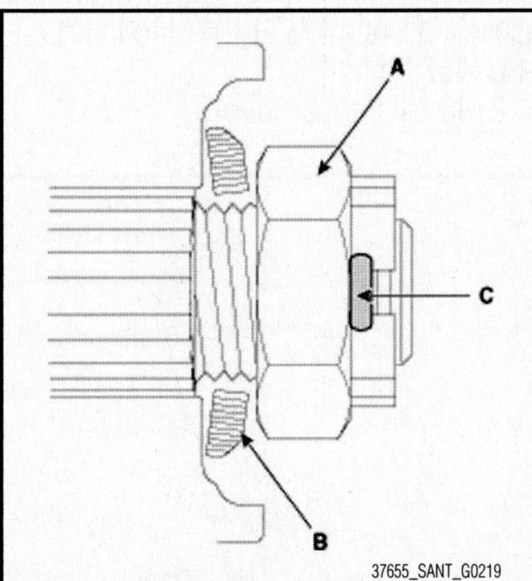

Fig. 544 The washer (B) should be assembled with convex surface outward when installing the castle nut (A) and split pin (C)

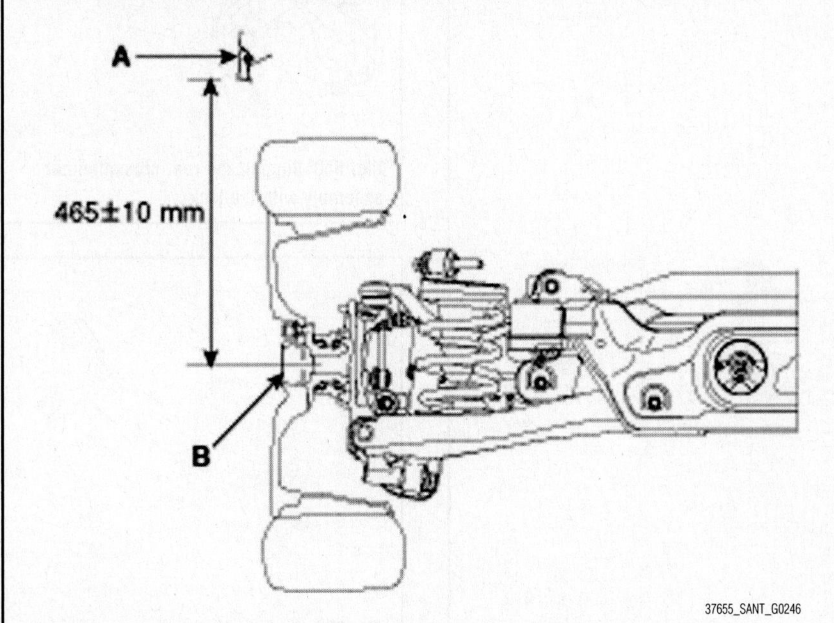

Fig. 543 Distance between the wheel housing molding (A) and hub assembly (B)

LOWER ARM

REMOVAL & INSTALLATION

See Figures 545 through 547.

1. Before servicing the vehicle, refer to the Precautions Section.

2. Raise and safely support the vehicle.

3. Remove the rear wheels and tires.

⁂ WARNING

Be careful not to damage the hub bolts when removing the rear wheel and tire.

4. Remove the mounting bolt of the rear lower arm and the rear carrier, while supporting the lower arm with a jack. Loosen the mounting bolt of the cross member and the rear lower arm.

5. Remove the spring, the upper pad and the lower pad.

6. Remove the lower arm.

To install:

7. Pre-tighten the mounting bolt of the cross member and the rear lower arm.

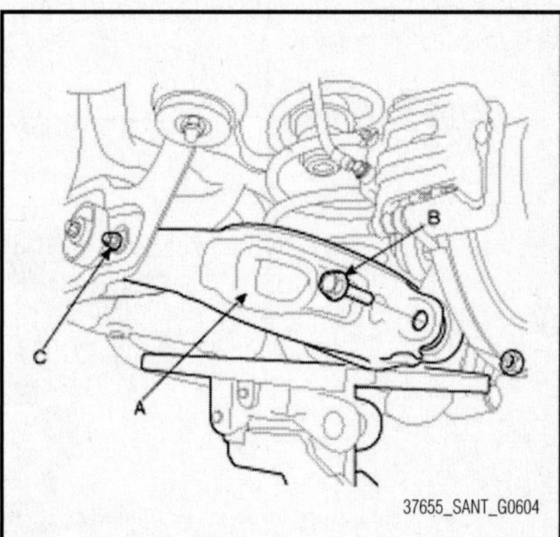

Fig. 545 Remove the mounting bolt (B) of the rear lower arm (A) and the rear carrier, while supporting the lower arm with a jack. Loosen the mounting bolt (C) of the cross member and the rear lower arm

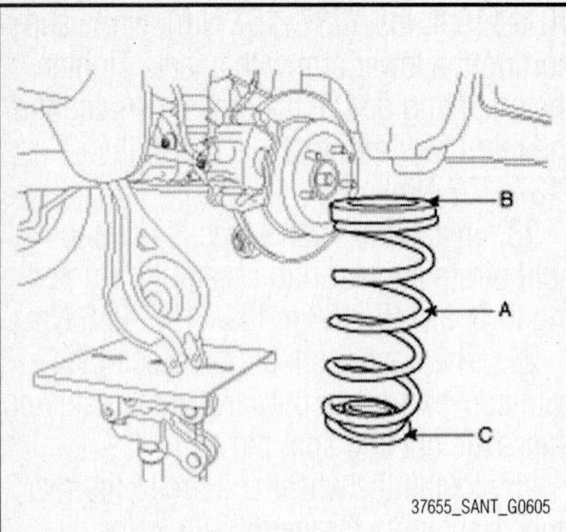

Fig. 546 Remove the spring (A), the upper pad (B) and the lower pad (C)

8. Install the spring, the upper pad and the lower pad.
9. Install the mounting bolt of the rear lower arm and the rear carrier with a specified torque, while supporting the lower arm with a jack as shown in the illustration. Tighten the mounting bolt and nuts of the cross member and the rear lower arm to 101–116 ft. lbs. (137–157 Nm).

➡After checking the distance [(18.31 in., plus or minus 0.39 in. (465mm, plus or minus 10mm)] between the wheel housing garnish and the hub assembly as shown in the illustration, tighten the mounting bolts and nuts of rear chassis part to the specified torque.

10. Install wheel and tire.

37655_SANT_G0602

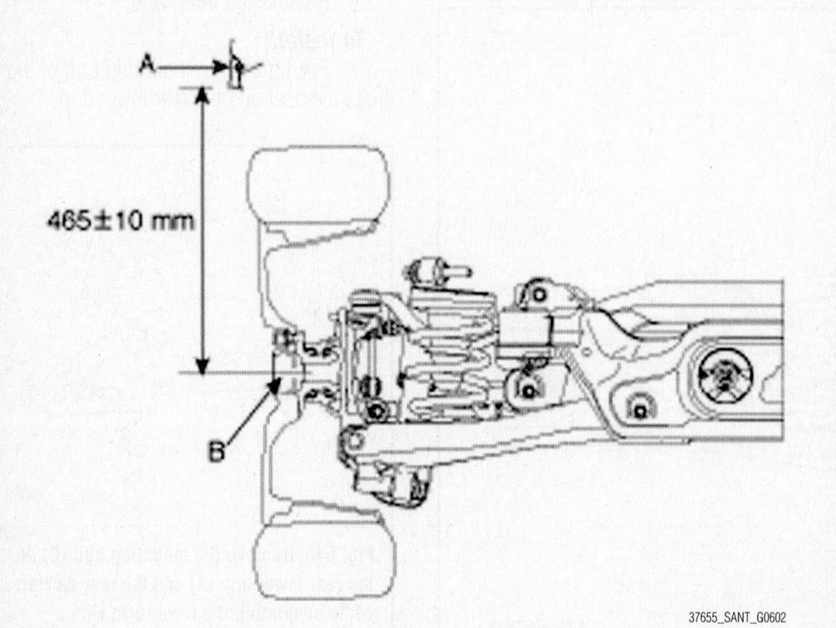

Fig. 547 Check the distance between the wheel housing garnish (A) and the hub assembly (B)

SHOCK ABSORBER

REMOVAL & INSTALLATION

See Figure 548.

1. Before servicing the vehicle, refer to the Precautions Section.
2. Support the vehicle under the lower control arm.
3. Remove the rear wheel.
4. Remove the upper shock absorber upper nut.
5. Remove the lower shock absorber nut.
6. Remove the shock absorber.

To install:

7. Install the shock absorber. Tighten the upper and lower nuts to 101–116 ft. lbs. (137–157 Nm).
8. Install the rear wheel.

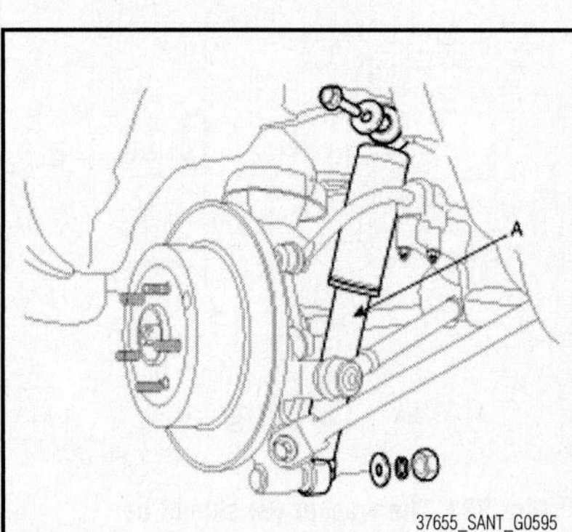

Fig. 548 Rear shock absorber removal and installation

STABILIZER BAR

REMOVAL & INSTALLATION

See Figures 549 through 557.

1. Before servicing the vehicle, refer to the Precautions Section.
2. Remove the rear wheel.
3. Support the vehicle under the lower control arm.
4. Remove the rear shock absorber.
5. Remove the brake caliper mounting bolts, and then support the brake caliper assembly with wire.

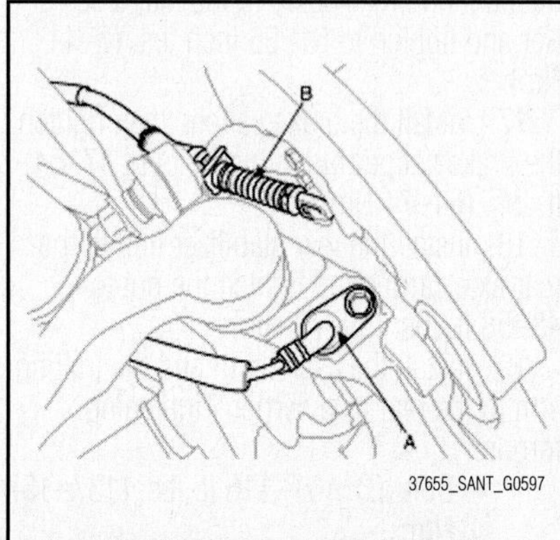

Fig. 549 Remove the wheel speed (A) sensor and the parking brake cable (B)

Fig. 550 Support the rear crossmember assembly with the jack

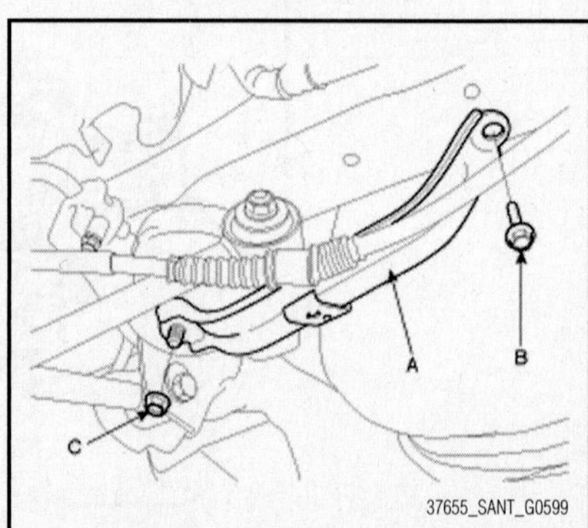

Fig. 551 Remove the rear cross member stay (A) and mounting bolt (B) and nut (C)

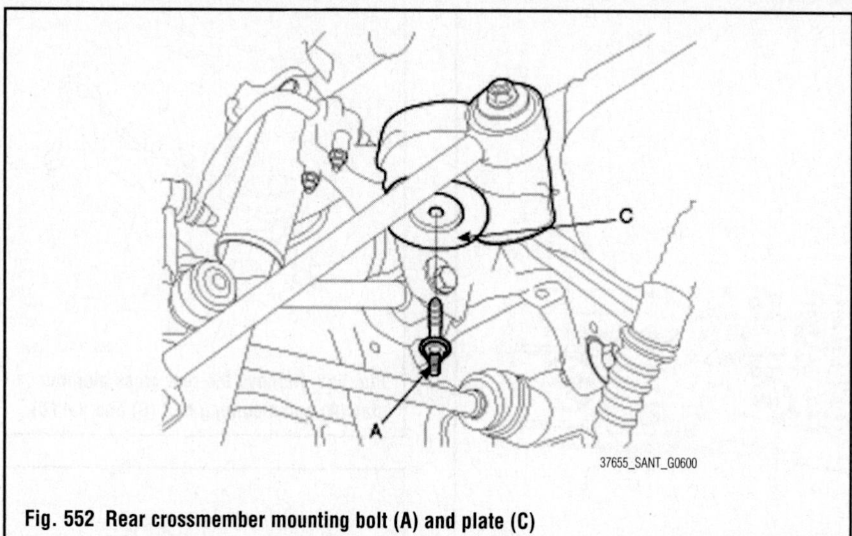

Fig. 552 Rear crossmember mounting bolt (A) and plate (C)

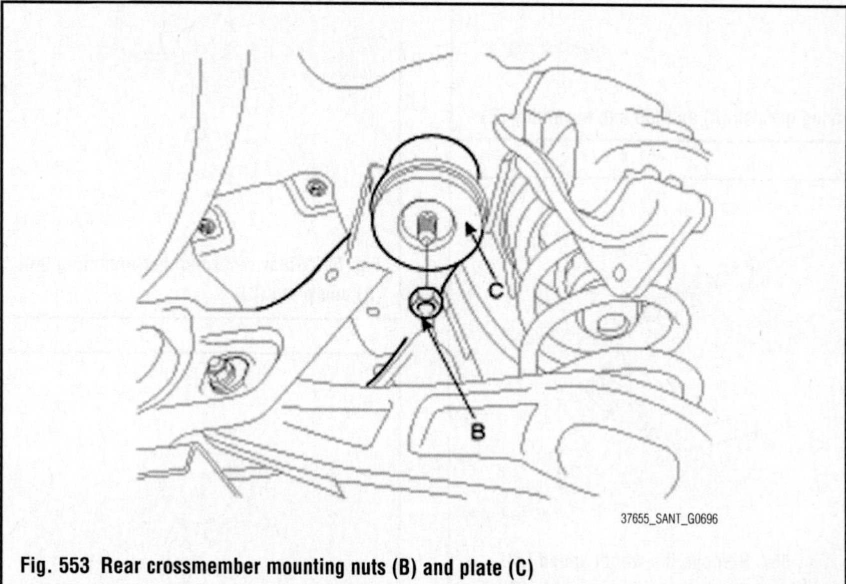

Fig. 553 Rear crossmember mounting nuts (B) and plate (C)

6. Remove the wheel speed sensor and the parking brake cable from the rear axle carrier.

7. Remove the propeller shaft. Refer to Propeller Shaft Removal & Installation in Drive Train.

8. Support the rear crossmember assembly with the jack.

9. Remove the rear cross member stay and mounting bolt and nut.

10. Remove the rear crossmember mounting bolts, nuts and plate.

11. Remove the rear cross member.

12. Remove the rear stabilizer bar link and bracket.

13. Remove the rear stabilizer bar.

To install:

14. Install the bushing on the stabilizer bar.

15. Bring clamp of stabilizer bar into contact with bushing.

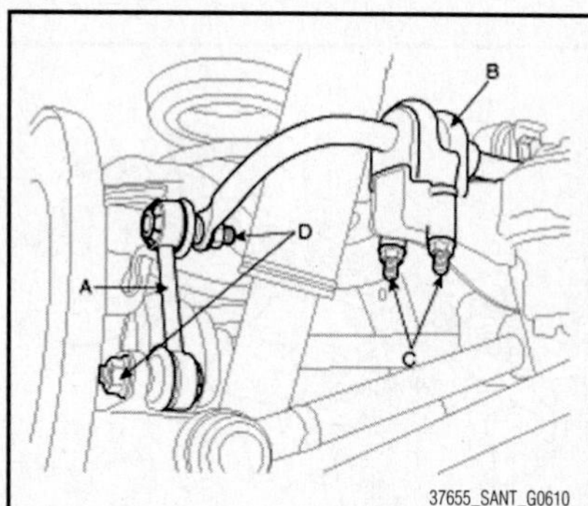

Fig. 554 Stabilizer link (A), bracket (B), bracket nuts (C) and link nut(D)

16. Install the bracket on the bushing.

17. Install the rear stabilizer bar to the rear cross member.

18. Install the rear stabilizer bar bracket and link:

- Tighten nuts (C) to 25–32 ft. lbs. (34–44 Nm)
- Tighten nuts (D) to 43–58 ft. lbs. (59–79 Nm)

19. Support the rear cross member assembly with the jack.

20. Install the rear cross member mounting bolts, nuts and plate. Tighten the bolts and nuts to 116–130 ft. lbs. (157–176 Nm).

21. Install the rear cross member stay. Tighten the bolts and nuts to 51–65 ft. lbs. (69–83 Nm).

22. After matching a matchmark on the rubber coupling and rear differential companion, Install the propeller shaft mounting bolts and tighten to 36–51 ft. lbs. (49–69 Nm).

23. Install the wheel speed sensor and the parking brake cable to the rear axle carrier. Tighten the bolt to 60–96 inch lbs. (6.9–10.8 Nm).

24. Install the brake caliper, and tighten the brake caliper mounting bolts to 47–54 ft. lbs. (64–74 Nm).

25. Support the rear lower arm.

➡After checking the distance [(18.31 in., plus or minus 0.39 in. (465mm, plus or minus 10mm)] between the wheel housing garnish and the hub assembly as shown in the illustration, tighten the mounting bolts and nuts of rear chassis part to the specified torque.

26. Install the shock absorber. Tighten the upper nut to 101–116 ft. lbs. (137–157 Nm). Tighten the lower nut to 101–116 ft. lbs. (137–157 Nm).

27. Install the rear wheel.

TRAILING ARM

REMOVAL & INSTALLATION

See Figures 555 and 556.

1. Before servicing the vehicle, refer to the Precautions Section.

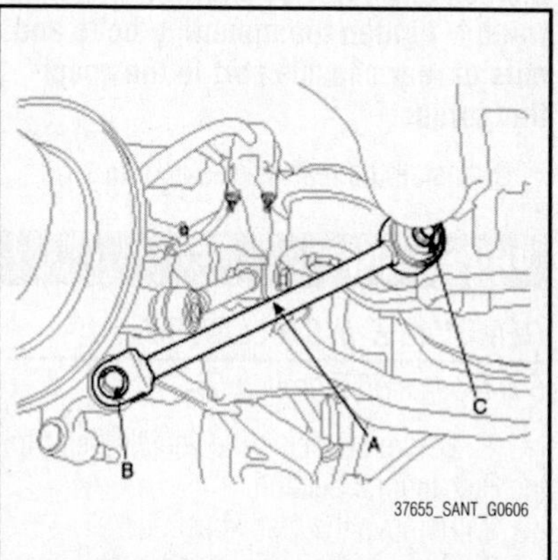

Fig. 555 Trailing arm (A) to rear carrier

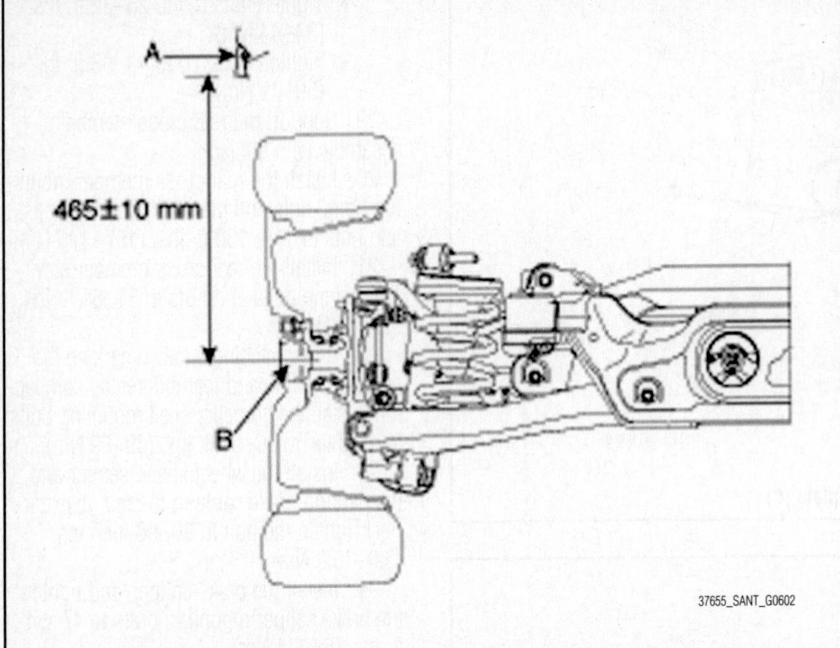

Fig. 556 Check the distance between the wheel housing garnish (A) and the hub assembly (B)

2. Raise and safely support the vehicle.

3. Remove the rear wheels and tires.

WARNING

Be careful not to damage the hub bolts when removing the rear wheel and tire.

4. Remove the trailing arm from the rear carrier.

To install:

5. Install the trailing arm to the rear carrier. Tighten bolt and nut to 101–116 ft. lbs. (137–157 Nm).

➡After checking the distance [(18.31 inches, plus or minus 0.39 in. (465mm, plus or minus 10mm)] between the wheel housing garnish and the hub assembly as shown in the illustration, tighten the mounting bolts and nuts of rear chassis part to the specified torque.

6. Install the rear wheels and tires.

UPPER ARM

REMOVAL & INSTALLATION

See Figures 557 through 564.

1. Before servicing the vehicle, refer to the Precautions Section.

2. Remove the rear wheel.

3. Support the vehicle under the lower control arm.

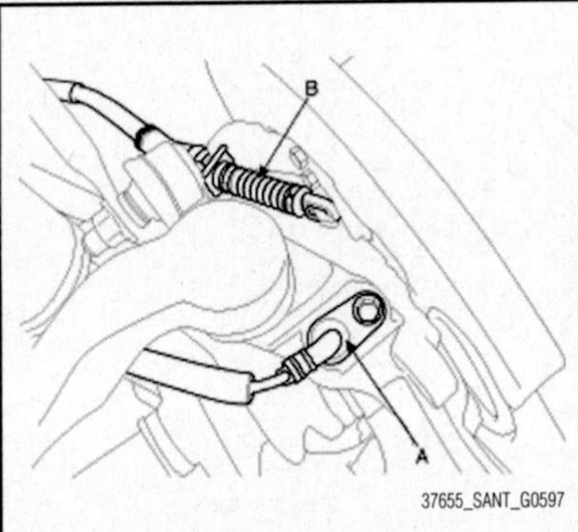

Fig. 557 Remove the wheel speed (A) sensor and the parking brake cable (B)

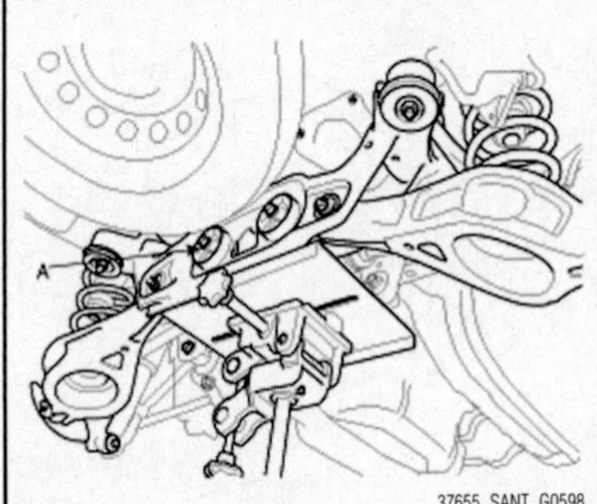

Fig. 558 Support the rear cross member assembly with the jack

4. Remove the rear shock absorber.

5. Remove the brake caliper mounting bolts, and then support the brake caliper assembly with wire.

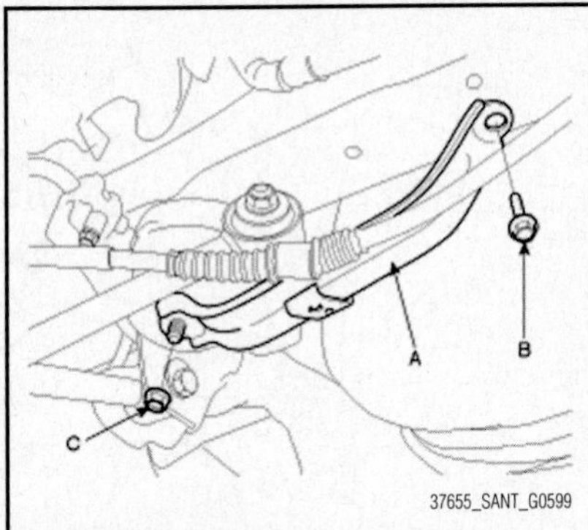

Fig. 559 Remove the rear cross member stay (A) and mounting bolt (B) and nut (C)

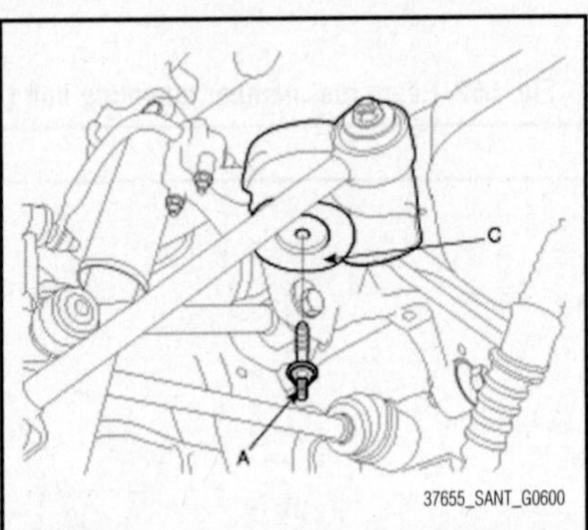

Fig. 560 Rear crossmember mounting bolt (A) and plate (C)

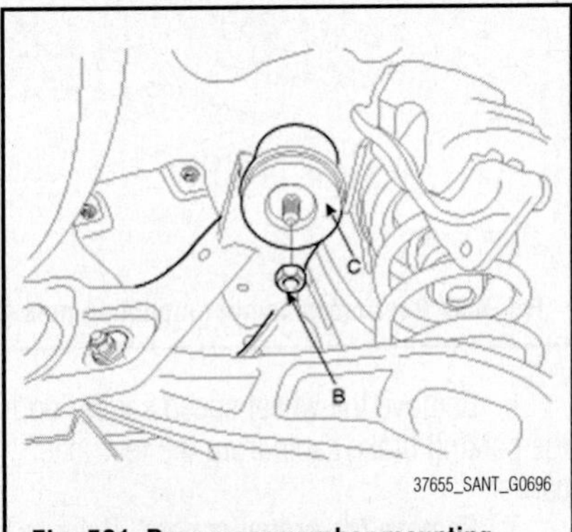

Fig. 561 Rear crossmember mounting nuts (B) and plate (C)

6. Remove the wheel speed sensor and the parking brake cable from the rear axle carrier.

7. Remove the propeller shaft. Refer to Propeller Shaft Removal & Installation in the Drive Train section.

8. Support the rear cross member assembly with the jack.

9. Remove the rear cross member stay and mounting bolt and nut.

10. Remove the rear crossmember mounting bolts, nuts and plate.

11. Remove the rear crossmember.

12. Remove the split pin and the castle nut from the rear upper arm ball joint, and

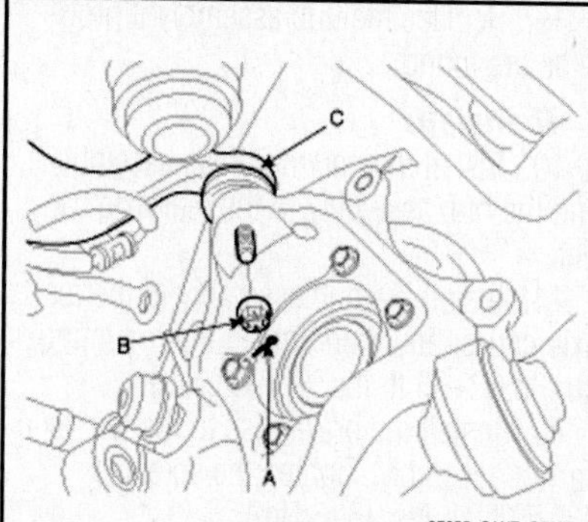

Fig. 562 Remove the split pin (A) and the castle nut (B) from the rear upper arm ball joint (C),

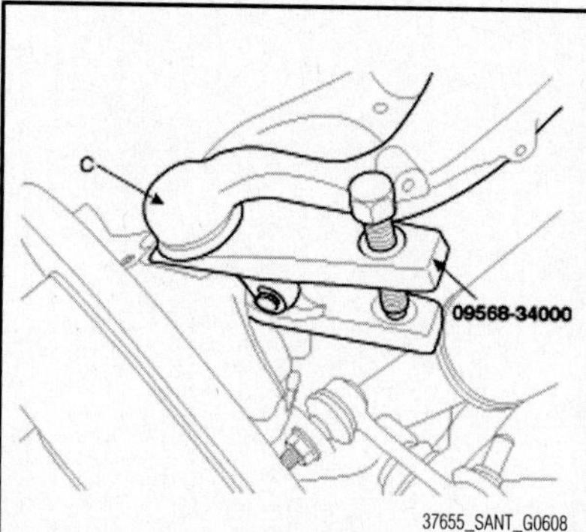

Fig. 563 Remove the rear upper arm ball joint (C) by using the special tool (09568-34000)

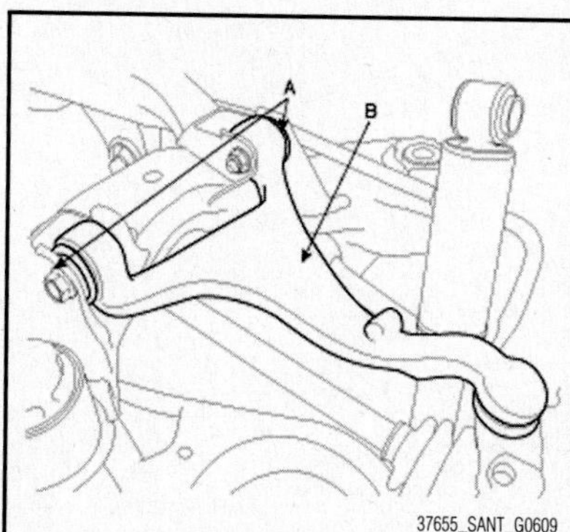

Fig. 564 Remove the mounting bolts (A) of the rear upper arm and the cross member, then remove the rear upper arm (B)

then remove the rear upper arm ball joint by using the special tool (09568-34000).

13. Remove the mounting bolts of the rear upper arm and the cross member, then remove the rear upper arm.

To install:

14. Install the rear upper arm, and mounting bolts to the cross member. Tighten the bolts to 72–87 ft. lbs. (98–118 Nm).

➡After checking the distance [(18.31 in., plus or minus 0.39 in. (465mm, plus or minus 10mm)] between the wheel housing garnish and the hub assembly as shown in the illustration, tighten the mounting bolts and nuts of rear chassis part to the specified torque.

15. Install the split pin and the castle nut to the rear upper arm ball joint. Tighten the nut to 58–72 ft. lbs. (78–98 Nm).

16. Support the rear cross member assembly with the jack.

17. Install the rear cross member mounting bolts, nuts and plate. Tighten the bolts and nuts to 116–130 ft. lbs. 157–176 Nm).

18. Install the rear cross member stay and mounting bolt and nut. Tighten the bolts and nuts to 51–65 ft. lbs. (69–83 Nm).

19. After matching a match mark on the rubber coupling and rear differential companion, Install the propeller shaft mounting bolts. Tighten the bolts to 36–51 ft. lbs. (49–69 Nm).

20. Install the wheel speed sensor and the parking brake cable to the rear axle carrier. Tighten the bolt to 60–96 inch lbs. (7–11 Nm).

21. Install the brake caliper, and tighten the brake caliper mounting bolts to 47–54 ft. lbs. (64–74 Nm).

22. Support the rear lower arm.

23. Install the shock absorber. Tighten the upper nut to 101–116 ft. lbs. (137–157 Nm). Tighten the lower nut to 101–116 ft. lbs. (137–157 Nm).

24. Install the rear wheel.

WHEEL HUB & BEARING

REMOVAL & INSTALLATION

See Figures 565 through 567.

1. Before servicing the vehicle, refer to the Precautions Section.
2. Remove the brake disc rotor from the rear axle carrier assembly.
3. Remove the hub assembly mounting bolts from the rear axle carrier.
4. Remove the hub assembly and the parking brake assembly from the rear axle carrier.

➡Do not disassemble the hub assembly.

5. Check the hub for cracks and the splines for wear.
6. Check the brake disc for scoring and damage.
7. Check the rear axle carrier for cracks
8. Check the bearing for cracks or damage.

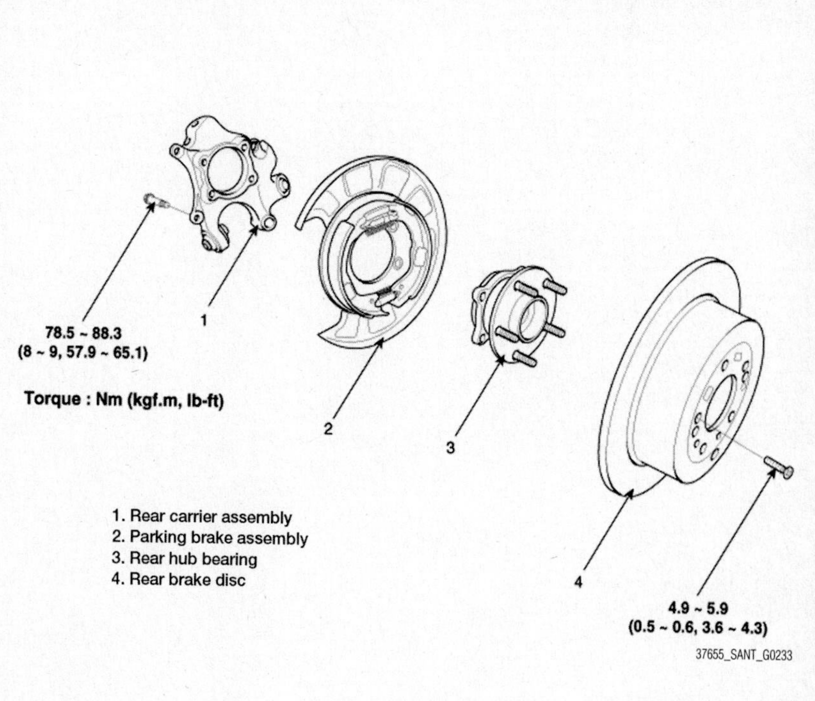

Fig. 565 Rear hub components

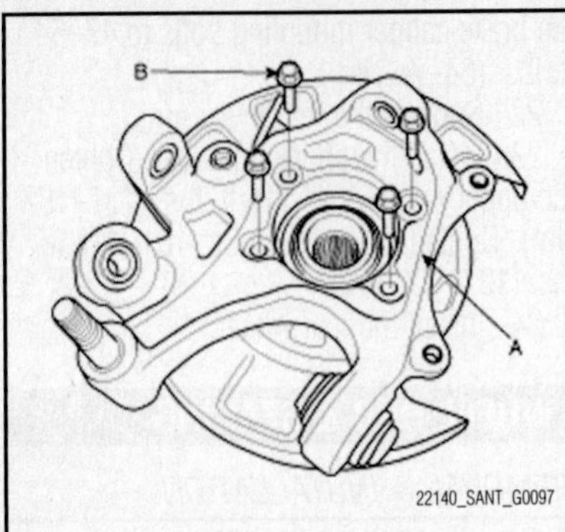

Fig. 566 Remove the hub assembly mounting bolts (B) from the rear axle carrier (A)

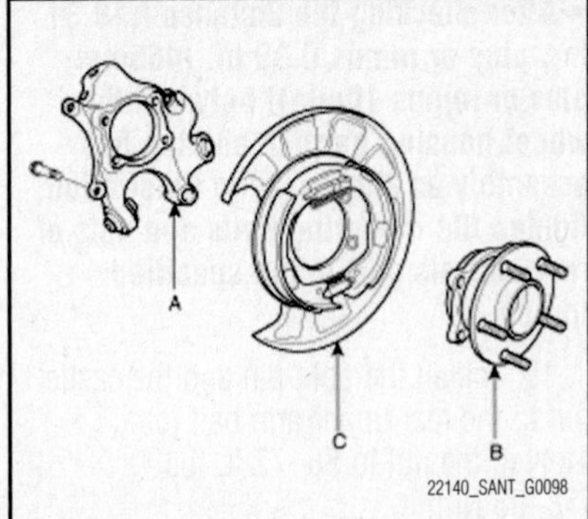

Fig. 567 Remove the hub assembly (B) and the parking brake assembly (C) from the rear axle carrier (A)

9. Replace the hub assembly if problems are found.

To install:

10. Install the parking brake assembly and the hub assembly to the rear axle carrier.

11. Install the hub assembly to the rear axle carrier and then tighten the mounting bolt to 58–65 ft. lbs. (79–88 Nm).

12. Install the brake disc to the rear axle carrier assembly. Tighten the screw to 43–52 inch lbs. (5–6 Nm).

ADJUSTMENT

The wheel bearings are sealed units and are not adjustable.

HYUNDAI

Sonata

7

SPECIFICATIONS AND MAINTENANCE CHARTS

ENGINE AND VEHICLE IDENTIFICATION

Engine							Model Year	
Code ①	Liters (cc)	Cu. In.	Cyl.	Fuel Sys.	Engine Type	Eng. Mfg.	Code ②	Year
C	2.4 (2359)	143.90	I4	MPFI	DOHC	Hyundai	9	2009
F	3.3 (3342)	203.86	V6	MPFI	DOHC	Hyundai	A	2010

MPFI: Multi-Point Fuel Injection

DOHC: Double Overhead Camshafts

① 8th digit of VIN

② 10th digit of VIN

37655_SONA_C0001

GENERAL ENGINE SPECIFICATIONS

All measurements are given in inches.

Year	Model	Engine Displacement Liters	Engine Series VIN	Net Horsepower @ rpm	Net Torque @ rpm (ft. lbs.)	Bore x Stroke (in.)	Com-pression Ratio	Oil Pressure @ rpm
2009	Sonata GLS	2.4	C	175@6000	168@4000	3.464 x 3.819	10.5:1	NA
	Sonata SE/Limited	3.3	F	249@6000	229@4500	3.622 x 3.299	10.4:1	18.8@1000
2010	Sonata GLS	2.4	C	175@6000	168@4000	3.464 x 3.819	10.5:1	NA
	Sonata SE/Limited	3.3	F	249@6000	229@4500	3.622 x 3.299	10.4:1	18.8@1000

37655_SONA_C0002

GASOLINE ENGINE TUNE-UP SPECIFICATIONS

Year	Engine Displacement Liters	Engine VIN	Spark Plug Gap (in.)	Ignition Timing (deg.)		Fuel Pump (psi)	Idle Speed (rpm)		Valve Clearance	
				MT	AT		MT	AT	In.	Ex.
2009	2.4	C	0.039-0.043	①	①	50.0-51.5	②	②	HYD	HYD
	3.3	D/F	0.039-0.043	①	①	54.3-55.8	②	②	HYD	HYD
2010	2.4	C	0.039-0.043	①	①	50.0-51.5	②	②	HYD	HYD
	3.3	D/F	0.039-0.043	①	①	54.3-55.8	②	②	HYD	HYD

Follow the figures on the label if they differ from those in this chart.

HYD: Hydraulic

① Ignition timing is computer controlled and is not adjustable

② Idle speed is maintained by the Electronic Control Module (ECM)

37655_SONA_C0003

CAPACITIES

Year	Model	Engine Displacement Liters	Engine VIN	Engine Oil with Filter (qts.)	Transmission (pts.) Manual	Transmission (pts.) Auto. ①	Fuel Tank (gal.)	Cooling System (qts.)
2009	Sonata	2.4	C	4.2	4.6	16.4	18.5	8.2
	Sonata	3.3	F	6.3	—	11.5	17.7	9.4
2010	Sonata	2.4	C	4.5	3.8	16.4	18.5	7.6
	Sonata	3.3	F	5.5	—	11.5	17.7	9.4

NA: Not Available

NOTE: All capacities are approximate. Add fluid gradually and check to be sure a proper fluid level is obtained.

① Drain and refill

37655_SONA_C0004

FLUID SPECIFICATIONS

Year	Model	Engine Displacement Liters	Engine ID/VIN	Engine Oil	Auto. Trans.	Manual Trans.	Power Steering Fluid	Brake Master Cylinder
2009	Sonata	2.4	C	5W-20	①	②	PSF-3	③
	Sonata	3.3	F	5W-20	①	②	PSF-3	③
2010	Sonata	2.4	C	5W-20	①	②	PSF-3	③
	Sonata	3.3	F	5W-20	①	②	PSF-3	③

DOT: Department Of Transportation

① DIAMOND ATF SP-III, SK ATF SP-III

② GENUINE PART MTF 75W/85 (API GL - 4)

③ DOT 3, DOT 4, or equivalent

37655_SONA_C0005

VALVE SPECIFICATIONS

Year	Engine Displacement Liters	Engine VIN	Seat Angle (deg.)	Face Angle (deg.)	Spring Test Pressure (lbs. @ in.)	Spring Installed Height (in.)	Stem-to-Guide Clearance (in.) Intake	Stem-to-Guide Clearance (in.) Exhaust	Stem Diameter (in.) Intake	Stem Diameter (in.) Exhaust
2009	2.4	C	44.75-45.10	45.25-45.75	85.1-90.4 @1.024	NA	0.0008-0.0019	0.0012-0.0021	0.2151-0.2157	0.2149-0.2153
	3.3	D/F	44.75-45.20	45.25-45.75	90.4-96.2 @0.953	NA	0.0008-0.0019	0.0012-0.0021	0.2151-0.2157	0.2149-0.2153
2010	2.4	C	44.75-45.10	45.25-45.75	85.1-90.4 @1.024	NA	0.0008-0.0019	0.0012-0.0021	0.2151-0.2157	0.2149-0.2153
	3.3	D/F	44.75-45.20	45.25-45.75	90.4-96.2 @0.953	NA	0.0008-0.0019	0.0012-0.0021	0.2151-0.2157	0.2149-0.2153

NA: Not Available

37655_SONA_C0006

CAMSHAFT AND BEARING SPECIFICATIONS CHART

All measurements are given in inches.

Year	Engine Displ. Liters	Engine ID/VIN	Journal Dia.	Brg. Oil Clearance	Shaft End-play	Runout	Journal Bore	Lobe Height Intake	Lobe Height Exhaust
2009	2.4	C	①	②	0.0039-0.0086	NA	NA	1.7244	1.7716
	3.3	D/F	③	④	0.0008-0.0071	NA	NA	1.8228	1.8031
2010	2.4	C	①	②	0.0039-0.0086	NA	NA	1.7244	1.7716
	3.3	D/F	③	④	0.0008-0.0071	NA	NA	1.8228	1.8031

NA: Not Available

① Intake No. 1 is 1.1811 inch
Intake No. 2, 3, 4, 5 are 0.9449 inch
Exhaust No.1 is 1.4173 inch
Exhaust No. 2, 3, 4, 5 are 0.9449 inch

② Intake No. 1 is 0.0008-0.0022 inch
Intake No. 2, 3, 4, 5 are 0.0017-0.0032 inch
Exhaust No. 1 is 0.0000-0.0012 inch
Exhaust No. 2, 3, 4, 5 are 0.0017-0.0032 inch

③ Intake No. 1 is 1.1009-1.1016 inch
Intake No. 2, 3, 4 are 0.9430-0.9437 inch
Exhaust No.1 is 1.1009-1.1016 inch
Exhaust No. 2, 3, 4 are 0.9430-0.9437 inch

④ Intake No. 1 is 0.0011-0.0022 inch
Intake No. 2, 3, 4 are 0.0012-0.0026 inch
Exhaust No.1 is 0.0011-0.0022 inch
Exhaust No. 2, 3, 4 are 0.0012-0.0026 inch

37655_SONA_C0007

CRANKSHAFT AND CONNECTING ROD SPECIFICATIONS

All measurements are given in inches.

Year	Engine Displacement Liters	Engine VIN	Crankshaft Main Brg. Journal Dia.	Crankshaft Main Brg. Oil Clearance	Crankshaft Shaft End-play	Crankshaft Thrust on No.	Connecting Rod Journal Diameter	Connecting Rod Oil Clearance	Connecting Rod Side Clearance
2009	2.4	C	2.0449-2.0456	0.0007-0.0014	0.0027-0.0098	3	1.8879-1.8886	0.0009-0.0016	0.0039-0.0100
	3.3	D/F	2.7142-2.7149	0.0008-0.0016	0.0039-0.0110	3	2.1635-2.1642	0.0015-0.0022	0.0039-0.0098
2010	2.4	C	2.0449-2.0456	0.0010-0.0019	0.0027-0.0098	3	1.8879-1.8886	0.0011-0.0018	0.0039-0.0100
	3.3	D	2.7142-2.7149	0.0008-0.0016	0.0039-0.0110	3	2.1635-2.1642	0.0015-0.0022	0.0039-0.0098

37655_SONA_C0008

PISTON AND RING SPECIFICATIONS

All measurements are given in inches.

Year	Engine Displ. Liters	Engine VIN	Piston Clearance	Ring Gap Top Compression	Ring Gap Bottom Compression	Ring Gap Oil Control	Ring Side Clearance Top Compression	Ring Side Clearance Bottom Compression	Ring Side Clearance Oil Control
2009	2.4	C	0.0005-0.0013	0.0059-0.0118	0.0118-0.0204	0.0078-0.0275	0.0019-0.0031	0.0015-0.0031	0.0023-0.0051
	3.3	D/F	0.0008-0.0016	0.0067-0.0126	0.0126-0.0185	0.0078-0.0275	0.0016-0.0031	0.0012-0.0027	0.0024-0.0059
2010	2.4	C	0.0005-0.0013	0.0059-0.0118	0.0118-0.0204	0.0078-0.0275	0.0019-0.0031	0.0015-0.0031	0.0023-0.0051
	3.3	D/F	0.0008-0.0016	0.0067-0.0126	0.0126-0.0185	0.0078-0.0275	0.0016-0.0031	0.0012-0.0027	0.0024-0.0059

NA: Not Available

37655_SONA_C0009

TORQUE SPECIFICATIONS

All readings in ft. lbs.

Year	Engine Displacement Liters	Engine VIN	Cylinder Head Bolts	Main Bearing Bolts	Rod Bearing Bolts	Crankshaft Damper Bolts	Flywheel Bolts	Manifold Intake	Manifold Exhaust	Spark Plugs	Oil Pan Drain Plug
2009	2.4	C	①	②	③	123-130	87-94	14-17	④	15-22	29-33
	3.3	F	⑤	⑥	③	210-224	53-56	14-17	29-33	15-22	25-33
2010	2.4	C	①	②	③	123-130	87-94	14-17	④	15-22	29-33
	3.3	F	⑤	⑥	③	210-224	53-56	14-17	29-33	15-22	25-33

① Step 1: 25 ft. lbs.
Step 2: Plus 90 degrees
Step 3: Plus 90 degrees

② Step 1: 22 ft. lbs.
Step 2: Plus 120 degrees

③ Step 1: 15 ft. lbs.
Step 2: Plus 90 degrees

④ Exhaust manifold heat protector bolt and stay bolt (M8): 6-9 ft. lbs.
Exhaust manifold nut: 29-33 ft. lbs.
Exhaust manifold stay bolt (M10): 31-40 ft. lbs.

⑤ Step 1: 29 ft. lbs.
Step 2: Plus 120 degrees
Step 3: Plus 90 degrees

⑥ M11 bolts (inner) Step 1: 36 ft. lbs.
M11 bolts (inner) Step 2: Plus 90 degrees
M8 bolts (outer) Step 1: 15 ft. lbs.
M8 bolts (outer) Step 2: Plus 120 degrees
M8 bolts (side): 22-23 ft. lbs.

37655_SONA_C0010

WHEEL ALIGNMENT

Year	Model		Caster Range (+/-Deg.)	Caster Preferred Setting (Deg.)	Camber Range (+/-Deg.)	Camber Preferred Setting (Deg.)	Toe-in (Deg.)
2009	Sonata	Front	0.50	+4.80	0.50	0.00	0.00 +/- 0.20
		Rear	—	—	0.50	-0.50	0.20 +/- 0.20
2010	Sonata	Front	0.50	+4.80	0.50	-0.00	0.00 +/- 0.20
		Rear	—	—	0.50	-0.50	0.20 +/- 0.20

37655_SONA_C0011

TIRE, WHEEL AND BALL JOINT SPECIFICATIONS

Year	Model	OEM Tires Standard	OEM Tires Optional	Tire Pressures (psi) Front	Tire Pressures (psi) Rear	Wheel Size	Ball Joint Inspection	Lug Nut Torque (ft. lbs.)
2009	Sonata	P215/60R16	P215/55R17	32	32	6.5J×16 6.5J×17	①	65-80
2010	Sonata	P215/60R16	P215/55R17	32	32	6.5J×16 6.5J×17	①	65-80

① Replace if any measurable movement is found.

37655_SONA_C0012

BRAKE SPECIFICATIONS

All measurements in inches unless noted

Year	Model		Brake Disc Original Thickness	Brake Disc Minimum Thickness	Brake Disc Maximum Runout	Brake Drum Diameter Original Inside Diameter	Brake Drum Diameter Max. Wear Limit	Brake Drum Diameter Maximum Machine Diameter	Minimum Lining Thickness	Brake Caliper Bracket Bolts (ft. lbs.)	Brake Caliper Mounting Bolts (ft. lbs.)
2009	Sonata (2.4L)	F	1.024	0.961	0.002	—	—	—	0.120-0.160	59-74	16-23
		R	0.390	0.330	0.002	—	—	—	0.120	59-74	16-23
	Sonata (3.3L)	F	1.100	1.040	0.002	—	—	—	0.120-0.160	59-74	16-23
		R	0.390	0.330	0.002	—	—	—	0.120	59-74	16-23
2010	Sonata (2.4L)	F	1.024	0.961	0.002	—	—	—	0.120-0.160	59-74	16-23
		R	0.390	0.330	0.002	—	—	—	0.120	59-74	16-23
	Sonata (3.3L)	F	1.100	1.040	0.002	—	—	—	0.120-0.160	59-74	16-23
		R	0.390	0.330	0.002	—	—	—	0.120	59-74	16-23

37655_SONA_C0013

SCHEDULED MAINTENANCE INTERVALS

HYUNDAI—Sonata

TO BE SERVICED	TYPE OF SERVICE	VEHICLE MILEAGE INTERVAL (x1000)												
		7.5	15	22.5	30	37.5	45	52.5	60	67.5	75	82.5	90	97.5
Engine oil & filter	R	✓	✓	✓	✓	✓	✓	✓	✓	✓	✓	✓	✓	✓
Brake pads, calipers & rotors	S/I	✓	✓	✓	✓	✓	✓	✓	✓	✓	✓	✓	✓	✓
Driveshafts & boots	S/I		✓		✓		✓		✓		✓		✓	
Air cleaner filter ①	S/I	✓	✓	✓	✓	✓	✓	✓	✓	✓	✓	✓	✓	✓
A/C refrigerant	S/I		✓		✓		✓		✓		✓		✓	
Brake fluid	S/I				✓								✓	
Engine coolant ②	R								✓				✓	
Fuel hose, vapor hose & fuel filter cap	S/I	✓	✓	✓	✓	✓	✓	✓	✓	✓	✓	✓	✓	✓
Spark plugs: Platinum (Iridium) 100,000 mile replacement	R													✓
Suspension mounting bolts and nuts, upper and lower ball joint.	S/I	✓	✓	✓	✓	✓	✓	✓	✓	✓	✓	✓	✓	✓
Drive belts	S/I				✓				✓				✓	
Exhaust pipe connections, muffler & suspension bolts	S/I		✓		✓		✓		✓		✓		✓	
Valve clearance ③	S/I								✓					
Electronic throttle control	S/I							✓						
Brake hoses & lines	S/I		✓		✓		✓		✓		✓		✓	
Rear brake discs, linings & parking brake	S/I	✓	✓	✓	✓	✓	✓	✓	✓	✓	✓	✓	✓	✓
Steering gear rack, linkage & boots	S/I	✓	✓	✓	✓	✓	✓	✓	✓	✓	✓	✓	✓	✓
Fuel tank air filter ④	S/I		✓		✓		✓		✓		✓		✓	
Power steering belt ond hoses	S/I		✓		✓		✓		✓		✓		✓	
Fuel filter	R					✓					✓		✓	
Fuel lines & connections	S/I	✓	✓	✓	✓	✓	✓	✓	✓	✓	✓	✓	✓	✓
Manual transaxle fluid	S/I				✓				✓				✓	
Automatic transaxle fluid	S/I		✓		✓		✓		✓		✓		✓	
Climate control air filter ⑤	S/I	✓	✓	✓	✓	✓	✓	✓	✓	✓	✓	✓	✓	✓
Crankcase ventilation hoses	S/I				✓				✓				✓	

R: Replace S/I: Service or Inspect

① Replace every 30,000 miles

② Replace every 24 months or 30,000 miles after 60,000 miles

③ Adjust every 60,000 miles

④ Replace every 30,000 miles

⑥ Replace every 12 months or 10,000 miles.

FREQUENT OPERATION MAINTENANCE (SEVERE SERVICE)

If a vehicle is operated under any of the following conditions it is considered severe service:

- **Repeatedly driving short distance of less than 5miles (8km) in normal temperature or less than 10miles (16km) in freezing temperature**
- **Extensive engine idling or low speed driving for long distance**
- **Driving on rough, dusty, muddy, unpaved, graveled or salt- spread roads**
- **Driving in areas using salt or other corrosive materials or in very cold weather**
- **Driving in heavy traffic area over 90°F (32°C)**
- **Driving on uphill, downhill, or mountain road**
- **Towing a Trailer, or using a camper, or roof rack**
- **Driving as a patrol car, taxi, other commercial use or vehicle towing**
- **Driving over 100 MPH (170 Km/h)**
- **Frequently driving in stop-and-go conditions**

Oil & oil filter: change every 3,000 miles.

Brake pads, calipers & rotors: service or inspect more frequently

Brake linings, parking brake: service or inspect more frequently

Driveshaft boots: service or inspect every 7500 miles

Steering gear box, linkage & boots, lower and upper ball joints: service or inspect more frequently

Air cleaner filter: replace more frequently

Automatic transaxle fluid: replace every 30,000 miles.

Manual transaxle fluid: replace every 60,000 miles.

Climate control air filter: replace more frequently

Spark plugs: replace more frequently

37655_SONA_C0014

BRAKES — INFORMATION AND PRECAUTIONS

ANTI-LOCK SYSTEMS

- Certain components within the ABS system are not intended to be serviced or repaired individually.
- Do not use rubber hoses or other parts not specifically specified for and ABS system. When using repair kits, replace all parts included in the kit. Partial or incorrect repair may lead to functional problems and require the replacement of components.
- Lubricate rubber parts with clean, fresh brake fluid to ease assembly. Do not use shop air to clean parts; damage to rubber components may result.
- Use only DOT 3 brake fluid from an unopened container.
- If any hydraulic component or line is removed or replaced, it may be necessary to bleed the entire system.
- A clean repair area is essential. Always clean the reservoir and cap thoroughly before removing the cap. The slightest amount of dirt in the fluid may plug an orifice and impair the system function. Perform repairs after components have been thoroughly cleaned; use only denatured alcohol to clean components. Do not allow ABS components to come into contact with any substance containing mineral oil; this includes used shop rags.
- The Anti-Lock control unit is a microprocessor similar to other computer units in the vehicle. Ensure that the ignition switch is **OFF** before removing or installing controller harnesses. Avoid static electricity discharge at or near the controller.
- If any arc welding is to be done on the vehicle, the control unit should be unplugged before welding operations begin.

DISC AND DRUM SYSTEMS

**** CAUTION**

Dust and dirt accumulating on brake parts during normal use may contain asbestos fibers from production or aftermarket brake linings. Breathing excessive concentrations of asbestos fibers can cause serious bodily harm. Exercise care when servicing brake parts. Do not sand or grind brake lining unless equipment used is designed to contain the dust residue. Do not clean brake parts with compressed air or by dry brushing. Cleaning should be done by dampening the brake components with a fine mist of water, then wiping the brake components clean with a dampened cloth. Dispose of cloth and all residue containing asbestos fibers in an impermeable container with the appropriate label. Follow practices prescribed by the Occupational Safety and Health Administration (OSHA) and the Environmental Protection Agency (EPA) for the handling, processing, and disposing of dust or debris that may contain asbestos fibers.

BRAKES — BLEEDING THE BRAKE SYSTEM

BLEEDING PROCEDURE

BLEEDING PROCEDURE

See Figure 1.

1. Before servicing the vehicle, refer to the Precautions Section.

**** WARNING**

The reservoir on the master cylinder must be at the MAX (upper) level mark at the start of bleeding procedure and checked after bleeding each brake caliper. Add fluid as required.

2. Make sure the brake fluid level in the reservoir is at the MAX (upper) level line.
3. Have someone slowly pump the brake pedal several times, and then apply steady pressure.

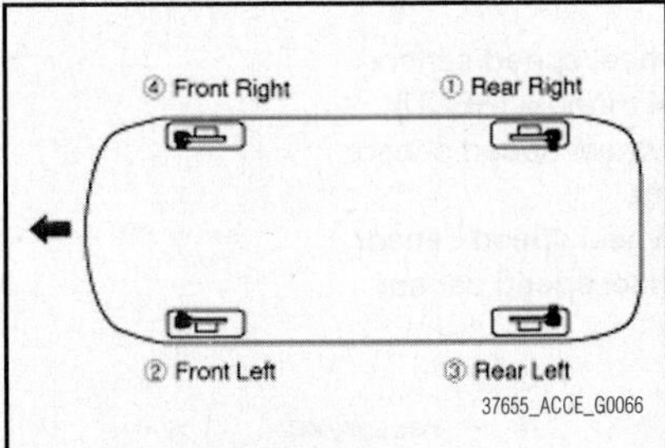

Fig. 1 Brake bleeding sequence

4. Loosen the right-rear brake bleed screw to allow air to escape from the system. Then tighten the bleed screw securely.
5. Repeat the procedure for each wheel in the sequence shown below until air bubbles no longer appear in the fluid.
6. Refill the master cylinder reservoir to the MAX (upper) level line.

BLEEDING THE ABS SYSTEM

This procedure should be followed to ensure adequate bleeding of air and the filling of the ABS unit, the brake lines, and the master cylinder with brake fluid.

1. Before servicing the vehicle, refer to the Precautions Section.
2. Remove the reservoir cap and fill the brake reservoir with brake fluid.

**** WARNING**

If there is any brake fluid on any painted surface, wash it off immediately.

➡When pressure bleeding, do not depress the brake pedal. Recommended brake fluid: DOT3 or DOT4.

3. Connect a clear plastic tube to the wheel cylinder bleeder plug and insert the other end of the tube into a clear plastic bottle that is half filled with clean brake fluid.
4. Connect the Hi-Scan Pro® to the data link connector located underneath the dash panel.
5. Select and operate according to the instructions on the Hi-Scan Pro® screen.

**** CAUTION**

You must obey the maximum operating time of the ABS motor with the Hi-Scan Pro® to prevent the motor pump from burning.

6. Select Hyundai vehicle diagnosis.
7. Select vehicle name.
8. Select Anti-Lock Brake system.
9. Select air bleeding mode.
10. Press "YES" to operate motor pump and solenoid valve.

**** WARNING**

Wait 60 seconds before operating the air bleeding or damage to the motor may occur.

11. Wait 60 seconds before operating the air bleeding.
12. Pump the brake pedal several times, and then loosen the bleeder screw until fluid starts to run out without bubbles. Then, close the bleeder screw.
13. Repeat until there are no more bubbles in the fluid for each wheel.

SPEED SENSORS

REMOVAL & INSTALLATION

Front

See Figures 2 through 4.

1. Before servicing the vehicle, refer to the Precautions Section.
2. Remove the front wheel speed sensor mounting bolt.
3. Remove the front wheel guard.
4. Disconnect the wheel speed sensor connector.
5. Remove the front wheel speed sensor.

To install:

6. Installation is the reverse of the removal procedure.

Rear

See Figures 2, 5 and 6.

1. Before servicing the vehicle, refer to the Precautions Section.
2. Remove the rear wheel speed sensor mounting bolt.
3. Remove the rear seat side pad then disconnect the rear wheel speed sensor connector.

To install:

4. Installation is the reverse of the removal procedure.

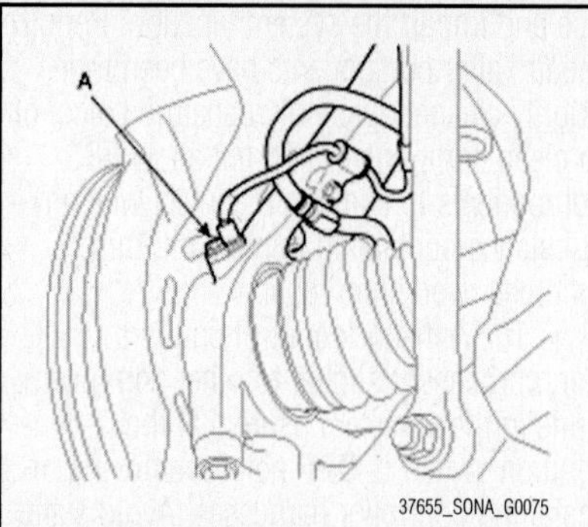

Fig. 3 Remove the front wheel speed sensor mounting bolt (A)

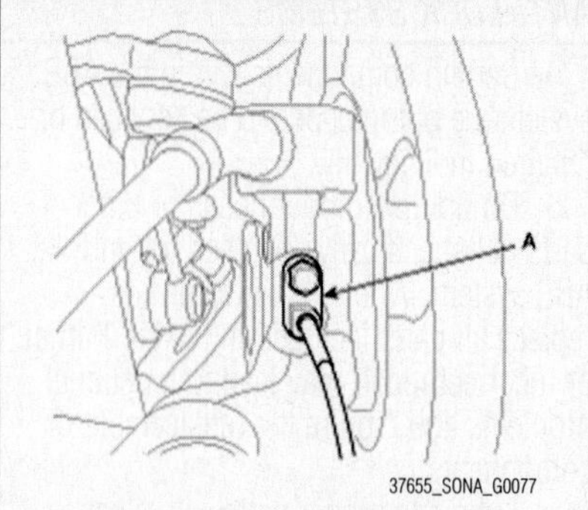

Fig. 6 Remove the rear wheel speed sensor mounting bolt (A)

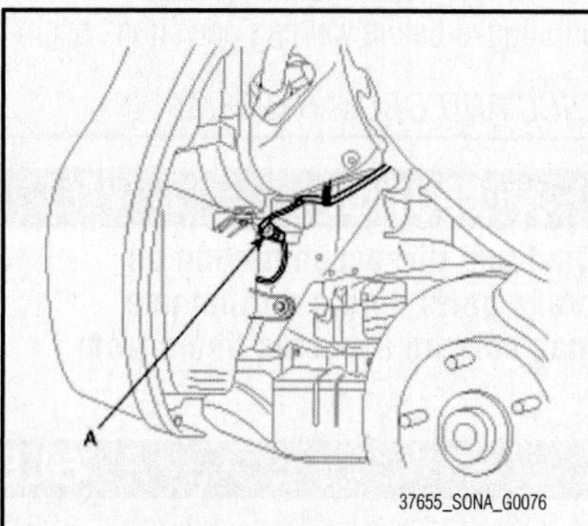

Fig. 4 Front wheel speed sensor and connector (A)

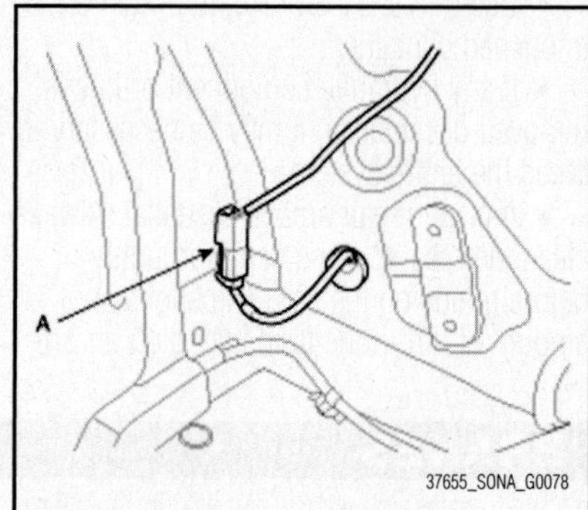

Fig. 7 Rear wheel speed sensor and connector (A)

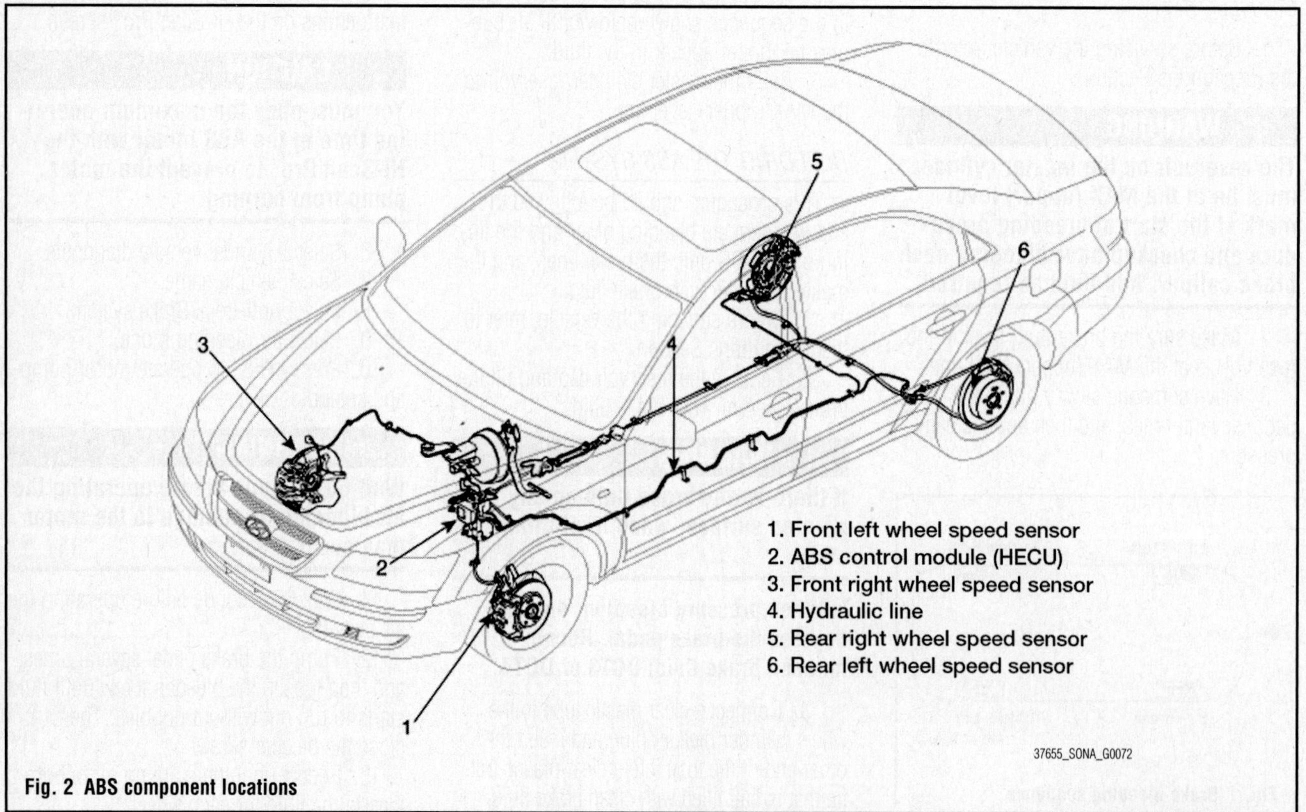

Fig. 2 ABS component locations

BRAKE CALIPER

REMOVAL & INSTALLATION

See Figures 7 through 9.

1. Before servicing the vehicle, refer to the Precautions Section.
2. Disconnect the negative battery cable.
3. Raise and safely support the vehicle.
4. Remove the front wheel and tire from front hub.

✻✻ WARNING

Be careful not to damage the hub bolts.

5. Remove the brake hose bolt and guide rod bolts from the caliper assembly.
6. Remove the caliper assembly.
7. Remove the brake hose from the caliper.

To install:

8. Install the brake hose.
9. Using the SST (09581-11000) or equivalent tool, push in the piston so that the caliper will fit over the pads. Make sure that the piston boot is in position to prevent damaging it when pivoting the caliper down.
10. Install the caliper assembly.

✻✻ WARNING

Be careful not to damage the piston pin boot.

11. Install the brake hose bolt and tighten to 18–22 ft. lbs. (25–29 Nm).
12. Install the guide rod bolts and tighten to 16–23 ft. lbs. (22–31 Nm).
13. Refill the master cylinder reservoir to the MAX line.
14. Bleed the brake system.
15. Install wheel and tire.

➡Engagement of the brake may require a greater pedal stroke immediately after the brake pads have been replaced as a set. Several applications of the brake will restore the normal pedal stroke.

16. After installation, check for leaks at hose and line joints or connections, and retighten if necessary.
17. Depress the brake pedal several times to make sure the brakes work, and then test-drive.

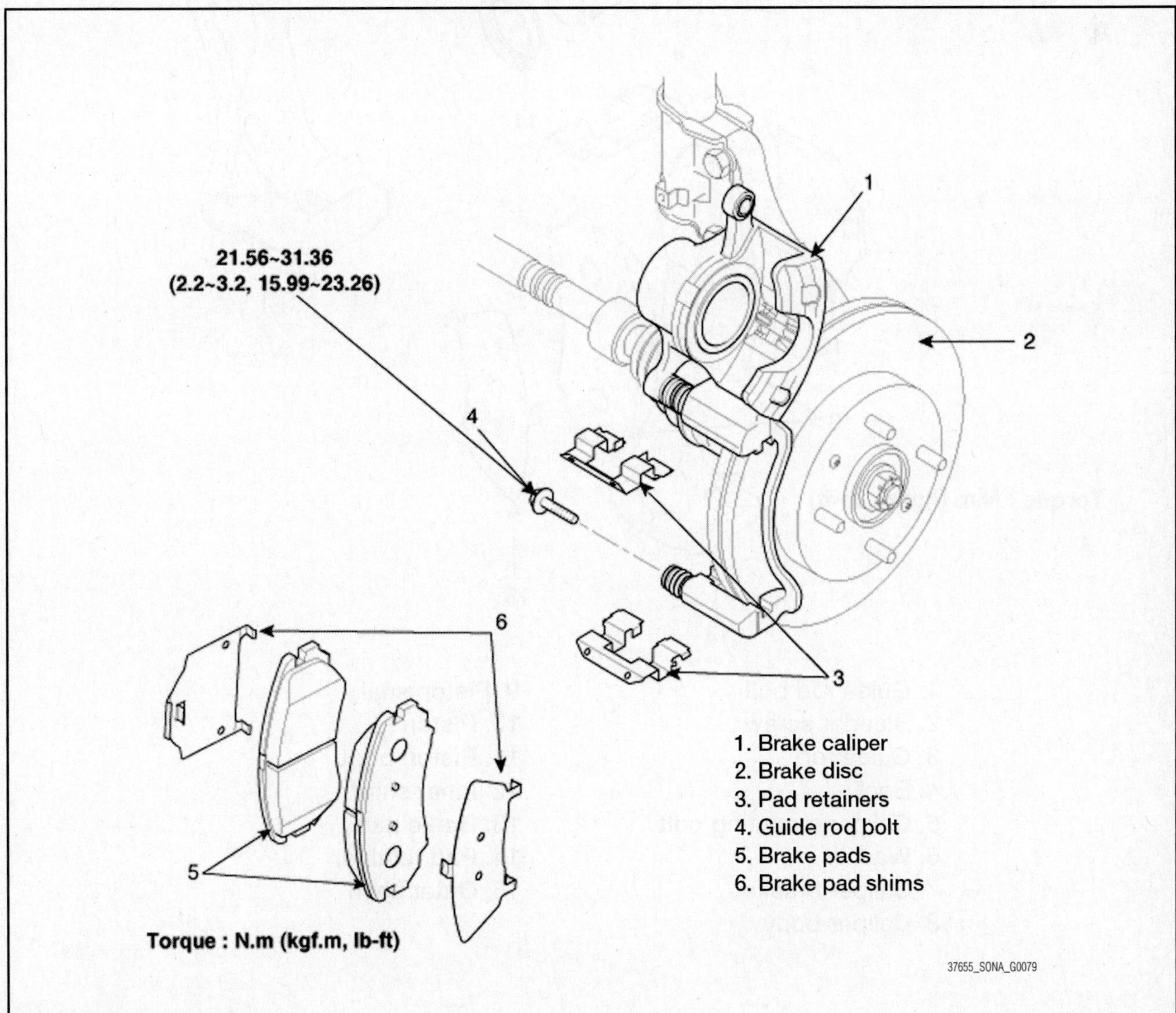

Fig. 7 Front disc brake components

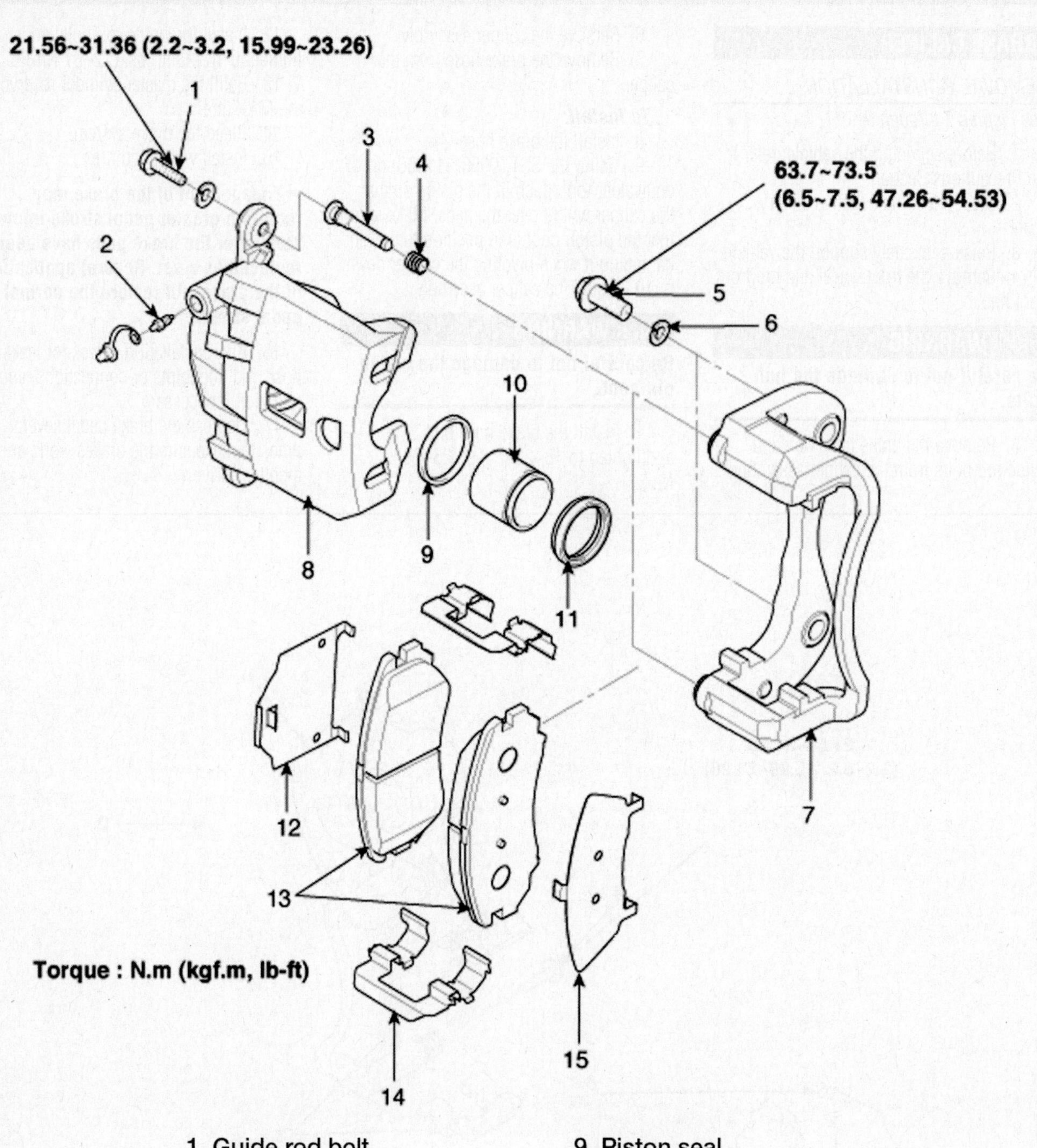

Fig. 8 Front disc brake—Exploded view

B
A
21.56~31.36 N.m
(2.2~3.2 kgf.m, 15.99~23.26 lb-ft)
37655_SONA_G0081

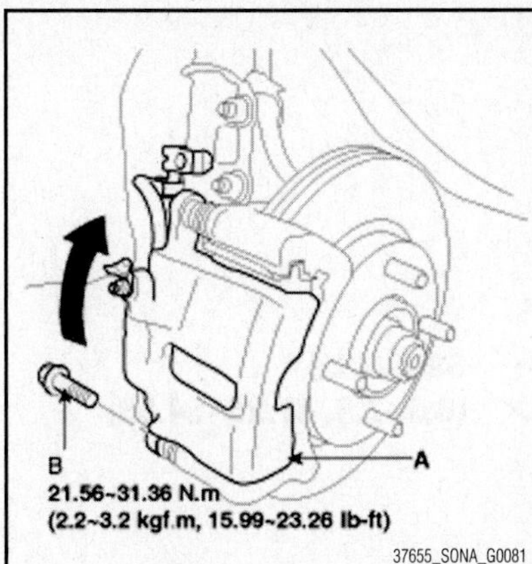

Fig. 9 Front guide rod bolts (B) and caliper assembly (A)

DISC BRAKE PADS

REMOVAL & INSTALLATION

See Figures 10 through 14.

1. Before servicing the vehicle, refer to the Precautions Section.
2. Disconnect the negative battery cable.
3. Raise and safely support the vehicle.
4. Remove the front wheel and tire from front hub.

WARNING

Be careful not to damage the hub bolts.

5. Remove the lower guide rod bolt from the caliper assembly.
6. Raise the caliper assembly and support it with a wire.
7. Remove the pad shim, pad retainer and pad assembly in the caliper bracket.

To install:

8. Install the pad retainers on the caliper bracket.
9. Check the foreign material at the pad shims and the back of the pads.

CAUTION

Contaminated brake discs or pads reduce stopping ability. Keep grease off the discs and pads.

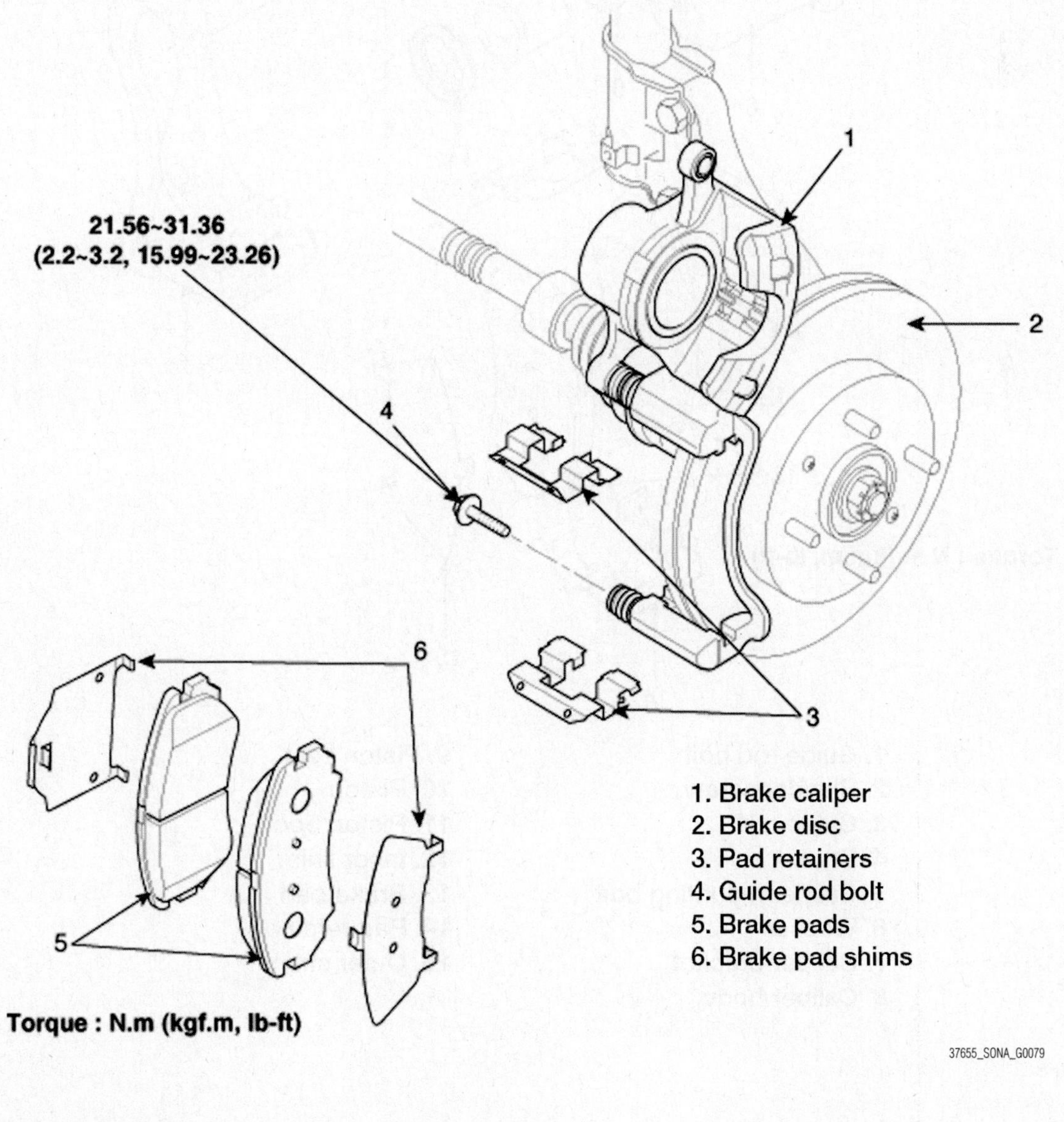

21.56~31.36
(2.2~3.2, 15.99~23.26)
1
2
3
4
5
6
1. Brake caliper
2. Brake disc
3. Pad retainers
4. Guide rod bolt
5. Brake pads
6. Brake pad shims
Torque : N.m (kgf.m, lb-ft)
37655_SONA_G0079

Fig. 10 Front disc brake components

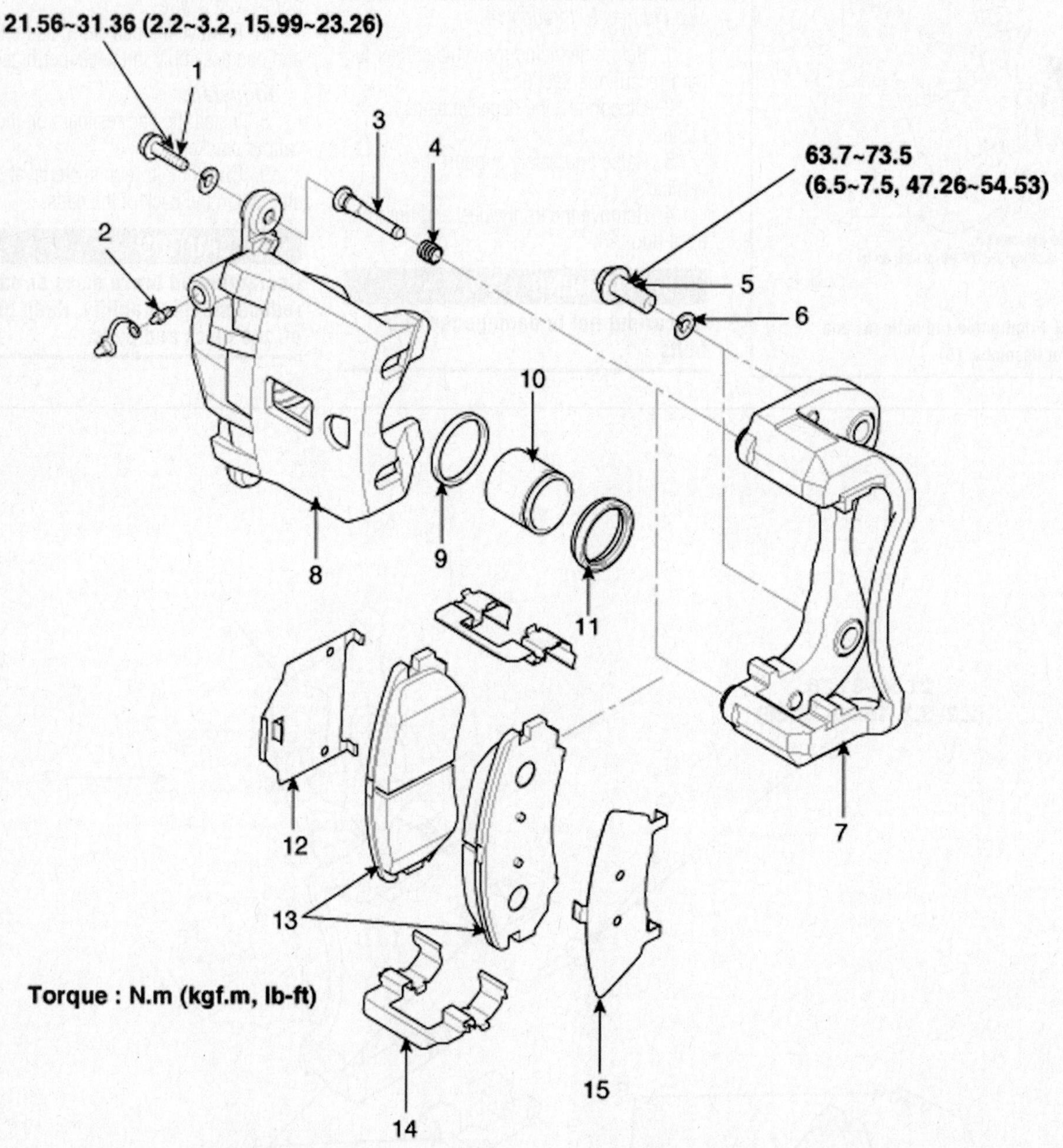

Fig. 11 Front disc brake—Exploded view

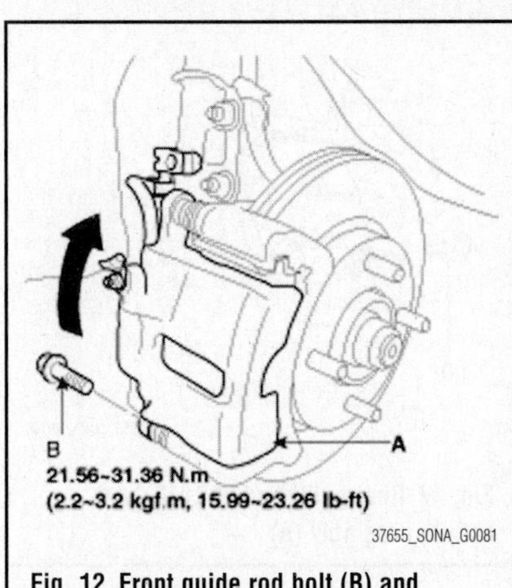

Fig. 12 Front guide rod bolt (B) and caliper assembly (A)

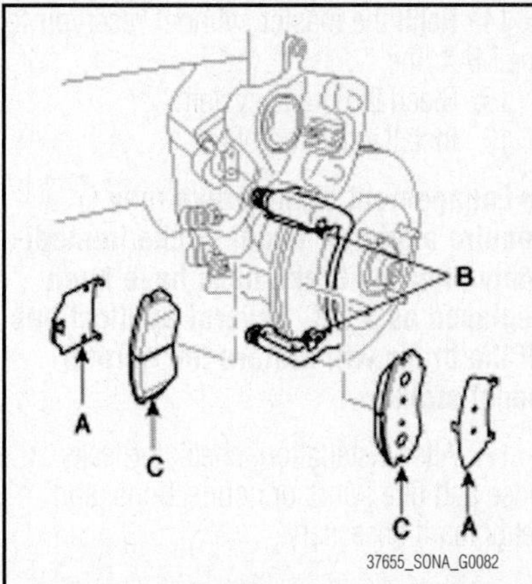

Fig. 13 Front pad shim (A), pad retainer (B) and pad assembly (C)

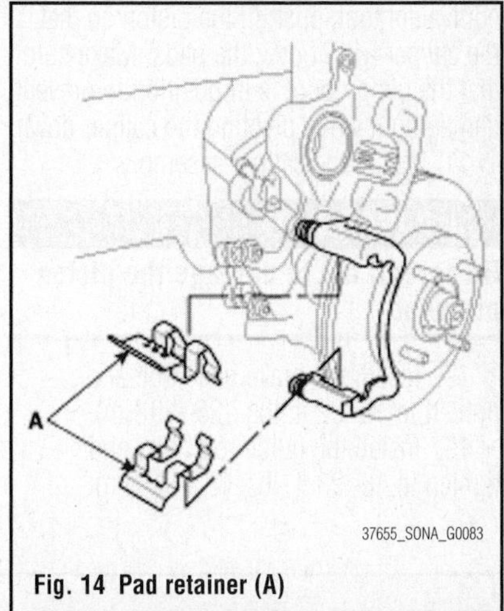

Fig. 14 Pad retainer (A)

10. Install the brake pads with the wear indicator on the inside.

CAUTION

If you are reusing the pads, always reinstall the brake pads in their original positions to prevent a momentary loss of braking efficiency.

11. Using the SST (09581-11000) or equivalent tool, push in the piston so that the caliper will fit over the pads. Make sure that the piston boot is in position to prevent damaging it when pivoting the caliper down.
12. Pivot caliper down into position. Being careful not to damage the pin boot, install the guide rod bolt and tighten to 16–24 ft. lbs. (22–32 Nm).
13. Install the wheel and tire.
14. Depress the brake pedal several times to make sure the brakes work, then test-drive.

CAUTION

Engagement of the brake may require a greater pedal stroke immediately after the brake pads have been replaced as a set. Several applications of the brake will restore the normal pedal stroke. Be sure to do this before driving the vehicle.

15. After installation, check for leaks at hose and line joints or connections, and retighten if necessary.

BRAKES — REAR DISC BRAKES

BRAKE CALIPER

REMOVAL & INSTALLATION

See Figures 15 through 17.

1. Before servicing the vehicle, refer to the Precautions Section.
2. Disconnect the negative battery cable.
3. Raise and safely support the vehicle.
4. Remove the rear wheel and tire from rear hub.

WARNING

Be careful not to damage the hub bolts.

5. Remove the brake hose bolts and guide rod bolt from the caliper assembly.
6. Remove the caliper assembly.
7. If replacing the caliper, remove the brake hose from the caliper.

To install:

8. Remove the brake hose from the caliper.

To install:

9. Install the brake hose.
10. Using the SST (09581-11000) or

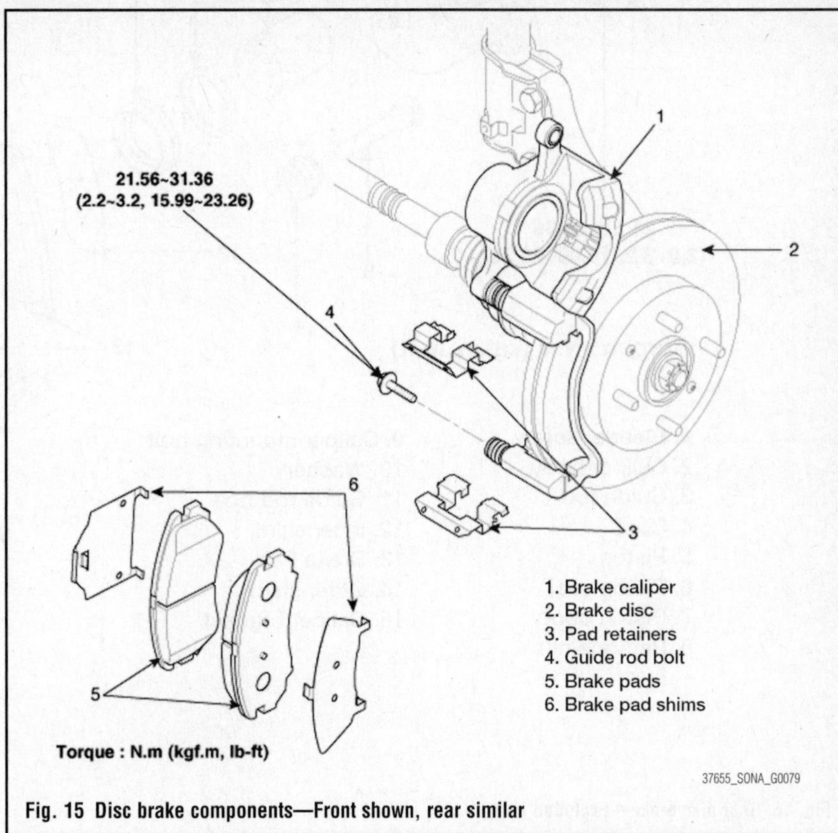

Fig. 15 Disc brake components—Front shown, rear similar

equivalent tool, push in the piston so that the caliper will fit over the pads. Make sure that the piston boot is in position to prevent damaging it when pivoting the caliper down.

11. Install the caliper assembly.

⁂ WARNING

Be careful not to damage the piston pin boot.

12. Install the brake hose bolt and tighten to 18–22 ft. lbs. (25–29 Nm).
13. Install the guide rod bolts and tighten to 16–23 ft. lbs. (22–31 Nm).
14. Refill the master cylinder reservoir to the MAX line.
15. Bleed the brake system.
16. Install wheel and tire.

➡Engagement of the brake may require a greater pedal stroke immediately after the brake pads have been replaced as a set. Several applications of the brake will restore the normal pedal stroke.

17. After installation, check for leaks at hose and line joints or connections, and retighten if necessary.

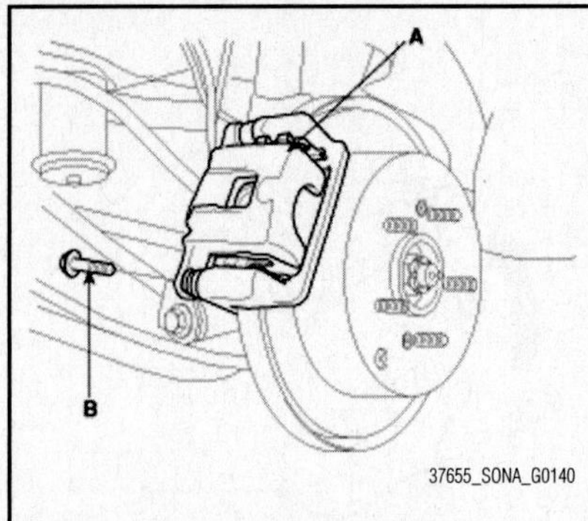

Fig. 17 Rear guide rod bolts (B) and caliper assembly (A)

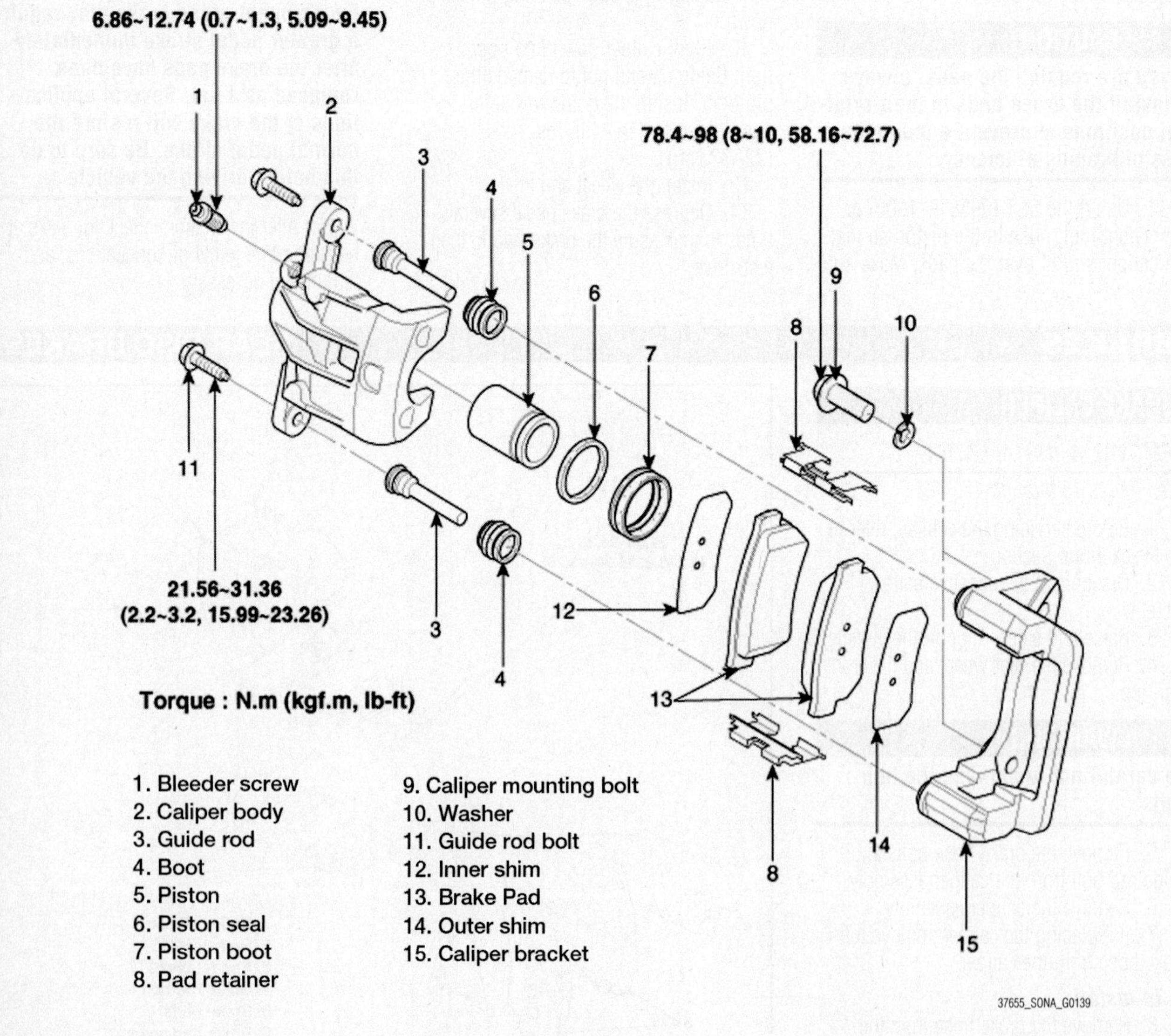

6.86~12.74 (0.7~1.3, 5.09~9.45)

78.4~98 (8~10, 58.16~72.7)

21.56~31.36 (2.2~3.2, 15.99~23.26)

Torque : N.m (kgf.m, lb-ft)

1. Bleeder screw
2. Caliper body
3. Guide rod
4. Boot
5. Piston
6. Piston seal
7. Piston boot
8. Pad retainer
9. Caliper mounting bolt
10. Washer
11. Guide rod bolt
12. Inner shim
13. Brake Pad
14. Outer shim
15. Caliper bracket

37655_SONA_G0139

Fig. 16 Rear disc brake—Exploded view

18. Depress the brake pedal several times to make sure the brakes work, and then test-drive.

DISC BRAKE PADS

REMOVAL & INSTALLATION

See Figures 18 through 22.

1. Before servicing the vehicle, refer to the Precautions Section.
2. Disconnect the negative battery cable.
3. Raise and safely support the vehicle.
4. Remove the rear wheel and tire from rear hub.

⁂ WARNING

Be careful not to damage the hub bolts.

5. Remove the lower guide rod bolt from the caliper assembly.
6. Raise the caliper assembly and support it with a wire.
7. Remove the pad shim, pad retainer and pad assembly in the caliper bracket.
8. Remove the caliper bracket.

To install:

9. Install the caliper bracket.
10. Install the pad retainers on the caliper bracket.
11. Check the foreign material at the pad shims and the back of the pads.

⁂ CAUTION

Contaminated brake discs or pads reduce stopping ability. Keep grease off the discs and pads.

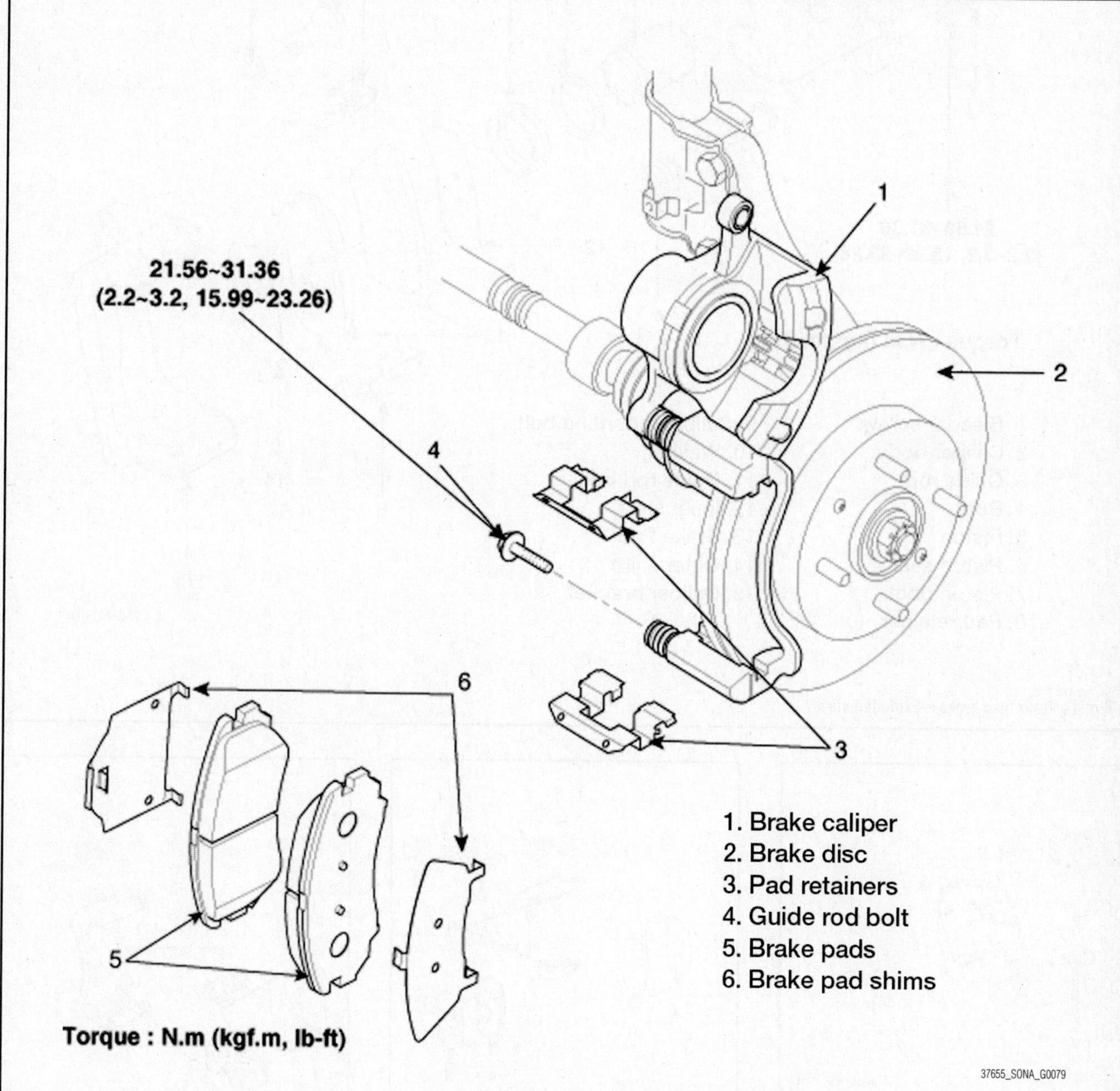

Fig. 18 Disc brake components—Front shown, rear similar

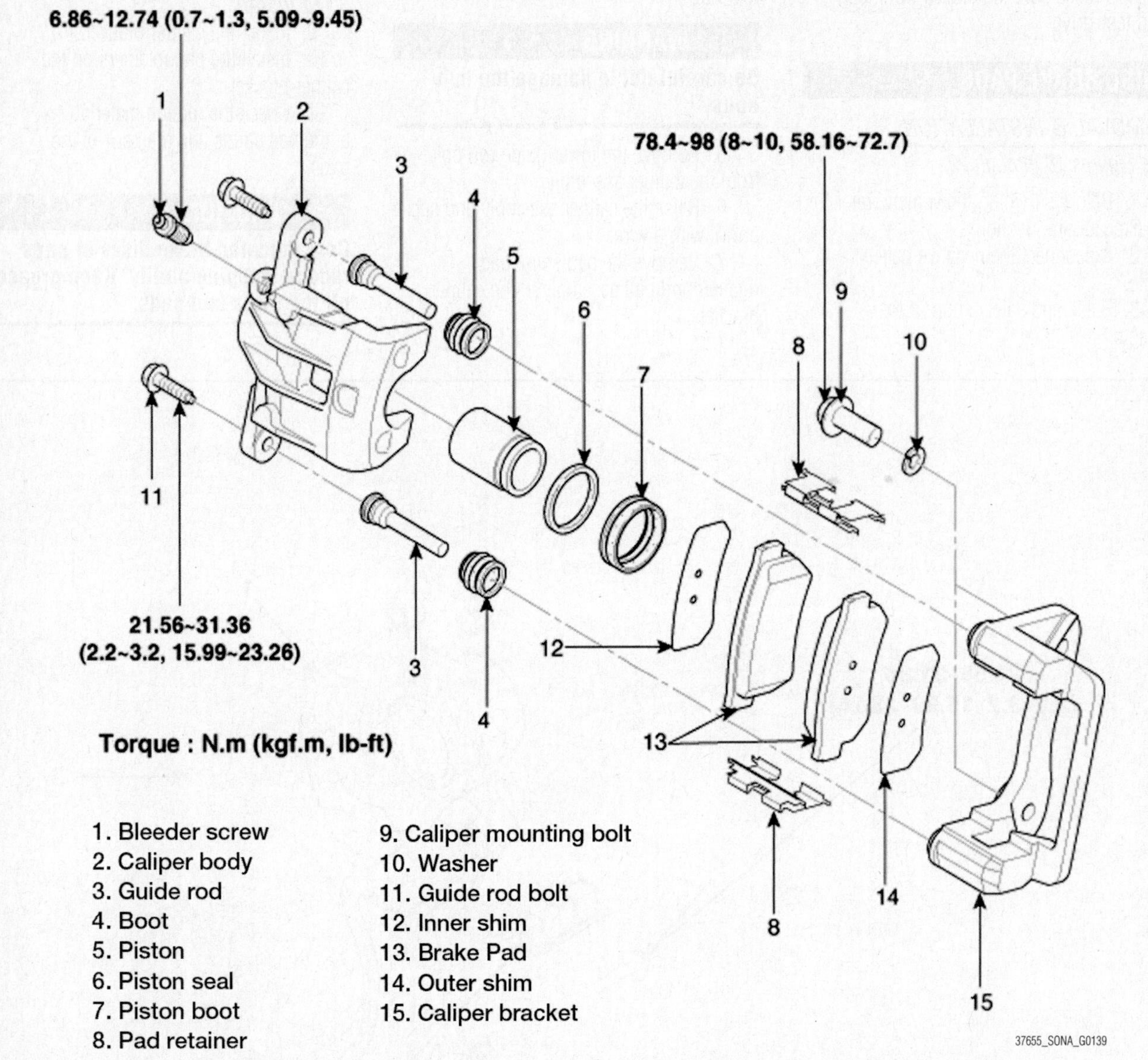

6.86~12.74 (0.7~1.3, 5.09~9.45)
78.4~98 (8~10, 58.16~72.7)
21.56~31.36
(2.2~3.2, 15.99~23.26)
Torque : N.m (kgf.m, lb-ft)
1. Bleeder screw
2. Caliper body
3. Guide rod
4. Boot
5. Piston
6. Piston seal
7. Piston boot
8. Pad retainer
9. Caliper mounting bolt
10. Washer
11. Guide rod bolt
12. Inner shim
13. Brake Pad
14. Outer shim
15. Caliper bracket
37655_SONA_G0139

Fig. 19 Rear disc brake—Exploded view

A
B
37655_SONA_G0140

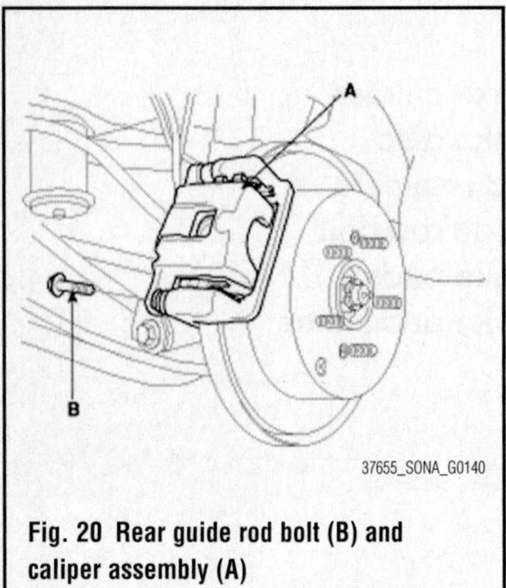

Fig. 20 Rear guide rod bolt (B) and caliper assembly (A)

B
A
C
C
A
37655_SONA_G0163

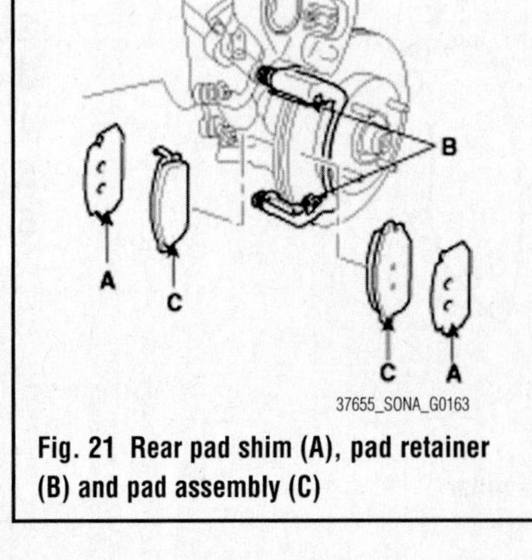

Fig. 21 Rear pad shim (A), pad retainer (B) and pad assembly (C)

A
37655_SONA_G0083

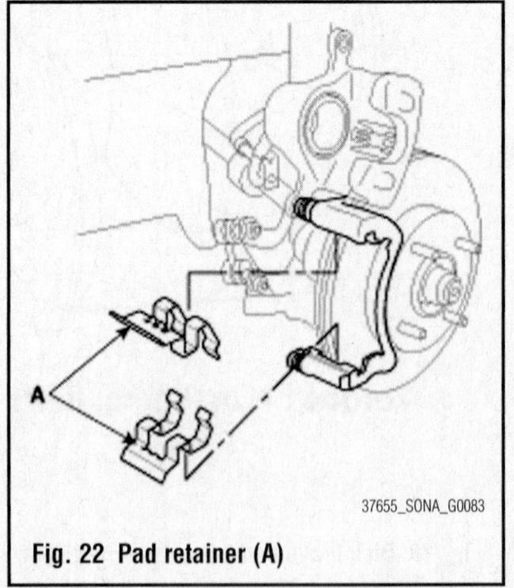

Fig. 22 Pad retainer (A)

⁂ CAUTION

If you are reusing the pads, always reinstall the brake pads in their original positions to prevent a momentary loss of braking efficiency.

12. Install the brake pads and pad shims correctly. Install the pad with the wear indicator on the inside.
13. Using the SST (09581-11000) or equivalent tool, push in the piston so that the caliper will fit over the pads. Make sure that the piston boot is in position to prevent damaging it when pivoting the caliper down.
14. Pivot caliper down into position. Being careful not to damage the pin boot, install the guide rod bolt and tighten to 16–24 ft. lbs. (22–32 Nm).
15. Install the wheel and tire.
16. Depress the brake pedal several times to make sure the brakes work, then test-drive.

⁂ CAUTION

Engagement of the brake may require a greater pedal stroke immediately after the brake pads have been replaced as a set. Several applications of the brake will restore the normal pedal stroke. Be sure to do this before driving the vehicle.

17. After installation, check for leaks at hose and line joints or connections, and retighten if necessary.

BRAKES — PARKING BRAKE

PARKING BRAKE CABLES

ADJUSTMENT

See Figure 23.

1. Before servicing the vehicle, refer to the Precautions Section.

➡After repairing the parking brake shoe, adjust the brake shoe clearance, and then adjust the parking brake lever stroke.

2. Raise and securely support the vehicle.
3. Pull the parking brake lever up one click.
4. Remove the floor console, if equipped.
5. Adjust the parking brake lever stroke by turning adjusting nut. Stroke: 8 clicks.
6. Release the parking brake lever completely.
7. Check if the parking brakes drag when the rear wheels are turned. Readjust if necessary until there is no drag from the parking brakes.
8. Check the proper operation of the parking brakes by fully applying the parking brakes.
9. Reinstall the floor console, if equipped.

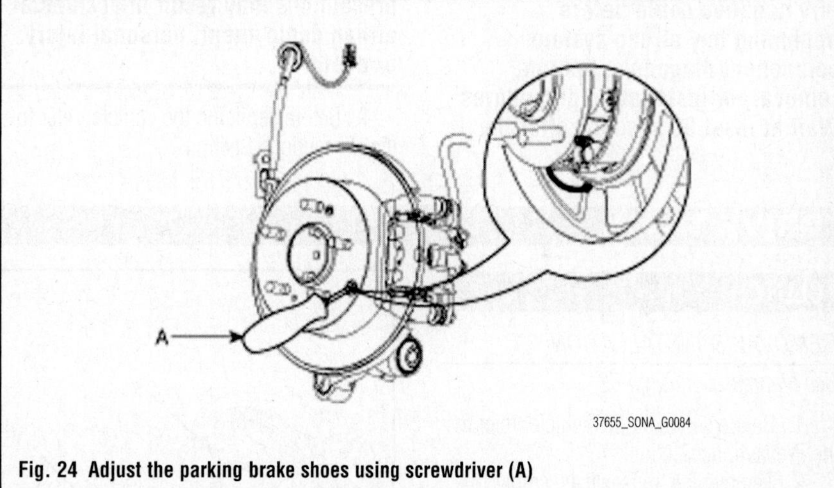

Fig. 24 Adjust the parking brake shoes using screwdriver (A)

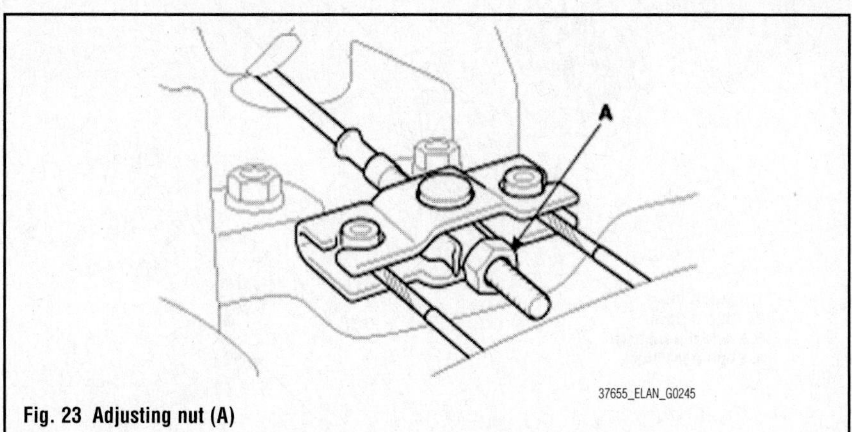

Fig. 23 Adjusting nut (A)

PARKING BRAKE SHOES

ADJUSTMENT

See Figure 24.

1. Before servicing the vehicle, refer to the Precautions Section.
2. Raise and securely support the vehicle.
3. Remove the rear tire and wheel.
4. Remove the plug from the disc.
5. Using a screwdriver, turn the adjusting wheel in the direction shown until a rotational force of 6.6 lbs. (29.4 N) is reached. Then turn the adjusting wheel back by 5 notches.
6. Install the rear tire and wheel after installing the plug.

CHASSIS ELECTRICAL

AIR BAG (SUPPLEMENTAL RESTRAINT SYSTEM)

GENERAL INFORMATION

⁂ CAUTION

These vehicles are equipped with an air bag system. The system must be disarmed before performing service on, or around, system components, the steering column, instrument panel components, wiring and sensors. Failure to follow the safety precautions and the disarming procedure could result in accidental air bag deployment, possible injury and unnecessary system repairs.

SERVICE PRECAUTIONS

⁂ CAUTION

Disconnect and isolate the battery negative cable before beginning any airbag system component diagnosis, testing, removal, or installation procedures. Wait at least 90 seconds after the ignition switch is turned off and the negative (-) terminal cable is disconnected from the battery before starting the operation. The SRS is equipped with a backup power source, so if work is started within 90 seconds after disconnecting the negative (-) terminal cable from the battery, the SRS may be deployed. Failure to disable the airbag system may result in accidental airbag deployment, personal injury, or death.

DISARMING THE SYSTEM

⁂ CAUTION

Before servicing components near or affected by the SRS (airbag) system, read and observe all SRS Service Precautions. Failure to observe all precautions may result in accidental airbag deployment, personal injury, or death.

1. Before servicing the vehicle, refer to the Precautions Section.
2. Disconnect and isolate the negative battery cable.
3. Wait 3 minutes for the system capacitor to discharge before performing any service.
4. Remove the ignition key from the vehicle.

ARMING THE SYSTEM

⁂ CAUTION

Before servicing components near or affected by the SRS (airbag) system, read and observe all SRS Service Precautions. Failure to observe all precautions may result in accidental airbag deployment, personal injury, or death.

1. Before servicing the vehicle, refer to the Precautions Section.
2. Reconnect the negative battery cable.
3. Turn the ignition switch to the **RUN** position.
4. Confirm proper system operation:
   a. Turn the ignition switch ON; the SRS indicator light should be turned on for about six seconds and then go off.

DRIVE TRAIN

CLUTCH

REMOVAL & INSTALLATION

See Figures 25 through 32.

1. Before servicing the vehicle, refer to the Precautions Section.
2. Remove the transaxle assembly.
3. Insert the special tool (09411-11000) in the clutch disc to prevent the disc from falling.
4. Loosen the bolts which attach the clutch cover to the flywheel in a star pattern. Loosen the bolts in succession, 1 or 2 turns at a time, to avoid bending the cover flange.

➡Do not clean the clutch disc or the release bearing with cleaning solvent.

5. Remove the clutch cover and disc.

To install:

6. Using the SST (09411-11000), install the disc.
7. Apply multipurpose grease (CASMOLY L9508) to the disc spline and the transmission input shaft spline.

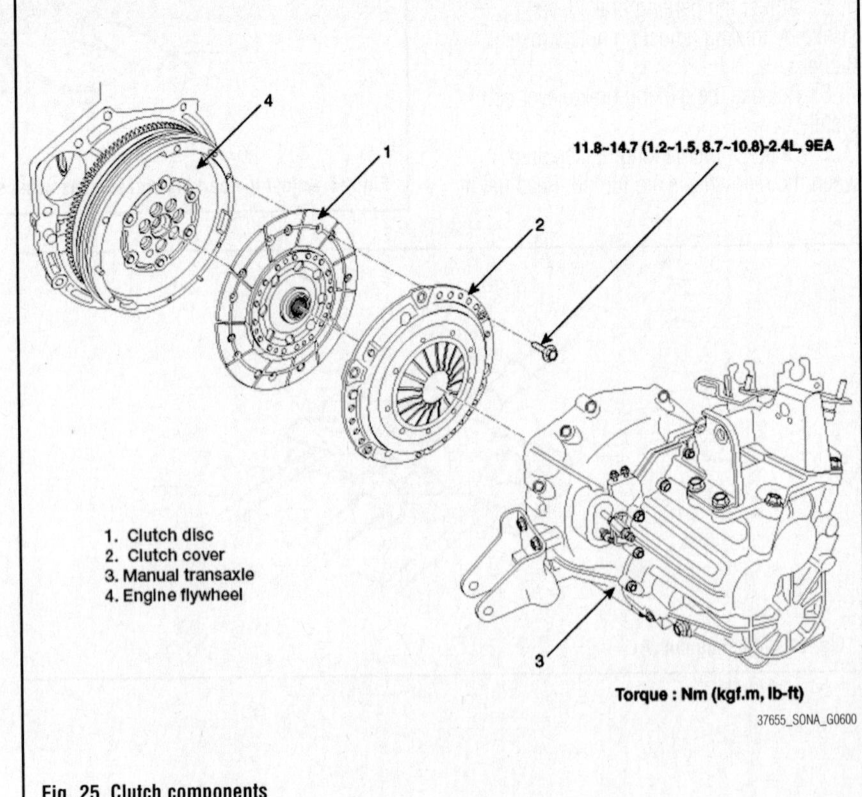

Fig. 25 Clutch components

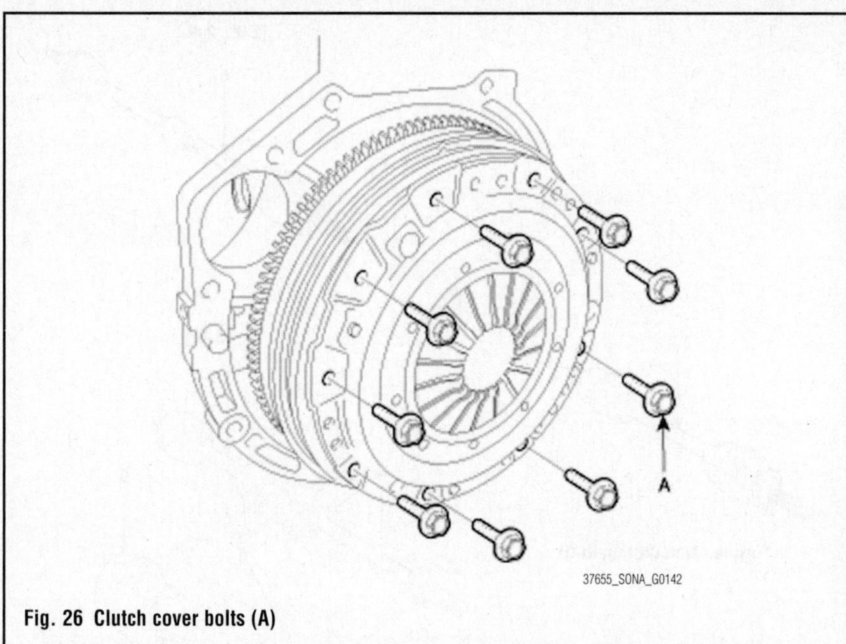

Fig. 26 Clutch cover bolts (A)

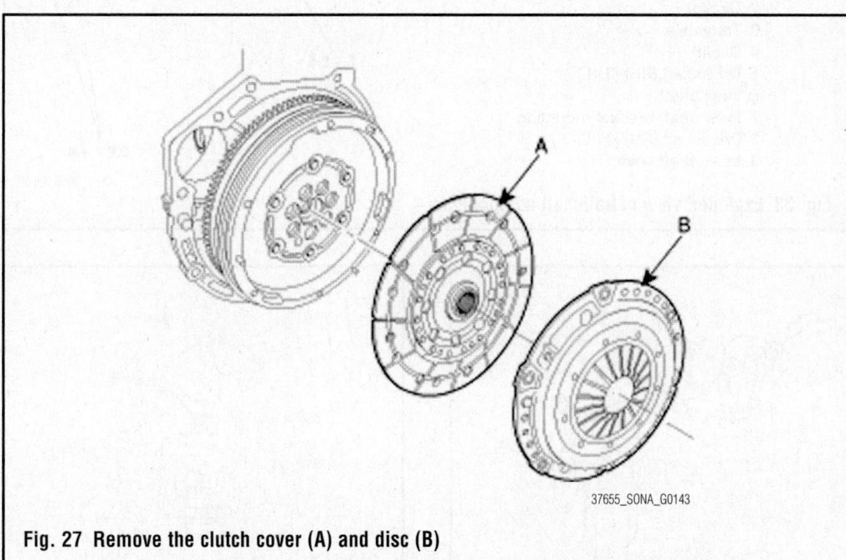

Fig. 27 Remove the clutch cover (A) and disc (B)

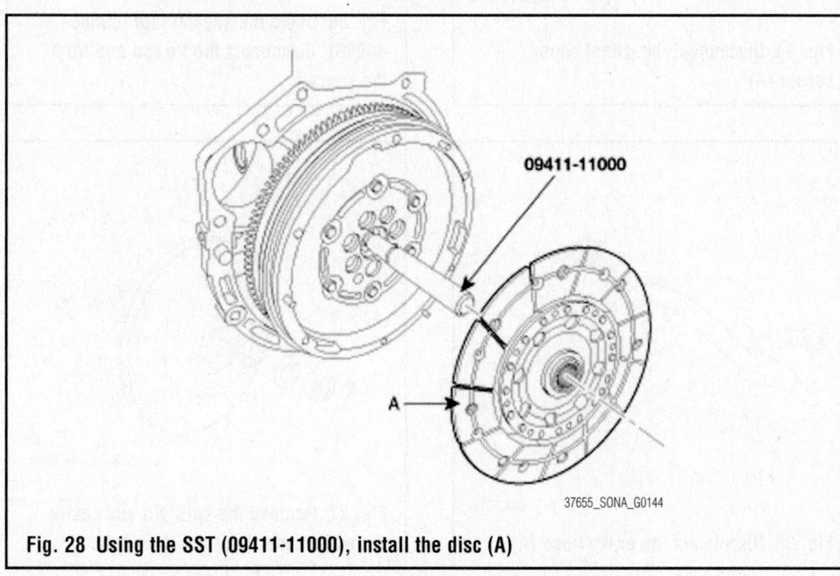

Fig. 28 Using the SST (09411-11000), install the disc (A)

➡The 'T/M SIDE' marked surface should face the transaxle.

8. Install the clutch cover bolts in a star pattern. Tighten the bolts in succession, 1 or 2 turns at a time, to 9–11 ft. lbs. (12–15 Nm).
9. Align the bearing to the release fork and then install it to the sleeve of the housing.

✻✻ CAUTION

Apply multipurpose grease (CASMOLY L9508) to the bearing sleeve, contact point of the release fork and the bushing inner surface.

10. Install the release lever to the release fork.

✻✻ CAUTION

If the transaxle assembly is installed to the engine without performing this step, the release bearing can be separated, as the release fork rotates freely.

11. Install the transaxle assembly to the engine.

ADJUSTMENTS

No adjustment is necessary

FRONT HALFSHAFTS

REMOVAL & INSTALLATION

See Figures 33 through 48.

➡Special service tools used for this procedure: SST (09568-4A000) and SST (09517-21500)

1. Before servicing the vehicle, refer to the Precautions Section.
2. Raise and safely support the vehicle.
3. Remove the front wheel and tire.
4. Disconnect the wheel speed sensor from the knuckle.
5. Disconnect the brake hose from the knuckle.
6. Remove the caliper assembly and suspend it with a wire.
7. Using the special tool (09568-4A000), disconnect the tie rod end from the knuckle.
8. Remove the split pin, and then remove castle nut and washer from the front hub.
9. Remove the bolts and disconnect the ball joint from the knuckle.
10. Using a plastic hammer, disconnect halfshaft from the axle hub.

Fig. 29 The 'T/M SIDE' marked surface should face the transaxle

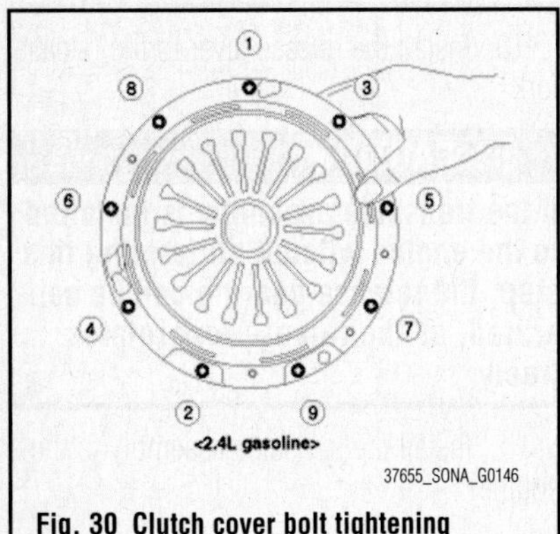

Fig. 30 Clutch cover bolt tightening sequence

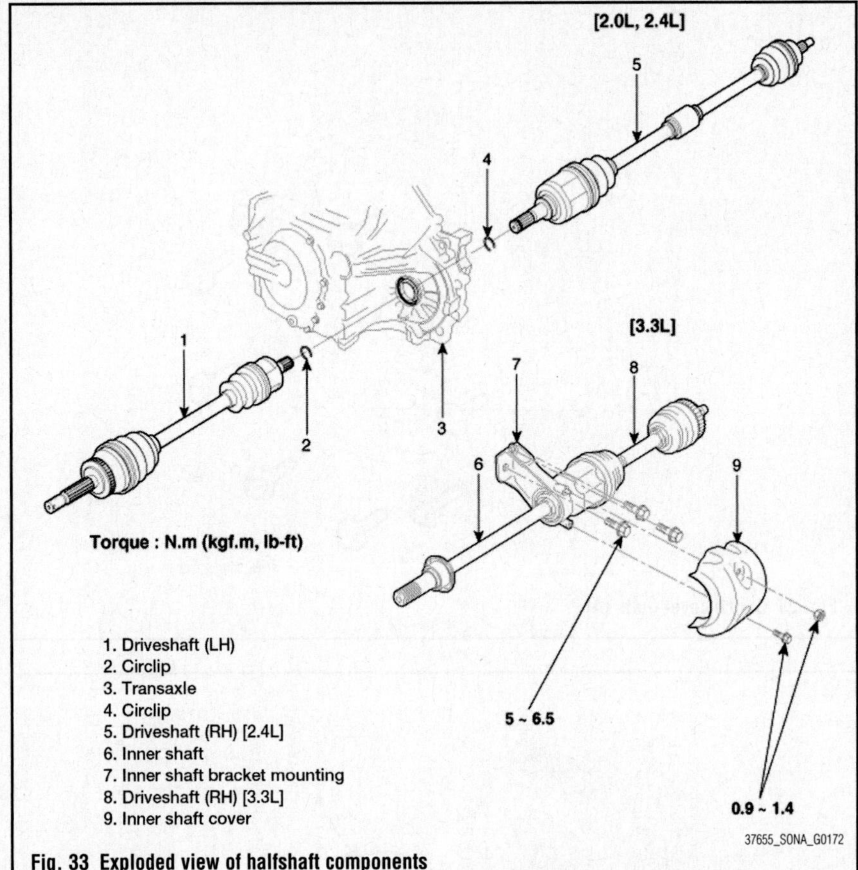

Fig. 33 Exploded view of halfshaft components

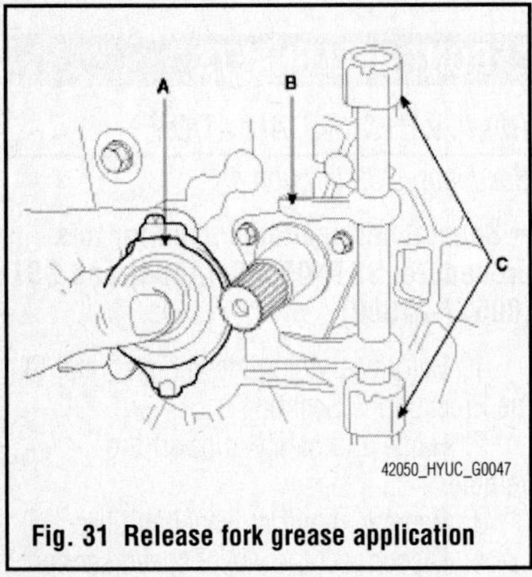

Fig. 31 Release fork grease application

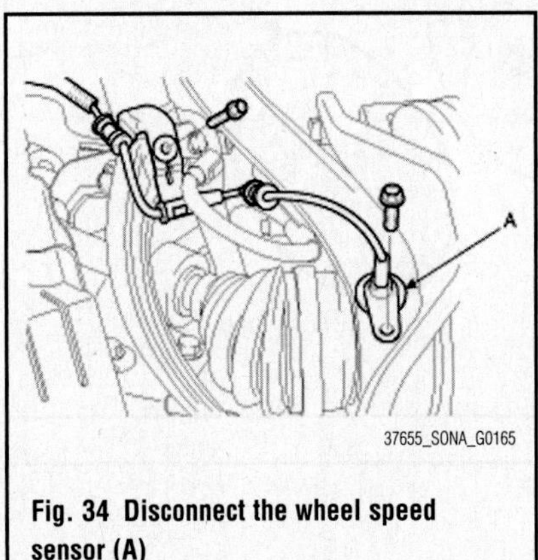

Fig. 34 Disconnect the wheel speed sensor (A)

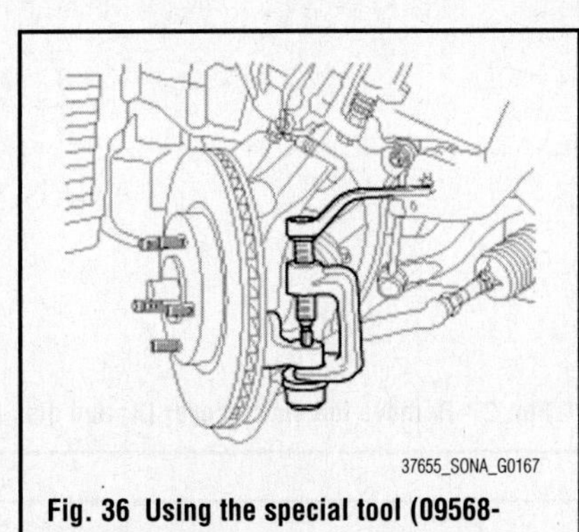

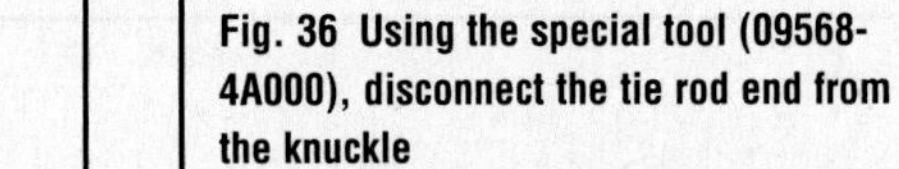

Fig. 36 Using the special tool (09568-4A000), disconnect the tie rod end from the knuckle

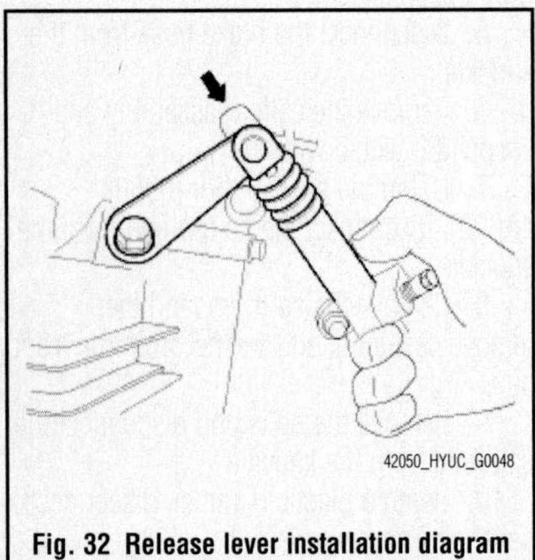

Fig. 32 Release lever installation diagram

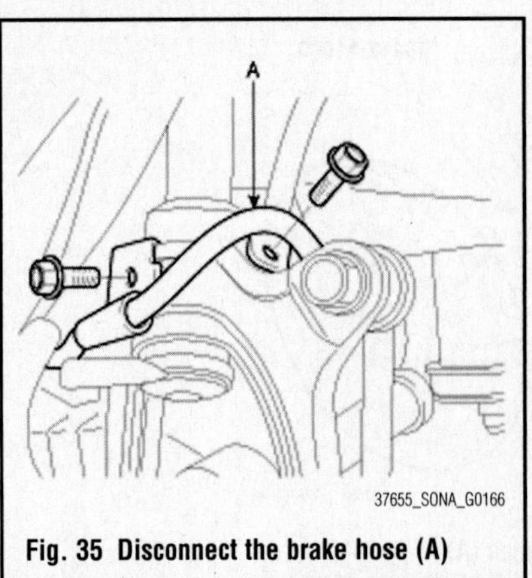

Fig. 35 Disconnect the brake hose (A)

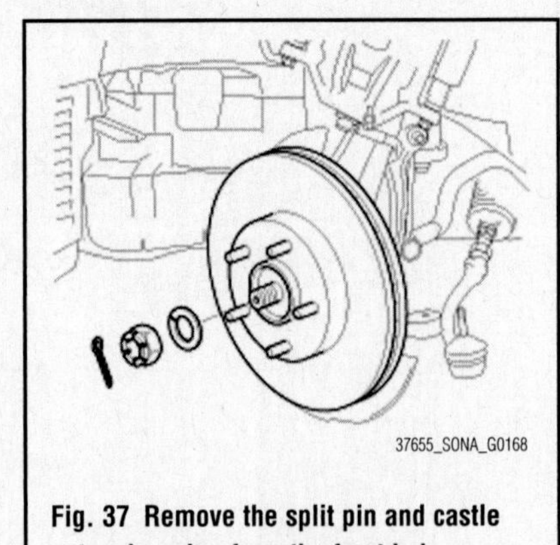

Fig. 37 Remove the split pin and castle nut and washer from the front hub

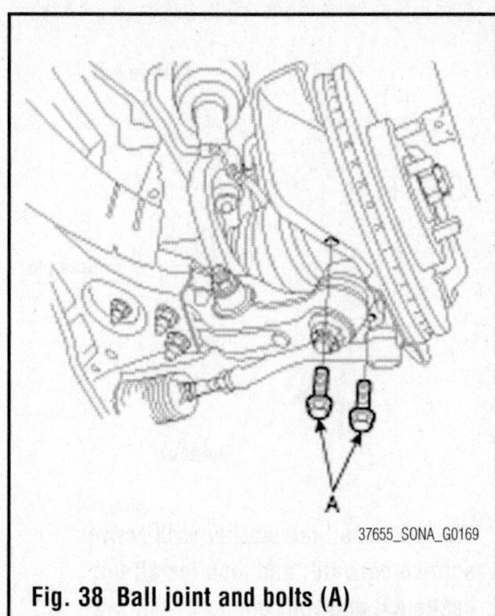

Fig. 38 Ball joint and bolts (A)

11. Insert a pry bar between the transaxle case and joint case, and separate the halfshaft from the transaxle case.

WARNING

Use a pry bar being careful not to damage the transaxle and joint. Do not insert the pry bar too deep, as this may cause damage to the oil seal. Do not pull the halfshaft by excessive force it may cause components inside the joint to dislodge resulting in a torn boot or a damaged bearing.

12. Pull out the halfshaft from the transaxle case.

➡Plug the hole of the transaxle case with the oil seal cap to prevent contamination. Support the halfshaft properly. Replace the retainer ring whenever the halfshaft is removed from the transaxle case.

➡While loosening the halfshaft nut, do not allow vehicle weight to be concentrated on the wheel bearing. If the vehicle moves, hold the wheel bearing using the special tool (09517-21500).

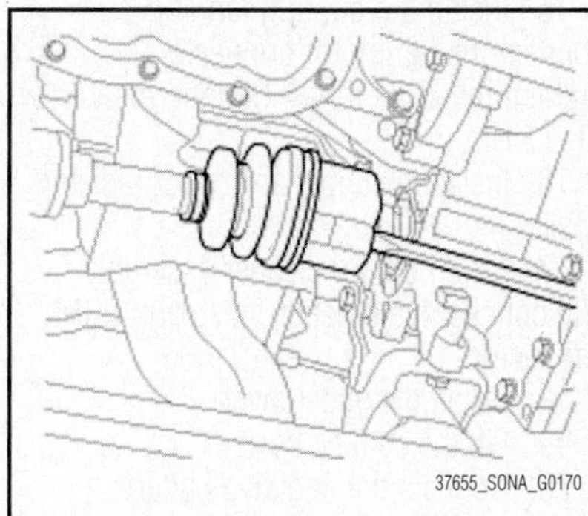

Fig. 39 Using a pry bar, remove the halfshaft from the transaxle

13. For vehicles with 3.3L engines, remove the right-hand halfshaft:

a. Remove the stabilizer link from the fork.

b. Remove the fork from the front lower arm.

c. Remove the fork from the front strut assembly.

d. Remove the inner shaft cover from the inner shaft bracket.

e. Remove the inner shaft bracket mounting bolts.

f. Remove the front halfshaft assembly with the inner shaft from the transaxle.

➡Do not try to disconnect the inner shaft from the halfshaft, because they cannot be disconnected once assembled. Do not reuse the halfshaft which is disassembled from the inner shaft.

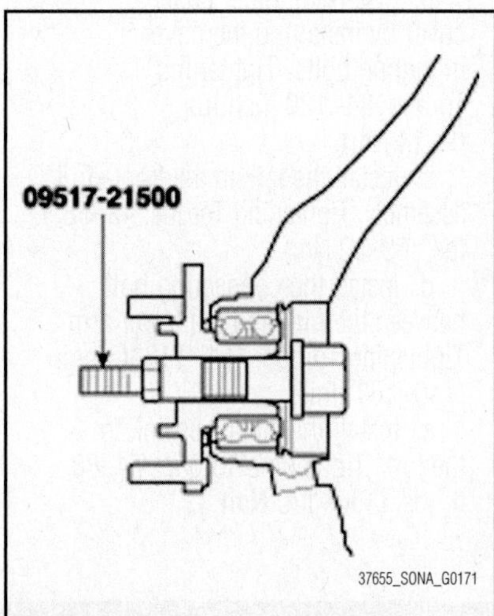

Fig. 40 If the vehicle moves, hold the wheel bearing using the special tool (09517-21500)

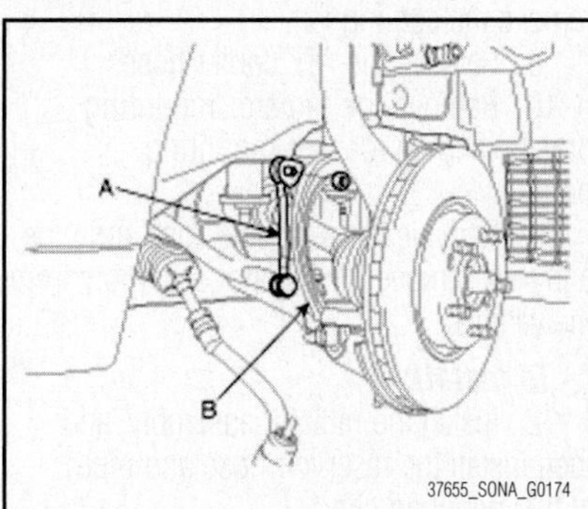

Fig. 41 Remove the stabilizer link (A) from the fork (B)

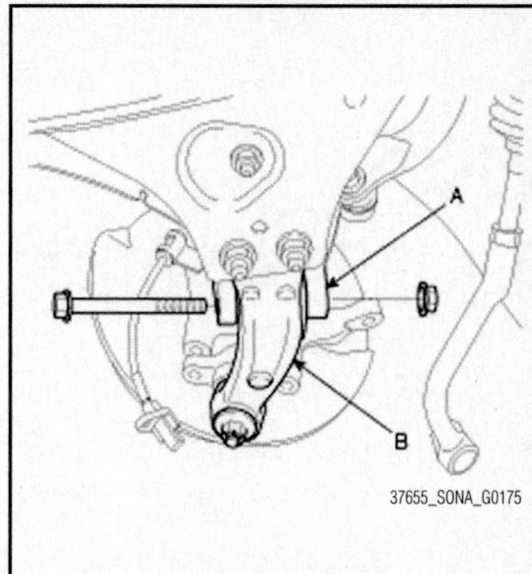

Fig. 42 Remove the fork (A) from the front lower arm (B)

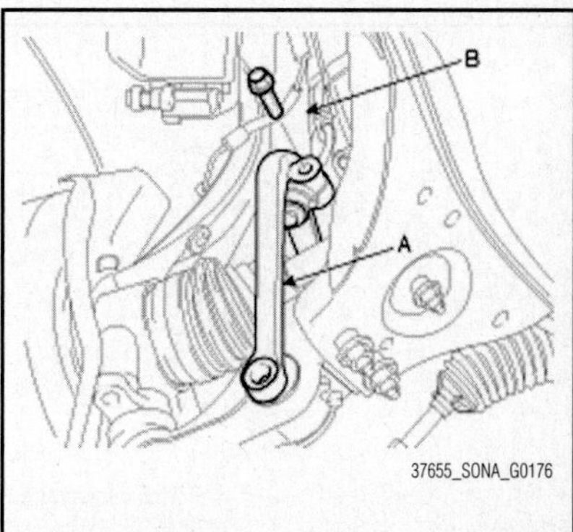

Fig. 43 Remove the fork (A) from the front strut assembly (B)

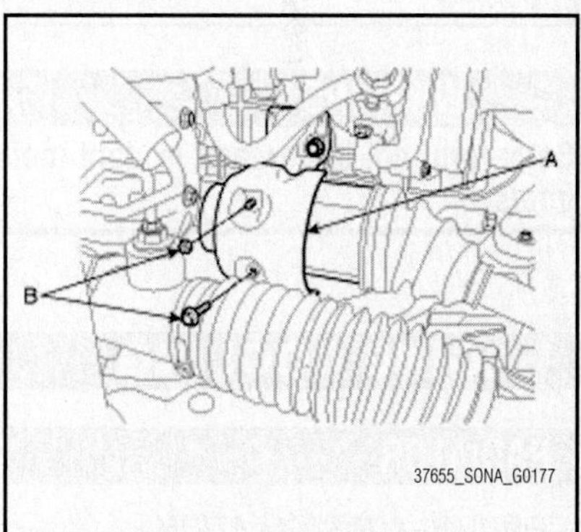

Fig. 44 Remove the inner shaft cover (A) from the inner shaft bracket (B)

To install:

14. Apply gear oil on the oil seal contacting surface of transaxle case and the halfshaft splines.

➡Replace circlip with a new one. Be careful to install the same kind of circlip.

15. Before installing the halfshaft, set the opening side of the circlip facing downward.

16. After installation, check that the halfshaft cannot be removed by hand.

17. Install the halfshaft to the knuckle.

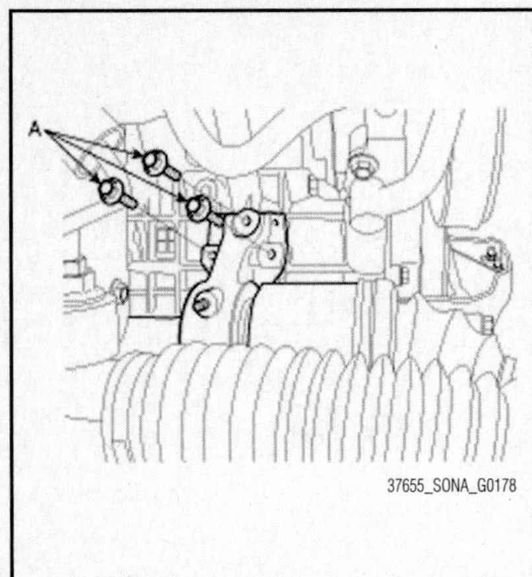

Fig. 45 Remove the inner shaft bracket mounting bolts (A)

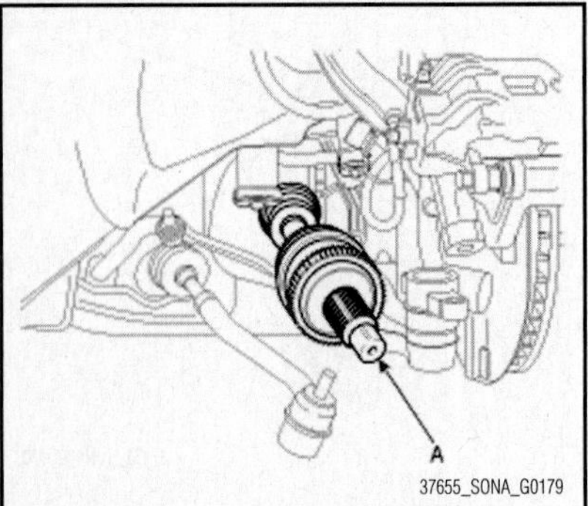

Fig. 46 Remove the front halfshaft assembly (A) with the inner shaft from the transaxle

WARNING

Be careful not to damage or dent the boots.

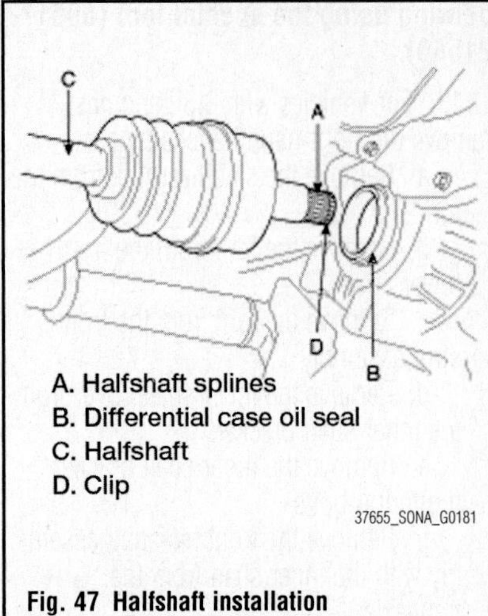

Fig. 47 Halfshaft installation

18. For vehicles with 3.3L engines, install the right-hand halfshaft:
    a. Install the inner shaft bracket mounting bolts. Tightening Torque: 29–36 ft. lbs. (39–49 Nm).
    b. Install the inner shaft cover by installing the cover mounting bolts. Tightening Torque: 84–120 inch lbs. (9–14 Nm).
    c. Install the fork to the front strut assembly. Tightening Torque: 43–58 ft. lbs. (60–80 Nm).
    d. Install the connecting bolt between the fork and the lower arm. Tightening Torque: 101–116 ft. lbs. (140–160 Nm).
    e. Install the stabilizer link to the fork. Tightening Torque: 74–88 ft. lbs. (100–120 Nm).

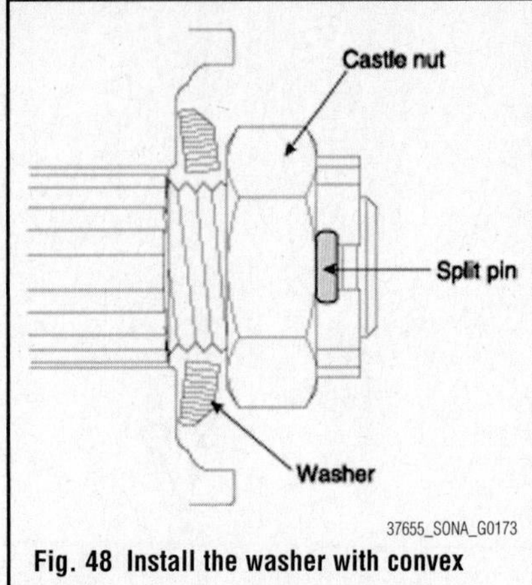

Fig. 48 Install the washer with convex surface outward, and then install the castle nut and split pin

19. Install the knuckle in the lower arm assembly. Tightening torque: 74–88 ft. lbs. (100–120 Nm).
20. Install the tie rod end in the knuckle. Tightening torque: 18–25 ft. lbs. (24–34 Nm).
21. Install the wheel speed sensor in the knuckle. Tightening torque: 70–86 inch lbs. (8–10 Nm).
22. Install the washer with convex surface outward, and then install the castle nut and split pin. Tightening torque: 148–207 ft. lbs. (200–280 Nm).
23. Install the wheel and tire.

CV-BOOTS INSPECTION

1. Check splines for wear.
2. Check boots for tears, water, foreign matter, or rust.

ENGINE COOLING

RADIATOR

REMOVAL & INSTALLATION

2.4L Engines

See Figures 49 through 52.

1. Before servicing the vehicle, refer to the Precautions Section.
2. Disconnect the negative battery cable.
3. Raise and safely support the vehicle.
4. Remove the undercover.
5. Drain the cooling system.
6. Disconnect the breather hose, ECM connector and remove the air cleaner assembly.
7. Remove the upper and lower radiator hoses.
8. Disconnect fan motor connector, and remove the cooling fan.
9. Remove the ATF cooler hoses.
10. Remove the radiator mounting bracket and condenser mounting bolts.
11. Remove the reservoir hose and pipe, and then remove the radiator assembly from the vehicle.

To install:

12. Install the radiator assembly, and then install the reservoir hose and pipe at the mounting clip.
13. Install the condenser mounting bolts and the radiator mounting bracket. Tighten the bolts to 78–95 inch lbs. (5–8 Nm).
14. Install the ATF cooler hoses.
15. Install the cooling fan and connect the fan motor connector. Tighten the bolts to 43–68 inch lbs. (9–11 Nm).
16. Install the upper and lower radiator hoses.
17. Install the air cleaner assembly, and connect the breather hose and ECM connector.
18. Install the undercover.
19. Fill the cooling system.
20. Connect the negative battery cable.
21. Check for leaks.
22. Recheck the coolant level.

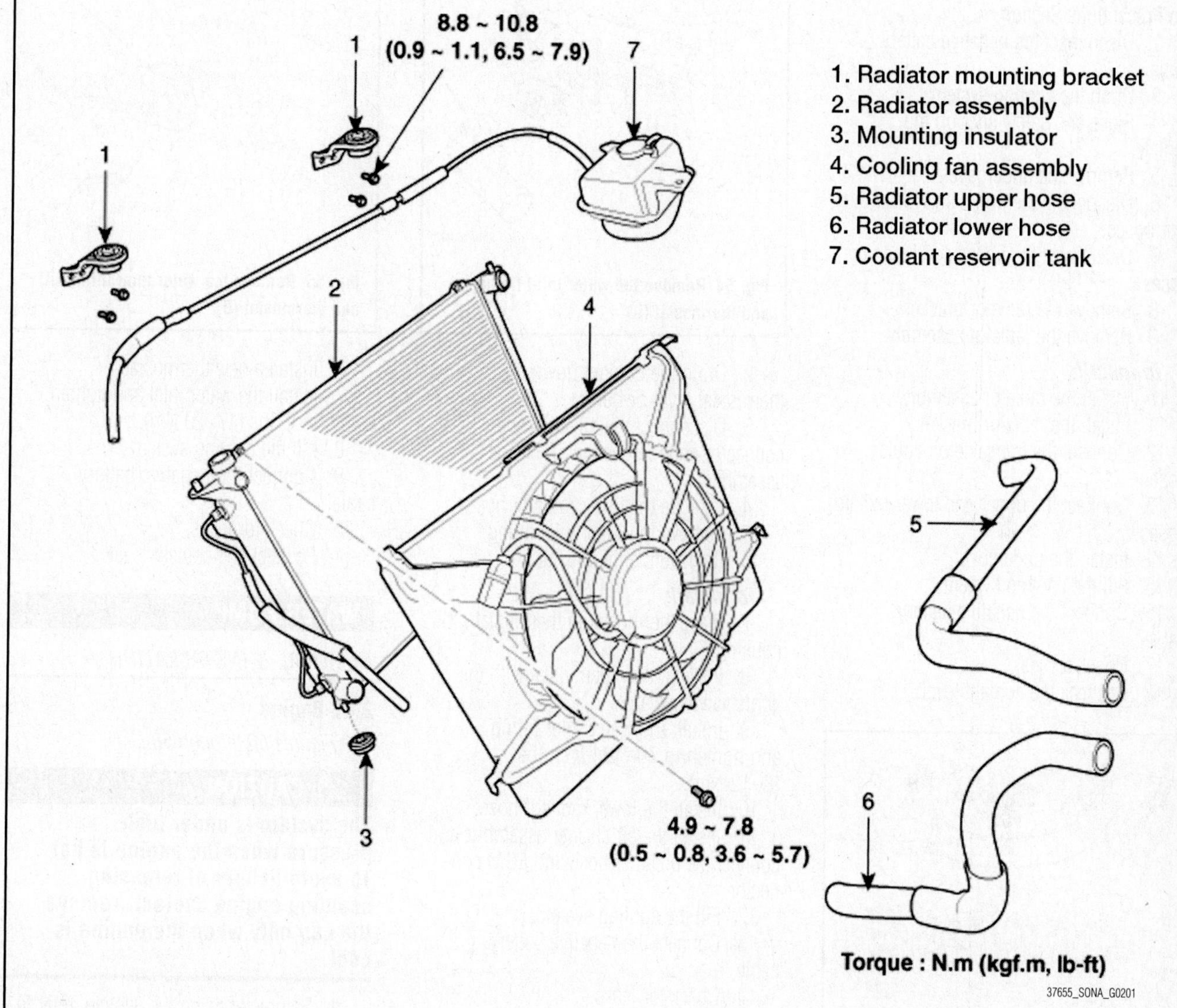

8.8 ~ 10.8
(0.9 ~ 1.1, 6.5 ~ 7.9)
1
7
1
2
4
5
6
3
4.9 ~ 7.8
(0.5 ~ 0.8, 3.6 ~ 5.7)
1. Radiator mounting bracket
2. Radiator assembly
3. Mounting insulator
4. Cooling fan assembly
5. Radiator upper hose
6. Radiator lower hose
7. Coolant reservoir tank
Torque : N.m (kgf.m, lb-ft)
37655_SONA_G0201

Fig. 49 Disconnect fan motor connector (A), and remove the cooling fan (B)

B
A
37655_SONA_G0197

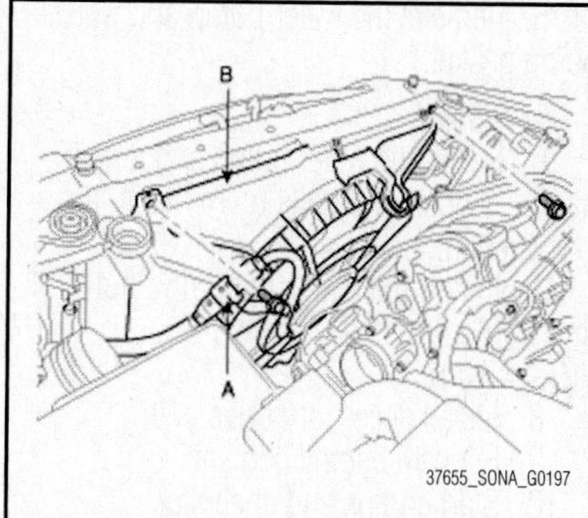

Fig. 50 Disconnect fan motor connector (A), and remove the cooling fan (B)

A
37655_SONA_G0198

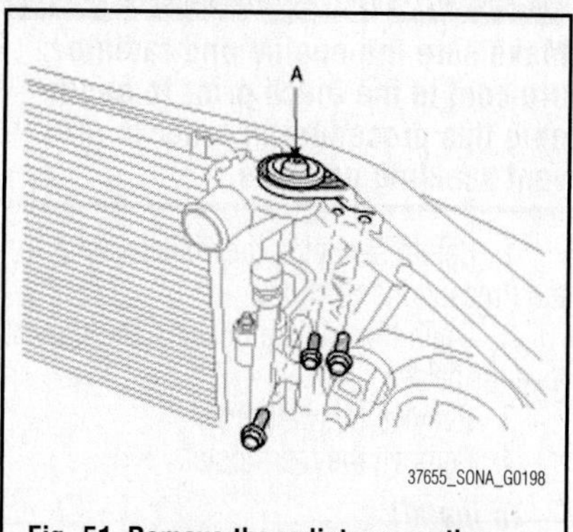

Fig. 51 Remove the radiator mounting bracket (A) and condenser mounting bolts

B
A
37655_SONA_G0199

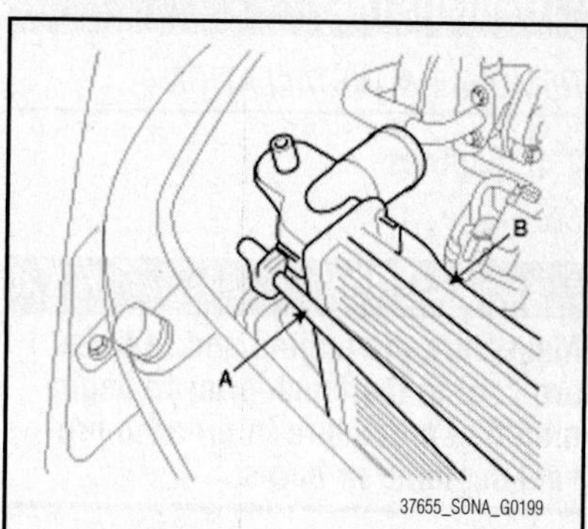

Fig. 52 Remove the reservoir hose and pipe (A) and radiator assembly (B)

3.3L Engines

See Figure 53.

1. Before servicing the vehicle, refer to the Precautions Section.
2. Disconnect the negative battery cable.
3. Drain the cooling system.
4. Raise and safely support the vehicle.
5. Remove the undercover.
6. Disconnect the upper and lower radiator hoses.
7. Disconnect the transaxle oil cooler hoses.
8. Remove the radiator bracket.
9. Remove the radiator assembly.

To install:

10. Install the radiator assembly.
11. Install the radiator bracket.
12. Connect the transaxle oil cooler hoses.
13. Connect the upper and lower radiator hoses.
14. Install the undercover.
15. Fill the cooling system.
16. Connect the negative battery cable.
17. Check for leaks.
18. Recheck the coolant level.

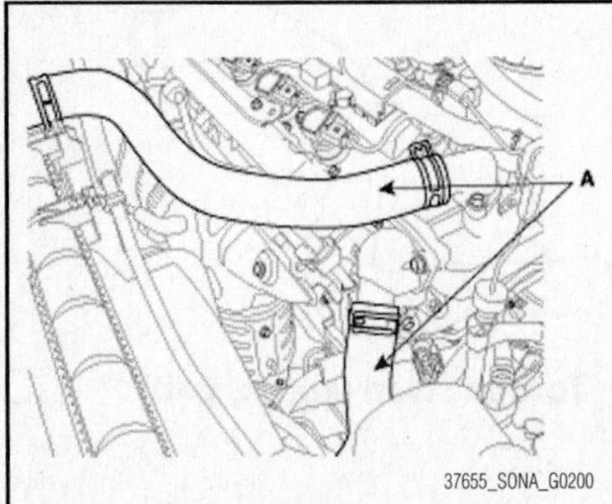

Fig. 53 Remove the radiator upper hose (A) and lower hose (B)

THERMOSTAT

REMOVAL & INSTALLATION

2.4L Engines

See Figure 54.

✱✱ CAUTION

Make sure the engine and radiator are cool to the touch prior to beginning this procedure in order to prevent scalding or burns.

1. Before servicing the vehicle, refer to the Precautions Section.

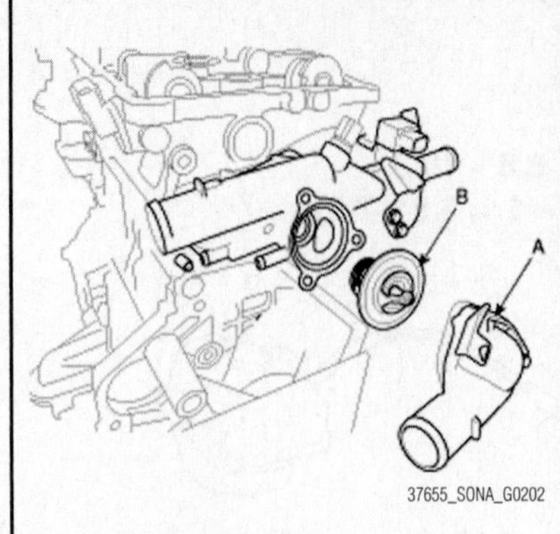

Fig. 54 Remove the water inlet fitting (A) and thermostat (B)

2. Drain the coolant down to thermostat level or below.
3. Disconnect the breather hose, ECM connector and remove the air cleaner assembly.
4. Remove the lower radiator hose.
5. Remove the water inlet fitting.
6. Remove the thermostat.

To install:

7. Place thermostat in thermostat housing.
8. Install a new thermostat with the jiggle valve upward.
9. Install the water inlet fitting and tighten to 70–104 inch lbs (8–12 Nm).
10. Install the lower radiator hose.
11. Install the air cleaner assembly, and connect the breather hose and ECM connector.
12. Fill the cooling system.
13. Connect the negative battery cable.
14. Check for leaks.
15. Recheck the coolant level.

3.3L Engines

See Figure 55.

✱✱ CAUTION

Make sure the engine and radiator are cool to the touch prior to beginning this procedure in order to prevent scalding or burns.

1. Before servicing the vehicle, refer to the Precautions Section.
2. Drain the coolant down to thermostat level or below.
3. Remove the water inlet.
4. Remove the thermostat.

To install:

5. Place the thermostat in thermostat housing with the jiggle valve upward.

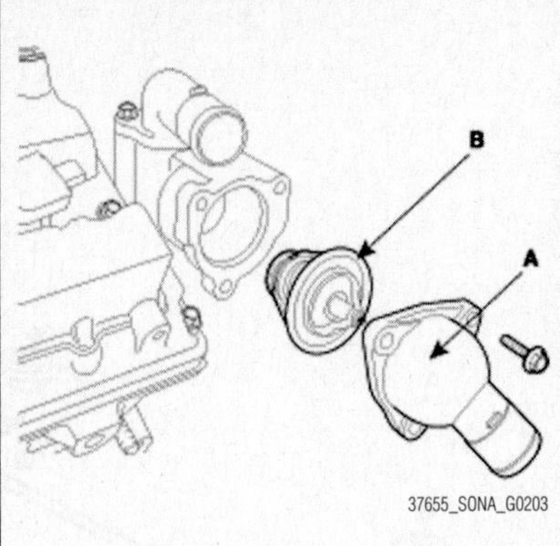

Fig. 55 Remove the water inlet fitting (A) and thermostat (B)

6. Install a new thermostat.
7. Install the water inlet and tighten to 12–14 ft. lbs (17–20 Nm).
8. Fill the cooling system.
9. Connect the negative battery cable.
10. Check for leaks.
11. Recheck the coolant level.

WATER PUMP

REMOVAL & INSTALLATION

2.4L Engine

See Figures 56 through 58.

✱✱ CAUTION

The system is under high pressure when the engine is hot. To avoid danger of releasing scalding engine coolant, remove the cap only when the engine is cool.

1. Before servicing the vehicle, refer to the Precautions Section.
2. Drain the engine coolant.
3. Remove the accessory drive belt.
4. Remove the water inlet pipe nut.
5. Remove the water pump and water pump gasket.

To install:

6. Install the water pump and a new gasket. Tighten the bolts to 14–17 ft. lbs. (19–24 Nm).
7. Install the water inlet pipe nut and tighten to 14–17 ft. lbs. (19–24 Nm).
8. Install accessory drive belt.
9. Fill with engine coolant.
10. Start engine and check for leaks.
11. Recheck engine coolant level.

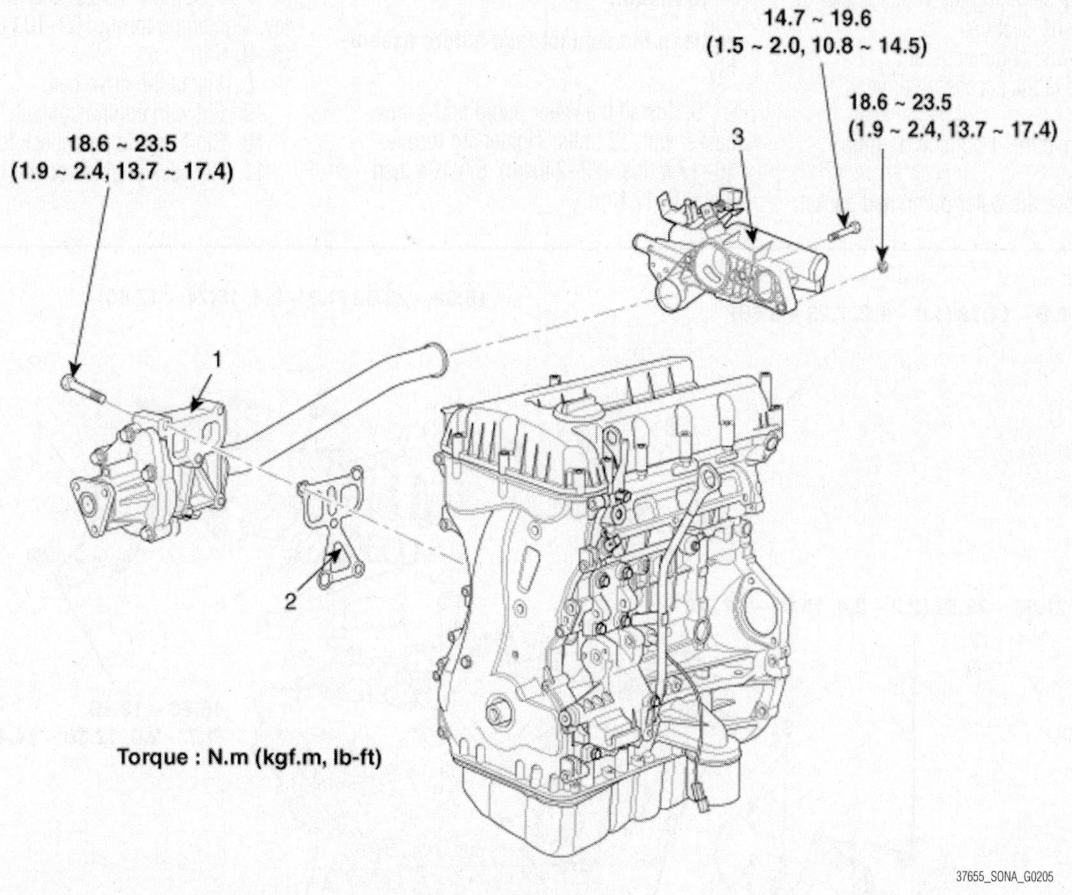

Fig. 56 Exploded view of the water pump assembly and related components—2.4L engine

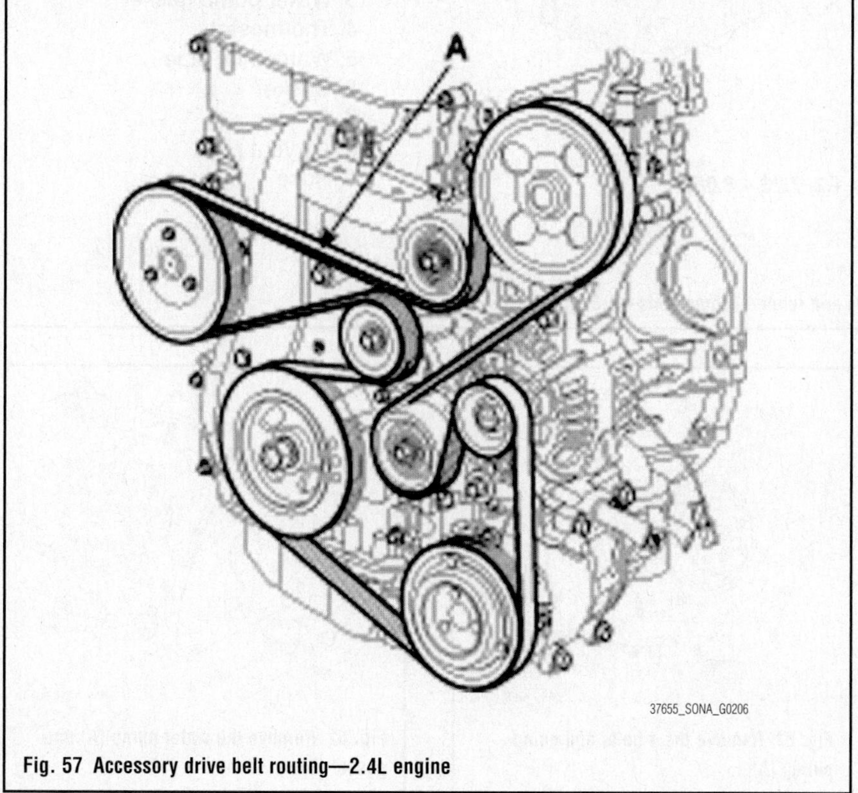

Fig. 57 Accessory drive belt routing—2.4L engine

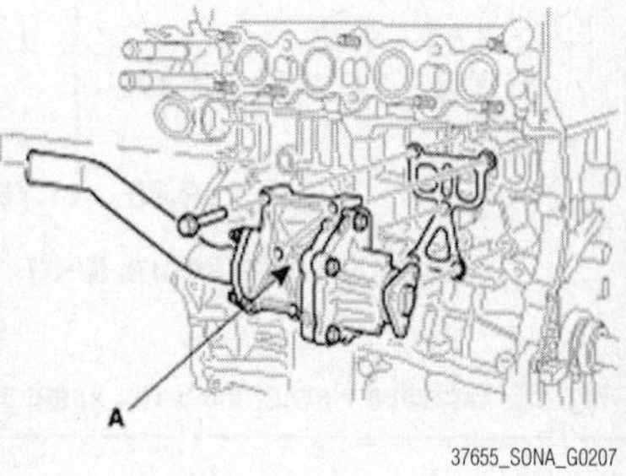

Fig. 58 Remove the water pump (A) and gasket

3.3L Engines

See Figures 59 through 62.

CAUTION

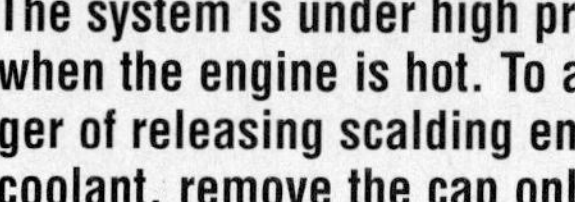

The system is under high pressure when the engine is hot. To avoid danger of releasing scalding engine coolant, remove the cap only when the engine is cool.

1. Before servicing the vehicle, refer to the Precautions Section.
2. Drain the engine coolant.
3. Remove the accessory drive belt.
4. Remove the 4 bolts and pump pulley.
5. Remove the water pump and gasket.

To install:

➡Clean the contact face before assembly.

6. Install the water pump and a new gasket with 12 bolts. Tightening torque: 16–17 ft. lbs. (22–24 Nm); 87–104 inch lbs. (10–12 Nm).
7. Install the 4 bolts and the pump pulley. Tightening torque: 69–104 inch lbs. (8–10 Nm).
8. Install the drive belt.
9. Fill with engine coolant.
10. Start engine and check for leaks.
11. Recheck engine coolant level.

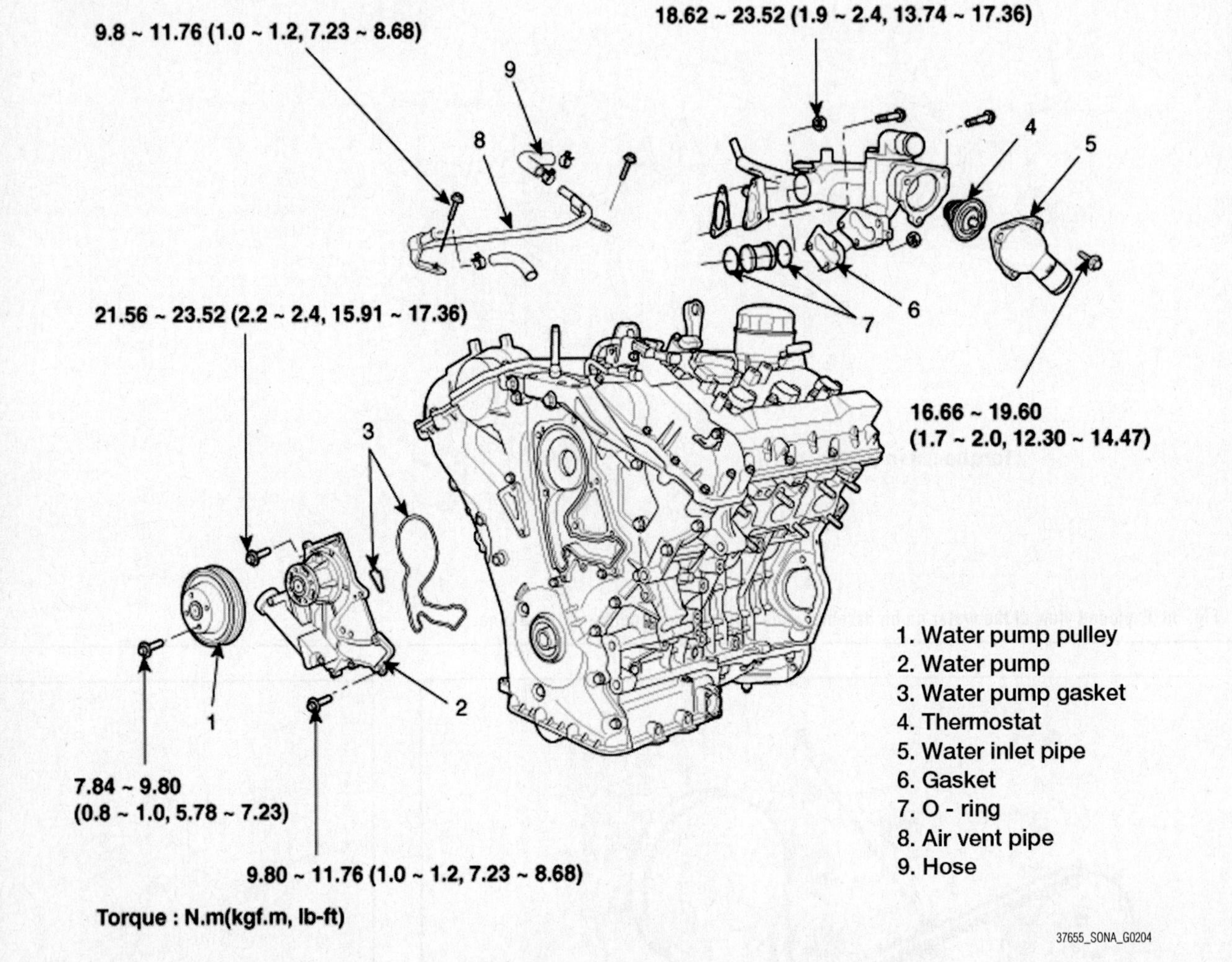

9.8 ~ 11.76 (1.0 ~ 1.2, 7.23 ~ 8.68)
18.62 ~ 23.52 (1.9 ~ 2.4, 13.74 ~ 17.36)
21.56 ~ 23.52 (2.2 ~ 2.4, 15.91 ~ 17.36)
16.66 ~ 19.60
(1.7 ~ 2.0, 12.30 ~ 14.47)
7.84 ~ 9.80
(0.8 ~ 1.0, 5.78 ~ 7.23)
9.80 ~ 11.76 (1.0 ~ 1.2, 7.23 ~ 8.68)
Torque : N.m(kgf.m, lb-ft)
1. Water pump pulley
2. Water pump
3. Water pump gasket
4. Thermostat
5. Water inlet pipe
6. Gasket
7. O - ring
8. Air vent pipe
9. Hose
37655_SONA_G0204

Fig. 59 Exploded view of the water pump assembly and related components—3.3L engine

A
22140_HYUN_G0032

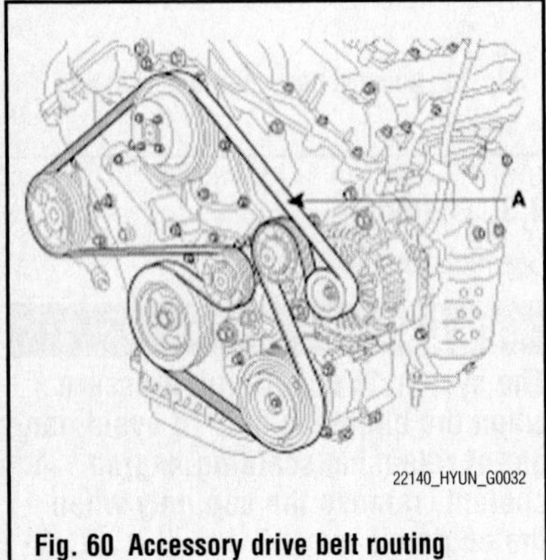

Fig. 60 Accessory drive belt routing

A
37655_AZER_G0182

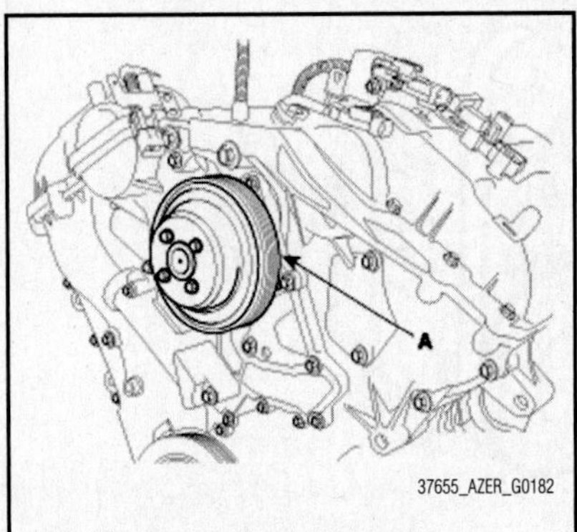

Fig. 61 Remove the 4 bolts and pump pulley (A)

B
A
37655_AZER_G0183

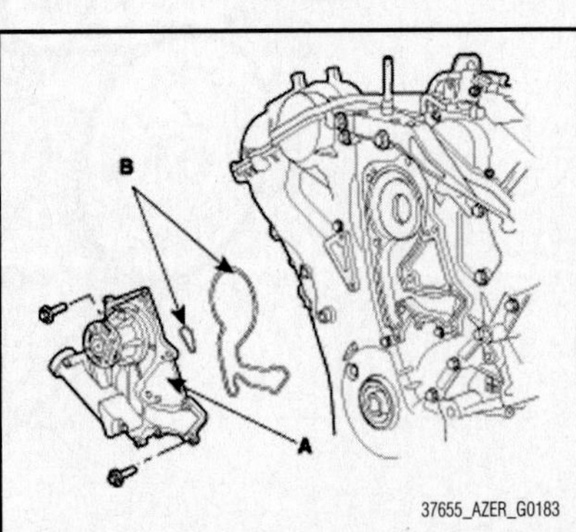

Fig. 62 Remove the water pump (A) and gasket (B)

ENGINE ELECTRICAL — CHARGING SYSTEM

ALTERNATOR

REMOVAL & INSTALLATION

See Figures 63 through 66.

1. Before servicing the vehicle, refer to the Precautions Section.
2. Disconnect the battery negative terminal first, then the positive terminal.
3. Disconnect the alternator connector, and remove the cable from alternator "B" terminal.
4. Remove the drive belt.
5. Pull out the through bolt and then remove the alternator.

To install:

6. Installation is the reverse of removal.

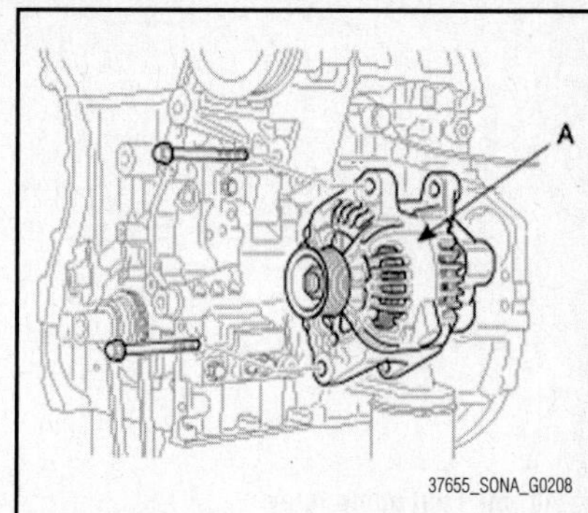

Fig. 65 Remove the alternator (A)–2.4L engine

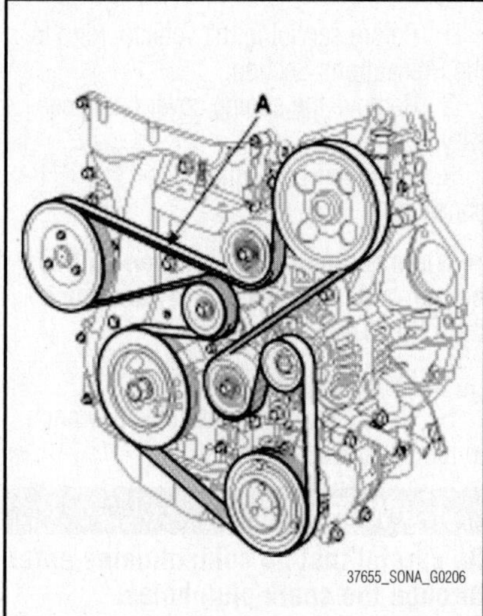

Fig. 63 Accessory drive belt (A) routing—2.4L engine

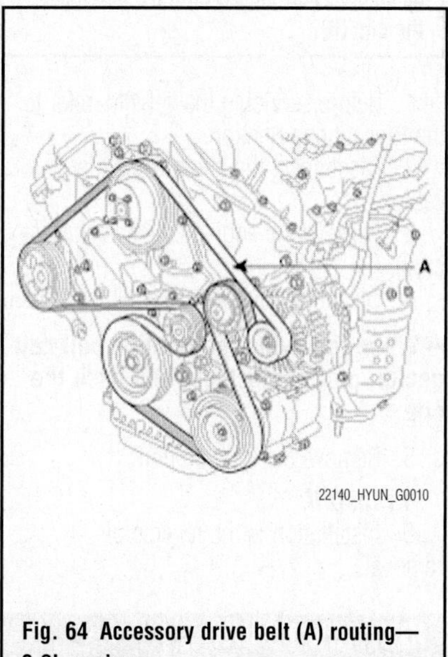

Fig. 64 Accessory drive belt (A) routing—3.3L engines

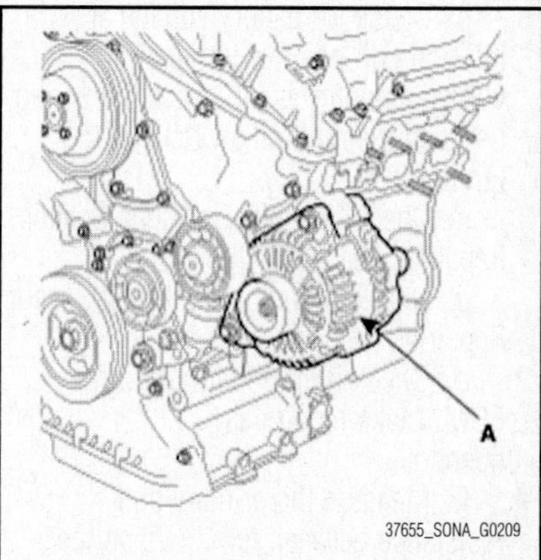

Fig. 67 Remove the alternator (A)—3.3L engine

ENGINE ELECTRICAL — IGNITION SYSTEM

FIRING ORDER

See Figures 67 and 68.

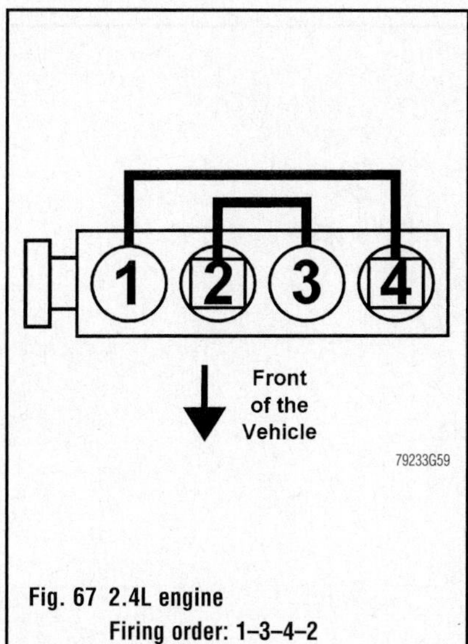

Fig. 67 2.4L engine
Firing order: 1–3–4–2

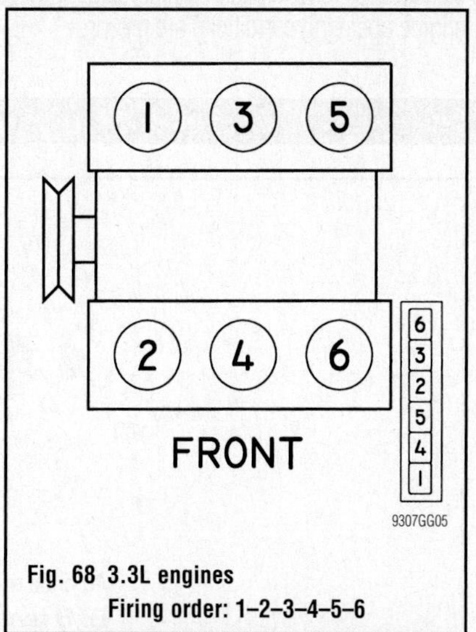

Fig. 68 3.3L engines
Firing order: 1–2–3–4–5–6

IGNITION COIL

TESTING

See Figure 69.

1. Before servicing the vehicle, refer to the Precautions Section.
2. Remove the ignition coil connector.
3. Remove the ignition coil.
4. Using a spark plug socket, remove the spark plug.
5. Install the spark plug to the ignition coil.
6. Ground the spark plug to the engine.
7. Check if spark occurs while engine is being cranked.

➡To prevent fuel being injected from injectors while the engine is being cranked, remove the fuel pump relay from the fuse box.

8. Crank the engine for no more than 5–10 seconds.
9. Inspect for spark at each of the spark plugs one at a time.
10. If the spark test fails:
    a. Check connection at ignition coil

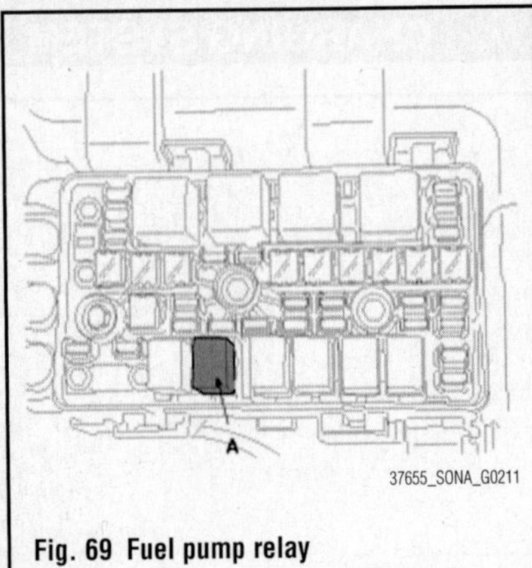

Fig. 69 Fuel pump relay

connector. If faulty, make necessary repairs.

b. Check for battery voltage at ignition coil positive terminal. If voltage is not present, check the wiring between the ignition switch and the ignition coil.

c. Check for faulty Camshaft Sensor. Repair as needed.

d. Check for faulty Crankshaft Sensor. Repair as needed.

e. Check ignition signal from ECM. Make repairs as needed.

f. Measure the primary coil resistance between terminals on the effected ignition coil. Standard value: 0.62 ohms plus or minus 10 percent. If out of specification, replace the ignition coil.

REMOVAL & INSTALLATION

See Figure 70.

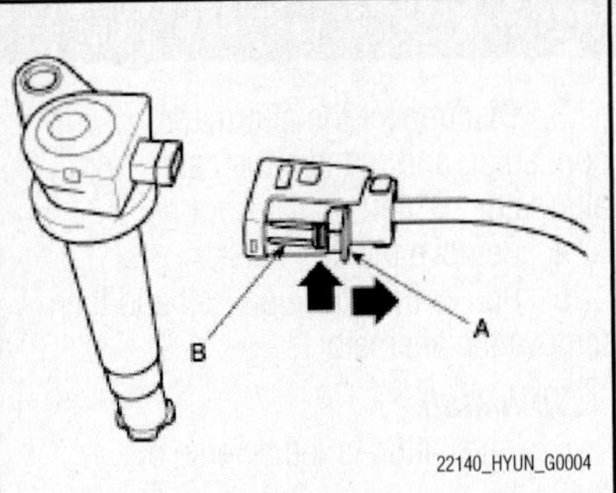

Fig. 70 When removing the ignition coil connector, pull the lock pin (A) and push the clip (B)

1. Before servicing the vehicle, refer to the Precautions Section.
2. Disconnect the negative battery cable.
3. Remove the engine cover (as necessary).
4. Disconnect the ignition coil connector.

➡When removing the ignition coil connector, pull the lock pin and push the clip.

5. Remove the ignition coil.

To install:

6. Installation is the reverse of removal.

IGNITION TIMING

ADJUSTMENT

Ignition timing is controlled by the electronic control ignition timing system. The standard reference ignition timing data for the engine operating conditions are pre-programmed in the memory of the Engine Control Module (ECM). The engine operating conditions (speed, load, warm-up condition, etc.) are detected by the various sensors. Based on these sensor signals and the ignition timing data, signals to interrupt the primary current are sent to the ECM. The ignition coil is activated, and timing is controlled.

SPARK PLUGS

REMOVAL & INSTALLATION

See Figure 70.

1. Before servicing the vehicle, refer to the Precautions Section.
2. Remove the engine cover (as necessary).
3. Disconnect the ignition coil connector.

➡When removing the ignition coil connector, pull the lock pin (A) and push the clip (B).

4. Remove the ignition coil.
5. Use a spark plug socket and wrench to remove the spark plugs.

⁂ WARNING

Be careful that no contaminates enter through the spark plug holes.

To install:

➡Check the electrode gap on the spark plugs before installation. Specification: 0.039–0.043 in. (1.0–1.1 mm).

6. To install, reverse the removal procedure. Tighten the spark plugs to 11 ft. lbs. (15 Nm).

ENGINE ELECTRICAL

STARTING SYSTEM

STARTER

REMOVAL & INSTALLATION

See Figure 71.

1. Before servicing the vehicle, refer to the Precautions Section.
2. Disconnect the negative battery cable.
3. Disconnect the starter cable from the B terminal on the solenoid, then disconnect the connector from the S terminal.
4. Remove the 2 bolts holding the starter.
5. Remove the starter.

To install:

6. Installation is the reverse of removal. Tighten starter motor bolts to 20–25 ft. lbs. (27–34 Nm).

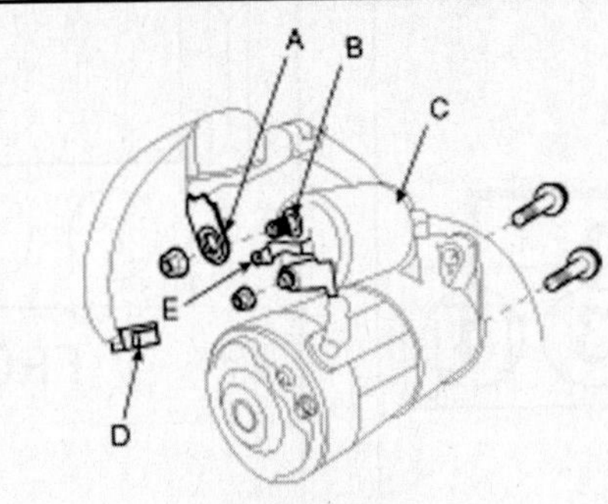

Fig. 71 Starter components

ENGINE MECHANICAL

➡Disconnecting the negative battery cable may interfere with the functions of the on board computer systems and may require the computer to undergo a relearning process, once the negative battery cable is reconnected.

ACCESSORY DRIVE BELTS

ACCESSORY BELT ROUTING

See Figures 72 and 73.

Refer to the accompanying illustrations.

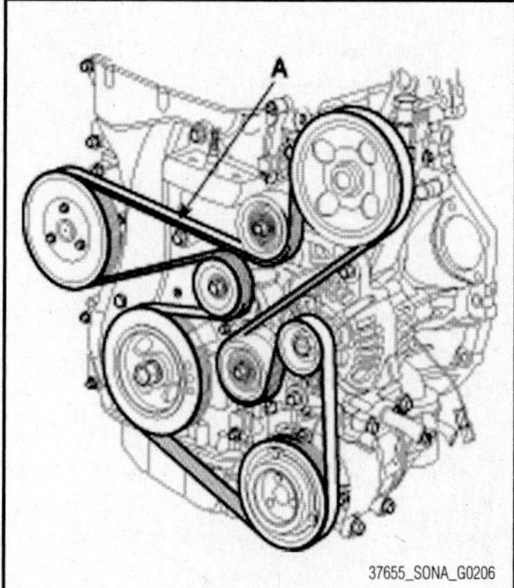

Fig. 72 Accessory drive belt (A) routing—2.4L engine

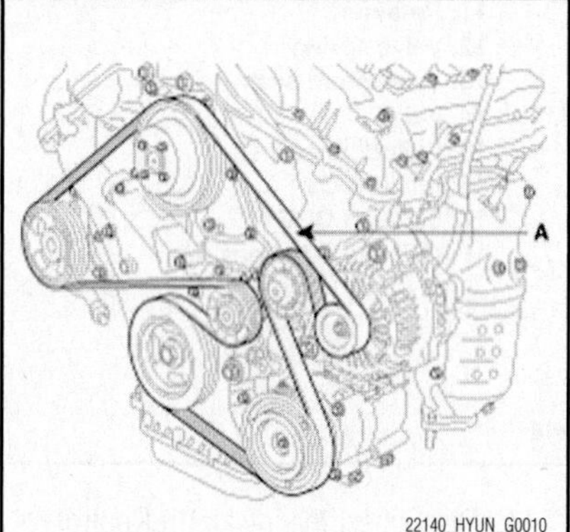

Fig. 73 Accessory drive belt (A) routing—3.3L engines

INSPECTION

Inspect the accessory drive belt for signs of glazing or cracking. A glazed belt will be perfectly smooth from slippage, while a good belt will have a slight texture of fabric visible. Cracks will usually start at the inner edge of the belt and run outward. All worn or damaged accessory drive belts should be replaced immediately.

ADJUSTMENT

The drive belt is tension controlled by a drive belt tensioner, no adjustment is possible or necessary.

REMOVAL & INSTALLATION

See Figures 72 and 73.

1. Before servicing the vehicle, refer to the Precautions Section.
2. Raise and support the vehicle.
3. Remove the engine splash shield.
4. Rotate the drive belt tensioner clockwise to release the drive belt tension.
5. Remove the drive belt from the alternator.
6. Slowly release the drive belt tensioner.
7. Remove the drive belt from the accessory drive pulleys.

To install:

8. Install the drive belt to the accessory drive pulley.
9. Rotate the drive belt tensioner clockwise.
10. Install the drive belt.
11. Ensure the drive belt is properly aligned and seated into the grooves of the accessory drive pulleys.
12. Slowly release the drive belt tensioner.
13. Install the engine splash shield.
14. Lower the vehicle.

CAMSHAFT AND VALVE LIFTERS

INSPECTION

See Figures 74 and 75.

1. Inspect the cam lobes:
   a. Using a micrometer, measure the cam lobe height.
   b. If the cam lobe height is less than specified, replace the camshaft.
2. Inspect the camshaft journal clearance:
   a. Clean the bearing caps and camshaft journals.
   b. Place the camshafts on the cylinder head.
   c. Lay a strip of Plastigage® across each of the camshaft journal.
   d. Install the bearing caps and tighten the bolts with specified torque.

➡Do not turn the camshaft.

   e. Remove the bearing caps.
   f. Measure the Plastigage® at its widest point.
   g. If the oil clearance is greater than specified, replace the camshaft. If necessary, replace the bearing caps and cylinder head as a set.
   h. Specifications are as follows:
      - No. 1 Journal: 0.0035 in. max
      - No. 2,3,4,5 Journal: 1.0047 in. max
      - Exhaust: 0.0047 in. max
   i. Completely remove the Plastigage®.
   j. Remove the camshafts.
3. Inspect the camshaft end play:
   a. Install the camshafts.
   b. Using a dial indicator, measure the end play while moving the camshaft back and forth.
   c. If the end play is greater than specified, replace the camshaft. If necessary, replace the bearing caps and cylinder head as a set.
   d. Specifications are as follows:
      - 0.0015–0.0062 in.
   e. Remove the camshafts.
4. Inspect the Continuous Variable Valve Timing (CVVT) assembly:
   a. Check that the CVVT assembly will not turn.
   b. Apply vinyl tape to all the parts except the one indicated by the arrow in the illustration.
   c. Wrap tape around the tip of the air gun and apply air of approx. 14 psi to the port of the camshaft. Perform this in order to release the lock pin for the maximum delay angle locking.

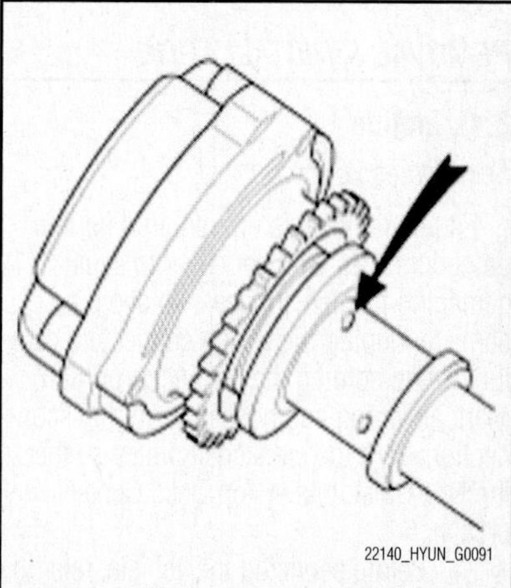

Fig. 74 Apply vinyl tape to the CVVT on all parts except the one indicated by the arrow

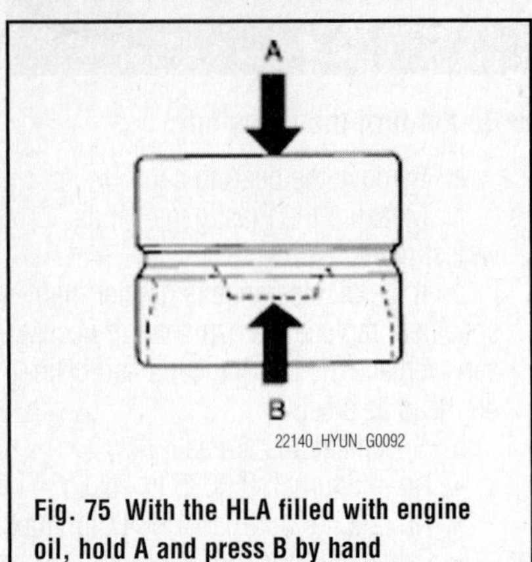

Fig. 75 With the HLA filled with engine oil, hold A and press B by hand

➡Wrap a shop rag around the CVVT as the oil may spray out when the air pressure is applied.

d. Under the condition of air pressure being applied, turn the CVVT assembly to the advance angle side with your hand.

- Depending on the air pressure, the CVVT assembly will turn to the advance side
- If air is leaking from the port and air pressure cannot be maintained, the locking pin will not release

5. Except the position where the lock pin meets at the maximum delay angle, let the CVVT assembly turn back and forth and check the movable range and that there is no disturbance.

a. The CVVT should move smoothly in the range of about 20°.

b. Turn the CVVT assembly with your hand and lock it at the maximum delay angle position.

6. Inspect the Hydraulic Lash Adjuster (HLA).

a. With the HLA filled with engine oil, hold A and press B by hand.

b. If B moves, replace the HLA.

REMOVAL & INSTALLATION

2.4L Engine

See Figures 76 through 92.

Engine removal is not required for this procedure. Use a fender cover to avoid damaging painted surfaces. To avoid damage, unplug the wiring connectors carefully while holding the connector portion. Mark all wiring and hoses to avoid misconnection. Turn the crankshaft pulley so that the No. 1 piston is at Top Dead Center (TDC).

1. Before servicing the vehicle, refer to the Precautions Section.
2. Disconnect the terminals from battery and remove the battery.

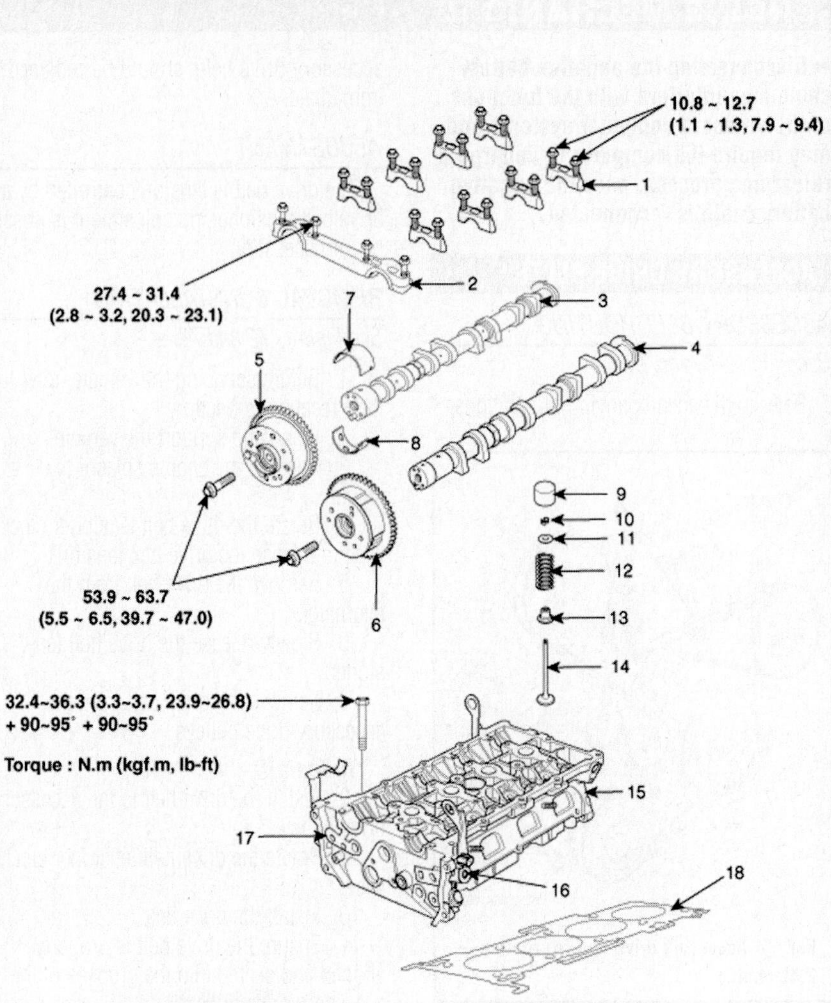

10.8 ~ 12.7 (1.1 ~ 1.3, 7.9 ~ 9.4)

27.4 ~ 31.4 (2.8 ~ 3.2, 20.3 ~ 23.1)

53.9 ~ 63.7 (5.5 ~ 6.5, 39.7 ~ 47.0)

32.4~36.3 (3.3~3.7, 23.9~26.8) + 90~95° + 90~95°

Torque : N.m (kgf.m, lb-ft)

1. Camshaft bearing cap
2. Camshaft front bearing cap
3. Exhaust camshaft
4. Intake camshaft
5. Exhaust CVVT assembly
6. Intake CVVT assembly
7. Exhaust camshaft upper bearing
8. Exhaust camshaft lower bearing
9. MLA
10. Retainer lock
11. Retainer
12. Valve spring
13. Valve stem seal
14. Valve
15. Cylinder head
16. Intake OCV
17. Exhaust OCV
18. Cylinder head gasket

37655_SONA_G0255

Fig. 76 Exploded view of cylinder head—2.4L engine

3. Remove the engine cover.
4. Remove the air duct.
5. Disconnect the breather hose.
6. Disconnect the ECM connector.
7. Remove the air cleaner assembly.
8. Remove the undercover.
9. Drain the engine coolant.
10. Remove the battery tray.
11. Remove the upper radiator hose and lower radiator hose.
12. Remove the heater hoses.
13. Disconnect the VIS connector, OPS connector, knock sensor connector and A/C switch connecter.
14. Disconnect the intake Oil Control Valve (OCV) connector.
15. Disconnect the injector connectors and ignition coil connectors.
16. Disconnect the ETC connector and MAP sensor connector.
17. Disconnect the VCM connector.
18. Disconnect the Camshaft Position sensor (CMP) connector, fuel hose, brake booster vacuum hose, and Purge Control Solenoid Valve (PCSV) hose.
19. Disconnect the PCSV connector, ECT connector, condenser connector, CKP sen-

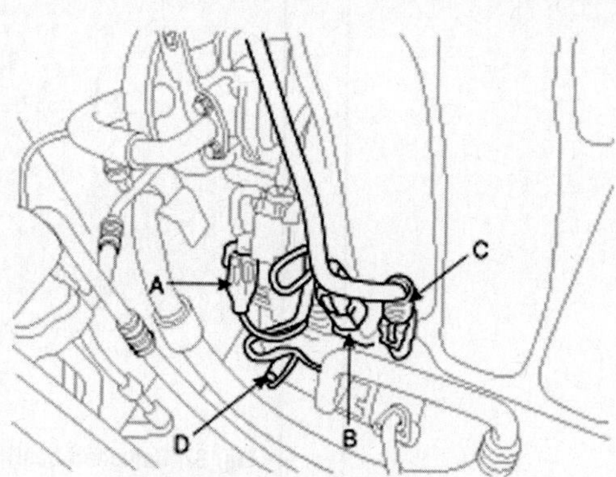

Fig. 77 Disconnect the VIS connector, OPS connector, knock sensor connector and A/C switch connecter

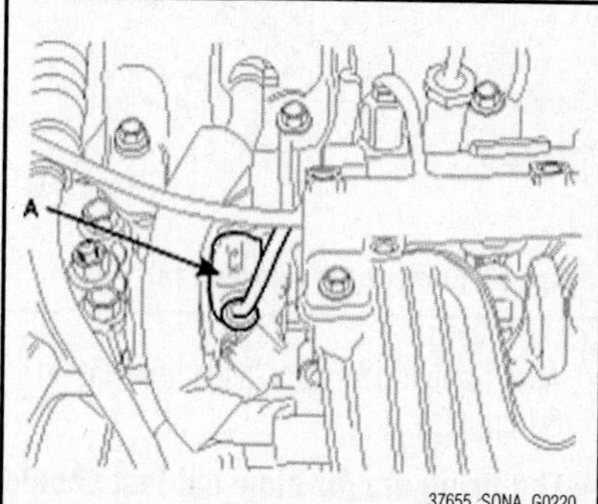

Fig. 78 Disconnect the intake OCV connector

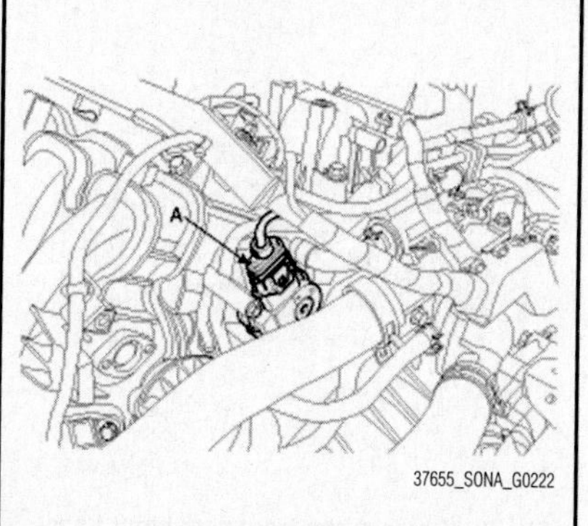

Fig. 80 Disconnect the VCM connector (A)

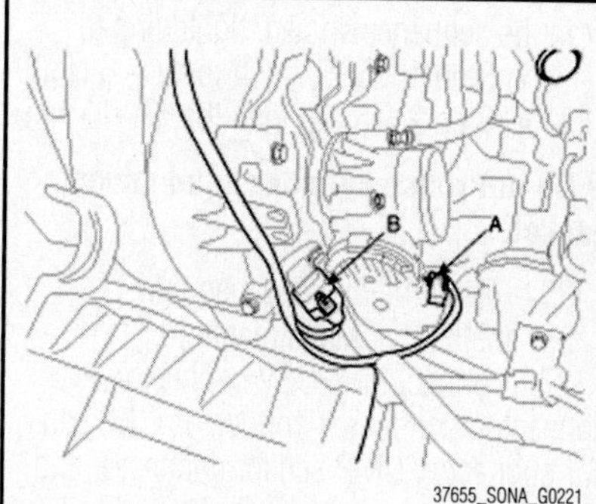

Fig. 79 Disconnect the ETC (A) and MAP sensor connectors (B)

sor connector, CMP sensor connector, and brake booster vacuum hose.

20. Disconnect the front and rear oxygen sensor connectors.
21. Remove the water temp control assembly.
22. Remove the timing chain. Refer to Timing Chain & Sprockets Removal & Installation.
23. Remove the intake and exhaust manifolds. Refer to Intake Manifold Removal & Installation, and Exhaust Manifold Removal & Installation.
24. Remove the intake and exhaust CVVT assembly.
25. Remove the camshaft:
    a. Remove the front cam shaft bearing cap.
26. Remove the exhaust cam shaft upper bearing.
27. Remove camshaft bearing caps, in the sequence shown.
28. Remove the camshaft.
29. Remove the exhaust camshaft lower bearing.

To install:

30. Install the camshafts:

➡Apply a light coat of engine oil on camshaft journals.

    a. Install the exhaust camshaft lower bearing.
    b. Install the camshafts.
    c. Install the exhaust camshaft upper bearing.
    d. Install the camshaft bearing caps in the tightening sequence shown, and tighten:
        - M6 bolts Step 1: 52 inch lbs. (6 Nm), Step 2: 95–113 inch lbs. (11–13 Nm)
        - M8 bolts Step 1: 11 ft. lbs. (15 Nm), Step 2: 20–23 ft. lbs. (28–31 Nm)

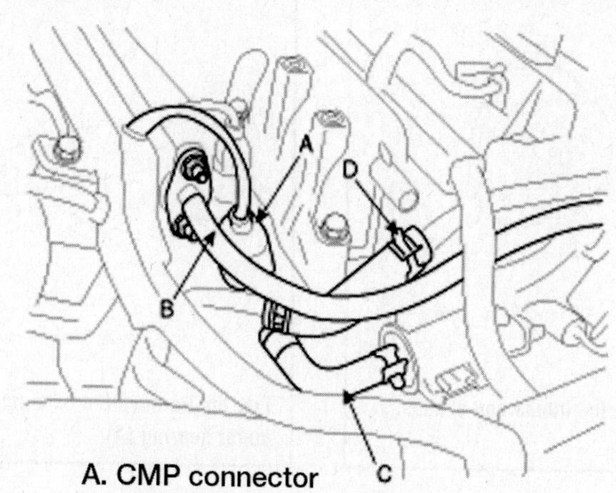

Fig. 81 Disconnect the CMP sensor connector, fuel hose, brake booster vacuum hose, and PCSV hose

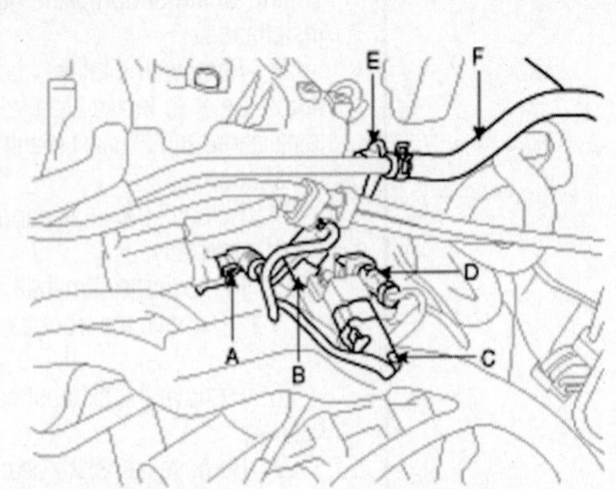

Fig. 82 Disconnect the PCSV connector, ECT connector, condenser connector, CKP sensor connector, CMP sensor connector, and brake booster vacuum hose

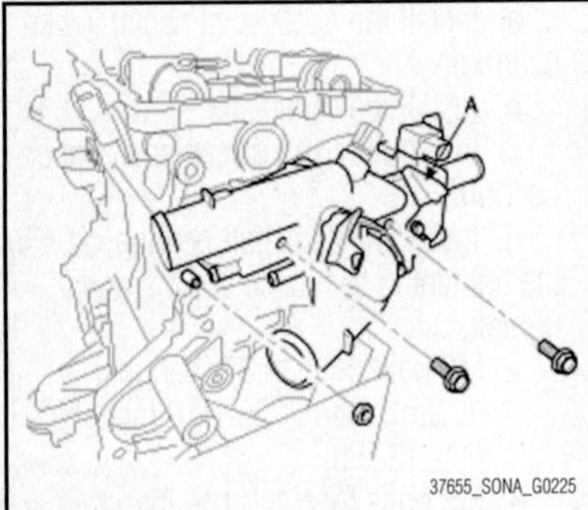

Fig. 83 Remove the water temp control assembly (A)

Fig. 85 Remove the front cam shaft bearing cap (A)

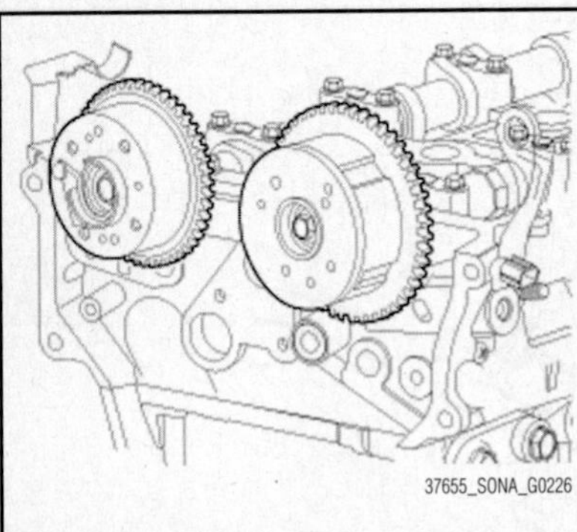

Fig. 84 Remove the intake and exhaust CVVT assembly

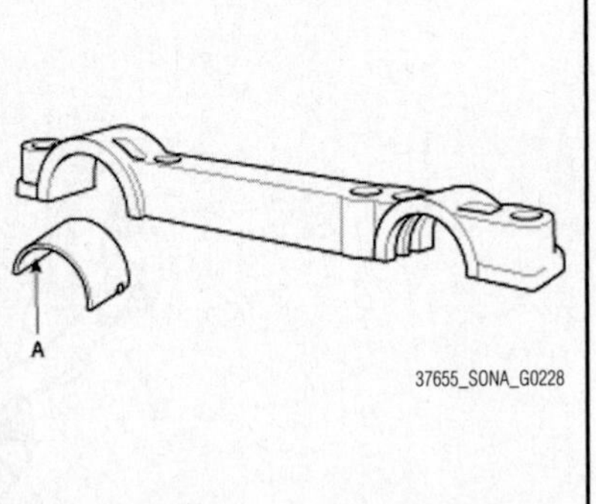

Fig. 86 Remove the exhaust cam shaft upper bearing (A)

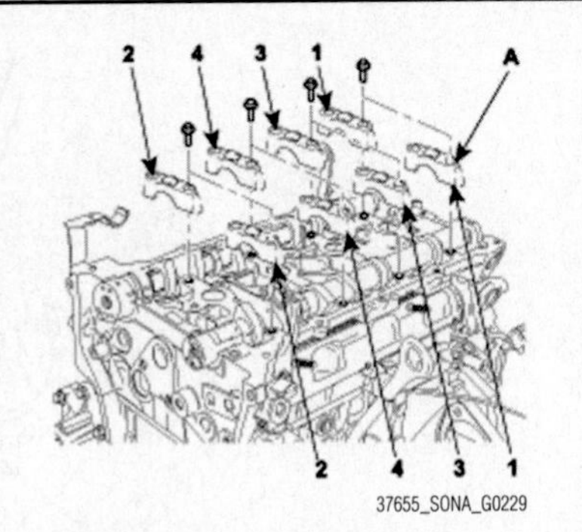

Fig. 87 Camshaft bearing cap removal sequence

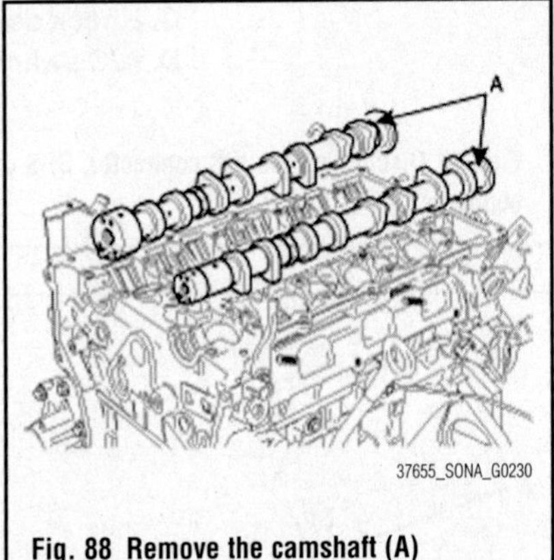

Fig. 88 Remove the camshaft (A)

31. Install the water temp control assembly, and tighten the bolts and nut to 14–17 ft. lbs. (19–24 Nm).

➡Assemble water temp control assembly and water inlet pipe to water pump assembly before nuts for assembling of water inlet pipe to be tightened.

➡Always use a new O-ring.

32. Install the timing chain.
33. Check and adjust valve clearance.
34. Install the cylinder head cover:
    a. Remove existing sealant and apply new sealant [Sealant: LOCTITE 5900H, Bead width: 0.1 in. (2.5mm)] as shown.

After applying sealant, assemble within 5 minutes.

➡The firing and/or blow out test should not be performed within 30 minutes after the cylinder head cover was assembled.

    b. Install the cylinder head cover bolts in the sequence shown, tightening to:
    - Step 1: 35–52 inch lbs. (4–6 Nm)
    - Step 2: 70–86 inch lbs. (8–10 Nm)

➡Do not reuse cylinder head cover gasket.

35. Install the exhaust manifold.
36. Install the intake manifold.
37. Connect the PCSV connector, ECT connector, condenser connector, CKP sensor connector, CMP sensor connector and brake booster vacuum hose.
38. Connect the front oxygen sensor connector and rear oxygen sensor connector.
39. Connect the brake booster vacuum hose, PCSV hose fuel hose and CMP sensor connector.
40. Connect the ETC connector and MAP sensor connector.
41. Connect the injector connectors and ignition coil connectors.
42. Connect the OCV connector.

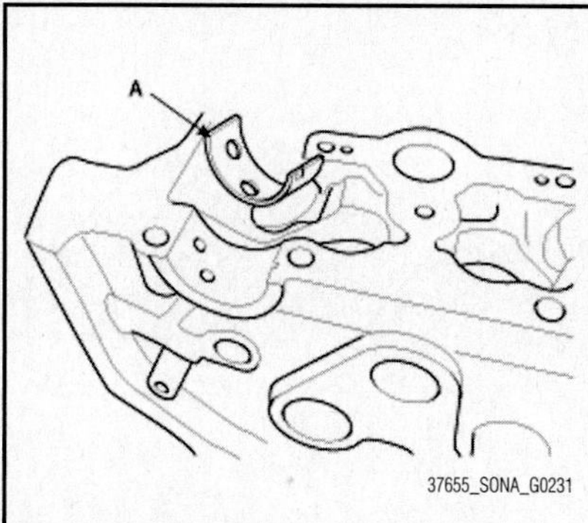

Fig. 89 Remove the exhaust camshaft lower bearing (A)

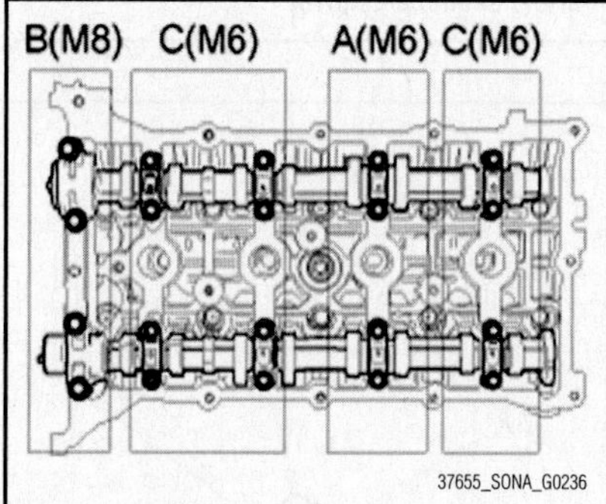

Fig. 90 Camshaft bearing cap Tightening order: Group A, Group B, Group C

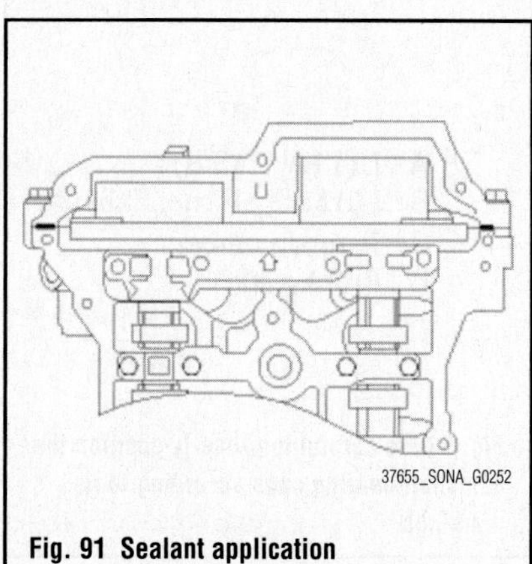

Fig. 91 Sealant application

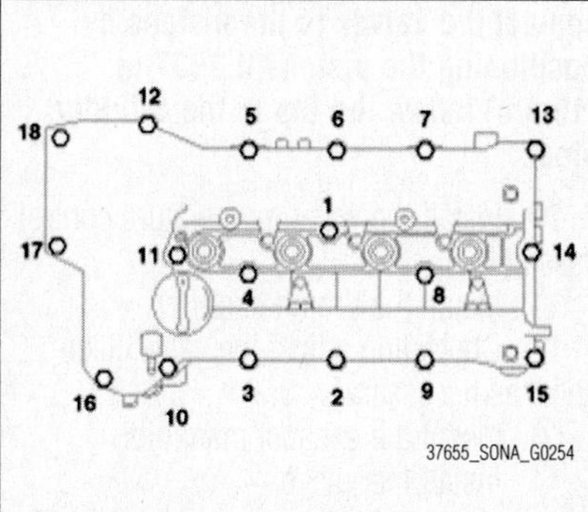

Fig. 92 Cylinder head bolt tightening sequence

43. Connect the VIS connector, OPS connector, knock sensor connector and A/C switch connecter.
44. Install the heater hoses.
45. Install the upper and lower radiator hoses.
46. Install the intake air hose and air cleaner assembly.
47. Install the engine cover.
48. Connect the negative battery cable.
49. Fill with engine coolant.
50. Start the engine and check for leaks.
51. Recheck the engine coolant level and oil level.

3.3L Engines

See Figures 93 through 98.

Use a fender cover to avoid damaging painted surfaces. To avoid damage, unplug the wiring connectors carefully while holding the connector portion. Mark all wiring and hoses to avoid misconnection. Turn the crankshaft pulley so that the No. 1 piston is at Top Dead Center (TDC).

1. Before servicing the vehicle, refer to the Precautions Section.
2. Remove the exhaust manifold. Refer to Exhaust Manifold Removal & Installation.
3. Remove the intake manifold. Refer to Intake Manifold Removal & Installation.
4. Remove the timing chain. Refer to Timing Chain & Sprockets Removal & Installation.
5. Remove the water temperature control assembly.
6. Remove the camshaft bearing cap.
7. Remove camshaft assembly.

To install:

8. Install the camshafts.
   a. Apply a light coat of engine oil on camshaft journals.
   b. Assemble the key groove of camshaft rear side to the same level of head top surface.
   c. Be careful to get the right bank, left bank, intake side, and exhaust side in the correct position before assembling.
9. Install the camshaft bearing caps in the sequence shown:

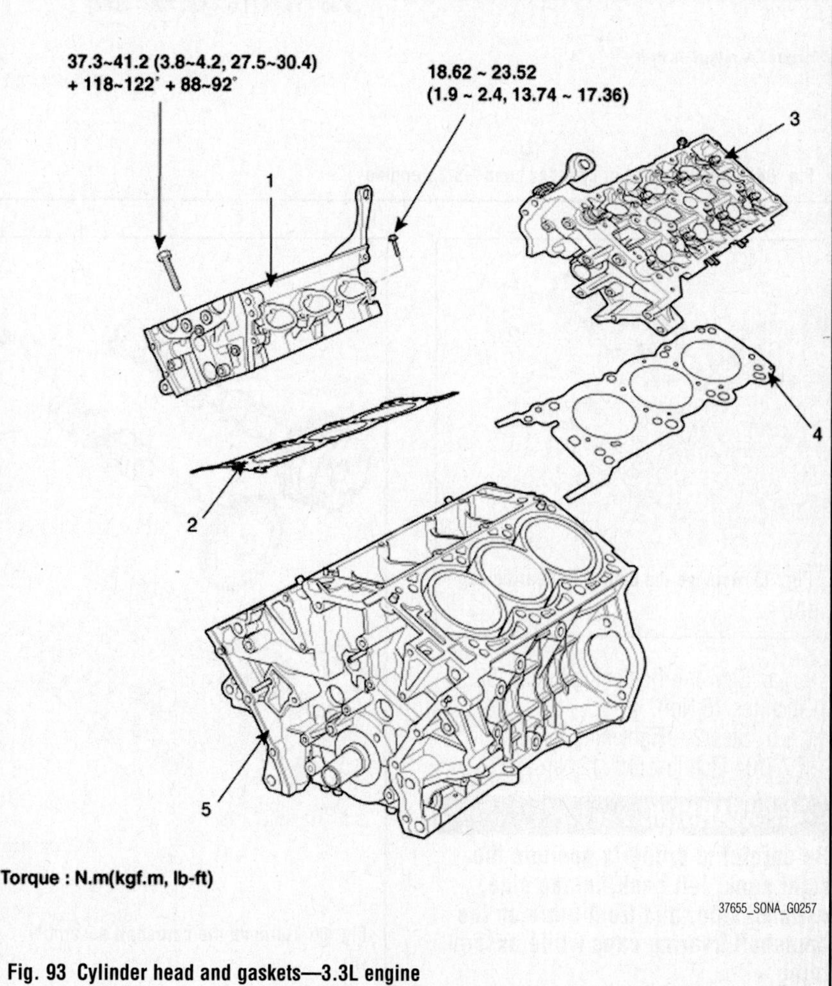

Fig. 93 Cylinder head and gaskets—3.3L engine

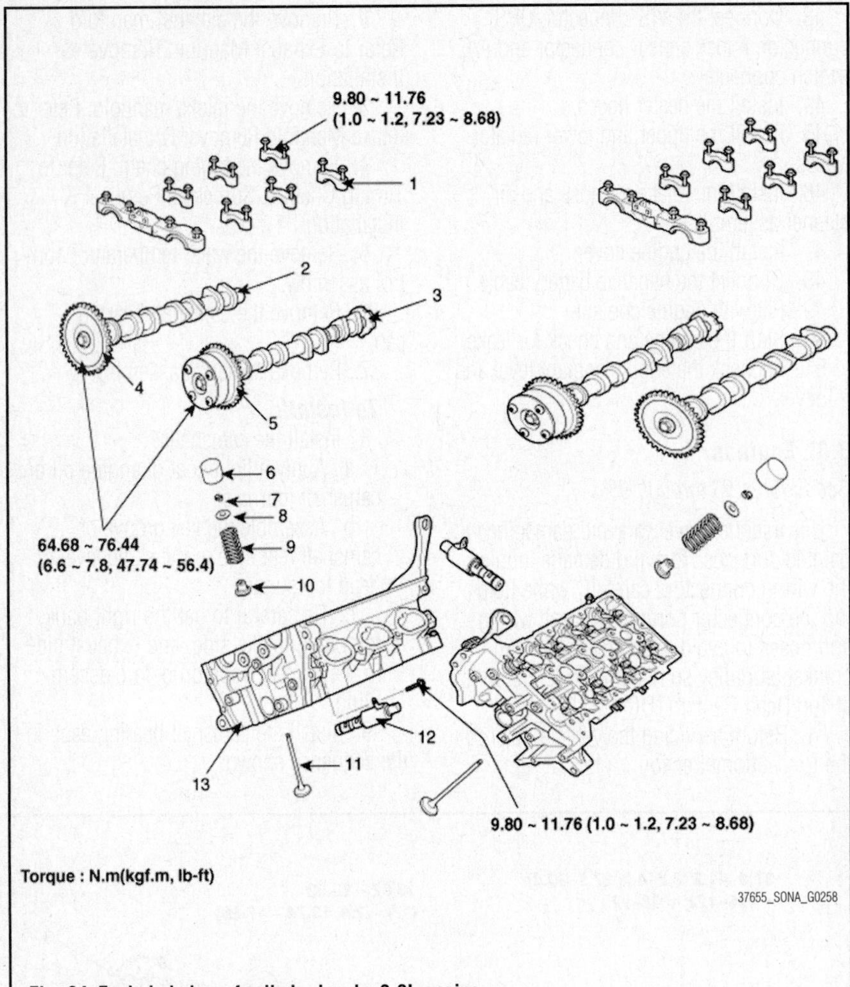

Fig. 94 Exploded view of cylinder head—3.3L engine

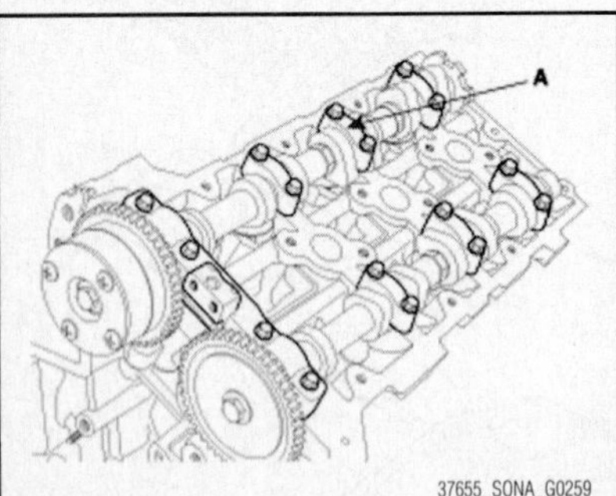

Fig. 95 Remove the camshaft bearing cap (A)

a. Step 1—Tightening torque: 52 inch lbs. (6 Nm).

b. Step 2—Tightening torque: 87–104 inch lbs. (10–12 Nm).

WARNING

Be careful to properly position the right bank, left bank, intake side, exhaust side, and front mark on the camshaft bearing caps while assembling.

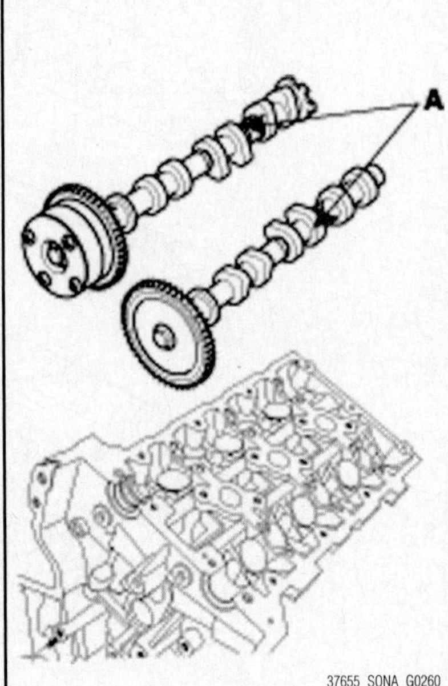

Fig. 96 Remove the camshaft assembly (A)

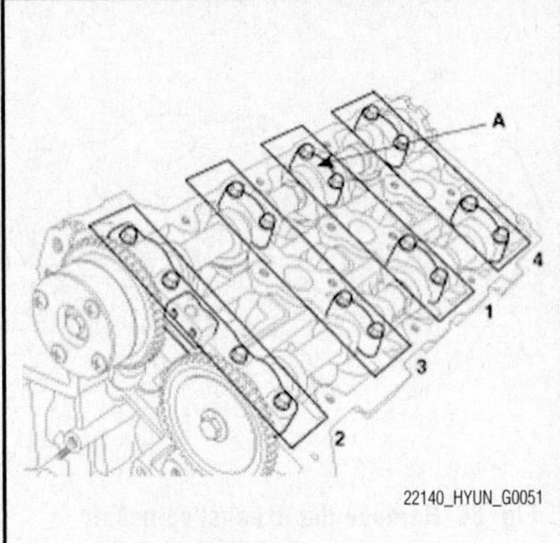

Fig. 97 Install the camshaft bearing caps in the sequence shown

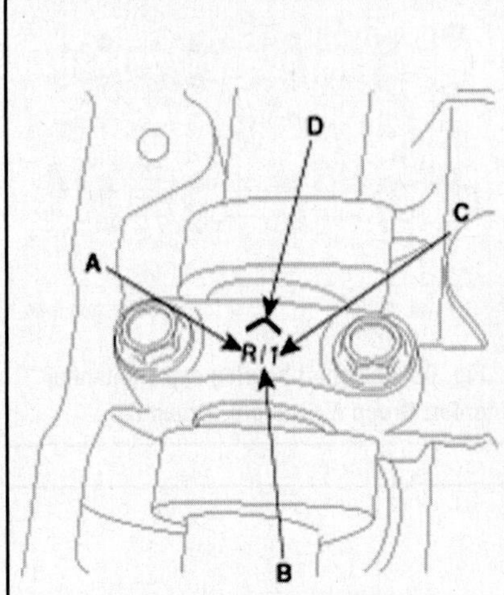

Fig. 98 Be careful to properly position the camshaft bearing caps according to its markings

➡Rotate the crankshaft so as not to contact the valves to the pistons by positioning the pistons 0.3937 in. (10mm) below the top of the cylinder block.

10. Install the water temperature control assembly.
11. Install the timing chain.
12. Check and adjust the valve clearance, as necessary.
13. Install the exhaust manifold.
14. Install the intake manifold.
15. Connect the negative battery cable.

16. Fill with engine coolant.
17. Start the engine and check for leaks.
18. Recheck the engine coolant level and oil level.

CATALYTIC CONVERTER

REMOVAL & INSTALLATION

See Figure 99.

No Removal & Installation procedure is given by the manufacturer. Refer to the accompanying illustrations for component locations.

CRANKSHAFT FRONT SEAL

REMOVAL & INSTALLATION

2.4L Engine

See Figures 101 and 102.

1. Before servicing the vehicle, refer to the Precautions Section.
2. Disconnect the negative battery cable.
3. Remove the right-hand front wheel.
4. Remove the right-hand side cover.
5. Remove the engine mounting bracket.

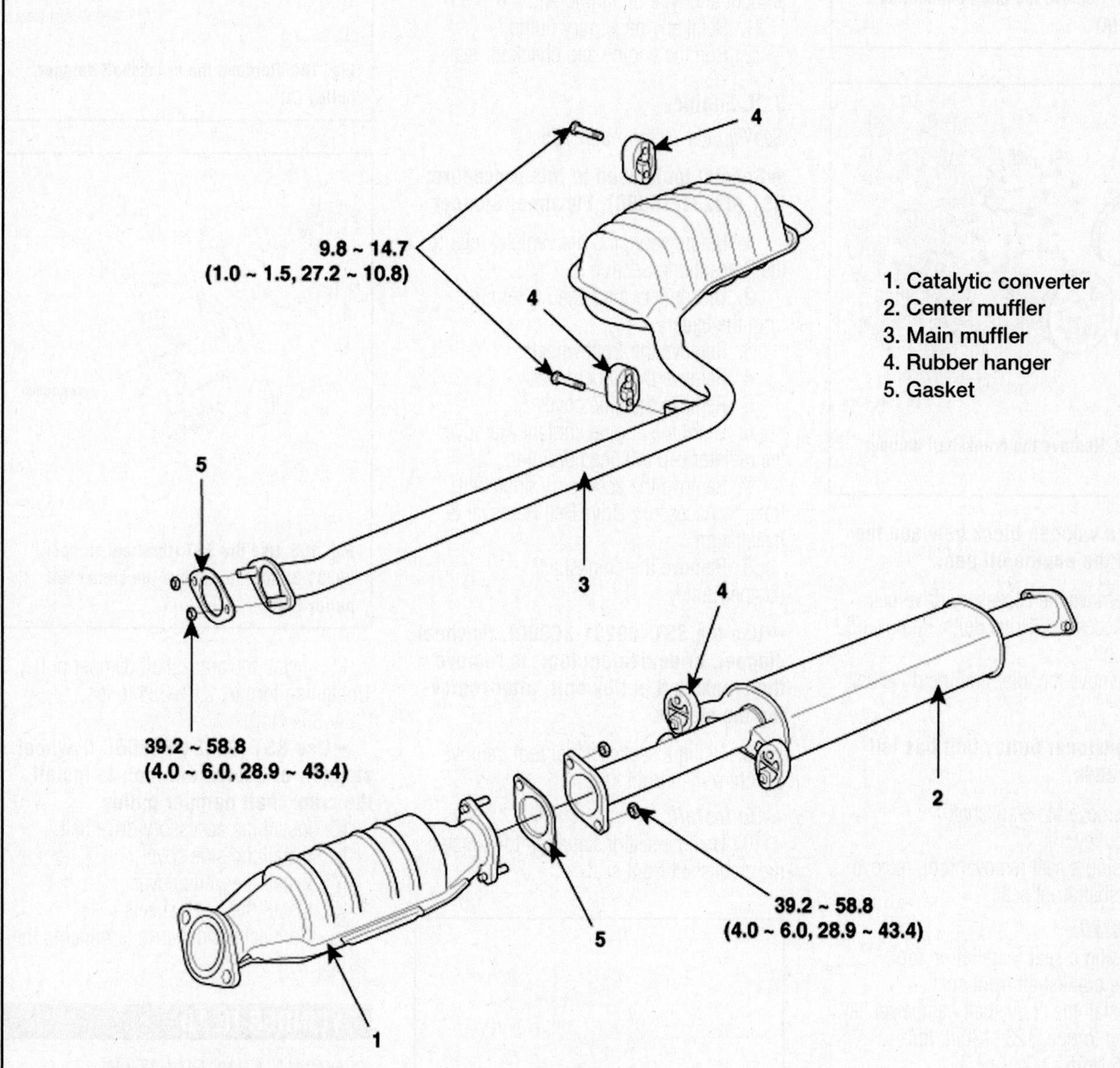

Fig. 99 Muffler components, including catalytic converter—2.4L engine

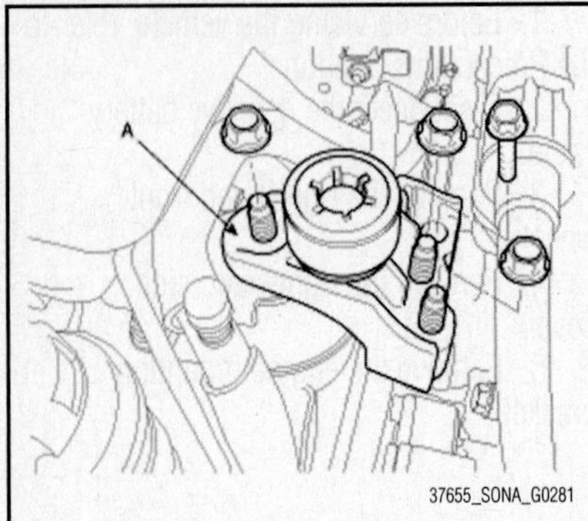

Fig. 101 Remove the engine mounting bracket (A)

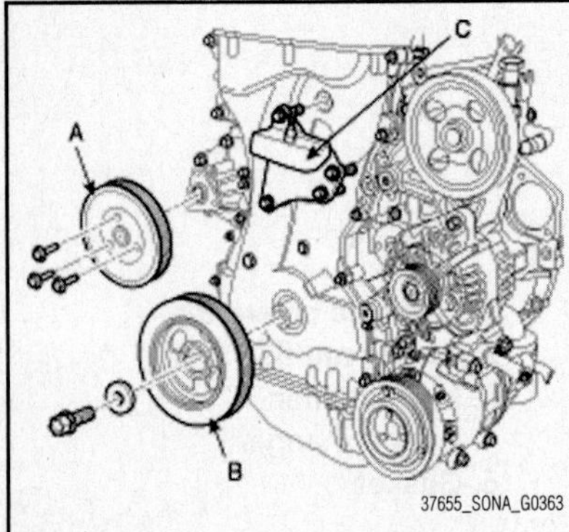

Fig. 102 Remove the crankshaft damper (B)

➡Place a wooden block between the jack and the engine oil pan.

6. Remove the accessory drive belt. Refer to Accessory Drive Belts, Removal & Installation.
7. Remove the idler and the drive belt tensioner.

➡The tensioner pulley bolt has left-hand threads.

8. Remove the crankshaft damper/pulley.
9. Using a seal removal tool, remove the crankshaft front seal.

To install:

10. Using a seal installation tool, install the crankshaft front seal.
11. Install the crankshaft damper/pulley. Tightening torque: 123–130 ft. lbs. (167–176 Nm).

➡Use the SST (09231-3K000), flywheel stopper, or equivalent tool to install the crankshaft damper bolt.

12. Install the accessory drive belt tensioner. Tightening torque: 40–47 ft. lbs. (54–64 Nm).
13. Install the idler. Tightening torque: 40–47 ft. lbs. (54–64 Nm).

➡The tensioner pulley bolt has left-hand threads.

14. Install the accessory drive belt.
15. Install the engine mounting bracket. Tightening torque: 47–62 ft. lbs. (64–83 Nm).
16. Install the right-hand side cover.
17. Install the right-hand front wheel.
18. Install the engine cover.
19. Connect the negative terminal to the battery.
20. Refill the engine with the proper amount and type of engine oil.
21. Refill any necessary fluids.
22. Run the engine and check for leaks.

3.3L Engine

See Figures 103 through 105.

➡Special tools used in this procedure: SST (09231-3C300), Flywheel Stopper

1. Before servicing the vehicle, refer to the Precautions Section.
2. Disconnect the negative terminal from the battery.
3. Remove the front wheels.
4. Remove the undercover.
5. Remove the side cover.
6. Drain the engine coolant. Remove the radiator cap to speed draining.
7. Remove the accessory drive belt. Refer to Accessory Drive Belt Removal & Installation.
8. Remove the crankshaft damper/pulley.

➡Use the SST (09231-3C300), flywheel stopper, or equivalent tool, to remove the crankshaft pulley bolt, after removing the starter.

9. Using a seal removal tool, remove the crankshaft front seal.

To install:

10. Using a seal installation tool, install the crankshaft front seal.

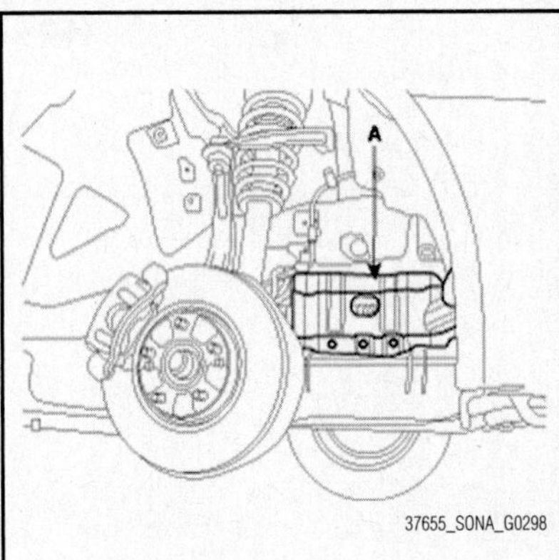

Fig. 103 Remove the side cover (A)

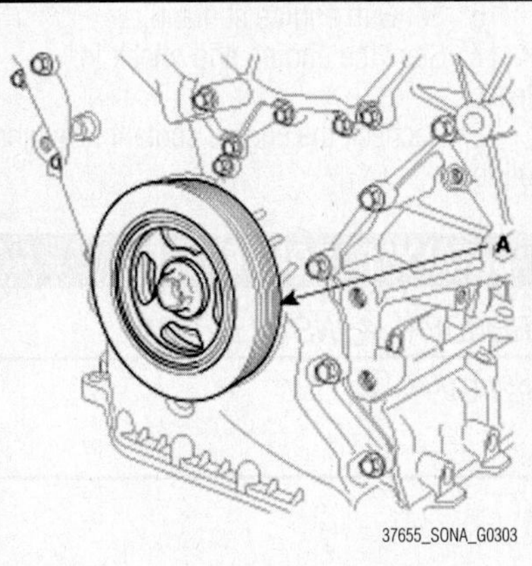

Fig. 104 Remove the crankshaft damper/pulley (A)

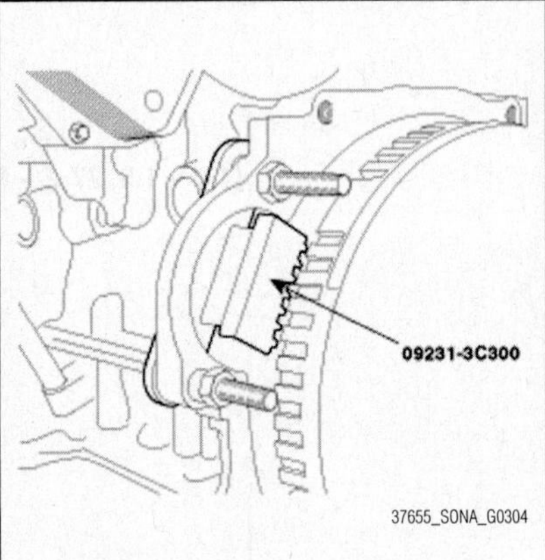

Fig. 105 Use the SST (flywheel stopper, 09231-3C300) to remove the crankshaft pulley bolt

11. Install the crankshaft damper pulley. Tightening torque: 210–224 ft. lbs. (284–304 Nm).

➡Use SST (09231-3C300), flywheel stopper, or equivalent tool, to install the crankshaft damper pulley.

12. Install the accessory drive belt.
13. Install the side cover.
14. Install the undercover.
15. Install the front wheels.
16. Connect the negative terminal to the battery.

CYLINDER HEAD

REMOVAL & INSTALLATION

2.4L Engines

See Figures 106 through 126.

Engine removal is not required for this procedure. Use a fender cover to avoid damaging painted surfaces. To avoid damaging the cylinder head, wait until the engine coolant temperature drops below

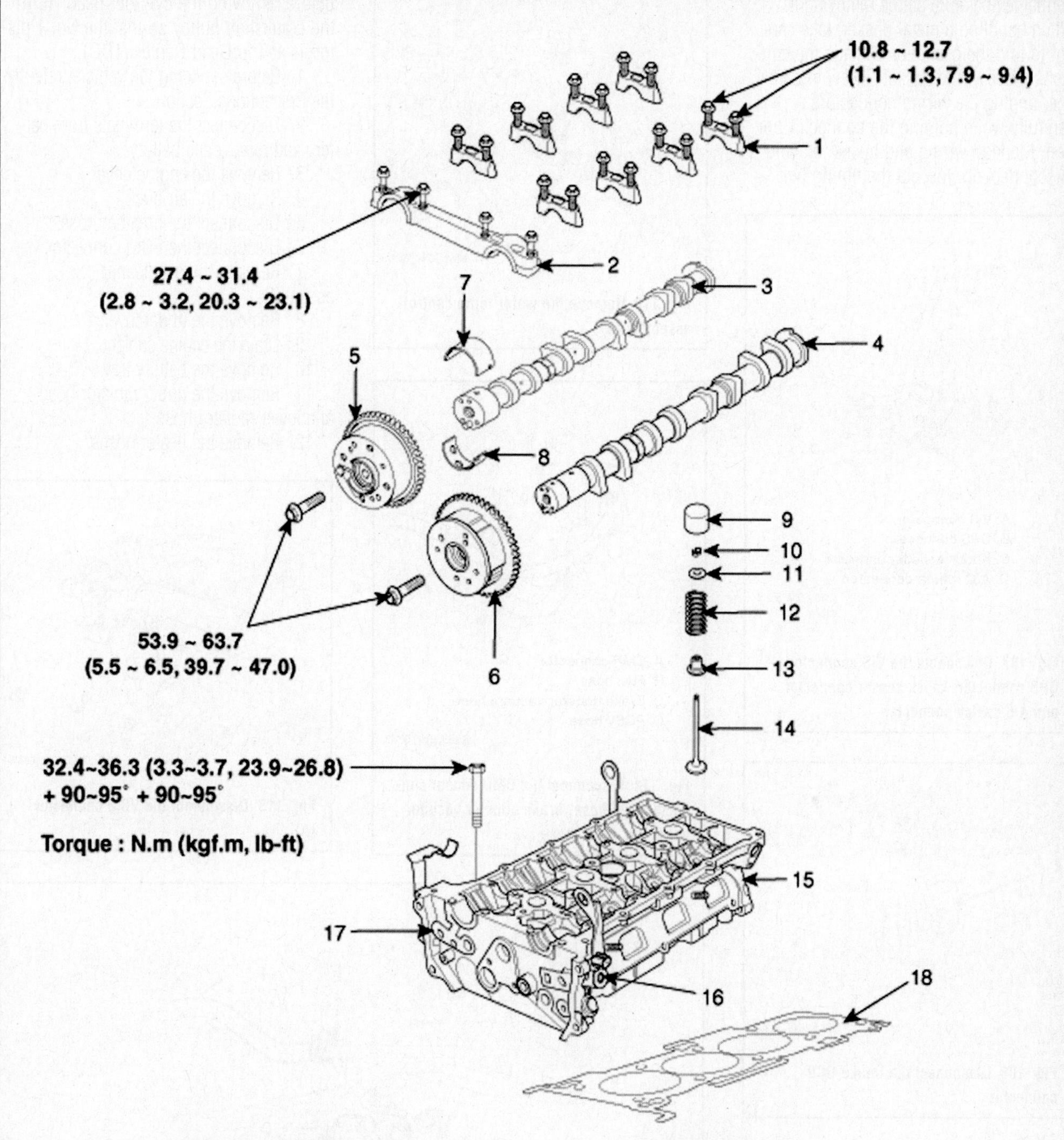

1. Camshaft bearing cap
2. Camshaft front bearing cap
3. Exhaust camshaft
4. Intake camshaft
5. Exhaust CVVT assembly
6. Intake CVVT assembly
7. Exhaust camshaft upper bearing
8. Exhaust camshaft lower bearing
9. MLA
10. Retainer lock
11. Retainer
12. Valve spring
13. Valve stem seal
14. Valve
15. Cylinder head
16. Intake OCV
17. Exhaust OCV
18. Cylinder head gasket

37655_SONA_G0255

Fig. 106 Exploded view of cylinder head—2.4L engine

normal temperature before removing it. When handling a metal gasket, take care not to fold the gasket or damage the contact surface of the gasket. To avoid damage, unplug the wiring connectors carefully while holding the connector portion. Mark all wiring and hoses to avoid misconnection. Inspect the timing belt before removing the cylinder head. Turn the crankshaft pulley so that the No. 1 piston is at Top Dead Center (TDC).

1. Before servicing the vehicle, refer to the Precautions Section.
2. Disconnect the terminals from battery and remove the battery.
3. Remove the engine cover.
4. Remove the air duct.
5. Disconnect the breather hose.
6. Disconnect the ECM connector.
7. Remove the air cleaner assembly.
8. Remove the undercover.
9. Drain the engine coolant.
10. Remove the battery tray.
11. Remove the upper radiator hose and lower radiator hose.
12. Remove the heater hoses.

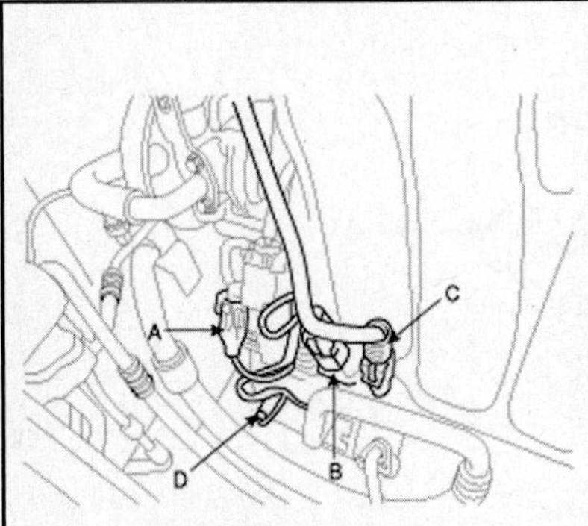

A. VIS connector
B. OPS connector
C. Knock sensor connector
D. A/C switch connector

37655_SONA_G0219

Fig. 107 Disconnect the VIS connector, OPS connector, knock sensor connector and A/C switch connecter

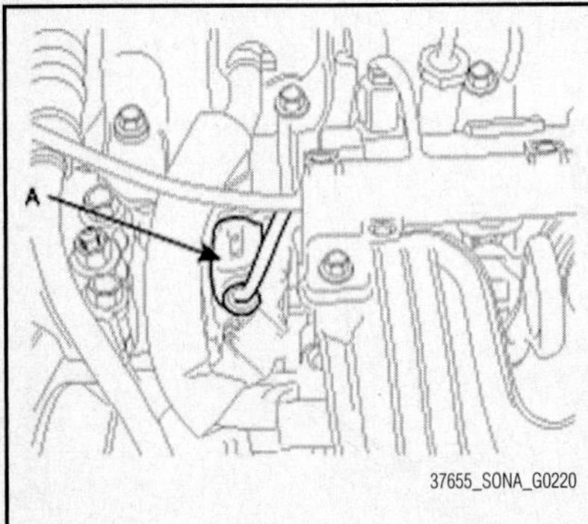

Fig. 108 Disconnect the intake OCV connector

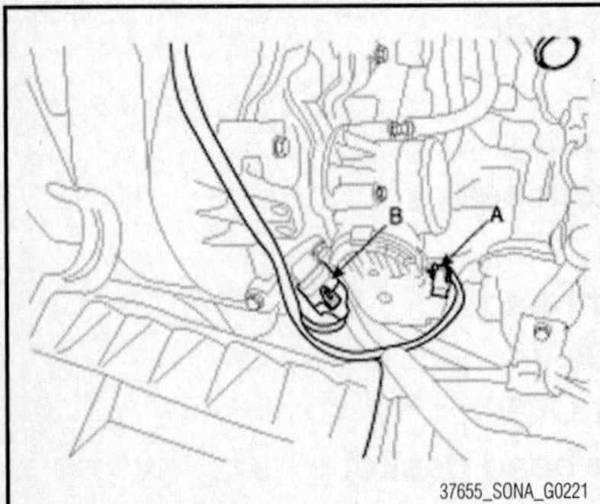

Fig. 109 Disconnect the ETC (A) and MAP sensor connectors (B)

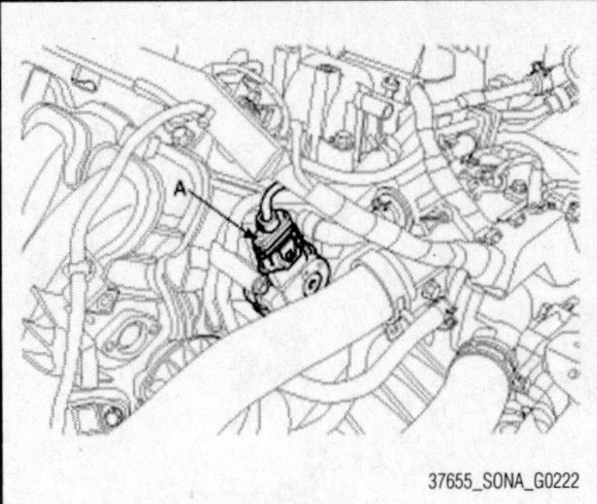

Fig. 110 Remove the water temp control assembly (A)

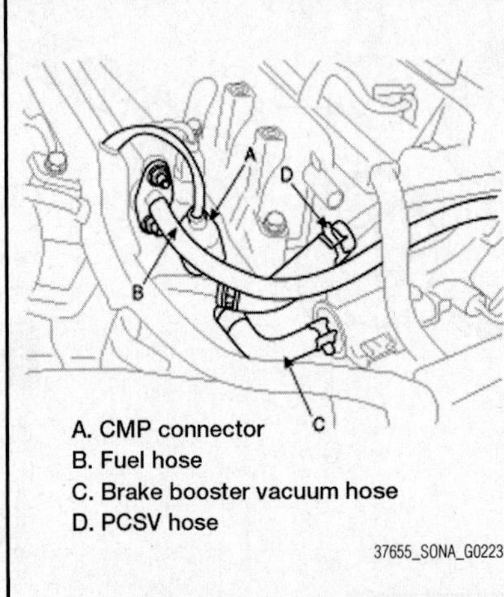

Fig. 111 Disconnect the CMP sensor connector, fuel hose, brake booster vacuum hose, and PCSV hose

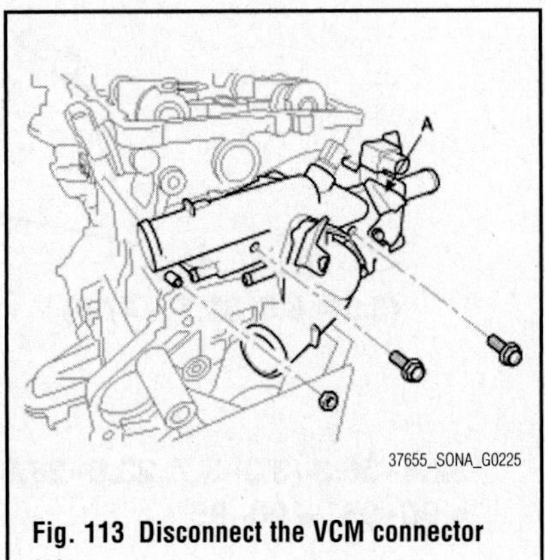

Fig. 113 Disconnect the VCM connector (A)

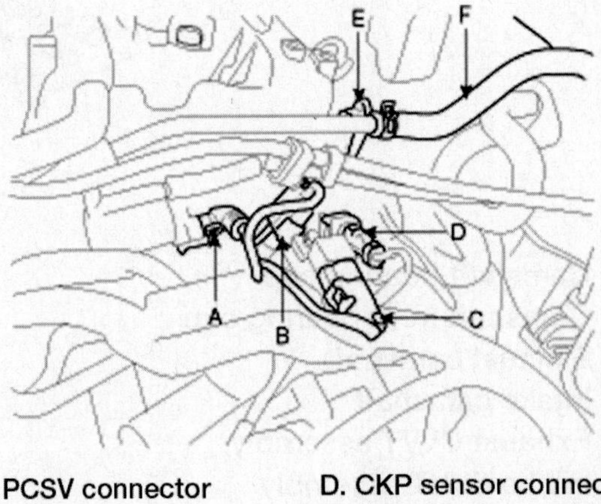

A. PCSV connector
B. ECT connector
C. Condenser connector
D. CKP sensor connector
E. CMP sensor connector
F. Brake booster vacuum hose

37655_SONA_G0224

Fig. 112 Disconnect the PCSV connector, ECT connector, condenser connector, CKP sensor connector, CMP sensor connector, and brake booster vacuum hose

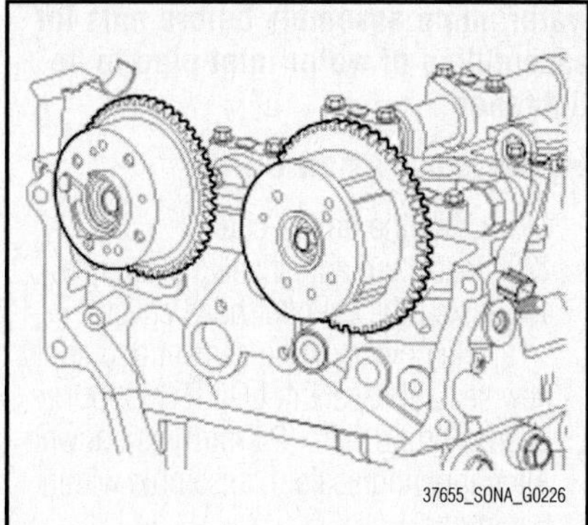

Fig. 114 Remove the intake and exhaust CVVT assembly

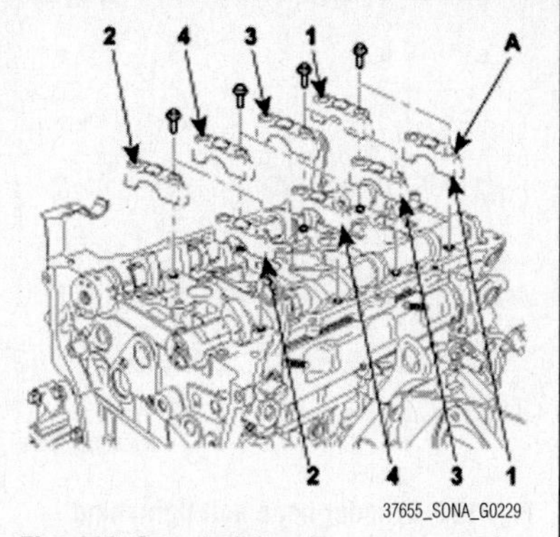

Fig. 117 Camshaft bearing cap removal sequence

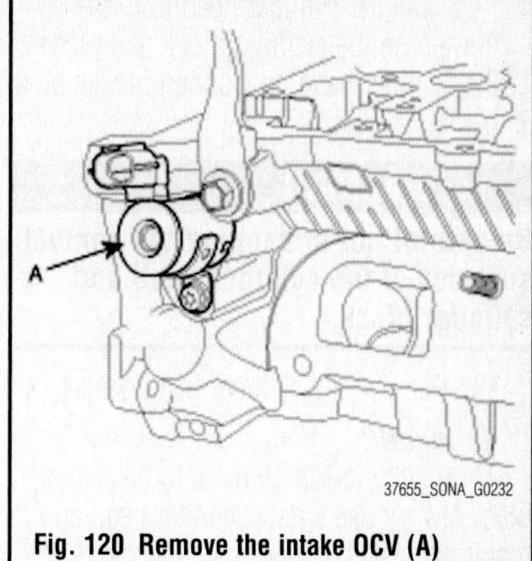

Fig. 120 Remove the intake OCV (A)

Fig. 115 Remove the front cam shaft bearing cap (A)

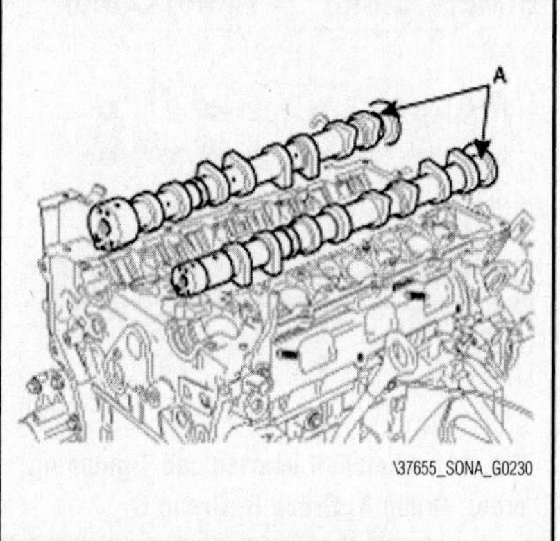

Fig. 118 Remove the camshaft (A)

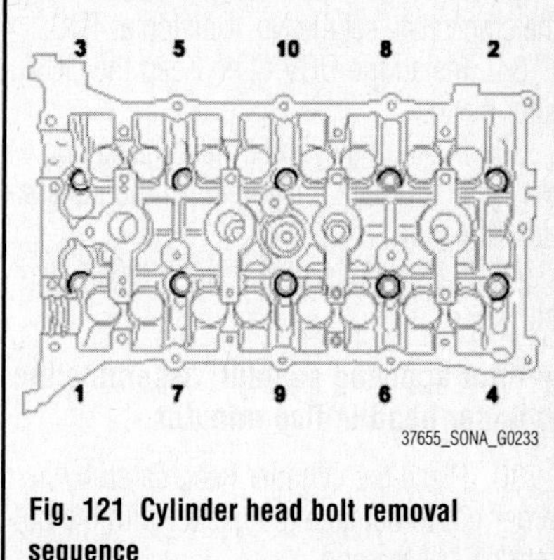

Fig. 121 Cylinder head bolt removal sequence

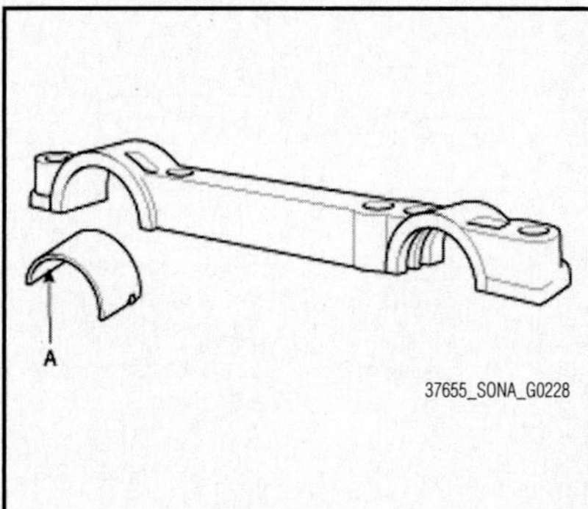

Fig. 116 Remove the exhaust cam shaft upper bearing (A)

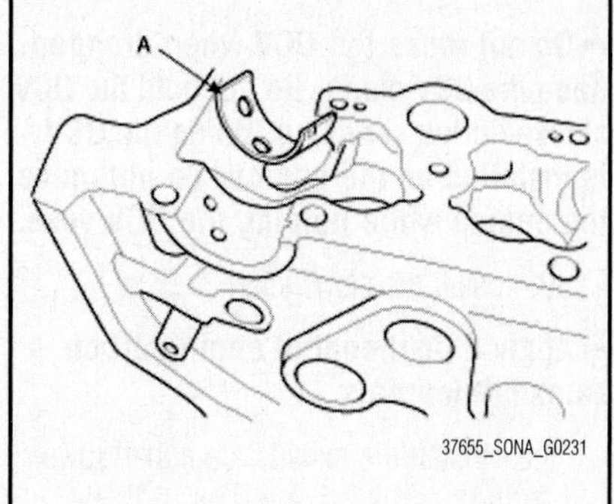

Fig. 119 Remove the exhaust camshaft lower bearing (A)

13. Disconnect the VIS connector, OPS connector, knock sensor connector and A/C switch connecter.
14. Disconnect the intake Oil Control Valve (OCV) connector.
15. Disconnect the injector connectors and ignition coil connectors.
16. Disconnect the ETC connector and MAP sensor connector.
17. Disconnect the VCM connector.
18. Disconnect the Camshaft Position sensor (CMP) connector, fuel hose, brake booster vacuum hose, and Purge Control Solenoid Valve (PCSV) hose.
19. Disconnect the PCSV connector, ECT connector, condenser connector, CKP sensor connector, CMP sensor connector, and brake booster vacuum hose.
20. Disconnect the front and rear oxygen sensor connectors.
21. Remove the water temp control assembly.
22. Remove the timing chain. Refer to Timing Chain & Sprockets Removal & Installation.
23. Remove the intake and exhaust manifolds. Refer to Intake Manifold Removal & Installation, and Exhaust Manifold Removal & Installation.
24. Remove the intake and exhaust CVVT assembly.
25. Remove the camshaft:
    a. Remove the front cam shaft bearing cap.
26. Remove the exhaust cam shaft upper bearing.
27. Remove camshaft bearing caps, in the sequence shown.
28. Remove the camshaft.
29. Remove the exhaust camshaft lower bearing.
30. Using a Torx® wrench, remove the intake OCV.
31. Remove the exhaust OCV.
32. Remove the cylinder head bolts, and then remove the cylinder head.
    a. Using triple square wrench, uniformly loosen and remove the 10 cylinder head bolts, in several passes, in the sequence shown.

⁂ WARNING

Head warpage or cracking could result from removing bolts in an incorrect order.

b. Lift the cylinder head from the dowels on the cylinder block and place the cylinder head on wooden blocks on a bench.

✻✻ WARNING

Be careful not to damage the contact surfaces of the cylinder head and cylinder block.

33. Remove the cylinder head gasket.

To install:

Thoroughly clean all parts to be assembled. Always use a new cylinder head and manifold gasket. Always use new cylinder head bolts. The cylinder head gasket is a metal gasket. Take care not to bend it. Rotate the crankshaft, set the No. 1 piston at TDC.

34. Install the OCV filter. Keep the OCV filter clean.

35. Install the cylinder head gasket on the cylinder block. Be careful of the installation direction.

Apply liquid gasket (Loctite 5900H), as shown.

➡After applying sealant, assemble the cylinder head in five minutes.

36. Place the cylinder head carefully in order not to damage the gasket with the bottom part of the end.

37. Install the cylinder head bolts:

a. Apply a light coat of engine oil on the threads and under the heads of the cylinder head bolts.

b. Using a hexagon wrench, install and tighten the 10 cylinder head bolts and plate washers, in several passes, in the sequence shown.

- Step 1: 24–27 ft. lbs. (32–36 Nm) plus 90°–95°
- Step 2: Additional 90°–95°

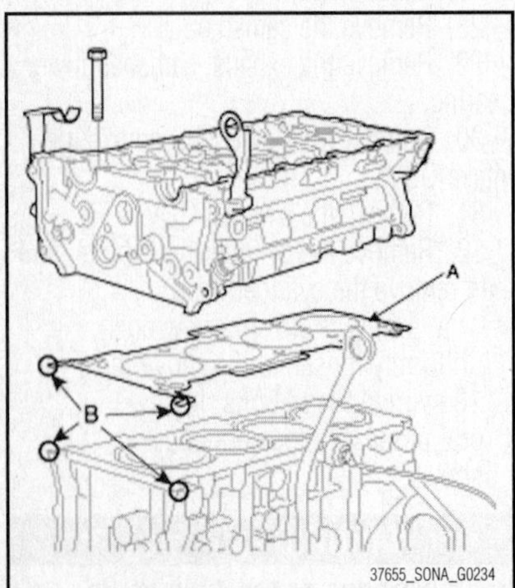

Fig. 122 Cylinder head gasket (A) and sealant locations (B)

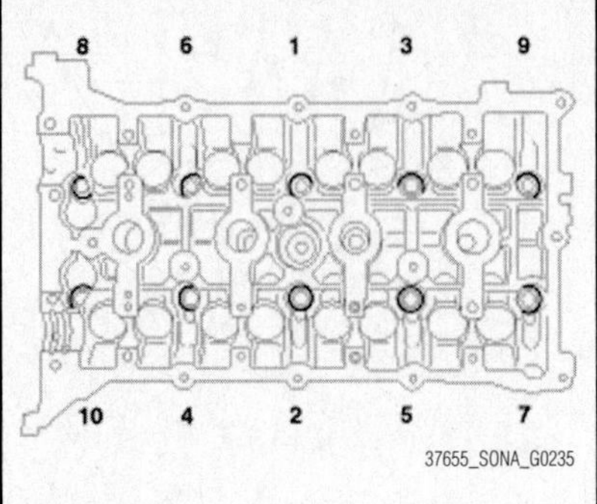

Fig. 123 Cylinder head bolt tightening sequence

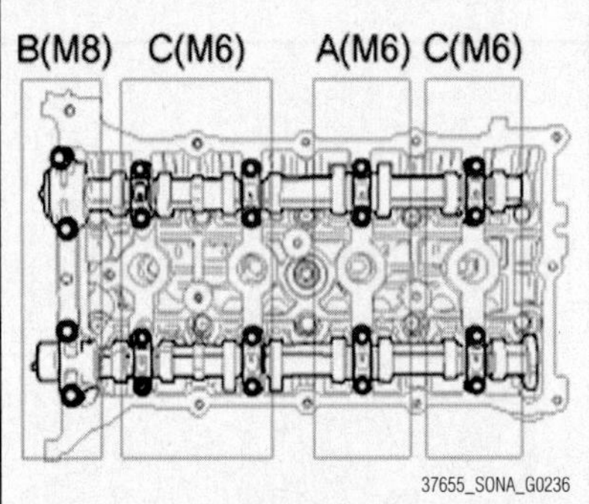

Fig. 127 Camshaft bearing cap Tightening order: Group A, Group B, Group C

38. Install the OCV. Tightening torque: 84–108 inch lbs. (10–12 Nm).

➡Do not reuse the OCV when dropped. Keep the OCV clean. Do not hold the OCV sleeve during servicing. When the OCV is installed on the engine, do not move the engine while holding the OCV yoke.

39. Install the camshafts:

➡Apply a light coat of engine oil on camshaft journals.

a. Install the exhaust camshaft lower bearing.

b. Install the camshafts.

c. Install the exhaust camshaft upper bearing.

d. Install the camshaft bearing caps in the tightening sequence shown, and tighten:

- M6 bolts Step 1: 52 inch lbs. (6 Nm), Step 2: 95–113 inch lbs. (11–13 Nm)
- M8 bolts Step 1: 11 ft. lbs. (15 Nm), Step 2: 20–23 ft. lbs. (28–31 Nm)

40. Install the water temp control assembly, and tighten the bolts and nut to 14–17 ft. lbs. (19–24 Nm).

➡Assemble water temp control assembly and water inlet pipe to water pump assembly before nuts for assembling of water inlet pipe to be tightened.

➡Always use a new O-ring.

41. Install the timing chain.

42. Check and adjust valve clearance.

43. Install the cylinder head cover:

a. Remove existing sealant and apply new sealant [Sealant: LOCTITE 5900H, Bead width: 0.1 in. (2.5mm)] as shown. After applying sealant, assemble within 5 minutes.

➡The firing and/or blow out test should not be performed within 30 minutes after the cylinder head cover was assembled.

b. Install the cylinder head cover bolts in the sequence shown, tightening to:

- Step 1: 35–52 inch lbs. (4–6 Nm)
- Step 2: 70–86 inch lbs. (8–10 Nm)

➡Do not reuse cylinder head cover gasket.

44. Install the exhaust manifold.

45. Install the intake manifold.

46. Connect the PCSV connector, ECT connector, condenser connector, CKP sensor connector, CMP sensor connector and brake booster vacuum hose.

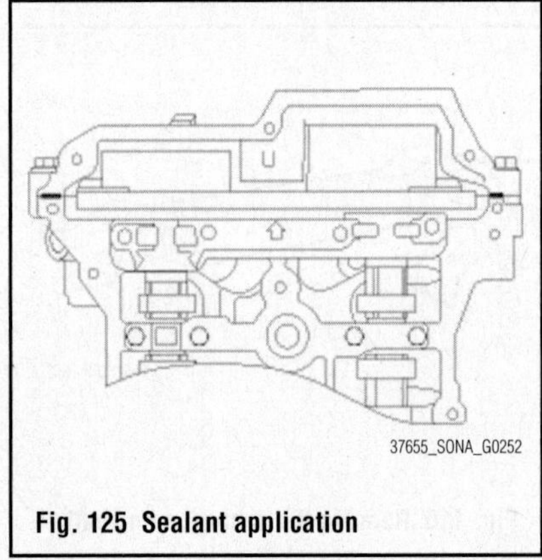

Fig. 125 Sealant application

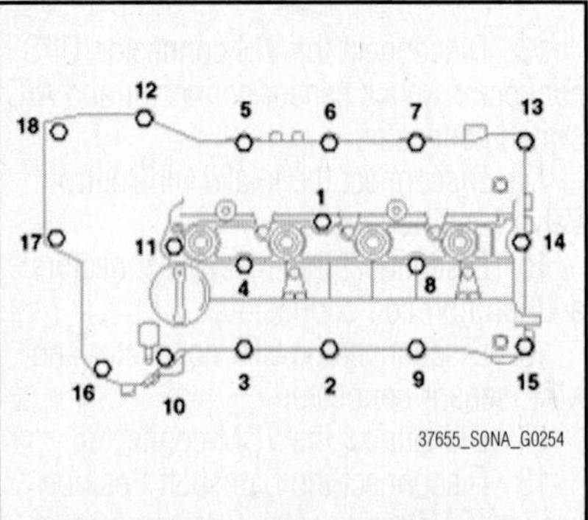

Fig. 126 Cylinder head bolt tightening sequence

47. Connect the front oxygen sensor connector and rear oxygen sensor connector.
48. Connect the brake booster vacuum hose, PCSV hose fuel hose and CMP sensor connector.
49. Connect the ETC connector and MAP sensor connector.
50. Connect the injector connectors and ignition coil connectors.
51. Connect the OCV connector.
52. Connect the VIS connector, OPS connector, knock sensor connector and A/C switch connecter.
53. Install the heater hoses.
54. Install the upper and lower radiator hoses.
55. Install the intake air hose and air cleaner assembly.
56. Install the engine cover.
57. Connect the negative battery cable.
58. Fill with engine coolant.
59. Start the engine and check for leaks.
60. Recheck the engine coolant level and oil level.

3.3L Engines

See Figures 127 through 141.

Use a fender cover to avoid damaging painted surfaces. To avoid damaging the cylinder head, wait until the engine coolant temperature drops below normal temperature before removing it. When handling a metal gasket, take care not to fold the gasket or damage the contact surface of the gasket. To avoid damage, unplug the wiring connectors carefully while holding the connector portion. Mark all wiring and hoses to avoid misconnection. Inspect the timing belt before removing the cylinder head. Turn the crankshaft pulley so that the No. 1 piston is at Top Dead Center (TDC).

1. Before servicing the vehicle, refer to the Precautions Section.
2. Remove the exhaust manifold. Refer to Exhaust Manifold Removal & Installation.
3. Remove the intake manifold. Refer to Intake Manifold Removal & Installation.
4. Remove the timing chain. Refer to Timing Chain & Sprockets Removal & Installation.

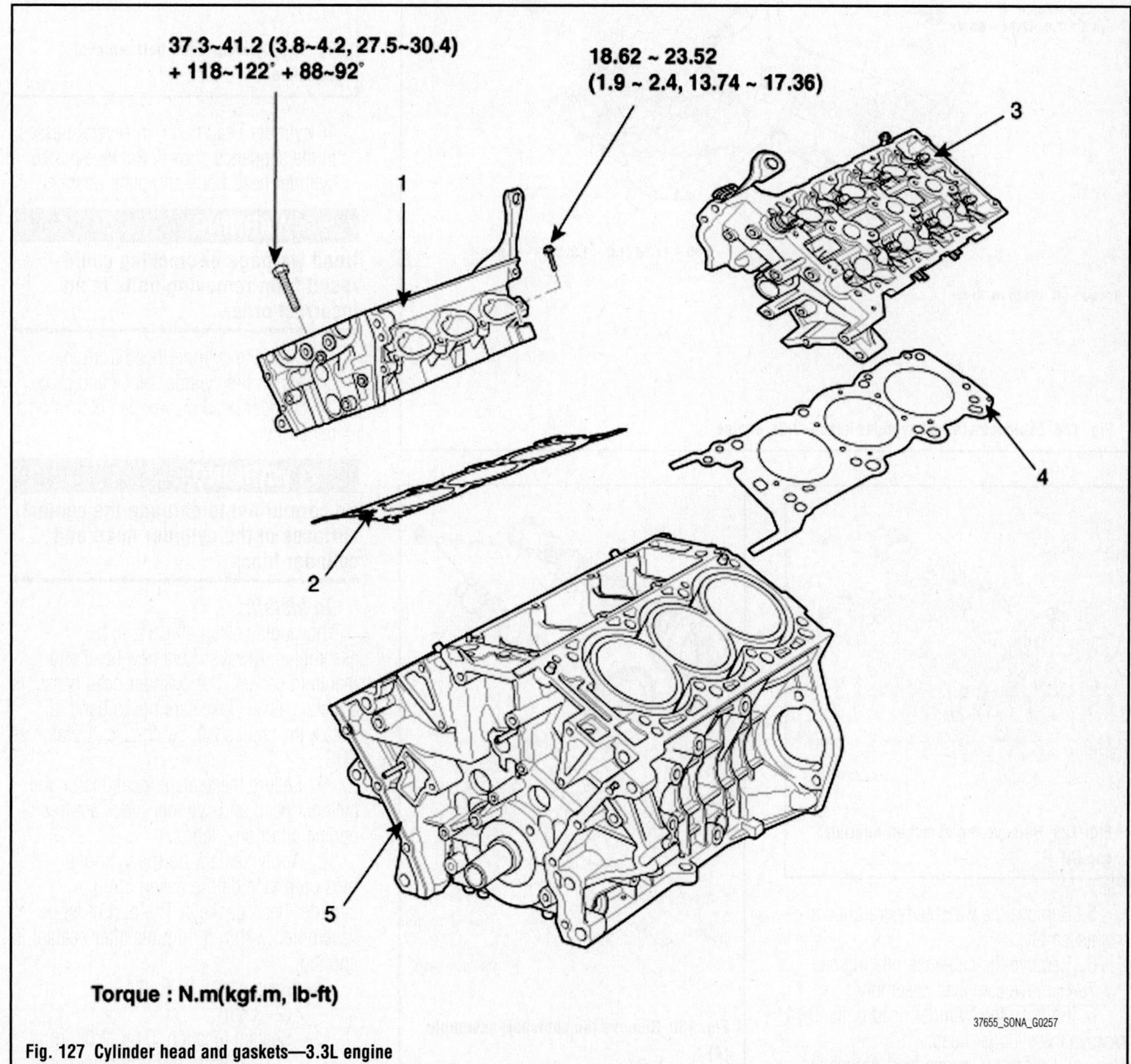

Fig. 127 Cylinder head and gaskets—3.3L engine

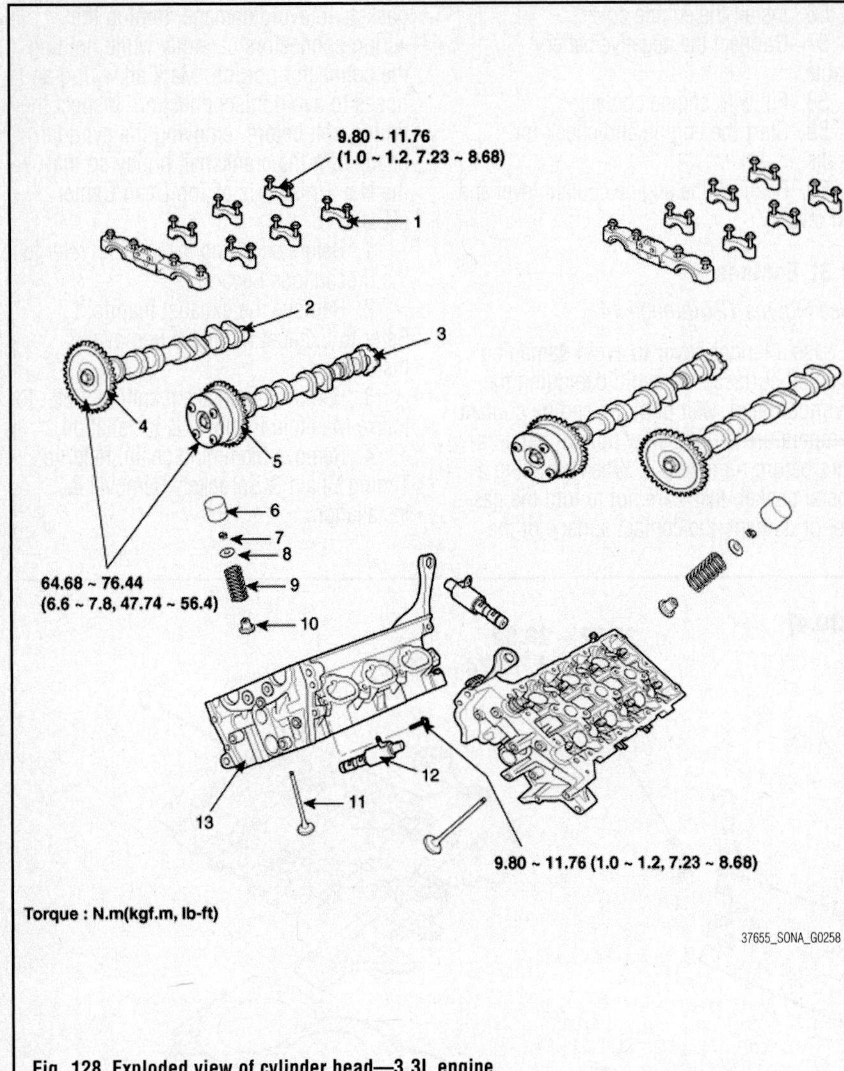

Fig. 128 Exploded view of cylinder head—3.3L engine

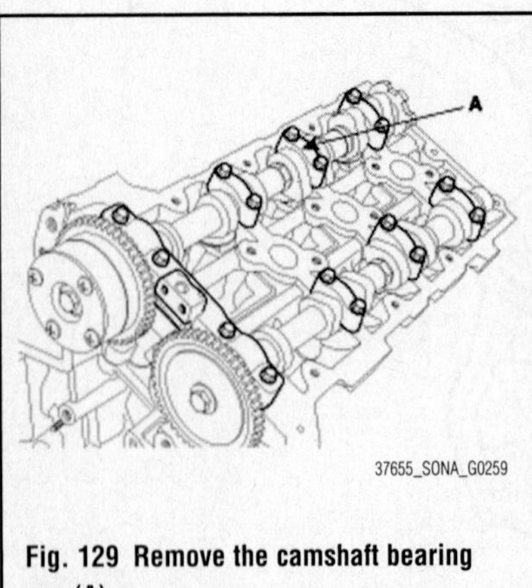

Fig. 129 Remove the camshaft bearing cap (A)

5. Remove the water temperature control assembly.
6. Remove the camshaft bearing cap.
7. Remove camshaft assembly.
8. Remove the cylinder head bolts, then remove the cylinder head:
   a. Uniformly loosen and remove the

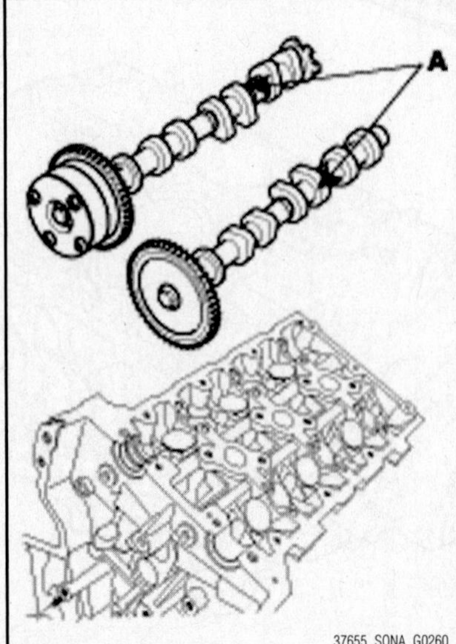

Fig. 130 Remove the camshaft assembly (A)

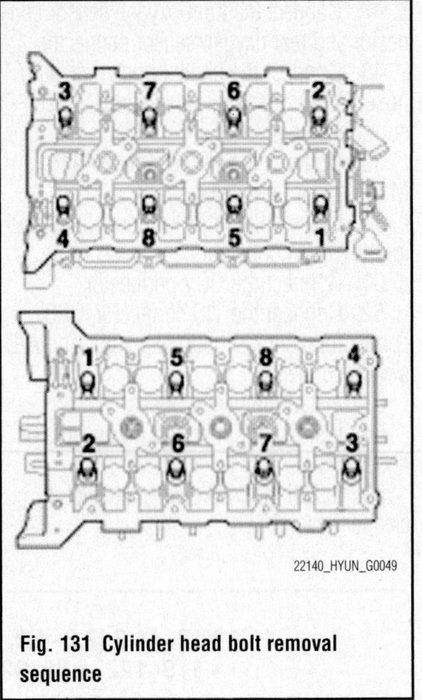

Fig. 131 Cylinder head bolt removal sequence

16 cylinder head bolts, in several passes, in the sequence shown. Remove the 16 cylinder head bolts and plate washers.

✲✲ WARNING

Head warpage or cracking could result from removing bolts in an incorrect order.

   b. Lift the cylinder head from the dowels on the cylinder block and place the cylinder head on wooden blocks on a bench.

✲✲ WARNING

Be careful not to damage the contact surfaces of the cylinder head and cylinder block.

To install:

Thoroughly clean all parts to be assembled. Always use a new head and manifold gasket. The cylinder head gasket is a metal gasket. Take care not to bend it. Rotate the crankshaft, set the No. 1 piston at TDC.

9. Ensure the sealant locations on the cylinder head and cylinder block are free of engine oil or any debris.
10. Apply sealant on the cylinder block top face before assembling cylinder head gaskets. The part must be assembled within 5 minutes after sealant is applied.
    - Bead width: 0.08–0.12 in. (2–3mm)
    - Sealant location: 0.04–0.06 in. (1.0–1.5mm) from block surface

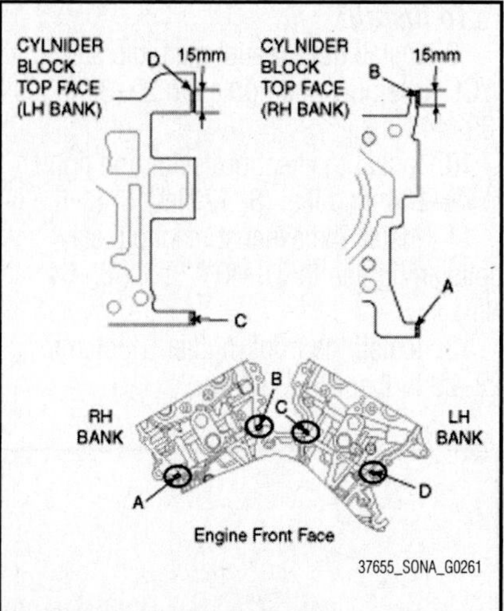

Fig. 132 Cylinder block top face sealant application locations

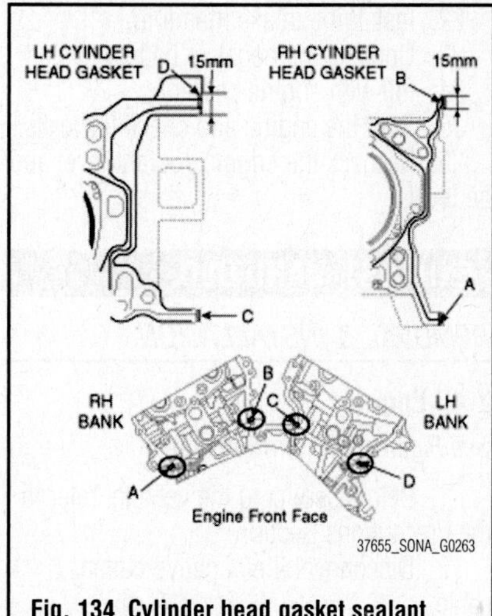

Fig. 134 Cylinder head gasket sealant application locations

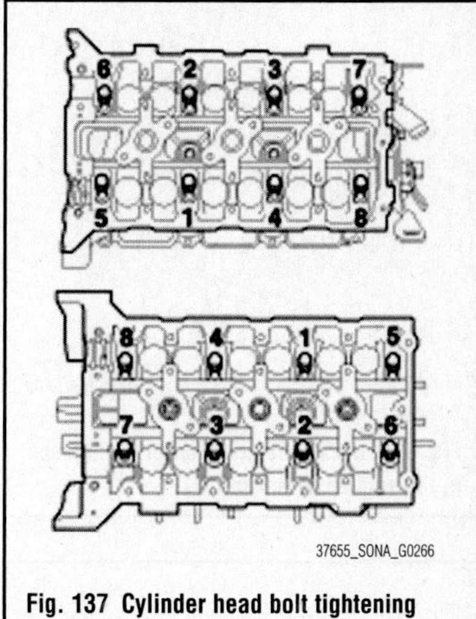

Fig. 137 Cylinder head bolt tightening sequence

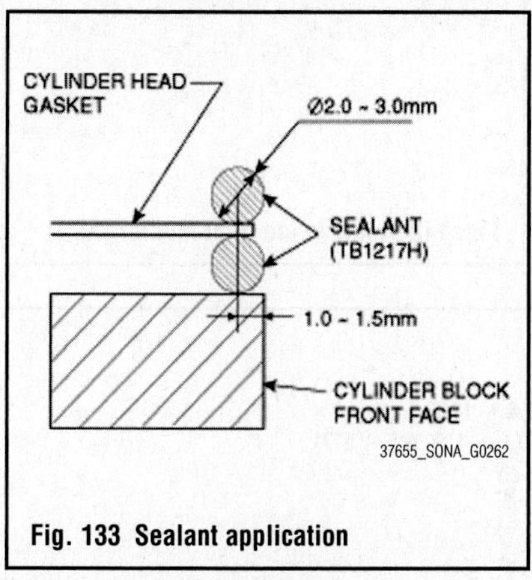

Fig. 133 Sealant application

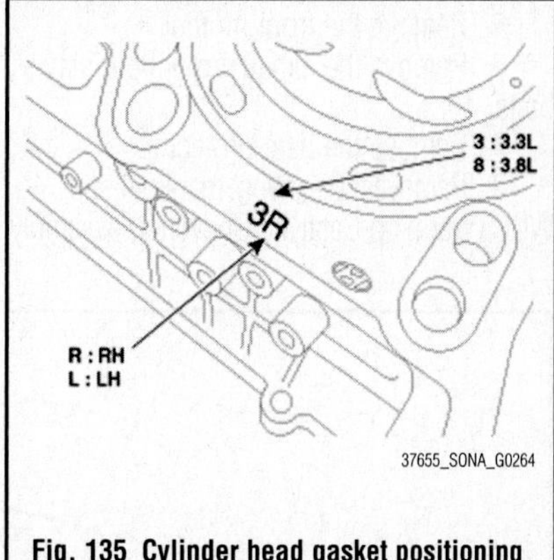

Fig. 135 Cylinder head gasket positioning

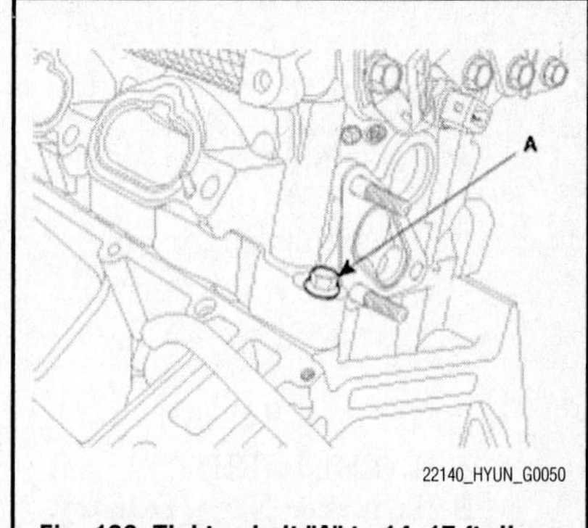

Fig. 138 Tighten bolt "A" to 14–17 ft. lbs. (19–24 Nm)

- Recommended sealant: Liquid sealant TB1217H

11. Apply sealant on cylinder head gaskets after assembling cylinder head gaskets on cylinder block.

The part must be assembled within 5 minutes after sealant was applied.

12. Be careful of installation direction.
13. Install the cylinder head. Remove any extruded sealant after assembling cylinder heads.
14. Place the cylinder head carefully in order not to damage the gasket with the bottom part of the end.
15. Install cylinder head bolts.
    a. Do not apply engine oil on the threads or under the heads of the cylinder head bolts.

➡Always use new cylinder head bolts.

b. Using SST (09221-4A000), install and tighten the cylinder head bolts and plate washers, in several passes, in the sequence shown.

- Step 1—Tightening torque: 28–30 ft. lbs. (37–41 Nm)

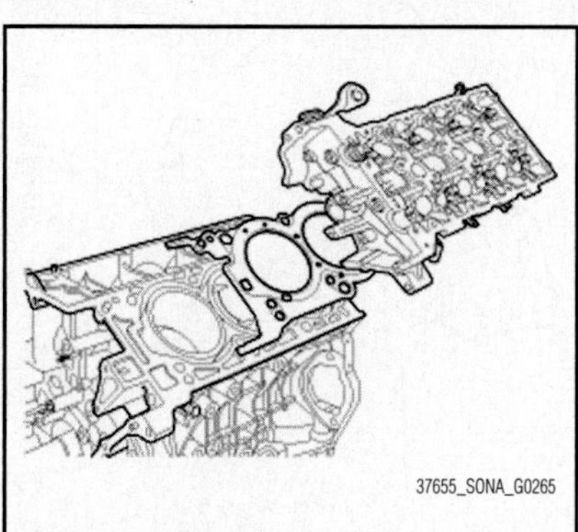

Fig. 136 Cylinder head and gasket installation

- Step 2—Tighten an additional: 120° plus or minus 2°
- Step 3—Tighten an additional: 90° plus or minus 2°

c. Tighten bolt "A" to 14–17 ft. lbs. (19–24 Nm).

16. Install the camshafts.
    a. Apply a light coat of engine oil on camshaft journals.
    b. Assemble the key groove of camshaft rear side to the same level of head top surface.
    c. Be careful to get the right bank, left bank, intake side, and exhaust side in the correct position before assembling.

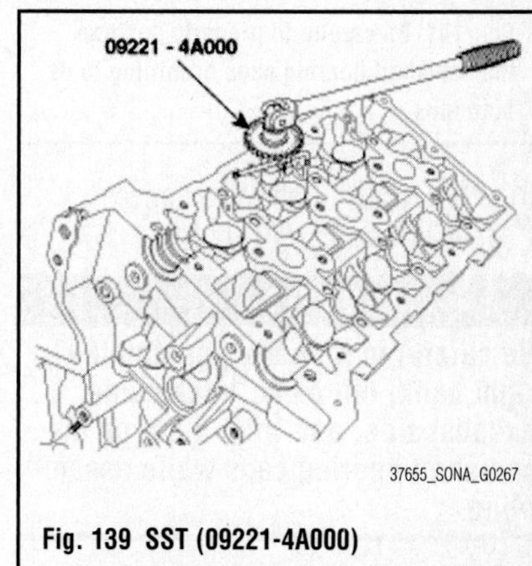

Fig. 139 SST (09221-4A000)

17. Install the camshaft bearing caps in the sequence shown:
    a. Step 1—Tightening torque: 52 inch lbs. (6 Nm).

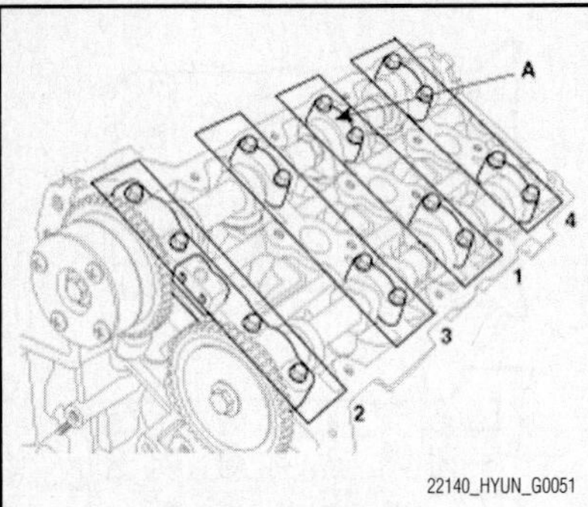

Fig. 140 Install the camshaft bearing caps in the sequence shown

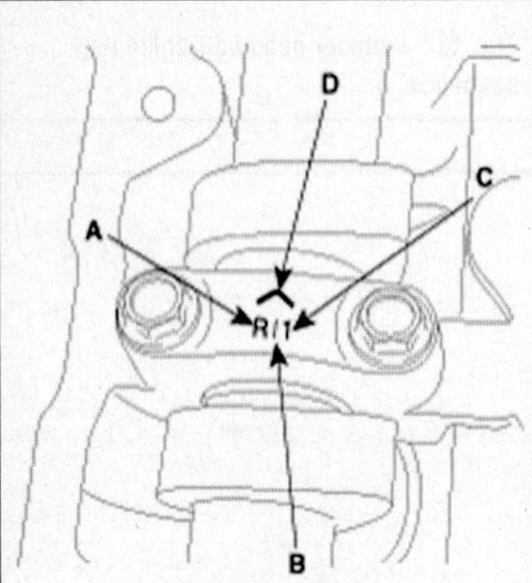

Fig. 141 Be careful to properly position the camshaft bearing caps according to its markings

b. Step 2—Tightening torque: 87–104 inch lbs. (10–12 Nm).

WARNING

Be careful to properly position the right bank, left bank, intake side, exhaust side, and front mark on the camshaft bearing caps while assembling.

➡Rotate the crankshaft so as not to contact the valves to the pistons by positioning the pistons 0.3937 in. (10mm) below the top of the cylinder block.

18. Install the water temperature control assembly.
19. Install the timing chain.
20. Check and adjust the valve clearance, as necessary.
21. Install the exhaust manifold.
22. Install the intake manifold.
23. Connect the negative battery cable.
24. Fill with engine coolant.
25. Start the engine and check for leaks.
26. Recheck the engine coolant level and oil level.

EXHAUST MANIFOLD

REMOVAL & INSTALLATION

2.4L Engines

See Figures 142 through 144.

1. Before servicing the vehicle, refer to the Precautions Section.
2. Disconnect the negative battery cable.
3. Remove the engine cover.
4. Disconnect the front and rear oxygen sensor connectors.
5. Remove the front muffler.
6. Remove the exhaust manifold stay bolts.
7. Remove the heat protector.
8. Remove the exhaust manifold and WCC (Warm up Catalytic Converter) assembly.

To install:

9. Install the exhaust manifold and WCC assembly and tighten to 29–33 ft. lbs. (39–44 Nm).
10. Install the heat protector and tighten to 70–104 inch lbs. (8–12 Nm).
11. Install the exhaust manifold stay bolts an tighten to 31–40 ft. lbs. (42–54 Nm).
12. Install the front muffler. Tighten to 22–29 ft. lbs. (29–39 Nm).

Fig. 143 Remove the front muffler (A)

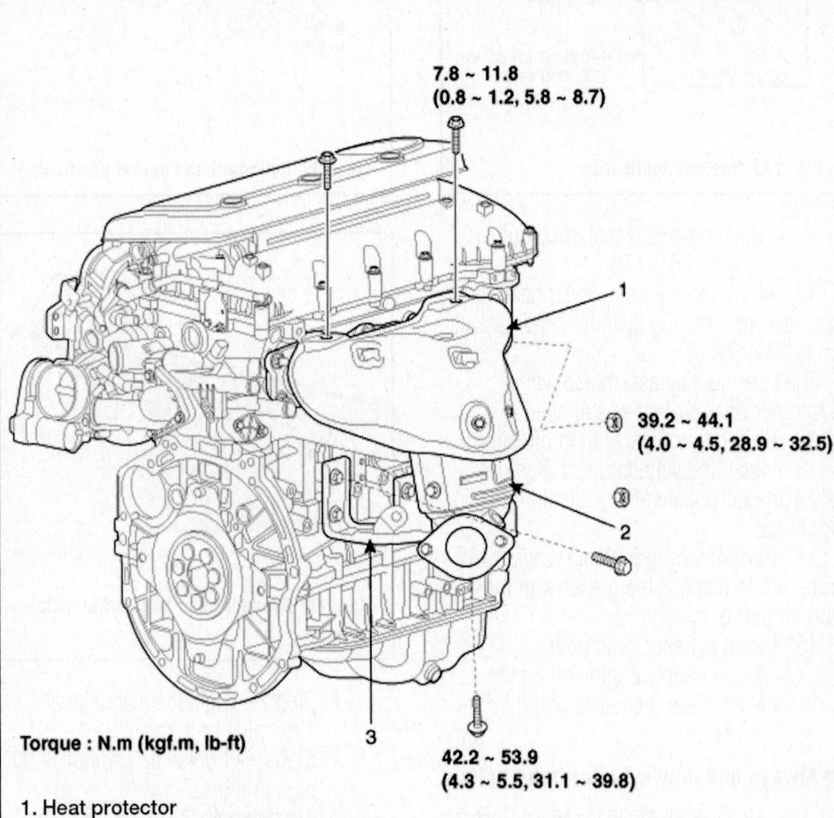

Fig. 142 Exhaust manifold and related components

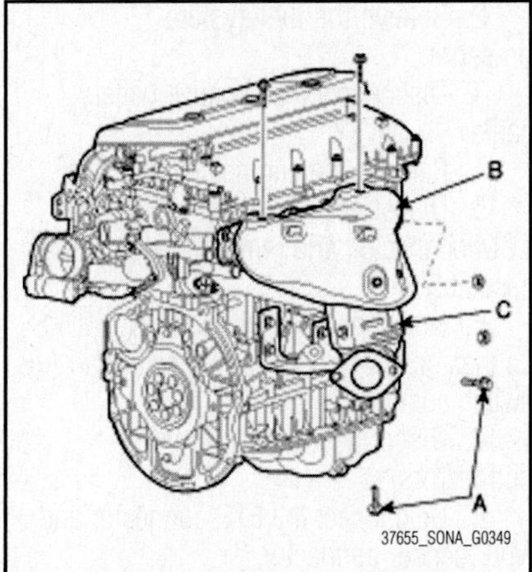

Fig. 144 Remove the exhaust manifold stay bolts (A), heat protector (B), and exhaust manifold and WCC assembly (C)

13. Install the front muffler and tighten to 29–43 ft. lbs. (39–59 Nm).
14. Connect the front and rear oxygen sensor connectors.
15. Install the engine cover.
16. Connect the negative battery cable.

3.3L Engines

See Figures 145 through 148.

1. Before servicing the vehicle, refer to the Precautions Section.
2. Disconnect the negative battery cable.
3. Remove the undercover.
4. Disconnect the left-hand and right-hand rear oxygen sensor connectors from bracket.
5. Remove the front muffler.
6. Remove the oil level gauge.
7. Disconnect the left-hand front oxygen sensor connector from bracket.
8. Remove the left-hand heat protector.
9. Remove the left-hand exhaust manifold.
10. Disconnect the right-hand front oxygen sensor connector from bracket.
11. Remove the right-hand heat protector.
12. Remove the right-hand exhaust manifold.

To install:

13. Install a new gasket and exhaust manifold. Tighten to 29–33 ft. lbs. (39–44 Nm).
14. Install heat protector. Tighten to 12–16 ft. lbs. (17–22 Nm).
15. Install front muffler. Tightening torque: 29–43 ft. lbs. (39–59 Nm).
16. Connect oxygen sensor connectors.
17. Install undercover.
18. Connect the negative battery cable.

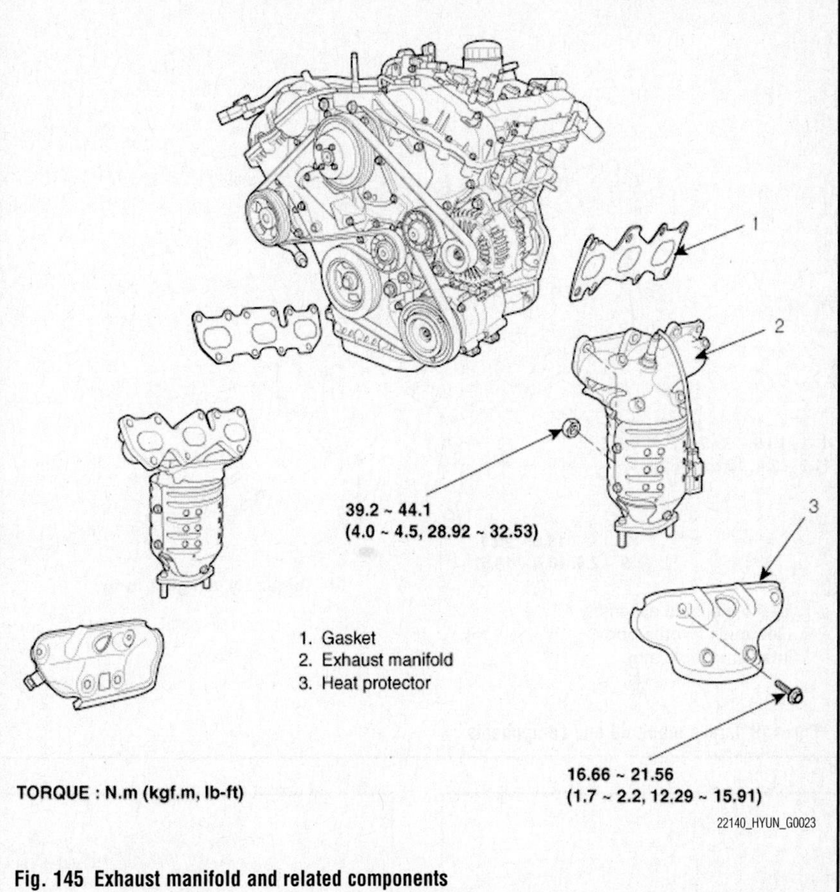

Fig. 145 Exhaust manifold and related components

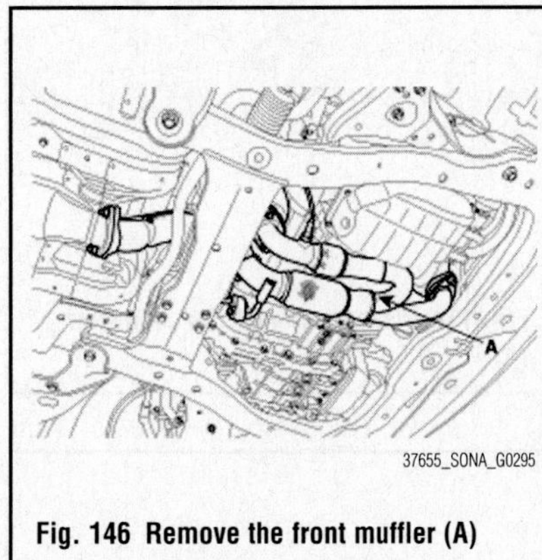

Fig. 146 Remove the front muffler (A)

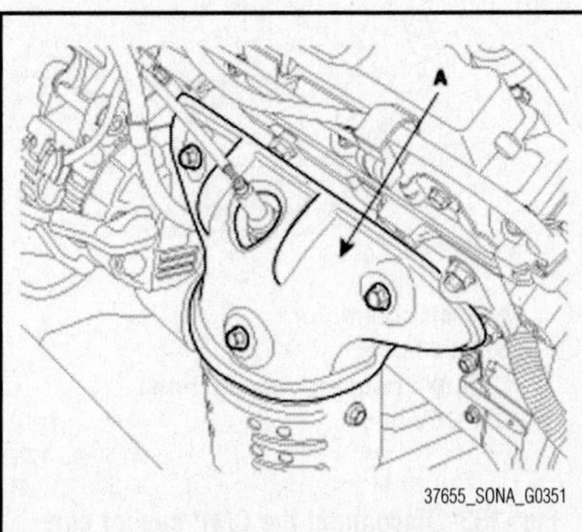

Fig. 147 Remove the left-hand heat protector (A)

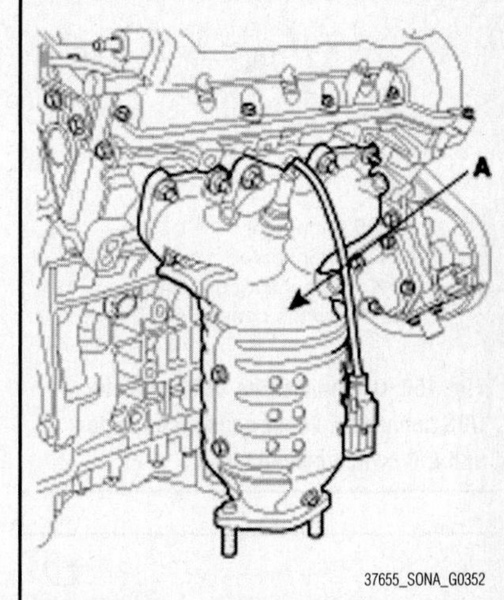

Fig. 148 Remove the right-hand exhaust manifold (A)

INTAKE MANIFOLD

REMOVAL & INSTALLATION

2.4L Engines

See Figures 149 through 155.

1. Before servicing the vehicle, refer to the Precautions Section.

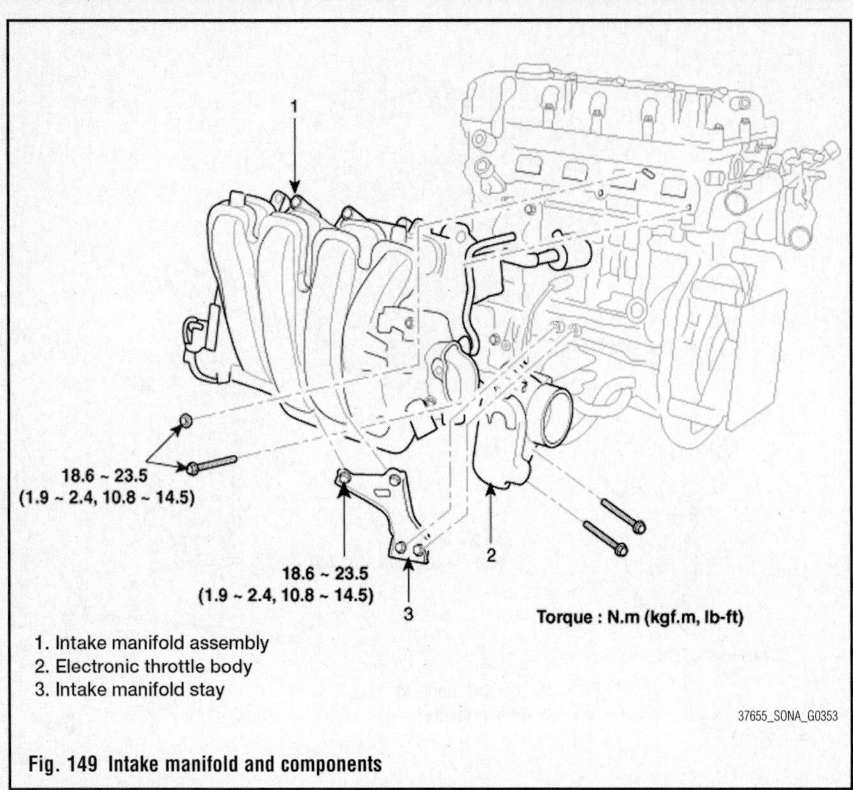

Fig. 149 Intake manifold and components

2. Relieve the fuel system pressure.

3. Disconnect the negative battery cable.

4. Remove the engine cover.

5. Disconnect the breather hose and ECM connector, and remove the air cleaner assembly.

6. Disconnect the VIS connector, OPS connector, knock sensor connector and A/C switch connecter.

7. Disconnect the injector connectors.

8. Disconnect the ETC connector and MAP sensor connector.

9. Disconnect the VCM connector.

10. Disconnect the Camshaft Position sensor (CMP) connector, fuel hose, brake booster vacuum hose, and Purge Control Solenoid Valve (PCSV) hose.

11. Remove the coolant hose from the throttle body.

12. Remove the PCV hose.

13. Remove the sensor connectors from the bracket and then remove the intake manifold stay.

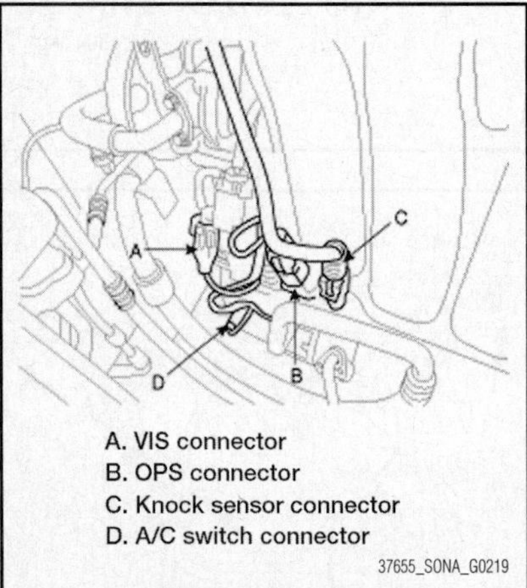

Fig. 150 Disconnect the VIS connector, OPS connector, knock sensor connector and A/C switch connecter

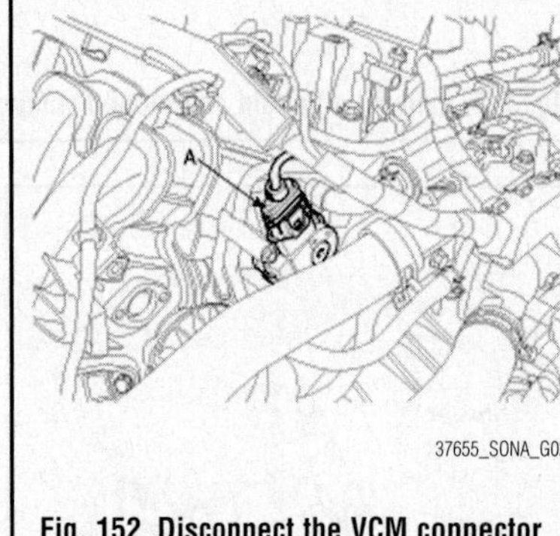

Fig. 152 Disconnect the VCM connector (A)

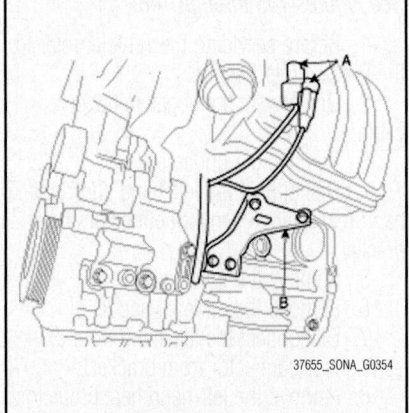

Fig. 154 Remove the sensor connectors (A) and then remove the intake manifold stay (B)

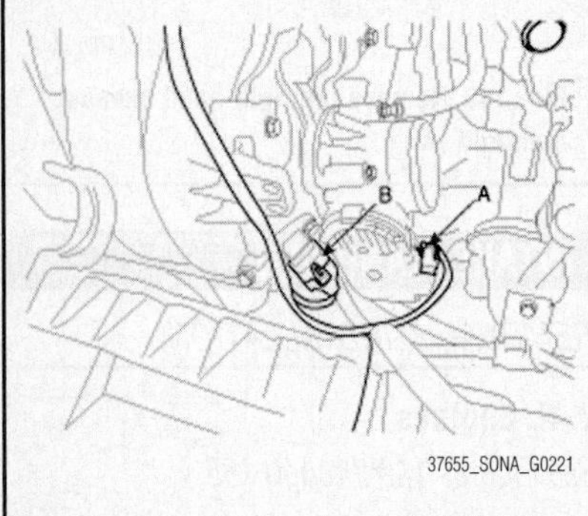

Fig. 151 Disconnect the ETC (A) and MAP sensor connectors (B)

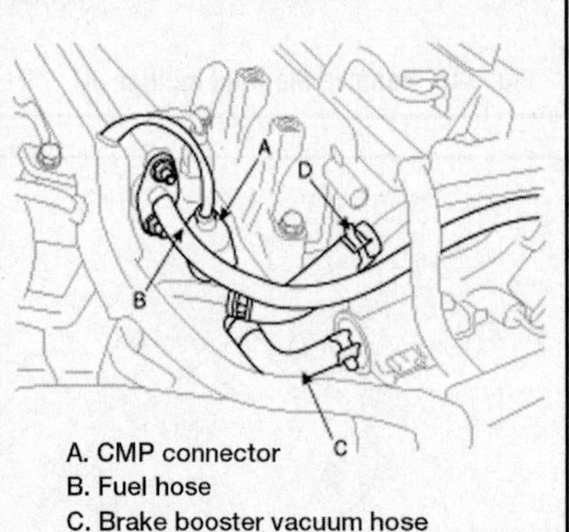
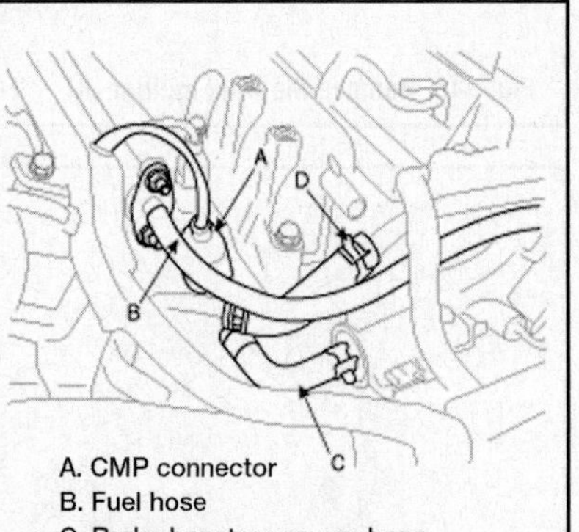

Fig. 153 Disconnect the CMP sensor connector, fuel hose, brake booster vacuum hose, and PCSV hose

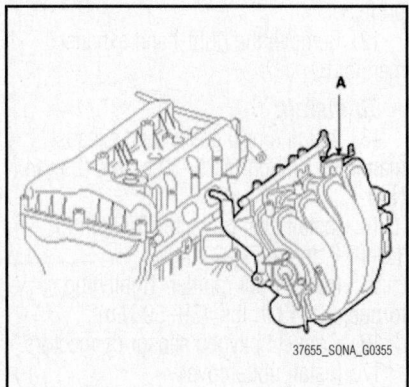

Fig. 155 Remove the intake manifold (A)

14. Remove the oil level gauge.
15. Remove the intake manifold.

To install:

16. Install the intake manifold. Tighten to 14–17 ft. lbs. (19–24 Nm).
17. Install the oil level gauge.
18. Install the intake manifold stay. Tighten to 14–17 ft. lbs. (19–24 Nm)
19. Connect the sensor connectors.
20. Install the PCV hose.
21. Install the coolant hose to the throttle body.
22. Connect the CMP connector, fuel hose, brake booster vacuum hose, and PCSV hose.
23. Connect the VCM connector.
24. Connect the ETC connector and MAP sensor connector.
25. Connect the injector connectors.
26. Connect the VIS connector, OPS connector, knock sensor connector and A/C switch connecter.
27. Install the air cleaner assembly, and connect the breather hose and ECM connector.
28. Install the engine cover.
29. Connect the negative battery cable.

3.3L Engines

See Figures 156 through 168.

1. Before servicing the vehicle, refer to the Precautions Section.
2. Relieve the fuel system pressure.
3. Disconnect the negative battery cable.
4. Disconnect AFS and breather hose.
5. Remove the air cleaner upper cover and intake hose.
6. Disconnect the right-hand oxygen sensor connector.
7. Disconnect the right-hand injector connector and ignition coil connector.
8. Disconnect the Electronic Throttle Control (ETC) connector and knock sensor connector.
9. Disconnect the Purge Control Solenoid Valve (PCSV) connector, Manifold Absolute Pressure (MAP) sensor connector, and PCSV hose.
10. Remove the ETC bracket.
11. Disconnect the water hoses from the ETC.
12. Disconnect the PCV hose.
13. Disconnect the brake vacuum hose.
14. Remove the surge tank stay.
15. Remove the connector bracket from surge tank.

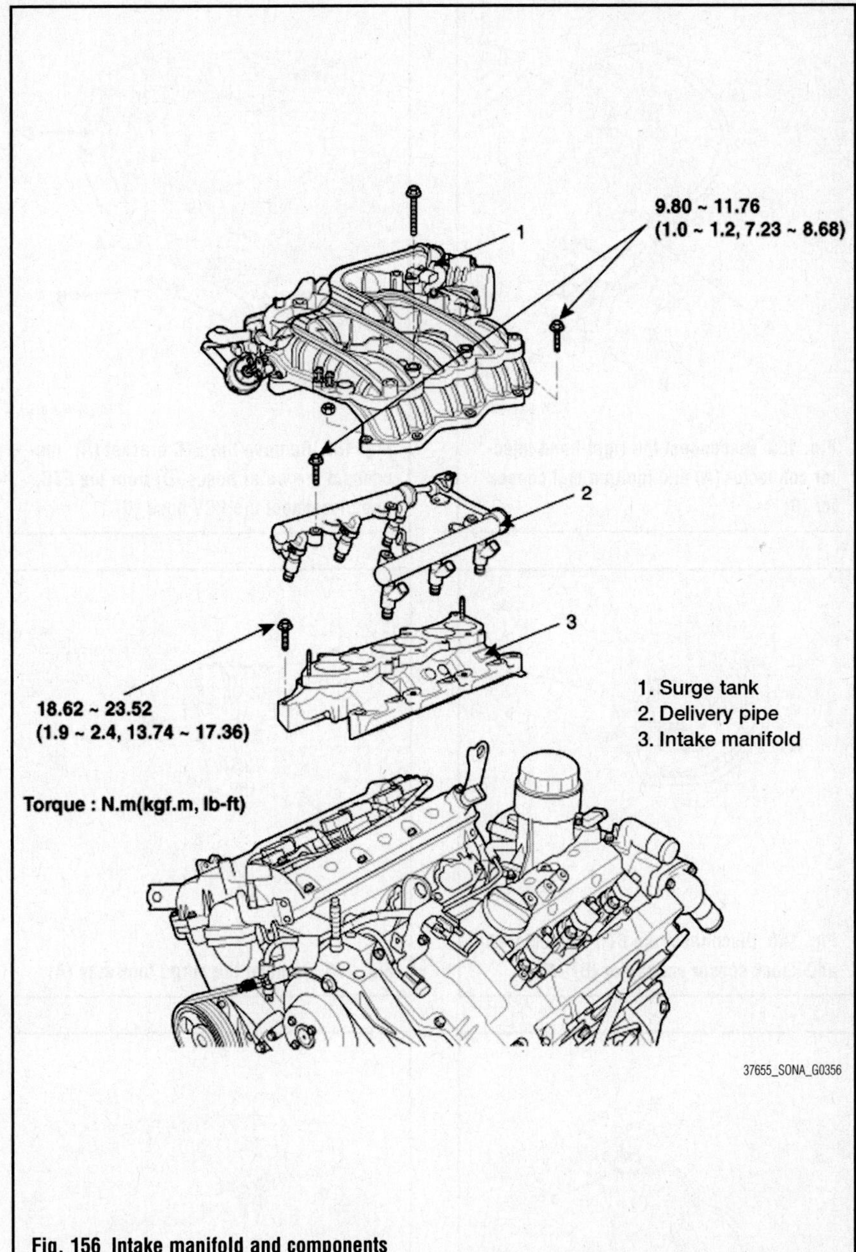

Fig. 156 Intake manifold and components

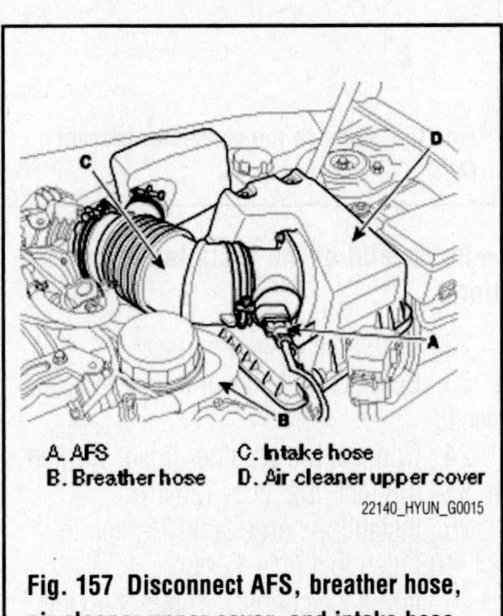

Fig. 157 Disconnect AFS, breather hose, air cleaner upper cover, and intake hose

Fig. 158 Disconnect the right-hand oxygen sensor connector (A)

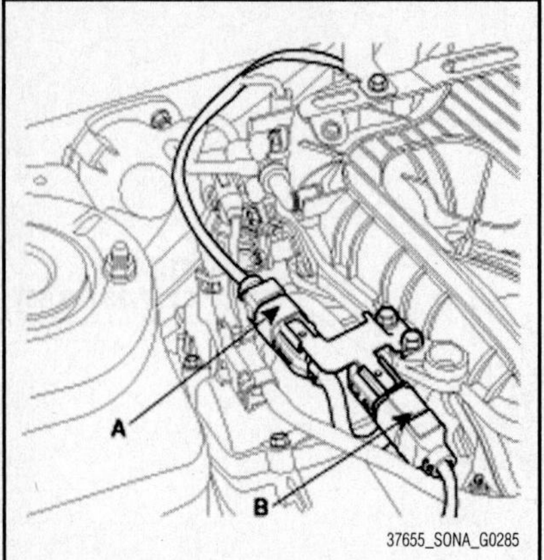

Fig. 159 Disconnect the right-hand injector connector (A) and ignition coil connector (B)

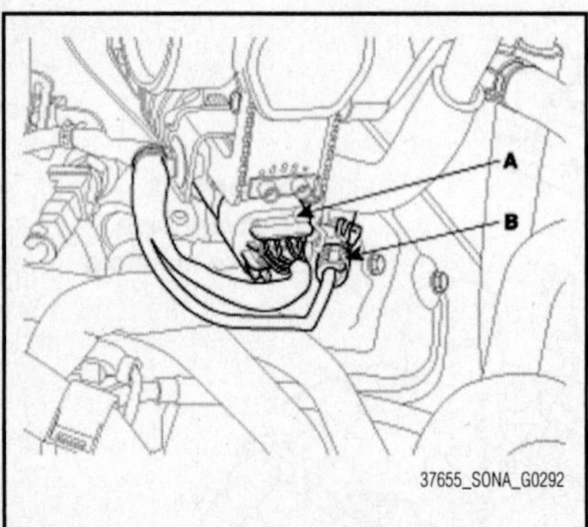

Fig. 160 Disconnect the ETC connector (A) and knock sensor connector (B)

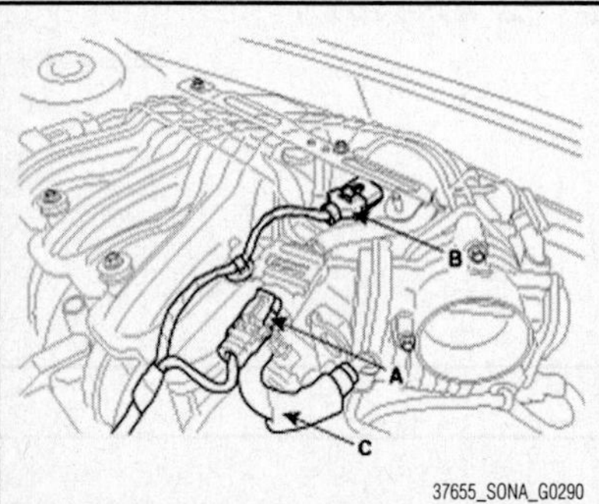
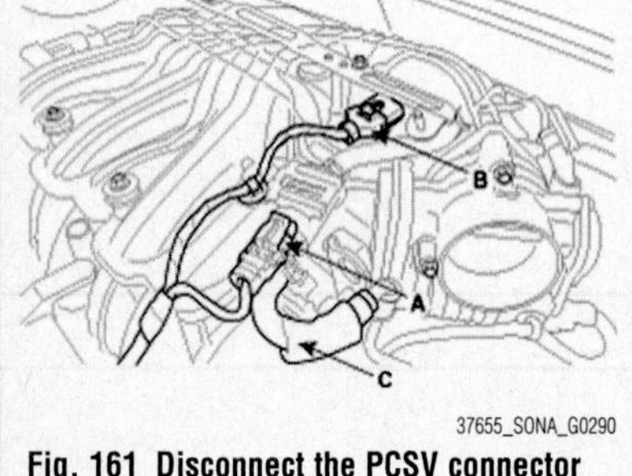

Fig. 161 Disconnect the PCSV connector (A), MAP sensor connector (B), and PCSV hose (C)

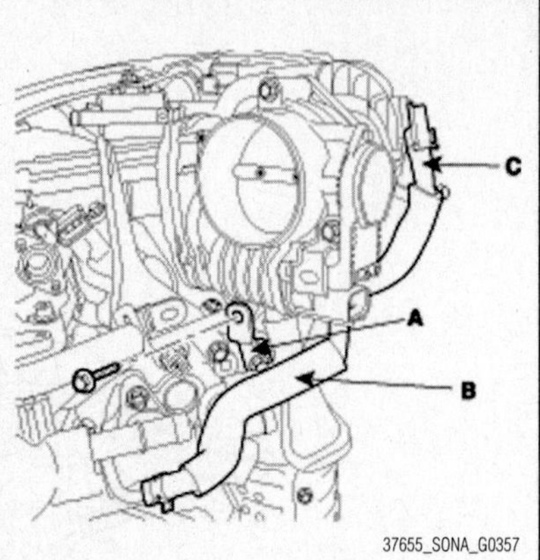

Fig. 162 Remove the ETC bracket (A), disconnect the water hoses (B) from the ETC, and disconnect the PCV hose (C)

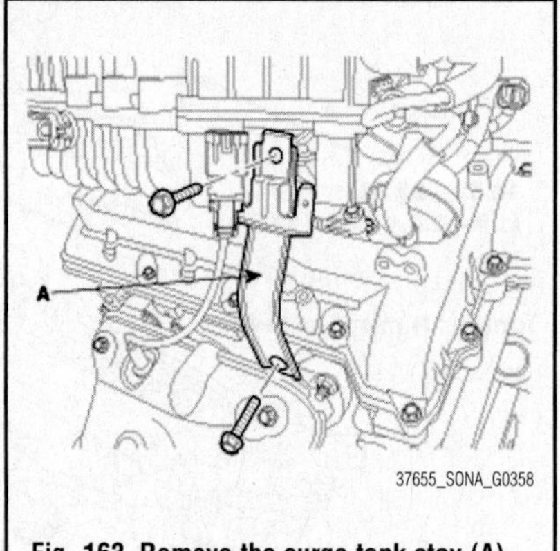

Fig. 163 Remove the surge tank stay (A)

Fig. 164 Remove the connector bracket (A)

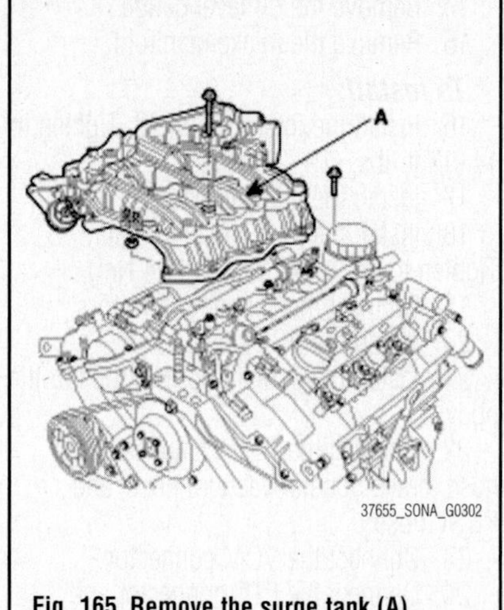

Fig. 165 Remove the surge tank (A)

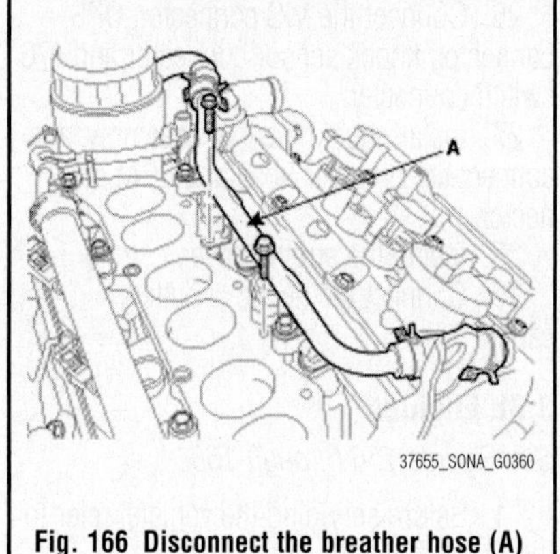

Fig. 166 Disconnect the breather hose (A)

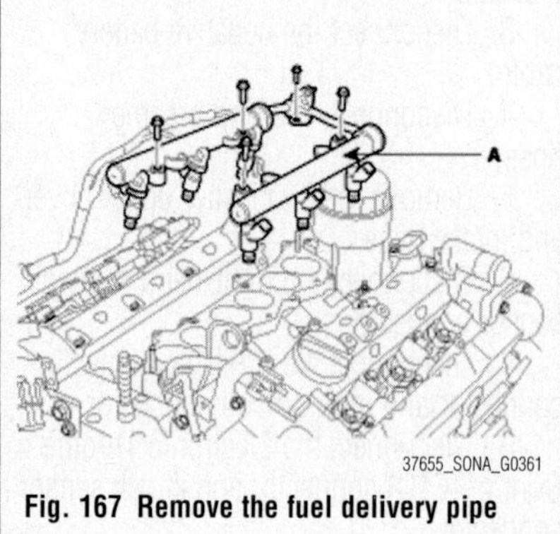

Fig. 167 Remove the fuel delivery pipe (A)

16. Remove the surge tank.
17. Disconnect the breather hose.
18. Disconnect the left-hand injector connector.
19. Remove the fuel delivery pipe.
20. Remove the intake manifold and gasket.

To install:

21. Install the intake manifold and a new gasket on the cylinder head. Tighten the bolts to 14–17 ft. lbs. (19–24 Nm).

➡Be careful of the installation direction.

22. Install the delivery pipe.
23. Connect the left-hand injector connector.
24. Connect the breather hose. Tighten to 84–108 inch lbs. (10–12 Nm).
25. Install the surge tank. Tighten to 84–108 inch lbs. (10–12 Nm).
26. Install the connector bracket on the surge tank. Tighten to 5–8 ft. lbs. (7–11 Nm).
27. Install the surge tank stay. Tighten to 20–23 ft. lbs. (27–31 Nm).
28. Connect the brake vacuum hose.
29. Connect the PCV hose.
30. Connect the water hoses to the ETC.
31. Install the ETC bracket. Tighten to 12–19 ft. lbs. (16–26 Nm).
32. Connect the ETC connector and the knock sensor connector.
33. Connect the PCSV connector, MAP sensor connector and PCSV hose.

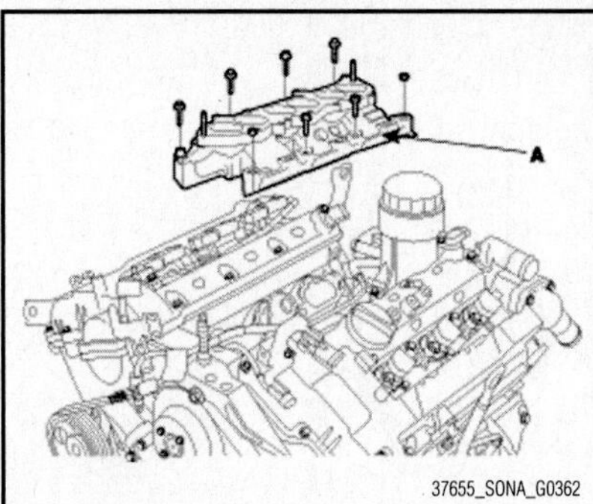

Fig. 168 Remove the intake manifold (A) and gasket

34. Connect the right-hand injector connector and ignition coil connector.
35. Connect the right-hand oxygen sensor connector.
36. Install the air cleaner upper cover and intake hose.
37. Connect the AFS and breather hose.
38. Connect the negative battery cable.

OIL PAN

REMOVAL & INSTALLATION

2.4L Engine

See Figures 169 through 171.

1. Before servicing the vehicle, refer to the Precautions Section.
2. Drain the engine oil.
3. Remove the oil pan bolts.
   a. Insert the tool between the oil pan and the ladder frame by tapping it with a plastic hammer in the direction of arrow.
   b. After tapping the tool with a plastic hammer along the direction of arrow around more than 2/3 edge of the oil pan, remove it from the ladder frame.

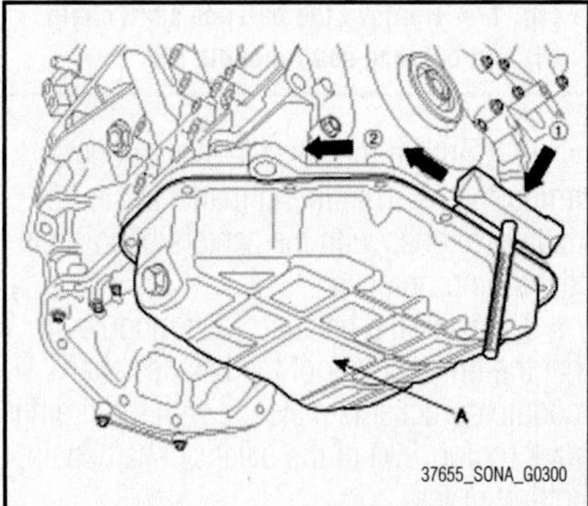

Fig. 169 Oil pan removal (A)—3.3L engine shown

➡Do not turn over the tool abruptly without tapping. It may result in damage to the tool. Be careful not to damage the contact surfaces of cylinder block and oil pan.

To install:

4. Using a gasket scraper, remove all the old packing material from the gasket surfaces.
5. Before assembling the oil pan, the liquid sealant Loctite 5900H or Three Bond® 1217H should be applied on oil pan. The part must be assembled within 5 minutes after the sealant was applied.

➡When applying sealant gasket, sealant must not be protruded into the inside of oil pan. To prevent leakage of oil, apply sealant gasket to the inner threads of the bolt holes.

6. Uniformly tighten the bolts in several passes. Tightening torque:
   - M8: 20–22 ft. lbs. (27–30 Nm)
   - M6: 87–104 inch lbs. (10–12 Nm)

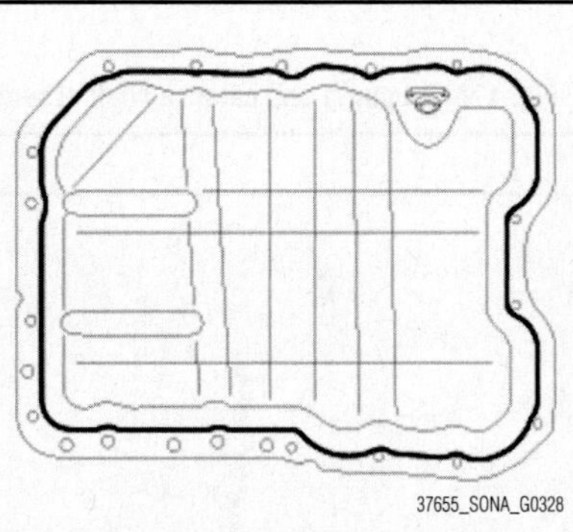

Fig. 170 Oil pan sealant application—2.4L engine

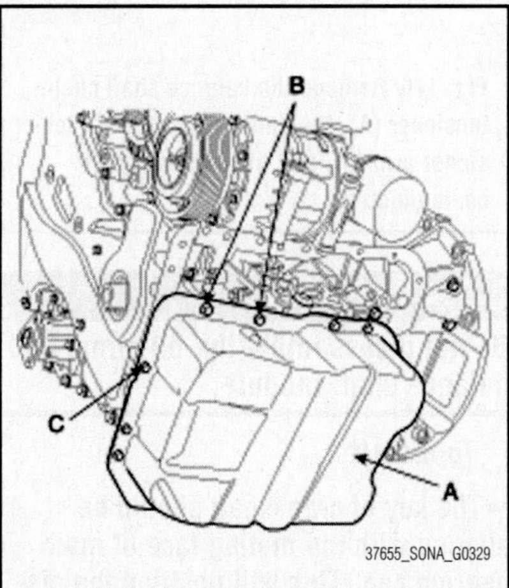

Fig. 171 Oil pan (A), M8 bolts (B), and M6 bolts (C)—2.4L engine

➡After assembly, wait at least 30 minutes before filling the engine with oil.

7. Fill engine with oil.

3.3L Engine

See Figures 172 through 174.

1. Before servicing the vehicle, refer to the Precautions Section.
2. Drain the engine oil.
3. Remove the lower oil pan bolts.
4. Insert the blade of SST (09215-3C000) between the upper oil pan and lower oil pan, and cut off applied sealer and removed lower oil pan.
   a. Insert the SST between the oil pan and the ladder frame by tapping it with a plastic hammer in the direction of arrow.
   b. After tapping the tool with a plastic hammer along the direction of arrow around more than 2/3 edge of the oil pan, remove it from the ladder frame.
   c. Do not turn over the tool abruptly without tapping. It may result in damage to the tool.
   d. Be careful not to damage the contact surfaces of upper oil pan and lower oil pan.

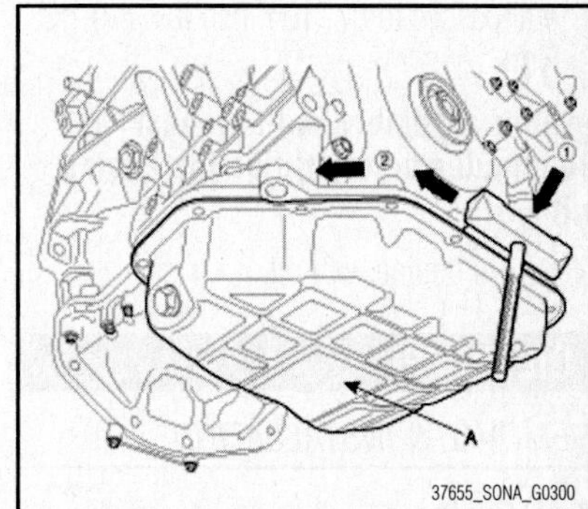

Fig. 172 Oil pan removal (A)—3.3L engine

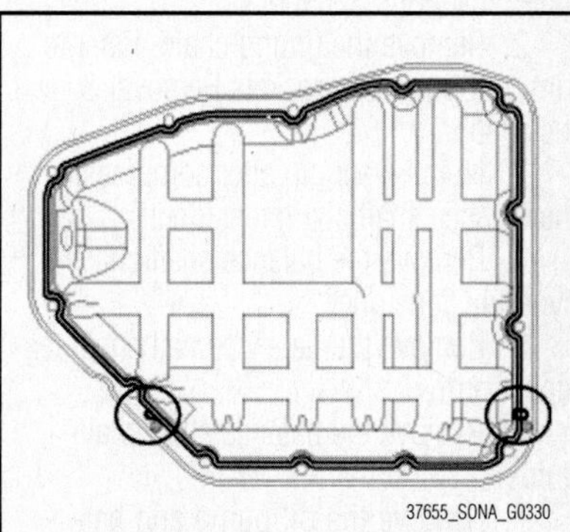

Fig. 173 Oil pan sealant application—3.3L engine

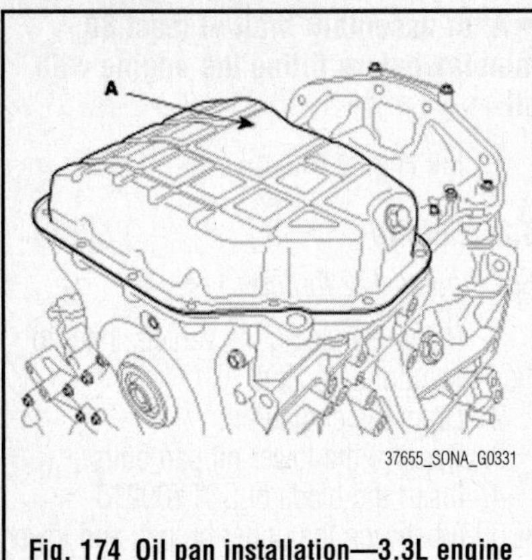

Fig. 174 Oil pan installation—3.3L engine

To install:

5. Install the lower oil pan.
   a. Using a gasket scraper, remove all the old packing material from the gasket surfaces.

➡Before assembling the oil pan, the liquid sealant TB 1217H should be applied on oil pan.

➡The part must be assembled within 5 minutes after the sealant was applied. Bead width: 0.1 in. (2.5mm), except marked area (*) to be 0.2 in. (5.0mm)

   b. Uniformly tighten the bolts in several passes to 87–104 inch lbs. (10–12 Nm).

➡After assembly, wait at least 30 minutes before filling the engine with oil.

6. Fill engine with oil.

OIL PUMP

REMOVAL & INSTALLATION

2.4L Engine

See Figures 175 through 181.

1. Before servicing the vehicle, refer to the Precautions Section.
2. Remove the timing chain. Refer to Timing Chain & Sprockets Removal & Installation.
3. Install a set pin after compressing the balance shaft chain tensioner.
4. Remove the balance shaft chain hydraulic tensioner.
5. Remove the balance shaft chain tensioner arm.
6. Remove the balance shaft chain guide.
7. Remove the oil pump and balance shaft module and balance shaft chain.

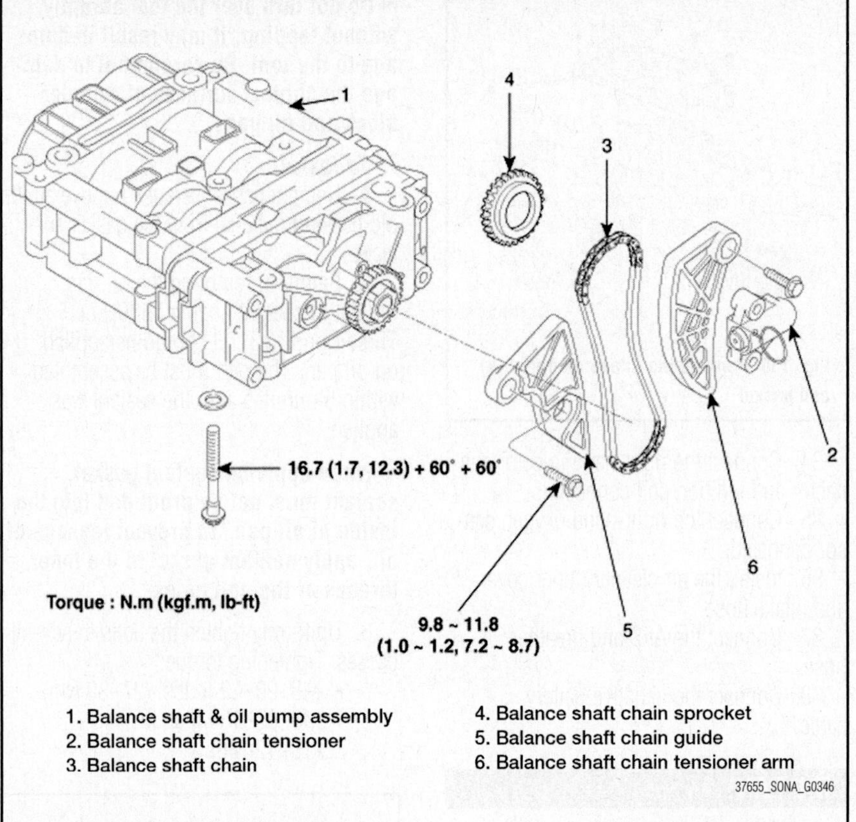

Fig. 175 Oil pump and balance shaft assembly

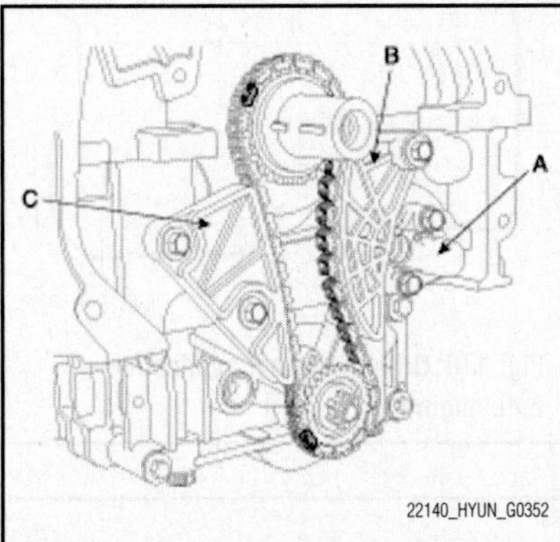

Fig. 176 Remove the balance shaft chain tensioner (A), the balance shaft chain tensioner arm (B), and the balance shaft chain guide (C)

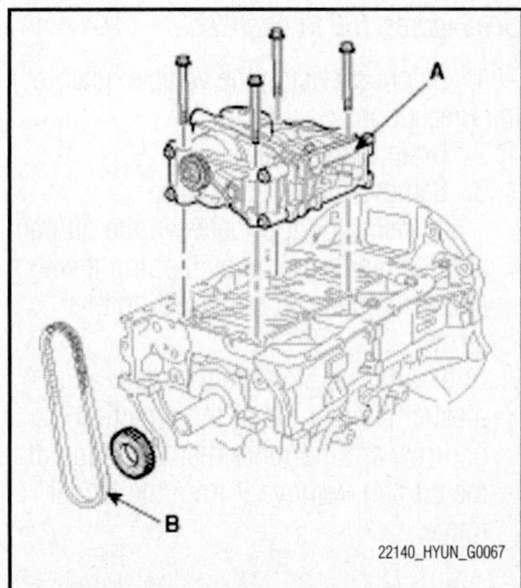

Fig. 177 Remove the balance shaft chain (B) and balance shaft module (A)

⁂ WARNING

Do not disassemble the oil pump and balance shaft module.

To install:

➡The key of crankshaft should be aligned with the mating face of main bearing cap. This will position the piston of No.1 cylinder at the top dead center on compression stroke.

8. Confirm the balance shaft module timing marks. The timing marks are to be visually aligned with the centers of adjacent cast timing notches.
9. Install the balance shaft module so that the timing mark of the balance shaft module sprocket is matched with the timing mark (color link) of the balance shaft chain. Bolting order:
   a. Assemble the bolts in order shown with seating torque 19 ft. lbs. (26 Nm).

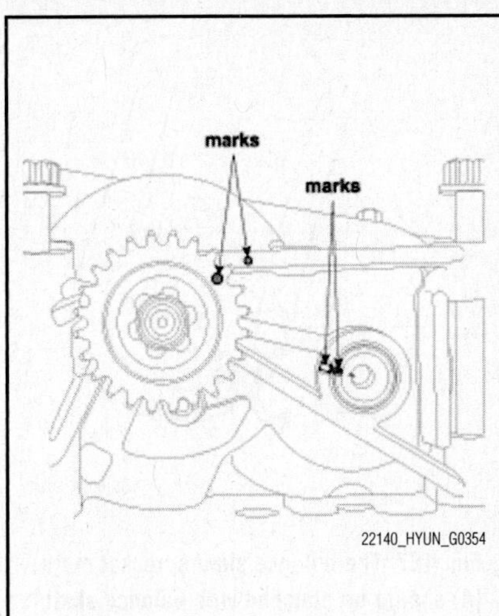

Fig. 178 The timing marks aligned as illustrated

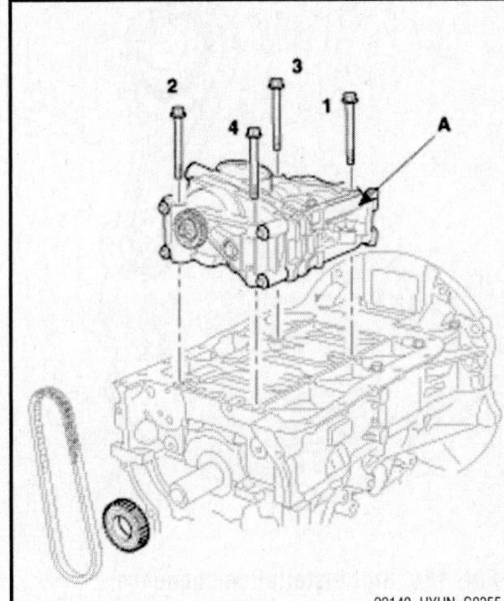

Fig. 179 Tighten the balance shaft module retaining bolts as illustrated

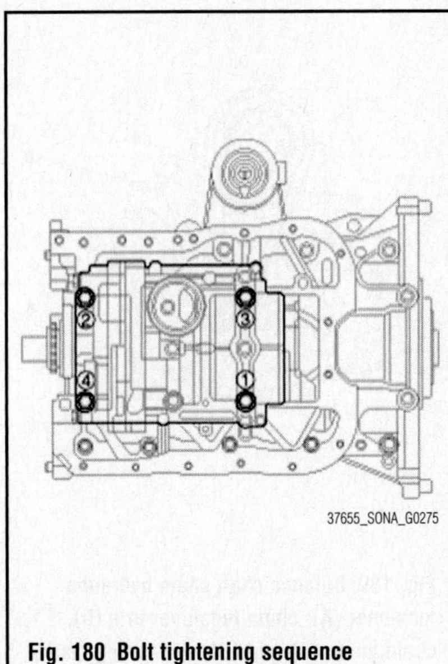

Fig. 180 Bolt tightening sequence

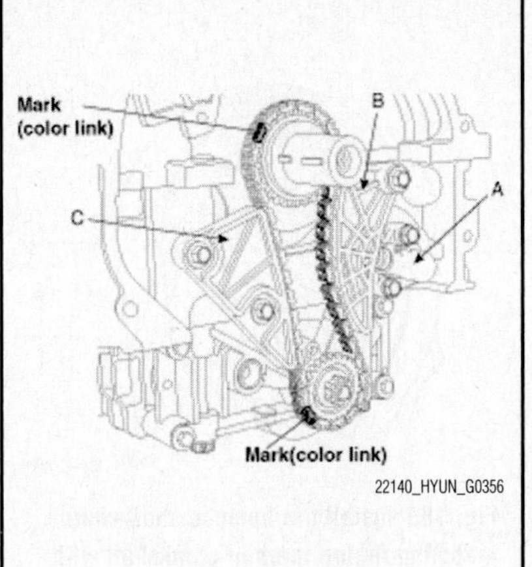

Fig. 181 Install the balance shaft chain guide (C), the balance shaft tensioner arm (B), and the balance shaft tensioner (A)

b. Unfasten the bolts as reverse bolting order (4-3-2-1).

c. Assemble the bolts in the specified order in same increments, as shown. Tightening torque: 12 ft. lbs. (17 Nm) plus 60° plus an additional 60°.

10. Install the balance shaft chain guide. Tightening torque: 87–104 inch lbs. (10–12 Nm).

11. Install the balance shaft chain tensioner arm. Tightening torque: 87–104 inch lbs. (10–12 Nm).

12. Install the balance shaft chain hydraulic tensioner. Tightening torque: 87–104 inch lbs. (10–12 Nm). Remove the set pin.

13. Confirm the timing marks.

14. Install the timing chain.

2.4L Engine—Revised

See Figures 182 through 189.

1. Before servicing the vehicle, refer to the Precautions Section.

2. Remove the timing chain. Refer to Timing Chain & Sprockets Removal & Installation.

➡Before removing the balance shaft chain, set the crankshaft key right direction to TDC position.

3. Install a set pin after compressing the balance shaft chain tensioner.

4. Remove the balance shaft chain hydraulic tensioner.

5. Remove the balance shaft chain tensioner arm.

6. Remove the balance shaft chain guide.

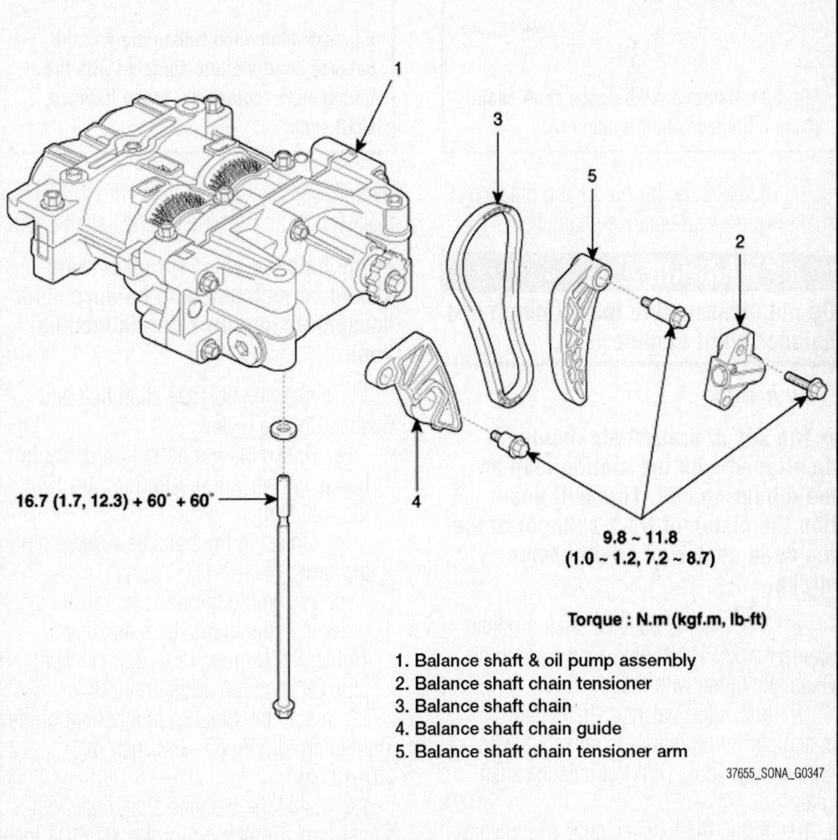

Fig. 182 Oil pump and balance shaft assembly—Revised

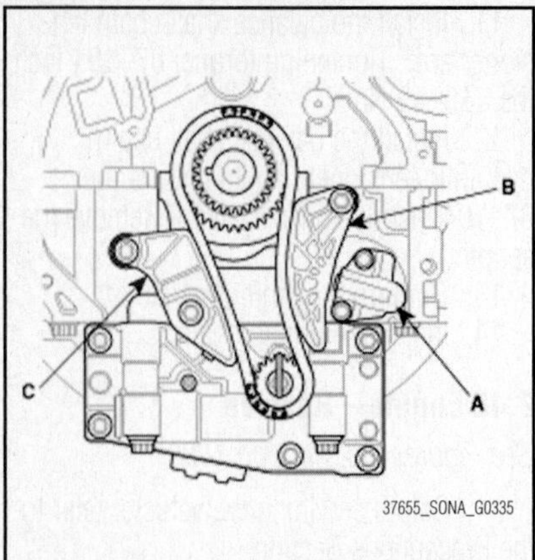

Fig. 183 Remove the balance shaft chain hydraulic tensioner (A), chain tensioner arm (B), and chain guide (C)

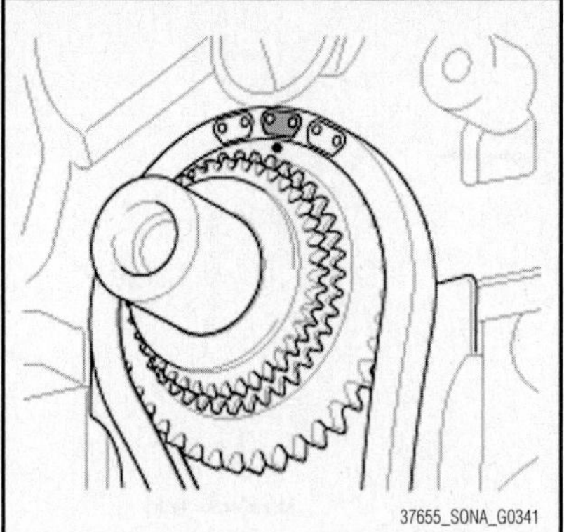

Fig. 185 Install the balance shaft chain matching timing mark of crankshaft with timing mark (color link) of balance shaft chain

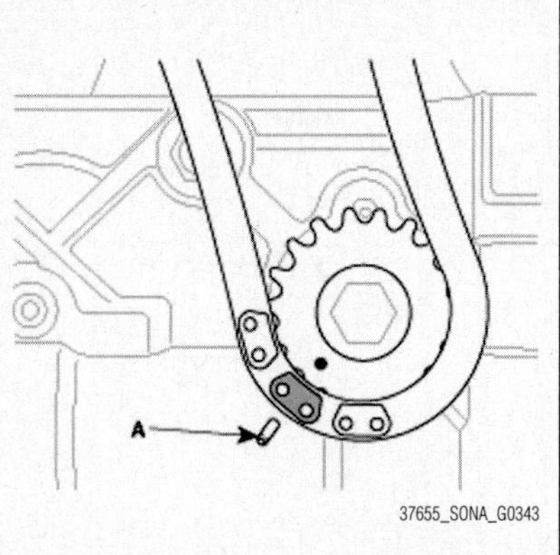

Fig. 187 The balance shaft sprocket mark (A) should be matched with balance shaft housing timing mark

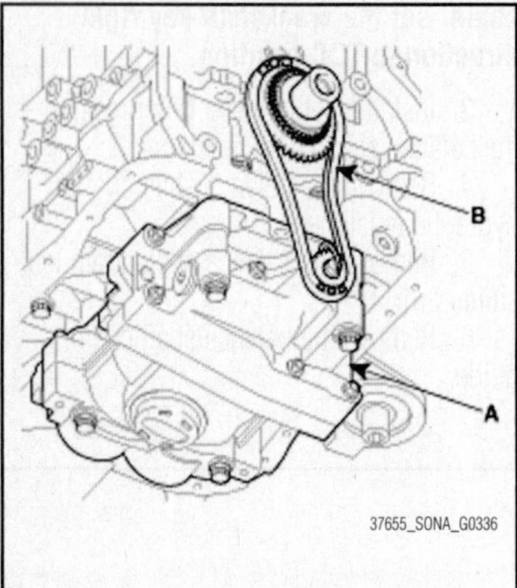

Fig. 184 Remove the balance shaft chain (B) and balance shaft module (A)

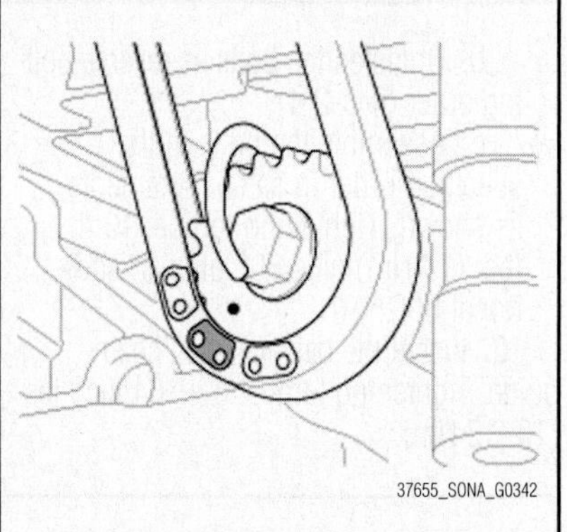

Fig. 186 Match the timing mark of the balance shaft module sprocket with the timing mark (color link) of the balance shaft chain

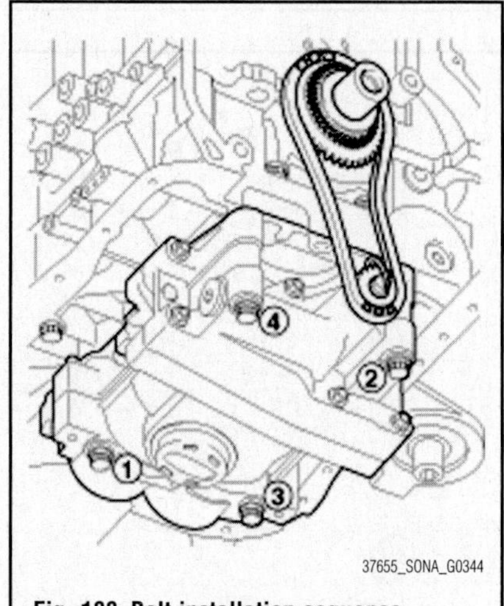

Fig. 188 Bolt installation sequence

7. Remove the oil pump and balance shaft module and balance shaft chain.

WARNING

Do not disassemble the oil pump and balance shaft module.

To install:

➡The key of crankshaft should be aligned with the mating face of main bearing cap. This will position the piston of No.1 cylinder at the top dead center on compression stroke.

8. Confirm the balance shaft module timing marks. The timing marks are to be visually aligned with the
9. Install the balance shaft chain matching timing mark of crankshaft with timing mark (color link) of balance shaft chain.
10. Install the balance shaft module so that the timing mark of the balance shaft module sprocket is matched with the timing mark (color link) of the balance shaft chain.

➡The balance shaft sprocket mark should be matched with balance shaft housing timing mark during installation.

11. Install the balance shaft bolt and washer. Bolting order:
   a. Assemble the bolts in order number as shown with seating torque 19 ft. lbs. (26 Nm).
   b. Unfasten the bolts as reverse bolting order (4-3-2-1).
   c. Assemble the bolts in specified order in same increments as shown. Tightening torque: 12 ft. lbs. (17 Nm) plus 60° plus an additional 60°.
12. Install the balance shaft chain guide. Tightening torque: 87–104 inch lbs. (10–12 Nm).
13. Install the balance shaft chain tensioner arm. Tightening torque: 87–104 inch lbs. (10–12 Nm).

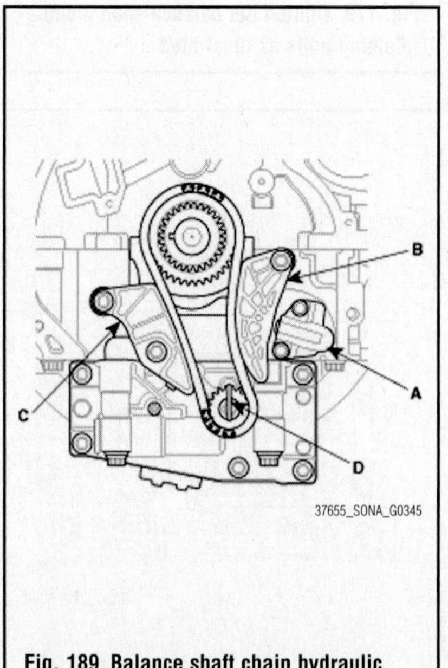

Fig. 189 Balance shaft chain hydraulic tensioner (A), chain tensioner arm (B), chain guide (C), pin (D), and timing marks

14. Install the balance shaft chain hydraulic tensioner then remove the stopper pin. Tightening torque: 87–104 inch lbs. (10–12 Nm). If installing a new balance shaft module, remove the pin.

15. Confirm the timing marks.

16. Install the timing chain.

3.3L Engines

See Figures 190 through 193.

1. Before servicing the vehicle, refer to the Precautions Section.
2. Remove the lower oil pan. Refer to Oil Pan Removal & Installation.
3. Remove the oil pump chain cover.
4. Remove the oil pump chain sprocket.
5. Remove the oil pump.

To install:

6. Install the oil pump using a new O-ring. Tighten the bolts to 14–17 ft. lbs. (20–24 Nm).

➡**Always use a new O-ring.**

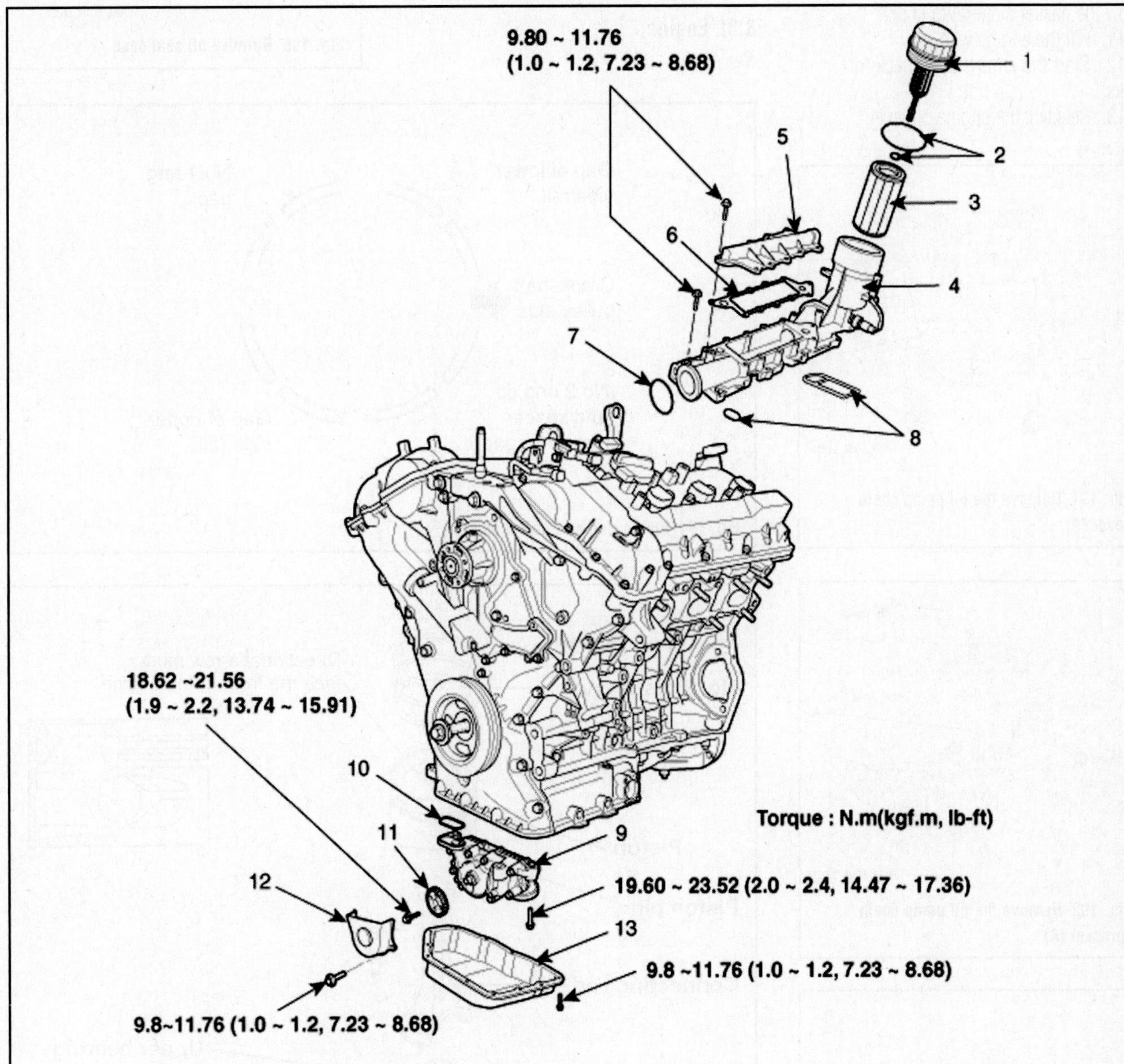

Fig. 190 Expanded view of lubrication system components

7. Install the oil pump sprocket and oil pump chain on the oil pump. Tightening torque: 14–16 ft. lbs. (19–22 Nm).
8. Install the oil pump chain cover. Tightening torque: 84–108 inch lbs. (10–12 Nm).
9. Install the lower oil pan.
10. After assembly, wait at least 30 minutes before filling the engine with oil to allow the gasket material to cure.
11. Fill the engine with oil.
12. Start the engine and check for leaks.
13. Recheck the engine oil level.

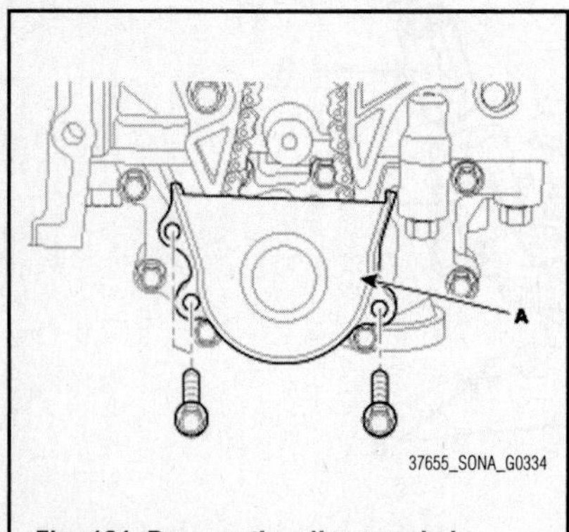

Fig. 191 Remove the oil pump chain cover (A)

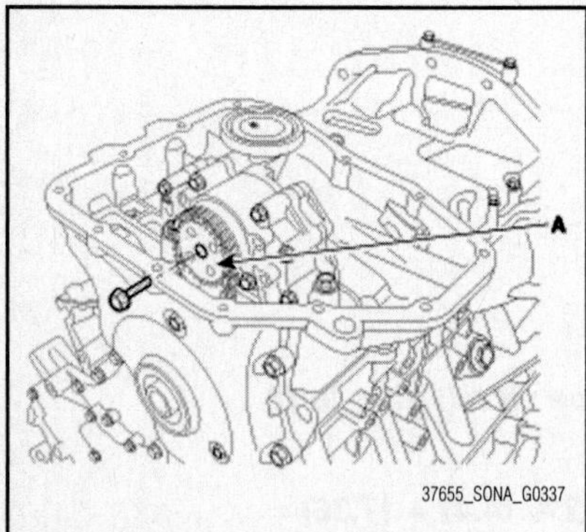

Fig. 192 Remove the oil pump chain sprocket (A)

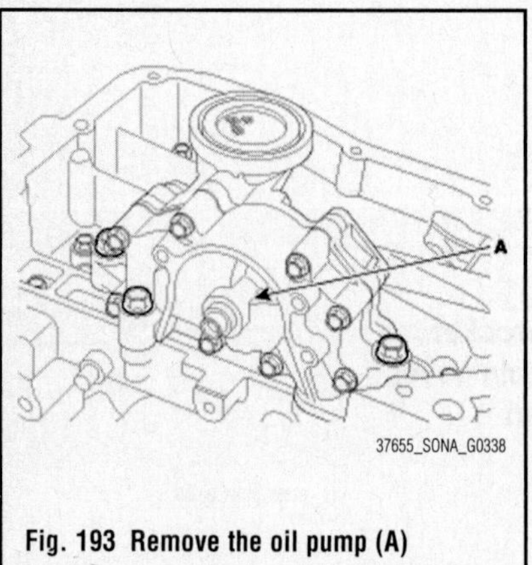

Fig. 193 Remove the oil pump (A)

PISTON AND RING

POSITIONING

See Figures 194 and 195.

REAR MAIN SEAL

REMOVAL & INSTALLATION

3.3L Engine

See Figure 196.

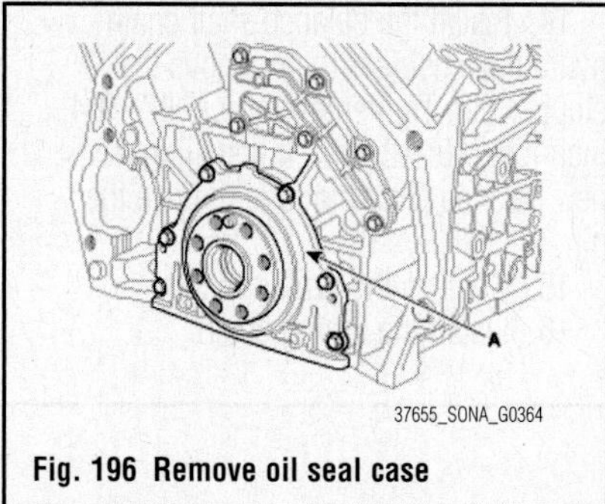

Fig. 196 Remove oil seal case

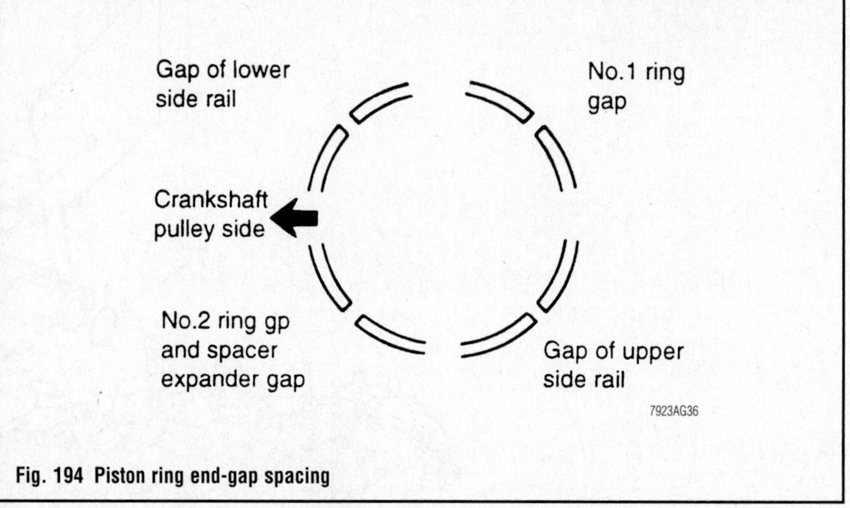

Fig. 194 Piston ring end-gap spacing

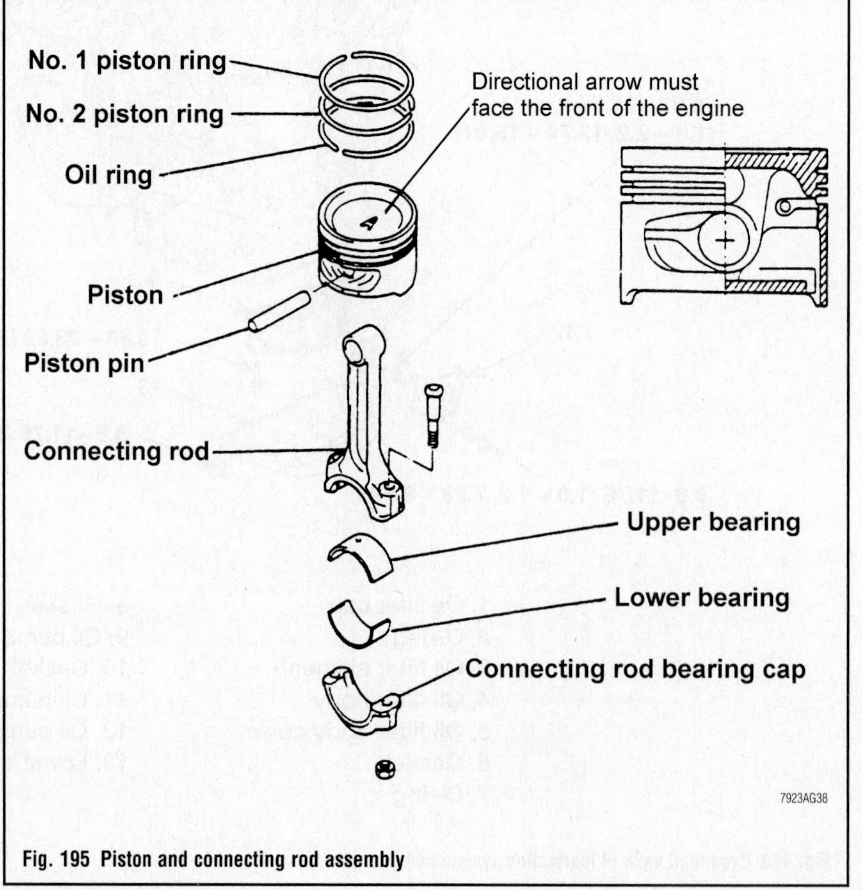

Fig. 195 Piston and connecting rod assembly

1. Before servicing the vehicle, refer to the Precautions Section.
2. Remove the transaxle.
3. Remove the drive plate. Refer to Flywheel/Drive Plate Removal & Installation.
4. Remove the oil seal case.
5. Remove the oil separator, if equipped.
6. Remove the oil seal.

To install:

7. Using seal a installer, install the oil seal in seal case.
8. Install the oil separator, if equipped.
9. Install the oil seal case and torque the case bolts to 84–108 inch lbs. (8–10 Nm).
10. Install drive plate.
11. Install the transaxle.

TIMING CHAIN & SPROCKETS

REMOVAL & INSTALLATION

2.4L Engine

See Figures 197 through 211.

➡Special tool used in this procedure: SST (09231-3K000), Flywheel Stopper

1. Before servicing the vehicle, refer to the Precautions Section.

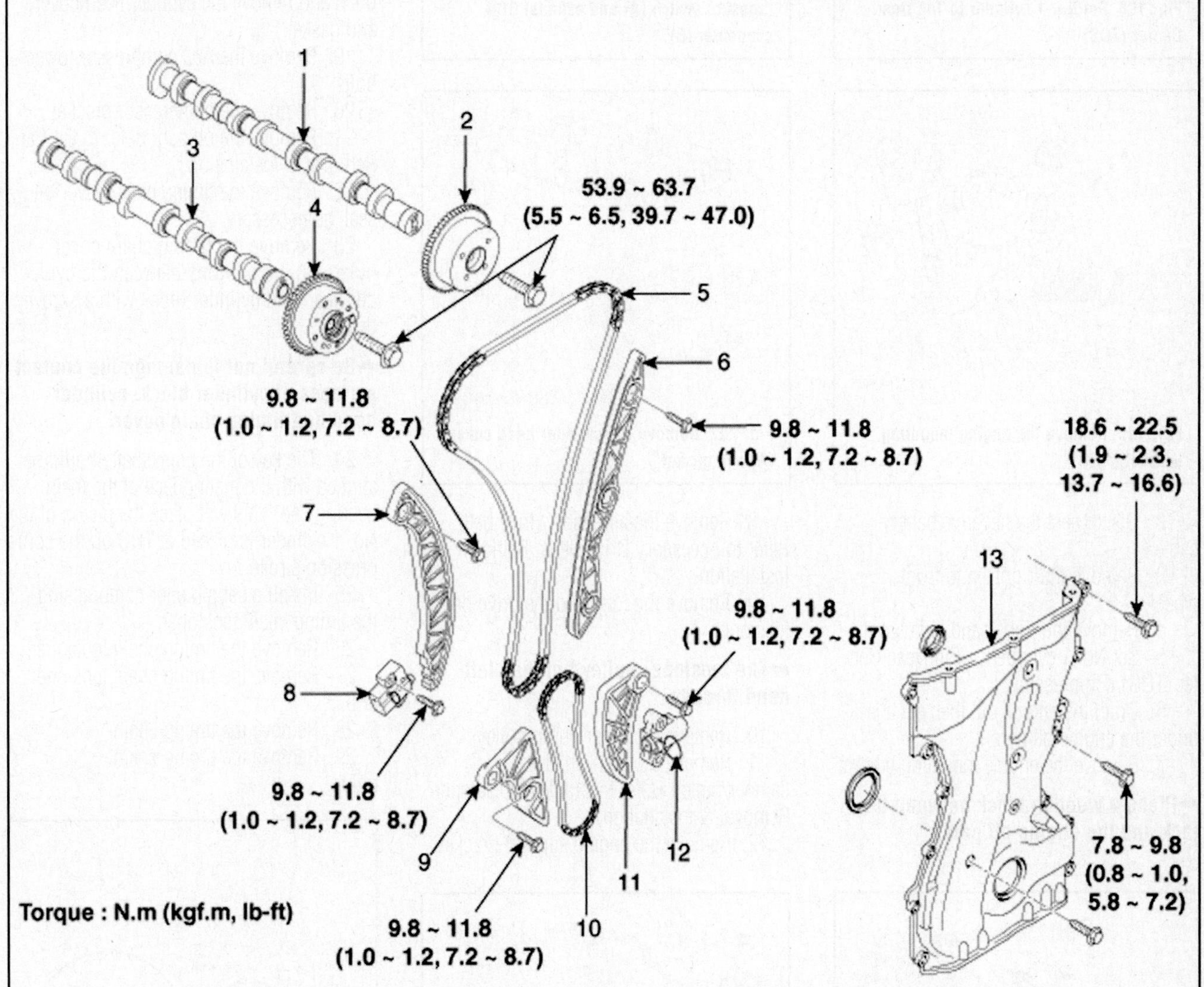

Fig. 197 Timing chain and components

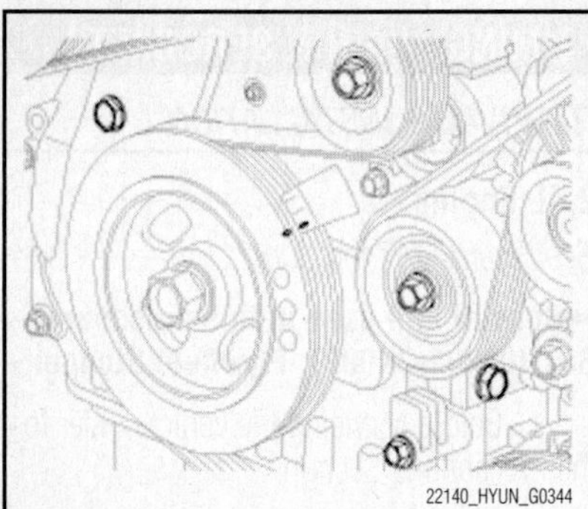

Fig. 198 Set No. 1 cylinder to Top Dead Center (TDC)

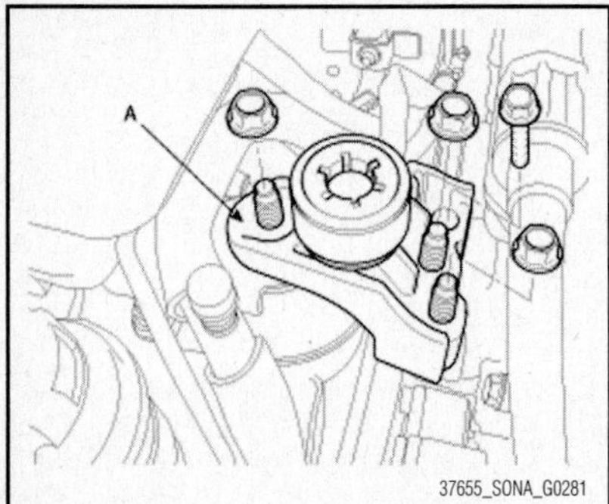

Fig. 199 Remove the engine mounting bracket (A)

2. Disconnect the negative battery cable.
3. Remove the right-hand front wheel.
4. Remove the right-hand side cover.
5. Set No. 1 cylinder to Top Dead Center (TDC)/compression.
6. Drain the engine oil, then set a jack under the engine oil pan.
7. Remove the engine mounting bracket.

➡Place a wooden block between the jack and the engine oil pan.

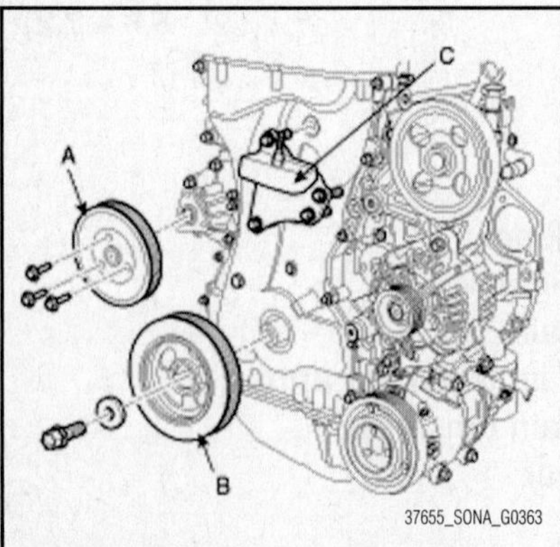

Fig. 200 Remove the water pump pulley (A), crankshaft damper (B), and engine support bracket (C)

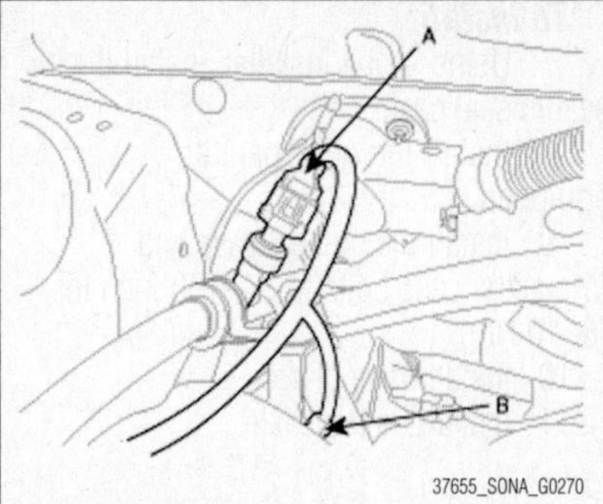

Fig. 201 Remove the power steering oil pressure switch (A) and exhaust OCV connector (B)

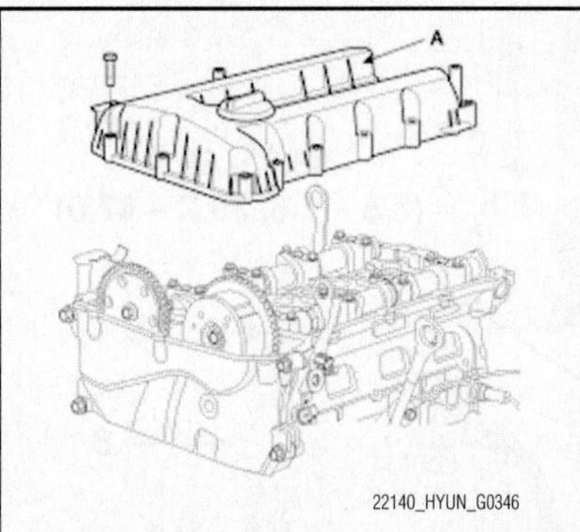

Fig. 202 Remove the cylinder head cover (A) and gasket

8. Remove the accessory drive belt. Refer to Accessory Drive Belts, Removal & Installation.
9. Remove the idler and the drive belt tensioner.

➡The tensioner pulley bolt has left-hand threads.

10. Remove the water pump pulley.
11. Remove the crankshaft damper/pulley. Refer to Crankshaft Damper Removal & Installation.
12. Remove the engine support bracket.

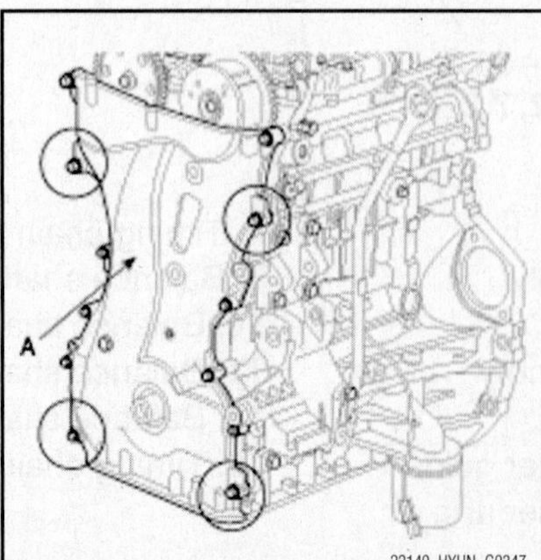

Fig. 203 Remove the timing chain cover (A)

13. Use the SST (09231-3K000), flywheel stopper, to remove the crankshaft pulley bolt, after removing the starter.
14. Disconnect the power steering oil pressure switch and exhaust OCV connector.
15. Remove the PCV hose and breather hose.
16. Disconnect the ignition coil connectors.
17. Remove the ignition coils.
18. Remove the cylinder head cover bolts and remove the cylinder head cover and gasket.
19. Remove the A/C compressor lower bolts.
20. Remove the compressor bracket.
21. Remove the oil pan. Refer to Oil Pan, Removal & Installation.
22. Remove the timing chain cover oil seal, as necessary.
23. Remove the timing chain cover by prying the portions between the cylinder head and cylinder block with a screwdriver.

➡Be careful not to damage the contact surfaces of cylinder block, cylinder head and timing chain cover.

24. The key of the crankshaft should be aligned with the mating face of the main bearing cap. This will place the piston of the No. 1 cylinder is placed at TDC on the compression stroke.
25. Install a set pin after compressing the timing chain tensioner.
26. Remove the timing chain tensioner.
27. Remove the timing chain tensioner arm.
28. Remove the timing chain.
29. Remove the timing chain guide.

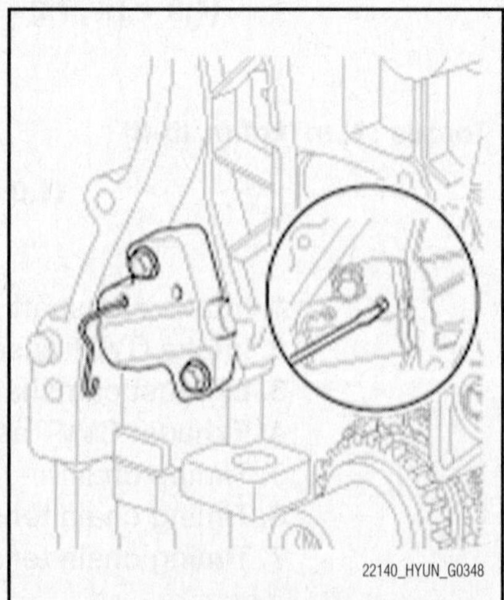

Fig. 204 Install a set pin after compressing the timing chain tensioner

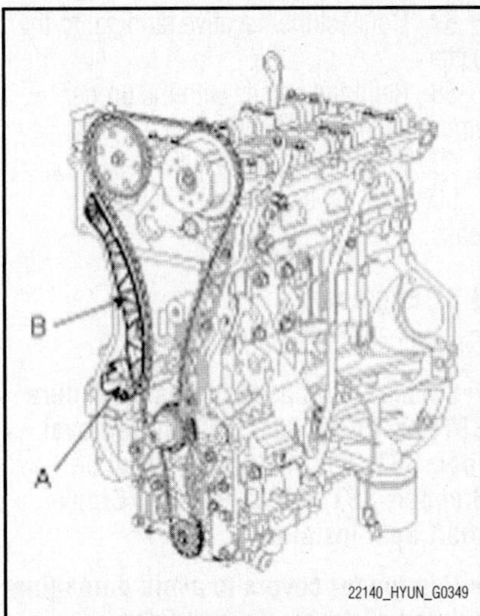

Fig. 205 Remove the timing chain tensioner (A) and the tensioner arm (B)

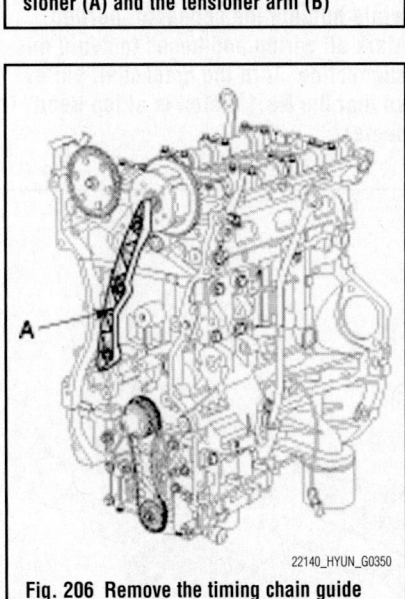

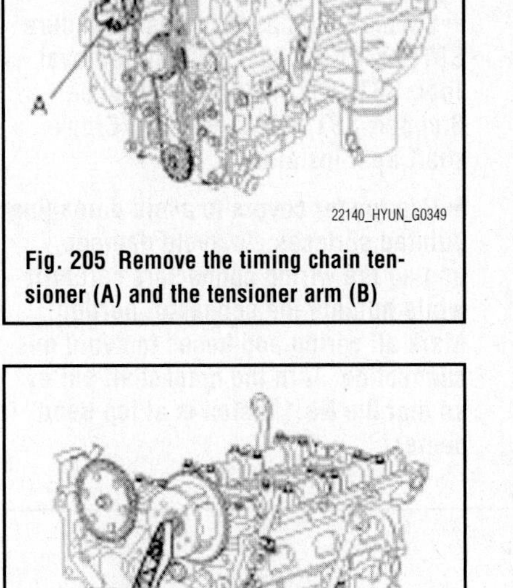

Fig. 206 Remove the timing chain guide (A)

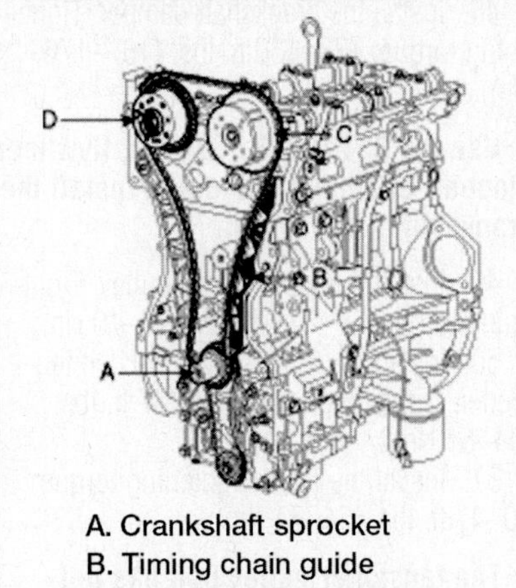

Fig. 207 Timing chain installation

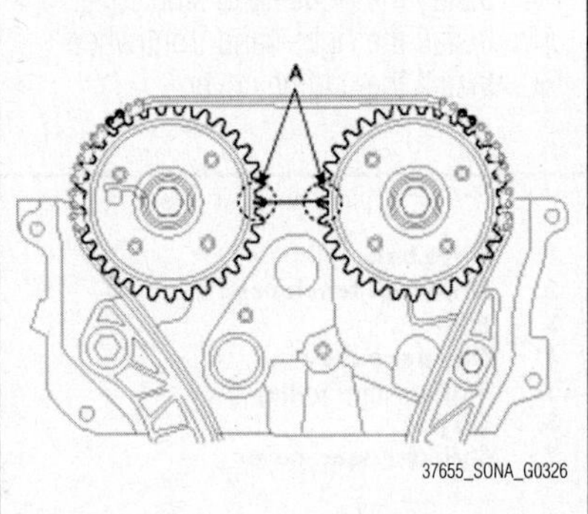

Fig. 208 Timing mark alignment

To install:

30. Set the crankshaft so that the key of the crankshaft is aligned with the mating surface of main bearing cap. Set the intake and exhaust camshaft assembly so that the TDC mark of the intake sprocket and the exhaust sprocket are aligned with the top surface of the cylinder head. This will place the piston on the No. 1 cylinder at TDC on the compression stroke.
31. Install the timing chain guide. Tightening torque: 87–104 inch lbs. (10–12 Nm).
32. Install the timing chain. To install the timing chain so that there is no slack between the camshaft and crankshaft, use the following procedure:
    a. Place the chain over the crankshaft sprocket (A), then the timing chain guide (B), then the intake CVVT assembly (C), then the exhaust CVVT assembly (D).
    b. The timing mark of each of the sprockets should be matched with the timing mark (color link) of the timing chain while installing the timing chain.
33. Install the timing chain tensioner arm. Tightening torque: 87–104 inch lbs. (10–12 Nm).
34. Install the timing chain auto tensioner and remove the set pin. Tightening torque: 87–104 inch lbs. (10–12 Nm).
35. After rotating the crankshaft 2 revolutions in a clockwise direction (viewed from the front), confirm that the timing marks are still aligned.
36. Install the timing chain cover.
    a. The sealant locations on the chain cover and on the counter parts (cylinder head, cylinder block, and ladder frame) must be free of engine oil and any debris.
    b. Before assembling the timing chain cover, the liquid sealant Loctite® 5900H or Three Bond® 1217H should be applied on the gap between the cylinder head and the cylinder block.

➡The parts must be assembled within 5 minutes after the sealant was applied. Use a bead width of 0.12 in. (3.0mm).

    c. Apply sealant on the timing chain cover. The part must be assembled within 5 minutes after the sealant was applied. The sealant should be applied in a continuous bead.
    d. The dowel pins on the cylinder block and the holes on the timing chain cover should be used as a reference in order to assemble the timing chain cover into exact position. Tightening torque:
        - M6 bolts: 69–87 inch lbs. (8–10 Nm)
        - M8 bolts: 14–17 ft. lbs. (19–23 Nm)
    e. Wait to start the engine for 30 minutes after the timing chain cover was assembled.
37. Install timing chain cover oil seal, as necessary, as follows:
    a. Apply engine oil to a new oil seal lip.
    b. Using SST (09214-3K000, 09231-H1100) and a hammer, tap in the oil seal.
38. Install oil pan. Refer to Oil Pan, Removal & Installation.
    a. Uniformly tighten the bolts in several passes.
    b. Tightening torque:
        - M8: 20–22 ft. lbs. (27–30 Nm)
        - M6: 87–104 inch lbs. (10–12 Nm)
    c. After assembly, wait at least 30 minutes before filling the engine with oil.
39. Install the air compressor bracket. Tightening torque: 14–15 ft. lbs. (20–24 Nm).
40. Install air compressor lower bolts. Tightening torque: 14–18 ft. lbs. (20–25 Nm).
41. Install the cylinder head cover.
    a. The hardening sealant located on the upper area between the timing chain

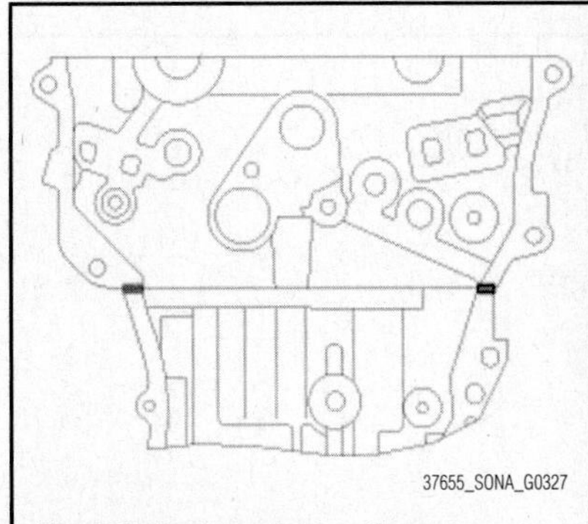

Fig. 209 Apply sealant on the gap between the cylinder head and the cylinder block

cover and the cylinder head should be removed before assembling the cylinder head cover.

b. After applying sealant, it should be assembled within 5 minutes. Apply in a bead width of 0.1 inch (2.5mm).

c. The firing of the engine should not be performed within 30 minutes after the cylinder head cover was assembled.

d. Install the cylinder head cover bolts.

- Step 1: 35–52 inch lbs. (4–6 Nm)
- Step 2: 69–87 inch lbs. (8–10 Nm)

➡Do not reuse the cylinder head cover gasket.

42. Install the ignition coils.
43. Connect the ignition coil connectors.
44. Install the PCV hose.
45. Install the breather hose.
46. Connect the power steering oil pressure switch connector and exhaust OCV connector.
47. Install the engine support bracket. Tightening torque:
    - M10 bolts: 29–33 ft. lbs. (39–44 Nm)
    - M8 bolts: 15–18 ft. lbs. (20–25 Nm)

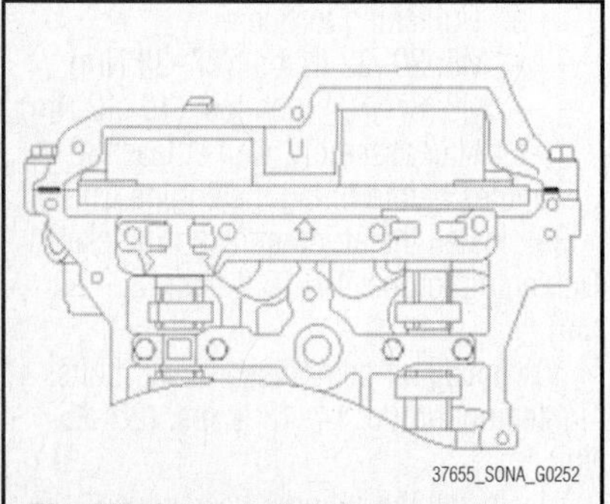

Fig. 210 Apply sealant on the gap between the cylinder head and the cylinder block

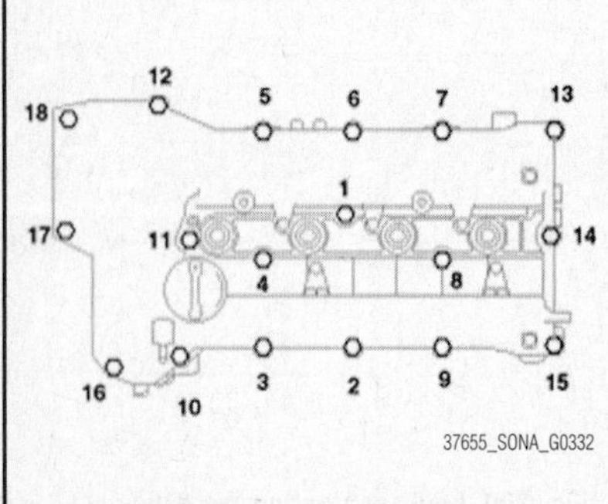

Fig. 211 Cylinder head cover bolt tightening sequence

48. Install the crankshaft damper. Tightening torque: 123–130 ft. lbs. (167–176 Nm).

➡Use the SST (09231-3K000), flywheel stopper, or equivalent tool to install the crankshaft damper bolt.

49. Install the water pump pulley. Tightening torque: 69–87 inch lbs. (8–10 Nm).
50. Install the accessory drive belt tensioner. Tightening torque: 40–47 ft. lbs. (54–64 Nm).
51. Install the idler. Tightening torque: 40–47 ft. lbs. (54–64 Nm).

➡The tensioner pulley bolt has left-hand threads.

52. Install the accessory drive belt.
53. Install the engine mounting bracket. Tightening torque: 47–62 ft. lbs. (64–83 Nm).
54. Install the right-hand side cover.
55. Install the right-hand front wheel.
56. Install the engine cover.
57. Connect the negative terminal to the battery.
58. Refill the engine with the proper amount and type of engine oil.
59. Refill any necessary fluids.
60. Run the engine and check for leaks.

3.3L Engines

See Figures 212 through 251.

➡Special tools used in this procedure: SST(09215-3C000), Oil Pan Removal Tool; SST (09231-3C300), Flywheel Stopper; SST (09231-3C100), Crankshaft Seal Installer

➡Use fender covers to avoid damaging painted surfaces. To avoid damage, unplug the wiring connectors carefully while holding the connector portion. Mark all wiring and hoses to avoid misconnection. Turn the crankshaft pulley so that the No.1 piston is at top dead center.

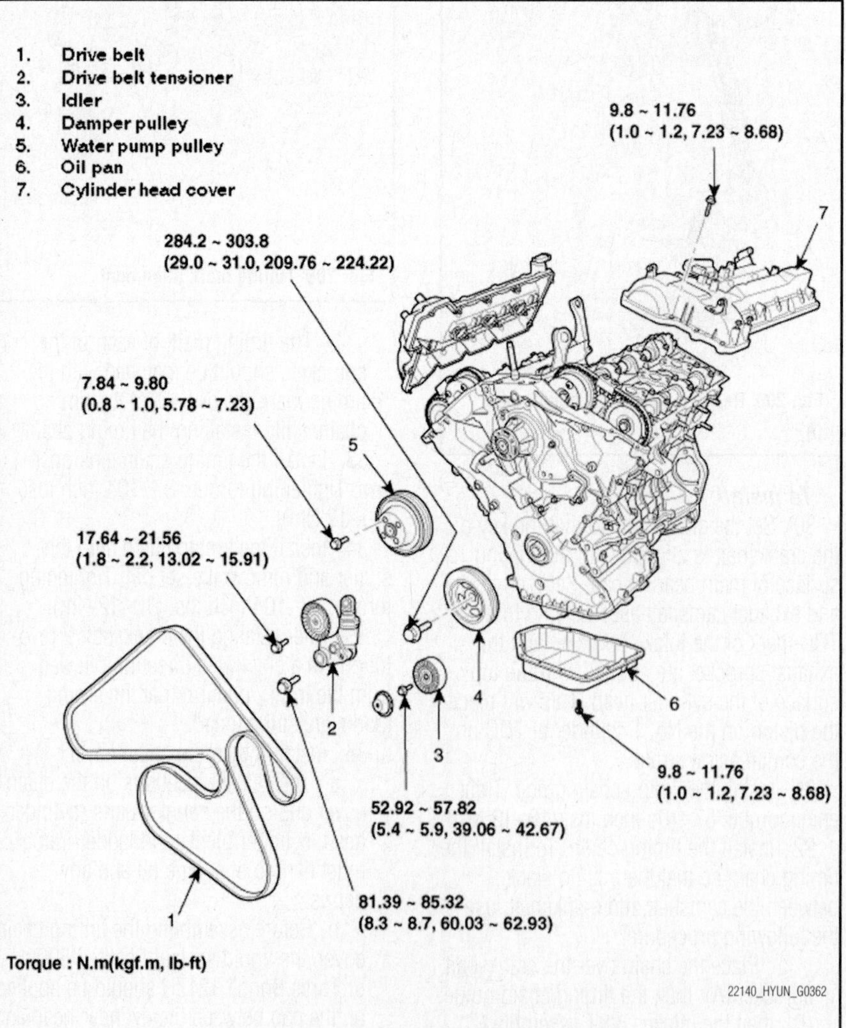

Fig. 212 Exploded view of cylinder head and accessory drive belt components

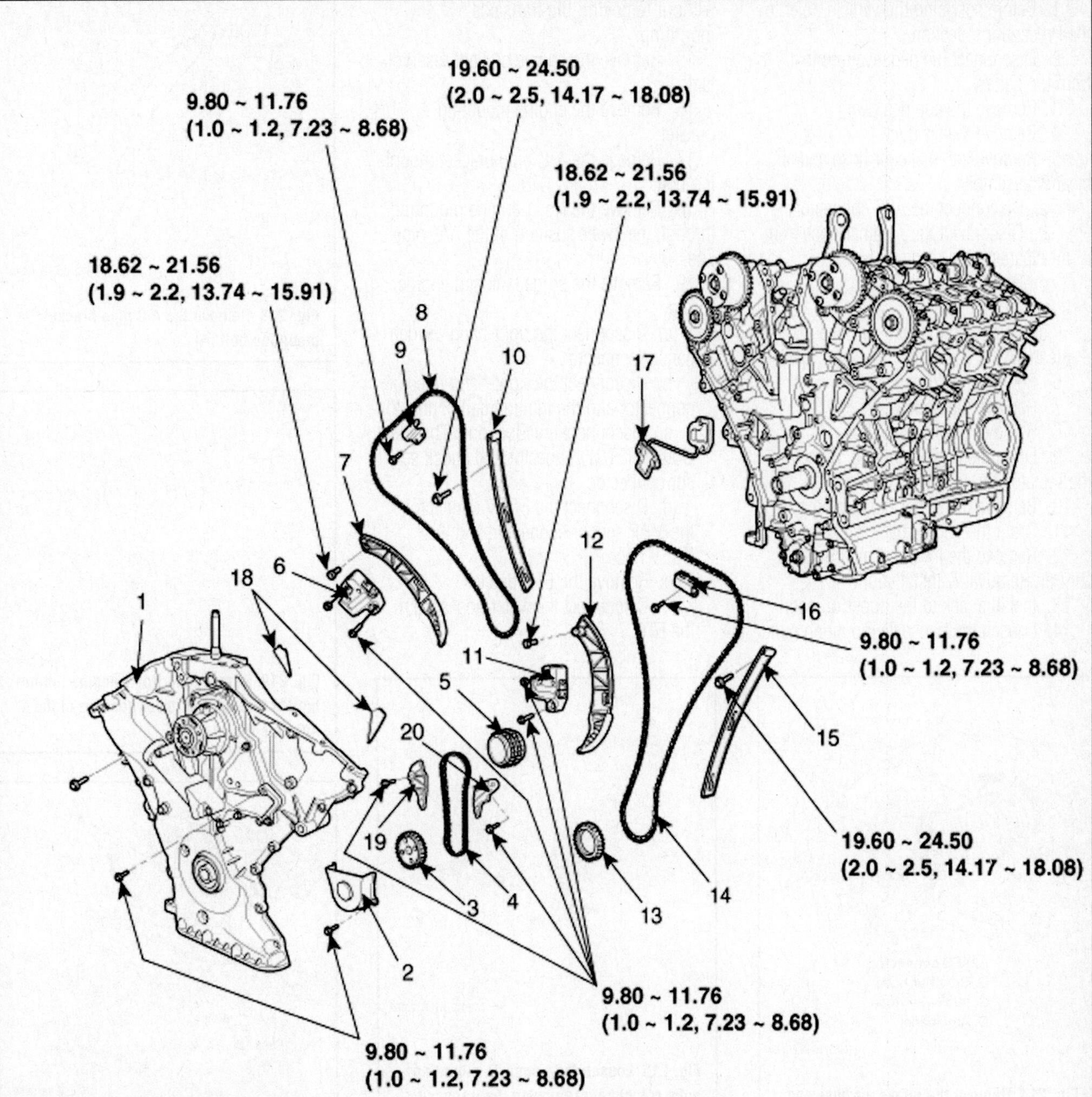

1. Timing chain cover
2. Oil pump chain cover
3. Oil pump sprocket
4. Oil pump chain
5. Crankshaft sprocket
6. Timing chain auto tensioner
7. Timing chain tensioner arm
8. Timing chain
9. Cam to cam guide
10. Timing chain guide
11. Timing chain auto tensioner
12. Timing chain tensioner arm
13. Crankshaft sprocket
14. Timing chain
15. Timing chain guide
16. Cam to cam guide
17. Tensioner adapter
18. Gasket
19. Oil pump chain guide
20. Oil pump tensioner assembly

22140_HYUN_G0363

Fig. 213 Exploded view of timing chain and related components

1. Before servicing the vehicle, refer to the Precautions Section.
2. Disconnect the negative terminal from the battery.
3. Remove the engine cover.
4. Remove the air duct.
5. Remove the intake air hose and air cleaner assembly:
   a. Disconnect the AFS connector.
   b. Disconnect the breather hose from air cleaner hose.
   c. Disconnect the ECM connector.
   d. Remove the intake air hose and air cleaner.
6. Remove the front wheels.
7. Remove the undercover.
8. Remove the side cover.
9. Drain the engine coolant. Remove the radiator cap to speed draining.
10. Remove the upper radiator hose.
11. Drain the engine oil.
12. Remove the lower oil pan. Refer to Oil Pan Removal & Installation.
13. Install a jack to the upper oil pan.
14. Loosen the transaxle mounting bolts without removing the transaxle mounting.
15. Remove the engine coolant reservoir tank.
16. Remove the engine mounting bracket.
17. Remove the A/C pipe bracket mounting bolt.
18. Remove the No. 1 engine mounting through the lower position of the A/C pipe line.
19. Remove the surge tank and engine wiring.
   a. Disconnect the right-hand oxygen sensor connector.
   b. Disconnect the right-hand injector connector and the ignition coil connector.
   c. Disconnect the Electronic Throttle Control (ETC) connector and knock sensor connector.
   d. Disconnect the PCSV connector, the MAP sensor connector, and the PCSV hose.
   e. Remove the ETC bracket.
   f. Disconnect the water hoses from the ETC.

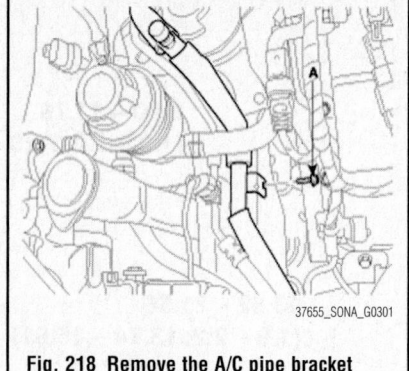

Fig. 218 Remove the A/C pipe bracket mounting bolt (A)

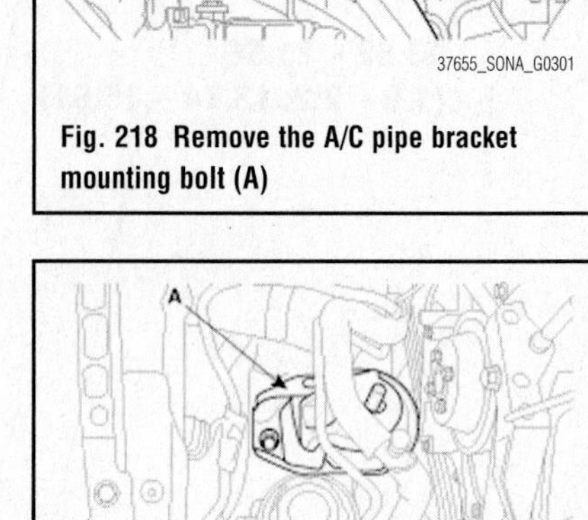

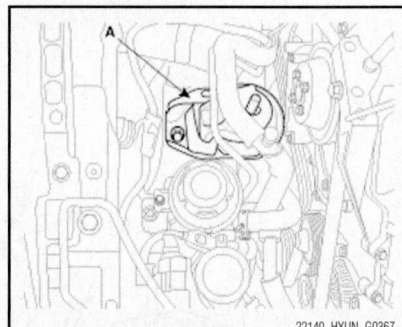

Fig. 219 Remove the No. 1 engine mounting (A) through the lower position of the A/C pipe line

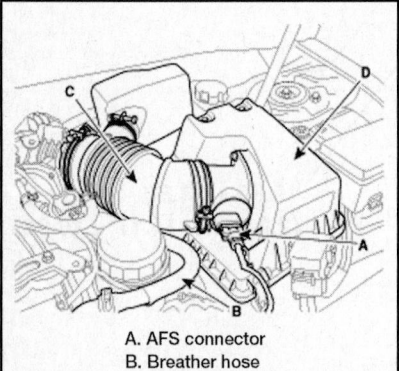

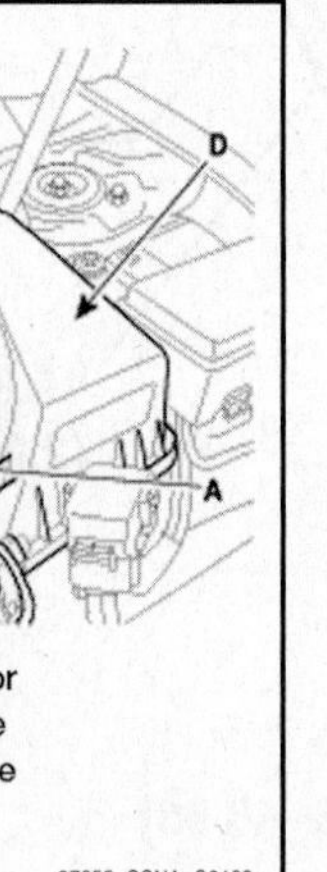

Fig. 214 Remove the intake air hose and air cleaner assembly

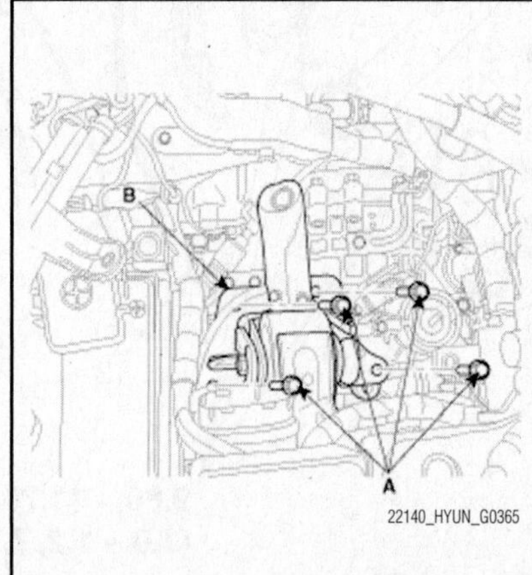

Fig. 216 Loosen the transaxle mounting bolts (A) without removing the transaxle mounting (B)

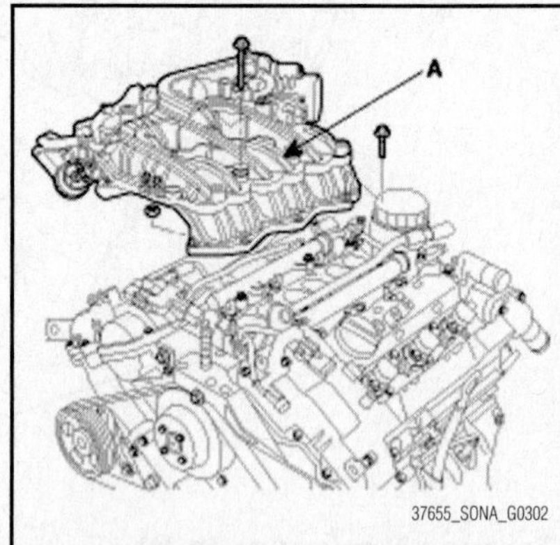

Fig. 220 Remove the surge tank (A)

Fig. 215 Remove the side cover (A)

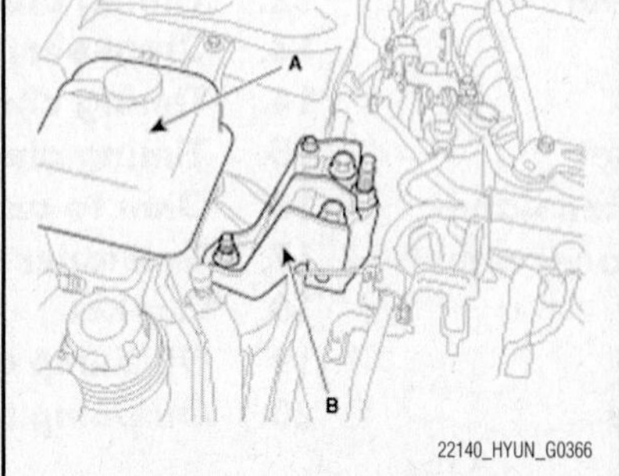

Fig. 217 Remove the engine coolant reservoir tank (A) and the engine mounting bracket (B)

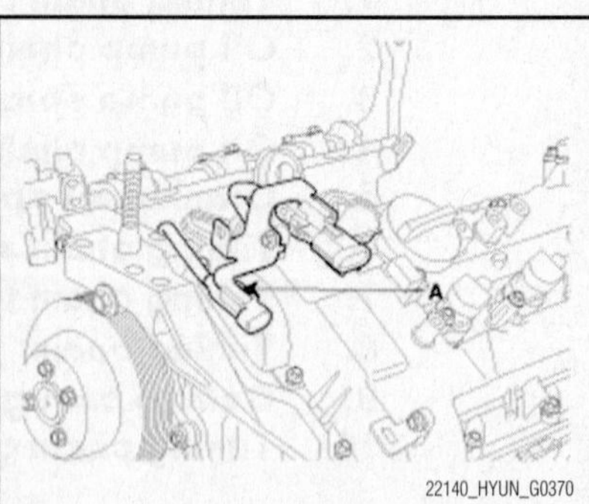

Fig. 221 Remove the connector bracket (A) from the left-hand cylinder head cover

g. Disconnect the PCV hose.

h. Disconnect the brake vacuum hose.

i. Remove the surge tank stay.

j. Remove the connector bracket from the surge tank.

k. Remove the surge tank.

➡Cover the inlet of intake manifold to prevent foreign materials from entering.

20. Remove the cylinder head cover.

a. Remove the connector bracket from the left-hand cylinder head cover.

b. Disconnect the right-hand ignition coil connector and condenser connector, and remove the wiring bracket.

c. Remove the left-hand and right-hand ignition coils.

d. Remove the left-hand and right-hand cylinder head cover.

➡Cover the top of engine head with a clean vinyl cover to prevent foreign materials from entering.

21. Set the No. 1 cylinder to TDC/compression:

a. Turn the crankshaft pulley and align its groove with the timing mark "T" of the lower timing chain cover.

⁂ WARNING

Do not rotate engine counterclockwise.

b. Check that the mark of the camshaft timing sprockets are in straight line on the cylinder head surface as shown in the illustration. If not, turn the crankshaft one revolution (360°). DO NOT rotate engine counterclockwise.

22. Remove the accessory drive belt. Refer to Accessory Drive Belt Removal & Installation.

23. Remove the crankshaft damper pulley. Refer to Crankshaft Balancer Removal & Installation.

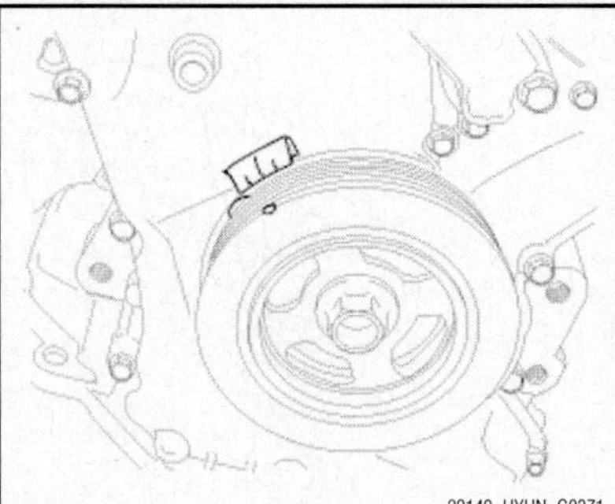

Fig. 222 Turn the crankshaft pulley and align its groove with the timing mark "T" of the lower timing chain cover

➡Use the SST (09231-3C300), flywheel stopper, or equivalent tool, to remove the crankshaft pulley bolt, after removing the starter.

24. Lift up the engine assembly by using the jack.

25. Remove the power steering pump.

26. Remove the air conditioner compressor.

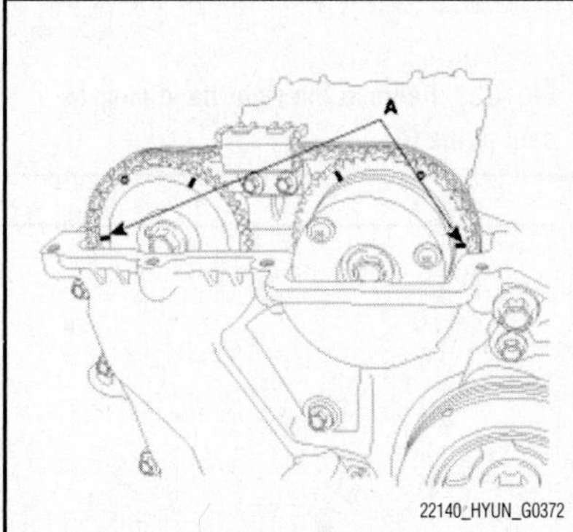

Fig. 223 Check that the mark (A) of the camshaft timing sprockets are in straight line on the cylinder head surface

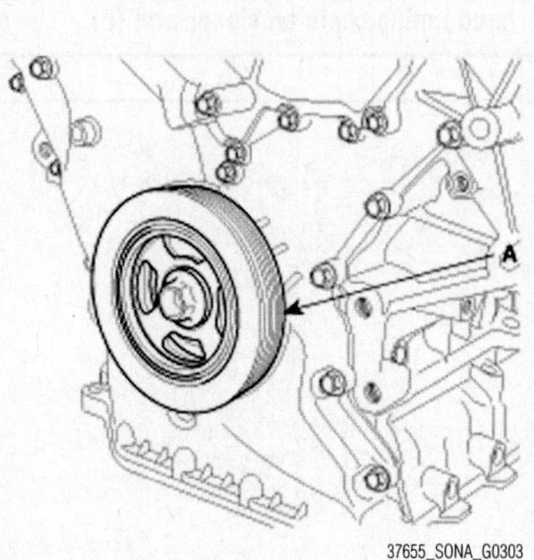

Fig. 224 Remove the crankshaft damper pulley (A)

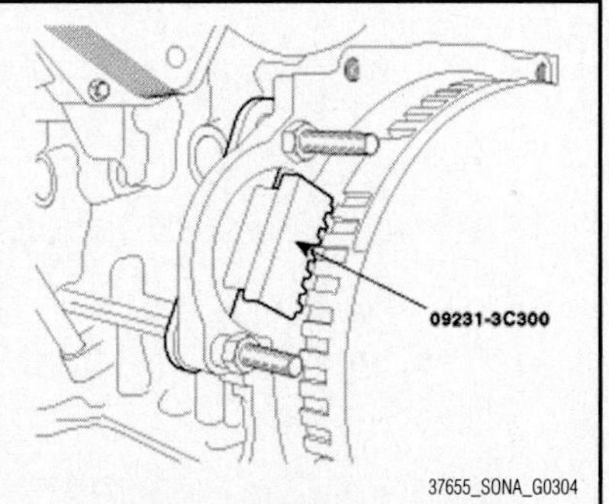

Fig. 225 Use the SST (flywheel stopper, 09231-3C300) to remove the crankshaft pulley bolt

27. Remove the alternator. Refer to Alternator Removal & Installation in the Engine Electrical section.

28. Remove the drive belt idler.

29. Remove the drive belt auto tensioner.

30. Remove water pump pulley.

31. Remove the timing chain cover.

⁂ WARNING

Be careful not to damage the contact surfaces of the cylinder block, cylinder head, or timing chain cover.

➡Before removing the timing chain, mark the right-hand/left-hand timing chain with an identification based on the location of the sprocket as the identification mark on the chain for TDC can be erased accidentally.

32. Install a set pin after compressing the right-hand timing chain tensioner.

33. Remove the right-hand cam-to-cam guide.

34. Remove the right-hand timing chain auto tensioner and the right-hand timing chain tensioner arm.

35. Remove the right-hand timing chain.

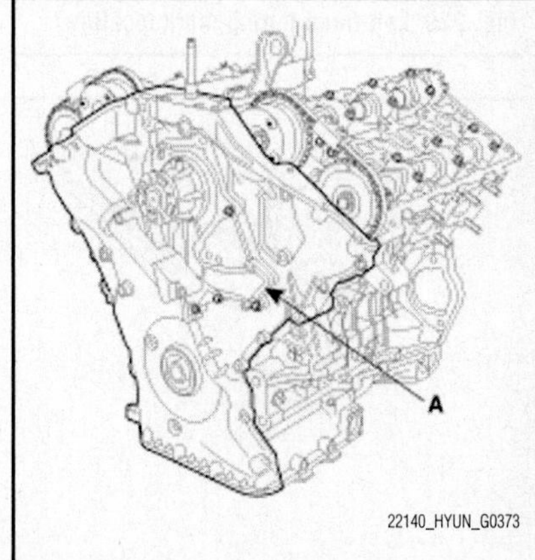

Fig. 226 Remove the timing chain cover (A)

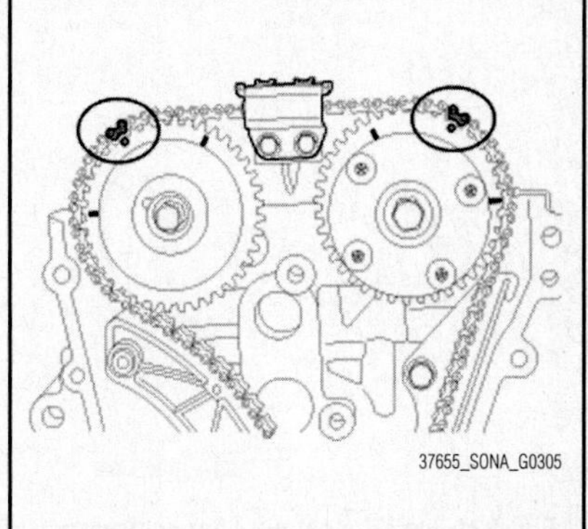

Fig. 227 Right-hand timing mark location

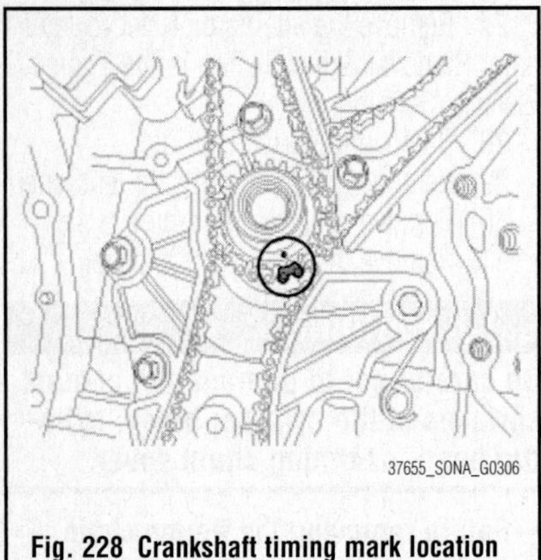
Fig. 228 Crankshaft timing mark location

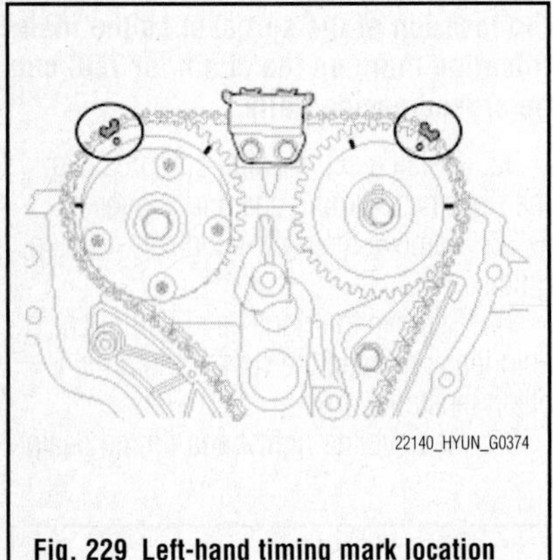
Fig. 229 Left-hand timing mark location

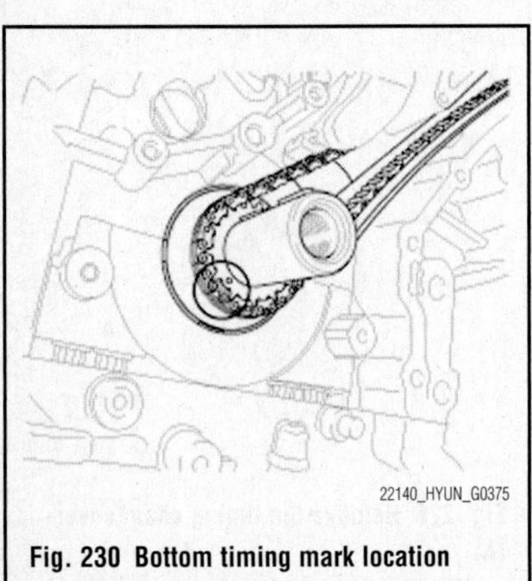
Fig. 230 Bottom timing mark location

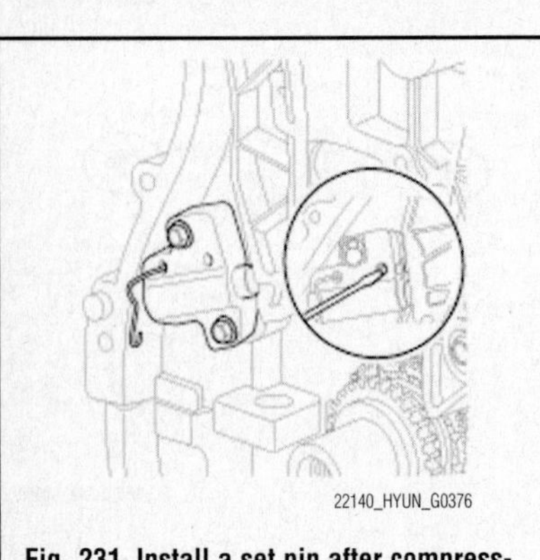
Fig. 231 Install a set pin after compressing the right-hand timing chain tensioner

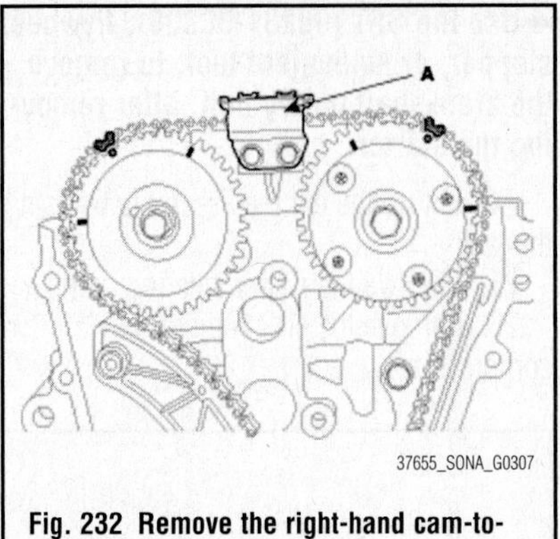

Fig. 232 Remove the right-hand cam-to-cam guide (A)

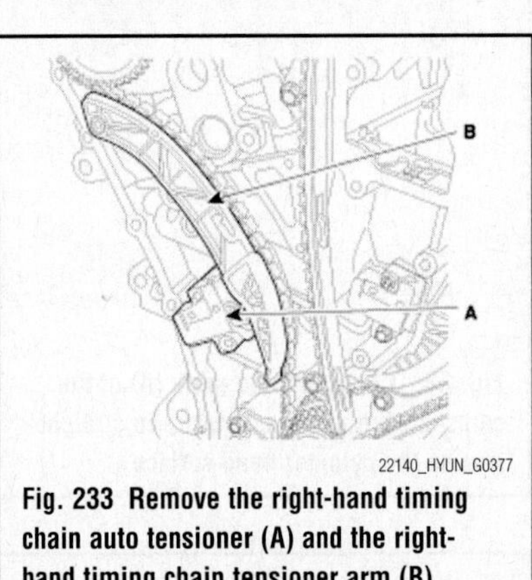

Fig. 233 Remove the right-hand timing chain auto tensioner (A) and the right-hand timing chain tensioner arm (B)

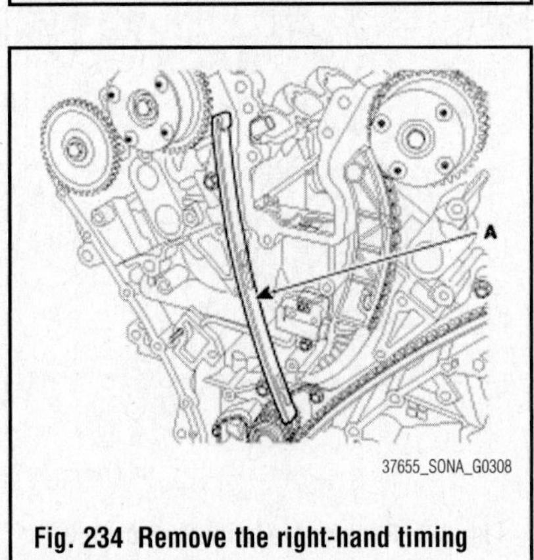

Fig. 234 Remove the right-hand timing chain guide (A)

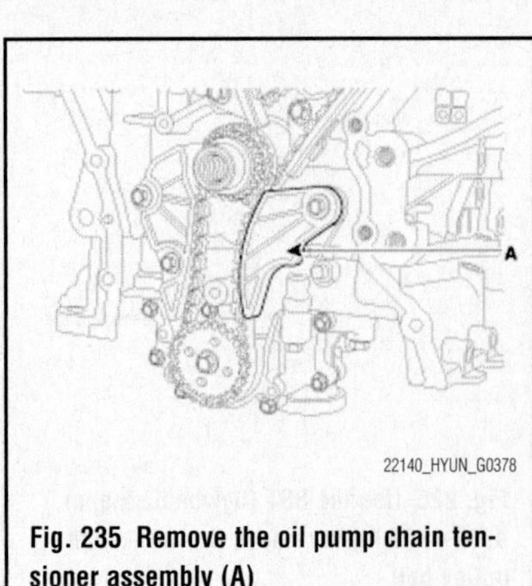

Fig. 235 Remove the oil pump chain tensioner assembly (A)

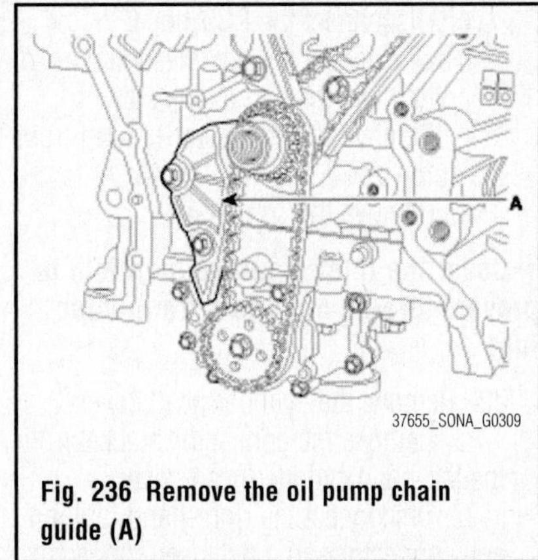

Fig. 236 Remove the oil pump chain guide (A)

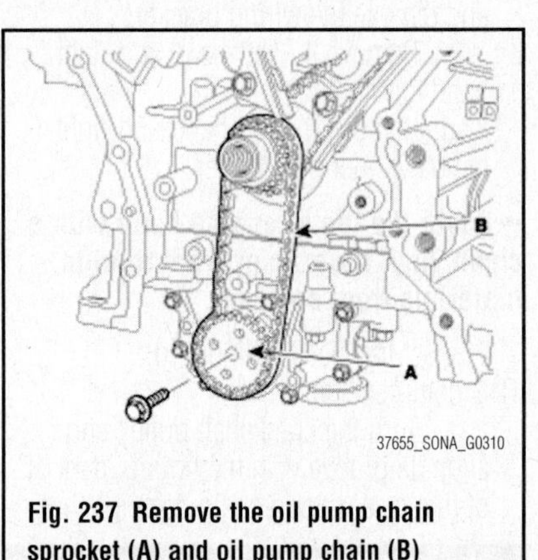

Fig. 237 Remove the oil pump chain sprocket (A) and oil pump chain (B)

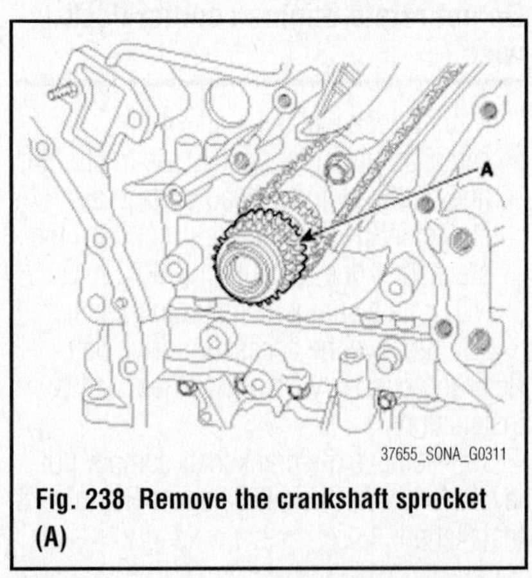

Fig. 238 Remove the crankshaft sprocket (A)

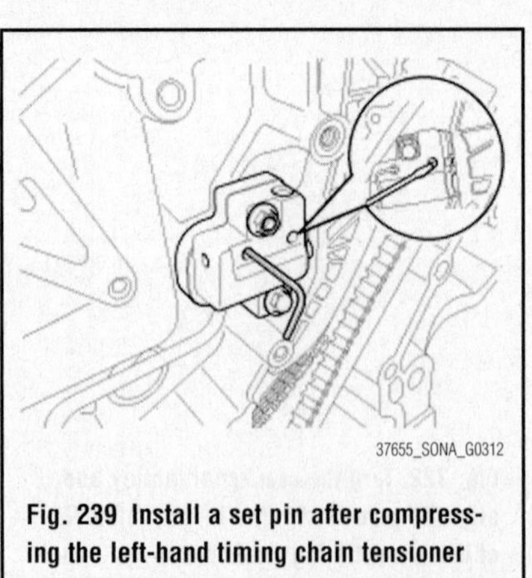
Fig. 239 Install a set pin after compressing the left-hand timing chain tensioner

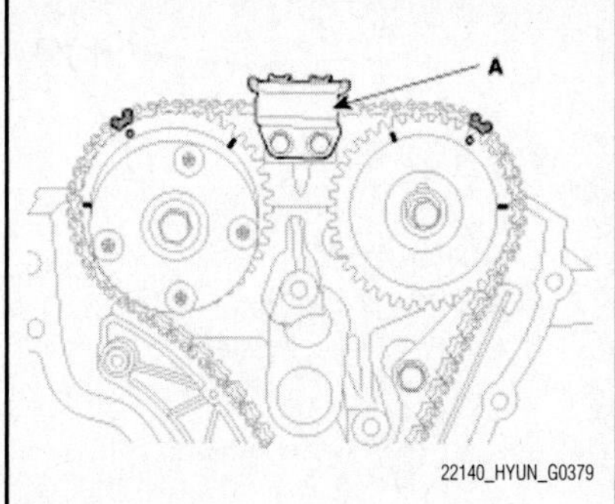

Fig. 240 Remove the left-hand cam-to-cam guide (A)

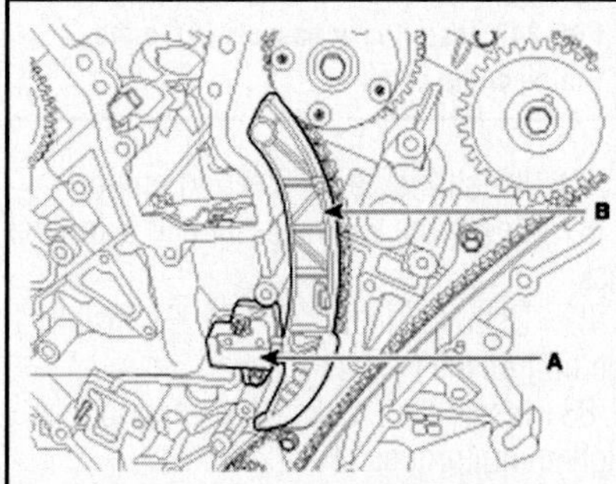

Fig. 241 Remove the left-hand timing chain auto tensioner (A) and the right-hand timing chain tensioner arm (B)

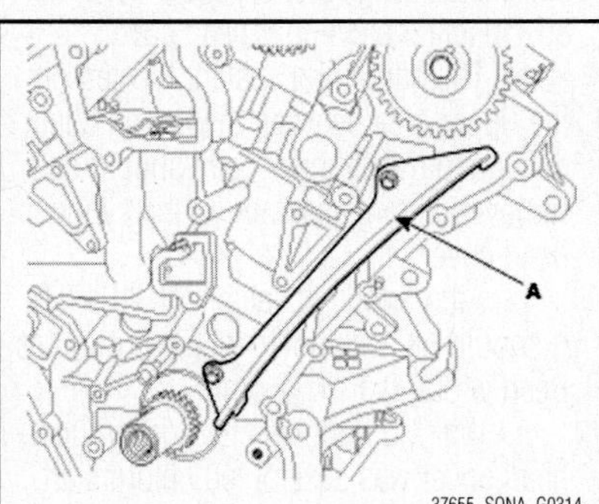

Fig. 242 Remove the left-hand timing chain guide (A)

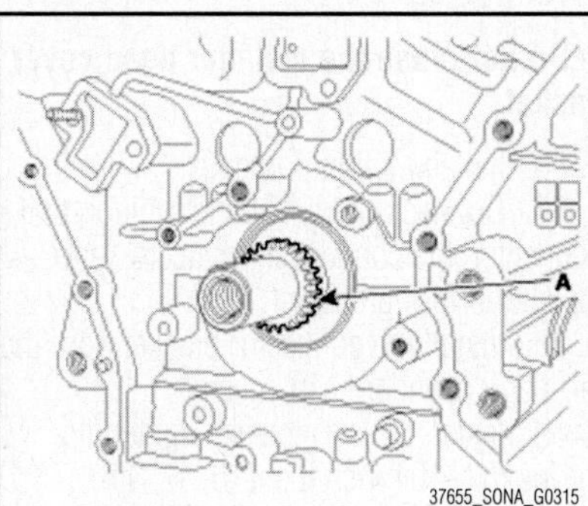

Fig. 243 Remove the crankshaft sprocket (A)—left-hand

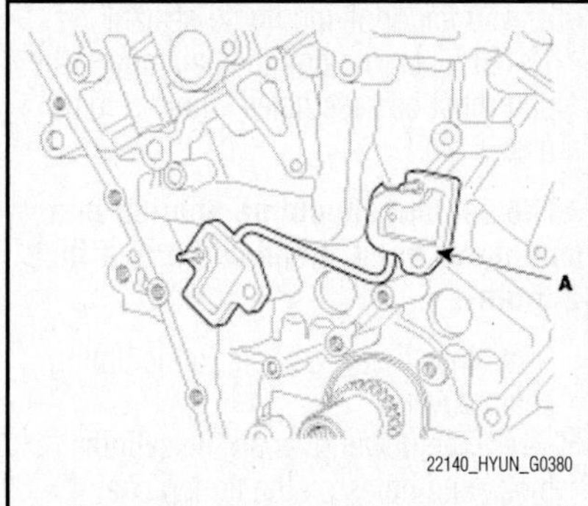

Fig. 244 Remove the tensioner adapter assembly (A)

36. Remove the right-hand timing chain guide.
37. Remove the oil pump chain cover.
38. Remove the oil pump chain tensioner assembly.
39. Remove the oil pump chain guide.
40. Remove the oil pump chain sprocket and oil pump chain.
41. Remove the crankshaft sprocket.
42. Install a set pin after compressing the left-hand timing chain tensioner.
43. Remove the left-hand cam-to-cam guide.
44. Remove the left-hand timing chain auto tensioner and left-hand timing chain tensioner arm.
45. Remove the left-hand timing chain.
46. Remove the left-hand timing chain guide.
47. Remove the crankshaft sprocket.
48. Remove the tensioner adapter assembly.

To install:

49. Check the camshaft sprocket and crankshaft sprocket for abnormal wear, cracks, or damage. Replace as necessary.
50. Inspect the tensioner arm and chain guide for abnormal wear, cracks, or damage. Replace as necessary.
51. Check that the tensioner piston moves smoothly when the ratchet pawl is released.
52. Install the jack to the upper oil pan.
53. The crankshaft key must be aligned with the timing mark of timing chain cover. Then, piston of No. 1 cylinder is placed at the TDC on the compression stroke.
54. Install the tensioner adapter assembly.
55. Install the crankshaft sprocket.
56. Install the left-hand timing chain guide. Tightening torque: 14–18 ft. lbs. (20–25 Nm).
57. Install the left-hand timing chain. To install the timing chain with no slack

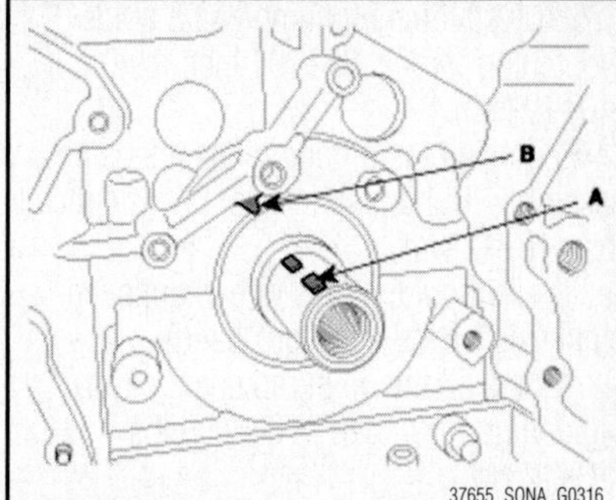

Fig. 245 The crankshaft key (A) must be aligned with the timing chain cover timing mark (B)

between the camshaft and crankshaft, use the following procedure:

a. Place the crankshaft sprocket on first, then the timing chain guide, then the exhaust camshaft sprocket, and finally the intake camshaft sprocket.
b. The timing mark of each of the sprockets should be matched with the timing mark (color link) of the timing chain when installing the timing chain.

58. Install the left-hand timing chain tensioner arm. Tightening torque: 14–16 ft. lbs. (19–22 Nm).
59. Install the chain tensioner. Tightening torque: 87–104 inch lbs. (10–12 Nm).
60. Install the left-hand cam-to-cam guide. Tightening torque: 87–104 inch lbs. (10–12 Nm).
61. Install the crankshaft sprocket.
62. Install the oil pump chain and the oil pump sprocket. Tightening torque: 14–16 ft. lbs. (19–22 Nm).
63. Install the right-hand timing chain guide. Tightening torque: 14–18 ft. lbs. (20–25 Nm).
64. Install the right-hand timing chain. To install the timing chain with no slack between the camshaft and the crankshaft, use the following procedure:

a. Place the chain on the crankshaft sprocket first, then the intake camshaft sprocket, then the exhaust camshaft sprocket.
b. The timing mark of each of the sprockets must be matched with the timing mark (color link) of timing chain when installing the timing chain.

65. Install the right-hand timing chain tensioner arm. Tightening torque: 14–16 ft. lbs. (19–22 Nm).
66. Install the right-hand timing chain auto tensioner. Tightening torque: 87–104 inch lbs. (10–12 Nm).
67. Install the right-hand cam-to-cam guide. Tightening torque: 87–104 inch lbs. (10–12 Nm).

68. Install the oil pump chain guide. Tightening torque: 87–104 inch lbs. (10–12 Nm).

69. Install the oil pump chain tensioner assembly. Tightening torque: 87–104 inch lbs. (10–12 Nm).

70. Pull out the pins of hydraulic tensioner (left-hand and right-hand).

71. Install the oil pump chain cover. Tightening torque: 87–104 inch lbs. (10–12 Nm).

72. After rotating the crankshaft 2 revolutions in a clockwise direction (viewed from front), confirm that the timing marks are aligned.

**** WARNING**

Always turn the crankshaft clockwise.

73. Install the timing chain cover.

a. The sealant locations on the chain cover and on the counter parts (cylinder head, cylinder block, and lower oil pan) must be free of engine oil and any debris.

b. Before assembling the timing chain cover, the liquid sealant TB 1217H should be applied on the gap between cylinder head and cylinder block.

➡The part must be assembled within 5 minutes after the sealant is applied. Use a bead width of 0.1 in. (2.5mm).

c. After applying liquid sealant TB1217H on the timing chain cover, the part must be assembled within 5 minutes.

➡The sealant should be applied in a continuous bead. Bead width: 0.1 inch (2.5mm).

d. Install new gaskets to the timing chain cover.

e. The dowel pins on the cylinder block and holes on the timing chain cover should be used as a reference in order to aid assembly of the timing chain cover into position.

f. Tighten the timing cover as shown in the illustration.

74. Wait 30 minutes after the timing chain cover was assembled before starting the engine.

75. Install the water pump pulley. Tightening torque: 69–87 inch lbs. (8–10 Nm).

76. Install the drive belt auto tensioner. Tightening torque:

- Bolt B: 60–63 ft. lbs. (81–85 Nm).
- Bolt C: 13–16 ft. lbs. (18–22 Nm).

77. Install the drive belt idler. Tightening torque: 39–43 ft. lbs. (53–58 Nm).

78. Install the alternator. Tightening torque: 20–25 ft. lbs. (26–33 Nm).

79. Install the air conditioner compressor.

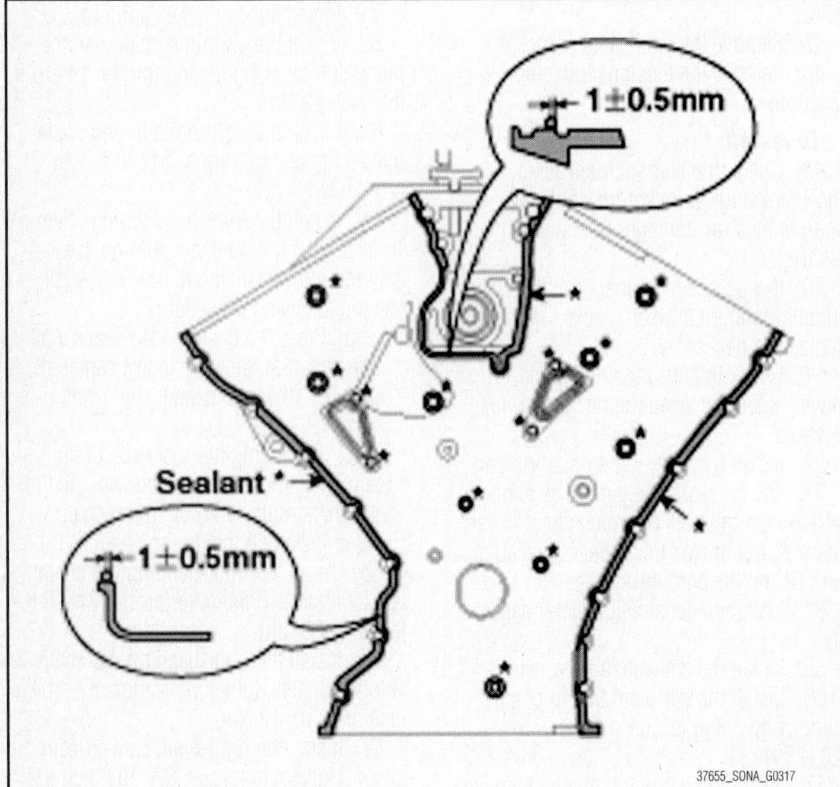

Fig. 246 Sealant application locations

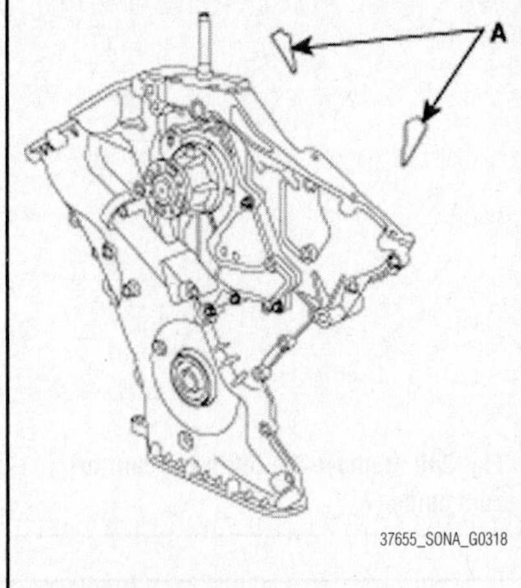

Fig. 247 Install new gaskets (A) to the timing chain cover

80. Install the power steering pump.

81. Lower the engine assembly using the jack.

82. Using SST (09231-3C100), install the timing chain cover oil seal.

83. Install the crankshaft damper pulley. Tightening torque: 210–224 ft. lbs. (284–304 Nm).

➡Use SST (09231-3C300), flywheel stopper, or equivalent tool, to install the crankshaft damper pulley.

84. Install the accessory drive belt.

85. Install the cylinder head cover.

a. The hardening sealant located on the upper area between the timing chain cover and the cylinder head should be removed before assembling the cylinder head cover.

b. After applying sealant (TB1217H), it should be assembled within 5 minutes. Bead width: 0.1 in. (2.5mm).

c. Wait 30 minutes after the cylinder head cover was assembled before starting the engine.

d. Install the cylinder head cover bolts as shown in the illustration. Tightening torque: 87–104 inch lbs. (10–12 Nm).

➡Do not reuse the cylinder head cover gasket.

86. Install the ignition coils.

87. Connect the right-hand ignition coil connector, the condenser connector, and install the wiring bracket.

88. Install the connector bracket from the left-hand cylinder head cover.

89. Install the surge tank. Tightening torque: 87–104 inch lbs. (10–12 Nm).

a. Install the connector bracket on the surge tank. Tightening torque: 61–96 inch lbs. (7–11 Nm).

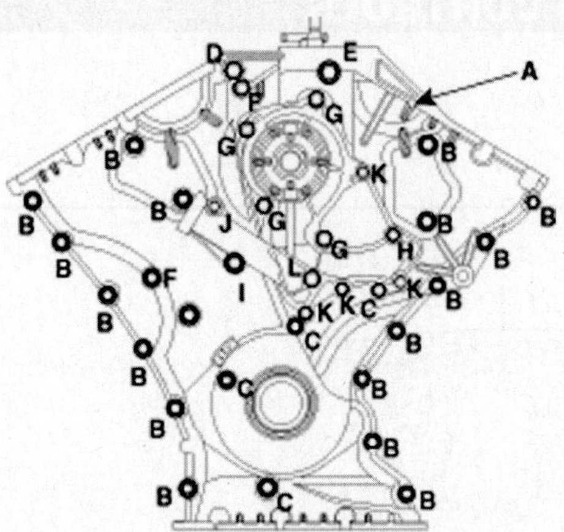

B (17): 14–16 ft. lbs. (19–22 Nm)
C (4): 87–104 inch lbs. (10–12 Nm)
D (1): 43–51 ft. lbs. (59–69 Nm)
E (1): 43–51 ft. lbs. (59–69 Nm)
F (2): 18–20 ft. lbs. (25–26 Nm)
G (4): 16–17 ft. lbs. (22–24 Nm)
H (1): 87–104 inch lbs. (10–12 Nm)
I (1): 87–104 inch lbs. (10–12 Nm)
J (1): 87–104 inch lbs. (10–12 Nm)
K (4): 87–104 inch lbs. (10–12 Nm)
L (1): 16–20 ft. lbs. (22–26 Nm)

37655_SONA_G0319

Fig. 248 Timing chain cover tightening sequence and tightening specifications

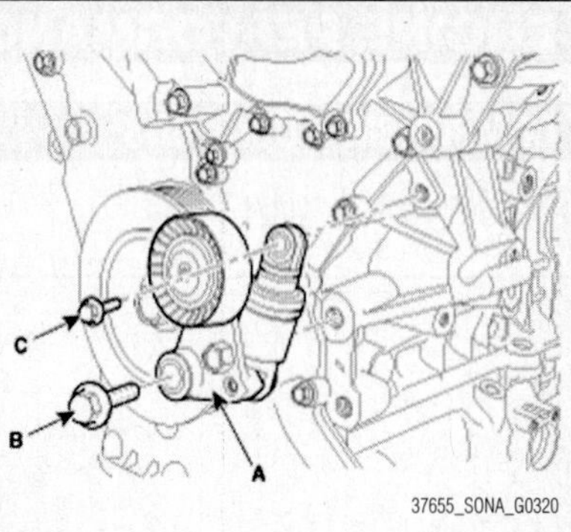

Fig. 249 Drive belt auto tensioner (A) and bolts (B, C)

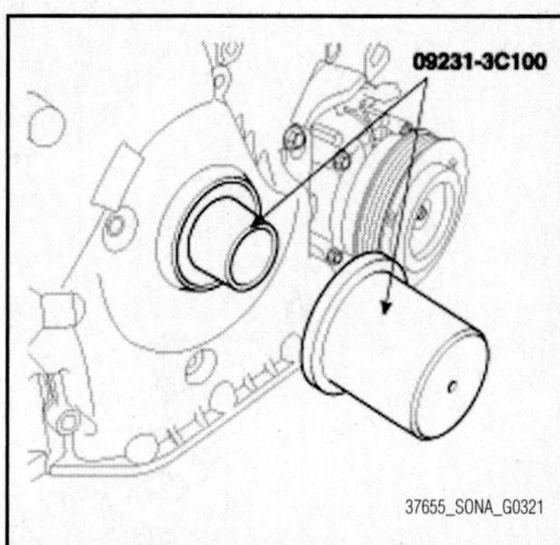

Fig. 250 Using SST (09231-3C100), install the timing chain cover oil seal

b. Install the surge tank stay. Tightening torque: 20–23 ft. lbs. (27–31 Nm).

90. Connect the brake vacuum hose.
91. Connect the PCV hose.
92. Connect the water hoses to the ETC.
93. Install the ETC bracket. Tightening torque: 12–19 ft. lbs. (16–26 Nm).
94. Connect the PCSV connector, the MAP sensor connector, and the PCSV hose.
95. Connect the ETC connector and the knock sensor connector.
96. Connect the right-hand injector connector and the ignition coil connector.
97. Connect the right-hand oxygen sensor connector.
98. Install the No. 1 engine mounting through the lower position of A/C pipe line. Tightening torque: 36–47 ft. lbs. (49–64 Nm).
99. Install the A/C pipe bracket mounting bolt.
100. Install the engine coolant reservoir tank.
101. Install the engine mounting bracket. Tightening torque: 47–61 ft. lbs. (64–83 Nm).
102. Install the transaxle mounting bolts. Tightening torque: 36–47 ft. lbs. (49–64 Nm).
103. Remove the jack from the upper oil pan.
104. Install the lower oil pan. Refer to Oil Pan, Removal & Installation. Uniformly tighten the bolts in several passes. Tightening torque: 87–104 inch lbs. (10–12 Nm).
105. Install the upper radiator hose.
106. Install the side cover.
107. Install the undercover.
108. Install the front wheels.
109. Install the intake air hose and air cleaner assembly.
    a. Install the intake air hose and air cleaner.
    b. Connect the ECM connector.
    c. Connect the breather hose to air cleaner hose.
    d. Connect the AFS connector.
    e. Install the air duct.
110. Install the engine cover.
111. Connect the negative terminal to the battery.
112. Refill the engine with the proper amount and type of engine oil.
113. Refill the radiator and reservoir tank with engine coolant.
114. Bleed the air from the cooling system.
115. Run the engine and check for leaks.

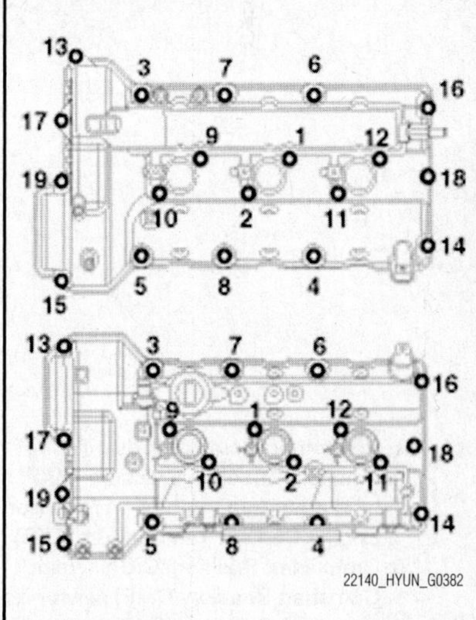

Fig. 251 Cylinder head cover bolt tightening sequence

VALVE LASH

ADJUSTMENT

All engines use hydraulic valve lash adjusters. Valve lash adjustments are not necessary or possible on these engines.

ENGINE PERFORMANCE & EMISSION CONTROLS

COMPONENT LOCATIONS

See Figures 252 through 254.

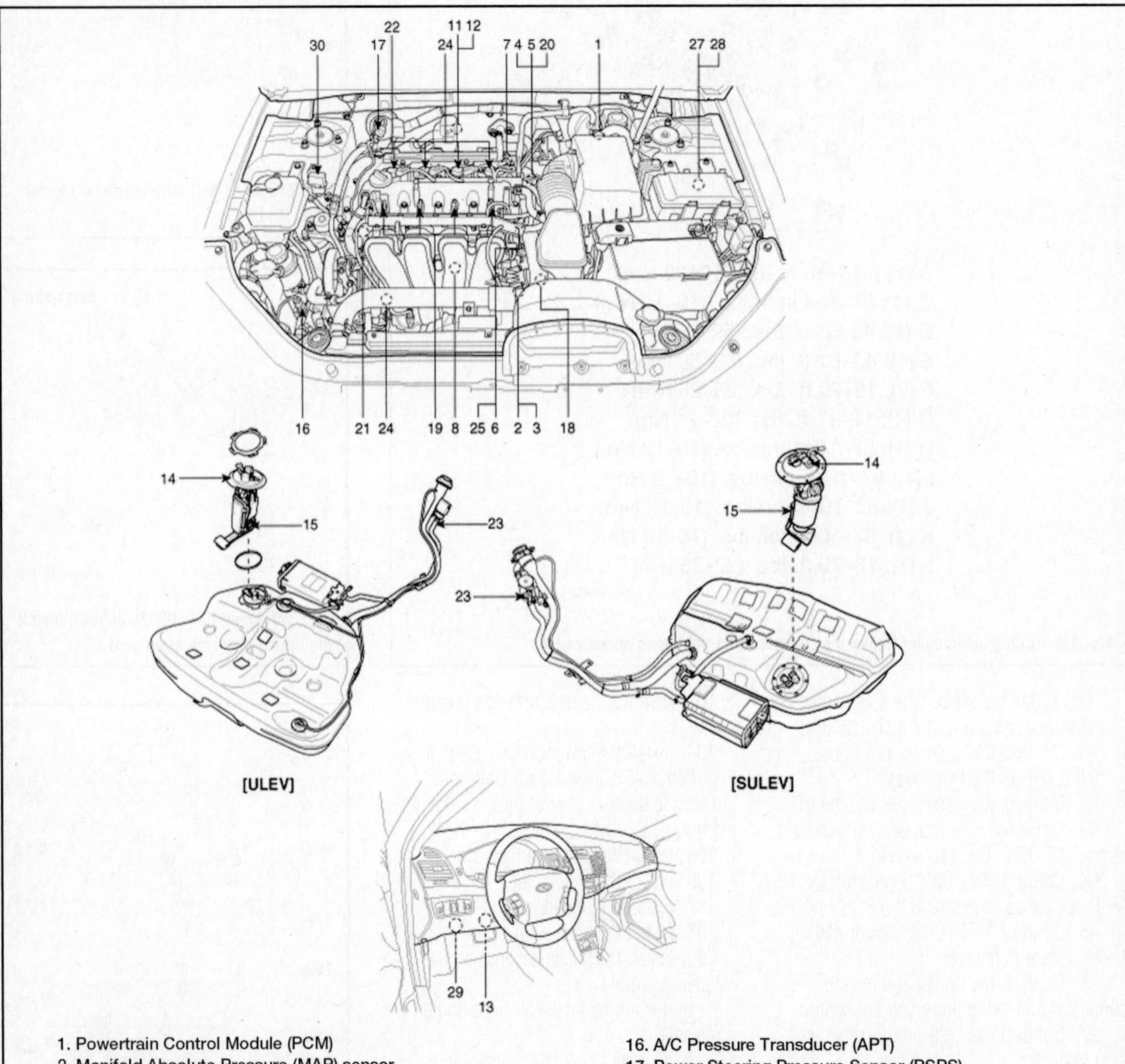

1. Powertrain Control Module (PCM)
2. Manifold Absolute Pressure (MAP) sensor
3. Intake Air Temperature (IAT) sensor (IAT)
4. Engine Coolant Temperature (ECT) sensor
5. Crankshaft Position (CKP) sensor
6. Camshaft Position (CMP) sensor No. 1 [Intake]
7. Camshaft Position (CMP) sensor No. 2 [Exhaust]
8. Knock Sensor (KS)
9. Heated Oxygen Sensor (HO2S) [Bank 1/Sensor 1] [ULEV]
10. Heated Oxygen Sensor (HO2S) [Bank 1/Sensor 2] [ULEV]
11. Heated Oxygen Sensor (HO2S) [Bank 1/Sensor 1] [SULEV]
12. Heated Oxygen Sensor (HO2S) [Bank 1/Sensor 2] [SULEV]
13. Accelerator Position (APP) sensor
14. Fuel Tank Pressure (FTP) sensor
15. Fuel Level Sensor (FLS)
16. A/C Pressure Transducer (APT)
17. Power Steering Pressure Sensor (PSPS)
18. ETC Module (Including TPS & ETC Motor)
19. Injector
20. Purge Control Solenoid Valve (PCSV)
21. CVVT Oil Control Valve (OCV) No. 1 [Intake]
22. CVVT Oil Control Valve (OCV) No. 2 [Exhaust]
23. Canister Close Valve (CCV)
24. Variable Intake Solenoid (VIS) valve
25. Variable Charge Motion Actuator (VCMA) [SULEV only]
26. Ignition coil
27. Main relay
28. Fuel pump relay
29. Data link connector (DLC)
30. Multi-purpose check connector

37655_SONA_G0375

Fig. 252 Underhood and instrument panel sensor locations—2.4L engine

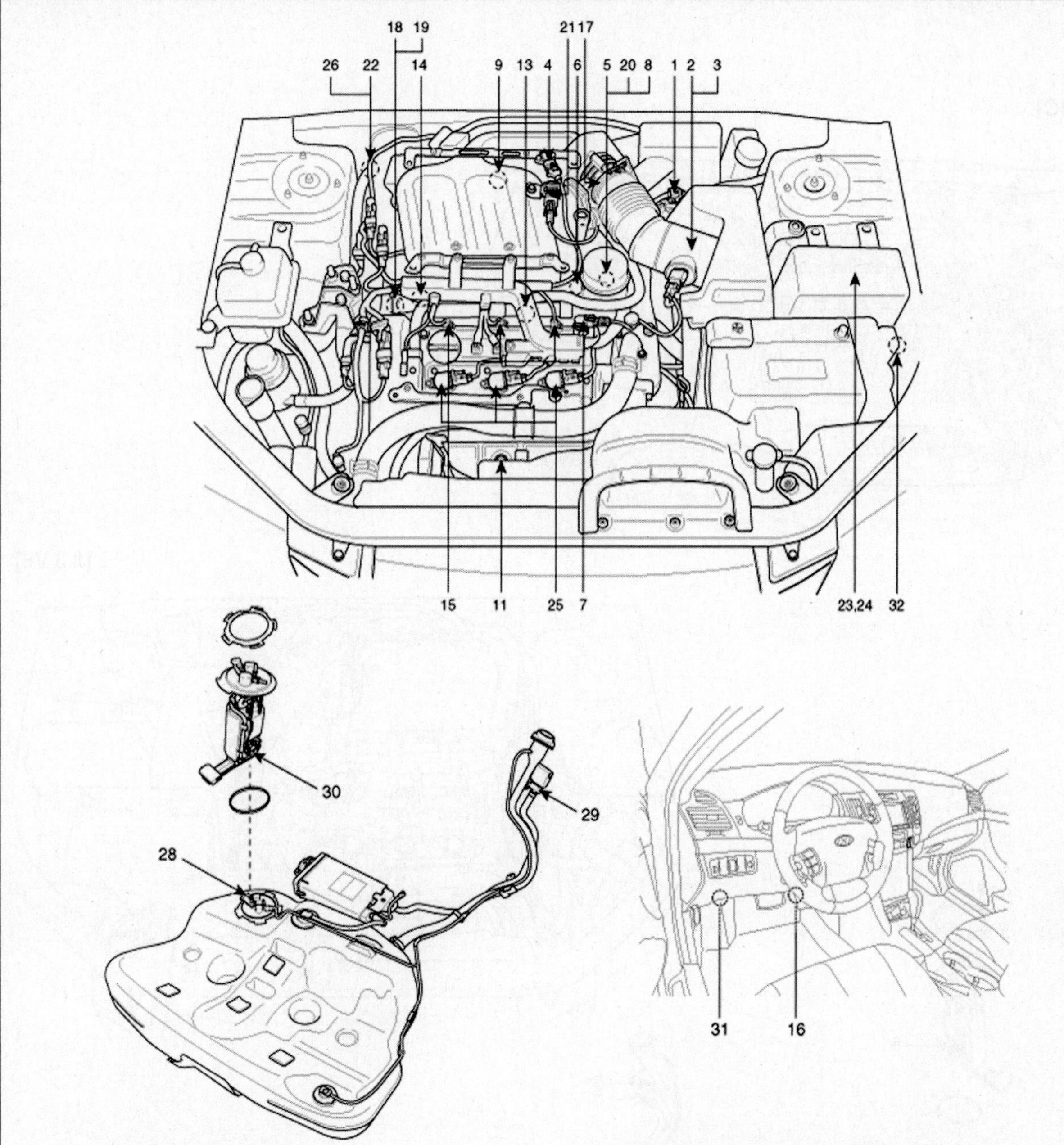

1. Engine Control Module (ECM)
2. Mass Air Flow (MAF) sensor
3. Intake Air Temperature (IAT) sensor
4. Manifold Absolute Pressure (MAP) sensor
5. Engine Coolant Temperature (ECT) sensor
6. Camshaft Position (CMP) sensor [Bank 1]
7. Camshaft Position (CMP) sensor [Bank 2]
8. Crankshaft Position (CKP) sensor
9. Heated Oxygen Sensor (HO2S) [Bank 1 / Sensor 1]
10. Heated Oxygen Sensor (HO2S) [Bank 1 / Sensor 2]
11. Heated Oxygen Sensor (HO2S) [Bank 2 / Sensor 1]
12. Heated Oxygen Sensor (HO2S) [Bank 2 / Sensor 2]
13. Knock Sensor (KS) No. 1
14. Knock Sensor (KS) No. 2
15. Injector
16. Accelerator Position (APP) sensor
17. ETC Module [Throttle Position Sensor (TPS) + ETC Motor]
18. CVVT Oil Control Valve (OCV) [Bank 1]
19. CVVT Oil Control Valve (OCV) [Bank 2]
20. CVVT Oil Temperature Sensor (OTS)
21. Purge Control Solenoid Valve (PCSV)
22. Variable Intake Solenoid (VIS) valve
23. Fuel Pump Relay
24. Main Relay
25. Ignition Coil
26. Power Steering Pressure Sensor (PSPS)
27. Wheel Speed Sensor (WSS)
28. Fuel Tank Pressure sensor (FTP)
29. Canister Close Valve (CCV)
30. Fuel Level Sensor (FLS)
31. Data Link Connector (DLC)
32. Multi-Purpose Check Connector

37655_SONA_G0376

Fig. 253 Underhood sensor locations—3.3L engine

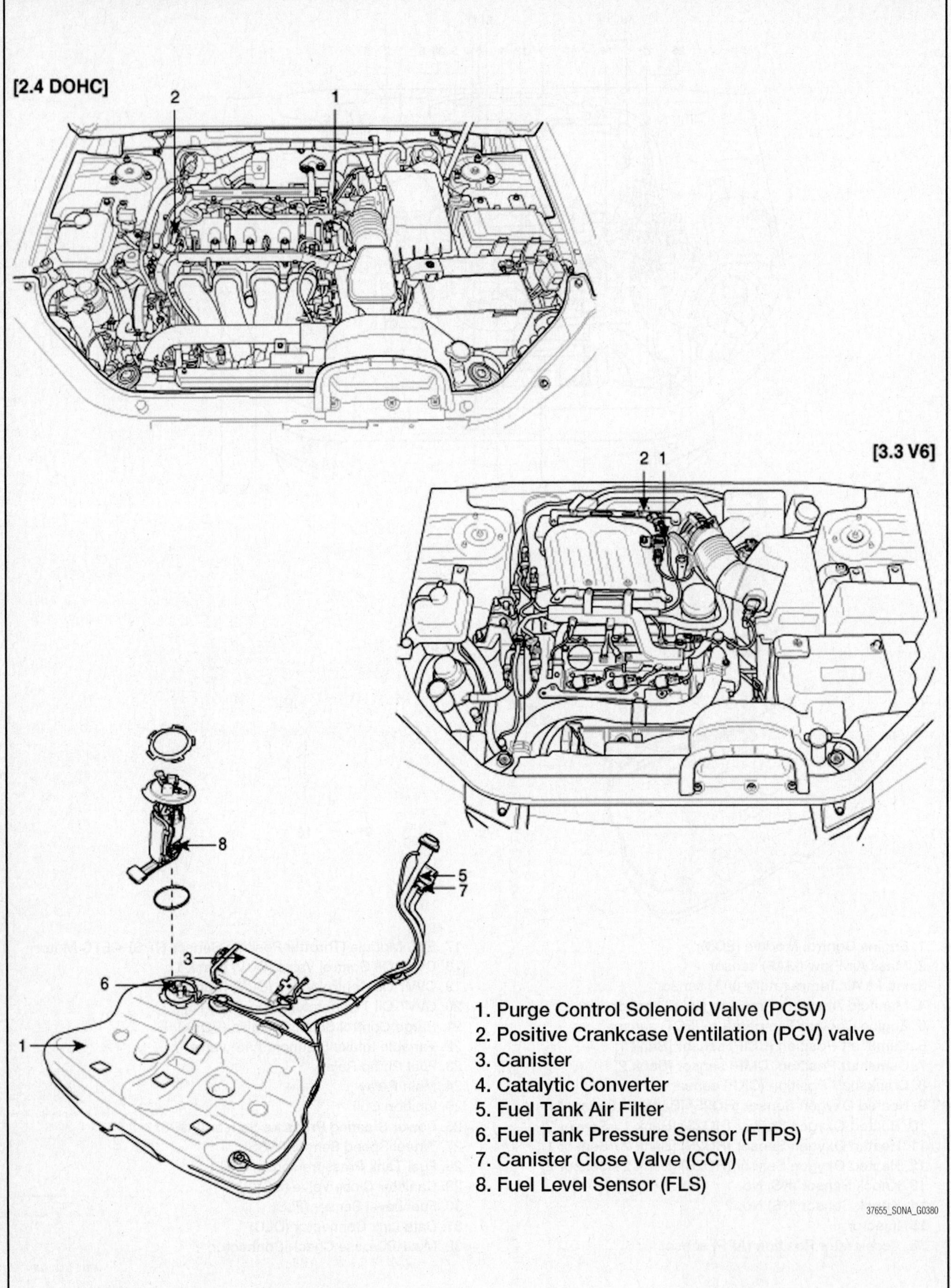

Fig. 254 Emission control system component locations

CAMSHAFT POSITION (CMP) SENSOR

LOCATION

See Figures 255 through 258.

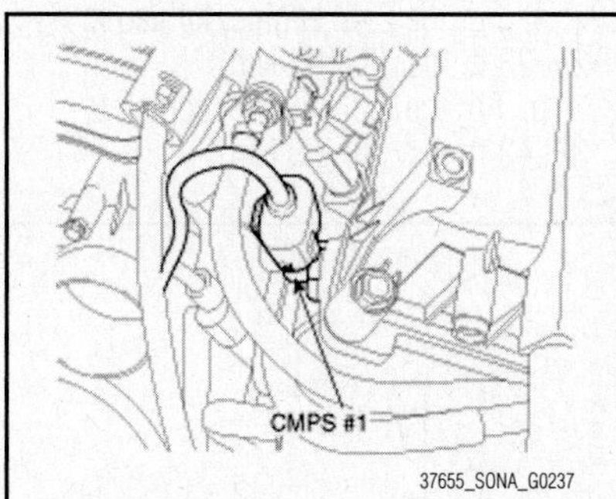

Fig. 255 Camshaft Position (CMP) sensor location—2.4L engine (No. 1)

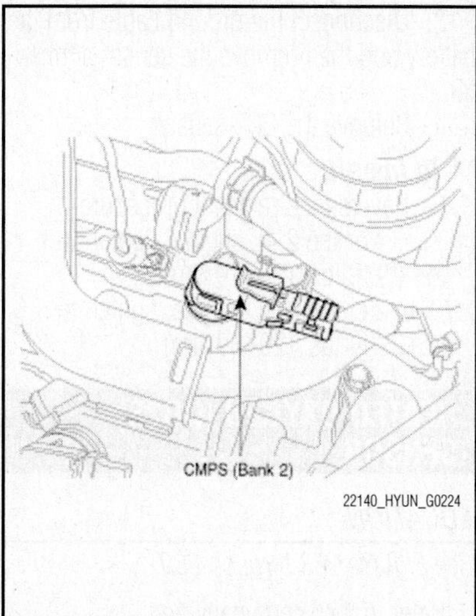

Fig. 258 Camshaft Position (CMP) sensor location—3.3L engine (Bank 2)

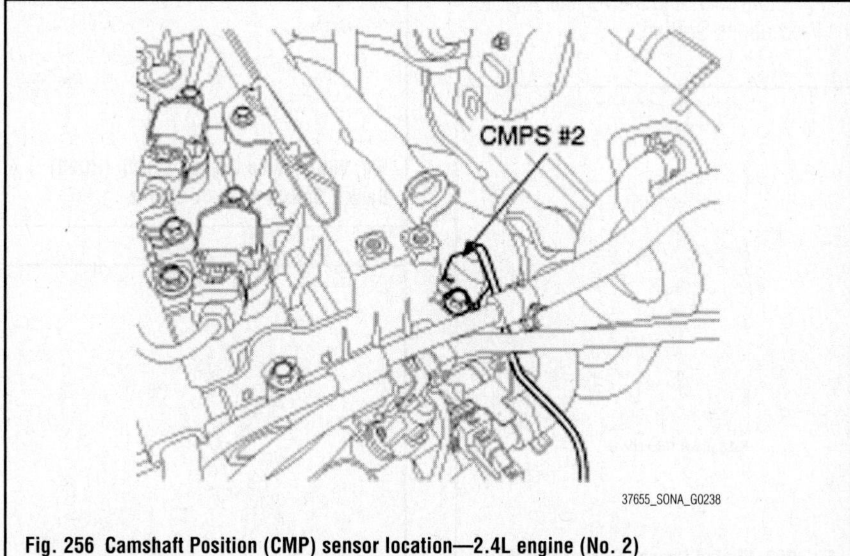

Fig. 256 Camshaft Position (CMP) sensor location—2.4L engine (No. 2)

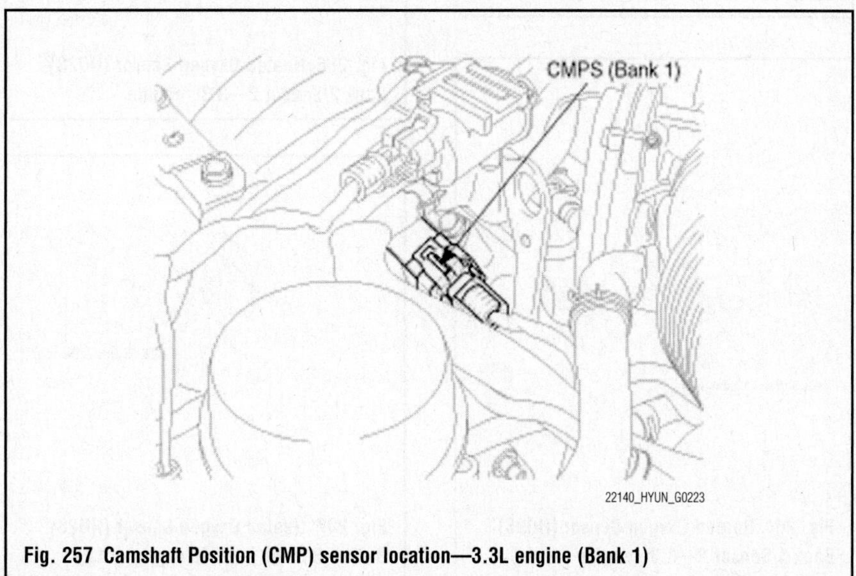

Fig. 257 Camshaft Position (CMP) sensor location—3.3L engine (Bank 1)

Refer to the accompanying illustrations.

REMOVAL & INSTALLATION

See Figures 255 through 258.

1. Before servicing the vehicle, refer to the Precautions Section.
2. Disconnect the negative battery cable.
3. Disconnect the connector from the CMP sensor.
4. Remove the CMP sensor.

To install:

5. Installation is the reverse of the removal procedure.
   a. For the 2.4L engine, tighten the bolts to 86–104 inch lbs. (10–12 Nm).
   b. For the 3.3L engine, tighten the bolts to 61–86 inch lbs. (7–10 Nm).

CRANKSHAFT POSITION (CKP) SENSOR

LOCATION

See Figures 259 and 260.

Refer to the accompanying illustrations.

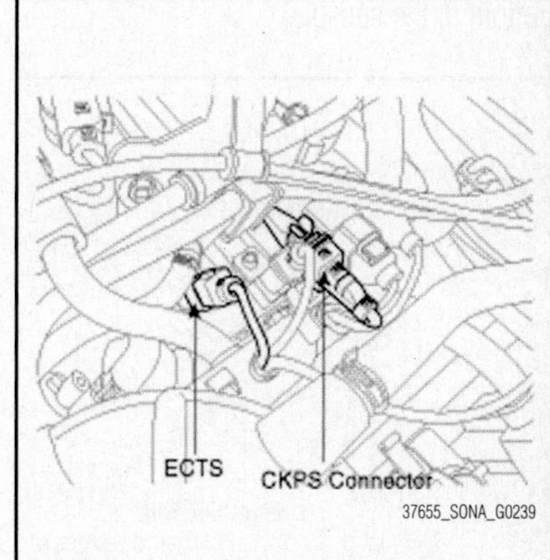

Fig. 259 Crankshaft Position (CKP) sensor location—2.4L engine

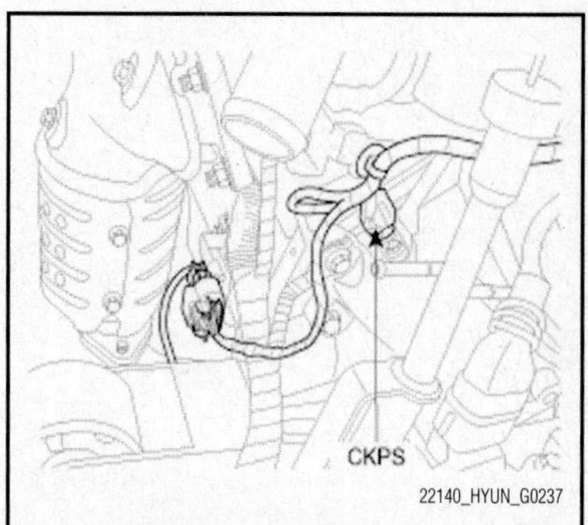

Fig. 260 Crankshaft Position (CKP) sensor location—3.3 L engine

REMOVAL & INSTALLATION

1. Before servicing the vehicle, refer to the Precautions Section.
2. Disconnect the negative battery cable.
3. Disconnect the connector from the sensor.
4. Remove the sensor from its mounting.

To install:

5. Installation is the reverse of the removal procedure.
   a. For the 2.4L engine, tighten the bolts to 86–104 inch lbs. (10–12 Nm).
   b. For the 3.3L engine, tighten the bolts to 61–86 inch lbs. (7–10 Nm).

ENGINE COOLANT TEMPERATURE (ECT) SENSOR

LOCATION

See Figures 261 and 262.

Refer to the accompanying illustrations.

REMOVAL & INSTALLATION

1. Before servicing the vehicle, refer to the Precautions Section.
2. Drain the coolant to a level below the bottom of the sensor.

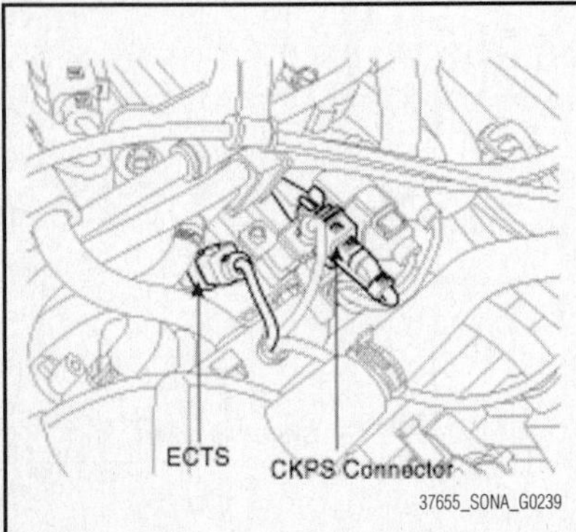

Fig. 261 Engine Coolant Temperature (ECT) sensor location—2.4L engine

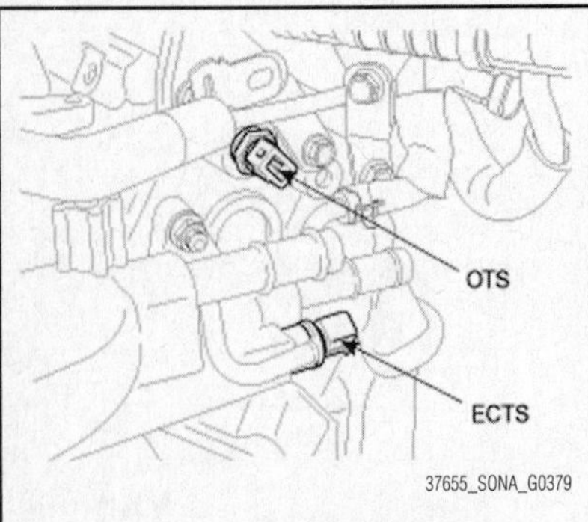

Fig. 262 Engine Coolant Temperature (ECT) sensor location—3.3L engine

3. Disconnect the ground cable from the battery and then remove the sensor connector.
4. Remove the ECT sensor.

To install:

5. Reverse the removal procedure.
   a. For the 2.4L engine, tighten to 29–39 ft. lbs. (22–29 Nm).
   b. For the 3.3L engine, tighten to 15–29 ft. lbs. (20–39 Nm).

HEATED OXYGEN (HO2S) SENSOR

LOCATION

See Figures 263 through 269.

Refer to the accompanying illustrations.

REMOVAL & INSTALLATION

1. Before servicing the vehicle, refer to the Precautions Section.

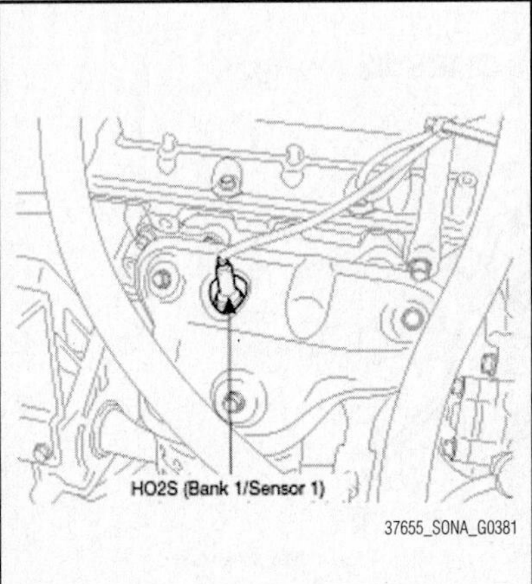

Fig. 263 Heated Oxygen Sensor (HO2S) Bank 1/Sensor 1—3.3L engine

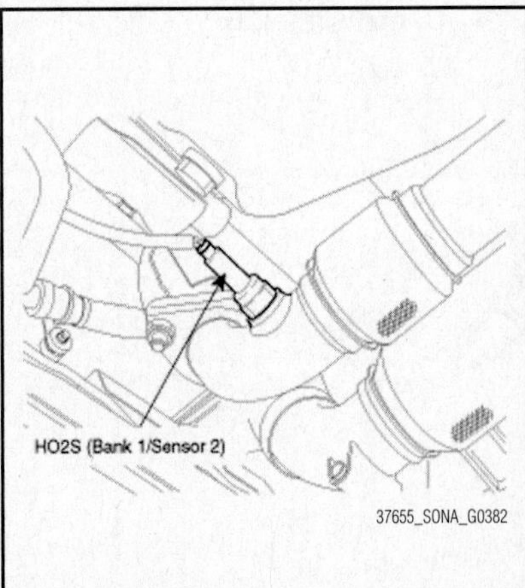

Fig. 264 Heated Oxygen Sensor (HO2S) Bank 1/Sensor 2—3.3L engine

2. Disconnect the electrical connector from the sensor.
3. Remove the oxygen sensor.

To install:

4. Installation is the reverse of the removal procedure.
   a. For the 2.4L engine, tighten to 25–33 ft. lbs. (34–44 Nm).
   b. For the 3.3L engine, tighten to 36–43 ft. lbs. (49–59 Nm).

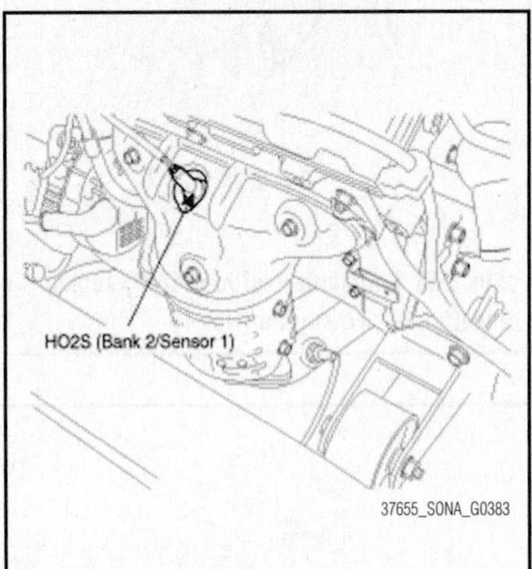

Fig. 265 Heated Oxygen Sensor (HO2S) Bank 2/Sensor 1—3.3L engine

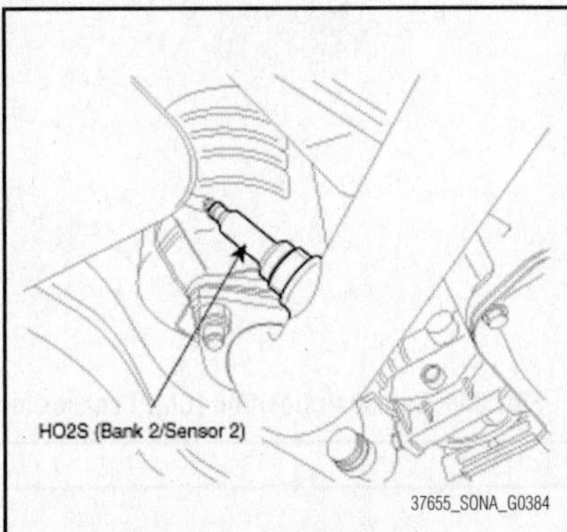

Fig. 266 Heated Oxygen Sensor (HO2S) Bank 2/Sensor 2—3.3L engine

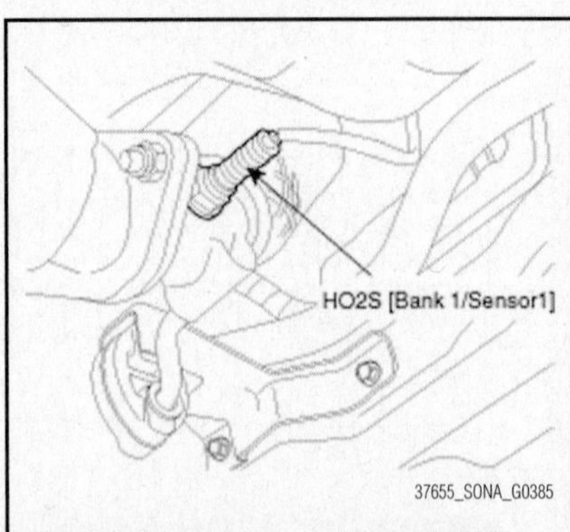

Fig. 267 Heated Oxygen Sensor (HO2S) Bank 1/Sensor 1—2.4L engine, ULEV

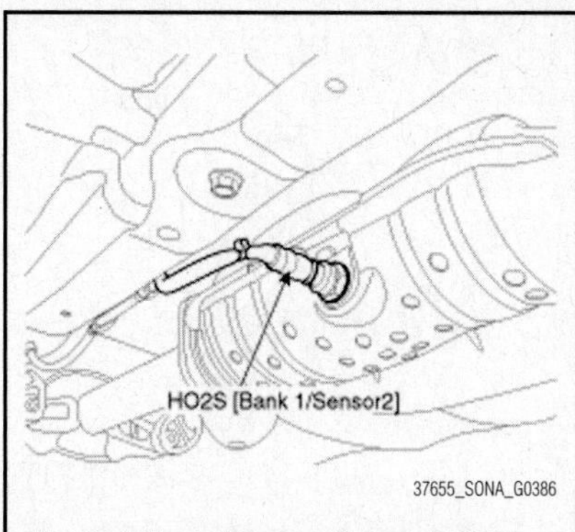

Fig. 268 Heated Oxygen Sensor (HO2S) Bank 1/Sensor 2—2.4L engine, ULEV

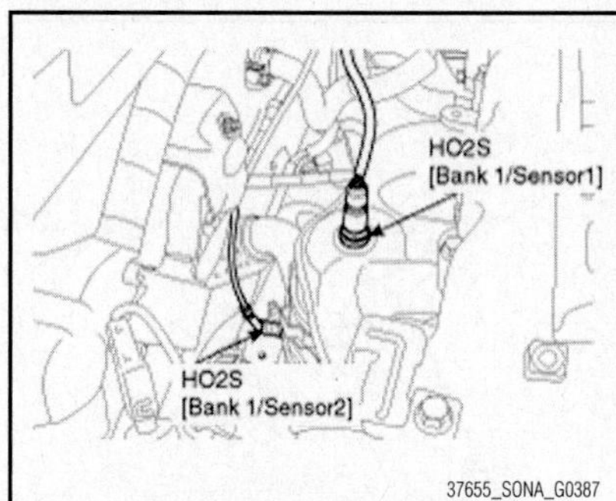

Fig. 269 Heated Oxygen Sensor (HO2S) Bank 1/Sensor 1 and Bank 1/Sensor 2—2.4L engine, SULEV

➡Apply anti-seize compound to the threaded portion of the sensor, prior to installation. Never apply anti-seize compound to the protector of the sensor.

INTAKE AIR TEMPERATURE (IAT) SENSOR

LOCATION

2.4L Engine

See Figure 270.

The Intake Air Temperature (IAT) sensor is mounted on the intake manifold behind the throttle actuator assembly. For the 2.4 Liter engine, the IAT is integrated inside the Manifold Absolute Pressure (MAP) sensor.

3.3L Engine

See Figure 271.

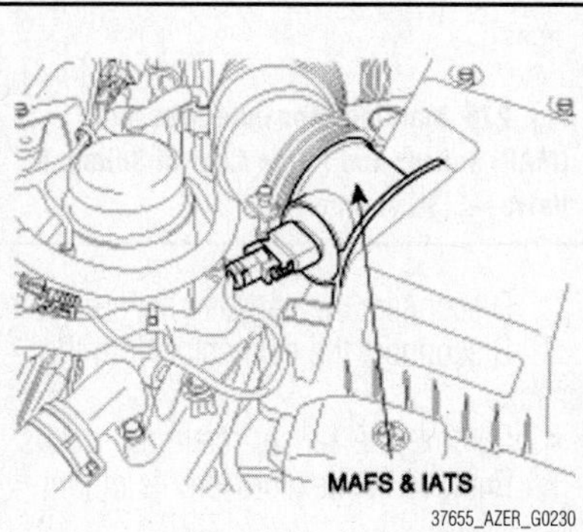

Fig. 271 Intake Air Temperature (IAT) sensor and Mass Air Flow (MAF) sensor—3.3L engine

The Intake Air Temperature (IAT) sensor is mounted in the intake air hose of the air cleaner assembly. For the 3.3 Liter engine, the IAT is integrated inside the Mass Air Flow (MAF) sensor.

REMOVAL & INSTALLATION

1. Before servicing the vehicle, refer to the Precautions Section.
2. Disconnect the negative battery cable.
3. Disconnect the connector from the sensor.
4. Remove the sensor retaining screws.
5. Remove the sensor from its mounting.

To install:

6. Installation is the reverse of the removal procedure.

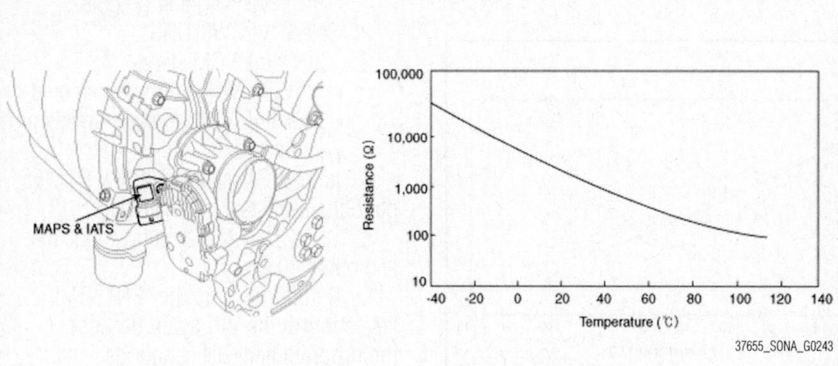

Fig. 270 Intake Air Temperature (IAT) sensor and Manifold Absolute Pressure (MAP) sensor—2.4L engine

KNOCK SENSOR (KS)

LOCATION

See Figures 272 and 273.

Refer to the accompanying illustrations.

REMOVAL & INSTALLATION

1. Before servicing the vehicle, refer to the Precautions Section.
2. Disconnect the negative battery cable.
3. Disconnect the sensor connector.
4. Remove the sensor from its mounting.

To install:

5. Installation is the reverse of the removal procedure.
6. Tighten the sensor to 14–17 ft. lbs. (19–24 Nm).

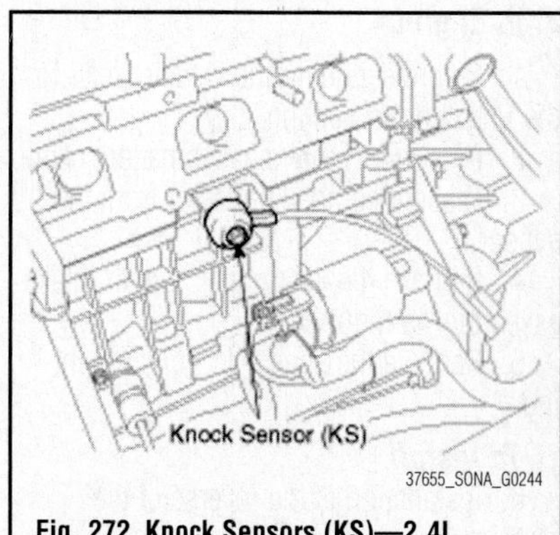

Fig. 272 Knock Sensors (KS)—2.4L engine

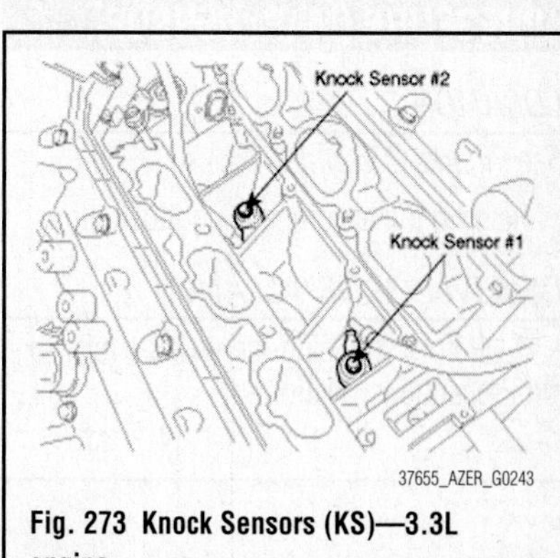

Fig. 273 Knock Sensors (KS)—3.3L engine

MASS AIR FLOW (MAF) SENSOR

LOCATION

3.3L Engine

See Figure 274.

The Mass Air Flow (MAF) sensor is mounted in the intake air hose of the air

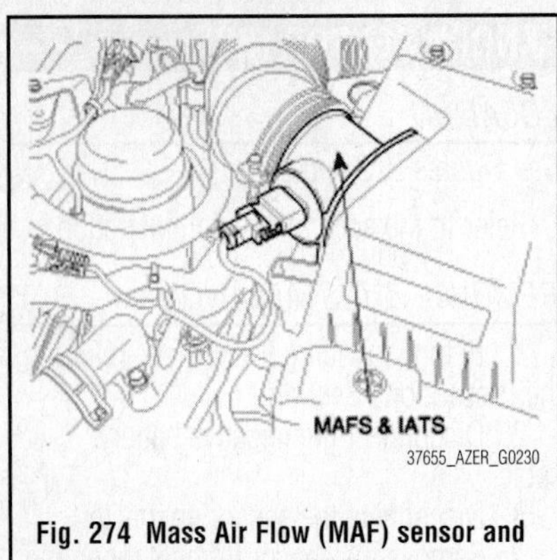

Fig. 274 Mass Air Flow (MAF) sensor and Intake Air Temperature (IAT) sensor

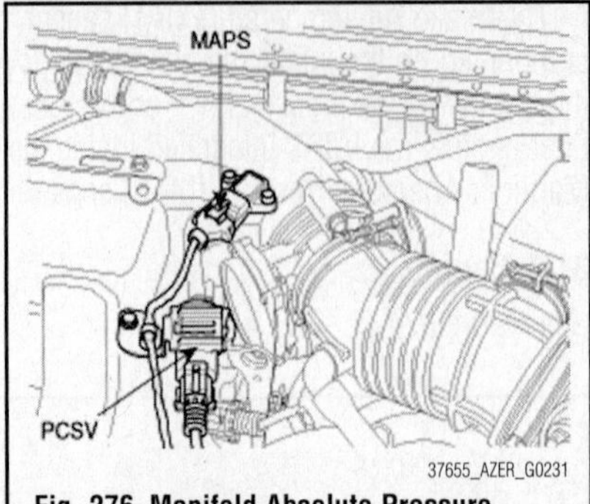

Fig. 276 Manifold Absolute Pressure (MAP) sensor and Purge Control Solenoid Valve—3.3L engine

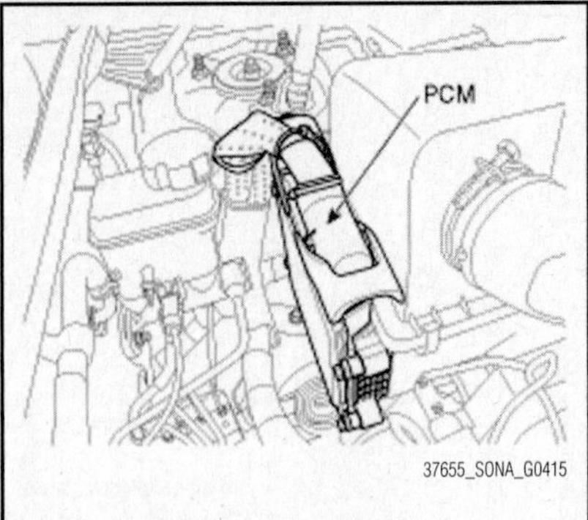

Fig. 277 Powertrain Control Module (ECM) location—2.4L engine

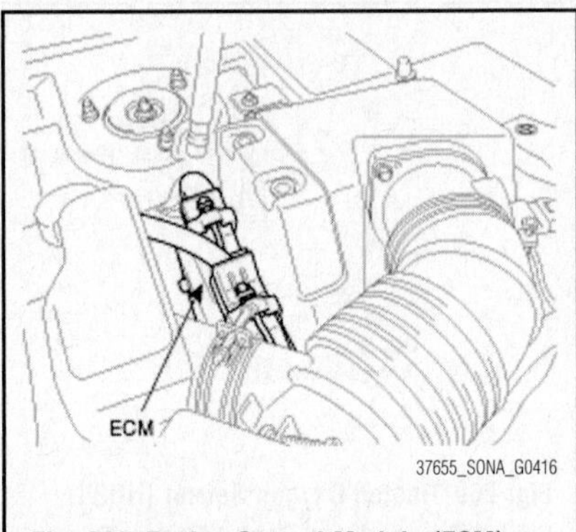

Fig. 278 Engine Control Module (ECM) location—3.3L engine

cleaner assembly. This sensor is combined with the Intake Air Temperature (IAT) sensor.

REMOVAL & INSTALLATION

3.3L Engine

1. Before servicing the vehicle, refer to the Precautions Section.
2. Disconnect the negative battery cable.
3. Disconnect the connector from the sensor.
4. Remove the air cleaner and air intake assembly, as required.
5. Remove the sensor from its mounting.

To install:

6. Installation is the reverse of the removal procedure.

MANIFOLD ABSOLUTE PRESSURE (MAP) SENSOR

LOCATION

See Figures 275 and 276.

Refer to the accompanying illustrations.

REMOVAL & INSTALLATION

1. Before servicing the vehicle, refer to the Precautions Section.
2. Disconnect the negative battery cable.
3. Disconnect the connector from the sensor.
4. Remove the sensor retaining screws.
5. Remove the sensor from its mounting.

To install:

6. Installation is the reverse of the removal procedure. Tighten the bolts to 78–104 inch lbs. (9–12 Nm).

POWERTRAIN CONTROL MODULE (PCM)

LOCATION

See Figures 277 and 278.

Refer to the accompanying illustrations.

REMOVAL & INSTALLATION

1. Turn ignition switch off.
2. Disconnect the battery (-) cable from the battery.
3. Remove the PCM/ECM cover and disconnect the PCM/ECM connector.
4. Unscrew the mounting bolts and remove the PCM/ECM from the air cleaner assembly.

To install:

5. Installation is the reverse of removal.

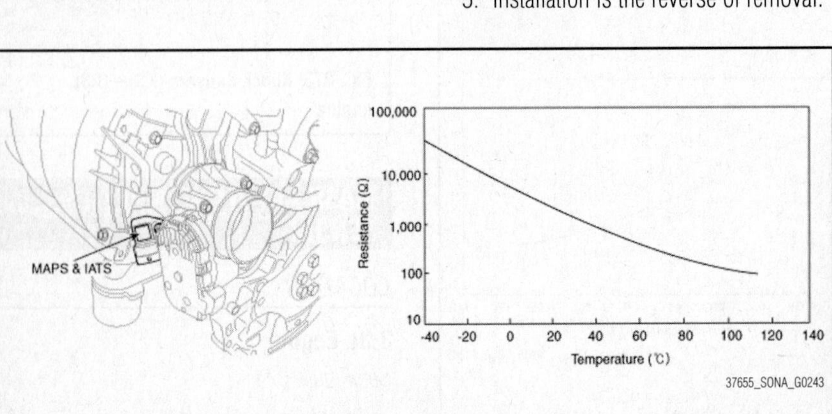

Fig. 275 Manifold Absolute Pressure (MAP) sensor—2.4L engine

6. Tighten retaining screws to 7–9 ft. lbs. (10–12 Nm).

Reset Procedure

➡When replacing a PCM, the VIN must be programmed in the PCM. If there is no VIN in PCM memory, the fault code (DTC P0630) is set.

⁂ WARNING

The programmed VIN cannot be changed. When writing the VIN, confirm the VIN carefully

1. Select "Vehicle" and "Engine".
2. Select "VIN WRITING".
3. Check the PCM status:
   - VIRGIN: VIN is not programmed
   - LEARNT: VIN has been already programmed
4. Is the PCM status "VIRGIN"? If YES, go to the next step. If NO, END.
5. Write the VIN with cursor, function and number keys.
6. Before pressing the "ENTER" key, confirm the VIN again because the programmed VIN cannot be changed.
7. After verifying the written VIN, press the "ENTER" key.

8. Turn the ignition switch OFF, and then turn ON.
9. Verify the programmed VIN in the PCM memory.

THROTTLE POSITION SENSOR (TPS)

LOCATION

See Figures 279 and 280.

Refer to the accompanying illustrations.

REMOVAL & INSTALLATION

1. Before servicing the vehicle, refer to the Precautions Section.
2. Disconnect the negative battery cable.
3. Disconnect the sensor connector.

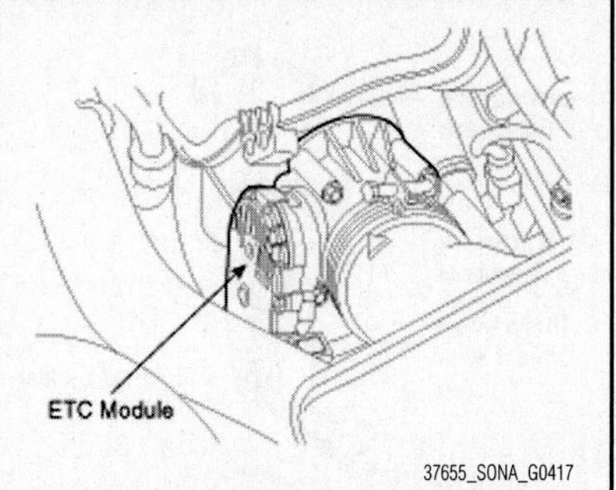

Fig. 279 Electronic Throttle Control (ETC) system, including the Throttle Position Sensor (TPS) location—2.4L engine

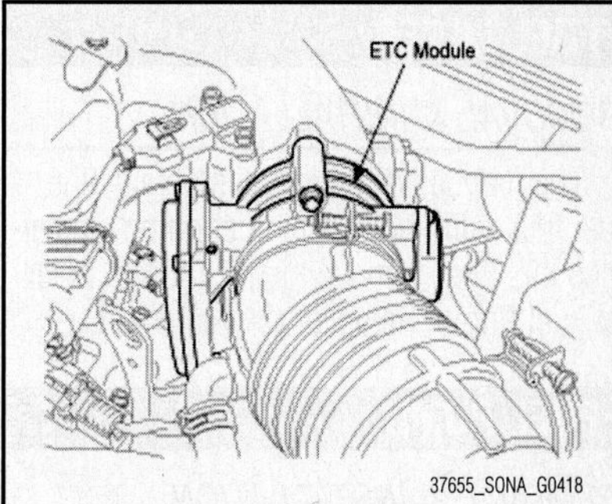

Fig. 280 Electronic Throttle Control (ETC) system, including the Throttle Position Sensor (TPS) location—3.3L engine

4. Remove the sensor retaining screws.
5. Remove the sensor from its mounting.

To install:

6. Installation is the reverse of the removal procedure.

FUEL — GASOLINE FUEL INJECTION SYSTEM

FUEL SYSTEM SERVICE PRECAUTIONS

Safety is the most important factor when performing not only fuel system maintenance but any type of maintenance. Failure to conduct maintenance and repairs in a safe manner may result in serious personal injury or death. Maintenance and testing of the vehicle's fuel system components can be accomplished safely and effectively by adhering to the following rules and guidelines.

- To avoid the possibility of fire and personal injury, always disconnect the negative battery cable unless the repair or test procedure requires that battery voltage be applied.
- Always relieve the fuel system pressure prior to disconnecting any fuel system component (injector, fuel rail, pressure regulator, etc.), fitting or fuel line connection. Exercise extreme caution whenever relieving fuel system pressure to avoid exposing skin, face and eyes to fuel spray. Please be advised that fuel under pressure may penetrate the skin or any part of the body that it contacts.
- Always place a shop towel or cloth around the fitting or connection prior to loosening to absorb any excess fuel due to spillage. Ensure that all fuel spillage (should it occur) is quickly removed from engine surfaces. Ensure that all fuel soaked cloths or towels are deposited into a suitable waste container.
- Always keep a dry chemical (Class B) fire extinguisher near the work area.
- Do not allow fuel spray or fuel vapors to come into contact with a spark or open flame.
- Always use a back-up wrench when loosening and tightening fuel line connection fittings. This will prevent unnecessary stress and torsion to fuel line piping.
- Always replace worn fuel fitting O-rings with new Do not substitute fuel hose or equivalent where fuel pipe is installed.

Before servicing the vehicle, make sure to also refer to the precautions in the beginning of this section as well.

RELIEVING FUEL SYSTEM PRESSURE

See Figures 281 through 283.

1. Before servicing the vehicle, refer to the Precautions Section.
2. Remove the rear seat cushion.
3. Remove the access panel.
4. Remove the fuel pump module connector.

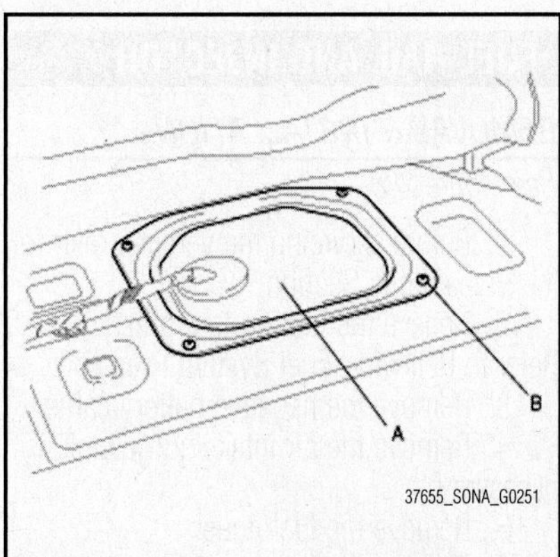

Fig. 281 Fuel pump cover assembly

5. Start the engine and allow it to run until it stalls.
6. Turn the ignition switch to the **OFF** position.
7. Disconnect the negative battery cable.
8. Attach the fuel pump harness connector.

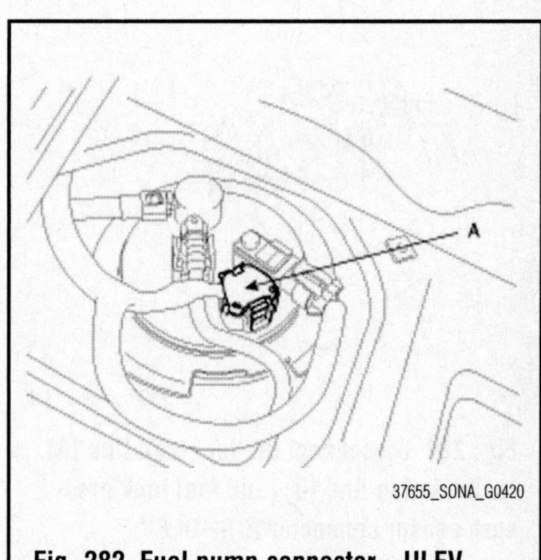

Fig. 282 Fuel pump connector—ULEV

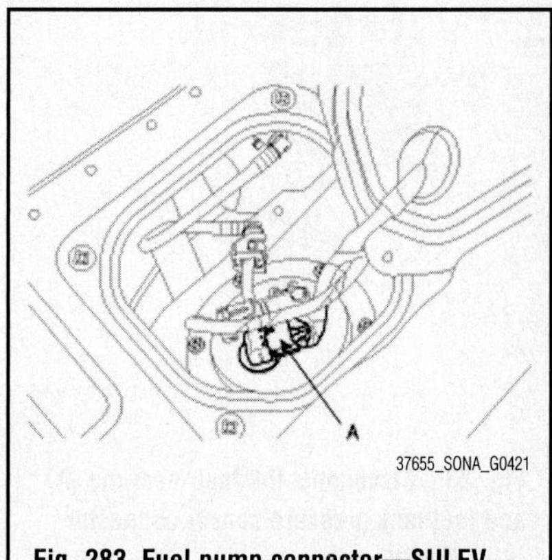

Fig. 283 Fuel pump connector—SULEV

FUEL FILTER

REMOVAL & INSTALLATION

The fuel delivery system integrates the fuel filter with the in-tank fuel pump. To service this filter, remove the fuel pump. Refer to Fuel Pump, Removal & Installation.

FUEL PUMP

REMOVAL & INSTALLATION

See Figures 284 through 287.

1. Before servicing the vehicle, refer to the Precautions Section.
2. Open the service cover.
3. Relieve the fuel system pressure. Refer to Relieving Fuel System Pressure.
4. Disconnect the fuel feed line, fuel tank pressure sensor connector, and recirculation line, as necessary.
5. For the ULEV system, unfasten the fuel pump locking ring with the appropriate service tool.
6. For the SULEV system, unscrew the fuel pump mounting bolts.
7. Remove the fuel pump assembly from the fuel tank.

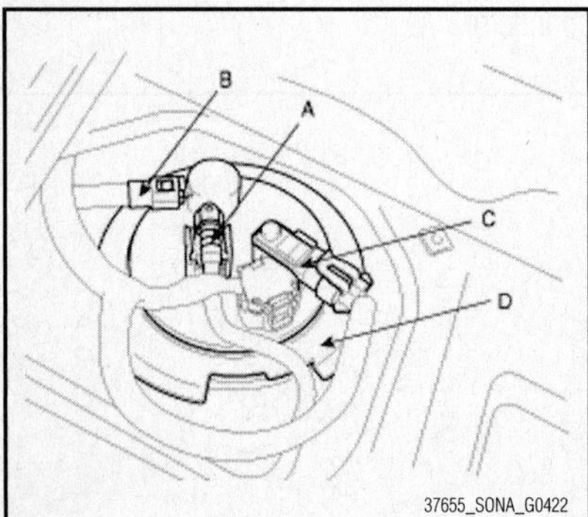

Fig. 284 Disconnect the fuel feed line (A), recirculation line (B), and fuel tank pressure sensor connector (C)—ULEV

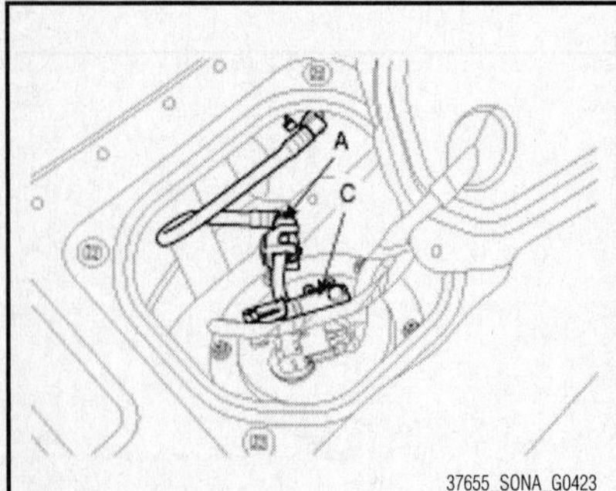

Fig. 285 Disconnect the fuel feed line (A) and fuel tank pressure sensor connector (C)—SULEV

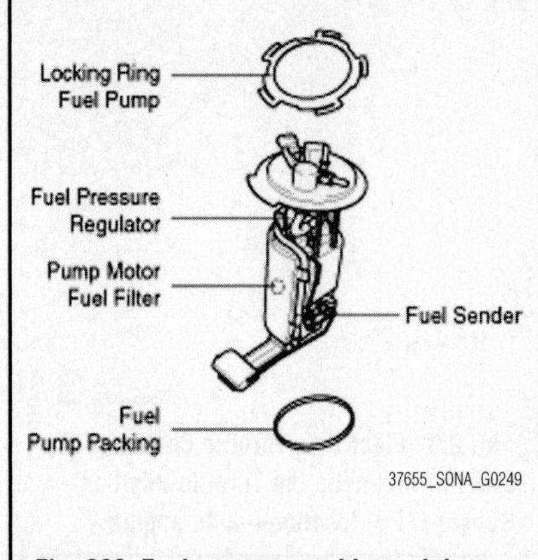

Fig. 286 Fuel pump assembly module—ULEV

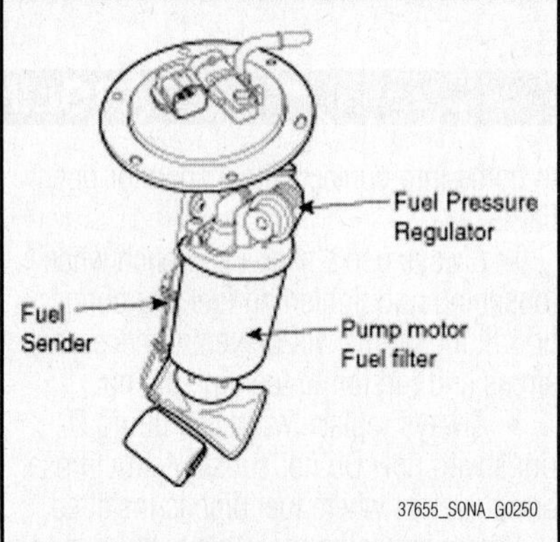

Fig. 287 Fuel pump assembly module—SULEV

To install:

8. Installation is the reverse of removal, noting the following:

➡Replace the Fuel Pump Locking Ring and Fuel Pump Plate Packing (A) with a new one when installing the fuel pump again.

9. Tighten the fuel pump mounting bolts/nuts to: 17–26 inch lbs. (2–3 Nm).
10. Connect the fuel feed line and canister hoses.

FUEL RAIL AND INJECTOR

REMOVAL & INSTALLATION

See Figure 288.

1. Before servicing the vehicle, refer to the Precautions Section.
2. Relieve the fuel system pressure. Refer to Relieving Fuel System Pressure.
3. Remove the negative battery cable.
4. Remove the air intake surge tank, if necessary.
5. Remove the fuel lines.
6. Remove the fuel injector connectors.

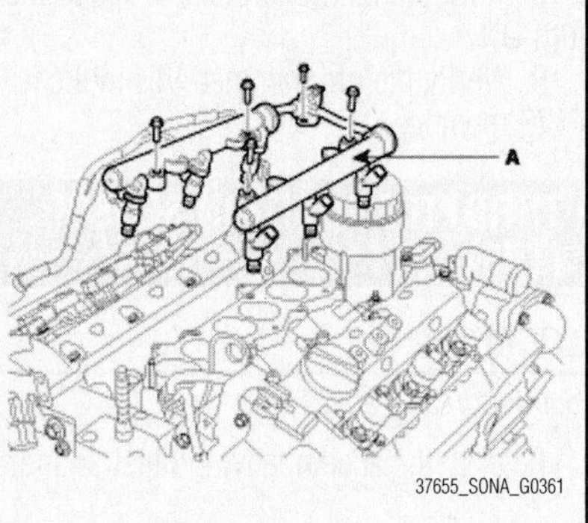

Fig. 288 Fuel supply manifold with injectors attached (A)—3.3L engine shown, 2.4L engine similar

7. Remove the fuel supply manifold with injectors attached.
8. Separate the injectors from the supply manifold.

To install:

9. Install the injectors to the fuel supply manifold using new O-rings.
10. Install the fuel supply manifold with injectors attached and torque the bolts:
    - 3.3L engine: 72–108 inch lbs. (9–12 Nm)
    - 2.4L engine: 15–19 ft. lbs. (20–25 Nm)
11. Install the fuel injector connectors.
12. Install the fuel lines.
13. Install the air intake surge tank, if removed.
14. Install the negative battery cable.
15. Start the engine and check for leaks.

FUEL TANK

REMOVAL & INSTALLATION

See Figures 289 through 298.

1. Before servicing the vehicle, refer to the Precautions Section.
2. Remove the rear seat cushion.

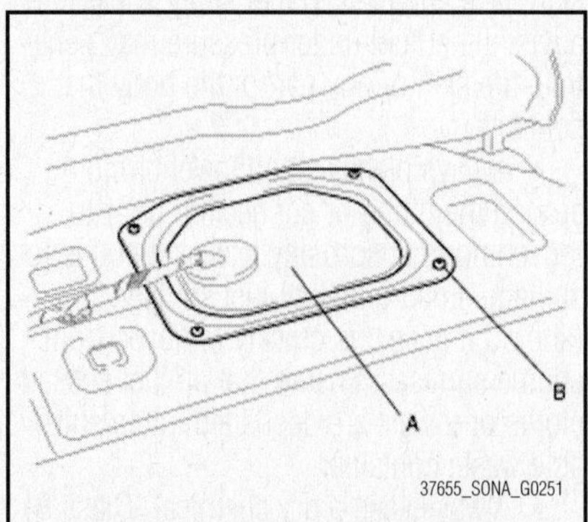

Fig. 289 Remove the service cover (A) under the back seat

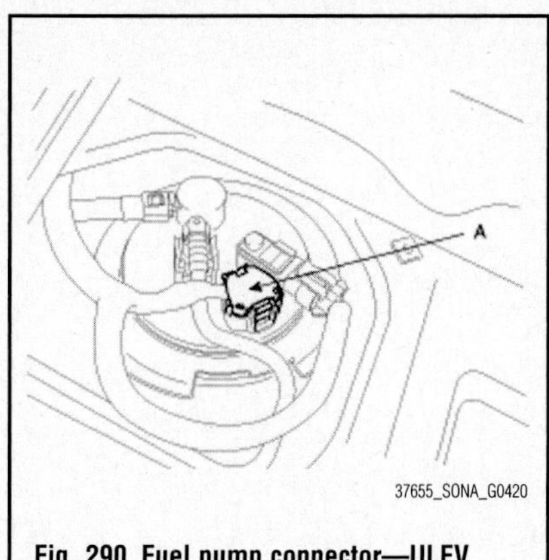

Fig. 290 Fuel pump connector—ULEV

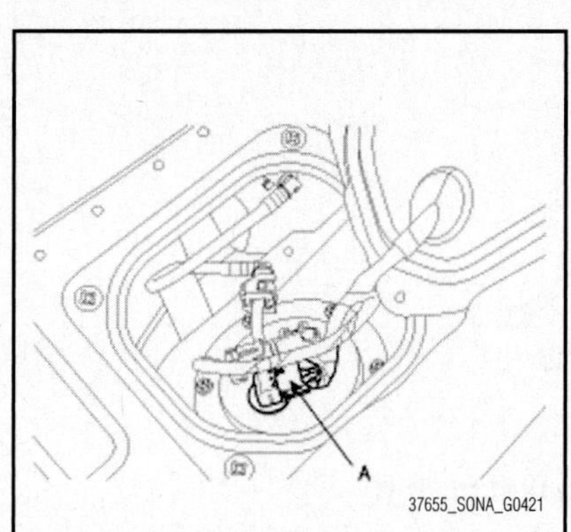

Fig. 291 Fuel pump connector—SULEV

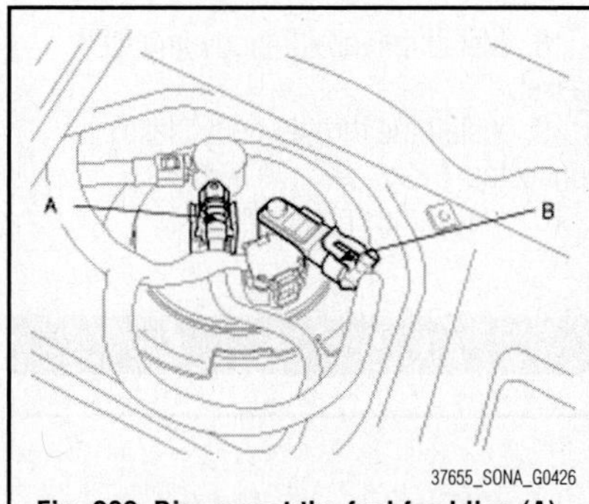

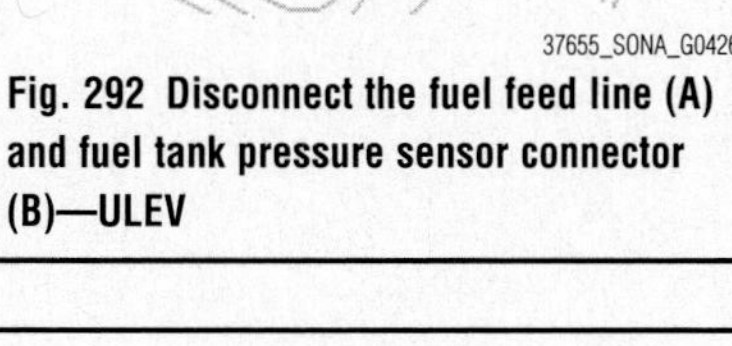

Fig. 292 Disconnect the fuel feed line (A) and fuel tank pressure sensor connector (B)—ULEV

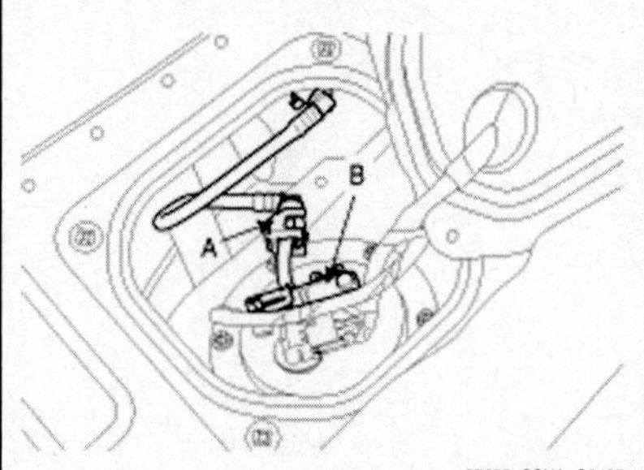

Fig. 293 Disconnect the fuel feed line (A) and fuel tank pressure sensor connector (B)—SULEV

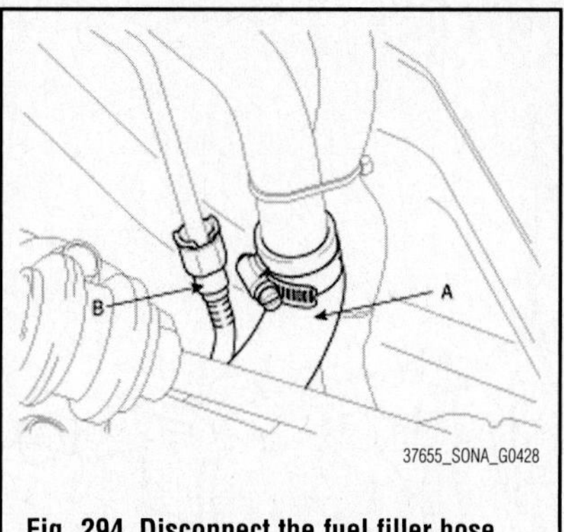

Fig. 294 Disconnect the fuel filler hose (A) and recirculation line (B)—ULEV

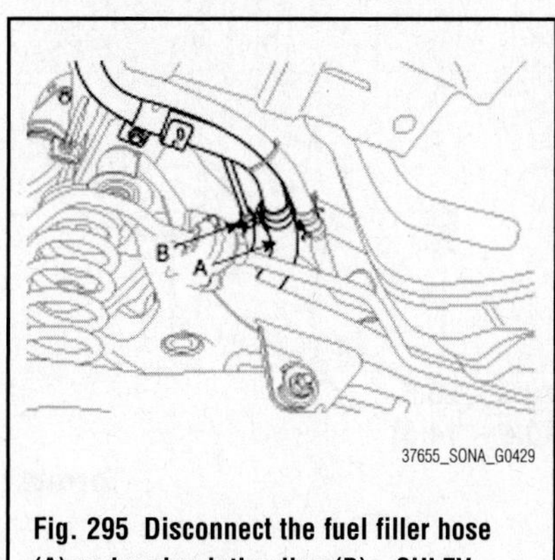

Fig. 295 Disconnect the fuel filler hose (A) and recirculation line (B)—SULEV

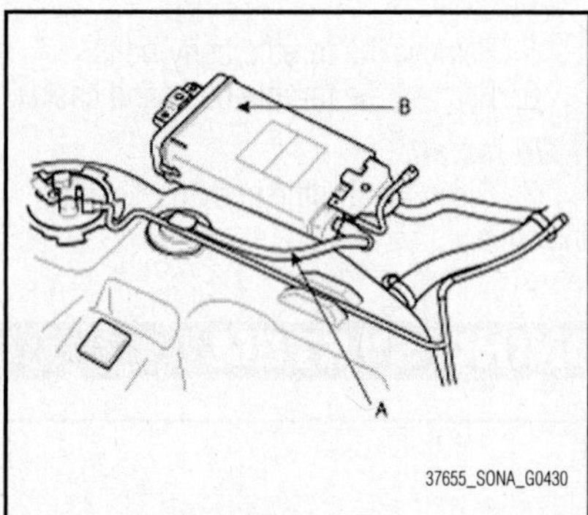

Fig. 296 Disconnect the canister hose (A) from the canister (B)—ULEV

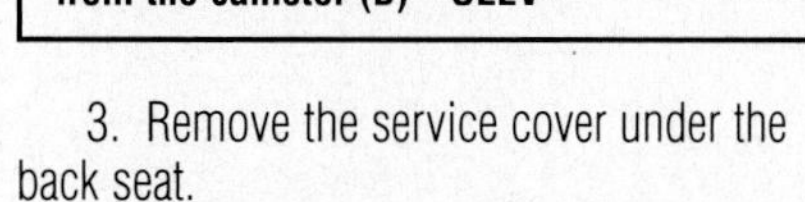

3. Remove the service cover under the back seat.

4. Disconnect the fuel pump connector.

5. Start the engine and wait until the fuel in the fuel line is exhausted.

6. After the engine stalls, turn the ignition switch to the OFF position.

7. Disconnect the negative battery connection.

8. Disconnect the fuel feed line and fuel tank pressure sensor connector.

9. Raise and safely support the vehicle.

10. Remove the center muffler and main muffler, as needed.

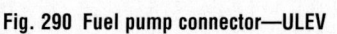

Fig. 297 Disconnect the canister hose (A)—SULEV

Fig. 298 Remove the fuel tank band mounting bolts (A) and fuel tank

11. Disconnect the fuel filler hose (A) and recirculation line (B).

12. Disconnect the canister hose (A) from the canister (B).

13. Remove the fuel tank band mounting bolts.

14. Remove the fuel tank.

To install:

15. Installation is the reverse of the removal.

16. Tighten the fuel tank mounting bolts:

- 2.4L engine: 29–40 ft. lbs. (39–54 Nm)
- 3.3L engine: 2–40 ft. lbs. (34–54 Nm)

IDLE SPEED

ADJUSTMENT

Idle speed is maintained by the Powertrain Control Module (PCM). No adjustment is necessary or possible.

THROTTLE BODY

REMOVAL & INSTALLATION

See Figure 299.

1. Before servicing the vehicle, refer to the Precautions Section.

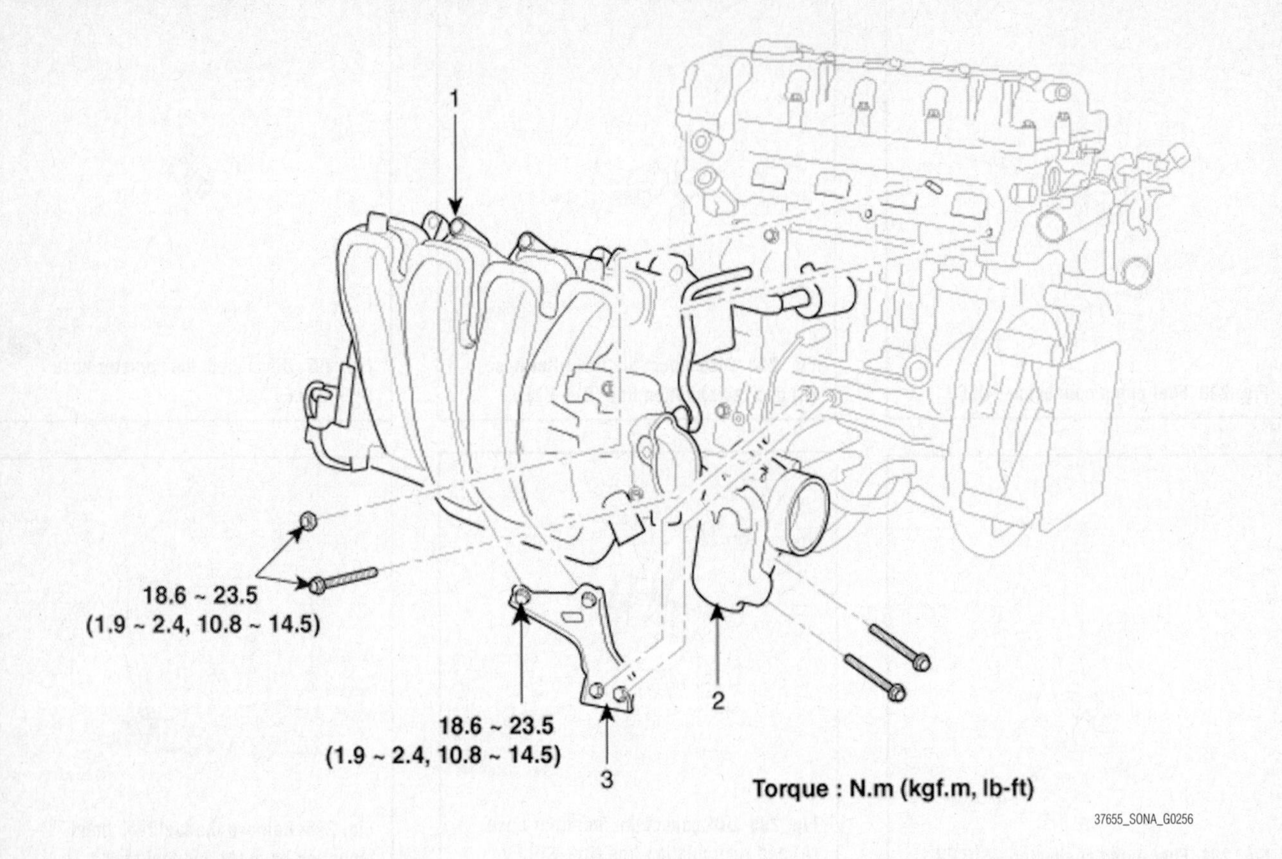

Fig. 299 Throttle body bolt locations—2.4L engine

2. Turn the ignition **OFF**.
3. Remove the engine cover, if necessary.
4. Remove the throttle body electrical connector.
5. Remove the throttle body bolts.
6. Remove the throttle body and gasket.

To install:

7. Clean the throttle body gasket mating surfaces.
8. Install the throttle body and NEW gasket.
9. Install the throttle body electrical connector.
10. Install the engine cover.

HEATING & AIR CONDITIONING SYSTEM

BLOWER MOTOR

REMOVAL & INSTALLATION

See Figures 300 and 301.

1. Before servicing the vehicle, refer to the Precautions Section.

CAUTION

Before servicing components near or affected by the SRS (air bag) system, read and observe all SRS Service Precautions. Refer to Supplemental Restraint System (SRS), in the Chassis Electrical section. Failure to observe all precautions may result in accidental airbag deployment, personal injury, or death. Refer to Airbag Removal & Installation.

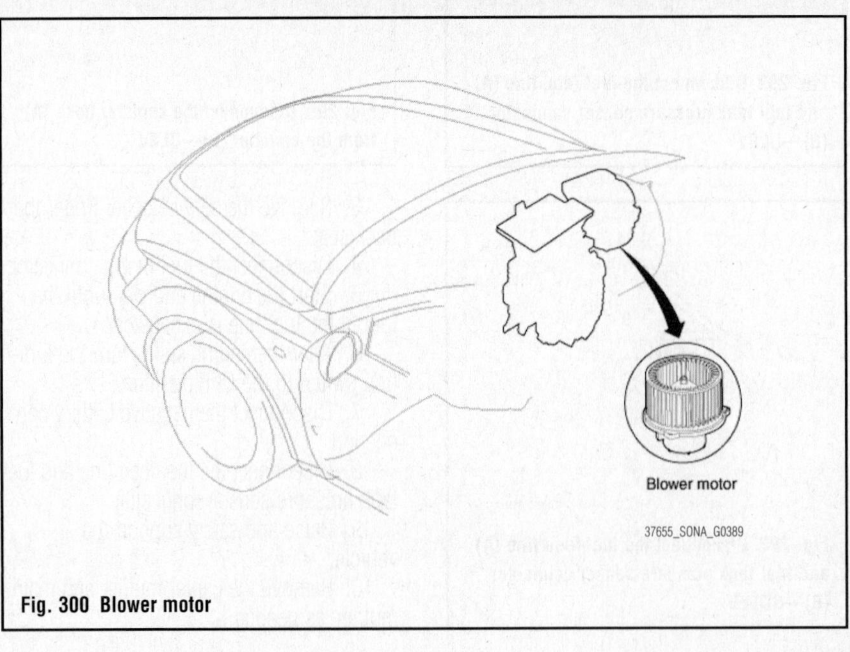

Fig. 300 Blower motor

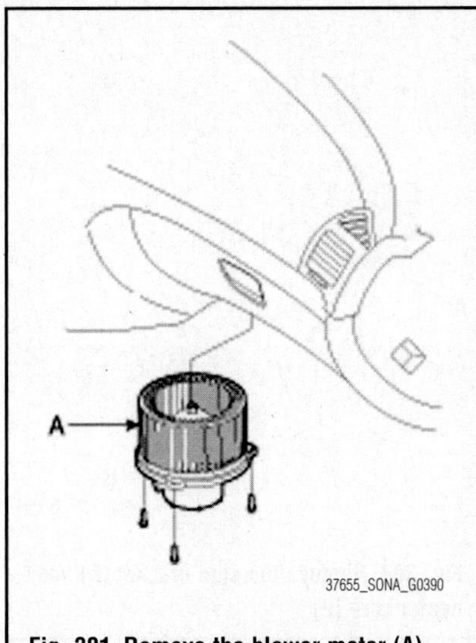

Fig. 301 Remove the blower motor (A)

2. Disconnect the negative battery cable and wait at least 3 minutes for the SRS memory to drain.
3. Remove the instrument panel under cover.
4. Disconnect the connector from the blower motor.
5. Remove the blower motor mounting screws.
6. Remove the blower motor.

To install:

7. Installation is the reverse order of removal.

➡Make sure that there is no air leaking out of the blower and duct joints.

HEATER UNIT/CORE

REMOVAL & INSTALLATION

See Figures 302 through 307.

1. Before servicing the vehicle, refer to the Precautions Section.

CAUTION

Before servicing components near or affected by the SRS (air bag) system, read and observe all SRS Service Precautions. Refer to Supplemental Restraint System (SRS), in the Chassis Electrical section. Failure to observe all precautions may result in accidental airbag deployment, personal injury, or death. Refer to Airbag Removal & Installation.

2. Disconnect the negative battery cable and wait at least 3 minutes for the SRS memory to drain.
3. Recover the refrigerant with a recovery/ recycling/ charging station.
4. When the engine is cool, drain the engine coolant from the radiator.
5. Remove the expansion valve cover after removing the nut.
6. Remove the expansion valve from the evaporator core after removing the nuts. Plug or cap the lines immediately after disconnecting them to avoid moisture and dust contamination.
7. Disconnect the inlet and outlet heater hoses from the heater unit.

WARNING

Engine coolant will run out when the hoses are disconnected; drain it into a clean drip pan. Be sure not to let coolant spill on electrical parts or painted surfaces. If any coolant spills, rinse it off immediately.

8. Remove the instrument panel.
9. Remove the cowl cross member.
10. Disconnect the connectors from the temperature control actuator, the mode control actuator and the evaporator temperature sensor, and then remove the mounting nut and the mounting bolts.
11. Remove the heater and evaporator unit after removing the mounting bolts.
12. Remove the blower unit from heater unit after removing fixing screws on the connected part.
13. Remove the side bracket and the heater core.
14. Be careful that inlet and outlet pipes are not to be bent during heater core removal, and pull out the heater core.

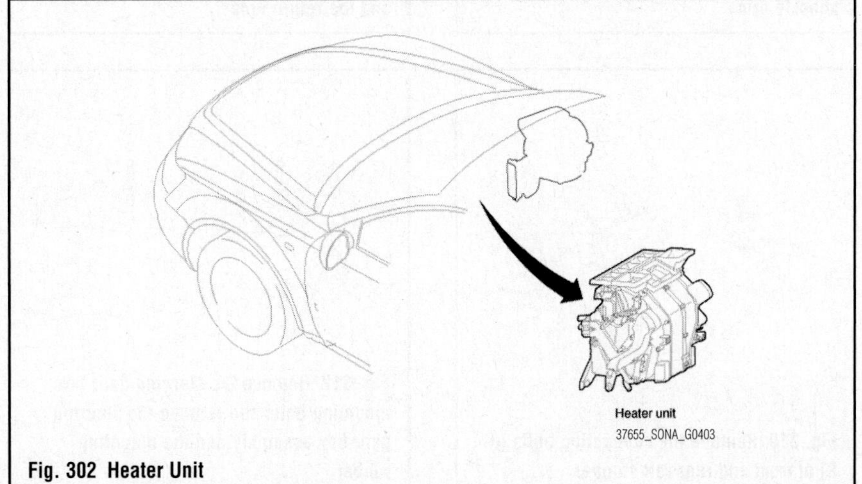

Fig. 302 Heater Unit

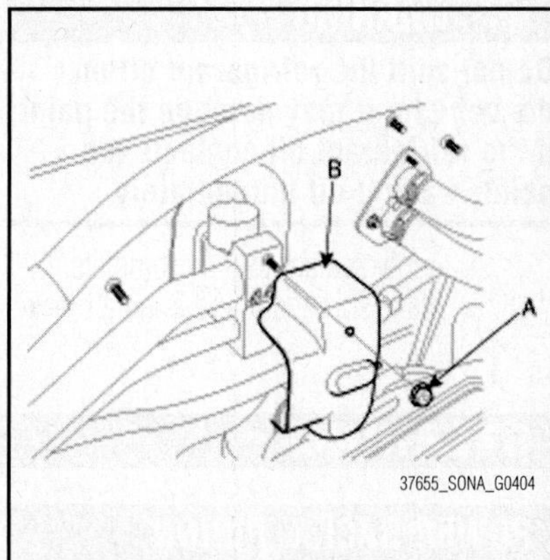

Fig. 303 Remove the expansion valve cover after removing the nut (A)

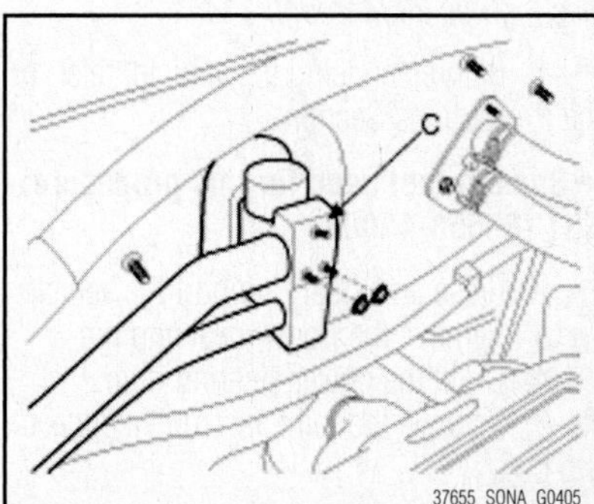

Fig. 304 Remove the expansion valve (C) from evaporator core after removing the nuts

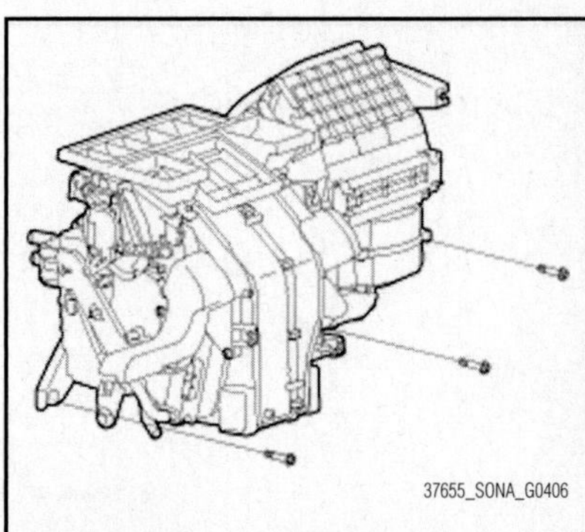

Fig. 305 Heater and evaporator unit and mounting bolts

To install:

15. Install in the reverse order of removal, noting the following:
    a. If you're installing a new evaporator, add refrigerant oil.
    b. Replace the O-rings with new ones at each fitting, and apply a thin coat of refrigerant oil before installing them. Be sure to use the right O-rings for R-134a to avoid leakage.
    c. Immediately after using the oil, replace the cap on the container, and seal it to avoid moisture absorption.

WARNING

Do not spill the refrigerant oil on the vehicle; it may damage the paint; if the refrigerant oil contacts the paint, wash it off immediately.

    d. Apply sealant to the grommets.
    e. Make sure that there is no air leakage.

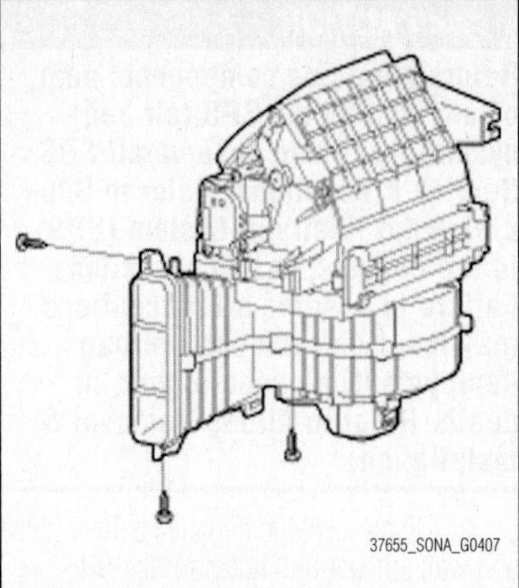

Fig. 306 Remove the blower unit from heater unit

    f. Charge the system and test its performance.
    g. Do not interchange the inlet and outlet heater hoses and install the hose clamps securely.

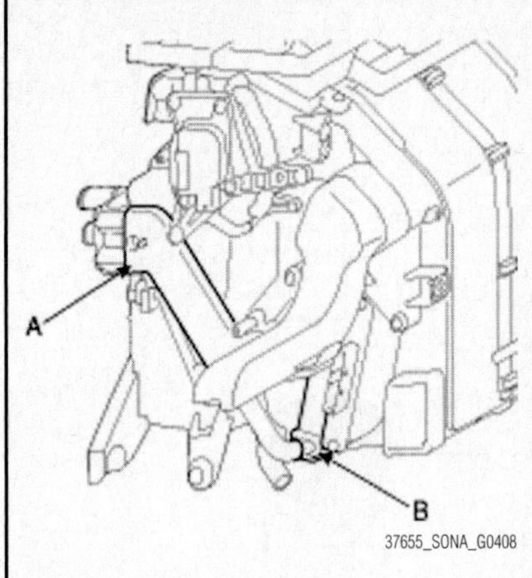

Fig. 307 Remove the side bracket (A) and heater core (B)

    h. Refill the cooling system with engine coolant.

STEERING

POWER RACK & PINION STEERING GEAR

REMOVAL & INSTALLATION

See Figures 308 through 314.

1. Before servicing the vehicle, refer to the Precautions Section.

➡Special tool used for this procedure: SST (09568-4A000)

2. Raise and safely support the vehicle.
3. Remove the front wheel and tire.
4. Drain the power steering fluid.
5. Remove the joint assembly connecting bolt.

WARNING

Keep the neutral-range to prevent the damage of the clock spring inner cable when you handle the steering wheel.

6. Using the special tool (09568-4A000), disconnect the tie rod end from the knuckle arm.

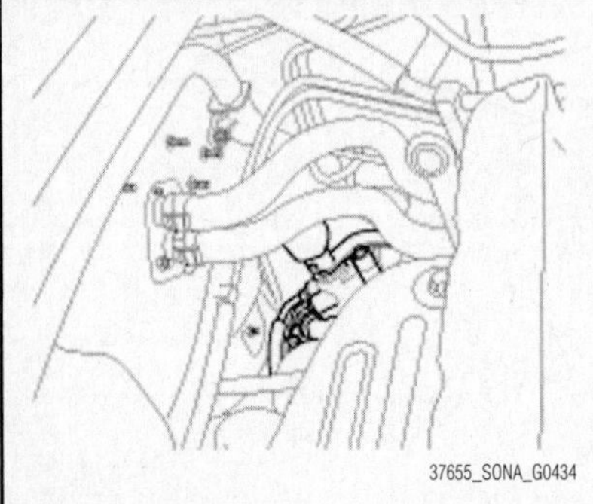

Fig. 308 Remove the joint assembly connecting bolt

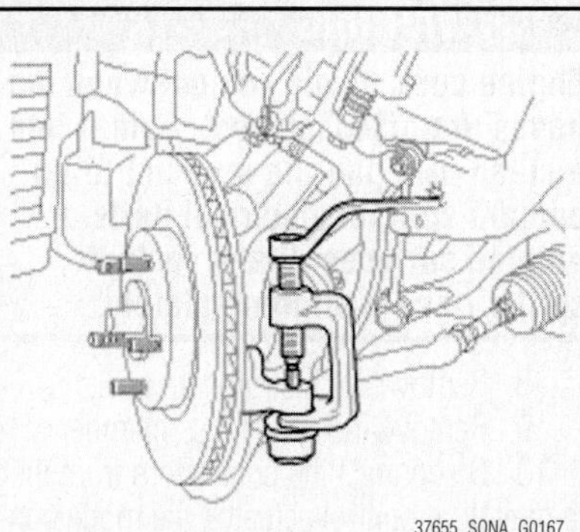

Fig. 309 Using the special tool (09568-4A000) disconnect the tie rod end from the knuckle arm

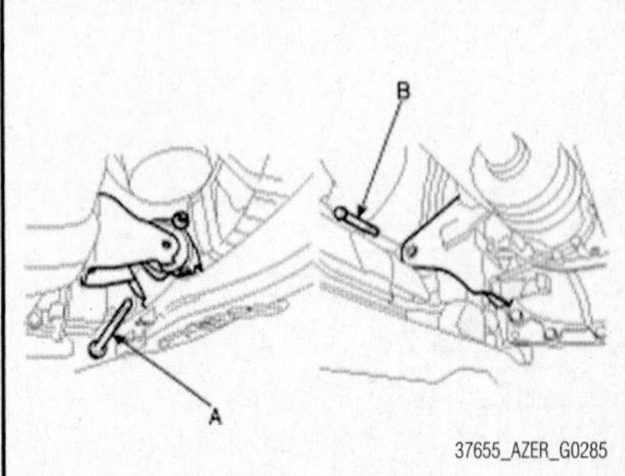

Fig. 310 Remove the connecting bolts (A, B) of front and rear roll stopper

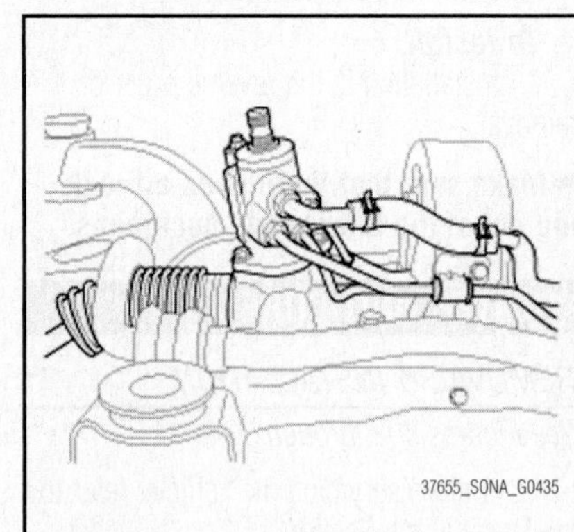

Fig. 311 Disconnect the pressure hose and the return tube

Fig. 312 Remove the steering gear box mounting bolts and remove the steering gear box assembly and the mounting rubber

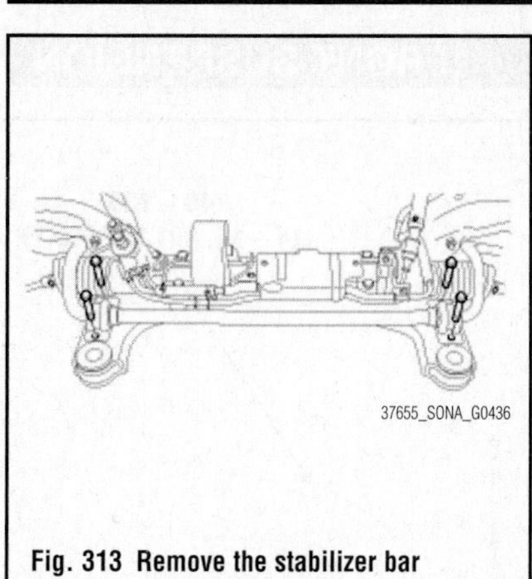

Fig. 313 Remove the stabilizer bar

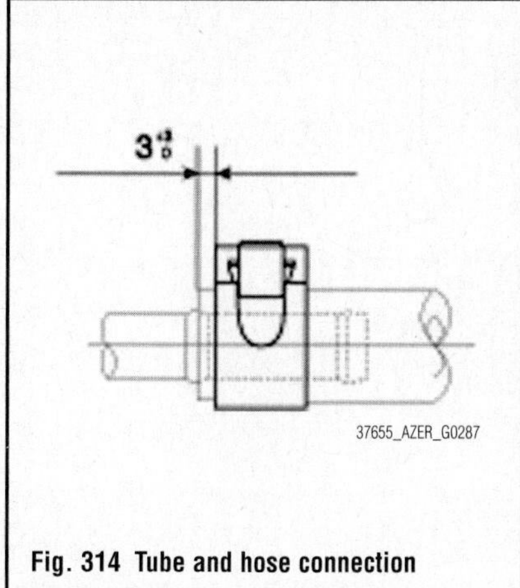

Fig. 314 Tube and hose connection

7. Remove the front fork and the knuckle ball joint from the front lower arm.
8. Remove the connecting bolts of front and rear roll stopper.
9. Remove the crossmember mounting bolts.
10. Disconnect the pressure hose and the return tube.
11. Remove the steering gear box mounting bolts and remove the steering gear box assembly and the mounting rubber.

⁂ WARNING

When removing the gear box, pull it out carefully and slowly to avoid damaging the boots.

12. Remove the stabilizer bar.

To install:

➡Be sure to connect between a tube and a hose as shown in the illustration.

13. Installation is the reverse of removal, noting the following tightening specifications:
    - Pressure hose to gear box: 9–13 ft. lbs. (12–18 Nm)
    - Return tube to gear box: 9–13 ft. lbs. (12–18 Nm)
    - Tie rod end lock nut: 36–40 ft. lbs. (50–55 Nm)
    - Pinion and valve assembly to self locking nut: 14–22 ft. lbs. (20–30 Nm)
    - Lock nut: 36–51 ft. lbs. (50–70 Nm)
    - Tie rod end self locking nut: 17–25 ft. lbs. (24–34 Nm)
    - Mounting bracket to crossmember: 43–58 ft. lbs. (60–80 Nm)
14. After installation, bleed the air in the power steering system.
15. Check front wheel alignment.

POWER STEERING PUMP

REMOVAL & INSTALLATION

See Figure 315.

1. Before servicing the vehicle, refer to the Precautions Section.
2. Loosen the bolt fixing the wiring bracket and move the wiring aside.
3. Remove the pressure hose from the oil pump.
4. Disconnect the suction hose from the suction connector and drain the fluid into a container.
5. Release the tension of the power steering drive belt by lifting the auto-tensioner pulley.
6. Remove the drive belt from the power steering oil pump pulley.
7. Loosen the power steering oil pump mounting bolt and the tension adjusting bolt.
8. Remove the steering oil pump assembly.

To install:

9. Installation is the reverse of the removal procedure.

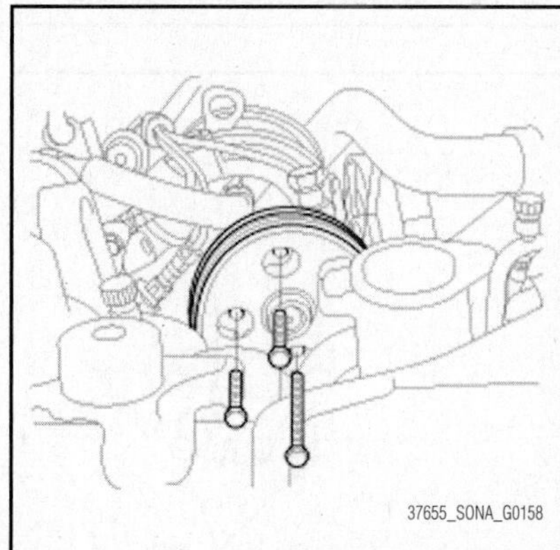

Fig. 315 Steering oil pump assembly bolt locations

➡Install the pressure hose being careful so that it does not twist and come in contact with other components.

10. Add power steering fluid (PSF-4).
11. Air bleed the system.

BLEEDING

⁂ CAUTION

The fluid level should be checked with the engine OFF to prevent injury from moving components. Use only PSF-4 power steering fluid, or equivalent. Do not overfill.

1. Before servicing the vehicle, refer to the Precautions Section.
2. Wipe the reservoir fill cap clean before removal.
3. Fill the pump fluid reservoir to the proper level. The fluid level should be within the "FILL RANGE" listed on the exterior of the reservoir when the fluid is at normal ambient temperature, approximately 70–80°F (21–27°C).
4. Let the fluid settle in the system for at least 2 minutes.
5. Start the engine and let it run for a few seconds. Then turn the engine **OFF**.
6. Add fluid if necessary. Repeat the above procedure until the fluid level remains constant after running the engine.
7. Raise the front wheels off the ground.
8. Start the engine. Slowly turn the steering wheel right and left, lightly contacting the wheel stops.
9. Turn the engine off. Check the fluid level and add power steering fluid if necessary.
10. Start the engine. Lower the vehicle and turn the steering wheel slowly from lock to lock.
11. Stop the engine. Check the fluid level and refill as required.
12. If the fluid is extremely foamy, allow the vehicle to stabilize a few minutes, then repeat the above procedure.

FLUID FILL PROCEDURE

1. Before servicing the vehicle, refer to the Precautions Section.
2. Wipe the reservoir fill cap clean before removal.
3. Fill the pump fluid reservoir to the proper level. The fluid level should be within the "FILL RANGE" listed on the exterior of the reservoir when the fluid is at normal ambient temperature, approximately 70–80°F (21–27°C).

CONTROL LINKS

REMOVAL & INSTALLATION

See Figure 316.

1. Before servicing the vehicle, refer to the Precautions Section.
2. Raise and safely support the vehicle.
3. Remove the front wheel and tire from the front hub.

⁂ WARNING

Be careful not to damage the hub bolts when removing the front wheel and tire.

4. Remove the stabilizer bar control link from the fork.

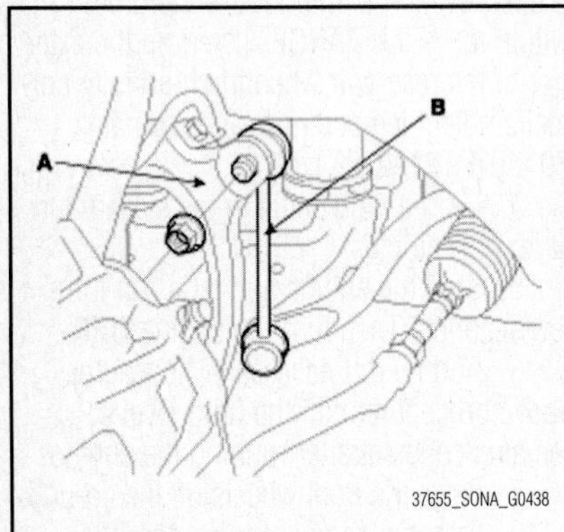

Fig. 316 Remove the stabilizer link (B) from the fork (A)

To install:

5. Install the stabilizer bar control link.
6. Tighten the stabilizer bar control link upper and lower mounting nuts:
    a. 72–87 ft. lbs. (100–120 Nm)
7. Install the front wheels, and lower the vehicle. Tighten the wheel nuts to: 65–80 ft. lbs. (90–110 Nm).

LOWER CONTROL ARM

REMOVAL & INSTALLATION

See Figures 317 through 320.

1. Before servicing the vehicle, refer to the Precautions Section.
2. Raise and safely support the vehicle.
3. Remove the front wheel and tire.
4. Remove the lower arm ball joint mounting bolts.
5. Remove the front lower arm fork mounting bolt from the fork.
6. Remove the lower arm mounting bolts.

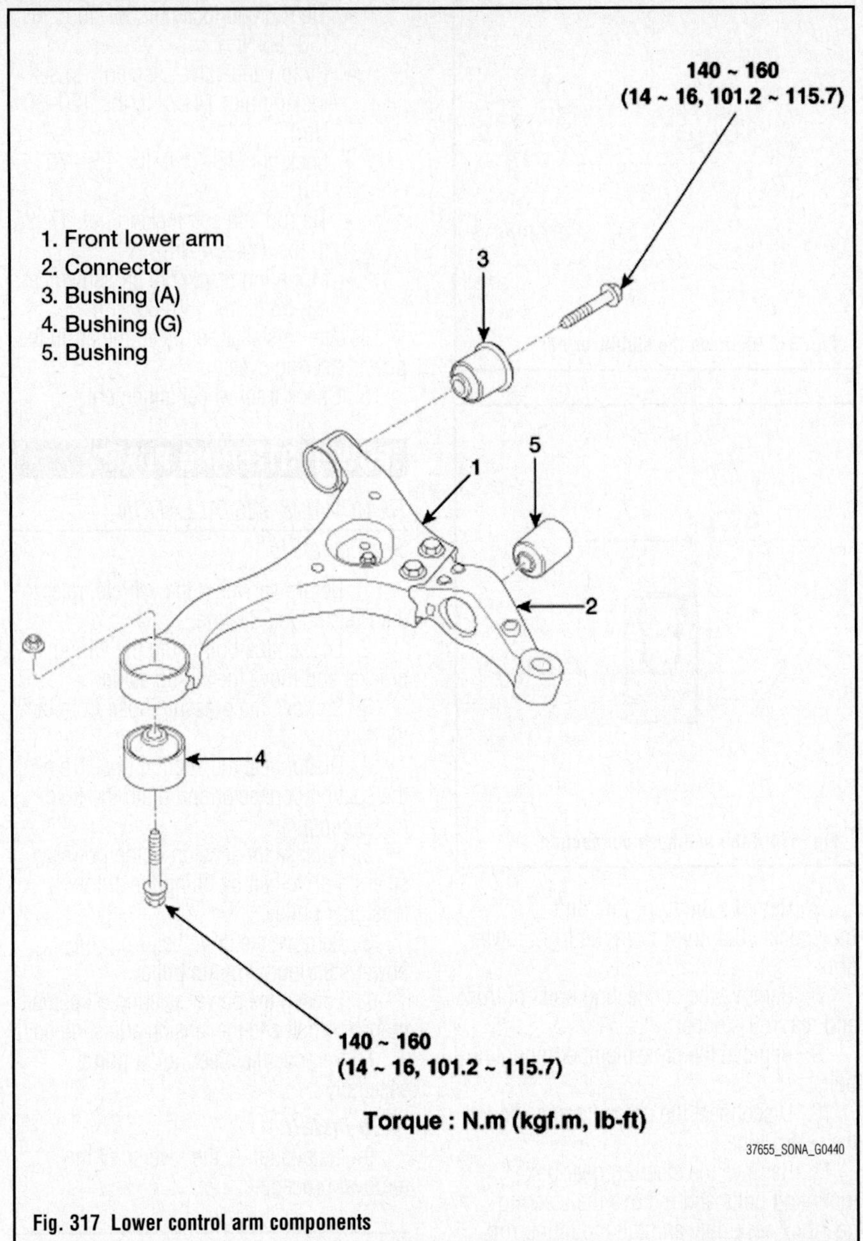

Fig. 317 Lower control arm components

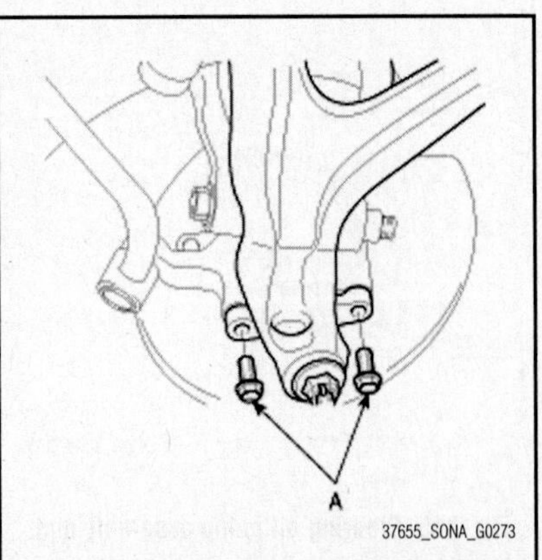

Fig. 318 Remove the lower arm ball joint mounting bolts (A)

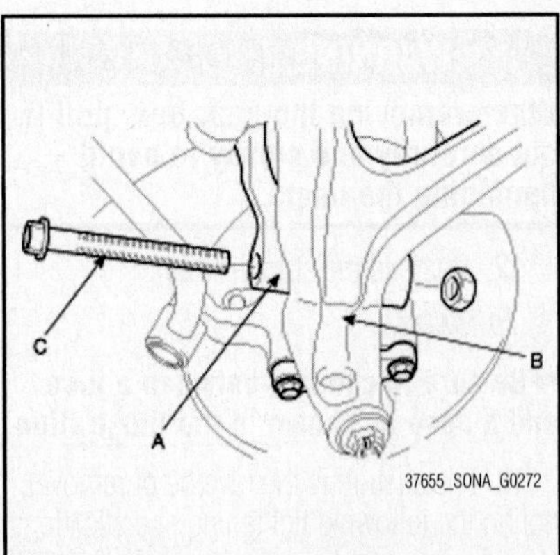

Fig. 319 Remove the front lower arm fork mounting bolt from the fork (A)

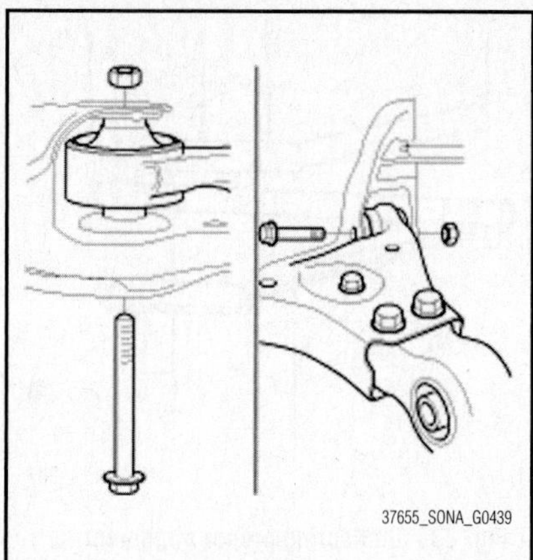

Fig. 320 Remove the lower arm mounting bolts (A, B)

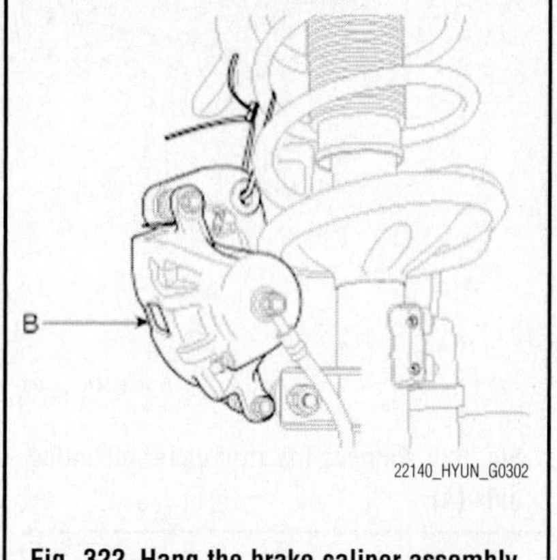

Fig. 322 Hang the brake caliper assembly (B) with a wire

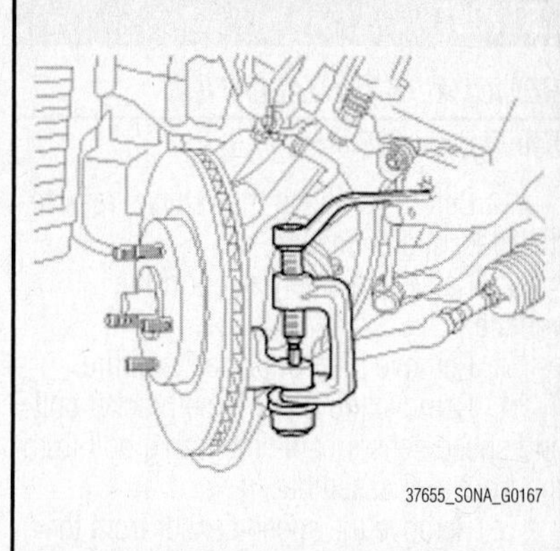

Fig. 324 Using the special tool (09568-4A000) disconnect the tie rod end from the knuckle arm

To install:

7. Install and tighten the lower arm mounting bolts and tighten to 101–116 ft. lbs. (140–160 Nm).
8. Install the lower arm fork mounting bolt to the fork and tighten to 101–116 ft. lbs. (140–160 Nm).
9. Install the lower arm ball joint mounting bolts and tighten to 72–87 ft. lbs. (100–120 Nm).
10. Install the front wheel and tire.
11. Check and/or adjust the wheel alignment.

STEERING KNUCKLE

REMOVAL & INSTALLATION

See Figures 321 through 331.

1. Before servicing the vehicle, refer to the Precautions Section.
2. Raise and safely support the vehicle.
3. Remove the front wheel and tire.

WARNING

Be careful not to damage the hub bolts when removing the front wheel and tire.

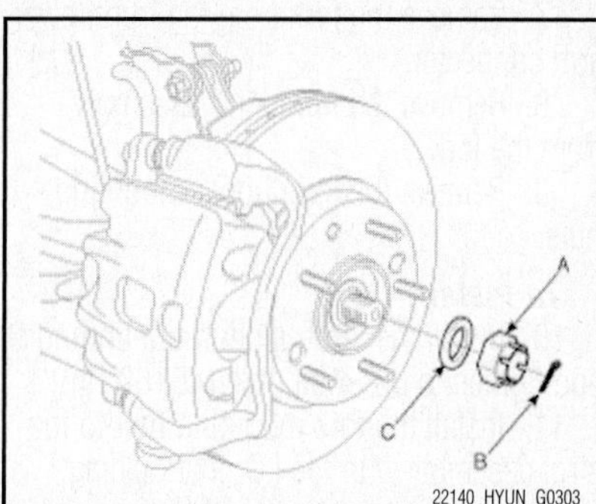

Fig. 323 Remove the split pin (B), the castle nut (A), and the washer (C) from the front hub

4. Remove the wheel speed sensor from the knuckle.
5. Disconnect the brake hose bracket from the knuckle.
6. Remove the brake caliper mounting bolts, and then hang the brake caliper assembly with a wire.

WARNING

Do not suspend the brake caliper assembly from the brake hose or damage may occur to the hose.

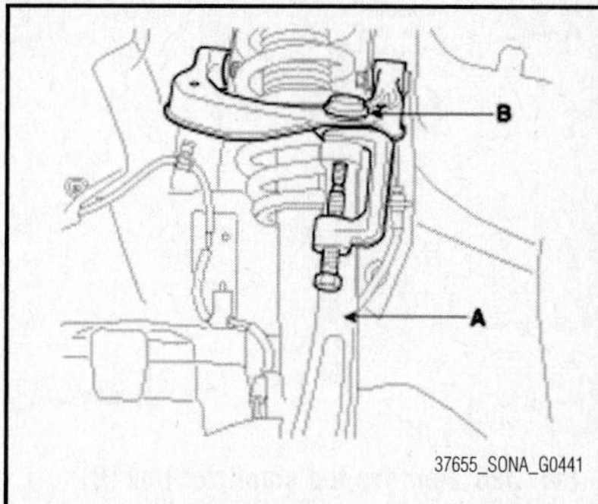

Fig. 325 Using the special tool (09568-4A000), disconnect the knuckle from the upper arm assembly (A)

7. Remove the split pin and castle nut from the front hub.
8. Remove the 2 bolts and disconnect the knuckle from the lower arm assembly.
9. Using a plastic hammer, disconnect the driveshaft from the axle hub.
10. Using the special tool (09568-4A000), disconnect the tie rod end from the knuckle.
11. Using a plastic hammer, disconnect the driveshaft from the axle hub.
12. Loosen the upper arm mounting nut but do not remove it.
13. Using the special tool (09568-4A000), disconnect the knuckle from the upper arm assembly.

To install:

14. Installation is the reverse of removal.
    a. Tighten lower arm to fork bolts to 101–116 ft. lbs. (140–160 Nm).
    b. Tighten lower arm mounting bolts to 101–116 ft. lbs. (140–160 Nm).
    c. Tighten lower ball joint mounting bolts to 72–87 ft. lbs. (100–120 Nm).

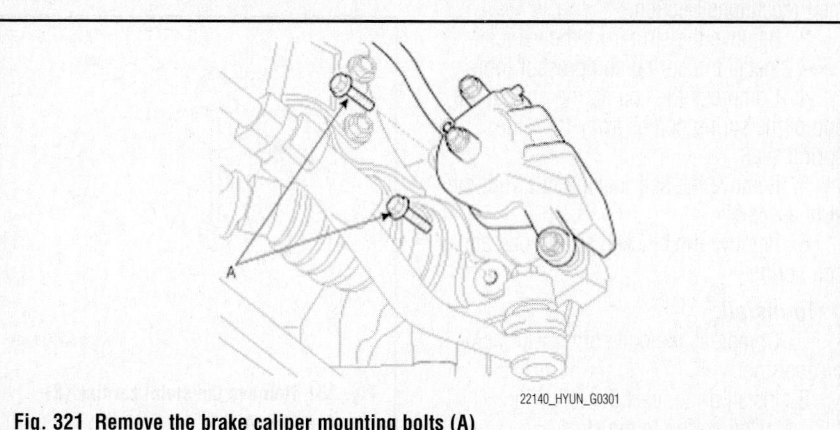

Fig. 321 Remove the brake caliper mounting bolts (A)

STRUT

REMOVAL & INSTALLATION

See Figures 326 through 329.

1. Before servicing the vehicle, refer to the Precautions Section.
2. Raise and safely support the vehicle.
3. Remove the front wheel and tire.
4. Remove the brake hose bracket bolt and speed sensor cable mounting bolt from the front axle assembly.
5. Remove the speed sensor from the knuckle.

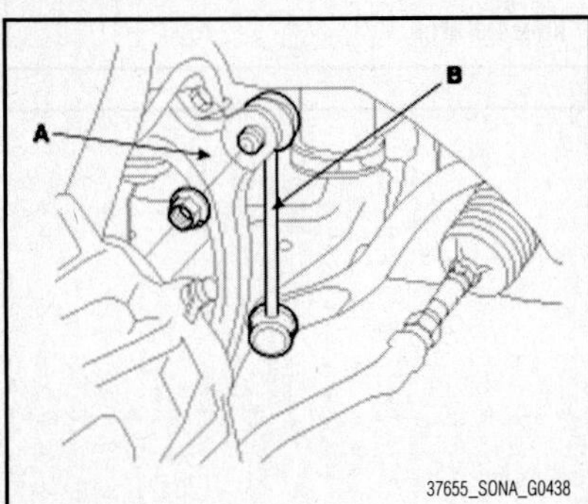

Fig. 326 Remove the stabilizer link (B) from the fork (A)

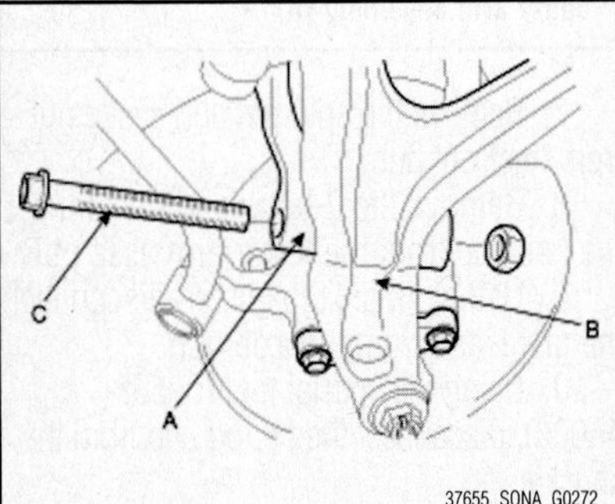

Fig. 327 Remove the fork (A) from the front lower arm connector (B)

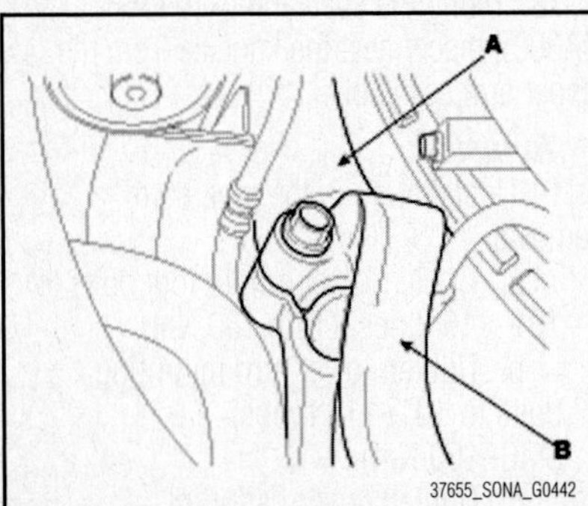

Fig. 328 Remove the front strut assembly (A) from the fork (B)

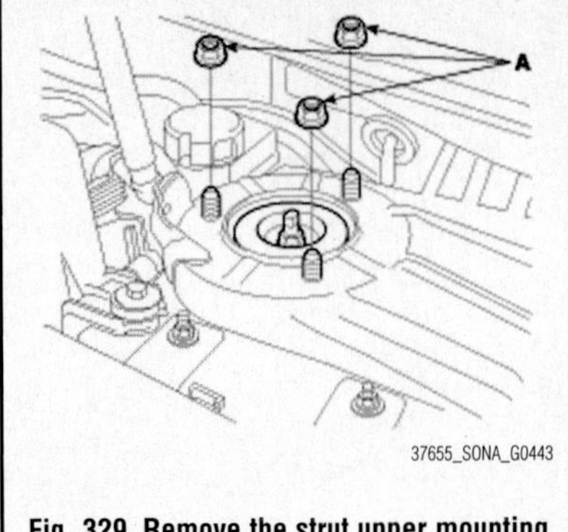

Fig. 329 Remove the strut upper mounting nuts (A)

6. Remove the front stabilizer link nut from the fork.
7. Remove the fork from the front lower arm connector.
8. Remove the front strut assembly from the fork.
9. Remove the strut upper mounting nuts.

To install:

10. Install the strut upper mounting nuts and tighten to 33–43 ft. lbs. (45–60 Nm).
11. Install the fork mounting bolt to the strut assembly with the I.D. mark facing outward and tighten to 43–58 ft. lbs. (60–80 Nm).
12. Install the mounting bolt to the fork and tighten to 101–116 ft. lbs. (140–160 Nm).
13. Install the front stabilizer link nut to the fork tighten to 72–87 ft. lbs. (100–120 Nm).
14. Install the speed sensor bolt.
15. Install the brake hose bracket to the fork and speed sensor cable mounting bolt to the axle assembly.
16. Install the wheel and the tire.

OVERHAUL

See Figure 330.

1. Before servicing the vehicle, refer to the Precautions Section.
2. Remove the strut from the vehicle.
3. Install the spring compressor tool.
4. Compress the coil spring so that the end of the spring comes away from the spring seat.
5. Remove the self-locking nut from the strut assembly.
6. Remove the bracket, spring pad and coil spring.

To install:

7. Compress the coil spring with compressor tool.
8. Install or connect the following:
   a. Coil spring to the strut

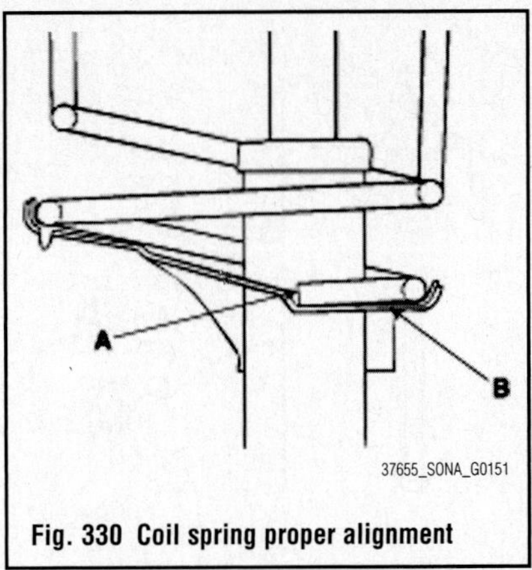

Fig. 330 Coil spring proper alignment

   b. Dust cover, upper spring pad, bushing and hand tighten the lock nut
9. Remove the compressor tool when the coil spring is properly aligned and torque the lock nut to 18 ft. lbs. (25 Nm).
10. Install the strut assembly.

STABILIZER BAR

REMOVAL & INSTALLATION

See Figures 331 through 333.

1. Before servicing the vehicle, refer to the Precautions Section.
2. Raise the front of the vehicle, and make sure it is securely supported.
3. Remove the front wheel and tire from the front hub.

⁂ WARNING

Be careful not to damage the hub bolts when removing the front wheel and tire.

4. Remove the stabilizer link from the fork.
5. Remove the two mounting bolts of the rear side of the subframe, supporting the subframe with a jack.
6. Remove the rear mounting bolts of subframe.

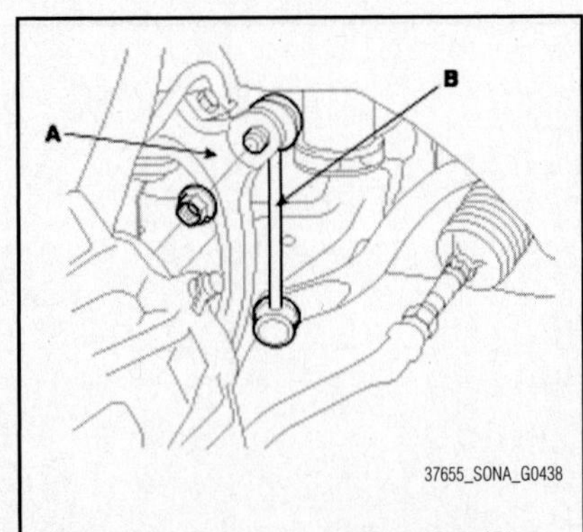

Fig. 331 Remove the stabilizer link (B) from the fork (A)

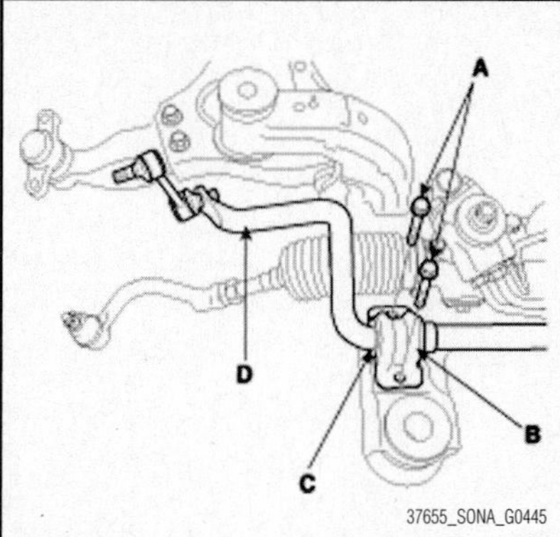

Fig. 332 Remove the mounting bolts (A), stabilizer bracket (B), bushing (C), and stabilizer bar (D)

7. Remove the stabilizer bracket and bushing.
8. Remove the stabilizer bar.

⁂ WARNING

Be careful not to damage the power steering pressure tube.

To install:

9. Installation is the reverse of the removal.
10. Tighten the stabilizer bracket bolts to: 32–39 ft. lbs. (45–55 Nm).
11. Tighten the stabilizer link upper to the fork: 72–87 ft. lbs. (100–120 Nm).
12. Tighten the wheel nuts to: 65–80 ft. lbs. (90–110 Nm).

UPPER CONTROL ARM

REMOVAL & INSTALLATION

See Figures 334 through 337.

1. Before servicing the vehicle, refer to the Precautions Section.
2. Support the lower control arm assembly with a floor jack.
3. Remove the front wheel.
4. Remove the upper arm ball joint self-locking nut and the split pin.
5. Remove the upper arm ball joint from the steering knuckle with Special Tool 09568-34000.
6. Remove the front strut assembly.

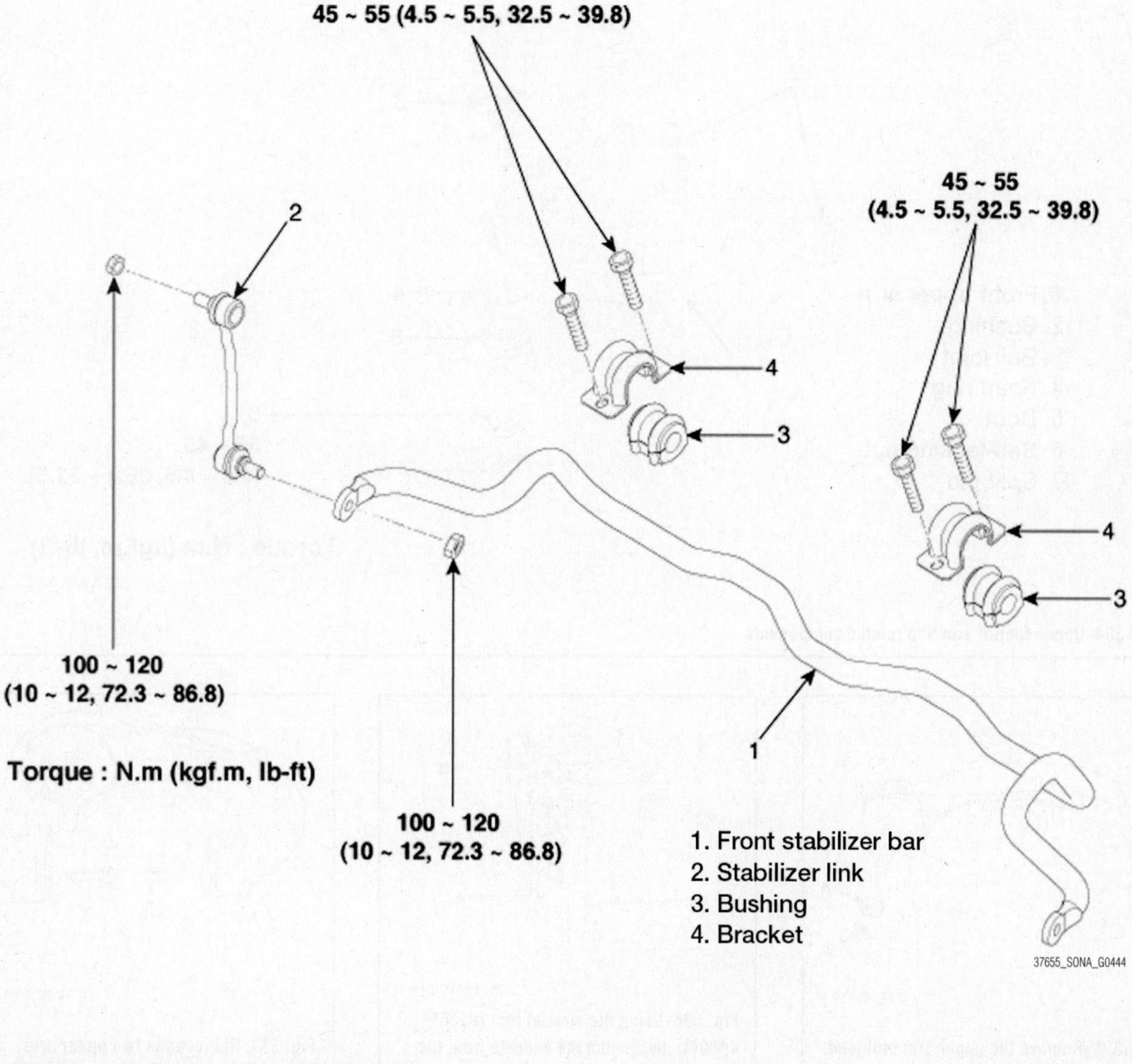

45 ~ 55 (4.5 ~ 5.5, 32.5 ~ 39.8)
45 ~ 55
(4.5 ~ 5.5, 32.5 ~ 39.8)
2
4
3
4
3
100 ~ 120
(10 ~ 12, 72.3 ~ 86.8)
Torque : N.m (kgf.m, lb-ft)
1
100 ~ 120
(10 ~ 12, 72.3 ~ 86.8)
1. Front stabilizer bar
2. Stabilizer link
3. Bushing
4. Bracket
37655_SONA_G0444

Fig. 333 View of stabilizer bar and related components

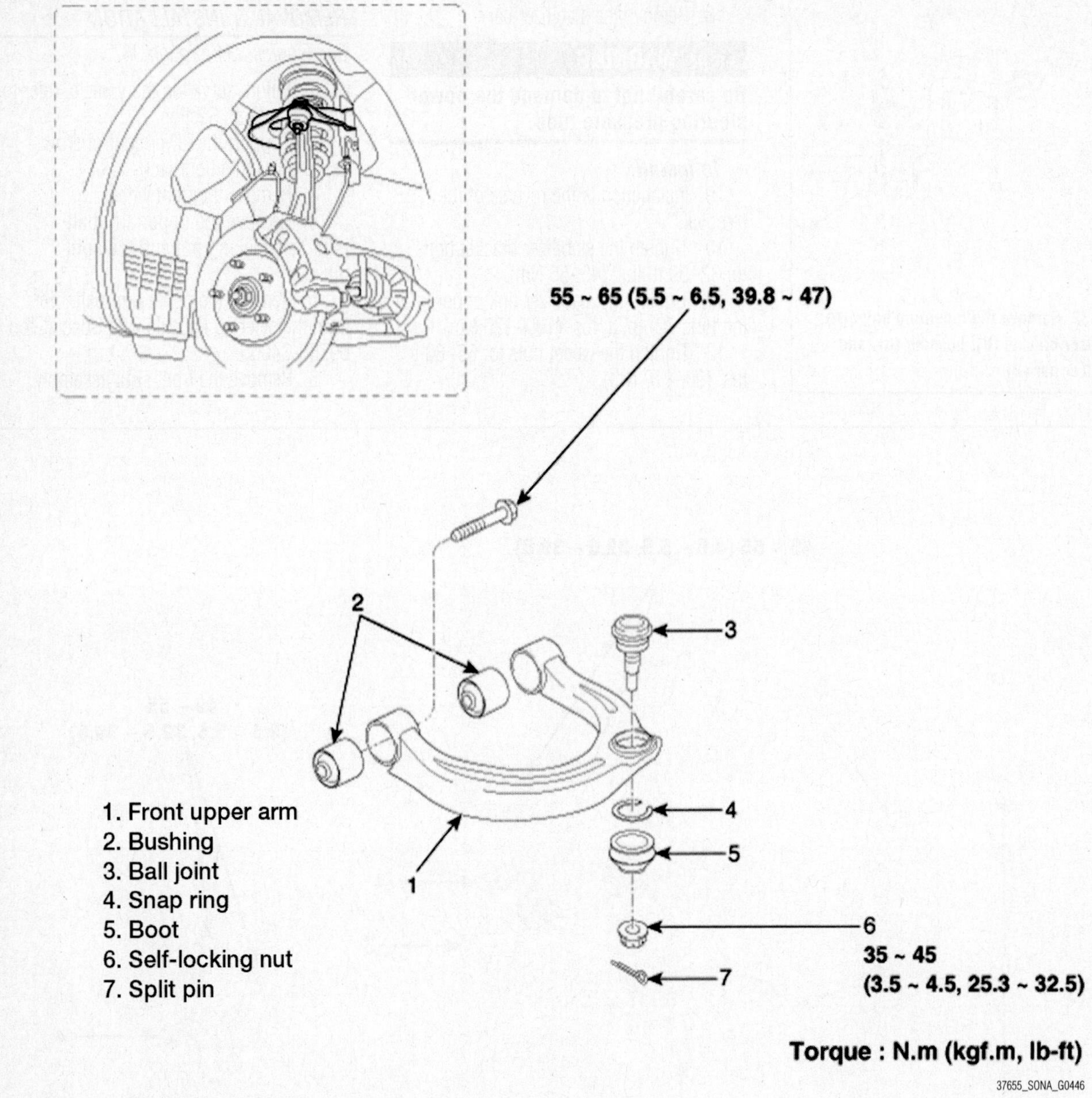

55 ~ 65 (5.5 ~ 6.5, 39.8 ~ 47)
2
3
1
4
5
6
35 ~ 45
(3.5 ~ 4.5, 25.3 ~ 32.5)
7
1. Front upper arm
2. Bushing
3. Ball joint
4. Snap ring
5. Boot
6. Self-locking nut
7. Split pin
Torque : N.m (kgf.m, lb-ft)
37655_SONA_G0446

Fig. 334 Upper control arm and related components

A
B
37655_SONA_G0447

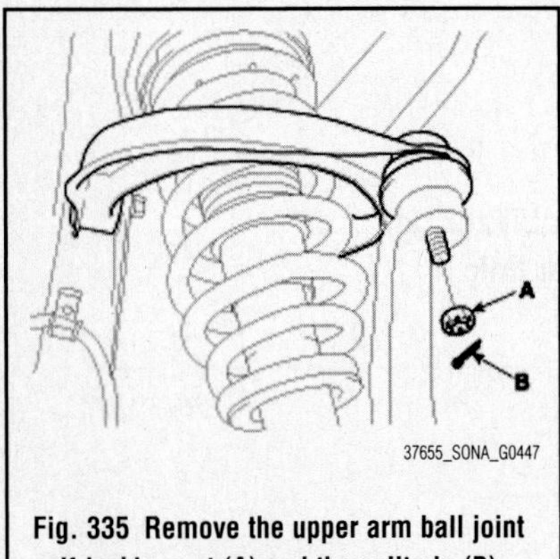

Fig. 335 Remove the upper arm ball joint self-locking nut (A) and the split pin (B)

B
A
37655_SONA_G0441

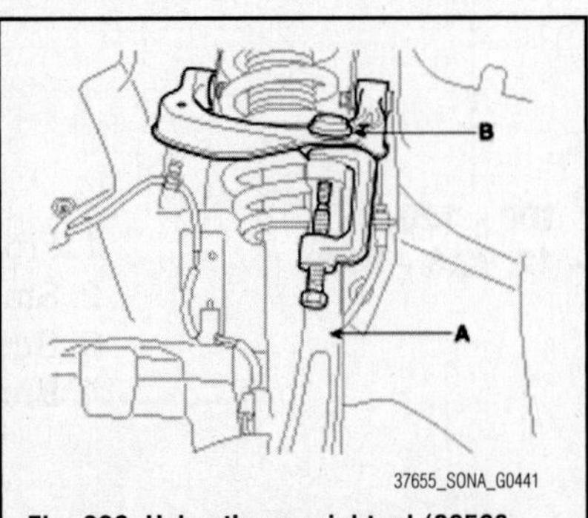

Fig. 336 Using the special tool (09568-4A000), disconnect the knuckle from the upper arm assembly (A)

A
37655_SONA_G0448

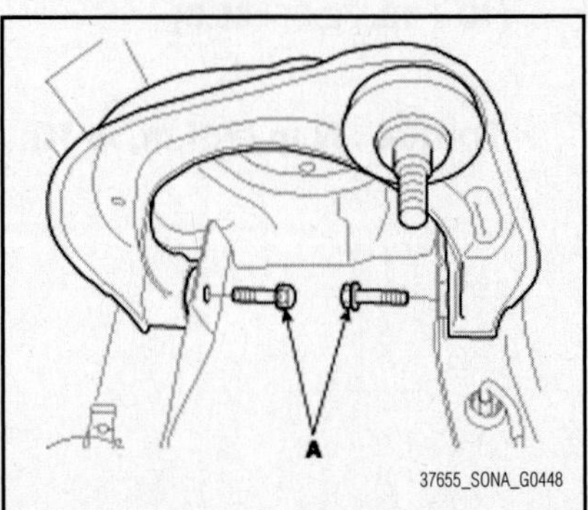

Fig. 337 Remove the two upper arm mounting bolts

7. Remove the two upper arm mounting bolts from the body.

To install:

8. Install the two upper arm mounting bolts to the body torque the bolts to 40–48 ft. lbs. (55–65 Nm).
9. Upper arm ball joint to the steering knuckle and torque the nut to 25–33 ft. lbs. (35–45 Nm).
10. Install a new split pin.
11. Install the front wheel.

WHEEL HUB & BEARING

REMOVAL & INSTALLATION

See Figures 338 through 344.

1. Before servicing the vehicle, refer to the Precautions Section.
2. Remove the steering knuckle. Refer to Steering Knuckle Removal & Installation.
3. Remove the brake disc from the front hub.
4. Remove the snap ring.
5. Using the special tool (09545-34100), disconnect the hub from the knuckle.
6. Using the special tools (09432-11000, 09545-34100), remove the wheel bearing inner race from the hub.
7. Using the special tools (09216-21600, 09216-22100), remove the wheel bearing outer race from the knuckle.

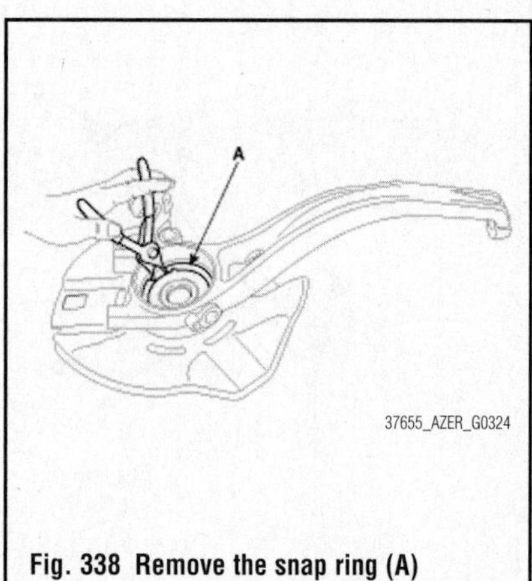

Fig. 338 Remove the snap ring (A)

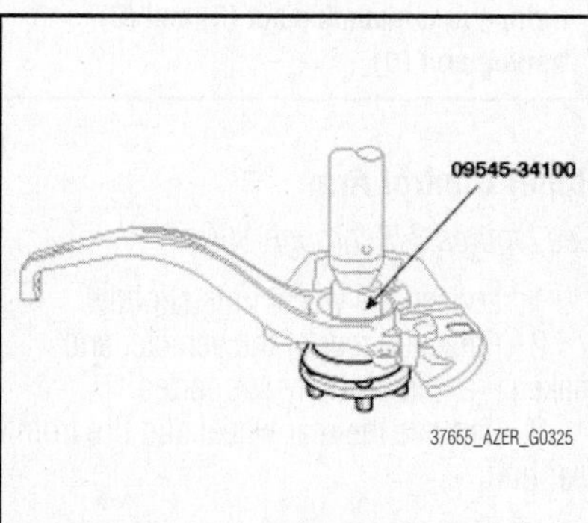

Fig. 339 Disconnect the hub from the knuckle

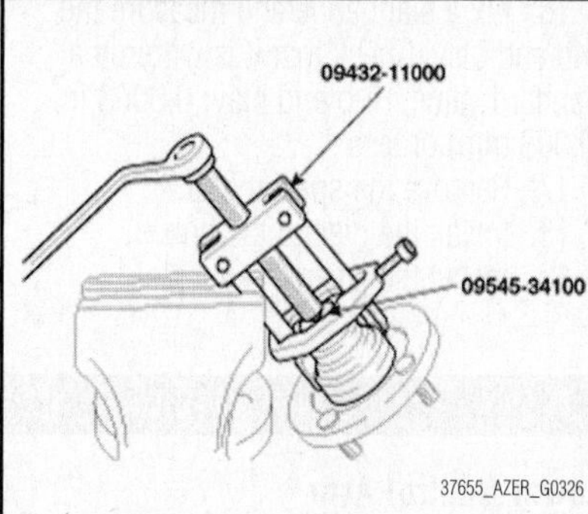

Fig. 340 Remove the wheel bearing inner race from the hub

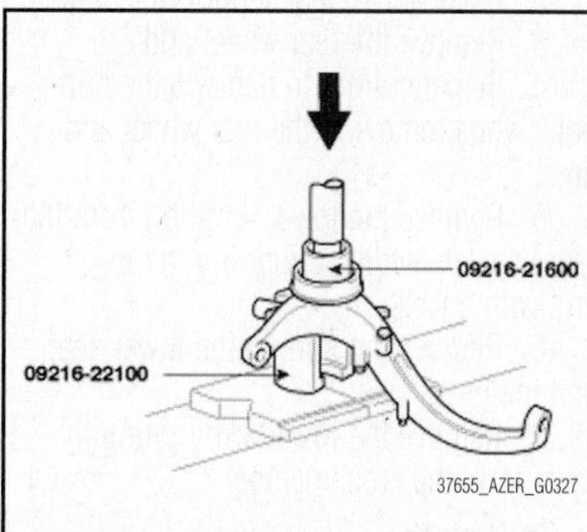

Fig. 341 Remove the wheel bearing outer race from the knuckle

To install:

8. Preparation for installation:
   a. Check the hub for cracks and the splines for wear.
   b. Check the oil seal for cracks or damage.
   c. Check the brake disc for scoring and damage.
   d. Check the steering knuckle for cracks.
   e. Check the bearing for cracks or damage.
9. Apply a thin coat of multi-purpose grease to the knuckle and bearing contact surface.
10. Using the special tool (09216-21100), press-in the bearing to the knuckle.

✲✲ WARNING

Do not press against the inner race of the wheel bearing because that can cause damage to the bearing assembly.

✲✲ WARNING

Always use a new bearing assembly.

11. Install the snap ring into the groove of the knuckle.
12. Using the special tool (09545-21100), press the hub on to the knuckle.

✲✲ WARNING

Do not press against the outer race of the wheel bearing because that can cause damage to the bearing assembly.

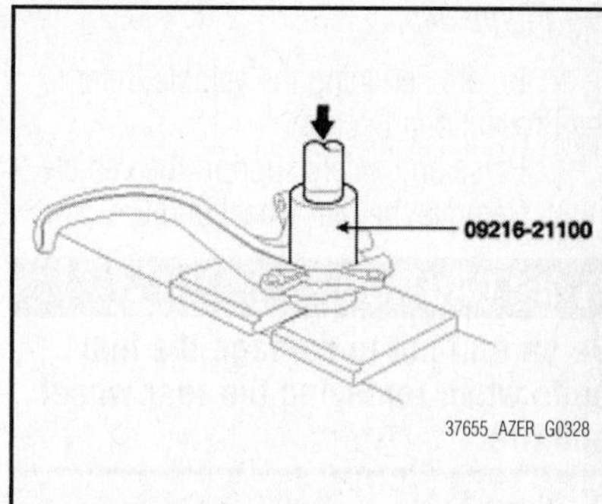

Fig. 342 Using the special tool (09216-21100), press-in the bearing to the knuckle

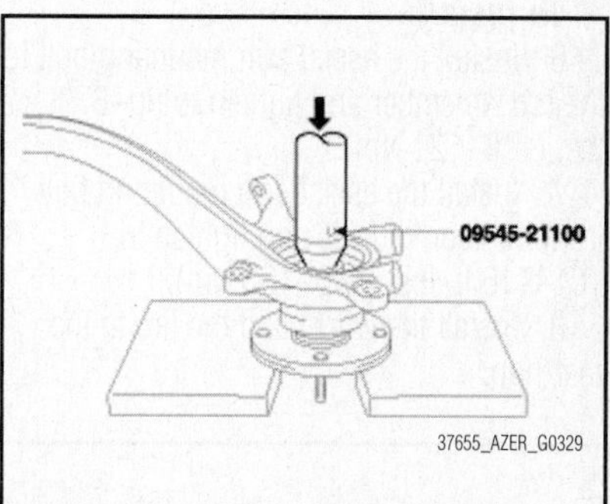

Fig. 343 Using the special tool (09545-21100), press the hub on to the knuckle

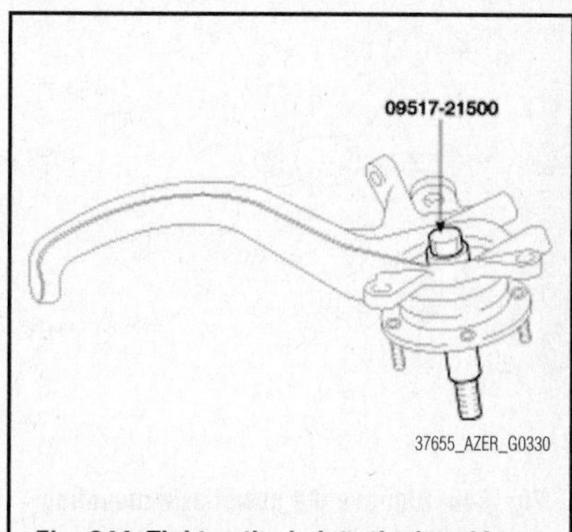

Fig. 344 Tighten the hub to the knuckle with the special tool (09517-21500)

13. Tighten the hub to the knuckle to 148 ft. lbs. (200 Nm) with the special tool (09517-21500).

14. Rotate the hub to seat the bearing.

15. Measure the wheel bearing starting torque: 16 ft. lbs. (2 Nm) or less.

16. Fix a dial gauge and measure the hub end play. Check that it is within the standard value. Hub end play: 0.0003 in. (0.008 mm) or less

17. Remove the special tool.

18. Install the disc to the hub.

19. Install the steering knuckle.

20. Check and/or adjust the wheel alignment.

ADJUSTMENT

The front wheel bearing is a sealed unit and is not adjustable.

SUSPENSION

REAR SUSPENSION

CONTROL ARMS/LINKS

REMOVAL & INSTALLATION

Rear Assist Arm

See Figure 345.

1. Before servicing the vehicle, refer to the Precautions Section.
2. Raise and safely support the vehicle.
3. Remove the rear wheel and tire.

CAUTION

Be careful not to damage the hub bolts when removing the rear wheel and tire.

4. Remove the assist arm mounting bolt from the rear axle.
5. Remove the assist arm mounting bolt from the crossmember.

To install:

6. Install the assist arm mounting bolt to the crossmember and tighten to 80–87 ft. lbs. (110–120 Nm).
7. Install the assist arm mounting bolt from the rear knuckle and tighten to 101–116 ft. lbs. (140–160 Nm).
8. Install the wheel and the tire to the rear hub.

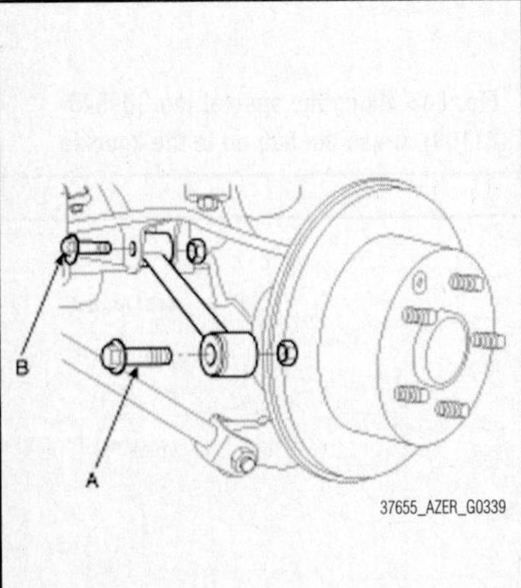

Fig. 345 Remove the assist arm mounting bolts (A, B) from the rear axle and the cross member

Lower Control Arm

See Figure 346.

1. Before servicing the vehicle, refer to the Precautions Section.
2. Raise and safely support the vehicle.
3. Remove the rear wheel and tire.
4. Be careful not to damage the hub bolts when removing the rear wheel and tire.
5. Remove the lower arm bolt from the rear knuckle, while supporting the lower arm with a jack.
6. Remove the spring, the lower seat, and the upper pad.
7. Remove the lower arm mounting bolts from the crossmember.

To install:

8. Install the lower arm mounting bolts to the cross member and tighten to 80–87 ft. lbs. (110–120 Nm).
9. Install the spring, the lower seat, and the upper pad.
10. Install the lower arm bolt to the rear knuckle and tighten to 101–116 ft. lbs. (140–160 Nm), while supporting the lower arm with a jack.
11. Install the wheel and the tire to the rear hub.

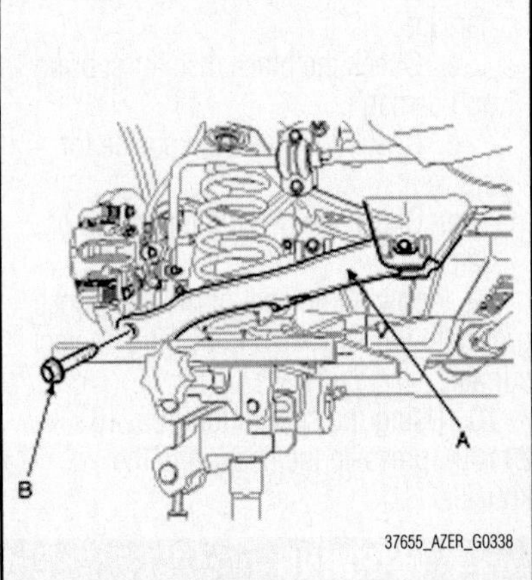

Fig. 346 Remove the lower arm bolt (B) from the rear knuckle, while supporting the lower arm (A)

Trailing Arm

See Figure 347.

1. Before servicing the vehicle, refer to the Precautions Section.
2. Remove the wheel speed sensor bracket.
3. Remove the trailing arm mounting nut from the rear axle assembly.
4. Remove the trailing arm mounting nut from the body.
5. Remove the trailing arm.

To install:

6. Installation is reverse of removal.
   a. Tighten trailing arm mounting nuts to 101–116 ft. lbs. (140–160 Nm).

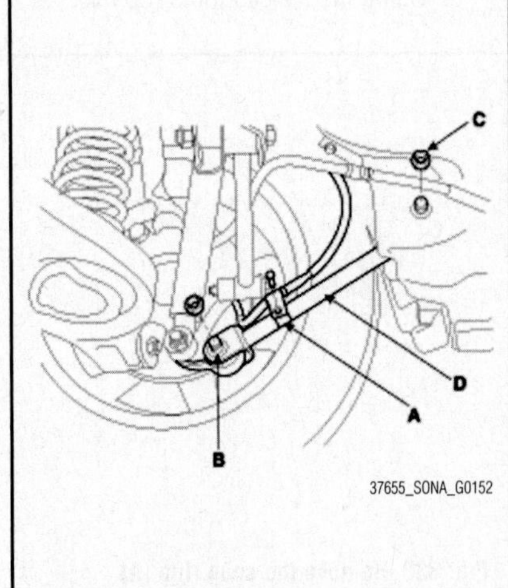

Fig. 347 Wheel speed sensor bracket (A), trailing arm mounting nut (B) and (C), trailing arm (D)

Upper Control Arm

See Figures 348 through 350.

1. Loosen the wheel nuts slightly.
2. Raise the rear of the vehicle, and make sure it is securely supported.
3. Remove the rear wheel and tire from rear hub.

➡Be careful not to damage the hub bolts when removing the rear wheel and tire.

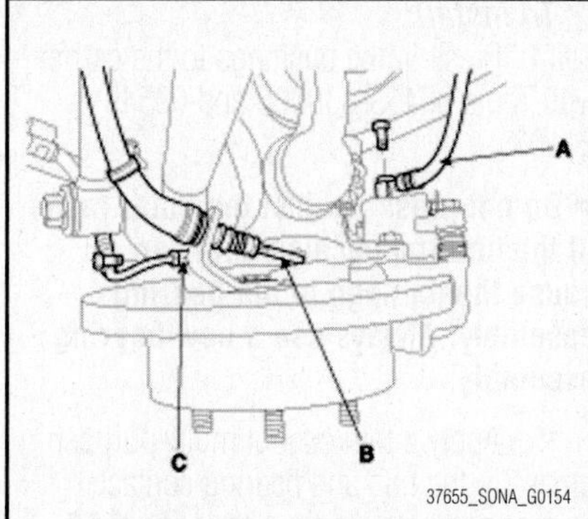

Fig. 348 Remove the brake hose (A), the parking brake cable (B) and the wheel speed sensor (C)

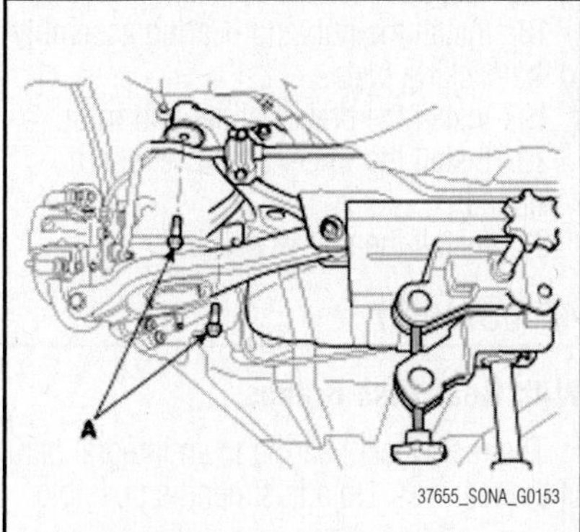

Fig. 349 Remove the four rear cross member mounting bolts (A)

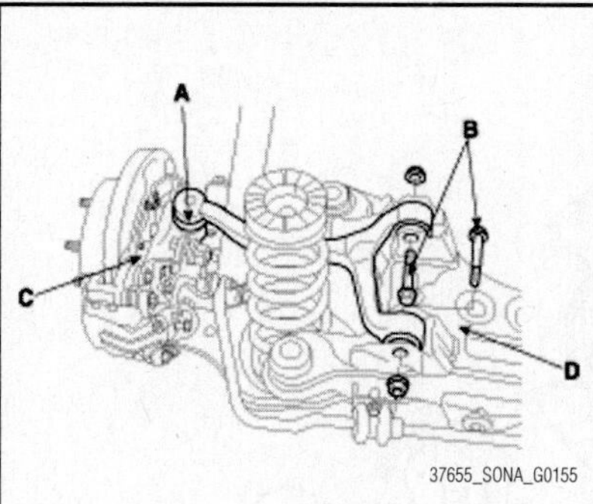

Fig. 350 Upper arm ball joint (A), rear axle assembly (C), upper arm mounting bolts (B) and the cross member (D)

4. Remove the brake hose, the parking brake cable and the wheel speed sensor.
5. Remove the trailing arm and the shock absorber from the rear axle assembly, supporting with a jack.
6. Remove the muffler.
7. Remove the four rear crossmember mounting bolts while supporting with a jack.
8. Remove the split pin and the upper arm ball joint self-locking nut.
9. Remove the upper arm ball joint from the rear axle assembly with special tool (09568-4A000).
10. Remove the two upper arm mounting bolts from the cross member.

To install:

11. Install the upper arm to the cross member with two mounting bolts and tighten to 73–87 ft. lbs. (100–120 Nm).
12. Install the upper arm ball joint self-locking nut to the rear axle assembly and tighten to 58–65 ft. lbs. (80–90 Nm) and install the split pin.
13. Install the cross member and tighten to 101–116 ft. lbs. (140–160 Nm).
14. Install the spring, the lower seat, and the upper pad.
15. Install the lower arm bolt to the axle assembly, while supporting with a jack.
16. Install the trailing arm and the shock absorber to the rear axle assembly and tighten to 101–116 ft. lbs. (140–160 Nm).
17. Install the brake hose, the parking brake cable and the wheel speed sensor.
18. Install the muffler.
19. Install the wheel and the tire to the rear hub.

SHOCK ABSORBER

REMOVAL & INSTALLATION

See Figures 351 and 352.

1. Before servicing the vehicle, refer to the Precautions Section.
2. Support the vehicle under the lower control arm.
3. Remove the rear wheel.
4. Remove the two rear shock absorber assembly mounting bolts.
5. Remove the rear shock absorber assembly nut from the rear knuckle, then remove the shock absorber assembly.
6. Disassemble the rubber bumper and the dust cover from the rear shock absorber.

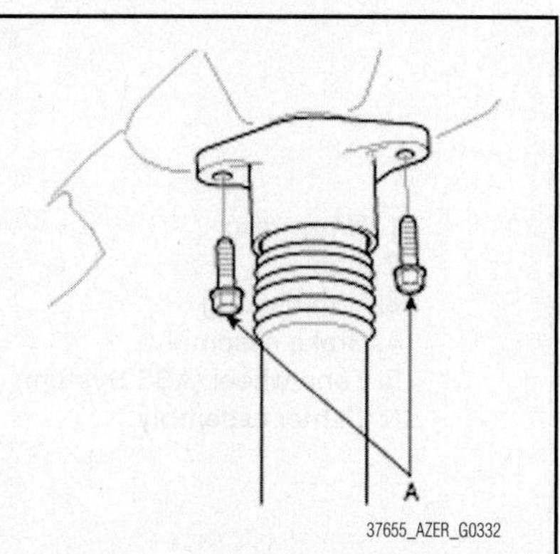

Fig. 351 Remove the two rear shock absorber assembly mounting bolts (A)

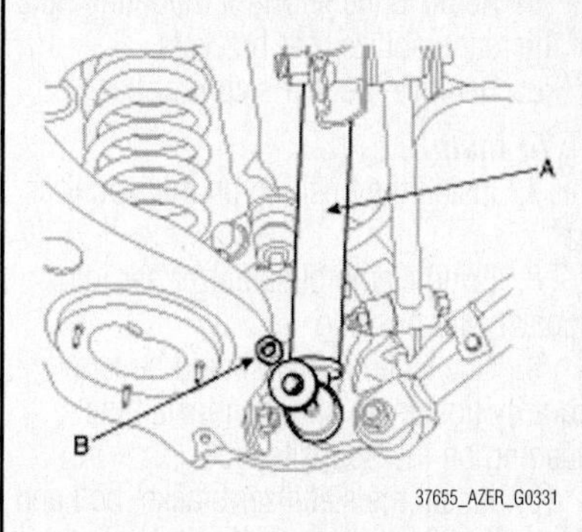

Fig. 352 Remove the rear shock absorber assembly nut (B) from the rear knuckle, then remove the shock absorber assembly (A)

To install:

7. Assembly the rubber bumper and the dust cover to the rear shock absorber, after pulling the rod of the rear shock absorber completely.
8. Install the shock absorber. Tighten the upper bolts to 36–47 ft. lbs. (50–65 Nm). 9. Tighten the lower nut to 101–116 ft. lbs. (140–160 Nm)
10. Install the rear wheel.

STABILIZER BAR

REMOVAL & INSTALLATION

See Figures 353.

1. Loosen the wheel nuts slightly.
2. Raise the rear of the vehicle, and make sure it is securely supported.
3. Remove the rear wheel and tire from rear hub.

➡Be careful not to damage the hub bolts when removing the rear wheel and tire.

4. Remove the left/right nuts of the rear stabilizer links.

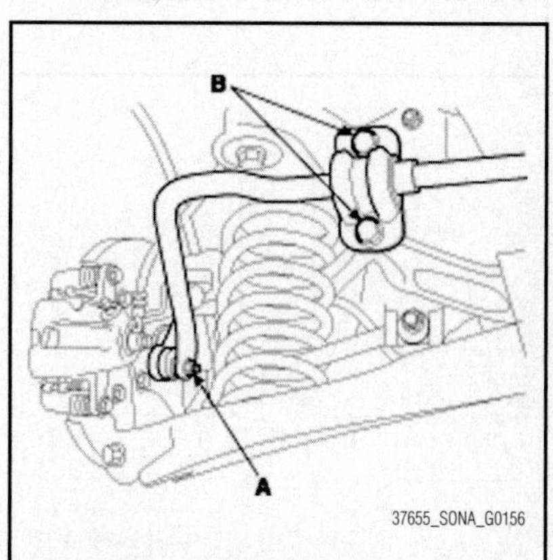

Fig. 353 Rear stabilizer links (A), rear stabilizer bar brackets bolts (B)

5. Remove the left/right mounting nuts of the rear stabilizer bar brackets.
6. Remove the rear stabilizer bar.

To install:

7. Install the bushing on the stabilizer bar.
8. Bring clamp of stabilizer bar into contact with bushing.
9. One side bracket should be temporarily tightened, and then install the bushing on the opposite side.
10. Install the stabilizer bracket bolt and tighten to 32–40 ft. lbs. (45–55 Nm).
11. Install the stabilizer link mounting nut and tighten to 25–32 ft. lbs. (35–45 Nm).
12. Repeat step 4 and 5 for the other side.
13. Install the wheel and the tire to the rear hub.

WHEEL HUB & BEARING

REMOVAL & INSTALLATION

See Figures 354 through 357.

➡Special tools 09453-33000B and 09545-21100 press adapters, are used in this procedure.

1. Before servicing the vehicle, refer to the Precautions Section.

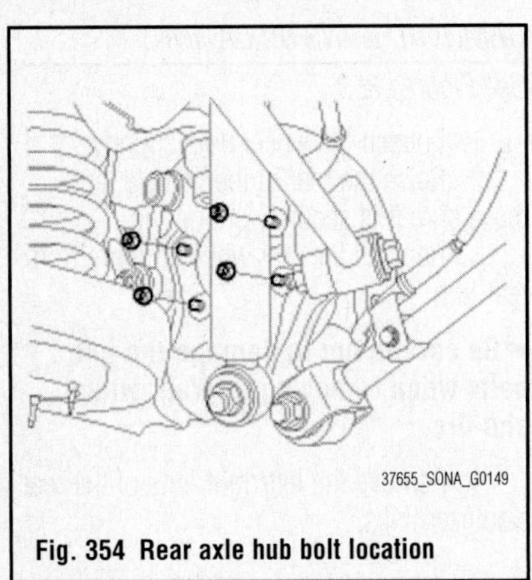

Fig. 354 Rear axle hub bolt location

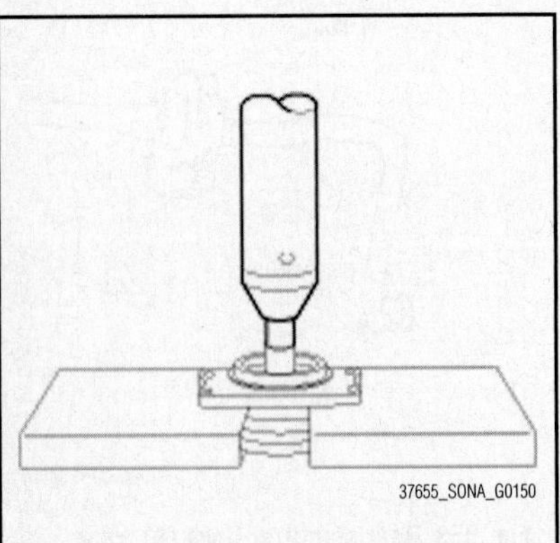

Fig. 355 Press out the rear axle hub

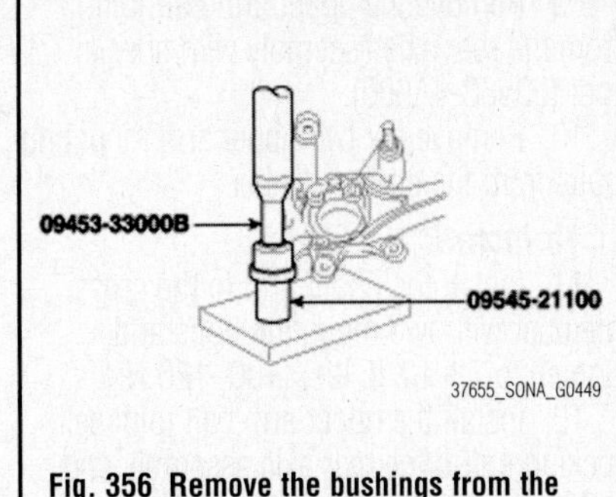

Fig. 356 Remove the bushings from the carrier

2. Release the parking brake.
3. Remove or disconnect the following:
4. Remove the rear wheel.
5. Remove the wheel speed sensor, if equipped.
6. Remove the brake caliper and rotor.
7. Remove the rear axle hub bolts.
8. Remove the tone wheel using puller.
9. Remove the carrier assembly.
10. Remove the nut, after un-staking it.
11. Press out the rear axle hub.
12. Remove the bearing inner race using puller.
13. Remove the bushings from the carrier with Tools 09453-33000B and 09545-21100.

To install:

14. Press in the bushings to the carrier with Tools 09453-33000B and 09545-21100.

➡Do not press against the outer race of the bearing because that can cause the damage to the bearing assembly. Always use a new bearing assembly.

15. Apply a thin coat of multi-purpose grease to the hub and bearing contact surface and press in the bearing to the hub.
16. Tighten the flange nut and stake the nut to meet the concave portion of the spindle.
17. Press in the tone wheel.
18. Install the hub and bearing assembly to the backing plate.
19. Install the brake caliper and rotor.
20. Install the wheel speed sensor, if equipped.
21. Install the rear wheel.

ADJUSTMENT

With Rear Disc Brakes

The rear wheel bearing is an integral part of the rear hub. No adjustment is possible.

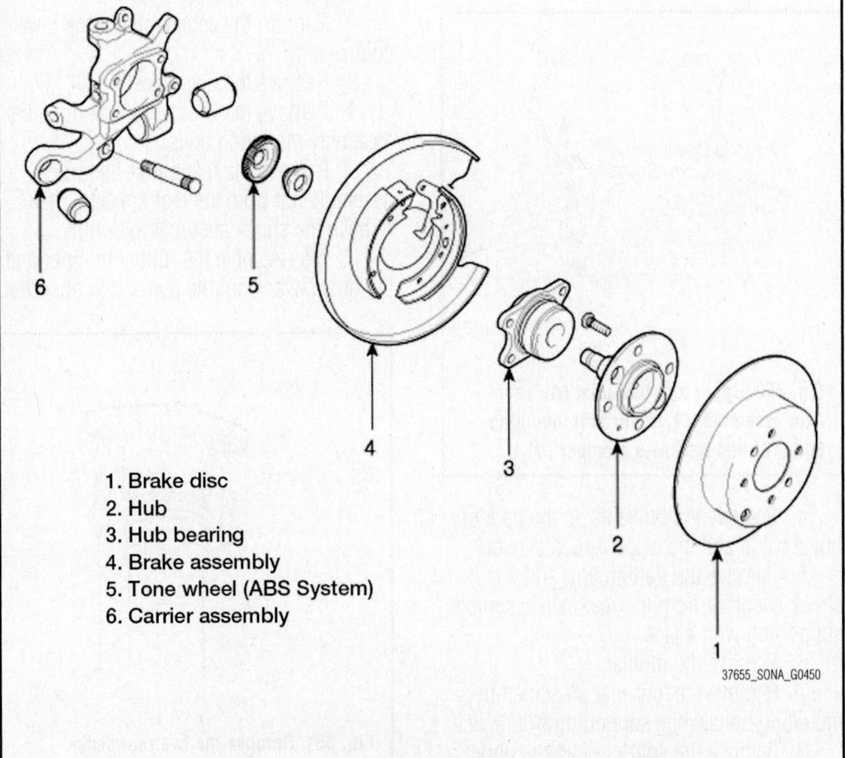

Fig. 357 Exploded view of the rear wheel bearing assembly—with disc brakes

HYUNDAI

Tucson

8

SPECIFICATIONS AND MAINTENANCE CHARTS

ENGINE AND VEHICLE IDENTIFICATION

Engine							Model Year	
Code ①	Liters (cc)	Cu. In.	Cyl.	Fuel Sys.	Engine Type	Eng. Mfg.	Code ②	Year
B	2.0 (1975)	120.52	I4	MPFI	DOHC	Hyundai	9	2009
D	2.7 (2656)	164.30	V6	MPFI	DOHC	Hyundai	A	2010
C	2.4 (2359)	143.90	I4	MPFI	DOHC	Hyundai		

MPFI: Multi-Point Fuel Injection

DOHC: Double Overhead Camshafts

① 8th digit of VIN

② 10th digit of VIN

37655_TUCS_C0001

GENERAL ENGINE SPECIFICATIONS

All measurements are given in inches.

Year	Model	Engine Displacement Liters	Engine Series VIN	Net Horsepower @ rpm	Net Torque @ rpm (ft. lbs.)	Bore x Stroke (in.)	Com-pression Ratio	Oil Pressure @ rpm
2009	Tucson	2.0	B	140@6000	136@4500	3.23 x 3.68	10.1:1	36@1500
	Tucson	2.7	D	173@6000	178@4000	3.41 x 2.95	10.0:1	NS
2010	Tucson	2.4	C	176@6000	168@4000	3.46 x 3.89	10.5:1	22@1500

NA: Not Supplied

37655_TUCS_C0002

GASOLINE ENGINE TUNE-UP SPECIFICATIONS

Year	Engine Displacement Liters	Engine VIN	Spark Plug Gap (in.)	Ignition Timing (deg.)		Fuel Pump (psi)	Idle Speed (rpm)		Valve Clearance (in.)	
				MT	AT		MT	AT	In.	Ex.
2009	2.0	B	0.039-0.043	①	①	49.8	②	②	0.0079	0.0110
	2.7	D	0.039-0.043	①	①	49.8	②	②	HYD	HYD
2010	2.4	C	0.039-0.043	①	①	46.9-52.6	②	②	0.0067-0.0090	0.0106-0.0129

NOTE: The Vehicle Emission Control Information label reflects specification changes made during production.

Follow the figures on the label if they differ from those in this chart.

HYD: Hydraulic

① Ignition timing is preset and cannot be adjusted

② Idle speed is maintained by the Electronic Control Module (ECM)

37655_TUCS_C0003

CAPACITIES

Year	Model	Engine Displacement Liters	Engine VIN	Engine Oil with Filter (qts.)	Transmission (pts.) Manual	Transmission (pts.) Auto. ①	Fuel Tank (gal.)	Cooling System (qts.)
2009	Tucson	2.0	B	4.2	4.5	16.4	17.7	6.4
	Tucson	2.7	D	4.8	—	16.4	17.7	8.2
2010	Tucson	2.4	C	4.9	3.8	15.0	14.5	7.2

NOTE: All capacities are approximate. Add fluid gradually and check to be sure a proper fluid level is obtained.

① Drain and refill

37655_TUCS_C0004

FLUID SPECIFICATIONS

Year	Model	Engine Displacement Liters	Engine ID/VIN	Engine Oil	Manual Trans.	Auto. Trans.	Drive Axle	Power Steering Fluid	Brake Master Cylinder	Cooling System
2009	Tucson	2.0	B	5W-20	①	②	③	PSF-3	④	⑤
	Tucson	2.7	D	5W-20	—	②	③	PSF-3	④	⑤
2010	Tucson	2.4	C	5W-20	①	⑥	③	NA	④	⑤

NA: Not Applicable

DOT: Department Of Transportation

① HYUNDAI GENUINE PART MTF 75W/90 (API GL-4)

② DIAMOND ATF SP-3, SK ATF SP-3

③ Hypoid gear oil (API GL-5, SAE 80W/90, SHELL SPIRAX AX or equivalent)

④ DOT 3, DOT 4, or equivalent

⑤ Ethylene glycol base for aluminum radiator

⑥ HYUNDAI GENUINE ATF SP-IV

37655_TUCS_C0005

VALVE SPECIFICATIONS

Year	Engine Displacement Liters	Engine VIN	Seat Angle (deg.)	Face Angle (deg.)	Spring Test Pressure (lbs. @ in.)	Spring Installed Height (in.)	Stem-to-Guide Clearance (in.) Intake	Stem-to-Guide Clearance (in.) Exhaust	Stem Diameter (in.) Intake	Stem Diameter (in.) Exhaust
2009	2.0	B	45	45	41.5 @1.535	1.535	0.0008-0.0019	0.0014-0.0026	0.2348-0.2354	0.2343-0.2348
	2.7	D	NA	45.0-45.5	48.4 @1.378	NS	0.0008-0.0020	0.0012-0.0026	0.2350-0.2354	0.2340-0.2350
2010	2.4	C	44.75-45.10	45.25-45.75	40.56-43.20 @1.3779	NS	0.0008-0.0018	0.0012-0.0021	0.2151-0.2157	0.2149-0.2153

NA: Not Supplied

37655_TUCS_C0006

CAMSHAFT AND BEARING SPECIFICATIONS CHART

All measurements are given in inches.

Year	Engine Displ. Liters	Engine ID/VIN	Journal Dia.	Brg. Oil Clearance	Shaft End-play	Runout	Journal Bore	Lobe Height Intake	Lobe Height Exhaust
2009	2.0	B	1.1023	0.0008-0.0024	0.0040-0.0080	NS	NS	1.7527-1.7566	1.7487-1.7527
	2.7	D	1.0222-1.0228	0.0007-0.0024	0.0039-0.0059	NS	NS	1.7303-1.7382	1.7303-1.7382
2010	2.4	C	①	②	0.0015-0.0062	NS	NS	1.7401	1.7716

NS: Not Supplied

① Intake journal No. 1: 1.811 inches

Intake journal No. 2,3,4,5: 0.9449 inches

Exhaust journal No. 1: 1.4173 inches

Exhaust journal No. 2,3,4,5: 0.9449 inches

② Intake journal No. 1: 0.0008-0.0022 inches

Intake journal No. 2,3,4,5: 0.0017- 0.0032 inches

Exhaust journal No. 1: 0.0012-0.0032 inches

Exhaust journal No. 2,3,4,5: 0.0017-0.0032 inches

37655_TUCS_C0007

CRANKSHAFT AND CONNECTING ROD SPECIFICATIONS

All measurements are given in inches.

Year	Engine Displacement Liters	Engine VIN	Crankshaft Main Brg. Journal Dia.	Crankshaft Main Brg. Oil Clearance	Crankshaft Shaft End-play	Crankshaft Thrust on No.	Connecting Rod Journal Diameter	Connecting Rod Oil Clearance	Connecting Rod Side Clearance
2009	2.0	B	2.2440	0.0011-0.0018	0.0023-0.0100	3	1.7700	0.0009-0.0016	0.0039-0.0100
	2.7	D	2.4402-2.4409	0.0002-0.0009	0.0028-0.0098	3	1.8891-1.8898	0.0007-0.0014	0.0039-0.0098
2010	2.4	C	2.0449	0.0007-0.0014	0.0027-0.0098	3	1.8879-1.8886	0.0012-0.0017	0.0039-0.0100

37655_TUCS_C0008

PISTON AND RING SPECIFICATIONS

All measurements are given in inches.

Year	Engine Displ. Liters	Engine VIN	Piston Clearance	Ring Gap Top Compression	Ring Gap Bottom Compression	Ring Gap Oil Control	Ring Side Clearance Top Compression	Ring Side Clearance Bottom Compression	Ring Side Clearance Oil Control
2009	2.0	B	0.0008-0.0016	0.0090-0.0149	0.0130-0.0189	0.0078-0.0236	0.0015-0.0031	0.0012-0.0027	NS
	2.7	D	0.0004-0.0012	0.0079-0.0138	0.0146-0.0205	0.0079-0.0276	0.0016-0.0031	0.0012-0.0028	NS
2010	2.4	C	0.0005-0.0013	0.0059-0.0118	0.0145-0.0204	0.0078-0.0275	0.0019-0.0031	0.0015-0.0031	0.0023-0.0059

NA: Not Supplied

37655_TUCS_C0009

TORQUE SPECIFICATIONS

All readings in ft. lbs.

Year	Engine Displacement Liters	Engine VIN	Cylinder Head Bolts	Main Bearing Bolts	Rod Bearing Bolts	Crankshaft Damper Bolts	Flywheel Bolts	Manifold Intake	Manifold Exhaust	Spark Plugs	Oil Pan Drain Plug
2009	2.0	B	①	②	37-39	125-133	88-95	12-17	32-40	15-22	30-33
	2.7	D	③	④	⑤	130-138	53-56	14-15	22-26	15-22	25-33
2010	2.4	C	⑥	⑦	⑧	123-130	87-94	14-17	36-40	15-22	25-33

NOTE: Dip main bearing bolts and crankshaft damper bolt in clean engine oil prior to tightening.

① Step 1: M10 bolts to 18 ft. lbs.; M12 bolts to 22 ft. lbs.
Step 2: Plus 60-65 degrees
Step 3: Plus 60-65 degrees

② Step 1: 20-24 ft. lbs.
Step 2: Plus 60-65 degrees

③ Step 1: 18 ft. lbs.
Step 2: Plus 58-62 degrees
Step 3: Plus 43-47 degrees

④ Step 1: M10 bolts to 20-24 ft. lbs.; M8 bolts to 10-14 ft. lbs.
Step 2: Plus 90-94 degrees

⑤ Step 1: 12-15 ft. lbs.
Step 2: Plus 90-94 degrees

⑥ Step 1: 24-27 ft. lbs.
Step 2: Plus 90-95 degrees
Step 3: Plus 90-95 degrees

⑦ Step 1: 11 ft. lbs.
Step 2: 20-23 ft. lbs.
Step 3: Plus 120-125 degrees

⑧ Step 1: 13-16 ft. lbs.
Step 2: Plus 88-92 degrees

37655_TUCS_C0010

WHEEL ALIGNMENT

Year	Model		Caster Range (+/-Deg.)	Caster Preferred Setting (Deg.)	Camber Range (+/-Deg.)	Camber Preferred Setting (Deg.)	Toe-in (Deg.)
2009	Tucson	Front	①	②	①	0.00	0 +/- 0.08
		Rear	—	—	①	③	④
2010	Tucson	Front	0.50	4.02	0.50	0.50	0 +/- 0.10
		Rear	—	—	0.50	-1.0	0.2 +/- 0.10

① +/- 30'

② 3 degrees 32'

③ -0 degrees 55'

④ 4.6 +3, -1

37655_TUCS_C0011

TIRE, WHEEL AND BALL JOINT SPECIFICATIONS

Year	Model	OEM Tires Standard	OEM Tires Optional	Tire Pressures (psi) Front	Tire Pressures (psi) Rear	Wheel Size	Ball Joint Inspection	Lug Nut Torque (ft. lbs.)
2009	Tucson	P215/65R16	P235/60R16	30	30	6.5J x 16	①	66-81
2010	Tucson	P225/60R17	P225/55R18	33	33	6.5J x 17,18	①	65-80

OEM: Original Equipment Manufacturer

PSI: Pounds Per Square Inch

① Replace the ball joint if too loose or if rotating torque exceeds specification: 4-21 inch lbs.

37655_TUCS_C0012

BRAKE SPECIFICATIONS

All measurements in inches unless noted

Year	Model		Brake Disc Original Thickness	Brake Disc Minimum Thickness	Brake Disc Maximum Runout	Brake Drum Diameter Original Inside Diameter	Brake Drum Diameter Max. Wear Limit	Brake Drum Diameter Maximum Machine Diameter	Minimum Lining Thickness	Brake Caliper Bracket Bolts (ft. lbs.)	Brake Caliper Mounting Bolts (ft. lbs.)
2009	Tucson	F	1.014	0.961	0.002	—	—	—	0.079	59-74	16-24
		R	0.394	0.315	0.001	—	—	—	0.079	59-74	16-24
2010	Tucson	F	①	②	0.0002	—	—	—	0.079	58-73	16-23
		R	0.390	0.330	0.0002	—	—	—	0.079	58-73	16-23

F: Front

R: Rear

① 2WD: 1.02 inches 4WD: 1.10 inches

② 2WD: 0.96 inches 4WD: 1.04 inches

37655_TUCS_C0013

SCHEDULED MAINTENANCE INTERVALS

HYUNDAI—TUCSON

TO BE SERVICED	TYPE OF SERVICE	VEHICLE MILEAGE INTERVAL (x1000)												
		7.5	15	22.5	30	37.5	45	52.5	60	67.5	75	82.5	90	97.5
Accessory drive belts	S/I								✓		✓		✓	
Air cleaner element (engine) ①	R	✓	✓	✓		✓	✓	✓	✓	✓	✓	✓		✓
Air conditioner system	S/I		✓		✓		✓		✓		✓		✓	
Brake lines, hoses, and connections	S/I		✓				✓		✓		✓		✓	
Brake pads, calipers, & rotors	S/I		✓		✓		✓		✓		✓		✓	
Brake fluid/clutch fluid	S/I				✓				✓				✓	
Cabin air filter ②	R		✓		✓		✓		✓		✓			
Crankcase ventilation hose, vapor hose, and fuel filter cap	S/I	✓	✓	✓	✓	✓	✓	✓	✓	✓	✓	✓	✓	✓
Driveshafts and CV-boots	S/I		✓		✓		✓		✓		✓		✓	
Electronic throttle control	S/I		✓		✓		✓		✓		✓		✓	
Engine oil and filter	R	✓	✓	✓	✓	✓	✓	✓	✓	✓	✓	✓	✓	✓
Exhaust pipe connections, muffler, and suspension bolts	S/I		✓		✓		✓		✓		✓		✓	
Fuel filter	R				✓				✓				✓	
Fuel tank air filter					✓				✓				✓	
Fuel lines, fuel hoses, and connections	S/I				✓				✓				✓	
Manual transaxle fluid	S/I					✓					✓			
Propeller Shaft	S/I		✓		✓		✓		✓		✓		✓	
Rear axle/transfer case oil ③	S/I					✓					✓			
Steering gear rack, linkage, and boots	S/I		✓		✓		✓		✓		✓		✓	
Suspension mounting bolts	S/I		✓		✓		✓		✓		✓		✓	

R: Replace S/I: Service or Inspect

① Replace air filter every 30,000 miles

② Replace cabin air filter every 30,000 miles

③ Transfer case oil and rear axle oil should be changed anytime they have been submerged in water.

Replace spark plugs every 105,000 miles

Inspect valve clearance for excessive noise, or engine vibration at 60,000 miles

Replace coolant at 120,000 miles, or 120 months and every 30,000 miles, or 24 months thereafter

FREQUENT OPERATION MAINTENANCE (SEVERE SERVICE)

If a vehicle is operated under any of the following conditions it is considered severe service:

- Extremely dusty areas.

- 50% or more of the vehicle operation is in 90°F (32°C) or higher temperatures, or constant operation in temperatures below 32°F (0°C).

- Prolonged idling (vehicle operation in stop and go traffic).

- Frequent short running periods (engine does not warm to normal operating temperatures).

- Police, taxi, delivery usage or trailer towing usage.

Automatic transaxle fluid: replace more frequently

Manual transaxle fluid: replace more frequently

Oil & oil filter: change every 3,750 miles.

Brake pads, calipers & rotors: service, or inspect more frequently

Driveshaft boots: service, or inspect every 7,500 miles

Rear axle/transfer case oil: replace every 75,000 miles

Air cleaner filter: replace more frequently

Cabin air cleaner filter: replace more frequently

Steering gear rack, linkage & boots: service, or inspect more frequently

Propeller shaft: Service, or inspect more frequently

Spark plugs: replace more frequently

37655_TUCS_C0014

BRAKES — INFORMATION AND PRECAUTIONS

ANTI-LOCK SYSTEMS

- Certain components within the ABS system are not intended to be serviced or repaired individually.
- Do not use rubber hoses or other parts not specifically specified for and ABS system. When using repair kits, replace all parts included in the kit. Partial or incorrect repair may lead to functional problems and require the replacement of components.
- Lubricate rubber parts with clean, fresh brake fluid to ease assembly. Do not use shop air to clean parts; damage to rubber components may result.
- Use only DOT 3 brake fluid from an unopened container.
- If any hydraulic component or line is removed or replaced, it may be necessary to bleed the entire system.
- A clean repair area is essential. Always clean the reservoir and cap thoroughly before removing the cap. The slightest amount of dirt in the fluid may plug an orifice and impair the system function. Perform repairs after components have been thoroughly cleaned; use only denatured alcohol to clean components. Do not allow ABS components to come into contact with any substance containing mineral oil; this includes used shop rags.
- The Anti-Lock control unit is a microprocessor similar to other computer units in the vehicle. Ensure that the ignition switch is **OFF** before removing or installing controller harnesses. Avoid static electricity discharge at or near the controller.
- If any arc welding is to be done on the vehicle, the control unit should be unplugged before welding operations begin.

DISC AND DRUM SYSTEMS

⁂ CAUTION

Dust and dirt accumulating on brake parts during normal use may contain asbestos fibers from production or aftermarket brake linings. Breathing excessive concentrations of asbestos fibers can cause serious bodily harm. Exercise care when servicing brake parts. Do not sand or grind brake lining unless equipment used is designed to contain the dust residue. Do not clean brake parts with compressed air or by dry brushing. Cleaning should be done by dampening the brake components with a fine mist of water, then wiping the brake components clean with a dampened cloth. Dispose of cloth and all residue containing asbestos fibers in an impermeable container with the appropriate label. Follow practices prescribed by the Occupational Safety and Health Administration (OSHA) and the Environmental Protection Agency (EPA) for the handling, processing, and disposing of dust or debris that may contain asbestos fibers.

BRAKES — BLEEDING THE BRAKE SYSTEM

BLEEDING PROCEDURE

BLEEDING THE ABS SYSTEM

See Figure 1.

This procedure should be followed to ensure adequate bleeding of air and the filling of the ABS unit, the brake lines, and the master cylinder with brake fluid.

1. Before servicing the vehicle, refer to the Precautions Section.

⁂ WARNING

The reservoir on the master cylinder must be at the MAX (upper) level mark at the start of bleeding procedure and checked after bleeding each brake caliper. Add fluid as required.

2. Remove the reservoir cap and fill the brake reservoir with brake fluid.

⁂ WARNING

If there is any brake fluid on any painted surface, wash it off immediately.

➡When pressure bleeding, do not depress the brake pedal. Recommended brake fluid: DOT3 or DOT4.

3. Connect a clear plastic tube to the wheel cylinder bleeder plug and insert the other end of the tube into a clear plastic bottle that is half filled with clean brake fluid.
4. Connect the scan tool to the data link connector located underneath the dash panel.
5. Select and operate according to the instructions on the scan tool screen.

⁂ CAUTION

You must obey the maximum operating time of the ABS motor with the scan tool to prevent the motor pump from burning.

   a. Select vehicle name.
   b. Select Anti-Lock Brake system.
   c. Select air bleeding mode.
   d. Press "OK" or "YES" to operate motor pump and solenoid valve.
   e. Wait 60 seconds before operating the air bleeding or damage to the motor may occur.
   f. Perform the air bleeding.
6. Pump the brake pedal several times, and then loosen the bleeder screw until fluid starts to run out without bubbles. Then, close the bleeder screw.
7. Repeat until there are no more bubbles in the fluid for each wheel.
8. Tighten the bleeder screw to 65–114 inch lbs. (7–13 Nm).

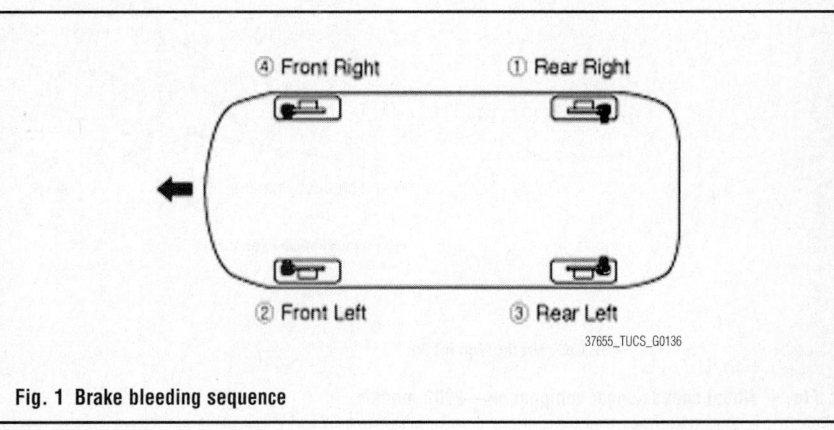

Fig. 1 Brake bleeding sequence

SPEED SENSORS

REMOVAL & INSTALLATION

2009 Models

Front

See Figures 2 through 4.

1. Before servicing the vehicle, refer to the Precautions Section.
2. Remove the front wheel speed sensor mounting bolt.
3. Remove the front wheel guard after removing the mud guard.
4. Disconnect the wheel speed sensor connector.
5. Remove the front wheel speed sensor.

A
37655_TUCS_G0143

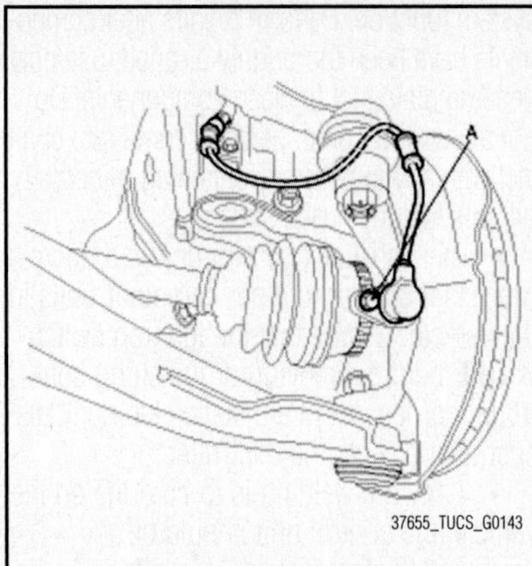

Fig. 2 Remove the front wheel speed sensor mounting bolt (A)

A
B
37655_TUCS_G0144

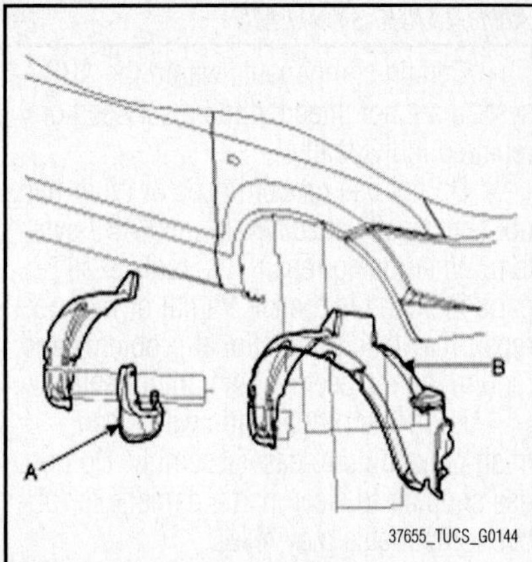

Fig. 3 Remove the front wheel guard (B) after removing the mud guard (A)

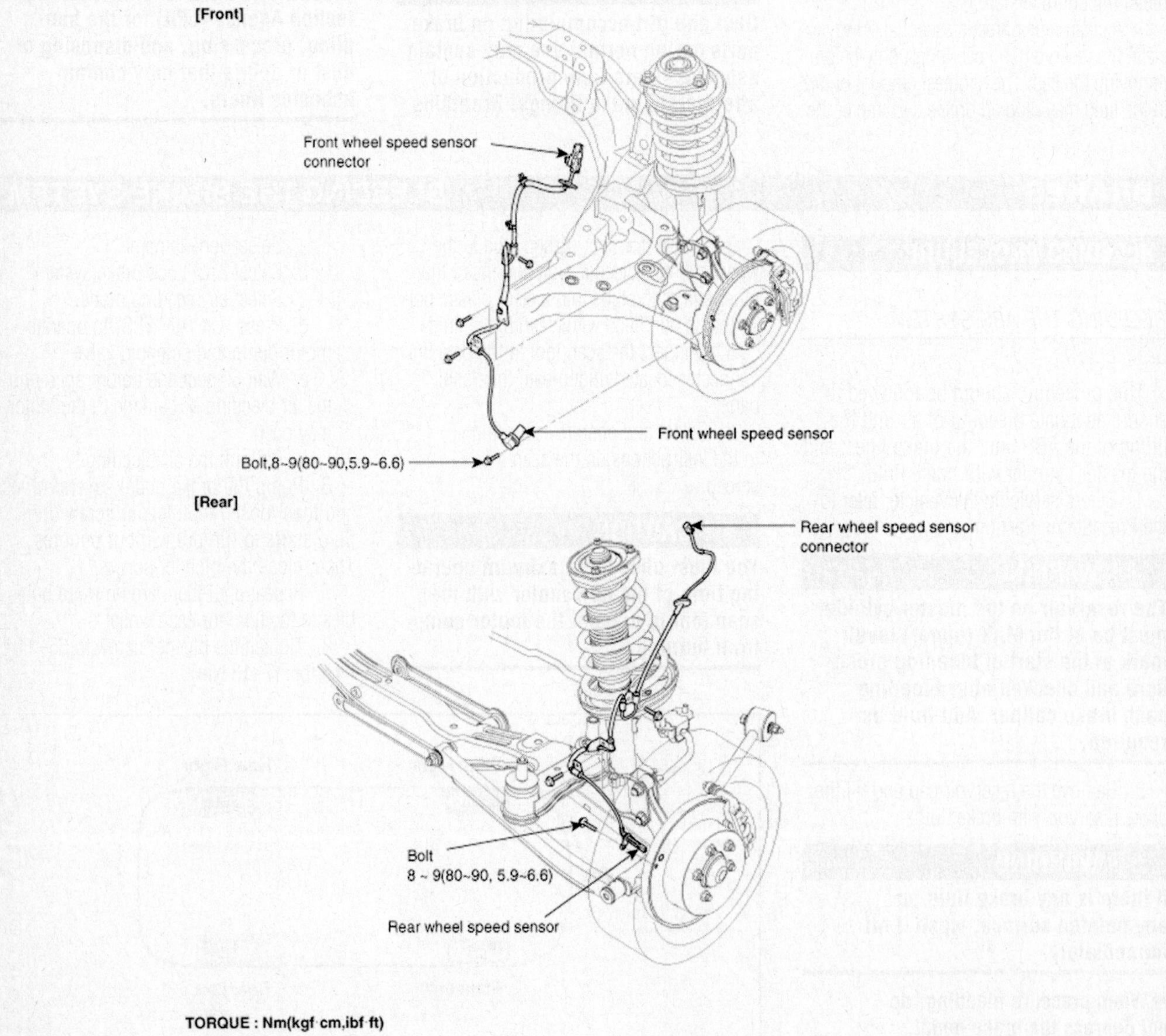

[Front]
Front wheel speed sensor connector
Front wheel speed sensor
Bolt,8~9(80~90,5.9~6.6)
[Rear]
Rear wheel speed sensor connector
Bolt
8 ~ 9(80~90, 5.9~6.6)
Rear wheel speed sensor
TORQUE : Nm(kgf·cm,lbf·ft)
22140_TUCS_G0028

Fig. 4 Wheel speed sensor components—2009 models

To install:

6. Installation is the reverse of the removal procedure. Tighten the sensor mounting bolt to 71–79 inch lbs. (8–9 Nm).

REAR

See Figures 4 and 5.

1. Before servicing the vehicle, refer to the Precautions Section.
2. Remove the rear wheel speed sensor mounting bolt.

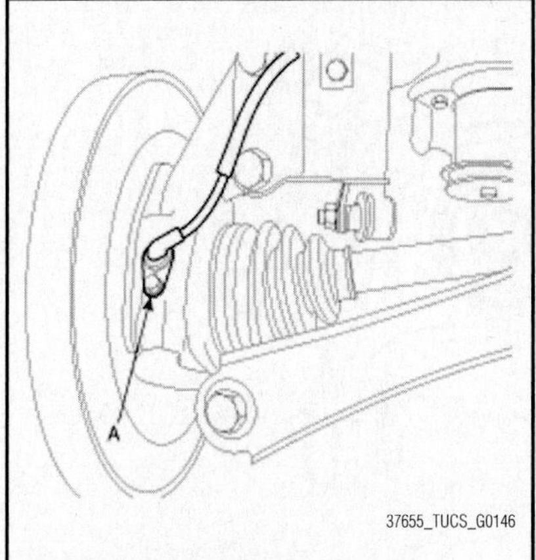

Fig. 5 Remove the rear wheel speed sensor mounting bolt (A)

3. Remove the rear seat side pad and disconnect the rear wheel speed sensor connector.

To install:

4. Installation is the reverse of the removal procedure. Tighten the sensor mounting bolt to 71–79 inch lbs. (8–9 Nm).

2010 Models

Front

See Figures 6 and 7.

1. Before servicing the vehicle, refer to the Precautions Section.
2. Remove the front wheel speed sensor mounting bolt.

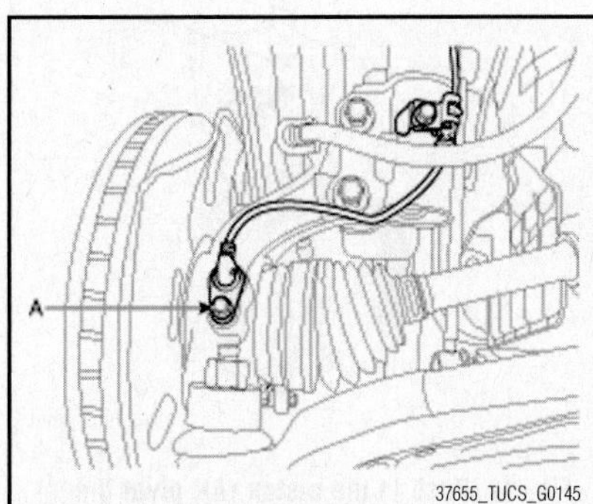

Fig. 6 Remove the front wheel speed sensor mounting bolt (A)

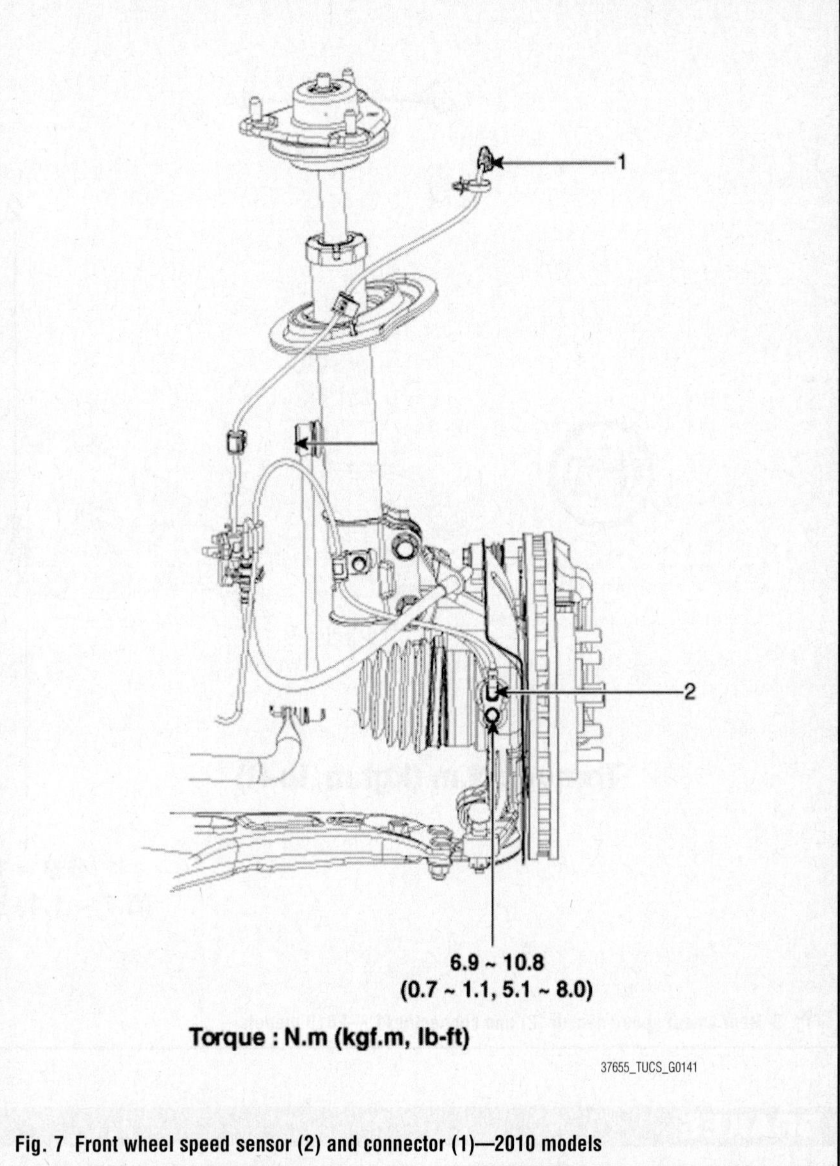

Fig. 7 Front wheel speed sensor (2) and connector (1)—2010 models

3. Remove the front wheel guard.
4. Disconnect the wheel speed sensor connector.
5. Remove the front wheel speed sensor.

To install:

6. Installation is the reverse of the removal procedure. Tighten the sensor mounting bolt to 61–96 inch lbs. (7–11 Nm).

Rear

See Figures 8 and 9.

1. Before servicing the vehicle, refer to the Precautions Section.
2. Remove the rear wheel speed sensor mounting bolt.
3. Remove the luggage side trim and disconnect the rear wheel speed sensor connector.

To install:

4. Installation is the reverse of the removal procedure. Tighten the sensor mounting bolt to 61–96 inch lbs. (7–11 Nm).

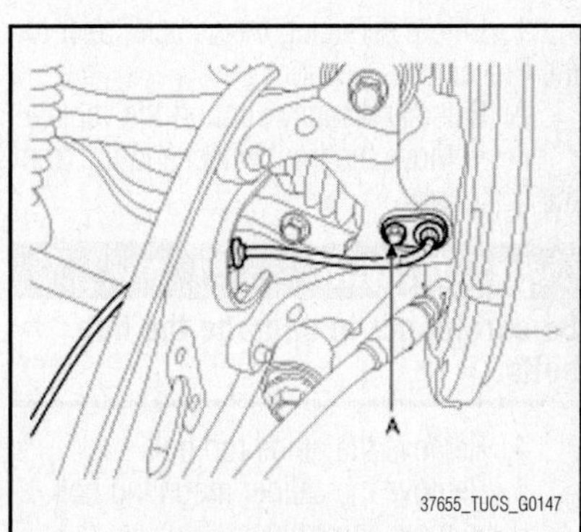

Fig. 8 Remove the rear wheel speed sensor mounting bolt (A)

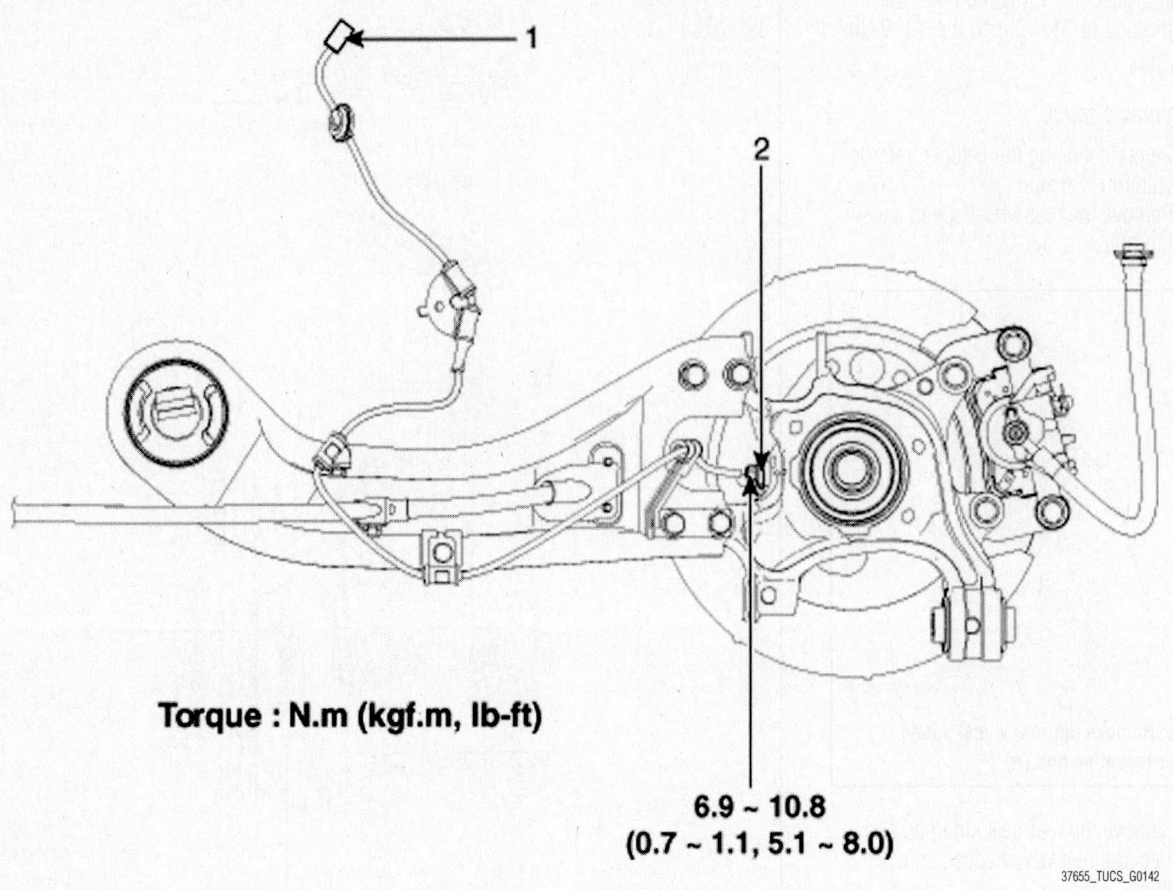

Fig. 9 Rear wheel speed sensor (2) and connector (1)—2010 models

BRAKES — FRONT DISC BRAKES

BRAKE CALIPER

REMOVAL & INSTALLATION

2009 Models

See Figures 10 and 11.

1. Before servicing the vehicle, refer to the Precautions Section.
2. Raise and safely support the vehicle.
3. Remove the front wheel and tire from the front hub.

⁂ WARNING

Be careful not to damage the hub bolts.

4. Remove the guide rod bolt.
5. Remove the caliper mounting bolt.
6. Remove the caliper.

To install:

7. Install the caliper.
8. Install the caliper mounting bolt and tighten to 48–55 ft. lbs. (65–75 Nm).
9. Install the guide rod bolt and tighten to 16–24 ft. lbs. (22–32 Nm).

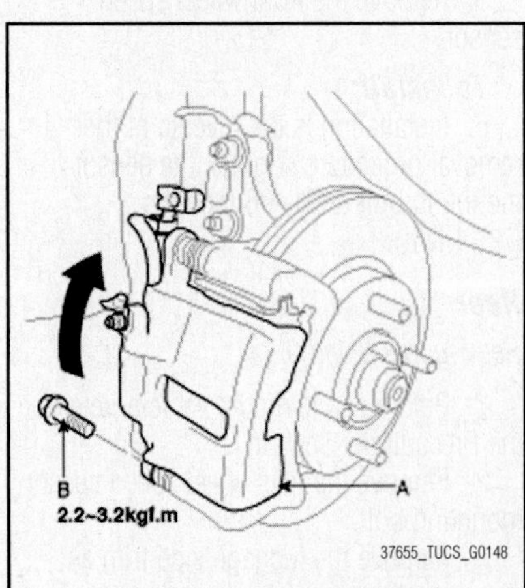

Fig. 10 Remove the guide rod bolt (B)

➡Insert the piston in the cylinder using the special tool (09581-11000).

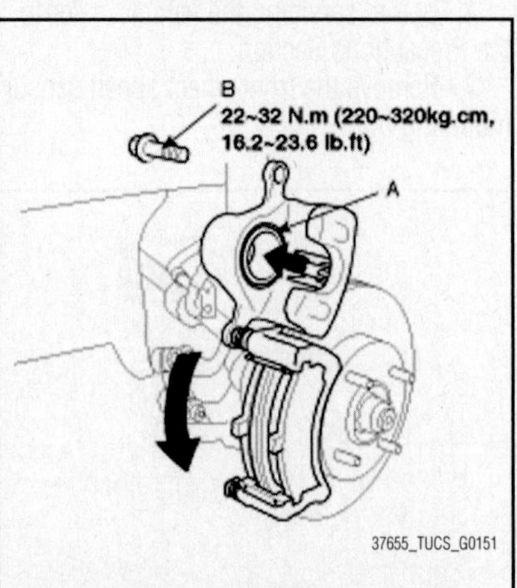

Fig. 11 Push in the piston (A), pivot the caliper into position, and install the guide rod bolt (B)

10. Depress the brake pedal several times to make sure the brakes work, then test-drive.

➡**Engagement of the brake may require a greater pedal stroke immediately after the brake pads have been replaced as a set. Several applications of the brake will restore the normal pedal stroke.**

11. After installation, check for leaks at hose and line joints or connections, and retighten if necessary.

2010 Models

See Figures 12 and 13.

1. Before servicing the vehicle, refer to the Precautions Section.
2. Raise and safely support the vehicle.
3. Remove the front wheel and tire from the front hub.

⁂ WARNING

Be careful not to damage the hub bolts.

4. Remove the hose eyebolt and caliper mounting bolts, then remove the front caliper assembly.

To install:

5. Installation is the reverse of removal.
   a. Tighten the brake hose to caliper bolt to 18–22 ft. lbs. (25–29 Nm).
   b. Tighten the caliper assembly to knuckle bolts to 58–72 ft. lbs. (79–98 Nm).
6. Use a SST (09581-11000) when installing the brake caliper assembly.
7. After installation, bleed the brake system.
8. Depress the brake pedal several times to make sure the brakes work, then test-drive.

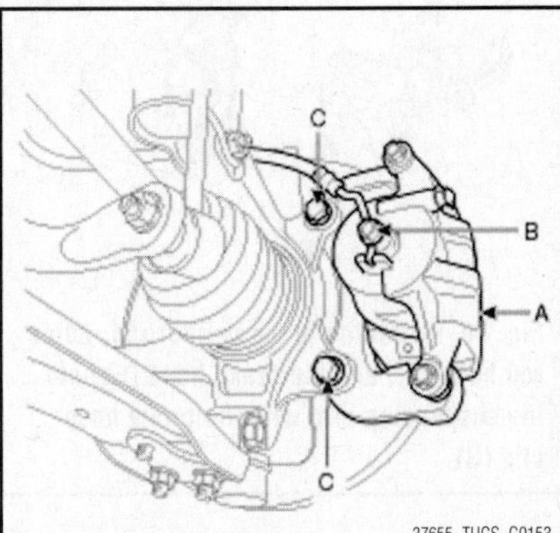

Fig. 12 Remove the hose eyebolt (B) and caliper mounting bolts (C), then remove the front caliper assembly (A)

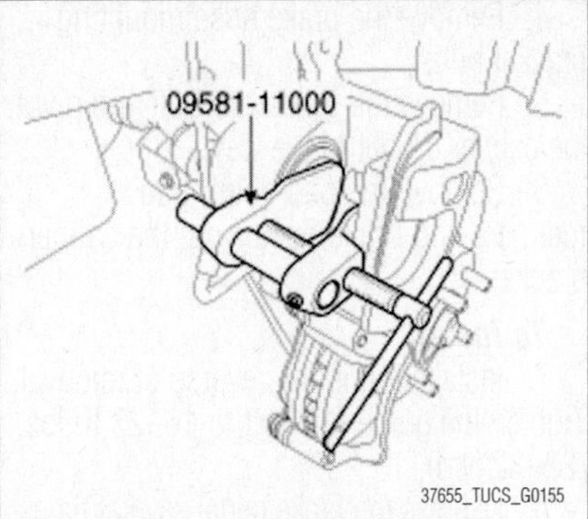

Fig. 13 Use a SST (09581-11000) when installing the brake caliper assembly

9. After installation, check for leaks at hose and line joints or connections, and retighten if necessary.

DISC BRAKE PADS

REMOVAL & INSTALLATION

2009 Models

See Figures 10, 11, 14 and 15.

1. Before servicing the vehicle, refer to the Precautions Section.
2. Raise and safely support the vehicle.
3. Remove the front wheel and tire from the front hub.

⁂ WARNING

Be careful not to damage the hub bolts.

4. Remove the guide rod bolt and support the caliper assembly with a piece of wire so that it does not hang from the brake hose.
5. Remove the pad shims, pad retainers, and pads.

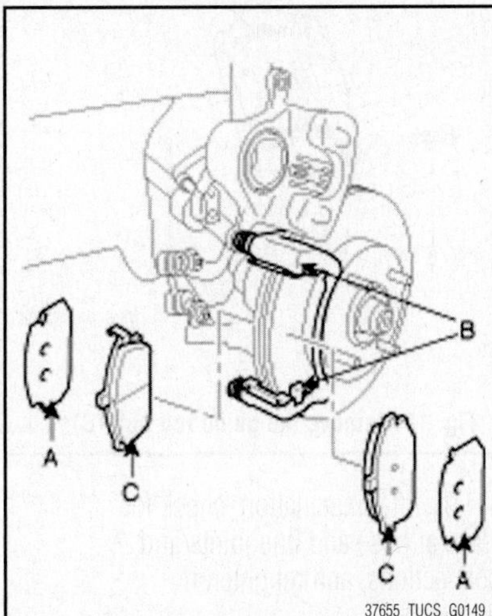

Fig. 14 Remove the pad shims (A), pad retainers (B), and pads (C)

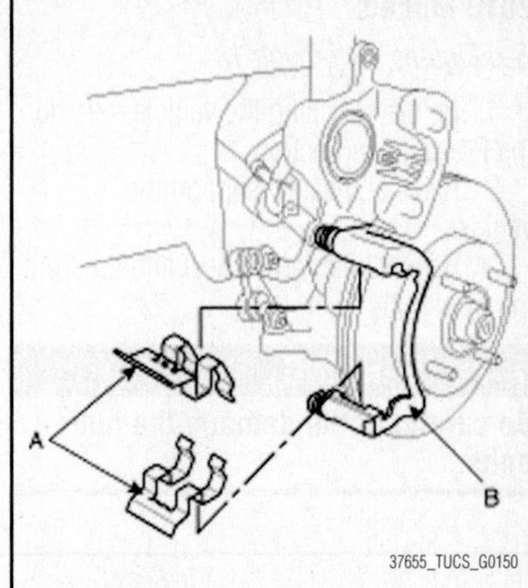

Fig. 15 Install the pad retainers (A) on the caliper bracket (B)

To install:

6. Install the pad retainers on the caliper bracket.
7. Check the foreign material at the pad shims and the back of the pads. Contaminated brake discs or pads reduce stopping ability. Keep grease off the discs and pads.
8. Install the brake pads and pad shims correctly. Install the pad with the wear indicator on the inside.

➡**If you are reusing the pads, always reinstall the brake pads in their original positions to prevent a momentary loss of braking efficiency.**

9. Push in the piston so that the caliper will fit over the pads. Make sure that the piston boot is in position to prevent damaging it when pivoting the caliper down.
10. Pivot the caliper down into position. Being careful not to damage the pin boot, install the guide rod bolt and tighten to 16–24 ft. lbs. (22–32 Nm).

➡**Insert the piston in the cylinder using the special tool (09581-11000).**

11. Depress the brake pedal several times to make sure the brakes work, then test-drive.

➡**Engagement of the brake may require a greater pedal stroke immediately after the brake pads have been replaced as a set. Several applications of the brake will restore the normal pedal stroke.**

12. After installation, check for leaks at hose and line joints or connections, and retighten if necessary.

2010 Models

See Figures 16 through 18.

1. Before servicing the vehicle, refer to the Precautions Section.
2. Raise and safely support the vehicle.
3. Remove the front wheel and tire from the front hub.

WARNING

Be careful not to damage the hub bolts.

4. Remove the brake hose mounting bracket bolt.
5. Remove the guide rod bolt and pivot the caliper up out of the way.
6. Remove the pad shim, pad retainers and brake pads from the caliper bracket.

To install:

7. Installation is the reverse of removal. Tighten the guide rod bolt to 16–23 ft. lbs. (22–32 Nm).
8. Depress the brake pedal several times to make sure the brakes work, then test-drive.

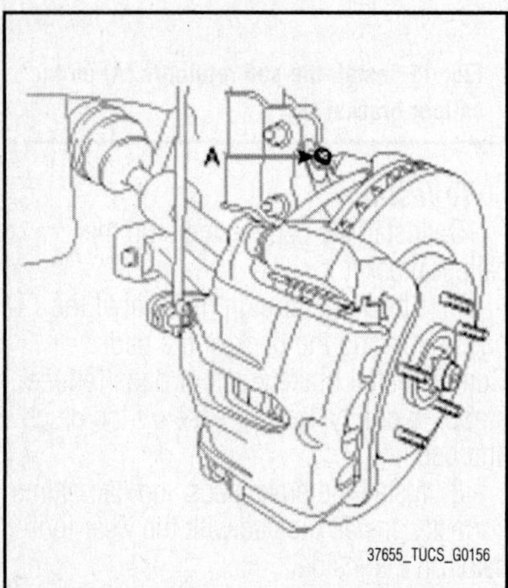

Fig. 16 Remove the brake hose mounting bracket bolt (A)

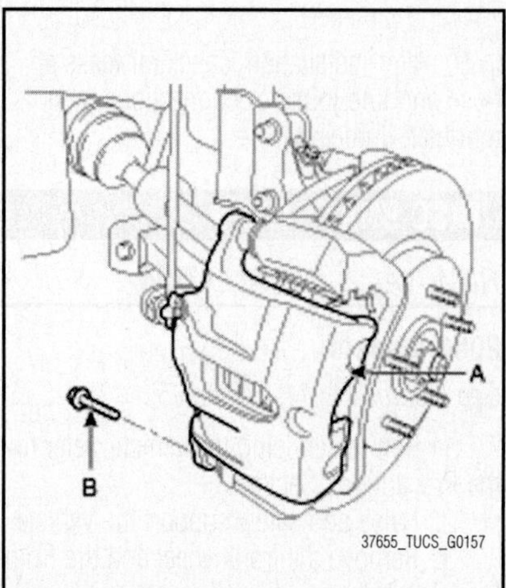

Fig. 17 Remove the guide rod bolt (B) and pivot the caliper (A) up out of the way

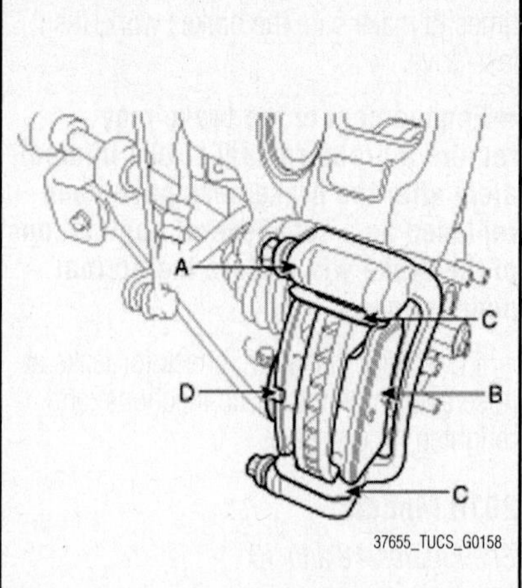

Fig. 18 Remove the pad shim (D), pad retainers (C) and brake pads (B) from the caliper bracket (A)

➡Engagement of the brake may require a greater pedal stroke immediately after the brake pads have been replaced as a set. Several applications of the brake will restore the normal pedal stroke.

9. After installation, check for leaks at hose and line joints or connections, and retighten if necessary.

BRAKES — REAR DISC BRAKES

BRAKE CALIPER

REMOVAL & INSTALLATION

2009 Models

See Figures 19 and 20.

1. Before servicing the vehicle, refer to the Precautions Section.
2. Raise and safely support the vehicle.
3. Remove the rear wheel and tire from the hub.
4. Release the parking brake.
5. Remove the guide rod bolt and remove the caliper assembly.

To install:

6. Install the brake caliper.
7. Install and torque the guide rod bolts to 16–24 ft. lbs. (22–32 Nm).
8. Install the brake hose onto the suspension arm with the brake hose clip.

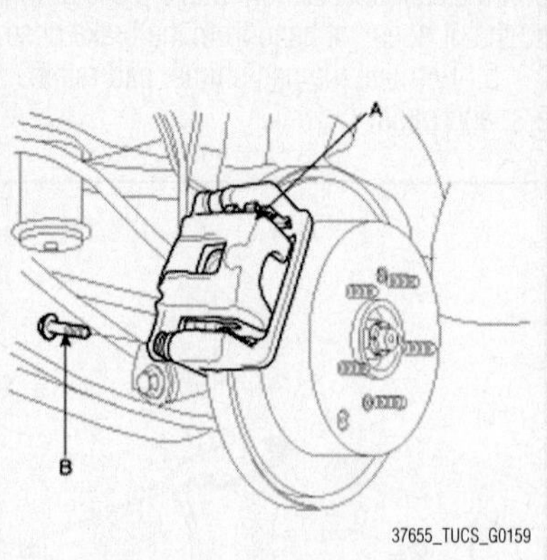

Fig. 19 Remove the guide rod bolt (B)

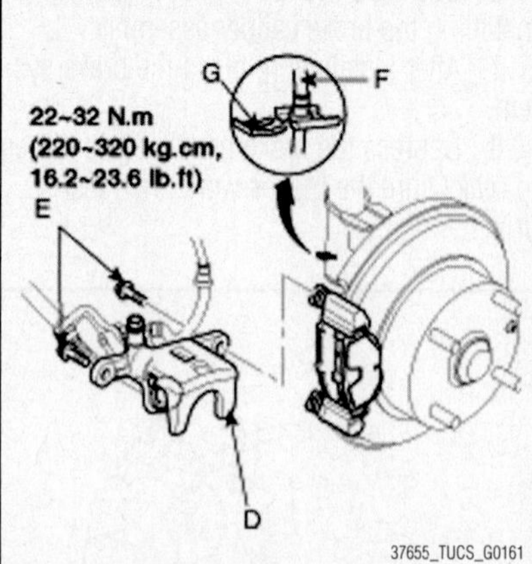

Fig. 20 Install the brake caliper (D), guide rod bolts (E), and the brake hose (F) onto the suspension arm with the brake hose clip (G)

9. After installation, check for leaks at hose and line joints and connections, and retighten if necessary.
10. Depress the brake pedal several times to make sure the brakes work, then test-drive.

2010 Models

See Figures 21 and 22.

1. Before servicing the vehicle, refer to the Precautions Section.

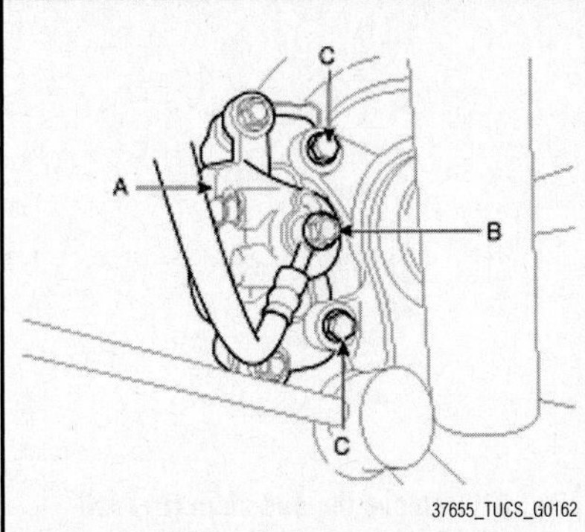

Fig. 21 Remove the hose eyebolt (B) and caliper mounting bolts (C), then remove the rear caliper assembly (A)

2. Raise and safely support the vehicle.
3. Remove the rear wheel and tire from the hub.
4. For 2WD vehicles, remove the rear shock absorber.
5. For 2WD vehicles, remove the rear upper arm.
6. Remove the hose eyebolt and caliper mounting bolts, then remove the rear caliper assembly.

To install:

7. Installation is the reverse of removal, noting the following:
    a. Use a SST (09581-11000) when installing the brake caliper assembly.
8. Install the brake hose to caliper and tighten the bolts to 18–22 ft. lbs. (25–29 Nm).
9. Install the caliper assembly to carrier and tighten the bolts to 58–72 ft. lbs. (79–98 Nm).
10. Bleed the brake system.
11. After installation, check for leaks at hose and line joints and connections, and retighten if necessary.

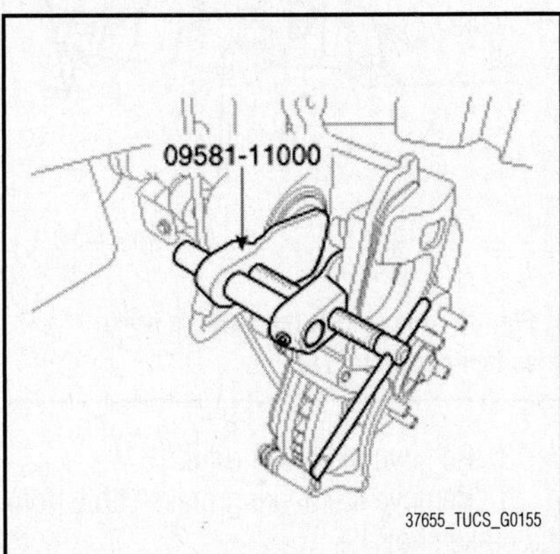

Fig. 22 Use a SST (09581-11000) when installing the brake caliper assembly

12. Depress the brake pedal several times to make sure the brakes work, then test-drive.

DISC BRAKE PADS

REMOVAL & INSTALLATION

2009 Models

See Figures 19, 20, 23 and 24.

1. Before servicing the vehicle, refer to the Precautions Section.
2. Raise and safely support the vehicle.
3. Remove the rear wheel and tire from the hub.
4. Release the parking brake.
5. Remove the guide rod bolt and support the caliper assembly with a piece of wire so that it does not hang from the brake hose.
6. Remove the pad shims, pad retainers, and pads.

To install:

7. Install the pad springs to the carrier.

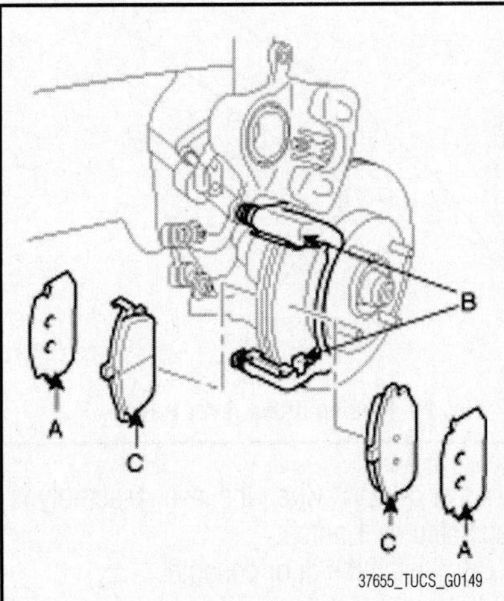

Fig. 23 Remove the pad shims (A), pad retainers (B), and pads (C)

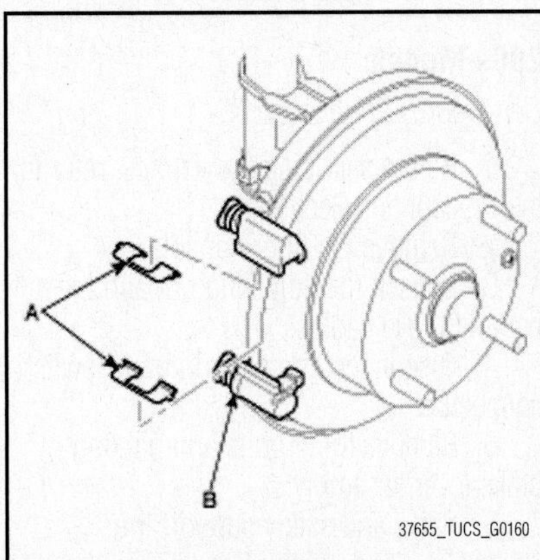

Fig. 24 Install the pad springs (A) to the carrier (B)

8. Check the foreign material at the pad shim and the back of the pads. Contaminated brake discs or pads reduce stopping ability. Keep grease off the discs and pads.
9. Insert the piston in the cylinder using the special tool (09581-11000).
10. Install the brake pads and pad shim on the caliper bracket. Install the inner pad with its wear indicator facing downward.

⁂ WARNING

If you are reusing the pads, always reinstall the brake pads in their original positions to prevent a momentary loss of braking efficiency.

11. Rotate the caliper piston clockwise into the cylinder, the align the cutout in the piston with the tab on the inner pad by turning the piston back. Lubricate the boot with rubber grease to avoid twisting the piston boot. If the piston boot is twisted, back it out so it is positioned properly.
12. Reposition the brake caliper.
13. Install and torque the guide rod bolts to 16–24 ft. lbs. (22–32 Nm).
14. Install the brake hose onto the suspension arm with the brake hose clip.
15. After installation, check for leaks at hose and line joints and connections, and retighten if necessary.
16. Depress the brake pedal several times to make sure the brakes work, then test-drive.

2010 Models

See Figures 25 through 27.

1. Before servicing the vehicle, refer to the Precautions Section.
2. Raise and safely support the vehicle.
3. Remove the rear wheel and tire from the hub.
4. For 2WD vehicles, remove the rear shock absorber.
5. For 2WD vehicles, remove the rear upper arm.
6. Remove the hose eyebolt and caliper mounting bolts, then remove the rear caliper assembly.
7. Remove the guide rod bolt and pivot the caliper up out of the way.
8. Remove the pad shim, pad retainers and brake pads in the caliper bracket.

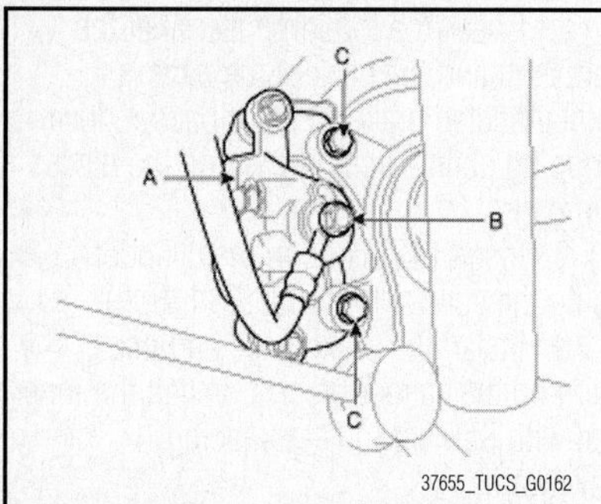

Fig. 25 Remove the hose eyebolt (B) and caliper mounting bolts (C), then remove the rear caliper assembly (A)

To install:

9. Installation is the reverse of removal.
10. Bleed the brake system.

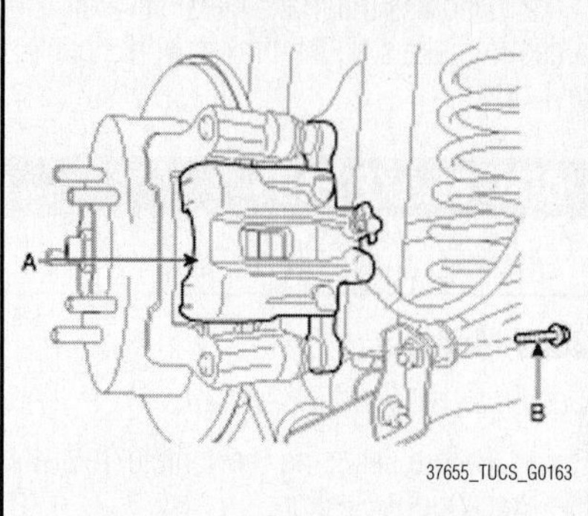

Fig. 26 Remove the guide rod bolt (B) and pivot the caliper (A) up out of the way

11. After installation, check for leaks at hose and line joints and connections, and retighten if necessary.

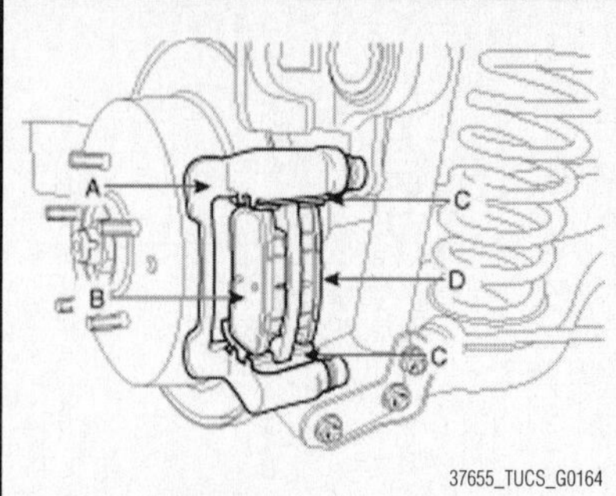

Fig. 27 Remove the pad shim (D), pad retainers (C) and brake pads (B) from the caliper bracket (A)

12. Depress the brake pedal several times to make sure the brakes work, then test-drive.

BRAKES — PARKING BRAKE

PARKING BRAKE CABLES

ADJUSTMENT

See Figures 28 and 29.

1. Before servicing the vehicle, refer to the Precautions Section.
2. Raise and safely support the vehicle.
3. Remove the floor console.
4. Adjust the parking brake lever stroke by turning the adjusting nut. Set the lever to 6 clicks.
5. Release the parking brake lever fully, and check that parking brakes do not drag when the rear wheels are turned. Readjust if necessary.
6. Make sure that the parking brakes are fully applied when the parking brake lever is pulled up fully.
7. The parking indicator lamp must be "OFF" when the lever assembly is released.

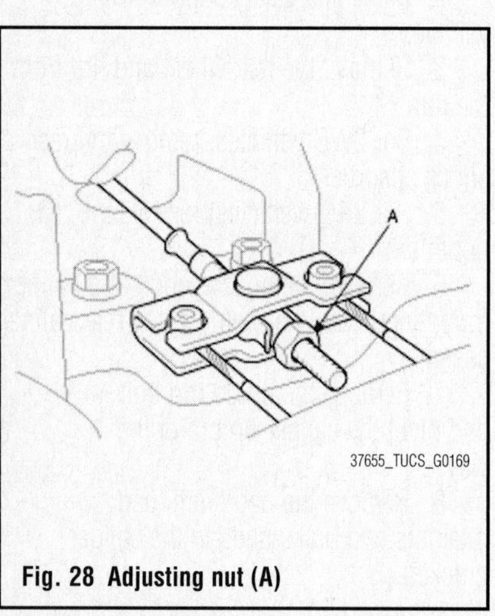

Fig. 28 Adjusting nut (A)

Pulled up with 196N (20kgf, 44lb)

A

40mm

37655_TUCS_G0174

Fig. 29 Parking brake lever travel

It must be "ON" when the lever assembly is operated by 1 notch.

8. Install the floor console.

PARKING BRAKE SHOES

REMOVAL & INSTALLATION

2009 Models

See Figures 30 through 33.

1. Before servicing the vehicle, refer to the Precautions Section.
2. Remove the floor console.
3. Loosen the adjusting nut and the parking brake cables.
4. Disconnect the parking brake switch connector.
5. Remove the 4 bolts and parking brake lever assembly.
6. Raise and safely support the vehicle.
7. Remove the rear wheel and tire assembly.

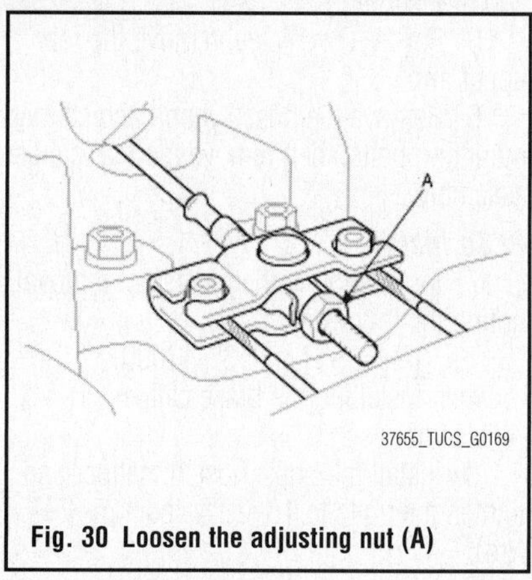

Fig. 30 Loosen the adjusting nut (A)

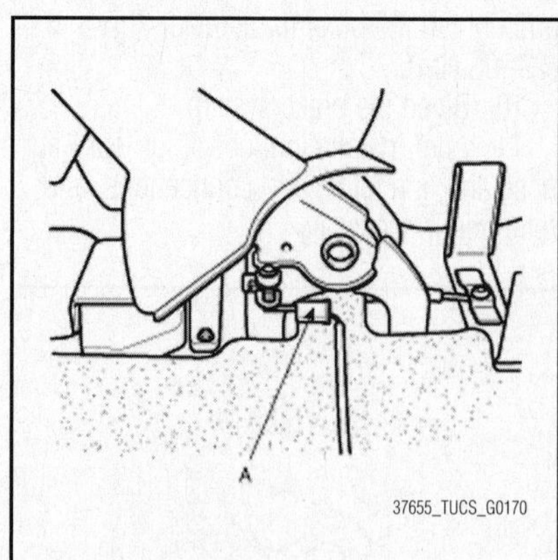

Fig. 31 Disconnect the parking brake switch connector (A)

8. Remove the brake rotor.
9. Remove the parking brake cable from the brake shoe.
10. Remove the parking brake cable retaining ring (B), from the parking brake cable (A).

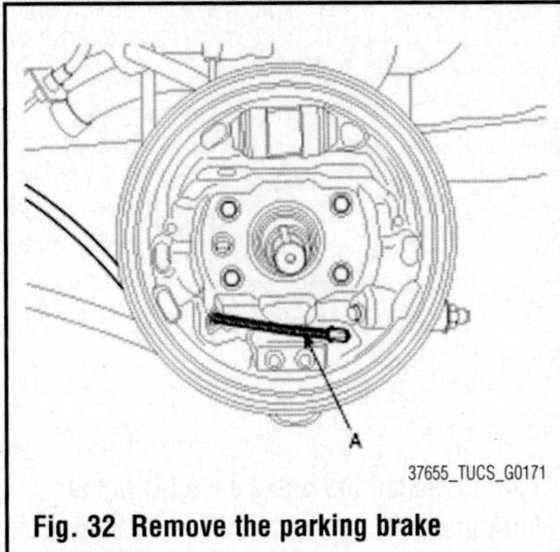

Fig. 32 Remove the parking brake cable (A)

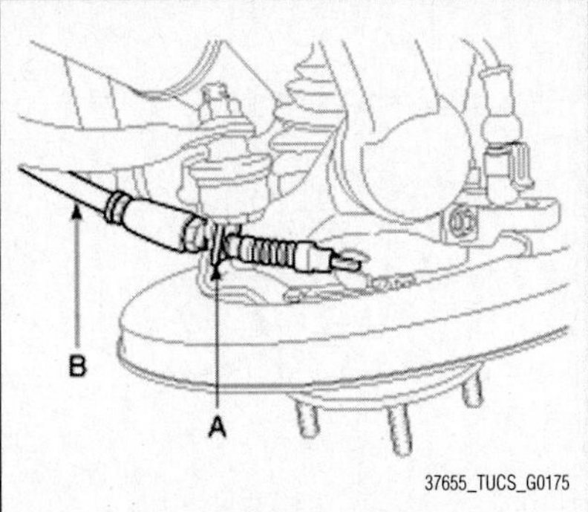

Fig. 34 Remove the parking brake cable (B) after removing the clip (A)

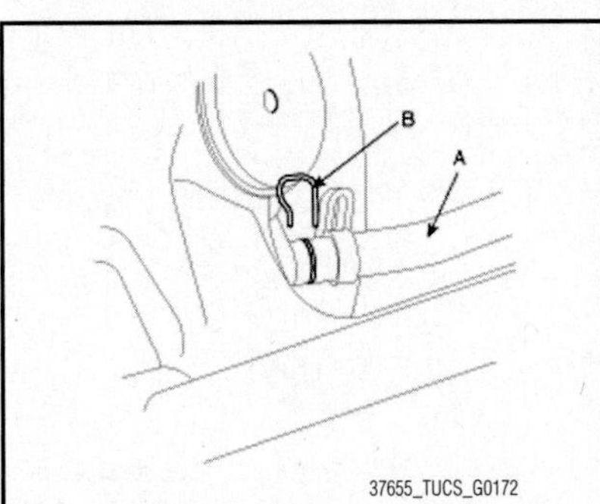

Fig. 33 Remove the parking brake cable retaining ring (B), from the parking brake cable (A)

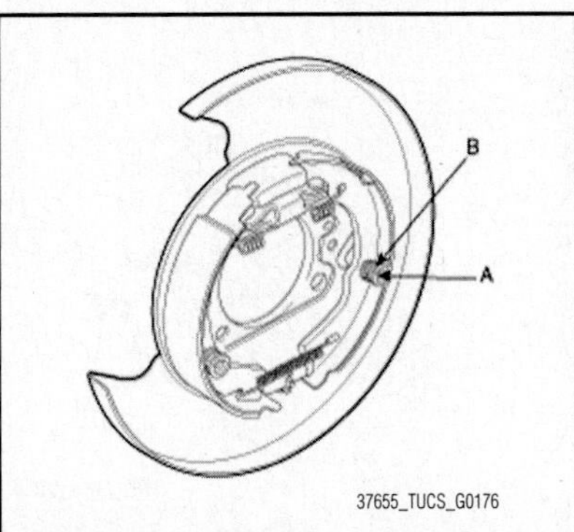

Fig. 35 Remove the shoe hold down pin (A) and the spring (B)

To install:

11. Install the removed parts in the reverse order of removal.
    a. Apply the specified grease (Multi-purpose grease SAE J310, NLGI No. 2) to each sliding parts of the ratchet plate or the ratchet pawl.
    b. After installing the parking brake cable adjuster, adjust the parking brake lever stroke.

2010 Models

2WD Vehicles

See Figures 34 through 36.

1. Before servicing the vehicle, refer to the Precautions Section.
2. Raise and safely support the vehicle.
3. Remove the rear wheel and tire assembly.
4. Remove the brake caliper and disc. Refer to Rear Disc Brakes, Caliper, Removal & Installation.
5. Remove the parking brake cable after removing the clip.
6. Remove the shoe hold down pin and the spring by pressing and rotating the spring.
7. Remove the adjuster assembly and the lower return spring.

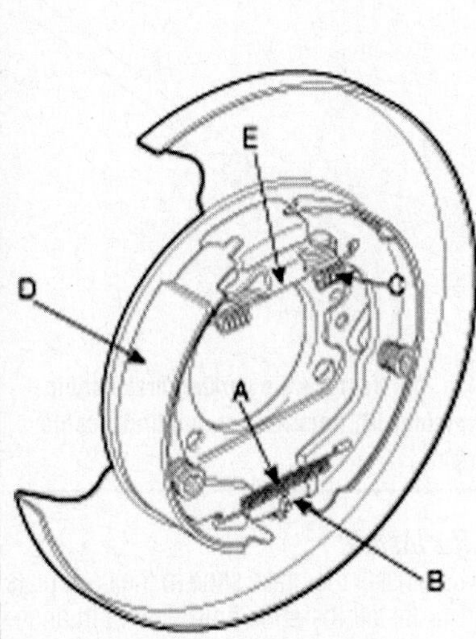

Fig. 36 Remove the adjuster assembly, lower return spring, upper return spring brake shoes, and operating lever assembly

8. Remove the upper return spring and the brake shoes.
9. Remove the operating lever assembly.

To install:

10. Install the operating lever assembly.
11. Install the upper return spring and the brake shoes.
12. Install the adjuster assembly and the lower return spring.
13. Install the shoe hold down pin and spring by pressing and rotating the spring.
14. Install the parking brake cable, then install the clip.
15. Install the rear brake disc, then adjust the rear brake shoe clearance.
    a. Remove the plug from the disc.
    b. Rotate the toothed wheel of adjuster with a screwdriver until the disc is not moving, and then return it by 4 notches in the opposite direction.
16. Install the brake caliper assembly.
17. Install the tire and wheel, after installing the plug on the disc.
18. If the parking brake shoe or the brake disc are replaced with a new one, perform the brake shoe bed-in procedure.
    a. Hand type: While operating the parking brake pedal to 15.4 ft. lbs. drive the vehicle 0.31 miles at a speed of 18.6 mph.
    b. Foot type: While operating the parking brake lever to 33 ft. lbs. drive the vehicle 0.31 miles at a speed of 18.6 mph.
    c. Repeat the above procedure three times.
19. After adjusting parking brake, check the following:
    a. Check that all parts move smoothly.
    b. The parking brake indicator lamp must be on after the parking pedal is worked and must be off after the pedal is released.

4WD Vehicles

See Figures 37 through 44.

1. Before servicing the vehicle, refer to the Precautions Section.
2. Raise and safely support the vehicle.
3. Remove the rear wheel and tire assembly.
4. Remove the brake caliper. Refer to Rear Disc Brakes, Caliper, Removal & Installation.

5. Remove the rear brake disc.
6. Remove the rear hub unit bearing.
7. Remove the shoe hold down pin and the spring by pushing the retainer spring and turning the pin.
8. Remove the adjuster assembly and the return spring.
9. Remove the parking brake cable from the brake shoe.
10. Remove the strut and the strut spring.
11. Remove the brake shoe.
12. Remove the parking brake cable retainer from the parking brake cable.

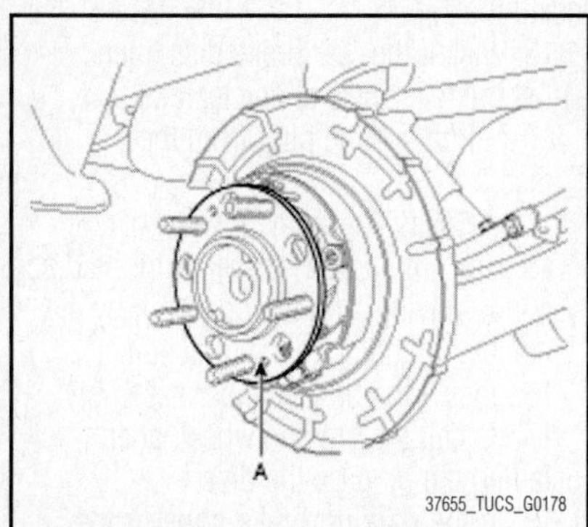

Fig. 37 Remove the rear hub unit bearing (A)

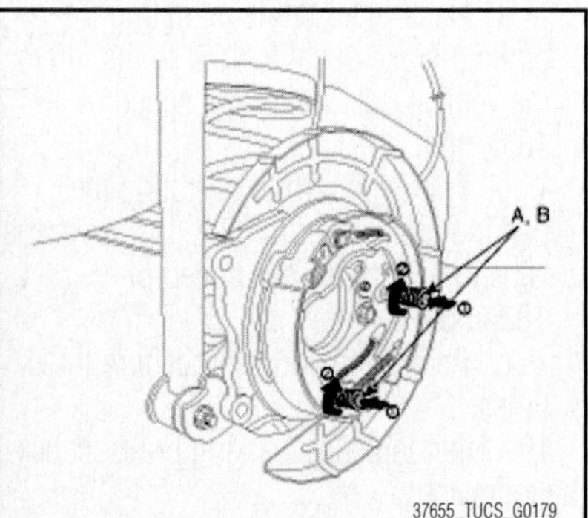

Fig. 38 Remove the shoe hold down pin (A) and the spring (B)

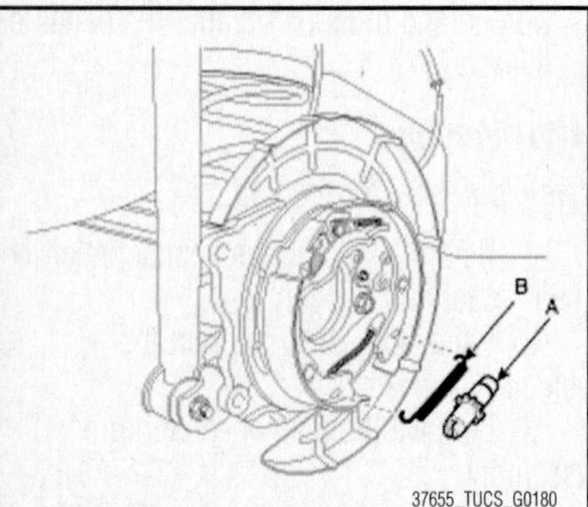

Fig. 39 Remove the adjuster assembly (A) and the return spring (B)

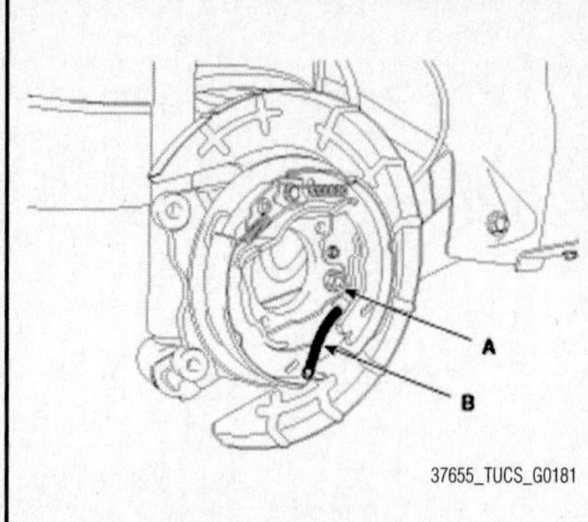

Fig. 40 Remove the parking brake cable (B) from the brake shoe (A)

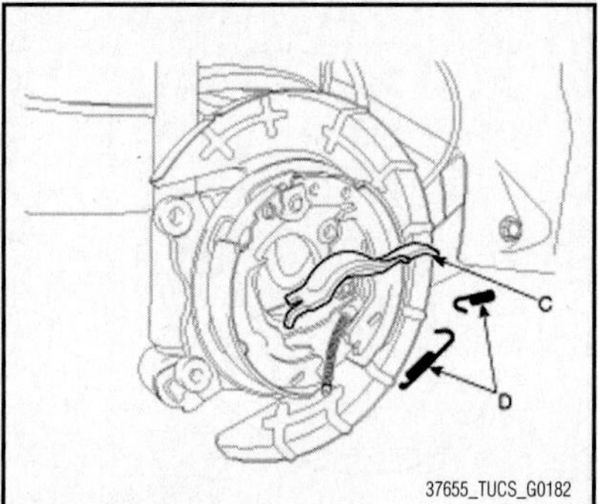

Fig. 41 Remove the strut (C) and the strut spring (D)

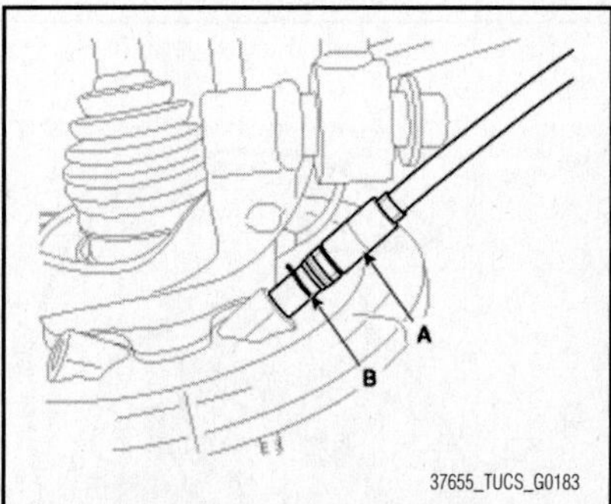

Fig. 42 Remove the parking brake cable retainer (B) from the parking brake cable (A)

To install:

13. Install the brake shoe to the back plate.
14. Install the shoe hold down pin and the spring by pushing the spring and turning the pins.
15. After installing the strut and upper return spring, install the adjuster assembly and the lower return spring.
16. Install the parking brake cable, then install the retainer.
17. Apply a coating of the specified grease (Multi-purpose grease SAE J310, NLGI No. 2) to all sliding parts of parking brake as shown.

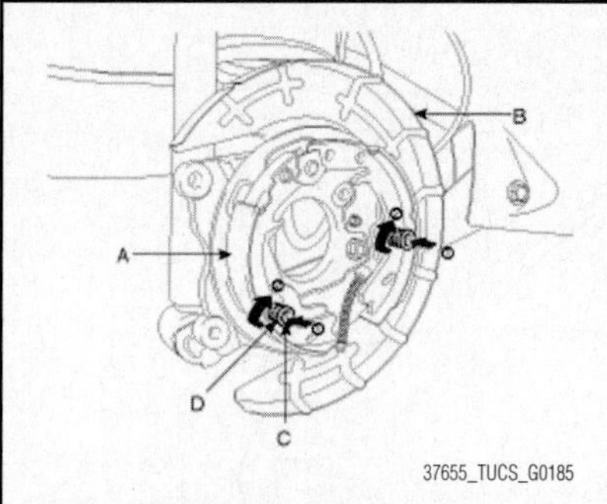

Fig. 43 Install the brake shoe (A) to the back plate (B) and install the shoe hold down pin (C) and the spring (D)

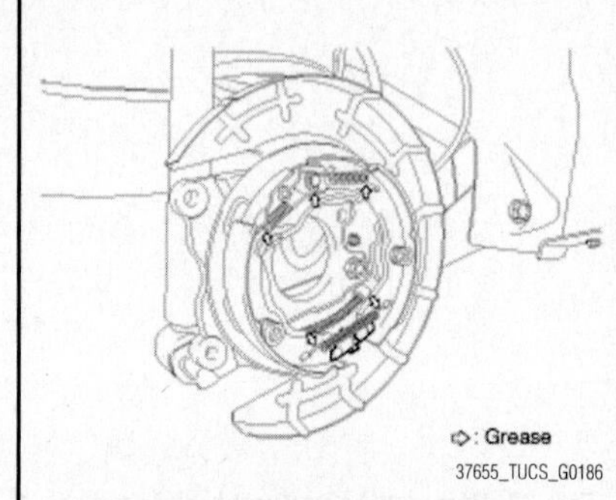

Fig. 44 Apply a coating of the specified grease (Multi-purpose grease SAE J310, NLGI No. 2) to all sliding parts of parking brake as shown

18. Install the rear brake disc, then adjust the rear brake shoe clearance.
    a. Remove the plug from the disc.
    b. Rotate the toothed wheel of adjuster with a screwdriver until the disc is not moving, and then return it by 4 notches in the opposite direction.
19. Install the brake caliper.
20. Install the tire and wheel.
21. If the parking brake shoe or the brake disc are replaced a newly one, perform the brake shoe bed-in procedure.
    a. Hand type: While operating the parking brake pedal to 15.4 ft. lbs. drive the vehicle 0.31 miles at a speed of 18.6 mph.
    b. Foot type: While operating the parking brake lever to 33 ft. lbs. drive the vehicle 0.31 miles at a speed of 18.6 mph.
    c. Repeat the above procedure more than three times.
    d. Check that all parts move smoothly.
    e. The parking brake indicator lamp must be on after the parking pedal is worked and must be off after the pedal is released.

GENERAL INFORMATION

✻✻ CAUTION

These vehicles are equipped with an air bag system. The system must be disarmed before performing service on, or around, system components, the steering column, instrument panel components, wiring and sensors. Failure to follow the safety precautions and the disarming procedure could result in accidental air bag deployment, possible injury and unnecessary system repairs.

SERVICE PRECAUTIONS

✻✻ CAUTION

Disconnect and isolate the battery negative cable before beginning any airbag system component diagnosis, testing, removal, or installation procedures. Wait at least 90 seconds after the ignition switch is turned off and the negative (-) terminal cable is disconnected from the battery before starting the operation. The SRS is equipped with a backup power source, so if work is started within 90 seconds after disconnecting the negative (-) terminal cable from the battery, the SRS may be deployed. Failure to disable the airbag system may result in accidental airbag deployment, personal injury, or death.

DISARMING THE SYSTEM

1. Before servicing the vehicle, refer to the Precautions Section.
2. Record the radio anti-theft code data. Remove the ignition key from the vehicle.
3. Disconnect the negative battery cable.
4. Wait at least 3 minutes for the system capacitor to discharge before performing any service.

ARMING THE SYSTEM

1. Before servicing the vehicle, refer to the Precautions Section.
2. Reconnect the negative battery cable.
3. To confirm proper system operation, turn the ignition switch to the ON position. The SRS indicator light will be lit for at least 6 seconds and then go off.

CLOCKSPRING CENTERING

See Figures 45 and 46.

1. Before servicing the vehicle, refer to the Precautions Section.
2. Disconnect and isolate the battery negative cable. Allow the system capacitor to discharge for 3 minutes before beginning any component service. This will disable the airbag system.

✻✻ CAUTION

Failure to disable the airbag system may result in accidental airbag deployment, personal injury, or death.

3. Remove the ignition key from the vehicle.
4. Connect the clockspring harness connector and horn harness connector to the clockspring.
5. Set the clockspring on neutral position and after turning the front wheels to the straight-ahead position, install the clockspring.

✻✻ CAUTION

Check connectors and protective tube for damage, and terminals for deformities. If even one abnormal point is discovered, replace the clockspring with a new one.

6. Connect the clockspring harness connector and the steering switch harness connector to the clockspring.
7. Set the center position by getting the marks between the clockspring and the cover into line.
8. Turn the clockspring clockwise to the stop and then 2–3 revolutions counterclockwise to match the marks.

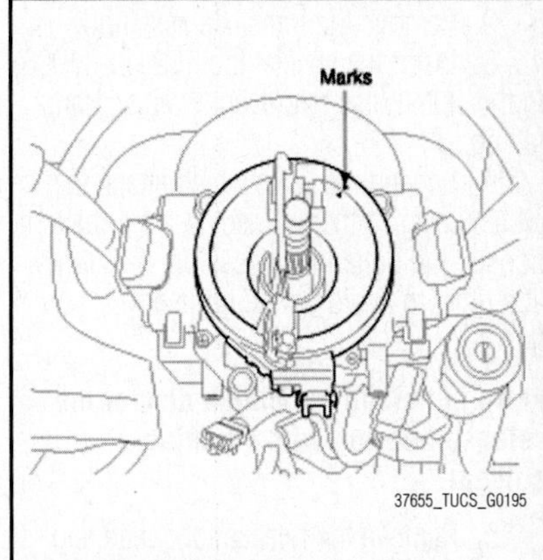

Fig. 45 Aligning the marks to center the clockspring—2009 models

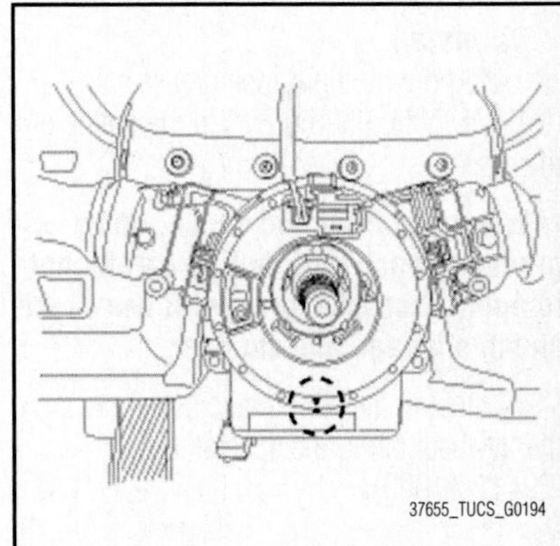

Fig. 46 Aligning the marks to center the clockspring–2010 models

DRIVE TRAIN

DRIVEN DISC & PRESSURE PLATE

REMOVAL & INSTALLATION

2009 Models

2.0L Engine

See Figures 47 through 51.

1. Before servicing the vehicle, refer to the Precautions Section.
2. Remove the transaxle assembly.
3. Insert the special tool (09411-11000) in the clutch disc to prevent the disc from falling.
4. Loosen the bolts which attach the clutch cover to the flywheel in a star pattern. Loosen the bolts in succession, 1–2 turns at a time, to avoid bending the cover flange.

➡Do not clean the clutch disc or the release bearing with cleaning solvent.

5. Remove the release fork shaft and bushing.
6. Remove clutch cover assembly and then driven disc and pressure plate.

To install:

7. Apply multipurpose grease (CASMOLY® L 9508 0.2gr) to the spline of the disc.

➡When installing the clutch, apply grease to each part, but be careful not to apply excessive grease. It can cause clutch slippage and shudder.

8. Install the clutch disc assembly to the flywheel using the special tool (09411-11000).

Fig. 47 Insert the special tool (09411-11000) in the clutch disc to prevent the disc from falling

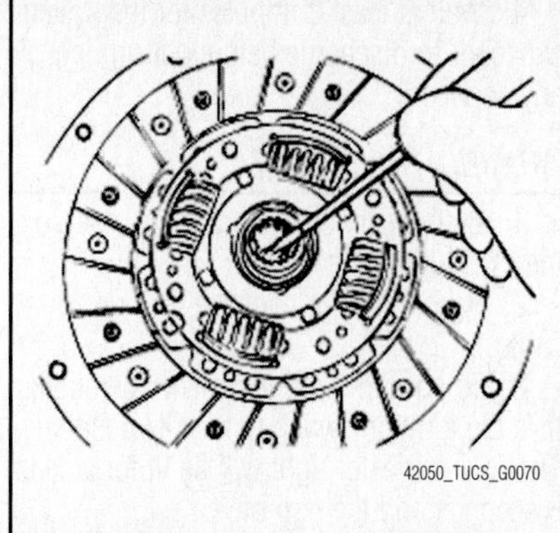

Fig. 48 Proper clutch greasing technique

9. Install the clutch cover assembly to the flywheel and temporarily tighten the bolts 1–2 steps at a time in a star pattern.
10. Tighten the clutch cover bolts to a final torque of 11–16 ft. lbs. (15–22 Nm).
11. Align the bearing to the release fork and then install it to the sleeve of the housing.
12. Apply multipurpose grease (CASMOLY® L9508) to the bearing sleeve, contact point of the release fork, and the bushing inner surface.
13. Install the release lever to the release fork.
14. Install the transaxle assembly to the engine.
15. Test the clutch for proper operation.

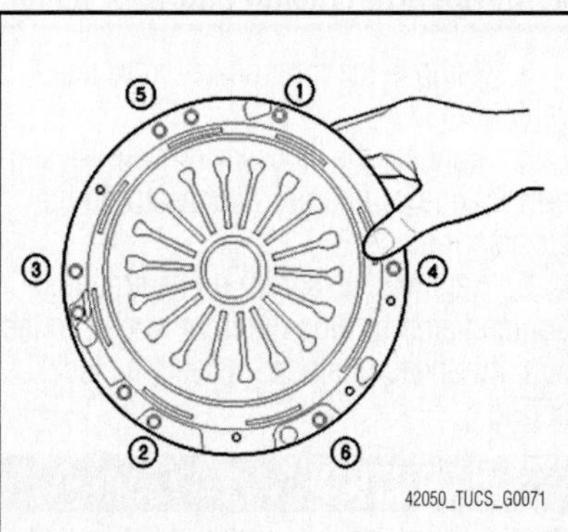

Fig. 49 Clutch cover bolt tightening sequence

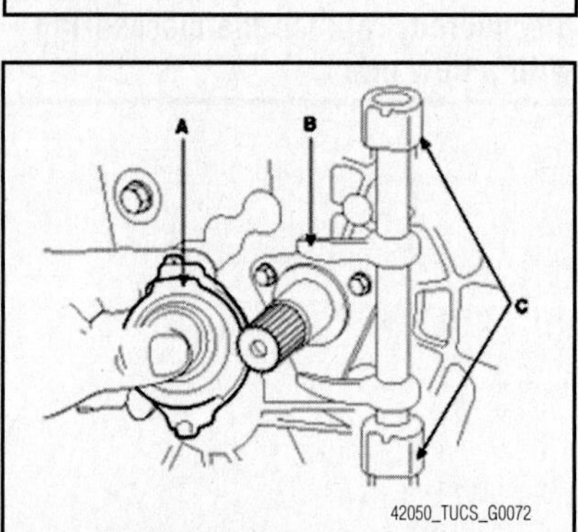

Fig. 50 Bearing sleeve (A), contact point of the release fork (B), and the bushing inner surface (C)

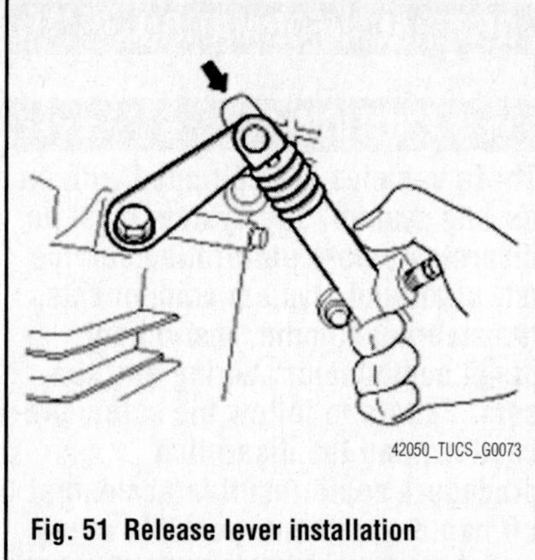

Fig. 51 Release lever installation

➡If the transaxle assembly is installed to the engine without installing the release lever to the release fork, the release bearing can be separated, as the release fork rotates freely.

2010 Models

See Figures 49, 52 through 54.

1. Before servicing the vehicle, refer to the Precautions Section.
2. Remove a transaxle assembly.
3. Remove the clutch cover bolts. To prevent them from being bent or twisted, loosen them in diagonal directions.
4. Remove the clutch cover and disc.

To install:

5. Using the SST (09411-11000), install the disc.

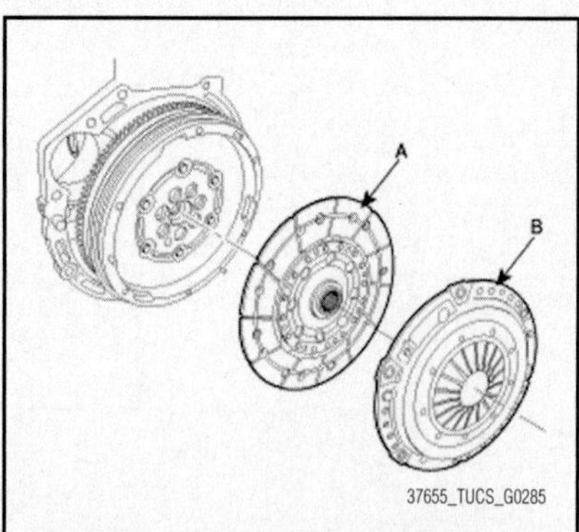

Fig. 52 Remove the clutch cover (A) and disc (B)

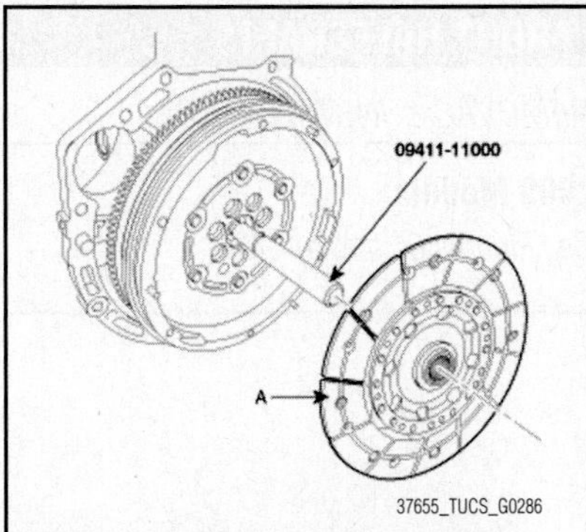

Fig. 53 Using the SST (09411-11000), install the disc (A)

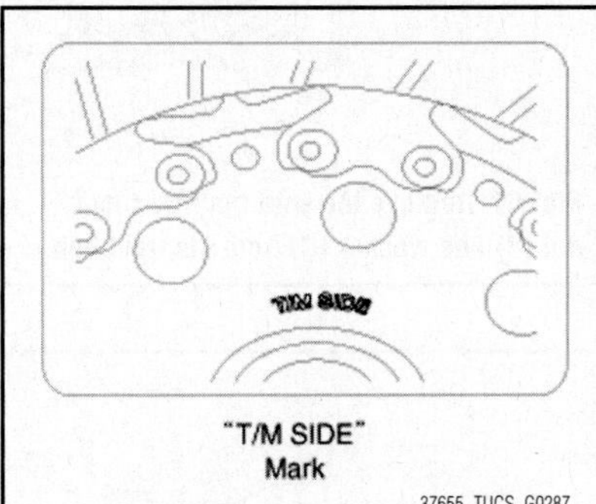

Fig. 54 The 'T/M SIDE' marked surface should face the transaxle

6. Apply grease (CASMOLY L9508, 0.2gr.) on a disc spline and transmission input shaft spline. The 'T/M SIDE' marked surface should face the transaxle.
7. Tighten the clutch cover.
8. Install the clutch cover assembly to the flywheel and temporarily tighten the bolts three steps at time in a star pattern.
9. Tighten the clutch cover bolts to a final torque of 18–26 ft. lbs. (25–34 Nm).

TRANSFER CASE ASSEMBLY

REMOVAL & INSTALLATION

2009 Models

2.0L Engine

1. Before servicing the vehicle, refer to the Precautions Section.
2. Remove the battery (-) terminal.
3. Raise and safely support the vehicle.
4. Remove the propeller shaft.
5. Remove the front muffler.
6. Remove the RH halfshaft.
7. Loosen the oil drain plug and drain the fluid.
8. After draining, re-tighten the oil drain plug to: 29–43 ft. lbs. (39–58 Nm).
9. Support the transfer assembly with a jack.
10. Remove the transfer case assembly by loosening the mounting bolts.
11. Remove the 2 transfer bracket mounting bolts together.
12. Remove the transfer case assembly.

To install:

➡Ensure that the transaxle is secured properly to the transaxle jack.

13. Position the transaxle onto a transaxle jack and secure the transaxle to the jack.
14. Install the transaxle into the vehicle.
15. Install the transfer case mounting bolts and tighten to: 45–49 ft. lbs. (61–66 Nm).
16. Fill the transfer case assembly with the proper type and amount of lubricant.
17. Connect the battery negative (-) cable.
18. Check for leaks.

2.7L Engine

See Figures 55 through 57.

The repair of the transfer assembly requires a special skill and furthermore an improper adjustment of the spacers may cause a severe noise and durability issue. The hypoid gear set is manufactured and controlled as a pair. Any replacement of the part is to be done as a pair, hypoid gear shaft assembly 47308-39200 and pinion shaft 47311-39000.

1. Remove the battery negative (-) cable.
2. Remove the air intake hose and air cleaner cover.
3. Disconnect the oxygen sensor connector left bank and right bank.
4. Remove the wheel and tire (RH).
5. Remove the engine side cover (RH).
6. Raise and safely support the vehicle.
7. Remove the lower arm ball joint mounting bolts.
8. Remove the lower arm ball joint from the front axle steering knuckle.
9. Remove the steering bar tie rod ball joint from the knuckle.
10. Remove the drive shaft from the transfer case assembly.
11. Remove the front exhaust pipe.
12. Drain the transfer oil through drain plug hole.
13. Remove the pinion case mounting bolts (A).
14. Remove the alternator drive belt using the tensioner and then alternator assembly by loosening the mounting bolts.

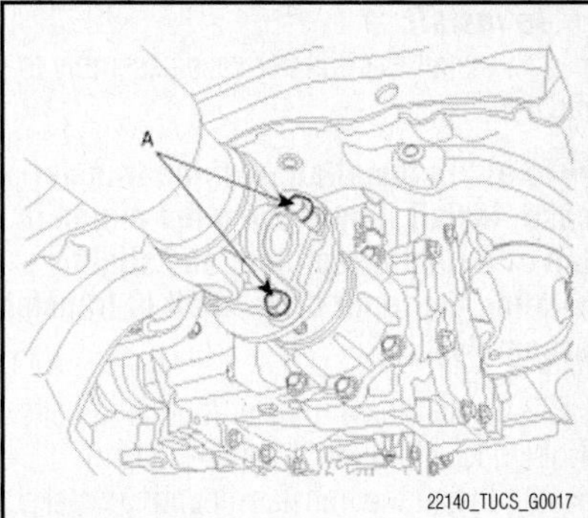

Fig. 55 Remove the pinion case mounting bolts (A)

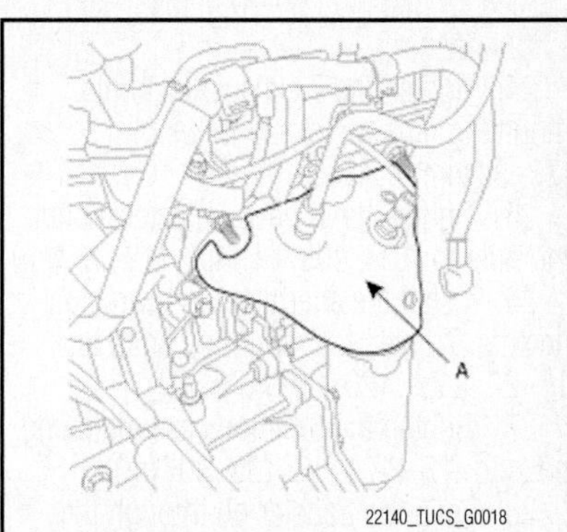

Fig. 56 Remove the heat protector (A) using a hexagonal socket

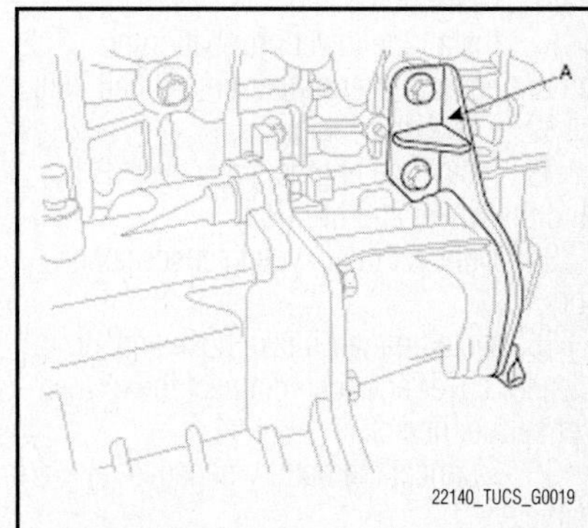

Fig. 57 Remove the transfer mounting bracket (A)

15. Remove the alternator wire terminal mounting nut and connector.
16. Remove the heat protector (A) using a hexagonal socket.
17. Remove the 7 exhaust manifold mounting nuts.
18. Remove the transfer mounting bracket (A).
19. Remove the 4 transfer mounting bolts.
20. Using a flat head screw driver, carefully remove the transfer case assembly from the transaxle.

To install:

21. Install the transfer case assembly to the transaxle.

➡To aid in installation of the transfer case, install it by moving the pinion in left and right directions, and slightly rotating the inner drive shaft of transfer assembly.

22. Install the 4 transfer mounting bolts. Tighten to: 45–49 ft. lbs. (61–66 Nm).
23. Install the transfer mounting bracket. Tighten to:
    a. 2 bolts to: 34–37 ft. lbs. (46–50 Nm).
    b. 3 bolts to: 17–20 ft. lbs. (24–28 Nm).
24. Tighten the 7 exhaust manifold mounting nuts to: 22–25 ft. lbs. (29–34 Nm).
25. Tighten the 3 heat protector mounting nuts to: 104–130 inch lbs. (12–15 Nm).
26. Install the alternator and wire connectors. Tighten the mounting bolts to: 15–22 ft. lbs. (20–29 Nm).
27. Tighten the 6 pinion case mounting bolts to: 15–22 ft. lbs. (36–39 Nm).
28. Refill the transfer oil through the filler plug.
29. Install the front exhaust pipe.
30. Install the drive shaft (RH) to the transfer assembly.
31. Install the lower arm ball joint mounting and steering bar tie rod ball joint.
32. Lower the vehicle.
33. Install the engine side cover (RH) and the wheel and tire (RH).
34. Connect the oxygen sensor connectors.
35. Install the air intake hose and air cleaner cover and then connect the oxygen sensor connector.
36. Connect the battery negative (-) cable.
37. Check for leaks.

2010 Models

See Figures 58 through 60.

1. Disconnect the (-) terminal from the battery.
2. Lift up the vehicle.
3. Remove the front muffler
4. Remove the propeller shaft bolts.
5. Matchmark propeller shaft.
6. Remove the right driveshaft (passenger side) from the transfer case.
7. Remove the roll bracket after removing bolts.
8. Remove the transfer case up and down mounting bolts.
9. Remove the transfer case with the lever after supporting the transfer case with a jack.

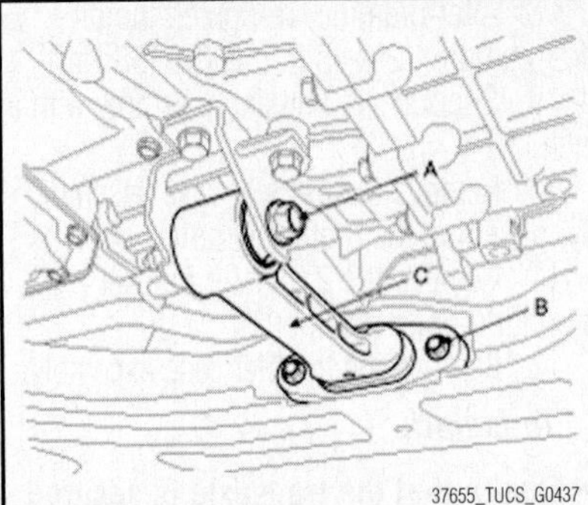

Fig. 58 Remove the roll bracket (C) after removing bolts (A,B)

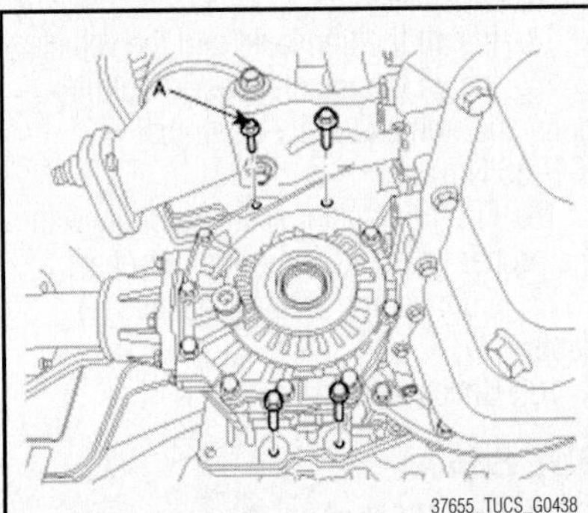

Fig. 59 Remove the transfer case up and down mounting bolts

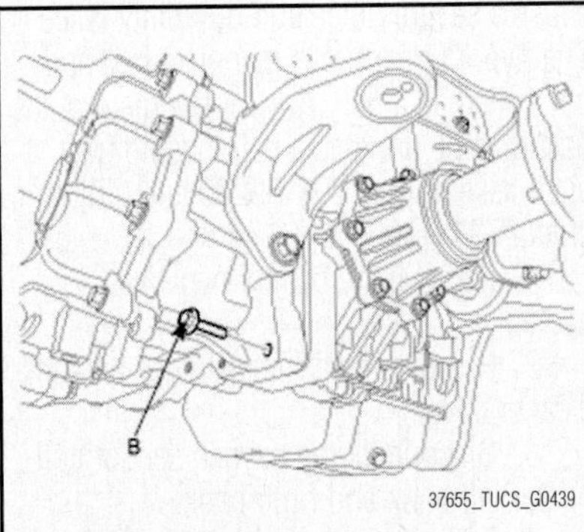

Fig. 60 Remove the transfer case lower mounting bolt

To install:

10. Installation is the reverse of removal.
    a. Tighten propeller shaft bolts to 36.2–50.6 ft. lbs. (49.0–68.6 Nm).
    b. Tighten roll bracket bolt (A) to 79.6–94.1 ft. lbs. (107.9–127.5 Nm).
    c. Tighten roll bracket bolt (B) to 36.2–47.0 ft. lbs. (49.0–63.7 Nm).
    d. Tighten up and down mounting bolts to 44.8–48.5 ft. lbs. (60.8–65.7 Nm).

FRONT HALFSHAFTS

REMOVAL & INSTALLATION

2009 Models

See Figures 61 through 70.

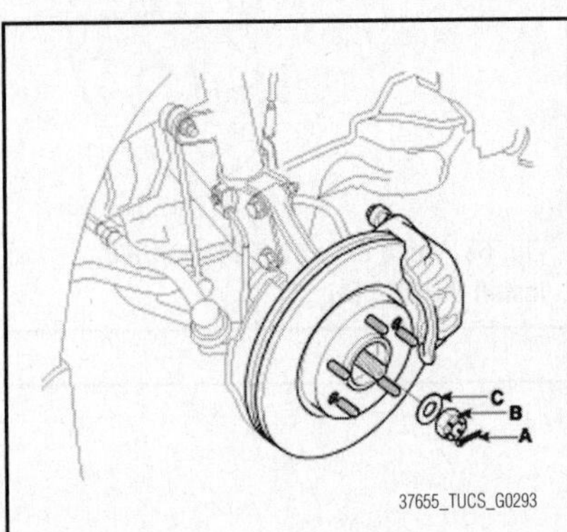

Fig. 61 Remove the split pin (A), castle nut (B) and washer (C) from the front hub

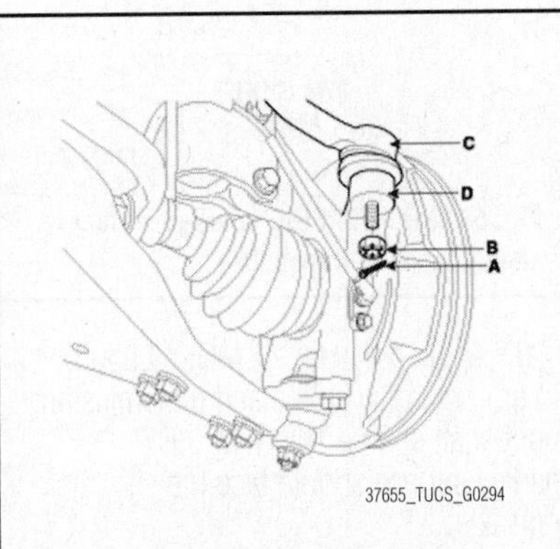

Fig. 62 Disconnect the tie rod end ball joint (C) from the knuckle (D), and remove the split pin (A) and castle nut (B)

1. Before servicing the vehicle, refer to the Precautions Section.
2. Remove the front wheels and tires from the front hub.
3. Remove the split pin, castle nut and washer from the front hub while applying the brake.
4. Remove the wheel speed sensor from the knuckle. Refer to ABS, Wheel Speed Sensor, Removal & Installation, in the Brake section.
5. Disconnect the tie rod end ball joint from the knuckle:
    a. Remove the split pin.
    b. Remove the castle nut.
    c. Disconnect the ball joint from knuckle using the special tool (09568-34000).

➡Apply a few drops of oil to the special tool (boot contact part).

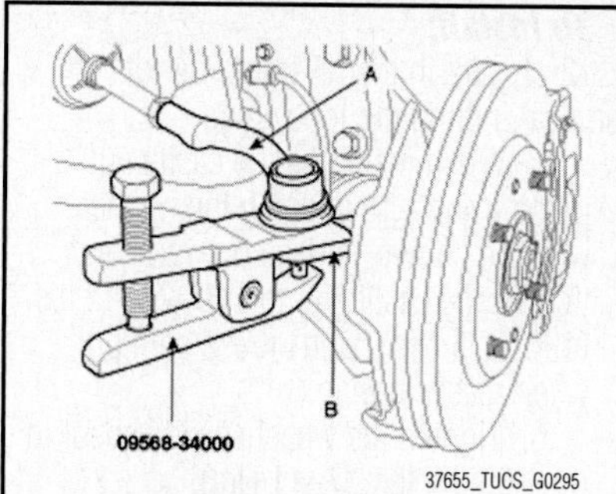

Fig. 63 Disconnect the ball joint from knuckle using the special tool (09568-34000)

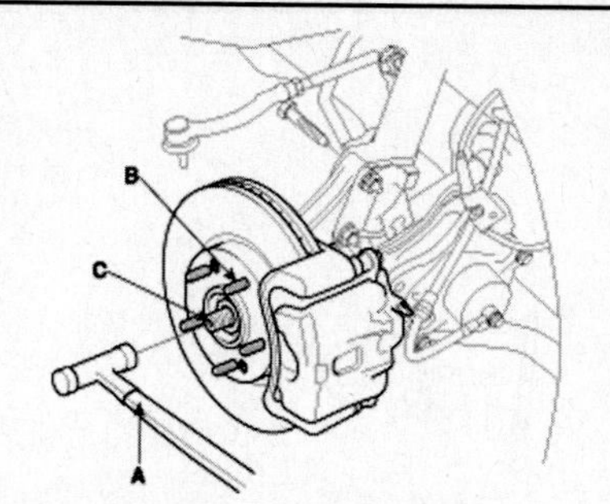

Fig. 66 Using a plastic hammer (A), disconnect the halfshaft (C) from the axle hub (B)

6. Remove the lower arm ball joint mounting bolts.

7. Using a plastic hammer, disconnect the halfshaft from the axle hub.

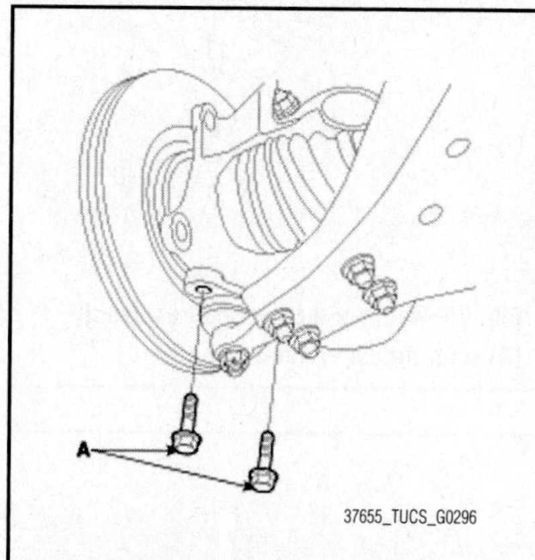

Fig. 64 Remove the lower arm ball joint mounting bolts (A)

8. Push the axle hub outward and separate the halfshaft from the axle hub.

9. Insert a pry bar between the transaxle case and joint case, and separate the halfshaft from the transaxle case.

➡Be careful not to damage the transaxle and joint. Do not insert the pry bar too deep, as this may damage the oil seal. Do not pull the halfshaft by excessive force. Plug the hole of the transaxle case with the oil seal cap to prevent contamination. Support the halfshaft properly. Replace the retainer ring whenever the halfshaft is removed from the transaxle case.

To install:

10. Apply gear oil on the halfshaft oil seal case contacting surface and transaxle case splines.

11. Before installing the halfshaft, position the opening side of the circlip facing downward.

12. After installation, check that the halfshaft cannot be removed by hand.

13. Install the ball joint into the knuckle.

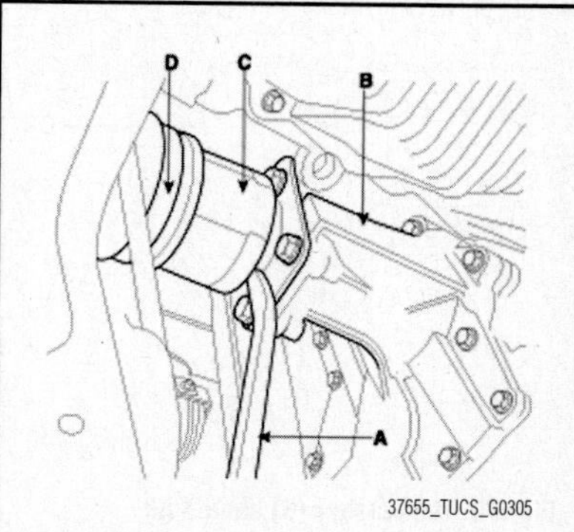

Fig. 67 Insert a pry bar (A) between the transaxle case (B) and joint case (C), and separate the halfshaft (D) from the transaxle case

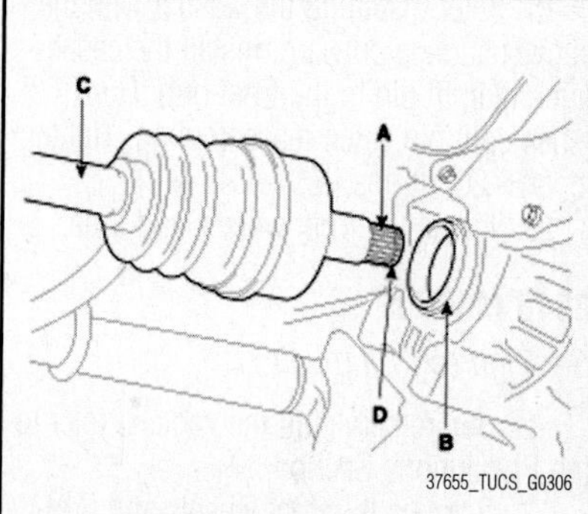

Fig. 68 Apply gear oil on the halfshaft oil seal case contacting surface (B) and transaxle case splines (A), position the opening side of the circlip (D) downward, and install the halfshaft (C)—2WD vehicles

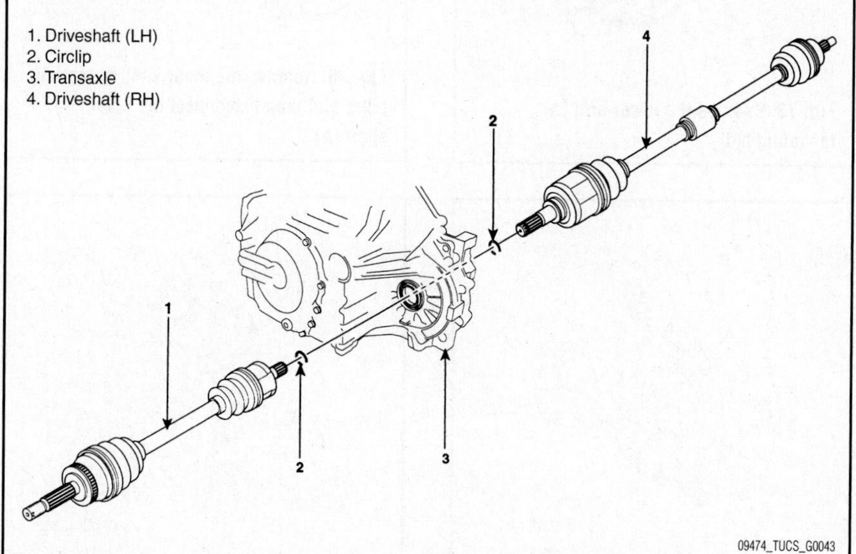

Fig. 65 Exploded view of the front halfshaft and related components—2009 models

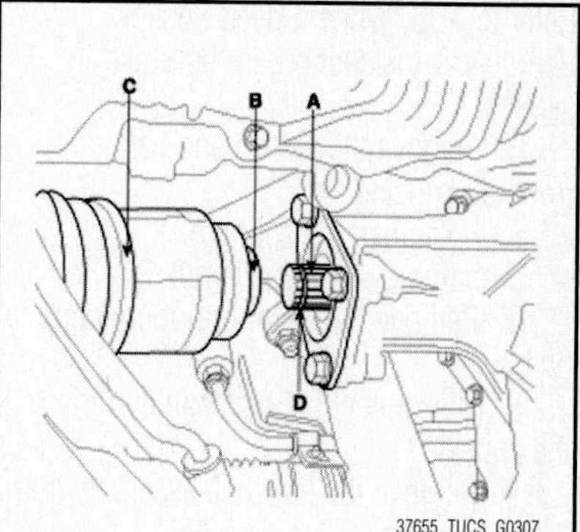

Fig. 69 Apply gear oil on the halfshaft oil seal case contacting surface (B) and transaxle case splines (A), position the opening side of the circlip (D) downward, and install the halfshaft (C)—4WD vehicles

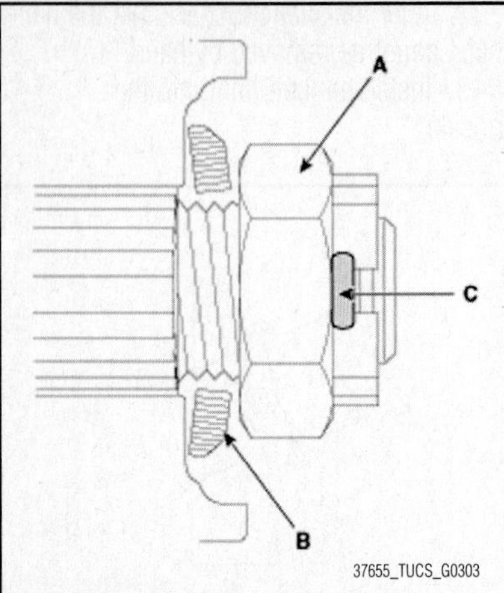

Fig. 70 The washer (B) should be assembled with convex surface outward when installing the castle nut (A) and split pin (C)

14. Install the lower arm mounting bolts.

15. After installing the washer with the convex surface outward, install the castle nut and split pin to the front hub. Don't reuse split pin when reassembling. Tighten to 148–207 ft. lbs. (200–280 Nm).

16. Install the front wheels and tires.

2010 Models

See Figures 71 through 77.

1. Before servicing the vehicle, refer to the Precautions Section.
2. Remove the front wheels and tires from the front hub.
3. Remove the caliper mounting bolts, and hang the caliper assembly to one side with a wire.
4. Remove the castle nut from the front hub while applying the brake.
5. Remove the wheel speed sensor. Refer to ABS, Wheel Speed Sensor, Removal & Installation, in the Brake section.
6. Remove the tie rod end ball joint from the knuckle:
   a. Remove the split pin.
   b. Remove the castle nut.
7. Remove the lower arm mounting bolt.
8. Disconnect the halfshaft end from the knuckle.
9. Remove the halfshaft assembly from the inner shaft.
10. Remove the inner shaft mounting bolts and then disconnect the inner shaft.
11. Insert a pry bar between the transaxle case and joint case and separate the halfshaft from the transaxle case.

Fig. 71 Remove the castle nut (A)

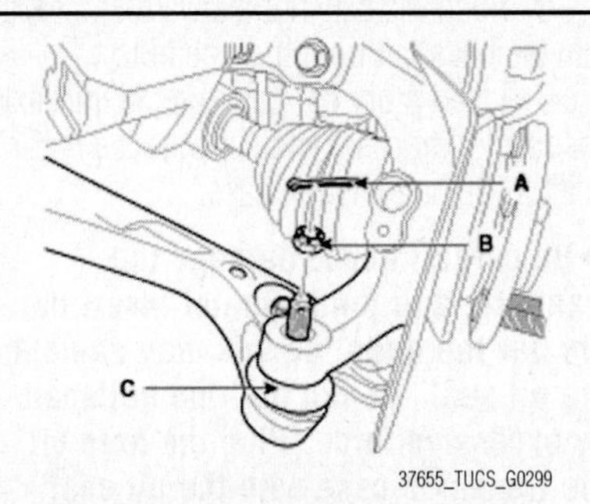

Fig. 72 Disconnect the tie rod end ball joint (C) from the knuckle (D), and remove the split pin (A) and castle nut (B)

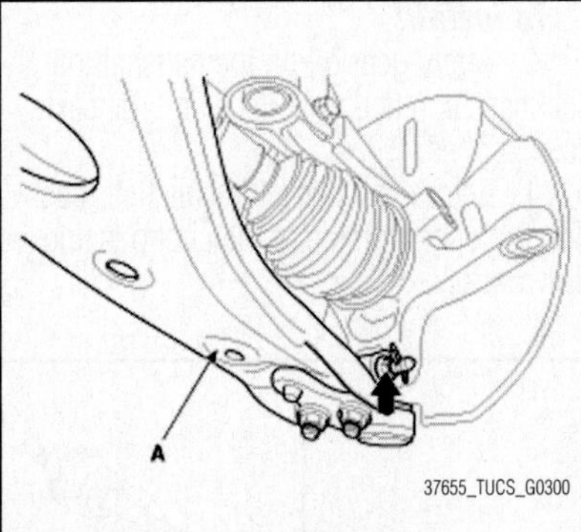

Fig. 73 Remove the lower arm (A) mounting bolt

Fig. 74 Disconnect the halfshaft end (A) from the knuckle

To install:

12. Install in the reverse order of removal, noting the following:
   a. When installing the castle nut to the front hub, position the washer with the convex surface outward, and tighten the castle nut to 145–203 ft. lbs. (196–275 Nm). Don't reuse split pin when reassembling.
   b. Tighten the wheel speed sensor to 61–95 inch lbs. (7–11 Nm).
   c. Tighten the tie rod end ball joint mounting nut to 25–33 ft. lbs. (34–44 Nm).
   d. Tighten the lower arm mounting bolt to 72–87 ft. lbs. (98–118 Nm).

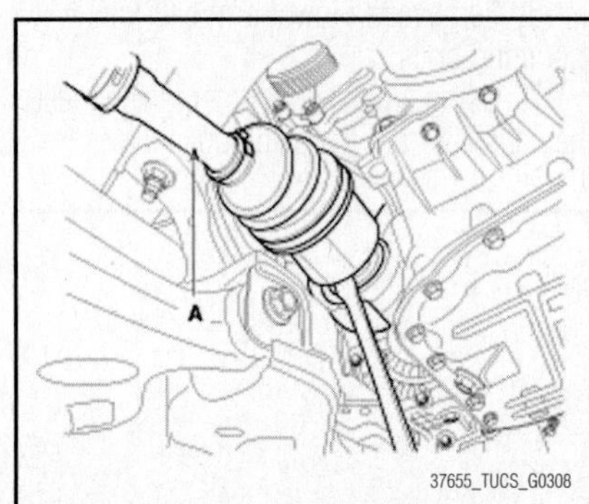

Fig. 75 Remove the halfshaft assembly (A) from the inner shaft

Fig. 76 Remove the inner shaft mounting bolts and then disconnect the inner shaft (A)

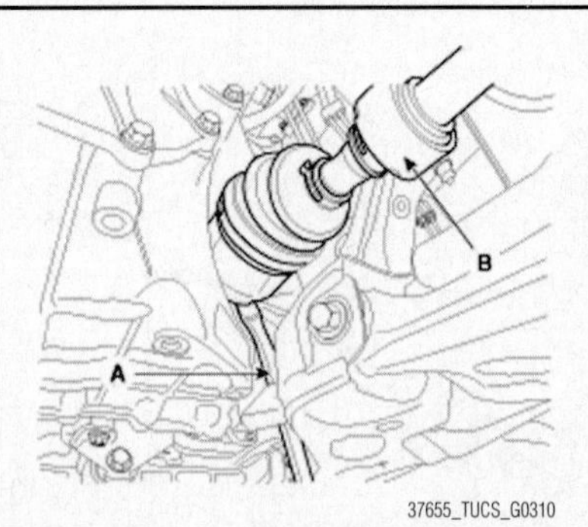

Fig. 77 Use a pry bar (A) and separate the halfshaft (B) from the transaxle case

➡Be careful not to damage the transaxle and joint. Do not insert the pry bar too deep, as this may damage the oil seal. Do not pull the halfshaft by excessive force. Plug the hole of the transaxle case with the oil seal cap to prevent contamination. Support the halfshaft properly. Replace the retainer ring whenever the halfshaft is removed from the transaxle case.

REAR DIFFERENTIAL

REMOVAL & INSTALLATION

2009 Models

See Figures 78 through 81.

1. Before servicing the vehicle, refer to the Precautions Section.
2. Drain the differential gear oil.
3. Remove the rear halfshaft. Refer to Halfshaft Removal & Installation.
4. Remove the propeller shaft. Refer to Propeller Shaft Removal & Installation.
5. Support the differential assembly with the jack.
6. Disconnect the coupling control connector.
7. After removing the differential mounting bolts, remove the differential.
8. After removing the cover bolts, remove the differential cover.

To install:

9. Apply liquid gasket to the differential cover.
10. Install the mounting bolts and tighten to 30–37 ft. lbs. (40–50 Nm).
11. Install the differential and tighten the mounting bolts to 66–89 ft. lbs. (90–120 Nm).
12. Using the transaxle jack, install the differential assembly.
13. Connect the coupling control connector.
14. Install the propeller shaft.
15. Install the rear halfshaft.

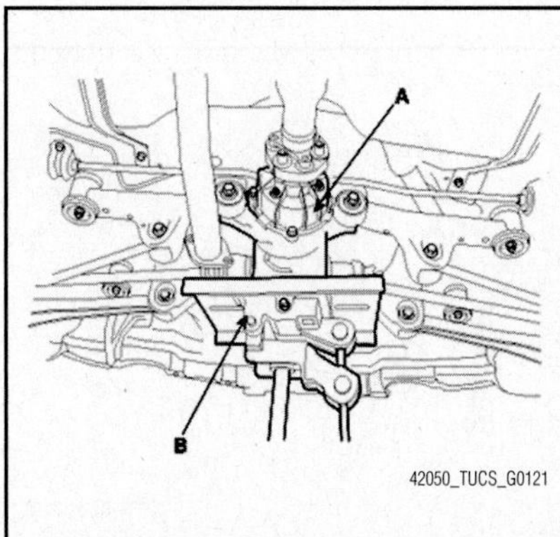

Fig. 78 Support the differential assembly (A) with the jack (B)

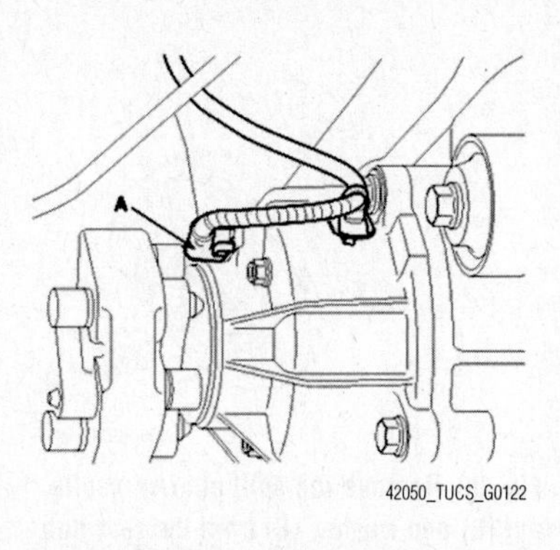

Fig. 79 Coupling control connector (A)

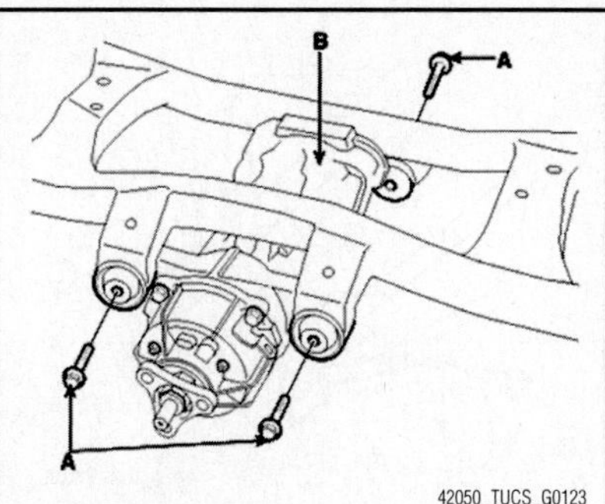

Fig. 80 After loosening the differential mounting bolts (A), remove the differential (B)

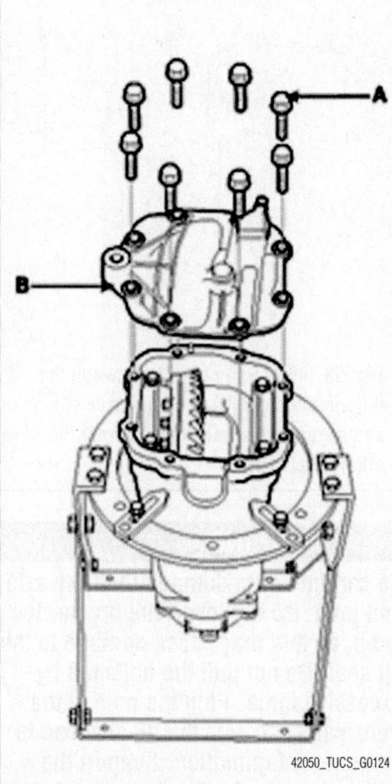

Fig. 81 After removing the cover bolts (A), remove the differential cover (B)

16. Fill the gear oil with Hypoid gear oil (GL-5, SAE 80W/90). Fill the reservoir to the plug hole (approx. 0.75–0.80L).

2010 Models

See Figures 82 through 85.

1. Before servicing the vehicle, refer to the Precautions Section.
2. Drain the differential gear oil.
3. Remove the rear halfshaft. Refer to Halfshaft Removal & Installation.
4. Remove the propeller shaft. Refer to Propeller Shaft Removal & Installation.

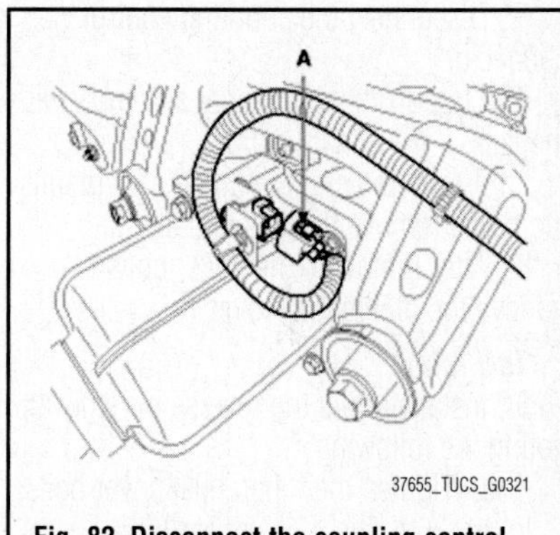

Fig. 82 Disconnect the coupling control connector (A)

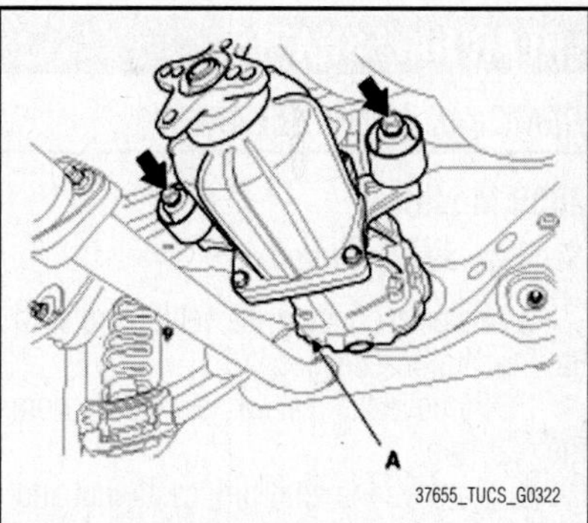

Fig. 83 Support the differential assembly (A) with the jack

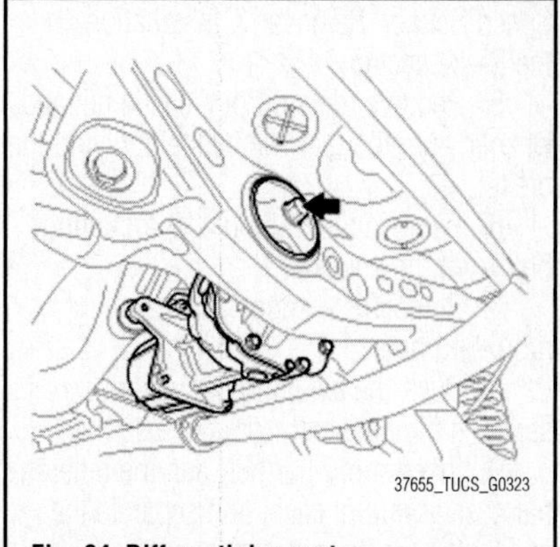

Fig. 84 Differential mount

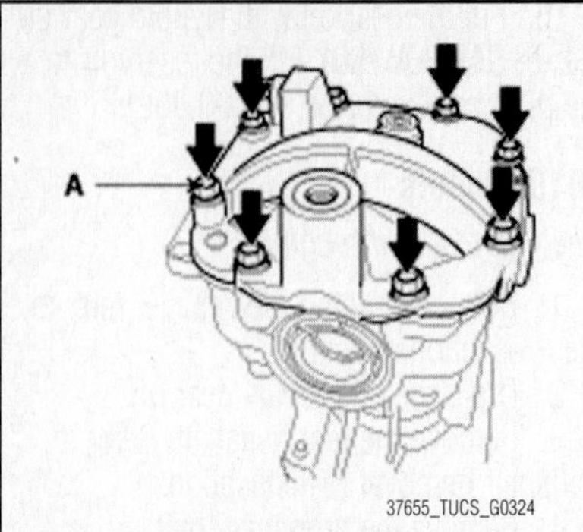

Fig. 85 After removing the cover bolts, remove the differential cover (A)

5. Disconnect the coupling control connector.
6. Support the differential assembly with the jack.
7. After removing the differential mounting bolts, remove the differential.
8. After removing the cover bolts, remove the differential cover.

To install:

9. Installation is the reverse of removal, noting the following:
    a. Tighten the differential cover bolts to 29–36 ft. lbs. (39–49 Nm).
    b. Tighten the differential mounting bolts to 51–65 ft. lbs. (69–88 Nm).

REAR HALFSHAFTS

REMOVAL & INSTALLATION

2009 Models

See Figures 86 through 88.

1. Before servicing the vehicle, refer to the Precautions Section.
2. Remove the rear wheel and tire from the rear hub.
3. Remove the split pin, castle nut and washer from the rear hub while applying the brake.
4. Remove the wheel speed sensor from the axle carrier. Refer to ABS, Wheel Speed Sensor, Removal & Installation, in the Brake section.
5. Remove the split pin, castle nut and washer from the rear hub while applying the brake.
6. Remove the trailing arm mounting bolt from the knuckle.
7. Remove the suspension arm mounting nuts.
8. Push the axle hub outward and separate the halfshaft from the axle hub.
9. Insert a pry bar between the differential case and joint case, and separate the halfshaft from the differential case.

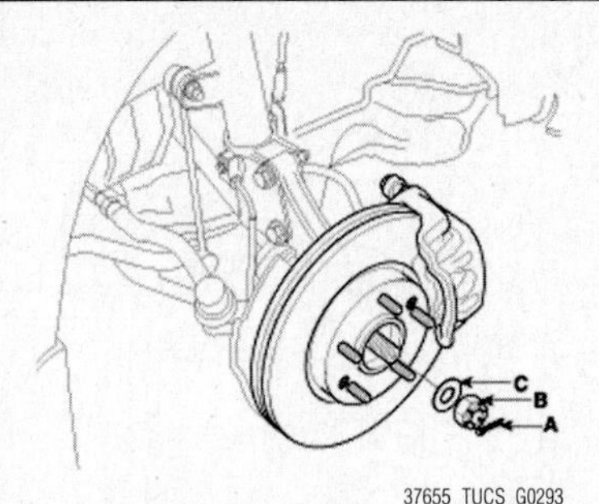

Fig. 86 Remove the split pin (A), castle nut (B) and washer (C) from the rear hub while applying the brake

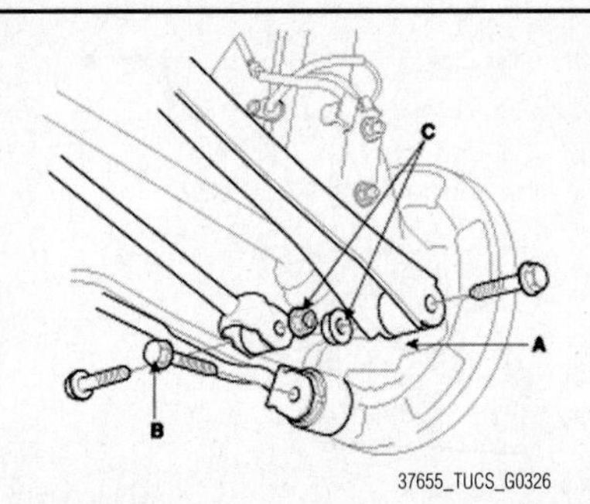

Fig. 87 Remove the trailing arm mounting bolt (B) from the knuckle (A) and remove the suspension arm mounting nuts (C)

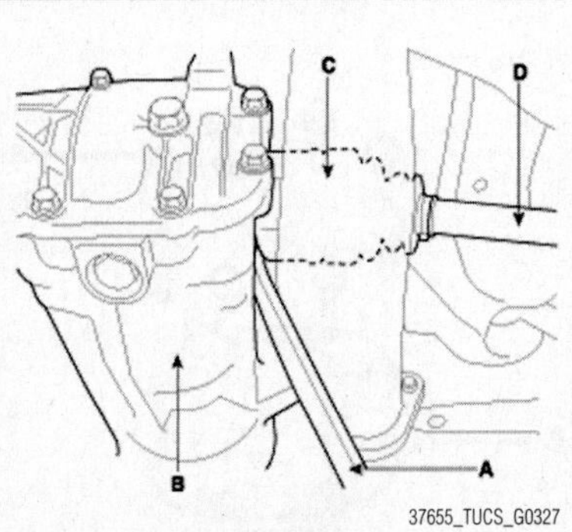

Fig. 88 Insert a pry bar (A) between the differential case (B) and joint case (C), and separate the halfshaft (D) from the differential case

WARNING

Be careful not to damage the transaxle and joint. Do not insert the pry bar too deep, as this may cause damage to the oil seal. Do not pull the halfshaft by excessive force. Plug the hole of the transaxle case with the oil seal cap to prevent contamination. Support the halfshaft properly. Replace the retainer ring whenever the halfshaft is removed from the transaxle case.

To install:

10. Apply gear oil on the halfshaft differential case contacting surface and halfshaft splines.
11. Before installing the halfshaft, set the opening side of the circlip facing downward.
12. After installation, check that the halfshaft cannot be removed by hand.
13. Install the BJ. Into the knuckle.
14. Install the suspension arm mounting nuts and tighten to 104–118 ft. lbs. (140–160 Nm).
15. Install the trailing arm mounting bolt and tighten to 74–89 ft. lbs. (100–120 Nm). and.
16. Install the washer, castle nut and split pin and tighten to 148–207 ft. lbs. (200–280 Nm).
17. Install the wheel speed sensor.
18. Install the rear wheel and tire.

2010 Models

See Figures 89 through 95.

1. Before servicing the vehicle, refer to the Precautions Section.
2. Remove the rear wheel and tire from the rear hub.
3. Remove the brake caliper mounting bolts, and then position the brake caliper assembly aside with a wire.
4. Remove castle nut from the rear hub while applying the break.
5. Remove the rear brake lining assembly.
6. Remove the parking brake cable from the brake shoe and remove the parking brake cable retaining ring. Refer to Parking Brake Shoe Removal & Installation in the Brake section.
7. Remove the wheel speed sensor from the knuckle.
8. Remove the assist arm from the rear axle carrier.
9. Remove the trailing arm from the rear axle carrier.

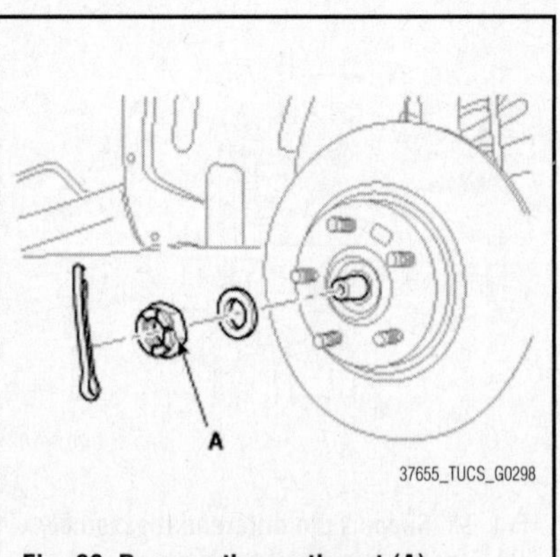

Fig. 89 Remove the castle nut (A)

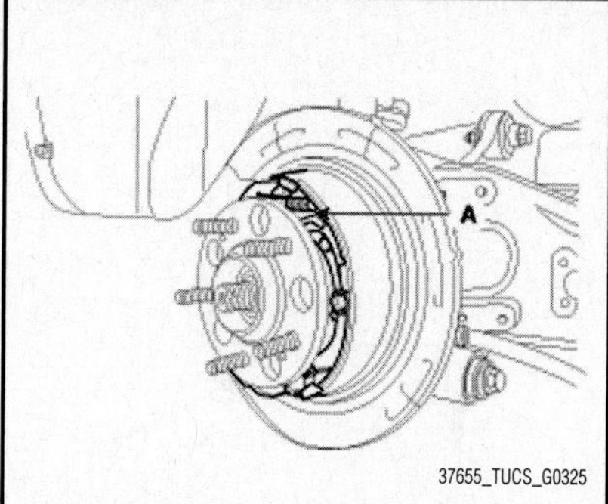

Fig. 90 Remove the rear brake lining assembly (A)

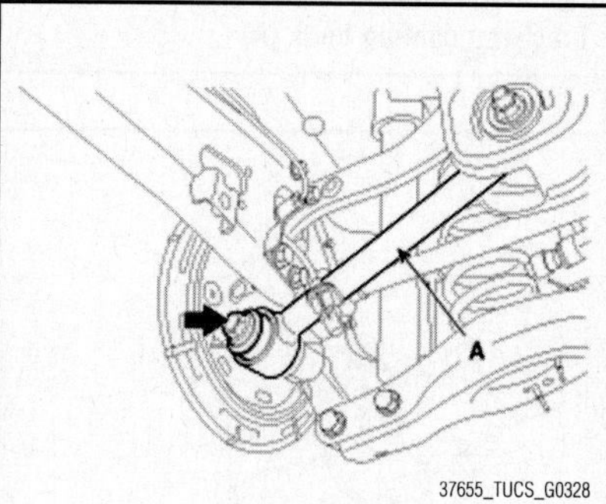

Fig. 91 Remove the assist arm (A) from the rear axle carrier

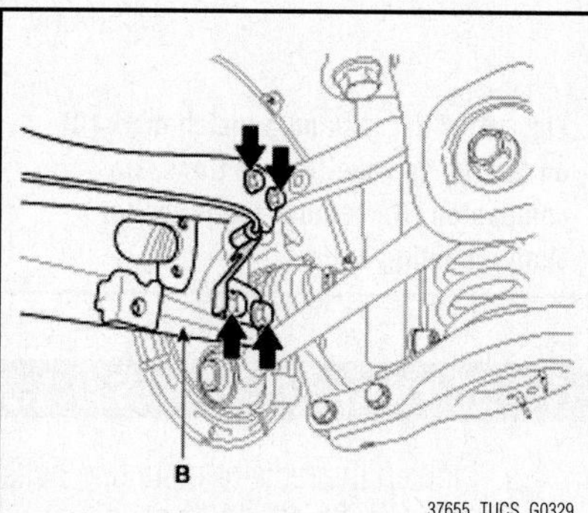

Fig. 92 Remove the trailing arm (B) from the rear axle carrier

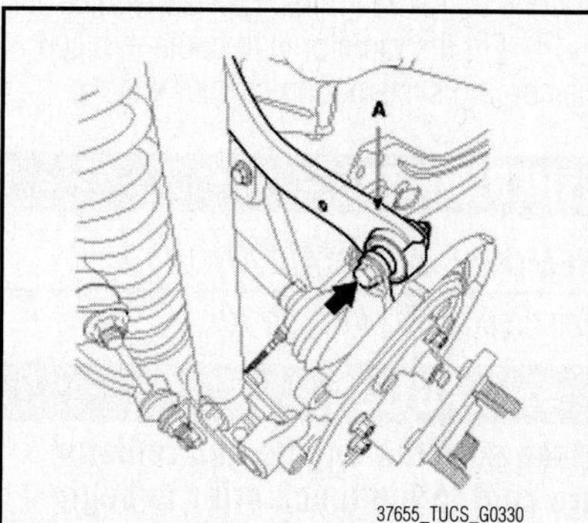

Fig. 93 Remove the upper arm (A) from the rear axle carrier

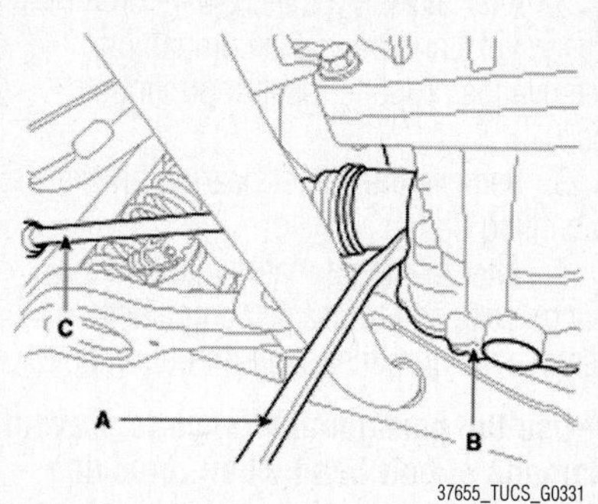

Fig. 94 Insert a pry bar (A) between the differential case and joint case, and separate the halfshaft (B) from the differential case

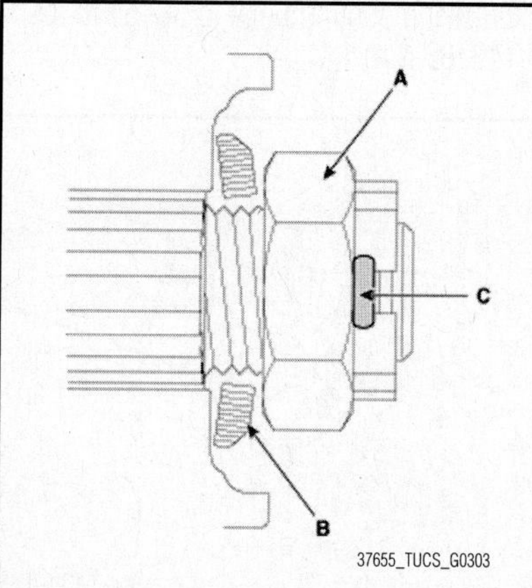

Fig. 95 The washer (B) should be assembled with convex surface outward when installing the castle nut (A) and split pin (C)

10. Remove the upper arm from the rear axle carrier.
11. Push the rear axle carrier outward and separate the halfshaft from the axle hub.
12. Insert a pry bar between the differential case and joint case, and separate the halfshaft from the differential case.

WARNING

Be careful not to damage the differential and joint. Do not insert the pry bar too deep, as this may cause damage to the oil seal. Do not pull the halfshaft by excessive force. Plug the hole of the differential case with the oil seal cap to prevent contamination. Support the halfshaft t properly. Replace the retainer ring whenever the halfshaft is removed from the differential case.

To install:

13. Installation is the reverse of removal, noting the following:

a. Tighten the castle nut to 145–203 ft. lbs. (196–275 Nm).

b. Tighten the assist arm to 101–116 ft. lbs. (137–157 Nm).

c. Tighten the trailing arm to 25–40 ft. lbs. (34–54 Nm).

d. Tighten the upper arm to 72–87 ft. lbs. (98–118 Nm).

PROPELLER SHAFT

REMOVAL & INSTALLATION

2009 Models

See Figures 96 through 98.

1. Before servicing the vehicle, refer to the Precautions Section.
2. After making a match mark on the rubber coupling and rear differential companion, remove the propeller shaft mounting bolts.
3. Remove the center bearing bracket mounting bolts.
4. After making a matchmark on the flange yoke and transaxle companion, remove the propeller shaft mounting bolts.

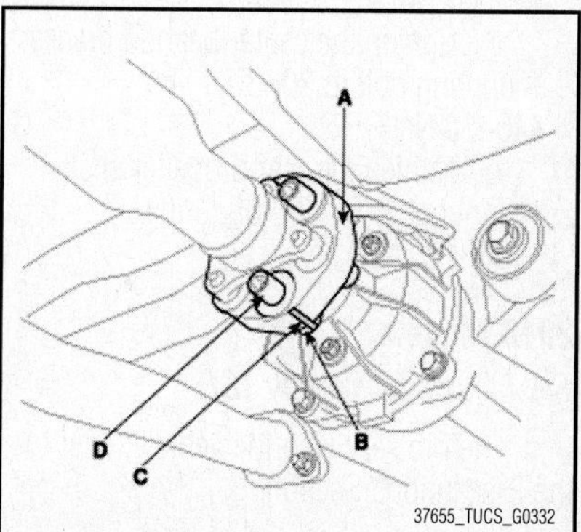

Fig. 96 After making a match mark (C) on the rubber coupling (A) and rear differential companion (B), remove the propeller shaft mounting bolts (D)

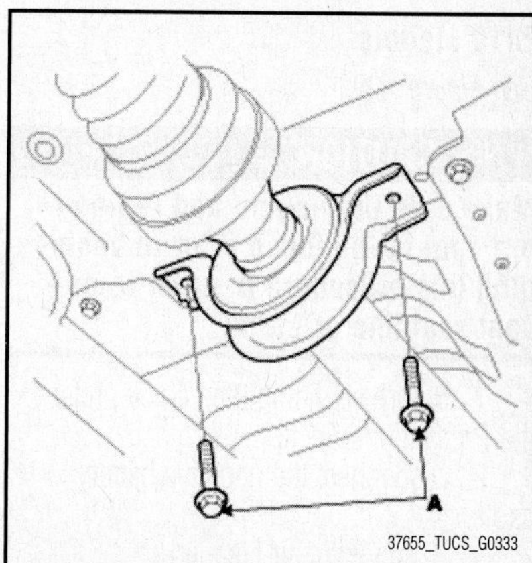

Fig. 97 Remove the center bearing bracket mounting bolts (A)

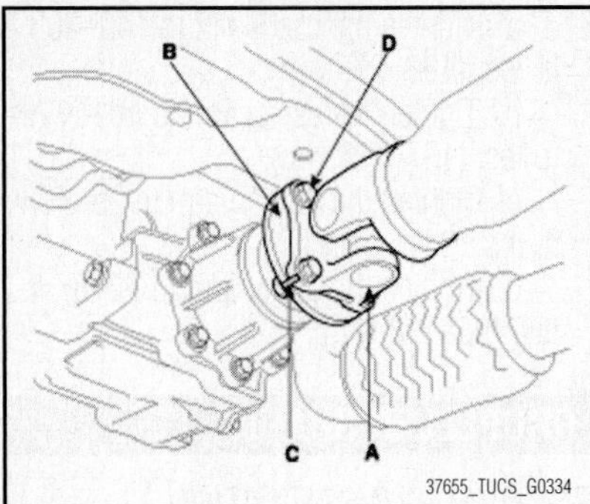

Fig. 98 After making a matchmark (C) on the flange yoke (A) and transaxle companion (B), remove the propeller shaft mounting bolts (D)

To install:

5. Installation is the reverse of removal, noting the following:
   a. Install according to match mark of the transaxle companion (or rear differential companion) and propeller shaft.
   b. Tighten the front propeller shaft mounting bolt to 37–44 ft. lbs. (50–60 Nm).
   c. Tighten the center bearing bracket mounting bolt to 30–37 ft. lbs. (40–50 Nm).
   d. Tighten the rear propeller shaft mounting bolt to 74–89 ft. lbs. (100–120 Nm).

2010 Models

See Figures 99 through 101.

1. Before servicing the vehicle, refer to the Precautions Section.
2. After making a match mark on the flange yoke and transaxle companion, remove the propeller shaft mounting bolts.
3. Remove the center bearing bracket mounting bolts.
4. After making a match mark on the flange yoke and transaxle companion, remove the propeller shaft mounting bolts.

➡Use the hexagonal wrench to prevent damage of bolt head when removing bolts.

To install:

5. Installation is the reverse of removal, noting the following:
   a. Tighten the propeller shaft mounting bolts and the center bearing bracket mounting bolts to 36–51 ft. lbs. (49–67 Nm).

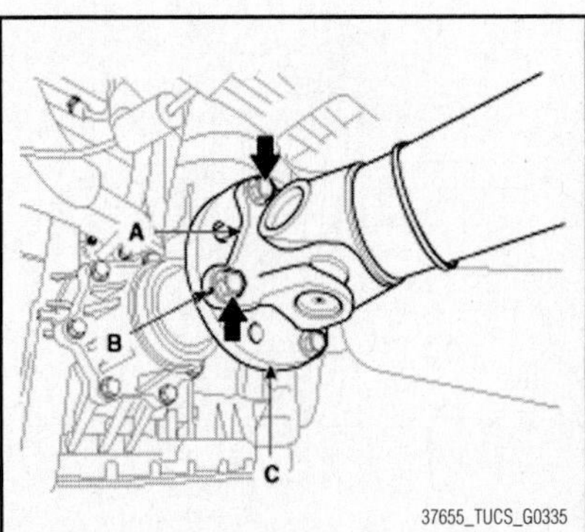

Fig. 99 After making a match mark (B) on the flange yoke (A) and transaxle companion (C), remove the propeller shaft mounting bolts

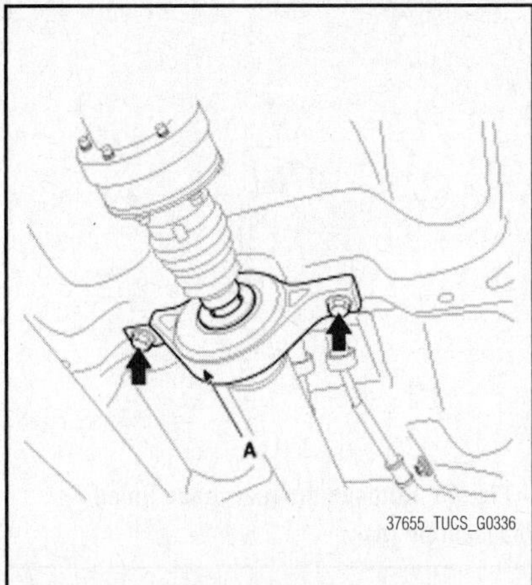

Fig. 100 Remove the center bearing bracket mounting bolts (A)

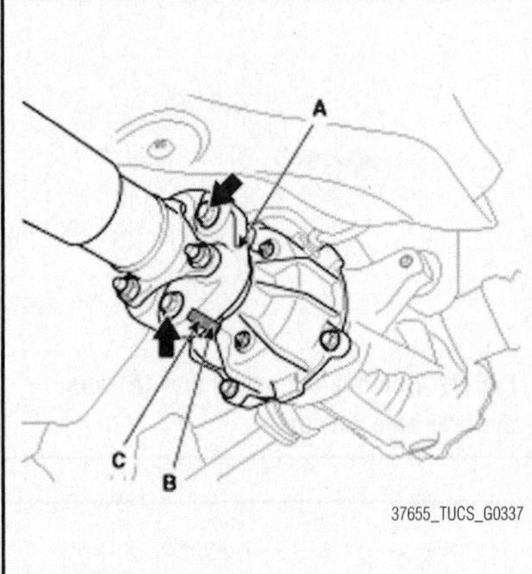

Fig. 101 After making a match mark (C) on the flange yoke (A) and transaxle companion (B), remove the propeller shaft mounting bolts

ENGINE COOLING

RADIATOR

REMOVAL & INSTALLATION

2010 Models

See Figure 102.

✲✲ CAUTION

Make sure the engine and radiator are cool to the touch prior to beginning this procedure in order to prevent scalding or burns.

1. Before servicing the vehicle, refer to the Precautions Section.
2. Disconnect the negative battery cable.
3. Remove the air cleaner assembly.
4. Remove the undercover.
5. Drain the cooling system. Remove the radiator cap to speed draining.
6. Remove the radiator upper hose (A) and lower hose (B).
7. Disconnect the ATF cooler hoses (A/T only).
8. Disconnect the fan motor connector (A) and then remove the cooling fan assembly.
9. Remove the front bumper.
10. Disconnect the over flow hose (A) from the radiator.
11. Remove the cover (A) and the air guard (B) and then separate the condenser from the radiator.
12. Remove the radiator mounting bracket (A) and then pull the radiator upper from vehicle.

To install:

13. Installation is in the reverse order of removal, noting the following:
    a. Tighten the radiator mounting bolts to 78–95 inch lbs. (9–11 Nm).
    b. Tighten the cooling fan mounting bolts to 78–95 inch lbs. (9–11 Nm).
    c. Tighten the undercover mounting bolts to 15–18 ft. lbs. (20–25 Nm).
14. Fill the radiator with coolant, bleed the cooling system, and check for leaks.

THERMOSTAT

REMOVAL & INSTALLATION

See Figures 103 through 105.

✲✲ CAUTION

Make sure the engine and radiator are cool to the touch prior to beginning this procedure in order to prevent scalding or burns.

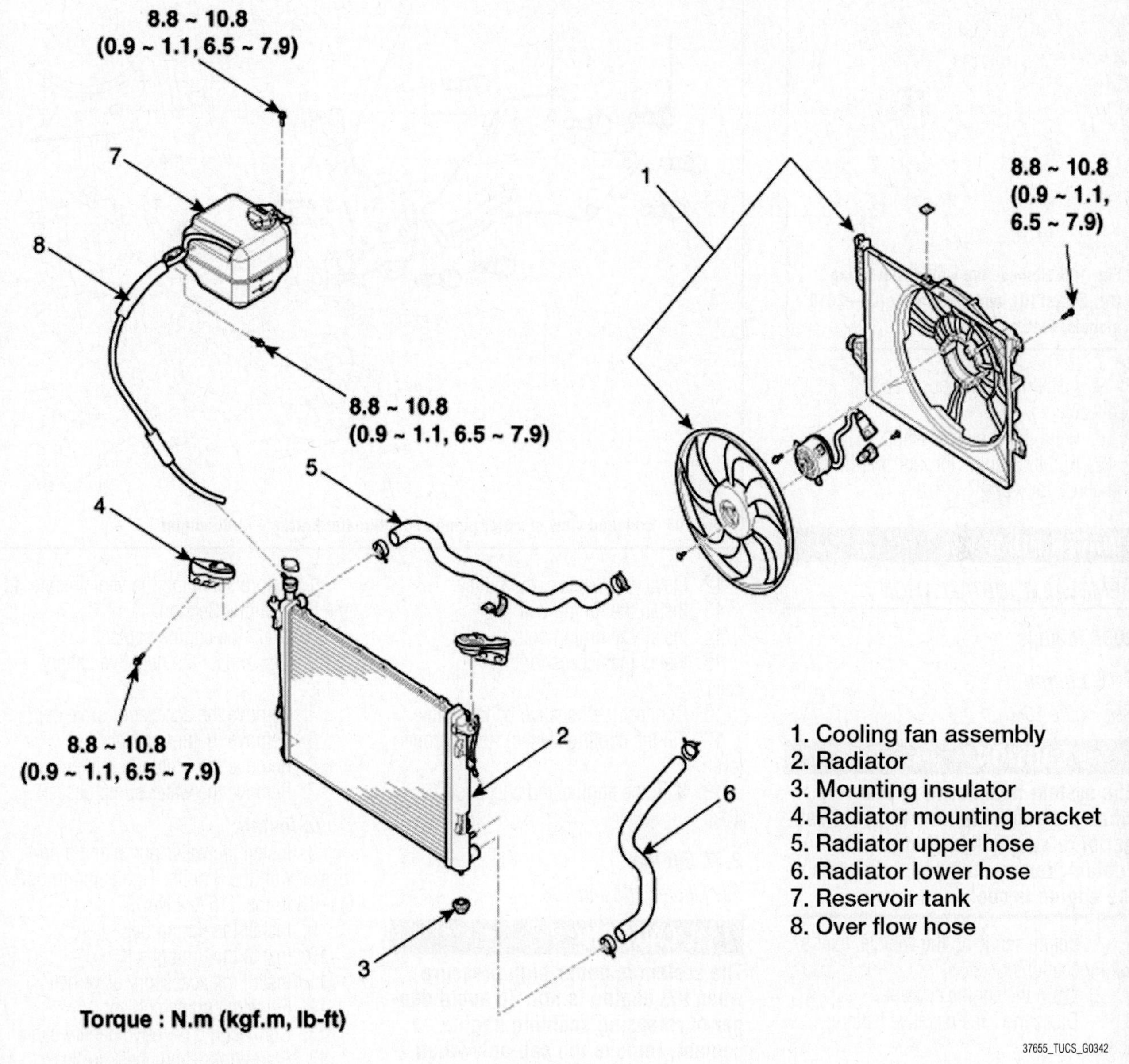

8.8 ~ 10.8
(0.9 ~ 1.1, 6.5 ~ 7.9)
7
8
1
8.8 ~ 10.8
(0.9 ~ 1.1,
6.5 ~ 7.9)
8.8 ~ 10.8
(0.9 ~ 1.1, 6.5 ~ 7.9)
5
4
8.8 ~ 10.8
(0.9 ~ 1.1, 6.5 ~ 7.9)
2
6
3
Torque : N.m (kgf.m, lb-ft)
1. Cooling fan assembly
2. Radiator
3. Mounting insulator
4. Radiator mounting bracket
5. Radiator upper hose
6. Radiator lower hose
7. Reservoir tank
8. Over flow hose
37655_TUCS_G0342

Fig. 102 Radiator components

1. Before servicing the vehicle, refer to the Precautions Section.
2. Drain the coolant down to thermostat level or below.
3. Remove the lower hose, as applicable.
4. Remove the water inlet and gasket.
5. Remove the thermostat.

To install:

6. Install the thermostat with the jiggle valve upward.
7. Install a new gasket to the thermostat.
8. Install the water inlet. Tighten the bolt to: 9–14 ft. lbs. (15–20 Nm).

A
37655_TUCS_G0339

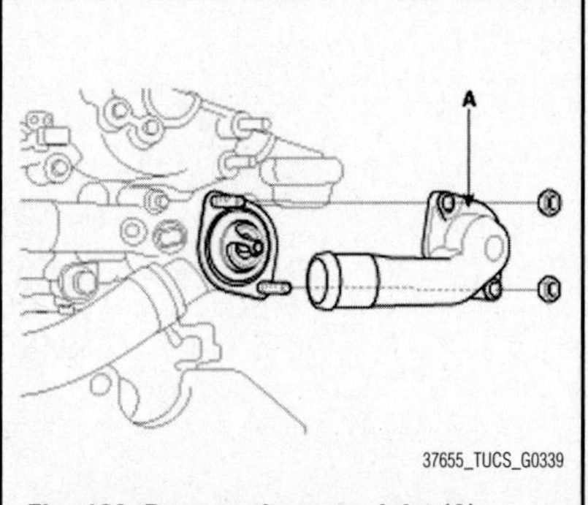

Fig. 103 Remove the water inlet (A), gasket and thermostat—2009 models, 2.0L engine

B
A
37655_TUCS_G0340

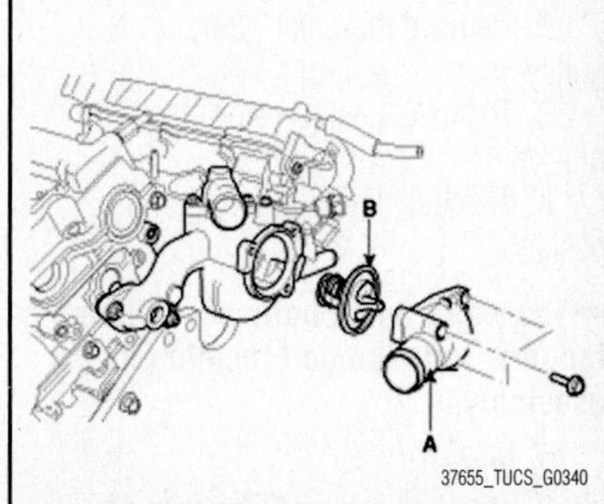

Fig. 104 Remove the water inlet (A) and thermostat (B)—2009 models, 2.7L engine

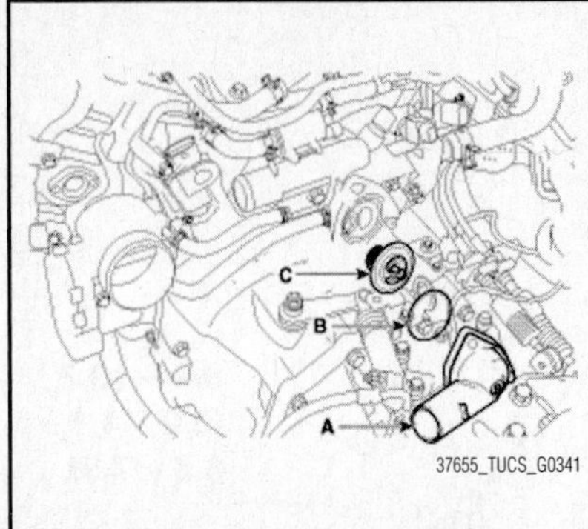

Fig. 105 Remove the water inlet fitting (A), gasket (B), and thermostat (C)—2010 models, 2.4L engine

9. Install the lower hose, as applicable.
10. Refill the engine coolant.
11. Run the engine through warm-up and check for leaks.

WATER PUMP

REMOVAL & INSTALLATION

2009 Models

2.0L Engine

See Figure 106.

⁂ CAUTION

The system is under high pressure when the engine is hot. To avoid danger of releasing scalding engine coolant, remove the cap only when the engine is cool.

1. Before servicing the vehicle, refer to the Precautions Section.
2. Drain the engine coolant.
3. Disconnect the negative battery cable.
4. Remove the accessory drive belts.
5. Remove the timing belt.
6. Remove the timing belt idler.
7. Remove the water pump pulley.
8. Remove the alternator bracket.
9. Remove the water pump and gasket.

➡The water pump bolts are different lengths. Note the bolt location for assembly.

To install:

10. Install the water pump with a new gasket. Tighten the bolts 15–20 ft. lbs. (20–27 Nm).
11. Install the alternator bracket.

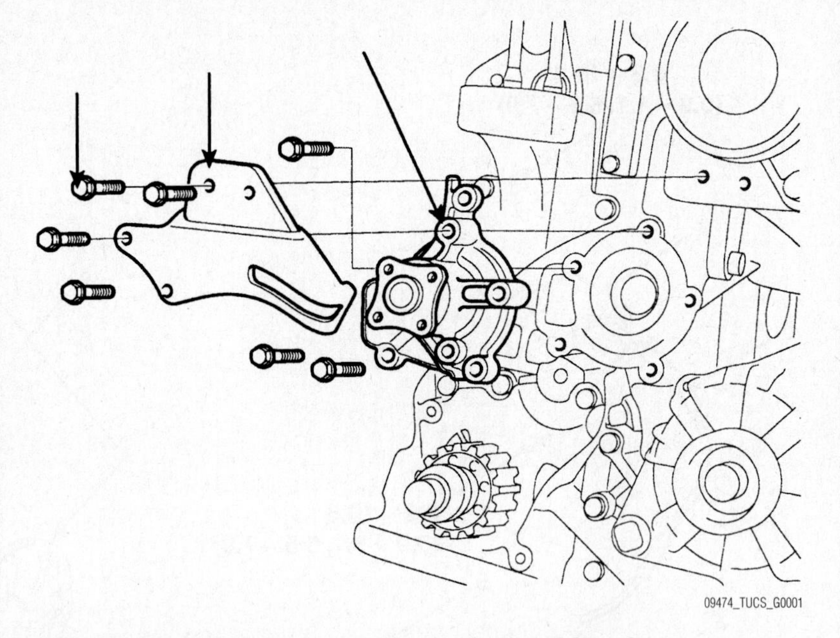

Fig. 106 Exploded view of water pump and alternator bracket—2.0L engine

12. Install the water pump pulley.
13. Install the timing belt idler.
14. Install the timing belt.
15. Install the accessory drive belts.
16. Connect the negative battery cable.
17. Fill the cooling system to the proper level.
18. Start the engine and check for leaks.

2.7L Engine

See Figures 107 and 108.

⁂ CAUTION

The system is under high pressure when the engine is hot. To avoid danger of releasing scalding engine coolant, remove the cap only when the engine is cool.

1. Before servicing the vehicle, refer to the Precautions Section.
2. Drain the engine coolant.
3. Disconnect the negative battery cable.
4. Remove the accessory drive belt.
5. Remove the timing belt.
6. Remove the timing belt idler.
7. Remove the water pump gasket.

To install:

8. Install the water pump and a new gasket with the 8 bolts. Tightening torque: 11–16 ft. lbs. (15–22 Nm).
9. Install the timing belt idler.
10. Install the timing belt.
11. Install the accessory drive belt.
12. Fill with engine coolant.
13. Connect the negative battery cable.
14. Start engine and check for leaks.
15. Recheck engine coolant level.

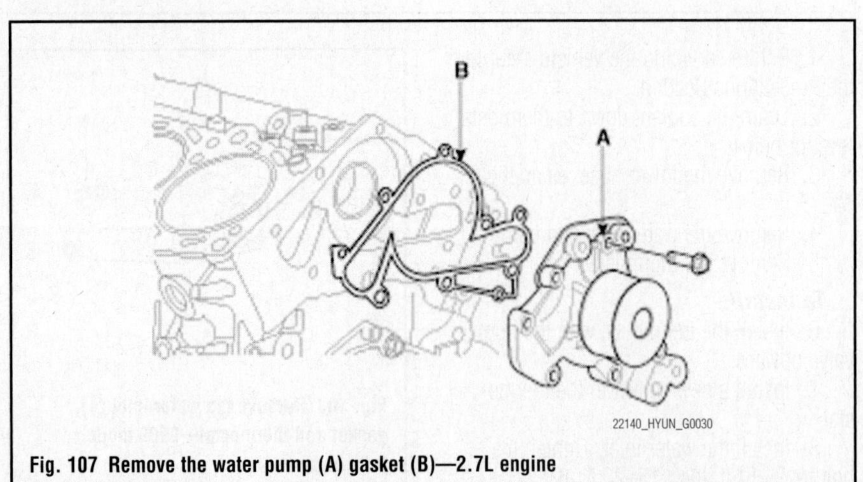

Fig. 107 Remove the water pump (A) gasket (B)—2.7L engine

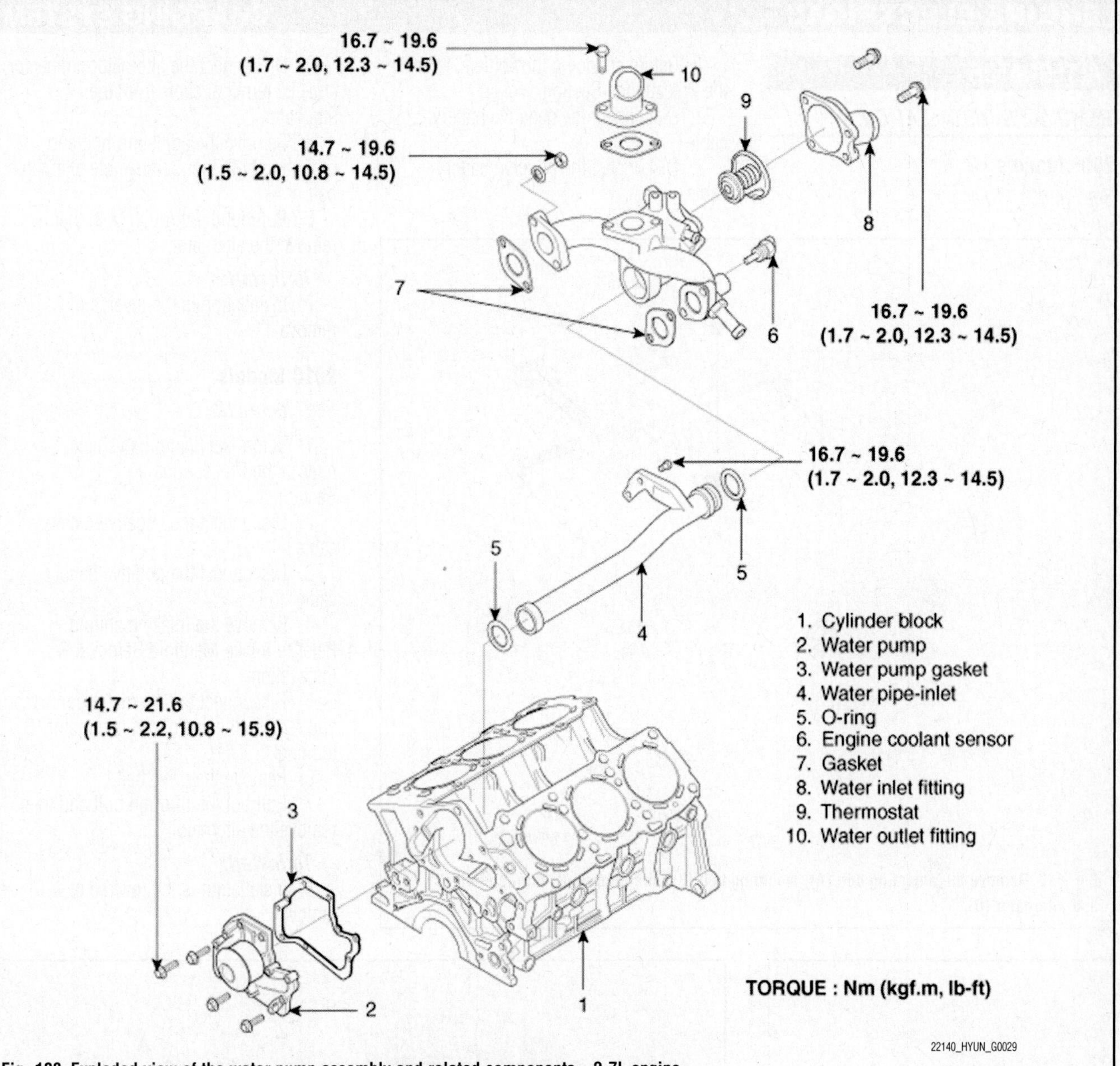

16.7 ~ 19.6
(1.7 ~ 2.0, 12.3 ~ 14.5)
10
14.7 ~ 19.6
(1.5 ~ 2.0, 10.8 ~ 14.5)
9
8
7
6
16.7 ~ 19.6
(1.7 ~ 2.0, 12.3 ~ 14.5)
16.7 ~ 19.6
(1.7 ~ 2.0, 12.3 ~ 14.5)
5
5
4
1. Cylinder block
2. Water pump
3. Water pump gasket
4. Water pipe-inlet
5. O-ring
6. Engine coolant sensor
7. Gasket
8. Water inlet fitting
9. Thermostat
10. Water outlet fitting
14.7 ~ 21.6
(1.5 ~ 2.2, 10.8 ~ 15.9)
3
2
1
TORQUE : Nm (kgf.m, lb-ft)
22140_HYUN_G0029

Fig. 108 Exploded view of the water pump assembly and related components—2.7L engine

2010 Models

See Figure 109.

B
A
37655_TUCS_G0338

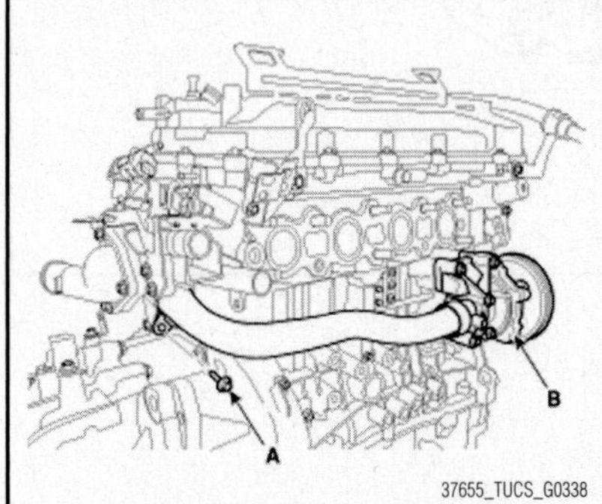

Fig. 109 Remove the water inlet pipe bolt (A), water pump (B) and the gasket

**** CAUTION**

The system is under high pressure when the engine is hot. To avoid danger of releasing scalding engine coolant, remove the cap only when the engine is cool.

1. Before servicing the vehicle, refer to the Precautions Section.
2. Drain the engine coolant.
3. Disconnect the negative battery cable.
4. Remove the drive belt.
5. Remove the exhaust manifold heat protector.
6. Remove the water inlet pipe bolt.
7. Remove the water pump and the gasket.

To install:

8. Installation is the reverse of removal, noting the following:
    a. Tighten the water inlet pipe bolt to 86–104 inch lbs. (10–12 Nm).
    b. Tighten the water pump bolts to 14–17 ft. lbs. (19–24 Nm).
    c. Tighten the exhaust manifold heat protector bolts to 70–104 inch lbs. (8–12 Nm).
9. Refill engine coolant, bleed the cooling system and check for leaks.

ALTERNATOR

REMOVAL & INSTALLATION

2009 Models

See Figures 110 and 111.

1. Before servicing the vehicle, refer to the Precautions Section.
2. Disconnect the negative battery cable.
3. Disconnect the positive battery cable.
4. Disconnect the alternator connector and "B" terminal cable from the alternator.
5. Remove the adjusting bolt and mounting bolt, then remove the alternator belt.
6. Pull out the through bolt, then remove the alternator.

To install:

7. Installation is the reverse of removal.

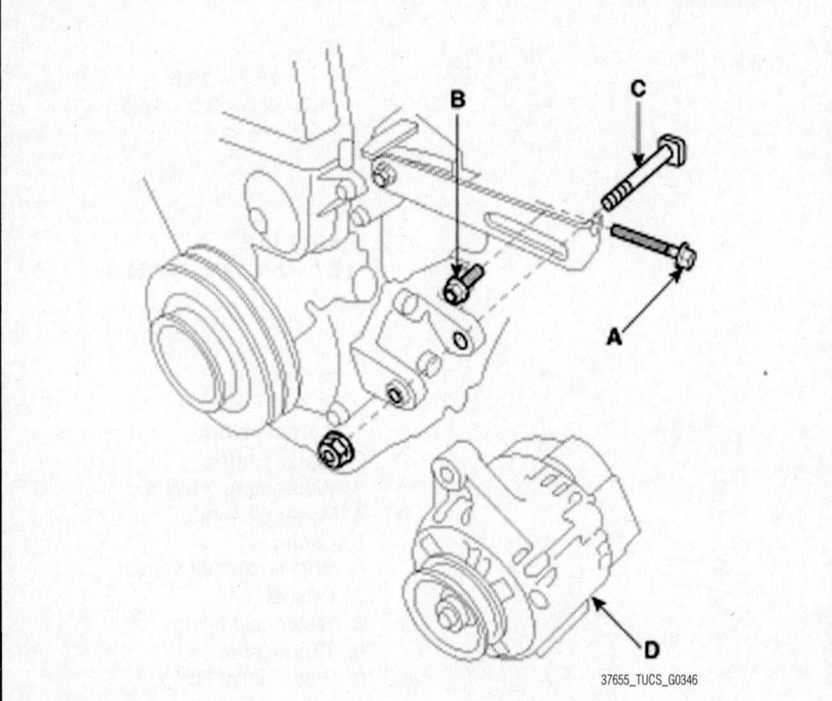

Fig. 110 Remove the adjusting bolt (A), mounting bolt (B), alternator belt, through bolt (C), and alternator (D)

2010 Models

See Figure 112.

1. Before servicing the vehicle, refer to the Precautions Section.
2. Disconnect the negative battery cable.
3. Disconnect the positive battery cable.
4. Remove the intake manifold. Refer to Intake Manifold Removal & Installation.
5. Disconnect the alternator connector and remove the cable from alternator "B" terminal.
6. Remove the drive belt.
7. Pull out the through bolt and then remove the alternator.

To install:

8. Installation is the reverse of removal.

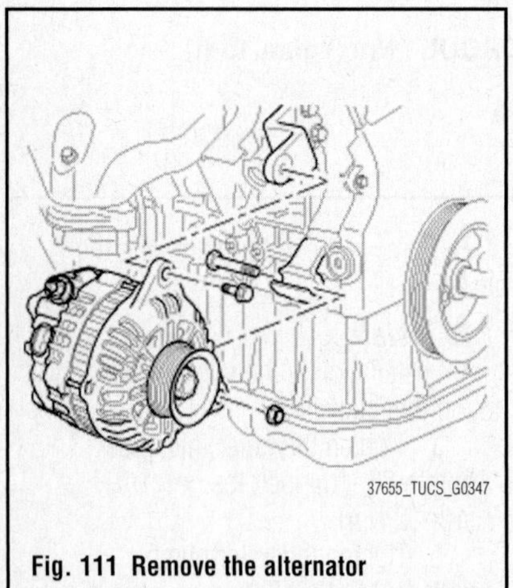

Fig. 111 Remove the alternator

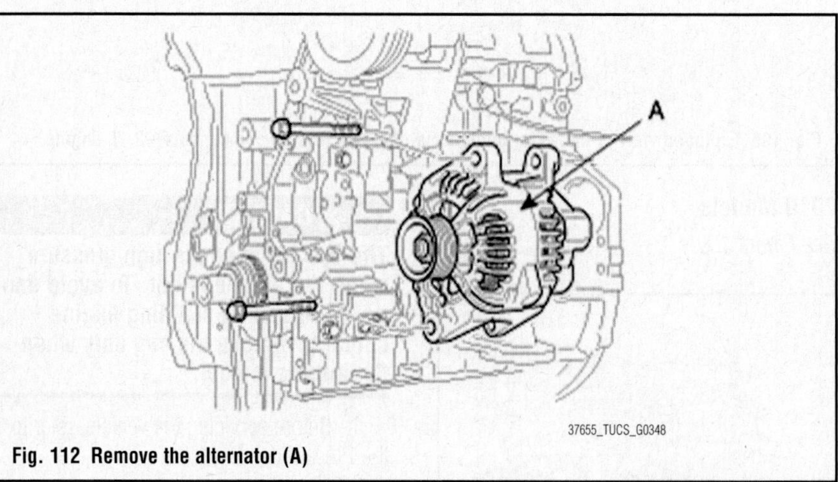

Fig. 112 Remove the alternator (A)

FIRING ORDER

2.0L and 2.4L engines firing order: 1-3-4-2

2.7L engine firing order: 1-2-3-4-5-6

IGNITION COIL

REMOVAL & INSTALLATION

See Figures 113 through 115.

1. Before servicing the vehicle, refer to the Precautions Section.
2. Remove the engine cover, if necessary.
3. Disconnect the spark plug cables, as applicable.
4. Disconnect the ignition coil connectors.
5. Remove the ignition coils.

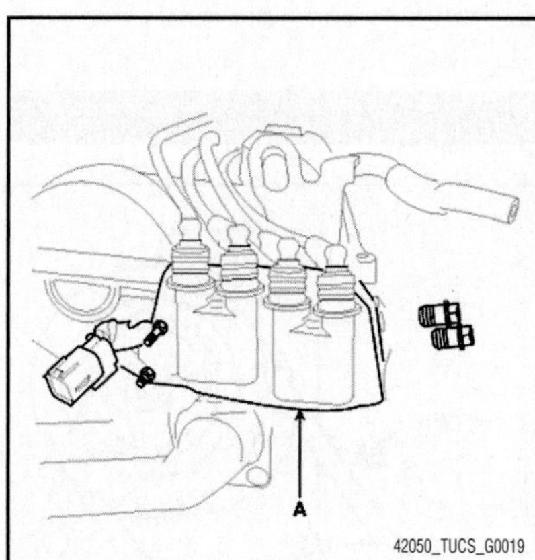

Fig. 113 Remove the ignition coils (A)—2009 models, 2.0L engine

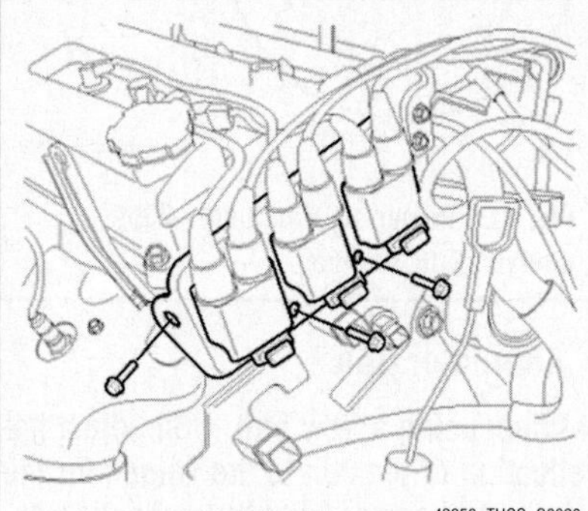

Fig. 114 Remove the ignition coils—2009 models, 2.7L engine

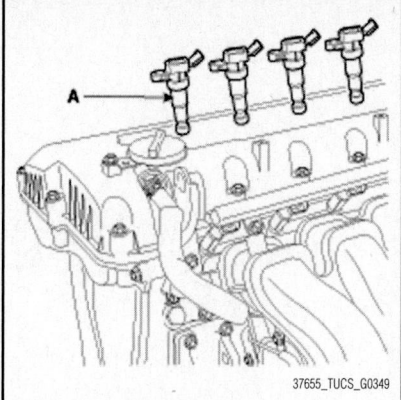

Fig. 115 Remove the ignition coils—2010 models, 2.4L engine

To install:

6. Installation is the reverse of removal.

IGNITION TIMING

ADJUSTMENT

Ignition timing is controlled by the electronic control ignition timing system. The standard reference ignition timing data for the engine operating conditions are pre-programmed in the memory of the Engine Control Module (ECM). No adjustment is necessary.

SPARK PLUGS

REMOVAL & INSTALLATION

2009 Models

2.0L Engine

1. Before servicing the vehicle, refer to the Precautions Section.
2. Remove the spark plug cables.

⁂ WARNING

When removing the spark plug cable, pull on the spark plug cable boot (not the cable), as it may be damaged.

3. Using a spark plug socket, remove the spark plugs.

⁂ WARNING

Be careful that no contaminants enter through the spark plug holes.

4. Check the electrode gap of the new spark plugs. Specification: 0.039–0.043 in. (1.0–1.1mm).

To install:

5. Installation is the reverse of the removal procedure.
6. Tighten the spark plugs to 15–22 ft. lbs. (20–30 Nm).

2.7L Engine

1. Before servicing the vehicle, refer to the Precautions Section.
2. Remove the engine cover.
3. Disconnect the Variable Intake System (VIS) actuator connectors and the fuel injector connectors.
4. Remove the accelerator cable.
5. Remove surge tank sub assembly.
6. Remove the spark plug cable.
7. Remove the spark plug.
8. Check the electrode gap of the new spark plugs. Specification: 0.039–0.043 in. (1.0–1.1mm).

To install:

9. Installation is the reverse of the removal procedure.
10. Tighten the spark plugs to 15–22 ft. lbs. (20–30 Nm).

2010 Models

1. Before servicing the vehicle, refer to the Precautions Section.
2. Remove the engine cover (as necessary).
3. Remove the ignition coil.
4. Use a spark plug socket and wrench to remove the spark plugs.

⁂ WARNING

Be careful that no contaminates enter through the spark plug holes.

➡Check the electrode gap on the spark plugs before installation. Specification: 0.039–0.043 inch (1.0–1.1 mm).

To install:

5. To install, reverse the removal procedure.

STARTER

REMOVAL & INSTALLATION

See Figure 116.

1. Before servicing the vehicle, refer to the Precautions Section.
2. Remove the negative battery cable and wait at least 3 minutes
3. Disconnect the starter cable from the B terminal on the solenoid.
4. Disconnect the connector from the S terminal.
5. Remove the 2 bolts holding the starter.
6. Remove the starter motor assembly.

To install:

7. Installation is the reverse of the removal procedure.
8. Tighten the starter motor bolts to 20–25 ft. lbs. (27–34 Nm).

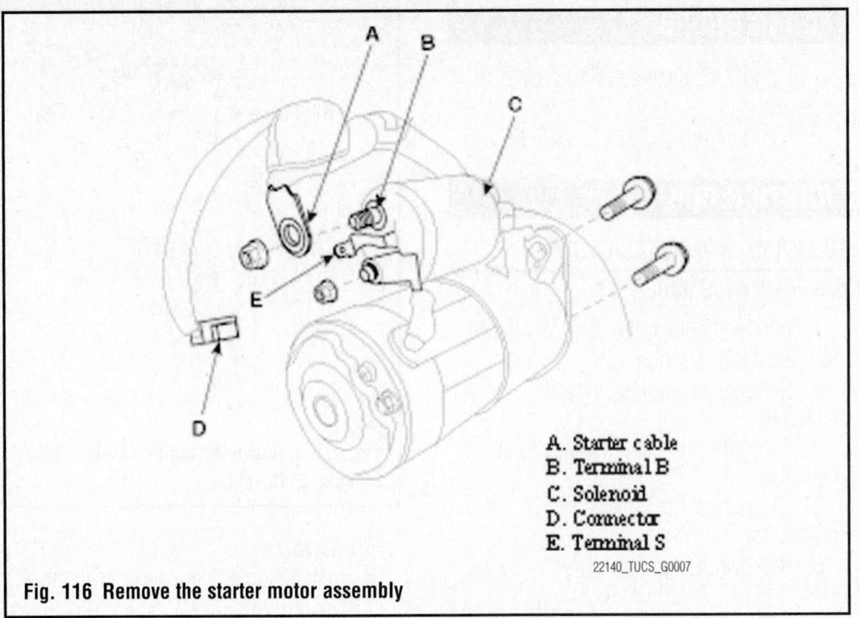

Fig. 116 Remove the starter motor assembly

ENGINE MECHANICAL

➡Disconnecting the negative battery cable may interfere with the functions of the on board computer systems and may require the computer to undergo a relearning process, once the negative battery cable is reconnected.

ACCESSORY DRIVE BELTS

ACCESSORY BELT ROUTING

See Figures 117 through 121.

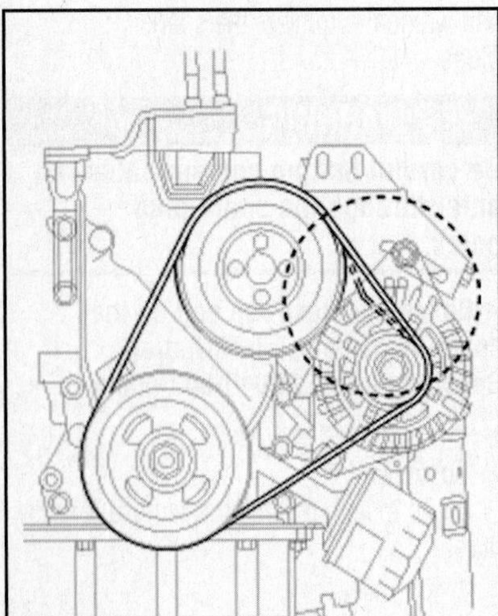

Fig. 117 Alternator belt—2009 models, 2.0L engine

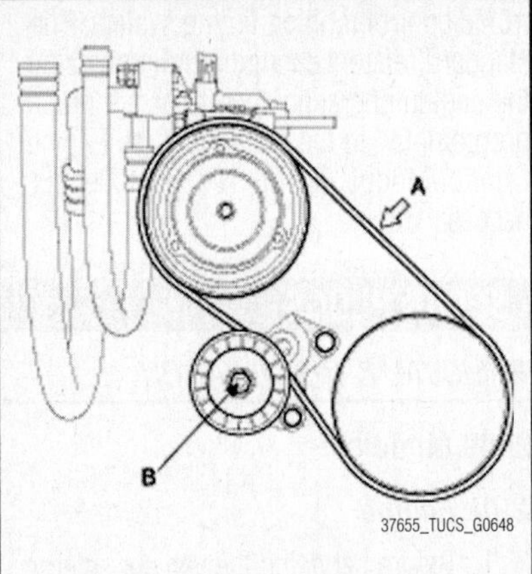

Fig. 118 A/C belt—2009 models, 2.0L engine

INSPECTION

2009 Models

2.0L Engine

See Figures 117 through 119, 122 and 123.

Inspect the accessory drive belt for signs of glazing or cracking. A glazed belt will be perfectly smooth from slippage, while a good belt will have a slight texture of fabric visible. Cracks will usually start at the inner edge of the belt and run outward. All worn or damaged accessory drive belts should be replaced immediately.

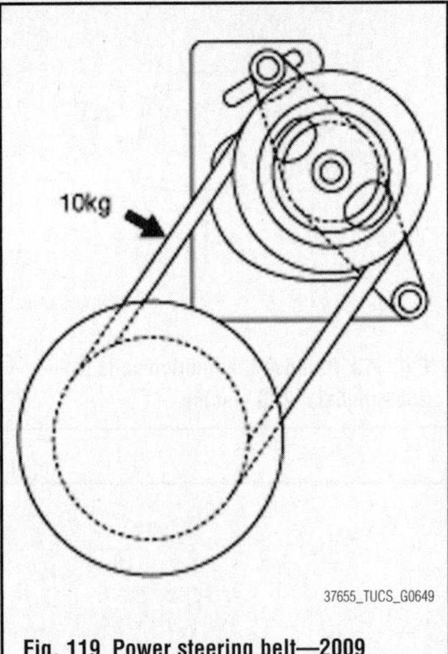

Fig. 119 Power steering belt—2009 models, 2.0L engine

Alternator Belt

➡When using a new belt, first adjust the deflection or tension to the values for the new belt, then readjust the deflection or tension to the values for the used belt after running engine for five minutes.

1. Deflection method: Apply a force of 22 lbs (98 N), and measure the deflection between the alternator and crankshaft pulley:
   - Used Belt: 0.20–0.23 in. (5.0–6.0 mm)

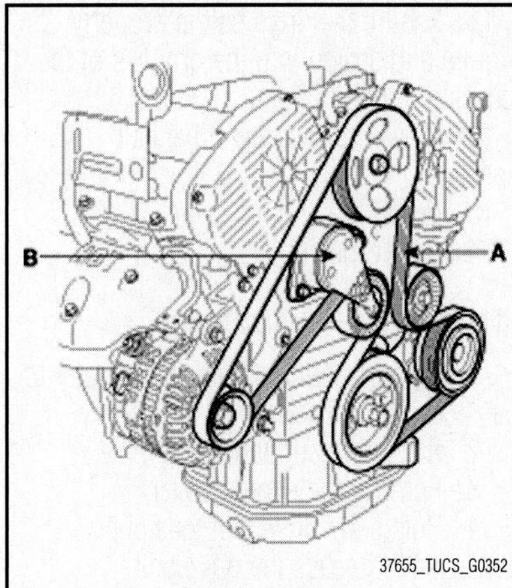

Fig. 120 Accessory drive belt routing, showing drive belt (A) and tensioner (B)—2009 models, 2.7L engine

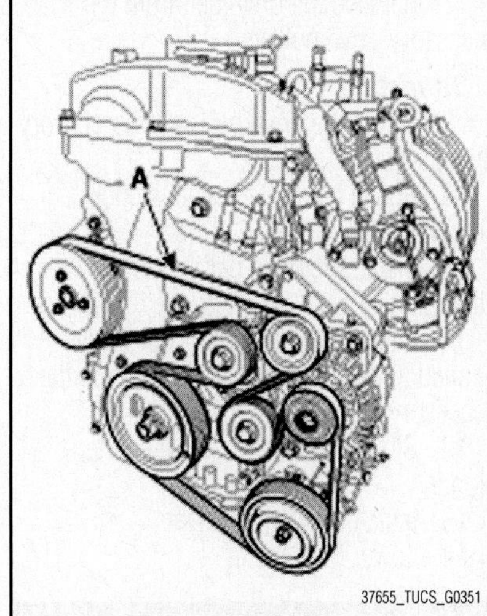

Fig. 121 Accessory drive belt routing—2010 models, 2.4L engine

- New Belt: 0.16–0.20 in. (4.0–5.0 mm)

Power Steering Belt

2. Press the V belt, applying a pressure of 22 lbs. (98N) at the specified point and measure the deflection to confirm that it is within the standard value:
   - New belt: 0.04–0.43 in. (8.8–11.0 mm)
   - Used belt: 0.49–0.56 in. (12.5–14.3 mm)
3. If the belt deflection is beyond the standard value, adjust the belt tension as follows:
   a. Loosen the bolt adjusting the power steering "V" belt tension.
   b. Put a bar or equivalent, between the bracket and the oil pump and adjust the tension so that the belt deflection is within the standard value.

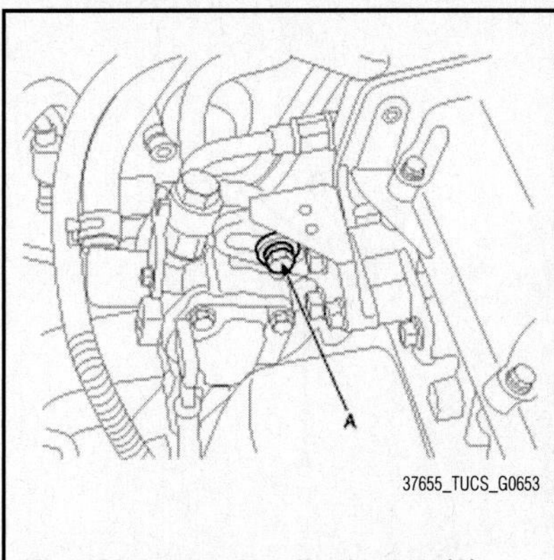

Fig. 122 Loosen the adjusting bolt (A)

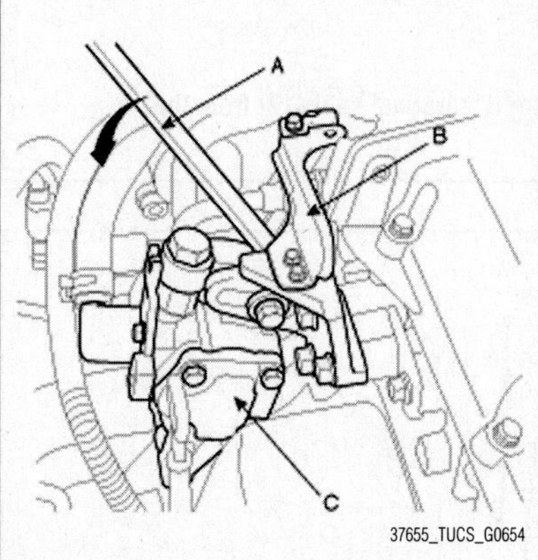

Fig. 123 Put a bar (A) or equivalent, between the bracket (B) and the oil pump (C)

   c. Tighten the bolt adjusting the power steering "V" belt tension.
   d. Check the belt deflection and adjust it again if necessary.
4. After turning the V belt in the normal rotation direction more than once, recheck the belt deflection.

A/C Belt

5. Only BETA ENG: Apply a force of 22 lbs. (98N), and measure the deflection at the midpoint (A) between the air condition compressor and crankshaft pulley:
   - Used belt: 0.24–0.28 in. (6.0–7.0mm)
   - New belt: 0.20–0.22 in. (5.0–5.5mm)

2.7L Engine

Inspect the accessory drive belt for signs of glazing or cracking. A glazed belt will be perfectly smooth from slippage, while a good belt will have a slight texture of fabric visible. Cracks will usually start at the inner edge of the belt and run outward. All worn or damaged accessory drive belts should be replaced immediately.

2010 Models—2.4L Engine

Inspect the accessory drive belt for signs of glazing or cracking. A glazed belt will be perfectly smooth from slippage, while a good belt will have a slight texture of fabric visible. Cracks will usually start at the inner edge of the belt and run outward. All worn or damaged accessory drive belts should be replaced immediately.

ADJUSTMENT

2009 Models

2.0L Engine

See Figure 124.

1. Before servicing the vehicle, refer to the Precautions Section.
2. Loosen the adjusting bolt and the lock bolt, as necessary.
3. Move the alternator to obtain the proper belt tension, then retighten the nuts.
4. Recheck the deflection or tension of the belt.

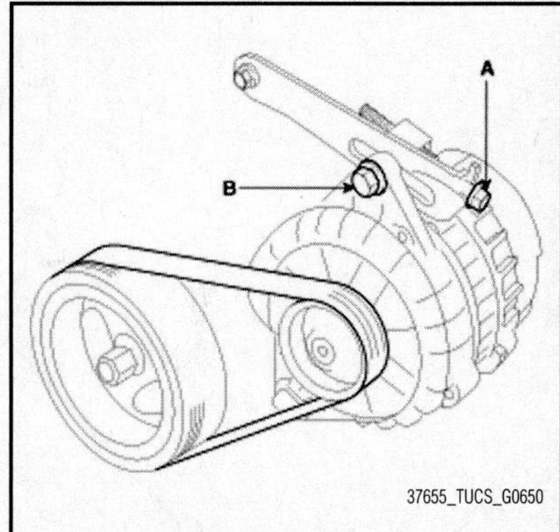

Fig. 124 Loosen the adjusting bolt (A) and the lock bolt (B)—Alternator belt shown, power steering belt and A/C belt similar

2.7L Engine

The drive belt tension is provided through a drive belt tensioner.

2010 Models—2.4L Engine

The drive belt tension is provided through a drive belt tensioner.

REMOVAL & INSTALLATION

2009 Models

2.0L Engine

See Figures 125 and 126.

1. Before servicing the vehicle, refer to the Precautions Section.

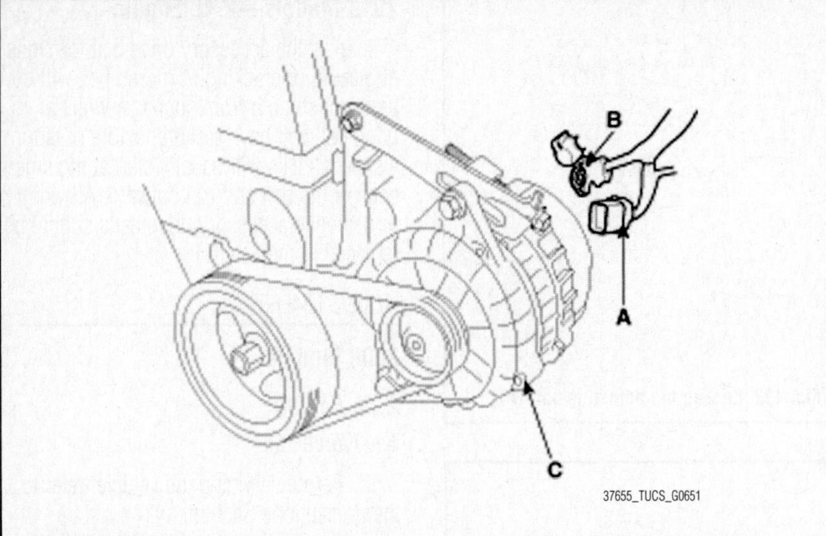

Fig. 125 Disconnect the alternator connector (A) and "B" terminal cable (B) from the alternator (C)

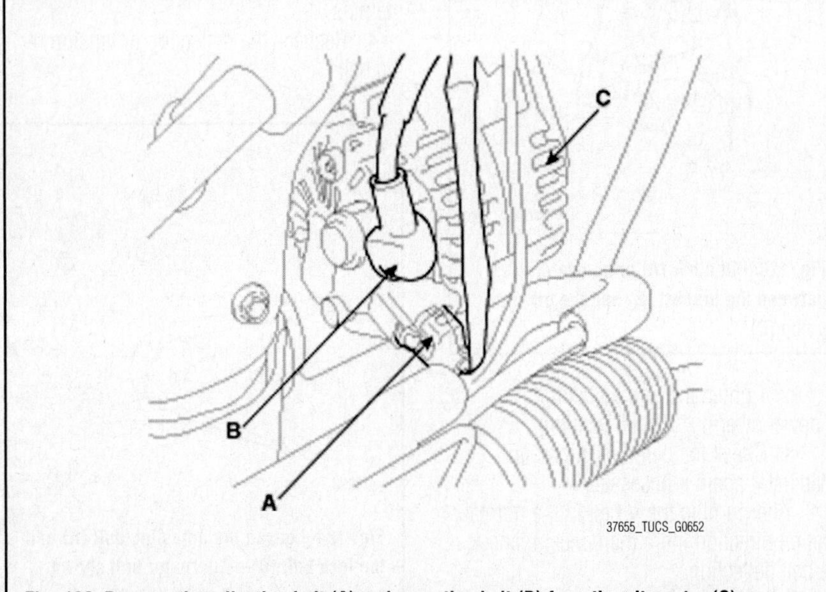

Fig. 126 Remove the adjusting bolt (A) and mounting bolt (B) from the alternator (C)

2. Disconnect the battery negative terminal first, then the positive terminal.
3. Disconnect the alternator connector and "B" terminal cable from the alternator.
4. Remove the adjusting bolt and mounting bolt, then remove the alternator belt.

To install:

5. Installation is the reverse of removal.
6. Adjust the alternator belt tension after installation.

2.7L Engine

1. Before servicing the vehicle, refer to the Precautions Section.
2. Raise and support the vehicle.
3. Remove the engine cover.
4. Rotate the drive belt tensioner clockwise to release the drive belt tension.
5. Remove the drive belt from the alternator.
6. Slowly release the drive belt tensioner.
7. Remove the drive belt from the accessory drive pulleys.

To install:

8. Install the drive belt to the accessory drive pulley.
9. Rotate the drive belt tensioner clockwise.
10. Install the drive belt to the alternator.
11. Ensure the drive belt is properly aligned and seated into the grooves of the accessory drive pulleys.
12. Slowly release the drive belt tensioner.
13. Install the engine cover.
14. Lower the vehicle.

2010 Models—2.4L Engine

1. Before servicing the vehicle, refer to the Precautions Section.
2. Raise and support the vehicle.
3. Remove the engine cover.
4. Rotate the drive belt tensioner clockwise to release the drive belt tension.
5. Remove the drive belt from the alternator.
6. Slowly release the drive belt tensioner.
7. Remove the drive belt from the accessory drive pulleys.

To install:

8. Install the drive belt to the accessory drive pulley.
9. Rotate the drive belt tensioner clockwise.
10. Install the drive belt to the alternator.
11. Ensure the drive belt is properly aligned and seated into the grooves of the accessory drive pulleys.
12. Slowly release the drive belt tensioner.
13. Install the engine cover.
14. Lower the vehicle.

CAMSHAFT AND VALVE LIFTERS

INSPECTION

2009 Models

2.0L Engine

1. Inspect the camshaft lobes. Using a micrometer, measure the cam lobe height.
    a. Standard value: Intake = 1.7566 inch (44.618mm), Exhaust = 1.7527 inch (44.518mm).
    b. Limit value: Intake = 1.7527 inch (44.518mm), Exhaust = 1.7487 inch (44.418mm).
2. If the cam lobe height is less than the minimum limit value, replace the camshaft.
3. Inspect the camshaft journal clearance.
    a. Clean the bearing caps and camshaft journals.

b. Place the camshafts on the cylinder head.
c. Lay a strip of Plastigage® across each of the camshaft journals.
d. Install the bearing caps.

➡**Do not turn the camshaft.**

e. Remove the bearing caps.
f. Measure the Plastigage® at its widest point. The bearing oil clearance should be within specification.

- Standard value: 0.0008–0.0024 inch (0.020–0.061mm)
- Limit value: 0.0039 inch (0.1mm)

g. If the oil clearance is greater than the maximum limit, replace the camshaft. If necessary, replace the bearing caps and cylinder head as a set.
4. Completely remove the Plastigage®.
5. Remove the camshafts.
6. Inspect camshaft end play.
a. Install the camshafts.
b. Using a dial indicator, measure the end play while moving the camshaft back and forth.
c. Camshaft end play standard value: 0.004–0.008 inch (0.1–0.2mm).
d. If the end play is greater than maximum limit, replace the camshaft. If necessary, replace the bearing caps and cylinder head as a set.

2.7L Engine

1. Inspect the camshaft lobes. Using a micrometer, measure the cam lobe height.
a. Standard value: Intake = 1.7303–1.7382 inch (43.95–44.15mm), Exhaust = 1.7303–1.7382 inch (43.95–44.15mm).
b. Limit value: Intake = 1.7527 inch (44.518mm), Exhaust = 1.7487 inch (44.418mm).
2. If the cam lobe height is less than the minimum limit value, replace the camshaft.
3. Inspect the camshaft journals. Using a micrometer, measure the journal diameter.
a. Standard value: 1.0222–1.0228 inch (25.964–25.980mm)
4. If the journal diameter is not as specified, check the oil clearance.
5. Inspect camshaft bearings.
a. Check that bearing for flaking and scoring. If the bearings are damaged, replace the bearing caps and cylinder head as a set.
6. Inspect the camshaft journal clearance.
a. Clean the bearing caps and camshaft journals.
b. Place the camshafts on the cylinder head.
c. Lay a strip of Plastigage® across each of the camshaft journals.
d. Install the bearing caps.

➡**Do not turn the camshaft.**

e. Remove the bearing caps.
f. Measure the Plastigage® at its widest point. The bearing oil clearance should be within specification.

- Standard value: 0.0008–0.0024 inch (0.020–0.061mm)
- Limit value: 0.0039 inch (0.1mm)

g. If the oil clearance is greater than the maximum limit, replace the camshaft. If necessary, replace the bearing caps and cylinder head as a set.
7. Completely remove the Plastigage®.
8. Remove the camshafts.
9. Inspect camshaft end play.
a. Install the camshafts.
b. Using a dial indicator, measure the end play while moving the camshaft back and forth.
c. Camshaft end play standard value: 0.004–0.0059 inch (0.1–0.15mm).
d. If the end play is greater than maximum limit, replace the camshaft. If necessary, replace the bearing caps and cylinder head as a set.

2010 Models—2.4L Engine

➡**For camshaft specifications, refer to Camshaft and Bearing Specifications in Specifications Chart.**

1. Inspect the cam lobes.
a. Using a micrometer, measure the cam lobe height.
b. If the cam lobe height is less than specified, replace the camshaft.
2. Inspect the camshaft journal clearance.
a. Clean the bearing caps and camshaft journals.
b. Place the camshafts on the cylinder head.
c. Lay a strip of Plastigage® across each of the camshaft journal.
d. Install the bearing caps and tighten the bolts with specified torque.

➡**Do not turn the camshaft.**

e. Remove the bearing caps.
f. Measure the Plastigage® at its widest point.
g. If the oil clearance is greater than specified, replace the camshaft. If necessary, replace the bearing caps and cylinder head as a set.
h. Completely remove the Plastigage®.
i. Remove the camshafts.
3. Inspect the camshaft end play.
a. Install the camshafts.
b. Using a dial indicator, measure the end play while moving the camshaft back and forth.
c. If the end play is greater than specified, replace the camshaft. If necessary, replace the bearing caps and cylinder head as a set.
d. Remove the camshafts.

REMOVAL & INSTALLATION

2009 Models

2.0L Engine

See Figures 127 through 131.

➡**Use fender covers to avoid damaging painted surfaces. To avoid damage, unplug the wiring connectors carefully while holding the connector portion. Mark all wiring and hoses to avoid misconnection.**

1. Before servicing the vehicle, refer to the Precautions Section.
2. Disconnect the negative battery cable.
3. Remove the air duct.
4. Drain the cooling system.
5. Remove the engine cover.
6. Remove the air cleaner assembly.
7. Remove the accessory drive belts. Refer to Accessory Drive Belt Removal & Installation.
8. Remove the cylinder head cover.
9. Remove the timing belt. Refer to Timing Belt Removal & Installation.
10. Remove the camshaft sprocket.
11. Remove the camshaft timing chain auto tensioner.
12. Remove the camshaft bearing caps, camshaft timing chain, and camshafts.

To install:

13. Install the mechanical lash adjusters in their original positions.
14. Install the intake and exhaust camshafts with the timing chain aligned as shown.

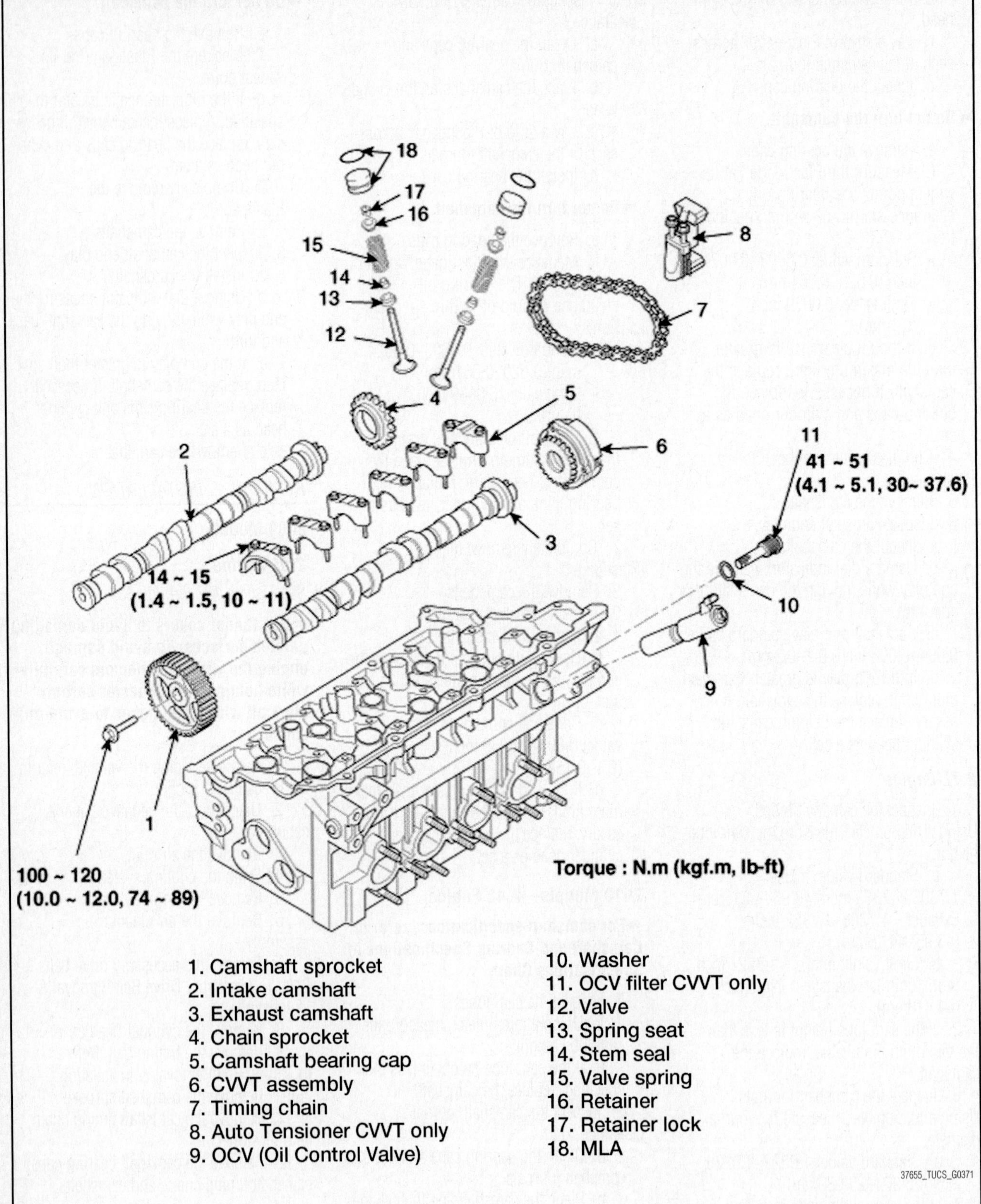

Fig. 127 Exploded view of camshafts and related components—2.0L engine

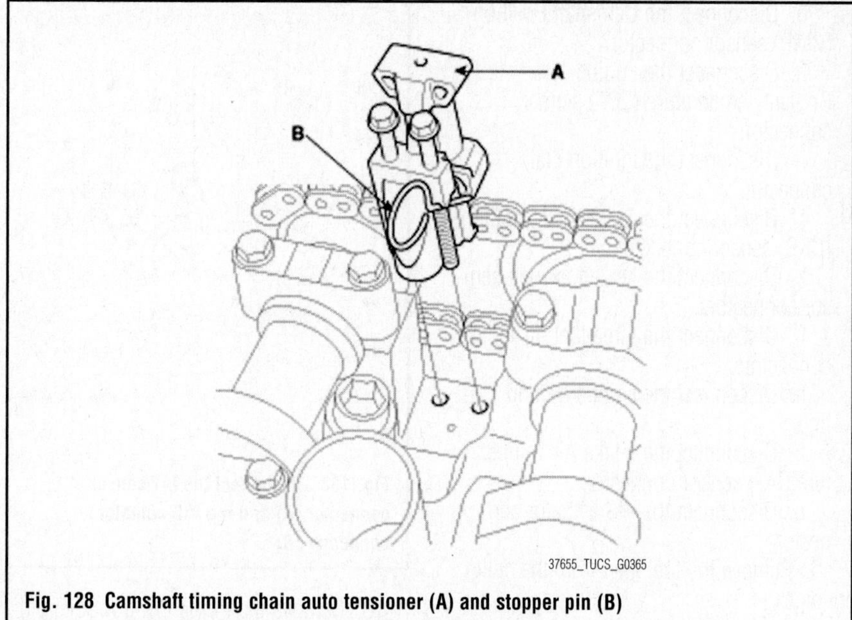

Fig. 128 Camshaft timing chain auto tensioner (A) and stopper pin (B)

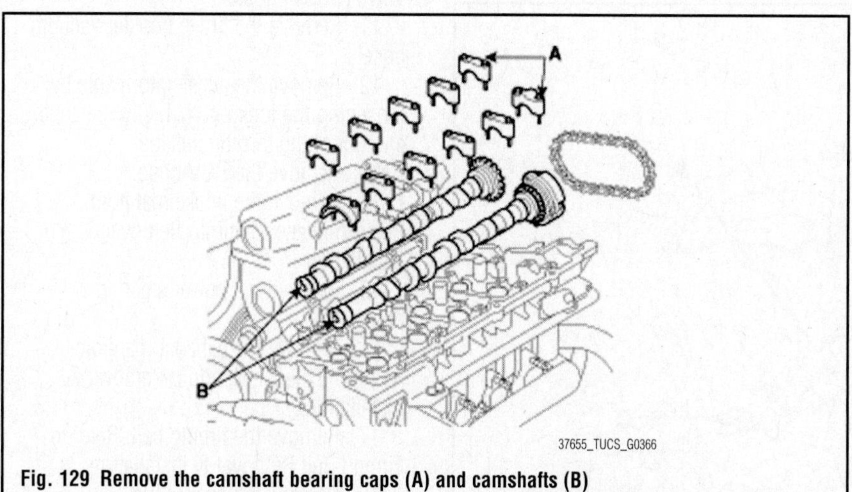

Fig. 129 Remove the camshaft bearing caps (A) and camshafts (B)

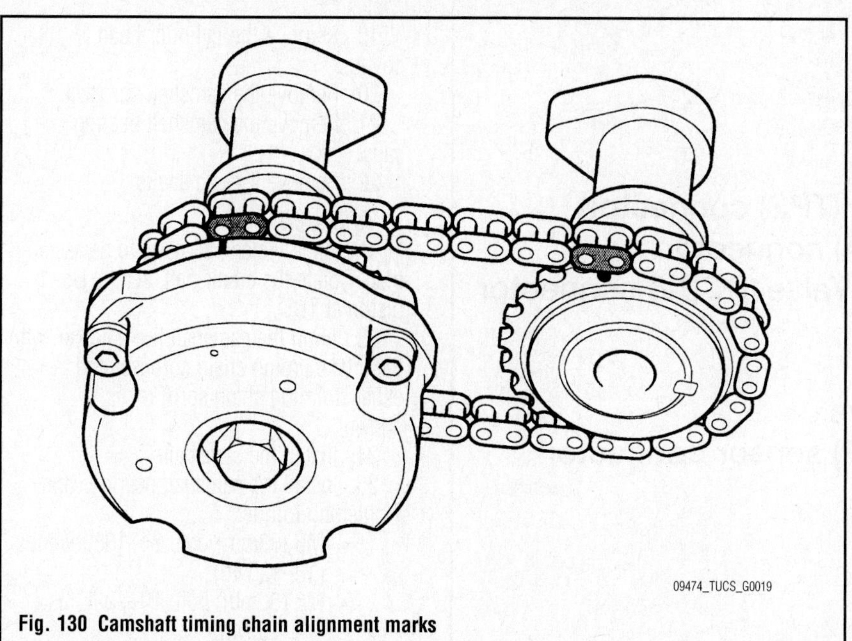

Fig. 130 Camshaft timing chain alignment marks

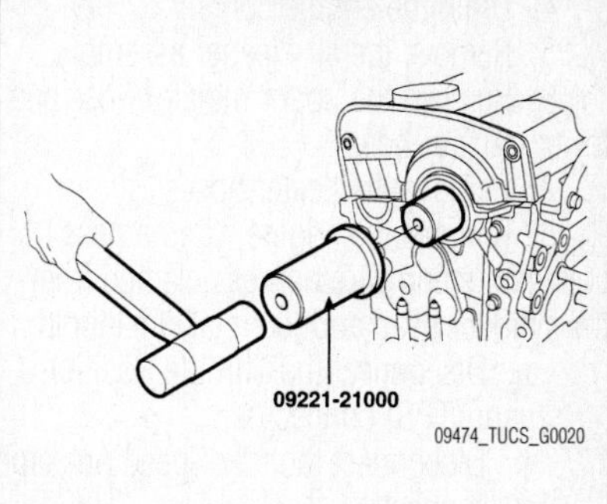

Fig. 131 Camshaft bearing oil seal install tool

15. Install the camshaft bearing caps. Tighten to 10–11 ft. lbs. (14–15 Nm).
16. Install the timing chain auto tensioner. Tighten to 72–89 inch lbs. (8–10 Nm).
17. Using Special Tool 09221-21000 seal installer, install the camshaft bearing oil seal.
18. Install the camshaft sprocket and torque the bolt to 74–89 ft. lbs. (100–120 Nm).
19. Install the timing belt.
20. Install the cylinder head cover.
21. Install the accessory drive belts.
22. Install the air cleaner assembly.
23. Install the engine cover.
24. Install the air duct.
25. Connect the negative battery cable.
26. Refill the cooling system to the correct level.
27. Refill the engine oil to the correct level.
28. Start the engine and check for leaks.

2.7L Engine

See Figures 132 through 138.

Engine removal is not required for this procedure. Use Fender cover to avoid damaging painted surfaces. To avoid damage, unplug the wiring connectors carefully while holding the connector portion. Mark all wiring and hoses to avoid misconnection. Inspect the timing belt before removing. Turn the crankshaft pulley so that the No. 1 piston is at Top Dead Center (TDC).

1. Before servicing the vehicle, refer to the Precautions Section.
2. Remove the air duct.
3. Disconnect the negative battery cable.

4. Drain the engine coolant.
5. Remove the air cleaner assembly.
6. Remove the upper radiator hose and lower radiator hose.
7. Remove the heater hoses.
8. Remove the engine wire harness connectors and wire harness clamps from the cylinder head and the intake manifold:
   a. Disconnect the Throttle Position Sensor (TPS) connector.
   b. Disconnect the Idle Speed Actuator (ISA) connector.
   c. Disconnect the Purge Control Solenoid Valve (PCSV) connector.
   d. Disconnect the VIS actuator connector.
   e. Disconnect the fuel injector connector.
   f. Disconnect the knock sensor connector.
   g. Disconnect the Camshaft Position (CMP) sensor connector.
   h. Disconnect the Engine Coolant Temperature (ECT) sensor connector.
   i. Disconnect the ignition coil connector.
   j. Disconnect the Crankshaft Position (CKP) sensor connector.
   k. Disconnect the heated oxygen sensor connectors.
   l. Disconnect the three fuel injector connectors.
   m. Disconnect the engine ground line.
   n. Disconnect the Intake Air Temperature (IAT) sensor connector.
   o. Disconnect the VIS actuator connector.
9. Remove the fuel inlet from the delivery pipe.

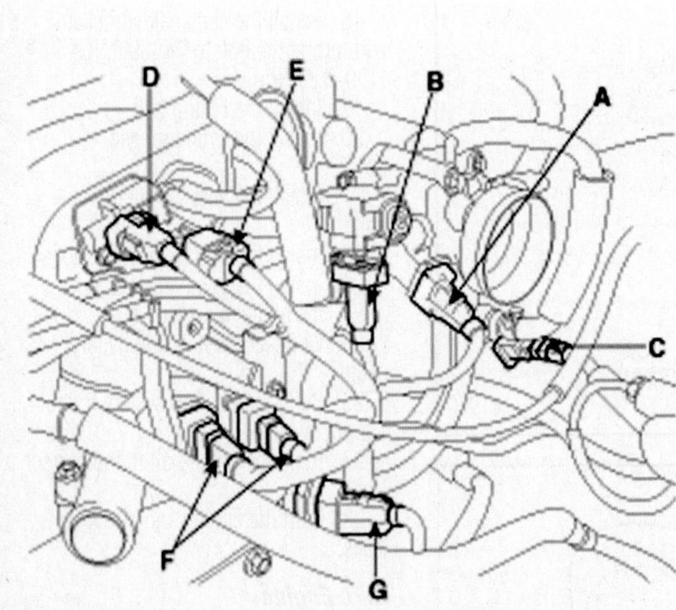

Fig. 132 Engine wiring connectors

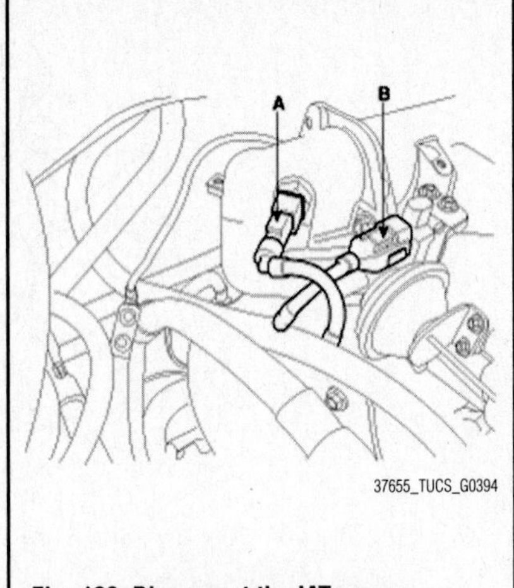

Fig. 133 Disconnect the IAT sensor connector (A) and the VIS actuator connector (B)

10. Remove the Purge Control Solenoid Valve (PCSV) hose.
11. Remove the brake booster vacuum hose.
12. Remove the accelerator cable by loosening the locknut, then slip the cable end out of the throttle linkage.
13. Remove the PCV hose.
14. Remove the intake manifold. Refer to Intake Manifold Removal & Installation.
15. Remove the power steering pump.
16. Remove the exhaust manifold. Refer to Exhaust Manifold Removal & Installation.
17. Remove the timing belt. Refer to Timing Belt Removal & Installation.
18. Remove the spark plug cables.
19. Remove the cylinder head covers.
20. Remove the camshaft sprocket.
21. Remove the camshaft bearing caps.
22. Remove the camshafts.

To install:

Thoroughly clean all parts to be assembled. Rotate the crankshaft, set the No. 1 piston at TDC.

23. Align the camshaft timing chain with the intake timing chain sprocket and exhaust timing chain sprocket as shown.
24. Install the camshafts.
25. Install the camshaft bearing caps. Tightening torque:
    - M6 (38mm) bolt: 84–108 inch lbs. (10–12 Nm)
    - M6 (50mm) bolt: 10–12 ft. lbs. (14–16 Nm)

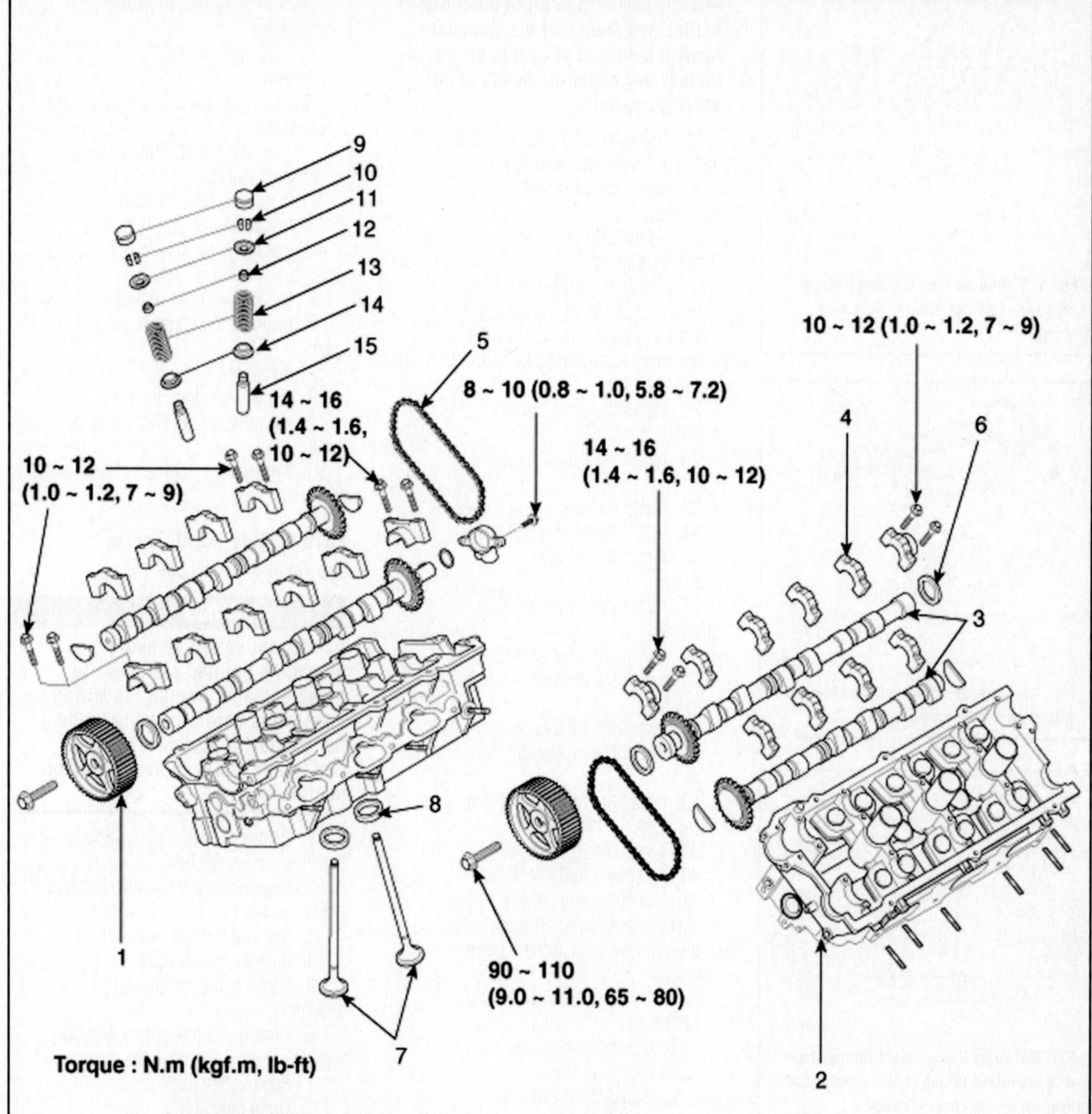

1. Camshaft sprocket
2. Cylinder head
3. Camshaft
4. Camshaft bearing cap
5. Timing chain
6. Oil seal
7. Valve
8. Valve seat
9. HLA
10. Valve spring retainer lock
11. Valve spring retainer
12. Valve stem seal
13. Valve spring
14. Valve spring seat
15. Valve guide

37655_TUCS_G0370

Fig. 134 Exploded view of camshafts and related components—2.7L engine

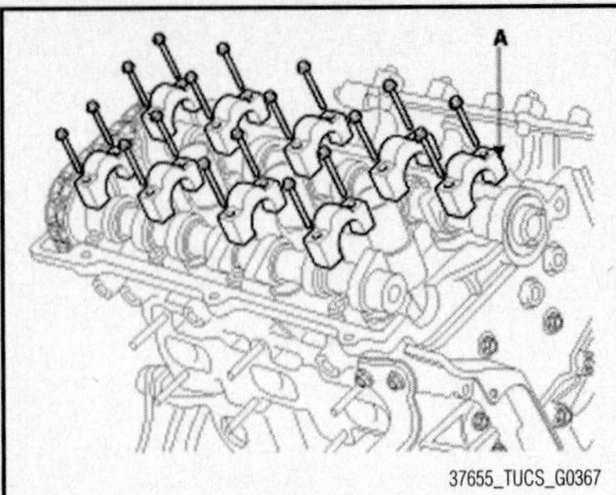

Fig. 135 Remove the camshaft bearing caps (A)—left side shown, right side similar

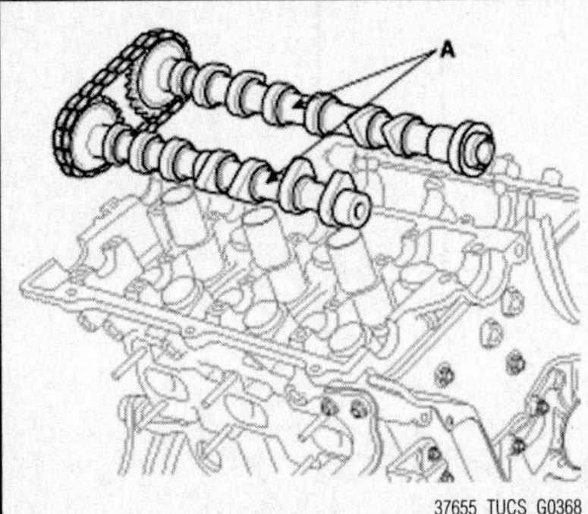

Fig. 136 Remove the camshafts (A)—left side shown, right side similar

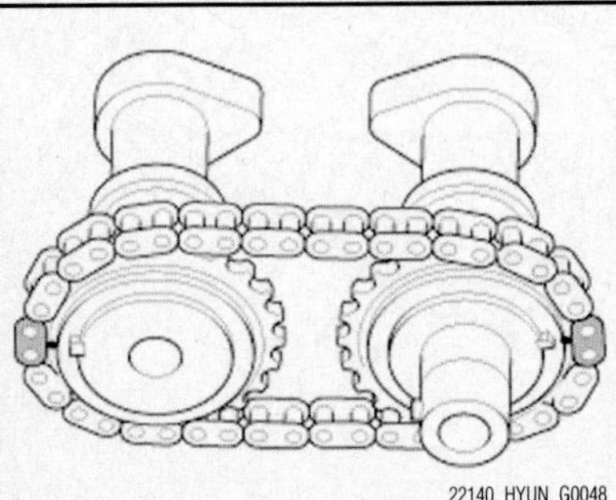

Fig. 137 Align the camshaft timing chain with the intake timing chain sprocket and exhaust timing chain sprocket

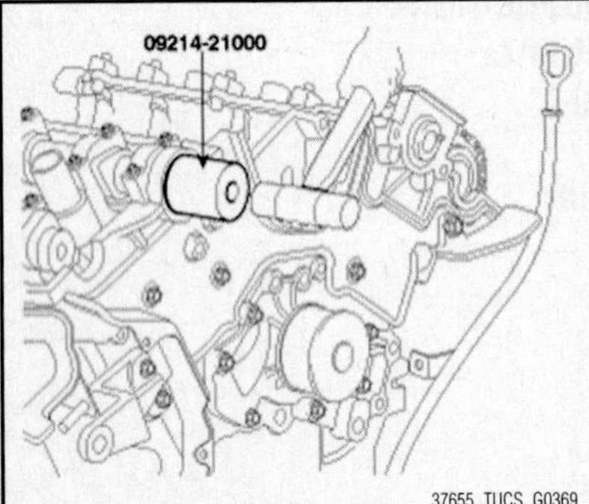

Fig. 138 Using the SST (09214-21000), install the camshaft bearing oil seal

➡Apply new engine oil to the thrust portion and journal of the camshafts. Apply a light coat of engine oil on the threads and under the heads of the bearing cap bolts.

26. Using the SST (09214-21000), install the camshaft bearing oil seal.
27. Install the camshaft sprocket:
    a. Temporarily install the camshaft sprocket bolts.
    b. Hold the hexagonal head wrench portion of the camshaft with a wrench, and tighten the camshaft sprocket bolts to 65–80 ft. lbs. (88–110 Nm).
28. Install the semi-circular packing.
29. Install the cylinder head cover.
30. Install the spark plug cables.
31. Install the timing belt.
32. Install the exhaust manifold.
33. Install the power steering pump.
34. Install the intake manifold.
35. Install the exhaust manifold.
36. Install the PCV hose.
37. Install the accelerator cable.
38. Install the brake booster vacuum hose.
39. Install the PCSV hose.
40. Install the fuel inlet hose.
41. Install the engine wire harness connectors and wire harness clamps to the cylinder head and the intake manifold:
    a. Connect the VIS actuator connector.
    b. Connect the IAT sensor connector.
    c. Connect the ground cable.
    d. Connect three fuel injector connectors.
    e. Connect the heated oxygen sensor connectors.
    f. Connect the crankshaft position sensor connector.
    g. Connect the ignition coil connectors.
    h. Connect the engine coolant temperature sensor connector.
    i. Connect the camshaft position sensor connector.
    j. Connect the knock sensor connector.
    k. Connect the injector connector.
    l. Connect the VIS actuator connector.
    m. Connect the PCSV connector.
    n. Connect the idle speed actuator connector.
    o. Connect the throttle position sensor connector.
42. Install the heater hoses.
43. Install the radiator hoses.
44. Install the air cleaner assembly.
45. Install the engine cover.
46. Connect the negative battery cable.
47. Install the air duct.
48. Fill with engine coolant.
49. Start the engine and check for leaks.
50. Recheck the engine coolant level and oil level.

2010 Models—2.4L Engine

See Figures 139 through 147.

WARNING

Use fender covers to avoid damaging painted surfaces. To avoid damage, unplug the wiring connectors carefully while holding the connector portion. Mark all wiring and hoses to avoid misconnection.

1. Before servicing the vehicle, refer to the Precautions Section.
2. Disconnect the negative battery cable.
3. Remove the engine cover.
4. Remove the air duct.
5. Remove the air cleaner assembly.
6. Disconnect the battery positive terminal and then remove the battery.
7. Remove the undercover.
8. Drain the cooling system.
9. Remove the radiator hoses.
10. Disconnect the wiring connectors, harness clamps from the engine:
    a. Disconnect the OCV connector.
    b. Disconnect the VIS connector.
    c. Disconnect the oil pressure switch.
    d. Disconnect the knock sensor connector.
    e. Disconnect the A/C compressor switch connector.
    f. Disconnect the alternator connector.
    g. Disconnect the injector connectors.

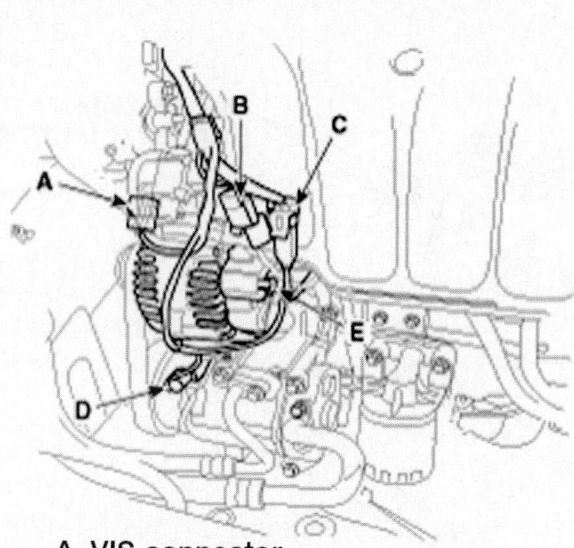

Fig. 139 Disconnect the VIS connector, oil pressure switch, knock sensor connector, A/C compressor switch connector, and alternator connector

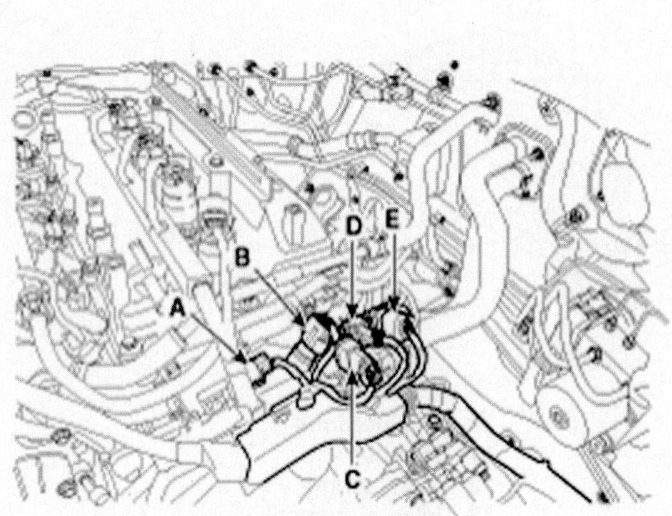

Fig. 140 Disconnect the PCSV connector, ECT sensor connector, condenser connector, CKP sensor connector, and oxygen sensor connector

h. Disconnect the ignition coil connectors.

i. Disconnect the intake Camshaft Position (CMP) sensor connector.

j. Disconnect the exhaust CMP connector.

k. Disconnect the exhaust OCV connector.

l. Disconnect the VCM connector.

m. Disconnect the ETC connector.

n. Disconnect the Manifold Absolute Pressure (MAP) and Intake Air Temperature (IAT) connector.

o. Disconnect the Purge Control Solenoid Valve (PCSV) connector.

p. Disconnect the Engine Coolant Temperature (ECT) sensor connector.

q. Disconnect the condenser connector.

r. Disconnect the Crankshaft Position (CKP) sensor connector.

s. Disconnect the oxygen sensor connector.

11. Disconnect the fuel hose, brake booster vacuum hose and heater hoses.
12. Disconnect the throttle body coolant hoses.
13. Remove the water inlet pipe mounting bolt.
14. Disconnect the oil cooler hoses and then remove the water temperature control assembly.
15. Remove the timing chain. Refer to Timing Chain Removal & Installation.
16. Remove the intake and exhaust manifold. Refer to Intake Manifold and Exhaust Manifold Removal & Installation.
17. Remove the intake and exhaust CVVT assembly.
18. Remove the front camshaft bearing cap.
19. Remove the exhaust camshaft upper bearing.
20. Remove the camshaft bearing caps, in the sequence shown.
21. Remove the camshaft.
22. Remove the exhaust camshaft lower bearing.

To install:

23. Apply a light coat of engine oil on camshaft journals.
24. Install the exhaust camshaft lower bearing.
25. Install the camshafts.

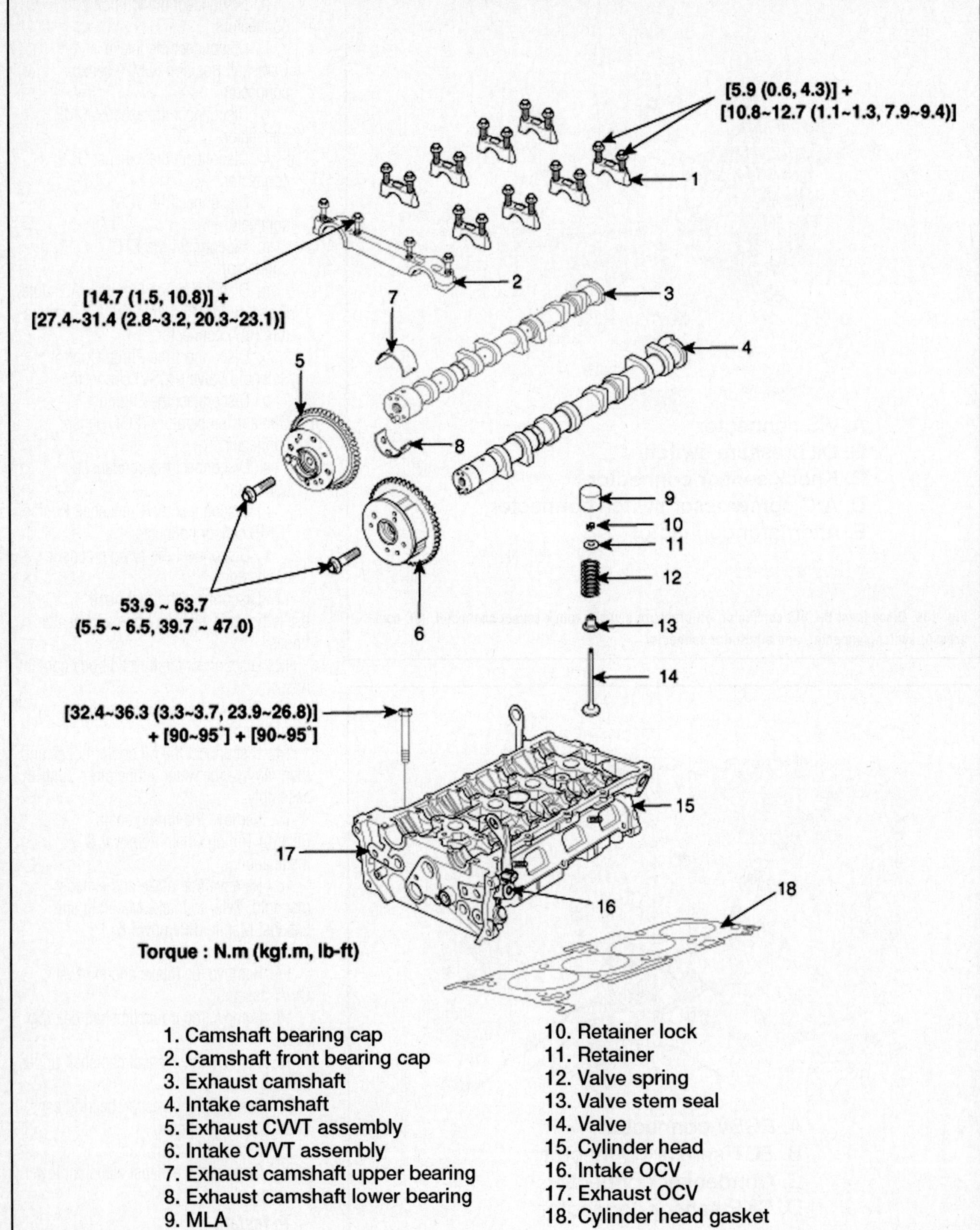

Fig. 141 Camshafts and related components

Fig. 142 Remove the front camshaft bearing cap (A)

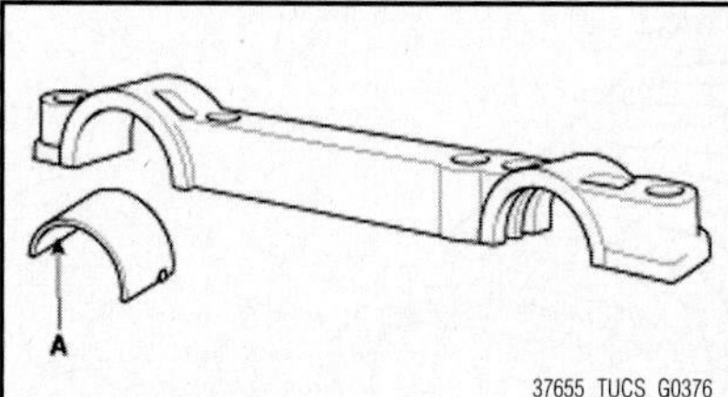

Fig. 143 Remove the exhaust camshaft upper bearing (A)

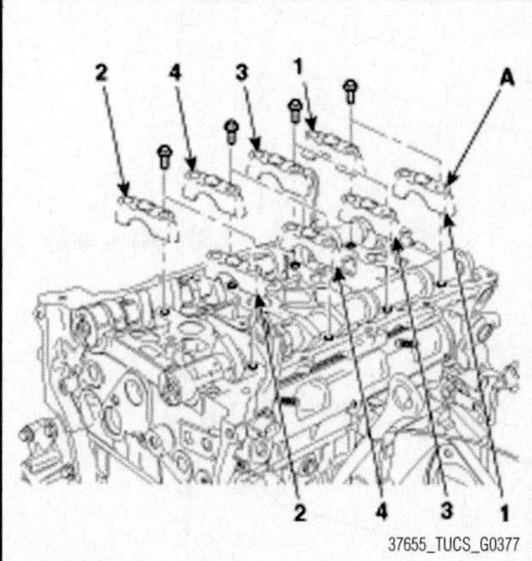

Fig. 144 Remove the camshaft bearing caps (A)

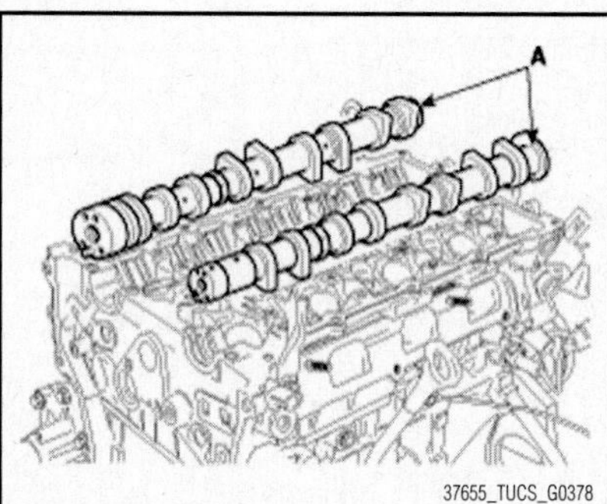

Fig. 145 Remove the camshaft (A)

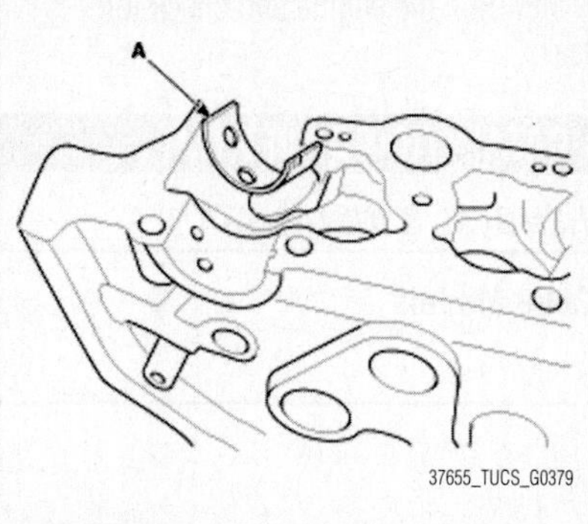

Fig. 146 Remove the exhaust cam shaft lower bearing (A)

26. Install the exhaust camshaft upper bearing.
27. Install camshaft bearing caps in their proper locations. Tightening torque:
    a. Step 1:
    - M6: 52 inch lbs. (6 Nm)
    - M8: 11 ft. lbs. (15 Nm)
    b. Step 2
    - M6: 95–113 inch lbs. (11–13 Nm)
    - M8: 20–23 ft. lbs. (28–31 Nm)
28. Install the water temperature control assembly and then connect the oil cooler hoses. Tighten the bolts to 11–15 ft. lbs. (15–20 Nm) and the nuts to 14–17 ft. lbs. (19–24 Nm).

➡Attach the water temp control assembly and water inlet pipe to the water pump assembly before tightening the water inlet pipe bolts. Always use a new O-ring.

29. Install the water inlet pipe mounting bolt and tighten to 86–104 inch lbs. (10–12 Nm).
30. Install the timing chain.
31. Check and adjust valve clearance.
32. Install the cylinder head cover.
33. Install the intake and exhaust manifolds.
34. Connect the throttle body coolant hoses.
35. Connect the fuel hose, the brake booster vacuum hose and heater hoses.
36. Connect the wiring connectors and harness clamps:
    a. Connect the OCV connector.
    b. Connect the VIS connector.
    c. Connect the oil pressure switch.
    d. Connect the knock sensor connector.
    e. Connect the A/C compressor switch connector.
    f. Connect the alternator connector.
    g. Connect the injector connectors.
    h. Connect the ignition coil connectors.
    i. Connect the intake CMP sensor connector.
    j. Connect the exhaust CMP sensor connector.
    k. Connect the exhaust OCV connector.
    l. Connect the VCM connector.
    m. Connect the ETC connector.
    n. Connect the MAP and IAT connectors.
    o. Connect the PCSV connector.
    p. Connect the ECT sensor connector.
    q. Connect the condenser connector.
    r. Connect the CKP sensor connector.
    s. Connect the oxygen sensor connector.
37. Install the radiator hoses.
38. Install the undercover and tighten the mounting bolts to 15–18 ft. lbs. (20–25 Nm).
39. Install the battery and connect the battery positive terminal.
40. Install the air cleaner assembly.
41. Install the air duct.

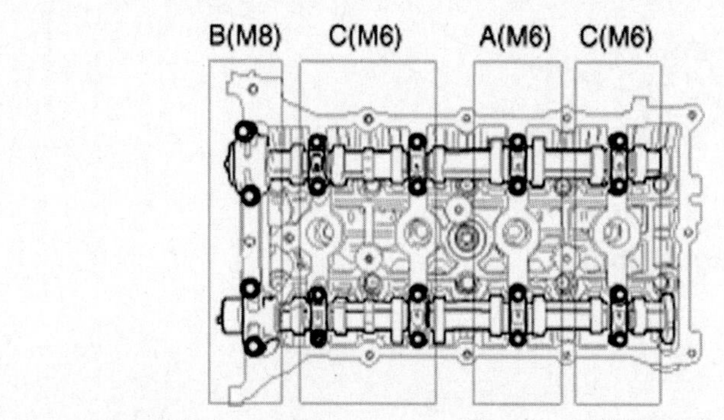

Fig. 147 Tightening order: Group A, Group B, Group C

42. Install the engine cover.
43. Connect the battery negative terminal.
44. Refill the engine oil.
45. Inspect for fuel leakage.
46. Refill radiator and reservoir tank with engine coolant.
47. Bleed air from the cooling system.
48. Run the engine and check for leaks.

CATALYTIC CONVERTER

REMOVAL & INSTALLATION

2009 Models

2.0L Engine

See Figure 148.

2.7L Engine

See Figure 149.

2010 Models—2.4L Engine

See Figure 150.

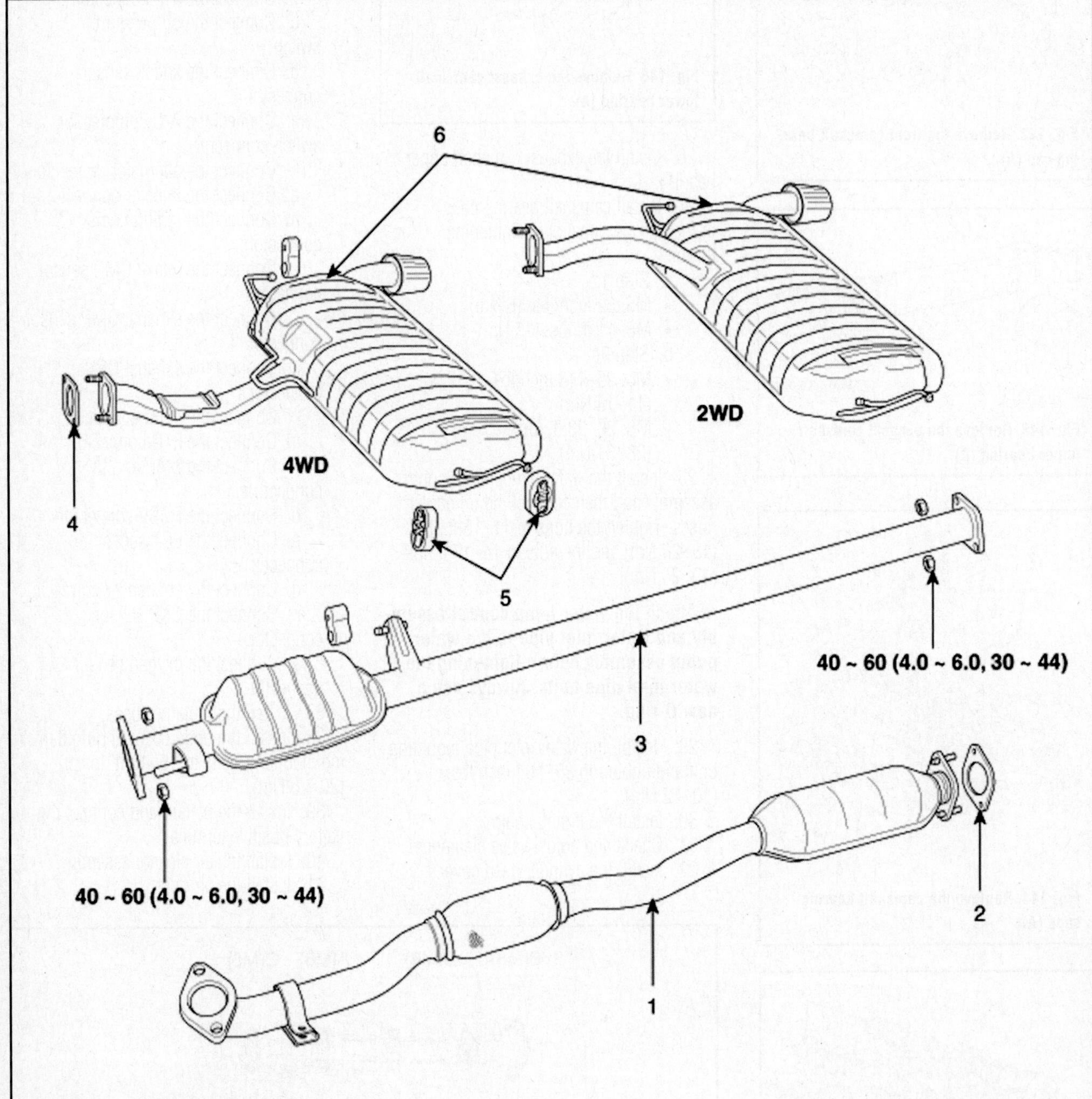

Fig. 148 Front catalytic converter location

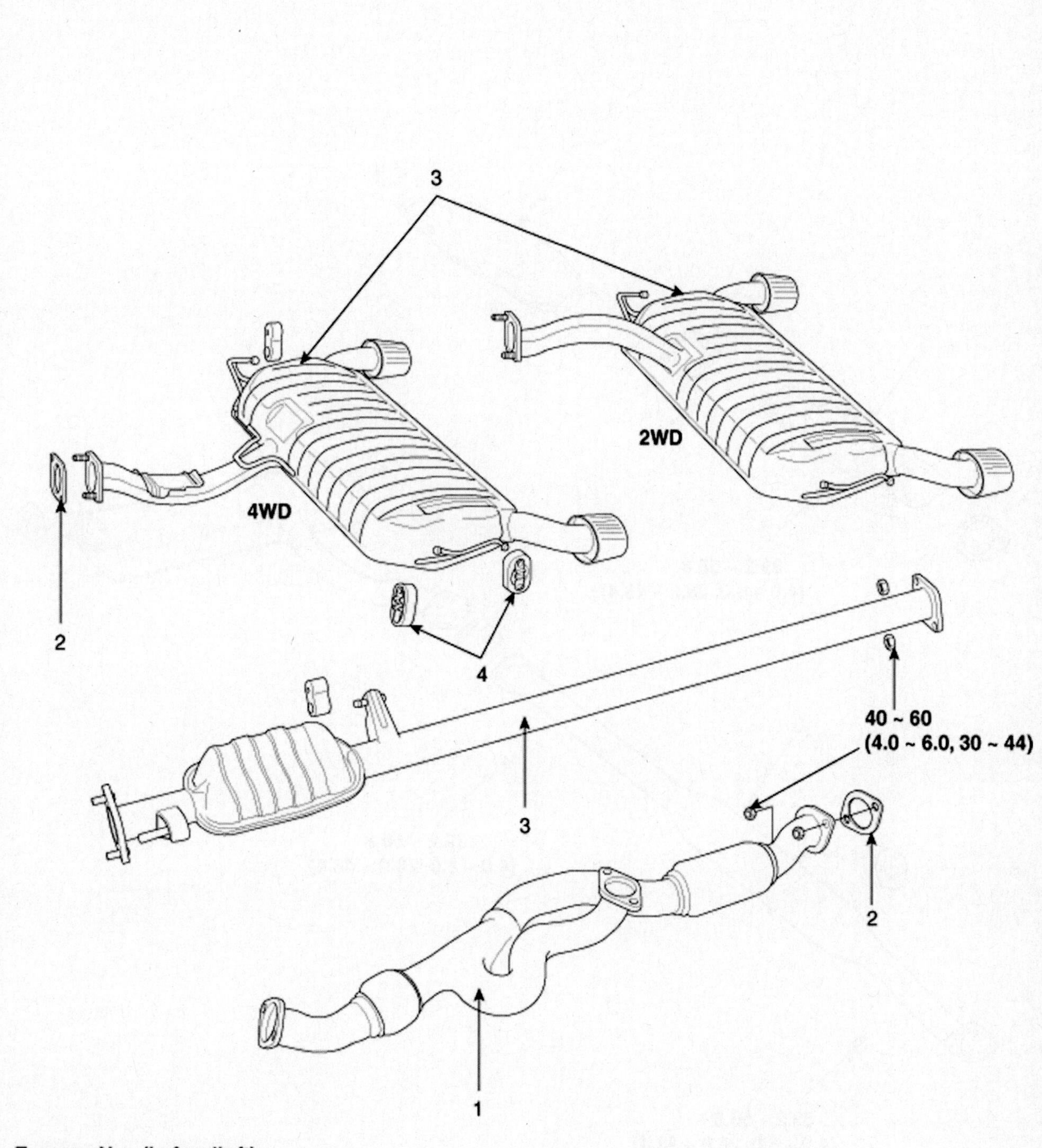

Fig. 149 Front catalytic converter location

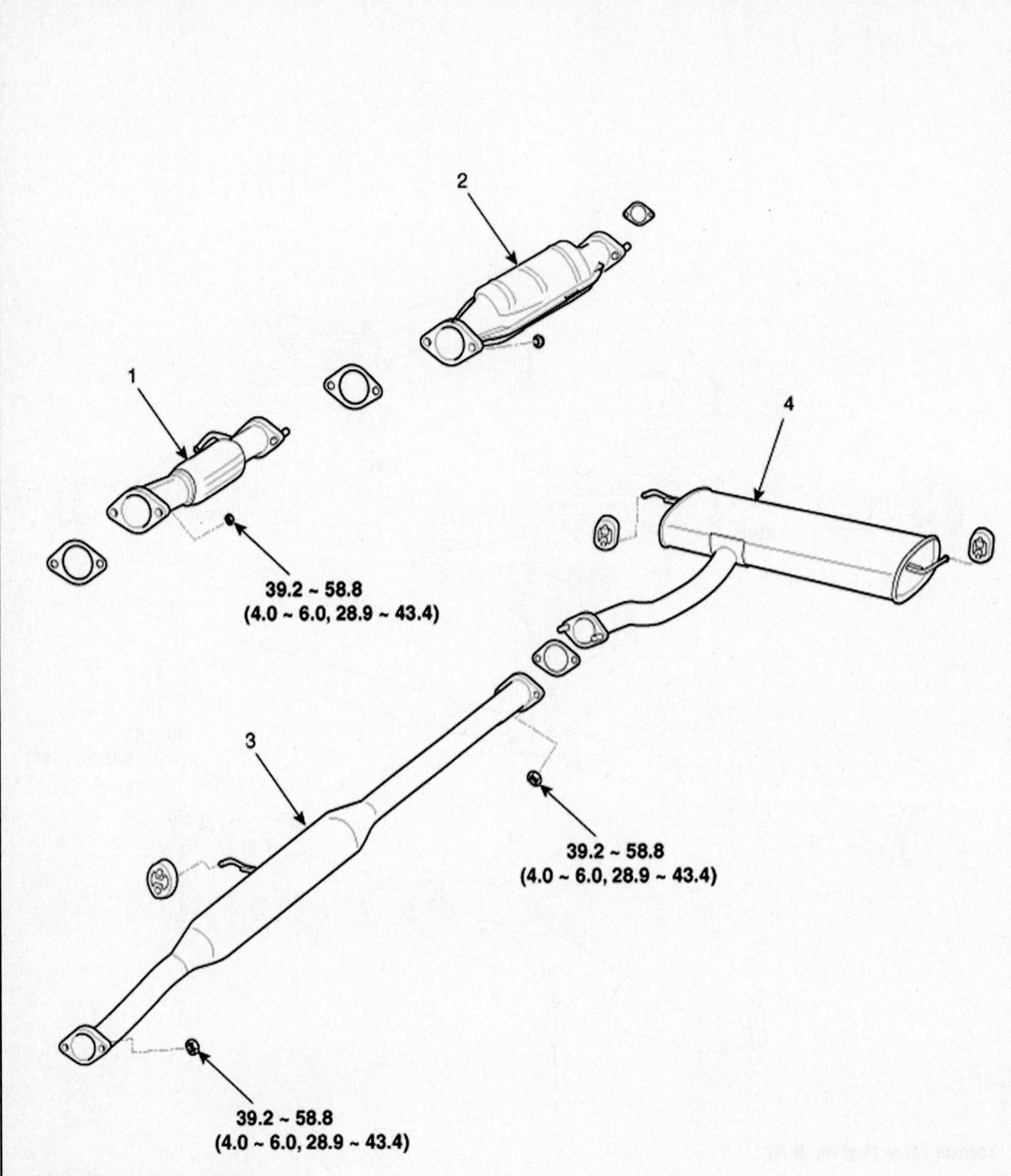

Fig. 150 Front catalytic converter location

CRANKSHAFT FRONT SEAL

REMOVAL & INSTALLATION

See Figure 151.

1. Before servicing the vehicle, refer to the Precautions Section.
2. Remove or disconnect the following:
   a. Negative battery cable
   b. Accessory drive belts
   c. Front cover
   d. Front crankshaft seal

To install:

3. Install the front crankshaft seal using Special Tool 09214–33000 seal installer
4. Install or connect the following:
   a. Front covers
   b. Accessory drive belts
   c. Negative battery cable
5. Start the engine and check for leaks.

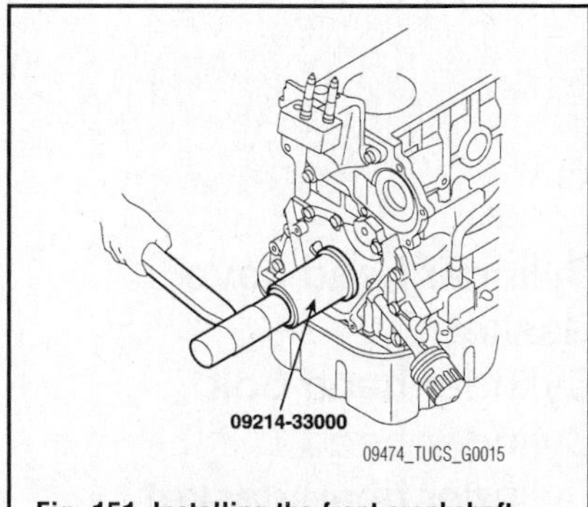

Fig. 151 Installing the front crankshaft seal—2.0L engine, 2.7L, and 2.4L engine similar

CYLINDER HEAD

REMOVAL & INSTALLATION

2009 Models

2.0L Engine

See Figures 152 through 155.

⁂ WARNING

Use fender covers to avoid damaging painted surfaces. To avoid damaging the cylinder head, wait until the engine coolant temperature drops below normal temperature before removing it. When handling a metal gasket, take care not to fold the gasket or damage the contact surface of the gasket. To avoid damage, unplug the wiring connectors carefully while holding the connector portion.

➡Mark all wiring and hoses to avoid misconnection. Inspect the timing belt before removing the cylinder head. Turn the crankshaft pulley so that the No. 1 piston is at top dead center.

1. Before servicing the vehicle, refer to the Precautions Section.
2. Disconnect the negative battery cable.
3. Remove the air duct.
4. Drain the cooling system.
5. Relieve the fuel system pressure. Refer to Relieving Fuel System Pressure in the Fuel System section.
6. Remove the engine cover.
7. Remove the air cleaner assembly.
8. Remove the upper and lower radiator hoses.
9. Remove the heater hoses.
10. Remove the engine wiring harnesses connectors and clamps from the cylinder head and intake manifold:
    a. Disconnect the Oil Control Valve (OCV) connector.
    b. Disconnect the oil temperature sensor connector.
    c. Disconnect the Engine Coolant Temperature (ECT) sensor connector.
    d. Disconnect the ignition coil connector.
    e. Disconnect the throttle position sensor connector.
    f. Disconnect the Idle Speed Actuator (ISA) connector.
    g. Disconnect the Camshaft Position (CMP) sensor connector.
    h. Disconnect the fuel injector connectors.
    i. Disconnect the knock sensor connector.
    j. Disconnect the ground cable from the intake manifold.
    k. Disconnect the front heated oxygen sensor connector.
    l. Disconnect the Purge Control Solenoid Valve (PCSV) connector.
11. Remove the fuel supply hose.
12. Remove the Purge Control Solenoid Valve (PCSV) hose.
13. Remove the brake booster vacuum hose.
14. Remove the throttle cable.
15. Remove the power steering pump and mounting bracket.
16. Remove the spark plug cables.
17. Remove the PCV hose.
18. Remove the cylinder head cover.
19. Remove the timing belt. Refer to Timing Belt Removal & Installation.
20. Remove the exhaust manifold. Refer to Exhaust Manifold Removal & Installation.
21. Remove the intake manifold. Refer to Intake Manifold Removal & Installation.

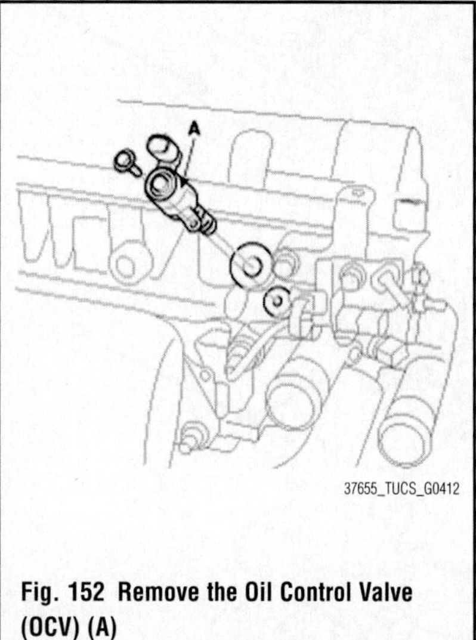

Fig. 152 Remove the Oil Control Valve (OCV) (A)

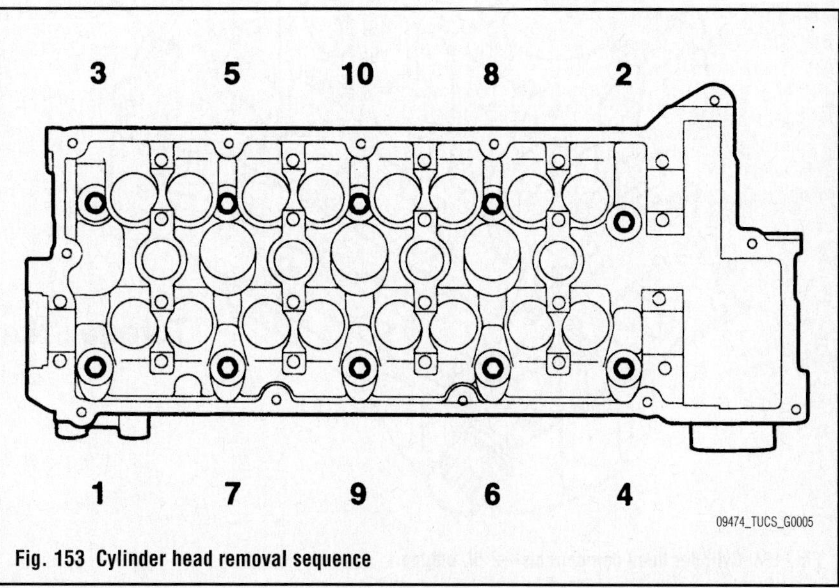

Fig. 153 Cylinder head removal sequence

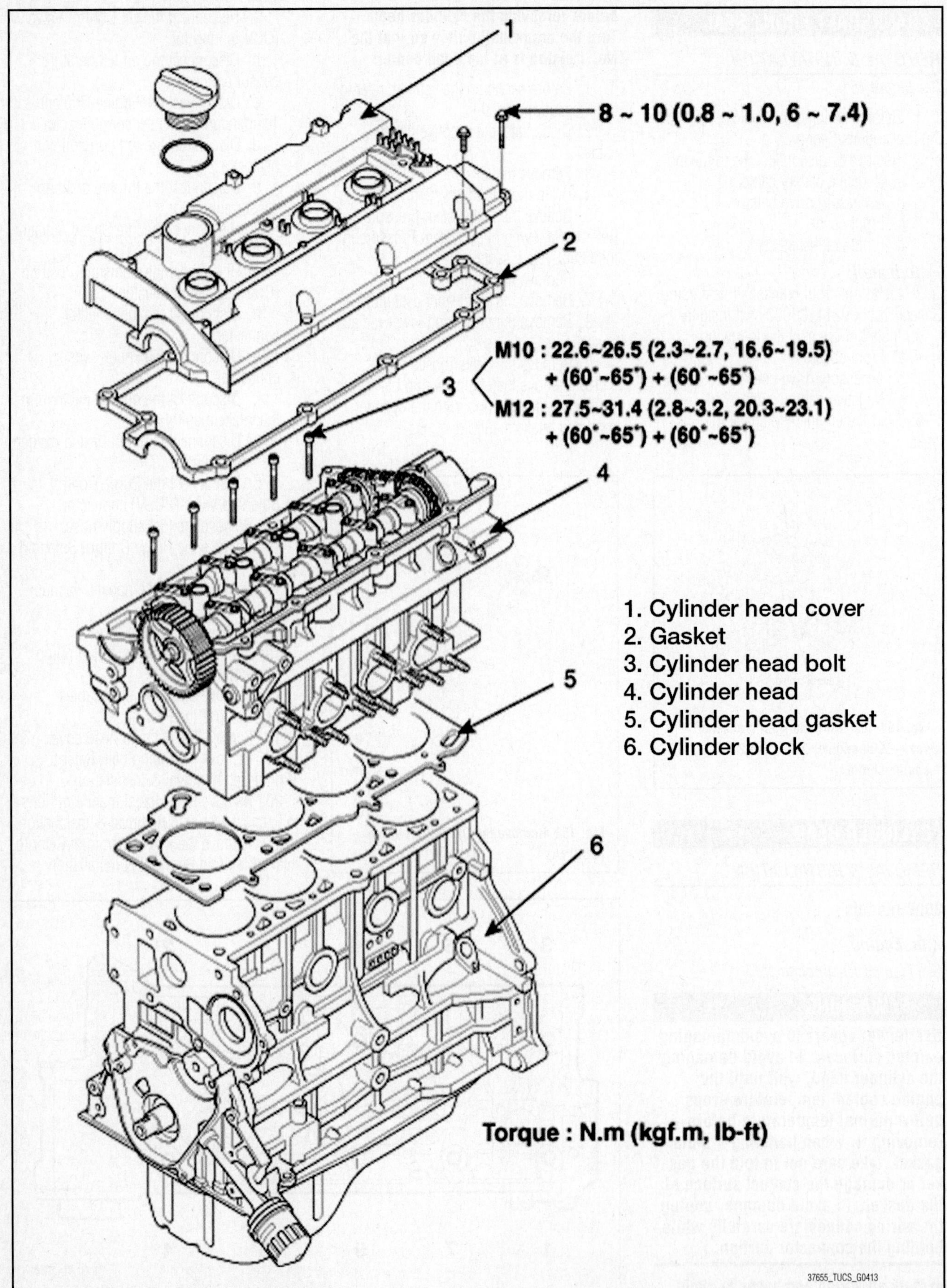

Fig. 154 Cylinder head components—2.0L engine

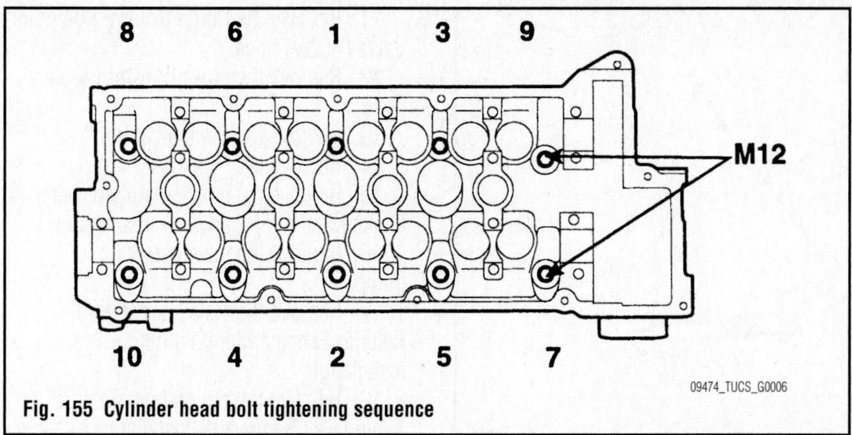

Fig. 155 Cylinder head bolt tightening sequence

22. Remove the camshaft sprocket, camshaft timing chain auto tensioner, bearing caps, and camshafts. Refer to Camshaft and Valve Lifters, Removal & Installation.
23. Remove the Oil Control Valve (OCV).
24. Remove the OCV filter.
25. Remove the coolant hose from coolant pipe.
26. Remove the cylinder head mounting bolts and cylinder head:
    a. Using 8mm and 10mm hexagon wrench, uniformly loosen and remove the 10 cylinder head bolts, in several passes, in the sequence shown. Remove the 10 cylinder head bolts and plate washers.

**** WARNING**

Head warpage or cracking could result from removing bolts in an incorrect order.

    b. Lift the cylinder head from the dowels on the cylinder block and place the cylinder head on wooden blocks on a bench.

**** WARNING**

Be careful not to damage the contact surfaces of the cylinder head and cylinder block.

To install:

➡Thoroughly clean all parts to be assembled. Always use a new head and manifold gasket. The cylinder head gasket is a metal gasket. Take care not to bend it. Rotate the crankshaft, set the No. 1 piston at TDC.

27. Install a new gasket on the cylinder block. Be careful of the installation direction.
28. Place the cylinder head carefully in order not to damage the gasket with the bottom part of the end.
29. Apply a light coat of engine oil to the mounting bolts and tighten in sequence as shown:
    a. M10 bolts:
        - Step 1: 17–20 ft. lbs. (23–27 Nm)
        - Step 2: Plus 60–65 degrees
        - Step 3: Plus an additional 60–65 degrees
    b. M12 bolts:
        - Step 1: 20–23 ft. lbs. (28–31 Nm)
        - Step 2: Plus 60–65 degrees
        - Step 3: Plus an additional 60–65 degrees
30. Install the OCV filter and tighten to 30–38 ft. lbs. (41–51 Nm). Use a new OCV filter gasket.
31. Install the OCV and tighten to 84–108 inch lbs. (10–12 Nm).

➡Do not reuse the OCV if dropped. Keep the OCV clean. Do not hold the OCV sleeve during servicing. When the OCV is installed on the engine, do not move the engine by holding the OCV yoke.

32. Install the camshafts, bearing caps, timing chain auto tensioner, and sprocket.
33. Install the timing belt.
34. Install the cylinder head cover and new gasket.
35. Install the intake manifold.
36. Install the exhaust manifold.
37. Install the PCV.
38. Install the spark plug wires.
39. Install the power steering pump mounting bracket. Tighten to 26–37 ft. lbs. (35–50 Nm).
40. Install the power steering pump.
41. Install the throttle cable.
42. Install the brake booster hose.
43. Install the PCSV hose.
44. Install the fuel supply hose.
45. Install the engine wiring harnesses to the cylinder head and intake manifold:
    a. Connect the PCSV connector.
    b. Connect the front heated oxygen sensor connector.
    c. Connect the ground cable to the intake manifold.
    d. Connect the knock sensor connector.
    e. Connect the fuel injector connectors.
    f. Connect the CMP sensor connector.
    g. Connect the ISA connector.
    h. Connect the throttle position sensor connector.
    i. Connect the ignition coil connector.
    j. Connect the ECT sensor connector.
    k. Connect the oil temperature sensor connector.
    l. Connect the Oil Control Valve (OCV) connector.
46. Install the heater hoses.
47. Install the radiator hoses.
48. Install the air intake assembly.
49. Install the engine cover.
50. Connect the negative battery cable.
51. Fill the engine with coolant to the correct level.
52. Start the engine and check for leaks.

2.7L Engine

See Figures 156 through 160.

1. Before servicing the vehicle, refer to the Precautions Section.
2. Drain the cooling system.
3. Relieve the fuel system pressure. Refer to Relieving Fuel System Pressure in the Fuel System section.
4. Disconnect the negative battery cable.
5. Remove the engine cover.
6. Remove the air intake assembly.
7. Remove the upper and lower radiator hoses.
8. Remove the heater hoses.
9. Remove the engine wiring harnesses from the cylinder head and intake manifold:
    a. Disconnect the Throttle Position Sensor (TPS) connector.
    b. Disconnect the Idle Speed Actuator (ISA) connector.
    c. Disconnect the Purge Control Solenoid Valve (PCSV) connector.
    d. Disconnect the VIS actuator connector.
    e. Disconnect the fuel injector connector.
    f. Disconnect the knock sensor connector.
    g. Disconnect the Camshaft Position (CMP) sensor connector.

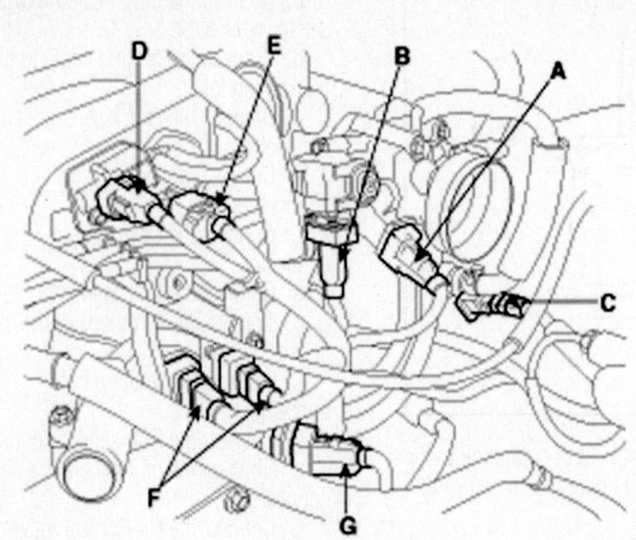

Fig. 156 Engine wiring connectors

h. Disconnect the Engine Coolant Temperature (ECT) sensor connector.

i. Disconnect the ignition coil connector.

j. Disconnect the Crankshaft Position (CKP) sensor connector.

k. Disconnect the heated oxygen sensor connectors.

l. Disconnect the three fuel injector connectors.

m. Disconnect the engine ground line.

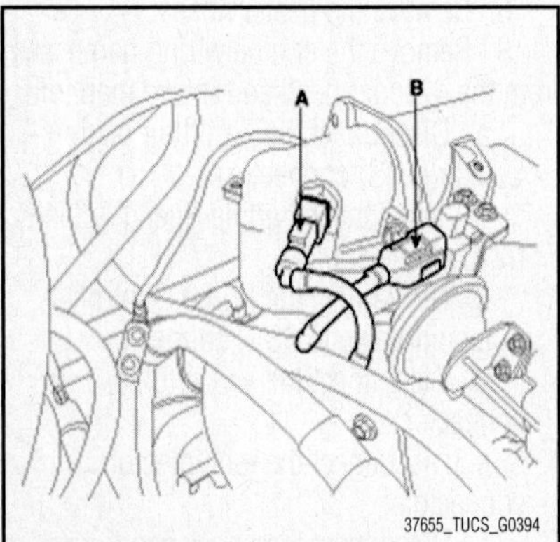

Fig. 157 Disconnect the IAT sensor connector (A) and the VIS actuator connector (B)

n. Disconnect the Intake Air Temperature (IAT) sensor connector.

o. Disconnect the VIS actuator connector.

10. Remove the fuel supply hoses.

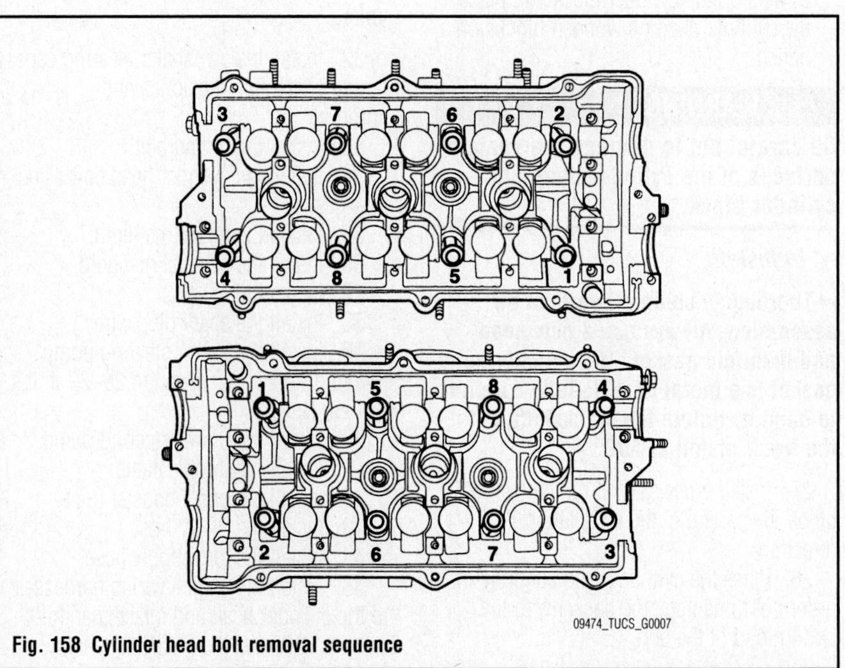

Fig. 158 Cylinder head bolt removal sequence

11. Remove the Purge Control Solenoid Valve (PCSV) hose.

12. Remove the brake booster vacuum hose.

13. Remove the throttle cable.

14. Remove the PCV hose.

15. Remove the intake manifold. Refer to Intake Manifold Removal & Installation.

16. Remove the power steering pump.

17. Remove the exhaust manifold. Refer to Exhaust Manifold Removal & Installation.

18. Remove the timing belt. Refer to Timing Belt Removal & Installation.

19. Remove the spark plug cable.

20. Remove the cylinder head covers.

21. Remove the camshaft bearing caps and camshafts. Refer to Camshaft and Valve Lifters, Removal & Installation.

22. Remove the timing belt rear cover.

23. Remove the water temperature control assembly and water pipe.

24. Loosen the cylinder head bolts in several passes in the sequence shown.

WARNING

Head warping or cracking could result from removing bolts in an incorrect order.

25. Remove the cylinder bolts and plate washer.

26. Remove the cylinder heads.

To install:

27. Install the cylinder head gaskets on the cylinder block.

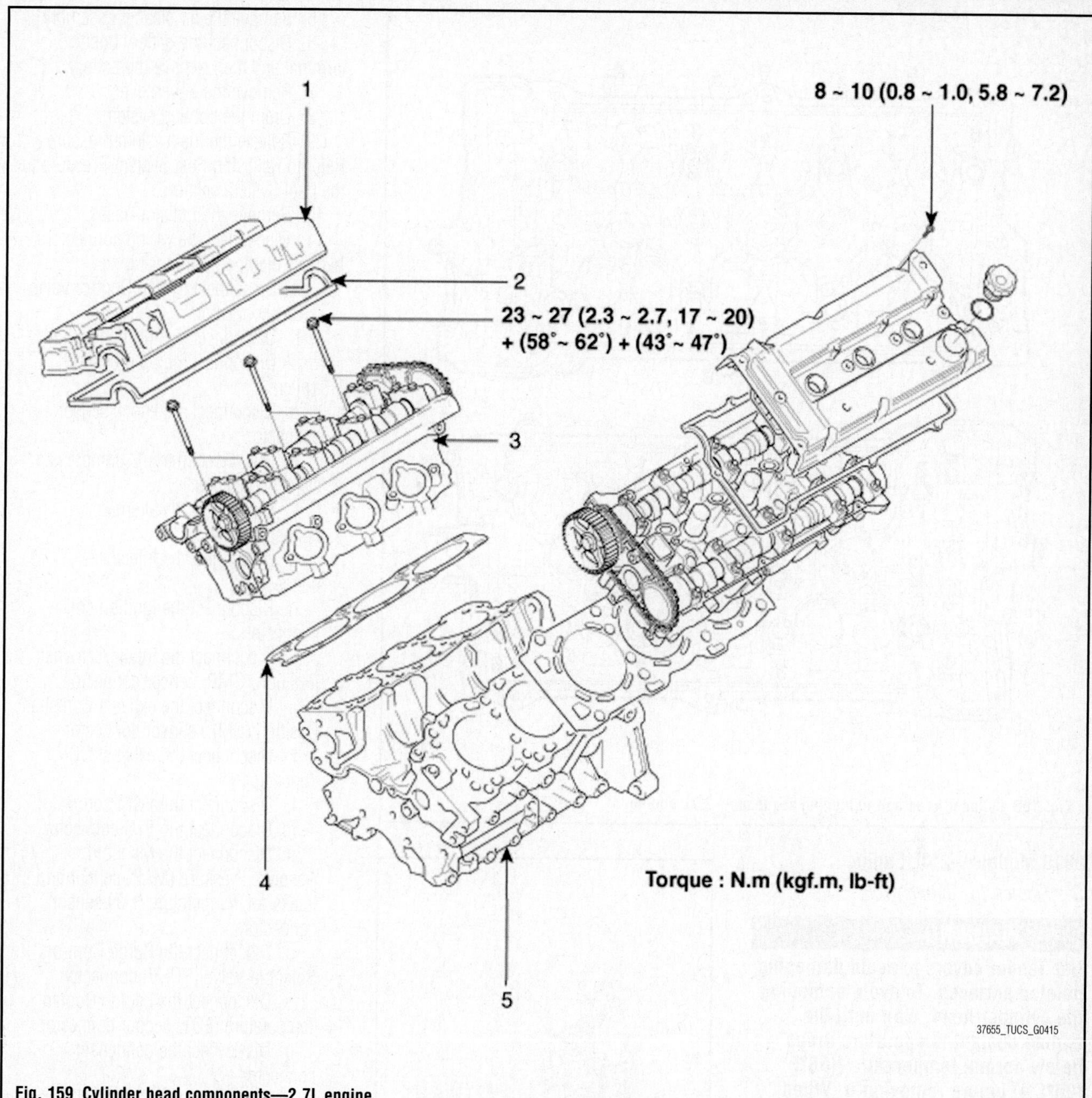

Fig. 159 Cylinder head components—2.7L engine

➡Be careful of the installation direction.

28. Apply a light coat if engine oil on the threads and under the heads of the cylinder head bolts.

29. Install the plate washer to the cylinder head bolt.

30. Install and uniformly tighten the cylinder head bolts on each cylinder head in several passes and in the sequence shown, then repeat for the other side, as shown.

- Step 1 = Tighten to 17–20 ft. lbs. (23–27 Nm)
- Step 2 = Plus 60 degrees
- Step 3 = Plus 45 degrees

31. Install the water pipe and water temperature control assembly. Tighten to 11–14 ft. lbs. (15–20 Nm)

32. Install the timing belt rear cover. Tighten to 84–108 inch lbs. (10–12 Nm)

33. Install the camshafts and camshaft bearing caps.

34. Install the cylinder head covers and gaskets.

35. Install the spark plug cable.

36. Install the timing belt.

37. Install the exhaust manifold.

38. Install the power steering pump.

39. Install the intake manifold.

40. Install the PCV hose.

41. Install the throttle cable.

42. Install the brake booster vacuum hose.

43. Install the PCSV hose.

44. Install the fuel supply hose.

45. Install the engine wiring harnesses to the cylinder head and intake manifold.

46. Install the heater hoses.

47. Install the upper and lower radiator hoses.

48. Install the air intake assembly.

49. Install the engine cover.

50. Connect the negative battery cable.

51. Refill the engine with coolant to the correct level.

52. Start the engine and check for leaks.

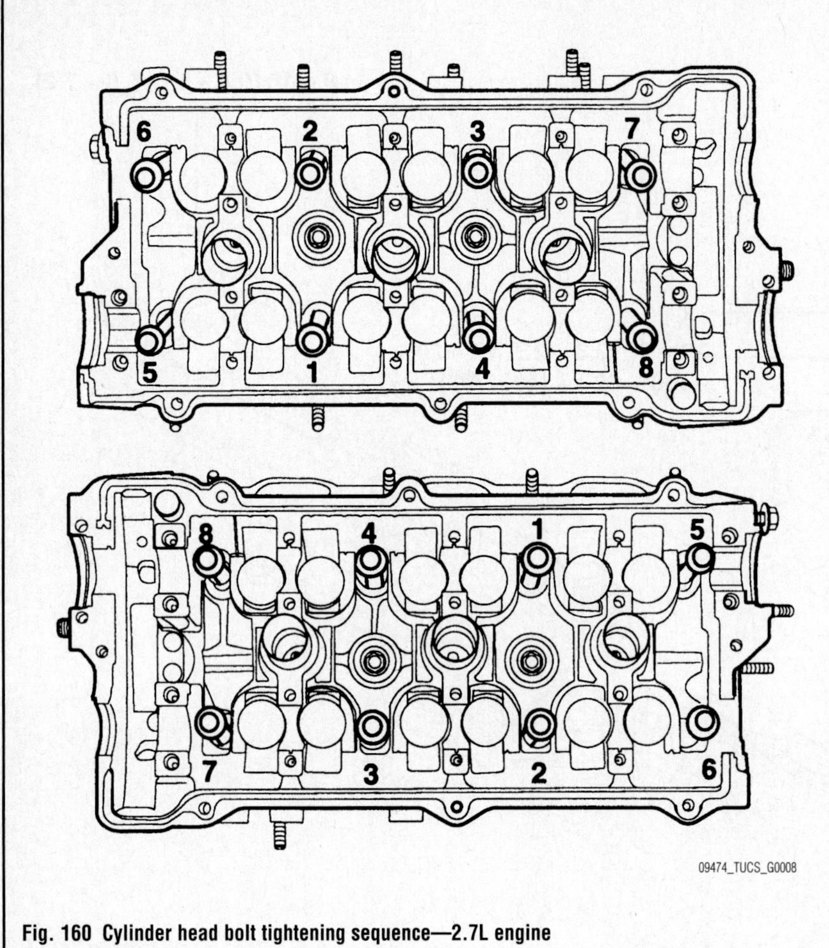

Fig. 160 Cylinder head bolt tightening sequence—2.7L engine

2010 Models—2.4L Engine

See Figures 161 through 168.

⁂ WARNING

Use fender covers to avoid damaging painted surfaces. To avoid damaging the cylinder head, wait until the engine coolant temperature drops below normal temperature [(68°F (20°C)]) before removing it. When handling a metal gasket, take care not to fold the gasket or damage the contact surface of the gasket. To avoid damage, unplug the wiring connectors carefully while holding the connector portion.

➡Mark all wiring and hoses to avoid misconnection. Turn the crankshaft pulley so that the No. 1 piston is at top dead center.

1. Before servicing the vehicle, refer to the Precautions Section.
2. Disconnect the negative battery cable.
3. Remove the engine cover.
4. Remove the air duct.

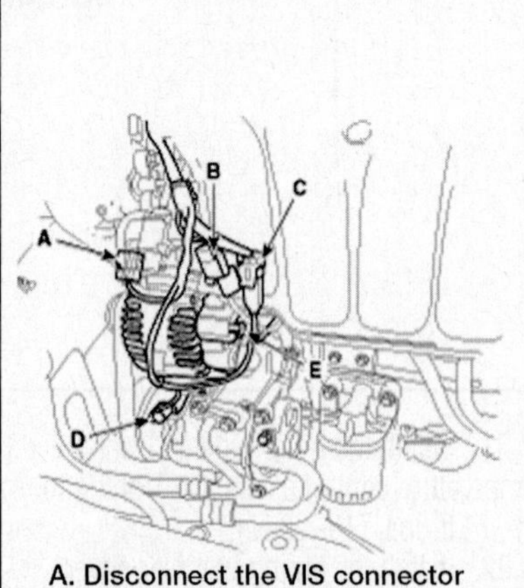

Fig. 161 Disconnect the VIS connector, the oil pressure switch, the knock sensor connector, the A/C compressor switch connector and the alternator connector

5. Remove the air cleaner assembly.
6. Disconnect the battery positive terminal and then remove the battery.
7. Remove the undercover.
8. Drain the cooling system.
9. Relieve the fuel system pressure. Refer to Relieving Fuel System Pressure in the Fuel System section.
10. Remove the radiator hoses.
11. Disconnect the wiring connectors, harness clamps from the engine.
    a. Disconnect the Oil Control Valve (OCV) connector.
    b. Disconnect the VIS connector.
    c. Disconnect the oil pressure switch.
    d. Disconnect the knock sensor connector.
    e. Disconnect the A/C compressor switch connector.
    f. Disconnect the alternator connector.
    g. Disconnect the injector connectors.
    h. Disconnect the ignition coil connectors.
    i. Disconnect the intake Camshaft Position (CMP) sensor connector.
    j. Disconnect the exhaust Camshaft Position (CMP) sensor connector.
    k. Disconnect the exhaust OCV connector.
    l. Disconnect the VCM connector.
    m. Disconnect the ETC connector.
    n. Disconnect the Manifold Absolute Pressure (MAP) sensor and Intake Air Temperature (IAT) sensor connectors.
    o. Disconnect the Purge Control Solenoid Valve (PCSV) connector.
    p. Disconnect the Engine Coolant Temperature (ECT) sensor connector.
    q. Disconnect the condenser connector.
    r. Disconnect the Crankshaft Position (CKP) sensor connector.
    s. Disconnect the oxygen sensor connector.
12. Disconnect the fuel hose, brake booster vacuum hose and heater hoses.
13. Disconnect the throttle body coolant hoses.
14. Remove the water inlet pipe mounting bolt.
15. Disconnect the oil cooler hoses and then remove the water temperature control assembly.
16. Remove the timing chain. Refer to Timing Chain Removal & Installation.
17. Remove the intake and exhaust manifold. Refer to Intake Manifold and Exhaust Manifold Removal & Installation.

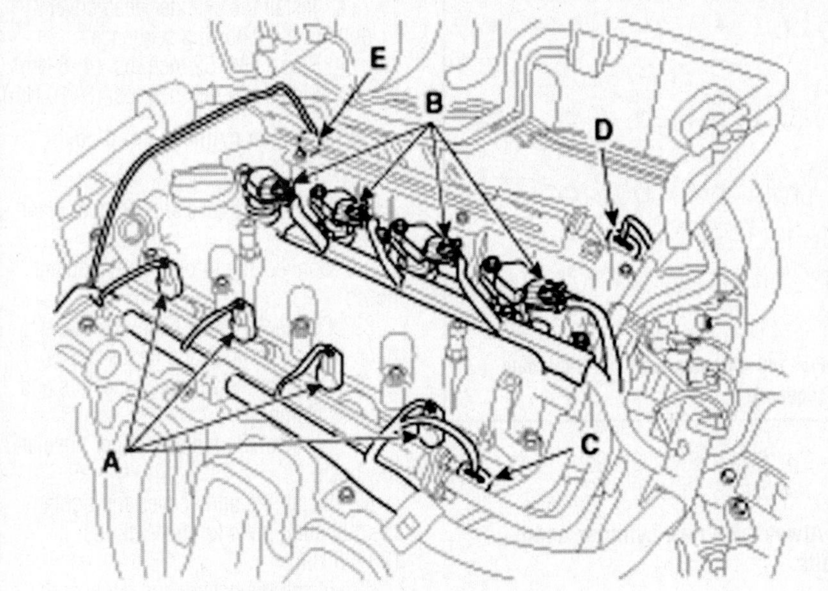

Fig. 162 Disconnect the injector connectors, ignition coil connectors, intake and exhaust CMP sensor connectors, and the exhaust OCV connector

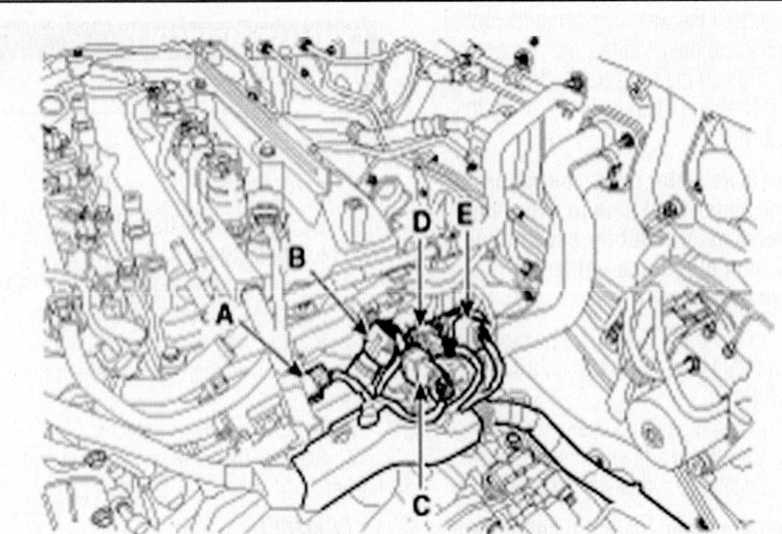

Fig. 163 Disconnect the PCSV connector, ECT sensor connector, condenser connector, CKP sensor connector, and the oxygen sensor connector

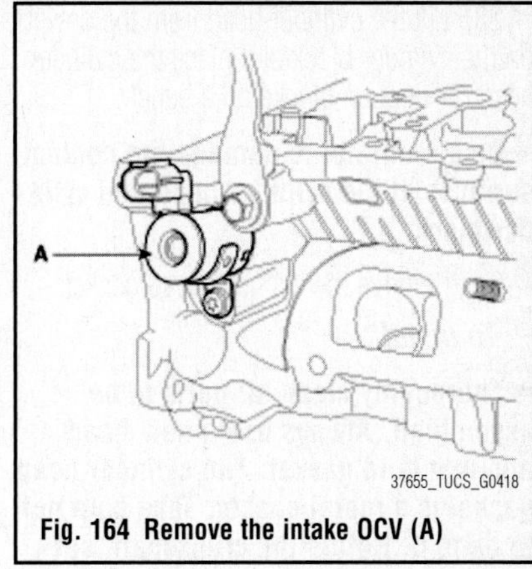

Fig. 164 Remove the intake OCV (A)

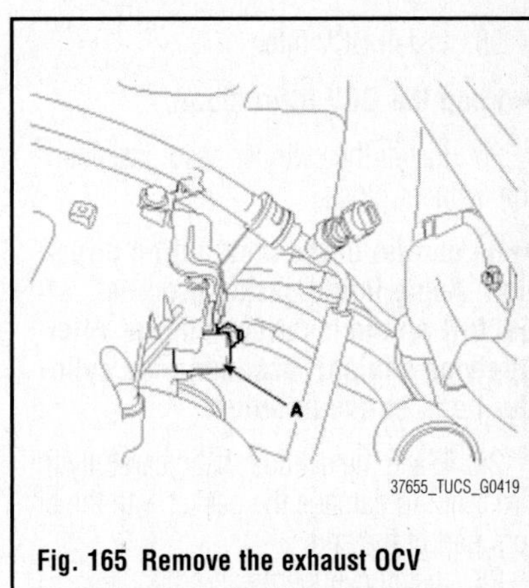

Fig. 165 Remove the exhaust OCV

18. Remove the intake and exhaust CVVT assembly.
19. Remove the camshaft bearings, bearing caps, and camshaft. Refer to Camshaft and Valve Lifters, Removal & Installation.
20. Remove the intake OCV.
21. Remove the exhaust OCV.
22. Using a triple square wrench, uniformly loosen and remove the 10 cylinder head bolts, in several passes, in the sequence shown. Head warpage or cracking could result from removing bolts in an incorrect order.

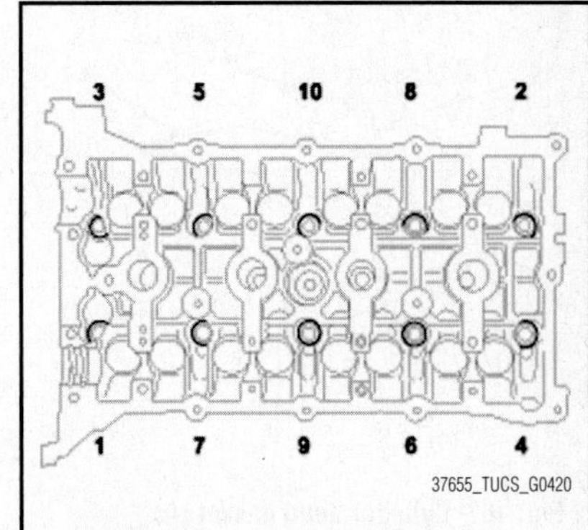

Fig. 166 Cylinder head bolt removal sequence

23. Lift the cylinder head from the dowels on the cylinder block and place the cylinder head on wooden blocks on a bench.

➡Be careful not to damage the contact surfaces of the cylinder head and cylinder block.

24. Remove the cylinder head gasket.

To install:

➡Thoroughly clean all parts to be assembled. Always use a new head and manifold gasket. The cylinder head gasket is a metal gasket. Take care not to bend it. Rotate the crankshaft, set the No.1 piston at TDC.

25. Install OCV filter.

➡Keep the OCV filter clean.

26. Install the cylinder head gasket on the cylinder block.

➡Be careful of the installation direction. Apply liquid gasket (Loctite® 5900H) on the locations shown. After applying sealant, assemble the cylinder head in five minutes.

27. Place the cylinder head carefully in order not to damage the gasket with the bottom part of the end.

28. Install cylinder head bolts.

a. Apply a light coat if engine oil on the threads and under the heads of the cylinder head bolts.

b. Using hexagon wrench, install and tighten the 10 cylinder head bolts and plate washers, in several passes, in the sequence shown. Tightening torque:

- Step 1: 24–27 ft. lbs. (33–36 Nm)

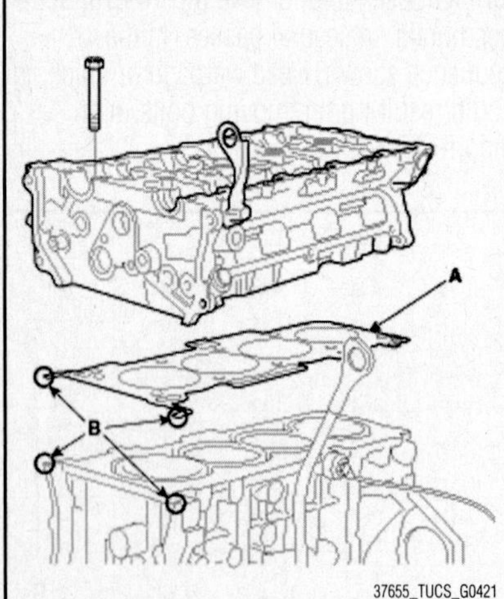

Fig. 167 Cylinder head gasket (A) installation, and sealant application locations (B)

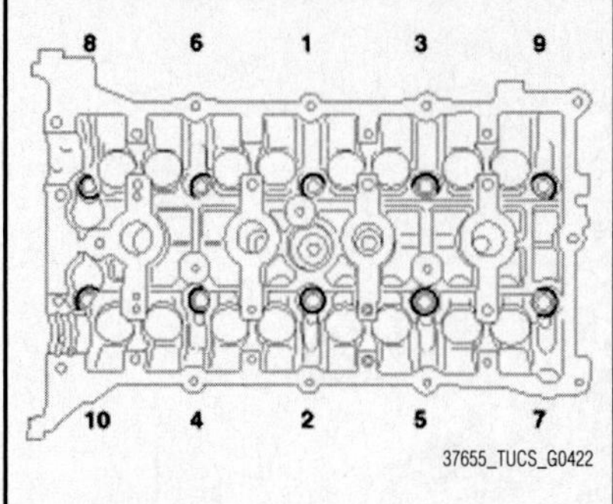

Fig. 168 Cylinder head bolt tightening sequence

- Step 2: Plus 90°–95°
- Step 3: Plus 90°–95°

➡Always use new cylinder head bolts.

29. Install the intake OCV and tighten to 86–104 inch lbs. (10–12 Nm).

30. Install the exhaust OCV and tighten to 86–104 inch lbs. (10–12 Nm).

➡Do not reuse the OCV if dropped. Keep the OCV filter clean. Do not hold the OCV sleeve during servicing. When the OCV is installed on the engine, do not move the engine with holding the OCV yoke.

31. Install the camshafts.

32. Install the water temperature control assembly and then connect the oil cooler hoses. Tighten the bolts to 11–15 ft. lbs. (15–20 Nm) and the nuts to 14–17 ft. lbs. (19–24 Nm).

➡Assemble water temp control assembly and water inlet pipe to water pump assembly before bolt for assembling of water inlet pipe to be tightened. Always use a new O-ring.

33. Install the water inlet pipe mounting bolt and tighten to 86–104 inch lbs. (10–12 Nm).

34. Install the timing chain.

35. Check and adjust valve clearance.

36. Install the cylinder head cover.

➡The hardening sealant located on the upper area between timing chain cover and cylinder head should be removed before assembling cylinder head cover. After applying sealant, it should be assembled within 5 minutes. Bead width: 0.1 in. (2.5mm), Sealant: LOCTITE® 5900H

➡Running the engine or performing a pressure test should not be performed within 30 minutes of assembly

a. Install the cylinder head cover bolts in the following sequence.

- Step 1: 35–52 inch lbs. (4–6 Nm)
- Step 2: 70–86 inch lbs. (8–10 Nm)

➡Do not reuse cylinder head cover gasket.

37. Install the intake and exhaust manifold.

38. Connect the throttle body coolant hoses.

39. Connect the fuel hose, the brake booster vacuum hose and heater hoses.

40. Connect the wiring connectors and harness clamps.

41. Connect the radiator upper hose and low hose.

42. Install the undercover and tighten the mounting bolts to 15–18 ft. lbs. (20–25 Nm).

43. Install the battery and connect the battery positive terminal.

44. Install the air cleaner assembly.

45. Install the air duct.

46. Install the engine cover.

47. Connect the battery negative terminal.

48. Refill engine oil.

49. Inspect for fuel leakage.

50. Refill radiator and reservoir tank with engine coolant.

51. Bleed air from the cooling system.

EXHAUST MANIFOLD

REMOVAL & INSTALLATION

2009 Models

See Figures 169 and 170.

1. Before servicing the vehicle, refer to the Precautions Section.

2. Disconnect the negative battery cable.

3. Remove the undercover.

4. Remove the front exhaust pipe.

5. Disconnect the oxygen sensor connector.

6. Remove the heat protector.

7. Remove the exhaust manifold and gasket.

To install:

8. Installation is the reverse of the removal procedure.

9. Observe the following torques:

a. 2.0L Engine

- Exhaust manifold: 32–40 ft. lbs. (43–55 Nm)
- Manifold heat shield: 12–16 ft. lbs. (17–22 Nm)

b. 2.7L Engine

- Exhaust manifold: 22–26 ft. lbs. (30–35 Nm)

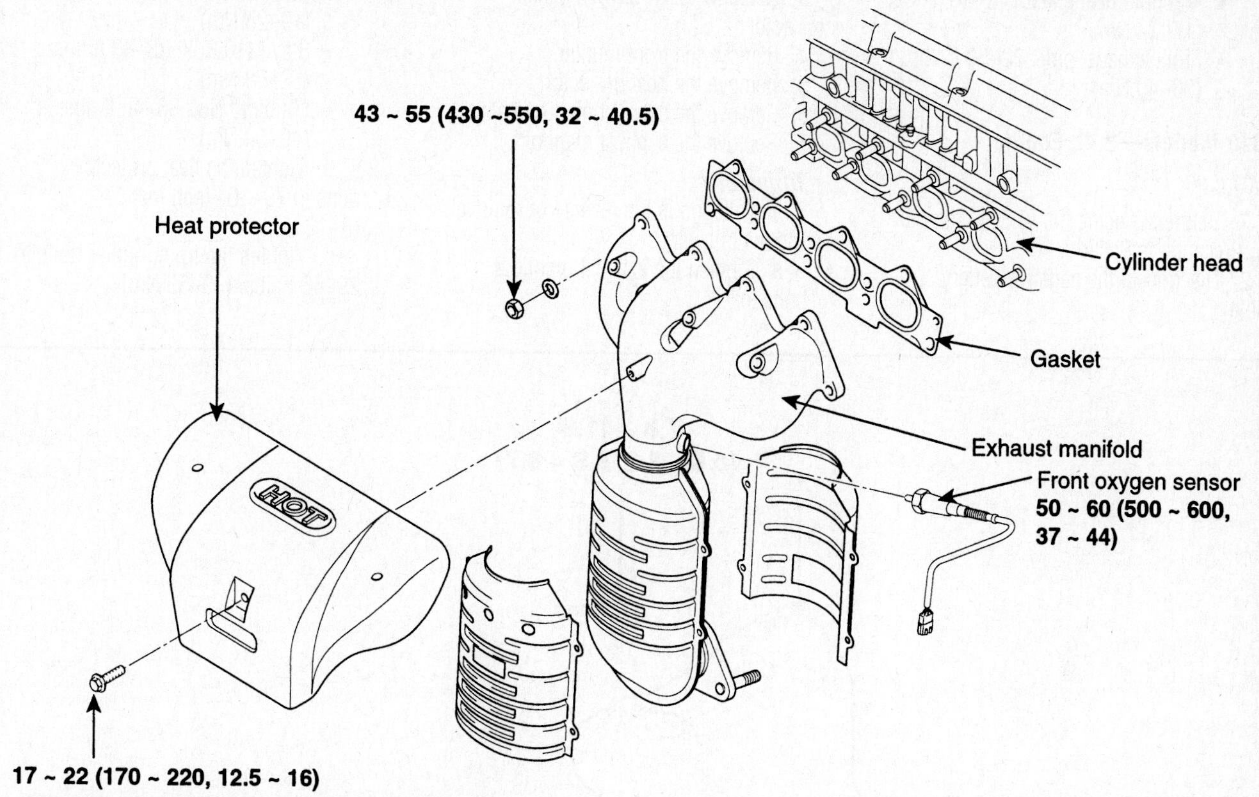

Fig. 169 Exhaust manifold exploded view—2.0L engine

09474_TUCS_G0013

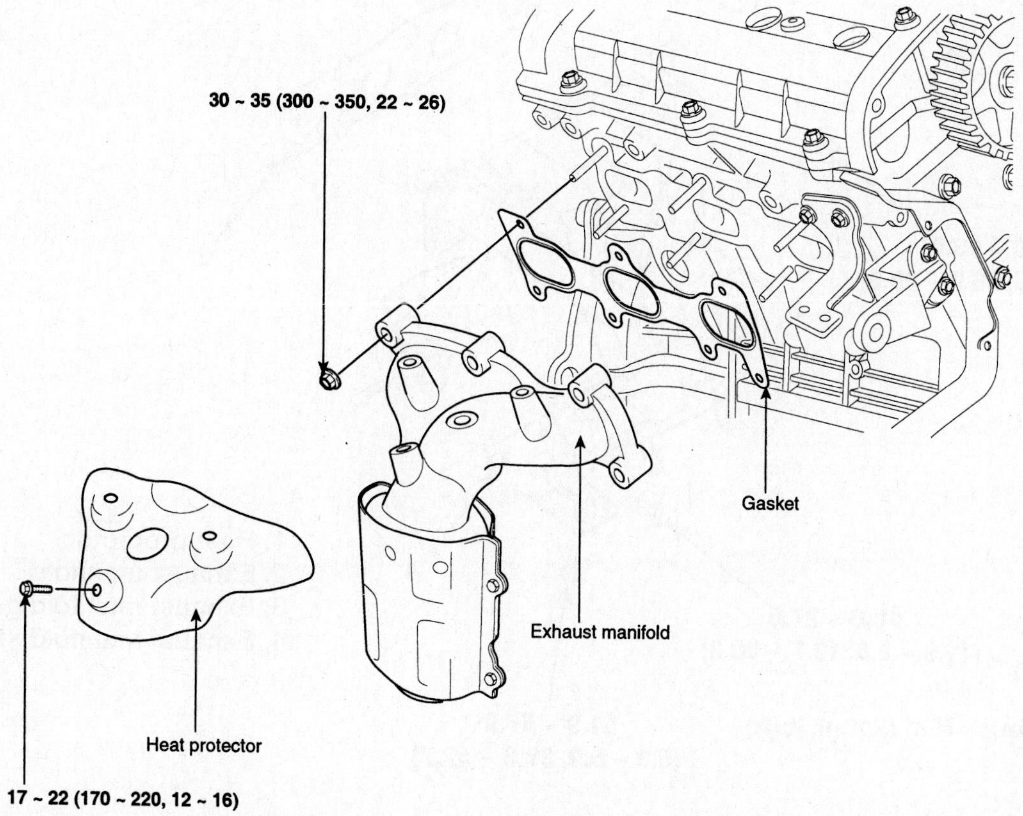

Fig. 170 Exhaust manifold exploded view—2.7L engine

09474_TUCS_G0014

- Manifold heat shield: 12–16 ft. lbs. (17–22 Nm)
- Front exhaust pipe: 22–30 ft. lbs. (30–40 Nm)

2010 Models—2.4L Engine

See Figure 171.

1. Before servicing the vehicle, refer to the Precautions Section.
2. Disconnect the negative battery cable.
3. Disconnect the oxygen sensor connector.
4. Remove the front muffler.
5. Remove the heat protector.
6. Remove the exhaust manifold stay.
7. Remove the exhaust manifold.

To install:

8. Installation is the reverse of removal, noting the following:
   a. Tighten the exhaust manifold bolts to:
      - Stay M8 Bolt: 14–20 ft. lbs. (19–28 Nm)
      - Stay M10 Bolt: 38–43 ft. lbs. (52–58 Nm)
      - Manifold Nut: 36–40 ft. lbs. (49–54 Nm)
   b. Tighten the heat protector bolts to 70–104 inch lbs. (8–12 Nm).
   c. Tighten the front muffler bolts to 29–43 ft. lbs. (39–59 Nm).

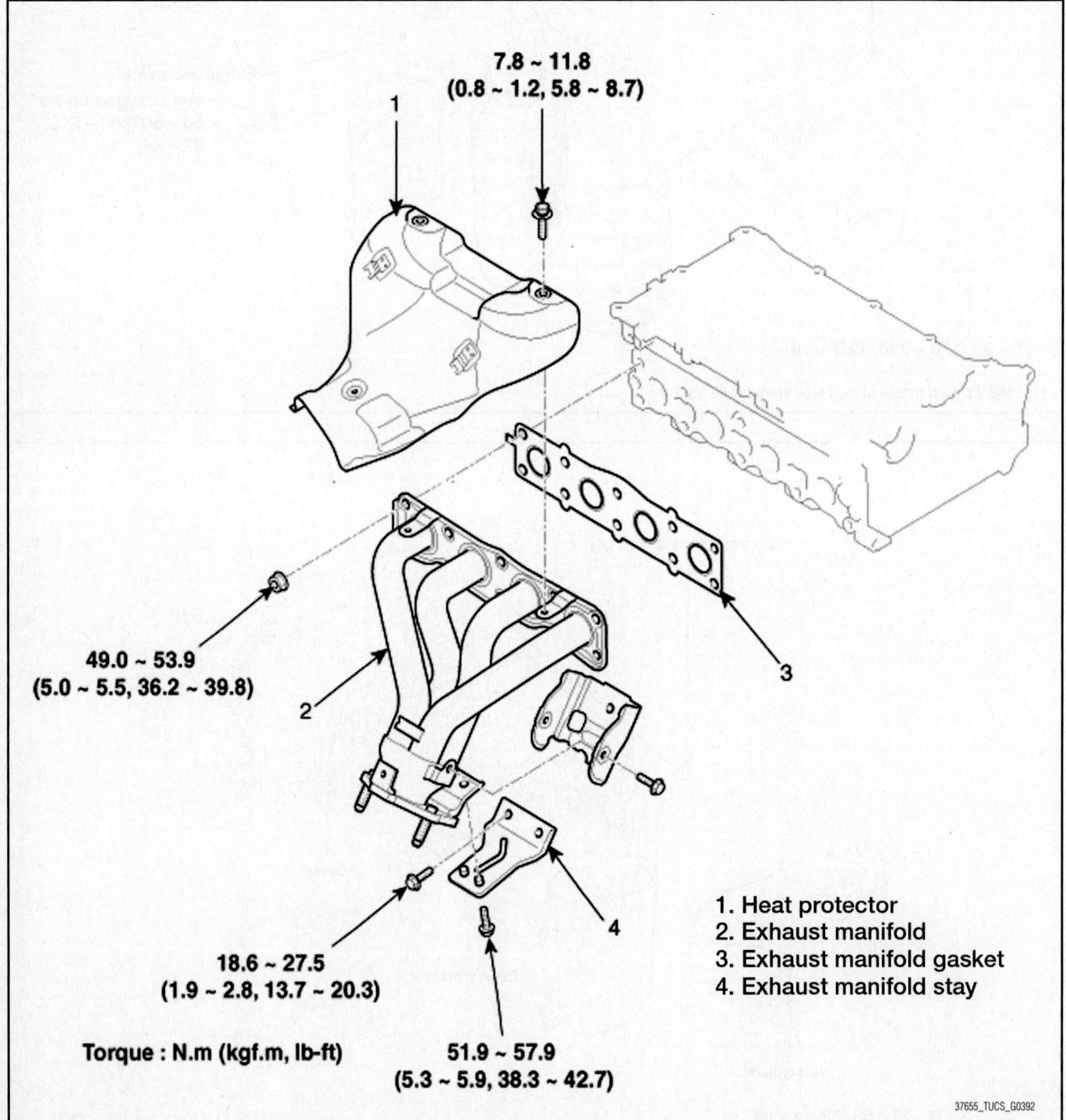

Fig. 171 Exhaust manifold exploded view

INTAKE MANIFOLD

REMOVAL & INSTALLATION

2009 Models

2.0L Engine

See Figure 172.

1. Before servicing the vehicle, refer to the Precautions Section.
2. Relieve the fuel system pressure. Refer to Relieving Fuel System Pressure.
3. Drain the cooling system.
4. Disconnect the negative battery cable.
5. Remove the engine cover.
6. Disconnect the TPS and ISA connectors.
7. Remove the PCV and breather hoses.
8. Disconnect the throttle cable.
9. Remove the fuel rail.
10. Remove the heater hoses and brake vacuum hose.
11. Remove the intake manifold stay.
12. Remove the intake manifold and gasket.

To install:

13. Installation is the reverse of the removal procedure.
14. Tighten the intake manifold nuts to 12–17 ft. lbs. (16–23 Nm).

2.7L Engine

See Figures 173 through 178.

1. Before servicing the vehicle, refer to the Precautions Section.
2. Relieve the fuel system pressure. Refer to Relieving Fuel System Pressure.
3. Drain the cooling system.
4. Disconnect the negative battery cable.
5. Remove the engine cover.
6. Remove the air cleaner hose.
7. Disconnect the throttle cable.
8. Disconnect the TPS and ISA connectors.
9. Disconnect the VIS actuator connector.
10. Disconnect the injector connector.
11. Disconnect the PCSV connector.
12. Disconnect the PCSV hose.
13. Disconnect the brake booster vacuum hose.
14. Disconnect the PCV hose.
15. Disconnect the IAT sensor connector.

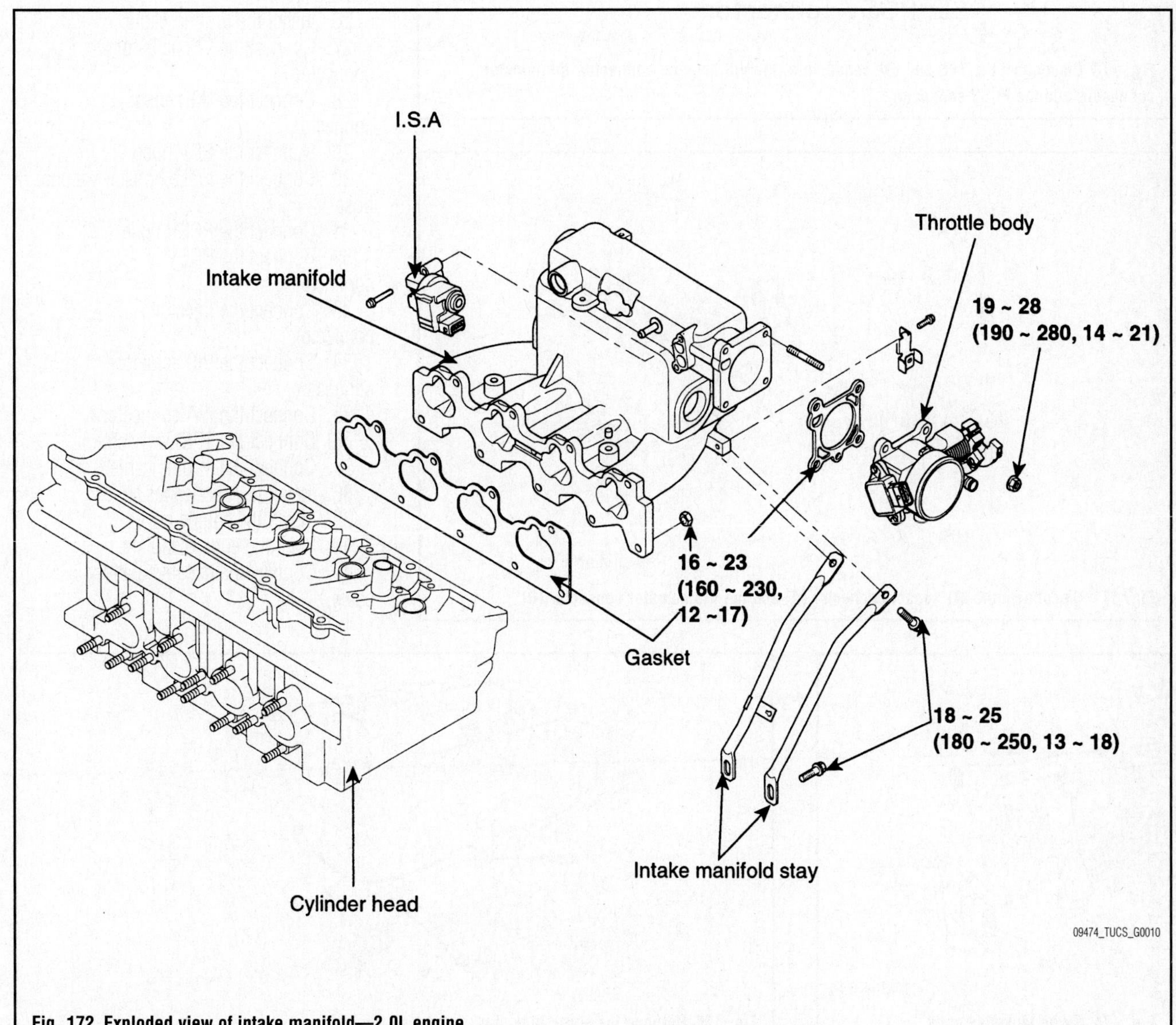

Fig. 172 Exploded view of intake manifold—2.0L engine

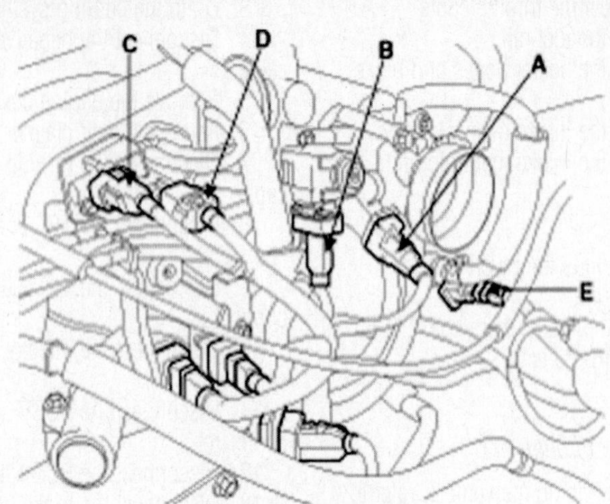

Fig. 173 Disconnect the TPS and ISA connectors, the VIS actuator connector, the injector connector, and the PCSV connector

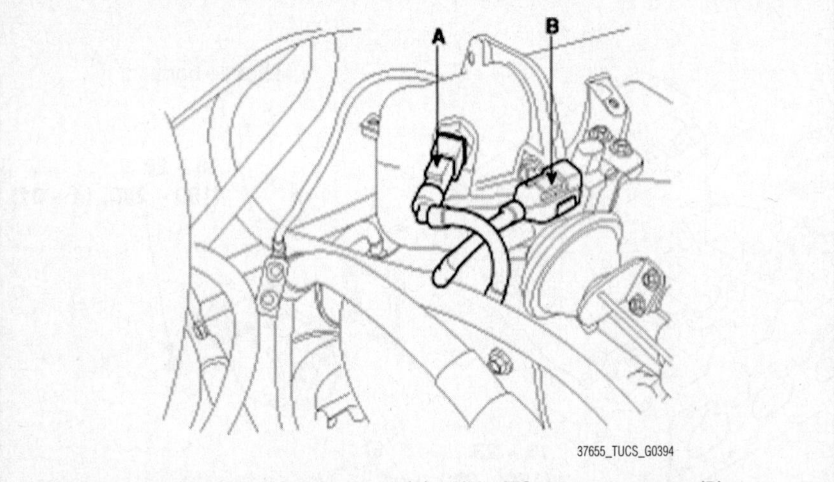

Fig. 174 Disconnect the IAT sensor connector (A) and the VIS actuator connector (B)

16. Disconnect the VIS actuator connector.
17. Disconnect the ground cable from the surge tank assembly.
18. Remove the surge tank stay.
19. Remove the surge tank assembly.
20. Remove the injector assembly.
21. Remove the intake manifold and gasket.

To install:

22. Install the intake manifold and gasket. Tighten the bolts in sequence to 14–15 ft. lbs. (19–21 Nm).
23. Install the injector assembly.
24. Install the surge tank assembly and tighten to 11–15 ft. lbs. (15–20 Nm).
25. Install the surge tank stay and tighten to 11–15 ft. lbs. (15–20 Nm).
26. Install the ground cable.
27. Connect the VIS actuator connector.
28. Connect the IAT sensor connector.
29. Connect the PCV hose.
30. Connect the brake booster vacuum hose.
31. Connect the PCSV hose.
32. Connect the PCSV connector.
33. Connect the injector connector.
34. Connect the VIS actuator connector.
35. Connect the ISA connector.
36. Connect the TPS connector.
37. Connect the throttle cable.
38. Install the air cleaner hose.
39. Install the engine cover.
40. Refill the engine coolant.
41. Connect the negative battery cable.

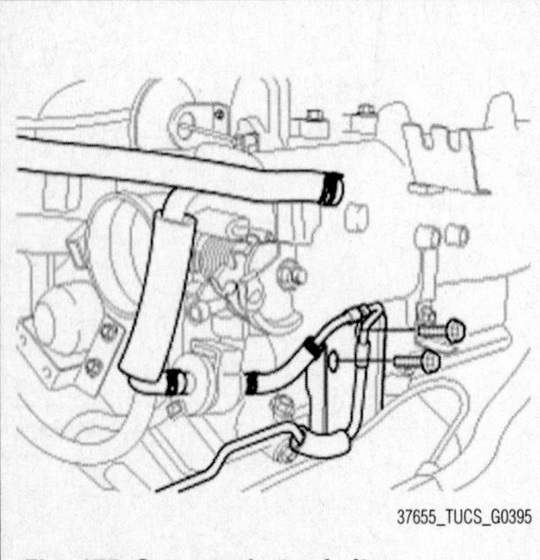

Fig. 175 Surge tank stay bolts

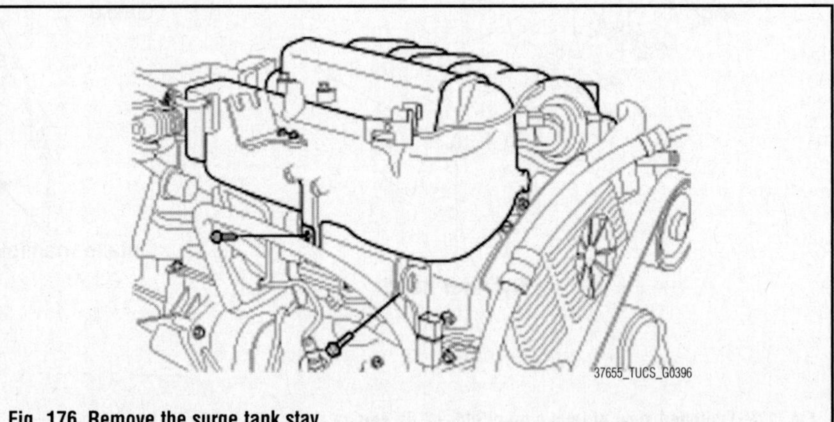

Fig. 176 Remove the surge tank stay

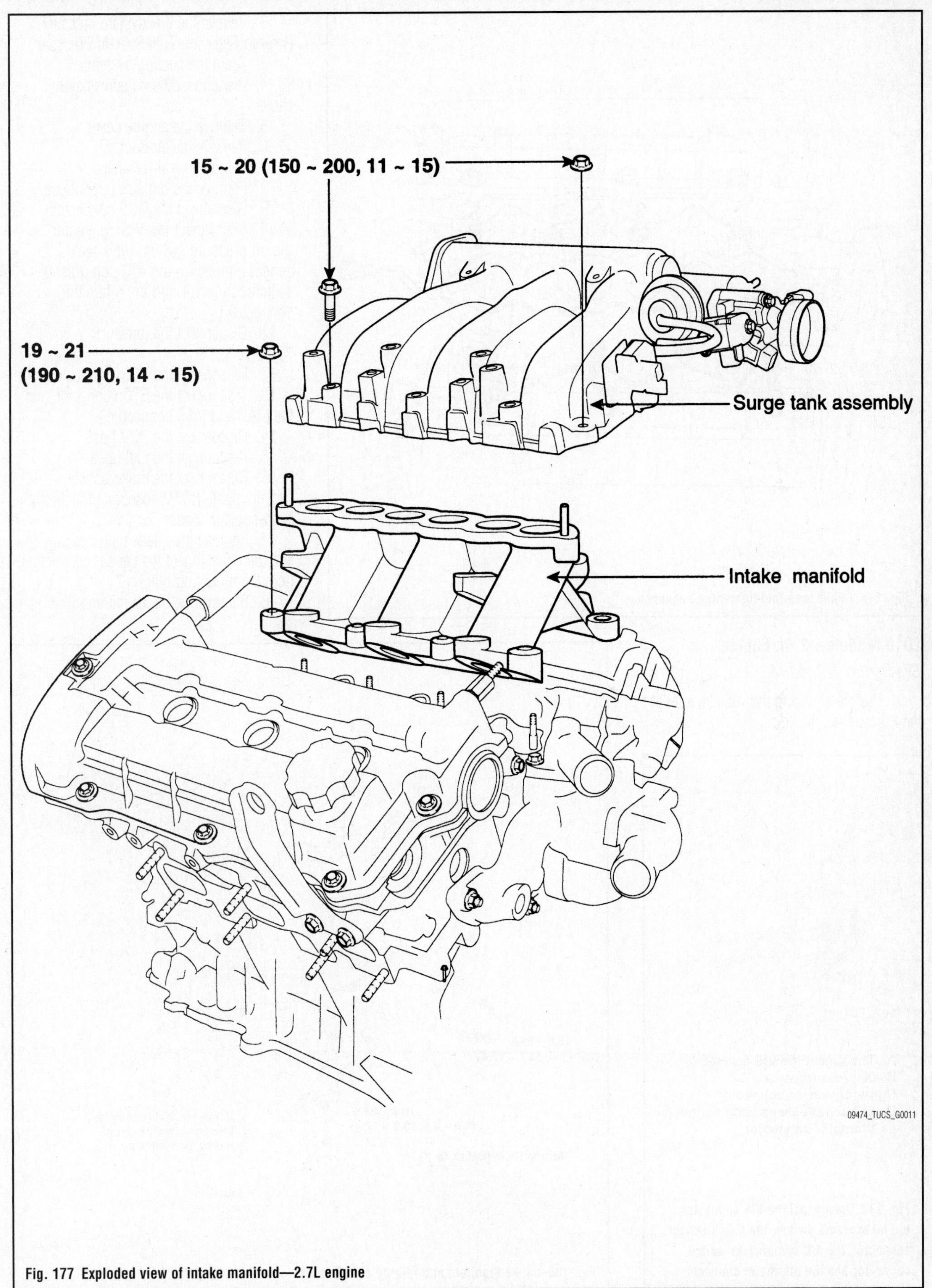

Fig. 177 Exploded view of intake manifold—2.7L engine

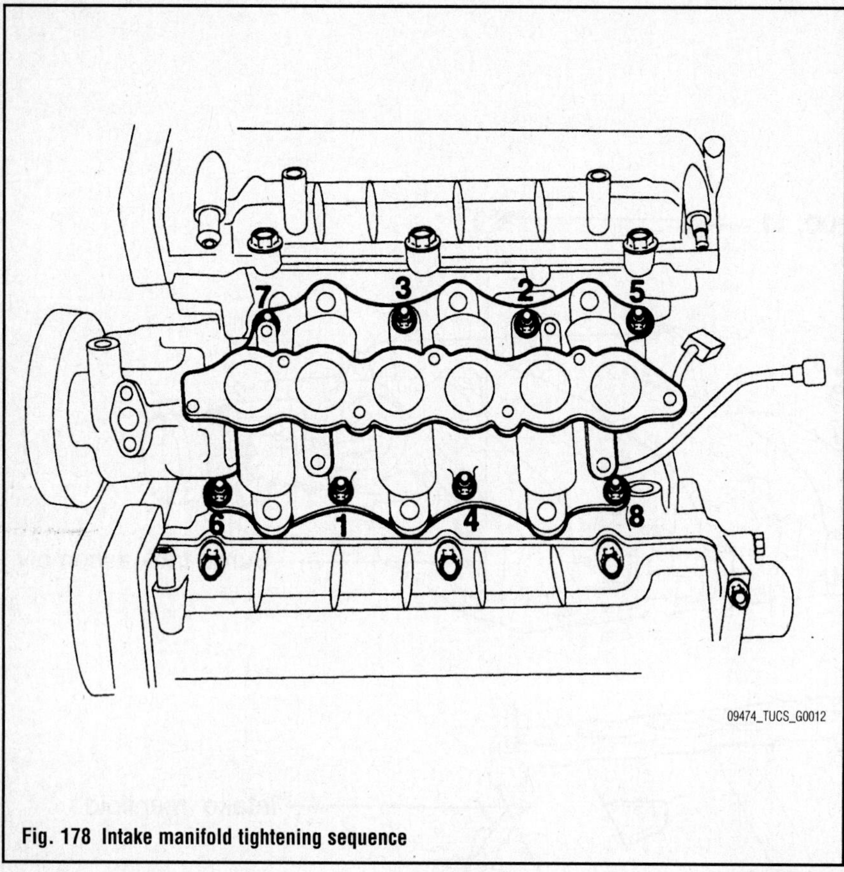

Fig. 178 Intake manifold tightening sequence

2. Relieve the fuel system pressure. Refer to Relieving Fuel System Pressure.
3. Drain the cooling system.
4. Disconnect the negative battery cable.
5. Remove the engine cover.
6. Remove the air duct.
7. Remove the air cleaner.
8. Remove the radiator upper hose.
9. Disconnect the OCV connector.
10. Disconnect the VIS connector, the oil pressure switch, the knock sensor connector, the A/C compressor switch connector and the alternator connector.
11. Disconnect the injector connectors.
12. Disconnect the VCM connector.
13. Disconnect the ETC connector, and the MAP and IATS connector.
14. Disconnect the PCV hose.
15. Disconnect the fuel hose.
16. Disconnect the brake booster vacuum hose, PCSV hose, and the throttle body coolant hoses.
17. Remove the intake manifold stay (A) and then disconnect the sensor connectors from the mounting bracket.
18. Remove the intake manifold.

2010 Models—2.4L Engine

See Figures 179 and 180.

1. Before servicing the vehicle, refer to the Precautions Section.

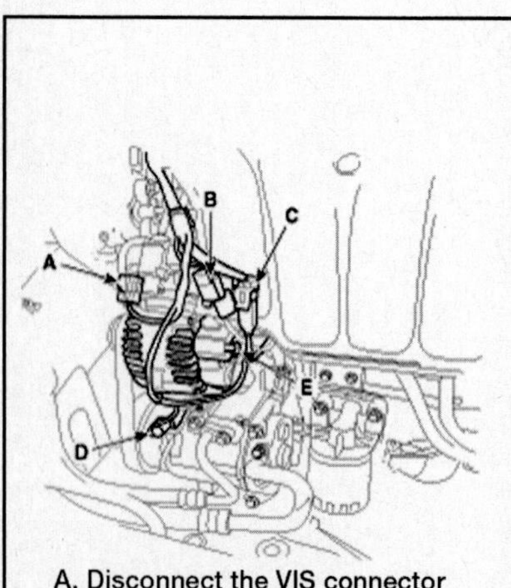

Fig. 179 Disconnect the VIS connector, the oil pressure switch, the knock sensor connector, the A/C compressor switch connector and the alternator connector

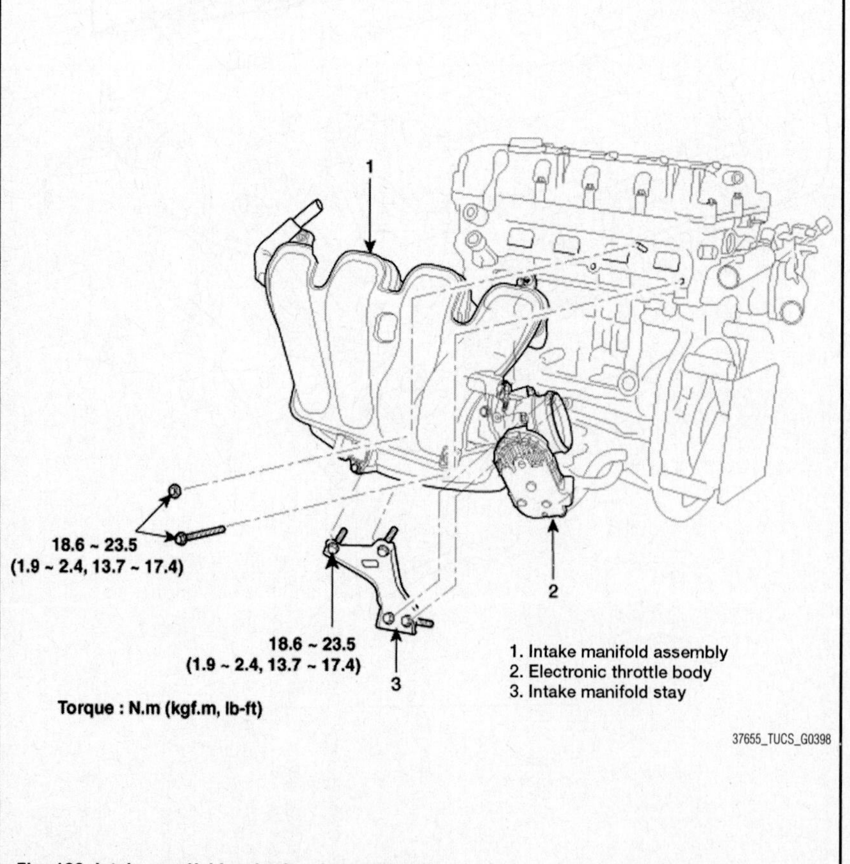

Fig. 180 Intake manifold and related components

To install:

19. Installation is the reverse of removal, noting the following:
    a. Tighten the intake manifold and intake manifold stay nuts and bolts to 14–17 ft. lbs. (19–24 Nm).

OIL PAN

REMOVAL & INSTALLATION

2009 Models

See Figures 181 and 182.

1. Before servicing the vehicle, refer to the Precautions Section.
2. Drain the engine oil.
3. Remove the front exhaust pipe—2.7L engine.
4. Remove the oil pan bolts.
5. Using the SST (09215-3C000), remove the oil pan.
    a. Insert the SST between the oil pan and the ladder frame by tapping it with a plastic hammer in the direction of arrow.
    b. After tapping the SST with a plastic hammer along the direction of arrow around more than ⅔ of the edge of the oil pan, remove it from the ladder frame.

⁂ WARNING

Do not turn over the SST abruptly without tapping or damage may occur to the SST or oil pan.

To install:

6. Using a razor blade and gasket scraper, carefully remove all the old packing material from the gasket surfaces.
7. Check that the mating surfaces are clean and dry before applying liquid gasket.
    a. Apply liquid gasket as an even bead, centered between the edges of the mating surface. Use liquid gasket: TB1217H or equivalent. Apply a bead ⅛ inch (3mm) wide to the oil pan.
    b. To prevent leakage of oil, apply liquid gasket to the inner threads of the bolt holes.

➡Do not install the parts if 5 minutes or more have elapsed since applying the liquid gasket. Instead, reapply liquid gasket after removing the residue.

8. Install the oil pan with the bolts. Uniformly tighten the bolts in several passes. Tightening torque: 84–108 inch lbs. (10–12 Nm).
9. Install the front exhaust pipe—2.7L engine.

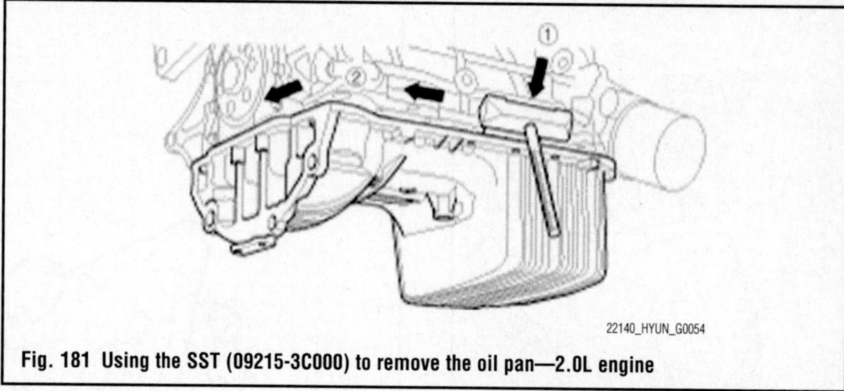

Fig. 181 Using the SST (09215-3C000) to remove the oil pan—2.0L engine

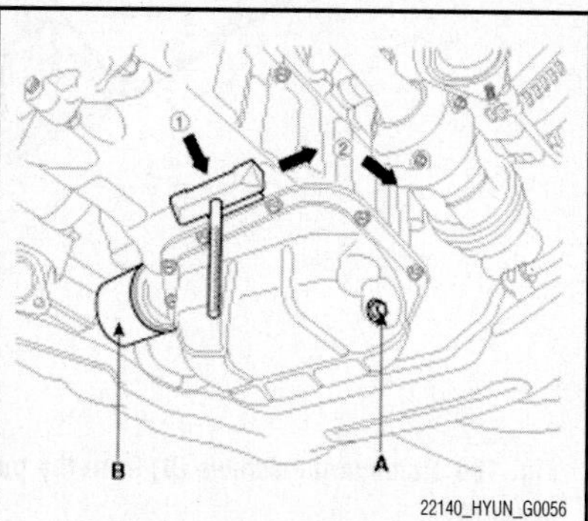

Fig. 182 Using the SST (09215-3C000) to remove the oil pan—2.7L engine

➡After assembly, wait at least 30 minutes before filling the engine with oil.

10. Fill with engine oil.
11. Run engine and check for leaks.

2010 Models—2.4L Engine

See Figures 183 through 185.

1. Before servicing the vehicle, refer to the Precautions Section.
2. Disconnect the negative battery cable.
3. Drain the engine oil.
4. Insert the blade of SST (09215-3C000) between the ladder frame and oil pan. Cut off applied sealer and remove the lower oil pan.

➡Insert the SST between the oil pan and the ladder frame by tapping it with a plastic hammer in the direction of arrow. After tapping the SST with a plastic hammer along the direction of arrow around more than ⅔ edge of the oil pan, remove it from the ladder frame. Do not use the SST as a prybar. Hold the tool in position (on gasket line) and tap in with a light hammer. Be careful not to damage the contact surfaces of ladder frame and lower oil pan.

To install:

5. Using a gasket scraper, remove all the old packing material from the gasket surfaces.
6. Before assembling the oil pan, the liquid sealant Loctite® 5900H should be applied on oil pan. The part must be assembled within 5 minutes after the sealant was applied. Bead width: 0.12 in. (3.0mm)

➡When applying sealant gasket, sealant must not protrude into the inside of the oil pan. To prevent leakage of oil, apply sealant gasket on the inner threads of the bolt holes.

7. Install oil pan. Uniformly tighten the bolts in several passes. Tightening torque:
    - M9 (B): 22–25 ft. lbs. (30–34 Nm)
    - M6 (C): 86–104 inch lbs. (10–12 Nm)

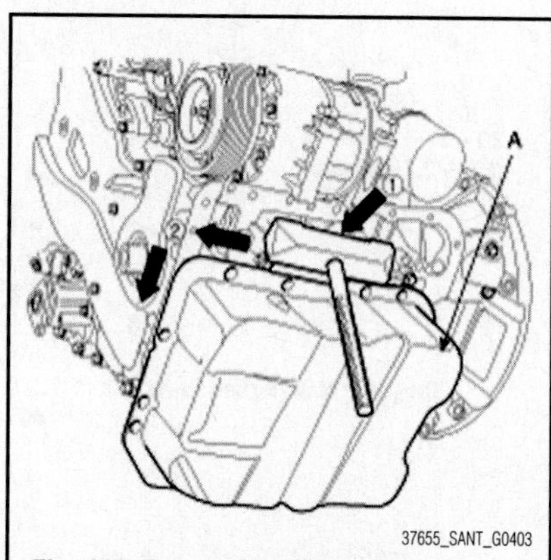

Fig. 183 Remove the oil pan (A)

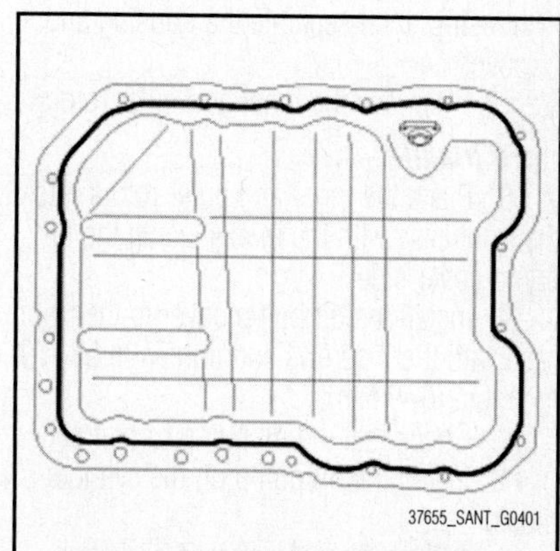

Fig. 184 Oil pan sealant application

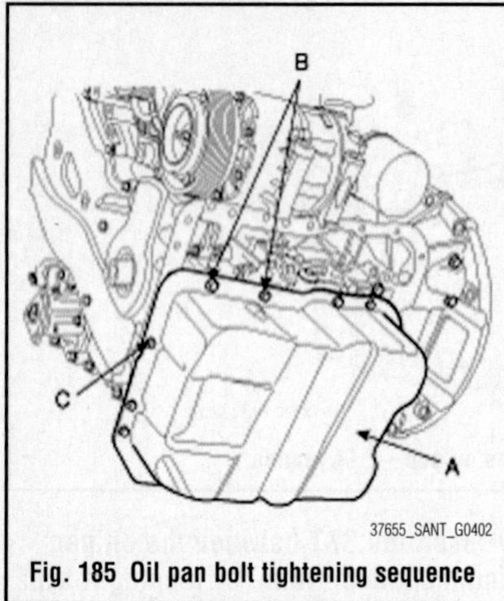

Fig. 185 Oil pan bolt tightening sequence

➡After assembly, wait at least 30 minutes before filling the engine with oil. Always use a new drain bolt gasket.

8. Refill the engine oil.
9. Connect the negative battery cable.

OIL PUMP

REMOVAL & INSTALLATION

2009 Models

2.0L Engine

See Figures 186 through 189.

1. Before servicing the vehicle, refer to the Precautions Section.
2. Drain the engine oil.
3. Remove the drive belts. Refer to Accessory Drive Belt Removal & Installation.
4. Turn the crankshaft and align the white groove on the crankshaft pulley with the pointer on the lower cover.
5. Remove the timing belt. Refer to Timing Belt Removal & Installation.
6. Remove the oil pan. Refer to Oil Pan Removal & Installation.
7. Remove the front case and oil pump:
   a. Remove the screws from the pump housing, then separate the housing and cover.
   b. Remove the inner and outer rotors.

To install:

8. Place the inner and outer rotors into the front case with the marks facing the oil pump cover side.
9. Install the oil pump cover to the front case with the 7 screws and tighten to 53–79 inch lbs. (6–9 Nm).
10. Check that the oil pump turns freely.
11. Install the oil pump on the cylinder block.
    a. Place a new front case gasket on the cylinder block.

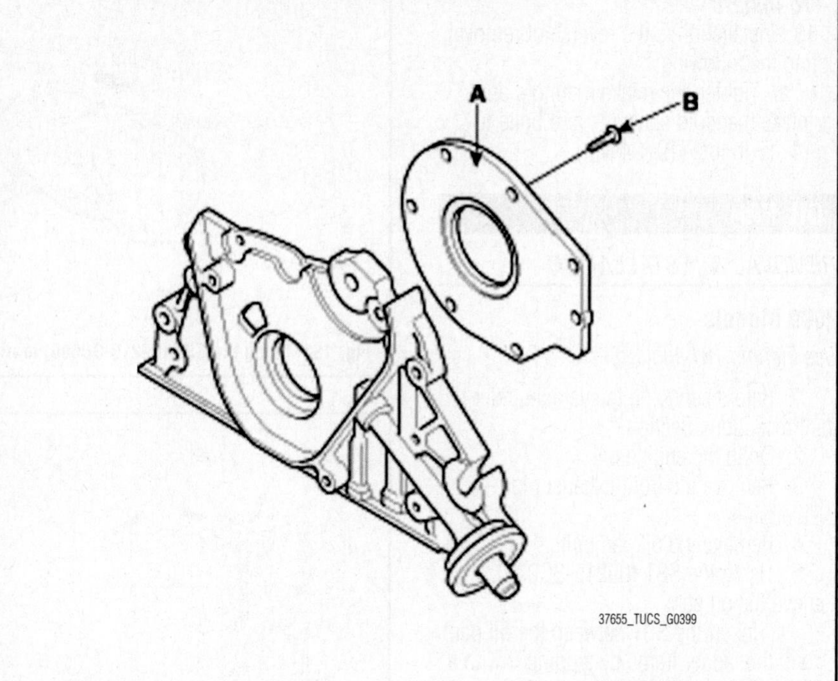

Fig. 186 Remove the screws (B) from the pump housing, then separate the housing and cover (A)

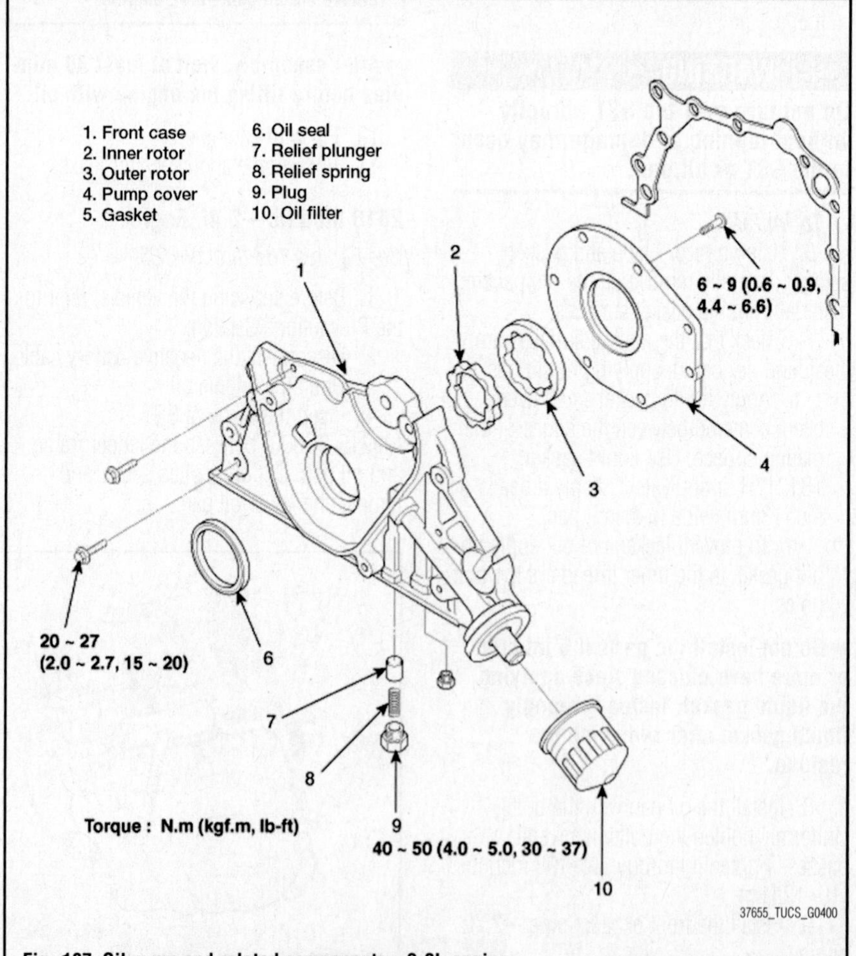

Fig. 187 Oil pump and related components—2.0L engine

b. Apply engine oil to the lip of the oil pump seal. Then, install the oil pump onto the crankshaft.

c. When the pump is in place, clean any excess grease off the crankshaft and check that the oil seal lip is not distorted.

d. Install the oil pump bolts in the correct position as illustrated. Tightening torque: 15–20 ft. lbs. (20–27 Nm).

12. Apply a light coat of oil to the seal lip.
13. Using the SST (09231-23100), install the oil seal.
14. Install the oil pan and oil screen.
15. Ensure that the crankshaft aligns with the white groove on the crankshaft pulley and the pointer on the lower cover.
16. Install the timing belt.
17. Install the accessory drive belts.
18. Fill with engine oil.
19. Check for leaks.

2.7L Engine

See Figures 190 through 199.

1. Before servicing the vehicle, refer to the Precautions Section.
2. Disconnect the negative battery cable.
3. Remove the right front wheel.
4. Remove the right-hand side cover.
5. Remove the front exhaust pipe.
6. Remove the alternator from the engine. Refer to Alternator Removal & Installation in the Engine Electrical section.
7. Turn the crankshaft and align the white groove on the crankshaft pulley with the pointer on the lower cover.
8. Remove the timing belt. Refer to Timing Belt Removal & Installation.
9. Remove the lower oil pan. Refer to Oil Pan Removal & Installation.
10. Remove the oil screen:

a. Remove the 2 bolts.

b. Remove the oil screen and gasket.

11. Remove the upper oil pan.
12. Remove the oil pump case:

a. Remove the screws from the pump housing, then separate the housing and cover.

b. Remove the inner and outer rotors.

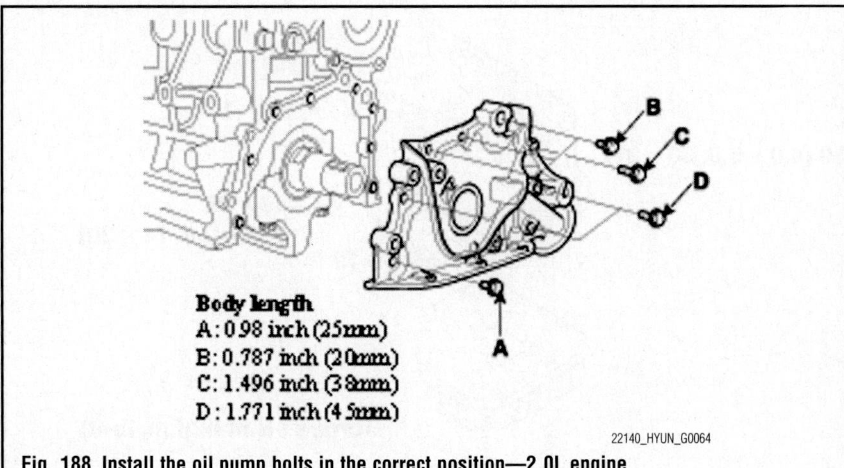

Fig. 188 Install the oil pump bolts in the correct position—2.0L engine

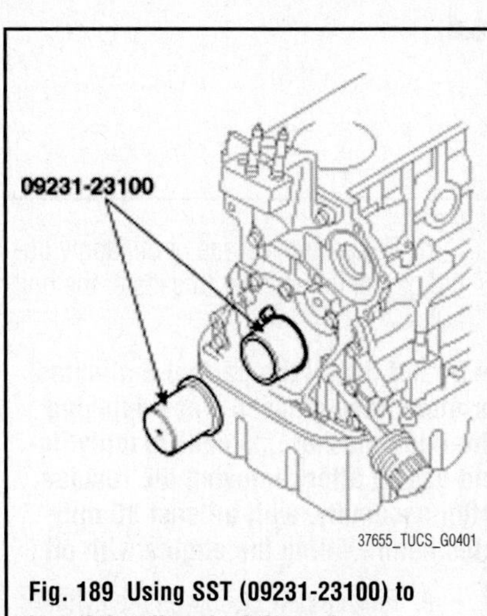

Fig. 189 Using SST (09231-23100) to install the oil seal—2.0L engine

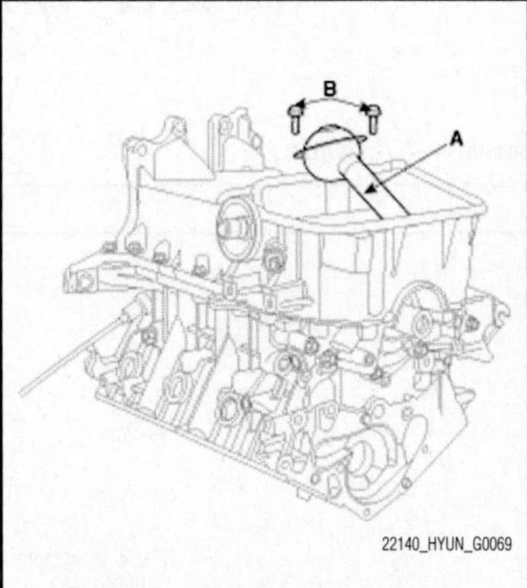

Fig. 190 Remove the 2 bolts (B) and oil screen (A)

Fig. 191 Remove the upper oil pan (A)

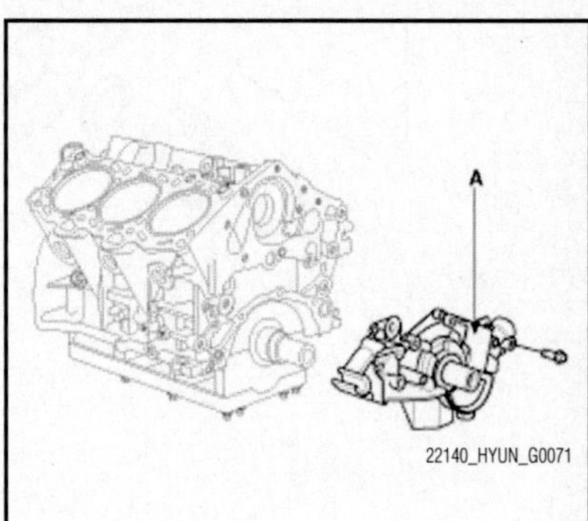

Fig. 192 Remove the upper oil pump case (A)

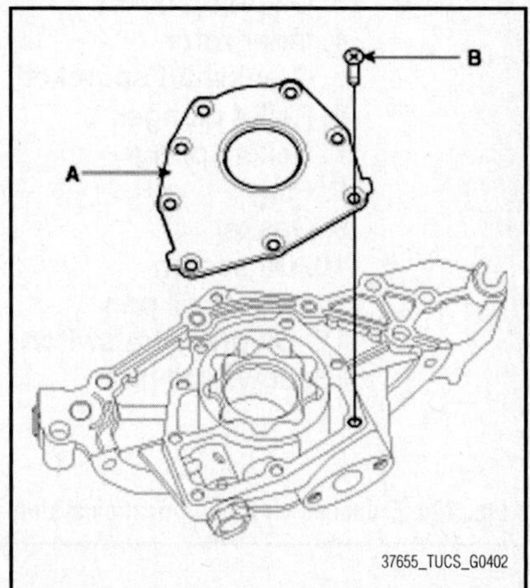

Fig. 193 Remove the screws (B) from the pump housing, then separate the housing and cover (A)

To install:

13. Place the inner and outer rotors into the front case with the marks facing the oil pump cover side.
14. Install the oil pump cover to front case with the 8 screws. Tightening torque: 72–106 inch lbs. (8–12 Nm).

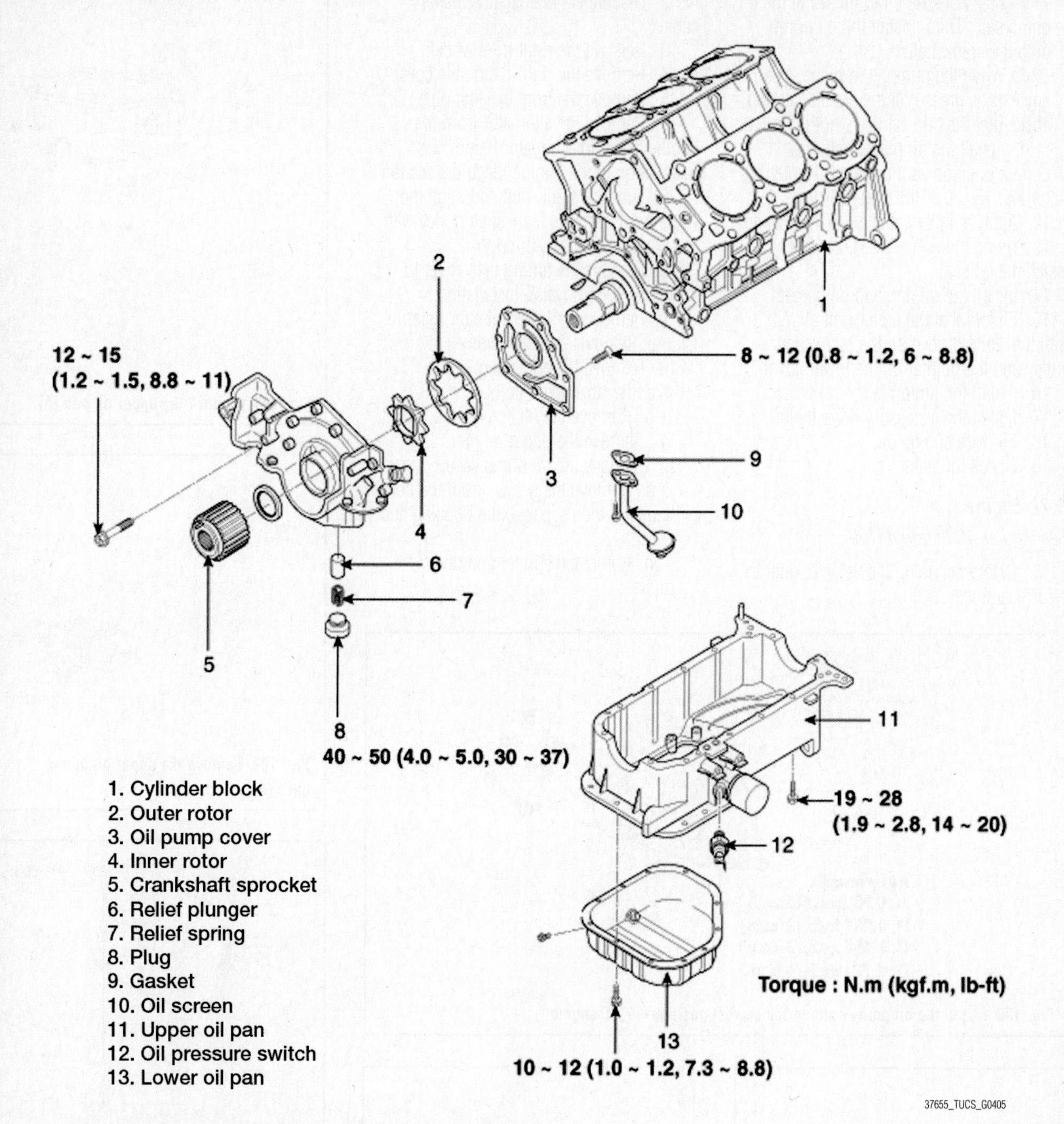

Fig. 194 Expanded view of lubrication system components—2.7L engine

15. Check that the oil pump turns freely.
16. Install the oil pump on the cylinder block.
   a. Using a razor blade and gasket scraper, remove all the old liquid gasket from the gasket surfaces and sealing grooves.
   b. Using a non-residue solvent, clean both sealing surfaces.
   c. Apply liquid gasket to the oil pump as shown in the illustration. Use liquid gasket MS 721-40A or equivalent.

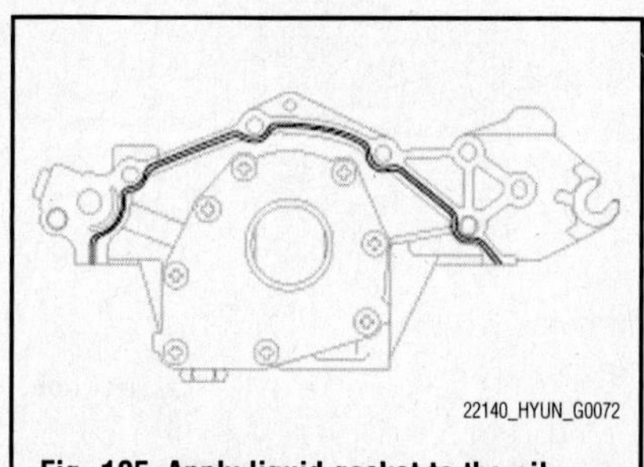

Fig. 195 Apply liquid gasket to the oil pump as shown

   d. To prevent leakage of oil, apply liquid gasket to the inner threads of the bolt holes.

➡Do not install the parts if 5 minutes or more have elapsed since applying the liquid gasket. Instead, reapply liquid gasket after removing the residue. After assembly, wait at least 30 minutes before filling the engine with oil.

   e. Place a new O-ring on the cylinder block.

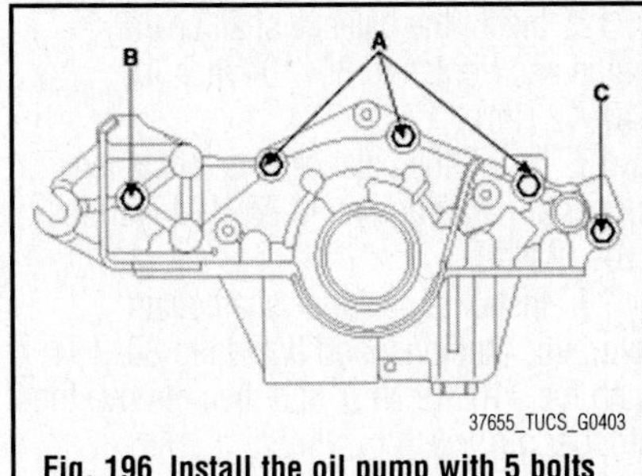

Fig. 196 Install the oil pump with 5 bolts

Bolt	Size	Number
A	8 × 25	3
B	8 × 35	1
C	8 × 45	1

37655_TUCS_G0404

Fig. 197 Oil pump bolts

f. Engage the spline teeth of the oil pump drive gear with large teeth of the crankshaft, and slide the oil pump on the crankshaft.

g. Install the oil pump with 5 bolts. Uniformly tighten the bolts in several passes. Tightening torque: 9–11 ft. lbs. (12–15 Nm).

17. Apply a light coat of oil to the seal lip.
18. Using the special tool (09214-33000), install the oil seal.
19. Install the upper oil pan:

a. Using a razor blade and gasket scraper, remove all the old packing material from the gasket surfaces.

➡Check that the mating surfaces are clean and dry before applying liquid gasket.

b. Install the oil pan with the 17 bolts. Uniformly tighten the bolts in several passes using the pattern illustrated. Tightening torque:

- Bolts 1–15: 14–17 ft. lbs. (19–24 Nm)
- Bolts 16–17: 48–61 inch lbs. (5–7 Nm)

c. To prevent leakage of oil, apply liquid gasket to the inner threads of the bolt holes.

➡Do not install the parts if 5 minutes or more have elapsed since applying the liquid gasket. Instead, reapply liquid gasket after removing the residue. After assembly, wait at least 30 minutes before filling the engine with oil.

20. Install the oil screen with a new gasket and the 2 bolts. Tightening torque: 11–16 ft. lbs. (15–22 Nm).
21. Install the lower oil pan.
22. Install the timing belt.
23. Install the accessory drive belt.
24. Install the alternator.
25. Install the front exhaust pipe.
26. Remove the right-hand side cover.
27. Install the right front wheel.
28. Fill the engine with oil.
29. Start the engine and check for leaks.
30. Recheck the engine oil level.

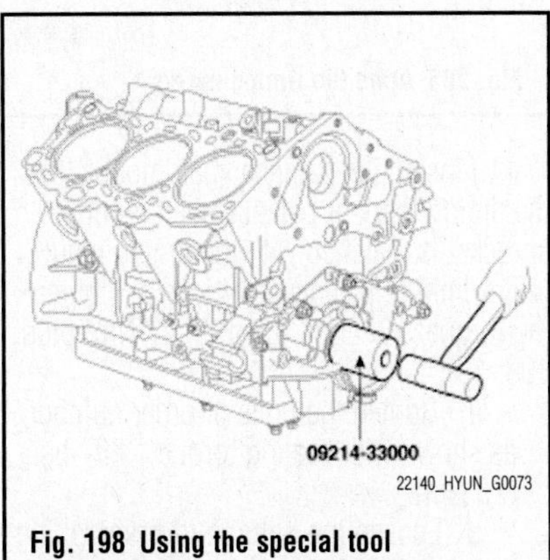

Fig. 198 Using the special tool (09214-33000) to install the oil seal

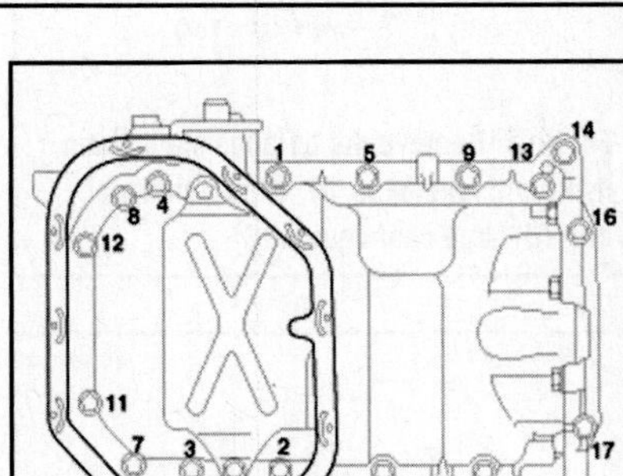

Fig. 199 Oil pan with the 17 bolts. Uniformly tighten the bolts in several passes

2010 Models—2.4L Engine

See Figures 200 through 205.

1. Before servicing the vehicle, refer to the Precautions Section.
2. Disconnect the negative battery cable.
3. Remove the timing chain. Refer to Timing Chain Removal & Installation.

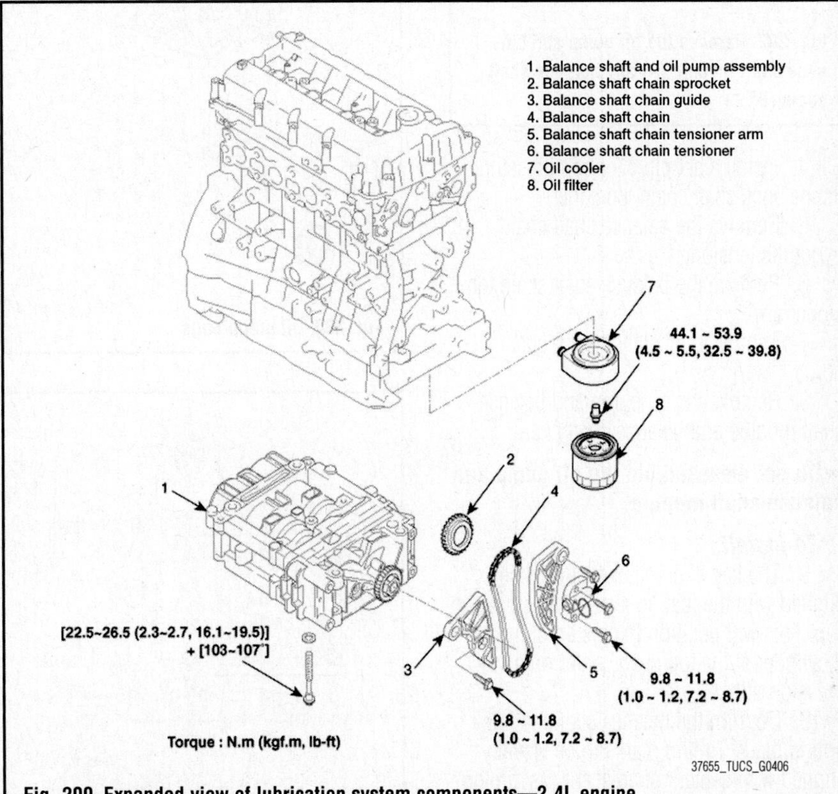

Fig. 200 Expanded view of lubrication system components—2.4L engine

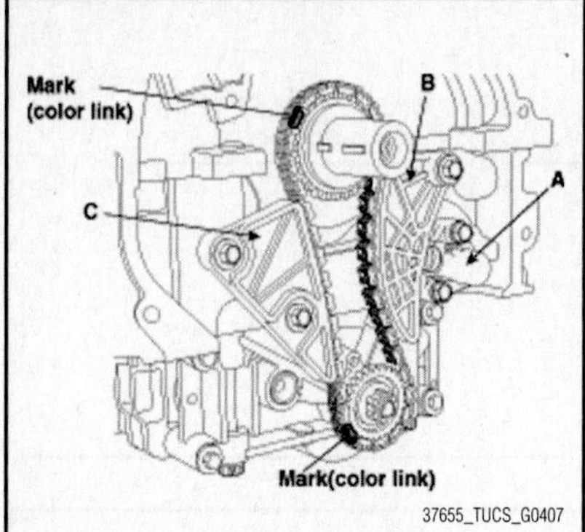

Fig. 201 Remove the balance shaft chain hydraulic tensioner (A), chain tensioner arm (B), and chain guide (C)

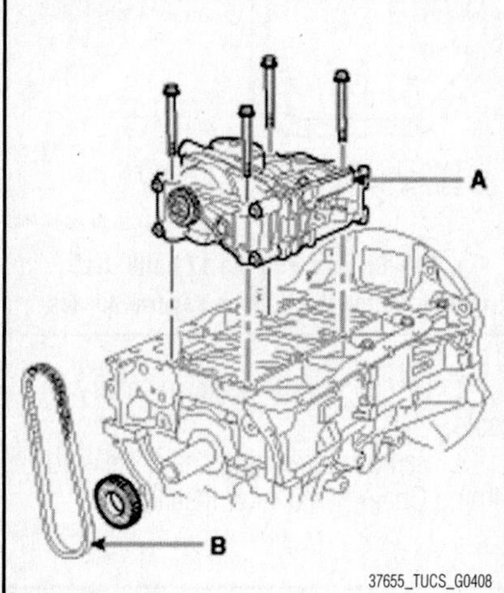

Fig. 202 Remove the oil pump and balance shaft module (A) and balance shaft chain (B)

4. Install a set pin after compressing the balance shaft chain tensioner.
5. Remove the balance shaft chain hydraulic tensioner.
6. Remove the balance shaft chain tensioner arm.
7. Remove the balance shaft chain guide.
8. Remove the oil pump and balance shaft module and balance shaft chain.

➡Do not disassemble the oil pump and balance shaft module.

To install:

9. The key of crankshaft should be aligned with the mating face of main bearing cap. This will position the piston of the No. 1 cylinder at the top dead center on compression stroke.
10. Confirm the balance shaft module timing mark. Timing marks to be visually aligned with centers of adjacent cast timing notches.

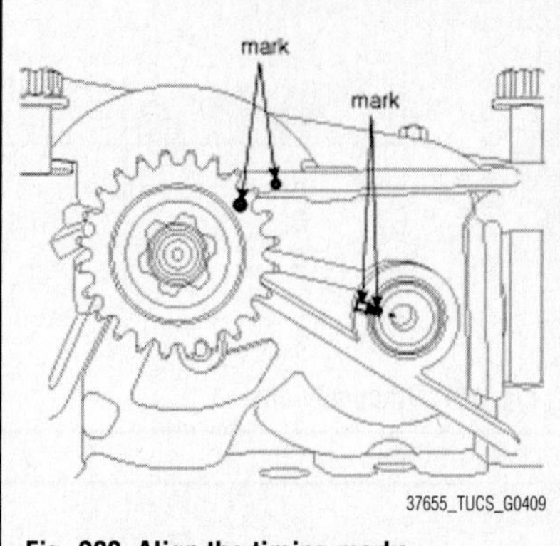

Fig. 203 Align the timing marks

11. Install the balance shaft module that the timing mark of balance shaft module sprocket is matched with the timing mark (color link) of balance shaft chain. Tightening torque: 16–20 ft. lbs. (23–27 Nm) plus 103–107°.
    a. Tighten the bolts in order number as shown with seating torque 19 ft. lbs. (26 Nm).
    b. Loosen the bolts in reverse bolting order (4-3-2-1).
    c. Tighten the bolts as shown.

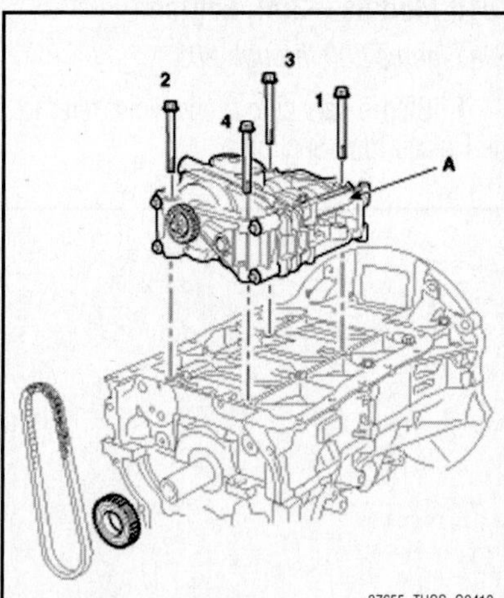

Fig. 204 Oil pump bolts

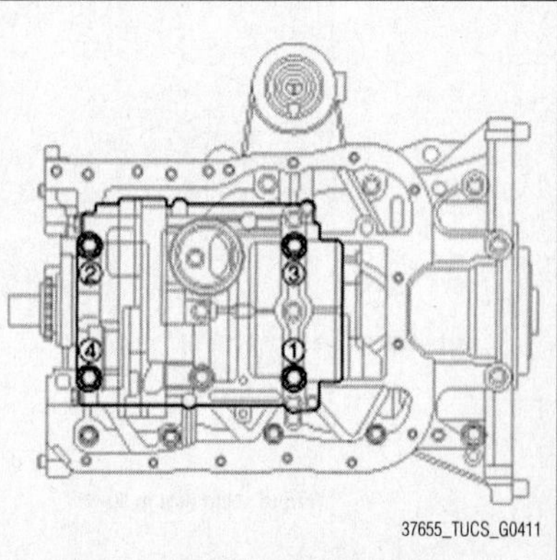

Fig. 205 Oil pump bolt installation

12. Install the balance shaft chain guide and tighten to 86–104 inch lbs. (10–12 Nm).
13. Install the balance shaft chain tensioner arm and tighten to 86–104 inch lbs. (10–12 Nm).
14. Install the balance shaft chain hydraulic tensioner and tighten to 86–104 inch lbs. (10–12 Nm), and then remove the stopper pin.
15. Confirm the timing marks.
16. Install the timing chain.
17. Connect the negative battery cable.

INSPECTION

See Figures 206 through 208.

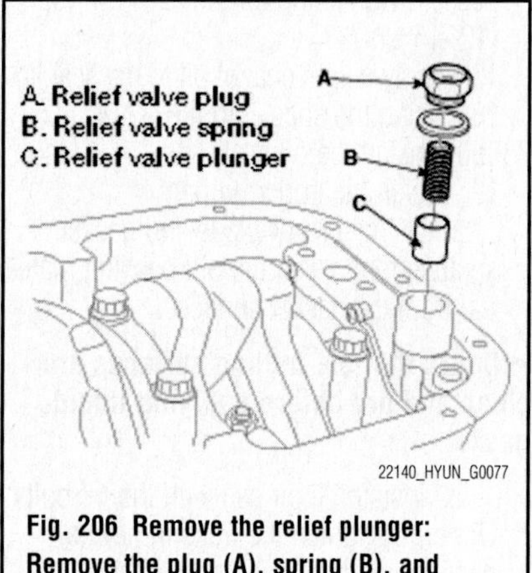

Fig. 206 Remove the relief plunger: Remove the plug (A), spring (B), and relief plunger (C)

1. Before servicing the vehicle, refer to the Precautions Section.
2. Remove the relief plunger: Remove the plug, spring, and relief plunger.
3. Inspect the relief plunger.
    a. Coat the plunger with engine oil
    b. Check that it falls smoothly into the plunger hole by its own weight.
    c. If it does not, replace the relief plunger.
    d. If necessary, replace the front case.
4. Inspect the relief valve spring.
5. Inspect for a distorted or broken relief valve spring.
6. Inspect the rotor side clearance.
    a. Using a feeler gauge and precision straight edge, measure the clearance between the rotors and precision straight edge.
    b. If the side clearance is greater than maximum, replace the rotors as a set. If necessary, replace the front case.

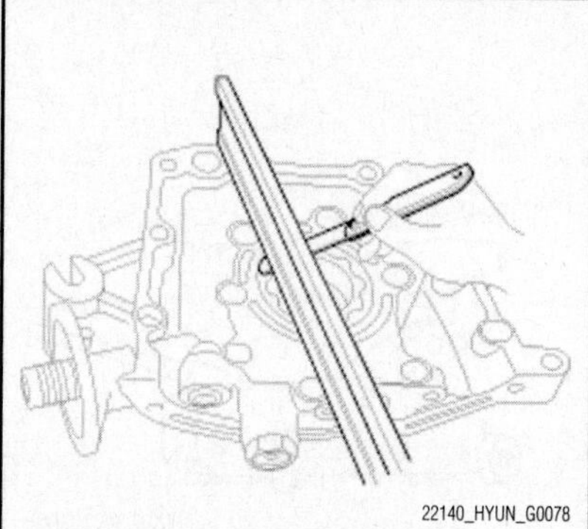

Fig. 207 Using a feeler gauge and precision straight edge to measure the side clearance of the oil pump

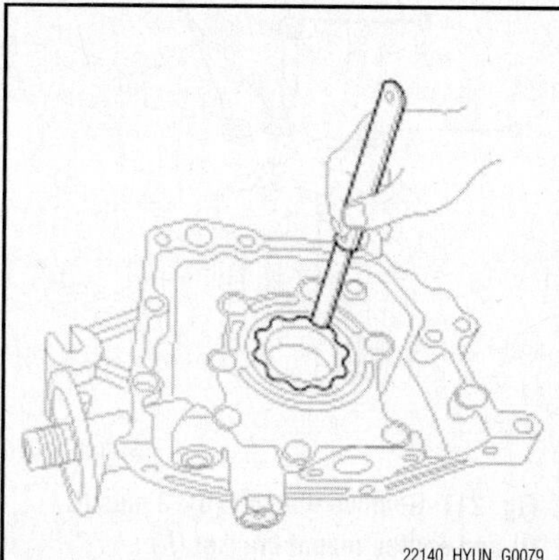

Fig. 208 Using a feeler gauge, measure the tip clearance between the inner and outer rotor tips of the oil pump

7. Inspect the rotor tip clearance.
   a. Using a feeler gauge, measure the tip clearance between the inner and outer rotor tips.
   b. If the tip clearance is greater than specified, replace the rotors as a set.
8. Inspect the rotor body clearance.
   a. Using a feeler gauge, measure the clearance between the outer rotor and body.
   b. If the body clearance is greater than specified, replace the rotors as a set. If necessary, replace the front case.
9. Install the relief plunger and spring into the front case hole.
10. Install the plug.

PISTON AND RING

POSITIONING

See Figures 209 through 211.

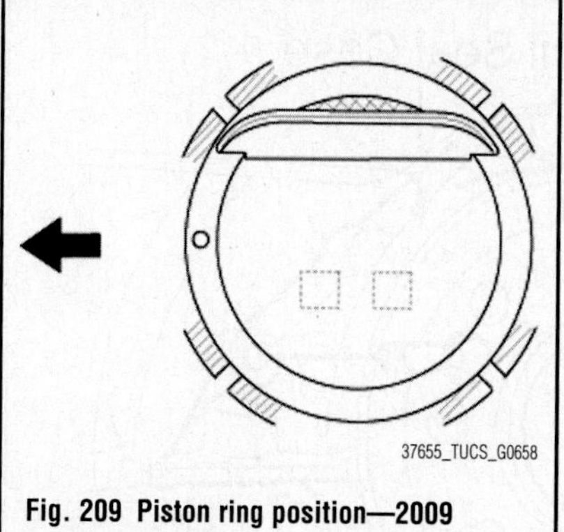

Fig. 209 Piston ring position—2009 models, 2.0L engine

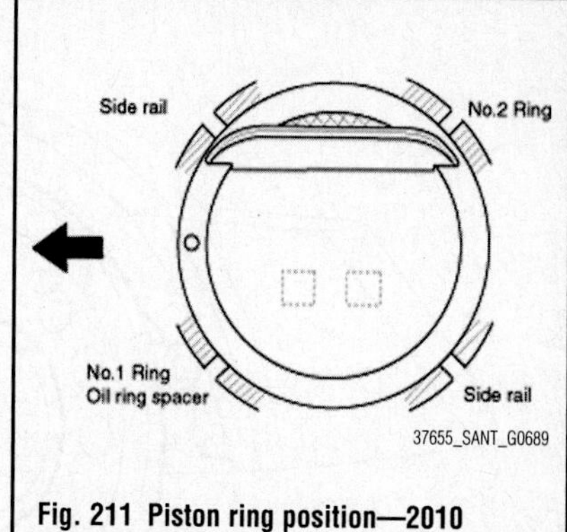

Fig. 211 Piston ring position—2010 models, 2.4L engines

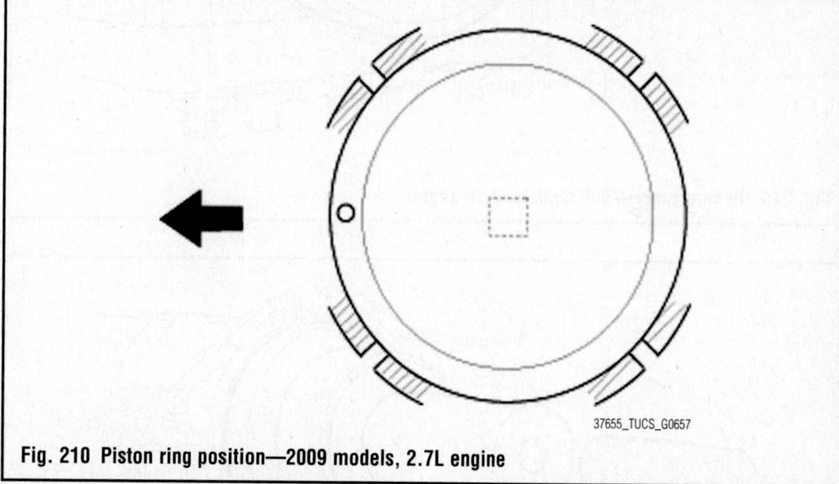

Fig. 210 Piston ring position—2009 models, 2.7L engine

REAR MAIN SEAL

REMOVAL & INSTALLATION

See Figures 212 through 214.

1. Before servicing the vehicle, refer to the Precautions Section.
2. Remove the flywheel. Refer to Flywheel Removal & Installation.
3. Using an appropriate removal tool, remove the rear main seal.

To install:

4. Installation is the reverse of removal.

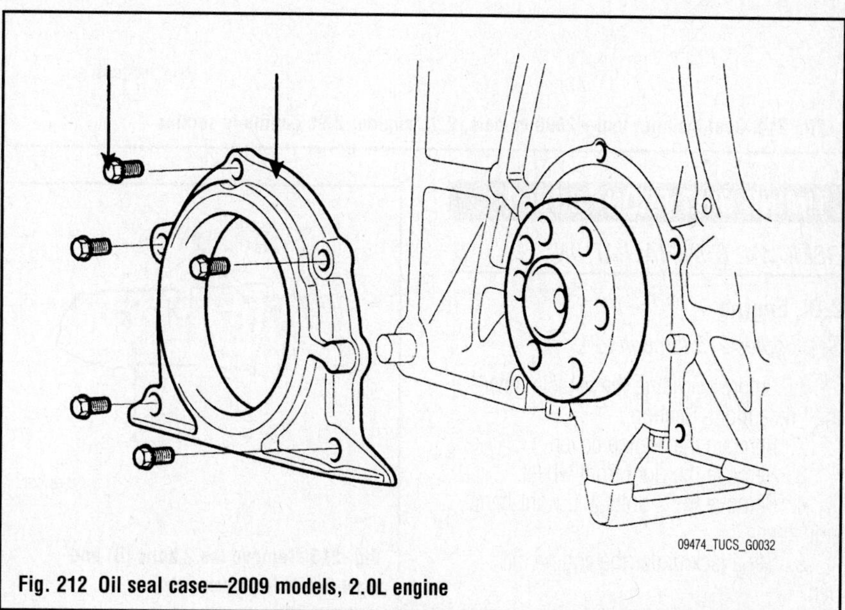

Fig. 212 Oil seal case—2009 models, 2.0L engine

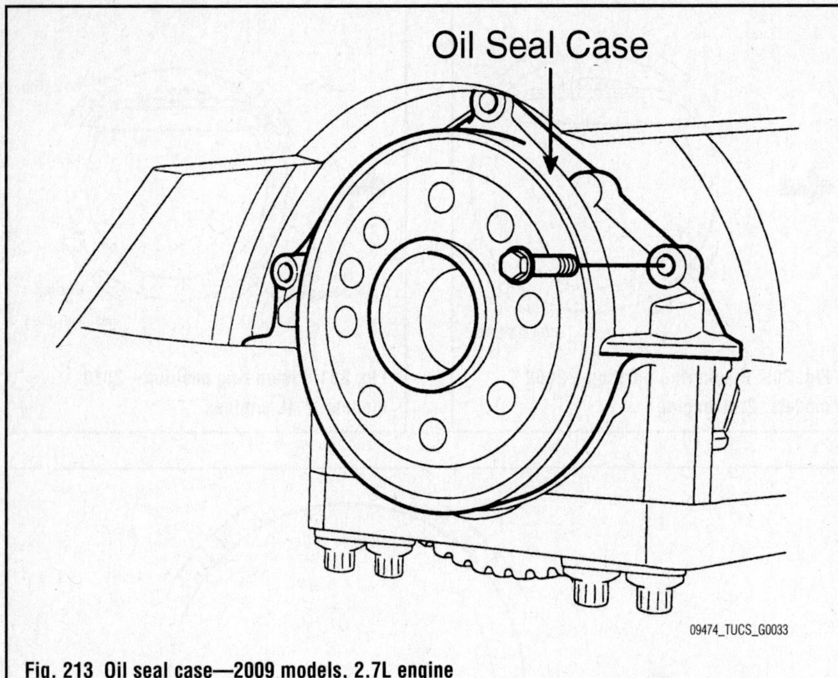

Fig. 213 Oil seal case—2009 models, 2.7L engine

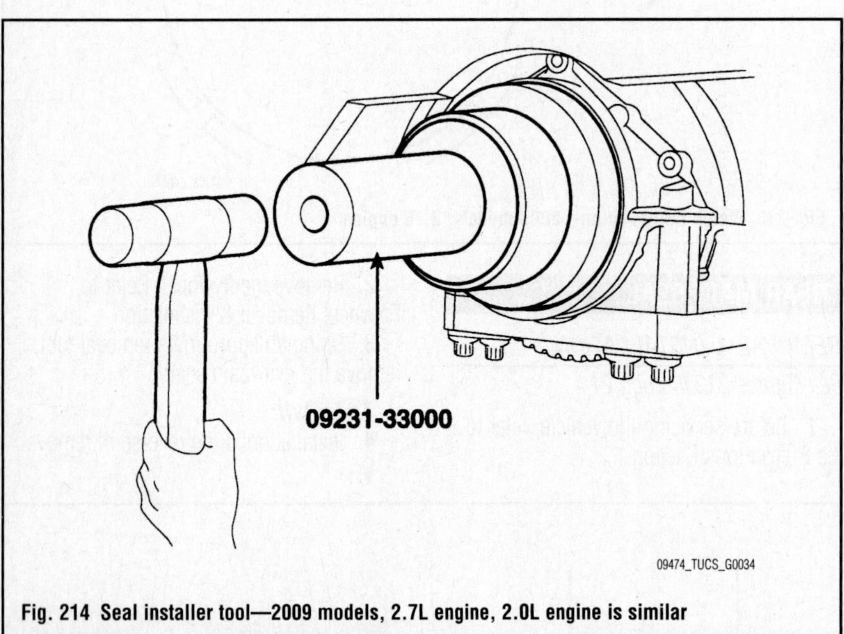

Fig. 214 Seal installer tool—2009 models, 2.7L engine, 2.0L engine is similar

TIMING BELT FRONT COVER

REMOVAL & INSTALLATION

2.0L Engine

See Figures 215 through 221.

1. Before servicing the vehicle, refer to the Precautions Section.
2. Remove the engine cover.
3. Remove the right front wheel.
4. Remove the 2 bolts and right-hand side cover.
5. Set a jack under the engine oil pan.

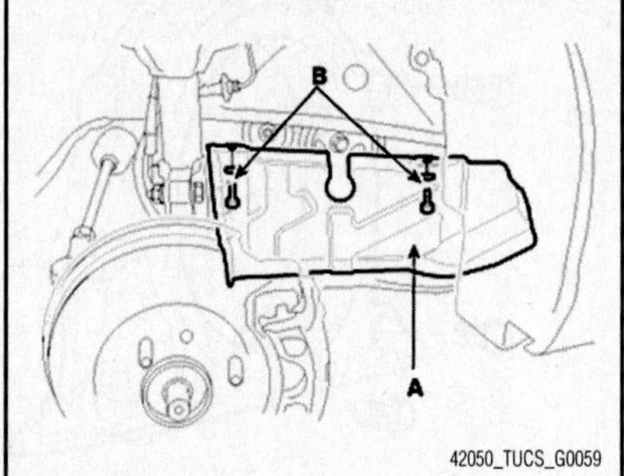

Fig. 215 Remove the 2 bolts (B) and right-hand side cover (A)

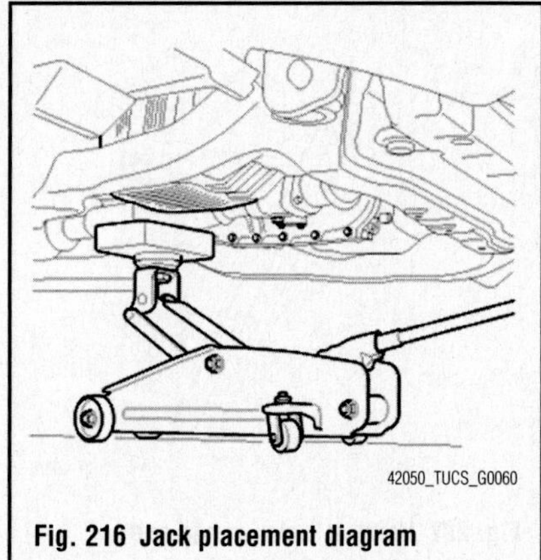

Fig. 216 Jack placement diagram

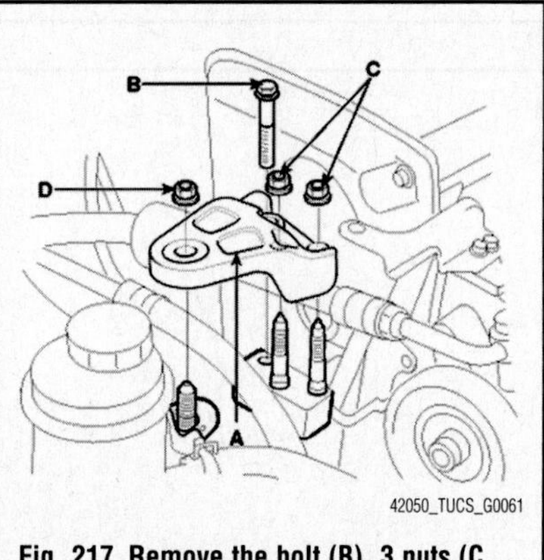

Fig. 217 Remove the bolt (B), 3 nuts (C, D) and engine mount bracket (A)

➡Place a wooden block between the jack and engine oil pan.

6. Remove the bolt, nuts and engine mounting bracket.
7. Remove the bolt and stay plate.
8. Temporarily loosen the water pump pulley bolts.
9. Remove alternator belt.
10. Remove air compressor belt.
11. Remove power steering belt.
12. Remove the bolts and water pump pulley.

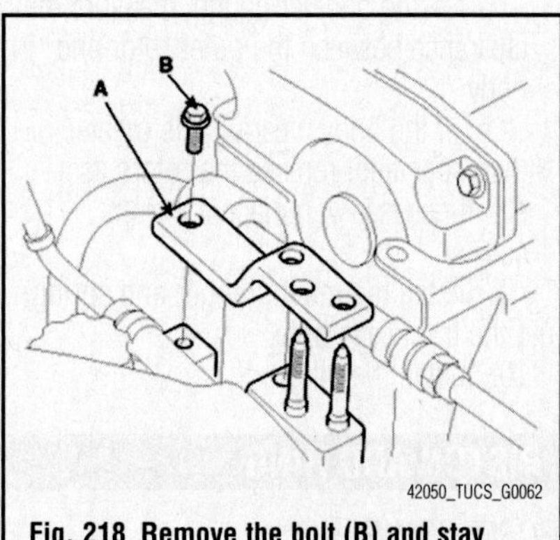

Fig. 218 Remove the bolt (B) and stay plate (A)

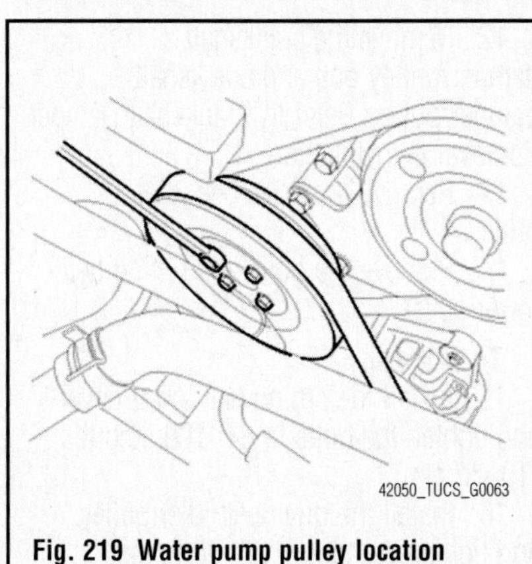
Fig. 219 Water pump pulley location

13. Remove the 4 bolts and timing belt upper cover.
14. Turn the crankshaft pulley, and align its groove with timing mark "T" of the timing belt cover.
15. Remove the crankshaft pulley. Refer to Crankshaft Damper Removal & Installation.
16. Remove the crankshaft flange.
17. Remove the bolts and timing belt lower cover.

To install:

18. Install timing belt lower cover and secure with the 5 bolts. Tighten to 72–84 inch lbs. (8–10 Nm).

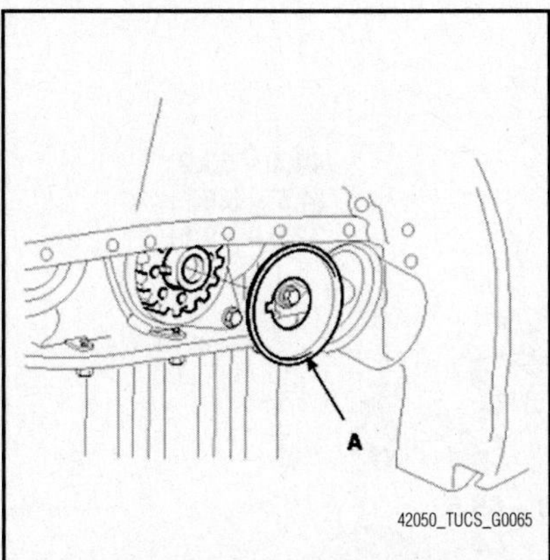

Fig. 220 Remove the crankshaft flange (A)

19. Install crankshaft flange.
20. Install crankshaft pulley and secure with the crankshaft pulley bolt. Tighten to 125–133 ft. lbs. (170–180 Nm).
21. Install the timing belt upper cover and secure with the 4 bolts. Tighten to 84 inch lbs. (10 Nm).
22. Install the crankshaft flange and pulley and tighten the crankshaft pulley bolt to 125–133 ft. lbs. (170–180 Nm).
23. Install the upper timing belt cover.
24. Install the water pump pulley.
25. Install the power steering belt.
26. Install the A/C compressor belt.
27. Install the alternator belt.
28. Install the engine mounting bracket:
    a. Install the stay plate and tighten the bolt to 32–40 ft. lbs. (43–55 Nm).
    b. Install the engine mounting bracket and tighten the bolt to 37–48 ft. lbs. (50–60 Nm) and the nuts to 44–59 ft. lbs. (60–80 Nm).
29. Lower the jack and remove from the vehicle.
30. Making sure all timing and alignment marks match, install the drive belt, the idler, and the tensioner.
31. Install the right side cover.
32. Install the right front tire and wheel.
33. Install the engine cover.

2.7L Engine

See Figures 222 through 228.

1. Before servicing the vehicle, refer to the Precautions Section.

➡Engine removal is not required for this procedure.

2. Remove the engine cover.
3. Disconnect the negative battery cable.
4. Remove the right front wheel.
5. Remove the bolts and right-hand side cover.

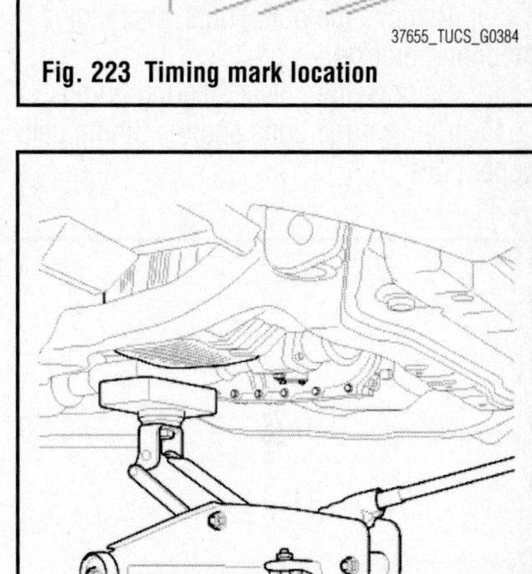
Fig. 223 Timing mark location

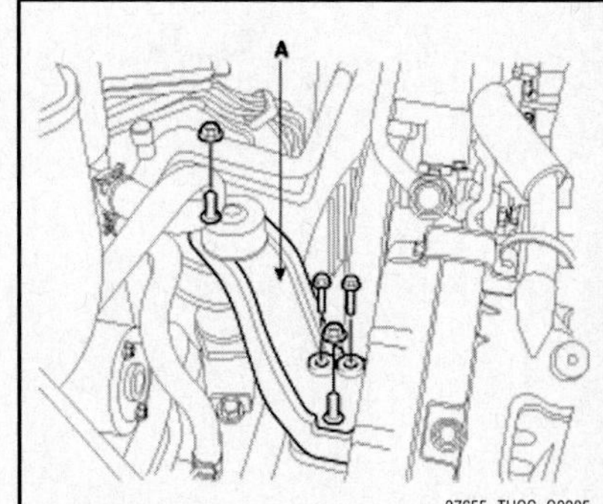
Fig. 224 Jack placement

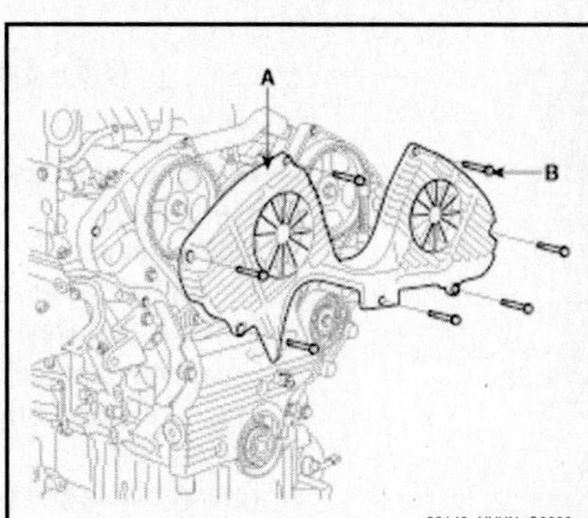

Fig. 226 Remove the 7 bolts (B) and timing belt upper cover (A)

Fig. 225 Remove the bolts, nuts, and engine mounting bracket (A)

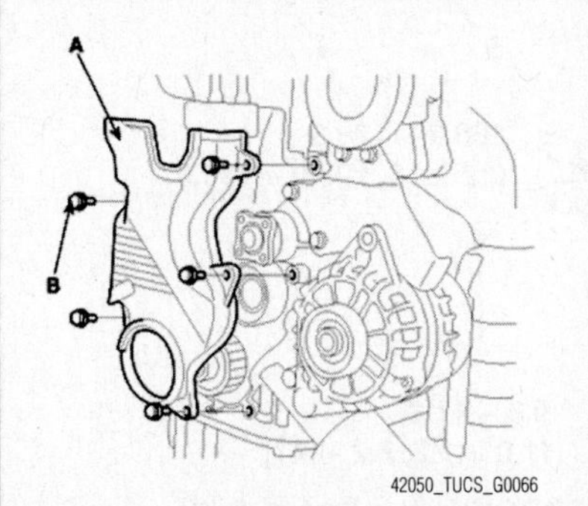

Fig. 221 Remove the 5 bolts (B) and timing belt lower cover (A)

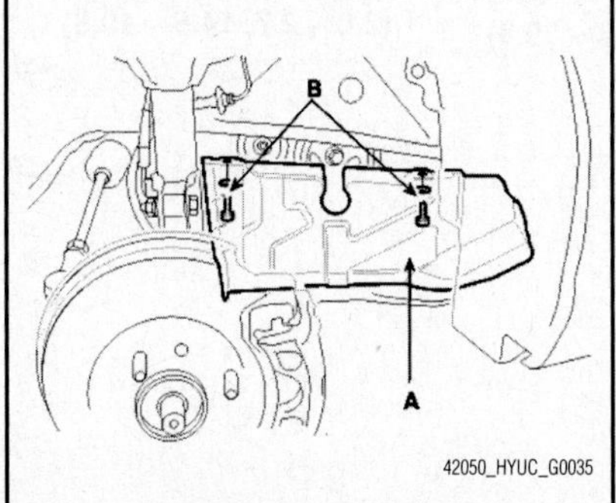

Fig. 222 Remove the 2 bolts (B) and right-hand side cover (A)

6. Turn the crankshaft pulley, and align its groove with timing mark "T" of the timing belt cover.

➡Always turn the crankshaft clockwise.

7. Remove the auto-tensioner and the drive belt. Refer to Accessory Drive Belt Removal & Installation.

8. Secure a jack to the engine oil pan.

➡Place wooden block between the jack and engine oil pan.

9. Remove the bolts, nuts, and engine mounting bracket.

10. Remove the power steering pump.

11. Remove the bolts and the timing belt upper cover.

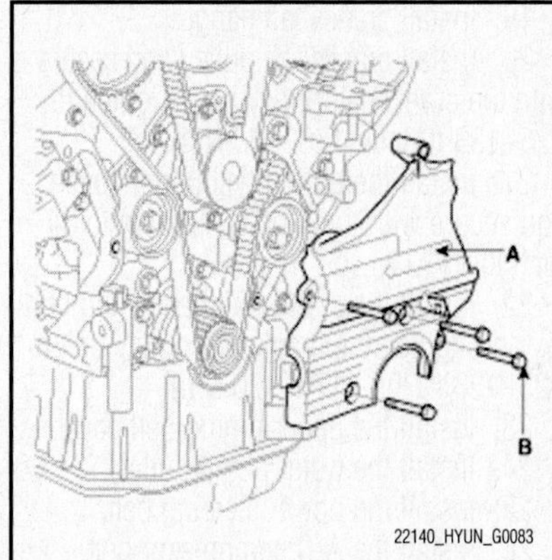

Fig. 227 Remove the 4 bolts (B) and timing belt lower cover (A)

12. Remove the crankshaft damper/pulley bolt and crankshaft damper/pulley. Refer to Crankshaft Damper Removal & Installation.

13. Remove the drive belt idler pulley.

14. Remove the bolts and timing belt lower cover.

To install:

15. Install the timing belt lower cover and tighten the bolts to 84–108 inch lbs. (10–12 Nm).

16. Install the drive belt idler pulley and tighten the bolt to 25–40 ft. lbs. (35–55 Nm).

17. Install the crankshaft pulley and

9.8 ~ 11.8 (1.0 ~ 1.2, 7.2 ~ 8.7)
44.1 ~ 53.9 (4.5 ~ 5.5, 32.5 ~ 39.8)
50 ~ 60 (5.0 ~ 6.0, 36 ~ 43)
58.8 ~ 68.6 (6.0 ~ 7.0, 43.4 ~ 50.6)
34.3 ~ 53.9 (3.5 ~ 5.5, 25.3 ~ 39.8)
19.6 ~ 26.5 (2.0 ~ 2.7, 14.5 ~ 19.5)
166.7 ~ 176.5 (17 ~ 18, 123.0 ~ 130.2)
9.8 ~ 11.8 (1.0 ~ 1.2, 7.2 ~ 8.7)

1. Tensioner arm assembly
2. Idler pulley
3. Auto tensioner
4. Timing belt lower cover
5. Crankshaft pulley
6. Drive belt idler pulley
7. Engine support bracket
8. Timing belt
9. Timing belt upper cover

TORQUE : Nm (kgf.m, lb-ft)

22140_HYUN_G0080

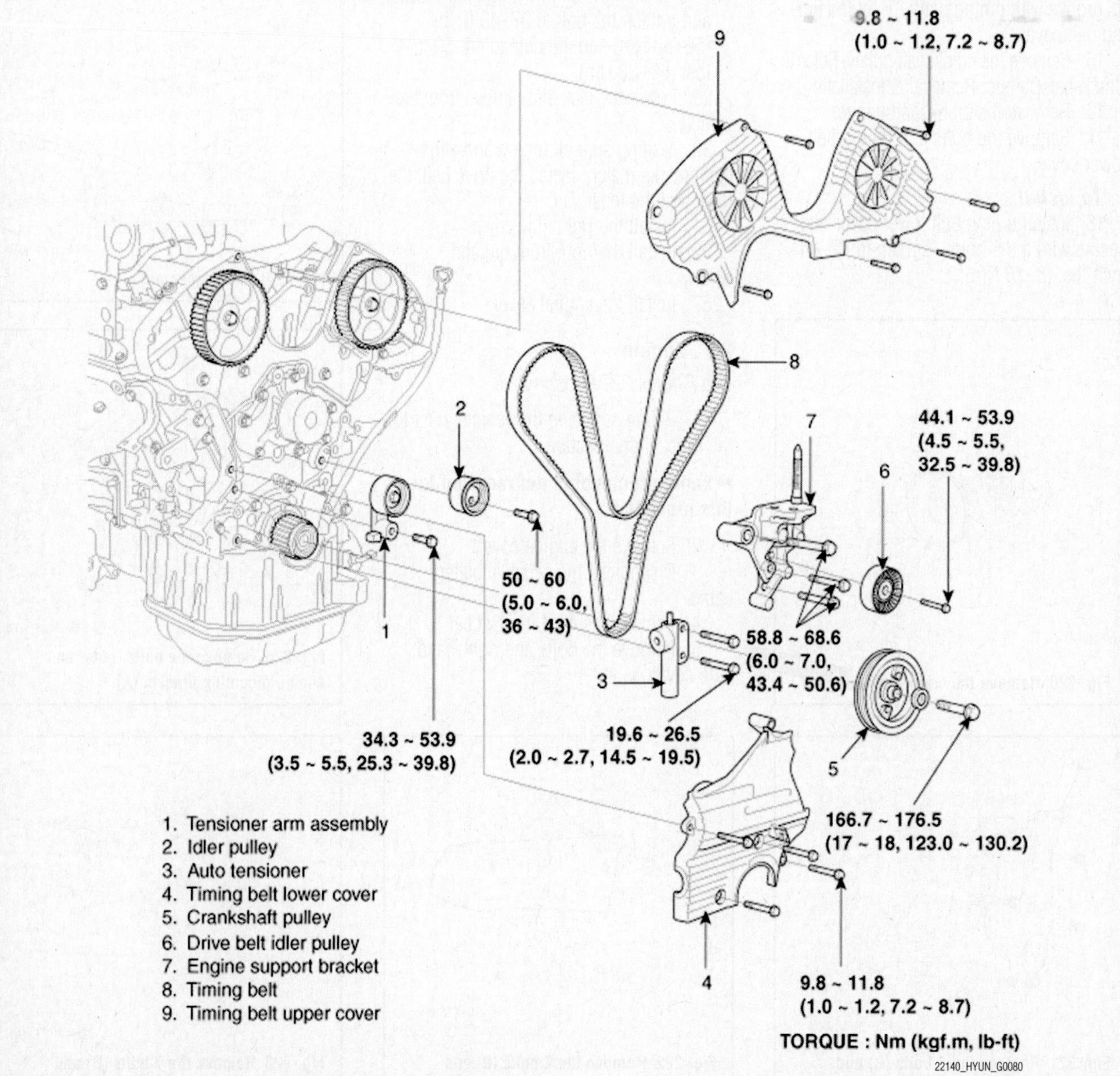

Fig. 228 Timing belt system and related components—2.7L engine

tighten the mounting bolt to 130–138 ft. lbs. (180–190 Nm). Make sure that crankshaft sprocket pin fits the small hole in the pulley.

18. Install the upper timing belt cover.
19. Install the power steering pump.
20. Install the drive belt and belt tensioner.
21. Install the engine mounting bracket and tighten the nuts and bolts to 44–59 ft. lbs. (60–80 Nm).
22. Install the right side cover.
23. Install the right front tire and wheel.
24. Install the engine cover.
25. Connect the negative battery cable.

TIMING BELT & SPROCKETS

REMOVAL & INSTALLATION

2.0L Engine

See Figures 229 through 231.

1. Before servicing the vehicle, refer to the Precautions Section.
2. Remove the timing belt front cover. Refer to Timing Belt Front Cover, Removal & Installation.
3. Remove the timing belt tensioner and timing belt.

➡If the timing belt is to be reused, make an arrow indicating the turning direction to make sure that the belt is reinstalled in the same direction as before.

4. Remove the bolt and timing belt idler.

To install:

5. Install the idler pulley and tighten the bolt to: 32–40 ft. lbs. (43–55 Nm).
6. Install the timing belt tensioner loosely enough for the adjuster to rotate. Make sure that the stopper of the base is leaning against the lower sealing cap on the cylinder head.

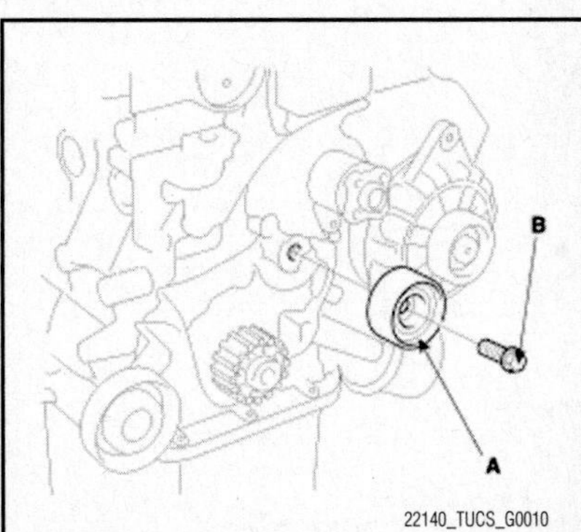

Fig. 229 Remove the bolt (B) and timing belt idler (A)

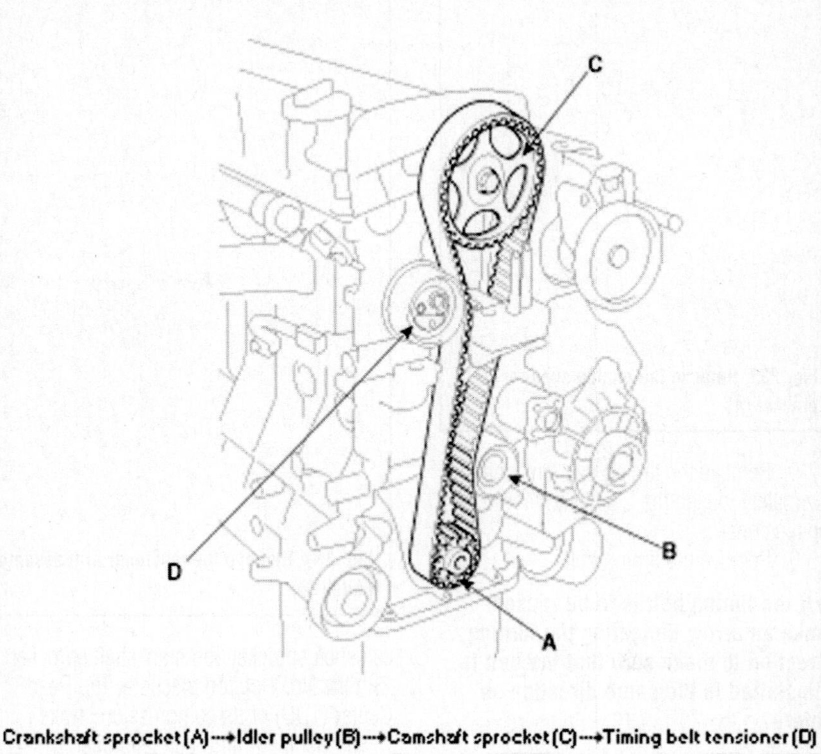

Fig. 230 Install the timing belt in the order illustrated—2.0L engine

7. Install the timing belt so there is no slack in the order illustrated.

➡The tensioner can be installed after the timing belt.

8. Check the alignment of the timing marks on each sprocket.
9. Remove the pin fixing the tensioner arm.
10. Using a hex wrench, turn the adjuster counterclockwise to make the indicator on the arm (A) align at the center of the base notch.

WARNING

Do not rotate the adjuster clockwise. It will result in damage to the auto tensioner.

11. Tightening the tensioner bolt so the indicator does not move. Tightening torque: 17–21 ft. lbs. (23–28 Nm).
12. Turn the crankshaft 2 revolutions in the operating direction (clockwise) and check that the indicator is in the center of the base.
13. If the indicator is not located at the center of the base, slacken the bolt and repeat the above procedure.
14. Install the timing belt front cover. Refer to Timing Belt Front Cover, removal & installation.

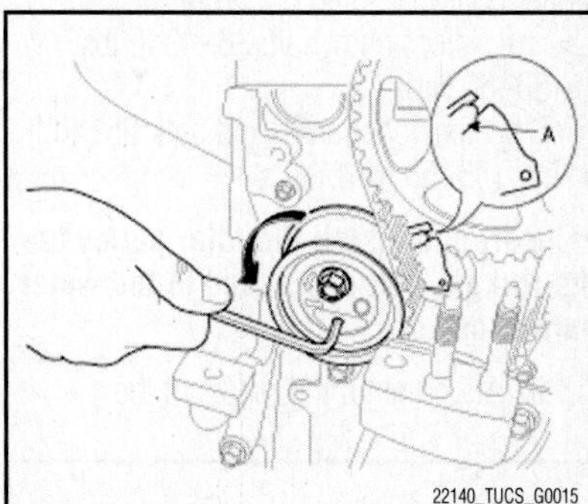

Fig. 231 Using a hex wrench, turn the adjuster counterclockwise to align the indicator on the arm (A) to the center of the base notch—2.0L engine

2.7L Engine

See Figures 232 through 240.

1. Before servicing the vehicle, refer to the Precautions Section.
2. Remove the timing belt front covers (upper and lower). Refer to Timing Belt Front Cover Removal & Installation.
3. Remove the engine support bracket.
4. Check that timing marks of the camshaft timing pulleys and cylinder head covers are aligned. If not, turn the crankshaft 1 revolution (360°).

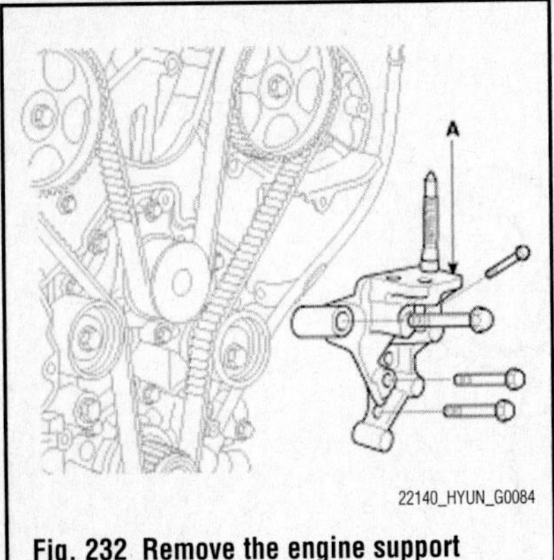

Fig. 232 Remove the engine support bracket (A)

5. Remove the timing belt tensioner. Alternately loosen the 2 bolts and remove the tensioner.

6. Remove the timing belt.

➡**If the timing belt is to be reused, make an arrow indicating the turning direction to make sure that the belt is reinstalled in the same direction as before.**

7. Remove the tensioner arm assembly and timing belt idler pulley.

To install:

8. Install the idler pulley and the tensioner pulley. Tightening torque:
   a. Idler pulley bolt: 36–43 ft. lbs. (50–60 Nm).
   b. Tensioner arm fixed bolt: 25–40 ft. lbs. (35–55 Nm).

➡**Insert and install the idler pulley to the roll pin that is pressed in the water pump boss.**

9. Align the timing marks of the camshaft sprocket and crankshaft sprocket with the No. 1 piston placed at Top Dead Center (TDC) of its compression stroke.

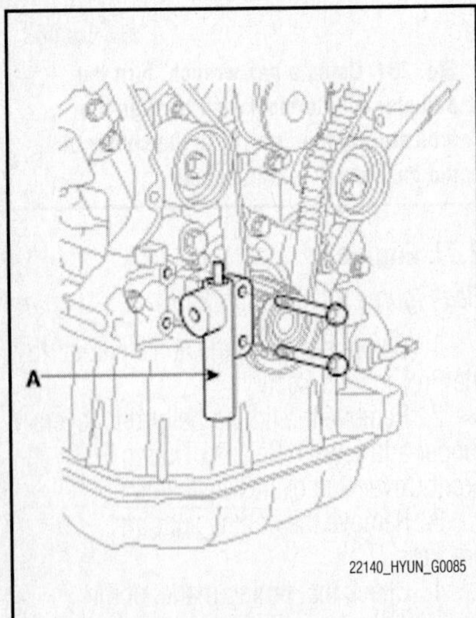

Fig. 233 Alternately loosen the 2 bolts and remove the tensioner (A)

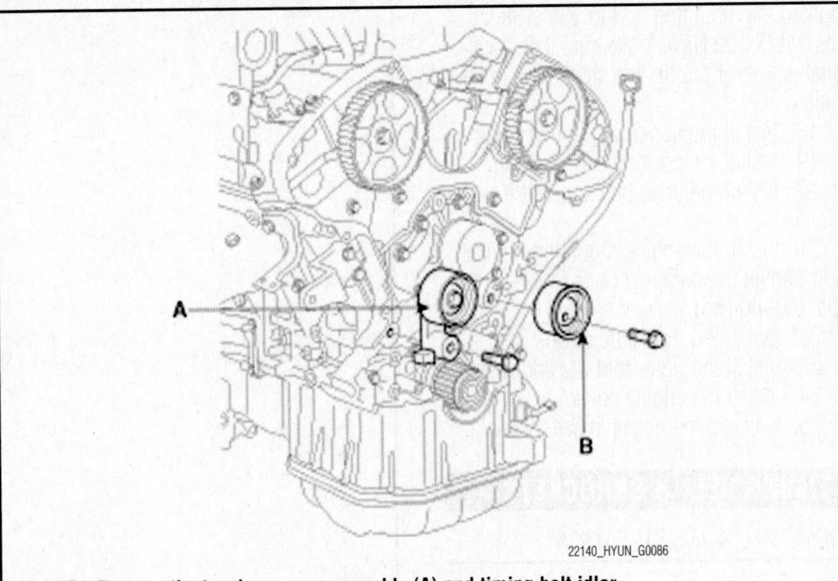

Fig. 234 Remove the tensioner arm assembly (A) and timing belt idler pulley (B)

10. Set the timing belt tensioner:
   a. Using a press, slowly press in the push rod.
   b. Align the holes of the push rod and housing.
   c. Pass a set pin through the holes to keep the setting position of the push rod.
   d. Release the press.

11. Install the timing belt tensioner:
   a. Temporarily install the tensioner with the 2 bolts.
   b. Alternately tighten the 2 bolts.

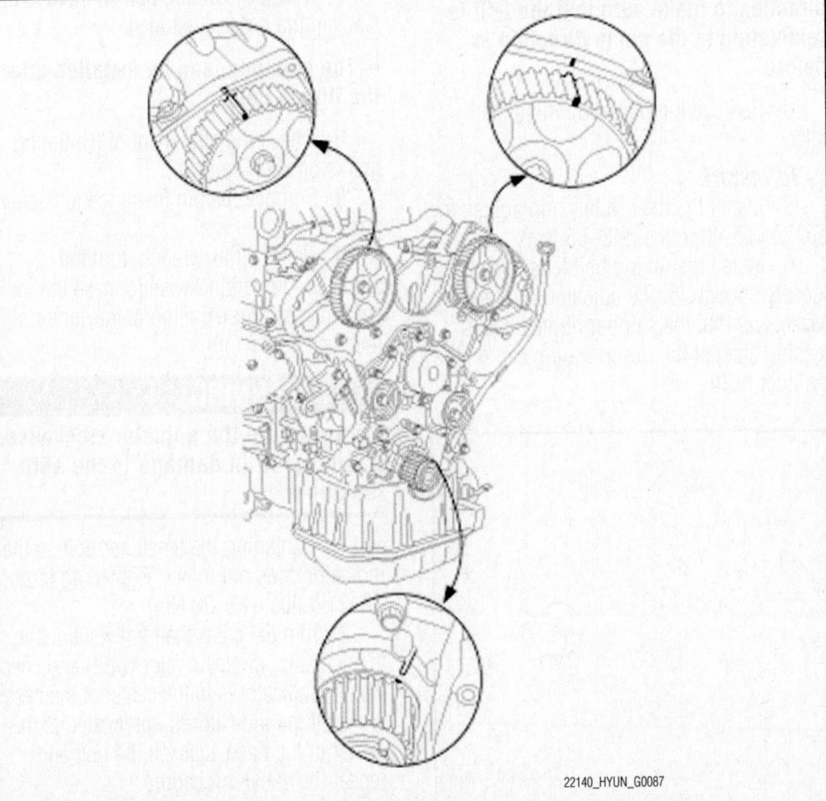

Fig. 235 Align the timing marks of the camshaft sprocket and crankshaft sprocket with the No. 1 piston placed at Top Dead Center (TDC) of its compression stroke

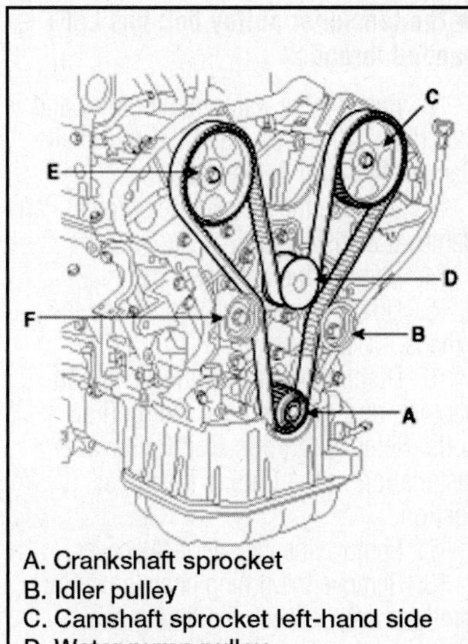

A. Crankshaft sprocket
B. Idler pulley
C. Camshaft sprocket left-hand side
D. Water pump pulley
E. Camshaft sprocket right-hand side
F. Tensioner pulley

37655_TUCS_G0434

Fig. 236 Timing belt installation

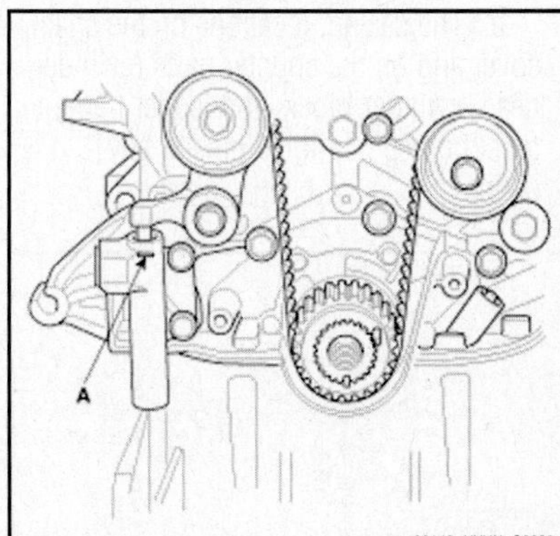

Fig. 237 Remove the set pin (A) from the tensioner

Tightening torque: 14–20 ft. lbs. (20–27 Nm).

12. Install the timing belt:
    a. Remove any oil or water on the sprockets, and keep them clean.
    b. Install the timing belt in this order:
    - Crankshaft sprocket
    - Idler pulley
    - Camshaft sprocket left-hand side
    - Water pump pulley
    - Camshaft sprocket right-hand side
    - Tensioner pulley
13. Remove the set pin (A) from the tensioner.
14. Check the timing belt tensioner:
    a. Rotate the crankshaft 2 turns clockwise and measure the projected length of the auto tensioner at TDC (No. 1 compression stroke) after 5 minutes.
    b. As illustrated in the example, the projected length of the timing belt tensioner should be 0.27–0.31 in. (7–9mm).
15. Install the engine support bracket. Tightening torque of the following bolts (see illustration):
    a. B: 43–51 ft. lbs. (60–70 Nm).
    b. C: 11–16 ft. lbs. (15–22 Nm).
16. Install the timing belt front covers (upper and lower).

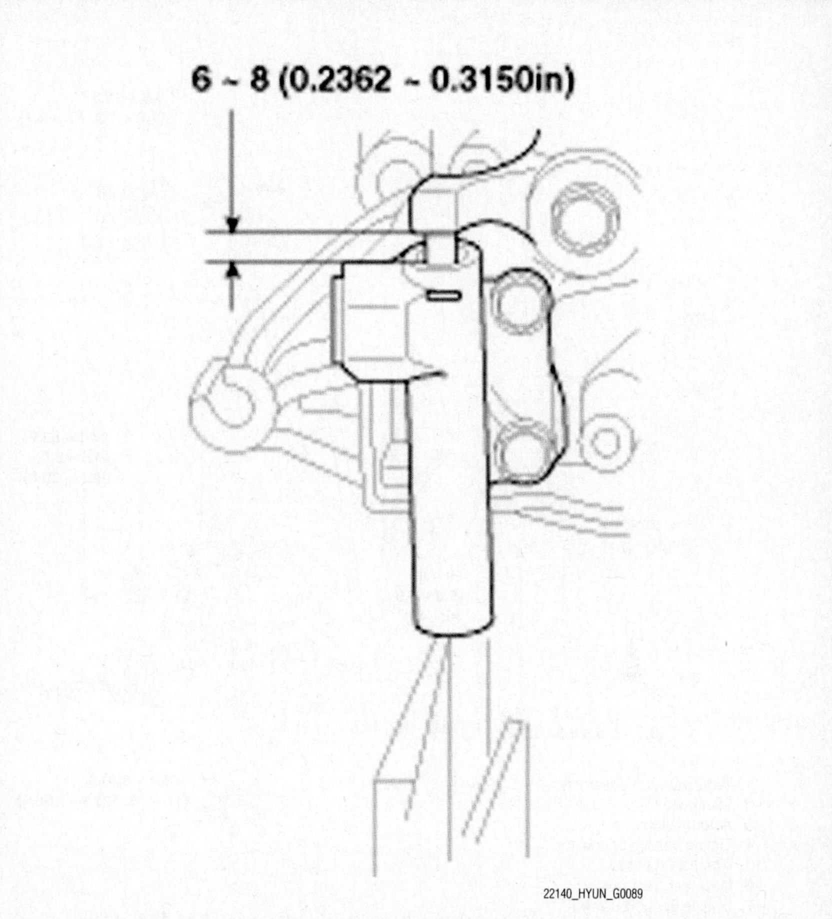

Fig. 238 The projected length of the timing belt tensioner should be 0.27–0.31 inch (7–9mm)

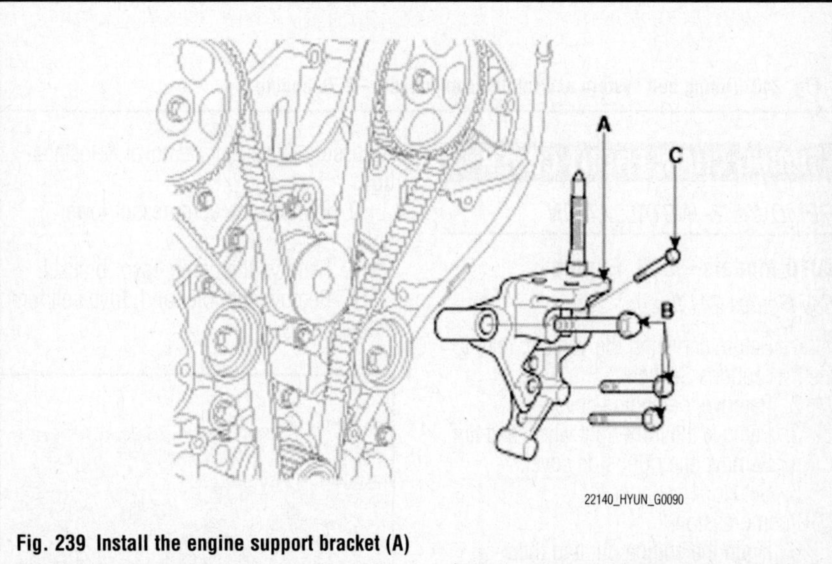

Fig. 239 Install the engine support bracket (A)

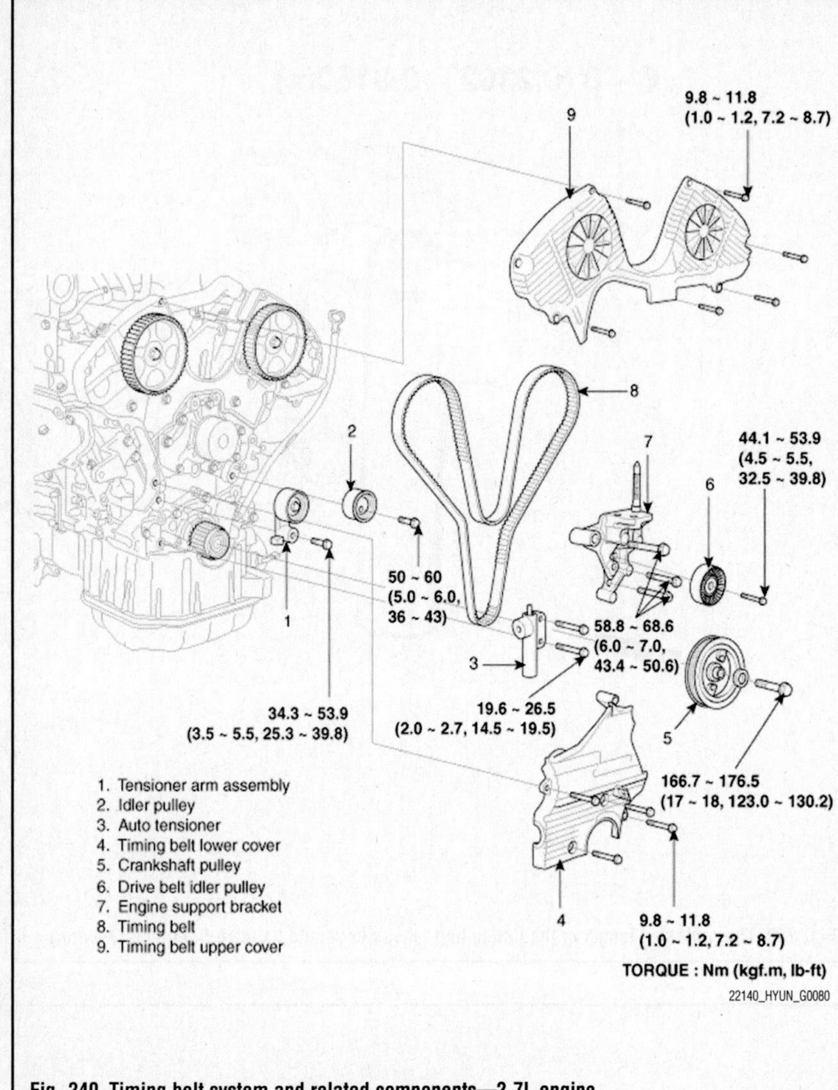

Fig. 240 Timing belt system and related components—2.7L engine

TIMING CHAIN FRONT COVER

REMOVAL & INSTALLATION

2010 Models—2.4L Engine

See Figures 241 through 248.

1. Before servicing the vehicle, refer to the Precautions Section.
2. Remove the engine cover.
3. Remove the front right wheel and tire.
4. Remove the right side cover.
5. Set No. 1 cylinder to TDC/compression.
6. Drain the engine oil, and then position a jack to the oil pan.

➡Place wooden block between the jack and engine oil pan.

7. Disconnect the ground cable and then remove the engine mounting bracket.
8. Remove the drive belt. Refer to Accessory Drive Belt Removal & Installation.
9. Remove the compressor lower bolts.
10. Remove the compressor bracket.
11. Remove the idler and drive belt tensioner pulley.

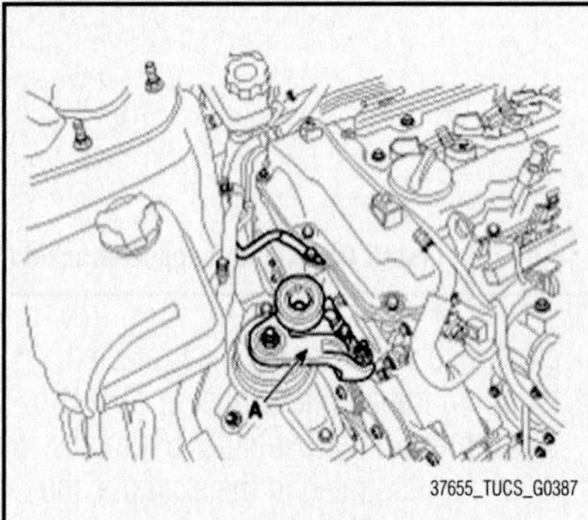

Fig. 241 Remove the engine mounting bracket (A)

➡The tensioner pulley bolt has LEFT handed threads.

12. Remove the water pump pulley and the crankshaft pulley. Refer to Crankshaft Damper Removal & Installation.
13. Remove the oil pan. Refer to Oil Pan, Removal & Installation.
14. Remove the breather hose.
15. Disconnect the PCV hose and the exhaust OCV connector.
16. Disconnect the ignition coil connectors, and then remove the ignition coils. Refer to Ignition Coil Removal & Installation in the Engine Electrical section.
17. Remove the cylinder head cover.
18. Remove the timing chain cover by gently prying between the timing chain cover and the cylinder block.

➡Be careful not to damage the contact surfaces of cylinder block, cylinder head and timing chain cover.

To install:

19. Install the timing chain cover:
    a. The sealant locations on the chain cover and on the counter parts (cylinder head, cylinder block, and ladder frame) must be free of engine oil and any debris.

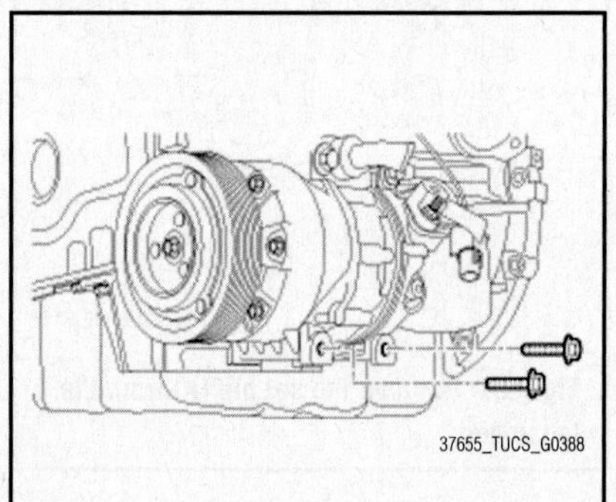

Fig. 242 Remove the compressor lower bolts

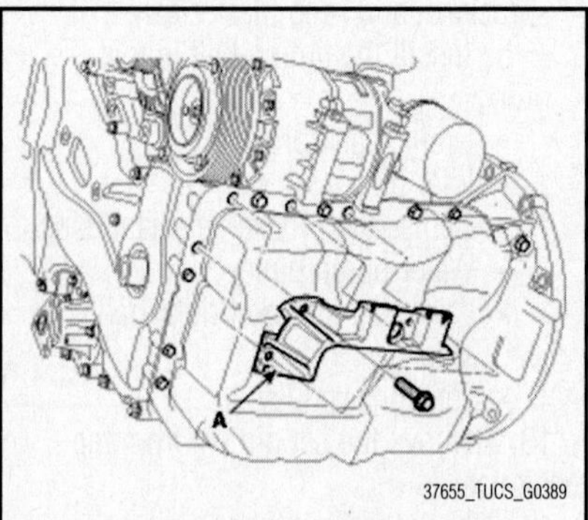

Fig. 243 Remove the compressor bracket (A)

Fig. 244 Remove the idler (A) and drive belt tensioner pulley (B)

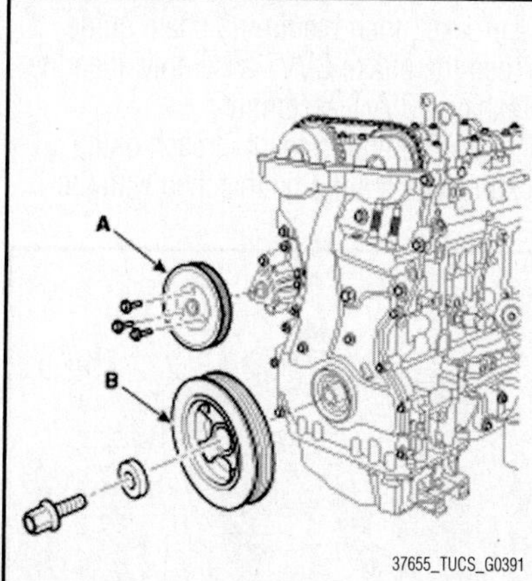

Fig. 245 Remove the water pump pulley (A), crankshaft pulley (B)

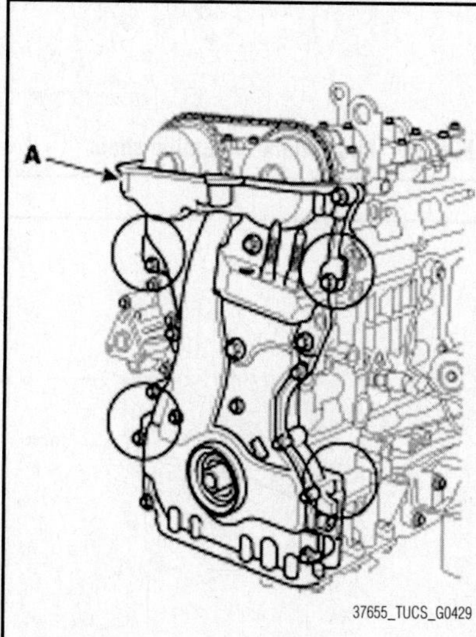

Fig. 246 Remove the timing chain cover (A)

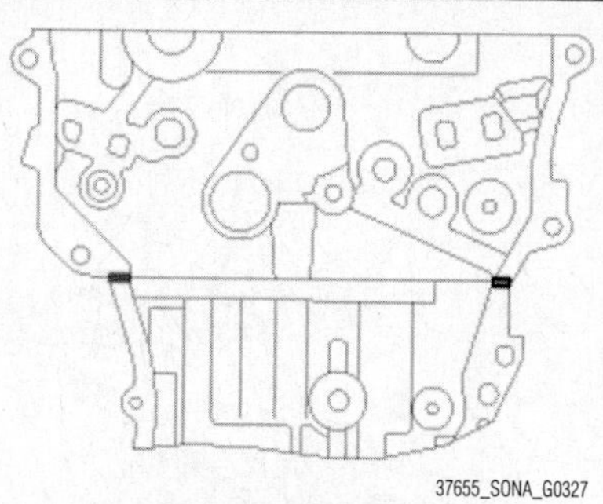

Fig. 247 Apply sealant on the gap between the cylinder head and the cylinder block

b. Before assembling the timing chain cover, the liquid sealant Loctite® 5900H or Three Bond® 1217H should be applied on the gap between the cylinder head and the cylinder block.

➡The parts must be assembled within 5 minutes after the sealant was applied. Use a bead width of 0.12 in. (3.0mm).

c. Apply sealant on the timing chain cover. The part must be assembled within 5 minutes after the sealant was applied. The sealant should be applied in a continuous bead.

d. The dowel pins on the cylinder block and the holes on the timing chain cover should be used as a reference in order to assemble the timing chain cover into exact position. Tightening torque:

Fig. 248 Timing chain cover sealant application

- 6x25: 70–86 inch lbs. (8–10 Nm)
- 8x28: 14–17 ft. lbs. (19–23 Nm)
- 10x45: 29–33 ft. lbs. (39–44 Nm)
- 10x40: 29–33 ft. lbs. (39–44 Nm)

e. Wait to start the engine for 30 minutes after the timing chain cover was assembled.

20. Install the oil pan.
21. Install the cylinder head cover.
22. Install the breather hose.
23. Connect the PCV hose and exhaust OCV connector.
24. Install the ignition coils and then connect the ignition coil connectors.
25. Install the crankshaft pulley and tighten to 123–130 ft. lbs. (167–176 Nm).
26. Install the water pump pulley and tighten to 70–86 inch lbs. (8–10 Nm).
27. Install the drive belt tensioner pulley and tighten to 40–47 ft. lbs. (54–64 Nm).
28. Install the idler and tighten to 40–47 ft. lbs. (54–64 Nm).
29. Install the air compressor bracket and tighten to 14–15 ft. lbs. (20–24 Nm).
30. Install the air compressor lower bolts and tighten to 18–24 ft. lbs. (20–33 Nm).
31. Install the drive belt.
32. Install the engine mounting bracket, tighten the bolts to 47–62 ft. lbs. (64–83 Nm), and then connect the ground cable.
33. Install the right side cover.
34. Install the right front tire and wheel.
35. Install the engine cover.
36. Connect the negative battery cable.

TIMING CHAIN & SPROCKETS

REMOVAL & INSTALLATION

2010 Models—2.4L Engine

See Figures 249 through 260.

1. Before servicing the vehicle, refer to the Precautions Section.
2. Remove the engine cover.
3. Remove the front right wheel and tire.
4. Remove the right side cover.
5. Set No. 1 cylinder to TDC/compression.
6. Drain the engine oil, and then position a jack to the oil pan.

➡Place wooden block between the jack and engine oil pan.

7. Disconnect the ground cable and then remove the engine mounting bracket.

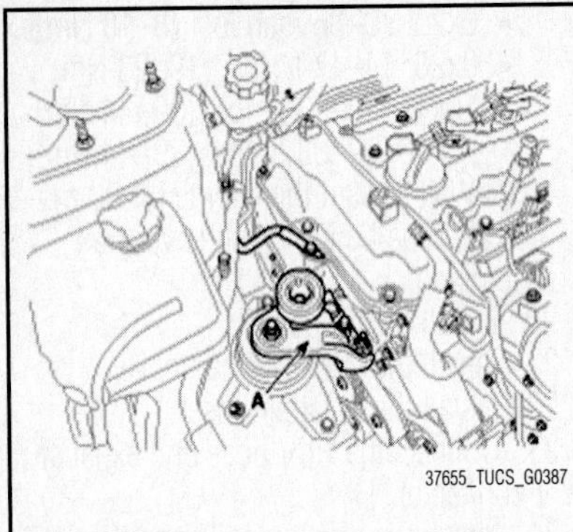

Fig. 249 Remove the engine mounting bracket (A)

8. Remove the drive belt. Refer to Accessory Drive Belt Removal & Installation.
9. Remove the compressor lower bolts.
10. Remove the compressor bracket.
11. Remove the idler and drive belt tensioner pulley.

➡**The tensioner pulley bolt has LEFT handed threads.**

12. Remove the water pump pulley and the crankshaft pulley. Refer to Crankshaft Damper Removal & Installation.
13. Remove the oil pan. Refer to Oil Pan, Removal & Installation.
14. Remove the breather hose.

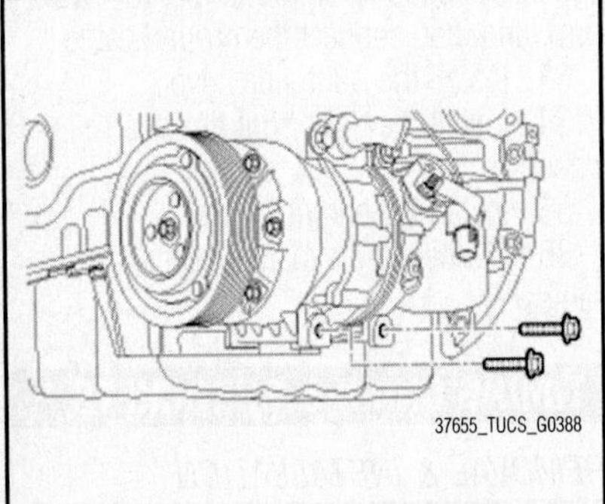

Fig. 250 Remove the compressor lower bolts

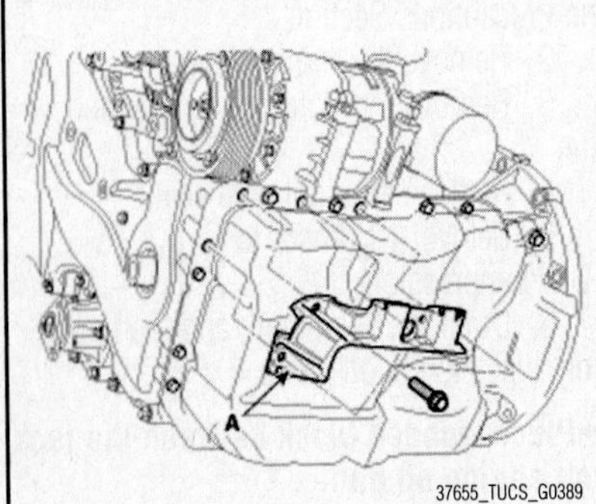

Fig. 251 Remove the compressor bracket (A)

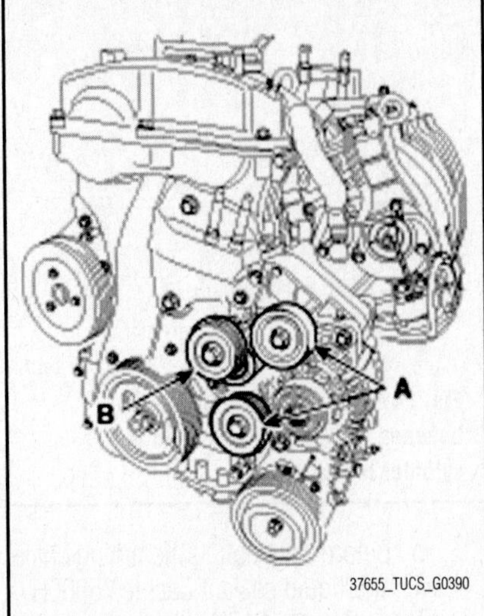

Fig. 252 Remove the idler (A) and drive belt tensioner pulley (B)

Fig. 253 Remove the water pump pulley (A), crankshaft pulley (B)

15. Disconnect the PCV hose and the exhaust OCV connector.
16. Disconnect the ignition coil connectors, and then remove the ignition coils. Refer to Ignition Coil Removal & Installation in the Engine Electrical section.
17. Remove the cylinder head cover.
18. Remove the timing chain cover. Refer to Timing Chain Front Cover Removal & Installation.
19. The key of the crankshaft should be aligned with the mating face of the main bearing cap. This will place the piston of the No. 1 cylinder is placed at TDC on the compression stroke.
20. Before removing the timing chain, matchmark the timing chain with the sprocket because the identification mark on the chain for TDC (Top Dead Center) can be erased.
21. Install a set pin after compressing the timing chain tensioner.
22. Remove the timing chain tensioner.
23. Remove the timing chain tensioner arm.
24. Remove the timing chain.
25. Remove the timing chain guide.

To install:

26. Set the crankshaft so that the key of the crankshaft is aligned with the mating surface of main bearing cap. Set the intake and exhaust camshaft assembly so that the TDC mark of the intake sprocket and the exhaust sprocket are aligned with the top surface of the cylinder head. This will place the piston on the No. 1 cylinder at TDC on the compression stroke.
27. Install the timing chain guide. Tightening torque: 86–104 inch lbs. (10–12 Nm).
28. Install the timing chain. To install the timing chain so that there is no slack between the camshaft and crankshaft, use the following procedure:
   a. Place the chain over the crankshaft sprocket, then the timing chain guide, then the intake CVVT assembly, then the exhaust CVVT assembly.
   b. The timing mark of each of the sprockets should be matched with the

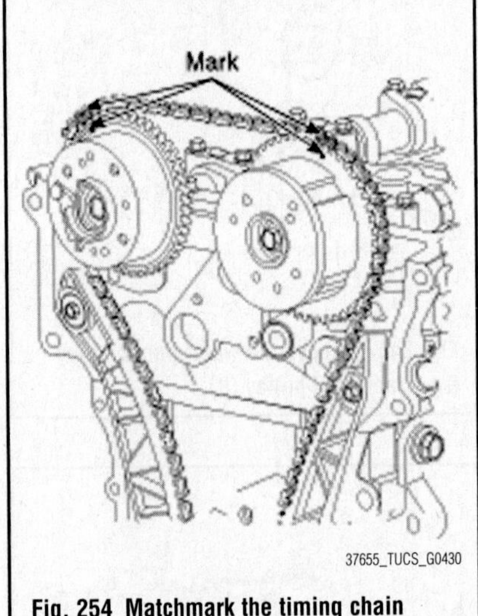

Fig. 254 Matchmark the timing chain

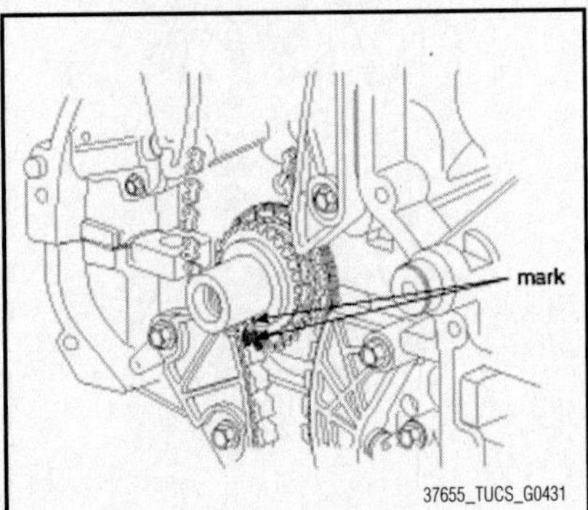

Fig. 255 Matchmark the timing chain with the sprocket

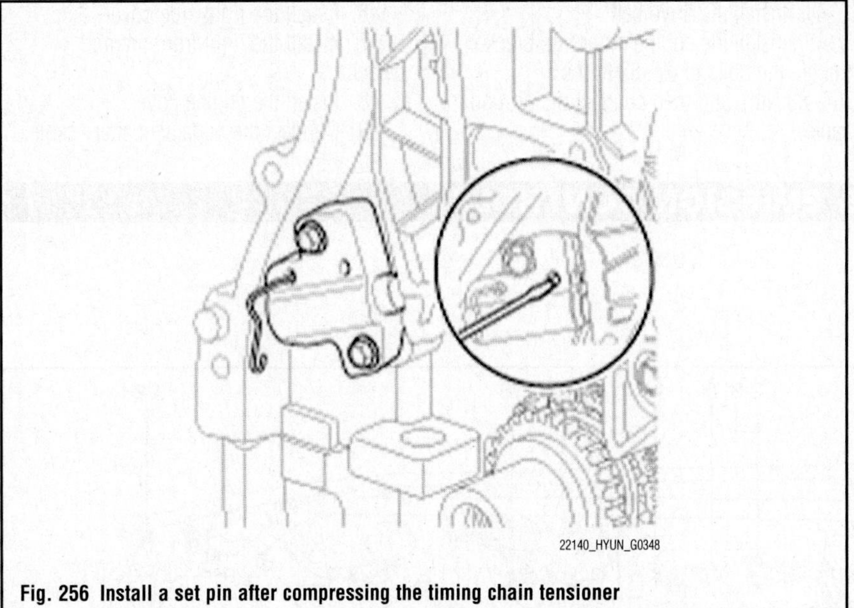

Fig. 256 Install a set pin after compressing the timing chain tensioner

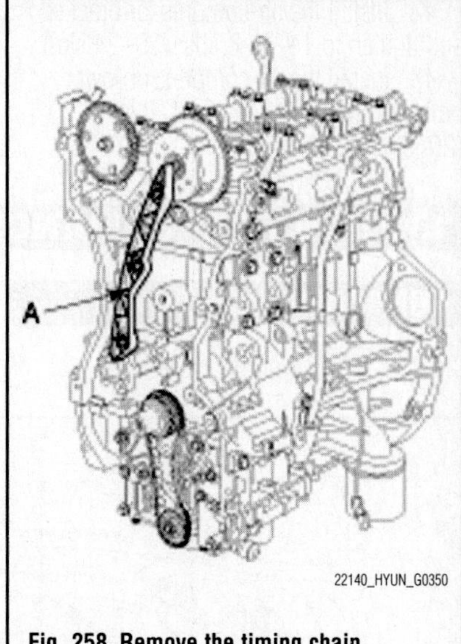

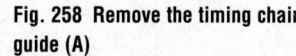

Fig. 258 Remove the timing chain guide (A)

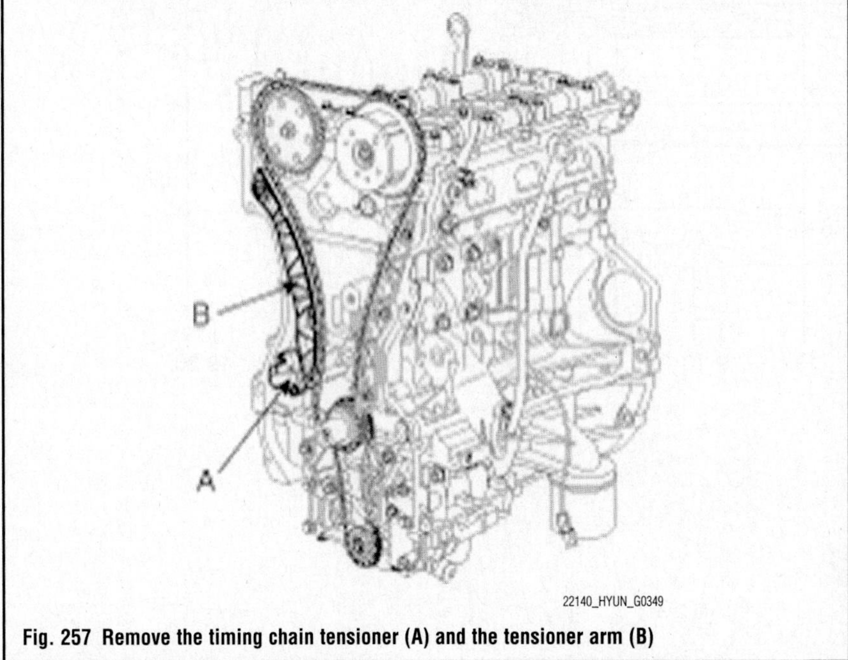

Fig. 257 Remove the timing chain tensioner (A) and the tensioner arm (B)

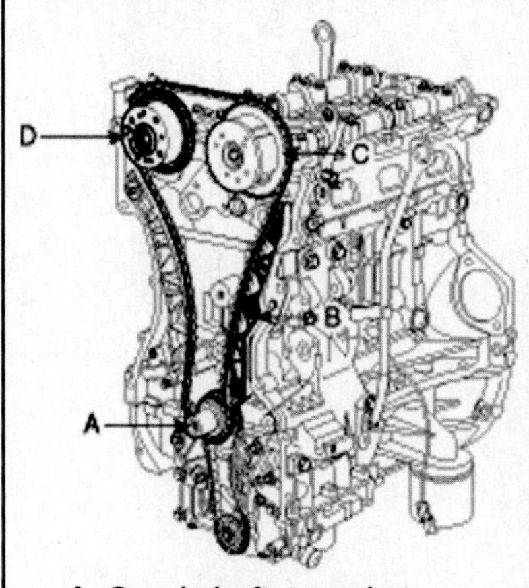

Fig. 259 Place the chain over the crankshaft sprocket (A), then the timing chain guide (B), then the intake CVVT assembly (C), then the exhaust CVVT assembly (D)

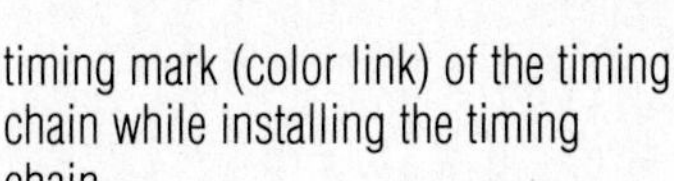
timing mark (color link) of the timing chain while installing the timing chain.

29. Install the timing chain tensioner arm. Tightening torque: 86–104 inch lbs. (10–12 Nm).

30. Install the timing chain auto tensioner and remove the set pin. Tightening torque: 86–104 inch lbs. (10–12 Nm).

31. After rotating the crankshaft 2 revolutions in a clockwise direction (viewed from the front), confirm that the timing marks are still aligned.

32. Install the timing chain cover:

33. Install the oil pan.

34. Install the cylinder head cover.

35. Install the breather hose.

36. Connect the PCV hose and exhaust OCV connector.

37. Install the ignition coils and then connect the ignition coil connectors.

38. Install the crankshaft pulley and tighten to 123–130 ft. lbs. (167–176 Nm).

39. Install the water pump pulley and tighten to 70–86 inch lbs. (8–10 Nm).

40. Install the drive belt tensioner pulley and tighten to 40–47 ft. lbs. (54–64 Nm).

41. Install the idler and tighten to 40–47 ft. lbs. (54–64 Nm).

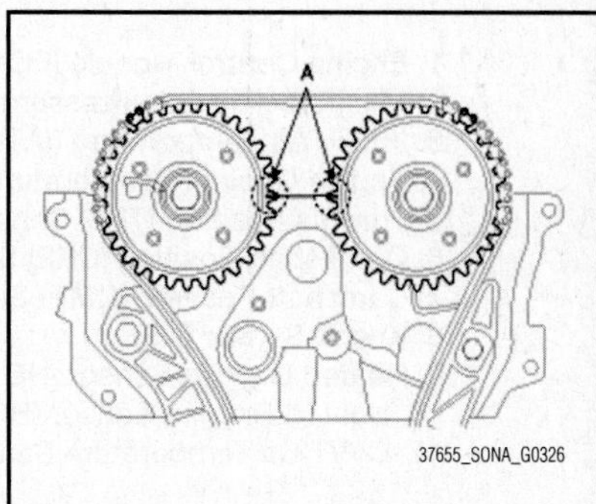

Fig. 260 Timing mark alignment

42. Install the air compressor bracket and tighten to 14–15 ft. lbs. (20–24 Nm).
43. Install the air compressor lower bolts and tighten to 18–24 ft. lbs. (20–33 Nm).
44. Install the drive belt.
45. Install the engine mounting bracket, tighten the bolts to 47–62 ft. lbs. (64–83 Nm), and then connect the ground cable.
46. Install the right side cover.
47. Install the right front tire and wheel.
48. Install the engine cover.
49. Connect the negative battery cable.

ENGINE PERFORMANCE & EMISSION CONTROLS

COMPONENT LOCATIONS

See Figures 261 through 266.

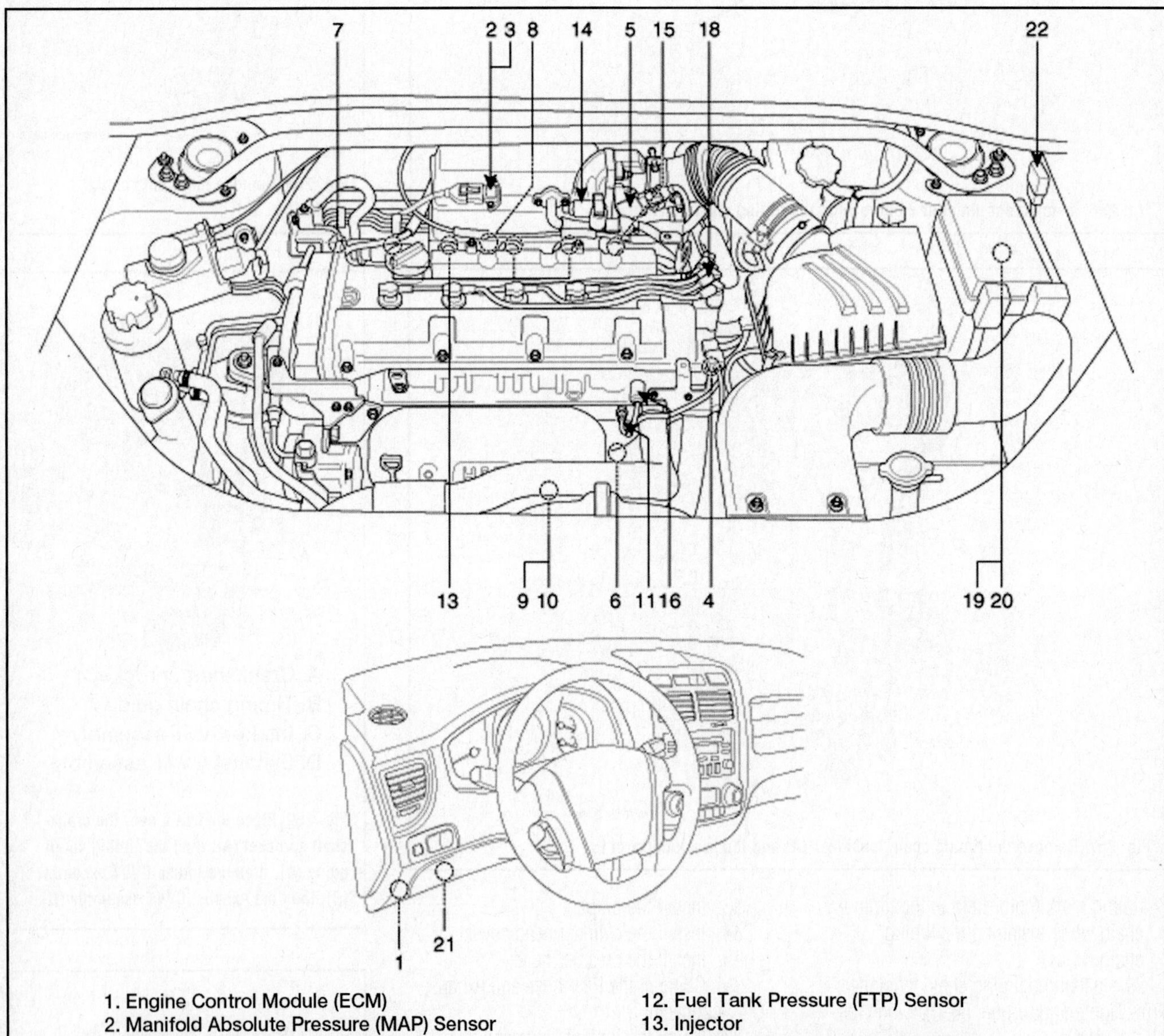

Fig. 261 Underhood and related sensor locations—2009 models, 2.0L engine

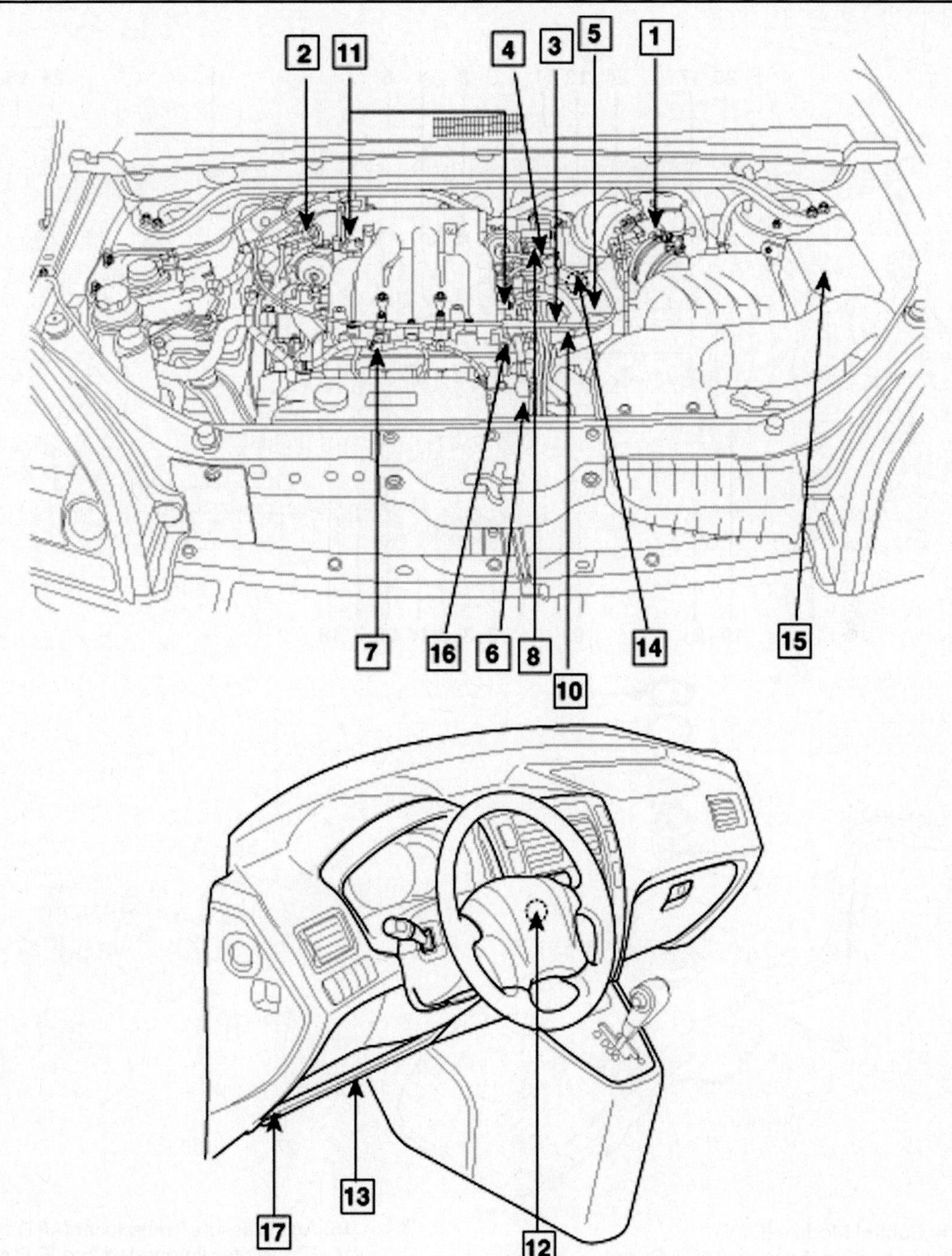

1. Mass Air Flow (MAF) Sensor [With CVVT]
2. Intake Air Temperature (IAT) Sensor
3. Engine Coolant Temperature (ECT) Sensor
4. Throttle Position Sensor (TPS)
5. Camshaft Position (CMP) Sensor
6. Crankshaft Position (CKP) Sensor
7. Injector
8. Idle Speed Control Actuator (ISCA)
9. Vehicle Speed Sensor (VSS)
10. Knock Sensor
11. VIS Control solenoid valve
12. Ignition Switch
13. ECM
14. Purge Control Solenoid Valve (PCSV)
15. Main Relay
16. Ignition Coil
17. DLC (Diagnostic Link Connector)
18. Heated Oxygen Sensor (Bank1, Sensor1)
19. Heated Oxygen Sensor (Bank1, Sensor2)
20. Heated Oxygen Sensor (Bank2, Sensor1)
21. Heated Oxygen Sensor (Bank2, Sensor2)

37655_TUCS_G0501

Fig. 262 Underhood and related sensor locations—2009 models, 2.7L engine

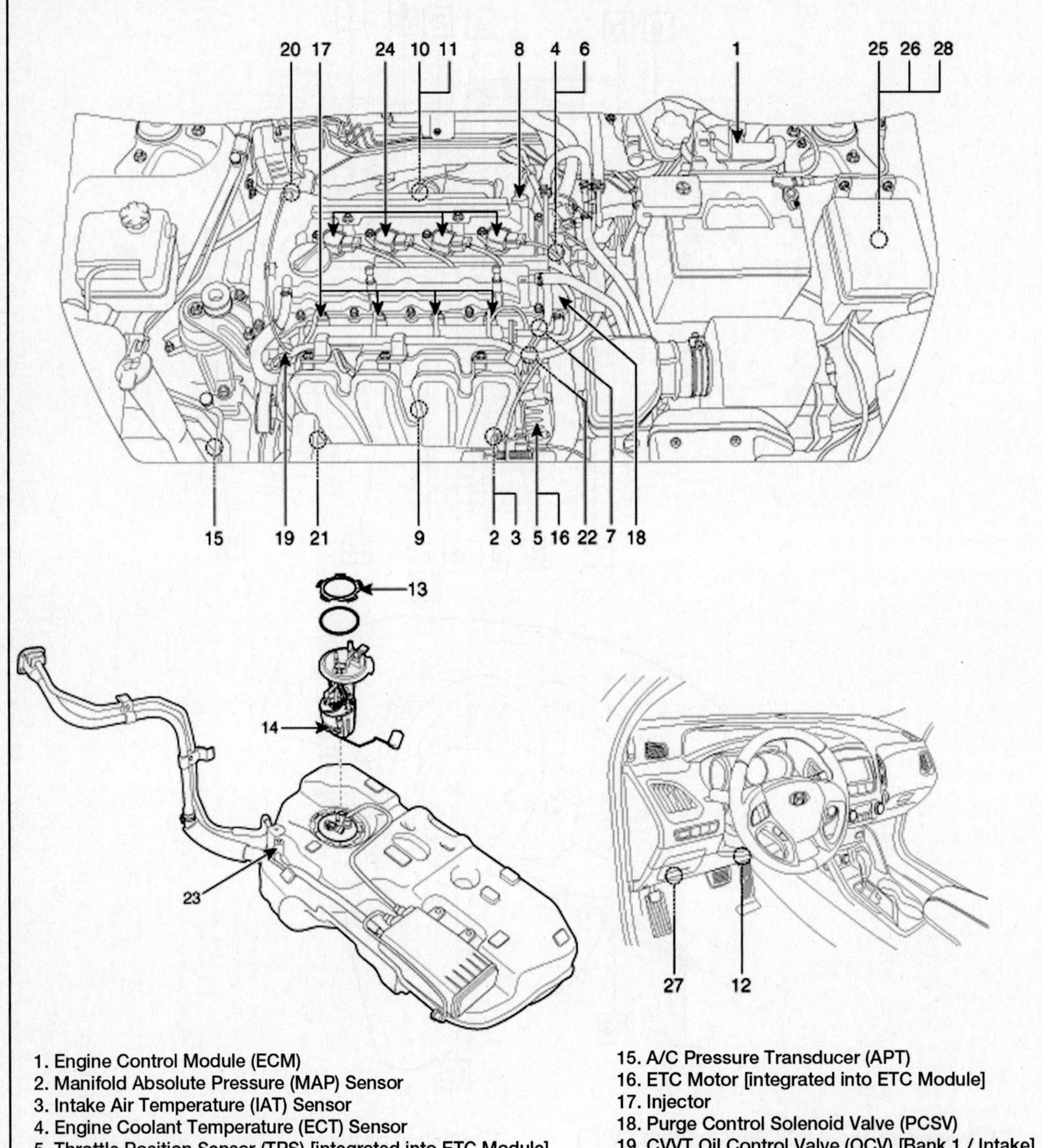

1. Engine Control Module (ECM)
2. Manifold Absolute Pressure (MAP) Sensor
3. Intake Air Temperature (IAT) Sensor
4. Engine Coolant Temperature (ECT) Sensor
5. Throttle Position Sensor (TPS) [integrated into ETC Module]
6. Crankshaft Position (CKP) Sensor
7. Camshaft Position (CMP) Sensor [Bank 1 / Intake]
8. Camshaft Position (CMP) Sensor [Bank 1 / Exhaust]
9. Knock Sensor (KS)
10. Heated Oxygen Sensor (HO2S) [Bank 1 / Sensor 1]
11. Heated Oxygen Sensor (HO2S) [Bank 1 / Sensor 2]
12. Accelerator Position (APP) Sensor
13. Fuel Tank Pressure (FTP) Sensor
14. Fuel Level Sensor (FLS)
15. A/C Pressure Transducer (APT)
16. ETC Motor [integrated into ETC Module]
17. Injector
18. Purge Control Solenoid Valve (PCSV)
19. CVVT Oil Control Valve (OCV) [Bank 1 / Intake]
20. CVVT Oil Control Valve (OCV) [Bank 1 / Exhaust]
21. Variable Intake Solenoid (VIS) Valve
22. Variable Charge Motion Actuator (VCMA)
23. Canister Close Valve (CCV)
24. Ignition Coil
25. Main Relay
26. Fuel Pump Relay
27. Data Link Connector (DLC) [16-Pin]
28. Multi-Purpose Check Connector [20-Pin]

37655_TUCS_G0502

Fig. 263 Underhood and related sensor locations—2010 models, 2.4L engine

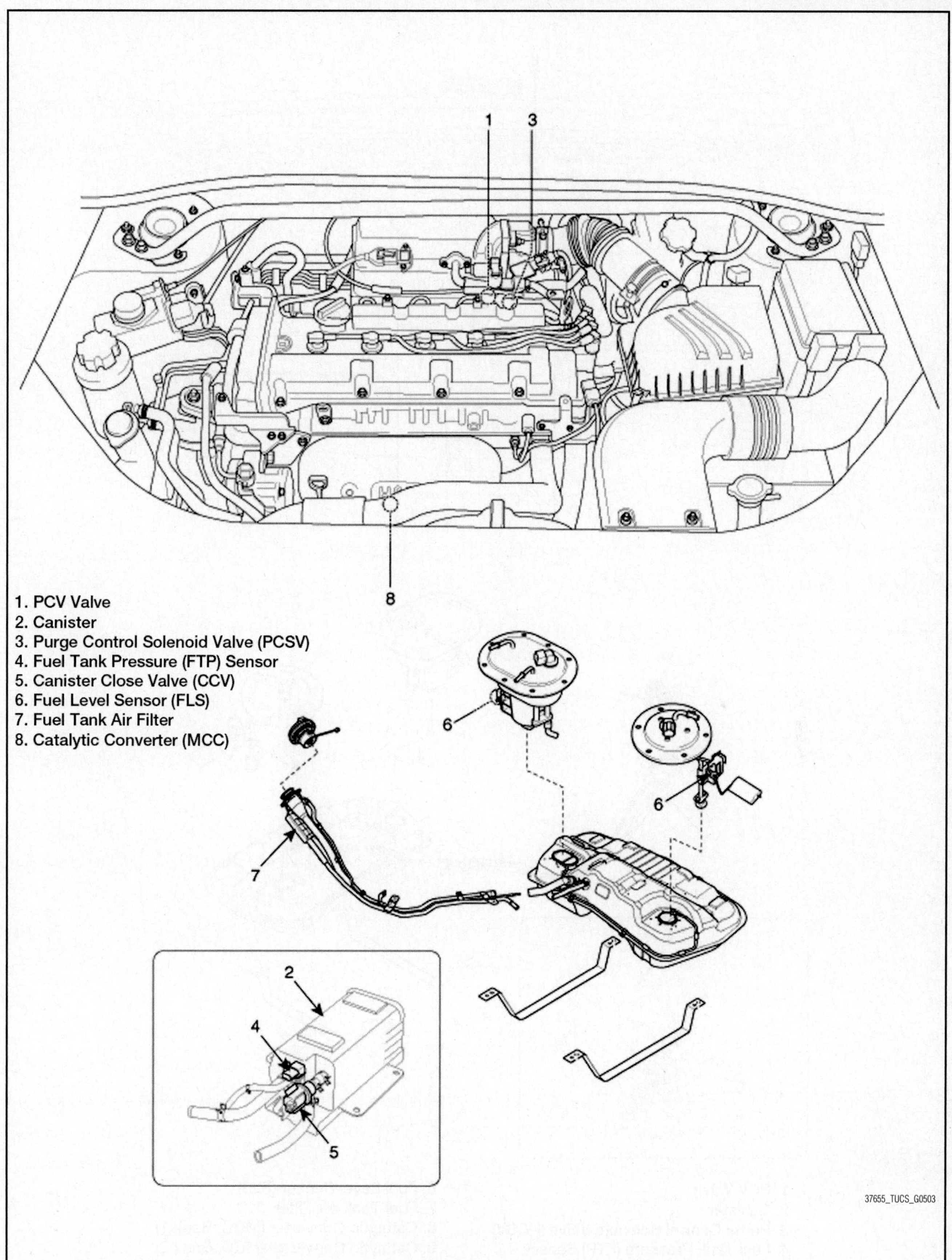

Fig. 264 Emission Control System component locations—2009 models, 2.0L engine

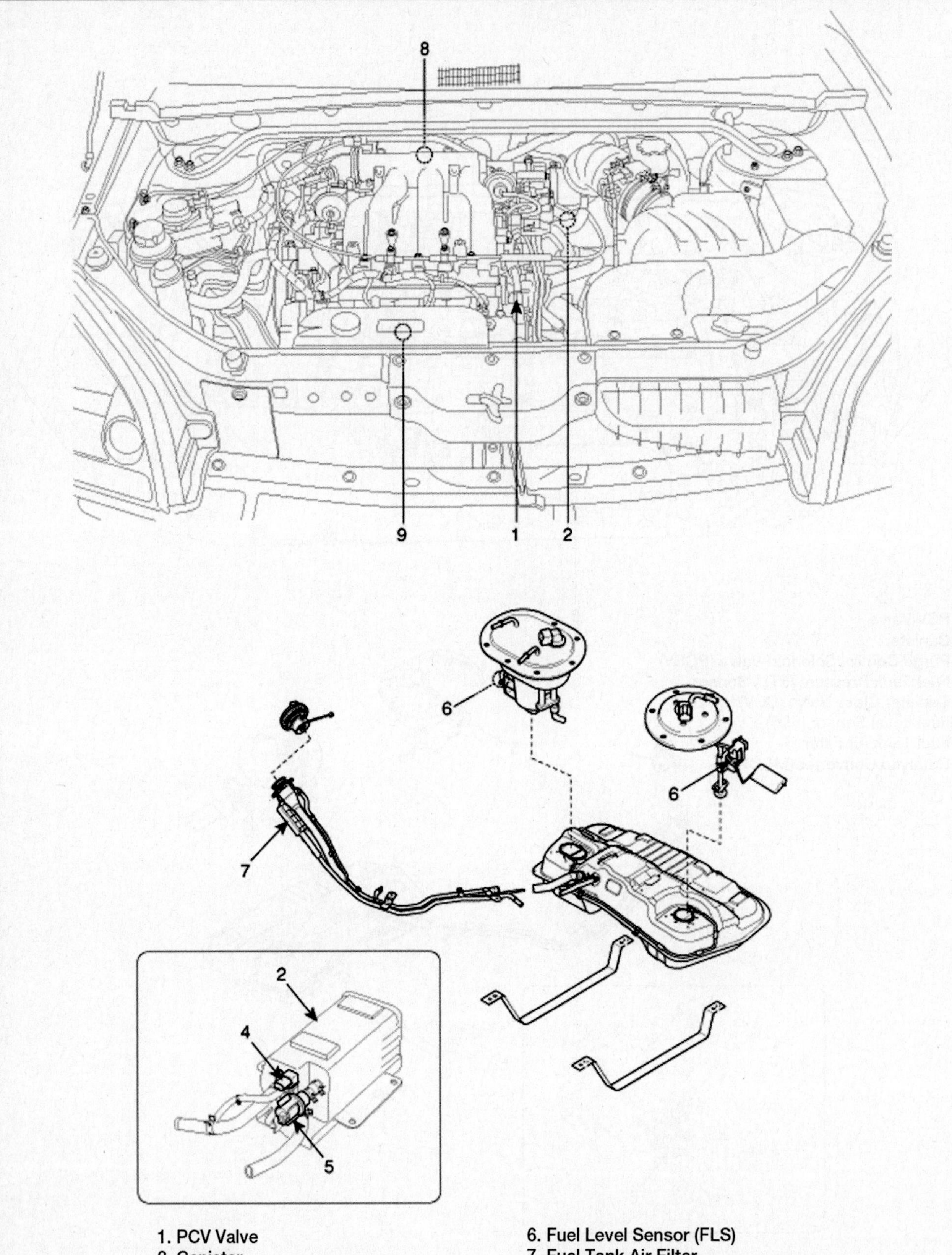

Fig. 265 Emission Control System component locations—2009 models, 2.7L engine

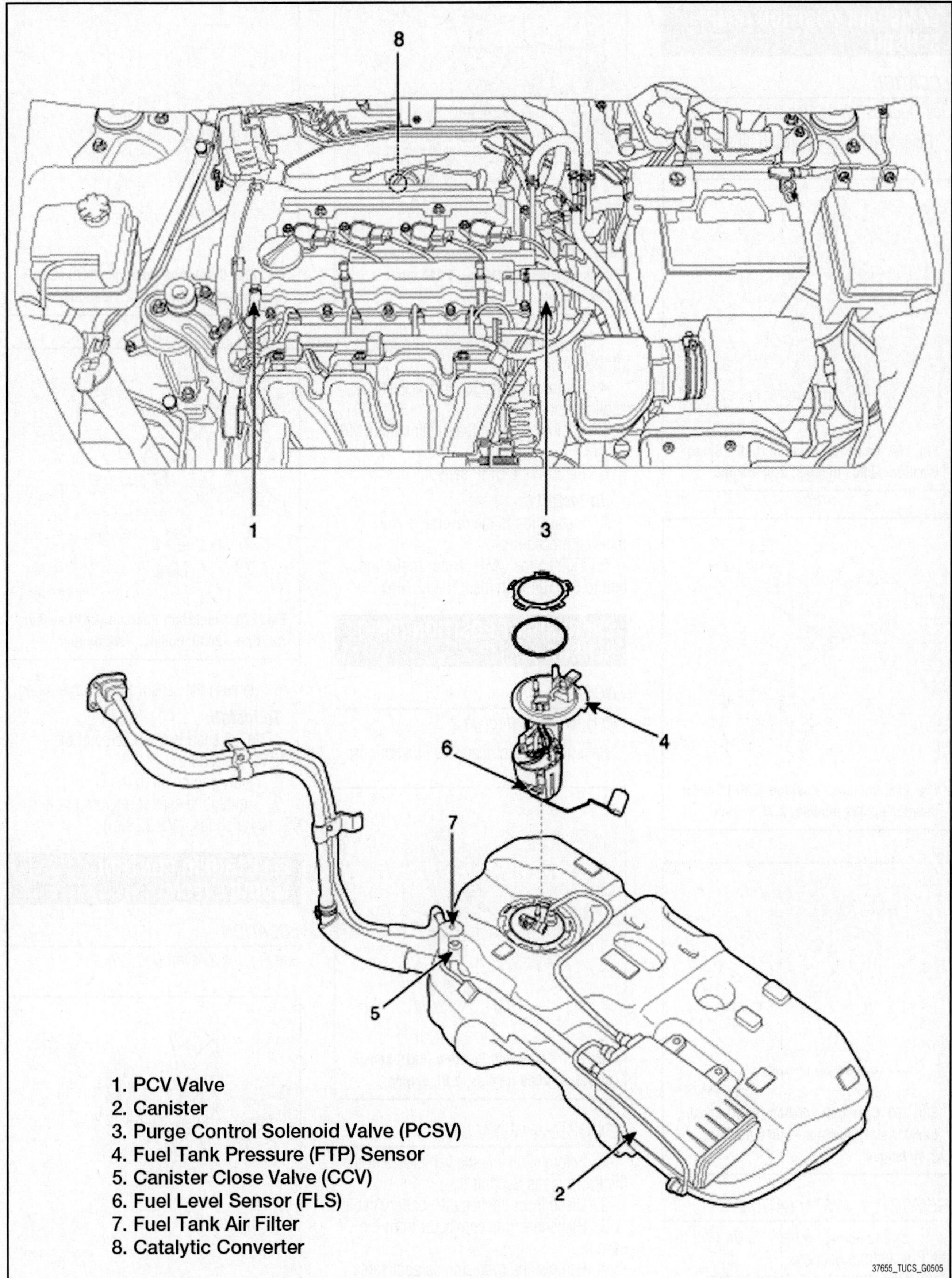

Fig. 266 Emission Control System component locations—2010 models, 2.4L engine

CAMSHAFT POSITION (CMP) SENSOR

LOCATION

See Figures 267 through 270.

Refer to the accompanying illustrations.

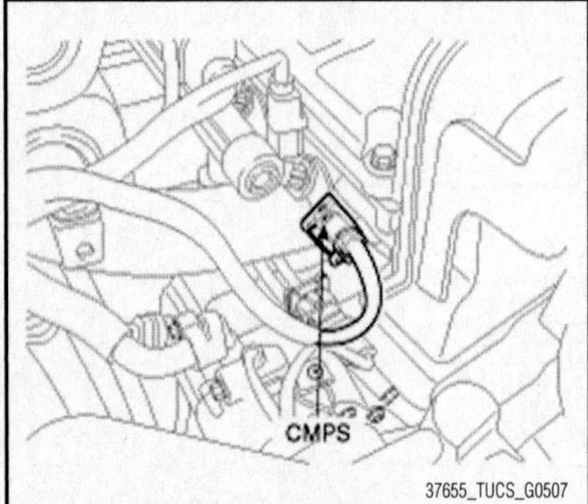

Fig. 267 Camshaft Position (CMP) Sensor location—2009 models, 2.0L engine

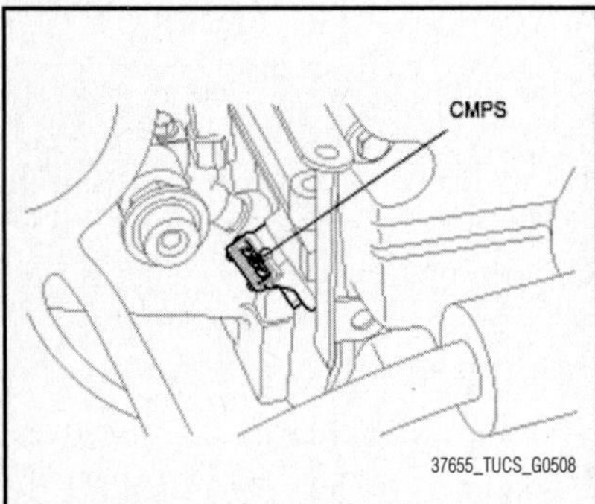

Fig. 268 Camshaft Position (CMP) Sensor location—2009 models, 2.7L engine

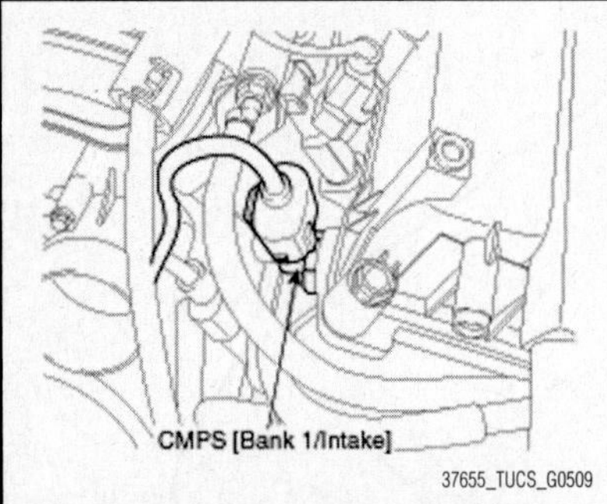

Fig. 269 Camshaft Position (CMP) Sensor Bank 1/Intake location—2010 models, 2.4L engine

REMOVAL & INSTALLATION

1. Before servicing the vehicle, refer to the Precautions Section.
2. Disconnect the negative battery cable.

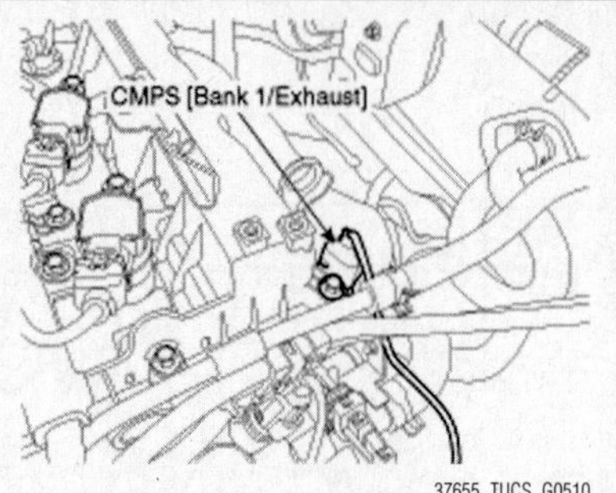

Fig. 270 Camshaft Position (CMP) Sensor Bank 1/Exhaust location—2010 models, 2.4L engine

3. Disconnect the connector from the CMP sensor.
4. Remove the hanger and protector, as applicable.
5. Remove the bolt that retains the CMP sensor.
6. Remove the CMP sensor.

To install:

7. Installation is the reverse of the removal procedure.
8. Tighten the CMP sensor mounting bolt to 86–104 inch lbs. (10–12 Nm).

CRANKSHAFT POSITION (CKP) SENSOR

LOCATION

See Figures 271 through 273.

Refer to the accompanying illustrations.

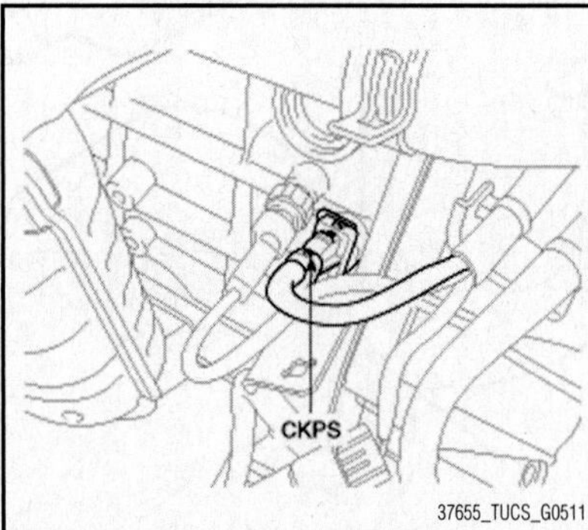

Fig. 271 Crankshaft Position (CKP) sensor location—2009 models, 2.0L engine

REMOVAL & INSTALLATION

1. Before servicing the vehicle, refer to the Precautions Section.
2. Disconnect the negative battery cable.
3. Disconnect the connector from the sensor.
4. Remove the protector, as applicable.
5. Remove the bolt that retains the sensor in place.

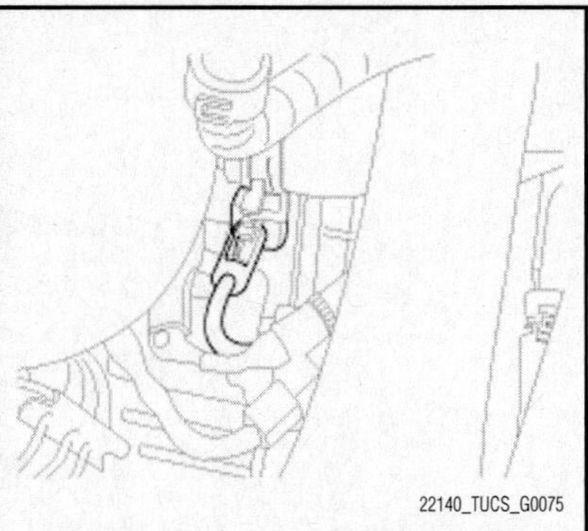

Fig. 272 Crankshaft Position (CKP) sensor location—2009 models, 2.7L engine

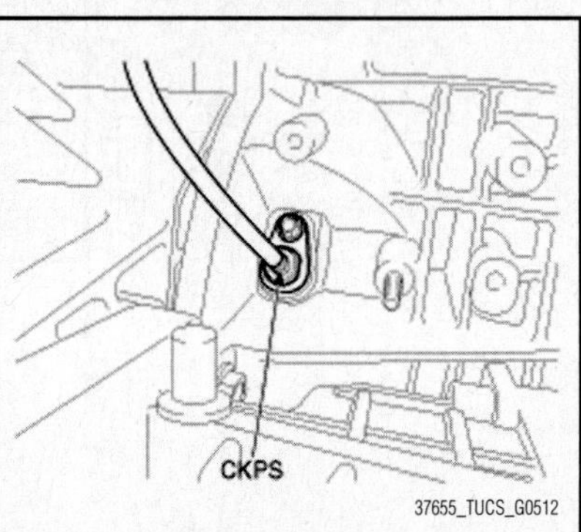

Fig. 273 Crankshaft Position (CKP) sensor location—2010 models, 2.4L engine

6. Remove the sensor from its mounting.

To install:

7. Installation is the reverse of the removal procedure.
8. Apply engine oil to the O-ring.
9. Tighten the sensor retaining bolt to: 86–104 inch lbs. (10–12 Nm).

ELECTRONIC CONTROL MODULE (ECM)

LOCATION

See Figures 274 through 276.

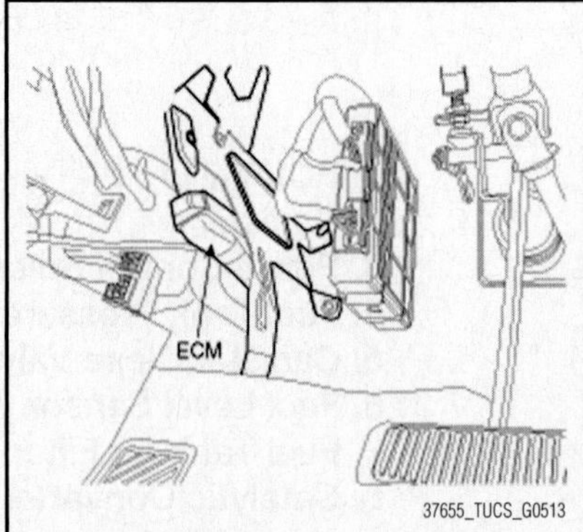

Fig. 274 ECM/PCM location—2009 models, 2.0L engine

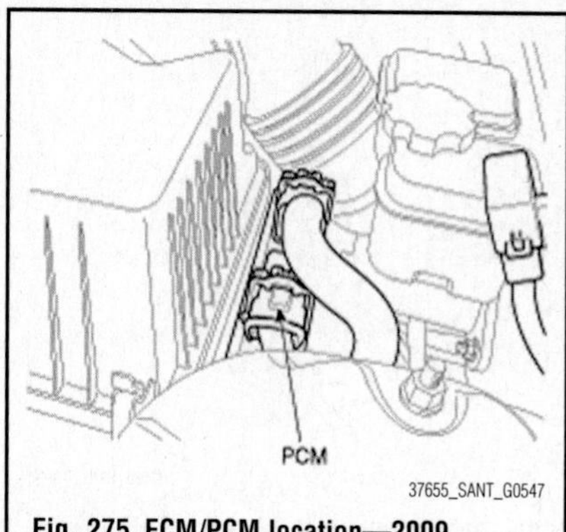

Fig. 275 ECM/PCM location—2009 models, 2.7L engine

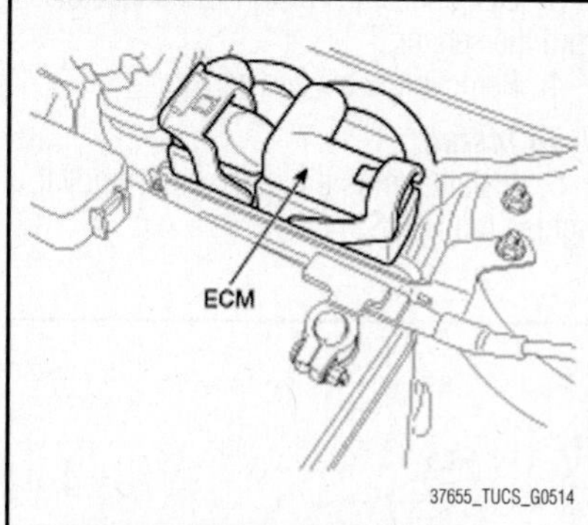

Fig. 276 ECM/PCM location—2010 models, 2.4L engine

REMOVAL & INSTALLATION

1. Before servicing the vehicle, refer to the Precautions Section.
2. Turn ignition switch **OFF**.
3. Disconnect the negative battery cable from the battery.
4. Disconnect the ECM connector.
5. For 2.4L engine, remove the battery.
6. Remove the ECM mounting bolts/nuts and remove the ECM from the vehicle.

To install:

7. Installation is the reverse of the removal.
8. Tighten the ECM mounting bolts to: 70–104 inch lbs. (8–12 Nm).

WARNING

When replacing the ECM, be careful to use the right part number, as damage to the injection system could occur.

RESET

2009 Models

➡Reprogramming of the VIN will require a scan tool.

➡When replacing an ECM, the VIN must be programmed in the ECM. If there is no VIN in ECM memory, the fault code (DTC P0630) is set.

➡The programmed VIN cannot be changed. When writing the VIN, confirm the VIN carefully.

1. Select "Vehicle" and "Engine".
2. Select "VIN WRITING".
3. Check the ECM status.

➡VIRGIN: VIN is not programmed. LEARNT: VIN has been already programmed

4. Is the PCM status "VIRGIN"? If yes, go to next step. If no, end.
5. Write the VIN with cursor, function and number keys.

➡Before pressing the "ENTER" key, confirm the VIN again because the programmed VIN cannot be changed.

6. After verifying the written VIN, press the "ENTER" key.
7. Turn the ignition switch OFF, and then turn ON.
8. Verify the programmed VIN in the ECM memory.

2010 Models

➡Reprogramming of the VIN will require a scan tool.

➡When replacing an ECM, the VIN must be programmed in the ECM. If there is no VIN in ECM memory, the fault code (DTC P0630) is set.

➡The programmed VIN cannot be changed. When writing the VIN, confirm the VIN carefully.

1. Select "VIN Writing" function in "Vehicle S/W Management".
2. Select "Write VIN" in "ID Register".

➡Before inputting the VIN, confirm the VIN again because the programmed VIN cannot be changed.

3. Input the VIN.
4. Turn the ignition switch OFF, then back ON.

ENGINE COOLANT TEMPERATURE (ECT) SENSOR

LOCATION

See Figures 277 through 279.

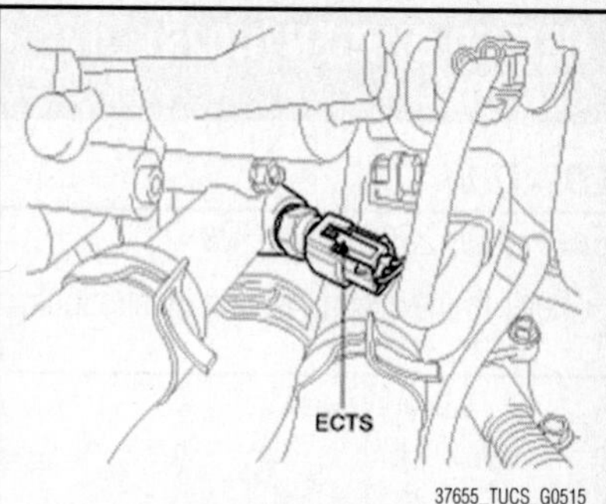

Fig. 277 Engine Coolant Temperature (ECT) Sensor location—2009 models, 2.0L engine

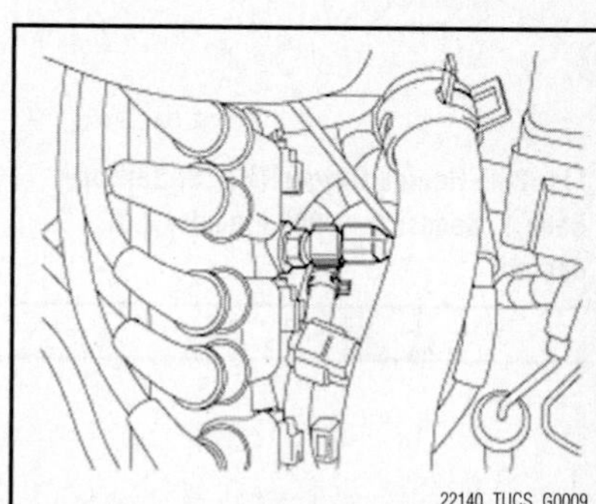

Fig. 278 Engine Coolant Temperature (ECT) sensor location—2009 models, 2.7L engine

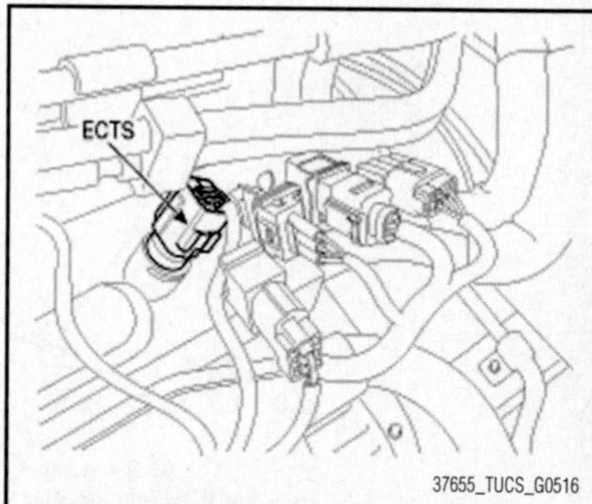

Fig. 279 Engine Coolant Temperature (ECT) Sensor location—2010 models, 2.4L engine

REMOVAL & INSTALLATION

1. Before servicing the vehicle, refer to the Precautions Section.
2. Drain the coolant to a level below the bottom of the sensor.
3. Turn ignition switch **OFF**.
4. Disconnect the negative battery cable from the battery.
5. Disconnect the ECT connector.
6. Remove the fixing clip, as necessary.
7. Remove the ECT.

To install:

8. Installation is the reverse of the removal.

HEATED OXYGEN (HO2S) SENSOR

LOCATION

See Figures 280 through 288.

Refer to the accompanying illustrations.

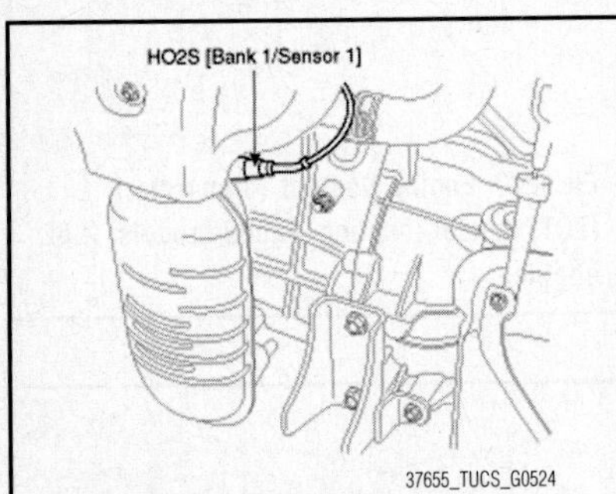

Fig. 280 Heated Oxygen (HO2S) Sensor—Bank 1, Sensor 1—2009 models, 2.0L engine

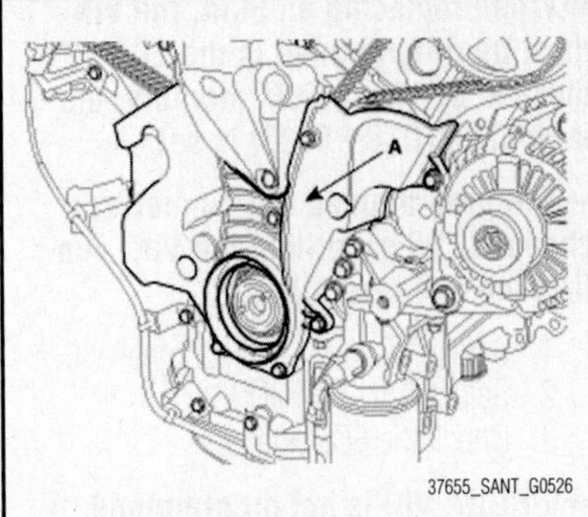

Fig. 282 Heated Oxygen (HO2S) sensor—Bank 1, Sensor 1—2.7L engine

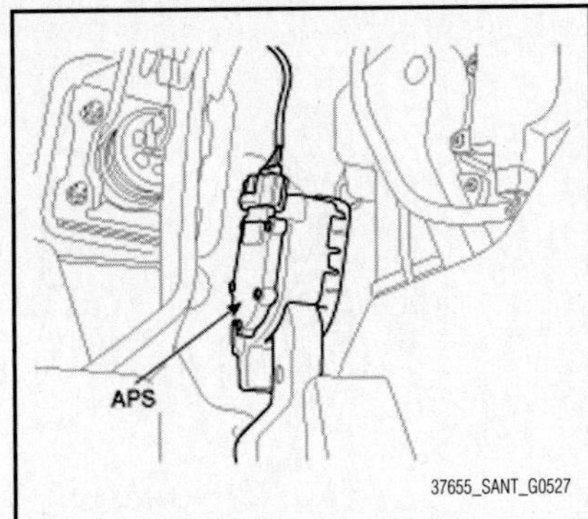

Fig. 283 Heated Oxygen (HO2S) sensor—Bank 1, Sensor 2—2.7L engine

REMOVAL & INSTALLATION

1. Before servicing the vehicle, refer to the Precautions Section.
2. Remove the console assembly, if necessary (2010 models, Bank 1, Sensor 2, ULEV).
3. Disconnect the electrical connector from the sensor.
4. Remove the oxygen sensor.

To install:

5. Installation is the reverse of removal. Tighten the HO2S:

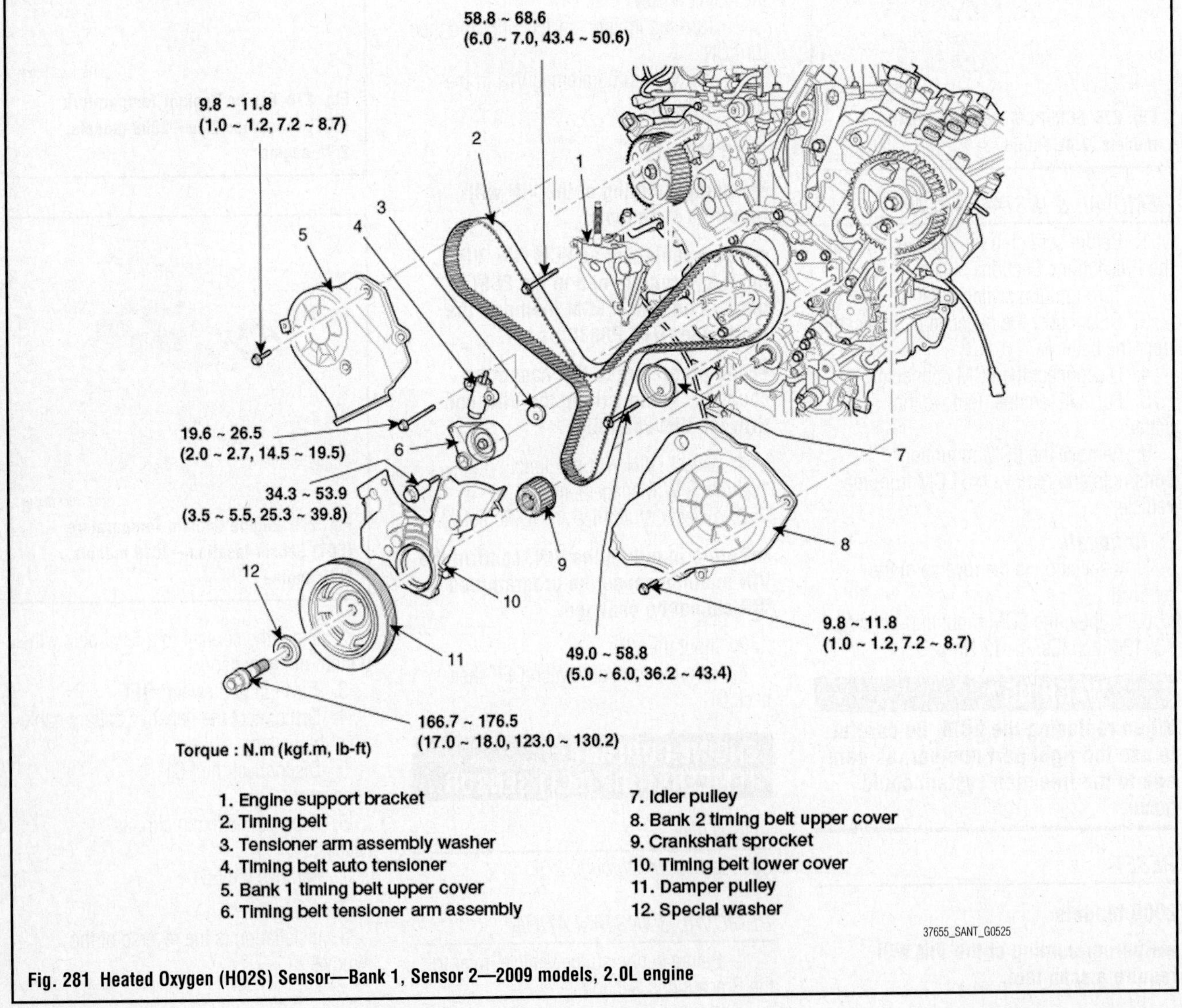

Fig. 281 Heated Oxygen (HO2S) Sensor—Bank 1, Sensor 2—2009 models, 2.0L engine

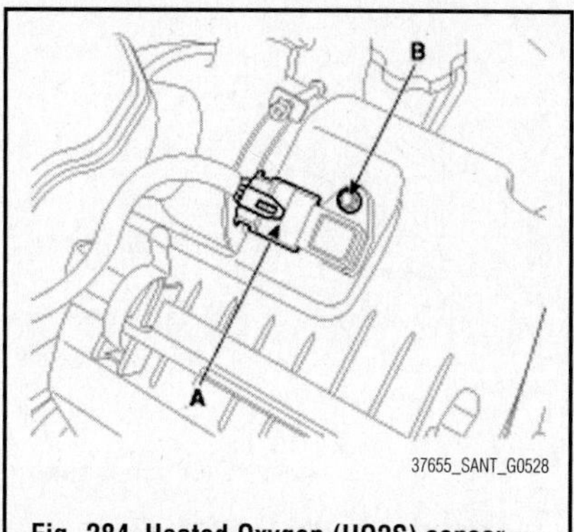

Fig. 284 Heated Oxygen (HO2S) sensor—Bank 2, Sensor 1—2.7L engine

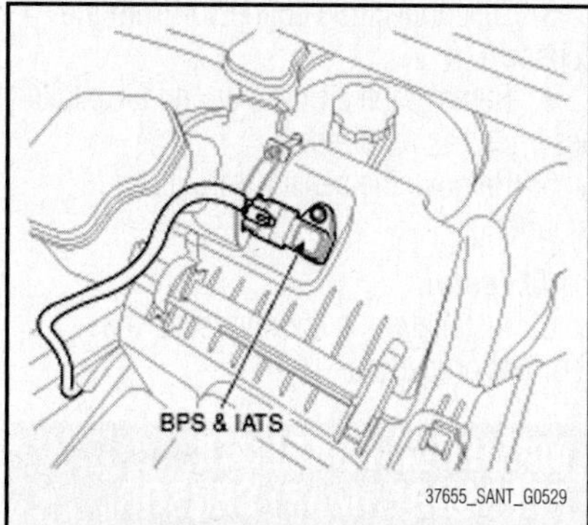

Fig. 285 Heated Oxygen (HO2S) sensor—Bank 2, Sensor 2—2.7L engine

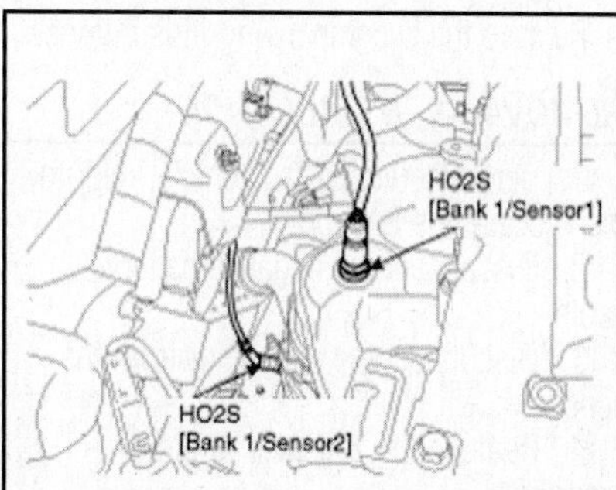

Fig. 286 Heated Oxygen (HO2S) sensor—2010 models, 2.4L engine, SULEV

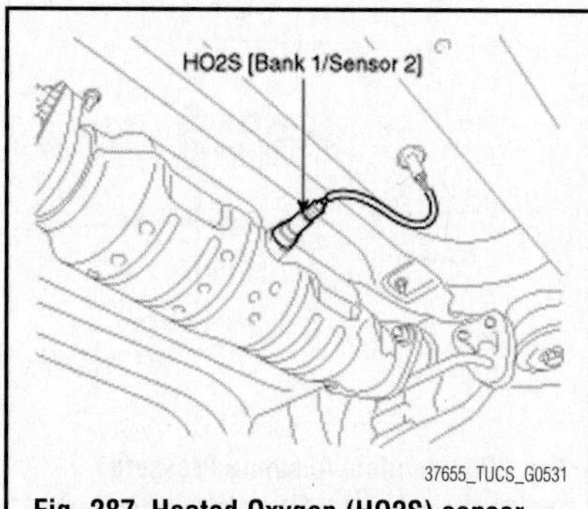

Fig. 287 Heated Oxygen (HO2S) sensor—Bank 1, Sensor 1—2010 models, 2.4L engine, ULEV

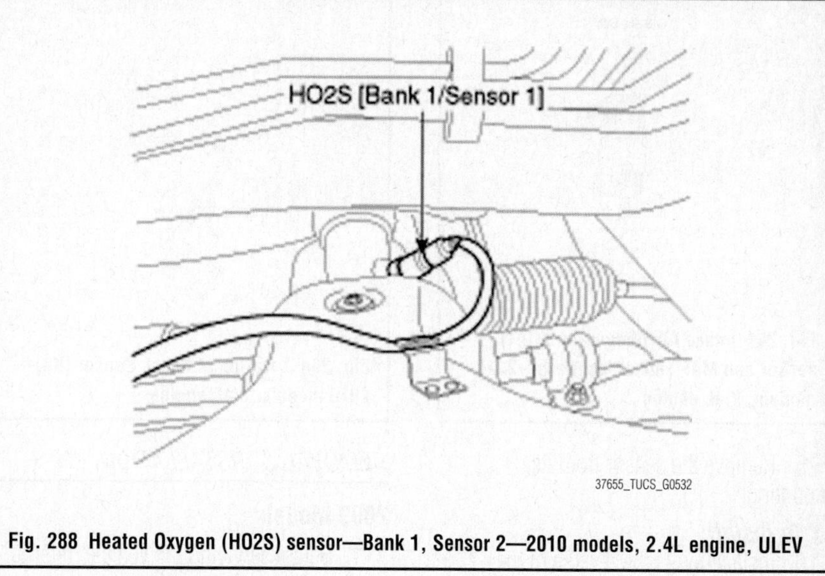

Fig. 288 Heated Oxygen (HO2S) sensor—Bank 1, Sensor 2—2010 models, 2.4L engine, ULEV

- 2.0L, 2.7L and 2.4L SULEV engines: 29–36 ft. lbs. (39–49 Nm)
- 2.4L ULEV engine: 33–36 ft. lbs. (44–49 Nm)

➡Apply anti-seize compound to the threaded portion of the sensor, prior to installation. Never apply anti-seize compound to the protector of the sensor.

INTAKE AIR TEMPERATURE (IAT) SENSOR

LOCATION

See Figures 289 through 291.

Refer to the accompanying illustrations.

REMOVAL & INSTALLATION

1. Before servicing the vehicle, refer to the Precautions Section.
2. Disconnect the negative battery cable.
3. Disconnect the connector from the sensor.
4. Remove the sensor retaining screws.

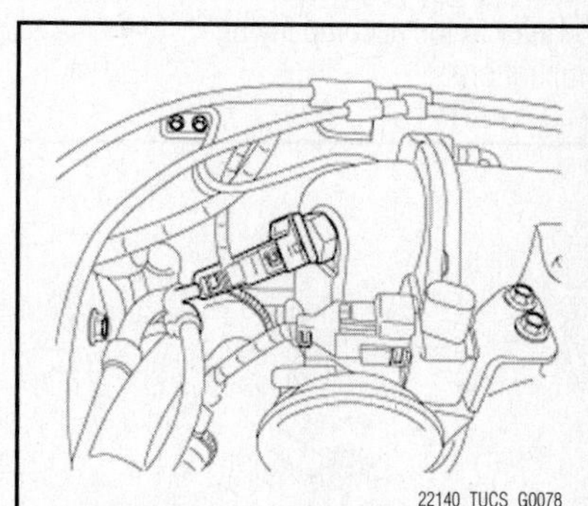

Fig. 290 Intake Air Temperature (IAT) sensor location—2009 models, 2.7L engine

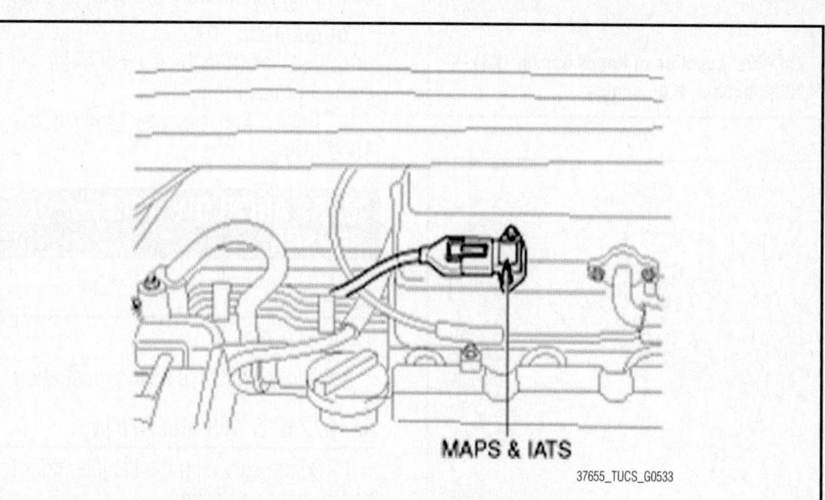

Fig. 289 Intake Air Temperature (IAT) sensor and MAP sensor locations—2009 models, 2.0L engine

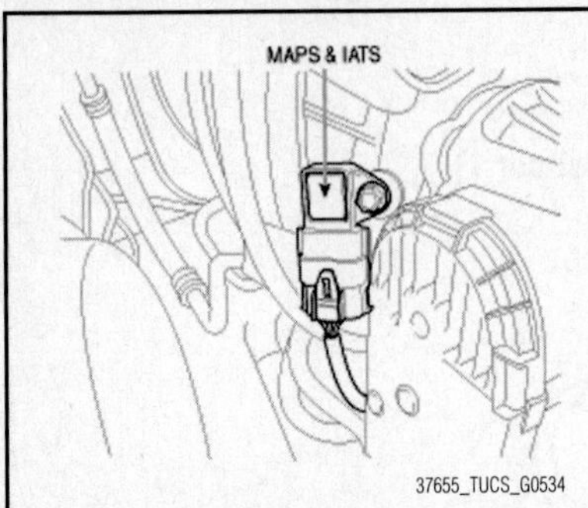

Fig. 291 Intake Air Temperature (IAT) sensor and MAP sensor locations—2010 models, 2.4L engine

5. Remove the sensor from its mounting.

To install:

6. Installation is the reverse of the removal procedure.

KNOCK SENSOR (KS)

LOCATION

See Figures 292 through 294.

Refer to the accompanying illustrations.

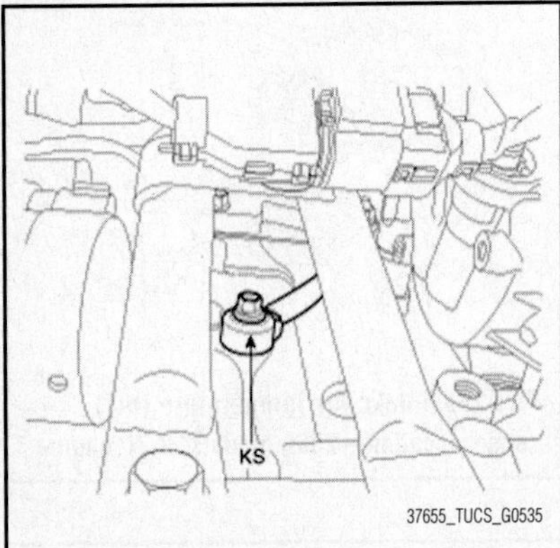

Fig. 292 Location of Knock Sensor (KS)—2009 models, 2.0L engine

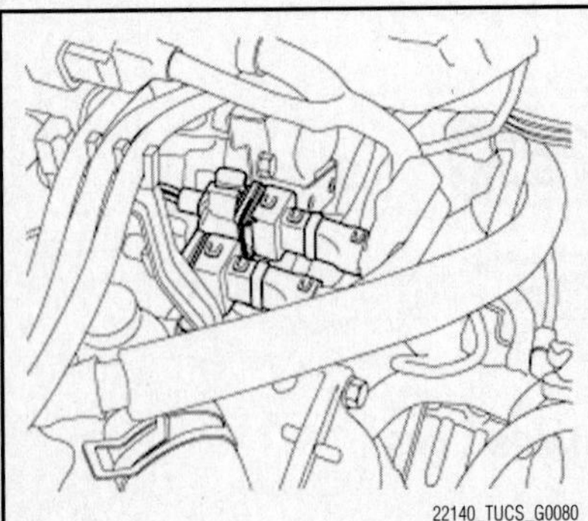

Fig. 293 Location of Knock Sensor (KS)—2009 models, 2.7L engine

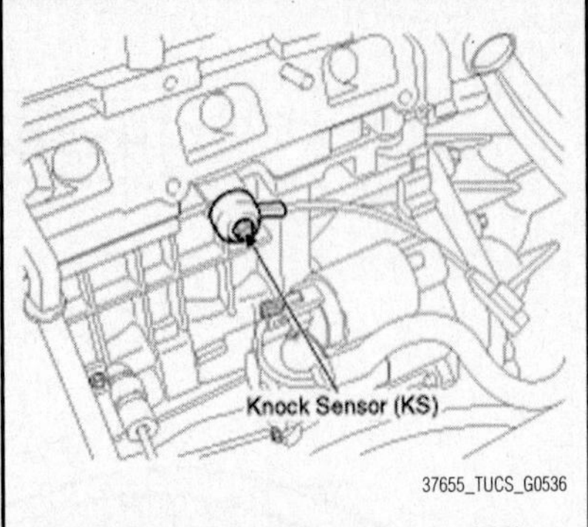

Fig. 294 Location of Knock Sensor (KS)—2010 models, 2.4L engine

REMOVAL & INSTALLATION

2009 Models

1. Before servicing the vehicle, refer to the Precautions Section.
2. Disconnect the negative battery cable.
3. Disconnect the sensor connector.
4. Remove the sensor from its mounting.

To install:

5. Installation is the reverse of the removal procedure.
6. Tighten the sensor to 12–17 ft. lbs. (16–24 Nm).

2010 Models

1. Before servicing the vehicle, refer to the Precautions Section.
2. Turn the ignition switch OFF and disconnect the battery negative cable.
3. Drain the engine coolant.
4. Remove the radiator upper hose.
5. Disconnect the knock sensor connector.
6. Remove the intake manifold.
7. Remove the installation bolt, and then remove the sensor from the cylinder block.

To install:

8. Installation is the reverse of the removal procedure.
9. Tighten the sensor to 14–17 ft. lbs. (19–24 Nm).

MASS AIR FLOW (MAF) SENSOR

LOCATION

See Figure 295.

Refer to the accompanying illustration.

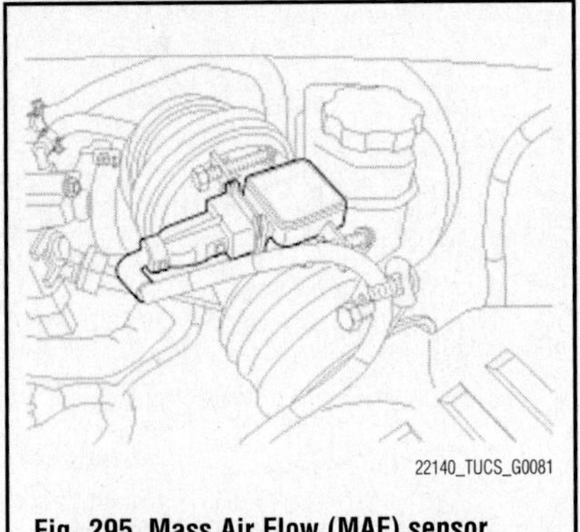

Fig. 295 Mass Air Flow (MAF) sensor location—2009 models, 2.7L engine

REMOVAL & INSTALLATION

1. Before servicing the vehicle, refer to the Precautions Section.
2. Disconnect the negative battery cable.
3. Disconnect the connector from the sensor.
4. Remove the air cleaner and air intake assembly, as required.
5. Remove the sensor from its mounting.

To install:

6. Installation is the reverse of the removal procedure.

MANIFOLD ABSOLUTE PRESSURE (MAP) SENSOR

LOCATION

See Figures 296 and 297.

Refer to the accompanying illustrations.

REMOVAL & INSTALLATION

1. Before servicing the vehicle, refer to the Precautions Section.
2. Disconnect the negative battery cable.
3. Disconnect the connector from the sensor.
4. Remove the sensor from its mounting.

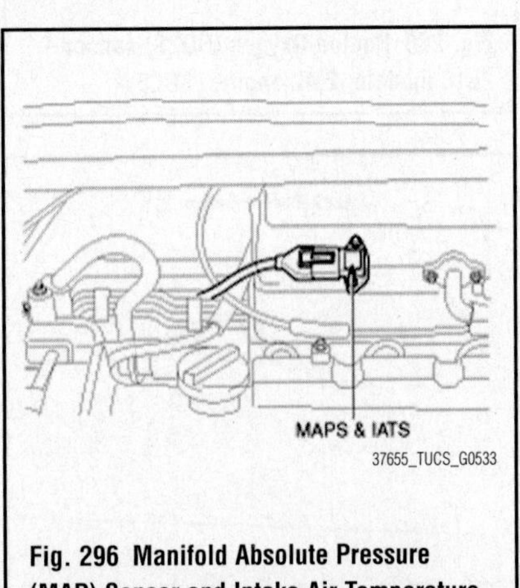

Fig. 296 Manifold Absolute Pressure (MAP) Sensor and Intake Air Temperature (IAT) sensor locations—2009 models, 2.0L engine

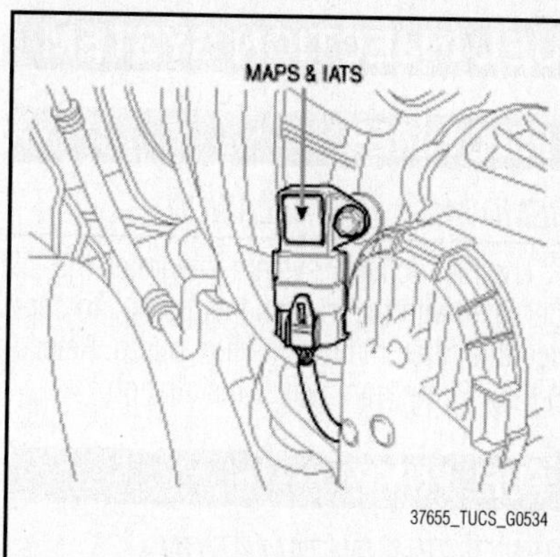

Fig. 297 Manifold Absolute Pressure (MAP) Sensor and Intake Air Temperature (IAT) sensor locations—2010 models, 2.4L engine

To install:

5. Installation is the reverse of the removal procedure. Tighten the mounting bolt to 86–104 inch lbs. (10–12 Nm).

THROTTLE POSITION SENSOR (TPS)

LOCATION

See Figures 298 through 300.

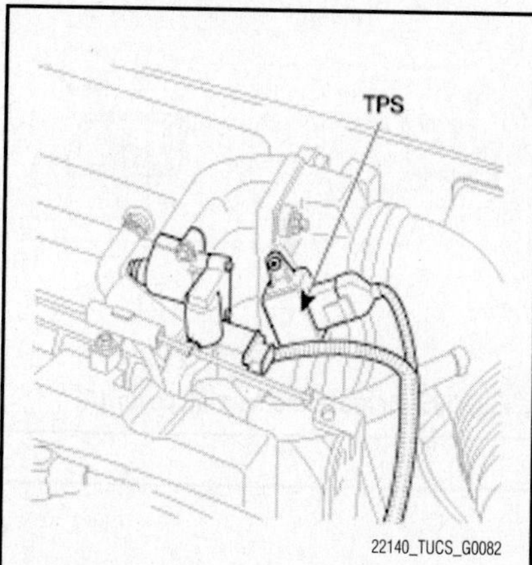

Fig. 298 Throttle Position Sensor (TPS) location—2009 models, 2.0L engine

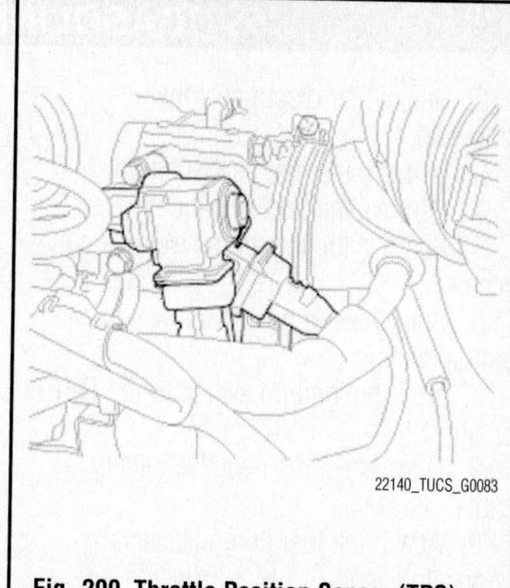

Fig. 299 Throttle Position Sensor (TPS) location—2009 models, 2.7L engine

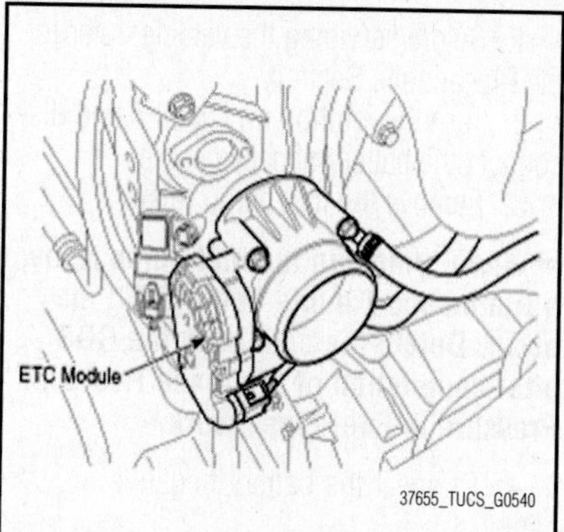

Fig. 300 Throttle Position Sensor (TPS) location—2010 models, 2.4L engine

Refer to the accompanying illustrations.

REMOVAL & INSTALLATION

2009 Models

1. Before servicing the vehicle, refer to the Precautions Section.
2. Disconnect the negative battery cable.
3. Disconnect the sensor connector.
4. Remove the sensor retaining screws.
5. Remove the sensor from its mounting.

To install:

6. Installation is the reverse of the removal procedure.

2010 Models—2.4L Engine

See Figure 301.

1. Before servicing the vehicle, refer to the Precautions Section.
2. Turn the ignition switch OFF and disconnect the battery negative cable.
3. Remove the resonator and the air intake hose.
4. Disconnect the ETC module connector.
5. Disconnect the coolant hoses.
6. Remove the installation bolts, and then remove the ETC module from the engine.

To install:

7. Installation is reverse of removal. Tighten the ETC module mounting bolt to 70–104 inch lbs. (8–12 Nm)

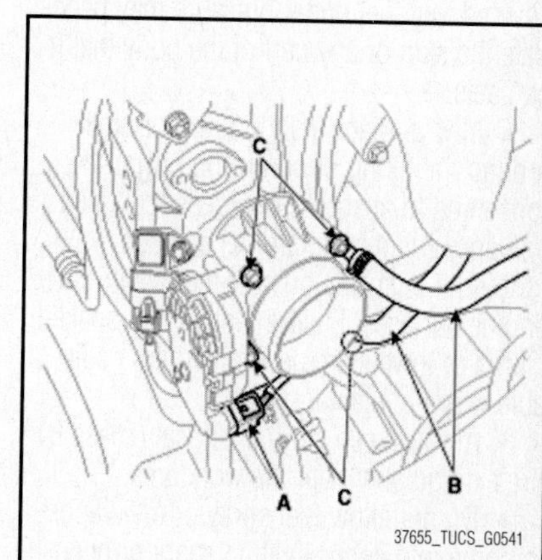

Fig. 301 Disconnect the ETC module connector (A) and coolant hoses (B), and remove the ETC module installation bolts (D)

FUEL SYSTEM SERVICE PRECAUTIONS

Safety is the most important factor when performing not only fuel system maintenance but any type of maintenance. Failure to conduct maintenance and repairs in a safe manner may result in serious personal injury or death. Maintenance and testing of the vehicle's fuel system components can be accomplished safely and effectively by adhering to the following rules and guidelines.

- To avoid the possibility of fire and personal injury, always disconnect the negative battery cable unless the repair or test procedure requires that battery voltage be applied.
- Always relieve the fuel system pressure prior to disconnecting any fuel system component (injector, fuel rail, pressure regulator, etc.), fitting or fuel line connection. Exercise extreme caution whenever relieving fuel system pressure to avoid exposing skin, face and eyes to fuel spray. Please be advised that fuel under pressure may penetrate the skin or any part of the body that it contacts.
- Always place a shop towel or cloth around the fitting or connection prior to loosening to absorb any excess fuel due to spillage. Ensure that all fuel spillage (should it occur) is quickly removed from engine surfaces. Ensure that all fuel soaked cloths or towels are deposited into a suitable waste container.
- Always keep a dry chemical (Class B) fire extinguisher near the work area.
- Do not allow fuel spray or fuel vapors to come into contact with a spark or open flame.
- Always use a back-up wrench when loosening and tightening fuel line connection fittings. This will prevent unnecessary stress and torsion to fuel line piping.
- Always replace worn fuel fitting O-rings with new Do not substitute fuel hose or equivalent where fuel pipe is installed.

Before servicing the vehicle, make sure to also refer to the precautions in the beginning of this section as well.

RELIEVING FUEL SYSTEM PRESSURE

2009 Models

1. Before servicing the vehicle, refer to the Precautions Section.
2. Remove or disconnect the following:
3. Remove the rear seat cushion.
4. Remove the access panel.
5. Remove the fuel pump module connector.
6. Start the engine and allow it to run until it stalls.
7. Turn the ignition switch to the **OFF**-position.
8. Disconnect the negative battery cable.
9. Attach the fuel pump harness connector.

2010 Models

See Figure 304.

1. Before servicing the vehicle, refer to the Precautions Section.
2. Turn the ignition switch OFF and disconnect the battery negative cable.
3. Remove the fuel pump relay.

➡When removing the fuel pump relay, a Diagnostic Trouble Code (DTC) may occur. Delete the code with the GDS after completion of "Release Residual Pressure in Fuel Line" work.

4. Connect the battery negative cable.
5. Start the engine and let idle, and then turn the ignition switch OFF after the engine has stopped on its own.
6. Disconnect the battery negative cable, and then install the fuel pump relay.
7. Connect the battery negative cable.
8. Delete the Diagnostic Trouble Code (DTC) related the fuel pump relay with the GDS.

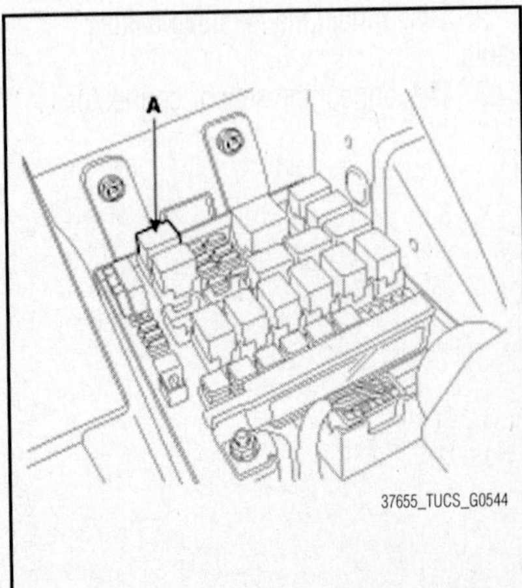

Fig. 304 Remove the fuel pump relay (A)

FUEL FILTER

REMOVAL & INSTALLATION

The fuel delivery system integrates the fuel filter with the in-tank fuel pump. To service this filter, remove the fuel pump. Refer to Fuel Pump, removal & installation.

FUEL PUMP

REMOVAL & INSTALLATION

2009 Models

See Figures 305 through 307.

1. Before servicing the vehicle, refer to the Precautions Section.
2. Remove the rear seat.
3. Remove the fuel pump service cover, disconnect the fuel pump connector, and relieve the fuel system pressure. Refer to Relieving Fuel System Pressure.
4. Disconnect the fuel feed tube quick-connector, and the suction tube quick-connector.

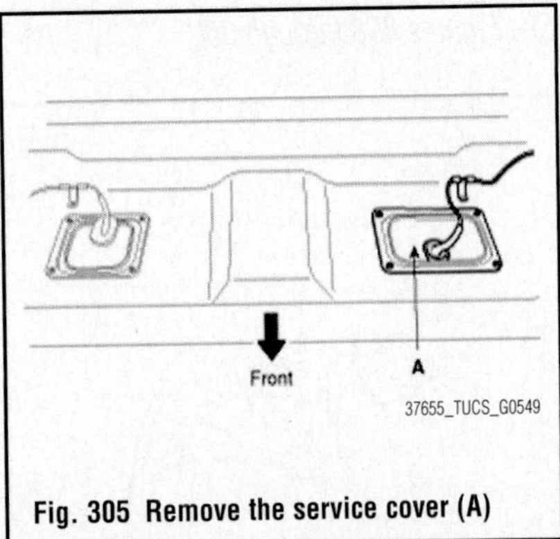

Fig. 305 Remove the service cover (A)

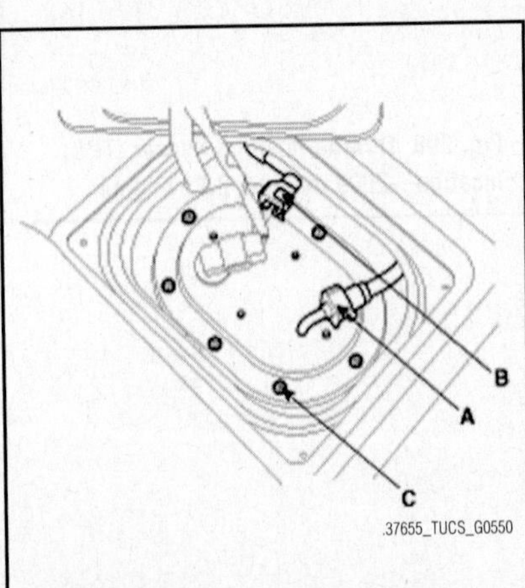

Fig. 306 Disconnect the fuel feed tube quick-connector (A) and the suction tube quick-connector (B) and remove the bolts (C)

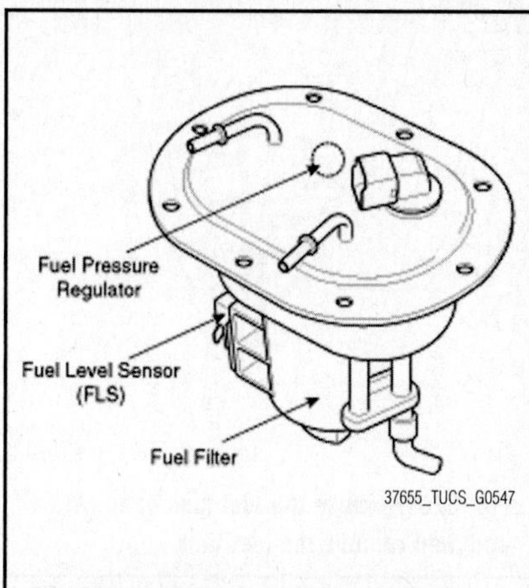

Fig. 307 Fuel pump assembly module

5. Remove the fuel pump installation bolts and remove the fuel pump assembly.

To install:

6. Installation is reverse of removal.
7. Tighten the fuel pump mounting bolts to 17–26 inch lbs. (2–3 Nm).

2010 Models

See Figures 308 through 311.

1. Before servicing the vehicle, refer to the Precautions Section.

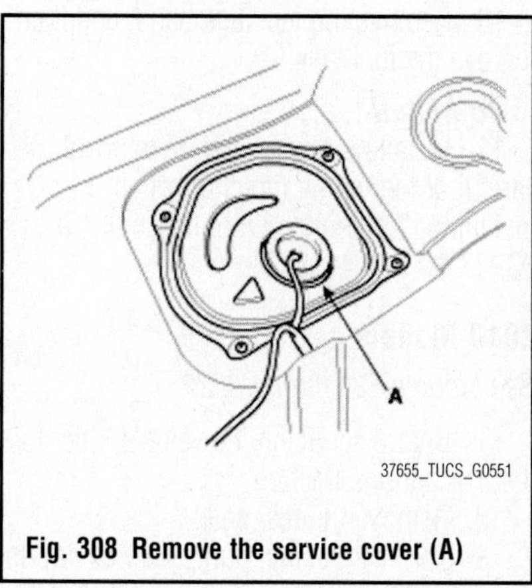

Fig. 308 Remove the service cover (A)

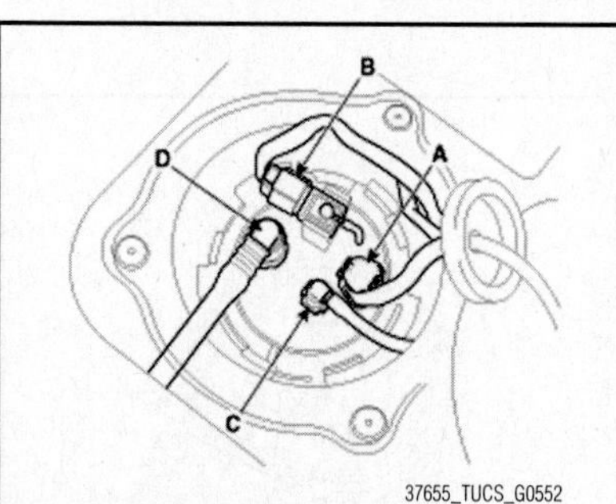

Fig. 309 Disconnect the fuel pump connector (A), fuel tank pressure sensor connector (B), fuel feed tube quick connector (C) and the vapor tube quick-connector (D)

2. Remove the rear seat.
3. Remove the fuel pump service cover, disconnect the fuel pump connector, and relieve the fuel system pressure. Refer to Relieving Fuel System Pressure.
4. Disconnect the fuel pump connector and the fuel tank pressure sensor connector.
5. Disconnect the fuel feed tube quick connector and the vapor tube quick-connector.
6. Remove locking ring with SST (No. 09310-2S200).
7. Remove the fuel pump from the fuel tank.

To install:

8. Installation is reverse of removal. Be careful of fuel pump direction when installing.

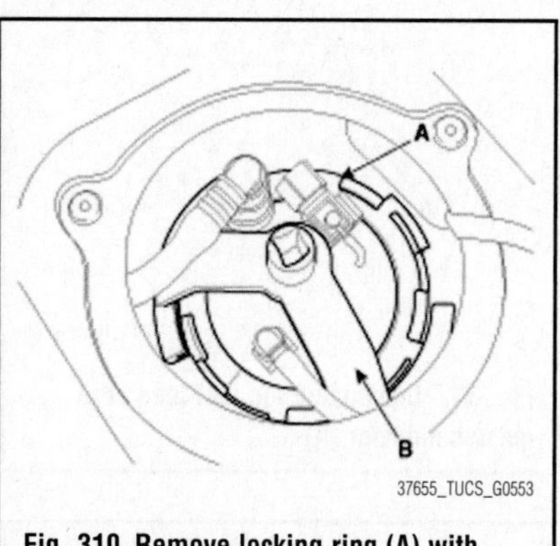

Fig. 310 Remove locking ring (A) with SST (No. 09310-2S200)

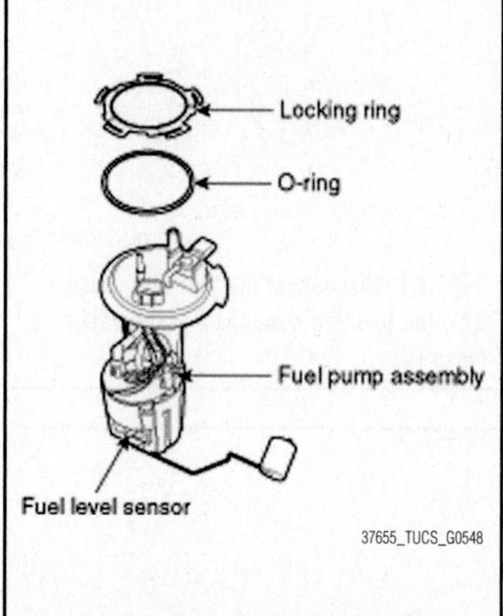

Fig. 311 Fuel pump assembly module

FUEL RAIL AND INJECTOR

REMOVAL & INSTALLATION

2009 Models

See Figures 312 and 313.

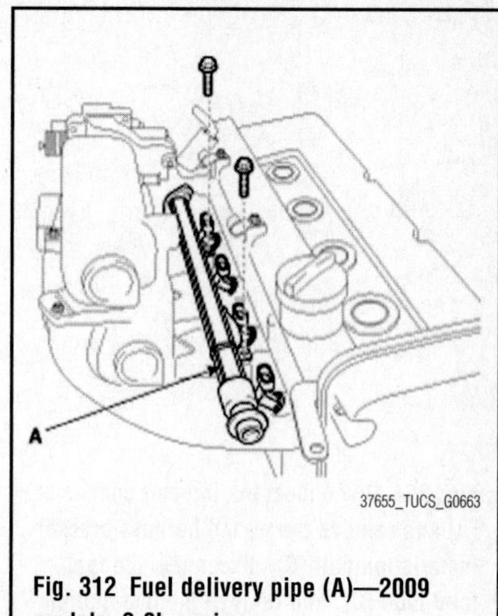

Fig. 312 Fuel delivery pipe (A)—2009 models, 2.0L engine

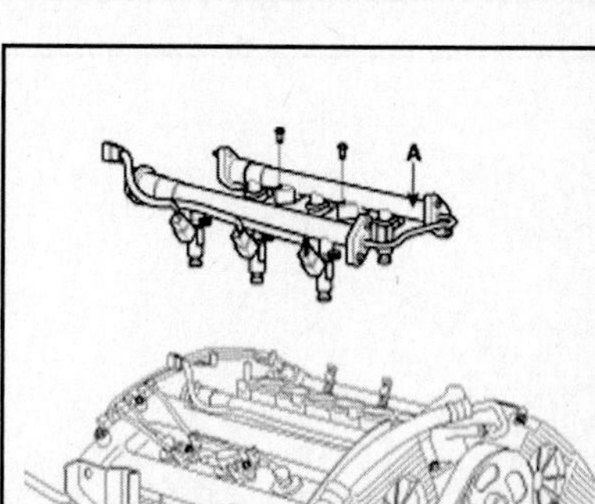

Fig. 313 Fuel injector assembly (A)—2009 models, 2.7L engine

1. Before servicing the vehicle, refer to the Precautions Section.
2. Relieve the fuel system pressure. Refer to Relieving Fuel System Pressure.
3. Disconnect the negative battery cable.
4. Remove the intake manifold. Refer to Intake Manifold Removal & Installation in the Engine Mechanical section.
5. Remove the fuel rail.

To install:

6. Installation is the reverse of removal.
7. Start the engine and check for leaks and for proper operation.

2010 Models

See Figures 314 and 315.

1. Before servicing the vehicle, refer to the Precautions Section.
2. Relieve the fuel system pressure. Refer to Relieving Fuel System Pressure.
3. Disconnect the negative battery cable.
4. Disconnect the injector connector.

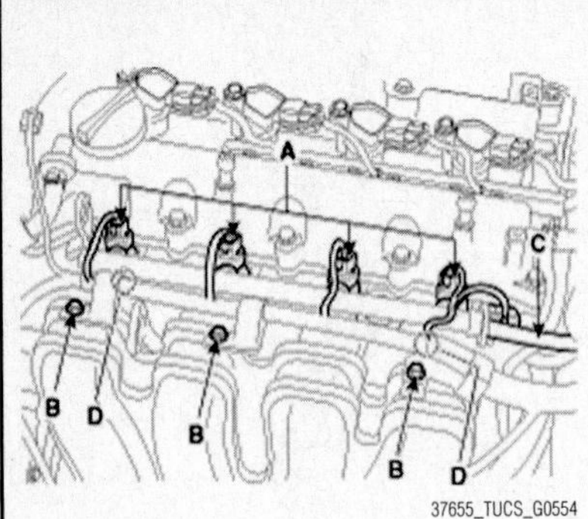

Fig. 314 Disconnect the injector connector (A) and remove the wiring harness bracket installation bolt (B), disconnect the fuel feed tube (C), and remove the installation bolt (D)

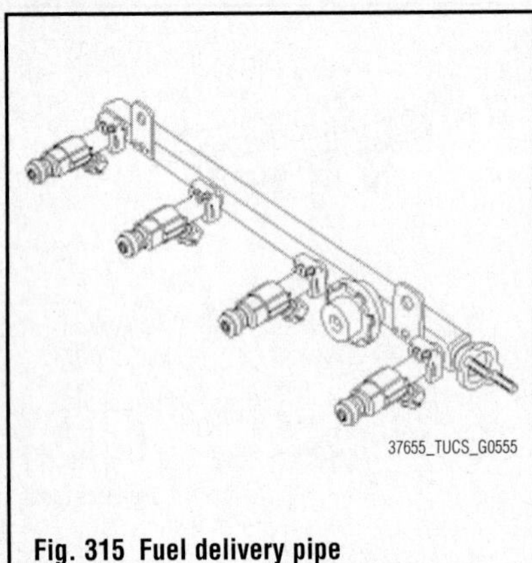

Fig. 315 Fuel delivery pipe

5. Remove the wiring harness bracket installation bolt.
6. Remove the installation nut, and then disconnect the fuel feed tube.
7. Remove the installation bolt, and then remove the delivery pipe and injector assembly from the engine.

To install:

8. Installation is reverse of removal, noting the following:
   a. Tighten the installation bolt to 14–17 ft. lbs. (19–24 Nm)
   b. Tighten the installation nut-to-fuel feed tube to 70–86 ft. lbs. (8–10 Nm)

FUEL TANK

REMOVAL & INSTALLATION

2009 Models

See Figures 316 through 320.

1. Before servicing the vehicle, refer to the Precautions Section.
2. Remove the rear seat.
3. Remove the fuel pump service cover, disconnect the fuel pump connector, and

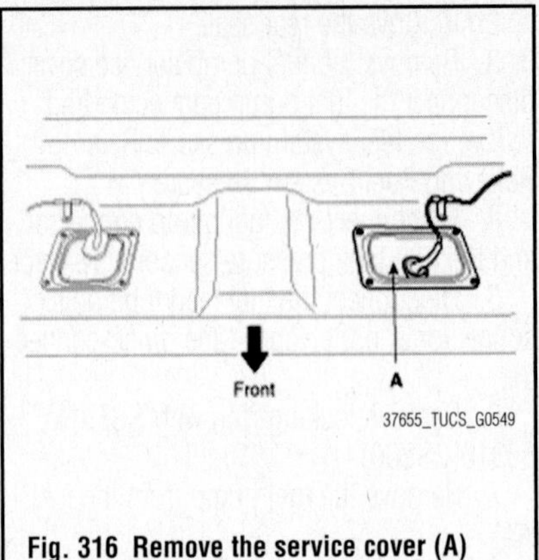

Fig. 316 Remove the service cover (A)

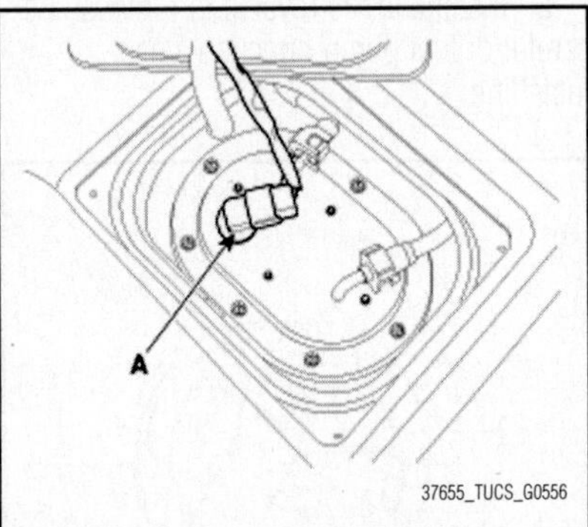

Fig. 317 Disconnect the fuel feed tube quick-connector (A)

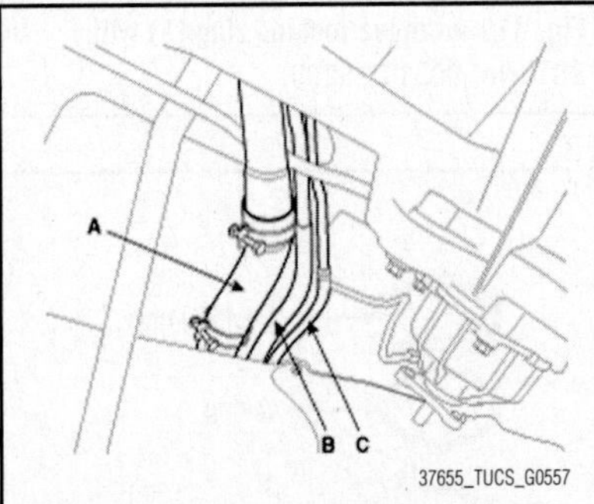

Fig. 318 Disconnect the fuel filler hose (A), the leveling hose (B) and the vapor hose (C)

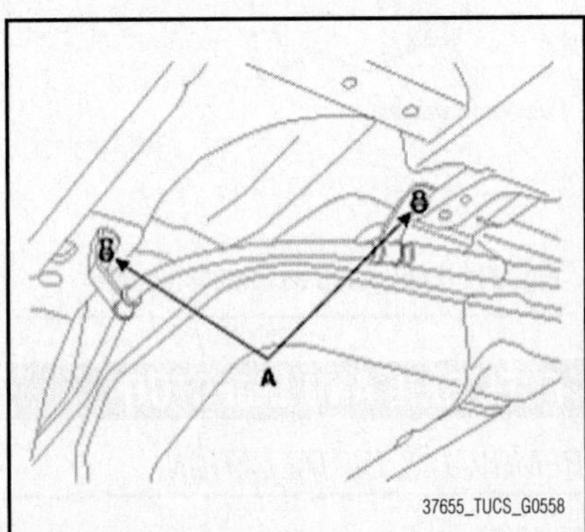

Fig. 319 Remove the four parking brake installation bolts (A)

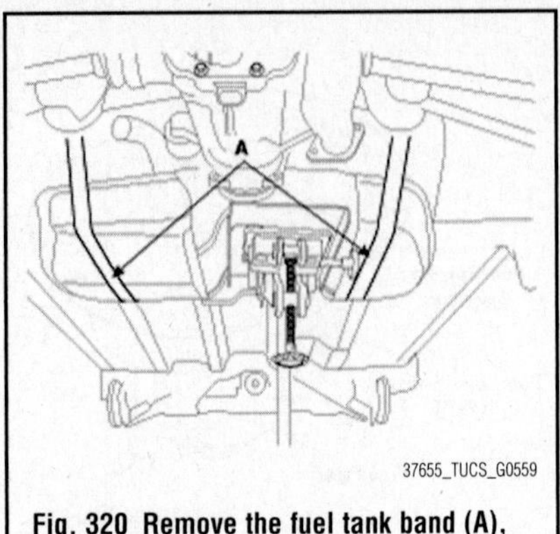

Fig. 320 Remove the fuel tank band (A), and then remove the fuel tank

relieve the fuel system pressure. Refer to Relieving Fuel System Pressure.

4. Disconnect the fuel feed tube quick-connector.
5. Lift the vehicle and support the fuel tank with a jack.
6. Remove the front and main muffler.
7. For 4WD vehicles, remove the propeller shaft.
8. Disconnect the fuel filler hose, the leveling hose and the vapor hose.
9. Remove the four parking brake installation bolts.
10. Remove the fuel tank band, and then remove the fuel tank.

To install:

11. Installation is reverse of removal. Be careful of fuel pump direction when installing. Tighten the mounting bolts to 33–43 ft. lbs. (44–59 Nm).

2010 Models

See Figures 321 through 326.

1. Before servicing the vehicle, refer to the Precautions Section.
2. Remove the rear seat.
3. Remove the fuel pump service cover, disconnect the fuel pump connector, and

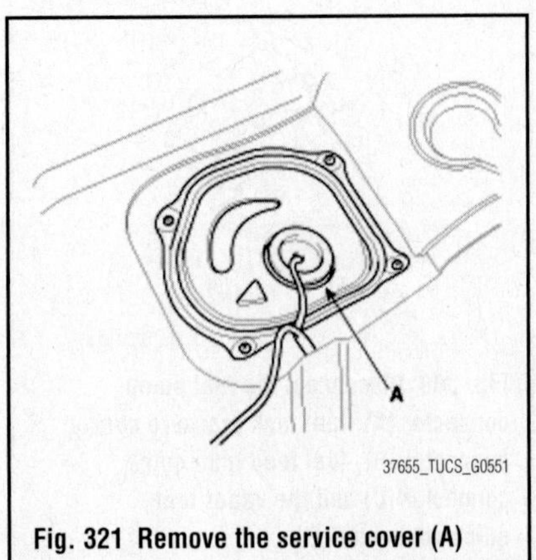

Fig. 321 Remove the service cover (A)

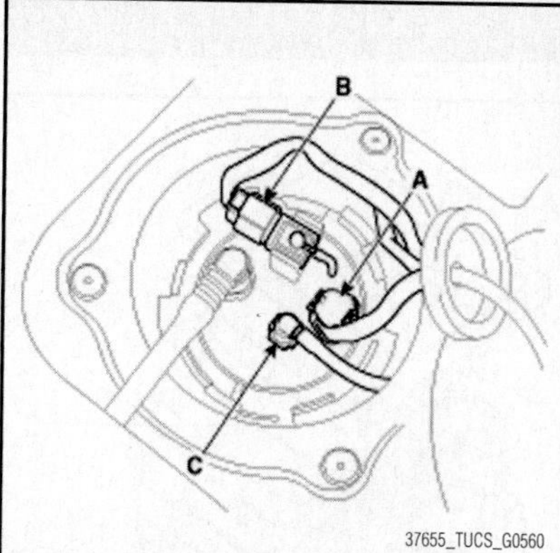

Fig. 322 Disconnect the fuel pump connector (A), fuel tank pressure sensor connector (B), and the fuel feed tube quick connector (C)

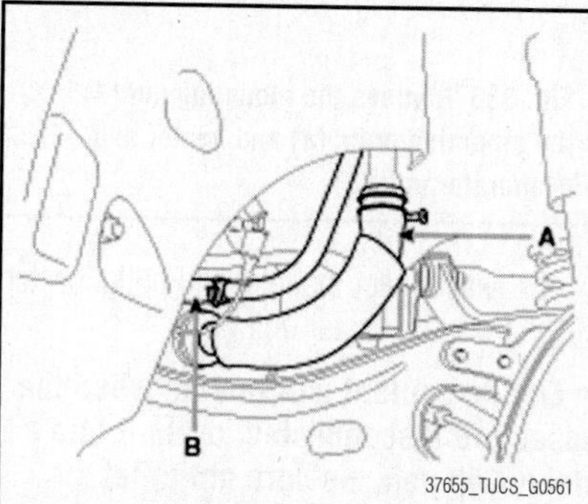

Fig. 323 Disconnect the fuel filler hose (A) and the ventilation hose (B)

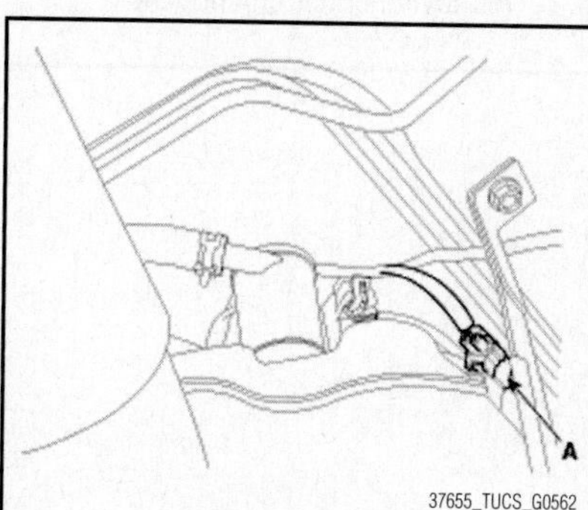

Fig. 324 Disconnect the canister close valve connector (A)

relieve the fuel system pressure. Refer to Relieving Fuel System Pressure.

4. Disconnect the fuel pump connector and the fuel tank pressure sensor connector.
5. Disconnect the fuel feed tube quick connector and the vapor tube quick-connector.
6. Remove the left rear wheel and tire.
7. Lift the vehicle and support the fuel tank with a jack.

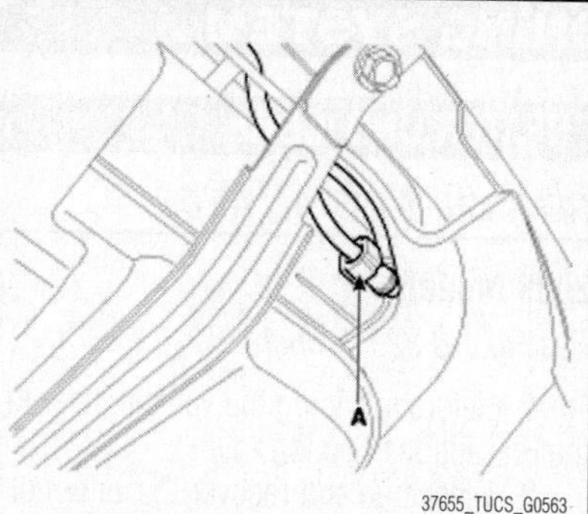

Fig. 325 Disconnect the vapor tube quick-connector (A)

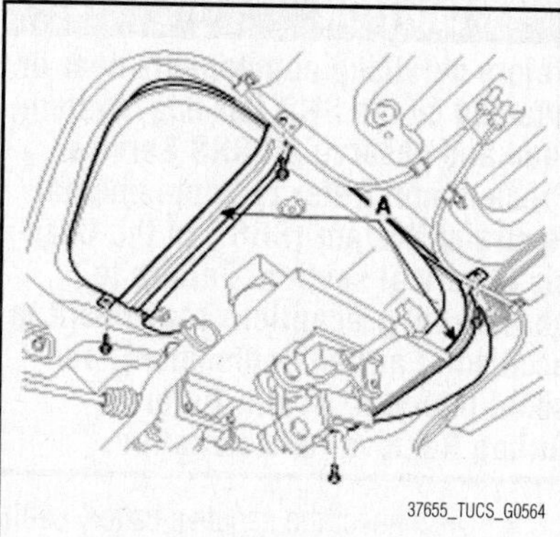

Fig. 326 Disconnect the vapor tube quick-connector (A)

8. Remove the center muffler assembly.
9. For 4WD vehicles, remove the propeller shaft.
10. Disconnect the fuel filler hose and the ventilation hose.
11. Disconnect the canister close valve connector.
12. Disconnect the vapor tube quick-connector.
13. Disconnect the vapor tube quick-connector.

➡**When removing the fuel tank, the fuel tank must be tilted to prevent interfering with coupling.**

To install:

14. Installation is reverse of removal. Be careful of fuel pump direction when installing. Tighten the mounting bolts to 29–40 ft. lbs. (39–54 Nm).

IDLE SPEED

ADJUSTMENT

Idle speed is maintained by the Electronic Control Module (ECM). No adjustment is necessary or possible.

THROTTLE BODY

REMOVAL & INSTALLATION

2009 Models

1. Before servicing the vehicle, refer to the Precautions Section.
2. Turn the ignition **OFF**.
3. Remove the engine cover.
4. Remove the throttle body electrical connector.
5. Remove the throttle body bolts.
6. Remove the throttle body and gasket.

To install:

7. Clean the throttle body gasket mating surfaces.
8. Install the throttle body and NEW gasket.
9. Install the throttle body bolts and tighten to 21 ft. lbs. (28 Nm).
10. Install the throttle body electrical connector.
11. Install the engine cover.

2010 Models

See Figure 327.

1. Before servicing the vehicle, refer to the Precautions Section.
2. Turn the ignition switch OFF and disconnect the battery negative cable.
3. Remove the resonator and the air intake hose.
4. Disconnect the ETC module connector.
5. Disconnect the coolant hoses.
6. Remove the installation bolts, and then remove the ETC module from the engine.

To install:

7. Installation is reverse of removal. Tighten the ETC module mounting bolt to 70–104 inch lbs. (8–12 Nm)

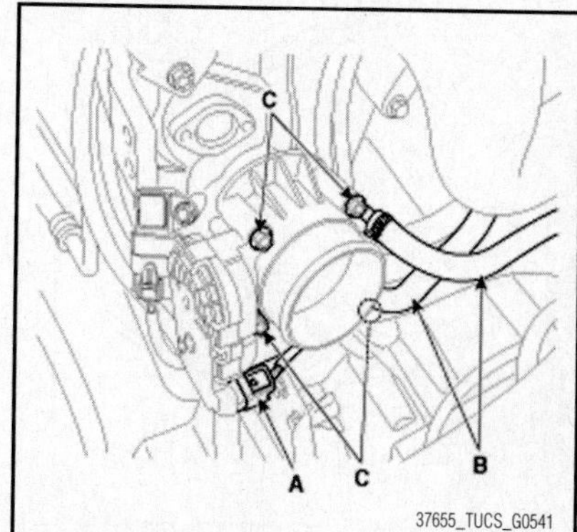

Fig. 327 Disconnect the ETC module connector (A) and coolant hoses (B), and remove the ETC module installation bolts (D)

HEATING & AIR CONDITIONING SYSTEM

BLOWER MOTOR

REMOVAL & INSTALLATION

See Figure 328.

1. Before servicing the vehicle, refer to the Precautions Section.

CAUTION

Before servicing components near or affected by the SRS (air bag) system, read and observe all SRS Service Precautions. Refer to Supplemental Restraint System (SRS), in the Chassis Electrical section. Failure to observe all precautions may result in accidental airbag deployment, personal injury, or death. Refer to Airbag Removal & Installation.

2. Disconnect the negative battery cable and wait at least 3 minutes for the SRS memory to drain.
3. Remove the instrument panel lower section.
4. Disconnect the blower motor connector.
5. Remove the mounting screws and the blower motor.

To install:

6. Installation is the reverse of the removal procedure.

HEATER UNIT/CORE

REMOVAL & INSTALLATION

2009 Models

See Figures 329 through 331.

1. Before servicing the vehicle, refer to the Precautions Section.
2. Discharge and recover the air conditioning system refrigerant.
3. Drain the engine coolant into a clean container for reuse.

CAUTION

Before servicing components near or affected by the SRS (air bag) system, read and observe all SRS Service Precautions. Refer to Supplemental Restraint System (SRS), in the Chassis Electrical section. Failure to observe all precautions may result in accidental airbag deployment, personal injury, or death. Refer to Airbag Removal & Installation.

4. Disconnect the negative battery cable and wait at least 3 minutes for the SRS memory to drain.
5. Remove the bolts and the expansion valve from the evaporator core. Plug or cap the lines immediately after disconnecting them to avoid moisture and dust contamination.

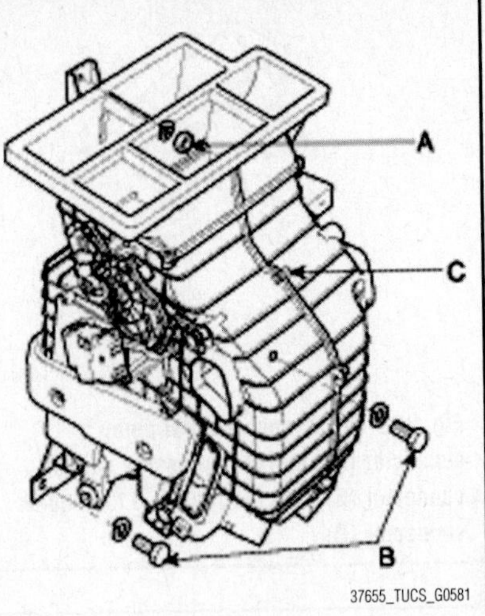

Fig. 330 Remove the mounting nut (A), the mounting bolts (B) and heater and evaporator unit (C)

6. Disconnect the inlet and outlet heater hoses from the heater unit.

➡Engine coolant will run out when the hoses are disconnected; drain it into a clean drip pan. Be sure not to let coolant spill on electrical parts or painted surfaces. If any coolant spills, rinse it off immediately.

7. Remove the instrument panel.

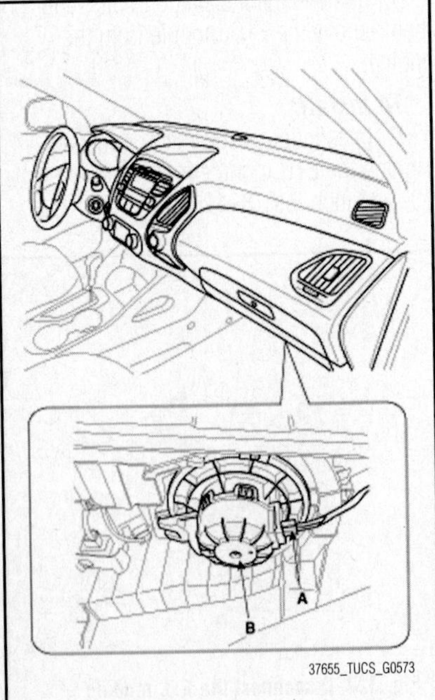

Fig. 328 Blower motor (B) and connector (A)

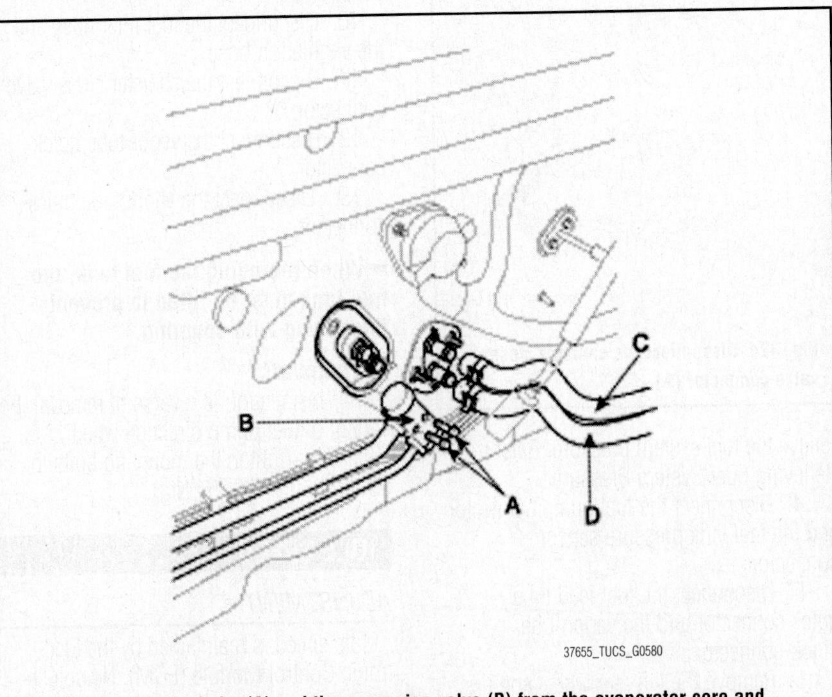

Fig. 329 Remove the bolts (A) and the expansion valve (B) from the evaporator core and disconnect the inlet (C) and outlet (D) heater hoses

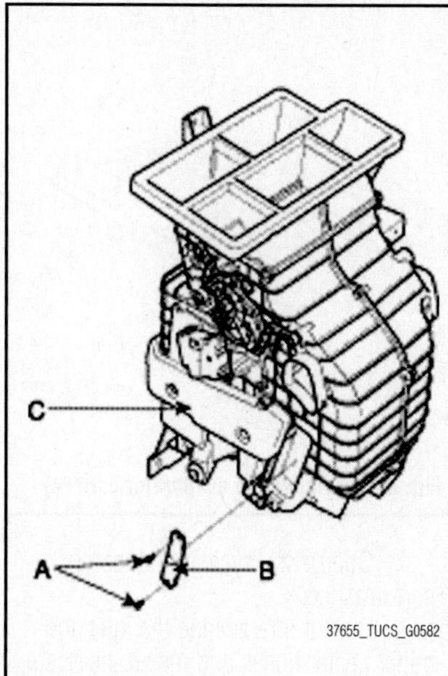

Fig. 331 Remove the self-tapping screws (A), the cover (B) and the side bracket (C)

8. Remove the instrument panel cross-member.

9. Remove the mounting nut, the mounting bolts and heater and evaporator unit.

10. Remove the self-tapping screws and the cover.

11. Remove the side bracket.

12. Remove the clip and lower cover.

➡**Be careful not to bend the inlet and outlet pipes during heater core and evaporator core removal.**

To install:

13. Install the heater core and evaporator core in the reverse order of removal, noting the following:

a. If you're installing a new evaporator, add refrigerant oil.

b. Replace the O-rings with new ones at each fitting, and apply a thin coat of refrigerant oil before installing them. Be sure to use the right O-rings for R-134a to avoid leakage. Immediately after using the oil, replace the cap on the container, and seal it to avoid moisture absorption.

➡**Do not spill the refrigerant oil on the vehicle ; it may damage the paint ; if the refrigerant oil contacts the paint, wash it off immediately.**

c. Apply sealant to the grommets.

d. Make sure that there is no air leakage.

e. Charge the system, and test its performance.

f. Do not interchange the inlet and outlet heater hoses and install the hose clamps securely.

g. Refill the cooling system with engine coolant.

2010 Models

See Figures 332 through 337.

1. Before servicing the vehicle, refer to the Precautions Section.

2. Discharge and recover the air conditioning system refrigerant.

3. Drain the engine coolant into a clean container for reuse.

CAUTION

Before servicing components near or affected by the SRS (air bag) system, read and observe all SRS Service Precautions. Refer to Supplemental Restraint System (SRS), in the Chassis Electrical section. Failure to observe all precautions may result in accidental airbag deployment, personal injury, or death. Refer to Airbag Removal & Installation.

4. Disconnect the negative battery cable and wait at least 3 minutes for the SRS memory to drain.

5. Remove the expansion valve cover.

6. Remove the bolts and the expansion valve from the evaporator core. Plug or cap the lines immediately after disconnecting them to avoid moisture and dust contamination.

7. Disconnect the inlet and outlet heater hoses from the heater unit.

➡**Engine coolant will run out when the hoses are disconnected; drain it into a clean drip pan. Be sure not to let coolant spill on electrical parts or painted surfaces. If any coolant spills, rinse it off immediately.**

8. Remove the instrument panel.

9. Disconnect the connectors from the temperature control actuator, the mode control actuator and the evaporator temperature sensor.

10. Remove the cowl cross bar assembly.

11. Remove the heater and blower unit after removing the mounting bolts.

12. Remove the blower unit from heater unit after removing the 3 screws.

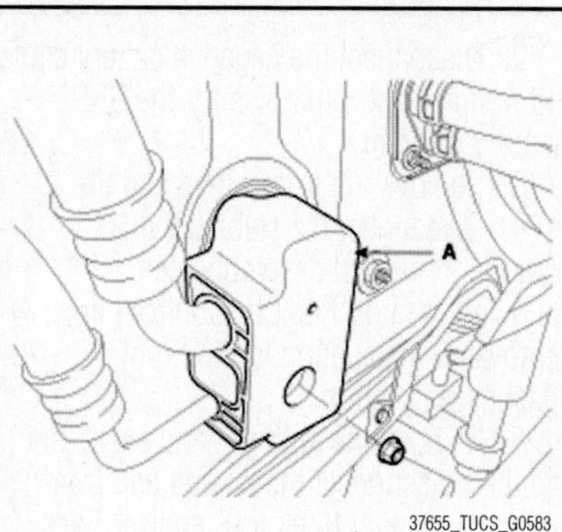

Fig. 332 Remove the expansion valve cover (A)

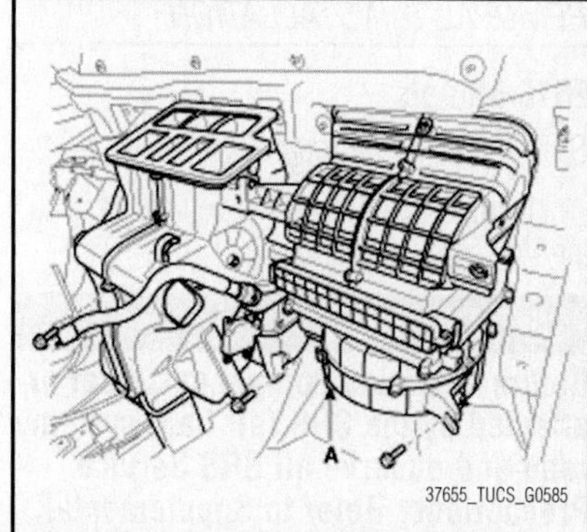

Fig. 334 Remove the heater and blower unit (A) after removing the 3 mounting bolts

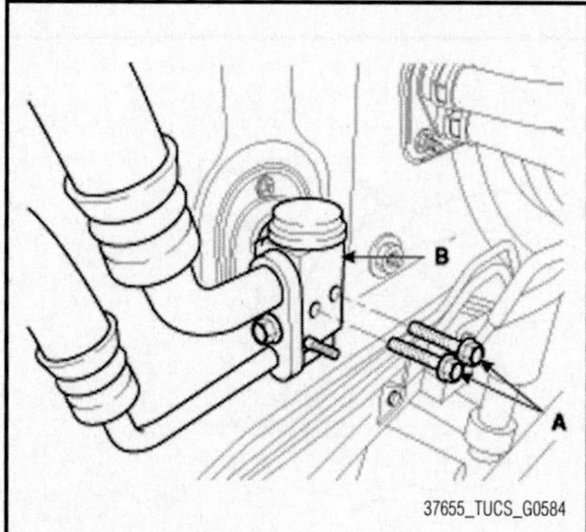

Fig. 333 Remove the bolts (A) and the expansion valve (B)

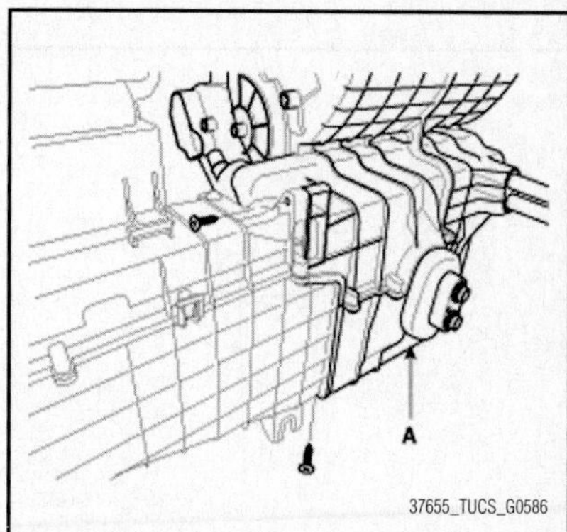

Fig. 335 Remove the blower unit (B) from heater unit after removing the 3 screws

13. Remove the mounting screw and then remove the heater core cover.

14. Remove the heater unit lower case mount screw and then remove the heater unit lower case.

15. Remove the evaporator core.

➡Be careful that the inlet and outlet pipe are not bent during heater core removal, and pull out the heater core.

To install:

16. Installation is the reverse order of removal, noting the following:

a. Tighten the expansion valve bolts to 68–104 ft. lbs. (8–12 Nm).

b. If you're installing a new evaporator, add refrigerant oil (ND-OIL8).

c. Replace the O-rings with new ones at each fitting, and apply a thin coat of refrigerant oil before installing them. Be sure to use the right O-rings for R-134a to avoid leakage. Immediately after using the oil, replace the cap on the container, and seal it to avoid moisture absorption.

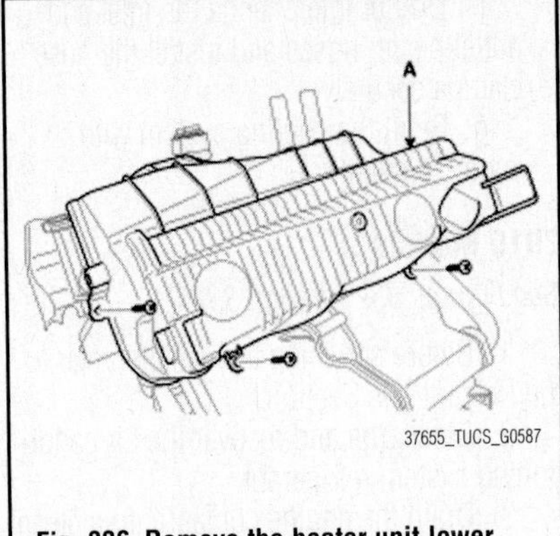

Fig. 336 Remove the heater unit lower case (A)

➡Do not spill the refrigerant oil on the vehicle ; it may damage the paint ; if the refrigerant oil contacts the paint, wash it off immediately

d. Apply sealant to the grommets.

e. Make sure that there is no air leakage.

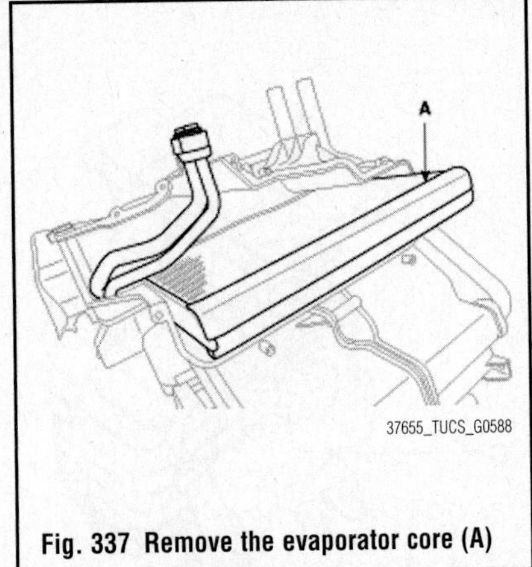

Fig. 337 Remove the evaporator core (A)

f. Charge the system and test its performance.

g. Do not interchange the inlet and outlet heater hoses and install the hose clamps securely.

h. Refill the cooling system with engine coolant

STEERING

ELECTRIC STEERING GEAR BOX

REMOVAL & INSTALLATION

2010 Models

See Figures 338 through 346.

1. Before servicing the vehicle, refer to the Precautions Section.

⁂ CAUTION

Before servicing components near or affected by the SRS (air bag) system, read and observe all SRS Service Precautions. Refer to Supplemental Restraint System (SRS), in the Chassis Electrical section. Failure to observe all precautions may result in accidental airbag deployment, personal injury, or death. Refer to Airbag Removal & Installation.

2. Disconnect the negative battery cable and wait at least 3 minutes for the SRS memory to drain.

3. Remove the front wheel and tire.

4. Disconnect the stabilizer link from the front strut assembly after removing the nut. Refer to Control Links, Removal & Installation in the Front Suspension section.

5. Remove the split pin and castle nut and then disconnect the tie-rod end from the front knuckle. Refer to Stabilizer Bar Removal & Installation in the Front Suspension section.

6. Remove the bolt and nut and then remove the lower arm.

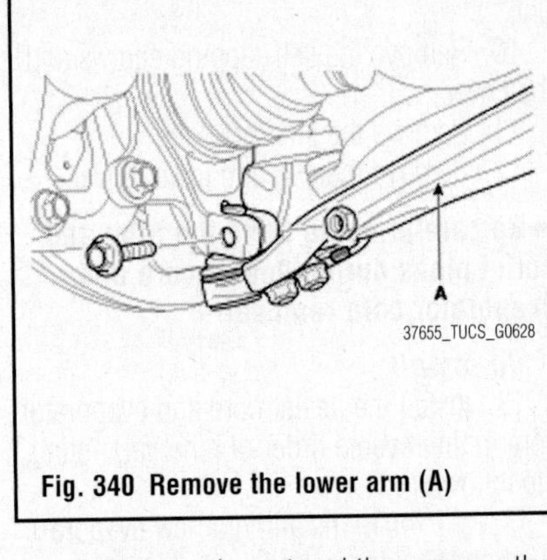

Fig. 340 Remove the lower arm (A)

7. Remove the nut and then remove the dust cover.

8. Remove the bolt and then disconnect the universal joint assembly from the pinion of the steering gear box.

⁂ WARNING

Lock the steering wheel in the straight ahead position to prevent the damage of the clock spring inner cable when you handle the steering wheel.

9. Remove the bolt and nut and then remove the roll rod stopper.

10. Disconnect the muffler rubber hanger.

11. Remove the bolts and nuts and then remove the subframe.

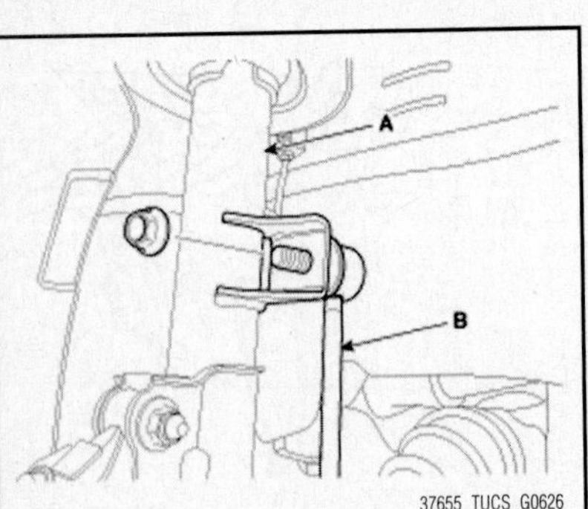

Fig. 338 Disconnect the stabilizer link (B) from the front strut assembly (A)

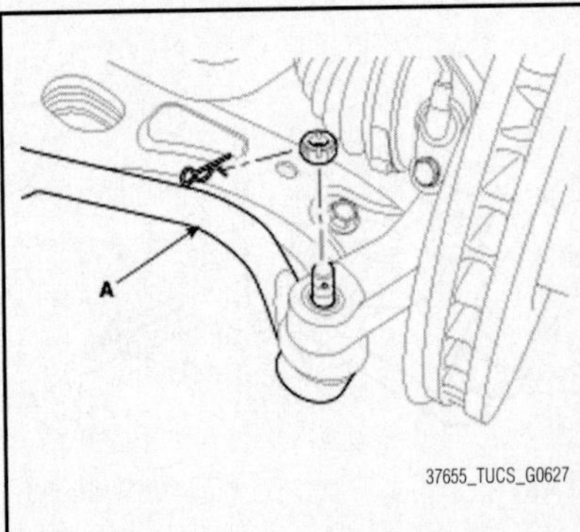

Fig. 339 Remove the split pin and castle nut and then disconnect the tie-rod end (A)

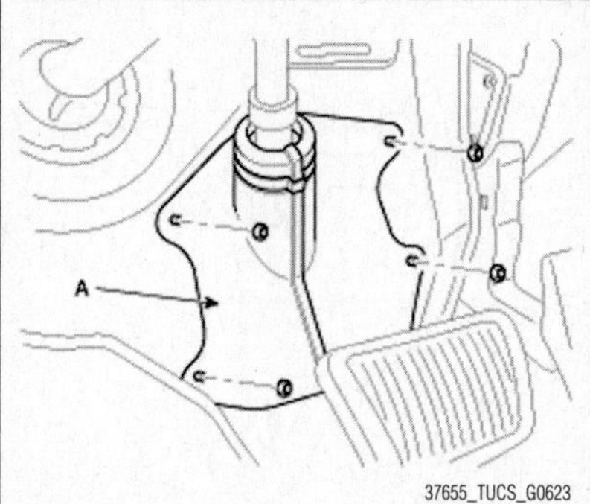

Fig. 341 Remove the nut and the dust cover (A)

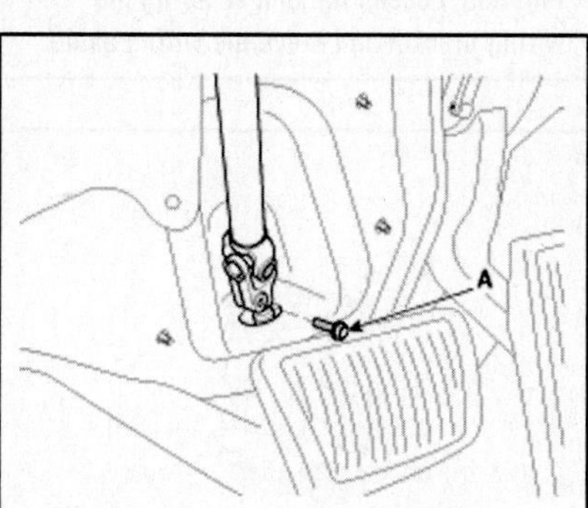

Fig. 342 Remove the bolt (A) and then disconnect the universal joint assembly

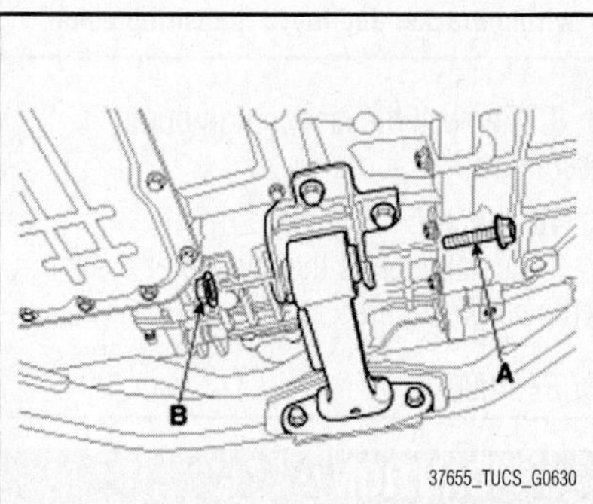

Fig. 343 Remove the bolt (A) and nut (B) and then remove the roll rod stopper

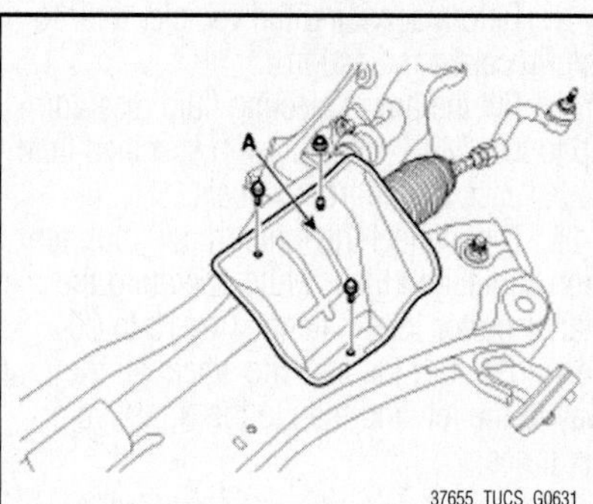

Fig. 344 Remove the bolt and nut and then remove the protector (A)

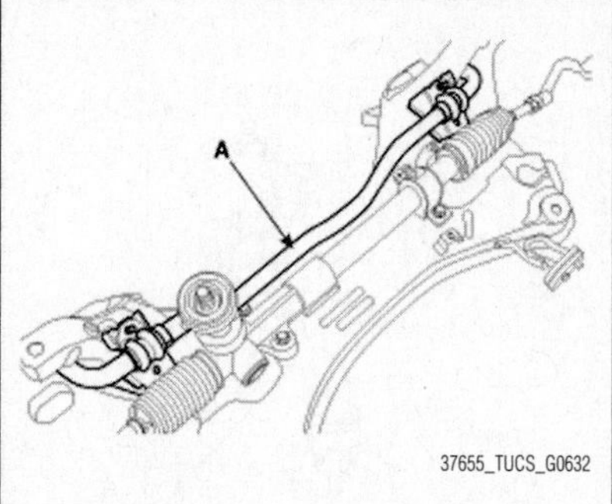

Fig. 345 Remove the stabilizer (A) from the sub frame

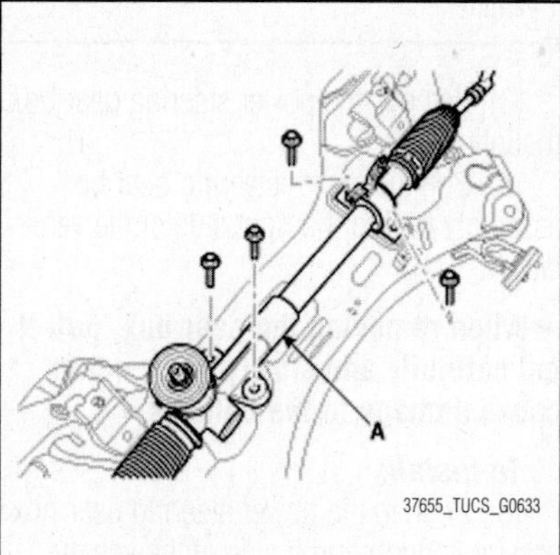

Fig. 346 Remove the bolt and then remove the steering gear box (A)

12. Remove the bolt and nut and then remove the protector.
13. Remove the bolt and then remove the stabilizer from the sub frame. Refer to Stabilizer Bar Removal & Installation in the Front Suspension section.
14. Remove the bolt and then remove the steering gear box.

To install:

15. Installation is the reverse of the removal, noting the following:
    a. Tighten the steering gear box bolt to 43–58 ft. lbs. (59–79 Nm).
    b. Tighten the stabilizer bar bolts to 33–40 ft. lbs. (44–54 Nm).
    c. Tighten the subframe bolts and nuts to 130–145 ft. lbs. (177–196 Nm).
    d. Tighten the roll rod stopper bolts and nuts to 80–94 ft. lbs. (108–128 Nm).
    e. Tighten the universal joint assembly bolt to 22–25 ft. lbs. (29–34 Nm).
    f. Tighten the lower arm bolts and nuts to 72–87 ft. lbs. (98–118 Nm).
    g. Tighten the tie-rod end nut to 25–33 ft. lbs. (34–44 Nm).
    h. Tighten the stabilizer link nut to 72–87 ft. lbs. (98–118 Nm).

POWER RACK & PINION STEERING GEAR

REMOVAL & INSTALLATION

2009 Models

See Figures 347 through 352.

1. Before servicing the vehicle, refer to the Precautions Section.
2. Disconnect the cover fixing clip on the universal joint indoor driver side, and loosen the noise covers.
3. Loosen the universal joint and the gear box mounting bolt and disconnect the universal joint from the gear box.

➡Keep the neutral-range to prevent the damage of the clock spring inner cable when you handle the steering wheel.

4. Raise and safely support the vehicle.
5. Remove the front wheels and tires.
6. After removing the split pin, disconnect the tie rod from the knuckle by using the special tool (09568-34000).

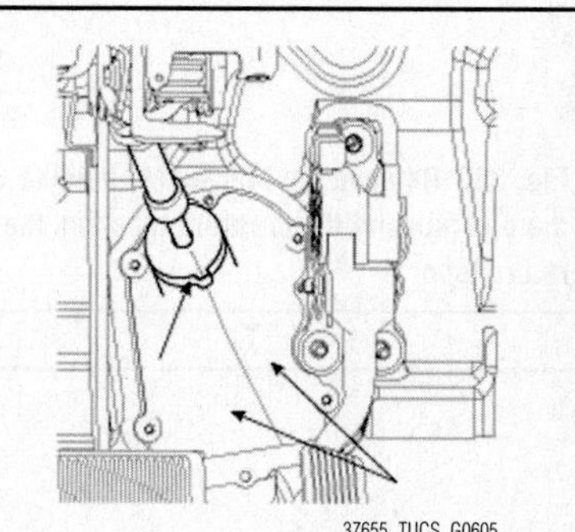

Fig. 347 Disconnect the cover fixing clip on the universal joint indoor driver side, and loosen the noise covers

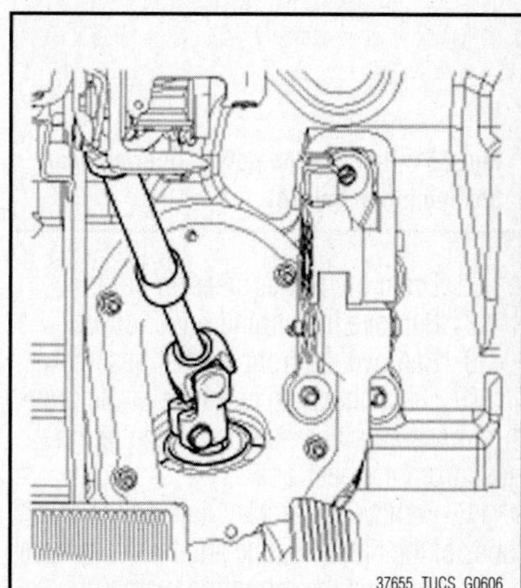

Fig. 348 Loosen the universal joint and the gear box mounting bolt and disconnect the universal joint from the gear box

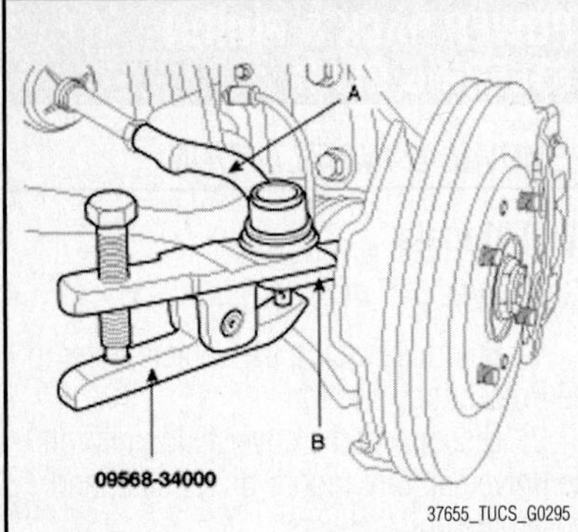

Fig. 349 Disconnect the tie rod (A) from the knuckle (B) by using the special tool (09568-34000)

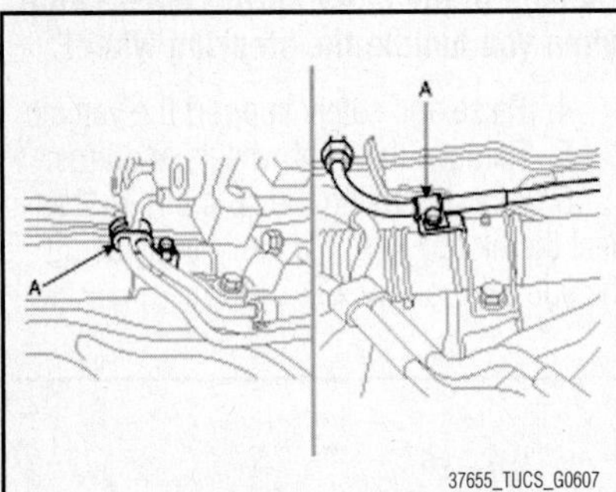

Fig. 350 Remove the bracket (A) holding the end tubes of the pressure tube and the return tube

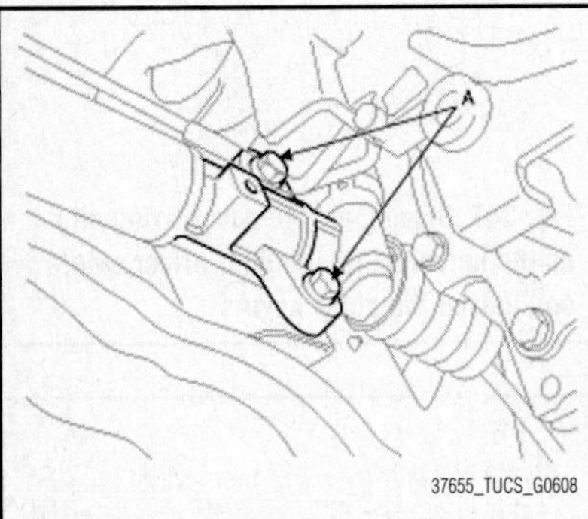

Fig. 351 Remove the power steering gear box mounting bolt (A)

7. Drain the power steering fluid.
8. Remove the engine under cover.
9. Remove the front muffler assembly.
10. Disconnect the end tube of the pressure hose and the end hose of the return hose from the gear box.
11. Remove the bracket holding the end tubes of the pressure tube and the return tube.
12. Remove the mounting clamp of power steering gear box, and also remove the clamp holding the pressure tube and the return tube.

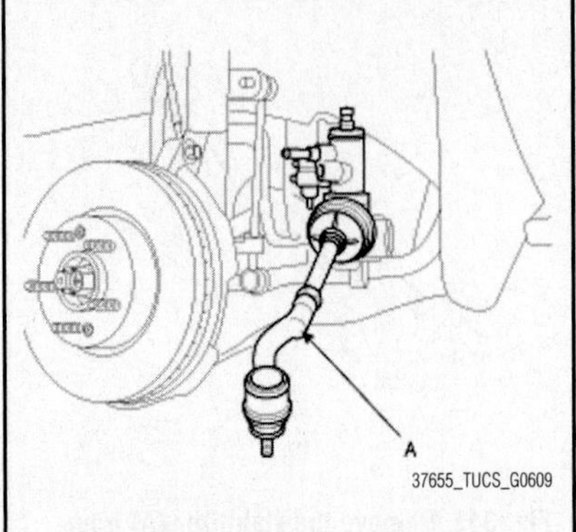

Fig. 352 Pull the power steering gear box assembly (A) toward the right side of the vehicle

13. Remove the power steering gear box mounting bolt.
14. Pull the power steering gear box assembly toward the right side of the vehicle.

➡**When removing the gear box, pull it out carefully and slowly so as not to cause damage to the bellows.**

To install:

15. Push in the power steering gear box assembly on the right side of the vehicle.
16. Install the dust cover mounting plate.
17. Connect the dust cover to its mounting plate with a new strap.
18. Connect the steering gear box assembly to the universal joint assembly by using the bolt.
19. The remainder of installation is the reverse of removal.
20. After installation, air bleed the system.

POWER STEERING PUMP

REMOVAL & INSTALLATION

2009 Models

See Figures 353 and 354.

1. Before servicing the vehicle, refer to the Precautions Section.
2. Loosen the bolt securing the wiring bracket and move the wiring aside.
3. Remove the pressure hose from the oil pump.
4. Disconnect the suction hose from the suction connector and drain the fluid into a container.
5. Loosen the tension adjusting bolt on the power steering belt.
6. Remove the belt from the power steering oil pump pulley.
7. Loosen the power steering oil pump mounting bolt and the tension adjusting bolt.

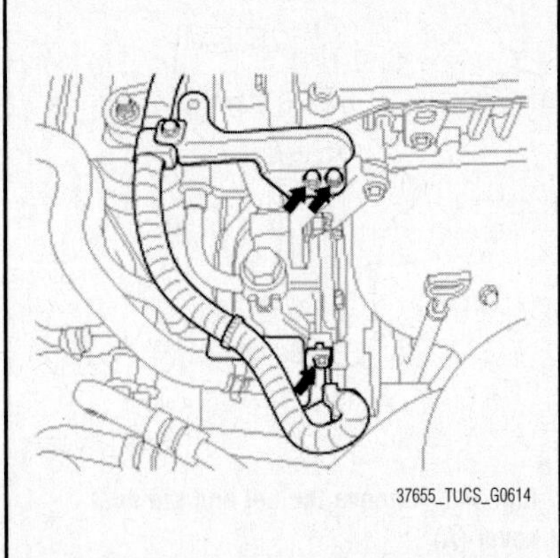

Fig. 353 Loosen the bolt securing the wiring bracket and move the wiring aside

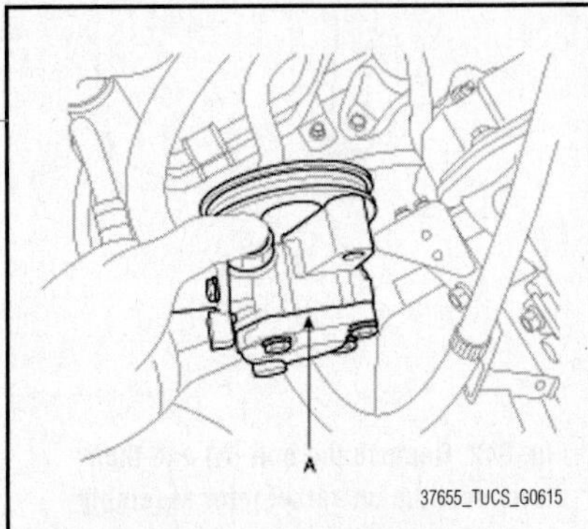

Fig. 354 Loosen the bolt securing the wiring bracket and move the wiring aside

8. Remove the steering oil pump assembly.

To install:

9. Installation is the reverse of the removal procedure.

BLEEDING

CAUTION

The fluid level should be checked with the engine OFF to prevent injury from moving components.

1. Before servicing the vehicle, refer to the Precautions Section.
2. Fill the power steering fluid reservoir up to the "MAX" position with specified fluid.
3. Jack up the front wheels.
4. Disconnect the ignition coil high tension cable, and then, while operating the starter motor intermittently (for 15 to 20 seconds), turn the steering wheel all the way to the left and then to the right five or six times.

➡**When bleeding fluid, replenish with the fluid so that the level does not fall below the bottom of the filter.**

➡If air bleeding is done while the vehicle is idling, the air will be broken up and absorbed into the fluid. Be sure to do the bleeding only while cranking.

5. Connect the high tension cable, and then start the engine (idling).
6. Turn the steering wheel to the left and then to the right, until there are no air bubbles in the oil reservoir.

➡Do not hold the steering wheel turned all the way to either side for more than ten seconds.

7. Confirm that the fluid is not milky and that the level is between "MAX" and "MIN" mark on the reservoir.
8. Check that there is a little change in the fluid level when the steering wheel is turned left and right.

a. If the fluid level varies 5mm (0.2 in.) or more, bleed the system again.
b. If the fluid level suddenly rises after stopping the engine, further bleeding is required.
c. Incomplete bleeding will produce a chattering sound in the pump and noise in the flow control valve, and lead to decreased durability of the pump.

SUSPENSION

FRONT SUSPENSION

CONTROL LINKS

REMOVAL & INSTALLATION

2009 Models

See Figure 355.

1. Before servicing the vehicle, refer to the Precautions Section.
2. Raise and safely support the vehicle.
3. Remove the front wheel and tire from the front hub.

⁂ WARNING

Be careful not to damage the hub bolts when removing the front wheel and tire.

4. Remove the stabilizer bar control link by removing the upper and lower mounting nuts.
5. Remove the stabilizer bar control link.

To install:

6. Install the stabilizer bar control link.
7. Tighten the stabilizer bar control link upper and lower mounting nuts to 74–89 ft. lbs. (100–120 Nm).
8. Install the front wheels. Tighten the wheel lug nuts to 66–81 ft. lbs. (90–110 Nm).

2010 Models

See Figure 356.

1. Before servicing the vehicle, refer to the Precautions Section.

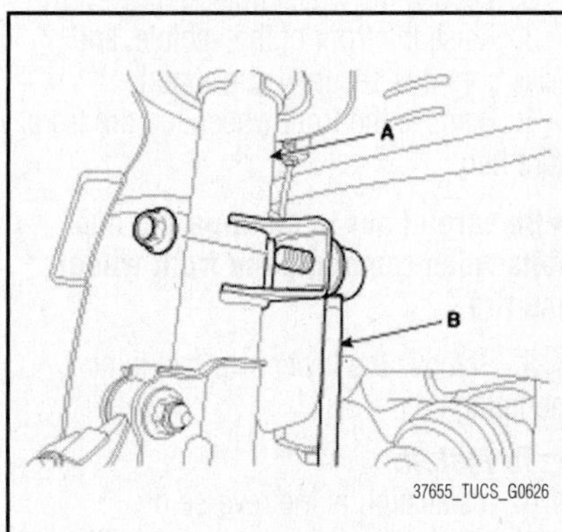

Fig. 356 Disconnect the stabilizer link (B) from the front strut assembly (A)

2. Remove the front wheel and tire.
3. Disconnect the stabilizer link from the front strut assembly after removing the nut.

To install:

4. Installation is the reverse of removal. Tighten the nut to 72–87 ft. lbs. (98–118).

LOWER BALL JOINT

REMOVAL & INSTALLATION

2009 Models

See Figures 357 through 359.

1. Before servicing the vehicle, refer to the Precautions Section.

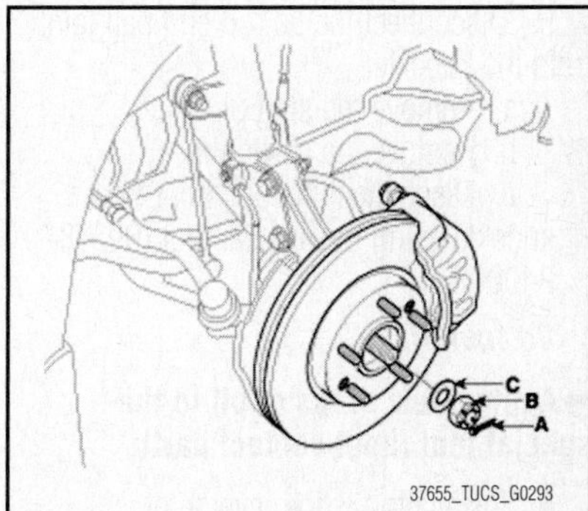

Fig. 357 Remove the split pin (A), castle nut (B) and washer (C) from the front hub

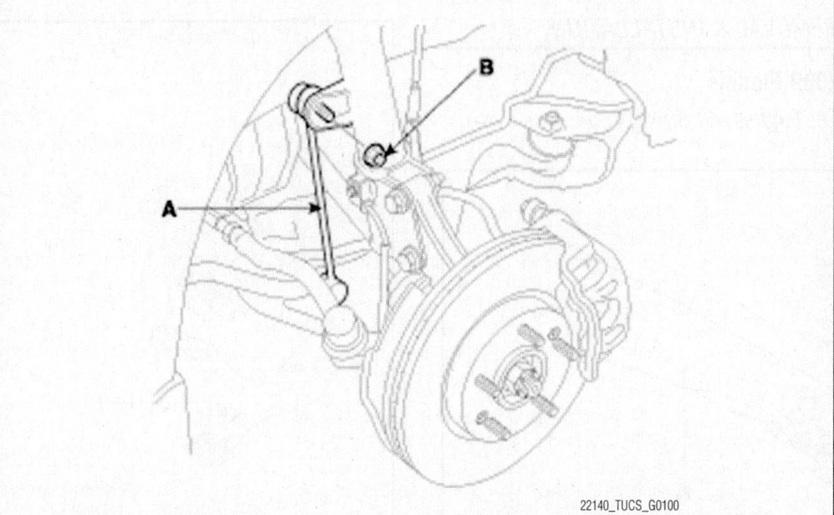

Fig. 355 Remove the stabilizer bar control link (A) by removing the upper (B) and lower mounting nuts

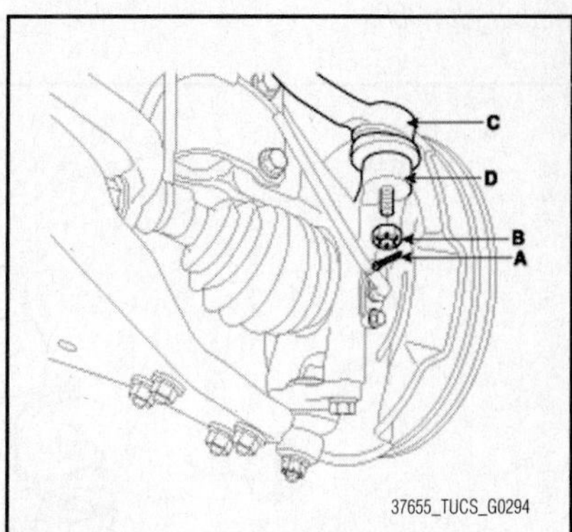

Fig. 358 Disconnect the tie rod end ball joint (C) from the knuckle (D), and remove the split pin (A) and castle nut (B)

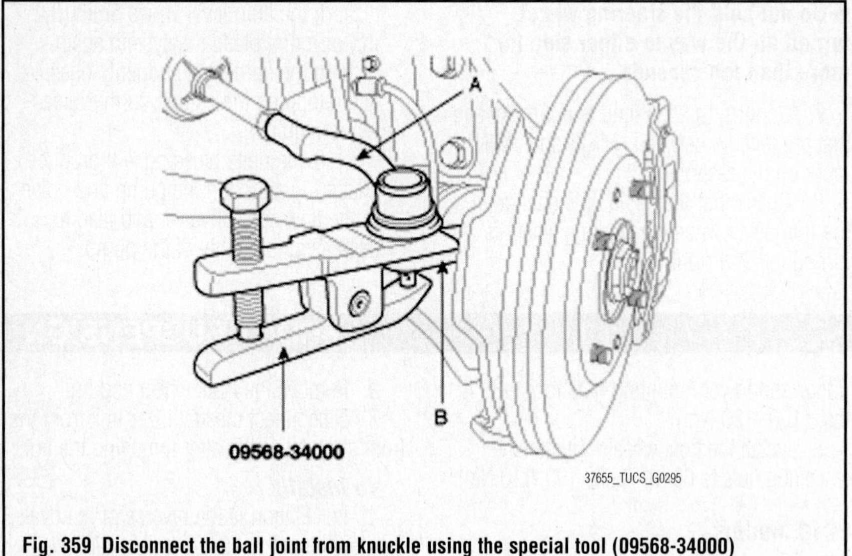

Fig. 359 Disconnect the ball joint from knuckle using the special tool (09568-34000)

2. Remove the front wheels and tires from the front hub.
3. Remove the split pin, castle nut and washer from the front hub while applying the brake.
4. Remove the wheel speed sensor from the knuckle. Refer to ABS, Wheel Speed Sensor, Removal & Installation, in the Brake section.
5. Disconnect the tie rod end ball joint from the knuckle:
   a. Remove the split pin.
   b. Remove the castle nut.
   c. Disconnect the ball joint from knuckle using the special tool (09568-34000).

To install:

➡Apply a few drops of oil to the special tool (boot contact part).

6. Installation is the reverse of removal.

2010 Models

See Figure 360.

1. Before servicing the vehicle, refer to the Precautions Section.
2. Loosen the wheel nuts slightly.
3. Raise the front of the vehicle, and make sure it is securely supported.
4. Remove the front wheel and tire from front hub.

➡Be careful not to damage the hub bolts when removing the front wheel and tire.

5. Remove the lower arm ball joint mounting bolt.

To install:

6. Installation is the reverse of removal.
7. Tighten bolts to 72–87 ft. lbs. (98–117 Nm)

LOWER CONTROL ARM

REMOVAL & INSTALLATION

2009 Models

See Figures 361 through 363.

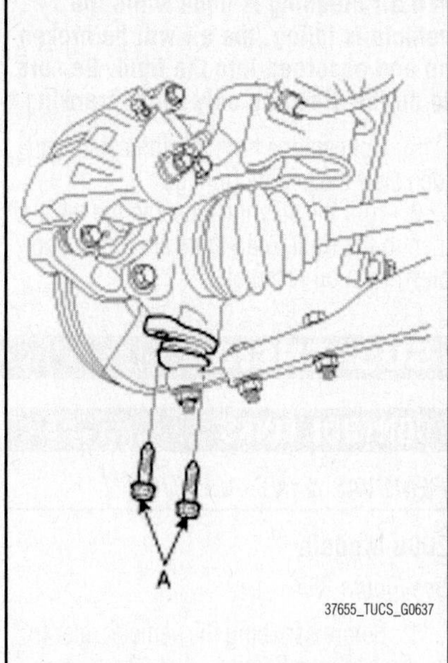

Fig. 361 Remove the lower arm ball joint mounting bolts (A)

1. Before servicing the vehicle, refer to the Precautions Section.
2. Remove the front wheels and tires.
3. Remove the lower arm ball joint mounting bolts.
4. Remove the lower arm mounting bolts.

To install:

5. Install the lower arm mounting bolts and tighten the front bolt to 74–89 ft. lbs. (100–120 Nm). Tighten the rear bolt to 103–118 ft. lbs. (140–160 Nm). Install the lower arm ball joint mounting bolts and tighten to 74–89 ft. lbs. (100–120 Nm).
6. Install the front wheels and tires.

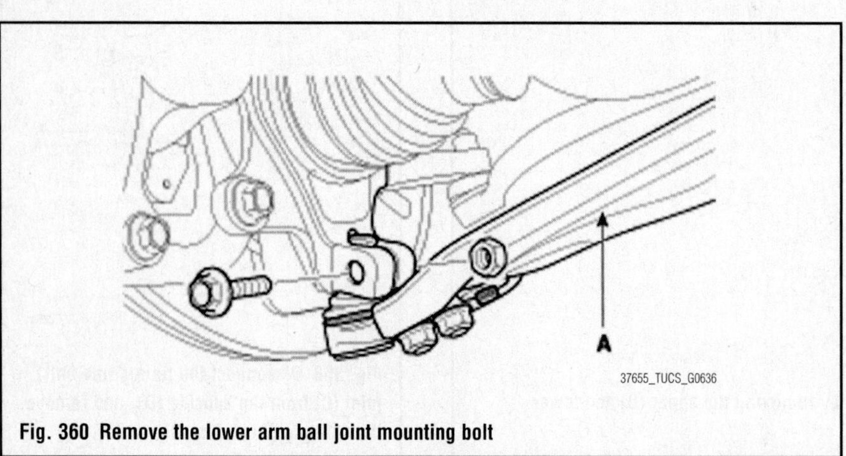

Fig. 360 Remove the lower arm ball joint mounting bolt

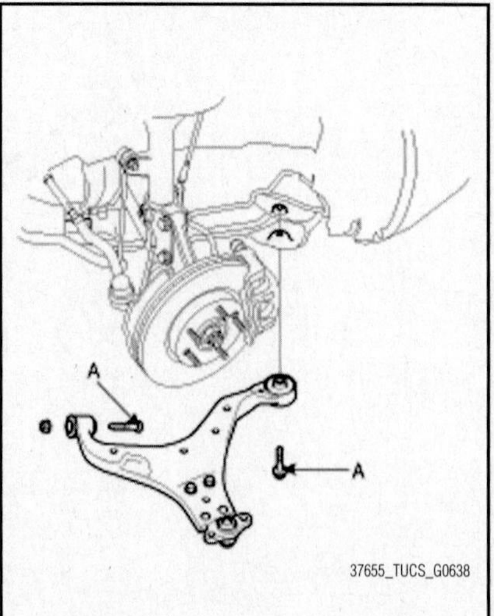

Fig. 362 Remove the lower arm mounting bolts (A)

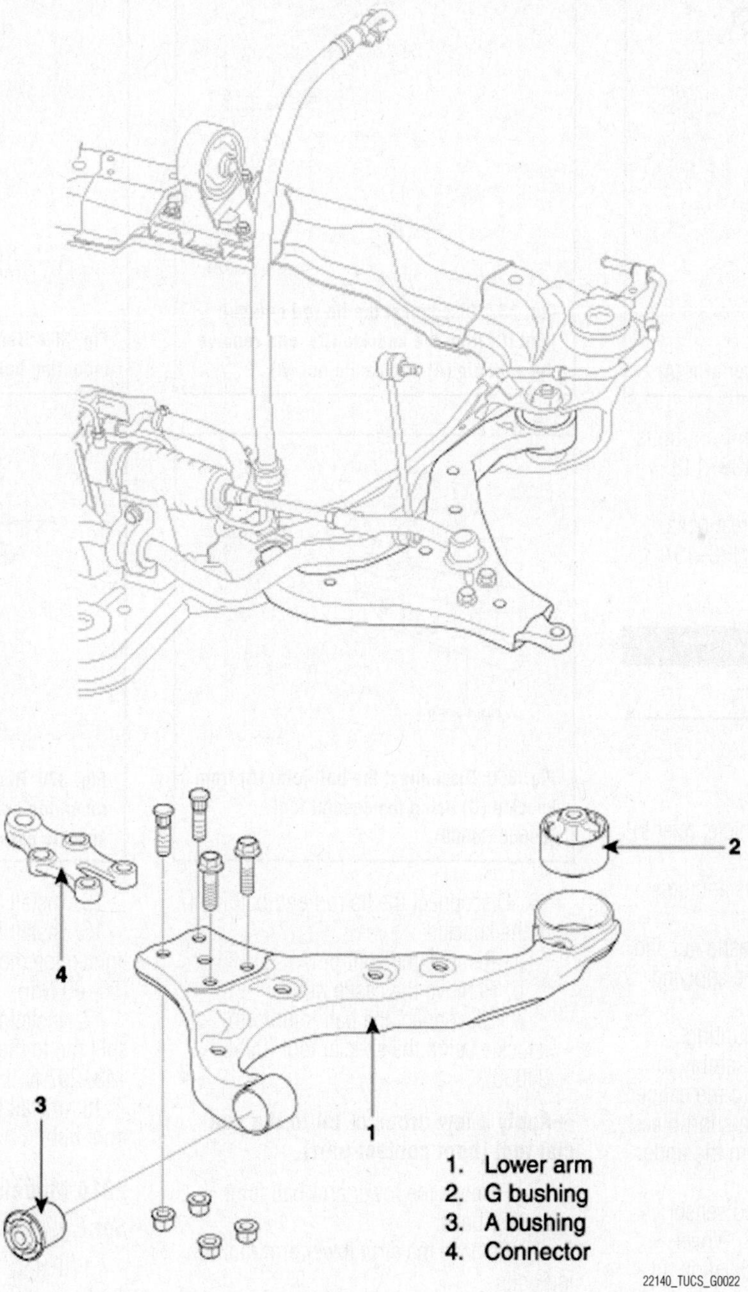

1. Lower arm
2. G bushing
3. A bushing
4. Connector

22140_TUCS_G0022

Fig. 363 View of front lower control arm assembly and related components

2010 Models

See Figures 364 and 365.

1. Before servicing the vehicle, refer to the Precautions Section.
2. Remove the front wheel and tire.
3. Remove the bolt and nut and then remove the lower arm.
4. Remove the bolts and nuts and then remove the front lower arm.

To install:

5. Installation is the reverse of the removal, noting the following:

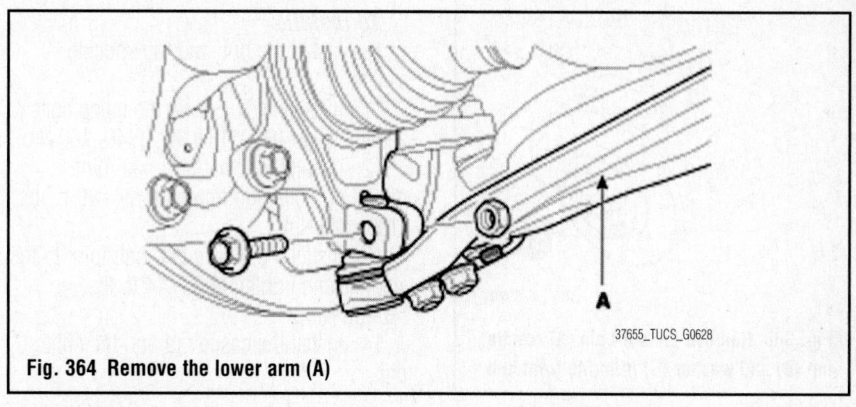

Fig. 364 Remove the lower arm (A)

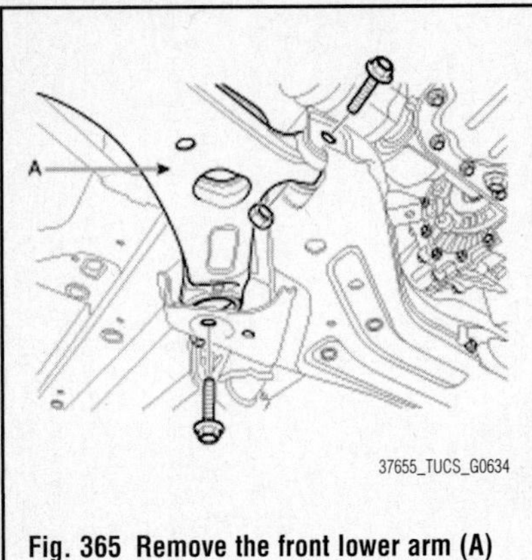

Fig. 365 Remove the front lower arm (A)

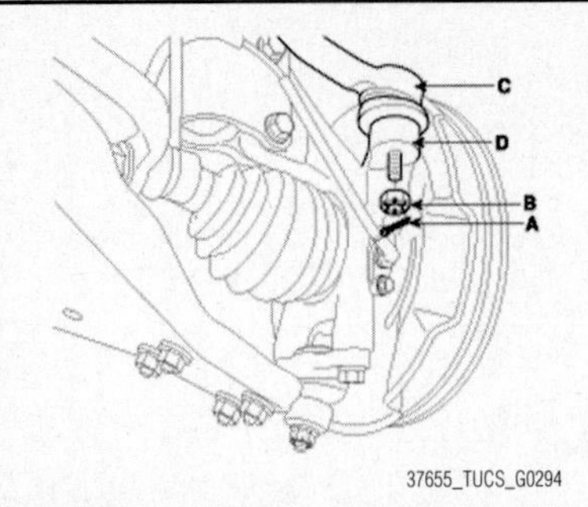

Fig. 367 Disconnect the tie rod end ball joint (C) from the knuckle (D), and remove the split pin (A) and castle nut (B)

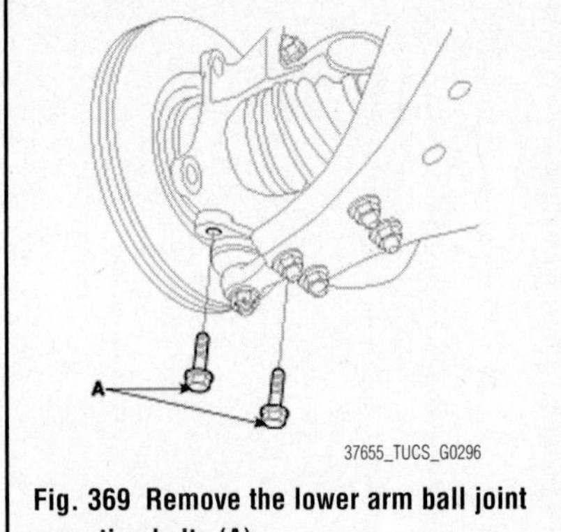

Fig. 369 Remove the lower arm ball joint mounting bolts (A)

a. Tighten the lower arm front bolts and nuts to 72–87 ft. lbs. (98–118 Nm).

b. Tighten the lower arm rear bolts and nuts to 101–116 ft. lbs. (138–157 Nm).

STEERING KNUCKLE

REMOVAL & INSTALLATION

2009 Models

See Figures 366 through 370.

1. Before servicing the vehicle, refer to the Precautions Section.
2. Remove the front wheels and tires from the front hub.
3. Remove the split pin, castle nut and washer from the front hub while applying the brake.
4. Remove the caliper mounting bolts, and hang the caliper assembly to one side. To prevent damage to the caliper assembly or brake hose, use a short piece of wire to hang the caliper from the undercarriage.
5. Remove the wheel speed sensor from the knuckle. Refer to ABS, Wheel Speed Sensor, Removal & Installation, in the Brake section.

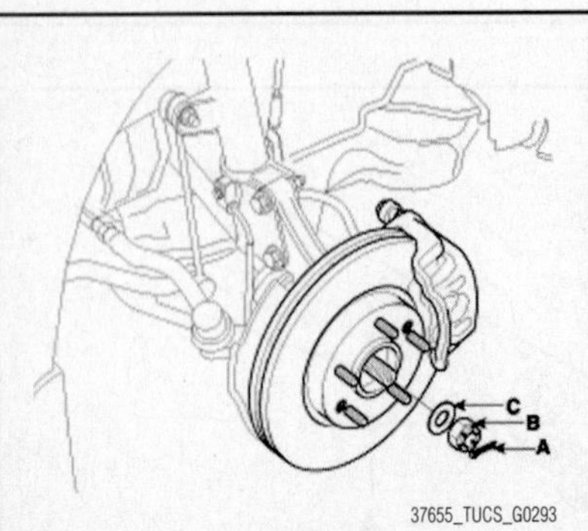

Fig. 366 Remove the split pin (A), castle nut (B) and washer (C) from the front hub

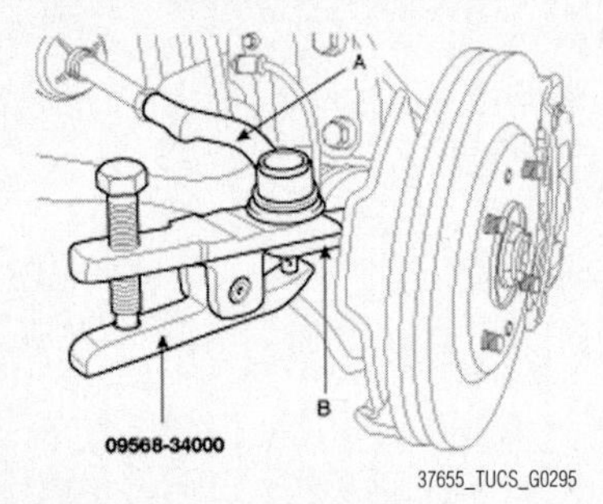

Fig. 368 Disconnect the ball joint (C) from knuckle (D) using the special tool (09568-34000)

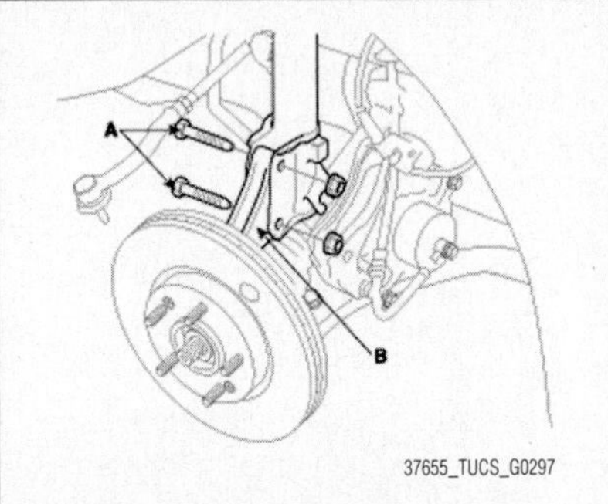

Fig. 370 Remove the strut lower arm mounting bolts (A) and the hub and the knuckle assembly (B)

6. Disconnect the tie rod end ball joint from the knuckle:
   a. Remove the split pin.
   b. Remove the castle nut.
   c. Disconnect the ball joint from knuckle using the special tool (09568-34000).

➡Apply a few drops of oil to the special tool (boot contact part).

7. Remove the lower arm ball joint mounting bolts.
8. Remove the strut lower arm mounting bolts.
9. Remove the hub and the knuckle assembly. Be careful not to damage the boot and rotor teeth.

To install:

10. Install the hub and the knuckle assembly.
11. Install the strut lower mounting bolts and tighten to 103–118 ft. lbs. (140–160 Nm).
12. Install the lower arm ball joint mounting bolts and tighten to 74–89 ft. lbs. (100–120 Nm).
13. Install the tie rod end ball joint to the knuckle and tighten to 33–44 ft. lbs. (45–60 Nm).
14. Install the castle nut and the split pin.
15. Install the wheel speed sensor.
16. Install the brake caliper, and then tighten the mounting bolts to 37–44 ft. lbs. (50–60 Nm).
17. Install the washer, castle nut and split pin to the front hub and tighten to 148–207 ft. lbs. (200–280 Nm).
18. Install the front wheel and tire on the front hub.

2010 Models

See Figures 371 through 376.

1. Before servicing the vehicle, refer to the Precautions Section.

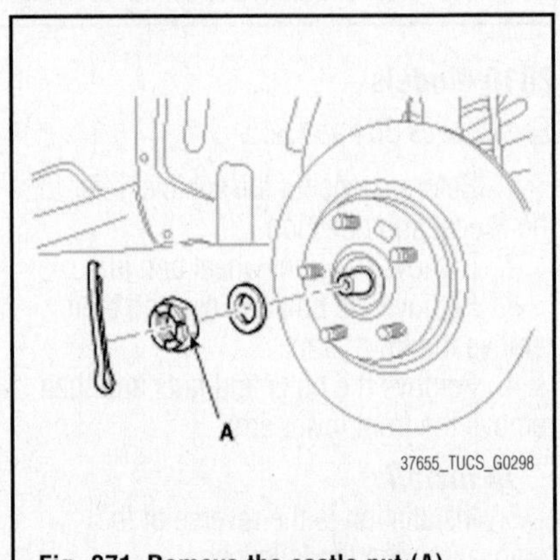

Fig. 371 Remove the castle nut (A)

2. Remove the front wheels and tires from the front hub.

3. Remove the caliper mounting bolts, and hang the caliper assembly to one side with a wire.

4. Remove the castle nut from the front hub while applying the brake.

5. Remove the wheel speed sensor. Refer to ABS, Wheel Speed Sensor, Removal & Installation, in the Brake section.

6. Remove the tie rod end ball joint from the knuckle:

a. Remove the split pin.

b. Remove the castle nut.

7. Remove the lower arm mounting bolts.

8. Disconnect the halfshaft end from the knuckle.

9. Remove the strut mounting bolts and then remove the hub and knuckle assembly.

To install:

10. Installation is the reverse of removal, noting the following:

a. Tighten the strut lower mounting bolts to 101–116 ft. lbs. (137–157 Nm).

b. Tighten the lower arm mounting bolts to 72–87 ft. lbs. (98–118 Nm).

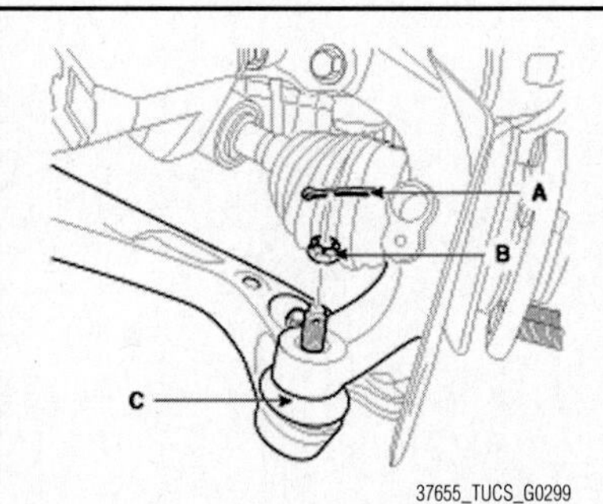

Fig. 372 Disconnect the tie rod end ball joint (C) from the knuckle (D), and remove the split pin (A) and castle nut (B)

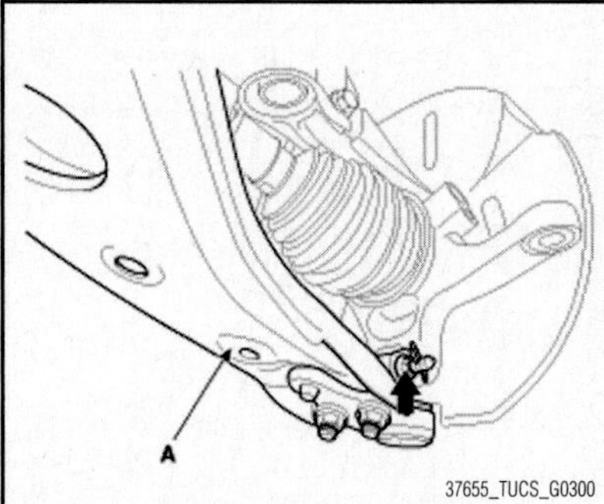

Fig. 373 Remove the lower arm (A) mounting bolts

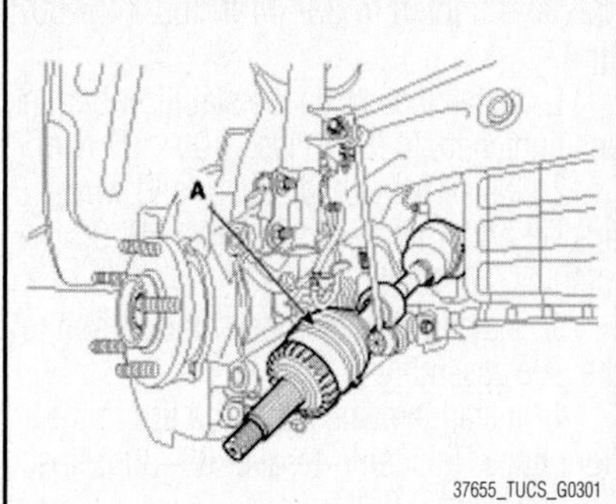

Fig. 374 Disconnect the halfshaft end (A) from the knuckle

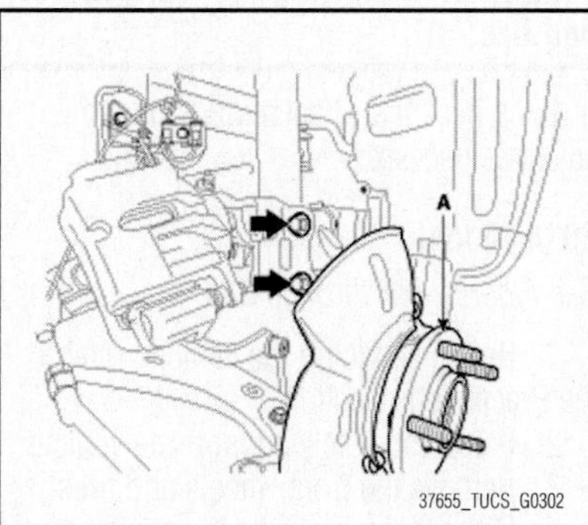

Fig. 375 Remove the strut mounting bolts and the hub and knuckle assembly (A)

c. Tighten the tie rod end ball joint castle nut to 25–33 ft. lbs. (34–44 Nm).

d. Tighten the front hub castle nut to 145–203 ft. lbs. (196–275 Nm).

e. The washer should be assembled with convex surface outward when installing the castle nut and split pin. Don't reuse split pin when reassembling.

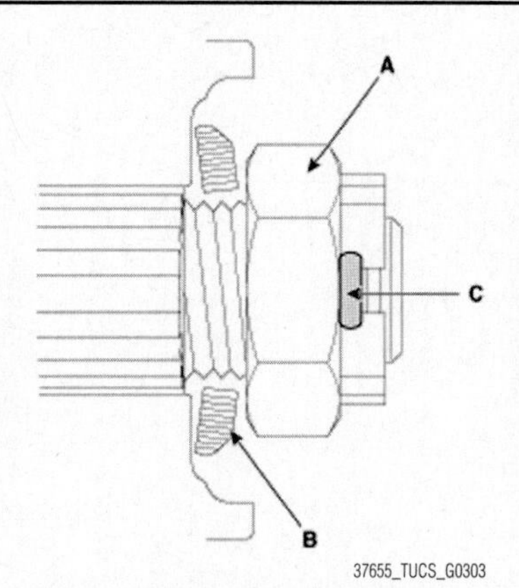

Fig. 376 The washer (B) should be assembled with convex surface outward when installing the castle nut (A) and split pin (C)

STRUTS

REMOVAL & INSTALLATION

2009 Models

See Figures 377 through 381.

1. Before servicing the vehicle, refer to the Precautions Section.

2. Raise and safely support the vehicle.

3. Remove the front wheels and tires.

4. Remove the brake hose bracket and speed sensor cable mounting bolt from the strut assembly.

5. Remove the speed sensor cable mounting bolt and speed sensor. Refer to Wheel Speed Sensors, Removal & Installation in the Brake section.

6. Remove the nut from the stabilizer bar link.

7. Remove the upper strut mounting nuts.

8. Remove the strut lower mounting bolts and then remove the strut assembly.

To install:

9. Install the strut assembly and tighten the lower mounting bolts to 103–118 ft. lbs. (140–160 Nm).

10. Install the strut upper mounting

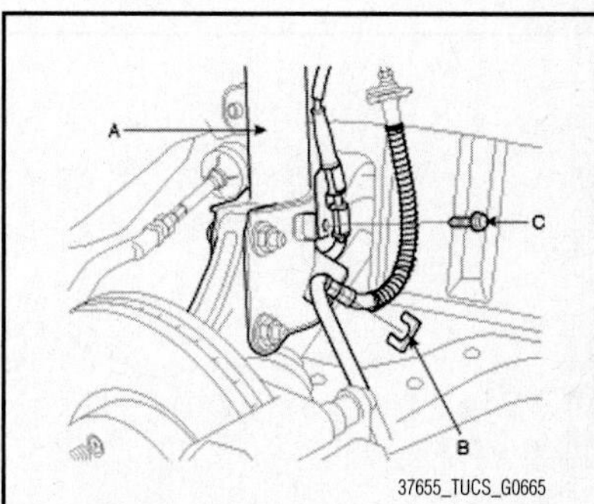

Fig. 377 Remove the brake hose bracket (B) and speed sensor cable mounting bolt (C) from the strut assembly (A)

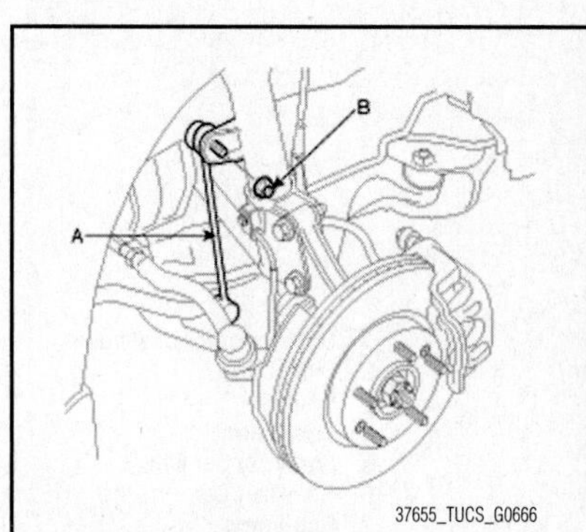

Fig. 378 Remove the nut (B) from the stabilizer bar link (A)

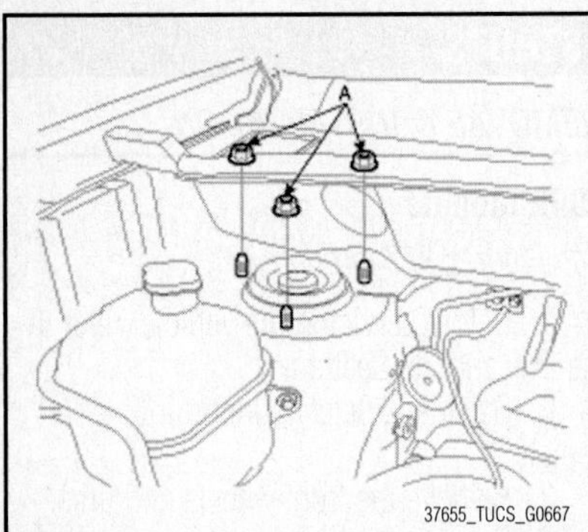

Fig. 379 Remove the upper strut mounting nuts (A)

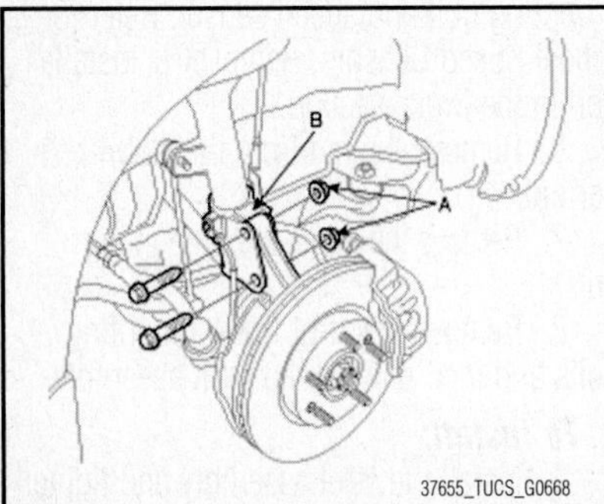

Fig. 380 Remove the strut lower mounting bolts (A) and then remove the strut assembly (B)

nuts and tighten to 33–44 ft. lbs. (45–60 Nm).

11. Install the nut on the stabilizer bar link and tighten to 74–89 ft. lbs. (100–120 Nm).

12. Install the speed sensor and wire bracket bolts. Tightening torque: 60–96 inch lbs. (7–11 Nm).

13. Install the brake hose bracket bolt to the axle assembly.

14. Install the wheel and the tire to the front hub. Tightening torque: 65–80 ft. lbs. (90–110 Nm).

⁂ WARNING

Be careful not to damage the hub bolts when installing the front wheel and tire.

15. Check the wheel alignment and adjust as necessary.

2010 Models

See Figures 382 through 384.

1. Before servicing the vehicle, refer to the Precautions Section.
2. Raise and safely support the vehicle.
3. Remove the front wheels and tires.
4. Remove the brake hose and wheel speed sensor bracket from the front strut assembly by removing the mounting bolts.
5. Disconnect the stabilizer link from the front strut assembly after removing the nut. Refer to Control Links, Removal & Installation in the Front Suspension section.
6. Disconnect the front strut assembly from the knuckle by removing the bolt and nut.
7. Remove the front strut assembly.

To install:

8. Installation is the reverse of removal, noting the following:
   a. Tighten the front strut mounting nuts to 33–43 ft. lbs. (44–59 Nm).

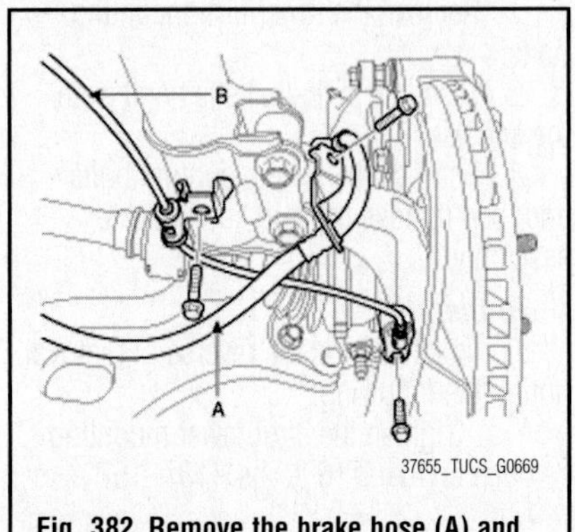

Fig. 382 Remove the brake hose (A) and wheel speed sensor bracket (B)

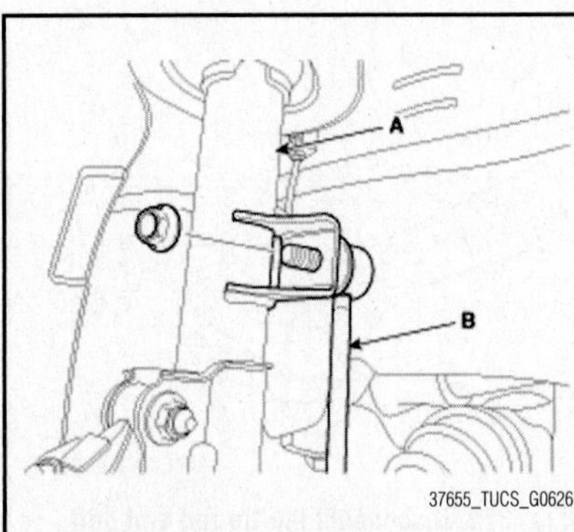

Fig. 383 Disconnect the stabilizer link (B) from the front strut assembly (A)

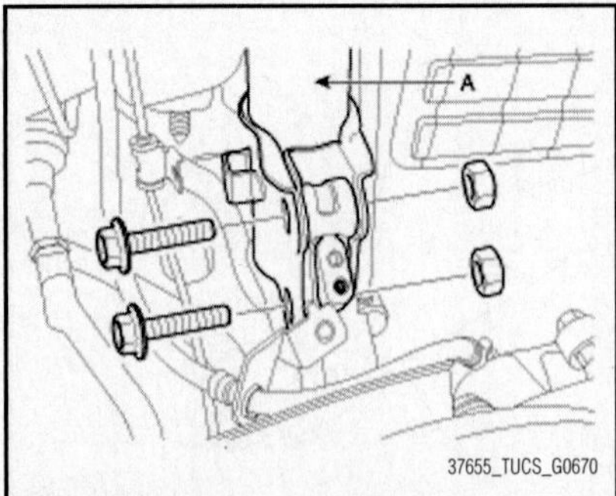

Fig. 384 Disconnect the front strut assembly (A) from the knuckle by removing the bolt and nut

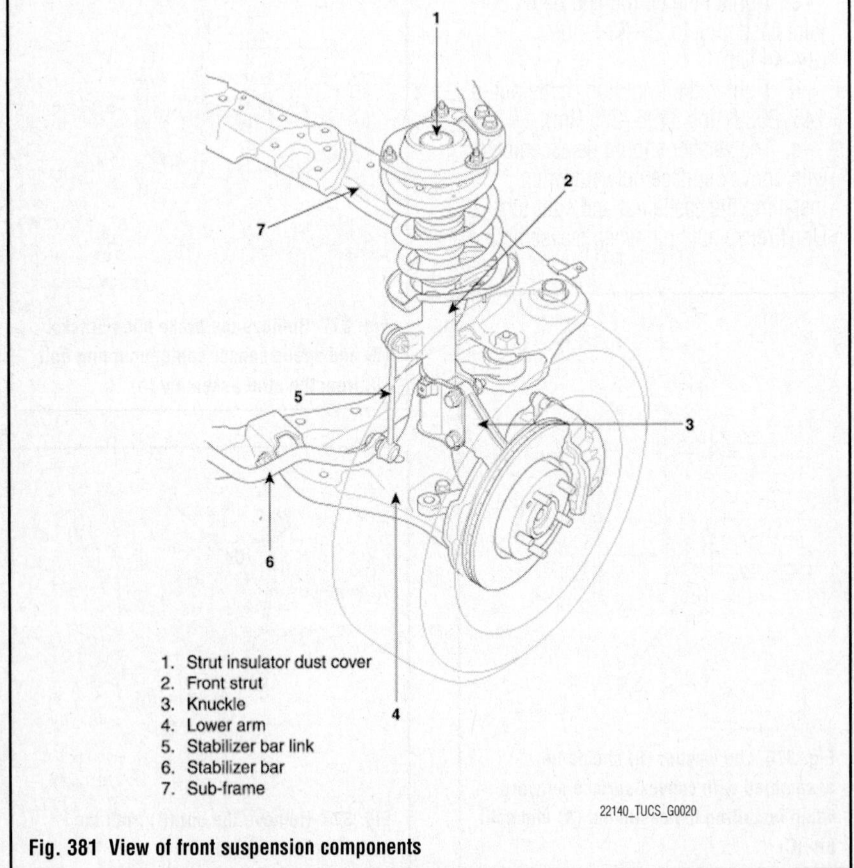

Fig. 381 View of front suspension components

b. Install the front strut assembly to the knuckle and tighten the nut and bolt to 101–116 ft. lbs. (137–157 Nm).

c. Tighten the stabilizer link nut to 72–87 ft. lbs. (98–118 Nm).

OVERHAUL

See Figure 385.

1. Before servicing the vehicle, refer to the Precautions Section.
2. Remove the strut from the vehicle and attach Service Tool 09546-2600 or another suitable spring compressor.
3. Compress the coil spring.
4. Remove the self-locking nut from the strut.
5. Remove the insulator, spring seat, coil spring, and dust cover.
6. Inspect the insulator for wear and damage, replace as required.
7. Check the rubber parts for damage or deterioration, replace as required.
8. Install the spring lower pad so that the protrusions fit the holes in the spring lower seat.
9. Compress the spring, using the spring compressor tool.
10. Install the compressed spring over the shock absorber.

➡ There are 2 color identification marks on the coil spring, one indicates model option and the other indicates load classification. Install the coil spring with the identification mark directed toward the steering knuckle.

11. After fully extending the piston rod, install the spring upper seat and insulator assembly.
12. After correctly seating the upper and lower ends of the coil spring in the upper and lower spring grooves, tighten the NEW self-locking nut temporarily.
13. Remove the spring compression tool. Tighten the self-locking nut to 43–51 ft. lbs. (60–70 Nm).
14. Install the strut to the vehicle.
15. Check the wheel alignment and adjust as necessary.

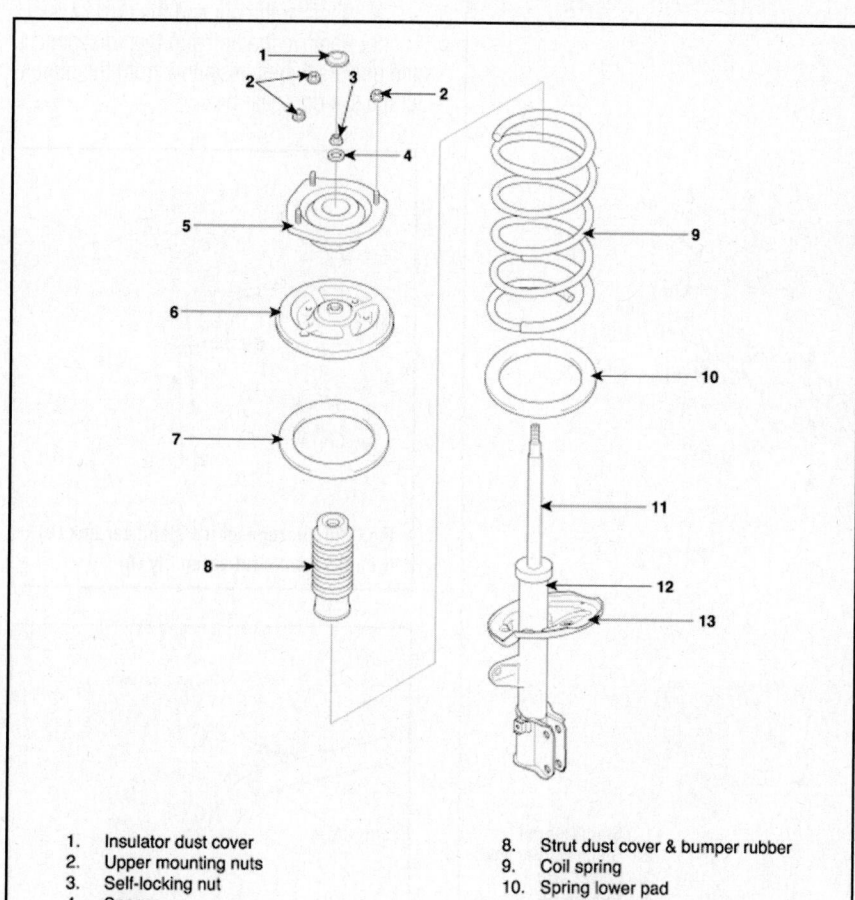

1. Insulator dust cover
2. Upper mounting nuts
3. Self-locking nut
4. Spacer
5. Insulator
6. Spring upper seat
7. Spring upper pad
8. Strut dust cover & bumper rubber
9. Coil spring
10. Spring lower pad
11. Piston rod
12. Strut assembly
13. Spring lower seat

22140_TUCS_G0021

Fig. 385 Exploded view of strut components

STABILIZER BAR

REMOVAL & INSTALLATION

2009 Models

See Figures 386 through 389.

1. Before servicing the vehicle, refer to the Precautions Section.
2. Raise the front of the vehicle, and make sure it is securely supported.
3. Remove the front wheel and tire from the front hub.

⁂ WARNING

Be careful not to damage the hub bolts when removing the front wheel and tire.

4. Remove the nut and stabilizer bar control link. Refer to Control Links Removal & Installation.
5. Remove the control link on the opposite side in the same way.
6. Remove the rear mounting bolts of the sub-frame.
7. Remove the stabilizer bolts, bracket and bushing.

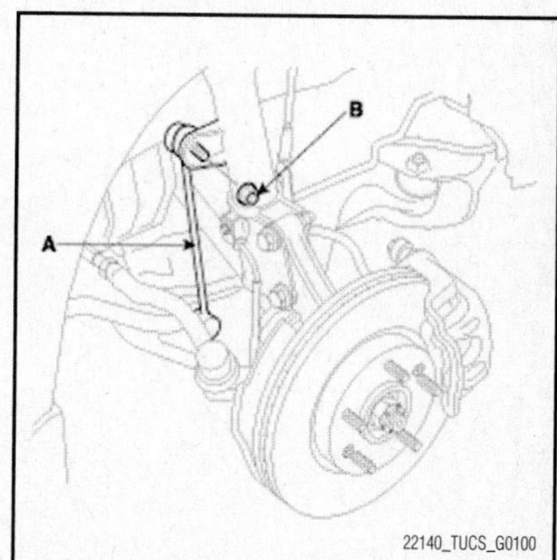

Fig. 386 Remove the stabilizer bar control link (A) by removing the upper (B) and lower mounting nuts

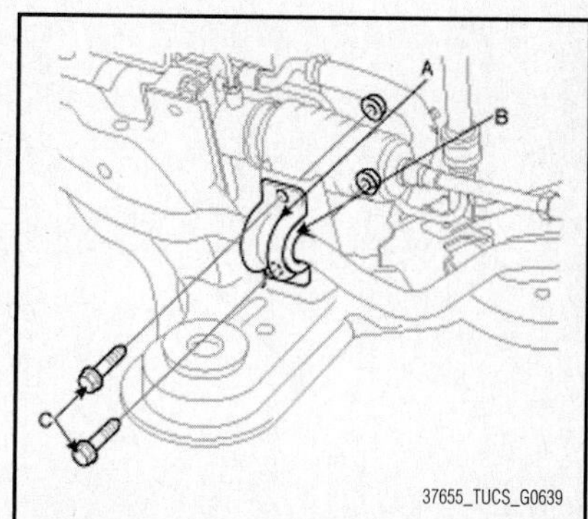

Fig. 387 Remove the stabilizer bolts (C), bracket (A) and bushing (B)

8. Remove the stabilizer bracket and bushing on the opposite side in the same way.

9. Remove the stabilizer bar.

⁂ WARNING

Be careful not to damage the power steering pressure tube.

To install:

10. Install the bushing on the stabilizer bar. Bring the clamp of the stabilizer bar into contact with the bushing.

11. Install the bracket on the bushing.

12. After tightening the bolts of the bushing bracket temporarily, install the bushing bracket on the opposite side and tighten to 37–48 ft. lbs. (50–65 Nm).

13. Install the rear mounting bolts of sub-frame.

14. Install the nut on the stabilizer bar link and tighten to 74–89 ft. lbs. (100–120 Nm).

15. Install the front wheel and tire.

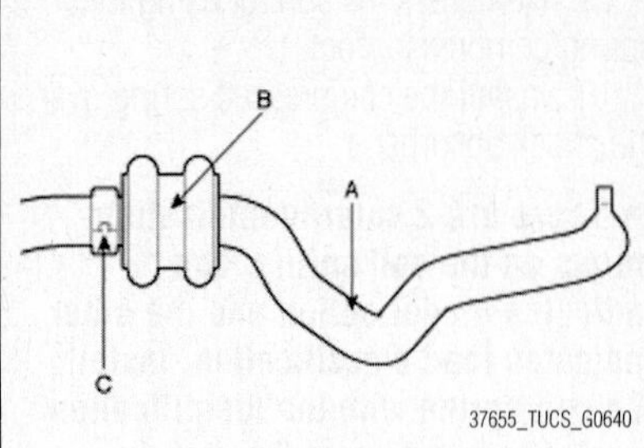

Fig. 389 Install the bushing (B) on the stabilizer bar (A) and bring the clamp (C) into contact with bushing

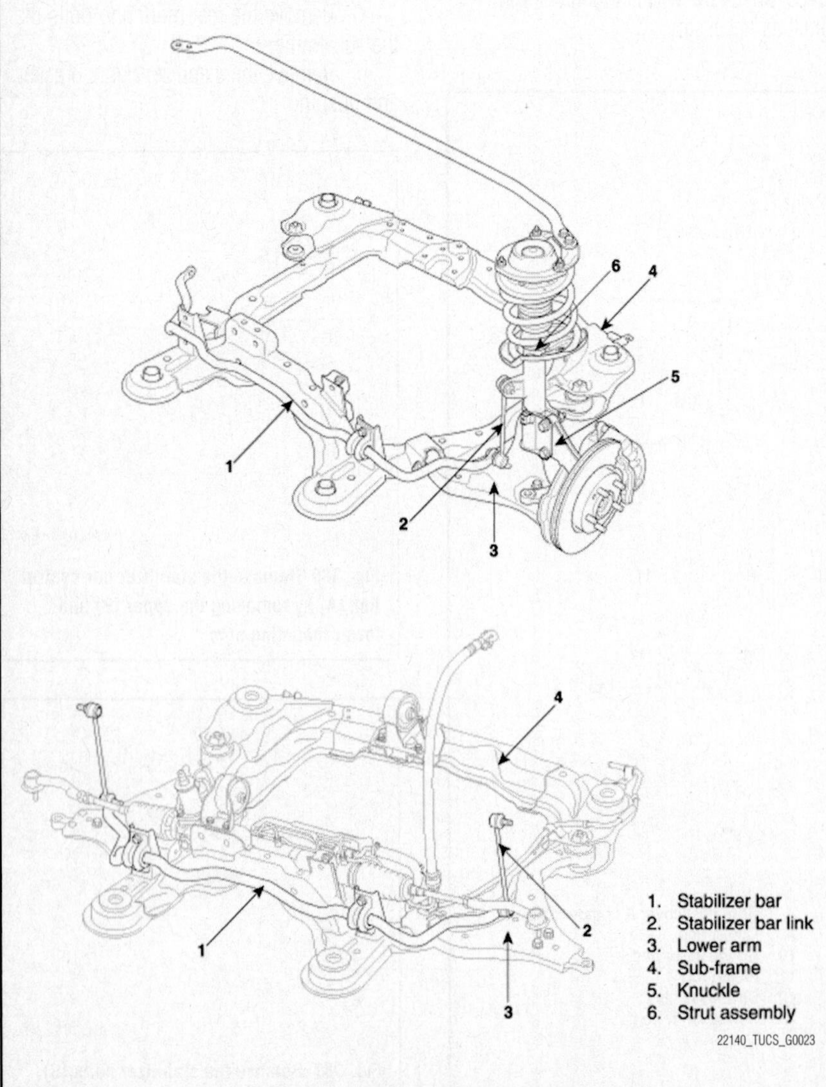

Fig. 388 View of stabilizer bar and related front suspension components

2010 Models

See Figures 390 through 397.

1. Before servicing the vehicle, refer to the Precautions Section.

⁂ CAUTION

Before servicing components near or affected by the SRS (air bag) system, read and observe all SRS Service Precautions. Refer to Supplemental Restraint System (SRS), in the Chassis Electrical section. Failure to observe all precautions may result in accidental airbag deployment, personal injury, or death. Refer to Airbag Removal & Installation.

2. Disconnect the negative battery cable and wait at least 3 minutes for the SRS memory to drain.

3. Remove the front wheel and tire.

4. Disconnect the stabilizer link from the front strut assembly after removing the nut.

5. Remove the split pin and castle nut and then disconnect the tie-rod end from the front knuckle.

6. Remove the bolt and nut and then remove the lower arm.

7. Remove the nut and the dust cover.

8. Remove the bolt and then disconnect the universal joint assembly from the pinion of the steering gear box.

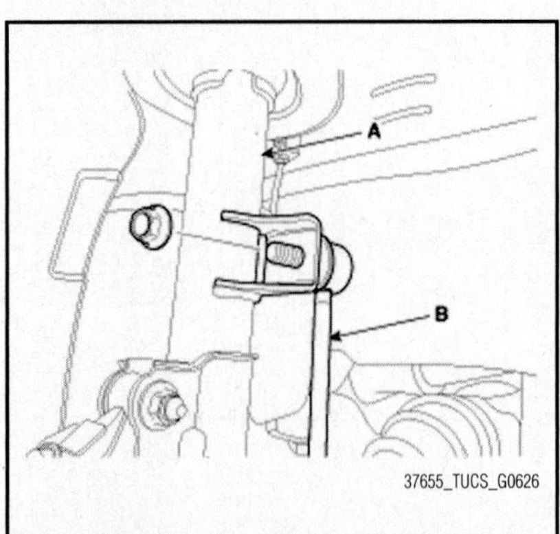

Fig. 390 Disconnect the stabilizer link (B) from the front strut assembly (A)

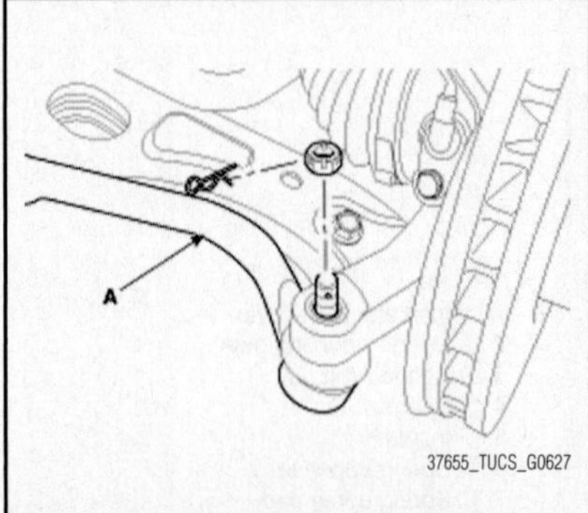

Fig. 391 Remove the split pin and castle nut and disconnect the tie-rod end (A)

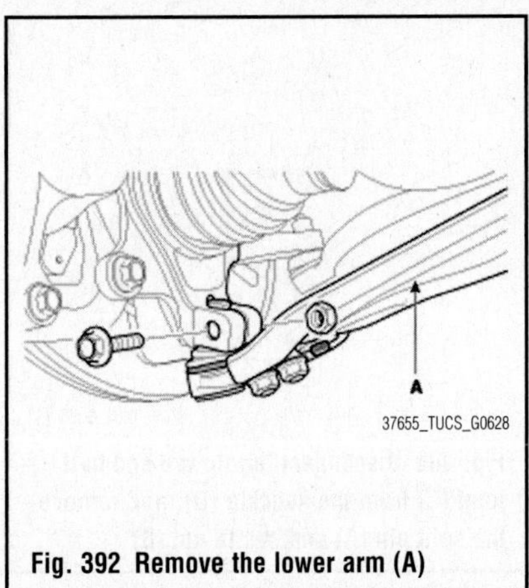

Fig. 392 Remove the lower arm (A)

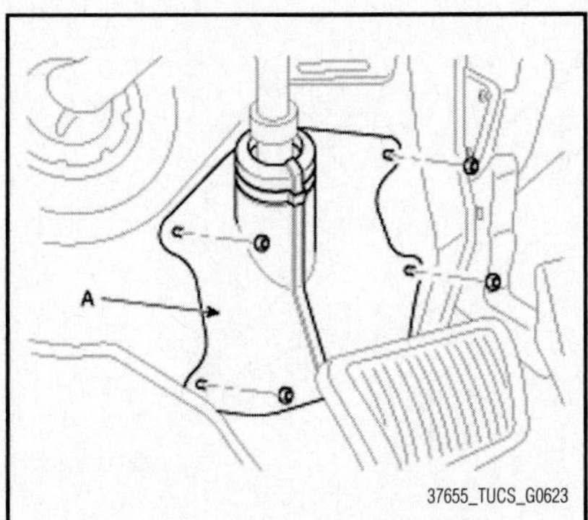

Fig. 393 Remove the nut and the dust cover (A)

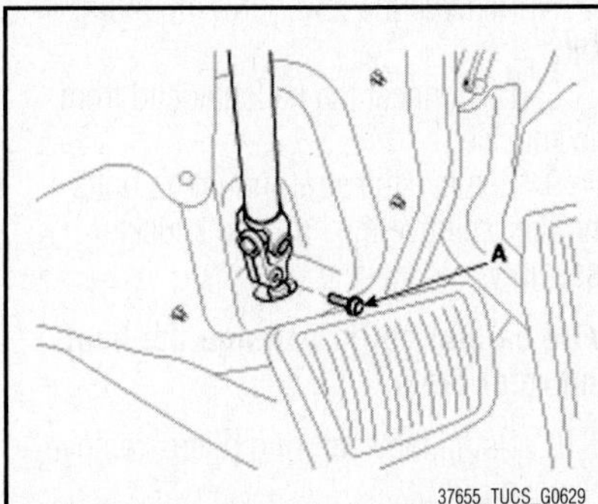

Fig. 394 Remove the bolt (A) and then disconnect the universal joint assembly

⁂ WARNING

Lock the steering wheel in the straight ahead position to prevent the damage of the clock spring inner cable when you handle the steering wheel.

9. Remove the bolt and nut and then remove the roll rod stopper.
10. Disconnect the muffler rubber hanger.
11. Remove the bolts and nuts and then remove the subframe.

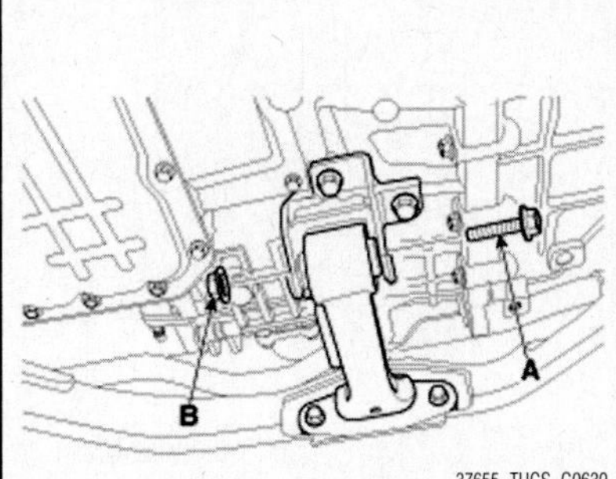

Fig. 395 Remove the bolt (A) and nut (B) and then remove the roll rod stopper

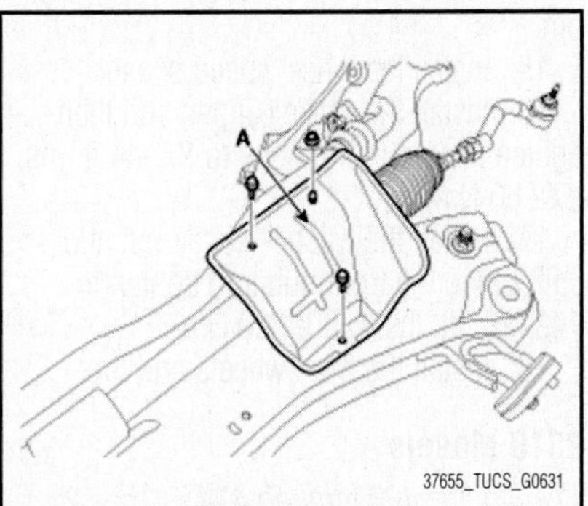

Fig. 396 Remove the bolt and nut and then remove the protector (A)

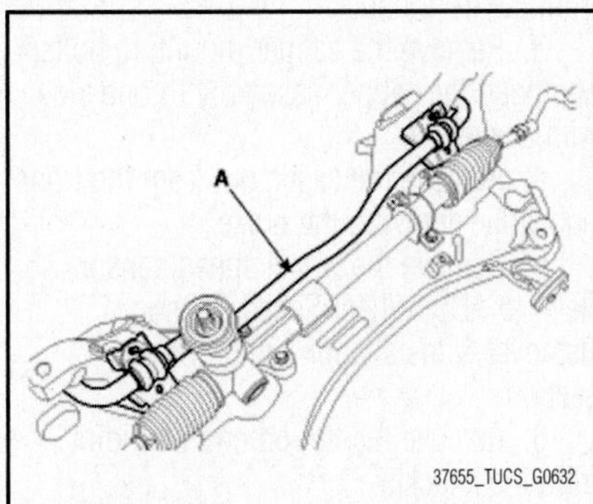

Fig. 397 Remove the stabilizer (A) from the sub frame

12. Remove the bolt and nut and then remove the protector.
13. Remove the bolt and then remove the stabilizer from the sub frame.

To install:

14. Installation is the reverse of the removal, noting the following:
    a. Tighten the stabilizer bar bolts to 33–40 ft. lbs. (44–54 Nm).
    b. Tighten the subframe bolts and nuts to 130–145 ft. lbs. (177–196 Nm).
    c. Tighten the roll rod stopper bolts and nuts to 80–94 ft. lbs. (108–128 Nm).
    d. Tighten the universal joint assembly bolt to 22–25 ft. lbs. (29–34 Nm).
    e. Tighten the lower arm bolts and nuts to 72–87 ft. lbs. (98–118 Nm).
    f. Tighten the tie-rod end nut to 25–33 ft. lbs. (34–44 Nm).
    g. Tighten the stabilizer link nut to 72–87 ft. lbs. (98–118 Nm).

WHEEL HUB & BEARING

REMOVAL & INSTALLATION

2009 Models

See Figures 398 through 402.

1. Before servicing the vehicle, refer to the Precautions Section.
2. Remove the front wheels and tires from the front hub.
3. Remove the split pin, castle nut and washer from the front hub while applying the brake.

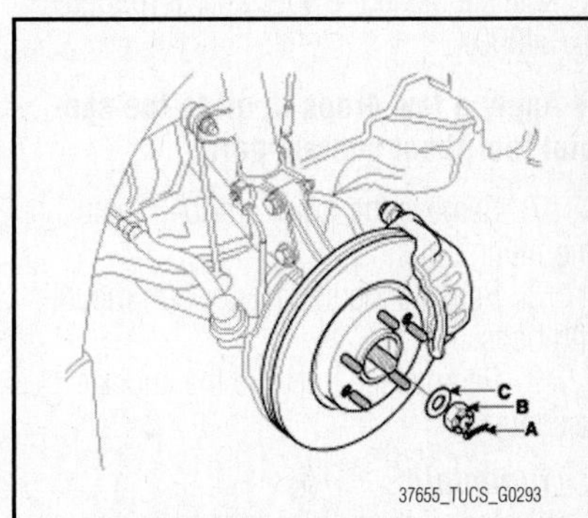

Fig. 398 Remove the split pin (A), castle nut (B) and washer (C) from the front hub

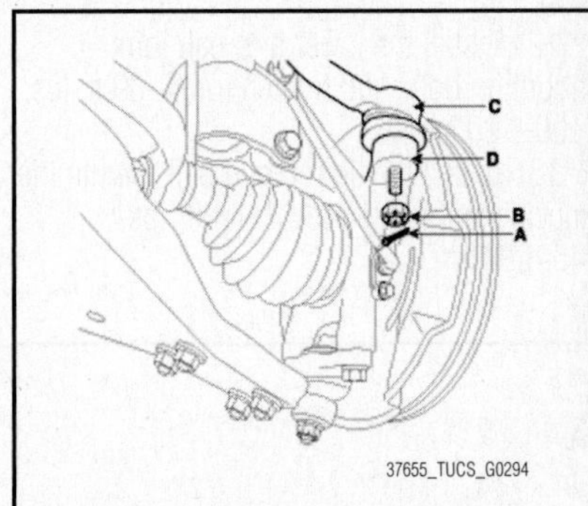

Fig. 399 Disconnect the tie rod end ball joint (C) from the knuckle (D), and remove the split pin (A) and castle nut (B)

4. Remove the caliper mounting bolts, and hang the caliper assembly to one side with a wire.
5. Remove the wheel speed sensor from the knuckle. Refer to ABS, Wheel

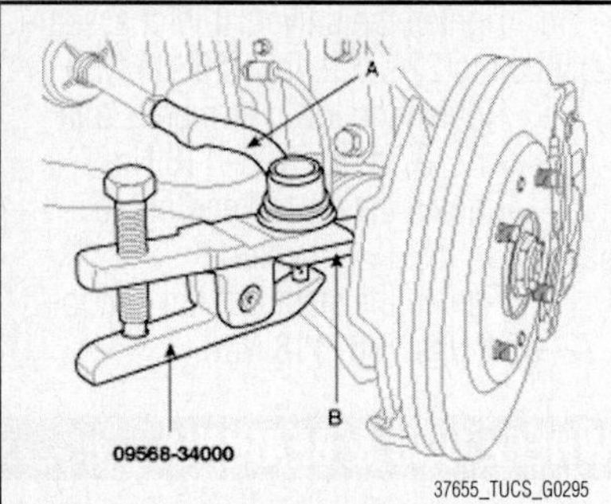

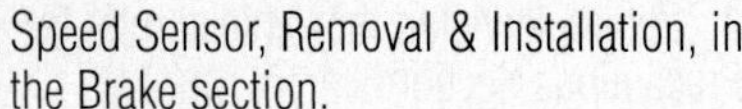
Fig. 400 Disconnect the ball joint from knuckle using the special tool (09568-34000)

Speed Sensor, Removal & Installation, in the Brake section.

6. Disconnect the tie rod end ball joint from the knuckle:
   a. Remove the split pin.
   b. Remove the castle nut.
   c. Disconnect the ball joint from knuckle using the special tool (09568-34000).

➡**Apply a few drops of oil to the special tool (boot contact part).**

7. Remove the lower arm ball joint mounting bolts.
8. Remove the strut lower arm mounting bolts.
9. Remove the hub and the knuckle assembly.

To install:

10. Install the hub and the knuckle assembly.
11. Install the strut lower mounting bolts and tighten to 103–118 ft. lbs. (140–160 Nm).
12. Install the lower arm ball joint mounting bolts and tighten to 74–89 ft. lbs. (100–120 Nm).
13. Install the tie rod end ball joint to the knuckle and tighten to 33–44 ft. lbs. (45–60 Nm).

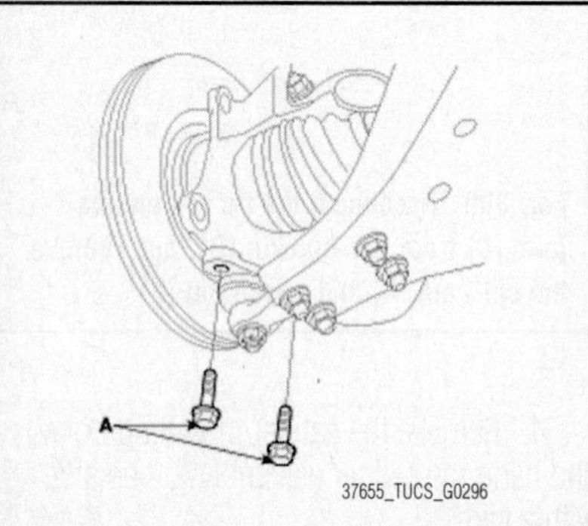

Fig. 401 Remove the lower arm ball joint mounting bolts (A)

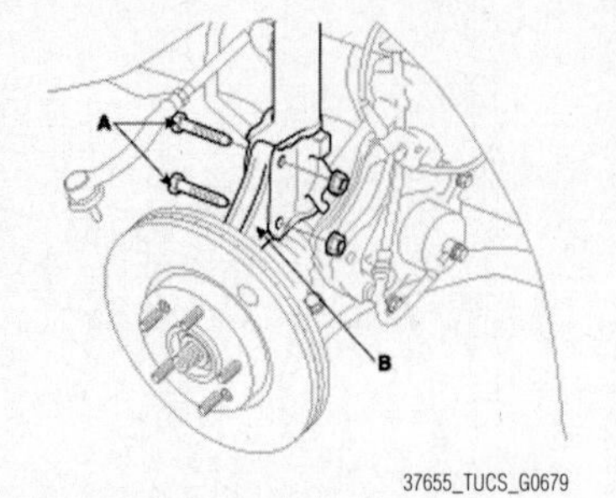

Fig. 402 Remove the strut lower arm mounting bolts (A) and the hub and the knuckle assembly (B)

14. Install the castle nut and split pin.
15. Install the wheel speed sensor.
16. Install the brake caliper, and then tighten the mounting bolts to 37–44 ft. lbs. (50–60 Nm).
17. Install the washer, castle nut and split pin to the front hub and tighten to 148–207 ft. lbs. (200–280 Nm).
18. Install the front wheels and tires.

2010 Models

See Figures 403 through 414.

1. Before servicing the vehicle, refer to the Precautions Section.
2. Remove the front wheels and tires from the front hub.
3. Remove the caliper mounting bolts, and hang the caliper assembly to one side with a wire.
4. Remove the castle nut from the front hub while applying the brake.
5. Remove the wheel speed sensor. Refer to ABS, Wheel Speed Sensor, Removal & Installation, in the Brake section.
6. Remove the tie rod end ball joint from the knuckle:
   a. Remove the split pin.
   b. Remove the castle nut.

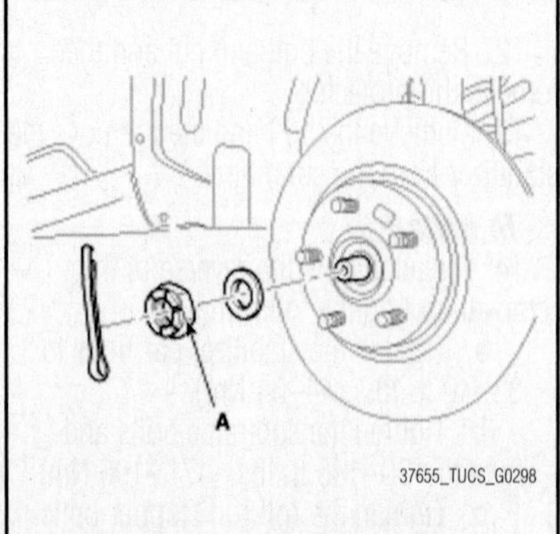

Fig. 403 Remove the castle nut (A)

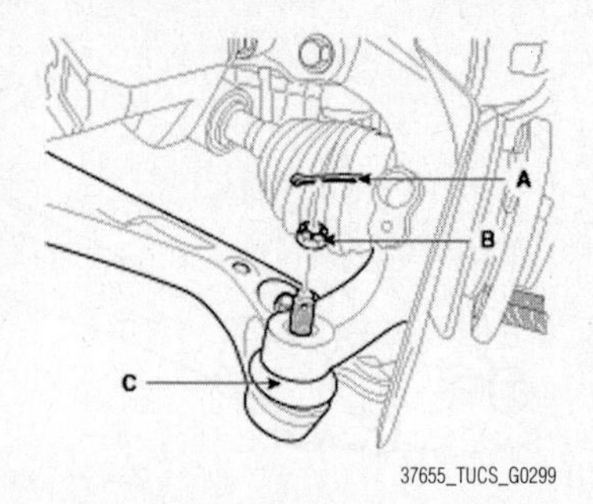

Fig. 404 Disconnect the tie rod end ball joint (C) from the knuckle (D), and remove the split pin (A) and castle nut (B)

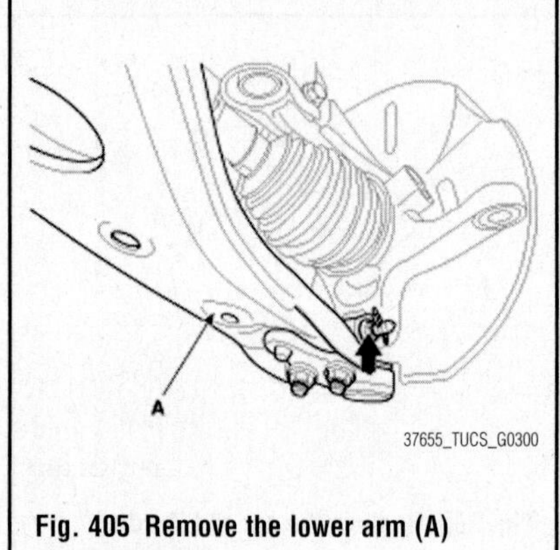

Fig. 405 Remove the lower arm (A) mounting bolt

7. Remove the lower arm mounting bolt.
8. Disconnect the halfshaft end from the knuckle.
9. Remove the strut mounting bolts and then remove the hub and knuckle assembly.

➡**Be careful not to damage the boot and rotor teeth.**

10. Using the snap ring pliers, remove the snap ring.
11. Remove the hub assembly from the knuckle assembly.

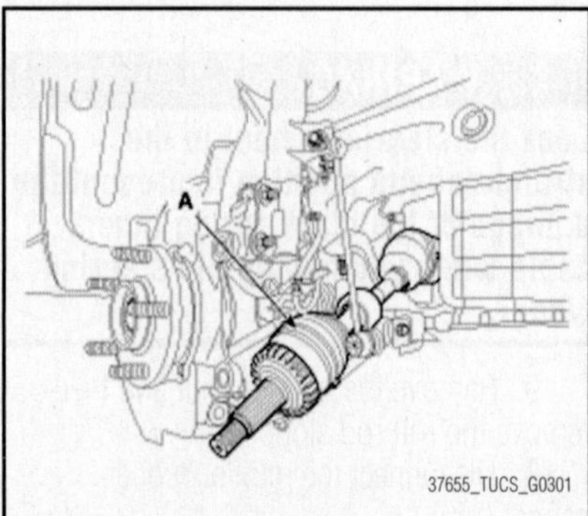

Fig. 406 Disconnect the halfshaft end (A) from the knuckle

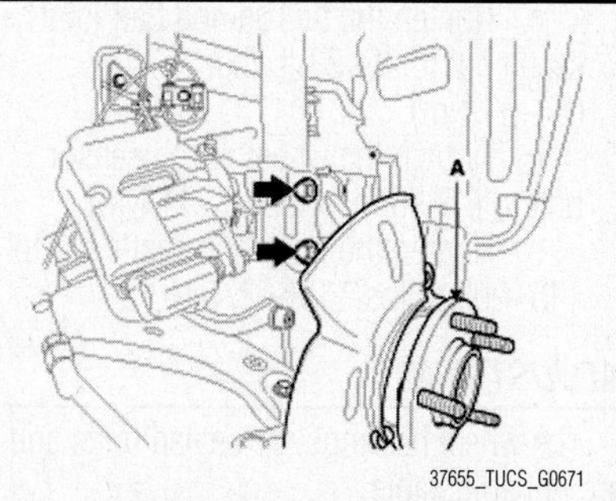

Fig. 407 Remove the strut mounting bolts and then remove the hub and knuckle assembly (A)

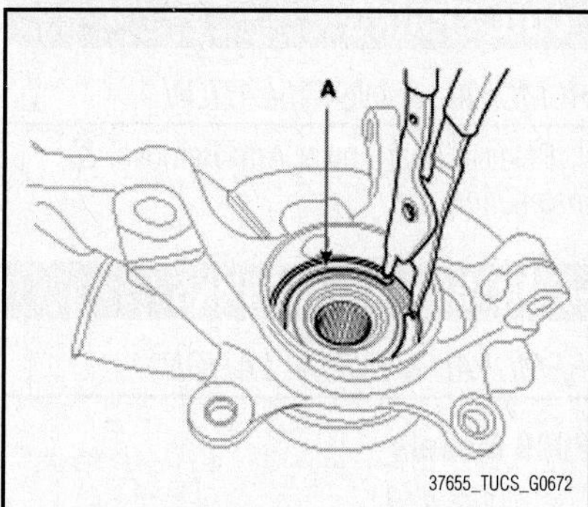

Fig. 408 Using the snap ring pliers, remove the snap ring (A)

a. Install the front knuckle assembly on press.

b. Lay a suitable adapter upon the hub assembly shaft.

12. Remove the dust cover.

13. Remove the hub bearing inner race from the hub assembly.

a. Install a suitable tool for removing the hub bearing inner race on the hub assembly.

b. Lay the hub assembly and tool upon a suitable adapter.

c. Lay a suitable adapter upon the hub assembly shaft.

d. Remove the hub bearing inner race from the hub assembly by using press.

14. Remove the hub bearing outer race from the knuckle assembly.

a. Lay the hub assembly upon a suitable adapter.

b. Lay a suitable adapter upon the hub bearing outer race.

c. Remove the hub bearing outer race from the knuckle assembly by using press.

15. Replace hub bearing with a new one.

To install:

16. Install the dust cover.

17. Install the hub bearing to the knuckle assembly.

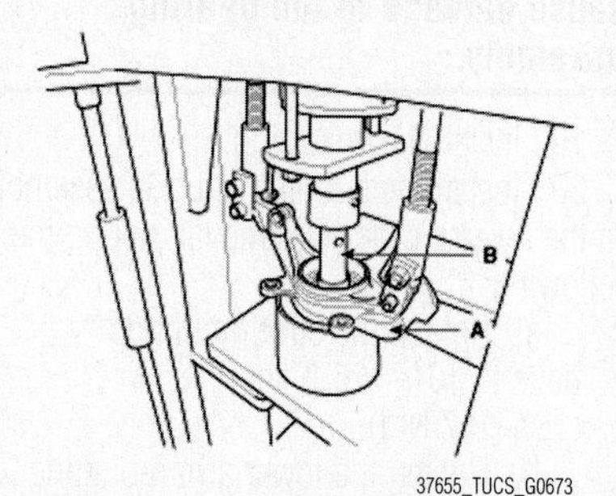

Fig. 409 Install the front knuckle assembly (A) on press and lay a suitable adapter (B) upon the hub assembly shaft

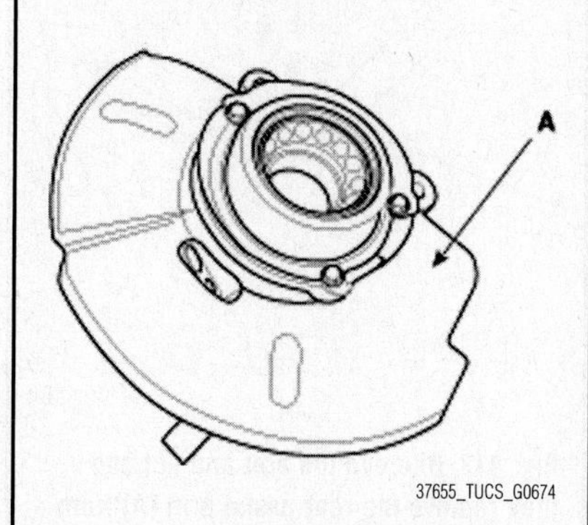

Fig. 410 Remove the dust cover (A)

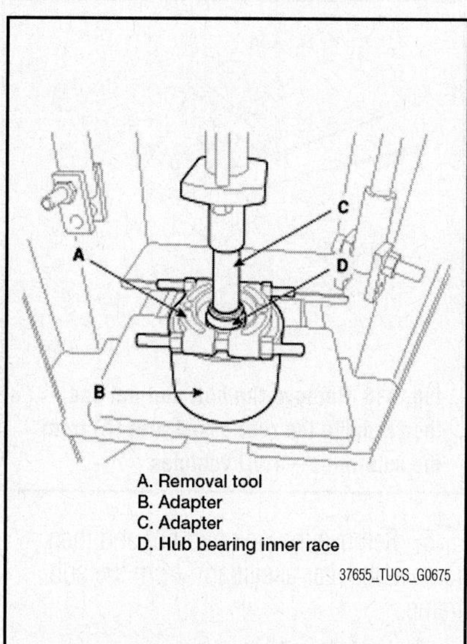

Fig. 411 Remove the hub bearing inner race from the hub assembly

a. Lay the knuckle assembly on the press.

b. Lay a new hub bearing upon the knuckle assembly.

c. Lay a suitable adapter upon the hub bearing.

d. Install the hub bearing to the knuckle assembly by using press.

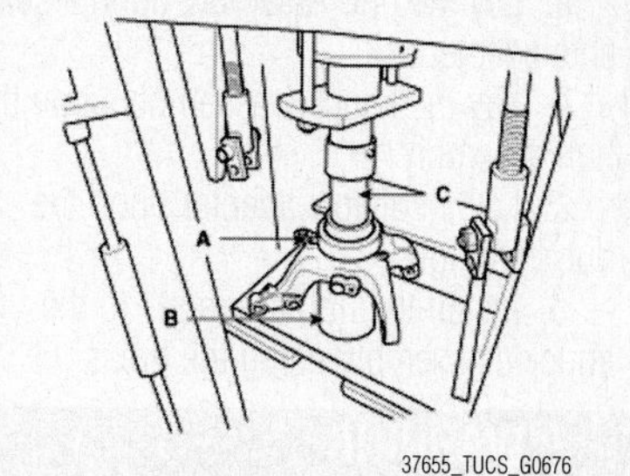

Fig. 412 Lay the hub assembly (A) upon a suitable adapter (B) and lay a suitable adapter (C) upon the hub bearing outer race

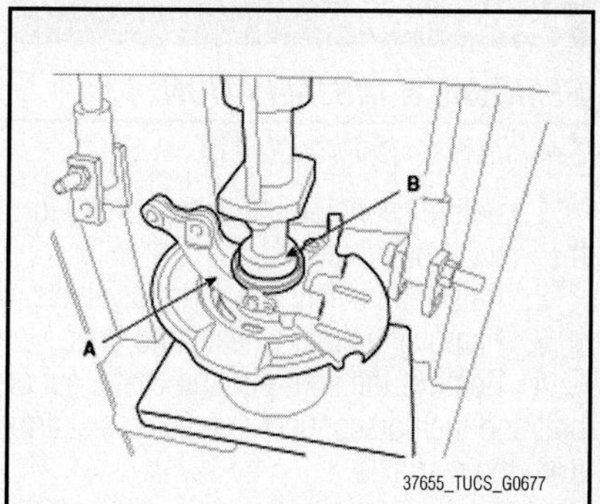

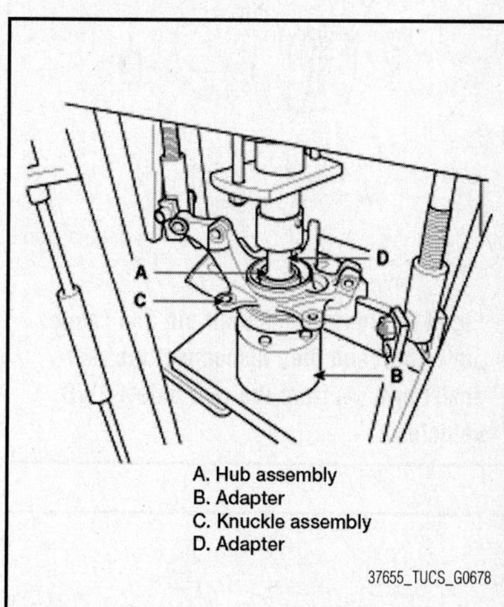

Fig. 413 Lay the knuckle assembly (A) on the press and a new hub bearing on the knuckle, and install a suitable adapter (B) upon the hub bearing

Fig. 414 Install the hub assembly to the knuckle assembly

➡Do not press against the inner race of the hub bearing because that can cause damage to the bearing assembly. Always use a new wheel bearing assembly.

18. Install the hub assembly to the knuckle assembly.

a. Lay the hub assembly upon a suitable adapter.

b. Lay the knuckle assembly upon the hub assembly.

c. Lay a suitable adapter upon the hub bearing.

d. Install the hub assembly to the knuckle assembly using the press.

WARNING

Do not press against the inner race of the hub bearing because that can cause damage to the bearing assembly.

19. Install the snap ring.
20. Install the hub and knuckle assembly in the reverse order of removal, noting the following:

a. Tighten the strut mounting bolts to 101–116 ft. lbs. (137–157 Nm).

b. Tighten the lower arm mounting bolt to 72–87 ft. lbs. (98–118 Nm).

c. Tighten the tie rod end ball joint castle nut to 25–33 ft. lbs. (34–44 Nm).

d. Tighten the wheel speed sensor bolt to 61–95 inch lbs. (7–11 Nm).

e. Tighten the front hub castle nut to 145–203 ft. lbs. (196–275 Nm).

ADJUSTMENT

The wheel bearings are sealed units and are not adjustable.

SUSPENSION — REAR SUSPENSION

ASSIST ARM

REMOVAL & INSTALLATION

See Figures 415 through 418.

1. Before servicing the vehicle, refer to the Precautions Section.
2. Raise and safely support the vehicle.
3. Remove the rear wheels and tires.
4. Remove the split pin and castle nut or bolt and then disconnect the rear assist arm from the rear axle.

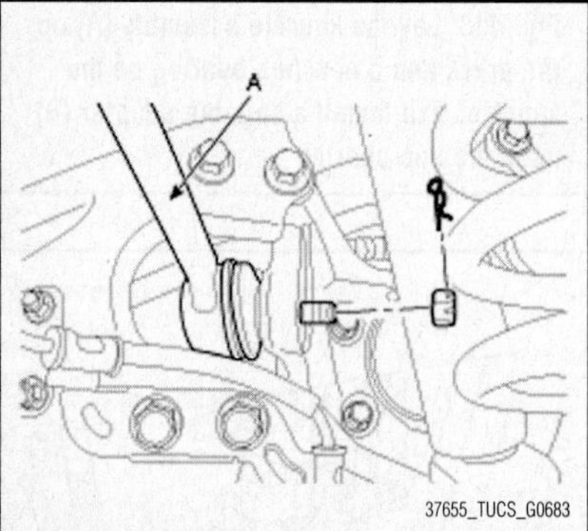

Fig. 415 Remove the split pin and castle nut or bolt and then disconnect the rear assist arm (A) from the rear axle—2WD vehicles

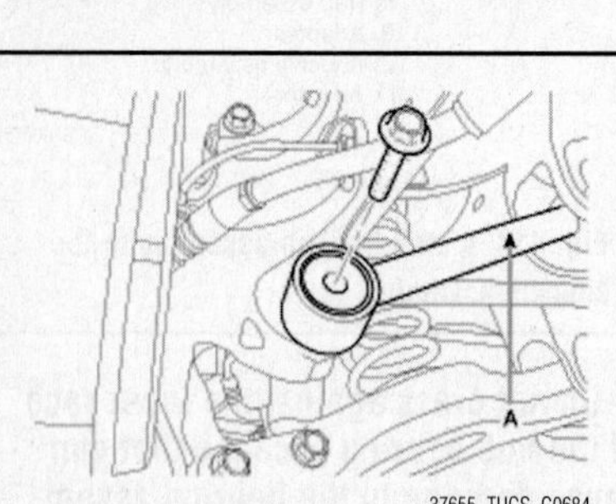

Fig. 416 Remove the split pin and castle nut or bolt and then disconnect the rear assist arm (A) from the rear axle—4WD vehicles

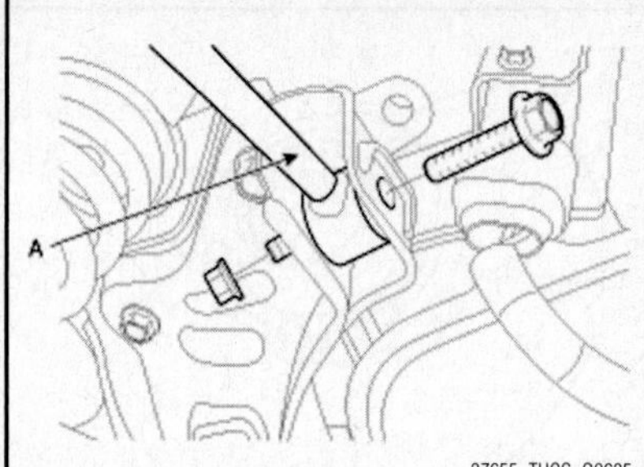

Fig. 417 Remove the bolt and nut and then remove the rear assist arm (A) from the sub frame—2WD vehicles

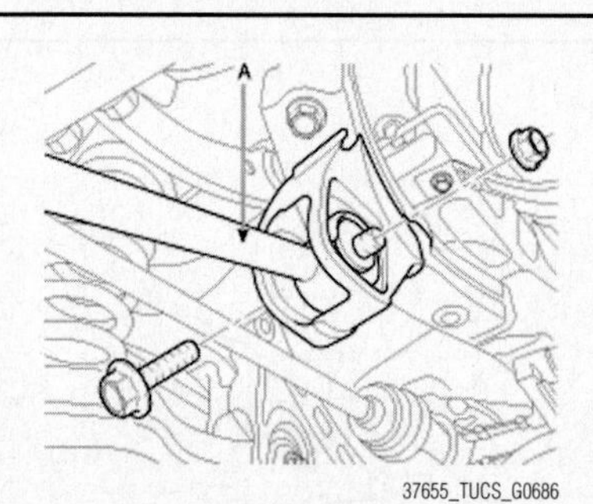

Fig. 418 Remove the bolt and nut and then remove the rear assist arm (A) from the sub frame—4WD vehicles

5. Remove the bolt and nut and then remove the rear assist arm from the sub frame.

To install:

6. Installation is the reverse of removal, noting the following:

a. Tighten the rear assist arm nut and bolt to 80–87 ft. lbs. (108–118 Nm).

b. Tighten the rear assist arm castle nut/bolt to:

- 2WD vehicles: 33–40 ft. lbs. (44–54 Nm).
- 4WD vehicles: 101–116 ft. lbs. (137–157 Nm).

COIL SPRING

REMOVAL & INSTALLATION

Refer to Rear Lower Arm Removal & Installation.

CONTROL ARMS/LINKS

REMOVAL & INSTALLATION

2009 Models

See Figure 419.

1. Before servicing the vehicle, refer to the Precautions Section.
2. Raise and safely support the vehicle.
3. Remove the rear wheels and tires.

WARNING

Be careful not to damage the hub bolts when removing the rear wheel and tire.

4. Remove the upper and lower nuts of the rear stabilizer control links.
5. Remove the rear stabilizer control link from the stabilizer bar assembly.

To install:

6. Install the rear stabilizer control link and nuts to the stabilizer bar assembly.

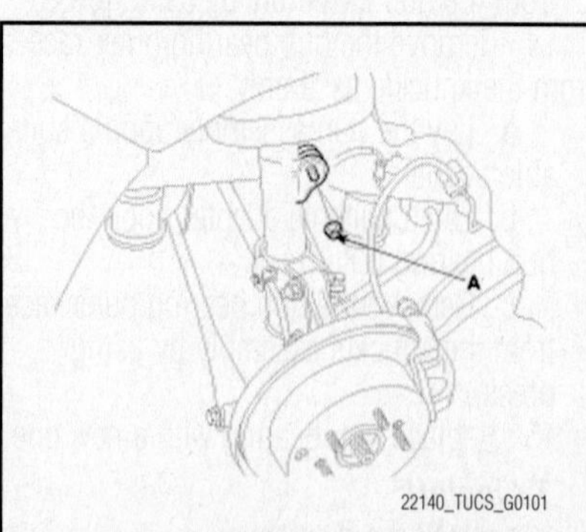

Fig. 419 Remove the upper (A) and lower nuts of the rear stabilizer control links

Tightening torque: 74–89 ft. lbs. (100–120 Nm).

7. Repeat appropriate steps for the other side.

8. Install the wheel and the tire to the rear hub. Tightening torque: 65–80 ft. lbs. (90–110 Nm).

2010 Models

See Figure 420.

1. Before servicing the vehicle, refer to the Precautions Section.
2. Raise and safely support the vehicle.
3. Remove the rear wheel and tire.
4. Remove the nut and then remove the rear stabilizer link with the rear lower arm.

To install:

5. Installation is the reverse of removal, noting the following:
   a. Tighten the rear stabilizer link mounting bolt to 72–87 ft. lbs. (98–118 Nm).

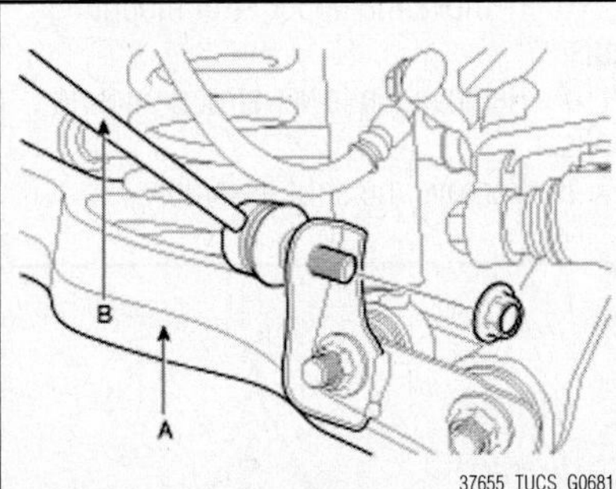

Fig. 420 Remove the nut and then remove the rear stabilizer link (B) with the rear lower arm (A)

LOWER ARM

REMOVAL & INSTALLATION

See Figures 421 through 422.

1. Before servicing the vehicle, refer to the Precautions Section.
2. Raise and safely support the vehicle.
3. Remove the rear wheels and tires.

⁂ WARNING

Be careful not to damage the hub bolts when removing the rear wheel and tire.

4. Remove the nut and then remove the rear stabilizer link from the rear lower arm.
5. Remove the bolt and nut and then remove the rear lower arm from the rear axle.
6. Remove the bolt and nut and then remove the rear lower arm from the subframe.

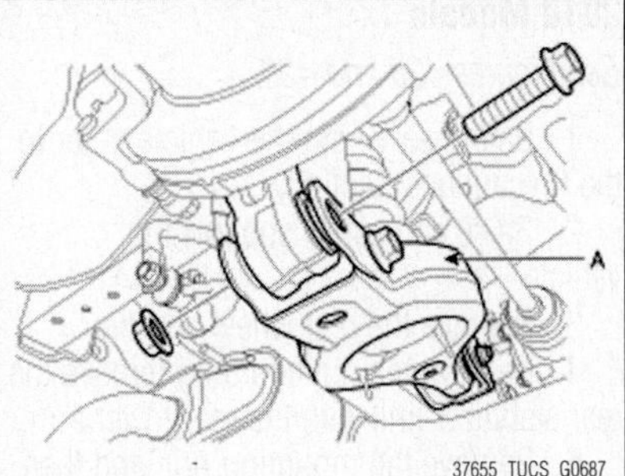

Fig. 421 Remove the bolt and nut and then remove the rear lower arm (A) from the rear axle

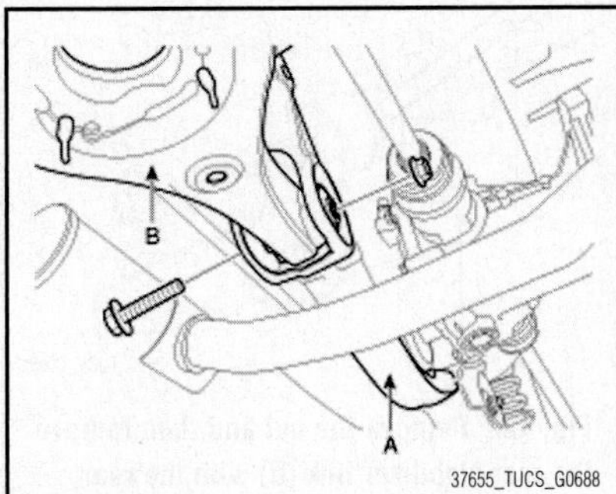

Fig. 422 Remove the bolt and nut and then remove the rear lower arm (B) from the subframe (A)

To install:

7. Installation is the reverse of removal, noting the following:
   a. Tighten the rear stabilizer link-to-subframe mounting nuts and bolts to 80–87 ft. lbs. (108–118 Nm).
   b. Tighten the rear stabilizer link-to-rear axle mounting nut and bolt to 101–116 ft. lbs. (137–157 Nm).
   c. Tighten the rear stabilizer link-to-rear lower arm mounting nut to 72–87 ft. lbs. (98–118 Nm).

SHOCK ABSORBER

REMOVAL & INSTALLATION

See Figures 423 and 424.

1. Before servicing the vehicle, refer to the Precautions Section.
2. Raise and safely support the vehicle.
3. Remove the rear wheels and tires.
4. Remove the bolt and nut and then remove the shock absorber from the rear axle.
5. Remove the shock absorber mounting bolts).

To install:

6. Installation is the reverse of removal, noting the following:

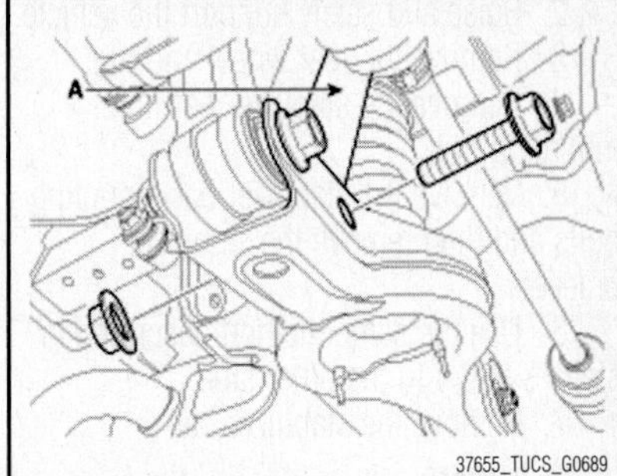

Fig. 423 Remove the bolt and nut and then remove the rear shock absorber (A) from the rear axle

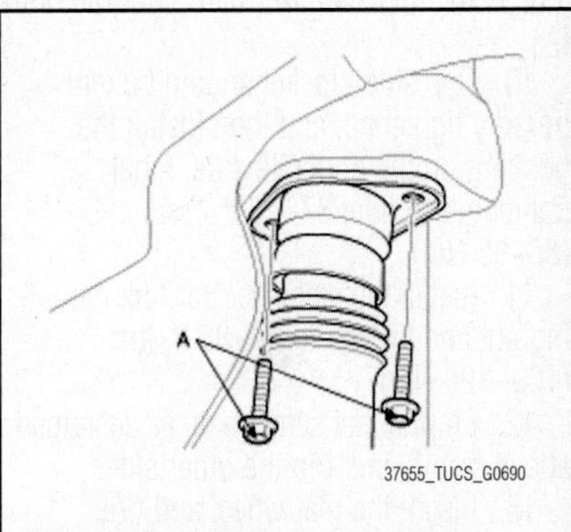

Fig. 424 Remove the shock absorber mounting bolts (A)

   a. Tighten the shock absorber mounting bolts to 36–47 ft. lbs. (49–64 Nm).
   b. Tighten the shock absorber-to-rear axle mounting nut and bolt to 101–116 ft. lbs. (137–157 Nm).

STABILIZER BAR

REMOVAL & INSTALLATION

2009 Models

See Figures 425 through 427.

1. Before servicing the vehicle, refer to the Precautions Section.

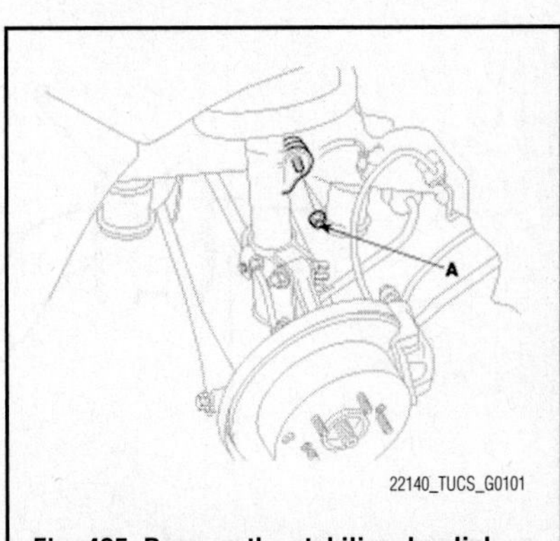

Fig. 425 Remove the stabilizer bar link mounting nut (A)

2. Raise and safely support the vehicle.
3. Remove the rear wheel and tire.
4. Remove the stabilizer bar link mounting nut.
5. Remove the stabilizer bar mounting bolts and then remove the stabilizer bracket.
6. Use the same method described in steps 3 and 4 to the other side.
7. Remove the stabilizer bar.

To install:

8. Install the bushing on the stabilizer bar. Bring the stabilizer bar clamp into contact with the bushing.
9. Install the stabilizer bracket (B) and then install the stabilizer bar mounting bolts (A).
10. One side bracket should be temporarily tightened, and then install the bushing on the opposite side. Final tightening torque: 37–48 ft. lbs. (50–65 Nm).
11. Install the stabilizer bar link mounting nut and tighten to 74–89 ft. lbs. (100–120 Nm).
12. Employ the same manner described above step 3 and 4 to the other side.
13. Install the rear wheel and tire.

2010 Models

See Figures 428 and 429.

1. Before servicing the vehicle, refer to the Precautions Section.
2. Raise and safely support the vehicle.
3. Remove the rear wheel and tire.
4. Remove the nut and then remove the rear stabilizer link with the rear lower arm.
5. Remove the mounting bolt and then remove the stabilizer bar with the sub frame.

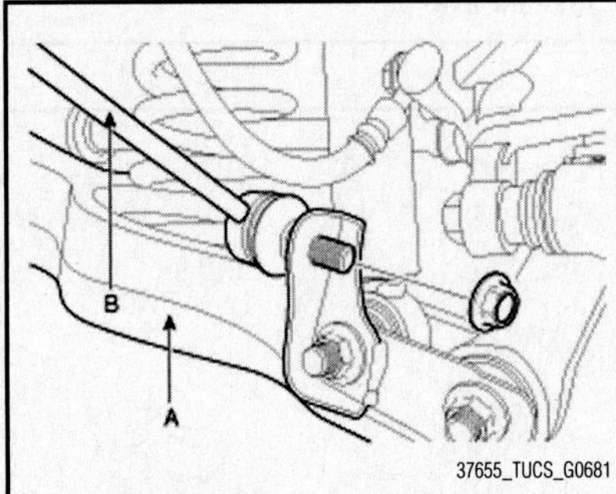

Fig. 428 Remove the nut and then remove the rear stabilizer link (B) with the rear lower arm (A)

To install:

6. Installation is the reverse of removal, noting the following:
   a. Tighten the stabilizer bar mounting bolt to 33–40 ft. lbs. (44–54 Nm).
   b. Tighten the rear stabilizer link mounting bolt to 72–87 ft. lbs. (98–118 Nm).

STRUTS

REMOVAL & INSTALLATION

See Figures 430 through 433.

1. Before servicing the vehicle, refer to the Precautions Section.
2. Raise and safely support the vehicle.
3. Remove the rear wheel and tire.
4. Remove the rear speed sensor. Refer to Wheel Speed Sensor Removal & Installation in the Brake section.
5. Remove the stabilizer bar control link mounting nut.
6. Remove the upper strut mounting nuts.
7. Remove the lower strut mounting bolts.
8. Remove the strut assembly.

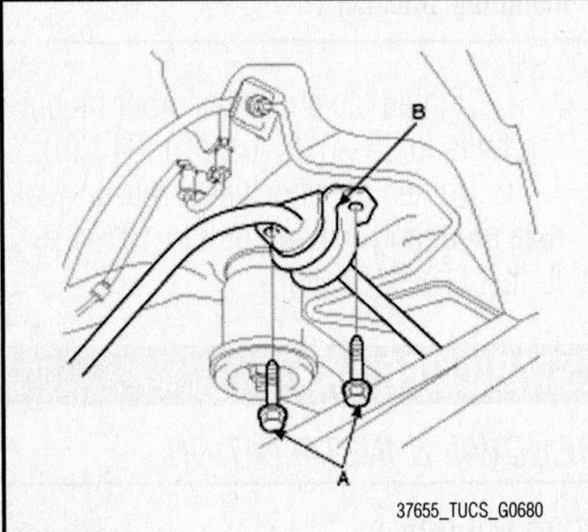

Fig. 426 Remove the stabilizer bar mounting bolts (A) and then remove the stabilizer bracket (B)

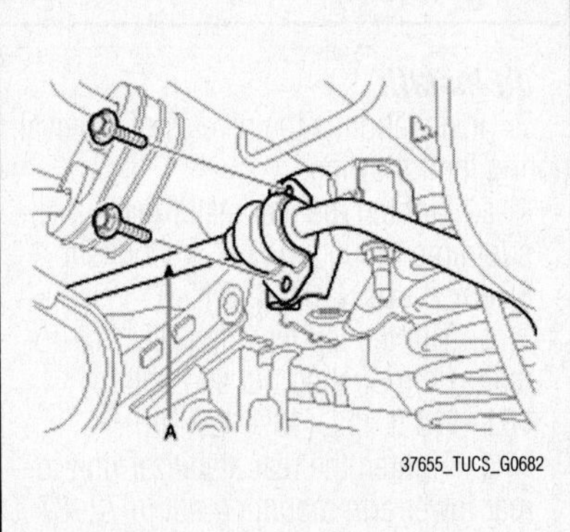

Fig. 429 Remove the stabilizer bar (A) with the sub frame

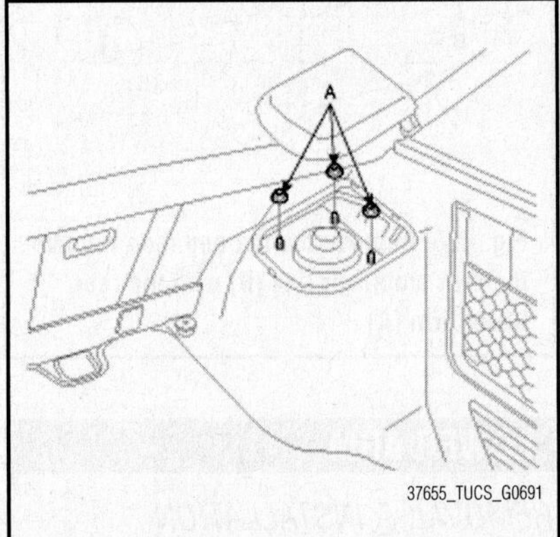

Fig. 430 Remove the upper strut mounting nuts (A)

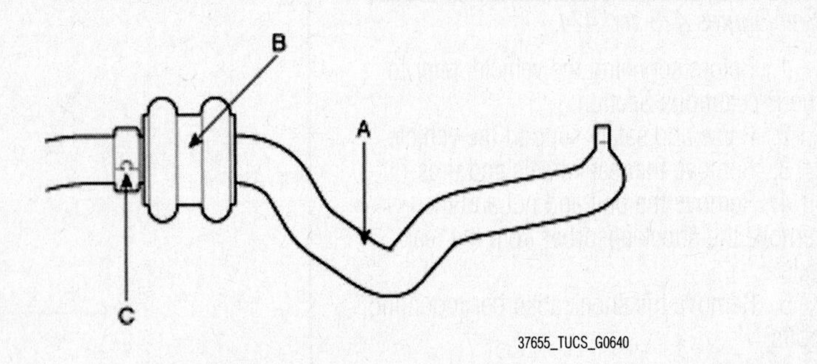

Fig. 427 Install the bushing (B) on the stabilizer bar (A) and bring the clamp (C) into contact with bushing

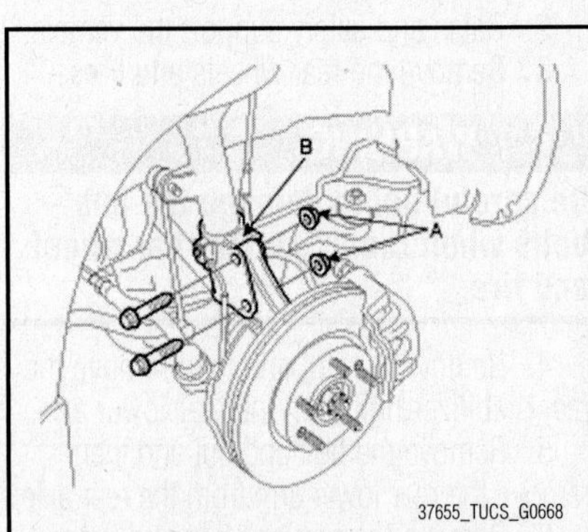

Fig. 431 Remove the lower strut mounting bolts (A) and the strut assembly (B)

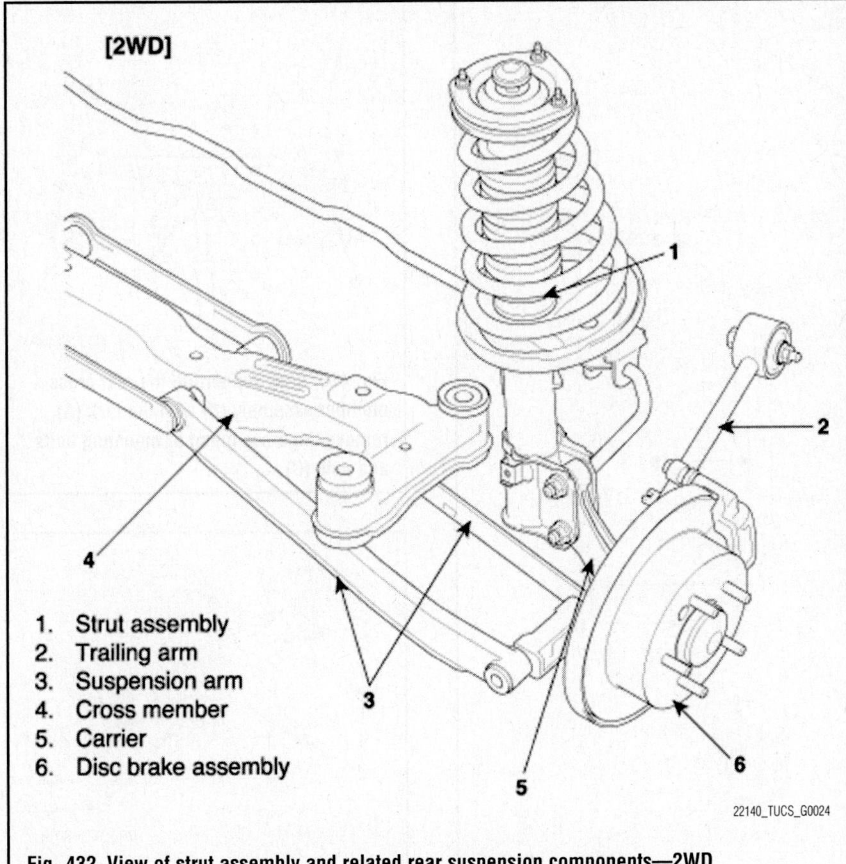

Fig. 432 View of strut assembly and related rear suspension components—2WD

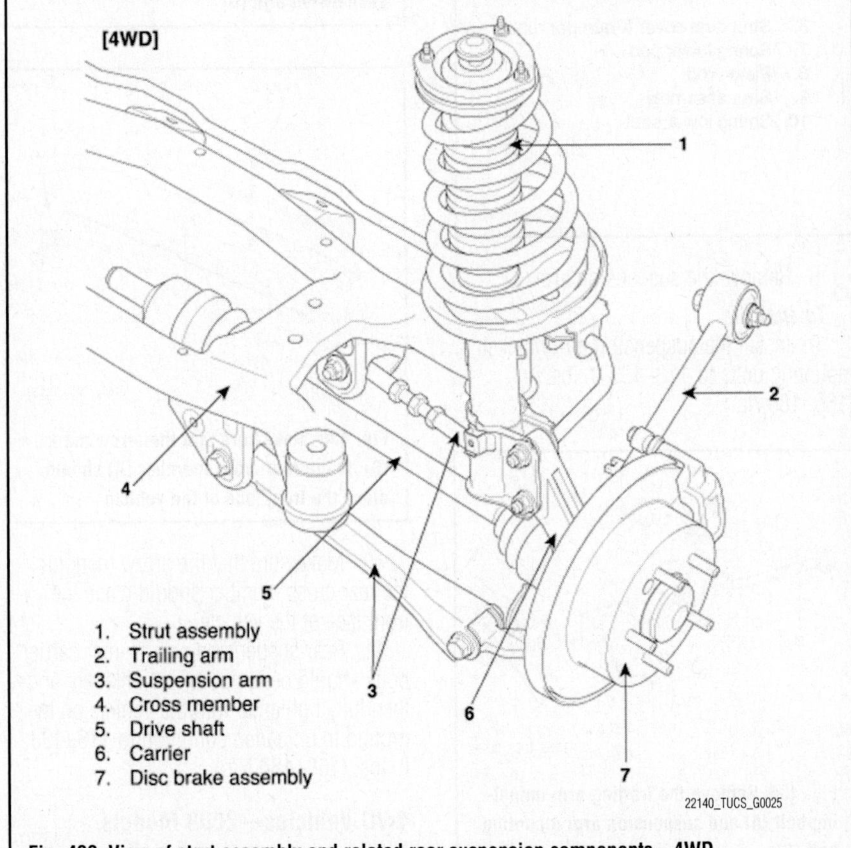

Fig. 433 View of strut assembly and related rear suspension components—4WD

To install:

9. Install the strut assembly and tighten the lower mounting bolts to 103–118 ft. lbs. (140–160 Nm)
10. Install the upper strut mounting bolts. Tighten to 22–30 ft. lbs. (30–40 Nm)
11. Install the stabilizer bar link mounting nut. Tighten to 74–89 ft. lbs. (100–120 Nm)
12. Install the speed sensor. Tighten the mounting bolt to 60–96 inch lbs. (7–11 Nm).
13. Install the rear wheel. Tighten the lug nuts to 66–81 ft. lbs. (90–110 Nm).
14. Check the alignment and adjust as necessary.

OVERHAUL

See Figure 434.

1. Before servicing the vehicle, refer to the Precautions Section.
2. Remove the strut from the vehicle and attach Service Tool 09546-2600 or another suitable spring compressor.
3. Compress the coil spring.
4. Remove the self-locking nut from the strut.
5. Remove the insulator, spring seat, coil spring, and dust cover.
6. Inspect the insulator for wear and damage, replace as required.
7. Check the rubber parts for damage or deterioration, replace as required.

To reassemble:

8. Install the spring lower pad so that the protrusions fit the holes in the spring lower seat.
9. Compress the spring, using the spring compressor tool.
10. Install the compressed spring over the shock absorber.

➡There are 2 color identification marks on the coil spring, one indicates model option and the other indicates load classification. Install the coil spring with the identification mark directed toward the steering knuckle.

11. After fully extending the piston rod, install the spring upper seat and insulator assembly.
12. After correctly seating the upper and lower ends of the coil spring in the upper and lower spring grooves, tighten the NEW self-locking nut temporarily.
13. Remove the spring compression tool. Tighten the self-locking nut to 43–51 ft. lbs. (60–70 Nm).
14. Install the strut to the vehicle.
15. Check the wheel alignment and adjust as necessary.

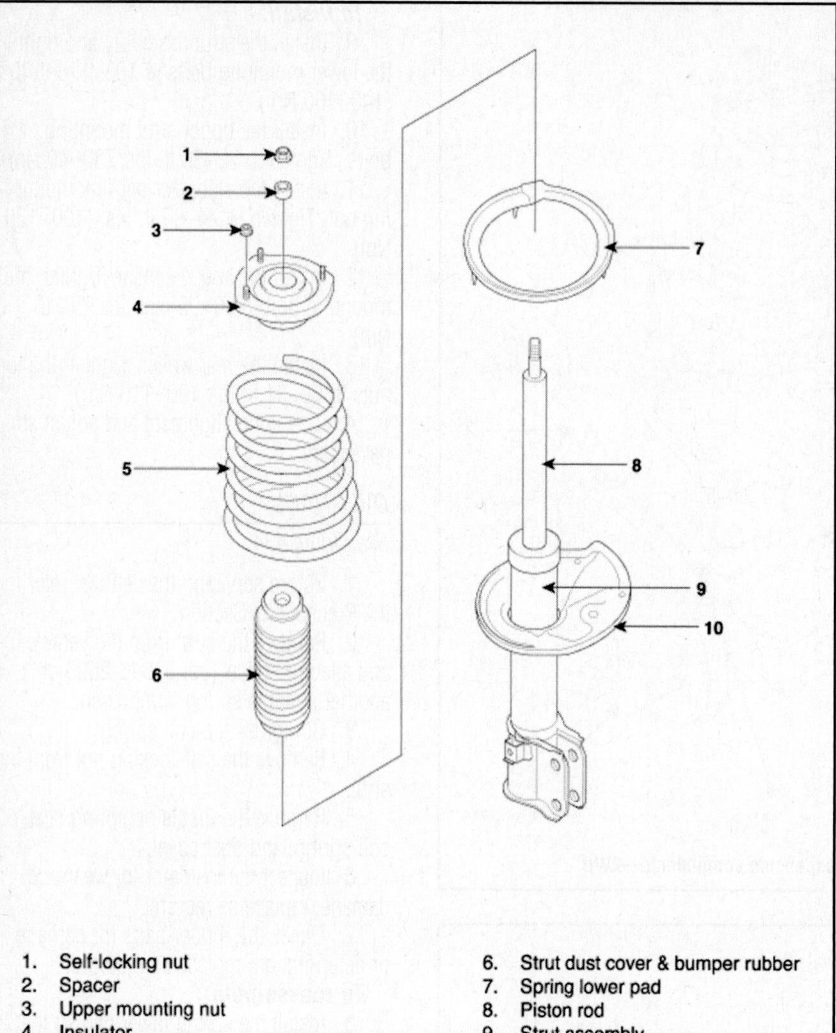

Fig. 434 Exploded view of rear strut assembly

SUSPENSION ARM

REMOVAL & INSTALLATION

2WD Vehicles—2009 Models

See Figures 435 through 438.

1. Before servicing the vehicle, refer to the Precautions Section.
2. Raise and safely support the vehicle.
3. Remove the rear wheel and tire.
4. Remove the trailing arm mounting bolt and suspension arm mounting bolt.
5. Remove the opposite side trailing arm mounting bolt and suspension arm mounting bolt.
6. After supporting the rear cross member assembly with the jack, remove the cross member mounting bolts and nuts.
7. Remove the suspension arm bracket mounting bolts.
8. Remove the suspension arm.

To install:

9. Install the suspension arm bracket mounting bolts to 118–133 ft. lbs. (160–180 Nm).

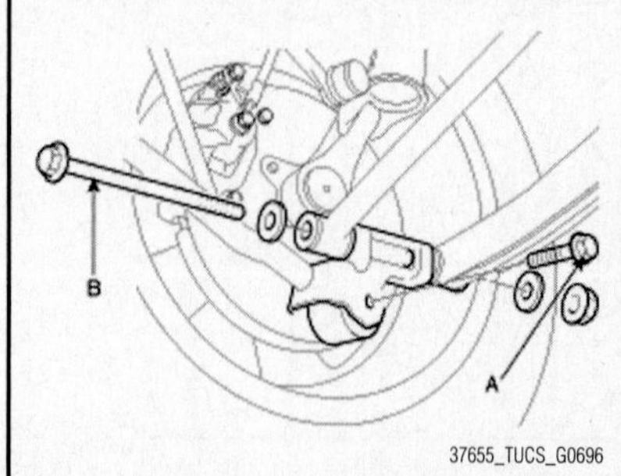

Fig. 435 Remove the trailing arm mounting bolt (A) and suspension arm mounting bolt (B)

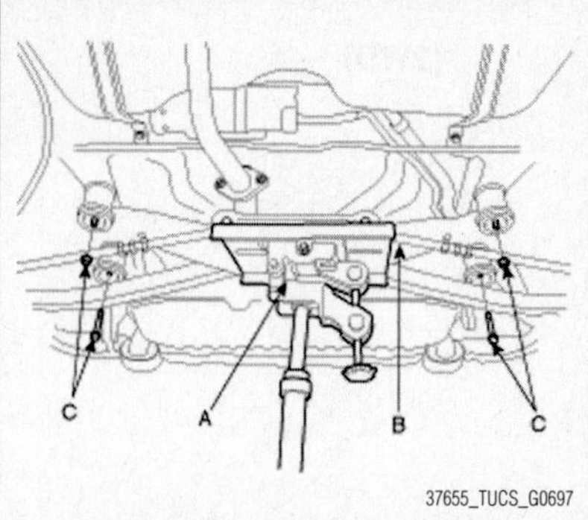

Fig. 436 After supporting the rear cross member assembly (B) with the jack (A), remove the cross member mounting bolts and nuts (C)

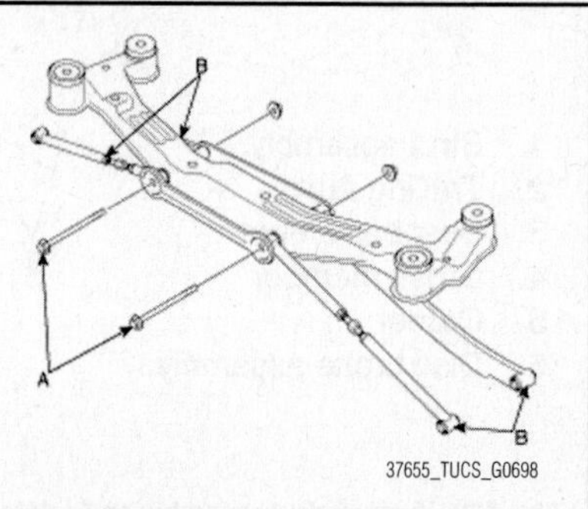

Fig. 437 Remove the suspension arm bracket mounting bolts (A) and the suspension arm (B)

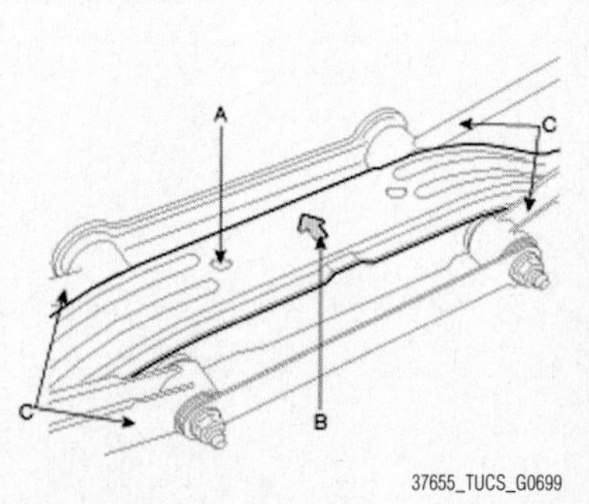

Fig. 438 Make sure that the arrow mark (B) on the rear cross member (A) should place the front face of the vehicle

10. Make sure that the arrow mark on the rear cross member should place the front face of the vehicle.
11. Rear suspension arm-to-rear carrier bolts should be temporarily tightened, and then fully tightened with the vehicle on the ground in unloaded condition to 118–133 ft. lbs. (160–180 Nm).

4WD Vehicles—2009 Models

See Figures 439 through 442.

1. Before servicing the vehicle, refer to the Precautions Section.
2. Raise and safely support the vehicle.
3. Remove the rear wheel and tire.
4. Remove the muffler.
5. Remove the suspension arm mounting bolts.
6. Remove the opposite side suspension mounting bolts.
7. Remove the coupling control connector.

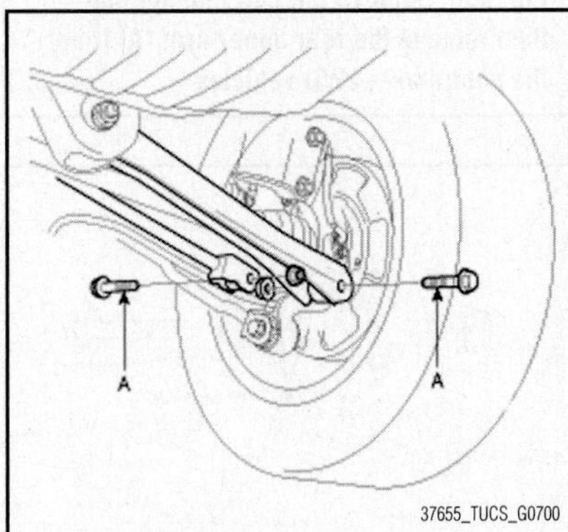

Fig. 439 Remove the suspension arm mounting bolts (A)

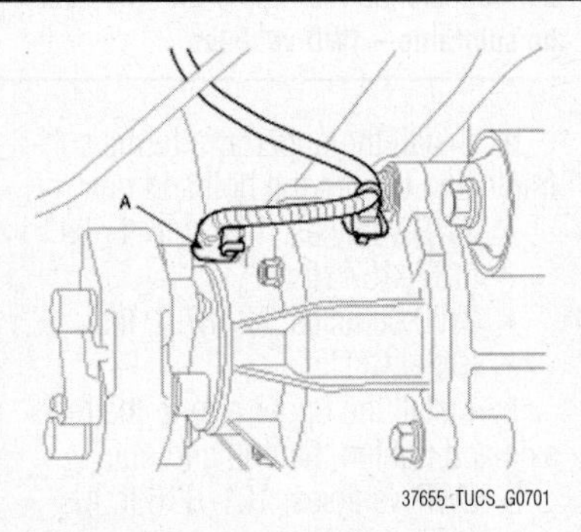

Fig. 440 Remove the coupling control connector (A)

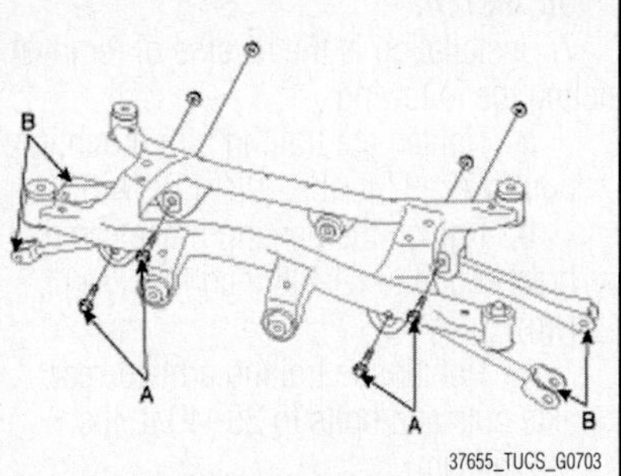

Fig. 442 Remove the suspension arm bracket mounting bolts (A) and the suspension arm (B)

8. After supporting the rear cross member assembly with a jack, remove the cross member mounting bolts and nuts.
9. Remove the propeller shaft. Refer to Propeller Shaft Removal & Installation in the Drive Train section.
10. Remove the rear differential from the cross member. Refer to Differential Removal & Installation in the Drive Train section.
11. Remove the suspension arm bracket mounting bolts.
12. Remove the suspension arm.

To install:

13. Install the suspension arm bracket mounting bolts and tighten to 103–118 ft. lbs. (140–160 Nm).
14. Install the rear differential on the cross member and tighten to 59–89 ft. lbs. (90–120 Nm).
15. Install the propeller shaft.
16. After supporting the rear cross member assembly with the jack, install the crossmember mounting bolts and nuts and tighten to 74–89 ft. lbs. (100–120 Nm).
17. Install the coupling control connector.
18. Rear suspension arm-to-rear carrier bolts should be temporarily tightened, and then fully tightened with the vehicle on the ground in unloaded condition to 103–118 ft. lbs. (140–160 Nm).

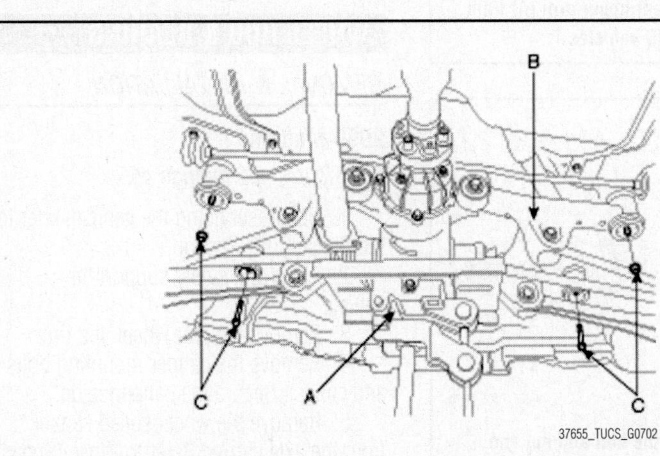

Fig. 441 After supporting the rear cross member assembly (B) with a jack (A), remove the cross member mounting bolts and nuts (C)

TRAILING ARM

REMOVAL & INSTALLATION

2009 Models

See Figure 443.

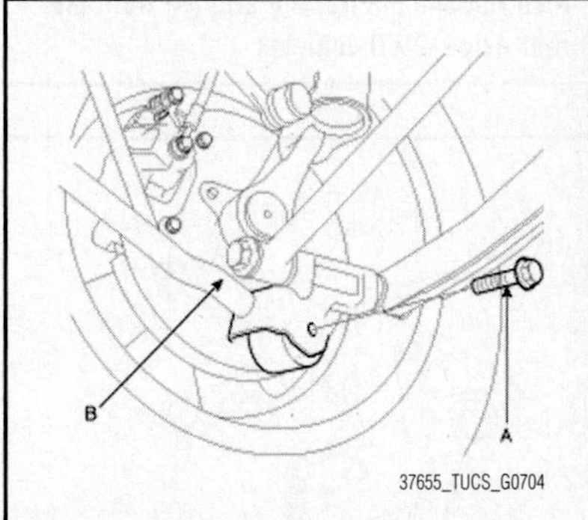

Fig. 443 Remove the trailing arm mounting bolts (A) and the trailing arm (B)

1. Before servicing the vehicle, refer to the Precautions Section.
2. Raise and safely support the vehicle.
3. Remove the rear wheel and tire.
4. Remove the trailing arm mounting bolts.
5. Remove the bracket mounting bolt, nut of the vehicle side.
6. Remove the trailing arm.

To install:

7. The trailing arm mounting bolts should be temporarily tightened, and then fully tightened with the vehicle on the ground in unloaded condition to 74–89 ft. lbs. (100–120 Nm).
8. Install the trailing arm bracket mounting bolt and nut and tighten to 74–89 ft. lbs. (100–120 Nm).

2010 Models

See Figures 444 through 447.

1. Before servicing the vehicle, refer to the Precautions Section.
2. Raise and safely support the vehicle.
3. Remove the rear wheel and tire.
4. Remove the bolts and nuts and then remove the trailing arm from the rear axle.
5. Remove the parking brake cable bracket bolt and height sensor bracket bolt.
6. Remove the mounting bolt and then remove the trailing arm from the frame.

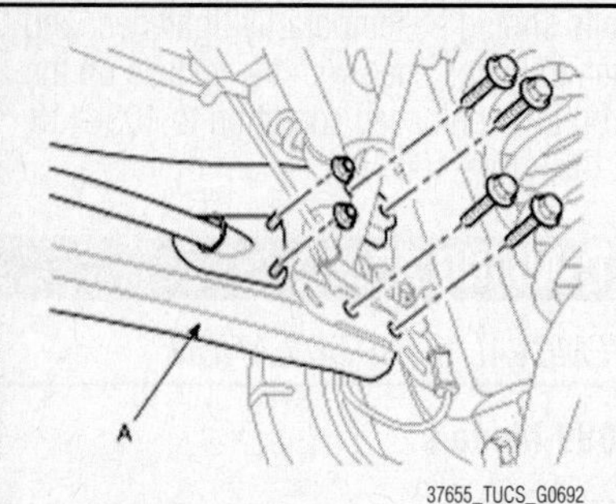

Fig. 444 Remove the bolts and nuts and then remove the trailing arm (A) from the rear axle—2WD vehicles

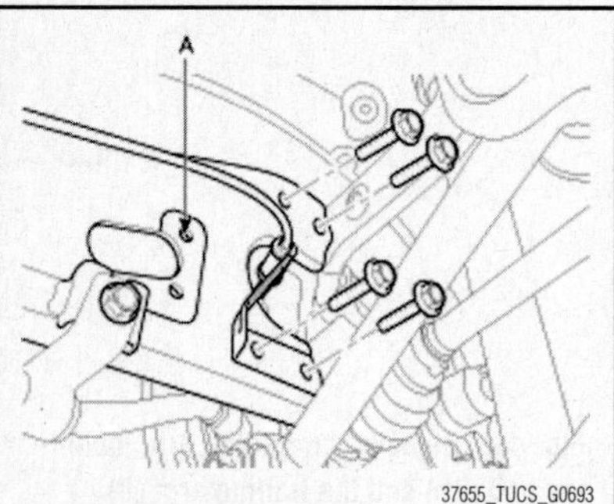

Fig. 445 Remove the bolts and nuts and then remove the trailing arm (A) from the rear axle—4WD vehicles

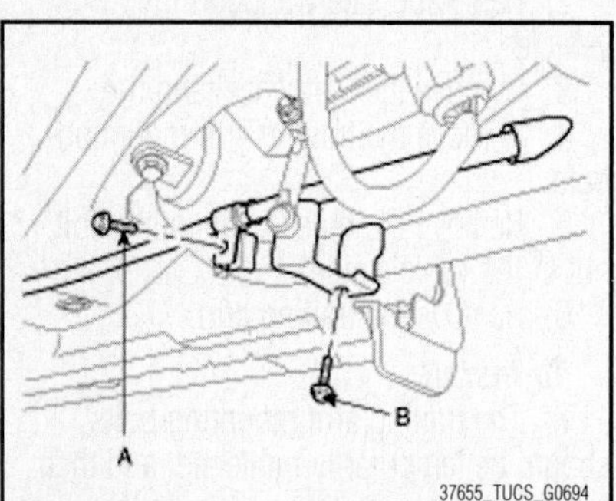

Fig. 446 Remove the parking brake cable bracket bolt (A) and height sensor bracket bolt (B)

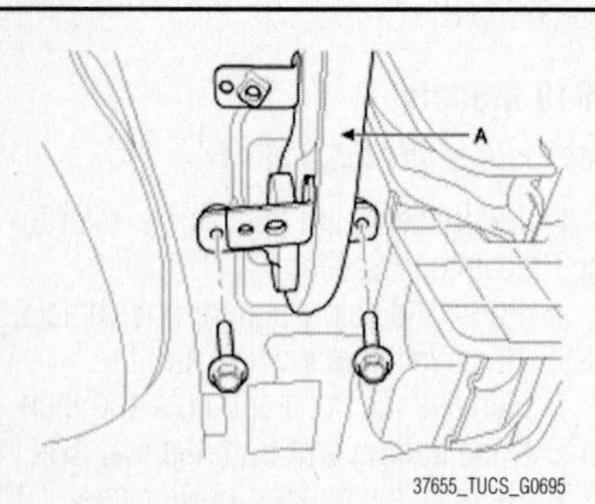
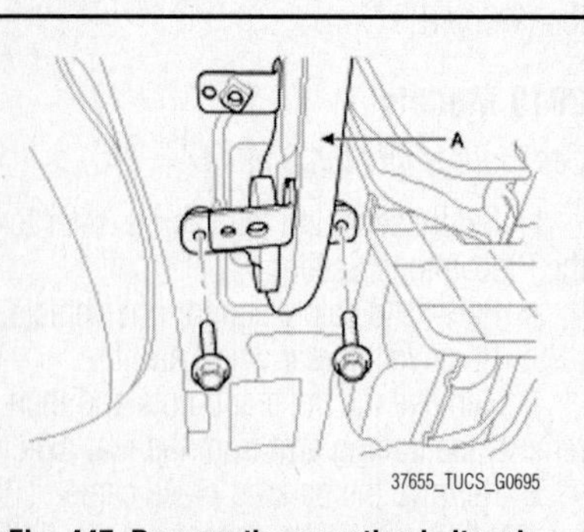

Fig. 447 Remove the mounting bolt and then remove the trailing arm (A) from the frame

To install:

7. Installation is the reverse of removal, noting the following:
   a. Tighten the trailing arm mounting bolt to 72–87 ft. lbs. (98–118 Nm).
   b. Tighten the parking brake cable bracket bolt to 61–96 inch lbs. (7–11 Nm).
   c. Tighten the trailing arm-to-rear axle nuts and bolts to 25–40 ft. lbs. (34–54 Nm).

UPPER ARM

REMOVAL & INSTALLATION

See Figures 448 through 451.

1. Before servicing the vehicle, refer to the Precautions Section.
2. Raise and safely support the vehicle.
3. Remove the rear wheel and tire.
4. Remove the bolt and nut and then remove the rear upper arm from the rear axle.
5. Remove the bolt and nut and then remove the rear upper arm from the subframe.

To install:

6. Installation is the reverse of removal, noting the following:

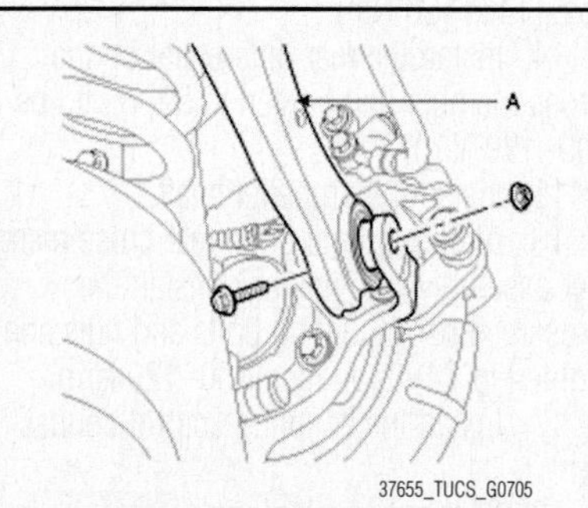

Fig. 448 Remove the bolt and nut and then remove the rear upper arm (A) from the rear axle—2WD vehicles

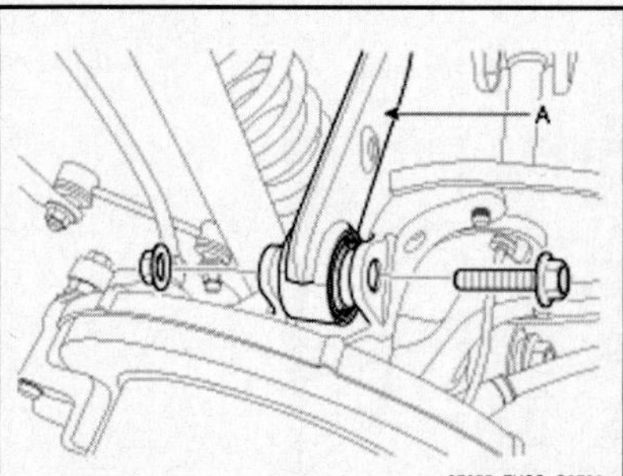

Fig. 449 Remove the bolt and nut and then remove the rear upper arm (A) from the rear axle—4WD vehicles

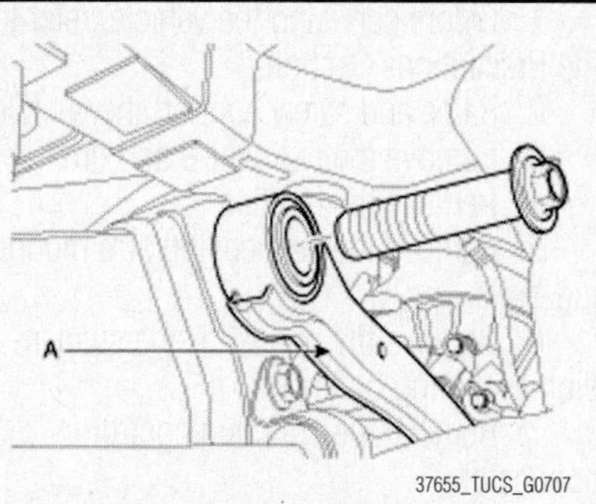

Fig. 450 Remove the bolt and nut and then remove the rear upper arm (A) from the subframe—2WD vehicles

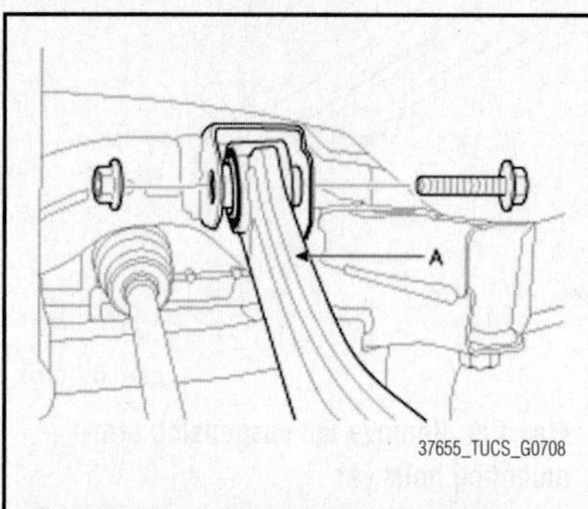

Fig. 451 Remove the bolt and nut and then remove the rear upper arm (A) from the subframe—4WD vehicles

   a. Install the upper arm to the subframe and tighten the bolt and nut:
      - 2WD vehicles: 101–116 ft. lbs. (137–157 Nm)
      - 2WD vehicles: 72–87 ft. lbs. (98–118 Nm)
   b. Install the upper arm to the rear axle and tighten the bolt and nut:
      - 2WD vehicles: 101–116 ft. lbs. (137–157 Nm)
      - 2WD vehicles: 72–87 ft. lbs. (98–118 Nm)

WHEEL HUB & BEARING

REMOVAL & INSTALLATION

2009 Models

See Figures 452 through 455.

1. Before servicing the vehicle, refer to the Precautions Section.
2. Raise and safely support the vehicle.
3. Remove the rear wheel and tire.
4. Remove the caliper mounting bolts and caliper. Secure the caliper aside.
5. Remove the wheel speed sensor from the axle carrier. Refer to Wheel Speed Sensor Removal & Installation in the Brake section.

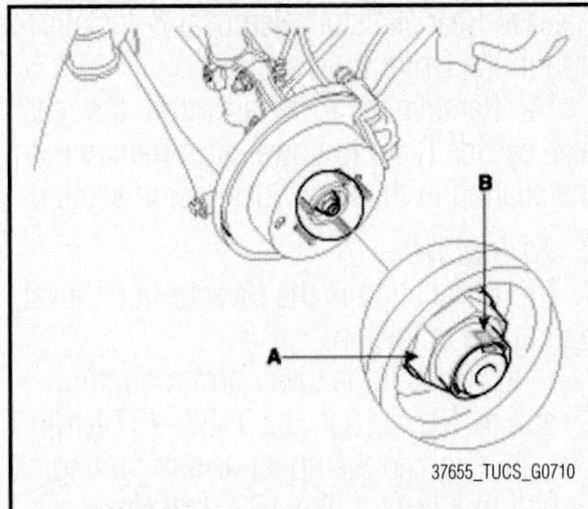

Fig. 452 Spread out the groove (B) on the flange nut (A)

6. Remove the brake disc from the hub.
7. Remove the hubcap.
8. Remove the hub bearing flange nut:
   a. Using a flat-tipped screwdriver, spread out the groove on the flange nut.
   b. Remove the hub bearing flange nut.
9. Remove the rear hub washer and rear hub assembly.

WARNING

Be careful not to disassemble the rear hub assembly. (For vehicles equipped with ABS.)

WARNING

Care must be taken not to scratch or damage the teeth of the rotor. The rotor must never be dropped. If the teeth of the rotor are chipped, it results in deformation of the rotor. It will make it impossible to detect the wheel rotation speed accurately and to operate the system normally.

10. Remove the rear dust cover mounting bolts and then remove the rear parking brake assembly.

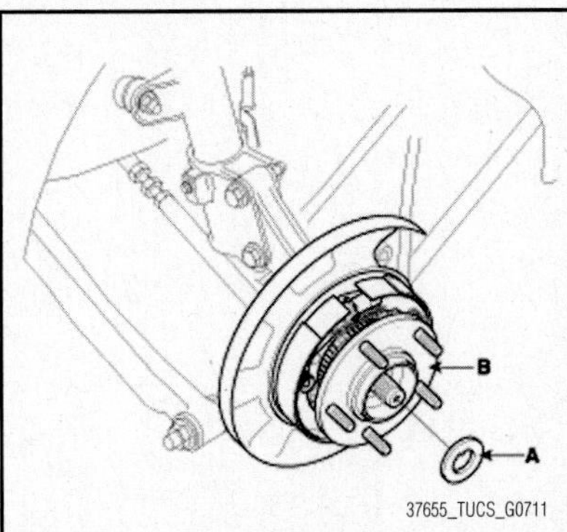

Fig. 453 Remove the rear hub washer (A) and rear hub assembly (B)

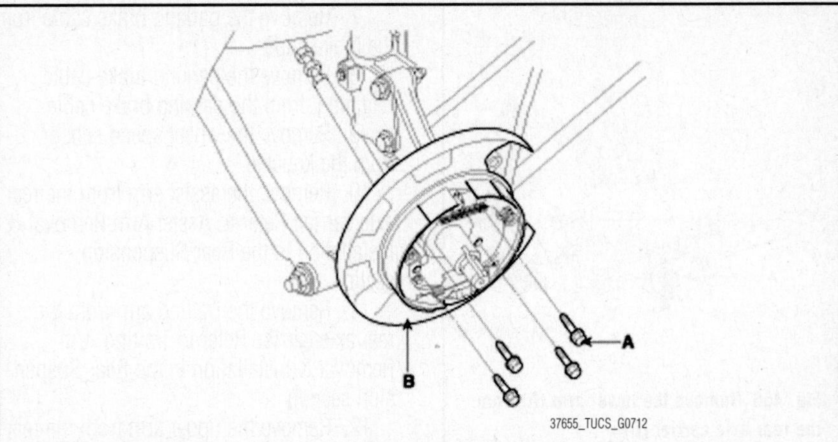

Fig. 454 Remove the rear dust cover mounting bolts (A) and then remove the rear parking brake assembly (B)

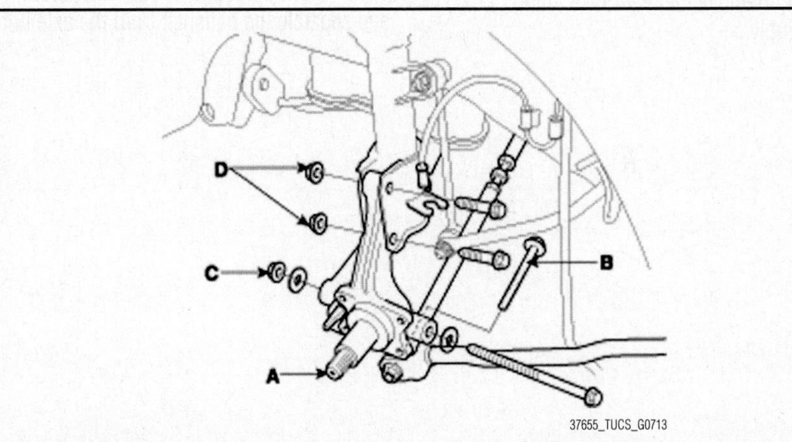

Fig. 455 Remove the rear axle carrier (A), trailing arm mounting bolt (B), suspension arm and strut mounting nuts (C, D)

11. Remove the rear axle carrier.
   a. Remove the trailing arm mounting bolt.
   b. Remove the suspension arm mounting nut.
   c. Remove the strut mounting nuts.

To install:

12. Install the rear axle carrier.
   a. Tighten the strut mounting nuts to 103–118 ft. lbs. (140–160 Nm).
   b. Tighten the suspension arm mounting nut to 118–133 ft. lbs. (160–180 Nm).
   c. Tighten the trailing arm mounting bolts to 74–89 ft. lbs. (100–120 Nm).

➡Replace the self-locking nut with new ones after removal.

13. Install the rear dust cover and then tighten the mounting bolts to 37–44 ft. lbs. (50–60 Nm).
14. Install the hub assembly and hub washer.
15. After tightening the hub bearing flange nut to 148–192 ft. lbs. (200–260 Nm), caulk the concave portion of the spindle by crimping the nut.

➡Replace the flange nut with new ones after removal.

16. Install a new hub cap.
17. Install the rear speed sensor.
18. Install the brake disc to the hub and tighten the brake disc mounting screw to 44–53 inch lbs. (5–6 Nm).
19. Install the brake caliper and tighten the mounting bolt to 37–44 ft. lbs. (50–60 Nm).
20. Install the rear wheel and tire.

2010 Models

See Figure 456.

1. Before servicing the vehicle, refer to the Precautions Section.
2. Raise and safely support the vehicle.
3. Remove the rear wheel and tire.
4. Remove the caliper mounting bolts and caliper. Secure the caliper aside.

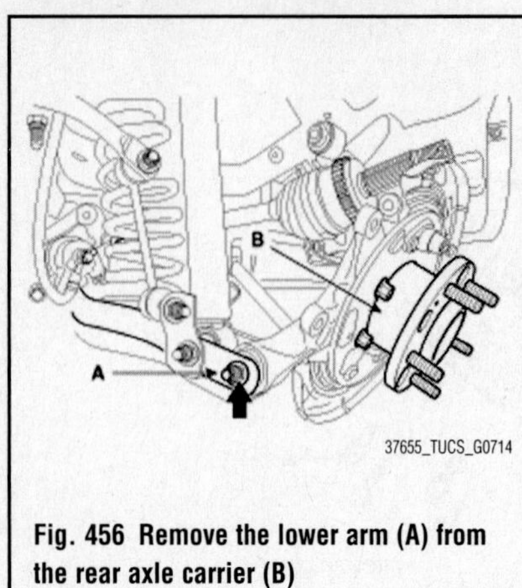

Fig. 456 Remove the lower arm (A) from the rear axle carrier (B)

5. Remove castle nut from the front hub while applying the break.
6. Remove the rear brake lining assembly.
7. Remove the parking brake cable from the brake shoe.
8. Remove the parking brake cable retaining, from the parking brake cable.
9. Remove the wheel speed sensor from the knuckle.
10. Remove the assist arm from the rear axle carrier. Refer to Assist Arm Removal & Installation in the Rear Suspension section.
11. Remove the trailing arm from the rear axle carrier. Refer to Trailing Arm Removal & Installation in the Rear Suspension section.
12. Remove the upper arm from the rear axle carrier. Refer to Upper Arm Removal & Installation in the Rear Suspension section.
13. Push the rear axle carrier outward and separate the halfshaft from the axle hub. Refer to Rear Halfshaft Removal & Installation in the Drive Train section.
14. Remove the lower arm from the rear axle carrier. Refer to Lower Arm Removal & Installation in the Rear Suspension section.

To install:

15. Installation is the reverse of removal, noting the following:
   a. Tighten the lower arm mounting bolt to 101–116 ft. lbs. (137–157 Nm).
   b. Tighten the upper arm mounting bolt to 72–87 ft. lbs. (98–118 Nm).
   c. Tighten the trailing arm mounting bolt to 25–40 ft. lbs. (34–54 Nm).
   d. Tighten the assist arm mounting bolt to 101–116 ft. lbs. (137–157 Nm).

ADJUSTMENT

The wheel bearings are sealed units and are not adjustable.

HYUNDAI

Veracruz

9

SPECIFICATIONS AND MAINTENANCE CHARTS

ENGINE AND VEHICLE IDENTIFICATION

Engine							Model Year	
Code ①	Liters (cc)	Cu. In.	Cyl.	Fuel Sys.	Engine Type	Eng. Mfg.	Code ②	Year
C	3.8 (3778)	230.55	V6	MPFI	DOHC	Hyundai	9	2009
							10	2010

MPFI: Multi-Point Fuel Injection

DOHC: Double Overhead Camshafts

① 8th digit of VIN

37655_VERA_C0001

GENERAL ENGINE SPECIFICATIONS

All measurements are given in inches.

Year	Model	Engine Displacement Liters	Engine Series VIN	Net Horsepower @ rpm	Net Torque @ rpm (ft. lbs.)	Bore x Stroke (in.)	Com-pression Ratio	Oil Pressure @ rpm
2009	Veracruz	3.8	C	260@6000	257@4500	3.780 x 3.425	10.4:1	18.8@1000
2010	Veracruz	3.8	C	260@6000	257@4500	3.780 x 3.425	10.4:1	18.8@1000

37655_VERA_C0002

GASOLINE ENGINE TUNE-UP SPECIFICATIONS

Year	Engine Displacement Liters	Engine VIN	Spark Plug Gap (in.)	Ignition Timing (deg.)		Fuel Pump (psi)	Idle Speed (rpm)		Valve Clearance	
				MT	AT		MT	AT	In.	Ex.
2009	3.8	C	0.039-0.043	①	①	54.3-55.8	②	②	HYD	HYD
2010	3.8	C	0.039-0.043	①	①	54.3-55.8	②	②	HYD	HYD

NOTE: The Vehicle Emission Control Information label reflects specification changes made during production. Follow the figures on the label if they differ from those in this chart.

HYD: Hydraulic

① Ignition timing is preset and cannot be adjusted

② Idle speed is maintained by the Electronic Control Module (ECM)

37655_VERA_C0003

CAPACITIES

Year	Model	Engine Displacement Liters	Engine VIN	Engine Oil with Filter (qts.)	Transmission (pts.) Manual	Auto. ①	Fuel Tank (gal.)	Cooling System (qts.)
2009	Veracruz	3.8	C	6.0	NA	23.0	20.6	9.0-11.1
2010	Veracruz	3.8	C	6.0	NA	23.0	20.6	9.0-11.1

NOTE: All capacities are approximate. Add fluid gradually and check to be sure a proper fluid level is obtained.

⑭① Not Applicable

37655_VERA_C0004

FLUID SPECIFICATIONS

Year	Model	Engine Displacement Liters	Engine ID/VIN	Engine Oil	Manual Trans.	Auto. Trans.	Drive Axle	Power Steering Fluid	Brake Master Cylinder	Cooling System
2009	Veracruz	3.8	C	5W-20	NA	①	②	PSF-3	DOT 3	③
2010	Veracruz	3.8	C	5W-20	NA	①	②	PSF-3	DOT 3	③

DOT: Department Of Transportation

① HYUNDAI GENUINE PART TFF ATF T-IV JWS-3309, Mobil ATF 3309

② Hypoid gear oil (API GL-5, SAE 75W/90)

③ Ethylene glycol base for aluminum radiator

NA: Not Applicable

37655_VERA_C0005

VALVE SPECIFICATIONS

Year	Engine Displacement Liters	Engine VIN	Seat Angle (deg.)	Face Angle (deg.)	Spring Test Pressure (lbs. @ in.)	Spring Installed Height (in.)	Stem-to-Guide Clearance (in.) Intake	Stem-to-Guide Clearance (in.) Exhaust	Stem Diameter (in.) Intake	Stem Diameter (in.) Exhaust
2009	3.8	C	44.75-45.20	45.25-45.75	90.4-96.2 @0.953	NS	0.0008-0.0019	0.0012-0.0021	0.2151-0.2157	0.2149-0.2153
2010	3.8	C	44.75-45.20	45.25-45.75	90.4-96.2 @0.953	NS	0.0008-0.0019	0.0012-0.0021	0.2151-0.2157	0.2149-0.2153

NS: Not Supplied

37655_VERA_C0006

CAMSHAFT AND BEARING SPECIFICATIONS CHART

All measurements are given in inches.

Year	Engine Displ. Liters	Engine ID/VIN	Journal Dia.	Brg. Oil Clearance	Shaft End-play	Runout	Journal Bore	Lobe Height Intake	Lobe Height Exhaust
2009	3.8	C	①	②	0.0008-0.0071	NS	NS	1.8425	1.8031
2010	3.8	C	①	②	0.0008-0.0071	NS	NS	1.8425	1.8031

NS: Not Supplied

① For LH and RH camshafts:
Intake No. 1 is 1.1009-1.1015 inch
Intake No. 2, 3, 4 are 0.9430-0.9437 inch
Exhaust No.1 is 1.1009-1.1015 inch
Exhaust No. 2, 3, 4 are 0.9430-0.9437 inch

② For LH and RH camshafts:
Intake No. 1 is 0.0011-0.0022 inch
Intake No. 2, 3, 4 are 0.0012-0.0026 inch
Exhaust No.1 is 0.0011-0.0022 inch
Exhaust No. 2, 3, 4 are 0.0012-0.0026 inch

37655_VERA_C0007

CRANKSHAFT AND CONNECTING ROD SPECIFICATIONS

All measurements are given in inches.

Year	Engine Displacement Liters	Engine VIN	Crankshaft Main Brg. Journal Dia.	Crankshaft Main Brg. Oil Clearance	Crankshaft Shaft End-play	Crankshaft Thrust on No.	Connecting Rod Journal Diameter	Connecting Rod Oil Clearance	Connecting Rod Side Clearance
2009	3.8	C	2.7142-2.7149	0.0008-0.0016	0.0039-0.0110	3	2.1635-2.1642	0.0015-0.0022	0.0039-0.0098
2010	3.8	C	2.7142-2.7149	0.0008-0.0016	0.0039-0.0110	3	2.1635-2.1642	0.0015-0.0022	0.0039-0.0098

37655_VERA_C0008

PISTON AND RING SPECIFICATIONS

All measurements are given in inches.

Year	Engine Displ. Liters	Engine VIN	Piston Clearance	Ring Gap Top Compression	Ring Gap Bottom Compression	Ring Gap Oil Control	Ring Side Clearance Top Compression	Ring Side Clearance Bottom Compression	Ring Side Clearance Oil Control
2009	3.8	C	0.0012-0.0020	0.0067-0.0126	0.0126-0.0185	0.0078-0.0275	0.0012-0.0027	0.0012-0.0027	0.0024-0.0059
2010	3.8	C	0.0012-0.0020	0.0067-0.0126	0.0126-0.0185	0.0078-0.0275	0.0012-0.0027	0.0012-0.0027	0.0024-0.0059

37655_VERA_C0009

TORQUE SPECIFICATIONS

All readings in ft. lbs.

Year	Engine Displacement Liters	Engine VIN	Cylinder Head Bolts	Main Bearing Bolts	Rod Bearing Bolts	Crankshaft Damper Bolts	Flywheel Bolts	Manifold Intake	Manifold Exhaust	Spark Plugs	Oil Pan Drain Plug
2009	3.8	C	①	②	③	210-224	53-56	14-17	29-33	15-22	25-33
2010	3.8	C	①	②	③	210-224	53-56	14-17	29-33	15-22	25-33

NOTE: Dip main bearing bolts and crankshaft damper bolt in clean engine oil prior to tightening.

① Step 1: 29 ft. lbs.
Step 2: Plus 120 degrees
Step 3: Plus 90 degrees

② M11 bolts (inner) Step 1: 36 ft. lbs.
M11 bolts (inner) Step 2: Plus 90 degrees
M8 bolts (outer) Step 1: 15 ft. lbs.
M8 bolts (outer) Step 2: Plus 120 degrees
M8 bolts (side): 22-23 ft. lbs.

③ 15 ft. lbs., plus 90 degrees

37655_VERA_C0010

WHEEL ALIGNMENT

Year	Model		Caster Range (+/-Deg.)	Caster Preferred Setting (Deg.)	Camber Range (+/-Deg.)	Camber Preferred Setting (Deg.)	Toe-in (Deg.)
2009	Veracruz	Front	0.50	4.33	0.50	-0.50	0+/-0.20
		Rear	NA	NA	0.50	-1.00	0.20+/-0.20
2010	Veracruz	Front	0.50	4.33	0.50	-0.50	0+/-0.20
		Rear	NA	NA	0.50	-1.00	0.20+/-0.20

NA: Not Applicable

37655_VERA_C0011

TIRE, WHEEL AND BALL JOINT SPECIFICATIONS

Year	Model	OEM Tires		Tire Pressures (psi)		Wheel	Ball Joint	Lug Nut Torque
		Standard	Optional	Front	Rear	Size	Inspection	(ft. lbs.)
2009	Veracruz	P245/65R17 P245/60R18	NA	30	30	7.0J x 17 7.0J x 18	①	65-80
2010	Veracruz	P245/65R17 P245/60R18	NA	30	30	7.0J x 17 7.0J x 18	①	65-80

NA: Not Applicable

① Replace the ball joint if rotating torque exceeds specification: 13-31 inch lbs.

37655_VERA_C0012

BRAKE SPECIFICATIONS

All measurements in inches unless noted

Year	Model		Brake Disc			Brake Drum Diameter			Minimum	Brake Caliper	
			Original Thickness	Minimum Thickness	Maximum Runout	Original Inside Diameter	Max. Wear Limit	Maximum Machine Diameter	Lining Thickness	Bracket Bolts (ft. lbs.)	Mounting Bolts (ft. lbs.)
2009	Veracruz	F	1.100	1.040	0.001	—	—	—	NS	54-62	19-28
		R	0.470	0.410	0.001	—	—	—		47-54	16-23
2010	Veracruz	F	1.100	1.040	0.001	—	—	—	NS	54-62	19-28
		R	0.470	0.410	0.001	—	—	—		47-54	16-23

F: Front

R: Rear

NS: Not Supplied

37655_VERA_C0013

SCHEDULED MAINTENANCE INTERVALS

HYUNDAI—VERACRUZ

TO BE SERVICED	TYPE OF SERVICE	VEHICLE MILEAGE INTERVAL (x1000)												
		7.5	15	22.5	30	37.5	45	52.5	60	67.5	75	82.5	90	97.5
Accessory drive belts	S/I				✓				✓				✓	
Air cleaner element (engine)	R				✓				✓				✓	
Air conditioner system	S/I		✓		✓		✓		✓		✓		✓	
Automatic transaxle fluid & filter	R	No service necessary												
Brake fluid, lines, hoses, and connections	S/I		✓		✓		✓		✓		✓		✓	
Brake pads, calipers, & rotors	S/I		✓		✓		✓		✓		✓		✓	
Cabin air filter	R	Every 12 months or 10,000 miles												
Chassis and body fasteners	S/I		✓		✓		✓		✓		✓		✓	
Crankcase ventilation hose, vapor hose, and fuel filter cap	S/I				✓				✓					
Driveshafts and CV-boots	S/I		✓		✓		✓		✓		✓		✓	
Electronic throttle control	S/I		✓		✓		✓		✓		✓		✓	
Engine coolant	R								✓					
Engine oil and filter	R	✓	✓	✓	✓	✓	✓	✓	✓	✓	✓	✓	✓	✓
Exhaust pipe connections, muffler, and suspension bolts	S/I		✓		✓		✓		✓		✓		✓	
Fuel filter	R					✓					✓			
Fuel lines, fuel hoses, and connections	S/I				✓				✓				✓	
Fuel tank air filter	S/I		✓				✓				✓			
Power steering fluid	S/I	✓	✓	✓	✓	✓	✓	✓	✓	✓	✓	✓	✓	✓
Power steering pump, belt, hoses	S/I		✓		✓		✓		✓		✓		✓	
Propeller shaft (AWD)	S/I		✓		✓		✓		✓		✓		✓	
Rear axle oil (AWD)	S/I	Inspect every 48 months or 40,000 miles												
Rear parking brake	S/I				✓				✓				✓	
Spark plugs (Iridium coated) 100,000 mile replacement	R													
Steering gear rack, linkage,	S/I		✓		✓		✓		✓		✓		✓	
Transfer case oil (AWD)	S/I	Inspect every 48 months or 40,000 miles												
Upper and lower arm ball joints	S/I		✓		✓		✓		✓		✓		✓	
Vacuum hose	S/I	✓	✓	✓	✓	✓	✓	✓	✓	✓	✓	✓	✓	✓
Valve clearance	S/I								✓					

R: Replace S/I: Service or Inspect

Replace coolant at 60,000 miles and every 30, 000 miles thereafter.

Fuel tank air filter: replace every 30,000 miles.

FREQUENT OPERATION MAINTENANCE (SEVERE SERVICE)

If a vehicle is operated under any of the following conditions it is considered severe service:

- Extremely dusty areas.

- 50% or more of the vehicle operation is in 90°F (32°C) or higher temperatures, or constant operation in temperatures below 32°F (0°C).

- Prolonged idling (vehicle operation in stop and go traffic).

- Frequent short running periods (engine does not warm to normal operating temperatures).

- Police, taxi, delivery usage or trailer towing usage.

Automatic transaxle fluid: replace every 60,000 miles

Oil & oil filter: change every 3,750 miles.

Brake pads, calipers & rotors: service, or inspect more frequently

Driveshaft boots: service, or inspect every 7,500 miles

Rear axle/transfer case oil: replace every 75,000 miles

Air cleaner filter: replace more frequently

Cabin air cleaner filter: replace more frequently

Steering gear rack, linkage & boots: service, or inspect more frequently

Propeller shaft: Inspect every 7,500 miles

Spark plugs: replace more frequently

Transfer case oil (AWD): replace every 80,000 miles

Rear axle oil (AWD): replace every 80,000 miles

37655_VERA_C0014

BRAKES — INFORMATION AND PRECAUTIONS

ANTI-LOCK SYSTEMS

• Certain components within the ABS system are not intended to be serviced or repaired individually.

• Do not use rubber hoses or other parts not specifically specified for and ABS system. When using repair kits, replace all parts included in the kit. Partial or incorrect repair may lead to functional problems and require the replacement of components.

• Lubricate rubber parts with clean, fresh brake fluid to ease assembly. Do not use shop air to clean parts; damage to rubber components may result.

• Use only DOT 3 brake fluid from an unopened container.

• If any hydraulic component or line is removed or replaced, it may be necessary to bleed the entire system.

• A clean repair area is essential. Always clean the reservoir and cap thoroughly before removing the cap. The slightest amount of dirt in the fluid may plug an orifice and impair the system function. Perform repairs after components have been thoroughly cleaned; use only denatured alcohol to clean components. Do not allow ABS components to come into contact with any substance containing mineral oil; this includes used shop rags.

• The Anti-Lock control unit is a microprocessor similar to other computer units in the vehicle. Ensure that the ignition switch is **OFF** before removing or installing controller harnesses. Avoid static electricity discharge at or near the controller.

• If any arc welding is to be done on the vehicle, the control unit should be unplugged before welding operations begin.

DISC AND DRUM SYSTEMS

⁂ CAUTION

Dust and dirt accumulating on brake parts during normal use may contain asbestos fibers from production or aftermarket brake linings. Breathing excessive concentrations of asbestos fibers can cause serious bodily harm. Exercise care when servicing brake parts. Do not sand or grind brake lining unless equipment used is designed to contain the dust residue. Do not clean brake parts with compressed air or by dry brushing. Cleaning should be done by dampening the brake components with a fine mist of water, then wiping the brake components clean with a dampened cloth. Dispose of cloth and all residue containing asbestos fibers in an impermeable container with the appropriate label. Follow practices prescribed by the Occupational Safety and Health Administration (OSHA) and the Environmental Protection Agency (EPA) for the handling, processing, and disposing of dust or debris that may contain asbestos fibers.

BRAKES — BLEEDING THE BRAKE SYSTEM

BLEEDING PROCEDURE

BLEEDING THE ABS SYSTEM

This procedure should be followed to ensure adequate bleeding of air and the filling of the ABS unit, the brake lines, and the master cylinder with brake fluid.

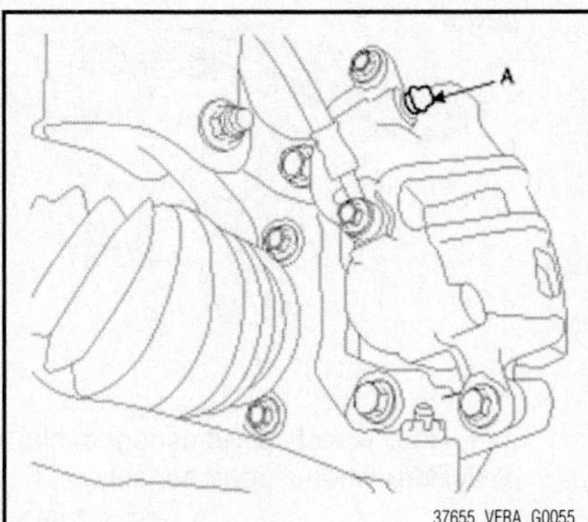

Fig. 1 Loosen the right-rear brake bleed screw (A)

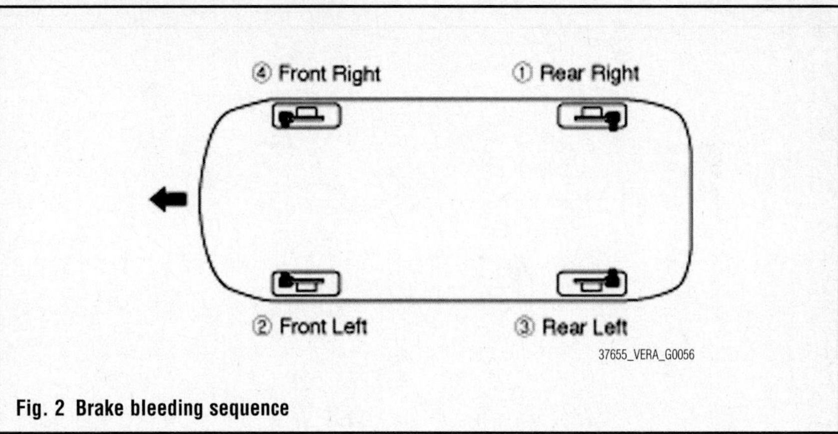

Fig. 2 Brake bleeding sequence

1. Before servicing the vehicle, refer to the Precautions Section.
2. Make sure the brake fluid in the reservoir is at the MAX (upper) level line.
3. Have someone slowly pump the brake pedal several times, and then apply pressure.
4. Loosen the right-rear brake bleed screw to allow air to escape from the system. Then tighten the bleed screw securely.
5. Repeat the procedure for wheel in the sequence shown below unit air bubbles no longer appear in the fluid.
6. Tighten the bleeder screw to 61–113 inch. lbs. (7–13 Nm).
7. Refill the master cylinder reservoir to MAX (upper) level line.

SPEED SENSORS

REMOVAL & INSTALLATION

Front

See Figures 3 and 4.

1. Before servicing the vehicle, refer to the Precautions Section.

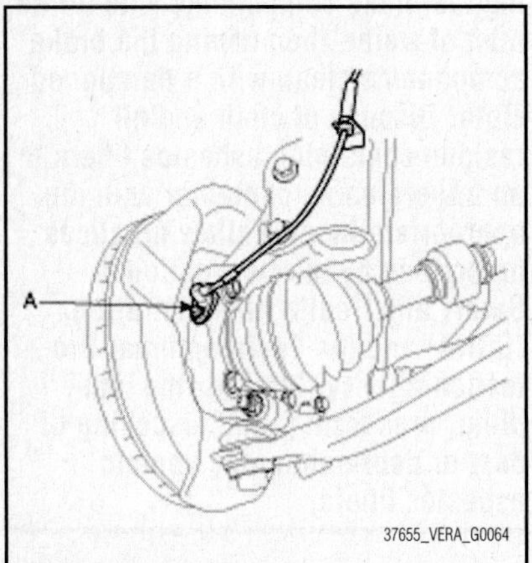

Fig. 3 Remove the front wheel speed sensor mounting bolt (A)

2. Remove the front wheel speed sensor mounting bolt.
3. Remove the front wheel speed sensor bracket.
4. Remove the front wheel guard.
5. Remove the front wheel speed sensor after disconnecting the connector.

To install:

6. Installation is the reverse of the removal procedure.

Rear

See Figures 5 and 6.

1. Before servicing the vehicle, refer to the Precautions Section.
2. Remove the rear halfshaft. Refer to Rear Halfshafts, Removal & Installation in the Drive Train section.
3. Disconnect the rear wheel speed sensor connector.
4. Remove the hub assembly mounting bolts.
5. Remove the rear wheel speed sensor.

To install:

6. Installation is the reverse of the removal procedure.

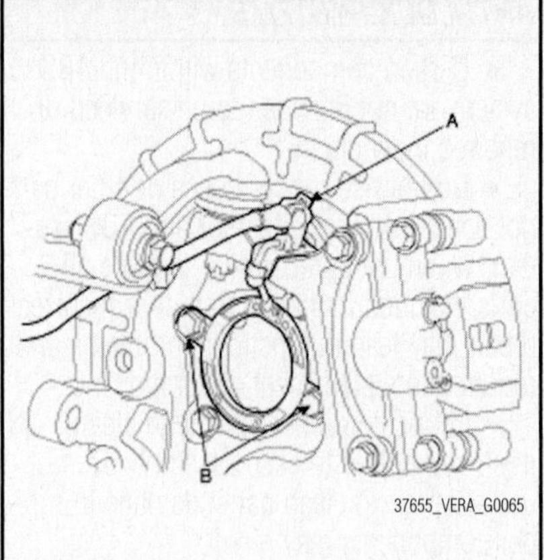

Fig. 5 Disconnect the connector (A) and remove the hub assembly mounting bolts (B)

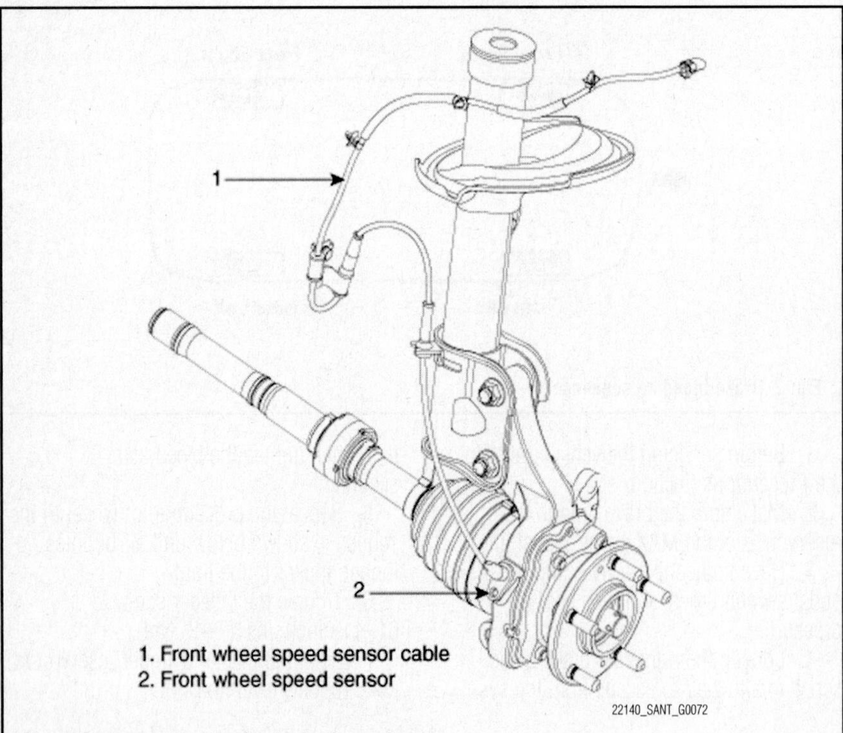

Fig. 4 Components of the front wheel speed sensor

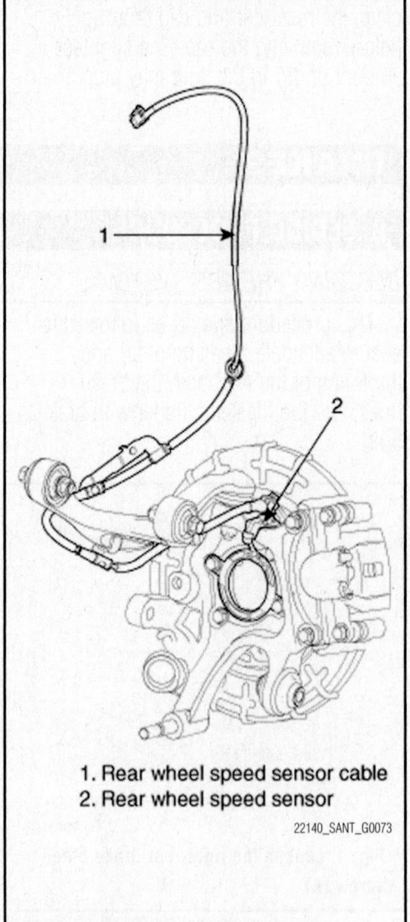

Fig. 6 Components of the rear wheel speed sensor

BRAKE CALIPER

REMOVAL & INSTALLATION

See Figures 7 through 9.

➡**Special Service Tool used for this procedure: SST 09581-11000**

1. Before servicing the vehicle, refer to the Precautions Section.
2. Remove the wheels and tires.
3. Remove the split pin and castle nut.
4. Remove the hose eye-bolt and the brake hose from the caliper.

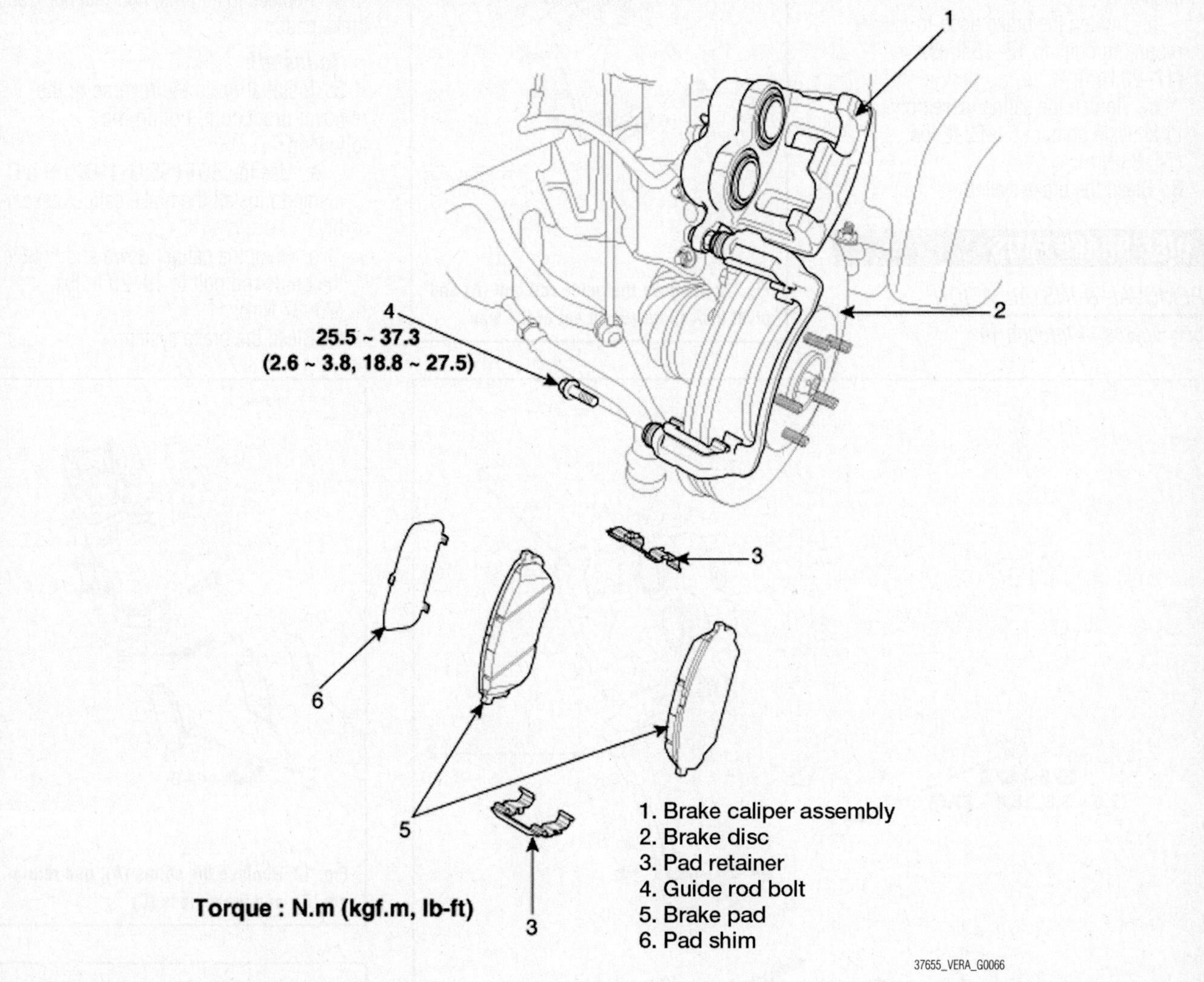

1
2
3
4
25.5 ~ 37.3
(2.6 ~ 3.8, 18.8 ~ 27.5)
6
5
3
Torque : N.m (kgf.m, lb-ft)
1. Brake caliper assembly
2. Brake disc
3. Pad retainer
4. Guide rod bolt
5. Brake pad
6. Pad shim
37655_VERA_G0066

Fig. 7 Front disc brake—Exploded view

B
A
37655_VERA_G0067

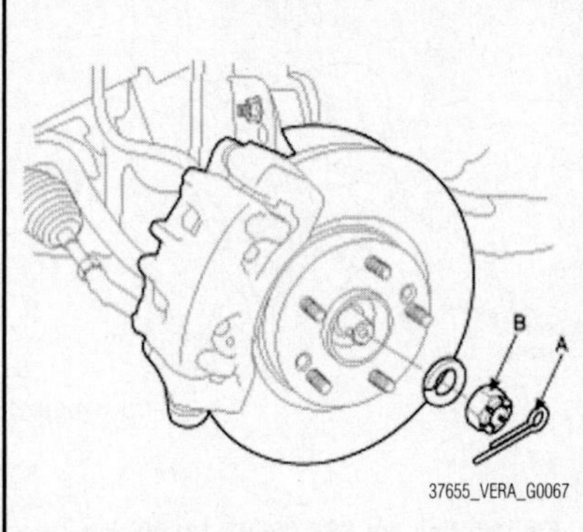

Fig. 8 Remove the split pin (A) and castle nut (B)

B
A
37655_VERA_G0068

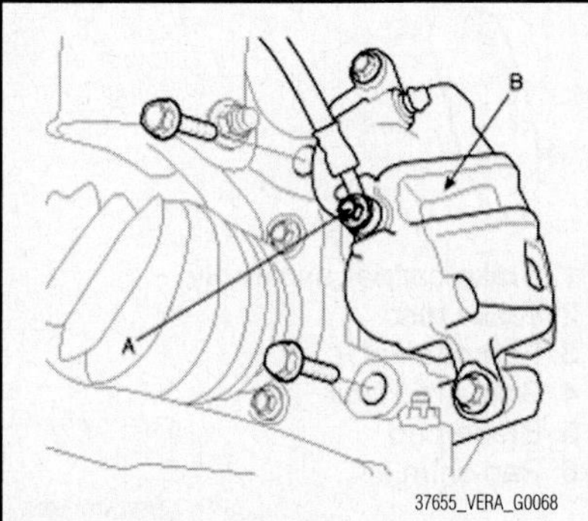

Fig. 9 Remove the hose eye-bolt (A), caliper mounting bolts, and front caliper (B)

09581
-11000
37655_VERA_G0069

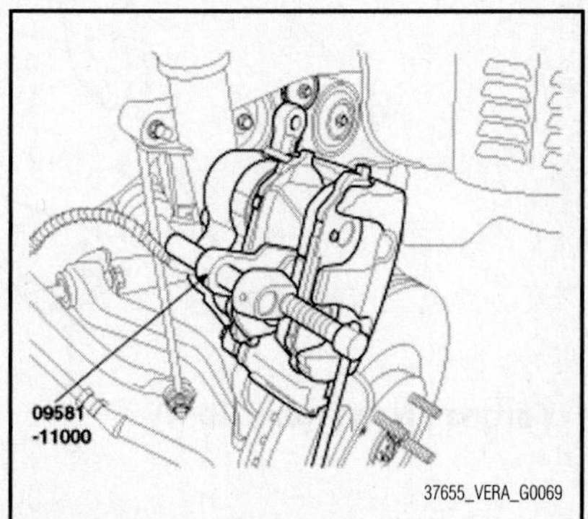

Fig. 10 Use the SST 09581-11000 or a C-clamp to install the brake caliper assembly

5. Remove the caliper mounting bolts.
6. Remove the caliper assembly.

To install:

7. Installation is the reverse of the removal procedure, noting the following:
   a. Use the SST 09581-11000 or a C-clamp to install the brake caliper assembly.
   b. Tighten the brake hose-to-caliper mounting bolts to 12–15 ft. lbs. (17–20 Nm).
   c. Tighten the caliper assembly-to-knuckle bolts to 54–62 ft. lbs. (74–83 Nm).
8. Bleed the brake system.

DISC BRAKE PADS

REMOVAL & INSTALLATION

See Figures 11 through 14.

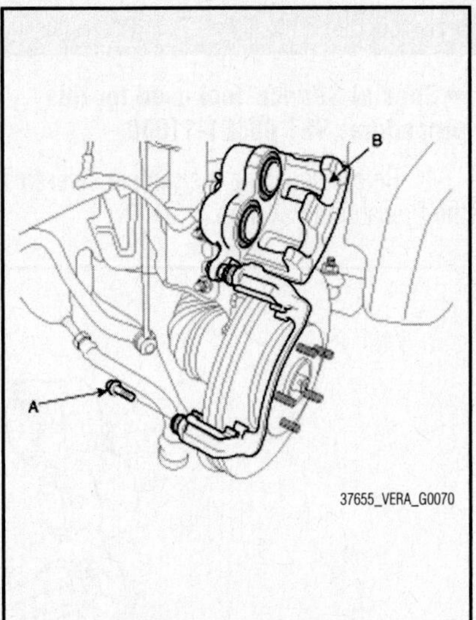

Fig. 12 Remove the guide rod bolt (A) and pivot the caliper (B) up out of the way

➡Special Service Tool used for this procedure: SST 09581-11000

1. Before servicing the vehicle, refer to the Precautions Section.
2. Remove the wheels and tires.
3. Remove the guide rod bolt and pivot the caliper up out of the way.
4. Replace the shims, pad retainers, and brake pads.

To install:

5. Installation is the reverse of the removal procedure, noting the following:
   a. Use the SST 09581-11000 or a C-clamp to install the brake caliper assembly.
   b. Pivot the caliper down and tighten the guide rod bolt to 19–28 ft. lbs. (26–37 Nm).
6. Bleed the brake system.

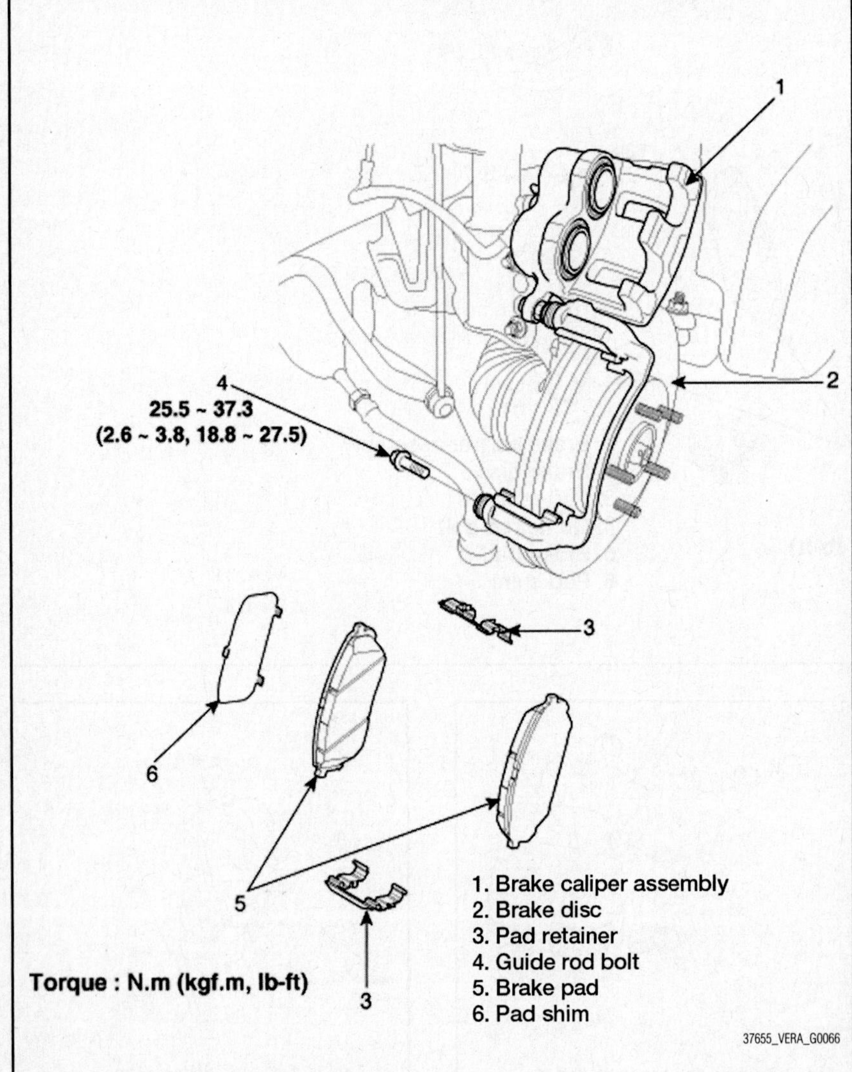

Fig. 11 Front disc brake—Exploded view

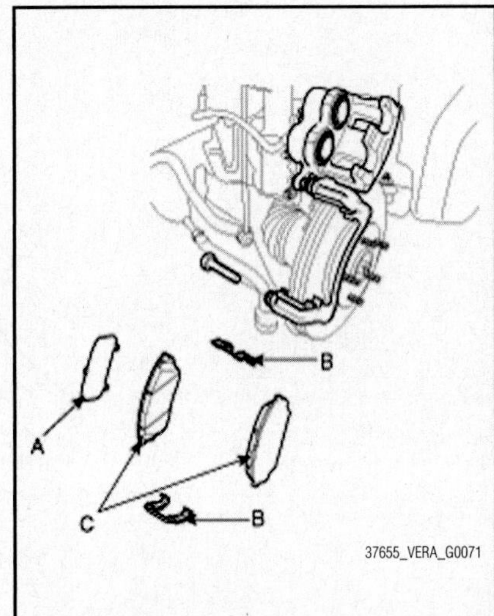

Fig. 13 Replace the shims (A), pad retainers (B), and brake pads (C)

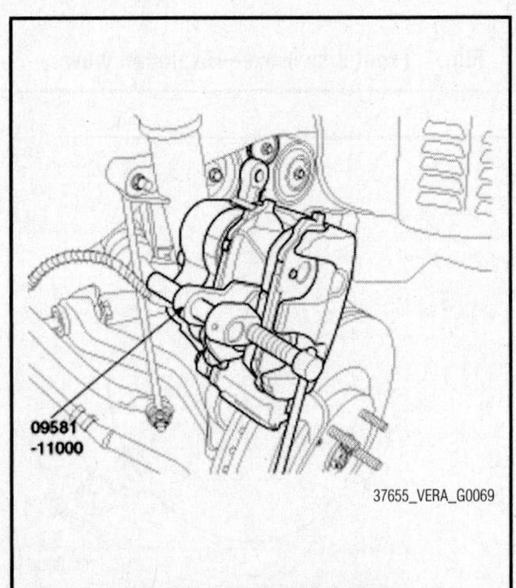

Fig. 14 Use the SST 09581-11000 or a C-clamp to install the brake caliper assembly

BRAKE CALIPER

REMOVAL & INSTALLATION

See Figures 15 through 18.

➡Special Service Tool used for this procedure: SST 09581-11000

1. Before servicing the vehicle, refer to the Precautions Section.
2. Remove the wheels and tires.
3. Remove the split pin and castle nut.
4. Remove the hose eye-bolt and the brake hose from the caliper.
5. Remove the caliper mounting bolts.
6. Remove the caliper assembly.

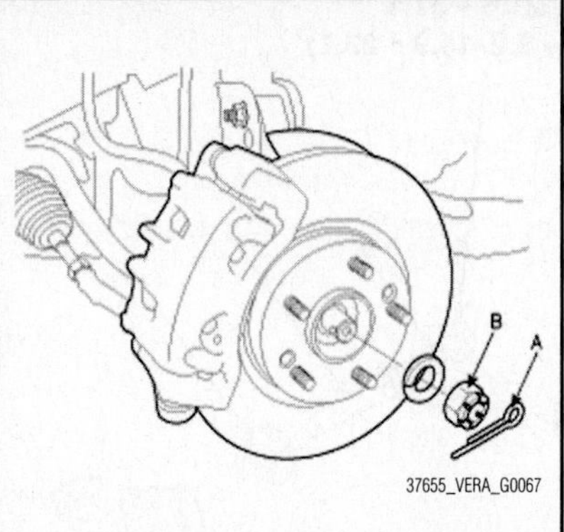

Fig. 16 Remove the split pin (A) and castle nut (B)

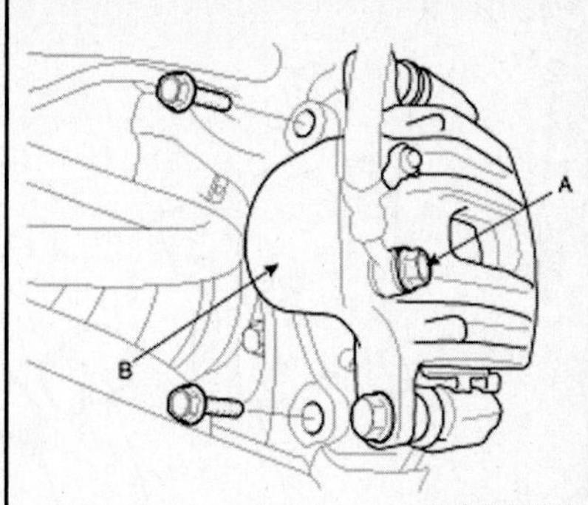

Fig. 17 Remove the hose eye-bolt (A) and rear caliper (B)

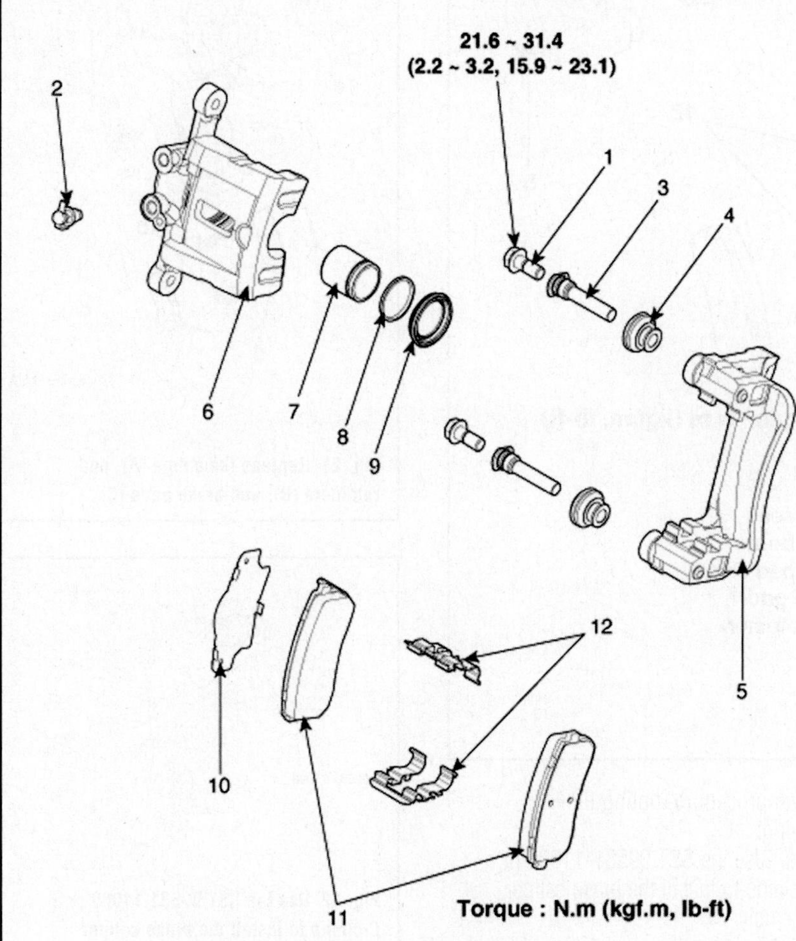

Fig. 15 Rear disc brake—Exploded view

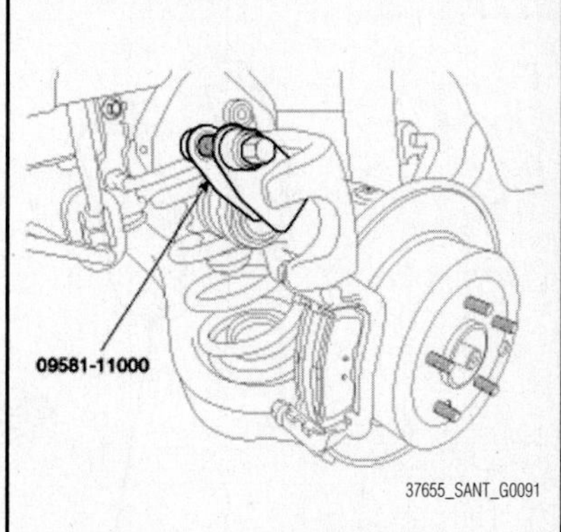

Fig. 18 Use the SST 09581-11000 or a C-clamp to install the brake caliper assembly

To install:

7. Installation is the reverse of the removal procedure, noting the following:
    a. Use the SST 09581-11000 or a C-clamp to install the brake caliper assembly.
    b. Tighten the brake hose-to-caliper mounting bolts to 12–15 ft. lbs. (17–20 Nm).
    c. Tighten the caliper assembly-to-knuckle bolts to 47–54 ft. lbs. (64–74 Nm)
8. Bleed the brake system.

DISC BRAKE PADS

REMOVAL & INSTALLATION

See Figures 19 through 22.

➡Special Service Tool used for this procedure: SST 09581-11000

1. Before servicing the vehicle, refer to the Precautions Section.

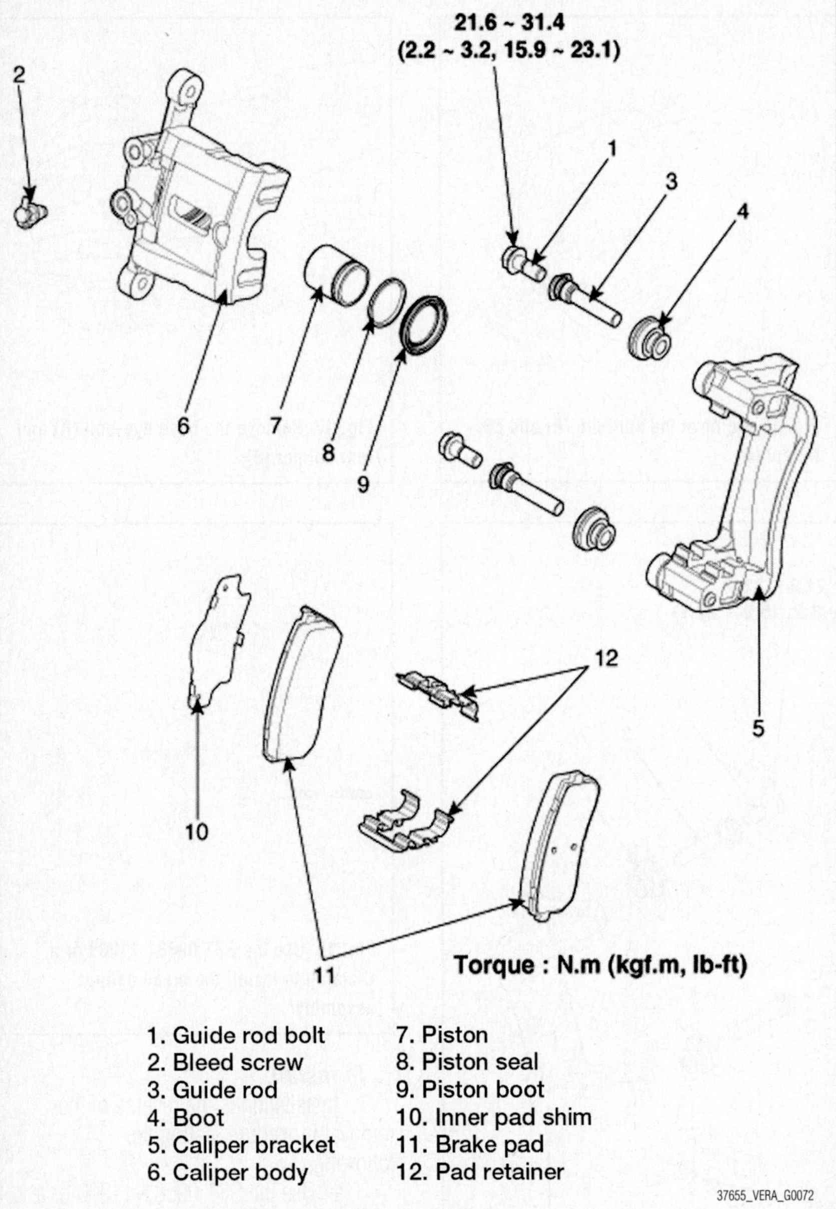

21.6 ~ 31.4
(2.2 ~ 3.2, 15.9 ~ 23.1)
1
2
3
4
5
6
7
8
9
10
11
12
Torque : N.m (kgf.m, lb-ft)
1. Guide rod bolt
2. Bleed screw
3. Guide rod
4. Boot
5. Caliper bracket
6. Caliper body
7. Piston
8. Piston seal
9. Piston boot
10. Inner pad shim
11. Brake pad
12. Pad retainer
37655_VERA_G0072

Fig. 19 Rear disc brake—Exploded view

A
37655_VERA_G0073

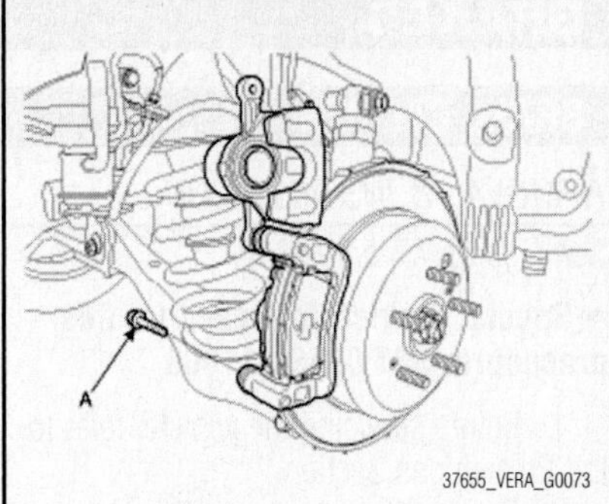

Fig. 20 Remove the guide rod bolt (A) and pivot the caliper up out of the way

A
B
C
37655_VERA_G0074

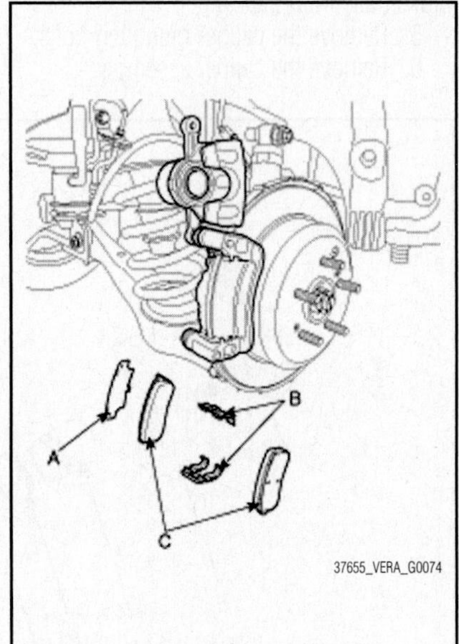

Fig. 21 Replace the shims (A), pad retainers (B), and brake pads (C)

09581-11000
37655_SANT_G0091

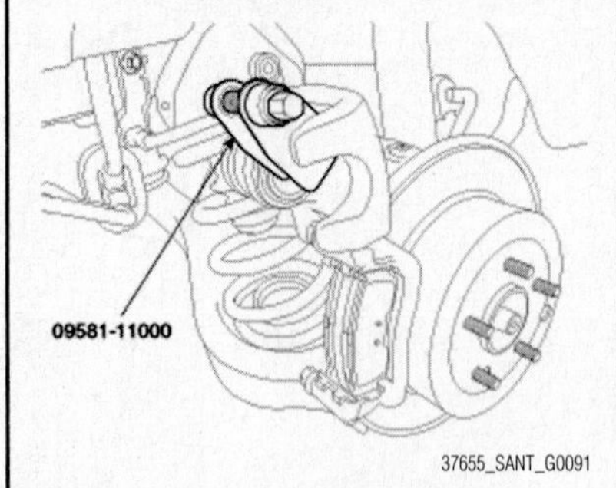

Fig. 22 Use the SST 09581-11000 or a C-clamp to install the brake caliper assembly

2. Remove the wheels and tires.
3. Remove the guide rod bolt and pivot the caliper up out of the way.
4. Replace the shims, pad retainers, and brake pads.

To install:

5. Installation is the reverse of the removal procedure, noting the following:
   a. Use the SST 09581-11000 or a C-clamp to install the brake caliper assembly.
   b. Tighten the guide rod bolt to 16–23 ft. lbs. (22–31 Nm).

PARKING BRAKE CABLES

ADJUSTMENT

See Figure 23.

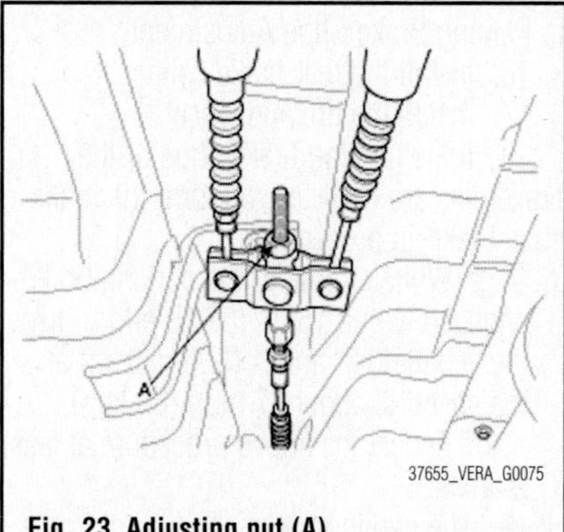

Fig. 23 Adjusting nut (A)

1. Before servicing the vehicle, refer to the Precautions Section.

➡**The parking brake adjustment must be carried out after adjusting the rear shoe.**

2. Operate the parking brake lever through a full stoke 3 times to position the cables.
3. Turn the adjusting nut so that parking brake pedal stroke is to be 3–4 in. (88–98mm) when applying a force of approx. 44 lbs. 196 N).

➡**Check the clearance. Check securely that the brake is not dragging.**

PARKING BRAKE SHOES

REMOVAL & INSTALLATION

See Figures 24 through 26.

1. Before servicing the vehicle, refer to the Precautions Section.
2. Raise and safely support the vehicle.
3. Remove the rear wheel and tire assembly.
4. Remove the rear brake caliper. Refer to Rear Disc Brake Caliper Removal & Installation.
5. Remove the clip and the parking brake cable.
6. Remove the shoe hold down pin and spring by pressing and rotating the spring.
7. Remove the adjuster assembly and the lower return spring.
8. Remove the upper return spring and the brake shoes.
9. Remove the operating lever assembly.

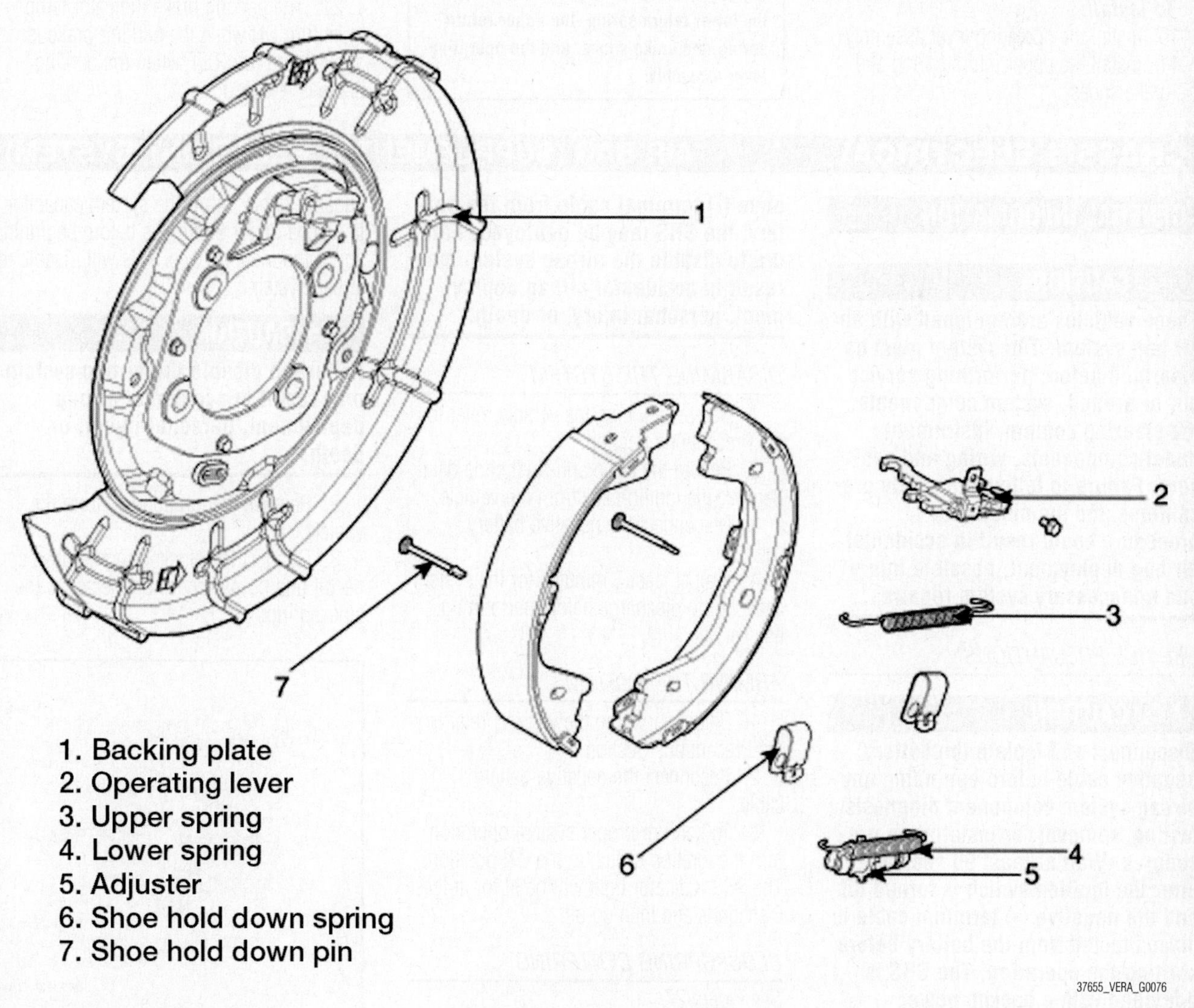

1. Backing plate
2. Operating lever
3. Upper spring
4. Lower spring
5. Adjuster
6. Shoe hold down spring
7. Shoe hold down pin

37655_VERA_G0076

Fig. 24 Parking brake components

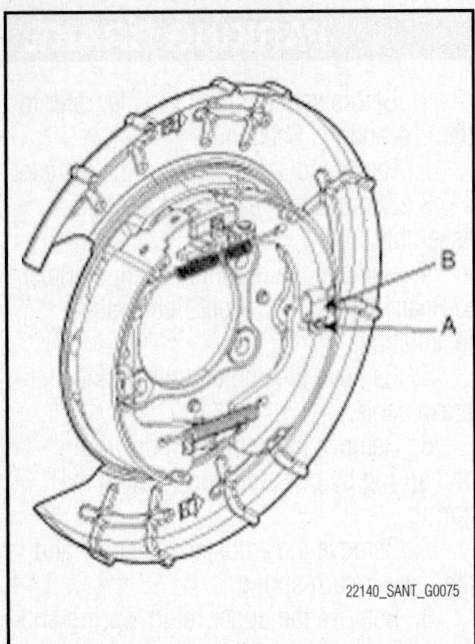

Fig. 25 Remove the shoe hold down pin (A) and spring (B) by pressing and rotating the spring

To install:

10. Install the operating lever assembly.
11. Install the upper return spring and the brake shoes.

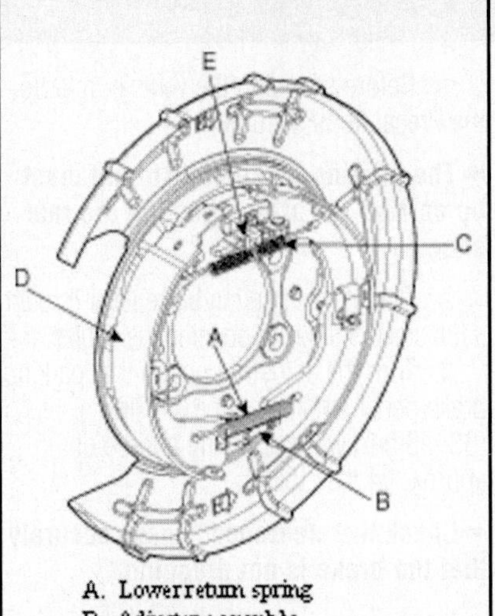

Fig. 26 Remove the adjuster assembly, the lower return spring, the upper return spring, the brake shoes, and the operating lever assembly

12. Install the adjuster assembly and the lower return spring.
13. Install the shoe hold down pin and spring by pressing and rotating the spring.
14. Install the parking brake cable and the clip.
15. Install the rear brake disc, then adjust the rear brake shoe clearance. Refer to Parking Brake Shoe Adjustment.
16. Install the rear brake caliper.
17. Install the tire and wheel.
18. If the parking brake shoe or the brake disc are replaced, perform the brake shoe brake-in procedure:
    a. While operating the parking brake engaged with a 15 lb. (69 N) effort, drive the vehicle 0.31 miles (500 meters) at the speed of about 37 mph (60 kph).
    b. Repeat the above procedure at least 2 times.
19. The parking brake should hold on a 30 percent grade.
20. Ensure that all parts move smoothly.
21. The parking brake indicator lamp must turn ON when the parking brake is applied and turn OFF when the parking brake is released.

CHASSIS ELECTRICAL

AIR BAG (SUPPLEMENTAL RESTRAINT SYSTEM)

GENERAL INFORMATION

**** CAUTION**

These vehicles are equipped with an air bag system. The system must be disarmed before performing service on, or around, system components, the steering column, instrument panel components, wiring and sensors. Failure to follow the safety precautions and the disarming procedure could result in accidental air bag deployment, possible injury and unnecessary system repairs.

SERVICE PRECAUTIONS

**** CAUTION**

Disconnect and isolate the battery negative cable before beginning any airbag system component diagnosis, testing, removal, or installation procedures. Wait at least 90 seconds after the ignition switch is turned off and the negative (-) terminal cable is disconnected from the battery before starting the operation. The SRS is equipped with a backup power source, so if work is started within 90 seconds after disconnecting the negative (-) terminal cable from the battery, the SRS may be deployed. Failure to disable the airbag system may result in accidental airbag deployment, personal injury, or death.

DISARMING THE SYSTEM

1. Before servicing the vehicle, refer to the Precautions Section.
2. Record the radio anti-theft code data. Remove the ignition key from the vehicle.
3. Disconnect the negative battery cable.
4. Wait at least 3 minutes for the system capacitor to discharge before performing any service.

ARMING THE SYSTEM

1. Before servicing the vehicle, refer to the Precautions Section.
2. Reconnect the negative battery cable.
3. To confirm proper system operation, turn the ignition switch to the ON position. The SRS indicator light will be lit for at least 6 seconds and then go off.

CLOCKSPRING CENTERING

See Figure 27.

1. Disconnect and isolate the battery negative cable. Allow the system capacitor to discharge for 3 minutes before beginning any component service. This will disable the airbag system.

**** CAUTION**

Failure to disable the airbag system may result in accidental airbag deployment, personal injury, or death.

2. Remove the ignition key from the vehicle.
3. Connect the clockspring harness connector and horn harness connector to the clockspring.

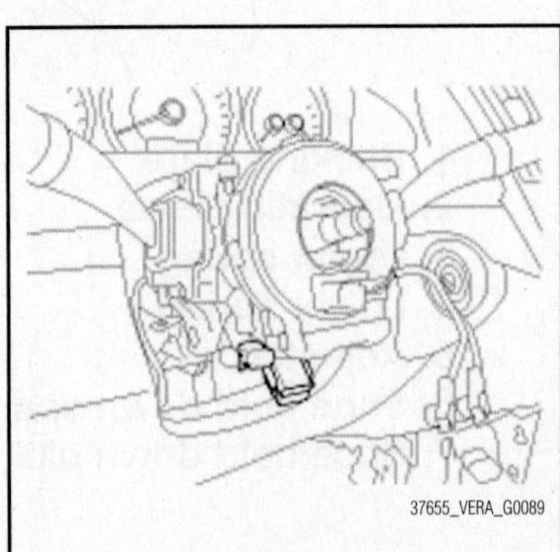

Fig. 27 Clockspring component

4. Set the clockspring on neutral position and, after turning the front wheels to the straight-ahead position, install the clockspring.

a. Check connectors and protective tube for damage, and terminals for deformities.

b. If even one abnormal point is discovered, replace the clockspring with a new one.

5. Connect the clockspring harness connector and the steering switch harness connector to the clockspring.

6. Set the center position by getting the marks between the clockspring and the cover into line.

7. Set the center position by setting the marks between the clock spring and the cover into line. Make an array the mark by turning the clock spring clockwise to the stop and then 3 revolutions counterclockwise.

DRIVE TRAIN

DIFFERENTIAL

REMOVAL & INSTALLATION

See Figures 28 through 31.

1. Before servicing the vehicle, refer to the Precautions Section.
2. Drain the differential gear oil.
3. Remove the rear halfshaft.
4. Remove the propeller shaft. Refer to Propeller Shaft Removal & Installation.
5. Support the differential assembly with the jack.
6. Disconnect the coupling control connector.
7. Remove the differential mounting bolts and the differential.

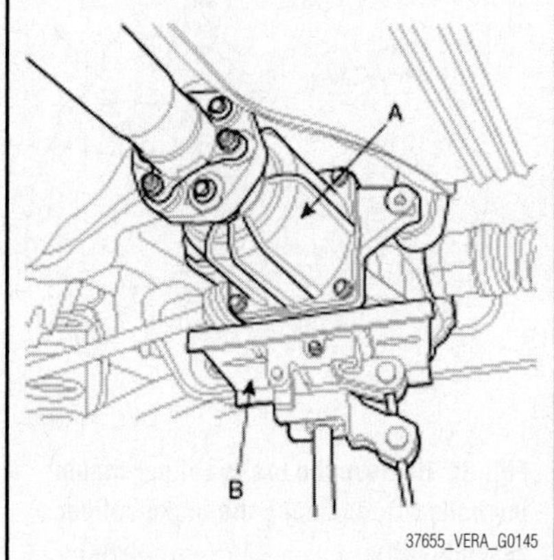

Fig. 28 Support the differential assembly (A) with the jack (B)

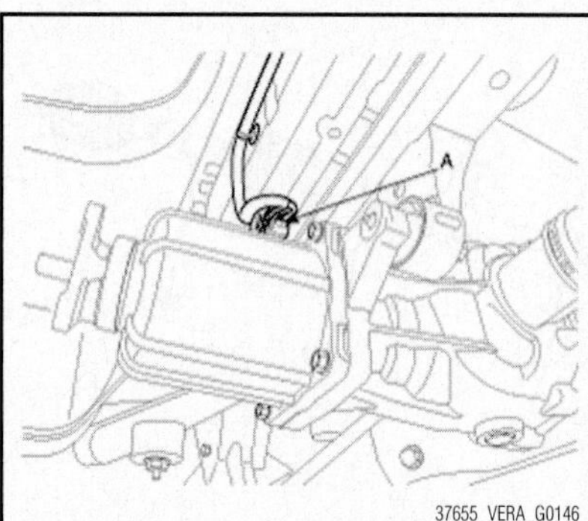

Fig. 29 Disconnect the coupling control connector (A)

8. Remove the cover bolts and the differential cover.

To install:

9. Apply liquid gasket to the differential cover.
10. Install the mounting bolts and tighten to 29–36 ft. lbs. (39–49 Nm).

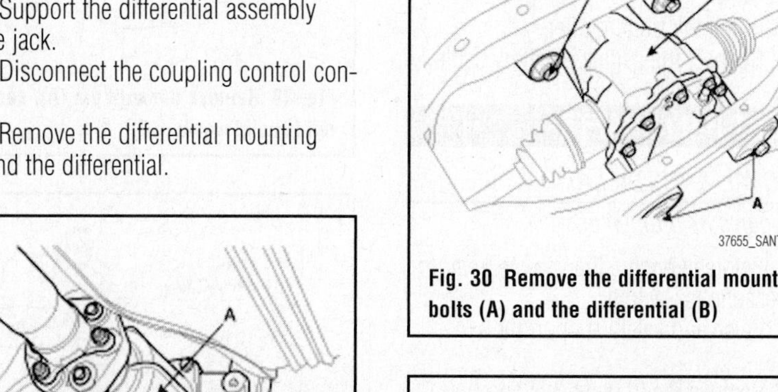

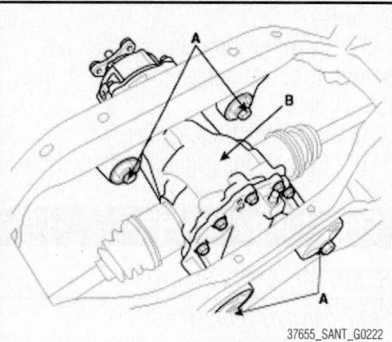

Fig. 30 Remove the differential mounting bolts (A) and the differential (B)

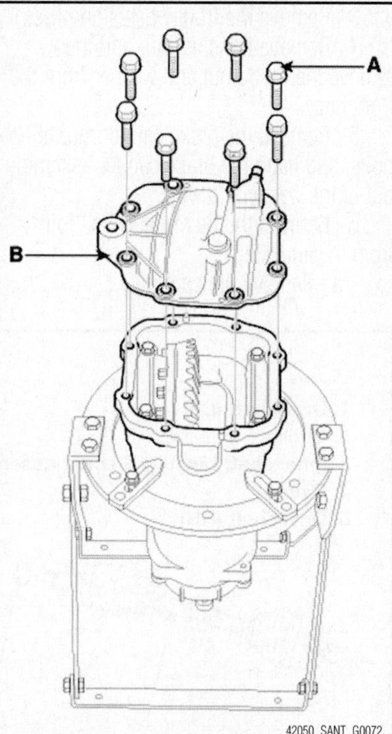

Fig. 31 Remove the cover bolts (A) and the differential cover (B)

11. Install the differential and tighten the mounting bolts to: 51–65 ft. lbs. (69–88 Nm).
12. Using the transaxle jack, install the differential assembly.
13. Connect the coupling control connector.
14. Install the propeller shaft.
15. Install the rear halfshaft.
16. Fill the gear oil with the proper amount and type of gear oil.

TRANSFER CASE ASSEMBLY

REMOVAL & INSTALLATION

See Figures 32 through 36.

1. Before servicing the vehicle, refer to the Precautions Section.
2. Disconnect the negative terminal from the battery.

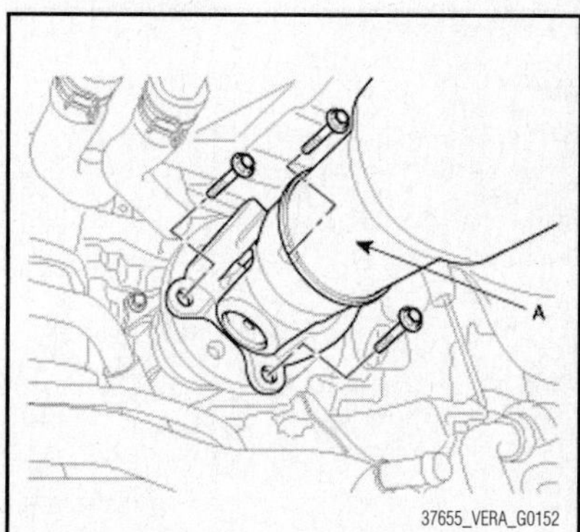

Fig. 32 Disconnect the propeller shaft (A) by removing the three bolts

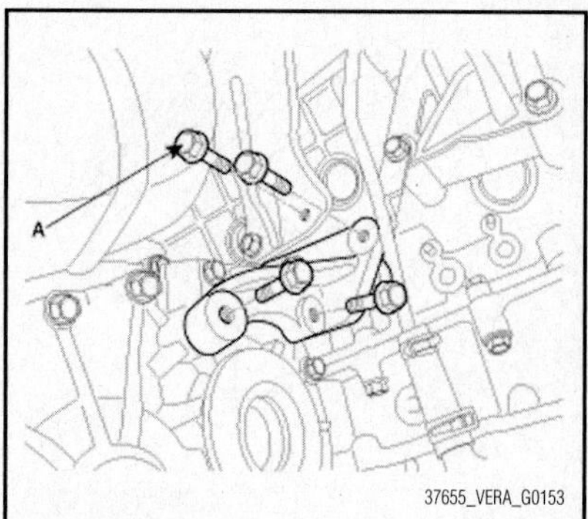

Fig. 33 Remove the four transfer case bracket mounting bolts (A)

3. Raise and safely support the vehicle.
4. Remove the front muffler.
5. Disconnect the propeller shaft by removing the three bolts.
6. Disconnect the right halfshaft from the transfer case.
7. Drain the fluid by removing the oil drain plug.
8. After completing draining the fluid, install the oil drain plug and tighten to 29–43 ft. lbs. (40–60 Nm).
9. Remove the transfer case bracket mounting bolts.
10. Remove the transfer case assembly mounting bolts (6 total).
11. Remove the transfer case assembly.

To install:

➡Use a new O-ring (47354-39300).

12. Temporarily install the transfer case assembly to the transaxle and then tighten the six mounting bolts to 45–49 ft. lbs. (62–67 Nm).
13. Install the four transfer case bracket mounting bolts and tighten to 34–37 ft. lbs. (47–51 Nm).
14. Remove the filler plug.
15. After refilling the fluid amount of the specified, tighten the filler plug. Specification: API GL-5, SAE 75W/90, Quantity: Approx 0.8L.
16. Install the right halfshaft to the transfer case.
17. Install the propeller shaft to the transfer case and tighten to 36–51 ft. lbs. (50–70 Nm).
18. Install the front muffler.
19. Connect the negative battery cable.

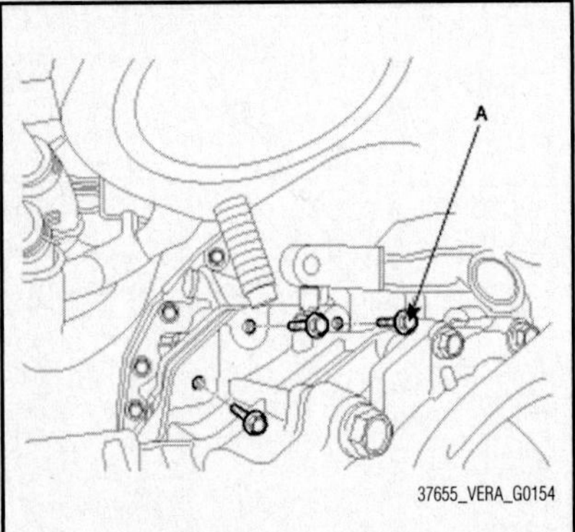

Fig. 34 Remove the transfer case assembly mounting bolts (A)

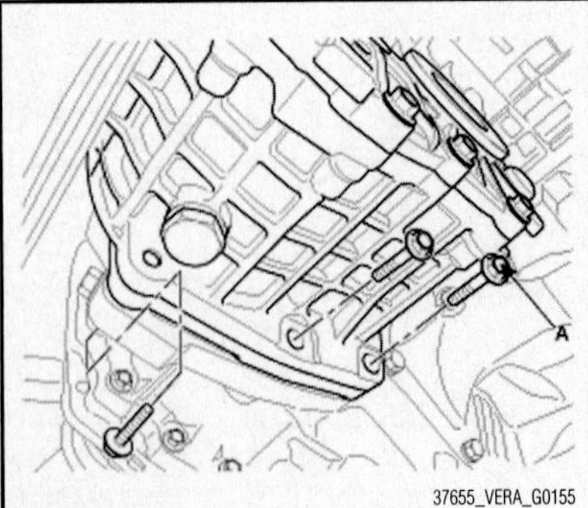

Fig. 35 Remove the remaining transfer case assembly mounting bolts (A)

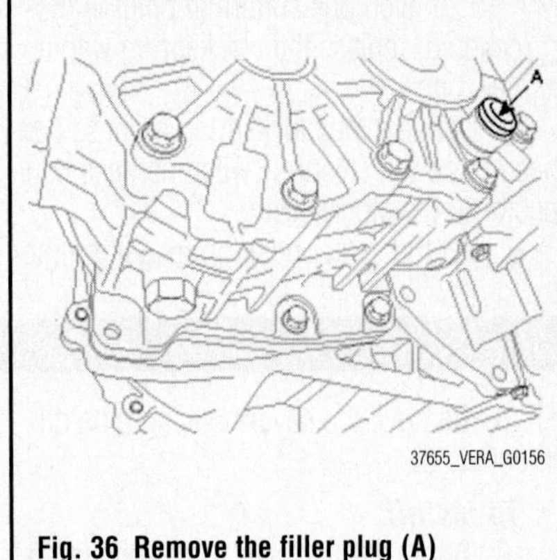

Fig. 36 Remove the filler plug (A)

FRONT HALFSHAFTS

REMOVAL & INSTALLATION

See Figures 37 through 50.

1. Before servicing the vehicle, refer to the Precautions Section.
2. Raise and safely support the vehicle.
3. Remove the front wheels and tires.
4. Remove the split pin, and then remove the castle nut and washer from the front hub.
5. Remove the brake caliper mounting bolts, and hang the brake caliper assembly out of the way with a wire.
6. Remove the tie rod end ball joint from the knuckle:
   a. Remove the split pin.
   b. Remove the castle nut.
   c. Disconnect the ball joint from knuckle using the SST (09568-4A000). Apply a few drops of oil to the boot contact part of the SST.
7. Remove the split pin and the lower arm mounting bolt from the knuckle.
8. Using a plastic hammer, disconnect the halfshaft from the axle hub.
9. Push the axle hub outward and separate the halfshaft from the axle hub.

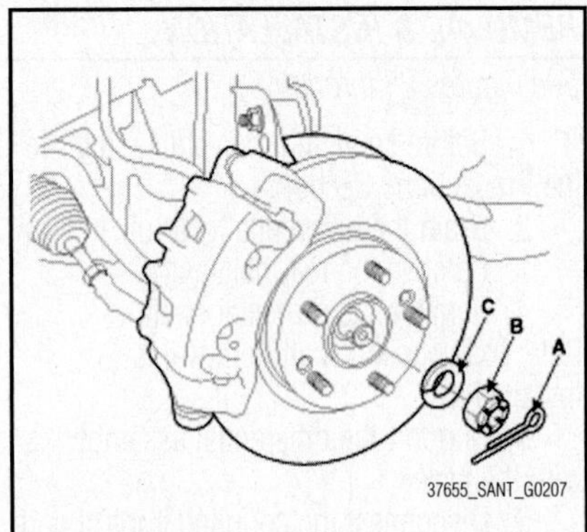

Fig. 38 Remove the split pin (A), castle nut (B) and washer (C)

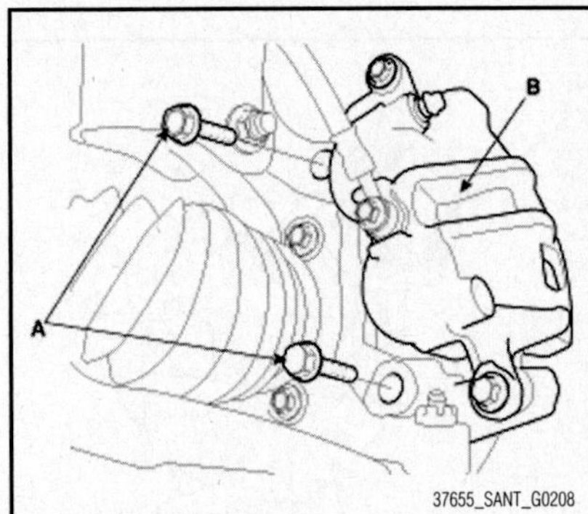

Fig. 39 Remove the brake caliper mounting bolts (A), and hang the brake caliper assembly (B)

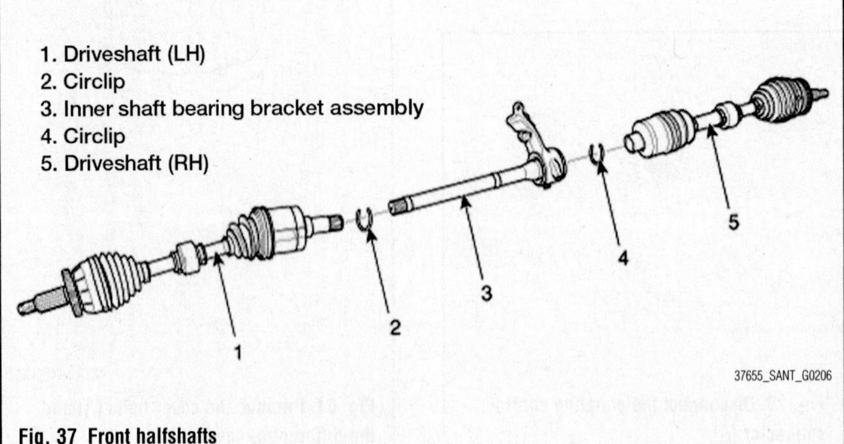

Fig. 37 Front halfshafts

10. Remove the dust cover (A).

11. Remove the mounting bolts of inner shaft bearing bracket assembly.

12. Separate the driveshaft from the transaxle case.

❊❊ WARNING

Do not use a pry bar to remove the driveshaft, as this may cause damage to the dust cover and dust seal.

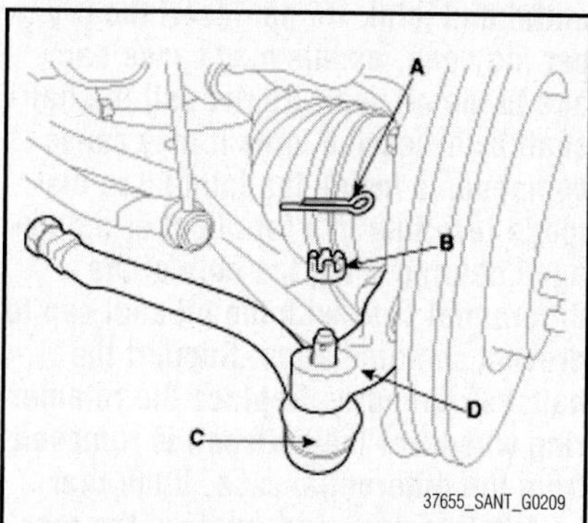

Fig. 40 Remove the split pin (A) and castle nut (B) and disconnect the ball joint (C) from knuckle (D)

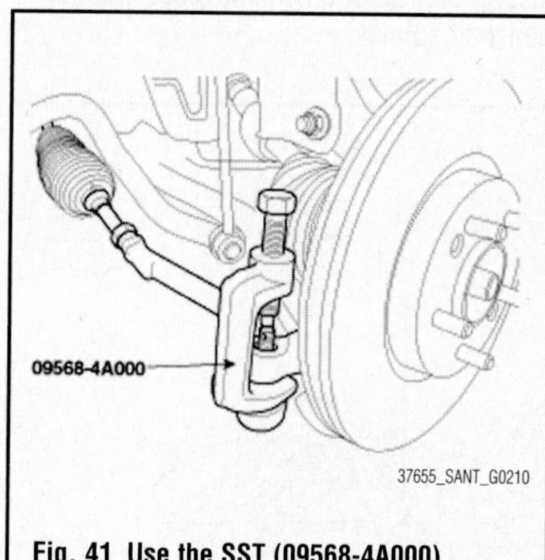

Fig. 41 Use the SST (09568-4A000)

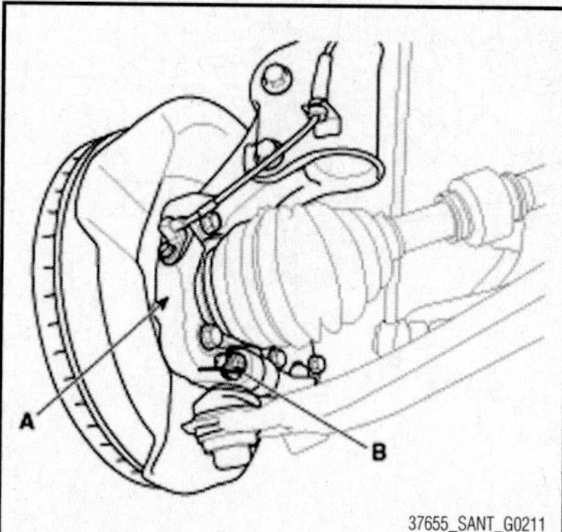

Fig. 42 Remove the split pin and the lower arm mounting bolt (B) from the knuckle (A)

13. Remove the driveshaft after striking a pipe added on the joint case with a hammer to prevent damage.

❊❊ WARNING

Do not pull the driveshaft by excessive force it may cause components inside the joint kit to dislodge resulting in a torn boot or a damaged bearing. Plug the hole of the transaxle case with the oil seal cap to prevent contamination. Support the driveshaft properly.

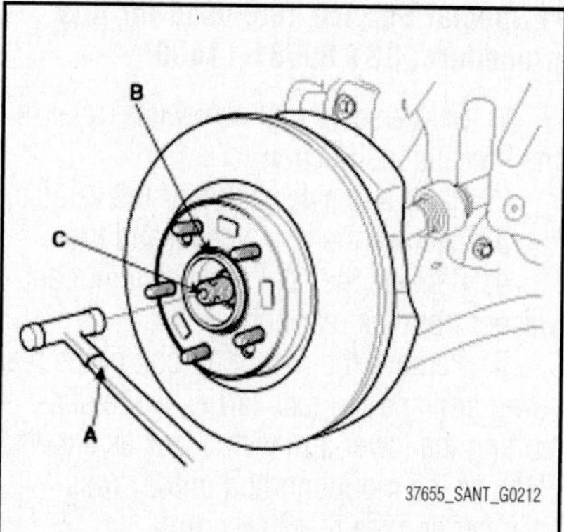

Fig. 43 Using a plastic hammer (A), disconnect halfshaft (C) from the axle hub (B)

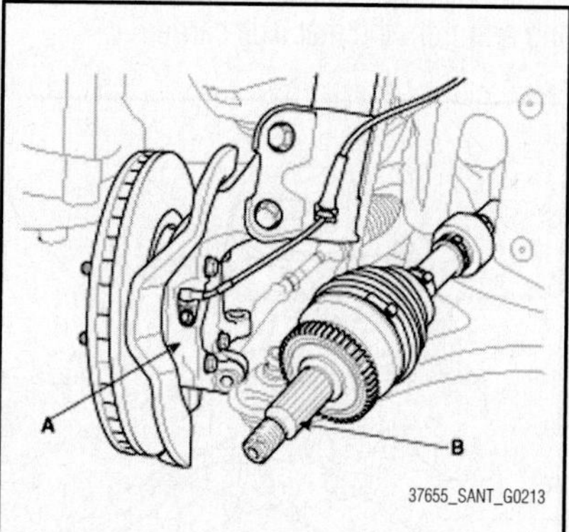

Fig. 44 Push the axle hub (A) outward and separate the halfshaft (B) from the axle hub (A)

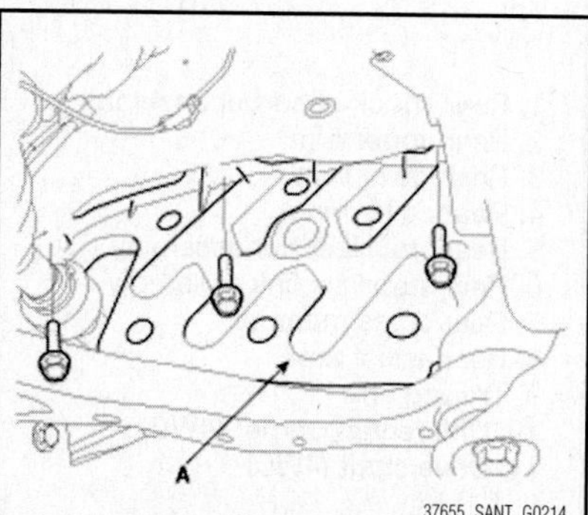

Fig. 45 Remove the dust cover (A)—Right-hand side

14. Replace the retainer ring whenever the driveshaft is removed from the transaxle case.

To install:

15. Apply gear oil on the oil seal contacting surface of transaxle case and the halfshaft splines.

16. Before installing the halfshaft, set the opening side of the circlip facing downward.

17. After installation, check that the halfshaft cannot be removed by hand.

18. Install the inner shaft bearing bracket

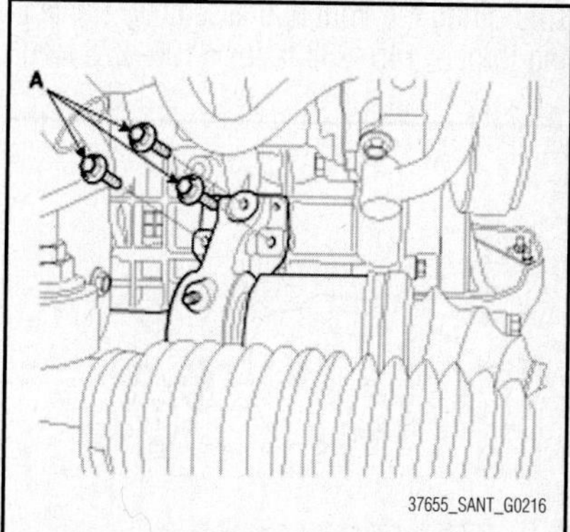

Fig. 46 Remove the inner shaft bearing bracket assembly mounting bolts (A)

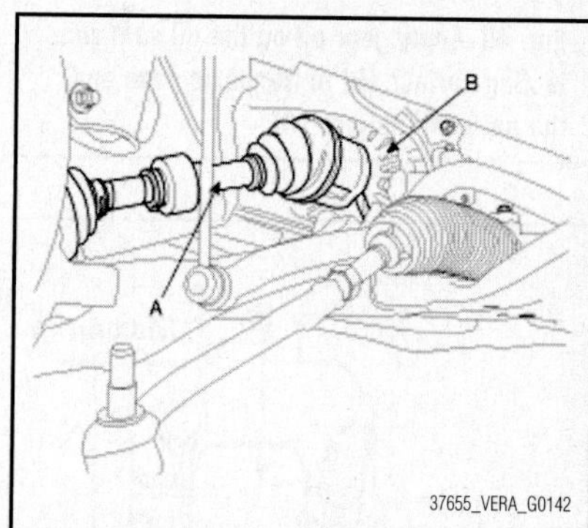

Fig. 47 Separate the driveshaft (A) from the transaxle case (B)

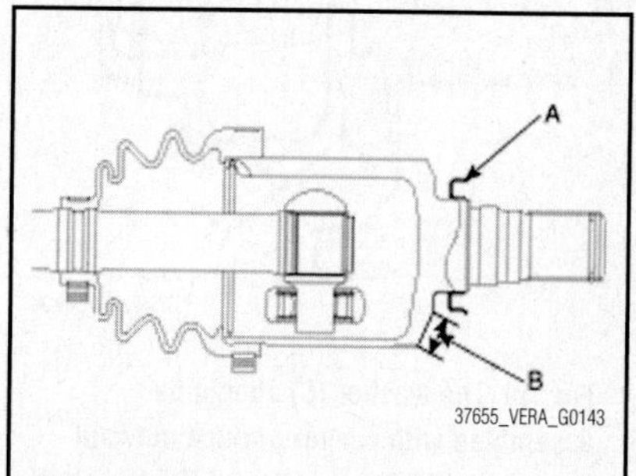

Fig. 48 Dust cover (A) and joint case (B)

assembly and tighten the mounting bolts to 36–51 ft. lbs. (49–69 Nm).

19. Install the dust cover and tighten to 70–104 inch lbs. (8–12 Nm).

20. Install the halfshaft to the axle hub.

21. Install the lower arm mounting bolt and the split pin to the knuckle. Tightening torque: 72–87 ft. lbs. (98–118 Nm).

22. Install the tie rod end ball joint to the knuckle.

23. Install the castle nut and the split pin. Tightening torque: 17–25 ft. lbs. (24–33 Nm).

24. Install the brake caliper and tighten the mounting bolts to 54–62 ft. lbs. (74–83 Nm).

25. Install the washer, castle nut and split pin to the front hub assembly. Tightening torque: 145–203 ft. lbs. (196–275 Nm).

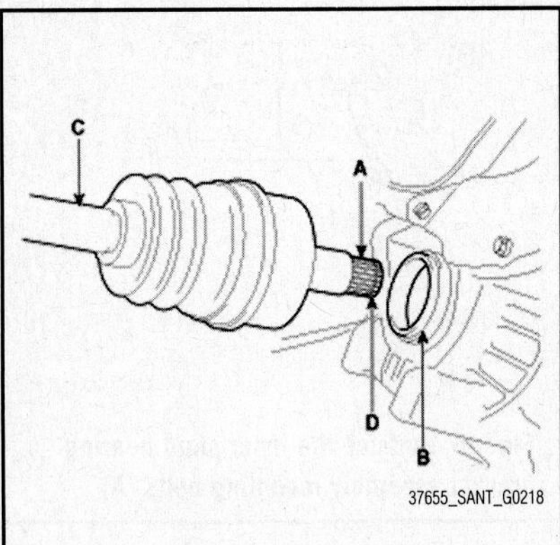

Fig. 49 Apply gear oil on the oil seal contacting surface (B) of transaxle case and the halfshaft splines (A)

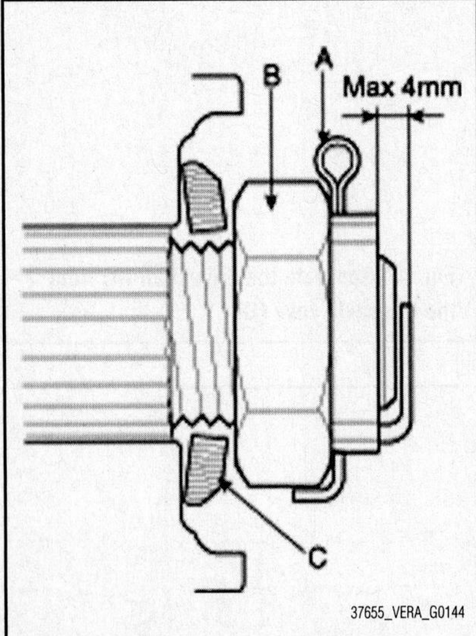

Fig. 50 The washer (C) should be assembled with convex surface outward when installing the castle nut (B) and split pin (A)

➡The washer should be assembled with convex surface outward when installing the castle nut and split pin.

26. Install the wheel and tire to the front hub. Tightening torque: 65–80 ft. lbs. (88–108 Nm).

REAR HALFSHAFTS

REMOVAL & INSTALLATION

See Figures 51 through 63.

➡Special Service Tool used for this procedure: SST 09581-11000

1. Before servicing the vehicle, refer to the Precautions Section.
2. Raise and safely support the vehicle.
3. Remove the rear wheels and tires.
4. Remove the split pin, castle nut and washer from the rear hub.
5. Remove the mounting bolt of the rear lower arm and the rear carrier, while supporting the lower arm with a jack as shown. Remove the mounting bolt of the crossmember and the rear lower arm.
6. Remove the spring, the upper pad and the lower pad.
7. Remove the rear shock absorber.
8. Remove the assist arm and the trailing arm from the rear axle carrier.

➡Remove the rear assist arm ball joint by using the SST (09568-4A000).

9. Remove the rear stabilizer link from the rear axle carrier.
10. Push the rear axle carrier outward and separate the halfshaft from the axle hub.
11. Insert a pry bar between the differential case and joint case, and separate the halfshaft from the differential case.

➡Be careful not to damage the differential and joint. Do not insert the pry bar too deep, as this may cause damage to the oil seal. Do not pull the halfshaft by excessive force it may cause components inside the joint kit to dislodge resulting in a torn boot or a damaged bearing. Plug the hole of the differential case with the oil seal cap to prevent contamination. Support the halfshaft properly. Replace the retainer ring whenever the halfshaft is removed from the differential case. If the rear halfshaft is damaged, replace the rear halfshaft assembly.

To install:

12. Apply gear oil on the oil seal contacting surface of differential case and the halfshaft splines.

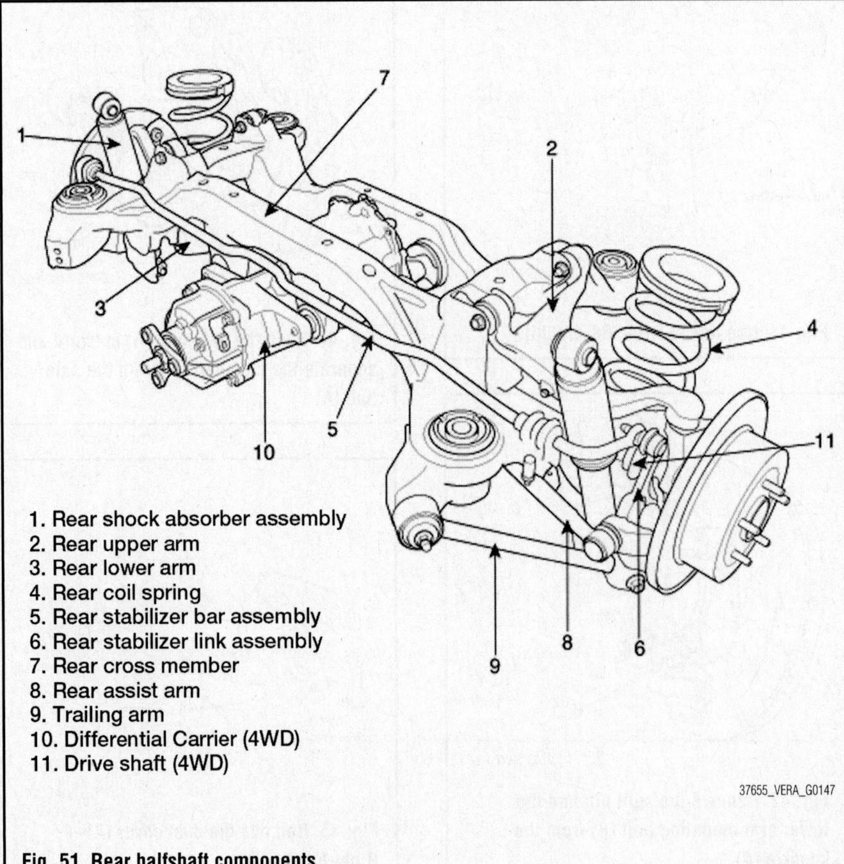

Fig. 51 Rear halfshaft components

13. Before installing the halfshaft, set the opening side of the circlip facing downward.

14. After installation, check that the halfshaft cannot be removed by hand.

15. Install the halfshaft to the rear axle carrier assembly.

16. Install the rear stabilizer link to the rear axle carrier and tighten the nut to 43–58 ft. lbs. (59–79 Nm).

17. Install the assist arm and the trailing arm to the rear axle carrier. Tightening torque:

- Bolt (C): 101–116 ft. lbs. (137–157 Nm)
- Nut (D): 72–87 ft. lbs. (98–118 Nm)
- Nut (E): 101–116 ft. lbs. (137–157 Nm)
- Bolt (F): 101–116 ft. lbs. (137–157 Nm)

➡Before tightening the chassis components, measure the distance between the wheel housing molding and the hub assembly, as shown. Specified distance: 18.31 inches, plus or minus 0.39 in. (465mm, plus or minus 10 mm)

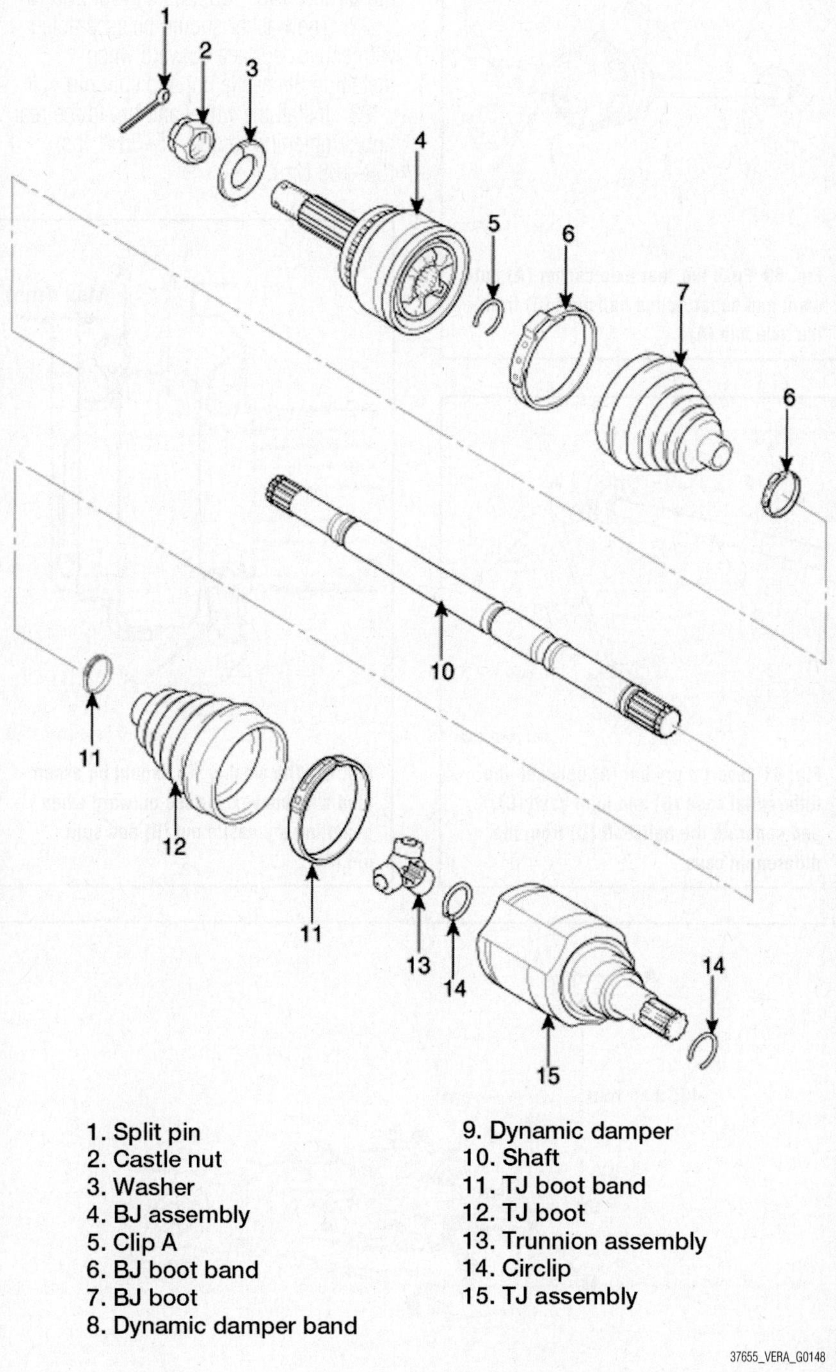

1
2
3
4
5
6
7
6
10
11
12
11
13
14
14
15
1. Split pin
2. Castle nut
3. Washer
4. BJ assembly
5. Clip A
6. BJ boot band
7. BJ boot
8. Dynamic damper band
9. Dynamic damper
10. Shaft
11. TJ boot band
12. TJ boot
13. Trunnion assembly
14. Circlip
15. TJ assembly
37655_VERA_G0148

Fig. 52 Rear halfshaft

C
B
A
37655_SANT_G0223

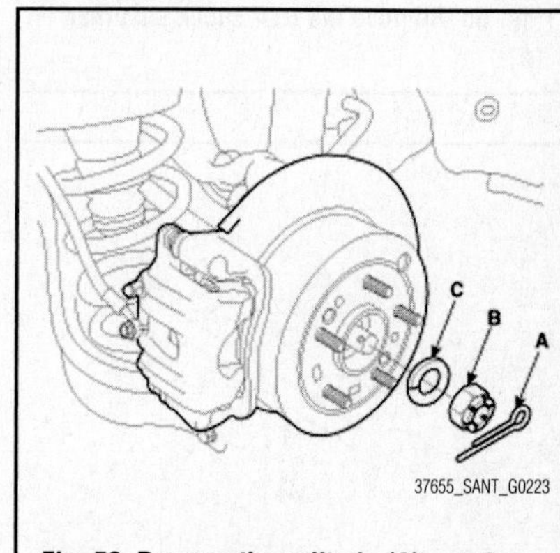

Fig. 53 Remove the split pin (A), castle nut (B) and washer (C)

C
B
A
37655_VERA_G0149

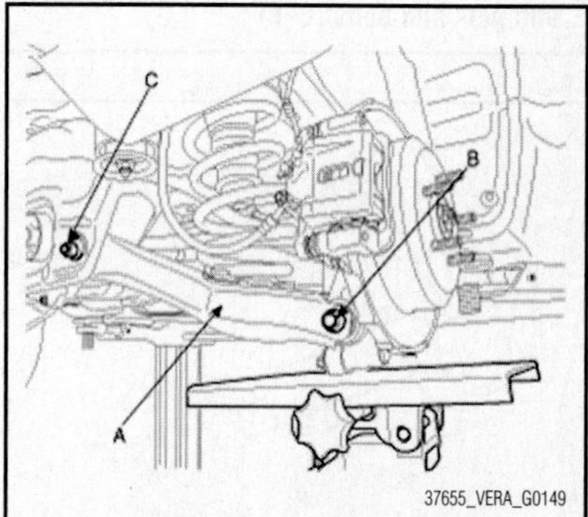

Fig. 54 Remove the rear lower arm (A) and the rear carrier mounting bolt (B), and the crossmember and the rear lower arm mounting bolt (C)

B
A
C
37655_VERA_G0150

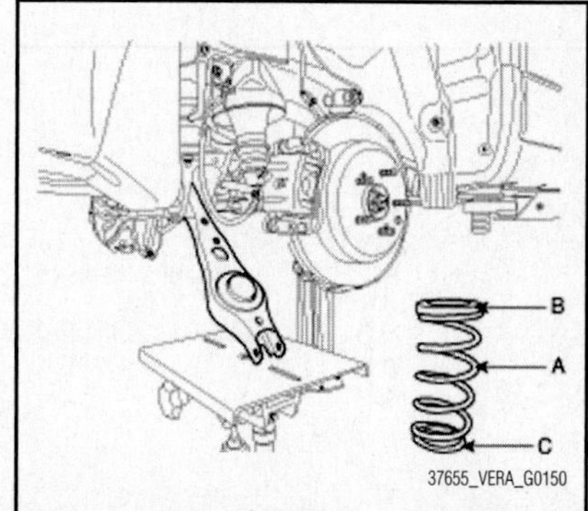

Fig. 55 Remove the spring (A), the upper pad (B) and the lower pad (C)

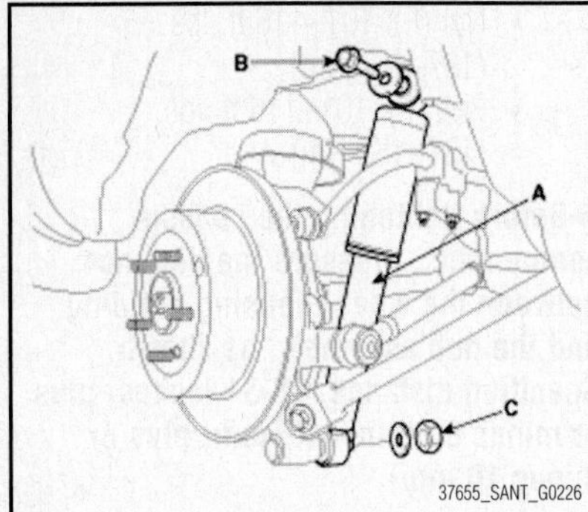

Fig. 56 Remove the rear shock absorber (A)

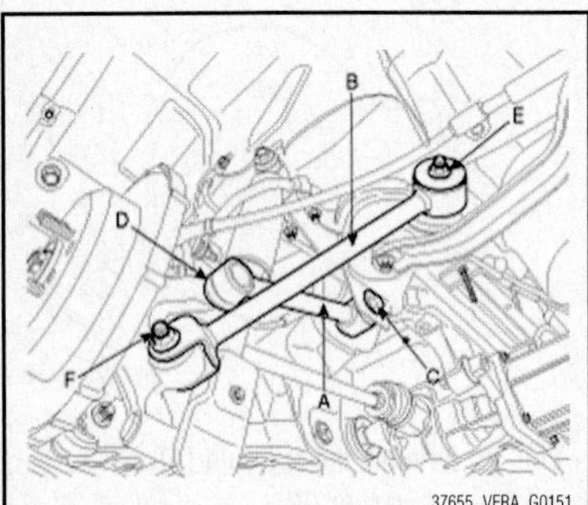

Fig. 57 Assist arm (A), trailing arm (B), and nuts and bolts (C–F)

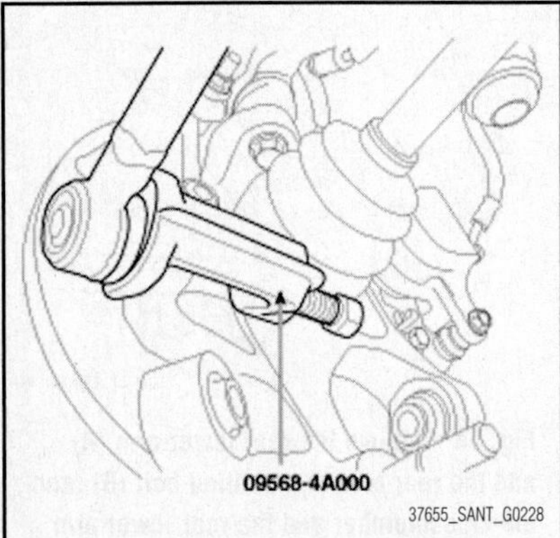

Fig. 58 Remove the rear assist arm ball joint by using the SST (09568-4A000)

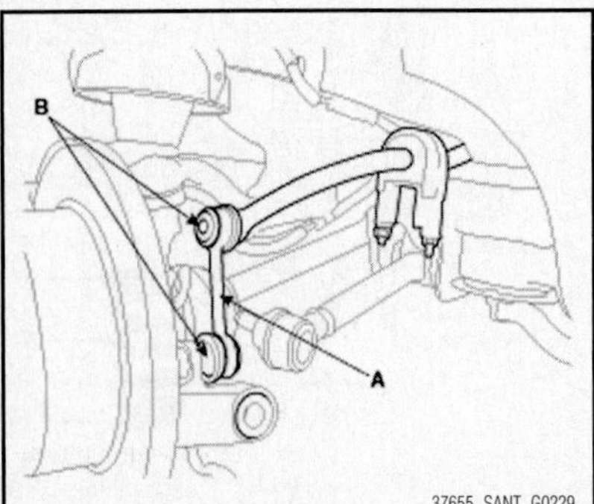

Fig. 59 Remove the rear stabilizer link (A)

18. Install the rear shock absorber. Tightening torque:
    - Bolt (B): 101–116 ft. lbs. (137–157 Nm)
    - Nut (C): 72–87 ft. lbs. (98–118 Nm)

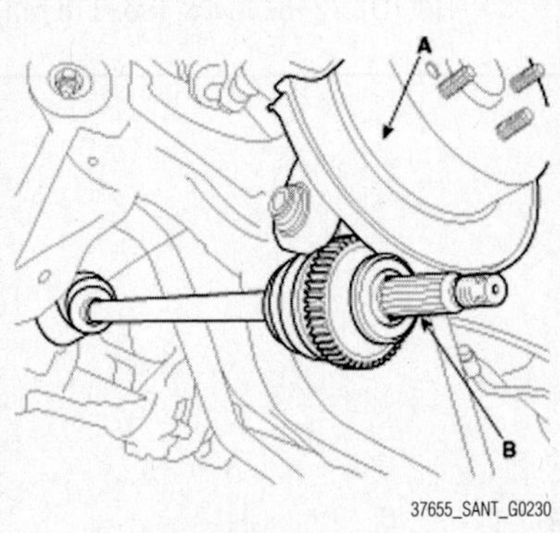

Fig. 60 Push the rear axle carrier (A) outward and separate the halfshaft (B) from the axle hub (A)

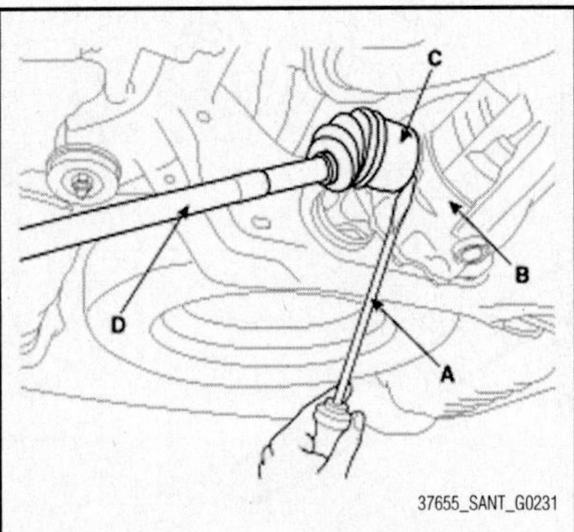

Fig. 61 Insert a pry bar (A) between the differential case (B) and joint case (C), and separate the halfshaft (D) from the differential case

19. Install the spring, the upper pad and the lower pad.
20. Install the mounting bolt of the rear lower arm and the rear carrier and tighten to 101–116 ft. lbs. (137–157 Nm), while supporting the lower arm with a jack. Tighten the mounting bolt of the cross member and the rear lower arm to 101–116 ft. lbs. (137–157 Nm).
21. Install the washer, castle nut and split pin to the rear hub assembly. Tightening torque: 145–188 ft. lbs. (196–255 Nm)
22. The washer should be assembled with convex surface outward when installing the castle nut and split pin.
23. Install the wheel and tire to the rear hub. Tightening torque: 65–80 ft. lbs. (88–108 Nm).

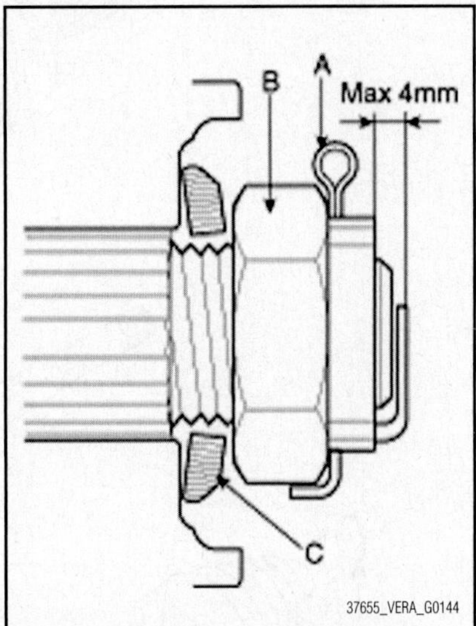

Fig. 63 The washer (C) should be assembled with convex surface outward when installing the castle nut (B) and split pin (A)

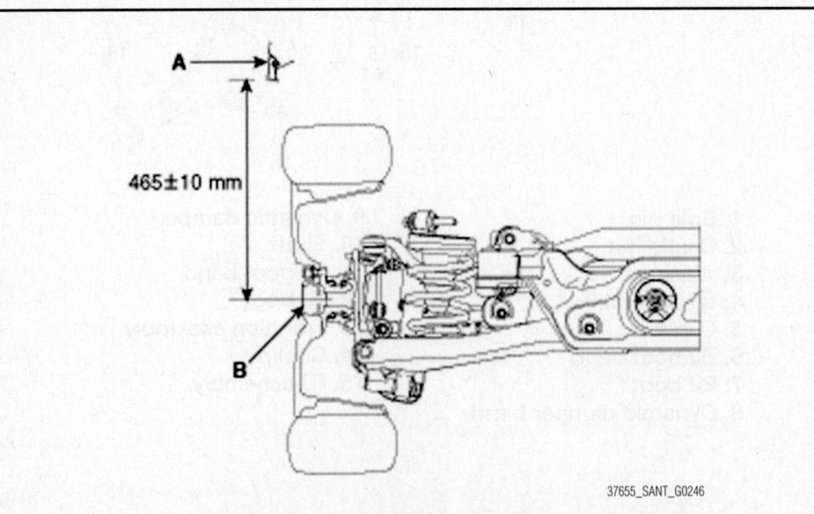

Fig. 62 Distance between the wheel housing molding (A) and hub assembly (B)

PROPELLER SHAFT

REMOVAL & INSTALLATION

See Figures 64 through 66.

1. Before servicing the vehicle, refer to the Precautions Section.
2. After making a match mark on the rubber coupling and rear differential companion, remove the propeller shaft mounting bolts.
3. Remove the center bearing bracket mounting bolts.
4. After making a match mark on the flange yoke and transaxle companion, remove the propeller shaft mounting bolts. Use the hexagonal wrench to prevent damage of bolt head when removing bolts.

To install:

5. Installation is the reverse of removal, noting the following:
    a. Tighten the propeller shaft mounting bolts to 36–51 ft. lbs. (49–69 Nm).
    b. Tighten the center bearing bracket mounting bolts to 29–36 ft. lbs. (39–49 Nm).

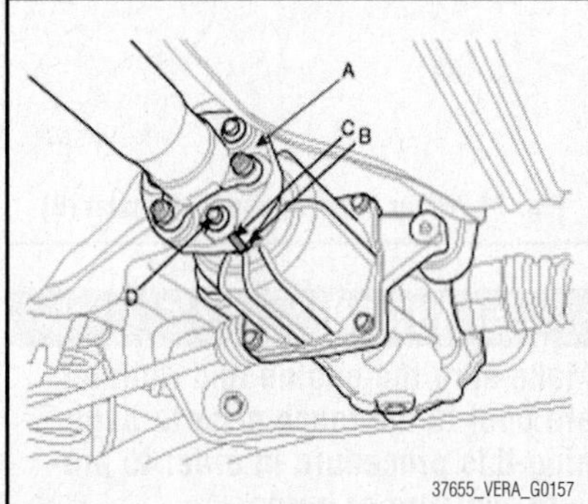

Fig. 64 After making a match mark (C) on the rubber coupling (A) and rear differential companion (B), remove the propeller shaft mounting bolts (D)

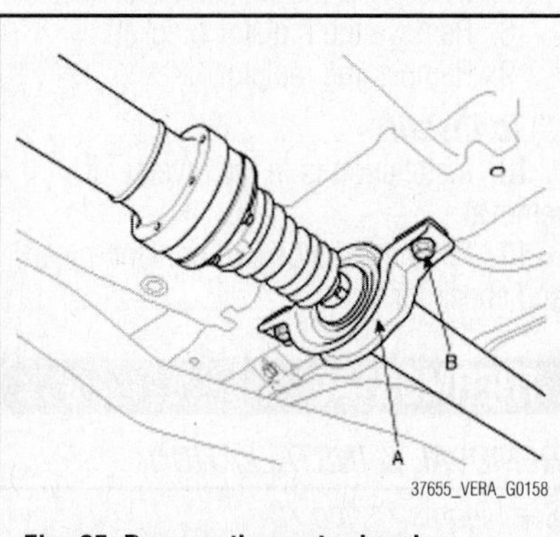

Fig. 65 Remove the center bearing bracket (A) mounting bolts (B)

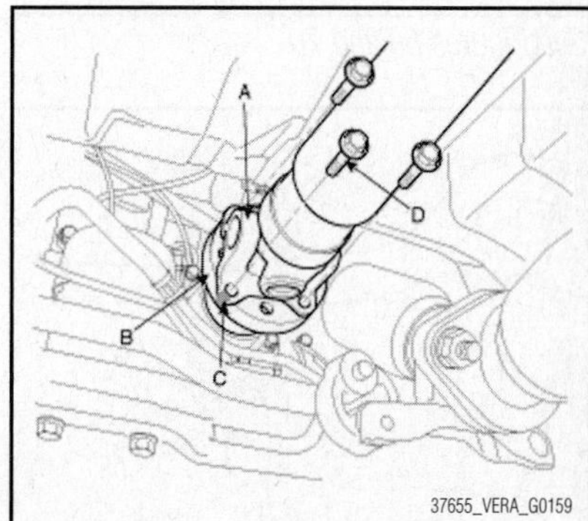

Fig. 66 After making a match mark (C) on the flange yoke (A) and transaxle companion (B), remove the propeller shaft mounting bolts (D)

ENGINE COOLING

ENGINE FAN

REMOVAL & INSTALLATION

See Figures 67 and 68.

1. Before servicing the vehicle, refer to the Precautions Section.
2. Disconnect the negative battery cable.
3. Remove the air duct.
4. Drain the coolant. Remove the radiator cap to speed draining.
5. Remove the radiator upper and lower hoses.
6. Disconnect transaxle oil cooler hoses.
7. Disconnect the fan connectors.
8. Remove the cooling fan.

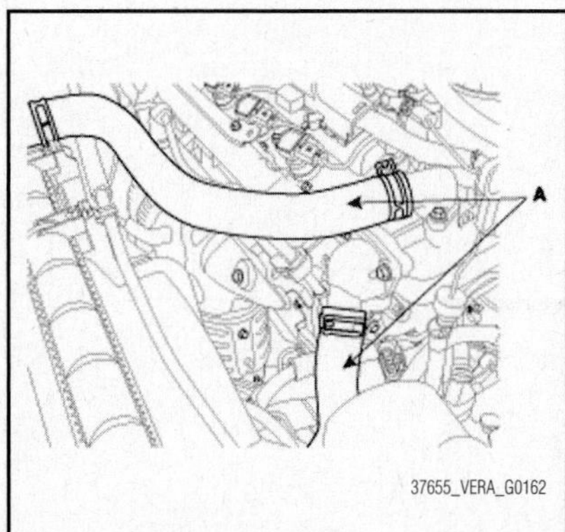

Fig. 67 Radiator upper and lower hose (A)

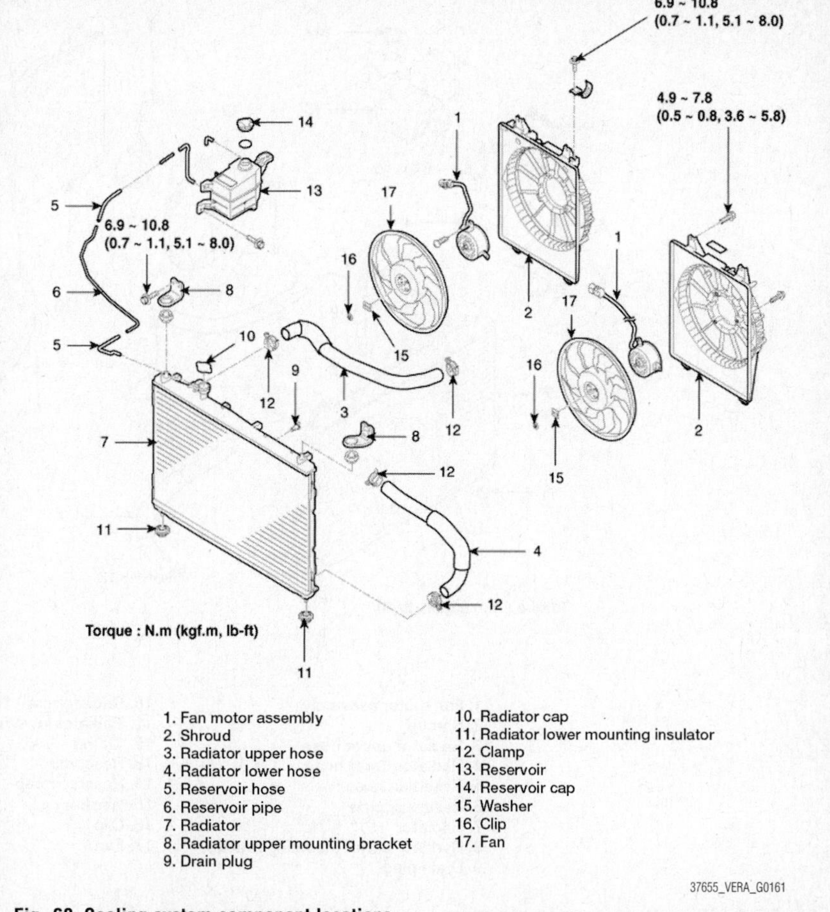

Fig. 68 Cooling system component locations

To install:

9. Installation is in the reverse of removal.
10. Fill the radiator with coolant, bleed, and check for leaks.

RADIATOR

REMOVAL & INSTALLATION

See Figures 69 and 70.

A
37655_VERA_G0162

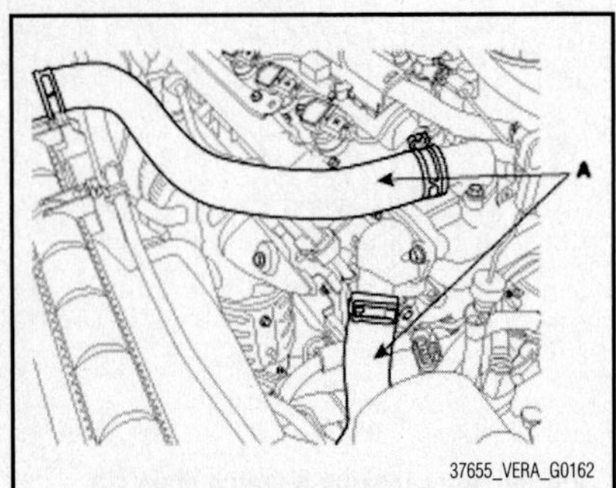

Fig. 69 Radiator upper and lower hose (A)

1. Before servicing the vehicle, refer to the Precautions Section.
2. Disconnect the negative battery cable.
3. Remove the air duct.
4. Drain the coolant.
5. Remove the radiator upper and lower hoses.
6. Disconnect the transaxle oil cooler hoses.
7. Disconnect the radiator fan connectors.
8. Remove the radiator bracket.
9. Remove the radiator.

To install:

10. Installation is in the reverse of removal.
11. Fill the radiator with coolant, bleed, and check for leaks.

THERMOSTAT

REMOVAL & INSTALLATION

See Figures 71 and 72.

B
A
37655_SANT_G0273

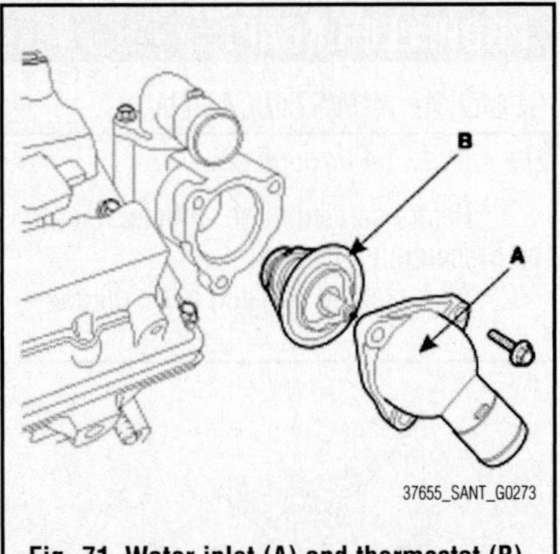

Fig. 71 Water inlet (A) and thermostat (B)

⁂ CAUTION

Make sure the engine and radiator are cool to the touch prior to beginning this procedure in order to prevent scalding or burns.

1. Before servicing the vehicle, refer to the Precautions Section.

6.9 ~ 10.8
(0.7 ~ 1.1, 5.1 ~ 8.0)
4.9 ~ 7.8
(0.5 ~ 0.8, 3.6 ~ 5.8)
6.9 ~ 10.8
(0.7 ~ 1.1, 5.1 ~ 8.0)
Torque : N.m (kgf.m, lb-ft)
1. Fan motor assembly
2. Shroud
3. Radiator upper hose
4. Radiator lower hose
5. Reservoir hose
6. Reservoir pipe
7. Radiator
8. Radiator upper mounting bracket
9. Drain plug
10. Radiator cap
11. Radiator lower mounting insulator
12. Clamp
13. Reservoir
14. Reservoir cap
15. Washer
16. Clip
17. Fan
37655_VERA_G0161

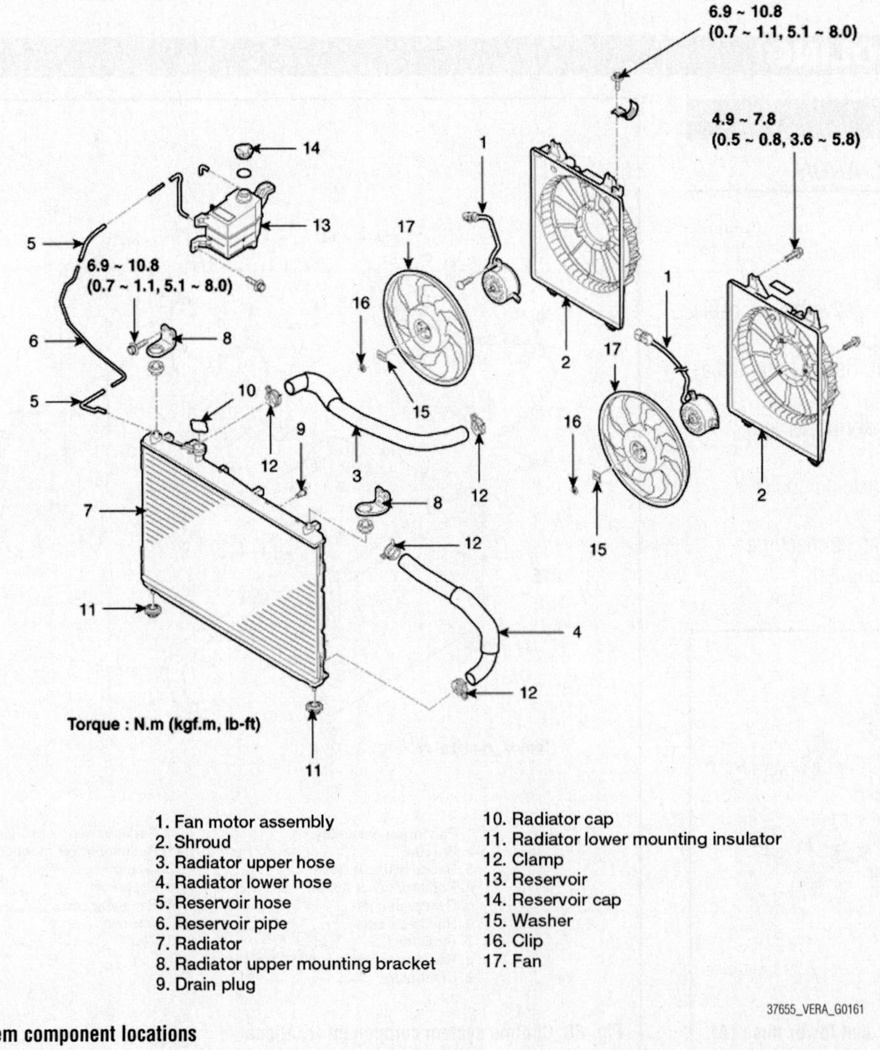

Fig. 70 Cooling system component locations

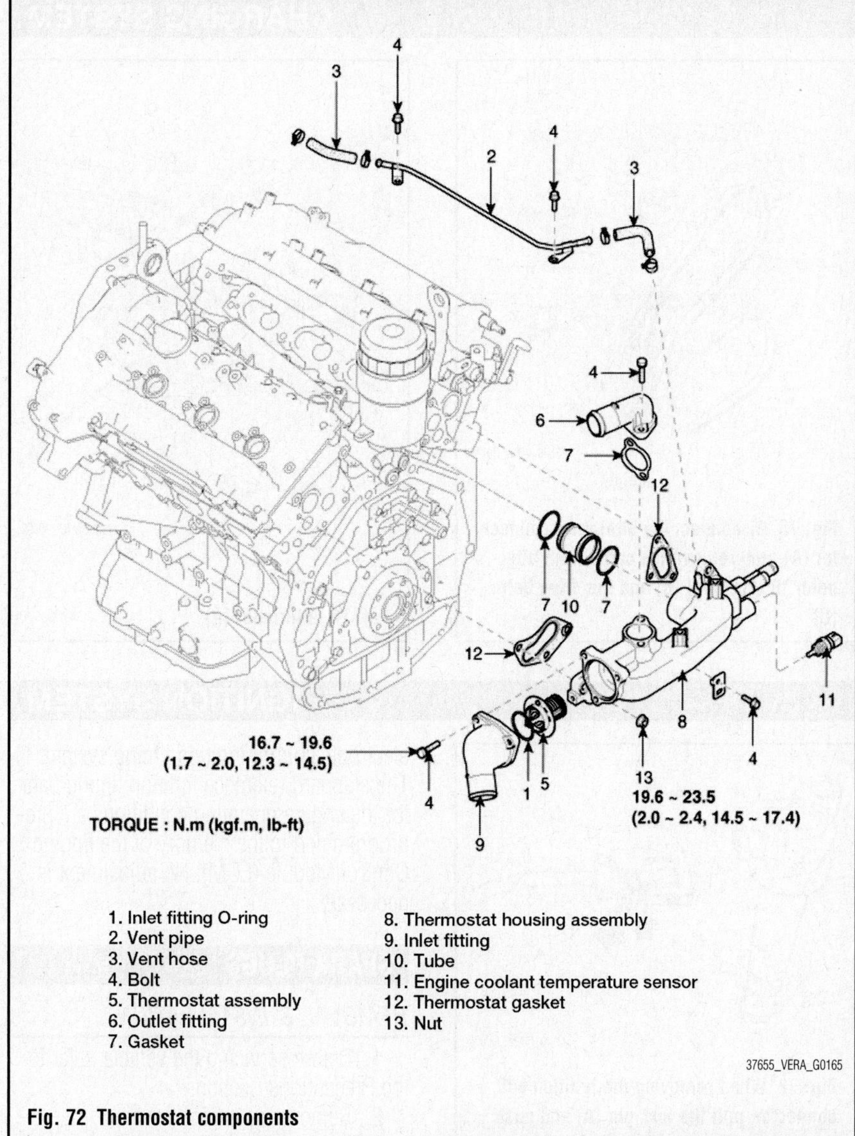

Fig. 72 Thermostat components

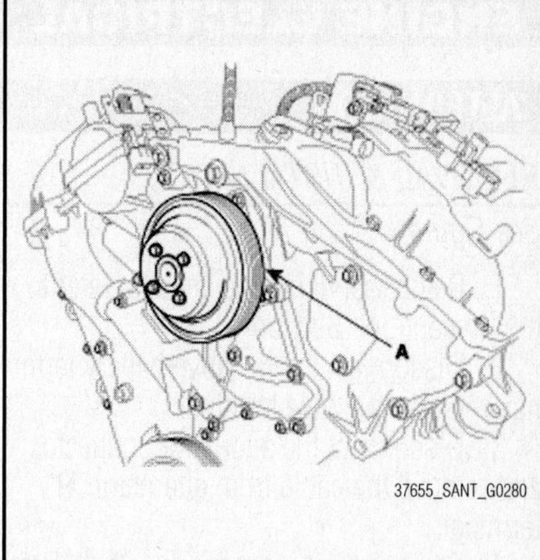

Fig. 73 Remove the 4 bolts and pump pulley (A)

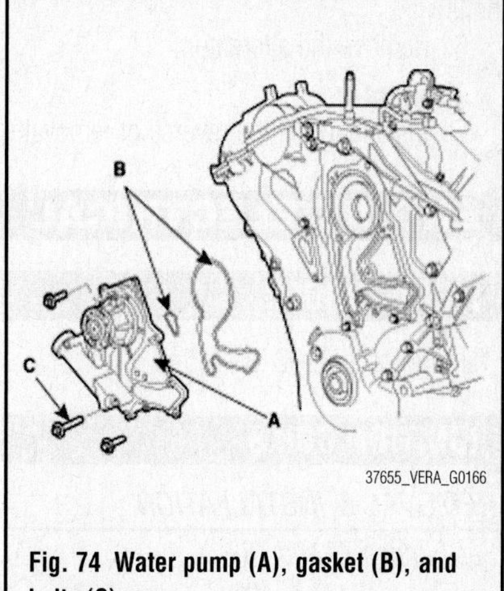

Fig. 74 Water pump (A), gasket (B), and bolts (C)

2. Drain the coolant down to below thermostat level.
3. Remove the water inlet and gasket.
4. Remove the thermostat.

To install:

5. Install a new thermostat. Install the thermostat with the jiggle valve upward.
6. Tighten the water inlet mounting bolts to 12–15 ft. lbs. (17–20 Nm).
7. Refill the coolant and check for leaks.

WATER PUMP

REMOVAL & INSTALLATION

See Figures 73 and 74.

CAUTION

The system is under high pressure when the engine is hot. To avoid danger of releasing scalding engine coolant, remove the cap only when the engine is cool.

1. Before servicing the vehicle, refer to the Precautions Section.
2. Drain the engine coolant.
3. Remove the accessory drive belt. Refer to Accessory Drive Belt Removal & Installation in the Engine Mechanical section.
4. Remove the 4 bolts and pump pulley.
5. Remove the water pump and gasket.

To install:

➡Clean the contact face before assembly.

6. Install the water pump and a new gasket. Always use new bolts and gaskets. Tighten the bolts to 16–17 ft. lbs. (22–24 Nm); 87–104 inch lbs. (10–12 Nm). Apply sealant (Three Bond 13868 or 13860) where shown.
7. Install the 4 bolts and the pump pulley. Tighten to 70–86 inch lbs. (8–10 Nm).
8. Install the accessory drive belt.
9. Fill with engine coolant.
10. Start engine and check for leaks.
11. Recheck engine coolant level.

ENGINE ELECTRICAL — CHARGING SYSTEM

ALTERNATOR

REMOVAL & INSTALLATION

See Figures 75 and 76.

1. Before servicing the vehicle, refer to the Precautions Section.
2. Disconnect the negative battery terminal, then the positive terminal.
3. Disconnect the alternator connector, and remove the cable from alternator "B" terminal.
4. Remove the accessory drive belt. Refer to Accessory Drive Belts, Removal & Installation, in the Engine Mechanical section.
5. Remove the alternator mounting bolts.
6. Remove the alternator.

To install:

7. Installation is the reverse of removal.

Fig. 75 Disconnect the alternator connector (A), and remove the cable from alternator "B" terminal (B) and the drive belt (C)

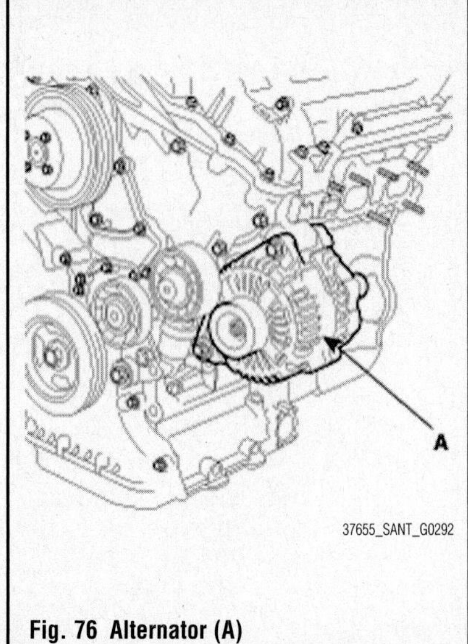

Fig. 76 Alternator (A)

ENGINE ELECTRICAL — IGNITION SYSTEM

FIRING ORDER

Firing order: 1–2–3–4–5–6

IGNITION COILS

REMOVAL & INSTALLATION

See Figures 77 and 78.

1. Before servicing the vehicle, refer to the Precautions Section.
2. Remove the engine cover.
3. Disconnect the ignition coil connectors.

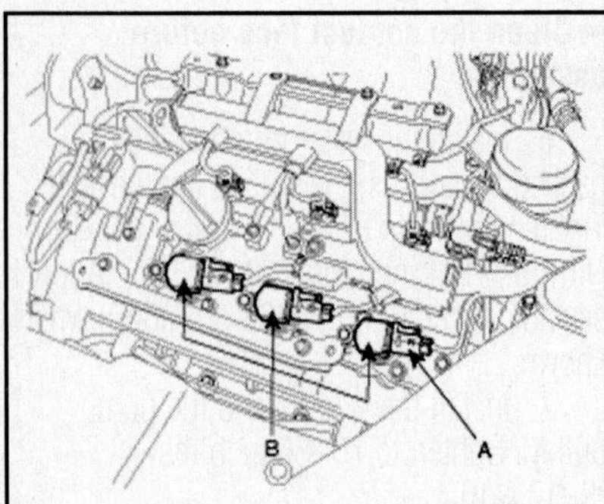

Fig. 77 Disconnect the ignition coil connectors (A) similar

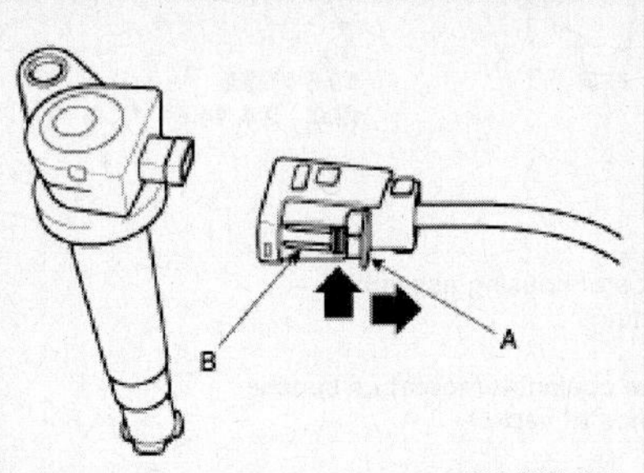

Fig. 78 When removing the ignition coil connector, pull the lock pin (A) and push the clip (B)

➡When removing the ignition coil connector, pull the lock pin and push the clip.

4. Remove the ignition coil.

To install:

5. Installation is the reverse of the removal procedure.

IGNITION TIMING

ADJUSTMENT

Ignition timing is controlled by the electronic control ignition timing system. The standard reference ignition timing data for the engine operating conditions are pre-programmed in the memory of the Engine Control Module (ECM). No adjustment is necessary.

SPARK PLUGS

REMOVAL & INSTALLATION

1. Before servicing the vehicle, refer to the Precautions Section.
2. Remove the engine cover.
3. Remove the ignition coil.
4. Use a spark plug socket and wrench to remove the spark plugs.

WARNING

Be careful that no contaminates enter through the spark plug holes.

➡Check the electrode gap on the spark plugs before installation. Specification: 0.039–0.043 inch (1.0–1.1 mm).

To install:

5. To install, reverse the removal procedure.

STARTER

REMOVAL & INSTALLATION

See Figures 79 through 81.

1. Before servicing the vehicle, refer to the Precautions Section.
2. Disconnect the negative battery cable.
3. Remove the front roll stopper under mounting bolts and disconnect the electronic controlled mounting solenoid connector.

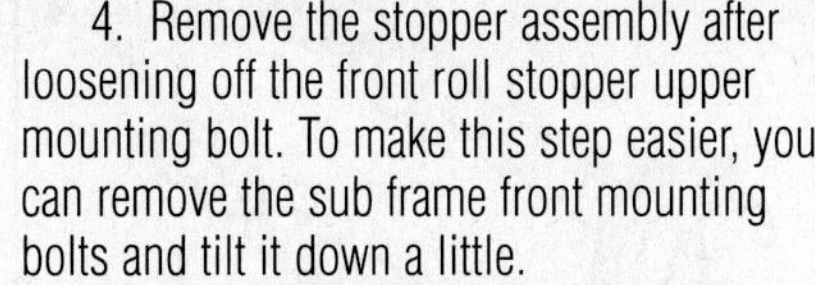

4. Remove the stopper assembly after loosening off the front roll stopper upper mounting bolt. To make this step easier, you can remove the sub frame front mounting bolts and tilt it down a little.
5. Remove the front engine bracket.
6. Remove the starter cover by loosening the two bolts and the one nut.
7. Disconnect the starter cable from the B terminal on the solenoid, then disconnect the connector from the S terminal.
8. Remove the starter mounting bolts.
9. Remove the starter.

To install:

10. Installation is the reverse of removal, noting the following:
    a. Tighten the starter mounting bolts (D) to 36–47 ft. lbs. (49–64 Nm)
    b. Tighten the starter cover bolts (A) to 78–12 inch lbs. (9–14 Nm) and nut (B) to 35–52 12 inch lbs. (4–6 Nm).
    c. Tighten the front roll stopper under mounting bolts to 36–47 ft. lbs. (49–64 Nm)

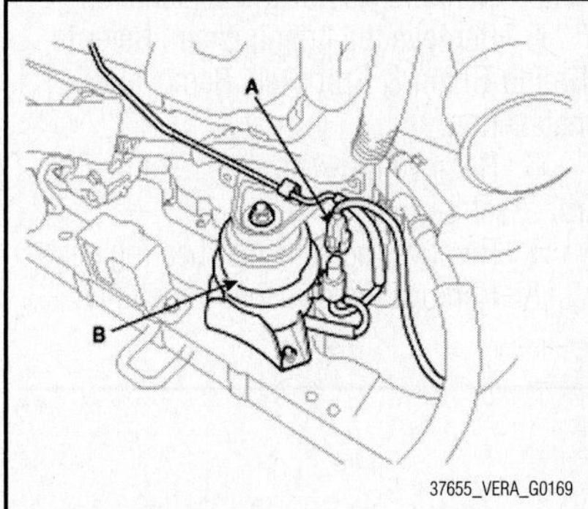

Fig. 79 Remove the front roll stopper under mounting (B) bolts and disconnect the electronic controlled mounting solenoid connector (A)

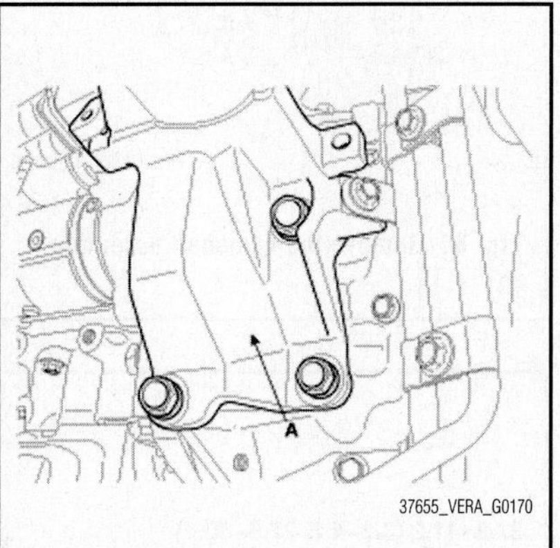

Fig. 80 Remove the front engine bracket (A)

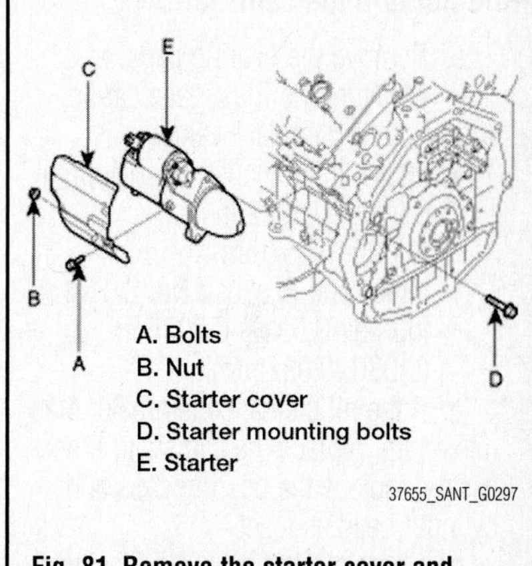

Fig. 81 Remove the starter cover and starter

ENGINE MECHANICAL

➡Disconnecting the negative battery cable may interfere with the functions of the on board computer systems and may require the computer to undergo a relearning process, once the negative battery cable is reconnected.

ACCESSORY DRIVE BELTS

ACCESSORY BELT ROUTING

See Figure 82.

INSPECTION

Inspect the accessory drive belt for signs of glazing or cracking. A glazed belt will be perfectly smooth from slippage, while a good belt will have a slight texture of fabric visible. Cracks will usually start at the inner edge of the belt and run outward. All worn or damaged accessory drive belts should be replaced immediately.

ADJUSTMENT

The drive belt tension is provided through a drive belt tensioner.

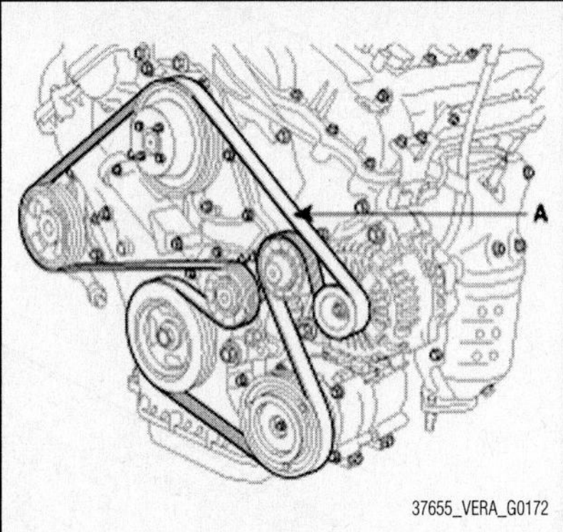

Fig. 82 Accessory drive belt routing

REMOVAL & INSTALLATION

1. Before servicing the vehicle, refer to the Precautions Section.
2. Raise and support the vehicle.
3. Remove the engine splash shield.
4. Rotate the drive belt tensioner clockwise to release the drive belt tension.
5. Remove the drive belt from the alternator.
6. Slowly release the drive belt tensioner.
7. Remove the drive belt from the accessory drive pulleys.

To install:

8. Install the drive belt to the accessory drive pulley.
9. Rotate the drive belt tensioner clockwise.
10. Install the drive belt to the alternator.
11. Ensure the drive belt is properly aligned and seated into the grooves of the accessory drive pulleys.
12. Slowly release the drive belt tensioner.
13. Install the engine splash shield.
14. Lower the vehicle.

CAMSHAFT AND VALVE LIFTERS

INSPECTION

➡For camshaft specifications, refer to Camshaft and Bearing Specifications in Specifications Chart.

1. Inspect the cam lobes.
   a. Using a micrometer, measure the cam lobe height. Standard value:
   - Intake: 1.8425 in. (46.8mm)
   - Exhaust: 1.8031 in. (45.8mm)

   b. If the cam lobe height is less than standard, replace the camshaft.
2. Inspect the camshaft journal clearance.
   a. Clean the bearing caps and camshaft journals.
   b. Place the camshafts on the cylinder head.
   c. Lay a strip of Plastigage®across each of the camshaft journal.
   d. Install the bearing caps.

➡Do not turn the camshaft.

   e. Remove the bearing caps.
   f. Measure the Plastigage® at its widest point. Standard value:
   - Intake and Exhaust: No. 1 journal: 0.0008–0.0022 in. (0.020–0.057mm)
   - Intake and Exhaust: No. 2, 3, 4 journal: 0.0012–0.0026 in. (0.030–0.067mm)

   g. If the oil clearance is greater than maximum, replace the camshaft. If necessary, replace the bearing caps and cylinder head as a set.
   h. Completely remove the Plastigage®.
   i. Remove the camshafts.
3. Inspect the camshaft end play.
   a. Install the camshafts.
   b. Using a dial indicator, measure the end play while moving the camshaft back and forth. Standard value:
   - 0.0008–0.0071 in. (0.02–0.18mm)

   c. If the end play is greater than maximum, replace the camshaft. If necessary, replace the bearing caps and cylinder head as a set.
   d. Remove the camshafts.

REMOVAL & INSTALLATION

See Figures 83 through 88.

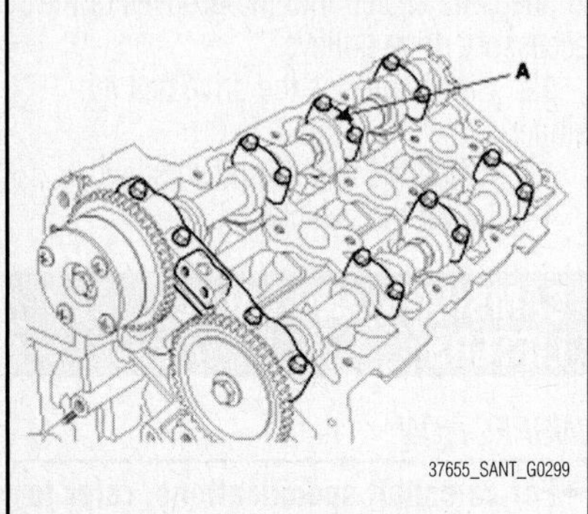

Fig. 83 Remove the camshaft bearing caps (A)

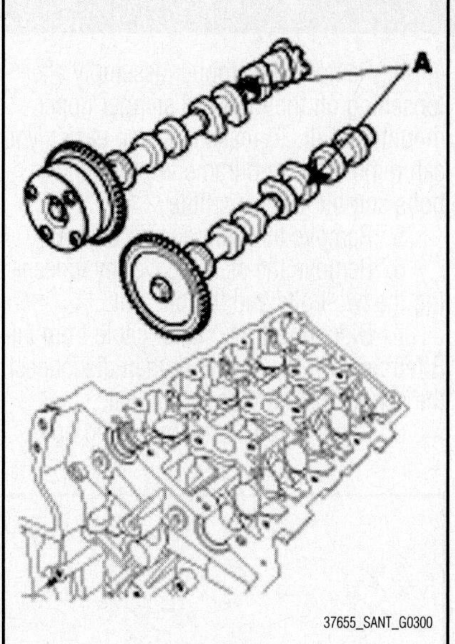

Fig. 84 Remove the camshaft assembly (A)

Engine removal is required for this procedure. Use fender covers to avoid damaging painted surfaces. To avoid damage, unplug the wiring connectors carefully while holding the connector portion. Mark all wiring and hoses to avoid misconnection. Inspect the timing belt before removing. Turn the crankshaft pulley so that the No. 1 piston is at Top Dead Center (TDC).

1. Before servicing the vehicle, refer to the Precautions Section.
2. Disconnect the negative battery cable.
3. Remove the exhaust manifold. Refer to Exhaust Manifold Removal & Installation.
4. Remove the intake manifold. Refer to Intake Manifold Removal & Installation.
5. Remove the timing chain. Refer to Timing Chain & Sprockets Removal & Installation.
6. Remove the water temperature control assembly.
7. Remove the camshaft bearing caps.
8. Remove the camshaft assembly.

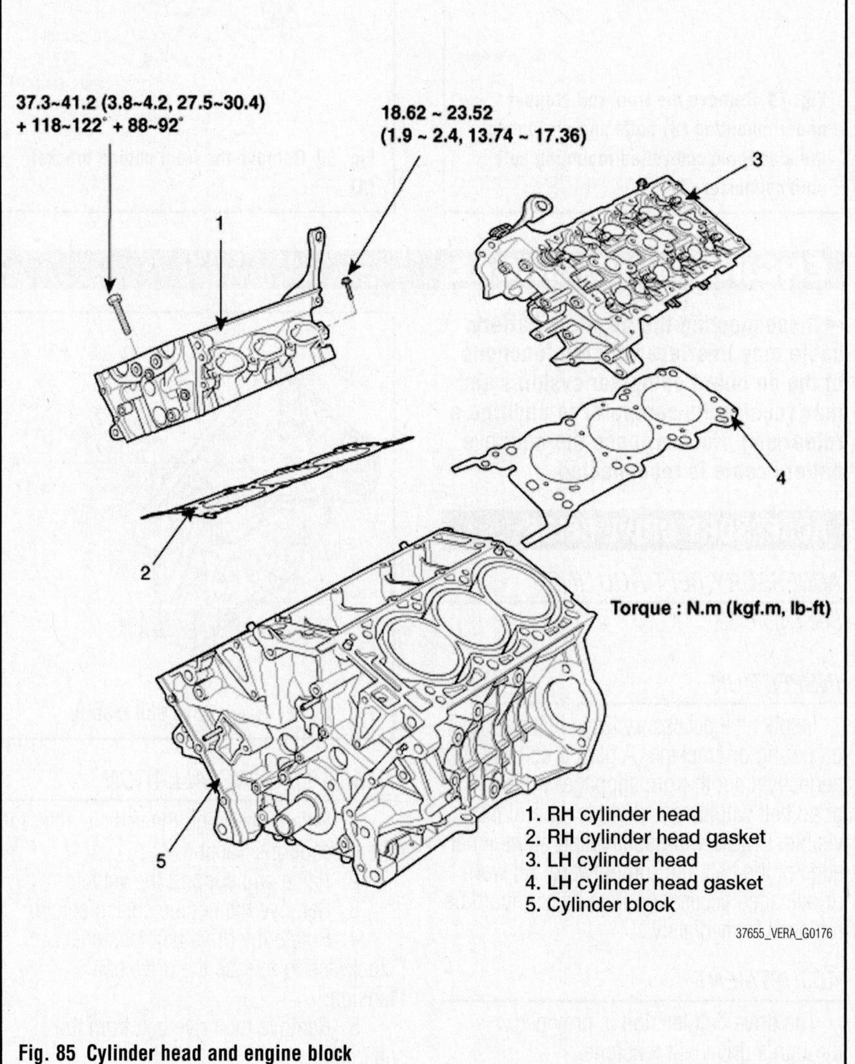

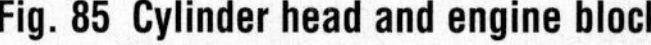

Fig. 85 Cylinder head and engine block

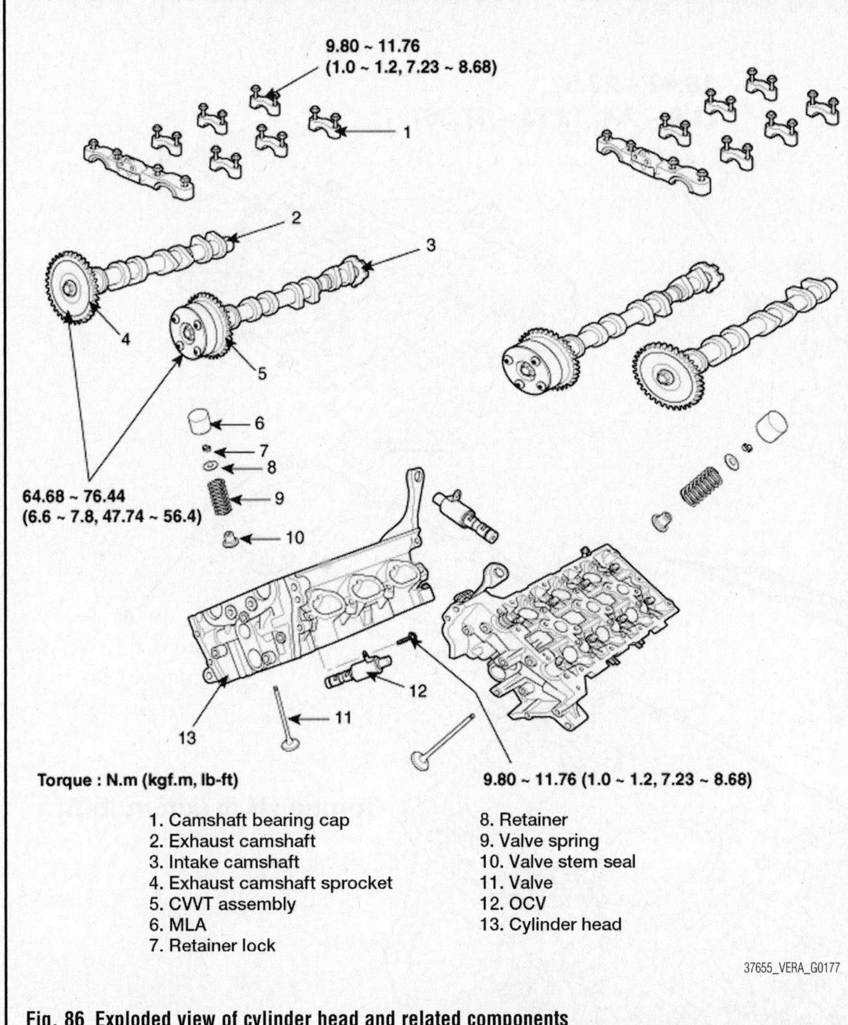

Fig. 86 Exploded view of cylinder head and related components

A. L(LH); R (RH)
B. I (Intake); None (Exhaust)
C. Journal number
D. Front mark

22140_HYUN_G0052

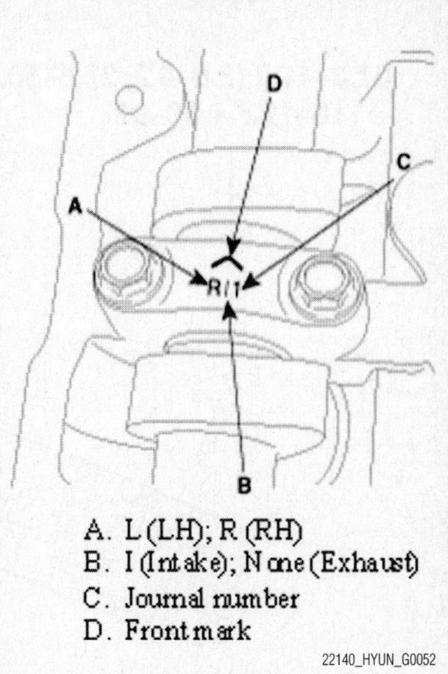

Fig. 88 Be careful to properly position the camshaft bearing caps according to its markings

To install:

Thoroughly clean all parts to be assembled. Rotate the crankshaft, set the No. 1 piston at TDC.

9. Install the camshafts.
   a. Apply a light coat of engine oil on camshaft journals.
   b. Assemble the key groove of camshaft rear side to the same level of head top surface.
   c. Be careful to get the right bank, left bank, intake side, and exhaust side in the correct position before assembling.
10. Install the camshaft bearing caps in the sequence shown:
   a. Step 1—Tighten to 52 inch lbs. (6 Nm).
   b. Step 2—Tighten to 86–104 inch lbs. (10–12 Nm).

⁂ WARNING

Be sure to properly position the right bank, left bank, intake side, exhaust side, and front mark on the camshaft bearing caps while assembling.

⁂ WARNING

Rotate the crankshaft so as not to contact the valves to the pistons by positioning the pistons 0.3937 in. (10mm) from the top of the cylinder block.

11. Install the water temperature control assembly.

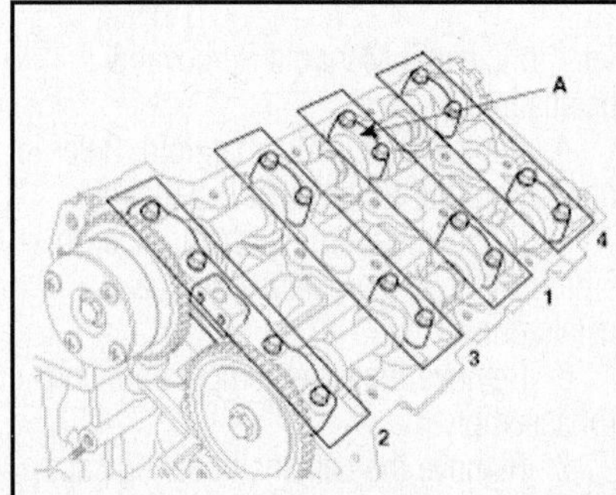

Fig. 87 Install the camshaft bearing caps in the sequence shown

12. Install the timing chain.
13. Check and adjust the valve clearance, as necessary.
14. Install the intake manifold.
15. Install the exhaust manifold.
16. Connect the negative battery cable.
17. Fill with engine coolant.
18. Start the engine and check for leaks.
19. Recheck the engine coolant level and oil level.

CRANKSHAFT FRONT SEAL

REMOVAL & INSTALLATION

1. Before servicing the vehicle, refer to the Precautions Section.
2. Remove the crankshaft pulley. Refer to Crankshaft Damper Removal & Installation.
3. Using a seal removal tool, remove, the crankshaft front seal.

To install:

4. Using a seal installation tool, install the crankshaft front seal.
5. Install the crankshaft pulley.
6. Start the engine and check for leaks.

CYLINDER HEAD

REMOVAL & INSTALLATION

See Figures 89 through 100.

➡Special Service Tool used for this procedure: SST 09221-4A000

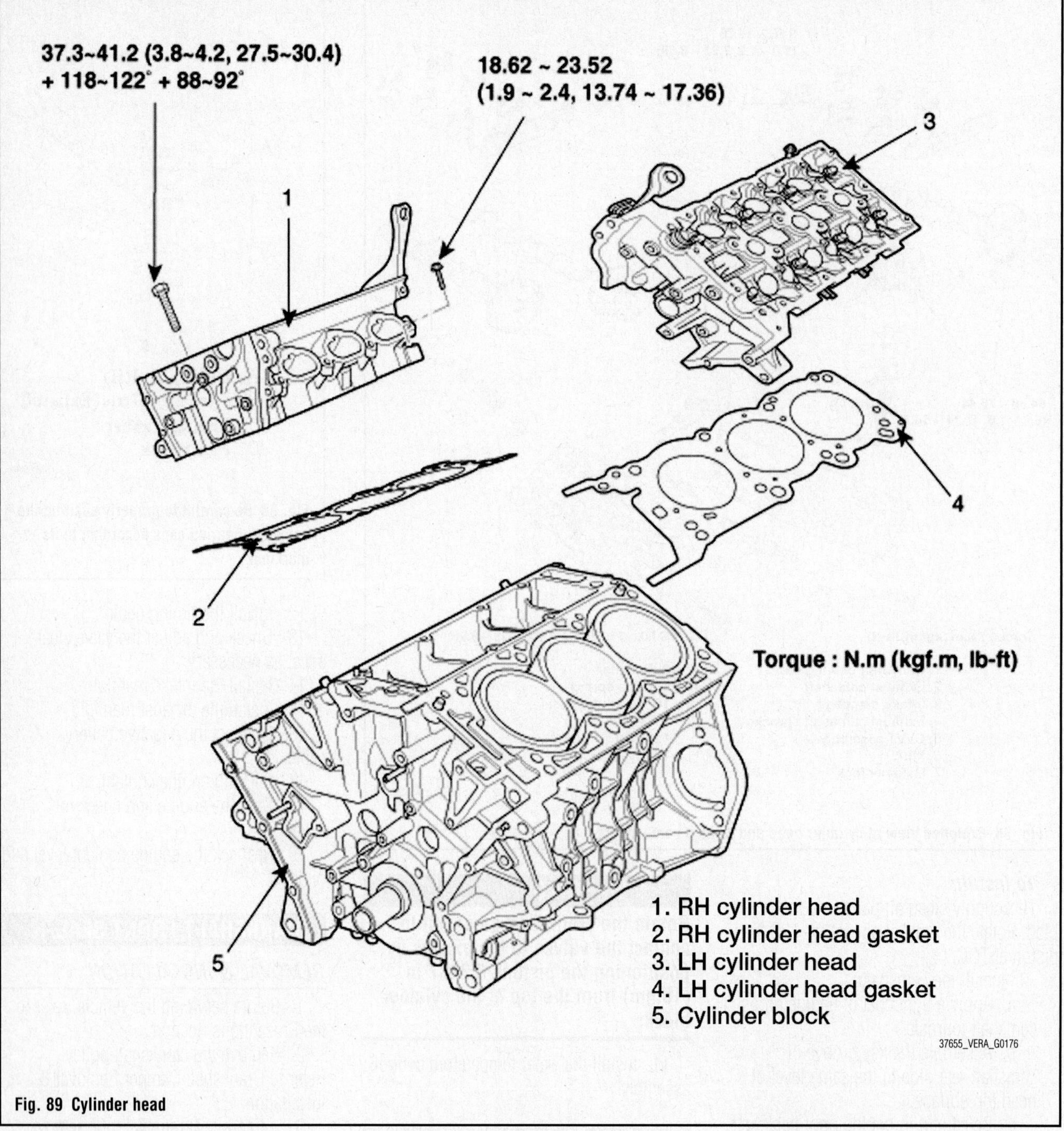

Fig. 89 Cylinder head

Engine removal is required for this procedure. Use fender covers to avoid damaging painted surfaces. To avoid damage, unplug the wiring connectors carefully while holding the connector portion. Mark all wiring and hoses to avoid misconnection. Inspect the timing belt before removing. Turn the crankshaft pulley so that the No. 1 piston is at Top Dead Center (TDC).

1. Before servicing the vehicle, refer to the Precautions Section.
2. Disconnect the negative battery cable.
3. Remove the exhaust manifold. Refer to Exhaust Manifold Removal & Installation.
4. Remove the intake manifold. Refer to Intake Manifold Removal & Installation.
5. Remove the timing chain. Refer to Timing Chain & Sprockets Removal & Installation.
6. Remove the water temperature control assembly.
7. Remove the camshaft bearing caps.
8. Remove the camshaft assembly.
9. Uniformly loosen and remove the 16 cylinder head bolts, in several passes, in the sequence shown.

➡Head warpage or cracking could result from removing bolts in an incorrect order.

10. Lift the cylinder head from the dowels on the cylinder block and place the cylinder head on wooden blocks on a bench.

➡Be careful not to damage the contact surfaces of the cylinder head and cylinder block.

To install:

Thoroughly clean all parts to be assembled. Always use a new head and manifold

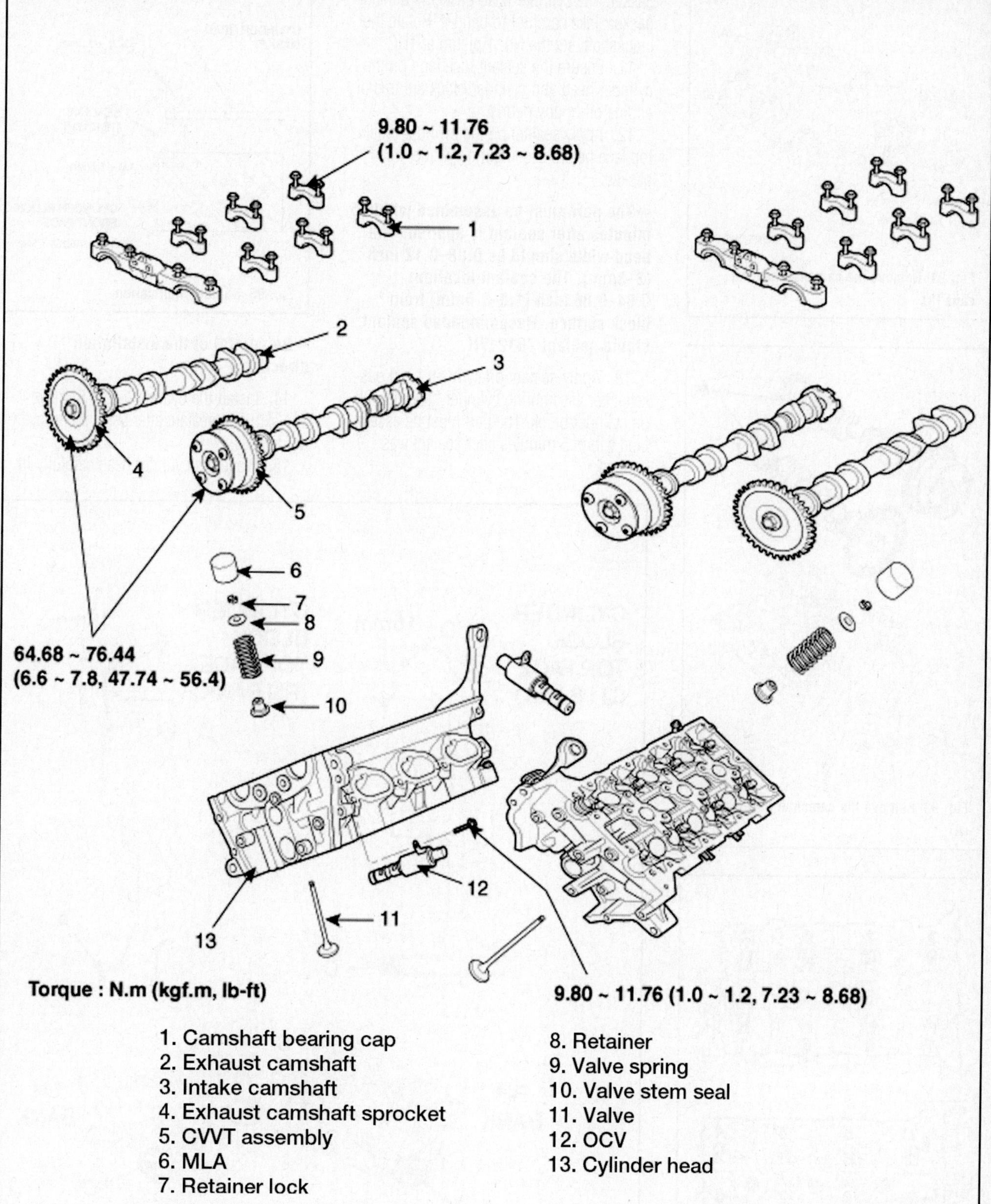

Fig. 90 Camshafts and related components

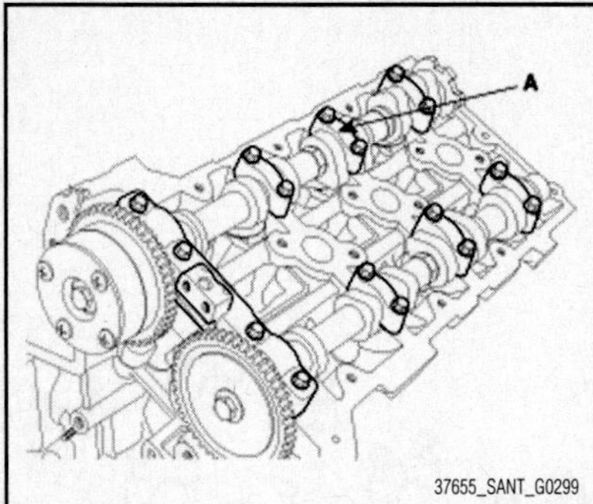

Fig. 91 Remove the camshaft bearing caps (A)

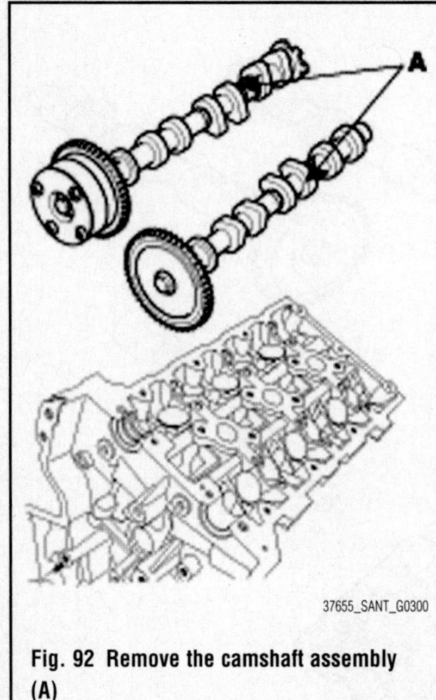

Fig. 92 Remove the camshaft assembly (A)

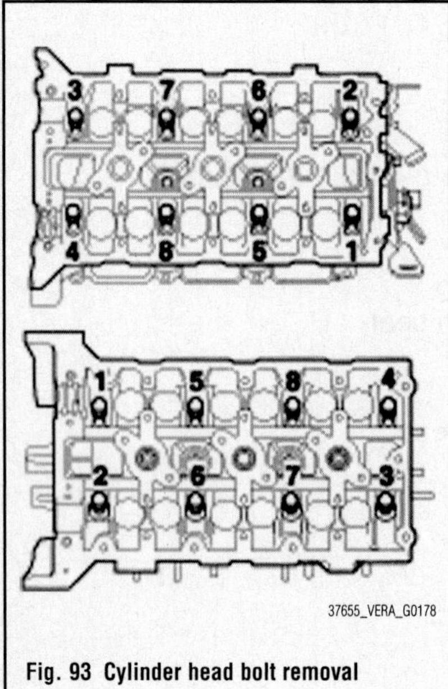

Fig. 93 Cylinder head bolt removal sequence

gasket. The cylinder head gasket is a metal gasket. Take care not to bend it. Rotate the crankshaft, set the No. 1 piston at TDC.

11. Ensure the sealant locations on the cylinder head and cylinder block are free of engine oil or any debris.

12. Apply sealant on the cylinder block top face before assembling cylinder head gaskets.

➡The part must be assembled within 5 minutes after sealant is applied. The bead width should be 0.08–0.12 inch (2–3mm). The sealant location: 0.04–0.06 inch (1.0–1.5mm) from block surface. Recommended sealant: Liquid sealant TB1217H.

13. Apply sealant on cylinder head gaskets after assembling cylinder head gaskets on cylinder block. The part must be assembled within 5 minutes after sealant was applied.

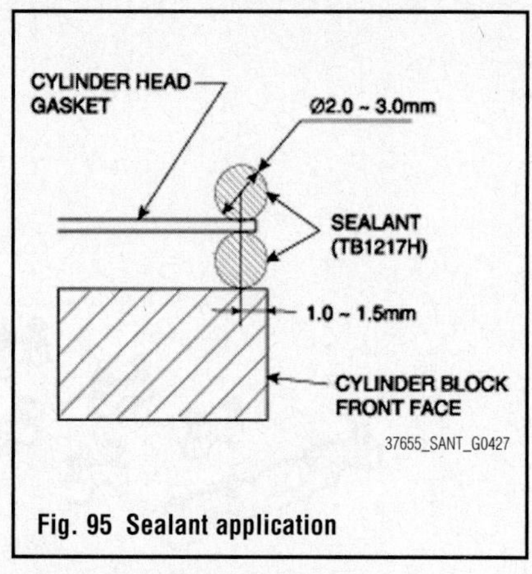

Fig. 95 Sealant application

➡Be careful of the installation direction.

14. Install the cylinder head. Remove any extruded sealant after assembling cylinder heads.

15. Place the cylinder head carefully in

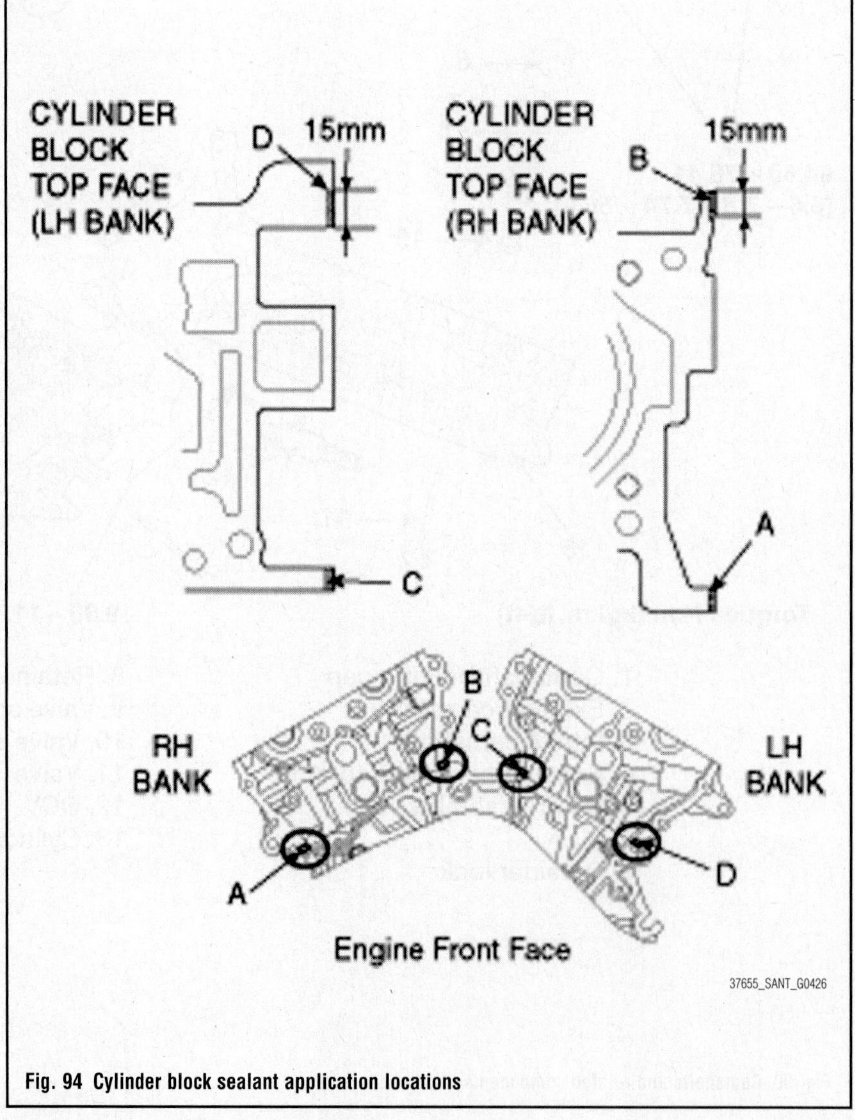

Fig. 94 Cylinder block sealant application locations

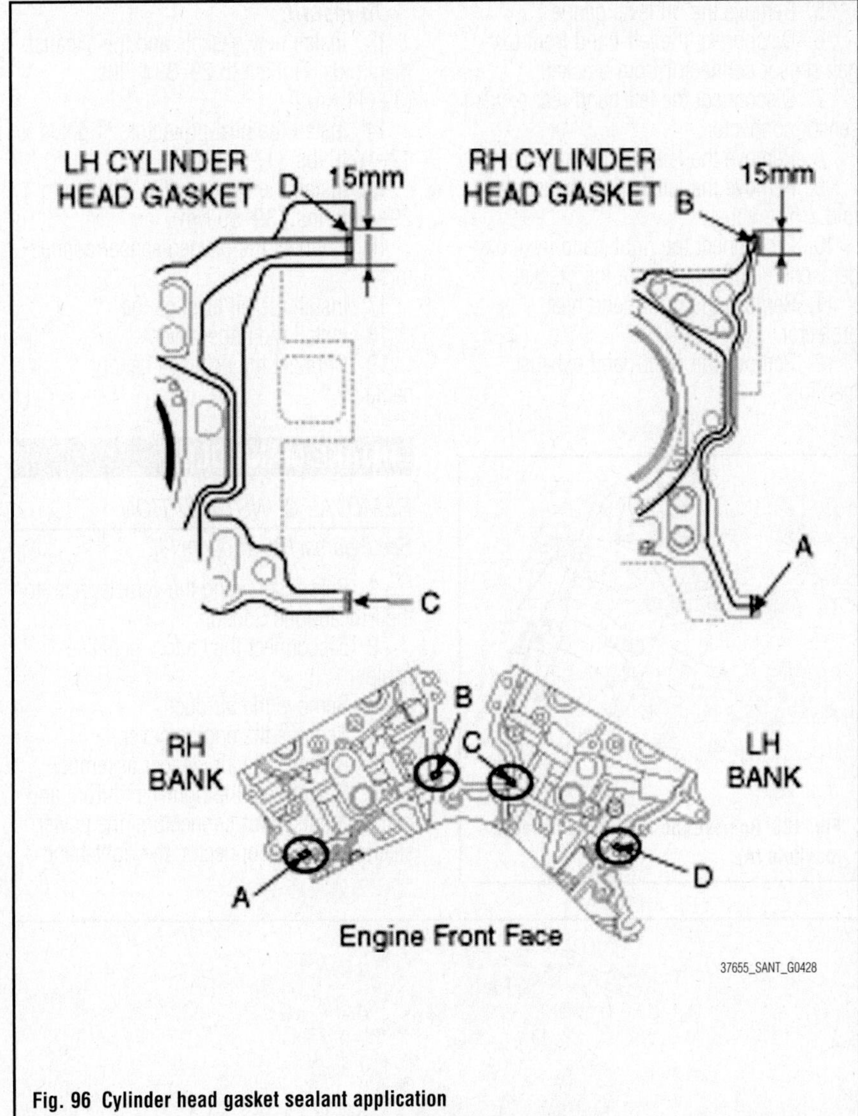

Fig. 96 Cylinder head gasket sealant application

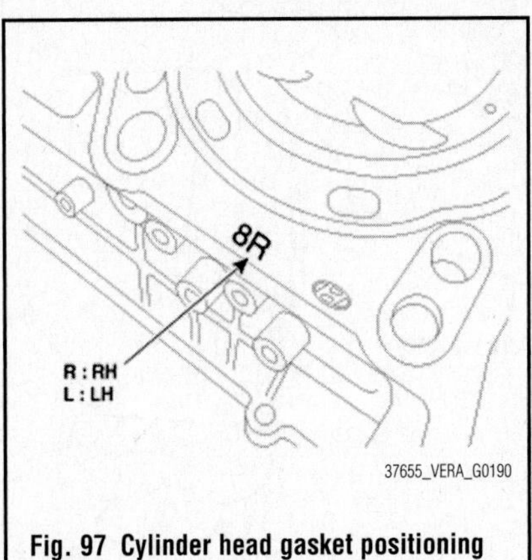

Fig. 97 Cylinder head gasket positioning

order not to damage the gasket with the bottom part of the end.

16. Install cylinder head bolts.
    a. Do not apply engine oil on the threads or under the heads of the cylinder head bolts.
    b. Using SST (09221-4A000), install and tighten the cylinder head bolts and plate washers, in several passes, in the sequence shown.
        - Step 1—Tighten to 28–30 ft. lbs. (37–41 Nm)
        - Step 2—Tighten an additional: 120° plus or minus 2°
        - Step 3—Tighten an additional: 90° plus or minus 2°
    c. Tighten bolt "A" to: 14–17 ft. lbs. (19–24 Nm).

➡Always use new cylinder head bolts.

17. Install the CVVT and camshaft sprocket and tighten to 48–56 ft. lbs. (65–76 Nm).

➡Install camshaft-inlet to dowel pin of CVVT assembly. At this time, do not install to oil hole of camshaft-inlet. Hold the hexagonal head wrench portion of the camshaft with a vise, and install the bolt and CVVT assembly. Do not rotate CVVT assembly when camshaft is installed to dowel pin of CVVT assembly.

18. Install the camshafts and bearing caps.

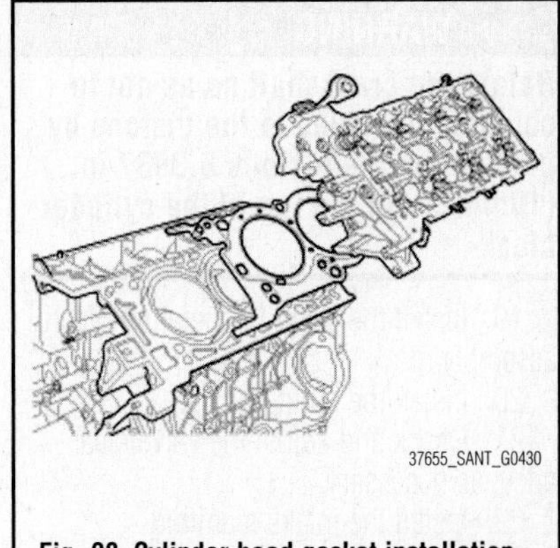

Fig. 98 Cylinder head gasket installation

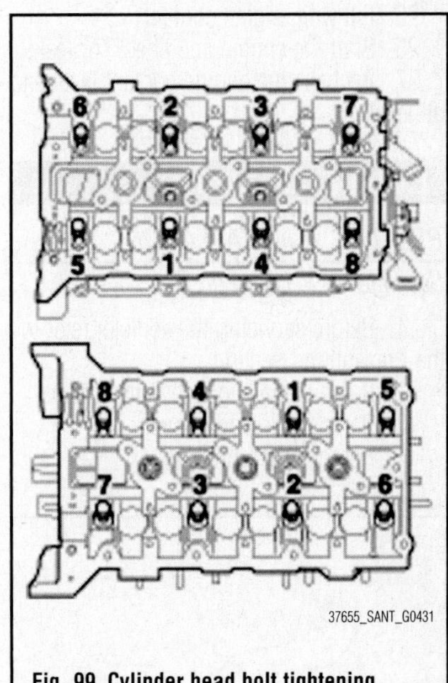

Fig. 99 Cylinder head bolt tightening sequence

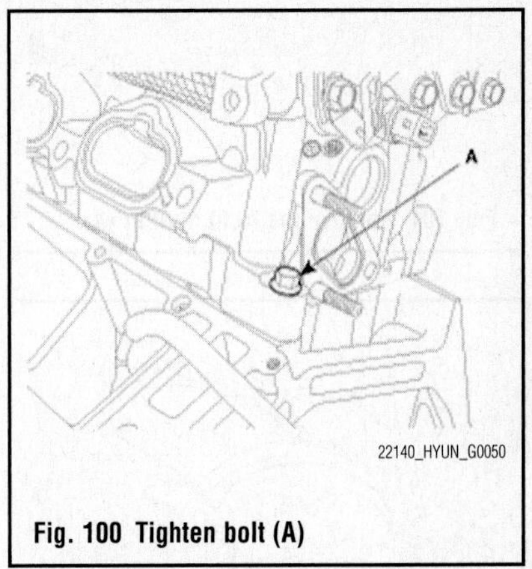

Fig. 100 Tighten bolt (A)

✲✲ WARNING

Rotate the crankshaft so as not to contact the valves to the pistons by positioning the pistons 0.3937 in. (10mm) from the top of the cylinder block.

19. Install the water temperature control assembly.
20. Install the timing chain.
21. Check and adjust the valve clearance, as necessary.
22. Install the intake manifold.
23. Install the exhaust manifold.
24. Connect the negative battery cable.
25. Fill with engine coolant.
26. Start the engine and check for leaks.
27. Recheck the engine coolant level and oil level.

EXHAUST MANIFOLD

REMOVAL & INSTALLATION

See Figures 101 through 104.

1. Before servicing the vehicle, refer to the Precautions Section.
2. Remove the negative battery cable.
3. Remove the undercover.
4. Remove the front muffler.
5. Remove the oil level gauge.
6. Disconnect the left-hand front oxygen sensor connector from bracket.
7. Disconnect the left-hand rear oxygen sensor connector.
8. Remove the left-hand heat protector.
9. Remove the left-hand exhaust manifold.
10. Disconnect the right-hand front oxygen sensor connector from the bracket.
11. Remove the right-hand heat protector.
12. Remove the right-hand exhaust manifold.

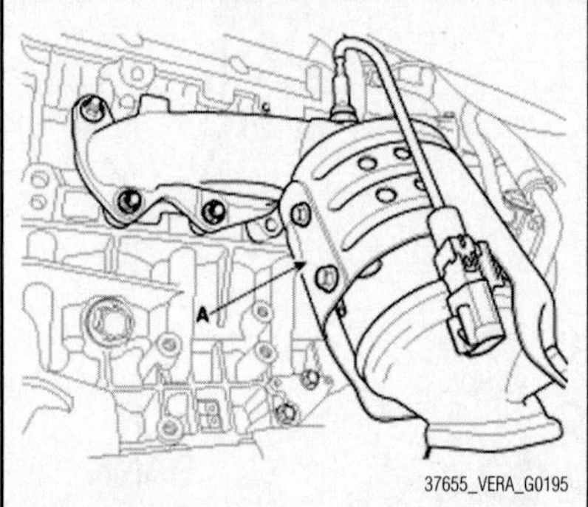

Fig. 103 Remove the right-hand exhaust manifold (A)

To install:

13. Install new gaskets and the exhaust manifolds. Tighten to 29–33 ft. lbs. (39–44 Nm).
14. Install the heat protectors. Tighten to 12–16 ft. lbs. (17–22 Nm).
15. Install the front muffler. Tighten to 29–43 ft. lbs. (39–59 Nm).
16. Connect the oxygen sensor connectors.
17. Install the oil level gauge.
18. Install the undercover.
19. Connect the negative battery cable.

INTAKE MANIFOLD

REMOVAL & INSTALLATION

See Figures 105 through 112.

1. Before servicing the vehicle, refer to the Precautions Section.
2. Disconnect the battery negative cable.
3. Remove the air duct.
4. Remove the engine cover.
5. Remove the air cleaner assembly.
6. Disconnect the right-hand front and rear oxygen sensor connectors, the power steering sensor connector, the right-hand

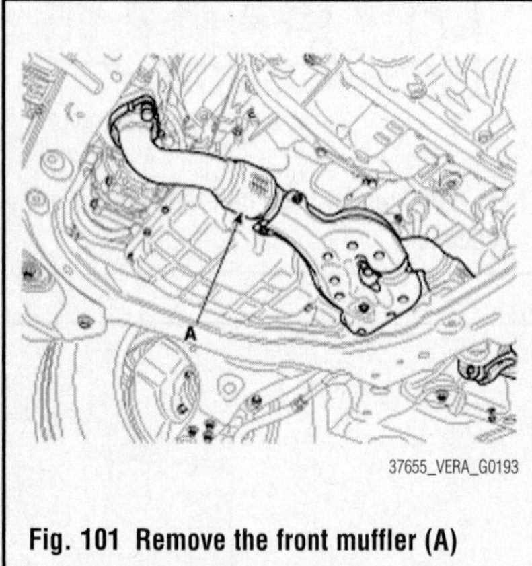

Fig. 101 Remove the front muffler (A)

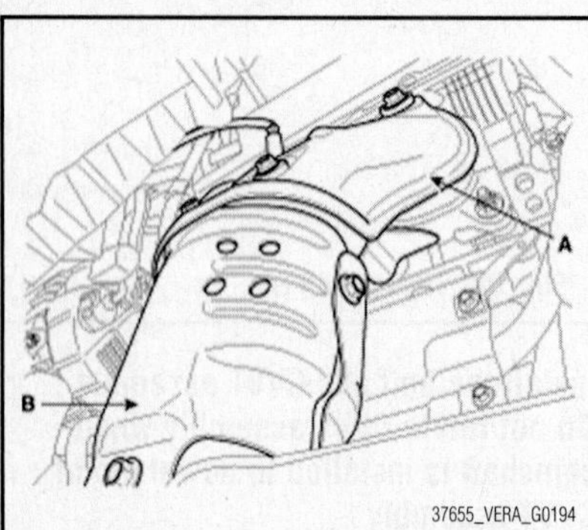

Fig. 102 Remove the left-hand heat protector (A) and exhaust manifold (B)

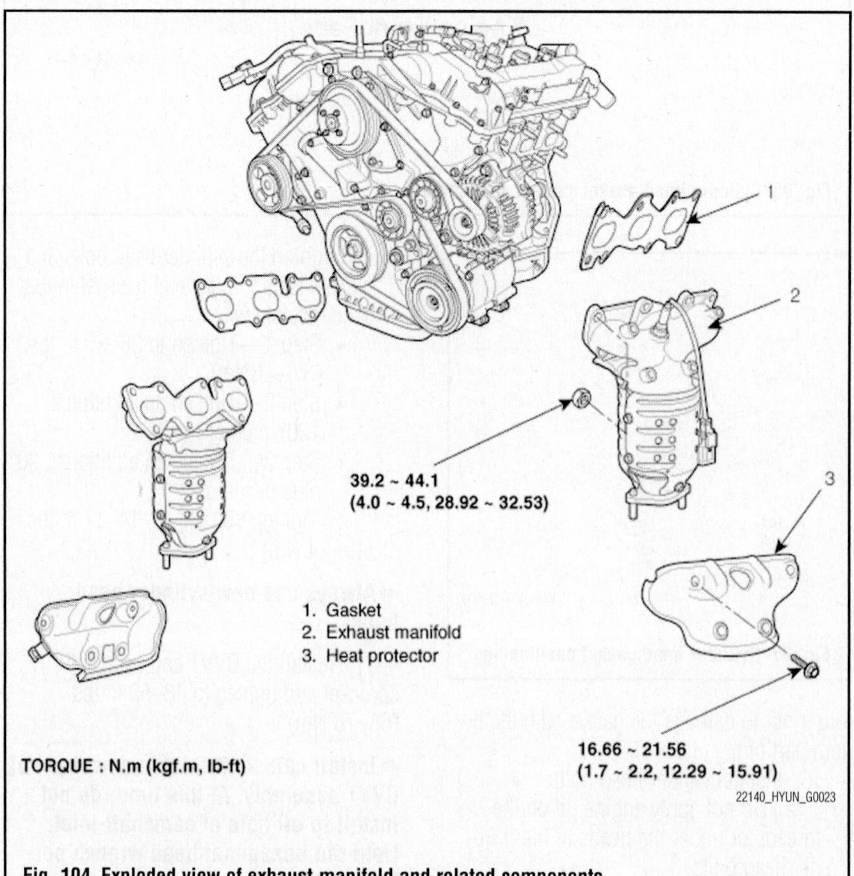

Fig. 104 Exploded view of exhaust manifold and related components

injector harness connector, and the Variable Intake Solenoid (VIS) valve connector.

7. Disconnect the Purge Control Solenoid Valve (PCSV) connector, MAP Manifold Absolute Pressure (MAP) sensor connector, and the PCSV hose.

8. Disconnect the Electronic Throttle Control (ETC) connector and knock sensor connector.

9. Disconnect the water hoses from the ETC and disconnect the Positive Crankcase Ventilation (PCV) hose.

10. Disconnect the brake vacuum hose.

11. Remove the surge tank stays.

12. Remove the surge tank.

13. Disconnect the breather pipe assembly.

14. Disconnect the left-hand injector harness connectors.

15. Remove the intake manifold and gasket.

To install:

16. Install the intake manifold assembly with a new gasket to the cylinder head assembly. Tighten the bolts in two steps:

- Step 1 (a–h): 35–52 inch lbs. (4–6 Nm)
- Step 2 (1–8): and tighten to 14–17 ft. lbs. (19–24 Nm)

17. Install the delivery pipe.

18. Connect the left-hand injector connector.

19. Connect the breather pipe assembly and tighten to 86–104 inch lbs. (10–12 Nm).

20. Install the surge tank and tighten the mounting bolts:

- Long bolt (Qty: 1): 86–104 inch lbs. (10–12 Nm)
- Short bolts (Qty: 3), Nuts (Qty: 2): 14–17 ft. lbs. (19–24 Nm)

21. Install the surge tank stays and tighten:

- Engine front side: 20–23 ft. lbs. (27–31 Nm)
- Engine rear side: 14–17 ft. lbs. (19–24 Nm)

22. Connect the brake vacuum hose.

23. Connect the water hoses to the ETC.

24. Connect the ETC connector and knock sensor connector.

25. Connect the PCSV connector, MAP sensor connector, and the PCSV hose.

26. Connect the right-hand front and rear oxygen sensor connectors, the power steering sensor connector, the right-hand

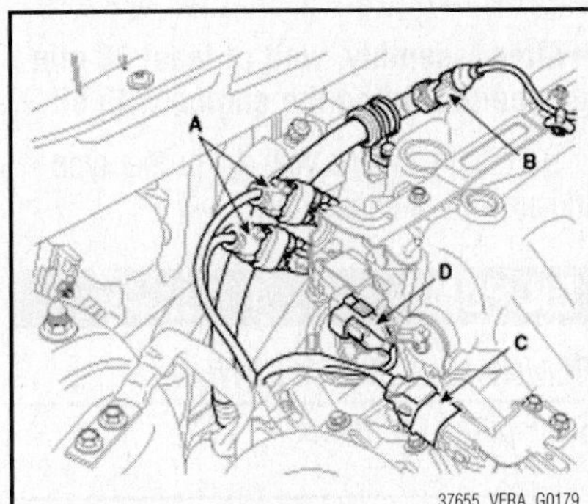

Fig. 105 Disconnect the oxygen sensor connectors (A), power steering sensor connector (B), right-hand injector harness connector (C), and VIS connector (D)

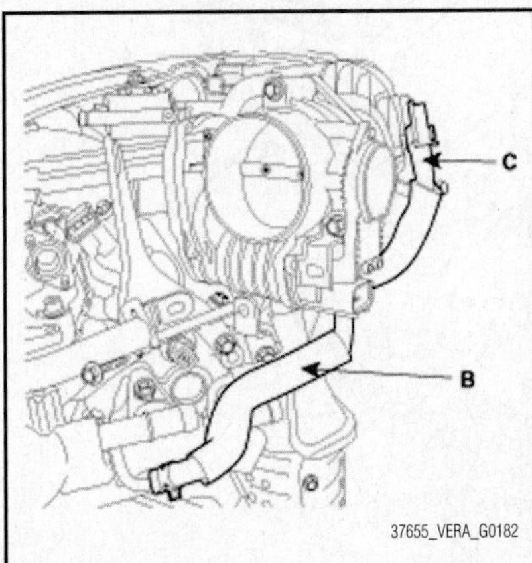

Fig. 106 Disconnect the water hoses (B) from ETC and PCV (C) hose

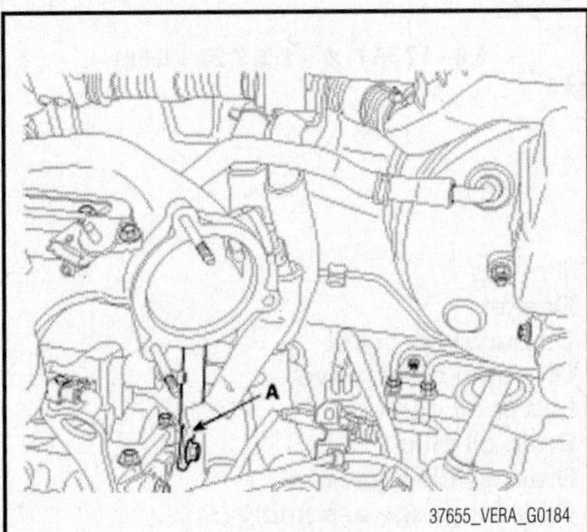

Fig. 107 Remove the surge tank stay (A)

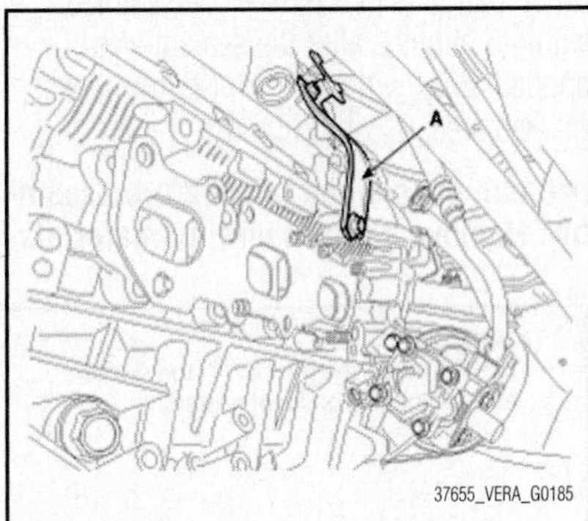

Fig. 108 Remove the remaining surge tank stay (A)

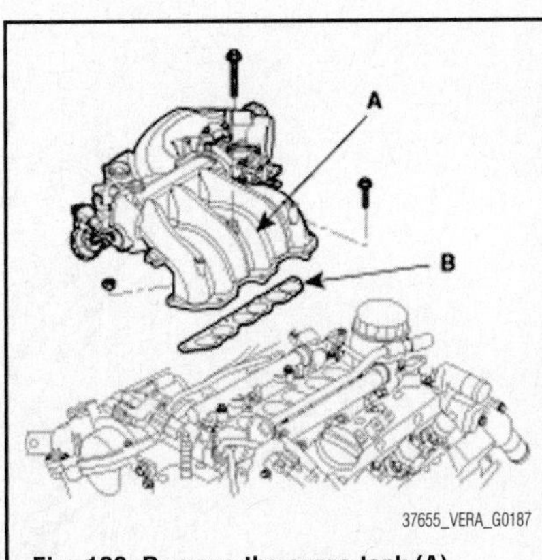

Fig. 109 Remove the surge tank (A)

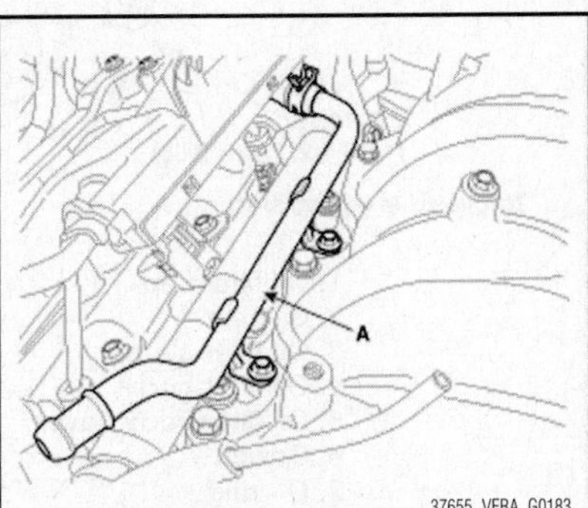

Fig. 110 Disconnect the breather pipe assembly (A)

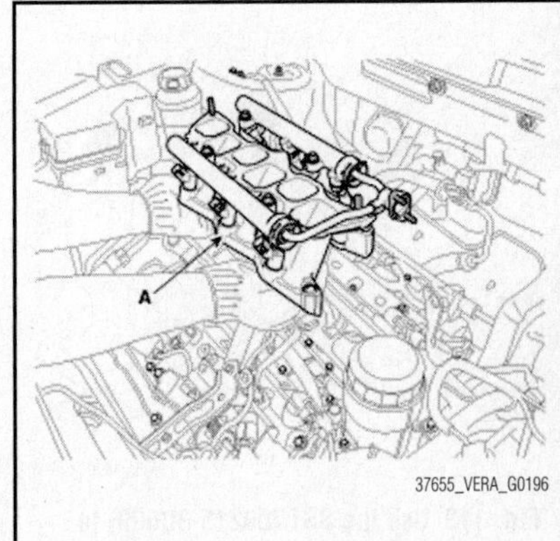

Fig. 111 Remove the intake manifold (A)

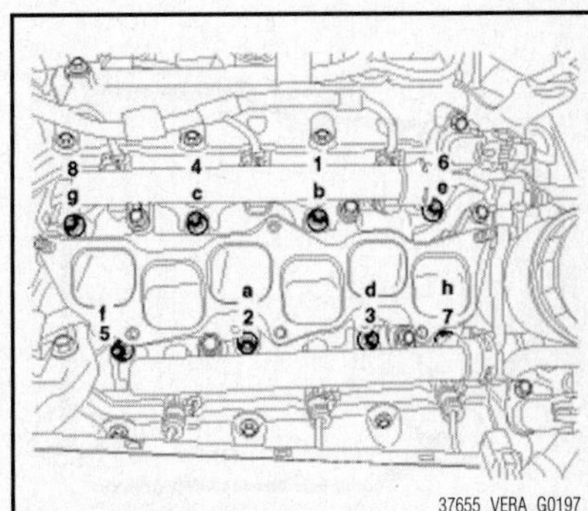

Fig. 112 Intake manifold bolt tightening sequence

injector harness connector, and the VIS solenoid valve connector.

27. Install the air cleaner assembly.
28. Install the air duct.
29. Connect the negative battery cable.
30. Install the engine cover.
31. Start the engine and check for proper operation.

OIL PAN

REMOVAL & INSTALLATION

See Figures 113 and 114.

➡Special Service Tool used for this procedure: SST 09215-3C000

1. Before servicing the vehicle, refer to the Precautions Section.
2. Disconnect the negative battery cable.
3. Drain the engine oil.
4. Remove the oil pan bolts.
5. Using the SST (09215-3C000), remove the oil pan.
    a. Insert the SST between the oil pan and the ladder frame by tapping it with a plastic hammer in the direction of arrow.
    b. After tapping the SST with a plastic hammer along the direction of arrow around more than ⅔ of the edge of the oil pan, remove it from the ladder frame.

⁂ WARNING

Do not turn over the SST abruptly without tapping. Damage may occur to the SST or the oil pan.

➡Be careful not to damage the contact surfaces of upper oil pan and lower oil pan.

To install:

6. Using a gasket scraper, carefully remove all the old packing material from the gasket surfaces.
7. Before assembling the oil pan, the liquid sealant TB1217H should be applied on oil pan. The part must be assembled within 5 minutes after the sealant was applied. Bead width: 0.1 in. (2.5mm), except marked area: 0.2 in. (5.0mm)

➡Clean the sealing face before assembly. Remove harmful foreign materials on the sealing face before applying sealant. When applying sealant gasket, sealant must not be protrude into the inside of oil pan. To prevent leakage of oil, apply sealant gasket to the inner threads of the bolt holes. After assembly, wait at least 30 minutes before filling the engine with oil.

8. Install the oil pan. Uniformly tighten the bolts in several passes to 86–104 inch lbs. (10–12 Nm).

➡After assembly, wait at least 30 minutes before filling the engine with oil.

9. Fill the engine with the proper type and amount of engine oil.

OIL PUMP

REMOVAL & INSTALLATION

See Figures 115 through 119.

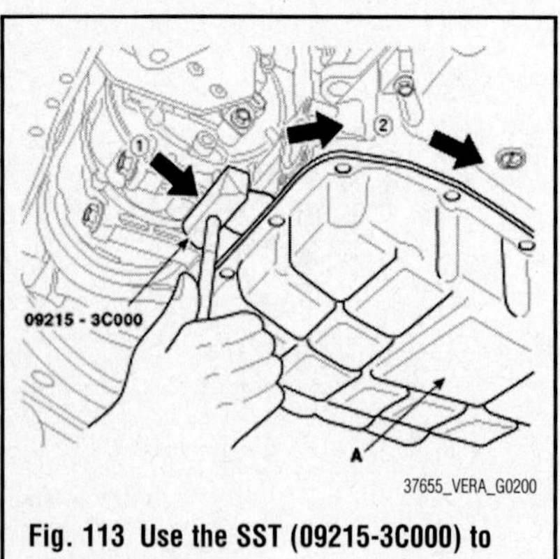

Fig. 113 Use the SST (09215-3C000) to remove the oil pan

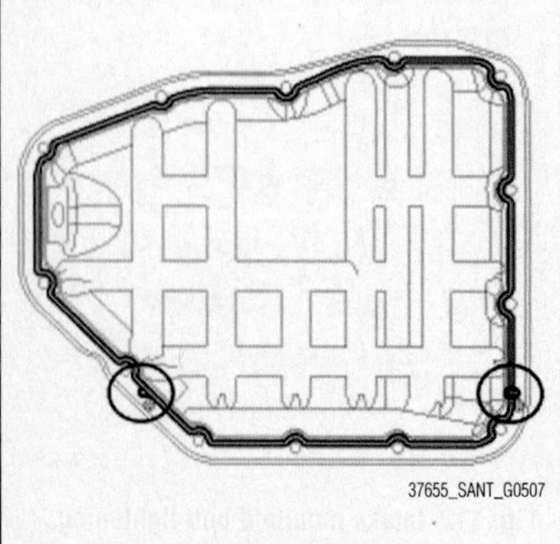

Fig. 114 Oil pan sealant application

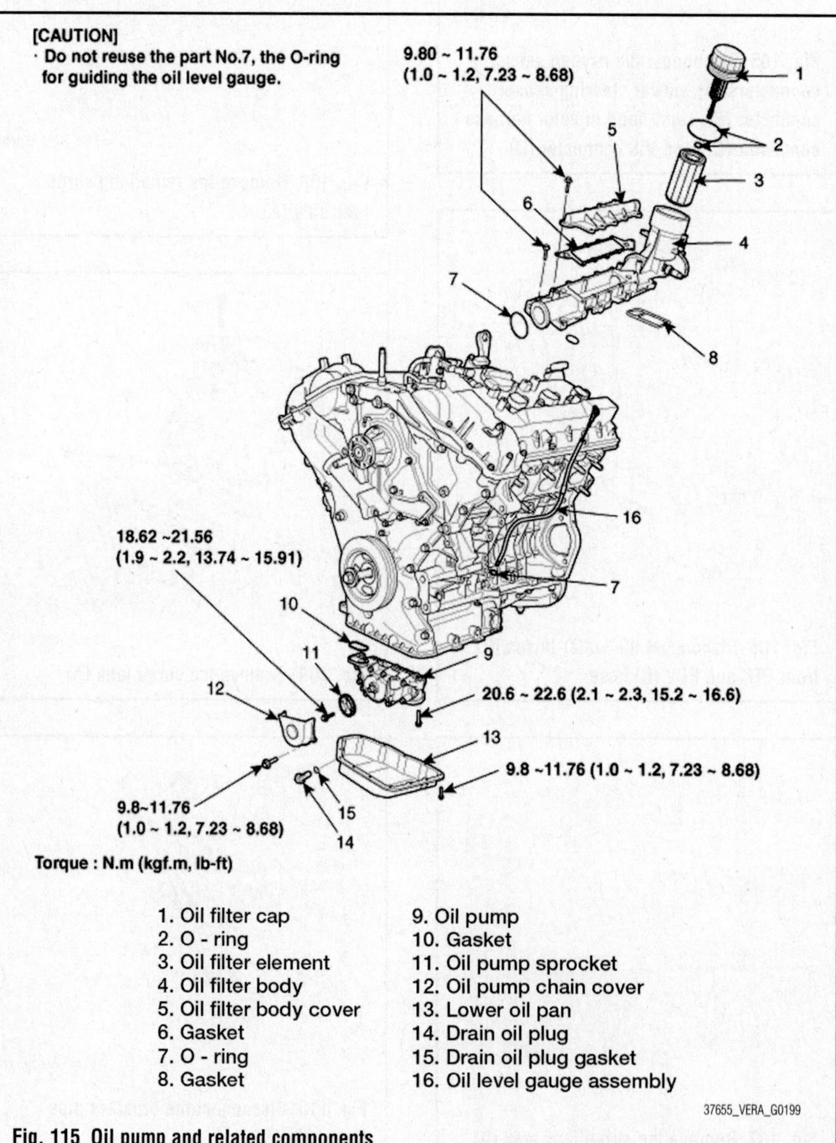

Fig. 115 Oil pump and related components

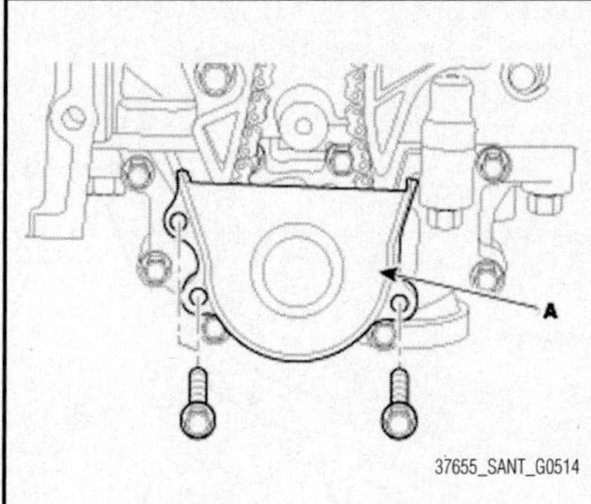

Fig. 116 Remove the oil pump chain cover (A)

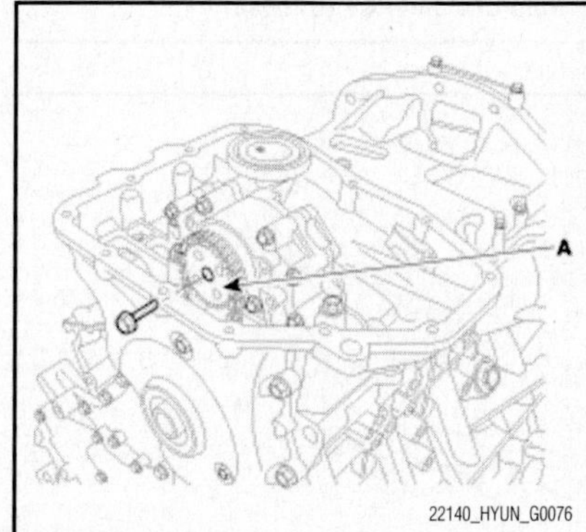

Fig. 117 Remove the oil pump chain sprocket (A)

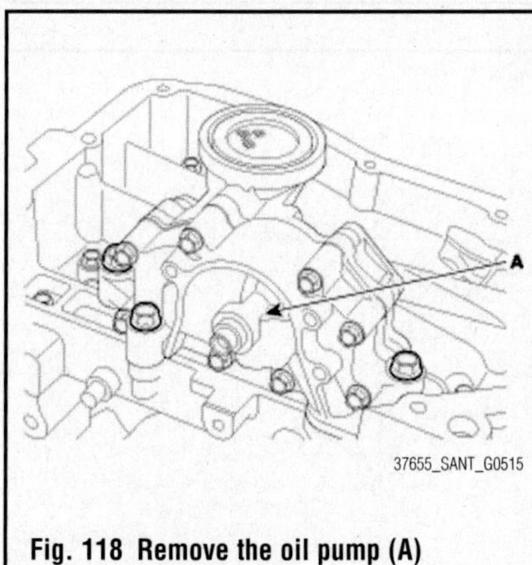

Fig. 118 Remove the oil pump (A)

1. Before servicing the vehicle, refer to the Precautions Section.
2. Remove the lower oil pan. Refer to Oil Pan Removal & Installation.
3. Remove the oil pump chain cover.
4. Remove the oil pump chain sprocket.
5. Remove the oil pump.

To install:

6. Install the oil pump and a new O-ring. Tighten the bolts to 15–17 ft. lbs. (21–23 Nm).

➡Always use a new O-ring.

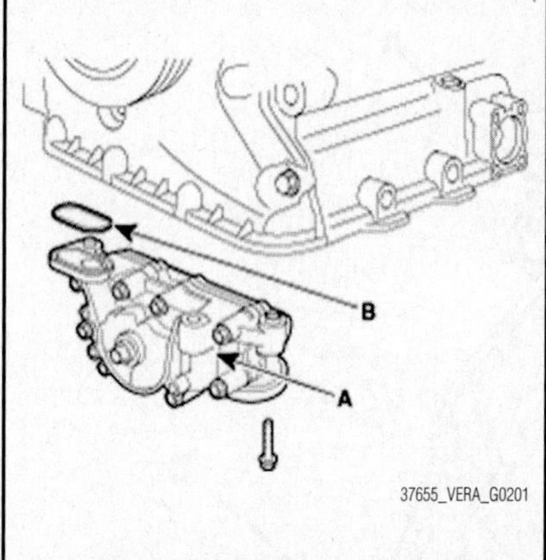

Fig. 119 Install the oil pump (A) and new O-ring (B)

7. Install the oil pump sprocket and oil pump chain on the oil pump. Tighten to 14–16 ft. lbs. (19–22 Nm).
8. Install the oil pump chain cover. Tighten to 86–104 inch lbs. (10–12 Nm).
9. Install the lower oil pan.
10. After assembly, wait at least 30 minutes before filling the engine with oil to allow the gasket material to cure.
11. Fill the engine with oil.
12. Start the engine and check for leaks.
13. Recheck the engine oil level.

PISTON AND RING

POSITIONING

See Figure 120.

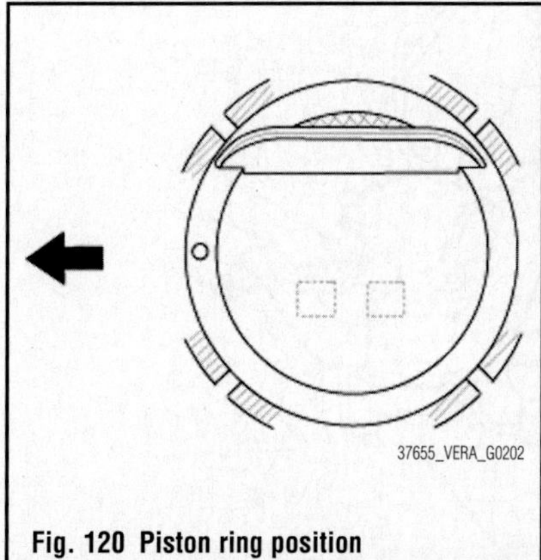

Fig. 120 Piston ring position

REAR MAIN SEAL

REMOVAL & INSTALLATION

See Figure 121.

➡Special Service Tool used for this procedure: SST 09231-3C200, 09231-H1100

1. Before servicing the vehicle, refer to the Precautions Section.

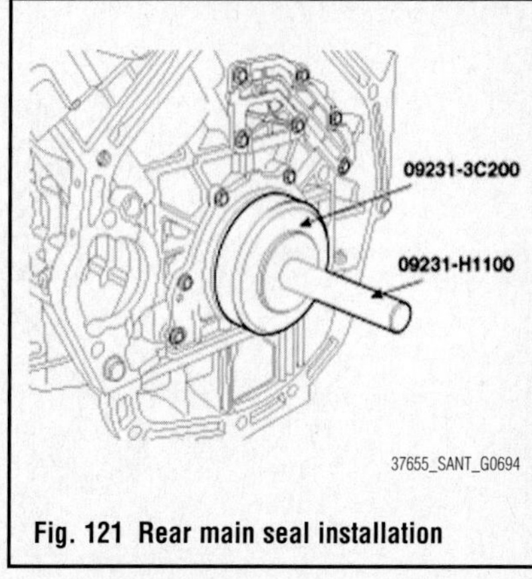

Fig. 121 Rear main seal installation

2. Remove the drive plate. Refer to Flexplate/Drive Plate Removal & Installation.
3. Using a seal removal tool, remove the rear main seal.

To install:

4. Using SST (09231-3C200, 09231-H1100), or appropriate installation tool, tap in the oil seal until its surface is flush with the rear oil seal retainer edge.

TIMING CHAIN FRONT COVER

REMOVAL & INSTALLATION

See Figures 122 through 142.

➡Special Service Tools used for this procedure: SST 09231-3C300 and 09231-3C100

➡Use fender covers to avoid damaging painted surfaces. To avoid damage, unplug the wiring connectors carefully while holding the connector portion. Mark all wiring and hoses to avoid mis-connection. Turn the crankshaft pulley so that the No. 1 piston is at top dead center.

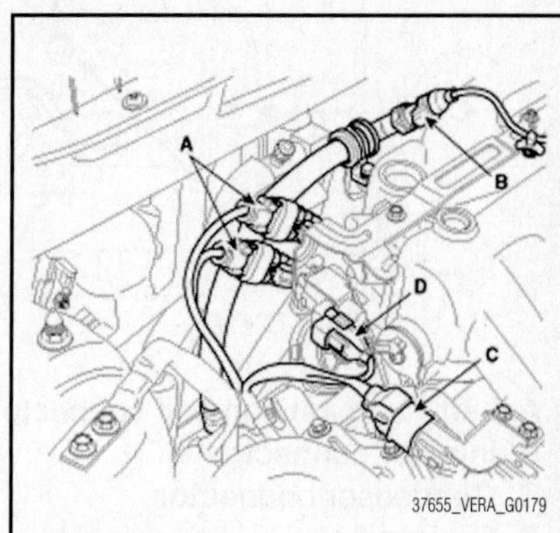

Fig. 122 Disconnect the oxygen sensor connectors (A), power steering sensor connector (B), right-hand injector harness connector (C), and VIS connector (D)

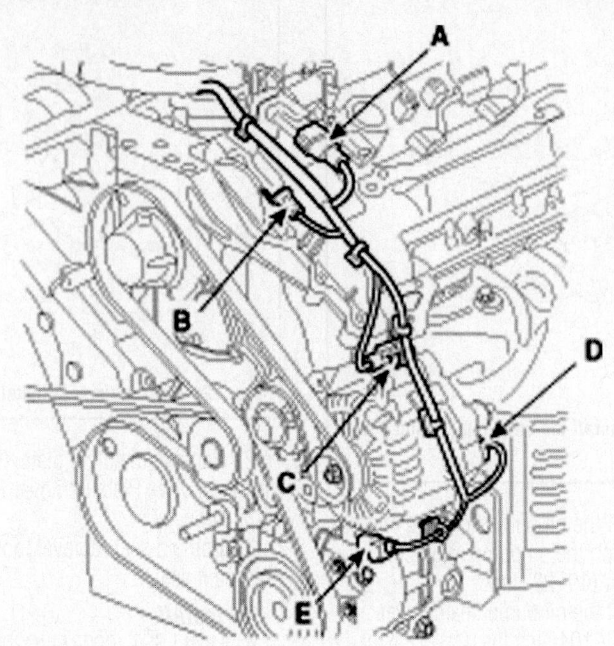

Fig. 123 Disconnect the OCV connector, knock sensor connector, left-hand front oxygen sensor connector, alternator connector, and the air compressor connector

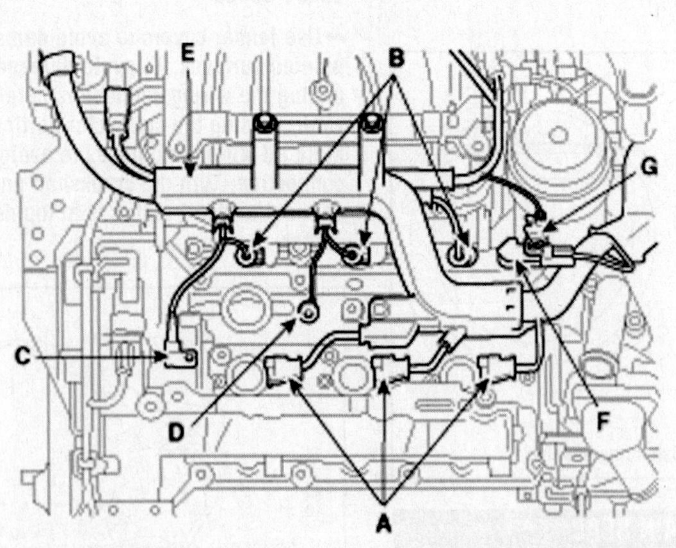

Fig. 124 Disconnect the left-hand ignition coil connector, injector connector, condenser connector, and the ground, and remove the wiring harness protector

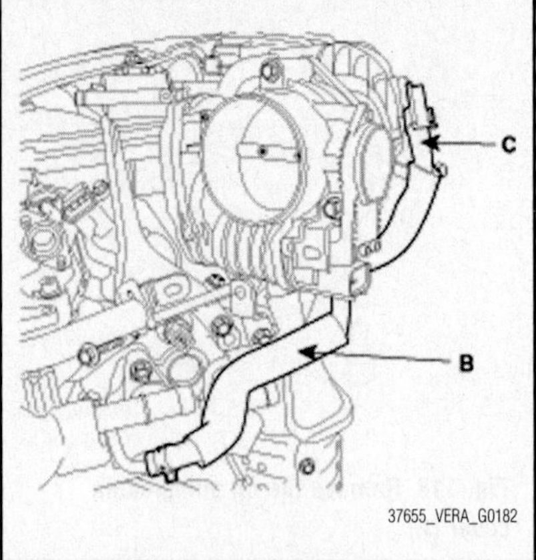

Fig. 125 Disconnect the water hoses (B) from ETC and PCV (C) hose

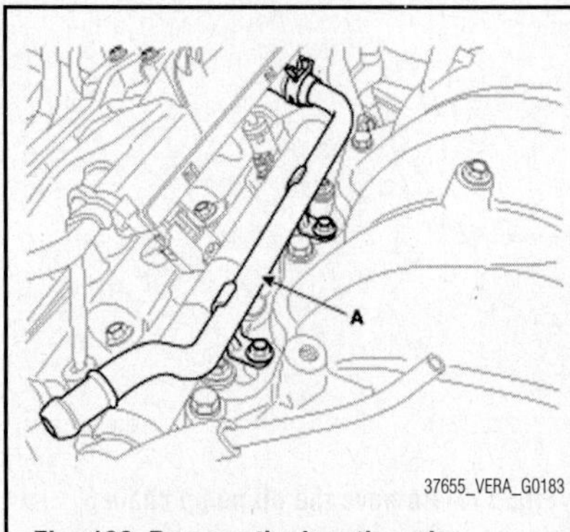

Fig. 126 Remove the breather pipe assembly (A)

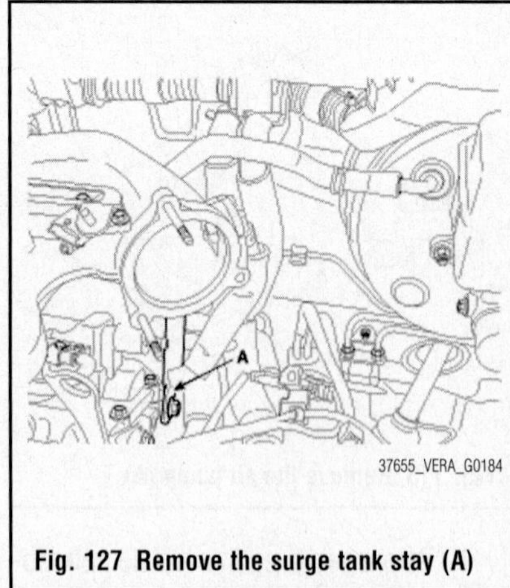

Fig. 127 Remove the surge tank stay (A)

1. Before servicing the vehicle, refer to the Precautions Section.
2. Disconnect the battery negative cable.
3. Remove the air duct.
4. Remove the engine cover.
5. Remove the intake air hose and air cleaner assembly.
6. Remove the right front wheel.
7. Remove the undercover.
8. Remove the side cover.

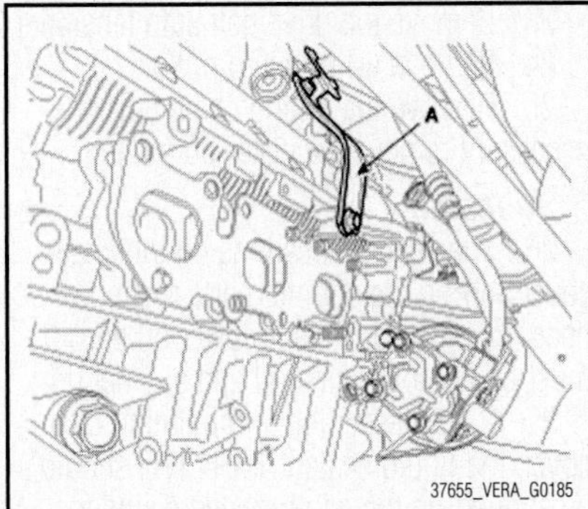

Fig. 128 Remove the remaining surge tank stay (A)

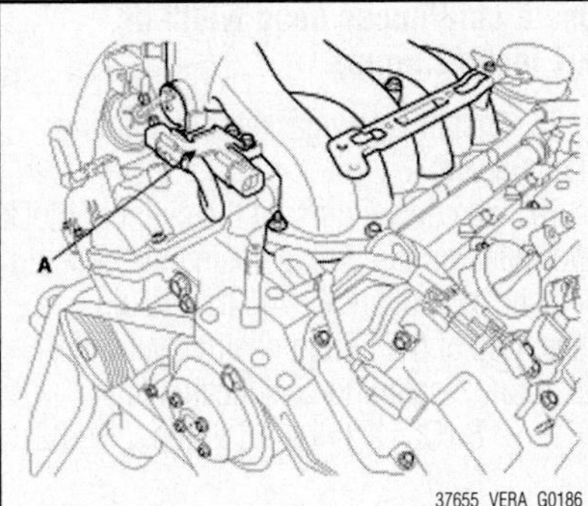

Fig. 129 Remove the surge tank connector bracket (A)

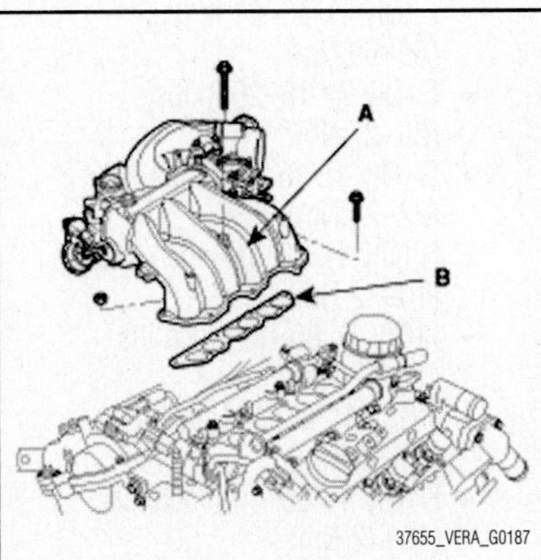

Fig. 130 Remove the surge tank (A)

9. Drain the engine coolant.
10. Drain the engine oil.
11. Disconnect the ground cable and remove the power steering hose mounting bolt.
12. Remove the surge tank and engine wiring.
 a. Disconnect the right-hand front and rear oxygen sensor connectors, the power steering sensor connector, the right-hand injector harness connector, and the Variable Intake Solenoid (VIS) valve connector.

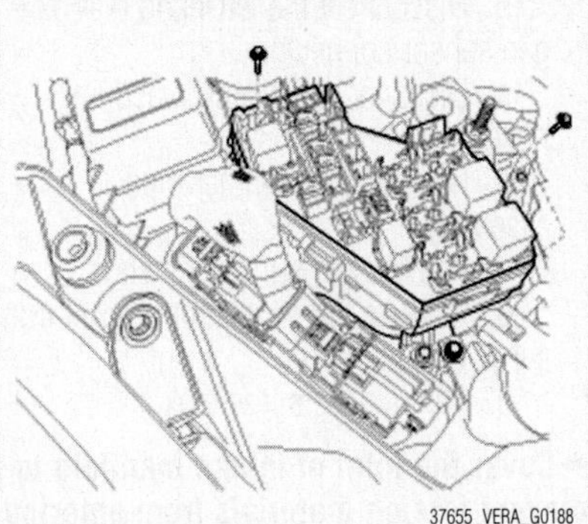

Fig. 131 Remove the PCM and relay box cover and remove the relay box mounting bolts

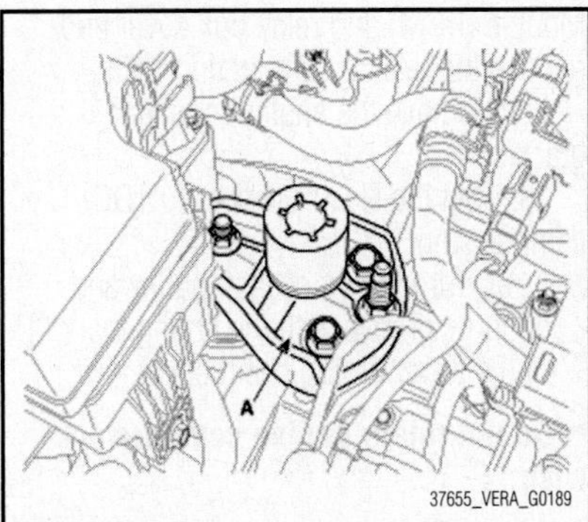

Fig. 132 Remove the engine mounting bracket (A)

 b. Disconnect the right-hand ignition coil connector.
 c. Disconnect the Oil Control Valve (OCV) connector and the knock sensor connector.
 d. Disconnect the left-hand front oxygen sensor connector.
 e. Disconnect the alternator connector and the air compressor connector.
 f. Disconnect the left-hand ignition coil connector, the injector connector, the condenser connector, and the ground, and remove the wiring harness protector.
 g. Disconnect the right-hand Camshaft Position (CMP) Sensor and the Oil Temperature Sensor (OTS) connector.
 h. Disconnect the Purge Control Solenoid Valve (PCSV) connector,

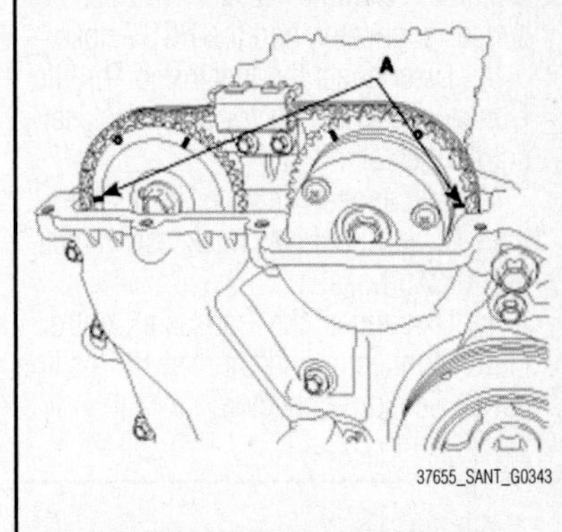

Fig. 134 Check that the mark (A) of the camshaft timing sprockets are in straight line on the cylinder head surface

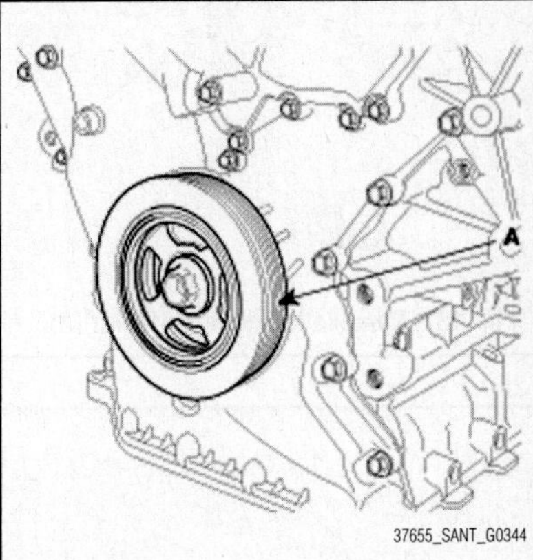

Fig. 135 Remove the crankshaft damper pulley (A)

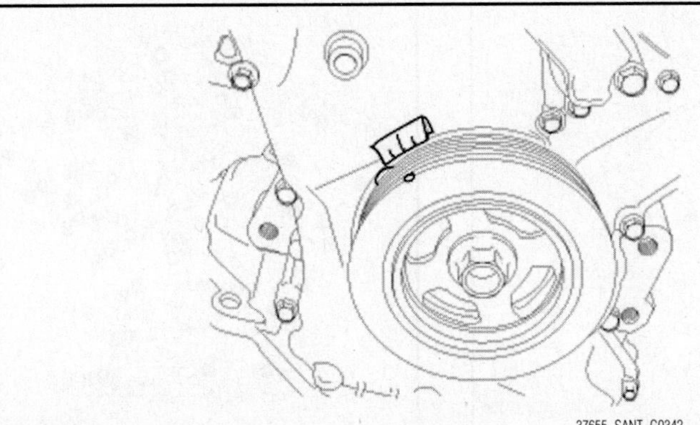

Fig. 133 Turn the crankshaft pulley and align its groove with the timing mark "T" of the lower timing chain cover

Manifold Absolute Pressure (MAP) sensor connector, and the PCSV hose.

i. Disconnect the Electronic Throttle Control (ETC) connector and knock sensor connector.

j. Disconnect the water hoses from the ETC and Positive Crankcase Ventilation (PCV) hose.

k. Disconnect the Transaxle Control (TCU) connector and Crankshaft Position (CKP) Sensor connector.

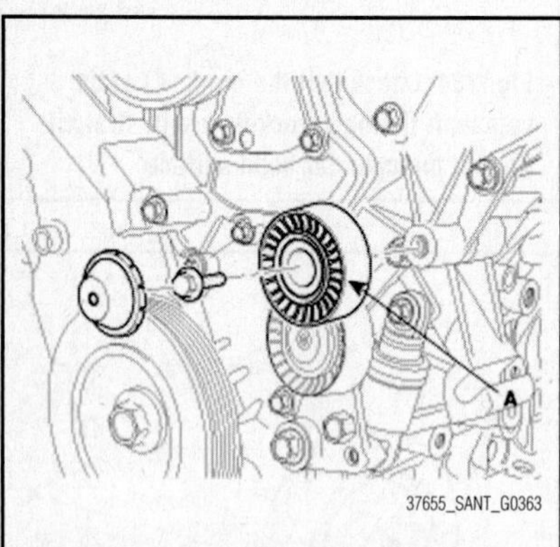

Fig. 137 Remove the drive belt idler (A)

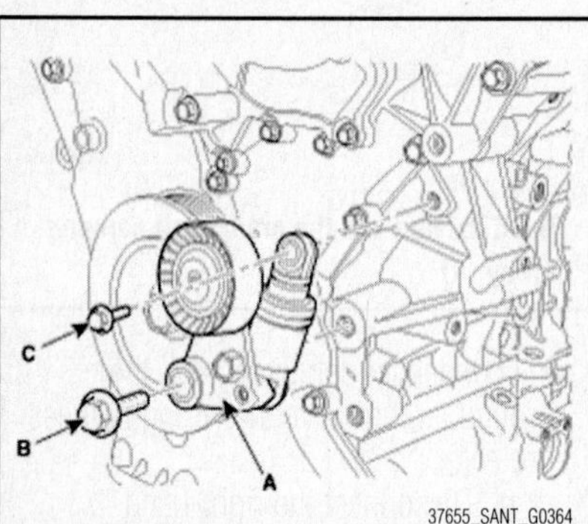

Fig. 138 Remove the drive belt auto tensioner (A) and bolts (B, C)

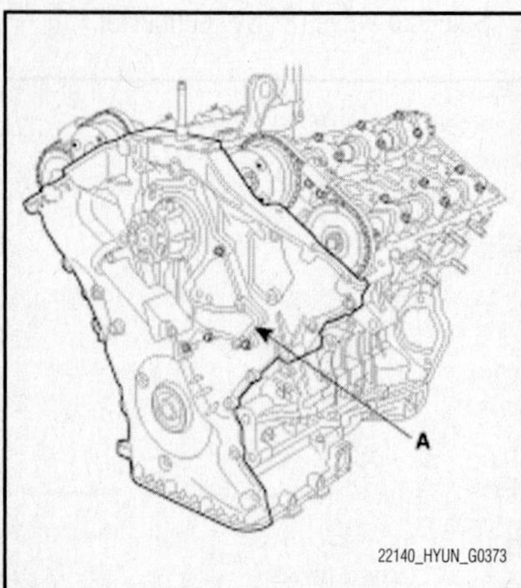

Fig. 139 Remove the timing chain cover (A)

l. Disconnect the left-hand rear oxygen sensor connector.

m. Disconnect the brake vacuum hose.

n. Remove the breather pipe assembly.

o. Remove the surge tank stays.

p. Remove the connector bracket from surge tank.

q. Remove the surge tank.

➡Cover the inlet of intake manifold to prevent foreign materials from entering.

13. Remove the cylinder head cover.
14. Remove the lower oil pan. Refer to Oil Pan Removal & Installation.
15. Install a jack to the upper oil pan.
16. Remove the Powertrain Control Module (PCM) and relay box cover and remove the relay box mounting bolts.
17. Remove the engine mounting bracket.
18. Set the No. 1 cylinder to TDC/compression:

a. Turn the crankshaft pulley and align its groove with the timing mark "T" of the lower timing chain cover.

➡Do not rotate engine counterclockwise.

b. Check that the mark of the camshaft timing sprockets are in straight line on the cylinder head surface as shown. If not, turn the crankshaft one revolution (360°).

19. Remove the drive belt. Refer to Accessory Drive Belt Removal & Installation.
20. Using SST (09231-3C300) remove the crankshaft damper pulley. Refer to Crankshaft Damper Removal & Installation.
21. Lift up the engine assembly using the jack.
22. Remove the drive belt idler.

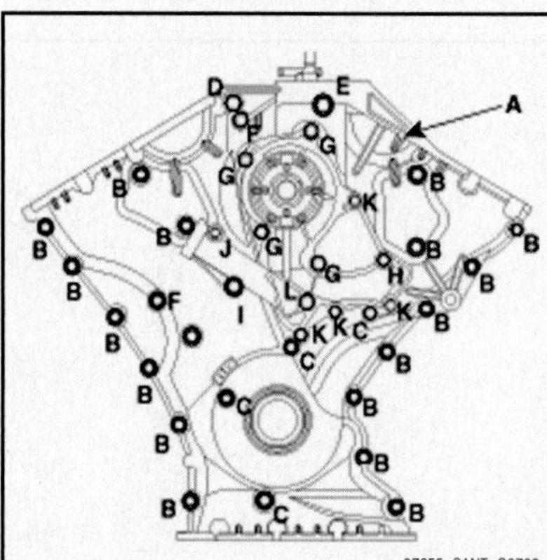

Fig. 140 Timing chain cover bolt tightening sequence

23. Remove the drive belt auto tensioner.
24. Remove water pump pulley.
25. Remove the timing chain cover. If necessary remove the water pump first.

To install:

26. The sealant locations on the chain cover and on the counter parts (cylinder head, cylinder block, and lower oil pan) must be free of engine oil and any debris.
27. Before assembling the timing chain cover, the liquid sealant TB 1217H should be applied on the gap between cylinder head and cylinder block.

➡The part must be assembled within 5 minutes after the sealant is applied. Use a continuous bead width of 0.1 inch (2.5mm).

28. Install a new gasket to the timing chain cover.
29. The dowel pins on the cylinder block and holes on the timing chain cover should be used as a reference in order to aid assembly of the timing chain cover into position. Tightening specifications:
    - B (Qty 17): 14–16 ft. lbs. (19–22 Nm)
    - C (Qty 4): 86–104 inch lbs. (10–12 Nm)
    - D (Qty 2): 43–51 ft. lbs. (59–69 Nm)
    - E (Qty 1): 43–51 ft. lbs. (59–69 Nm)
    - F (Qty 2): 18–20 ft. lbs. (25–27 Nm)
    - G (Qty 4): 16–17 ft. lbs. (22–24 Nm)
    - H (Qty 1): 86–104 inch lbs. (10–12 Nm)
    - I (Qty 1): 86–104 inch lbs. (10–12 Nm)
    - J (Qty 1): 86–104 inch lbs. (10–12 Nm)
    - K (Qty 4): 86–104 inch lbs. (10–12 Nm)
    - L (Qty 1): 16–20 ft. lbs. (22–27 Nm), New bolt
30. Wait 30 minutes after the timing chain cover was assembled before starting the engine.
31. Install the water pump pulley. Tighten to 69–87 inch lbs. (8–10 Nm).
32. Install the drive belt auto tensioner. Tighten to Bolt (B): 60–63 ft. lbs. (81–85 Nm). Bolt (C): 13–16 ft. lbs. (18–22 Nm).
33. Install the drive belt idler. Tighten to 39–43 ft. lbs. (53–58 Nm).
34. Lower the engine assembly using the jack.
35. Using SST (09231-3C100), install the timing chain cover oil seal.

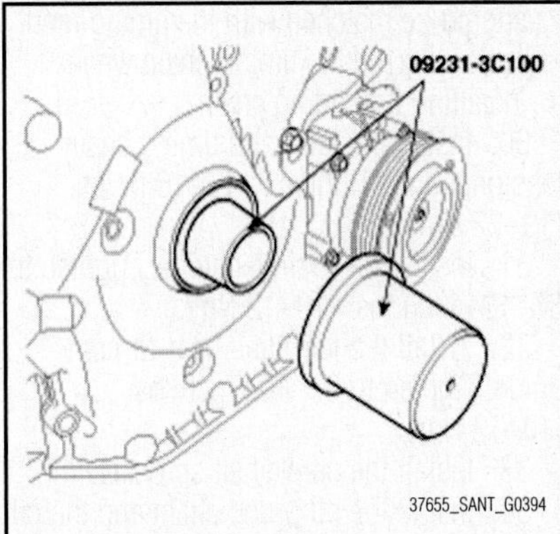

Fig. 141 Using SST (09231-3C100), install timing chain cover oil seal

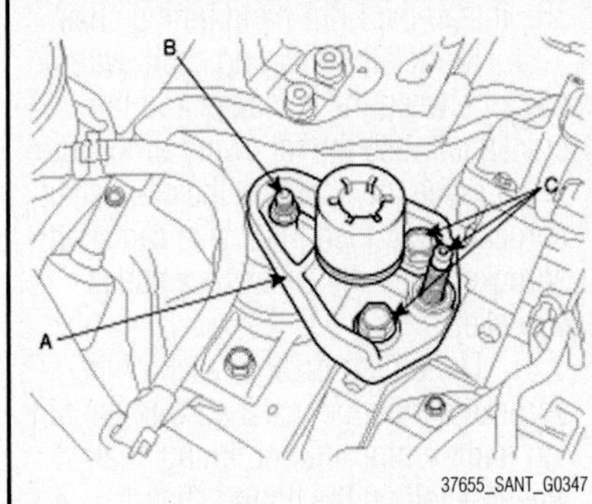

Fig. 142 Install the engine mounting bracket (A) and bolts (B, C)

36. Using SST (09231-3C300) install the crankshaft damper pulley and tighten to 210–224 ft. lbs. (284–304 Nm).
37. Install the drive belt.
38. Install the cylinder head cover.
39. Install the surge tank and wiring connectors:
  a. Install the surge tank and tighten to 86–104 inch lbs. (10–12 Nm).
  b. Install the connector bracket to the surge tank and tighten to 60–95 inch lbs. (7–11 Nm).
  c. Install the surge tank stays and tighten to 20–23 ft. lbs. (27–31 Nm).
  d. Install the breather pipe assembly.
  e. Connect the brake vacuum hose.
  f. Connect the left-hand rear oxygen sensor connector.
  g. Connect the TCU connector and CKP sensor connector.
  h. Connect the water hoses to the ETC and PCV hose.
  i. Connect the ETC connector and knock sensor connector.
  j. Connect the PCSV connector, MAP sensor connector, and the PCSV hose.
  k. Connect the right-hand CMP sensor and the OTS connector.
  l. Install the wiring harness protector.
  m. Connect the left-hand ignition coil connector, the injector connector, the condenser connector, and the ground.
  n. Connect the alternator connector and the air compressor connector.
  o. Connect the left-hand front oxygen sensor connector.
  p. Connect the OCV connector and the knock sensor connector.
  q. Connect the right-hand ignition coil connector.
  r. Connect the right-hand front and rear oxygen sensor connectors, the power steering sensor connector, the right-hand injector harness connector, and the VIS solenoid valve connector.
40. Connect the ground cable and tighten the power steering hose mounting bolt.
41. Install the engine mounting bracket. Tightening torque:
  - Bolt "B": 58–72 ft. lbs. (79–98 Nm)
  - Bolt "C": 47–62 ft. lbs. (64–83 Nm)
42. Tighten the relay box mounting bolts and install the PCM and relay box cover.
43. Remove the jack from the upper oil pan.
44. Install the lower oil pan.
45. Install the side cover and tighten to 78–95 inch lbs. (9–11 Nm).
46. Install the undercover and tighten to 78–95 inch lbs. (9–11 Nm).
47. Install the right front wheel.
48. Install the intake air hose and air cleaner assembly.
49. Install the engine cover.
50. Install the air duct.
51. Connect the battery negative cable.
52. Refill engine with engine oil.
53. Refill radiator and reservoir tank with engine coolant.
54. Bleed air from the cooling system.
55. Run the engine and check for leaks.

TIMING CHAIN & SPROCKETS

REMOVAL & INSTALLATION

See Figures 143 through 149.

➡Use fender covers to avoid damaging painted surfaces. To avoid damage, unplug the wiring connectors carefully while holding the connector portion. Mark all wiring and hoses to avoid misconnection. Turn the crankshaft pulley so that the No. 1 piston is at top dead center.

1. Before servicing the vehicle, refer to the Precautions Section.
2. Disconnect the battery negative cable.

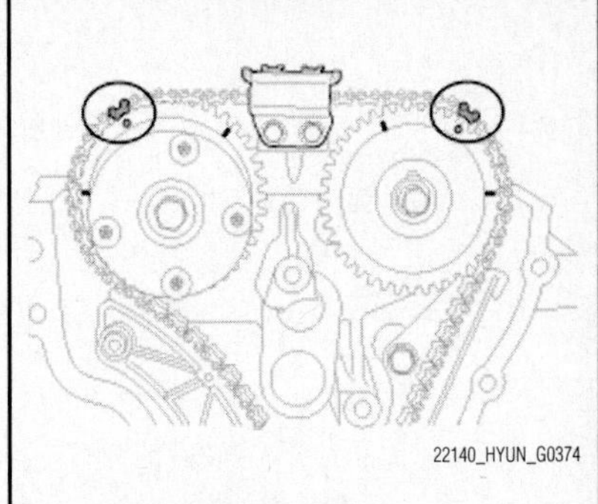

Fig. 143 Timing marks on chain and camshaft sprockets illustrated

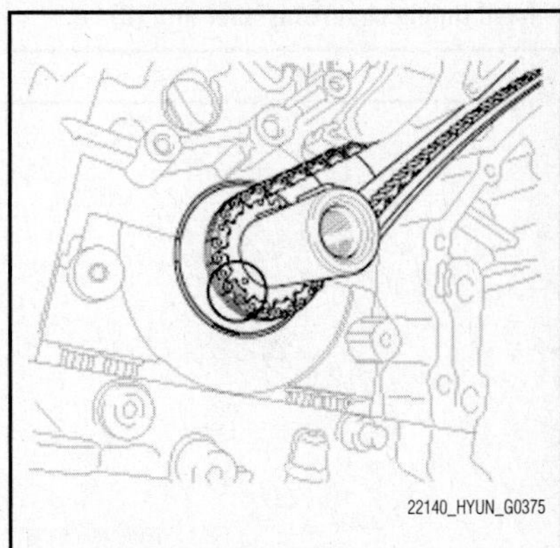

Fig. 144 Timing marks on chain and crankshaft sprocket illustrated

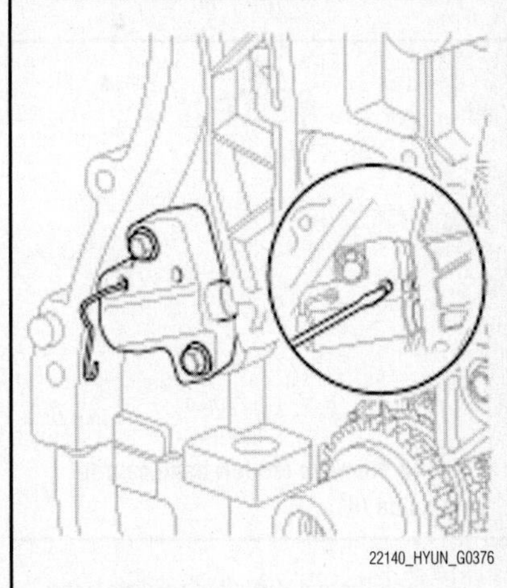

Fig. 145 Install a set pin after compressing the right-hand timing chain tensioner

3. Remove the timing chain cover. Refer to Timing Chain Front Cover Removal & Installation.

➡Before removing the timing chain, mark the right-hand/left-hand timing chain with an identification based on the location of the sprocket as the identification mark on the chain for the TDC can be erased accidentally.

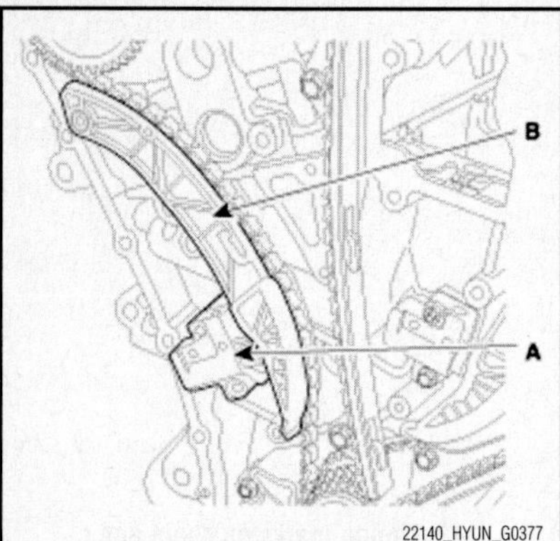

Fig. 146 Remove the right-hand timing chain auto tensioner (A) and the right-hand timing chain tensioner arm (B)

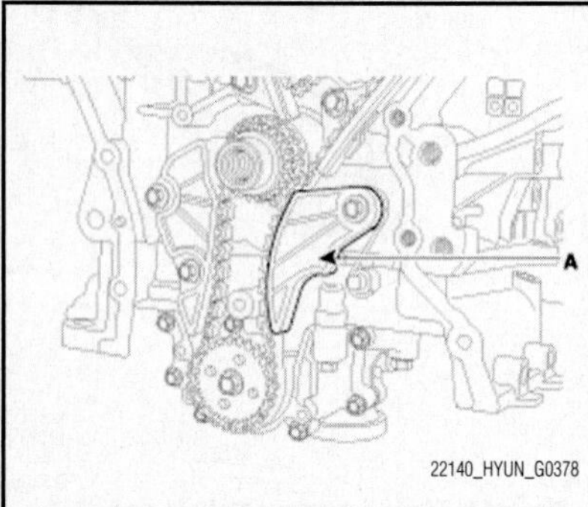

Fig. 147 Remove the oil pump chain tensioner assembly (A)

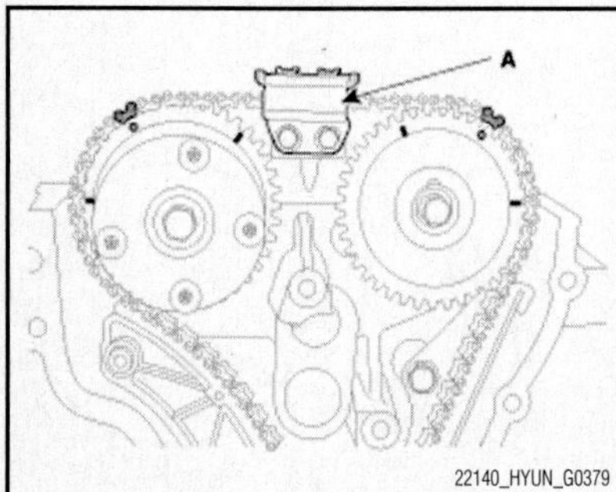

Fig. 148 Remove the left-hand cam-to-cam guide (A)

4. Install a set pin after compressing the right-hand timing chain tensioner.
5. Remove the right-hand cam-to-cam guide.
6. Remove the right-hand timing chain auto tensioner and the right-hand timing chain tensioner arm.
7. Remove the right-hand timing chain.
8. Remove the right-hand timing chain guide.
9. Remove the oil pump chain cover.
10. Remove the oil pump chain tensioner assembly.

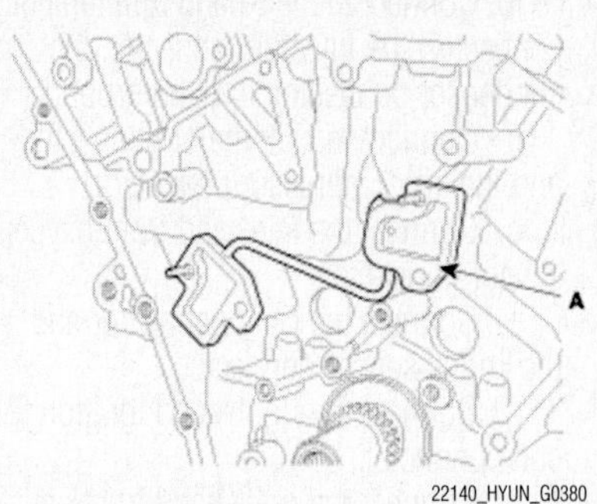

Fig. 149 Remove the tensioner adapter assembly (A)

11. Remove the oil pump chain guide.
12. Remove the oil pump chain sprocket and oil pump chain.
13. Remove the crankshaft sprocket that drives the oil pump.
14. Install a set pin after compressing the left-hand timing chain tensioner.
15. Remove the left-hand cam-to-cam guide.
16. Remove the left-hand timing chain auto tensioner and left-hand timing chain tensioner arm.
17. Remove the left-hand timing chain.
18. Remove the left-hand timing chain guide.
19. Remove the crankshaft sprocket.
20. Remove the tensioner adapter assembly.

To install:

21. Check the camshaft sprocket and crankshaft sprocket for abnormal wear, cracks, or damage. Replace as necessary.
22. Inspect the tensioner arm and chain guide for abnormal wear, cracks, or damage. Replace as necessary.
23. Check that the tensioner piston moves smoothly when the ratchet pawl is released.
24. Install the jack to the upper oil pan.
25. The key of crankshaft must be aligned with the timing mark of timing chain cover. Then, piston of No. 1 cylinder is placed at the TDC on the compression stroke.
26. Install the tensioner adapter assembly.
27. Install the crankshaft sprocket.
28. Install the left-hand timing chain guide. Tighten to 14–18 ft. lbs. (20–25 Nm).
29. Install the left-hand timing chain.
    a. To install the timing chain with no slack between the camshaft and crankshaft, use the following procedure.
    b. Place the crankshaft sprocket on first, then the timing chain guide, then the exhaust camshaft sprocket, and finally the intake camshaft sprocket.
    c. The timing mark of each sprockets should be matched with the timing mark (color link) of the timing chain when installing the timing chain.
30. Install the left-hand timing chain tensioner arm. Tighten to 14–16 ft. lbs. (19–22 Nm).
31. Install the chain tensioner. Tighten to 87–104 inch lbs. (10–12 Nm).
32. Install the left-hand cam-to-cam guide. Tighten to 87–104 inch lbs. (10–12 Nm).
33. Install the crankshaft sprocket.
34. Install the oil pump chain and the oil pump sprocket. Tighten to 14–16 ft. lbs. (19–22 Nm).
35. Install the right-hand timing chain guide. Tighten to 14–18 ft. lbs. (20–25 Nm).
36. Install the right-hand timing chain.
    a. To install the timing chain with no slack between the camshaft and the crankshaft, use the following procedure.
    b. Place the chain on the crankshaft sprocket first, then the intake camshaft sprocket, then the exhaust camshaft sprocket.
    c. The timing mark of each of the sprockets must be matched with the timing mark (color link) of timing chain when installing the timing chain.
37. Install the right-hand timing chain tensioner arm. Tighten to 14–16 ft. lbs. (19–22 Nm).
38. Install the right-hand timing chain auto tensioner. Tighten to 87–104 inch lbs. (10–12 Nm).
39. Install the right-hand cam-to-cam guide. Tighten to 87–104 inch lbs. (10–12 Nm).
40. Install the oil pump chain guide. Tighten to 87–104 inch lbs. (10–12 Nm).
41. Install the oil pump chain tensioner assembly. Tighten to 87–104 inch lbs. (10–12 Nm).
42. Pull out the pins of hydraulic tensioner (left-hand and right-hand).
43. Install the oil pump chain cover. Tighten to 87–104 inch lbs. (10–12 Nm).
44. After rotating the crankshaft 2 revolutions in a clockwise direction (viewed from front), confirm that the timing marks are aligned.

✲✲ WARNING

Always turn the crankshaft clockwise.

45. Install the timing chain cover.
46. Connect the battery negative cable.
47. Refill engine with engine oil.
48. Refill radiator and reservoir tank with engine coolant.
49. Bleed air from the cooling system.
50. Run the engine and check for leaks.

ENGINE PERFORMANCE & EMISSION CONTROLS

CAMSHAFT POSITION (CMP) SENSOR

LOCATION

See Figures 152 and 153.

Refer to the accompanying illustrations.

REMOVAL & INSTALLATION

1. Before servicing the vehicle, refer to the Precautions Section.
2. Disconnect the negative battery cable.
3. Disconnect the CMP sensor connector.
4. Remove the CMP sensor mounting bolt.
5. Remove the CMP sensor.

To install:

6. Installation is the reverse of the removal procedure.
7. Tighten the bolt that retains the CMP sensor to 61–86 inch lbs. (7–10 Nm).

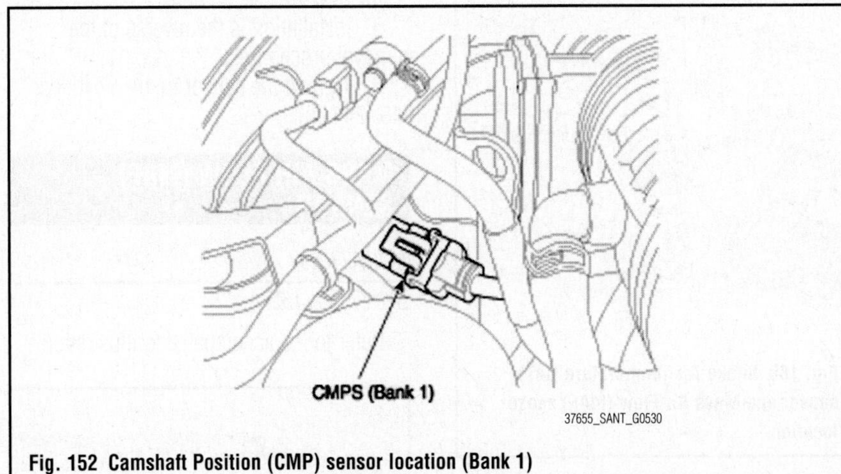

Fig. 152 Camshaft Position (CMP) sensor location (Bank 1)

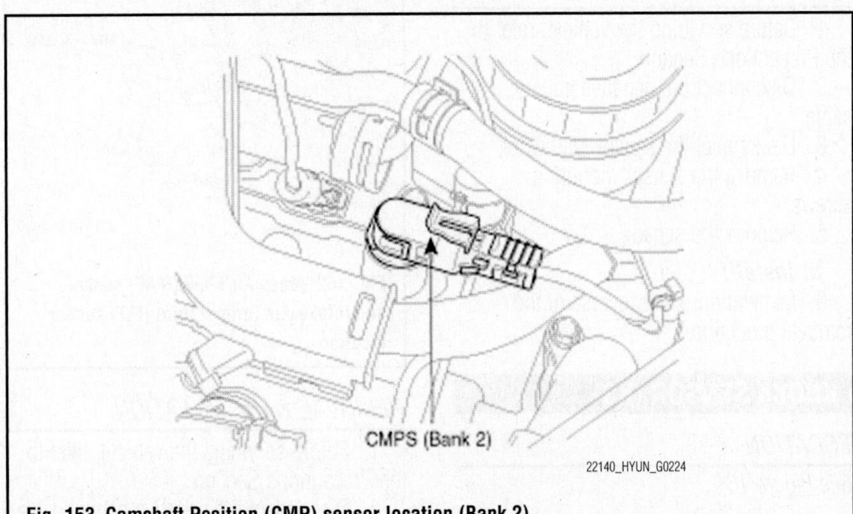

Fig. 153 Camshaft Position (CMP) sensor location (Bank 2)

CRANKSHAFT POSITION (CKP) SENSOR

LOCATION

See Figure 154.

Refer to the accompanying illustration.

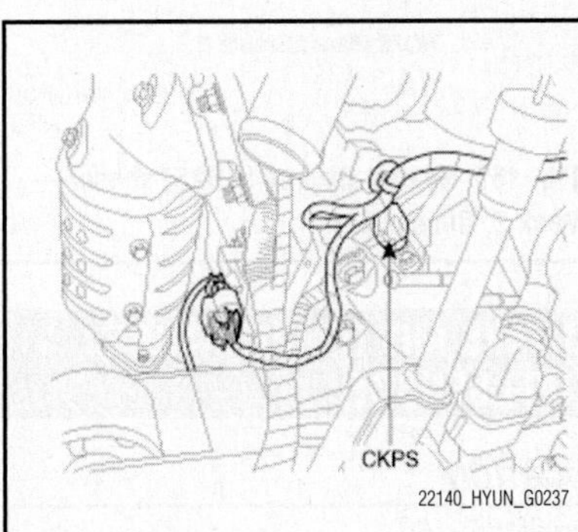

Fig. 154 Crankshaft Position (CKP) sensor location

REMOVAL & INSTALLATION

1. Before servicing the vehicle, refer to the Precautions Section.
2. Disconnect the negative battery cable.
3. Disconnect the CKP sensor connector.
4. Remove the CKP sensor mounting bolt.
5. Remove the sensor.

To install:

6. Installation is the reverse of the removal procedure.
7. Tighten the sensor retaining bolt to 61–86 inch lbs. (7–10 Nm).

ENGINE COOLANT TEMPERATURE (ECT) SENSOR

LOCATION

See Figure 155.

Refer to the accompanying illustration.

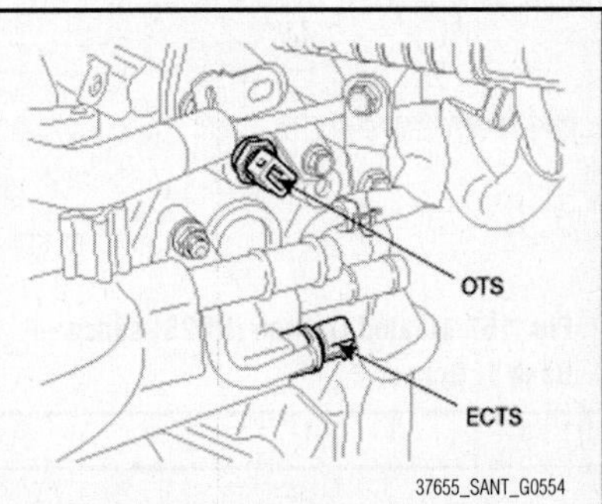

Fig. 155 Engine Coolant Temperature (ECT) sensor and CVVT Oil Temperature sensor (OTS) location

REMOVAL & INSTALLATION

1. Before servicing the vehicle, refer to the Precautions Section.
2. Turn ignition switch OFF and disconnect the negative battery cable.
3. Remove the air cleaner assembly, as necessary.
4. Disconnect the ECT sensor connector.
5. Remove the ECT sensor.

To install:

6. Installation is the reverse of removal. Apply engine coolant to the O-ring. Tighten the mounting bolts to 15–29 ft. lbs. (20–39 Nm).

HEATED OXYGEN (HO2S) SENSOR

LOCATION

See Figures 156 through 159.

Refer to the accompanying illustrations.

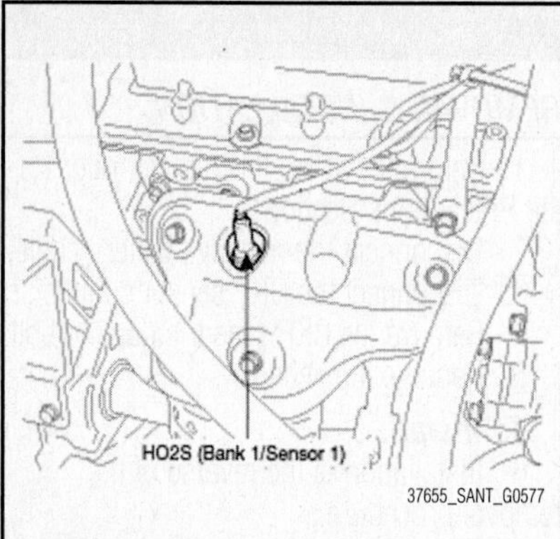

Fig. 156 Heated Oxygen (HO2S) sensor—Bank 1, Sensor 1

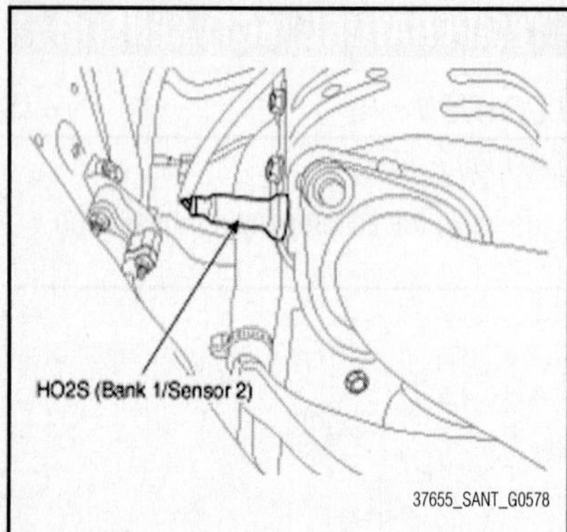

Fig. 157 Heated Oxygen (HO2S) sensor—Bank 1, Sensor 2

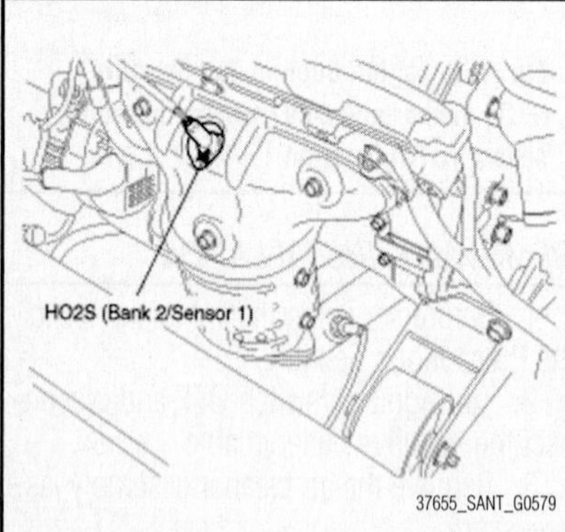

Fig. 158 Heated Oxygen (HO2S) sensor—Bank 2, Sensor 1

REMOVAL & INSTALLATION

1. Before servicing the vehicle, refer to the Precautions Section.
2. Disconnect the negative battery cable.
3. Disconnect the oxygen sensor connector.
4. Remove the oxygen sensor.

To install:

5. Installation is the reverse of removal. Tighten the HO2S to 29–36 ft. lbs. (39–49 Nm).

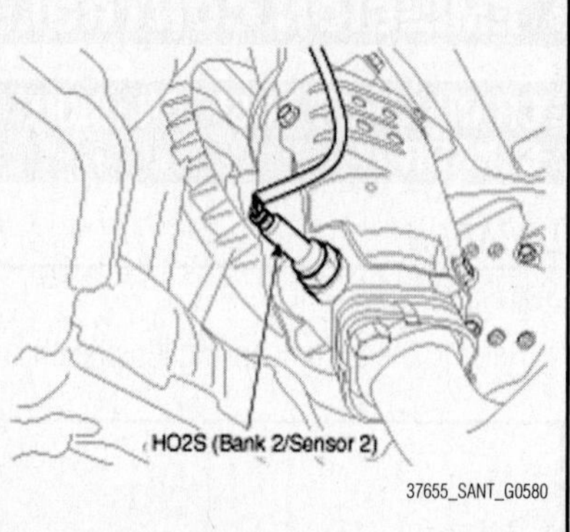

Fig. 159 Heated Oxygen (HO2S) sensor—Bank 2, Sensor 2

INTAKE AIR TEMPERATURE (IAT) SENSOR

LOCATION

See Figure 160.

Refer to the accompanying illustration.

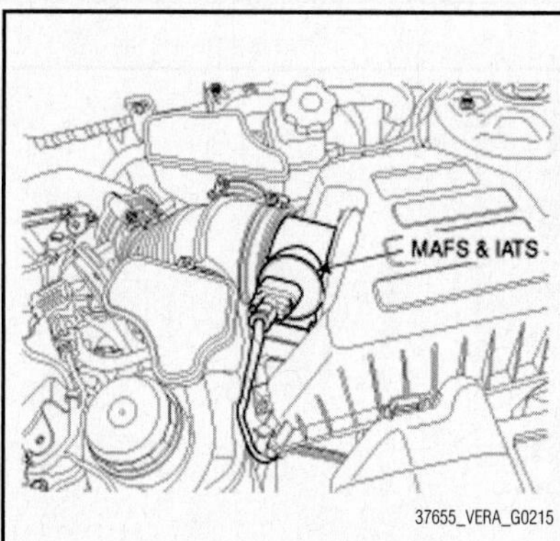

Fig. 160 Intake Air Temperature (IAT) sensor and Mass Air Flow (MAF) sensor location

REMOVAL & INSTALLATION

1. Before servicing the vehicle, refer to the Precautions Section.
2. Disconnect the negative battery cable.
3. Disconnect the sensor connector.
4. Remove the sensor mounting screws.
5. Remove the sensor.

To install:

6. Installation is the reverse of the removal procedure.

KNOCK SENSOR (KS)

LOCATION

See Figure 161.

Refer to the accompanying illustration.

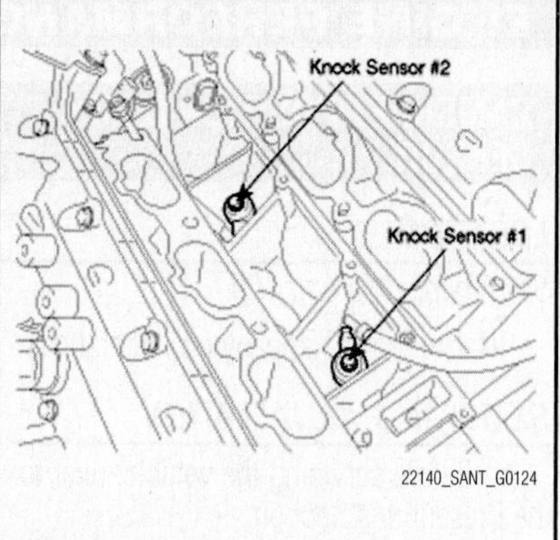

Fig. 161 Knock Sensor (KS) No. 1 and No. 2 locations

REMOVAL & INSTALLATION

1. Before servicing the vehicle, refer to the Precautions Section.
2. Disconnect the negative battery cable.
3. Disconnect the sensor connector.
4. Remove the sensor from its mounting.

To install:

5. Installation is the reverse of the removal procedure.
6. Tighten the sensor to 14–17 ft. lbs. (19–24 Nm).

MASS AIR FLOW (MAF) SENSOR

LOCATION

See Figure 162.

Refer to the accompanying illustration.

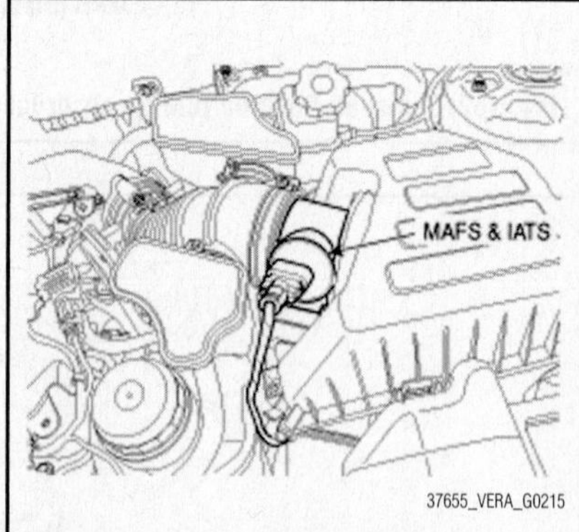

Fig. 162 Mass Air Flow (MAF) sensor and Intake Air Temperature (IAT) sensor location

REMOVAL & INSTALLATION

1. Before servicing the vehicle, refer to the Precautions Section.
2. Disconnect the negative battery cable.

3. Disconnect the Mass Air Flow (MAF) sensor connector.
4. Remove the MAF sensor.

To install:

5. Installation is the reverse of removal. Tighten the MAF sensor installation bolt to 35–52 inch lbs. (4–6 Nm).

MANIFOLD ABSOLUTE PRESSURE (MAP) SENSOR

LOCATION

See Figure 163.

Refer to the accompanying illustration.

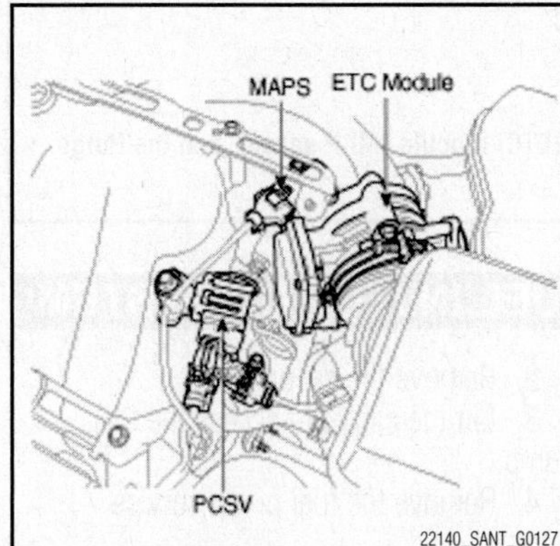

Fig. 163 Location of the MAP sensor, Purge Control Solenoid Valve (PCSV), and the Electronic Throttle Control (ETC) module

REMOVAL & INSTALLATION

1. Before servicing the vehicle, refer to the Precautions Section.
2. Disconnect the negative battery cable.
3. Disconnect the connector from the Manifold Absolute Pressure (MAP) sensor.
4. Remove the MAP sensor retaining screws.
5. Remove the MAP sensor.

To install:

6. Installation is the reverse of the removal procedure. Tighten the MAP sensor installation bolt to: 78–104 inch lbs. (9–12 Nm).

POWERTRAIN CONTROL MODULE (PCM)

LOCATION

See Figure 164.

Refer to the accompanying illustration.

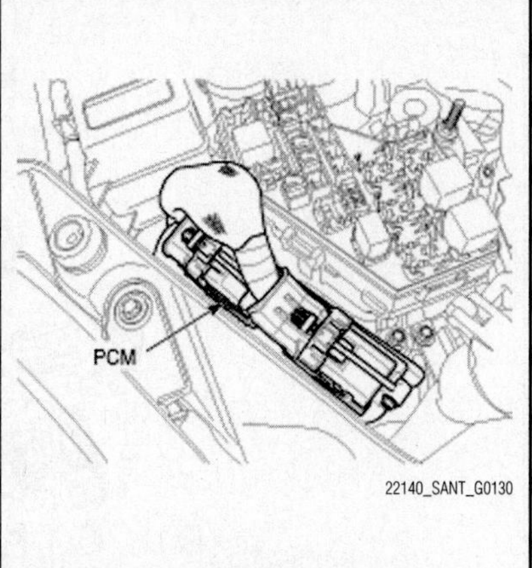

Fig. 164 Powertrain Control Module (PCM) location

REMOVAL & INSTALLATION

See Figures 165 and 166.

➡If the vehicle is equipped with immobilizer, perform "KEY TEACHING" procedure together.

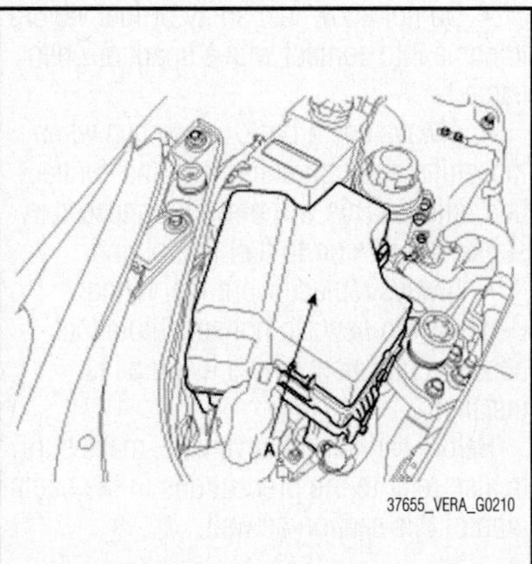

Fig. 165 Remove the cover of the PCM and relay box (A)

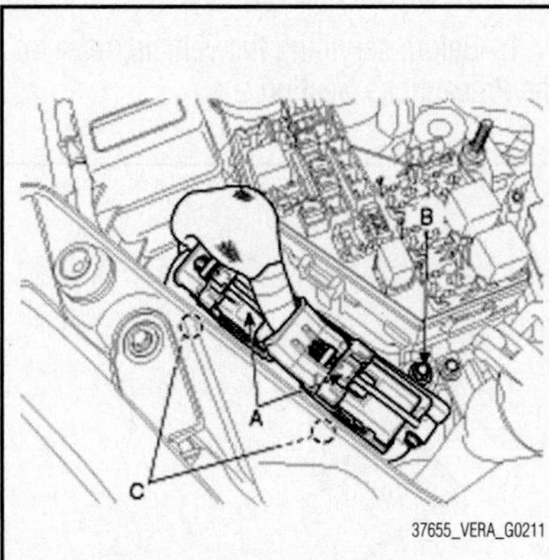

Fig. 166 Disconnect the PCM connectors (A) and remove the mounting bolt (B), nuts (C), and the PCM

➡When replacing a PCM, the VIN must be programmed in the PCM. If there is no VIN in the PCM memory, the fault code (DTC P0630) is set.

1. Before servicing the vehicle, refer to the Precautions Section.
2. Turn the ignition switch **OFF**.
3. Disconnect the negative battery cable from the battery.
4. Remove the cover of the PCM and relay box.
5. Disconnect the PCM connectors.
6. Remove the PCM bracket mounting bolt and nuts.
7. Remove the PCM.

To install:

8. Installation is the reverse of the removal procedure.
9. Tighten the PCM and bracket mounting bolts to: 86–104 inch lbs. (10–12 Nm).
10. Perform the "Key Teaching" procedure, if necessary.
11. Input the Vehicle Identification Number (VIN). Refer to Reset procedure.

RESET PROCEDURE

➡Reprogramming of the VIN will require a scan tool.

➡When replacing an ECM, the VIN must be programmed in the ECM. If there is no VIN in ECM memory, the fault code (DTC P0630) is set.

➡The programmed VIN cannot be changed. When writing the VIN, confirm the VIN carefully.

1. Select "Vehicle" and "Engine".
2. Select "VIN WRITING".
3. Check the PCM status.

➡VIRGIN: VIN is not programmed. LEARNT: VIN has been already programmed

4. Is the PCM status "VIRGIN"? If yes, go to next step. If no, end.
5. Write the VIN with cursor, function and number keys.

➡Before pressing the "ENTER" key, confirm the VIN again because the programmed VIN cannot be changed.

6. After verifying the written VIN, press the "ENTER" key.
7. Turn the ignition switch OFF, and then turn ON.
8. Verify the programmed VIN in the ECM memory.

THROTTLE POSITION SENSOR (TPS)

LOCATION

See Figure 167.

The Throttle Position Sensor (TPS) is part of the Electronic Throttle Control (ETC) module.

REMOVAL & INSTALLATION

1. Before servicing the vehicle, refer to the Precautions Section.
2. Disconnect the negative battery cable.
3. Disconnect the Electronic Throttle Control (ETC) module connector.
4. Remove the retaining screws.
5. Remove the module.

To install:

6. Installation is the reverse of the removal procedure.

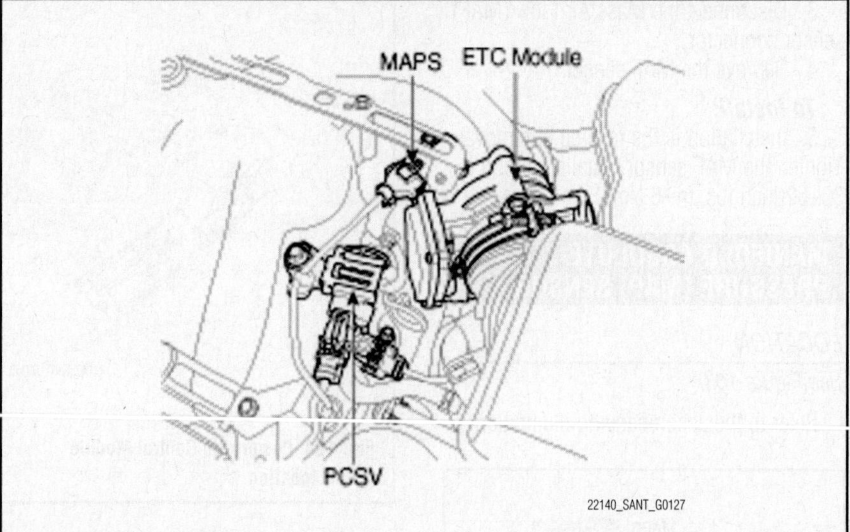

Fig. 167 Location of the Electronic Throttle Control (ETC) module, MAP sensor, and the Purge Control Solenoid Valve (PCSV)

FUEL — GASOLINE FUEL INJECTION SYSTEM

FUEL SYSTEM SERVICE PRECAUTIONS

Safety is the most important factor when performing not only fuel system maintenance but any type of maintenance. Failure to conduct maintenance and repairs in a safe manner may result in serious personal injury or death. Maintenance and testing of the vehicle's fuel system components can be accomplished safely and effectively by adhering to the following rules and guidelines.

- To avoid the possibility of fire and personal injury, always disconnect the negative battery cable unless the repair or test procedure requires that battery voltage be applied.
- Always relieve the fuel system pressure prior to disconnecting any fuel system component (injector, fuel rail, pressure regulator, etc.), fitting or fuel line connection. Exercise extreme caution whenever relieving fuel system pressure to avoid exposing skin, face and eyes to fuel spray. Please be advised that fuel under pressure may penetrate the skin or any part of the body that it contacts.
- Always place a shop towel or cloth around the fitting or connection prior to loosening to absorb any excess fuel due to spillage. Ensure that all fuel spillage (should it occur) is quickly removed from engine surfaces. Ensure that all fuel soaked cloths or towels are deposited into a suitable waste container.
- Always keep a dry chemical (Class B) fire extinguisher near the work area.
- Do not allow fuel spray or fuel vapors to come into contact with a spark or open flame.
- Always use a back-up wrench when loosening and tightening fuel line connection fittings. This will prevent unnecessary stress and torsion to fuel line piping.
- Always replace worn fuel fitting O-rings with new Do not substitute fuel hose or equivalent where fuel pipe is installed.

Before servicing the vehicle, make sure to also refer to the precautions in the beginning of this section as well.

RELIEVING FUEL SYSTEM PRESSURE

See Figures 168 through 170.

1. Before servicing the vehicle, refer to the Precautions Section.

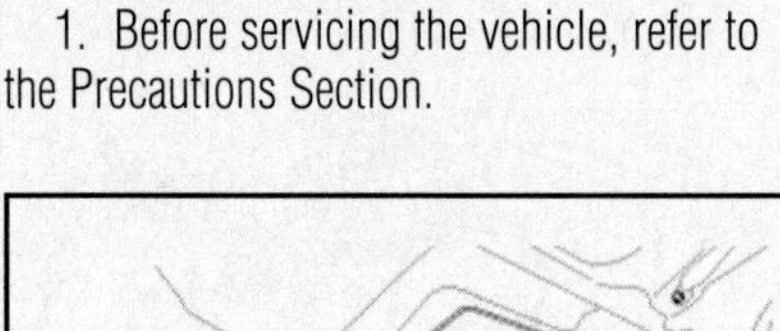

Fig. 168 Lift the carpet to access the fuel pump (A)

2. Remove the second seat.
3. Lift the carpet to access the fuel pump.
4. Remove the fuel pump service cover.

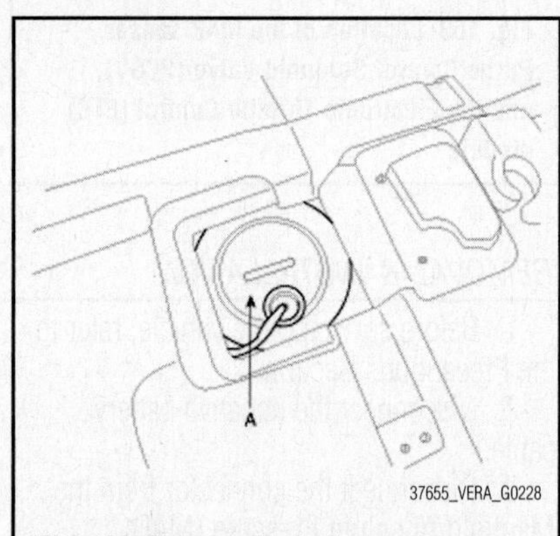

Fig. 169 Remove the fuel pump service cover (A)

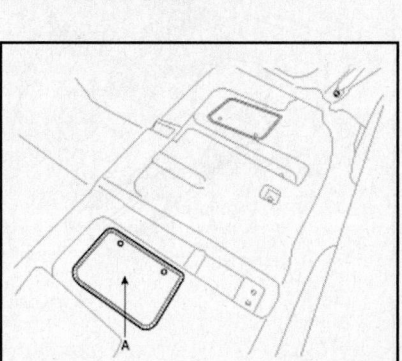

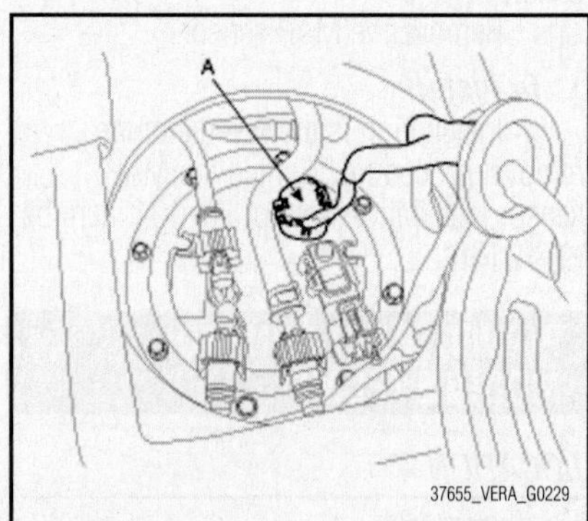

Fig. 170 Disconnect the fuel pump connector (A)

5. Disconnect the fuel pump connector.
6. Start the engine and wait until fuel in fuel line is exhausted.
7. After engine stops, turn the ignition switch off and disconnect the negative battery cable.

FUEL FILTER

REMOVAL & INSTALLATION

See Figures 171 through 176.

1. Before servicing the vehicle, refer to the Precautions Section.
2. Relieve the fuel system pressure. Refer to Relieving Fuel System Pressure.
3. Remove the fuel pump. Refer to Fuel Pump Removal & Installation.
4. Disconnect the fuel sender wiring connector and the electric pump wiring connector.

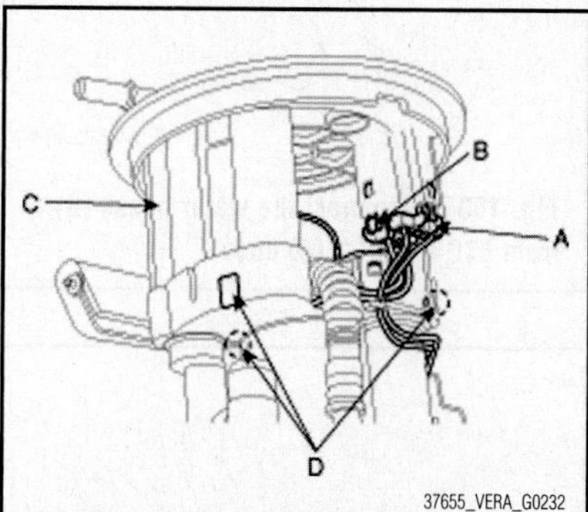

Fig. 171 Disconnect the fuel sender wiring connector (A) and the electric pump wiring connector (B) and separate the flange assembly (C) from the fuel pump assembly by detaching the three hooks (D)

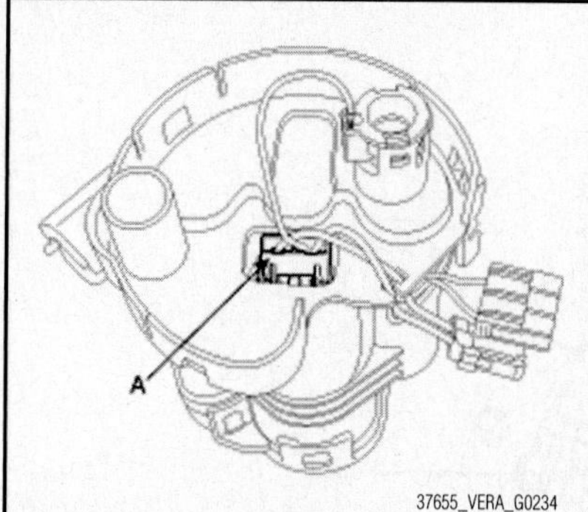

Fig. 173 Disconnect the electric pump wiring connector (A)

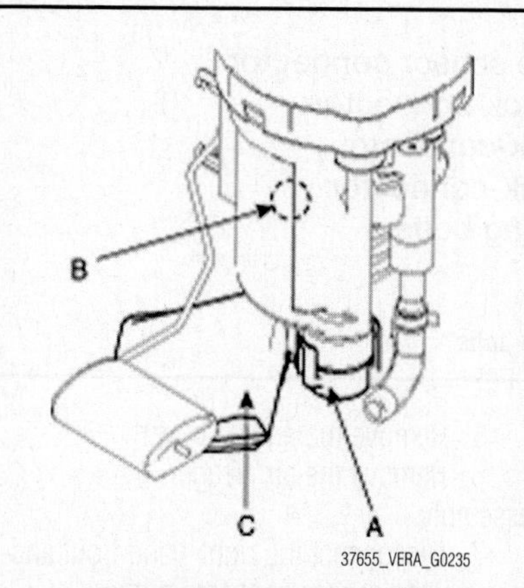

Fig. 174 Remove the cover (A) and the electric pump (B) and the pre-filter (C)

5. Separate the flange assembly from the fuel pump assembly by carefully detaching the hooks.
6. Separate the assist pump with hose by carefully detaching the hooks.
7. Disconnect the electric pump wiring connector.

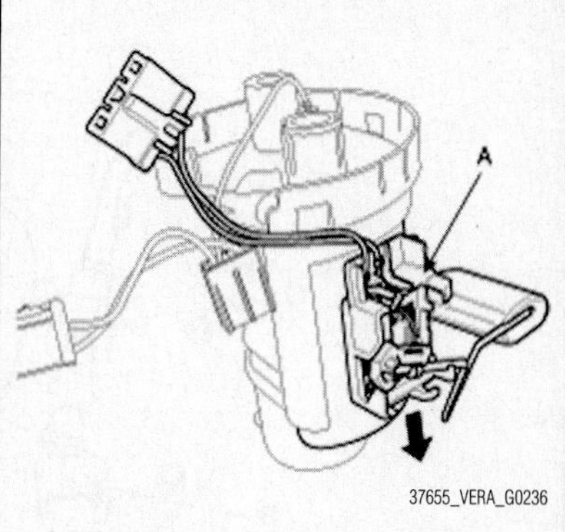

Fig. 175 Push the sender assembly downward, and then separate it from the fuel filter assembly

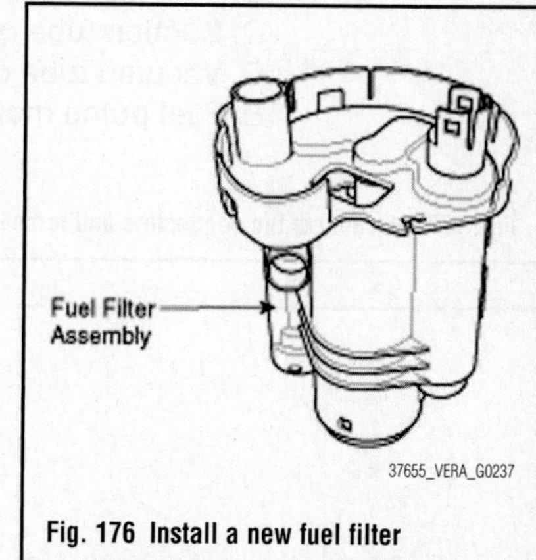

Fig. 176 Install a new fuel filter

8. Remove the cover.
9. Remove the electric pump and the pre-filter from the fuel filter.
10. Push the sender assembly downward, and then separate it from the fuel filter assembly.

To install:

11. Install a new fuel filter.
12. Assemble the fuel pump assembly. Reverse the disassembly steps, above.

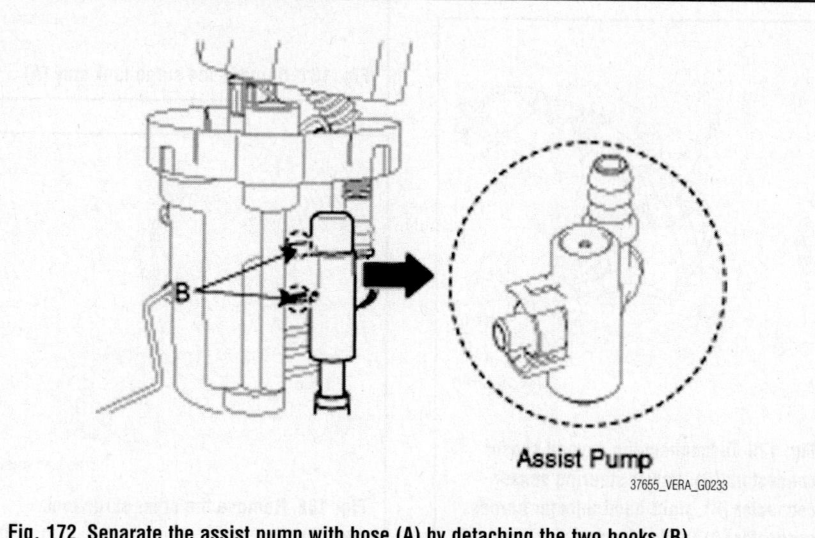

Fig. 172 Separate the assist pump with hose (A) by detaching the two hooks (B)

FUEL PUMP

REMOVAL & INSTALLATION

See Figures 177 and 178.

1. Before servicing the vehicle, refer to the Precautions Section.
2. Relieve the fuel system pressure. Refer to Relieving Fuel System Pressure.
3. Disconnect the fuel tank pressure sensor connector.
4. Disconnect the fuel feed tube quick-connector, the suction tube quick-connector, and the vacuum tube quick-connector.
5. Remove the fuel pump mounting bolts.

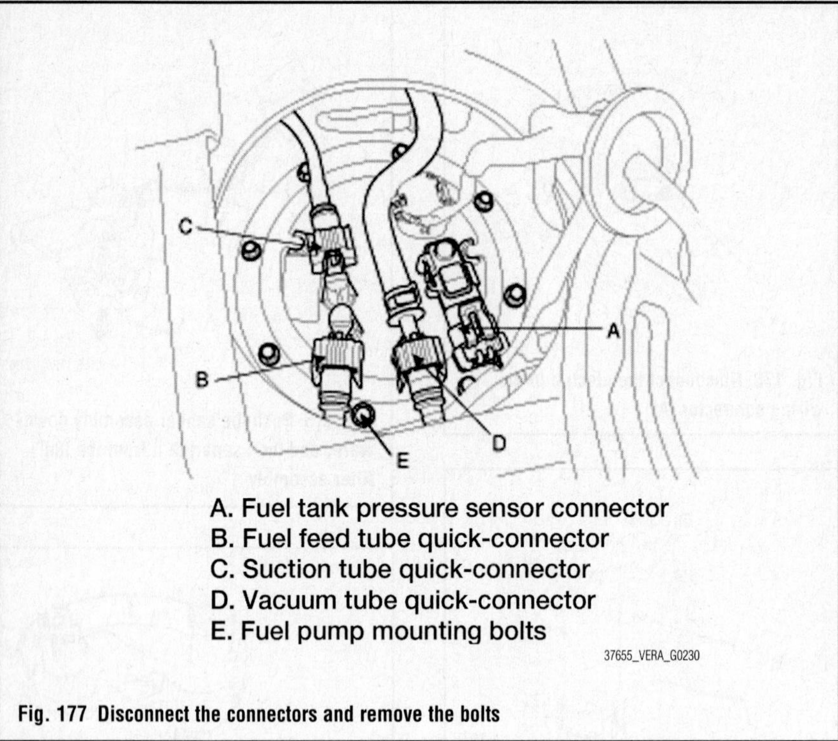

Fig. 177 Disconnect the connectors and remove the bolts

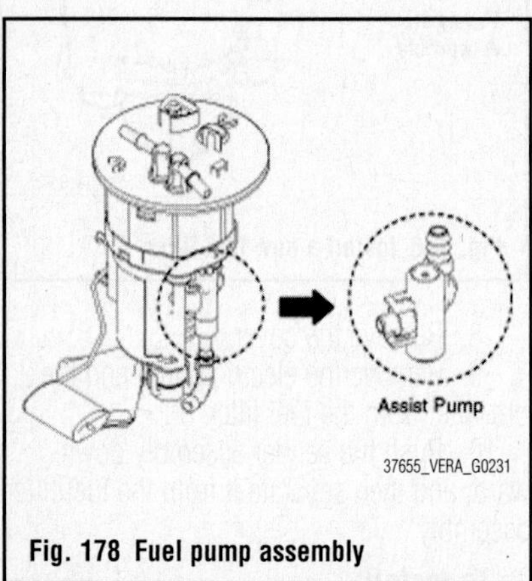

Fig. 178 Fuel pump assembly

6. Remove the fuel pump.

To install:

7. Installation is the reverse of removal. Tighten the fuel pump mounting bolts to 17–26 inch lbs. (2–3 Nm).

FUEL RAIL AND INJECTOR

REMOVAL & INSTALLATION

See Figures 179 through 185.

1. Before servicing the vehicle, refer to the Precautions Section.
2. Relieve the fuel system pressure. Refer to Relieving Fuel System Pressure.
3. Turn the ignition switch OFF and disconnect the battery negative cable.
4. Remove the air duct.
5. Remove the engine cover.
6. Remove the air cleaner assembly.
7. Disconnect the right-hand front and rear oxygen sensor connectors, the power steering sensor connector, the right-hand injector harness connector, and the Variable Intake Solenoid (VIS) valve connector.
8. Disconnect the Purge Control Solenoid Valve (PCSV) connector, Manifold Absolute Pressure (MAP) sensor connector, and the PCSV hose.
9. Disconnect the Electronic Throttle Control (ETC) connector and knock sensor connector.

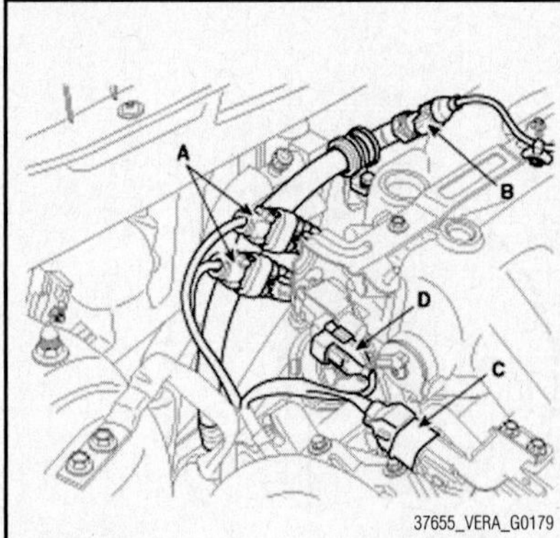

Fig. 179 Disconnect the oxygen sensor connectors (A), power steering sensor connector (B), right-hand injector harness connector (C), and VIS connector (D)

10. Disconnect the water hoses from the ETC and disconnect the Positive Crankcase Ventilation (PCV) hose.
11. Disconnect the brake vacuum hose.
12. Remove the surge tank stays.
13. Remove the surge tank.
14. Disconnect the breather pipe assembly.
15. Disconnect the left-hand injector harness connectors.
16. Remove the fuel rail and injectors.

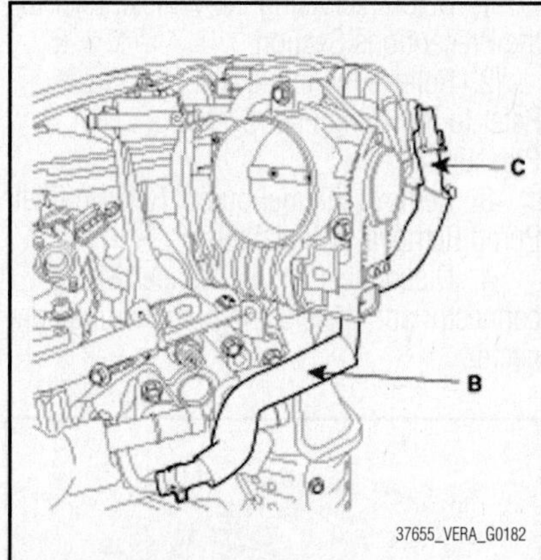

Fig. 180 Disconnect the water hoses (B) from ETC and PCV (C) hose

Fig. 181 Remove the surge tank stay (A)

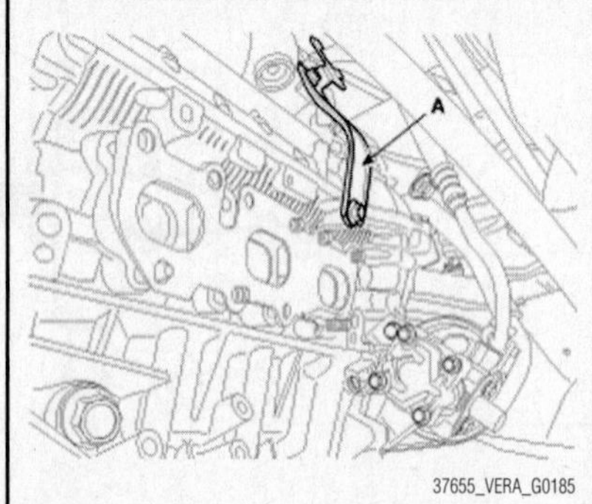

Fig. 182 Remove the other surge tank stay (A)

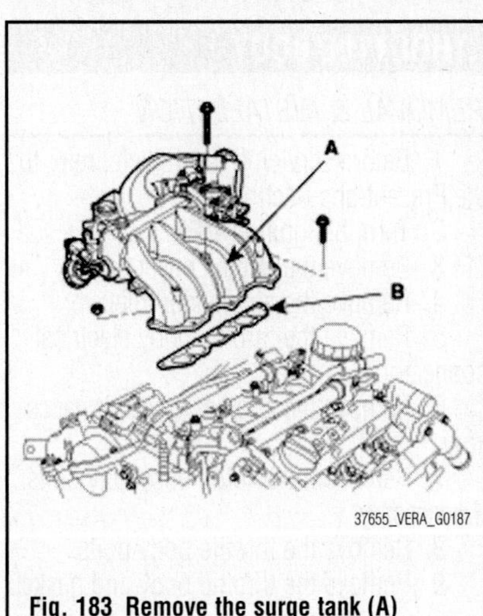

Fig. 183 Remove the surge tank (A)

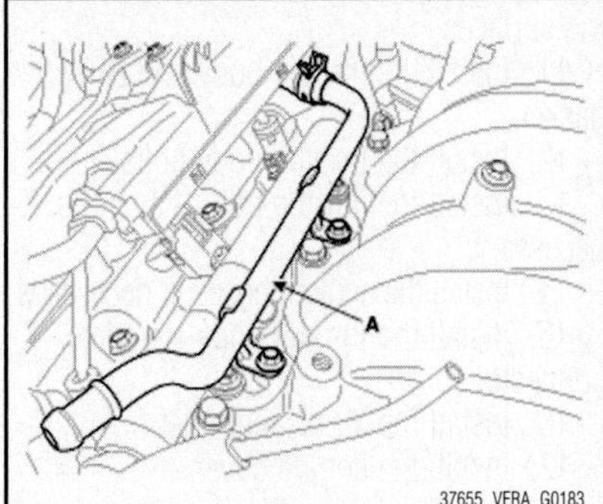

Fig. 184 Disconnect the breather pipe assembly (A)

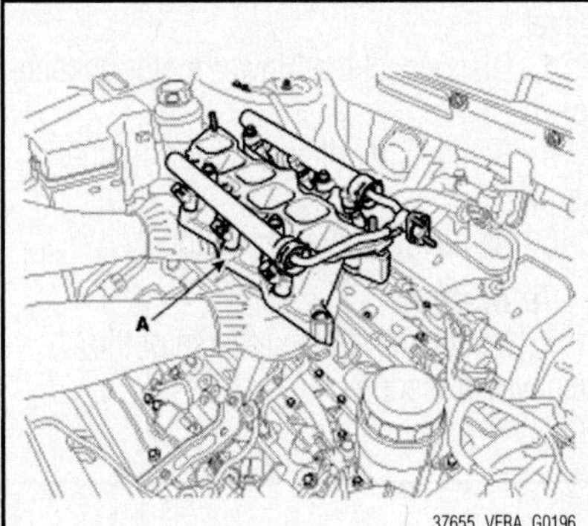

Fig. 185 Remove the fuel rail and injectors

To install:

17. Install the fuel rail and injectors.

18. Connect the left-hand injector connector.

19. Connect the breather pipe assembly and tighten to 86–104 inch lbs. (10–12 Nm).

20. Install the surge tank and tighten the mounting bolts:

- Long bolt (Qty: 1): 86–104 inch lbs. (10–12 Nm)
- Short bolts (Qty: 3), Nuts (Qty: 2): 14–17 ft. lbs. (19–24 Nm)

21. Install the surge tank stays and tighten:

- Engine front side: 20–23 ft. lbs. (27–31 Nm)
- Engine rear side: 14–17 ft. lbs. (19–24 Nm)

22. Connect the brake vacuum hose.

23. Connect the water hoses to the ETC.

24. Connect the ETC connector and knock sensor connector.

25. Connect the PCSV connector, MAP sensor connector, and the PCSV hose.

26. Connect the right-hand front and rear oxygen sensor connectors, the power steering sensor connector, the right-hand injector harness connector, and the VIS solenoid valve connector.

27. Install the air cleaner assembly.

28. Install the air duct.

29. Connect the negative battery cable.

30. Install the engine cover.

31. Start the engine and check for proper operation.

FUEL TANK

REMOVAL & INSTALLATION

See Figures 186 through 191.

1. Before servicing the vehicle, refer to the Precautions Section.

2. Relieve the fuel system pressure. Refer to Relieving Fuel System Pressure.

3. Disconnect the fuel tank pressure sensor connector.

4. Open the sub fuel sender service cover. Refer to Sub Fuel Sender Removal & Installation.

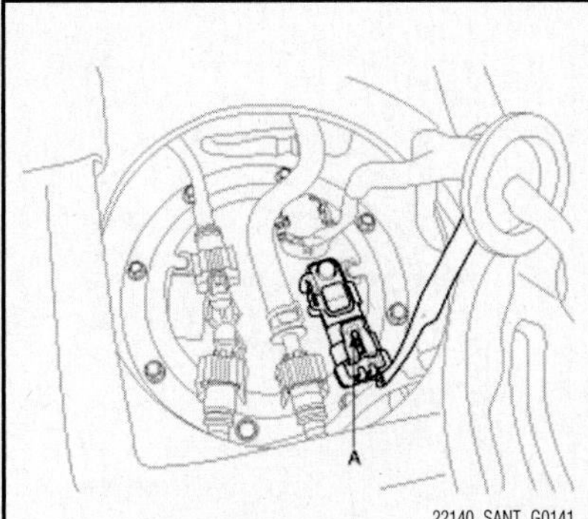

Fig. 186 Disconnect the fuel tank pressure sensor connector (A)

5. Disconnect the sub fuel sender connector and the canister close valve connector.

6. Lift the vehicle and remove the muffler assembly.

7. For 4WD vehicles, remove the propeller shaft. Refer to Propeller Shaft Removal & Installation in the Drive Train section.

8. Support the fuel tank with a jack.

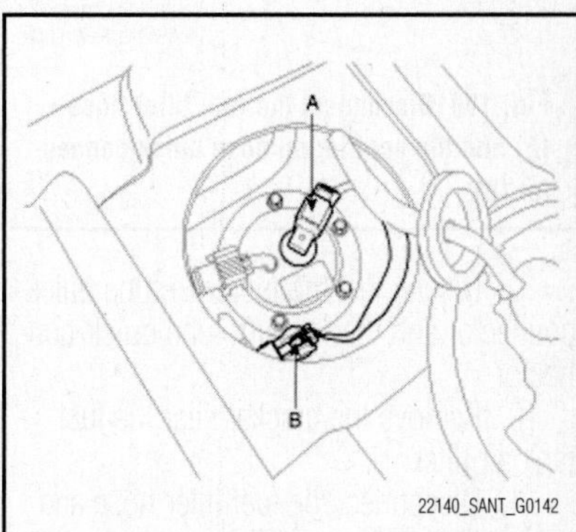

Fig. 187 Disconnect the sub fuel sender connector and the canister close valve connector (B)

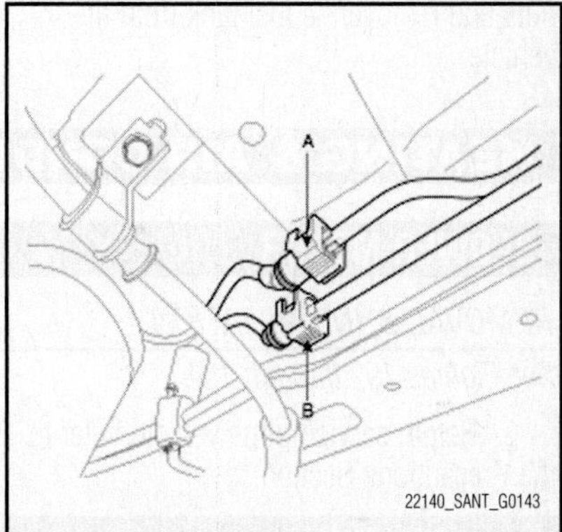

Fig. 188 Disconnect the fuel feed tube quick-connector (A) and the vacuum tube quick-connector (B)

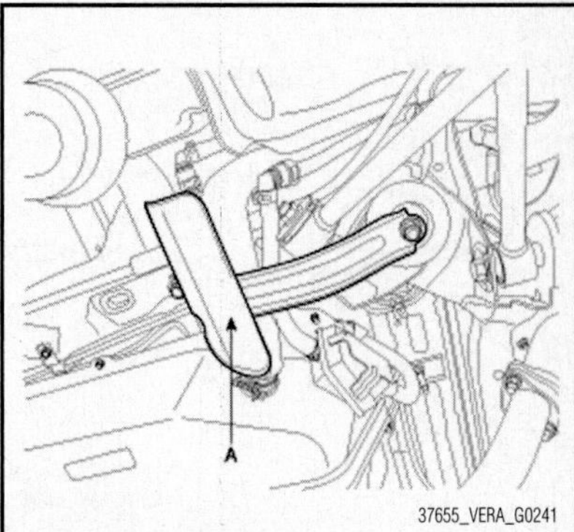

Fig. 189 Remove the bracket (A) near the fuel tank air filter

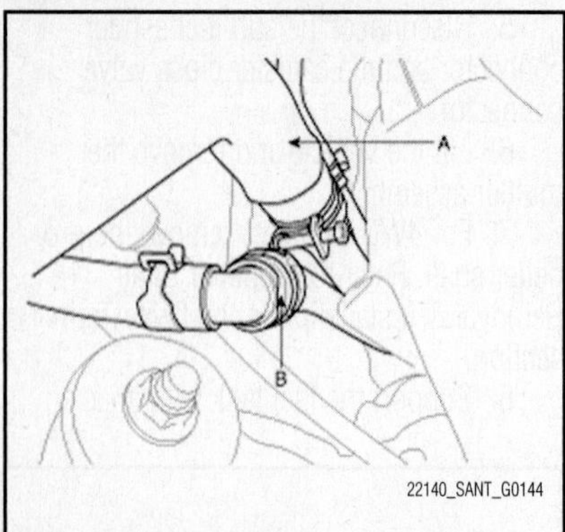

Fig. 190 Disconnect the fuel filler hose (A) and the ventilation hose quick-connector (B)

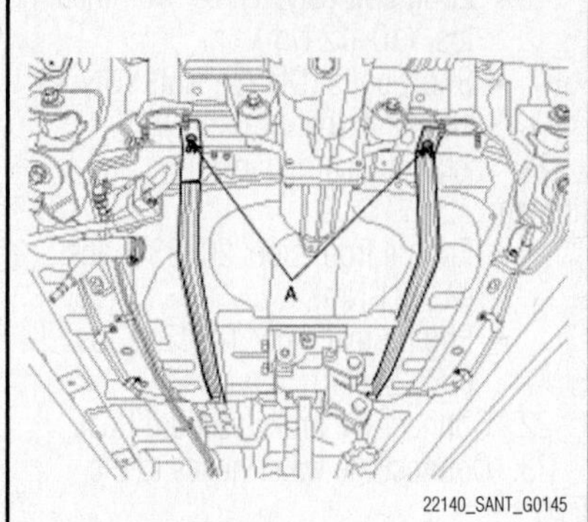

Fig. 191 Remove the fuel tank bank mounting nuts (A)

9. Disconnect the fuel feed tube quick-connector and the vacuum tube quick-connector.
10. Remove the bracket near the fuel tank air filter.
11. Disconnect the fuel filler hose and the ventilation hose quick-connector.
12. Disconnect the leveling tube quick-connector.
13. Remove the fuel tank cover.
14. Remove the fuel tank bank mounting nuts and remove the fuel tank from the vehicle.

To install:

15. Installation is the reverse of the removal procedure. Tighten the fuel tank band mounting nuts to: 29–40 ft. lbs. (39–54 Nm).

IDLE SPEED

ADJUSTMENT

Idle speed is maintained by the Powertrain Control Module (PCM). No adjustment is necessary or possible.

THROTTLE BODY

REMOVAL & INSTALLATION

1. Before servicing the vehicle, refer to the Precautions Section.
2. Turn the ignition **OFF**.
3. Remove the engine cover.
4. Remove the air cleaner inlet.
5. Remove the throttle body electrical connector.
6. Remove the coolant hoses, if necessary.
7. Remove the throttle body stay bolts, if necessary.
8. Remove the throttle body bolts.
9. Remove the throttle body and gasket.

To install:

10. Clean the throttle body gasket mating surfaces.
11. Install the throttle body and NEW gasket.
12. Install the throttle body bolts.
13. Install the throttle body stay bolts, if necessary.
14. Install the coolant hoses, if necessary.
15. Install the throttle body electrical connector.
16. Install the air cleaner inlet.
17. Install the engine cover.

HEATING & AIR CONDITIONING SYSTEM

BLOWER MOTOR

REMOVAL & INSTALLATION

See Figures 192 through 194.

1. Before servicing the vehicle, refer to the Precautions Section.

⁂ CAUTION

Before servicing components near or affected by the SRS (air bag) system, read and observe all SRS Service Precautions. Refer to Supplemental Restraint System (SRS), in the Chassis Electrical section. Failure to observe all precautions may result in accidental airbag deployment, personal injury, or death.

2. Disconnect the negative battery cable and wait at least 3 minutes for the SRS memory to drain.
3. Remove the instrument panel lower panel.
4. Disconnect the blower motor connector.
5. Remove the blower motor mounting screws.
6. Remove the blower motor.

To install:

7. Installation is the reverse of the removal procedure.

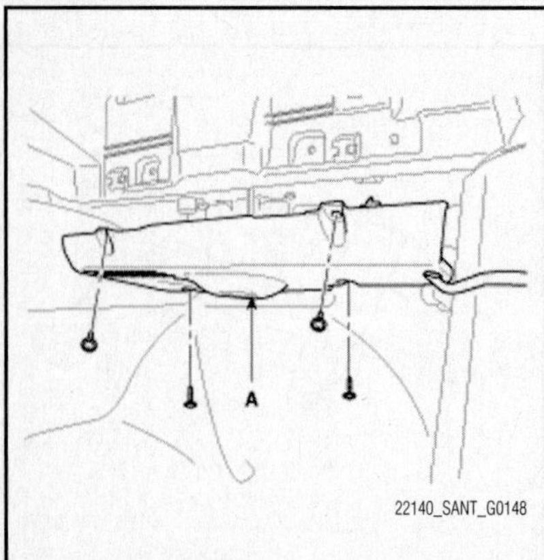

Fig. 192 Remove the instrument panel lower panel (A)

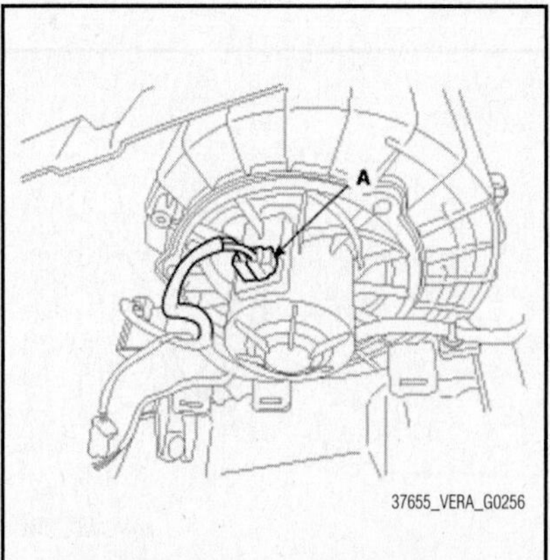

Fig. 193 Disconnect the blower motor connector (A)

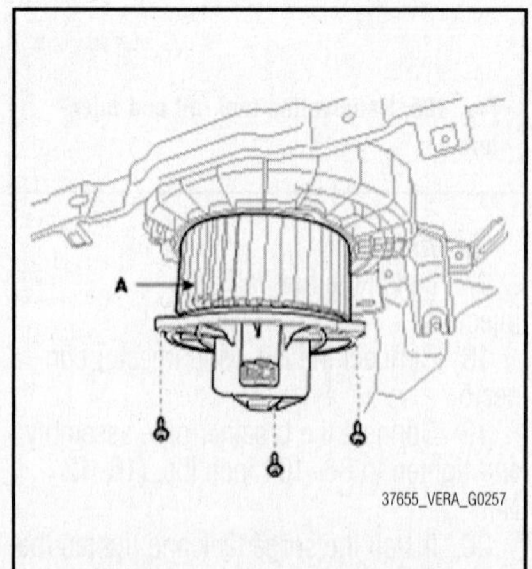

Fig. 194 Remove the blower motor (A)

HEATER UNIT/CORE

REMOVAL & INSTALLATION

See Figures 195 through 201.

1. Before servicing the vehicle, refer to the Precautions Section.
2. Discharge and recover the air conditioning system refrigerant.
3. Drain the engine coolant.

CAUTION

Before servicing components near or affected by the SRS (air bag) system, read and observe all SRS Service Precautions. Refer to Supplemental Restraint System (SRS), in the Chassis Electrical section. Failure to observe all precautions may result in accidental airbag deployment, personal injury, or death.

4. Disconnect the negative battery cable and wait at least 3 minutes for the SRS memory to drain.
5. Remove the bolts and expansion valve from the evaporate core and plug the lines.
6. Disconnect the heater hoses from the heater unit.
7. Remove the instrument panel.
8. Remove the cowl cross bar assembly.
9. Disconnect the connectors from the temperature control actuator, the mode control actuator and the evaporator temperature sensor.
10. Remove the heater and blower unit after removing the mounting nuts.
11. Remove the blower unit from heater unit after removing the screws.
12. Remove the heater core cover.
13. Remove the heater core from the heater unit.
14. Remove the heater unit cover after removing the screws.
15. Remove the evaporator core.
16. Be careful that the inlet and outlet pipe are not bent during heater core removal, and pull out the heater core.

To install:

17. Installation is the reverse of the removal procedure. Tighten the expansion valve bolts to 70–104 inch lbs. (8–12 Nm).
18. Observe the following:

- If you're installing a new evaporator, add refrigerant oil (ND-OIL8).
- Replace the O-rings with new ones at each fitting, and apply a thin coat of refrigerant oil before installing them. Be sure to use the right O-rings for R-134a to avoid leakage.
- Immediately after using the oil, replace the cap on the container,

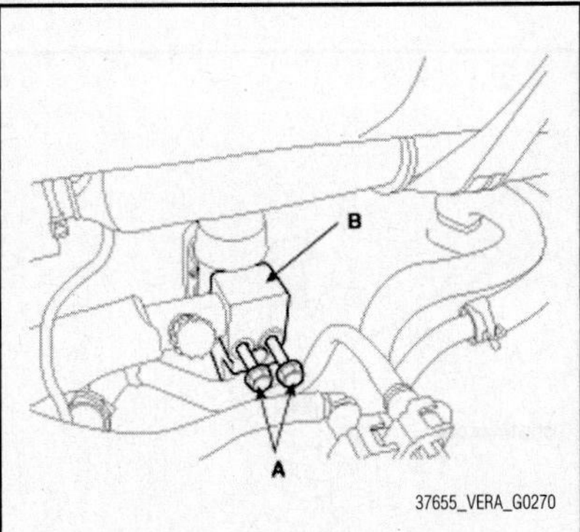

Fig. 195 Remove the bolts (A) and the expansion valve (B)

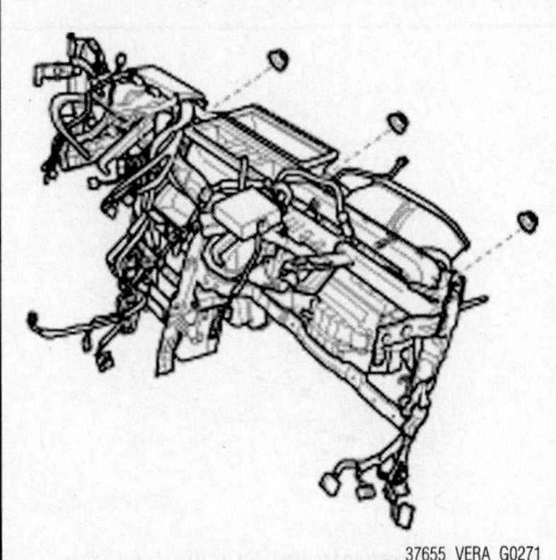

Fig. 196 Remove the heater and blower unit after removing the 3 mounting nuts

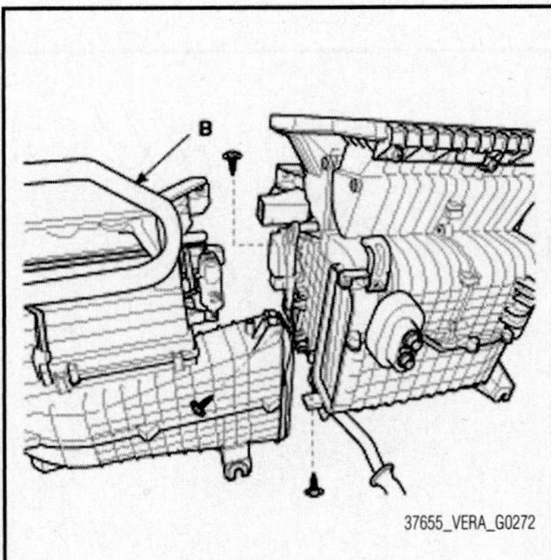

Fig. 197 Remove the blower unit (B) from heater unit after removing the 3 screws

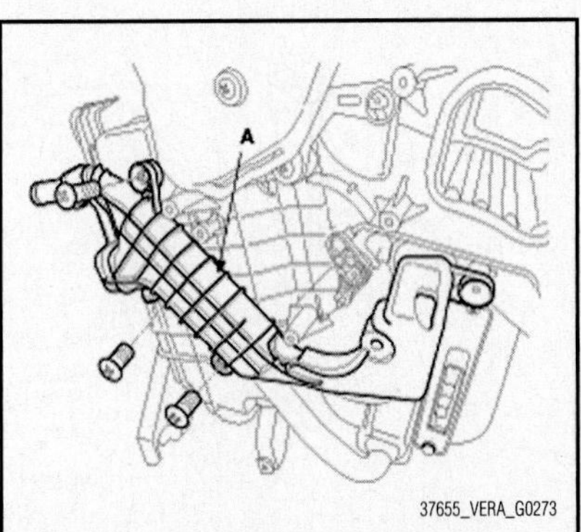

Fig. 198 Remove the heater core cover (A)

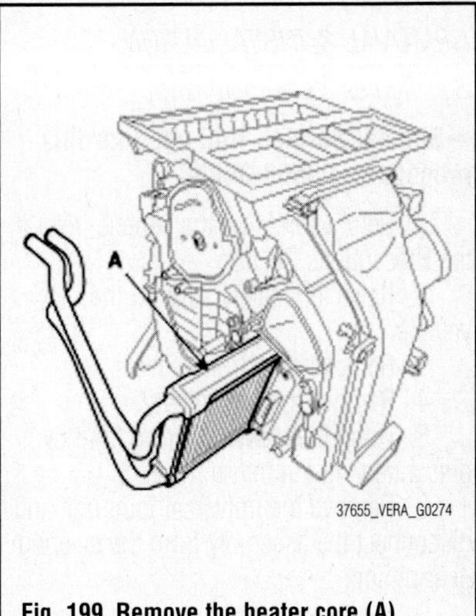

Fig. 199 Remove the heater core (A)

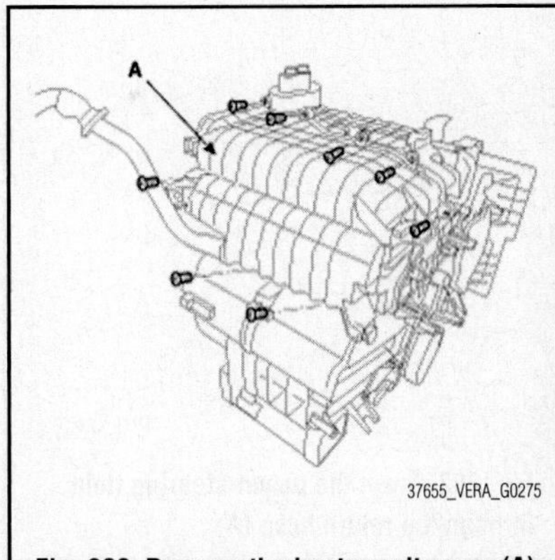

Fig. 200 Remove the heater unit cover (A)

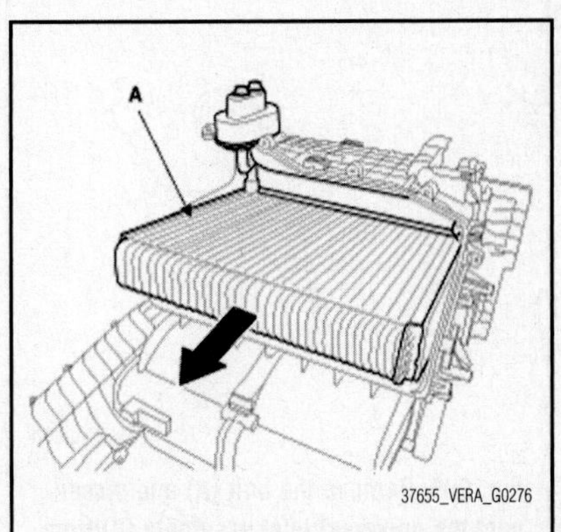

Fig. 201 Remove the evaporator core (A)

and seal it to avoid moisture absorption.

- Apply sealant to the grommets.
- Make sure that there is no air leakage.
- Charge the system and test its performance.
- Do not interchange the inlet and outlet heater hoses and install the hose clamps securely.

WARNING

Do not spill the refrigerant oil on the vehicle ; it may damage the paint ; if the refrigerant oil contacts the paint, wash it off immediately.

STEERING

POWER RACK & PINION STEERING GEAR

REMOVAL & INSTALLATION

See Figures 202 through 215.

➡Special Service Tool used for this procedure: 09568-34000

1. Before servicing the vehicle, refer to the Precautions Section.
2. Raise and safely support the vehicle.
3. Remove both front wheels.
4. Remove the undercover.
5. Drain the power steering fluid by disconnecting the return hose.
6. Remove the universal joint bolt and disconnect the assembly from the steering gear pinion.

WARNING

Keep the steering gear in the neutral-range to prevent the damage of the clock spring inner cable when you handle the steering wheel.

7. Disconnect the pressure tube from the power steering pump by removing the eye-bolt.
8. Disconnect the stabilizer link from the strut assembly by removing the nut.
9. Remove the split pin and castle nut.
10. Disconnect the tie-rod end from the knuckle using a SST (09568-34000).
11. Separate the knuckle from the lower arm by removing the bolt and nut.
12. Remove the muffler rubber hanger.
13. Remove the front and rear roll stopper mounting bolts.
14. Remove the sub-frame and stay.
15. Remove the heat protector.
16. Remove the pressure and return tube from the steering gear box.

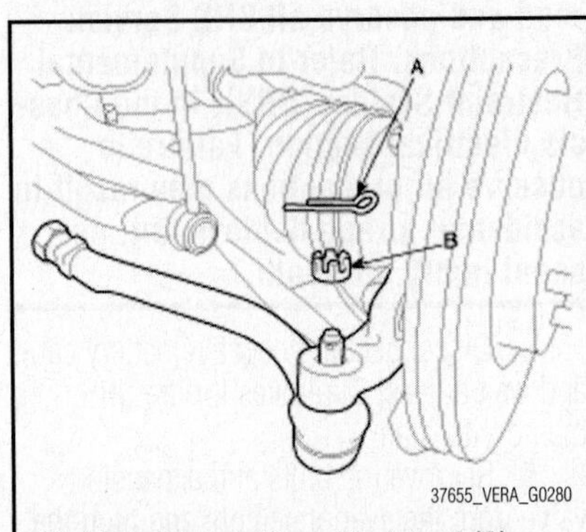

Fig. 206 Remove the split pin and (A) castle nut (B)

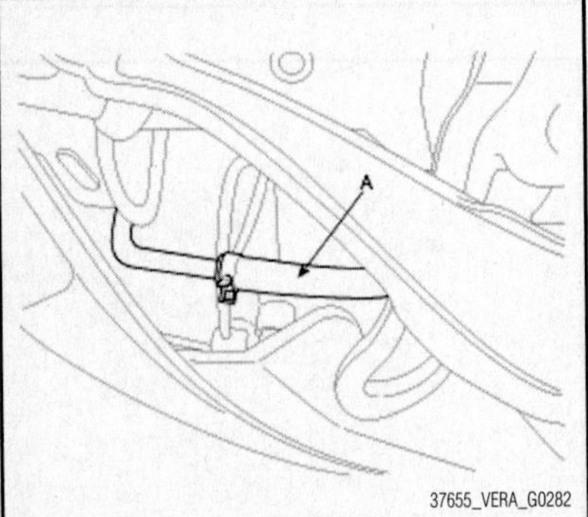

Fig. 202 Drain the power steering fluid through the return hose (A)

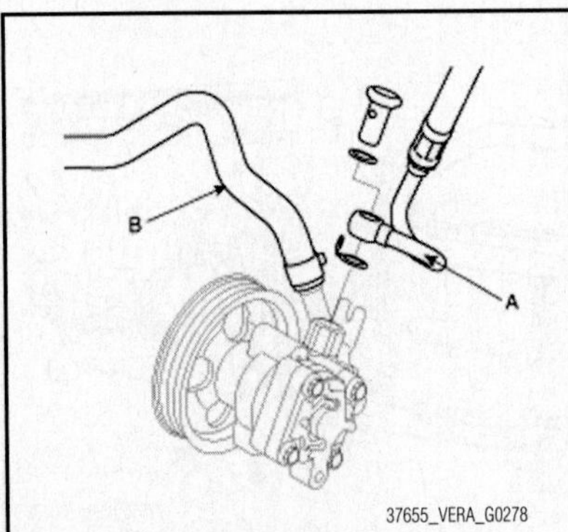

Fig. 204 Disconnect the pressure tube (A) from the power steering pump

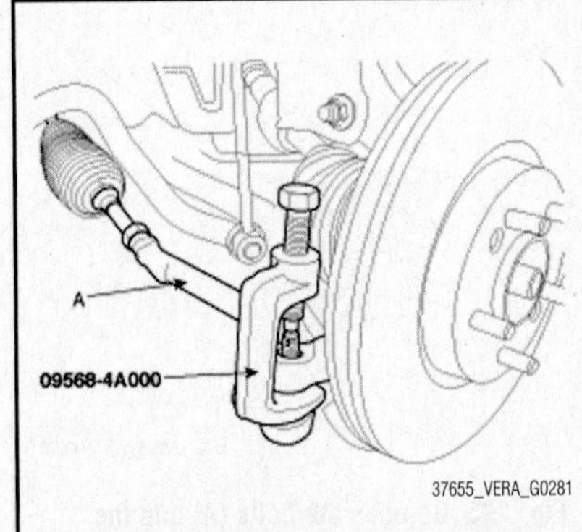

Fig. 207 Disconnect the tie-rod end (A) from the knuckle using SST (09568-34000)

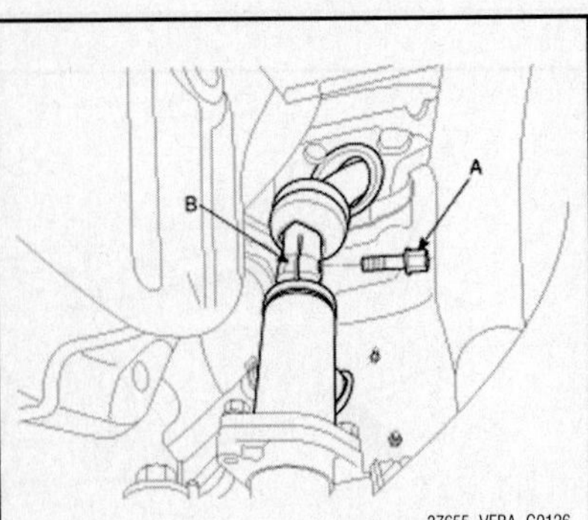

Fig. 203 Remove the bolt (A) and disconnect the universal joint assembly (B) from the steering gear pinion

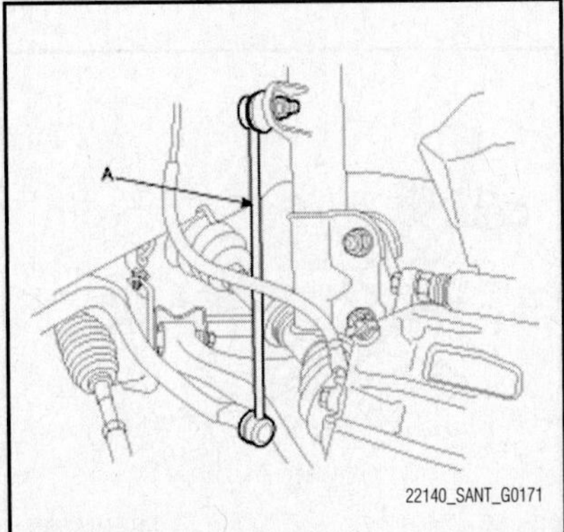

Fig. 205 Disconnect the stabilizer link (A) from the strut assembly

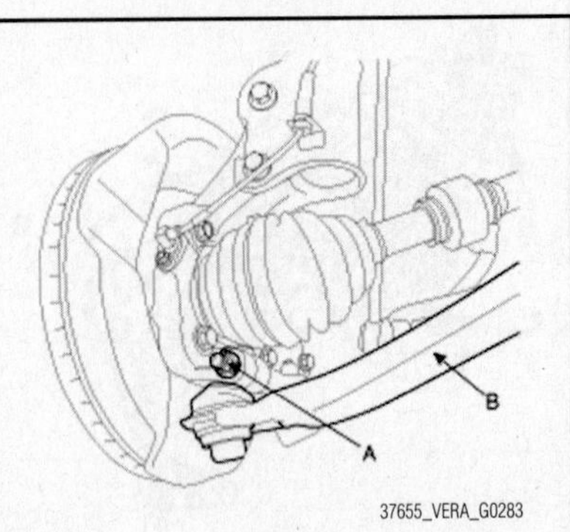

Fig. 208 Separate the knuckle from the lower arm (B) by removing the bolt and nut (A)

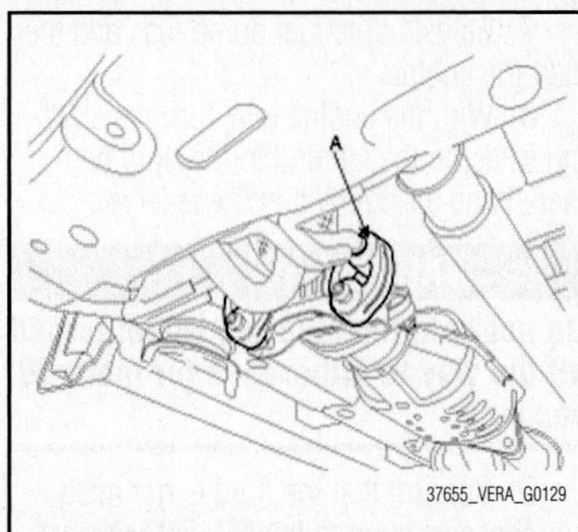

Fig. 209 Remove the muffler hanger rubber (A)

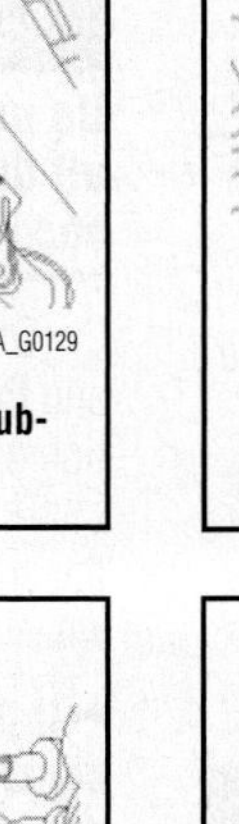

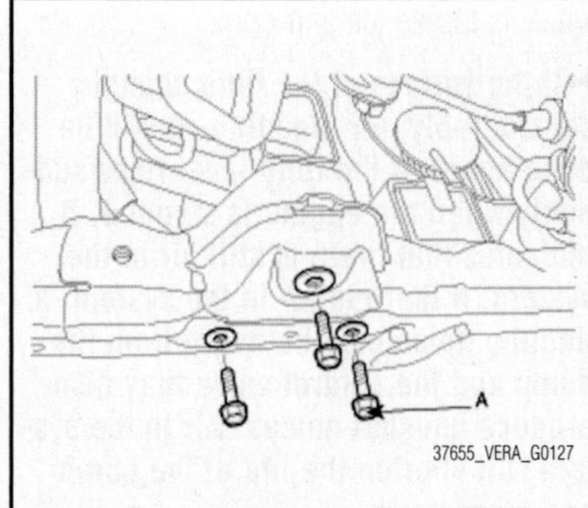

Fig. 210 Remove the front roll stopper mounting bolts (A)

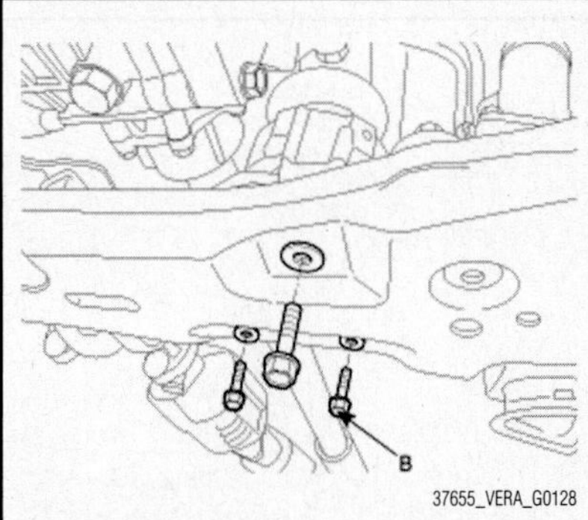

Fig. 211 Remove the rear roll stopper mounting bolts (B)

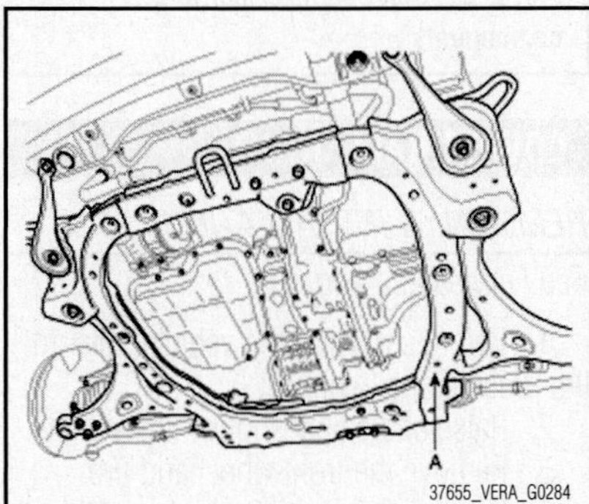

Fig. 212 Remove the sub-frame (A) and stay

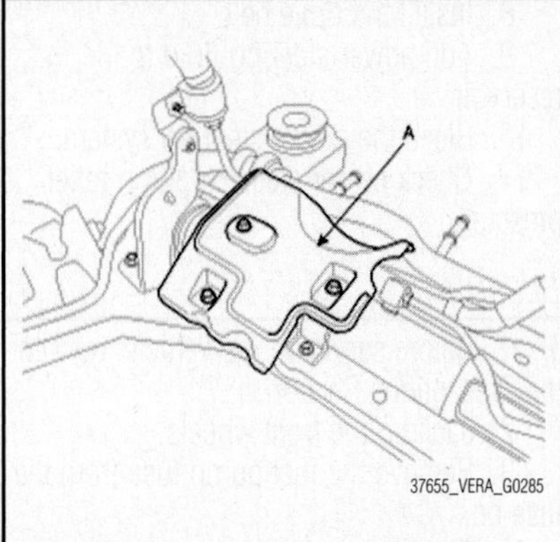

Fig. 213 Remove the heat protector (A)

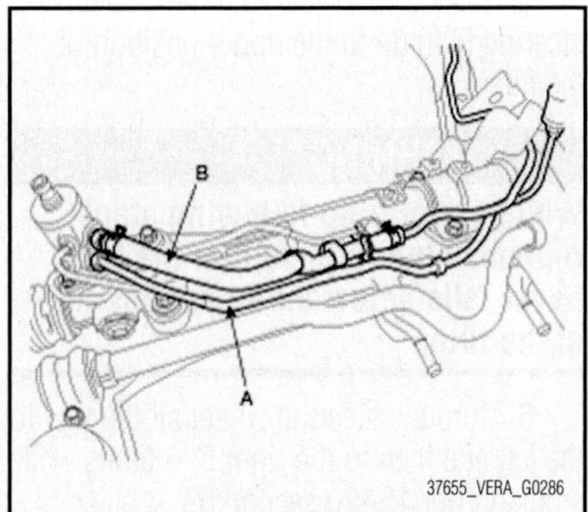

Fig. 214 Remove the pressure (A) and return tube (B) from the steering gear box

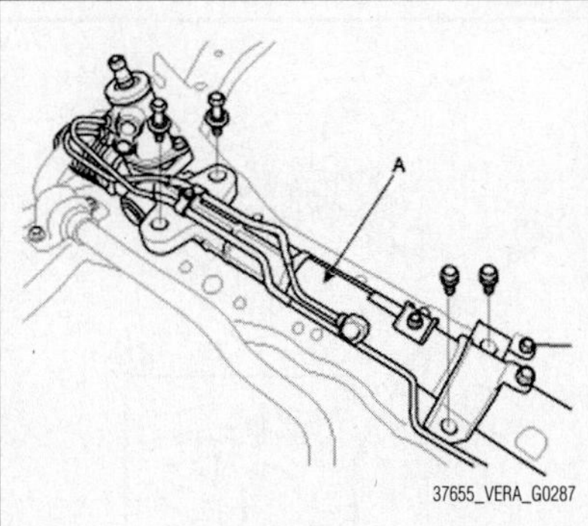

Fig. 215 Remove the steering gear box (A)

17. Remove the steering gear box from the sub-frame after removing the mounting bolts.

To install:

18. Install the steering gear box to the sub-frame and tighten the mounting bolts to 65–80 ft. lbs. (90–110 Nm).
19. Install the pressure tube and return tube to the steering gear box and then tighten the flare nut to 9–13 ft. lbs. (12–18 Nm).
20. Install the heat protector.
21. Install the sub-frame and stay and tighten the mounting bolts and nuts:
    - Sub-frame mounting bolt: 101–116 ft. lbs. (140–160 Nm)
    - Sub-frame stay mounting bolt and nuts: 51–65 ft. lbs. (70–90 Nm)
22. Install the front and rear roll stopper mounting bolts and tighten to 36–47 ft. lbs. (50–65 Nm).
23. Install the muffler rubber hanger.
24. Connect the lower arm with the knuckle and tighten the bolt and nut to 72–87 ft. lbs. (100–120 Nm).
25. Connect the tie-rod end with the knuckle and install the castle nut and split pin and tighten to 17–25 ft. lbs. (24–34 Nm).
26. Connect the stabilizer link with the front strut assembly and tighten the nut to 72–87 ft. lbs. (100–120 Nm).
27. Connect the pressure tube to power steering pump and tighten the eye bolt to 40–47 ft. lbs. (55–65 Nm).
28. Connect the universal joint assembly with the steering column assembly and then tighten the bolt to 22–25 ft. lbs. (30–35 Nm).
29. Connect the power steering return hose.
30. Install the undercover.
31. Install the front wheel and tire.
32. Add power steering fluid to reservoir.
33. Bleed the power steering system.
34. Check and adjust the front wheel alignment.

POWER STEERING PUMP

REMOVAL & INSTALLATION

See Figures 216 and 217.

1. Before servicing the vehicle, refer to the Precautions Section.
2. Remove the drive belt.

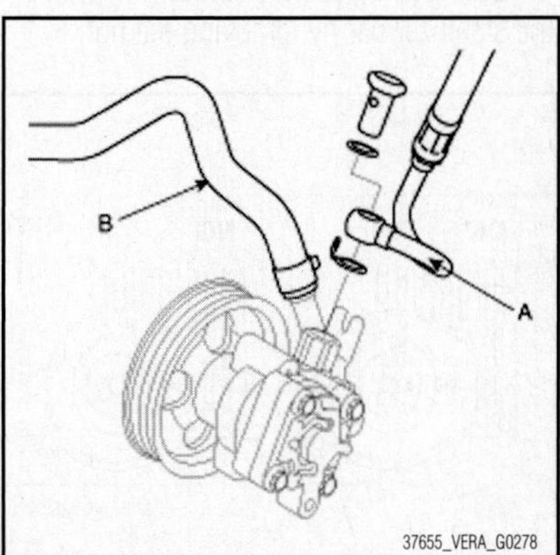

Fig. 216 Disconnect the pressure tube (A) and return hose (B) from the power steering pump

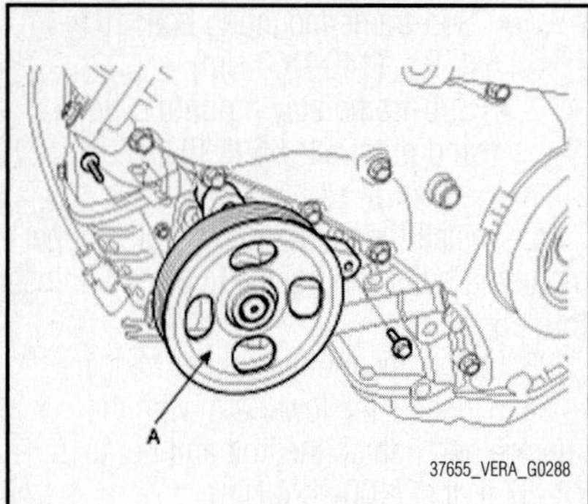

Fig. 217 Remove the power steering pump (A)

3. Disconnect the pressure tube and return hose from the power steering pump.
4. Remove the mounting bolts and remove the power steering pump.

To install:

5. Install the power steering pump and tighten the mounting bolts to 14–20 ft. lbs. (35—50 Nm).
6. Connect the suction hose with suction pipe.
7. Connect the pressure tube to power steering pump and tighten the eye bolt to 40–47 ft. lbs. (55–65 Nm).
8. Install the drive belt.
9. Add power steering fluid to the reservoir.
10. Bleed the power steering system.
11. Check power steering pump relief pressure.

BLEEDING

1. Before servicing the vehicle, refer to the Precautions Section.
2. Jack up the front wheels.
3. Remove the fuel pump fuse from the fuse box.
4. Start the engine and wait for the engine to stall.
5. Fill the reservoir with the power steering fluid up to the upper position of the filter.

WARNING

While performing following steps, replenish the fluid so that the level never falls below the lower position of the filter.

6. Turn the steering wheel all the way to the left and then to the right 5–6 times while cranking (for 15–20 seconds).
7. Reinstall the fuel pump fuse and then start the engine.
8. With the engine idle, turn the steering wheel to the left and to the right until there is no air bubbles in the reservoir.

WARNING

Do not hold the steering wheel turned all the way to either side for more 10 seconds.

9. Confirm that the fluid is not milky, and the level is up to the COLD MAX position specified on the level gauge.
10. Confirm that there is little change in the surface of the fluid when the steering wheel is turned left and right.

➡If the surface of the fluid changes considerably, air bleeding should be done again. If the fluid level rises suddenly when the engine is stopped, it indicates that there is still air in the system. If there is air in the system, a jingling noise may be heard from the pump and the control valve may also produce unusual noises. Air in the system will shorten the life of the pump and other parts.

SUSPENSION — FRONT SUSPENSION

CONTROL LINKS

REMOVAL & INSTALLATION

See Figures 218 through 220.

1. Before servicing the vehicle, refer to the Precautions Section.
2. Raise and safely support the vehicle.
3. Remove the both front wheels.
4. Remove the stabilizer link from the strut assembly by removing the nut.
5. Disconnect the stabilizer link from the stabilizer bar by removing the nut.

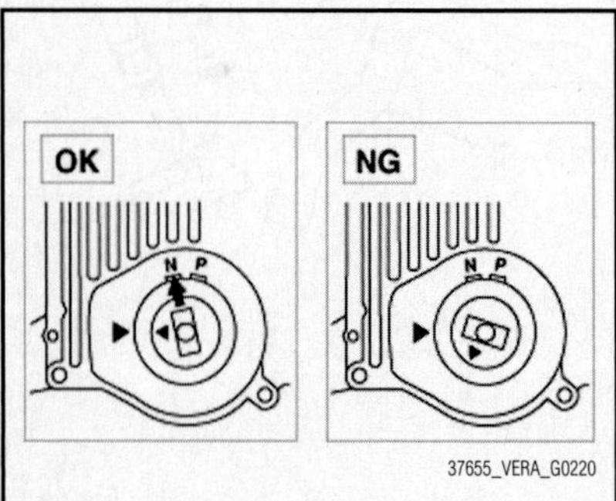

Fig. 218 Remove the stabilizer link (A) from the strut assembly

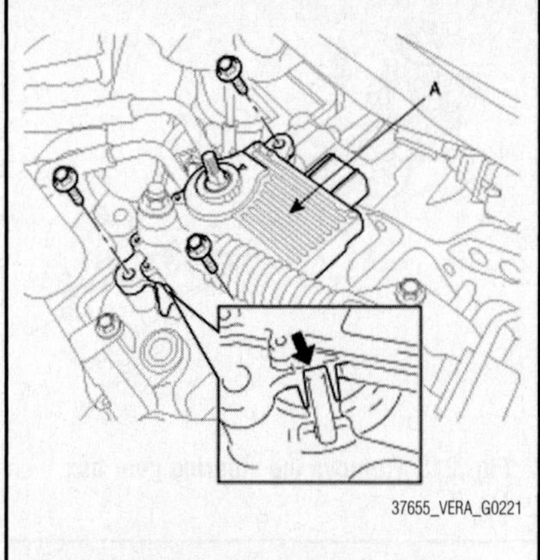

Fig. 219 Remove the stabilizer link (A) from the strut assembly

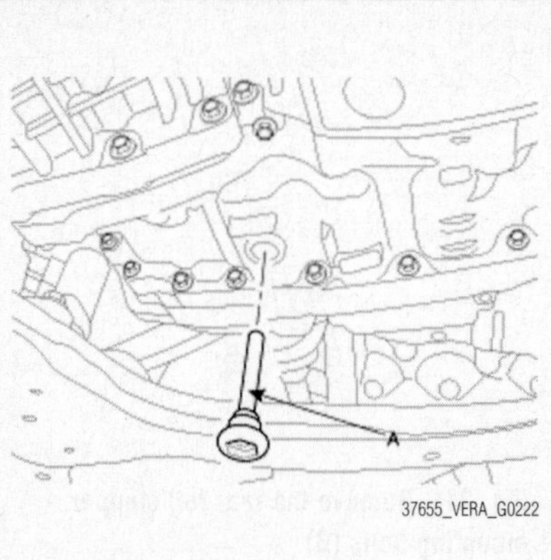

Fig. 220 Front stabilizer bar and components

To install:

6. Connect the stabilizer link to the stabilizer bar and tighten the nut to 72–87 ft. lbs. (100–120 Nm).
7. Connect the stabilizer link to the front strut assembly and tighten the nut to 72–87 ft. lbs. (100–120 Nm).
8. Install the front wheels.
9. Add power steering fluid to reservoir.
10. Bleed the power steering system.
11. Check and adjust the front wheel alignment.

LOWER BALL JOINT

REMOVAL & INSTALLATION

See Figures 221 and 222.

1. Before servicing the vehicle, refer to the Precautions Section.
2. Raise and safely support the vehicle.
3. Remove the front wheel and tire.
4. Remove the split pin, bolt, and nut.
5. Disconnect the lower arm ball joint from the knuckle.

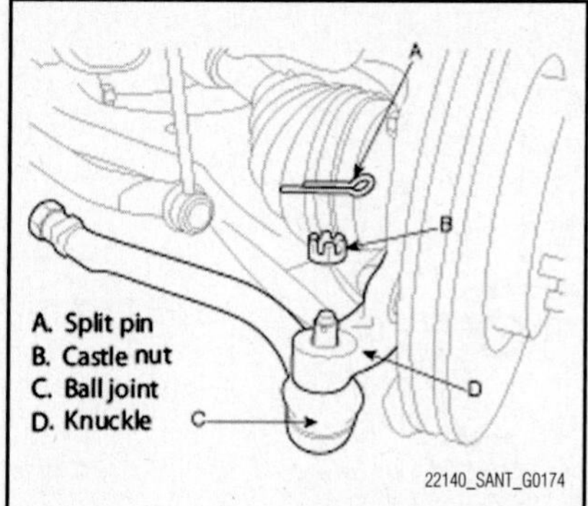

Fig. 221 Remove the split pin and castle nut. Disconnect the ball joint from the knuckle

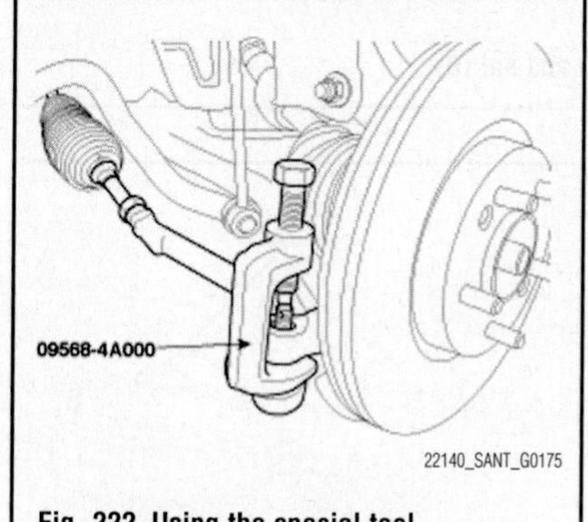

Fig. 222 Using the special tool (09568-4A000) to remove the ball joint

To install:

6. Connect the lower arm ball joint to the knuckle and tighten the bolt and nut to 72–87 ft. lbs. (100–120 Nm).
7. Install the split pin to the bolt.
8. Install the front wheel and tire.

LOWER CONTROL ARM

REMOVAL & INSTALLATION

See Figures 221, 223 and 224.

1. Before servicing the vehicle, refer to the Precautions Section.

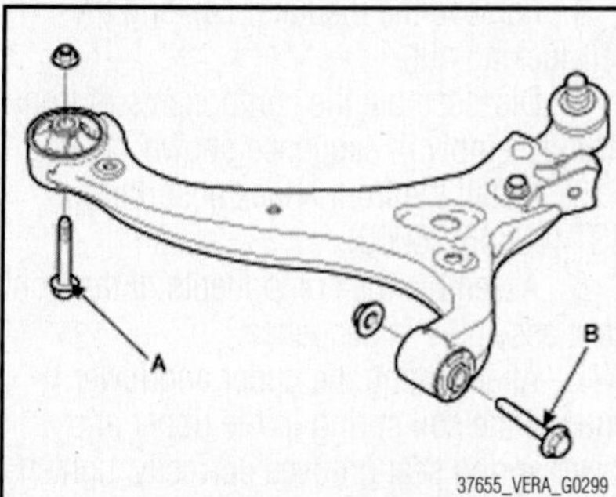

Fig. 223 Remove the bolts and nut (A, B) and remove the lower arm from the sub-frame

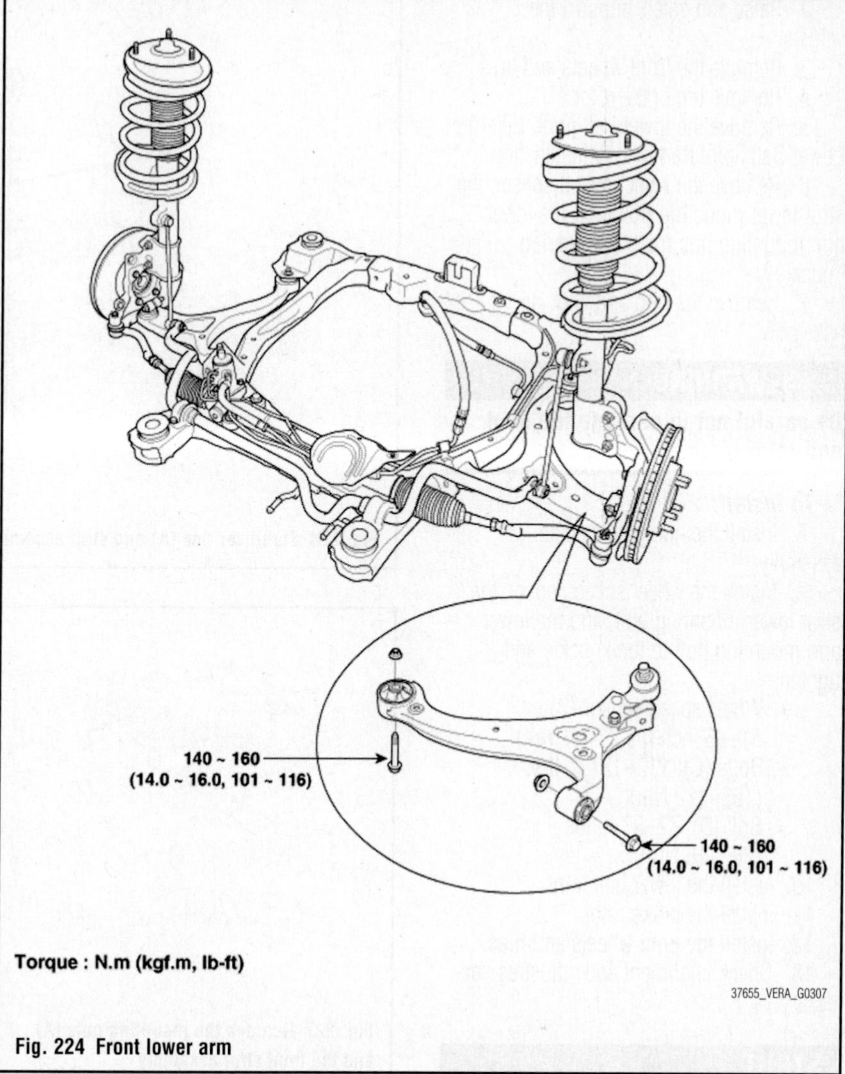

Fig. 224 Front lower arm

2. Remove the front wheel and tire.
3. Remove the split pin, bolt, and nut.
4. Disconnect the lower arm ball joint from the knuckle. Refer to Lower Ball Joint Removal & Installation.
5. Remove the bolts and nut and remove the lower arm from the sub-frame.

To install:

6. Install the front lower arm to the sub-frame and tighten the bolts and nuts to 101–116 ft. lbs. (140–160 Nm).
7. Connect the lower arm ball joint with the knuckle and tighten the bolt and nut to 72–87 ft. lbs. (100–120 Nm).
8. Install the split pin to the bolt.
9. Install the front wheel and tire.

STEERING KNUCKLE

REMOVAL & INSTALLATION

See Figure 225.

1. Before servicing the vehicle, refer to the Precautions Section.

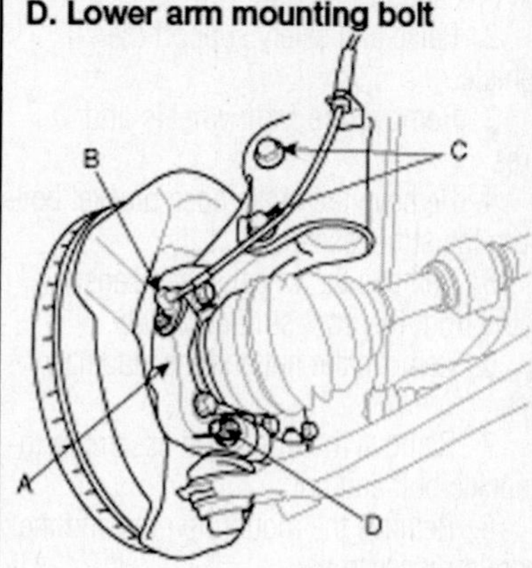

Fig. 225 Remove the wheel speed sensor (B), the strut lower mounting bolt (C), and the lower arm mounting bolt (D) from the steering knuckle (A)

2. Raise and safely support the vehicle.
3. Remove the front wheels and tires.
4. Remove the brake rotor.
5. Remove the lower ball joint. Refer to Lower Ball Joint Removal & Installation.
6. Remove the wheel speed sensor, the strut lower mounting bolt, and the lower arm mounting bolt from the steering knuckle.
7. Remove the hub and knuckle assembly.

⁂ WARNING

Be careful not to damage the boot and rotor teeth.

To install:

8. Install the hub and knuckle assembly.
9. Install the wheel speed sensor, the strut lower mounting bolt, and the lower arm mounting bolt to the knuckle and tighten:
   - Wheel speed sensor (B): 61–96 inch lbs. (7–11 Nm)
   - Bolts (C): 112–127 ft. lbs. (152–172 Nm)
   - Bolt (D): 72–87 ft. lbs. (98–118 Nm)
10. Install the lower ball joint.
11. Install the brake rotor.
12. Install the front wheels and tires.
13. Check alignment and adjust as necessary.

STRUT

REMOVAL & INSTALLATION

See Figures 226 and 227.

1. Before servicing the vehicle, refer to the Precautions Section.
2. Raise and safely support the vehicle.
3. Remove the front wheels and tires.
4. Remove the brake hose bracket bolts from the strut assembly.
5. Remove the wheel speed sensor cable from the front strut assembly.
6. Remove the nut and the stabilizer link.
7. Remove the front strut assembly-to-knuckle bolt and nut.
8. Remove the mounting nuts and the front strut assembly.

To install:

9. Install the front strut assembly and tighten the mounting nuts to 33–43 ft. lbs. (45–60 Nm).
10. Install the front strut assembly bolt and nut to the front knuckle and tighten to 112–127 ft. lbs. (155–175 Nm).
11. Install the front stabilizer link to the strut assembly and tighten the nut to 72–87 ft. lbs. (100–120 Nm).
12. Install the wheel speed sensor cable to the front strut assembly.
13. Install the wheel and the tire to the front hub. Tighten to 65–80 ft. lbs. (90–110 Nm).

⁂ WARNING

Be careful not to damage the hub bolts when installing the front wheel and tire.

14. Check the wheel alignment and adjust as necessary.

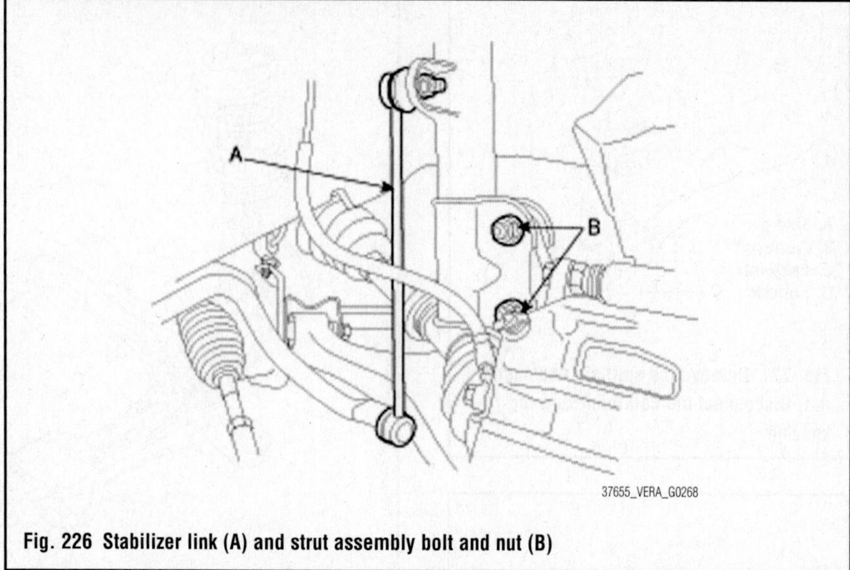

Fig. 226 Stabilizer link (A) and strut assembly bolt and nut (B)

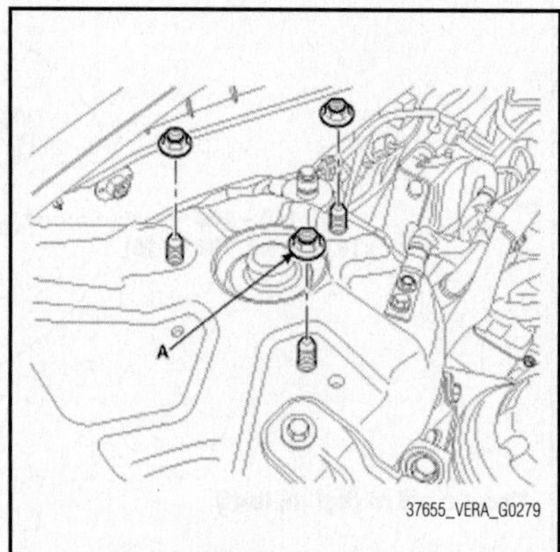

Fig. 227 Remove the mounting nuts (A) and the front strut assembly

OVERHAUL

See Figures 228 and 229.

➡Special Service Tool used for this procedure: 09546-26000

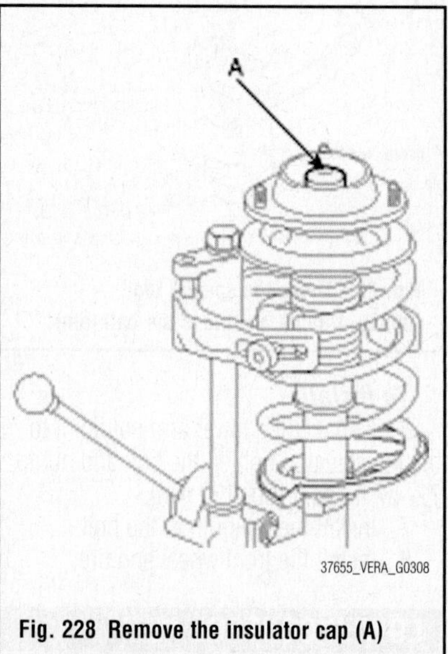

Fig. 228 Remove the insulator cap (A)

1. Before servicing the vehicle, refer to the Precautions Section.
2. Compress the coil spring with a SST (09546-26000). Do not compress the spring more than necessary.
3. Remove the insulator cap and the self-locking nut.
4. Disassemble the components of front strut assembly in sequence shown.
5. Install the front shock absorber to SST (09546-26000).
6. Assemble the components of the front strut assembly in sequence.
7. After seating the upper and lower ends of the coil spring in the upper and lower spring seat grooves correctly, tighten new self-locking nut to 43–51 ft. lbs. (60–70 Nm).
8. Install the insulator cap.
9. Remove the special service tool.

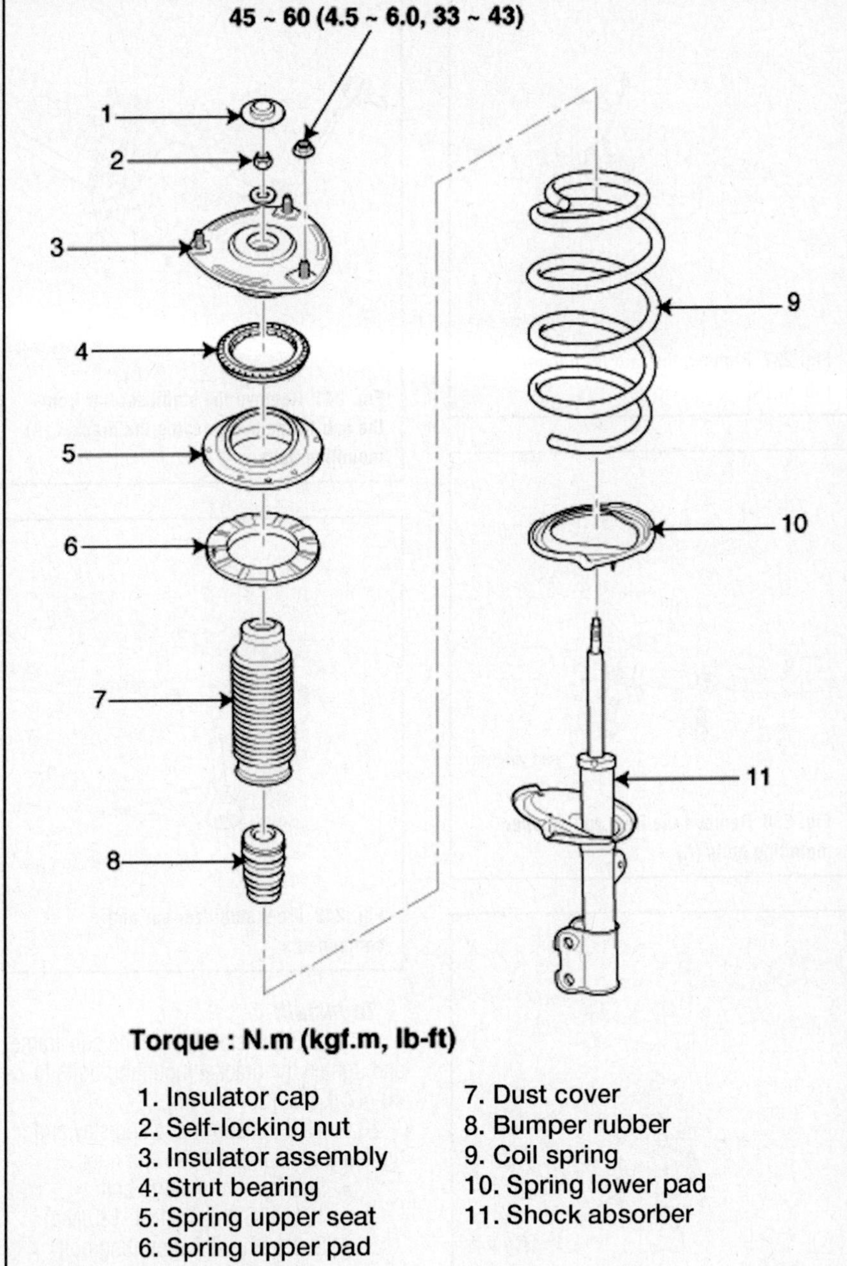

Fig. 229 Front strut components

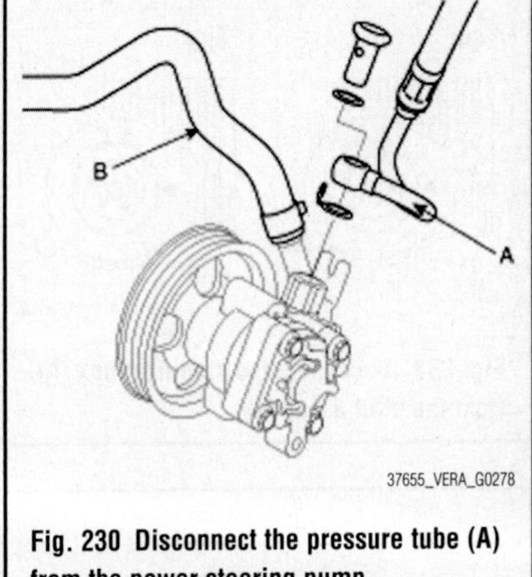

Fig. 230 Disconnect the pressure tube (A) from the power steering pump

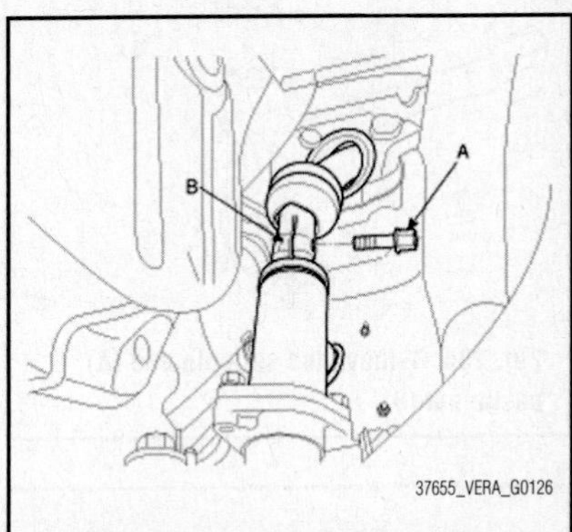

Fig. 231 Remove the bolt (A) and disconnect the universal joint assembly (B) from the steering gear pinion

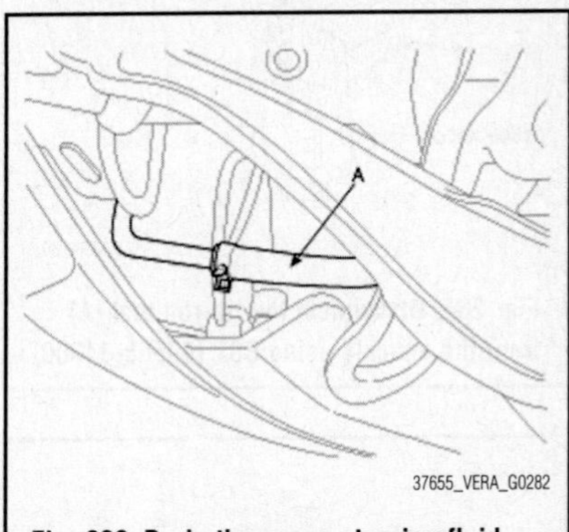

Fig. 232 Drain the power steering fluid through the return hose (A)

STABILIZER BAR

REMOVAL & INSTALLATION

See Figures 230 through 242.

➡Special Service Tool used for this procedure: 09568-34000

1. Before servicing the vehicle, refer to the Precautions Section.
2. Raise and safely support the vehicle.
3. Remove both front wheels.
4. Disconnect the pressure tube from the power steering pump by removing the eye-bolt.
5. Remove the universal joint bolt and disconnect the assembly from the steering gear pinion.

⁂ WARNING

Keep the steering gear in the neutral-range to prevent the damage of the clock spring inner cable when you handle the steering wheel.

6. Drain the power steering fluid by disconnecting the return hose.
7. Disconnect the stabilizer link from the strut assembly by removing the nut.
8. Remove the split pin and castle nut.
9. Disconnect the tie-rod end from the knuckle using a SST (09568-34000).
10. Separate the knuckle from the lower arm ball joint by removing the bolt and nut.
11. Remove the muffler rubber hanger.
12. Remove the front and rear roll stopper mounting bolts.
13. Remove the sub-frame and stay.
14. Remove the stabilizer bar from the sub-frame by removing the bracket mounting bolts.

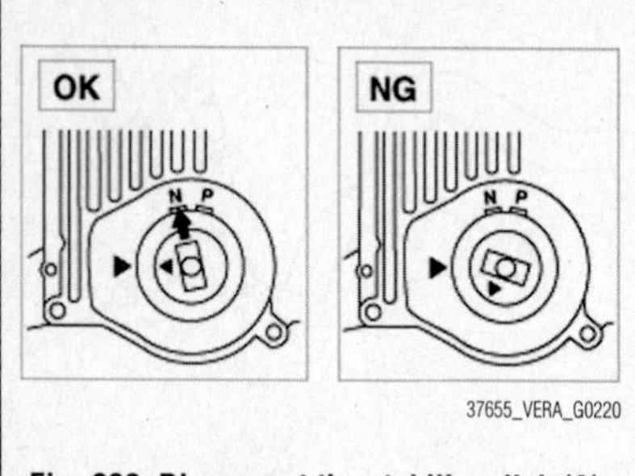

Fig. 233 Disconnect the stabilizer link (A) from the strut assembly

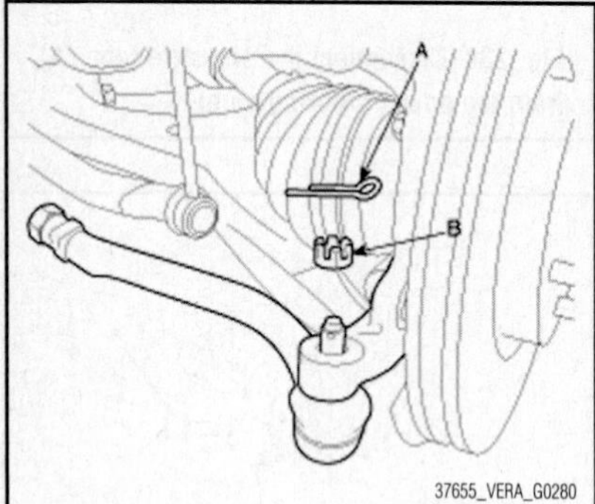

Fig. 234 Remove the split pin and (A) castle nut (B)

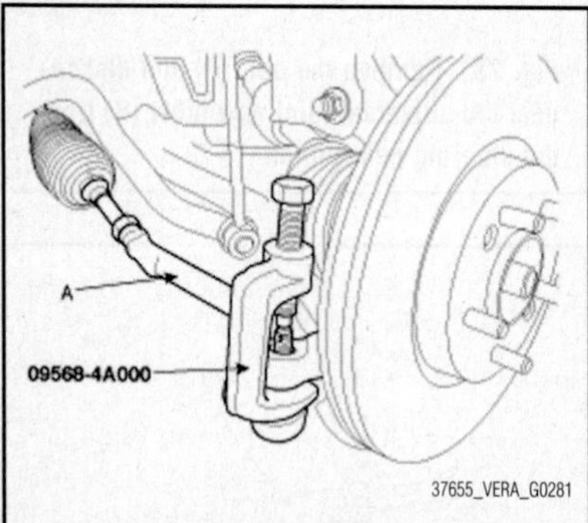

Fig. 235 Disconnect the tie-rod end (A) from the knuckle using SST (09568-34000)

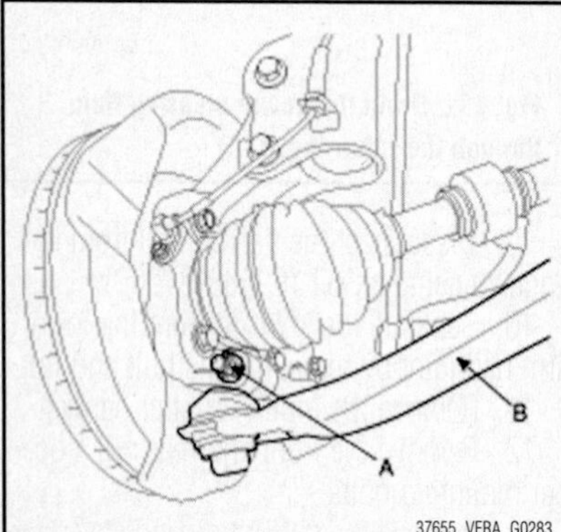

Fig. 236 Separate the knuckle from the lower arm (B) by removing the bolt and nut (A)

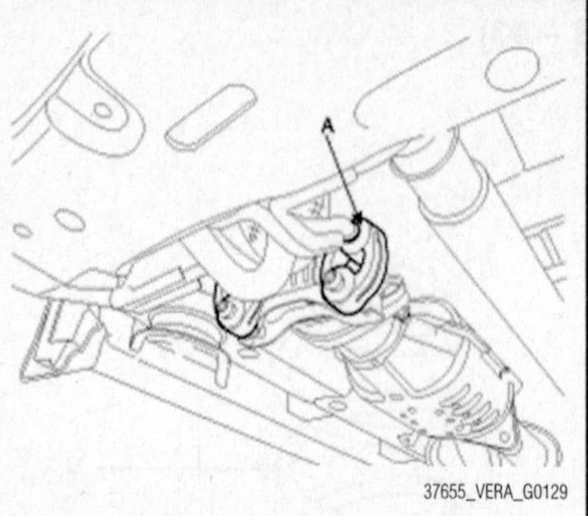

Fig. 237 Remove the muffler hanger rubber (A)

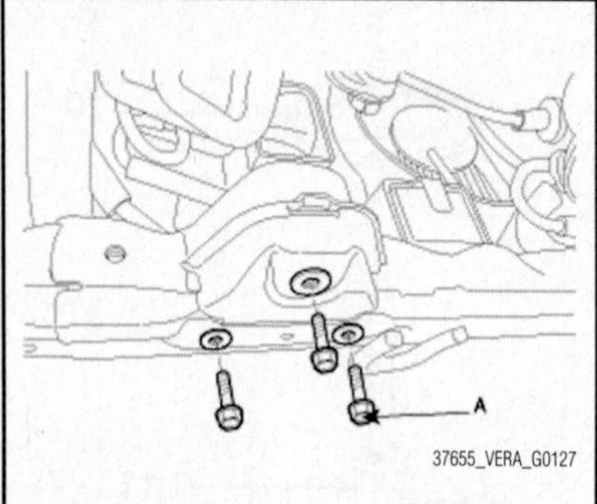

Fig. 238 Remove the front roll stopper mounting bolts (A)

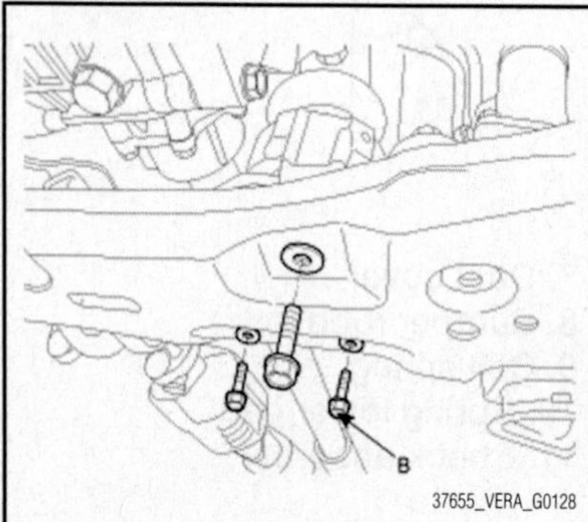

Fig. 239 Remove the rear roll stopper mounting bolts (B)

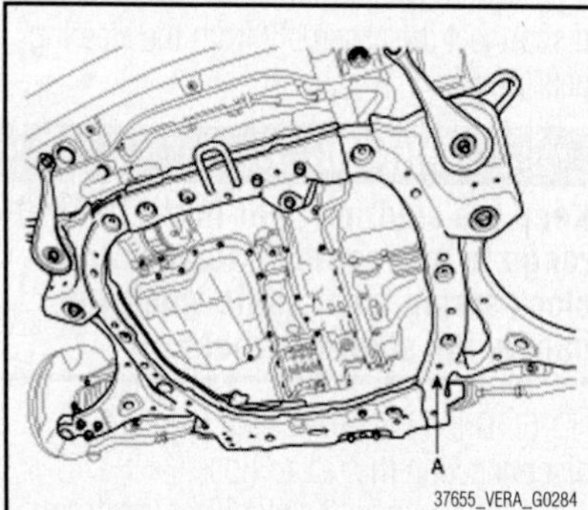

Fig. 240 Remove the sub-frame (A) and stay

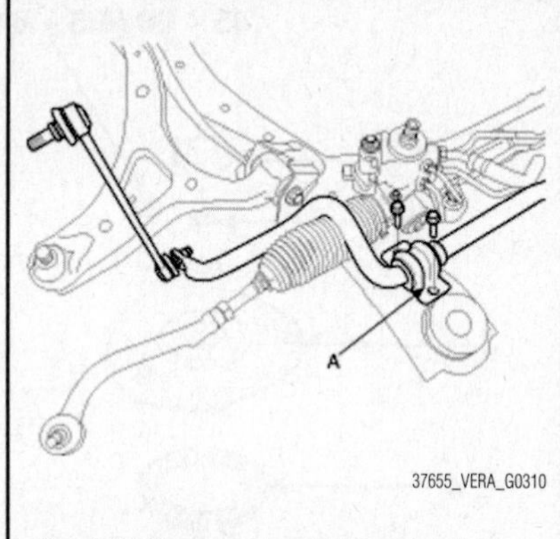

Fig. 241 Remove the stabilizer bar from the sub-frame by removing the bracket (A) mounting bolts

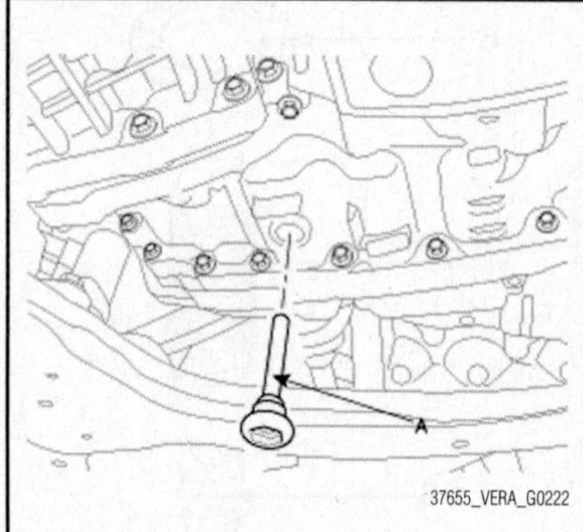

Fig. 242 Front stabilizer bar and components

To install:

15. Install the stabilizer to the sub-frame and tighten the bracket mounting bolts to 40–47 ft. lbs. (50–65 Nm).
16. Install the sub-frame and stay and tighten the mounting bolts and nuts:
    - Sub-frame mounting bolt: 101–116 ft. lbs. (140–160 Nm)
    - Sub-frame stay mounting bolt and nuts: 51–65 ft. lbs. (70–90 Nm)
17. Install the front and rear roll stopper mounting bolts and tighten to 36–47 ft. lbs. (50–65 Nm).
18. Install the muffler rubber hanger.
19. Connect the lower arm with the knuckle and tighten the bolt and nut to 72–87 ft. lbs. (100–120 Nm).
20. Connect the tie-rod end with the knuckle and install the castle nut and split pin and tighten to 17–25 ft. lbs. (24–34 Nm).
21. Connect the stabilizer link with the front strut assembly and tighten the nut to 72–87 ft. lbs. (100–120 Nm).
22. Connect the universal joint assembly with the steering column assembly and

then tighten the bolt to 22–25 ft. lbs. (30–35 Nm).

23. Connect the power steering return hose.

24. Connect the pressure tube to power steering pump and tighten the eye bolt to 40–47 ft. lbs. (55–65 Nm).

25. Install the front wheel and tire.

26. Add power steering fluid to reservoir.

27. Bleed the power steering system.

28. Check and adjust the front wheel alignment.

WHEEL HUB & BEARING

REMOVAL & INSTALLATION

See Figures 243 through 247.

1. Before servicing the vehicle, refer to the Precautions Section.

2. Remove the brake disc from the knuckle assembly.

3. Remove the hub bearing mounting bolts from the knuckle.

4. Remove the hub bearing and the dust cover from the knuckle.

➡Do not disassemble the hub bearing.

To install:

5. Check the hub for cracks and the splines for wear.

6. Check the brake disc for scoring and damage.

7. Check the knuckle for cracks.

8. Check the bearing for cracks or damage.

9. Install the dust cover to the knuckle and tighten the mounting bolt to 61–96 inch lbs. (7–11 Nm).

10. Install the hub bearing to the knuckle and then tighten the mounting bolt to 58–72 ft. lbs. (79–98 Nm).

11. Install the brake disc to the knuckle assembly and tighten the screws to 43–52 inch lbs. (5–6 Nm).

ADJUSTMENT

The wheel bearings are sealed units and are not adjustable.

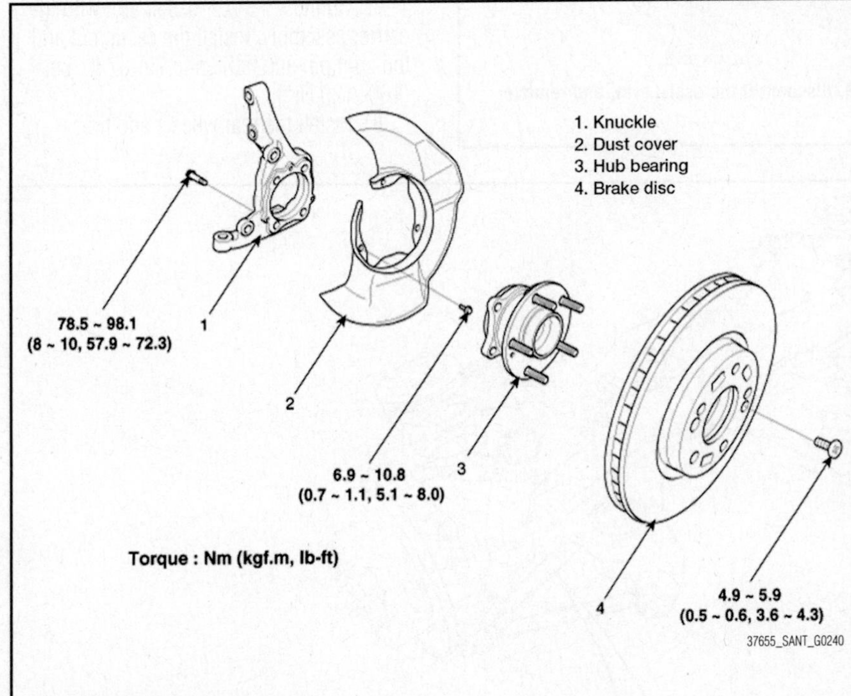

Fig. 243 Front hub components

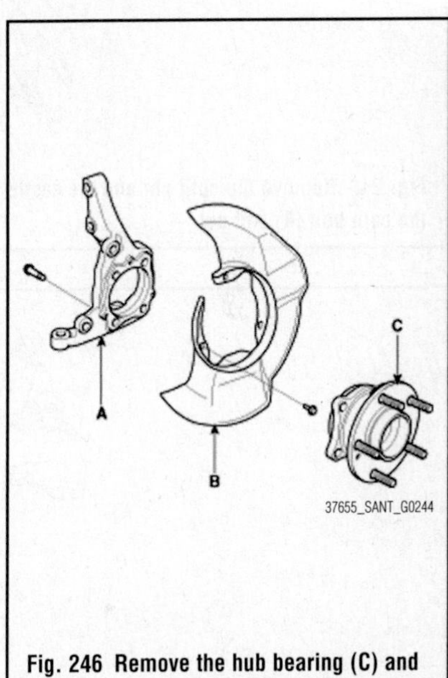

Fig. 246 Remove the hub bearing (C) and the dust cover (B) from the knuckle (A)

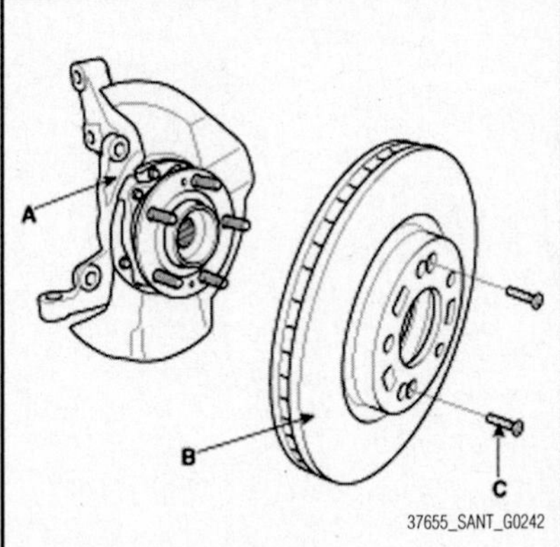

Fig. 244 Remove the brake disc (B) from the knuckle assembly (A)

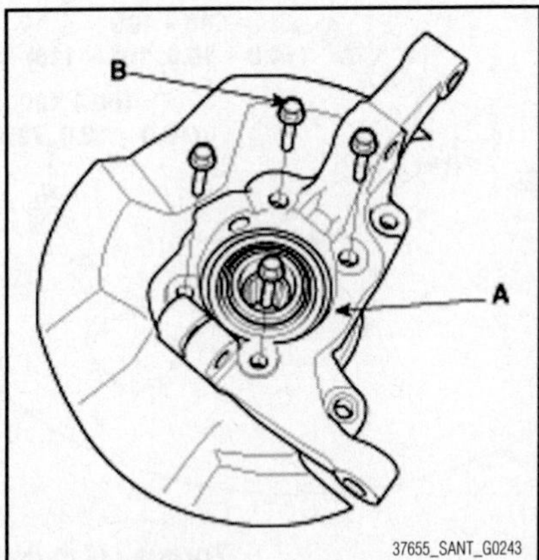

Fig. 245 Remove the hub bearing mounting bolts (B) from the knuckle (A)

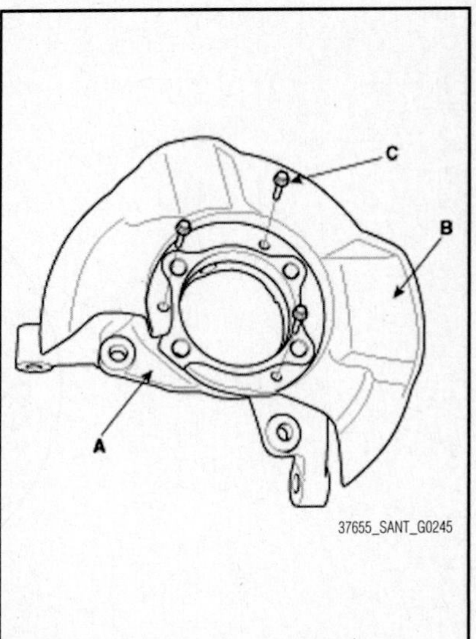

Fig. 247 Install the dust cover (B) to the knuckle (A) and tighten the mounting bolt (C)

ASSIST ARM

REMOVAL & INSTALLATION

See Figures 248 and 249.

1. Before servicing the vehicle, refer to the Precautions Section.
2. Raise and safely support the vehicle.
3. Remove the rear wheels and tires.

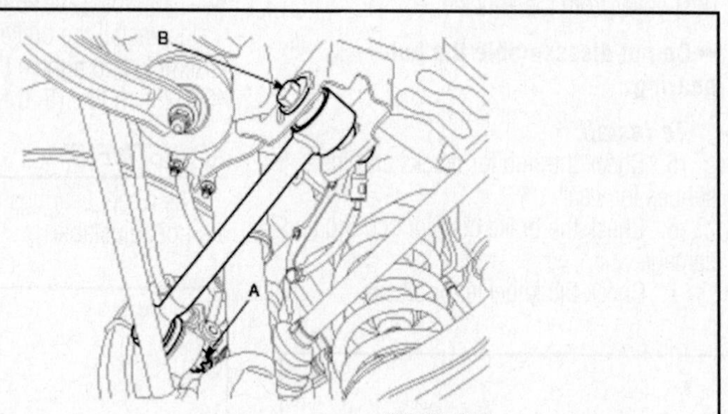

Fig. 248 Remove the split pin and the castle nut (A), disconnect the assist arm, and remove the cam bolt (B) and nut

⁂ WARNING

Be careful not to damage the hub bolts when removing the rear wheel and tire.

4. Remove the split pin and the castle nut and then disconnect the rear assist arm from the carrier assembly.
5. Remove the rear assist arm from the crossmember by removing the cam bolt and nut.

To install:

6. Connect the rear assist arm with the crossmember and tighten the cam bolt and nut to 101–116 ft. lbs. (140–160 Nm).
7. Connect the rear assist arm with the carrier assembly, install the castle nut and the split pin and tighten to 72–87 ft. lbs. (100–120 Nm).
8. Install the rear wheels and tires.

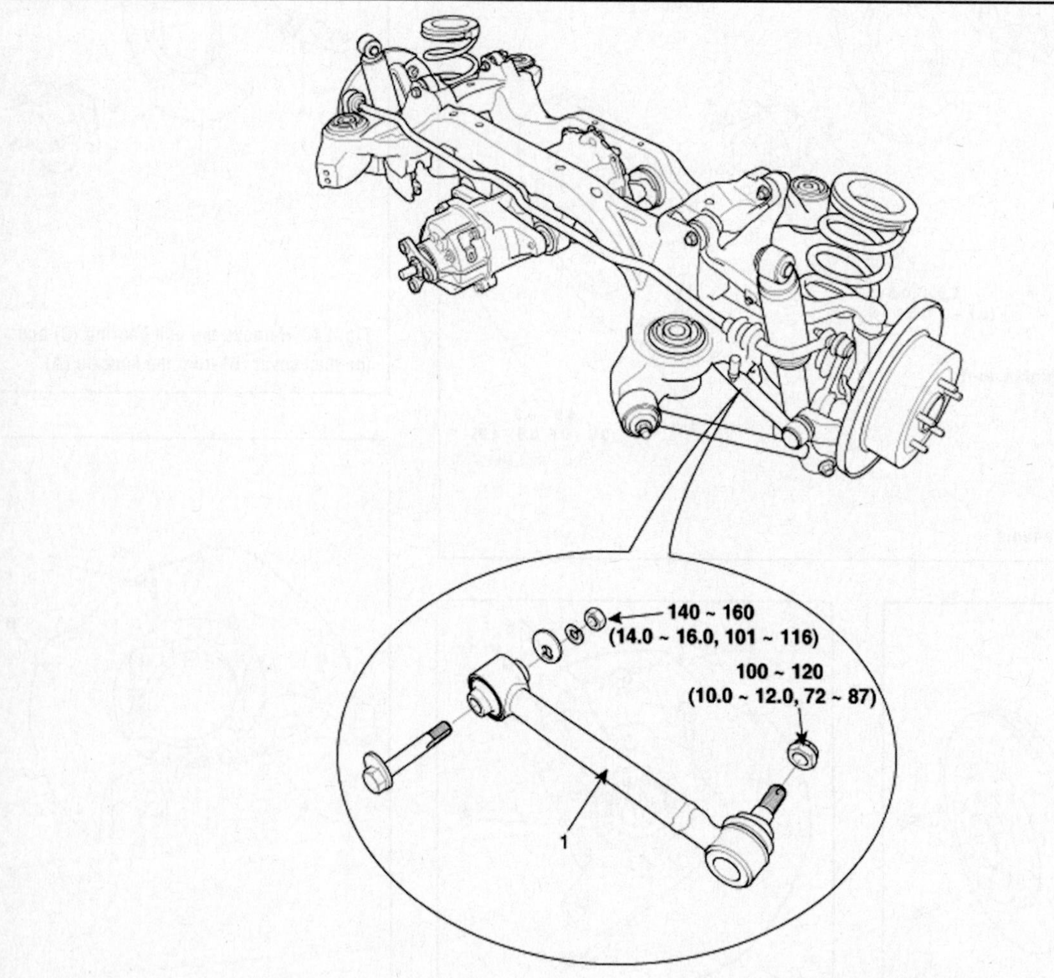

Fig. 249 Rear assist arm (1)

COIL SPRING

REMOVAL & INSTALLATION

Refer to Rear Lower Arm Removal & Installation.

CONTROL LINKS

REMOVAL & INSTALLATION

See Figures 250 and 251.

A
B
37655_VERA_G0321

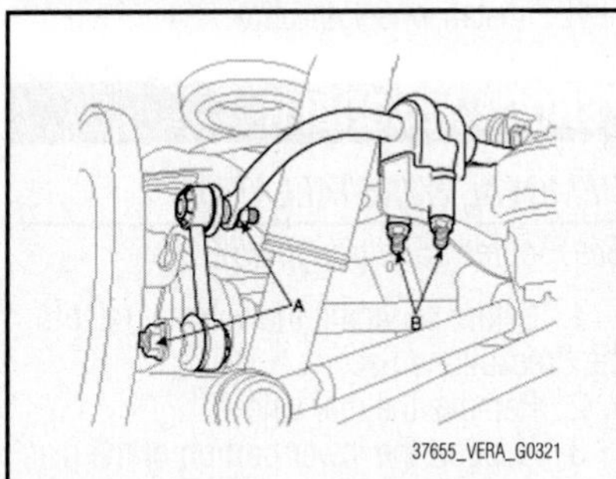

Fig. 250 Remove the nuts (A) and remove the rear stabilizer control links

1. Before servicing the vehicle, refer to the Precautions Section.
2. Raise and safely support the vehicle.
3. Remove the rear wheels and tires.
4. Remove the nuts and remove the rear stabilizer control links.

To install:

5. Install the rear stabilizer control links and tighten the nuts to 43–58 ft. lbs. (60–80 Nm).
6. Install the wheels.

LOWER ARM

REMOVAL & INSTALLATION

See Figures 252 through 254.

1. Before servicing the vehicle, refer to the Precautions Section.
2. Raise and safely support the vehicle.
3. Remove the rear wheels and tires.
4. Support the lower portion of the rear lower arm with the jack.
5. Temporarily loosen the nut holding the crossmember to the rear lower arm.

B
A
37655_VERA_G0315

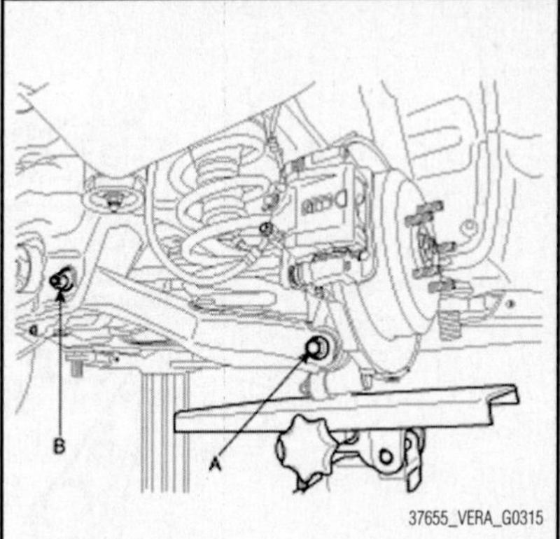

Fig. 252 Temporarily loosen the nut (B) holding the crossmember to the rear lower arm, and remove the bolt and nut (A) holding the rear lower arm to the carrier assembly

6. Remove the bolt and nut holding the rear lower arm to the carrier assembly.
7. Lower the jack and then remove the coil spring and the spring pad.

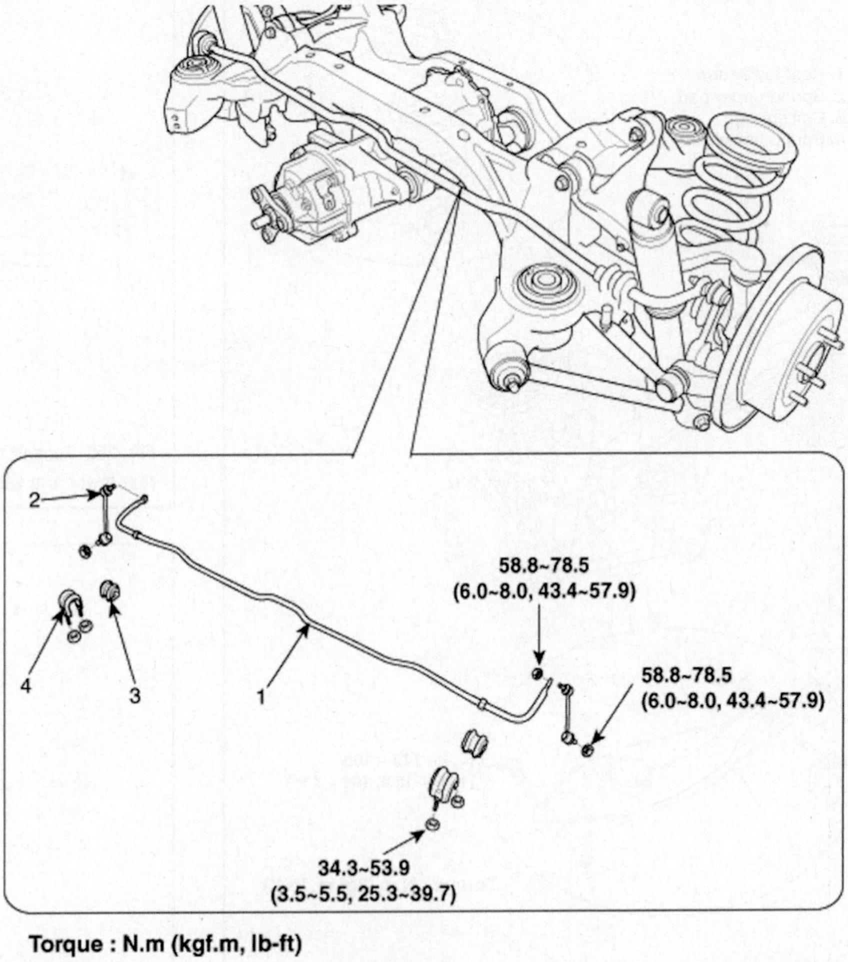

2
4
3
1
58.8~78.5
(6.0~8.0, 43.4~57.9)
58.8~78.5
(6.0~8.0, 43.4~57.9)
34.3~53.9
(3.5~5.5, 25.3~39.7)
Torque : N.m (kgf.m, lb-ft)
37655_SANT_G0596

Fig. 251 Rear stabilizer bar components

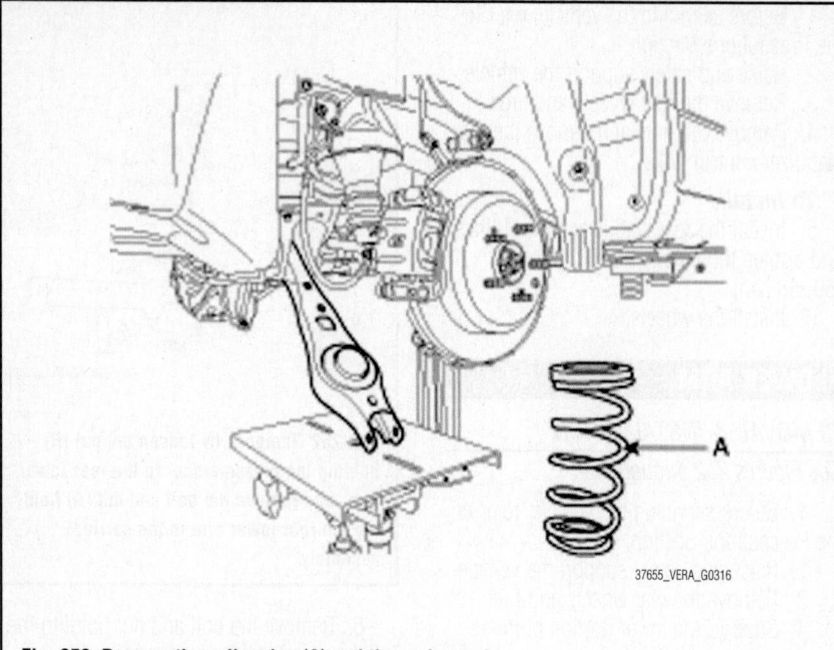

Fig. 253 Remove the coil spring (A) and the spring pad

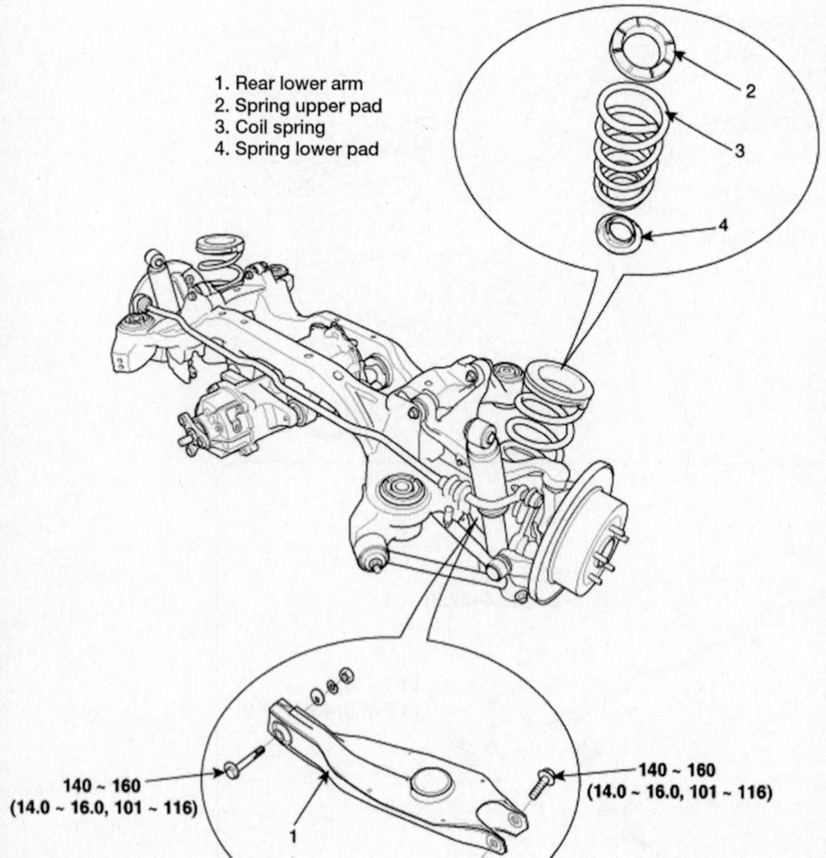

1. Rear lower arm
2. Spring upper pad
3. Coil spring
4. Spring lower pad

140 ~ 160
(14.0 ~ 16.0, 101 ~ 116)

140 ~ 160
(14.0 ~ 16.0, 101 ~ 116)

Torque : N.m (kgf.m, lb-ft)

37655_VERA_G0317

Fig. 254 Rear lower arm components

8. Remove the rear lower arm from the crossmember by removing the cam bolt.

To install:

9. Connect the rear lower arm with the cross member by installing the cam bolt.
10. Install the coil spring and pad and support the lower portion of the rear lower arm with a jack.
11. Connect the rear lower arm with the carrier assembly and tighten the bolt and nut to 101–116 ft. lbs. (140–160 Nm).
12. Install wheel and tire.

SHOCK ABSORBER

REMOVAL & INSTALLATION

See Figures 255 through 257.

1. Before servicing the vehicle, refer to the Precautions Section.
2. Remove the rear wheel.
3. Support the lower portion of the rear lower arm with a jack.

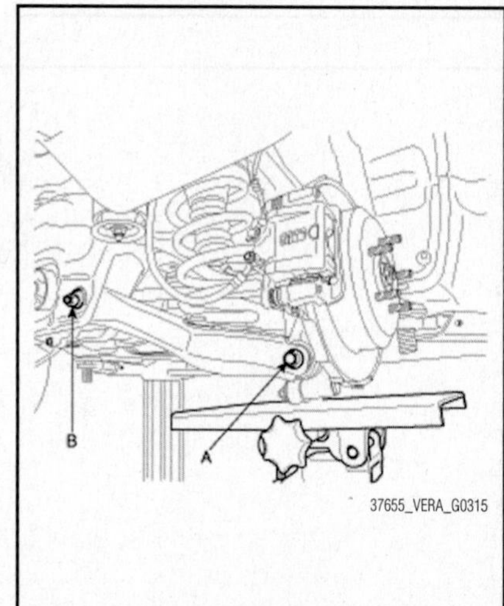

Fig. 255 Support the lower portion of the rear lower arm with the jack

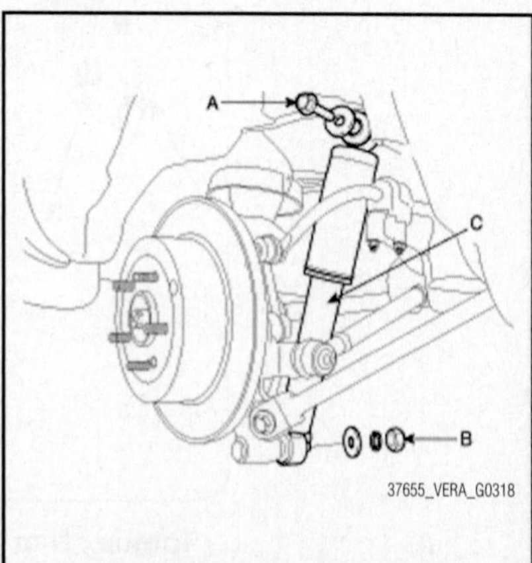

Fig. 256 Rear shock absorber (C), bolt (A), and nut (B)

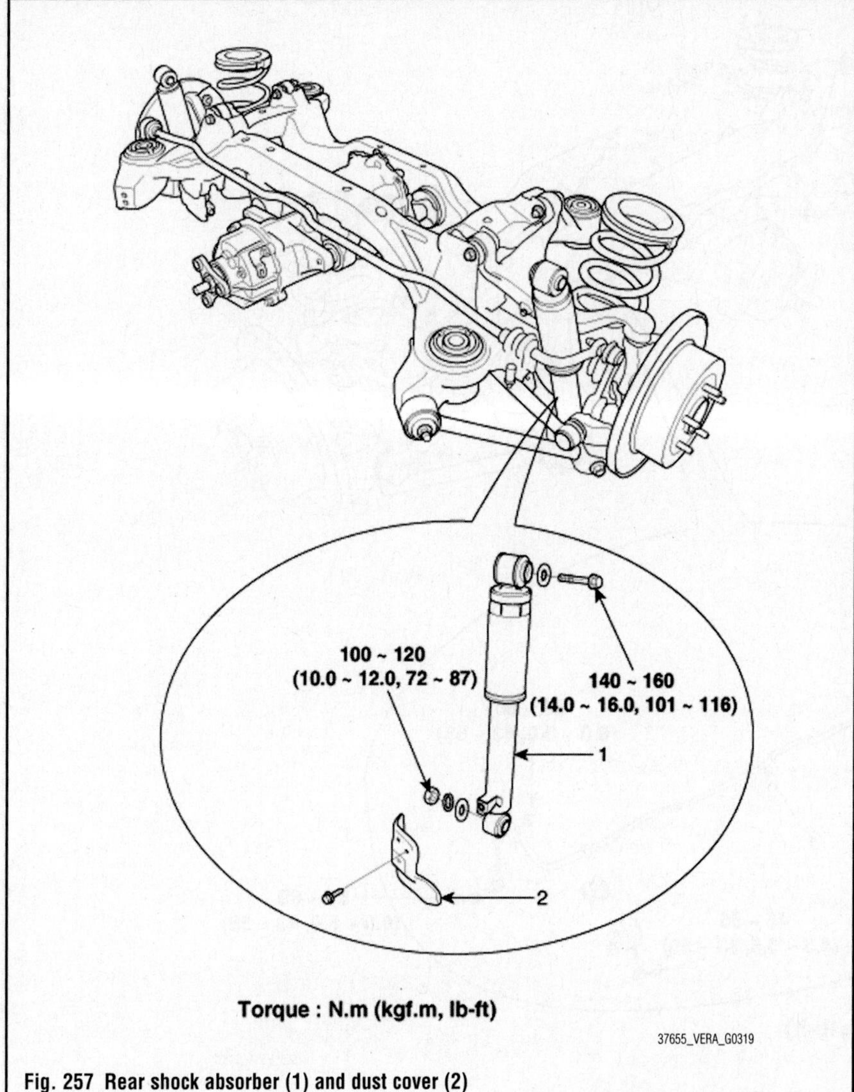

Fig. 257 Rear shock absorber (1) and dust cover (2)

4. Remove the shock absorber bolt and nut.
5. Remove the shock absorber.

To install:

6. Install the shock absorber. Tighten the upper and lower nuts to 101–116 ft. lbs. (137–157 Nm).
7. Install the rear wheel.

STABILIZER BAR

REMOVAL & INSTALLATION

See Figures 258 through 263.

1. Before servicing the vehicle, refer to the Precautions Section.
2. Remove the rear wheel.
3. Remove the wheel speed sensor and the parking brake cable from the rear axle carrier.
4. Remove the brake caliper mounting bolts, and then support the brake caliper assembly with wire.
5. Support the lower portion of the rear lower arm with a jack.
6. Remove the rear shock absorber.
7. Remove the center and main muffler.
8. Remove the propeller shaft, as applicable. Refer to Propeller Shaft Removal & Installation in Drive Train.

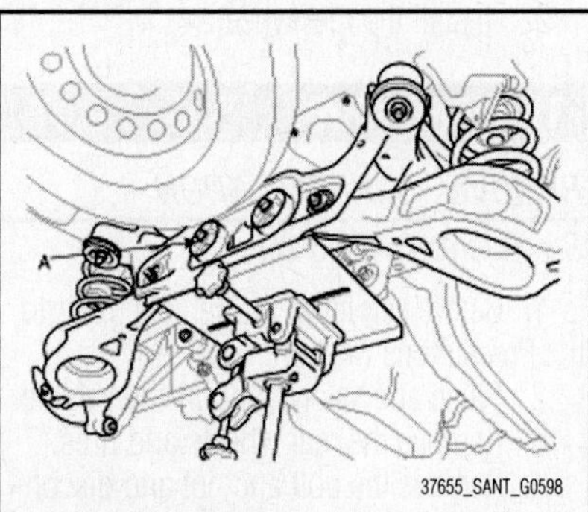

Fig. 258 Support the rear crossmember assembly with the jack

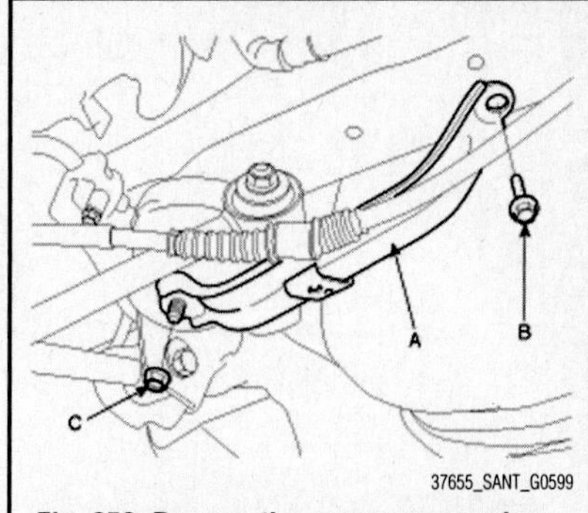

Fig. 259 Remove the rear cross member stay (A) and mounting bolt (B) and nut (C)

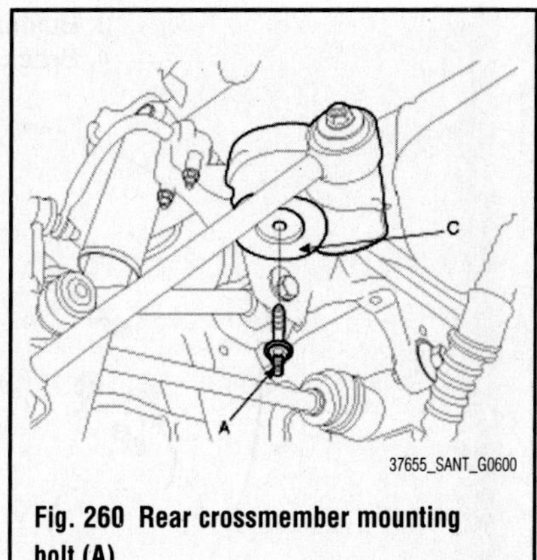

Fig. 260 Rear crossmember mounting bolt (A)

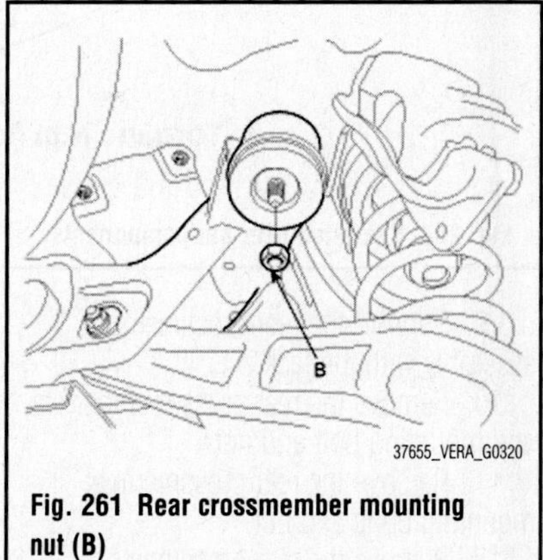

Fig. 261 Rear crossmember mounting nut (B)

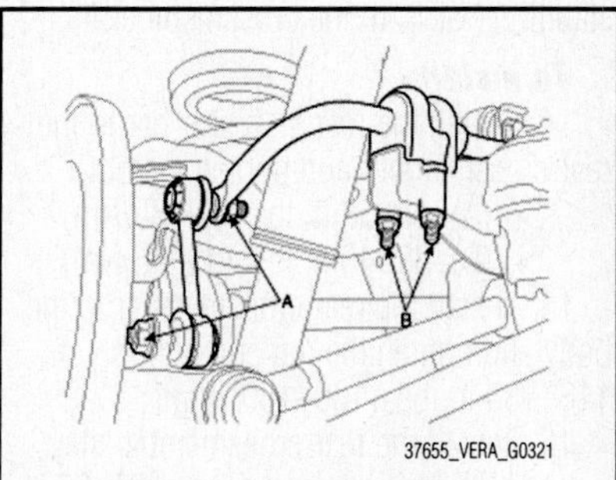

Fig. 262 Remove nuts (A, B) and remove the rear stabilizer bar from the crossmember

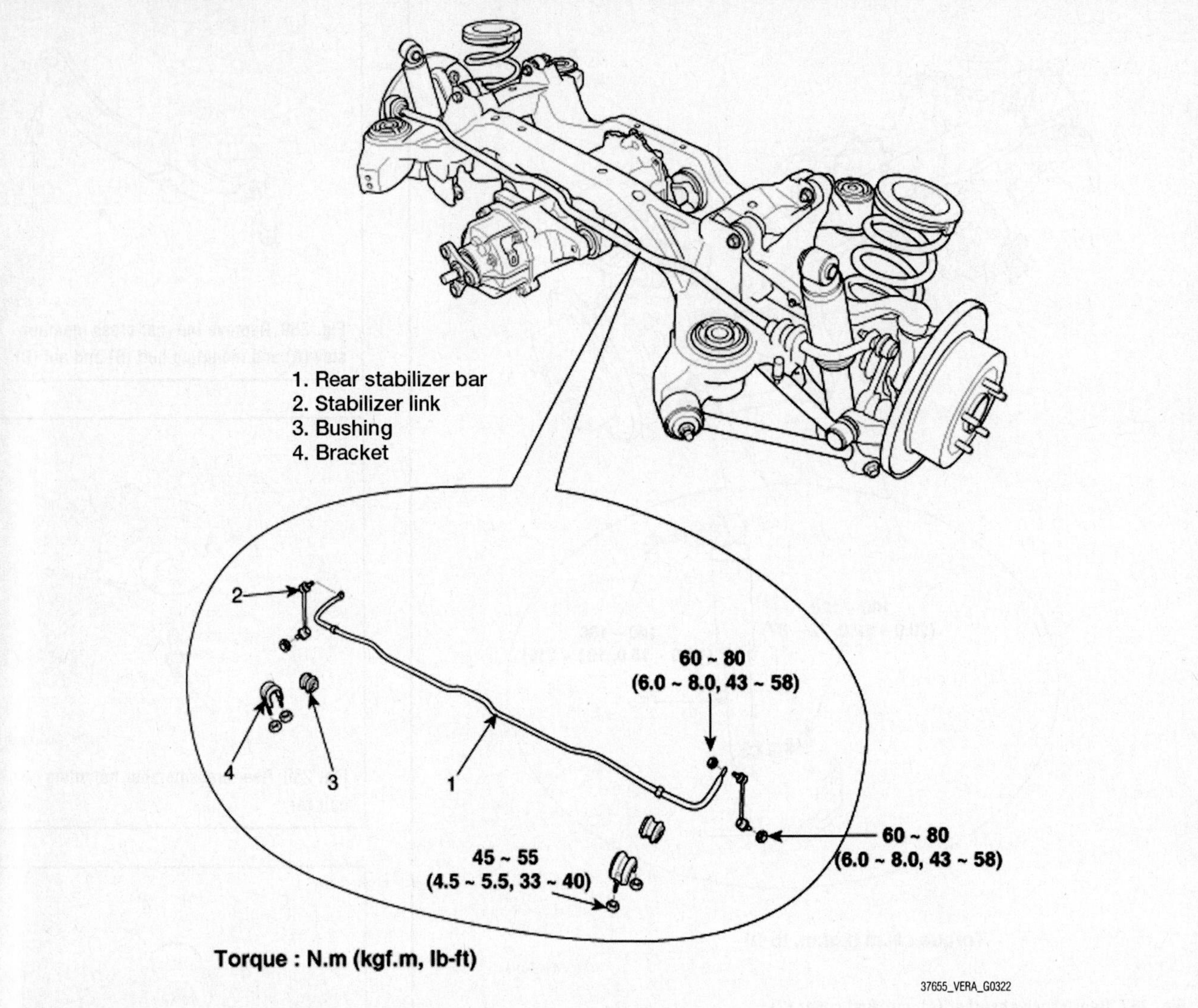

1. Rear stabilizer bar
2. Stabilizer link
3. Bushing
4. Bracket

60 ~ 80 (6.0 ~ 8.0, 43 ~ 58)

60 ~ 80 (6.0 ~ 8.0, 43 ~ 58)

45 ~ 55 (4.5 ~ 5.5, 33 ~ 40)

Torque : N.m (kgf.m, lb-ft)

37655_VERA_G0322

Fig. 263 Rear stabilizer bar components

9. Support the rear crossmember assembly with the jack.
10. Remove the rear cross member stay and mounting bolt and nut.
11. Remove the rear crossmember mounting bolts and nut.
12. Remove the rear crossmember.
13. Remove nuts and remove the rear stabilizer bar from the crossmember.

To install:

14. Install the rear stabilizer bar to the rear crossmember and tighten the nuts:
    - (A): 43–58 ft. lbs. (60–80 Nm)
    - (B): 33–40 ft. lbs. (45–55 Nm)
15. Install the rear crossmember to the body and tighten the nut and bolt to 116–130 ft. lbs. (160–180 Nm).
16. Install the rear crossmember stay and tighten the bolts and nuts to 51–65 ft. lbs. (70–90 Nm).
17. Install the propeller shaft.
18. Install the center and main muffler.
19. Install the shock absorber and tighten the bolt to 101–116 ft. lbs. (140–160 Nm).
20. Install the brake caliper, and tighten the brake caliper mounting bolts to 36–43 ft. lbs. (50–60 Nm).
21. Install the wheel speed sensor and the parking brake cable to the rear axle carrier.
22. Install the rear wheel.

TRAILING ARM

REMOVAL & INSTALLATION

See Figures 264 and 265.

1. Before servicing the vehicle, refer to the Precautions Section.
2. Raise and safely support the vehicle.
3. Remove the rear wheels and tires.
4. Remove the bolt and nut and disconnect the trailing arm from the carrier assembly.
5. Remove the nut and disconnect the trailing arm from the crossmember.

To install:

6. Install the trailing arm to the crossmember and tighten the nut to 101–116 ft. lbs. (140–160 Nm).

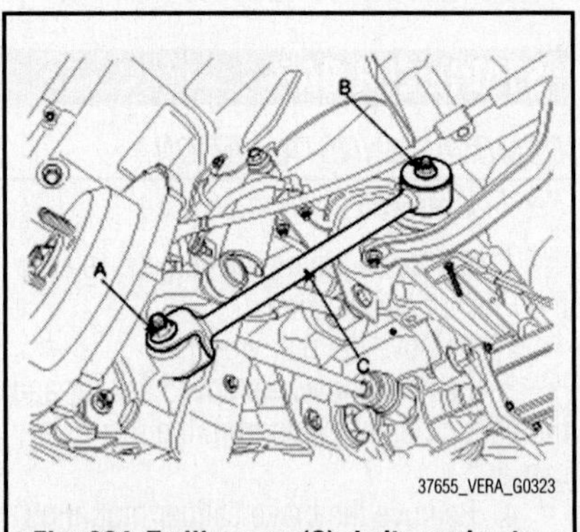

Fig. 264 Trailing arm (C), bolts, and nuts (A, B)

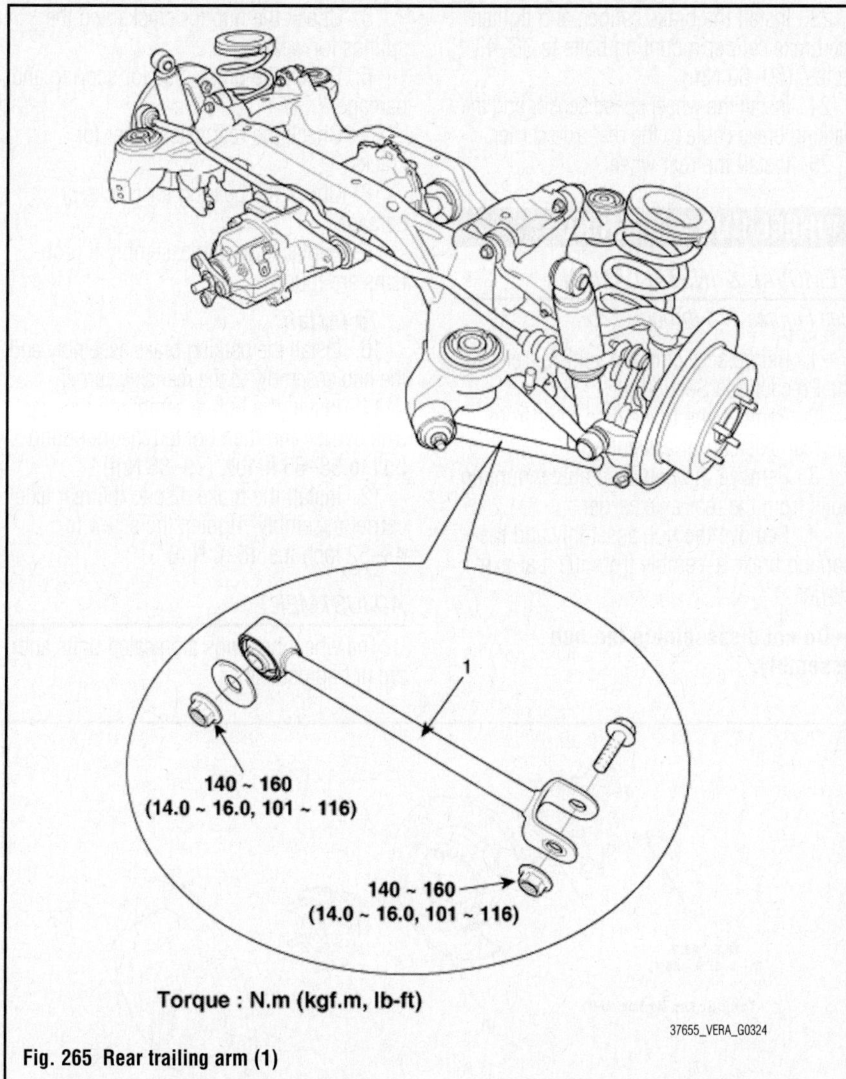

Fig. 265 Rear trailing arm (1)

7. Install the trailing arm to the rear carrier and tighten bolt and nut to 101–116 ft. lbs. (140–160 Nm).
8. Install the rear wheels and tires.

UPPER ARM

REMOVAL & INSTALLATION

See Figures 266 through 272.

➡Special Service Tool used for this procedure: 09568-34000

1. Before servicing the vehicle, refer to the Precautions Section.
2. Remove the rear wheel.
3. Remove the wheel speed sensor and the parking brake cable from the rear axle carrier.
4. Remove the brake caliper mounting bolts, and then support the brake caliper assembly with wire.
5. Support the lower portion of the rear lower arm with a jack.
6. Remove the rear shock absorber.
7. Remove the center and main muffler.
8. Remove the propeller shaft, as applicable. Refer to Propeller Shaft Removal & Installation in Drive Train.
9. Support the rear crossmember assembly with the jack.
10. Remove the rear cross member stay and mounting bolt and nut.
11. Remove the rear crossmember mounting bolts and nut.
12. Remove the rear crossmember.
13. Remove the split pin and the castle nut from the rear upper arm ball joint.
14. Remove the rear upper arm ball joint from the rear carrier assembly using SST (09568-34000).
15. Remove the mounting bolts and nuts and remove the rear upper arm.

To install:

16. Install the rear upper arm to the crossmember and tighten the bolts and nuts to 72–87 ft. lbs. (100–120 Nm).
17. Connect the rear upper arm ball joint with the carrier assembly and install the castle nut and split pin. Tighten to 58–72 ft. lbs. (80–100 Nm).
18. Install the rear crossmember to the body and tighten the nut and bolt to 116–130 ft. lbs. (160–180 Nm).
19. Install the rear crossmember stay and tighten the bolts and nuts to 51–65 ft. lbs. (70–90 Nm).
20. Install the propeller shaft.
21. Install the center and main muffler.
22. Install the shock absorber and tighten the bolt to 101–116 ft. lbs. (140–160 Nm).

Fig. 266 Support the rear crossmember assembly with the jack

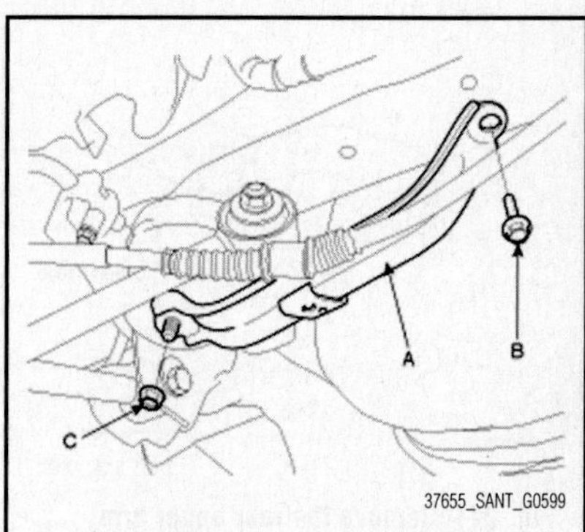

Fig. 267 Remove the rear cross member stay (A) and mounting bolt (B) and nut (C)

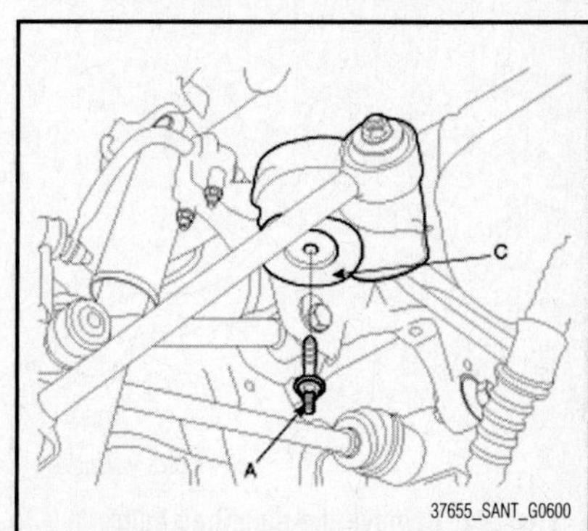

Fig. 268 Rear crossmember mounting bolt (A)

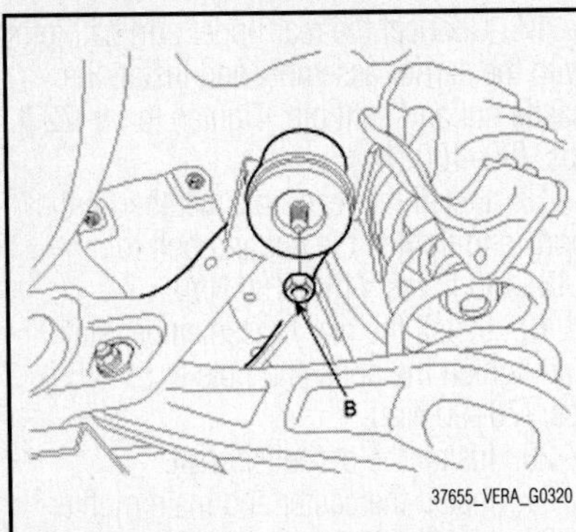

Fig. 269 Rear crossmember mounting nut (B)

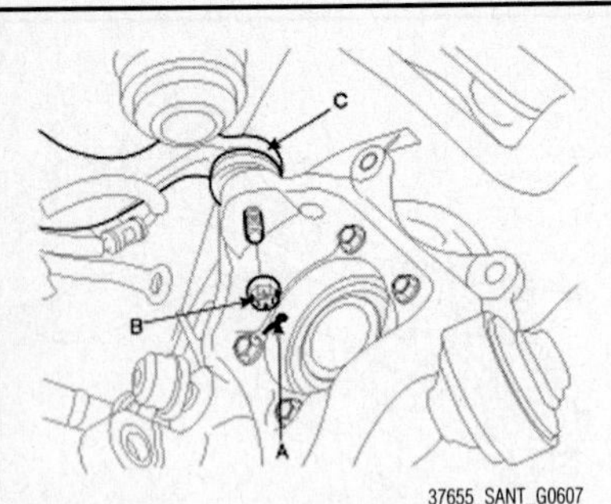

Fig. 270 Remove the split pin (A) and the castle nut (B) from the rear upper arm ball joint (C),

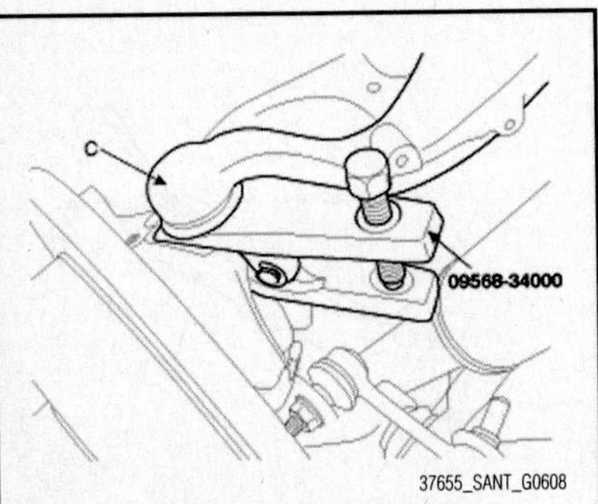

Fig. 271 Remove the rear upper arm ball joint (C) by using the special tool (09568-34000)

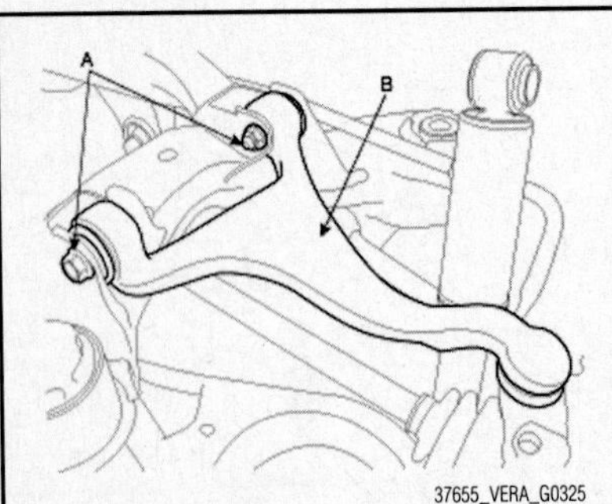

Fig. 272 Remove the mounting bolts and nuts (A) and remove the rear upper arm (B)

23. Install the brake caliper, and tighten the brake caliper mounting bolts to 36–43 ft. lbs. (50–60 Nm).

24. Install the wheel speed sensor and the parking brake cable to the rear axle carrier.

25. Install the rear wheel.

WHEEL HUB & BEARING

REMOVAL & INSTALLATION

See Figures 273 through 275.

1. Before servicing the vehicle, refer to the Precautions Section.
2. Remove the brake disc rotor from the rear axle carrier assembly.
3. Remove the hub assembly mounting bolts from the rear axle carrier.
4. Remove the hub assembly and the parking brake assembly from the rear axle carrier.

➡Do not disassemble the hub assembly.

5. Check the hub for cracks and the splines for wear.
6. Check the brake disc for scoring and damage.
7. Check the rear axle carrier for cracks
8. Check the bearing for cracks or damage.
9. Replace the hub assembly if problems are found.

To install:

10. Install the parking brake assembly and the hub assembly to the rear axle carrier.
11. Install the hub assembly to the rear axle carrier and then tighten the mounting bolt to 58–65 ft. lbs. (79–88 Nm).
12. Install the brake disc to the rear axle carrier assembly. Tighten the screw to 43–52 inch lbs. (5–6 Nm).

ADJUSTMENT

The wheel bearings are sealed units and are not adjustable.

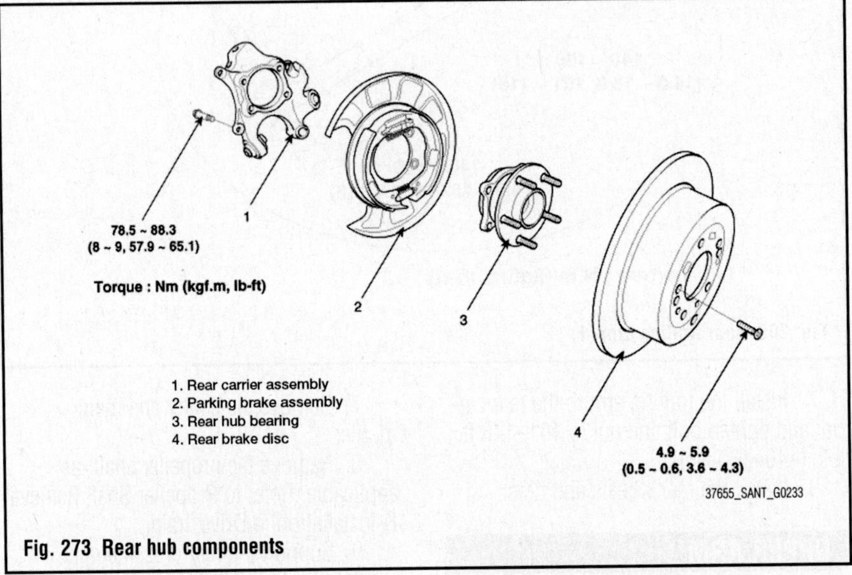

Fig. 273 Rear hub components

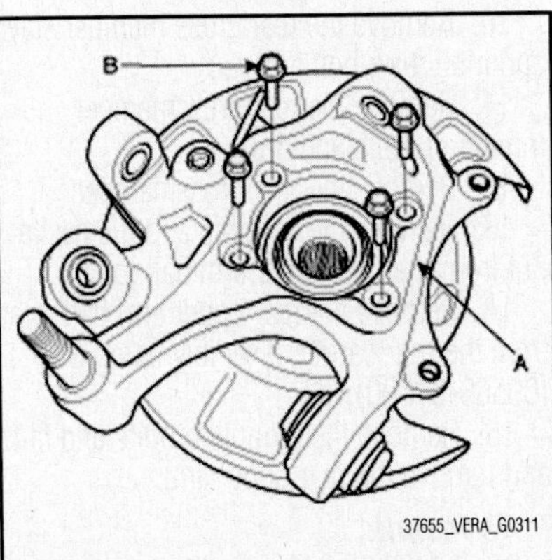

Fig. 274 Remove the hub assembly mounting bolts (B) from the rear axle carrier (A)

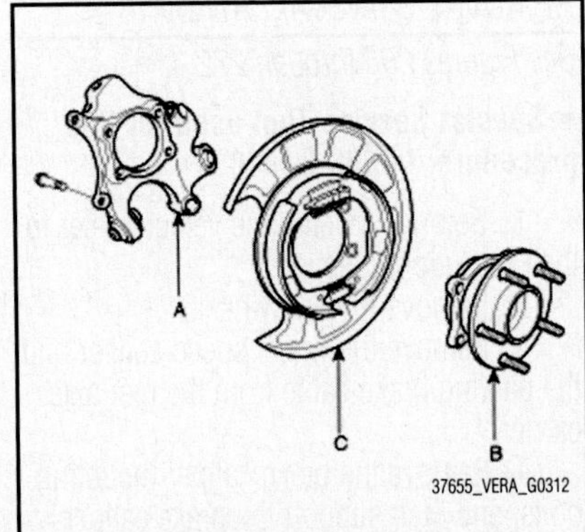

Fig. 275 Remove the hub assembly (B) and the parking brake assembly (C) from the rear axle carrier (A)

HYUNDAI

Diagnostic Trouble Codes

10

DIAGNOSTIC TROUBLE CODES

OBD II VEHICLE APPLICATIONS

HYUNDAI

Accent
2009–2010
- 1.6L I4 MPFI (DOHC) .... VIN C

Azera
2009–2010
- 3.3L V6 MPFI (DOHC) .... VIN D
- 3.8L V6 MPFI (DOHC) .... VIN F

Elantra
2009–2010
- 2.0L I4 MPFI (DOHC) .... VIN D/E

Genesis Coupe
2010
- 2.0L I4 MPFI (DOHC) .... VIN D
- 3.8L V6 MPFI (DOHC) .... VIN H

Genesis Sedan
2009–2010
- 3.8L V6 MPFI (DOHC) .... VIN E
- 4.6L V8 MPFI (DOHC) .... VIN F

Santa Fe
2009–2010
- 2.7L V6 MPFI (DOHC) .... VIN D
- 3.3L V6 MPFI (DOHC) .... VIN E

Sonata
2009–2010
- 2.4L I4 MPFI (DOHC) .... VIN C
- 3.3L V6 MPFI (DOHC) .... VIN F

Tucson
2009–2010
- 2.0L I4 MPFI (DOHC) .... VIN B
- 2.4L I4 MPFI (DOHC) .... VIN C
- 2.7L V6 MPFI (DOHC) .... VIN D

Veracruz
2009–2010
- 3.8L V6 MPFI (DOHC) .... VIN C

HYUNDAI REFERENCE INFORMATION

OBD II TROUBLE CODE LIST

To use this information, first read and record All codes in memory along with Freeze Frame data. *If a PCM Reset function is done prior to recording this data,* All *codes and freeze frame data are lost!*

Look up the appropriate trouble code in the list on the following pages. The left hand column includes the code number, the number of trips to set the code (e.g., **1T or 2T**), the year, model description, and type of OBD II Monitor that failed (e.g., **CCM or O2S**). This data can be used to determine how to drive a vehicle after a repair in order to validate the repair has been completed.

The **(N/MIL)** designator in the left hand column indicates the trouble code does not turn on the Malfunction Indicator Lamp or MIL. The **(STS Lamp)** indicator in the left column indicates a code that turns on the Service Transmission Soon lamp. This code may or may not turn "on" the MIL.

OBD II Trouble Code List (P0XXX Codes)

DTC	Trouble Code Title, Conditions & Possible Causes
DTC: P0011 **2T PCM, MIL: Yes** **Year:** 2009, 2010 **Model:** Accent, Azera, Elantra, Genesis, Genesis Coupe, Santa Fe, Sonata, Tucson, Veracruz **Engine:** 1.6L L4 VIN C, 2.0L L4 VIN B, 2.0L L4 VIN D, 2.0L L4 VIN E, 2.4L L4 VIN B, 2.4L L4 VIN C, 2.7L V6 VIN D, 3.3L V6 VIN D, 3.3L V6 VIN E, 3.3L V6 VIN F, 3.5L V6 VIN G, 3.8L V6 VIN C, 3.8L V6 VIN E, 3.8L V6 VIN F, 3.8L V6 VIN H, 4.6L V8 VIN F **Transmission:** All	**'A' Camshaft Position - Timing Over-Advanced or System Performance (Bank 1):** The CVVT (Continuously Variable Valve Timing) system is installed to the chain sprocket of the intake camshaft. This system controls the intake camshaft to provide the optimal valve timing. The PCM controls the Oil Control Valve(OCV), based on the signals output from mass air flow, throttle position and engine coolant temperature. The CVVT controller regulates the intake camshaft angle using oil pressure through the OCV. PCM detects CAM phasing average rate while cam signal is normally generating. PCM determines that a fault exists and a DTC is stored while vehicle is tip - in and out driving for 5 minutes. **Possible Causes:** • Oil level and condition • OCV • CVVT
DTC: P0012 **2T PCM, MIL: Yes** **Year:** 2009, 2010 **Model:** Accent, Azera, Genesis, Genesis Coupe, Santa Fe, Sonata, Veracruz **Engine:** 1.6L L4 VIN C, 2.7L V6 VIN D, 3.3L V6 VIN D, 3.3L V6 VIN E, 3.3L V6 VIN F, 3.8L V6 VIN C, 3.8L V6 VIN E, 3.8L V6 VIN F, 3.8L V6 VIN H **Transmission:** All	**'A' Camshaft Position - Timing Over-Retarded (Bank 1):** The CVVT (Continuously Variable Valve Timing) system is installed to the chain sprocket of the intake camshaft. This system controls the intake camshaft to provide the optimal valve timing. The PCM controls the Oil Control Valve(OCV), based on the signals output from mass air flow, throttle position and engine coolant temperature. The CVVT controller regulates the intake camshaft angle using oil pressure through the OCV. PCM monitors CAM phaser error while CMP signal is normally generating and vehicle is driving in 2000 ~ 3000rpm .If the CAM phaser does not move although PCM commands OCV duty cycle PCM determines that a fault exists and a DTC is stored. **Possible Causes:** • Oil level and condition • OCV • CVVT
DTC: P0014 **2T PCM, MIL: Yes** **Year:** 2009, 2010 **Model:** Genesis, Genesis Coupe, Santa Fe, Sonata, Tucson **Engine:** 2.0L L4 VIN D, 2.4L L4 VIN B, 2.4L L4 VIN C, 3.5L V6 VIN G, 3.8L V6 VIN E, 3.8L V6 VIN H, 4.6L V8 VIN F **Transmission:** All	**"B" Camshaft Position -Timing Over-Advanced or System Performance (Bank 1):** The CVVT (Continuously Variable Valve Timing) system is installed to the chain sprocket of the intake camshaft. This system controls the intake camshaft to provide the optimal valve timing. The PCM controls the Oil Control Valve(OCV), based on the signals output from mass air flow, throttle position and engine coolant temperature. The CVVT controller regulates the intake camshaft angle using oil pressure through the OCV. PCM detects CAM phasing average rate while cam signal is normally generating. PCM determines that a fault exists and a DTC is stored while vehicle is tip - in and out driving for 5 minutes. **Possible Causes:** • Oil level and condition • OCV • CVVT
DTC: P0015 **2T PCM, MIL: Yes** **Year:** 2009, 2010 **Model:** Genesis, Genesis Coupe **Engine:** 3.8L V6 VIN E, 3.8L V6 VIN H **Transmission:** All	**"B" Camshaft Position -Timing Over-Retarded (Bank 1):** Enable Conditions • CAM signal is normally generating • Vehicle is on driving (2000 ~ 3000RPM) for 5 minutes Threshold Value Case 1 • 5 CAD < Cam Actual Position < 50 CAD • Duty Cycle > 90% or Duty Cycle < 10% Case 2 • Cam Position error > 15 CAD (Difference between Actual Position and Desire Position is more than 15°) • Timing Counter > 80 Diagnosis Time • Continuous (within 5min.) **Possible Causes:** • Engine Oil • OCV stuck • CVVT stuck

DTC	Trouble Code Title, Conditions & Possible Causes
DTC: P0016 **2T CCM, MIL: Yes** **Year:** 2009, 2010 **Model:** Genesis, Genesis Coupe **Engine:** 2.0L L4 VIN D, 3.8L V6 VIN E, 3.8L V6 VIN H, 4.6L V8 VIN F **Transmission:** All	**Crankshaft Position- Camshaft Position Correlation (Bank 1 Sensor "A"):** Monitor camshaft position in the full retard condition or during CVVT control. Camshaft switching out of 109 to 141 degrees in full retard position, 70 to 140 degrees CRK during CVVT control. **Possible Causes:** • Abnormal installation of camshaft • Abnormal installation of crankshaft • Abnormal installation of tone wheel
DTC: P0016 **2T PCM, MIL: Yes** **Year:** 2009, 2010 **Model:** Accent, Azera, Elantra, Santa Fe, Sonata, Tucson, Veracruz **Engine:** 1.6L L4 VIN C, 2.0L L4 VIN B, 2.0L L4 VIN D, 2.0L L4 VIN E, 2.4L L4 VIN B, 2.4L L4 VIN C, 2.7L V6 VIN D, 3.3L V6 VIN D, 3.3L V6 VIN E, 3.3L V6 VIN F, 3.5L V6 VIN G, 3.8L V6 VIN C, 3.8L V6 VIN F **Transmission:** All	**Crankshaft Position - Camshaft Position Correlation (Bank 1 Sensor A):** The CVVT (Continuously Variable Valve Timing) system is installed to the chain sprocket of the intake camshaft. This system controls the intake camshaft to provide the optimal valve timing. The PCM controls the Oil Control Valve(OCV), based on the signals output from mass air flow, throttle position and engine coolant temperature. The CVVT controller regulates the intake camshaft angle using oil pressure through the OCV. PCM monitors timing misalignment while no active faults are present and fully warmed up engine oil at idle. If the timing is misaligned PCM determines that a fault exists and a DTC is stored. **Possible Causes:** • Oil level and condition • OCV • CVVT • Timing mark alignment
DTC: P0017 **2T PCM, MIL: Yes** **Year:** 2009, 2010 **Model:** Genesis, Genesis Coupe, Santa Fe, Sonata, Tucson **Engine:** 2.0L L4 VIN D, 2.4L L4 VIN B, 2.4L L4 VIN C, 3.5L V6 VIN G, 3.8L V6 VIN E, 3.8L V6 VIN H, 4.6L V8 VIN F **Transmission:** All	**Crankshaft Position- Camshaft Position Correlation (Bank 1 Sensor B):** Deviation between CKPS and CMPS(EX-CMPS Bank 1) is bigger than the threshold value **Possible Causes:** • Poor connection • Contamination of Oil / Clog of Oil path • CKPS, CMPS • OCV • CVVT
DTC: P0018 **2T PCM, MIL: Yes** **Year:** 2009, 2010 **Model:** Azera, Genesis, Genesis Coupe, Santa Fe, Sonata, Veracruz **Engine:** 2.7L V6 VIN D, 3.3L V6 VIN D, 3.3L V6 VIN E, 3.3L V6 VIN F, 3.5L V6 VIN G, 3.8L V6 VIN C, 3.8L V6 VIN E, 3.8L V6 VIN F, 3.8L V6 VIN H, 4.6L V8 VIN F **Transmission:** All	**Crankshaft Position- Camshaft Position Correlation (Bank 2 Sensor A):** Deviation between CKPS and CMPS (IN CMPS Bank 2) is bigger than the threshold value **Possible Causes:** • Poor connection • Contamination of Oil / Clog of Oil path • CKPS, CMPS • OCV • CVVT
DTC: P0019 **2T PCM, MIL: Yes** **Year:** 2009, 2010 **Model:** Genesis, Genesis Coupe, Santa Fe **Engine:** 3.5L V6 VIN G, 3.8L V6 VIN E, 3.8L V6 VIN H, 4.6L V8 VIN F **Transmission:** All	**Crankshaft Position- Camshaft Position Correlation (Bank 2 Sensor B):** Deviation between CKPS and CMPS (EX CMPS Bank 2) is bigger than the threshold value **Possible Causes:** • Poor connection • Contamination of Oil / Clog of Oil path • CKPS, CMPS • OCV • CVVT
DTC: P0021 **2T PCM, MIL: Yes** **Year:** 2009, 2010 **Model:** Azera, Genesis, Genesis Coupe, Santa Fe, Sonata, Veracruz **Engine:** 2.7L V6 VIN D, 3.3L V6 VIN D, 3.3L V6 VIN E, 3.3L V6 VIN F, 3.5L V6 VIN G, 3.8L V6 VIN C, 3.8L V6 VIN E, 3.8L V6 VIN F, 3.8L V6 VIN H, 4.6L V8 VIN F **Transmission:** All	**'A' Camshaft Position- Timing Over-Advanced or System Performance (Bank 2):** delay between the target angle and the real angle over 10 times • Time after | target VVT angle –VVT angle | > 6 deg(2 sec) • Time after engine start > 10 ~ 60 sec • Coolant temperature : 0~120°C • Engine oil temperature : 0~130°C • 1000 < rpm < 6600rpm **Possible Causes:** • Poor connection • Contamination of Oil / Clog of Oil path • OCV • CVVT

DTC	Trouble Code Title, Conditions & Possible Causes
DTC: P0022 **2T CCM, MIL: Yes** **Year:** 2009, 2010 **Model:** Azera, Genesis, Genesis Coupe, Santa Fe, Sonata, Veracruz **Engine:** 2.7L V6 VIN D, 3.3L V6 VIN D, 3.3L V6 VIN E, 3.3L V6 VIN F, 3.8L V6 VIN C, 3.8L V6 VIN E, 3.8L V6 VIN F, 3.8L V6 VIN H **Transmission:** All	**"A" Camshaft Position- Timing Over Retarded (Bank 2):** Determines if the phaser is stuck or has a steady state error. Off sets available. Cam velocity below threshold at 15 CAD/s. **Possible Causes:** • Engine oil • OCV • CVVT stuck • Faulty PCM
DTC: P0024 **2T PCM, MIL: Yes** **Year:** 2009, 2010 **Model:** Genesis, Genesis Coupe, Santa Fe **Engine:** 3.5L V6 VIN G, 3.8L V6 VIN E, 3.8L V6 VIN H, 4.6L V8 VIN F **Transmission:** All	**'B' Camshaft Position- Timing Over-Advanced or System Performance (Bank 2):** Enable Conditions • CAM signal is normally generating • Accelerate and decelerate more than 10 times within 5 minutes – while driving Threshold value • Cam phasing is abnormally fast or slow **Possible Causes:** • Excessive phasing system leakage • Binding Oil pressure • (ex. Blockage in OCV filter) • Faulty OCV
DTC: P0025 **2T PCM, MIL: Yes** **Year:** 2009, 2010 **Model:** Genesis, Genesis Coupe **Engine:** 3.8L V6 VIN E, 3.8L V6 VIN H **Transmission:** All	**'B' Camshaft Position- Timing Over-Retarded (Bank 2):** CAM phaser does not move although ECM commands OCV duty cycle **Possible Causes:** • Engine Oil • OCV stuck • CVVT stuck
DTC: P0026 **2T PCM, MIL: Yes** **Year:** 2009, 2010 **Model:** Azera, Genesis, Genesis Coupe, Santa Fe, Sonata, Veracruz **Engine:** 2.7L V6 VIN D, 3.3L V6 VIN D, 3.3L V6 VIN E, 3.3L V6 VIN F, 3.8L V6 VIN C, 3.8L V6 VIN E, 3.8L V6 VIN F, 3.8L V6 VIN H **Transmission:** All	**Intake Valve Control Solenoid Circuit Range/Performance (Bank 1):** Determines if oil control valve is stuck. Valve cleaning not in progress. CAM Actual Position is too high or low and Difference between Cam Actual Position and Desired Position is higher than 20 degrees **Possible Causes:** • Oil pressure loss • OCV seizure • Faulty PCM
DTC: P0027 **2T PCM, MIL: Yes** **Year:** 2009, 2010 **Model:** Genesis, Genesis Coupe **Engine:** 3.8L V6 VIN E, 3.8L V6 VIN H **Transmission:** All	**Exhaust Valve Control Solenoid Circuit Range/Performance (Bank 1):** CAM Actual Position is too high or low and Difference between Cam Actual Position and Desired Position is higher than 20 degrees **Possible Causes:** • Oil Pressure Loss • OCV seizure
DTC: P0028 **2T CCM, MIL: Yes** **Year:** 2009, 2010 **Model:** Azera, Genesis, Genesis Coupe, Santa Fe, Sonata, Veracruz **Engine:** 2.7L V6 VIN D, 3.3L V6 VIN D, 3.3L V6 VIN E, 3.3L V6 VIN F, 3.8L V6 VIN C, 3.8L V6 VIN E, 3.8L V6 VIN F, 3.8L V6 VIN H **Transmission:** All	**Intake Valve Control Solenoid Circuit Range/Performance (Bank 2):** Determines if oil control valve is stuck. Valve cleaning not in progress. CAM Actual Position is too high or low and Difference between Cam Actual Position and Desired Position is higher than 20 degrees **Possible Causes:** • Oil pressure loss • OCV seizure

DTC	Trouble Code Title, Conditions & Possible Causes
DTC: P0029 **2T PCM, MIL: Yes** **Year:** 2009, 2010 **Model:** Genesis, Genesis Coupe **Engine:** 3.8L V6 VIN E, 3.8L V6 VIN H **Transmission:** All	**Exhaust Valve Control Solenoid Circuit Range/Performance (Bank 2):** Determines if oil control valve is stuck. Valve cleaning not in progress. CAM Actual Position is too high or low and Difference between Cam Actual Position and Desired Position is higher than 20 degrees **Possible Causes:** • Oil Pressure Loss • OCV seizure
DTC: P0030 **2T CCM, MIL: Yes** **Year:** 2009, 2010 **Model:** Accent, Azera, Elantra, Genesis, Genesis Coupe, Santa Fe, Sonata, Tucson, Veracruz **Engine:** 1.6L L4 VIN C, 2.0L L4 VIN B, 2.0L L4 VIN D, 2.0L L4 VIN E, 2.4L L4 VIN B, 2.4L L4 VIN C, 2.7L V6 VIN D, 3.3L V6 VIN D, 3.3L V6 VIN E, 3.3L V6 VIN F, 3.5L V6 VIN G, 3.8L V6 VIN C, 3.8L V6 VIN E, 3.8L V6 VIN F, 3.8L V6 VIN H, 4.6L V8 VIN F **Transmission:** All	**HO2S-11 Heater Circuit Malfunction:** Engine started, engine runtime over 3 minutes, and the PCM determined the resistance of the HO2S heater was more than a calculated amount. **Possible Causes:** • HO2S heater control circuit is open or shorted to ground • HO2S heater control circuit is shorted to system power (B+) • HO2S heater is damaged or has failed • PCM has failed
DTC: P0031 **2T CCM, MIL: Yes** **Year:** 2009, 2010 **Model:** Accent, Azera, Elantra, Genesis, Genesis Coupe, Santa Fe, Sonata, Tucson, Veracruz **Engine:** 1.6L L4 VIN C, 2.0L L4 VIN B, 2.0L L4 VIN D, 2.0L L4 VIN E, 2.4L L4 VIN B, 2.4L L4 VIN C, 2.7L V6 VIN D, 3.3L V6 VIN D, 3.3L V6 VIN E, 3.3L V6 VIN F, 3.5L V6 VIN G, 3.8L V6 VIN C, 3.8L V6 VIN E, 3.8L V6 VIN F, 3.8L V6 VIN H, 4.6L V8 VIN F **Transmission:** All	**O2 Sensor Heater Circuit Low (Bank 1/Sensor 1):** Heater check, low. Open or short circuit. **Possible Causes:** • Open in battery and control circuit • Short to ground in control circuit (pin 48 to 36) • Faulty HO2S heater • Faulty PCM
DTC: P0032 **2T CCM, MIL: Yes** **Year:** 2009, 2010 **Model:** Accent, Azera, Elantra, Genesis, Genesis Coupe, Santa Fe, Sonata, Tucson, Veracruz **Engine:** 1.6L L4 VIN C, 2.0L L4 VIN B, 2.0L L4 VIN D, 2.0L L4 VIN E, 2.4L L4 VIN B, 2.4L L4 VIN C, 2.7L V6 VIN D, 3.3L V6 VIN D, 3.3L V6 VIN E, 3.3L V6 VIN F, 3.5L V6 VIN G, 3.8L V6 VIN C, 3.8L V6 VIN E, 3.8L V6 VIN F, 3.8L V6 VIN H, 4.6L V8 VIN F **Transmission:** All	**O2 Sensor Heater Circuit High (Bank 1/Sensor 1):** Heater check, high. Short circuit. **Possible Causes:** • Short to battery in control circuit • Faulty HO2S heater • Faulty PCM

DTC	Trouble Code Title, Conditions & Possible Causes
DTC: P0034 **2T PCM, MIL: Yes** **Year:** 2010 **Model:** Genesis Coupe **Engine:** 2.0L L4 VIN D **Transmission:** All	**Turbocharger/Supercharger Bypass :** Enable Conditions • 10V< Battery voltage <16V Threshold Value • ECM power stage diagnosis Diagnostic Time • 1sec **Possible Causes:** • Short to ground in signal harness • Poor connection or damaged harness • Faulty RCV
DTC: P0035 **2T PCM, MIL: Yes** **Year:** 2010 **Model:** Genesis Coupe **Engine:** 2.0L L4 VIN D **Transmission:** All	**Turbocharger/Supercharger Bypass Valve Control Circuit High:** Enable Conditions • 10V< Battery voltage <16V Threshold Value • IG ON Diagnostic Time • 1sec **Possible Causes:** • Poor connection or damaged harness • short to battery in control harness • Faulty RCV
DTC: P0036 **2T CCM, MIL: Yes** **Year:** 2009, 2010 **Model:** Accent, Azera, Elantra, Genesis, Genesis Coupe, Santa Fe, Sonata, Tucson, Veracruz **Engine:** 1.6L L4 VIN C, 2.0L L4 VIN B, 2.0L L4 VIN D, 2.0L L4 VIN E, 2.4L L4 VIN B, 2.4L L4 VIN C, 2.7L V6 VIN D, 3.3L V6 VIN D, 3.3L V6 VIN E, 3.3L V6 VIN F, 3.5L V6 VIN G, 3.8L V6 VIN C, 3.8L V6 VIN E, 3.8L V6 VIN F, 3.8L V6 VIN H, 4.6L V8 VIN F **Transmission:** All	**O2 Sensor Heater Control Circuit Bank 1/Sensor 2):** Check heater current. Internal resistance above threshold temperature (exhaust temperature, heater power). **Possible Causes:** • Contaminated, deteriorated or aged sensor • Heater resistance out of range • Faulty HO2S heater • Faulty PCM • Misplaced, bent, loose or corroded terminals
DTC: P0037 **2T CCM, MIL: Yes** **Year:** 2009, 2010 **Model:** Accent, Azera, Elantra, Genesis, Genesis Coupe, Tucson, Veracruz **Engine:** 1.6L L4 VIN C, 2.0L L4 VIN B, 2.0L L4 VIN D, 2.0L L4 VIN E, 2.4L L4 VIN C, 2.7L V6 VIN D, 3.3L V6 VIN D, 3.8L V6 VIN C, 3.8L V6 VIN E, 3.8L V6 VIN F, 3.8L V6 VIN H, 4.6L V8 VIN F **Transmission:** All	**O2 Sensor Heater Circuit Low (Bank 1/Sensor 2):** Heater check, low. Open or short circuit. **Possible Causes:** • Open in battery and control circuit • Short to ground in control circuit • Faulty HO2S heater • Faulty PCM

DTC	Trouble Code Title, Conditions & Possible Causes
DTC: P0038 **2T CCM, MIL: Yes** **Year:** 2009, 2010 **Model:** Accent, Azera, Elantra, Genesis, Genesis Coupe, Santa Fe, Sonata, Tucson, Veracruz **Engine:** 1.6L L4 VIN C, 2.0L L4 VIN B, 2.0L L4 VIN D, 2.0L L4 VIN E, 2.4L L4 VIN B, 2.4L L4 VIN C, 2.7L V6 VIN D, 3.3L V6 VIN D, 3.3L V6 VIN E, 3.3L V6 VIN F, 3.5L V6 VIN G, 3.8L V6 VIN C, 3.8L V6 VIN E, 3.8L V6 VIN F, 3.8L V6 VIN H, 4.6L V8 VIN F **Transmission:** All	**O2 Sensor Heater Circuit High (Bank 1/Sensor 2):** Heater check, high. Short circuit. **Possible Causes:** • Short to battery in control circuit • Faulty HO2S heater • Faulty PCM
DTC: P0049 **2T PCM, MIL: Yes** **Year:** 2010 **Model:** Genesis Coupe **Engine:** 2.0L L4 VIN D **Transmission:** All	**Turbocharger/Supercharger Turbine Overspeed:** Enable Conditions • - Threshold Value • Turbocharger speed > 172000rpm Diagnostic Time • 5 sec **Possible Causes:** • Detected for following DTC check • Air leak in intake (Turbocharger to ETC) • Restriction in intake(Air Filter to Turbocharger) • WGT/RCV Control Air line check • Faulty RCV • Faulty PUT(Boost sensor) • Faulty Turbocharger
DTC: P0050 **2T CCM, MIL: Yes** **Year:** 2009, 2010 **Model:** Azera, Genesis, Genesis Coupe, Santa Fe, Sonata, Tucson, Veracruz **Engine:** 2.7L V6 VIN D, 3.3L V6 VIN D, 3.3L V6 VIN E, 3.3L V6 VIN F, 3.5L V6 VIN G, 3.8L V6 VIN C, 3.8L V6 VIN E, 3.8L V6 VIN F, 3.8L V6 VIN H, 4.6L V8 VIN F **Transmission:** All	**OHO2S Heater Control Circuit High (Bank 2/Sensor 1):** Evaluate O2 sensor element temperature via measuring element resistance. Sensor preheating and full heating phases finished. **Possible Causes:** • Related fuse blown or missing • Heater control circuit open or short • Power supply circuit open or short • Contact resistance in connectors • Faulty HO2S
DTC: P0051 **2T CCM, MIL: Yes** **Year:** 2009, 2010 **Model:** Azera, Genesis, Genesis Coupe, Santa Fe, Sonata, Tucson, Veracruz **Engine:** 2.7L V6 VIN D, 3.3L V6 VIN D, 3.3L V6 VIN E, 3.3L V6 VIN F, 3.5L V6 VIN G, 3.8L V6 VIN C, 3.8L V6 VIN E, 3.8L V6 VIN F, 3.8L V6 VIN H, 4.6L V8 VIN F **Transmission:** All	**OHO2S Heater Circuit Low (Bank 2/Sensor 1):** Short circuit to ground on front HO2S heater line. Battery voltage above 10 volts. **Possible Causes:** • Related fuse blown or missing • Open or short to ground in power supply or control harness • Contact resistance in connectors • Faulty HO2S

DTC	Trouble Code Title, Conditions & Possible Causes
DTC: P0052 **2T CCM, MIL: Yes** **Year:** 2009, 2010 **Model:** Azera, Genesis, Genesis Coupe, Santa Fe, Sonata, Tucson, Veracruz **Engine:** 2.7L V6 VIN D, 3.3L V6 VIN D, 3.3L V6 VIN E, 3.3L V6 VIN F, 3.5L V6 VIN G, 3.8L V6 VIN C, 3.8L V6 VIN E, 3.8L V6 VIN F, 3.8L V6 VIN H, 4.6L V8 VIN F **Transmission:** All	**OHO2S Heater Circuit High (Bank 2/Sensor 1):** Open or short circuit to battery line on front HO2S heater line. Battery voltage above 10 volts. **Possible Causes:** • Open or short to battery in control harness • Contact resistance in connectors • Faulty HO2S
DTC: P0056 **2T CCM, MIL: Yes** **Year:** 2009, 2010 **Model:** Azera, Genesis, Genesis Coupe, Santa Fe, Sonata, Tucson, Veracruz **Engine:** 2.7L V6 VIN D, 3.3L V6 VIN D, 3.3L V6 VIN E, 3.3L V6 VIN F, 3.5L V6 VIN G, 3.8L V6 VIN C, 3.8L V6 VIN E, 3.8L V6 VIN F, 3.8L V6 VIN H, 4.6L V8 VIN F **Transmission:** All	**HO2S Heater Control Circuit (Bank 2/Sensor 2):** Evaluate O2 sensor element temperature via measuring element resistance. Sensor preheating and full heating phases finished. **Possible Causes:** • Related fuse blown or missing • Heater control circuit open or short • Power supply circuit open or short • Contact resistance in connections • Faulty HO2S
DTC: P0057 **2T CCM, MIL: Yes** **Year:** 2009, 2010 **Model:** Azera, Genesis, Genesis Coupe, Santa Fe, Sonata, Tucson, Veracruz **Engine:** 2.7L V6 VIN D, 3.3L V6 VIN D, 3.3L V6 VIN E, 3.3L V6 VIN F, 3.5L V6 VIN G, 3.8L V6 VIN C, 3.8L V6 VIN E, 3.8L V6 VIN F, 3.8L V6 VIN H, 4.6L V8 VIN F **Transmission:** All	**HO2S Heater Circuit Low (Bank 2/Sensor 2):** Check short circuit to ground on rear HO2S heater line. **Possible Causes:** • Related fuse blown or missing • Open or short to ground in power supply or control harness • Contact resistance in connections • Faulty HO2S
DTC: P0058 **2T CCM, MIL: Yes** **Year:** 2009, 2010 **Model:** Azera, Genesis, Genesis Coupe, Santa Fe, Sonata, Tucson, Veracruz **Engine:** 2.7L V6 VIN D, 3.3L V6 VIN D, 3.3L V6 VIN E, 3.3L V6 VIN F, 3.5L V6 VIN G, 3.8L V6 VIN C, 3.8L V6 VIN E, 3.8L V6 VIN F, 3.8L V6 VIN H, 4.6L V8 VIN F **Transmission:** All	**HO2S Heater Circuit High (Bank 2/Sensor 2):** Check short circuit to ground on rear HO2S heater line. **Possible Causes:** • Open or short to battery in control harness • Contact resistance in connections • Faulty HO2S
DTC: P0068 **2T PCM, MIL: Yes** **Year:** 2009, 2010 **Model:** Accent, Genesis, Genesis Coupe **Engine:** 1.6L L4 VIN C, 2.0L L4 VIN D, 4.6L V8 VIN F **Transmission:** All	**MAP(MAF)-Throttle Position Sensor Correlation:** Enable Conditions • Correction factor for secondary load > 1.2 or Correction factor for secondary load < 0.8 Threshold Value • Time for secondary load adaptation > 300s Diagnostic Time • 1sec **Possible Causes:** • Poor connection • TPS • MAFS • ECM/PCM

DTC	Trouble Code Title, Conditions & Possible Causes
DTC: P0071 **2T PCM, MIL: Yes** **Year:** 2010 **Model:** Genesis Coupe **Engine:** 2.0L L4 VIN D **Transmission:** All	**Ambient Air Temperature Circuit Range / Performance:** Enable Conditions Case1 • modeled Temp. AAT < 190.5 °C • change of IAT(between two measurements) <1.5 °C • change of AAT(between two measurements)<1.5 °C • 79.5 °C < ECT Temp. < 105.0 °C Case2 • 1500rpm < engine speed < 4000rpm • 35kg/h <mass air flow < 200kg/h • pressure quotient(PUT/AMP) <1.5 • 43.5mph < Vehicle speed < 81mph • Failure not detected for DTCs Case3 • Time after start < 180 sec • change of IAT (between two measurements) <1.5 °C • change of AAT (between two measurements) <5.25 °C • engine off time > 450 min • ECT temp.(at engine start) - ECT temp.(at engine stop) > 40 °C • ECT temp. < 34 °C • Failure not detected for DTCs Threshold Value Case1 • measured AAT - modeled AAT <=-15 °C (hot engine) Case2 • measured AAT - modeled AAT >= 15 °C (hot engine) Case3 • measured AAT - modeled AAT >= 12 °C (cold engine) **Possible Causes:** • Faulty AAT
DTC: P0072 **2T PCM, MIL: Yes** **Year:** 2010 **Model:** Genesis Coupe **Engine:** 2.0L L4 VIN D **Transmission:** All	**Ambient Air Temperature Circuit Low:** Enable Conditions • 10V < Battery voltage < 16V • Time after start > 0 sec • Vehicle speed >0km/h Threshold Value • Sensor voltage < 0.024V Diagnostic Time • 1sec **Possible Causes:** • Short to ground in signal harness • Poor connection or damaged harness • Faulty AAT
DTC: P0073 **2T PCM, MIL: Yes** **Year:** 2010 **Model:** Genesis Coupe **Engine:** 2.0L L4 VIN D **Transmission:** All	**Ambient Air Temperature Circuit High:** Enable Conditions • 10V < Battery voltage < 16V • Time after start > 0 sec • Vehicle speed >0km/h Threshold Value • Sensor voltage < 0.024V Diagnostic Time • 1sec **Possible Causes:** • Poor connection or damaged harness • short to battery in control harness • Open in ground circuit • Open in signal circuit • Faulty AAT

DTC	Trouble Code Title, Conditions & Possible Causes
DTC: P0075 **2T PCM, MIL: Yes** **Year:** 2009, 2010 **Model:** Accent, Genesis **Engine:** 1.6L L4 VIN C, 4.6L V8 VIN F **Transmission:** All	**Intake Valve Control Solenoid Circuit (Bank 1):** Enable Conditions • Threshold Value • Disconnected Diagnostic Time • Continuous **Possible Causes:** • Poor connection • Open or Short to ground in power circuit • Open in control circuit • OCV • ECM/PCM
DTC: P0076 **2T CCM, MIL: Yes** **Year:** 2009, 2010 **Model:** Accent, Azera, Elantra, Genesis, Genesis Coupe, Santa Fe, Sonata, Tucson, Veracruz **Engine:** 1.6L L4 VIN C, 2.0L L4 VIN B, 2.0L L4 VIN D, 2.0L L4 VIN E, 2.4L L4 VIN B, 2.4L L4 VIN C, 2.7L V6 VIN D, 3.3L V6 VIN D, 3.3L V6 VIN E, 3.3L V6 VIN F, 3.5L V6 VIN G, 3.8L V6 VIN C, 3.8L V6 VIN E, 3.8L V6 VIN F, 3.8L V6 VIN H, 4.6L V8 VIN F **Transmission:** All	**Intake Valve Control Solenoid Circuit Low (Bank 1):** PCM sets the code if it detects that the intake valve control solenoid control circuit is short to ground. Electrical check. **Possible Causes:** • Short to ground in control circuit • Contact resistance in connectors • Faulty intake valve control solenoid
DTC: P0077 **2T CCM, MIL: Yes** **Year:** 2009, 2010 **Model:** Accent, Azera, Elantra, Genesis, Genesis Coupe, Santa Fe, Sonata, Tucson, Veracruz **Engine:** 1.6L L4 VIN C, 2.0L L4 VIN B, 2.0L L4 VIN D, 2.0L L4 VIN E, 2.4L L4 VIN B, 2.4L L4 VIN C, 2.7L V6 VIN D, 3.3L V6 VIN D, 3.3L V6 VIN E, 3.3L V6 VIN F, 3.5L V6 VIN G, 3.8L V6 VIN C, 3.8L V6 VIN E, 3.8L V6 VIN F, 3.8L V6 VIN H, 4.6L V8 VIN F **Transmission:** All	**Intake Valve Control Solenoid Circuit High (Bank 1):** PCM sets the code if it detects that the OCV control circuit is open or short to battery. Electrical check. **Possible Causes:** • Open or short to battery in control circuit • Contact resistance in connectors • Faulty intake valve control solenoid
DTC: P0078 **2T PCM, MIL: Yes** **Year:** 2009, 2010 **Model:** Genesis **Engine:** 4.6L V8 VIN F **Transmission:** All	**Exhaust Valve Control Solenoid Circuit (Bank 1):** Enable Conditions • Threshold value • Open Diagnosis Time • Continuous **Possible Causes:** • Poor Connection • Open or Short to ground in power circuit • Open in control circuit • OCV

DTC	Trouble Code Title, Conditions & Possible Causes
DTC: P0079 **2T PCM, MIL: Yes** **Year:** 2009, 2010 **Model:** Genesis, Genesis Coupe, Santa Fe, Sonata, Tucson **Engine:** 2.0L L4 VIN B, 2.0L L4 VIN D, 2.4L L4 VIN B, 2.4L L4 VIN C, 3.5L V6 VIN G, 3.8L V6 VIN E, 3.8L V6 VIN H, 4.6L V8 VIN F **Transmission:** All	**Exhaust Valve Control Solenoid Circuit Low (Bank 1):** Enable Conditions • No disabling Faults Present • Engine Running • 11V ≤Battery Voltage ≤16V Threshold value • Short to ground or open circuit Diagnosis Time • Continuous (More than 5 seconds failure for every 10 seconds test) **Possible Causes:** • Poor Connection • Open in Power circuit • Open or short to ground in Control Circuit • OCV • ECM
DTC: P0080 **2T PCM, MIL: Yes** **Year:** 2009, 2010 **Model:** Genesis, Genesis Coupe, Santa Fe, Sonata, Tucson **Engine:** 2.0L L4 VIN B, 2.0L L4 VIN D, 2.4L L4 VIN B, 2.4L L4 VIN C, 3.5L V6 VIN G, 3.8L V6 VIN E, 3.8L V6 VIN H, 4.6L V8 VIN F **Transmission:** All	**Exhaust Valve Control Solenoid Circuit High(Bank 1):** Enable Conditions • No disabling Faults Present • Engine Running • 11V ≤Battery Voltage ≤16V Threshold value • Short to battery Diagnosis Time • Continuous (More than 5 seconds failure for every 10 seconds test) **Possible Causes:** • Poor Connection • Short to battery in Control Circuit • OCV • ECM
DTC: P0081 **2T PCM, MIL: Yes** **Year:** 2009, 2010 **Model:** Genesis **Engine:** 4.6L V8 VIN F **Transmission:** All	**Intake Valve Control Solenoid Circuit (Bank 2):** Enable Conditions • - Threshold value • Open Diagnosis Time • Continuous **Possible Causes:** • Poor Connection • Open or Short to ground in power circuit • Open in control circuit • OCV
DTC: P0082 **2T CCM, MIL: Yes** **Year:** 2009, 2010 **Model:** Azera, Genesis, Genesis Coupe, Santa Fe, Sonata, Veracruz **Engine:** 2.7L V6 VIN D, 3.3L V6 VIN D, 3.3L V6 VIN E, 3.3L V6 VIN F, 3.5L V6 VIN G, 3.8L V6 VIN C, 3.8L V6 VIN E, 3.8L V6 VIN F, 3.8L V6 VIN H, 4.6L V8 VIN F **Transmission:** All	**Intake Valve Control Solenoid Circuit Low (Bank 2):** Detects a short to ground or open circuit of VCPD bank 1 intake circuit output. No disabling faults present. Engine running. Enable time delay equal to or greater than 0.5 second. **Possible Causes:** • Poor connection • Open in power circuit • Open or short to ground in control circuit • OCV • Faulty PCM

DTC	Trouble Code Title, Conditions & Possible Causes
DTC: P0083 **2T CCM, MIL: Yes** **Year:** 2009, 2010 **Model:** Azera, Genesis, Genesis Coupe, Santa Fe, Sonata, Veracruz **Engine:** 2.7L V6 VIN D, 3.3L V6 VIN D, 3.3L V6 VIN E, 3.3L V6 VIN F, 3.5L V6 VIN G, 3.8L V6 VIN C, 3.8L V6 VIN E, 3.8L V6 VIN F, 3.8L V6 VIN H, 4.6L V8 VIN F **Transmission:** All	**Intake Valve Control Solenoid Circuit High (Bank 2):** Detects a short to battery of VCPD bank 1 intake circuit output. No disabling faults present. Engine running. Enable time delay equal to or greater than 0.5 second. **Possible Causes:** • poor connection • Short to battery in control circuit • OCV • Faulty PCM
DTC: P0084 **2T PCM, MIL: Yes** **Year:** 2009, 2010 **Model:** Genesis **Engine:** 4.6L V8 VIN F **Transmission:** All	**Exhaust Valve Control Solenoid Circuit (Bank 2):** Enable Conditions Threshold value • Open Diagnosis Time • Continuous **Possible Causes:** • Poor Connection • Open or Short to ground in power circuit • Open in control circuit • OCV
DTC: P0085 **2T PCM, MIL: Yes** **Year:** 2009, 2010 **Model:** Genesis, Genesis Coupe, Santa Fe **Engine:** 3.5L V6 VIN G, 3.8L V6 VIN E, 3.8L V6 VIN H, 4.6L V8 VIN F **Transmission:** All	**Exhaust Valve Control Solenoid Circuit Low (Bank 2):** Enable Conditions • - Threshold value • Short circuit to ground Diagnosis Time • Continuous **Possible Causes:** • Poor Connection • Open or Short to ground in power circuit • Open in control circuit • OCV
DTC: P0086 **2T PCM, MIL: Yes** **Year:** 2009, 2010 **Model:** Genesis, Genesis Coupe, Santa Fe **Engine:** 3.5L V6 VIN G, 3.8L V6 VIN E, 3.8L V6 VIN H, 4.6L V8 VIN F **Transmission:** All	**Exhaust Valve Control Solenoid Circuit High (Bank 2):** Enable Conditions • No disabling Faults Present • Engine Running • 11V ≤Battery Voltage ≤16V Threshold value • Short to battery Diagnosis Time • Continuous (More than 5 seconds failure for every 10 seconds test **Possible Causes:** • Poor Connection • Short to battery in Control Circuit • OCV • ECM
DTC: P0101 **2T CCM, MIL: Yes** **Year:** 2009, 2010 **Model:** Azera, Elantra, Genesis, Genesis Coupe, Santa Fe, Sonata, Tucson, Veracruz **Engine:** 2.0L L4 VIN D, 2.0L L4 VIN E, 2.7L V6 VIN D, 3.3L V6 VIN D, 3.3L V6 VIN E, 3.3L V6 VIN F, 3.8L V6 VIN C, 3.8L V6 VIN E, 3.8L V6 VIN F, 3.8L V6 VIN H, 4.6L V8 VIN F **Transmission:** All	**Mass Airflow or Volume Airflow Sensor Performance:** Engine started, engine running at idle speed, and the PCM detected the MAF sensor signal was less than 0.5 volt, or with the engine speed more than 3000 rpm, the MAF sensor signal was more than 4.5 volts. **Possible Causes:** • MAF sensor signal circuit is shorted to ground • MAF sensor signal circuit is shorted to VREF or system power • MAF sensor ground circuit is open between sensor and ground • MAF sensor is contaminated (dirty), damaged or has failed • PCM has failed

DTC	Trouble Code Title, Conditions & Possible Causes
DTC: P0102 **2T CCM, MIL: Yes** **Year:** 2009, 2010 **Model:** Azera, Elantra, Genesis, Genesis Coupe, Santa Fe, Sonata, Tucson, Veracruz **Engine:** 2.0L L4 VIN D, 2.0L L4 VIN E, 2.7L V6 VIN D, 3.3L V6 VIN D, 3.3L V6 VIN E, 3.3L V6 VIN F, 3.8L V6 VIN C, 3.8L V6 VIN E, 3.8L V6 VIN F, 3.8L V6 VIN H, 4.6L V8 VIN F **Transmission:** All	**Mass or Volume Airflow Sensor Circuit Low Input:** Engine started, engine runtime over 5 seconds, and the PCM detected the MAF sensor signal was less than 0.5 volt during the test. **Possible Causes:** • MAF sensor signal circuit is open or shorted to ground • MAF sensor power (VREF) circuit is open or shorted to ground • MAF sensor is damaged or has failed • PCM has failed
DTC: P0103 **2T CCM, MIL: Yes** **Year:** 2009, 2010 **Model:** Azera, Elantra, Genesis, Genesis Coupe, Santa Fe, Sonata, Tucson, Veracruz **Engine:** 2.0L L4 VIN D, 2.0L L4 VIN E, 2.7L V6 VIN D, 3.3L V6 VIN D, 3.3L V6 VIN E, 3.3L V6 VIN F, 3.8L V6 VIN C, 3.8L V6 VIN E, 3.8L V6 VIN F, 3.8L V6 VIN H, 4.6L V8 VIN F **Transmission:** All	**Volume Airflow Sensor Circuit High Input:** Engine runtime over 5 seconds, and the PCM detected the MAF sensor input was out of range high. **Possible Causes:** • MAF sensor signal circuit is open between the sensor and PCM • MAF sensor signal circuit is shorted to VREF or system power • MAF sensor ground circuit is open between sensor and ground • MAF sensor is damaged or has failed • PCM has failed
DTC: P0105 **2T PCM, MIL: Yes** **Year:** 2009, 2010 **Model:** Azera, Genesis, Genesis Coupe, Sonata, Veracruz **Engine:** 3.3L V6 VIN D, 3.3L V6 VIN F, 3.8L V6 VIN C, 3.8L V6 VIN E, 3.8L V6 VIN F, 3.8L V6 VIN H **Transmission:** All	**Manifold Absolute Pressure/Barometric Pressure Circuit:** Enable Conditions • No Disabling Fault Present • Shutdown time > 20 minutes • Engine running Threshold value • The difference between the signal at key-on and the signal at engine start < 0.5 kPa Diagnosis Time • For 3 seconds out of 5 seconds **Possible Causes:** • Faulty MAPS
DTC: P0106 **2T CCM, MIL: Yes** **Year:** 2009, 2010 **Model:** Accent, Azera, Genesis, Genesis Coupe, Santa Fe, Sonata, Tucson, Veracruz **Engine:** 1.6L L4 VIN C, 2.0L L4 VIN B, 2.0L L4 VIN D, 2.4L L4 VIN B, 2.4L L4 VIN C, 2.7L V6 VIN D, 3.3L V6 VIN D, 3.3L V6 VIN E, 3.3L V6 VIN F, 3.5L V6 VIN G, 3.8L V6 VIN C, 3.8L V6 VIN E, 3.8L V6 VIN F, 3.8L V6 VIN H **Transmission:** All	**Manifold Absolute Pressure/Barometric Pressure Circuit Range/Performance:** Map sensor output voltage is out of the threshold value. Rationality check. Coolant temperature above 19.4 degrees F. **Possible Causes:** • Poor connection • Open or short in MAP sensor circuit • Faulty MAPS • Faulty PCM

DTC	Trouble Code Title, Conditions & Possible Causes
DTC: P0107 **2T CCM, MIL: Yes** **Year:** 2009, 2010 **Model:** Accent, Azera, Genesis, Genesis Coupe, Santa Fe, Sonata, Tucson, Veracruz **Engine:** 1.6L L4 VIN C, 2.0L L4 VIN B, 2.0L L4 VIN D, 2.4L L4 VIN B, 2.4L L4 VIN C, 2.7L V6 VIN D, 3.3L V6 VIN D, 3.3L V6 VIN E, 3.3L V6 VIN F, 3.5L V6 VIN G, 3.8L V6 VIN C, 3.8L V6 VIN E, 3.8L V6 VIN F, 3.8L V6 VIN H **Transmission:** All	**Manifold Absolute Pressure/Barometric Pressure Circuit Low Input:** Map sensor output voltage is lower than threshold value. Low voltage check. Coolant temperature above 19.4 degrees F. **Possible Causes:** • Poor connection • Open or short to ground in MAP sensor circuit • Faulty MAPS • Faulty PCM
DTC: P0108 **2T CCM, MIL: Yes** **Year:** 2009, 2010 **Model:** Accent, Azera, Genesis, Genesis Coupe, Santa Fe, Sonata, Tucson, Veracruz **Engine:** 1.6L L4 VIN C, 2.0L L4 VIN B, 2.0L L4 VIN D, 2.4L L4 VIN B, 2.4L L4 VIN C, 2.7L V6 VIN D, 3.3L V6 VIN D, 3.3L V6 VIN E, 3.3L V6 VIN F, 3.5L V6 VIN G, 3.8L V6 VIN C, 3.8L V6 VIN E, 3.8L V6 VIN F, 3.8L V6 VIN H **Transmission:** All	**Manifold Absolute Pressure/Barometric Pressure Circuit High Input:** Map sensor output voltage is higher than 4.9 volts. High voltage check. Coolant temperature above 19.4 degrees F. **Possible Causes:** • Poor connection • Open or short to battery in MAP sensor circuit • Faulty MAPS • Faulty PCM
DTC: P0109 **2T PCM** **Year:** 2009, 2010 **Model:** Azera, Genesis, Genesis Coupe, Santa Fe, Sonata, Veracruz **Engine:** 2.7L V6 VIN D, 3.3L V6 VIN D, 3.3L V6 VIN E, 3.3L V6 VIN F, 3.8L V6 VIN C, 3.8L V6 VIN E, 3.8L V6 VIN F, 3.8L V6 VIN H **Transmission:** All	**Manifold Absolute Pressure/Barometric Pressure Circuit Intermittent:** Enable Conditions • Engine running • $\mid \Delta APS \mid < 5\%$ • Engine Speed > 800rpm Threshold value • MAP_stable –MAP_current | >10 % Diagnosis Time • - MIL On Condition • NO MIL ON(DTC only) **Possible Causes:** • Connecting condition • Open or short to ground in power circuit • Open or short to ground in signal circuit • MAPS • ECM
DTC: P0110 **2T CCM, MIL: Yes** **Year:** 2009, 2010 **Model:** Azera, Genesis, Genesis Coupe, Santa Fe, Sonata, Veracruz **Engine:** 2.7L V6 VIN D, 3.3L V6 VIN D, 3.3L V6 VIN E, 3.3L V6 VIN F, 3.8L V6 VIN C, 3.8L V6 VIN E, 3.8L V6 VIN F, 3.8L V6 VIN H **Transmission:** All	**Intake Air Temperature Sensor Circuit Malfunction:** Key on or engine running, and the PCM detected an unexpected voltage condition on the IAT sensor signal circuit during the test. **Possible Causes:** • IAT sensor signal circuit is open or shorted to ground • IAT sensor ground circuit is open between sensor and the PCM • IAT sensor signal circuit is shorted to VREF or system power • IAT sensor is damaged or has failed • PCM has failed

DTC	Trouble Code Title, Conditions & Possible Causes
DTC: P0111 **2T CCM, MIL: Yes** **Year:** 2009, 2010 **Model:** Accent, Azera, Elantra, Genesis, Genesis Coupe, Santa Fe, Sonata, Tucson, Veracruz **Engine:** 1.6L L4 VIN C, 2.0L L4 VIN B, 2.0L L4 VIN D, 2.0L L4 VIN E, 2.4L L4 VIN B, 2.4L L4 VIN C, 2.7L V6 VIN D, 3.3L V6 VIN D, 3.3L V6 VIN E, 3.3L V6 VIN F, 3.5L V6 VIN G, 3.8L V6 VIN C, 3.8L V6 VIN E, 3.8L V6 VIN F, 3.8L V6 VIN H, 4.6L V8 VIN F **Transmission:** All	**Intake Air Temperature Sensor 1 Circuit Range/Performance:** If the sensor is out of specification, a code is set. Output voltage is monitored. Engine coolant is above 167 degrees F. Vehicle speed is above 30MPH for more than 60 seconds. Vehicle speed is below 7mph for more than 30 seconds. **Possible Causes:** • Poor connection • Faulty IATS • Faulty PCM
DTC: P0112 **2T CCM, MIL: Yes** **Year:** 2009, 2010 **Model:** Accent, Azera, Elantra, Genesis, Genesis Coupe, Santa Fe, Sonata, Tucson, Veracruz **Engine:** 1.6L L4 VIN C, 2.0L L4 VIN B, 2.0L L4 VIN D, 2.0L L4 VIN E, 2.4L L4 VIN B, 2.4L L4 VIN C, 2.7L V6 VIN D, 3.3L V6 VIN D, 3.3L V6 VIN E, 3.3L V6 VIN F, 3.5L V6 VIN G, 3.8L V6 VIN C, 3.8L V6 VIN E, 3.8L V6 VIN F, 3.8L V6 VIN H, 4.6L V8 VIN F **Transmission:** All	**Intake Air Temperature Sensor Circuit High Input:** Key on or engine running, and the PCM detected the IAT sensor indicated more than 4.96 volts during the CCM test. **Possible Causes:** • IAT sensor signal circuit is open between sensor and the PCM • IAT sensor ground circuit is open between sensor and the PCM • IAT sensor signal circuit is shorted to VREF or system power • IAT sensor is damaged or has failed • PCM has failed
DTC: P0113 **2T CCM, MIL: Yes** **Year:** 2009, 2010 **Model:** Accent, Azera, Elantra, Genesis, Genesis Coupe, Santa Fe, Sonata, Tucson, Veracruz **Engine:** 1.6L L4 VIN C, 2.0L L4 VIN B, 2.0L L4 VIN D, 2.0L L4 VIN E, 2.4L L4 VIN B, 2.4L L4 VIN C, 2.7L V6 VIN D, 3.3L V6 VIN D, 3.3L V6 VIN E, 3.3L V6 VIN F, 3.5L V6 VIN G, 3.8L V6 VIN C, 3.8L V6 VIN E, 3.8L V6 VIN F, 3.8L V6 VIN H, 4.6L V8 VIN F **Transmission:** All	**Intake Air Temperature Sensor Circuit Low Input:** Key on or engine runtime over 5 seconds, and the PCM detected the IAT sensor indicated less than 0.20 volt during the CCM test. **Possible Causes:** • IAT sensor signal circuit is shorted to ground • IAT sensor is damaged or has failed • PCM has failed
DTC: P0115 **2T CCM, MIL: Yes** **Year:** 2009, 2010 **Model:** Accent, Azera, Genesis, Genesis Coupe, Santa Fe, Sonata, Veracruz **Engine:** 1.6L L4 VIN C, 2.7L V6 VIN D, 3.3L V6 VIN D, 3.3L V6 VIN E, 3.3L V6 VIN F, 3.8L V6 VIN C, 3.8L V6 VIN E, 3.8L V6 VIN F, 3.8L V6 VIN H **Transmission:** All	**Engine Coolant Temperature Sensor Circuit Malfunction:** Key on or engine running, and the PCM detected an unexpected voltage condition on the ECT sensor signal circuit during the test. **Possible Causes:** • ECT sensor signal circuit is open or shorted ground • ECT sensor ground circuit is open between sensor and PCM • ECT sensor signal circuit is shorted to VREF or system power • ECT sensor is damaged or has failed • PCM has failed

DTC	Trouble Code Title, Conditions & Possible Causes
DTC: P0116 **2T CCM, MIL: Yes** **Year:** 2009, 2010 **Model:** Accent, Azera, Elantra, Genesis, Genesis Coupe, Santa Fe, Sonata, Tucson, Veracruz **Engine:** 1.6L L4 VIN C, 2.0L L4 VIN B, 2.0L L4 VIN D, 2.0L L4 VIN E, 2.4L L4 VIN B, 2.4L L4 VIN C, 2.7L V6 VIN D, 3.3L V6 VIN D, 3.3L V6 VIN E, 3.3L V6 VIN F, 3.5L V6 VIN G, 3.8L V6 VIN C, 3.8L V6 VIN E, 3.8L V6 VIN F, 3.8L V6 VIN H, 4.6L V8 VIN F **Transmission:** All	**Engine Coolant Temperature Sensor Performance:** Engine started, engine runtime over20 minutes, and the PCM detected the ECT sensor signal tailed more than 68°F from the model curve stored in memory (could be an intermittent fault). **NOTE: Check for a possible problem related to the Cooling system.** **Possible Causes:** • ECT sensor circuit is open or shorted ground (intermittent fault) • ECT sensor ground circuit is open (an intermittent fault) • ECT sensor has drifted out of calibration or has failed • PCM has failed
DTC: P0117 **2T CCM, MIL: Yes** **Year:** 2009, 2010 **Model:** Accent, Azera, Elantra, Genesis, Genesis Coupe, Santa Fe, Sonata, Tucson, Veracruz **Engine:** 1.6L L4 VIN C, 2.0L L4 VIN B, 2.0L L4 VIN D, 2.0L L4 VIN E, 2.4L L4 VIN B, 2.4L L4 VIN C, 2.7L V6 VIN D, 3.3L V6 VIN D, 3.3L V6 VIN E, 3.3L V6 VIN F, 3.5L V6 VIN G, 3.8L V6 VIN C, 3.8L V6 VIN E, 3.8L V6 VIN F, 3.8L V6 VIN H, 4.6L V8 VIN F **Transmission:** All	**Engine Coolant Temperature Sensor Low Input:** Key on or engine runtime over 5 seconds, and the PCM detected that the ECT sensor input was less than 0.20 volt during the CCM test. **Possible Causes:** • ECT sensor signal circuit is shorted to ground • ECT sensor is damaged or has failed • PCM has failed
DTC: P0118 **2T CCM, MIL: Yes** **Year:** 2009, 2010 **Model:** Accent, Azera, Elantra, Genesis, Genesis Coupe, Santa Fe, Sonata, Tucson, Veracruz **Engine:** 1.6L L4 VIN C, 2.0L L4 VIN B, 2.0L L4 VIN D, 2.0L L4 VIN E, 2.4L L4 VIN B, 2.4L L4 VIN C, 2.7L V6 VIN D, 3.3L V6 VIN D, 3.3L V6 VIN E, 3.3L V6 VIN F, 3.5L V6 VIN G, 3.8L V6 VIN C, 3.8L V6 VIN E, 3.8L V6 VIN F, 3.8L V6 VIN H, 4.6L V8 VIN F **Transmission:** All	**Engine Coolant Temperature Sensor High Input:** Key on or engine running, and the PCM detected the ECT sensor indicated more than 4.96 volts during the CCM test. **Possible Causes:** • ECT sensor signal circuit is open between sensor and the PCM • ECT sensor ground circuit is open between sensor and PCM • ECT sensor signal circuit is shorted to VREF or system power • ECT sensor is damaged or has failed • PCM has failed
DTC: P0119 **2T CCM, MIL: Yes** **Year:** 2009, 2010 **Model:** Genesis, Genesis Coupe, Santa Fe, Sonata, Tucson **Engine:** 2.0L L4 VIN B, 2.0L L4 VIN D, 2.4L L4 VIN B, 2.4L L4 VIN C, 2.7L V6 VIN D, 4.6L V8 VIN F **Transmission:** All	**Engine Coolant Temperature Sensor Circuit Malfunction:** Engine started, engine runtime over 5 seconds, and the PCM detected an intermittent loss of the ECT sensor input during the test. **Possible Causes:** • ECT sensor signal circuit is open (an intermittent fault) • ECT sensor ground circuit is open (an intermittent fault) • ECT sensor signal circuit is shorted to VREF (intermittent fault) • ECT sensor is damaged or has failed (an intermittent fault) • PCM has failed

DTC	Trouble Code Title, Conditions & Possible Causes
DTC: P0121 **1T PCM, MIL: Yes** **Year:** 2009, 2010 **Model:** Accent, Elantra, Genesis, Genesis Coupe, Santa Fe, Sonata, Tucson **Engine:** 1.6L L4 VIN C, 2.0L L4 VIN B, 2.0L L4 VIN D, 2.0L L4 VIN E, 2.4L L4 VIN B, 2.4L L4 VIN C, 2.7L V6 VIN D, 3.5L V6 VIN G, 4.6L V8 VIN F **Transmission:** All	**Throttle/Pedal Position Sensor/Switch 'A' Circuit Range/Performance:** Enable Conditions • Engine speed > 480rpm • ECT > 75C • Engine load < 95% Threshold value • I TPS1 Position – TPS2 Position I > 6.3% Diagnosis Time • 0.5 sec **Possible Causes:** • Poor Connection • TPS • ECM
DTC: P0122 **2T CCM, MIL: Yes** **Year:** 2009, 2010 **Model:** Accent, Azera, Elantra, Genesis, Genesis Coupe, Santa Fe, Sonata, Tucson, Veracruz **Engine:** 1.6L L4 VIN C, 2.0L L4 VIN B, 2.0L L4 VIN D, 2.0L L4 VIN E, 2.4L L4 VIN B, 2.4L L4 VIN C, 2.7L V6 VIN D, 3.3L V6 VIN D, 3.3L V6 VIN E, 3.3L V6 VIN F, 3.5L V6 VIN G, 3.8L V6 VIN C, 3.8L V6 VIN E, 3.8L V6 VIN F, 3.8L V6 VIN H, 4.6L V8 VIN F **Transmission:** All	**Throttle Position Sensor 'A' Circuit Low Input:** Key on or engine running and the PCM detected the TP sensor 'A' signal was less than 0.170-0.200 volt during the CCM test. **Possible Causes:** • TP sensor signal circuit is shorted to ground • TP sensor VREF circuit is open or shorted to ground • TP sensor is damaged or has failed • PCM has failed
DTC: P0123 **2T CCM, MIL: Yes** **Year:** 2009, 2010 **Model:** Accent, Azera, Elantra, Genesis, Genesis Coupe, Santa Fe, Sonata, Tucson, Veracruz **Engine:** 1.6L L4 VIN C, 2.0L L4 VIN B, 2.0L L4 VIN D, 2.0L L4 VIN E, 2.4L L4 VIN B, 2.4L L4 VIN C, 2.7L V6 VIN D, 3.3L V6 VIN D, 3.3L V6 VIN E, 3.3L V6 VIN F, 3.5L V6 VIN G, 3.8L V6 VIN C, 3.8L V6 VIN E, 3.8L V6 VIN F, 3.8L V6 VIN H, 4.6L V8 VIN F **Transmission:** All	**Throttle Position Sensor 'A' Circuit High Input:** Key on or engine running and the PCM detected the TP sensor signal was more than 4.60-4.80 volts during the CCM test. **Possible Causes:** • TP sensor signal circuit is open • TP sensor ground circuit is open • TP sensor signal circuit is shorted to VREF or system power • TP sensor is damaged or has failed\ • PCM has failed
DTC: P0124 **2T CCM, MIL: Yes** **Year:** 2009, 2010 **Model:** Accent **Engine:** 1.6L L4 VIN C **Transmission:** All	**Throttle/Pedal Position Sensor/Switch "A" Circuit Intermittent:** Rationality check. Rate of change in throttle angle 0.1221 percent. Engine speed 600 rpm. Coolant temperature 167 degrees F. **Possible Causes:** • Poor connection • TPS • Faulty ECM/PCM

DTC	Trouble Code Title, Conditions & Possible Causes
DTC: P0125 **2T ECT, MIL: Yes** **Year:** 2009, 2010 **Model:** Azera, Elantra, Genesis, Genesis Coupe, Santa Fe, Sonata, Tucson, Veracruz **Engine:** 2.0L L4 VIN B, 2.0L L4 VIN D, 2.0L L4 VIN E, 2.4L L4 VIN B, 2.4L L4 VIN C, 2.7L V6 VIN D, 3.3L V6 VIN D, 3.3L V6 VIN E, 3.3L V6 VIN F, 3.5L V6 VIN G, 3.8L V6 VIN C, 3.8L V6 VIN E, 3.8L V6 VIN F, 3.8L V6 VIN H **Transmission:** All	**Insufficient Coolant Temperature for Closed Loop:** DTC P0116, P0117 and P0118 not set, engine run time from 6-8 minutes, and the PCM detected the ECT signal did not reach the closed loop temperature of at least 68°F. **Possible Causes:** • Check for low coolant level or incorrect coolant mixture • Cooling system component failure (thermostat stuck open) • ECT sensor is out of calibration ("skewed") or it has failed • PCM has failed
DTC: P0128 **2T ECT, MIL: Yes** **Year:** 2009, 2010 **Model:** Accent, Azera, Elantra **Engine:** 1.6L L4 VIN C, 2.0L L4 VIN D, 2.0L L4 VIN E, 3.3L V6 VIN D, 3.8L V6 VIN F **Transmission:** All	**Thermostat Malfunction Detected:** ECT sensor input less than 40°F at startup, engine started, and the PCM detected the ECT sensor did not reach 167°F after a normal warm up period had expired during the CCM Rationality test. **Possible Causes:** • Check the operation of the thermostat (it may be stuck open) • ECT sensor is out-of-calibration or skewed • Inspect for low coolant level or for an incorrect coolant mixture
DTC: P0128 **2T PCM, MIL: Yes** **Year:** 2009, 2010 **Model:** Genesis, Genesis Coupe, Santa Fe, Sonata, Tucson, Veracruz **Engine:** 2.0L L4 VIN B, 2.0L L4 VIN D, 2.4L L4 VIN B, 2.4L L4 VIN C, 2.7L V6 VIN D, 3.3L V6 VIN E, 3.3L V6 VIN F, 3.5L V6 VIN G, 3.8L V6 VIN C, 3.8L V6 VIN E, 3.8L V6 VIN H, 4.6L V8 VIN F **Transmission:** All	**Coolant Thermostat (Coolant Temperature below Thermostat Regulating Temperature):** Enable Conditions • Fuel-cut off phase <20% • Low load phase < 50% • 11V < Battery voltage < 16V • Intake air temperature decrease compared to Start Intake air temperature > -10°C(-14°F) • -10°C(14°F) < Coolant temperature at start < 54.4°C(131°F) • -10°C(14°F) < Intake air temperature at start • No high engine speed(4800rpm) with vehicle stopped • No relevant failure • Ambient temperature > -10°C(-14°F) • Percentage of high vehicle speed phase (Vehicle speed > 87 mph) < 90% Threshold Value • Measured coolant temp. < 74°C(165°F) When modeled coolant temp. > 85°C(185°F) Diagnostic Time • 10~30 min. depending on coolant temperature at start and driving pattern **Possible Causes:** • Faulty cooling system • Faulty ECT sensor • Poor connection or damaged harness
DTC: P0130 **2T CCM, MIL: Yes** **Year:** 2009, 2010 **Model:** Accent, Genesis, Genesis Coupe, Tucson **Engine:** 1.6L L4 VIN C, 2.0L L4 VIN B, 2.0L L4 VIN D, 2.7L V6 VIN D, 4.6L V8 VIN F **Transmission:** All	**HO2S-11 (Bank 1 Sensor 1) Circuit Malfunction:** Engine started, engine running in closed loop, and the PCM detected one of the following "failure" conditions existed: - HO2S signal was too high (more than 1.0 volt at idle speed) - HO2S signal was fixed from 350-600 mv - HO2S switch time from rich-to-lean or lean-to -rich was too long - HO2S input fixed at mid-range (from 350-550 mv) **Possible Causes:** • HO2S signal circuit is open between the sensor and the PCM • HO2S signal circuit is shorted to sensor or chassis ground • HO2S signal circuit is shorted to VREF or system power (B+) • HO2S is damaged, contaminated or it has failed • PCM has failed

DTC	Trouble Code Title, Conditions & Possible Causes
DTC: P0131 **2T CCM, MIL: Yes** **Year:** 2009, 2010 **Model:** Accent, Azera, Elantra, Genesis, Genesis Coupe, Santa Fe, Sonata, Tucson, Veracruz **Engine:** 1.6L L4 VIN C, 2.0L L4 VIN B, 2.0L L4 VIN D, 2.0L L4 VIN E, 2.4L L4 VIN B, 2.4L L4 VIN C, 2.7L V6 VIN D, 3.3L V6 VIN D, 3.3L V6 VIN E, 3.3L V6 VIN F, 3.5L V6 VIN G, 3.8L V6 VIN C, 3.8L V6 VIN E, 3.8L V6 VIN F, 3.8L V6 VIN H, 4.6L V8 VIN F **Transmission:** All	**HO2S-11 (Bank 1 Sensor 1) Circuit Low Input:** Engine started, engine running at cruise speed in closed loop for 2 minutes, and the PCM detected the HO2S-11 signal indicated less than 0.16 volt during the CCM test. **Possible Causes:** • Low fuel pressure, fuel filter restricted or fuel injectors plugged • HO2S signal circuit is shorted to ground (an intermittent fault) • HO2S may be contaminated or it has failed • HO2S heater is damaged or has failed • PCM has failed
DTC: P0132 **2T CCM, MIL: Yes** **Year:** 2009, 2010 **Model:** Accent, Azera, Elantra, Genesis, Genesis Coupe, Santa Fe, Sonata, Tucson, Veracruz **Engine:** 1.6L L4 VIN C, 2.0L L4 VIN B, 2.0L L4 VIN D, 2.0L L4 VIN E, 2.4L L4 VIN B, 2.4L L4 VIN C, 2.7L V6 VIN D, 3.3L V6 VIN D, 3.3L V6 VIN E, 3.3L V6 VIN F, 3.5L V6 VIN G, 3.8L V6 VIN C, 3.8L V6 VIN E, 3.8L V6 VIN F, 3.8L V6 VIN H, 4.6L V8 VIN F **Transmission:** All	**HO2S-11 (Bank 1 Sensor 1) Circuit High Input:** Engine started, engine running in closed loop at cruise speed for 2 minutes, and the PCM detected the HO2S-11 signal indicated more than 1.20 volts during the CCM test. **Possible Causes:** • Fuel pressure regulator leaking or fuel injectors leaking • HO2S signal circuit is shorted to the heater power circuit • HO2S may be contaminated or it has failed • HO2S heater is damaged or has failed • PCM has failed
DTC: P0133 **2T O2S, MIL: Yes** **Year:** 2009, 2010 **Model:** Accent, Azera, Elantra, Genesis, Genesis Coupe **Engine:** 1.6L L4 VIN C, 2.0L L4 VIN D, 2.0L L4 VIN E, 3.3L V6 VIN D, 3.8L V6 VIN E, 3.8L V6 VIN F, 3.8L V6 VIN H, 4.6L V8 VIN F **Transmission:** All	**HO2S-11 (Bank 1 Sensor 1) Slow Response:** DTC P0135 not set, engine started, engine running at idle speed in closed loop, and PCM detected the HO2S-11 response time to switch from rich-to-lean or from lean-to-rich was over one second. **Possible Causes:** • HO2S signal circuit is open or shorted to ground • HO2S element is contaminated or it has failed • HO2S heater is damaged or has failed • Intake air leaks, exhaust manifold leaks or PCV system leaks • MAF sensor out of calibration (it may be dirty or contaminated)
DTC: P0133 **2T O2S, MIL: Yes** **Year:** 2009, 2010 **Model:** Santa Fe, Sonata, Tucson, Veracruz **Engine:** 2.0L L4 VIN B, 2.4L L4 VIN B, 2.4L L4 VIN C, 2.7L V6 VIN D, 3.3L V6 VIN E, 3.3L V6 VIN F, 3.5L V6 VIN G, 3.8L V6 VIN C **Transmission:** All	**HO2S-11 (Bank 1 Sensor 1) Slow Response:** DTC P0135 not set, engine started, engine running at idle speed in closed loop, and PCM detected the HO2S-11 response time to switch from rich-to-lean or from lean-to-rich was over one second. **Possible Causes:** • HO2S signal circuit is open or shorted to ground • HO2S element is contaminated or it has failed • HO2S heater is damaged or has failed • Intake air leaks, exhaust manifold leaks or PCV system leaks • MAF sensor out of calibration (it may be dirty or contaminated) • TSB 02-360010 (3/02) contains information for this code
DTC: P0134 **2T O2S, MIL: Yes** **Year:** 2009, 2010 **Model:** Santa Fe, Sonata, Tucson, Veracruz **Engine:** 2.0L L4 VIN B, 2.7L V6 VIN D, 3.3L V6 VIN E, 3.3L V6 VIN F, 3.5L V6 VIN G, 3.8L V6 VIN C **Transmission:** All	**HO2S-11 (Bank 1 Sensor 1) No Activity Detected:** DTC P0135 not set, engine started, engine at idle speed and running in closed loop, and the PCM detected the HO2S-11 signal remained fixed from 400-550 mv for more than 1 minute during the CCM test. **Possible Causes:** • HO2S signal circuit is open or shorted to ground • HO2S element is contaminated or it has failed • HO2S heater is damaged or has failed • PCM has failed • TSB 02-360010 (3/02) contains information about this code

DTC	Trouble Code Title, Conditions & Possible Causes
DTC: P0134 **2T O2S, MIL: Yes** **Year:** 2009, 2010 **Model:** Accent, Azera, Genesis, Genesis Coupe **Engine:** 1.6L L4 VIN C, 2.0L L4 VIN D, 3.3L V6 VIN D, 3.8L V6 VIN E, 3.8L V6 VIN F, 3.8L V6 VIN H, 4.6L V8 VIN F **Transmission:** All	**HO2S-11 (Bank 1 Sensor 1) No Activity Detected:** DTC P0135 not set, engine started, engine at idle speed and running in closed loop, and the PCM detected the HO2S-11 signal remained fixed from 400-550 mv for more than 1 minute during the CCM test. **Possible Causes:** • HO2S signal circuit is open or shorted to ground • HO2S element is contaminated or it has failed • HO2S heater is damaged or has failed • PCM has failed
DTC: P0135 **2T O2S HTR1, MIL: Yes** **Year:** 2009, 2010 **Model:** Accent, Santa Fe, Sonata, Tucson **Engine:** 1.6L L4 VIN C, 2.4L L4 VIN B, 2.4L L4 VIN C **Transmission:** All	**HO2S-11 (Bank 1 Sensor 1) Heater Circuit Malfunction:** Engine started, engine running in closed loop at cruise speed, and the PCM detected the HO2S-11 heater current was less than 0.2 amps, or that is was more than 3.5 amps during the test period. **Possible Causes:** • HO2S heater control circuit is open or shorted to ground • HO2S heater control circuit is shorted to power • HO2S heater power circuit is open (check fuse in Engine J/B) • HO2S heater is damaged or has failed • PCM has failed
DTC: P0136 **2T CCM, MIL: Yes** **Year:** 2009, 2010 **Model:** Accent, Elantra, Genesis, Genesis Coupe, Santa Fe, Sonata, Tucson **Engine:** 1.6L L4 VIN C, 2.0L L4 VIN B, 2.0L L4 VIN D, 2.0L L4 VIN E, 2.4L L4 VIN B, 2.4L L4 VIN C, 2.7L V6 VIN D, 4.6L V8 VIN F **Transmission:** All	**HO2S-12 (Bank 1 Sensor 2) Circuit Malfunction:** Engine started, running in closed loop, and the PCM detected an unexpected high voltage on the HO2S circuit; or the HO2S signal was fixed at mid-range (350-550 mv) or not switching properly. **Possible Causes:** • HO2S signal circuit is open between the sensor and the PCM • HO2S signal circuit is shorted to sensor or chassis ground • HO2S signal circuit is shorted to VREF or system power (B+) • HO2S is damaged, contaminated or it has failed • PCM has failed
DTC: P0137 **2T CCM, MIL: Yes** **Year:** 2009, 2010 **Model:** Accent, Azera, Elantra, Genesis, Genesis Coupe, Santa Fe, Sonata, Tucson, Veracruz **Engine:** 1.6L L4 VIN C, 2.0L L4 VIN B, 2.0L L4 VIN D, 2.0L L4 VIN E, 2.4L L4 VIN B, 2.4L L4 VIN C, 2.7L V6 VIN D, 3.3L V6 VIN D, 3.3L V6 VIN E, 3.3L V6 VIN F, 3.5L V6 VIN G, 3.8L V6 VIN C, 3.8L V6 VIN E, 3.8L V6 VIN F, 3.8L V6 VIN H, 4.6L V8 VIN F **Transmission:** All	**HO2S-12 (Bank 1 Sensor 2) Circuit Low Input:** Engine started, engine running at cruise speed in closed loop for 2 minutes, and the PCM detected the HO2S-12 signal indicated less than 0.16 volt during the CCM test. **Possible Causes:** • Low fuel pressure, fuel filter restricted or fuel injectors plugged • HO2S signal circuit is shorted to ground (an intermittent fault) • HO2S may be contaminated or it has failed • HO2S heater is damaged or has failed • PCM has failed
DTC: P0138 **2T CCM, MIL: Yes** **Year:** 2009, 2010 **Model:** Accent, Azera, Elantra, Genesis, Genesis Coupe, Santa Fe, Sonata, Tucson, Veracruz **Engine:** 1.6L L4 VIN C, 2.0L L4 VIN B, 2.0L L4 VIN D, 2.0L L4 VIN E, 2.4L L4 VIN B, 2.4L L4 VIN C, 2.7L V6 VIN D, 3.3L V6 VIN D, 3.3L V6 VIN E, 3.3L V6 VIN F, 3.5L V6 VIN G, 3.8L V6 VIN C, 3.8L V6 VIN E, 3.8L V6 VIN F, 3.8L V6 VIN H, 4.6L V8 VIN F **Transmission:** All	**HO2S-12 (Bank 1 Sensor 2) Circuit High Input:** Engine started, engine running in closed loop at cruise speed for 2 minutes, and the PCM detected the HO2S-12 signal indicated more than 1.20 volts during the CCM test. **Possible Causes:** • Fuel pressure regulator leaking or fuel injectors leaking • HO2S signal circuit is shorted to the heater power circuit • HO2S may be contaminated or it has failed • HO2S heater is damaged or has failed • PCM has failed

DTC	Trouble Code Title, Conditions & Possible Causes
DTC: P0139 **2T CCM, MIL: Yes** **Year:** 2009, 2010 **Model:** Accent, Azera, Elantra, Genesis, Genesis Coupe, Santa Fe, Sonata, Tucson, Veracruz **Engine:** 1.6L L4 VIN C, 2.0L L4 VIN B, 2.0L L4 VIN D, 2.0L L4 VIN E, 2.4L L4 VIN B, 2.4L L4 VIN C, 2.7L V6 VIN D, 3.3L V6 VIN D, 3.3L V6 VIN E, 3.3L V6 VIN F, 3.8L V6 VIN C, 3.8L V6 VIN E, 3.8L V6 VIN F, 3.8L V6 VIN H, 4.6L V8 VIN F **Transmission:** All	**HO2S-12 (Bank 1 Sensor 2) Slow Response:** DTC P0141 not set, engine started, engine running at idle speed in closed loop, and PCM detected the HO2S-12 response time to switch from rich-to-lean or from lean-to-rich was over one second. **Possible Causes:** • HO2S signal circuit is open or shorted to ground • HO2S element is contaminated or it has failed • HO2S heater is damaged or has failed • Intake air leaks, exhaust manifold leaks or PCV system leaks • MAF sensor out of calibration (it may be dirty or contaminated)
DTC: P0140 **2T O2S, MIL: Yes** **Year:** 2009, 2010 **Model:** Accent, Azera, Elantra, Genesis, Genesis Coupe, Santa Fe, Sonata, Tucson, Veracruz **Engine:** 1.6L L4 VIN C, 2.0L L4 VIN B, 2.0L L4 VIN D, 2.0L L4 VIN E, 2.4L L4 VIN B, 2.4L L4 VIN C, 2.7L V6 VIN D, 3.3L V6 VIN D, 3.3L V6 VIN E, 3.3L V6 VIN F, 3.5L V6 VIN G, 3.8L V6 VIN C, 3.8L V6 VIN E, 3.8L V6 VIN F, 3.8L V6 VIN H, 4.6L V8 VIN F **Transmission:** All	**HO2S-12 (Bank 1 Sensor 2) Circuit No Activity:** DTC P0141 not set, engine started, engine at idle speed and running in closed loop, and the PCM detected the HO2S-12 signal remained fixed from 400-550 mv for more than 1 minute during the CCM test. **Possible Causes:** • HO2S signal circuit is open or shorted to ground • HO2S element is contaminated, damaged or has failed • PCM has failed
DTC: P0141 **2T O2S HTR1, MIL: Yes** **Year:** 2009, 2010 **Model:** Accent **Engine:** 1.6L L4 VIN C **Transmission:** All	**HO2S-12 (Bank 1 Sensor 2) Heater Circuit Malfunction:** Engine started, engine running in closed loop at cruise speed, and the PCM detected the HO2S-12 heater current was less than 0.2 amps, or that is was more than 3.5 amps during the test period. **Possible Causes:** • HO2S heater control circuit shorted to ground or system power • HO2S heater power circuit is open (check fuse in Engine J/B) • HO2S heater is damaged or has failed • PCM has failed
DTC: P0150 **2T CCM, MIL: Yes** **Year:** 2009, 2010 **Model:** Genesis, Tucson **Engine:** 2.7L V6 VIN D, 4.6L V8 VIN F **Transmission:** All	**HO2S-21 (Bank 2 Sensor 1) Circuit Malfunction:** Engine started, engine running in closed loop, and the PCM detected one of the following "failure" conditions were present: - The HO2S signal was too high (more than 1.0 volt) - The HO2S signal was fixed between 350-600 mv - The HO2S switch time was too long - The HO2S signal fixed at mid-range (350-550 mv) **Possible Causes:** • HO2S signal circuit is open between the sensor and the PCM • HO2S signal circuit is shorted to sensor or chassis ground • HO2S signal circuit is shorted to VREF or system power (B+) • HO2S is damaged, contaminated or it has failed • PCM has failed
DTC: P0151 **2T O2S2, MIL: Yes** **Year:** 2009, 2010 **Model:** Azera, Genesis, Genesis Coupe, Santa Fe, Sonata, Tucson, Veracruz **Engine:** 2.7L V6 VIN D, 3.3L V6 VIN D, 3.3L V6 VIN E, 3.3L V6 VIN F, 3.5L V6 VIN G, 3.8L V6 VIN C, 3.8L V6 VIN E, 3.8L V6 VIN F, 3.8L V6 VIN H, 4.6L V8 VIN F **Transmission:** All	**HO2S Circuit Low Voltage (Bank 2 Sensor 1):** Engine started, engine running in closed loop at a speed over 3 mph, and the PCM detected the average ratio between the HO2S Actual and maximum allowed frequency during 100 Lambda cycles was more than the Threshold value (e.g., 0.66 Hz) during the test. **Possible Causes:** • Poor connection • Short to ground in harness • Faulty PCM • HO2S has deteriorated, is contaminated or has failed

DTC	Trouble Code Title, Conditions & Possible Causes
DTC: P0152 **2T CCM, MIL: Yes** **Year:** 2009, 2010 **Model:** Azera, Genesis, Genesis Coupe, Sonata, Tucson, Veracruz **Engine:** 2.7L V6 VIN D, 3.3L V6 VIN D, 3.3L V6 VIN F, 3.8L V6 VIN C, 3.8L V6 VIN E, 3.8L V6 VIN F, 3.8L V6 VIN H, 4.6L V8 VIN F **Transmission:** All	**HO2S-21 (Bank 2 Sensor 1) Circuit High Input:** Engine started, engine running in closed loop at cruise speed, and the PCM detected the HO2S-21 signal was more 1.20 volts in the test. **Possible Causes:** • Fuel pressure regulator leaking or fuel injectors leaking • HO2S signal circuit is shorted to the heater power circuit • HO2S may be contaminated or it has failed • HO2S heater is damaged or has failed • PCM has failed
DTC: P0153 **2T O2S, MIL: Yes** **Year:** 2009, 2010 **Model:** Azera, Genesis, Genesis Coupe, Santa Fe, Sonata, Tucson, Veracruz **Engine:** 2.7L V6 VIN D, 3.3L V6 VIN D, 3.3L V6 VIN E, 3.3L V6 VIN F, 3.5L V6 VIN G, 3.8L V6 VIN C, 3.8L V6 VIN E, 3.8L V6 VIN F, 3.8L V6 VIN H, 4.6L V8 VIN F **Transmission:** All	**HO2S-21 (Bank 2 Sensor 1) Slow Response:** DTC P0155 not set, engine started, engine running at idle speed in closed loop, and PCM detected the HO2S-12 response time to switch from rich-to-lean or from lean-to-rich was over one second. **Possible Causes:** • HO2S signal circuit is open or shorted to ground • HO2S element is contaminated or it has failed • HO2S heater is damaged or has failed • Intake air leaks, exhaust manifold leaks or PCV system leaks • MAF sensor out of calibration (it may be dirty or contaminated)
DTC: P0154 **2T O2S, MIL: Yes** **Year:** 2009, 2010 **Model:** Azera, Genesis, Genesis Coupe, Santa Fe, Sonata, Tucson, Veracruz **Engine:** 2.7L V6 VIN D, 3.3L V6 VIN D, 3.3L V6 VIN E, 3.3L V6 VIN F, 3.5L V6 VIN G, 3.8L V6 VIN C, 3.8L V6 VIN E, 3.8L V6 VIN F, 3.8L V6 VIN H, 4.6L V8 VIN F **Transmission:** All	**HO2S-21 (Bank 2 Sensor 1) No Activity Detected:** DTC P0141 not set, engine started, engine at idle speed and running in closed loop, and the PCM detected the HO2S-21 signal remained fixed from 400-550 mv for more than 1 minute during the CCM test. **Possible Causes:** • HO2S signal circuit is open or shorted to ground • HO2S element is contaminated or it has failed • HO2S heater is damaged or has failed • PCM has failed
DTC: P0156 **2T CCM, MIL: Yes** **Year:** 2009, 2010 **Model:** Genesis, Tucson **Engine:** 2.7L V6 VIN D, 4.6L V8 VIN F **Transmission:** All	**HO2S-22 (Bank 2 Sensor 2) Circuit Malfunction:** Engine started, engine running in closed loop, and the PCM detected one of the following "failure" conditions were present: - The HO2S signal was too high (more than 1.0 volt) - The HO2S signal was fixed between 350-600 mv - The HO2S switch time was too long - The HO2S signal fixed at mid-range (350-550 mv) **Possible Causes:** • HO2S signal circuit is open between the sensor and the PCM • HO2S signal circuit is shorted to sensor or chassis ground • HO2S signal circuit is shorted to VREF or system power (B+) • HO2S is damaged, contaminated or it has failed • PCM has failed
DTC: P0157 **2T CCM, MIL: Yes** **Year:** 2009, 2010 **Model:** Azera, Genesis, Genesis Coupe, Santa Fe, Sonata, Tucson, Veracruz **Engine:** 2.7L V6 VIN D, 3.3L V6 VIN D, 3.3L V6 VIN E, 3.3L V6 VIN F, 3.5L V6 VIN G, 3.8L V6 VIN C, 3.8L V6 VIN E, 3.8L V6 VIN F, 3.8L V6 VIN H, 4.6L V8 VIN F **Transmission:** All	**HO2S Circuit Low Voltage (Bank 2 Sensor 2):** The signal voltage of the front or rear sensor changes the rear circuit voltage specification when air fuel ratio is rich, a DTC is set. Out of range low failure (ground short open circuit). **Possible Causes:** • Poor connection • Short to ground in HO2S circuit • Faulty HO2S • Faulty PCM

DTC	Trouble Code Title, Conditions & Possible Causes
DTC: P0158 **2T CCM, MIL: Yes** **Year:** 2009, 2010 **Model:** Azera, Genesis, Genesis Coupe, Santa Fe, Sonata, Tucson, Veracruz **Engine:** 2.7L V6 VIN D, 3.3L V6 VIN D, 3.3L V6 VIN E, 3.3L V6 VIN F, 3.5L V6 VIN G, 3.8L V6 VIN C, 3.8L V6 VIN E, 3.8L V6 VIN F, 3.8L V6 VIN H, 4.6L V8 VIN F **Transmission:** All	**HO2S Circuit High Voltage (Bank 2 Sensor 2):** The signal voltage is higher than 1.2 volts after open in circuit. Out of range high failure. **Possible Causes:** • Poor connection • Short to battery in HO2S circuit • Faulty HO2S • Faulty PCM
DTC: P0159 **2T O2S, MIL: Yes** **Year:** 2009, 2010 **Model:** Azera, Genesis, Genesis Coupe, Santa Fe, Sonata, Tucson, Veracruz **Engine:** 2.7L V6 VIN D, 3.3L V6 VIN D, 3.3L V6 VIN E, 3.3L V6 VIN F, 3.8L V6 VIN C, 3.8L V6 VIN E, 3.8L V6 VIN F, 3.8L V6 VIN H, 4.6L V8 VIN F **Transmission:** All	**HO2S-22 (Bank 2 Sensor 2) Rationality Check:** DTC P0141 not set, engine started, engine running at idle speed in closed loop, and PCM detected the HO2S-22 response time to switch from rich-to-lean or from lean-to-rich was over one second. **Possible Causes:** • HO2S signal circuit is open or shorted to ground • HO2S element is contaminated or it has failed • HO2S heater is damaged or has failed • Intake air leaks, exhaust manifold leaks or PCV system leaks • MAF sensor out of calibration (it may be dirty or contaminated)
DTC: P0160 **2T O2S, MIL: Yes** **Year:** 2009, 2010 **Model:** Azera, Genesis, Genesis Coupe, Santa Fe, Tucson, Veracruz **Engine:** 2.7L V6 VIN D, 3.3L V6 VIN D, 3.3L V6 VIN E, 3.5L V6 VIN G, 3.8L V6 VIN C, 3.8L V6 VIN E, 3.8L V6 VIN F, 3.8L V6 VIN H, 4.6L V8 VIN F **Transmission:** All	**HO2S-22 (Bank 2 Sensor 2) No Activity Detected:** DTC P0141 not set, engine started, engine at idle speed and running in closed loop, and the PCM detected the HO2S-22 signal remained fixed at 550 mv or less for more than 1 minute during the CCM test. **Possible Causes:** • HO2S signal circuit is open or shorted to ground • HO2S element is contaminated or it has failed • HO2S heater is damaged or has failed • PCM has failed
DTC: P0170 **2T Fuel, MIL: Yes** **Year:** 2009, 2010 **Model:** Elantra, Genesis Coupe, Santa Fe, Sonata, Tucson **Engine:** 2.0L L4 VIN B, 2.0L L4 VIN D, 2.0L L4 VIN E, 2.4L L4 VIN B, 2.4L L4 VIN C, 2.7L V6 VIN D **Transmission:** All	**Fuel Trim Too Rich or Too Lean (Bank 1):** Engine started, engine running at cruise speed in closed loop for 3-5 minutes, and PCM the detected the Fuel system control was too rich or too lean under these conditions in the Fuel System Monitor test. **Possible Causes:** • Air leaks after the MAF sensor, or in the EGR or PCV system • Base engine "mechanical" fault affecting one or more cylinders • Exhaust leaks located in front of the A/FS or HO2S location • Fuel control sensor is out of calibration (i.e., ECT, IAT or MAP) • Fuel delivery system supplying too little fuel during cruise or idle periods (e.g., faulty fuel pump or dirty, restricted fuel filter) • Fuel injector (one or more) dirty or pressure regulator has failed • HO2S is contaminated, deteriorated or it has failed • Vehicle driven low on fuel or until it ran out of fuel

DTC	Trouble Code Title, Conditions & Possible Causes
DTC: P0171 **2T Fuel, MIL: Yes** **Year:** 2009, 2010 **Model:** Accent, Azera, Elantra, Genesis, Genesis Coupe, Santa Fe, Sonata, Tucson, Veracruz **Engine:** 1.6L L4 VIN C, 2.0L L4 VIN B, 2.0L L4 VIN D, 2.0L L4 VIN E, 2.7L V6 VIN D, 3.3L V6 VIN D, 3.3L V6 VIN E, 3.3L V6 VIN F, 3.5L V6 VIN G, 3.8L V6 VIN C, 3.8L V6 VIN E, 3.8L V6 VIN F, 3.8L V6 VIN H **Transmission:** All	**Fuel System Too Lean (Bank 1):** Engine started, engine running at cruise speed for 3-5 minutes in closed loop, and the PCM detected the Fuel system was too lean (i.e., it was beyond a calibrated value stored in the PCM memory). **Possible Causes:** • Air leaks after the MAF sensor, or in the EGR or PCV system • Base engine "mechanical" fault affecting one or more cylinders • Exhaust leaks located in front of the A/FS or HO2S location • Fuel control sensor is out of calibration (i.e., ECT, IAT or MAP) • Fuel delivery system supplying too little fuel during cruise or idle periods (e.g., faulty fuel pump or dirty, restricted fuel filter) • Fuel injector (one or more) dirty or pressure regulator has failed • HO2S is contaminated, deteriorated or it has failed • Vehicle driven low on fuel or until it ran out of fuel
DTC: P0172 **2T Fuel, MIL: Yes** **Year:** 2009, 2010 **Model:** Accent, Azera, Elantra, Genesis, Genesis Coupe, Santa Fe, Sonata **Engine:** 1.6L L4 VIN C, 2.0L L4 VIN D, 2.0L L4 VIN E, 2.7L V6 VIN D, 3.3L V6 VIN D, 3.3L V6 VIN E, 3.3L V6 VIN F, 3.5L V6 VIN G, 3.8L V6 VIN E, 3.8L V6 VIN F, 3.8L V6 VIN H **Transmission:** All	**Fuel System Too Rich (Bank 1):** Engine started, engine running at cruise speed for 3-5 minutes in closed loop, and the PCM detected the Fuel system was too rich (i.e., it was beyond a calibrated value stored in the PCM memory). **Possible Causes:** • Base engine "mechanical" fault affecting one or more cylinders • EVAP system component has failed or canister fuel saturated • Exhaust leaks located in front of the HO2S location • Fuel control sensor is out of calibration (i.e., ECT, IAT or MAF) • Fuel delivery system supplying too much fuel during cruise or idle periods (e.g., faulty fuel pump, or faulty pressure regulator) • Fuel injector(s) is leaking or stuck partially open (one or more) • HO2S is contaminated, deteriorated or it has failed
DTC: P0173 **2T Fuel, MIL: Yes** **Year:** 2009 **Model:** Tucson **Engine:** 2.7L V6 VIN D **Transmission:** All	**Fuel Trim Too Rich or Too Lean (Bank 2):** Engine running in closed loop, and the PCM detected the Fuel system was too rich or too lean during two or more consecutive trips. **Possible Causes:** • Base engine "mechanical" fault affecting one or more cylinders • EVAP system component has failed or canister fuel saturated • Exhaust leaks located in front of the HO2S location • Fuel control sensor is out of calibration (i.e., ECT, IAT or MAF) • Fuel delivery system supplying too much fuel during cruise or idle periods (e.g., faulty fuel pump, or faulty pressure regulator) • Fuel injector(s) is leaking or stuck partially open (one or more) • HO2S is contaminated, deteriorated or it has failed
DTC: P0174 **2T Fuel, MIL: Yes** **Year:** 2009, 2010 **Model:** Azera, Genesis, Genesis Coupe, Santa Fe, Sonata, Tucson, Veracruz **Engine:** 2.7L V6 VIN D, 3.3L V6 VIN D, 3.3L V6 VIN E, 3.3L V6 VIN F, 3.5L V6 VIN G, 3.8L V6 VIN C, 3.8L V6 VIN E, 3.8L V6 VIN F, 3.8L V6 VIN H **Transmission:** All	**Fuel System Too Lean (Bank 2):** Engine started, engine running at cruise speed for 3-5 minutes in closed loop, and the PCM detected the Fuel system was too lean (i.e., it was beyond a calibrated value stored in the PCM memory). **Possible Causes:** • Air leaks after the MAF sensor, or in the EGR or PCV system • Base engine "mechanical" fault affecting one or more cylinders • Exhaust leaks located in front of the A/FS or HO2S location • Fuel control sensor is out of calibration (i.e., ECT, IAT or MAP) • Fuel delivery system supplying too little fuel during cruise or idle periods (e.g., faulty fuel pump or dirty, restricted fuel filter) • Fuel injector (one or more) dirty or pressure regulator has failed • HO2S is contaminated, deteriorated or it has failed • Vehicle driven low on fuel or until it ran out of fuel

DTC	Trouble Code Title, Conditions & Possible Causes
DTC: P0175 **2T Fuel, MIL: Yes** **Year:** 2009, 2010 **Model:** Azera, Genesis, Genesis Coupe, Santa Fe, Sonata, Tucson, Veracruz **Engine:** 2.7L V6 VIN D, 3.3L V6 VIN D, 3.3L V6 VIN E, 3.3L V6 VIN F, 3.5L V6 VIN G, 3.8L V6 VIN C, 3.8L V6 VIN E, 3.8L V6 VIN F, 3.8L V6 VIN H **Transmission:** All	**Fuel System Too Rich (Bank 2):** Engine started, engine running at cruise speed for 3-5 minutes in closed loop, and the PCM detected the Fuel system was too rich (i.e., it was beyond a calibrated value stored in the PCM memory). **Possible Causes:** • Base engine "mechanical" fault affecting one or more cylinders • EVAP system component has failed or canister fuel saturated • Exhaust leaks located in front of the HO2S location • Fuel control sensor is out of calibration (i.e., ECT, IAT or MAF) • Fuel delivery system supplying too much fuel during cruise or idle periods (e.g., faulty fuel pump, or faulty pressure regulator) • Fuel injector(s) is leaking or stuck partially open (one or more) • HO2S is contaminated, deteriorated or it has failed
DTC: P0196 **2T CCM, MIL: Yes** **Year:** 2009, 2010 **Model:** Azera, Elantra, Genesis, Genesis Coupe, Santa Fe, Sonata, Tucson, Veracruz **Engine:** 2.0L L4 VIN B, 2.0L L4 VIN D, 2.0L L4 VIN E, 2.7L V6 VIN D, 3.3L V6 VIN D, 3.3L V6 VIN E, 3.3L V6 VIN F, 3.5L V6 VIN G, 3.8L V6 VIN C, 3.8L V6 VIN E, 3.8L V6 VIN F, 3.8L V6 VIN H **Transmission:** All	**Engine Oil Temperature Sensor Range/Performance:** Stuck oil temperature sensor signal or unusual low or high signal. Condition 1 (signal high or low), engine coolant temperature more than 158 degrees F and oil temperature less than 68 degrees F. Condition 2 (signal high or low), engine coolant temperature less than 158 degrees F and oil temperature above 212 degrees F. Condition 3 (stuck signal) engine coolant temperature less than 104 degrees F. **Possible Causes:** • Contact resistance in connectors • faulty OTS
DTC: P0197 **2T CCM, MIL: Yes** **Year:** 2009, 2010 **Model:** Azera, Elantra, Genesis, Genesis Coupe, Santa Fe, Sonata, Tucson, Veracruz **Engine:** 2.0L L4 VIN B, 2.0L L4 VIN D, 2.0L L4 VIN E, 2.7L V6 VIN D, 3.3L V6 VIN D, 3.3L V6 VIN E, 3.3L V6 VIN F, 3.5L V6 VIN G, 3.8L V6 VIN C, 3.8L V6 VIN E, 3.8L V6 VIN F, 3.8L V6 VIN H **Transmission:** All	**Engine Oil temperature Sensor Low Input:** Signal voltage lower than the possible range of a properly operating OTS. Voltage range check. Engine coolant temperature less than 212 degrees F. Oil temperature above 309 degrees F. **Possible Causes:** • Short circuit to ground • Contact resistance in connectors • faulty OTS
DTC: P0198 **2T CCM, MIL: Yes** **Year:** 2009, 2010 **Model:** Azera, Elantra, Genesis, Genesis Coupe, Santa Fe, Sonata, Tucson, Veracruz **Engine:** 2.0L L4 VIN B, 2.0L L4 VIN D, 2.0L L4 VIN E, 2.7L V6 VIN D, 3.3L V6 VIN D, 3.3L V6 VIN E, 3.3L V6 VIN F, 3.5L V6 VIN G, 3.8L V6 VIN C, 3.8L V6 VIN E, 3.8L V6 VIN F, 3.8L V6 VIN H **Transmission:** All	**Engine Oil Temperature Sensor High Input:** Signal voltage higher than the possible range of a properly operating OTS. Voltage range check. Five minutes after engine start if engine coolant temperature less than 14 degrees F. Oil temperature minus 33 degrees F. **Possible Causes:** • Open circuit to battery • Contact resistance in connectors • faulty OTS
DTC: P0201 **2T CCM, MIL: Yes** **Year:** 2009, 2010 **Model:** Accent, Genesis **Engine:** 1.6L L4 VIN C, 4.6L V8 VIN F **Transmission:** All	**Fuel Injector 1 Circuit Malfunction:** Engine running and the PCM detected an unexpected voltage on the Fuel Injector 1 control circuit during the Component Monitor test. **Possible Causes:** • Injector 1 control circuit open or grounded • Injector 1 control circuit shorted to system power • Injector 1 power (B+) circuit open • Injector 1 damaged or has failed • PCM has failed

DTC	Trouble Code Title, Conditions & Possible Causes
DTC: P0202 **2T CCM, MIL: Yes** **Year:** 2009, 2010 **Model:** Accent, Genesis **Engine:** 1.6L L4 VIN C, 4.6L V8 VIN F **Transmission:** All	**Fuel Injector 2 Circuit Malfunction:** Engine running and the PCM detected an unexpected voltage on the Fuel Injector control circuit during the Component Monitor test. **Possible Causes:** • Injector 2 control circuit open or grounded • Injector 2 control circuit shorted to system power • Injector 2 power (B+) circuit open • Injector 2 damaged or has failed • PCM has failed
DTC: P0203 **2T CCM, MIL: Yes** **Year:** 2009, 2010 **Model:** Accent, Genesis **Engine:** 1.6L L4 VIN C, 4.6L V8 VIN F **Transmission:** All	**Fuel Injector 3 Circuit Malfunction:** Engine running and the PCM detected an unexpected voltage on the Fuel Injector control circuit during the Component Monitor test. **Possible Causes:** • Injector 3 control circuit open or grounded • Injector 3 control circuit shorted to system power • Injector 3 power (B+) circuit open • Injector 3 damaged or has failed • PCM has failed
DTC: P0204 **2T CCM, MIL: Yes** **Year:** 2009, 2010 **Model:** Accent, Genesis **Engine:** 1.6L L4 VIN C, 4.6L V8 VIN F **Transmission:** All	**Fuel Injector 4 Circuit Malfunction:** Engine running and the PCM detected an unexpected voltage on the Fuel Injector control circuit during the Component Monitor test. **Possible Causes:** • Injector 4 control circuit open or grounded • Injector 4 control circuit shorted to system power • Injector 4 power (B+) circuit open • Injector 4 damaged or has failed • PCM has failed
DTC: P0205 **2T CCM, MIL: Yes** **Year:** 2009, 2010 **Model:** Genesis **Engine:** 4.6L V8 VIN F **Transmission:** All	**Fuel Injector 5 Circuit Malfunction:** Engine running and the PCM detected an unexpected voltage on the Fuel Injector control circuit during the Component Monitor test. **Possible Causes:** • Injector 5 control circuit open or grounded • Injector 5 control circuit shorted to system power • Injector 5 power (B+) circuit open • Injector 5 damaged or has failed • PCM has failed
DTC: P0206 **2T CCM, MIL: Yes** **Year:** 2009, 2010 **Model:** Genesis **Engine:** 4.6L V8 VIN F **Transmission:** All	**Fuel Injector 6 Circuit Malfunction:** Engine running and the PCM detected an unexpected voltage on the Fuel Injector control circuit during the Component Monitor test. **Possible Causes:** • Injector 6 control circuit open or grounded • Injector 6 control circuit shorted to system power • Injector 6 power (B+) circuit open • Injector 6 damaged or has failed • PCM has failed
DTC: P0207 **2T CCM, MIL: Yes** **Year:** 2009, 2010 **Model:** Genesis **Engine:** 4.6L V8 VIN F **Transmission:** All	**Cylinder 7 Injector Circuit Malfunction:** Engine running and the PCM detected the identified fuel injector control circuit signal was more than the upper limit, or that it was less than the lower limit, or that no control signal was present. **Possible Causes:** • Main relay power supply circuit to the injector is open • Fuel injector 6 control circuit is open or shorted to ground • Fuel injector 6 is damaged or has failed • Injector "driver" circuit in the PCM is damaged or has failed
DTC: P0208 **2T CCM, MIL: Yes** **Year:** 2009, 2010 **Model:** Genesis **Engine:** 4.6L V8 VIN F **Transmission:** All	**Cylinder 8 Injector Circuit Malfunction:** Engine running and the PCM detected the identified fuel injector control circuit signal was more than the upper limit, or that it was less than the lower limit, or that no control signal was present. **Possible Causes:** • Main relay power supply circuit to the injector is open • Fuel injector 6 control circuit is open or shorted to ground • Fuel injector 6 is damaged or has failed • Injector "driver" circuit in the PCM is damaged or has failed

DTC	Trouble Code Title, Conditions & Possible Causes
DTC: P0217 **2T CCM, MIL: Yes** **Year:** 2009, 2010 **Model:** Azera, Genesis, Genesis Coupe, Santa Fe, Sonata, Veracruz **Engine:** 2.7L V6 VIN D, 3.3L V6 VIN D, 3.3L V6 VIN E, 3.3L V6 VIN F, 3.8L V6 VIN C, 3.8L V6 VIN E, 3.8L V6 VIN F, 3.8L V6 VIN H **Transmission:** All	**Engine Coolant Over Temperature Condition:** This diagnostic introduces a delay and also looks out for excessive engine loads. Once the delay period passes and excessive loads were not experienced, the diagnostic checks whether the undefaulted coolant temperature has exceeded a maximum threshold in order to make a pass/fail determination. No engine running status. No disabling faults present. Coolant sensor within range. Coolant temperature equal or greater than 122 degrees F. IAT equal or greater than 95 degrees F. **Possible Causes:** • Poor connection • Lack of engine coolant • Water pump problems • ECTS • Faulty PCM
DTC: P0219 **1T PCM** **Year:** 2009, 2010 **Model:** Genesis **Engine:** 4.6L V8 VIN F **Transmission:** All	**Engine Overspeed Condition:** Enable Conditions • Coolant sensor is normal • No disabling faults present(DTCs related to MAFS/MAPS, catalyst, fuel system or engine oil temperature sensor) • Coolant Temperature at startup < 45°C(113 °F) • Engine running state • Coolant temperature > 50°C(122 °F) • Intake air temperature < 35°C(95 °F) Threshold value • Coolant temperature ≥110°C (230 °F) (Average airflow< 30 g/sec and filtered airflow< 50 g/sec) Diagnosis Time • Once per driving cycle (about 2 minutes) **Possible Causes:** • ECTS
DTC: P0221 **2T CCM, MIL: Yes** **Year:** 2009, 2010 **Model:** Genesis, Genesis Coupe, Santa Fe, Sonata, Tucson **Engine:** 2.0L L4 VIN D, 2.4L L4 VIN B, 2.4L L4 VIN C, 3.5L V6 VIN G, 4.6L V8 VIN F **Transmission:** All	**Throttle/Pedal Position Sensor/Switch "B" Circuit Range/Performance:** Plausibility check between TPS1 and TPS2. No engine start mode. No TSP adaptation request. No relevant failure. **Possible Causes:** • Poor connection or damaged harness • Air leakage in intake system • Faulty TPS2
DTC: P0222 **2T CCM, MIL: Yes** **Year:** 2009, 2010 **Model:** Azera, Genesis, Genesis Coupe, Santa Fe, Sonata, Tucson, Veracruz **Engine:** 2.0L L4 VIN D, 2.4L L4 VIN B, 2.4L L4 VIN C, 2.7L V6 VIN D, 3.3L V6 VIN D, 3.3L V6 VIN E, 3.3L V6 VIN F, 3.5L V6 VIN G, 3.8L V6 VIN C, 3.8L V6 VIN E, 3.8L V6 VIN F, 3.8L V6 VIN H, 4.6L V8 VIN F **Transmission:** All	**Throttle/Pedal Position Sensor/Switch "B" Circuit Low Input:** The DTC is recorded if the output voltage of the TPS 1 is lower than threshold value (Vtps1 less than or equal to 0.2 volt). TPS 1 low input. **Possible Causes:** • Poor connection • Open or short to ground in TPS circuit • Faulty TPS • Faulty PCM
DTC: P0223 **2T CCM, MIL: Yes** **Year:** 2009, 2010 **Model:** Azera, Genesis, Genesis Coupe, Santa Fe, Sonata, Tucson, Veracruz **Engine:** 2.0L L4 VIN D, 2.4L L4 VIN B, 2.4L L4 VIN C, 2.7L V6 VIN D, 3.3L V6 VIN D, 3.3L V6 VIN E, 3.3L V6 VIN F, 3.5L V6 VIN G, 3.8L V6 VIN C, 3.8L V6 VIN E, 3.8L V6 VIN F, 3.8L V6 VIN H, 4.6L V8 VIN F **Transmission:** All	**Throttle/Pedal Position Sensor/Switch "B" Circuit High Input:** The DTC is recorded if the output voltage of the TPS 1 higher than threshold value (Vtps1 greater than or equal to 4.85 volts, load value, EV less than 70 percent) when TPS 2 (Vtps2 less than or equal to 2.5 volts) is normal. TPS 1 high input. **Possible Causes:** • Poor connection • Open or short to ground in TPS circuit • Faulty TPS • Faulty PCM

DTC	Trouble Code Title, Conditions & Possible Causes
DTC: P0230 **1T CCM, MIL: Yes** **Year:** 2009, 2010 **Model:** Accent, Azera, Elantra, Genesis, Genesis Coupe, Santa Fe, Sonata, Tucson, Veracruz **Engine:** 1.6L L4 VIN C, 2.0L L4 VIN B, 2.0L L4 VIN D, 2.0L L4 VIN E, 2.4L L4 VIN B, 2.4L L4 VIN C, 2.7L V6 VIN D, 3.3L V6 VIN D, 3.3L V6 VIN E, 3.3L V6 VIN F, 3.5L V6 VIN G, 3.8L V6 VIN C, 3.8L V6 VIN E, 3.8L V6 VIN F, 3.8L V6 VIN H, 4.6L V8 VIN F **Transmission:** All	**Fuel Pump Circuit Malfunction:** Key on, and then the PCM detected an unexpected voltage condition on the fuel pump circuit through the fuel pump monitoring input. **Possible Causes:** • Fuel pump control circuit is open or shorted to ground • Fuel pump relay power circuit from ignition switch is open • Fuel pump relay is damaged or has failed • PCM has failed
DTC: P0231 **2T CCM, MIL: Yes** **Year:** 2009, 2010 **Model:** Accent **Engine:** 1.6L L4 VIN C **Transmission:** All	**Electric Fuel Pump Relay Open Or Short Circuit:** Circuit continuity check, high. **Possible Causes:** • Poor connection • Short to power in control circuit • Fuel pump relay • Faulty ECM/PCM
DTC: P0232 **2T CCM, MIL: Yes** **Year:** 2009, 2010 **Model:** Accent **Engine:** 1.6L L4 VIN C **Transmission:** All	**Electric Fuel Pump Relay Short Circuit:** Circuit continuity check, low. **Possible Causes:** • Poor connection • Short to ground in control circuit • Fuel pump relay • Faulty ECM/PCM
DTC: P0234 **2T PCM, MIL: Yes** **Year:** 2010 **Model:** Genesis Coupe **Engine:** 2.0L L4 VIN D **Transmission:** All	**Turbocharger/Supercharger Overboost Condition:** Enable Conditions • charge pressure >530hPa (0.53 bar) • Failure not detected for DTCs Threshold Value • measured pressure up throttle >1400 . . . 2200hPa (1.4 bar . . . 2.2 bar) Diagnostic Time • 2 sec **Possible Causes:** • Inspect Air cleaner • intake side leakage • intake side Blocked • Inspect waste gate valve and Recirculation valve (RCV) control hose blocked /leak /assemble improperly. • Faulty Boost Sensor

DTC	Trouble Code Title, Conditions & Possible Causes
DTC: P0236 **2T PCM, MIL: Yes** **Year:** 2010 **Model:** Genesis Coupe **Engine:** 2.0L L4 VIN D **Transmission:** All	**Turbocharger/Supercharger Boost Sensor 'A' Circuit Range/Performance:** Enable Conditions Case1 Turbocharger boost sensor out of range (engine off) • engine off time > 10 sec • Vehicle speed <0.6 mph • 0.8V < PUT voltage < 4.9V • 0.1V < MAP voltage < 4.85V • 0.1V < pressure up throttle sensor voltage < 4.7V • engine operating state off Case2 Turbocharger boost sensor out of range (engine on) • engine operating state out of start • throttle position > -1°TPS(modeled throttle position setpoint) • 0.35 < Pressure quotient(Manifold pressure/Ambient Pressure < 1 Threshold Value Case1 Turbocharger boost sensor out of range (engine off) • turbocharger boost pressure-manifold air pressure > 100hPa • barometric pressure-turbocharger boost > 100hPa • barometric pressure-manifold air pressure < 100hPa Case2 Turbocharger boost sensor out of range (engine on) • barometric pressure-pressure up throttle > 500hPa • pressure up throttle-pressure up throttle(full load) < 0hPa • barometric pressure-pressure up throttle during full load > 2716hPa Diagnostic Time • immediately **Possible Causes:** • Boost Pressure Sensor(PUT)
DTC: P0237 **2T PCM, MIL: Yes** **Year:** 2010 **Model:** Genesis Coupe **Engine:** 2.0L L4 VIN D **Transmission:** All	**Turbocharger/Supercharger Boost Sensor 'A' Circuit Low:** Enable Conditions • 10V < Battery voltage <16V Threshold Value • PUT voltage < 0.92 V Diagnostic Time • 1 sec **Possible Causes:** • Short to ground in signal harness • Poor connection or damaged harness • Faulty PUT
DTC: P0238 **2T PCM, MIL: Yes** **Year:** 2010 **Model:** Genesis Coupe **Engine:** 2.0L L4 VIN D **Transmission:** All	**Turbocharger/Supercharger Boost Sensor 'A' Circuit High:** Enable Conditions • 10V < Battery voltage <16V Threshold Value • PUT voltage > 4.7 V Diagnostic Time • 1 sec **Possible Causes:** • Poor connection • Short to power in signal circuit • Open in ground circuit

DTC	Trouble Code Title, Conditions & Possible Causes
DTC: P0244 **2T PCM, MIL: Yes** **Year:** 2010 **Model:** Genesis Coupe **Engine:** 2.0L L4 VIN D **Transmission:** All	**Turbocharger/Supercharger Wastegate Sol. 'A' Range/Performance:** Enable Conditions • engine operating state out start Threshold Value case 1 • 100hPa < pressure charge air adaptation< -100hPa case 2 • 280hPa < pressure up throttle adaptation< -280hPa Diagnostic Time • immediately **Possible Causes:** • Waste gate valve
DTC: P0245 **2T PCM, MIL: Yes** **Year:** 2010 **Model:** Genesis Coupe **Engine:** 2.0L L4 VIN D **Transmission:** All	**Turbocharger/Supercharger Wastegate Sol. 'A' Low:** Enable Conditions • Waste gate PWM < 98.44% Threshold Value • ECM power outage diagnosis Diagnostic Time • 1 sec **Possible Causes:** • Poor connection • Short to power in signal circuit • Open in ground circuit
DTC: P0246 **2T PCM, MIL: Yes** **Year:** 2010 **Model:** Genesis Coupe **Engine:** 2.0L L4 VIN D **Transmission:** All	**Turbocharger/Supercharger Wastegate Sol. 'A' High:** Enable Conditions Case 1 • WGT PWM > 1.56% Case 2 • 1.56% < WGT PWM < 98.44% Threshold Value • WGT pressure >1400 . . . 2200hPa Diagnostic Time • 1 sec **Possible Causes:** • Open or short to battery in control harness. • Poor connection or damaged harness
DTC: P0261 **2T CCM, MIL: Yes** **Year:** 2009, 2010 **Model:** Accent, Azera, Elantra, Genesis, Genesis Coupe, Santa Fe, Sonata, Tucson, Veracruz **Engine:** 1.6L L4 VIN C, 2.0L L4 VIN B, 2.0L L4 VIN D, 2.0L L4 VIN E, 2.4L L4 VIN B, 2.4L L4 VIN C, 2.7L V6 VIN D, 3.3L V6 VIN D, 3.3L V6 VIN E, 3.3L V6 VIN F, 3.5L V6 VIN G, 3.8L V6 VIN C, 3.8L V6 VIN E, 3.8L V6 VIN F, 3.8L V6 VIN H, 4.6L V8 VIN F **Transmission:** All	**Cylinder 1- Injector Circuit Low:** The PCM sets the DTC if the control circuit is shorted to ground. Driver stage check. **Possible Causes:** • Open in power supply harness • Short to ground in control harness • Contact resistance in connectors • Faulty injector

DTC	Trouble Code Title, Conditions & Possible Causes
DTC: P0262 **2T CCM, MIL: Yes** **Year:** 2009, 2010 **Model:** Accent, Azera, Elantra, Genesis, Genesis Coupe, Santa Fe, Sonata, Tucson, Veracruz **Engine:** 1.6L L4 VIN C, 2.0L L4 VIN B, 2.0L L4 VIN D, 2.0L L4 VIN E, 2.4L L4 VIN B, 2.4L L4 VIN C, 2.7L V6 VIN D, 3.3L V6 VIN D, 3.3L V6 VIN E, 3.3L V6 VIN F, 3.5L V6 VIN G, 3.8L V6 VIN C, 3.8L V6 VIN E, 3.8L V6 VIN F, 3.8L V6 VIN H, 4.6L V8 VIN F **Transmission:** All	**Cylinder 1- Injector Circuit High:** The PCM sets the DTC if the control circuit is open or shorted to battery voltage. Driver stage check. **Possible Causes:** • Open or short to battery control harness • Contact resistance in connectors • Faulty injector
DTC: P0264 **2T CCM, MIL: Yes** **Year:** 2009, 2010 **Model:** Accent, Azera, Elantra, Genesis, Genesis Coupe, Santa Fe, Sonata, Tucson, Veracruz **Engine:** 1.6L L4 VIN C, 2.0L L4 VIN B, 2.0L L4 VIN D, 2.0L L4 VIN E, 2.4L L4 VIN B, 2.4L L4 VIN C, 2.7L V6 VIN D, 3.3L V6 VIN D, 3.3L V6 VIN E, 3.3L V6 VIN F, 3.5L V6 VIN G, 3.8L V6 VIN C, 3.8L V6 VIN E, 3.8L V6 VIN F, 3.8L V6 VIN H, 4.6L V8 VIN F **Transmission:** All	**Cylinder 2- Injector Circuit Low:** The PCM sets the DTC if the control circuit is shorted to ground. Driver stage check. **Possible Causes:** • Open in power supply harness • Short to ground in control harness • Contact resistance in connectors • Faulty injector
DTC: P0265 **2T CCM, MIL: Yes** **Year:** 2009, 2010 **Model:** Accent, Azera, Elantra, Genesis, Genesis Coupe, Santa Fe, Sonata, Tucson, Veracruz **Engine:** 1.6L L4 VIN C, 2.0L L4 VIN B, 2.0L L4 VIN D, 2.0L L4 VIN E, 2.4L L4 VIN B, 2.4L L4 VIN C, 2.7L V6 VIN D, 3.3L V6 VIN D, 3.3L V6 VIN E, 3.3L V6 VIN F, 3.5L V6 VIN G, 3.8L V6 VIN C, 3.8L V6 VIN E, 3.8L V6 VIN F, 3.8L V6 VIN H, 4.6L V8 VIN F **Transmission:** All	**Cylinder 2- Injector Circuit High:** The PCM sets the DTC if the control circuit is open or shorted to battery voltage. Driver stage check. **Possible Causes:** • Open or short to battery control harness • Contact resistance in connectors • Faulty injector
DTC: P0267 **2T CCM, MIL: Yes** **Year:** 2009, 2010 **Model:** Accent, Azera, Elantra, Genesis, Genesis Coupe, Santa Fe, Sonata, Tucson, Veracruz **Engine:** 1.6L L4 VIN C, 2.0L L4 VIN B, 2.0L L4 VIN D, 2.0L L4 VIN E, 2.4L L4 VIN B, 2.4L L4 VIN C, 2.7L V6 VIN D, 3.3L V6 VIN D, 3.3L V6 VIN E, 3.3L V6 VIN F, 3.5L V6 VIN G, 3.8L V6 VIN C, 3.8L V6 VIN E, 3.8L V6 VIN F, 3.8L V6 VIN H, 4.6L V8 VIN F **Transmission:** All	**Cylinder 3- Injector Circuit Low:** The PCM sets the DTC if the control circuit is shorted to ground. Driver stage check. **Possible Causes:** • Open in power supply harness • Short to ground in control harness • Contact resistance in connectors • Faulty injector

DTC	Trouble Code Title, Conditions & Possible Causes
DTC: P0268 **2T CCM, MIL: Yes** **Year:** 2009, 2010 **Model:** Accent, Azera, Elantra, Genesis, Genesis Coupe, Santa Fe, Sonata, Tucson, Veracruz **Engine:** 1.6L L4 VIN C, 2.0L L4 VIN B, 2.0L L4 VIN D, 2.0L L4 VIN E, 2.4L L4 VIN B, 2.4L L4 VIN C, 2.7L V6 VIN D, 3.3L V6 VIN D, 3.3L V6 VIN E, 3.3L V6 VIN F, 3.5L V6 VIN G, 3.8L V6 VIN C, 3.8L V6 VIN E, 3.8L V6 VIN F, 3.8L V6 VIN H, 4.6L V8 VIN F **Transmission:** All	**Cylinder 3- Injector Circuit High:** The PCM sets the DTC if the control circuit is open or shorted to battery voltage. Driver stage check. **Possible Causes:** • Open or short to battery control harness • Contact resistance in connectors • Faulty injector
DTC: P0270 **2T CCM, MIL: Yes** **Year:** 2009, 2010 **Model:** Accent, Azera, Elantra, Genesis, Genesis Coupe, Santa Fe, Sonata, Tucson, Veracruz **Engine:** 1.6L L4 VIN C, 2.0L L4 VIN B, 2.0L L4 VIN D, 2.0L L4 VIN E, 2.4L L4 VIN B, 2.4L L4 VIN C, 2.7L V6 VIN D, 3.3L V6 VIN D, 3.3L V6 VIN E, 3.3L V6 VIN F, 3.5L V6 VIN G, 3.8L V6 VIN C, 3.8L V6 VIN E, 3.8L V6 VIN F, 3.8L V6 VIN H, 4.6L V8 VIN F **Transmission:** All	**Cylinder 4- Injector Circuit Low:** The PCM sets the DTC if the control circuit is shorted to ground. Driver stage check. **Possible Causes:** • Open in power supply harness • Short to ground in control harness • Contact resistance in connectors • Faulty injector
DTC: P0271 **2T CCM, MIL: Yes** **Year:** 2009, 2010 **Model:** Accent, Azera, Elantra, Genesis, Genesis Coupe, Santa Fe, Sonata, Tucson, Veracruz **Engine:** 1.6L L4 VIN C, 2.0L L4 VIN B, 2.0L L4 VIN D, 2.0L L4 VIN E, 2.4L L4 VIN B, 2.4L L4 VIN C, 2.7L V6 VIN D, 3.3L V6 VIN D, 3.3L V6 VIN E, 3.3L V6 VIN F, 3.5L V6 VIN G, 3.8L V6 VIN C, 3.8L V6 VIN E, 3.8L V6 VIN F, 3.8L V6 VIN H, 4.6L V8 VIN F **Transmission:** All	**Cylinder 4- Injector Circuit High:** The PCM sets the DTC if the control circuit is open or shorted to battery voltage. Driver stage check. **Possible Causes:** • Open or short to battery control harness • Contact resistance in connectors • Faulty injector
DTC: P0273 **2T CCM, MIL: Yes** **Year:** 2009, 2010 **Model:** Azera, Genesis, Genesis Coupe, Santa Fe, Sonata **Engine:** 2.7L V6 VIN D, 3.3L V6 VIN D, 3.3L V6 VIN E, 3.3L V6 VIN F, 3.5L V6 VIN G, 3.8L V6 VIN E, 3.8L V6 VIN F, 3.8L V6 VIN H, 4.6L V8 VIN F **Transmission:** All	**Cylinder 5- Injector Circuit Low:** The PCM sets the DTC if the control circuit is shorted to ground. Driver stage check. **Possible Causes:** • Open in power supply harness • Short to ground in control harness • Contact resistance in connectors • Faulty injector

DTC	Trouble Code Title, Conditions & Possible Causes
DTC: P0274 **2T CCM, MIL: Yes** **Year:** 2009, 2010 **Model:** Azera, Genesis, Genesis Coupe, Santa Fe, Sonata, Tucson, Veracruz **Engine:** 2.7L V6 VIN D, 3.3L V6 VIN D, 3.3L V6 VIN E, 3.3L V6 VIN F, 3.5L V6 VIN G, 3.8L V6 VIN C, 3.8L V6 VIN E, 3.8L V6 VIN F, 3.8L V6 VIN H, 4.6L V8 VIN F **Transmission:** All	**Cylinder 5- Injector Circuit High:** The PCM sets the DTC if the control circuit is open or shorted to battery voltage. Driver stage check. **Possible Causes:** • Open or short to battery control harness • Contact resistance in connectors • Faulty injector
DTC: P0276 **2T CCM, MIL: Yes** **Year:** 2009, 2010 **Model:** Azera, Genesis, Genesis Coupe, Santa Fe, Sonata, Tucson, Veracruz **Engine:** 2.7L V6 VIN D, 3.3L V6 VIN D, 3.3L V6 VIN E, 3.3L V6 VIN F, 3.5L V6 VIN G, 3.8L V6 VIN C, 3.8L V6 VIN E, 3.8L V6 VIN F, 3.8L V6 VIN H, 4.6L V8 VIN F **Transmission:** All	**Cylinder 6- Injector Circuit Low:** The PCM sets the DTC if the control circuit is shorted to ground. Driver stage check. **Possible Causes:** • Open in power supply harness • Short to ground in control harness • Contact resistance in connectors • Faulty injector
DTC: P0277 **2T CCM, MIL: Yes** **Year:** 2009, 2010 **Model:** Azera, Genesis, Genesis Coupe, Santa Fe, Sonata, Tucson, Veracruz **Engine:** 2.7L V6 VIN D, 3.3L V6 VIN D, 3.3L V6 VIN E, 3.3L V6 VIN F, 3.5L V6 VIN G, 3.8L V6 VIN C, 3.8L V6 VIN E, 3.8L V6 VIN F, 3.8L V6 VIN H, 4.6L V8 VIN F **Transmission:** All	**Cylinder 6- Injector Circuit High:** The PCM sets the DTC if the control circuit is open or shorted to battery voltage. Driver stage check. **Possible Causes:** • Open or short to battery control harness • Contact resistance in connectors • Faulty injector
DTC: P0279 **2T PCM, MIL: Yes** **Year:** 2009, 2010 **Model:** Genesis **Engine:** 4.6L V8 VIN F **Transmission:** All	**Cylinder 7 Injector Circuit Low:** Enable Conditions • - Threshold value • Short to ground Diagnosis Time • 5 sec **Possible Causes:** • Poor connection • Open or short to ground in power harness • Open or short to ground in control harness • Injector • ECM

DTC	Trouble Code Title, Conditions & Possible Causes
DTC: P0280 **2T PCM, MIL: Yes** **Year:** 2009, 2010 **Model:** Genesis **Engine:** 4.6L V8 VIN F **Transmission:** All	**Cylinder 7 - Injector Circuit High:** Enable Conditions • - Threshold value • Short to battery Diagnosis Time • 5 sec **Possible Causes:** • Poor connection • Short to battery in harness • Injector • ECM
DTC: P0282 **2T PCM, MIL: Yes** **Year:** 2009, 2010 **Model:** Genesis **Engine:** 4.6L V8 VIN F **Transmission:** All	**Cylinder 8 - Injector Circuit Low:** Enable Conditions • - Threshold value • Short to ground Diagnosis Time • 5 sec **Possible Causes:** • Poor connection • Open or short to ground in power harness • Open or short to ground in control harness • Injector • ECM
DTC: P0283 **2T PCM, MIL: Yes** **Year:** 2009, 2010 **Model:** Genesis **Engine:** 4.6L V8 VIN F **Transmission:** All	**Cylinder 8 - Injector Circuit High:** Enable Conditions • - Threshold value • Short to battery Diagnosis Time • 5 sec **Possible Causes:** • Poor connection • Short to battery in harness • Injector • ECM

DTC	Trouble Code Title, Conditions & Possible Causes
DTC: P0299 **T PCM** **Year:** 2010 **Model:** Genesis Coupe **Engine:** 2.0L L4 VIN D **Transmission:** All	**Turbocharger/Supercharger Underboost:** Enable Conditions Case1 • engine speed > 2016 ~ 3008 rpm • ambient pressure > 600 hpa • Fuel tank level> 8.6 % • ECT Temp. > -9.75 °C • Pressure quotient(Manifold pressure/Ambient Pressure) > 1.2 • Failure not detected for DTCs Case2 • Wastegate PWM < 20.31 % • mass air flow > 70| kg/h • engine speed > 2016 ~ 3008 rpm • ambient pressure > 600 hpa • Fuel tank level > 8.6 % • ECT Temp.> -9.75 °C • Pressure quotient(Manifold pressure/Ambient Pressure) > 1.2 • Failure not detected for DTCs Threshold Value Case1 • measured pressure up throttle< 600 ~ 1700 hpa(charge air pressure too low) • pressure up throttle deviation > 100 hpa Case2 • pressure up throttle-ambient pressure < 40 ~151.41 hpa(basic charge air pressure too low) Diagnostic Time • 2.2 sec **Possible Causes:** • Inspect Air cleaner • Inspect intake air line • Inspect vacuum hose • Faulty RCV • Faulty Wastegate
DTC: P0300 **2T MISFIRE, MIL: Yes** **Year:** 2009, 2010 **Model:** Accent, Azera, Elantra, Genesis, Genesis Coupe, Santa Fe, Sonata, Tucson, Veracruz **Engine:** 1.6L L4 VIN C, 2.0L L4 VIN B, 2.0L L4 VIN D, 2.0L L4 VIN E, 2.4L L4 VIN B, 2.4L L4 VIN C, 2.7L V6 VIN D, 3.3L V6 VIN D, 3.3L V6 VIN E, 3.3L V6 VIN F, 3.5L V6 VIN G, 3.8L V6 VIN C, 3.8L V6 VIN E, 3.8L V6 VIN F, 3.8L V6 VIN H, 4.6L V8 VIN F **Transmission:** All	**Random/Multiple Misfire Detected:** Engine started, vehicle driven to over 3 mph at an engine speed of 400-3500 rpm, and the PCM detected irregular CKP sensor signals indicating a misfire condition in two or more cylinders was present during the 200-revolution or 1000-revolution Misfire Monitor test. **NOTE: If the misfire is severe, the MIL will flash on/off on the 1st trip!** **Possible Causes:** • Ignition system or fuel metering fault in 2 or more cylinders • Fuel pressure too low or too high, fuel supply contaminated • CKP/CMP signals, EVAP canister saturated, EGR valve stuck

DTC	Trouble Code Title, Conditions & Possible Causes
DTC: P0301 **2T MISFIRE, MIL: Yes** **Year:** 2009, 2010 **Model:** Accent, Azera, Elantra, Genesis, Genesis Coupe, Santa Fe, Sonata, Tucson, Veracruz **Engine:** 1.6L L4 VIN C, 2.0L L4 VIN B, 2.0L L4 VIN D, 2.0L L4 VIN E, 2.4L L4 VIN B, 2.4L L4 VIN C, 2.7L V6 VIN D, 3.3L V6 VIN D, 3.3L V6 VIN E, 3.3L V6 VIN F, 3.5L V6 VIN G, 3.8L V6 VIN C, 3.8L V6 VIN E, 3.8L V6 VIN F, 3.8L V6 VIN H, 4.6L V8 VIN F **Transmission:** All	**Cylinder 1 Misfire Detected:** Engine started, vehicle driven to over 3 mph at an engine speed of 400-3500 rpm, and the PCM detected irregular CKP sensor signals indicating a misfire condition in one cylinder was present during the 200-revolution or 1000-revolution Misfire Detection Monitor test. **NOTE: If the misfire is severe, the MIL will flash on/off on the 1st trip!** **Possible Causes:** • Base engine mechanical fault that affects only one cylinder • Fuel metering fault that affects only one cylinder • EGR valve is stuck open or the PCV system has a vacuum leak • Ignition system fault (i.e., a coil) that affects only one cylinder
DTC: P0302 **2T MISFIRE, MIL: Yes** **Year:** 2009, 2010 **Model:** Accent, Azera, Elantra, Genesis, Genesis Coupe, Santa Fe, Sonata, Tucson, Veracruz **Engine:** 1.6L L4 VIN C, 2.0L L4 VIN B, 2.0L L4 VIN D, 2.0L L4 VIN E, 2.4L L4 VIN B, 2.4L L4 VIN C, 2.7L V6 VIN D, 3.3L V6 VIN D, 3.3L V6 VIN E, 3.3L V6 VIN F, 3.5L V6 VIN G, 3.8L V6 VIN C, 3.8L V6 VIN E, 3.8L V6 VIN F, 3.8L V6 VIN H, 4.6L V8 VIN F **Transmission:** All	**Cylinder 2 Misfire Detected:** Engine started, vehicle driven to over 3 mph at an engine speed of 400-3500 rpm, and the PCM detected irregular CKP sensor signals indicating a misfire condition in one cylinder was present during the 200-revolution or 1000-revolution Misfire Detection Monitor test. **NOTE: If the misfire is severe, the MIL will flash on/off on the 1st trip!** **Possible Causes:** • Base engine mechanical fault that affects only one cylinder • Fuel metering fault that affects only one cylinder • EGR valve is stuck open or the PCV system has a vacuum leak • Ignition system fault (i.e., a coil) that affects only one cylinder
DTC: P0303 **2T MISFIRE, MIL: Yes** **Year:** 2009, 2010 **Model:** Accent, Azera, Elantra, Genesis, Genesis Coupe, Santa Fe, Sonata, Tucson, Veracruz **Engine:** 1.6L L4 VIN C, 2.0L L4 VIN B, 2.0L L4 VIN D, 2.0L L4 VIN E, 2.4L L4 VIN B, 2.4L L4 VIN C, 2.7L V6 VIN D, 3.3L V6 VIN D, 3.3L V6 VIN E, 3.3L V6 VIN F, 3.5L V6 VIN G, 3.8L V6 VIN C, 3.8L V6 VIN E, 3.8L V6 VIN F, 3.8L V6 VIN H, 4.6L V8 VIN F **Transmission:** All	**Cylinder 3 Misfire Detected:** Engine started, vehicle driven to over 3 mph at an engine speed of 400-3500 rpm, and the PCM detected irregular CKP sensor signals indicating a misfire condition in one cylinder was present during the 200-revolution or 1000-revolution Misfire Detection Monitor test. **NOTE: If the misfire is severe, the MIL will flash on/off on the 1st trip!** **Possible Causes:** • Base engine mechanical fault that affects only one cylinder • Fuel metering fault that affects only one cylinder • EGR valve is stuck open or the PCV system has a vacuum leak • Ignition system fault (i.e., a coil) that affects only one cylinder
DTC: P0304 **2T MISFIRE, MIL: Yes** **Year:** 2009, 2010 **Model:** Accent, Azera, Elantra, Genesis, Genesis Coupe, Santa Fe, Sonata, Tucson, Veracruz **Engine:** 1.6L L4 VIN C, 2.0L L4 VIN B, 2.0L L4 VIN D, 2.0L L4 VIN E, 2.4L L4 VIN B, 2.4L L4 VIN C, 2.7L V6 VIN D, 3.3L V6 VIN D, 3.3L V6 VIN E, 3.3L V6 VIN F, 3.5L V6 VIN G, 3.8L V6 VIN C, 3.8L V6 VIN E, 3.8L V6 VIN F, 3.8L V6 VIN H, 4.6L V8 VIN F **Transmission:** All	**Cylinder 4 Misfire Detected:** Engine speed from 400-3500 rpm, VSS input over 3 mph, and the PCM detected irregular CKP inputs indicating a misfire condition in one cylinder during the 200-revolution or 1000-revolution test period. **NOTE: If the misfire is severe, the MIL will flash on/off on the 1st trip!** **Possible Causes:** • Base engine mechanical fault that affects only one cylinder • Fuel metering fault that affects only one cylinder • EGR valve is stuck open or the PCV system has a vacuum leak • Ignition system fault (i.e., a coil) that affects only one cylinder

DTC	Trouble Code Title, Conditions & Possible Causes
DTC: P0305 **2T MISFIRE, MIL: Yes** **Year:** 2009, 2010 **Model:** Azera, Genesis, Genesis Coupe, Santa Fe, Sonata, Tucson, Veracruz **Engine:** 2.7L V6 VIN D, 3.3L V6 VIN D, 3.3L V6 VIN E, 3.3L V6 VIN F, 3.5L V6 VIN G, 3.8L V6 VIN C, 3.8L V6 VIN E, 3.8L V6 VIN F, 3.8L V6 VIN H, 4.6L V8 VIN F **Transmission:** All	**Cylinder 5 Misfire Detected:** Engine started, vehicle driven to over 3 mph at an engine speed of 400-3500 rpm, and the PCM detected irregular CKP sensor signals indicating a misfire condition in one cylinder was present during the 200-revolution or 1000-revolution Misfire Detection Monitor test. **NOTE: If the misfire is severe, the MIL will flash on/off on the 1st trip!** **Possible Causes:** • Base engine mechanical fault that affects only one cylinder • Fuel metering fault that affects only one cylinder • EGR valve is stuck open or the PCV system has a vacuum leak • Ignition system fault (i.e., a coil) that affects only one cylinder
DTC: P0306 **2T MISFIRE, MIL: Yes** **Year:** 2009, 2010 **Model:** Azera, Genesis, Genesis Coupe, Santa Fe, Sonata, Tucson, Veracruz **Engine:** 2.7L V6 VIN D, 3.3L V6 VIN D, 3.3L V6 VIN E, 3.3L V6 VIN F, 3.5L V6 VIN G, 3.8L V6 VIN C, 3.8L V6 VIN E, 3.8L V6 VIN F, 3.8L V6 VIN H, 4.6L V8 VIN F **Transmission:** All	**Cylinder 6 Misfire Detected:** Engine started, vehicle driven to over 3 mph at an engine speed of 400-3500 rpm, and the PCM detected irregular CKP sensor signals indicating a misfire condition in one cylinder was present during the 200-revolution or 1000-revolution Misfire Detection Monitor test. **NOTE: If the misfire is severe, the MIL will flash on/off on the 1st trip!** **Possible Causes:** • Base engine mechanical fault that affects only one cylinder • Fuel metering fault that affects only one cylinder • EGR valve is stuck open or the PCV system has a vacuum leak • Ignition system fault (i.e., a coil) that affects only one cylinder
DTC: P0307 **T PCM, MIL: Yes** **Year:** 2009, 2010 **Model:** Genesis **Engine:** 4.6L V8 VIN F **Transmission:** All	**Cylinder 7 - Misfire Detected:** Case1 Threshold Value • Misfire rate > 3.25% Diagnostic Time • 1000 rev MIL ON Condition • 2 driving cycle Case2 Threshold Value • Misfire rate > 2.4~11.7% Diagnostic Time • 200 rev MIL ON Condition • Immediately (Blink) **Possible Causes:** • Poor connection • Ignition system • Fuel system • Intake/exhaust air system • Ignition timing • Injector

DTC	Trouble Code Title, Conditions & Possible Causes
DTC: P0308 **T PCM, MIL: Yes** **Year:** 2009, 2010 **Model:** Genesis **Engine:** 4.6L V8 VIN F **Transmission:** All	**Cylinder 8 - Misfire Detected:** Case1 Threshold Value • Misfire rate > 3.25% Diagnostic Time • 1000 rev MIL ON Condition • 2 driving cycle Case2 Threshold Value • Misfire rate > 2.4~11.7% Diagnostic Time • 200 rev MIL ON Condition • Immediately (Blink) **Possible Causes:** • Poor connection • Ignition system • Fuel system • Intake/exhaust air system • Ignition timing • Injector
DTC: P0315 **2T PCM** **Year:** 2009, 2010 **Model:** Genesis, Genesis Coupe, Santa Fe, Sonata, Tucson, Veracruz **Engine:** 2.0L L4 VIN B, 2.0L L4 VIN D, 2.4L L4 VIN B, 2.4L L4 VIN C, 2.7L V6 VIN D, 3.5L V6 VIN G, 3.8L V6 VIN C, 3.8L V6 VIN E, 3.8L V6 VIN H **Transmission:** All	**Camshaft Position (CMP) Sensor Error:** • 10% ≤Engine load < 90% • 2000 rpm ≤engine speed ≤4000 rpm • TEC(Tooth Error Correction) RPM stability timer > 10sec • 0°C(32°F) < coolant temperature < 110°C(230°F) • Not active disabling faults **Possible Causes:** • Loosened CKPS • Target wheel • PCM
DTC: P0315 **2T CCM, MIL: Yes** **Year:** 2009, 2010 **Model:** Azera, Elantra, Santa Fe, Sonata **Engine:** 2.0L L4 VIN D, 2.0L L4 VIN E, 2.7L V6 VIN D, 3.3L V6 VIN D, 3.3L V6 VIN E, 3.3L V6 VIN F, 3.8L V6 VIN F **Transmission:** All	**Segment Time Acquisition Incorrect:** A misfire induces a decrease in the engine speed and causes a variation in the segment period. Monitor segment time adaptation. **Possible Causes:** • Improperly installed target wheel • Contact resistance in connectors
DTC: P0325 **2T CCM, MIL: Yes** **Year:** 2009, 2010 **Model:** Azera, Elantra, Genesis, Genesis Coupe, Santa Fe, Sonata, Tucson, Veracruz **Engine:** 2.0L L4 VIN B, 2.0L L4 VIN D, 2.0L L4 VIN E, 2.7L V6 VIN D, 3.3L V6 VIN D, 3.3L V6 VIN E, 3.3L V6 VIN F, 3.5L V6 VIN G, 3.8L V6 VIN C, 3.8L V6 VIN E, 3.8L V6 VIN F, 3.8L V6 VIN H **Transmission:** All	**Knock Sensor 1 Circuit Malfunction:** Engine started, engine running, and the PCM detected an unexpected voltage condition on the Knock Sensor 1 circuit. **Possible Causes:** • KS signal circuit is open, shorted to ground or to system power • KS ground circuit is open between sensor and PCM • Knock sensor is damaged or it has failed • PCM has failed

DTC	Trouble Code Title, Conditions & Possible Causes
DTC: P0326 **2T CCM, MIL: Yes** **Year:** 2009, 2010 **Model:** Accent, Azera, Genesis, Genesis Coupe, Santa Fe, Sonata, Tucson, Veracruz **Engine:** 1.6L L4 VIN C, 2.0L L4 VIN D, 2.4L L4 VIN B, 2.4L L4 VIN C, 2.7L V6 VIN D, 3.3L V6 VIN D, 3.3L V6 VIN E, 3.3L V6 VIN F, 3.5L V6 VIN G, 3.8L V6 VIN C, 3.8L V6 VIN E, 3.8L V6 VIN F, 3.8L V6 VIN H **Transmission:** All	**Knock Sensor Circuit Malfunction (Bank 1):** Engine speed from 1000-2200 rpm, ECT sensor signal more than 104°F, engine load more than 2 ms, and the PCM detected the Knock sensor signal was out of range at a calculated engine speed. **Possible Causes:** • Knock sensor signal circuit open or shorted to ground • Knock sensor signal circuit shorted to VREF or system power • Knock sensor is damaged or has failed • PCM has failed
DTC: P0327 **2T CCM, MIL: Yes** **Year:** 2009, 2010 **Model:** Accent, Genesis **Engine:** 1.6L L4 VIN C, 4.6L V8 VIN F **Transmission:** All	**Knock Sensor 1 Circuit Low Input:** Engine speed greater than 2600 rpm. **Possible Causes:** • Poor connection • Open or short to ground in signal circuit • Knock sensor is damaged or has failed • PCM/ECM has failed
DTC: P0328 **2T CCM, MIL: Yes** **Year:** 2009, 2010 **Model:** Accent, Genesis **Engine:** 1.6L L4 VIN C, 4.6L V8 VIN F **Transmission:** All	**Knock Sensor 1 Circuit High Input:** Coolant temperature more than 104 degrees F. **Possible Causes:** • Poor connection • Short to power in signal circuit • Knock sensor is damaged or has failed • PCM/ECM has failed
DTC: P032C **2T PCM, MIL: Yes** **Year:** 2009, 2010 **Model:** Genesis **Engine:** 4.6L V8 VIN F **Transmission:** All	**Knock Sensor 3 Circuit Low Input :** Enable Conditions • ECT > 40C • Engine speed > 2400rpm • Engine load > 35% Threshold value • Normalized reference level < lower limit, f(rpm) Diagnosis Time • Continuous **Possible Causes:** • Poor connection • Open or short to ground in signal circuit • Knock sensor
DTC: P032D **2T PCM, MIL: Yes** **Year:** 2009, 2010 **Model:** Genesis **Engine:** 4.6L V8 VIN F **Transmission:** All	**Knock Sensor 3 Circuit High Input:** Enable Conditions • ECT > 40C • Engine speed > 2400rpm • Engine load > 35% Threshold value • Normalized reference level > upper limit, f(rpm) Diagnosis Time • Continuous **Possible Causes:** • Poor connection • Short to power in signal circuit • Knock sensor

DTC	Trouble Code Title, Conditions & Possible Causes
DTC: P0330 **2T CCM, MIL: Yes** **Year:** 2009, 2010 **Model:** Azera, Genesis, Genesis Coupe, Santa Fe, Sonata, Tucson, Veracruz **Engine:** 2.7L V6 VIN D, 3.3L V6 VIN D, 3.3L V6 VIN E, 3.3L V6 VIN F, 3.5L V6 VIN G, 3.8L V6 VIN C, 3.8L V6 VIN E, 3.8L V6 VIN F, 3.8L V6 VIN H **Transmission:** All	**Knock Sensor 2 Circuit Malfunction:** Engine started, engine running, and the PCM detected an unexpected voltage condition on the Knock Sensor 2 circuit. **Possible Causes:** • KS signal circuit is open, shorted to ground or to system power • KS ground circuit is open between sensor and PCM • Knock sensor is damaged or it has failed • PCM has failed
DTC: P0331 **2T CCM, MIL: Yes** **Year:** 2009, 2010 **Model:** Azera, Genesis, Genesis Coupe, Santa Fe, Sonata, Veracruz **Engine:** 2.7L V6 VIN D, 3.3L V6 VIN D, 3.3L V6 VIN E, 3.3L V6 VIN F, 3.5L V6 VIN G, 3.8L V6 VIN C, 3.8L V6 VIN E, 3.8L V6 VIN F, 3.8L V6 VIN H **Transmission:** All	**Knock Sensor 2 Circuit Range/Performance (Bank 2):** Signal short. Pressure in intake manifold is normal. Engine speed is equal to or less than 1600 rpm. **Possible Causes:** • Poor connection • Short in harness • Faulty knock sensor • Faulty PCM
DTC: P0332 **2T PCM, MIL: Yes** **Year:** 2009, 2010 **Model:** Genesis **Engine:** 4.6L V8 VIN F **Transmission:** All	**Knock Sensor 2 Circuit Low (Bank 2):** Enable Conditions • ECT > 40C • Engine speed > 2400rpm • Engine load > 35% Threshold value • Normalized reference level < lower limit, f(rpm) Diagnosis Time • Continuous **Possible Causes:** • Poor connection • Open or short to ground in signal circuit • Knock sensor
DTC: P0333 **2T PCM, MIL: Yes** **Year:** 2009, 2010 **Model:** Genesis **Engine:** 4.6L V8 VIN F **Transmission:** All	**Knock Sensor 2 Circuit High (Bank 2):** Enable Conditions • ECT > 40C • Engine speed > 1600rpm • Engine load > 35% Threshold value • Normalized reference level > upper limit, f(rpm) Diagnosis Time • Continuous **Possible Causes:** • Poor connection • Short to power in signal circuit • Knock sensor

DTC	Trouble Code Title, Conditions & Possible Causes
DTC: P0335 **1T CCM, MIL: Yes** **Year:** 2009, 2010 **Model:** Accent, Azera, Elantra, Genesis, Genesis Coupe, Santa Fe, Sonata, Tucson, Veracruz **Engine:** 1.6L L4 VIN C, 2.0L L4 VIN B, 2.0L L4 VIN D, 2.0L L4 VIN E, 2.4L L4 VIN B, 2.4L L4 VIN C, 2.7L V6 VIN D, 3.3L V6 VIN D, 3.3L V6 VIN E, 3.3L V6 VIN F, 3.5L V6 VIN G, 3.8L V6 VIN C, 3.8L V6 VIN E, 3.8L V6 VIN F, 3.8L V6 VIN H, 4.6L V8 VIN F **Transmission:** All	**Crankshaft Position Sensor 'A' Circuit Malfunction:** Engine cranking for over 2 seconds, CMP sensor signals detected, and the PCM did not receive any CKP sensor signals during the test. **Possible Causes:** • CKP (Magnetic) sensor signal (+) or (-) circuit is open or shorted to ground between the sensor and the PCM • CKP (Magnetic) sensor signal is damaged or has failed • PCM has failed
DTC: P0336 **2T PCM, MIL: Yes** **Year:** 2009, 2010 **Model:** Accent, Azera, Genesis, Genesis Coupe, Santa Fe, Sonata, Tucson, Veracruz **Engine:** 1.6L L4 VIN C, 2.0L L4 VIN D, 2.4L L4 VIN B, 2.4L L4 VIN C, 2.7L V6 VIN D, 3.3L V6 VIN D, 3.3L V6 VIN E, 3.3L V6 VIN F, 3.5L V6 VIN G, 3.8L V6 VIN C, 3.8L V6 VIN E, 3.8L V6 VIN F, 3.8L V6 VIN H, 4.6L V8 VIN F **Transmission:** All	**Crankshaft Position Sensor 'A' Circuit Range/Performance:** Enable Conditions • IG "ON", Cranking or engine-off during driving • No DTC related to CAM • Camshaft position sensor state change Threshold value • Implausible signal counter > 32 times Diagnosis Time • Continuous **Possible Causes:** • Poor connection • Noise • Short in harness • Target wheel • ECM
DTC: P0337 **2T CCM, MIL: Yes** **Year:** 2009, 2010 **Model:** Accent **Engine:** 1.6L L4 VIN C **Transmission:** All	**Crankshaft Position Sensor "A" Circuit Low Input:** If the output voltage of the CKPS remains low for more than two seconds. When the change of the CMPS output voltage is zero, the PCM determines a fault and stores a code. Change in output voltage (delta sign Vckp) is monitored. **Possible Causes:** • Poor connection • Open or short to ground in CKPS circuit • Faulty CKPS • Faulty PCM
DTC: P0338 **2T CCM, MIL: Yes** **Year:** 2009, 2010 **Model:** Accent **Engine:** 1.6L L4 VIN C **Transmission:** All	**Crankshaft Position Sensor "A" Circuit High Input:** If the output voltage of the CKPS remains high for more than two seconds. When the change of the CMPS output voltage is zero, the PCM determines a fault and stores a code. Change in output voltage (delta sign Vckp) is monitored. **Possible Causes:** • Poor connection • Open or short to ground in CKPS circuit • Faulty CKPS • Faulty PCM
DTC: P0339 **2T CCM, MIL: Yes** **Year:** 2009, 2010 **Model:** Accent **Engine:** 1.6L L4 VIN C **Transmission:** All	**Crankshaft Position Sensor "A" Circuit:** Signal check. Edge counter of camshaft position sensor 8. **Possible Causes:** • Poor connection • Open or short in signal circuit • CKPS • Faulty ECM/PCM

DTC	Trouble Code Title, Conditions & Possible Causes
DTC: P033C **2T PCM, MIL: Yes** **Year:** 2009, 2010 **Model:** Genesis **Engine:** 4.6L V8 VIN F **Transmission:** All	**Knock Sensor 4 Circuit Low Input:** Enable Conditions • ECT > 40C • Engine speed > 2400rpm • Engine load > 35% Threshold value • Normalized reference level > upper limit, f(rpm) Diagnosis Time • Continuous **Possible Causes:** • Poor connection • Open or short to ground in signal circuit • Knock sensor
DTC: P033D **2T PCM, MIL: Yes** **Year:** 2009, 2010 **Model:** Genesis **Engine:** 4.6L V8 VIN F **Transmission:** All	**Knock Sensor 4 Circuit High Input:** Enable Conditions • ECT > 40C • Engine speed > 1600rpm • Engine load > 35% Threshold value • Normalized reference level > upper limit, f(rpm) Diagnosis Time • Continuous **Possible Causes:** • Poor connection • Short to power in signal circuit • Knock sensor
DTC: P0340 **2T PCM, MIL: Yes** **Year:** 2009, 2010 **Model:** Accent, Azera, Elantra, Genesis, Genesis Coupe, Santa Fe, Sonata, Tucson, Veracruz **Engine:** 1.6L L4 VIN C, 2.0L L4 VIN B, 2.0L L4 VIN D, 2.0L L4 VIN E, 2.4L L4 VIN B, 2.4L L4 VIN C, 2.7L V6 VIN D, 3.3L V6 VIN D, 3.3L V6 VIN E, 3.3L V6 VIN F, 3.5L V6 VIN G, 3.8L V6 VIN C, 3.8L V6 VIN E, 3.8L V6 VIN F, 3.8L V6 VIN H, 4.6L V8 VIN F **Transmission:** All	**Camshaft Position Sensor 'A' Circuit (Bank 1 Or Single Sensor):** Enable Conditions • Reference mark found • No CKP sensor error • No 1 tooth off error Threshold value • Abnormal phase edges (High or Low) (No signal counter > 2 times) Diagnosis Time • Continuous **Possible Causes:** • Poor connection • Open in harness • CMPS(B1-Intake) • ECM
DTC: P0341 **2T PCM, MIL: Yes** **Year:** 2009, 2010 **Model:** Accent, Azera, Genesis, Genesis Coupe, Santa Fe, Sonata, Tucson, Veracruz **Engine:** 1.6L L4 VIN C, 2.0L L4 VIN D, 2.4L L4 VIN B, 2.4L L4 VIN C, 2.7L V6 VIN D, 3.3L V6 VIN D, 3.3L V6 VIN E, 3.3L V6 VIN F, 3.5L V6 VIN G, 3.8L V6 VIN C, 3.8L V6 VIN E, 3.8L V6 VIN F, 3.8L V6 VIN H, 4.6L V8 VIN F **Transmission:** All	**Camshaft Position Sensor 'A' Circuit Range/Performance (Bank 1 Or Single Sensor):** Enable Conditions • Reference mark found • No CKP sensor error • No 1 tooth off error Threshold value • Implausible signal counter > 10 times Diagnosis Time • Continuous **Possible Causes:** • Poor connection • Short in harness • electrical noise • Target wheel • CMPS(B1-Intake) • ECM

DTC	Trouble Code Title, Conditions & Possible Causes
DTC: P0342 **2T CCM, MIL: Yes** **Year:** 2009 **Model:** Accent **Engine:** 1.6L L4 VIN C **Transmission:** All	**Camshaft Position Sensor Low Input:** Engine started, engine speed over 600 rpm, and the PCM detected invalid or irregular CMP signals during the CCM test. **Possible Causes:** • CMP (Hall) sensor signal circuit is open or shorted (intermittent) • CMP (Hall) sensor is damaged or has failed • PCM has failed
DTC: P0343 **2T CCM, MIL: Yes** **Year:** 2009, 2010 **Model:** Accent **Engine:** 1.6L L4 VIN C **Transmission:** All	**Camshaft Position Sensor High Input:** Engine started, engine speed over 600 rpm, and the PCM did not detect any CMP signals within 200 revolutions during the CCM test. **Possible Causes:** • CMP (Hall) sensor signal circuit is open or shorted to ground • CMP (Hall) sensor ground circuit is open • CMP (Hall) sensor power circuit is open (check fuse in the I/P) • CMP (Hall) sensor is damaged or has failed • PCM has failed
DTC: P0345 **2T PCM, MIL: Yes** **Year:** 2009, 2010 **Model:** Genesis **Engine:** 4.6L V8 VIN F **Transmission:** All	**Camshaft Position Sensor 'A' Circuit (Bank 2):** Enable Conditions • Reference mark found • No CKP sensor error • No 1 tooth off error Threshold value • Abnormal phase edges (High or Low) • No signal counter > 2 times Diagnosis Time • Continuous **Possible Causes:** • Poor connection • Open in harness • CMPS(B2-Intake) • ECM
DTC: P0346 **2T CCM, MIL: Yes** **Year:** 2009, 2010 **Model:** Azera, Genesis, Genesis Coupe, Santa Fe, Sonata, Veracruz **Engine:** 2.7L V6 VIN D, 3.3L V6 VIN D, 3.3L V6 VIN E, 3.3L V6 VIN F, 3.5L V6 VIN G, 3.8L V6 VIN C, 3.8L V6 VIN E, 3.8L V6 VIN F, 3.8L V6 VIN H, 4.6L V8 VIN F **Transmission:** All	**Camshaft Position Sensor "A" Circuit Range/Performance (Bank 2):** Engine running and the PCM detected the CMP sensor signal was above a threshold value stored in memory during the CCM test. **Possible Causes:** • Poor connection • Open or Short in harness • Electrical noise • Target wheel • CMPS • PCM has failed
DTC: P0350 **2T CCM, MIL: Yes** **Year:** 2009 **Model:** Tucson **Engine:** 2.7L V6 VIN D **Transmission:** All	**Ignition Coil Primary or Secondary Circuit Malfunction:** Engine started, engine running, and the PCM detected an unexpected voltage condition on the Ignition Coil primary or secondary circuit during the CCM test. **Possible Causes:** • Ignition Coil 1, 2 or 3 primary/secondary circuit open or shorted • Ignition Coil power circuit is open (test power from the relay) • Ignition Coil 1, 2 or 3 is damaged or has failed • PCM has failed

DTC	Trouble Code Title, Conditions & Possible Causes
DTC: P0351 **2T CCM, MIL: Yes** **Year:** 2009, 2010 **Model:** Azera, Genesis, Genesis Coupe, Santa Fe, Sonata, Tucson, Veracruz **Engine:** 2.7L V6 VIN D, 3.3L V6 VIN D, 3.3L V6 VIN E, 3.3L V6 VIN F, 3.5L V6 VIN G, 3.8L V6 VIN C, 3.8L V6 VIN E, 3.8L V6 VIN F, 3.8L V6 VIN H **Transmission:** All	**Ignition Coil 'A' Circuit Malfunction:** Engine started, engine running, and the PCM detected an unexpected voltage condition on the Ignition Coil 'A' primary circuit. **Possible Causes:** • Ignition Coil 'A' primary circuit is open or shorted to ground • Ignition Coil 'A' power circuit is open (test power from I/P fuse) • Ignition Coil 'A' is damaged or has failed • PCM has failed
DTC: P0352 **2T CCM, MIL: Yes** **Year:** 2009, 2010 **Model:** Azera, Genesis, Genesis Coupe, Santa Fe, Sonata, Tucson, Veracruz **Engine:** 2.7L V6 VIN D, 3.3L V6 VIN D, 3.3L V6 VIN E, 3.3L V6 VIN F, 3.5L V6 VIN G, 3.8L V6 VIN C, 3.8L V6 VIN E, 3.8L V6 VIN F, 3.8L V6 VIN H **Transmission:** All	**Ignition Coil 'B' Circuit Malfunction:** Engine started, engine running, and the PCM detected an unexpected voltage condition on the Ignition Coil 'B' primary circuit. **Possible Causes:** • Ignition Coil 'B' primary circuit is open or shorted to ground • Ignition Coil 'B' power circuit is open (test power from I/P fuse) • Ignition Coil 'B' is damaged or has failed • PCM has failed
DTC: P0353 **2T CCM, MIL: Yes** **Year:** 2009, 2010 **Model:** Azera, Genesis, Genesis Coupe, Santa Fe, Sonata, Tucson, Veracruz **Engine:** 2.7L V6 VIN D, 3.3L V6 VIN D, 3.3L V6 VIN E, 3.3L V6 VIN F, 3.5L V6 VIN G, 3.8L V6 VIN C, 3.8L V6 VIN E, 3.8L V6 VIN F, 3.8L V6 VIN H **Transmission:** All	**Ignition Coil 'C' Circuit Malfunction:** Engine started, engine running, and the PCM detected an unexpected voltage condition on the Ignition Coil 'C' primary circuit. **Possible Causes:** • Ignition Coil 'C' primary circuit is open or shorted to ground • Ignition Coil 'C' power circuit is open (test power from I/P fuse) • Ignition Coil 'C' is damaged or has failed • PCM has failed
DTC: P0354 **2T CCM, MIL: Yes** **Year:** 2009, 2010 **Model:** Azera, Genesis, Genesis Coupe, Santa Fe, Sonata, Tucson, Veracruz **Engine:** 2.7L V6 VIN D, 3.3L V6 VIN D, 3.3L V6 VIN E, 3.3L V6 VIN F, 3.5L V6 VIN G, 3.8L V6 VIN C, 3.8L V6 VIN E, 3.8L V6 VIN F, 3.8L V6 VIN H **Transmission:** All	**Ignition Coil 'D' Circuit Malfunction:** Engine started, engine running, and the PCM detected an unexpected voltage condition on the Ignition Coil 'D' primary circuit. **Possible Causes:** • Ignition Coil 'D' primary circuit is open or shorted to ground • Ignition Coil 'D' power circuit is open (test power from I/P fuse) • Ignition Coil 'D' is damaged or has failed • PCM has failed
DTC: P0355 **2T CCM, MIL: Yes** **Year:** 2009, 2010 **Model:** Azera, Genesis, Genesis Coupe, Santa Fe, Sonata, Tucson, Veracruz **Engine:** 2.7L V6 VIN D, 3.3L V6 VIN D, 3.3L V6 VIN E, 3.3L V6 VIN F, 3.5L V6 VIN G, 3.8L V6 VIN C, 3.8L V6 VIN E, 3.8L V6 VIN F, 3.8L V6 VIN H **Transmission:** All	**Ignition Coil 'E' Circuit Malfunction:** Engine started, engine running, and the PCM detected an unexpected voltage condition on the Ignition Coil 'E' primary circuit. **Possible Causes:** • Ignition Coil 'E' primary circuit is open or shorted to ground • Ignition Coil 'E' power circuit is open (test power from I/P fuse) • Ignition Coil 'E' is damaged or has failed • PCM has failed

DTC	Trouble Code Title, Conditions & Possible Causes
DTC: P0356 **2T CCM, MIL: Yes** **Year:** 2009, 2010 **Model:** Azera, Genesis, Genesis Coupe, Santa Fe, Sonata, Tucson, Veracruz **Engine:** 2.7L V6 VIN D, 3.3L V6 VIN D, 3.3L V6 VIN E, 3.3L V6 VIN F, 3.5L V6 VIN G, 3.8L V6 VIN C, 3.8L V6 VIN E, 3.8L V6 VIN F, 3.8L V6 VIN H **Transmission:** All	**Ignition Coil 'F' Circuit Malfunction:** Engine started, engine running, and the PCM detected an unexpected voltage condition on the Ignition Coil 'F' primary circuit. **Possible Causes:** • Ignition Coil 'F' primary circuit is open or shorted to ground • Ignition Coil 'F' power circuit is open (test power from I/P fuse) • Ignition Coil 'F' is damaged or has failed • PCM has failed
DTC: P0365 **2T PCM, MIL: Yes** **Year:** 2009, 2010 **Model:** Genesis, Genesis Coupe, Sonata, Tucson **Engine:** 2.0L L4 VIN D, 2.4L L4 VIN C, 4.6L V8 VIN F **Transmission:** All	**Camshaft Position Sensor B Circuit Malfunction (Bank 1):** Enable Conditions • Reference mark found • No CKP sensor error • No 1 tooth off error Threshold value • Abnormal phase edges (High or Low) (No signal counter > 2 times) Diagnosis Time • Continuous **Possible Causes:** • Poor connection • Open in harness • CMPS(B1-Exhaust) • ECM
DTC: P0366 **2T PCM, MIL: Yes** **Year:** 2009, 2010 **Model:** Genesis, Genesis Coupe, Santa Fe, Sonata, Tucson **Engine:** 2.0L L4 VIN D, 2.4L L4 VIN B, 2.4L L4 VIN C, 3.5L V6 VIN G, 3.8L V6 VIN E, 3.8L V6 VIN H, 4.6L V8 VIN F **Transmission:** All	**Camshaft Position Sensor 'B' Circuit Range/Performance (Bank 1):** Enable Conditions • Reference mark found • No CKP sensor error • No 1 tooth off error Threshold value • Implausible signal counter > 10 times Diagnosis Time • Continuous **Possible Causes:** • Poor connection • Short in harness • electrical noise • Target wheel • CMPS(B1-Exhaust) • ECM
DTC: P0390 **2T PCM, MIL: Yes** **Year:** 2009, 2010 **Model:** Genesis **Engine:** 4.6L V8 VIN F **Transmission:** All	**Camshaft Position Sensor 'B' Circuit (Bank 2):** Enable Conditions • Reference mark found • No CKP sensor error • No 1 tooth off error Threshold value • Abnormal phase edges (High or Low) (No signal counter > 2 times) Diagnosis Time • Continuous **Possible Causes:** • Poor connection • Open in harness • CMPS(B2-Exhaust) • ECM

DTC	Trouble Code Title, Conditions & Possible Causes
DTC: P0391 **2T PCM, MIL: Yes** **Year:** 2009, 2010 **Model:** Genesis, Genesis Coupe, Santa Fe **Engine:** 3.5L V6 VIN G, 3.8L V6 VIN E, 3.8L V6 VIN H, 4.6L V8 VIN F **Transmission:** All	**Camshaft Position Sensor 'B' Circuit Range/Performance (Bank 2):** Enable Conditions • Reference mark found • No CKP sensor error • No 1 tooth off error Threshold value • Implausible signal counter > 10 times Diagnosis Time • Continuous **Possible Causes:** • Poor connection • Short in harness • electrical noise • Target wheel • CMPS(B2-Exhaust) • ECM
DTC: P0420 **2T CCM, MIL: Yes** **Year:** 2009, 2010 **Model:** Accent, Azera, Elantra, Genesis, Genesis Coupe, Santa Fe, Sonata, Tucson, Veracruz **Engine:** 1.6L L4 VIN C, 2.0L L4 VIN B, 2.0L L4 VIN D, 2.0L L4 VIN E, 2.4L L4 VIN B, 2.4L L4 VIN C, 2.7L V6 VIN D, 3.3L V6 VIN D, 3.3L V6 VIN E, 3.3L V6 VIN F, 3.5L V6 VIN G, 3.8L V6 VIN C, 3.8L V6 VIN E, 3.8L V6 VIN F, 3.8L V6 VIN H, 4.6L V8 VIN F **Transmission:** All	**Catalyst Efficiency Below Normal (Bank 1):** Engine started, vehicle driven at a speed of 45-60 mph for 8-10 minutes in closed loop, and the PCM detected the rear HO2S-12 switch rate was similar to the front HO2S-11 switch rate for over 3 seconds under these conditions during the Catalyst Monitor test. **Possible Causes:** • Air leaks at the exhaust manifold or in the exhaust pipes • Catalytic converter is contaminated, damaged or has failed • Front HO2S is older (aged) than the rear HO2S (HO2S is lazy) • Front HO2S or rear HO2S is contaminated with fuel or moisture
DTC: P0430 **2T CAT, MIL: Yes** **Year:** 2009, 2010 **Model:** Azera, Genesis, Genesis Coupe, Santa Fe, Sonata, Tucson, Veracruz **Engine:** 2.7L V6 VIN D, 3.3L V6 VIN D, 3.3L V6 VIN E, 3.3L V6 VIN F, 3.5L V6 VIN G, 3.8L V6 VIN C, 3.8L V6 VIN E, 3.8L V6 VIN F, 3.8L V6 VIN H, 4.6L V8 VIN F **Transmission:** All	**Catalyst Efficiency Below Normal (Bank 2):** Engine started, vehicle driven at a speed of 45-60 mph for 8 minutes in closed loop, and the PCM detected the rear HO2S-22 switch rate was similar to the front HO2S-21 switch rate for over 3 seconds. **Possible Causes:** • Air leaks at the exhaust manifold or in the exhaust pipes • Catalytic converter is contaminated, damaged or has failed • Front HO2S is older (aged) than the rear HO2S (HO2S is lazy) • Front HO2S or rear HO2S is contaminated with fuel or moisture
DTC: P0441 **2T EVAP, MIL: Yes** **Year:** 2009, 2010 **Model:** Azera, Elantra, Genesis, Genesis Coupe, Santa Fe, Sonata, Tucson, Veracruz **Engine:** 2.0L L4 VIN B, 2.0L L4 VIN D, 2.0L L4 VIN E, 2.4L L4 VIN B, 2.4L L4 VIN C, 2.7L V6 VIN D, 3.3L V6 VIN D, 3.3L V6 VIN E, 3.3L V6 VIN F, 3.5L V6 VIN G, 3.8L V6 VIN C, 3.8L V6 VIN E, 3.8L V6 VIN F, 3.8L V6 VIN H **Transmission:** All	**EVAP System Incorrect Purge Flow (Stuck Open):** Engine started, vehicle driven at a speed of 35-40 for 5-10 minutes under light engine load conditions, then with the Purge solenoid commanded "on" and then "off", the PCM detected the Purge solenoid valve remained "open" during the EVAP Monitor flow test. **NOTE: This is a "functionality" test of the EVAP system (flow test).** **Possible Causes:** • Charcoal canister is damaged, clogged or restricted • Purge solenoid control circuit open, shorted to ground or power • Purge solenoid power circuit is open (check the relay or fuse) • Purge valve vacuum line is clogged or contains water or debris • Fuel filler cap loose, cross-threaded, incorrect part or damaged • Fuel tank or fuel tank sender assembly 'O' ring is leaking • Fuel tank vapor line(s) blocked, damaged or disconnected • PCM has failed

DTC	Trouble Code Title, Conditions & Possible Causes
DTC: P0442 **2T EVAP, MIL: Yes** **Year:** 2009, 2010 **Model:** Accent, Azera, Elantra, Genesis, Genesis Coupe, Santa Fe, Sonata, Tucson, Veracruz **Engine:** 1.6L L4 VIN C, 2.0L L4 VIN B, 2.0L L4 VIN D, 2.0L L4 VIN E, 2.4L L4 VIN B, 2.4L L4 VIN C, 2.7L V6 VIN D, 3.3L V6 VIN D, 3.3L V6 VIN E, 3.3L V6 VIN F, 3.5L V6 VIN G, 3.8L V6 VIN C, 3.8L V6 VIN E, 3.8L V6 VIN F, 3.8L V6 VIN H, 4.6L V8 VIN F **Transmission:** All	**EVAP System Small Leak (0.040") Detected:** ECT sensor signal less than158°F at startup, IAT sensor signal more than 9.05°F, system voltage more than 10.9 volts engine runtime 15-20 minutes at cruise speed, then returned to idle speed, VSS indicating 0 mph, load value 2.2 ms, canister load factor less than 4.0, fuel tank pressure less than 0.5" Hg, then after the Idle Control system and Fuel Trim had stabilized, the PCM detected a fuel vapor leak (as small as 0.040") in the EVAP system during the EVAP Leak Test. **Possible Causes:** • Fuel filler cap damaged, cross-threaded or loosely installed • Small leaks or cuts present in the EVAP vapor hoses/lines • EVAP purge valve is damaged or has failed • PCM has failed
DTC: P0444 **2T CCM, MIL: Yes** **Year:** 2009, 2010 **Model:** Accent, Azera, Elantra, Genesis, Genesis Coupe, Santa Fe, Sonata, Tucson, Veracruz **Engine:** 1.6L L4 VIN C, 2.0L L4 VIN B, 2.0L L4 VIN D, 2.0L L4 VIN E, 2.4L L4 VIN B, 2.4L L4 VIN C, 2.7L V6 VIN D, 3.3L V6 VIN D, 3.3L V6 VIN E, 3.3L V6 VIN F, 3.5L V6 VIN G, 3.8L V6 VIN C, 3.8L V6 VIN E, 3.8L V6 VIN F, 3.8L V6 VIN H, 4.6L V8 VIN F **Transmission:** All	**EVAP Emission System- Purge Control Valve Circuit Open:** Engine running. Checking output signals from PCSV every 10 seconds, under detecting condition. **Possible Causes:** • Poor connection • Open or short to ground in harness • PCVS • PCM
DTC: P0445 **2T CCM, MIL: Yes** **Year:** 2009, 2010 **Model:** Azera, Elantra, Genesis, Genesis Coupe, Santa Fe, Sonata, Tucson, Veracruz **Engine:** 2.0L L4 VIN B, 2.0L L4 VIN D, 2.0L L4 VIN E, 2.4L L4 VIN B, 2.4L L4 VIN C, 2.7L V6 VIN D, 3.3L V6 VIN D, 3.3L V6 VIN E, 3.3L V6 VIN F, 3.5L V6 VIN G, 3.8L V6 VIN C, 3.8L V6 VIN E, 3.8L V6 VIN F, 3.8L V6 VIN H **Transmission:** All	**EVAP Purge Solenoid Circuit Malfunction (Shorted):** Engine started, engine running at idle speed, and the PCM detected an unexpected "high" voltage condition on the EVAP Purge solenoid circuit as the solenoid was commanded "on" and "off" in the test. **Possible Causes:** • Purge solenoid control circuit is shorted to system power • Purge control solenoid is damaged or has failed (short circuit) • PCM has failed
DTC: P0446 **2T CCM, MIL: Yes** **Year:** 2009, 2010 **Model:** Accent **Engine:** 1.6L L4 VIN C **Transmission:** All	**EVAP Emission System- Vent Control Circuit:** CCV stuck open. Time after engine start greater than 600 seconds. Idle speed controller activated. Coolant temperature less than 12 degrees F. **Possible Causes:** • Poor connection • CCV • ECM/PCM

DTC	Trouble Code Title, Conditions & Possible Causes
DTC: P0447 **2T CCM, MIL: Yes** **Year:** 2009, 2010 **Model:** Azera, Elantra, Genesis, Genesis Coupe, Santa Fe, Sonata, Tucson, Veracruz **Engine:** 2.0L L4 VIN B, 2.0L L4 VIN D, 2.0L L4 VIN E, 2.4L L4 VIN B, 2.4L L4 VIN C, 2.7L V6 VIN D, 3.3L V6 VIN D, 3.3L V6 VIN E, 3.3L V6 VIN F, 3.5L V6 VIN G, 3.8L V6 VIN C, 3.8L V6 VIN E, 3.8L V6 VIN F, 3.8L V6 VIN H, 4.6L V8 VIN F **Transmission:** All	**EVAP Emission System- Vent Control Circuit Open:** Detects a short to ground or open circuit on vent valve output circuit. No disabling faults present. Engine running. **Possible Causes:** • Poor connection • Open or short in power circuit • Open or short in control circuit • CCV • ECM/PCM
DTC: P0448 **2T CCM, MIL: Yes** **Year:** 2009, 2010 **Model:** Azera, Elantra, Genesis, Genesis Coupe, Santa Fe, Sonata, Tucson, Veracruz **Engine:** 2.0L L4 VIN B, 2.0L L4 VIN D, 2.0L L4 VIN E, 2.4L L4 VIN B, 2.4L L4 VIN C, 2.7L V6 VIN D, 3.3L V6 VIN D, 3.3L V6 VIN E, 3.3L V6 VIN F, 3.5L V6 VIN G, 3.8L V6 VIN C, 3.8L V6 VIN E, 3.8L V6 VIN F, 3.8L V6 VIN H, 4.6L V8 VIN F **Transmission:** All	**EVAP Emission System- Vent Control Circuit Shorted:** Detects a short to battery on vent valve output circuit. No disabling faults present. Engine running. **Possible Causes:** • Poor connection • Short to battery in CCV circuit • CCV • ECM/PCM
DTC: P0449 **2T CCM, MIL: Yes** **Year:** 2009, 2010 **Model:** Accent, Elantra, Genesis, Genesis Coupe, Santa Fe, Sonata, Tucson **Engine:** 1.6L L4 VIN C, 2.0L L4 VIN B, 2.0L L4 VIN D, 2.0L L4 VIN E, 2.4L L4 VIN B, 2.4L L4 VIN C, 2.7L V6 VIN D, 4.6L V8 VIN F **Transmission:** All	**EVAP Emission System- Vent Valve/Solenoid Circuit:** The PCM measures pressure in the fuel tank, by means of a sensor during all engine operating conditions, except start and stop. If pressure is lower than threshold (less than 1.6L volts), a DTC is set. Monitoring CCV stuck closed. **Possible Causes:** • Canister air filter contamination • Faulty CCV
DTC: P0450 **2T CCM, MIL: Yes** **Year:** 2009, 2010 **Model:** Accent **Engine:** 1.6L L4 VIN C **Transmission:** All	**EVAP Pressure Sensor Circuit Malfunction:** Engine started, engine running, vehicle not moving, VSV for the EVAP Vapor sensor commanded "on", and the PCM detected an unexpected voltage condition on the EVAP Pressure sensor circuit. **Possible Causes:** • Pressure sensor signal circuit is open or shorted to ground • Pressure sensor signal circuit is shorted to VREF or power • Pressure sensor power (VREF) circuit is open • Pressure sensor is damaged or has failed • PCM has failed

DTC	Trouble Code Title, Conditions & Possible Causes
DTC: P0451 **2T CCM, MIL: Yes** **Year:** 2009, 2010 **Model:** Accent, Azera, Elantra, Genesis, Genesis Coupe, Santa Fe, Sonata, Tucson, Veracruz **Engine:** 1.6L L4 VIN C, 2.0L L4 VIN B, 2.0L L4 VIN D, 2.0L L4 VIN E, 2.4L L4 VIN B, 2.4L L4 VIN C, 2.7L V6 VIN D, 3.3L V6 VIN D, 3.3L V6 VIN E, 3.3L V6 VIN F, 3.5L V6 VIN G, 3.8L V6 VIN C, 3.8L V6 VIN E, 3.8L V6 VIN F, 3.8L V6 VIN H, 4.6L V8 VIN F **Transmission:** All	**EVAP Pressure Sensor Performance:** Engine started, engine running with the vehicle not moving, VSV for the EVAP Vapor sensor commanded "on", and the PCM detected the EVAP pressure sensor signal was not plausible during the test. **NOTE: This condition (code) can be due to a fuel sloshing condition.** **Possible Causes:** • Pressure sensor vacuum hoses loose or damaged • Pressure sensor is damaged or out-of-calibration • VSV for the EVAP pressure sensor is damaged or has failed • PCM has failed
DTC: P0452 **2T CCM, MIL: Yes** **Year:** 2009, 2010 **Model:** Accent, Azera, Elantra, Genesis, Genesis Coupe, Santa Fe, Sonata, Tucson, Veracruz **Engine:** 1.6L L4 VIN C, 2.0L L4 VIN B, 2.0L L4 VIN D, 2.0L L4 VIN E, 2.4L L4 VIN B, 2.4L L4 VIN C, 2.7L V6 VIN D, 3.3L V6 VIN D, 3.3L V6 VIN E, 3.3L V6 VIN F, 3.5L V6 VIN G, 3.8L V6 VIN C, 3.8L V6 VIN E, 3.8L V6 VIN F, 3.8L V6 VIN H, 4.6L V8 VIN F **Transmission:** All	**EVAP Pressure Sensor Circuit Low Input:** Engine started, engine runtime over 5 seconds, and the PCM detected an unexpected "low" voltage condition on the EVAP Pressure sensor circuit during the CCM test. **Possible Causes:** • Pressure sensor signal circuit is shorted to ground • Pressure sensor power (VREF) circuit is open • Pressure sensor is damaged or has failed • PCM has failed
DTC: P0453 **2T CCM, MIL: Yes** **Year:** 2009, 2010 **Model:** Accent, Azera, Elantra, Genesis, Genesis Coupe, Santa Fe, Sonata, Tucson, Veracruz **Engine:** 1.6L L4 VIN C, 2.0L L4 VIN B, 2.0L L4 VIN D, 2.0L L4 VIN E, 2.4L L4 VIN B, 2.4L L4 VIN C, 2.7L V6 VIN D, 3.3L V6 VIN D, 3.3L V6 VIN E, 3.3L V6 VIN F, 3.5L V6 VIN G, 3.8L V6 VIN C, 3.8L V6 VIN E, 3.8L V6 VIN F, 3.8L V6 VIN H, 4.6L V8 VIN F **Transmission:** All	**EVAP Pressure Sensor Circuit High Input:** Engine started, engine runtime over 5 seconds, and the PCM detected an unexpected "low" voltage condition on the EVAP Pressure sensor circuit during the CCM test. **Possible Causes:** • Pressure sensor signal circuit is shorted to power • Pressure sensor ground circuit open between sensor and PCM • Pressure sensor is damaged or has failed • PCM has failed

DTC	Trouble Code Title, Conditions & Possible Causes
DTC: P0454 **2T CCM, MIL: Yes** **Year:** 2009, 2010 **Model:** Accent, Azera, Elantra, Genesis, Genesis Coupe, Santa Fe, Sonata, Tucson, Veracruz **Engine:** 1.6L L4 VIN C, 2.0L L4 VIN B, 2.0L L4 VIN D, 2.0L L4 VIN E, 2.4L L4 VIN B, 2.4L L4 VIN C, 2.7L V6 VIN D, 3.3L V6 VIN D, 3.3L V6 VIN E, 3.3L V6 VIN F, 3.5L V6 VIN G, 3.8L V6 VIN C, 3.8L V6 VIN E, 3.8L V6 VIN F, 3.8L V6 VIN H **Transmission:** All	**EVAP Emission System- Pressure Sensor Intermittent:** The PCM measures pressure stability in the fuel tank, by means of a sensor for a predetermined duration. If fluctuation is larger than predetermined threshold a DTC is set. Sensor signal noise check. **Possible Causes:** • Contact resistance in connectors • Faulty FTPS
DTC: P0455 **2T EVAP, MIL: Yes** **Year:** 2009, 2010 **Model:** Accent, Azera, Elantra, Genesis, Genesis Coupe, Santa Fe, Sonata, Tucson, Veracruz **Engine:** 1.6L L4 VIN C, 2.0L L4 VIN B, 2.0L L4 VIN D, 2.0L L4 VIN E, 2.4L L4 VIN B, 2.4L L4 VIN C, 2.7L V6 VIN D, 3.3L V6 VIN D, 3.3L V6 VIN E, 3.3L V6 VIN F, 3.5L V6 VIN G, 3.8L V6 VIN C, 3.8L V6 VIN E, 3.8L V6 VIN F, 3.8L V6 VIN H, 4.6L V8 VIN F **Transmission:** All	**EVAP System Large Leak (0.080") Detected:** DTC P0443 not set, engine started, ECT sensor more than 185°F, IAT sensor from 14-122°F, fuel level from 25-75%, engine running at cruise speed, and the PCM detected a large change in the fuel tank pressure (due to a large leak) during the EVAP Monitor leak test. **Possible Causes:** • Canister vent (CV) solenoid may be stuck in open position • EVAP canister tube, EVAP canister purge outlet tube or EVAP return tube disconnected or cracked, or canister is damaged • EVAP canister purge valve stuck closed, or canister damaged • Fuel filler cap missing, loose (not tightened) or the wrong part • Fuel vapor hoses/tubes blocked or restricted, or fuel vapor control valve tube or fuel vapor vent valve assembly blocked • Fuel tank pressure (FTP) sensor has failed (mechanical fault) • Fuel tank control valve is contaminated, damaged or has failed
DTC: P0456 **2T CCM, MIL: Yes** **Year:** 2009, 2010 **Model:** Accent, Azera, Elantra, Genesis, Genesis Coupe, Santa Fe, Sonata, Tucson, Veracruz **Engine:** 1.6L L4 VIN C, 2.0L L4 VIN B, 2.0L L4 VIN D, 2.0L L4 VIN E, 2.4L L4 VIN B, 2.4L L4 VIN C, 2.7L V6 VIN D, 3.3L V6 VIN D, 3.3L V6 VIN E, 3.3L V6 VIN F, 3.5L V6 VIN G, 3.8L V6 VIN C, 3.8L V6 VIN E, 3.8L V6 VIN F, 3.8L V6 VIN H, 4.6L V8 VIN F **Transmission:** All	**EVAP Emission System- Leak Detected (Very Small):** Monitoring the tank pressure sensor (DTP) signal with under pressure. **Possible Causes:** • Leakage in EVAP system line • Faulty CCV, PCSV or FTPS
DTC: P0457 **2T EVAP, MIL: Yes** **Year:** 2009, 2010 **Model:** Accent, Genesis **Engine:** 1.6L L4 VIN C, 4.6L V8 VIN F **Transmission:** All	**EVAP System Leak (Fuel Tank Cap) Detected:** DTC P0443 not set, engine started, ECT sensor more than 185°F, IAT sensor from 14-122°F, fuel level from 25-75%, and the PCM detected a large change in fuel tank pressure during the EVAP System Monitor leak test. **Possible Causes:** • Canister vent (CV) solenoid may be stuck in open position • EVAP canister tube, EVAP canister purge outlet tube or EVAP return tube disconnected or cracked, or canister is damaged • Fuel filler cap missing, loose (not tightened) or the wrong part • Fuel tank pressure (FTP) sensor has failed (mechanical fault) • Fuel tank control valve is contaminated, damaged or has failed
DTC: P0458 **2T CCM, MIL: Yes** **Year:** 2009, 2010 **Model:** Accent, Genesis **Engine:** 1.6L L4 VIN C, 4.6L V8 VIN F **Transmission:** All	**Evaporative Emission System Purge Control valve Circuit Low:** Circuit continuity check, low. **Possible Causes:** • Poor connection • Short to ground in control circuit • PCSV • Faulty ECM/PCM

DTC	Trouble Code Title, Conditions & Possible Causes
DTC: P0459 **2T CCM, MIL: Yes** **Year:** 2009, 2010 **Model:** Accent, Genesis **Engine:** 1.6L L4 VIN C, 4.6L V8 VIN F **Transmission:** All	**Evaporative Emission System Purge Control valve Circuit High:** Circuit continuity check, high. **Possible Causes:** • Poor connection • Short to power in control circuit • PCSV • Faulty ECM/PCM
DTC: P0461 **2T CCM, MIL: Yes** **Year:** 2009, 2010 **Model:** Accent, Azera, Genesis, Genesis Coupe, Santa Fe, Sonata, Tucson, Veracruz **Engine:** 1.6L L4 VIN C, 2.0L L4 VIN D, 2.4L L4 VIN B, 2.4L L4 VIN C, 2.7L V6 VIN D, 3.3L V6 VIN D, 3.3L V6 VIN E, 3.3L V6 VIN F, 3.5L V6 VIN G, 3.8L V6 VIN C, 3.8L V6 VIN E, 3.8L V6 VIN F, 3.8L V6 VIN H, 4.6L V8 VIN F **Transmission:** All	**Fuel Level Sensor "A" Circuit Range/Performance:** Filtered and unfiltered signal of fuel sensor are monitored. **Possible Causes:** • Poor connection • Faulty fuel level sensor • Faulty PCM
DTC: P0462 **2T CCM** **Year:** 2009, 2010 **Model:** Accent, Azera, Genesis, Genesis Coupe, Santa Fe, Sonata, Tucson, Veracruz **Engine:** 1.6L L4 VIN C, 2.0L L4 VIN D, 2.4L L4 VIN B, 2.4L L4 VIN C, 2.7L V6 VIN D, 3.3L V6 VIN D, 3.3L V6 VIN E, 3.3L V6 VIN F, 3.5L V6 VIN G, 3.8L V6 VIN C, 3.8L V6 VIN E, 3.8L V6 VIN F, 3.8L V6 VIN H, 4.6L V8 VIN F **Transmission:** All	**Fuel Level Sensor "A" Circuit Low Input:** If the sensor output voltage is higher than 4.96 volts or output voltage is less than 0.039 volt for 300 seconds, when the engine speed is not 0 miles. Output voltage (VFLS) is monitored. **Possible Causes:** • Poor connection • Faulty fuel level sensor • Short to battery in fuel level (FLS) circuit • Faulty fuel level circuit (FLS) • Faulty PCM
DTC: P0463 **2T CCM** **Year:** 2009, 2010 **Model:** Accent, Azera, Genesis, Genesis Coupe, Santa Fe, Sonata, Tucson, Veracruz **Engine:** 1.6L L4 VIN C, 2.0L L4 VIN D, 2.4L L4 VIN B, 2.4L L4 VIN C, 2.7L V6 VIN D, 3.3L V6 VIN D, 3.3L V6 VIN E, 3.3L V6 VIN F, 3.5L V6 VIN G, 3.8L V6 VIN C, 3.8L V6 VIN E, 3.8L V6 VIN F, 3.8L V6 VIN H, 4.6L V8 VIN F **Transmission:** All	**Fuel Level Sensor "A" Circuit high Input:** If the sensor output voltage is less than 0.02 volt, or output voltage is less than 0.039 volt for 300 seconds, when the engine speed is not 0 miles. Output voltage (VFLS) is monitored. **Possible Causes:** • Short to battery in fuel level (FLS) circuit • Faulty fuel level sensor • Faulty PCM

DTC	Trouble Code Title, Conditions & Possible Causes
DTC: P0464 **2T CCM** **Year:** 2009, 2010 **Model:** Azera, Genesis, Genesis Coupe, Santa Fe, Sonata, Tucson, Veracruz **Engine:** 2.0L L4 VIN D, 2.4L L4 VIN B, 2.4L L4 VIN C, 2.7L V6 VIN D, 3.3L V6 VIN D, 3.3L V6 VIN E, 3.3L V6 VIN F, 3.5L V6 VIN G, 3.8L V6 VIN C, 3.8L V6 VIN E, 3.8L V6 VIN F, 3.8L V6 VIN H **Transmission:** All	**Fuel Level Sensor "A" Circuit Intermittent:** Check signal for fluctuation. The ECM sets the DTC if the fuel level signal is higher than the threshold value (signal fluctuation greater than 50 percent). **Possible Causes:** • Contact resistance in connectors • Short to battery in fuel level (FLS) circuit
DTC: P0480 **2T CCM, MIL: Yes** **Year:** 2009, 2010 **Model:** Azera, Genesis, Genesis Coupe, Santa Fe, Sonata, Veracruz **Engine:** 2.7L V6 VIN D, 3.3L V6 VIN D, 3.3L V6 VIN E, 3.3L V6 VIN F, 3.5L V6 VIN G, 3.8L V6 VIN C, 3.8L V6 VIN E, 3.8L V6 VIN F, 3.8L V6 VIN H **Transmission:** All	**Fan 1 Control Circuit Malfunction:** This will detect a short to ground, to battery or open circuit of fan relay output. Fault information provided by an output driver chip. No disabling faults present. Engine running. Enable time delay equal or greater than 0.5 seconds. **Possible Causes:** • Poor connection • Open in power circuit to cooling fan • Open or short in control circuit to PCM • Faulty fan relay • Faulty cooling fan module • Faulty PCM
DTC: P0481 **2T CCM, MIL: Yes** **Year:** 2009, 2010 **Model:** Genesis, Genesis Coupe, Santa Fe, Veracruz **Engine:** 2.7L V6 VIN D, 3.3L V6 VIN E, 3.5L V6 VIN G, 3.8L V6 VIN C, 3.8L V6 VIN E, 3.8L V6 VIN H **Transmission:** All	**Fan 2 Control Circuit Malfunction:** This will detect a short to ground, to battery or open circuit of fan relay output. Fault information provided by an output driver chip. No disabling faults present. Engine running. Enable time delay equal or greater than 0.5 seconds. **Possible Causes:** • Poor connection • Open in power circuit to cooling fan • Open or short in control circuit to PCM • Faulty fan relay
DTC: P0496 **2T CCM, MIL: Yes** **Year:** 2009, 2010 **Model:** Accent, Genesis **Engine:** 1.6L L4 VIN C, 4.6L V8 VIN F **Transmission:** All	**Evaporative Emission System High Purge Flow:** Fuel tank pressure behavior (canister purge valve stuck). Time after engine start 600 seconds. Idle speed controller activated. Mixture adaptation activated. Coolant temperature at start 11.88 degrees F. Tank ventilation must be active for 10 seconds. **Possible Causes:** • Leakage at the fuel evaporative system • PCSV • Faulty ECM/PCM
DTC: P0497 **2T CCM, MIL: Yes** **Year:** 2009, 2010 **Model:** Accent, Genesis **Engine:** 1.6L L4 VIN C, 4.6L V8 VIN F **Transmission:** All	**Evaporative Emission System Low Purge Flow:** Fuel tank pressure behavior (canister purge valve stuck). Time after engine start 600 seconds. Idle speed controller activated. Mixture adaptation activated. Coolant temperature at start 11.88 degrees F. Tank ventilation must be active for 10 seconds. **Possible Causes:** • Clog in the fuel evaporative system • PCSV • Faulty ECM/PCM
DTC: P0498 **2T CCM, MIL: Yes** **Year:** 2009, 2010 **Model:** Accent **Engine:** 1.6L L4 VIN C **Transmission:** All	**Evaporative Emission System Vent Valve Control Circuit Low:** Circuit continuity check, low. **Possible Causes:** • Poor connection • Short to ground in control circuit • CCV • Faulty ECM/PCM

DTC	Trouble Code Title, Conditions & Possible Causes
DTC: P0499 **2T CCM, MIL: Yes** **Year:** 2009, 2010 **Model:** Accent **Engine:** 1.6L L4 VIN C **Transmission:** All	**Evaporative Emission System Vent Valve Control Circuit High:** Circuit continuity check, high. **Possible Causes:** • Poor connection • Short to power in control circuit • CCV • Faulty ECM/PCM
DTC: P0501 **2T CCM, MIL: Yes** **Year:** 2009, 2010 **Model:** Accent, Azera, Elantra, Genesis, Genesis Coupe, Santa Fe, Sonata, Tucson, Veracruz **Engine:** 1.6L L4 VIN C, 2.0L L4 VIN D, 2.0L L4 VIN E, 2.7L V6 VIN D, 3.3L V6 VIN D, 3.3L V6 VIN E, 3.3L V6 VIN F, 3.8L V6 VIN C, 3.8L V6 VIN E, 3.8L V6 VIN F, 3.8L V6 VIN H, 4.6L V8 VIN F **Transmission:** All	**Vehicle Speed Sensor Circuit Range/Performance:** Vehicle driven with the engine speed over 2000 rpm in Drive, engine load from 2-3 ms, and the PCM did not detect any VSS signals. **Possible Causes:** • VSS signal circuit from the sensor to the I/P Cluster to the PCM is open, shorted to ground or to system power • VSS (Magnetic) signal (+) or (-) circuit is open or shorted • VSS (Magnetic) is damaged or has failed • PCM has failed
DTC: P0502 **2T PCM, MIL: Yes** **Year:** 2009, 2010 **Model:** Genesis **Engine:** 4.6L V8 VIN F **Transmission:** All	**Vehicle Speed Sensor 'A' Circuit Low Input:** Enable Conditions • Vehicle speed > 0mph Threshold Value • Old and New value are same for 10sec Diagnostic Time • 10sec **Possible Causes:** • Poor connection • Open or Short in harness • ESC(ESP) control unit • ECM
DTC: P0503 **2T PCM, MIL: Yes** **Year:** 2009, 2010 **Model:** Genesis **Engine:** 4.6L V8 VIN F **Transmission:** All	**Vehicle Speed Sensor 'A' Intermittent/Erratic/High:** Enable Conditions • Vehicle speed raw signal > 168mph **Possible Causes:** • Poor connection • Short in harness • ESC(ESP) control unit • ECM
DTC: P0504 **2T CCM, MIL: Yes** **Year:** 2009, 2010 **Model:** Azera, Genesis, Genesis Coupe, Santa Fe, Sonata **Engine:** 2.0L L4 VIN D, 2.4L L4 VIN C, 2.7L V6 VIN D, 3.3L V6 VIN D, 3.3L V6 VIN E, 3.3L V6 VIN F, 3.5L V6 VIN G, 3.8L V6 VIN E, 3.8L V6 VIN F, 3.8L V6 VIN H, 4.6L V8 VIN F **Transmission:** All	**Brake Switch "A"/"B" Correlation (1):** Comparing two brake signals during driving. Case 1: Engine works. Vehicle speed sensor is normal. Case 2: Engine works. Vehicle speed sensor is normal. Vehicle speed is over 20 kph, for at least 1 second. **Possible Causes:** • Poor connection • Open or short • Faulty PCM

DTC	Trouble Code Title, Conditions & Possible Causes
DTC: P0504 **2T CCM, MIL: Yes** **Year:** 2009, 2010 **Model:** Azera, Genesis, Genesis Coupe, Sonata, Tucson, Veracruz **Engine:** 2.0L L4 VIN D, 2.4L L4 VIN C, 3.3L V6 VIN D, 3.3L V6 VIN F, 3.8L V6 VIN C, 3.8L V6 VIN E, 3.8L V6 VIN F, 3.8L V6 VIN H, 4.6L V8 VIN F **Transmission:** All	**Brake Switch "A"/"B" Correlation (2):** Plausibility check between brake light switch and brake test switch. Engine running. Time between brake light switch and brake test switch do not correlate longer than 10 seconds. **Possible Causes:** • Open or short circuit in harness • Poor connection or damaged harness • Faulty brake warning lamp or brake test switch
DTC: P0505 **2T CCM, MIL: Yes** **Year:** 2009, 2010 **Model:** Accent, Elantra, Sonata, Tucson **Engine:** 1.6L L4 VIN C, 2.0L L4 VIN B, 2.0L L4 VIN D, 2.0L L4 VIN E, 2.4L L4 VIN C, 2.7L V6 VIN D, 3.3L V6 VIN F **Transmission:** All	**Idle Speed Control System (Mechanical) Fault:** Engine started, engine running at hot idle speed in closed loop for 30 seconds, and the PCM detected the difference between the Actual and the Target idle speed was more than 300 higher or lower than a calibrated amount in memory during the CCM Rationality test. **Possible Causes:** • Stepper motor Coil A1 or A2 circuit is open or shorted to ground • Stepper motor Coil B1 or B2 circuit is open or shorted to ground • Stepper motor coil circuit(s) shorted to system power (B+) • Stepper motor power circuit is open (check power at MFI relay) • Stepper motor is damaged or has failed • PCM has failed
DTC: P0506 **2T CCM, MIL: Yes** **Year:** 2009, 2010 **Model:** Accent, Azera, Elantra, Genesis, Genesis Coupe, Santa Fe, Sonata, Tucson, Veracruz **Engine:** 1.6L L4 VIN C, 2.0L L4 VIN B, 2.0L L4 VIN D, 2.0L L4 VIN E, 2.4L L4 VIN B, 2.4L L4 VIN C, 2.7L V6 VIN D, 3.3L V6 VIN D, 3.3L V6 VIN E, 3.3L V6 VIN F, 3.5L V6 VIN G, 3.8L V6 VIN C, 3.8L V6 VIN E, 3.8L V6 VIN F, 3.8L V6 VIN H, 4.6L V8 VIN F **Transmission:** All	**Idle Speed Lower Than Expected:** Engine started, engine running at idle speed under these conditions: IAT sensor less than 114°F during the last ignition cycle, Long Term fuel trim from -8% to +8%, ECT sensor more than 176°F, IAT sensor more than 14°F, system voltage over 10.0 volts, and the PCM detected the Actual idle speed was over 200 rpm lower than the Target idle speed for 10 seconds during the CCM Rationality test. Power steering pressure switch signal indicating "off", engine load less than 40%, IAT sensor more than 14°F, and the PCM detected the Actual idle speed was over 120 rpm lower than the Target idle speed for 10 seconds during the CCM Rationality test. **Possible Causes:** • ISC motor "open" circuit is open or shorted to ground • ISC motor "close" circuit is open or shorted to ground • ISC motor is damaged or has failed • PCM is damaged
DTC: P0507 **2T CCM, MIL: Yes** **Year:** 2009, 2010 **Model:** Accent, Azera, Elantra, Genesis, Genesis Coupe, Santa Fe, Sonata, Tucson, Veracruz **Engine:** 1.6L L4 VIN C, 2.0L L4 VIN B, 2.0L L4 VIN D, 2.0L L4 VIN E, 2.4L L4 VIN B, 2.4L L4 VIN C, 2.7L V6 VIN D, 3.3L V6 VIN D, 3.3L V6 VIN E, 3.3L V6 VIN F, 3.5L V6 VIN G, 3.8L V6 VIN C, 3.8L V6 VIN E, 3.8L V6 VIN F, 3.8L V6 VIN H, 4.6L V8 VIN F **Transmission:** All	**Idle Speed Higher Than Expected:** Engine at idle speed in closed loop, and under these conditions: Condition 1 - IAT sensor input less than 114°F during last drive cycle, Long Term fuel trim from -8% to +8%, ECT sensor input more than 176°F, IAT sensor input more than 14°F, system voltage over 10.0 volts, and the PCM detected the Actual idle speed was more than 200 rpm higher than the Target idle speed for 10 seconds. Condition 2 - Power steering pressure switch signal indicating "off", engine load less than 40%, IAT sensor input more than 14°F, and the PCM detected the Actual idle speed was more than 120 rpm higher than the Target idle speed for 10 seconds. **Possible Causes:** • ISC motor control circuit(s) open or shorted to ground • ISC motor is damaged or has failed • PCM is damaged

DTC	Trouble Code Title, Conditions & Possible Causes
DTC: P050B **2T PCM, MIL: Yes** **Year:** 2009, 2010 **Model:** Azera, Genesis, Genesis Coupe, Santa Fe, Sonata, Veracruz **Engine:** 3.3L V6 VIN D, 3.3L V6 VIN E, 3.3L V6 VIN F, 3.5L V6 VIN G, 3.8L V6 VIN C, 3.8L V6 VIN E, 3.8L V6 VIN F, 3.8L V6 VIN H **Transmission:** All	**Cold Start Ignition Timing Performance:** Enable Conditions • After engine overnight soaking • Vehicle is not rapidly accelerating or decelerating. • 11V< Battery Voltage < 16V • Engine is running. • NO DTC related to CKPS, Ignition coil, and Misfire Threshold value • The actual spark timing > the commanded spark timing + 15° or < the commanded spark timing -15° Diagnosis Time • Within 1 minute after cold-starting **Possible Causes:** • Faulty Ignition Coil • Faulty PCM
DTC: P0532 **2T CCM** **Year:** 2009, 2010 **Model:** Accent, Azera, Genesis, Genesis Coupe, Santa Fe, Sonata, Tucson, Veracruz **Engine:** 1.6L L4 VIN C, 2.0L L4 VIN D, 2.4L L4 VIN B, 2.4L L4 VIN C, 2.7L V6 VIN D, 3.3L V6 VIN D, 3.3L V6 VIN E, 3.3L V6 VIN F, 3.5L V6 VIN G, 3.8L V6 VIN C, 3.8L V6 VIN E, 3.8L V6 VIN F, 3.8L V6 VIN H, 4.6L V8 VIN F **Transmission:** All	**A/C Refrigerant Pressure Sensor "A" Circuit Low Input:** Detects sensor signal short to low voltage. Engine works. Sensor output 0.05 volt. **Possible Causes:** • Poor connection • Open in power circuit • Open or short to ground in signal circuit • Faulty A/C pressure sensor • Faulty PCM
DTC: P0533 **2T CCM** **Year:** 2009, 2010 **Model:** Accent, Azera, Genesis, Genesis Coupe, Santa Fe, Sonata, Tucson, Veracruz **Engine:** 1.6L L4 VIN C, 2.0L L4 VIN D, 2.4L L4 VIN B, 2.4L L4 VIN C, 2.7L V6 VIN D, 3.3L V6 VIN D, 3.3L V6 VIN E, 3.3L V6 VIN F, 3.5L V6 VIN G, 3.8L V6 VIN C, 3.8L V6 VIN E, 3.8L V6 VIN F, 3.8L V6 VIN H, 4.6L V8 VIN F **Transmission:** All	**A/C Refrigerant Pressure Sensor "A" Circuit High Input:** Detects sensor signal short to high voltage. Engine works. Sensor output 4.65 volts. **Possible Causes:** • Poor connection • Open in signal circuit open • Open in ground circuit • Faulty A/C pressure sensor • Faulty PCM
DTC: P0551 **2T CCM, MIL: Yes** **Year:** 2009, 2010 **Model:** Genesis Coupe, Sonata, Tucson **Engine:** 2.0L L4 VIN D, 2.4L L4 VIN C, 2.7L V6 VIN D **Transmission:** All	**Power Steering Pressure Sensor/Switch Circuit Range/Performance:** If a power steering switch signal is ON when the engine speed is more than 2500 rpm, load value is greater than 55 percent and engine coolant temperature is above 50 degrees F, the DTC will set. Signal of power steering pressure switch is monitored. **Possible Causes:** • Poor connection • Faulty power steering switch • Open or short in power steering switch • Faulty PCM
DTC: P0552 **2T CCM, MIL: Yes** **Year:** 2009, 2010 **Model:** Genesis, Santa Fe, Sonata, Veracruz **Engine:** 2.4L L4 VIN C, 2.7L V6 VIN D, 3.3L V6 VIN F, 3.8L V6 VIN C, 3.8L V6 VIN E **Transmission:** All	**Power Steering Pressure Sensor/Switch Circuit Low Input:** Detects sensor signal short to low voltage. Engine works. Sensor output 0.25 volt. **Possible Causes:** • Poor connection • Open in power circuit • Open or short to ground in signal circuit • Faulty P/S pressure sensor • Faulty PCM

DTC	Trouble Code Title, Conditions & Possible Causes
DTC: P0553 **2T CCM, MIL: Yes** **Year:** 2009, 2010 **Model:** Genesis, Santa Fe, Sonata, Veracruz **Engine:** 2.4L L4 VIN C, 2.7L V6 VIN D, 3.3L V6 VIN F, 3.8L V6 VIN C, 3.8L V6 VIN E **Transmission:** All	**Power Steering Pressure Sensor/Switch Circuit High Input:** Detects sensor signal short to low voltage. Engine works. Sensor output 4.65 volts. **Possible Causes:** • Poor connection • Short in signal circuit • Open in ground circuit • Faulty P/S pressure sensor • Faulty PCM
DTC: P0560 **2T CCM, MIL: Yes** **Year:** 2009, 2010 **Model:** Accent, Elantra, Genesis Coupe, Santa Fe, Sonata, Tucson **Engine:** 1.6L L4 VIN C, 2.0L L4 VIN B, 2.0L L4 VIN D, 2.0L L4 VIN E, 2.4L L4 VIN B, 2.4L L4 VIN C, 2.7L V6 VIN D **Transmission:** All	**Battery Backup Line Circuit Malfunction:** Engine runtime over 4 minutes and the PCM did not detect any system voltage on the Battery Backup circuit for 5 seconds. **Possible Causes:** • Battery backup circuit to the PCM is open • Battery backup fuse to the PCM is open or missing • Battery backup circuit to the PCM has high resistance • PCM has failed
DTC: P0561 **2T CCM** **Year:** 2009, 2010 **Model:** Accent **Engine:** 1.6L L4 VIN C **Transmission:** All	**System Voltage Unstable:** Engine runtime over 4 minutes, and the PCM detected the system voltage rapidly changed its value by more than 3 volts. **NOTE: If the Battery Backup circuit is open, the vehicle will not run.** **Possible Causes:** • Charging system problem (charging voltage interrupted) • Backup voltage circuit to the PCM open (intermittent fault) • PCM has failed
DTC: P0562 **2T CCM, MIL: Yes** **Year:** 2009, 2010 **Model:** Accent, Azera, Elantra, Genesis, Genesis Coupe, Santa Fe, Sonata, Tucson, Veracruz **Engine:** 1.6L L4 VIN C, 2.0L L4 VIN B, 2.0L L4 VIN D, 2.0L L4 VIN E, 2.4L L4 VIN B, 2.4L L4 VIN C, 2.7L V6 VIN D, 3.3L V6 VIN D, 3.3L V6 VIN E, 3.3L V6 VIN F, 3.5L V6 VIN G, 3.8L V6 VIN C, 3.8L V6 VIN E, 3.8L V6 VIN F, 3.8L V6 VIN H, 4.6L V8 VIN F **Transmission:** All	**System Voltage Low Input:** Engine started engine running at idle or cruise speed, and the PCM detected an unexpected low voltage condition on the ignition circuit. **Possible Causes:** • Ignition system voltage circuit is open • Generator is damaged or has failed (generator output too low) • PCM has failed
DTC: P0563 **2T CCM, MIL: Yes** **Year:** 2009, 2010 **Model:** Accent, Azera, Elantra, Genesis, Genesis Coupe, Santa Fe, Sonata, Tucson, Veracruz **Engine:** 1.6L L4 VIN C, 2.0L L4 VIN B, 2.0L L4 VIN D, 2.0L L4 VIN E, 2.4L L4 VIN B, 2.4L L4 VIN C, 2.7L V6 VIN D, 3.3L V6 VIN D, 3.3L V6 VIN E, 3.3L V6 VIN F, 3.5L V6 VIN G, 3.8L V6 VIN C, 3.8L V6 VIN E, 3.8L V6 VIN F, 3.8L V6 VIN H, 4.6L V8 VIN F **Transmission:** All	**System Voltage High Input:** Engine started engine running at idle or cruise speed, and the PCM detected an unexpected high voltage condition on the ignition circuit. **Possible Causes:** • Generator is damaged or has failed (generator output to high) • PCM has failed

DTC	Trouble Code Title, Conditions & Possible Causes
DTC: P0564 **2T CCM, MIL: Yes** **Year:** 2009, 2010 **Model:** Genesis Coupe, Santa Fe, Sonata, Tucson, Veracruz **Engine:** 2.0L L4 VIN D, 2.4L L4 VIN B, 2.4L L4 VIN C, 3.3L V6 VIN F, 3.5L V6 VIN G, 3.8L V6 VIN C **Transmission:** All	**Cruise Control Multifunction Input "A" Circuit:** Invalid voltage range check. A DTC code is set for the following conditions. Check SET/COAST switch stuck. Check RES/ACC switch stuck. **Possible Causes:** • Open or short in harness • Poor connection or damaged harness • Faulty cruise control remote control switch
DTC: P0571 **2T CCM, MIL: Yes** **Year:** 2009, 2010 **Model:** Azera, Genesis, Genesis Coupe, Santa Fe, Sonata, Veracruz **Engine:** 2.7L V6 VIN D, 3.3L V6 VIN D, 3.3L V6 VIN E, 3.3L V6 VIN F, 3.8L V6 VIN C, 3.8L V6 VIN E, 3.8L V6 VIN F, 3.8L V6 VIN H **Transmission:** All	**Brake Switch "A" Circuit:** PCM detects brake light input signal when the vehicle stops. VSS is normal. Vehicle speed 0 mph, during one second or more. **Possible Causes:** • Poor connection • Open or short to ground in signal circuit • Faulty PCM
DTC: P0600 **2T CCM, MIL: Yes** **Year:** 2009 **Model:** Tucson **Engine:** 2.0L L4 VIN B **Transmission:** All	**CAN Communication Bus:** CAN message transfer incorrect. **Possible Causes:** • Open or short in CAN line • Contact resistance in connectors • Faulty PCM
DTC: P0601 **1T PCM, MIL: Yes** **Year:** 2009, 2010 **Model:** Azera, Genesis, Genesis Coupe, Santa Fe, Sonata, Veracruz **Engine:** 2.7L V6 VIN D, 3.3L V6 VIN D, 3.3L V6 VIN E, 3.3L V6 VIN F, 3.8L V6 VIN C, 3.8L V6 VIN E, 3.8L V6 VIN F, 3.8L V6 VIN H **Transmission:** All	**PCM (Internal Controller) Checksum Error:** Key on or engine running for 1 second, and the PCM detected an internal checksum data error during the initial Self-Test. **Possible Causes:** • Clear the trouble codes and retest for this trouble code. If the same trouble code resets, the PCM has failed and must be replaced to repair this problem.
DTC: P0602 **2T CCM, MIL: Yes** **Year:** 2009, 2010 **Model:** Azera, Genesis, Genesis Coupe, Santa Fe, Sonata, Veracruz **Engine:** 2.7L V6 VIN D, 3.3L V6 VIN D, 3.3L V6 VIN E, 3.3L V6 VIN F, 3.8L V6 VIN C, 3.8L V6 VIN E, 3.8L V6 VIN F, 3.8L V6 VIN H **Transmission:** All	**EEPROM Programming Error:** Check internal CPU **Possible Causes:** • Faulty PCM
DTC: P0604 **2T CCM, MIL: Yes** **Year:** 2009, 2010 **Model:** Accent, Azera, Genesis, Genesis Coupe, Santa Fe, Sonata, Veracruz **Engine:** 1.6L L4 VIN C, 2.7L V6 VIN D, 3.3L V6 VIN D, 3.3L V6 VIN E, 3.3L V6 VIN F, 3.8L V6 VIN C, 3.8L V6 VIN E, 3.8L V6 VIN F, 3.8L V6 VIN H, 4.6L V8 VIN F **Transmission:** All	**Internal Control Module Random Access Memory (RAM) Error:** Check internal CPU. **Possible Causes:** • Faulty PCM

DTC	Trouble Code Title, Conditions & Possible Causes
DTC: P0605 **1T PCM, MIL: Yes** **Year:** 2009, 2010 **Model:** Elantra, Genesis, Genesis Coupe, Santa Fe, Sonata, Tucson **Engine:** 2.0L L4 VIN B, 2.0L L4 VIN D, 2.0L L4 VIN E, 2.4L L4 VIN B, 2.4L L4 VIN C, 2.7L V6 VIN D, 3.5L V6 VIN G, 4.6L V8 VIN F **Transmission:** All	**PCM (Internal Controller) ROM Error:** Key on for 1 second, and the PCM detected an internal ROM error occurred during the initial Self-Test. **Possible Causes:** • Clear the trouble codes and retest for this trouble code. If the same trouble code resets, the PCM has failed and must be replaced to repair this problem.
DTC: P0606 **2T CCM, MIL: Yes** **Year:** 2009, 2010 **Model:** Azera, Elantra, Genesis, Genesis Coupe, Santa Fe, Sonata, Veracruz **Engine:** 2.0L L4 VIN D, 2.0L L4 VIN E, 2.4L L4 VIN C, 2.7L V6 VIN D, 3.3L V6 VIN D, 3.3L V6 VIN E, 3.3L V6 VIN F, 3.5L V6 VIN G, 3.8L V6 VIN C, 3.8L V6 VIN E, 3.8L V6 VIN F, 3.8L V6 VIN H, 4.6L V8 VIN F **Transmission:** All	**ECM Processor (ECU-Self Test Failed):** Controller error. No electrical fault of the front HO2S. **Possible Causes:** • Faulty PCM
DTC: P061B **2T CCM** **Year:** 2009, 2010 **Model:** Azera, Genesis, Genesis Coupe, Santa Fe, Sonata, Veracruz **Engine:** 2.7L V6 VIN D, 3.3L V6 VIN D, 3.3L V6 VIN E, 3.3L V6 VIN F, 3.5L V6 VIN G, 3.8L V6 VIN C, 3.8L V6 VIN E, 3.8L V6 VIN F, 3.8L V6 VIN H, 4.6L V8 VIN F **Transmission:** All	**Internal Control Module Torque Calculation Performance:** Desired torque error. **Possible Causes:** • Faulty PCM
DTC: P0620 **1T PCM, MIL: Yes** **Year:** 2010 **Model:** Santa Fe, Tucson **Engine:** 2.4L L4 VIN B, 2.4L L4 VIN C, 3.5L V6 VIN G **Transmission:** All	**Alternator Control Circuit:** Enable Conditions • Desired alternator voltage duty cycle ≥3.12%, ≤94.5% • Engine speed >240rpm • Battery voltage <16V, >10.73V Threshold Value • Open or Short circuit Diagnostic Time • Continuous **Possible Causes:** • Open or Short in harness • Poor connection or damaged harness
DTC: P0624 **2T CCM** **Year:** 2009, 2010 **Model:** Accent **Engine:** 1.6L L4 VIN C **Transmission:** All	**Fuel Cap Lamp Control Circuit:** Circuit continuity check, (high, low, or open). **Possible Causes:** • Poor connection • Open or short • Instrument cluster • Faulty ECM/PCM

DTC	Trouble Code Title, Conditions & Possible Causes
DTC: P0625 **2T CCM** **Year:** 2009, 2010 **Model:** Elantra, Genesis, Genesis Coupe, Santa Fe, Sonata, Tucson **Engine:** 2.0L L4 VIN D, 2.0L L4 VIN E, 2.4L L4 VIN B, 2.4L L4 VIN C, 4.6L V8 VIN F **Transmission:** All	**Alternator Field "F" Terminal Circuit Low:** Electrical check. **Possible Causes:** • Short to battery in harness • Poor connection or damaged harness
DTC: P0626 **2T CCM** **Year:** 2009, 2010 **Model:** Elantra, Genesis, Genesis Coupe, Santa Fe, Sonata, Tucson **Engine:** 2.0L L4 VIN D, 2.0L L4 VIN E, 2.4L L4 VIN B, 2.4L L4 VIN C, 4.6L V8 VIN F **Transmission:** All	**Alternator Field "F" Terminal Circuit High:** Electrical check. Time after ignition ON, 1 second. Engine speed 0. No main relay error. **Possible Causes:** • Open or short to ground in harness • Faulty charging system
DTC: P0630 **2T CCM, MIL: Yes** **Year:** 2009, 2010 **Model:** Accent, Azera, Elantra, Genesis, Genesis Coupe, Santa Fe, Sonata, Tucson, Veracruz **Engine:** 1.6L L4 VIN C, 2.0L L4 VIN B, 2.0L L4 VIN D, 2.0L L4 VIN E, 2.4L L4 VIN B, 2.4L L4 VIN C, 2.7L V6 VIN D, 3.3L V6 VIN D, 3.3L V6 VIN E, 3.3L V6 VIN F, 3.5L V6 VIN G, 3.8L V6 VIN C, 3.8L V6 VIN E, 3.8L V6 VIN F, 3.8L V6 VIN H, 4.6L V8 VIN F **Transmission:** All	**VIN Not Programmed Or Incompatible- ECM/PCMECM:** PCM internal check. Enable condition, ignition ON. VIN does not exist in boot area. **Possible Causes:** • PCM is new and has not yet been programmed
DTC: P0638 **2T CCM, MIL: Yes** **Year:** 2009, 2010 **Model:** Azera, Genesis, Genesis Coupe, Santa Fe, Sonata, Tucson, Veracruz **Engine:** 2.0L L4 VIN D, 2.4L L4 VIN B, 2.4L L4 VIN C, 2.7L V6 VIN D, 3.3L V6 VIN D, 3.3L V6 VIN E, 3.3L V6 VIN F, 3.5L V6 VIN G, 3.8L V6 VIN C, 3.8L V6 VIN E, 3.8L V6 VIN F, 3.8L V6 VIN H, 4.6L V8 VIN F **Transmission:** All	**Throttle Actuator Control Range/Performance:** ETS position control malfunction. Battery voltage more than 5 volts. **Possible Causes:** • Throttle stuck • Open in motor circuit • Faulty motor • Faulty PCM
DTC: P0641 **2T CCM, MIL: Yes** **Year:** 2009, 2010 **Model:** Azera, Genesis, Genesis Coupe, Santa Fe, Sonata, Veracruz **Engine:** 2.7L V6 VIN D, 3.3L V6 VIN D, 3.3L V6 VIN E, 3.3L V6 VIN F, 3.5L V6 VIN G, 3.8L V6 VIN C, 3.8L V6 VIN E, 3.8L V6 VIN F, 3.8L V6 VIN H **Transmission:** All	**Sensor Reference Voltage "A" Circuit Open:** Sensor reference voltage check. Ignition ON. **Possible Causes:** • Short in sensor power supply line • Faulty PCM

DTC	Trouble Code Title, Conditions & Possible Causes
DTC: P0642 **2T CCM** **Year:** 2009, 2010 **Model:** Accent, Genesis, Genesis Coupe, Santa Fe, Sonata, Tucson **Engine:** 1.6L L4 VIN C, 2.0L L4 VIN D, 2.4L L4 VIN B, 2.4L L4 VIN C, 4.6L V8 VIN F **Transmission:** All	**Sensor Reference Voltage "A" Circuit High:** Electrical check. Ignition ON. APS2 voltage 5.5 volts, for at least 1 second. **Possible Causes:** • Open or short to battery in power circuit • Poor connection or damaged harness • Faulty ECM
DTC: P0643 **2T CCM** **Year:** 2009, 2010 **Model:** Accent, Genesis Coupe, Santa Fe, Sonata, Tucson **Engine:** 1.6L L4 VIN C, 2.0L L4 VIN D, 2.4L L4 VIN B, 2.4L L4 VIN C **Transmission:** All	**Sensor Reference Voltage "A" Circuit Low:** Electrical check. Ignition ON. APS2 voltage 0.7 volt, for at least 1 second. **Possible Causes:** • Open or short to ground in power circuit • Poor connection or damaged harness • Faulty ECM
DTC: P0645 **2T CCM** **Year:** 2009, 2010 **Model:** Accent, Genesis **Engine:** 1.6L L4 VIN C, 4.6L V8 VIN F **Transmission:** All	**A/C Clutch Relay Control Circuit:** DTC is set if the PCM detects that the relay line is open or shorted to ground or battery line. Circuit continuity check. **Possible Causes:** • Open in battery and control circuit • Short to ground in control circuit • Short to battery in control circuit • Faulty A/C relay • Faulty PCM
DTC: P0646 **2T CCM** **Year:** 2009, 2010 **Model:** Accent, Azera, Genesis, Genesis Coupe, Santa Fe, Sonata, Veracruz **Engine:** 1.6L L4 VIN C, 2.0L L4 VIN D, 2.4L L4 VIN C, 2.7L V6 VIN D, 3.3L V6 VIN D, 3.3L V6 VIN E, 3.3L V6 VIN F, 3.5L V6 VIN G, 3.8L V6 VIN C, 3.8L V6 VIN E, 3.8L V6 VIN F, 3.8L V6 VIN H, 4.6L V8 VIN F **Transmission:** All	**A/C Clutch Relay Control Circuit Low:** Detects circuit short to low voltage. No DTC exists. Engine works. After 0.5 seconds. **Possible Causes:** • Poor connection • Open or short to ground in A/C relay circuit • Faulty A/C relay • Faulty PCM
DTC: P0647 **2T CCM** **Year:** 2009, 2010 **Model:** Accent, Azera, Genesis, Genesis Coupe, Santa Fe, Sonata, Veracruz **Engine:** 1.6L L4 VIN C, 2.0L L4 VIN D, 2.4L L4 VIN C, 2.7L V6 VIN D, 3.3L V6 VIN D, 3.3L V6 VIN E, 3.3L V6 VIN F, 3.5L V6 VIN G, 3.8L V6 VIN C, 3.8L V6 VIN E, 3.8L V6 VIN F, 3.8L V6 VIN H, 4.6L V8 VIN F **Transmission:** All	**A/C Clutch Relay Control Circuit High:** Detects circuit short to high voltage. No DTC exists. Engine works. After 0.5 seconds. **Possible Causes:** • Poor connection • Short to power in A/C relay circuit • Faulty A/C relay • Faulty PCM

DTC	Trouble Code Title, Conditions & Possible Causes
DTC: P0650 **2T CCM** **Year:** 2009, 2010 **Model:** Accent, Azera, Elantra, Genesis, Genesis Coupe, Santa Fe, Sonata, Tucson, Veracruz **Engine:** 1.6L L4 VIN C, 2.0L L4 VIN B, 2.0L L4 VIN D, 2.0L L4 VIN E, 2.4L L4 VIN B, 2.4L L4 VIN C, 2.7L V6 VIN D, 3.3L V6 VIN D, 3.3L V6 VIN E, 3.3L V6 VIN F, 3.5L V6 VIN G, 3.8L V6 VIN C, 3.8L V6 VIN E, 3.8L V6 VIN F, 3.8L V6 VIN H, 4.6L V8 VIN F **Transmission:** All	**Malfunction Indicator Lamp (MIL) Control Circuit:** DTC is set if the PCM detects that the MIL line is open or shorted to ground or battery line. Driver stage check. **Possible Causes:** • Open or short between MIL and PCM • Contact resistance in connectors • Burned out MIL bulb
DTC: P0651 **2T CCM, MIL: Yes** **Year:** 2009, 2010 **Model:** Azera, Genesis, Santa Fe, Sonata, Veracruz **Engine:** 2.7L V6 VIN D, 3.3L V6 VIN D, 3.3L V6 VIN E, 3.3L V6 VIN F, 3.8L V6 VIN C, 3.8L V6 VIN E, 3.8L V6 VIN F **Transmission:** All	**Sensor reference Voltage "B" Circuit Open:** Sensor reference voltage check. Key ON. **Possible Causes:** • Short in sensor power supply line • Faulty PCM
DTC: P0652 **2T CCM, MIL: Yes** **Year:** 2009, 2010 **Model:** Genesis, Genesis Coupe, Santa Fe, Sonata, Tucson **Engine:** 2.0L L4 VIN D, 2.4L L4 VIN B, 2.4L L4 VIN C, 4.6L V8 VIN F **Transmission:** All	**Sensor Reference Voltage "B" Circuit Low:** Electrical check. Ignition ON. APS2 voltage 0.7 volt, for at least 0.04 second. **Possible Causes:** • Open or short to ground in power circuit • Poor connection or damaged harness • Faulty ECM
DTC: P0653 **2T CCM, MIL: Yes** **Year:** 2009, 2010 **Model:** Genesis Coupe, Santa Fe, Sonata, Tucson **Engine:** 2.0L L4 VIN D, 2.4L L4 VIN B, 2.4L L4 VIN C **Transmission:** All	**Sensor Reference Voltage "B" Circuit High:** Electrical check. Ignition ON. TPS voltage 5.5 volts, for at least 0.04 second. **Possible Causes:** • Open or short to ground in power circuit • Poor connection or damaged harness • Faulty ECM
DTC: P0660 **2T CCM, MIL: Yes** **Year:** 2009, 2010 **Model:** Azera, Genesis, Santa Fe, Sonata, Veracruz **Engine:** 2.7L V6 VIN D, 3.3L V6 VIN D, 3.3L V6 VIN E, 3.3L V6 VIN F, 3.5L V6 VIN G, 3.8L V6 VIN C, 3.8L V6 VIN E, 3.8L V6 VIN F, 4.6L V8 VIN F **Transmission:** All	**Intake Manifold Tuning Valve Control Circuit/Open (Bank 1):** Signal low, high. **Possible Causes:** • Poor connection • Open or short in VIS circuit • Faulty VIS • Faulty PCM
DTC: P0661 **2T CCM** **Year:** 2009, 2010 **Model:** Genesis, Santa Fe, Sonata, Tucson **Engine:** 2.4L L4 VIN B, 2.4L L4 VIN C, 2.7L V6 VIN D, 4.6L V8 VIN F **Transmission:** All	**Intake Manifold Tuning Valve Control Circuit Low (Bank 1) Solenoid Type:** DTC is set if the ECM detects that the valve control circuit is shorted to ground. Driver stage check. **Possible Causes:** • Open in power supply harness • Short to ground in control harness • Contact resistance in connectors • Faulty valve

DTC	Trouble Code Title, Conditions & Possible Causes
DTC: P0662 **2T CCM** **Year:** 2009, 2010 **Model:** Genesis, Santa Fe, Sonata, Tucson **Engine:** 2.4L L4 VIN B, 2.4L L4 VIN C, 2.7L V6 VIN D, 4.6L V8 VIN F **Transmission:** All	**Intake Manifold Tuning Valve Control Circuit High (Bank 1) Solenoid Type:** DTC is set if the ECM detects that the valve control circuit is open or shorted to battery voltage. Driver stage check. **Possible Causes:** • Open or short to battery in control harness • Contact resistance in connectors • Faulty valve
DTC: P0663 **2T PCM, MIL: Yes** **Year:** 2009, 2010 **Model:** Santa Fe **Engine:** 2.7L V6 VIN D, 3.5L V6 VIN G **Transmission:** All	**Intake Manifold Tuning Valve Control Circuit/Open (Bank 2):** Enable Conditions • After 0.5 sec under conditions below • Engine works • 11V ≤Battery voltage ≤16V Threshold value • Open or short Diagnosis Time • Continuous (More than 5 seconds failure for every 10 seconds test) **Possible Causes:** • Poor connection • Open or short in VIS #2 circuit • Faulty VIS #2 • Faulty PCM
DTC: P0664 **2T CCM** **Year:** 2009 **Model:** Tucson **Engine:** 2.7L V6 VIN D **Transmission:** All	**Intake Manifold Tuning Valve Control Circuit High (Bank 2) Solenoid Type:** DTC is set if the ECM detects that the valve control circuit is shorted to ground. Driver stage check. **Possible Causes:** • Open in power supply harness • Short to ground in control harness • Contact resistance in connectors • Faulty valve
DTC: P0665 **2T CCM** **Year:** 2009, 2010 **Model:** Santa Fe, Tucson **Engine:** 2.7L V6 VIN D, 3.3L V6 VIN E, 3.5L V6 VIN G **Transmission:** All	**Intake Manifold Tuning Valve Control Circuit High (Bank 1) Solenoid Type:** DTC is set if the ECM detects that the valve control circuit is open or shorted to battery voltage. Driver stage check. **Possible Causes:** • Open or short to battery in control harness • Contact resistance in connectors • Faulty valve
DTC: P0685 **2T CCM** **Year:** 2009, 2010 **Model:** Azera, Genesis, Genesis Coupe, Sonata, Veracruz **Engine:** 3.3L V6 VIN D, 3.3L V6 VIN F, 3.8L V6 VIN C, 3.8L V6 VIN E, 3.8L V6 VIN F, 3.8L V6 VIN H **Transmission:** All	**ECM/PCM Power Relay Control Circuit/Open:** Engine running. Ignition voltage less than or equal to 11 volts. **Possible Causes:** • poor connection • Open or short to in control circuit • Main relay • PCM

DTC	Trouble Code Title, Conditions & Possible Causes
DTC: P0698 **1T PCM, MIL: Yes** **Year:** 2009, 2010 **Model:** Genesis, Genesis Coupe, Santa Fe, Sonata, Tucson **Engine:** 2.0L L4 VIN D, 2.4L L4 VIN B, 2.4L L4 VIN C, 4.6L V8 VIN F **Transmission:** All	**Sensor Reference Voltage 'C' Circuit Low:** Case 1 DTC Strategy • Short to ground Enable Conditions • IG "ON" Threshold Value • APS power supply voltage_1 < 0.7 V Case 2 DTC Strategy • Electrical check Enable Conditions • IG "ON" Threshold Value • 0.7 V ≤APS power supply voltage_1 < 4.5 V Fail Safe • Forced limited power mode : The ECM limits opening angle of the throttle valve to max. 50% and engine torque to a certain pre-determined value. • After idle recognition, the ECM uses APS2 signal to calculate the current opening angle of the throttle valve.. **Possible Causes:** • Short to ground in power circuit • Poor connection or damaged harness • Faulty APS
DTC: P0699 **1T PCM, MIL: Yes** **Year:** 2009, 2010 **Model:** Genesis Coupe, Santa Fe, Sonata, Tucson **Engine:** 2.0L L4 VIN D, 2.4L L4 VIN B, 2.4L L4 VIN C **Transmission:** All	**Sensor Reference Voltage 'C' Circuit High:** Enable Conditions • IG "ON" Threshold Value • APS1 > 5.5V Diagnostic Time • 0.1sec. Fail Safe • Forced limited power mode : The ECM limits opening angle of the throttle valve to max. 50% and engine torque to a certain pre-determined value. • After idle recognition, the ECM uses APS2 signal to calculate the current opening angle of the throttle valve. **Possible Causes:** • Short to battery in Power circuit • Poor connection or damaged harness • Faulty APS
DTC: P06A4 **1T PCM, MIL: Yes** **Year:** 2009, 2010 **Model:** Genesis Coupe, Santa Fe, Sonata, Tucson **Engine:** 2.0L L4 VIN D, 2.4L L4 VIN B, 2.4L L4 VIN C **Transmission:** All	**Sensor Reference Voltage 'D' Circuit Low:** Case 1 DTC Strategy • Short to Ground Enable Conditions • IG "ON" Threshold Value • APT/PSPS power supply voltage < 0.7 V Case 2 DTC Strategy • Electrical check Enable Conditions • IG "ON" Threshold Value • 0.7 V ≤APT/DTP power supply voltage < 4.5 V Diagnostic Time • 0.1 sec. **Possible Causes:** • Short to ground in Power circuit • Poor connection or damaged harness • Faulty APT/DTP

DTC	Trouble Code Title, Conditions & Possible Causes
DTC: P06A5 **1T PCM, MIL: Yes** **Year:** 2009, 2010 **Model:** Genesis Coupe, Santa Fe, Sonata, Tucson **Engine:** 2.0L L4 VIN D, 2.4L L4 VIN B, 2.4L L4 VIN C **Transmission:** All	**Sensor Reference Voltage 'D' Circuit High:** Enable Conditions • IG "ON" Threshold Value • APT/PSPS > 5.5V Diagnostic Time • 0.1sec. **Possible Causes:** • Short to battery in Power circuit • Poor connection or damaged harness • Faulty APT/PSPS
DTC: P0700 **2T CCM, MIL: Yes** **Year:** 2009, 2010 **Model:** Accent, Elantra, Genesis, Genesis Coupe, Tucson **Engine:** 1.6L L4 VIN C, 2.0L L4 VIN B, 2.0L L4 VIN D, 2.0L L4 VIN E, 2.7L V6 VIN D, 3.8L V6 VIN E, 3.8L V6 VIN H, 4.6L V8 VIN F **Transmission:** All	**Transmission Control System Signal:** Key on or engine running, and the PCM received a signal from the TCM that indicating an internal problem with the TCM had occurred. **Possible Causes:** • Clear the trouble codes and retest for this trouble code. If the same trouble code resets, the TCM may have failed.
DTC: P0704 **T ECM** **Year:** 2009, 2010 **Model:** Genesis Coupe, Santa Fe, Sonata, Tucson **Engine:** 2.0L L4 VIN D, 2.4L L4 VIN B, 2.4L L4 VIN C, 2.7L V6 VIN D **Transmission:** All	**Clutch Switch Input Circuit Malfunction:** Threshold Value • Open / Short **Possible Causes:** • Open or Short in signal circuit • Poor connection or damaged harness • Faulty clutch switch

OBD II Trouble Code List (P1XXX Codes)

DTC	Trouble Code Title, Conditions & Possible Causes
DTC: P1106 **2T CCM** **Year:** 2009, 2010 **Model:** Azera, Genesis, Genesis Coupe, Santa Fe, Sonata, Veracruz **Engine:** 2.7L V6 VIN D, 3.3L V6 VIN D, 3.3L V6 VIN E, 3.3L V6 VIN F, 3.8L V6 VIN C, 3.8L V6 VIN E, 3.8L V6 VIN F, 3.8L V6 VIN H **Transmission:** All	**Manifold Absolute Pressure Sensor Circuit Short- Intermittent High Input:** This code detects an intermittent short to high in either the signal circuit or the MAP sensor. **Possible Causes:** • Poor connection • Short to battery in signal circuit • Open in ground circuit • Faulty MAPS • Faulty PCM
DTC: P1107 **2T CCM** **Year:** 2009, 2010 **Model:** Azera, Genesis, Genesis Coupe, Santa Fe, Sonata, Veracruz **Engine:** 2.7L V6 VIN D, 3.3L V6 VIN D, 3.3L V6 VIN E, 3.3L V6 VIN F, 3.8L V6 VIN C, 3.8L V6 VIN E, 3.8L V6 VIN F, 3.8L V6 VIN H **Transmission:** All	**Manifold Absolute Pressure Sensor Circuit Short- Intermittent Low Input:** This code detects an intermittent short to high in either the signal circuit or the MAP sensor. **Possible Causes:** • Poor connection • Open or short to ground in the power circuit • Open or short to ground in the signal circuit • Faulty MAPS • Faulty PCM

DTC	Trouble Code Title, Conditions & Possible Causes
DTC: P1111 **2T CCM** **Year:** 2009, 2010 **Model:** Azera, Genesis, Genesis Coupe, Santa Fe, Sonata, Veracruz **Engine:** 2.7L V6 VIN D, 3.3L V6 VIN D, 3.3L V6 VIN E, 3.3L V6 VIN F, 3.8L V6 VIN C, 3.8L V6 VIN E, 3.8L V6 VIN F, 3.8L V6 VIN H **Transmission:** All	**Intake Air Temperature Sensor Circuit Short- Intermittent High Input:** This code detects a continuous short to high in either the signal circuit or the sensor. **Possible Causes:** • Poor connection • Open or short in signal circuit • Open in ground circuit • Faulty IATS • Faulty PCM
DTC: P1112 **2T CCM** **Year:** 2009, 2010 **Model:** Azera, Genesis, Genesis Coupe, Santa Fe, Sonata, Veracruz **Engine:** 2.7L V6 VIN D, 3.3L V6 VIN D, 3.3L V6 VIN E, 3.3L V6 VIN F, 3.8L V6 VIN C, 3.8L V6 VIN E, 3.8L V6 VIN F, 3.8L V6 VIN H **Transmission:** All	**Intake Air Temperature Sensor Circuit Short- Intermittent Low Input:** This code detects a continuous short to high in either the signal circuit or the sensor. **Possible Causes:** • Poor connection • Short to ground in the signal circuit • Open in ground circuit • Faulty IATS • Faulty PCM
DTC: P1114 **2T CCM** **Year:** 2009, 2010 **Model:** Azera, Genesis, Genesis Coupe, Santa Fe, Sonata, Veracruz **Engine:** 2.7L V6 VIN D, 3.3L V6 VIN D, 3.3L V6 VIN E, 3.3L V6 VIN F, 3.8L V6 VIN C, 3.8L V6 VIN E, 3.8L V6 VIN F, 3.8L V6 VIN H **Transmission:** All	**Engine Coolant temperature Sensor Circuit- Intermittent Low Input:** This code detects an intermittent short to ground in the signal circuit or the sensor. **Possible Causes:** • Poor connection • Short to ground in signal circuit • Open in ground circuit • Faulty ECTS • Faulty PCM
DTC: P1115 **2T CCM** **Year:** 2009, 2010 **Model:** Azera, Genesis, Genesis Coupe, Santa Fe, Sonata, Veracruz **Engine:** 2.7L V6 VIN D, 3.3L V6 VIN D, 3.3L V6 VIN E, 3.3L V6 VIN F, 3.8L V6 VIN C, 3.8L V6 VIN E, 3.8L V6 VIN F, 3.8L V6 VIN H **Transmission:** All	**Engine Coolant temperature Sensor Circuit- Intermittent High Input:** This code detects an intermittent open or short to battery in the signal circuit or the sensor. **Possible Causes:** • Poor connection • Open or short to battery in signal circuit • Open in ground circuit • Faulty ECTS • Faulty PCM
DTC: P1295 **2T CCM** **Year:** 2009, 2010 **Model:** Azera, Genesis, Genesis Coupe, Santa Fe, Sonata, Veracruz **Engine:** 2.7L V6 VIN D, 3.3L V6 VIN D, 3.3L V6 VIN E, 3.3L V6 VIN F, 3.8L V6 VIN C, 3.8L V6 VIN E, 3.8L V6 VIN F, 3.8L V6 VIN H **Transmission:** All	**Electronic Throttle Control (ETC) System Malfunction- Power Management:** This code is set is there is a problem in the power management system. Ignition ON. **Possible Causes:** • TPS malfunction • TPS malfunction plus MAFS malfunction • MAP malfunction plus TPS malfunction • Faulty PCM
DTC: P1501 **T PCM** **Year:** 2009, 2010 **Model:** Genesis **Engine:** 3.8L V6 VIN E **Transmission:** All	**Battery Temperature Too High Diagnostic :** Enable Conditions • Engine running state • No error related to LIN communication Threshold value • Battery temperature from battery sensor > 70°C Diagnosis Time • Continuous **Possible Causes:** • Battery temperature high

DTC	Trouble Code Title, Conditions & Possible Causes
DTC: P1502 **T PCM** **Year:** 2009, 2010 **Model:** Genesis **Engine:** 3.8L V6 VIN E **Transmission:** All	**Battery Not Charging Diagnostic:** Enable Conditions • Engine running state • No error related to LIN communication Threshold value • Battery charging current < 10A Diagnosis Time • Continuous **Possible Causes:** • Poor connection • Open or short in charging system • Faulty charging system • Faulty ECM
DTC: P1505 **2T CCM, MIL: Yes** **Year:** 2009, 2010 **Model:** Accent, Elantra, Tucson **Engine:** 1.6L L4 VIN C, 2.0L L4 VIN B, 2.0L L4 VIN D, 2.0L L4 VIN E, 2.7L V6 VIN D **Transmission:** All	**Idle Charge Actuator Signal Low of Coil #1:** The PCM sets a DTC if the ICAV (open) control circuit is open or short to ground. Driver stage check. **Possible Causes:** • Open or short to ground in harness • Contact resistance in connectors • Faulty ICA valve
DTC: P1506 **2T CCM, MIL: Yes** **Year:** 2009, 2010 **Model:** Accent, Elantra, Tucson **Engine:** 1.6L L4 VIN C, 2.0L L4 VIN B, 2.0L L4 VIN D, 2.0L L4 VIN E, 2.7L V6 VIN D **Transmission:** All	**Idle Charge Actuator Signal High of Coil #1:** The PCM sets a DTC if the ICAV (open) control circuit is short to battery. Driver stage check. **Possible Causes:** • Short to battery in harness • Contact resistance in connectors • Faulty ICA valve
DTC: P1507 **2T CCM, MIL: Yes** **Year:** 2009, 2010 **Model:** Accent, Elantra, Tucson **Engine:** 1.6L L4 VIN C, 2.0L L4 VIN B, 2.0L L4 VIN D, 2.0L L4 VIN E, 2.7L V6 VIN D **Transmission:** All	**Idle Charge Actuator Signal Low of Coil #2:** The PCM sets a DTC if the ICAV (open) control circuit is open or short to ground. Driver stage check. **Possible Causes:** • Open or short to ground in harness • Contact resistance in connectors • Faulty ICA valve
DTC: P1508 **2T CCM, MIL: Yes** **Year:** 2009, 2010 **Model:** Accent, Elantra, Tucson **Engine:** 1.6L L4 VIN C, 2.0L L4 VIN B, 2.0L L4 VIN D, 2.0L L4 VIN E, 2.7L V6 VIN D **Transmission:** All	**Idle Charge Actuator Signal High of Coil #2:** The PCM sets a DTC if the ICAV (open) control circuit is short to battery. Driver stage check. **Possible Causes:** • Short to battery in harness • Contact resistance in connectors • Faulty ICA valve
DTC: P1523 **2T CCM** **Year:** 2009, 2010 **Model:** Azera, Genesis, Genesis Coupe, Santa Fe, Sonata, Veracruz **Engine:** 2.7L V6 VIN D, 3.3L V6 VIN D, 3.3L V6 VIN E, 3.3L V6 VIN F, 3.8L V6 VIN C, 3.8L V6 VIN E, 3.8L V6 VIN F, 3.8L V6 VIN H **Transmission:** All	**Electronic Throttle Control (ETC) System Malfunction- Throttle Valve Stuck:** This code is set when the throttle fails to return to unpowered default position when power to the ETC motor is turned off. Fault set for failure to return to default position within a specified time. Throttle actuation previous mode not OFF. Throttle actuator mode is OFF. ETC power control mode equal normal. TPS1 and 2 equal normal. Sensor supply voltage equals normal. **Possible Causes:** • Carbon in throttle • Broken throttle return spring • Throttle sticky • Throttle icy • Faulty PCM

DTC	Trouble Code Title, Conditions & Possible Causes
DTC: P1550 **2T CCM** **Year:** 2009, 2010 **Model:** Accent **Engine:** 1.6L L4 VIN C **Transmission:** All	**Knock Sensor Evaluation IC:** Circuit continuity check, pulse test. **Possible Causes:** • Poor connection • Open or short in control circuit • Faulty knock sensor • Faulty PCM
DTC: P1560 **2T CCM** **Year:** 2009, 2010 **Model:** Accent **Engine:** 1.6L L4 VIN C **Transmission:** All	**Knock Control SPI (Serial Port Interface) Check:** SPI communication check. **Possible Causes:** • Poor connection • Faulty ECM/PCM
DTC: P161B **2T CCM, MIL: Yes** **Year:** 2009, 2010 **Model:** Azera, Genesis, Genesis Coupe, Santa Fe, Sonata, Veracruz **Engine:** 2.7L V6 VIN D, 3.3L V6 VIN D, 3.3L V6 VIN E, 3.3L V6 VIN F, 3.8L V6 VIN C, 3.8L V6 VIN E, 3.8L V6 VIN F, 3.8L V6 VIN H **Transmission:** All	**PCM Internal Error- Torque Calculating:** This code is set if delivered torque is grossly different from the desired torque. **Possible Causes:** • Intake air leakage • Faulty ETS system • Clogged exhaust system • Faulty PCM

OBD II Trouble Code List (P2XXX Codes)

DTC	Trouble Code Title, Conditions & Possible Causes
DTC: P2004 **2T PCM, MIL: Yes** **Year:** 2009, 2010 **Model:** Santa Fe, Sonata, Tucson **Engine:** 2.4L L4 VIN B, 2.4L L4 VIN C **Transmission:** All	**Intake Manifold Runner Control Stuck Open (Bank 1):** Enable Conditions • Port position set-point : Opened for more than 1 sec. • No relevant DTCs • 11V < Battery Voltage < 16V Threshold value • Flap position > Closed stop position - 0.28V Diagnosis Time • 0.2 sec. **Possible Causes:** • Short to ground in Closing Control Circuit • Open in Opening Control Circuit • Carbon or foreign objects • Poor connection or damaged harness • Faulty VCMA
DTC: P2006 **2T PCM, MIL: Yes** **Year:** 2009, 2010 **Model:** Santa Fe, Sonata, Tucson **Engine:** 2.4L L4 VIN B, 2.4L L4 VIN C **Transmission:** All	**Intake Manifold Runner Control Stuck Closed (Bank 1):** Enable Conditions • Port position set-point : Opened for more than 1 sec. • No relevant DTCs • 11V < Battery Voltage < 16V Threshold value • Flap position < Opened stop position + 0.28V Diagnosis Time • 0.2 sec. **Possible Causes:** • Short to ground in Opening Control Circuit • Open in closing Control Circuit • Carbon or foreign objects • Poor connection or damaged harness • Faulty VCMA

DTC	Trouble Code Title, Conditions & Possible Causes
DTC: P2008 **2T PCM, MIL: Yes** **Year:** 2009, 2010 **Model:** Santa Fe, Sonata, Tucson **Engine:** 2.4L L4 VIN B, 2.4L L4 VIN C **Transmission:** All	**Intake Manifold Runner Control Circuit/Open (Bank 1):** Enable Conditions • Engine running • Battery Voltage > 10V Threshold value • ECU power stage diagnosis error Diagnosis Time • 0.1 sec. **Possible Causes:** • Poor connection or damaged harness • Faulty VCMA • Faulty PCM
DTC: P2009 **2T PCM, MIL: Yes** **Year:** 2010 **Model:** Santa Fe **Engine:** 3.5L V6 VIN G **Transmission:** All	**Intake Manifold Runner Control Circuit Low (Bank 1):** Enable Conditions • After 0.5 sec under conditions below • Engine works • 11V ≤Battery voltage ≤16V Threshold value • The VCM commend line open or short to ground Diagnosis Time • Continuous (within 9.88sec.) **Possible Causes:** • Poor connection • Open or short to ground in VCMA commend circuit • Faulty VCMA • Faulty ECM

DTC	Trouble Code Title, Conditions & Possible Causes
DTC: P200A **2T PCM, MIL: Yes** **Year:** 2009, 2010 **Model:** Santa Fe, Sonata, Tucson **Engine:** 2.4L L4 VIN B, 2.4L L4 VIN C, 3.5L V6 VIN G **Transmission:** All	**Intake Manifold Runner Performance (Bank1):** Case 1 DTC Strategy • Opened stop position out of range Enable Conditions • IG KEY ON • Open position stop reached for more than 0.3sec. • No relevant DTCs • 11V < Battery voltage < 16V Threshold Value • Signal : < 0.70V or > 1.73V Case 2 DTC Strategy • Closed stop position out of range Enable Conditions • IG KEY ON • Close position stop reached for more than 0.3sec. • No relevant DTCs • 11V < Battery voltage < 16V Threshold Value • Signal : < 3.90V or > 4.83V Case 3 DTC Strategy • Stuck in range Enable Conditions • IG KEY ON • Port set-point change (Closed -> Opened or Opened -> Closed) for more than 1sec. • Valid learnt Opened / Closed stop position • No relevant DTCs • 11V < Battery voltage < 16V Threshold Value • Signal > Opened stop position + 0.28V • Signal < Closed stop position - 0.28V Case 4 DTC Strategy • Opened stop position exceeded Enable Conditions • IG KEY ON • Port set-point change (Closed -> Opened or Opened -> Closed) for more than 1sec. • Valid learnt Opened / Closed stop position • No relevant DTCs • 11V < Battery voltage < 16V Threshold Value • Signal < Opened stop position - 0.29V Case 1, 2 Diagnostic Time • Immediate Case 3, 4 Diagnostic Time • 0.3 sec. **Possible Causes:** • Carbon or foreign objects • Poor connection or damaged harness • Faulty VCMA
DTC: P2010 **2T PCM, MIL: Yes** **Year:** 2010 **Model:** Santa Fe **Engine:** 3.5L V6 VIN G **Transmission:** All	**Intake Manifold Runner Control Circuit High(Bank 1):** Enable Conditions • After 0.5 sec under conditions below • Engine works • 11V ≤Battery voltage ≤16V Threshold value • The VCM commend line short to battery Diagnosis Time • Continuous (within 9.88sec.) **Possible Causes:** • Poor connection • Short to battery in VCMA commend circuit • Faulty VCMA • Faulty ECM

DTC	Trouble Code Title, Conditions & Possible Causes
DTC: P2016 **2T PCM, MIL: Yes** **Year:** 2009, 2010 **Model:** Santa Fe, Sonata, Tucson **Engine:** 2.4L L4 VIN B, 2.4L L4 VIN C, 3.5L V6 VIN G **Transmission:** All	**Intake Manifold Runner Position Sensor/Switch Circuit Low(Bank 1):** Enable Conditions • IG KEY ON • 11V < Battery voltage < 16V Threshold value • Position signal > 4.85V Diagnosis Time • 0.12 sec. **Possible Causes:** • Short to battery or Open in Signal Circuit • Open in Ground Circuit • Poor connection or damaged harness • Faulty PCM
DTC: P2017 **2T PCM, MIL: Yes** **Year:** 2009, 2010 **Model:** Santa Fe, Sonata, Tucson **Engine:** 2.4L L4 VIN B, 2.4L L4 VIN C, 3.5L V6 VIN G **Transmission:** All	**Intake Manifold Runner Position Sensor/Switch Circuit High(Bank 1):** Enable Conditions • IG KEY ON • 11V < Battery voltage < 16V Threshold value • Position signal < 0.1V Diagnosis Time • 0.12 sec. **Possible Causes:** • Open in Power Circuit • Short to ground in Signal Circuit • Poor connection or damaged harness • Faulty VCMA
DTC: P2021 **2T PCM, MIL: Yes** **Year:** 2010 **Model:** Santa Fe **Engine:** 3.5L V6 VIN G **Transmission:** All	**Intake Manifold Runner Valve Position Sensor Circuit Low:** Enable Conditions • Ignition Voltage ≥10V Threshold value • The valve position sensor voltage < 0.15V Diagnosis Time • Continuous (within 15.6sec.) **Possible Causes:** • Poor connection • Open or short to ground in power circuit • Open or short to ground in signal circuit • Faulty VPS • Faulty ECM
DTC: P2022 **2T PCM, MIL: Yes** **Year:** 2010 **Model:** Santa Fe **Engine:** 3.5L V6 VIN G **Transmission:** All	**Intake Manifold Runner Valve Position Sensor Circuit High:** Enable Conditions • Ignition Voltage ≥10V Threshold value • The valve position sensor voltage > 4.8V Diagnosis Time • Continuous (within 15.6sec.) **Possible Causes:** • Poor connection • Open or short to ground in power circuit • Open or short to ground in signal circuit • Faulty VPS • Faulty ECM

DTC	Trouble Code Title, Conditions & Possible Causes
DTC: P2065 **2T PCM, MIL: Yes** **Year:** 2009, 2010 **Model:** Azera, Genesis, Genesis Coupe, Santa Fe, Veracruz **Engine:** 2.7L V6 VIN D, 3.3L V6 VIN D, 3.3L V6 VIN E, 3.5L V6 VIN G, 3.8L V6 VIN C, 3.8L V6 VIN E, 3.8L V6 VIN F, 3.8L V6 VIN H **Transmission:** All	**Fuel Level Sensor 'B' Circuit:** Enable Conditions • Engine Running • Ignition Voltage > 11V Threshold value • The difference between fuel level sender A and B < 2% Diagnosis Time • Continuous(More than 5 sec. failure for every 10 sec. test) **Possible Causes:** • Poor connection • Shorted signal circuits • Short in signal circuit • Faulty Fuel Level sender A/B • Faulty PCM
DTC: P2066 **2T PCM, MIL: Yes** **Year:** 2009, 2010 **Model:** Azera, Genesis, Genesis Coupe, Santa Fe, Veracruz **Engine:** 2.0L L4 VIN D, 2.7L V6 VIN D, 3.3L V6 VIN D, 3.3L V6 VIN E, 3.5L V6 VIN G, 3.8L V6 VIN C, 3.8L V6 VIN E, 3.8L V6 VIN F, 3.8L V6 VIN H, 4.6L V8 VIN F **Transmission:** All	**Fuel Level Sensor 'B' Performance:** Enable Conditions • Engine Running • Fuel Level Fault Not Present • Ignition Voltage > 11V Threshold value • Current fuel level sender's signal - Previous fuel level sender's signal ≤10%(After Present Odometer - Previous Odometer ≥170 km(105.6 mile)) **Possible Causes:** • Poor connection • Faulty Fuel Level sender A
DTC: P2067 **2T PCM, MIL: Yes** **Year:** 2009, 2010 **Model:** Azera, Genesis, Genesis Coupe, Santa Fe, Veracruz **Engine:** 2.0L L4 VIN D, 2.7L V6 VIN D, 3.3L V6 VIN D, 3.3L V6 VIN E, 3.5L V6 VIN G, 3.8L V6 VIN C, 3.8L V6 VIN E, 3.8L V6 VIN F, 3.8L V6 VIN H, 4.6L V8 VIN F **Transmission:** All	**Fuel Level Sensor 'B' Circuit Low:** Enable Conditions • Engine Running • Ignition Voltage > 11V Threshold value • Raw fuel level sender signal < 0.9 % Diagnosis Time • Continuous(More than 5 sec. failure for every 10 sec. test) **Possible Causes:** • Poor connection • Open or short to ground in signal circuit • Faulty Fuel Level sender A • Faulty PCM
DTC: P2068 **2T PCM, MIL: Yes** **Year:** 2009, 2010 **Model:** Azera, Genesis, Genesis Coupe, Santa Fe, Veracruz **Engine:** 2.0L L4 VIN D, 2.7L V6 VIN D, 3.3L V6 VIN D, 3.3L V6 VIN E, 3.5L V6 VIN G, 3.8L V6 VIN C, 3.8L V6 VIN E, 3.8L V6 VIN F, 3.8L V6 VIN H, 4.6L V8 VIN F **Transmission:** All	**Fuel Level Sensor 'B' Circuit High:** Enable Conditions • Engine Running • Ignition Voltage > 11V Threshold value • Raw fuel level sensor signal > 43 % Diagnosis Time • Continuous(More than 5 sec. failure for every 10 sec. test) **Possible Causes:** • Poor connection • Short to battery in signal Circuit • Faulty Fuel Level sender A • Faulty PCM

DTC	Trouble Code Title, Conditions & Possible Causes
DTC: P2069 **2T PCM, MIL: Yes** **Year:** 2010 **Model:** Genesis Coupe **Engine:** 2.0L L4 VIN D **Transmission:** All	**Fuel Level Sensor 'B' Circuit Intermittent:** Enable Conditions • Vehicle Speed > 22 mph during 20 sec. • Wheel speed gradient < 1.8 /1000 during 20 sec. • No relevant DTCs. • 11< Battery voltage <16 Threshold Value • FL measured - FL filtered value > 50 % Diagnostic Time • 10 sec **Possible Causes:** • Poor connection or damaged harness • Faulty Fuel Level Sender "A"
DTC: P2096 **2T CCM, MIL: Yes** **Year:** 2009, 2010 **Model:** Accent, Azera, Elantra, Genesis, Genesis Coupe, Santa Fe, Sonata, Tucson, Veracruz **Engine:** 1.6L L4 VIN C, 2.0L L4 VIN D, 2.0L L4 VIN E, 2.4L L4 VIN B, 2.4L L4 VIN C, 2.7L V6 VIN D, 3.3L V6 VIN D, 3.3L V6 VIN E, 3.3L V6 VIN F, 3.5L V6 VIN G, 3.8L V6 VIN C, 3.8L V6 VIN E, 3.8L V6 VIN F, 3.8L V6 VIN H, 4.6L V8 VIN F **Transmission:** All	**Post Catalyst Fuel Trim System Too Lean (Bank 1):** Case 1: Monitoring deviation of fuel trim control (long term). No relevant failure. Long term fuel trim active. Case 2: Monitoring deviation of fuel trim control (short term). No relevant failure. Short term fuel trim active. Current engine speed less than 500 rpm. Current mass air flow less than 400mg/rev. Current lambda correction mean value less than 4 percent. **Possible Causes:** • Three way catalytic converter (TWC) • Rear HO2S
DTC: P2097 **2T CCM, MIL: Yes** **Year:** 2009, 2010 **Model:** Accent, Azera, Elantra, Genesis, Genesis Coupe, Santa Fe, Sonata, Tucson, Veracruz **Engine:** 1.6L L4 VIN C, 2.0L L4 VIN D, 2.0L L4 VIN E, 2.4L L4 VIN B, 2.4L L4 VIN C, 2.7L V6 VIN D, 3.3L V6 VIN D, 3.3L V6 VIN E, 3.3L V6 VIN F, 3.5L V6 VIN G, 3.8L V6 VIN C, 3.8L V6 VIN E, 3.8L V6 VIN F, 3.8L V6 VIN H, 4.6L V8 VIN F **Transmission:** All	**Post Catalyst Fuel Trim System Too Rich (Bank 1):** Case 1: Monitoring deviation of fuel trim control (long term). No relevant failure. Long term fuel trim active. Case 2: Monitoring deviation of fuel trim control (short term). No relevant failure. Short term fuel trim active. Current engine speed less than 500 rpm. Current mass air flow less than 400mg/rev. Current lambda correction mean value less than 4 percent. **Possible Causes:** • Three way catalytic converter (TWC) • Rear HO2S
DTC: P2098 **2T CCM, MIL: Yes** **Year:** 2009, 2010 **Model:** Azera, Genesis, Genesis Coupe, Santa Fe, Sonata, Veracruz **Engine:** 2.7L V6 VIN D, 3.3L V6 VIN D, 3.3L V6 VIN E, 3.3L V6 VIN F, 3.5L V6 VIN G, 3.8L V6 VIN C, 3.8L V6 VIN E, 3.8L V6 VIN F, 3.8L V6 VIN H, 4.6L V8 VIN F **Transmission:** All	**Post Catalyst Fuel Trim System Too Lean (Bank 2):** Case 1: Monitoring deviation of fuel trim control (long term). No relevant failure. Long term fuel trim active. Case 2: Monitoring deviation of fuel trim control (short term). No relevant failure. Short term fuel trim active. Current engine speed less than 500 rpm. Current mass air flow less than 400mg/rev. Current lambda correction mean value less than 4 percent. **Possible Causes:** • Three way catalytic converter (TWC) • Rear HO2S

DTC	Trouble Code Title, Conditions & Possible Causes
DTC: P2099 **2T CCM, MIL: Yes** **Year:** 2009, 2010 **Model:** Azera, Genesis, Genesis Coupe, Santa Fe, Sonata, Veracruz **Engine:** 2.7L V6 VIN D, 3.3L V6 VIN D, 3.3L V6 VIN E, 3.3L V6 VIN F, 3.5L V6 VIN G, 3.8L V6 VIN C, 3.8L V6 VIN E, 3.8L V6 VIN F, 3.8L V6 VIN H, 4.6L V8 VIN F **Transmission:** All	**Post Catalyst Fuel Trim System Too Rich (Bank 2):** Case 1: Monitoring deviation of fuel trim control (long term). No relevant failure. Long term fuel trim active. Case 2: Monitoring deviation of fuel trim control (short term). No relevant failure. Short term fuel trim active. Current engine speed less than 500 rpm. Current mass air flow less than 400mg/rev. Current lambda correction mean value less than 4 percent. **Possible Causes:** • Three way catalytic converter (TWC) • Rear HO2S
DTC: P2101 **2T CCM** **Year:** 2009, 2010 **Model:** Genesis, Genesis Coupe, Santa Fe, Sonata, Tucson **Engine:** 2.0L L4 VIN D, 2.4L L4 VIN B, 2.4L L4 VIN C, 4.6L V8 VIN F **Transmission:** All	**Throttle Actuator Control Motor Circuit Range/Performance:** Hardware check. Battery voltage 9 volts. ECU power stage error. **Possible Causes:** • Poor connection or damaged harness • Faulty ETC motor
DTC: P2104 **2T CCM, MIL: Yes** **Year:** 2009, 2010 **Model:** Azera, Genesis, Genesis Coupe, Santa Fe, Sonata, Tucson, Veracruz **Engine:** 2.0L L4 VIN D, 2.4L L4 VIN B, 2.4L L4 VIN C, 2.7L V6 VIN D, 3.3L V6 VIN D, 3.3L V6 VIN E, 3.3L V6 VIN F, 3.5L V6 VIN G, 3.8L V6 VIN C, 3.8L V6 VIN E, 3.8L V6 VIN F, 3.8L V6 VIN H, 4.6L V8 VIN F **Transmission:** All	**Electronic Throttle Control (ETC) System Malfunction- Forced Idle:** This code is set if the system is in forced idle mode. Ignition ON. **Possible Causes:** • Faulty AFS • Faulty AFS plus brake • Faulty AFS plus vehicle speed sensor • Faulty AFS plus brake plus vehicle speed sensor • Faulty PCM
DTC: P2105 **2T CCM, MIL: Yes** **Year:** 2009, 2010 **Model:** Azera, Genesis, Genesis Coupe, Santa Fe, Sonata, Tucson, Veracruz **Engine:** 2.0L L4 VIN D, 2.4L L4 VIN B, 2.4L L4 VIN C, 2.7L V6 VIN D, 3.3L V6 VIN D, 3.3L V6 VIN E, 3.3L V6 VIN F, 3.5L V6 VIN G, 3.8L V6 VIN C, 3.8L V6 VIN E, 3.8L V6 VIN F, 3.8L V6 VIN H **Transmission:** All	**Electronic Throttle Control (ETC) System Malfunction- Forced Engine Shutdown:** This code is set if the system is in forced engine shutdown mode. Ignition ON. **Possible Causes:** • Faulty AFS plus MAPS plus ETS • Faulty PCM
DTC: P2106 **1T PCM, MIL: Yes** **Year:** 2009, 2010 **Model:** Genesis Coupe, Sonata, Tucson, Veracruz **Engine:** 2.0L L4 VIN D, 2.4L L4 VIN C, 3.3L V6 VIN F, 3.8L V6 VIN C **Transmission:** All	**Throttle Actuator Control System-Force Limited Power:** • Additional DTC when TPS sensor electrical error or plausibility of TPS_1 and TPS_2 error detected **Possible Causes:** • ETC system malfunction • (P0121,P0122,P0123,P0221,P0222,P0223,P0642,P0643, • P0698,P0699,P2122,P2123,P2127,P2128 or P2138)

DTC	Trouble Code Title, Conditions & Possible Causes
DTC: P2106 **2T CCM, MIL: Yes** **Year:** 2009, 2010 **Model:** Azera, Genesis, Genesis Coupe, Santa Fe **Engine:** 2.4L L4 VIN B, 2.7L V6 VIN D, 3.3L V6 VIN D, 3.3L V6 VIN E, 3.5L V6 VIN G, 3.8L V6 VIN E, 3.8L V6 VIN F, 3.8L V6 VIN H, 4.6L V8 VIN F **Transmission:** All	**Electronic Throttle Control (ETC) System Malfunction- Forced Limited Power:** This code is set if the system is in forced limited power mode. Ignition ON. **Possible Causes:** • Faulty APS • Faulty APS + Brake • Faulty APS + vehicle speed sensor • Faulty APS + vehicle speed sensor + brake • Faulty PCM
DTC: P2110 **1T PCM, MIL: Yes** **Year:** 2009, 2010 **Model:** Genesis Coupe, Santa Fe, Sonata, Tucson **Engine:** 2.0L L4 VIN D, 2.4L L4 VIN B, 2.4L L4 VIN C, 3.5L V6 VIN G **Transmission:** All	**Throttle Actuator Control System-Forced Limited RPM:** • Additional DTC when PVS single error or ETC hardware check error(P2101, 2118) or TPS start check adaptation(P2119, P0638) or TPS reference voltage(P0652, P0653) or PVS reference voltage(P0642, P0643,P0698,P0699) or PVS ratio error(P2138) or both TPS_1 and TPS_2 error or Level 2 monitoring error detected **Possible Causes:** • ETC system malfunction
DTC: P2118 **2T CCM, MIL: Yes** **Year:** 2009, 2010 **Model:** Genesis, Genesis Coupe **Engine:** 2.0L L4 VIN D, 4.6L V8 VIN F **Transmission:** All	**Throttle Actuator Control Motor Current Range/Performance/Throttle Actuator Control Motor Circuit Open:** Vb open. Motor relay ON. Voltage to detect circuit open less than or equal to 4.0 volts. **Possible Causes:** • Poor connection • Open in ETS relay circuit • Faulty ETS relay/fuse • Faulty PCM
DTC: P2118 **2T CCM, MIL: Yes** **Year:** 2009, 2010 **Model:** Genesis, Genesis Coupe, Santa Fe, Sonata, Tucson **Engine:** 2.0L L4 VIN D, 2.4L L4 VIN B, 2.4L L4 VIN C, 4.6L V8 VIN F **Transmission:** All	**Throttle Actuator Control Motor Circuit Range/Performance/Throttle Actuator Control Motor Circuit Low:** Motor circuit low. Ignition switch ON. **Possible Causes:** • Poor connection • Short to ground in ETS motor circuit • Faulty ETS motor • Faulty PCM
DTC: P2118 **2T CCM, MIL: Yes** **Year:** 2009, 2010 **Model:** Genesis, Genesis Coupe, Santa Fe, Sonata, Tucson **Engine:** 2.0L L4 VIN D, 2.4L L4 VIN B, 2.4L L4 VIN C, 4.6L V8 VIN F **Transmission:** All	**Throttle Actuator Control Motor Circuit Range/Performance/Throttle Actuator Control Motor Circuit High:** Motor circuit High. Ignition switch ON. **Possible Causes:** • Poor connection • Short to battery in ETS motor circuit • Faulty ETS motor • Faulty PCM
DTC: P2119 **2T CCM, MIL: Yes** **Year:** 2009, 2010 **Model:** Genesis, Genesis Coupe, Sonata, Tucson **Engine:** 2.0L L4 VIN D, 2.4L L4 VIN C, 4.6L V8 VIN F **Transmission:** All	**Throttle Actuator Control Module Performance/Throttle Actuator Control System Stuck Closed (IG OFF):** Valve stuck closed (#1). Ignition switch OFF. TPS output as throttle valve is closed less than 0.025 volt. **Possible Causes:** • Poor connector • Faulty throttle valve • Faulty ETS motor • Faulty PCM

DTC	Trouble Code Title, Conditions & Possible Causes
DTC: P2122 **2T CCM, MIL: Yes** **Year:** 2009, 2010 **Model:** Azera, Genesis, Genesis Coupe, Santa Fe, Sonata, Tucson, Veracruz **Engine:** 2.0L L4 VIN D, 2.4L L4 VIN B, 2.4L L4 VIN C, 2.7L V6 VIN D, 3.3L V6 VIN D, 3.3L V6 VIN E, 3.3L V6 VIN F, 3.5L V6 VIN G, 3.8L V6 VIN C, 3.8L V6 VIN E, 3.8L V6 VIN F, 3.8L V6 VIN H, 4.6L V8 VIN F **Transmission:** All	**Throttle/Pedal Position Sensor/Switch "D" Circuit Low Input:** Accelerator position sensor (APS1) low input. ETS/PCM communication is normal. Output voltage of APS1 is less than 0.2 volt. **Possible Causes:** • Poor connector • Faulty APS1 • Open or short in APS1 circuit • Faulty PCM
DTC: P2123 **2T CCM, MIL: Yes** **Year:** 2009, 2010 **Model:** Azera, Genesis, Genesis Coupe, Santa Fe, Sonata, Tucson, Veracruz **Engine:** 2.0L L4 VIN D, 2.4L L4 VIN B, 2.4L L4 VIN C, 2.7L V6 VIN D, 3.3L V6 VIN D, 3.3L V6 VIN E, 3.3L V6 VIN F, 3.5L V6 VIN G, 3.8L V6 VIN C, 3.8L V6 VIN E, 3.8L V6 VIN F, 3.8L V6 VIN H, 4.6L V8 VIN F **Transmission:** All	**Throttle/Pedal Position Sensor/Switch "D" Circuit High Input:** Accelerator position sensor (APS1) high input. ETS/PCM communication is normal. Output voltage of APS1 is equal to or greater than 4.9 volts. Output voltage of APS2 is less than 4.1 volts. **Possible Causes:** • Poor connector • Faulty APS1 • Open or short in APS1 circuit • Faulty PCM
DTC: P2127 **2T CCM, MIL: Yes** **Year:** 2009, 2010 **Model:** Azera, Genesis, Genesis Coupe, Santa Fe, Sonata, Tucson, Veracruz **Engine:** 2.0L L4 VIN D, 2.4L L4 VIN B, 2.4L L4 VIN C, 2.7L V6 VIN D, 3.3L V6 VIN D, 3.3L V6 VIN E, 3.3L V6 VIN F, 3.5L V6 VIN G, 3.8L V6 VIN C, 3.8L V6 VIN E, 3.8L V6 VIN F, 3.8L V6 VIN H, 4.6L V8 VIN F **Transmission:** All	**Throttle/Pedal Position Sensor/Switch "E" Circuit Low Input:** Accelerator position sensor (APS2) low input. ETS/PCM communication is normal. Output voltage of APS2 is less than 0.2 volt. **Possible Causes:** • Poor connection • Faulty APS2 • Open or short in APS2 circuit • Faulty PCM
DTC: P2128 **2T CCM, MIL: Yes** **Year:** 2009, 2010 **Model:** Azera, Genesis, Genesis Coupe, Santa Fe, Sonata, Tucson, Veracruz **Engine:** 2.0L L4 VIN D, 2.4L L4 VIN B, 2.4L L4 VIN C, 2.7L V6 VIN D, 3.3L V6 VIN D, 3.3L V6 VIN E, 3.3L V6 VIN F, 3.5L V6 VIN G, 3.8L V6 VIN C, 3.8L V6 VIN E, 3.8L V6 VIN F, 3.8L V6 VIN H, 4.6L V8 VIN F **Transmission:** All	**Throttle/Pedal Position Sensor/Switch "E" Circuit High Input:** Accelerator position sensor (APS2) high input. ETS/PCM communication is normal. Output voltage of APS2 is greater than or equal to 4.9 volts. Output voltage of ASP1 is less than 4.1 volts. **Possible Causes:** • Poor connection • Faulty APS2 • Open or short in APS2 circuit • Faulty PCM

DTC	Trouble Code Title, Conditions & Possible Causes
DTC: P2135 **2T CCM, MIL: Yes** **Year:** 2009, 2010 **Model:** Azera, Genesis, Genesis Coupe, Santa Fe, Sonata, Veracruz **Engine:** 2.4L L4 VIN B, 2.7L V6 VIN D, 3.3L V6 VIN D, 3.3L V6 VIN E, 3.3L V6 VIN F, 3.5L V6 VIN G, 3.8L V6 VIN C, 3.8L V6 VIN E, 3.8L V6 VIN F, 3.8L V6 VIN H **Transmission:** All	**Throttle/Pedal Position Sensor/Switch "A"/"B" Voltage Correlation:** Determines if TPS #1 disagrees with TPS #2. Ignition "ON". **Possible Causes:** • Poor connection • Open or short in TPS circuit • Faulty TPS • Faulty PCM
DTC: P2138 **2T CCM, MIL: Yes** **Year:** 2009, 2010 **Model:** Azera, Genesis, Genesis Coupe, Santa Fe, Sonata, Tucson, Veracruz **Engine:** 2.0L L4 VIN D, 2.4L L4 VIN C, 2.7L V6 VIN D, 3.3L V6 VIN D, 3.3L V6 VIN E, 3.3L V6 VIN F, 3.5L V6 VIN G, 3.8L V6 VIN C, 3.8L V6 VIN E, 3.8L V6 VIN F, 3.8L V6 VIN H, 4.6L V8 VIN F **Transmission:** All	**Throttle/Pedal Position Sensor/Switch "D/E" Voltage Correlation:** Monitoring abnormal APS. Output voltage of APS1: 0.2 to 4.9 volts. Output voltage of APS2: 0.2 to 4.9 volts. Ignition switch ON. **Possible Causes:** • Poor connection • Faulty APS • Faulty PCM
DTC: P2159 **2T CCM, MIL: Yes** **Year:** 2009, 2010 **Model:** Genesis Coupe, Santa Fe, Sonata, Tucson **Engine:** 2.0L L4 VIN D, 2.4L L4 VIN B, 2.4L L4 VIN C, 3.5L V6 VIN G **Transmission:** All	**Vehicle Speed Sensor "B" Range/Performance:** Plausibility check. Enabling conditions are as follows: engine speed greater than 2100 rpm, air mass flow greater than 0.44 g/rev, no fuel injection shut off, coolant temperature 140 degrees F. **Possible Causes:** • Open or short in harness • Poor connection or damaged harness • VSS
DTC: P2173 **2T CCM, MIL: Yes** **Year:** 2009, 2010 **Model:** Azera, Genesis, Genesis Coupe, Santa Fe, Sonata, Veracruz **Engine:** 2.7L V6 VIN D, 3.3L V6 VIN D, 3.3L V6 VIN E, 3.3L V6 VIN F, 3.8L V6 VIN C, 3.8L V6 VIN E, 3.8L V6 VIN F, 3.8L V6 VIN H **Transmission:** All	**Electronic Throttle Control (ETC) System Malfunction- High Air Flow Detected:** The engine airflow measurements are not based on throttle position. They are compared with throttle position based on estimated air flow. If measured air flow is much higher, the throttle body may not be throttling the engine. Engine running. Throttle actuation mode is not off. MAP sensor is not failed. MAF sensor is not failed. IAT sensor is not failed. **Possible Causes:** • Air leakage between TPS and MAFS • Faulty throttle body • Faulty PCM
DTC: P2187 **2T CCM, MIL: Yes** **Year:** 2009, 2010 **Model:** Azera, Genesis, Genesis Coupe, Santa Fe, Sonata, Tucson, Veracruz **Engine:** 2.0L L4 VIN D, 2.4L L4 VIN B, 2.4L L4 VIN C, 2.7L V6 VIN D, 3.3L V6 VIN D, 3.3L V6 VIN E, 3.3L V6 VIN F, 3.5L V6 VIN G, 3.8L V6 VIN C, 3.8L V6 VIN E, 3.8L V6 VIN F, 3.8L V6 VIN H, 4.6L V8 VIN F **Transmission:** All	**System Too Lean At Idle (Additive) (Bank 1):** Engine coolant temperature 140 degrees F. Intake air temperature 140 degrees F. System voltage greater than 11 volts. Closed loop active. **Possible Causes:** • Sensors related to fuel trim • Intake system • Fuel pressure • Faulty PCM

DTC	Trouble Code Title, Conditions & Possible Causes
DTC: P2188 **2T CCM, MIL: Yes** **Year:** 2009, 2010 **Model:** Azera, Genesis, Genesis Coupe, Santa Fe, Sonata, Tucson, Veracruz **Engine:** 2.0L L4 VIN D, 2.4L L4 VIN B, 2.4L L4 VIN C, 2.7L V6 VIN D, 3.3L V6 VIN D, 3.3L V6 VIN E, 3.3L V6 VIN F, 3.5L V6 VIN G, 3.8L V6 VIN C, 3.8L V6 VIN E, 3.8L V6 VIN F, 3.8L V6 VIN H, 4.6L V8 VIN F **Transmission:** All	**System Too Rich At Idle (Bank 1):** Fuel trim limit. Coolant temperature greater than 158 degrees F. Intake air temperature less than 176 degrees F. Throttle angle less than 60 percent. Integrated air mass greater than 10 grams. Closed loop control enabled. No transient control phase. No canister purge phase. Engine speed less than 920 rpm. **Possible Causes:** • Faulty ignition system • EVAP PCSV malfunction • Faulty fuel injectors • Leak in exhaust system • Faulty MAP, TPS, ECTS • Faulty front HO2S • Faulty PCM
DTC: P2189 **2T CCM, MIL: Yes** **Year:** 2009, 2010 **Model:** Azera, Genesis, Genesis Coupe, Santa Fe, Sonata, Veracruz **Engine:** 2.7L V6 VIN D, 3.3L V6 VIN D, 3.3L V6 VIN E, 3.3L V6 VIN F, 3.5L V6 VIN G, 3.8L V6 VIN C, 3.8L V6 VIN E, 3.8L V6 VIN F, 3.8L V6 VIN H, 4.6L V8 VIN F **Transmission:** All	**System Too Lean At Idle (additive) Bank 2:** Fuel trim idle condition (option limits exceeded). **Possible Causes:** • Sensors related to fuel trim • Intake system • Fuel pressure • Faulty PCM
DTC: P2190 **2T CCM, MIL: Yes** **Year:** 2009, 2010 **Model:** Azera, Genesis, Genesis Coupe, Santa Fe, Sonata, Veracruz **Engine:** 2.7L V6 VIN D, 3.3L V6 VIN D, 3.3L V6 VIN E, 3.3L V6 VIN F, 3.5L V6 VIN G, 3.8L V6 VIN C, 3.8L V6 VIN E, 3.8L V6 VIN F, 3.8L V6 VIN H, 4.6L V8 VIN F **Transmission:** All	**System Too Rich At Idle (additive) Bank 2:** Fuel trim idle condition (option limits exceeded). **Possible Causes:** • Sensors related to fuel trim • Intake system • Fuel pressure • Faulty PCM
DTC: P2191 **2T CCM, MIL: Yes** **Year:** 2009, 2010 **Model:** Genesis, Genesis Coupe, Santa Fe, Sonata, Tucson **Engine:** 2.0L L4 VIN D, 2.4L L4 VIN B, 2.4L L4 VIN C, 4.6L V8 VIN F **Transmission:** All	**System Too Lean At Higher Load (Multiple) (Bank 1):** Fuel trim limit. Coolant temperature greater than 158 degrees F. Intake air temperature less than 176 degrees F. Throttle angle less than 60 percent. Integrated air mass greater than 10 grams. Closed loop control enabled. No transient control phase. No canister purge phase. Air mass1 40 to 80 kg/h. Air mass2 greater than 100 kg/h. **Possible Causes:** • Faulty ignition system • EVAP PCSV malfunction • Faulty fuel injectors • Leak in exhaust system • Faulty MAP, TPS, ECTS • Faulty front HO2S • Faulty PCM
DTC: P2192 **2T CCM, MIL: Yes** **Year:** 2009, 2010 **Model:** Genesis, Genesis Coupe, Santa Fe, Sonata, Tucson **Engine:** 2.0L L4 VIN D, 2.4L L4 VIN B, 2.4L L4 VIN C, 4.6L V8 VIN F **Transmission:** All	**System Too Rich At Higher Load (Bank 1):** Fuel trim limit. Coolant temperature greater than 158 degrees F. Intake air temperature less than 176 degrees F. Throttle angle less than 60 percent. Integrated air mass greater than 10 grams. Closed loop control enabled. No transient control phase. No canister purge phase. Engine load1 30 to 55 percent. Engine load2 greater than 70 percent. **Possible Causes:** • Faulty ignition system • EVAP PCSV malfunction • Faulty fuel injectors • Leak in exhaust system • Faulty MAP, TPS, ECTS • Faulty front HO2S • Faulty PCM

DTC	Trouble Code Title, Conditions & Possible Causes
DTC: P2193 **2T PCM, MIL: Yes** **Year:** 2009, 2010 **Model:** Genesis **Engine:** 4.6L V8 VIN F **Transmission:** All	**System Too Lean at Higher Load (Bank 2):** Enable Conditions • 12.3%< load < 65% • Engine speed > 500rpm • ECT > 70C • IAT < 80C Threshold value • Long Term FT (multiplicative) > 1.23 Diagnosis Time • 10 sec **Possible Causes:** • Air leakage • Improper fuel pressure • PCV valve stuck • Clogging of injector
DTC: P2194 **2T PCM, MIL: Yes** **Year:** 2009, 2010 **Model:** Genesis **Engine:** 4.6L V8 VIN F **Transmission:** All	**System Too Rich at Higher Load (Bank 2):** Enable Conditions • 12.3%< load < 65% • Engine speed > 500rpm • ECT > 70C • IAT < 80C Threshold value • Long Term FT (multiplicative) < 0.77 Diagnosis Time • 10 sec **Possible Causes:** • Blocking of intake system • Fuel leakage in injector • Improper fuel pressure
DTC: P2195 **2T CCM, MIL: Yes** **Year:** 2009, 2010 **Model:** Azera, Elantra, Genesis, Genesis Coupe, Santa Fe, Sonata, Tucson, Veracruz **Engine:** 2.0L L4 VIN D, 2.0L L4 VIN E, 2.4L L4 VIN B, 2.4L L4 VIN C, 2.7L V6 VIN D, 3.3L V6 VIN D, 3.3L V6 VIN E, 3.3L V6 VIN F, 3.5L V6 VIN G, 3.8L V6 VIN C, 3.8L V6 VIN E, 3.8L V6 VIN F, 3.8L V6 VIN H **Transmission:** All	**HO2S Signal Stuck Lean (Bank 1 Sensor 1):** Sensor characteristic line shifted to lean. No relevant failure. No misfire detected. Fuel trim control active. **Possible Causes:** • Contact resistance in connectors • Faulty HO2S
DTC: P2196 **2T CCM, MIL: Yes** **Year:** 2009, 2010 **Model:** Azera, Elantra, Genesis, Genesis Coupe, Santa Fe, Sonata, Tucson, Veracruz **Engine:** 2.0L L4 VIN D, 2.0L L4 VIN E, 2.4L L4 VIN B, 2.4L L4 VIN C, 2.7L V6 VIN D, 3.3L V6 VIN D, 3.3L V6 VIN E, 3.3L V6 VIN F, 3.5L V6 VIN G, 3.8L V6 VIN C, 3.8L V6 VIN E, 3.8L V6 VIN F, 3.8L V6 VIN H **Transmission:** All	**HO2S Signal Stuck Rich (Bank 1 Sensor 1):** Sensor characteristic line shifted to lean. No relevant failure. No misfire detected. Fuel trim control active. **Possible Causes:** • Contact resistance in connectors • Faulty HO2S

DTC	Trouble Code Title, Conditions & Possible Causes
DTC: P2197 **2T CCM, MIL: Yes** **Year:** 2009, 2010 **Model:** Azera, Genesis, Genesis Coupe, Santa Fe, Sonata, Veracruz **Engine:** 2.7L V6 VIN D, 3.3L V6 VIN D, 3.3L V6 VIN E, 3.3L V6 VIN F, 3.5L V6 VIN G, 3.8L V6 VIN C, 3.8L V6 VIN E, 3.8L V6 VIN F, 3.8L V6 VIN H **Transmission:** All	**HO2S Signal Stuck Lean (Bank 2 Sensor 1):** Determines if O2 sensor indicates lean exhaust while in power enrichment. Sensor not in cooled status flag. Not in transient conditions status flag. Device control not active. Engine running. Minimum air flow present is equal or greater than 2 g/s. Engine coolant warm (140 degrees F. Above conditions met for at least 1.5L seconds. **Possible Causes:** • Poor connection • Faulty HO2S • Faulty PCM
DTC: P2198 **2T CCM, MIL: Yes** **Year:** 2009, 2010 **Model:** Azera, Genesis, Genesis Coupe, Santa Fe, Sonata, Veracruz **Engine:** 2.7L V6 VIN D, 3.3L V6 VIN D, 3.3L V6 VIN E, 3.3L V6 VIN F, 3.5L V6 VIN G, 3.8L V6 VIN C, 3.8L V6 VIN E, 3.8L V6 VIN F, 3.8L V6 VIN H **Transmission:** All	**HO2S Signal Stuck Rich (Bank 2 Sensor 1):** Determines if O2 sensor indicates rich exhaust while in decal fuel cut off (DFCO). Sensor not in cooled status flag. Not in transient conditions status flag. Device control not active. Engine running. Minimum air flow present is equal or greater than 2 g/s. Ignition voltage equal to or greater than 10 volts. Fuel reduction not active. Engine running long enough (more than 60 seconds). Engine coolant warm (140 degrees F. Above conditions met for at least 1.5L seconds. **Possible Causes:** • Poor connection • Faulty HO2S • Faulty PCM
DTC: P2226 **2T CCM, MIL: Yes** **Year:** 2009, 2010 **Model:** Accent **Engine:** 1.6L L4 VIN C **Transmission:** All	**Barometric Pressure Circuit:** Rationality check. **Possible Causes:** • Clog at sensing hole • Faulty ECM
DTC: P2227 **2T CCM, MIL: Yes** **Year:** 2009, 2010 **Model:** Accent, Genesis Coupe, Santa Fe **Engine:** 1.6L L4 VIN C, 2.0L L4 VIN D, 3.5L V6 VIN G **Transmission:** All	**Barometric Pressure Circuit Range/Performance:** Rationality check. **Possible Causes:** • Clog at sensing hole • Faulty ECM
DTC: P2228 **2T CCM, MIL: Yes** **Year:** 2009, 2010 **Model:** Accent, Genesis Coupe, Santa Fe **Engine:** 1.6L L4 VIN C, 2.0L L4 VIN D, 3.5L V6 VIN G **Transmission:** All	**Barometric Pressure Circuit Low Input:** Signal check low. **Possible Causes:** • Faulty ECM
DTC: P2229 **2T CCM, MIL: Yes** **Year:** 2009, 2010 **Model:** Accent, Genesis Coupe, Santa Fe **Engine:** 1.6L L4 VIN C, 2.0L L4 VIN D, 3.5L V6 VIN G **Transmission:** All	**Barometric Pressure Circuit High Input:** Signal check, high. **Possible Causes:** • Faulty ECM
DTC: P2231 **2T CCM, MIL: Yes** **Year:** 2009, 2010 **Model:** Elantra **Engine:** 2.0L L4 VIN D, 2.0L L4 VIN E **Transmission:** All	**HO2S Signal Circuit Shorted To Heater Circuit (Bank 1 Sensor 1):** Front HO2S signal monitoring. Exhaust temperature greater than 752 degrees F. No relevant failure. Amplitude of forced lambda simulation less than 0.05. Period time of forced lambda simulation less than 2.55 seconds. Current engine speed less than 500 rpm. Current mass air flow less than 400 mg/rev. **Possible Causes:** • Contact resistance in connectors • Interference in HO2S

DTC	Trouble Code Title, Conditions & Possible Causes
DTC: P2232 **2T CCM, MIL: Yes** **Year:** 2009, 2010 **Model:** Accent **Engine:** 1.6L L4 VIN C **Transmission:** All	**HO2S Signal Circuit Shorted To Heater Circuit (Bank 1 Sensor 2):** Rationality check. **Possible Causes:** • Poor connection • Short to power in signal circuit • B1S2 • Faulty ECM
DTC: P2237 **2T CCM, MIL: Yes** **Year:** 2009, 2010 **Model:** Elantra, Santa Fe, Sonata, Tucson **Engine:** 2.0L L4 VIN D, 2.0L L4 VIN E, 2.4L L4 VIN B, 2.4L L4 VIN C **Transmission:** All	**HO2S Pumping Current Circuit/Open Bank 1, Sensor 1:** Open circuit of front HO2S circuit. No relevant failure. **Possible Causes:** • Contact resistance in connectors • Open or short to ground in HO2S circuit • Front HO2S sensor
DTC: P2243 **2T CCM, MIL: Yes** **Year:** 2009, 2010 **Model:** Elantra, Santa Fe, Sonata, Tucson **Engine:** 2.0L L4 VIN D, 2.0L L4 VIN E, 2.4L L4 VIN B, 2.4L L4 VIN C **Transmission:** All	**HO2S Reference Voltage Circuit/Open Bank 1, Sensor 1:** Open circuit of front HO2S circuit. No relevant failure. **Possible Causes:** • Contact resistance in connectors • Open or short to ground in HO2S circuit • Front HO2S sensor
DTC: P2251 **2T CCM, MIL: Yes** **Year:** 2009, 2010 **Model:** Elantra, Santa Fe, Sonata, Tucson **Engine:** 2.0L L4 VIN D, 2.0L L4 VIN E, 2.4L L4 VIN B, 2.4L L4 VIN C **Transmission:** All	**HO2S Reference Ground Circuit/Open Bank 1, Sensor 1:** Open circuit of front HO2S circuit. No pump current malfunction. **Possible Causes:** • Contact resistance in connectors • Open or short to ground in HO2S circuit • Front HO2S sensor
DTC: P2261 **2T PCM, MIL: Yes** **Year:** 2010 **Model:** Genesis Coupe **Engine:** 2.0L L4 VIN D **Transmission:** All	**Turbocharger/Supercharger Bypass Valve - Mechanical:** Enable Conditions Case1 • time after valve close request > 500ms • Failure not detected for following DTCs • Stuck open Case2 • time after valve close request > 500ms • pedal value gradient < -97.7%/s • maximum pressure up throttle deviation < 100hPa • Failure not detected for following DTCs • Stuck close Threshold Value Case1 • pressure up throttle mean value > modeled pressure up throttle with closed • pressure up throttle mean value < modeled pressure up throttle with opened valve + 50hPa Case2 • 6.5hPa*s > modeled pressure up throttle value > 6.5hPa*s **Possible Causes:** • RCV stuck

DTC	Trouble Code Title, Conditions & Possible Causes
DTC: P2270 **2T CCM, MIL: Yes** **Year:** 2009, 2010 **Model:** Azera, Elantra, Genesis, Genesis Coupe, Santa Fe, Sonata, Tucson, Veracruz **Engine:** 2.0L L4 VIN D, 2.0L L4 VIN E, 2.4L L4 VIN B, 2.4L L4 VIN C, 2.7L V6 VIN D, 3.3L V6 VIN D, 3.3L V6 VIN E, 3.3L V6 VIN F, 3.5L V6 VIN G, 3.8L V6 VIN C, 3.8L V6 VIN E, 3.8L V6 VIN F, 3.8L V6 VIN H, 4.6L V8 VIN F **Transmission:** All	**HO2S Signal Stuck Lean (Bank 1 Sensor 2):** Plausibility check during shift of lambda set point to rich from lean. No fuel cut off. No full load phase. No fuel trim error detected. Delay time to start diagnosis: 13 to 30 seconds. No relevant failure. **Possible Causes:** • Three way catalytic converter (TWC) • Air leakage in exhaust system • Faulty rear HO2S sensor
DTC: P2271 **2T CCM, MIL: Yes** **Year:** 2009, 2010 **Model:** Azera, Elantra, Genesis, Genesis Coupe, Santa Fe, Sonata, Tucson, Veracruz **Engine:** 2.0L L4 VIN D, 2.0L L4 VIN E, 2.4L L4 VIN B, 2.4L L4 VIN C, 2.7L V6 VIN D, 3.3L V6 VIN D, 3.3L V6 VIN E, 3.3L V6 VIN F, 3.5L V6 VIN G, 3.8L V6 VIN C, 3.8L V6 VIN E, 3.8L V6 VIN F, 3.8L V6 VIN H, 4.6L V8 VIN F **Transmission:** All	**O2 Signal Stuck Rich (Bank 1/2 Sensor 1):** Plausibility check during shift of lambda set point to rich from lean. No fuel cut off. No full load phase. No fuel trim error detected. Delay time to start diagnosis: 13 to 30 seconds. No relevant failure. **Possible Causes:** • Three way catalytic converter (TWC) • Air leakage in exhaust system • Faulty rear HO2S sensor
DTC: P2272 **2T CCM, MIL: Yes** **Year:** 2009, 2010 **Model:** Azera, Genesis, Genesis Coupe, Santa Fe, Sonata, Veracruz **Engine:** 2.7L V6 VIN D, 3.3L V6 VIN D, 3.3L V6 VIN E, 3.3L V6 VIN F, 3.5L V6 VIN G, 3.8L V6 VIN C, 3.8L V6 VIN E, 3.8L V6 VIN F, 3.8L V6 VIN H, 4.6L V8 VIN F **Transmission:** All	**HO2S Signal Stuck Lean (Bank 2 Sensor 2):** Determines if O2 sensor indicates lean exhaust while in power enrichment mode. Sensor not in cooled status flag. Not in transient conditions status flag. Device control not active. Engine running. Minimum air flow present is equal or greater than 2 g/s. Ignition voltage equal to or greater than 10 volts. Fuel reduction not active. Engine running long enough (more than 60 seconds). Engine coolant warm (140 degrees F. Above conditions met for at least 2.5 seconds. **Possible Causes:** • Poor connection • Faulty HO2S • Faulty PCM
DTC: P2273 **2T CCM, MIL: Yes** **Year:** 2009, 2010 **Model:** Azera, Genesis, Genesis Coupe, Santa Fe, Sonata, Veracruz **Engine:** 2.7L V6 VIN D, 3.3L V6 VIN D, 3.3L V6 VIN E, 3.3L V6 VIN F, 3.5L V6 VIN G, 3.8L V6 VIN C, 3.8L V6 VIN E, 3.8L V6 VIN F, 3.8L V6 VIN H, 4.6L V8 VIN F **Transmission:** All	**HO2S Signal Stuck Rich (Bank 2 Sensor 2):** Determines if O2 sensor indicates rich exhaust while in decal fuel cut off (DFCO). Sensor not in cooled status flag. Not in transient conditions status flag. Device control not active. Engine running. Minimum air flow present is equal or greater than 2 g/s. Ignition voltage equal to or greater than 10 volts. Fuel reduction not active. Engine running long enough (more than 60 seconds). Engine coolant warm (140 degrees F. Above conditions met for at least 2.0L seconds. **Possible Causes:** • Poor connection • Faulty HO2S • Faulty PCM

DTC	Trouble Code Title, Conditions & Possible Causes
DTC: P2297 **2T PCM, MIL: Yes** **Year:** 2009, 2010 **Model:** Santa Fe, Sonata, Tucson **Engine:** 2.4L L4 VIN B, 2.4L L4 VIN C **Transmission:** All	**O2 Sensor Out of Range During Deceleration - (Bank 1 Sensor 1):** Enable Conditions • Fuel cut-off • Wide range sensor pump current valid • Sensor heater controlled heating • Exhaust gas temp. > 300°C (572°F) • 10V < Battery voltage < 16V Threshold value • Signal < 3.1V Diagnosis Time • 6 sec. After fuel cut-off **Possible Causes:** • Poor connection or damaged harness • Front HO2S
DTC: P2414 **2T CCM, MIL: Yes** **Year:** 2009, 2010 **Model:** Elantra, Santa Fe, Sonata, Tucson **Engine:** 2.0L L4 VIN D, 2.0L L4 VIN E, 2.4L L4 VIN B, 2.4L L4 VIN C **Transmission:** All	**HO2S Exhaust Sample Error Bank 1 Sensor 1:** Sensor not mounted. Plausibility check in part load or full load conditions. Sensor tip temperature 1202 degrees F. Part load or full load. No relevant failure. **Possible Causes:** • Incorrect installation of HO2S sensor • Contact resistance in connectors
DTC: P2422 **2T PCM, MIL: Yes** **Year:** 2009, 2010 **Model:** Genesis, Genesis Coupe, Santa Fe, Sonata, Veracruz **Engine:** 3.3L V6 VIN F, 3.5L V6 VIN G, 3.8L V6 VIN C, 3.8L V6 VIN E, 3.8L V6 VIN H **Transmission:** All	**Evaporative Emission System Vent Valve Stuck Closed:** Enable Conditions • 10 V < Battery voltage < 16 V • Barometric pressure > 72 kPa (0.72 bar) • Engine coolant temperature at startup - Intake air temperature at startup < 6.7°C(12 °F) • Engine coolant temperature at startup: 4.5 ~ 35°C(40 ~ 95 °F) • Intake air temperature at startup: 4.5 ~ 35°C(40 ~ 95 °F) • Fuel level: 15 ~ 85 % Threshold value • Fuel tank's vacuum at purging > a prescribed threshold Diagnosis Time • Continuous **Possible Causes:** • Faulty Canister Close Valve • Clogging of canister air filter
DTC: P2422 **2T CCM, MIL: Yes** **Year:** 2009, 2010 **Model:** Azera, Santa Fe **Engine:** 2.7L V6 VIN D, 3.3L V6 VIN D, 3.3L V6 VIN E, 3.8L V6 VIN F **Transmission:** All	**Evaporative Emission System Canister Clogging:** Clogging is monitored. Purge duty greater than or equal to 60 percent. Canister close valve OFF. Canister close valve greater than 1600 rpm. Load value 20 to 85 percent. **Possible Causes:** • Blockage of canister filter • Faulty CCV • Faulty canister • Faulty fuel cut valve
DTC: P2501 **T PCM** **Year:** 2010 **Model:** Santa Fe, Tucson **Engine:** 2.4L L4 VIN B, 2.4L L4 VIN C, 3.5L V6 VIN G **Transmission:** All	**EPMD Voltage Rationality:** Enable Conditions • IG "ON" Threshold value • There is difference between battery voltage at power supply terminal of ECM and the voltage from battery senor. Diagnosis Time • Continuous **Possible Causes:** • Open or short in harness • Faulty battery sensor • Poor connection

DTC	Trouble Code Title, Conditions & Possible Causes
DTC: P2502 **T PCM** **Year:** 2009, 2010 **Model:** Genesis, Santa Fe, Tucson **Engine:** 2.4L L4 VIN B, 2.4L L4 VIN C, 3.5L V6 VIN G, 3.8L V6 VIN E **Transmission:** All	**Battery Voltage Rationality:** Enable Conditions • Engine running state • No error related to LIN communication Threshold value • Battery voltage at power supply terminal of ECM – voltage from battery sensor > 3V Diagnosis Time • Continuous **Possible Causes:** • Open or short in harness • Faulty battery sensor • Poor connection
DTC: P2503 **T PCM** **Year:** 2010 **Model:** Santa Fe, Tucson **Engine:** 2.4L L4 VIN B, 2.4L L4 VIN C, 3.5L V6 VIN G **Transmission:** All	**No Battery Charge:** Enable Conditions • Engine running state • No LIN Bus error / No Plausibility Error related to smart sensor / No Alternator Error Threshold value • Battery charging current < Calibration • State of Charge(SOC) < Calibration Diagnosis Time • Continuous **Possible Causes:** • Poor connection • Open or short in charging system • Faulty charging system • Faulty PCM
DTC: P2507 **1T PCM, MIL: Yes** **Year:** 2009, 2010 **Model:** Azera, Genesis, Genesis Coupe, Santa Fe, Sonata, Veracruz **Engine:** 2.7L V6 VIN D, 3.3L V6 VIN D, 3.3L V6 VIN E, 3.3L V6 VIN F, 3.8L V6 VIN C, 3.8L V6 VIN E, 3.8L V6 VIN F, 3.8L V6 VIN H **Transmission:** All	**ECM/PCM Power Input Signal Low:** DTC Strategy • Monitor the battery power input line Threshold value • Open or short to ground in line Diagnosis Time • Continuous (More than 5 sec. failure for every 10 sec. test) **Possible Causes:** • Poor connection • Open or short to ground in line • Faulty PCM
DTC: P2610 **2T CCM, MIL: Yes** **Year:** 2009, 2010 **Model:** Azera, Genesis, Genesis Coupe, Santa Fe, Sonata, Tucson, Veracruz **Engine:** 2.0L L4 VIN D, 2.4L L4 VIN B, 2.4L L4 VIN C, 2.7L V6 VIN D, 3.3L V6 VIN D, 3.3L V6 VIN E, 3.3L V6 VIN F, 3.5L V6 VIN G, 3.8L V6 VIN C, 3.8L V6 VIN E, 3.8L V6 VIN F, 3.8L V6 VIN H, 4.6L V8 VIN F **Transmission:** All	**ECM/PCM Internal Engine Off Timer Performance:** The LPC SPI diagnostic allows the low power counter to count down and simultaneously enables a test timer to run for a calibrated length of time and then compares the lapsed time recorded by the counter to make a pass/fail determination. Engine running. Enough time (10 seconds). Battery voltage 8 volts. No memory failure. **Possible Causes:** • Faulty PCM
DTC: P2626 **2T CCM, MIL: Yes** **Year:** 2009, 2010 **Model:** Elantra, Santa Fe, Sonata, Tucson **Engine:** 2.0L L4 VIN D, 2.0L L4 VIN E, 2.4L L4 VIN B, 2.4L L4 VIN C **Transmission:** All	**HO2S Pumping Current Trim Circuit Open Bank 1 Sensor 1:** Check open circuit of front HO2S sensor. **Possible Causes:** • Contact resistance in connectors • Open or short to ground in HO2S circuit • Faulty canister • Faulty front HO2S sensor

DTC	Trouble Code Title, Conditions & Possible Causes
DTC: P2A00 **2T CCM, MIL: Yes** **Year:** 2009, 2010 **Model:** Azera, Genesis, Genesis Coupe, Santa Fe, Sonata, Veracruz **Engine:** 2.7L V6 VIN D, 3.3L V6 VIN D, 3.3L V6 VIN E, 3.3L V6 VIN F, 3.5L V6 VIN G, 3.8L V6 VIN C, 3.8L V6 VIN E, 3.8L V6 VIN F, 3.8L V6 VIN H **Transmission:** All	**O2 Sensor Not Ready (Bank 1 Sensor 1):** Detects loss of O2 ready status, which would lead to open loop fueling operation, a default mode. Engine running. Ignition ON. DFCO not present too long (less than 15 seconds). No disabling faults present. All of the above for at least 20 seconds. **Possible Causes:** • Poor connection • Faulty HO2S • Faulty PCM
DTC: P2A01 **2T O2S2, MIL: Yes** **Year:** 2009, 2010 **Model:** Azera, Genesis, Genesis Coupe, Santa Fe, Sonata, Veracruz **Engine:** 2.7L V6 VIN D, 3.3L V6 VIN D, 3.3L V6 VIN E, 3.3L V6 VIN F, 3.5L V6 VIN G, 3.8L V6 VIN C, 3.8L V6 VIN E, 3.8L V6 VIN F, 3.8L V6 VIN H **Transmission:** All	**HO2S Circuit Range/Performance (Bank 1 / Sensor 2):** Enable Conditions • After engine warming-up • Deceleration fuel cut-off state • No other disabling faults Threshold value • The average time for voltage drop > approx. 0.3 seconds Diagnosis Time • During deceleration fuel cut-off **Possible Causes:** • Poor connection • Faulty HO2S • Faulty PCM
DTC: P2A03 **2T CCM, MIL: Yes** **Year:** 2009, 2010 **Model:** Azera, Genesis, Genesis Coupe, Santa Fe, Sonata, Veracruz **Engine:** 2.7L V6 VIN D, 3.3L V6 VIN D, 3.3L V6 VIN E, 3.3L V6 VIN F, 3.5L V6 VIN G, 3.8L V6 VIN C, 3.8L V6 VIN E, 3.8L V6 VIN F, 3.8L V6 VIN H **Transmission:** All	**O2 Sensor Not Ready (Bank 1 Sensor 2):** Detects loss of O2 ready status, which would lead to open loop fueling operation, a default mode. Engine running. Ignition ON. DFCO not present too long (less than 15 seconds). No disabling faults present. All of the above for at least 20 seconds. **Possible Causes:** • Poor connection • Faulty HO2S • Faulty PCM
DTC: P2A04 **2T O2S2, MIL: Yes** **Year:** 2009, 2010 **Model:** Azera, Genesis, Genesis Coupe, Santa Fe, Sonata, Veracruz **Engine:** 2.7L V6 VIN D, 3.3L V6 VIN D, 3.3L V6 VIN E, 3.3L V6 VIN F, 3.5L V6 VIN G, 3.8L V6 VIN C, 3.8L V6 VIN E, 3.8L V6 VIN F, 3.8L V6 VIN H **Transmission:** All	**HO2S Circuit Range/Performance (Bank 2 / Sensor 2):** Enable Conditions • After engine warming-up • Deceleration fuel cut-off state • No other disabling faults Threshold value • The average time for voltage drop > approx. 0.3 seconds Diagnosis Time • During deceleration fuel cut-off **Possible Causes:** • Poor connection • Faulty HO2S • Faulty PCM

OBD II Trouble Code List (U0XXX Codes)

DTC	Trouble Code Title, Conditions & Possible Causes
DTC: U0001 **2T PCM, MIL: Yes** **Year:** 2009, 2010 **Model:** Accent, Azera, Elantra, Genesis, Genesis Coupe, Santa Fe, Sonata, Tucson, Veracruz **Engine:** 1.6L L4 VIN C, 2.0L L4 VIN B, 2.0L L4 VIN D, 2.0L L4 VIN E, 2.4L L4 VIN B, 2.4L L4 VIN C, 2.7L V6 VIN D, 3.3L V6 VIN D, 3.3L V6 VIN E, 3.3L V6 VIN F, 3.5L V6 VIN G, 3.8L V6 VIN C, 3.8L V6 VIN E, 3.8L V6 VIN F, 3.8L V6 VIN H, 4.6L V8 VIN F **Transmission:** All	**CAN Communication Malfunction:** Enable Conditions • Engine Run Time ≥2sec. • Ignition Voltage ≥11V Threshold value • CAN communication error Diagnosis Time • Continuous **Possible Causes:** • CAN BUS • CAN communication module component
DTC: U0101 **2T PCM, MIL: Yes** **Year:** 2009, 2010 **Model:** Accent, Elantra, Genesis, Genesis Coupe, Santa Fe, Tucson **Engine:** 1.6L L4 VIN C, 2.0L L4 VIN D, 2.0L L4 VIN E, 2.4L L4 VIN B, 2.4L L4 VIN C, 3.8L V6 VIN E, 3.8L V6 VIN H, 4.6L V8 VIN F **Transmission:** All	**Lost Communication With TCM:** Enable Conditions • Engine Run Time ≥2sec. • Ignition Voltage ≥11V Threshold value • CAN communication error with TCM Diagnosis Time • Continuous **Possible Causes:** • TCM faulty • CAN communication • line between ECM and TCM
DTC: U0122 **T PCM** **Year:** 2009, 2010 **Model:** Genesis **Engine:** 4.6L V8 VIN F **Transmission:** All	**CAN TCS Communication:** Enable Conditions • B+ > 10.8V • Time after Ignition switch on = 0.5 sec Threshold value • Duration of communication lost with TCS > 2 sec Diagnosis Time • Continuous MIL On Condition • No MIL **Possible Causes:** • TCS faulty • CAN communication line between ECM and TCS

OBD II Trouble Code List (U1XXX Codes)

DTC	Trouble Code Title, Conditions & Possible Causes
DTC: U1111 **T PCM** **Year:** 2009, 2010 **Model:** Genesis, Santa Fe, Tucson **Engine:** 2.4L L4 VIN B, 2.4L L4 VIN C, 3.5L V6 VIN G, 3.8L V6 VIN E **Transmission:** All	**Battery Sensor fault detected by ECU:** Enable Conditions • Engine Run Time ≥2sec. • Ignition Voltage ≥11V Threshold value • Battery sensor error Diagnosis Time • Continuous **Possible Causes:** • Faulty battery sensor

DTC	Trouble Code Title, Conditions & Possible Causes
DTC: U1112 **T PCM** **Year:** 2009, 2010 **Model:** Genesis, Santa Fe **Engine:** 3.5L V6 VIN G, 3.8L V6 VIN E **Transmission:** All	**LIN Communication Error :** Enable Conditions • Engine Run Time ≥2sec. • Ignition Voltage ≥11V Threshold value • LIN communication error Diagnosis Time • Continuous **Possible Causes:** • Open or short in LIN communication line

KIA

Amanti

11

SPECIFICATIONS AND MAINTENANCE CHARTS

ENGINE AND VEHICLE IDENTIFICATION

Engine							Model Year	
Code ①	Liters (cc)	Cu. In.	Cyl.	Fuel Sys.	Engine Type	Eng. Mfg.	Code ②	Year
5	3.8 (3778)	230.55	6	MPFI	DOHC	KIA	9	2009

MPFI: Multi-Point Fuel Injection

DOHC: Dual Overhead Camshafts

① 8th Digit of VIN

② 10th Digit of VIN

37655_AMAN_C0001

GENERAL ENGINE SPECIFICATIONS

Year	Engine Displacement Liters	Engine ID/VIN	Net Horsepower @ rpm	Net Torque @ rpm (ft. lbs.)	Bore x Stroke (in.)	Com-pression Ratio	Oil Pressure @ rpm
2009	3.8	G6DA/5	200@5500	220@3500	3.78 x 3.43	10.4:1	18.8@1000

37655_AMAN_C0002

GASOLINE ENGINE TUNE-UP SPECIFICATIONS

Year	Engine Displacement Liters	Engine ID/VIN	Spark Plugs Gap (in.)	Ignition Timing (deg.)		Fuel Pump (psi)	Idle Speed (rpm)		Valve Clearance	
				MT	AT		MT	AT	In.	Ex.
2009	3.8	G6DA/5	0.039-0.043	—	①	54.3-55.8	—	550-750	②	③

HYD: Hydraulic Valve Lifters

B: Before Top Dead Center

① 10 degrees, plus or minus 5 degrees

② 0.0039-0.01185 degrees

③ 0.0078-0.0157 degrees

37655_AMAN_C0003

CAPACITIES

Year	Model	Engine Displacement Liters	Engine ID/VIN	Engine Oil with Filter	Transmission (pts.) 5–Spd	Transmission (pts.) Auto.	Fuel Tank (gal.)	Cooling System (qts.)
2009	Amanti	3.8	G6DA/5	5.49	—	23.04	18.4	9.19

NOTE: Add fluid gradually and check to be sure a proper fluid level is obtained.

37655_AMAN_C0004

FLUID SPECIFICATIONS

Year	Model	Engine Displ. Liters (VIN)	Engine Oil	Auto. Trans.	Power Steering Fluid	Brake Master Cylinder	Cooling System
2009	Amanti	3.8 (5)	①	②	PSF III	DOT 3 or 4	③

DOT: Department Of Transpotation

① If 5W-20 is not available, 5W-30 can be used

② Diamond ATF SP-III or SK ATF SP III

③ Ethlyene glycol base for aluminium radiator

37655_AMAN_C0014

VALVE SPECIFICATIONS

Year	Engine Displacement Liters	Engine ID/VIN	Seat Angle (deg.)	Face Angle (deg.)	Spring Test Pressure (lbs. @ in.)	Spring Free Height (in.)	Stem-to-Guide Clearance (in.) Intake	Stem-to-Guide Clearance (in.) Exhaust	Stem Diameter (in.) Intake	Stem Diameter (in.) Exhaust
2009	3.8	G6DA/5	44.75-45.20	45.25-45.75	NA	1.7267	0.00078-0.00185	0.00118-0.00212	0.2151-0.2157	0.2149-0.2153

NA: Not Available

37655_AMAN_C0007

CAMSHAFT AND BEARING SPECIFICATIONS

All measurements are given in inches unless noted.

Year	Engine Displacement Liters	Engine ID/VIN	Camshaft Cam Height Intake	Cam Height Exhaust	Shaft End-play	Bearing Cap Torque (ft. lbs.) Outer	Inner
2009	3.8	G6DA/5	1.8425	1.8031	0.0008-0.0071	① 7.23 - 8.68	① 7.23 - 8.68

① Step 1: 7.23 ft. lbs.
Step 2: 8.68 ft. lbs.

37655_AMAN_C0005

CRANKSHAFT AND CONNECTING ROD SPECIFICATIONS

All measurements are given in inches.

Year	Engine Displacement Liters	Engine ID/VIN	Crankshaft Main Brg. Journal Dia.	Main Brg. Oil Clearance	Shaft End-play	Thrust on No.	Connecting Rod Journal Diameter	Oil Clearance	Side Clearance
2009	3.8	G6DA/5	2.7142-2.7149	0.0008-0.0016	0.0039-0.0110	3	2.2834-2.2842	0.0015-0.0022	0.0039-0.0098

37655_AMAN_C0006

PISTON AND RING SPECIFICATIONS

Year	Engine Displacement Liters	Engine ID/VIN	Piston Clearance	Ring Gap Top Compression	Bottom Compression	Oil Control	Ring Side Clearance Top Compression	Bottom Compression	Oil Control
2009	3.8	G6DA/5	0.0012-0.0020	0.0067-0.0126	0.0126-0.0185	0.0078-0.0275	0.0012-0.0027	0.0012-0.0027	0.0024-0.0059

NA: Not Available

37655_AMAN_C0008

TORQUE SPECIFICATIONS

All readings in ft. lbs.

Year	Engine Displacement Liters	Engine ID/VIN	Cylinder Head Bolts	Main Bearing Bolts	Rod Bearing Bolts	Crankshaft Damper Bolts	Flywheel Bolts	Manifold Intake	Manifold Exhaust	Spark Plugs	Oil Pan Drain Plug
2009	3.8	G6DA/5	①	②	③	210-224	53-56	④	29-33	18-22	25-33

① Bolts (16)

Step 1: 27.5-30.4 ft. lbs.

Step 2: Plus 118-122 degrees

Step 3: Plus 88-92 degrees

Bolt (1) 13.74-17.36 ft. lbs.

② Inner cap bolt (M11): 36.16 ft. lbs., plus 90 degrees

Outer cap bolt (M8): 14.46 ft. lbs., plus 120 degrees

Side cap bolt: 21.70-23.14 ft. lbs.Plus 90 degrees

③ 13.0-15.9 ft. lbs., plus 88-92 degrees

④ Bolts: 19.5-23.1 ft. lbs. Nuts: 13.74-17.36 ft. lbs.

37655_AMAN_C0009

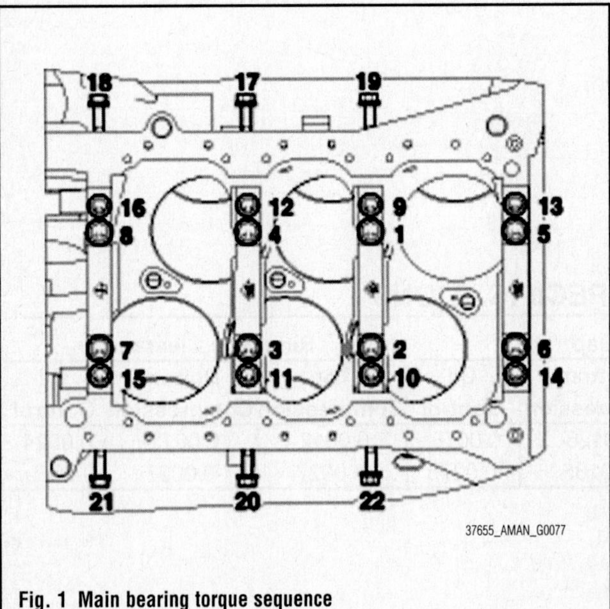

Fig. 1 Main bearing torque sequence

WHEEL ALIGNMENT

Year	Model		Caster Range (+/-Deg.)	Caster Preferred Setting (Deg.)	Camber Range (+/-Deg.)	Camber Preferred Setting (Deg.)	Toe-in (in.)
2009	Amanti	F	0.5	4.53	0.5	0	0 +/- 0.02
		R	—	—	0.5	-0.1	0.02 +/- 0.02

37655_AMAN_C0010

TIRE, WHEEL AND BALL JOINT SPECIFICATIONS

Year	Model	OEM Tires Standard	OEM Tires Optional	Tire Pressures (psi) Front	Tire Pressures (psi) Rear	Wheel Size	Ball Joint Inspection	Lug Nut Torque (ft. lbs.)
2009	Amanti	P225/60R16	NA	31	30	6.5Jx16	NA	67-82

NA: Not Available

OEM: Original Equipment Manufacturer

PSI: Pounds Per Square Inch

37655_AMAN_C0011

BRAKE SPECIFICATIONS

All measurements in inches unless noted

Year	Model		Brake Disc Original Thickness	Brake Disc Minimum Thickness	Brake Disc Maximum Run-out	Brake Drum Diameter Original Inside Diameter	Brake Drum Diameter Max. Wear Limit	Brake Drum Diameter Maximum Machine Diameter	Minimum Lining Thickness Front	Minimum Lining Thickness Rear	Brake Caliper Bracket Bolts (ft. lbs.)	Brake Caliper Mounting Bolts (ft. lbs.)
2009	Amanti	F	1.100	NA	NA	—	—	—	0.043	—	60-72	16-23
		R	0.390	NA	NA	—	—	—	—	0.390	58-72	16-23

NA: Not Available

F: Front

R: Rear

37655_AMAN_C0012

SCHEDULED MAINTENANCE INTERVALS

Kia Amanti

TO BE SERVICED	TYPE OF SERVICE	VEHICLE MILEAGE INTERVAL (x1000)												
		7.5	15	22.5	30	37.5	45	52.5	60	67.5	75	82.5	90	97.5
Accessory drive belts	S/I			✓			✓			✓			✓	
Air cleaner filter	R		✓		✓		✓		✓		✓		✓	
Air conditioner system filter	S/I	Inspect and replace every 10,000 miles or 10 months												
Brake lines, hoses and connections	S/I		✓		✓		✓		✓		✓		✓	
Chassis and body fasteners	T	✓	✓	✓	✓	✓	✓	✓	✓	✓	✓	✓	✓	✓
Cooling system hoses and coolant level	S/I		✓		✓		✓		✓		✓		✓	
CV-joint boots	S/I				✓				✓				✓	
Engine coolant	R				✓				✓				✓	
Engine oil and filter	R	✓	✓	✓	✓	✓	✓	✓	✓	✓	✓	✓	✓	✓
Exhaust system heat shields	S/I				✓				✓				✓	
Front and rear brakes	S/I				✓				✓				✓	
Front ball joints	S/I				✓				✓				✓	
Fuel filter	R								✓					
Fuel lines and hoses	S/I				✓				✓				✓	
Locks and hinges	L	✓	✓	✓	✓	✓	✓	✓	✓	✓	✓	✓	✓	✓
Spark plugs	R				✓				✓				✓	
Steering operation and linkage	S/I				✓				✓				✓	
Timing belt	R								✓					

R: Replace S/I: Service or Inspect L: Lubricate T: Tighten

FREQUENT OPERATION MAINTENANCE (SEVERE SERVICE)

If a vehicle is operated under any of the following conditions it is considered severe service

- Towing a trailer or using a camper or car-top carrier

- Repeated short trips of less than 5 miles in temperatures below freezing, or trips of less than 10 miles in any temperature

- Prolonged idling (vehicle operation in stop and go traffic).

- Operating on rough, muddy, unpaved, dusty or salt-covered roads.

- Police, taxi, delivery usage or trailer towing usage.

- Driving in extremely hot (over 90°F) conditions

Oil & oil filter: change every 5000 miles or 5 months, whichever occurs first.

Air cleaner filter: inspect every 15,000 miles or 15 months and replace everything 30,000 miles or 30 months, whichever occur

Fuel system hoses (California models only): replace every 105,000 miles

Emission system hoses (non-CA models): inspect every 55,000 or 55 months, whichever occurs first

Emission system hoses (CA models): inspect every 60,000 miles or 60 months, which occurs first

Front and rear brakes: inspect every 15,000 miles or 15 months, whichever occurs first

Chassis and body fasteners: tighten every 15,000 miles or 15 months, whichever occurs first

Locks and hinges: lubricate every 5000 miles or 5 months, whichever occurs first

37655_AMAN_C0013

BRAKES — INFORMATION AND PRECAUTIONS

ANTI-LOCK SYSTEMS

- Certain components within the ABS system are not intended to be serviced or repaired individually.
- Do not use rubber hoses or other parts not specifically specified for and ABS system. When using repair kits, replace all parts included in the kit. Partial or incorrect repair may lead to functional problems and require the replacement of components.
- Lubricate rubber parts with clean, fresh brake fluid to ease assembly. Do not use shop air to clean parts; damage to rubber components may result.
- Use only DOT 3 brake fluid from an unopened container.
- If any hydraulic component or line is removed or replaced, it may be necessary to bleed the entire system.
- A clean repair area is essential. Always clean the reservoir and cap thoroughly before removing the cap. The slightest amount of dirt in the fluid may plug an orifice and impair the system function. Perform repairs after components have been thoroughly cleaned; use only denatured alcohol to clean components. Do not allow ABS components to come into contact with any substance containing mineral oil; this includes used shop rags.
- The Anti-Lock control unit is a microprocessor similar to other computer units in the vehicle. Ensure that the ignition switch is **OFF** before removing or installing controller harnesses. Avoid static electricity discharge at or near the controller.
- If any arc welding is to be done on the vehicle, the control unit should be unplugged before welding operations begin.

DISC AND DRUM SYSTEMS

⁂ CAUTION

Dust and dirt accumulating on brake parts during normal use may contain asbestos fibers from production or aftermarket brake linings. Breathing excessive concentrations of asbestos fibers can cause serious bodily harm. Exercise care when servicing brake parts. Do not sand or grind brake lining unless equipment used is designed to contain the dust residue. Do not clean brake parts with compressed air or by dry brushing. Cleaning should be done by dampening the brake components with a fine mist of water, then wiping the brake components clean with a dampened cloth. Dispose of cloth and all residue containing asbestos fibers in an impermeable container with the appropriate label. Follow practices prescribed by the Occupational Safety and Health Administration (OSHA) and the Environmental Protection Agency (EPA) for the handling, processing, and disposing of dust or debris that may contain asbestos fibers.

BRAKES — BLEEDING THE BRAKE SYSTEM

BLEEDING PROCEDURE

BLEEDING PROCEDURE

See Figures 2 and 3.

⁂ WARNING

Clean, high quality brake fluid is essential to the safe and proper operation of the brake system. You should always buy the highest quality brake fluid that is available. If the brake fluid becomes contaminated, drain and flush the system, then refill the master cylinder with new fluid. Never reuse any brake fluid. Any brake fluid that is removed from the system should be discarded. Also, do not allow any brake fluid to come in contact with a painted surface; it will damage the paint.

⁂ CAUTION

Brake fluid contains polyglycol ethers and polyglycols. Avoid contact with the eyes and wash your hands thoroughly after handling brake fluid. If you do get brake fluid in your eyes, flush your eyes with clean, running water for 15 minutes. If eye irritation persists, or if you have taken brake fluid internally, IMMEDIATELY seek medical assistance.

1. Remove the reservoir cap and fill the brake reservoir with brake fluid.
2. Connect a vinyl tube to the wheel cylinder bleeder screw and insert the other end of the tube in a clear container.
3. Slowly depress the brake pedal several times.
4. While depressing the brake pedal fully, loosen the bleeder screw until fluid runs out. Then close the bleeder screw and release the brake pedal.
5. Repeat these steps until there are no more bubbles in the fluid escaping to the clear container.
6. Tighten the bleeder screw to specification: 60–84 inch lbs. (7–9 Nm).

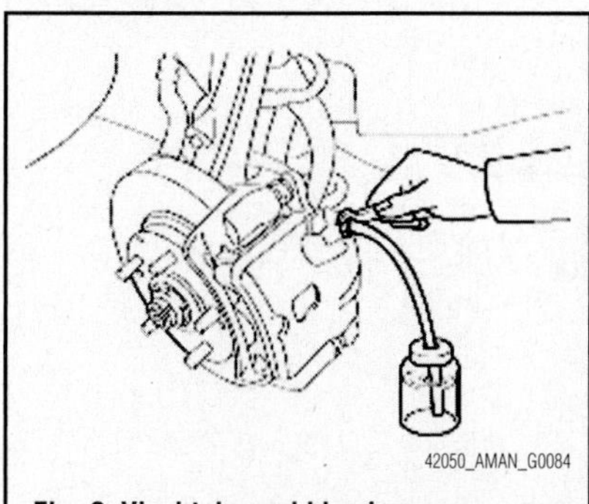

Fig. 2 Vinyl tube and bleeder screw

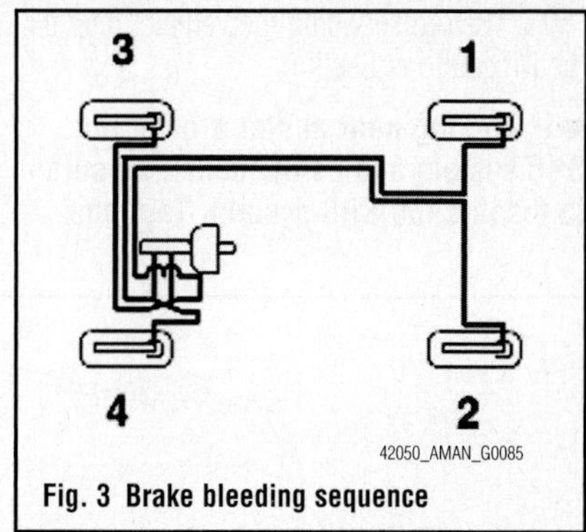

Fig. 3 Brake bleeding sequence

7. Repeat the above procedure for each wheel in the sequence shown in the illustration.

BLEEDING THE ABS SYSTEM

⁂ CAUTION

Brake fluid contains polyglycol ethers and polyglycols. Avoid contact with the eyes and wash your hands thoroughly after handling brake fluid. If you do get brake fluid in your eyes, flush your eyes with clean, running water for 15 minutes. If eye irritation persists, or if you have taken brake fluid internally, IMMEDIATELY seek medical assistance.

⁂ WARNING

Clean, high quality brake fluid is essential to the safe and proper operation of the brake system. You should always buy the highest quality brake fluid that is available. If the brake fluid becomes contaminated, drain and flush the system, then refill the master cylinder with new fluid. Never reuse any brake fluid. Any brake fluid that is removed from the system should be discarded. Also, do not allow any brake fluid to come in contact with a painted surface; it will damage the paint.

1. Remove the reservoir cap and fill the brake reservoir with brake fluid.
2. Connect a clear plastic tube to the wheel cylinder bleeder plug and insert the other end of the tube into a clear plastic bottle.
3. Connect the Hi-Scan (Pro) to the Data Link Connector located underneath the dash panel.
4. Select and operate according to the instructions on the Hi-Scan (Pro) screen:
    - Select KIA vehicle diagnosis
    - Select vehicle name
    - Select Anti-Lock Brake system

⁂ CAUTION

You must obey the maximum operating time of the ABS motor with the Hi-Scan (Pro) to prevent the motor pump from burning.

- Select air bleeding mode
- Press **YES** to operate motor pump and solenoid valve
- Wait 60 seconds before operating the air bleeding, otherwise may damage the motor

5. Pump the brake pedal several times, and then loosen the bleeder screw until fluid starts to run out without bubbles. Then close the bleeder screw.
6. Repeat step 5 until there are no more bubbles in the fluid for each wheel.
7. Tighten the bleeder screw to specification: 60–79 inch lbs. (7–9 Nm).
8. After completion of the repair or correction of the problem, erase the stored fault codes using the clear key on the Hi-Scan (Pro).
9. Disconnect the Hi-Scan (Pro).
10. Fill the brake reservoir with the proper amount of brake fluid.

BRAKES — ANTI-LOCK BRAKE SYSTEM (ABS)

WHEEL SPEED SENSORS

REMOVAL & INSTALLATION

Front Sensor

See Figure 4.

1. Before servicing the vehicle, refer to the Precautions Section.

➡If working near and/or around the SRS system and components, be sure to disable the SRS system. Tape the negative battery cable with insulating tape. Always disconnect the negative battery cable first.

⁂ CAUTION

To avoid personal injury when working on vehicles equipped with an air bag, the negative battery cable must be disconnected and at least three minutes must elapse before working on the system. Failure to do so may result in deployment of the air bag.

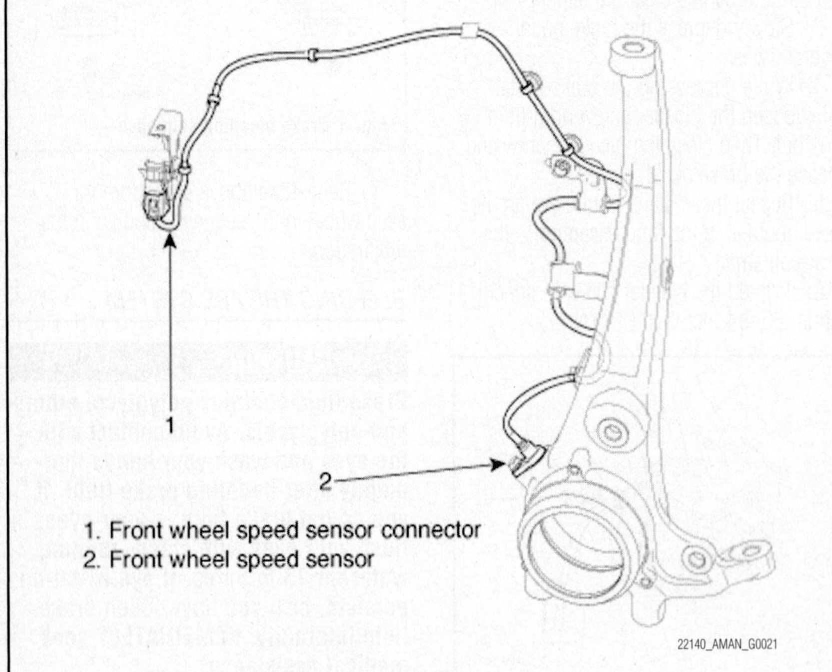

Fig. 4 Front wheel speed sensor and related components

2. Remove the battery top cover, if equipped.
3. Disconnect the negative battery cable. Tape the cable with insulating tape.
4. Raise and support the vehicle safely.
5. Remove the tire and wheel assembly.
6. Remove the wheel speed sensor mounting screw.

➡Remove the rotor to gain access to the wheel sensor mounting bolt.

7. Pull the sensor out, being careful to turn it as little as possible. Do not pull on the sensor harness.
8. Disconnect the wheel speed sensor electrical connector.
9. Remove the harness from its mount.

To install:

10. Inspect the sensor O-ring, replace as required.
11. Before installing the sensor, be certain there are no foreign materials, like iron fragments adhering to:
    - The pick-up part of the sensor
    - The inside of the sensor mounting hole
    - The rotor mounting surface
12. Apply a thin coat of a suitable grease to the wheel sensor O-ring and mounting hole.
13. Tighten the sensor mounting bolt to specification: 72–84 inch lbs. (8–10 Nm).
14. Continue the installation in the reverse order of the removal procedure.

Rear Sensor

See Figure 5.

1. Before servicing the vehicle, refer to the Precautions Section.

➡If working near and/or around the SRS system and components, be sure to disable the SRS system. Tape the negative battery cable with insulating tape. Always disconnect the negative battery cable first.

⁂ CAUTION

To avoid personal injury when working on vehicles equipped with an air bag, the negative battery cable must be disconnected and at least three minutes must elapse before working on the system. Failure to do so may result in deployment of the air bag.

2. Remove the battery top cover, if equipped.
3. Disconnect the negative battery cable. Tape the cable with insulating tape.
4. Raise and support the vehicle safely.
5. Remove the tire and wheel assembly.
6. Remove the wheel speed sensor mounting screw.

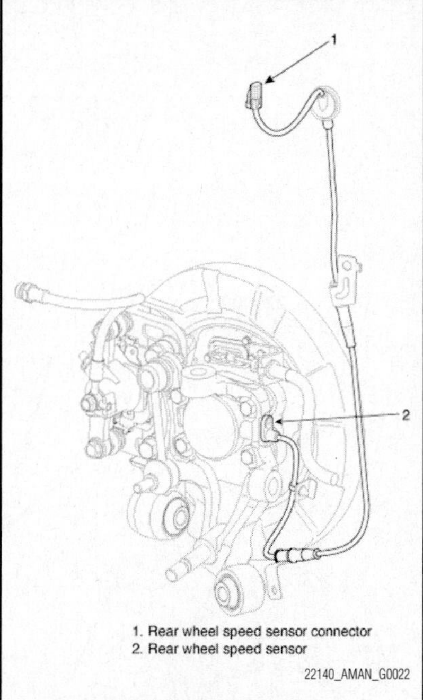

Fig. 5 Rear wheel speed sensor and related components

7. Pull the sensor out, being careful to turn it as little as possible. Do not pull on the sensor harness.
8. Disconnect the wheel speed sensor electrical connector.
9. Remove the harness from its mount.

To install:

10. Inspect the sensor O-ring, replace as required.
11. Before installing the sensor, be certain there are no foreign materials, like iron fragments adhering to:
    - The pick-up part of the sensor
    - The inside of the sensor mounting hole
    - The rotor mounting surface.
12. Apply a thin coat of a suitable grease to the wheel sensor O-ring and mounting hole.
13. Tighten the sensor mounting bolt to specification: 67–83 inch lbs. (8–10 Nm).
14. Continue the installation in the reverse order of the removal procedure.

BRAKES — FRONT DISC BRAKES

BRAKE CALIPER

REMOVAL & INSTALLATION

See Figure 6.

1. Before servicing the vehicle, refer to the Precautions Section.

➡If working near and/or around the SRS system and components, be sure to disable the SRS system. Tape the negative battery cable with insulating tape. Always disconnect the negative battery cable first.

⁂ CAUTION

To avoid personal injury when working on vehicles equipped with an air bag, the negative battery cable must be disconnected and at least three minutes must elapse before working on the system. Failure to do so may result in deployment of the air bag.

2. Remove the battery top cover, if equipped.
3. Disconnect the negative battery cable. Tape the cable with insulating tape.
4. Remove or disconnect the following:
    - Wheel
    - Brake hose at the caliper
    - Caliper mounting bolts
    - Caliper

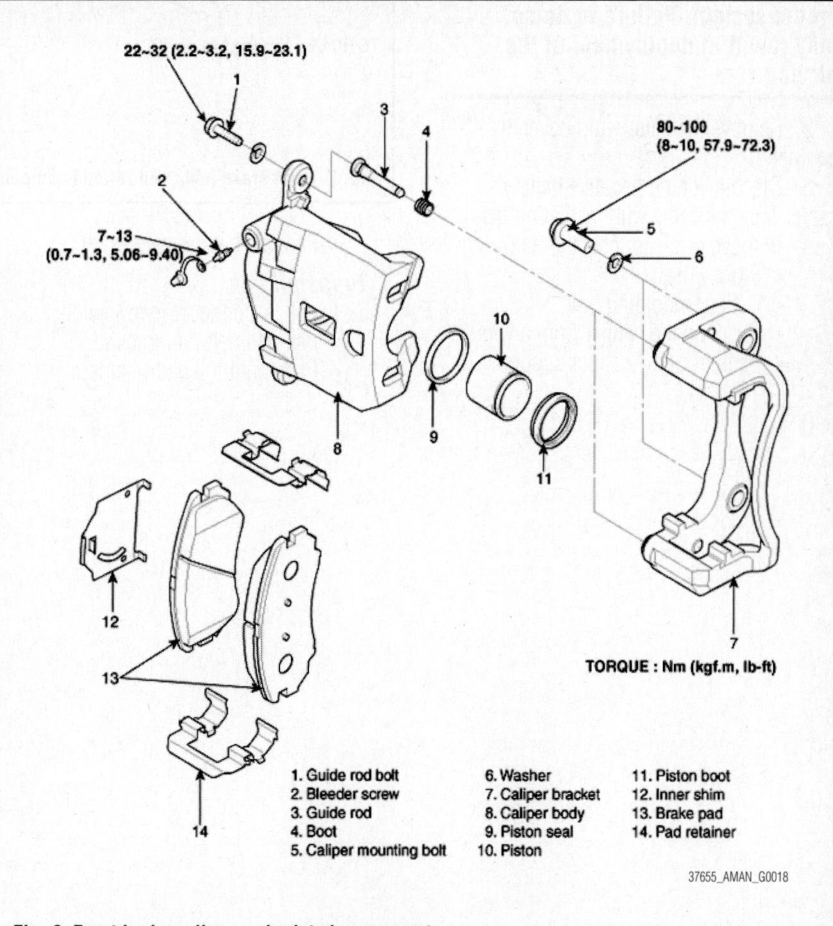

Fig. 6 Front brake caliper and related components

To install:

5. Install or connect the following:
   - Caliper. Tighten the mounting bolts to specification.
   - Brake line to the caliper. Torque the brake line bolt to specification.
6. Bleed the system.
7. Install the wheel.

DISC BRAKE PADS

REMOVAL & INSTALLATION

See Figure 7.

1. Before servicing the vehicle, refer to the Precautions Section.

➡If working near and/or around the SRS system and components, be sure to disable the SRS system. Tape the negative battery cable with insulating tape. Always disconnect the negative battery cable first.

⁂ CAUTION

To avoid personal injury when working on vehicles equipped with an air bag, the negative battery cable must be disconnected and at least three minutes must elapse before working on the system. Failure to do so may result in deployment of the air bag.

2. Remove the battery top cover, if equipped.
3. Disconnect the negative battery cable. Tape the cable with insulating tape.
4. Remove or disconnect the following:
   - Front wheel
   - Caliper mounting bolt
   - Suspend the caliper from a wire
   - Pads from the caliper support
   - Pad retainers, if necessary

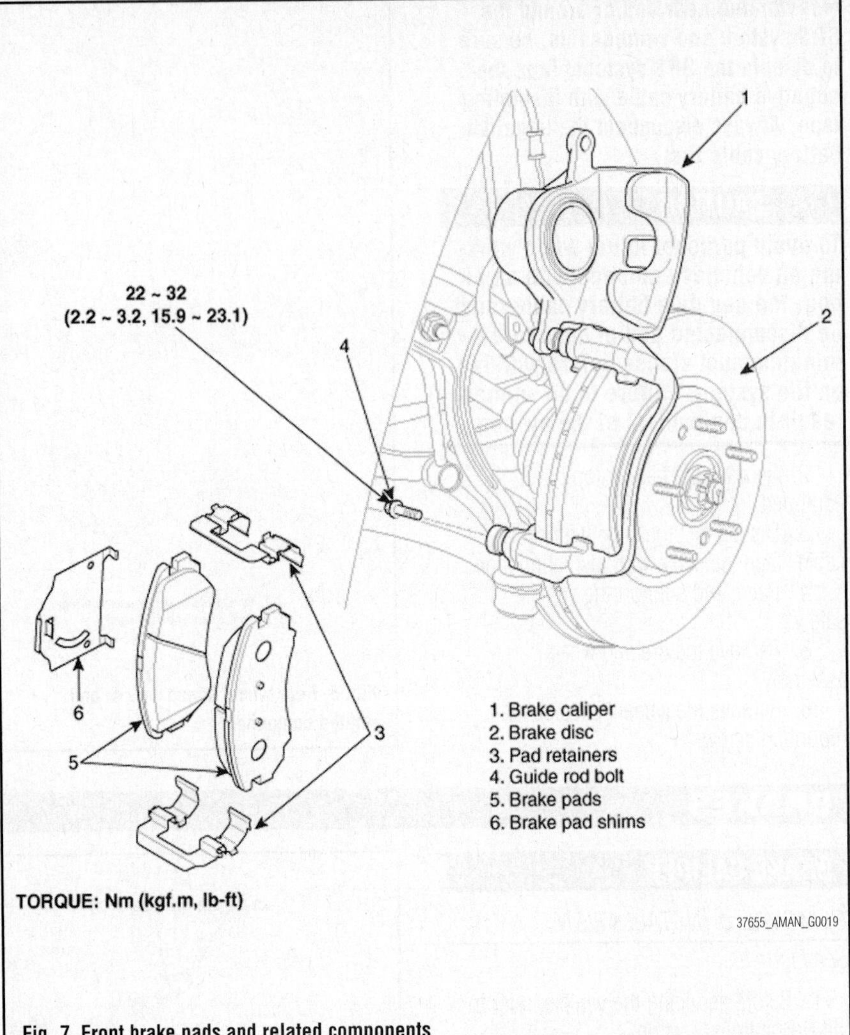

Fig. 7 Front brake pads and related components

To install:

5. Install or connect the following:
   - Pad retainers, if removed
   - Pads onto the pad retainers
6. Compress the caliper piston using a C-clamp. Rotate the caliper downward and install the mounting bolt. Tighten to specification.
7. Install the wheel.

BRAKE CALIPER

REMOVAL & INSTALLATION

See Figure 8.

1. Before servicing the vehicle, refer to the Precautions Section.

➡If working near and/or around the SRS system and components, be sure to disable the SRS system. Tape the negative battery cable with insulating tape. Always disconnect the negative battery cable first.

CAUTION

To avoid personal injury when working on vehicles equipped with an air bag, the negative battery cable must be disconnected and at least three minutes must elapse before working on the system. Failure to do so may result in deployment of the air bag.

2. Remove the battery top cover, if equipped.
3. Disconnect the negative battery cable. Tape the cable with insulating tape.
4. Release the parking brake.
5. Remove or disconnect the following:
   - Wheel
   - Brake line at the caliper
   - Caliper mounting bolts
   - Caliper

To install:

6. Install or connect the following:
   - Caliper onto its mounting
   - Mounting bolts. Torque the bolts to specification.
   - Brake line to the caliper. Torque the brake line bolt to specification.
7. Bleed the system.
8. Install the wheel.

DISC BRAKE PADS

REMOVAL & INSTALLATION

See Figure 8.

1. Before servicing the vehicle, refer to the Precautions Section.

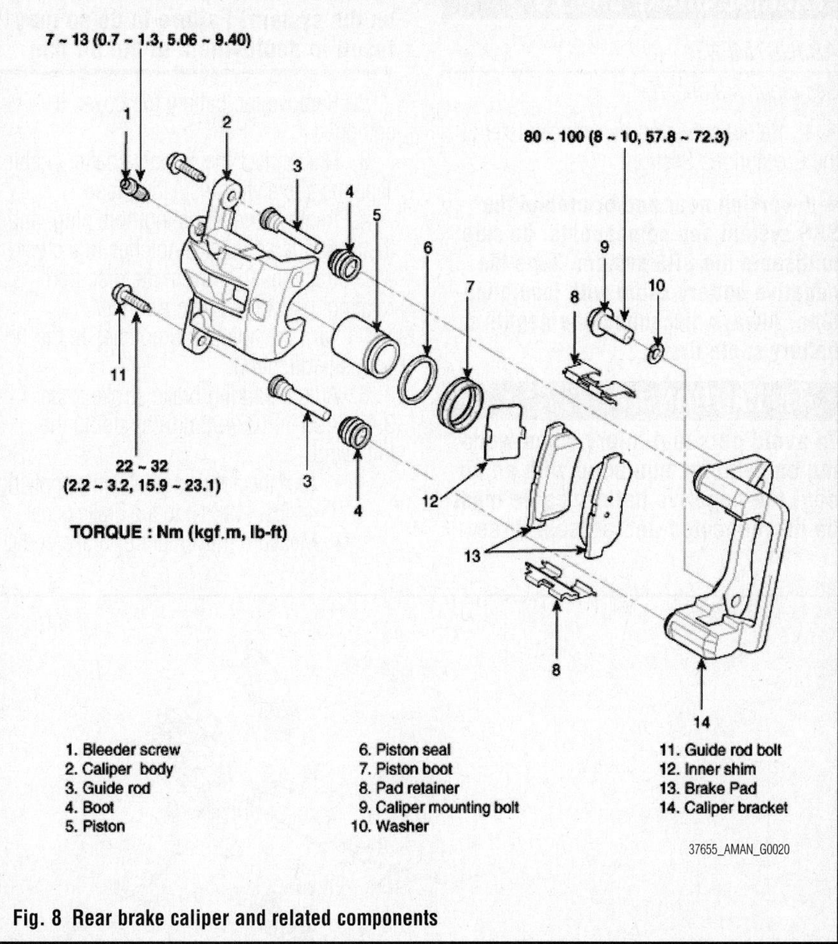

Fig. 8 Rear brake caliper and related components

➡If working near and/or around the SRS system and components, be sure to disable the SRS system. Tape the negative battery cable with insulating tape. Always disconnect the negative battery cable first.

CAUTION

To avoid personal injury when working on vehicles equipped with an air bag, the negative battery cable must be disconnected and at least three minutes must elapse before working on the system. Failure to do so may result in deployment of the air bag.

2. Remove the battery top cover, if equipped.
3. Disconnect the negative battery cable. Tape the cable with insulating tape.
4. Remove or disconnect the following:
   - Rear wheel
   - Caliper mounting bolt
   - Suspend the caliper from a wire
   - Pads from the caliper support
   - Pad retainers, if necessary

To install:

5. Install or connect the following:
   - Pad retainers, if removed
   - Pads onto the pad retainers
6. Compress the caliper piston using a C-clamp. Rotate the caliper downward and install the mounting bolt. Tighten to 16–24 ft. lbs. (22–32 Nm).
7. Install the wheel.

PARKING BRAKE CABLES

ADJUSTMENT

See Figures 9 and 10.

1. Before servicing the vehicle, refer to the Precautions Section.

➡If working near and/or around the SRS system and components, be sure to disable the SRS system. Tape the negative battery cable with insulating tape. Always disconnect the negative battery cable first.

⁂ CAUTION

To avoid personal injury when working on vehicles equipped with an air bag, the negative battery cable must be disconnected and at least three minutes must elapse before working on the system. Failure to do so may result in deployment of the air bag.

2. Remove the battery top cover, if equipped.
3. Disconnect the negative battery cable. Tape the cable with insulating tape.
4. Remove the adjusting hole plug, and then turn the adjuster 5 notches in a clockwise direction. To prevent the disc from rotating, use a flat tip screwdriver.
5. Turn the adjuster 5 notches in the opposite direction.
6. Adjust parking brake stroke to be 3.47–3.90 in. (88–98mm) by doing the following:
   - Turn the adjusting nut while applying 44 lbs. (196 N) to the brake pedal
   - Measure movement of brake pedal
   - Depress and release parking brake pedal 3 times and check the stroke again
7. After adjusting the parking brake stroke, raise the rear of vehicle with a jack.
8. Check that the rear brakes do not drag by turning the rear wheel when the parking brake lever is released.

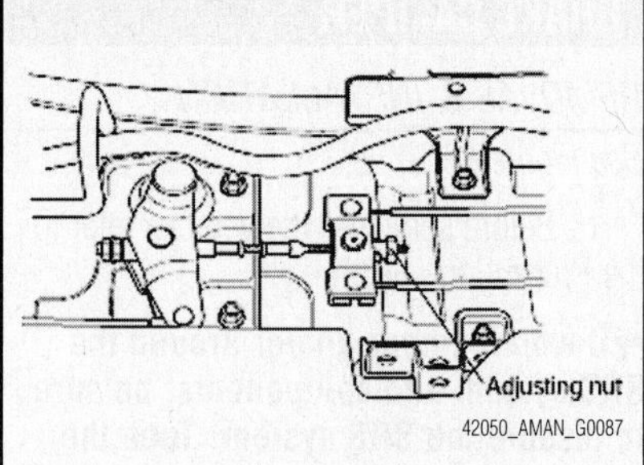

Fig. 10 Parking brake adjusting nut and pin

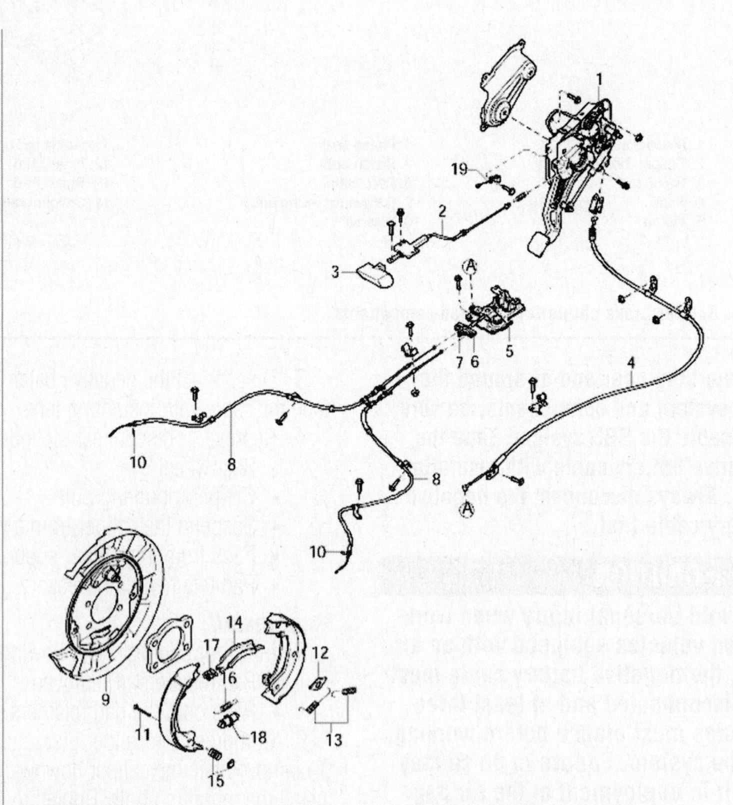

Fig. 9 Parking brake assembly and related components

PARKING BRAKE SHOES

REMOVAL & INSTALLATION

See Figures 11 and 12.

1. Before servicing the vehicle, refer to the Precautions Section.

➡If working near and/or around the SRS system and components, be sure to disable the SRS system. Tape the negative battery cable with insulating tape. Always disconnect the negative battery cable first.

⁂ CAUTION

To avoid personal injury when working on vehicles equipped with an air bag, the negative battery cable must be disconnected and at least three minutes must elapse before working on the system. Failure to do so may result in deployment of the air bag.

2. Remove the battery top cover, if equipped.
3. Disconnect the negative battery cable. Tape the cable with insulating tape.
4. Remove rear disc brake caliper assembly

➡Before removing the brake disc rotor, make a chalk marking by the bolts to aid in reassembly.

5. Remove the rotor.
6. Remove the shoe hold spring by turning the pin to coincide with hole of spring cap.

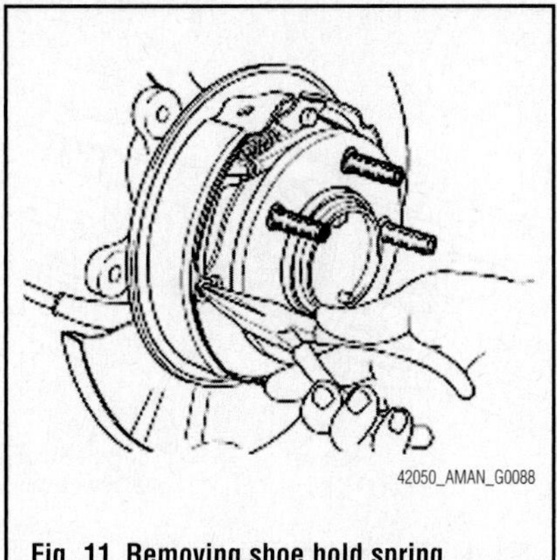

Fig. 11 Removing shoe hold spring

7. Remove the return spring.
8. Disconnect the cable end from the trailing shoe.
9. Remove the parking brake shoes.

➡**Complete the removal and installation on one side of the vehicle at a time using the others side as a reference.**

To install:

10. Install the parking brake cable to the operating lever.
11. Apply brake grease to return spring and areas of movement on the mechanism, but do not apply grease to the brake shoe material.
12. Install the upper return spring and brake shoes.
13. Turn the adjuster in counter clockwise direction and install.
14. Install the lower return spring.
15. Install the shoe hold spring with a pliers.
16. Install the disc brake and then align the mark while tightening the screw.

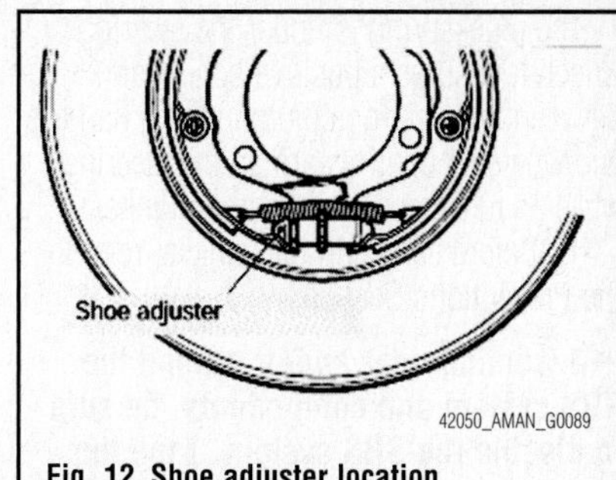

Fig. 12 Shoe adjuster location

ADJUSTMENT

See Figure 12.

The shoe adjuster changes the gap between the shoe and the drum.

1. Turn the adjuster clockwise to narrow the shoe gap
2. Turn the adjuster counterclockwise to widen the shoe gap.

CHASSIS ELECTRICAL

AIR BAG (SUPPLEMENTAL RESTRAINT SYSTEM)

GENERAL INFORMATION

⁂ CAUTION

These vehicles are equipped with an air bag system. The system must be disarmed before performing service on, or around, system components, the steering column, instrument panel components, wiring and sensors. Failure to follow the safety precautions and the disarming procedure could result in accidental air bag deployment, possible injury and unnecessary system repairs.

SERVICE PRECAUTIONS

⁂ CAUTION

Disconnect and isolate the battery negative cable before beginning any airbag system component diagnosis, testing, removal, or installation procedures. Wait at least 90 seconds after the ignition switch is turned off and the negative (-) terminal cable is disconnected from the battery before starting the operation. The SRS is equipped with a backup power source, so if work is started within 90 seconds after disconnecting the negative (-) terminal cable from the battery, the SRS may be deployed. Failure to disable the airbag system may result in accidental airbag deployment, personal injury, or death.

DISARMING THE SYSTEM

1. Before servicing the vehicle, refer to the Precautions Section.
2. Disconnect and isolate the negative battery cable. Wait 3 minutes for the system capacitor to discharge before performing any service.

➡**Wait at least 3 minutes before working on the vehicle. The air bag system is designed to retain enough power to deploy the air bag for a short time after the battery has been disconnected.**

ARMING THE SYSTEM

1. After repairs are complete, connect the negative battery cable. Turn the ignition switch to the **ON** position and check that the air bag warning light blinks as it would for normal operation.

CLOCKSPRING CENTERING

See Figures 13 and 14.

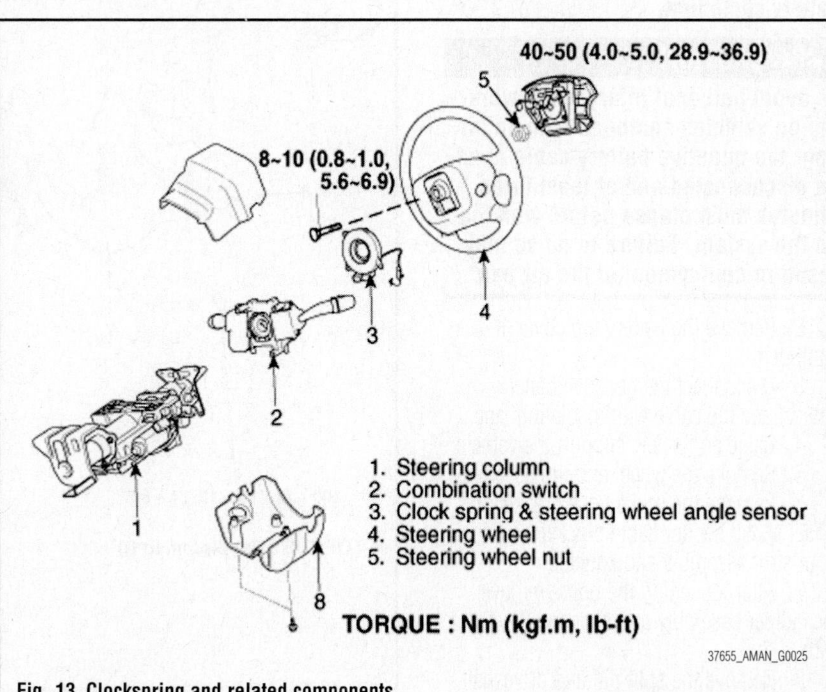

Fig. 13 Clockspring and related components

The clockspring is under the steering wheel. It ensures a positive connection between the steering column wiring harness and whatever controls are on the steering wheel, and especially the airbag igniter.

1. Before servicing the vehicle, refer to the Precautions Section.

➡If working near and/or around the SRS system and components, be sure to disable the SRS system. Tape the negative battery cable with insulating tape. Always disconnect the negative battery cable first.

⁂ CAUTION

To avoid personal injury when working on vehicles equipped with an air bag, the negative battery cable must be disconnected and at least three minutes must elapse before working on the system. Failure to do so may result in deployment of the air bag.

2. Remove the battery top cover, if equipped.
3. Disconnect the negative battery cable. Tape the cable with insulating tape.
4. Remove ignition key from vehicle
5. Remove the airbag module.
6. Set the clockspring in the center position. Make certain front wheels are in the straight-ahead position.
7. Make sure the mating mark of the clockspring is properly aligned.

⁂ CAUTION

If the mating mark is not properly aligned, the steering wheel may not completely rotate during a turn, or the flat cable within the clockspring may be broken causing an obstruction of the normal operation of the SRS and possibly lead to a serious injury to the driver of the vehicle.

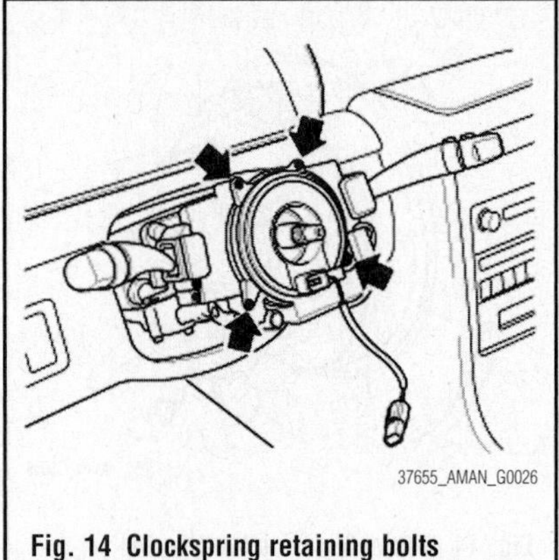

Fig. 14 Clockspring retaining bolts

DRIVE TRAIN

FRONT AXLE SHAFT, BEARING & SEAL

REMOVAL & INSTALLATION

See Figure 15.

1. Before servicing the vehicle, refer to the Precautions Section.

➡If working near and/or around the SRS system and components, be sure to disable the SRS system. Tape the negative battery cable with insulating tape. Always disconnect the negative battery cable first.

⁂ CAUTION

To avoid personal injury when working on vehicles equipped with an air bag, the negative battery cable must be disconnected and at least three minutes must elapse before working on the system. Failure to do so may result in deployment of the air bag.

2. Remove the battery top cover, if equipped.
3. Disconnect the negative battery cable. Tape the cable with insulating tape.
4. Raise and safely support the vehicle.
5. Remove the wheel and tire assembly.
6. Remove the brake hose bracket and wheel speed sensor cable bracket from the front strut assembly and knuckle.
7. After loosening the bolts, remove the caliper assembly and suspend it with wire.
8. Remove the split pin and driveshaft castle nut from the front hub.
9. Remove the 2 bolts and disconnect the knuckle from the lower arm assembly.
10. Using the special tool (09568-4A000), disconnect the tie rod end from the knuckle.
11. Using a plastic hammer, disconnect the driveshaft from the axle hub.
12. Loosen the upper arm mounting nut but do not remove it.
13. Using the special tool (09568-4A000),

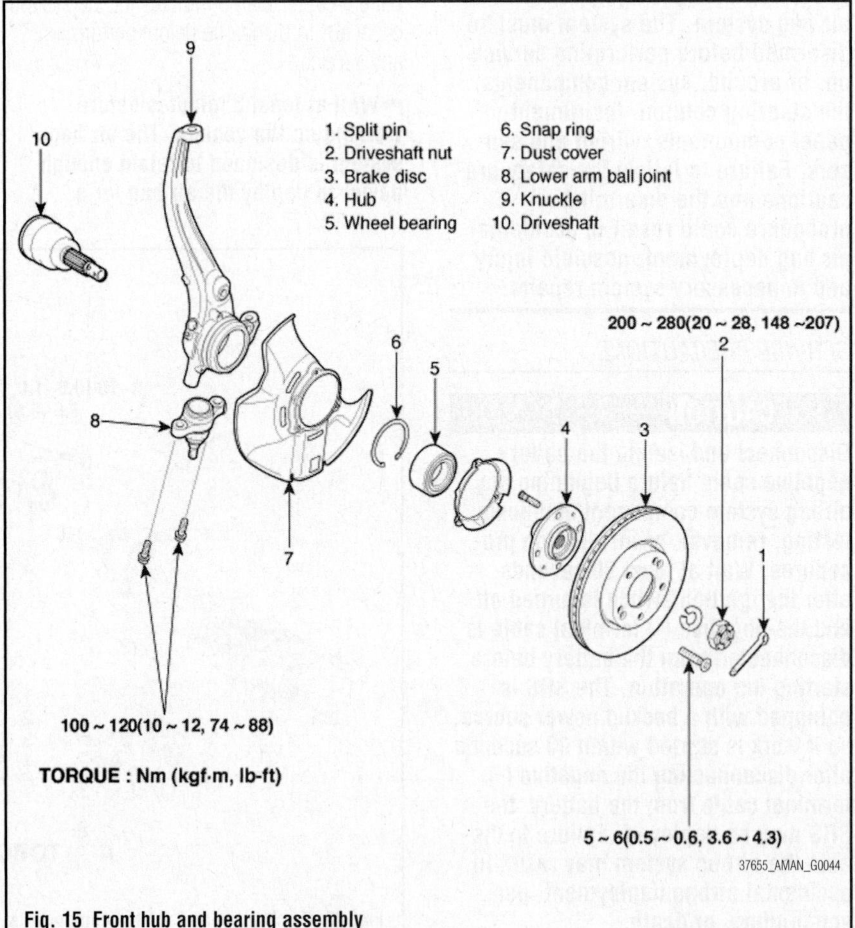

Fig. 15 Front hub and bearing assembly

disconnect the knuckle from the upper arm assembly.

14. Remove the front axle and knuckle together.

To install:

15. Installation is the reverse of removal.

FRONT HALFSHAFTS

REMOVAL & INSTALLATION

See Figure 16.

1. Before servicing the vehicle, refer to the Precautions Section.

➡If working near and/or around the SRS system and components, be sure to disable the SRS system. Tape the negative battery cable with insulating tape. Always disconnect the negative battery cable first.

✲✲ CAUTION

To avoid personal injury when working on vehicles equipped with an air bag, the negative battery cable must be disconnected and at least three minutes must elapse before working on the system. Failure to do so may result in deployment of the air bag.

2. Remove the battery top cover, if equipped.
3. Disconnect the negative battery cable. Tape the cable with insulating tape.
4. Raise and support the vehicle safely.
5. Remove the wheel and tire assembly.
6. Remove the spilt pin and driveshaft castle nut and washer from the front hub.
7. Using the special tool (09568-4A000), disconnect the tie rod end (A) from the knuckle.
8. Remove the 2 bolts and disconnect the knuckle from the lower arm assembly.
9. Remove the brake hose bracket and wheel speed sensor cable bracket from the front strut assembly and knuckle.
10. Using a plastic hammer, disconnect the driveshaft from the axle assembly.
11. Removing the left hand driveshaft (A) from the transaxle by using a pry bar (C) as shown.
12. For the right hand driveshaft, do the following:
    a. Remove the stabilizer link from the fork.
    b. Remove the fork from the front lower arm.

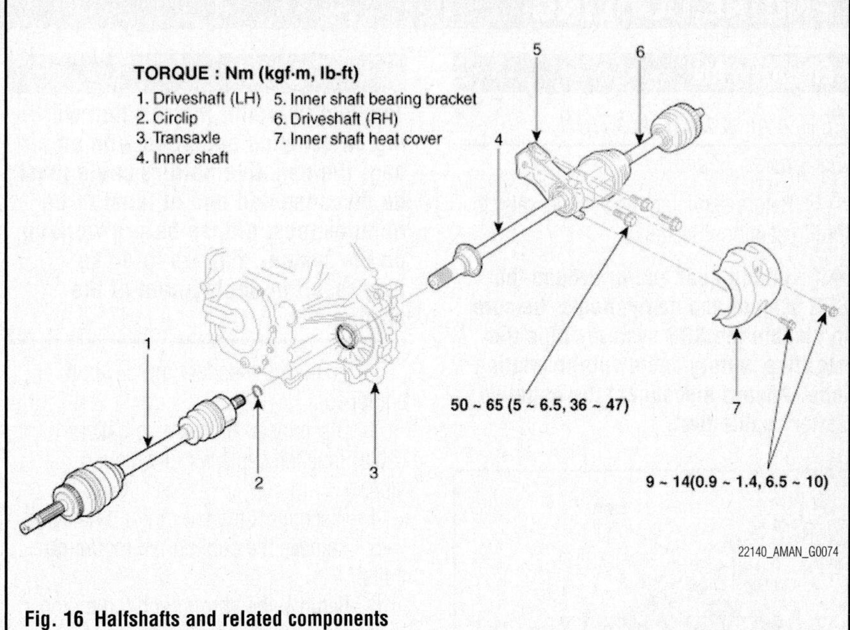

Fig. 16 Halfshafts and related components

    c. Remove the fork from the front strut assembly.
    d. Remove the inner shaft heat over and the heat cover mounting bolts.
    e. Remove the inner shaft bracket mounting bolts.
    f. Remove the front driveshaft assembly with the inner shaft from the transaxle.

✲✲ CAUTION

Do not try to disconnect the inner shaft from the driveshaft. Because they cannot be disconnected once assembled. Do not reuse the driveshaft which is disassembled from the inner shaft.

13. Using the special tool (09432-11000), remove the tone wheel.

To install:

✲✲ CAUTION

Replace the circlip with new ones after removal.

14. Apply gear oil on the drive shaft splines and the contacting surface of differential case oil seal.
15. After installation, check if the drive shaft cannot be removed.
16. For the right hand driveshaft, do the following:
    a. Install the inner shaft bearing bracket mounting bolt (A). Tightening Torque: 36–47 ft. lbs. (50–65 Nm).
    b. Install the inner shaft heat cover by installing the heat cover mounting bolts. Tightening Torque: 6.5–10 ft. lbs. (9–14 Nm).
    c. Install the fork to the front strut assembly. Tightening Torque: 44–59 ft. lbs. (60–80 Nm).
    d. Install the connecting bolt between the fork and the lower arm. Tightening Torque: 101–118 ft. lbs. (140–160 Nm).
    e. Install the stabilizer link to the fork. Tightening Torque: 74–88 ft. lbs. (100–120 Nm).
17. Install the drive shaft into the front axle assembly.
18. Install the knuckle in the lower arm assembly and tighten the bolts. Tightening Torque: 74–88 ft. lbs. (100–120 Nm).
19. Install the tie rod end in the knuckle. Tightening Torque: 18–25 ft. lbs. (24–34 Nm)
20. Install the brake hose bracket and wheel speed sensor cable bracket to the front strut assembly and knuckle.
21. After installing the washer with convex surface outward, install the castle nut and the spilt pin. Tightening Torque: 148–207 ft. lbs. (200–280 Nm).
22. Install the wheel and tire assembly.

ENGINE COOLING

ENGINE FAN

REMOVAL & INSTALLATION

See Figures 17 and 18.

1. Before servicing the vehicle, refer to the Precautions Section.

➡**If working near and/or around the SRS system and components, be sure to disable the SRS system. Tape the negative battery cable with insulating tape. Always disconnect the negative battery cable first.**

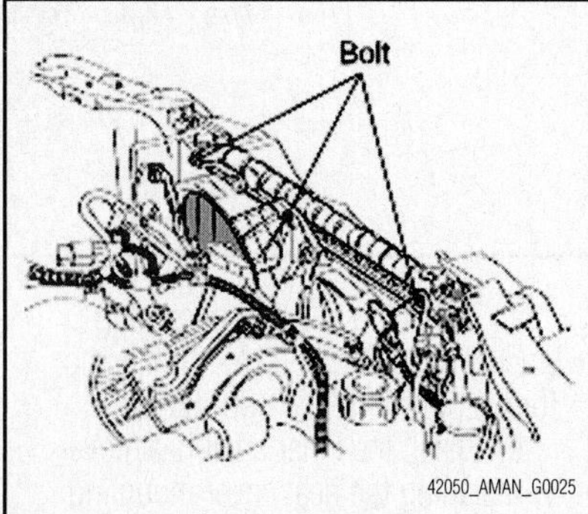

Fig. 17 Cooling fan assembly bolt locations

⁂ CAUTION

To avoid personal injury when working on vehicles equipped with an air bag, the negative battery cable must be disconnected and at least three minutes must elapse before working on the system. Failure to do so may result in deployment of the air bag.

2. Remove the battery top cover, if equipped.
3. Disconnect the negative battery cable. Tape the cable with insulating tape.
4. Disconnect the fan motor connector.
5. Remove the cooling fan mounting bolts.
6. Remove the shroud and fan motor assembly.
7. Remove the fan mounting clip to remove the fan.

To install:

8. Installation is the reverse of removal.
9. After installation, make sure there are no unusual noises or excessive vibration when the fan is rotating.

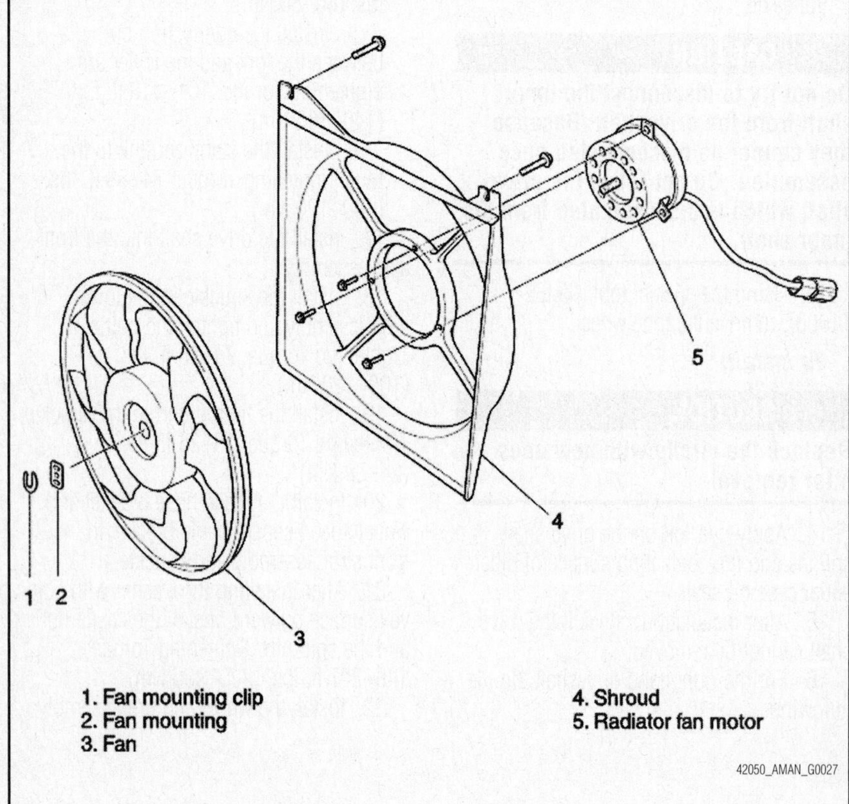

Fig. 18 Cooling fan motor assembly and related components

RADIATOR

REMOVAL & INSTALLATION

1. Before servicing the vehicle, refer to the Precautions Section.

➡**If working near and/or around the SRS system and components, be sure to disable the SRS system. Tape the negative battery cable with insulating tape. Always disconnect the negative battery cable first.**

⁂ CAUTION

To avoid personal injury when working on vehicles equipped with an air bag, the negative battery cable must be disconnected and at least three minutes must elapse before working on the system. Failure to do so may result in deployment of the air bag.

2. Remove the battery top cover, if equipped.
3. Disconnect the negative battery cable. Tape the cable with insulating tape.
4. Drain the engine coolant.
5. Remove the air duct.
6. Disconnect radiator upper and lower hoses.
7. Disconnect transaxle oil cooler hoses.
8. Disconnect the radiator fan connector.
9. Remove the radiator bracket.
10. Remove the radiator.

To install:

11. Install the radiator.
12. Install the radiator bracket.
13. Reconnect the radiator fan connector.
14. Connect transaxle oil cooler hoses.
15. Connect radiator upper and lower hoses.
16. Install the air duct.
17. Fill with engine coolant.
18. Start engine and check for leaks.
19. Recheck engine coolant level.

THERMOSTAT

REMOVAL & INSTALLATION

See Figure 19.

⁂ CAUTION

Never open, service or drain the radiator or cooling system when hot; serious burns can occur from the steam and hot coolant. Also, when draining engine coolant, keep in mind that cats and dogs are attracted

to ethylene glycol antifreeze and could drink any that is left in an uncovered container or in puddles on the ground. This will prove fatal in sufficient quantities. Always drain coolant into a sealable container. Coolant should be reused unless it is contaminated or is several years old.

1. Before servicing the vehicle, refer to the Precautions Section.

➡**If working near and/or around the SRS system and components, be sure to disable the SRS system. Tape the negative battery cable with insulating tape. Always disconnect the negative battery cable first.**

⁂ CAUTION

To avoid personal injury when working on vehicles equipped with an air bag, the negative battery cable must be disconnected and at least three minutes must elapse before working on the system. Failure to do so may result in deployment of the air bag.

2. Remove the battery top cover, if equipped.
3. Disconnect the negative battery cable. Tape the cable with insulating tape.
4. Drain the engine coolant. Be sure to properly dispose of used coolant.
5. Remove the thermostat housing retaining bolts.
6. Remove the housing.
7. Remove the thermostat.
8. Discard the gasket.

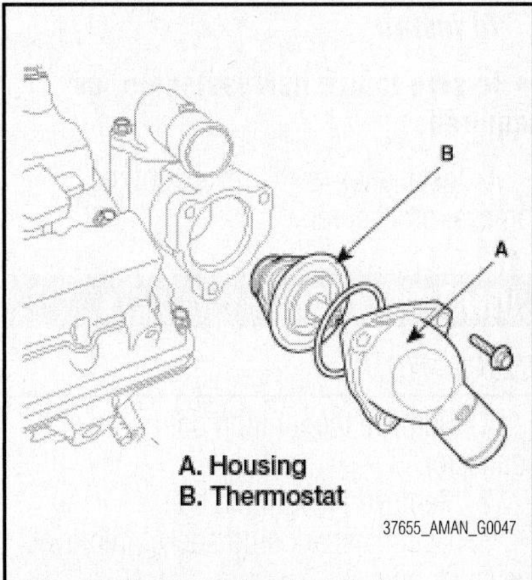

Fig. 19 Thermostat and related components

To install:

➡**Be sure to use new fasteners, as required.**

9. Installation is the reverse of the removal procedure.
10. Be sure to use a new gasket.
11. Be sure to install the thermostat with the jiggle valve upward.
12. Fill the cooling system with the proper grade and type coolant.
13. Start the engine and check for leaks.
14. Correct as required.

WATER PUMP

REMOVAL & INSTALLATION

See Figure 20.

1. Before servicing the vehicle, refer to the Precautions Section.

➡**If working near and/or around the SRS system and components, be sure to disable the SRS system. Tape the negative battery cable with insulating tape. Always disconnect the negative battery cable first.**

⁂ CAUTION

To avoid personal injury when working on vehicles equipped with an air bag, the negative battery cable must be disconnected and at least three minutes must elapse before working on the system. Failure to do so may result in deployment of the air bag.

2. Remove the battery top cover, if equipped.
3. Disconnect the negative battery cable. Tape the cable with insulating tape.
4. Drain the engine coolant.

⁂ WARNING

System is under high pressure when the engine is hot. To avoid danger of releasing scalding engine coolant, remove the cap only when the engine is cool.

5. Remove the accessory drive belt.
6. Remove the 4 bolts and pump pulley.
7. Remove the water pump and gasket.

To install:

8. Installation is the reverse order of the removal.
9. Tightening torque for water pump bolts: 7 ft. lbs. (8 Nm)
10. Refill the engine coolant to the correct level.
11. Start the engine and check for leaks.

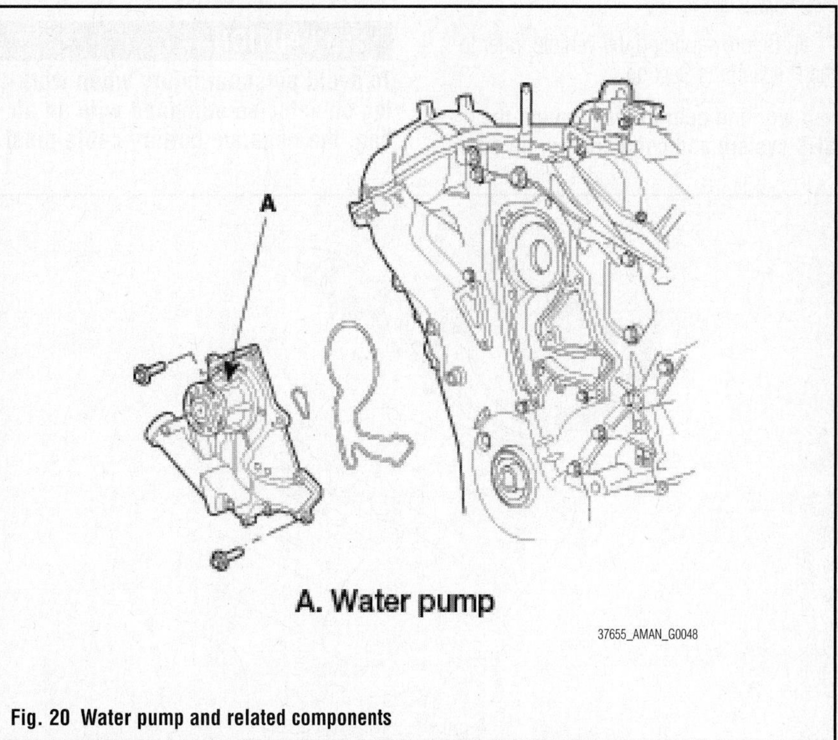

Fig. 20 Water pump and related components

ENGINE ELECTRICAL — CHARGING SYSTEM

ALTERNATOR

REMOVAL & INSTALLATION

See Figure 21.

1. Before servicing the vehicle, refer to the Precautions Section.

➡**If working near and/or around the SRS system and components, be sure to disable the SRS system. Tape the negative battery cable with insulating tape. Always disconnect the negative battery cable first.**

⁂ CAUTION

To avoid personal injury when working on vehicles equipped with an air bag, the negative battery cable must be disconnected and at least three minutes must elapse before working on the system. Failure to do so may result in deployment of the air bag.

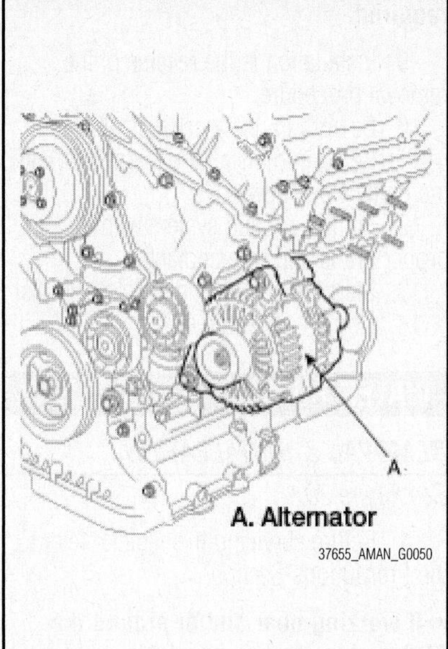

Fig. 21 Alternator and related components

2. Remove the battery top cover, if equipped.
3. Disconnect the negative battery cable. Tape the cable with insulating tape.
4. Remove or disconnect the following:
   - Alternator wiring harness connectors
   - Accessory drive belt
   - Alternator support bracket
   - Alternator

To install:

5. Install the alternator. Tighten the mounting bolts as follows:
   a. Mounting bracket bolts: Tighten to 15–18 ft. lbs. (20–25 Nm)
   b. Support bracket bolt: Tighten to 13–16 ft. lbs. (18–22 Nm)
6. Install or connect the following:
   - Accessory drive belt
   - Alternator wiring harness connectors
   - Negative battery cable

ENGINE ELECTRICAL — IGNITION SYSTEM

IGNITION COIL

REMOVAL & INSTALLATION

See Figure 22.

1. Before servicing the vehicle, refer to the Precautions Section.

➡**If working near and/or around the SRS system and components, be sure to disable the SRS system. Tape the negative battery cable with insulating tape. Always disconnect the negative battery cable first.**

⁂ CAUTION

To avoid personal injury when working on vehicles equipped with an air bag, the negative battery cable must be disconnected and at least three minutes must elapse before working on the system. Failure to do so may result in deployment of the air bag.

2. Remove the battery top cover, if equipped.
3. Disconnect the negative battery cable. Tape the cable with insulating tape.
4. Remove the engine cover.
5. Disconnect the electrical connector.
6. Remove the coil from its mounting.

To install:

➡**Be sure to use new fasteners, as required.**

7. Installation is the reverse of the removal procedure.

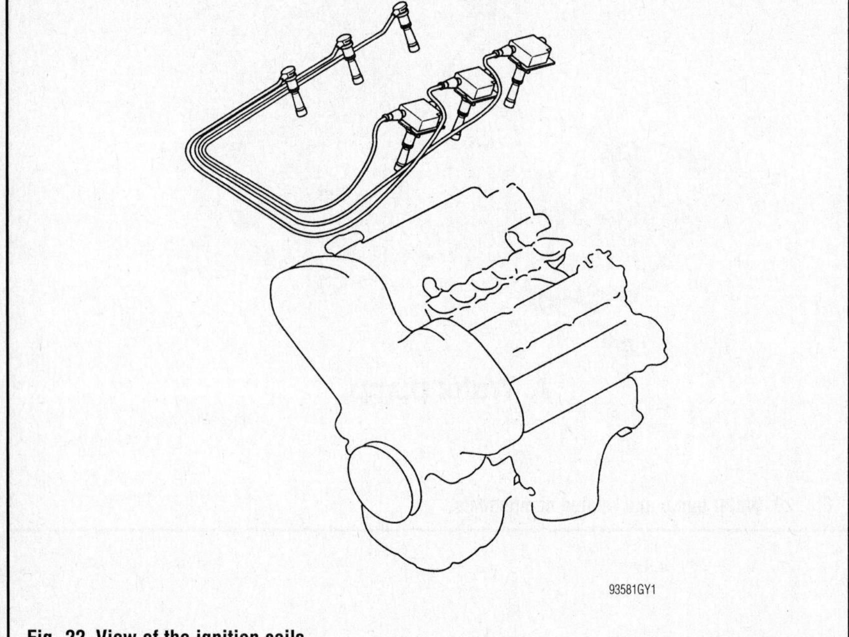

Fig. 22 View of the ignition coils

IGNITION TIMING

INSPECTION

1. Remove the ignition coil connector.
2. Remove the ignition coil.
3. Using a spark plug socket, remove the spark plug.
4. Install the spark plug to the ignition coil.
5. Ground the spark plug to the engine.
6. Check if spark occurs while engine is being cranked.

➡To prevent fuel being injected from injectors while the engine is being cranked, remove the fuel pump relay from the fuse box. Crank the engine for no more than 5–10 seconds.

7. Inspect all the spark plugs.
8. Using a spark plug socket, install the spark plug.
9. Install the ignition coil.
10. Reconnect the ignition coil connector.

ADJUSTMENT

The ignition timing is controlled by the Powertrain Control Module (PCM). No adjustment is necessary.

SPARK PLUGS

REMOVAL & INSTALLATION

1. Before servicing the vehicle, refer to the Precautions Section.

➡If working near and/or around the SRS system and components, be sure to disable the SRS system. Tape the negative battery cable with insulating tape. Always disconnect the negative battery cable first.

⁂ CAUTION

To avoid personal injury when working on vehicles equipped with an air bag, the negative battery cable must be disconnected and at least three minutes must elapse before working on the system. Failure to do so may result in deployment of the air bag.

2. Remove the battery top cover, if equipped.
3. Disconnect the negative battery cable. Tape the cable with insulating tape.
4. Remove the ignition coil.
5. Remove the spark plug from its mounting.

To install:

➡Be sure to use new fasteners, as required.

6. Installation is the reverse of the removal procedure.
7. Tighten the spark plug to specification.

ENGINE ELECTRICAL

STARTING SYSTEM

STARTER

REMOVAL & INSTALLATION

See Figure 23.

1. Before servicing the vehicle, refer to the Precautions Section.

➡If working near and/or around the SRS system and components, be sure to disable the SRS system. Tape the negative battery cable with insulating tape. Always disconnect the negative battery cable first.

⁂ CAUTION

To avoid personal injury when working on vehicles equipped with an air bag, the negative battery cable must be disconnected and at least three minutes must elapse before working on the system. Failure to do so may result in deployment of the air bag.

2. Remove the battery top cover, if equipped.
3. Disconnect the negative battery cable. Tape the cable with insulating tape.
4. Raise and safely support the vehicle, as required.
5. Remove or disconnect the following:
   - Starter electrical connectors
   - Starter motor

To install:

6. Install or connect the following:
   - Starter motor. Tighten mounting bolt and nut to 33–40 ft. lbs. (45–55 Nm).
   - Starter electrical connectors. Tighten nut to 20–25 ft. lbs. (27–34 Nm).
   - Speedometer and shift cables
   - Negative battery cable

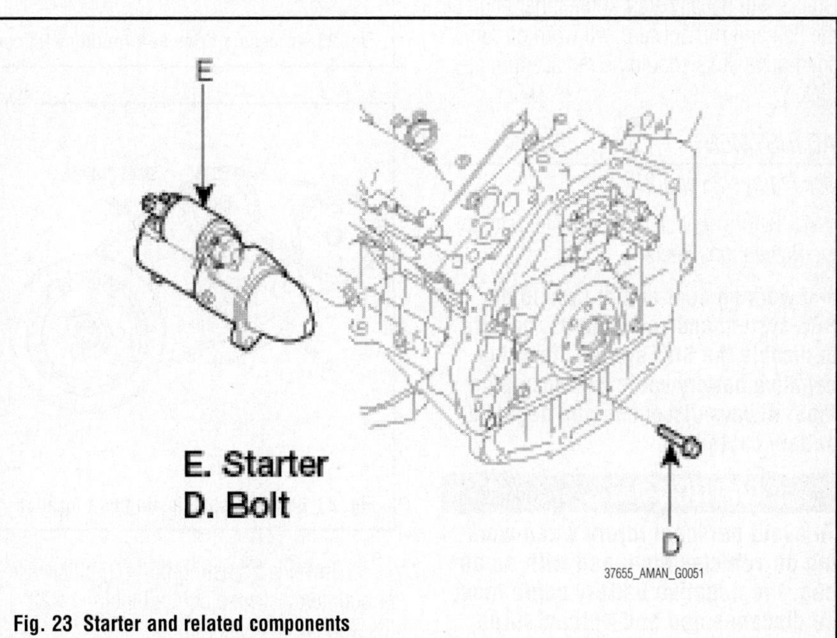

Fig. 23 Starter and related components

ENGINE MECHANICAL

ACCESSORY DRIVE BELTS

ACCESSORY BELT ROUTING

See Figure 24.

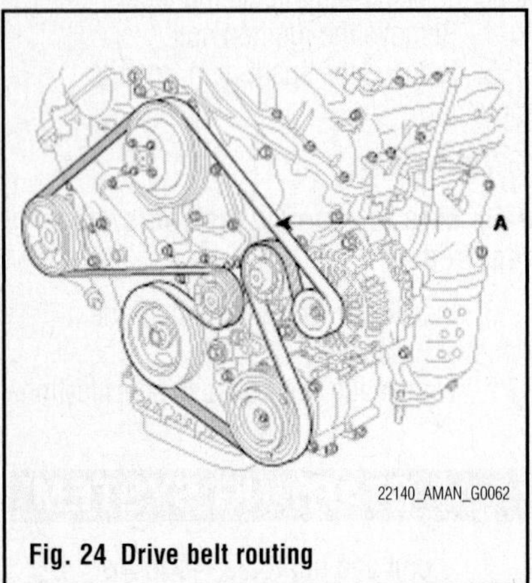

Fig. 24 Drive belt routing

INSPECTION

Inspect the drive belt for signs of glazing or cracking. A glazed belt will be perfectly smooth from slippage, while a good belt will have a slight texture of fabric visible. Cracks will usually start at the inner edge of the belt and run outward. All worn or damaged drive belts should be replaced immediately.

ADJUSTMENT

See Figures 25 and 26.

1. Before servicing the vehicle, refer to the Precautions Section.

➡If working near and/or around the SRS system and components, be sure to disable the SRS system. Tape the negative battery cable with insulating tape. Always disconnect the negative battery cable first.

✲✲ CAUTION

To avoid personal injury when working on vehicles equipped with an air bag, the negative battery cable must be disconnected and at least three minutes must elapse before working on the system. Failure to do so may result in deployment of the air bag.

2. Remove the battery top cover, if equipped.
3. Disconnect the negative battery cable. Tape the cable with insulating tape.
4. Use a tensioner gauge to measure the adjustment of the accessory drive belt.

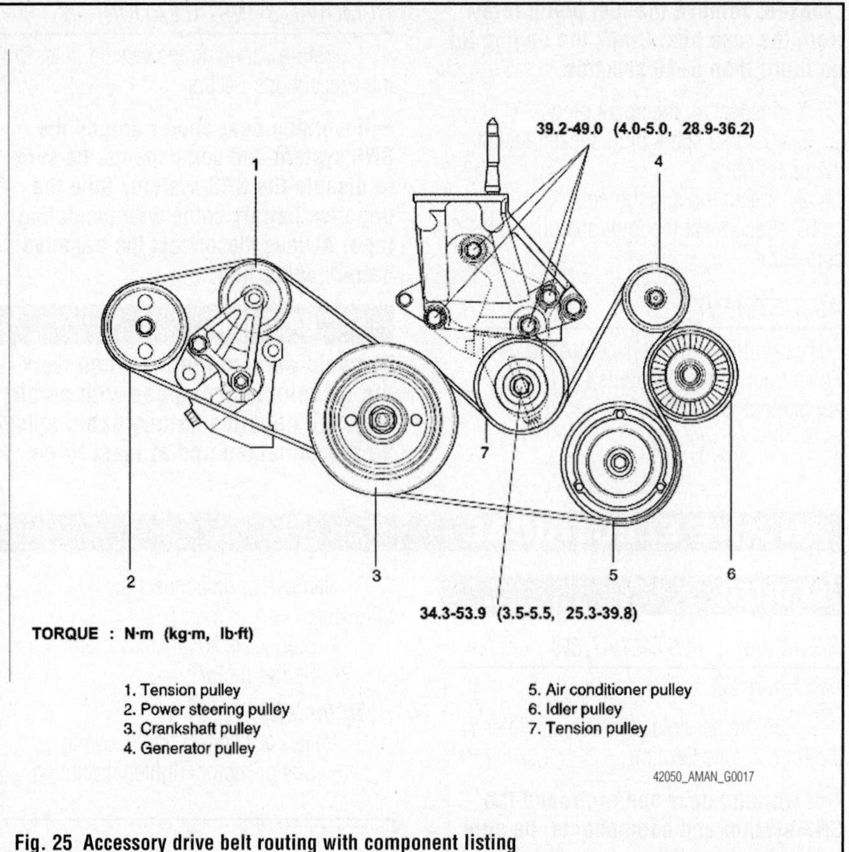

Fig. 25 Accessory drive belt routing with component listing

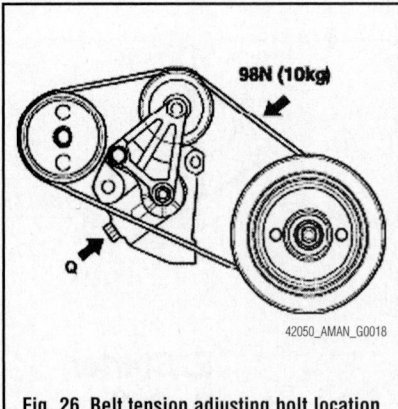

Fig. 26 Belt tension adjusting bolt location

5. Turn the adjusting bolt (Q) clockwise or counterclockwise until a tension of 22 lbs. (98 N) is measured.

REMOVAL & INSTALLATION

See Figures 24 through 26.

1. Before servicing the vehicle, refer to the Precautions Section.

➡If working near and/or around the SRS system and components, be sure to disable the SRS system. Tape the negative battery cable with insulating tape. Always disconnect the negative battery cable first.

✲✲ CAUTION

To avoid personal injury when working on vehicles equipped with an air bag, the negative battery cable must be disconnected and at least three minutes must elapse before working on the system. Failure to do so may result in deployment of the air bag.

2. Remove the battery top cover, if equipped.
3. Disconnect the negative battery cable. Tape the cable with insulating tape.
4. Release tension from drive belt by turning the belt tension adjusting bolt (Q).
5. Remove drive belt from the vehicle.

To install:

6. Installation is the reverse of the removal procedure.
7. Be sure that the belt is securely installed around all pulleys.
8. Rotate the crankshaft several times clockwise to equalize belt tension between the pulleys.
9. Adjust the drive belt tension.

CAMSHAFT AND VALVE LIFTERS

INSPECTION

See Figure 27.

1. Check the camshaft journals for wear. If the journals are badly worn out, replace the camshaft.
2. Check the cam lobes for damage. If the lobe is damaged or excessively worn out, replace the camshaft.
3. Measure the cam height with a micrometer. The specifications for cam lobe height:
   a. Intake 1.8425 in. (46.8mm).
   b. Exhaust 1.8031 in. (45.8mm).
4. Check the cam surface for abnormal wear or damage, and replace if necessary.
5. Check each bearing for damage. If the bearing surface is excessively damaged, replace the cylinder head assembly or camshaft bearing cap, as necessary.

➡Camshaft end play should be within specification: 0.0008–0.0071 in. (0.02–0.18mm)

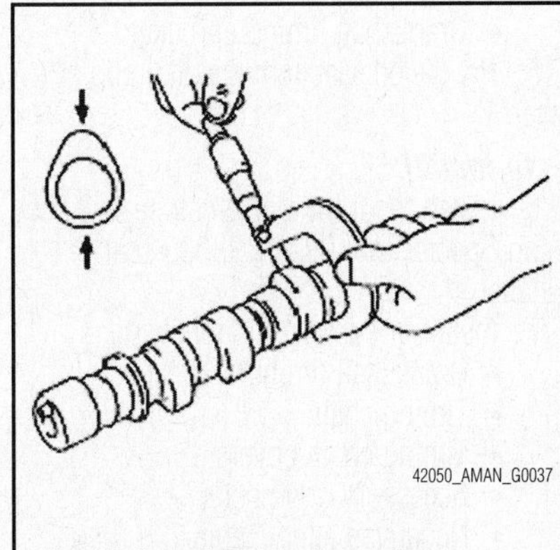

Fig. 27 Camshaft lobe inspection

REMOVAL & INSTALLATION

See Figures 28 through 30.

1. Before servicing the vehicle, refer to the Precautions Section.

➡If working near and/or around the SRS system and components, be sure to disable the SRS system. Tape the negative battery cable with insulating tape. Always disconnect the negative battery cable first.

CAUTION

To avoid personal injury when working on vehicles equipped with an air bag, the negative battery cable must be disconnected and at least three minutes must elapse before working on the system. Failure to do so may result in deployment of the air bag.

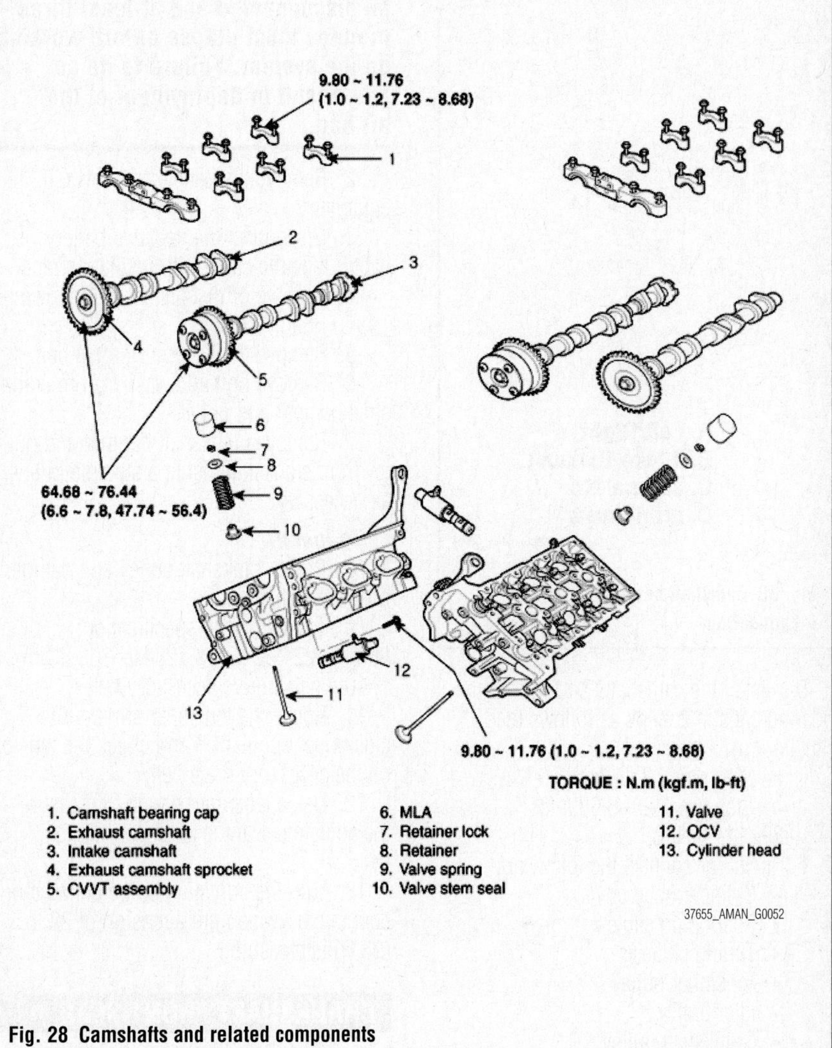

Fig. 28 Camshafts and related components

2. Remove the battery top cover, if equipped.
3. Disconnect the negative battery cable. Tape the cable with insulating tape.
4. Remove or disconnect the following:
   - Engine cover
   - Intake manifold
   - Power steering pulley
   - A/C pulley
   - Crankshaft pulley
   - Idler pulley
   - Tensioner pulley
   - Timing chain
   - Spark plug cables
   - Cylinder head cover
   - Camshaft sprockets
   - Camshaft bearing caps
   - Camshafts

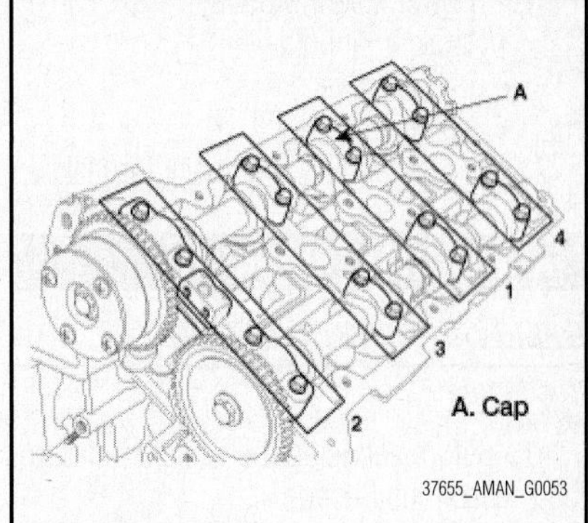

Fig. 29 Camshaft bearing cap tightening sequence

➡Keep all valvetrain components in order for assembly.

To install:

➡Be sure to use new fasteners, as required.

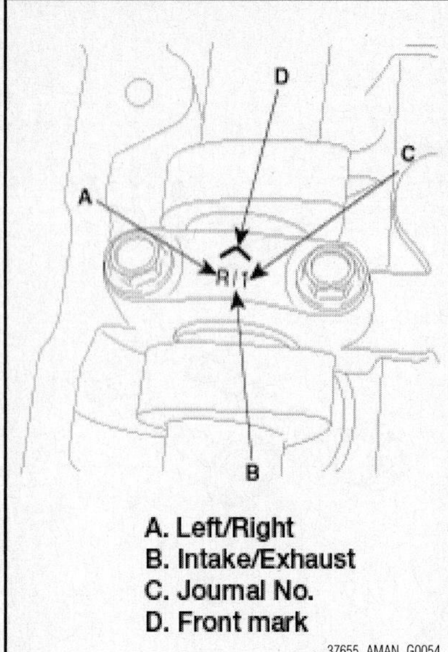

Fig. 30 Camshaft bearing cap identification

5. Install the camshafts and tighten the bearing caps in 2 steps as follows (see illustration):
    a. Step one 4.3 ft. lbs. (5.9 Nm).
    b. Step two 7.23–8.68 ft lbs. (9.80–11.76 Nm).
6. Install or connect the following:
    - Cylinder head cover
    - Spark plug cables
    - Timing chain
    - Tensioner pulley
    - Idler pulley
    - Crankshaft pulley
    - A/C pulley
    - Power steering pulley
    - Intake manifold
    - Engine cover
    - Negative battery
7. Start the engine and check for leaks.

CRANKSHAFT DAMPER

REMOVAL & INSTALLATION

See Figure 33.

1. Before servicing the vehicle, refer to the Precautions Section.

➡If working near and/or around the SRS system and components, be sure to disable the SRS system. Tape the negative battery cable with insulating tape. Always disconnect the negative battery cable first.

✲✲ CAUTION

To avoid personal injury when working on vehicles equipped with an air bag, the negative battery cable must be disconnected and at least three minutes must elapse before working on the system. Failure to do so may result in deployment of the air bag.

2. Remove the battery top cover, if equipped.
3. Disconnect the negative battery cable. Tape the cable with insulating tape.
4. Remove or disconnect the following:
    - Engine cover
5. Remove the accessory drive belt.
6. Remove bolt and washer from crankshaft damper and pulley.
7. Remove crankshaft damper and pulley from crankshaft using a suitable puller, as necessary.

To install:

8. Press crankshaft pulley and damper onto crankshaft end.
9. Torque bolt to specification: 209.76–224.22 ft. lbs. (284.2–303.8 Nm).
10. Install drive accessory belt.
11. Rotate the tensioner arm counterclockwise (about 14°) and check the tension on the accessory drive belt.
12. Use a tensioner gauge to measure the adjustment of the accessory drive belt.
13. Turn the adjusting bolt clockwise or counterclockwise until a tension of 22 lb. (98 N) is measured.

CRANKSHAFT FRONT SEAL

REMOVAL & INSTALLATION

See Figure 34.

1. Before servicing the vehicle, refer to the Precautions Section.

➡If working near and/or around the SRS system and components, be sure to disable the SRS system. Tape the negative battery cable with insulating tape. Always disconnect the negative battery cable first.

✲✲ CAUTION

To avoid personal injury when working on vehicles equipped with an air bag, the negative battery cable must be disconnected and at least three minutes must elapse before working on the system. Failure to do so may result in deployment of the air bag.

2. Remove the battery top cover, if equipped.
3. Disconnect the negative battery cable. Tape the cable with insulating tape.
4. Remove or disconnect the following:
    - Accessory drive belts
    - Timing chain covers
    - Timing chain
    - Crankshaft timing sprocket
5. Pry the oil seal from the oil pump case.

To install:

6. Install the front crankshaft seal using Special Tool 09214-33000 seal installer.
7. Install or connect the following:
    - Crankshaft timing sprocket
    - Timing chain
    - Timing chain covers
    - Accessory drive belts
    - Negative battery cable

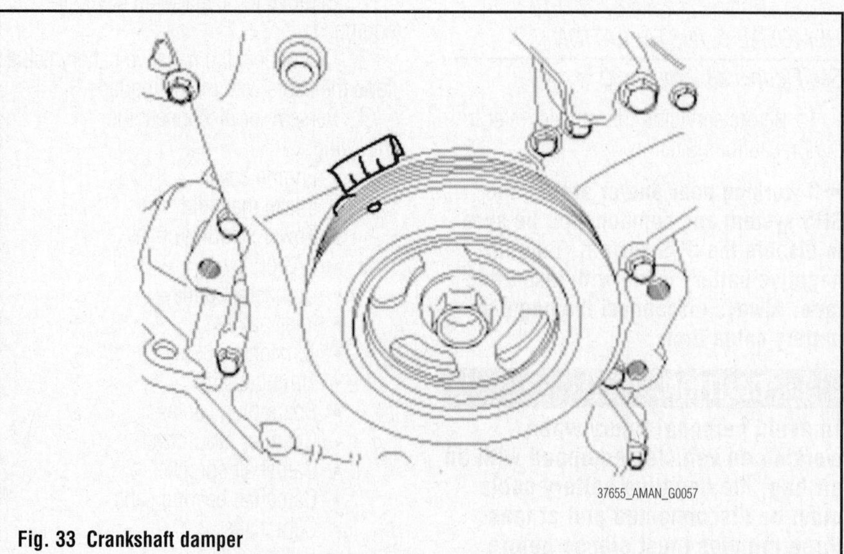

Fig. 33 Crankshaft damper

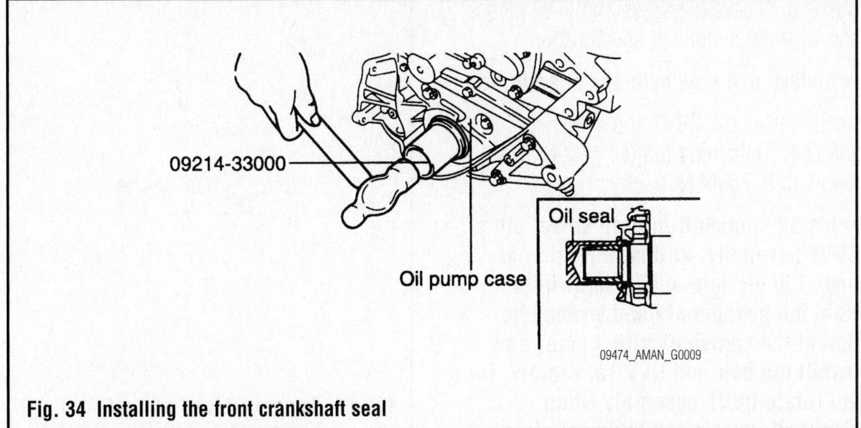

Fig. 34 Installing the front crankshaft seal

8. Start the engine and check for leaks.

CYLINDER HEAD

REMOVAL & INSTALLATION

See Figures 29 and 30, 35 through 40.

1. Before servicing the vehicle, refer to the Precautions Section.

➡If working near and/or around the SRS system and components, be sure to disable the SRS system. Tape the negative battery cable with insulating tape. Always disconnect the negative battery cable first.

CAUTION

To avoid personal injury when working on vehicles equipped with an air bag, the negative battery cable must be disconnected and at least three minutes must elapse before working on the system. Failure to do so may result in deployment of the air bag.

2. Remove the battery top cover, if equipped.
3. Disconnect the negative battery cable. Tape the cable with insulating tape.
4. Turn the crankshaft pulley so that the No. 1 piston is at top dead center.

➡Engine removal is required for this procedure.

5. Remove exhaust manifold.
6. Remove intake manifold.
7. Remove timing chain.
8. Remove water temperature control assembly.
9. Remove camshaft bearing cap.
10. Remove camshaft assembly.
11. Remove cylinder head bolts, then remove cylinder head as shown:
    - Uniformly loosen and remove the 16 cylinder head bolts, in several passes, in the sequence shown. Remove the 16 cylinder head bolts and plate washers.

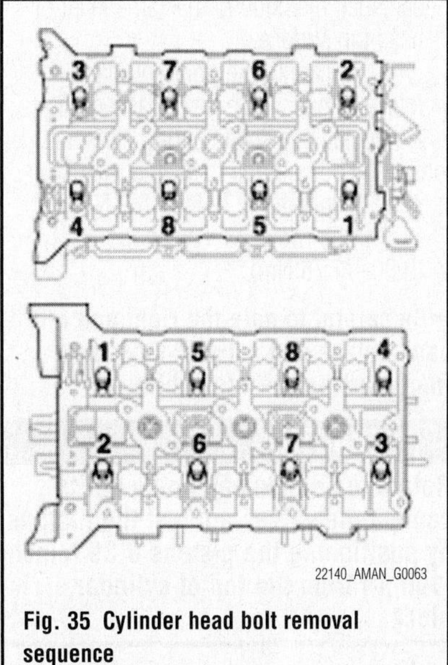

Fig. 35 Cylinder head bolt removal sequence

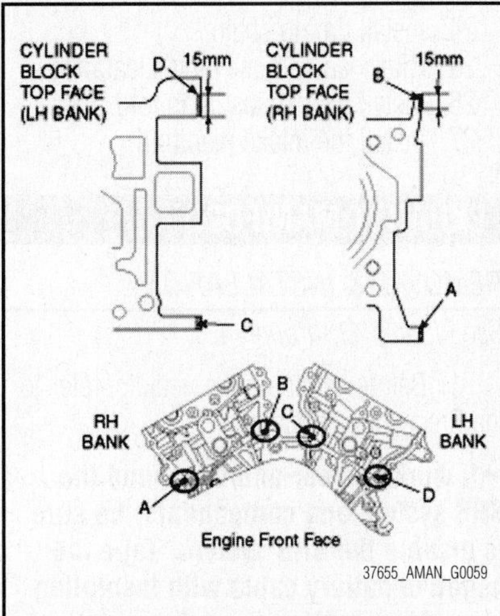

Fig. 36 Cylinder head gasket installation 1 of 4

CAUTION

Head warpage or cracking could result from removing bolts in an incorrect order.

- Lift the cylinder head from the dowels on the cylinder block and place the cylinder head on wooden blocks on a bench.

CAUTION

Be careful not to damage the contact surfaces of the cylinder head and cylinder block.

To install:

12. Thoroughly clean all parts to be assembled.
13. Always use a new head and manifold gasket.
14. The cylinder head gasket is a metal gasket. Take care not to bend it.

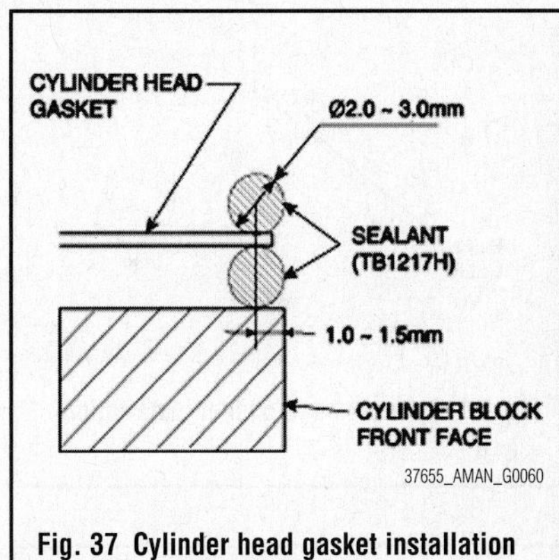

Fig. 37 Cylinder head gasket installation 2 of 4

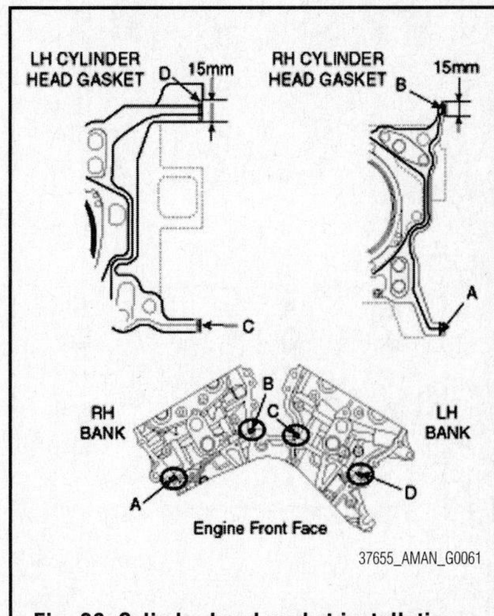

Fig. 38 Cylinder head gasket installation 3 of 4

15. Rotate the crankshaft, set the No.1 piston at TDC.
16. Install the cylinder head.

➡Apply sealant on cylinder block top face before assembling cylinder head gaskets. The part must be assembled within 5 minutes after sealant was applied.

➡Be careful of the installation direction.

➡Remove the extruded sealant after assembling cylinder heads.

17. Place the cylinder head carefully to avoid damaging the gasket.
18. Install cylinder head bolts.

➡Do not apply engine oil on the threads and under the heads of the cylinder head bolts.

19. Using SST(09221-4A000), install and tighten the cylinder head bolts and plate washers, in several passes, in the sequence shown. Tighten bolts to specification.

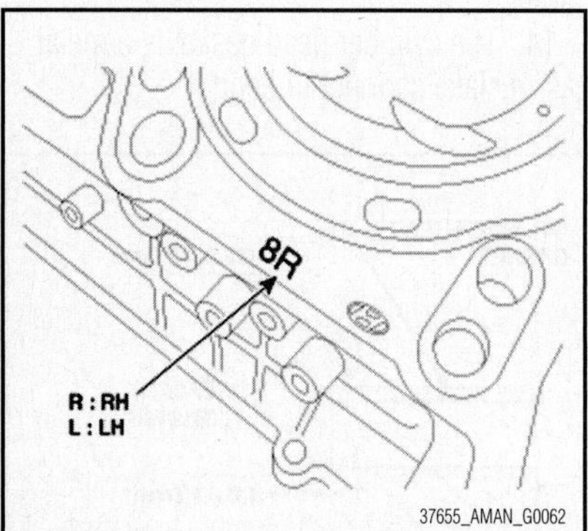

Fig. 39 Cylinder head gasket installation 4 of 4

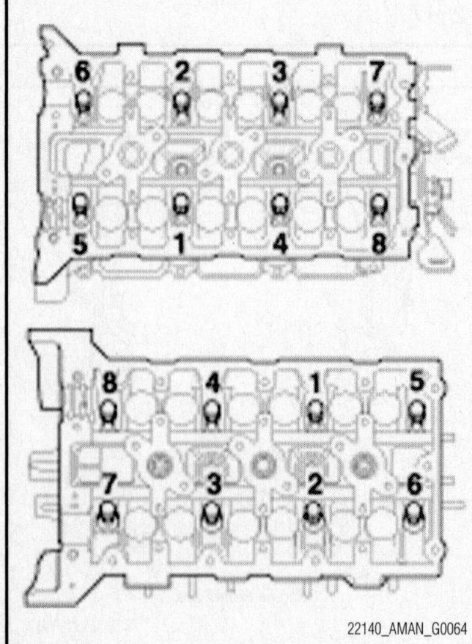

Fig. 40 Cylinder head bolt tightening sequence

➡Always use new cylinder head bolts.

20. Install the CVVT and camshaft sprocket. Tightening torque: 47.74–56.4ft. lbs. (64.68–76.44 Nm)

➡Install camshaft-inlet to dowel pin of CVVT assembly. At this time, do not install to oil hole of camshaft-inlet. Hold the hexagonal head wrench portion of the camshaft with a vise, and install the bolt and CVVT assembly. Do not rotate CVVT assembly when camshaft is installed to dowel pin of CVVT assembly.

21. Install camshafts.
    a. Apply a light coat of engine oil on camshaft journals.
    b. Assemble the key groove of camshaft rear side to the same level of head top surface.
    c. Be careful the right, left bank, intake, exhaust side before assembling.
22. Install camshaft bearing caps and torque the bolts in the following order:
    a. Step one 4.3 ft. lbs. (5.9 Nm).
    b. Step two 7.23–8.68 ft lbs. (9.80–11.76 Nm).

➡Be careful to note the right and left bank; intake and exhaust side; and front mark before assembling.

✲✲ CAUTION

Rotate the crankshaft so as not to contact the valves against the pistons by positioning the pistons 0.3937inch (10mm) from the top of cylinder block.

23. Install water temperature control assembly.
24. Install timing chain.
25. Check and adjust valve clearance.
26. Install the exhaust manifold.
27. Install the intake manifold.

EXHAUST MANIFOLD

REMOVAL & INSTALLATION

See Figures 41 through 43.

1. Before servicing the vehicle, refer to the Precautions Section.

➡If working near and/or around the SRS system and components, be sure to disable the SRS system. Tape the negative battery cable with insulating tape. Always disconnect the negative battery cable first.

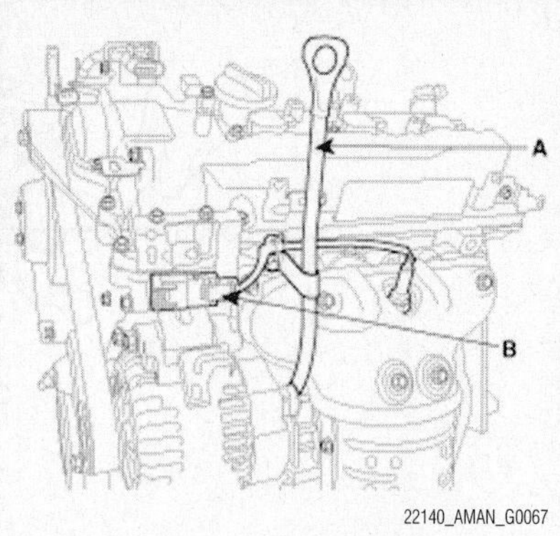

Fig. 41 Remove oil level gauge(A) and disconnect LH front oxygen sensor connector(B).

✲✲ CAUTION

To avoid personal injury when working on vehicles equipped with an air bag, the negative battery cable must be disconnected and at least three minutes must elapse before working on the system. Failure to do so may result in deployment of the air bag.

2. Remove the battery top cover, if equipped.
3. Disconnect the negative battery cable. Tape the cable with insulating tape.
4. Remove undercover.
5. Disconnect LH,RH rear oxygen sensor connector from bracket.
6. Remove front muffler.
7. Remove oil level gauge(A).
8. Disconnect LH front oxygen sensor connector(B) from bracket.
9. Remove LH heat protector.
10. Remove LH exhaust manifold.
11. Disconnect RH front oxygen sensor connector from bracket.
12. Remove RH heat protector.
13. Remove RH exhaust manifold.

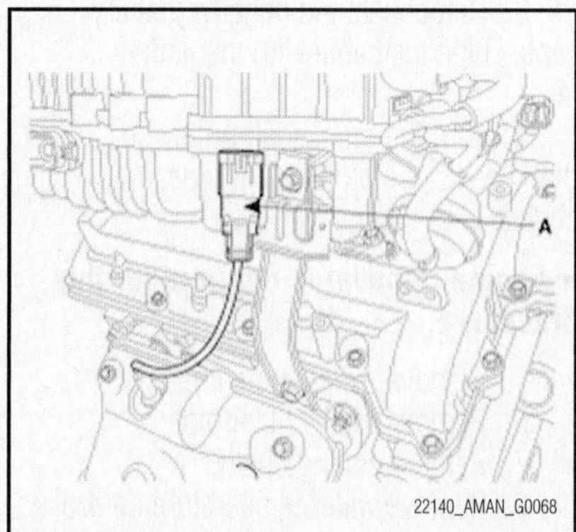

Fig. 42 Disconnect RH front oxygen sensor connector (A).

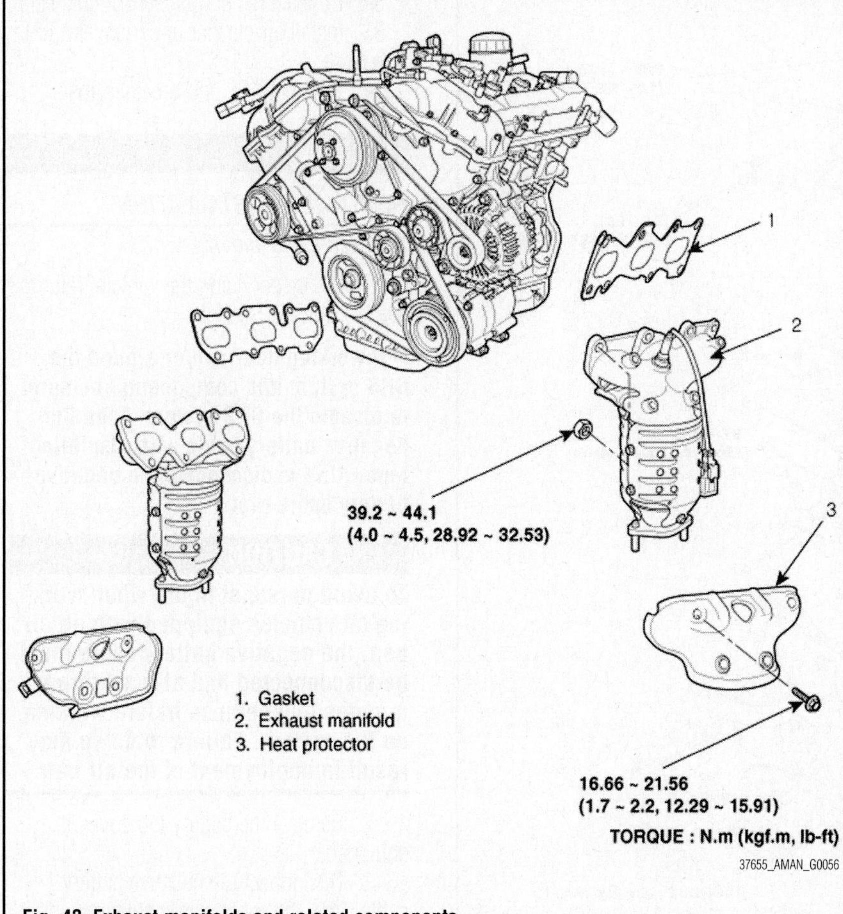

Fig. 43 Exhaust manifolds and related components

To install:

14. Install new gasket and exhaust manifold. Tightening torque: 29–33 ft. lbs. (40–44 Nm)
15. Install heat protectors.
16. Install front muffler.
17. Connect oxygen sensor connectors.
18. Install undercover.

FLEXPLATE

REMOVAL & INSTALLATION

1. Before servicing the vehicle, refer to the Precautions Section.

➡If working near and/or around the SRS system and components, be sure to disable the SRS system. Tape the negative battery cable with insulating tape. Always disconnect the negative battery cable first.

⁂ CAUTION

To avoid personal injury when working on vehicles equipped with an air bag, the negative battery cable must be disconnected and at least three minutes must elapse before working on the system. Failure to do so may result in deployment of the air bag.

2. Remove the battery top cover, if equipped.
3. Disconnect the negative battery cable. Tape the cable with insulating tape.
4. Remove the transaxle.
5. Remove the flexplate retaining bolts.
6. Remove the flexplate from the engine.
7. Replace if gears are damaged by starter engagement.

To install:

8. Reverse the removal procedure to install the flexplate.
9. Replace flexplate retaining bolts and tighten in a cross pattern to specification.

INTAKE MANIFOLD

REMOVAL & INSTALLATION

See Figures 44 and 45.

1. Before servicing the vehicle, refer to the Precautions Section.

➡If working near and/or around the SRS system and components, be sure to disable the SRS system. Tape the negative battery cable with insulating tape. Always disconnect the negative battery cable first.

⁂ CAUTION

To avoid personal injury when working on vehicles equipped with an air bag, the negative battery cable must be disconnected and at least three minutes must elapse before working on the system. Failure to do so may result in deployment of the air bag.

2. Remove the battery top cover, if equipped.
3. Disconnect the negative battery cable. Tape the cable with insulating tape.
4. Disconnect AFS and breather hose.
5. Remove air cleaner body and intake hose.
6. Disconnect RH oxygen sensor connector.
7. Disconnect RH injector connector and ignition coil connector.
8. Disconnect PCSV connector, MAP sensor connector and PCSV hose.
9. Disconnect ETC connector and knock sensor connector.
10. Disconnect water hoses from ETC.
11. Disconnect PCV hose.
12. Disconnect brake vacuum hose.
13. Remove surge tank stay.
14. Remove connector bracket from surge tank or connectors(2EA).
15. Remove surge tank.
16. Disconnect breather Pipe assembly.
17. Disconnect LH injector connector.
18. Remove intake manifold and gasket.

To install:

19. Install intake manifold and new gasket on the cylinder head. Tightening torque is done in the following steps:
    - 1: 4 ft. lbs. (6 Nm)
    - 2: 14–17 ft. lbs. (19–24 Nm)
    - 3: Repeat 2nd step twice

➡Be careful of the installation direction.

- a—h: 1st step order
- 1—8: 2nd step order

20. Install delivery pipe.
21. Connect LH injector connector.
22. Connect breather pipe assembly. Tightening torque: 7–9 ft. lbs. (10–12 Nm).
23. Install surge tank. Tightening torque: 7–9 ft. lbs. (10–12 Nm)
24. Install connector bracket on the surge tank.
25. Install surge tank stay.
26. Connect brake vacuum hose.

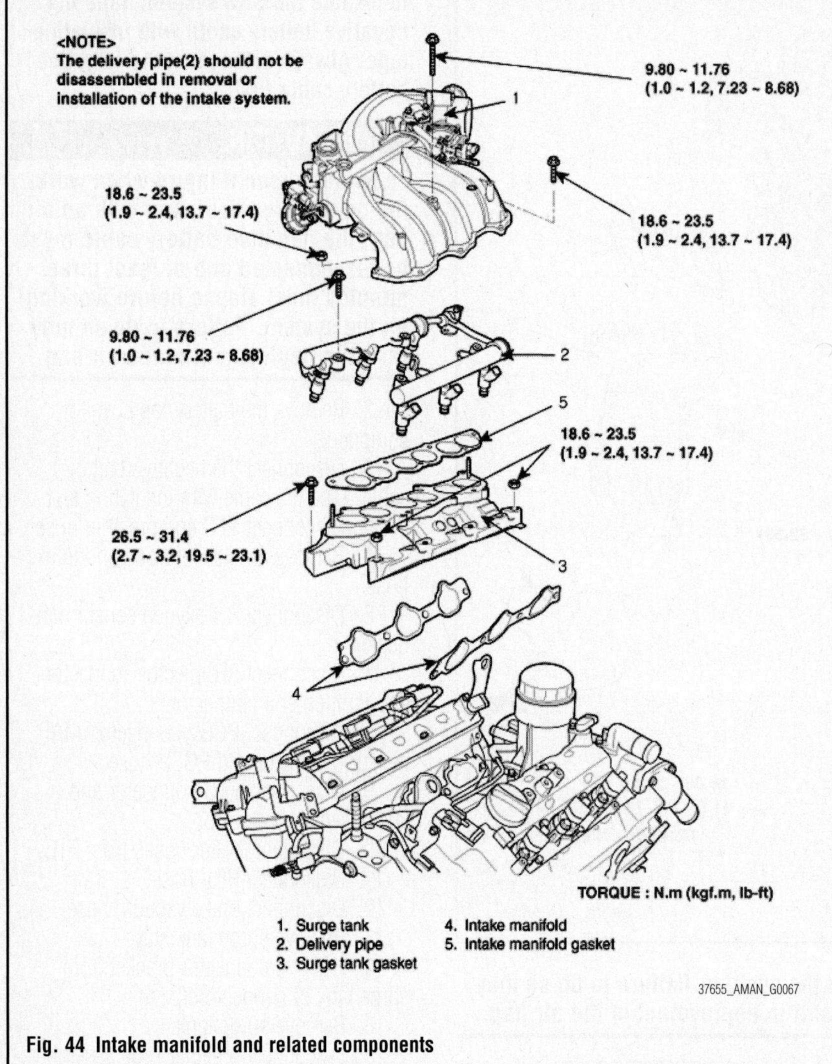

Fig. 44 Intake manifold and related components

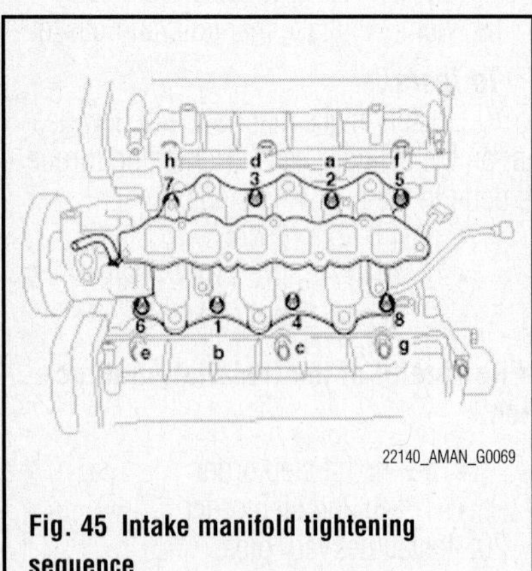

Fig. 45 Intake manifold tightening sequence

27. Connect PCV hose.
28. Connect water hoses to ETC.
29. Connect ETC connector and knock sensor connector.
30. Connect PCSV connector, MAP sensor connector and PCSV hoe.
31. Connect RH injector connector and ignition coil connector.
32. Connect RH oxygen sensor connector.
33. Install air cleaner upper cover and intake hose.
34. Connect AFS and breather hose.

OIL PAN

REMOVAL & INSTALLATION

See Figures 46 and 47.

1. Before servicing the vehicle, refer to the Precautions Section.

➡If working near and/or around the SRS system and components, be sure to disable the SRS system. Tape the negative battery cable with insulating tape. Always disconnect the negative battery cable first.

CAUTION

To avoid personal injury when working on vehicles equipped with an air bag, the negative battery cable must be disconnected and at least three minutes must elapse before working on the system. Failure to do so may result in deployment of the air bag.

2. Remove the battery top cover, if equipped.
3. Disconnect the negative battery cable. Tape the cable with insulating tape.
4. Raise and support the vehicle safely.
5. Drain the engine oil.
6. Remove the oil pan retaining bolts.
7. Using tool 09215-3C000, or equivalent, remove the oil pan.

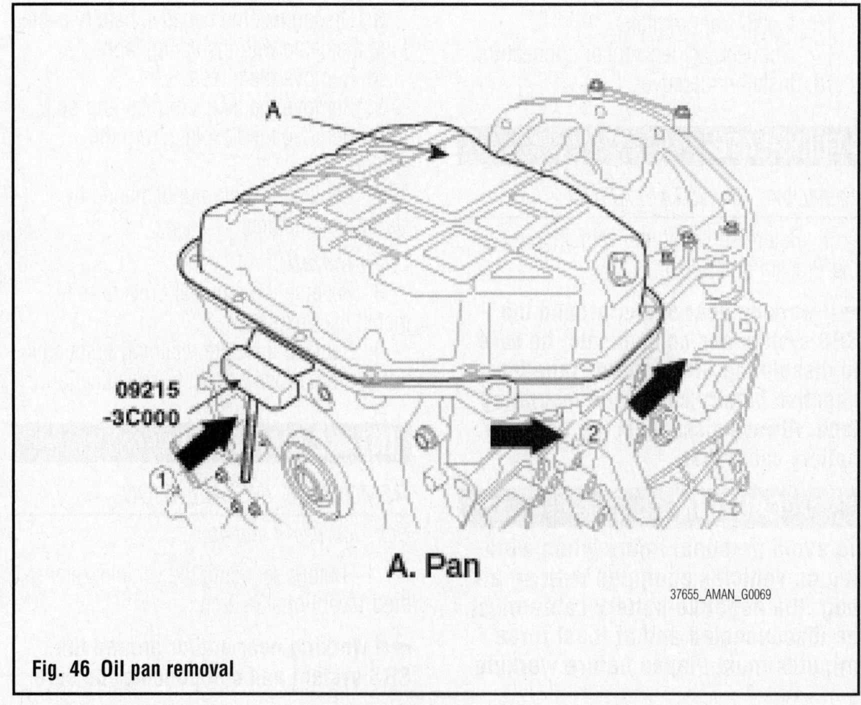

Fig. 46 Oil pan removal

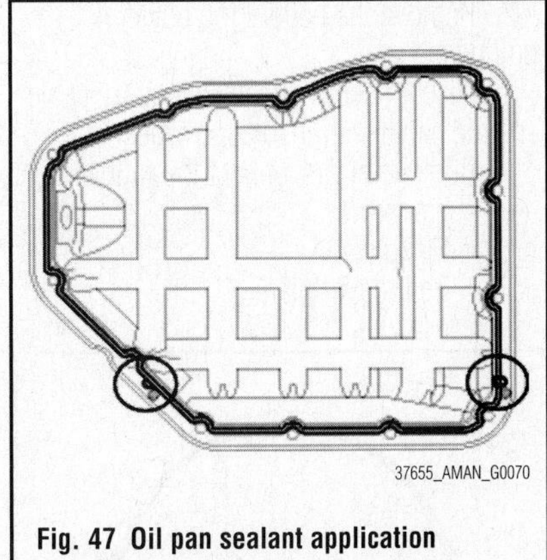

Fig. 47 Oil pan sealant application

➡**Be sure not to damage the contact surfaces of the upper and lower oil pan.**

To install:

8. Clean all gasket surfaces of the oil pans and cylinder block.
9. Apply sealant (TB1217H) to the upper oil pan. Component must be installed within five minutes after sealant application.
10. Bead with should be 0.1 inch, except marked area (see illustration) should be 0.2 inch.
11. Install lower oil pan. Tighten bolts to specification uniformly and in several passes.
12. Fill the engine with clean oil.
13. Start the vehicle and check for leaks.

OIL PUMP

REMOVAL & INSTALLATION

See Figure 48.

1. Before servicing the vehicle, refer to the Precautions Section.

➡**If working near and/or around the SRS system and components, be sure to disable the SRS system. Tape the negative battery cable with insulating tape. Always disconnect the negative battery cable first.**

⁂ CAUTION

To avoid personal injury when working on vehicles equipped with an air bag, the negative battery cable must be disconnected and at least three minutes must elapse before working on the system. Failure to do so may result in deployment of the air bag.

2. Remove the battery top cover, if equipped.
3. Disconnect the negative battery cable. Tape the cable with insulating tape.
4. Drain engine oil.

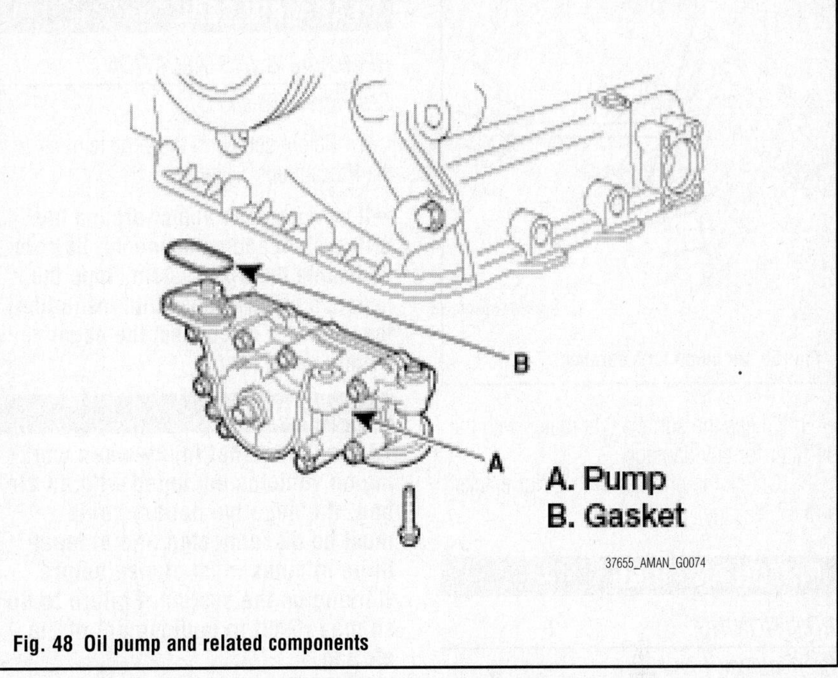

Fig. 48 Oil pump and related components

5. Using SST(09215-3C000) remove lower oil pan.

➡**Be careful not to damage the contact surfaces of upper oil pan and lower oil pan.**

6. Remove oil pump chain cover.
7. Remove oil pump chain sprocket.
8. Remove oil pump.

To install:

9. Install oil pump. Tightening torque: 15–17 ft. lbs. (21–23 Nm).

➡**Always use a new O-ring.**

10. Install oil pump sprocket and oil pump chain on the oil pump. Tightening torque: 14–16 ft. lbs. (19–22 Nm).
11. Install oil pump chain cover. Tightening torque: 7–9 ft. lbs. (10–12 Nm).
12. Install lower oil pan as follows:
    a. Clean the sealing face before assembling two parts.
    b. Remove harmful foreign materials on the sealing face before applying sealant.
    c. When applying sealant gasket, sealant must not be protrude into the inside of oil pan.
    d. To prevent leakage of oil, apply sealant gasket to the inner threads of the bolt holes.
13. Install lower oil pan.
14. Uniformly tighten the bolts in several passes. Tightening torque: 7–9 ft. lbs. (10–12 Nm).

➡**After assembly, wait at least 30 minutes before filling the engine with oil.**

INSPECTION

See Figures 49 and 50.

1. With the oil pump removed, visually check the parts of the oil pump case for cracks or damage.
2. Assemble the rotor on the oil pump and then check the following clearances with a feeler gauge:
    a. Body clearance: 0.0039–0.0071 inch. (0.100–0.181mm).
    b. Side clearance: 0.0016–0.0037 inch. (0.040–0.095mm).
    c. Oil pump tip clearance: 0.0024–0.0071 inch. (0.06–0.18mm).
3. Check the relief plunger for smooth operation.
4. Check the relief spring for deformation or a break.

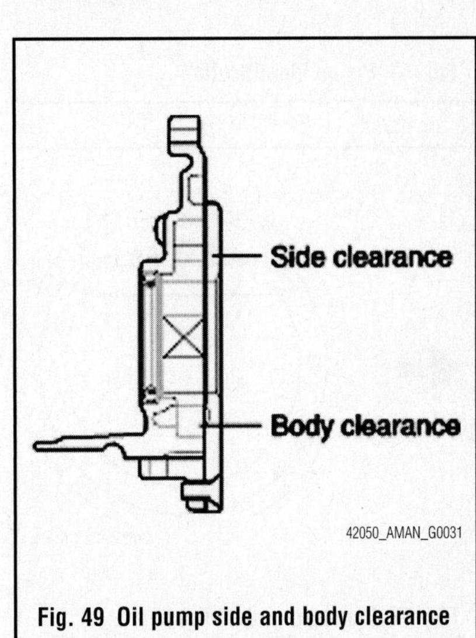

Fig. 49 Oil pump side and body clearance

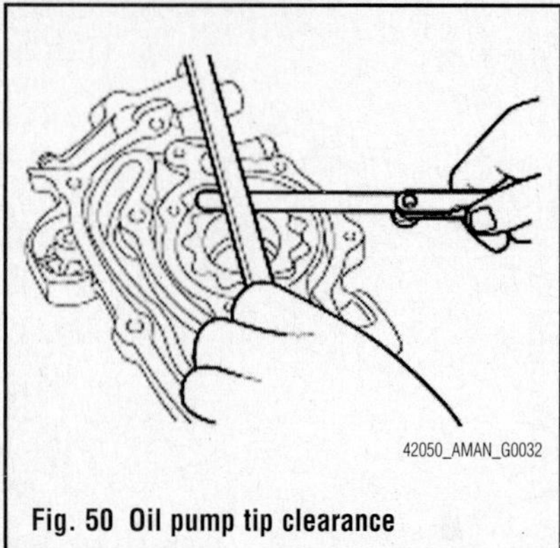

Fig. 50 Oil pump tip clearance

5. Check the surface that mates with the oil filter for any damage.
6. Check the oil filter bracket for cracks and oil leakage.

PISTON AND RING

POSITIONING

See Figures 51 and 52.

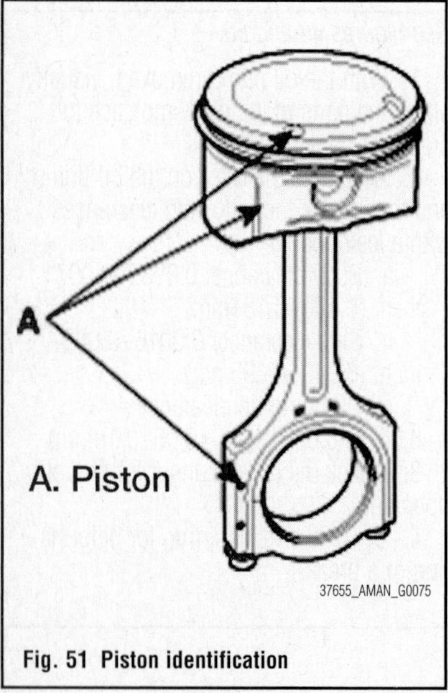

Fig. 51 Piston identification

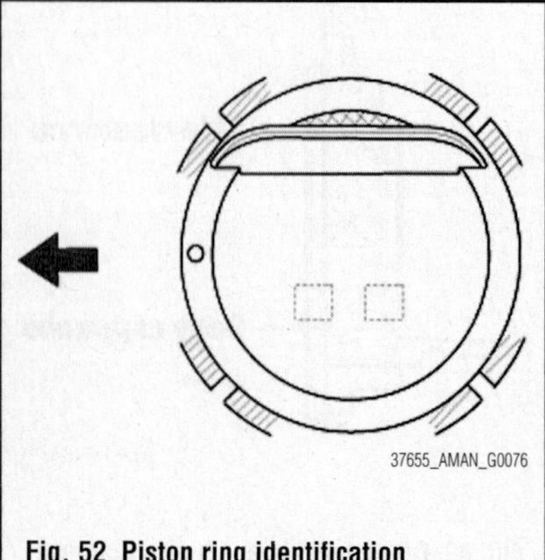

Fig. 52 Piston ring identification

REAR MAIN SEAL

REMOVAL & INSTALLATION

See Figures 53 and 54.

1. Before servicing the vehicle, refer to the Precautions Section.

➡If working near and/or around the SRS system and components, be sure to disable the SRS system. Tape the negative battery cable with insulating tape. Always disconnect the negative battery cable first.

⁂ CAUTION

To avoid personal injury when working on vehicles equipped with an air bag, the negative battery cable must be disconnected and at least three minutes must elapse before working on the system. Failure to do so may result in deployment of the air bag.

2. Remove the battery top cover, if equipped.
3. Disconnect the negative battery cable. Tape the cable with insulating tape.
4. Remove or disconnect the following:
   - Transaxle
   - Flexplate
   - Rear cover plate

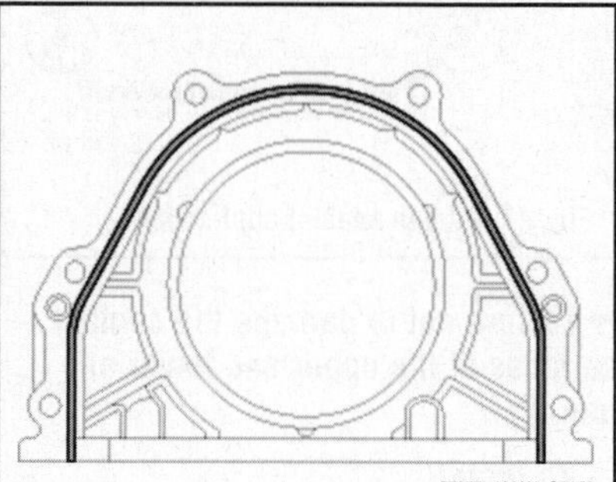

Fig. 54 Rear case sealant application location

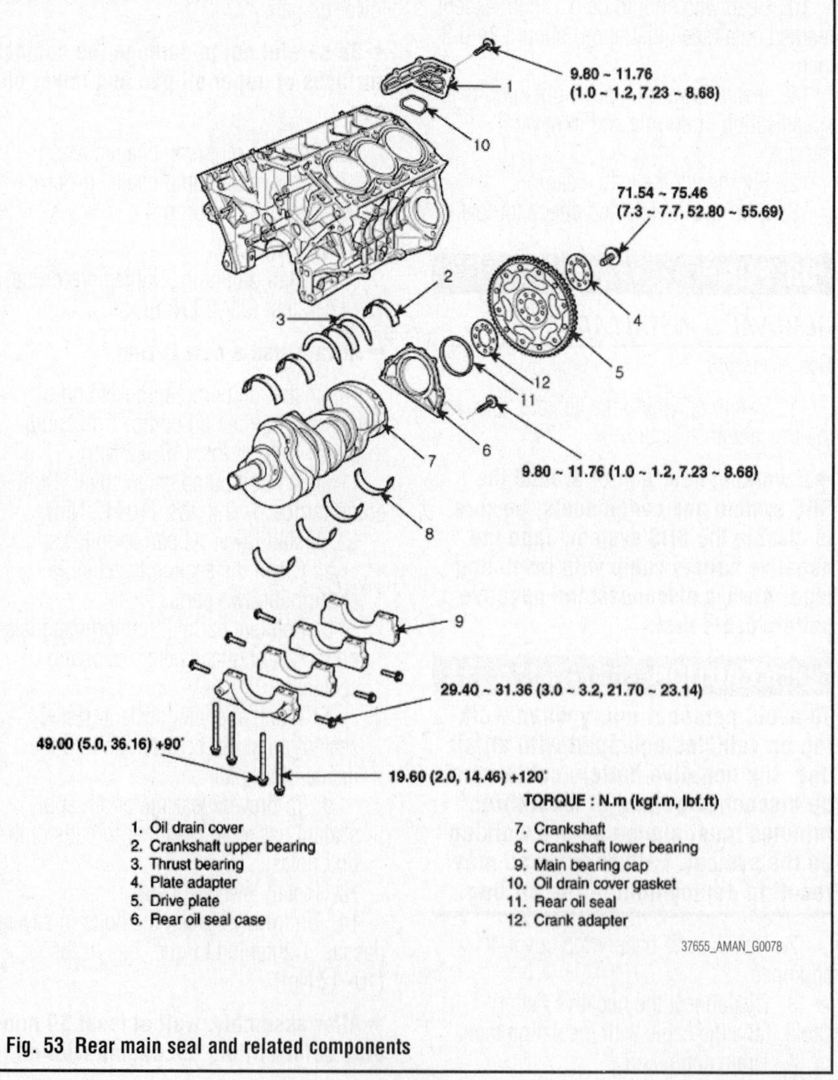

Fig. 53 Rear main seal and related components

- Oil seal case
- Oil seal

To install:

5. Install the oil seal to the oil seal case, using Seal Installer tool 09231-33000.
6. Apply a silicone sealant to the oil seal case as shown and Torque the case bolts to 8–9 ft. lbs. (10–12 Nm).
7. Install or connect the following:
   - Rear plate
   - Flexplate
   - Transaxle

TIMING CHAIN FRONT COVER

REMOVAL & INSTALLATION

See Figures 55 through 69.

1. Before servicing the vehicle, refer to the Precautions Section.

➡If working near and/or around the SRS system and components, be sure to disable the SRS system. Tape the negative battery cable with insulating tape. Always disconnect the negative battery cable first.

⁂ CAUTION

To avoid personal injury when working on vehicles equipped with an air bag, the negative battery cable must be disconnected and at least three minutes must elapse before working on the system. Failure to do so may result in deployment of the air bag.

2. Remove the battery top cover, if equipped.
3. Disconnect the negative battery cable. Tape the cable with insulating tape.
4. Remove the engine cover.
5. Remove the intake air hose and air cleaner assembly.

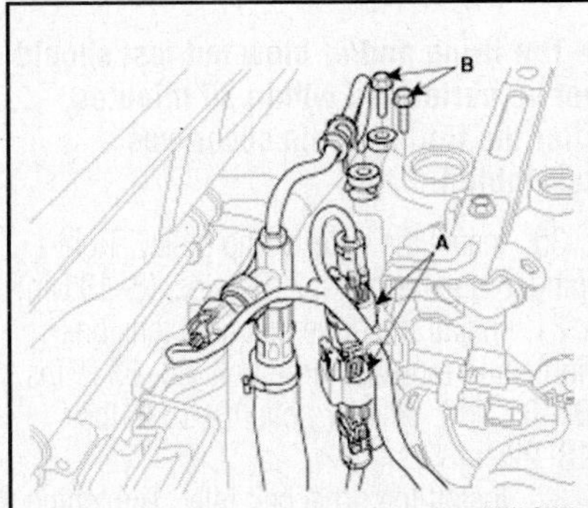

Fig. 55 Disconnect the RH oxygen sensor connector (A) and loosen the power steering hose mounting bolts (B)

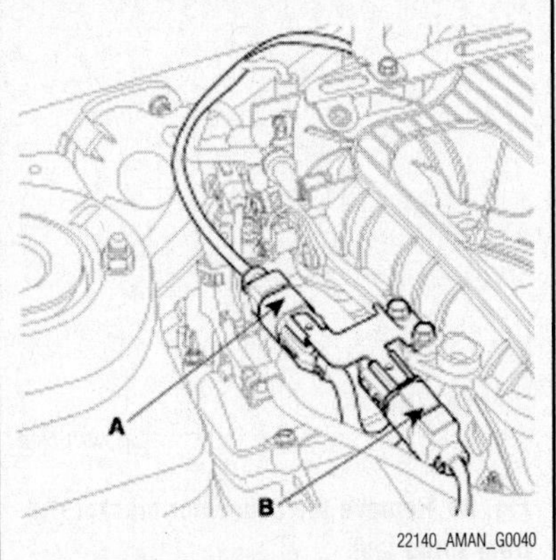

Fig. 56 Disconnect the RH injector connector (A) and ignition coil connector (B)

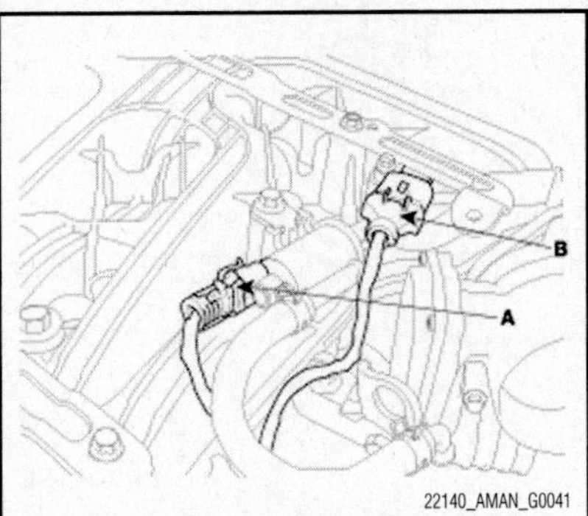

Fig. 57 Disconnect the PCSV connector (A), MAP sensor connector (B) and PCSV hose

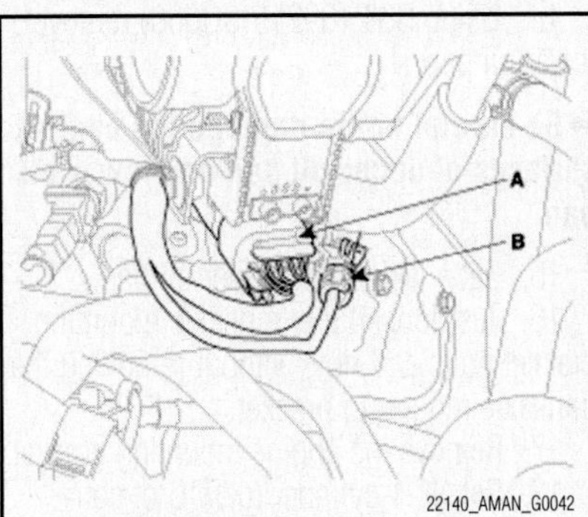

Fig. 58 Disconnect the ETC connector (A) and knock sensor connector (B)

6. Remove the RH front wheel.
7. Remove the undercover.
8. Remove the side cover.
9. Loosen the drain plug and drain the engine coolant.
10. Drain the engine oil.
11. Loosen the power steering oil cooler return pipe mounting bolt.
12. Remove the surge tank by the following:
    - Disconnect the RH oxygen sensor connector (A) and loosen the power steering hose mounting bolts (B).
    - Disconnect the RH injector connector (A) and ignition coil connector (B).
    - Disconnect the PCSV connector (A), MAP sensor connector (B) and PCSV hose.

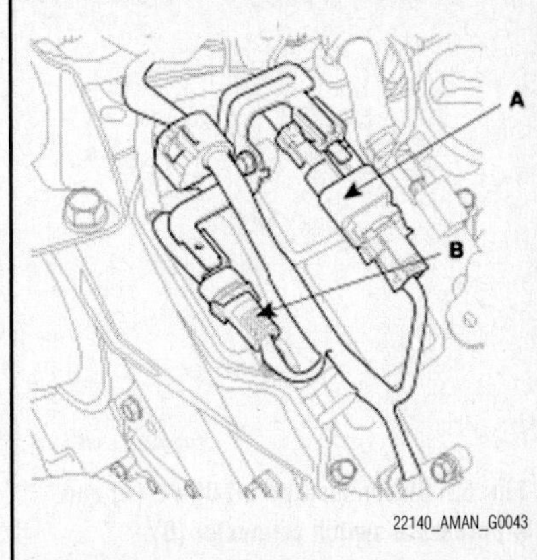

Fig. 59 Disconnect the OCV connector (A) and knock sensor connector (B)

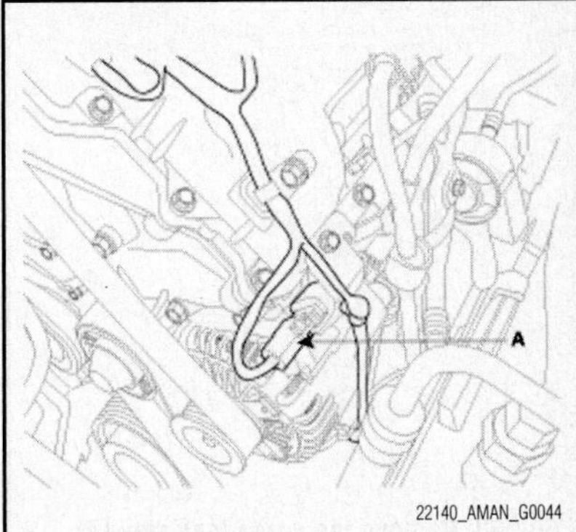

Fig. 60 Disconnect the LH front oxygen sensor connector (A)

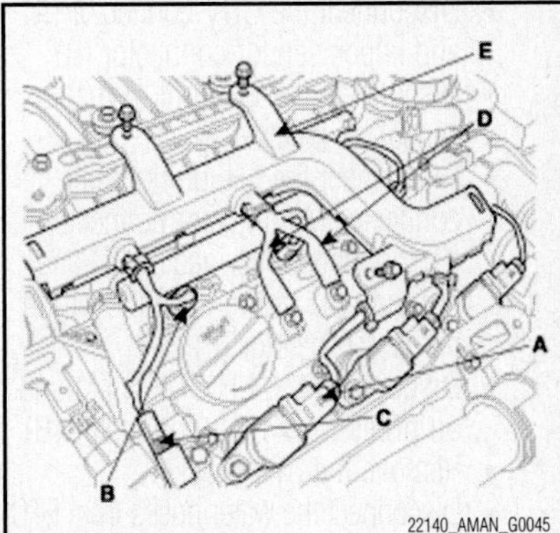

Fig. 61 Disconnect the LH ignition coil connector (A), injector connector (B), condenser connector (C) and ground (D), and remove the wiring harness protector (E)

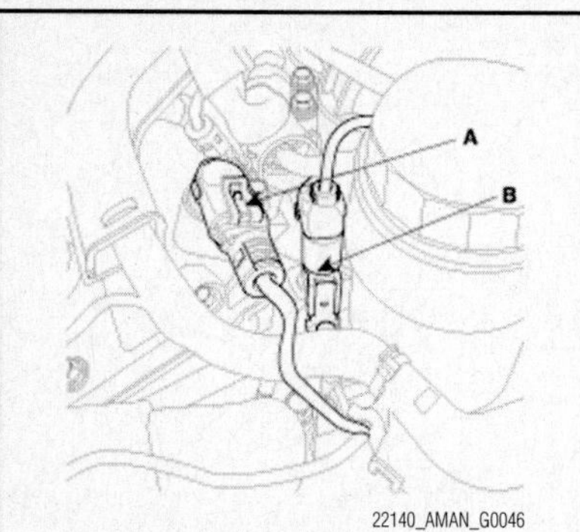

Fig. 62 Disconnect the LH CMPS (A) and oil pressure switch connector (B)

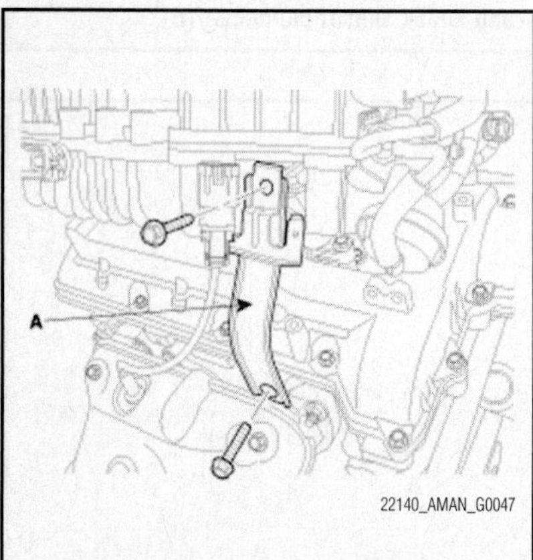

Fig. 63 Remove the surge tank stay (A)

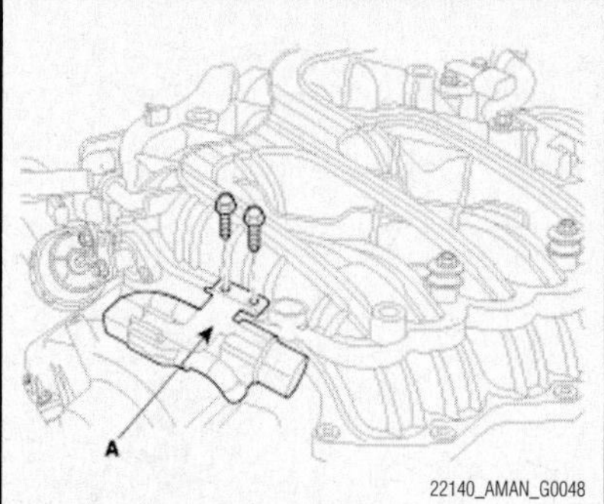

Fig. 64 Remove the connector bracket (A) from surge tank

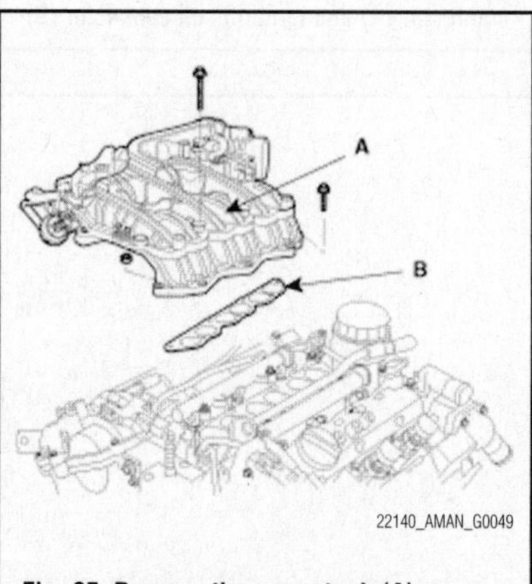

Fig. 65 Remove the surge tank (A)

- Disconnect the ETC connector (A) and knock sensor connector (B).
- Disconnect the OCV connector (A) and knock sensor connector (B).
- Disconnect the LH front oxygen sensor connector (A).
- Disconnect the LH ignition coil connector (A), injector connector (B), condenser connector (C) and ground (D), and remove the wiring harness protector (E).
- Disconnect the LH CMPS (A) and oil pressure switch connector (B)
- Remove the ETC bracket
- Disconnect the water hoses from ETC
- Disconnect the PCV hose
- Disconnect the brake vacuum hose
- Remove the surge tank stay (A)
- Remove the connector bracket (A) from surge tank
- Remove the surge tank (A).

13. Remove the cylinder head covers by the following:
    - Remove the connector bracket (A) from LH cylinder head cover
    - Disconnect the RH ignition coil connector, condenser connector and remove the wiring bracket.
    - Remove the LH, RH ignition coil
    - Remove the LH, RH cylinder head cover
14. Using SST (09215-3C000) remove lower oil pan.

➡Be careful not to damage the contact surfaces of upper oil pan and lower oil pan.

15. Set a jack to the upper oil pan.
16. Just loosen the transaxle mounting bracket bolts and nuts without removing the transaxle mounting bracket.
17. Remove the engine mounting bracket.
18. Set No.1 cylinder to TDC of compression stroke.
    a. Turn the crankshaft pulley and align its groove with the timing mark "T" of the lower timing chain cover.

➡Do not rotate engine counterclockwise.

    b. Check that the mark (A) of the camshaft timing sprockets are in straight line on the cylinder head surface as shown in the illustration. If not, turn the crankshaft one revolution (360°).
19. Remove the drive belt.
20. Using SST (09231-3C300) remove the crankshaft damper pulley.
21. Lift up the engine assembly to using the jack.
22. Remove the power steering pump.
23. Remove the alternator.
24. Remove the drive belt idler.
25. Remove the drive belt auto tensioner.
26. Remove the water pump pulley.
27. Remove the timing chain cover. If necessary remove the water pump first.

To install:

28. Install the timing chain cover by the following:
    - The sealant locations on chain cover and on counter parts (cylinder head, cylinder block, and lower oil pan) must be free of engine oil and ETC.
    - Before assembling the timing chain cover, the liquid sealant TB1217H should be applied on the gap between cylinder head and cylinder block.
    - After applying liquid sealant TB1217H on the timing chain cover, the part must be assembled within 5 minutes after sealant was applied.
    - Install the new gasket to the timing chain cover.
    - The dowel pins on the cylinder block and holes on the timing chain cover should be used as a reference in order to assemble the timing chain cover to be in exact position.
29. Tightening torques for timing chain cover are as follows:
    - B (17 ea): 14–16 ft. lbs. (19–22 Nm)
    - C (4 ea): 7–9 ft. lbs. (10–12 Nm)
    - D, E: (1 ea): 43–51 ft. lbs. (59–69 Nm)
    - F (2 ea): 19 ft. lbs. (26 Nm)
    - G (4 ea): 17 ft. lbs. (22 Nm)
    - H, I, J, K (1 ea): 7–9 ft. lbs. (10–12 Nm)
    - L (1 ea): 16–20 ft. lbs. (22–26 Nm)–New bolt

➡The firing and/or blow out test should not be performed within 30 minutes after the timing chain cover was assembled.

30. Install the water pump pulley. Tightening torque for bolts: 7–9 ft. lbs. (8–10 Nm)
31. Install the drive belt auto tensioner. Tightening torques: large bolt: 60–63 ft. lbs. (81–85 Nm); smaller bolt: 13–16 ft. lbs. (18–22 Nm)
32. Install the drive belt idler. Tightening torque: 39–43 ft. lbs. (53–58 Nm)
33. Install the alternator. Tightening torque: 20–25 ft. lbs. (26–33 Nm)
34. Install the power steering pump.

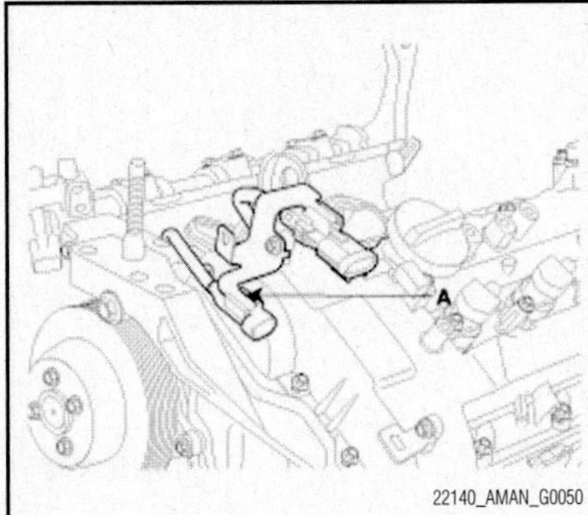

Fig. 66 Remove the connector bracket (A) from LH cylinder head cover

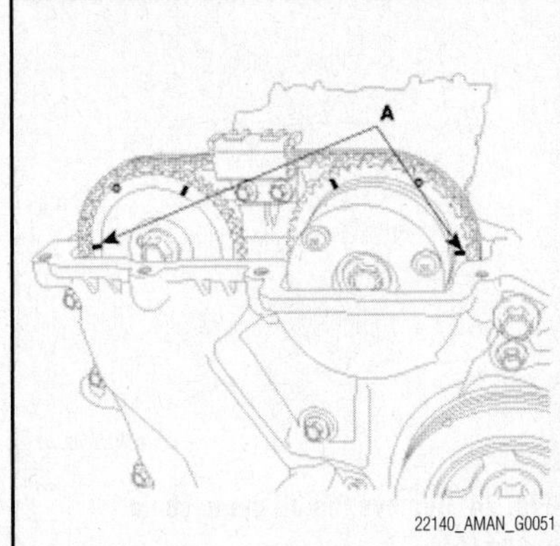

Fig. 67 Check that the mark (A) of the camshaft timing sprockets are in straight line on the cylinder head

35. Lower the engine assembly by using the jack.

36. Using SST (09231-3C100), install timing chain cover oil seal.

37. Using SST (09231-3C300) install the crankshaft damper pulley. Tightening torque: 210–224 ft. lbs. (284–304 Nm)

38. Install the drive belt.

39. After putting belt on auto tensioner pulley, release the auto tensioner pulley slowly.

40. Install the cylinder head cover by the following:

- The hardening sealant located on the upper area between timing chain cover and cylinder head should be removed before assembling cylinder head cover.
- After applying sealant (TB1217H), it should be assembled within 5 minutes.

➡The firing and/or blow out test should not be performed within 30 minutes after the cylinder head cover was assembled.

- Install the cylinder head cover bolts as shown. Tightening

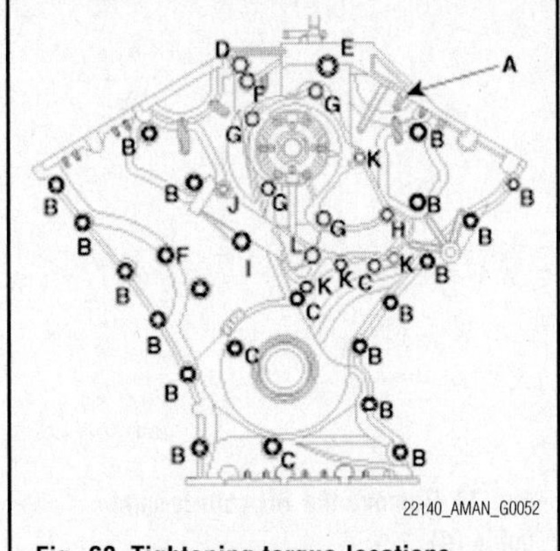

Fig. 68 Tightening torque locations

torque: 7.23–8.68ft. lbs. (9.80–11.76 Nm)
- Install the ignition coil
- Connect the RH ignition coil connector, the condenser connector and install the wiring bracket
- Install the connector bracket to the LH cylinder head cove.

41. Install the surge tank and wiring connectors by the following:

- Install the surge tank. Tightening torque: 7–9 ft. lbs. (10–12 Nm)
- Install the connector bracket to the surge tank. Tightening torque: 5–8 ft. lbs. (7–11 Nm)
- Install the surge tank stay. Tightening torque: 20–23 ft. lbs. (27–31 Nm)
- Connect the brake vacuum hose
- Connect the PCV hose (C)
- Connect the water hoses to the ETC
- Install the ETC bracket
- Connect the LH CMPS and oil pressure switch connector
- Install the wiring harness protector and connect the LH ignition coil connector, injector connector, condenser connector and ground
- Connect the LH front oxygen sensor connector
- Connect the OCV connector and knock sensor connector
- Connect the ETC connector and knock sensor connector
- Connect the PCSV connector, MAP sensor connector and PCSV hose
- Connect the RH injector connector and ignition coil connector
- Connect the RH oxygen sensor connector and tighten the power steering hose mounting bolts

42. Install the engine mounting bracket. Tightening torque: Side bolt: 65–80 ft. lbs. (88–108 Nm); other two bolts and one nut: 43–58 ft. lbs. (59–79 Nm)

43. Install the transaxle mounting bracket bolts and nuts. Tightening torque: 43–58 ft. lbs. (59–79 Nm)

44. Remove the jack from the upper oil pan.

45. Install the lower oil pan by the following:

- Using a gasket scraper, remove all the old packing material from the gasket surfaces
- Before assembling the oil pan, the liquid sealant TB1217H should be applied on oil pan. The part must be assembled within 5 minutes after the sealant was applied.

⁂ CAUTION

Be sure to do the following:

a. Make clean the sealing face before assembling two parts.

b. Remove harmful foreign matters on the sealing face before applying sealant.

c. When applying sealant gasket, sealant must not be protruded into the inside of oil pan.

d. To prevent leakage of oil, apply sealant gasket to the inner threads of the bolt holes.

- Install the lower oil pan. Tightening torque: 7–9 ft. lbs. (10–12 Nm)

46. Tighten the power steering oil cooler return pipe mounting bolt.

47. Install the side cover and the under-cover. Tightening torque: 6–8 ft. lbs. (9–11 Nm)

48. Install the RH front wheel.

49. Install the intake air hose and air cleaner assembly as follows:

- Install the intake air hose and air cleaner body.

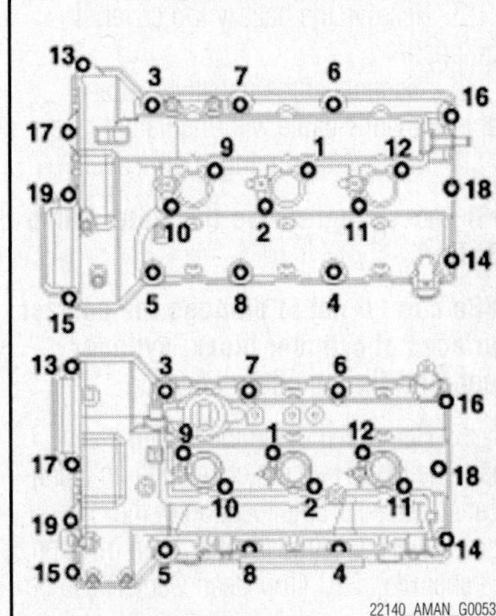

Fig. 69 Install the cylinder head cover bolts as shown

- Connect the breather hose to the air cleaner hose.
- Connect the AFS connector.

50. Install the engine cover.
51. Connect the battery negative cable.
52. Refill engine with engine oil.
53. Refill radiator and reservoir tank with engine coolant.
54. Bleed air from the cooling system.
    a. Start engine and let it run until it warms up. (until the radiator fan operates 3 or 4 times.)
    b. Turn Off the engine. Check the level in the radiator, add coolant if needed. This will allow trapped air to be removed from the cooling system.
    c. Put radiator cap on tightly, then run the engine again and check for leaks.

TIMING CHAIN & SPROCKETS

REMOVAL & INSTALLATION

See Figures 70 through 77.

1. Before servicing the vehicle, refer to the Precautions Section.

➡If working near and/or around the SRS system and components, be sure to disable the SRS system. Tape the negative battery cable with insulating tape. Always disconnect the negative battery cable first.

⁂ CAUTION

To avoid personal injury when working on vehicles equipped with an air bag, the negative battery cable must be disconnected and at least three minutes must elapse before working on the system. Failure to do so may result in deployment of the air bag.

2. Remove the battery top cover, if equipped.
3. Disconnect the negative battery cable. Tape the cable with insulating tape.
4. Remove the timing chain cover.

➡If necessary remove the water pump (B) first.

➡Be careful not to damage the contact surfaces of cylinder block, cylinder head and timing chain cover.

5. Before removing the timing chain, mark the RH/LH timing chain with an identification based on the location of the sprocket because the identification mark on the chain for TDC (Top Dead Center) can be erased.
6. Install a set pin after compressing the timing chain tensioner.

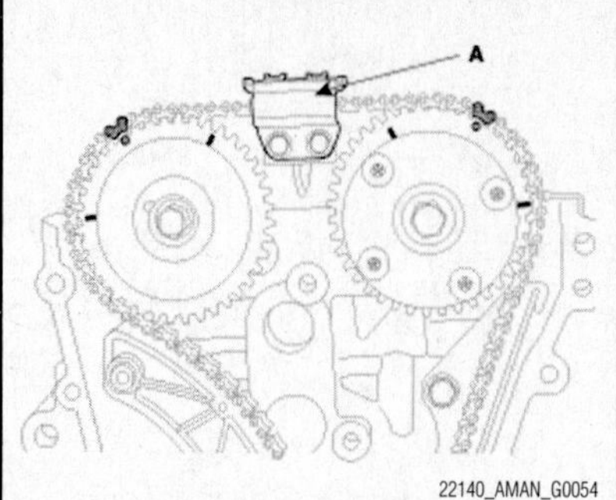

Fig. 70 Remove the RH cam-to-cam guide (A)

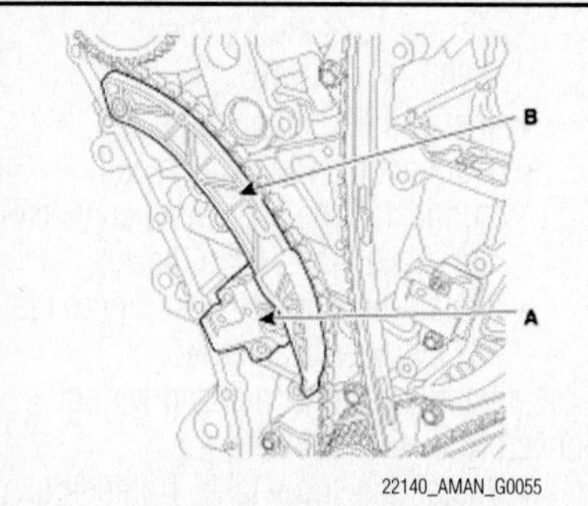

Fig. 71 Remove the RH timing chain auto tensioner (A) and RH timing chain tensioner arm (B)

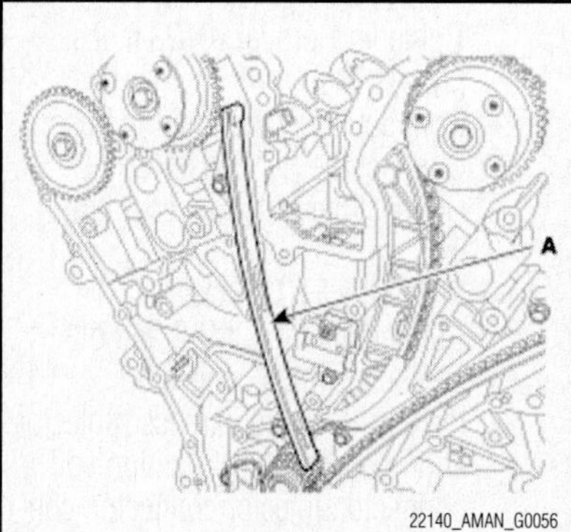

Fig. 72 Remove the RH timing chain guide (A)

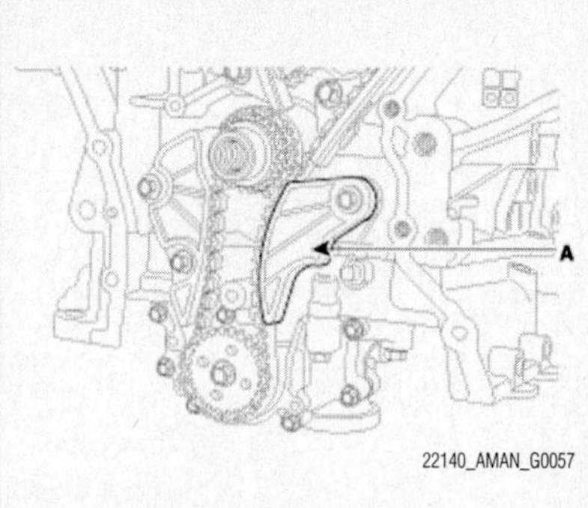

Fig. 73 Remove the oil pump chain tensioner assembly (A)

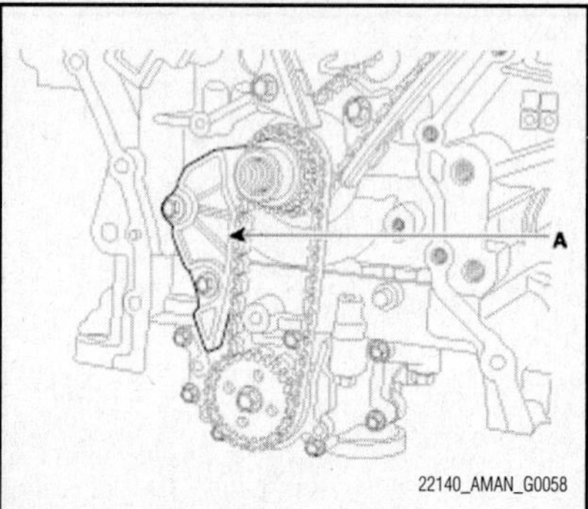

Fig. 74 Remove the oil pump chain guide (A)

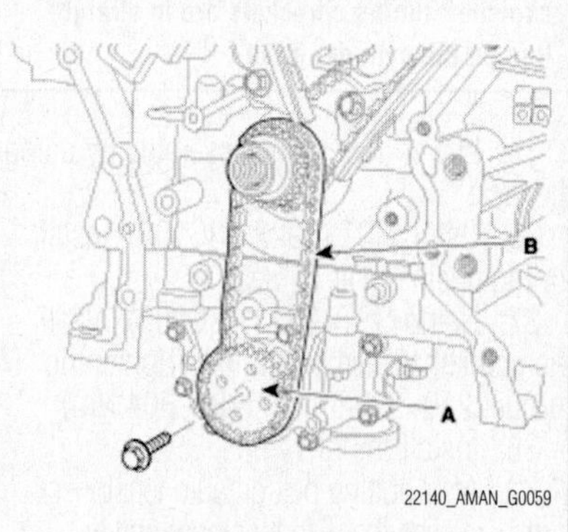

Fig. 75 Remove the oil pump chain sprocket (A) and oil pump chain (B)

7. Remove the RH cam-to-cam guide (A).
8. Remove the RH timing chain auto tensioner (A) and RH timing chain tensioner arm (B).
9. Remove the RH timing chain.
10. Remove the RH timing chain guide (A).
11. Remove the oil pump chain cover.
12. Remove the oil pump chain tensioner assembly (A).
13. Remove the oil pump chain guide (A).
14. Remove the oil pump chain sprocket (A) and oil pump chain (B).
15. Remove the crankshaft sprocket (A) (Oil pump & RH camshaft drive).
16. Install a set pin after compressing the LH timing chain tensioner.
17. Remove the LH cam-to-cam guide.
18. Remove the LH timing chain auto tensioner and LH timing chain tensioner arm.
19. Remove the LH timing chain.
20. Remove the LH timing chain guide.
21. Remove the crankshaft sprocket (LH camshaft drive).
22. Remove the tensioner adapter assembly.

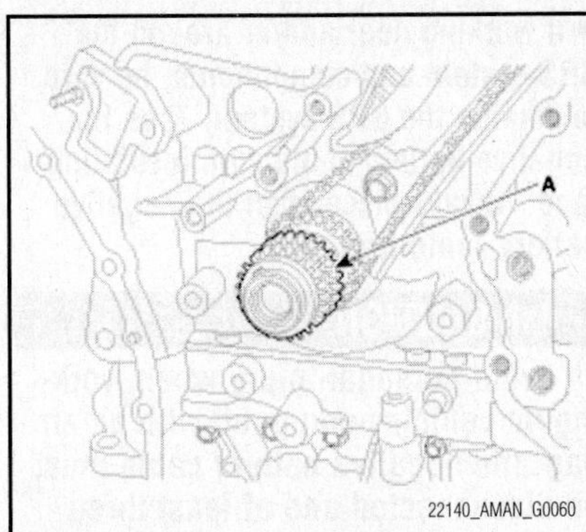

Fig. 76 Remove the crankshaft sprocket (A)

To install:

23. Check the camshaft sprocket and crankshaft sprocket for abnormal wear, cracks, or damage. Replace as necessary.
24. Inspect the tensioner arm and chain guide for abnormal wear, cracks, or damage. Replace as necessary.
25. Check that the tensioner piston moves smoothly when the ratchet pawl is released with thin rod.
26. Install the jack to the upper oil pan.
27. The key (A) of crankshaft should be aligned with the timing mark (B) of timing chain cover. As a result of this, the piston of No.1 cylinder is placed at the top dead center on compression stroke.
28. Install the tensioner adapter assembly.
29. Install the crankshaft sprocket (LH camshaft drive).
30. Install the LH timing chain guide. Tightening torque: 15–18 ft. lbs. (20–25 Nm)
31. Install LH timing chain. To install the timing chain with no slack between each shaft (cam, crank), follow the procedure below:
    - Crankshaft sprocket
    - Timing chain guide
    - Exhaust camshaft sprocket
    - Intake camshaft sprocket

➡The timing mark of each sprocket should be matched with timing mark (color link) of the timing chain when installing the timing chain.

32. Install the LH timing chain tensioner arm. Tightening torque: 14–16 ft. lbs. (19–22 Nm).
33. Install the LH chain tensioner. Tightening torque: 7–9 ft. lbs. (10–12 Nm).
34. Install the LH cam-to-cam guide. Tightening torque: 7–9 ft. lbs. (10–12 Nm).
35. Install the crankshaft sprocket (Oil pump & RH camshaft drive).
36. Install the oil pump chain and oil pump sprocket. Tightening torque: 15 ft. lbs. (20 Nm)
37. Install the RH timing chain guide. Tightening torque: 14–18 ft. lbs. (20–25 Nm)
38. Install the RH timing chain. To install the timing chain with no slack between each shaft (cam, crank), follow the procedure below:
    - Crankshaft sprocket
    - Intake camshaft sprocket
    - Exhaust camshaft sprocket

➡The timing mark of each sprocket should be matched with timing mark (color link) of timing chain at installing timing chain.

39. Install the RH timing chain tensioner arm. Tightening torque: 14–16 ft. lbs. (19–22 Nm)
40. Install the RH timing chain auto tensioner. Tightening torque: 7–9 ft. lbs. (10–12 Nm).
41. Install the RH cam-to-cam guide. Tightening torque: 7–9 ft. lbs. (10–12 Nm).

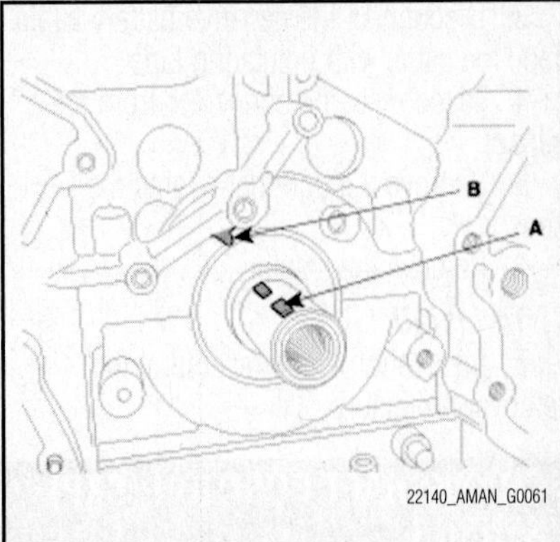

Fig. 77 The key (A) of crankshaft should be aligned with the timing mark (B) of timing chain cover

42. Install the oil pump chain guide. Tightening torque: 7–9 ft. lbs. (10–12 Nm).
43. Install the oil pump chain tensioner assembly. Tightening torque: 7–9 ft. lbs. (10–12 Nm).
44. Pull out the pins of hydraulic tensioners (LH & RH).
45. Install the oil pump chain cover. Tightening torque: 7–9 ft. lbs. (10–12 Nm).
46. After rotating crankshaft 2 revolutions in regular direction (clockwise viewed from front), confirm the timing mark.

➡Always turn the crankshaft clockwise.

47. Install the timing chain cover

VALVE LASH

ADJUSTMENT

This vehicle uses hydraulic valve lash adjusters. Valve lash adjustments are not necessary.

ENGINE PERFORMANCE & EMISSION CONTROLS

CAMSHAFT POSITION (CMP) SENSOR

LOCATION

The two Camshaft Position (CMP) sensors are installed on engine head cover of bank 1 and 2 and uses a target wheel installed on the camshaft.

REMOVAL & INSTALLATION

See Figure 78.

1. Before servicing the vehicle, refer to the Precautions Section.

➡If working near and/or around the SRS system and components, be sure to disable the SRS system. Tape the negative battery cable with insulating tape. Always disconnect the negative battery cable first.

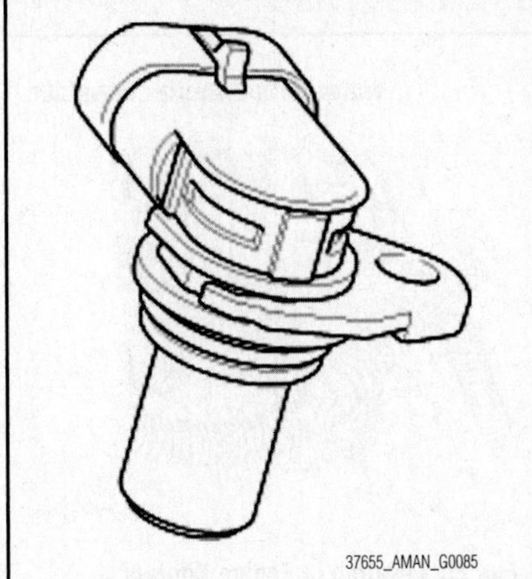

Fig. 78 Camshaft Position (CMP) sensor

⁂ CAUTION

To avoid personal injury when working on vehicles equipped with an air bag, the negative battery cable must be disconnected and at least three minutes must elapse before working on the system. Failure to do so may result in deployment of the air bag.

2. Remove the battery top cover, if equipped.

3. Disconnect the negative battery cable. Tape the cable with insulating tape.
4. Disconnect the connector from the sensor.
5. Remove the bolt that retains the sensor.
6. Remove the sensor.

To install:

7. Installation is the reverse of the removal procedure.

CRANKSHAFT POSITION (CKP) SENSOR

LOCATION

The Crankshaft Position (CKP) Sensor is installed on transaxle housing.

REMOVAL & INSTALLATION

See Figure 79.

1. Before servicing the vehicle, refer to the Precautions Section.

➡If working near and/or around the SRS system and components, be sure to disable the SRS system. Tape the negative battery cable with insulating tape. Always disconnect the negative battery cable first.

⁂ CAUTION

To avoid personal injury when working on vehicles equipped with an air bag, the negative battery cable must be disconnected and at least three minutes must elapse before working on the system. Failure to do so may result in deployment of the air bag.

2. Remove the battery top cover, if equipped.

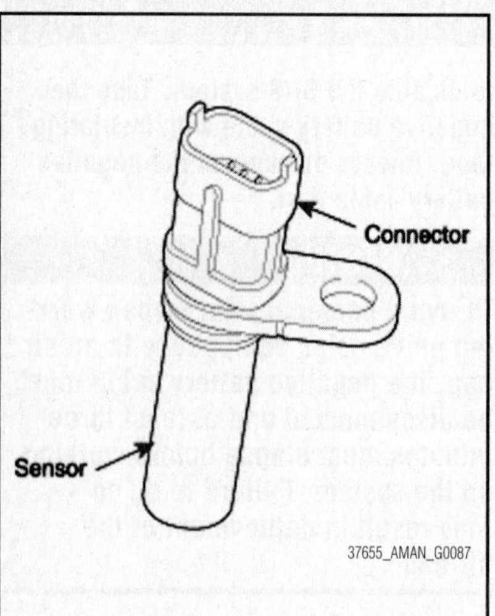

Fig. 79 Crankshaft Position (CKP) sensor

3. Disconnect the negative battery cable. Tape the cable with insulating tape.
4. Disconnect the connector from the sensor.
5. Remove the bolt that retains the sensor in place.
6. Remove the sensor from its mounting.

To install:

7. Installation is the reverse of the removal procedure.
8. Clearance between the sensor and the sensor wheel should be 0.020–0.059 inch.

ENGINE COOLANT TEMPERATURE (ECT) SENSOR

LOCATION

The Engine Coolant Temperature (ECT) Sensor is located in the engine coolant passage of the cylinder head for detecting the engine coolant temperature.

REMOVAL & INSTALLATION

See Figure 80.

⁂ CAUTION

Never open, service or drain the radiator or cooling system when hot; serious burns can occur from the steam and hot coolant. Also, when draining engine coolant, keep in mind that cats and dogs are attracted to ethylene glycol antifreeze and could drink any that is left in an uncovered container or in puddles on the ground. This will prove fatal in sufficient quantities. Always drain coolant into a sealable container. Coolant should be reused unless it is contaminated or is several years old.

1. Before servicing the vehicle, refer to the Precautions Section.

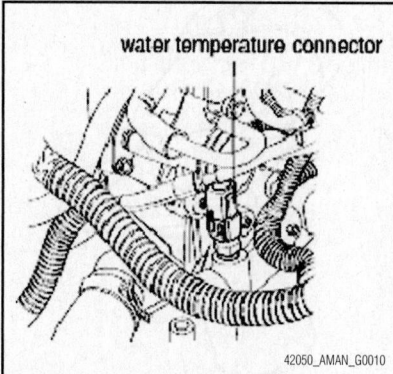

water temperature connector

42050_AMAN_G0010

Fig. 80 Location of Engine Coolant Temperature (ECT) sensor

➡If working near and/or around the SRS system and components, be sure to disable the SRS system. Tape the negative battery cable with insulating tape. Always disconnect the negative battery cable first.

⁂ CAUTION

To avoid personal injury when working on vehicles equipped with an air bag, the negative battery cable must be disconnected and at least three minutes must elapse before working on the system. Failure to do so may result in deployment of the air bag.

2. Remove the battery top cover, if equipped.
3. Disconnect the negative battery cable. Tape the cable with insulating tape.
4. Drain the engine coolant.
5. Remove the electrical connector from the sensor.
6. Remove the Engine Coolant Temperature (ECT) sensor.

To install:

7. Apply sealant to sensor threads. Install the sensor and tighten to 15–29 ft. lbs. (20–39 Nm).
8. Connect the coolant sensor to the harness.
9. Connect the ground cable of battery.
10. Refill the coolant.

HEATED OXYGEN SENSOR (HO2S)

LOCATION

The Heated Oxygen Sensor (HO2S) is installed on upstream and downstream of the Manifold Catalyst Converter (MCC).

REMOVAL & INSTALLATION

1. Before servicing the vehicle, refer to the Precautions Section.

➡If working near and/or around the SRS system and components, be sure to disable the SRS system. Tape the negative battery cable with insulating tape. Always disconnect the negative battery cable first.

⁂ CAUTION

To avoid personal injury when working on vehicles equipped with an air bag, the negative battery cable must be disconnected and at least three minutes must elapse before working

on the system. Failure to do so may result in deployment of the air bag.

2. Remove the battery top cover, if equipped.
3. Disconnect the negative battery cable. Tape the cable with insulating tape.
4. Raise and support the vehicle, as necessary.
5. Disconnect the electrical connector from the sensor.
6. Remove the oxygen sensor.

To install:

7. Installation is the reverse of the removal procedure.

➡Apply anti-seize compound to the threaded portion of the sensor, prior to installation. Never apply anti-seize compound to the protector of the sensor.

INTAKE AIR TEMPERATURE (IAT) SENSOR

LOCATION

See Figure 82.

The Intake Air Temperature Sensor (IATS) is installed inside the Mass Air Flow (MAF) sensor.

REMOVAL & INSTALLATION

See Figure 82.

1. Before servicing the vehicle, refer to the Precautions Section.

➡If working near and/or around the SRS system and components, be sure to disable the SRS system. Tape the negative battery cable with insulating tape. Always disconnect the negative battery cable first.

⁂ CAUTION

To avoid personal injury when working on vehicles equipped with an air bag, the negative battery cable must be disconnected and at least three minutes must elapse before working on the system. Failure to do so may result in deployment of the air bag.

2. Remove the battery top cover, if equipped.
3. Disconnect the negative battery cable. Tape the cable with insulating tape.
4. Disconnect the connector from the sensor.

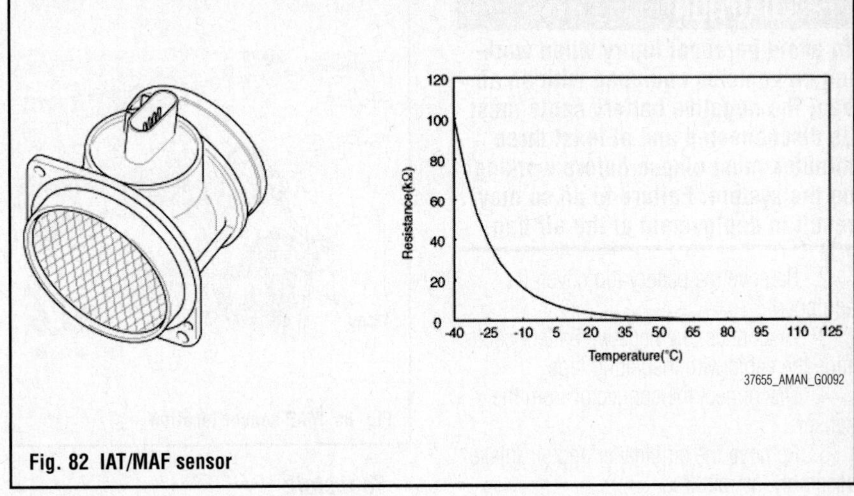

Fig. 82 IAT/MAF sensor

5. Remove the sensor retaining screws, as required.
6. Remove the air cleaner and air intake assembly, as required.
7. Remove the sensor from its mounting.

To install:

8. Installation is the reverse of the removal procedure.

KNOCK SENSOR (KS)

LOCATION

The Knock Sensor (KS) consists of two sensors which are installed inside the V-valley of the cylinder block.

REMOVAL & INSTALLATION

See Figure 83.

1. Before servicing the vehicle, refer to the Precautions Section.

➡If working near and/or around the SRS system and components, be sure to disable the SRS system. Tape the negative battery cable with insulating tape. Always disconnect the negative battery cable first.

⁂ CAUTION

To avoid personal injury when working on vehicles equipped with an air bag, the negative battery cable must be disconnected and at least three minutes must elapse before working on the system. Failure to do so may result in deployment of the air bag.

2. Remove the battery top cover, if equipped.
3. Disconnect the negative battery cable. Tape the cable with insulating tape.
4. Remove the intake manifold.

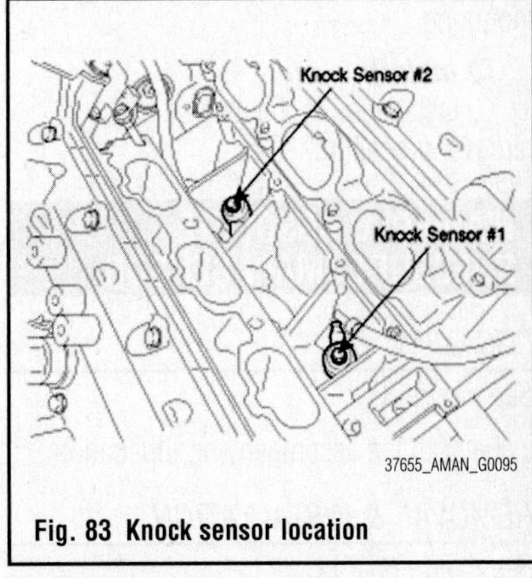

Fig. 83 Knock sensor location

5. Remove the sensor from its mounting.

To install:

6. Installation is the reverse of the removal procedure.
7. Tighten the sensor to 11–18 ft. lbs.

MASS AIR FLOW (MAF) SENSOR

LOCATION

See Figure 82.

The Mass Air Flow (MAF) Sensor is a located in between the air cleaner and the throttle body.

REMOVAL & INSTALLATION

See Figure 82.

1. Before servicing the vehicle, refer to the Precautions Section.

➡If working near and/or around the SRS system and components, be sure to disable the SRS system. Tape the negative battery cable with insulating tape. Always disconnect the negative battery cable first.

⁂ CAUTION

To avoid personal injury when working on vehicles equipped with an air bag, the negative battery cable must be disconnected and at least three minutes must elapse before working on the system. Failure to do so may result in deployment of the air bag.

2. Remove the battery top cover, if equipped.
3. Disconnect the negative battery cable. Tape the cable with insulating tape.
4. Disconnect the connector from the sensor.
5. Remove the air cleaner and air intake assembly, as required.
6. Remove the sensor from its mounting.

To install:

7. Installation is the reverse of the removal procedure.

MANIFOLD ABSOLUTE PRESSURE (MAP) SENSOR

LOCATION

See Figure 85.

Refer to the accompanying illustration.

REMOVAL & INSTALLATION

See Figure 85.

1. Before servicing the vehicle, refer to the Precautions Section.

➡If working near and/or around the SRS system and components, be sure to disable the SRS system. Tape the negative battery cable with insulating tape. Always disconnect the negative battery cable first.

⁂ CAUTION

To avoid personal injury when working on vehicles equipped with an air bag, the negative battery cable must be disconnected and at least three minutes must elapse before working on the system. Failure to do so may result in deployment of the air bag.

2. Remove the battery top cover, if equipped.
3. Disconnect the negative battery cable. Tape the cable with insulating tape.
4. Disconnect the connector from the sensor.
5. Remove the sensor retaining screws.
6. Remove the sensor from its mounting.

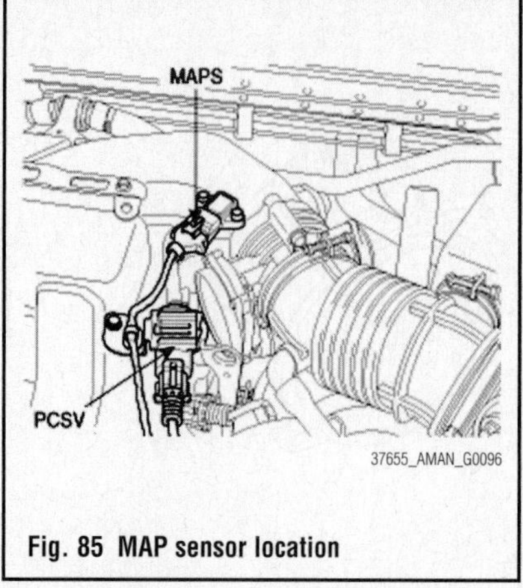

Fig. 85 MAP sensor location

To install:

7. Installation is the reverse of the removal procedure.

POSITIVE CRANKCASE VENTILATION (PCV) VALVE

LOCATION

See Figure 86.

Refer to the accompanying illustration.

REMOVAL & INSTALLATION

See Figure 86.

1. Before servicing the vehicle, refer to the Precautions Section.

➡If working near and/or around the SRS system and components, be sure to disable the SRS system. Tape the negative battery cable with insulating tape. Always disconnect the negative battery cable first.

⁂ CAUTION

To avoid personal injury when working on vehicles equipped with an air bag, the negative battery cable must be disconnected and at least three minutes must elapse before working on the system. Failure to do so may result in deployment of the air bag.

2. Remove the battery top cover, if equipped.
3. Disconnect the negative battery cable. Tape the cable with insulating tape.
4. Disconnect the vacuum hose.
5. Remove the valve from its mounting.

To install:

➡Be sure to use new fasteners, as required.

6. Installation is the reverse of the removal procedure.

VEHICLE SPEED SENSOR (VSS)

LOCATION

See Figure 87.

Refer to the accompanying illustration.

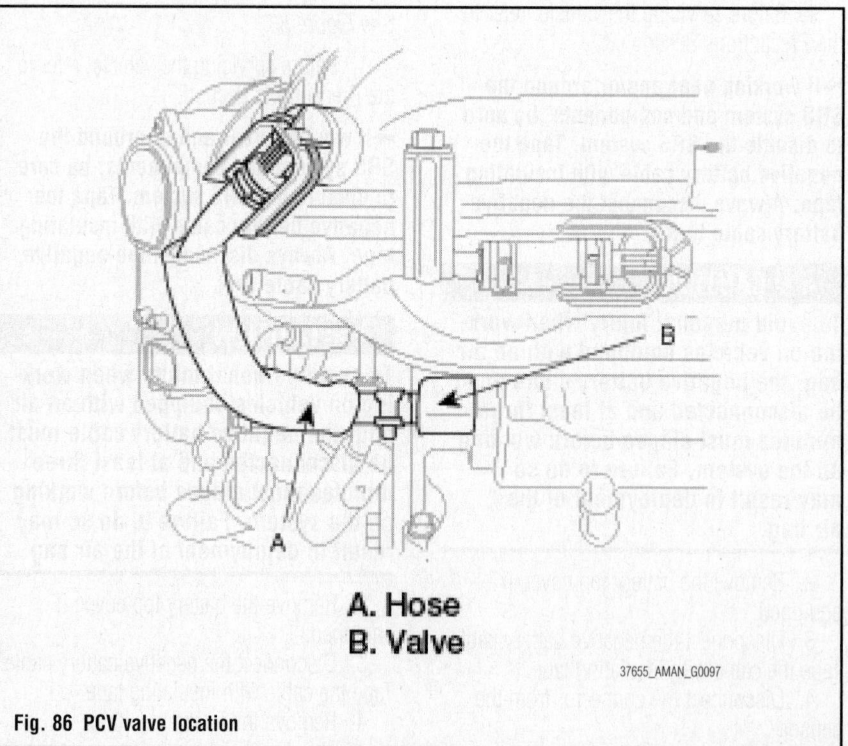

Fig. 86 PCV valve location

REMOVAL & INSTALLATION

See Figure 87.

At this time the manufacturer does not provide removal and installation procedures for this component. The following procedure is a guideline and may differ from the vehicle you are servicing.

1. Before servicing the vehicle, refer to the Precautions Section.

➡If working near and/or around the SRS system and components, be sure to disable the SRS system. Tape the negative battery cable with insulating tape. Always disconnect the negative battery cable first.

✱✱ CAUTION

To avoid personal injury when working on vehicles equipped with an air bag, the negative battery cable must be disconnected and at least three minutes must elapse before working on the system. Failure to do so may result in deployment of the air bag.

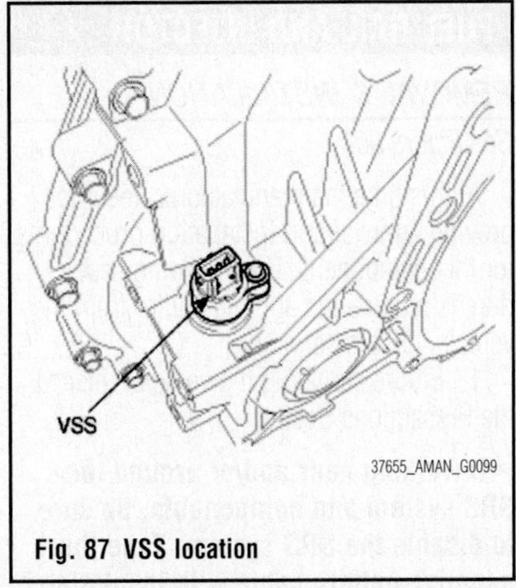

Fig. 87 VSS location

2. Remove the battery top cover, if equipped.
3. Disconnect the negative battery cable. Tape the cable with insulating tape.
4. Raise and support the vehicle, as necessary.
5. Remove the necessary components to gain access to the sensor.
6. Disconnect the sensor electrical connector.
7. Remove the sensor from its mounting.

To install:

➡Be sure to use new fasteners, as required.

8. Installation is the reverse of the removal procedure.

FUEL — GASOLINE FUEL INJECTION SYSTEM

FUEL SYSTEM SERVICE PRECAUTIONS

Safety is the most important factor when performing not only fuel system maintenance, but any type of maintenance. Failure to conduct maintenance and repairs in a safe manner may result in serious personal injury or death. Work on a vehicle's fuel system components can be accomplished safely and effectively by adhering to the following rules and guidelines.

- To avoid the possibility of fire and personal injury, always disconnect the negative battery cable unless the repair or test procedure requires that battery voltage be applied.
- Always relieve the fuel system pressure prior to disconnecting any fuel system component (injector, fuel rail, pressure regulator, etc.) fitting or fuel line connection. Exercise extreme caution whenever relieving fuel system pressure to avoid exposing skin, face and eyes to fuel spray. Please be advised that fuel under pressure may penetrate the skin or any part of the body that it contacts.
- Always place a shop towel or cloth around the fitting or connection prior to loosening to absorb any excess fuel due to spillage. Ensure that all fuel spillage is quickly removed from engine surfaces. Ensure that all fuel-soaked cloths or towels are deposited into a flame-proof waste container with a lid.
- Always keep a dry chemical (Class B) fire extinguisher near the work area.
- Do not allow fuel spray or fuel vapors to come into contact with a spark or open flame.
- Always use a second wrench when loosening or tightening fuel line connection fittings. This will prevent unnecessary stress and torsion on fuel piping. Always follow the proper torque specifications.
- Always replace worn fuel fitting O-rings with new ones. Do not substitute fuel hose where rigid pipe is installed.

FUEL SYSTEM PRESSURE

RELIEVING

See Figure 88.

1. Before servicing the vehicle, refer to the Precautions Section.
2. Open the service cover in the trunk area.
3. Disconnect the fuel pump wiring harness.
4. Start the engine and allow it to stall.
5. Turn the ignition switch to the **OFF** position.

Fig. 88 Fuel hose locations beneath the service cover

6. Reconnect the electrical connections after fuel system repairs are completed.

➡If working near and/or around the SRS system and components, be sure to disable the SRS system. Tape the negative battery cable with insulating tape. Always disconnect the negative battery cable first.

✱✱ CAUTION

To avoid personal injury when working on vehicles equipped with an air bag, the negative battery cable must be disconnected and at least three minutes must elapse before working on the system. Failure to do so may result in deployment of the air bag.

FUEL FILTER

REMOVAL & INSTALLATION

The fuel filter is part of the fuel pump assembly located in the fuel tank.

FUEL PUMP

REMOVAL & INSTALLATION

See Figure 89.

1. Before servicing the vehicle, refer to the Precautions Section.

➡If working near and/or around the SRS system and components, be sure to disable the SRS system. Tape the negative battery cable with insulating tape. Always disconnect the negative battery cable first.

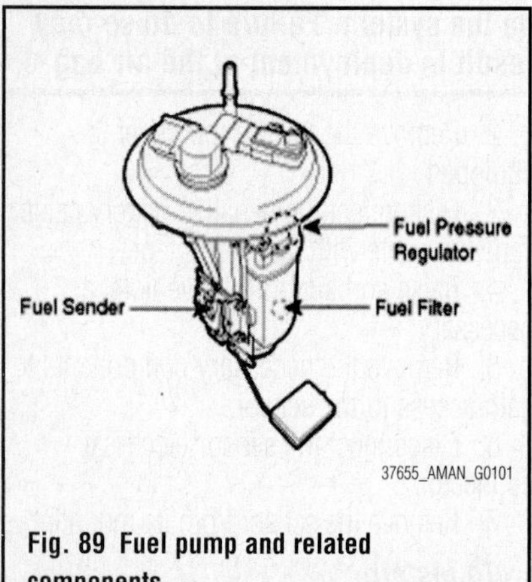

Fig. 89 Fuel pump and related components

CAUTION

To avoid personal injury when working on vehicles equipped with an air bag, the negative battery cable must be disconnected and at least three minutes must elapse before working on the system. Failure to do so may result in deployment of the air bag.

2. Remove the battery top cover, if equipped.
3. Disconnect the negative battery cable. Tape the cable with insulating tape.
4. Relieve the fuel system pressure. Gain access to the service cover located in the trunk.
5. Disconnect the fuel pump electrical connector.
6. Start the engine and wait until the fuel in the fuel line is exhausted.
7. After the engine stalls, turn the ignition switch to the **OFF** position and disconnect the negative (−) terminal from the battery.

➡Be sure to reduce the fuel pressure before disconnecting the fuel feed hose, otherwise fuel may spill out.

8. Disconnect the fuel supply and return hoses from the top of the fuel pump assembly.
9. Remove the mounting bolts and remove the fuel pump assembly from the fuel tank.

To install:

10. Installation is the reverse of removal.
11. Fill the tank with fuel and check for proper fuel pump operation.

FUEL PRESSURE REGULATOR

REMOVAL & INSTALLATION

The fuel pressure regulator is built in to the fuel pump.

FUEL RAIL AND INJECTOR

REMOVAL & INSTALLATION

See Figure 90.

At this time the manufacturer does not provide removal and installation procedures for this component. The following procedure is a guideline and may differ from the vehicle you are servicing.

1. Before servicing the vehicle, refer to the Precautions Section.

➡If working near and/or around the SRS system and components, be sure to disable the SRS system. Tape the negative battery cable with insulating tape. Always disconnect the negative battery cable first.

CAUTION

To avoid personal injury when working on vehicles equipped with an air bag, the negative battery cable must be disconnected and at least three minutes must elapse before working on the system. Failure to do so may result in deployment of the air bag.

2. Remove the battery top cover, if equipped.
3. Disconnect the negative battery cable. Tape the cable with insulating tape.
4. Relieve the fuel system pressure.
5. Drain the cooling system.
6. Remove or disconnect the following:
   - Engine cover
   - Air intake assembly
7. Remove the following engine wiring harnesses:
   a. Crankshaft angle sensor
   b. Camshaft angle sensor
   c. Fuel injector harness

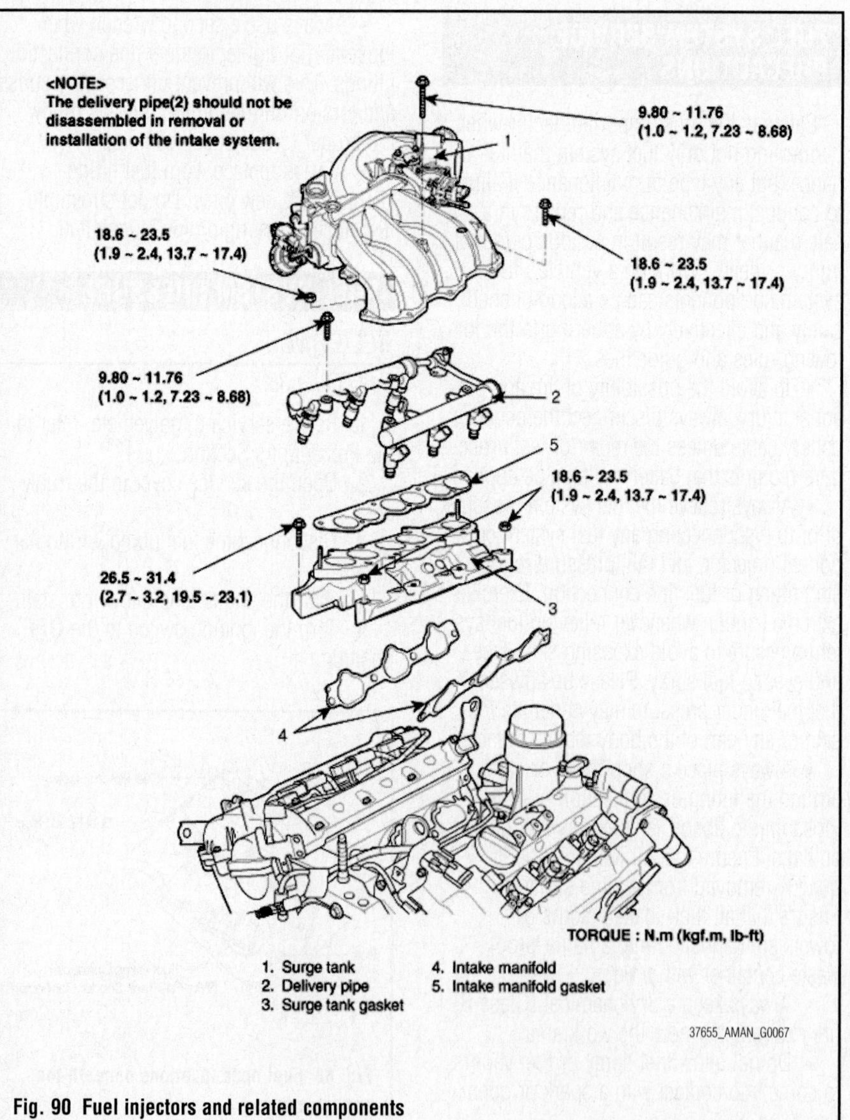

Fig. 90 Fuel injectors and related components

➡When disconnecting the injector, lift the fuel supply hose and injector assembly upward. Then unscrew the mounting bolt to lift the fuel supply hose. Reinstall the fuel hose and injector assembly after disconnecting the injector harness.

d. Power steering switch
e. Variable intake motor connector
f. Accelerator position sensor
g. Throttle position sensor
h. Electronic throttle system (ETS) motor
i. Limp-home connector
j. Purge solenoid valve connector

8. Remove or disconnect the following:
    - Vacuum hoses and heater hoses between the intake manifold and cylinder head cover
    - Engine wiring harness bracket
    - EGR valve hose and bracket
    - Surge tank assembly
    - Fuel supply hose
    - Fuel injector assembly
    - Fuel injector

To install:

➡**Be sure to use new fasteners, as required.**

9. Installation is the reverse of the removal procedure.

FUEL TANK

REMOVAL & INSTALLATION

See Figure 91.

CAUTION

Observe all applicable safety precautions when working around fuel. Whenever servicing the fuel system, always work in a well ventilated area. Do not allow fuel spray or vapors to come in contact with a spark or open flame. Keep a dry chemical fire extinguisher near the work area. Always keep fuel in a container specifically designed for fuel storage; also, always properly seal fuel containers to avoid the possibility of fire or explosion.

1. Before servicing the vehicle, refer to the Precautions Section.

➡**If working near and/or around the SRS system and components, be sure to disable the SRS system. Tape the negative battery cable with insulating tape. Always disconnect the negative battery cable first.**

CAUTION

To avoid personal injury when working on vehicles equipped with an air bag, the negative battery cable must be disconnected and at least three minutes must elapse before working on the system. Failure to do so may result in deployment of the air bag.

2. Remove the battery top cover, if equipped.
3. Disconnect the negative battery cable. Tape the cable with insulating tape.
4. Drain the fuel from the fuel tank to an acceptable level.
5. Relieve the fuel system pressure. Gain access to the service cover located in the trunk.
6. Disconnect the fuel pump electrical connector.
7. Start the engine and wait until the fuel in the fuel line is exhausted.
8. After the engine stalls, turn the ignition switch to the **OFF** position and disconnect the negative (–) terminal from the battery.

➡**Be sure to reduce the fuel pressure before disconnecting the fuel feed hose, otherwise fuel may spill out.**

9. Disconnect the fuel supply and return hoses from the top of the fuel pump assembly.
10. Raise and support the vehicle safely.
11. Properly support the fuel tank using a suitable jack.
12. Remove the center and main mufflers.
13. Remove the fuel tank cover.
14. Remove the fuel filler hose, the fuel hose (connecting the fuel tank with the canister) and the fuel leveling hose.
15. Remove the fuel tank band mounting bolts and remove the fuel tank.

To install:

16. Installation is the reverse of removal.
17. Tighten fuel tank mounting bolts to 29–40 ft. lbs. (39–54 Nm).
18. Fill the tank with fuel and check for proper fuel pump operation.

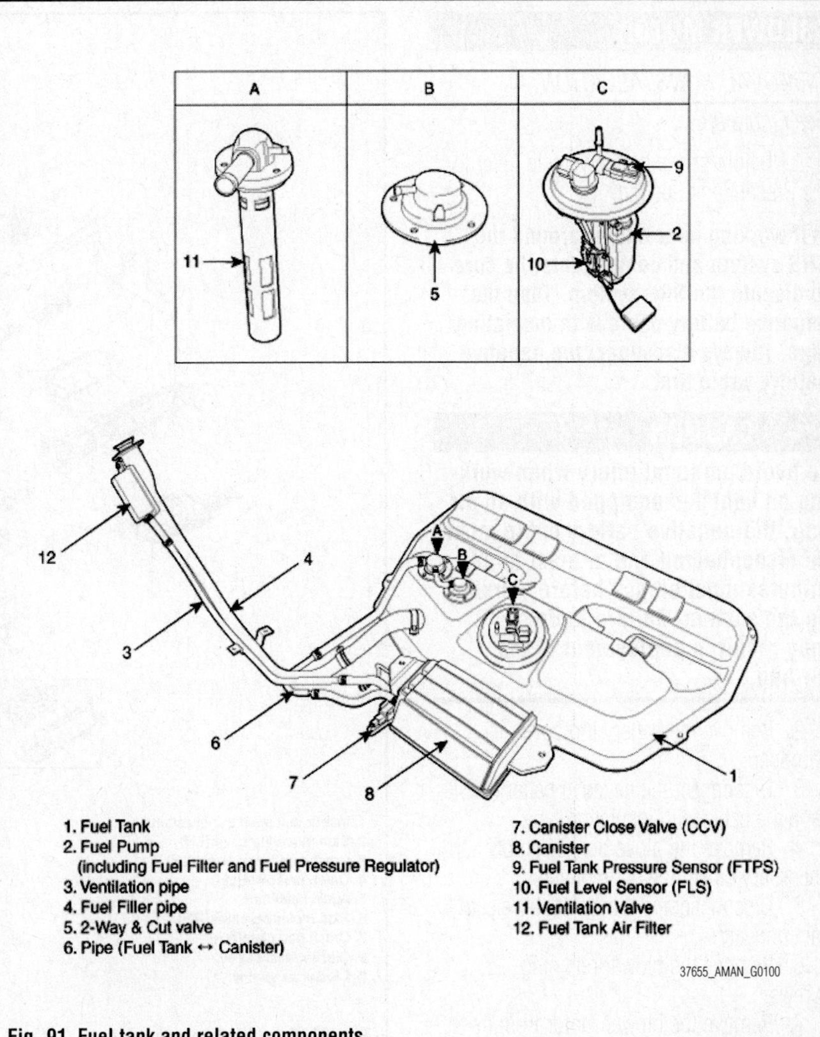

Fig. 91 Fuel tank and related components

IDLE SPEED

ADJUSTMENT

Idle speed is maintained by the Powertrain Control Module (PCM). No adjustment is necessary or possible.

THROTTLE BODY

REMOVAL & INSTALLATION

See Figure 92.

At this time the manufacturer does not provide removal and installation procedures for this component. The following procedure is a guideline and may differ from the vehicle you are servicing.

1. Before servicing the vehicle, refer to the Precautions Section.

➡If working near and/or around the SRS system and components, be sure to disable the SRS system. Tape the negative battery cable with insulating tape. Always disconnect the negative battery cable first.

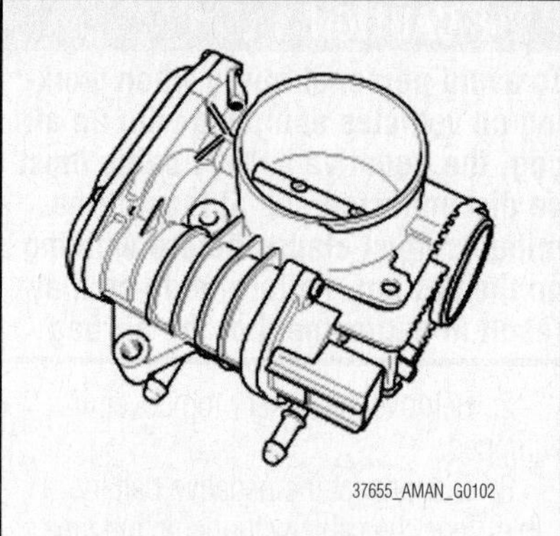

Fig. 92 Throttle body assembly

⁂ CAUTION

To avoid personal injury when working on vehicles equipped with an air bag, the negative battery cable must be disconnected and at least three minutes must elapse before working on the system. Failure to do so may result in deployment of the air bag.

2. Remove the battery top cover, if equipped.
3. Remove the necessary components in order to gain access to the throttle body assembly.
4. Disconnect the electrical connectors.
5. Remove the throttle body assembly from its mounting.

To install:

➡Be sure to use new fasteners, as required.

6. Installation is the reverse of the removal procedure.

HEATING & AIR CONDITIONING SYSTEM

BLOWER MOTOR

REMOVAL & INSTALLATION

See Figure 93.

1. Before servicing the vehicle, refer to the Precautions Section.

➡If working near and/or around the SRS system and components, be sure to disable the SRS system. Tape the negative battery cable with insulating tape. Always disconnect the negative battery cable first.

⁂ CAUTION

To avoid personal injury when working on vehicles equipped with an air bag, the negative battery cable must be disconnected and at least three minutes must elapse before working on the system. Failure to do so may result in deployment of the air bag.

2. Remove the battery top cover, if equipped.
3. Disconnect the negative battery cable. Tape the cable with insulating tape.
4. Remove the glove box assembly and necessary crash pad components.
5. Disconnect the blower motor electrical connector.
6. Remove the blower retaining screws.
7. Remove the blower motor from its mounting.

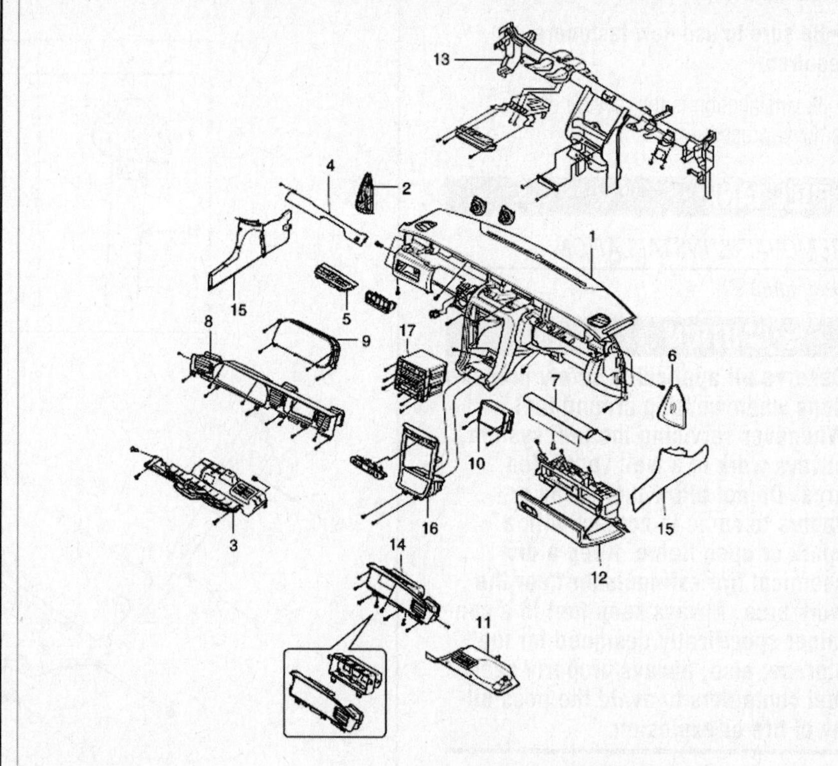

Fig. 93 Crash pad components

To install:

➡**Be sure to use new fasteners, as required.**

8. Installation is the reverse of the removal procedure.

HEATER CORE

REMOVAL & INSTALLATION

1. Before servicing the vehicle, refer to the Precautions Section.

➡**If working near and/or around the SRS system and components, be sure to disable the SRS system. Tape the negative battery cable with insulating tape. Always disconnect the negative battery cable first.**

CAUTION

To avoid personal injury when working on vehicles equipped with an air bag, the negative battery cable must be disconnected and at least three minutes must elapse before working on the system. Failure to do so may result in deployment of the air bag.

2. Remove the battery top cover, if equipped.
3. Disconnect the negative battery cable. Tape the cable with insulating tape.
4. Remove the heater unit.
5. Remove the heater core from the heater unit.

To install:

➡**Be sure to use new fasteners, as required.**

6. Installation is the reverse of the removal procedure.

HEATER UNIT

REMOVAL & INSTALLATION

See Figure 94.

At this time the manufacturer does not provide removal and installation procedures for this component, refer to the illustration as required.

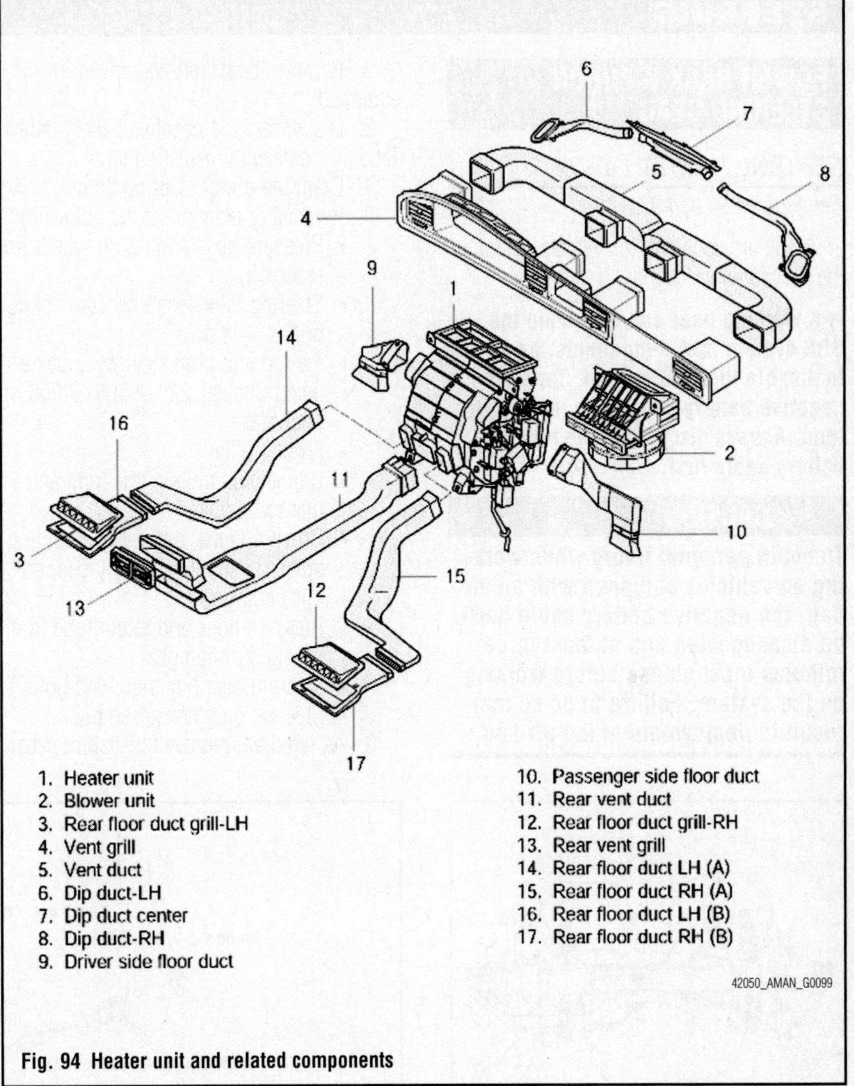

Fig. 94 Heater unit and related components

1. Before servicing the vehicle, refer to the Precautions Section.

➡**If working near and/or around the SRS system and components, be sure to disable the SRS system. Tape the negative battery cable with insulating tape. Always disconnect the negative battery cable first.**

CAUTION

To avoid personal injury when working on vehicles equipped with an air bag, the negative battery cable must be disconnected and at least three minutes must elapse before working on the system. Failure to do so may result in deployment of the air bag.

2. Remove the battery top cover, if equipped.
3. Disconnect the negative battery cable. Tape the cable with insulating tape.
4. Drain the cooling system.
5. Properly discharge the air conditioning system.

STEERING

POWER RACK & PINION STEERING GEAR

REMOVAL & INSTALLATION

See Figures 96 and 97.

1. Before servicing the vehicle, refer to the Precautions Section.

➡If working near and/or around the SRS system and components, be sure to disable the SRS system. Tape the negative battery cable with insulating tape. Always disconnect the negative battery cable first.

⁂ CAUTION

To avoid personal injury when working on vehicles equipped with an air bag, the negative battery cable must be disconnected and at least three minutes must elapse before working on the system. Failure to do so may result in deployment of the air bag.

2. Remove the battery top cover, if equipped.
3. Disconnect the negative battery cable. Tape the cable with insulating tape.
4. Drain the power steering fluid.
5. Remove or disconnect the following:
   - Pressure hose and return tube from reservoir
   - Steering joint assembly connecting bolt
   - Tie rod end from steering knuckle using Special Tool 09568-34000 or equivalent
   - Front muffler
   - Connecting bolts of the front and rear roll stopper
   - Cross-member assembly mounting bolts. Be sure to properly support this component.
   - Pressure hose and return tube from rack and pinion gear
   - Steering gear box mounting bolts
   - Steering gear box assembly
6. As required, remove the stabilizer bar.

➡If the vehicle is equipped with electronic power steering, you will notice the EPS solenoid valve mounted on the steering box assembly.

To install:

7. When installing the mounting rubber, align the projection of the mounting rubber with the indentation in the cross-member.
8. Installation is the reverse of removal.
9. Check the wheel alignment and adjust as necessary.

POWER STEERING PUMP

REMOVAL & INSTALLATION

See Figure 98.

1. Before servicing the vehicle, refer to the Precautions Section.

➡If working near and/or around the SRS system and components, be sure to disable the SRS system. Tape the negative battery cable with insulating

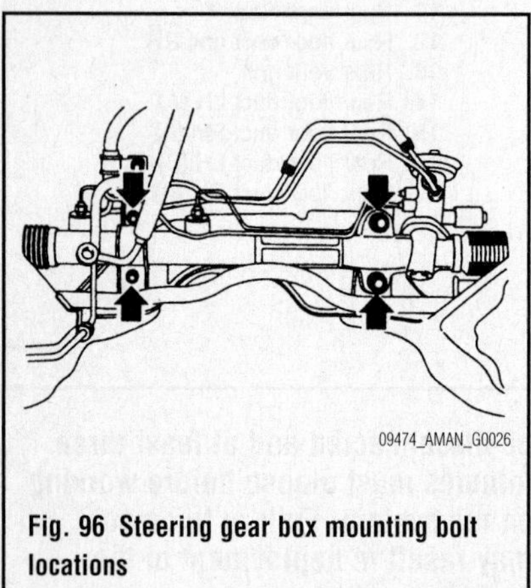

Fig. 96 Steering gear box mounting bolt locations

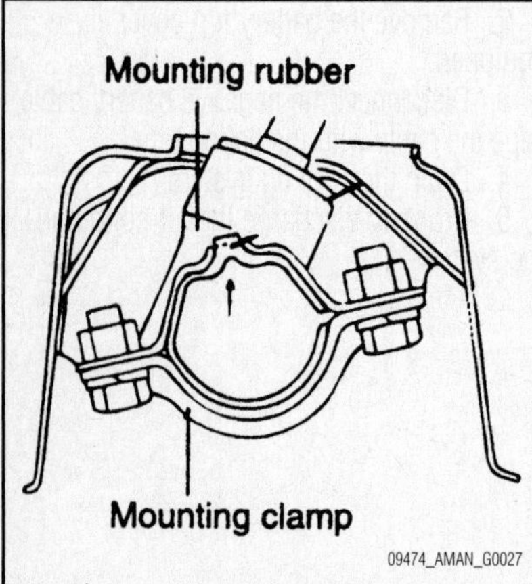

Fig. 97 Align the mounting rubber with the cross-member indentation

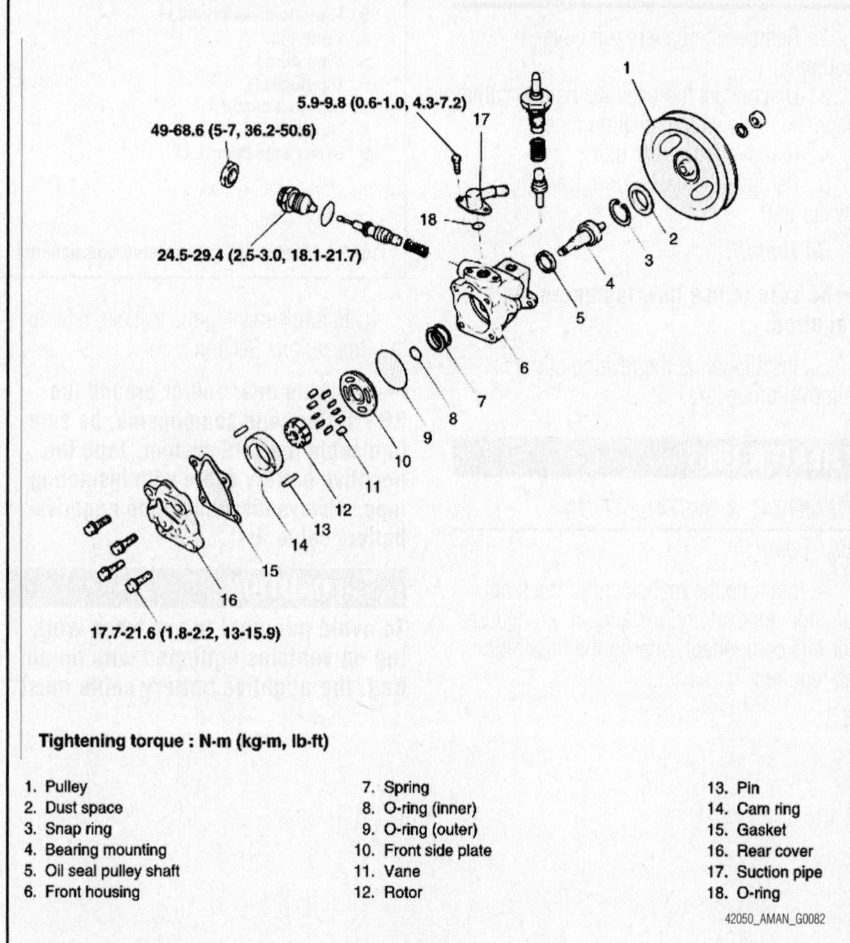

Fig. 98 Power steering pump and related components

tape. Always disconnect the negative battery cable first.

⁂ CAUTION

To avoid personal injury when working on vehicles equipped with an air bag, the negative battery cable must be disconnected and at least three minutes must elapse before working on the system. Failure to do so may result in deployment of the air bag.

2. Remove the battery top cover, if equipped.
3. Disconnect the negative battery cable. Tape the cable with insulating tape.
4. Remove the pressure hose from the power steering pump.

➡When assembling, use a new O-ring.

5. Disconnect the suction hose from the suction connector and drain the fluid into a container.

⁂ CAUTION

When removing the suction hose, cover the alternator with vinyl to keep oil from damaging that component.

6. Remove the drive belt.
7. Loosen the tension adjusting bolt to remove the drive belt.
8. Remove the power steering pump mounting bolts and disconnect the pressure switch connector.

To install:

9. Install the power steering pump to the power steering pump bracket.
10. Install the suction hose.
11. Install the ribbed V-belt and adjust its tension.
12. Connect the pressure hose to the power steering pump, and the suction hose to the oil reservoir.

➡Install the hoses so that they are not twisted and they do not come in contact with any other parts.

13. Replenish the reservoir.
14. Bleed the system.
15. Check the power steering pump pressure.

BLEEDING

1. Before servicing the vehicle, refer to the Precautions Section.
2. With engine off, turn the steering wheel fully to the right and left several times.

➡Do not allow the fluid level in the reservoir tank to go below the MIN level line. Check and add fluid as needed.

3. Run the engine at idle speed. Turn the steering wheel fully to the right and then fully to the left. Hold for about three seconds. Check for fluid leakage.
4. Repeat the above step several times at three second intervals.

➡Do not hold the steering wheel in the locked position for more than ten seconds.

5. Check for air bubbles or cloudy fluid. If found, repeat the bleeding procedure.
6. Stop the engine and check the fluid level. Fill as required.

SUSPENSION

FRONT SUSPENSION

COIL SPRING

REMOVAL & INSTALLATION

See Figure 99.

1. Before servicing the vehicle, refer to the Precautions Section.

➡If working near and/or around the SRS system and components, be sure to disable the SRS system. Tape the negative battery cable with insulating tape. Always disconnect the negative battery cable first.

⁂ CAUTION

To avoid personal injury when working on vehicles equipped with an air bag, the negative battery cable must be disconnected and at least three minutes must elapse before working on the system. Failure to do so may result in deployment of the air bag.

2. Remove the battery top cover, if equipped.
3. Disconnect the negative battery cable. Tape the cable with insulating tape.
4. Remove the strut from the vehicle and install a spring compressor.
5. Remove the front cap.
6. Compress the coil spring so that the end of the spring comes away from the spring seat.
7. Remove or disconnect the following:
   - Upper strut mounting nut
   - Insulator
   - Upper spring seat
   - Compressed spring from the strut
   - Spring from the spring compressor

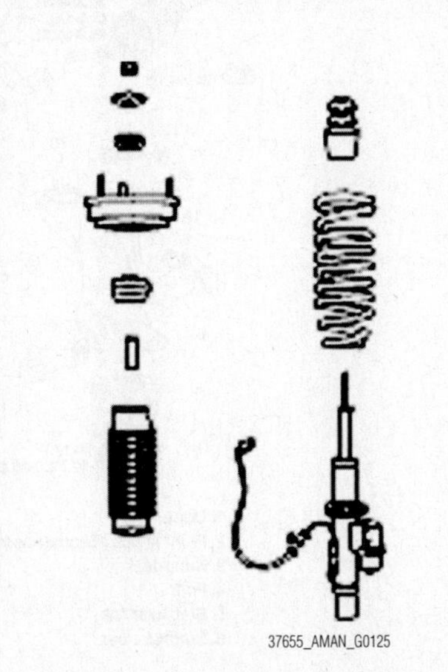

Fig. 99 Front coil spring and related components

⁂ CAUTION

Do not use an impact gun.

To install:

8. Compress the spring and install it on the strut.
9. Install or connect the following:
   - Upper spring seat
   - Insulator
   - Upper strut mount
   - Strut to the vehicle
10. Check and/or adjust the wheel alignment.

LOWER BALL JOINT

REMOVAL & INSTALLATION

See Figures 100 through 102.

1. Before servicing the vehicle, refer to the Precautions Section.

➡If working near and/or around the SRS system and components, be sure to disable the SRS system. Tape the negative battery cable with insulating tape. Always disconnect the negative battery cable first.

✻✻ CAUTION

To avoid personal injury when working on vehicles equipped with an air bag, the negative battery cable must be disconnected and at least three minutes must elapse before working on the system. Failure to do so may result in deployment of the air bag.

2. Remove the battery top cover, if equipped.

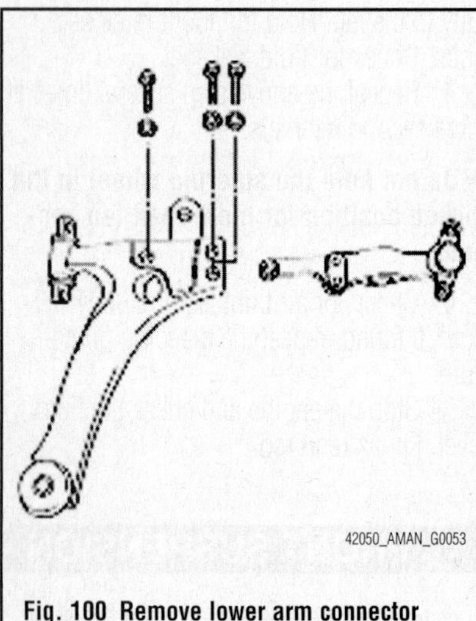

Fig. 100 Remove lower arm connector

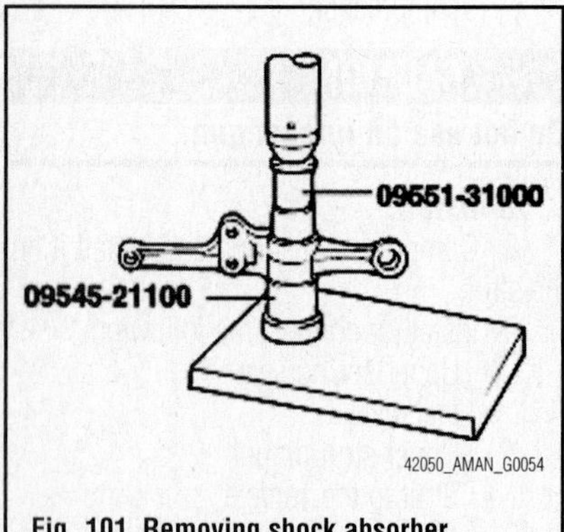

Fig. 101 Removing shock absorber mounting bushing

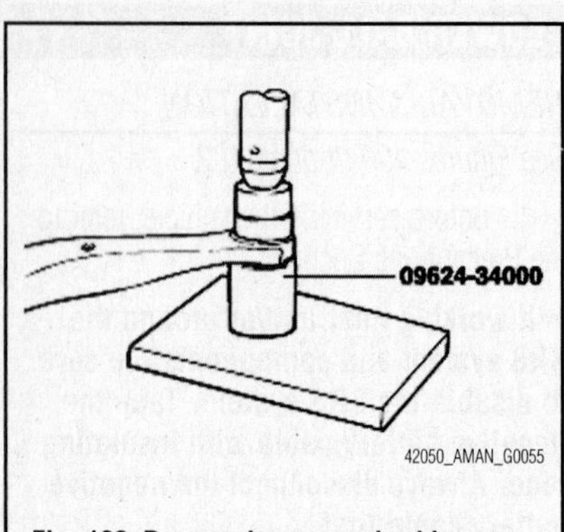

Fig. 102 Remove lower arm bushing

3. Disconnect the negative battery cable. Tape the cable with insulating tape.
4. Raise and safely support the vehicle. Remove the wheel and tire.
5. Remove the lower arm connector from the lower arm.
6. Using the special tools(09551-31000, 09545-21100), remove the shock absorber mounting bushing.
7. Using the special tools (09624-34000), remove the lower arm bushing (G).
8. Check the ball joint for rotating Torque:
   - If a crack is noted in the dust cover, replace the ball joint assembly
   - Measure the lower ball joint for rotating torque
   - Check that it is according to specification: 2–9 inch lbs. (0.2–1 Nm).

LOWER CONTROL ARM

REMOVAL & INSTALLATION

See Figure 103.

1. Before servicing the vehicle, refer to the Precautions Section.

➡If working near and/or around the SRS system and components, be sure to disable the SRS system. Tape the negative battery cable with insulating tape. Always disconnect the negative battery cable first.

✻✻ CAUTION

To avoid personal injury when working on vehicles equipped with an air bag, the negative battery cable must be disconnected and at least three minutes must elapse before working on the system. Failure to do so may result in deployment of the air bag.

2. Remove the battery top cover, if equipped.
3. Disconnect the negative battery cable. Tape the cable with insulating tape.
4. Raise and safely support the vehicle. Remove the wheel and tire.

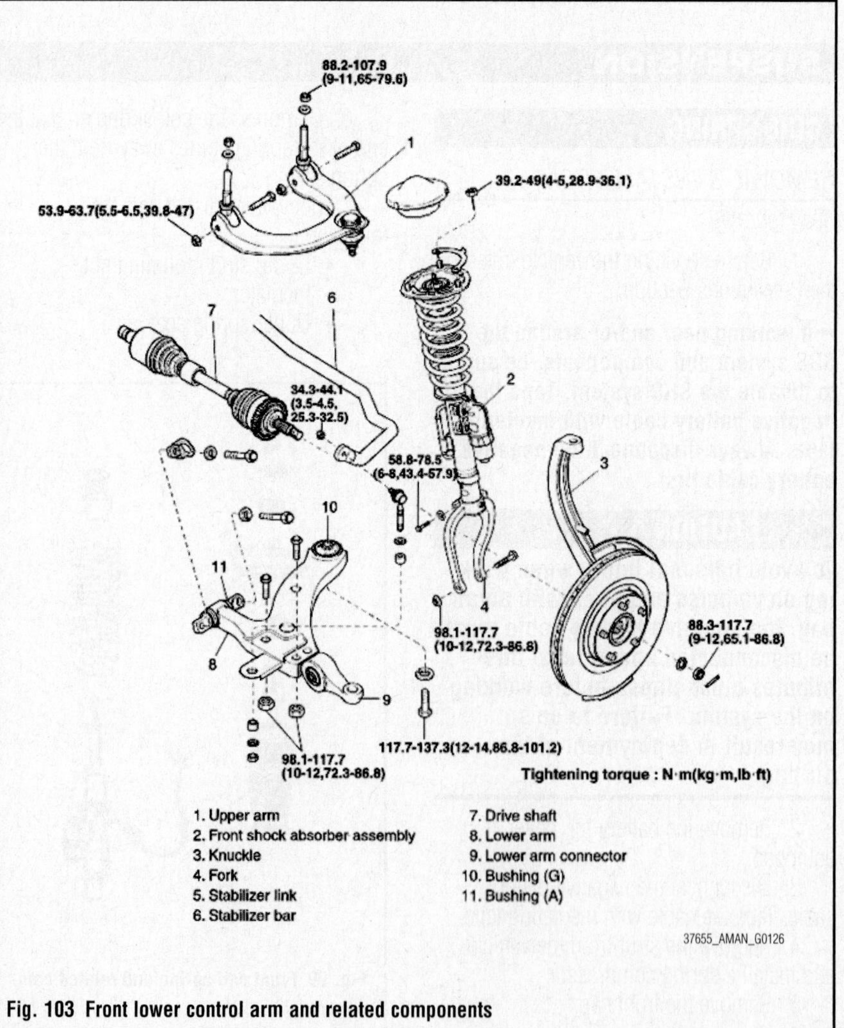

Fig. 103 Front lower control arm and related components

5. Loosen the ball joint nut, but do not remove.

6. Using Special Tool 09455-21000 or equivalent, disconnect the lower arm ball joint from the lower arm connector.

7. Remove the ball joint assembly.

8. Remove the fork to lower arm connector mounting bolt.

9. Remove the stabilizer link from the lower control arm.

10. Remove the lower control arm mounting bolts.

11. Remove the lower control arm.

To install:

12. Install or connect the following:
   - Lower control arm and torque the mounting bolts to 72–87 ft. lbs. (98–118 Nm).
   - Tighten the rear bushing bolt to 87–101 ft. lbs. (118–137 Nm).
   - Stabilizer bar link
   - Fork to lower arm connector mounting bolt and tighten to 72–87 ft. lbs. (98–118 Nm).
   - Ball joint assembly
   - Front wheel

13. Check and/or adjust the wheel alignment.

MACPHERSON STRUT

REMOVAL & INSTALLATION

See Figure 104.

1. Before servicing the vehicle, refer to the Precautions Section.

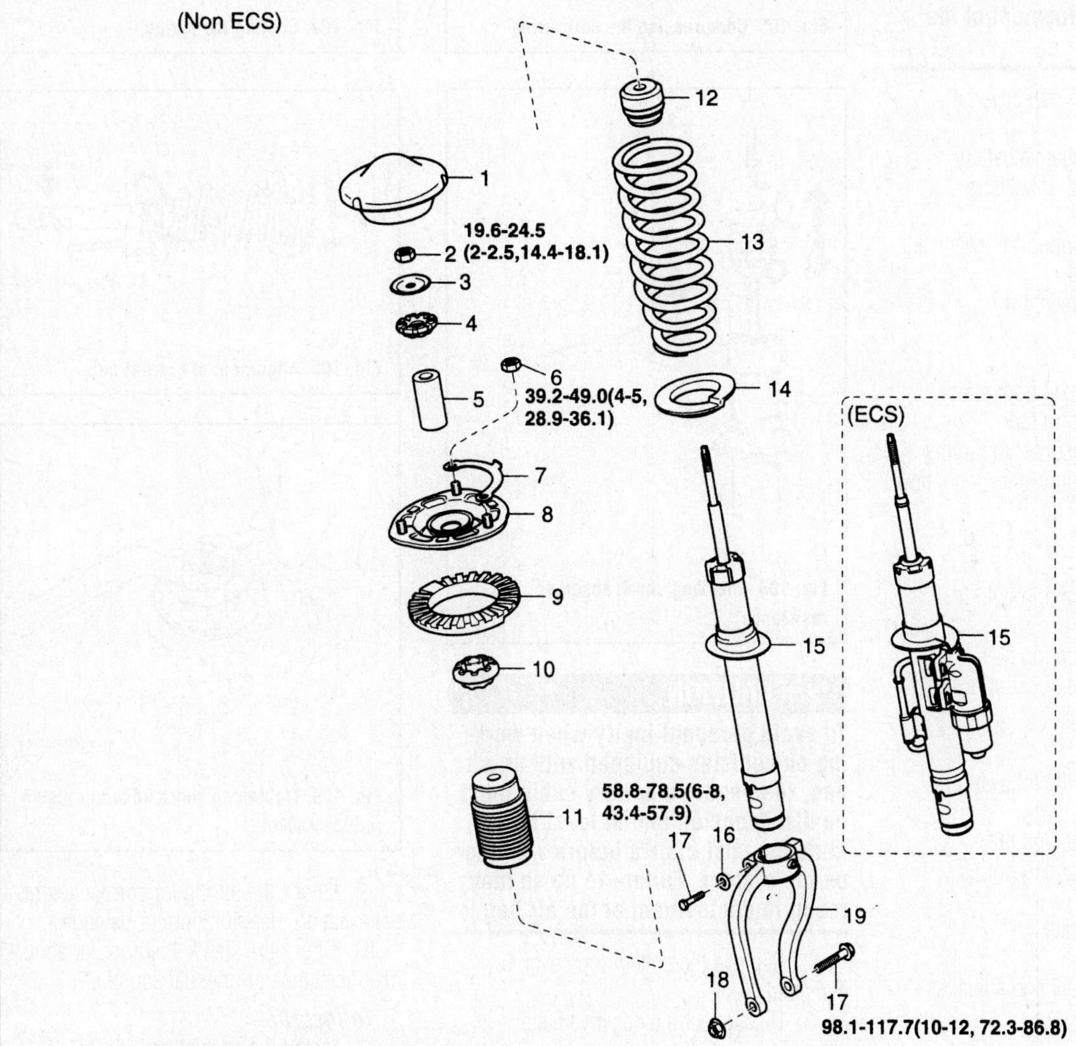

Fig. 104 Front strut and related components

➡If working near and/or around the SRS system and components, be sure to disable the SRS system. Tape the negative battery cable with insulating tape. Always disconnect the negative battery cable first.

⁂ CAUTION

To avoid personal injury when working on vehicles equipped with an air bag, the negative battery cable must be disconnected and at least three minutes must elapse before working on the system. Failure to do so may result in deployment of the air bag.

2. Remove the battery top cover, if equipped.
3. Disconnect the negative battery cable. Tape the cable with insulating tape.
4. Raise and safely support the vehicle, as required.
5. Remove or disconnect the following:
   - Front wheel
   - Brake hose bracket from shock absorber mounting fork
   - Lower shock absorber mounting fork/lower arm connector mounting bolt
   - Mounting fork from the shock absorber
   - ECS wiring mounting bolt, if equipped
   - Upper strut mounting nuts
6. Push the axle assembly upward and remove the strut assembly.

To install:

7. Install or connect the following:
   - Strut assembly
   - Upper strut mounting nut
   - ECS wiring mounting bolt, if equipped
   - Brake hose bracket
   - Front wheel
8. Check the alignment and adjust as necessary.

OVERHAUL

See Figures 105 through 109.

1. Before servicing the vehicle, refer to the Precautions Section.

➡If working near and/or around the SRS system and components, be sure to disable the SRS system. Tape the negative battery cable with insulating tape. Always disconnect the negative battery cable first.

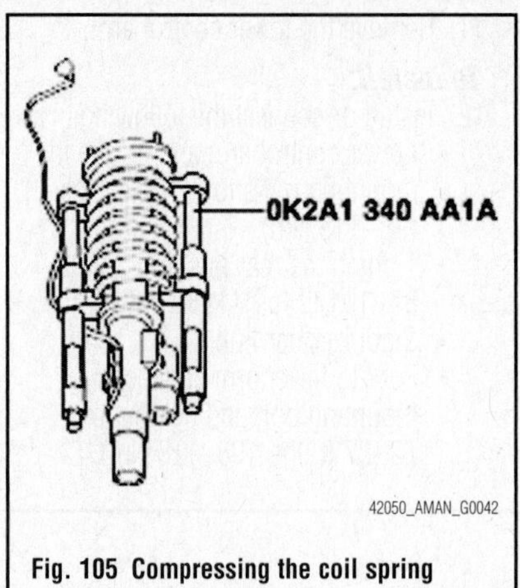

Fig. 105 Compressing the coil spring

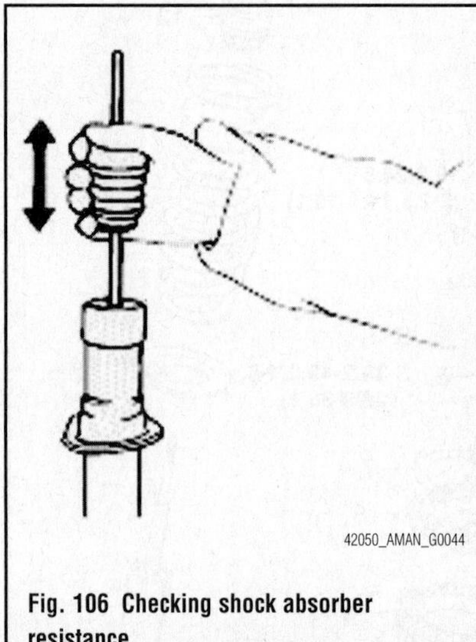

Fig. 106 Checking shock absorber resistance

⁂ CAUTION

To avoid personal injury when working on vehicles equipped with an air bag, the negative battery cable must be disconnected and at least three minutes must elapse before working on the system. Failure to do so may result in deployment of the air bag.

2. Remove the battery top cover, if equipped.
3. Disconnect the negative battery cable. Tape the cable with insulating tape.
4. With the MacPherson Strut assembly removed, attach a special tool (0K2A1 341 AA1A) to the coil spring.
5. Compress the coil spring until there is only a little tension on the strut.
6. Remove the self-locking nut at the top end of shock absorber.
7. Remove the bracket, spring pad, and coil spring.
8. Check the rubber parts for damage or deterioration.

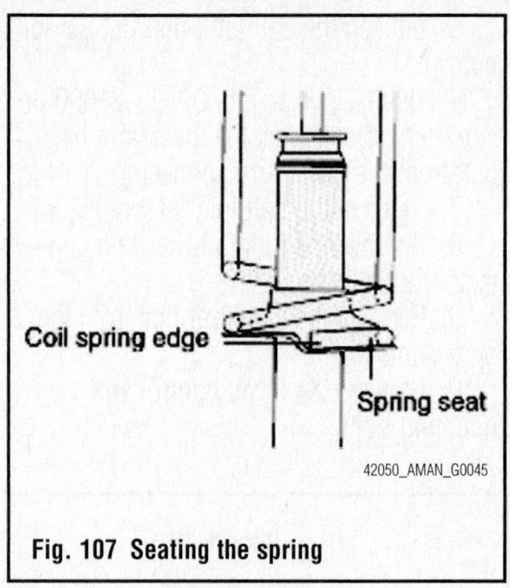

Fig. 107 Seating the spring

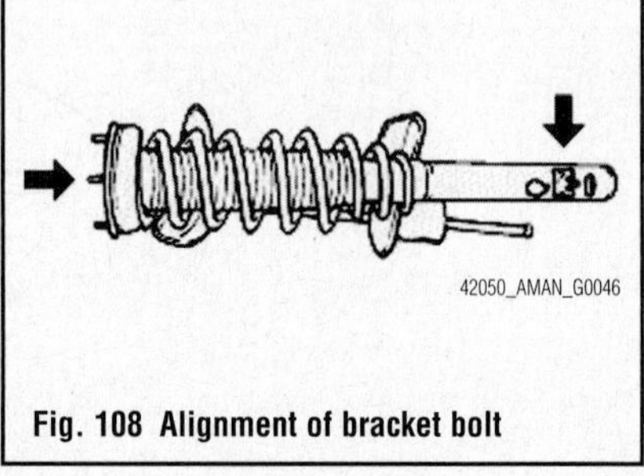

Fig. 108 Alignment of bracket bolt

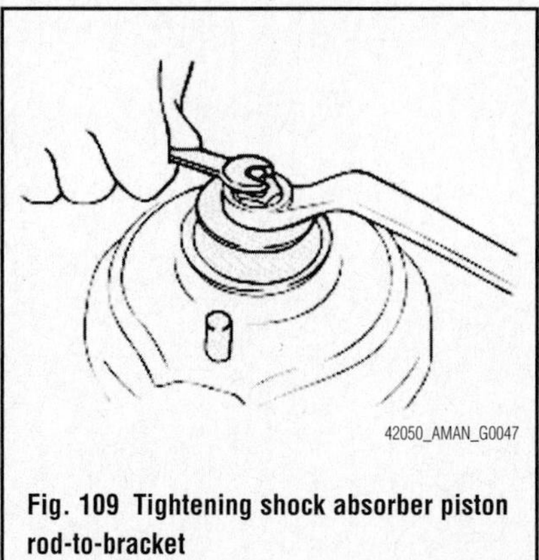

Fig. 109 Tightening shock absorber piston rod-to-bracket

9. Check the spring for correct height, deformation, deterioration, or damage.
10. Check the shock absorber for abnormal resistance or unusual sounds.

To install:

11. Install the special tool (0K2A1 341 AA1A) and compress the coil spring. After spring is fully compressed, install it on the shock absorber assembly.
12. After seating the dust cover, upper spring pad, bushings, and bracket, tighten the new self-locking nut temporarily.
13. Position the upper and lower ends of the coil spring in the upper spring pad and lower spring seat grooves correctly.
14. Place the bracket to align the bracket

bolt with the projection of the fork bracket in a straight line.

15. Remove the special tool (0K2A1 341 AA1A) from the coil spring.

16. While holding the piston rod, tighten the new self-locking nut to specification: 15–18 ft. lbs. (20–25 Nm).

STEERING KNUCKLE

REMOVAL & INSTALLATION

See Figure 110.

1. Before servicing the vehicle, refer to the Precautions Section.

➡If working near and/or around the SRS system and components, be sure to disable the SRS system. Tape the negative battery cable with insulating tape. Always disconnect the negative battery cable first.

✻✻ CAUTION

To avoid personal injury when working on vehicles equipped with an air bag, the negative battery cable must be disconnected and at least three minutes must elapse before working on the system. Failure to do so may result in deployment of the air bag.

2. Remove the battery top cover, if equipped.
3. Disconnect the negative battery cable. Tape the cable with insulating tape.
4. Raise and support the vehicle safely.
5. Remove the front wheel and tire.
6. Disconnect the wheel speed sensor from the knuckle. Remove the caliper assembly and suspend it with wire.
7. Remove the ABS sensor and wire.
8. Using the special tool (09568-34000), disconnect the tie rod end from the knuckle:
   - Be sure to tie the special tool to a nearby part with cord.
   - Loosen the nut but do not remove it.
9. Remove the 2 bolts and disconnect the ball joint from the knuckle.
10. Remove the brake disc from the knuckle.
11. Remove the split pin and drive shaft castle nut from the front hub.
12. Using a plastic hammer, disconnect the drive shaft from the axle hub.
13. Loosen the upper arm mounting nut, but do not remove it.
14. Using the special tool (09568-34000), disconnect the upper arm from the knuckle
15. Remove the front axle and knuckle together.

To install:

16. Installation is the reverse of the removal procedure.

STABILIZER BAR

REMOVAL & INSTALLATION

See Figures 111 through 113.

1. Before servicing the vehicle, refer to the Precautions Section.

➡If working near and/or around the SRS system and components, be sure to disable the SRS system. Tape the negative battery cable with insulating tape. Always disconnect the negative battery cable first.

✻✻ CAUTION

To avoid personal injury when working on vehicles equipped with an air bag, the negative battery cable must be disconnected and at least three minutes must elapse before working on the system. Failure to do so may result in deployment of the air bag.

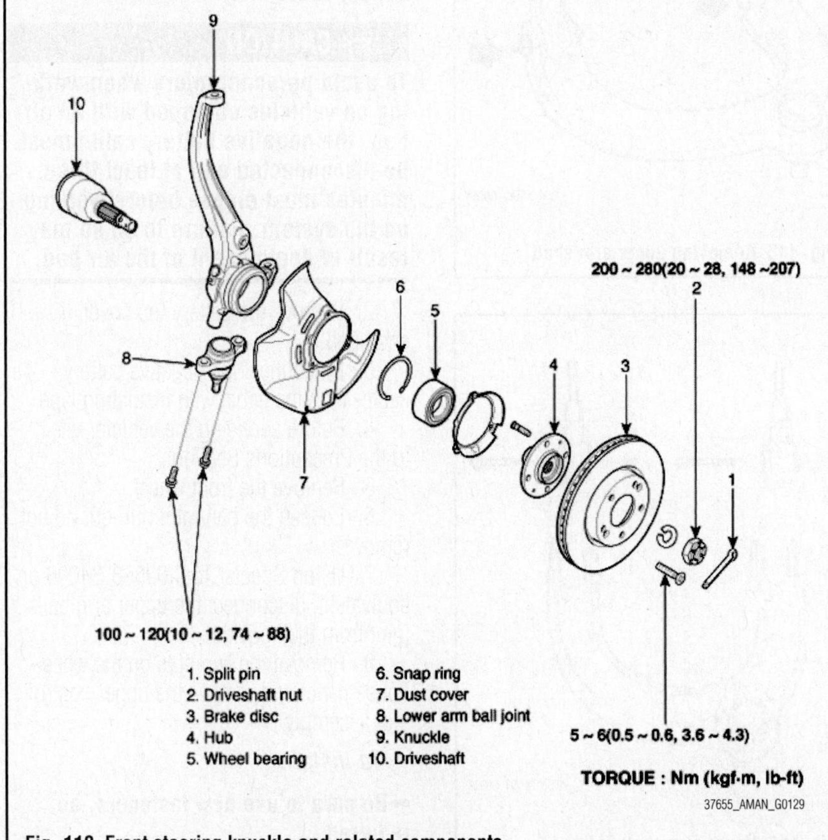

Fig. 110 Front steering knuckle and related components

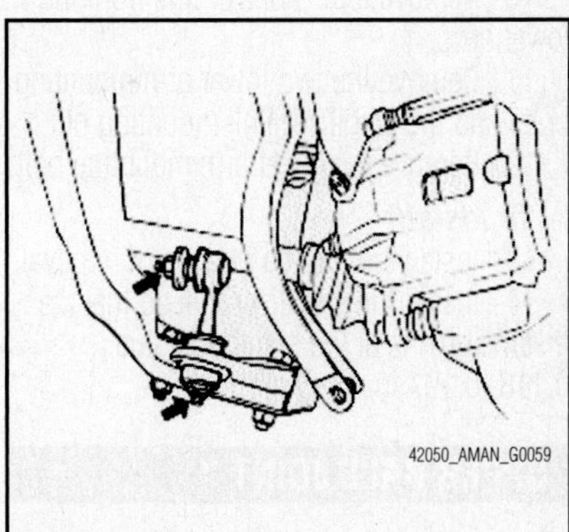

Fig. 111 Disconnecting the stabilizer bar from the lower arm

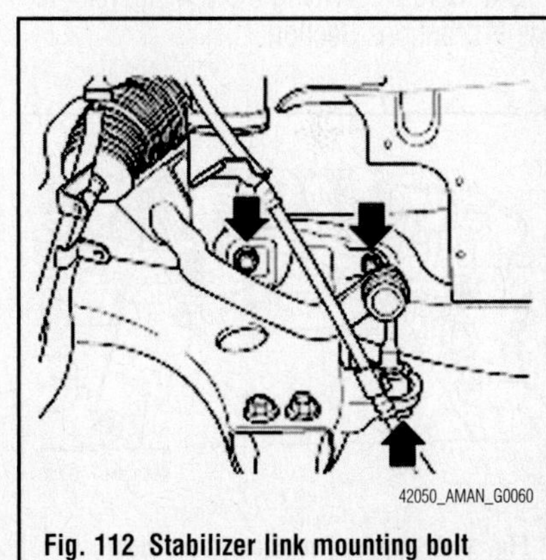

Fig. 112 Stabilizer link mounting bolt

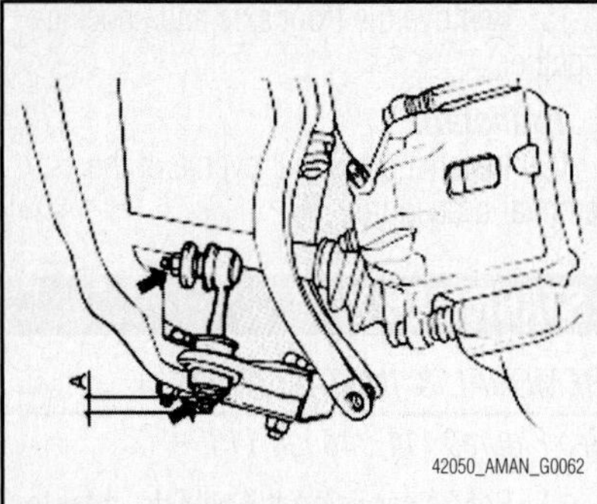

Fig. 113 Standard value for stabilizer link installation

2. Remove the battery top cover, if equipped.
3. Disconnect the negative battery cable. Tape the cable with insulating tape.
4. Raise and support the vehicle safely.
5. Remove the front wheel and tire.
6. Loosen the ball joint nut, but do not remove it.
7. Using the special tools (09455-21000), disconnect the lower arm ball joint from the lower arm connector.
8. Remove ball joint assembly.
9. Remove the fork to lower arm connector mounting bolt.
10. Remove the stabilizer link from the lower arm.
11. Remove the two lower arm mounting bolts and the stabilizer link mounting nut.
12. Remove the lower arm mounting bolt.

To install:

13. Installation is the reverse of removal.
14. Install the stabilizer link so that the distance (A) is at the standard value 0.118–0.197 in. (3–5 mm).

UPPER BALL JOINT

REMOVAL & INSTALLATION

See Figures 114 through 119.

1. Before servicing the vehicle, refer to the Precautions Section.

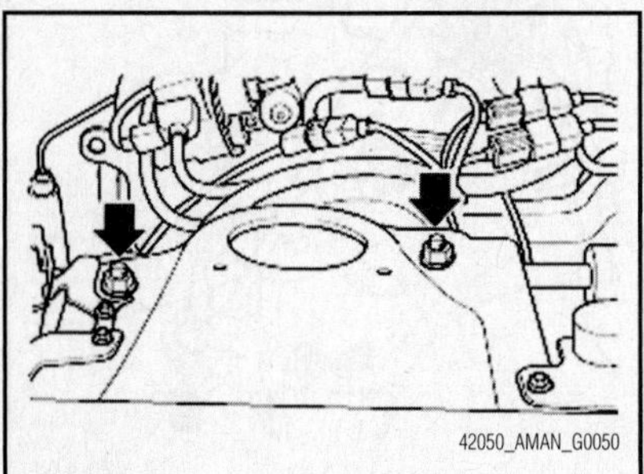

Fig. 114 Upper arm wheelhouse nuts

➡If working near and/or around the SRS system and components, be sure to disable the SRS system. Tape the negative battery cable with insulating tape. Always disconnect the negative battery cable first.

⁂ CAUTION

To avoid personal injury when working on vehicles equipped with an air bag, the negative battery cable must be disconnected and at least three minutes must elapse before working on the system. Failure to do so may result in deployment of the air bag.

2. Remove the battery top cover, if equipped.
3. Disconnect the negative battery cable. Tape the cable with insulating tape.
4. Remove the wheel and tire.
5. Loosen the ball joint nut, but do not remove it.
6. Using the special tool (09568-34000), disconnect the upper arm ball joint from the knuckle.

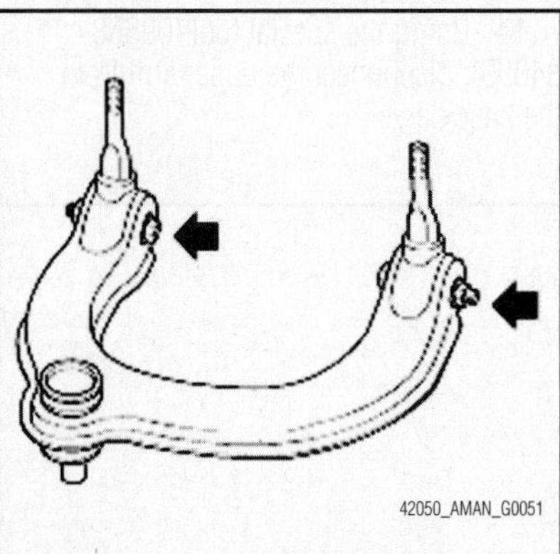

Fig. 115 Removing upper arm shaft

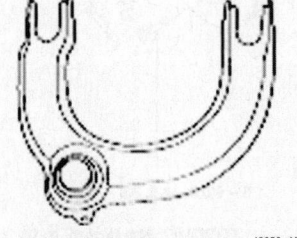

Fig. 116 Upper arm shaft bolts

7. Remove 2 nuts on the wheelhouse panel and remove the upper arm assembly.
8. Remove the upper arm shaft.

To install:

9. Installation is the reverse of the removal procedure.
10. Tighten the wheelhouse panel nuts to 65–80 ft. lbs. (88–108 Nm).
11. Check the ball joint for rotating Torque:
   - If there is a crack in the dust cover, replace it and add grease.
   - Mount the self-locking nut on the ball joint, and then measure the ball joint rotating Torque.
   - Check that it is according to specification: 13–22 inch lbs. (2–3 Nm).

UPPER CONTROL ARM

REMOVAL & INSTALLATION

See Figure 117.

1. Before servicing the vehicle, refer to the Precautions Section.

➡If working near and/or around the SRS system and components, be sure to disable the SRS system. Tape the negative battery cable with insulating tape. Always disconnect the negative battery cable first.

⁂ CAUTION

To avoid personal injury when working on vehicles equipped with an air bag, the negative battery cable must be disconnected and at least three minutes must elapse before working on the system. Failure to do so may result in deployment of the air bag.

2. Remove the battery top cover, if equipped.
3. Disconnect the negative battery cable. Tape the cable with insulating tape.
4. Before servicing the vehicle, refer to the Precautions Section.
5. Remove the front wheel.
6. Loosen the ball joint nut, but do not remove.
7. Using Special Tool 09568-34000 or equivalent, disconnect the upper arm ball joint from the knuckle.
8. Remove the two nuts on the wheelhouse panel and remove the upper control arm assembly.

To install:

➡Be sure to use new fasteners, as required.

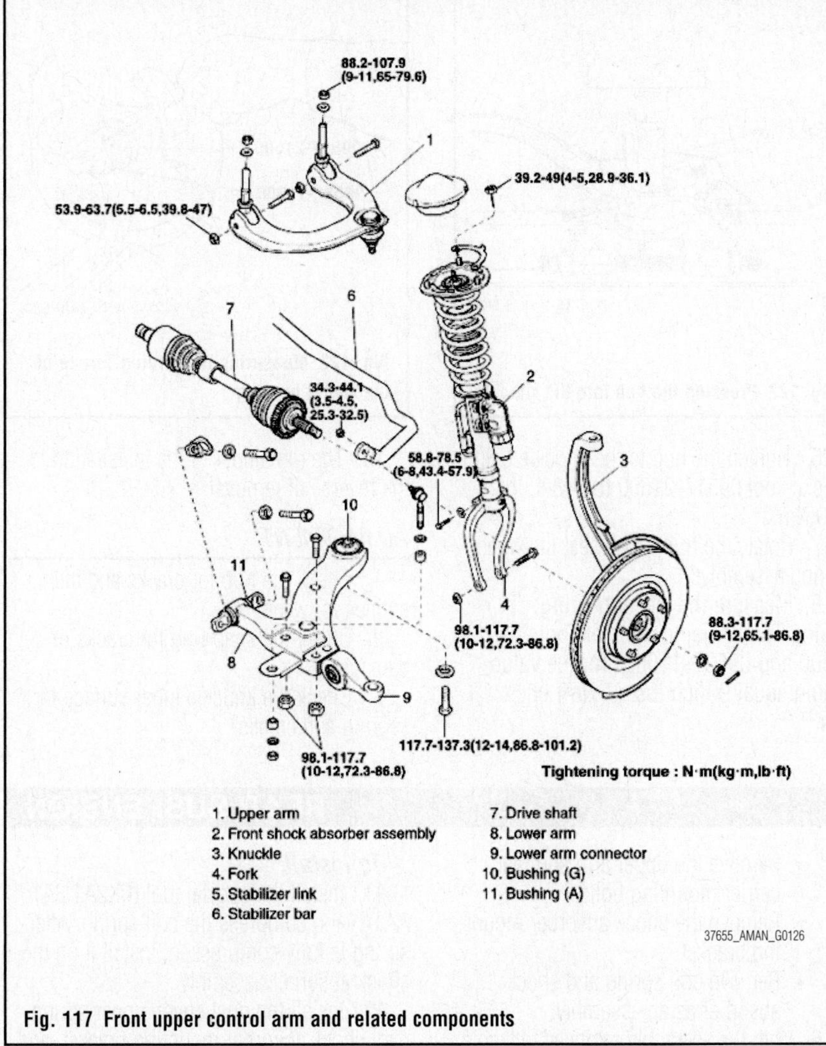

Fig. 117 Front upper control arm and related components

9. Installation is the reverse of the removal procedure.

10. Tighten the wheelhouse panel nuts to 65–80 ft. lbs. (88–108 Nm).

WHEEL HUB & BEARING

REMOVAL & INSTALLATION

See Figures 118 through 123.

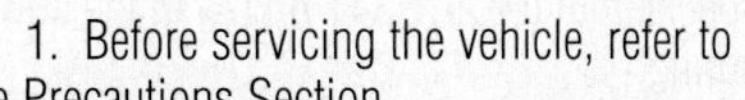

1. Before servicing the vehicle, refer to the Precautions Section.

➡If working near and/or around the SRS system and components, be sure to disable the SRS system. Tape the negative battery cable with insulating tape. Always disconnect the negative battery cable first.

⁂ CAUTION

To avoid personal injury when working on vehicles equipped with an air bag, the negative battery cable must be disconnected and at least three minutes must elapse before working on the system. Failure to do so may result in deployment of the air bag.

2. Remove the battery top cover, if equipped.

3. Disconnect the negative battery cable. Tape the cable with insulating tape.

4. raise and safely support the vehicle.

5. Remove or disconnect the following:
   - Front wheel
   - Brake caliper
   - Wheel speed sensor, if equipped
   - Tie rod end from the knuckle
   - Lower ball joint from the knuckle
   - Brake disc
   - Spindle nut
   - Halfshaft from the axle hub, using a plastic hammer if necessary.
   - Upper control arm from the knuckle
   - Front axle and knuckle as an assembly

6. Remove the snap ring from the knuckle assembly.

7. Using Special Tool 0K-130-331-AA0A or equivalent slide hammer, disconnect the hub from the knuckle.

8. Using special Tool 09455-21000 and 09545-34100 or equivalent gear puller, remove the wheel bearing inner race from the hub.

9. Remove the dust cover.

10. Using Special Tool 09216-21600 and 09216-22100, remove the wheel bearing outer race from the knuckle.

To install:

11. Apply a thin coat of grease to the knuckle and bearing contact surfaces.

12. Using Special Tool 09216-21100, press the bearing into the knuckle.

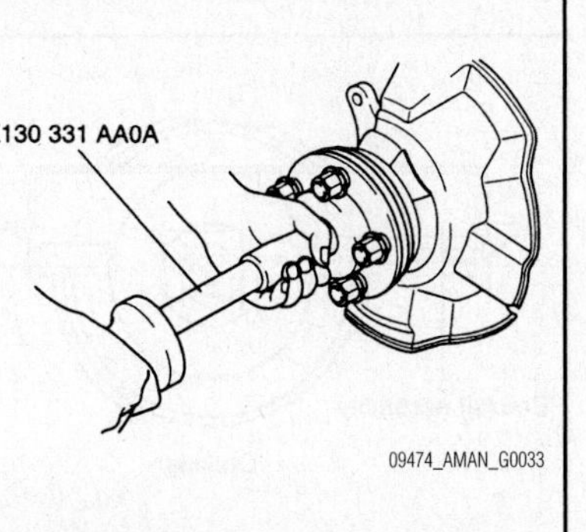

Fig. 118 Disconnect the hub assembly from the knuckle

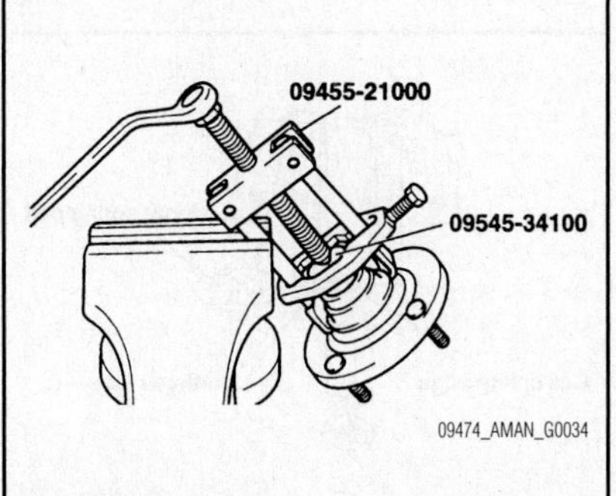

Fig. 119 Remove the wheel bearing inner race from the hub

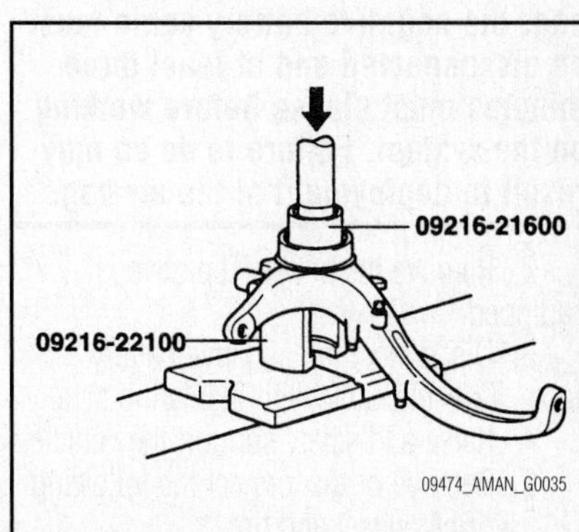

Fig. 120 Press the outer race from the knuckle

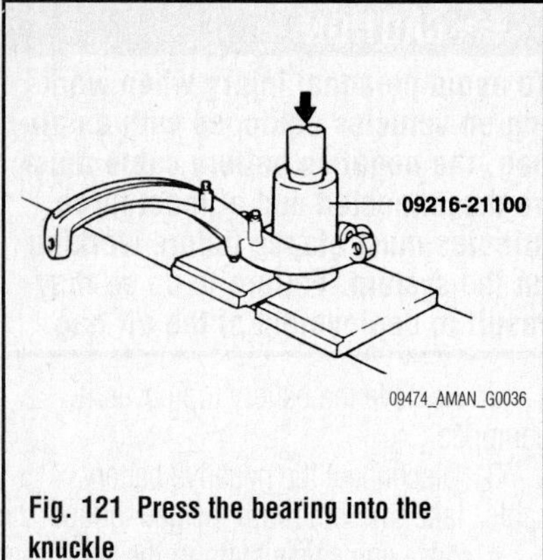

Fig. 121 Press the bearing into the knuckle

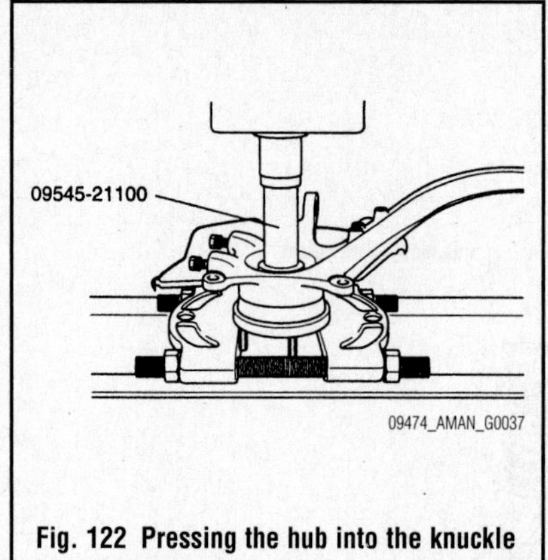

Fig. 122 Pressing the hub into the knuckle

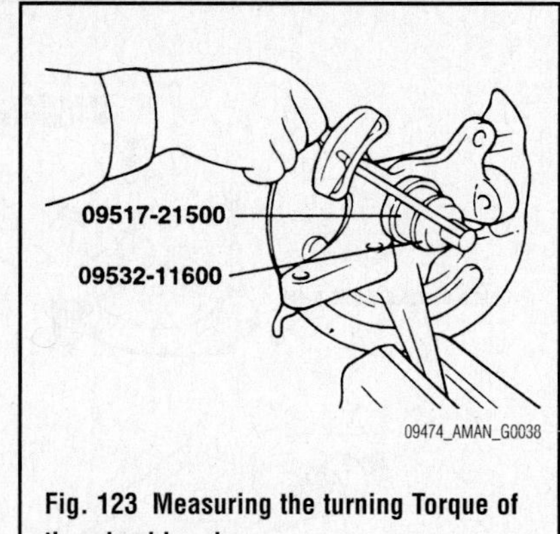

Fig. 123 Measuring the turning Torque of the wheel bearing

13. Install the snap ring into the groove of the knuckle.
14. Install the dust cover.
15. Using Special Tool 09545-21100, press the hub onto the knuckle.

⁂ WARNING

Do not press against the outer race of the wheel bearing. Damage to the bearing assembly could occur.

16. Tighten the hub to the knuckle using Special Tool 09517-21500 to 148 ft. lbs. (200 Nm).
17. Rotate the hub to the seat the wheel bearing assembly.
18. Measure the wheel bearing turning Torque using Special Tools 0951-21500 and 09532-11600. Torque value should equal 9 inch lbs. (1 Nm) or less.
19. The remainder of the installation is the reverse of removal.

ADJUSTMENT

1. Check the hub for cracks and the splines for wear.
2. Check the snap ring for cracks or damage.
3. Check the knuckle inner surface for scoring and cracks.

SUSPENSION — REAR SUSPENSION

COIL SPRING

REMOVAL & INSTALLATION

See Figures 124 and 125.

1. Before servicing the vehicle, refer to the Precautions Section.

➡If working near and/or around the SRS system and components, be sure to disable the SRS system. Tape the negative battery cable with insulating tape. Always disconnect the negative battery cable first.

⁂ CAUTION

To avoid personal injury when working on vehicles equipped with an air bag, the negative battery cable must be disconnected and at least three minutes must elapse before working on the system. Failure to do so may result in deployment of the air bag.

2. Remove the battery top cover, if equipped.
3. Disconnect the negative battery cable. Tape the cable with insulating tape.
4. Raise and safely support the vehicle.
5. Remove or disconnect the following:
   - Rear wheel and tire
   - Remove the shock absorber lower mounting bolts.
   - Remove the upper arm and rear carrier mounting bolts.
   - Remove the shock absorber mounting bracket.
   - Remove coil spring and shock absorber as an assembly.
6. With the assembly removed, attach a special tool (0K2A1 341 AA1A) to the coil spring.
7. Compress the coil spring until there is only a little tension on the strut.
8. Remove the self-locking nut at the top end of shock absorber.
9. Remove the shock absorber mounting bracket, dust cover, spring upper pad.
10. Remove the special tool (0K2A1 341 AA1A) and then remove the coil spring.

To install:

11. Install the special tool (0K2A1 341 AA1A) and compress the coil spring. After spring is fully compressed, install it on the shock absorber assembly.
12. Install the dust cover, upper spring pad, shock absorber mounting bracket, and washer
13. Tighten the new self-locking nut temporarily.
14. Position the upper and lower ends of the coil spring in the upper spring pad and lower spring seat grooves correctly.
15. Install so that the upper spring pad is fit in the shock absorber mounting bracket correctly and that the spring urethane tube is located down.

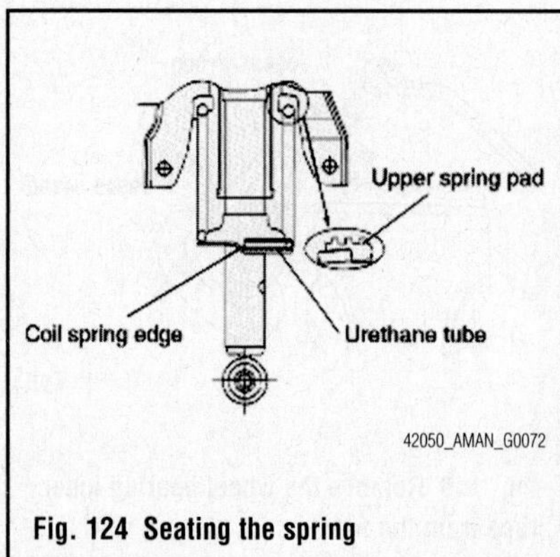

Fig. 124 Seating the spring

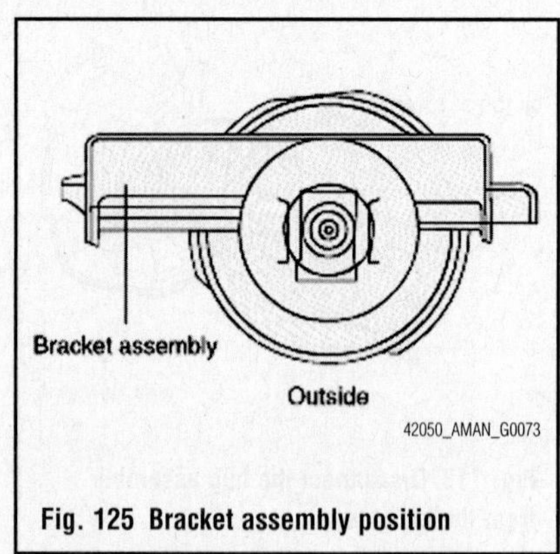

Fig. 125 Bracket assembly position

16. When the position of the bracket assembly is as shown in the illustration, tighten the new self-locking nut.
17. Remove the special tool.
18. While holding the piston rod, tighten the new self-locking nut to specification: 15–18 ft. lbs. (20–25 Nm).
19. Continue the installation of the coil spring and shock absorber assembly in the reverse order of the removal procedure.
20. Install or connect the following:
    - Shock absorber mounting bracket tightening to 72–87 ft. lbs. (98–118 Nm).
    - Upper arm and rear carrier mounting bolts to 72–87 ft. lbs. (98–118 Nm).
    - Shock absorber lower mounting bolts 58–65 ft. lb. (79–88 Nm).
    - Rear wheel and tire tightening lug nuts to 67–82 ft .lbs. (91–112 Nm).

CONTROL ARMS/LINKS

REMOVAL & INSTALLATION

See Figures 126 and 127.

At this time the manufacturer does not provide removal and installation procedures

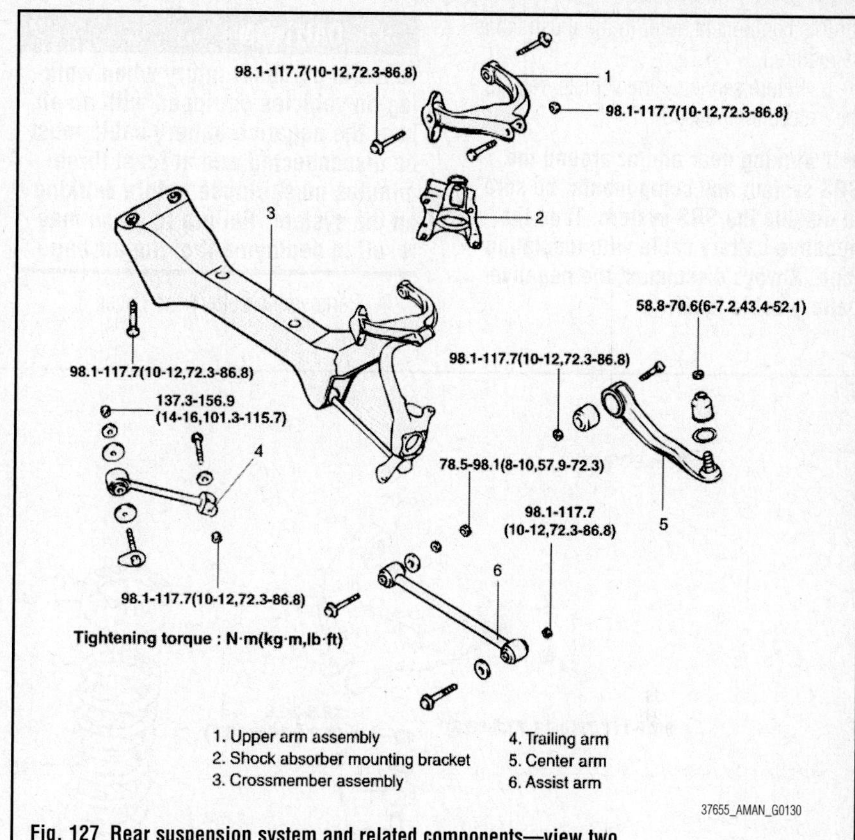

Fig. 127 Rear suspension system and related components—view two

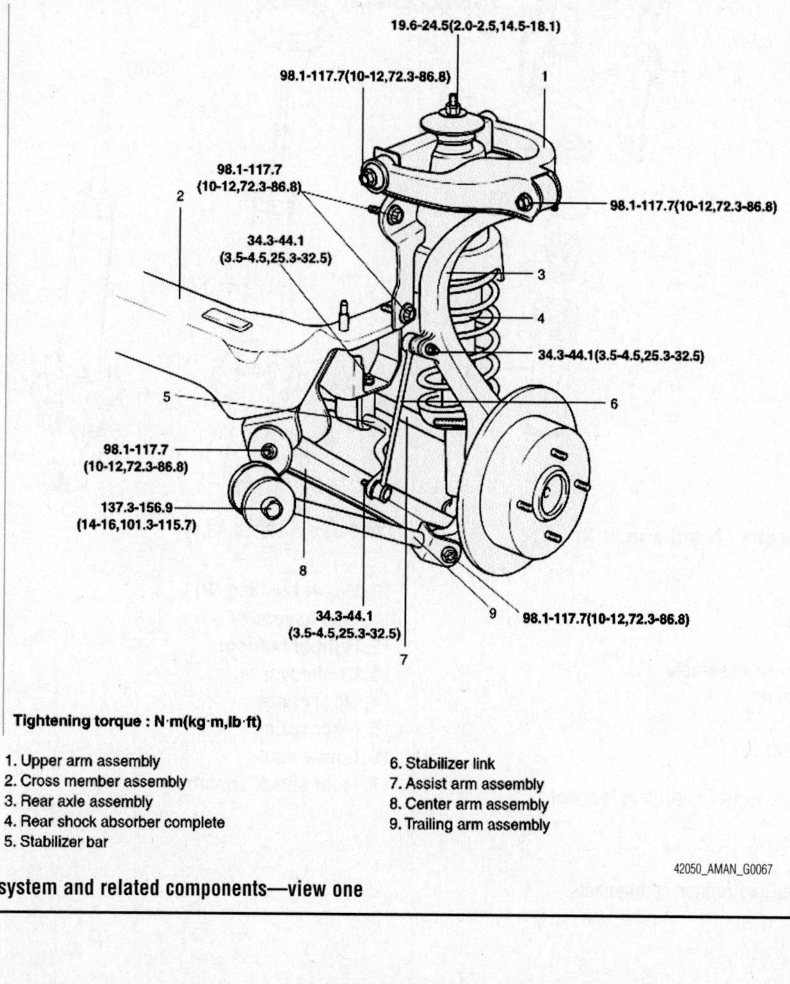

19.6-24.5(2.0-2.5,14.5-18.1)
98.1-117.7(10-12,72.3-86.8)
98.1-117.7 (10-12,72.3-86.8)
98.1-117.7(10-12,72.3-86.8)
34.3-44.1 (3.5-4.5,25.3-32.5)
34.3-44.1(3.5-4.5,25.3-32.5)
98.1-117.7 (10-12,72.3-86.8)
137.3-156.9 (14-16,101.3-115.7)
34.3-44.1 (3.5-4.5,25.3-32.5)
98.1-117.7(10-12,72.3-86.8)

Tightening torque : N·m(kg·m,lb·ft)

1. Upper arm assembly
2. Cross member assembly
3. Rear axle assembly
4. Rear shock absorber complete
5. Stabilizer bar
6. Stabilizer link
7. Assist arm assembly
8. Center arm assembly
9. Trailing arm assembly

42050_AMAN_G0067

Fig. 126 Rear suspension system and related components—view one

for this component, refer to the illustrations as required.

1. Before servicing the vehicle, refer to the Precautions Section.

➡If working near and/or around the SRS system and components, be sure to disable the SRS system. Tape the negative battery cable with insulating tape. Always disconnect the negative battery cable first.

✲✲ CAUTION

To avoid personal injury when working on vehicles equipped with an air bag, the negative battery cable must be disconnected and at least three minutes must elapse before working on the system. Failure to do so may result in deployment of the air bag.

2. Remove the battery top cover, if equipped.

3. Disconnect the negative battery cable. Tape the cable with insulating tape.

SHOCK ABSORBER

REMOVAL & INSTALLATION

See Figure 128.

1. Before servicing the vehicle, refer to the Precautions Section.

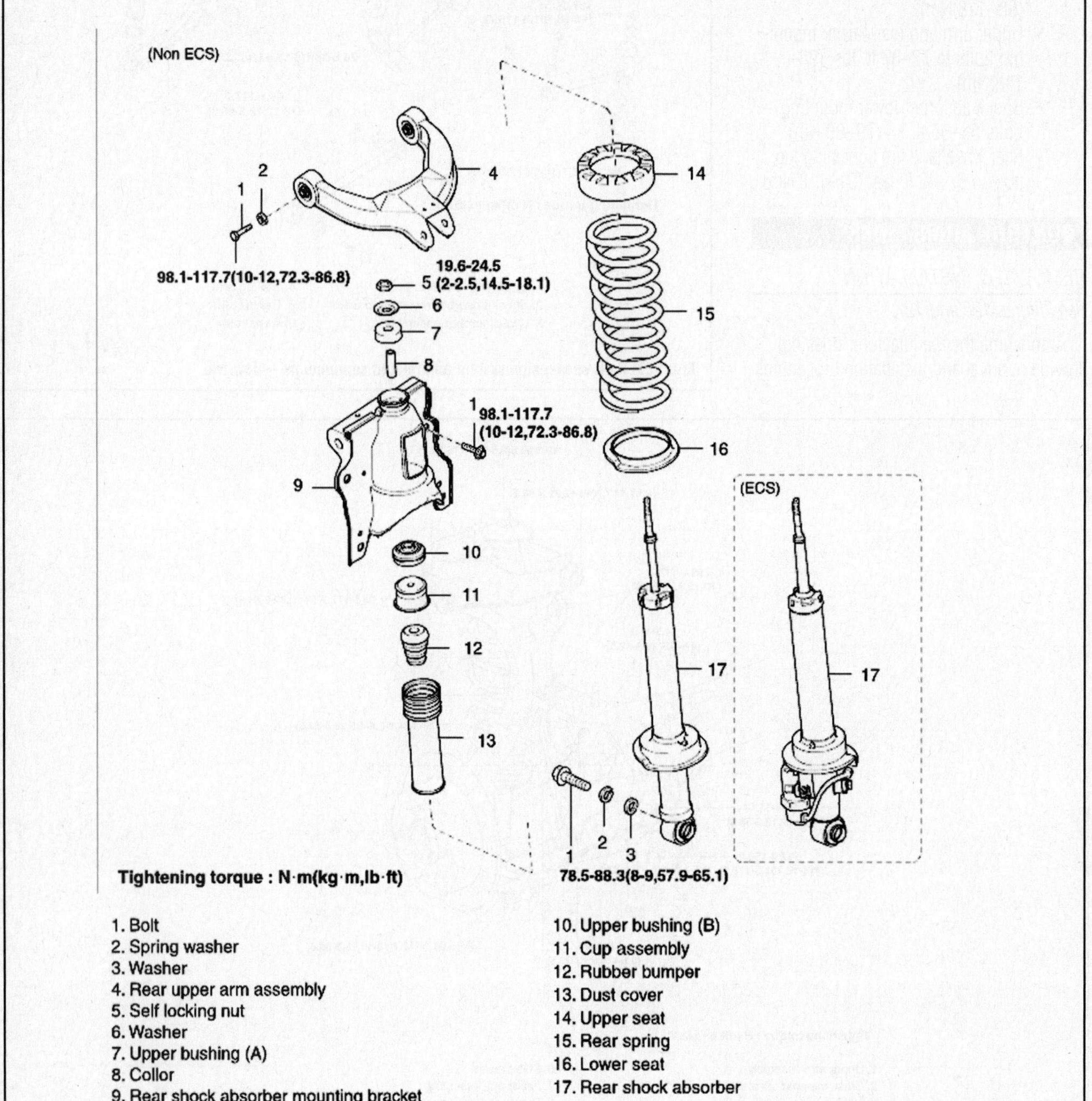

Fig. 128 Rear shock absorber and related components

➡**If working near and/or around the SRS system and components, be sure to disable the SRS system. Tape the negative battery cable with insulating tape. Always disconnect the negative battery cable first.**

⁂ CAUTION

To avoid personal injury when working on vehicles equipped with an air bag, the negative battery cable must be disconnected and at least three minutes must elapse before working on the system. Failure to do so may result in deployment of the air bag.

2. Remove the battery top cover, if equipped.
3. Disconnect the negative battery cable. Tape the cable with insulating tape.
4. Raise and safely support the vehicle.
5. Support the rear axle assembly, as required.
6. Disconnect the lower mounting bolts.
7. Remove the upper arm and rear carrier mounting bolt.
8. Remove the mounting bracket.
9. Remove the component from the vehicle.

To install:

10. When installing the rear shock absorber, be sure to clear the connecting surface.
11. Installation is the reverse of removal.
12. Install or connect the following:
    - Shock absorber mounting bracket tightening to 72–87 ft. lbs. (98–118 Nm).
    - Upper arm and rear carrier mounting bolts to 72–87 ft. lbs. (98–118 Nm).
    - Shock absorber lower mounting bolts 58–65 ft. lb. (79–88 Nm).
    - Rear wheel and tire tightening lug nuts to 67–82 ft .lbs. (91–112 Nm).

WHEEL HUB & BEARING

REMOVAL & INSTALLATION

See Figures 129 through 133.

1. Before servicing the vehicle, refer to the Precautions Section.

➡**If working near and/or around the SRS system and components, be sure to disable the SRS system. Tape the negative battery cable with insulating tape. Always disconnect the negative battery cable first.**

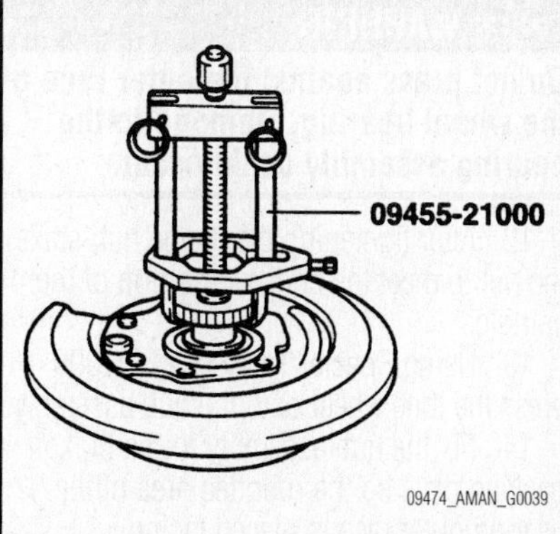

Fig. 129 Removing tone wheel from the hub assembly

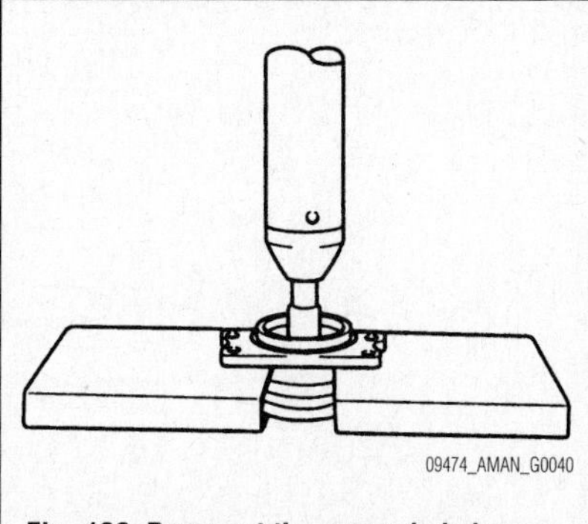

Fig. 130 Press out the rear axle hub

⁂ CAUTION

To avoid personal injury when working on vehicles equipped with an air bag, the negative battery cable must be disconnected and at least three minutes must elapse before working on the system. Failure to do so may result in deployment of the air bag.

2. Remove the battery top cover, if equipped.

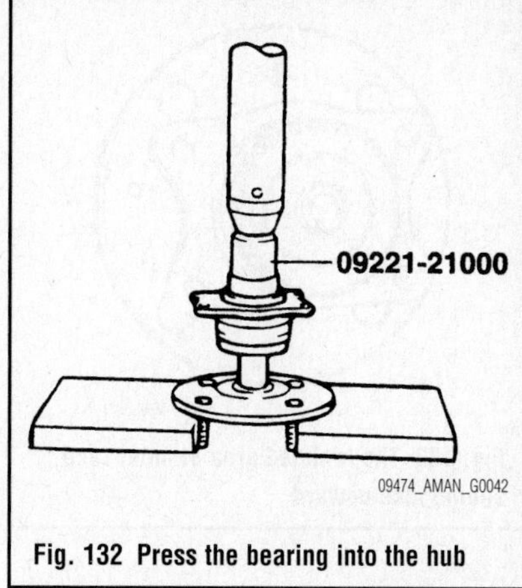

Fig. 132 Press the bearing into the hub

3. Disconnect the negative battery cable. Tape the cable with insulating tape.
4. Release the parking brake.
5. Remove or disconnect the following:
    - Rear wheel
    - Wheel speed sensor, if equipped
    - Caliper assembly
    - Parking brake assembly
    - Rear axle hub mounting bolts
    - Hub assembly
6. Using Special Tool 09455-21000, remove the tone wheel from the hub assembly.
7. Remove the flange nut from the hub assembly.
8. While supporting the flange area of the bearing outer race, press out the axle hub.
9. Using Special Tool 09544-21000, remove the bearing inner race from the axle hub.

To install:

10. Apply a thin coat of grease to the knuckle and bearing contact surfaces.
11. Using Special Tool 09221-21000, press the bearing into the hub.

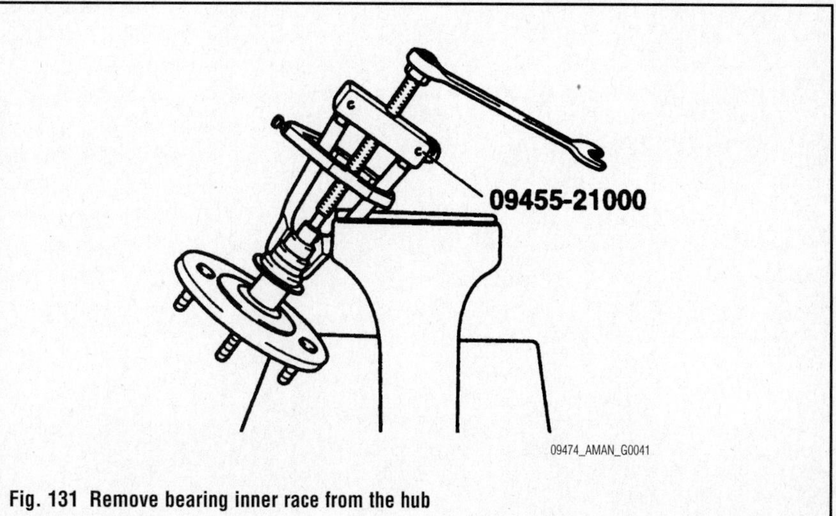

Fig. 131 Remove bearing inner race from the hub

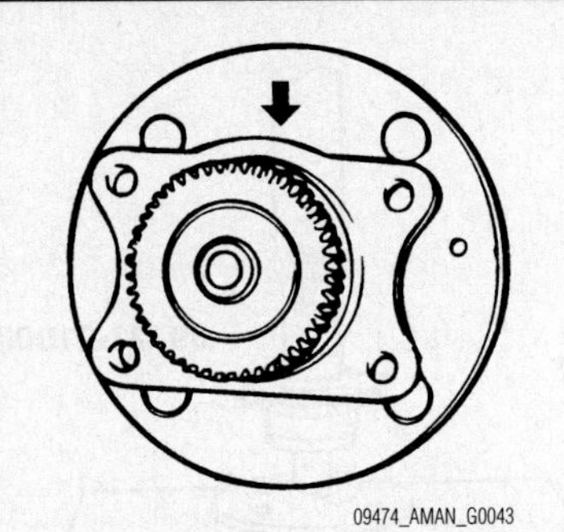

Fig. 133 The rounded area of outer race should face upward

WARNING

Do not press against the outer race of the wheel bearing. Damage to the bearing assembly could occur.

12. After tightening the flange nut, stake the nut to meet the concave portion of the spindle.
13. Using Special Tool 09221-21000, press the tone wheel onto the hub assembly.
14. Fix the hub assembly to the brake backing plate so the rounded area of the bearing outer race is placed facing upward.
15. Install the hub mounting bolts and tighten to 59–74 ft. lbs. (100–120 Nm).
16. Rotate the hub to seat the bearing.
17. Using a spring balance, measure the wheel bearing turning Torque. Torque should be 9 inch lbs. (1 Nm) or less.
18. The remainder of the installation is the reverse of removal.

ADJUSTMENT

If bearings are worn, replace with new sealed bearings.

KIA

Borrego

12

SPECIFICATIONS AND MAINTENANCE CHARTS

ENGINE AND VEHICLE IDENTIFICATION

Engine							Model Year	
Code ①	Liters (cc)	Cu. In.	Cyl.	Fuel Sys.	Engine Type	Eng. Mfg.	Code ②	Year
1	3.8 (3778)	230.5	6	MPFI	DOHC	KIA	9	2009
2	4.6 (4627)	282.4	8	MPFI	DOHC	KIA	A	2010

MPFI: Multi-Point Fuel Injection

DOHC: Double Overhead Camshafts

① 8th digit of VIN

② 10th digit of VIN

37655_BORR_C0001

GENERAL ENGINE SPECIFICATIONS

Year	Engine Displacement Liters	Engine VIN	Net Horsepower @ rpm	Net Torque @ rpm (ft. lbs.)	Bore x Stroke (in.)	Com-pression Ratio	Oil Pressure @ rpm
2009	3.8	1	276@6000	267@4400	3.7795x3.4252	10.4:1	18.77@1000
	4.6	2	337@6000	323@3500	3.6220x3.4252	10.4:1	22.76@1000
2010	3.8	1	276@6000	267@4400	3.7795x3.4252	10.4:1	18.77@1000
	4.6	2	337@6000	323@3500	3.6220x3.4252	10.4:1	22.76@1000

37655_BORR_C0002

ENGINE TUNE-UP SPECIFICATIONS

Year	Engine Displacement Liters	Engine VIN	Spark Plug Gap (in.)	Ignition Timing (deg.)		Fuel Pump (psi)	Idle Speed (rpm)		Valve Clearance	
				MT	AT		MT	AT	Intake	Exhaust
2009	3.8	1	0.039-0.043	—	6-16B	55	—	①	②	③
	4.6	2	0.039-0.043	—	④	55	—	500-700	HYD	HYD
2010	3.8	1	0.039-0.043	—	6-16B	55	—	①	②	③
	4.6	2	0.039-0.043	—	④	55	—	500-700	HYD	HYD

NOTE: The Vehicle Emission Control Information label often reflects specification changes made during production. The label figures must be used if they differ from those in this chart.

B: Before Top Dead Center

HYD: Hydraulic

① N and P range with AC off: 720 +/- 100. D range with AC off: 570 +/- 100.
N and P range AC on: 720 +/- 100. D range AC on: 630 +/- 100

② 0.0067- 0.0090 inch, limit 0.0039-0.0118 inch

③ 0.01060067- 0.0129 inch, limit 0.0078-0.0157 inch

④ N and D range: 0 +/- 10B. D range: 10 +/- 10B.

37655_BORR_C0003

CAPACITIES

Year	Model	Engine Displacement Liters	Engine VIN	Engine Oil with Filter	Transaxle (pts.) Manual	Transaxle (pts.) Auto.	Fuel Tank (gal.)	Cooling System (qts.)
2009	Borrego	3.8	1	5.49	—	① ②	20.6	12.0
		4.6	2	4.75	—	① ②	20.6	14.79
2010	Borrego	3.8	1	5.49	—	① ②	20.6	12.0
		4.6	2	4.75	—	① ②	20.6	14.79

① A5SR2 transaxle: 21.14 pints. 6HP26 transaxle: not available

② Transfer case: 3.17- 3.59 pints

37655_BORR_C0004

FLUID SPECIFICATIONS

Year	Model	Engine Displ. Liters (VIN)	Engine Oil	Auto. Trans.	Drive Axle Front	Drive Axle Rear	Transfer Case	Power Steering Fluid	Brake Master Cylinder	Cooling System
2009	Borrego	3.8 (1)	①	②	③	③	④	PSF 3	DOT 3	NA
		4.6 (2)	①	⑤	③	③	④	PSF 3	DOT 3	NA
2010	Borrego	3.8 (1)	①	②	③	③	④	PSF 3	DOT 3	NA
		4.6 (2)	①	⑤	③	③	④	PSF 3	DOT 3	NA

DOT: Department Of Transpotation

NA: Not Available

① 5W-20/GF4&SM

② APOLLOIL ATF RED-1K or Kia Genuine Red-1

③ Hypoid gear oil (API GL-5, SAE 90)

④ Part time unit: ATF DEXRO. Full time unit: ATF Mobile LT

④ Ethlyene glycol base for aluminium radiator

⑤ Shell M-1375.4 and ZF-lifegardfluid 6

37655_BORR_C0014

VALVE SPECIFICATIONS

Year	Engine Displacement Liters	Engine VIN	Seat Angle (deg.)	Face Angle (deg.)	Maximum out of Square (in.)	Spring Free Length (in.)	Stem-to-Guide Clearance (in.) Intake	Stem-to-Guide Clearance (in.) Exhaust	Stem Diameter (in.) Intake	Stem Diameter (in.) Exhaust
2009	3.8	1	44.75-45.20	45.25-45.75	①	1.7267	0.00078-0.00185	0.00118-0.00212	0.2151-0.2157	0.2149-0.2153
	4.6	2	44.75-45.10	45.25-45.75	①	1.8819	0.0008-0.0018	0.0012-0.0021	0.2348-0.2354	0.2346-0.2350
2010	3.8	1	44.75-45.20	45.25-45.75	①	1.7267	0.00078-0.00185	0.00118-0.00212	0.2151-0.2157	0.2149-0.2153
	4.6	2	44.75-45.10	45.25-45.75	①	1.8819	0.0008-0.0018	0.0012-0.0021	0.2348-0.2354	0.2346-0.2350

① Less than 1.5 degrees

37655_BORR_C0005

CAMSHAFT AND BEARING SPECIFICATIONS

All measurements are given in inches unless noted.

Year	Engine Displacement Liters	Engine VIN	Cam Height Intake	Camshaft Cam Height Exhaust	Shaft End-play	Bearing Cap Torque (ft. lbs.)
2009	3.8	1	①	1.8031	②	③
	4.6	2	NA	NA	④	⑤
2010	3.8	1	①	1.8031	②	③
	4.6	2	NA	NA	④	⑤

NA: Not Available

① 1.8228- 1.8425

② 0.0008- 0.0071

③ 7.2- 8.7 ft. lbs. (9.8- 11.8 Nm)

④ RH- intake: 0.0047- 0.0086. RH- exhaust, LH intake and exhaust: 0.0047- 0.0086.

⑤ 10.1- 10.8 ft. lbs. (13.7- 14.7 Nm)

37655_BORR_C0013

CRANKSHAFT AND CONNECTING ROD SPECIFICATIONS

All measurements are given in inches.

Year	Engine Displacement Liters	Engine VIN	Crankshaft Main Brg. Journal Dia.	Crankshaft Main Brg. Oil Clearance	Crankshaft Shaft End-play	Crankshaft Thrust on No.	Connecting Rod Journal Diameter	Connecting Rod Oil Clearance	Connecting Rod Side Clearance
2009	3.8	1	2.7142-2.7149	0.0008-0.0016	0.0039-0.0110	3	2.2834-2.8420	0.0015-0.0022	0.0039-0.0098
	4.6	2	2.5583-2.5591	0.0002-0.0009	0.0039-0.0110	3	2.1654-2.1661	0.0007-0.0014	0.0039-0.0118
2010	3.8	1	2.7142-2.7149	0.0008-0.0016	0.0039-0.0110	3	2.2834-2.8420	0.0015-0.0022	0.0039-0.0098
	4.6	2	2.5583-2.5591	0.0002-0.0009	0.0039-0.0110	3	2.1654-2.1661	0.0007-0.0014	0.0039-0.0118

37655_BORR_C0007

PISTON AND RING SPECIFICATIONS

All measurements are given in inches.

Year	Engine Displacement Liters	Engine VIN	Piston Clearance	Ring Gap Top Compression	Ring Gap Bottom Compression	Ring Gap Oil Control	Ring Side Clearance Top Compression	Ring Side Clearance Bottom Compression	Ring Side Clearance Oil Control
2009	3.8	1	0.0012-0.0020	0.0067-0.0126	0.0126-0.0185	0.0078-0.0275	0.0012-0.0027	0.0012-0.0027	0.0024-0.0059
	4.6	2	0.0018-0.0026	0.0067-0.0126	0.0146-0.0205	0.0079-0.0276	0.0016-0.0031	0.0016-0.0031	0.0024-0.0059
2010	3.8	1	0.0012-0.0020	0.0067-0.0126	0.0126-0.0185	0.0078-0.0275	0.0012-0.0027	0.0012-0.0027	0.0024-0.0059
	4.6	2	0.0018-0.0026	0.0067-0.0126	0.0146-0.0205	0.0079-0.0276	0.0016-0.0031	0.0016-0.0031	0.0024-0.0059

37655_BORR_C0006

TORQUE SPECIFICATIONS

All readings in ft. lbs.

Year	Engine Displacement Liters	Cylinder Head Bolts	Main Bearing Bolts	Rod Bearing Bolts	Crankshaft Damper Bolts	Flywheel Bolts	Manifold Intake	Manifold Exhaust	Spark Plugs	Oil Pan Drain Plug
2009	3.8	①	②	③	209.8-224.2	52.8-55.7	④	28.9-32.6	15-21	⑤
	4.6	⑥	⑦	⑧	289.3-296.6	72.3-79.6	14.5-19.5	36-2-39.8	18.1-21.7	NA
2010	3.8	①	②	③	209.8-224.2	52.8-55.7	④	28.9-32.6	15-21	⑤
	4.6	⑥	⑦	⑧	289.3-296.6	72.3-79.6	14.5-19.5	36-2-39.8	18.1-21.7	NA

NA: Not Available

① 27.5-30.4 ft. lbs., plus 118-122 degrees, plus an additional 88-92 degrees (for 16 bolts). 13.7-17.4 ft. lbs (for 1 bolt).

② Inner cap bolt (M11): 36.2 ft. lbs., plus 90 degrees. Outer cap bolt (M8): 14.5 ft. lbs., plus 120 degrees. Side cap bolts (M8): 21.7-23.1 ft. lbs.

③ 13.0-15.9 ft. lbs., plus 88-92 degrees

④ Nut: 13.7-17.4 ft. lbs. Bolt: 19.5-23.1 ft. lbs.

⑤ Bolt cap: 25.3-32.5 ft. lbs. Cover bolt: 7.2-8.7 ft. lbs.

⑥ Long bolts (20): 23.9-26.8 ft. lbs., plus 88-92 degrees, plus an additional 118-122 degrees. Flange bolt (4): 23.9-26.8 ft. lbs.

⑦ Main bearing (20): 27.5-30.4 ft. lbs., plus 120 degrees. Flange (15): 15.9-18.8 ft. lbs.

⑧ 16.6-19.5 ft. lbs., plus 98-102 degrees

37655_BORR_C0008

WHEEL ALIGNMENT

Year	Model		Caster Range (+/-Deg.)	Caster Preferred Setting (Deg.)	Camber Range (+/-Deg.)	Camber Preferred Setting (Deg.)	Toe-in (in.)
2009	Borrego	F	3.80 +/- 0.5	NA	-0.50 +/- 0.5	—	①
		R	—	—	-1.00 +/- 0.5	—	②
2010	Borrego	F	3.80 +/- 0.5	NA	-0.50 +/- 0.5	—	①
		R	—	—	-1.00 +/- 0.5	—	②

NA: Not Available

① Total 0.15 +/- 0.15. Individual 0.075 +/- 0.075

② Total 0.15 +/- 0.15. Individual 0.075 +/- 0.075

37655_BORR_C0009

TIRE, WHEEL AND BALL JOINT SPECIFICATIONS

Year	Model	OEM Tires Standard	OEM Tires Optional	Tire Pressures (psi) Front	Tire Pressures (psi) Rear	Wheel Size	Ball Joint Inspection	Lug Nut Torque (ft. lbs.)
2009	Borrego	245/70 R17	265/60 R18	32	32	①	②	65.1-79.6
2010	Borrego	245/70 R17	265/60 R18	32	32	①	②	65.1-79.6

OEM: Original Equipment Manufacturer

PSI: Pounds Per Square Inch

① 7.5Jx17 or 7.5Jx18

② Replace if any measureable movent is found.

37655_BORR_C0010

BRAKE SPECIFICATIONS

All measurements in inches unless noted

Year	Model		Brake Disc Original Thickness	Brake Disc Minimum Thickness	Brake Disc Maximum Run-out	Minimum Lining Thickness	Brake Caliper Bracket Bolts (ft. lbs.)	Brake Caliper Mounting Bolts (ft. lbs.)
2009	Borrego	Front	1.100	NA	NA	0.0787	57.9-72.3	15.9-23.1
		Rear	0.510	NA	NA	0.0787	47.0-54.2	15.9-23.1
2010	Borrego	Front	1.100	NA	NA	0.0787	57.9-72.3	15.9-23.1
		Rear	0.510	NA	NA	0.0787	47.0-54.2	15.9-23.1

NA: Not Available

F: Front

R: Rear

37655_BORR_C0011

SCHEDULED MAINTENANCE INTERVALS

Kia - Borrego

TO BE SERVICED	TYPE OF SERVICE	VEHICLE MILEAGE INTERVAL (x1000)												
		7.5	15	22.5	30	37.5	45	52.5	60	67.5	75	82.5	90	97.5
Accessory drive belts	S/I	✓	✓	✓	✓	✓	✓	✓	✓	✓	✓	✓	✓	✓
Air cleaner filter	R		✓		✓		✓		✓		✓		✓	
Air conditioner system filter	S/I	Inspect and replace every 10,000 miles or 10 months												
Brake lines, hoses and connections	S/I		✓		✓		✓		✓		✓		✓	
Chassis and body fasteners	T	✓	✓	✓	✓	✓	✓	✓	✓	✓	✓	✓	✓	✓
Cooling system hoses and coolant level	S/I		✓		✓		✓		✓		✓		✓	
CV-joint boots	S/I				✓				✓				✓	
Engine coolant	R								✓					
Engine oil and filter	R	✓	✓	✓	✓	✓	✓	✓	✓	✓	✓	✓	✓	✓
Exhaust system heat shields	S/I				✓				✓				✓	
Front and rear brakes	S/I	✓	✓	✓	✓	✓	✓	✓	✓	✓	✓	✓	✓	✓
Front and rear differentials	S/I	✓	✓	✓	✓	✓	✓	✓	✓	✓	✓	✓	✓	✓
Front ball joints	S/I				✓				✓				✓	
Fuel filter	R					✓					✓			
Fuel lines and hoses	S/I	✓	✓	✓	✓	✓	✓	✓	✓	✓	✓	✓	✓	✓
Locks and hinges	L	✓	✓	✓	✓	✓	✓	✓	✓	✓	✓	✓	✓	✓
Spark plugs (Iridium coated)	R	Replace every 100,000 miles or 120 months												
Steering operation and linkage	S/I	✓	✓	✓	✓	✓	✓	✓	✓	✓	✓	✓	✓	✓
Transfer case	S/I					✓					✓			
Timing belt	R								✓					

R: Replace S/I: Service or Inspect L: Lubricate T: Tighten

FREQUENT OPERATION MAINTENANCE (SEVERE SERVICE)

If a vehicle is operated under any of the following conditions it is considered severe service

- Towing a trailer or using a camper or car-top carrier

- Repeated short trips of less than 5 miles in temperatures below freezing, or trips of less than 10 miles in any temperature

- Prolonged idling (vehicle operation in stop and go traffic).

- Operating on rough, muddy, unpaved, dusty or salt-covered roads.

- Police, taxi, delivery usage or trailer towing usage.

- Driving in extremely hot (over 90°F) conditions

Oil & oil filter: change every 5000 miles or 5 months, whichever occurs first.

Air cleaner filter: inspect every 15,000 miles or 15 months and replace everything 30,000 miles or 30 months, whichever occurs fir:

Fuel system hoses: inspect/replace every 105,000 miles

Emission system hoses: inspect every 60,000 miles or 60 months, which occurs first

Front and rear brakes: inspect every 15,000 miles or 15 months, whichever occurs first

Chassis and body fasteners: tighten every 15,000 miles or 15 months, whichever occurs first

Locks and hinges: lubricate every 5000 miles or 5 months, whichever occurs first

37655_BORR_C0012

BRAKES — INFORMATION AND PRECAUTIONS

ANTI-LOCK SYSTEMS

- Certain components within the ABS system are not intended to be serviced or repaired individually.
- Do not use rubber hoses or other parts not specifically specified for and ABS system. When using repair kits, replace all parts included in the kit. Partial or incorrect repair may lead to functional problems and require the replacement of components.
- Lubricate rubber parts with clean, fresh brake fluid to ease assembly. Do not use shop air to clean parts; damage to rubber components may result.
- Use only DOT 3 brake fluid from an unopened container.
- If any hydraulic component or line is removed or replaced, it may be necessary to bleed the entire system.
- A clean repair area is essential. Always clean the reservoir and cap thoroughly before removing the cap. The slightest amount of dirt in the fluid may plug an orifice and impair the system function. Perform repairs after components have been thoroughly cleaned; use only denatured alcohol to clean components. Do not allow ABS components to come into contact with any substance containing mineral oil; this includes used shop rags.
- The Anti-Lock control unit is a microprocessor similar to other computer units in the vehicle. Ensure that the ignition switch is **OFF** before removing or installing controller harnesses. Avoid static electricity discharge at or near the controller.
- If any arc welding is to be done on the vehicle, the control unit should be unplugged before welding operations begin.

DISC AND DRUM SYSTEMS

**** CAUTION**

Dust and dirt accumulating on brake parts during normal use may contain asbestos fibers from production or aftermarket brake linings. Breathing excessive concentrations of asbestos fibers can cause serious bodily harm. Exercise care when servicing brake parts. Do not sand or grind brake lining unless equipment used is designed to contain the dust residue. Do not clean brake parts with compressed air or by dry brushing. Cleaning should be done by dampening the brake components with a fine mist of water, then wiping the brake components clean with a dampened cloth. Dispose of cloth and all residue containing asbestos fibers in an impermeable container with the appropriate label. Follow practices prescribed by the Occupational Safety and Health Administration (OSHA) and the Environmental Protection Agency (EPA) for the handling, processing, and disposing of dust or debris that may contain asbestos fibers.

BRAKES — BLEEDING THE BRAKE SYSTEM

BLEEDING PROCEDURE

BLEEDING PROCEDURE

See Figures 1 through 3.

The ABS brake system is bled in the usual fashion with no special procedures required.

**** WARNING**

When bleeding the brakes, note the following:

- Do not reuse the drained fluid
- Always use Genuine DOT 3 or DOT 4 Brake Fluid. Using a non-Genuine DOT3 or DOT 4 brake fluid can cause corrosion and decrease the life of the system
- Make sure no dirt of other foreign matter is allowed to contaminate the brake fluid
- Do not spill brake fluid on the vehicle, it may damage the paint; if brake fluid does contact the paint, wash it off immediately with water
- The reservoir on the master cylinder must be at the MAX (upper) level mark at the start of bleeding procedure and checked after bleeding each brake caliper. Add fluid as required

1. Before servicing the vehicle, refer to the Precautions Section.
2. Make sure the brake fluid level in the master cylinder fluid reservoir is at the MAX (upper) level line.
3. Have someone slowly pump the brake pedal several times, and then apply steady pressure
4. Loosen the right-rear brake bleed screw to allow air to escape from the system. Then tighten the bleed screw securely
5. Repeat the procedure for each wheel in the sequence shown below until air bubbles no longer appear in the fluid
6. Refill the master cylinder reservoir to the MAX (upper) level line

BLEEDING THE ABS SYSTEM

See Figures 4 through 6.

The ABS brake system is bled in the usual fashion with no special procedures required.

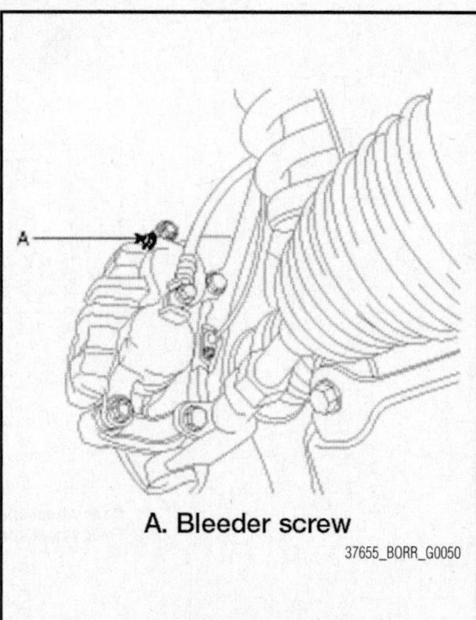

Fig. 1 Front caliper bleed screw location

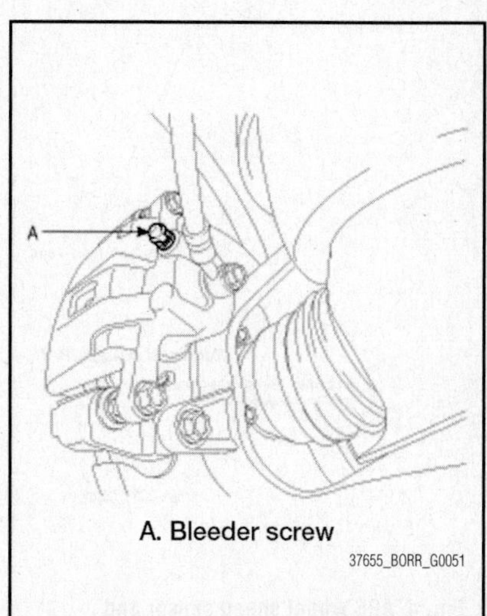

Fig. 2 Rear caliper bleed screw location

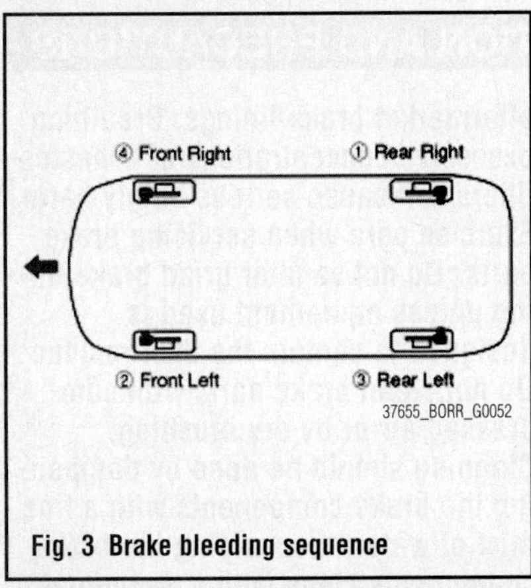

Fig. 3 Brake bleeding sequence

⁂ WARNING

When bleeding the brakes, note the following:

- Do not reuse the drained fluid
- Always use Genuine DOT 3 or DOT 4 Brake Fluid. Using a non-Genuine DOT3 or DOT 4 brake fluid can cause corrosion and decrease the life of the system
- Make sure no dirt of other foreign matter is allowed to contaminate the brake fluid
- Do not spill brake fluid on the vehicle, it may damage the paint; if brake fluid does contact the paint, wash it off immediately with water
- The reservoir on the master cylinder must be at the MAX (upper) level mark at the start of bleeding procedure and checked after bleeding each brake caliper. Add fluid as required

1. Before servicing the vehicle, refer to the Precautions Section.
2. Make sure the brake fluid level in the master cylinder fluid reservoir is at the MAX (upper) level line.
3. Have someone slowly pump the brake pedal several times, and then apply steady pressure
4. Loosen the right-rear brake bleed screw to allow air to escape from the system. Then tighten the bleed screw securely
5. Repeat the procedure for each wheel in the sequence shown below until air bubbles no longer appear in the fluid
6. Refill the master cylinder reservoir to the MAX (upper) level line

BRAKES — ANTI-LOCK BRAKE SYSTEM (ABS)

WHEEL SPEED SENSORS

REMOVAL & INSTALLATION

See Figures 4 and 5.

1. Before servicing the vehicle, refer to the Precautions Section.

➡If working near and/or around the SRS system and components, be sure to disable the SRS system. Tape the negative battery cable with insulating tape. Always disconnect the negative battery cable first.

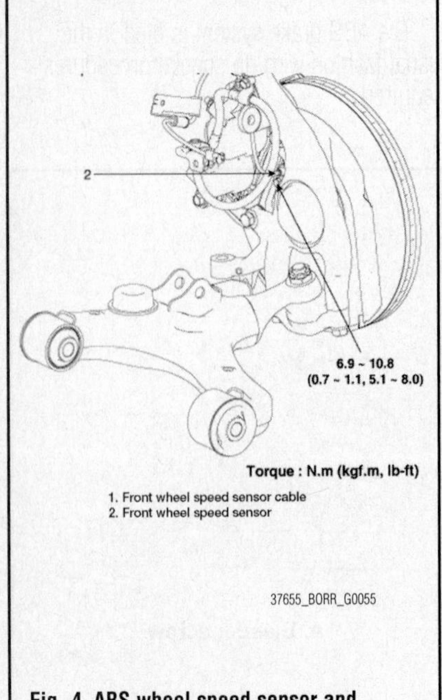

Fig. 4 ABS wheel speed sensor and related components—front

⁂ CAUTION

To avoid personal injury when working on vehicles equipped with an air bag, the negative battery cable must be disconnected and at least three minutes must elapse before working on the system. Failure to do so may result in deployment of the air bag.

2. Remove the battery top cover, if equipped.
3. Disconnect the negative battery cable. Tape the cable with insulating tape.

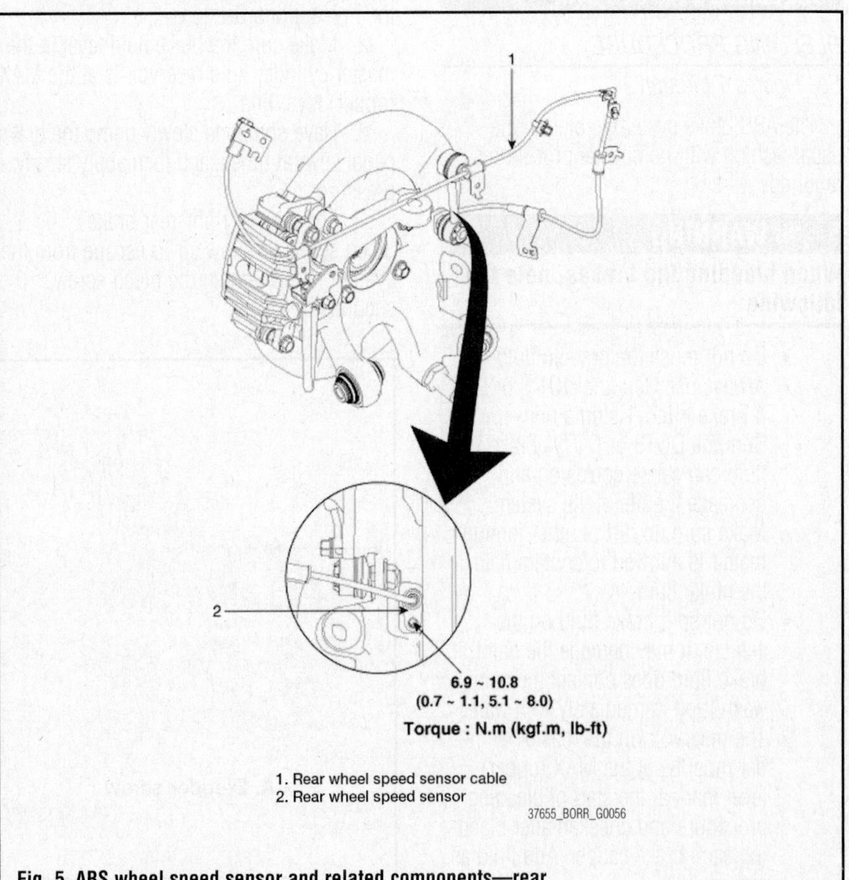

Fig. 5 ABS wheel speed sensor and related components—rear

4. Raise and safely support the vehicle.
5. Remove the sensor mounting bolt.
6. Disconnect the electrical connector.
7. Remove the sensor bracket(s).
8. Remove the wheel guard.
9. Remove the component from its mounting.

To install:

➡Be sure to use new fasteners, as required.

10. Installation is the reverse of the removal procedure.
11. Be sure to reprogram the necessary systems.
12. Reset/Initialize the sunroof, if equipped.

BRAKES — FRONT DISC BRAKES

BRAKE CALIPER

REMOVAL & INSTALLATION

See Figure 6.

1. Before servicing the vehicle, refer to the Precautions Section.

➡If working near and/or around the SRS system and components, be sure to disable the SRS system. Tape the negative battery cable with insulating tape. Always disconnect the negative battery cable first.

⁂ CAUTION

To avoid personal injury when working on vehicles equipped with an air bag, the negative battery cable must be disconnected and at least three minutes must elapse before working on the system. Failure to do so may result in deployment of the air bag.

2. Remove the battery top cover, if equipped.
3. Disconnect the negative battery cable. Tape the cable with insulating tape.
4. Drain the brake fluid to an acceptable level.
5. Raise and safely support the vehicle.
6. Remove the tire and wheel assemblies.
7. Disconnect and plug the brake line.
8. Remove the caliper retaining bolts.

➡Do not allow the caliper to hang by the brake hose. Position it to the side using mechanics wire.

9. Remove the caliper from its mounting.

To install:

➡Be sure to use new fasteners, as required.

10. Installation is the reverse of the removal procedure.
11. Be sure to reprogram the necessary systems.
12. Reset/Initialize the sunroof, if equipped.
13. Fill the master cylinder with the proper grade and type fluid.
14. Bleed the brake system.
15. Check for leaks, correct as required.

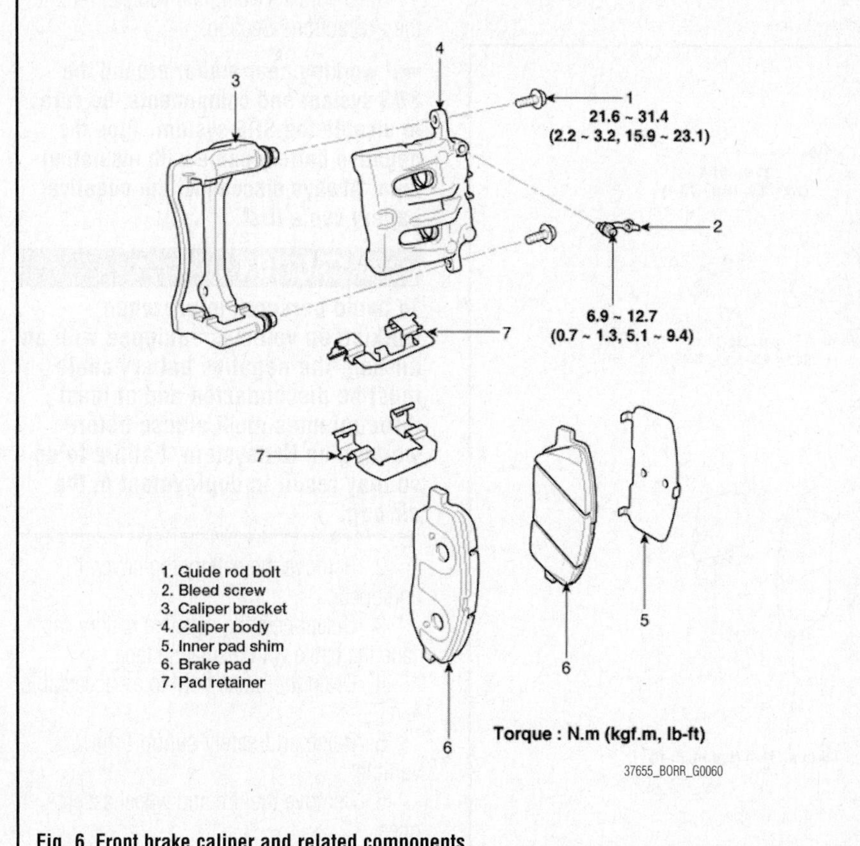

Fig. 6 Front brake caliper and related components

DISC BRAKE PADS

REMOVAL & INSTALLATION

See Figure 6.

1. Before servicing the vehicle, refer to the Precautions Section.

➡If working near and/or around the SRS system and components, be sure to disable the SRS system. Tape the negative battery cable with insulating tape. Always disconnect the negative battery cable first.

⁂ CAUTION

To avoid personal injury when working on vehicles equipped with an air bag, the negative battery cable must be disconnected and at least three minutes must elapse before working on the system. Failure to do so may result in deployment of the air bag.

2. Remove the battery top cover, if equipped.
3. Disconnect the negative battery cable. Tape the cable with insulating tape.
4. Drain the brake fluid to an acceptable level.
5. Raise and safely support the vehicle.
6. Remove the tire and wheel assemblies.

7. Remove the caliper guide bolt and pivot the caliper up out of the way.

➡Do not allow the caliper to hang by the brake hose. Position it to the side using mechanics wire.

8. Remove the pad shim, pad retainers and brake pads in the caliper bracket.

To install:

➡Be sure to use new fasteners, as required.

9. Installation is the reverse of the removal procedure.
10. Be sure to reprogram the necessary systems.
11. Reset/Initialize the sunroof, if equipped.
12. Fill the master cylinder with the proper grade and type fluid.
13. Bleed the brake system, as required.
14. Check for leaks, correct as required.

BRAKES — REAR DISC BRAKES

BRAKE CALIPER

REMOVAL & INSTALLATION

See Figure 7.

1. Before servicing the vehicle, refer to the Precautions Section.

➡If working near and/or around the SRS system and components, be sure to disable the SRS system. Tape the negative battery cable with insulating tape. Always disconnect the negative battery cable first.

✻✻ CAUTION

To avoid personal injury when working on vehicles equipped with an air bag, the negative battery cable must be disconnected and at least three minutes must elapse before working on the system. Failure to do so may result in deployment of the air bag.

2. Remove the battery top cover, if equipped.
3. Disconnect the negative battery cable. Tape the cable with insulating tape.
4. Drain the brake fluid to an acceptable level.
5. Raise and safely support the vehicle.
6. Remove the tire and wheel assemblies.
7. Disconnect and plug the brake line.
8. Remove the caliper retaining bolts.

➡Do not allow the caliper to hang by the brake hose. Position it to the side using mechanics wire.

9. Remove the caliper from its mounting.

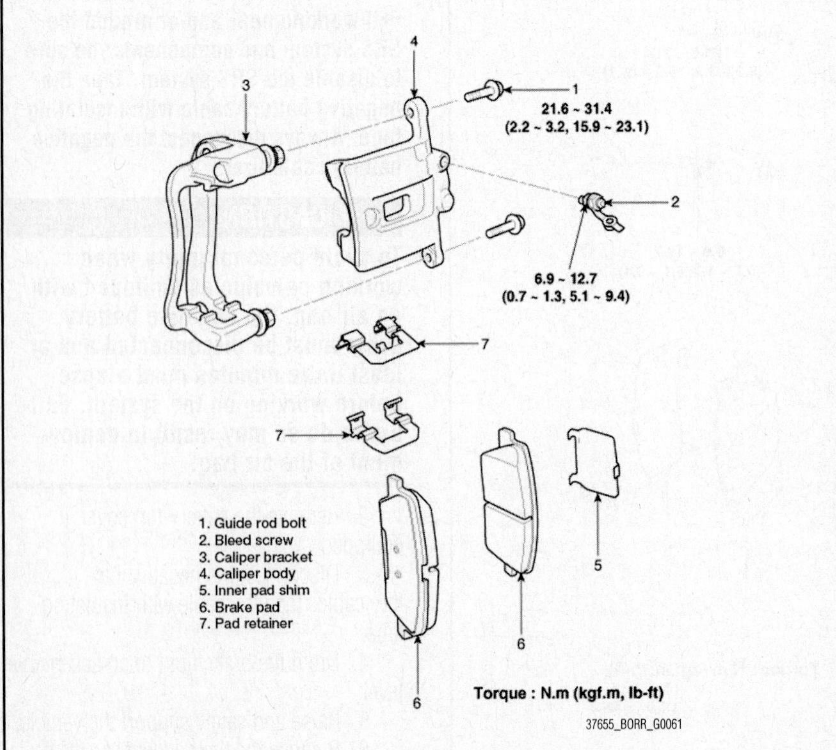

Fig. 7 Rear brake caliper and related components

To install:

➡Be sure to use new fasteners, as required.

10. Installation is the reverse of the removal procedure.
11. Be sure to reprogram the necessary systems.
12. Reset/Initialize the sunroof, if equipped.
13. Fill the master cylinder with the proper grade and type fluid.
14. Bleed the brake system.
15. Check for leaks, correct as required.

DISC BRAKE PADS

REMOVAL & INSTALLATION

See Figure 7.

1. Before servicing the vehicle, refer to the Precautions Section.

➡If working near and/or around the SRS system and components, be sure to disable the SRS system. Tape the negative battery cable with insulating tape. Always disconnect the negative battery cable first.

✻✻ CAUTION

To avoid personal injury when working on vehicles equipped with an air bag, the negative battery cable must be disconnected and at least three minutes must elapse before working on the system. Failure to do so may result in deployment of the air bag.

2. Remove the battery top cover, if equipped.
3. Disconnect the negative battery cable. Tape the cable with insulating tape.
4. Drain the brake fluid to an acceptable level.
5. Raise and safely support the vehicle.
6. Remove the tire and wheel assemblies.

7. Remove the caliper guide bolt and pivot the caliper up out of the way.

➡**Do not allow the caliper to hang by the brake hose. Position it to the side using mechanics wire.**

8. Remove the pad shim, pad retainers and brake pads in the caliper bracket.

To install:

➡**Be sure to use new fasteners, as required.**

9. Installation is the reverse of the removal procedure.
10. Be sure to reprogram the necessary systems.
11. Reset/Initialize the sunroof, if equipped.
12. Fill the master cylinder with the proper grade and type fluid.
13. Bleed the brake system, as required.
14. Check for leaks, correct as required.

BRAKES — PARKING BRAKE

PARKING BRAKE CABLES

ADJUSTMENT

At this time the manufacturer does not provide adjustment procedures for this component.

PARKING BRAKE SHOES

REMOVAL & INSTALLATION

See Figure 8.

1. Before servicing the vehicle, refer to the Precautions Section.

➡**If working near and/or around the SRS system and components, be sure to disable the SRS system. Tape the negative battery cable with insulating tape. Always disconnect the negative battery cable first.**

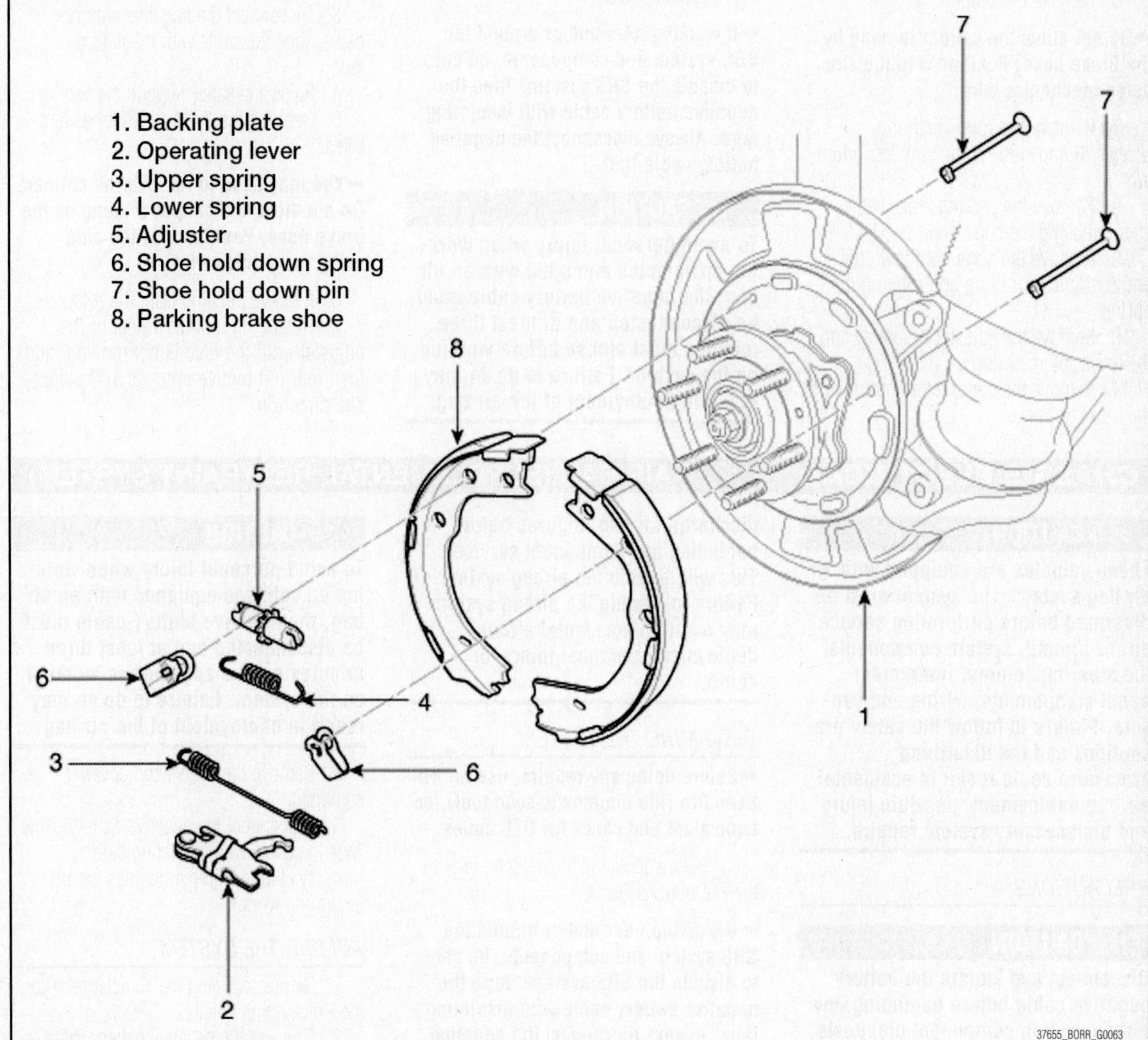

Fig. 8 Rear parking brake shoes and related components

⁂ CAUTION

To avoid personal injury when working on vehicles equipped with an air bag, the negative battery cable must be disconnected and at least three minutes must elapse before working on the system. Failure to do so may result in deployment of the air bag.

2. Remove the battery top cover, if equipped.
3. Disconnect the negative battery cable. Tape the cable with insulating tape.
4. Raise and safely support the vehicle.
5. Remove the tire and wheel assemblies.
6. Remove the caliper.

➡Do not allow the caliper to hang by the brake hose. Position it to the side using mechanics wire.

7. Remove the rotor retaining screws. Remove the rotor from its mounting.
8. Remove the parking brake cable, after removing the clip.
9. Remove the shoe holddown pin and spring by pressing and rotating the spring.
10. Remove the adjuster assembly and the lower return spring.
11. Remove the upper return spring and the brake shoes.

To install:

➡Be sure to use new fasteners, as required.

12. Installation is the reverse of the removal procedure.
13. Be sure to reprogram the necessary systems.
14. Reset/Initialize the sunroof, if equipped.
15. Adjust the parking brake.
16. Fill the master cylinder with the proper grade and type fluid.
17. Bleed the brake system, as required.
18. Check for leaks, correct as required.

ADJUSTMENT

See Figure 9.

1. Before servicing the vehicle, refer to the Precautions Section.

➡If working near and/or around the SRS system and components, be sure to disable the SRS system. Tape the negative battery cable with insulating tape. Always disconnect the negative battery cable first.

⁂ CAUTION

To avoid personal injury when working on vehicles equipped with an air bag, the negative battery cable must be disconnected and at least three minutes must elapse before working on the system. Failure to do so may result in deployment of the air bag.

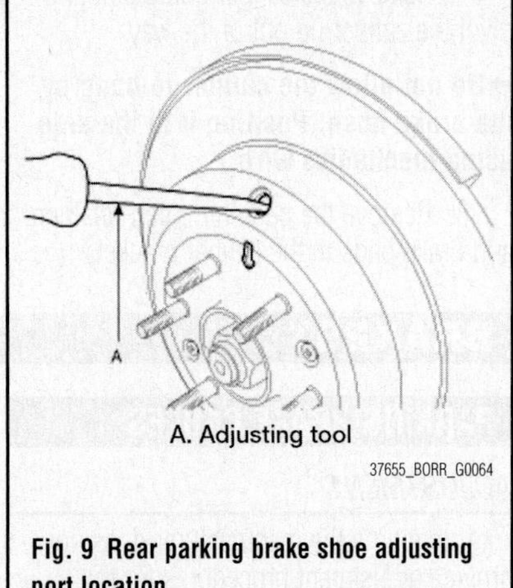

Fig. 9 Rear parking brake shoe adjusting port location

2. Remove the battery top cover, if equipped.
3. Disconnect the negative battery cable. Tape the cable with insulating tape.
4. Raise and safely support the vehicle.
5. Remove the tire and wheel assemblies.

➡You may have to remove the caliper. Do not allow the caliper to hang by the brake hose. Position it to the side using mechanics wire.

6. Remove the plug from the rotor.
7. Rotate the toothed wheel of the adjuster until the disc is not moving, and then return it by five notches in the opposite direction.

CHASSIS ELECTRICAL

AIR BAG (SUPPLEMENTAL RESTRAINT SYSTEM)

⁂ CAUTION

These vehicles are equipped with an air bag system. The system must be disarmed before performing service on, or around, system components, the steering column, instrument panel components, wiring and sensors. Failure to follow the safety precautions and the disarming procedure could result in accidental air bag deployment, possible injury and unnecessary system repairs.

PRECAUTIONS

⁂ CAUTION

Disconnect and isolate the battery negative cable before beginning any airbag system component diagnosis, testing, removal, or installation procedures. Allow system capacitor to discharge for two minutes before beginning any component service. This will disable the airbag system. Failure to disable the airbag system may result in accidental airbag deployment, personal injury, or death.

DISARMING THE SYSTEM

➡Before doing any repairs, use the Hi-Scan Pro (Kia diagnostic scan tool), or equivalent and check for DTC codes.

1. Before servicing the vehicle, refer to the Precautions Section.

➡If working near and/or around the SRS system and components, be sure to disable the SRS system. Tape the negative battery cable with insulating tape. Always disconnect the negative battery cable first.

⁂ CAUTION

To avoid personal injury when working on vehicles equipped with an air bag, the negative battery cable must be disconnected and at least three minutes must elapse before working on the system. Failure to do so may result in deployment of the air bag.

2. Remove the battery top cover, if equipped.
3. Disconnect the negative battery cable. Tape the cable with insulating tape.
4. Wait at least three minutes before beginning work.

ARMING THE SYSTEM

1. Before servicing the vehicle, refer to the Precautions Section.
2. Connect the negative battery cable.
3. Turn the ignition switch **ON** .

4. Verify that the air bag indicator illuminates for about six seconds, then goes off.

CLOCKSPRING CENTERING

See Figure 10.

➡Be sure that the front wheels are in the straight ahead position. Set the center position by getting marks between the clockspring and the cover into line. Make an array the mark (><) by turning the clockspring to the stop and then 2.4 revolutions counterclockwise. If the mating mark of the clockspring is not properly aligned, the steering wheel may not completely rotate during a turn, or the flat cable within the clockspring may be severed, obstructing normal operation of the SRS and possibly leading to serious injury to the vehicle's driver.

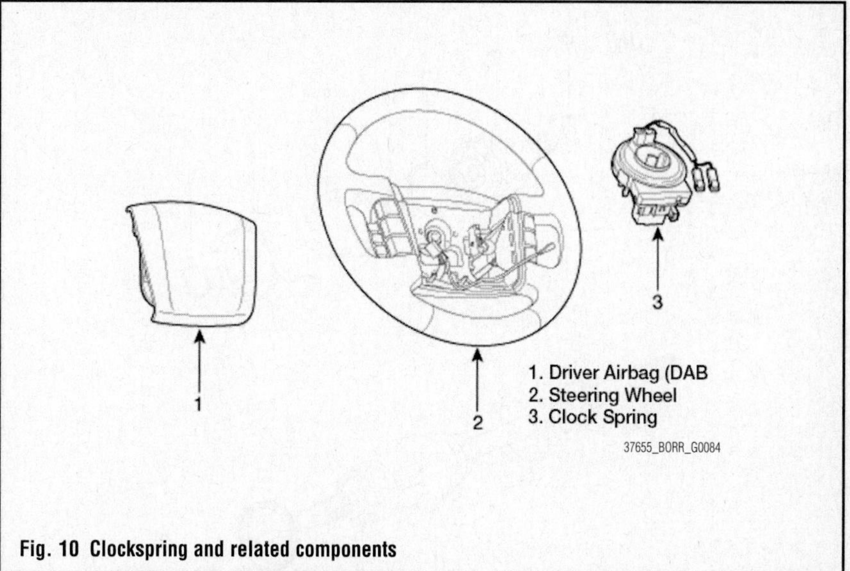

Fig. 10 Clockspring and related components

DRIVE TRAIN

FRONT HALFSHAFT

REMOVAL & INSTALLATION

See Figures 11 through 13.

1. Before servicing the vehicle, refer to the Precautions Section.

➡If working near and/or around the SRS system and components, be sure to disable the SRS system. Tape the negative battery cable with insulating tape. Always disconnect the negative battery cable first.

⁂ CAUTION

To avoid personal injury when working on vehicles equipped with an air bag, the negative battery cable must be disconnected and at least three minutes must elapse before working on the system. Failure to do so may result in deployment of the air bag.

2. Remove the battery top cover, if equipped.
3. Disconnect the negative battery cable. Tape the cable with insulating tape.
4. Raise and support the vehicle safely.
5. Remove the tire and wheel assemblies.
6. Remove the front hub nut.
7. Remove the brake caliper mounting bolts and position the caliper to the side. Do not allow the caliper to hang by the brake hose.
8. Remove the tie rod end ball joint from the knuckle.
9. Remove the wheel speed sensor, strut lower mounting bolt and the lower arm mounting bolt from the knuckle.
10. Disconnect the halfshaft from the knuckle.
11. Remove the inner shaft mounting bolts and then disconnect the inner shaft.
12. Disconnect the left shaft from the differential carrier.

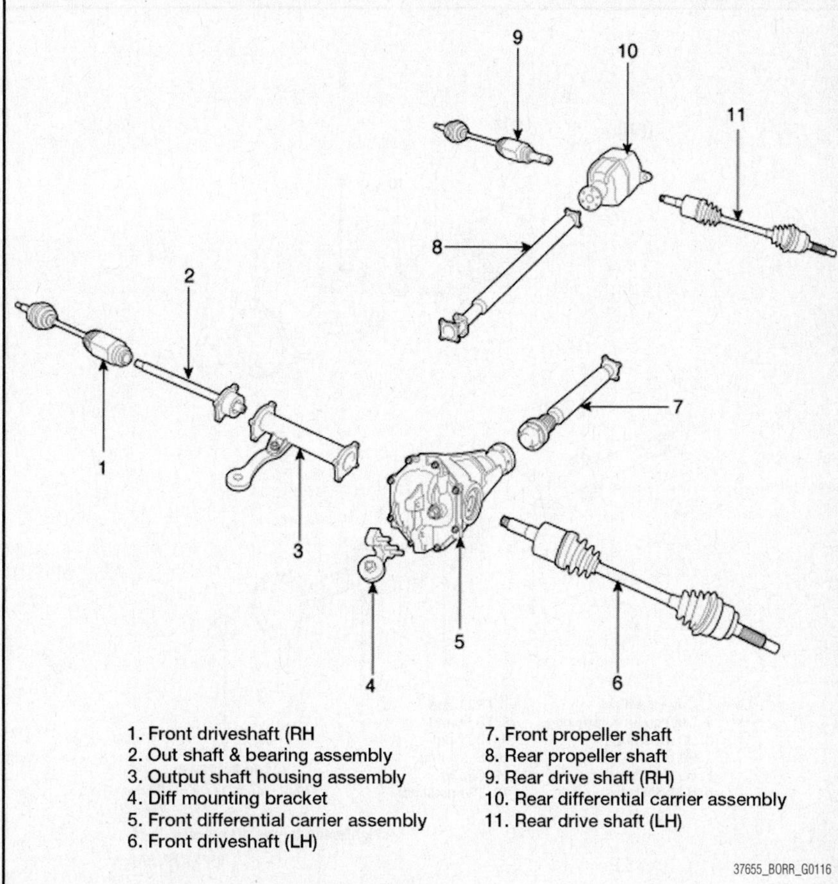

Fig. 11 Front halfshafts and related components

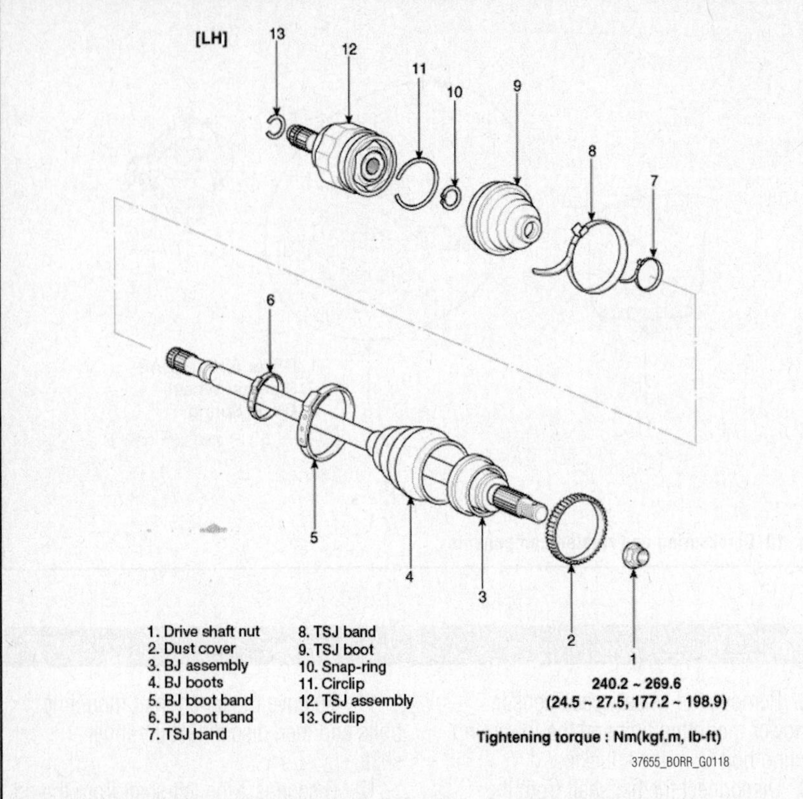

Fig. 12 Front halfshaft and related components—left

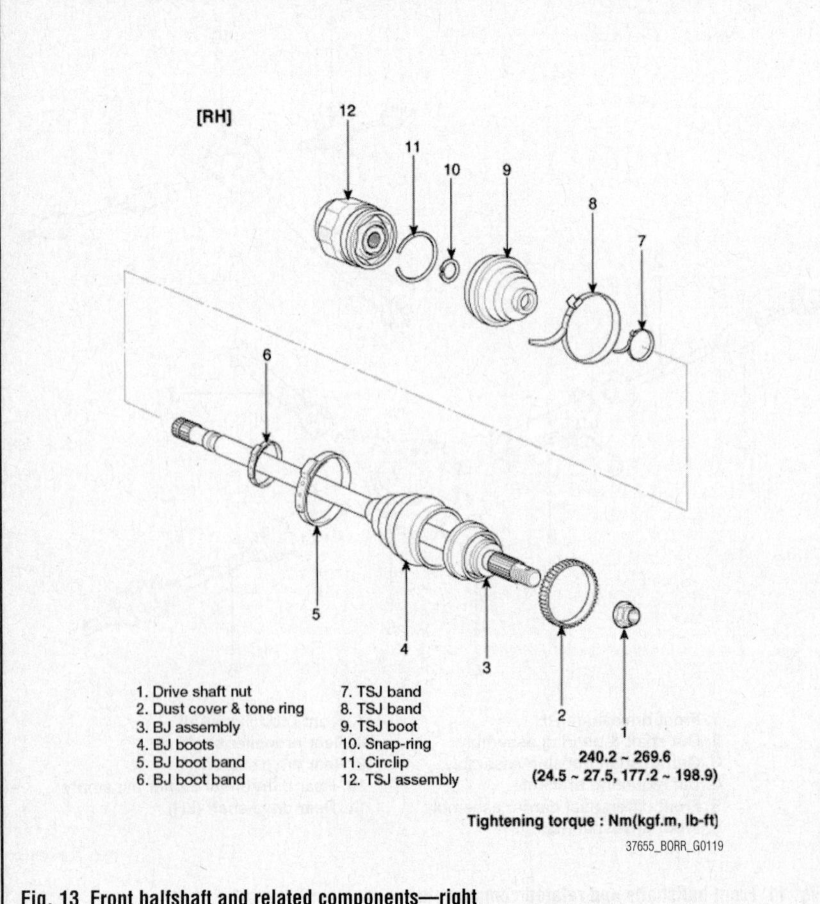

Fig. 13 Front halfshaft and related components—right

To install:

➡Be sure to use new fasteners, as required.

13. Installation is the reverse of the removal procedure.
14. Be sure to reprogram the necessary systems.
15. Reset/Initialize the sunroof, if equipped.

CV-BOOTS INSPECTION

1. Before servicing the vehicle, refer to the Precautions Section.
2. Check the driveshaft boots for damage and deterioration.
   - Raise and support the vehicle safely
   - Rotate axle and inspect for cracked or ripped CV boot material on inner and outer CV joints on both sides of vehicle
3. Replace boot if damaged or deteriorated.

REAR HALFSHAFT

REMOVAL & INSTALLATION

See Figure 14.

1.Before servicing the vehicle, refer to the Precautions Section.

➡If working near and/or around the SRS system and components, be sure to disable the SRS system. Tape the negative battery cable with insulating tape. Always disconnect the negative battery cable first.

⁂ CAUTION

To avoid personal injury when working on vehicles equipped with an air bag, the negative battery cable must be disconnected and at least three minutes must elapse before working on the system. Failure to do so may result in deployment of the air bag.

2. Remove the battery top cover, if equipped.
3. Disconnect the negative battery cable. Tape the cable with insulating tape.
4. Raise and safely support the vehicle.
5. Remove the tire and wheel assemblies.
6. Remove the hub nut.
7. Remove the strut mounting bolt.
8. Remove the assist arm and the trailing arm from the rear axle carrier.
9. Remove the rear assist arm ball joint using special tool 09568-4A000, or equivalent.
10. Remove the stabilizer link from the axle carrier.
11. Remove the caliper mounting bolts

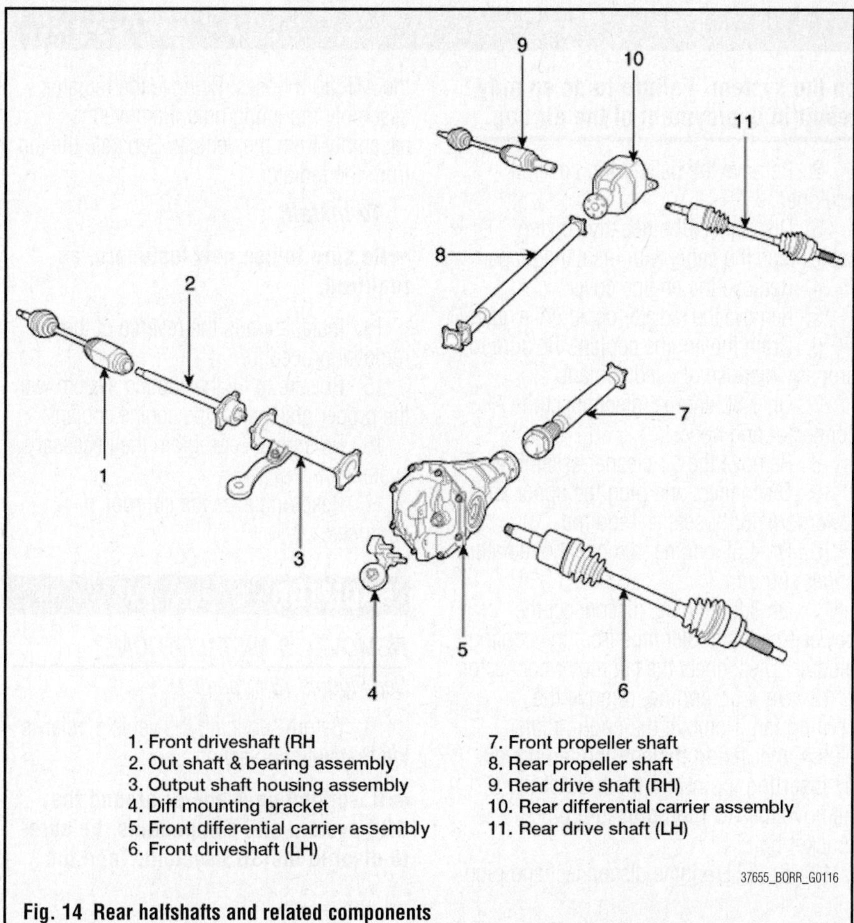

Fig. 14 Rear halfshafts and related components

and position the caliper to the side. Do not allow the caliper to hang by the brake line.

12. Remove the wheel speed sensor and the parking brake cable from the axle carrier.
13. Disconnect the rear shaft end from the knuckle.
14. Insert a pry bar between the differential case and the joint case, and separate the halfshaft from the case.

To install:

➡**Be sure to use new fasteners, as required.**

15. Installation is the reverse of the removal procedure.
16. Be sure to reprogram the necessary systems.
17. Reset/Initialize the sunroof, if equipped.

TRANSFER CASE ASSEMBLY

REMOVAL & INSTALLATION

1. Before servicing the vehicle, refer to the Precautions Section.

➡**If working near and/or around the SRS system and components, be sure to disable the SRS system. Tape the negative battery cable with insulating tape. Always disconnect the negative battery cable first.**

CAUTION

To avoid personal injury when working on vehicles equipped with an air bag, the negative battery cable must be disconnected and at least three minutes must elapse before working on the system. Failure to do so may result in deployment of the air bag.

2. Remove the battery top cover, if equipped.
3. Disconnect the negative battery cable. Tape the cable with insulating tape.
4. Raise and support the vehicle safely.
5. Drain the transmission fluid. Be sure to properly dispose of used fluid.
6. Matchmark and remove the front driveshaft, 4WD vehicles.
7. Remove the front and rear mufflers.
8. Matchmark and remove the driveshaft.
9. Properly support the transmission using a suitable support tool.

➡**Be careful, the fluid pan is plastic.**

10. Disconnect the oxygen sensor electrical connector.
11. Remove the exhaust manifold stay bolt.
12. Remove the crossmember assembly.
13. Remove the insulator support bracket (transmission mount).
14. Disconnect the shift cable.
15. Disconnect the 4WD ECU connector or EMC connector.
16. Be sure that the transfer case is properly supported. Remove the transfer case retaining bolts.
17. Carefully remove the transfer case from the vehicle, by lowering the support jack.

To install:

➡**Be sure to use new fasteners, as required.**

18. Installation is the reverse of the removal procedure.
19. Be sure to reprogram the necessary systems.
20. Reset/Initialize the sunroof, if equipped.
21. Refill the transfer case with the proper grade and type fluid.

ENGINE COOLING

ENGINE FAN

REMOVAL & INSTALLATION

See Figures 15 and 16.

1. Before servicing the vehicle, refer to the Precautions Section.

➡If working near and/or around the SRS system and components, be sure to disable the SRS system. Tape the negative battery cable with insulating tape. Always disconnect the negative battery cable first.

⁂ CAUTION

To avoid personal injury when working on vehicles equipped with an air bag, the negative battery cable must be disconnected and at least three minutes must elapse before working on the system. Failure to do so may result in deployment of the air bag.

2. Remove the battery top cover, if equipped.
3. Disconnect the negative battery cable. Tape the cable with insulating tape.
4. Remove the engine cover.
5. Remove the radiator upper grille guard.
6. Drain the engine coolant. Be sure to properly dispose of used coolant.
7. On 3.8L engine, disconnect the AFS connector and hose.
8. Remove the air cleaner assembly.
9. Disconnect and plug the upper and lower radiator hoses, as required.
10. On 4.6L engine, remove the radiator upper shroud.
11. On 3.8L engine, disconnect the power steering cooler pipe from the cooling module. Disconnect the fan motor connector.
12. On 4.6L engine, remove the cooling fan. Remove the cooling fan clutch after fixing the cooling fan pulley by inserting a pin into the hole of it. Remove the AC high pressure pipe bracket bolt.
13. On 3.8L engine disconnect and plug the ATF fluid hoses. Remove the radiator assembly mounting nuts. Remove the assembly from the vehicle. Separate the fan from the radiator.

To install:

➡Be sure to use new fasteners, as required.

14. Installation is the reverse of the removal procedure.
15. Be sure to fill the cooling system with the proper grade and type engine coolant.
16. Be sure to reprogram the necessary systems.
17. Reset/Initialize the sunroof, if equipped.

RADIATOR

REMOVAL & INSTALLATION

See Figures 16 through 18.

1. Before servicing the vehicle, refer to the Precautions Section.

➡If working near and/or around the SRS system and components, be sure to disable the SRS system. Tape the

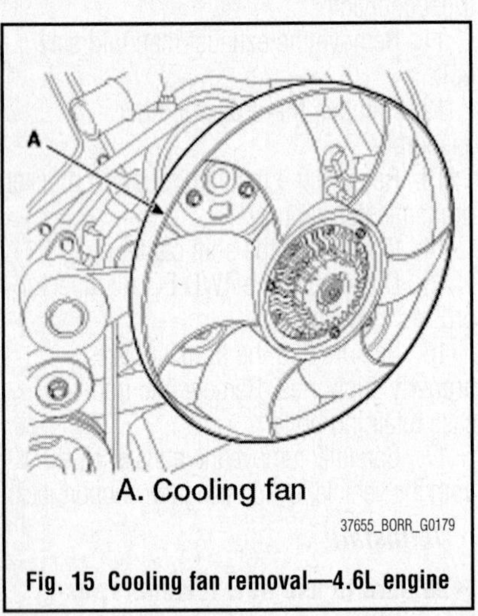

Fig. 15 Cooling fan removal—4.6L engine

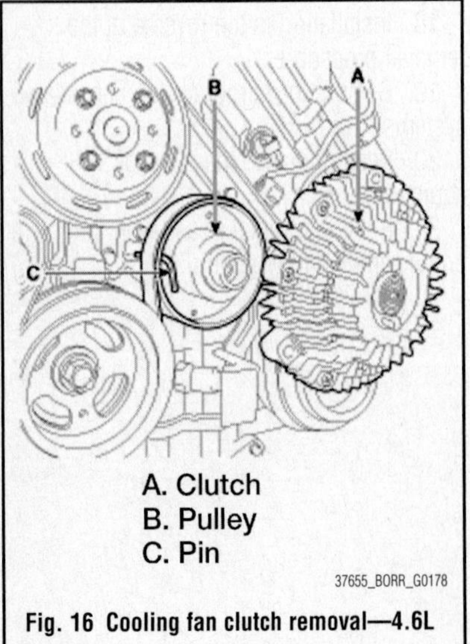

Fig. 16 Cooling fan clutch removal—4.6L engine

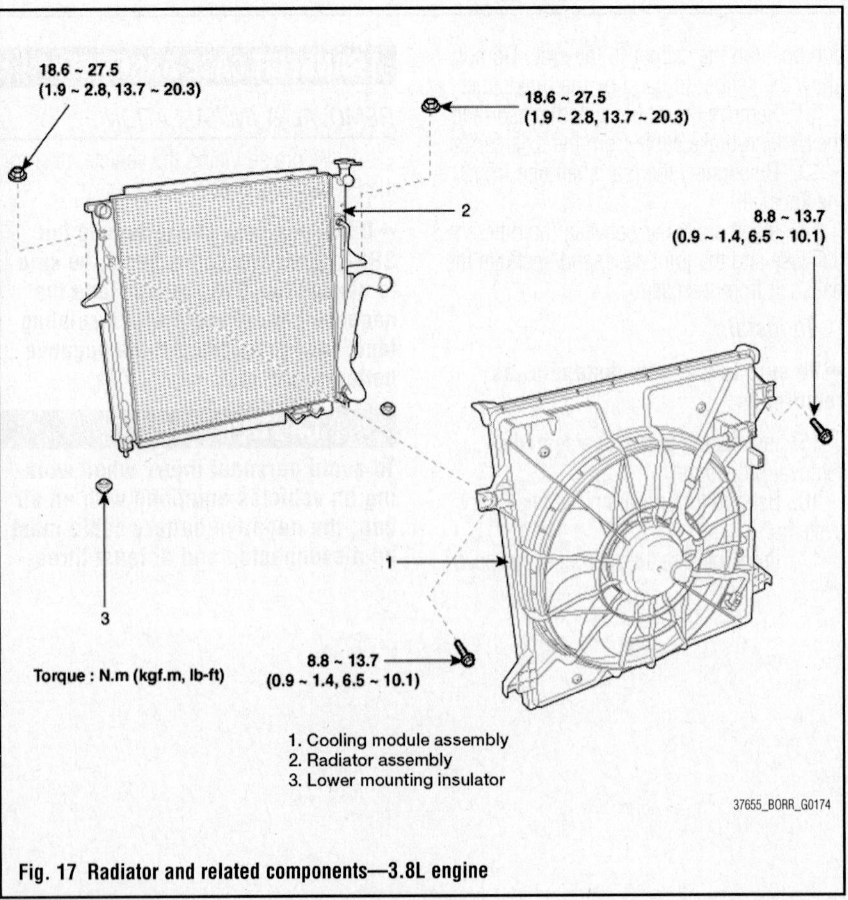

Fig. 17 Radiator and related components—3.8L engine

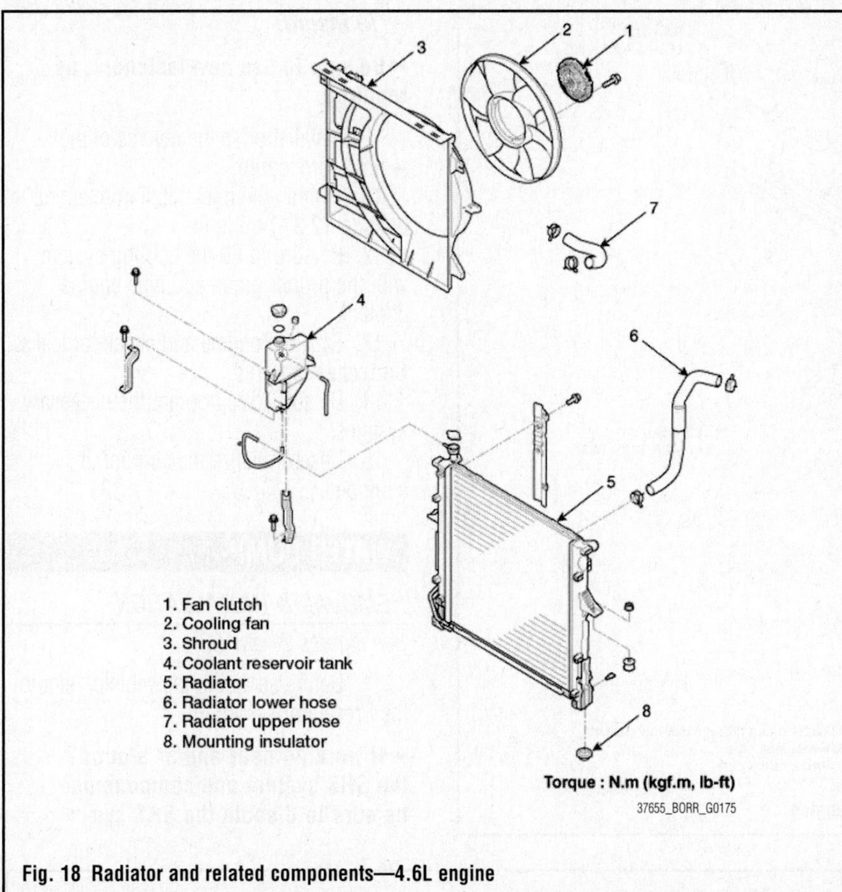

Fig. 18 Radiator and related components—4.6L engine

negative battery cable with insulating tape. Always disconnect the negative battery cable first.

⁂ CAUTION

To avoid personal injury when working on vehicles equipped with an air bag, the negative battery cable must be disconnected and at least three minutes must elapse before working on the system. Failure to do so may result in deployment of the air bag.

2. Remove the battery top cover, if equipped.
3. Disconnect the negative battery cable. Tape the cable with insulating tape.
4. Remove the engine cover.
5. Remove the radiator upper grille guard.
6. Drain the engine coolant. Be sure to properly dispose of used coolant.
7. On 3.8L engine, disconnect the AFS connector and hose.
8. Remove the air cleaner assembly.
9. Disconnect and plug the upper and lower radiator hoses.
10. On 4.6L engine, remove the radiator upper shroud.
11. On 3.8L engine, disconnect the power steering cooler pipe from the cooling module. Disconnect the fan motor connector.
12. On 4.6L engine, remove the cooling fan. Remove the cooling fan clutch after fixing the cooling fan pulley by inserting a pin into the hole of it. Remove the AC high pressure pipe bracket bolt.
13. Disconnect and plug the ATF fluid hoses.
14. On 4.6L engine, remove the power steering oil cooler bracket bolts.
15. Remove the radiator assembly mounting nuts. Remove the assembly from the vehicle.
16. On 3.8L engine, separate the fan from the radiator.

To install:

➡Be sure to use new fasteners, as required.

17. Installation is the reverse of the removal procedure.
18. Be sure to fill the cooling system with the proper grade and type engine coolant.
19. Be sure to reprogram the necessary systems.
20. Reset/Initialize the sunroof, if equipped.

THERMOSTAT

REMOVAL & INSTALLATION

See Figures 19 and 20.

1. Before servicing the vehicle, refer to the Precautions Section.

➡If working near and/or around the SRS system and components, be sure to disable the SRS system. Tape the negative battery cable with insulating tape. Always disconnect the negative battery cable first.

⁂ CAUTION

To avoid personal injury when working on vehicles equipped with an air bag, the negative battery cable must be disconnected and at least three minutes must elapse before working on the system. Failure to do so may result in deployment of the air bag.

2. Remove the battery top cover, if equipped.
3. Disconnect the negative battery cable. Tape the cable with insulating tape.
4. Remove the engine cover, as required.
5. Drain the cooling system, below the thermostat. Be sure to properly dispose of used engine coolant.
6. On 4.6L engine, remove the throttle body assembly.
7. Disconnect the radiator hose.
8. Remove the water inlet retaining bolts. Remove the water inlet. Discard the gasket/O-ring.
9. Remove the thermostat from its mounting.

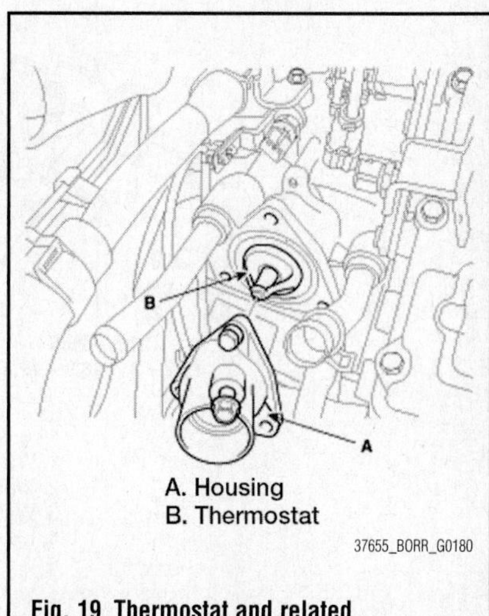

Fig. 19 Thermostat and related components—3.8L engine

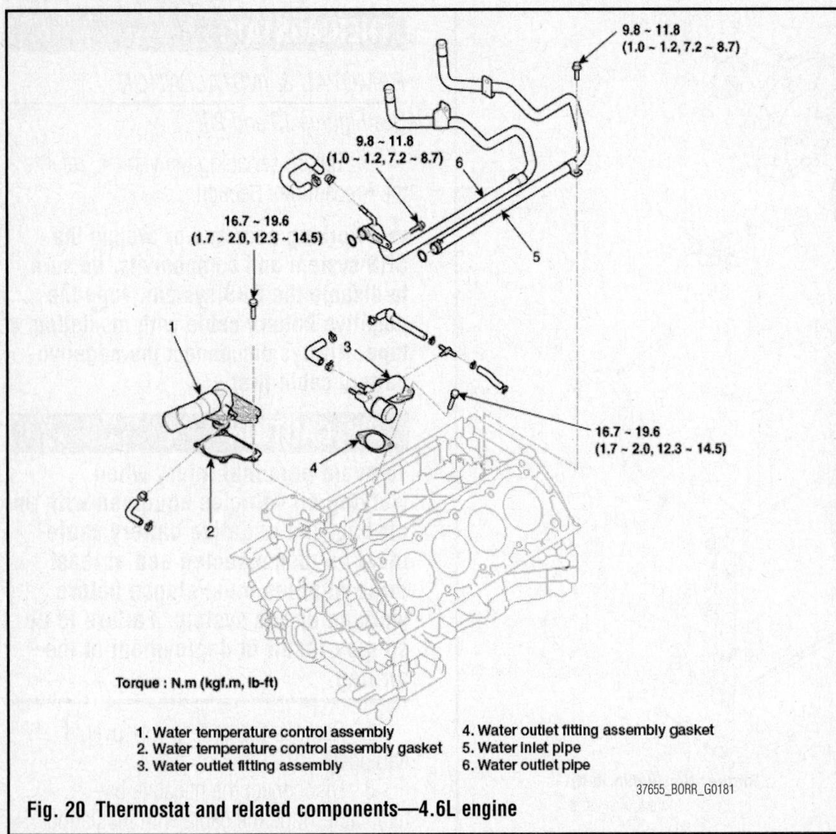

9.8 ~ 11.8
(1.0 ~ 1.2, 7.2 ~ 8.7)
9.8 ~ 11.8
(1.0 ~ 1.2, 7.2 ~ 8.7)
6
16.7 ~ 19.6
(1.7 ~ 2.0, 12.3 ~ 14.5)
5
1
3
16.7 ~ 19.6
(1.7 ~ 2.0, 12.3 ~ 14.5)
4
2
Torque : N.m (kgf.m, lb-ft)
1. Water temperature control assembly
2. Water temperature control assembly gasket
3. Water outlet fitting assembly
4. Water outlet fitting assembly gasket
5. Water inlet pipe
6. Water outlet pipe
37655_BORR_G0181
Fig. 20 Thermostat and related components—4.6L engine

To install:

➡Be sure to use new fasteners, as required.

10. Installation is the reverse of the removal procedure.
11. Tighten the inlet housing retaining bolts to 12.3–14.5 ft. lbs.
12. Be sure to fill the cooling system with the proper grade and type engine coolant.
13. Start the engine and check for leaks. Correct as required.
14. Be sure to reprogram the necessary systems.
15. Reset/Initialize the sunroof, if equipped.

WATER PUMP

REMOVAL & INSTALLATION

See Figures 21 and 22.

1. Before servicing the vehicle, refer to the Precautions Section.

➡If working near and/or around the SRS system and components, be sure to disable the SRS sys-

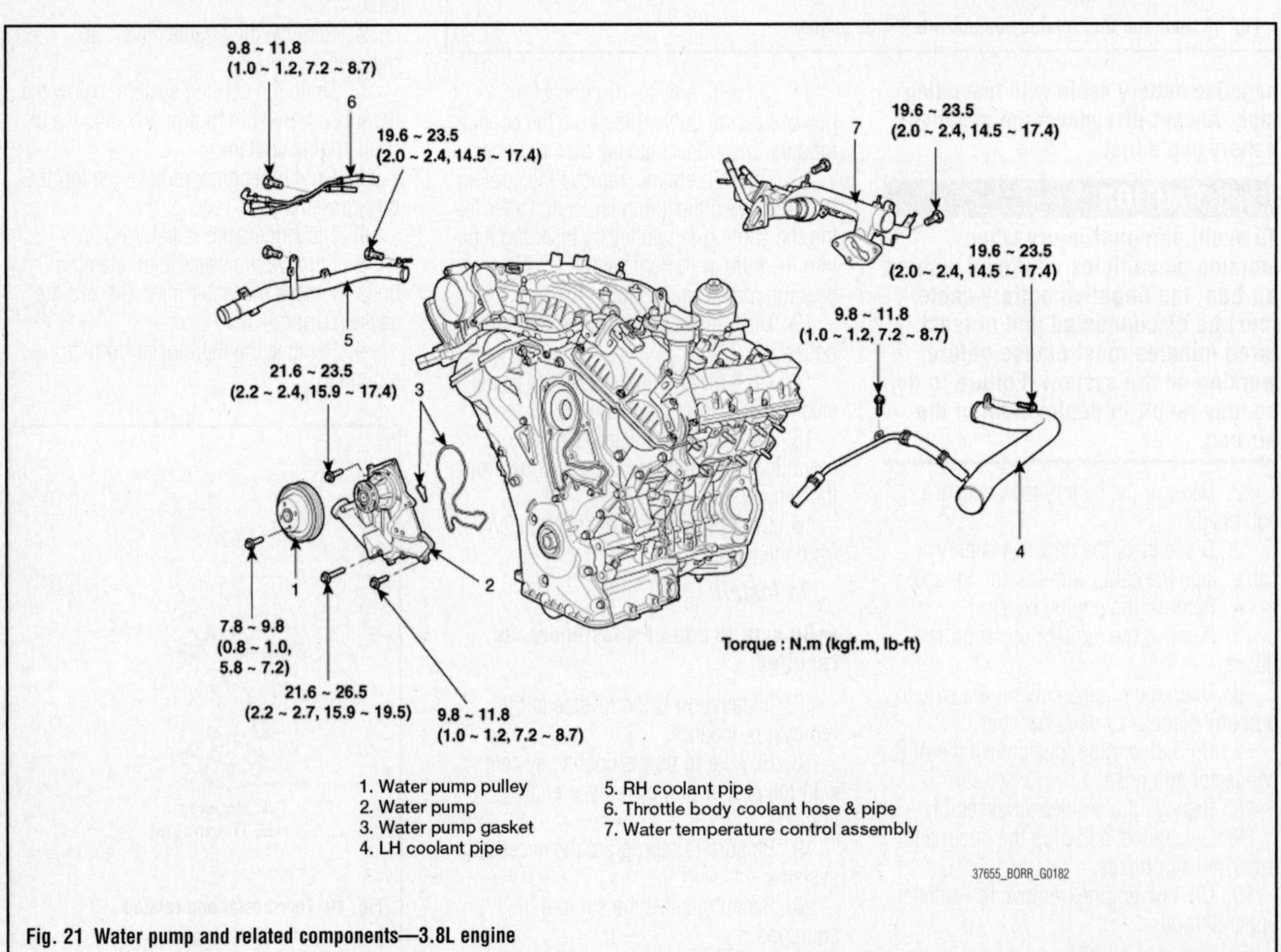

9.8 ~ 11.8
(1.0 ~ 1.2, 7.2 ~ 8.7)
6
19.6 ~ 23.5
(2.0 ~ 2.4, 14.5 ~ 17.4)
7
19.6 ~ 23.5
(2.0 ~ 2.4, 14.5 ~ 17.4)
19.6 ~ 23.5
(2.0 ~ 2.4, 14.5 ~ 17.4)
9.8 ~ 11.8
(1.0 ~ 1.2, 7.2 ~ 8.7)
5
21.6 ~ 23.5
(2.2 ~ 2.4, 15.9 ~ 17.4)
3
4
1
2
7.8 ~ 9.8
(0.8 ~ 1.0,
5.8 ~ 7.2)
21.6 ~ 26.5
(2.2 ~ 2.7, 15.9 ~ 19.5)
9.8 ~ 11.8
(1.0 ~ 1.2, 7.2 ~ 8.7)
Torque : N.m (kgf.m, lb-ft)
1. Water pump pulley
2. Water pump
3. Water pump gasket
4. LH coolant pipe
5. RH coolant pipe
6. Throttle body coolant hose & pipe
7. Water temperature control assembly
37655_BORR_G0182
Fig. 21 Water pump and related components—3.8L engine

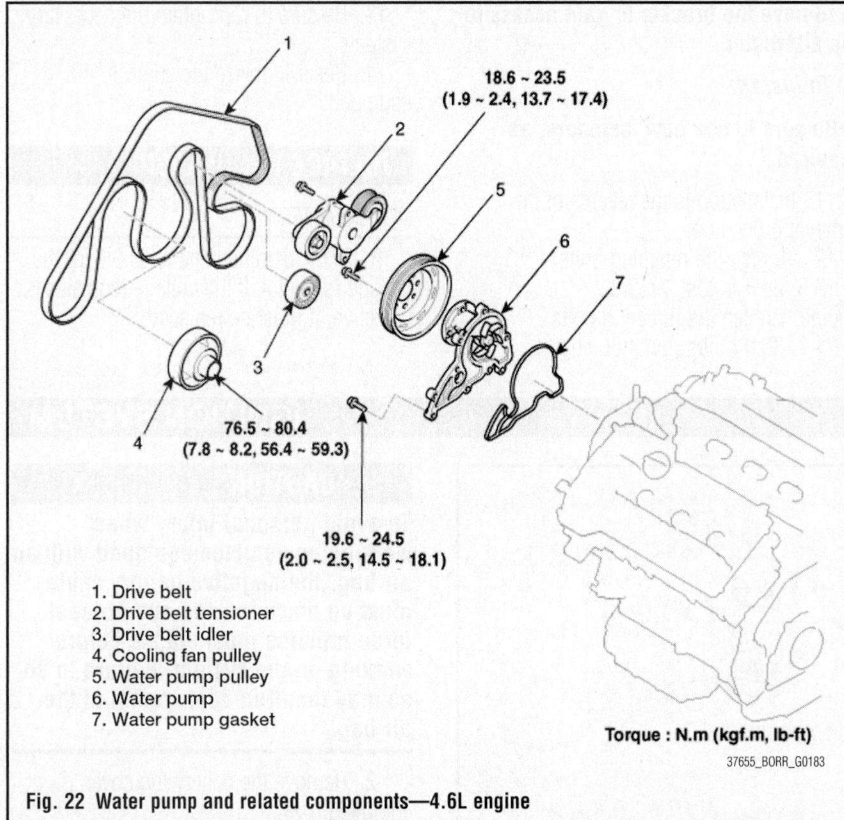

Fig. 22 Water pump and related components—4.6L engine

tem. Tape the negative battery cable with insulating tape. Always disconnect the negative battery cable first.

⁂ CAUTION

To avoid personal injury when working on vehicles equipped with an air bag, the negative battery cable must be disconnected and at least three minutes must elapse before working on the system. Failure to do so may result in deployment of the air bag.

2. Remove the battery top cover, if equipped.

3. Disconnect the negative battery cable. Tape the cable with insulating tape.

4. Drain the cooling system. Be sure to properly dispose of used coolant.

5. On 4.6L engine, remove the cooling fan and cooling fan clutch.

6. Remove the drive belt. On 4.6L engine, remove the tensioner.

7. Remove the water pump pulley.

➡On 4.6L engine the retaining bolt is left hand thread. To remove the bolt screw it clockwise.

8. Remove the water pump retaining bolts. Remove the water pump from its mounting. Discard the gasket.

To install:

➡Be sure to use new fasteners, as required.

9. Installation is the reverse of the removal procedure.

10. Be sure to fill the cooling system with the proper grade and type coolant.

11. Start the engine and check for leaks. Correct, as required.

12. Be sure to reprogram the necessary systems.

13. Reset/Initialize the sunroof, if equipped.

ALTERNATOR

REMOVAL & INSTALLATION

See Figures 23 and 24.

1. Before servicing the vehicle, refer to the Precautions Section.

➡If working near and/or around the SRS system and components, be sure to disable the SRS system. Tape the negative battery cable with insulating tape. Always disconnect the negative battery cable first.

⁂ CAUTION

To avoid personal injury when working on vehicles equipped with an air bag, the negative battery cable must be disconnected and at least three minutes must elapse before working on the system. Failure to do so may result in deployment of the air bag.

2. Remove the battery top cover, if equipped.

3. Disconnect the negative battery cable. Tape the cable with insulating tape.

4. Remove the engine cover.

5. Remove the radiator grille upper guard.

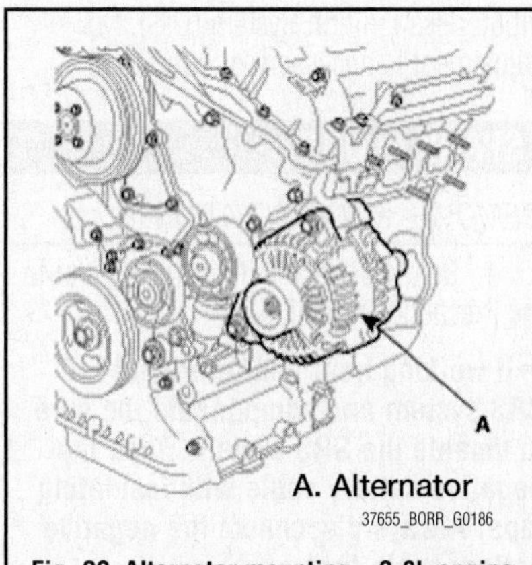

Fig. 23 Alternator mounting—3.8L engine

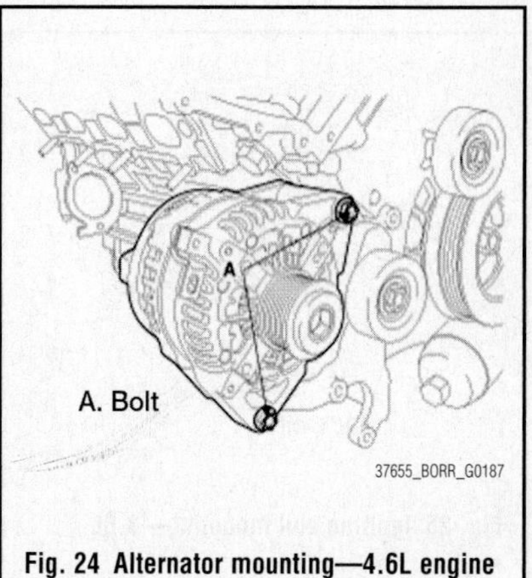

Fig. 24 Alternator mounting—4.6L engine

6. Remove the air cleaner assembly.
7. Remove the drive belt.
8. On 4.6L engine, remove the oil level gauge. Remove the power steering pump without disconnecting the hoses. Position the pump to the side. Do not allow the component to hang by the hoses.
9. Disconnect the alternator electrical connectors.
10. Remove the alternator mounting bolts. Remove the alternator from the vehicle.

➡On 4.6L engine it may be necessary to remove the bracket to gain access to the alternator.

To install:

➡Be sure to use new fasteners, as required.

11. Installation is the reverse of the removal procedure.
12. Tighten the retaining bolts to 19.5–24.6 ft. lbs., on 3.8L engine. Tighten the retaining bolts to 21.7–30.4 ft. lbs., on 4.6L engine.
13. Be sure to reprogram the necessary systems.
14. Reset/Initialize the sunroof, if equipped.

VOLTAGE REGULATOR

ADJUSTMENT

There is no adjustment possible for the voltage regulator. If the voltage regulator is defective, it must be replaced.

ENGINE ELECTRICAL — IGNITION SYSTEM

IGNITION COIL

REMOVAL & INSTALLATION

See Figures 25 and 26.

1. Before servicing the vehicle, refer to the Precautions Section.

➡If working near and/or around the SRS system and components, be sure to disable the SRS system. Tape the negative battery cable with insulating tape. Always disconnect the negative battery cable first.

✲✲ CAUTION

To avoid personal injury when working on vehicles equipped with an air bag, the negative battery cable must be disconnected and at least three minutes must elapse before working on the system. Failure to do so may result in deployment of the air bag.

2. Remove the battery top cover, if equipped.
3. Disconnect the negative battery cable. Tape the cable with insulating tape.
4. Remove the engine cover.
5. Disconnect the coil connector.
6. Remove the coil from its mounting.

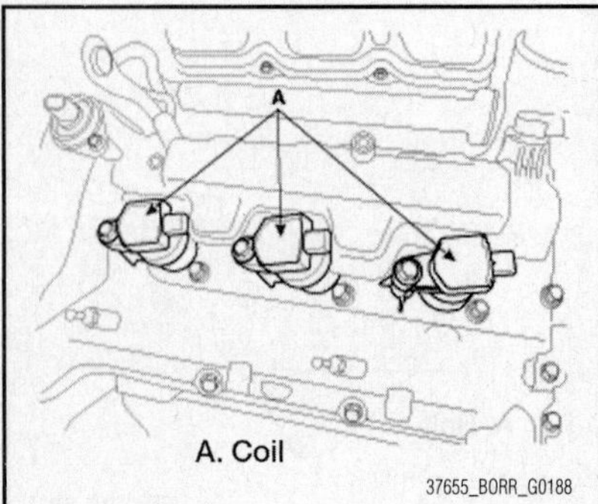
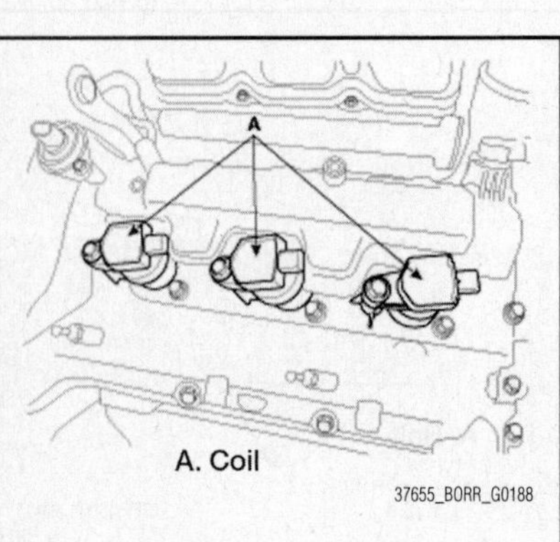

Fig. 25 Ignition coil mounting—3.8L engine

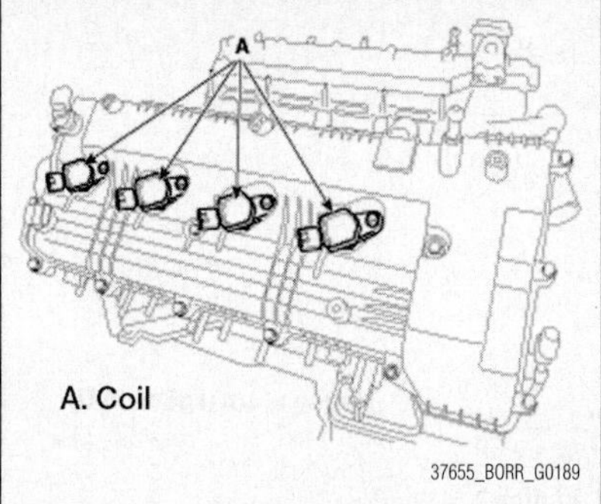

Fig. 26 Ignition coil mounting—4.6L engine

To install:

➡Be sure to use new fasteners, as required.

7. Installation is the reverse of the removal procedure.
8. Be sure to reprogram the necessary systems.
9. Reset/Initialize the sunroof, if equipped.

IGNITION TIMING

ADJUSTMENT

These vehicles are equipped with a Distributorless Ignition System (DIS). No adjustment is necessary or possible.

SPARK PLUGS

REMOVAL & INSTALLATION

1. Before servicing the vehicle, refer to the Precautions Section.

➡If working near and/or around the SRS system and components, be sure to disable the SRS system. Tape the negative battery cable with insulating tape. Always disconnect the negative battery cable first.

✲✲ CAUTION

To avoid personal injury when working on vehicles equipped with an air bag, the negative battery cable must be disconnected and at least three minutes must elapse before working on the system. Failure to do so may result in deployment of the air bag.

2. Remove the battery top cover, if equipped.
3. Disconnect the negative battery cable. Tape the cable with insulating tape.
4. Remove the engine cover.
5. Disconnect the coil connector.
6. Remove the coil from its mounting.
7. Use spark plug wrench to remove the spark plug (s) from its mounting.

✲✲ WARNING

Do not let any dirt or debris get into the engine through the spark plugs holes while the plugs are removed.

To install:

➡Be sure to use new fasteners, as required.

8. Installation is the reverse of the removal procedure.
9. Be sure to reprogram the necessary systems.
10. Reset/Initialize the sunroof, if equipped.

INSPECTION

Check spark plugs for the following:

1. Broken insulator
2. Worn electrode
3. Carbon deposits
4. Damaged or broken gasket
5. Condition of the porcelain insulator at the tip of the spark plug.
6. Check the electrode gap.

ENGINE ELECTRICAL

STARTER

REMOVAL & INSTALLATION

See Figures 27 and 28.

1. Before servicing the vehicle, refer to the Precautions Section.

➡If working near and/or around the SRS system and components, be sure to disable the SRS system. Tape the negative battery cable with insulating tape. Always disconnect the negative battery cable first.

⁂ CAUTION

To avoid personal injury when working on vehicles equipped with an air bag, the negative battery cable must be disconnected and at least three minutes must elapse before working on the system. Failure to do so may result in deployment of the air bag.

2. Remove the battery top cover, if equipped.
3. Disconnect the negative battery cable. Tape the cable with insulating tape.

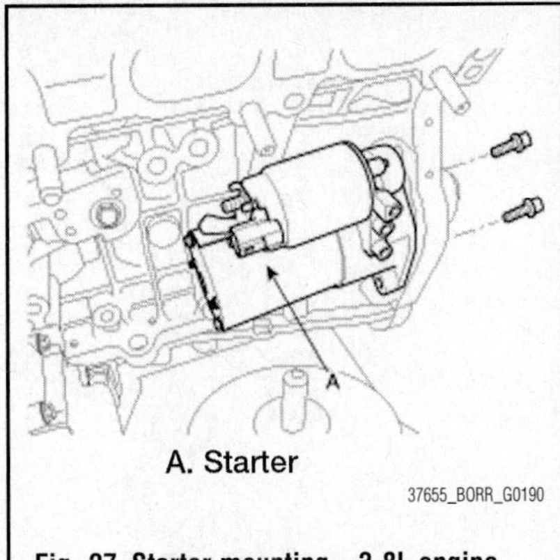

Fig. 27 Starter mounting—3.8L engine

4. On 3.8L engine, remove the left hand engine mounting bracket. Raise the engine slightly, using the proper jack, to gain access to the side of the engine. Remove the engine support bracket and the starter protector.
5. Disconnect the starter electrical connectors. Disconnect the positive battery cable, at the starter.
6. Remove the retaining bolts. Remove the starter from the vehicle.

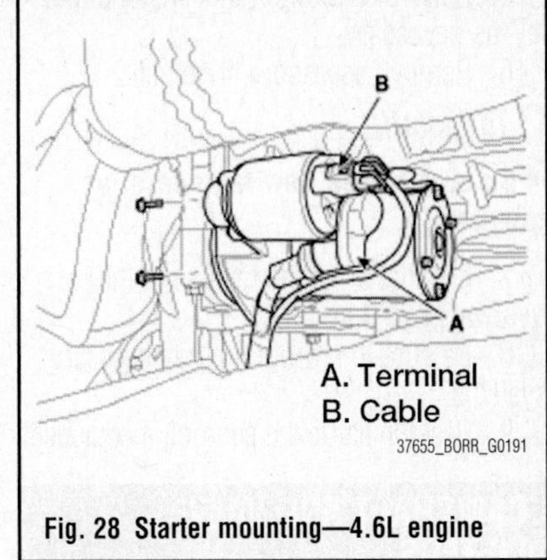

Fig. 28 Starter mounting—4.6L engine

To install:

➡Be sure to use new fasteners, as required.

7. Installation is the reverse of the removal procedure.
8. Tighten the starter retaining bolts to 36.2–47.0 ft. lbs.
9. Be sure to reprogram the necessary systems.
10. Reset/Initialize the sunroof, if equipped.

ENGINE MECHANICAL

ACCESSORY DRIVE BELTS

ACCESSORY BELT ROUTING

See Figures 29 and 30.

INSPECTION

Inspect the drive belt for signs of glazing or cracking. A glazed belt will be perfectly smooth from slippage, while a good belt will have a slight texture of fabric visible. Cracks will usually start at the inner edge of the belt and run outward. All worn or damaged drive belts should be replaced immediately.

ADJUSTMENT

At this time the manufacturer does not provide service information for this component.

REMOVAL & INSTALLATION

See Figures 37 and 38.

At this time the manufacturer does not provide removal and installation procedures for this component. The following procedure is a guideline and may differ from the vehicle you are servicing.

1. Before servicing the vehicle, refer to the Precautions Section.

➡If working near and/or around the SRS system and components, be sure to disable the SRS system. Tape the negative battery cable with insulating tape. Always disconnect the negative battery cable first.

⁂ CAUTION

To avoid personal injury when working on vehicles equipped with an air bag, the negative battery cable must be disconnected and at least three minutes must elapse before working on the system. Failure to do so may result in deployment of the air bag.

2. Remove the battery top cover, if equipped.
3. Disconnect the negative battery cable. Tape the cable with insulating tape.

A. Drive belt
37655_BORR_G0184

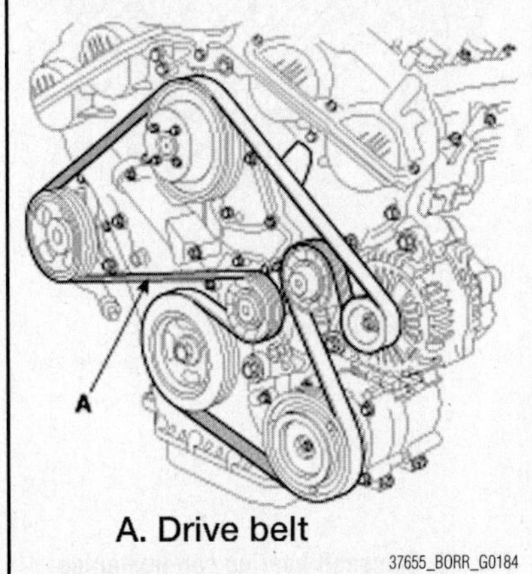

Fig. 29 Drive belt routing—3.8L engine

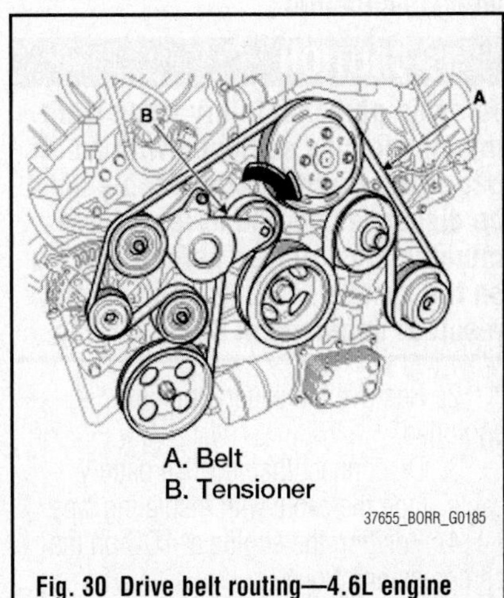

Fig. 30 Drive belt routing—4.6L engine

4. Remove the necessary components to gain access to the drive belt.

5. Relieve tension on the tensioner pulley, as necessary.

6. Remove accessory drive belt.

To install:

➡**Be sure to use new fasteners, as required.**

7. Installation is the reverse of the removal procedure.

8. Be sure to reprogram the necessary systems.

9. Reset/Initialize the sunroof, if equipped.

CAMSHAFT AND VALVE LIFTERS

INSPECTION

1. Check the camshaft journals for wear. If the journals are badly worn out, replace the camshaft.

2. Check the cam lobes for damage. If the lobe is damaged or excessively worn out, replace the camshaft.

3. Check the cam surface for abnormal wear or damage, and replace if necessary.

4. Check each bearing for damage. If the bearing surface is excessively damaged, replace the cylinder head assembly or camshaft bearing cap, as necessary.

REMOVAL & INSTALLATION

3.8L Engine

See Figures 31 through 33.

1. Before servicing the vehicle, refer to the Precautions Section.

➡**If working near and/or around the SRS system and components, be sure to disable the SRS system. Tape the negative battery cable with insulating tape. Always disconnect the negative battery cable first.**

CAUTION

To avoid personal injury when working on vehicles equipped with an air bag, the negative battery cable must be disconnected and at least three minutes must elapse before working on the system. Failure to do so may result in deployment of the air bag.

2. Remove the battery top cover, if equipped.

3. Disconnect the negative battery cable. Tape the cable with insulating tape.

4. Position the engine at TDC on the compression stroke.

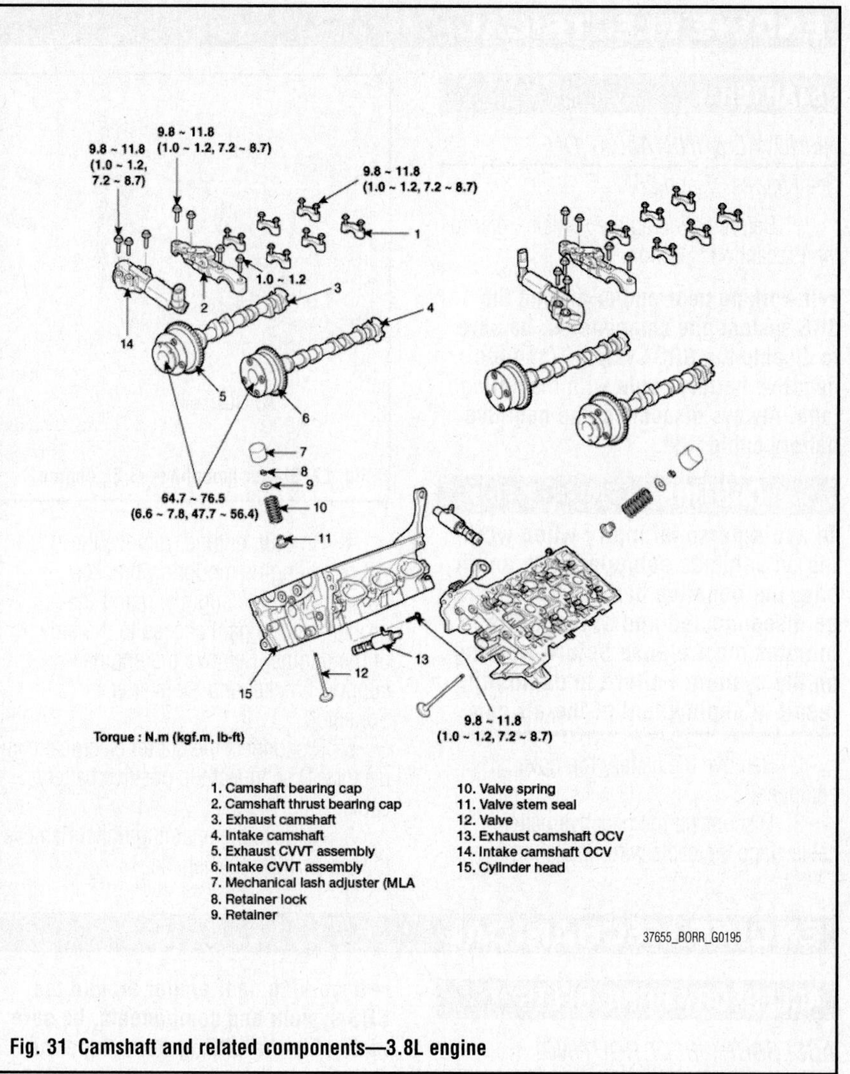

Fig. 31 Camshaft and related components—3.8L engine

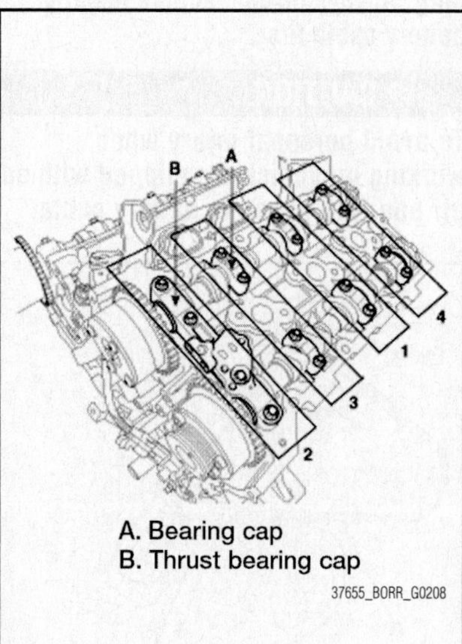

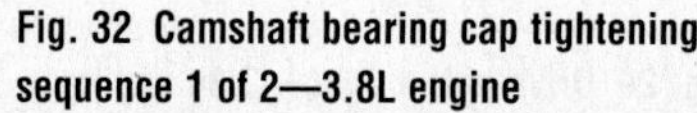

Fig. 32 Camshaft bearing cap tightening sequence 1 of 2—3.8L engine

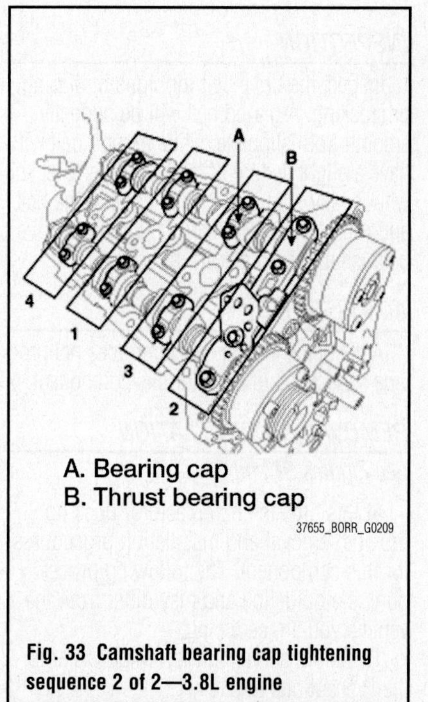

Fig. 33 Camshaft bearing cap tightening sequence 2 of 2—3.8L engine

5. Drain the engine coolant. Be sure to properly dispose of used coolant.

6. Properly relieve the fuel system pressure.

7. Drain the engine oil. Be sure to properly dispose of used oil.

8. Remove the timing chain.

9. Remove the mounting bolts and remove the water temperature control assembly, as necessary

10. Remove the intake manifold, as required.

11. Remove the exhaust manifolds, as required.

12. Remove the valve covers.

13. Remove the exhaust camshaft OCA valve (both sides).

14. Remove the camshaft bearing retaining cap bolts, and thrust bearing cap retaining bolts.

15. Remove the camshafts.

To install:

➡Be sure to use new fasteners, as required.

16. Installation is the reverse of the removal procedure.

17. Install the camshafts.

18. Tighten the bearing caps and thrust bearing caps to specification and in two steps.

19. Be sure to reprogram the necessary systems.

20. Reset/Initialize the sunroof, if equipped.

4.6L Engine

See Figures 34 through 37.

At this time the manufacturer does not provide removal and installation procedures for this component. The following procedure is a guideline and may differ from the vehicle you are servicing.

1. Before servicing the vehicle, refer to the Precautions Section.

➡If working near and/or around the SRS system and components, be sure to disable the SRS system. Tape the negative battery cable with insulating tape. Always disconnect the negative battery cable first.

⁂ CAUTION

To avoid personal injury when working on vehicles equipped with an air bag, the negative battery cable must be disconnected and at least three minutes must elapse before working on the system. Failure to do so may result in deployment of the air bag.

2. Remove the battery top cover, if equipped.

3. Disconnect the negative battery cable. Tape the cable with insulating tape.

4. Position the engine at TDC on the compression stroke.

5. Remove the engine cover.

6. Remove the air cleaner assembly.

7. Drain the engine coolant. Be sure to properly dispose of used coolant.

8. Properly relieve the fuel system pressure.

9. Drain the engine oil. Be sure to properly dispose of used oil.

10. Disconnect the heater hoses, fuel hose and PCVS hose.

11. Remove the cooling fan.

12. Remove the cooling fan clutch.

➡Remove the cooling fan clutch after fixing the cooling fan pulley by inserting a pin into the hole of it.

13. Remove the drive belt. Remove the drive belt tensioner.

14. Disconnect the water temperature sensor connector, CVVT oil control valve connectors, ECT module connector, PCVS connector, injector connectors, ignition coil connectors and the wiring harness protector.

15. Disconnect the knock sensor connectors, camshaft position sensor connectors, oxygen sensor connectors, condenser connector, variable intake system solenoid valve connector and the wiring harness protector.

16. Disconnect the oil pressure switch connector, CVVT oil valve connectors and the AC compressor switch connector.

17. Remove the oil level gauge.

18. Disconnect the oxygen sensor connector and remove the sensor connector bracket.

19. As necessary, remove the exhaust manifold heat protectors and remove the exhaust manifolds.

20. Disconnect the PCV hose, the PCVS hose and the water hoses from the intake manifold.

21. Remove the intake manifold.

22. Remove the knock sensors. Remove the ignition coils.

23. Remove the valve cover retaining bolts. Remove the valve covers.

24. Remove the water temperature sensor wiring assembly and bracket.

25. Remove the upper timing chain upper covers. Remove the oil seals from the covers, as necessary.

26. Remove the oil control valves.

27. Remove the cam to cam guides.

28. Align the timing mark of the camshaft sprocket.

29. Remove the alternator.

30. Remove the water outlet pipe and water inlet pipe.

31. Remove the water temperature control assembly and the water outlet fitting assembly.

32. Remove the timing chain tensioners.

33. Remove the camshaft bearing caps and the thrust bearing caps.

34. Remove the camshafts.

35. As required, remove the roller finger follower and hydraulic valve lifter assembly.

To install:

➡Be sure to use new fasteners, as required.

36. Installation is the reverse of the removal procedure.

37. Install the camshafts.

38. Tighten the bearing caps and thrust bearing caps to specification and in two steps. Step one specification: 5.1–5.8 ft. lbs. Step two specification: 10.1–10.8 ft. lbs.

39. Be sure to reprogram the necessary systems.

40. Reset/Initialize the sunroof, if equipped.

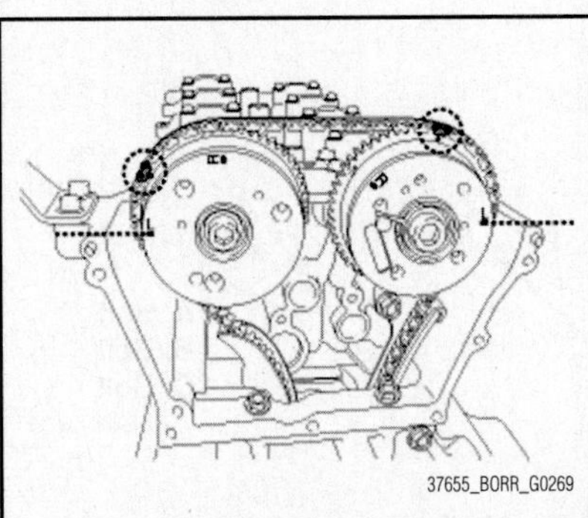

Fig. 34 Left camshaft timing mark alignment—4.6L engine

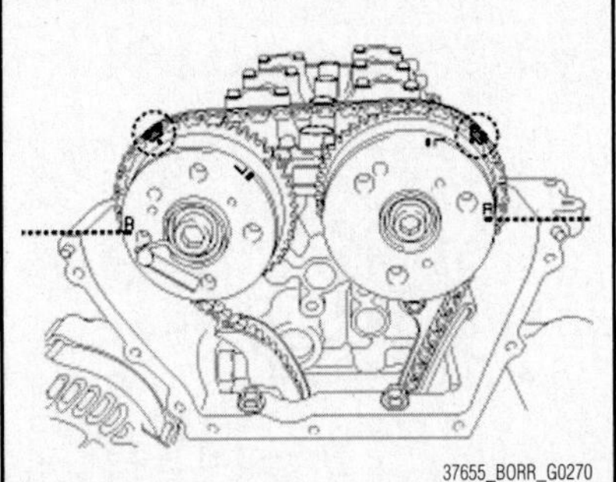

Fig. 35 Right camshaft timing mark alignment—4.6L engine

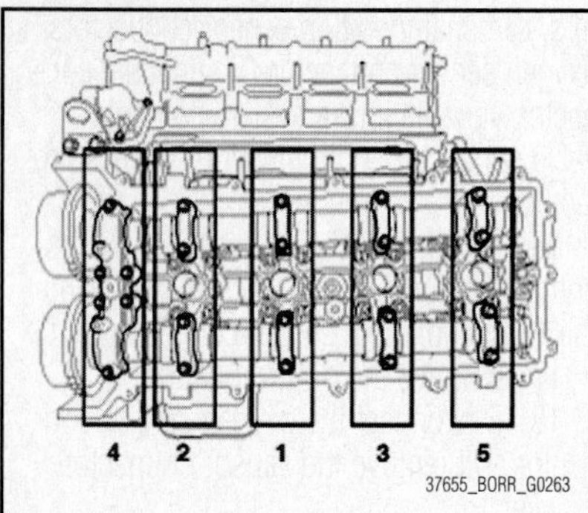

Fig. 36 Camshaft bolt tightening sequence (step one)—4.6L engine

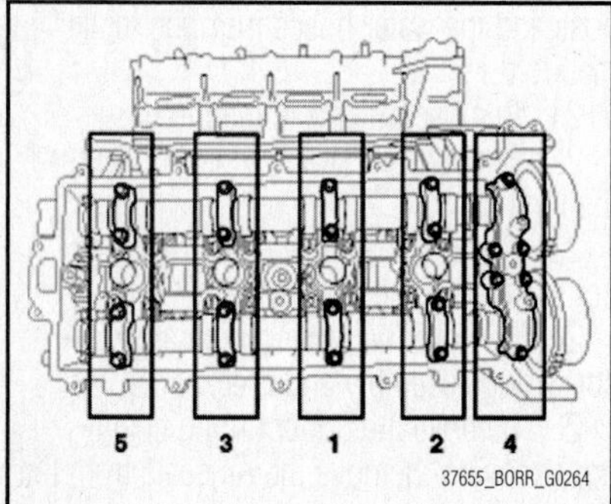

Fig. 37 Camshaft bolt tightening sequence (step two)—4.6L engine

CRANKSHAFT DAMPER

REMOVAL & INSTALLATION

See Figure 38.

At this time the manufacturer does not provide removal and installation procedures for this component, refer to the illustration as required.

1. Before servicing the vehicle, refer to the Precautions Section.

➡If working near and/or around the SRS system and components, be sure to disable the SRS system. Tape the negative battery cable with insulating tape. Always disconnect the negative battery cable first.

⁂ CAUTION

To avoid personal injury when working on vehicles equipped with an air bag, the negative battery cable must be disconnected and at least three minutes must elapse before working on the system. Failure to do so may result in deployment of the air bag.

2. Remove the battery top cover, if equipped.

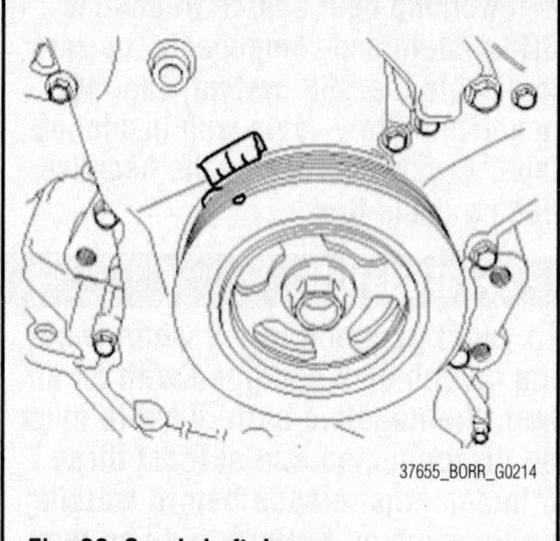

Fig. 38 Crankshaft damper

3. Disconnect the negative battery cable. Tape the cable with insulating tape.
4. Be sure to reprogram the necessary systems.
5. Reset/Initialize the sunroof, if equipped.

CRANKSHAFT FRONT SEAL

REMOVAL & INSTALLATION

See Figure 39.

At this time the manufacturer does not provide removal and installation procedures for this component, refer to the illustration as required.

1. Before servicing the vehicle, refer to the Precautions Section.

➡If working near and/or around the SRS system and components, be sure to disable the SRS system. Tape the negative battery cable with insulating tape. Always disconnect the negative battery cable first.

⁂ CAUTION

To avoid personal injury when working on vehicles equipped with an air bag, the negative battery cable must be disconnected and at least three minutes must elapse before working on the system. Failure to do so may result in deployment of the air bag.

2. Remove the battery top cover, if equipped.
3. Disconnect the negative battery cable. Tape the cable with insulating tape.
4. Remove the crankshaft damper.
5. Be sure to reprogram the necessary systems.
6. Reset/Initialize the sunroof, if equipped.

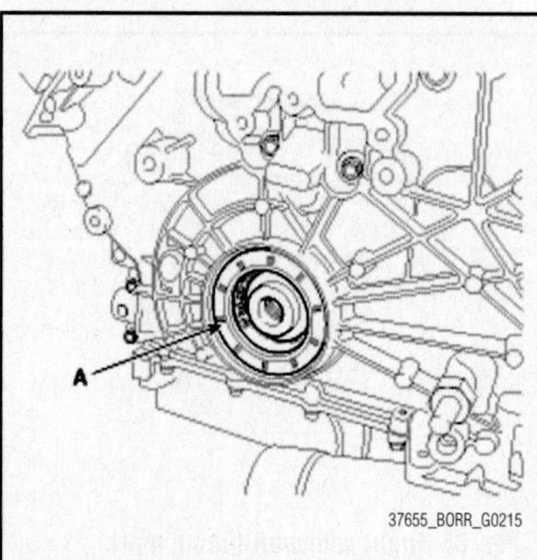

Fig. 39 Crankshaft front seal

CYLINDER HEAD

REMOVAL & INSTALLATION

3.8L Engine

See Figures 40 through 48.

1. Before servicing the vehicle, refer to the Precautions Section.

➡If working near and/or around the SRS system and components, be sure to disable the SRS system. Tape the negative battery cable with insulating tape. Always disconnect the negative battery cable first.

⁂ CAUTION

To avoid personal injury when working on vehicles equipped with an air bag, the negative battery cable must be disconnected and at least three minutes must elapse before working on the system. Failure to do so may result in deployment of the air bag.

2. Remove the battery top cover, if equipped.
3. Disconnect the negative battery cable. Tape the cable with insulating tape.
4. Position the engine at TDC on the compression stroke.

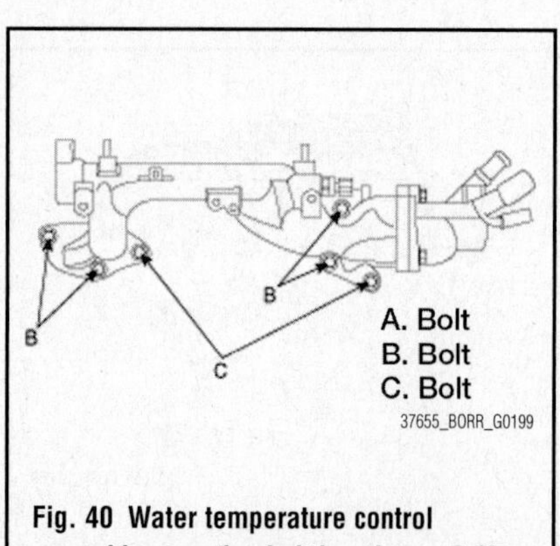

Fig. 40 Water temperature control assembly mounting bolt locations—3.8L engine

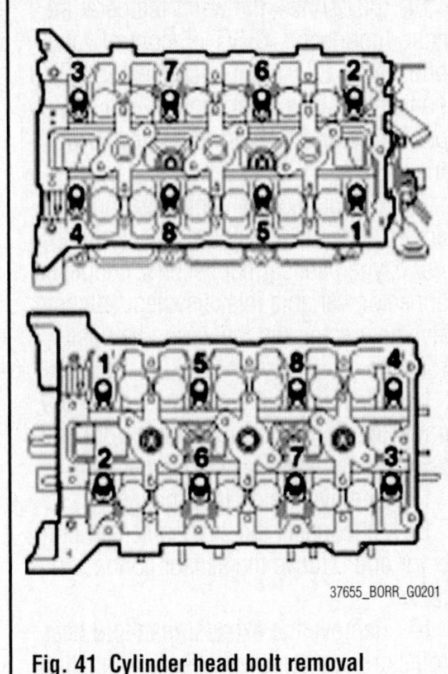

Fig. 41 Cylinder head bolt removal sequence—3.8L engine

5. Drain the engine coolant. Be sure to properly dispose of used coolant.

6. Properly relieve the fuel system pressure.

7. Drain the engine oil. Be sure to properly dispose of used oil.

8. Remove the timing chain.

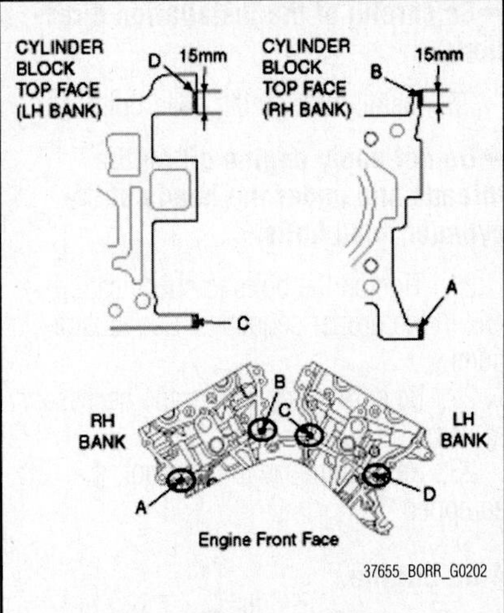

Fig. 43 Cylinder head sealant application 1 of 3—3.8L engine

9. Remove the mounting bolts and remove the water temperature control assembly.

10. Remove the intake manifold.

11. Remove the exhaust manifolds.

12. Remove the valve covers.

13. Remove the exhaust camshaft OCA valve (both sides).

14. Remove the camshaft bearing retaining cap bolts, and thrust bearing cap retaining bolts.

15. Remove the camshafts.

16. Remove the cylinder head retaining bolts, in the proper sequence and in several passes.

17. Remove the cylinder head from its mounting.

To install:

➡Be sure to use new fasteners, as required.

18. Installation is the reverse of the removal procedure.

19. The sealant locations on the head and block must be free of dirt and engine oil. Apply sealant on the block top surface before assembling the cylinder head gaskets. The component must be installed within five minutes after the sealant was applied. Bead width 0.078–0.118 inch. Sealant locations 0.039–0.059 inch from block surface. Apply sealant on the cylinder head gaskets after assembling the

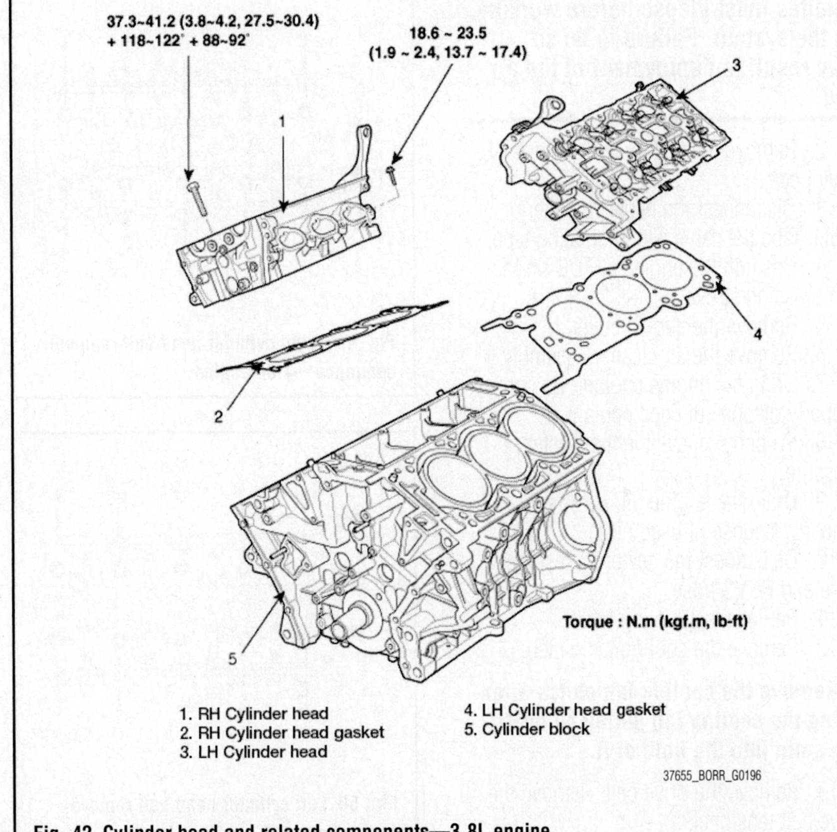

Fig. 42 Cylinder head and related components—3.8L engine

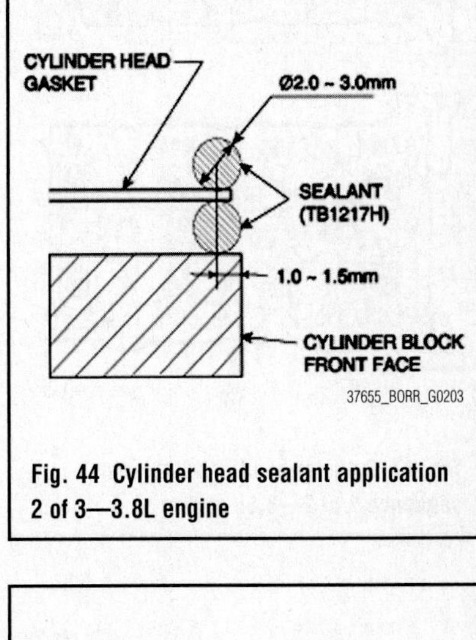

Fig. 44 Cylinder head sealant application 2 of 3—3.8L engine

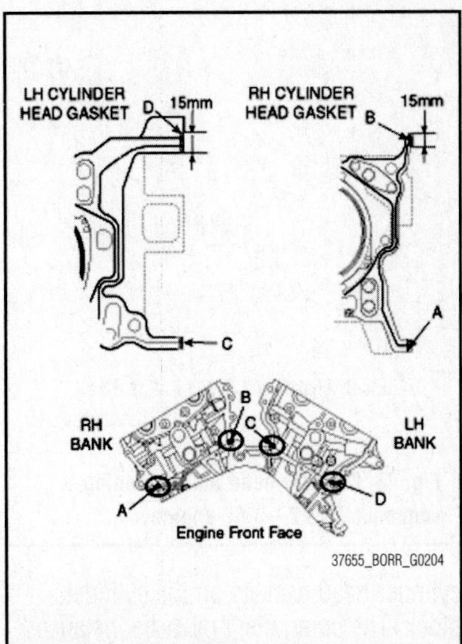

Fig. 45 Cylinder head sealant application 3 of 3—3.8L engine

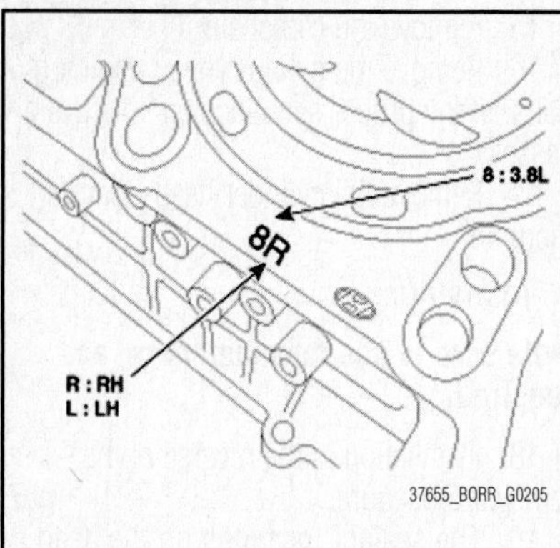

Fig. 46 Cylinder head gasket identification—3.8L engine

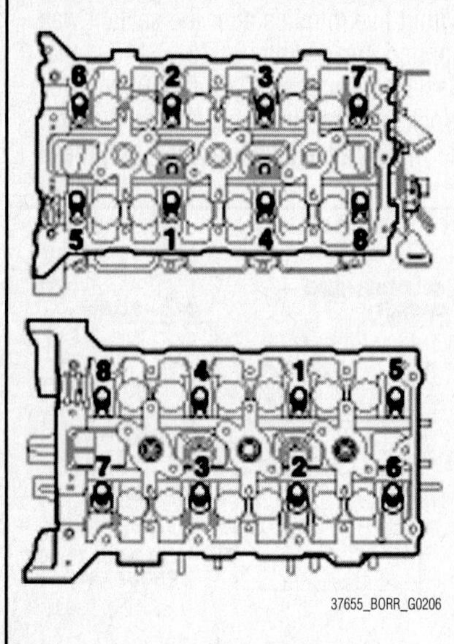

Fig. 47 Cylinder head bolt tightening sequence 1 of 2—3.8L engine

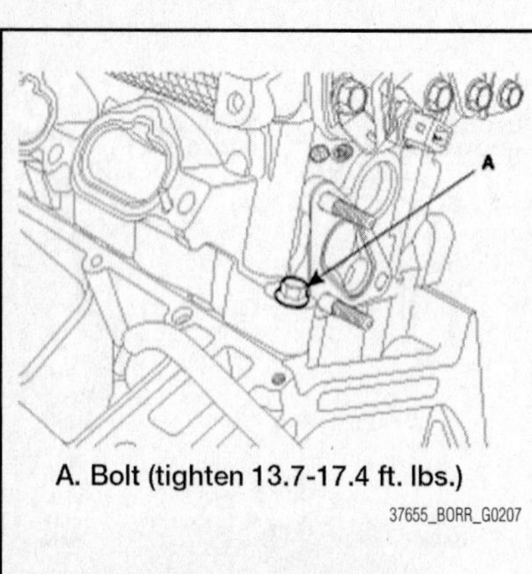

A. Bolt (tighten 13.7-17.4 ft. lbs.)

Fig. 48 Cylinder head bolt tightening sequence 2 of 2—3.8L engine

cylinder head gaskets on the cylinder block. The component must be installed within five minutes after the sealant was applied. See illustrations.

➡Be careful of the installation direction.

20. Install the cylinder head bolts.

➡Do not apply engine oil on the threads and under the heads of the cylinder head bolts.

21. Tighten the bolts to specification and in the proper sequence. See illustrations.
22. Be sure to reprogram the necessary systems.
23. Reset/Initialize the sunroof, if equipped.

4.6L Engine

See Figures 49 through 55.

1. Before servicing the vehicle, refer to the Precautions Section.

➡If working near and/or around the SRS system and components, be sure to disable the SRS system. Tape the negative battery cable with insulating tape. Always disconnect the negative battery cable first.

⁂ CAUTION

To avoid personal injury when working on vehicles equipped with an air bag, the negative battery cable must be disconnected and at least three minutes must elapse before working on the system. Failure to do so may result in deployment of the air bag.

2. Remove the battery top cover, if equipped.
3. Disconnect the negative battery cable. Tape the cable with insulating tape.
4. Position the engine at TDC on the compression stroke.
5. Remove the engine cover.
6. Remove the air cleaner assembly.
7. Drain the engine coolant. Be sure to properly dispose of used coolant.
8. Properly relieve the fuel system pressure.
9. Drain the engine oil. Be sure to properly dispose of used oil.
10. Disconnect the heater hoses, fuel hose and PCVS hose.
11. Remove the cooling fan.
12. Remove the cooling fan clutch.

➡Remove the cooling fan clutch after fixing the cooling fan pulley by inserting a pin into the hole of it.

13. Remove the drive belt. Remove the drive belt tensioner.
14. Disconnect the water temperature sensor connector, CVVT oil control valve connectors, ECT module connector, PCVS connector, injector connectors, ignition coil connectors and the wiring harness protector.
15. Disconnect the knock sensor connectors, camshaft position sensor connectors, oxygen sensor connectors, condenser connector, variable intake system solenoid valve connector and the wiring harness protector.
16. Disconnect the oil pressure switch connector, CVVT oil valve connectors and the AC compressor switch connector.
17. Remove the oil level gauge.
18. Disconnect the oxygen sensor connector and remove the sensor connector bracket.
19. Remove the exhaust manifold heat protectors and remove the exhaust manifolds.
20. Disconnect the PCV hose, the PCVS hose and the water hoses from the intake manifold.
21. Remove the intake manifold.
22. Remove the knock sensors. Remove the ignition coils.
23. Remove the valve cover retaining bolts. Remove the valve covers.

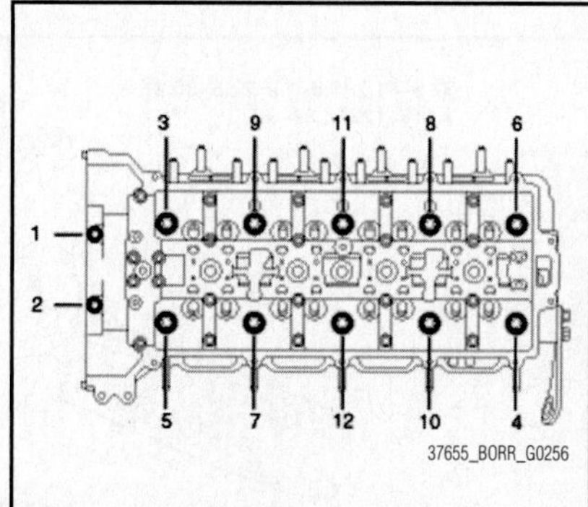

Fig. 49 Right cylinder head bolt removal sequence—4.6L engine

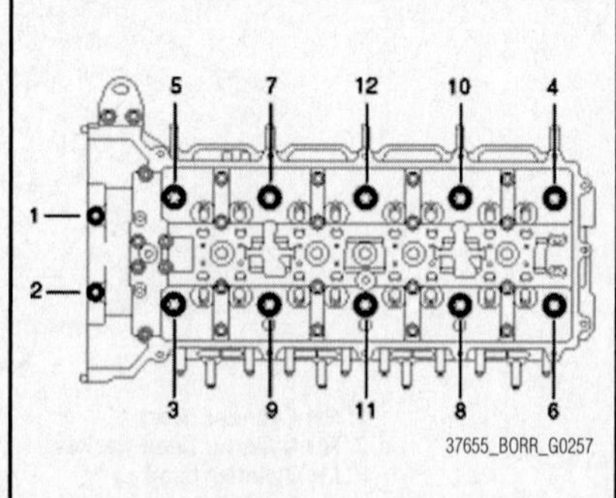

Fig. 50 Left cylinder head bolt removal sequence—4.6L engine

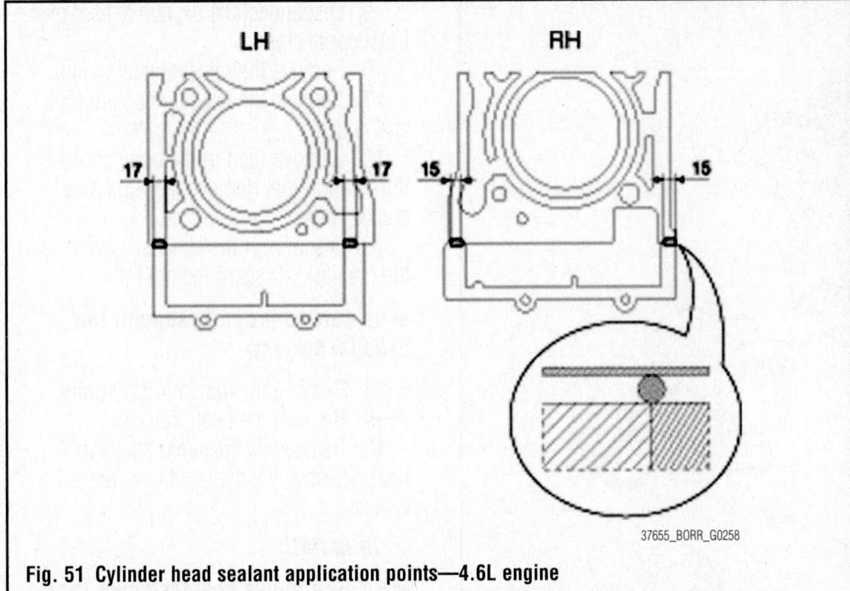

Fig. 51 Cylinder head sealant application points—4.6L engine

24. Remove the water temperature sensor wiring assembly and bracket.

25. Remove the upper timing chain upper covers. Remove the oil seals from the covers, as necessary.

26. Remove the oil control valves.

27. Remove the cam to cam guides.

28. Align the timing mark of the camshaft sprocket.

29. Remove the alternator.

30. Remove the water outlet pipe and water inlet pipe.

31. Remove the water temperature control assembly and the water outlet fitting assembly.

32. Remove the timing chain tensioners.

33. Remove the camshaft bearing caps and the thrust bearing caps.

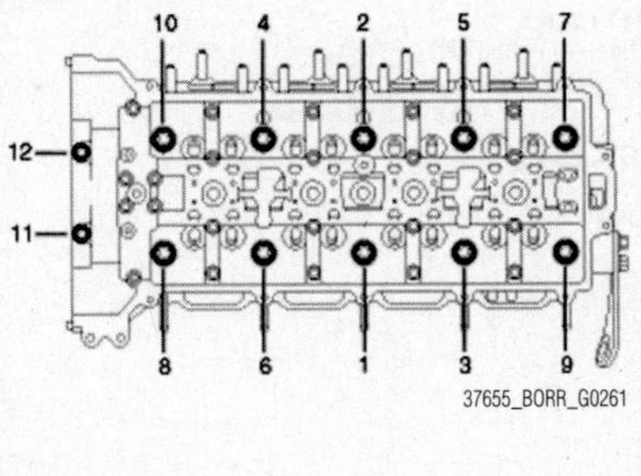

Fig. 52 Right cylinder head bolt tightening sequence—4.6L engine

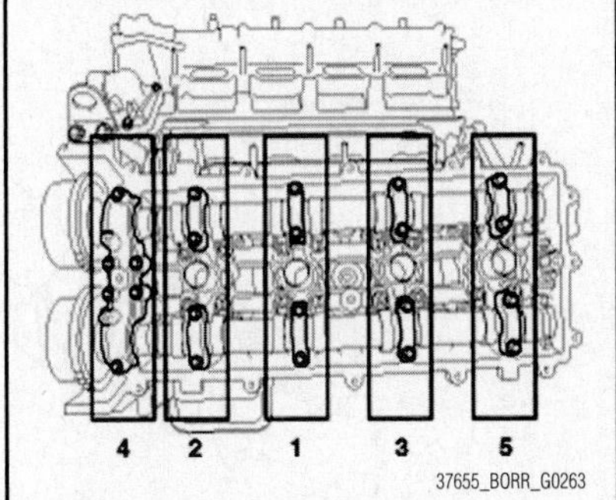

Fig. 54 Camshaft bolt tightening sequence (step one)—4.6L engine

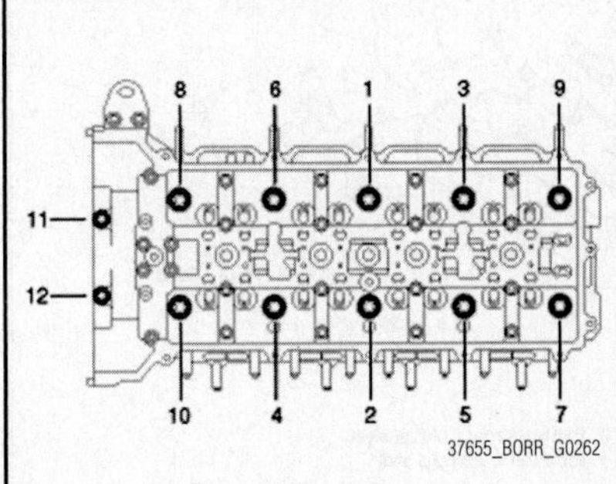

Fig. 53 Left cylinder head bolt tightening sequence—4.6L engine

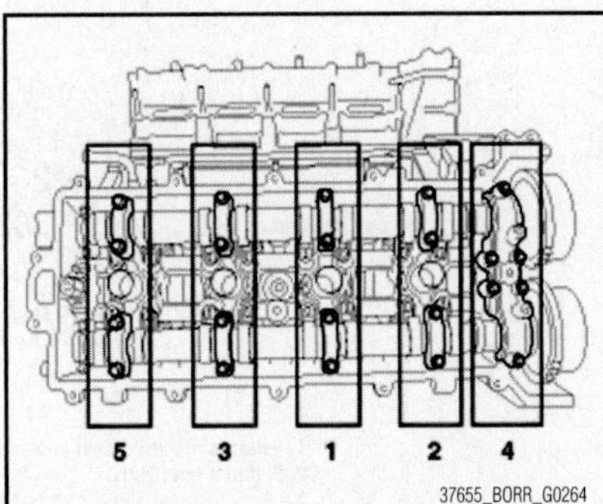

Fig. 55 Camshaft bolt tightening sequence (step two)—4.6L engine

34. Remove the camshafts.

35. As required, remove the roller finger follower and hydraulic valve lifter assembly.

36. Remove the cylinder head retaining bolts. See illustration for proper removal sequence. Remove bolts in several passes

37. Remove the cylinder heads from the engine.

To install:

➡Be sure to use new fasteners, as required.

38. Installation is the reverse of the removal procedure.

➡The sealant locations on the head gasket, block and timing chain lower cover must be free of engine oil and debris.

39. Before assembling the head gasket, liquid sealant part number TB 1217H, or equivalent, should be applied on the gap between the block and the timing chain lower case cover. The component must be installed within five minutes after the sealant was applied. Bead width 0.1378–0.1772 inch. Sealant locations 0.0591–0.0984 inch from the timing chain lower case inner surface.

40. Apply sealant on the head gaskets after assembling the gaskets on the cylinder block. The component must be installed within five minutes after the sealant was applied. Remove the extruded sealant, as necessary after assembling the heads.

41. Install the cylinder head bolts.

➡Do not apply engine oil on the threads and under the heads of the cylinder head bolts.

42. Tighten the bolts to specification and in the proper sequence. See illustrations.

43. Install the camshafts.

44. Tighten the bearing caps and thrust bearing caps to specification and in two steps. Step one specification: 5.1–5.8 ft. lbs. Step two specification: 10.1–10.8 ft. lbs.

45. Continue the installation in the reverse order of the removal procedure.

46. Be sure to reprogram the necessary systems.

47. Reset/Initialize the sunroof, if equipped.

EXHAUST MANIFOLD

REMOVAL & INSTALLATION

3.8L Engine

See Figure 56.

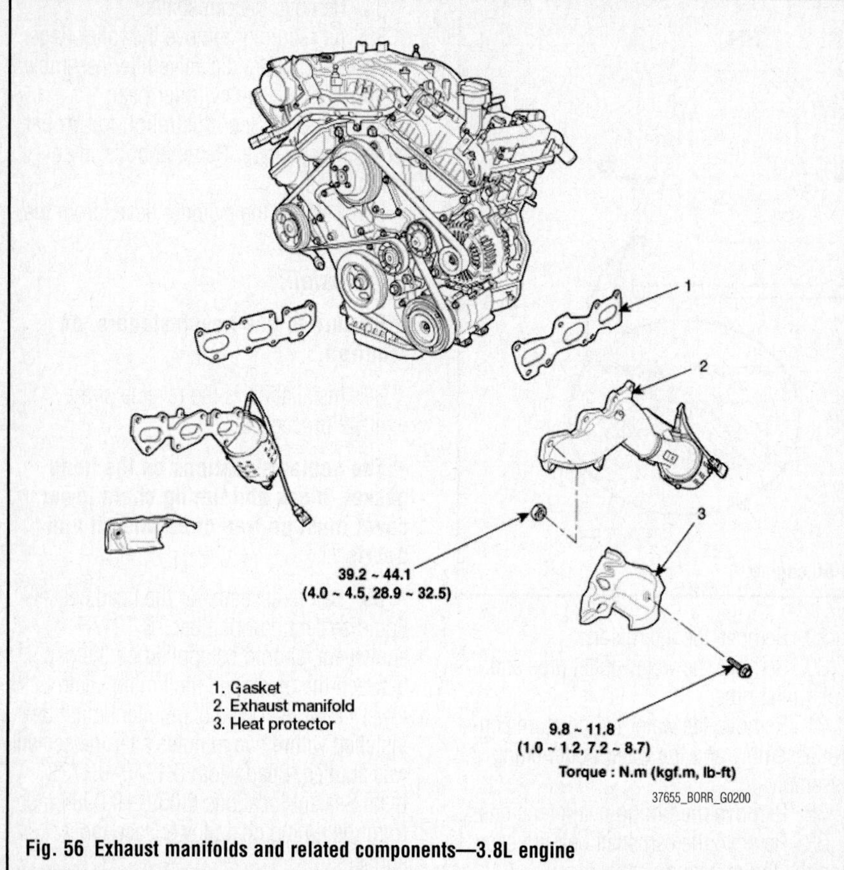

Fig. 56 Exhaust manifolds and related components—3.8L engine

1. Before servicing the vehicle, refer to the Precautions Section.

➡**If working near and/or around the SRS system and components, be sure to disable the SRS system. Tape the negative battery cable with insulating tape. Always disconnect the negative battery cable first.**

CAUTION

To avoid personal injury when working on vehicles equipped with an air bag, the negative battery cable must be disconnected and at least three minutes must elapse before working on the system. Failure to do so may result in deployment of the air bag.

2. Remove the battery top cover, if equipped.
3. Disconnect the negative battery cable. Tape the cable with insulating tape.
4. Remove the engine cover.
5. Drain the engine coolant. Be sure to properly dispose of used coolant.
6. Remove the radiator grille upper guard.
7. Remove the air cleaner assembly.
8. Disconnect the upper and lower radiator hoses.
9. Disconnect the oxygen sensor electrical connectors.
10. Remove the oil level gauge tube.
11. Remove the left side coolant pipe and hoses.
12. Remove the throttle body coolant hoses and pipe. Remove the right side coolant pipe.
13. Disconnect the exhaust system from the manifold. Discard the gasket.

➡**Be sure to properly support the exhaust system.**

14. Remove the manifold stay bolts.
15. Remove the heat protector.
16. Remove the manifold retaining bolts. Remove the manifold from the vehicle.

To install:

➡**Be sure to use new fasteners, as required.**

17. Installation is the reverse of the removal procedure.
18. Be sure to reprogram the necessary systems.
19. Reset/Initialize the sunroof, if equipped.

4.6L Engine

See Figures 57 through 59.

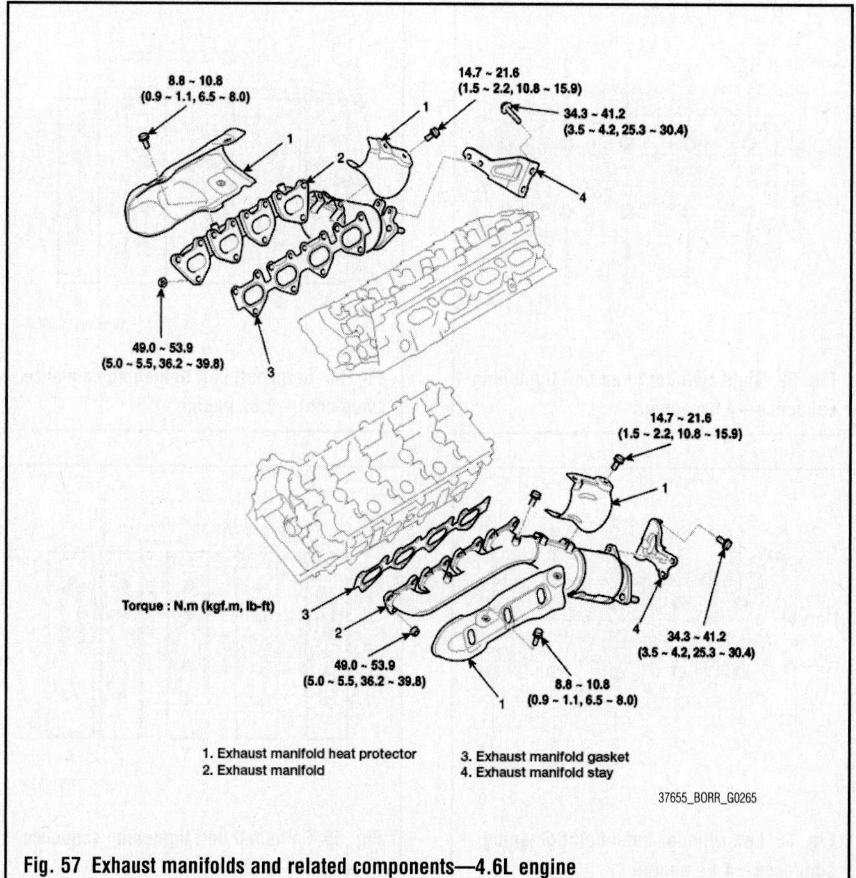

Fig. 57 Exhaust manifolds and related components—4.6L engine

1. Before servicing the vehicle, refer to the Precautions Section.

➡**If working near and/or around the SRS system and components, be sure to disable the SRS system. Tape the negative battery cable with insulating tape. Always disconnect the negative battery cable first.**

⁂ CAUTION

To avoid personal injury when working on vehicles equipped with an air bag, the negative battery cable must be disconnected and at least three minutes must elapse before working on the system. Failure to do so may result in deployment of the air bag.

2. Remove the battery top cover, if equipped.
3. Disconnect the negative battery cable. Tape the cable with insulating tape.
4. Remove the engine cover.
5. Remove the air cleaner assembly.
6. Remove the oil level gauge.
7. Disconnect the oxygen sensor connector and remove the sensor connector bracket.
8. Properly support the exhaust system, as required. Disconnect the exhaust manifold at the exhaust flange.

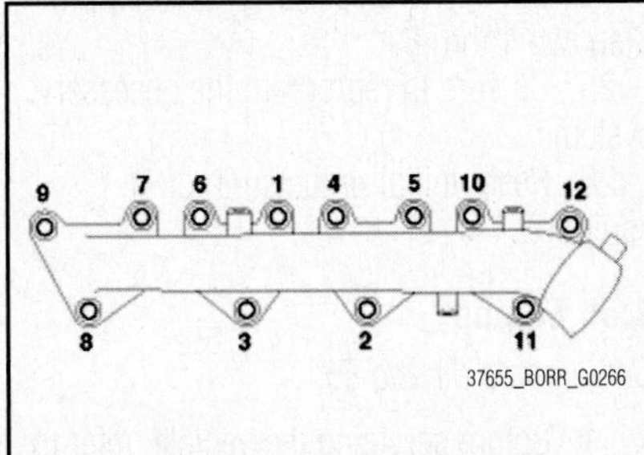

Fig. 58 Left exhaust manifold bolt tightening sequence—4.6L engine

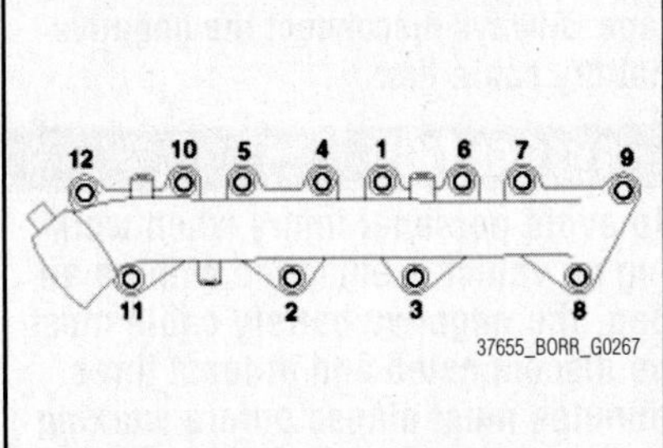

Fig. 59 Right exhaust manifold bolt tightening sequence—4.6L engine

9. Remove the left and right exhaust manifold stay(s).
10. Remove the heat protectors.
11. Remove the manifold retaining bolts.
12. Remove the manifold from the engine.

To install:

➡**Be sure to use new fasteners, as required.**

13. Installation is the reverse of the removal procedure.
14. The TOP mark on the gasket must face the exhaust manifold.
15. Install the retaining bolts. Tighten to specification and in the proper sequence.
16. Be sure to reprogram the necessary systems.
17. Reset/Initialize the sunroof, if equipped.

FLEXPLATE

REMOVAL & INSTALLATION

See Figures 60 and 61.

At this time the manufacturer does not provide removal and installation procedures for this component. The following procedure is a guideline and may differ from the vehicle you are servicing.

1. Before servicing the vehicle, refer to the Precautions Section.

➡**If working near and/or around the SRS system and components, be sure to disable the SRS system. Tape the negative battery cable with insulating tape. Always disconnect the negative battery cable first.**

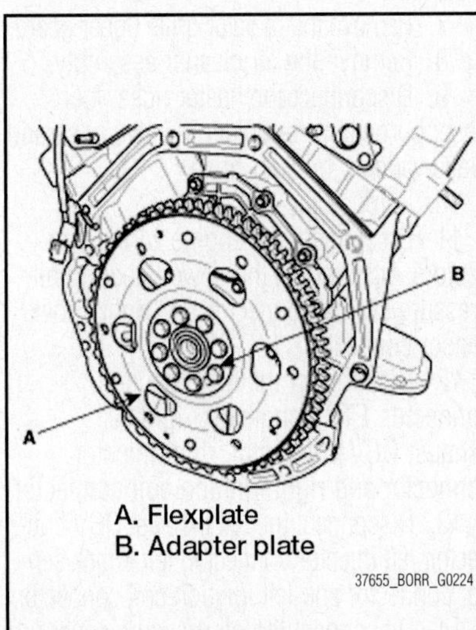

Fig. 60 Flywheel and related components—3.8L engine

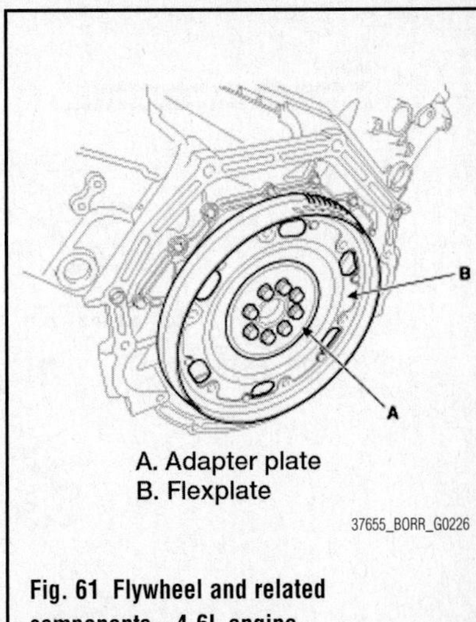

Fig. 61 Flywheel and related components—4.6L engine

⁂ CAUTION

To avoid personal injury when working on vehicles equipped with an air bag, the negative battery cable must be disconnected and at least three minutes must elapse before working on the system. Failure to do so may result in deployment of the air bag.

2. Remove the battery top cover, if equipped.
3. Disconnect the negative battery cable. Tape the cable with insulating tape.
4. Raise and safely support the vehicle.
5. Remove the transmission.
6. Matchmark the flywheel.
7. Remove the flywheel retaining bolts.
8. Remove the flywheel from its mounting.

To install:

➡**Be sure to use new fasteners, as required.**

9. Installation is the reverse of the removal procedure.
10. Be sure to reprogram the necessary systems.
11. Reset/Initialize the sunroof, if equipped.

INTAKE MANIFOLD

REMOVAL & INSTALLATION

3.8L Engine

See Figures 62 and 63.

1. Before servicing the vehicle, refer to the Precautions Section.

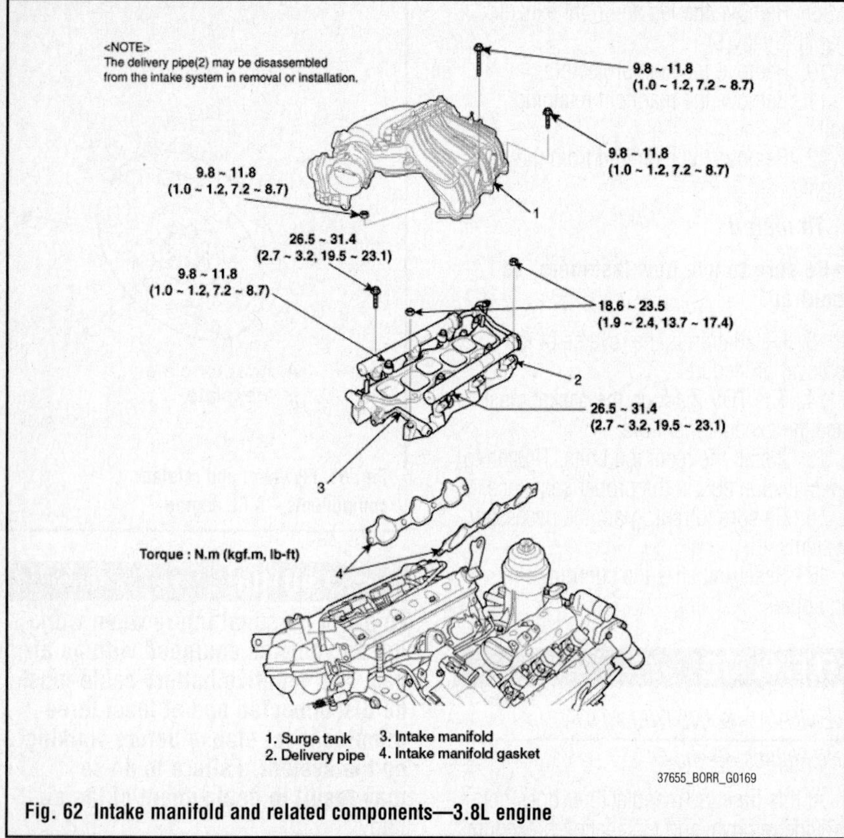

Fig. 62 Intake manifold and related components—3.8L engine

➡**If working near and/or around the SRS system and components, be sure to disable the SRS system. Tape the negative battery cable with insulating tape. Always disconnect the negative battery cable first.**

✲✲ CAUTION

To avoid personal injury when working on vehicles equipped with an air bag, the negative battery cable must be disconnected and at least three minutes must elapse before working on the system. Failure to do so may result in deployment of the air bag.

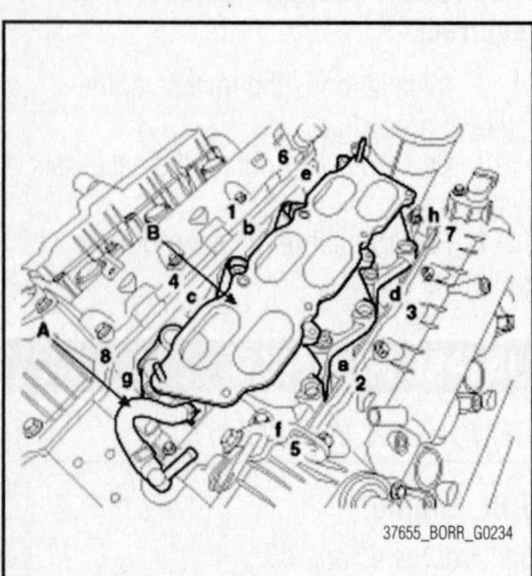

Fig. 63 Intake manifold bolt tightening sequence—3.8L engine

2. Remove the battery top cover, if equipped.
3. Disconnect the negative battery cable. Tape the cable with insulating tape.
4. Drain the cooling system. Be sure to properly dispose of used coolant.
5. Remove the engine cover.
6. Properly relieve the fuel system pressure.
7. Remove the radiator grille upper guard.
8. Remove the air cleaner assembly.
9. Disconnect the heater hose, fuel hose, purge control solenoid valve hose and brake vacuum booster hose.
10. Remove the fuel pipe.
11. Disconnect the engine wiring connectors. Disconnect the power steering oil pressure switch connector and right knock sensor connector.
12. Disconnect the MAP sensor connector, ETC connector, right exhaust OCV connector, right injector connector and right ignition coil connector.
13. Disconnect the left exhaust OCV connector, left injector connector, left knock sensor connector and left ignition coil connector.
14. Disconnect the oil pressure sensor connector, left exhaust camshaft CMP sensor and left oxygen sensor.
15. Disconnect the left intake camshaft CMP sensor connector, right intake camshaft CMP sensor connector and condenser connector.
16. Disconnect the VIS solenoid valve connector, PCSV connector, right oxygen sensor connector and right exhaust camshaft CMP sensor connector.
17. Disconnect the water temperature sensor connector and the oil temperature sensor connector.
18. Disconnect the throttle body coolant hoses and breather hoses.
19. Remove the surge tank stay bolts. Remove the surge tank.
20. Disconnect the right ignition coil connector. Disconnect the water vent hose.
21. Remove the intake manifold.

➡**Be sure to drain the coolant, before removing the manifold. If any coolant drained from the cylinder head vent hole has entered the intake port can lead to serious engine trouble.**

To install:

➡**Be sure to use new fasteners, as required.**

22. Installation is the reverse of the removal procedure.
23. Be sure to check for proper gasket identification (LH and/or RH).
24. Tighten the retaining bolts to specification and in two steps. Step one a thru h. Step two 1 thru 8.
25. Be sure to reprogram the necessary systems.
26. Reset/Initialize the sunroof, if equipped.

4.6L Engine

See Figures 64 and 65.

1. Before servicing the vehicle, refer to the Precautions Section.

➡**If working near and/or around the SRS system and components, be sure to disable the SRS system. Tape the negative battery cable with insulating tape. Always disconnect the negative battery cable first.**

✲✲ CAUTION

To avoid personal injury when working on vehicles equipped with an air bag, the negative battery cable must be disconnected and at least three minutes must elapse before working on the system. Failure to do so may result in deployment of the air bag.

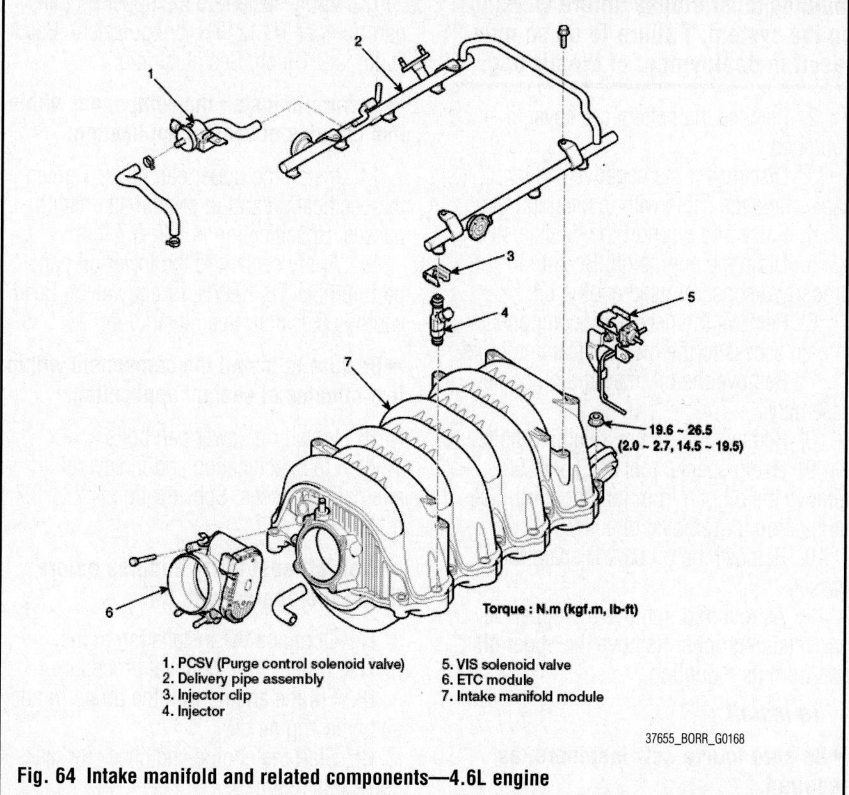

Fig. 64 Intake manifold and related components—4.6L engine

2. Remove the battery top cover, if equipped.
3. Disconnect the negative battery cable. Tape the cable with insulating tape.
4. Remove the engine cover.
5. Remove the air cleaner assembly.
6. Properly relieve the fuel system pressure.
7. Drain the cooling system. Be sure to properly dispose of used coolant.

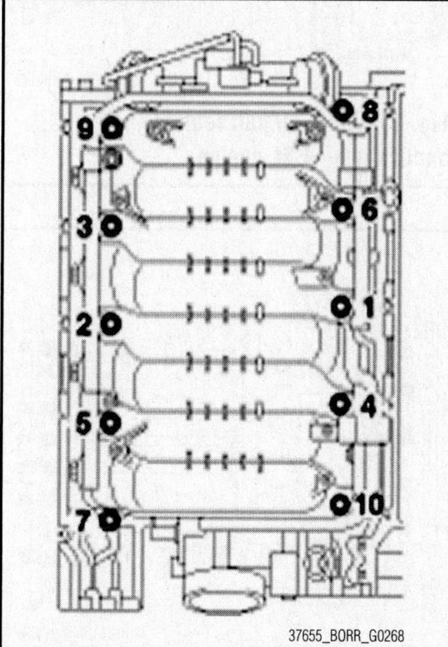

Fig. 65 Intake manifold bolt tightening sequence—4.6L engine

8. Disconnect the fuel hose, PCSV hose and ECT module connector.
9. Disconnect the PCSV connector, injector connectors, variable intake system solenoid valve connector and wiring harness protector.
10. Disconnect the injector connectors, PCV hose, PCSV hose and water hoses from the intake manifold.
11. Remove the intake manifold retaining bolts.
12. Remove the intake manifold. Discard the gasket.

To install:

➡Be sure to use new fasteners, as required.

13. Installation is the reverse of the removal procedure.
14. Be sure to use new gaskets. Tighten the retaining bolts to specification and in the proper sequence.
15. Be sure to reprogram the necessary systems.
16. Reset/Initialize the sunroof, if equipped.

OIL PAN

REMOVAL & INSTALLATION

3.8L Engine

See Figures 66 and 67.

At this time the manufacturer does not provide removal and installation procedures for this component. The following procedure is a guideline and may differ from the vehicle you are servicing.

1. Before servicing the vehicle, refer to the Precautions Section.

➡If working near and/or around the SRS system and components, be sure to disable the SRS system. Tape the negative battery cable with insulating tape. Always disconnect the negative battery cable first.

⁂ CAUTION

To avoid personal injury when working on vehicles equipped with an air bag, the negative battery cable must be disconnected and at least three minutes must elapse before working on the system. Failure to do so may result in deployment of the air bag.

2. Remove the battery top cover, if equipped.
3. Disconnect the negative battery cable. Tape the cable with insulating tape.
4. Raise and support the vehicle safely.

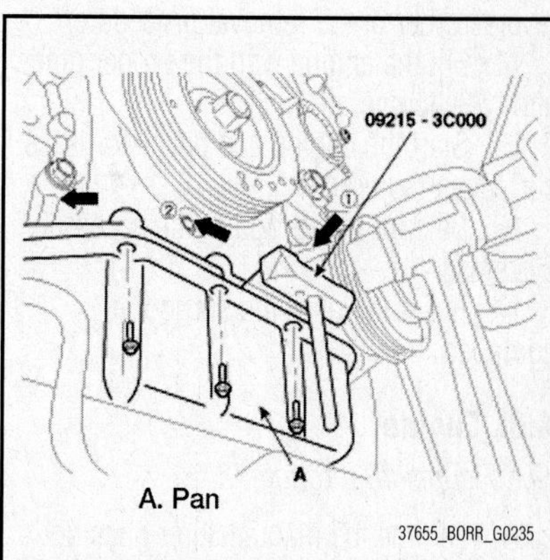

Fig. 66 Lower oil pan removal direction—3.8L engine

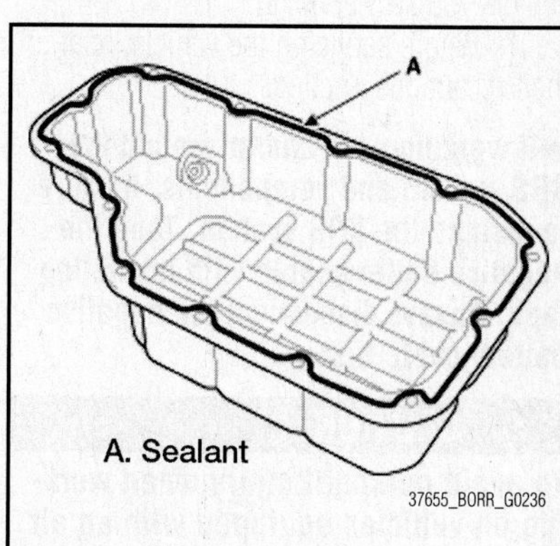

Fig. 67 Lower oil pan sealant application—3.8L engine

5. Drain the engine oil. Be sure to properly dispose of used engine oil.

6. Remove the necessary components to gain access to the oil pan retaining bolts.

7. Remove the oil pan retaining bolts.

8. Using special tool 09215-3C00, remove the oil pan from its mounting. See illustration for removal direction.

9. Remove the oil pan. Discard the gasket.

To install:

➡Be sure to use new fasteners, as required.

10. Installation is the reverse of the removal procedure.

11. Apply sealant to the oil pan part number TB 1217H, or equivalent. Bead width is 0.1 inch.

➡Be sure to install the component within five minutes of sealant application.

12. Install the pan bolts. Tighten to specification and in several alternating passes.

➡Wait at least thirty minutes before filling the engine with oil.

13. Continue the installation in the reverse order of the removal procedure.

14. Fill the engine with the proper grade and type engine oil.

15. Start the engine and check for leaks, correct as required.

16. Be sure to reprogram the necessary systems.

17. Reset/Initialize the sunroof, if equipped.

4.6L Engine

See Figures 68 through 73.

At this time the manufacturer does not provide removal and installation procedures for this component. The following procedure is a guideline and may differ from the vehicle you are servicing.

1. Before servicing the vehicle, refer to the Precautions Section.

➡If working near and/or around the SRS system and components, be sure to disable the SRS system. Tape the negative battery cable with insulating tape. Always disconnect the negative battery cable first.

⁂ CAUTION

To avoid personal injury when working on vehicles equipped with an air bag, the negative battery cable must be disconnected and at least three minutes must elapse before working on the system. Failure to do so may result in deployment of the air bag.

2. Remove the battery top cover, if equipped.

3. Disconnect the negative battery cable. Tape the cable with insulating tape.

4. Raise and support the vehicle safely.

5. Drain the engine oil. Be sure to properly dispose of used engine oil.

6. Remove the necessary components to gain access to the oil pan retaining bolts.

7. Remove the oil filter and oil cooler assembly.

8. Remove the oil pan retaining bolts.

9. Using special tool 09215-3C00, remove the oil pan from its mounting. See illustration for removal direction.

10. Remove the oil pan. Discard the gasket.

11. As required, remove the upper oil pan retaining bolts. Remove the upper oil pan from its mounting.

To install:

➡Be sure to use new fasteners, as required.

12. Installation is the reverse of the removal procedure.

13. Apply sealant to the upper oil pan part number TB 1217H, or equivalent. Bead width is 0.1 inch. See illustration.

➡Be sure to install the component within five minutes of sealant application.

14. Install the upper pan bolts. Tighten to specification and in several alternating passes. Specification is 7.2–8.7 ft. lbs.

15. Apply sealant to the lower oil pan part number TB 1217H, or equivalent. Bead width is 0.1 inch. See illustration.

➡Be sure to install the component within five minutes of sealant application.

16. Install the lower pan bolts. Tighten to specification and in several alternating passes. Specification is 7.2–8.7 ft. lbs.

➡Wait at least thirty minutes before filling the engine with oil.

17. Continue the installation in the reverse order of the removal procedure.

18. Fill the engine with the proper grade and type engine oil.

19. Start the engine and check for leaks, correct as required.

20. Be sure to reprogram the necessary systems.

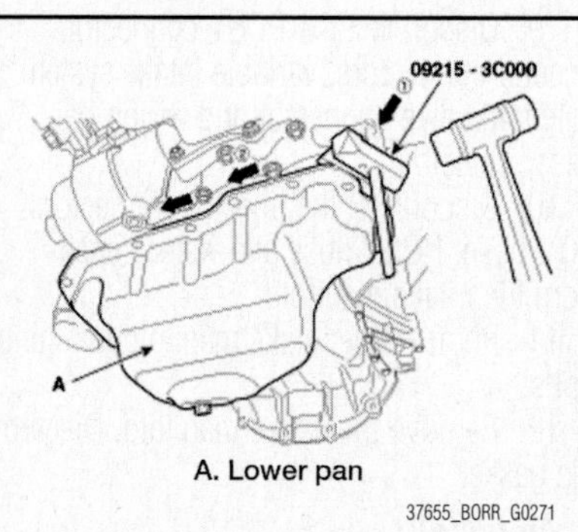

Fig. 68 Lower oil pan removal direction—4.6L engine

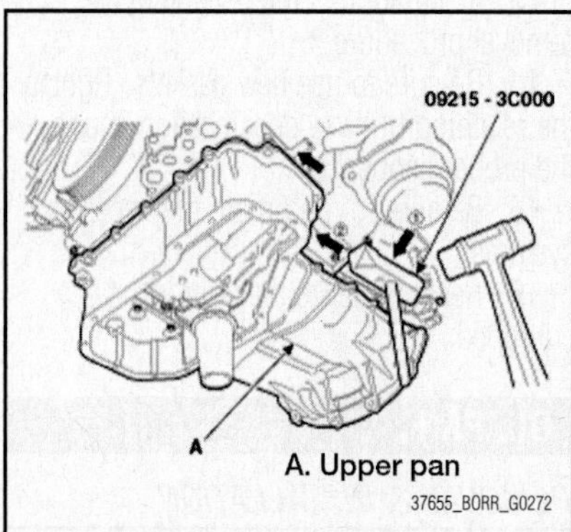

Fig. 69 Upper oil pan removal direction—4.6L engine

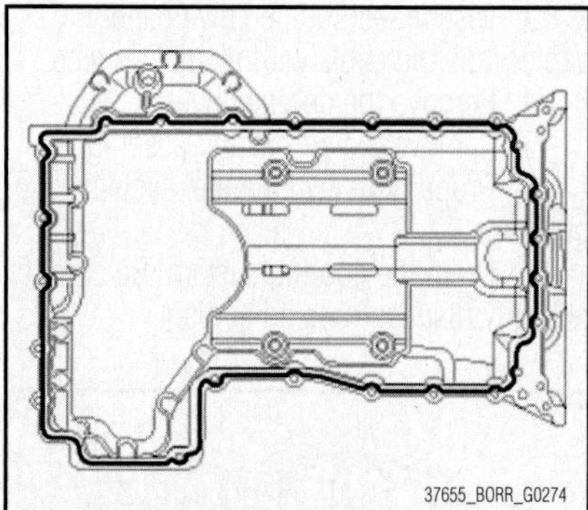

Fig. 70 Upper oil pan sealant application—4.6L engine

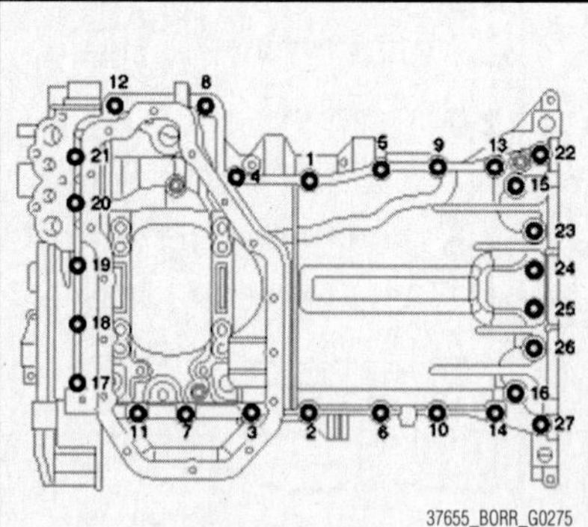

Fig. 71 Upper oil pan tightening sequence—4.6L engine

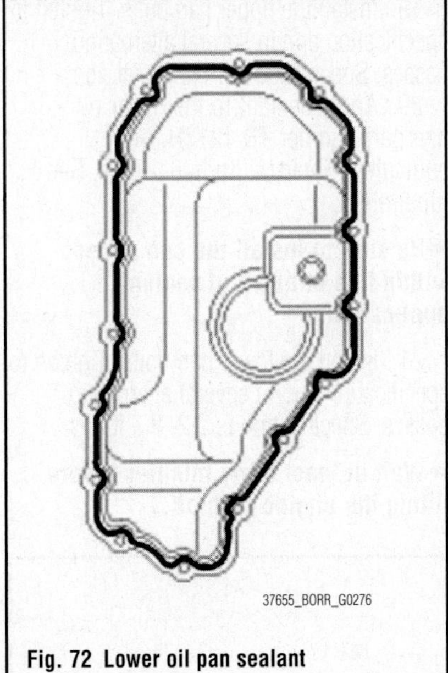

Fig. 72 Lower oil pan sealant application—4.6L engine

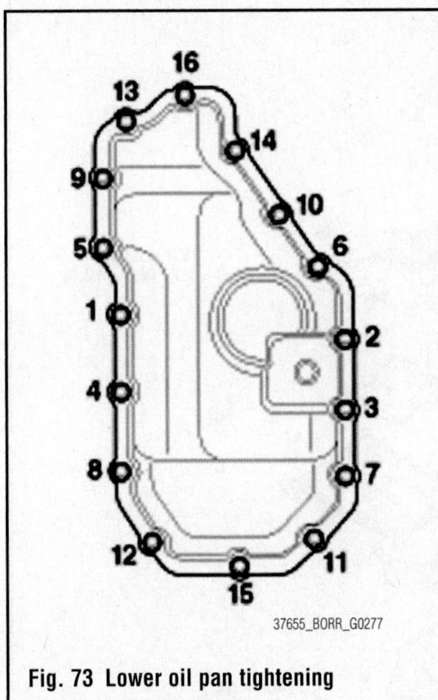

Fig. 73 Lower oil pan tightening sequence—4.6L engine

21. Reset/Initialize the sunroof, if equipped.

OIL PUMP

REMOVAL & INSTALLATION

3.8L Engine

See Figure 74.

1. Before servicing the vehicle, refer to the Precautions Section.

➡**If working near and/or around the SRS system and components, be sure to disable the SRS system. Tape the negative battery cable with insulating tape. Always disconnect the negative battery cable first.**

⁂ CAUTION

To avoid personal injury when working on vehicles equipped with an air bag, the negative battery cable must be disconnected and at least three minutes must elapse before working on the system. Failure to do so may result in deployment of the air bag.

2. Remove the battery top cover, if equipped.
3. Disconnect the negative battery cable. Tape the cable with insulating tape.
4. Raise and support the vehicle safely.
5. Drain the engine oil. Be sure to properly dispose of used engine oil.
6. Remove the necessary components to gain access to the oil pan retaining bolts.
7. Remove the oil pan retaining bolts.
8. Using special tool 09215-3C00, remove the oil pan from its mounting. See illustration for removal direction.
9. Remove the oil pump chain cover. Remove the chain sprocket.
10. Remove the oil pump retaining bolts.
11. Remove the oil pump from its mounting. Discard the O-ring.

To install:

➡**Be sure to use new fasteners, as required.**

12. Installation is the reverse of the removal procedure.
13. Install the oil pump to its mounting using a new O-ring. Tighten the retaining bolts to 13.7–15.9 ft. lbs.
14. Apply sealant to the oil pan part number TB 1217H, or equivalent. Bead width is 0.1 inch.

➡**Be sure to install the component within five minutes of sealant application.**

15. Install the pan bolts. Tighten to specification and in several alternating passes.

➡**Wait at least thirty minutes before filling the engine with oil.**

16. Fill the engine with the proper grade and type engine oil.
17. Start the engine and check for leaks, correct as required.
18. Be sure to reprogram the necessary systems.
19. Reset/Initialize the sunroof, if equipped.

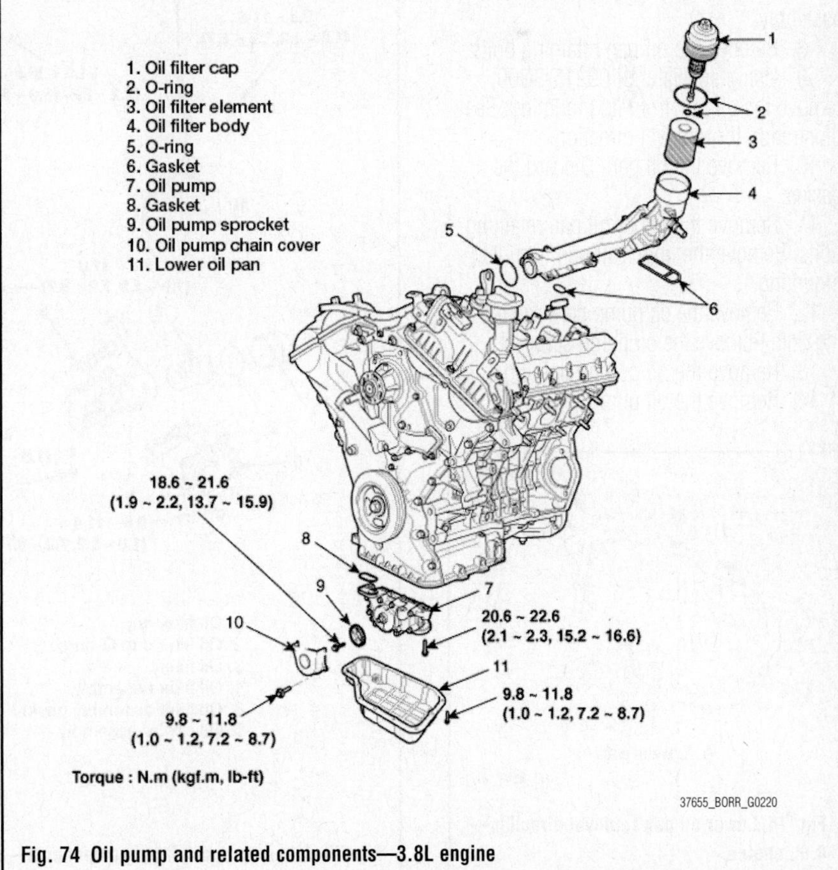

Fig. 74 Oil pump and related components—3.8L engine

4.6L Engine

See Figures 69, 75 and 76.

1. Before servicing the vehicle, refer to the Precautions Section.

➡If working near and/or around the SRS system and components, be sure to disable the SRS system. Tape the negative battery cable with insulating tape. Always disconnect the negative battery cable first.

⁂ CAUTION

To avoid personal injury when working on vehicles equipped with an air bag, the negative battery cable must be disconnected and at least three minutes must elapse before working on the system. Failure to do so may result in deployment of the air bag.

2. Remove the battery top cover, if equipped.
3. Disconnect the negative battery cable. Tape the cable with insulating tape.
4. Raise and support the vehicle safely.
5. Drain the engine oil. Be sure to properly dispose of used engine oil.
6. Remove the necessary components to gain access to the oil pan retaining bolts.
7. Remove the oil filter and oil cooler assembly.
8. Remove the oil pan retaining bolts.
9. Using special tool 09215-3C00, remove the oil pan from its mounting. See illustration for removal direction.
10. Remove the oil pan. Discard the gasket.
11. Remove the upper oil pan retaining bolts. Remove the upper oil pan from its mounting.
12. Remove the oil pump sprocket retaining bolt. Remove the oil pump sprocket.
13. Remove the oil pump retaining bolts.
14. Remove the oil pump.

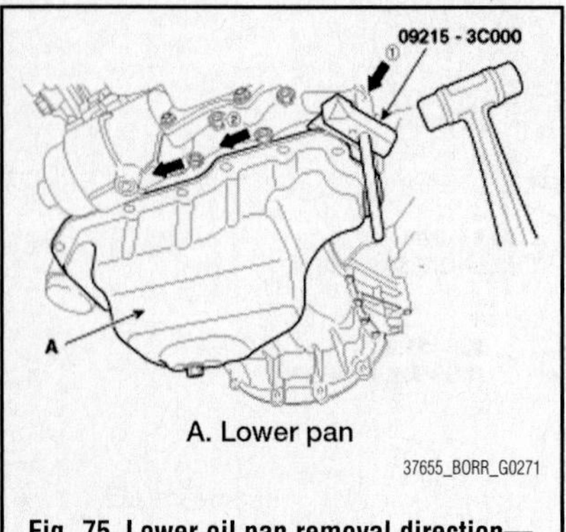

Fig. 75 Lower oil pan removal direction—4.6L engine

To install:

➡Be sure to use new fasteners, as required.

15. Installation is the reverse of the removal procedure.
16. Install the oil pump sprocket. Tighten the retaining bolt to 15.9–18.8 ft. lbs.
17. Install the oil pump assembly. Tighten the retaining bolts to 14.5–17.4 ft. lbs.
18. Apply sealant to the upper oil pan part number TB 1217H, or equivalent. Bead width is 0.1 inch. See illustration.

➡Be sure to install the component within five minutes of sealant application.

19. Install the upper pan bolts. Tighten to specification and in several alternating passes. Specification is 7.2–8.7 ft. lbs.
20. Apply sealant to the lower oil pan part number TB 1217H, or equivalent. Bead width is 0.1 inch. See illustration.

➡Be sure to install the component within five minutes of sealant application.

21. Install the lower pan bolts. Tighten to specification and in several alternating passes. Specification is 7.2–8.7 ft. lbs.

➡Wait at least thirty minutes before filling the engine with oil.

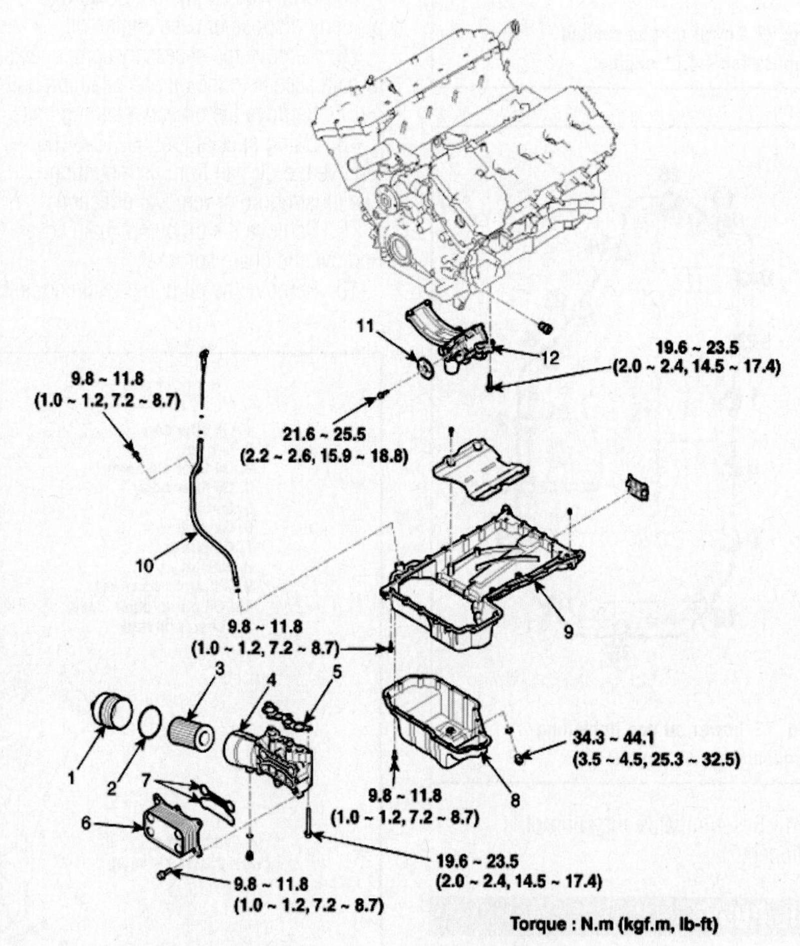

Torque : N.m (kgf.m, lb-ft)

1. Oil filter cap
2. Oil filter cap O-ring
3. Oil filter
4. Oil filter assembly
5. Oil filter assembly gasket
6. Oil cooler assembly
7. Oil cooler gasket
8. Lower oil pan
9. Upper oil pan
10. Oil level gauge
11. Oil pump sprocket
12. Oil pump assembly

37655_BORR_G0223

Fig. 76 Oil pump and related components—4.6L engine

22. Continue the installation in the reverse order of the removal procedure.
23. Fill the engine with the proper grade and type engine oil.
24. Start the engine and check for leaks, correct as required.
25. Be sure to reprogram the necessary systems.
26. Reset/Initialize the sunroof, if equipped.

INSPECTION

1. Before servicing the vehicle, refer to the Precautions Section.
2. With the oil pump removed, visually check the parts of the oil pump case for cracks or damage.
3. Check the relief plunger for smooth operation.
4. Check the relief spring for deformation or a break.
5. Check the surface that mates with the oil filter for any damage.
6. Check the oil filter bracket for cracks and oil leakage.
7. Replace the pump, as required.

PISTON AND RING

POSITIONING

See Figures 77 and 78.

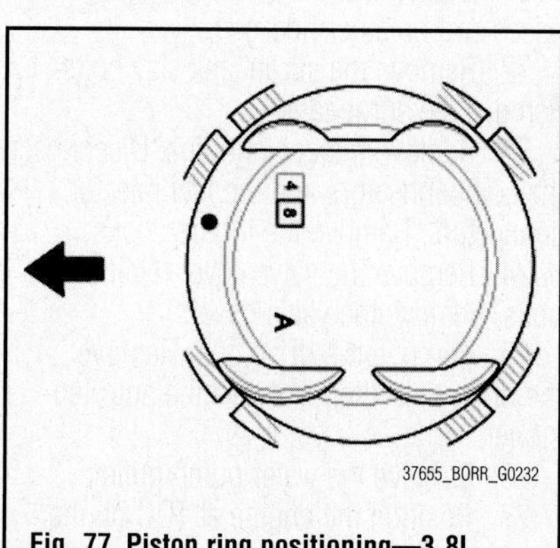

Fig. 77 Piston ring positioning—3.8L engine

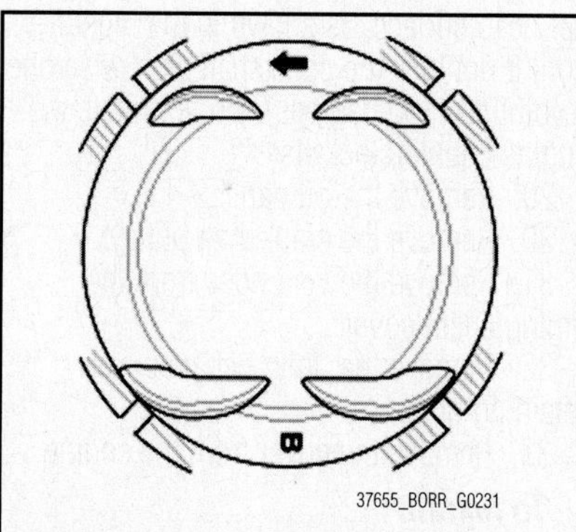

Fig. 78 Piston ring positioning—4.6L engine

REAR MAIN SEAL

REMOVAL & INSTALLATION

See Figures 79 and 80.

1. Before servicing the vehicle, refer to the Precautions Section.

➡If working near and/or around the SRS system and components, be sure to disable the SRS system. Tape the negative battery cable with insulating tape. Always disconnect the negative battery cable first.

✻✻ CAUTION

To avoid personal injury when working on vehicles equipped with an air bag, the negative battery cable must be disconnected and at least three minutes must elapse before working on the system. Failure to do so may result in deployment of the air bag.

2. Remove the battery top cover, if equipped.
3. Disconnect the negative battery cable. Tape the cable with insulating tape.
4. Remove the transmission assembly.
5. Remove the flexplate.
6. Remove the seal from its mounting. Clean the gasket mounting surfaces, as necessary.

To install:

➡Be sure to use new fasteners, as required.

7. Installation is the reverse of the removal procedure.
8. Be sure to reprogram the necessary systems.
9. Reset/Initialize the sunroof, if equipped.

ROCKER ARMS

REMOVAL & INSTALLATION

The engines covered are not equipped with rocker arms/shafts. The camshafts directly actuate the valve through a bucket type follower.

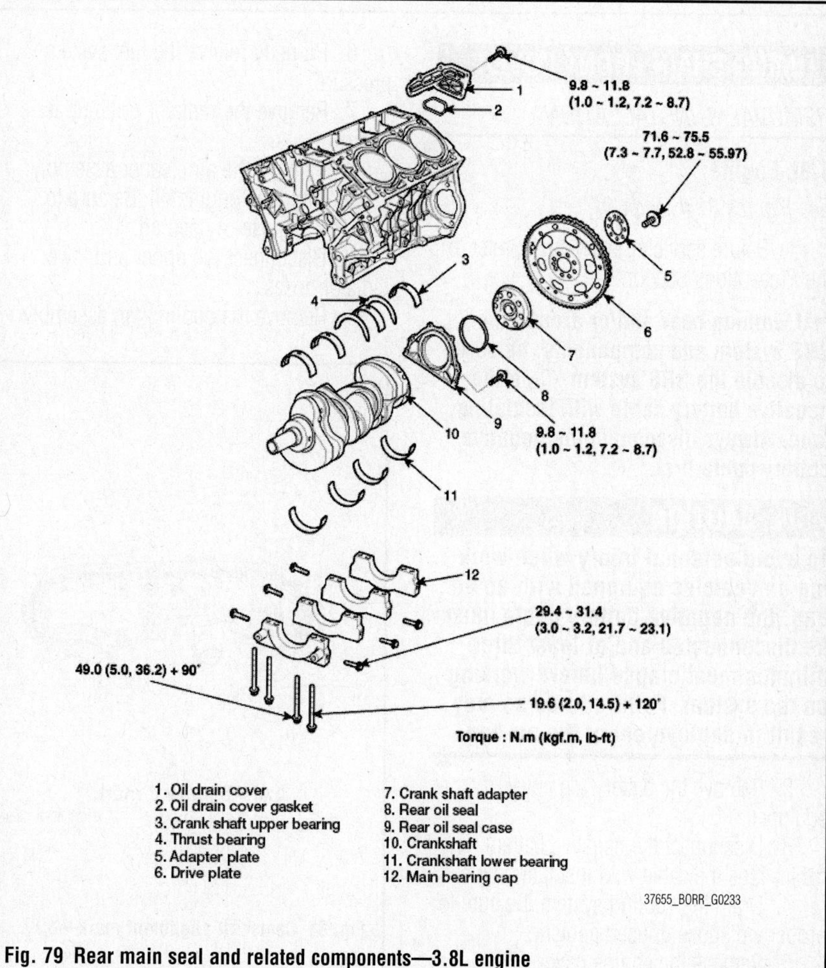

Fig. 79 Rear main seal and related components—3.8L engine

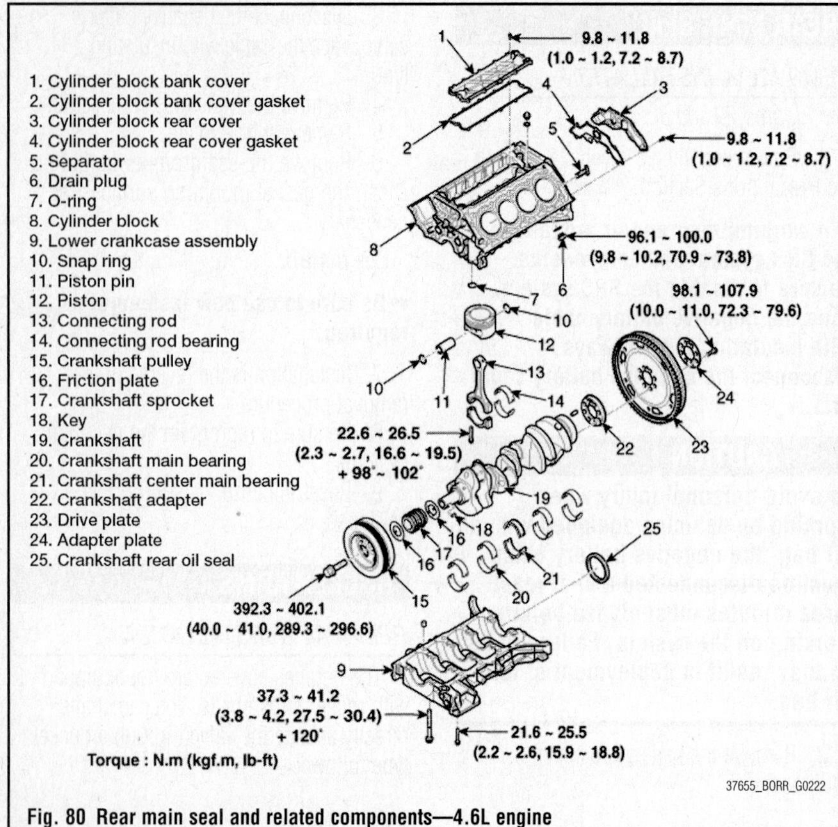

Fig. 80 Rear main seal and related components—4.6L engine

TIMING CHAIN FRONT COVER

REMOVAL & INSTALLATION

3.8L Engine

See Figures 81 through 87.

1. Before servicing the vehicle, refer to the Precautions Section.

➡If working near and/or around the SRS system and components, be sure to disable the SRS system. Tape the negative battery cable with insulating tape. Always disconnect the negative battery cable first.

⁂ CAUTION

To avoid personal injury when working on vehicles equipped with an air bag, the negative battery cable must be disconnected and at least three minutes must elapse before working on the system. Failure to do so may result in deployment of the air bag.

2. Remove the battery top cover, if equipped.
3. Disconnect the negative battery cable. Tape the cable with insulating tape.
4. Drain the cooling system. Be sure to properly dispose of used coolant.
5. Remove the engine cover.
6. Properly relieve the fuel system pressure.
7. Remove the radiator grille upper guard.
8. Remove the air cleaner assembly.
9. Drain the engine oil. Be sure to properly dispose of used oil.
10. Disconnect the upper and lower radiator hoses.
11. Remove the cooling fan assembly.

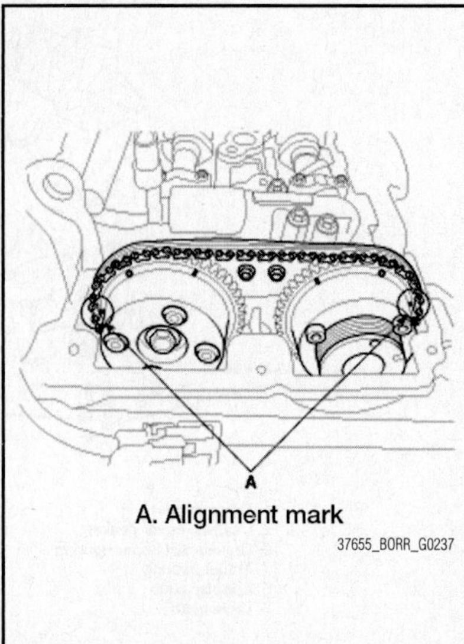

Fig. 81 Camshaft alignment mark—3.8L engine

12. Disconnect the heater hose, fuel hose, purge control solenoid valve hose and brake vacuum booster hose.
13. Remove the fuel pipe.
14. Disconnect the engine wiring connectors. Disconnect the power steering oil pressure switch connector and right knock sensor connector.
15. Disconnect the MAP sensor connector, ETC connector, right exhaust OCV connector, right injector connector and right ignition coil connector.
16. Disconnect the left exhaust OCV connector, left injector connector, left knock sensor connector and left ignition coil connector.
17. Disconnect the oil pressure sensor connector, left exhaust camshaft CMP sensor and left oxygen sensor.
18. Disconnect the left intake camshaft CMP sensor connector, right intake camshaft CMP sensor connector and condenser connector.
19. Disconnect the VIS solenoid valve connector, PCSV connector, right oxygen sensor connector and right exhaust camshaft CMP sensor connector.
20. Disconnect the water temperature sensor connector and the oil temperature sensor connector.
21. Disconnect the throttle body coolant hoses and breather hoses.
22. Remove the surge tank stay bolts. Remove the surge tank.
23. Remove the coolant pipe. Disconnect the coil connectors and the fuel injector connectors. Remove the ignition coils.
24. Remove the valve cover retaining bolts. Remove the valve covers.
25. Remove the drive belt. Remove the drive belt idler. Remove the auto tensioner.
26. Remove the water pump pulley.
27. Position the engine at TDC on the compression stroke.
28. Check that the mark of the camshaft timing sprockets are lined up on the cylinder head surface, as shown in the illustration. If not turn the crankshaft clockwise one revolution (360 degrees). Do not rotate the engine counterclockwise.
29. Remove the oil pan.
30. Remove the crankshaft pulley.
31. Remove the vent hose from the timing chain cover.
32. Remove the timing chain cover retaining bolts.
33. Remove the cover from the engine.

To install:

➡Be sure to use new fasteners, as required.

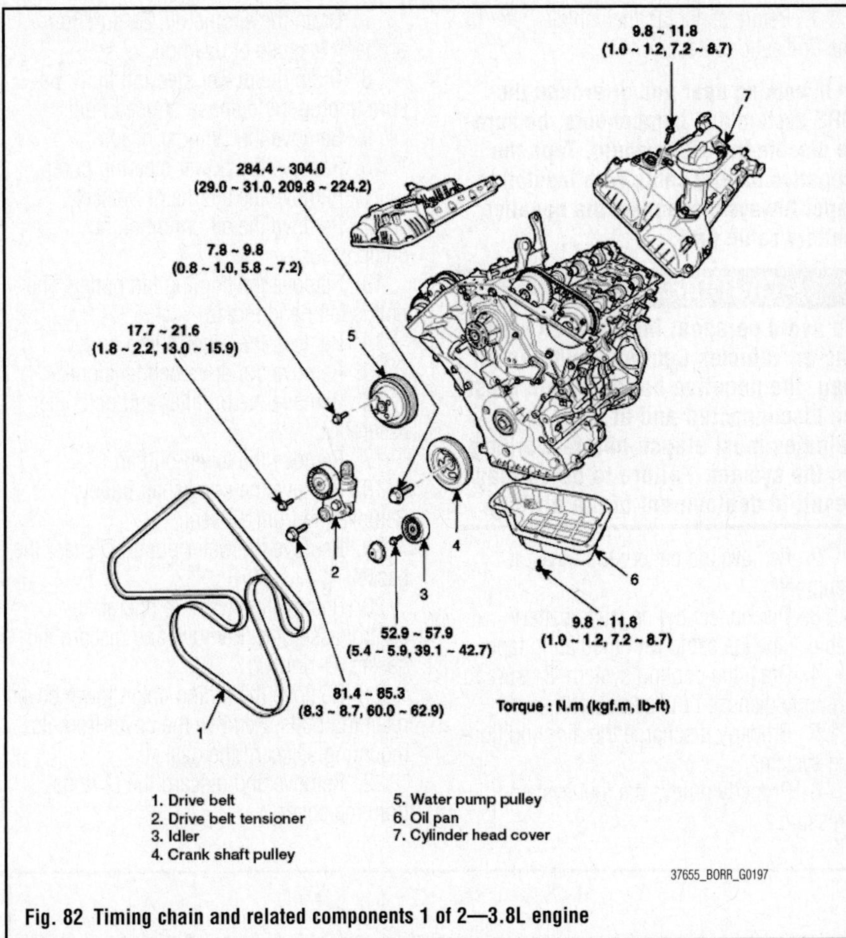

Fig. 82 Timing chain and related components 1 of 2—3.8L engine

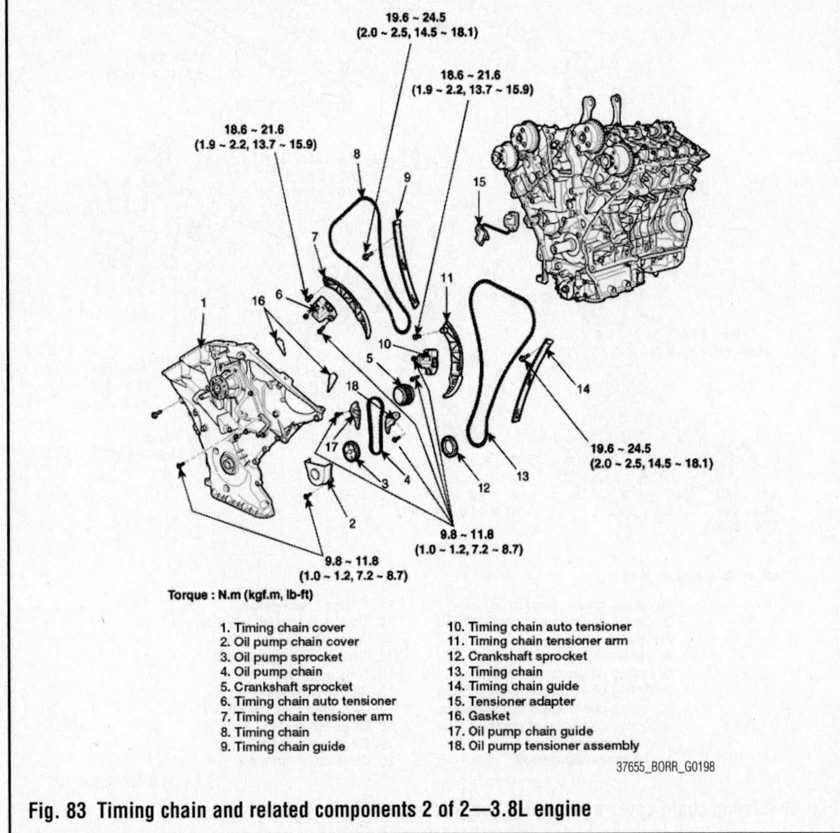

Fig. 83 Timing chain and related components 2 of 2—3.8L engine

34. Installation is the reverse of the removal procedure.

➡**Be sure that all sealant application surfaces are free of debris and engine oil.**

35. Before assembling the timing chain cover apply sealant part number TB 1217H, or equivalent on the gap between the cylinder head and cylinder block. See illustration.

➡**The part must be assembled within five minutes of sealant application.**

36. Apply sealant to the timing chain cover pan part number TB 1217H, or equivalent. Sealant should be applied in a continuous bead in each area indicated in the illustration. Install a new gasket to the cover.

➡**Be sure to install the component within five minutes of sealant application.**

➡**The dowel pins on the block and holes in the cover should be used as a reference when installing the cover.**

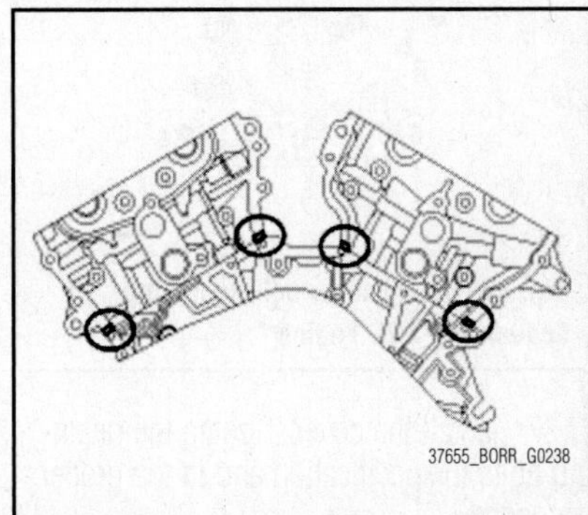

Fig. 84 Cylinder head/cylinder block sealant application points—3.8L engine

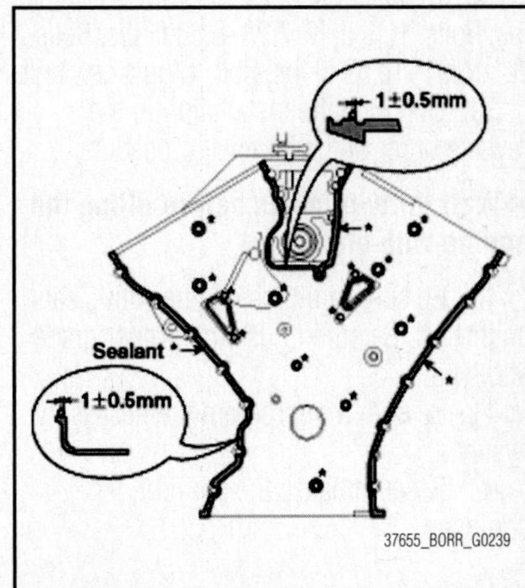

Fig. 85 Timing cover sealant application—3.8L engine

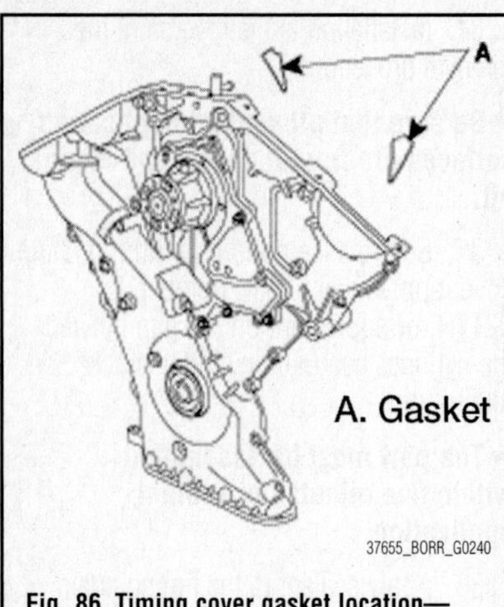

Fig. 86 Timing cover gasket location—3.8L engine

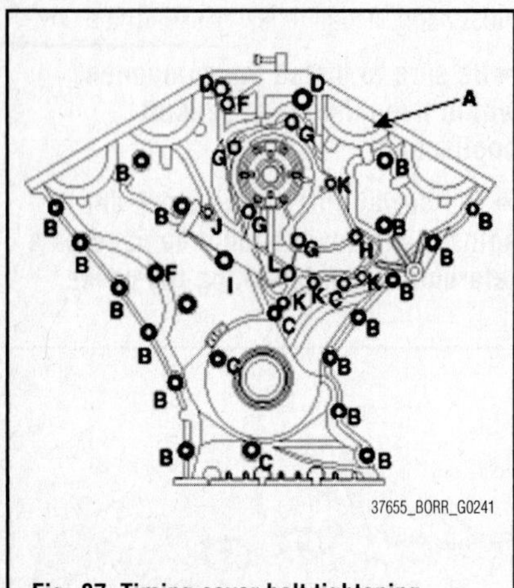

Fig. 87 Timing cover bolt tightening sequence—3.8L engine

37. Install the cover. Tighten the retaining bolts to specification and in the proper sequence.
38. Specification is as follows: Bolts "B" 13.74–15.91 ft. lbs. Bolts "C" 7.23–8.68 ft. lbs. Bolts "D" 43.40–50.63 ft. lbs. Bolts "F" 18.08–19.53 ft. lbs. Bolts "G" 15.91–17.36 ft. lbs. Bolts "H, I, J, K" 7.23–8.68 ft. lbs. Bolts "L" 15.91–19.53 ft. lbs. Bolt "L" is a new bolt.
39. Continue the installation in the reverse order of the removal procedure.

➡Wait thirty minutes before filling the engine with engine oil

40. Be sure to fill the engine with clean engine oil. Be sure to use the proper grade and type.
41. Be sure to reprogram the necessary systems.
42. Reset/Initialize the sunroof, if equipped.

4.6L Engine

See Figures 88 through 91.

1. Before servicing the vehicle, refer to the Precautions Section.

➡If working near and/or around the SRS system and components, be sure to disable the SRS system. Tape the negative battery cable with insulating tape. Always disconnect the negative battery cable first.

⁂ CAUTION

To avoid personal injury when working on vehicles equipped with an air bag, the negative battery cable must be disconnected and at least three minutes must elapse before working on the system. Failure to do so may result in deployment of the air bag.

2. Remove the battery top cover, if equipped.
3. Disconnect the negative battery cable. Tape the cable with insulating tape.
4. Drain the cooling system. Be sure to properly dispose of used coolant.
5. Properly discharge the air conditioning system.
6. Properly relieve the fuel system pressure.
7. Drain the engine oil. Be sure to properly dispose of used oil.
8. Drain the power steering fluid. Be sure to properly dispose of used fluid.
9. Remove the cylinder heads.
10. Remove the power steering pump.
11. Remove the alternator bracket.
12. Remove the air conditioning compressor.
13. Remove the cooling fan pulley. The bolt is left hand thread.
14. Remove the water pump pulley.
15. Remove the drive belt tensioner.
16. Remove the oil filter and oil cooler assembly.
17. Remove the lower oil pan.
18. Remove the crankshaft pulley. Remove the front oil seal.
19. Remove the water pump. Discard the gasket.
20. Remove the water temperature control assembly. Remove and discard the gasket (O-ring).
21. Remove the timing chain lower cover retaining bolts. Remove the cover from its mounting. Discard the gasket.
22. Remove and discard the O-rings from the cover.

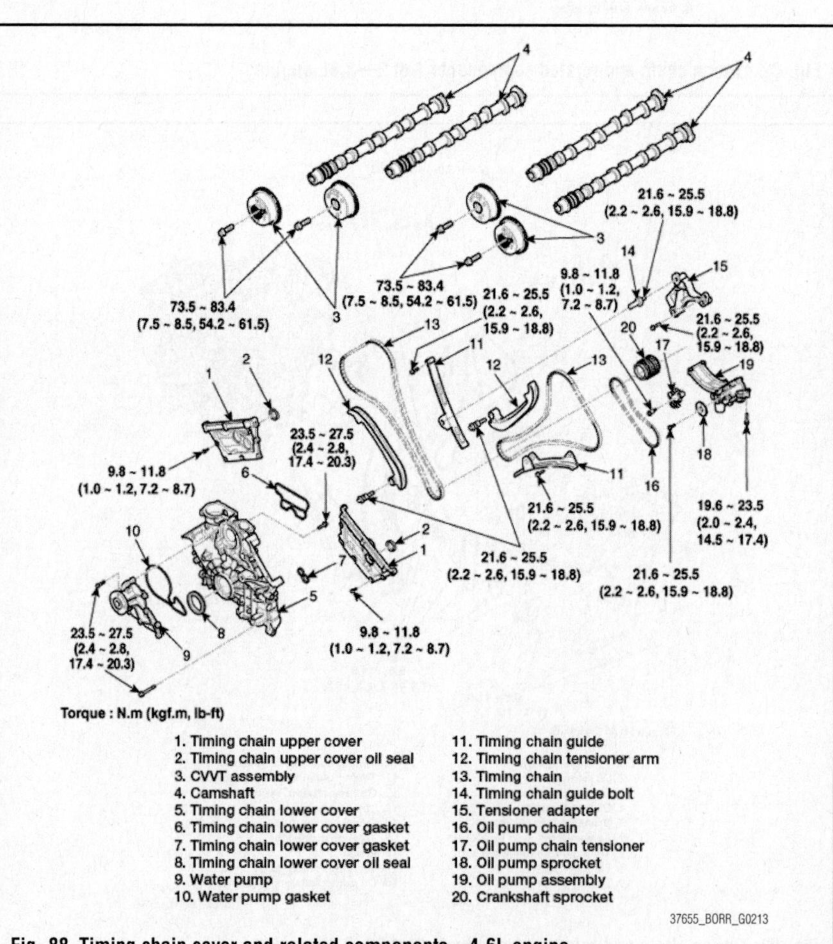

Fig. 88 Timing chain cover and related components—4.6L engine

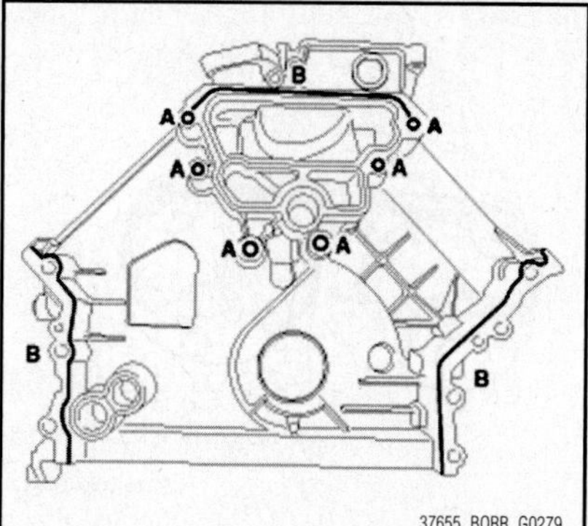

Fig. 89 Timing chain cover sealant application 1 of 2—4.6L engine

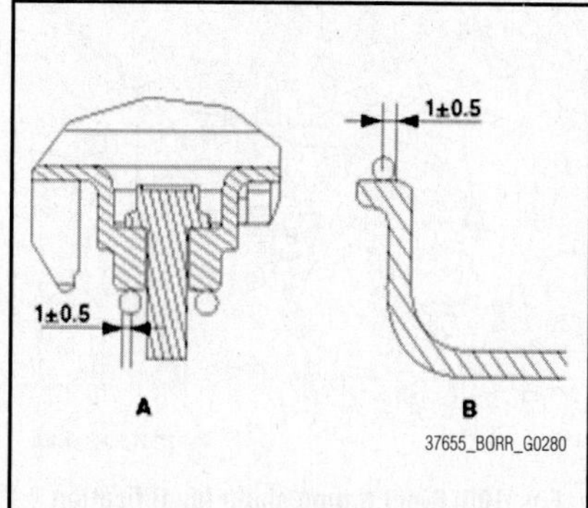

Fig. 90 Timing chain cover sealant application 2 of 2—4.6L engine

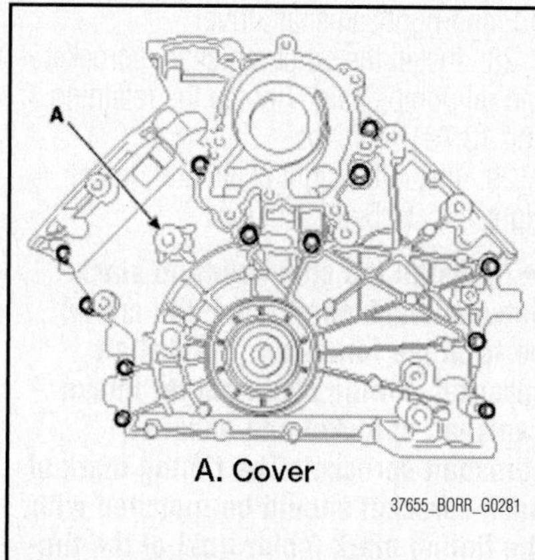

Fig. 91 Timing chain cover bolt location—4.6L engine

To install:

➡**Be sure to use new fasteners, as required.**

23. Installation is the reverse of the removal procedure.

➡**The sealant locations on the chain lower cover and on counter parts (head, block and lower pan) must be free of engine oil and debris.**

24. Install a new gasket to the timing chain lower cover.

25. Apply sealant part number TB 1217H or LT 5900H on the timing chain lower cover. The part must be assembled within five minutes of sealant application. See illustration for sealant bead information, bead width is 02.5 +/- 0.5mm.

➡**The dowel pins on the block and holes in the cover should be used as a reference when installing the cover.**

26. Install the cover. Tighten the retaining bolts to specification. Specification is 17.4–20.3 ft. lbs.

27. Continue the installation in the reverse order of the removal procedure.

➡**Wait thirty minutes before filling the engine with engine oil**

28. Be sure to fill the engine with clean engine oil. Be sure to use the proper grade and type.

29. Be sure to reprogram the necessary systems.

30. Reset/Initialize the sunroof, if equipped.

TIMING CHAIN & SPROCKETS

REMOVAL & INSTALLATION

3.8L Engine

See Figures 92 through 100.

1. Before servicing the vehicle, refer to the Precautions Section.

➡**If working near and/or around the SRS system and components, be sure to disable the SRS system. Tape the negative battery cable with insulating tape. Always disconnect the negative battery cable first.**

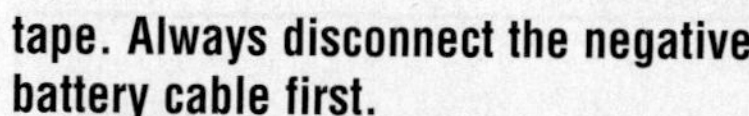

To avoid personal injury when working on vehicles equipped with an air bag, the negative battery cable must be disconnected and at least three minutes must elapse before working on the system. Failure to do so may result in deployment of the air bag.

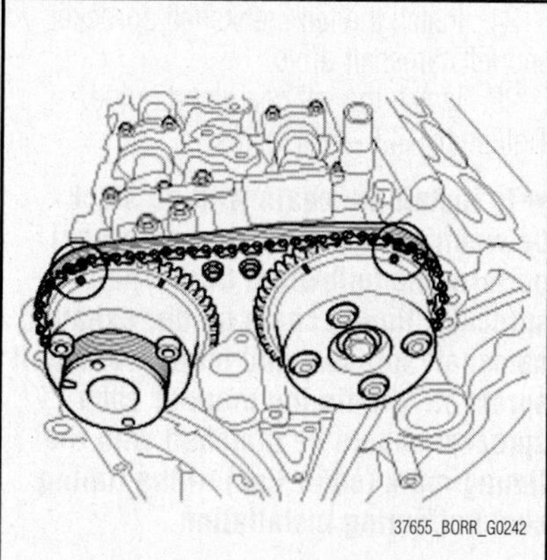

Fig. 92 Timing chain identification marks 1 of 4—3.8L engine

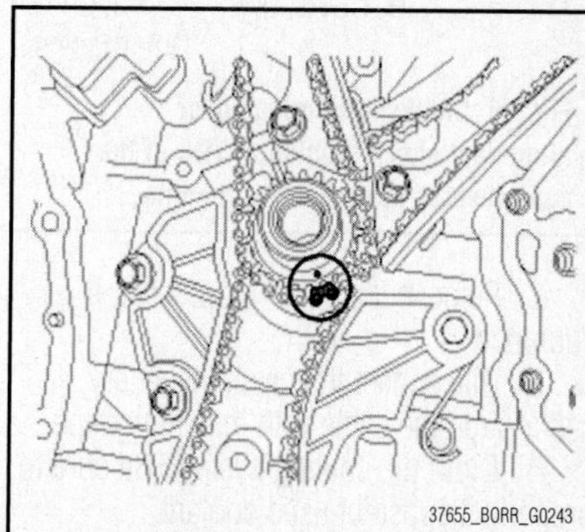

Fig. 93 Timing chain identification marks 2 of 4—3.8L engine

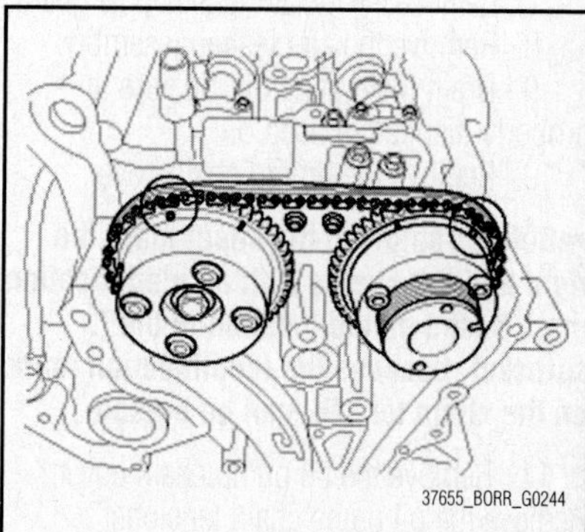

Fig. 94 Timing chain identification marks 3 of 4—3.8L engine

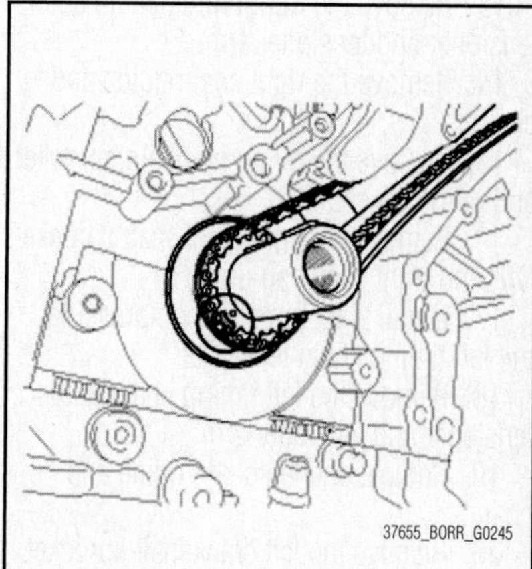

Fig. 95 Timing chain identification marks 4 of 4—3.8L engine

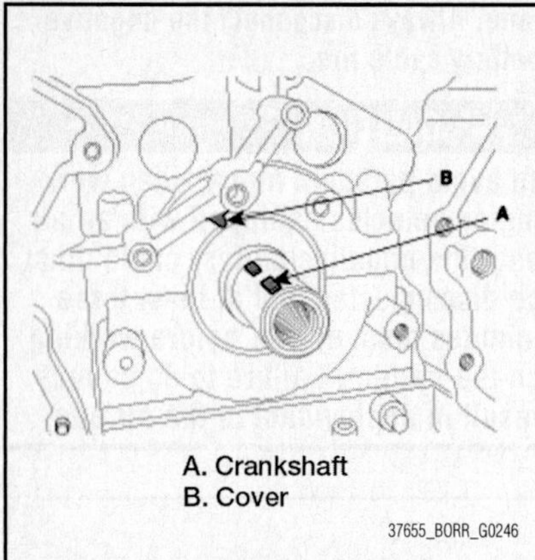

Fig. 96 Crankshaft/timing cover alignment check (engine at TDC of the compression stroke)—3.8L engine

2. Remove the battery top cover, if equipped.
3. Disconnect the negative battery cable. Tape the cable with insulating tape.
4. Drain the cooling system. Be sure to properly dispose of used coolant.
5. Remove the engine cover.
6. Properly relieve the fuel system pressure.
7. Remove the radiator grille upper guard.
8. Remove the air cleaner assembly.
9. Drain the engine oil. Be sure to properly dispose of used oil.
10. Remove the timing chain cover.

➡Before removing the chain mark the right and left chains with an identification mark based on the location of the sprocket, because the identification mark on the chain for TDC can be erased.

11. Remove the oil pump chain cover. Remove the oil pump chain tensioner assembly. Remove the oil pump chain guide.
12. Install a set pin after compressing the right timing chain tensioner.
13. Remove the right timing chain auto tensioner and tensioner arm.
14. Remove the right chain guide and chain.
15. Remove the oil pump chain sprocket and oil pump chain.
16. Remove the right crankshaft sprocket, O/P and right camshaft drive.
17. Install a set pin after compressing the left timing chain tensioner.
18. Remove the left timing chain auto tensioner and tensioner arm.
19. Remove the left chain guide and chain.
20. Remove the left crankshaft sprocket, O/P and left camshaft drive.
21. Remove the tensioner adapter assembly.

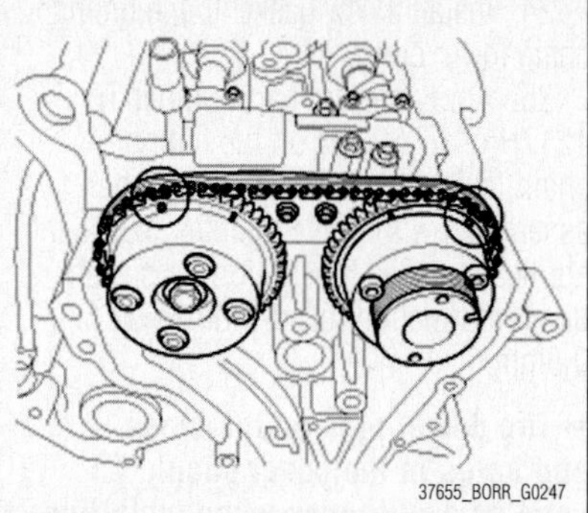

Fig. 97 Left timing chain identification marks 1 of 2—3.8L engine

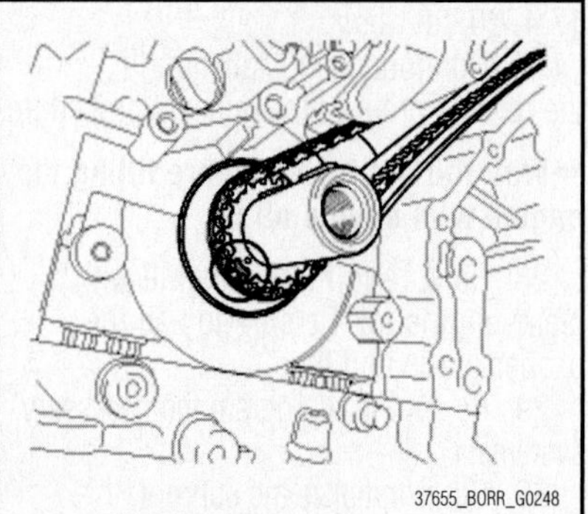

Fig. 98 Left timing chain identification marks 2 of 2—3.8L engine

To install:

➡Be sure to use new fasteners, as required.

22. Installation is the reverse of the removal procedure.

➡The key of the crankshaft should be aligned with the timing mark of the cover. As a result of this the piston of No. 1 cylinder is at TDC of the compression stroke.

23. Install the tensioner adapter assembly.
24. Install the left crankshaft sprocket and left camshaft drive.
25. Install the left chain and guide. Tighten to 14.5–18.1 ft. lbs.

➡To install the chain with no slack between each shaft (cam and crank) perform the following. Crankshaft sprocket, timing chain guide, exhaust camshaft sprocket and intake camshaft sprocket. The timing mark of each sprocket should be matched with the timing mark (color link) of the timing chain at/during installation.

26. Install the left tensioner arm and auto tensioner. Tighten tension retaining bolts to 7.2–8.7 ft. lbs. Tighten the arm retaining bolts to 13.7–15.9 ft. lbs.

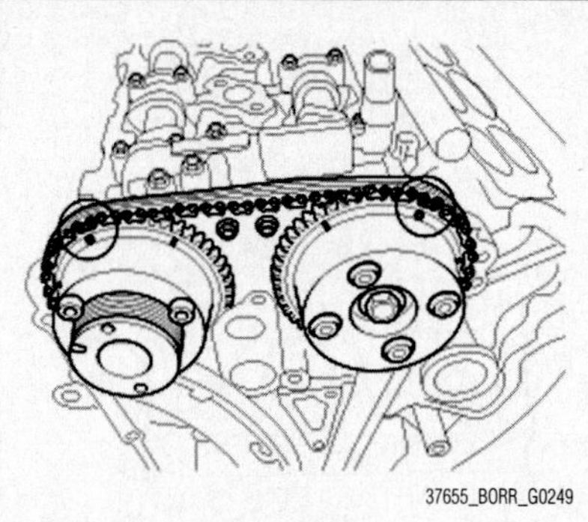

Fig. 99 Right timing chain identification marks 1 of 2—3.8L engine

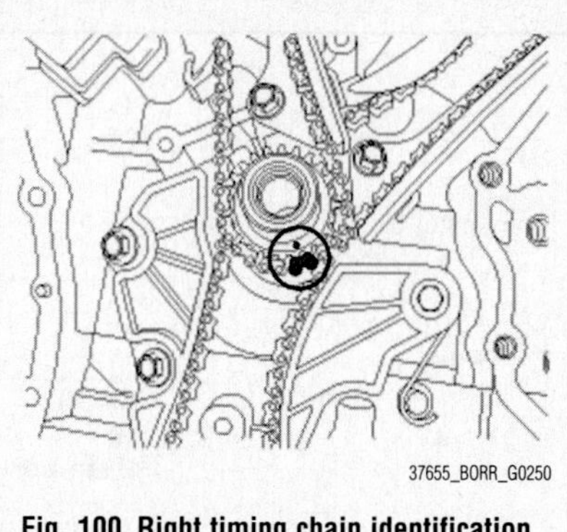

Fig. 100 Right timing chain identification marks 2 of 2—3.8L engine

27. Install the right crankshaft sprocket, O/P and right camshaft drive.
28. Install the oil pump chain sprocket and oil pump chain. Tighten the retaining bolt 13.7–15.9 ft. lbs.
29. Install the right chain and guide. Tighten to 14.5–18.1 ft. lbs.

➡To install the chain with no slack between each shaft (cam and crank) perform the following. Crankshaft sprocket, timing chain guide, intake camshaft sprocket and exhaust camshaft sprocket. The timing mark of each sprocket should be matched with the timing mark (color link) of the timing chain at/during installation.

30. Install the right tensioner arm and auto tensioner. Tighten tension retaining bolts to 7.2–8.7 ft. lbs. Tighten the arm retaining bolts to 13.7–15.9 ft. lbs.
31. Install the oil pump chain guide. Tighten the retaining bolt to 7.2–8.7 ft. lbs.
32. Install the oil pump chain tensioner assembly. Tighten the retaining bolts to 7.2–8.7 ft. lbs.
33. Pull out the pins of the hydraulic tensioners.
34. Install the oil pump chain cover. Tighten the retaining bolts to 7.8–8.7 ft. lbs.

35. Rotate the crankshaft two revolutions in regular direction (clockwise) as viewed from the front to confirm the timing mark. Never rotate the engine in the counterclockwise direction.

36. Continue the installation in the reverse order of the removal procedure.

➡Wait thirty minutes before filling the engine with engine oil

37. Be sure to fill the engine with clean engine oil. Be sure to use the proper grade and type.

38. Be sure to reprogram the necessary systems.

39. Reset/Initialize the sunroof, if equipped.

4.6L Engine

See Figures 101 through 103.

1. Before servicing the vehicle, refer to the Precautions Section.

➡If working near and/or around the SRS system and components, be sure to disable the SRS system. Tape the negative battery cable with insulating tape. Always disconnect the negative battery cable first.

⁂ CAUTION

To avoid personal injury when working on vehicles equipped with an air bag, the negative battery cable must be disconnected and at least three minutes must elapse before working on the system. Failure to do so may result in deployment of the air bag.

2. Remove the battery top cover, if equipped.
3. Disconnect the negative battery cable. Tape the cable with insulating tape.
4. Drain the cooling system. Be sure to properly dispose of used coolant.
5. Properly discharge the air conditioning system.
6. Properly relieve the fuel system pressure.
7. Drain the engine oil. Be sure to properly dispose of used oil.
8. Drain the power steering fluid. Be sure to properly dispose of used fluid.
9. Remove the cylinder heads.
10. Remove the lower oil pan.
11. Remove the timing chain cover.

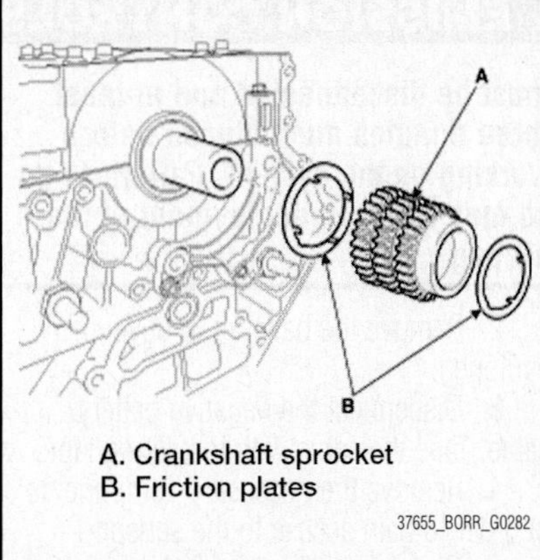

Fig. 101 Crankshaft sprocket and friction plate installation—4.6L engine

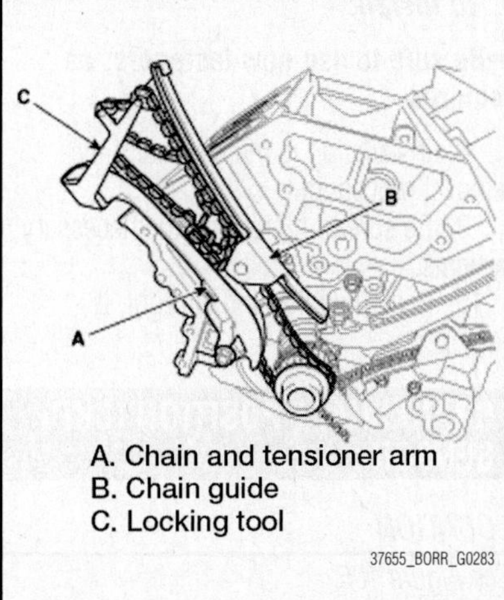

Fig. 102 Timing chain, chain locking tool and arm (side one)—4.6L engine

12. Remove the timing chain tensioner arm, guide and chain.
13. Insert a set pin after compressing the oil pump chain tensioner and then remove the oil pump sprocket and chain.
14. Remove the crankshaft sprocket and friction plates.
15. Remove the tensioner adapter and the timing chain guide bolts.

To install:

➡Be sure to use new fasteners, as required.

16. Installation is the reverse of the removal procedure.
17. Install the adapter and chain guide bolts. Tighten the bolts to 15.9–18.8 ft. lbs.

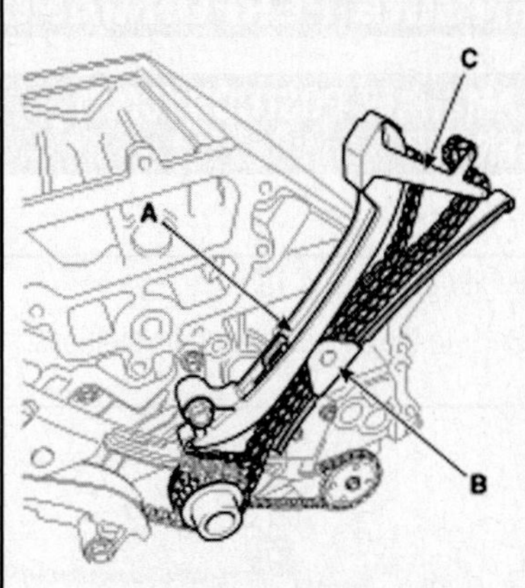

Fig. 103 Timing chain, chain locking tool and arm (side two)—4.6L engine

18. When installing the friction plates and crankshaft sprocket the mark on the friction plate should be placed top (engine is placed bottom up).
19. Install the oil pump chain with the pump sprocket and pump chain tensioner. Tighten the tensioner bolts to 7.2–8.7 ft. lbs. Tighten the sprocket bolt to 15.9–18.8 ft. lbs.
20. Install the chain with the tension arm, the guide and the chain locking tool. Tighten the chain tensioner arm bolts to 15.9–18.8 ft. lbs.
21. Continue the installation in the reverse order of the removal procedure.

➡Wait thirty minutes before filling the engine with engine oil

22. Be sure to fill the engine with clean engine oil. Be sure to use the proper grade and type.
23. Be sure to reprogram the necessary systems.
24. Reset/Initialize the sunroof, if equipped.

VALVE LASH

ADJUSTMENT

The valve lash on all engines is kept in adjustment hydraulically. No adjustment is necessary or possible.

ENGINE PERFORMANCE & EMISSION CONTROLS

CAMSHAFT POSITION (CMP) SENSOR

LOCATION

See Figures 104 and 105.

Refer to the accompanying illustrations.

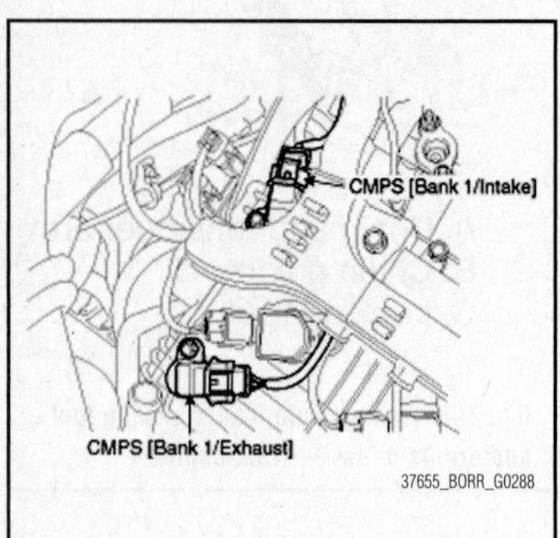

Fig. 104 CMP sensor location—bank 1

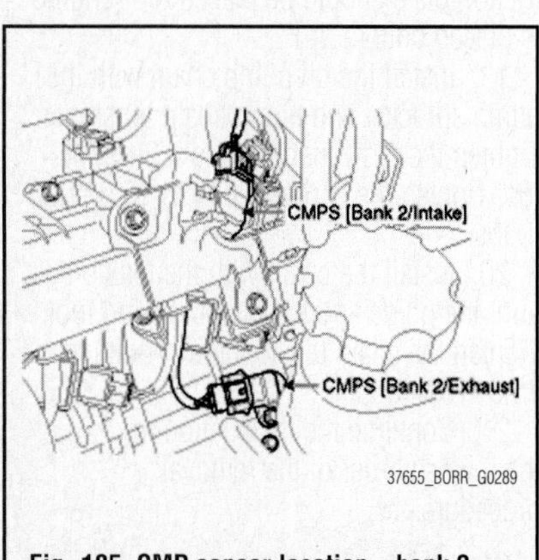

Fig. 105 CMP sensor location—bank 2

REMOVAL & INSTALLATION

At this time the manufacturer does not provide removal and installation procedures for this component. The following procedure is a guideline and may differ from the vehicle you are servicing.

1. Before servicing the vehicle, refer to the Precautions Section.

➡If working near and/or around the SRS system and components, be sure to disable the SRS system. Tape the negative battery cable with insulating tape. Always disconnect the negative battery cable first.

⁂ CAUTION

To avoid personal injury when working on vehicles equipped with an air bag, the negative battery cable must be disconnected and at least three minutes must elapse before working on the system. Failure to do so may result in deployment of the air bag.

2. Remove the battery top cover, if equipped.
3. Disconnect the negative battery cable. Tape the cable with insulating tape.
4. Remove the necessary components in order to gain access to the sensor.
5. Disconnect the electrical connector.
6. Remove the sensor from its mounting.
7. Discard the O-ring/gasket.

To install:

➡Be sure to use new fasteners, as required.

8. Installation is the reverse of the removal procedure.
9. Be sure to reprogram the necessary systems.
10. Reset/Initialize the sunroof, if equipped.

CRANKSHAFT POSITION (CKP) SENSOR

LOCATION

See Figure 106.

Refer to the accompanying illustration.

REMOVAL & INSTALLATION

At this time the manufacturer does not provide removal and installation procedures for this component. The following procedure is a guideline and may differ from the vehicle you are servicing.

1. Before servicing the vehicle, refer to the Precautions Section.

➡If working near and/or around the SRS system and components, be sure to disable the SRS system. Tape the negative battery cable with insulating tape. Always disconnect the negative battery cable first.

⁂ CAUTION

To avoid personal injury when working on vehicles equipped with an air bag, the negative battery cable must be disconnected and at least three minutes must elapse before working on the system. Failure to do so may result in deployment of the air bag.

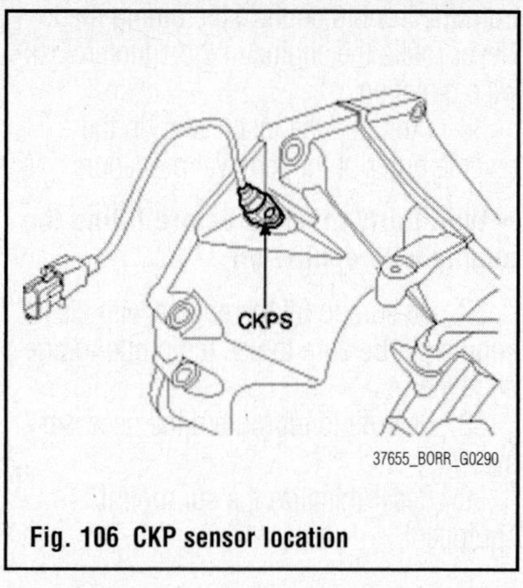

Fig. 106 CKP sensor location

2. Remove the battery top cover, if equipped.
3. Disconnect the negative battery cable. Tape the cable with insulating tape.
4. Remove the necessary components in order to gain access to the sensor.
5. Disconnect the electrical connector.
6. Remove the sensor from its mounting.
7. Discard the O-ring/gasket.

To install:

➡Be sure to use new fasteners, as required.

8. Installation is the reverse of the removal procedure.
9. Be sure to reprogram the necessary systems.
10. Reset/Initialize the sunroof, if equipped.

ELECTRONIC CONTROL MODULE (ECM)

LOCATION

See Figure 107.

Refer to the accompanying illustration.

REMOVAL & INSTALLATION

➡The VIN information must be programmed into a new ECM. Follow the directions on the diagnostic scan tool to perform this operation.

1. Before servicing the vehicle, refer to the Precautions Section.

➡If working near and/or around the SRS system and components, be sure to disable the SRS system. Tape the negative battery cable with insulating tape. Always disconnect the negative battery cable first.

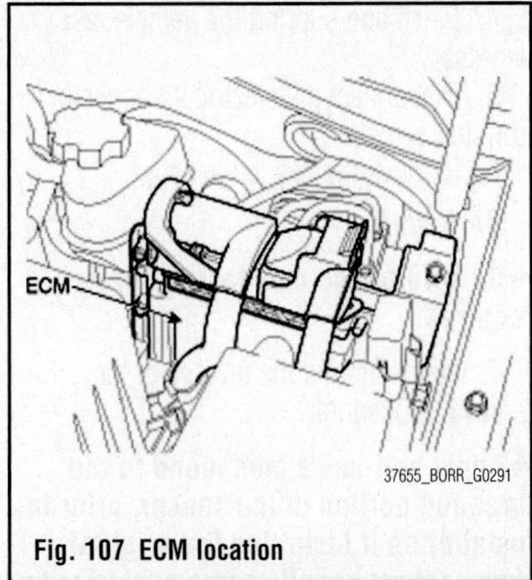

Fig. 107 ECM location

✲✲ CAUTION

To avoid personal injury when working on vehicles equipped with an air bag, the negative battery cable must be disconnected and at least three minutes must elapse before working on the system. Failure to do so may result in deployment of the air bag.

2. Remove the battery top cover, if equipped.
3. Disconnect the negative battery cable. Tape the cable with insulating tape.
4. Disconnect the ECM connector.
5. Remove the retaining bolts.
6. Remove the ECM from the bracket.

To install:

➡Be sure to use new fasteners, as required.

7. Installation is the reverse of the removal procedure.

➡If the vehicle is equipped with immobilizer system, perform the "key teaching" procedure, using the diagnostic scan tool. Follow the directions on the tool.

8. Be sure to reprogram the necessary systems.
9. Reset/Initialize the sunroof, if equipped.

ENGINE COOLANT TEMPERATURE (ECT) SENSOR

LOCATION

See Figures 108 and 109.

The Engine Coolant Temperature (ECT) Sensor is located in the engine coolant passage of the cylinder head for detecting the engine coolant temperature.

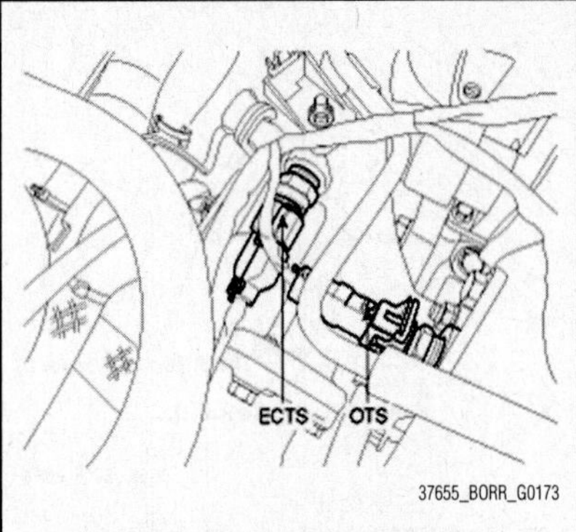

Fig. 108 Engine Coolant Temperature (ECT) sensor location—3.8L engine

REMOVAL & INSTALLATION

✲✲ CAUTION

Never open, service or drain the radiator or cooling system when hot; serious burns can occur from the steam and hot coolant. Also, when draining engine coolant, keep in mind that cats and dogs are attracted to ethylene glycol antifreeze and could drink any that is left in an uncovered container or in puddles on the ground. This will prove fatal in sufficient quantities. Always drain coolant into a sealable container.

At this time the manufacturer does not provide removal and installation procedures for this component. The following procedure is a guideline and may differ from the vehicle you are servicing.

1. Before servicing the vehicle, refer to the Precautions Section.

➡If working near and/or around the SRS system and components, be sure to disable the SRS system. Tape the negative battery cable with insulating tape. Always disconnect the negative battery cable first.

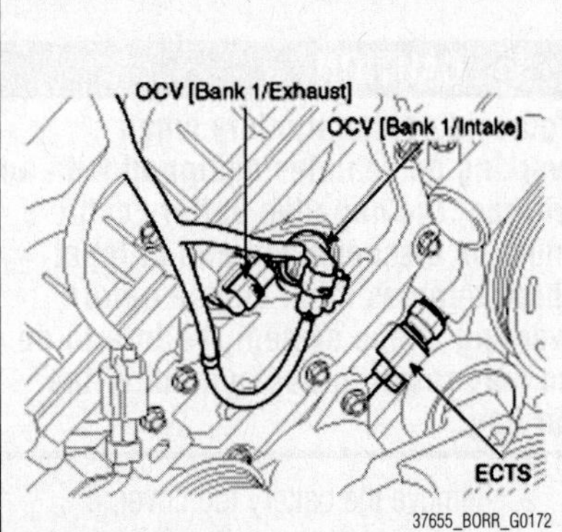

Fig. 109 Engine Coolant Temperature (ECT) sensor location—4.6L engine

✲✲ CAUTION

To avoid personal injury when working on vehicles equipped with an air bag, the negative battery cable must be disconnected and at least three minutes must elapse before working on the system. Failure to do so may result in deployment of the air bag.

2. Remove the battery top cover, if equipped.
3. Disconnect the negative battery cable. Tape the cable with insulating tape.
4. Drain the engine coolant, to an acceptable level.
5. Remove the electrical connector from the sensor.
6. Remove the Engine Coolant Temperature (ECT) sensor.

To install:

➡Be sure to use new fasteners, as required.

7. Installation is the reverse of the removal procedure.
8. refill the cooling system, as required. Be sure to use the proper grade and type engine coolant.
9. Be sure to reprogram the necessary systems.
10. Reset/Initialize the sunroof, if equipped.

HEATED OXYGEN (HO2S) SENSOR

LOCATION

See Figures 110 through 116.

Refer to the accompanying illustrations.

REMOVAL & INSTALLATION

At this time the manufacturer does not provide removal and installation procedures for this component. The following procedure is a guideline and may differ from the vehicle you are servicing.

1. Before servicing the vehicle, refer to the Precautions Section.

➡If working near and/or around the SRS system and components, be sure to disable the SRS system. Tape the negative battery cable with insulating tape. Always disconnect the negative battery cable first.

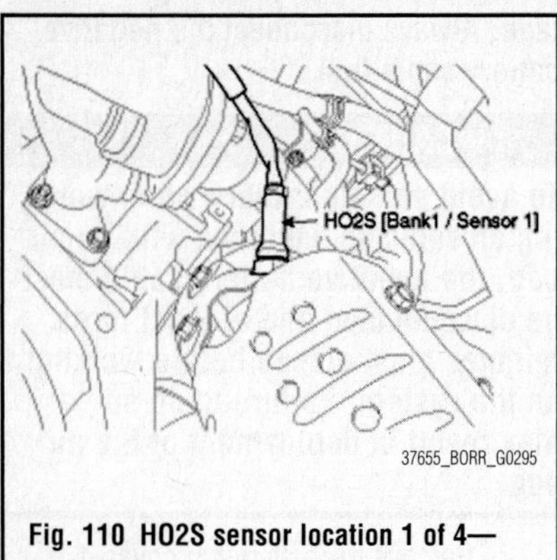

Fig. 110 HO2S sensor location 1 of 4—3.8L engine

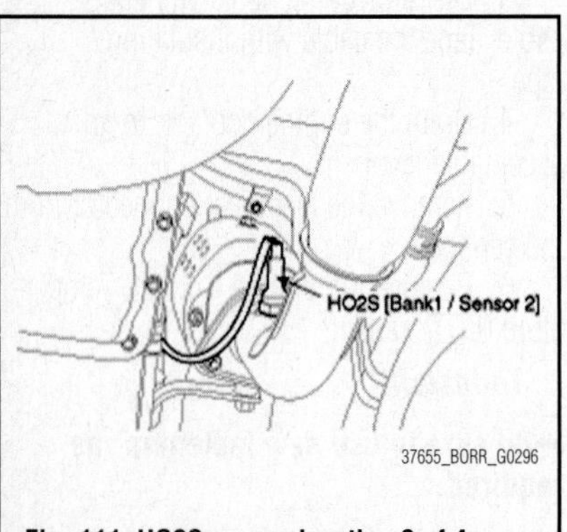

Fig. 111 HO2S sensor location 2 of 4—3.8L engine

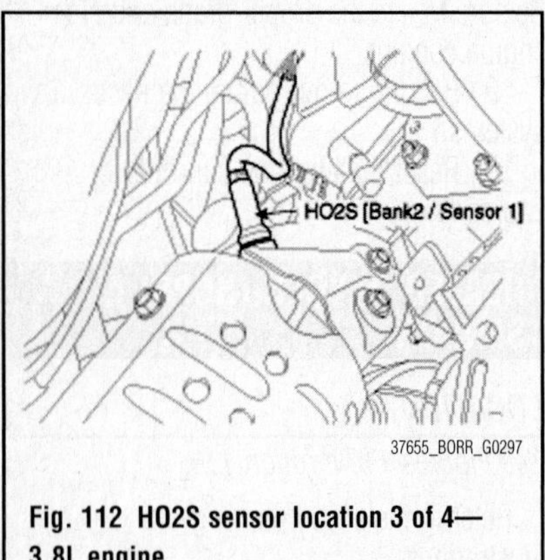

Fig. 112 HO2S sensor location 3 of 4—3.8L engine

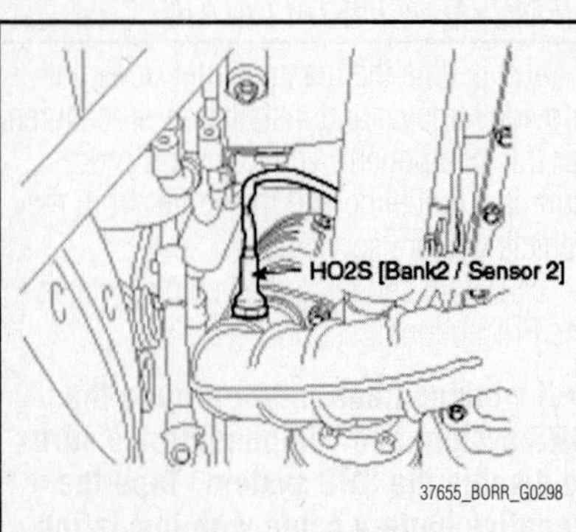

Fig. 113 HO2S sensor location 4 of 4—3.8L engine

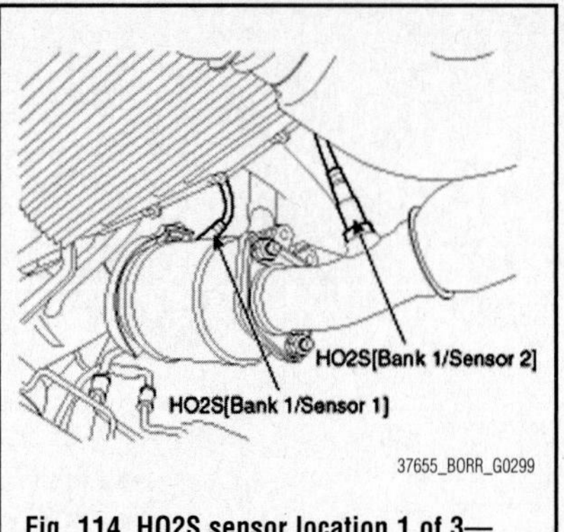

Fig. 114 HO2S sensor location 1 of 3—4.6L engine

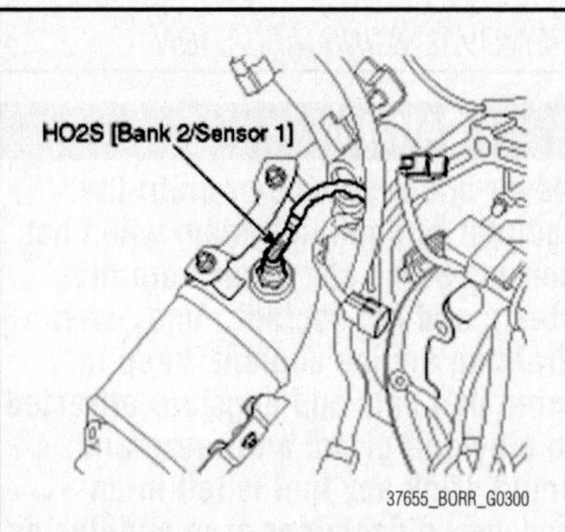

Fig. 115 HO2S sensor location 2 of 3—4.6L engine

HO2S [Bank 2/Sensor 2]

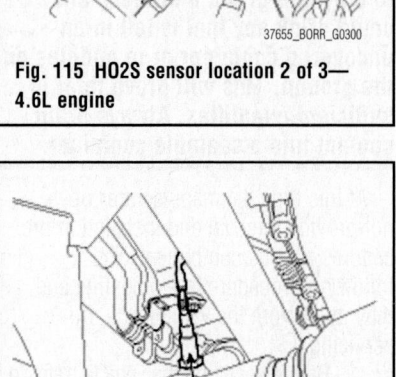

37655_BORR_G0301

Fig. 116 HO2S sensor location 3 of 3—4.6L engine

CAUTION

To avoid personal injury when working on vehicles equipped with an air bag, the negative battery cable must be disconnected and at least three minutes must elapse before working on the system. Failure to do so may result in deployment of the air bag.

2. Remove the battery top cover, if equipped.
3. Disconnect the negative battery cable. Tape the cable with insulating tape.
4. Raise and support the vehicle, as necessary.
5. Disconnect the electrical connector from the sensor.
6. Remove the oxygen sensor.

To install:

➡Be sure to use new fasteners, as required.

7. Installation is the reverse of the removal procedure.

➡Apply anti-seize compound to the threaded portion of the sensor, prior to installation if installing the original. New sensors usually come coated with anti-seize compound.. Never apply anti-seize compound to the protector of the sensor.

8. Be sure to reprogram the necessary systems.
9. Reset/Initialize the sunroof, if equipped.

INTAKE AIR TEMPERATURE (IAT) SENSOR

LOCATION

See Figures 117 and 118.

Refer to the accompanying illustrations.

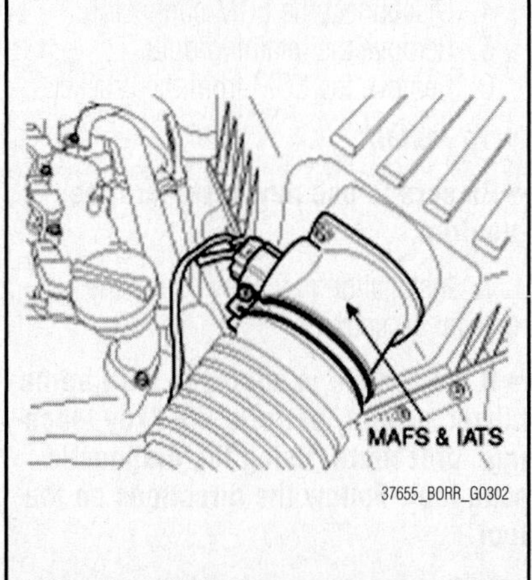

Fig. 117 IAT/MAFS sensor location—3.8L engine

REMOVAL & INSTALLATION

At this time the manufacturer does not provide removal and installation procedures for this component.

KNOCK SENSOR (KS)

LOCATION

See Figures 119 through 122.

Refer to the accompanying illustrations.

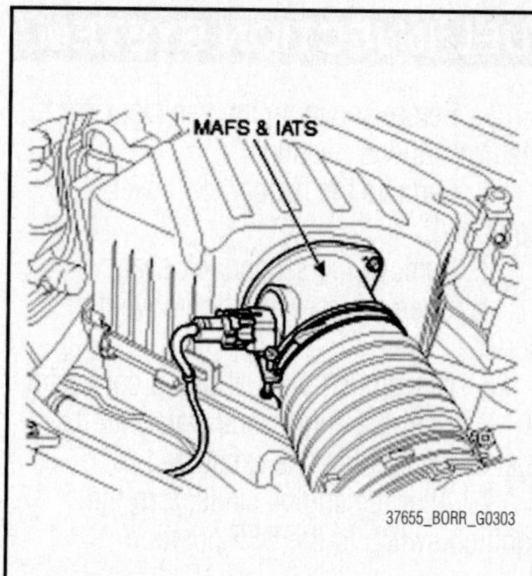

Fig. 118 IAT/MAFS sensor location—4.6L engine

Fig. 119 KS location 1 of 2—3.8L engine

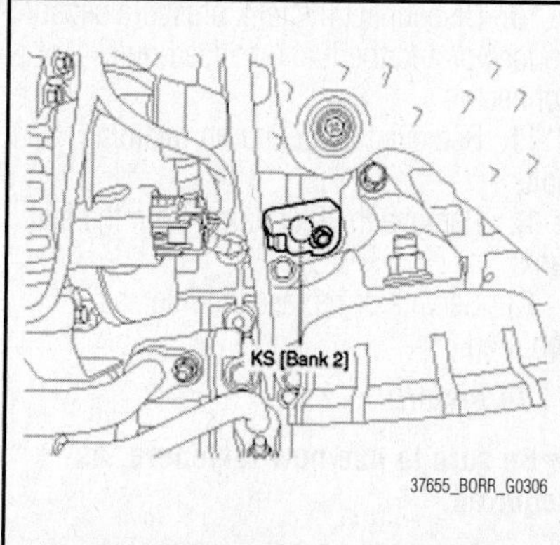

Fig. 120 KS location 2 of 2—3.8L engine

REMOVAL & INSTALLATION

At this time the manufacturer does not provide removal and installation procedures for this component, refer to the illustration as required.

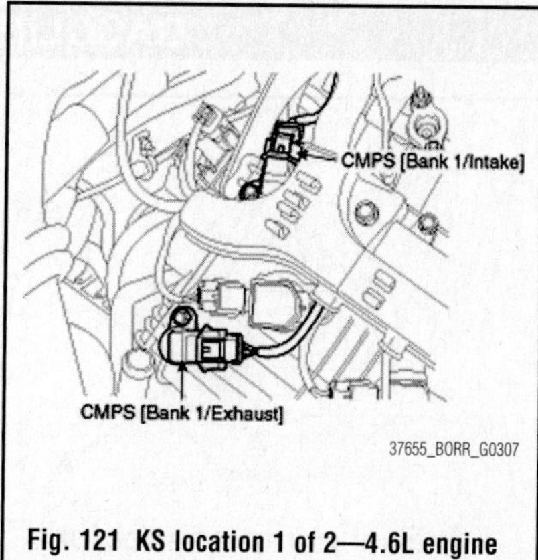

Fig. 121 KS location 1 of 2—4.6L engine

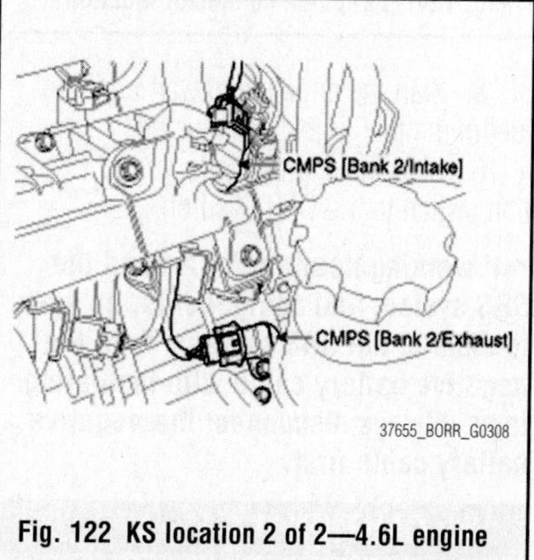

Fig. 122 KS location 2 of 2—4.6L engine

MASS AIR FLOW (MAF) SENSOR

LOCATION

This sensor is incorporated along with the IAT sensor, in the air cleaner assembly. Refer to IAT sensor.

MANIFOLD ABSOLUTE PRESSURE (MAP) SENSOR

LOCATION

See Figure 123

Refer to the accompanying illustration.

REMOVAL & INSTALLATION

At this time the manufacturer does not provide removal and installation procedures for this component. The following procedure is a guideline and may differ from the vehicle you are servicing.

1. Before servicing the vehicle, refer to the Precautions Section.

➡If working near and/or around the SRS system and components, be sure to disable the SRS system. Tape the negative battery cable with insulating tape. Always disconnect the negative battery cable first.

⁂ CAUTION

To avoid personal injury when working on vehicles equipped with an air bag, the negative battery cable must be disconnected and at least three minutes must elapse before working on the system. Failure to do so may result in deployment of the air bag.

2. Remove the battery top cover, if equipped.
3. Disconnect the negative battery cable. Tape the cable with insulating tape.
4. Remove the necessary components to gain access to the sensor retaining bolt.
5. Disconnect the electrical connector.
6. Remove the component from its mounting.

To install:

➡Be sure to use new fasteners, as required.

7. Installation is the reverse of the removal procedure.
8. Be sure to reprogram the necessary systems.
9. Reset/Initialize the sunroof, if equipped.

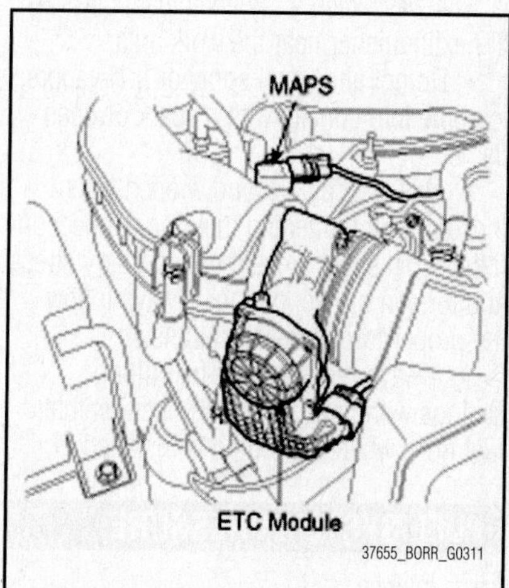

Fig. 123 MAP sensor location—3.8L engine

FUEL SYSTEM SERVICE PRECAUTIONS

Safety is the most important factor when performing not only fuel system maintenance, but any type of maintenance. Failure to conduct maintenance and repairs in a safe manner may result in serious personal injury or death. Work on a vehicle's fuel system components can be accomplished safely and effectively by adhering to the following rules and guidelines.

- To avoid the possibility of fire and personal injury, always disconnect the negative battery cable unless the repair or test procedure requires that battery voltage be applied.
- Always relieve the fuel system pressure prior to disconnecting any fuel system component (injector, fuel rail, pressure regulator, etc.) fitting or fuel line connection. Exercise extreme caution whenever relieving fuel system pressure to avoid exposing skin, face and eyes to fuel spray. Please be advised that fuel under pressure may penetrate the skin or any part of the body that it contacts.
- Always place a shop towel or cloth around the fitting or connection prior to loosening to absorb any excess fuel due to spillage. Ensure that all fuel spillage is quickly removed from engine surfaces. Ensure that all fuel-soaked cloths or towels are deposited into a flame-proof waste container with a lid.
- Always keep a dry chemical (Class B) fire extinguisher near the work area.
- Do not allow fuel spray or fuel vapors to come into contact with a spark or open flame.
- Always use a second wrench when loosening or tightening fuel line connection fittings. This will prevent unnecessary stress and torsion on fuel piping. Always follow the proper torque specifications.
- Always replace worn fuel fitting O-rings with new ones. Do not substitute fuel hose where rigid pipe is installed.

FUEL SYSTEM PRESSURE

RELIEVING

See Figure 124.

1. Before servicing the vehicle, refer to the Precautions Section.
2. Remove the second left seat.
3. Open the carpet and remove the service cover.
4. Disconnect the fuel pump connector.

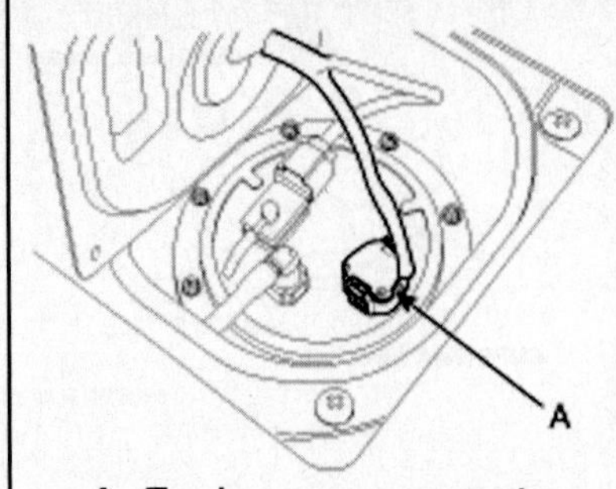

Fig. 124 Fuel pump connector location

5. Start the engine and wait until the fuel in the fuel lines is exhausted.
6. After the engine stalls, turn the ignition switch to the OFF position.

➡**If working near and/or around the SRS system and components, be sure to disable the SRS system. Tape the negative battery cable with insulating tape. Always disconnect the negative battery cable first.**

⁂ CAUTION

To avoid personal injury when working on vehicles equipped with an air bag, the negative battery cable must be disconnected and at least three minutes must elapse before working on the system. Failure to do so may result in deployment of the air bag.

7. Remove the battery top cover, if equipped.
8. Disconnect the negative battery cable. Tape the cable with insulating tape.

To install:

➡**Be sure to use new fasteners, as required.**

9. Installation is the reverse of the removal procedure.
10. Be sure to reprogram the necessary systems.
11. Reset/Initialize the sunroof, if equipped.

FUEL FILTER

REMOVAL & INSTALLATION

See Figure 125.

The fuel filter is an integral part of the fuel pump assembly. This component is located inside the fuel tank.

1. Before servicing the vehicle, refer to the Precautions Section.
2. Remove the battery top cover, if equipped.
3. Remove the second left seat.
4. Open the carpet and remove the service cover.
5. Disconnect the fuel pump connector.
6. Start the engine and wait until the fuel in the fuel lines is exhausted.
7. After the engine stalls, turn the ignition switch to the OFF position.

➡**If working near and/or around the SRS system and components, be sure to disable the SRS system. Tape the negative battery cable with insulating tape. Always disconnect the negative battery cable first.**

⁂ CAUTION

To avoid personal injury when working on vehicles equipped with an air bag, the negative battery cable must be disconnected and at least three minutes must elapse before working on the system. Failure to do so may result in deployment of the air bag.

8. Remove the battery top cover, if equipped.
9. Disconnect the negative battery cable. Tape the cable with insulating tape.
10. Disconnect the tank pressure sensor connector and the fuel tube feed quick connector.
11. Remove the fuel pump retaining bolts.
12. Remove the fuel pump from the fuel tank.
13. Service or replace the filter, as required.

To install:

➡**Be sure to use new fasteners, as required.**

14. Installation is the reverse of the removal procedure.
15. Be sure to reprogram the necessary systems.
16. Reset/Initialize the sunroof, if equipped.

FUEL PUMP

REMOVAL & INSTALLATION

See Figure 125.

1. Before servicing the vehicle, refer to the Precautions Section.

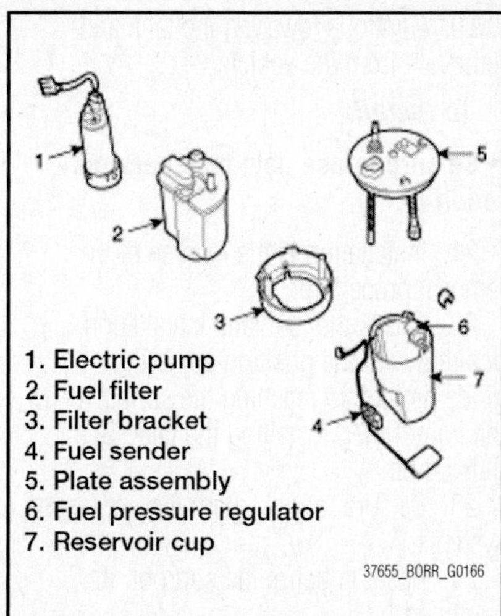

Fig. 125 Fuel pump, filter and related components

2. Remove the second left seat.
3. Open the carpet and remove the service cover.
4. Disconnect the fuel pump connector.
5. Start the engine and wait until the fuel in the fuel lines is exhausted.
6. After the engine stalls, turn the ignition switch to the OFF position.

➡If working near and/or around the SRS system and components, be sure to disable the SRS system. Tape the negative battery cable with insulating tape. Always disconnect the negative battery cable first.

⁂ CAUTION

To avoid personal injury when working on vehicles equipped with an air bag, the negative battery cable must be disconnected and at least three minutes must elapse before working on the system. Failure to do so may result in deployment of the air bag.

7. Remove the battery top cover, if equipped.
8. Disconnect the negative battery cable. Tape the cable with insulating tape.
9. Disconnect the tank pressure sensor connector and the fuel tube feed quick connector.
10. Remove the fuel pump retaining bolts.
11. Remove the fuel pump from the fuel tank.

To install:

➡Be sure to use new fasteners, as required.

12. Installation is the reverse of the removal procedure.
13. Be sure to reprogram the necessary systems.
14. Reset/Initialize the sunroof, if equipped.

FUEL RAIL AND INJECTOR

REMOVAL & INSTALLATION

See Figures 126 and 127.

At this time the manufacturer does not provide removal and installation procedures for this component. Before servicing this component be sure to properly relieve the fuel system pressure.

FUEL TANK

REMOVAL & INSTALLATION

See Figure 128.

1. Before servicing the vehicle, refer to the Precautions Section.
2. Drain the fuel tank to an acceptable level. Be sure to properly dispose of fuel.
3. Remove the second left seat.
4. Open the carpet and remove the service cover.
5. Disconnect the fuel pump connector.
6. Start the engine and wait until the fuel in the fuel lines is exhausted.
7. After the engine stalls, turn the ignition switch to the OFF position.

➡If working near and/or around the SRS system and components, be sure to disable the SRS system. Tape the negative battery cable with insulating tape. Always disconnect the negative battery cable first.

⁂ CAUTION

To avoid personal injury when working on vehicles equipped with an air bag, the negative battery cable must be disconnected and at least three minutes must elapse before working on the system. Failure to do so may result in deployment of the air bag.

8. Remove the battery top cover, if equipped.
9. Disconnect the negative battery cable. Tape the cable with insulating tape.
10. Disconnect the tank pressure sensor connector and the fuel tube feed quick connector.
11. raise and support the vehicle safely.

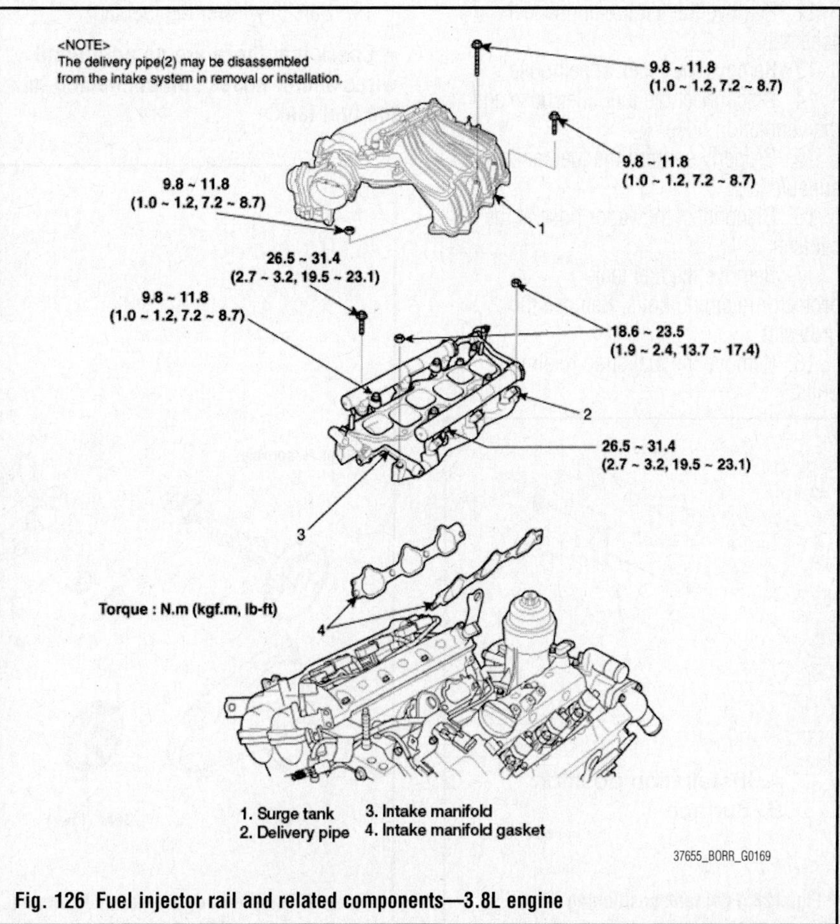

Fig. 126 Fuel injector rail and related components—3.8L engine

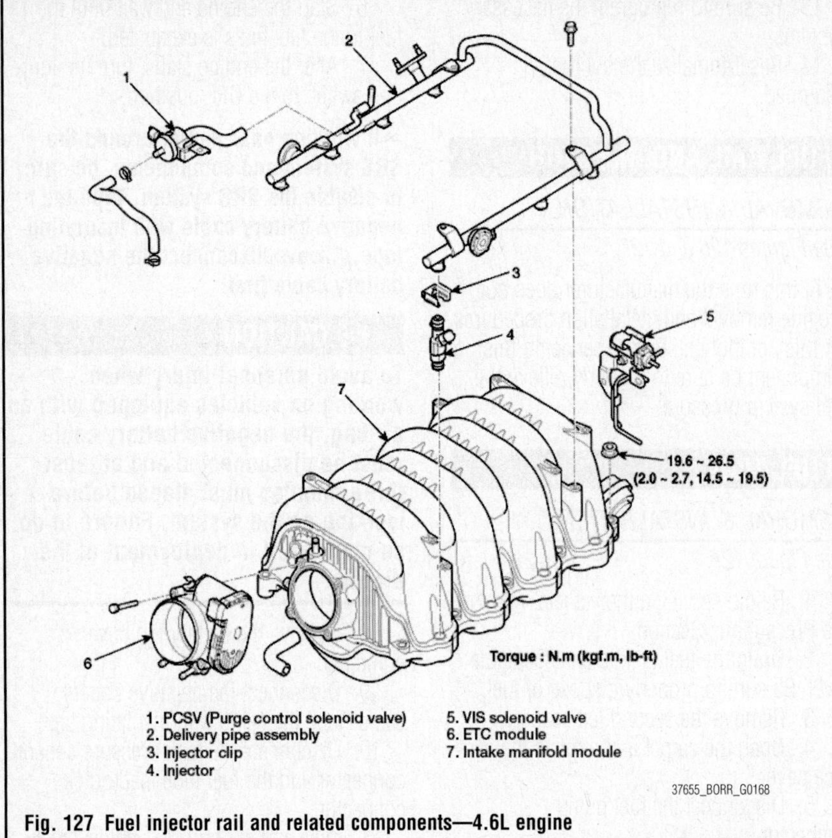

Fig. 127 Fuel injector rail and related components—4.6L engine

12. Remove the left tire and wheel assembly.

13. Remove the inner wheelhouse.

14. Disconnect the fuel filler hose and the ventilation hose.

15. Properly support the fuel tank with a suitable jack.

16. Disconnect the vapor hose at the canister.

17. Remove the fuel tank protector retaining bolts. Remove the protector.

18. Remove the fuel strap retaining bolts.

19. Carefully lower the fuel tank.

➡Check that there are no additional wires and/or hoses still connected to the fuel tank.

20. Continue lowering the tank and remove it from the vehicle.

To install:

➡Be sure to use new fasteners, as required.

21. Installation is the reverse of the removal procedure.

22. Check the fuel tank installation position with the position adjusting guide pin before reaching the surface of the frame when installing the tank. See illustration.

23. Be sure to reprogram the necessary systems.

24. Reset/Initialize the sunroof, if equipped.

IDLE SPEED

ADJUSTMENT

The idle speed is controlled by the PCM. It is not adjustable.

THROTTLE BODY

REMOVAL & INSTALLATION

See Figures 129 and 130.

1. Before servicing the vehicle, refer to the Precautions Section.

➡If working near and/or around the SRS system and components, be sure to disable the SRS system. Tape the

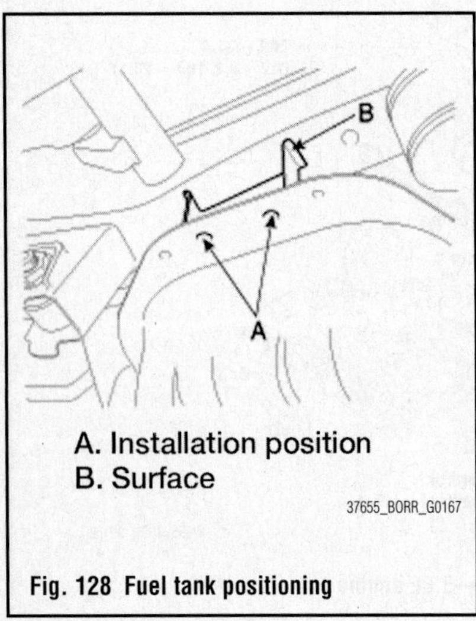

Fig. 128 Fuel tank positioning

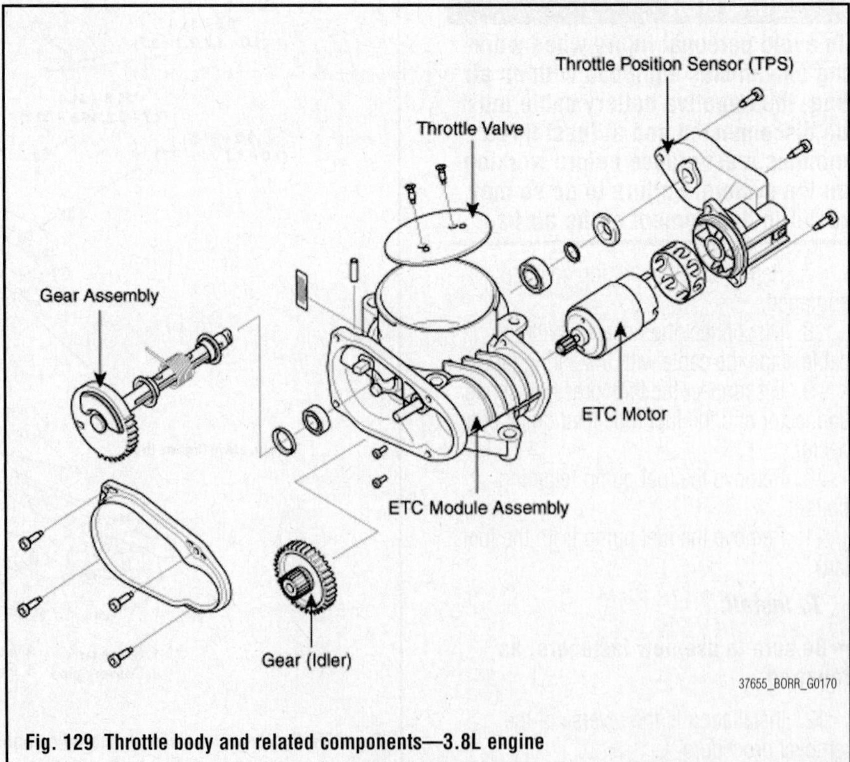

Fig. 129 Throttle body and related components—3.8L engine

negative battery cable with insulating tape. Always disconnect the negative battery cable first.

⁂ CAUTION

To avoid personal injury when working on vehicles equipped with an air bag, the negative battery cable must be disconnected and at least three minutes must elapse before working on the system. Failure to do so may result in deployment of the air bag.

2. Remove the battery top cover, if equipped.
3. Disconnect the negative battery cable. Tape the cable with insulating tape.
4. Remove the necessary components in order to gain access to the throttle body assembly.
5. Disconnect the electrical connectors.
6. Drain the cooling system to an acceptable level. Disconnect the hose.
7. Remove the throttle body assembly from its mounting.

To install:

➡Be sure to use new fasteners, as required.

8. Installation is the reverse of the removal procedure.
9. Be sure to reprogram the necessary systems.
10. Reset/Initialize the sunroof, if equipped.

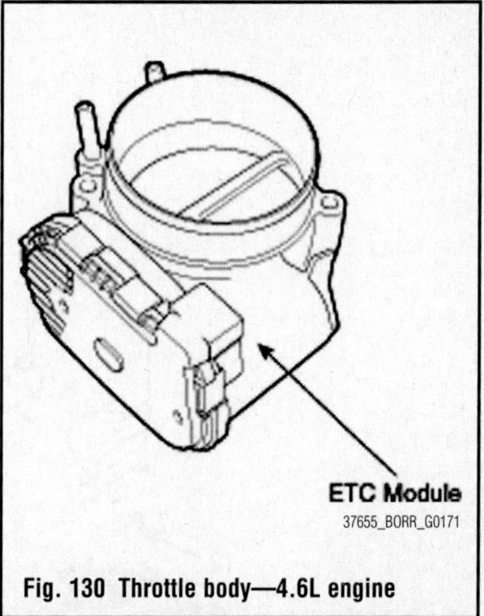

Fig. 130 Throttle body—4.6L engine

HEATING & AIR CONDITIONING SYSTEM

BLOWER MOTOR

REMOVAL & INSTALLATION

1. Before servicing the vehicle, refer to the Precautions Section.

➡If working near and/or around the SRS system and components, be sure to disable the SRS system. Tape the negative battery cable with insulating tape. Always disconnect the negative battery cable first.

⁂ CAUTION

To avoid personal injury when working on vehicles equipped with an air bag, the negative battery cable must be disconnected and at least three minutes must elapse before working on the system. Failure to do so may result in deployment of the air bag.

2. Remove the battery top cover, if equipped.
3. Disconnect the negative battery cable. Tape the cable with insulating tape.
4. Remove the instrument panel lower panel.
5. Disconnect the blower motor electrical connector.
6. Remove the blower motor retaining screws.
7. Remove the component from its mounting.

To install:

➡Be sure to use new fasteners, as required.

8. Installation is the reverse of the removal procedure.
9. Be sure to reprogram the necessary systems.
10. Reset/Initialize the sunroof, if equipped.

HEATER CORE

REMOVAL & INSTALLATION

See Figures 131 through 136.

1. Before servicing the vehicle, refer to the Precautions Section.

➡If working near and/or around the SRS system and components, be sure to disable the SRS system. Tape the negative battery cable with insulating tape. Always disconnect the negative battery cable first.

⁂ CAUTION

To avoid personal injury when working on vehicles equipped with an air bag, the negative battery cable must be disconnected and at least three minutes must elapse before working on the system. Failure to do so may result in deployment of the air bag.

2. Remove the battery top cover, if equipped.
3. Disconnect the negative battery cable. Tape the cable with insulating tape.
4. Properly discharge the air conditioning system.
5. Drain the engine coolant. Be sure to properly dispose of used coolant.

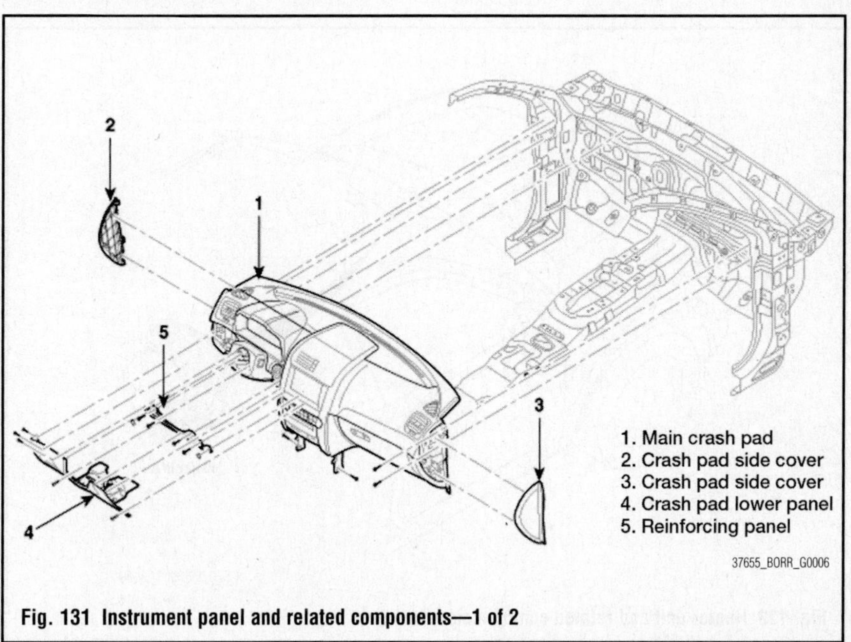

Fig. 131 Instrument panel and related components—1 of 2

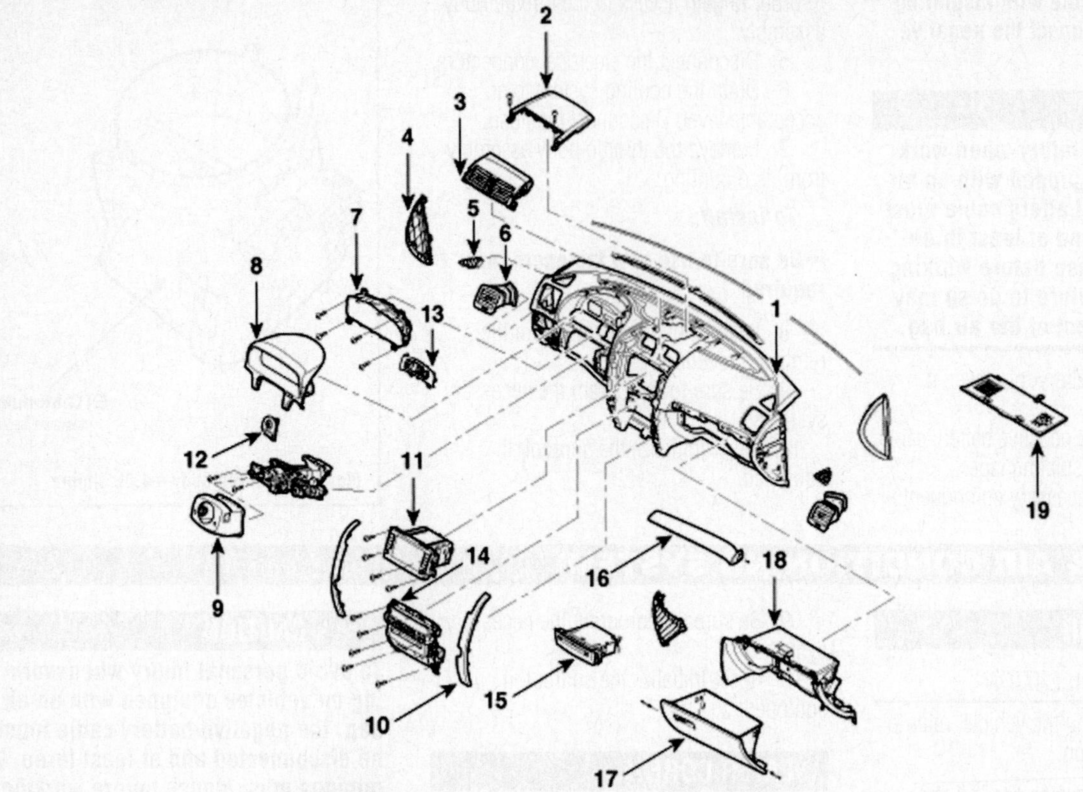

1. Crash pad
2. Speaker gill
3. Center air vent
4. Crash pad side cover
5. Crash pad side cover
6. Side air vent
7. Cluster
8. Cluster fascia panel
9. Shround
10. Center fascia side cover
11. Audio
12. Key switch
13. Switch assembly
14. Center fascia
15. DVD player
16. Crash pad side garnish
17. Glove box
18. Glove box housing
19 Glove box lower panel

37655_BORR_G0007

Fig. 132 Instrument panel and related components—2 of 2

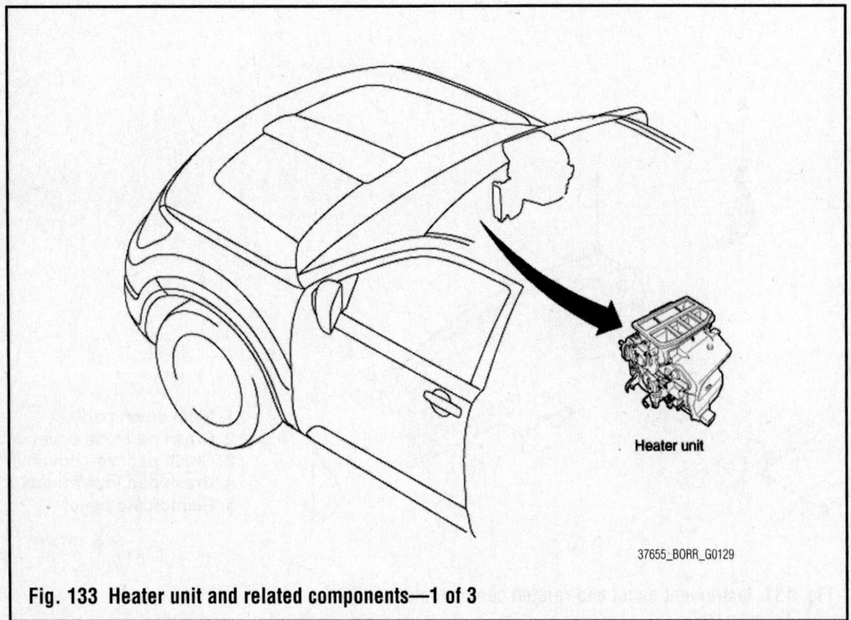

Fig. 133 Heater unit and related components—1 of 3

6. Remove the heater/blower unit assembly.
7. Remove the heater core cover.
8. Remove the heater core from its mounting.

To install:

➡Be sure to use new fasteners, as required.

9. Installation is the reverse of the removal procedure.
10. Be sure to fill the cooling system with the proper grade and type coolant.
11. Be sure to properly recharge the air conditioning system.
12. Be sure to use new O-rings, coated with clean refrigerant oil.
13. Be sure to reprogram the necessary systems.
14. Reset/Initialize the sunroof, if equipped.

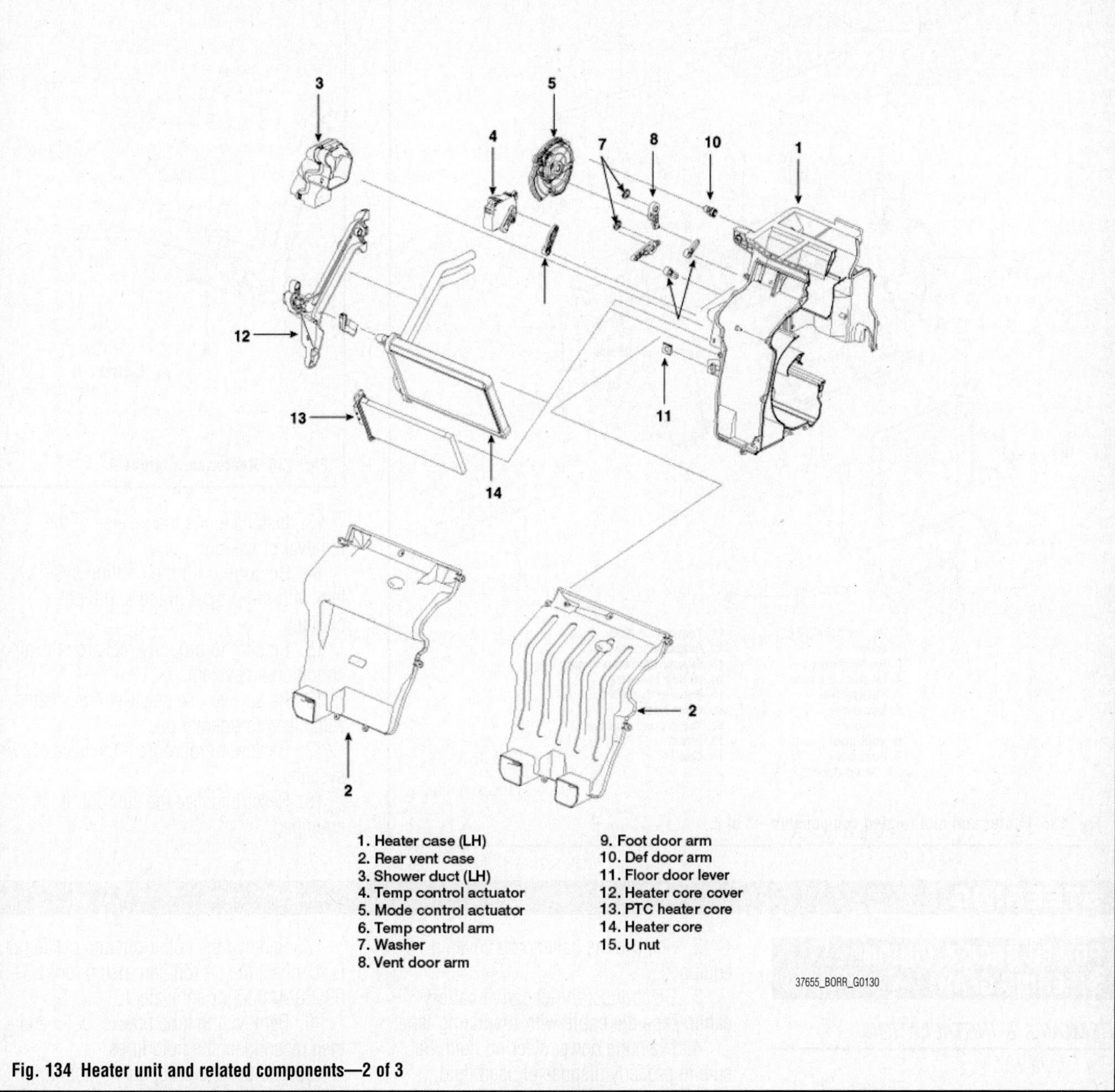

Fig. 134 Heater unit and related components—2 of 3

HEATER UNIT

REMOVAL & INSTALLATION

See Figures 131 through 136.

1. Before servicing the vehicle, refer to the Precautions Section.

➡If working near and/or around the SRS system and components, be sure to disable the SRS system. Tape the negative battery cable with insulating tape. Always disconnect the negative battery cable first.

⁂ CAUTION

To avoid personal injury when working on vehicles equipped with an air bag, the negative battery cable must be disconnected and at least three minutes must elapse before working on the system. Failure to do so may result in deployment of the air bag.

2. Remove the battery top cover, if equipped.
3. Disconnect the negative battery cable. Tape the cable with insulating tape.
4. Properly discharge the air conditioning system.
5. Drain the engine coolant. Be sure to properly dispose of used coolant.
6. Disconnect and plug the refrigerant lines at the expansion valve.
7. Disconnect and plug the heater hoses at the heater core.
8. Remove the crash pad (instrument panel).
9. Remove the cross bar assembly.
10. Disconnect the connectors from the temperature control actuator, mode control actuator and the evaporator temperature sensor.
11. Remove the heater/blower unit retaining screws.
12. Remove the heater/blower unit assembly.

To install:

➡Be sure to use new fasteners, as required.

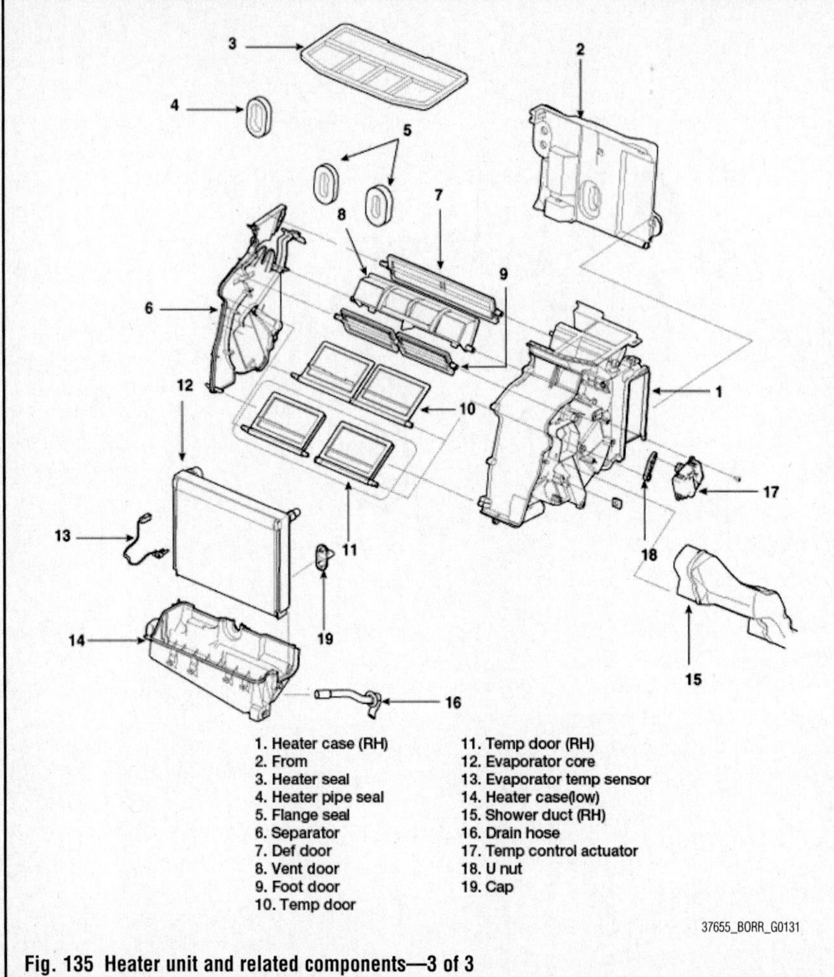

Fig. 135 Heater unit and related components—3 of 3

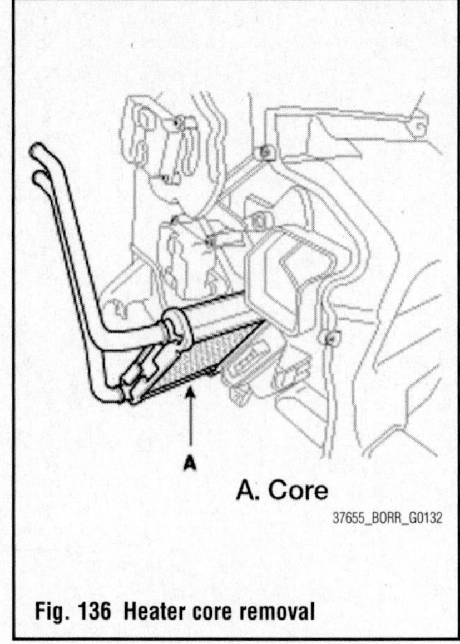

Fig. 136 Heater core removal

13. Installation is the reverse of the removal procedure.
14. Be sure to fill the cooling system with the proper grade and type coolant.
15. Be sure to properly recharge the air conditioning system.
16. Be sure to use new O-rings, coated with clean refrigerant oil.
17. Be sure to reprogram the necessary systems.
18. Reset/Initialize the sunroof, if equipped.

STEERING

POWER RACK & PINION STEERING GEAR

REMOVAL & INSTALLATION

See Figure 137.

1. Before servicing the vehicle, refer to the Precautions Section.

➡If working near and/or around the SRS system and components, be sure to disable the SRS system. Tape the negative battery cable with insulating tape. Always disconnect the negative battery cable first.

⁂ CAUTION

To avoid personal injury when working on vehicles equipped with an air bag, the negative battery cable must be disconnected and at least three minutes must elapse before working on the system. Failure to do so may result in deployment of the air bag.

2. Remove the battery top cover, if equipped.
3. Disconnect the negative battery cable. Tape the cable with insulating tape.
4. Drain the power steering fluid. Be sure to properly dispose of used fluid.
5. Raise and safely support the vehicle.
6. Remove the tire and wheel assemblies.

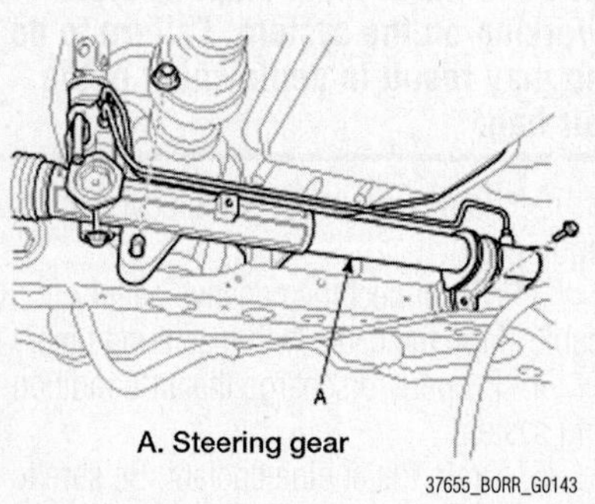

Fig. 137 Power steering gear and related components

7. Remove the cotter pin and castle nut. Disconnect the tie rod end, using tool SST 09568-4A000, or equivalent.
8. Remove the tube bracket bolts and then disconnect the fluid lines.
9. Remove the component mounting bolts.
10. Remove the steering gear from the vehicle.

To install:

➡Be sure to use new fasteners, as required.

11. Installation is the reverse of the removal procedure.
12. Fill the system with the proper grade and type fluid.
13. Bleed the system.
14. Check and adjust alignment, as required.
15. Be sure to reprogram the necessary systems.
16. Reset/Initialize the sunroof, if equipped.

POWER STEERING PUMP

REMOVAL & INSTALLATION

See Figure 138.

1. Before servicing the vehicle, refer to the Precautions Section.

➡**If working near and/or around the SRS system and components, be sure to disable the SRS system. Tape the negative battery cable with insulating tape. Always disconnect the negative battery cable first.**

⁂ CAUTION

To avoid personal injury when working on vehicles equipped with an air bag, the negative battery cable must be disconnected and at least three minutes must elapse before working on the system. Failure to do so may result in deployment of the air bag.

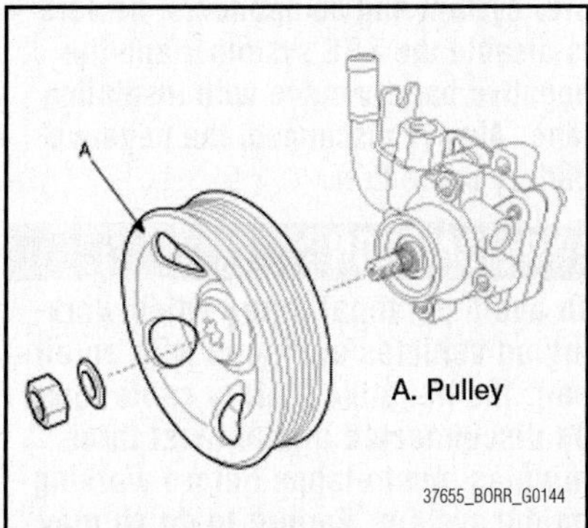

Fig. 138 Power steering pump and related components

2. Remove the battery top cover, if equipped.
3. Disconnect the negative battery cable. Tape the cable with insulating tape.
4. Drain the power steering fluid. Be sure to properly dispose of used fluid.
5. Disconnect and plug the fluid lines.
6. Remove the drive belt.
7. Remove the retaining bolts.
8. Remove the power steering pump from its mounting.

To install:

➡**Be sure to use new fasteners, as required.**

9. Installation is the reverse of the removal procedure.
10. Fill the system with the proper grade and type fluid.
11. Bleed the system.
12. Check and adjust alignment, as required.
13. Adjust the drive belt.
14. Be sure to reprogram the necessary systems.
15. Reset/Initialize the sunroof, if equipped.

BLEEDING

1. Before servicing the vehicle, refer to the Precautions Section.
2. Fill the fluid reservoir with the proper grade and type fluid. Fill to the "COLD/MAX" specification.

➡**Be sure that the fluid is within the "COLD/MAX" and "COLD/MIN" marks on the reservoir throughout this procedure.**

3. Raise the front of the vehicle and support it safely.
4. Crank the engine once or twice by turning the ignition key very quickly from the ON to the START position. Do not allow the engine to start.
5. Turn the steering wheel from lock to lock five or six times, within 15–20 seconds.
6. Start the engine and keep turning the steering wheel from lock to lock until air bubbles stop appearing in the reservoir with the engine at idle.
7. Check the color of the fluid and replenish as required.

➡**If the fluid level moves up and down when turning the steering wheel, the fluid overflows out of the reservoir when turning off the engine or the fluid has a white color, it indicates that all air bubbles have not been removed. Repeat the procedure again or until all air bubbles have been removed from the system.**

SUSPENSION

FRONT SUSPENSION

LOWER CONTROL ARM

REMOVAL & INSTALLATION

See Figure 139.

1. Before servicing the vehicle, refer to the Precautions Section.

➡**If working near and/or around the SRS system and components, be sure to disable the SRS system. Tape the negative battery cable with insulating tape. Always disconnect the negative battery cable first.**

⁂ CAUTION

To avoid personal injury when working on vehicles equipped with an air bag, the negative battery cable must be disconnected and at least three minutes must elapse before working on the system. Failure to do so may result in deployment of the air bag.

2. Remove the battery top cover, if equipped.
3. Disconnect the negative battery cable. Tape the cable with insulating tape.
4. Raise and support the vehicle safely.
5. Remove the tire and wheel assembly.
6. Properly support the component.
7. Remove the two retaining bolts.
8. Remove the two retaining bolts and nuts.
9. Carefully remove the component from its mounting.

To install:

➡**Be sure to use new fasteners, as required.**

10. Installation is the reverse of the removal procedure.
11. Check and adjust alignment, as required.
12. Tighten the bolts to 101–116 ft. lbs.
13. Tighten the bolts and nuts to 105–119 ft. lbs.

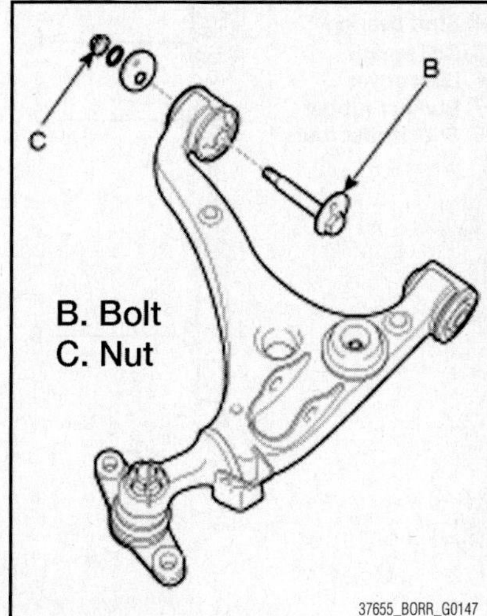

Fig. 139 Front lower control arm and related components

14. Be sure to reprogram the necessary systems.
15. Reset/Initialize the sunroof, if equipped.

MACPHERSON STRUT

REMOVAL & INSTALLATION

See Figure 140.

1. Before servicing the vehicle, refer to the Precautions Section.

➡If working near and/or around the SRS system and components, be sure to disable the SRS system. Tape the negative battery cable with insulating tape. Always disconnect the negative battery cable first.

⁂ CAUTION

To avoid personal injury when working on vehicles equipped with an air bag, the negative battery cable must be disconnected and at least three minutes must elapse before working on the system. Failure to do so may result in deployment of the air bag.

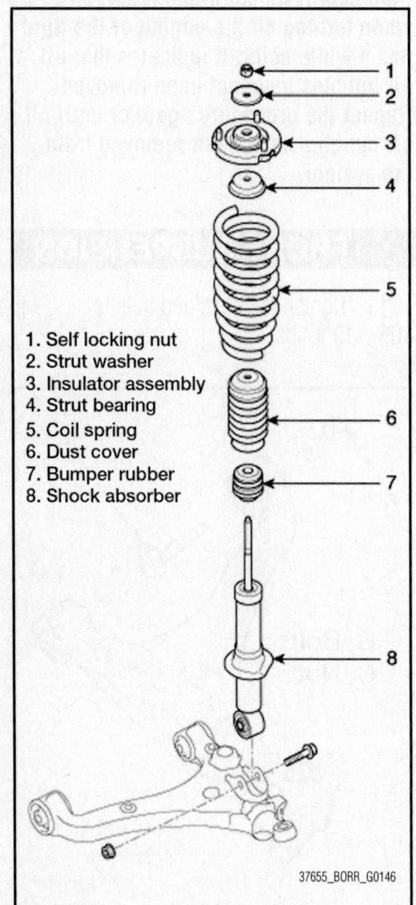

Fig. 140 Front strut and related components

2. Remove the battery top cover, if equipped.
3. Disconnect the negative battery cable. Tape the cable with insulating tape.
4. Raise and support the vehicle safely.
5. Remove the tire and wheel assembly.
6. Remove the lower strut retaining bolts and nuts.
7. Remove the upper strut nuts.
8. Remove the strut from its mounting.

To install:

➡Be sure to use new fasteners, as required.

9. Installation is the reverse of the removal procedure.
10. Tighten the lower strut bolts and nuts to 87–101 ft. lbs.
11. Tighten the upper strut retaining nuts to 33–40 ft. lbs.
12. Be sure to reprogram the necessary systems.
13. Reset/Initialize the sunroof, if equipped.

OVERHAUL

See Figure 140.

1. Before servicing the vehicle, refer to the Precautions Section.

➡If working near and/or around the SRS system and components, be sure to disable the SRS system. Tape the negative battery cable with insulating tape. Always disconnect the negative battery cable first.

⁂ CAUTION

To avoid personal injury when working on vehicles equipped with an air bag, the negative battery cable must be disconnected and at least three minutes must elapse before working on the system. Failure to do so may result in deployment of the air bag.

2. Remove the battery top cover, if equipped.
3. Disconnect the negative battery cable. Tape the cable with insulating tape.
4. Raise and support the vehicle safely.
5. Remove the tire and wheel assembly.
6. Remove the strut.
7. Position the strut in a suitable holding fixture.
8. Disassemble the strut. Refer to the illustration.

To install:

➡Be sure to use new fasteners, as required.

9. Installation is the reverse of the removal procedure.
10. Tighten the lower strut bolts and nuts to 87–101 ft. lbs.
11. Tighten the upper strut retaining nuts to 33–40 ft. lbs.
12. Be sure to reprogram the necessary systems.
13. Reset/Initialize the sunroof, if equipped.

STEERING KNUCKLE

REMOVAL & INSTALLATION

See Figure 141.

1. Before servicing the vehicle, refer to the Precautions Section.

➡If working near and/or around the SRS system and components, be sure to disable the SRS system. Tape the negative battery cable with insulating tape. Always disconnect the negative battery cable first.

⁂ CAUTION

To avoid personal injury when working on vehicles equipped with an air bag, the negative battery cable must be disconnected and at least three minutes must elapse before working on the system. Failure to do so may result in deployment of the air bag.

2. Remove the battery top cover, if equipped.
3. Disconnect the negative battery cable. Tape the cable with insulating tape.
4. Raise and support the vehicle safely.
5. Remove the tire and wheel assembly.
6. Remove the castle nut.
7. Remove the brake caliper and position it to the side. Do not allow the caliper to hang by the brake line.
8. Remove the tie rod end ball joint from the knuckle.
9. Remove the wheel speed sensor, strut lower mounting bolt and lower arm mounting bolts.
10. Remove the split pin.
11. Remove the castle nut.
12. Disconnect the ball joint from the knuckle assembly, using tool SST 09568-4A000, or equivalent.
13. Remove the component from the vehicle.

To install:

➡Be sure to use new fasteners, as required.

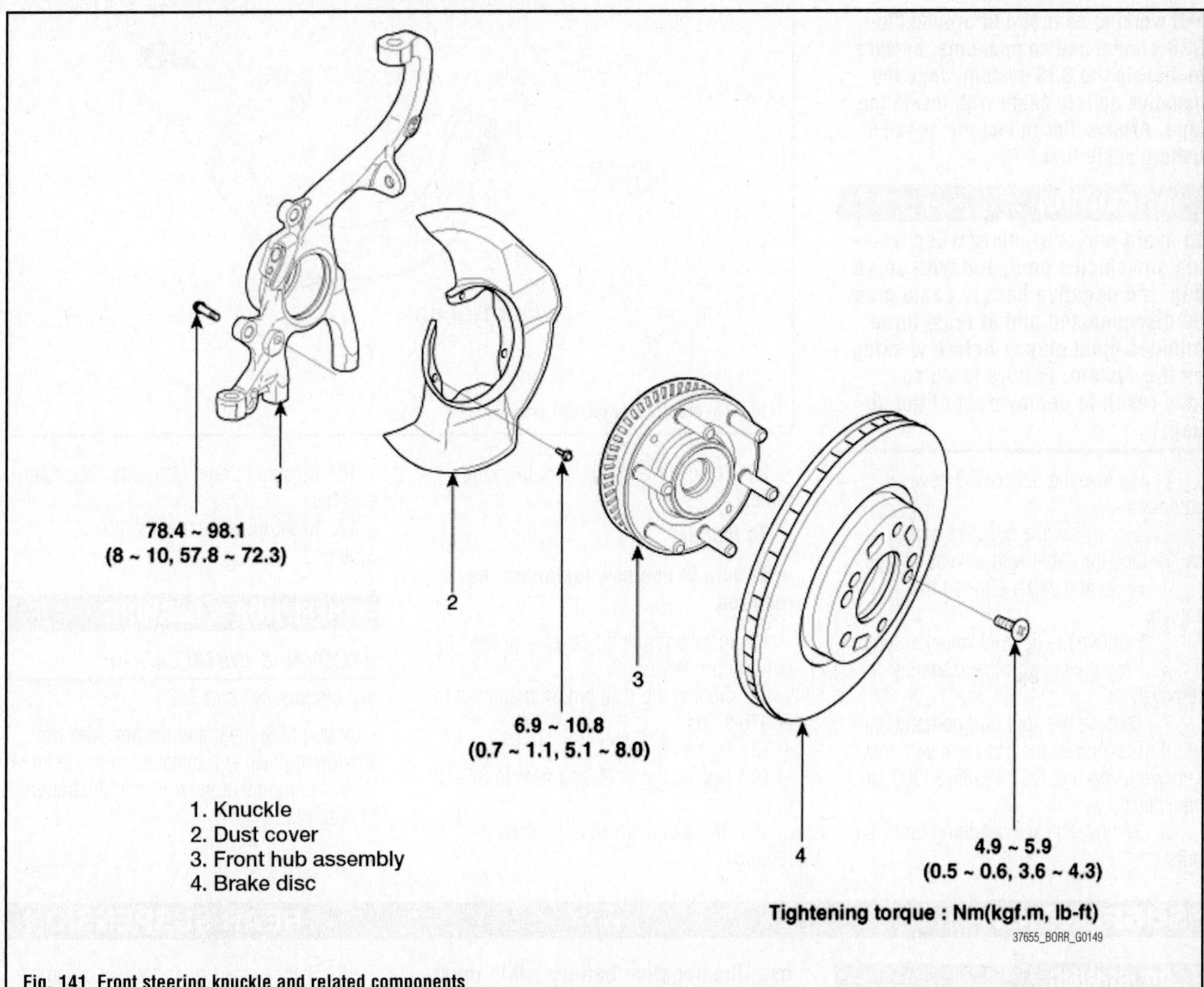

Fig. 141 Front steering knuckle and related components

14. Installation is the reverse of the removal procedure.
15. Be sure to reprogram the necessary systems.
16. Adjust alignment, as required.
17. Reset/Initialize the sunroof, if equipped.

STABILIZER BAR

REMOVAL & INSTALLATION

1. Before servicing the vehicle, refer to the Precautions Section.

➡**If working near and/or around the SRS system and components, be sure to disable the SRS system. Tape the negative battery cable with insulating tape. Always disconnect the negative battery cable first.**

⁂ CAUTION

To avoid personal injury when working on vehicles equipped with an air bag, the negative battery cable must be disconnected and at least three minutes must elapse before working on the system. Failure to do so may result in deployment of the air bag.

2. Remove the battery top cover, if equipped.
3. Disconnect the negative battery cable. Tape the cable with insulating tape.
4. Raise and support the vehicle safely.
5. Disconnect the stabilizer links at the knuckle.
6. Remove the stabilizer bar bushing bracket bolts.
7. Remove the bushings and brackets.
8. Remove the stabilizer bar from the vehicle.

To install:

➡**Be sure to use new fasteners, as required.**

9. Installation is the reverse of the removal procedure.
10. Tighten the link retaining nuts to 72–87 ft. lbs.
11. Tighten the bushing bracket bolts to 33–40 ft. lbs.
12. Be sure to reprogram the necessary systems.
13. Reset/Initialize the sunroof, if equipped.

UPPER BALL JOINT

REMOVAL & INSTALLATION

At this time the manufacturer does not provide removal and installation procedures for this component.

UPPER CONTROL ARM

REMOVAL & INSTALLATION

See Figure 142.

1. Before servicing the vehicle, refer to the Precautions Section.

➡If working near and/or around the SRS system and components, be sure to disable the SRS system. Tape the negative battery cable with insulating tape. Always disconnect the negative battery cable first.

⁂ CAUTION

To avoid personal injury when working on vehicles equipped with an air bag, the negative battery cable must be disconnected and at least three minutes must elapse before working on the system. Failure to do so may result in deployment of the air bag.

2. Remove the battery top cover, if equipped.
3. Disconnect the negative battery cable. Tape the cable with insulating tape.
4. Raise and safely support the vehicle.
5. Remove the tire and wheel assembly.
6. Properly support the assembly, for removal.
7. Remove the split pin, bolt and nut.
8. Disconnect the front arm with the knuckle using tool SST 09568-34000, or equivalent.
9. Remove the arm retaining bolts and nuts.

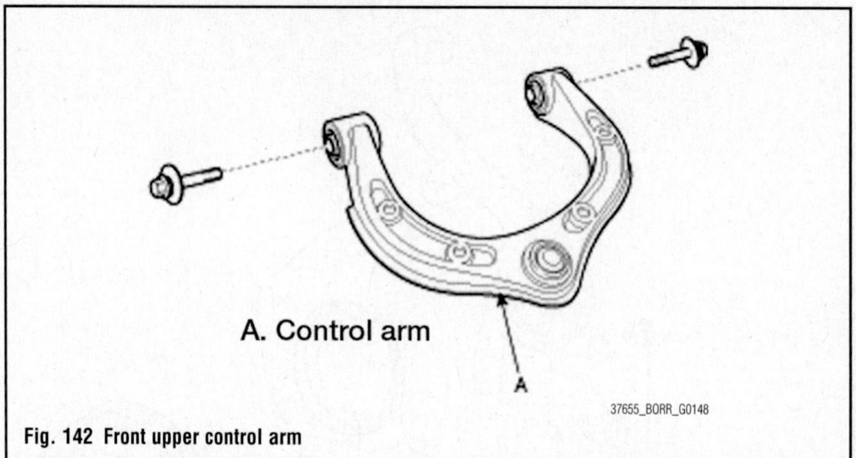

Fig. 142 Front upper control arm

10. Remove the component from the vehicle.

To install:

➡**Be sure to use new fasteners, as required.**

11. Installation is the reverse of the removal procedure.
12. Tighten the split pin, bolt and nut to 58–78 ft. lbs.
13. Tighten the nut to 65–87 ft. lbs.
14. Tighten the bolts and nuts to 87–101 ft. lbs.
15. Check and adjust alignment, as required.
16. Be sure to reprogram the necessary systems.
17. Reset/Initialize the sunroof, if equipped.

WHEEL HUB & BEARING

REMOVAL & INSTALLATION

See Figures 141 and 142.

At this time the manufacturer does not provide removal and installation procedures for this component, refer to the illustration as required.

SUSPENSION

REAR SUSPENSION

COIL SPRING

REMOVAL & INSTALLATION

At this time the manufacturer does not provide removal and installation procedures for this component.

CONTROL ARMS/LINKS

REMOVAL & INSTALLATION

Rear Assist Arm

See Figure 143.

1. Before servicing the vehicle, refer to the Precautions Section.

➡If working near and/or around the SRS system and components, be sure to disable the SRS system. Tape the negative battery cable with insulating tape. Always disconnect the negative battery cable first.

⁂ CAUTION

To avoid personal injury when working on vehicles equipped with an air bag, the negative battery cable must be disconnected and at least three minutes must elapse before working on the system. Failure to do so may result in deployment of the air bag.

2. Remove the battery top cover, if equipped.
3. Disconnect the negative battery cable. Tape the cable with insulating tape.
4. Raise and safely support the vehicle.

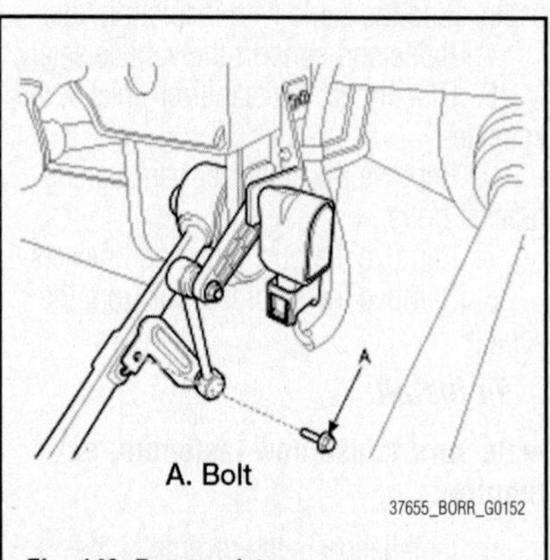

Fig. 143 Rear assist arm

5. Remove the tire and wheel assembly.
6. Remove the mounting bolt.
7. Using tool SST 09568-34000, or equivalent, remove the rear assist arm from the rear carrier.
8. Remove the bolt and then remove the rear assist arm from its mounting.

To install:

➡**Be sure to use new fasteners, as required.**

9. Installation is the reverse of the removal procedure.
10. Be sure to reprogram the necessary systems.
11. Reset/Initialize the sunroof, if equipped.

Rear Lower Arm

See Figure 144.

1. Before servicing the vehicle, refer to the Precautions Section.

➡If working near and/or around the SRS system and components, be sure to disable the SRS system. Tape the

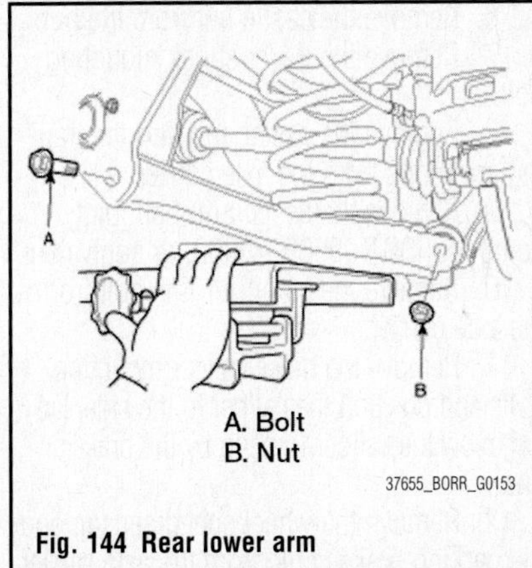

Fig. 144 Rear lower arm

negative battery cable with insulating tape. Always disconnect the negative battery cable first.

⁂ CAUTION

To avoid personal injury when working on vehicles equipped with an air bag, the negative battery cable must be disconnected and at least three minutes must elapse before working on the system. Failure to do so may result in deployment of the air bag.

2. Remove the battery top cover, if equipped.
3. Disconnect the negative battery cable. Tape the cable with insulating tape.
4. Raise and safely support the vehicle.
5. Remove the tire and wheel assembly.
6. Properly support the lower portion of the lower arm with a jack.
7. Remove the bolt.
8. Remove the nut.
9. Remove the arm from its mounting.

To install:

➡Be sure to use new fasteners, as required.

10. Installation is the reverse of the removal procedure.
11. Tighten the bolt and nut to 101–116 ft. lbs.
12. Be sure to reprogram the necessary systems.
13. Reset/Initialize the sunroof, if equipped.

Rear Upper Arm

See Figure 145.

1. Before servicing the vehicle, refer to the Precautions Section.

➡If working near and/or around the SRS system and components, be sure to disable the SRS system. Tape the negative battery cable with insulating tape. Always disconnect the negative battery cable first.

⁂ CAUTION

To avoid personal injury when working on vehicles equipped with an air bag, the negative battery cable must be disconnected and at least three minutes must elapse before working on the system. Failure to do so may result in deployment of the air bag.

2. Remove the battery top cover, if equipped.
3. Disconnect the negative battery cable. Tape the cable with insulating tape.
4. Raise and safely support the vehicle.
5. Remove the tire and wheel assembly.
6. Properly support the lower portion of the lower arm with a jack.
7. Remove the split pin, bolt and nut.
8. Disconnect the upper arm from the carrier using tool SST 09568-34000, or equivalent.
9. Disconnect the upper arm from the frame.
10. Remove the arm from its mounting.

To install:

➡Be sure to use new fasteners, as required.

11. Installation is the reverse of the removal procedure.
12. Reset/Initialize the sunroof, if equipped.

SHOCK ABSORBER

REMOVAL & INSTALLATION

1. Before servicing the vehicle, refer to the Precautions Section.

➡If working near and/or around the SRS system and components, be sure to disable the SRS system. Tape the negative battery cable with insulating tape. Always disconnect the negative battery cable first.

⁂ CAUTION

To avoid personal injury when working on vehicles equipped with an air bag, the negative battery cable must be disconnected and at least three minutes must elapse before working on the system. Failure to do so may result in deployment of the air bag.

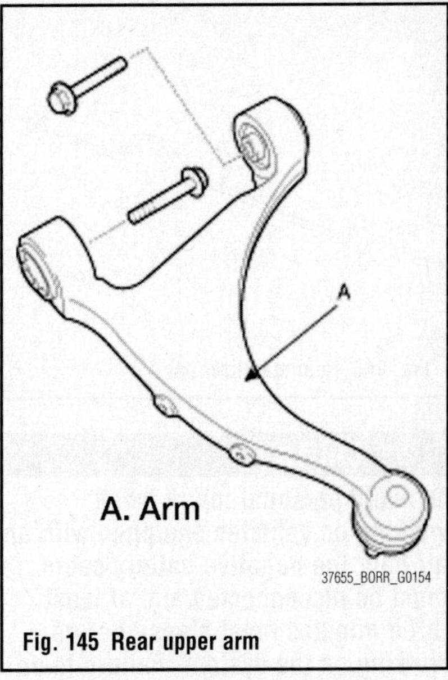

Fig. 145 Rear upper arm

2. Remove the battery top cover, if equipped.
3. Disconnect the negative battery cable. Tape the cable with insulating tape.
4. Raise and safely support the vehicle.
5. Remove the tire and wheel assembly.
6. Properly support the rear differential carrier assembly.
7. Remove the lower shock retaining bolt.
8. Remove the upper shock retaining nut.
9. Remove the shock from the vehicle.

To install:

➡Be sure to use new fasteners, as required.

10. Installation is the reverse of the removal procedure.
11. Tighten the lower bolt to 58–65 ft. lbs.
12. Tighten the upper nut to 58–65 ft. lbs.
13. Be sure to reprogram the necessary systems.
14. Reset/Initialize the sunroof, if equipped.

STABILIZER BAR

REMOVAL & INSTALLATION

See Figure 146.

1. Before servicing the vehicle, refer to the Precautions Section.

➡If working near and/or around the SRS system and components, be sure to disable the SRS system. Tape the negative battery cable with insulating tape. Always disconnect the negative battery cable first.

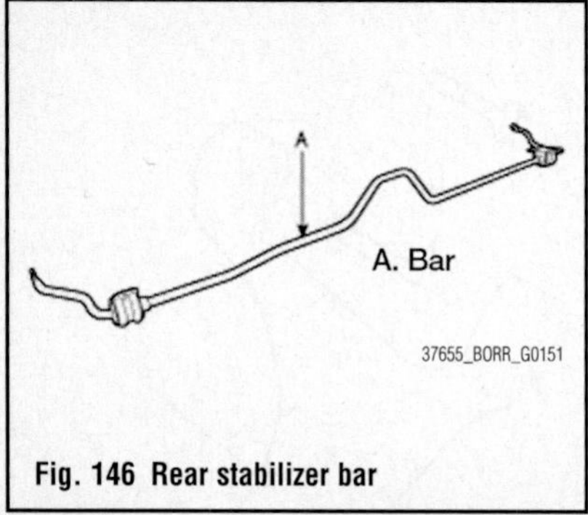

Fig. 146 Rear stabilizer bar

⁂ CAUTION

To avoid personal injury when working on vehicles equipped with an air bag, the negative battery cable must be disconnected and at least three minutes must elapse before working on the system. Failure to do so may result in deployment of the air bag.

2. Remove the battery top cover, if equipped.
3. Disconnect the negative battery cable. Tape the cable with insulating tape.
4. Raise and safely support the vehicle.
5. Remove the tire and wheel assembly.
6. Remove the fuel tank.
7. Remove the driveshaft.
8. Remove the stabilizer bar nut from the stabilizer link.
9. Remove the stabilizer bushing bolts. Remove the bushing brackets.
10. Remove the stabilizer bar from its mounting.

To install:

➡Be sure to use new fasteners, as required.

11. Installation is the reverse of the removal procedure.
12. Be sure to reprogram the necessary systems.
13. Reset/Initialize the sunroof, if equipped.

WHEEL HUB & BEARING

REMOVAL & INSTALLATION

See Figures 147 and 148.

1. Before servicing the vehicle, refer to the Precautions Section.

➡If working near and/or around the SRS system and components, be sure to disable the SRS system. Tape the negative battery cable with insulating tape. Always disconnect the negative battery cable first.

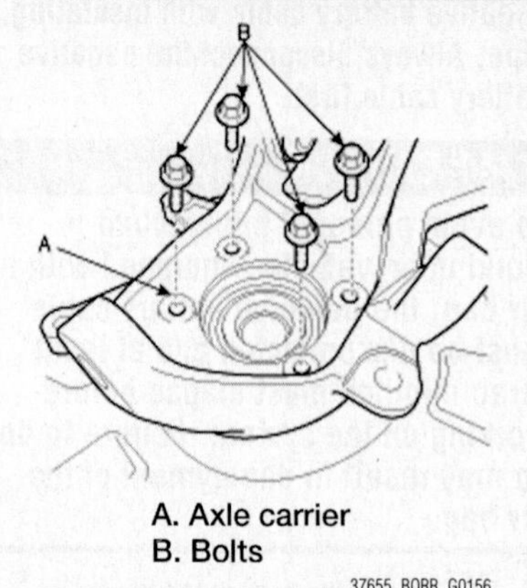

Fig. 147 Rear hub assembly mounting bolt locations

⁂ CAUTION

To avoid personal injury when working on vehicles equipped with an air bag, the negative battery cable must be disconnected and at least three minutes must elapse before working on the system. Failure to do so may result in deployment of the air bag.

2. Remove the battery top cover, if equipped.
3. Disconnect the negative battery cable. Tape the cable with insulating tape.
4. Raise and safely support the vehicle.
5. Remove the tire and wheel assembly.
6. Remove the castle nut from the hub.
7. Remove the lower shock mounting bolt.
8. Remove the assist arm and the trailing arm bolts from the axle carrier.
9. Remove the assist arm ball joint using tool SST 09568-4A000, or equivalent.
10. Remove the stabilizer link bolt from the axle carrier.
11. Remove the brake caliper mounting bolts and position the caliper to the side. Do not allow the caliper to hang by the brake line.
12. Remove the wheel speed sensor, and the parking brake cable from the axle carrier.
13. Disconnect the halfshaft from the knuckle.
14. Remove the cotter pin and castle nut from the rear of the upper arm ball joint. Remove the ball joint using tool SST 09568-34000, or equivalent.
15. Remove the hub assembly mounting bolts from the axle carrier.

To install:

➡Be sure to use new fasteners, as required.

16. Installation is the reverse of the removal procedure.
17. Be sure to reprogram the necessary systems.
18. Reset/Initialize the sunroof, if equipped.

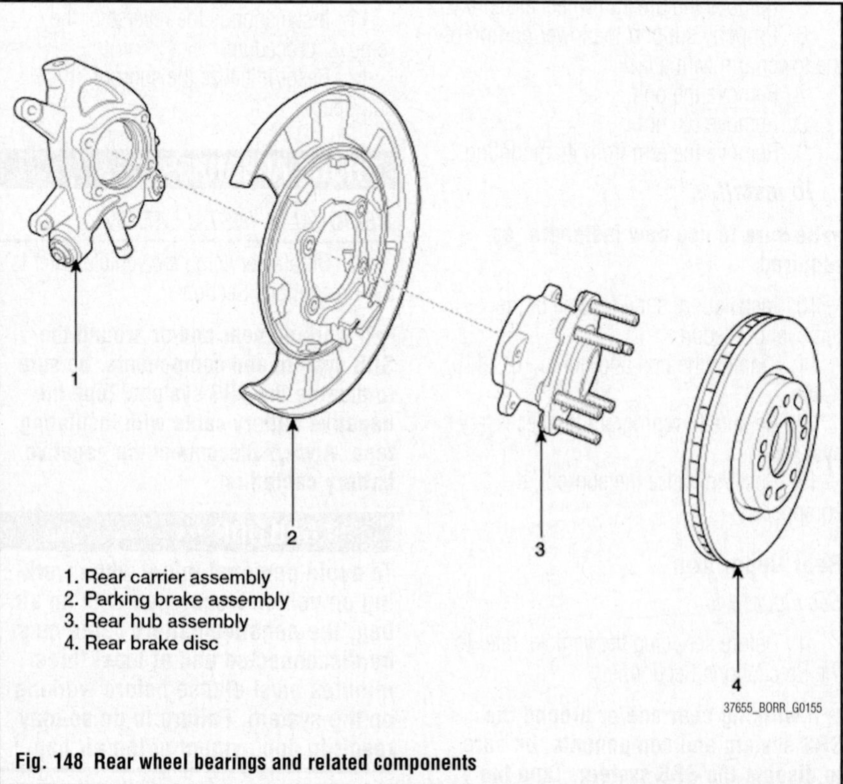

Fig. 148 Rear wheel bearings and related components

KIA

Forte

13

SPECIFICATIONS AND MAINTENANCE CHARTS

ENGINE AND VEHICLE IDENTIFICATION

Engine							Model Year	
Code ①	Liters (cc)	Cu. In.	Cyl.	Fuel Sys.	Engine Type	Eng. Mfg.	Code ②	Year
2	2.0 (1998)	121.9	4	EGI	DOHC	KIA	A	2010
3	2.4 (2359)	143.9	4	EGI	DOHC	KIA		

EGI: Electronic Gasoline Injection

DOHC: Double Overhead Camshafts

① 8th digit of VIN

② 10th digit of VIN

37655_FORT_C0001

GENERAL ENGINE SPECIFICATIONS

Year	Model	Engine Displacement Liters	Engine VIN	Net Horsepower @ rpm	Net Torque @ rpm (ft. lbs.)	Bore x Stroke (in.)	Com-pression Ratio	Oil Pressure @ rpm
2010	Forte	2.0	2	156@6200	144@4300	3.39x3.39	10.5:1	15.6@1000
		2.4	3	173@6000	168@4000	3.46x3.82	10.5:1	21.3@1000

37655_FORT_C0002

ENGINE TUNE-UP SPECIFICATIONS

Year	Engine Displacement Liters	Engine VIN	Spark Plug Gap (in.)	Fuel Pump (psi)	Idle Speed (rpm)		Valve Clearance	
					MT	AT	Intake	Exhaust
2010	2.0	2	0.039-0.043	49.8	550-750	520-720	0.006-0.009	0.011-0.013
	2.4	3	0.039-0.043	49.8	550-750	520-720	0.006-0.009	0.011-0.013

NOTE: The Vehicle Emission Control Information label often reflects specification changes made during production.

The label figures must be used if they differ from those in this chart.

B: Before top dead center

37655_FORT_C0003

CAPACITIES

Year	Model	Engine Displacement Liters	Engine VIN	Engine Oil with Filter	Transaxle (pts.) Manual	Transaxle (pts.) Auto.	Fuel Tank (gal.)	Cooling System (qts.)
2010	Forte	2.0	2	4.33	4.0	①	13.7	6.6
		2.4	3	4.76	3.7	②	13.7	6.6

① A4CF2 4 speed auto: 16, A5CF2 5 speed auto: 13.8

② A5CF2 5 speed auto 13.8

37655_FORT_C0004

FLUID SPECIFICATIONS

Year	Model	Engine Displ. Liters (VIN)	Engine Oil	Man. Trans.	Auto. Trans.	Transfer Case	Power Steering Fluid	Brake Master Cylinder	Cooling System
2010	Forte	2.0 (2)	5W-20	75W/85	①	NA	PSF - 3	DOT3 or 4	②
		2.4 (3)	5W-20	75W/85	①	NA	PSF - 3	DOT3 or 4	②

DOT: Department Of Transpotation

NA: Not Applicape

① GENUINE DIAMOND ATF SP-III

② Ethylene glycol

37655_FORT_C0005

VALVE SPECIFICATIONS

Year	Engine Displacement Liters	Engine VIN	Seat Angle (deg.)	Face Angle (deg.)	Maximum out of Square (in.)	Spring Free Length (in.)	Stem-to-Guide Clearance (in.) Intake	Stem-to-Guide Clearance (in.) Exhaust	Stem Diameter (in.) Intake	Stem Diameter (in.) Exhaust
2010	2.0	2	45.1-45.5	45.25-45.75	①	1.867	0.0008-0.0019	0.0012-0.0021	0.2151-0.2157	0.2149-0.2153
	2.4	3	45.1-45.5	45.25-45.75	①	1.867	0.0008-0.0019	0.0012-0.0021	0.2151-0.2157	0.2149-0.2153

NA: Not Available

① Less than 1.5 degrees

37655_FORT_C0006

CAMSHAFT AND BEARING SPECIFICATIONS

All measurements are given in inches.

Year	Engine Displacement Liters	Engine VIN	Journal Diameter	Brg. Oil Clearance	Shaft End-play	Runout	Journal Bore	Lobe Lift Intake	Lobe Lift Exhaust
2010	2.0	2	①	②	0.0015-0.0062	NA	NA	1.7244	1.7716
	2.4	3	③	④	0.0040-0.0079	NA	NA	1.7401	1.7716

NA: Not Available

① Intake No.1=1.1811, No.2-5=0.9449 Exhaust No.1=1.5748, No.2-5=0.9449

② Intake No.1=0.0008-0.0022, No.2-5=0.0017-0.0032, Exhaust No.1=0.0017-0.0032, No.2-5=0.0017-0.0032

③ Intake No.1=1.1811, No.2-5=0.9449 Exhaust No.1=1.4173, No.2-5=0.9449

④ Intake No.1=0.0008-0.0022, No.2-5=0.0012, Exhaust No.1=0.00-0.0012, No.2-5=0.0017-0.0032

37655_FORT_C0007

CRANKSHAFT AND CONNECTING ROD SPECIFICATIONS

All measurements are given in inches.

Year	Engine Displacement Liters	Engine VIN	Crankshaft Main Brg. Journal Dia.	Crankshaft Main Brg. Oil Clearance	Crankshaft Shaft End-play	Crankshaft Thrust on No.	Connecting Rod Journal Diameter	Connecting Rod Oil Clearance	Connecting Rod Side Clearance
2010	2.0	2	2.0449-2.0456	0.0007-0.0014	0.0027-0.0098	3	1.8879-1.8886	0.0007-0.0014	0.0039-0.0100
	2.4	3	2.0449-2.0456	0.0007-0.0014	0.0027-0.0098	3	1.8879-1.8886	0.0007-0.0014	0.0039-0.0100

37655_FORT_C0009

PISTON AND RING SPECIFICATIONS

All measurements are given in inches.

Year	Engine Displacement Liters	Engine VIN	Piston Clearance	Ring Gap Top Compression	Ring Gap Bottom Compression	Ring Gap Oil Control	Ring Side Clearance Top Compression	Ring Side Clearance Bottom Compression	Ring Side Clearance Oil Control
2010	2.0	2	0.0005-0.0013	0.0059-0.0118	0.0145-0.0204	0.0078-0.0275	0.0019-0.0031	0.0015-0.0031	0.0023-0.0059
	2.4	3	0.0005-0.0013	0.0059-0.0118	0.0145-0.0204	0.0078-0.0275	0.0019-0.0031	0.0015-0.0031	0.0023-0.0059

NA: Not Available

37655_FORT_C0008

TORQUE SPECIFICATIONS

All readings in ft. lbs.

Year	Engine Displacement Liters	Cylinder Head Bolts	Main Bearing Bolts	Rod Bearing Bolts	Crankshaft Damper Bolts	Flywheel Bolts	Manifold Intake	Manifold Exhaust	Spark Plugs	Oil Pan Drain Plug
2010	2.0	①	②	③	123-130	87-94	14-17	36-40	18-22	25-33
	2.4	①	②	③	123-130	87-94	14-17	36-40	18-22	25-33

① Step 1: 24-27 ft. lbs.
Step 2: Plus 90-95 degrees
Step 3: Plus 90-95 degrees

② Step 1: 10.8 ft. lbs.
Step 2: 20-23 ft. lbs.
Step 3: Plus 120-125 degrees

③ Step 1: 13-16 ft. lbs.
Step 2: plus88-92 degrees

③ M10 bolts; 17-20 ft. lbs. plus 60-65 degrees; plus an additional 6-65 degrees.
M12 bolts: 20.3-23.1 ft. lbs., plus 60-65 degrees, plus additional 60-65 degrees

④ 20-23 ft. lbs. plus 60 degrees

37655_FORT_C0010

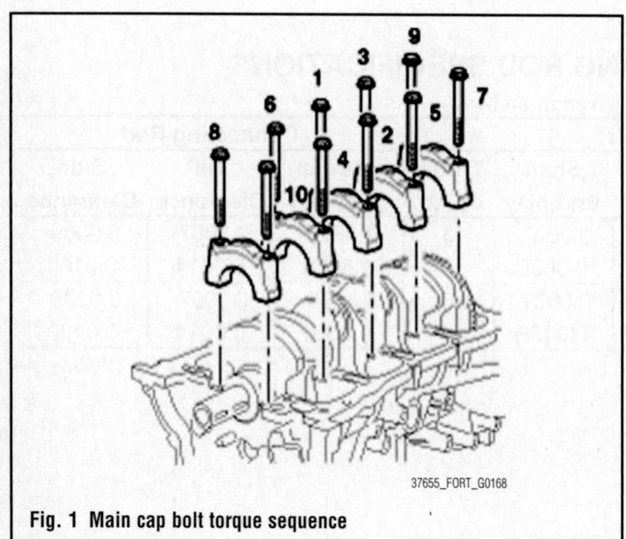

Fig. 1 Main cap bolt torque sequence

WHEEL ALIGNMENT

Year	Model		Caster Range (+/-Deg.)	Caster Preferred Setting (Deg.)	Camber Range (+/-Deg.)	Camber Preferred Setting (Deg.)	Toe-in (Deg.)
2010	Forte	F	0.5	+4.38	0.50	-0.64	0.1 +/- 0.2
		R	—	—	0.50	-1.50	0.4 +/- 0.2

37655_FORT_C0011

TIRE, WHEEL AND BALL JOINT SPECIFICATIONS

Year	Model	OEM Tires Standard	OEM Tires Optional	Tire Pressures (psi) Front	Tire Pressures (psi) Rear	Wheel Size	Ball Joint Inspection	Lug Nut Torque (ft. lbs.)
2010	Forte LX	185/70R15	195/65R15	30	30	5.0JX15 5.5JX15 Steel	①	65-80
	Forte EX	195/65R15	205/55R16	30	30	6.0JX16	①	65-80
	Forte SX	P215/45R17	—	30	30	7.0JX17	①	65-80
	Forte Koup EX	P205/55R16	—	30	30	6.0JX16	①	65-80
	Forte Koup SX	P215/45R17	—	30	30	7.0JX17	①	65-80

OEM: Original Equipment Manufacturer

PSI: Pounds Per Square Inch

STD: Standard

OPT: Optional

① Replace if any measureable movent is found.

37655_FORT_C0012

BRAKE SPECIFICATIONS

All measurements in inches unless noted

Year	Model		Brake Disc Original Thickness	Brake Disc Minimum Thickness	Brake Disc Maximum Run-out	Brake Drum Original Inside Diameter	Brake Drum Max. Wear Limit	Brake Drum Maximum Machine Diameter	Minimum Lining Thickness	Brake Caliper Bracket Bolts (ft. lbs.)	Brake Caliper Mounting Bolts (ft. lbs.)
2010	Forte 2.0	F	1.020	0.960	0.0016	—	—	—	0.079	58-72	16-23
		R	0.390	0.330	0.0019	6.61 ①	—	—	0.079	47-54	16-23
	Forte 2.4	F	1.100	1.040	0.0016	—	—	—	0.079	58-72	16-23
		R	0.390	0.330	0.0019	6.61 ①	—	—	0.079	47-54	16-23

F: Front

R: Rear

① Drum in hat parking b rake

37655_FORT_C0013

SCHEDULED MAINTENANCE INTERVALS

Kia - Forte

TO BE SERVICED	TYPE OF SERVICE	VEHICLE MILEAGE INTERVAL (x1000)												
		7.5	15	22.5	30	37.5	45	52.5	60	67.5	75	82.5	90	97.5
A/C system components and performance	S/I	✓	✓					✓	✓		✓	✓	✓	
Air fliter element	S/I	✓	✓					✓	✓		✓	✓	✓	
Cooling system	I	✓	✓		✓			✓	✓		✓	✓	✓	
Axle boots	I	✓	✓		✓			✓	✓		✓	✓	✓	
Battery condition	I	✓	✓		✓			✓	✓		✓	✓	✓	
Drive belts	I	✓	✓		✓			✓	✓		✓	✓	✓	
Brake/clutch fluid	I	✓	✓		✓			✓	✓		✓	✓	✓	
Brake operation and components	I	✓	✓		✓			✓	✓		✓	✓	✓	
Chassis and suspension fasteners	I	✓	✓		✓			✓	✓		✓	✓	✓	
Power steering components	I	✓	✓		✓			✓	✓		✓	✓	✓	
Locks and hinges	L	✓	✓		✓			✓	✓		✓	✓	✓	
Engine Oil and filter	R	✓	✓	✓	✓	✓	✓	✓	✓	✓	✓	✓	✓	✓
Rotate tires	S	✓	✓	✓	✓	✓	✓	✓	✓	✓	✓	✓	✓	✓
Cabin air filter	R		✓		✓			✓		✓		✓		
Air cleaner filter	R				✓				✓			✓		
Spark plugs (1.6L)	R				✓				✓			✓		
Fuel filter	I				✓									
Fuel tank (EVAP) filter	I				✓									
Automatic transmission fluid	I	Every 40,000 miles												
Manual transaxle fluid	I				✓				✓				✓	
Engine Coolant	R								✓				✓	
Timing belt (2.0L)													✓	

R: Replace S/I: Service or Inspect L: Lubricate T: Tighten

FREQUENT OPERATION MAINTENANCE (SEVERE SERVICE)

If a vehicle is operated under any of the following conditions it is considered severe service

- **Towing a trailer or using a camper or car-top carrier**
- **Repeated short trips of less than 5 miles in temperatures below freezing, or trips of less than 10 miles in any temperature**
- **Prolonged idling (vehicle operation in stop and go traffic).**
- **Operating on rough, muddy, unpaved, dusty or salt-covered roads.**
- **Police, taxi, delivery usage or trailer towing usage.**
- **Driving in extremely hot (over 90°F) conditions**

Oil & oil filter: change every 5000 miles or 5 months, whichever occurs first.

Air cleaner filter: inspect every 15,000 miles or 15 months and replace everything 30,000 miles or 30 months, whichever occurs fi

Front and rear brakes: inspect every 15,000 miles or 15 months, whichever occurs first

37655_FORT_C0014

BRAKES — INFORMATION AND PRECAUTIONS

ANTI-LOCK SYSTEMS

- Certain components within the ABS system are not intended to be serviced or repaired individually.
- Do not use rubber hoses or other parts not specifically specified for and ABS system. When using repair kits, replace all parts included in the kit. Partial or incorrect repair may lead to functional problems and require the replacement of components.
- Lubricate rubber parts with clean, fresh brake fluid to ease assembly. Do not use shop air to clean parts; damage to rubber components may result.
- Use only DOT 3 brake fluid from an unopened container.
- If any hydraulic component or line is removed or replaced, it may be necessary to bleed the entire system.
- A clean repair area is essential. Always clean the reservoir and cap thoroughly before removing the cap. The slightest amount of dirt in the fluid may plug an orifice and impair the system function. Perform repairs after components have been thoroughly cleaned; use only denatured alcohol to clean components. Do not allow ABS components to come into contact with any substance containing mineral oil; this includes used shop rags.
- The Anti-Lock control unit is a microprocessor similar to other computer units in the vehicle. Ensure that the ignition switch is **OFF** before removing or installing controller harnesses. Avoid static electricity discharge at or near the controller.
- If any arc welding is to be done on the vehicle, the control unit should be unplugged before welding operations begin.

DISC AND DRUM SYSTEMS

⁂ CAUTION

Dust and dirt accumulating on brake parts during normal use may contain asbestos fibers from production or aftermarket brake linings. Breathing excessive concentrations of asbestos fibers can cause serious bodily harm. Exercise care when servicing brake parts. Do not sand or grind brake lining unless equipment used is designed to contain the dust residue. Do not clean brake parts with compressed air or by dry brushing. Cleaning should be done by dampening the brake components with a fine mist of water, then wiping the brake components clean with a dampened cloth. Dispose of cloth and all residue containing asbestos fibers in an impermeable container with the appropriate label. Follow practices prescribed by the Occupational Safety and Health Administration (OSHA) and the Environmental Protection Agency (EPA) for the handling, processing, and disposing of dust or debris that may contain asbestos fibers.

BRAKES — BLEEDING THE BRAKE SYSTEM

BLEEDING PROCEDURE

BLEEDING PROCEDURE

⁂ CAUTION

Note the following:

- Do not reuse the drained fluid.
- Always use genuine DOT3/DOT4 brake Fluid.
- Using a non–genuine DOT3/DOT4 brake fluid can cause corrosion and decrease the life of the system.
- Make sure no dirt or other foreign matter is allowed to contaminate the brake fluid.
- Do not spill brake fluid on the vehicle, it may damage the paint; if brake fluid does contact the paint, wash it off immediately with water.
- The reservoir on the master cylinder must be at the MAX (upper) level mark at the start of bleeding procedure and checked after bleeding each brake caliper. Add fluid as required.

1. Make sure the brake fluid in the reservoir is at the MAX (upper) level line.
2. Have someone slowly pump the brake pedal several times, and then apply pressure.
3. Loosen the right–rear brake bleed screw to allow air to escape from the system. Then tighten the bleed screw securely.
4. Repeat the procedure for wheel in order until air bubbles no longer appear in the fluid. Bleeding order is right rear, left front, left rear, then right front.
5. Refill the master cylinder reservoir to MAX (upper) level line.

BLEEDING THE ABS SYSTEM

This procedure should be followed to ensure adequate bleeding of air and filling of the ABS unit, brake lines and master cylinder with brake fluid.

1. Remove the reservoir cap and fill the brake reservoir with brake fluid.

➡If there is any brake fluid on any painted surface, wash it off immediately.

⁂ CAUTION

When pressure bleeding, do not depress the brake pedal.

2. Connect a clear plastic tube to the wheel cylinder bleeder plug and insert the other end of the tube into a half filled clear plastic bottle.
3. Connect the GDS to the data link connector located underneath the dash panel.
4. Select and operate according to the instructions on the GDS screen.

⁂ CAUTION

You must obey the maximum operating time of the ABS motor with the GDS to prevent the motor pump from burning.

a. Select vehicle name.
b. Select Anti–Lock Brake system.
c. Select HCU air bleeding mode.
d. Press "OK" to operate motor pump and solenoid valve.
e. Wait 60 sec. before operating the air bleeding. If not, you may damage the motor.
f. Perform the air bleeding.

5. Pump the brake pedal several times, and then loosen the bleeder screw until fluid starts to run out without bubbles. Then close the bleeder screw.
6. Repeat step 5 until there are no more bubbles in the fluid for each wheel in order, right rear, left front, left rear, then right front.
7. Tighten the bleeder screw to 5.1–9.4ft. lbs. (7–13 Nm).

BRAKES — ANTI-LOCK BRAKE SYSTEM (ABS)

WHEEL SPEED SENSORS

REMOVAL & INSTALLATION

Front

See Figure 2.

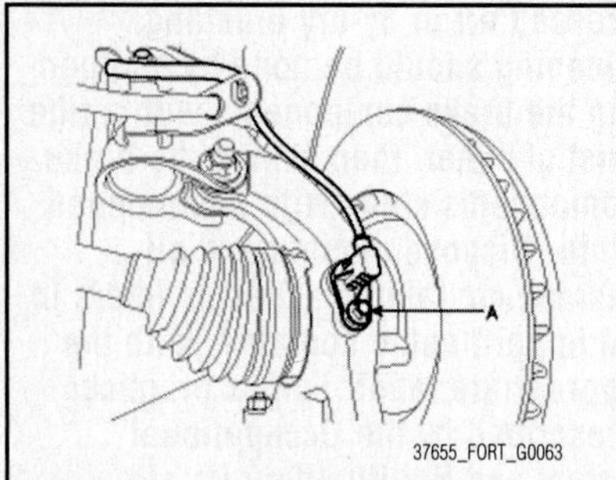

Fig. 2 Front wheel speed sensor mounting bolt (A)

1. Remove the front wheel speed sensor mounting bolt (A).
2. Remove the front wheel guard.
3. Disconnect the front wheel speed sensor connector. Remove the front wheel speed sensor.
4. Installation is the reverse of removal.
5. Tighten the front wheel speed sensor mounting bolt to 5–8 ft. lbs. (7–11 Nm).

Rear

See Figure 3.

➡Sensor is part of rear hub bearing assembly.

1. Remove the connector (A) after removing the rear wheel speed sensor clip.

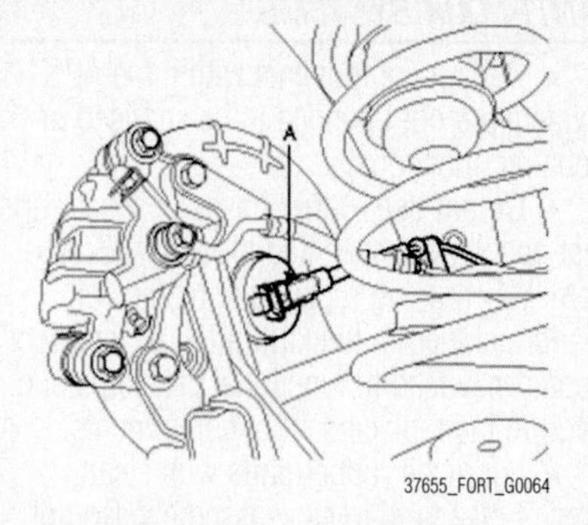

Fig. 3 Rear wheel speed sensor connector (A)

2. Remove the rear wheel speed sensor. See Rear Hub Bearing.
3. Installation is the reverse of removal.

BRAKES — FRONT DISC BRAKES

BRAKE CALIPER

REMOVAL & INSTALLATION

See Figure 4.

1. Raise and support vehicle.
2. Remove the wheel and tire.
3. Loosen the hose eye–bolt (B) and caliper bracket mounting bolts (C), then remove the front caliper assembly (A).
4. Installation is the reverse of removal.

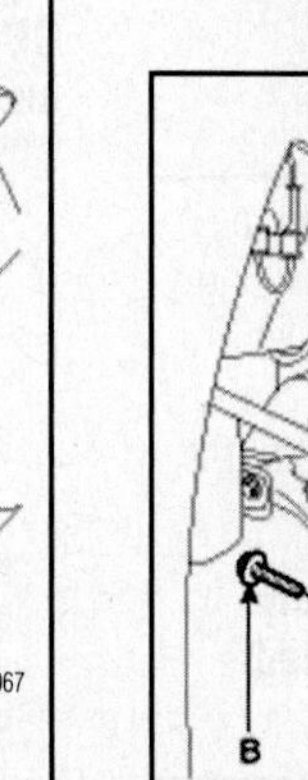

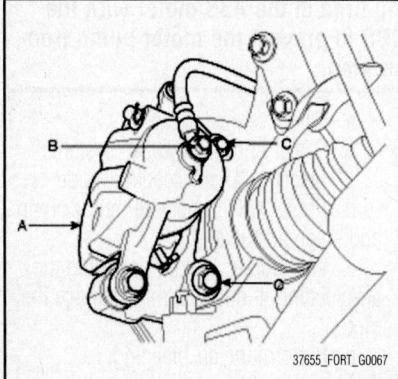

Fig. 4 Front caliper assembly (A), hose eye–bolt (B) and caliper bracket mounting bolts (C)

5. Tighten the front caliper bracket mounting bolt to 58–72 ft. lbs. (79–98 Nm).
6. Tighten the front caliper hose eye bolt to 18–22 ft. lbs. (25–30 Nm).
7. Tighten the wheel nuts to 65–80 ft. lbs. (88–108 Nm).

DISC BRAKE PADS

REMOVAL & INSTALLATION

See Figures 5 and 6.

1. Raise and support vehicle.

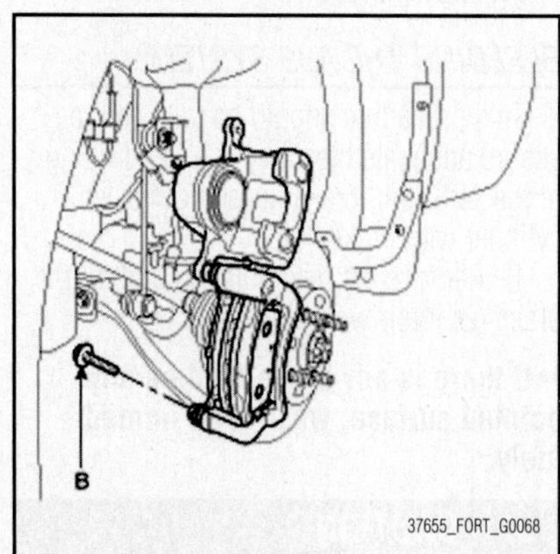

Fig. 5 Guide rod bolt (B)

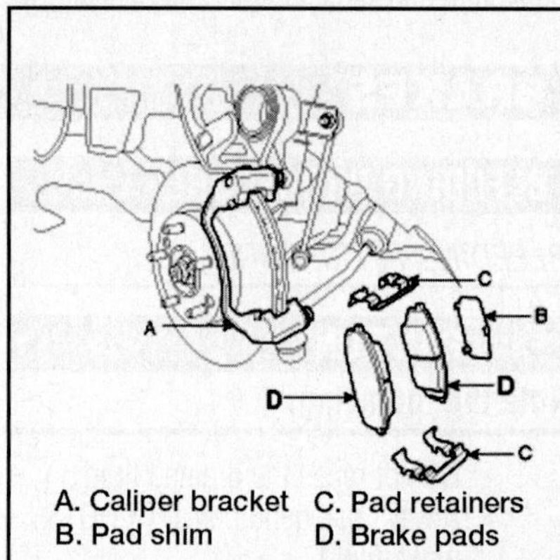

Fig. 6 Caliper bracket (A), pad shim (B), pad retainers (C) and brake pads (D)

2. Remove the front wheel and tire. Loosen the guide rod bolt (B) and pivot the caliper up out of the way.
3. Replace pad shim (B), pad retainers (C) and brake pads (B) in the caliper bracket (A).
4. Installation is the reverse of removal. Press back caliper piston to allow caliper to fit over pads.
5. Tighten the front caliper guide rod bolt to 16–23 ft. lbs. (22–31 Nm).

BRAKES — REAR DISC BRAKES

BRAKE CALIPER

REMOVAL & INSTALLATION

See Figure 7.

1. Raise and support vehicle.
2. Remove the wheel and tire.
3. Loosen the hose eye–bolt (B) and caliper bracket mounting bolts (C), then remove the front caliper assembly (A).
4. Installation is the reverse of removal.
5. Tighten the rear caliper bracket mounting bolt to 47–54 ft. lbs. (64–74 Nm).
6. Tighten the front caliper hose eye bolt to 18–22 ft. lbs. (25–30 Nm).
7. Tighten the wheel nuts to 65–80 ft. lbs. (88–108 Nm).

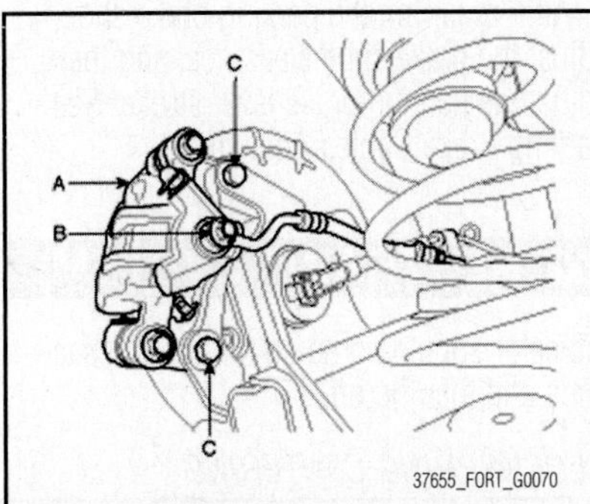

Fig. 7 Rear caliper assembly (A), hose eye–bolt (B) and caliper bracket mounting bolts (C)

DISC BRAKE PADS

REMOVAL & INSTALLATION

See Figures 8 and 9.

1. Raise and support vehicle.
2. Remove the wheel and tire. Loosen the caliper mounting bolts (A) and move the caliper up out of the way with hose attached.
3. Replace pad retainers (C) and brake pads (B) in the caliper bracket (A).
4. Installation is the reverse of removal. Press back caliper piston to allow caliper to fit over pads.
5. Tighten the front caliper mounting bolts to 16–23 ft. lbs. (22–31 Nm).

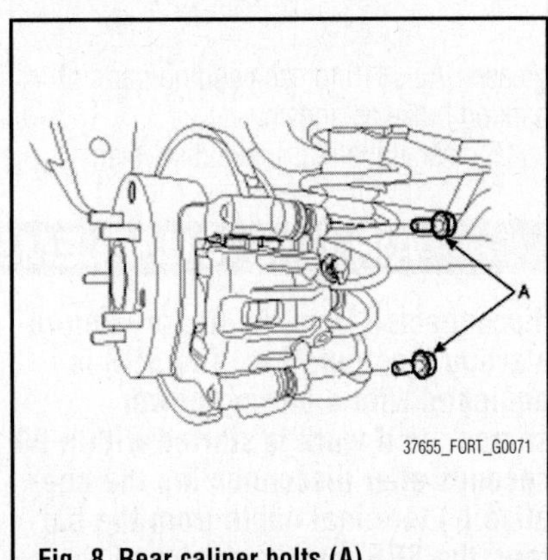

Fig. 8 Rear caliper bolts (A)

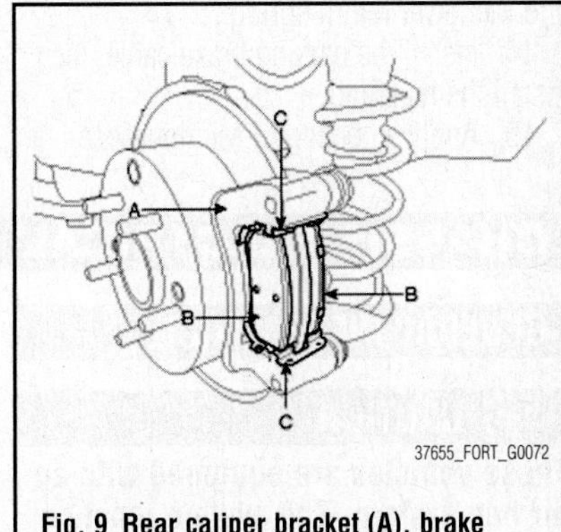

Fig. 9 Rear caliper bracket (A), brake pads (B) and pad retainers (C)

BRAKES — PARKING BRAKE

PARKING BRAKE CABLES

ADJUSTMENT

1. Block the front wheels, then raise the rear of the vehicle and make sure it is securely supported.
2. Pull the parking brake lever up 6–8 clicks with 44 pounds of force.
3. Make sure that the parking brakes are fully applied when the parking brake lever is pulled up fully.
4. Release the parking brake lever fully, and check that parking brakes do not drag when the rear wheels are turned.
5. If adjustment is necessary, check shoe adjustment first. See Parking Brake Shoes.
6. If shoe adjustment is complete and parking brake still needs adjustment, remove the console.
7. Tighten the adjusting nut until the parking brakes drag slightly when the rear wheels are turned.
8. Reinstall the console.

PARKING BRAKE SHOES

REMOVAL & INSTALLATION

See Figures 10 through 13.

1. Raise the vehicle, and make sure it is securely supported.
2. Remove the rear tire and wheel, then remove the brake caliper. See Brake Caliper.
3. Remove the brake Disc.
4. Remove the rear hub unit bearing.
5. Remove the shoe hold down pin (A) and the spring (B) by pushing the retainer spring and turning the pin.
6. Remove the adjuster assembly (A) and the return spring (B).
7. Remove the parking brake cable from the brake shoe.

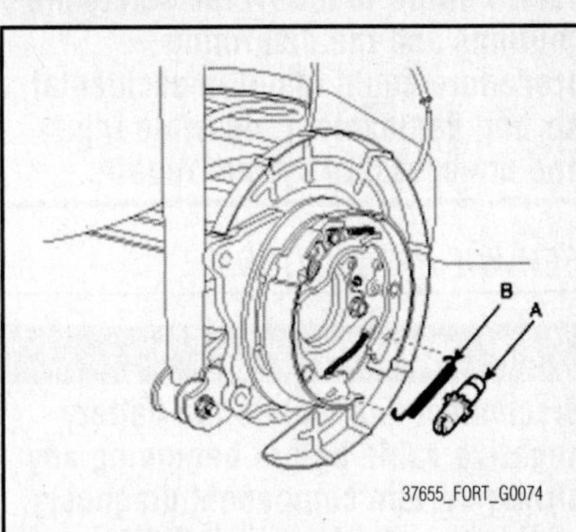

Fig. 11 Adjuster assembly (A) and return spring (B)

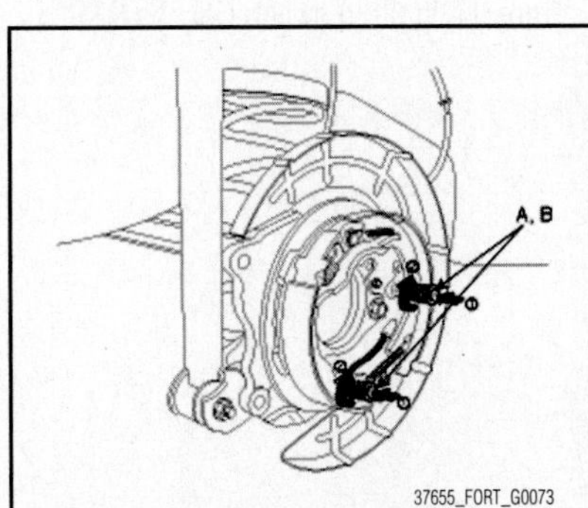

Fig. 10 Shoe hold down pin (A) and spring (B)

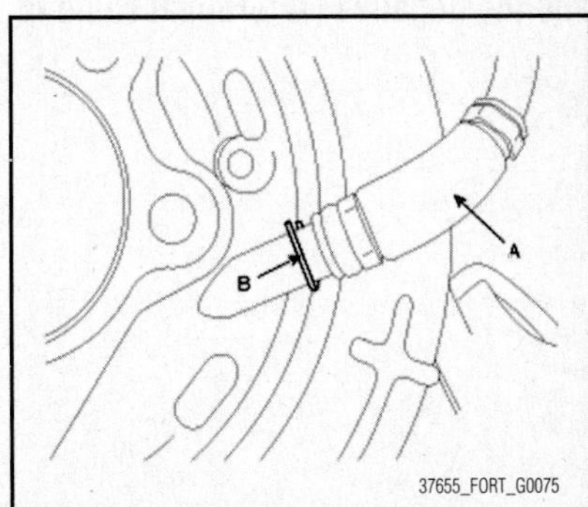

Fig. 12 Parking brake cable (A) and parking brake cable retaining (B)

8. Remove the strut and the strut spring.
9. Remove the brake shoe.
10. Remove the parking brake cable retaining (B), from the parking brake cable (A).

To install:

11. Install the brake shoe to the back plate.
12. Install the shoe hold down pin and the spring by pushing the retainer spring and turning the pins.
13. After installing the strut and upper return spring, install the adjuster assembly and the lower return spring.
14. Install the parking brake cable, then install the retaining.
15. Apply a coating of Multipurpose grease SAE J310 to each sliding parts of parking brake as shown.

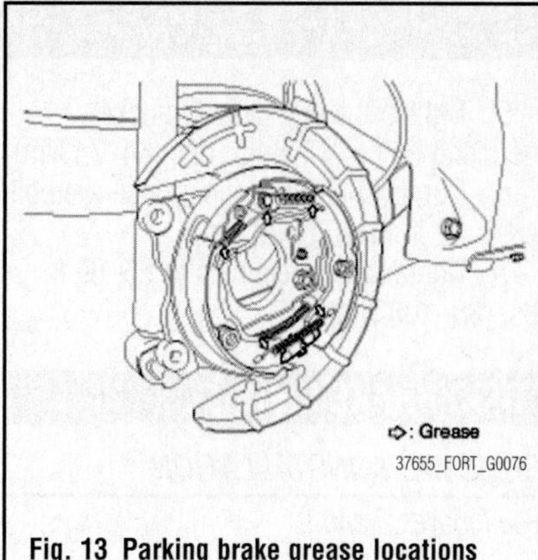

Fig. 13 Parking brake grease locations

16. Install the rear brake disc, then adjust the rear brake shoe clearance. See Adjustment.
17. Install the brake caliper. See Brake Caliper.
18. Install the tire and wheel.
19. Adjust the parking brake lever. See Parking Brake Cables Adjustment.

ADJUSTMENT

1. Remove the plug from the disc.
2. Rotate the toothed wheel of adjuster by a screw driver until the disc is not moving, and then return it by 3~5 notches in the opposite direction.

After repairing the parking brake shoe, adjust the brake shoe clearance, and then adjust the parking brake lever stroke. See Parking Brake Cables Adjustment.

CHASSIS ELECTRICAL

AIR BAG (SUPPLEMENTAL RESTRAINT SYSTEM)

GENERAL INFORMATION

CAUTION

These vehicles are equipped with an air bag system. The system must be disarmed before performing service on, or around, system components, the steering column, instrument panel components, wiring and sensors. Failure to follow the safety precautions and the disarming procedure could result in accidental air bag deployment, possible injury and unnecessary system repairs.

SERVICE PRECAUTIONS

CAUTION

Disconnect and isolate the battery negative cable before beginning any airbag system component diagnosis, testing, removal, or installation procedures. Wait at least 90 seconds after the ignition switch is turned off and the negative (-) terminal cable is disconnected from the battery before starting the operation. The SRS is equipped with a backup power source, so if work is started within 90 seconds after disconnecting the negative (-) terminal cable from the battery, the SRS may be deployed. Failure to disable the airbag system may result in accidental airbag deployment, personal injury, or death.

DISARMING THE SYSTEM

Turn the ignition switch OFF and disconnect the negative cable from the battery, and wait at least three minutes before beginning work.

➡See Engine Electrical, Battery System for battery disconnect/connect relearn procedures.

ARMING THE SYSTEM

After the vehicle is completely repaired, confirm the SRS airbag system is OK.

Turn the ignition switch ON, the SRS indicator should come on for about 6 seconds and then go off.

CLOCKSPRING CENTERING

See Figure 14.

Set the center position by getting marks between the clock spring and the cover into line. Ensure marks are correct by turning the clock spring clockwise to the stop and then 3 revolutions counterclockwise.

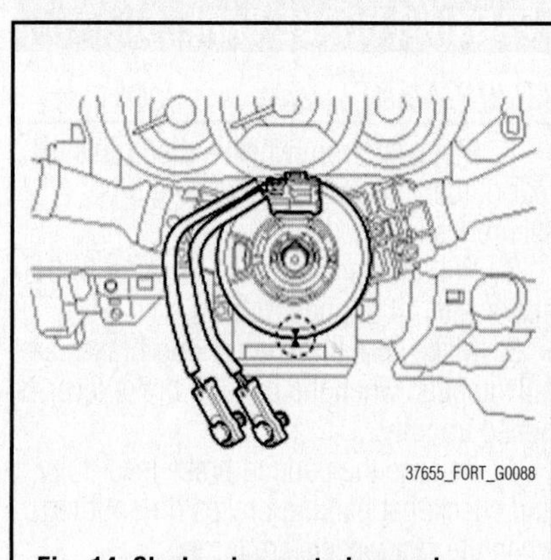

Fig. 14 Clockspring centering marks

DRIVE TRAIN

CLUTCH

REMOVAL & INSTALLATION

See Figures 15 and 16.

1. Remove the transaxle assembly.
2. Insert the special tool (09411–25000) in the clutch disc to prevent the disc from shifting.
3. Loosen the bolts which attach the clutch cover to the flywheel in a star pattern. Loosen the bolts in succession, one or two turns at a time, to avoid bending the cover.

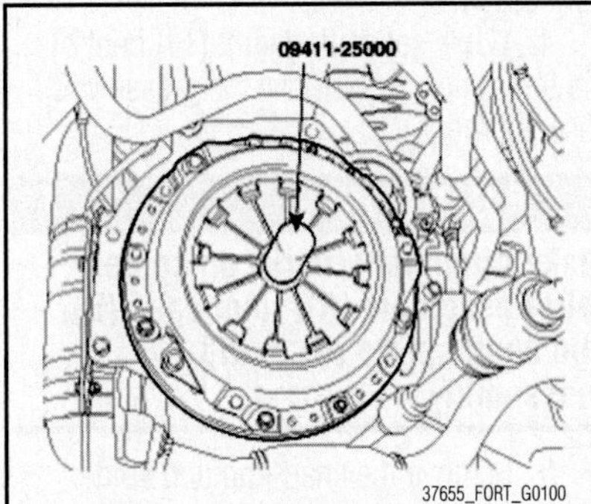

Fig. 15 Insert the special tool (09411–25000) in the clutch disc

To install:

4. Apply CASMOLY L 9508 multipurpose grease to the spline of the disc.

➡When installing the clutch, apply grease to each part, but be careful not to apply excessive grease. It can cause clutch slippage and vibration (shudder).

5. Temporarily install the clutch disc assembly to the flywheel using the special tool (09411–11000).
6. Tighten the bolts one or two steps at a time in a star pattern. Tighten to 11–16 ft. lbs. (15–22 Nm).
7. Remove the clutch disc guide (09411–11000).
8. Install the transaxle assembly to the engine. See Manual Transaxle.

ADJUSTMENTS

See Figure 17.

1. Measure the clutch pedal height (from the face of the pedal pad to the floorboard) and the clutch pedal clevis pin play (measured at the face of the pedal pad.)

➡If the clutch pedal height is lower than the standard value, loosen the bolt and adjust the push rod. After adjustment, adjust stopper bolt so that the clearance with pedal stopper becomes 0.5mm (0.02 in) to 1.0mm (0.04 in) and secure with lock nut.

2. If the clutch pedal freeplay and height is not within the standard value range, adjust as follows:
   a. Turn and adjust the bolt within the standard value, then secure by tightening the lock nut.
   b. Turn the push rod to agree with the standard value and then secure the push rod with the lock nut.

➡When adjusting the clutch pedal height or the clutch pedal play, be careful not to push the push rod toward the master cylinder.

   c. If the clutch pedal free play and the distance between the clutch pedal and the floor board, do not meet with the standard values when the clutch is disengaged, it may be the result of either air in the hydraulic system or a faulty clutch master cylinder.

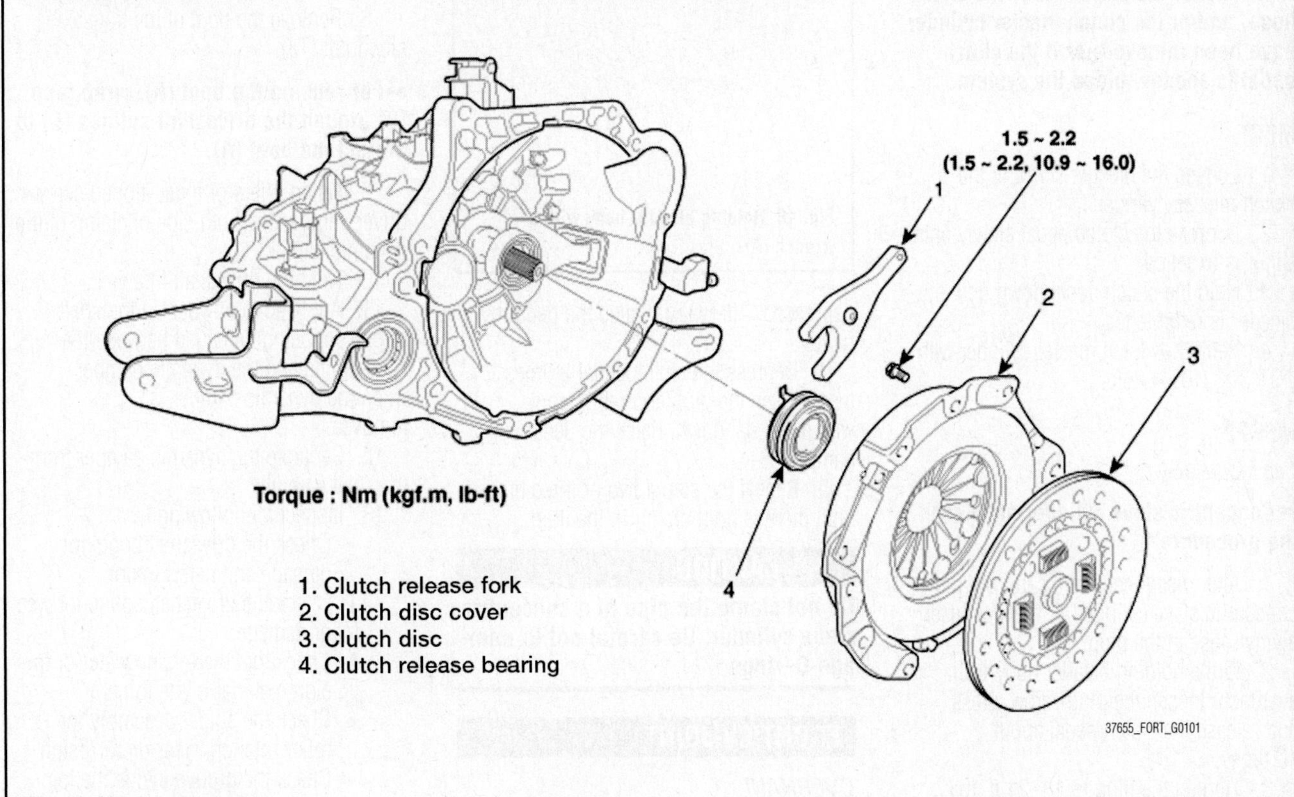

Fig. 16 Clutch components

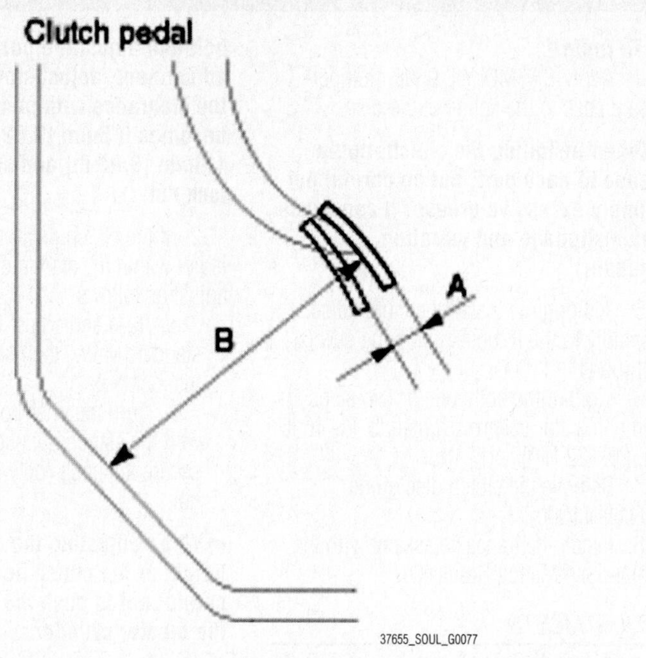

Fig. 17 Clutch pedal play height: Standard value (A) : 0.24–0.51 in. (6–13 mm) (B) : 7.19 in. (182.8 mm)

HYDRAULIC SYSTEM BLEEDING

BLEEDING PROCEDURE

➡Whenever the clutch tube, the clutch hose, and/or the clutch master cylinder have been removed, or if the clutch pedal is spongy, bleed the system.

M5CF

1. Loosen the bleeder screw at the clutch release cylinder.
2. Depress the clutch pedal slowly until all air is expelled.
3. Hold the clutch pedal down until the bleeder is retightened.
4. Refill the clutch master cylinder with DOT 3 or DOT 4 fluid.

M6GF2

See Figure 18.

➡Concentric slave cylinder air bleeding procedure

1. After disconnecting a cap from the concentric slave cylinder air bleeder, insert a vinyl hose in the plug.
2. While holding bleeder body with a wrench, loosen the plug screw, press and release the clutch pedal about 10 times.
3. Tighten the plug to 18–21 ft. lbs. (35–39 Nm) while the clutch pedal is depressed. Afterwards, raise the pedal by hand.

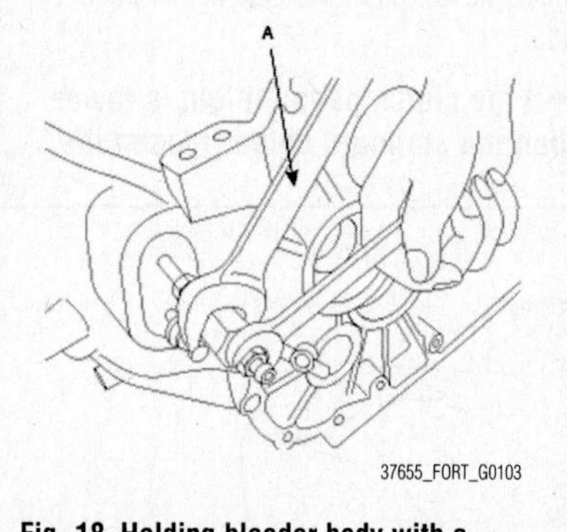

Fig. 18 Holding bleeder body with a wrench (A)

4. Depress the clutch pedal 3 times, then loosen the plug and retighten it with the pedal down. Raise the pedal by hand.
5. Repeat the step 4 two or three times, until there is no more air in the fluid.

⁂ CAUTION

Do not clamp the pipe of a concentric slave cylinder. Be careful not to damage O–rings.

FRONT CV-JOINT

OVERHAUL

See Figures 19 through 22.

⁂ CAUTION

Note the following:

- Do not disassemble the BJ assembly.
- Special grease must be applied to the driveshaft joint. Do not substitute with another type of grease.
- The boot band should be replaced with a new one.

1. Remove the circlip from the driveshaft spline.
2. Remove both boot bands from the transaxle side joint (TJ) case.
3. Pull out the boot from transaxle side joint case (B).
4. While separating joint (TJ) boot of the transaxle side, remove the grease in TJ case and boot.

⁂ CAUTION

Make alignment marks on spider roller assembly (A), joint case (B), and shaft spline (C) to aid reassembly.

5. Remove the snap ring and spider roller assembly from the shaft.
6. Remove the spider assembly (B) from the driveshaft (A) using the special tool (09495–33000).
7. Clean the spider assembly.
8. Remove the boot of the transaxle side joint (TJ).

➡For reusing the boot (A), wrap tape (B) around the driveshaft splines (C) to protect the boot (A).

9. Using pliers or a flat–tipped screwdriver, remove the both side of clamp of the dynamic damper.
10. Hold the driveshaft in a vice.
11. Apply soap powder on the shaft to prevent being damaged between the shaft spline and the dynamic damper when the dynamic damper is removed.
12. Separate the dynamic damper from the shaft carefully.
13. Inspect the following:
    - Check the driveshaft boots for damage and deterioration.
    - Check the driveshaft spline for wear or damage.
    - Check that there is no water or foreign material in the joint.
    - Check the spider assembly for roller rotation, wear or corrosion.
    - Check the groove inside the joint case for wear or corrosion.

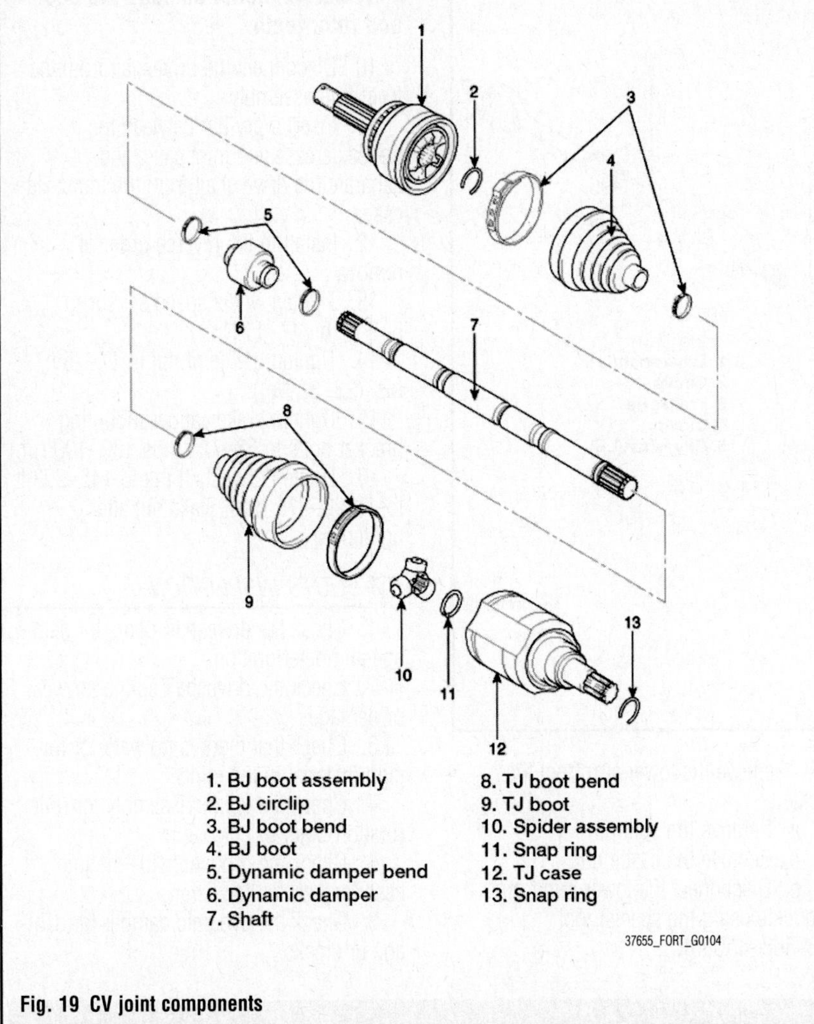

Fig. 19 CV joint components

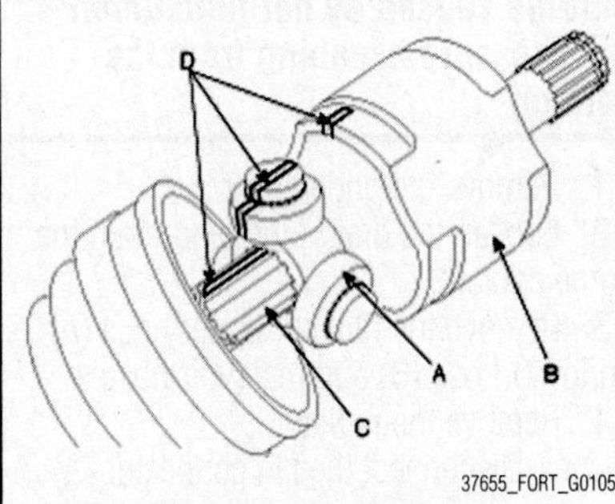

Fig. 20 Make alignment marks (D) on spider roller assembly (A), joint case (B), and shaft spline (C)

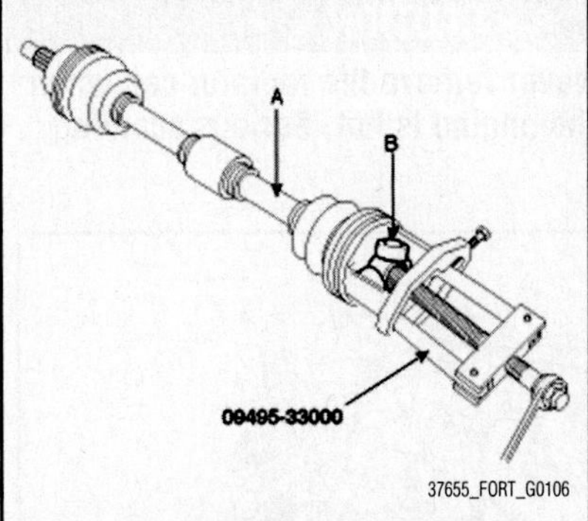

Fig. 21 Driveshaft (A), spider assembly (B) and special tool 09495–33000

- Check the dynamic damper for damage or cracks.

To assemble:

14. Wrap tape around the driveshaft spline (TJ) to prevent damage to the boots.
15. Apply grease to the joint boot on the side of the wheel and install the boot.
16. Install the clamp.
17. To install the dynamic damper, keep the shaft in a straight line and assemble the dynamic damper with the bands.
18. Assemble the transaxle side joint boot and bands.
19. Using the alignment marks (D) made during disassembly as a guide, install the spider assembly (A) and snap ring (B) on the driveshaft splines (C).

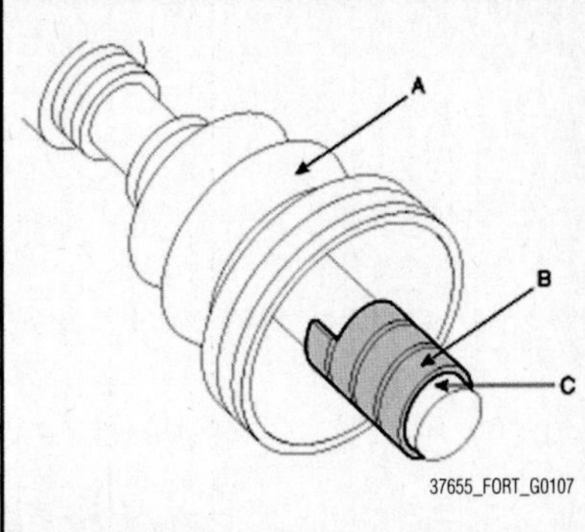

Fig. 22 Boot (A), tape (B) and driveshaft splines (C)

20. Add RBA grease to the joint boot. Quantity should be as much as it was removed during disassembly.
    Outer BJ joint 90 grams (Do not disassemble the BJ assembly)
    Inner TJ joint (Right side) 125 grams
    Inner UTJ joint (Left side) 210 grams
21. Install both boot bands.

FRONT HALFSHAFTS

REMOVAL & INSTALLATION

See Figure 23.

CAUTION

Note the following:

- Use a pry bar being careful not to damage the transaxle and joint.
- Do not insert the pry bar too deep, as this may cause damage to the oil seal.
- Do not pull the driveshaft by excessive force it may cause components inside the joint kit to dislodge resulting in a torn boot or a damaged bearing.
- Plug the hole of the transaxle case with the oil seal cap to prevent contamination.
- Support the driveshaft properly.
- Replace the retainer ring whenever the driveshaft is removed from the transaxle case

1. Loosen the wheel nuts slightly.
2. Raise And support the vehicle.
3. Remove the front wheel and tire.
4. Unstake the driveshaft lock nut using a chisel and hammer.
5. Apply brake and remove driveshaft nut from the front hub.
6. Remove the brake caliper mounting bracket bolts and caliper, and then support the brake caliper assembly with wire.
7. Remove the tie rod end ball joint from the knuckle.

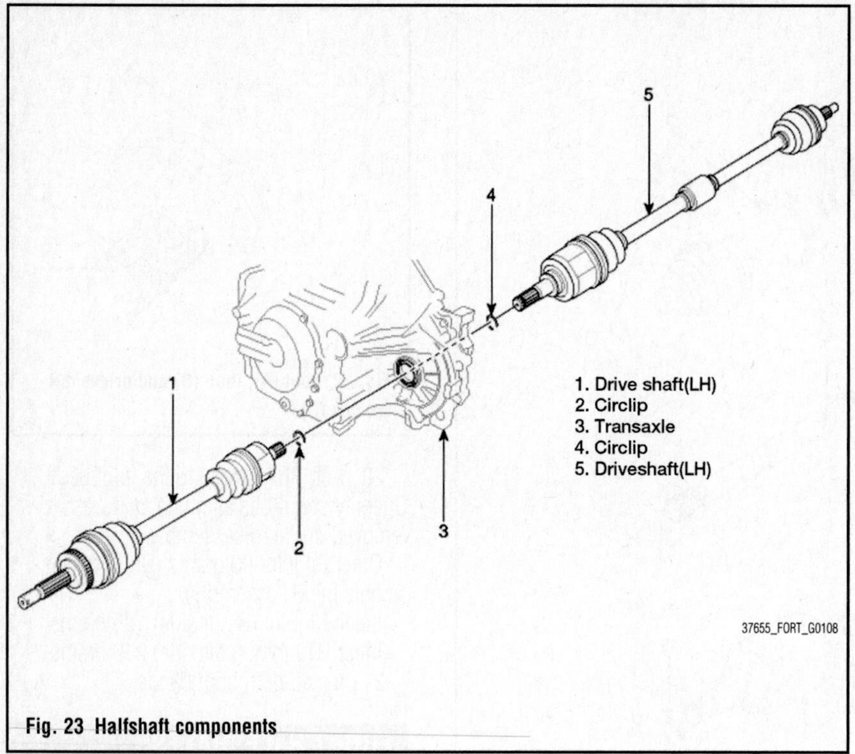

Fig. 23 Halfshaft components

a. Remove the split pin.
b. Remove the castle nut.
c. Disconnect the ball joint from knuckle using the special tool (09568–4A000).

8. Remove the wheel speed sensor.

9. Remove the lower arm from the knuckle.

a. Remove the split pin.
b. Remove the castle nut.
c. Disconnect the lower arm from knuckle using the special tool (09568–4A000).

➡Be careful not to damage the boot and rotor teeth.

10. Disconnect the driveshaft from the front hub assembly.
11. Insert a pry bar between the transaxle case and joint case, and separate the drive shaft from the transaxle case.
12. Install in the reverse order of removal.
13. Tighten wheel speed sensor to 5–8 ft. lbs. (7–11 Nm).
14. Tighten ball joint nut to 17–25 ft. lbs. (23–32 Nm).
15. Tighten brake caliper mounting bracket bolts to 58–72 ft. lbs. (80–100 Nm).
16. Tighten driveshaft nut to 145–203 ft. lbs. (196–275 Nm). Stake nut after tightening.

CV–BOOTS INSPECTION

1. Check the driveshaft boots for damage and deterioration.
2. Check the driveshaft spline for wear or damage.
3. Check that there is no water or foreign material in the joint.
4. Check the spider assembly for roller rotation, wear or corrosion.
5. Check the groove inside the joint case for wear or corrosion.
6. Check the dynamic damper for damage or cracks.

ENGINE COOLING

ENGINE FAN

REMOVAL & INSTALLATION

See Figures 24 through 26.

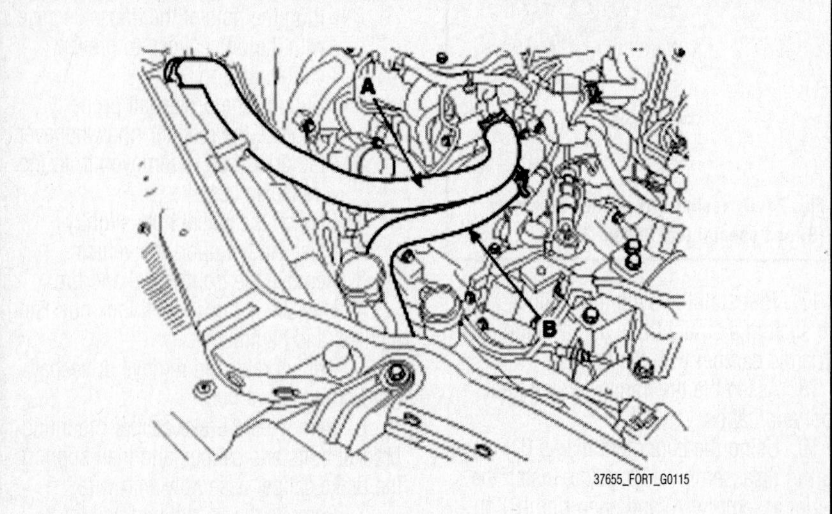

Fig. 24 Radiator upper hose (A) and lower hose (B)

CAUTION

Never remove the radiator cap when the engine is hot. Serious scalding could be caused by hot fluid under high pressure escaping from the radiator.

1. Remove the undercover.
2. Loosen the drain plug, and drain the engine coolant.
3. Remove the radiator upper hose (A) and lower hose (B) and reservoir hose.
4. Remove the radiator.
   a. Disconnect the fan connector (A) and remove the fan assembly (B).
5. Installation is reverse order of removal.
6. Tighten fan assembly mounting brackets bolts to 7–8 ft. lbs. (9–11 Nm).
7. Bleed air from the cooling system.
   a. Start engine and let it run until it warms up. (Until the radiator fan operates 3 or 4 times.)
   b. Turn off engine. Check the coolant level and add coolant if needed. This will allow trapped air to be removed from the cooling system.

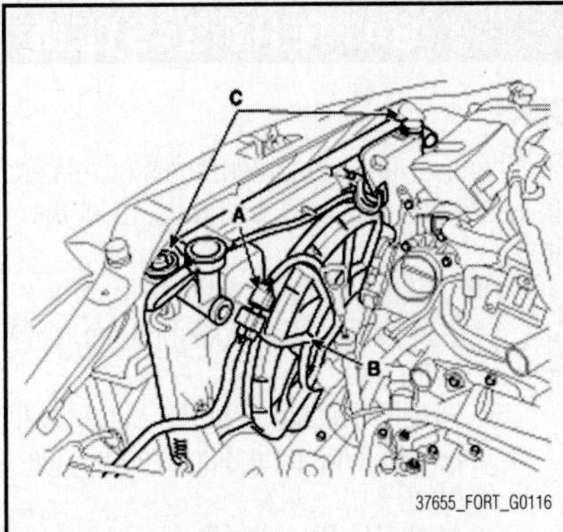

Fig. 25 Fan connector (A), fan assembly (B) and radiator upper mounting brackets (C)—Koup

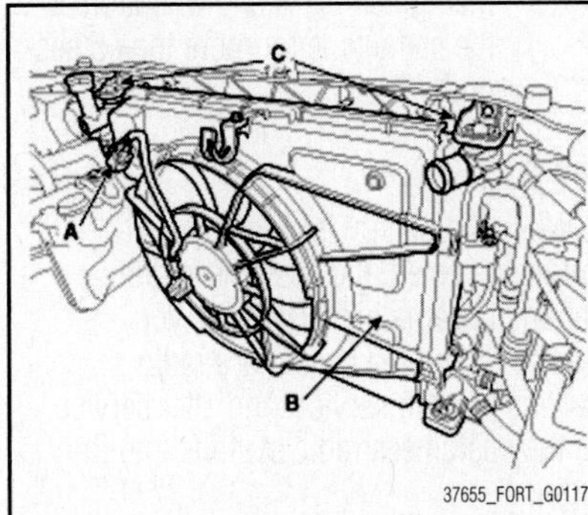

Fig. 26 Fan connector (A), fan assembly (B) and radiator upper mounting brackets (C)—Sedan

c. Put the radiator cap on tightly, then run engine again and check for leaks.

RADIATOR

REMOVAL & INSTALLATION

See Figures 24 through 26.

CAUTION

Never remove the radiator cap when the engine is hot. Serious scalding could be caused by hot fluid under high pressure escaping from the radiator.

1. Remove the undercover.
2. Loosen the drain plug, and drain the engine coolant.
3. Remove the radiator upper hose (A) and lower hose (B) and reservoir hose.
4. Remove the radiator.
   a. Disconnect the fan connector (A) and remove the fan assembly (B).
   b. Remove the radiator upper mounting brackets (C).
   c. After pulling back the condenser fixing bracket and then, remove the radiator assembly.
5. Installation is reverse order of removal.
6. Tighten radiator upper mounting brackets bolts to 7–8 ft. lbs. (9–11 Nm).
7. Tighten fan assembly mounting brackets bolts to 7–8 ft. lbs. (9–11 Nm).
8. Bleed air from the cooling system.
   a. Start engine and let it run until it warms up. (Until the radiator fan operates 3 or 4 times.)
   b. Turn off engine. Check the coolant level and add coolant if needed. This will allow trapped air to be removed from the cooling system.
   c. Put the radiator cap on tightly, then run engine again and check for leaks.

THERMOSTAT

REMOVAL & INSTALLATION

See Figure 27.

➡Removal of the thermostat would have an adverse effect, causing a lowering of cooling efficiency. Do not remove the thermostat, even if the engine tends to overheat.

1. Disconnect the battery terminals and remove the battery.

➡See Engine Electrical, Battery System for battery disconnect/connect relearn procedures.

2. Remove the engine cover.
3. Disconnect the ECM connector.
4. Disconnect the breather hose and then, remove the air duct and air cleaner assembly.
5. Drain engine coolant so its level is below thermostat.
6. Remove the radiator lower hose.
7. Remove water inlet fitting (A) and thermostat (B).
8. Installation is reverse order of removal.
9. Tighten thermostat housing bolts to 6–9 ft. lbs. (8–12 Nm).

➡Install the thermostat with the jiggle valve upward.

10. Fill the engine coolant.
11. Start the engine and check for leaks.
12. Recheck the coolant level.

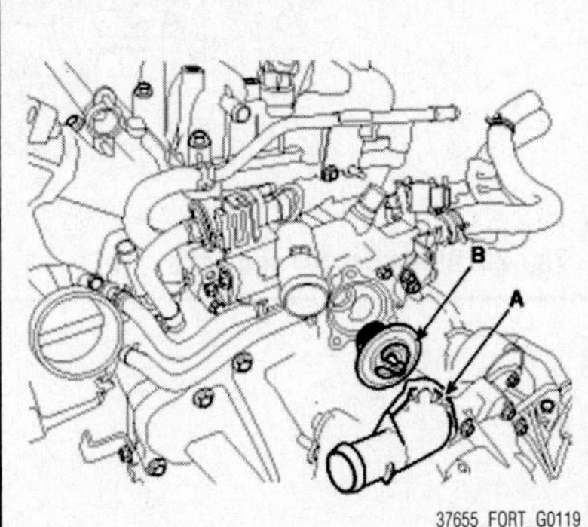

Fig. 27 Water inlet fitting (A) and thermostat (B)

WATER PUMP

REMOVAL & INSTALLATION

See Figure 28.

1. Loosen the drain plug, and then drain the engine coolant.

CAUTION

Never remove the radiator cap when the engine is hot. Serious scalding could be caused by hot fluid under high pressure escaping from the radiator.

2. Remove the drive belt.
3. Remove the water inlet pipe nut.
4. Remove the water pump (A) and the water pump gasket.
5. Installation is reverse order of removal with a new water pump gasket.
6. Tighten water pump bolts to 14–17 ft. lbs. (19–24 Nm).

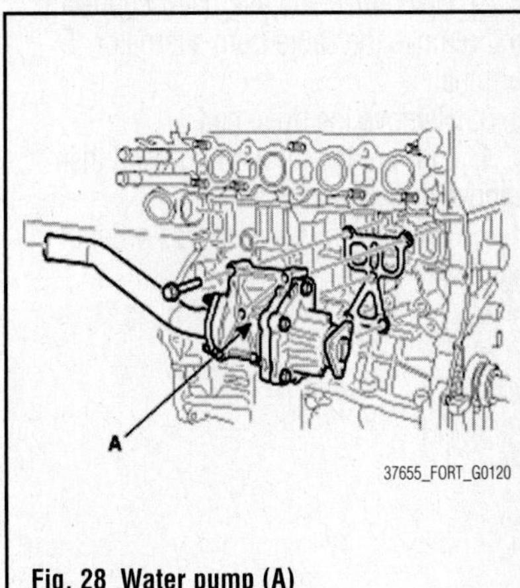

Fig. 28 Water pump (A)

BATTERY RECONNECT/RELEARN PROCEDURE

When reconnecting the battery cable after disconnecting, recharging battery after discharged or installing the memory fuse located on the driver's side panel after removing, be sure to reset the following systems.

Auto Up/Down Window

Whenever the battery is disconnected, discharged or the related fuse is replaced or reinstalled, reset the Auto up/down window system.

1. Turn the ignition switch to the ON position.
2. Pull up the power window switch in order that the window can close completely, and then keep pulling up the power switch for about 1 second.

Sunroof

Whenever the battery is disconnected, discharged or the related fuse is replaced or reinstalled, the sunroof system must be reset.

When To Initialize The Motor

- First operation the vehicle after manufacture.
- Initial value is erased or damaged because of short power electric discharge during operation
- After using the manual handle.
- After motor is replaced.

1. Check that the glass has been installed.
2. Push and hold the UP switch. The slide will move 5mm forward after 15 seconds.
3. After the slide has moved 5mm forward, release the switch, then push and hold the UP switch again.

➡If the motor initialization is successful, the sunroof should fully side open and close once.

4. When the sunroof is closed completely, release the switch to complete the initialization.

Trip Computer

When the battery is disconnected and reconnected, the set functions of the trip computer become initialized. You need to explain this to the customer.

Clock

Whenever the battery terminals or related fuses are disconnected, you must reset the time.

1. When the ignition switch is in the ACC or ON position, the clock buttons operate as follows:
   - HOUR: Pressing the "H" button will advance the time displayed by one hour.
   - MINUTE: Pressing the "M" button will advance the time displayed by one minute.
   - Display conversion: To change the 12 hour format to the 24 hour format, press the H and M button at the same time for more than 5 seconds.

Audio

When the battery is disconnected and reconnected, the customer's radio stations become initialized. So, you need to record the customer's radio stations prior to service, and after service, set the customer's radio stations into the audio.

ALTERNATOR

REMOVAL & INSTALLATION

See Figure 29.

1. Disconnect the battery negative terminal first, then the positive terminal.
2. Disconnect the alternator connector, and remove the cable from alternator "B" terminal.
3. Remove the drive belt.
4. Pull out the through bolt and then remove the alternator (A).
5. Installation is the reverse order of removal.

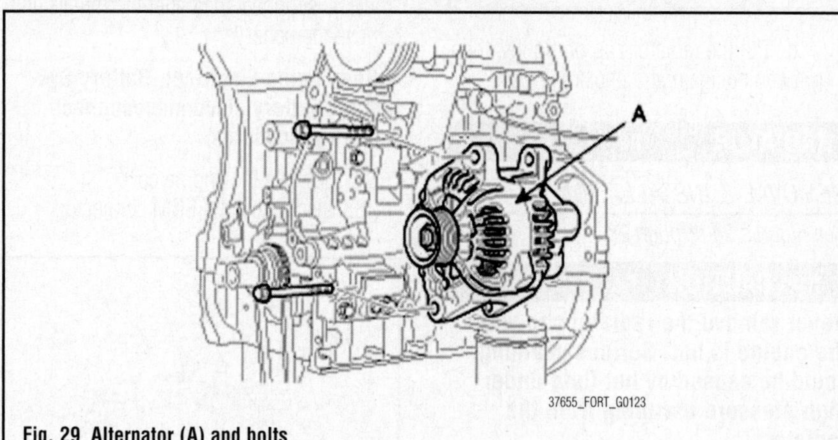

Fig. 29 Alternator (A) and bolts

ENGINE ELECTRICAL

IGNITION SYSTEM

FIRING ORDER

Ignition firing order is 1, 3, 4, 2.

IGNITION COIL

REMOVAL & INSTALLATION

See Figure 30.

1. Disconnect the ignition coil connectors (A), and then remove the ignition coils.
2. Installation is reverse order of removal.

IGNITION TIMING

ADJUSTMENT

Ignition timing is controlled by the electronic control ignition timing system.

The standard reference ignition timing data for the engine operating conditions are preprogrammed in the memory of the ECM (Engine Control Module).

The engine operating conditions (speed, load, warm–up condition, etc.) are detected by the various sensors. Based on these sensor signals and the ignition timing data, signals to interrupt the primary current are sent to the ECM. The ignition coil is activated, and timing is controlled.

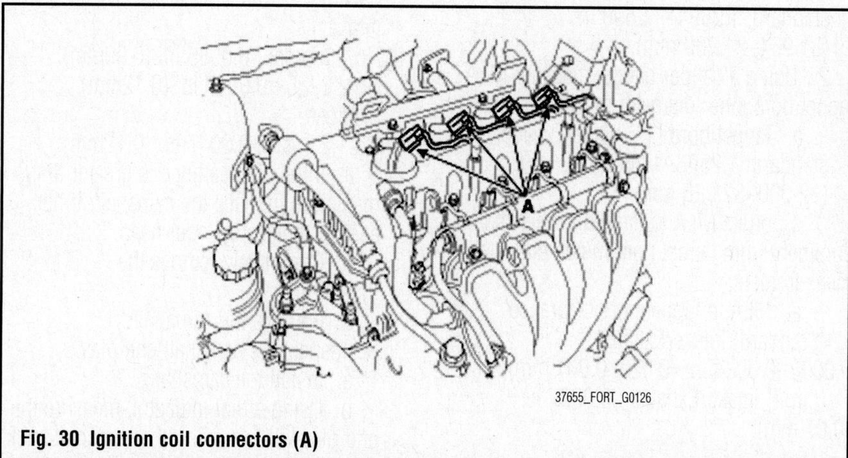

Fig. 30 Ignition coil connectors (A)

SPARK PLUGS

REMOVAL & INSTALLATION

See Figure 30.

1. Disconnect the ignition coil connectors (A), and then remove the ignition coils.
2. Using a spark plug socket, remove the spark plug.
3. Installation is reverse order of removal.

ENGINE ELECTRICAL

STARTING SYSTEM

STARTER

REMOVAL & INSTALLATION

See Figure 31.

1. Disconnect the battery negative cable.
2. Disconnect the starter cable (A) from the B terminal (B) on the solenoid (C), then disconnect the connector (D) from the S terminal (E).
3. Remove the 2 bolts holding the starter, then remove the starter.
4. Installation is the reverse of removal.
5. Connect the battery negative cable to the battery.

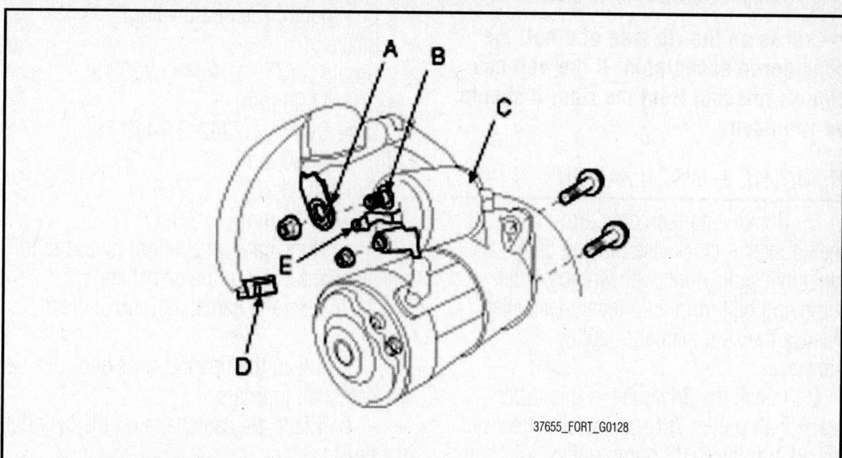

Fig. 31 Starter cable (A), B terminal (B), solenoid (C), connector (D) and S terminal (E)

ENGINE MECHANICAL

ACCESSORY DRIVE BELTS

ACCESSORY BELT ROUTING

See Figure 32.

Refer to the accompanying illustration.

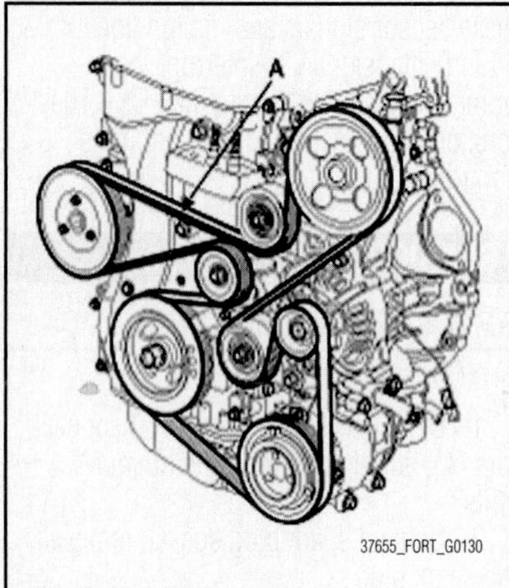

Fig. 32 Drive belt routing

INSPECTION

Visually check the belt for excessive wear, frayed cords etc. If any defect has been found, replace the drive belt.

➡Cracks on the rib side of a belt are considered acceptable. If the belt has chunks missing from the ribs, it should be replaced.

REMOVAL & INSTALLATION

1. Rotate auto tensioner arm in the counter – clockwise moving auto tensioner pulley bolt with wrench. After removing belt from auto tensioner pulley, release the auto tensioner pulley slowly.
2. Install the drive belt in this order: crankshaft pulley, A/C pulley, alternator pulley, idler pulley, P/S pump pulley, idler pulley, water pump pulley then tensioner pulley.
3. Rotate auto tensioner arm in the counter – clockwise moving auto tensioner pulley bolt with wrench. After putting belt on auto tensioner pulley, release the auto tensioner pulley slowly.

BALANCE SHAFT

REMOVAL & INSTALLATION

2.4L Engine

See Oil Pump.

CAMSHAFT AND VALVE LIFTERS

INSPECTION

Mechanical Lifter Assemblies (MLA)

1. Inspect the MLA.
   a. Using a micrometer, measure the MLA outside diameter.
   b. MLA O.D Intake/Exhaust standard: 1.2584–1.2590 in. (31.964–31.980 mm).
2. Using a caliper gauge, measure MLA tappet bore inner diameter of cylinder head.
   a. Tappet bore I.D. Intake/Exhaust standard: 1.2598–1.2608 in. (32.000–32.025 mm).
3. Subtract MLA outside diameter measurement from tappet bore inside diameter measurement.
   a. MLA to tappet bore clearance (Standard) Intake/Exhaust : 0.0008–0.0024 in. (0.020–0.061 mm).
   (Limit) Intake/Exhaust : 0.0027 in. (0.07 mm).

Camshaft

1. Inspect the cam lobes. Using a micrometer, measure the cam lobe height.
   Cam height (Standard value)
   Intake :
   Single CVVT : 1.7204–1.7283 in. (43.70–43.90 mm).
   Dual CVVT : 1.7362–1.7440 in. (44.10–44.30 mm).
   Exhaust : 1.7677–1.7756 in. (44.90–45.10 mm).
   a. If the cam lobe height is less than standard, replace the camshaft.
2. Inspect the camshaft journal clearance.
   a. Clean the bearing caps and camshaft journals.
   b. Place the camshafts on the cylinder head.
   c. Lay a strip of Plastigage®across each of the camshaft journal.
   d. Install the bearing caps.

⁂ CAUTION

Do not turn the camshaft.

   e. Remove the bearing caps.
   f. Measure the Plastigage®at its widest point.
   Bearing oil clearance (Standard value)
   Intake :
   No.1 journal : 0.00087–0.00224 in. (0.022–0.057 mm).
   No.2,3,4,5 : 0.00177–0.00323 in. (0.045–0.082 mm).
   Exhaust:
   Single CVVT
   No.1,2,3,4,5 : (0.00177–0.00323 in. (0.045–0.082 mm).
   Dual CVVT
   No.1 : 0–0.0012 in. (0–0.032 mm).
   No.2,3,4,5 : 0.00177–0.00323 in. (0.045–0.082 mm).
   (Limit)
   Intake:
   No.1 journal : 0.0035 in. (0.09mm).
   No.2,3,4,5 : 0.0047 in. (0.12mm).
   Exhaust :
   No.1,2,3,4,5 : 0.0047 in. (0.12mm).
   a. If the oil clearance is greater than maximum, replace the camshaft. If necessary, replace cylinder head.
   b. Completely remove the Plastigage®.
   c. Remove the camshafts.
3. Inspect the camshaft end play.
   a. Install the camshafts.
   b. Using a dial indicator, measure the end play while moving the camshaft back and forth.
   Camshaft end play
   (Standard value) : 0.0016–0.0062 in. (0.04–0.16 mm).
   (Limit) : 0.0071 in. (0.18 mm).
   a. If the end play is greater than maximum, replace the camshaft. If necessary, replace cylinder head.
   b. Remove the camshafts.

Exhaust Camshaft Bearing

See Figures 33 and 34.

1. Check the cylinder head bore mark.
   a. Mark indicates inside diameter of cylinder head bore for exhaust No. 1.

Fig. 33 Location of cylinder head bore mark

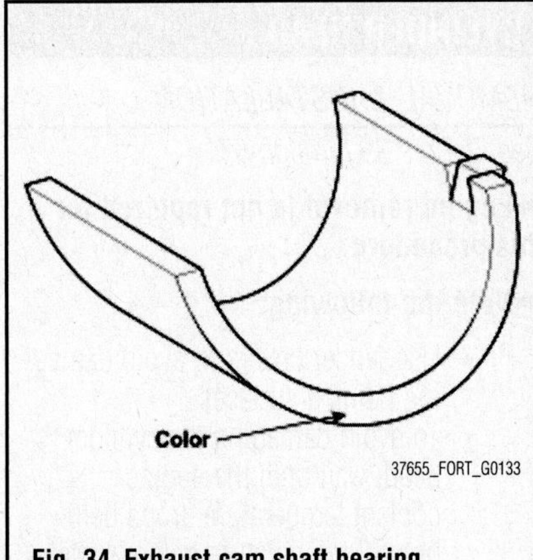

37655_FORT_G0133

Fig. 34 Exhaust cam shaft bearing identification mark

Mark A indicates (1.5748–1.5751 in. (40.000–40.008 mm).

Mark B indicates (1.5751–1.5754 in. (40.008–40.016 mm).

Mark C indicates (1.5754–1.5757 in. (40.016–40.024 mm).

2. Select camshaft bearing to match cylinder head bore. Color indicates size.

Cylinder head bore A uses Green bearing (1.996–2.000 in. (0.0785–0.0787 mm).

Cylinder head bore B uses No Color bearing (2.000–2.004 in. (0.0787–0.0788 mm).

Cylinder head bore C uses Black bearing (2.004–2.008 in. (0.0788–0.0790 mm).

Oil clearance should be: 0–0.0012 in. (0–0.032 mm).

CVVT Assembly

See Figures 35 and 36.

1. Inspect CVVT assembly.
   a. Check that the CVVT assembly will not turn.
   b. Apply vinyl tape to the retard hole EXCEPT the one indicated in the figure.
   c. Using an air gun, apply approx. 21psi (150kpa) to the retard hole of the camshaft. This is in order to release the lock pin.

➡Oil will be pushed from CVVT.

   d. With air applied, turn the CVVT assembly to the advance position as shown with your hand.

➡Depending on the air pressure, the CVVT assembly will turn to the advance position without applying force by hand. If there is too much air leakage from the port, the lock pin may not be released.

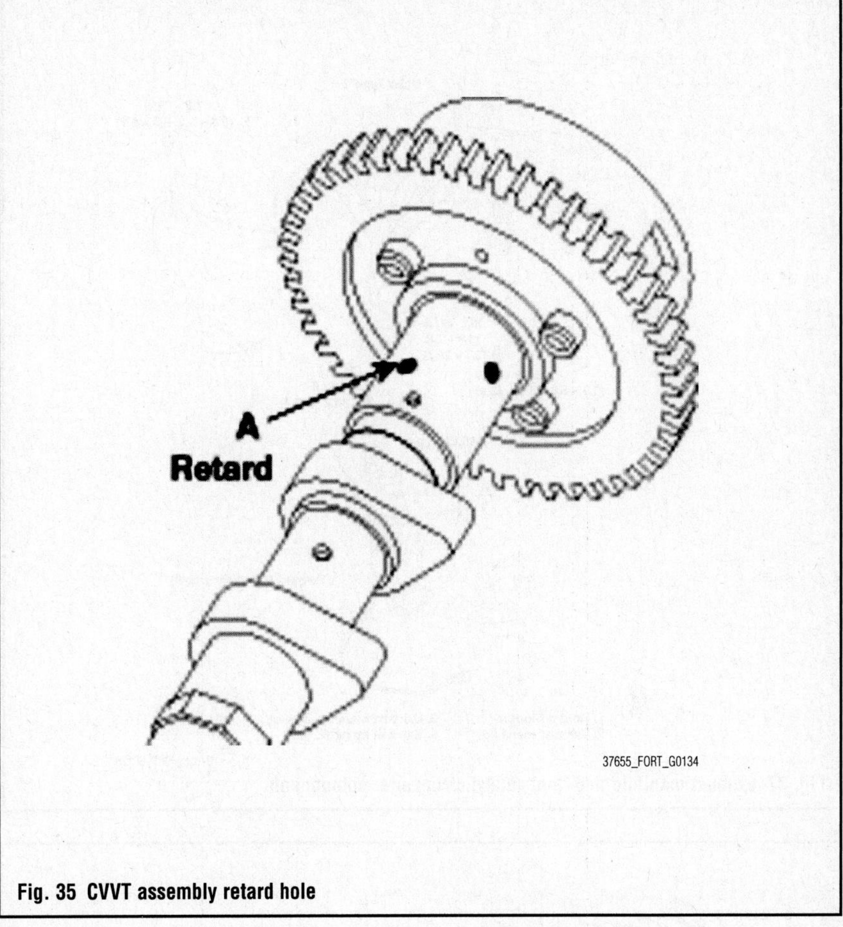

37655_FORT_G0134

Fig. 35 CVVT assembly retard hole

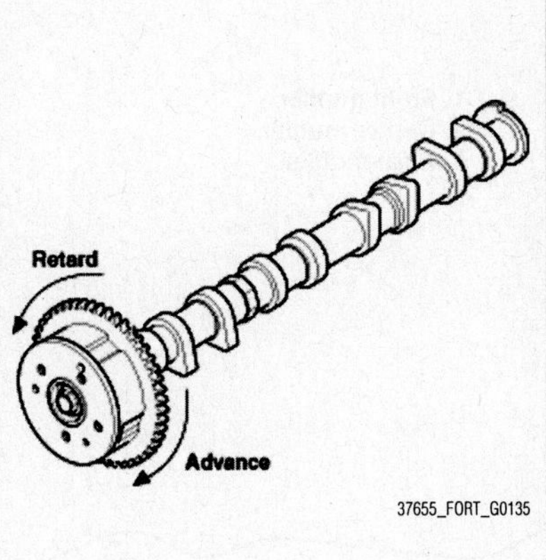

37655_FORT_G0135

Fig. 36 CVVT assembly advance and retard positions

   e. Once the lock pin is released, turn the CVVT assembly back and forth and check the movable range and smooth operation. It should move smoothly in a range from about 22.5° (Intake) / 20.0° (Exhaust)
   f. Turn the CVVT assembly with your hand and lock it at the maximum delay angle position (counter clockwise).

REMOVAL & INSTALLATION

See Cylinder Head.

CATALYTIC CONVERTER

REMOVAL & INSTALLATION

See Figures 37 and 38.

1. Remove exhaust manifold (SULEV only). See Exhaust Manifold.
2. Remove exhaust system.
3. Installation is reverse order of removal.

CRANKSHAFT DAMPER

REMOVAL & INSTALLATION

No separate procedure is available. See Timing Chain Cover.

CRANKSHAFT FRONT SEAL

REMOVAL & INSTALLATION

No separate procedure is available. See Timing Chain Cover.

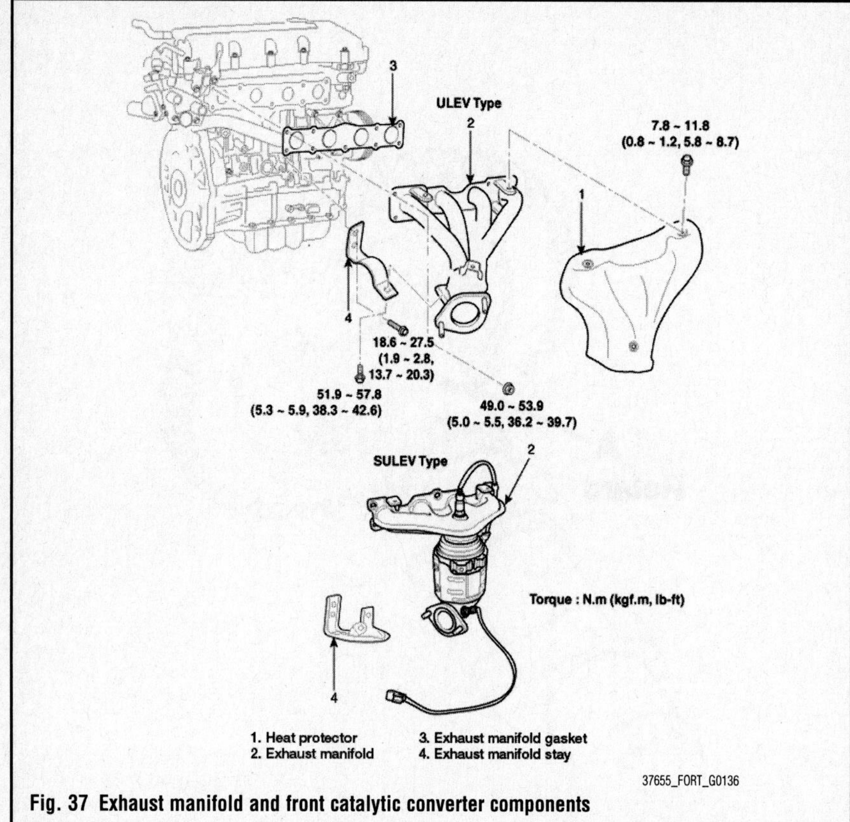

Fig. 37 Exhaust manifold and front catalytic converter components

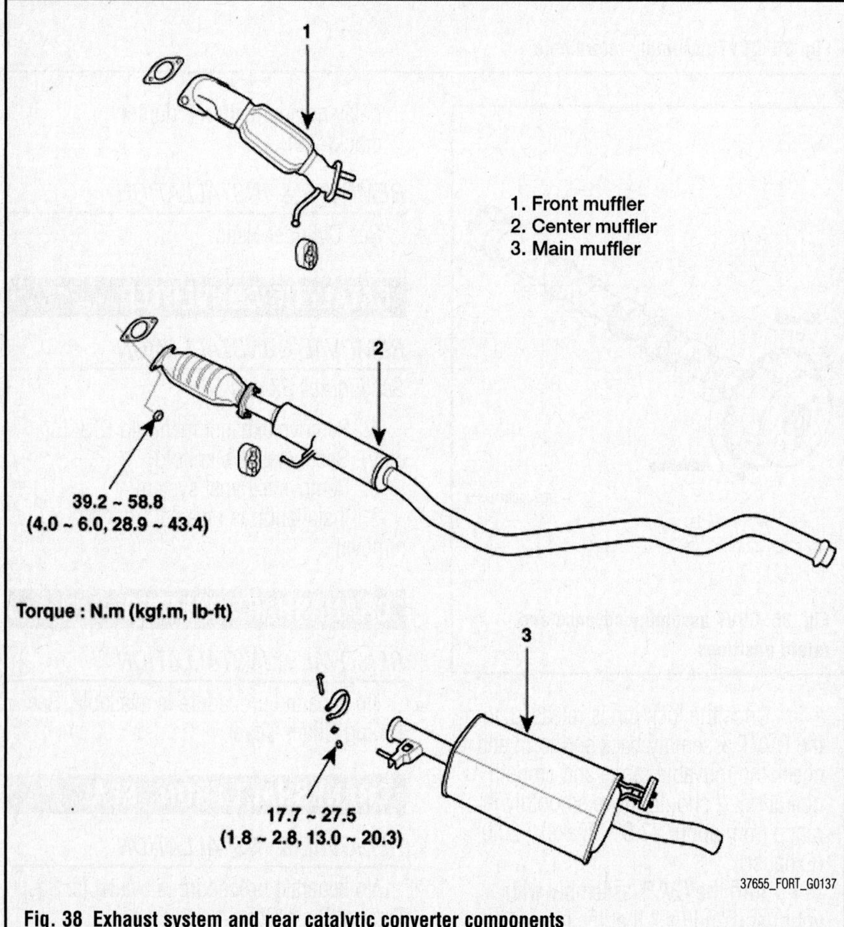

Fig. 38 Exhaust system and rear catalytic converter components

CYLINDER HEAD

REMOVAL & INSTALLATION

See Figures 39 through 57.

➡Engine removal is not required for this procedure.

➡Note the following:

- Use fender covers to avoid damaging painted surfaces.
- To avoid damaging the cylinder head, wait until the engine coolant temperature drops below normal temperature before removing it.
- When handling a metal gasket, take care not to fold the gasket or damage the contact surface of the gasket.
- To avoid damage, unplug the wiring connectors carefully while holding the connector portion.
- Mark all wiring and hoses to avoid misconnection.

1. Disconnect the battery terminals (A).
2. Remove the engine cover (B).
3. Disconnect the ECM connector (C).
4. Disconnect the breather hose and then, remove the air duct (D) and air cleaner assembly (E).
5. Raise and support vehicle. Remove the under covers.
6. Loosen the drain plug, and then drain the engine coolant.
7. Remove the radiator upper hose (A) and lower hose (B).
8. Remove the heater hoses (C).
9. Disconnect the brake booster vacuum hose (D).
10. Disconnect the wiring connectors and harness clamps from the engine.
    a. Disconnect the ETC connector (A) and knock sensor connector (B).
    b. Disconnect the PCSV connector (C).
    c. Disconnect the ECT connector (D).
    d. Disconnect the condenser connector (E).
    e. Disconnect the CKP sensor connector (F).
    f. Disconnect the oxygen sensor connector (G).
    g. Disconnect the VCM connector and MAP sensor connector. (SULEV type only)
    h. Disconnect the power steering fluid pressure switch connector (A).
    i. Disconnect the MAP sensor connector (B).
    j. Disconnect the OPS connector (C).

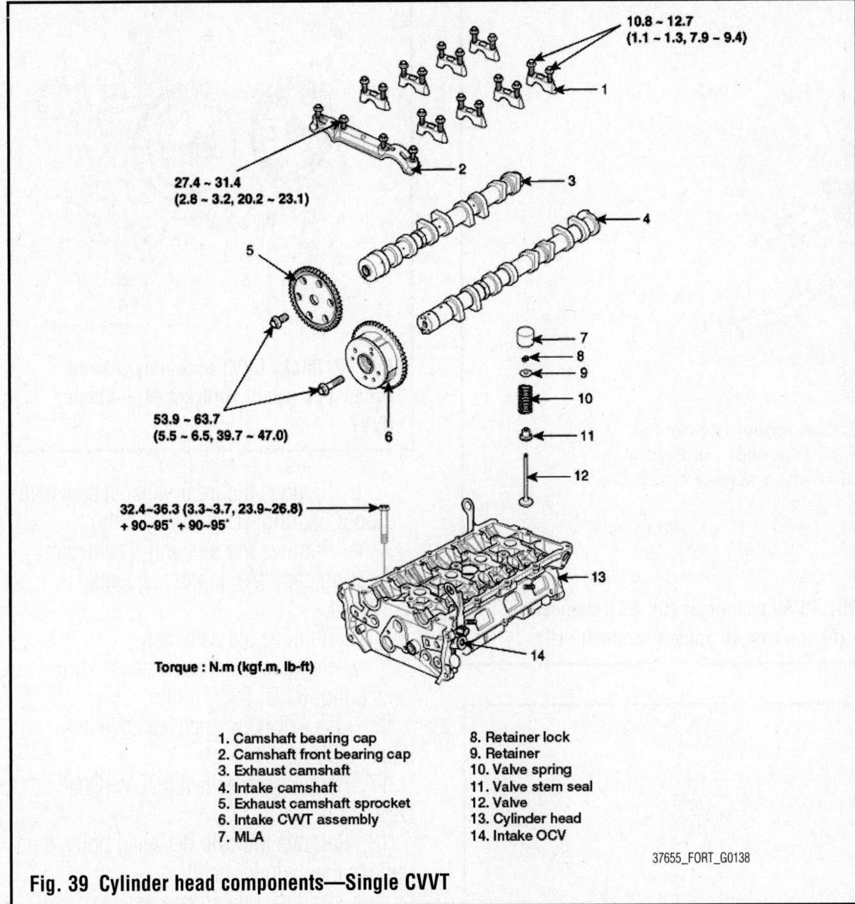

Fig. 39 Cylinder head components—Single CVVT

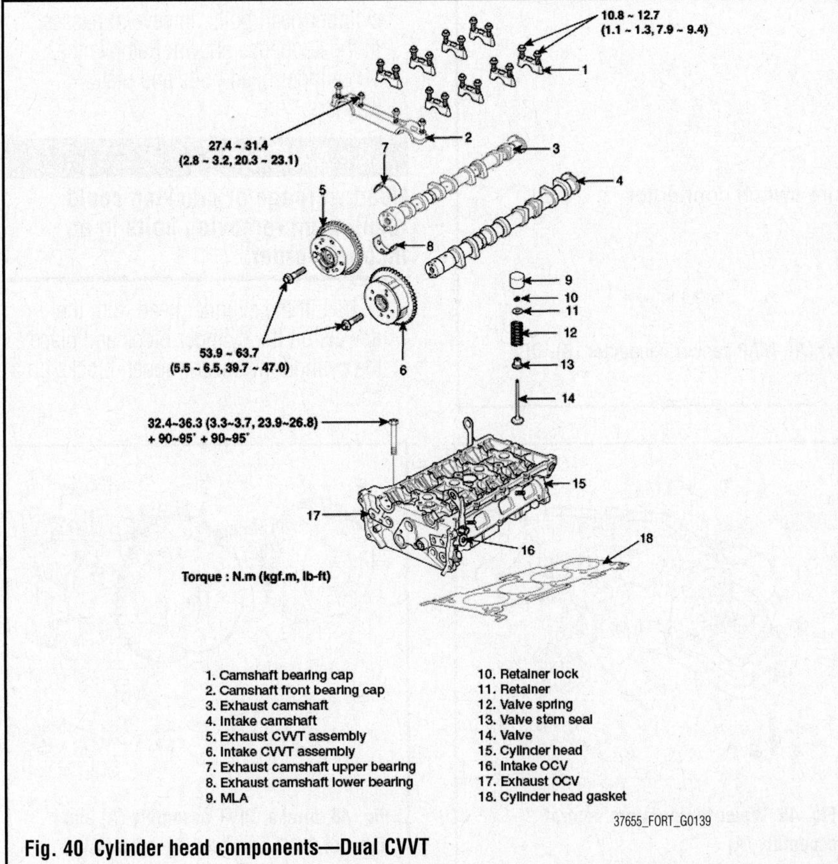

Fig. 40 Cylinder head components—Dual CVVT

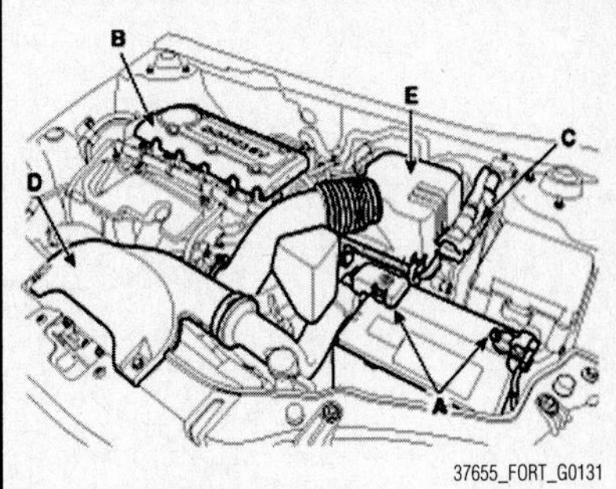

Fig. 41 Battery terminals (A), engine cover (B), ECM connector (C), air duct (D) and air cleaner assembly (E).

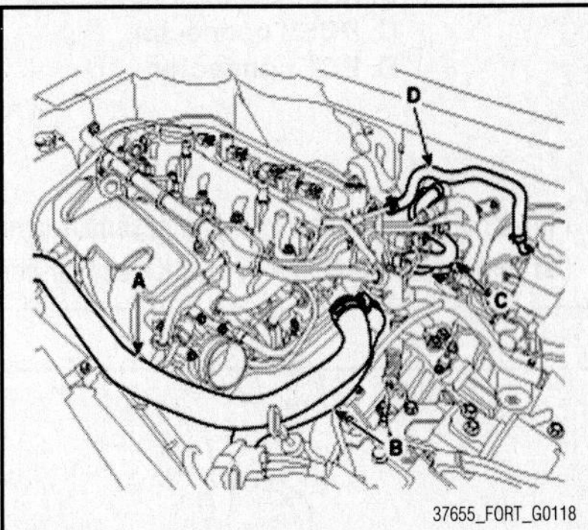

Fig. 42 Radiator upper hose (A), lower hose (B), heater hoses (C), and brake booster vacuum hose (D)

k. Disconnect the alternator connector (D) and 'B' terminal cable from the alternator.

l. Disconnect the A/C switch connector from the compressor.

m. Disconnect the VIS connector.

n. Disconnect the intake OCV connector (A).

o. Disconnect the CMP sensor connector (A) and fuel hose (B).

p. Disconnect the injector connectors (A) and ignition coil connectors (B).

11. Remove timing chain. See Timing Chain.

12. Remove the intake and exhaust manifold. See Intake Manifold and Exhaust Manifold.

13. Remove the water temperature control assembly (A).

14. Remove the intake CVVT assembly (A) and exhaust CVVT sprocket or camshaft sprocket (B).

➡When removing the sprocket bolt or CVVT assembly bolt, hold the camshaft by wrench at position A.

15. Remove the camshaft.

a. Remove the front camshaft bearing cap.

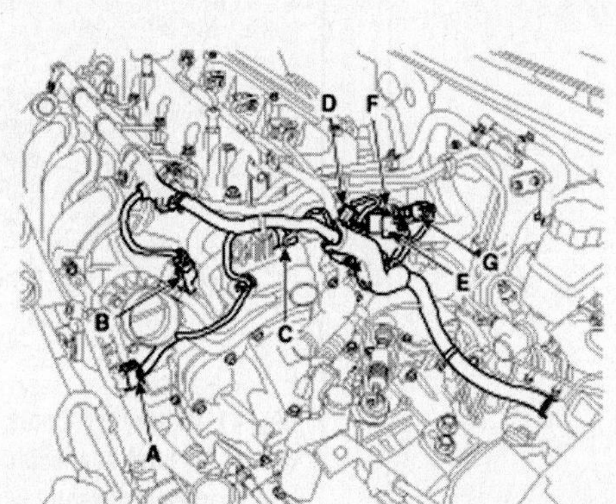

Fig. 43 ETC connector (A), knock sensor connector (B), PCSV connector (C), ECT connector (D), condenser connector (E), CKP sensor connector (F) and oxygen sensor connector (G)

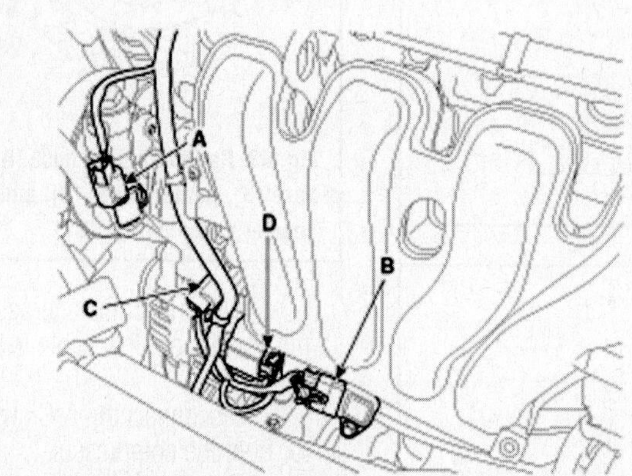

Fig. 44 Power steering fluid pressure switch connector (A), MAP sensor connector (B), OPS connector (C) and alternator connector (D)

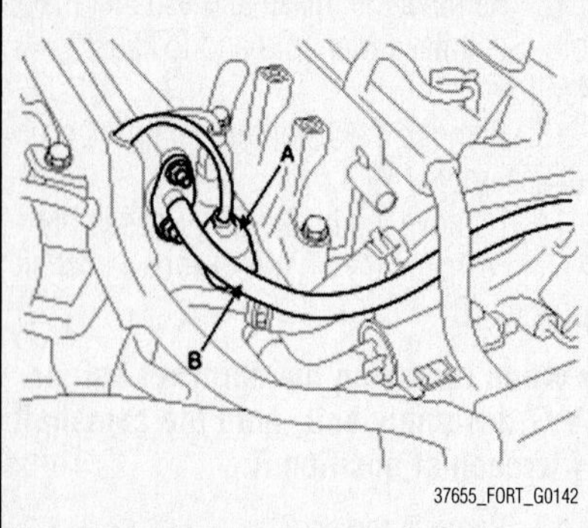

Fig. 45 CMP sensor connector (A) and fuel hose (B)

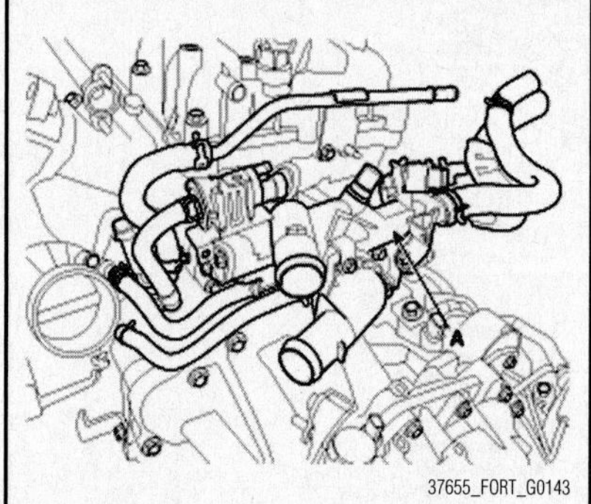

Fig. 46 Water temperature control assembly (A)

Fig. 47 Intake CVVT assembly (A) and exhaust camshaft sprocket (B)—Single CVVT

b. Remove the front exhaust camshaft upper bearing. (Dual CVVT only)

c. Remove the remaining camshaft bearing caps (A), in the sequence shown.

d. Remove the camshafts.

e. Remove the front exhaust lower bearing. (Dual CVVT only)

16. Use a Torx®wrench, remove the intake OCV.

17. Remove the exhaust OCV. (Dual CVVT only)

18. Remove the cylinder head bolts, then remove the cylinder head.

a. Using triple square wrench, uniformly loosen and remove the 10 cylinder head bolts, in several passes, in the sequence shown. Remove the 10 cylinder head bolts and plate washers.

✲✲ CAUTION

Head warpage or cracking could result from removing bolts in an incorrect order.

b. Lift the cylinder head from the dowels on the cylinder block and place the cylinder head on wooden blocks on a bench.

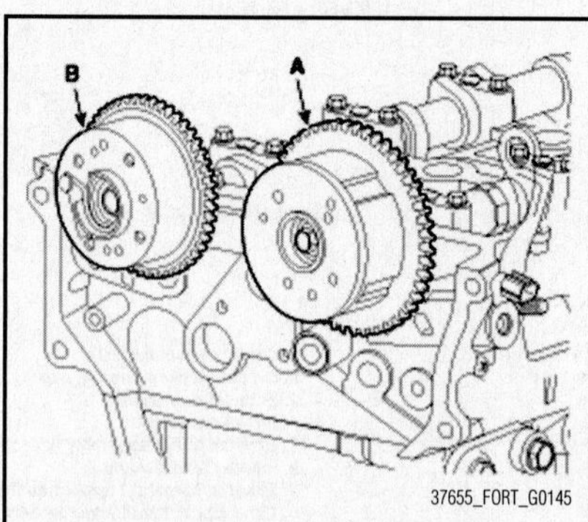

Fig. 48 Intake CVVT assembly (A) and exhaust CVVT sprocket (B)—Dual CVVT

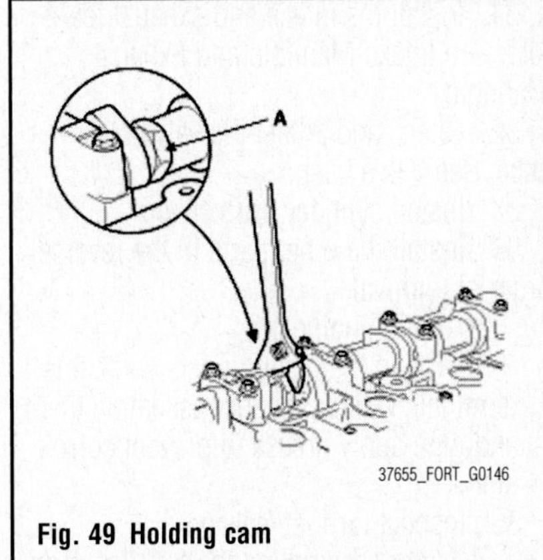

Fig. 49 Holding cam

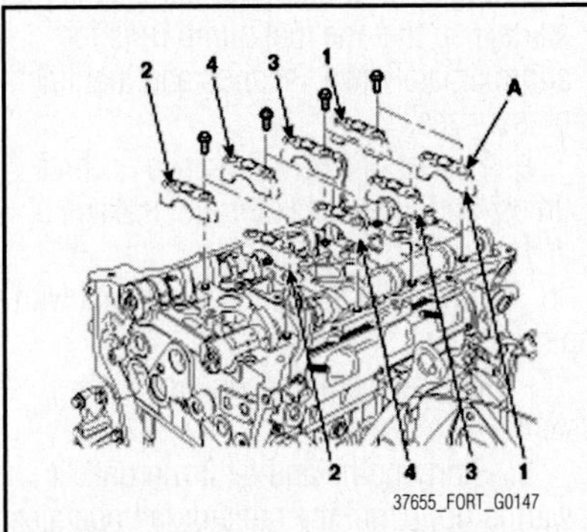

Fig. 50 Camshaft bearing caps (A) removal sequence

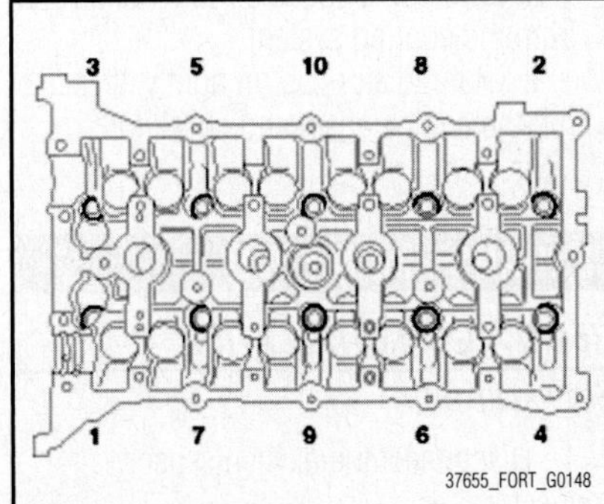

Fig. 51 Cylinder head bolt removal sequence

CAUTION

Be careful not to damage the contact surfaces of the cylinder head and cylinder block.

19. Remove the cylinder head gasket.

To install:

➡Note the following:

- Thoroughly clean all parts to be assembled.
- Always use a new head and manifold gasket.

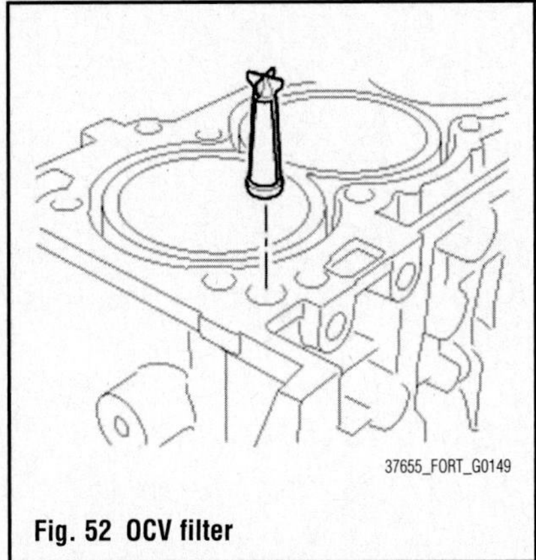

Fig. 52 OCV filter

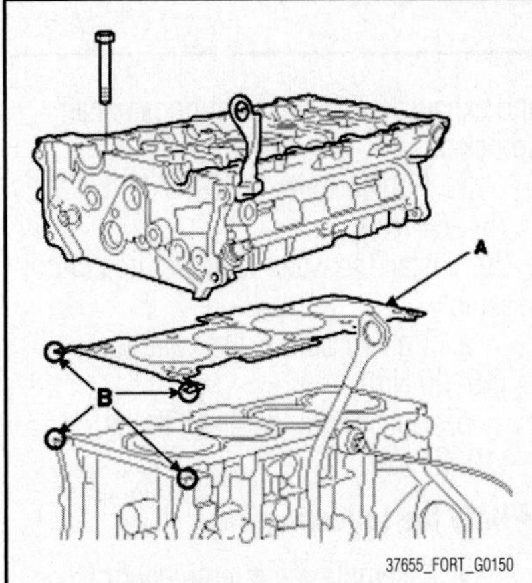

Fig. 53 Cylinder head gasket (A) and liquid gasket locations (B)

- The cylinder head gasket is a metal gasket. Take care not to bend it.
- Rotate the crankshaft, set the No.1 piston at TDC.

20. Install the OCV filter.

➡Keep the OCV filter clean.

21. Install the cylinder head gasket (A) on the cylinder block.

➡Note the following:

- Be careful of the installation direction.
- Apply liquid gasket (Loctite 5900H) on the edge of cylinder head gasket on both sides. (At position 'B')
- After applying sealant, assemble the cylinder head in five minutes.

22. Place the cylinder head carefully in order not to damage the gasket with the bottom part of the end.

CAUTION

Always use new cylinder head bolt.

23. Install cylinder head bolts.
   a. Apply a light coat if engine oil on the threads and under the heads of the cylinder head bolts.
   b. Using the SST (09221–4A000), tighten the cylinder head bolts and plate washers, in several passes, in the sequence shown.
   c. Tighten bolts to 24–27 ft. lbs. (32–36 Nm), then 90–95 degrees, then an additional 90–95 degrees.
24. Install the intake OCV (A).
   a. Tighten bolts to 7–9 ft. lbs. (10–12 Nm).
25. Install the exhaust OCV (A). (Dual CVVT only)
   a. Tighten bolts to 7–9 ft. lbs. (10–12 Nm).

➡Note the following:

- Do not reuse the OCV if dropped.
- Keep the OCV filter clean.
- Do not hold the OCV sleeve during servicing.
- When the OCV is installed on the engine, do not move the engine by holding the OCV yoke.

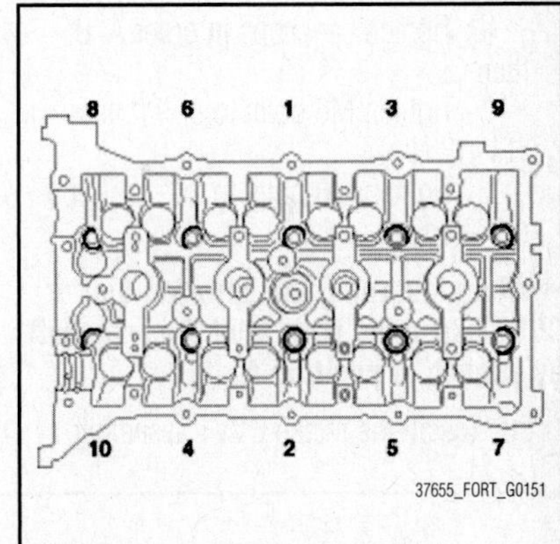

Fig. 54 Cylinder head bolt torque sequence

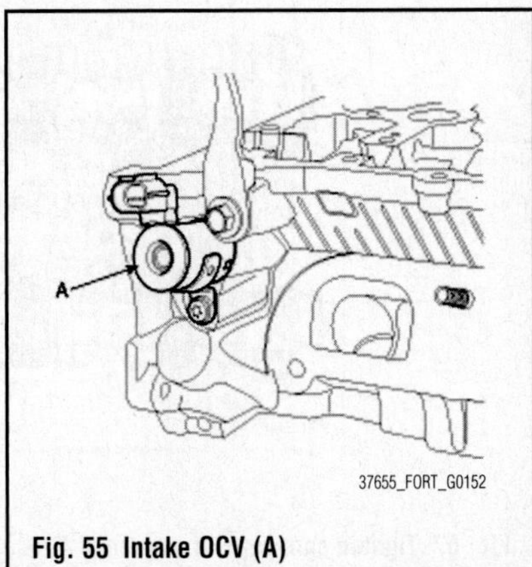

Fig. 55 Intake OCV (A)

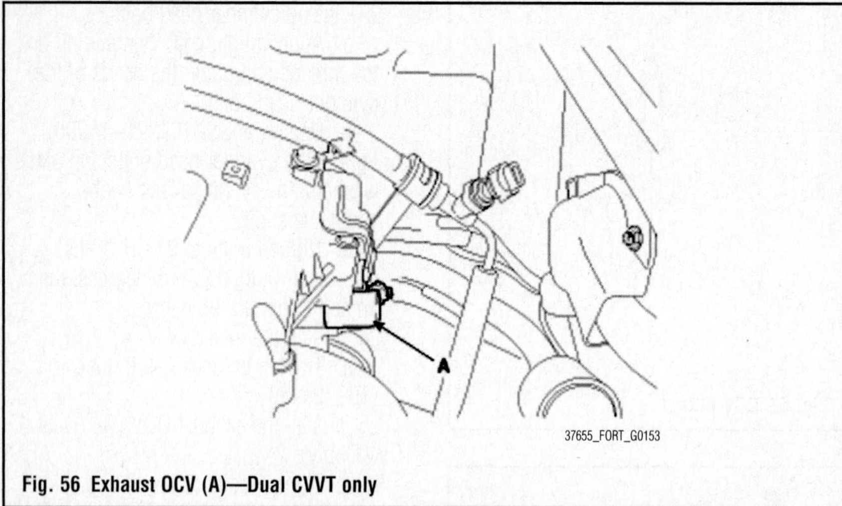

Fig. 56 Exhaust OCV (A)—Dual CVVT only

26. Install the camshafts.
   a. Apply a light coat of engine oil on camshaft journals.
   b. Install the front exhaust camshaft lower bearing. (Dual CVVT only)
   c. Install the camshafts.
   d. Install the front exhaust camshaft upper bearing to the front bearing cap. (Dual CVVT only)
27. Install camshaft bearing caps in their proper locations.
   a. Tighten cam caps in order A, B, then C.
   b. Tighten M6 bolts to 8–9 ft. lbs. (11–13 Nm).
   c. Tighten M8 bolts to 20–23 ft. lbs. (27–31 Nm).

➡When installing the sprocket bolt or CVVT assembly bolt, hold the camshaft by wrench at position A.

28. Install the intake CVVT assembly and exhaust CVVT sprocket or camshaft sprocket.
   a. Tighten bolts to 40–47 ft. lbs. (54–64 Nm).
29. Install the water temperature control assembly.
   a. Tighten bolt to 11–15 ft. lbs. (15–20 Nm).
   b. Tighten nut to 14–17 ft. lbs. (19–24 Nm).

➡Note the following:

- Assemble water temp control assembly and water inlet pipe to water pump assembly before nuts for assembling of water inlet pipe are tightened.
- Insert after wetting O–ring or inner surface of thermostat housing.
- Always use a new O–ring.

30. Install the timing chain. See Timing Chain.
31. Install the intake and exhaust manifold. See Intake Manifold and Exhaust Manifold.
32. Check and adjust the valve clearance. See Valve Lash.
33. Install cylinder head cover.
34. Install the other parts in the reverse order of removal.
   a. Refill engine oil.
   b. Clean the battery posts and cable terminals with sandpaper assemble them, and then apply grease to prevent corrosion.
35. Inspect for fuel leakage.
   a. After assembling the fuel line, turn on the ignition switch (do not operate the starter) so that the fuel pump runs for approximately two seconds and fuel line pressurizes.
   b. Repeat this operation two or three times, and then check for fuel leakage at any point in the fuel lines.
36. Refill radiator and reservoir tank with engine coolant.
37. Bleed air from the cooling system.
   a. Start engine and let it run until it warms up. (Until the radiator fan operates 3 or 4 times.)
   b. Turn off the engine. Check the level in the radiator, add coolant if needed. This will allow trapped air to be removed from the cooling system.
   c. Put radiator cap on tightly, then run the engine again and check for leaks.

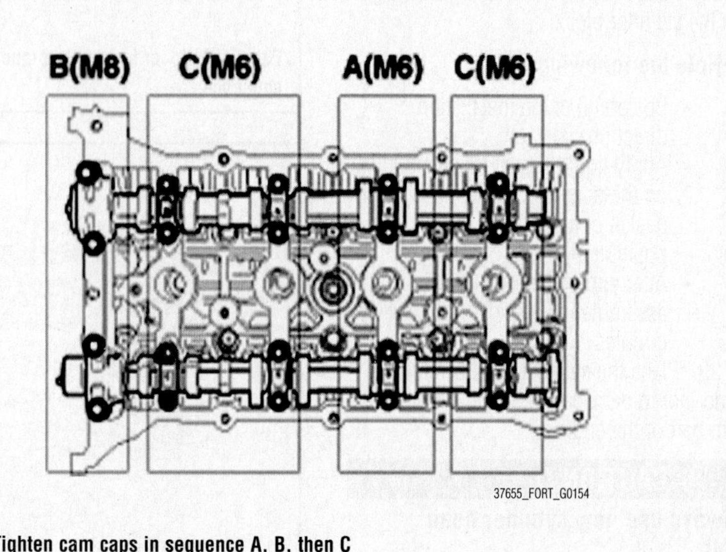

Fig. 57 Tighten cam caps in sequence A, B, then C

EXHAUST MANIFOLD

REMOVAL & INSTALLATION

See Figure 58.

1. Disconnect the oxygen sensor connector.
2. Remove the front muffler.
3. Remove the heat shield.
4. Remove the exhaust manifold stay.
5. Remove the exhaust manifold.
6. Installation is reverse order of removal.
   a. Tighten the exhaust manifold bolts to 36–40 ft. lbs. (49–54 Nm).
   b. Tighten the exhaust manifold stay M8 bolts to 14–20 ft. lbs. (19–28 Nm).
   c. Tighten the exhaust manifold stay M10 bolts to 38–43 ft. lbs. (52–58 Nm).
   d. Tighten the heat shield bolts to 6–9 ft. lbs. (8–12 Nm).
   e. Tighten the front muffler bolts to 29–43 ft. lbs. (39–59 Nm).

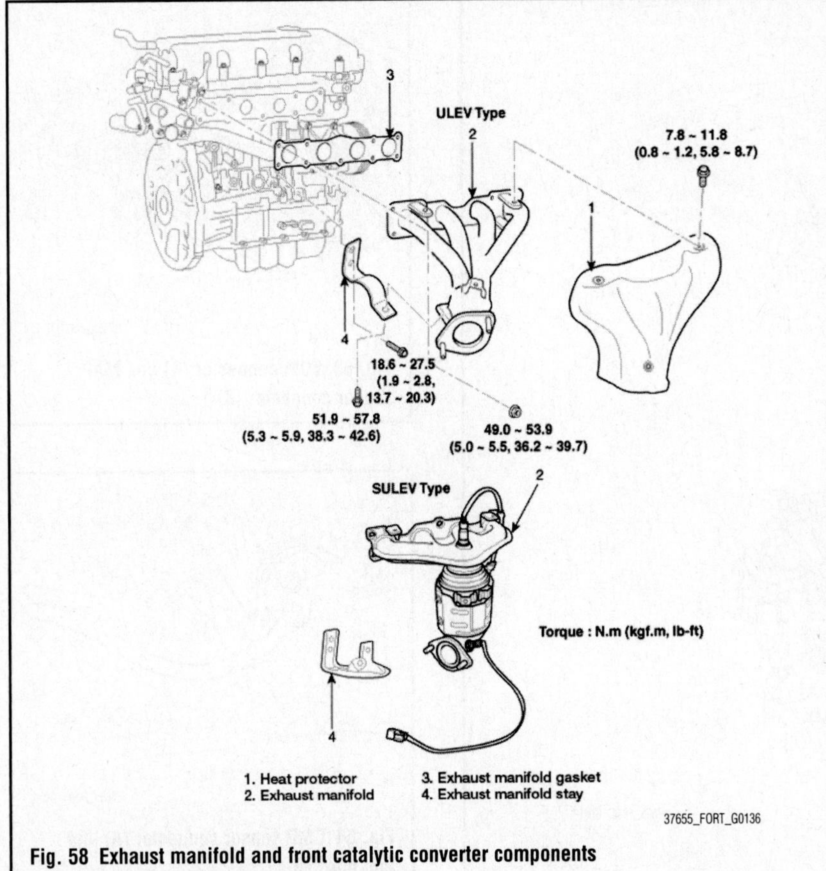

Fig. 58 Exhaust manifold and front catalytic converter components

FLYWHEEL/FLEXPLATE

REMOVAL & INSTALLATION

See Figure 59.

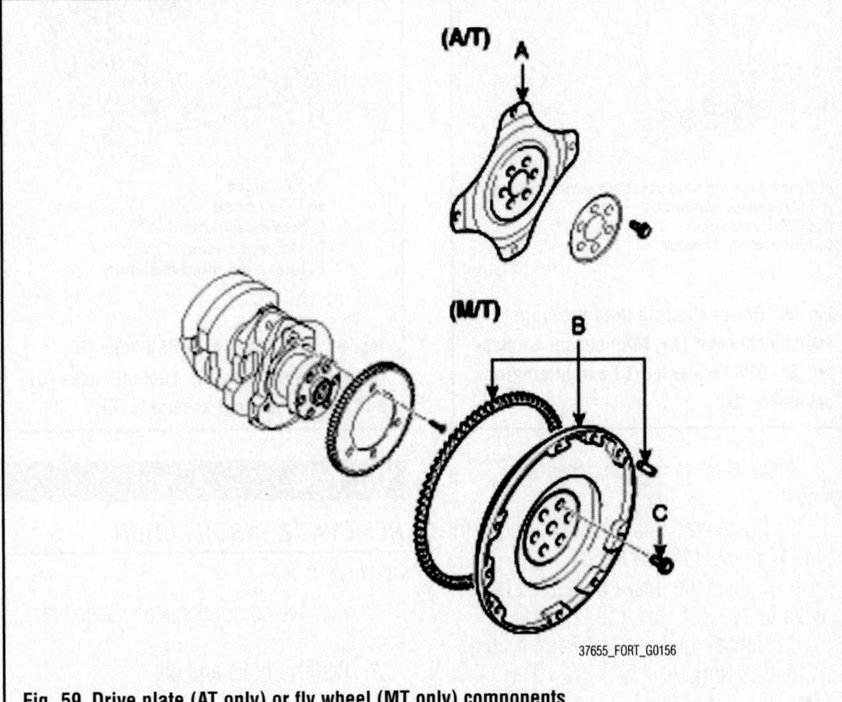

Fig. 59 Drive plate (AT only) or fly wheel (MT only) components

1. Remove the transaxle assembly.
2. Remove the clutch disc and pressure plate (MT only).
3. Remove the drive plate (AT only) or fly wheel (MT only).

To install:

➡Note the following:

- Always use new flywheel (drive plate) bolts.
- Apply Three bond 2403, Loctite 200 or 204 sealant to the screw part (8mm from the end of the bolt).
- Install and uniformly tighten the 7 bolts, in several passes.

4. Install the drive plate (AT only) or fly wheel (MT only).
   a. Tighten the bolts to 87–94 ft. lbs. (118–128 Nm).
5. Install the clutch disc and pressure plate (MT only).
6. Install the transaxle assembly.

INTAKE MANIFOLD

REMOVAL & INSTALLATION

See Figures 60 through 65.

1. Disconnect the battery terminals and remove the battery.
2. Remove the engine cover.
3. Disconnect the ECM connector.
4. Disconnect the breather hose and then, remove the air duct and air cleaner assembly.
5. Disconnect the wiring connectors and harness clamps from the intake manifold.
   a. Disconnect the ETC connector (A) and knock sensor connector (B).
   b. Disconnect the injector connectors (C).
   c. Remove the wiring harness mounting bolts (D).
   d. Disconnect the power steering fluid pressure switch connector (A).
   e. Disconnect the MAP sensor connector (B).
   f. Disconnect the OPS connector (C).
   g. Disconnect the alternator connector (D).
   h. Disconnect the A/C switch connector from the compressor.
   i. Disconnect the VIS connector.
   j. Disconnect the VCM connector (A) and MAP sensor connector (B).
   k. Disconnect the intake OCV connector.
   l. Disconnect the CMP sensor connector (A) and fuel hose (B).
6. Remove the intake manifold stay.

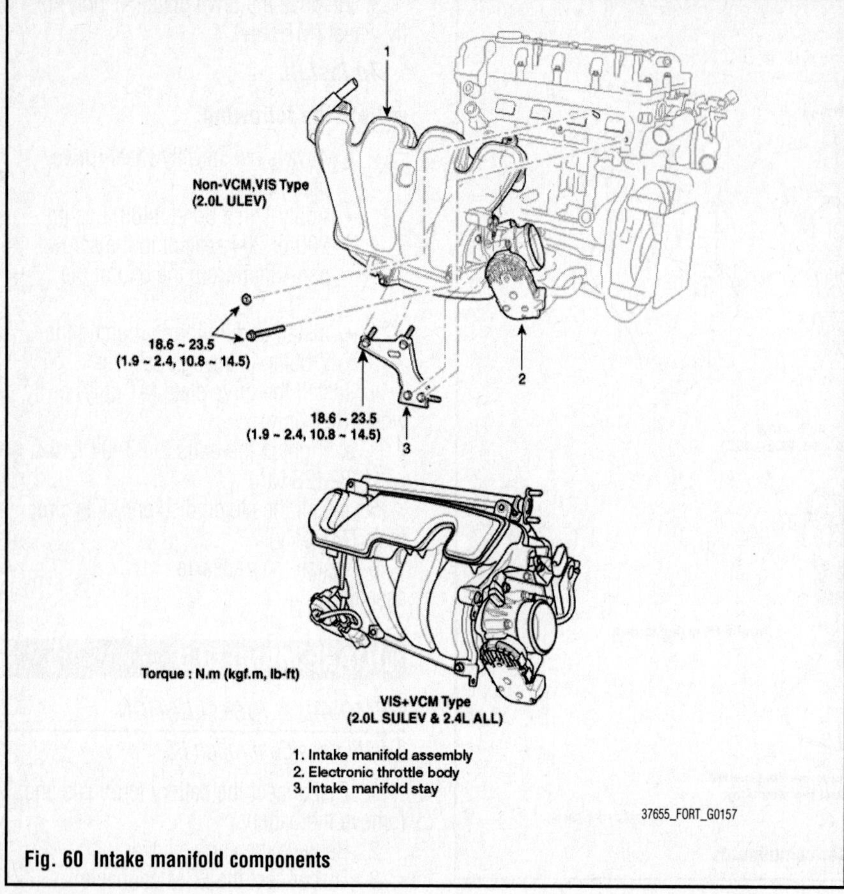

Fig. 60 Intake manifold components

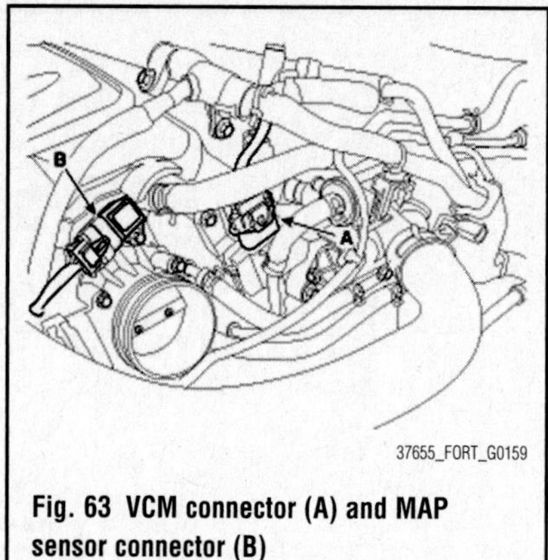

Fig. 63 VCM connector (A) and MAP sensor connector (B)

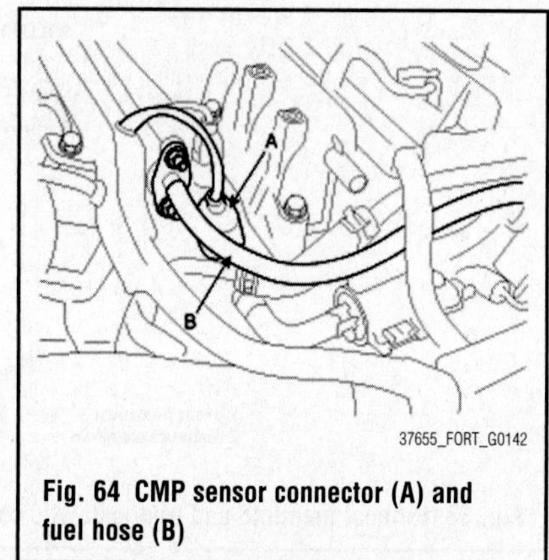

Fig. 64 CMP sensor connector (A) and fuel hose (B)

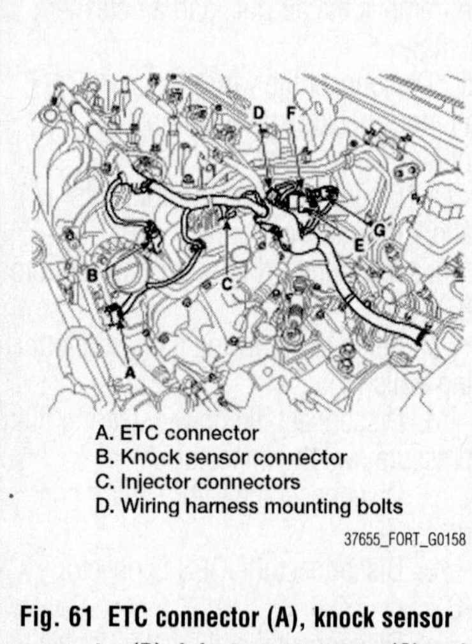

Fig. 61 ETC connector (A), knock sensor connector (B), injector connectors (C) and wiring harness mounting bolts (D)

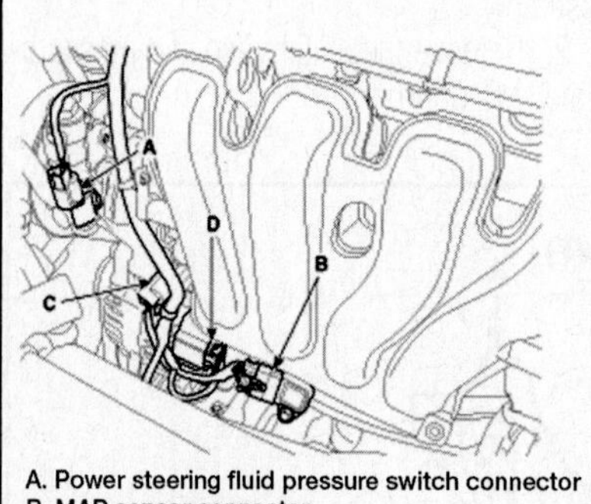

Fig. 62 Power steering fluid pressure switch connector (A), MAP sensor connector (B), OPS connector (C) and alternator connector (D)

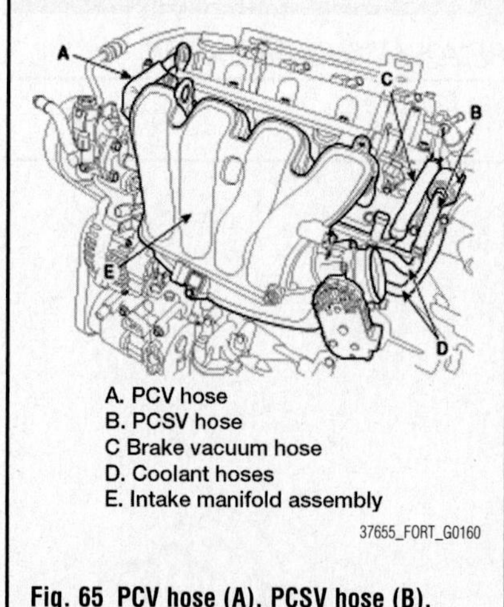

Fig. 65 PCV hose (A), PCSV hose (B), brake vacuum hose (C), coolant hoses (D) and intake manifold assembly (E).

7. Remove the intake manifold assembly.
   (1) Disconnect the PCV hose (A).
   (2) Disconnect the PCSV hose (B) and the brake vacuum hose (C).
   (3) Disconnect the coolant hoses (D) from the throttle body.
   (4) Remove the intake manifold assembly (E).

8. Installation is reverse order of removal.
   a. Tighten the intake manifold bolts to 14–17 ft. lbs. (19–24 Nm).
   b. Tighten the intake manifold stay bolts to 14–17 ft. lbs. (19–24 Nm).
   c. Tighten the bolts to 87–94 ft. lbs. (118–128 Nm).

OIL PAN

REMOVAL & INSTALLATION

See Figure 66.

1. Remove components as needed for access.
2. Remove bolts and oil pan.

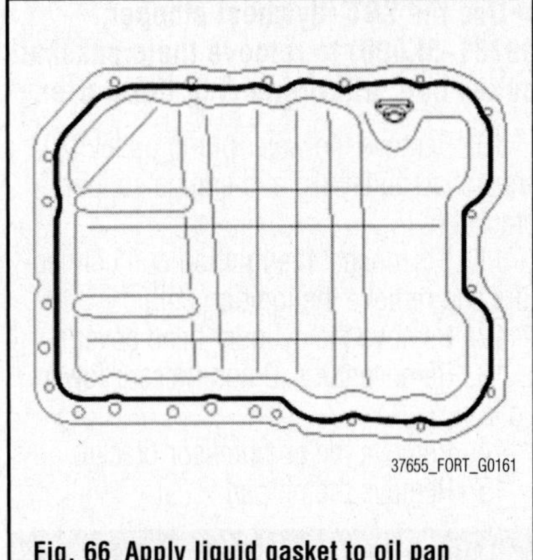

Fig. 66 Apply liquid gasket to oil pan

To install:

3. Using a razor blade and gasket scraper, remove all the old gasket material from the gasket surfaces.

➡Check that the mating surfaces are clean and dry before applying liquid gasket.

4. Apply liquid gasket LOCTITE 5900H or THREEBOND 1217H equivalent (MS721–40) as an even bead, centered between the edges of the mating surface.

➡Note the following:

- To prevent leakage of oil, apply liquid gasket to the inner threads of the bolt holes.
- Do not install the parts if five minutes or more have elapsed since applying the liquid gasket. Instead, reapply liquid gasket after removing the residue.
- After assembly, wait at least 30 minutes before filling the engine with oil.

5. Install the oil pan.
   a. Uniformly tighten the bolts in several passes.
   b. Tighten M6 bolts to 7–9 ft. lbs. (10–12 Nm).
   c. Tighten M8 bolts to 20–22 ft. lbs. (27–30 Nm).

OIL PUMP

REMOVAL & INSTALLATION

2.0L Engine

See Figure 67.

1. Remove the timing chain. (Refer to timing system in this group)
2. Remove the oil pump mechanical tensioner (B).
3. Remove the oil pump chain guide (D).

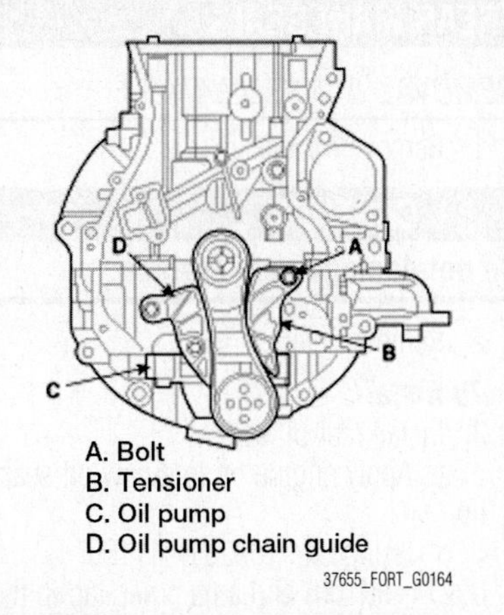

Fig. 67 Oil pump components—2.0L engine

4. Remove the bolts (A), oil pump (C) and oil pump chain.

To install:

5. The key of crankshaft should be aligned with the mating face of main bearing cap. As a result of this, the piston of No.1 cylinder is placed at the top dead center on compression stroke.
6. Assemble the crankshaft sprocket on the crankshaft with the front mark on the crankshaft sprocket outward.
7. Tighten the oil pump tensioner bolt after placing the tensioner spring on the dowel pin located in ladder frame, and then insert stopper pin to fix the tensioner.
   a. Tighten bolt to 7–9 ft. lbs. (9–12 Nm).
8. Assemble the oil pump chain on the crankshaft sprocket.
9. Install the oil pump assembly to engine while assembling chain on sprocket.
   a. Tighten bolts to 17–20 ft. lbs. (24–27 Nm).
10. Install the oil pump chain guide then remove the stopper pin.
    a. Tighten bolts to 7–9 ft. lbs. (9–12 Nm).
11. Install the timing chain.

2.4L Engine

See Figures 68 through 70.

1. Remove the timing chain. See Timing Chain.
2. Install a set pin after compressing the balance shaft chain tensioner.
3. Remove the balance shaft chain hydraulic tensioner (A).
4. Remove the balance shaft chain tensioner arm (B).

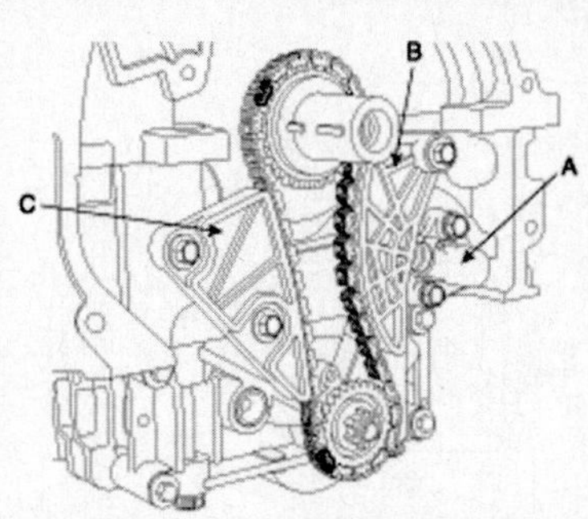

Fig. 68 Oil pump components—2.4L engine

5. Remove the balance shaft chain guide (C).
6. Remove the oil pump & balance shaft module and balance shaft chain.

⁂ CAUTION

Do not disassemble the oil pump & balance shaft module.

To install:

7. The key of crankshaft should be aligned with the mating face of main bearing cap. As a result of this, the piston of No.1 cylinder is placed at the top dead center on compression stroke.
8. Confirm the balance shaft module timing mark. Timing marks to be visually aligned with centers of adjacent cast timing notches.
9. Install balance shaft module that the timing mark of balance shaft module sprocket should be matched with the timing mark (color link) of balance shaft chain.
   a. Seat the bolts in sequence to 19 ft. lbs. (26 Nm).

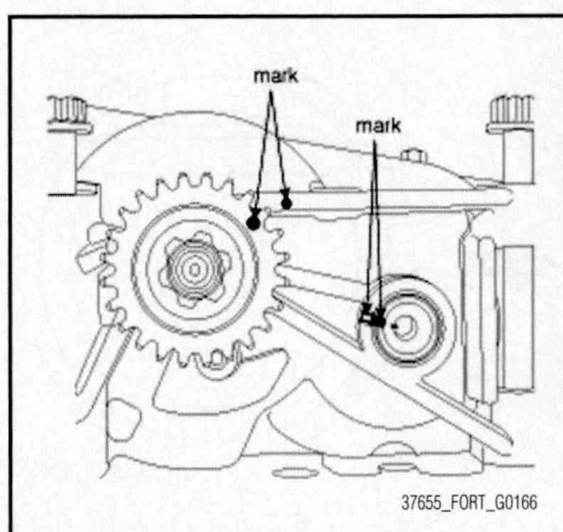

Fig. 69 Align balance shaft timing marks—2.4L engine

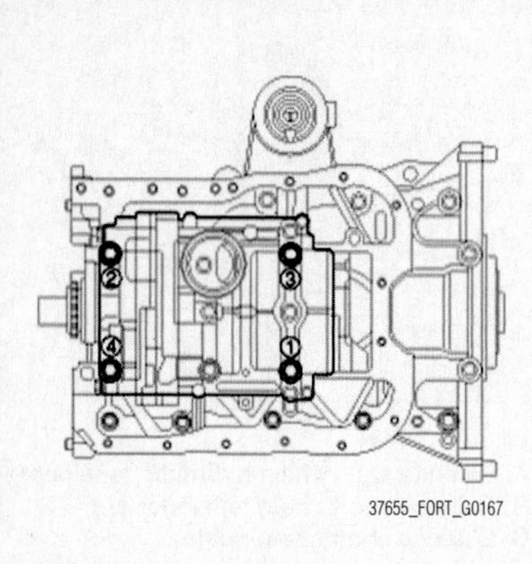

Fig. 70 Oil pump bolt torque sequence—2.4L engine

b. Unfasten the bolts in reverse order.

c. Tighten the bolts in sequence to 12 ft. lbs. (17 Nm). Tighten in sequence 60 degrees. Tighten in sequence an additional 60 degrees.

10. Install the balance shaft chain guide.

a. Tighten bolts to 7–9 ft. lbs. (9–12 Nm).

11. Install the balance shaft chain tensioner arm.

a. Tighten bolts to 7–9 ft. lbs. (9–12 Nm).

12. Install the balance shaft chain hydraulic tensioner then remove the stopper pin.

a. Tighten bolts to 7–9 ft. lbs. (9–12 Nm).

13. Confirm the timing marks.

14. Install the timing chain.

PISTON AND RING

POSITIONING

See Figure 71.

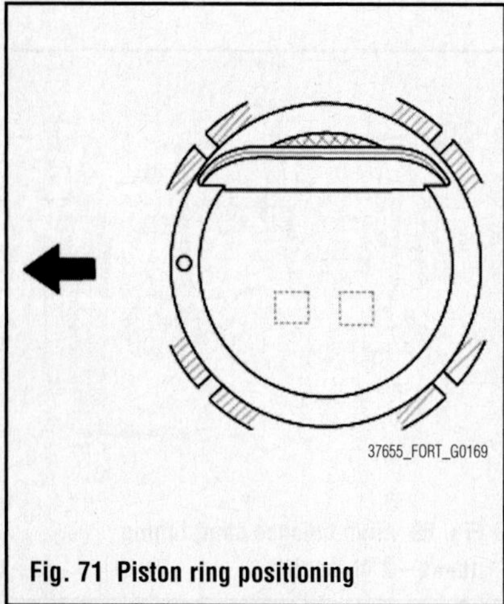

Fig. 71 Piston ring positioning

REAR MAIN SEAL

REMOVAL & INSTALLATION

1. Remove transaxle.

**** CAUTION**

Do not damage seal surface.

2. Remove seal.

To install:

3. Install rear oil seal.

a. Apply engine oil to a new oil seal lip.

b. Using SST (09231–H1100, 09214–3K100) and a hammer, tap in the oil seal until its surface is flush with the rear oil seal retainer edge.

TIMING CHAIN FRONT COVER

REMOVAL & INSTALLATION

See Figures 72 through 75.

1. Disconnect the battery negative cable.
2. Remove the engine cover.
3. Remove RH front wheel.
4. Remove RH side cover.
5. Set No.1 cylinder to TDC/compression.
6. Drain the engine oil, and then set a jack to the oil pan.

➡Place wooden block between the jack and engine oil pan.

7. Disconnect the ground line and then remove the engine mounting bracket.
8. Remove the drive belt.
9. Remove the idler and drive belt tensioner.

**** CAUTION**

Tensioner pulley bolt is left – handed screw.

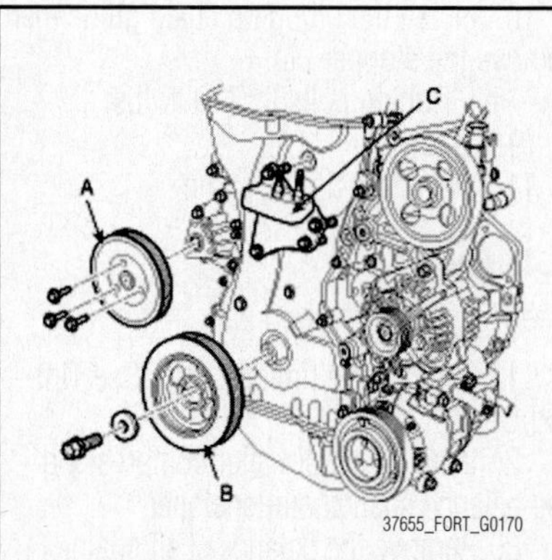

Fig. 72 Water pump pulley (A), crankshaft pulley (B) and engine support bracket (C)

➡Use the SST (flywheel stopper, 09231–3K000) to remove the crankshaft pulley bolt, after removing the starter.

10. Remove the water pump pulley (A), crankshaft pulley (B) and engine support bracket (C).
11. Disconnect the ignition coil connectors and remove the ignition coils.
12. Remove the cylinder head cover.
13. Remove the A/C compressor lower bolts.
14. Remove the compressor bracket.
15. Remove the oil pan.

**** CAUTION**

Be careful not to damage the contact surfaces of cylinder block and oil pan.

16. Remove the bolts and then the timing chain cover by gently prying the portions between the cylinder head and cylinder block.

**** CAUTION**

Be careful not to damage the contact surfaces of cylinder block, cylinder head and timing chain cover.

To install:

17. Install timing chain cover.

a. Using a gasket scraper, remove all the old sealer and gasket from the gasket surfaces.

b. All surfaces must be clean and dry.

c. Before assembling the timing chain cover, the liquid sealant Loctite 5900H or THREEBOND 1217H should be applied on the gap between cylinder head and cylinder block. The part must be assembled within 5 minutes after sealant was applied.

d. Apply liquid sealant Loctite 5900H on timing chain cover. The part must be assembled within 5 minutes after sealant was applied.

e. The dowel pins on the cylinder block and holes on the timing chain cover should be used as a reference in order to assemble the timing chain cover to exact position.

f. Tighten the M6 bolts to 6–7 ft. lbs. (8–10 Nm).

g. Tighten the M8 bolts to 14–17 ft. lbs. (19–23 Nm).

➡Allow 30 minutes setup time after the timing chain cover was assembled before adding oil or starting engine.

18. Install oil pan.

a. Using a gasket scraper, remove all the old sealer and gasket from the gasket surfaces.

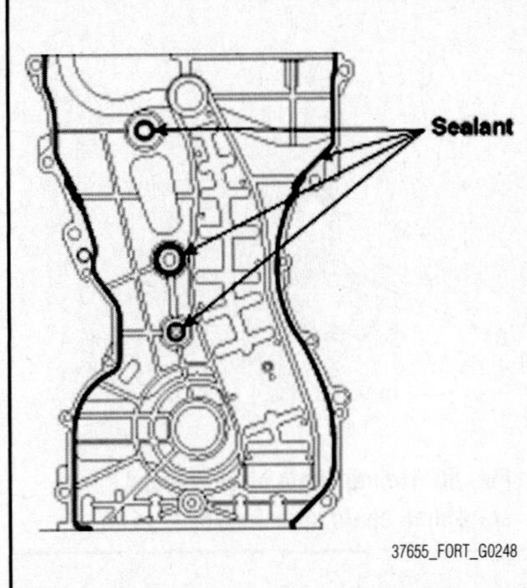

Fig. 73 Apply liquid sealant to timing chain cover

b. All surfaces must be clean and dry.

c. Apply liquid sealant Loctite 5900H to oil pan and the inner threads of the bolt holes. The part must be assembled within 5 minutes after the sealant was applied. Do not apply sealant so it extends into inside of pan.

d. Install oil pan.

➡**Uniformly tighten the bolts in several passes.**

e. Tighten the M6 bolts to 7–9 ft. lbs. (10–12 Nm).

f. Tighten the M8 bolts to 20–22 ft. lbs. (27–30 Nm).

➡**After assembly, wait at least 30 minutes before filling the engine with oil.**

19. Install the air compressor bracket.

a. Tighten the bolts to 14–17 ft. lbs. (19–23 Nm).

20. Install the compressor lower bolts.

a. Tighten the M8 bolts to 14–18 ft. lbs. (19–24 Nm).

➡**Do not reuse cylinder head cover gasket.**

21. Install cylinder head cover.

a. The hardening sealant located on the upper area between timing chain cover and cylinder head should be removed before assembling cylinder head cover.

b. Apply a fresh bead of sealant between timing chain cover and cylinder head. After applying sealant, it should be assembled within 5 minutes.

c. On 2.0L. install the cylinder head cover bolts in sequence, in two steps: Tighten the bolts to 3–4 ft. lbs. (4–6 Nm) then tighten the bolts to 6–7 ft. lbs. (8–10 Nm).

d. On 2.4L. install the cylinder head cover bolts in two steps: Tighten the

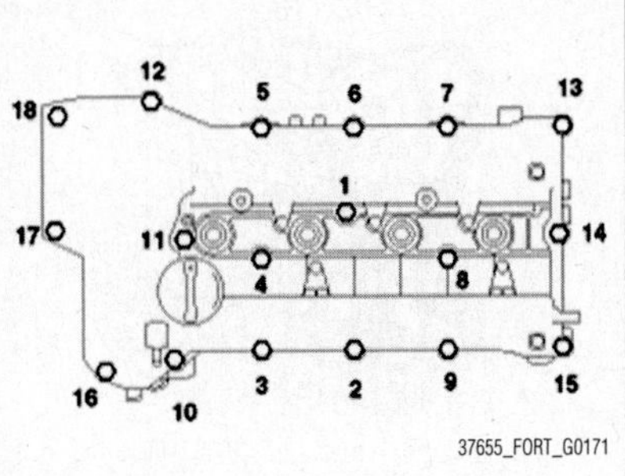

Fig. 74 Valve cover torque sequence—2.0L engine

bolts to 3–4 ft. lbs. (4–6 Nm) then tighten the bolts to 6–7 ft. lbs. (8–10 Nm).

➡**After assembly, wait at least 30 minutes before filling the engine with oil.**

22. Install the ignition coils and connect the ignition coil connectors.

23. Install the engine support bracket.

a. Tighten the M8 bolts to 15–18 ft. lbs. (20–25 Nm).

b. Tighten the M10 bolts to 29–33 ft. lbs. (39–44 Nm).

➡**Use the SST (flywheel stopper, 09231-3K000) to install the crankshaft pulley bolt.**

24. Install the crankshaft pulley.

a. Tighten the bolt to 123–130 ft. lbs. (167–176 Nm).

25. Install the water pump pulley.

a. Tighten the bolts to 6–7 ft. lbs. (8–10 Nm).

➡**Tensioner pulley bolt is left-handed screw.**

26. Install the drive belt tensioner and tensioner pulley.

a. Tighten the bolts to 40–47 ft. lbs. (54–64 Nm).

27. Install idler pulley.

a. Tighten the bolt to 40–47 ft. lbs. (54–64 Nm).

28. Install the drive belt in this order: crankshaft pulley, A/C pulley, alternator pulley, idler pulley, P/S pump pulley, idler pulley, water pump pulley then tensioner pulley.

a. Rotate auto tensioner arm in the counter – clockwise moving auto tensioner pulley bolt with wrench. After putting belt on auto tensioner pulley, release the auto tensioner pulley slowly.

29. Install the engine mounting bracket and then connect the ground cable.

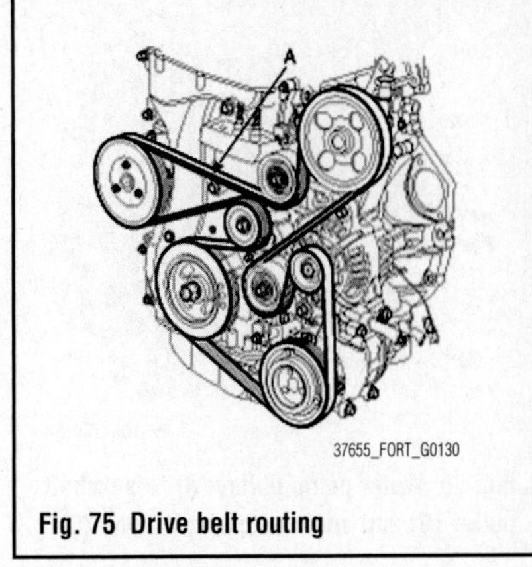

Fig. 75 Drive belt routing

a. Tighten the bolts to 47–62 ft. lbs. (64–83 Nm).

30. Install the RH side cover.

a. Tighten the bolts to 7–8 ft. lbs. (9–11 Nm).

31. Install the RH front wheel.

32. Install the engine cover.

33. Connect the battery negative cable.

TIMING CHAIN & SPROCKETS

REMOVAL & INSTALLATION

See Figures 76 through 83.

1. Disconnect the battery negative cable.
2. Remove the engine cover.
3. Remove RH front wheel.
4. Remove RH side cover.
5. Set No.1 cylinder to TDC/compression.
6. Drain the engine oil, and then set a jack to the oil pan.

➡**Place wooden block between the jack and engine oil pan.**

7. Disconnect the ground line and then remove the engine mounting bracket.
8. Remove the drive belt.
9. Remove the idler and drive belt tensioner.

CAUTION

Tensioner pulley bolt is left – handed screw.

➡**Use the SST (flywheel stopper, 09231-3K000) to remove the crankshaft pulley bolt, after removing the starter.**

10. Remove the water pump pulley (A), crankshaft pulley (B) and engine support bracket (C).
11. Disconnect the ignition coil connectors and remove the ignition coils.
12. Remove the cylinder head cover.

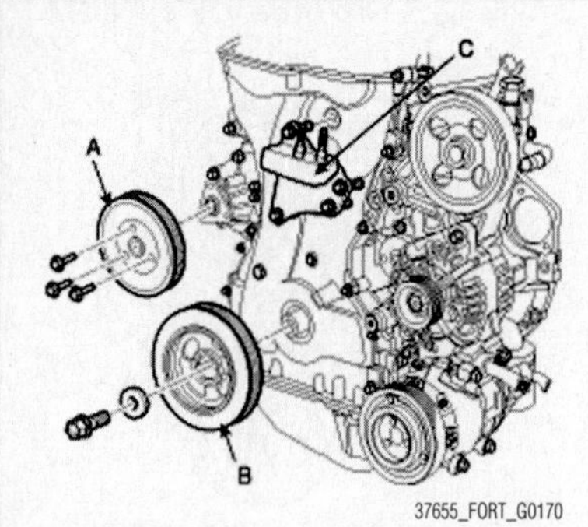

Fig. 76 Water pump pulley (A), crankshaft pulley (B) and engine support bracket (C)

13. Remove the A/C compressor lower bolts.
14. Remove the compressor bracket.
15. Remove the oil pan.

⁂ CAUTION

Be careful not to damage the contact surfaces of cylinder block and oil pan.

16. Remove the bolts and then the timing chain cover by gently prying the portions between the cylinder head and cylinder block.

⁂ CAUTION

Be careful not to damage the contact surfaces of cylinder block, cylinder head and timing chain cover.

17. The key of crankshaft should be aligned with the mating face of main bearing cap. As a result of this, the piston of No.1 cylinder is placed at the top dead center on compression stroke.

➡Before removing the timing chain, mark the timing chain and all the sprockets for TDC (Top Dead Center).

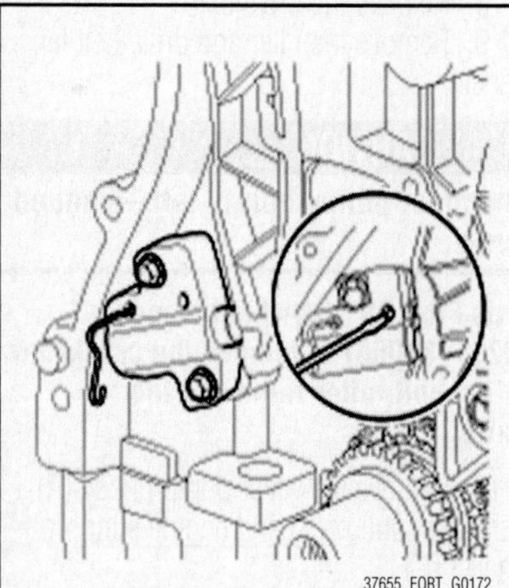

Fig. 77 Install a lock pin after compressing the timing chain tensioner

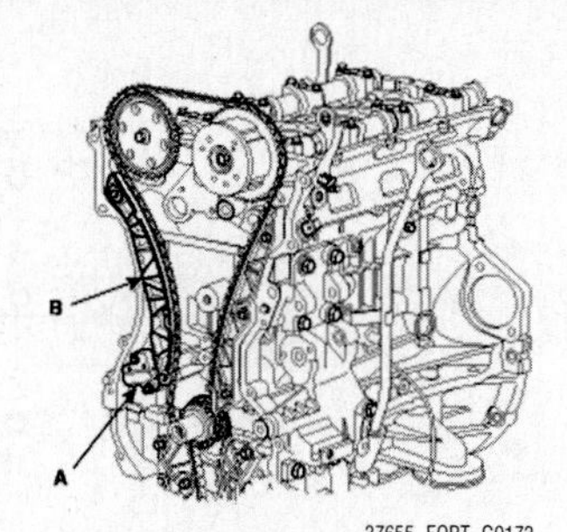

Fig. 78 Timing chain tensioner (A) and timing chain tensioner arm (B)—Single CVVT

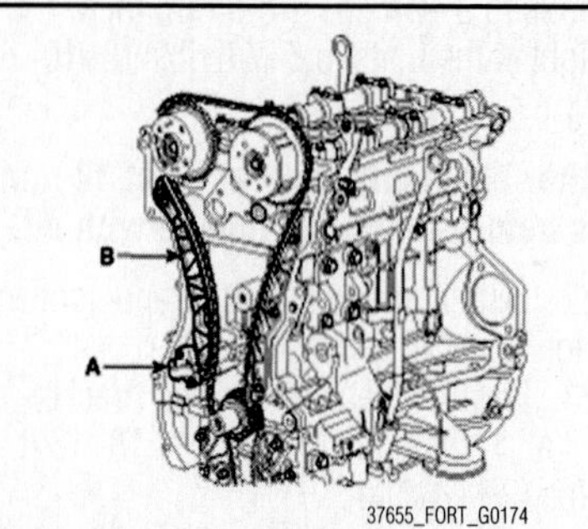

Fig. 79 Timing chain tensioner (A) and timing chain tensioner arm (B)—Dual CVVT

18. Install a lock pin after compressing the timing chain tensioner.
19. Remove the timing chain tensioner (A) and timing chain tensioner arm (B).
20. Remove the timing chain.
21. Remove the timing chain guide.
22. Remove the timing chain oil jet (A).
23. Remove the crankshaft chain sprocket (B).
24. Remove the oil pump chain.
25. Check the camshaft sprocket and crankshaft sprocket for abnormal wear, cracks, or damage. Replace as necessary.
26. Inspect the tensioner arm and chain guide for abnormal wear, cracks, or damage. Replace as necessary.
27. Check that the tensioner piston moves smoothly when the ratchet pawl is released with thin rod.
28. Check the idler for excessive oil leakage, abnormal rotation or vibration. Replace if necessary.
29. Check belt for maintenance and abnormal wear of V–ribbed part. Replace if necessary.
30. Check the pulleys for vibration in rotation, oil or dust deposit of V–ribbed part. Replace if necessary.

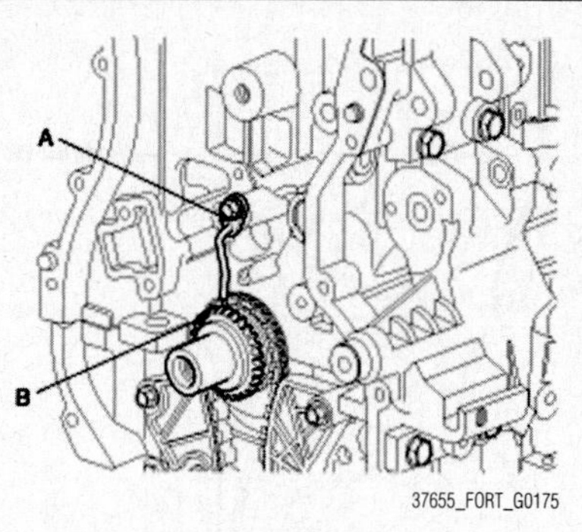

Fig. 80 Timing chain oil jet (A)and crankshaft chain sprocket (B)

To install:

31. Install the oil pump chain.
32. Install the crankshaft sprocket.
33. Install the timing chain oil jet.
34. Tighten bolt to 6–7 ft. lbs. (6–10 Nm).
35. Align crankshaft, intake and exhaust camshafts at TDC marks made during disassembly.
36. Install the timing chain guide (A).
37. Tighten bolt to 7–9 ft. lbs. (10–12 Nm).
38. Install the timing chain on the sprockets in the following order: crankshaft sprocket, timing chain guide, intake CVVT sprocket, exhaust CVVT sprocket or camshaft sprocket.
39. Check alignment of crankshaft, intake and exhaust camshafts at TDC marks. Reinstall chain if marks are not aligned.
40. Install timing chain tensioner arm.
41. Tighten bolt to 7–9 ft. lbs. (10–12 Nm).
42. Install timing chain auto tensioner and remove set pin.
43. Tighten bolt to 7–9 ft. lbs. (10–12 Nm).
44. After rotating crankshaft 2 revolutions in regular direction (clockwise viewed from front), recheck the timing marks.
45. Install timing chain cover.
    a. Using a gasket scraper, remove all the old sealer and gasket from the gasket surfaces.
    b. All surfaces must be clean and dry.
    c. Before assembling the timing chain cover, the liquid sealant Loctite 5900H or THREEBOND 1217H should be applied on the gap between cylinder head and cylinder block. The part must be assembled within 5 minutes after sealant was applied.
    d. Apply liquid sealant Loctite 5900H on timing chain cover. The part must be assembled within 5 minutes after sealant was applied.

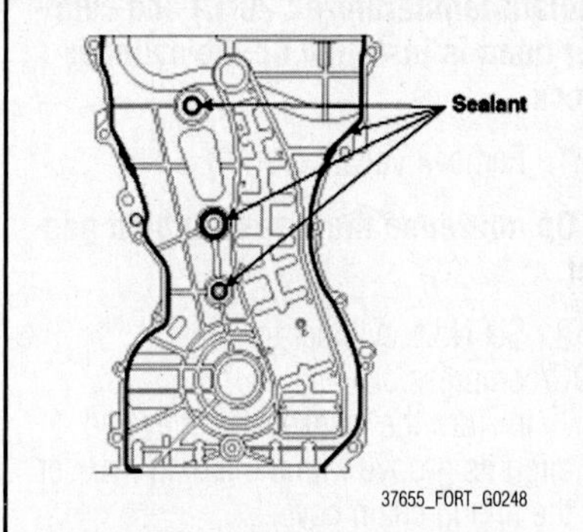

Fig. 81 Apply liquid sealant to timing chain cover

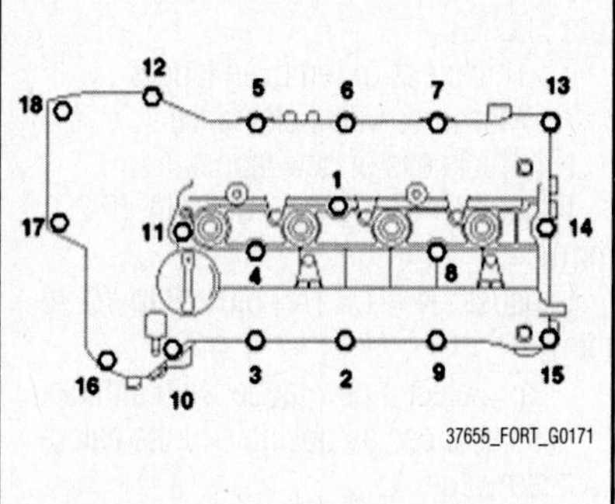

Fig. 82 Valve cover torque sequence—2.0L engine

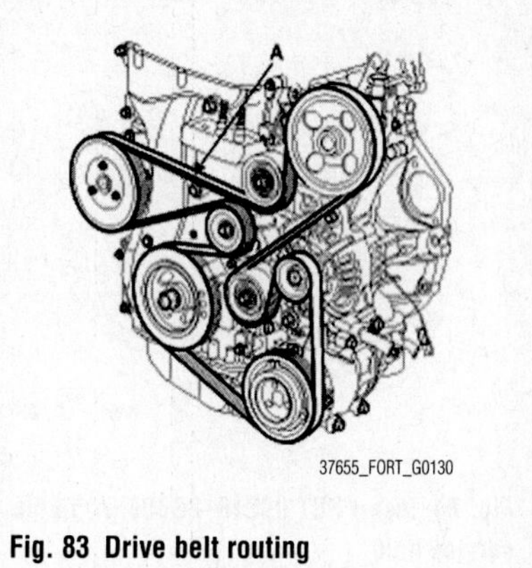

Fig. 83 Drive belt routing

e. The dowel pins on the cylinder block and holes on the timing chain cover should be used as a reference in order to assemble the timing chain cover to exact position.

f. Tighten the M6 bolts to 6–7 ft. lbs. (8–10 Nm).

g. Tighten the M8 bolts to 14–17 ft. lbs. (19–23 Nm).

➡Allow 30 minutes setup time after the timing chain cover was assembled before adding oil or starting engine.

46. Install oil pan.

a. Using a gasket scraper, remove all the old sealer and gasket from the gasket surfaces.

b. All surfaces must be clean and dry.

c. Apply liquid sealant Loctite 5900H to oil pan and the inner threads of the bolt holes. The part must be assembled within 5 minutes after the sealant was applied. Do not apply sealant so it extends into inside of pan.

d. Install oil pan.

➡Uniformly tighten the bolts in several passes.

e. Tighten the M6 bolts to 7–9 ft. lbs. (10–12 Nm).

f. Tighten the M8 bolts to 20–22 ft. lbs. (27–30 Nm).

➡After assembly, wait at least 30 minutes before filling the engine with oil.

47. Install the air compressor bracket.

a. Tighten the bolts to 14–17 ft. lbs. (19–23 Nm).

48. Install the compressor lower bolts.

a. Tighten the M8 bolts to 14–18 ft. lbs. (19–24 Nm).

➡Do not reuse cylinder head cover gasket.

49. Install cylinder head cover.

a. The hardening sealant located on the upper area between timing chain cover and cylinder head should be removed before assembling cylinder head cover.

b. Apply a fresh bead of sealant between timing chain cover and cylinder head. After applying sealant, it should be assembled within 5 minutes.

c. On 2.0L. install the cylinder head cover bolts in sequence, in two steps: Tighten the bolts to 3–4 ft. lbs. (4–6 Nm) then tighten the bolts to 6–7 ft. lbs. (8–10 Nm).

d. On 2.4L. install the cylinder head cover bolts in two steps: Tighten the bolts to 3–4 ft. lbs. (4–6 Nm) then tighten the bolts to 6–7 ft. lbs. (8–10 Nm).

➡After assembly, wait at least 30 minutes before filling the engine with oil.

50. Install the ignition coils and connect the ignition coil connectors.

51. Install the engine support bracket.

a. Tighten the M8 bolts to 15–18 ft. lbs. (20–25 Nm).

b. Tighten the M10 bolts to 29–33 ft. lbs. (39–44 Nm).

➡Use the SST (flywheel stopper, 09231-3K000) to install the crankshaft pulley bolt.

52. Install the crankshaft pulley.

a. Tighten the bolt to 123–130 ft. lbs. (167–176 Nm).

53. Install the water pump pulley.

a. Tighten the bolts to 6–7 ft. lbs. (8–10 Nm).

➡Tensioner pulley bolt is left-handed screw.

54. Install the drive belt tensioner and tensioner pulley.

a. Tighten the bolts to 40–47 ft. lbs. (54–64 Nm).

55. Install idler pulley.

a. Tighten the bolt to 40–47 ft. lbs. (54–64 Nm).

56. Install the drive belt in this order: crankshaft pulley, A/C pulley, alternator pulley, idler pulley, P/S pump pulley, idler pulley, water pump pulley then tensioner pulley.

a. Rotate auto tensioner arm in the counter – clockwise moving auto tensioner pulley bolt with wrench. After putting belt on auto tensioner pulley, release the auto tensioner pulley slowly.

57. Install the engine mounting bracket and then connect the ground cable.

a. Tighten the bolts to 47–62 ft. lbs. (64–83 Nm).

58. Install the RH side cover.

a. Tighten the bolts to 7–8 ft. lbs. (9–11 Nm).

59. Install the RH front wheel.

60. Install the engine cover.

61. Connect the battery negative cable.

VALVE LASH

ADJUSTMENT

See Figure 84.

1. Perform inspection, then adjust the intake and exhaust valve clearance if needed. See Inspection.

a. Set the No.1 cylinder to the TDC/compression position.

b. Put paint marks on the timing chain links (2 places) that meet with the timing marks of the intake, exhaust camshaft sprockets.

c. Remove the service hole bolt of the timing chain cover.

➡The bolt must not be reused once it has been assembled.

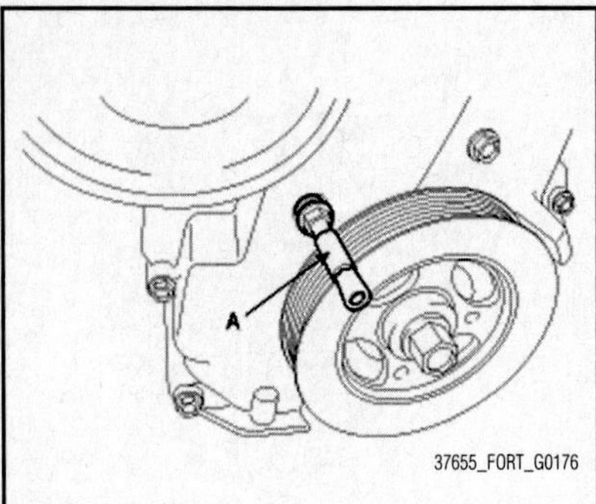

Fig. 84 Insert SST 09240–2G000 (A) in the service hole

d. Insert SST 09240–2G000 (A) in the service hole of the timing chain cover and release the ratchet.

e. Remove the front camshaft bearing cap.

f. Remove the exhaust camshaft bearing cap and exhaust camshaft.

g. Remove the intake camshaft bearing cap and intake camshaft.

⁂ CAUTION

When disconnect the timing chain from the camshaft timing sprocket, hold the timing chain.

h. Tie down timing chain so that it doesn't move.

⁂ CAUTION

Be careful not to drop anything inside timing chain cover.

i. Measure the thickness of the removed tappet using a micrometer.

j. Calculate the thickness of a new tappet so that the valve clearance comes within the specified value.

Valve clearance (Engine coolant temperature : 20°C)

T : Thickness of removed tappet

A : Measured valve clearance

N : Thickness of new tappet

Intake : N = T + [A – 0.0079 in. (0.20 mm)]

Exhaust : N = T + [A– 0.0118 in. (0.30 mm)]

a. Select a new tappet with a thickness as close as possible to the calculated value.

➡Shims are available in 47 size increments of 0.0006 in. (0.015 mm) from 0.118 in. (3.00 mm) to 0.1417 in. (3.600 mm).

b. Place a new tappet on the cylinder head.

c. Hold the timing chain, and place the intake camshaft and timing sprocket assembly.

d. Align the matchmarks on the timing chain and camshaft timing sprocket.

e. Install the exhaust camshaft.

f. Install the exhaust camshaft sprocket.

g. Install the front camshaft bearing cap. Tighten to 8.7–10 ft. lbs. (11.8–13.7 Nm).

h. Turn the crankshaft two turns in the operating direction (clockwise) and realign crankshaft sprocket and camshaft sprocket timing marks.

i. Recheck the valve clearance.

j. Install new service hole bolt. Tighten to 8.7–11 ft. lbs. (11.8–14.7 Nm).

INSPECTION

➡Inspect and adjust the valve clearance when the engine is cold (Engine coolant temperature : 20°C) and cylinder head is installed on the cylinder block.

1. Remove Valve cover.

➡Do not reuse the disassembled gasket.

2. Set No.1 cylinder to TDC/compression.

a. Turn the crankshaft pulley and align its groove with the timing mark of the timing chain cover.

b. Check that the marks of the camshaft timing sprockets are in straight line on the cylinder head surface. If not, turn the crankshaft one revolution (360°).

3. Inspect the valve clearance.

a. Check only the intake valves of the 1st and 2nd cylinders and exhaust valves of the 1st and 3rd cylinders for their clearance.

b. Using a thickness gauge, measure the clearance between the tappet and the base circle of camshaft.

c. Record the out–of–specification valve clearance measurements. They will be used later to determine the required tappet for adjusting.

Valve clearance specification (Engine coolant temperature : 20°C [68°F])

Intake : 0.0039–0.0118 in. (0.10–0.30 mm)

Exhaust : 0.0079–0.0157 in. (0.20–0.40 mm)

a. Turn the crankshaft pulley one revolution (clockwise 360°) and align its groove with timing mark of the timing chain cover.

b. Check the intake valves of the 3rd and 4th cylinders and exhaust valves of the 2nd and 4th cylinders for their clearance.

ENGINE PERFORMANCE & EMISSION CONTROLS

ACCELERATOR PEDAL POSITION (APP) SENSOR

LOCATION

See Figure 85.

Refer to the accompanying illustration.

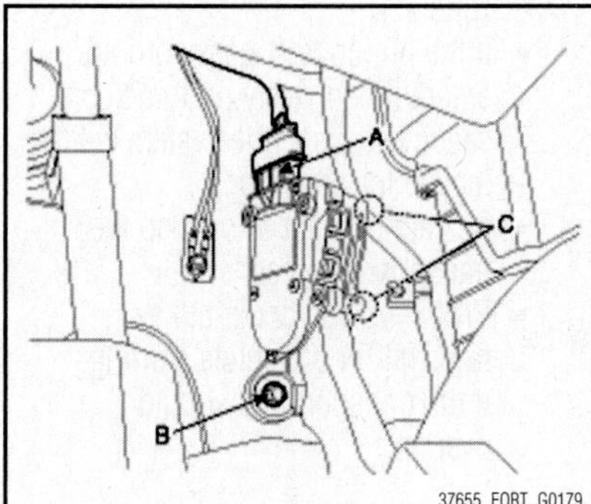

Fig. 85 APP sensor connector (A), mounting bolt (B) and nuts (C)

REMOVAL & INSTALLATION

See Figure 85.

1. Turn the ignition switch OFF and disconnect the negative (–) battery cable.
2. Disconnect the accelerator position sensor connector (A).
3. Remove the mounting bolt (B) and nuts (C), and then remove the accelerator pedal from the vehicle.
4. Installation is reverse of removal.
5. Tighten nuts and bolts to 9–12 ft. lbs. (13–16 Nm).

CAMSHAFT POSITION (CMP) SENSOR

LOCATION

See Figures 86 and 87.

Refer to the accompanying illustrations.

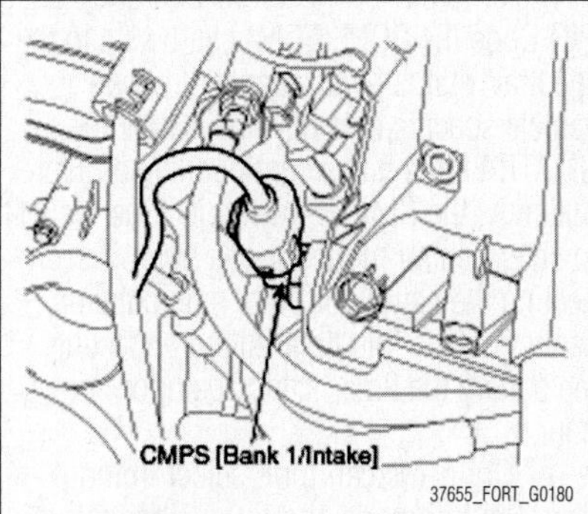

Fig. 86 Camshaft Position (CMP) Sensor [Bank 1 / Intake]

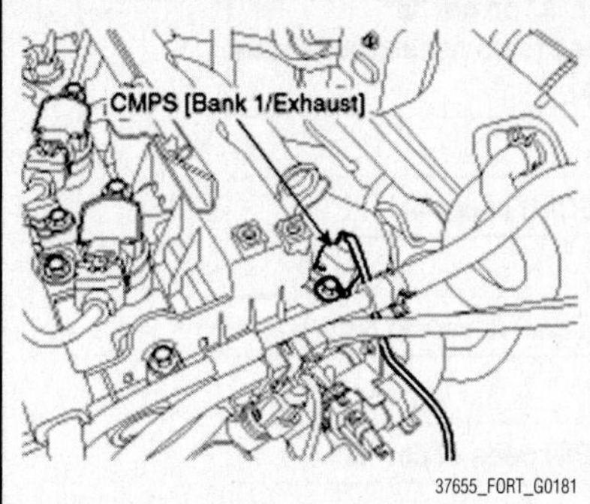

Fig. 87 Camshaft Position (CMP) Sensor [Bank 1 / Exhaust] [Except 2.0L ULEV]

REMOVAL & INSTALLATION

1. Remove harness connector.
2. Remove bolt and sensor.
3. Installation is reverse of removal.

CRANKSHAFT POSITION (CKP) SENSOR

LOCATION

See Figure 88.

Refer to the accompanying illustration.

REMOVAL & INSTALLATION

1. Remove harness connector.
2. Remove bolt and sensor.
3. Installation is reverse of removal.

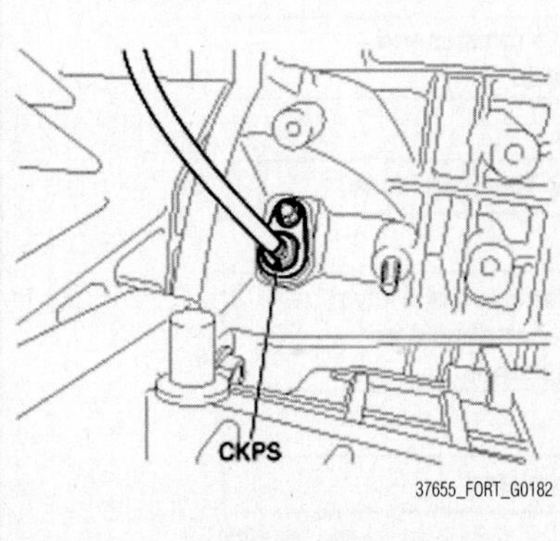

Fig. 88 Crankshaft Position (CKP) Sensor

ELECTRONIC CONTROL MODULE (ECM)

LOCATION

See Figure 89.

Refer to the accompanying illustration.

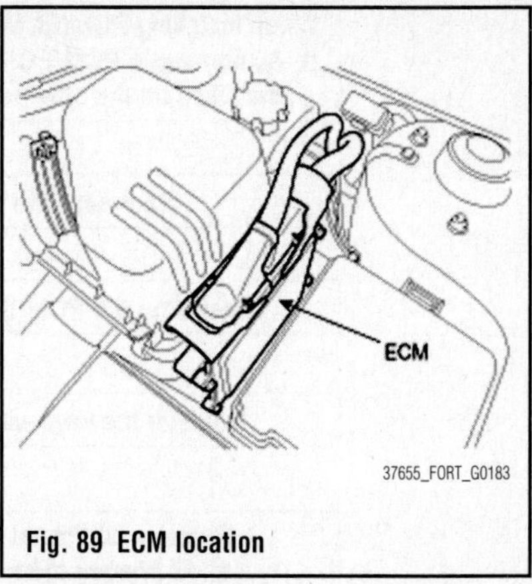

Fig. 89 ECM location

REMOVAL & INSTALLATION

⁂ CAUTION

If vehicle ECM has been replaced and is equipped with immobilizer, "Key Teaching" procedure must be performed to allow starting. See Reset.

1. Turn ignition switch OFF and disconnect the negative (–) battery cable.
2. Remove the air cleaner assembly.
3. Disconnect the ECM Connector (A).
4. Remove the mounting bolts, and then remove the ECM from the air cleaner assembly.
5. Installation is reverse of removal.

RESET

Immobilizer

See Figures 90 and 91.

The immobilizer system will disable the vehicle unless the proper ignition key is used, in addition to the currently available anti–theft systems such as car alarms, the immobilizer system aims to drastically reduce the rate of auto theft.

In case of a defective ECM, the unit has to be replaced with a "virgin" or "neutral" ECM. All keys have to be taught to the new ECM. Keys, which are not taught to the ECM, are invalid for the new ECM. See Key Teaching Procedure. The vehicle specific data have to be left unchanged due to the unique programming of transponder.

In case of a defective SMARTRA, it needs teaching the smartra. A new SMARTRA device replaces the old one and smartra need teaching.

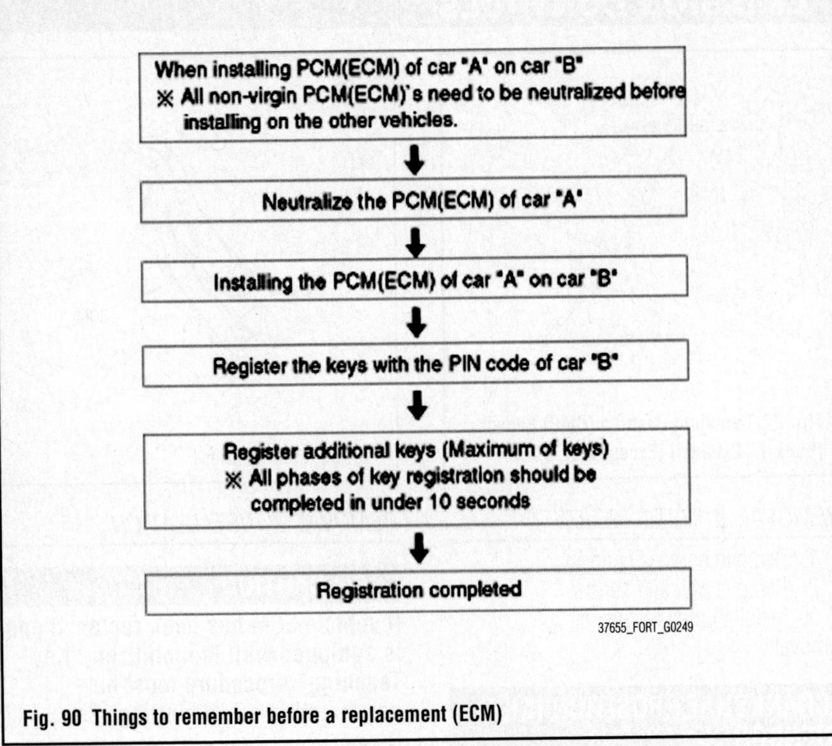

Fig. 90 Things to remember before a replacement (ECM)

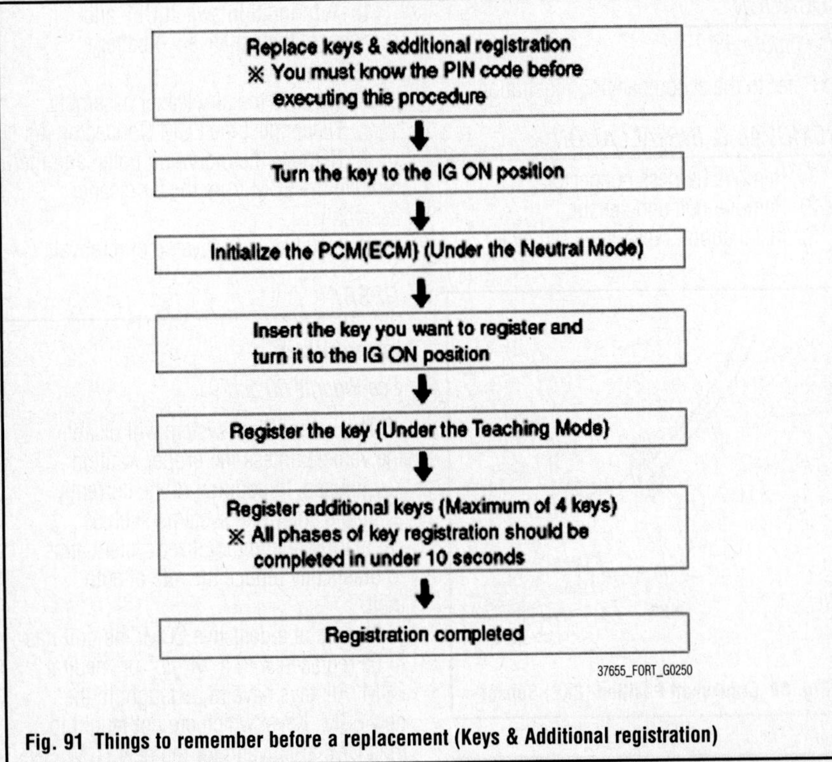

Fig. 91 Things to remember before a replacement (Keys & Additional registration)

➡Note the following:

- When there is only one key registered and you wish to register another key, you need to re–register the key which was already registered.
- When the key #1 is registered and master key #2 is not registered, Put the key #1 in the IG/ON or the start position and remove it. The engine can be started with the unregistered key #2.
(Note that key #2 must be used within 10 seconds of removing key #1)
- When the key #1 is registered and key #2 is not registered, put the unregistered master key #2 in the IG/ON or the start position. The engine cannot be started even with the registered key #1.
- When you inspect the immobilizer system, refer to the above paragraphs 1, 2 and 3. Always remember the 10 seconds zone.
- If the pin code & password are entered incorrectly on three consecutive inputs, the system will be locked for one hour.
- Be cautious not to overlap the transponder areas.
- Problems can occur at key registration or vehicle starting if the transponders should overlap.

Neutralizing of ECM

The PCM (ECM) can be set to the "neutral" status by a tester.

A valid ignition key is inserted and after ignition on is recorded, the PCM (ECM) requests the vehicle specific data from the tester. The communication messages are described at "Neutral Mode" After successfully receiving the data, the PCM (ECM) is neutralized.

The ECM remains locked. Neither the limp home mode nor the "twice ignition on" function, is accepted by the PCM (ECM).

The teaching of keys follows the procedure described for the virgin PCM (ECM). The vehicle specific data have to be unchanged due to the unique programming of the transponder. If data should be changed, new keys with a virgin transponder are requested.

This function is for neutralizing the PCM (ECM) and Key. Ex) when lost key, Neutralize the PCM (ECM) then teach keys.

(Refer to the Things to do when Key & PIN Code the PCM (ECM) can be set to the "neutral" status by a scanner. If wrong vehicle specific data have been sent to SMATRA three times continuously or intermittently, the SMATRA will reject the request to enter neutral mode for one hour. Disconnecting the battery or other manipulation cannot reduce this time. After connecting the battery the timer starts again for one hour.

1. Connect scan tool, select Immobilizer, Neutral mode and follow the instructions.

➡**Keys must be relearned after ECM replacement. See Key Teaching Procedure.**

Key Teaching Procedure

See Figure 92.

Key teaching must be done after replacing a defective PCM (ECM) or when providing additional keys to the vehicle owner.

The procedure starts with an PCM (ECM) request for vehicle specific data (PIN code: 6digits) from the tester. The "virgin" PCM (ECM) stores the vehicle specific data and the key teaching can be started. The "learnt" PCM (ECM) compares the vehicle specific data from the tester with the stored data. If the data are correct, the teaching can proceed.

If incorrect vehicle specific data have been sent to the PCM (ECM) three times, the PCM (ECM) will reject the request of key teaching for one hour. This time cannot be reduced by disconnecting the battery or any other manipulation. After reconnecting the battery, the timer starts again for one hour.

The key teaching is done by ignition on with the key and additional tester commands. The PCM (ECM) stores the relevant data in the EEPROM and in the transponder. Then the PCM (ECM) runs the authentication required for confirmation of the teaching process. The successful programming is then confirmed by a message to the tester.

If the key is already known to the PCM (ECM) from a previous teaching, the authentication will be accepted and the EEPROM data are updated. There is no changed transponder content (this is impossible for a learnt transponder).

The attempt to repeatedly teach a key, which has been taught already during the same teaching cycle, is recognized by the PCM (ECM). This rejects the key and a message is sent to the tester.

The PCM (ECM) rejects invalid keys, which are presented for teaching. A message is sent to the tester. The key can be invalid due to faults in the transponder or other reasons, which result from unsuccessful programming of data. If the PCM (ECM) detects different authenticators of a transponder and an PCM (ECM), the key is considered to be invalid.

The maximum number of taught keys is 8

If an error occurs during the Immobilizer Service Menu, the PCM (ECM) status remains unchanged and a specific fault code is stored.

If the PCM (ECM) status and the key status do not match for teaching of keys, the tester procedure will be stopped and a specific fault code will be stored at PCM (ECM).

1. Connect scan tool, select Immobilizer, Teaching mode and follow the instructions.

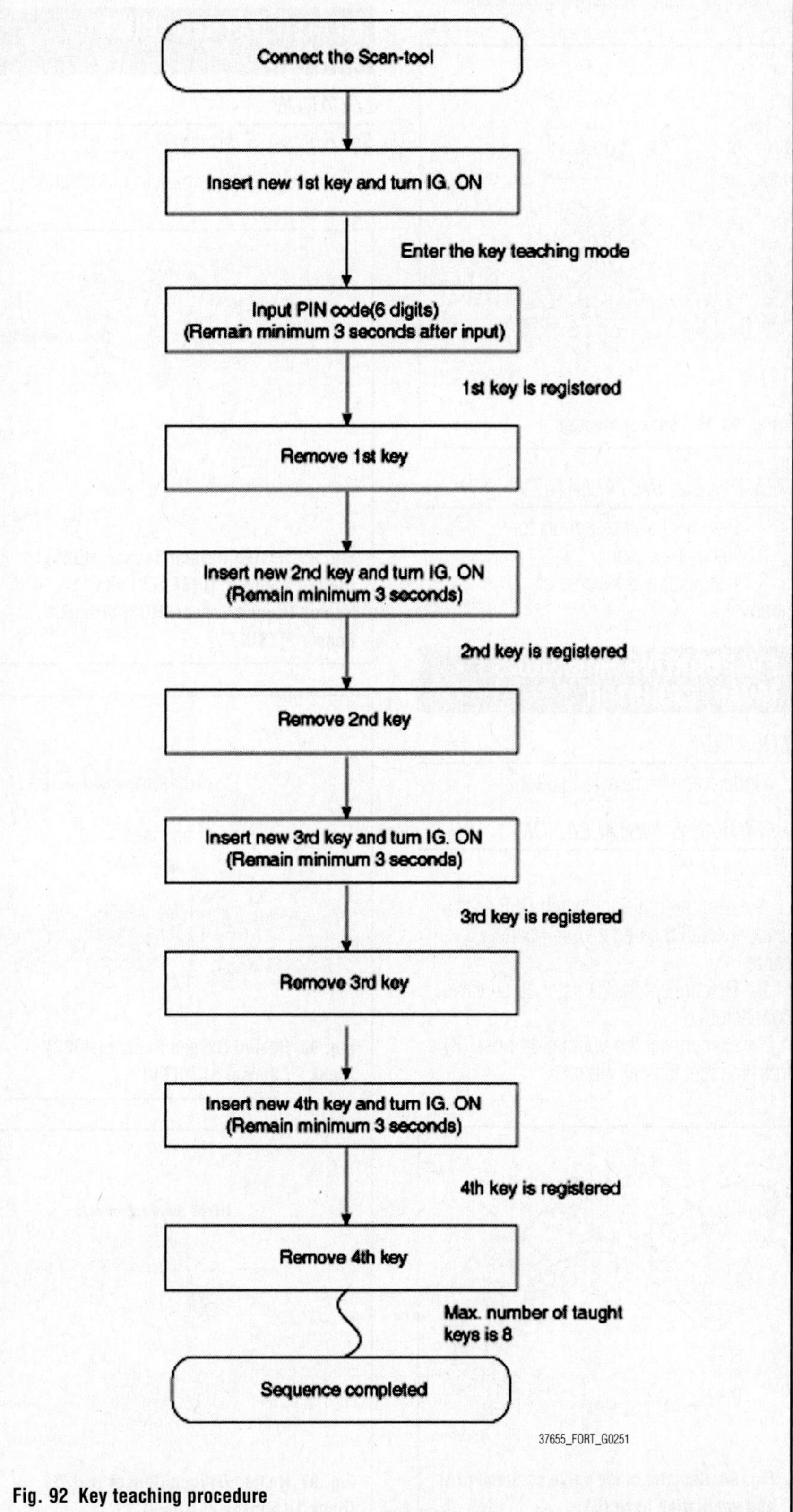

Fig. 92 Key teaching procedure

ENGINE COOLANT TEMPERATURE (ECT) SENSOR

LOCATION

See Figure 93.

Refer to the accompanying illustration.

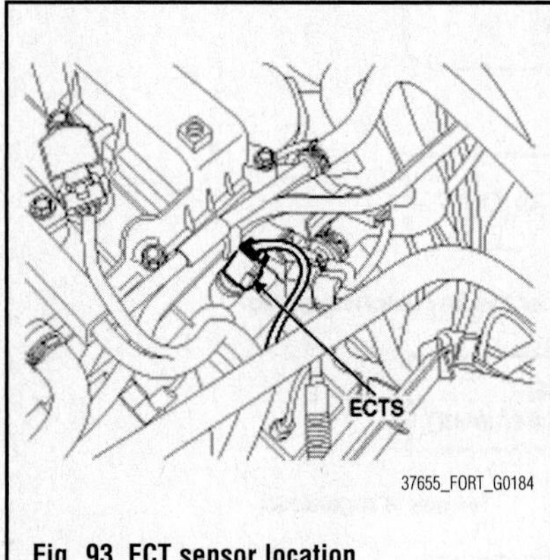

Fig. 93 ECT sensor location

REMOVAL & INSTALLATION

1. Remove harness connector.
2. Remove sensor.
3. Installation is reverse of removal.

EVAPORATIVE EMISSIONS (EVAP) CANISTER

LOCATION

Under vehicle, next to fuel tank.

REMOVAL & INSTALLATION

See Figure 94.

1. Turn the ignition switch OFF and disconnected the negative (–) battery cable.
2. Disconnect the canister close valve connector (A).
3. Disconnect the ventilation hose (B) from the fuel tank air filter.

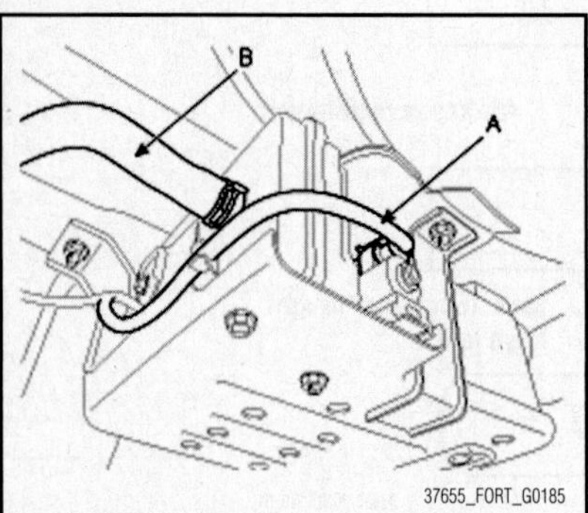

Fig. 94 Canister close valve connector (A) and ventilation hose (B)

4. Disconnect the vapor tube quick–connector and the vapor hose from the canister.
5. Remove the mounting bolts and nut, and then remove the canister from the vehicle.
6. Installation is reverse of removal.

HEATED OXYGEN (HO2S) SENSOR

LOCATION

See Figures 95 through 97.

Refer to the accompanying illustrations.

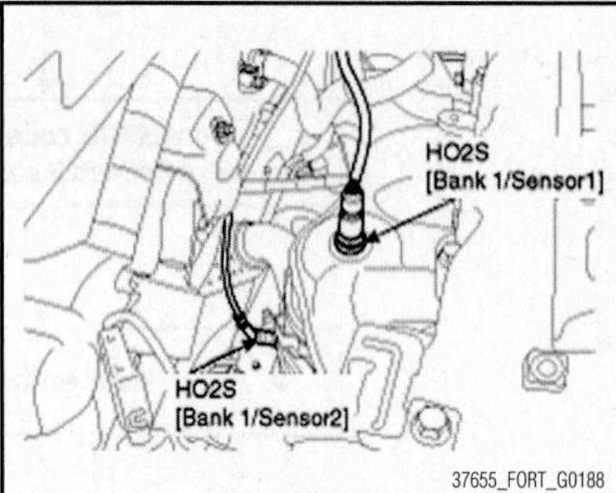

Fig. 95 Heated Oxygen Sensor (HO2S) [Bank 1 / Sensor 1] [SULEV] and 11. Heated Oxygen Sensor (HO2S) [Bank 1 / Sensor 2] [SULEV]

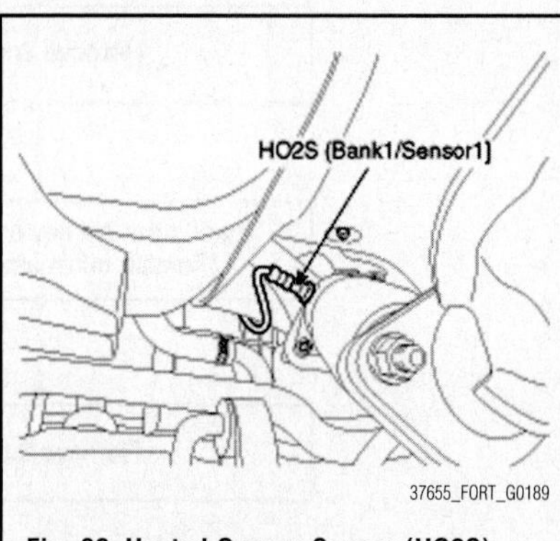

Fig. 96 Heated Oxygen Sensor (HO2S) [Bank 1 / Sensor 1] [ULEV]

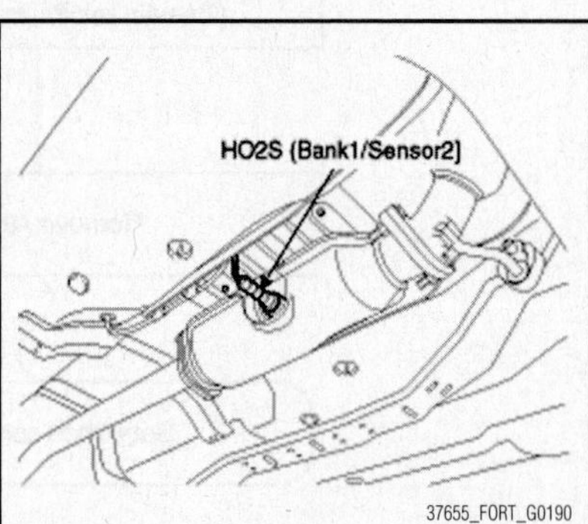

Fig. 97 Heated Oxygen Sensor (HO2S) [Bank 1 / Sensor 2] [ULEV]

REMOVAL & INSTALLATION

**** CAUTION**

Do not remove sensor unless exhaust system is cool.

1. Remove harness connector.
2. Remove sensor.
3. Installation is reverse of removal.
   a. Coat threads with anti–seize.

INTAKE AIR TEMPERATURE (IAT) SENSOR

LOCATION

See Figure 98.

Refer to the accompanying illustration.

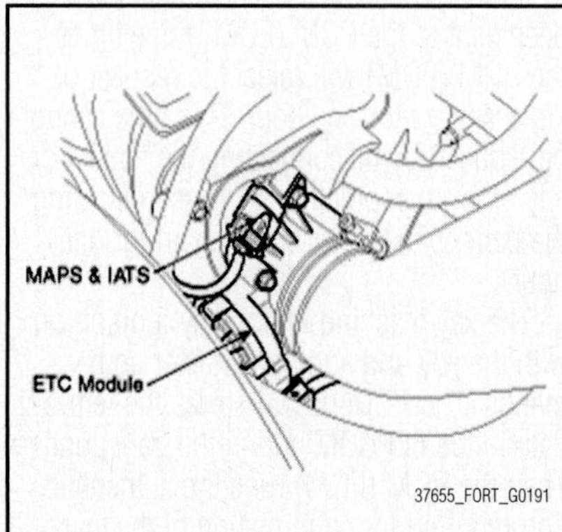

Fig. 98 Manifold Absolute Pressure Sensor (MAPS), Intake Air Temperature Sensor (IATS), Throttle Position Sensor (TPS) and ETC Motor

REMOVAL & INSTALLATION

Intake Air Temperature Sensor (IATS) is included inside Manifold Absolute Pressure Sensor. See Manifold Absolute Pressure (MAP) Sensor.

KNOCK SENSOR (KS)

LOCATION

See Figure 99.

Refer to the accompanying illustration.

REMOVAL & INSTALLATION

1. Remove harness connector.
2. Remove bolt.
3. Remove sensor.
4. Installation is reverse of removal. Tighten bolt to 13.4–13.7 ft. lbs. (19–24 Nm).

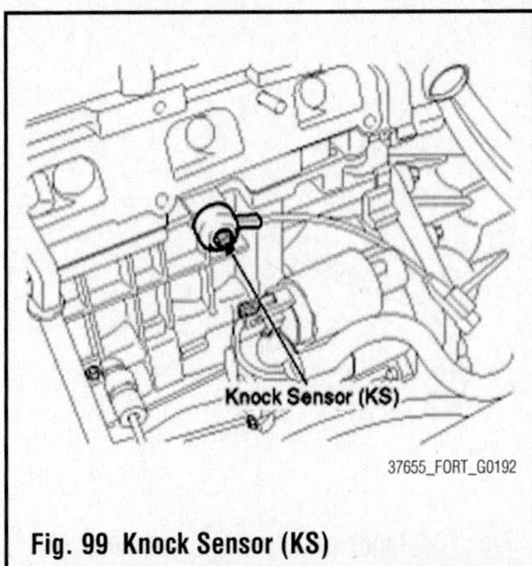

Fig. 99 Knock Sensor (KS)

MANIFOLD ABSOLUTE PRESSURE (MAP) SENSOR

LOCATION

See Figure 98.

Refer to the accompanying illustration.

REMOVAL & INSTALLATION

1. Remove harness connector.
2. Remove bolt.
3. Remove sensor.
4. Installation is reverse of removal.

OUTPUT SHAFT SPEED (OSS) SENSOR

LOCATION

On side of transaxle.

REMOVAL & INSTALLATION

See Figures 100 and 101.

1. Remove the battery terminal.
2. Remove the battery and battery tray.
3. Remove the air duct.
4. Remove the air cleaner assembly.
5. Remove the output speed sensor connector (A).
6. Remove the output speed sensor (A).

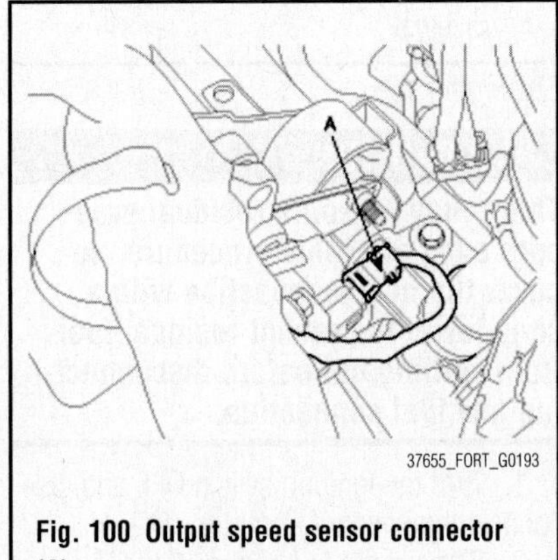

Fig. 100 Output speed sensor connector (A)

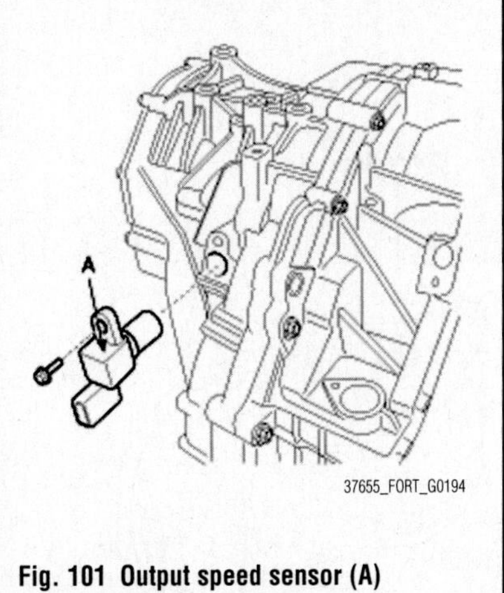

Fig. 101 Output speed sensor (A)

To install:

7. Install the new O–ring to the output shaft speed sensor.
8. Install the output speed sensor.
9. Tighten bolt to 7–8 ft. lbs. (10–12 Nm).

POSITIVE CRANKCASE VENTILATION (PCV) VALVE

LOCATION

See Figure 102.

Refer to the accompanying illustration.

REMOVAL & INSTALLATION

1. Disconnect the vapor hose.
2. Remove the PCV valve.
3. Installation is reverse of removal.

THROTTLE CONTROL ACTUATOR (TAC)

LOCATION

See Figure 98.

Refer to the accompanying illustration.

REMOVAL & INSTALLATION

See Figure 103.

Information is not available. Refer to the accompanying illustration for system components.

THROTTLE POSITION SENSOR (TPS)

LOCATION

See Figure 98.

Refer to the accompanying illustration.

REMOVAL & INSTALLATION

Information is not available. See Throttle Control Actuator (TAC).

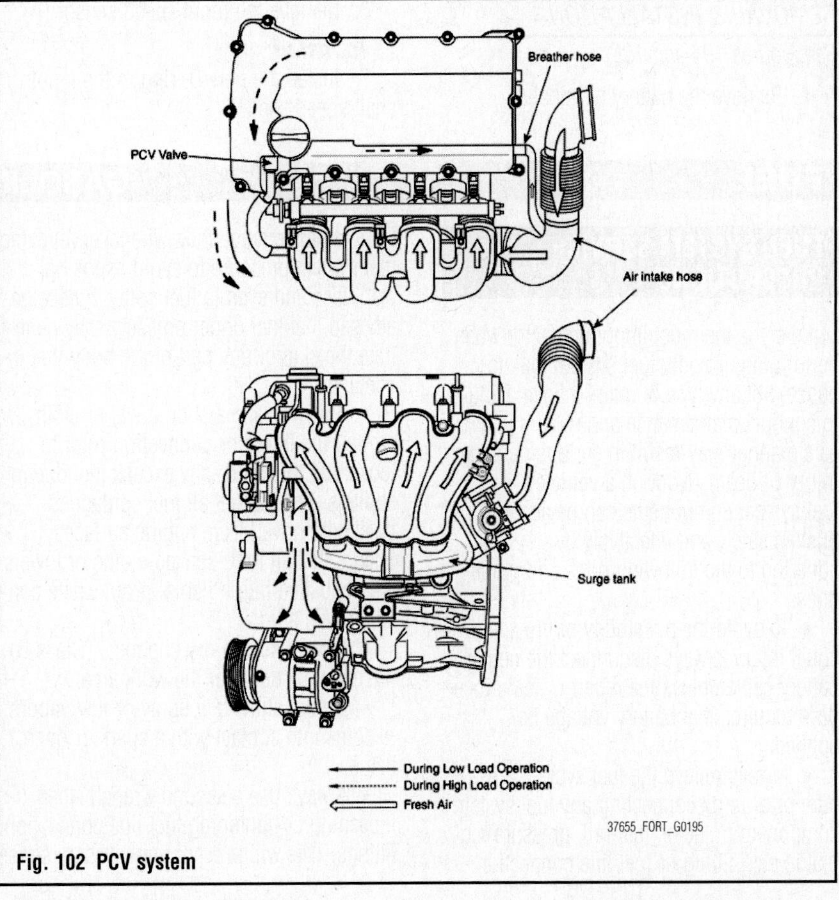

Fig. 102 PCV system

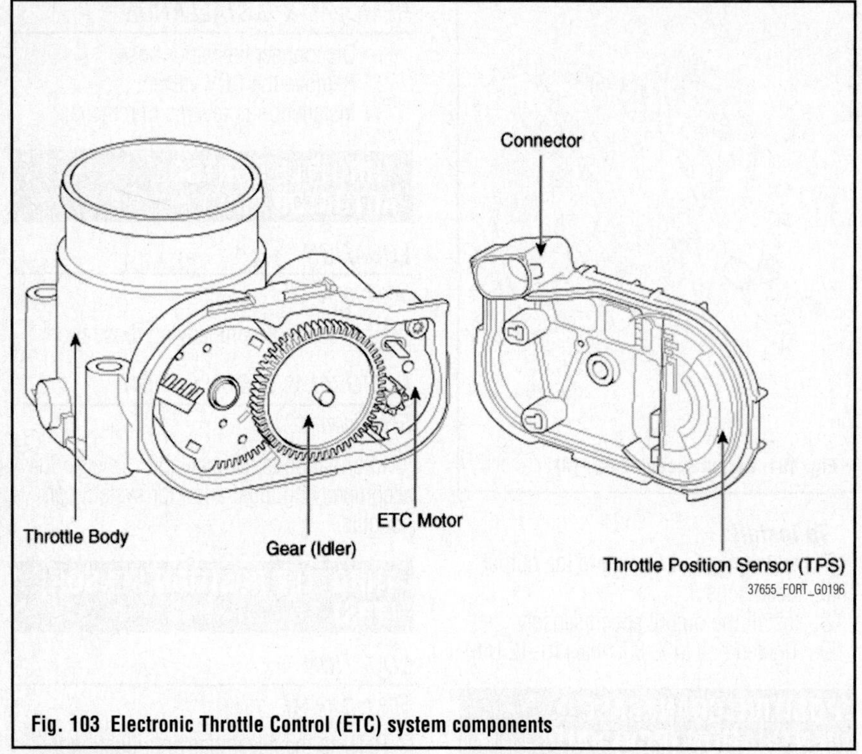

Fig. 103 Electronic Throttle Control (ETC) system components

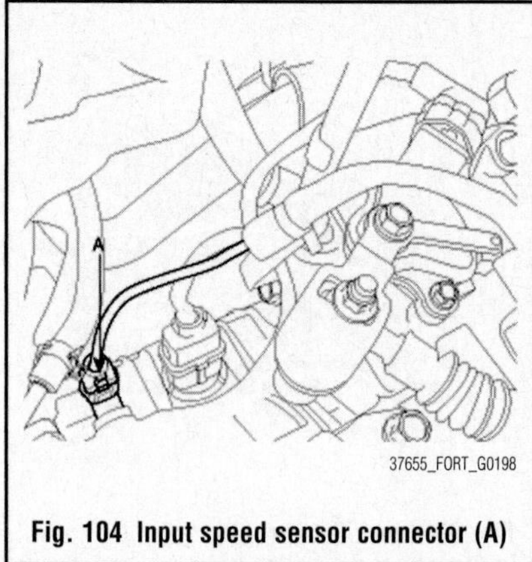

Fig. 104 Input speed sensor connector (A)

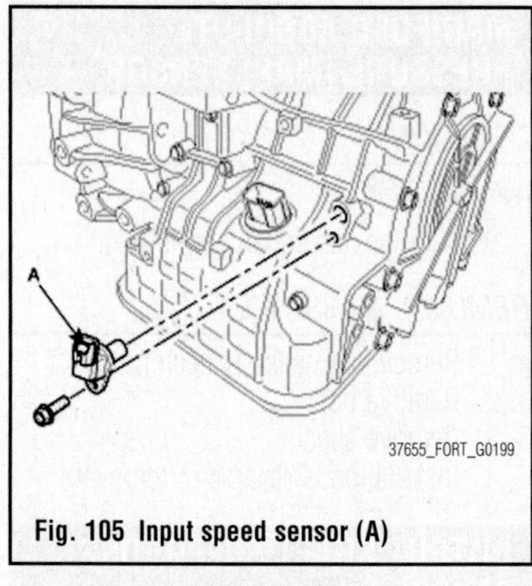

Fig. 105 Input speed sensor (A)

TURBINE SPEED SENSOR (TSS)

LOCATION

On side of transaxle.

REMOVAL & INSTALLATION

See Figures 104 and 105.

1. Remove the battery terminal.
2. Remove the battery and battery tray.
3. Remove the air duct.
4. Remove the air cleaner assembly.
5. Remove the input speed sensor connector (A).
6. Remove the input speed sensor (A).

To install:

7. Install the new O–ring to the input shaft speed sensor.
8. Install the input speed sensor.
9. Tighten bolt to 7–8 ft. lbs. (10–12 Nm).

FUEL

GASOLINE FUEL INJECTION SYSTEM

FUEL SYSTEM SERVICE PRECAUTIONS

Safety is the most important factor when performing not only fuel system maintenance, but any type of maintenance. Failure to conduct maintenance and repairs in a safe manner may result in serious personal injury or death. Work on a vehicle's fuel system components can be accomplished safely and effectively by adhering to the following rules and guidelines.

- To avoid the possibility of fire and personal injury, always disconnect the negative battery cable unless the repair or test procedure requires that battery voltage be applied.
- Always relieve the fuel system pressure prior to disconnecting any fuel system component (injector, fuel rail, pressure regulator, etc.) fitting or fuel line connection. Exercise extreme caution whenever relieving fuel system pressure to avoid exposing skin, face and eyes to fuel spray. Please be advised that fuel under pressure may penetrate the skin or any part of the body that it contacts.
- Always place a shop towel or cloth around the fitting or connection prior to loosening to absorb any excess fuel due to spillage. Ensure that all fuel spillage is quickly removed from engine surfaces. Ensure that all fuel–soaked cloths or towels are deposited into a flame–proof waste container with a lid.
- Always keep a dry chemical (Class B) fire extinguisher near the work area.
- Do not allow fuel spray or fuel vapors to come into contact with a spark or open flame.
- Always use a second wrench when loosening or tightening fuel line connection fittings. This will prevent unnecessary stress and torsion on fuel piping. Always follow the proper torque specifications.
- Always replace worn fuel fitting O–rings with new ones. Do not substitute fuel hose where rigid pipe is installed.

FUEL SYSTEM PRESSURE

RELIEVING

See Figure 106.

⁂ CAUTION

There may be some residual pressure even after this procedure, so cover the hose connection with a shop towel to prevent residual fuel from spilling out before disconnecting any fuel connection.

1. Turn the ignition switch OFF and disconnect the battery (–) cable.
2. Remove the fuel pump relay (A).

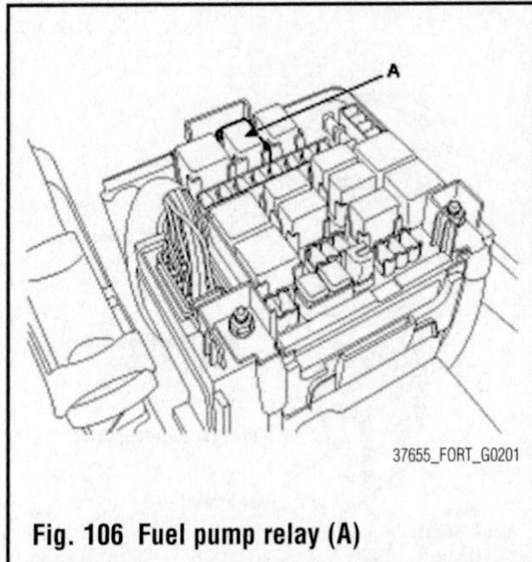

Fig. 106 Fuel pump relay (A)

➡When removing the fuel pump relay, a Diagnostic Trouble Code (DTC) may occur. Delete the code with the GDS after completion of this procedure.

3. Connect the battery (–) cable.
4. Start the engine and let idle, and then turn the ignition switch OFF after the engine has stopped on its own.
5. Disconnect the battery (–) cable, and then install the fuel pump relay (A).
6. Connect the battery (–) cable.
7. Delete the Diagnostic Trouble Code (DTC) related the fuel pump relay with the GDS.

FUEL FILTER

REMOVAL & INSTALLATION

See Figures 107 through 111.

1. Remove the fuel pump. See Fuel Pump.
2. Disconnect the electric pump wiring connector (A) and the fuel sender wiring connector (B).
3. Disconnect the electric pump wiring connector (A) from the pump.
4. Remove the fuel sender (B) with sliding it downward after releasing the latch (C).

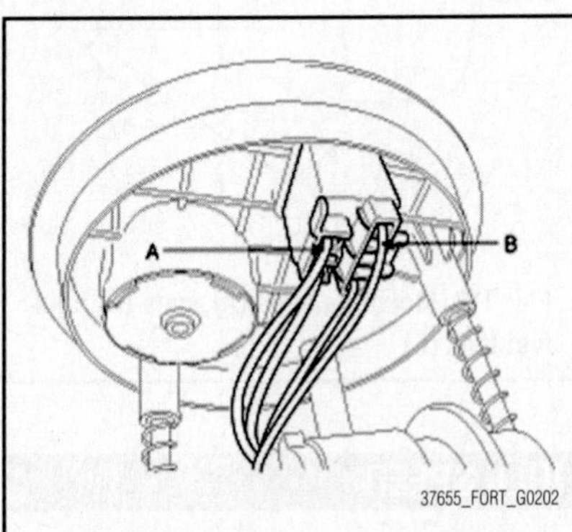

Fig. 107 Electric pump wiring connector (A) and fuel sender wiring connector (B)

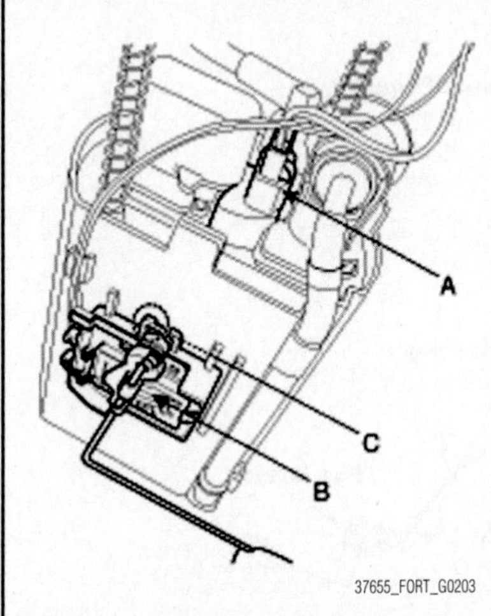

Fig. 108 Electric pump wiring connector (A), fuel sender (B) and latch (C)

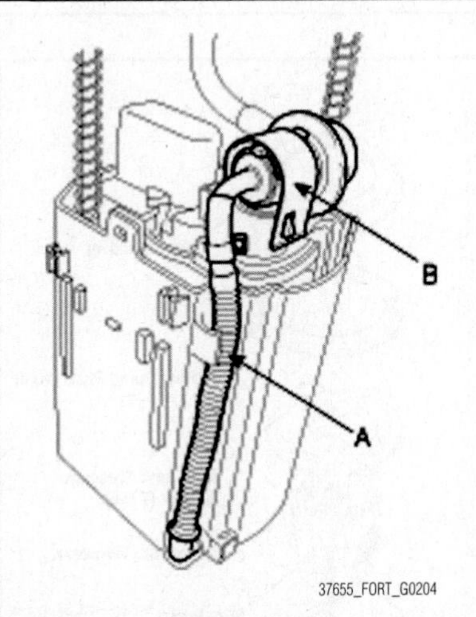

Fig. 109 Fuel pressure regulator & hose assembly (A) and cap (B)

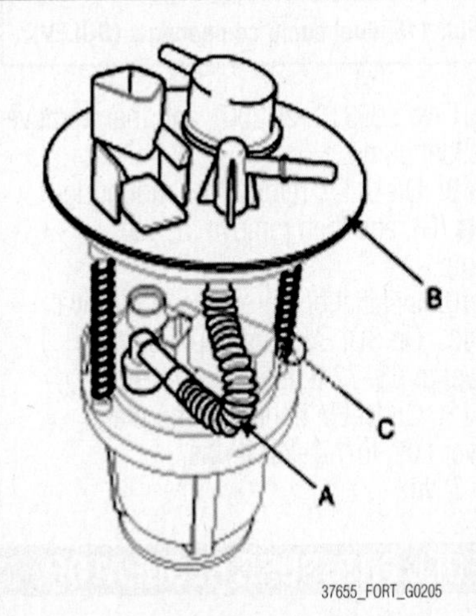

Fig. 110 Fuel feed tube (A), plate assembly (B) and cushion pipe fixing clip (C)

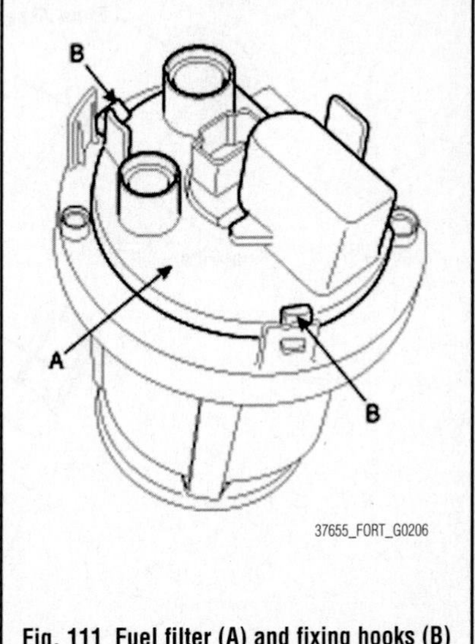

Fig. 111 Fuel filter (A) and fixing hooks (B)

5. Remove the fuel pressure regulator & hose assembly (A) after releasing the cap (B).
6. Remove the outer cover after releasing the three fixing hooks.
7. Remove the fuel feed tube (A) from the fuel filter after releasing the two fixing hooks.
8. Remove the plate assembly (B) after removing the cushion pipe fixing clip (C).
9. Extract the fuel filter (A) upward after releasing the two fixing hooks (B).
10. Extract the fuel filter two fixing hooks (B).
11. Installation is the reverse order of removal.

FUEL PUMP

REMOVAL & INSTALLATION

See Figures 112 through 116.

Fuel pump assembly includes fuel filter, pressure regulator, fuel sender and pump.

1. Release the residual pressure in fuel line. See Fuel Pump Pressure.
2. Remove the rear seat cushion.
3. Remove the service cover.
4. On SULEV, remove the rubber cover.
5. Disconnect the fuel pump connector (A) and the fuel tank pressure sensor connector (B).
6. Disconnect the fuel feed tube quick–connector (C) and the vapor tube quick–connector (D).
7. On ULEV, disconnect the vapor hose (E).
8. On SULEV, remove the fuel pump plate cover (F) with the special service tool

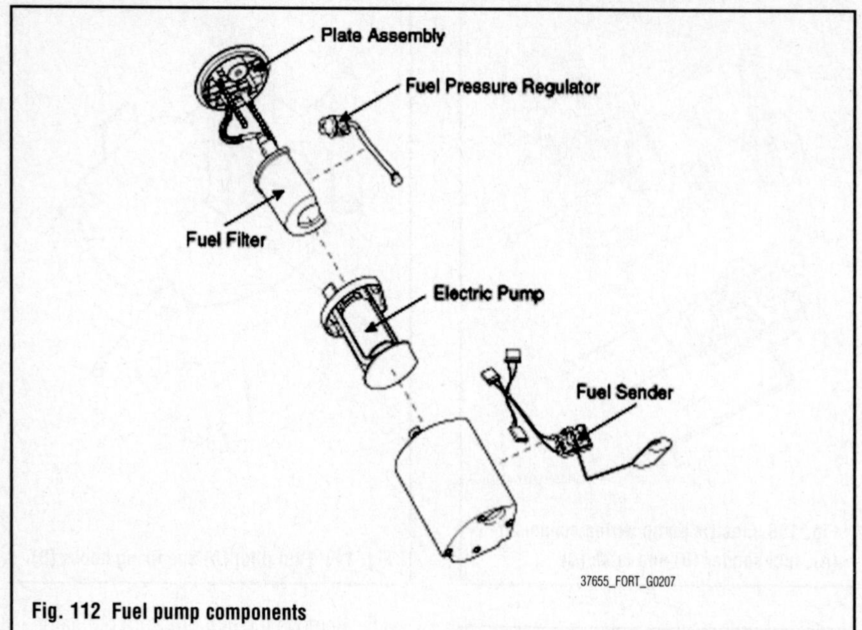

Fig. 112 Fuel pump components

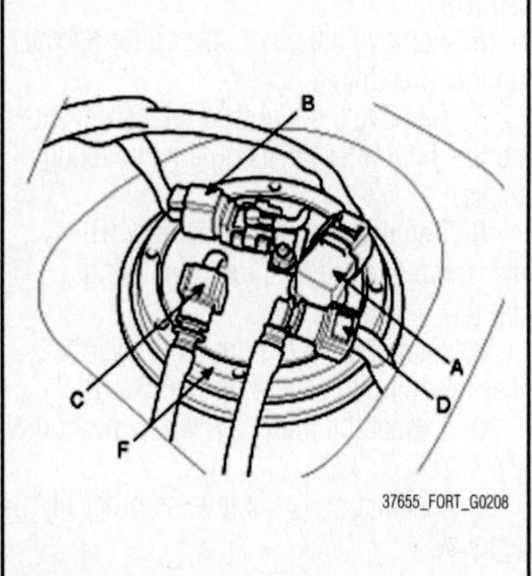

Fig. 113 Fuel pump connector (A), fuel tank pressure sensor connector (B), fuel feed tube quick–connector (C), vapor tube quick–connector (D) and fuel pump plate cover (F) (SULEV)

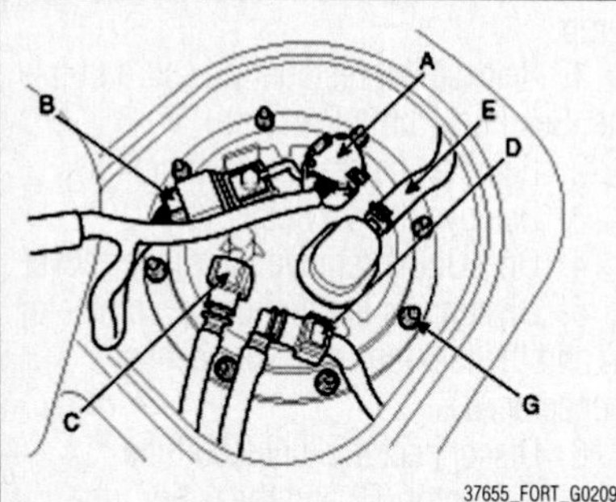

Fig. 114 Fuel pump connector (A), fuel tank pressure sensor connector (B), fuel feed tube quick–connector (C), vapor tube quick–connector (D) and mounting nuts (G) (ULEV)

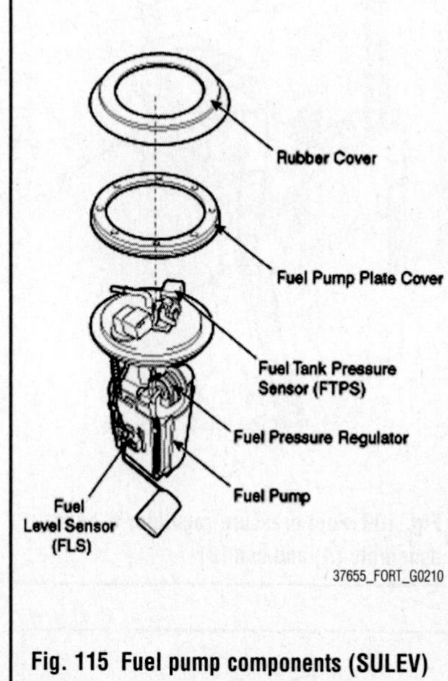

Fig. 115 Fuel pump components (SULEV)

(SST No.: 09310–2B200), and then remove the fuel pump.

9. On ULEV, remove the mounting nuts (G), and then remove the fuel pump.
10. Installation is reverse of removal.
11. On SULEV, tighten pump plate cover to 58–72 ft. lbs. (79–98 Nm).
12. On ULEV, tighten pump plate cover nuts to 1.4–2.2 ft. lbs. (2–3 Nm).

FUEL PRESSURE REGULATOR

REMOVAL & INSTALLATION

See Fuel Filter.

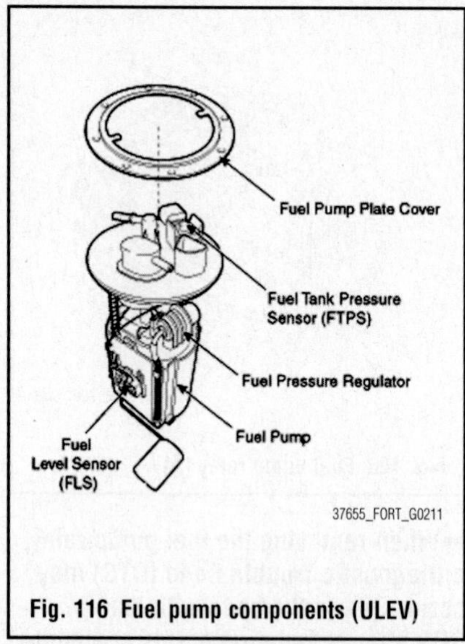

Fig. 116 Fuel pump components (ULEV)

FUEL TANK

REMOVAL & INSTALLATION

See Figure 117.

1. Remove the fuel pump. See Fuel Pump.
2. Lift the vehicle and support the fuel tank with a jack.
3. Remove the center muffler.
4. Disconnect the fuel filler hose.
5. Disconnect the vapor tube quick–connector and the vapor hose.
6. Remove the mounting bolts (A) and nuts (B), and then remove the fuel tank (C).
7. Installation is reverse of removal.
8. Tighten fuel tank bolts to 33–43 ft. lbs. (44–59 Nm).
9. Tighten fuel tank nuts to 29–40 ft. lbs. (40–54 Nm).

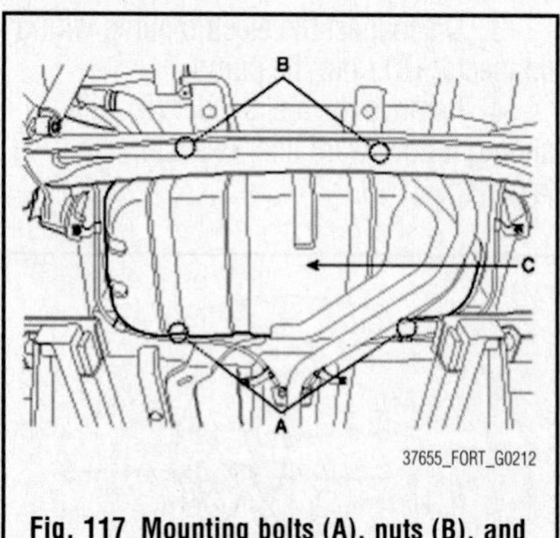

Fig. 117 Mounting bolts (A), nuts (B), and fuel tank (C)

IDLE SPEED

ADJUSTMENT

There is no adjustment for idle speed.

HEATING & AIR CONDITIONING SYSTEM

BLOWER MOTOR

REMOVAL & INSTALLATION

Removal requires removing heater unit. See Heater Unit.

HEATER CORE

REMOVAL & INSTALLATION

Removal requires removing heater unit. See Heater Unit.

HEATER UNIT

REMOVAL & INSTALLATION

See Figures 118 through 122.

1. Disconnect the negative (–) battery terminal.
2. Recover the refrigerant with a recovery/ recycling/ charging station.

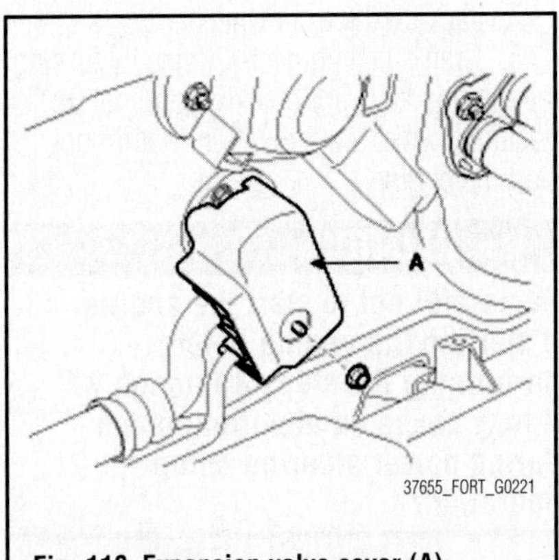

Fig. 118 Expansion valve cover (A)

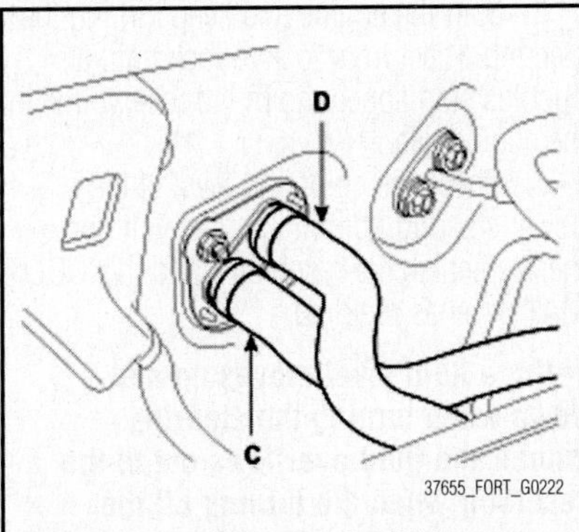

Fig. 119 Inlet (C) and outlet (D) heater hoses

3. When the engine is cool, drain the engine coolant from the radiator.
4. Remove the expansion valve cover (A).
5. Remove the bolts and the expansion valve from the evaporator core. Plug or cap the lines immediately after disconnecting them to avoid moisture and dust contamination.
6. Disconnect the inlet (C) and outlet (D) heater hoses from the heater unit.

✲✲ CAUTION

Engine coolant will spill when the hoses are disconnected; drain it into a clean drip pan. Be sure not to let coolant spill on electrical parts or painted surfaces. If any coolant spills, rinse it off immediately.

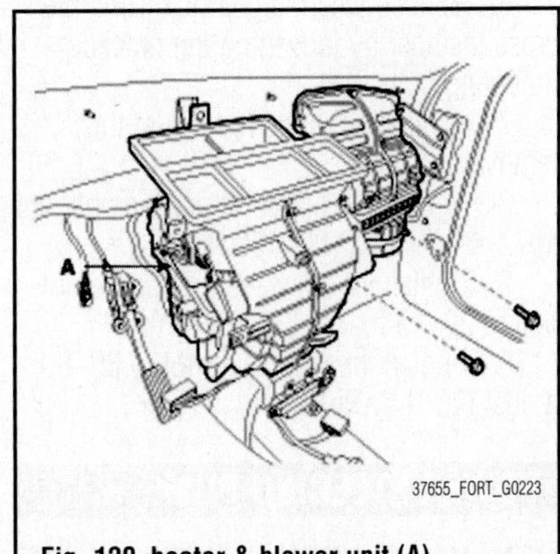

Fig. 120 heater & blower unit (A)

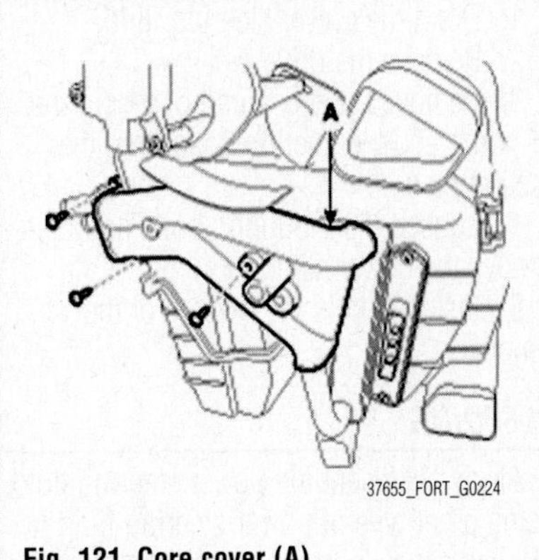

Fig. 121 Core cover (A)

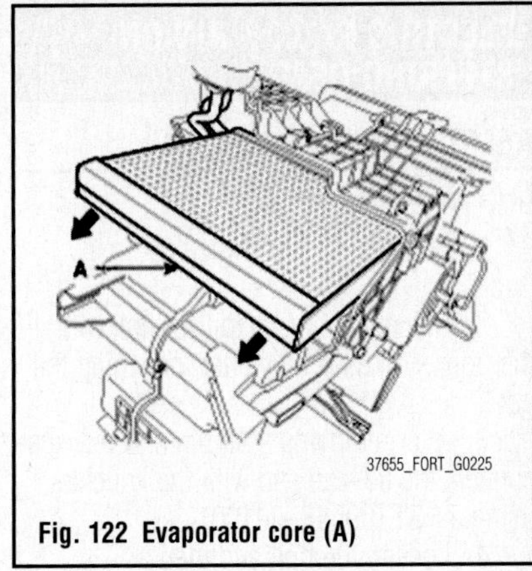

Fig. 122 Evaporator core (A)

7. Remove the instrument panel. See Instrument Panel.
8. Remove the cowl cross bar assembly.
9. Remove the heater & blower unit (A) after loosening mounting bolts.
10. Remove the blower unit from heater unit after loosening 3 screws.
11. Remove the cover (A).
12. Be careful that the inlet and outlet pipe are not bent during heater core removal, and pull out the heater core.
13. Remove the heater unit lower case.
14. Remove the evaporator core (A).
15. Installation is the reverse order of removal, and note these items :
    a. If you're installing a new evaporator, add refrigerant oil (ND–OIL8).
    b. Replace the O–rings with new ones at each fitting, and apply a thin coat of refrigerant oil before installing. Be sure to use the right O–rings for R–134a to avoid leakage.
    c. Immediately after using the oil, replace the cap on the container, and seal it to avoid moisture absorption.
    d. Do not spill the refrigerant oil on the vehicle ; it may damage paint ; if the refrigerant oil contacts the paint, wash off immediately.
    e. Apply sealant to the grommets.
    f. Make sure that there is no air leakage.
    g. Charge the system and test its performance.
    h. Do not interchange the inlet and outlet heater hoses and install the hose clamps securely.
    i. Refill the cooling system with engine coolant.

STEERING

ELECTRIC RACK & PINION STEERING GEAR

REMOVAL & INSTALLATION

See Figure 123.

1. Remove the front wheel & tire.
2. Disconnect the stabilizer link with the front strut assembly after loosening the nut.
3. After removing split pin and nut, disconnect the tie–rod end with the knuckle using a SST (09568–2J100).
4. Loosen the bolt and then disconnect the universal joint assembly from the pinion of the steering gear box.

➡Do not allow steering wheel to turn while rack is disconnected to prevent damage to clockspring.

5. Remove the cross member from the body by loosening the mounting bolts and nuts.
6. Remove steering gearbox from the cross member by loosening the bracket mounting bolts.
7. Installation is the reverse of the removal
8. Tighten the link to strut assembly nut to 72–87 ft. lbs. (100–120 Nm).
9. Tighten the universal joint assembly bolt to 22–25 ft. lbs. (30–35 Nm).
10. Tighten the wheel lug nut to 65–80 ft. lbs. (90–110 Nm).

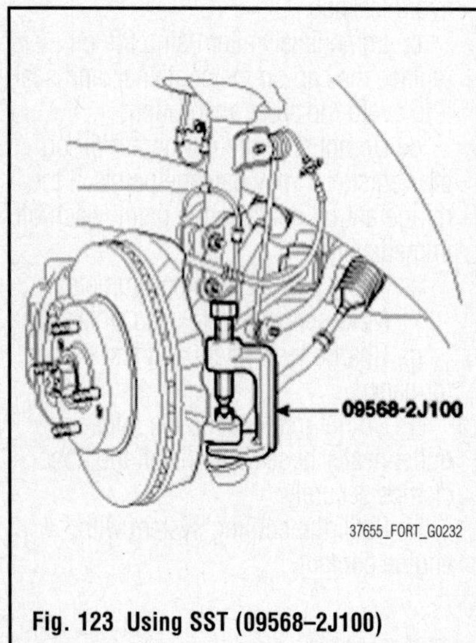

Fig. 123 Using SST (09568–2J100)

POWER RACK & PINION STEERING GEAR

REMOVAL & INSTALLATION

See Figure 123.

1. Remove the front wheel & tire.
2. Disconnect the stabilizer link with the front strut assembly after loosening the nut.
3. After removing split pin and nut, disconnect the tie–rod end with the knuckle using a SST (09568–2J100).
4. Loosen the bolt and then disconnect the universal joint assembly from the pinion of the steering gear box.

➡Do not allow steering wheel to turn while rack is disconnected to prevent damage to clockspring.

5. Remove the cross member from the body by loosening the mounting bolts and nuts.
6. Remove steering gearbox from the cross member by loosening the bracket mounting bolts.
7. Installation is the reverse of the removal
8. Tighten the link to strut assembly nut to 72–87 ft. lbs. (100–120 Nm).
9. Tighten the universal joint assembly bolt to 22–25 ft. lbs. (30–35 Nm).
10. Tighten the wheel lug nut to 65–80 ft. lbs. (90–110 Nm).

POWER STEERING PUMP

REMOVAL & INSTALLATION

See Figure 124.

1. Drain the power steering fluid.
2. Remove the drive belt.
3. Remove the pressure hose from the oil pump and the suction hose from the suction pipe.
4. Loosen the mounting bolts and then remove the power steering pump.
5. Installation is the reverse of the removal.

BLEEDING

Always use genuine power steering fluid. Using other type of power steering fluid or ATF can cause increased wear and poor steering in cold weather.

1. Fill the reservoir with the power steering fluid up to the level of 'COLD MAX' marked on the reservoir.

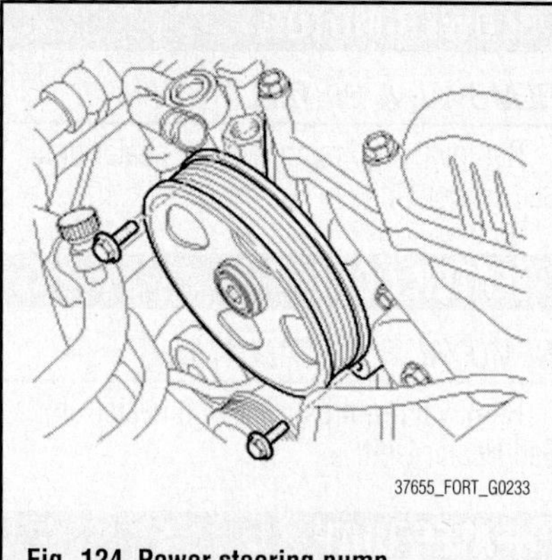

Fig. 124 Power steering pump

➡While conducting the following operations, keep replenishing the reservoir so that the fluid level can be always between the 'COLD MAX' and the 'COLD MIN' marked on the reservoir.

2. Jack up the front wheels.
3. Crank the engine 1–2times by turning the ignition key very quickly from the 'On' position to the 'Start' position, but do not start the engine.

⁂ CAUTION

Be careful not to start the engine. If starting the engine before performing the steps 3 through 4, it may cause an abnormal noise during power steering pump operation.

4. Turn the steering wheel from lock to lock 5–6 times for 15–20 seconds.
5. Start the engine and keep turning the steering wheel from lock to lock until air bubbles stop appearing in the reservoir with the engine idle.
6. Check the color and level of the power steering fluid in the reservoir and then replenish the reservoir up to the 'COLD MAX' level as required.

➡If the fluid level moves up and down when turning the steering wheel, the fluid overflows out of the reservoir when the turning off the engine or the fluid has white color, it indicates that air bubbles have not been removed sufficiently from the power steering system. Therefore, repeat the steps 5 through 6 as required.

LOWER BALL JOINT

REMOVAL & INSTALLATION

1. Remove the front wheel & tire.
2. Remove the lower arm ball joint mounting bolts.
3. Remove the lower arm from the knuckle.
    a. Remove the split pin.
    b. Remove the castle nut.
    c. Disconnect the lower arm from knuckle using the special tool (09568–4A000).

➡Be careful not to damage the boot and rotor teeth.

4. Installation is the reverse of removal.
5. Tighten the lower ball joint mounting bolts to 72–87 ft. lbs. (100–120 Nm).
6. Tighten the wheel lug nut to 65–80 ft. lbs. (90–110 Nm).

LOWER CONTROL ARM

REMOVAL & INSTALLATION

See Figure 125.

1. Remove the front wheel & tire.
2. Remove the lower arm ball joint mounting bolts.
3. Remove the lower arm mounting bolts.
4. Using the special tools (09214–32000 & 09216–211000), remove the bushing from the lower arm.

➡Separation force is over 800Kg

5. Apply soap solution to the following parts.
    a. Outer surface of the bushing.
    b. Inner surface of the lower bushing mounting part.
6. Using the special tools (09214–32000 & 09216–21100), install the busing on the lower arm. Remove old bushing on the lower arm.
7. Installation is the reverse of removal.
8. Tighten the lower arm ball joint mounting bolts to 72–87 ft. lbs. (100–120 Nm).
9. Tighten the horizontal inner control arm bolt to 72–87 ft. lbs. (100–120 Nm).
10. Tighten the vertical inner control arm bolt to 101–115 ft. lbs. (140–160 Nm).
11. Tighten the wheel lug nut to 65–80 ft. lbs. (90–110 Nm).

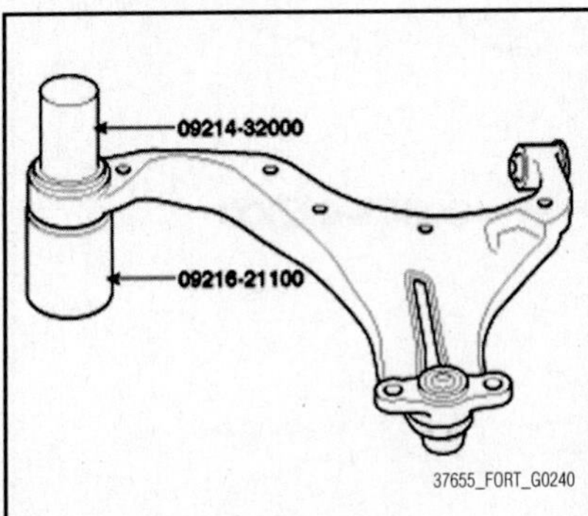

Fig. 125 Using special tools (09214–32000 & 09216–211000)

MACPHERSON STRUT

REMOVAL & INSTALLATION

1. Remove the front wheel & tire.
2. Remove the brake hose and the wheel speed sensor bracket from the front strut assembly by loosening the mounting bolts.
3. Disconnect the stabilizer link with the front strut assembly after loosening the nut.
4. Remove the top cap and loosen the strut mounting nuts.
5. Disconnect the front strut assembly with the knuckle by loosening the bolt & nut.
6. Installation is the reverse of removal.
7. Tighten the link to strut assembly nut to 72–87 ft. lbs. (100–120 Nm).
8. Tighten the strut assembly top nuts to 32–40 ft. lbs. (45–55 Nm).
9. Tighten the knuckle to strut assembly bolts to 144–202 ft. lbs. (137–160 Nm).
10. Tighten the wheel lug nut to 65–80 ft. lbs. (90–110 Nm).

OVERHAUL

See Figures 126 through 129.

1. Using the special tool (09546–26000), compress the coil spring (A) until there is only a little tension of the spring on the strut.
2. Remove the self–locking nut from the strut assembly.

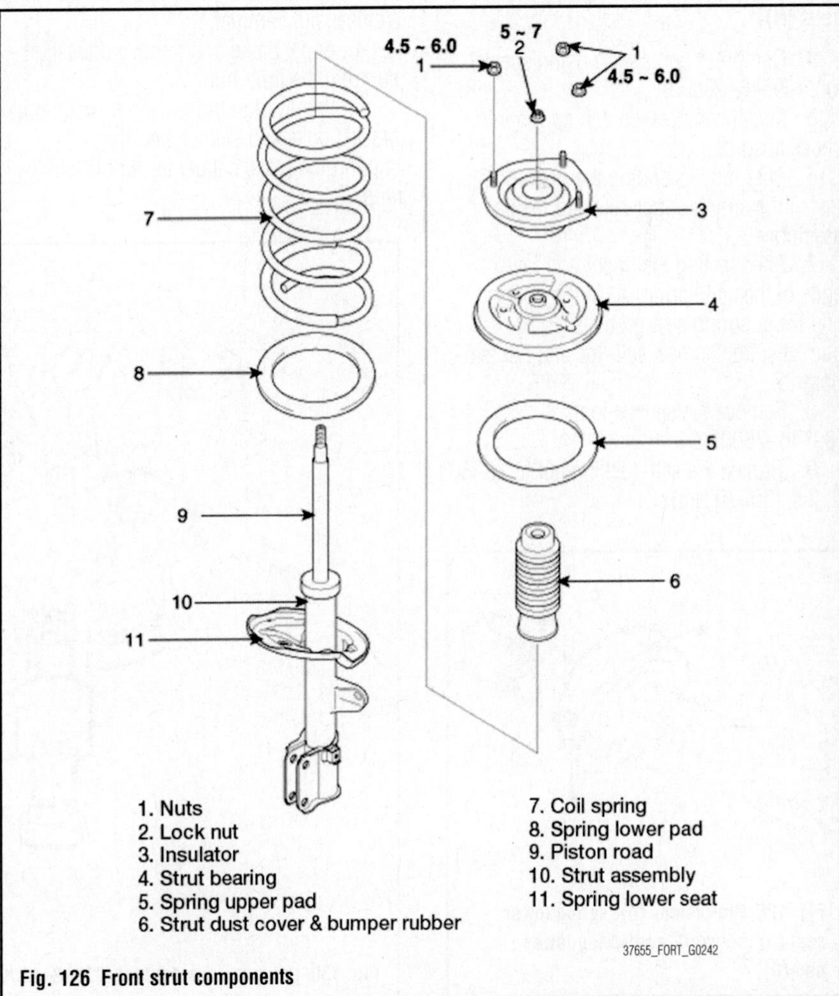

Fig. 126 Front strut components

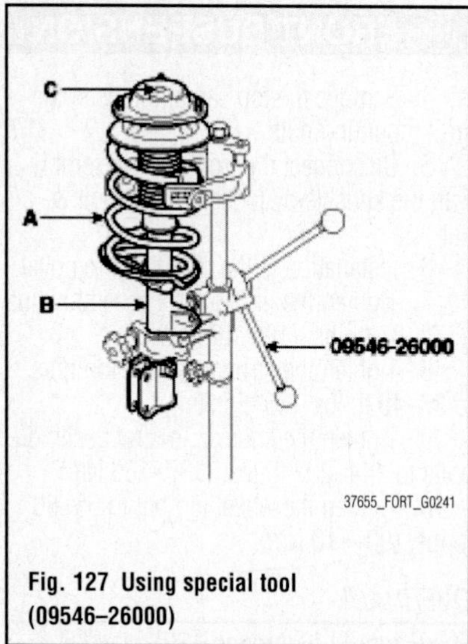

Fig. 127 Using special tool (09546–26000)

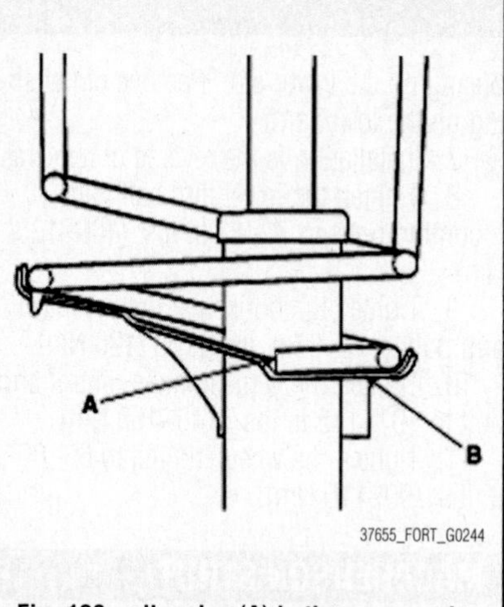

Fig. 129 coil spring (A) in the upper and lower spring seat grooves (B)

3. Remove the insulator, spring seat, coil spring and dust cover from the strut assembly.

To install:

➡Install the spring lower pad (D) so that the protrusions (A) fit in the holes (C) in the spring lower seat (B).

4. Compress coil spring using special tool (09546–26000).
5. Install compressed coil spring into shock absorber.
6. After fully extending the piston rod, install the spring upper seat and insulator assembly.
7. After seating the upper and lower ends of the coil spring (A) in the upper and lower spring seat grooves (B) correctly, tighten new self–locking nut temporarily.
8. Remove the special tool (09546–26000).
9. Tighten the self–locking nut to 36–50 ft. lbs. (50–70 Nm)..

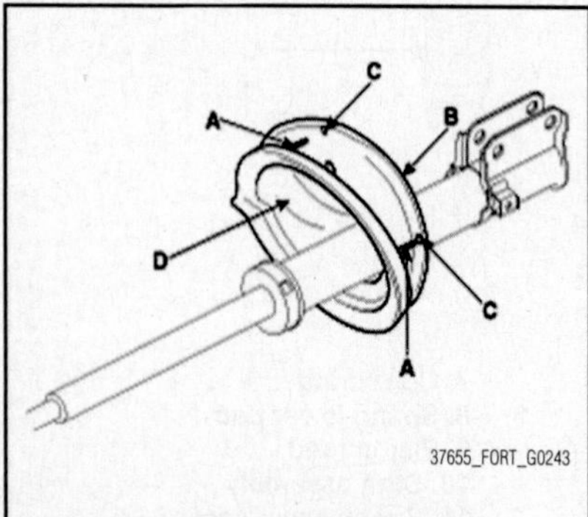

Fig. 128 Protrusions (A), spring lower seat (B), holes (C) and spring lower pad (D)

STEERING KNUCKLE

REMOVAL & INSTALLATION

See Figures 130 and 131.

1. Raise and support the vehicle.
2. Remove the front wheel and tire.
3. Unstake the driveshaft lock nut using a chisel and hammer.
4. Apply brake and remove driveshaft nut from the front hub.
5. Remove the brake caliper mounting bracket bolts and caliper, and then support the brake caliper assembly with wire.
6. Remove the tie rod end ball joint from the knuckle.
   a. Remove the split pin.
   b. Remove the castle nut.
   c. Disconnect the tie rod end ball joint from knuckle using the special tool (09568–4A000).
7. Remove the wheel speed sensor.
8. Remove the lower arm from the knuckle.
   a. Remove the split pin.
   b. Remove the castle nut.
   c. Disconnect the lower arm from knuckle using the special tool (09568–4A000).

➡Be careful not to damage the boot and rotor teeth.

9. Disconnect the driveshaft from the front hub assembly.
10. Loosen the strut mount bolts and then remove the knuckle assembly.
11. Installation is the reverse of removal.
12. Tighten the lower ball joint mounting bolts to 72–87 ft. lbs. (100–120 Nm).
13. Tighten the knuckle to strut assembly bolts to 144–202 ft. lbs. (137–160 Nm).
14. Tighten the tie rod end nut to 17–25 ft. lbs. (23–32 Nm).
15. Tighten the wheel lug nut to 65–80 ft. lbs. (90–110 Nm).

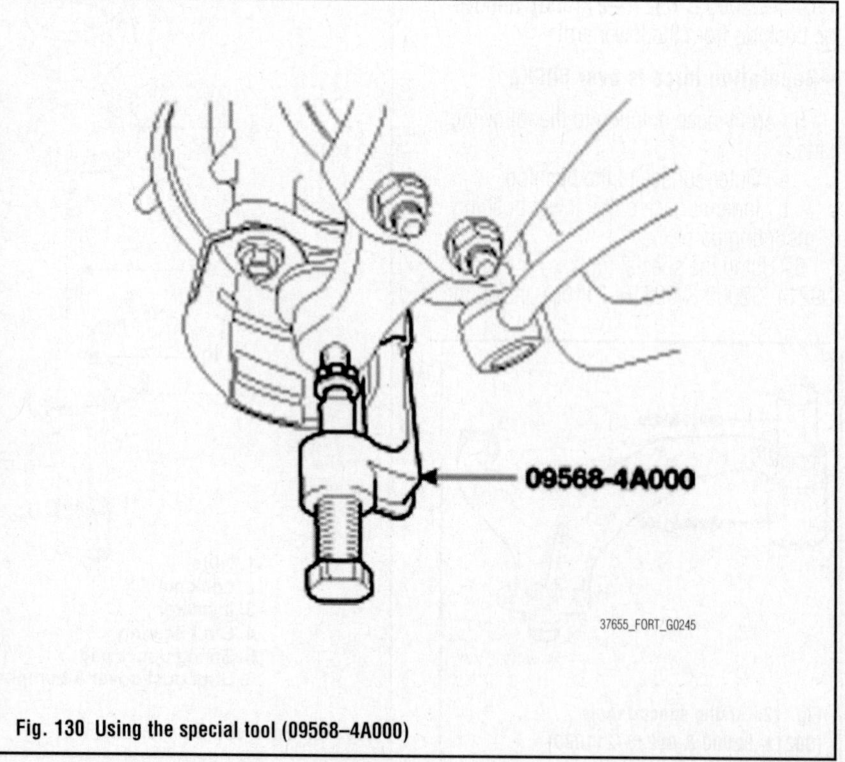

Fig. 130 Using the special tool (09568–4A000)

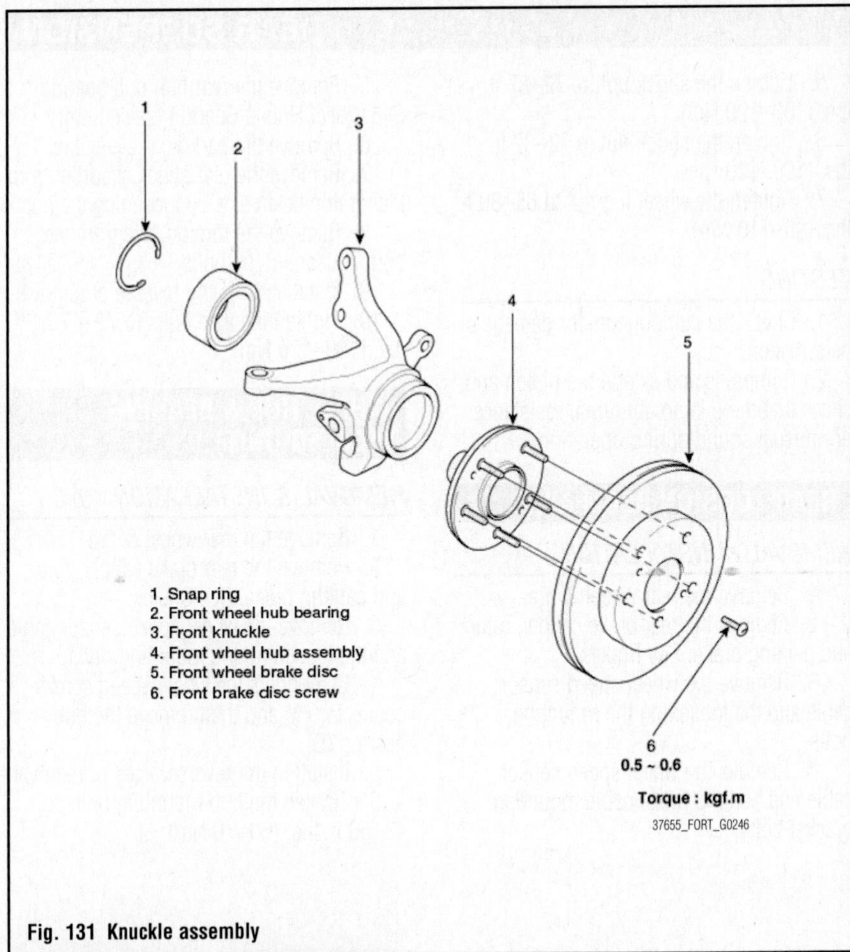

Fig. 131 Knuckle assembly

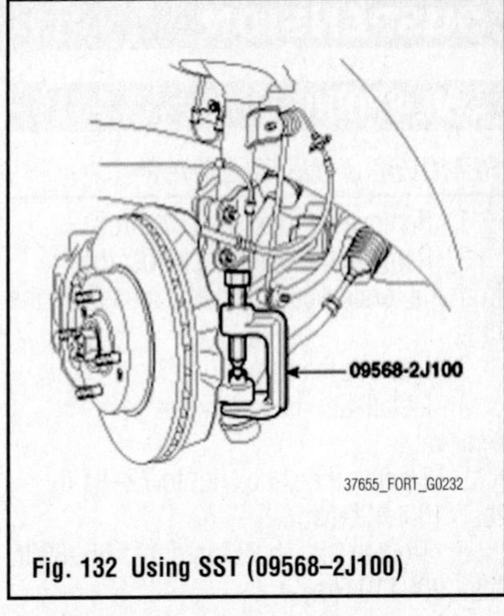

Fig. 132 Using SST (09568–2J100)

STABILIZER BAR

REMOVAL & INSTALLATION

See Figure 132.

1. Remove the front wheel & tire.
2. Disconnect the stabilizer link with the front strut assembly after loosening the nut.
3. Disconnect the tie–rod end with the knuckle using a SST (09568–2J100).
4. Remove the two bolts for lower arm ball joint.
5. Loosen the bolt and then disconnect the universal joint assembly from the pinion of the steering gear box.

CAUTION

Keep the neutral–range to prevent the damage of the clock spring inner cable when you handle the steering wheel.

6. Remove the cross member from the body by loosening the mounting bolts and nuts.
7. Remove the stabilizer from the cross member by loosening the bracket mounting bolts.
8. Installation is the reverse of removal.
9. Tighten the lower ball joint mounting bolts to 72–87 ft. lbs. (100–120 Nm).
10. Tighten the link to strut assembly nut to 72–87 ft. lbs. (100–120 Nm).
11. Tighten the stabilizer bracket bolts to 32–40 ft. lbs. (45–55 Nm).
12. Tighten the wheel lug nut to 65–80 ft. lbs. (90–110 Nm).

WHEEL BEARINGS

REMOVAL & INSTALLATION

1. Remove Steering Knuckle. See Steering Knuckle.
2. Remove the snap ring.
3. Remove bearing from back of knuckle.
4. Installation is the reverse of removal.

COIL SPRING

REMOVAL & INSTALLATION

1. Remove the rear wheel & tire.
2. Remove the rear shock absorber from the torsion beam axle by loosening the nut.
3. Remove the coil spring.
4. Installation is the reverse of removal.
5. Tighten the shock nut to 72–87 ft. lbs. (100–120 Nm).
6. Tighten the wheel lug nut to 65–80 ft. lbs. (90–110 Nm).

SHOCK ABSORBER

REMOVAL & INSTALLATION

1. Remove the rear wheel & tire.
2. Remove the rear shock absorber from the torsion beam axle by loosening the nut.
3. Remove the rear shock absorber from the frame by loosening the bolt.
4. Installation is the reverse of removal.
5. Tighten the shock bolt to 72–87 ft. lbs. (100–120 Nm).
6. Tighten the shock nut to 72–87 ft. lbs. (100–120 Nm).
7. Tighten the wheel lug nut to 65–80 ft. lbs. (90–110 Nm).

TESTING

1. Check the components for damage or deformation.
2. Compress and extend the piston and check that there is no abnormal resistance or unusual sound during operation.

REAR TORSION BEAM AXLE

REMOVAL & INSTALLATION

1. Remove the rear wheel & tire.
2. Remove the rear brake caliper, rotor and parking brake. See Brakes.
3. Remove the wheel speed sensor cable and the loosening the mounting bolts.
4. Remove the wheel speed sensor cable and parking brake cable mounting bracket bolts.
5. Remove the rear hub unit bearing. See Wheel Hub & Bearing (sealed unit).
6. Remove the parking brake cable.
7. Remove the rear shock absorber from the torsion beam axle by loosening the nut.
8. Remove the torsion axle from the body loosening the bolts.
9. Installation is the reverse of removal.
10. Tighten the axle bolts to 72–87 ft. lbs. (100–120 Nm).

WHEEL HUB & BEARING (SEALED UNIT)

REMOVAL & INSTALLATION

1. Remove the rear wheel & tire.
2. Remove the rear brake caliper, rotor and parking brake. See Brakes.
3. Remove the wheel speed sensor cable and the loosening the mounting bolts.
4. Disconnect the wheel speed sensor connector (A) and then remove the hub bearing (B).
5. Install in the reverse order of removal.
6. Tighten the hub mounting bolts to 43–50 ft. lbs. (60–70 Nm).

KIA

Optima

14

SPECIFICATIONS AND MAINTENANCE CHARTS

ENGINE AND VEHICLE IDENTIFICATION

Engine							Model Year	
Code ①	Liters (cc)	Cu. In.	Cyl.	Fuel Sys.	Engine Type	Eng. Mfg.	Code ②	Year
3	2.4 (2359)	144	4	MPFI	DOHC	KIA	9	2009
8	2.4 (2359)	144	4	MPFI	DOHC	KIA	A	2010
4	2.7 (2656)	163	6	MPFI	DOHC	KIA		

MPFI: Multi-Point Fuel Injection

DOHC: Double Overhead Camshafts

① 8th digit of VIN

② 10th digit of VIN

37655_OPTI_C0001

GENERAL ENGINE SPECIFICATIONS

Year	Model	Engine Displacement Liters	Engine VIN	Net Horsepower @ rpm	Net Torque @ rpm (ft. lbs.)	Bore x Stroke (in.)	Com-pression Ratio	Oil Pressure @ rpm
2009	Optima	2.4	3	149@6000	159@4500	3.41x3.94	10.0:1	43-57@3000
		2.7	4	178@6000	181@4000	3.413x2.952	10.4:1	18.77@1000
2010	Optima	2.4	8	175@6000	169@4000	3.464x3.819	10.5:1	21.34@1000
		2.7	4	194@6000	184@4500	3.413x2.952	10.4:1	18.77@1000

37655_OPTI_C0002

ENGINE TUNE-UP SPECIFICATIONS

Year	Engine Displacement Liters	Engine VIN	Spark Plug Gap (in.)	Ignition Timing (deg.)		Fuel Pump (psi)	Idle Speed (rpm)		Valve Clearance	
				MT	AT		MT	AT	Intake	Exhaust
2009	2.4	3	0.039-0.043	3-13B	3-13B	49.0-50.5	520-720	520-720	①	②
	2.7	4	0.039-0.043	—	-3-17B	55	—	550-750	①	②
2010	2.4	8	0.039-0.043	3-13B	3-13B	49.0-50.5	520-720	520-720	①	②
	2.7	4	0.039-0.043	—	-3-17B	55	—	550-750	①	②

NOTE: The Vehicle Emission Control Information label often reflects specification changes made during production.
The label figures must be used if they differ from those in this chart.

B: Before top dead center

HYD: Hydraulic

① 0.0039- 0.0118 inch

② 0.0078- 0.0157 inch

37655_OPTI_C0003

CAPACITIES

Year	Model	Engine Displacement Liters	Engine VIN	Engine Oil with Filter	Transaxle (pts.) Manual	Transaxle (pts.) Auto.	Fuel Tank (gal.)	Cooling System (qts.)
2009	Optima	2.4	3	4.23	4.6	16.4	16.4	7.40
		2.7	4	4.75	—	20.6	16.4	8.98
2010	Optima	2.4	8	4.23	4.6	16.4	16.4	7.40
		2.7	4	4.75	—	20.6	16.4	8.98

37655_OPTI_C0004

FLUID SPECIFICATIONS

Year	Model	Engine Displ. Liters (VIN)	Engine Oil	Man. Trans.	Auto. Trans.	Drive Axle Front	Drive Axle Rear	Power Steering Fluid	Brake Master Cylinder	Cooling System
2009	Optima	2.4 (3)	①	②	③	—	—	PSF 4	DOT 3	④
		2.7 (4)	⑤	②	③	—	—	PSF 4	DOT 3	④
2010	Optima	2.4 (8)	①	②	③	—	—	PSF 4	DOT 3	④
		2.7 (4)	⑤	②	③	—	—	PSF 4	DOT 3	④

DOT: Department Of Transpotation

① 5W-20. API SJ or SL ABOVE

② Hyundai genuine part MTF 75W/90 (API GL-4)

③ Diamond ATF SP-III or SK ATF SP III

④ Ethlyene glycol base for aluminium radiator

⑤ 5W-20, GF3 oil. API SJ or SL ABOVE

37655_OPTI_C0014

VALVE SPECIFICATIONS

Year	Engine Displacement Liters	Engine VIN	Seat Angle (deg.)	Face Angle (deg.)	Maximum out of Square (in.)	Spring Free Length (in.)	Stem-to-Guide Clearance (in.) Intake	Stem-to-Guide Clearance (in.) Exhaust	Stem Diameter (in.) Intake	Stem Diameter (in.) Exhaust
2009	2.4	3	44.75-45.10	45.25-45.75	①	1.8677	0.00078-0.00185	0.00118-0.00212	0.2151-0.2157	0.2149-0.2153
	2.7	4	NA	45.0-45.5	①	1.8425	0.0008-0.0020	0.0014-0.0026	0.2348-0.2354	0.2343-0.2348
2010	2.4	8	44.75-45.10	45.25-45.75	①	1.8677	0.00078-0.00185	0.00118-0.00212	0.2151-0.2157	0.2149-0.2153
	2.7	4	NA	45.0-45.5	①	1.8425	0.0008-0.0020	0.0014-0.0026	0.2348-0.2354	0.2343-0.2348

NA: Not Available

① Less than 1.5 degrees

37655_OPTI_C0005

CAMSHAFT AND BEARING SPECIFICATIONS

All measurements are given in inches unless noted.

Year	Engine Displacement Liters	Engine VIN	Camshaft Cam Height Intake	Cam Height Exhaust	Shaft End-play	Bearing Cap Torque (ft. lbs.)
2009	2.4	3	1.7401	1.7716	0.0015-0.0062	①
	2.7	4	1.7520	1.7520	0.0039-0.0079	②
2010	2.4	8	1.7401	1.7716	0.0015-0.0062	①
	2.7	4	1.7520	1.7520	0.0039-0.0079	②

① M6 bolts: 7.9-9.4 ft. lbs. M8 bolts: 20.3-23.1 ft. lbs.

② 6x38 bolts: 8.0-9.4 ft. lbs. 6x38 bolts: 15.2-18.8 ft. lbs.

37655_OPTI_C0013

CRANKSHAFT AND CONNECTING ROD SPECIFICATIONS

All measurements are given in inches.

Year	Engine Displacement Liters	Engine VIN	Crankshaft Main Brg. Journal Dia.	Main Brg. Oil Clearance	Shaft End-play	Thrust on No.	Connecting Rod Journal Diameter	Oil Clearance	Side Clearance
2009	2.4	3	2.0049-2.0456	0.0007-0.0014	0.0027-0.0098	3	2.0079-2.0086	0.0009-0.0016	0.0039-0.0098
	2.7	4	2.4402-2.4409	0.0002-0.0009	0.0028-0.0098	3	2.0079-2.0086	0.0007-0.0014	0.0039-0.0098
2010	2.4	8	2.0049-2.0456	0.0007-0.0014	0.0027-0.0098	3	2.0079-2.0086	0.0009-0.0016	0.0039-0.0098
	2.7	4	2.4402-2.4409	0.0002-0.0009	0.0028-0.0098	3	2.0079-2.0086	0.0007-0.0014	0.0039-0.0098

37655_OPTI_C0007

PISTON AND RING SPECIFICATIONS

All measurements are given in inches.

Year	Engine Displacement Liters	Engine VIN	Piston Clearance	Ring Gap Top Compression	Ring Gap Bottom Compression	Ring Gap Oil Control	Ring Side Clearance Top Compression	Ring Side Clearance Bottom Compression	Ring Side Clearance Oil Control
2009	2.4	3	NA	0.0059-0.0118	0.0118-0.0204	0.0078-0.0275	0.0019-0.0031	0.0015-0.0031	0.0023-0.0051
	2.7	4	NA	0.0059-0.0118	0.0118-0.0177	0.0078-0.0275	0.0016-0.0031	0.0012-0.0027	0.0024-0.0059
2010	2.4	8	NA	0.0059-0.0118	0.0118-0.0204	0.0078-0.0275	0.0019-0.0031	0.0015-0.0031	0.0023-0.0051
	2.7	4	NA	0.0059-0.0118	0.0118-0.0177	0.0078-0.0275	0.0016-0.0031	0.0012-0.0027	0.0024-0.0059

NA: Not Available

37655_OPTI_C0006

TORQUE SPECIFICATIONS

All readings in ft. lbs.

Year	Engine Displacement Liters	Cylinder Head Bolts	Main Bearing Bolts	Rod Bearing Bolts	Crankshaft Damper Bolts	Flywheel Bolts	Manifold Intake	Manifold Exhaust	Spark Plugs	Oil Pan Drain Plug
2009	2.4	①	②	③	122.9-130.1	86.8-93.9	13.7-17.4	④	15-21	⑤
	2.7	⑥	⑦	⑧	123.0-130.2	52.8-55.7	13.7-17.4	21.7-25.3	15-21	25.3-32.5
2010	2.4	①	②	③	122.9-130.1	86.8-93.9	13.7-17.4	④	15-21	⑤
	2.7	⑥	⑦	⑧	123.0-130.2	52.8-55.7	13.7-17.4	21.7-25.3	15-21	25.3-32.5

① 23.9-26.8 ft. lbs., plus 90-95 degrees, plus an additional 90-95 degrees

② 10.8 (20.3-23.1) ft. lbs., plus 120-125 degrees

③ 13.0-15.9 ft. lbs., plus 88-92 degrees

④ Nut: 28.9-43.4 ft. lbs. M8 stay bolt: 13.7-20.3 ft. lbs. M10 stay bolt: 31.1-39.8 ft. lbs.

⑤ M6x10 bolts: 7.2 ft. lbs. M8x103 bolts: 19.5-22.4 ft. lbs.

⑥ 16.6-19.5 ft. lbs., plus 58-62 degrees, plus an additional 43-47 degrees

⑦ M10 bolt: 19.5-23.9 ft. lbs., plus 90-95 degrees. M8 bolts 9.4-13.7 ft. lbs., plus 90-95 degrees

⑧ 13.0-15.9 ft. lbs., plus 90-94 degrees

37655_OPTI_C0008

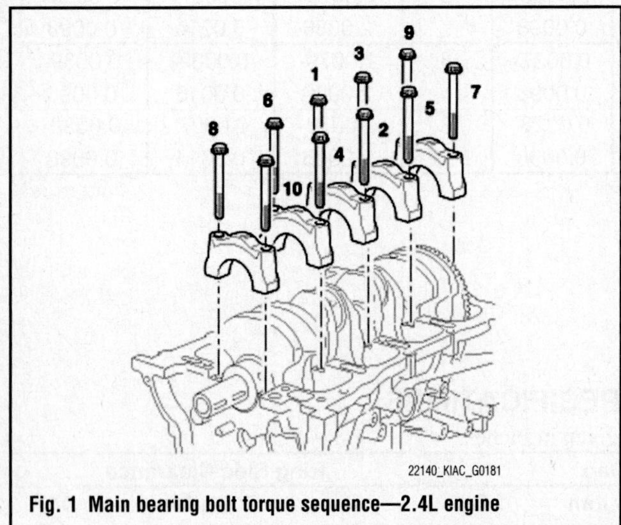

Fig. 1 Main bearing bolt torque sequence—2.4L engine

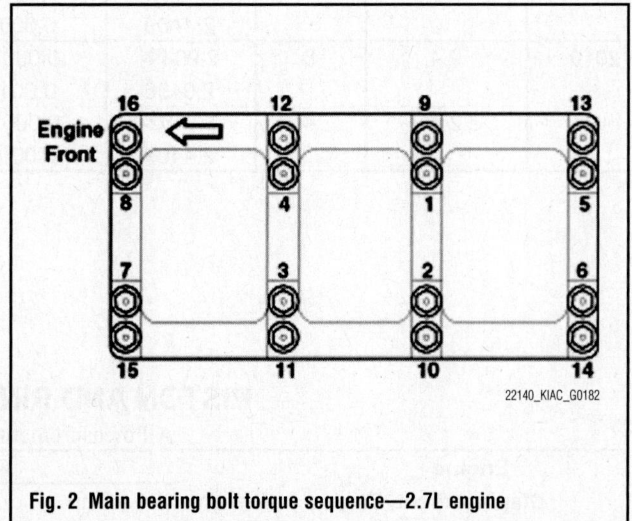

Fig. 2 Main bearing bolt torque sequence—2.7L engine

WHEEL ALIGNMENT

Year	Model		Caster Range (+/-Deg.)	Caster Preferred Setting (Deg.)	Camber Range (+/-Deg.)	Camber Preferred Setting (Deg.)	Toe-in (in.)
2009	Optima	F	①	+3.15	+/- 0.5	-0.5	②
		R	—	—	+/- 0.5	-1.0	②
2010	Optima	F	①	+3.15	+/- 0.5	-0.5	②
		R	—	—	+/- 0.5	-1.0	②

① To ground 4.75 degrees +/- 0.5 degrees

② Total 0 degrees +/- 0.2 degrees. Individual 0 degrees +/- 1 degree

37655_OPTI_C0009

TIRE, WHEEL AND BALL JOINT SPECIFICATIONS

Year	Model	OEM Tires Standard	OEM Tires Optional	Tire Pressures (psi) Front	Tire Pressures (psi) Rear	Wheel Size	Ball Joint Inspection	Lug Nut Torque (ft. lbs.)
2009	Optima	P205/60HR16	P215/50VR17 P215/50R17	①	①	②	③	65.1-79.6
2010	Optima	P205/60HR16	P215/50VR17 P215/50R17	①	①	②	③	65.1-79.6

OEM: Original Equipment Manufacturer

PSI: Pounds Per Square Inch

① Standard tire: 30. Optional tire: 33.

② Steel wheel: 6.5Jx16. Aluminum wheel: 6.5Jx16 or 6.5Jx17.

③ Replace if any measureable movent is found.

37655_OPTI_C0010

BRAKE SPECIFICATIONS

All measurements in inches unless noted

Year	Model		Brake Disc Original Thickness	Brake Disc Minimum Thickness	Brake Disc Maximum Run-out	Minimum Lining Thickness	Brake Caliper Bracket Bolts (ft. lbs.)	Brake Caliper Mounting Bolts (ft. lbs.)
2009	Optima	F	1.020	0.960	0.0016	0.430	57.9-72.3	15.9-23.1
		R	0.390	0.320	0.0039	0.590	36.1-43.4	15.9-23.1
2010	Optima	F	1.020	0.960	0.0016	0.430	57.9-72.3	15.9-23.1
		R	0.390	0.320	0.0039	0.590	36.1-43.4	15.9-23.1

F: Front

R: Rear

37655_OPTI_C0011

SCHEDULED MAINTENANCE INTERVALS

Kia - Optima

TO BE SERVICED	TYPE OF SERVICE	VEHICLE MILEAGE INTERVAL (x1000)												
		7.5	15	22.5	30	37.5	45	52.5	60	67.5	75	82.5	90	97.5
Accessory drive belts	S/I	✓	✓	✓	✓	✓	✓	✓	✓	✓	✓	✓	✓	✓
Air cleaner filter	R		✓		✓		✓		✓		✓		✓	
Air conditioner system filter	S/I	Inspect and replace every 10,000 miles or 10 months												
Brake lines, hoses and connections	S/I		✓		✓		✓		✓		✓		✓	
Chassis and body fasteners	T	✓	✓	✓	✓	✓	✓	✓	✓	✓	✓	✓	✓	✓
Cooling system hoses and coolant level	S/I		✓		✓		✓		✓		✓		✓	
CV-joint boots	S/I				✓				✓				✓	
Engine coolant	R				✓				✓				✓	
Engine oil and filter	R	✓	✓	✓	✓	✓	✓	✓	✓	✓	✓	✓	✓	✓
Exhaust system heat shields	S/I				✓				✓				✓	
Front and rear brakes	S/I	✓	✓	✓	✓	✓	✓	✓	✓	✓	✓	✓	✓	✓
Front ball joints	S/I				✓				✓				✓	
Fuel filter	R					✓					✓			
Fuel lines and hoses	S/I	✓	✓	✓	✓	✓	✓	✓	✓	✓	✓	✓	✓	✓
Locks and hinges	L	✓	✓	✓	✓	✓	✓	✓	✓	✓	✓	✓	✓	✓
Spark plugs	R								✓					
Steering operation and linkage	S/I	✓	✓	✓	✓	✓	✓	✓	✓	✓	✓	✓	✓	✓
Timing belt	R								✓					

R: Replace S/I: Service or Inspect L: Lubricate T: Tighten

FREQUENT OPERATION MAINTENANCE (SEVERE SERVICE)

If a vehicle is operated under any of the following conditions it is considered severe service

- Towing a trailer or using a camper or car-top carrier

- Repeated short trips of less than 5 miles in temperatures below freezing, or trips of less than 10 miles in any temperature

- Prolonged idling (vehicle operation in stop and go traffic).

- Operating on rough, muddy, unpaved, dusty or salt-covered roads.

- Police, taxi, delivery usage or trailer towing usage.

- Driving in extremely hot (over 90°F) conditions

Oil & oil filter: change every 5000 miles or 5 months, whichever occurs first.

Air cleaner filter: inspect every 15,000 miles or 15 months and replace everything 30,000 miles or 30 months, whichever occurs first

Fuel system hoses (California models only): replace every 105,000 miles

Emission system hoses (non-CA models): inspect every 55,000 or 55 months, whichever occurs first

Emission system hoses (CA models): inspect every 60,000 miles or 60 months, which occurs first

Front and rear brakes: inspect every 15,000 miles or 15 months, whichever occurs first

Chassis and body fasteners: tighten every 15,000 miles or 15 months, whichever occurs first

Locks and hinges: lubricate every 5000 miles or 5 months, whichever occurs first

37655_OPTI_C0012

BRAKES — INFORMATION AND PRECAUTIONS

ANTI-LOCK SYSTEMS

- Certain components within the ABS system are not intended to be serviced or repaired individually.
- Do not use rubber hoses or other parts not specifically specified for and ABS system. When using repair kits, replace all parts included in the kit. Partial or incorrect repair may lead to functional problems and require the replacement of components.
- Lubricate rubber parts with clean, fresh brake fluid to ease assembly. Do not use shop air to clean parts; damage to rubber components may result.
- Use only DOT 3 brake fluid from an unopened container.
- If any hydraulic component or line is removed or replaced, it may be necessary to bleed the entire system.
- A clean repair area is essential. Always clean the reservoir and cap thoroughly before removing the cap. The slightest amount of dirt in the fluid may plug an orifice and impair the system function. Perform repairs after components have been thoroughly cleaned; use only denatured alcohol to clean components. Do not allow ABS components to come into contact with any substance containing mineral oil; this includes used shop rags.
- The Anti-Lock control unit is a microprocessor similar to other computer units in the vehicle. Ensure that the ignition switch is **OFF** before removing or installing controller harnesses. Avoid static electricity discharge at or near the controller.
- If any arc welding is to be done on the vehicle, the control unit should be unplugged before welding operations begin.

DISC AND DRUM SYSTEMS

CAUTION

Dust and dirt accumulating on brake parts during normal use may contain asbestos fibers from production or aftermarket brake linings. Breathing excessive concentrations of asbestos fibers can cause serious bodily harm. Exercise care when servicing brake parts. Do not sand or grind brake lining unless equipment used is designed to contain the dust residue. Do not clean brake parts with compressed air or by dry brushing. Cleaning should be done by dampening the brake components with a fine mist of water, then wiping the brake components clean with a dampened cloth. Dispose of cloth and all residue containing asbestos fibers in an impermeable container with the appropriate label. Follow practices prescribed by the Occupational Safety and Health Administration (OSHA) and the Environmental Protection Agency (EPA) for the handling, processing, and disposing of dust or debris that may contain asbestos fibers.

BRAKES — BLEEDING THE BRAKE SYSTEM

BLEEDING PROCEDURE

BLEEDING PROCEDURE

See Figure 3.

WARNING

When bleeding the brakes, note the following:

- Do not reuse the drained fluid
- Always use Genuine DOT 3 or DOT 4 Brake Fluid. Using a non-Genuine DOT3 or DOT 4 brake fluid can cause corrosion and decrease the life of the system
- Make sure no dirt of other foreign matter is allowed to contaminate the brake fluid
- Do not spill brake fluid on the vehicle, it may damage the paint; if brake fluid does contact the paint, wash it off immediately with water
- The reservoir on the master cylinder must be at the MAX (upper) level mark at the start of bleeding procedure and checked after bleeding each brake caliper. Add fluid as required.

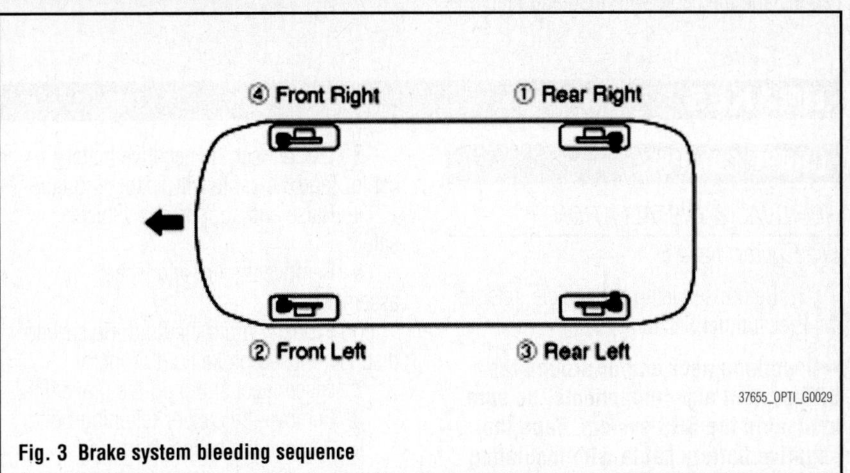

Fig. 3 Brake system bleeding sequence

1. Make sure the brake fluid level in the master cylinder fluid reservoir is at the MAX (upper) level line.
2. Have someone slowly pump the brake pedal several times, and then apply steady pressure
3. Loosen the right-rear brake bleed screw to allow air to escape from the system. Then tighten the bleed screw securely
4. Repeat the procedure for each wheel in the sequence shown below until air bubbles no longer appear in the fluid
5. Refill the master cylinder reservoir to the MAX (upper) level line

BLEEDING THE ABS SYSTEM

The ABS brake system is bled in the usual fashion with no special procedures required. Refer to the Bleeding Procedure above.

BRAKES — ANTI-LOCK BRAKE SYSTEM (ABS)

WHEEL SPEED SENSORS

REMOVAL & INSTALLATION

Front

1. Before servicing the vehicle, refer to the Precautions Section.

➡If working near and/or around the SRS system and components, be sure to disable the SRS system. Tape the negative battery cable with insulating tape. Always disconnect the negative battery cable first.

CAUTION

To avoid personal injury when working on vehicles equipped with an air bag, the negative battery cable must be disconnected and at least three minutes must elapse before working on the system. Failure to do so may result in deployment of the air bag.

2. Remove the battery top cover, if equipped.
3. Disconnect the negative battery cable. Tape the cable with insulating tape.
4. Raise and support the vehicle safely, as required.
5. Remove the front wheel speed sensor mounting bolt and cable mounting bolt.
6. Remove the front wheel speed sensor bracket.
7. Remove the front wheel guard.
8. Disconnect the front wheel speed sensor connector.
9. Remove the front wheel speed sensor.

To install:

➡Be sure to use new fasteners, as required.

10. Installation is the reverse of the removal procedure.

Rear

1. Before servicing the vehicle, refer to the Precautions Section.

➡If working near and/or around the SRS system and components, be sure to disable the SRS system. Tape the negative battery cable with insulating tape. Always disconnect the negative battery cable first.

CAUTION

To avoid personal injury when working on vehicles equipped with an air bag, the negative battery cable must be disconnected and at least three minutes must elapse before working on the system. Failure to do so may result in deployment of the air bag.

2. Remove the battery top cover, if equipped.
3. Disconnect the negative battery cable. Tape the cable with insulating tape.
4. Raise and support the vehicle safely, as required.
5. Remove the rear wheel speed sensor mounting bolt and cable mounting bolt.
6. Remove the rear wheel speed sensor bracket.
7. Disconnect the rear wheel speed sensor connector.
8. Remove the rear wheel speed sensor.

To install:

➡Be sure to use new fasteners, as required.

9. Installation is the reverse of the removal procedure.

BRAKES — FRONT DISC BRAKES

BRAKE CALIPER

REMOVAL & INSTALLATION

See Figures 4 and 5.

1. Before servicing the vehicle, refer to the Precautions Section.

➡If working near and/or around the SRS system and components, be sure to disable the SRS system. Tape the negative battery cable with insulating tape. Always disconnect the negative battery cable first.

CAUTION

To avoid personal injury when working on vehicles equipped with an air bag, the negative battery cable must be disconnected and at least three minutes must elapse before working on the system. Failure to do so may result in deployment of the air bag.

2. Remove the battery top cover, if equipped.
3. Disconnect the negative battery cable. Tape the cable with insulating tape.
4. Raise and support the vehicle safely.
5. Remove the tire and wheel assemblies.
6. Properly drain the fluid. Be sure to dispose of used brake fluid properly.
7. Disconnect and plug the brake line.
8. Remove the caliper retaining bolts.
9. Remove the caliper from its mounting.

➡On some vehicles it may be necessary to remove the pads and shims before removing the caliper retaining bolts.

10. Check the hoses and pin boots for damage and deterioration.
11. Remove the pad shims, pad retainers and pads.

To install:

➡Be sure to use new fasteners, as required.

12. Installation is the reverse of the removal procedure.
13. Tighten the caliper retaining bolts to specification.
14. Install the pad retainers.
15. Check the foreign material at the pad shims and the back of the pads.

CAUTION

Contaminated brake discs or pads reduce stopping ability. Keep grease off the discs and pads.

16. Install the brake pads and pad shims correctly. Install the pad with the wear indicator on the inside.

➡If you are reusing the pads, always reinstall the brake pads in their original positions to prevent a momentary loss of braking efficiency.

17. Push in the piston so that the caliper will fit over the pads. Make sure that the piston boot is in position to prevent damaging it when pivoting the caliper down.

➡Insert the piston in the cylinder using the special tool (09581-11000).

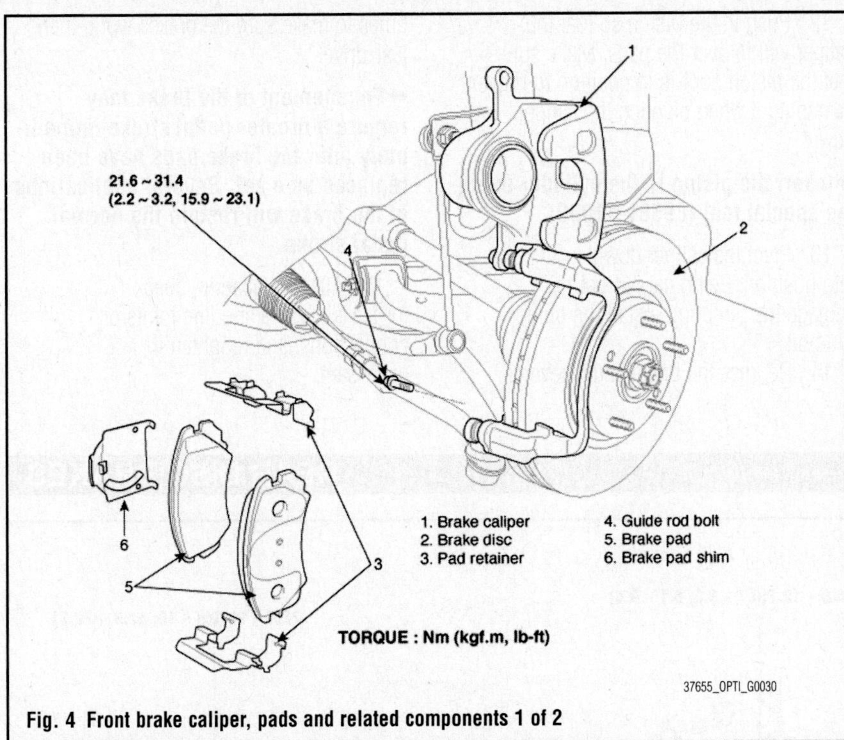

Fig. 4 Front brake caliper, pads and related components 1 of 2

18. Pivot the caliper down into position. Being careful not to damage the pin boot, install the guide rod bolt.

19. Depress the brake pedal several times to make sure the brakes work, then test-drive.

➡**Engagement of the brake may require a greater pedal stroke immediately after the brake pads have been replaced as a set. Several applications of the brake will restore the normal pedal stroke.**

20. After installation, check for leaks at hose and line joints or connections, and retighten if necessary.

DISC BRAKE PADS

REMOVAL & INSTALLATION

See Figures 4 and 5.

1. Before servicing the vehicle, refer to the Precautions Section.

➡**If working near and/or around the SRS system and components, be sure to disable the SRS system. Tape the negative battery cable with insulating tape. Always disconnect the negative battery cable first.**

CAUTION

To avoid personal injury when working on vehicles equipped with an air bag, the negative battery cable must be disconnected and at least three minutes must elapse before working on the system. Failure to do so may result in deployment of the air bag.

2. Remove the battery top cover, if equipped.

3. Disconnect the negative battery cable. Tape the cable with insulating tape.

4. Raise and support the vehicle safely.

5. Remove the tire and wheel assemblies.

6. Remove guide rod and the caliper up out of the way.

7. Check the hoses and pin boots for damage and deterioration.

8. Remove the pad shims, pad retainers and pads.

To install:

9. Install the pad retainers.

10. Check the foreign material at the pad shims and the back of the pads.

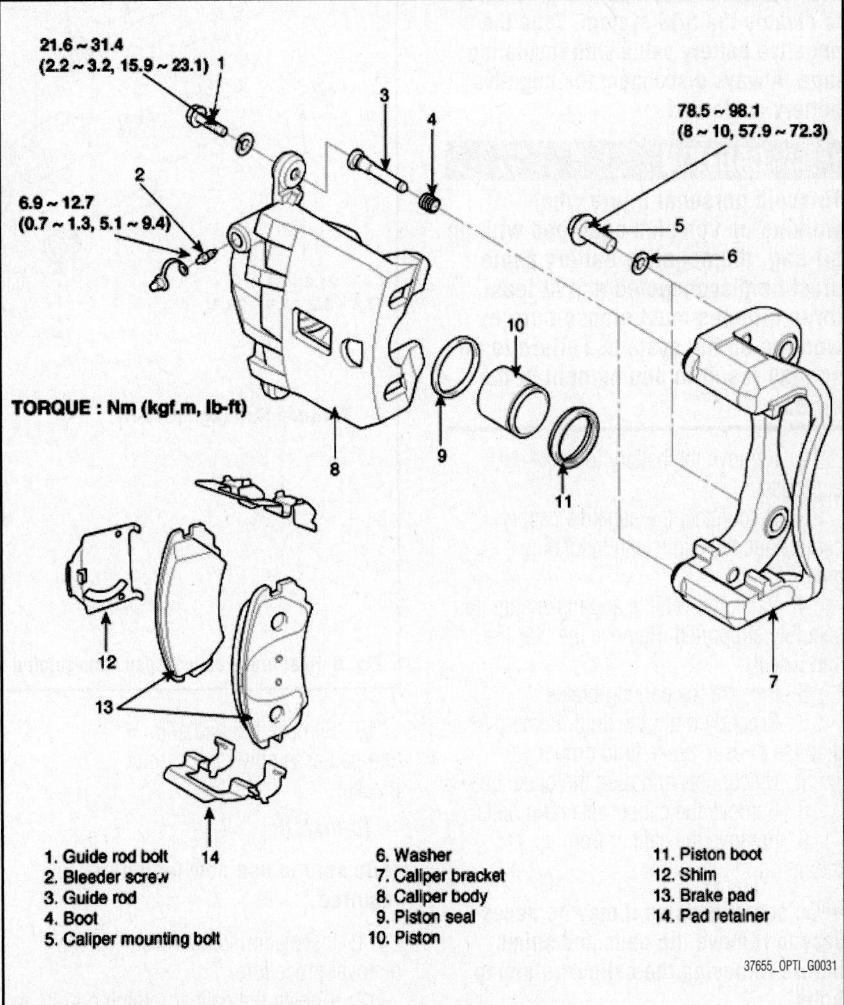

Fig. 5 Front brake caliper, pads, and related components 2 of 2

✻✻ CAUTION

Contaminated brake discs or pads reduce stopping ability. Keep grease off the discs and pads.

11. Install the brake pads and pad shims correctly. Install the pad with the wear indicator on the inside.

➡If you are reusing the pads, always reinstall the brake pads in their original positions to prevent a momentary loss of braking efficiency.

12. Push in the piston so that the caliper will fit over the pads. Make sure that the piston boot is in position to prevent damaging it when pivoting the caliper down.

➡Insert the piston in the cylinder using the special tool (09581-11000).

13. Pivot the caliper down into position. Being careful not to damage the pin boot, install the guide rod bolt.
14. Depress the brake pedal several times to make sure the brakes work, then test-drive.

➡Engagement of the brake may require a greater pedal stroke immediately after the brake pads have been replaced as a set. Several applications of the brake will restore the normal pedal stroke.

15. After installation, check for leaks at hose and line joints or connections, and retighten if necessary.

BRAKES — REAR DISC BRAKES

BRAKE CALIPER

REMOVAL & INSTALLATION

See Figure 6.

1. Before servicing the vehicle, refer to the Precautions Section.

➡If working near and/or around the SRS system and components, be sure to disable the SRS system. Tape the negative battery cable with insulating tape. Always disconnect the negative battery cable first.

✻✻ CAUTION

To avoid personal injury when working on vehicles equipped with an air bag, the negative battery cable must be disconnected and at least three minutes must elapse before working on the system. Failure to do so may result in deployment of the air bag.

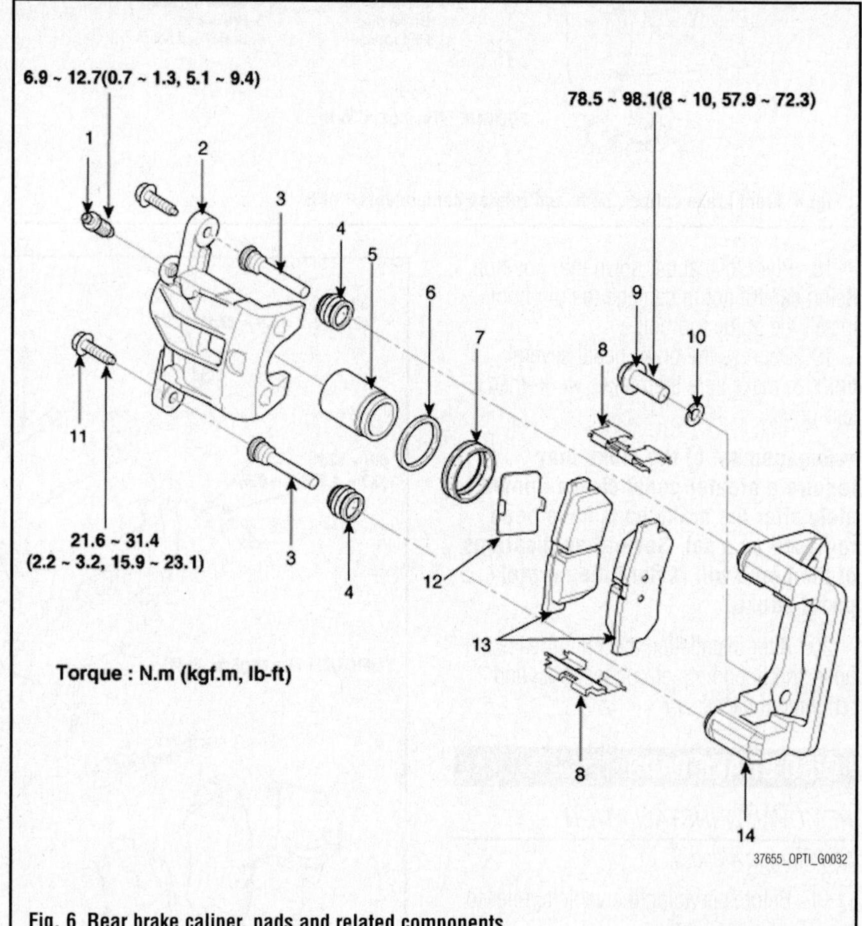

Fig. 6 Rear brake caliper, pads and related components

2. Remove the battery top cover, if equipped.
3. Disconnect the negative battery cable. Tape the cable with insulating tape.
4. Raise the vehicle and make sure it is securely supported. Remove the rear tire and wheel.
5. Release the parking brake.
6. Properly drain the fluid. Be sure to dispose of used brake fluid properly.
7. Disconnect and plug the brake line.
8. Remove the caliper retaining bolts.
9. Remove the caliper from its mounting.

➡On some vehicles it may be necessary to remove the pads and shims before removing the caliper retaining bolts.

10. Remove the pad shim and pad assembly from caliper bracket.

To install:

➡Be sure to use new fasteners, as required.

11. Installation is the reverse of the removal procedure.
12. Tighten the caliper retaining bolts to specification.
13. Install the pad retainers to the caliper.
14. Check the foreign material at the pad shim and the back of the pads.
15. Contaminated brake discs or pads reduce stopping ability. Keep grease off the discs and pads.
16. Install the brake pads and pad shim on the caliper bracket.
17. If you are reusing the pads, always reinstall the brake pads in their original

positions to prevent a momentary loss of braking efficiency. Push in the piston using SST (09581-11000) so that the caliper will fit over the pads. Make sure that the piston boot is in position to prevent damaging it when pivoting the caliper down.

18. Pivot caliper down into position. Being careful not to damage the pin boot, install the guide rod bolt and tighten it to the specified torque.

19. Install the brake caliper .

20. After installation, check for leaks at hose and line joints and connections, and retighten if necessary.

21. Depress the brake pedal several times to make sure the brakes work, then test-drive.

➡Engagement of the brake may require a greater pedal stroke immediately after the brake pads have been replaced as a set. Several applications of the brake pedal will restore the normal pedal stroke.

DISC BRAKE PADS

REMOVAL & INSTALLATION

See Figure 6.

1. Before servicing the vehicle, refer to the Precautions Section.

➡If working near and/or around the SRS system and components, be sure to disable the SRS system. Tape the negative battery cable with insulating tape. Always disconnect the negative battery cable first.

⁂ CAUTION

To avoid personal injury when working on vehicles equipped with an air bag, the negative battery cable must be disconnected and at least three minutes must elapse before working on the system. Failure to do so may result in deployment of the air bag.

2. Remove the battery top cover, if equipped.

3. Disconnect the negative battery cable. Tape the cable with insulating tape.

4. Raise the vehicle and make sure it is securely supported. Remove the rear tire and wheel.

5. Release the parking brake.

6. Remove the guide rod bolt.

7. Raise the caliper assembly, support it with a wire.

8. Remove the pad shim and pad assembly from caliper bracket.

To install:

9. Install the pad retainers to the caliper.

10. Check the foreign material at the pad shim and the back of the pads.

11. Contaminated brake discs or pads reduce stopping ability. Keep grease off the discs and pads.

12. Install the brake pads and pad shim on the caliper bracket.

13. If you are reusing the pads, always reinstall the brake pads in their original positions to prevent a momentary loss of braking efficiency. Push in the piston using SST (09581-11000) so that the caliper will fit over the pads. Make sure that the piston boot is in position to prevent damaging it when pivoting the caliper down.

14. Pivot caliper down into position. Being careful not to damage the pin boot, install the guide rod bolt and tighten it to the specified torque.

15. Install the brake caliper .

16. After installation, check for leaks at hose and line joints and connections, and retighten if necessary.

17. Depress the brake pedal several times to make sure the brakes work, then test-drive.

➡Engagement of the brake may require a greater pedal stroke immediately after the brake pads have been replaced as a set. Several applications of the brake pedal will restore the normal pedal stroke.

BRAKES — PARKING BRAKE

PARKING BRAKE CABLES

ADJUSTMENT

1. Block the front wheels, then raise the rear of the vehicle and make sure it is securely supported.

2. Make sure the parking brake arm on the rear brake caliper contacts the brake caliper pin.

3. Pull the parking brake lever up one click.

4. Remove the console.

5. Tighten the adjusting nuts until the parking brakes drag slightly when the rear wheels are turned.

6. Release the parking brake lever fully, and check that parking brakes do not drag when the rear wheels are turned. Readjust if necessary.

7. Make sure that the parking brakes are fully applied when the parking brake lever is pulled up fully.

8. Reinstall the console.

PARKING BRAKE SHOES

REMOVAL & INSTALLATION

See Figure 7.

1. Before servicing the vehicle, refer to the Precautions Section.

➡If working near and/or around the SRS system and components, be sure to disable the SRS system. Tape the negative battery cable with insulating tape. Always disconnect the negative battery cable first.

⁂ CAUTION

To avoid personal injury when working on vehicles equipped with an air bag, the negative battery cable must be disconnected and at least three minutes must elapse before working on the system. Failure to do so may result in deployment of the air bag.

2. Remove the battery top cover, if equipped.

3. Disconnect the negative battery cable. Tape the cable with insulating tape.

4. Remove the console.

5. Loosen the adjusting nut and the parking brake cables.

6. Disconnect the electrical connector of parking brake switch.

7. Remove the parking brake lever assembly by loosening the bolts.

8. Remove the wheel and tire.

9. Remove the brake shoe.

To install:

10. Install the removed parts in the reverse order of removal.

11. Apply a coating of the specified grease to each sliding parts of the ratchet plate or the ratchet pawl.

12. After installing the cable adjuster, adjust the parking brake lever stroke.

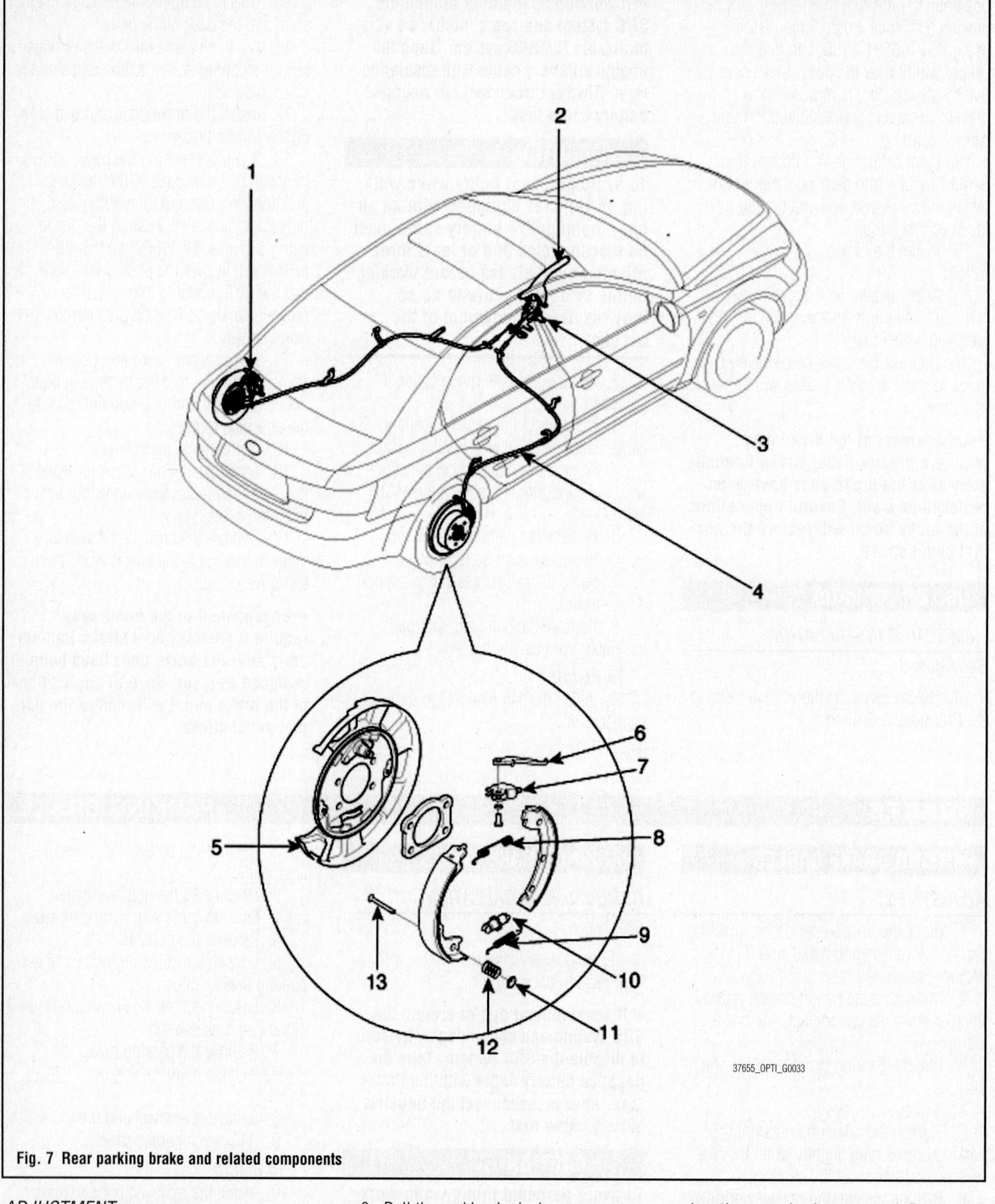

Fig. 7 Rear parking brake and related components

ADJUSTMENT

1. Block the front wheels, then raise the rear of the vehicle and make sure it is securely supported.
2. Make sure the parking brake arm on the rear brake caliper contacts the brake caliper pin.
3. Pull the parking brake lever up one click.
4. Remove the console.
5. Tighten the adjusting nuts until the parking brakes drag slightly when the rear wheels are turned.
6. Release the parking brake lever fully, and check that parking brakes do not drag when the rear wheels are turned. Readjust if necessary.
7. Make sure that the parking brakes are fully applied when the parking brake lever is pulled up fully.
8. Reinstall the console.

CHASSIS ELECTRICAL — AIR BAG (SUPPLEMENTAL RESTRAINT SYSTEM)

GENERAL INFORMATION

⁂ CAUTION

These vehicles are equipped with an air bag system. The system must be disarmed before performing service on, or around, system components, the steering column, instrument panel components, wiring and sensors. Failure to follow the safety precautions and the disarming procedure could result in accidental air bag deployment, possible injury and unnecessary system repairs.

PRECAUTIONS

⁂ CAUTION

Disconnect and isolate the battery negative cable before beginning any airbag system component diagnosis, testing, removal, or installation procedures. Allow system capacitor to discharge for two minutes before beginning any component service. This will disable the airbag system. Failure to disable the airbag system may result in accidental airbag deployment, personal injury, or death.

DISARMING THE SYSTEM

1. Before servicing the vehicle, refer to the Precautions Section.

➡If working near and/or around the SRS system and components, be sure to disable the SRS system. Tape the negative battery cable with insulating tape. Always disconnect the negative battery cable first.

⁂ CAUTION

To avoid personal injury when working on vehicles equipped with an air bag, the negative battery cable must be disconnected and at least three minutes must elapse before working on the system. Failure to do so may result in deployment of the air bag.

2. Turn the ignition switch to the **LOCK** position.
3. Remove the battery top cover, if equipped.
4. Disconnect the negative battery cable. Tape the cable with insulating tape.
5. Wait 3 minutes for the battery back-up power to discharge.

ARMING THE SYSTEM

1. Before servicing the vehicle, refer to the Precautions Section.
2. Connect the negative battery cable.
3. Turn the ignition switch **ON**.
4. Verify that the air bag indicator illuminates for about six seconds, then goes off.

CLOCKSPRING CENTERING

Prior to installing the clockspring, align the mating mark and "NEUTRAL" position indicator of the clockspring, and, after turning the front wheels to the straight-ahead position, install the clockspring to the column switch. If the mating mark of the clockspring is not properly aligned, the steering wheel may not completely rotate during a turn, or the flat cable within the clockspring may be severed, obstructing normal operation of the SRS and possibly leading to serious injury to the vehicle's driver.

DRIVE TRAIN

CLUTCH DRIVEN DISC & PRESSURE PLATE

REMOVAL & INSTALLATION

See Figures 8 and 9.

1. Before servicing the vehicle, refer to the Precautions Section.

➡If working near and/or around the SRS system and components, be sure to disable the SRS system. Tape the negative battery cable with insulating tape. Always disconnect the negative battery cable first.

⁂ CAUTION

To avoid personal injury when working on vehicles equipped with an air bag, the negative battery cable must be disconnected and at least three minutes must elapse before working on the system. Failure to do so may result in deployment of the air bag.

2. Remove the battery top cover, if equipped.
3. Disconnect the negative battery cable. Tape the cable with insulating tape.
4. Raise and safely support the vehicle.
5. Remove a transaxle assembly.
6. Remove the clutch cover bolts.
7. Remove the clutch cover and disc.

To install:

➡Be sure to use new fasteners, as required.

8. Using the SST (09411-11000) install the disc.

⁂ CAUTION

On 2.4L engine, replace a clutch cover and disc as a set. The 'T/M SIDE' marked surface should face the transaxle. If the surface faces the opposite side, there can be interference between a disc and a flywheel surface.

9. Tighten the clutch cover. Torque to 8.7–10.8 ft. lbs. (11.8–14.7 Nm).
10. Install transaxle assembly.
11. Continue the installation in the reverse order of the removal procedure.

ADJUSTMENTS

1. Measure the clutch pedal height (from the face of the pedal pad to the floorboard) and the clutch pedal free play (measured at the face of the pedal pad). Standard value is 0.25–0.5 inches (6–13 mm).
2. If the clutch pedal free play and height is not within the standard value range, adjust as follows:
   a. Turn and adjust the bolt within the standard value, then secure by tightening the lock nut.
   b. Turn the push rod to agree with the standard value and then secure the push rod with the lock nut.
3. If the clutch pedal free play and the distance between the clutch pedal and the floor board when the clutch is disengaged, do not meet with the standard values, it may be the result of either air in the hydraulic system or a faulty clutch master cylinder. Bleed the air or disassemble and inspect the master cylinder or clutch.

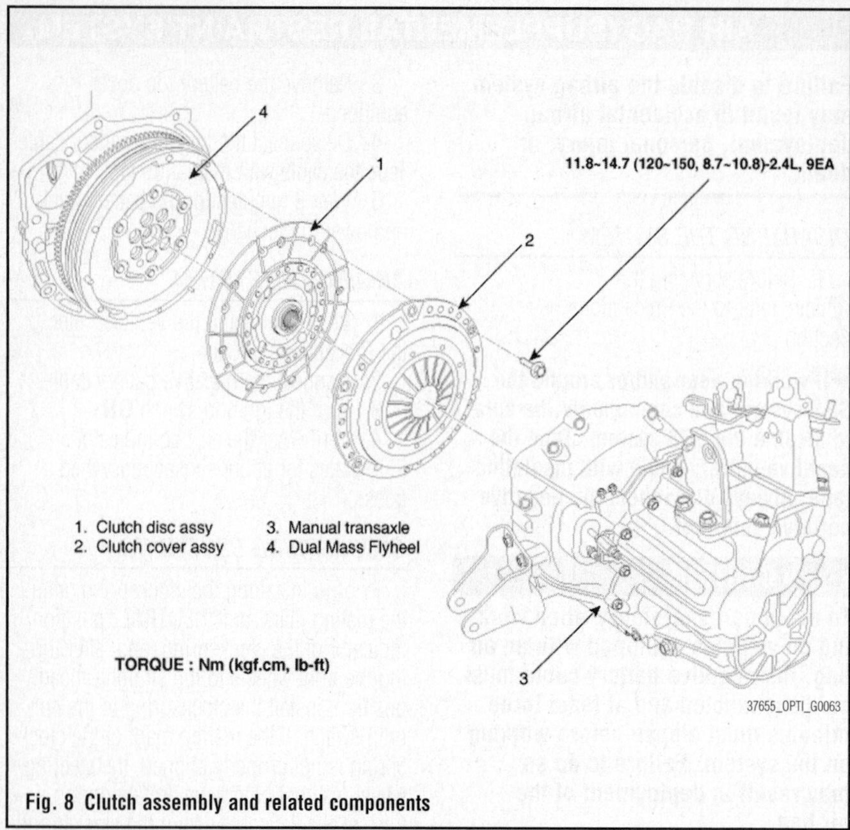

Fig. 8 Clutch assembly and related components

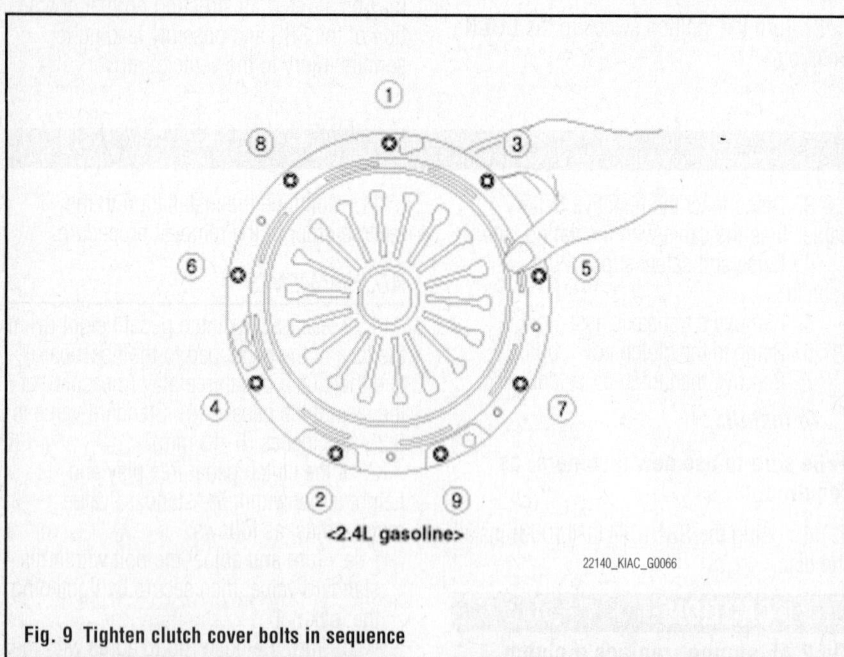

Fig. 9 Tighten clutch cover bolts in sequence

HYDRAULIC SYSTEM BLEEDING

BLEEDING PROCEDURE

Whenever the clutch tube, the clutch hose, and/or the clutch master cylinder have been removed, or if the clutch pedal is spongy, bleed the system.

⁂ CAUTION

Avoid mixing different brands of brake fluid. Use only SAE J1703 (DOT 3 or DOT 4).

1. Loosen the bleeder screw at the clutch release cylinder.
2. Depress the clutch pedal slowly until all air is expelled.
3. Hold the clutch pedal down until the bleeder is retightened.
4. Refill the clutch master cylinder with the specified fluid.

⁂ CAUTION

The rapidly-repeated operation of the clutch pedal may disrupt the release cylinder's position. During the bleeding operation, press the clutch pedal to the floor after it returns to the top of the stroke.

FRONT HALFSHAFT

REMOVAL & INSTALLATION

See Figure 10.

1. Before servicing the vehicle, refer to the Precautions Section.

➡If working near and/or around the SRS system and components, be sure to disable the SRS system. Tape the negative battery cable with insulating tape. Always disconnect the negative battery cable first.

⁂ CAUTION

To avoid personal injury when working on vehicles equipped with an air bag, the negative battery cable must be disconnected and at least three minutes must elapse before working on the system. Failure to do so may result in deployment of the air bag.

2. Remove the battery top cover, if equipped.
3. Disconnect the negative battery cable. Tape the cable with insulating tape.
4. Raise and safely support the vehicle.
5. Remove the front wheel and tire.
6. Remove the drain plug. Drain the transaxle oil.
7. Remove the split pin, the lock nut and the washer from the front hub.
8. Disconnect the tie rod end ball joint from the knuckle using the Special Tool (09568-34000) after removing the split pin and lock nut.
9. Remove the wheel speed sensor from the knuckle.
10. Remove the ball joint assembly mounting bolt from the knuckle.
11. On 2.7L engine (right side):
    a. Remove the heat protector.
    b. Remove the bearing and bracket assembly
12. Using a plastic hammer, disconnect the halfshaft from the axle hub.
13. Push the axle hub outward and separate the halfshaft from the axle hub.

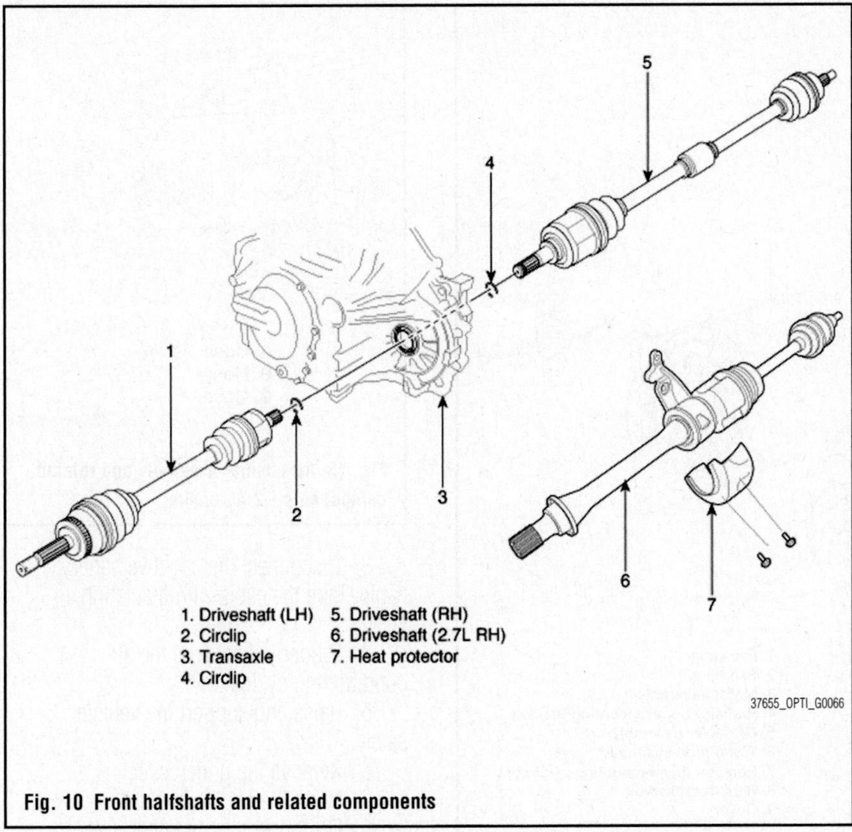

Fig. 10 Front halfshafts and related components

14. Insert a pry bar between the transaxle case and joint case, and separate the half-shaft from the transaxle case.
15. Pull out the halfshaft from the transaxle case.

To install:

16. Apply gear oil on the drive shaft splines and the contacting surface of differential case oil seal.
17. Before installing the halfshaft, set the opening side of the circlip facing downward.
18. After installation, check that the half-shaft cannot be removed by hand.
19. Install the halfshaft into the knuckle.
20. On 2.7L engine (right side):
    a. Install bearing and bracket assembly.
    b. Install heat protector.
21. Install the ball joint assembly mounting bolt to the knuckle. Torque to 72–86 ft. lbs. (100–120 Nm).
22. Install the tie rod end to the knuckle. Torque to 12–25 ft. lbs. (16–34 Nm).
23. Install the wheel speed sensor to the knuckle.
24. After installing the washer with convex surface outward, install the lock nut and the split pin.
25. Install the wheel and tire

ENGINE COOLING

ENGINE FAN

REMOVAL & INSTALLATION

See Figures 11 and 12.

1. Before servicing the vehicle, refer to the Precautions Section.

➡If working near and/or around the SRS system and components, be sure to disable the SRS system. Tape the negative battery cable with insulating tape. Always disconnect the negative battery cable first.

⁂ CAUTION

To avoid personal injury when working on vehicles equipped with an air bag, the negative battery cable must be disconnected and at least three minutes must elapse before working on the system. Failure to do so may result in deployment of the air bag.

2. Remove the battery top cover, if equipped.
3. Disconnect the negative battery cable. Tape the cable with insulating tape.
4. Drain the engine coolant.
5. Remove the upper radiator hose and lower radiator hose.
6. As required, disconnect the transaxle oil cooler hoses if equipped with automatic transaxle.
7. Remove the radiator protector, as required.
8. Disconnect the fan motor connector (s).

➡Some vehicles are equipped with two side-by-side fans.

9. Remove the cooling fan mounting bolts and remove cooling fan.

To install:

10. Install the cooling fan and mounting bolts.
11. Connect the fan motor connector (s).
12. Install the upper radiator hose and lower radiator hose.
13. Install the radiator protector, as required.
14. Install the ATF oil cooler hoses.
15. Fill with engine coolant.
16. Start engine and check for leaks.

RADIATOR

REMOVAL & INSTALLATION

2.4L Engine

See Figures 12 and 13.

1. Before servicing the vehicle, refer to the Precautions Section.

➡If working near and/or around the SRS system and components, be sure to disable the SRS system. Tape the negative battery cable with insulating tape. Always disconnect the negative battery cable first.

⁂ CAUTION

To avoid personal injury when working on vehicles equipped with an air bag, the negative battery cable must be disconnected and at least three minutes must elapse before working on the system. Failure to do so may result in deployment of the air bag.

2. Remove the battery top cover, if equipped.

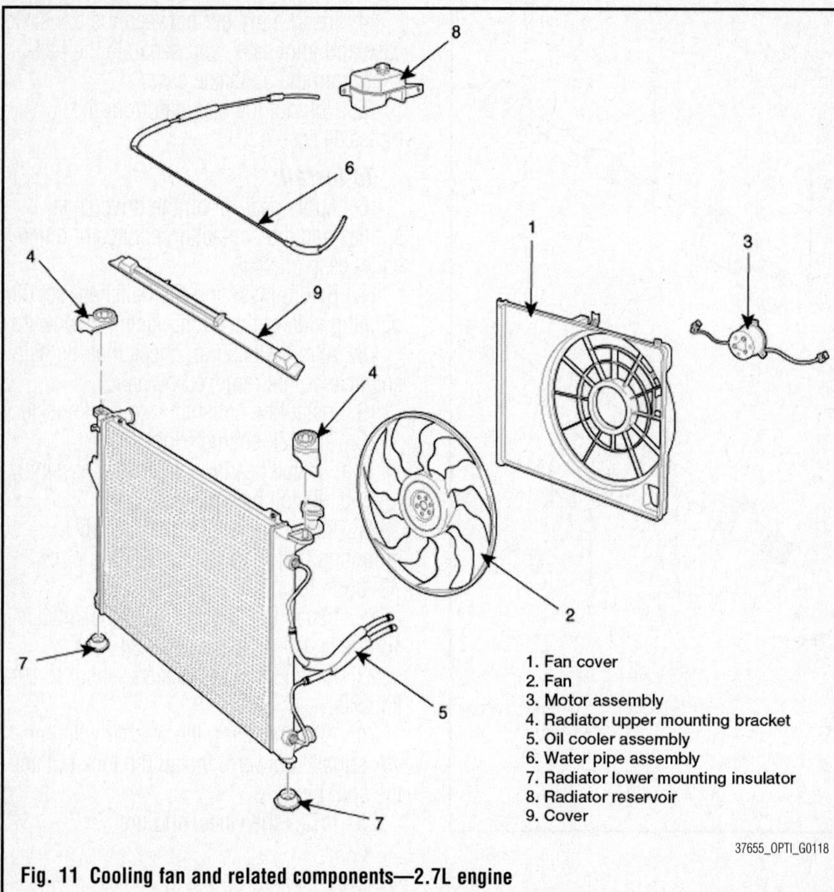

Fig. 11 Cooling fan and related components—2.7L engine

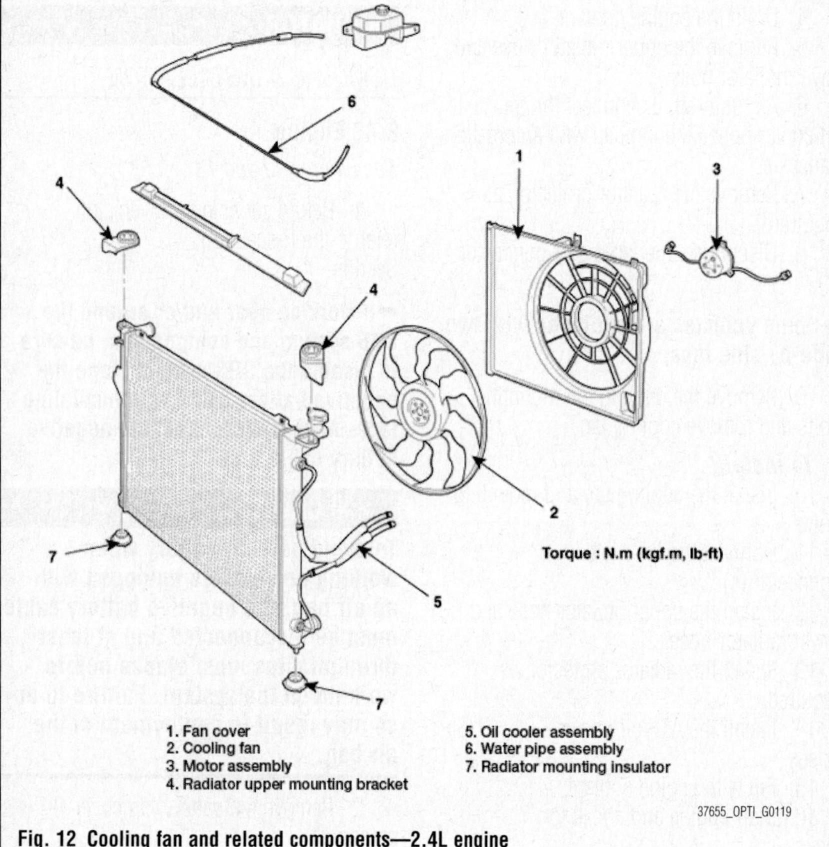

Fig. 12 Cooling fan and related components—2.4L engine

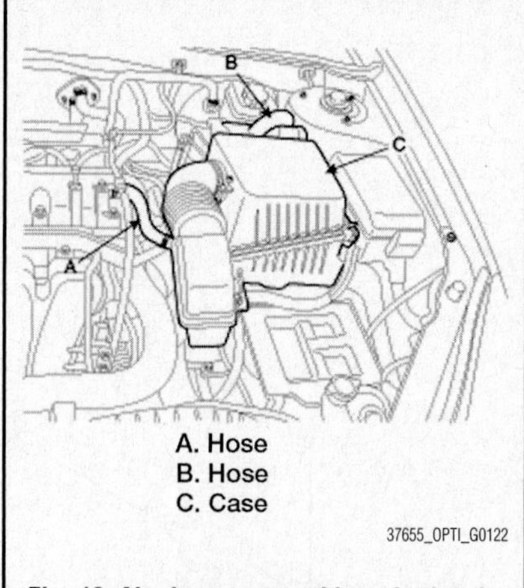

Fig. 13 Air cleaner assembly and related components—2.4L engine

3. Disconnect the negative battery cable. Tape the cable with insulating tape.
4. Properly discharge the air conditioning system.
5. Raise and support the vehicle safely.
6. Remove the undercover.
7. Drain the engine coolant.
8. Disconnect the breather hose and PCM connector. Remove the air cleaner assembly.
9. Remove the radiator protector.
10. Remove the upper radiator hose and lower radiator hose.
11. Remove the ATF oil cooler hoses if vehicle is equipped with automatic transaxle.
12. Disconnect the fan motor connector. Remove the cooling fan.
13. Remove the radiator mounting bolts.
14. Remove the engine coolant reservoir tank hoses.
15. Remove the radiator assembly along with the condenser from the vehicle.

To install:

➡Be sure to use new fasteners, as required.

16. Installation is the reverse of the removal procedure.
17. Properly recharge the air conditioning system.
18. Fill the cooling system with the proper grade and type engine coolant.
19. Start engine and check for leaks.

2.7L Engine

See Figure 18.

1. Before servicing the vehicle, refer to the Precautions Section.

➡If working near and/or around the SRS system and components, be sure to disable the SRS system. Tape the negative battery cable with insulating tape. Always disconnect the negative battery cable first.

⁂ CAUTION

To avoid personal injury when working on vehicles equipped with an air bag, the negative battery cable must be disconnected and at least three minutes must elapse before working on the system. Failure to do so may result in deployment of the air bag.

2. Remove the battery top cover, if equipped.
3. Disconnect the negative battery cable. Tape the cable with insulating tape.
4. Properly discharge the air conditioning system.
5. Drain the engine coolant.
6. Remove the air duct.
7. Remove the upper radiator hose and lower radiator hose.
8. Remove the ATF oil cooler hoses if vehicle is equipped with automatic transaxle.
9. Disconnect the fan motor connector.
10. Remove the radiator mounting bolts.
11. Disconnect the air conditioning tube hoses. Plug the lines.
12. Disconnect the engine coolant reservoir tank hoses.
13. Remove the radiator assembly along with the condenser from the vehicle.

To install:

➡Be sure to use new fasteners, as required.

14. Installation is the reverse of the removal procedure.
15. Properly recharge the air conditioning system.
16. Fill the cooling system with the proper grade and type engine coolant.
17. Start engine and check for leaks.

THERMOSTAT

REMOVAL & INSTALLATION

2.4L Engine

See Figures 13 and 14.

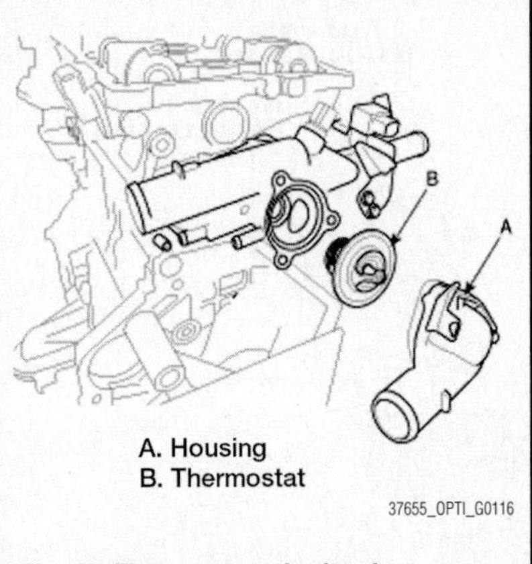

Fig. 14 Thermostat and related components—2.4L engine

1. Before servicing the vehicle, refer to the Precautions Section.

➡If working near and/or around the SRS system and components, be sure to disable the SRS system. Tape the negative battery cable with insulating tape. Always disconnect the negative battery cable first.

⁂ CAUTION

To avoid personal injury when working on vehicles equipped with an air bag, the negative battery cable must be disconnected and at least three minutes must elapse before working on the system. Failure to do so may result in deployment of the air bag.

2. Remove the battery top cover, if equipped.
3. Disconnect the negative battery cable. Tape the cable with insulating tape.
4. Drain the engine coolant so its level is below thermostat.
5. Remove the air cleaner assembly.
6. Remove the lower radiator hose.
7. Remove the thermostat housing retaining bolts.
8. Remove the housing.
9. Remove the thermostat from its mounting. Discard the gasket.

To install:

➡Be sure to use new fasteners, as required.

10. Installation is the reverse of the removal procedure.
11. Be sure to use a new gasket.
12. Tighten the housing retaining bolts to 10.8–15.9 ft. lbs (14.7–21.6 Nm).

2.7L Engine

See Figure 15.

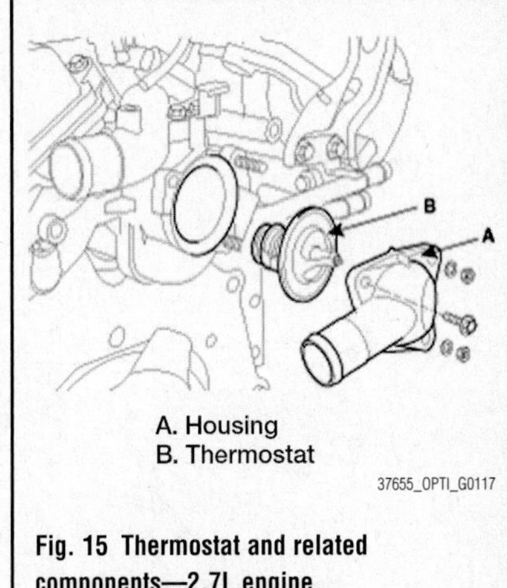

Fig. 15 Thermostat and related components—2.7L engine

1. Before servicing the vehicle, refer to the Precautions Section.

➡If working near and/or around the SRS system and components, be sure to disable the SRS system. Tape the negative battery cable with insulating tape. Always disconnect the negative battery cable first.

⁂ CAUTION

To avoid personal injury when working on vehicles equipped with an air bag, the negative battery cable must be disconnected and at least three minutes must elapse before working on the system. Failure to do so may result in deployment of the air bag.

2. Remove the battery top cover, if equipped.
3. Disconnect the negative battery cable. Tape the cable with insulating tape.
4. Drain the engine coolant so its level is below thermostat.
5. Remove the water inlet fitting, gasket and thermostat.

To install:

➡Be sure to use new fasteners, as required.

6. Installation is the reverse of the removal procedure.
7. Be sure to use a new gasket.
8. Tighten the housing retaining bolts to 12.30–14.47 ft. lbs (16.66–19.60 Nm).

WATER PUMP

REMOVAL & INSTALLATION

2.4L Engine

See Figure 16.

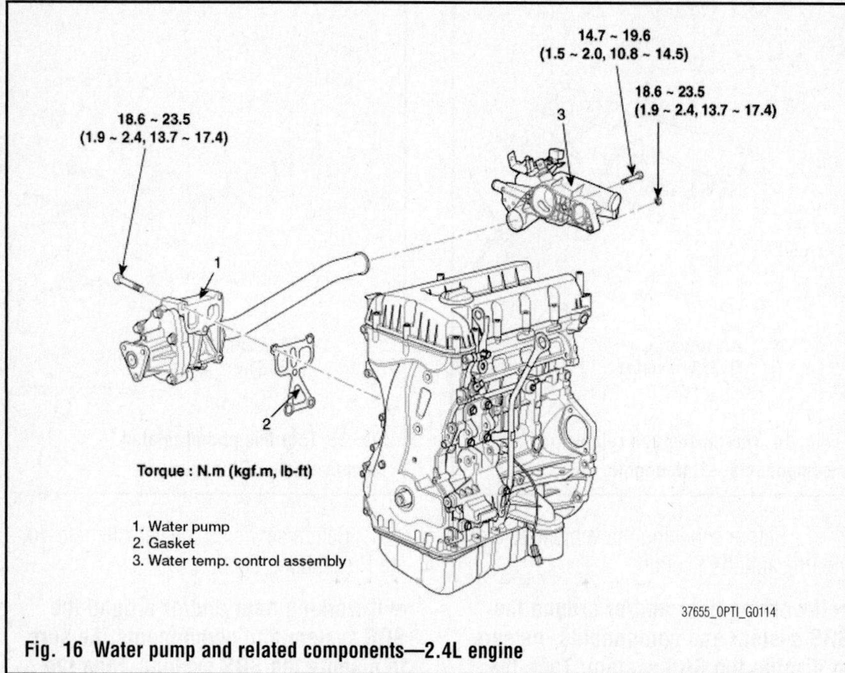

Fig. 16 Water pump and related components—2.4L engine

1. Before servicing the vehicle, refer to the Precautions Section.

➡**If working near and/or around the SRS system and components, be sure to disable the SRS system. Tape the negative battery cable with insulating tape. Always disconnect the negative battery cable first.**

⁂ CAUTION

To avoid personal injury when working on vehicles equipped with an air bag, the negative battery cable must be disconnected and at least three minutes must elapse before working on the system. Failure to do so may result in deployment of the air bag.

2. Remove the battery top cover, if equipped.
3. Disconnect the negative battery cable. Tape the cable with insulating tape.
4. Drain the engine coolant.
5. Remove drive belt.
6. Remove exhaust manifold protector.
7. Remove water inlet pipe nut.
8. Remove the water pump. Discard the gasket.

To install:

9. Install the water pump and a new gasket with the 5 bolts. Torque: 13.7– 17.4 ft. lbs. (18.6–23.5 Nm).
10. Install water inlet pipe nut. Torque: 13.7– 17.4 ft. lbs. (18.6–23.5 Nm).
11. Install exhaust manifold protector.
12. Install drive belt.
13. Fill with engine coolant.
14. Start engine and check for leaks.
15. Recheck engine coolant level.

2.7L Engine

See Figure 17.

1. Before servicing the vehicle, refer to the Precautions Section.

➡**If working near and/or around the SRS system and components, be sure to disable the SRS system. Tape the negative battery cable with insulating tape. Always disconnect the negative battery cable first.**

⁂ CAUTION

To avoid personal injury when working on vehicles equipped with an air bag, the negative battery cable must be disconnected and at least three minutes must elapse before working on the system. Failure to do so may result in deployment of the air bag.

2. Remove the battery top cover, if equipped.
3. Disconnect the negative battery cable. Tape the cable with insulating tape.
4. Drain the engine coolant.
5. Remove accessory drive belt.
6. Remove the timing belt.
7. Remove the water pump and gasket.

To install:

8. Install the water pump and a new gasket. Torque: 10.8–15.9 ft. lbs. (14.7 –21.6 Nm).
9. Install the timing belt.
10. Install accessory drive belt.
11. Fill with engine coolant.
12. Start engine and check for leaks.
13. Recheck engine coolant level.

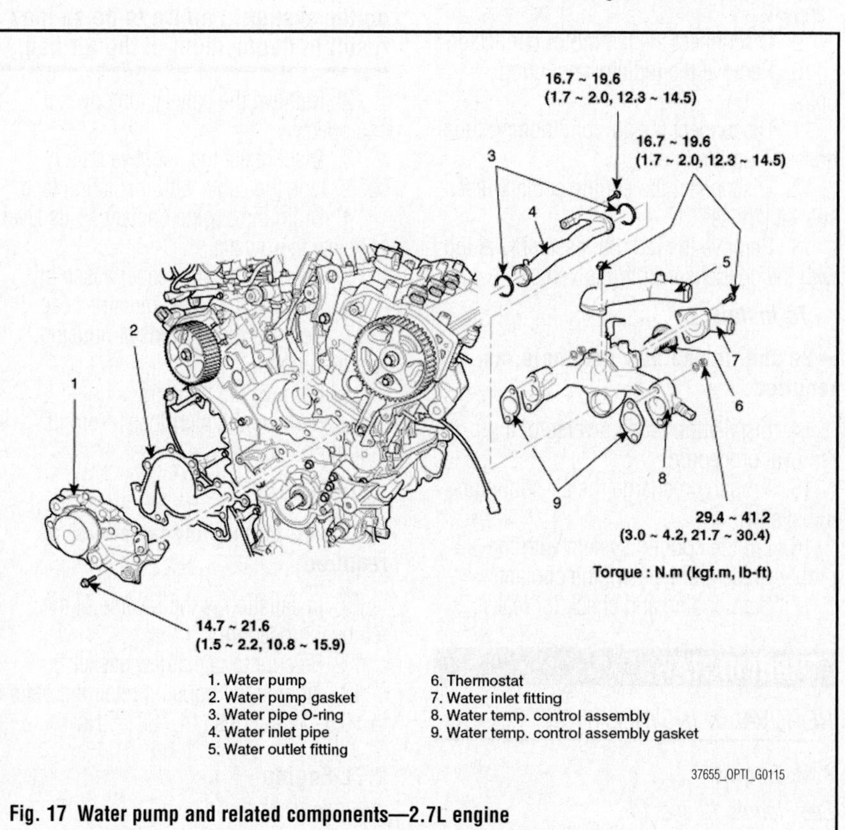

Fig. 17 Water pump and related components—2.7L engine

ENGINE ELECTRICAL — CHARGING SYSTEM

ALTERNATOR

REMOVAL & INSTALLATION

See Figures 18 and 19.

1. Before servicing the vehicle, refer to the Precautions Section.

➡**If working near and/or around the SRS system and components, be sure to disable the SRS system. Tape the negative battery cable with insulating tape. Always disconnect the negative battery cable first.**

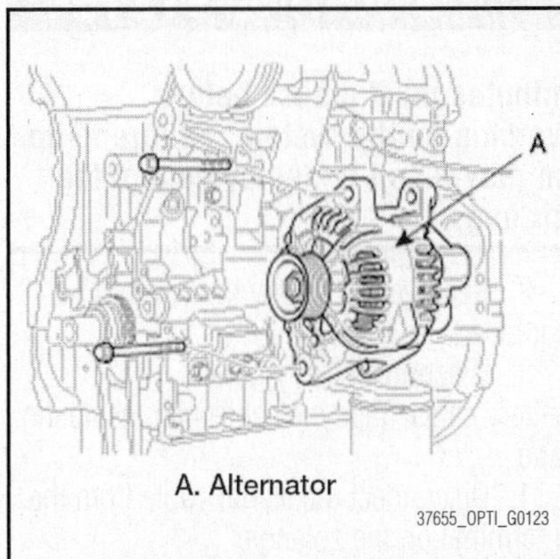

Fig. 18 Alternator mounting—2.4L engine

⁂ CAUTION

To avoid personal injury when working on vehicles equipped with an air bag, the negative battery cable must be disconnected and at least three minutes must elapse before working on the system. Failure to do so may result in deployment of the air bag.

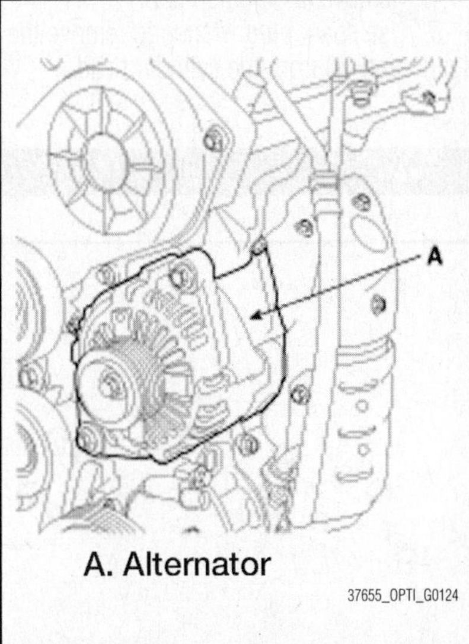

Fig. 19 Alternator mounting—2.7L engine

2. Remove the battery top cover, if equipped.
3. Disconnect the negative battery cable. Tape the cable with insulating tape.
4. Disconnect the positive battery terminal.
5. Disconnect the alternator connector, and remove the cable from alternator "B" terminal.
6. Remove the accessory drive belt.
7. Pull out the through bolt and then remove the alternator.

To install:

➡**Be sure to use new fasteners, as required.**

8. Installation is the reverse of the removal procedure.

ENGINE ELECTRICAL — IGNITION SYSTEM

IGNITION COIL

REMOVAL & INSTALLATION

See Figures 20 and 21.

1. Before servicing the vehicle, refer to the Precautions Section.

➡**If working near and/or around the SRS system and components, be sure to disable the SRS system. Tape the negative battery cable with insulating tape. Always disconnect the negative battery cable first.**

⁂ CAUTION

To avoid personal injury when working on vehicles equipped with an air bag, the negative battery cable must be disconnected and at least three minutes must elapse before working on the system. Failure to do so may result in deployment of the air bag.

2. Remove the battery top cover, if equipped.
3. Disconnect the negative battery cable. Tape the cable with insulating tape.

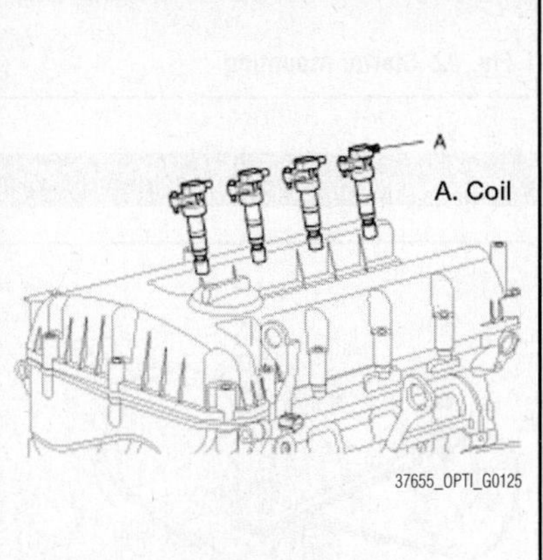

Fig. 20 Ignition coils and related components—2.4L engine

4. Remove the engine cover, 2.7L engine.
5. Disconnect the ignition coil connector.
6. Remove the ignition coil.

To install:

➡**Be sure to use new fasteners, as required.**

7. Installation is the reverse of the removal procedure.

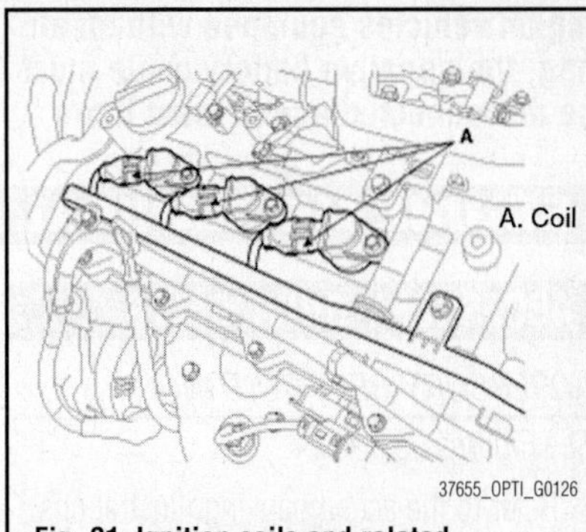

Fig. 21 Ignition coils and related components—2.7L engine

IGNITION TIMING

ADJUSTMENT

These vehicles are equipped with a Distributorless Ignition System (DIS). No adjustment is necessary or possible.

SPARK PLUGS

REMOVAL & INSTALLATION

1. Before servicing the vehicle, refer to the Precautions Section.

➡If working near and/or around the SRS system and components, be sure to disable the SRS system. Tape the negative battery cable with insulating tape. Always disconnect the negative battery cable first.

⁂ CAUTION

To avoid personal injury when working on vehicles equipped with an air bag, the negative battery cable must be disconnected and at least three minutes must elapse before working on the system. Failure to do so may result in deployment of the air bag.

2. Remove the battery top cover, if equipped.
3. Disconnect the negative battery cable. Tape the cable with insulating tape.
4. Remove the ignition coil.
5. Use spark plug wrench to remove the spark plug (s) from the cylinder head.

⁂ WARNING

Do not let any dirt or debris get into the engine through the spark plugs holes while when are removed.

To install:

6. Install the spark plug and tighten, to specification
7. Install the ignition coil (s).
8. Attach the spark plug wire (s) to the ignition coil.
9. Connect the negative battery cable.

ENGINE ELECTRICAL

STARTING SYSTEM

STARTER

REMOVAL & INSTALLATION

See Figure 22.

1. Before servicing the vehicle, refer to the Precautions Section.

➡If working near and/or around the SRS system and components, be sure to disable the SRS system. Tape the negative battery cable with insulating tape. Always disconnect the negative battery cable first.

⁂ CAUTION

To avoid personal injury when working on vehicles equipped with an air bag, the negative battery cable must be disconnected and at least three minutes must elapse before working on the system. Failure to do so may result in deployment of the air bag.

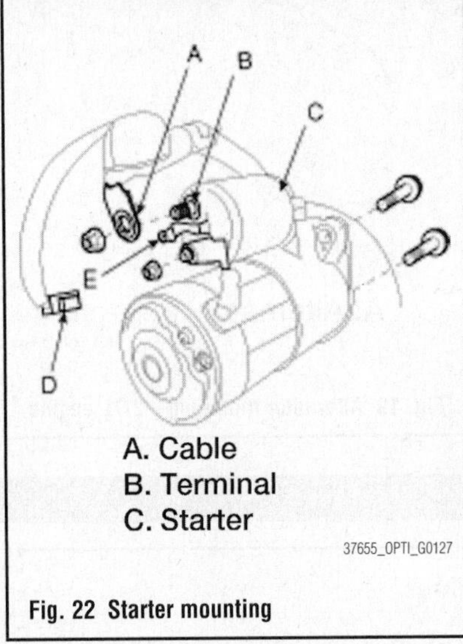

Fig. 22 Starter mounting

2. Remove the battery top cover, if equipped.
3. Disconnect the negative battery cable. Tape the cable with insulating tape.
4. Disconnect the starter cable from the B terminal on the solenoid.
5. Disconnect the connector from the S terminal.
6. Remove the 2 bolts holding the starter, then remove the starter.
7. Installation is the reverse of removal.

ENGINE MECHANICAL

ACCESSORY DRIVE BELTS

ACCESSORY BELT ROUTING

See Figures 23 and 24.

Refer to the accompanying illustrations for belt routing.

INSPECTION

Inspect the drive belt for signs of glazing or cracking. A glazed belt will be perfectly smooth from slippage, while a good belt will have a slight texture of fabric visible. Cracks will usually start at the inner edge of the belt and run outward. All worn or damaged drive belts should be replaced immediately.

ADJUSTMENT

The belt tension is maintained by an automatic tensioner. No adjustment is necessary or possible.

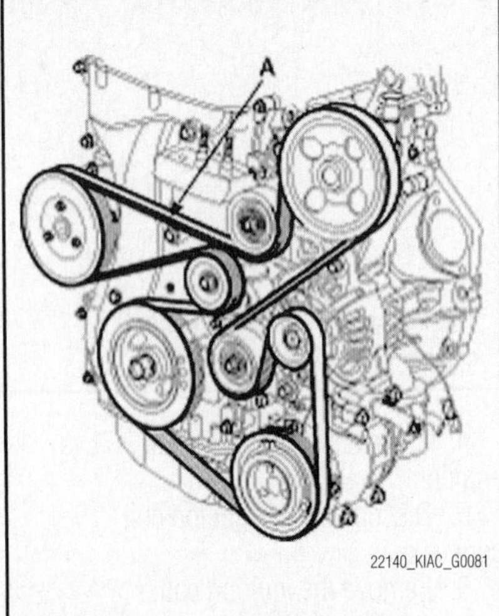

Fig. 23 Accessory drive belt routing—2.4L Engine

REMOVAL & INSTALLATION

See Figures 23 and 24.

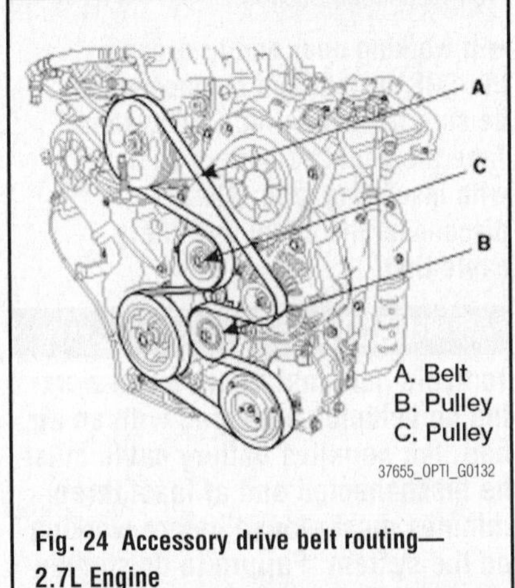

Fig. 24 Accessory drive belt routing—2.7L Engine

At this time the manufacturer does not provide removal and installation procedures for this component. The following procedure is a guideline and may differ from the vehicle you are servicing.

1. Before servicing the vehicle, refer to the Precautions Section.

➡If working near and/or around the SRS system and components, be sure to disable the SRS system. Tape the negative battery cable with insulating tape. Always disconnect the negative battery cable first.

⁂ CAUTION

To avoid personal injury when working on vehicles equipped with an air bag, the negative battery cable must be disconnected and at least three minutes must elapse before working on the system. Failure to do so may result in deployment of the air bag.

2. Remove the battery top cover, if equipped.
3. Disconnect the negative battery cable. Tape the cable with insulating tape.
4. Relieve tension on the tensioner pulley.
5. Remove accessory drive belt

To install:

➡Be sure to use new fasteners, as required.

6. Installation is the reverse of the removal procedure.

CAMSHAFT AND VALVE LIFTERS

INSPECTION

1. Check the camshaft journals for wear. If the journals are badly worn out, replace the camshaft.
2. Check the cam lobes for damage. If the lobe is damaged or excessively worn out, replace the camshaft.
3. Check the cam surface for abnormal wear or damage, and replace if necessary.
4. Check each bearing for damage. If the bearing surface is excessively damaged, replace the cylinder head assembly or camshaft bearing cap, as necessary.

REMOVAL & INSTALLATION

2.4L Engine

See Figures 25 through 36.

1. Before servicing the vehicle, refer to the Precautions Section.

➡If working near and/or around the SRS system and components, be sure to disable the SRS system. Tape the negative battery cable with insulating tape. Always disconnect the negative battery cable first.

⁂ CAUTION

To avoid personal injury when working on vehicles equipped with an air bag, the negative battery cable must be disconnected and at least three minutes must elapse before working on the system. Failure to do so may result in deployment of the air bag.

2. Remove the battery top cover, if equipped.
3. Disconnect the negative battery cable. Tape the cable with insulating tape.
4. Remove engine cover.
5. Remove air duct.
6. Remove the intake air hose and air cleaner assembly.
    a. Disconnect the AFS (B) connector.
    b. Disconnect the breather hose (C) from air cleaner hose.
    c. Disconnect the ECU connector.
    d. Remove the intake air hose and air cleaner (A).
7. Raise and support the vehicle safely. Remove front wheels, as necessary.
8. Remove lower cover.
9. Drain the engine coolant.
10. Remove the upper and lower radiator hoses.

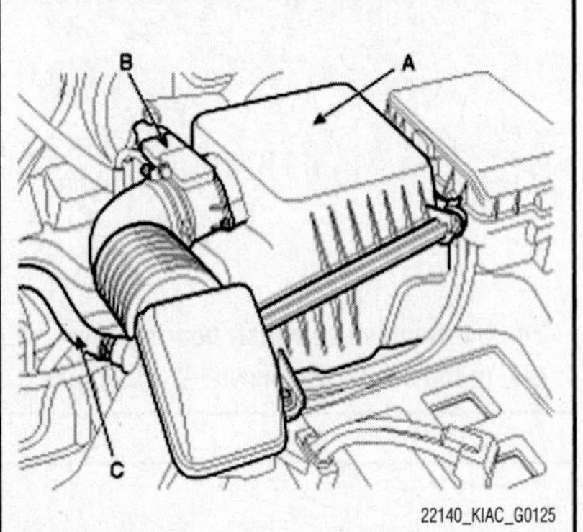

Fig. 25 Disconnect the AFS (B) connector, breather hose (C), and air cleaner (A)—2.4L engine

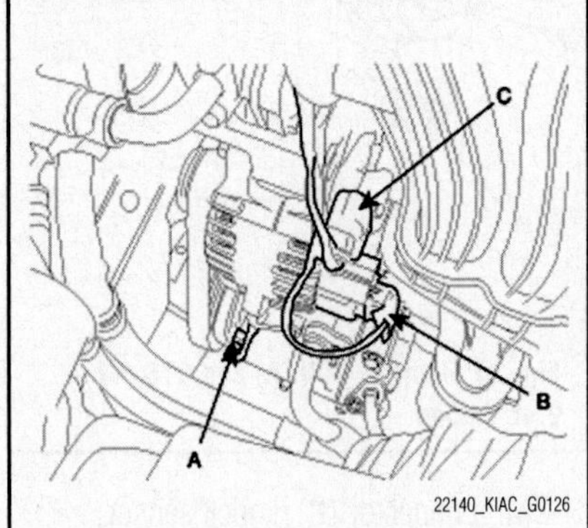

Fig. 26 Disconnect A/C switch (A), alternator connector (B), and oil pressure switch (C)—2.4L engine

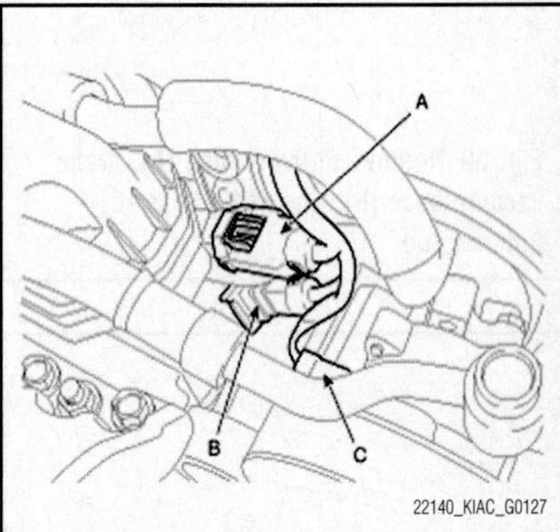

Fig. 27 Disconnect OCV connector (A) and OTS connector (B) and P/S switch connector (C)—2.4L engine

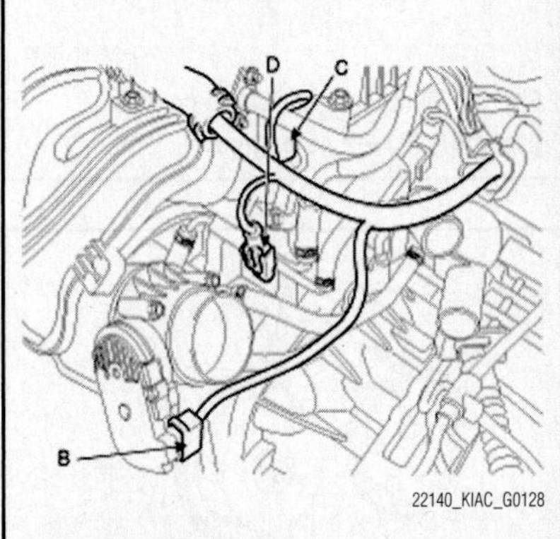

Fig. 28 Remove the ETC connector (B), CMP connector (C), knock sensor connector (D)—2.4L engine

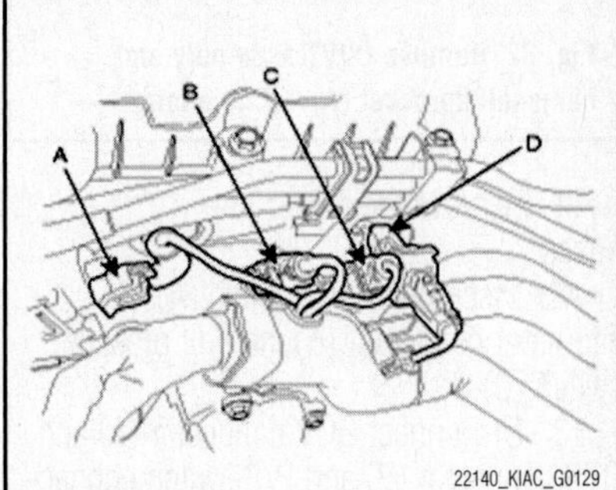

Fig. 29 Disconnect PCSV connector (A), WTS connector (B), condenser connector (C), and CKP sensor connector (D)—2.4L engine

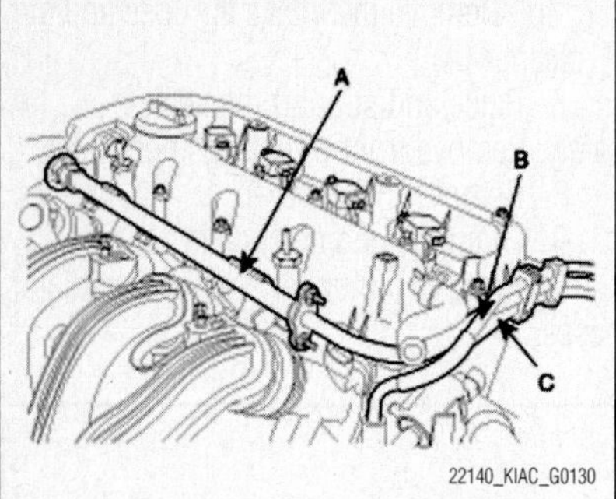

Fig. 30 Remove delivery pipe (A), brake vacuum hose (B), and PCSV hose (C)—2.4L engine

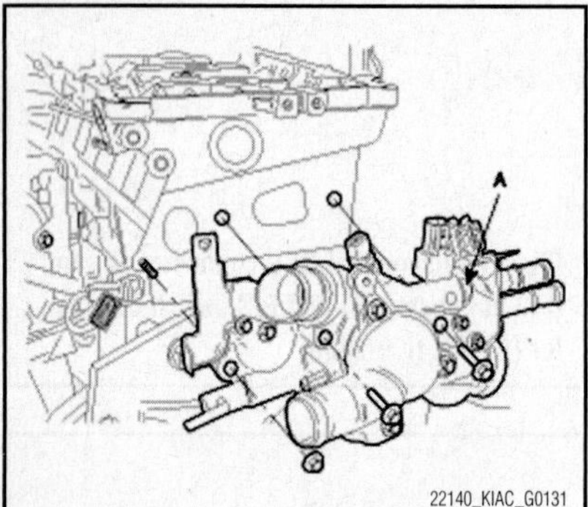

Fig. 31 Remove water temp control assembly (A)—2.4L engine

Fig. 32 Remove CVVT assembly and camshaft sprocket (A)—2.4L engine

11. Remove the heater hoses.
12. Disconnect A/C switch (A), alternator connector (B), and oil pressure switch (C).
13. Disconnect OCV connector (A) and OTS connector (B) and P/S switch connector (C).
14. Disconnect injector connectors.
15. Disconnect the engine wire harness connectors.
    a. Remove the ETC connector (B),

Fig. 33 Remove front camshaft bearing cap (A)—2.4L engine

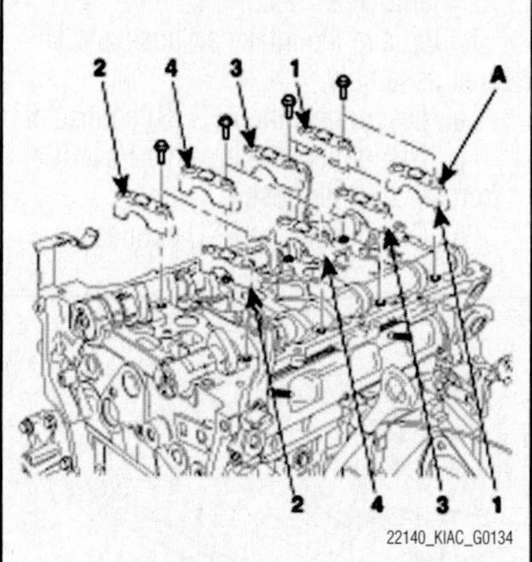

Fig. 34 Remove camshaft bearing cap (A), in the sequence shown—2.4L engine

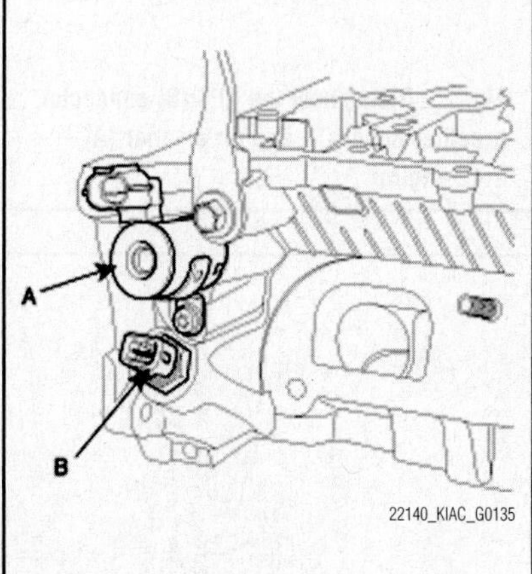

Fig. 35 Remove OCV (A) and OTS (B)—2.4L engine

CMP connector (C), knock sensor connector (D).
16. Disconnect ignition coil connectors.
17. Disconnect PCSV connector (A),

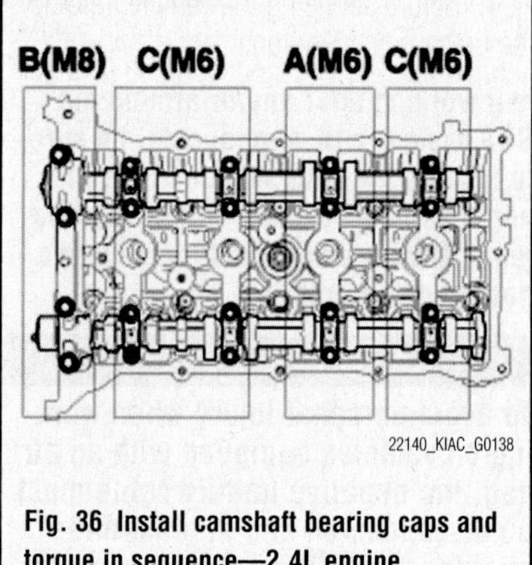

Fig. 36 Install camshaft bearing caps and torque in sequence—2.4L engine

WTS connector (B), condenser connector (C), and CKP sensor connector (D).
18. Remove delivery pipe (A), brake vacuum hose (B), and PCSV hose (C).
19. Remove water temp control assembly (A).
20. Remove intake manifold.
21. Remove exhaust manifold.
22. Remove timing chain.
23. Remove CVVT assembly and camshaft sprocket (A).
24. Remove camshaft.
    a. Remove front camshaft bearing cap (A).
    b. Remove camshaft bearing cap (A), in the sequence shown.
    c. Remove camshafts.
25. Remove OCV (A) and OTS (B), as necessary.

To install:

26. As required, install OCV and OTS. Torque: OCV: 7.23–8.67 ft. lbs. (9.8–11.76 Nm; OTS: 14.5–17.4 ft. lbs. (19.6–23.52 Nm).

CAUTION

Do not reuse the OCV when dropped. Keep the OCV clean. Do not hold the OCV sleeve during servicing. When the OCV is installed on the engine, do not move the engine with holding the OCV yoke.

27. Install the CVVT and camshaft sprocket. Torque: 39.7–47.0 ft. lbs. (53.9–63.7 Nm).

➡Hold the hexagonal head wrench portion of the camshaft with a vise, and install the bolt and CVVT assembly.

28. Install camshafts.

➡Apply a light coat of engine oil on camshaft journals.

29. Install camshaft bearing caps in their proper locations.
Tightening order:
- Group A
- Group B
- Group C.
- Torque: M6 bolts: 8.0–9.4 ft. lbs. (10.8–12.7 Nm; M8 bolts: 20.2–23.1 ft. lbs. (27.4–31.4 Nm).

30. Install timing chain.
31. Check and adjust valve clearance.
32. Install the exhaust manifold.
33. Install the intake manifold.
34. Install water temp control assembly. Torque: Bolt: 10.8–15.9 ft. lbs. (14.7–21.6 Nm; Nut: 14.5–19.5 ft. lbs. (19.6–26.5 Nm).

➡Assemble water temp control assembly and water inlet pipe to water pump assembly before nuts for assembling of water inlet pipe to be tightened. Insert after wetting O-ring or inner surface of thermostat housing. Always use a new O-ring.

35. Install delivery pipe, brake hose, and PCSV hose.
36. Install the PCSV connector, WTS connector, condenser connector, and CKP sensor connector.
37. Install ignition coil connector.
38. Install the engine wire harness connectors.
    a. Install the ETC connector, CMP connector, knock sensor connector.
39. Install the injector connectors.
40. Install the OCV connector, OTS connector, P/S switch connector.
41. Install the A/C switch, alternator connector, and oil pressure switch.
42. Install the upper and lower radiator hoses.
43. Install lower cover.
44. Install the intake air hose and air cleaner assembly.
    a. Install the AFS connector.
    b. Install the breather hose from air cleaner hose.
    c. Install the ECU connector.
    d. Install the intake air hose and air cleaner.
45. Install the engine cover.
46. Install the negative terminal to the battery.
47. Fill with engine coolant.
48. Start the engine and check for leaks.
49. Recheck engine coolant level and oil level.

2.7L Engine

See Figures 37 through 54.

1. Before servicing the vehicle, refer to the Precautions Section.

➡If working near and/or around the SRS system and components, be sure to disable the SRS system. Tape the negative battery cable with insulating tape. Always disconnect the negative battery cable first.

✻✻ CAUTION

To avoid personal injury when working on vehicles equipped with an air bag, the negative battery cable must be disconnected and at least three minutes must elapse before working on the system. Failure to do so may result in deployment of the air bag.

2. Remove the battery top cover, if equipped.
3. Disconnect the negative battery cable. Tape the cable with insulating tape.

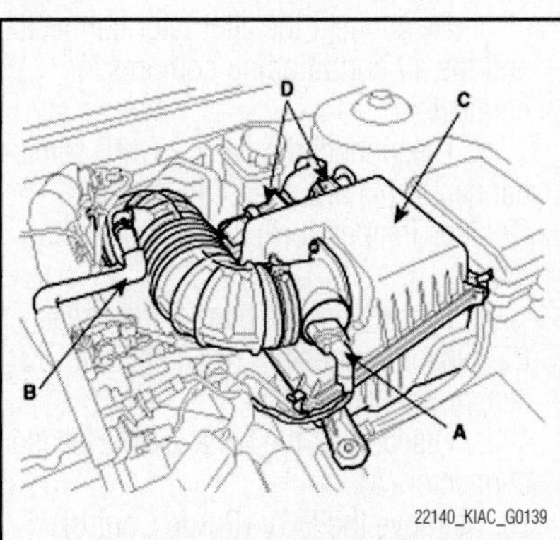

Fig. 37 Disconnect the MAF connector (A), breather hose (B), intake air hose and air cleaner assembly (C), and PCM connectors (D)—2.7L engine

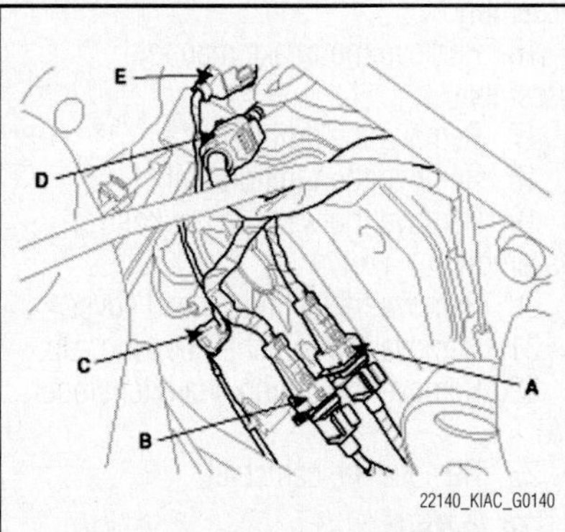

Fig. 38 Disconnect the engine wiring harness connectors—2.7L engine

4. Remove the air duct and the battery.
5. Remove the engine cover.
6. Remove the intake air hose and air cleaner assembly.
    a. Disconnect the MAF connector (A).

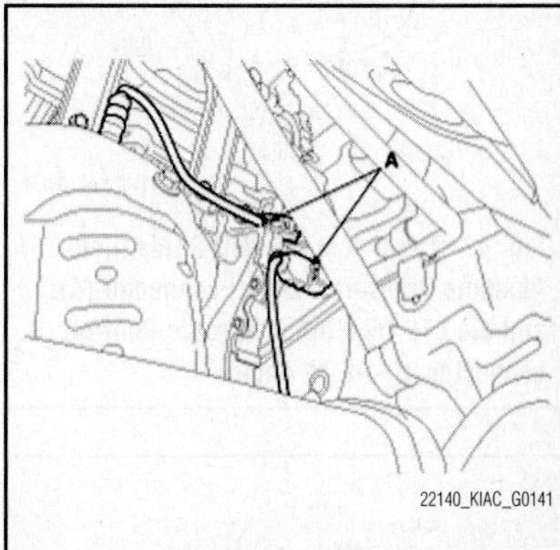

Fig. 39 Disconnect the bank 1 front/rear O2 sensor connectors (A)—2.7L engine

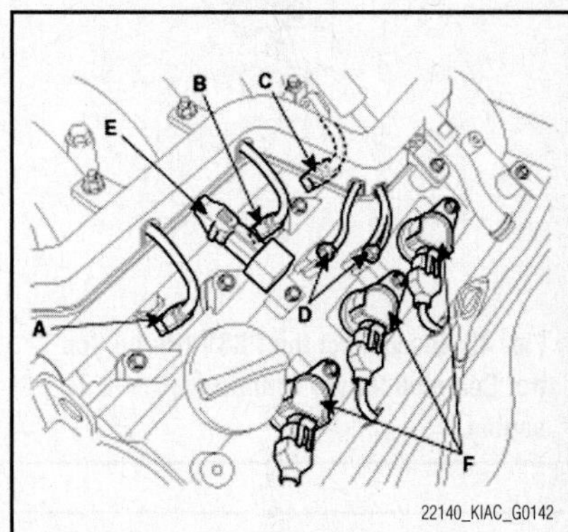

Fig. 40 Disconnect the injection connectors (A,B,C), the ground lines (D), the condenser connector (E) and the Ignition coil connectors (F)—2.7L engine

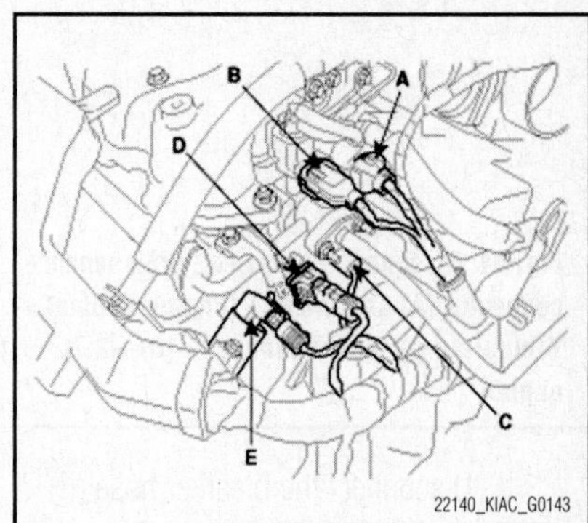

Fig. 41 Disconnect the injection harness connector (A), the No.2 VIS (Variable Induction System) connector (B), the No.1/No.2 OCV (Oil Control Valve) connectors (C,D) and the OTS (Oil Temperature Sensor) connector (E)—2.7L engine

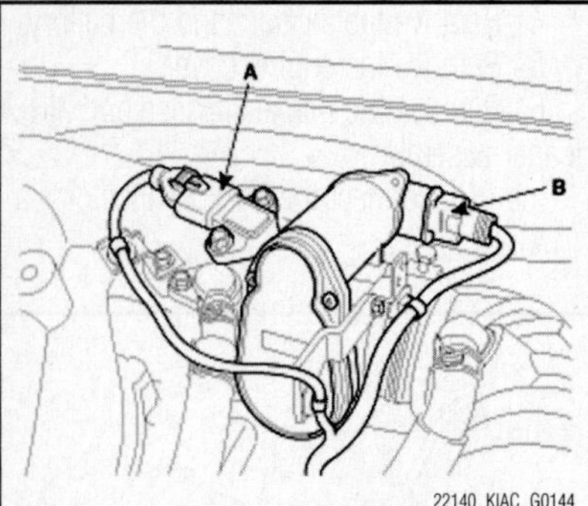

Fig. 42 Disconnect the MAPS (Manifold Absolute Pressure Sensor) connector (A), and the ETC (Electronic Throttle Control) connector (B)—2.7L engine

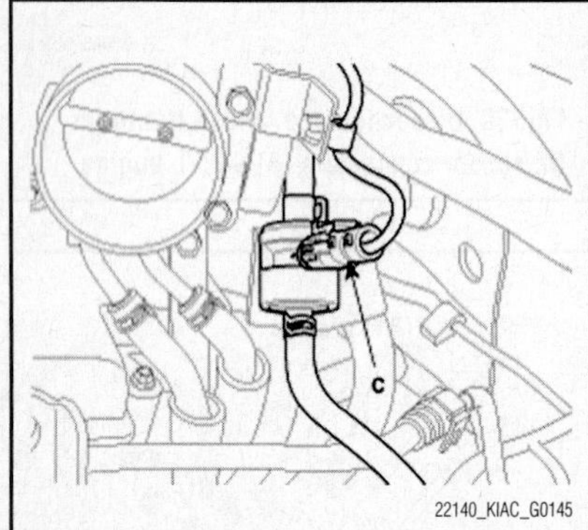

Fig. 43 Disconnect the PCSV (Purge Control Solenoid Valve) connector (C)—2.7L engine

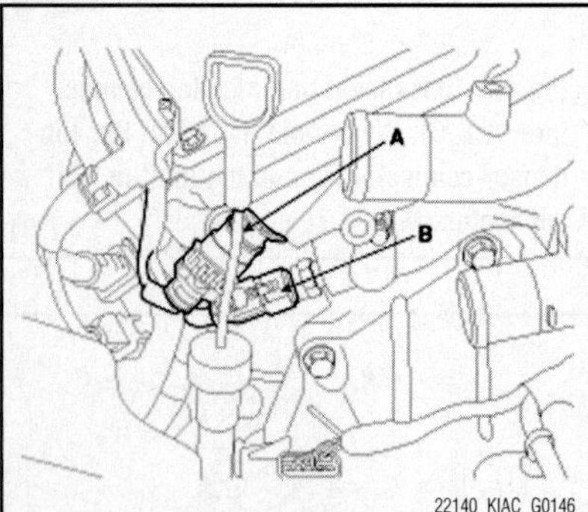

Fig. 44 Disconnect the bank 2 CMP sensor connector (A) and the ECT (Engine Coolant Temperature) sensor connector (B)—2.7L engine

b. Disconnect the breather hose (B) from air cleaner hose.

c. Remove the intake air hose and air cleaner assembly (C).

d. Disconnect the PCM connectors (D).

7. Remove the upper and lower radiator hoses.

8. Remove the fuel inlet hose from the delivery pipe.

9. Disconnect the engine wiring harness connectors.

a. Disconnect the No.1/No.2 knock sensor connectors (A,B), the oil pressure switch connector (C), the ignition coil harness (D) and the No.1 VIS (Variable Induction System) connector (E).

b. Disconnect the bank 1 front/rear O2 sensor connectors (A).

c. Disconnect the injection connectors (A,B,C), the ground lines (D), the condenser connector (E) and the Ignition coil connectors (F).

d. Disconnect the injection harness connector (A), the No.2 VIS (Variable Induction System) connector (B), the No.1/No.2 OCV (Oil Control Valve) connectors (C,D) and the OTS (Oil Temperature Sensor) connector (E).

e. Disconnect the MAPS (Manifold Absolute Pressure Sensor) connector (A), the ETC (Electronic Throttle Control) connector (B) and the PCSV (Purge Control Solenoid Valve) connector (C).

f. Disconnect the alternator connector and the air conditioning compressor connector.

g. Disconnect the bank 2 CMP sensor connector (A) and the ECT (Engine Coolant Temperature) sensor connector (B).

h. Disconnect the bank 2 front/rear O2 sensor connectors (A,B) and the CKP sensor connector (C).

i. Disconnect the bank 1 CMP sensor connector (A).

10. Remove the PCV (Purge Control Valve) hose.

11. Disconnect the brake vacuum hose.

12. Remove the heater hoses.

13. Remove the accessory drive belt.

14. Remove the power steering pump.

15. Remove the exhaust manifold assembly.

16. Remove the intake manifold assembly.

17. Remove the timing belt.

18. Remove the ignition coils.

19. Remove the water temp. control assembly.

20. Remove the cylinder head cover.

21. Remove the camshaft bearing cap.

22. Remove the timing chain tensioner (A).

23. Remove the camshaft.

To install:

24. Install the camshaft in the cylinder head assembly.

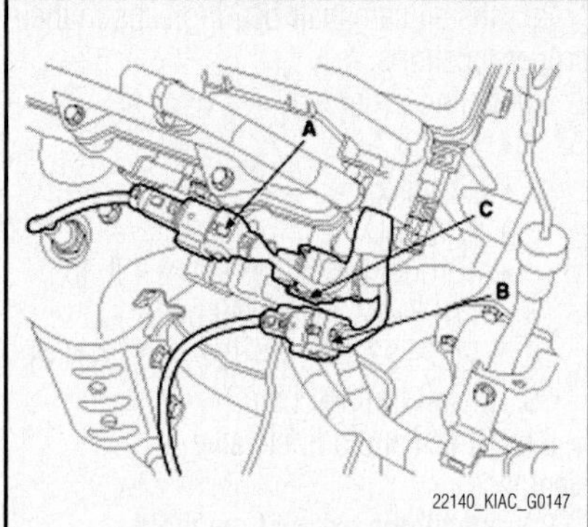

Fig. 45 Disconnect the bank 2 front/rear O2 sensor connectors (A,B) and the CKP sensor connector (C)—2.7L engine

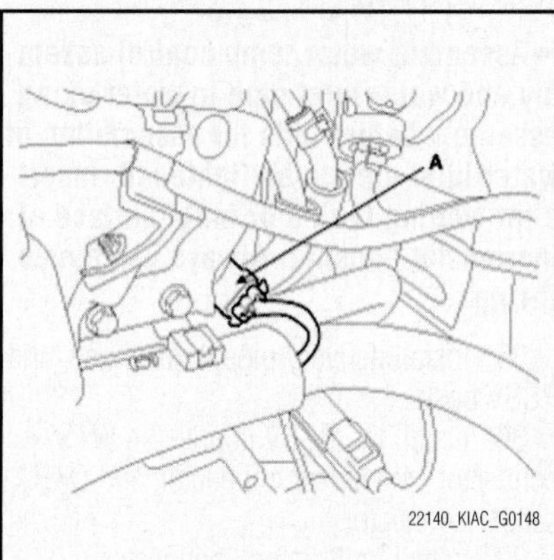

Fig. 46 Disconnect the bank 1 CMP sensor connector (A)—2.7L engine

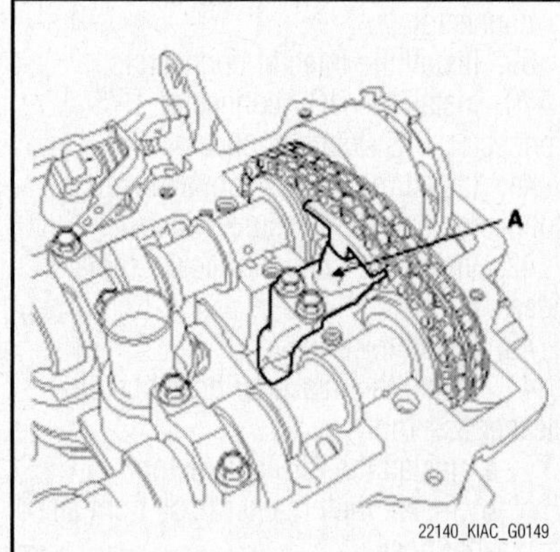

Fig. 47 Remove the timing chain tensioner (A)—2.7L engine

a. Align the timing mark of the camshaft timing chain.

CAUTION

Both timing marks should face upward in reassembly.

25. Install the timing chain tensioner.

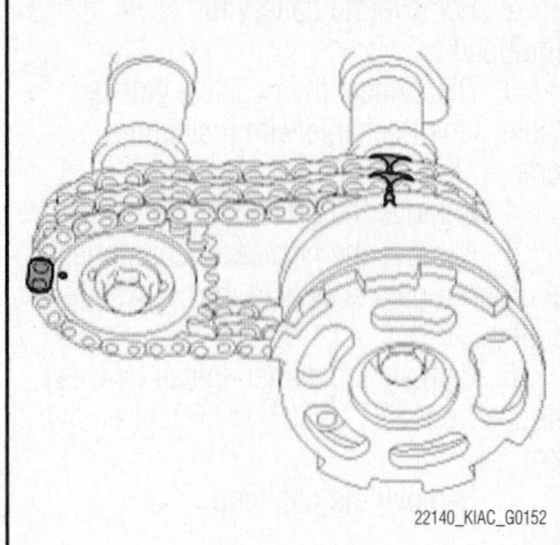
Fig. 48 Left hand camshaft chain timing mark—2.7L engine

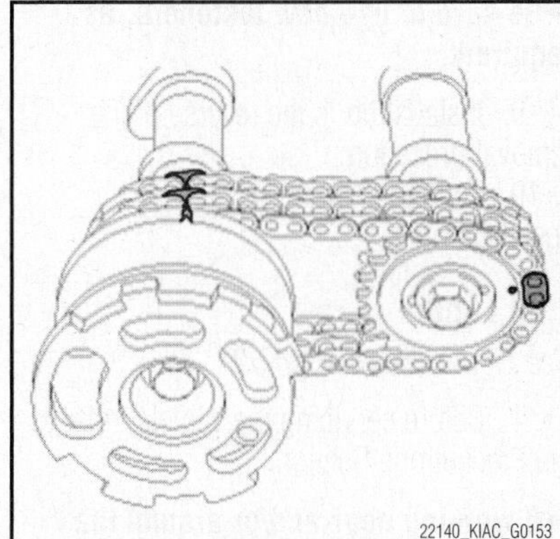
Fig. 49 Right hand camshaft chain timing mark—2.7L engine

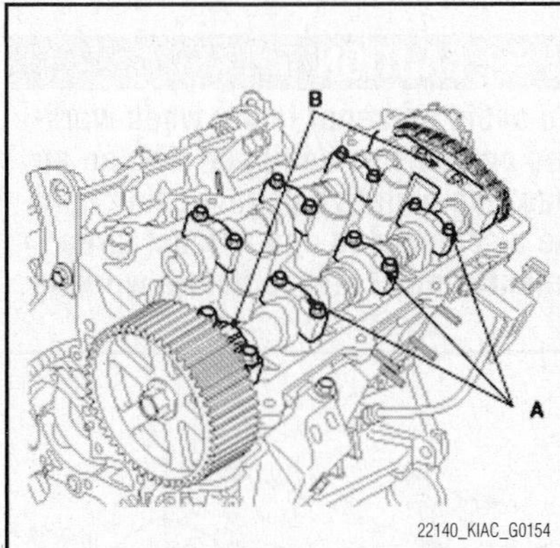

Fig. 50 Install the camshaft bearing caps—2.7L engine

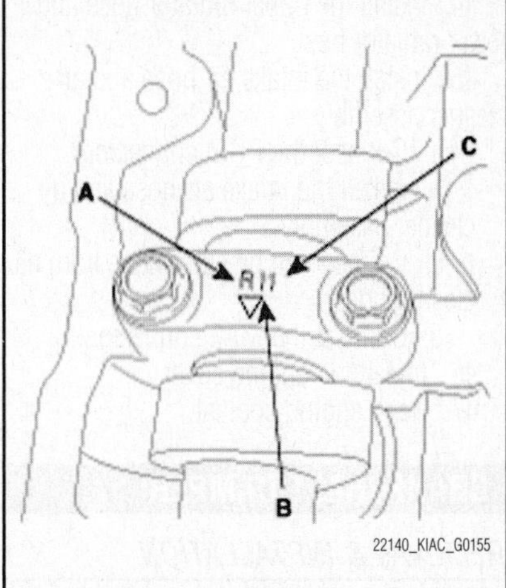

Fig. 51 When installing the bearing caps, check the marks—2.7L engine

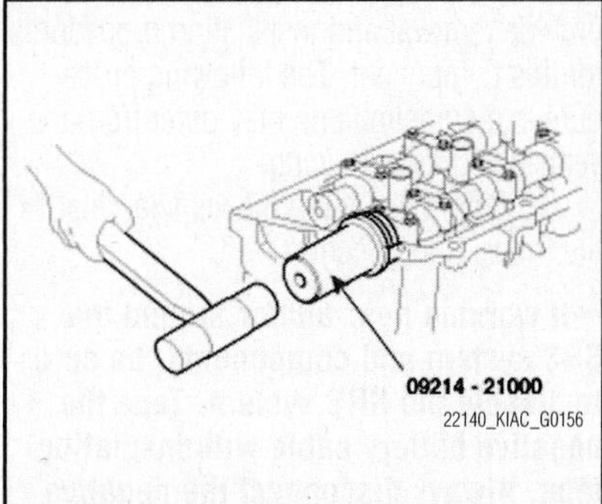

Fig. 52 Using the SST (09214-21000), install the camshaft oil seal—2.7L engine

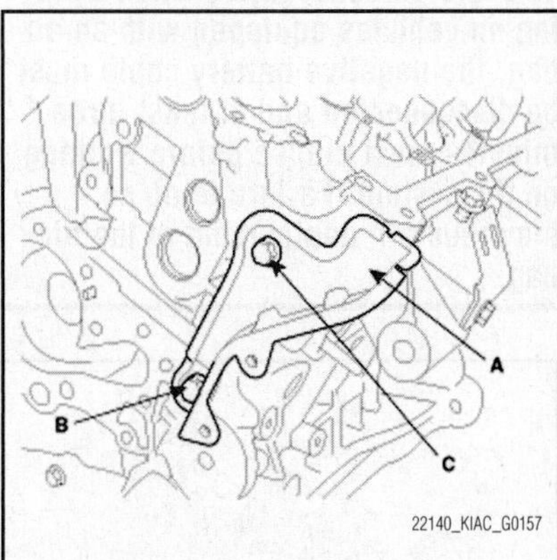

Fig. 53 Install the bank 1 timing belt rear cover—2.7L engine

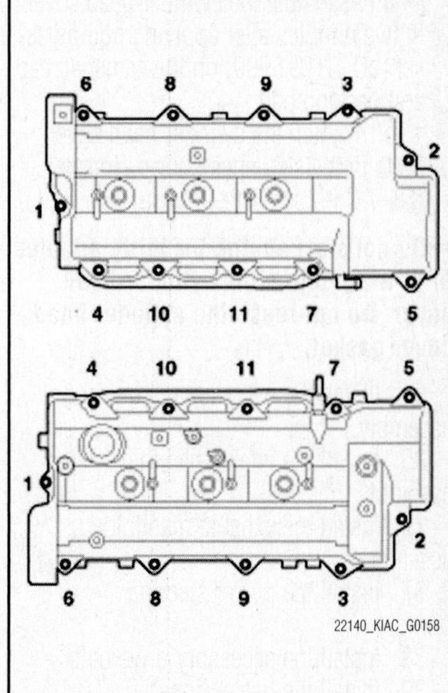

Fig. 54 Tighten the valve cover bolts in sequence—2.7L engine

a. Insert the set pin by pressing the timing chain tensioner.

b. Install the chain tensioner in the cylinder head assembly.

c. Remove the set pin from the tensioner after installing.

26. Install the camshaft bearing caps. Torque: Bearing cap bolt (A: 6_38): 8.0–9.4 ft. lbs. (10.8–12.7 Nm). Bearing cap bolt (B: 8_38): 15.2–18.8 ft. lbs. (20.6–22.5 Nm).

➡When installing the bearing caps, check the marks on them as shown below and install them in its proper position. Note the following designations as shown below:

- A: (LH/RH HEAD) L (LH), R (RH)
- B: (Intake/Exhaust) I (Intake), E (Exhaust)
- C: (Cap no.): 1,2,and 3

CAUTION

When installing the bearing caps, turn the crankshaft to place a piston in the middle of the block because interference between valves and pistons can occur.

27. Using the SST (09214-21000), install the camshaft oil seal.

➡Before installing, apply engine oil. The camshaft cap surface should adhere to the cylinder head assembly. Do not press an eccentric load.

28. Install the CKP sensor connector bracket.

29. Install the bank 2 timing belt rear cover.

30. Install the bank 1 timing belt rear cover.

➡The length of the bolt B is longer than that of the bolt C.

31. Install the timing belt.

32. Check and adjust the valve clearance.

33. Install the cylinder head cover.

a. Remove oil, dust or sealant on the upper surface of the cylinder before assembling cylinder head cover.

b. Assemble the cylinder head cover in five minutes after applying liquid gasket (LOCTITE 5900) on the camshaft cap and packing part.

c. Tighten the cylinder head cover bolts in the sequence shown. Torque: 5.8–7.2 ft. lbs. (7.8–9.8 Nm).

➡Do not start engine for thirty minutes after assembling the cylinder head cover. Do not reuse the cylinder head cover gasket.

34. Install the water temp. control assembly.
35. Install the intake manifold assembly.
36. Install the exhaust manifold assembly.
37. Install the power steering pump.
38. Install the accessory drive belt.
39. Install the heater hose.
40. Connect the brake vacuum hose.
41. Install the PCV (Positive Crankcase Ventilation) hose.
42. Connect the engine wiring harness connectors.
    a. Connect the bank 1 CMP sensor connector.
    b. Connect the bank 2 front/rear O2 sensor connectors and the CKP sensor connector.
    c. Connect the bank 2 CMP sensor connector and the WTS (Water Temperature Sensor) connector.
    d. Connect the alternator connector and the air conditioning compressor connector.
    e. Connect the MAPS (Manifold Absolute Pressure Sensor) connector, the ETC (Electronic Throttle Control) connector and the PCSV (Purge Control Solenoid Valve) connector.
    f. Connect the injection harness connector, the No.2 VIS (Variable Induction System) connector, the No.1/No.2 OCV (Oil Control Valve) connectors and the OTS (Oil Temperature Sensor) connector.
    g. Connect the injection connectors, the ground lines, the condenser connector and the Ignition coil connectors.
    h. Connect the bank 1 front/rear O2 sensor connectors.
    i. Connect the No.1/No.2 knock sensor connectors, the oil pressure switch connector, the ignition coil harness and the No.1 VIS (Variable Induction System) connector.
43. Install the fuel inlet hose from the delivery pipe.
44. Install the upper radiator hose and lower radiator hose.
45. Install the intake air hose and air cleaner assembly.
    a. Connect the PCM connectors.
    b. Install the intake air hose and air cleaner assembly.
    c. Connect the breather hose from air cleaner hose.
    d. Connect the MAF connector.
46. Install the engine cover.
47. Refill engine coolant.

CRANKSHAFT DAMPER

REMOVAL & INSTALLATION

2.4L Engine

See Figure 55.

At this time the manufacturer does not provide removal and installation procedures for this component. The following procedure is a guideline and may differ from the vehicle you are servicing.

1. Before servicing the vehicle, refer to the Precautions Section.

➡If working near and/or around the SRS system and components, be sure to disable the SRS system. Tape the negative battery cable with insulating tape. Always disconnect the negative battery cable first.

✻✻ CAUTION

To avoid personal injury when working on vehicles equipped with an air bag, the negative battery cable must be disconnected and at least three minutes must elapse before working on the system. Failure to do so may result in deployment of the air bag.

A. Pulley
B. Balancer
C. Bracket

37655_OPTI_G0139

Fig. 55 Crankshaft damper and related components—2.4L Engine

2. Remove the battery top cover, if equipped.
3. Disconnect the negative battery cable. Tape the cable with insulating tape.
4. Remove the drive belt.
5. Remove the necessary components in order to gain access to the crankshaft pulley.
6. Remove the starter. Install flywheel stopper tool SST 09231-3K000, or equivalent.
7. Remove the retaining bolt.
8. Remove the crankshaft balancer using the proper removal tool.

To install:

➡Be sure to use new fasteners, as required.

9. Installation is the reverse of the removal procedure.
10. Tighten the bolt to specification.

2.7L Engine

See Figures 56 through 59.

1. Before servicing the vehicle, refer to the Precautions Section.

➡If working near and/or around the SRS system and components, be sure to disable the SRS system. Tape the negative battery cable with insulating tape. Always disconnect the negative battery cable first.

✻✻ CAUTION

To avoid personal injury when working on vehicles equipped with an air bag, the negative battery cable must be disconnected and at least three minutes must elapse before working

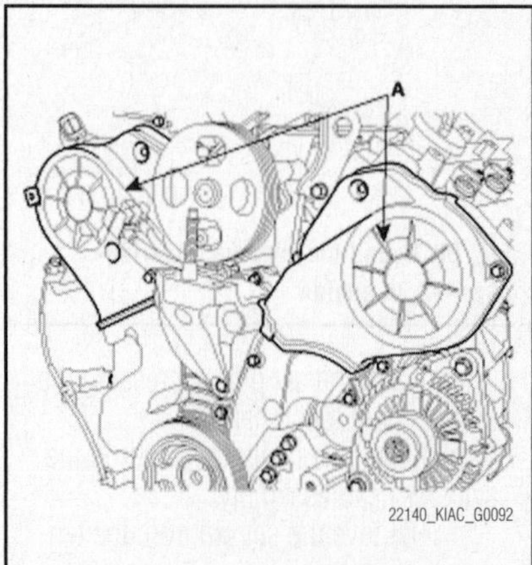

Fig. 56 Remove the timing belt upper covers (A)—2.7L engine

on the system. Failure to do so may result in deployment of the air bag.

2. Remove the battery top cover, if equipped.
3. Disconnect the negative battery cable. Tape the cable with insulating tape.
4. Remove the engine cover.
5. raise and safely support the

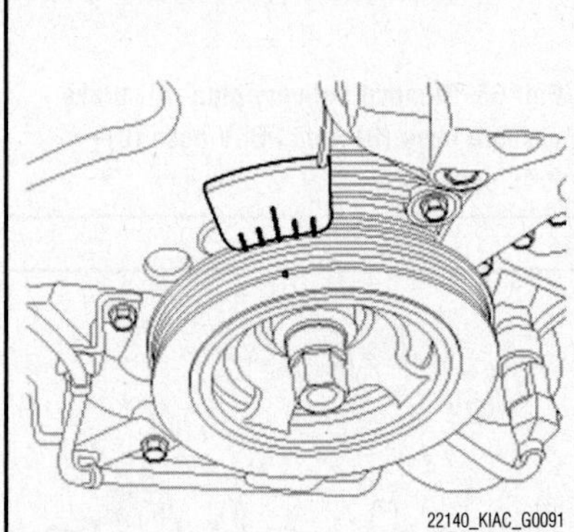

Fig. 57 Align the groove of the pulley with the timing mark of the timing belt cover—2.7L engine

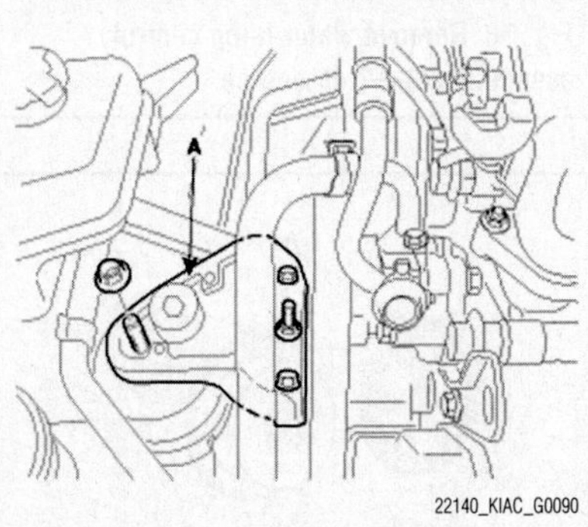

Fig. 58 Remove the engine mounting bracket (A)—2.7L engine

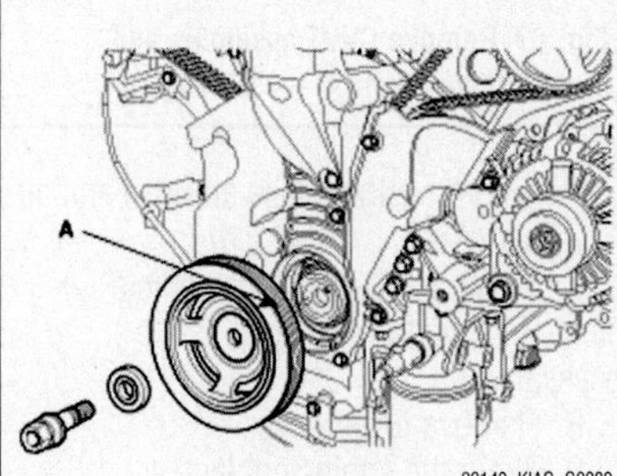

Fig. 59 Crankshaft damper (A) and related components—2.7L engine

vehicle. Remove the front right wheel and tire.
6. Remove the right side cover.
7. Remove the drive belt, the idler pulley and the tensioner pulley.

➡**In removing the drive belt, fix a tool in the auto tensioner pulley bolt and turn the bolt counter clockwise.**

8. Remove the timing belt upper covers (A).
9. Align the groove of the pulley with the timing mark of the timing belt cover by turning the crankshaft pulley clockwise. Check that the timing mark of the camshaft sprocket is aligned with that of the cylinder head cover. (No.1 cylinder piston at TDC)
10. Remove the engine mounting bracket.
   a. Support the engine oil pan with a jack.

CAUTION

Put a wooden or rubber block between the jack and the engine oil pan.

   b. Remove the engine mounting bracket (A).
11. Remove the crankshaft damper pulley (A).

To install:

12. Install the crankshaft damper pulley. Torque: 123.0–130.2 ft. lbs. (166.7–176.5 Nm).
13. Install the engine mounting bracket. Torque: 47.0–61.5 ft. lbs. (63.7–83.4 Nm).
14. Install the timing belt upper cover. Torque: 7.2–8.7 ft. lbs. (9.8–11.8 Nm).
15. Install the drive belt tensioner. Torque: 25.3–39.8 ft. lbs. (34.3–53.9 Nm).
16. Install the drive belt idler and the drive belt. Torque: 25.3–39.8 ft. lbs. (34.3–53.9 Nm).
17. Install the right side cover.
18. Install the front right wheel and tire.
19. Install the engine cover. Torque: 5.8–8.7 ft. lbs. (7.8–11.8 Nm).

CRANKSHAFT FRONT SEAL

REMOVAL & INSTALLATION

See Figures 55 and 59.

At this time the manufacturer does not provide removal and installation procedures for this component. The following procedure is a guideline and may differ from the vehicle you are servicing.

1. Before servicing the vehicle, refer to the Precautions Section.

➡**If working near and/or around the SRS system and components, be sure to disable the SRS system. Tape the negative battery cable with insulating tape. Always disconnect the negative battery cable first.**

CAUTION

To avoid personal injury when working on vehicles equipped with an air bag, the negative battery cable must be disconnected and at least three minutes must elapse before working on the system. Failure to do so may result in deployment of the air bag.

2. Remove the battery top cover, if equipped.
3. Disconnect the negative battery cable. Tape the cable with insulating tape.
4. Remove the crankshaft damper.
5. Carefully pry the seal from its mounting.

To install:

➡**Be sure to use new fasteners, as required.**

6. Installation is the reverse of the removal procedure.
7. Lubricate the seal lip with clean engine oil and push the seal slightly in by hand.
8. Install the seal using a seal installer. Install the seal until it is flush with the mating surface.
9. Start the engine and check for leaks.

CYLINDER HEAD

REMOVAL & INSTALLATION

2.4L Engine

See Figures 60 through 73.

1. Before servicing the vehicle, refer to the Precautions Section.

➡**If working near and/or around the SRS system and components, be sure to disable the SRS system. Tape the negative battery cable**

with insulating tape. Always disconnect the negative battery cable first.

⁂ CAUTION

To avoid personal injury when working on vehicles equipped with an air bag, the negative battery cable must be disconnected and at least three minutes must elapse before working on the system. Failure to do so may result in deployment of the air bag.

2. Remove the battery top cover, if equipped.
3. Disconnect the negative battery cable. Tape the cable with insulating tape.
4. Remove engine cover.
5. Remove air duct.
6. Remove the intake air hose and air cleaner assembly.

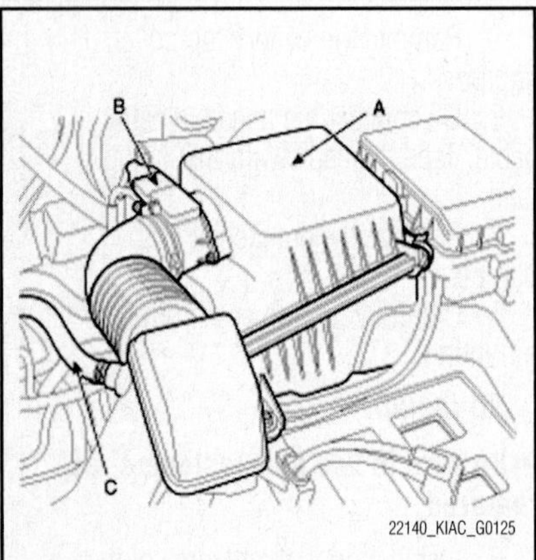

Fig. 60 Disconnect the AFS (B) connector, breather hose (C), and air cleaner (A)—2.4L engine

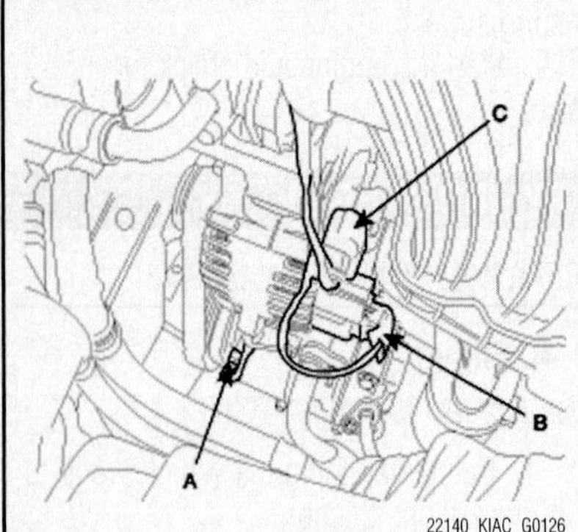

Fig. 61 Disconnect A/C switch (A), alternator connector (B), and oil pressure switch (C)—2.4L engine

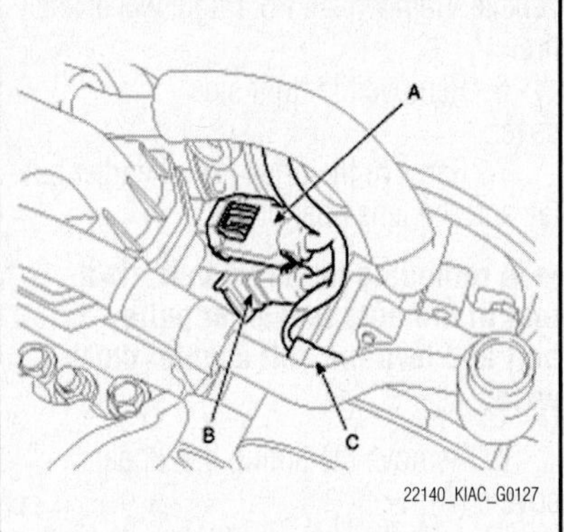

Fig. 62 Disconnect OCV connector (A) and OTS connector (B) and P/S switch connector (C)—2.4L engine

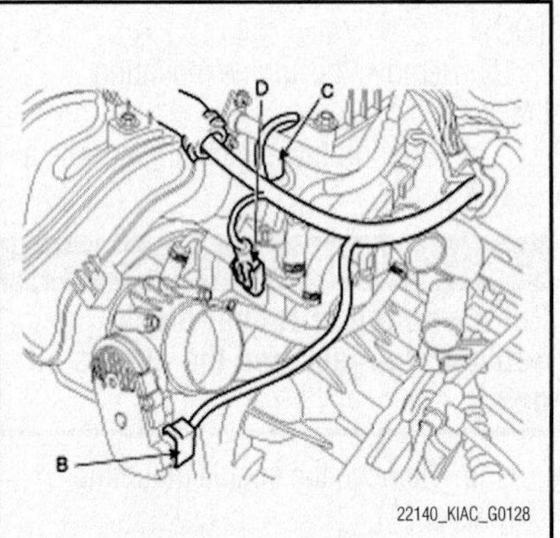

Fig. 63 Remove the ETC connector (B), CMP connector (C), knock sensor connector (D)—2.4L engine

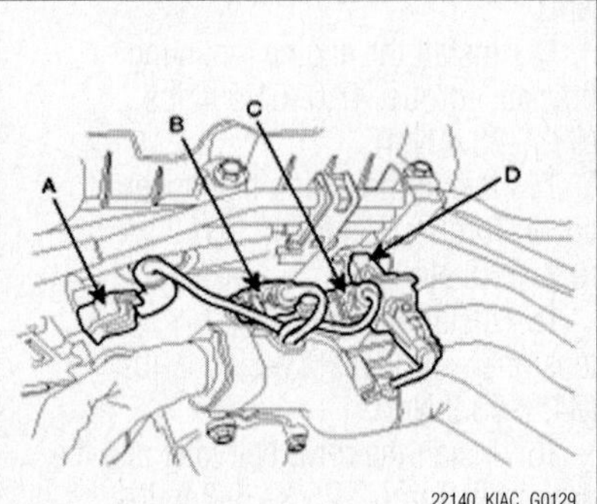

Fig. 64 Disconnect PCSV connector (A), WTS connector (B), condenser connector (C), and CKP sensor connector (D)—2.4L engine

a. Disconnect the AFS (B) connector.
b. Disconnect the breather hose (C) from air cleaner hose.
c. Disconnect the ECU connector.

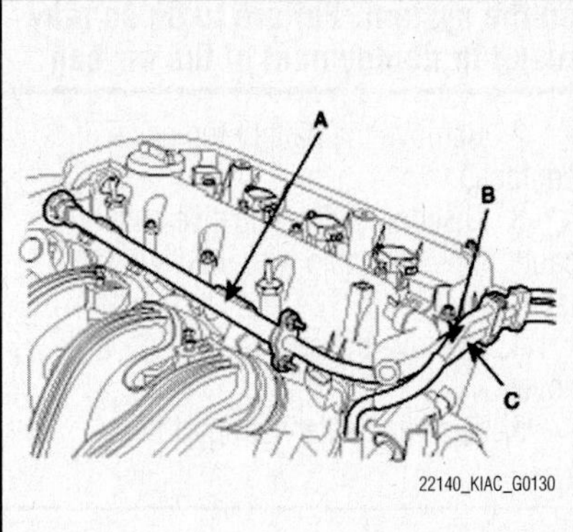

Fig. 65 Remove delivery pipe (A), brake vacuum hose (B), and PCSV hose (C)—2.4L engine

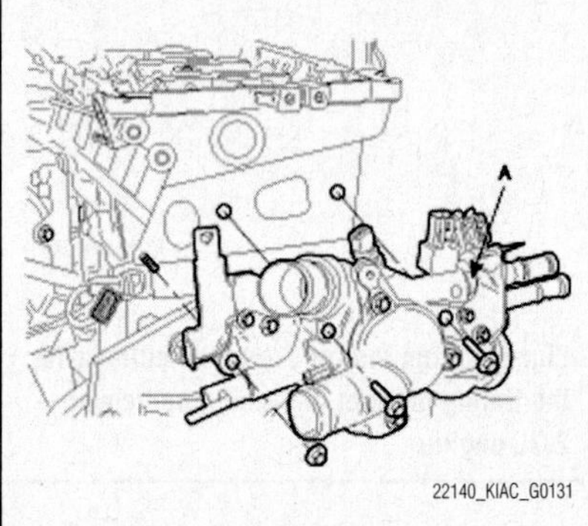

Fig. 66 Remove water temp control assembly (A)—2.4L engine

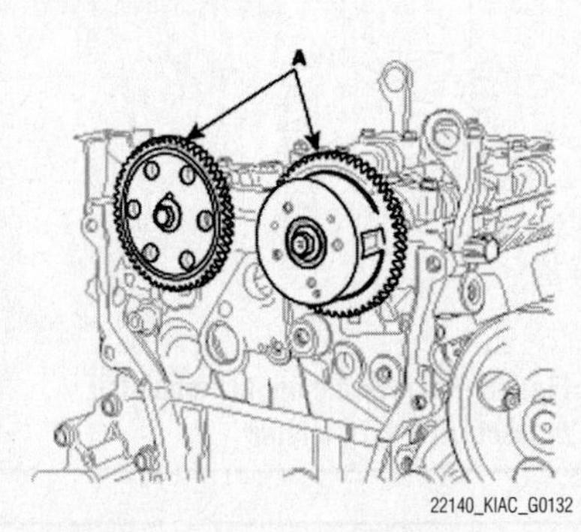

Fig. 67 Remove CVVT assembly and camshaft sprocket (A)—2.4L engine

d. Remove the intake air hose and air cleaner (A).
7. Raise and support the vehicle safely. Remove front wheels, as necessary.
8. Remove lower cover.
9. Drain the engine coolant.
10. Remove the upper and lower radiator hoses.
11. Remove the heater hoses.

Fig. 68 Remove front camshaft bearing cap (A)—2.4L engine

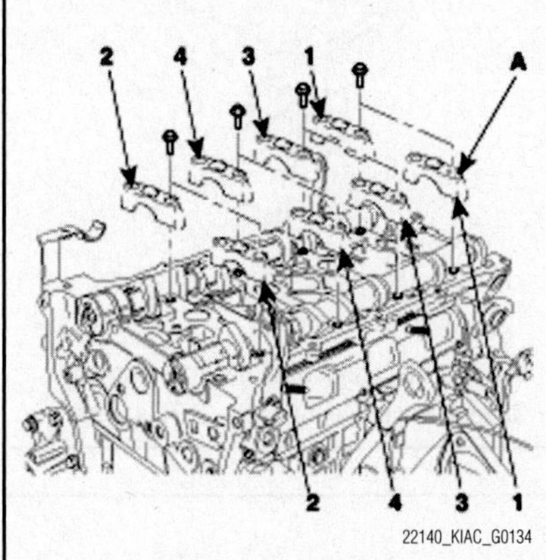

Fig. 69 Remove camshaft bearing cap (A), in the sequence shown—2.4L engine

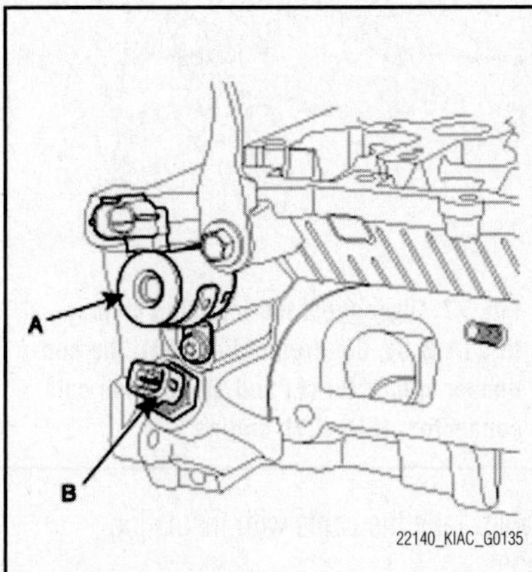

Fig. 70 Remove OCV (A) and OTS (B)—2.4L engine

12. Disconnect A/C switch (A), alternator connector (B), and oil pressure switch (C).

13. Disconnect OCV connector (A) and OTS connector (B) and P/S switch connector (C).

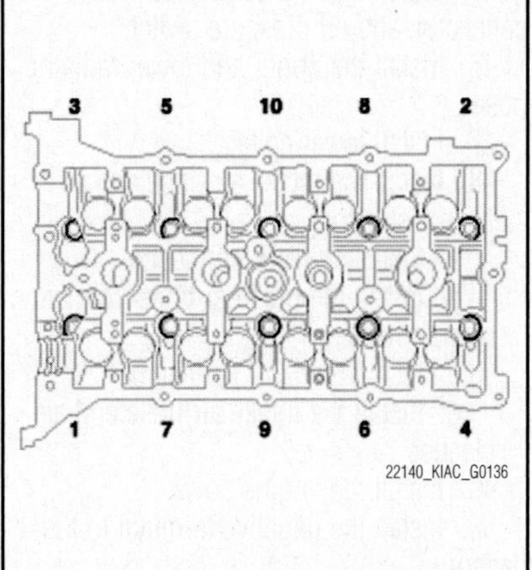

Fig. 71 Remove the cylinder head bolts in the sequence shown—2.4L engine

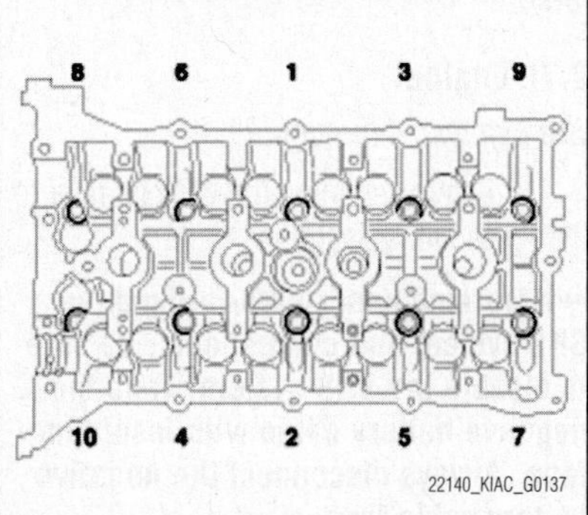

Fig. 72 Install cylinder head bolts in the sequence shown—2.4L engine

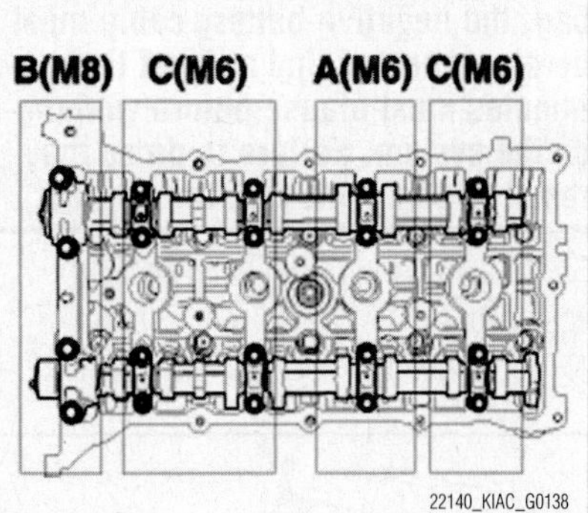

Fig. 73 Install camshaft bearing caps and torque in sequence—2.4L engine

14. Disconnect injector connectors.

15. Disconnect the engine wire harness connectors.

a. Remove the ETC connector (B), CMP connector (C), knock sensor connector (D).

16. Disconnect ignition coil connectors.

17. Disconnect PCSV connector (A), WTS connector (B), condenser connector (C), and CKP sensor connector (D).

18. Remove delivery pipe (A), brake vacuum hose (B), and PCSV hose (C).

19. Remove water temp control assembly (A).

20. Remove intake manifold.

21. Remove exhaust manifold.

22. Remove timing chain.

23. Remove CVVT assembly and camshaft sprocket (A).

24. Remove camshaft.

a. Remove front camshaft bearing cap (A).

b. Remove camshaft bearing cap (A), in the sequence shown.

c. Remove camshafts.

25. Remove OCV (A) and OTS (B).

26. Remove the cylinder head bolts, then remove the cylinder head.

a. Using triple square wrench, uniformly loosen and remove the 10 cylinder head bolts, in several passes, in the sequence shown. Remove the 10 cylinder head bolts and plate washers.

✲✲ CAUTION

Head warpage or cracking could result from removing bolts in an incorrect order.

b. Lift the cylinder head from the dowels on the cylinder block and place the cylinder head on wooden blocks on a bench.

To install:

27. Install the cylinder head gasket on the cylinder block.

➡Be careful of the installation direction.

28. Place the cylinder head carefully in order not to damage the gasket with the bottom part of the end.

29. Install cylinder head bolts.

a. Apply a light coat if engine oil on the threads and under the heads of the cylinder head bolts.

b. Install and tighten the 10 cylinder head bolts and plate washers, in several passes, in the sequence shown. Torque: 25.3 ft. lbs. (34.3 Nm) plus 90° plus an additional 90°.

➡Always use new cylinder head bolt.

30. Install OCV and OTS. Torque: OCV: 7.23–8.67 ft. lbs. (9.8–11.76 Nm; OTS: 14.5–17.4 ft. lbs. (19.6–23.52 Nm).

✲✲ CAUTION

Do not reuse the OCV when dropped. Keep the OCV clean. Do not hold the OCV sleeve during servicing. When the OCV is installed on the engine, do not move the engine with holding the OCV yoke.

31. Install the CVVT and camshaft sprocket. Torque: 39.7–47.0 ft. lbs. (53.9–63.7 Nm).

➡Hold the hexagonal head wrench portion of the camshaft with a vise, and install the bolt and CVVT assembly.

32. Install camshafts.

➡Apply a light coat of engine oil on camshaft journals.

33. Install camshaft bearing caps in their proper locations.
Tightening order:
- Group A
- Group B
- Group C.
- Torque: M6 bolts: 8.0–9.4 ft. lbs. (10.8–12.7 Nm; M8 bolts: 20.2–23.1 ft. lbs. (27.4–31.4 Nm).

34. Install timing chain.
35. Check and adjust valve clearance.
36. Install the exhaust manifold.
37. Install the intake manifold.
38. Install water temp control assembly. Torque: Bolt: 10.8–15.9 ft. lbs. (14.7–21.6 Nm; Nut: 14.5–19.5 ft. lbs. (19.6–26.5 Nm).

➡Assemble water temp control assembly and water inlet pipe to water pump assembly before nuts for assembling of water inlet pipe to be tightened. Insert after wetting O-ring or inner surface of thermostat housing. Always use a new O-ring.

39. Install delivery pipe, brake hose, and PCSV hose.
40. Install the PCSV connector, WTS connector, condenser connector, and CKP sensor connector.
41. Install ignition coil connector.
42. Install the engine wire harness connectors.
    a. Install the ETC connector, CMP connector, knock sensor connector.
43. Install the injector connectors.
44. Install the OCV connector, OTS connector, P/S switch connector.
45. Install the A/C switch, alternator connector, and oil pressure switch.
46. Install the upper and lower radiator hoses.
47. Install lower cover.
48. Install the intake air hose and air cleaner assembly.
    a. Install the AFS connector.
    b. Install the breather hose from air cleaner hose.
    c. Install the ECU connector.
    d. Install the intake air hose and air cleaner.
49. Install the engine cover.
50. Install the negative terminal to the battery.
51. Fill with engine coolant.
52. Start the engine and check for leaks.
53. Recheck engine coolant level and oil level.

2.7L Engine

See Figures 74 through 93.

1. Before servicing the vehicle, refer to the Precautions Section.

➡If working near and/or around the SRS system and components, be sure to disable the SRS system. Tape the negative battery cable with insulating tape. Always disconnect the negative battery cable first.

✲✲ CAUTION

To avoid personal injury when working on vehicles equipped with an air bag, the negative battery cable must be disconnected and at least three minutes must elapse before working on the system. Failure to do so may result in deployment of the air bag.

2. Remove the battery top cover, if equipped.
3. Disconnect the negative battery cable. Tape the cable with insulating tape.
4. Remove the air duct and the battery.
5. Remove the engine cover.
6. Remove the intake air hose and air cleaner assembly.
    a. Disconnect the MAF connector (A).
    b. Disconnect the breather hose (B) from air cleaner hose.

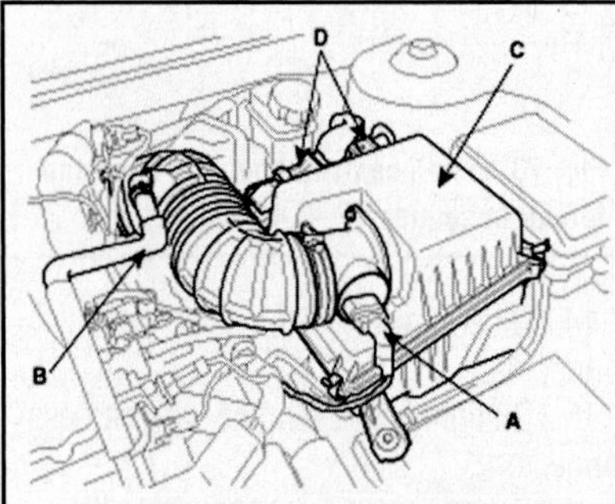

Fig. 74 Disconnect the MAF connector (A), breather hose (B), intake air hose and air cleaner assembly (C), and PCM connectors (D)—2.7L engine

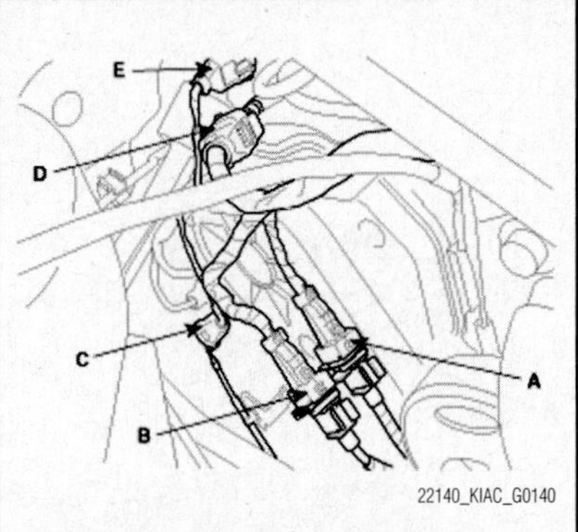

Fig. 75 Disconnect the engine wiring harness connectors—2.7L engine

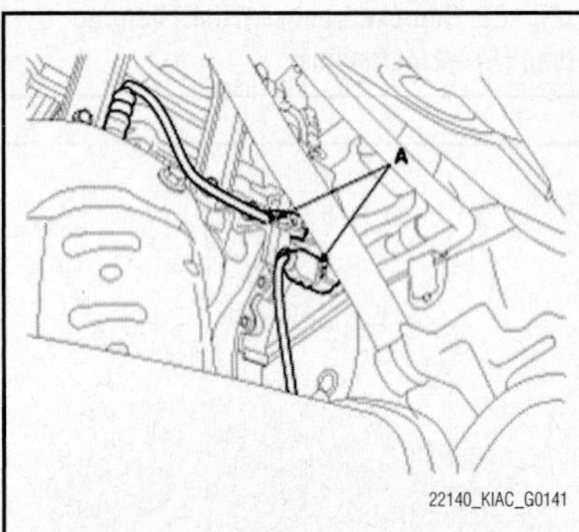

Fig. 76 Disconnect the bank 1 front/rear O2 sensor connectors (A)—2.7L engine

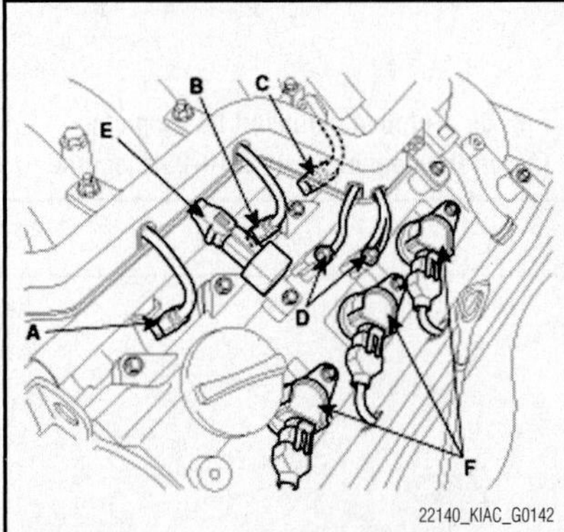

Fig. 77 Disconnect the injection connectors (A,B,C), the ground lines (D), the condenser connector (E) and the Ignition coil connectors (F)—2.7L engine

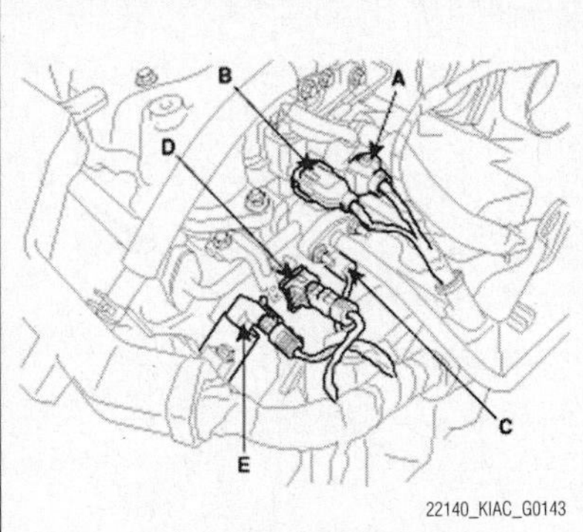

Fig. 78 Disconnect the injection harness connector (A), the No.2 VIS (Variable Induction System) connector (B), the No.1/No.2 OCV (Oil Control Valve) connectors (C,D) and the OTS (Oil Temperature Sensor) connector (E)—2.7L engine

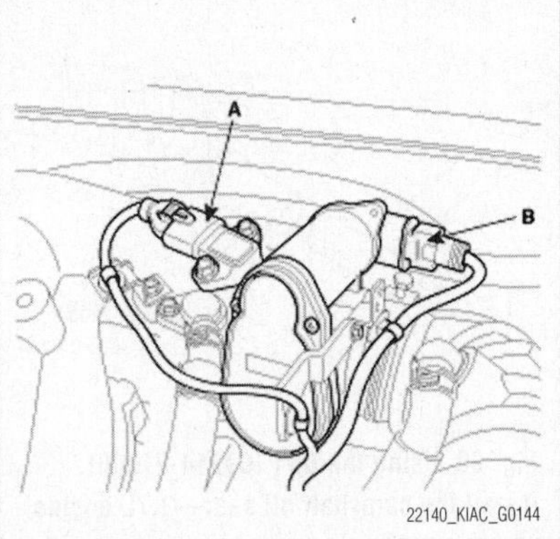

Fig. 79 Disconnect the MAPS (Manifold Absolute Pressure Sensor) connector (A), and the ETC (Electronic Throttle Control) connector (B)—2.7L engine

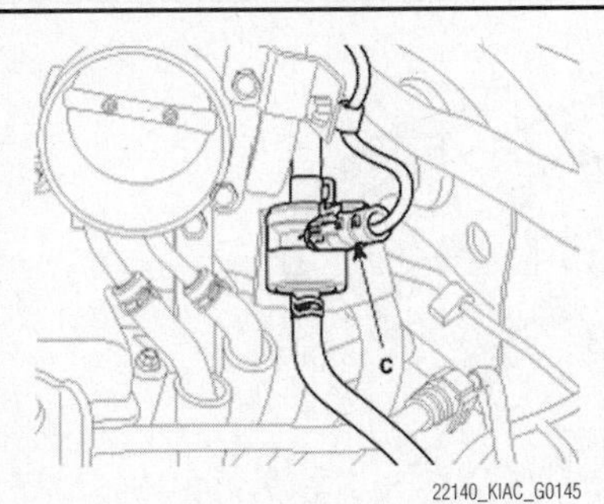

Fig. 80 Disconnect the PCSV (Purge Control Solenoid Valve) connector (C)—2.7L engine

c. Remove the intake air hose and air cleaner assembly (C).

d. Disconnect the PCM connectors (D).

7. Remove the upper and lower radiator hoses.

8. Remove the fuel inlet hose from the delivery pipe.

9. Disconnect the engine wiring harness connectors.

a. Disconnect the No.1/No.2 knock sensor connectors (A,B), the oil pressure switch connector (C), the ignition coil harness (D) and the No.1 VIS (Variable Induction System) connector (E).

b. Disconnect the bank 1 front/rear O2 sensor connectors (A).

c. Disconnect the injection connectors (A,B,C), the ground lines (D), the condenser connector (E) and the Ignition coil connectors (F).

d. Disconnect the injection harness connector (A), the No.2 VIS (Variable Induction System) connector (B), the No.1/No.2 OCV (Oil Control Valve) connectors (C,D) and the OTS (Oil Temperature Sensor) connector (E).

e. Disconnect the MAPS (Manifold Absolute Pressure Sensor) connector (A), the ETC (Electronic Throttle Control) connector (B) and the PCSV (Purge Control Solenoid Valve) connector (C).

f. Disconnect the alternator connector and the air conditioning compressor connector.

g. Disconnect the bank 2 CMP sensor connector (A) and the ECT (Engine Coolant Temperature) sensor connector (B).

h. Disconnect the bank 2 front/rear O2 sensor connectors (A,B) and the CKP sensor connector (C).

i. Disconnect the bank 1 CMP sensor connector (A).

10. Remove the PCV (Purge Control Valve) hose.

11. Disconnect the brake vacuum hose.

12. Remove the heater hoses.

13. Remove the accessory drive belt.

14. Remove the power steering pump.

15. Remove the exhaust manifold assembly.

16. Remove the intake manifold assembly.

17. Remove the timing belt.

18. Remove the ignition coils.

19. Remove the water temp. control assembly.

20. Remove the cylinder head cover.

21. Remove the camshaft bearing cap.

22. Remove the timing chain tensioner (A).

23. Remove the camshaft.

24. Remove the bank 1 timing belt rear cover.

25. Remove the bank 2 timing belt rear cover.

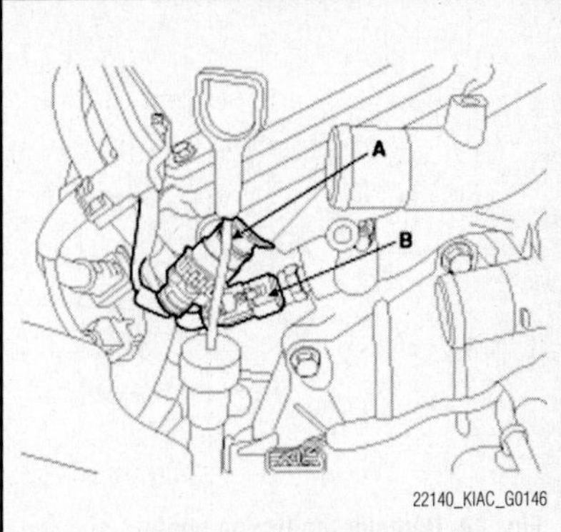

Fig. 81 Disconnect the bank 2 CMP sensor connector (A) and the ECT (Engine Coolant Temperature) sensor connector (B)—2.7L engine

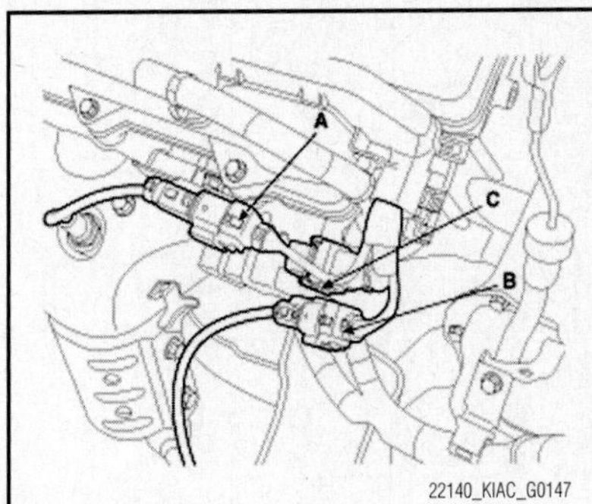

Fig. 82 Disconnect the bank 2 front/rear O2 sensor connectors (A,B) and the CKP sensor connector (C)—2.7L engine

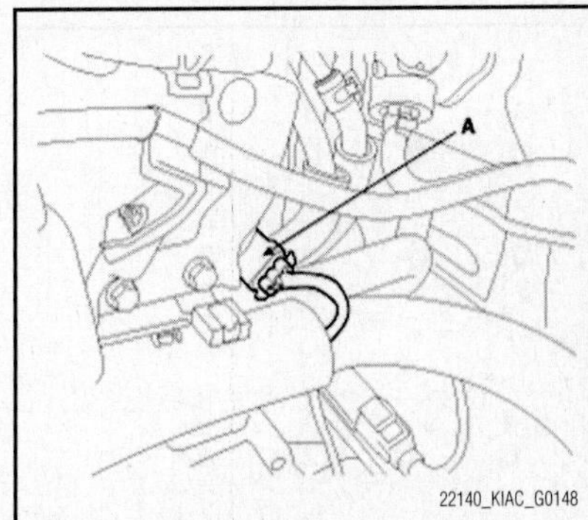

Fig. 83 Disconnect the bank 1 CMP sensor connector (A)—2.7L engine

26. Remove the CKP sensor connector bracket.

27. Remove the cylinder head assembly.

a. Remove the bolts in 2–3 steps in the sequence shown.

⁂ CAUTION

If the bolts are not removed as the order, the deformation of the head assembly can be occurred.

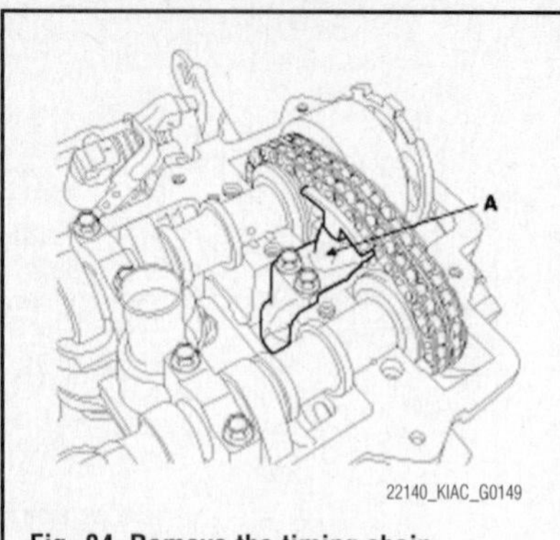

Fig. 84 Remove the timing chain tensioner (A)—2.7L engine

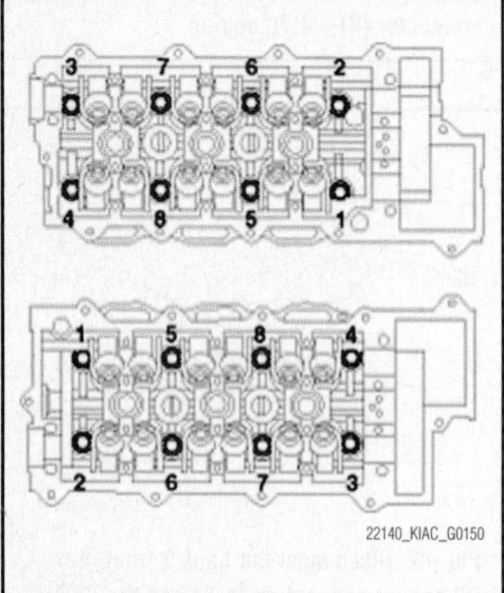

Fig. 85 Remove the bolts in the sequence shown–2.7L engine

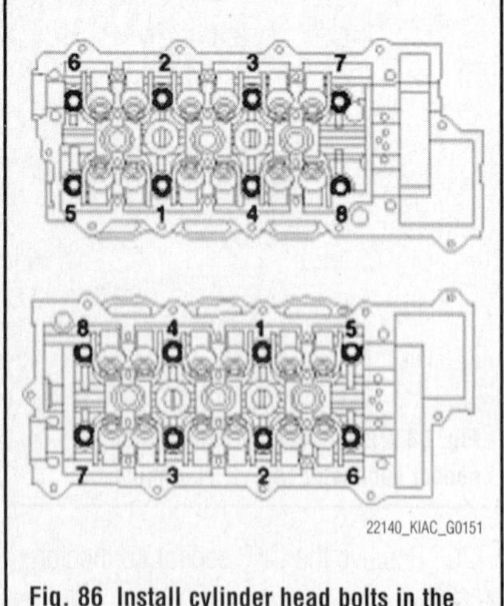

Fig. 86 Install cylinder head bolts in the sequence shown—2.7L engine

b. Put the cylinder head assembly on a wooden block after removal from the cylinder block.

To install:

28. Install the cylinder head (s) with new head gaskets.

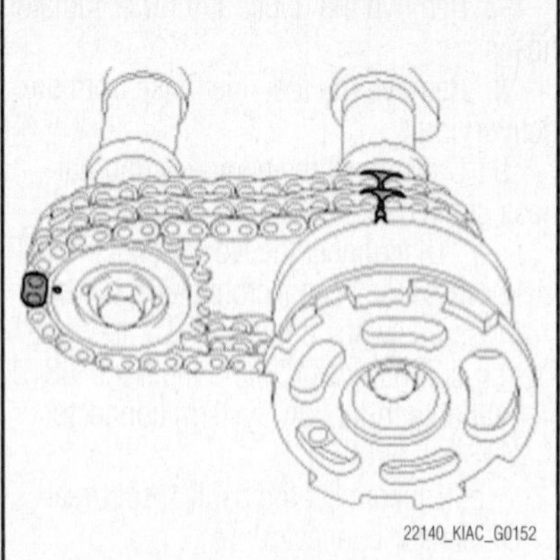

Fig. 87 Left hand camshaft chain timing mark—2.7L engine

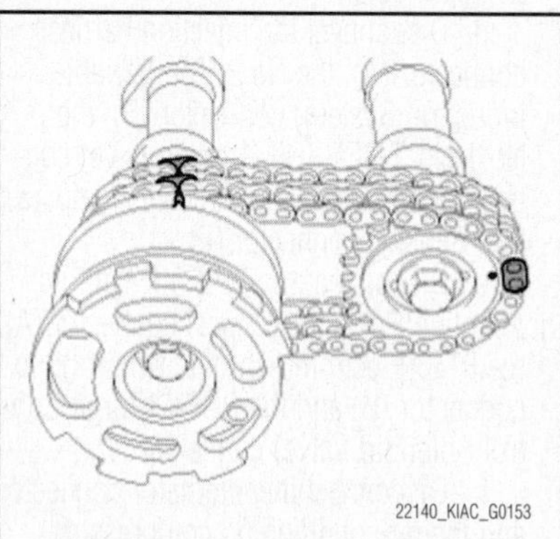

Fig. 88 Right hand camshaft chain timing mark—2.7L engine

✻✻ CAUTION

Ensure the LH/RH classification of the cylinder head gasket when installing.

29. Tighten the cylinder head bolts with the plain washers in several steps in the sequence shown.

➡**In assembling washers, the marked surface should face upward.**

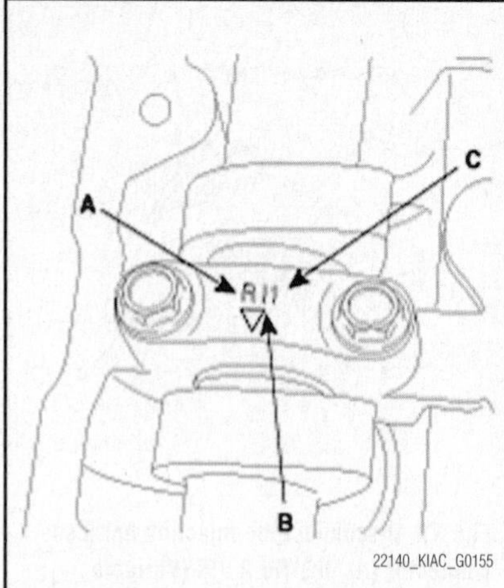

Fig. 89 When installing the bearing caps, check the marks—2.7L engine

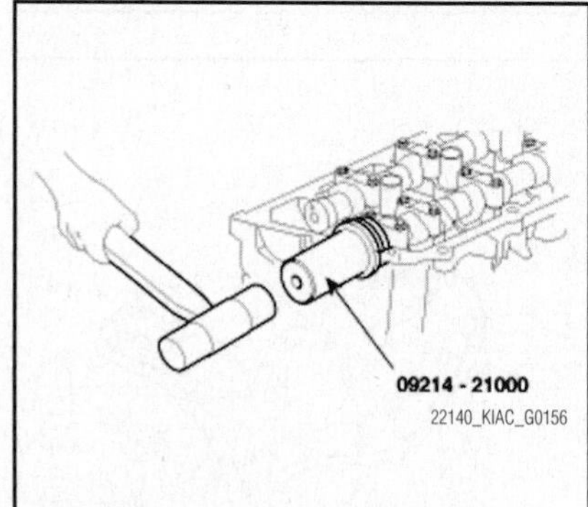

Fig. 90 Using the SST (09214-21000), install the camshaft oil seal—2.7L engine

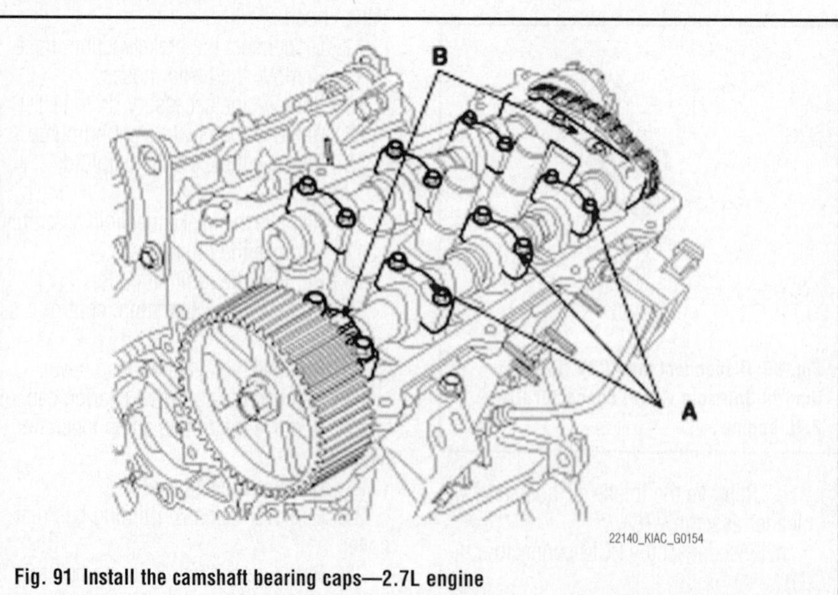

Fig. 91 Install the camshaft bearing caps—2.7L engine

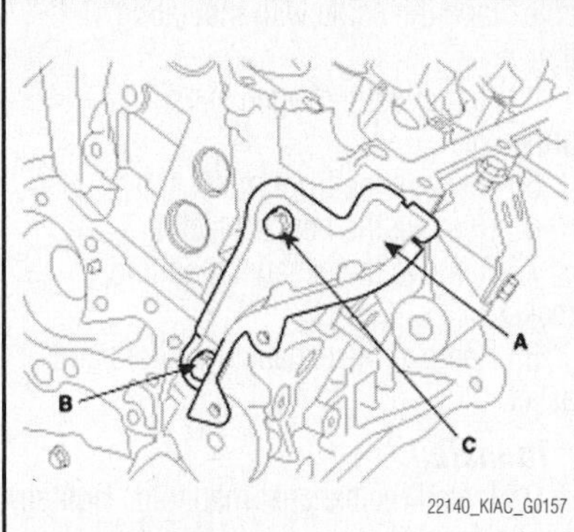

Fig. 92 Install the bank 1 timing belt rear cover—2.7L engine

30. Before installing the cylinder head bolts, apply engine oil on the thread of the bolts and the surface of the washers. Torque: 18.1 ft. lbs. (24.5 Nm) plus 60° plus an additional 45°.

➡Using the SST (09221-4A000), tighten the bolts which need to be tightened with the angular tightening method.

31. Install the CVVT assembly and camshaft chain sprocket with the dowel pin in the CVVT installed to the intake camshaft. Ensure that the pin will not be installed in the hole for oil feeding. Torque: 49.2–57.9 ft. lbs. (66.7–78.5 Nm).

➡After tightening the CVVT bolts, rotate the CVVT assembly housing counterclockwise by hand to seat the lock pin in the CVVT assembly in good position.

**** CAUTION**

Fix the hexagonal part of the camshaft in a vice when tightening the CVVT bolts. Do not fix the CVVT housing or sprocket in a vice.

32. Install the camshaft in the cylinder head assembly.

a. Align the timing mark of the camshaft timing chain.

**** CAUTION**

Both timing marks should face upward in reassembly.

33. Install the timing chain tensioner.

a. Insert the set pin by pressing the timing chain tensioner.

b. Install the chain tensioner in the cylinder head assembly.

c. Remove the set pin from the tensioner after installing.

34. Install the camshaft bearing caps. Torque: Bearing cap bolt (A: 6_38): 8.0–9.4 ft. lbs. (10.8–12.7 Nm). Bearing cap bolt (B: 8_38): 15.2–18.8 ft. lbs. (20.6–22.5 Nm).

➡When installing the bearing caps, check the marks on them as shown below and install them in its proper position. Note the following designations as shown below:

- A: (LH/RH HEAD) L (LH), R (RH)
- B: (Intake/Exhaust) I (Intake), E (Exhaust)
- C: (Cap no.): 1,2,and 3

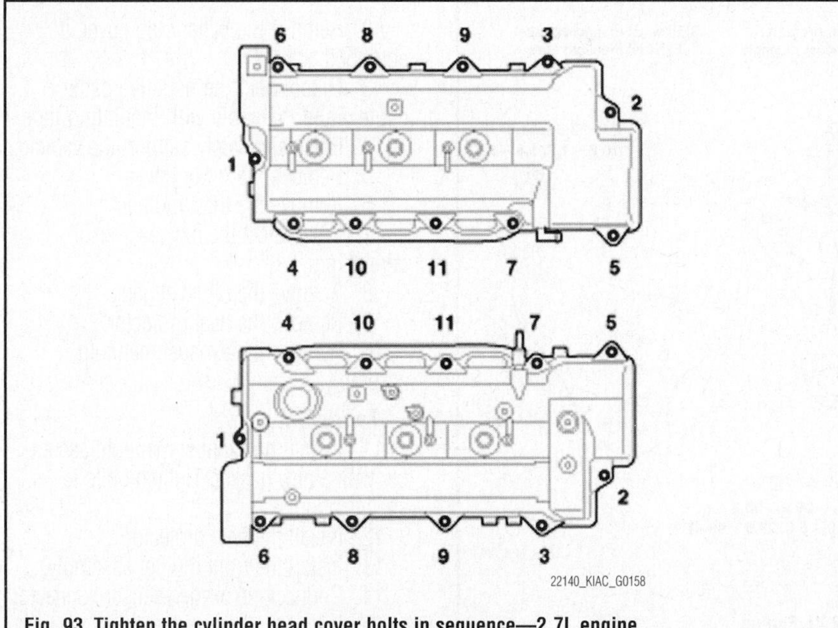

Fig. 93 Tighten the cylinder head cover bolts in sequence—2.7L engine

**** CAUTION**

When installing the bearing caps, turn the crankshaft to place a piston in the middle of the block because interference between valves and pistons can occur.

35. Using the SST (09214-21000), install the camshaft oil seal.

➡Before installing, apply engine oil. The camshaft cap surface should adhere to the cylinder head assembly. Do not press an eccentric load.

36. Install the CKP sensor connector bracket.

37. Install the bank 2 timing belt rear cover.

38. Install the bank 1 timing belt rear cover.

➡The length of the bolt B is longer than that of the bolt C.

39. Install the timing belt.

40. Check and adjust the valve clearance.

41. Install the cylinder head cover.

a. Remove oil, dust or sealant on the upper surface of the cylinder before assembling cylinder head cover.

b. Assemble the cylinder head cover in five minutes after applying liquid gasket (LOCTITE 5900) on the camshaft cap and packing part.

c. Tighten the cylinder head cover bolts in the sequence shown. Torque: 5.8–7.2 ft. lbs. (7.8–9.8 Nm).

➡Do not start engine for thirty minutes after assembling the cylinder head cover. Do not reuse the cylinder head cover gasket.

42. Install the water temp. control assembly.

43. Install the intake manifold assembly.

44. Install the exhaust manifold assembly.

45. Install the power steering pump.

46. Install the accessory drive belt.

47. Install the heater hose.

48. Connect the brake vacuum hose.

49. Install the PCV (Positive Crankcase Ventilation) hose.

50. Connect the engine wiring harness connectors.

a. Connect the bank 1 CMP sensor connector.

b. Connect the bank 2 front/rear O2 sensor connectors and the CKP sensor connector.

c. Connect the bank 2 CMP sensor connector and the WTS (Water Temperature Sensor) connector.

d. Connect the alternator connector and the air conditioning compressor connector.

e. Connect the MAPS (Manifold Absolute Pressure Sensor) connector, the ETC (Electronic Throttle Control) connector and the PCSV (Purge Control Solenoid Valve) connector.

f. Connect the injection harness connector, the No.2 VIS (Variable Induction System) connector, the No.1/No.2 OCV (Oil Control Valve) connectors and the OTS (Oil Temperature Sensor) connector.

g. Connect the injection connectors, the ground lines, the condenser connector and the Ignition coil connectors.

h. Connect the bank 1 front/rear O2 sensor connectors.

i. Connect the No.1/No.2 knock sensor connectors, the oil pressure switch connector, the ignition coil harness and the No.1 VIS (Variable Induction System) connector.

51. Install the fuel inlet hose from the delivery pipe.
52. Install the upper radiator hose and lower radiator hose.
53. Install the intake air hose and air cleaner assembly.

a. Connect the PCM connectors.

b. Install the intake air hose and air cleaner assembly.

c. Connect the breather hose from air cleaner hose.

d. Connect the MAF connector.

54. Install the engine cover.
55. Refill engine coolant.

EXHAUST MANIFOLD

REMOVAL & INSTALLATION

2.4L Engine

See Figure 94.

1. Before servicing the vehicle, refer to the Precautions Section.

➡If working near and/or around the SRS system and components, be sure to disable the SRS system. Tape the negative battery cable with insulating tape. Always disconnect the negative battery cable first.

✲✲ CAUTION

To avoid personal injury when working on vehicles equipped with an air bag, the negative battery cable must be disconnected and at least three minutes must elapse before working on the system. Failure to do so may result in deployment of the air bag.

2. Remove the battery top cover, if equipped.
3. Disconnect the negative battery cable. Tape the cable with insulating tape.
4. Remove the oxygen sensor connector.
5. Remove the front muffler.
6. Remove the heat protector.
7. Remove the exhaust manifold bracket.
8. Remove the exhaust manifold and gasket.

To install:

9. Install the exhaust manifold. Tighten bolts to specification.
10. Install the exhaust manifold bracket
11. Install the heat protector.
12. Install the front muffler.
13. Install the oxygen sensor connector.

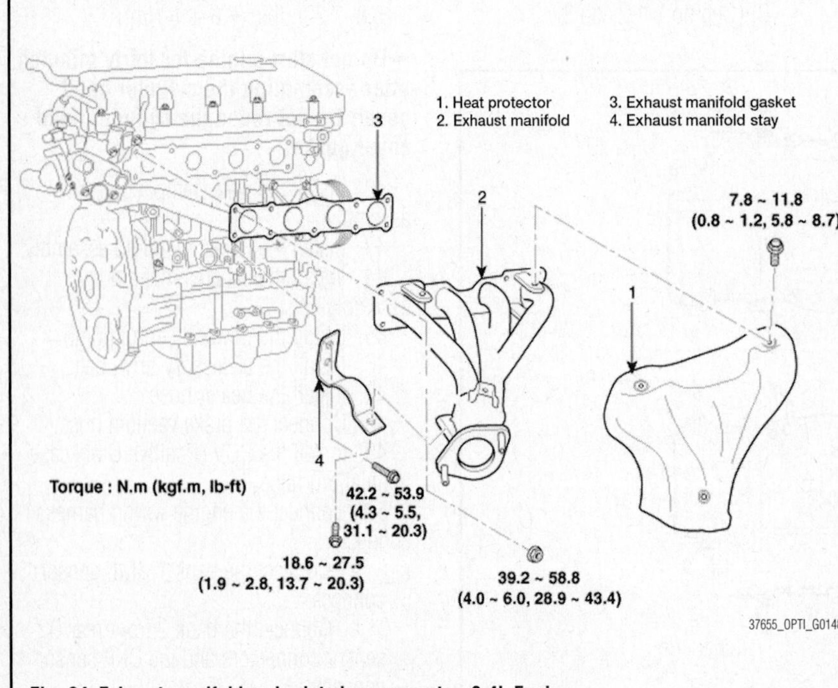

Fig. 94 Exhaust manifold and related components—2.4L Engine

2.7L Engine

See Figure 95.

1. Before servicing the vehicle, refer to the Precautions Section.

➡If working near and/or around the SRS system and components, be sure to disable the SRS system. Tape the negative battery cable with insulating tape. Always disconnect the negative battery cable first.

✲✲ CAUTION

To avoid personal injury when working on vehicles equipped with an air bag, the negative battery cable must be disconnected and at least three minutes must elapse before working on the system. Failure to do so may result in deployment of the air bag.

2. Remove the battery top cover, if equipped.
3. Disconnect the negative battery cable. Tape the cable with insulating tape.
4. Raise and safely support the vehicle.
5. Remove the lower cover.
6. Remove the front muffler.
7. Disconnect the oxygen sensor connectors.
8. Remove the oil level gauge.
9. Remove the heat protector.
10. Remove the exhaust manifold assembly.

To install:

11. Install the exhaust manifold assembly with a new gasket. Tighten bolts to specification.
12. Install the heat protector.
13. Install the front muffler assembly.
14. Connect the oxygen sensor connector.
15. Install the lower cover.

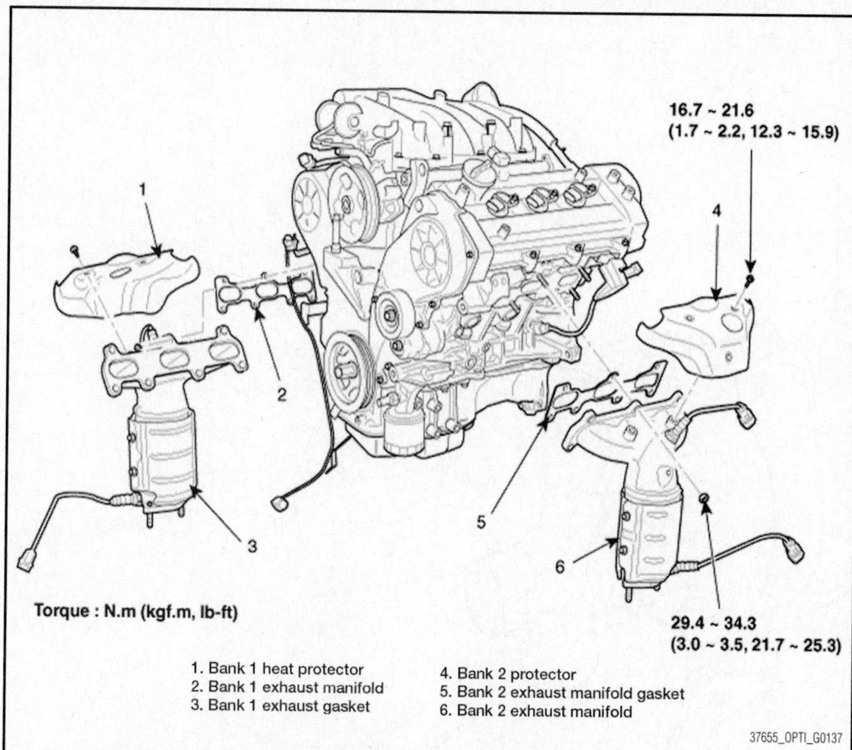

Fig. 95 Exhaust manifolds and related components—2.7L Engine

7. Remove the component from the vehicle.

To install:

➡Be sure to use new fasteners, as required.

8. Installation is the reverse of the removal procedure.

➡Apply sealant to the screw part (threads) of the bolt, when using new flywheel bolts.

9. Be sure to use new bolts. Tighten the bolts to specification and in a criss-cross pattern and in several passes.

INTAKE MANIFOLD

REMOVAL & INSTALLATION

2.4L Engine

See Figure 97.

1. Before servicing the vehicle, refer to the Precautions Section.

FLYWHEEL/FLEXPLATE

REMOVAL & INSTALLATION

See Figure 96.

At this time the manufacturer does not provide removal and installation procedures for this component. The following procedure is a guideline and may differ from the vehicle you are servicing.

1. Before servicing the vehicle, refer to the Precautions Section.

➡If working near and/or around the SRS system and components, be sure to disable the SRS system. Tape the negative battery cable with insulating tape. Always disconnect the negative battery cable first.

⁂ CAUTION

To avoid personal injury when working on vehicles equipped with an air bag, the negative battery cable must be disconnected and at least three minutes must elapse before working on the system. Failure to do so may result in deployment of the air bag.

2. Remove the battery top cover, if equipped.
3. Disconnect the negative battery cable. Tape the cable with insulating tape.

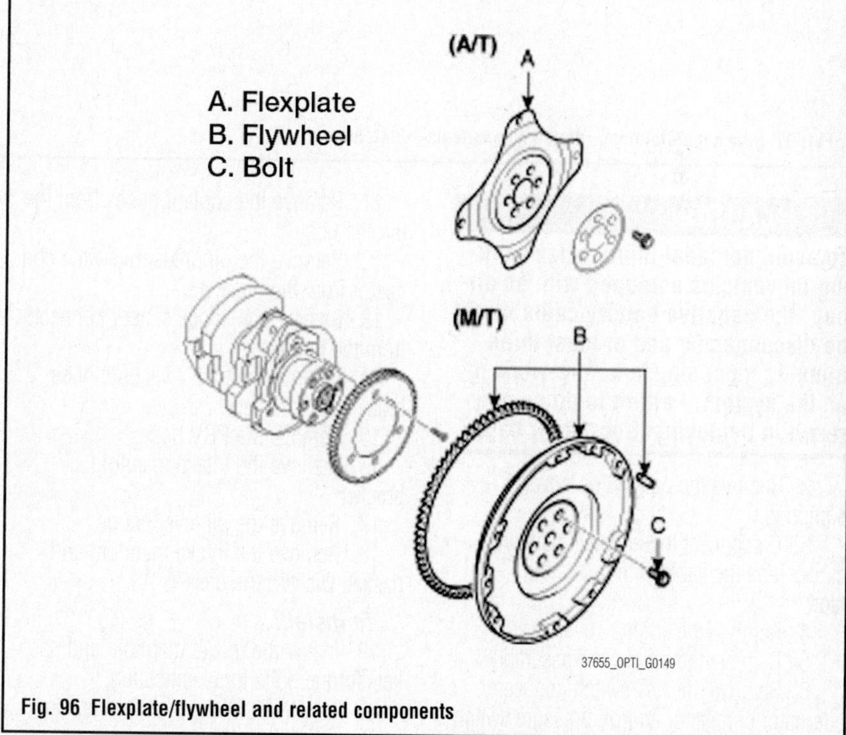

Fig. 96 Flexplate/flywheel and related components

4. Raise and safely support the vehicle.
5. Remove the transaxle.
6. Remove the flexplate/flywheel retaining bolts. Discard the bolts.

➡If working near and/or around the SRS system and components, be sure to disable the SRS system. Tape the negative battery cable with insulating tape. Always disconnect the negative battery cable first.

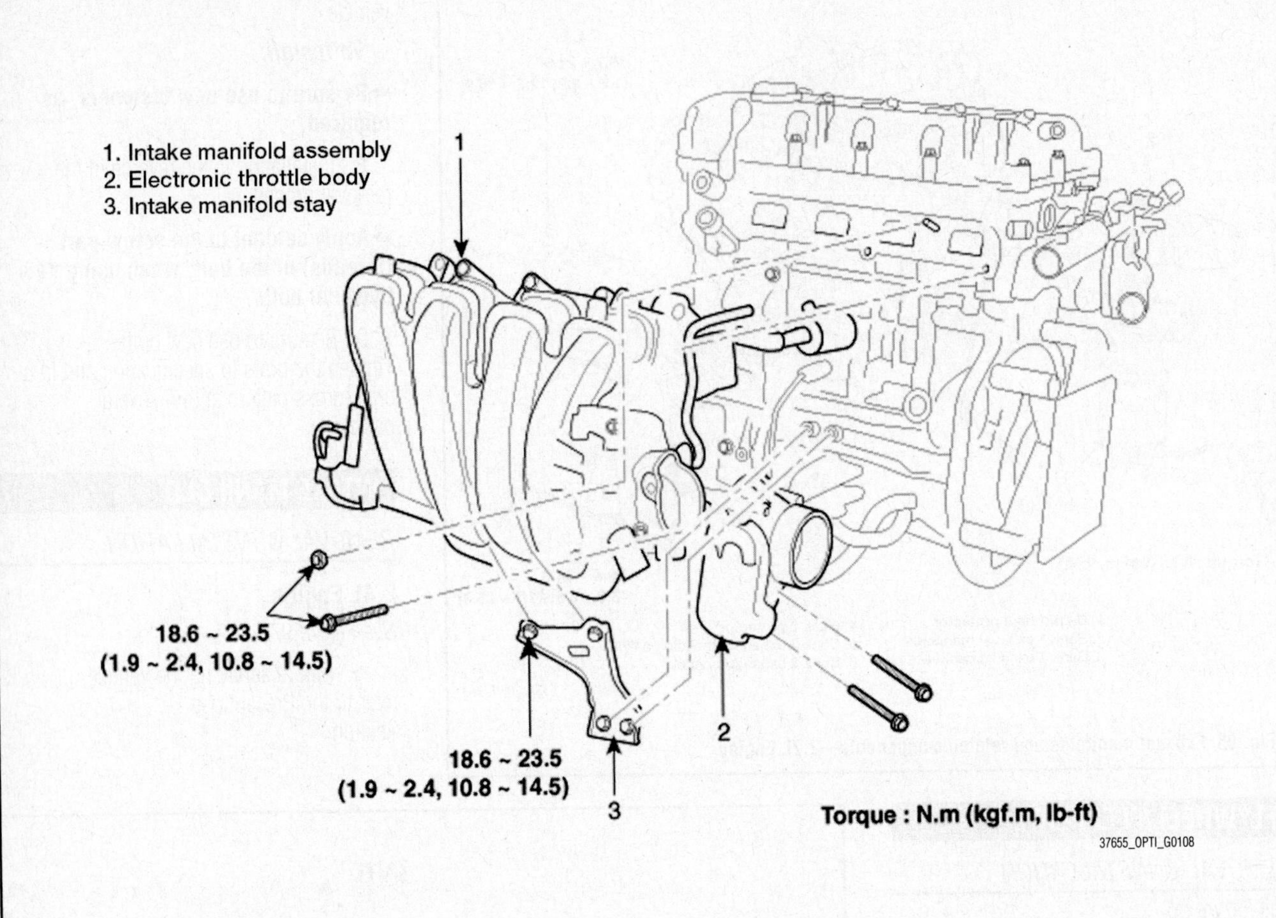

Fig. 97 Intake manifold and related components—2.4L engine

⁂ CAUTION

To avoid personal injury when working on vehicles equipped with an air bag, the negative battery cable must be disconnected and at least three minutes must elapse before working on the system. Failure to do so may result in deployment of the air bag.

2. Remove the battery top cover, if equipped.
3. Disconnect the negative battery cable. Tape the cable with insulating tape.
4. Remove the engine cover.
5. Remove the air cleaner assembly.
6. Remove the A/C switch connector, alternator connector and oil pressure switch connector.
7. Remove the OCV connector, OTS connector, P/S switch connector.
8. Remove the injector connector.
9. Disconnect the engine wire harness connectors.
10. Remove the fuel delivery pipe.
11. Remove the coolant hoses from the throttle body.
12. Remove the oil pressure switch connector from the bracket.
13. Remove the knock sensor connector from the bracket.
14. Remove the PCSV vacuum hose, brake vacuum hose.
15. Remove the PCV hose.
16. Remove the intake manifold bracket.
17. Remove the oil level gauge.
18. Remove the intake manifold and gasket. Discard the gasket.

To install:

19. Install the intake manifold and gasket. Torque bolts to specification.
20. Install the intake manifold bracket.
21. Install the PCV hose.
22. Install the PCSV vacuum hose, brake vacuum hose.
23. Install the knock sensor connector from the bracket.
24. Install the oil pressure switch connector from the bracket.
25. Install the coolant hoses from the throttle body.
26. Install the delivery pipe.
27. Install the engine wire harness connectors.
28. Install the injector connector.
29. Install Remove the OCV connector, OTS connector, P/S switch connector.
30. Install the A/C switch connector, alternator connector and oil pressure switch connector.
31. Install the air cleaner assembly.
32. Install the engine cover.

2.7L Engine

See Figures 98 through 112.

1. Before servicing the vehicle, refer to the Precautions Section.

➡If working near and/or around the SRS system and components, be sure to disable the SRS system. Tape the negative battery cable with insulating tape. Always disconnect the negative battery cable first.

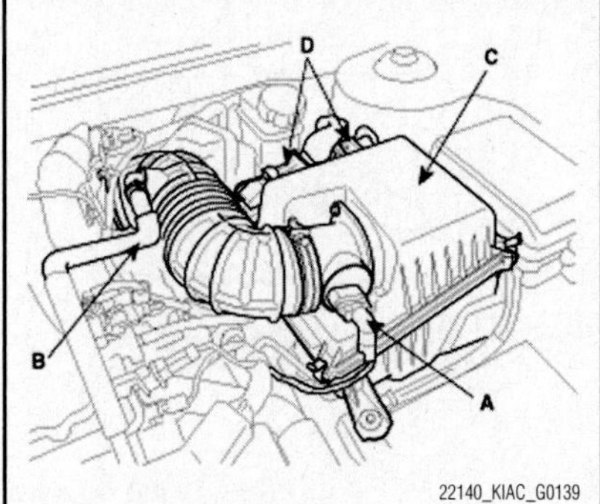

Fig. 98 Disconnect the MAF connector (A), breather hose (B), intake air hose and air cleaner assembly (C), and PCM connectors (D)—2.7L engine

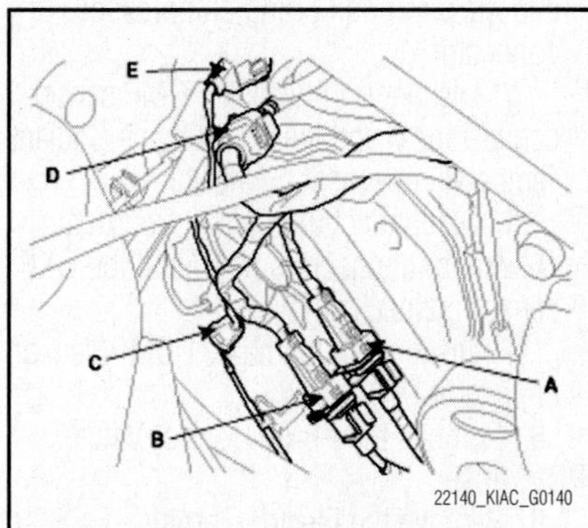

Fig. 99 Disconnect the engine wiring harness connectors—2.7L engine

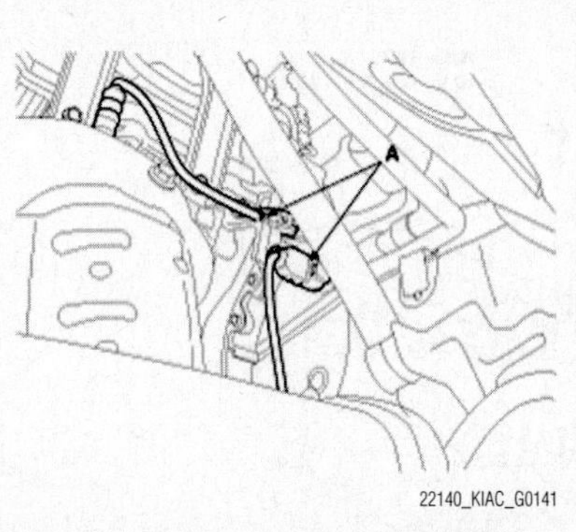

Fig. 100 Disconnect the bank 1 front/rear O2 sensor connectors (A)—2.7L engine

To avoid personal injury when working on vehicles equipped with an air bag, the negative battery cable must be disconnected and at least three minutes must elapse before working on the system. Failure to do so may result in deployment of the air bag.

2. Remove the battery top cover, if equipped.

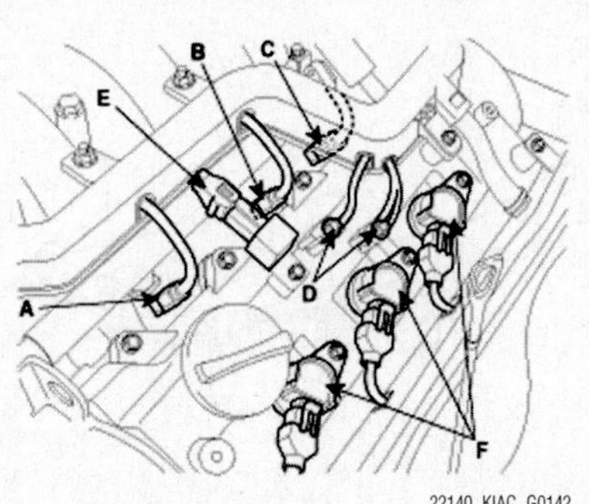

Fig. 101 Disconnect the injection connectors (A,B,C), the ground lines (D), the condenser connector (E) and the Ignition coil connectors (F)—2.7L engine

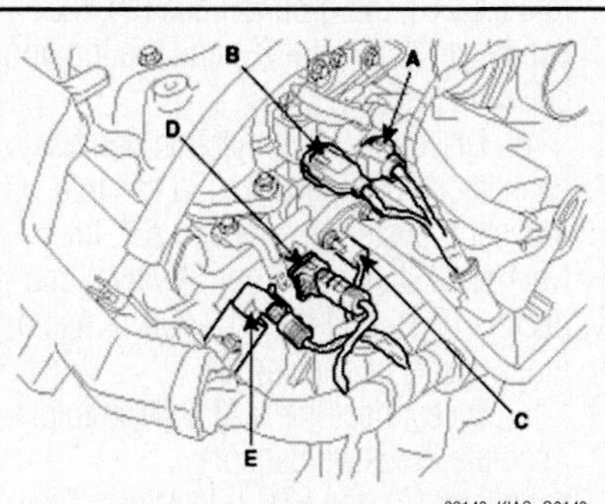

Fig. 102 Disconnect the injection harness connector (A), the No.2 VIS (Variable Induction System) connector (B), the No.1/No.2 OCV (Oil Control Valve) connectors (C,D) and the OTS (Oil Temperature Sensor) connector (E)—2.7L engine

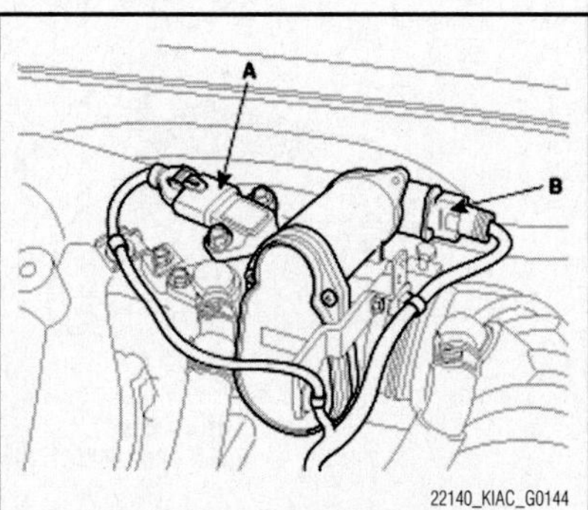

Fig. 103 Disconnect the MAPS (Manifold Absolute Pressure Sensor) connector (A), and the ETC (Electronic Throttle Control) connector (B)—2.7L engine

3. Disconnect the negative battery cable. Tape the cable with insulating tape.
4. Remove the air duct and the battery.
5. Remove the engine cover.
6. Remove the intake air hose and air cleaner assembly.
a. Disconnect the MAF connector (A).

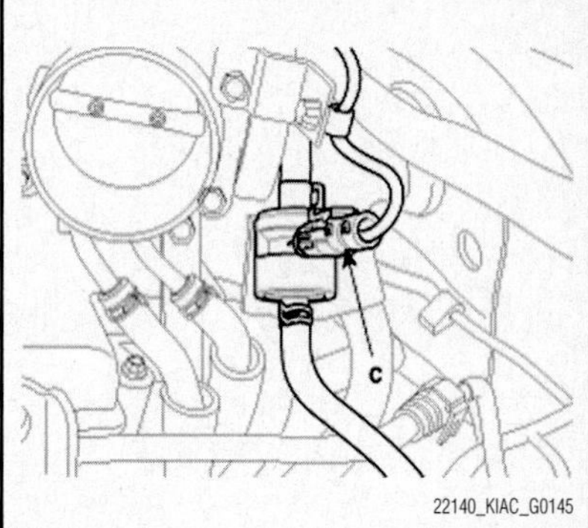

Fig. 104 Disconnect the PCSV (Purge Control Solenoid Valve) connector (C)—2.7L engine

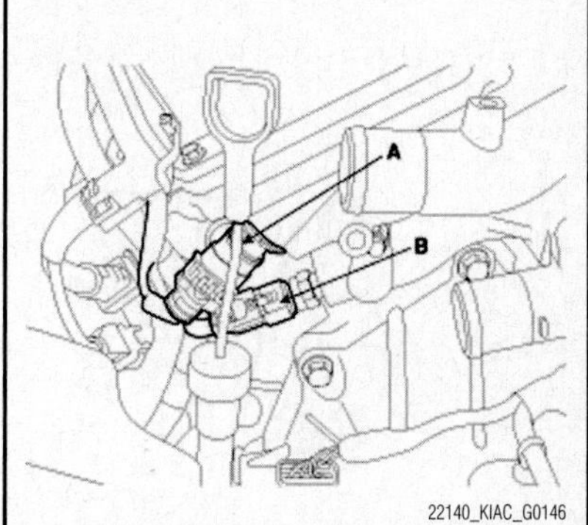

Fig. 105 Disconnect the bank 2 CMP sensor connector (A) and the ECT (Engine Coolant Temperature) sensor connector (B)—2.7L engine

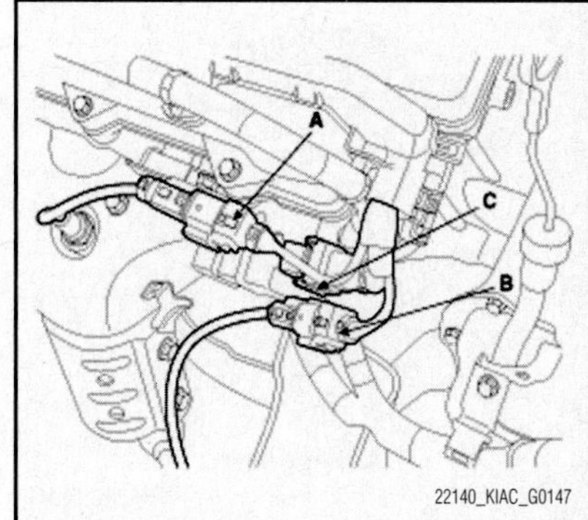

Fig. 106 Disconnect the bank 2 front/rear O2 sensor connectors (A,B) and the CKP sensor connector (C)—2.7L engine

b. Disconnect the breather hose (B) from air cleaner hose.
c. Remove the intake air hose and air cleaner assembly (C).
d. Disconnect the PCM connectors (D).
7. Disconnect the engine wiring harness connectors.

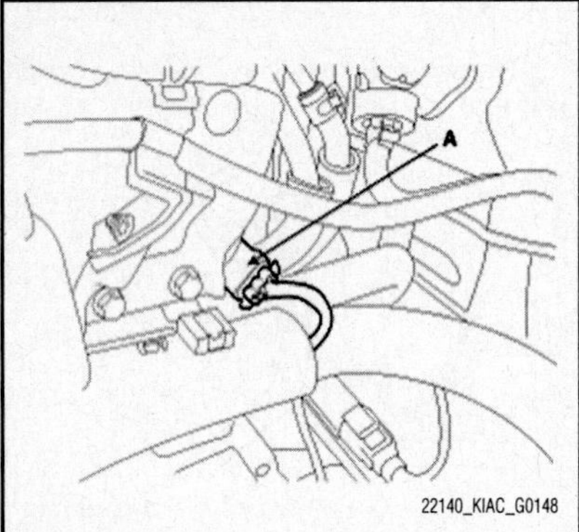

Fig. 107 Disconnect the bank 1 CMP sensor connector (A)—2.7L engine

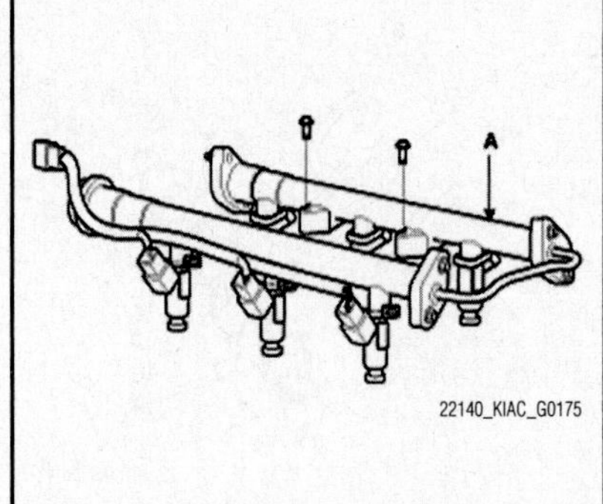

Fig. 110 Remove the fuel delivery pipe assembly (A)—2.7L engine

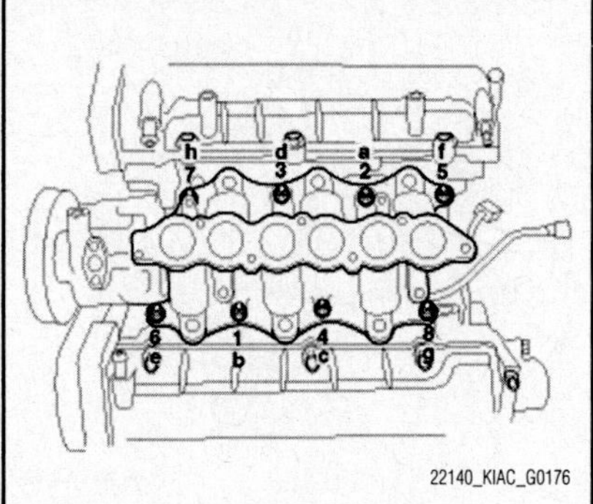

Fig. 112 Install the intake manifold assembly—2.7L engine

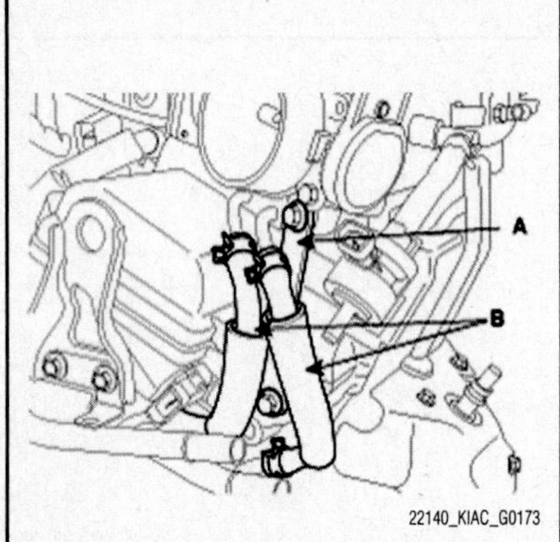

Fig. 108 Remove the Electric Throttle Control (ETC) bracket (A) and the cooling hoses (B)—2.7L engine

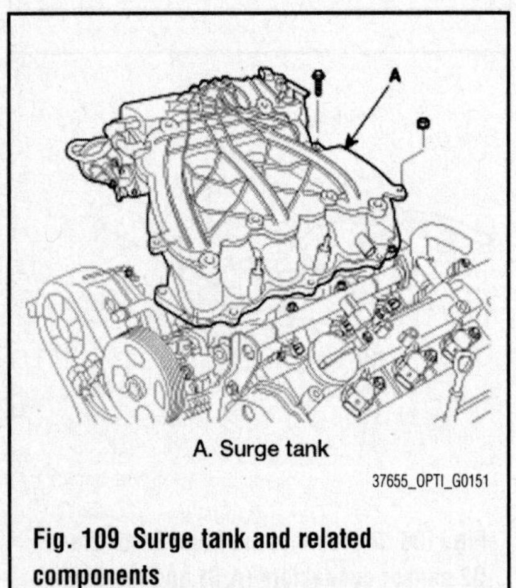

Fig. 109 Surge tank and related components

a. Disconnect the No.1/No.2 knock sensor connectors (A,B), the oil pressure switch connector (C), the ignition coil harness (D) and the No.1 VIS (Variable Induction System) connector (E).

b. Disconnect the bank 1 front/rear O2 sensor connectors (A).

c. Disconnect the injection connectors (A,B,C), the ground lines (D), the condenser connector (E) and the Ignition coil connectors (F).

d. Disconnect the injection harness connector (A), the No.2 VIS (Variable Induction System) connector (B), the No.1/No.2 OCV (Oil Control Valve) connectors (C,D) and the OTS (Oil Temperature Sensor) connector (E).

e. Disconnect the MAPS (Manifold Absolute Pressure Sensor) connector (A), the ETC (Electronic Throttle Control) connector (B) and the PCSV (Purge Control Solenoid Valve) connector (C).

f. Disconnect the alternator connector and the air conditioning compressor connector.

g. Disconnect the bank 2 CMP sensor connector (A) and the ECT (Engine Coolant Temperature) sensor connector (B).

h. Disconnect the bank 2 front/rear O2 sensor connectors (A,B) and the CKP sensor connector (C).

i. Disconnect the bank 1 CMP sensor connector (A).

8. Remove the Purge Control Valve (PCV) hose.

9. Remove the Electric Throttle Control (ETC) bracket (A) and the cooling hoses (B).

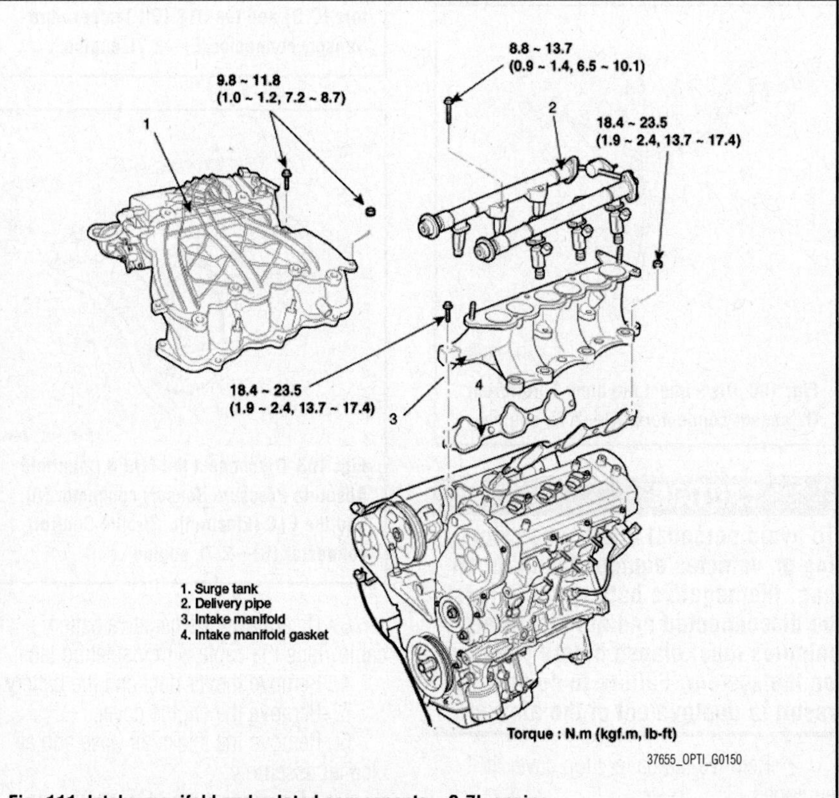

Fig. 111 Intake manifold and related components—2.7L engine

10. Disconnect the brake vacuum hose.
11. Remove the surge tank mounting bracket.
12. Remove the surge tank .
13. Remove the fuel delivery pipe assembly (A).
14. Remove the intake manifold assembly. Discard the gasket.

To install:

15. Install the intake manifold assembly with a new gasket to a cylinder head assembly. Tighten the bolts in two steps. Torque as follows:
    - Step 1: (a–h): 2.9–4.3 ft. lbs. (3.9–5.9 Nm)
    - Step 2: (1–8): 13.7–17.4 ft. lbs. (18.6–23.5 Nm)

CAUTION

When installing the gasket on the cylinder head, check the identification marks (LH/RH) to ensure correct installation.

16. Install the delivery pipe.
17. Connect the injector connectors.
18. Install the surge tank. Torque: 13.7–17.4 ft. lbs. (18.6–23.5 Nm)
19. Install the surge tank mounting bracket. Torque: 13.7–17.4 ft. lbs. (18.6–23.5 Nm)
20. Install the Electronic Throttle Control (ETC) system fixing bracket.
21. Connect the hoses and connectors.
22. Install the air cleaner assembly.
23. Install the engine cover.

OIL PAN

REMOVAL & INSTALLATION

2.4L Engine

At this time the manufacturer does not provide removal and installation procedures for this component.

2.7L Engine

See Figures 113 through 115.

1. Before servicing the vehicle, refer to the Precautions Section.

➡If working near and/or around the SRS system and components, be sure to disable the SRS system. Tape the negative battery cable with insulating tape. Always disconnect the negative battery cable first.

CAUTION

To avoid personal injury when working on vehicles equipped with an air bag, the negative battery cable must be disconnected and at least three minutes must elapse before working on the system. Failure to do so may result in deployment of the air bag.

2. Remove the battery top cover, if equipped.
3. Disconnect the negative battery cable. Tape the cable with insulating tape.
4. Raise and support the vehicle safely.
5. Drain the engine oil. Be sure to properly dispose of used oil.
6. Remove the right front tire and wheel assembly.
7. Remove the right front side cover.
8. Remove the front muffler.
9. Remove the alternator.
10. Remove the timing belt.
11. Remove the oil filter bracket.
12. Remove the lower oil pan retaining bolts.
13. Using tool SST 09215-3C00, remove

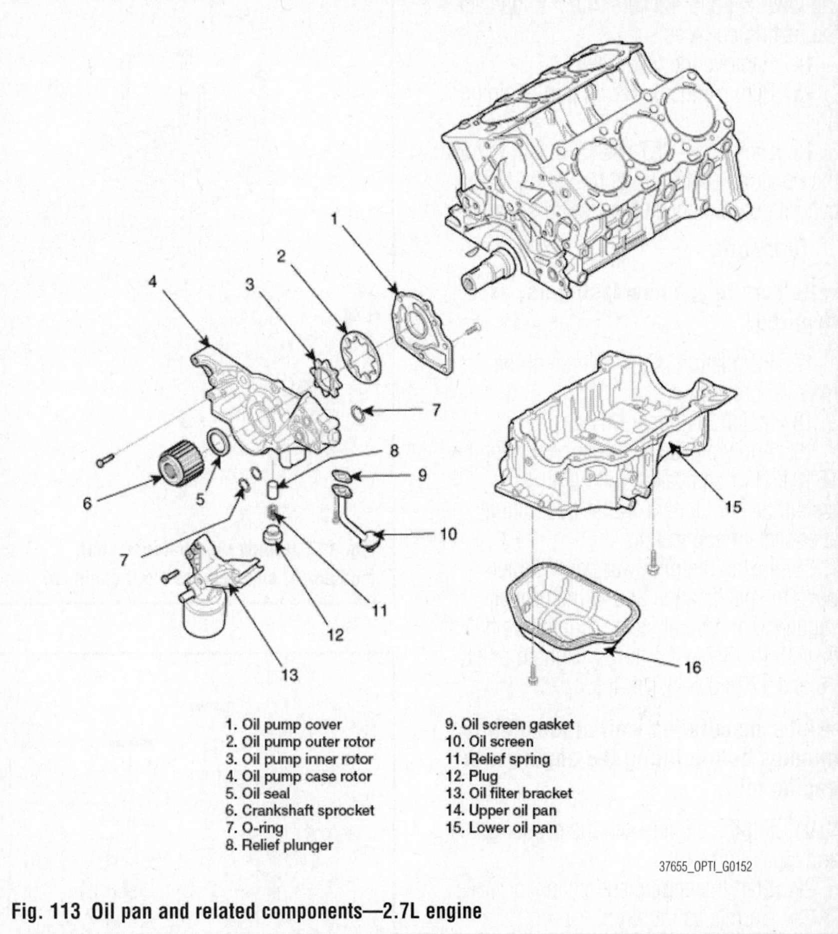

Fig. 113 Oil pan and related components—2.7L engine

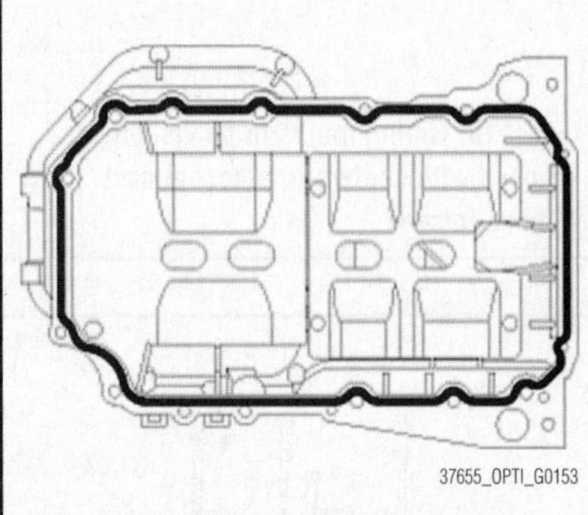

Fig. 114 Upper oil pan sealant application—2.7L engine

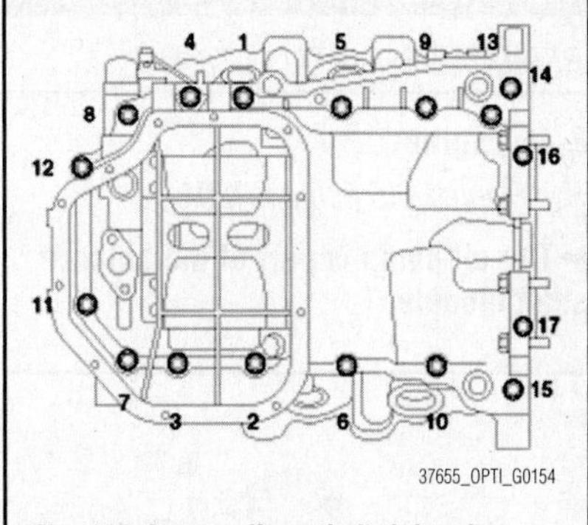

Fig. 115 Lower oil pan bolt tightening sequence—2.7L engine

the lower oil pan. Be careful not to damage the mating surfaces.

14. Remove the oil screen.
15. Remove the upper oil pan retaining bolts.
16. Using tool SST 09215-3C00, remove the upper oil pan from its mounting. Be careful not to damage the mating surfaces.

To install:

➡Be sure to use new fasteners, as required.

17. Installation is the reverse of the removal procedure.
18. Install the upper oil pan.
19. Apply a bead of sealant, part number TB1217H or equivalent on the oil pan. Install the component within five minutes after sealant application
20. Install the oil lower pan. Tighten bolts to specification and in the proper sequence in several steps. Tighten bolts 1 thru 15 to 13.7–17.4 ft. lbs. Tighten bolts 16 and 17 to 3.6–5.1 ft. lbs.

➡After installation wait at least thirty minutes before filling the engine with engine oil.

21. Fill the engine with the proper grade and type engine oil.
22. Start the engine and check for leaks.
23. Correct as required.

OIL PUMP

REMOVAL & INSTALLATION

2.4L Engine

See Figures 116 through 120.

➡The oil pump is part of the balance shaft module.

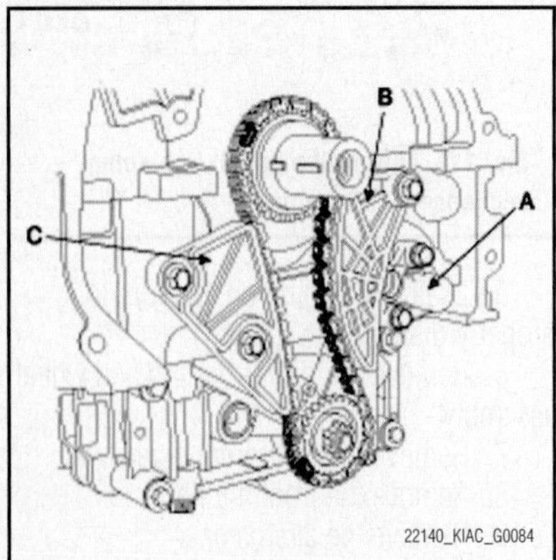

Fig. 116 Remove the balance shaft chain tensioner (A), balance shaft chain tensioner arm (B), and balance shaft chain guide (C)

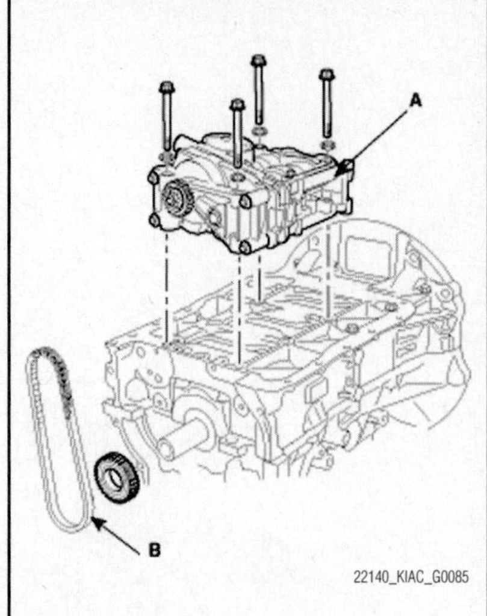

Fig. 117 Remove the balance shaft module (A) and balance shaft chain (B)

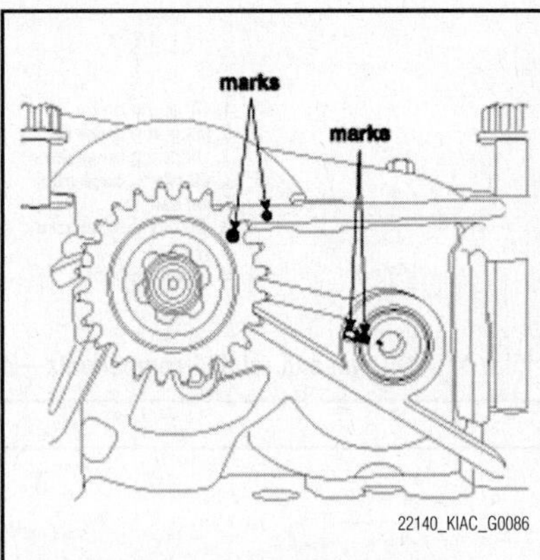

Fig. 118 Timing marks to be visually aligned with centers of adjacent cast timing notches

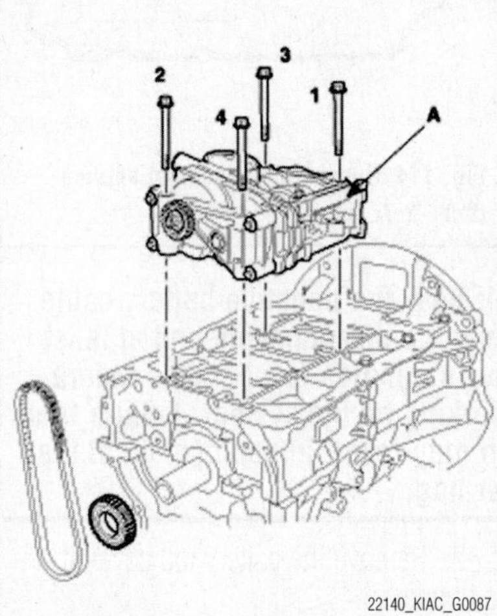

Fig. 119 Bolt tightening sequence and installation of balance shaft module (A)

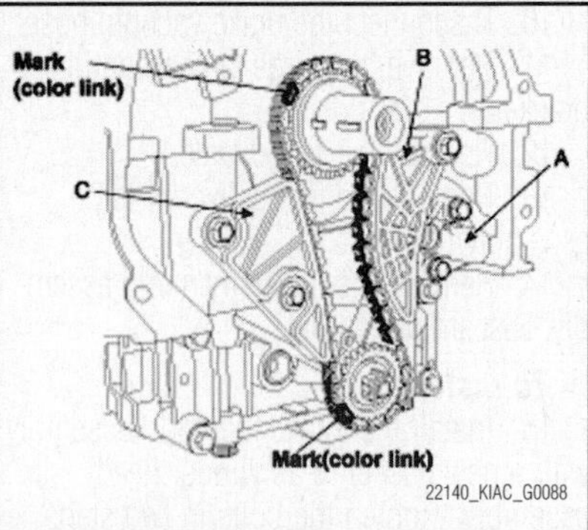

Fig. 120 Install the balance shaft chain guide (C), balance shaft tensioner arm (B)

1. Before servicing the vehicle, refer to the Precautions Section.

➡If working near and/or around the SRS system and components, be sure to disable the SRS system. Tape the negative battery cable with insulating tape. Always disconnect the negative battery cable first.

⁂ CAUTION

To avoid personal injury when working on vehicles equipped with an air bag, the negative battery cable must be disconnected and at least three minutes must elapse before working on the system. Failure to do so may result in deployment of the air bag.

2. Remove the battery top cover, if equipped.
3. Disconnect the negative battery cable. Tape the cable with insulating tape.
4. Remove the timing chain.
5. Install a set pin after compressing the balance shaft chain tensioner.
6. Remove the balance shaft chain tensioner (A).
7. Remove the balance shaft chain tensioner arm (B).
8. Remove the balance shaft chain guide (C).
9. Remove the balance shaft module (A) and balance shaft chain (B).

To install:

10. The key of crankshaft should be aligned with the mating face of main bearing cap. As a result of this, the piston of No.1 cylinder is placed at the top dead center on compression stroke.
11. Confirm the balance shaft module timing mark. Timing marks to be visually aligned with centers of adjacent cast timing notches.
12. Install balance shaft module (A) that the timing mark of balance shaft module

sprocket should be matched with the timing mark (color link) of balance shaft chain. Torque: 12.3 ft. lbs. (16.66Nm), then an additional 60° plus an additional 60°.

13. Install the balance shaft chain guide (C). Torque: 7.23–8.67 ft. lbs. (9.8–11.76 Nm).
14. Install the balance shaft tensioner arm (B). Torque: 7.23–8.67 ft. lbs. (9.8–11.76 Nm).
15. Install the balance shaft tensioner (A) and remove the set pin. Torque: 7.23–8.67 ft. lbs. (9.8–11.76 Nm).
16. Confirm the timing marks.
17. Install timing chain.

2.7L Engine

See Figure 121.

1. Before servicing the vehicle, refer to the Precautions Section.

➡If working near and/or around the SRS system and components, be sure to disable the SRS system. Tape the negative battery cable with insulating tape. Always disconnect the negative battery cable first.

⁂ CAUTION

To avoid personal injury when working on vehicles equipped with an air bag, the negative battery cable must be disconnected and at least three minutes must elapse before working on the system. Failure to do so may result in deployment of the air bag.

2. Remove the battery top cover, if equipped.
3. Disconnect the negative battery cable. Tape the cable with insulating tape.
4. Raise and support the vehicle safely.
5. Drain the engine oil. Be sure to properly dispose of used oil.

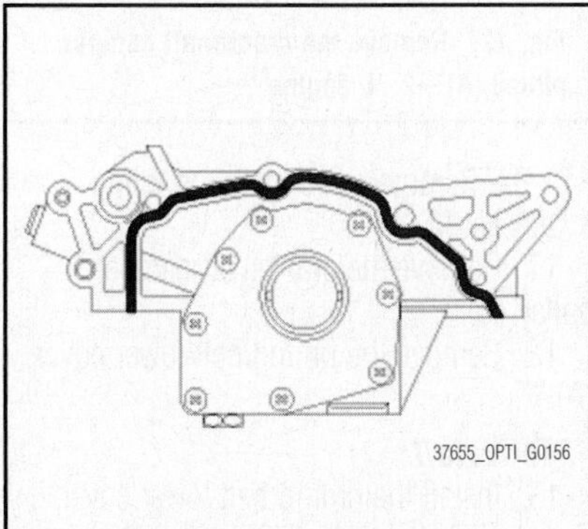

Fig. 121 Component sealant application—2.7L engine

6. Remove the right front tire and wheel assembly.
7. Remove the right front side cover.
8. Remove the front muffler.
9. Remove the alternator.
10. Remove the timing belt.
11. Remove the oil filter bracket.
12. Remove the lower oil pan retaining bolts.
13. Using tool SST 09215-3C00, remove the lower oil pan. Be careful not to damage the mating surfaces.
14. Remove the oil screen.
15. Remove the upper oil pan retaining bolts.
16. Using tool SST 09215-3C00, remove the upper oil pan from its mounting. Be careful not to damage the mating surfaces.
17. Remove the oil pump case.

To install:

➡Be sure to use new fasteners, as required.

18. Installation is the reverse of the removal procedure.
19. Apply a 0.0984 bead of sealant to the mating surfaces, see illustration.
20. Install the upper oil pan.
21. Apply a bead of sealant, part number TB1217H or equivalent on the oil pan. Install the component within five minutes after sealant application
22. Install the oil lower pan. Tighten bolts to specification and in the proper sequence in several steps. Tighten bolts 1 thru 15 to 13.7–17.4 ft. lbs. Tighten bolts 16 and 17 to 3.6–5.1 ft. lbs.

➡After installation wait at least thirty minutes before filling the engine with engine oil.

23. Fill the engine with the proper grade and type engine oil.
24. Start the engine and check for leaks.
25. Correct as required.

INSPECTION

1. Before servicing the vehicle, refer to the Precautions Section.
2. With the oil pump removed, visually check the parts of the oil pump case for cracks or damage.
3. Check the relief plunger for smooth operation.
4. Check the relief spring for deformation or a break.
5. Check the surface that mates with the oil filter for any damage.
6. Check the oil filter bracket for cracks and oil leakage.
7. Replace the pump, as required.

PISTON AND RING

POSITIONING

See Figures 122 and 123.

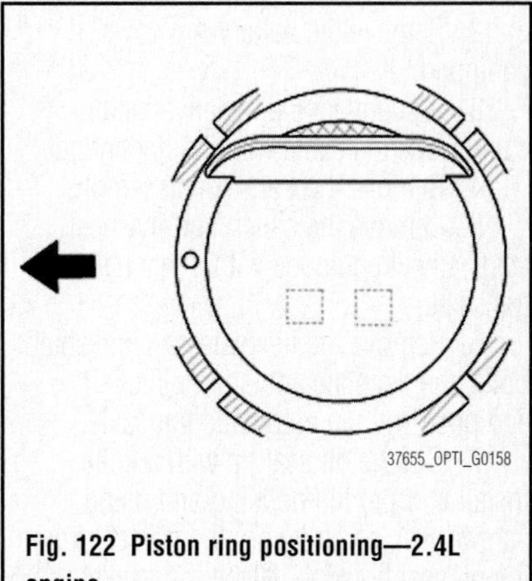

Fig. 122 Piston ring positioning—2.4L engine

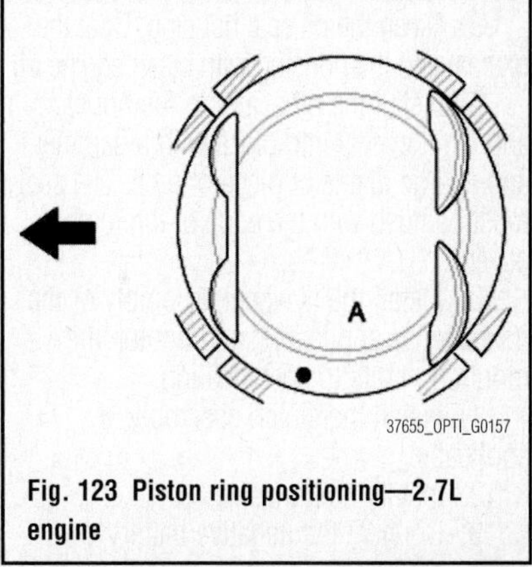

Fig. 123 Piston ring positioning—2.7L engine

REAR MAIN SEAL

REMOVAL & INSTALLATION

At this time the manufacturer does not provide removal and installation procedures for this component. The following procedure is a guideline and may differ from the vehicle you are servicing.

1. Before servicing the vehicle, refer to the Precautions Section.

➡If working near and/or around the SRS system and components, be sure to disable the SRS system. Tape the negative battery cable with insulating tape. Always disconnect the negative battery cable first.

✲✲ CAUTION

To avoid personal injury when working on vehicles equipped with an air bag, the negative battery cable must be disconnected and at least three minutes must elapse before working on the system. Failure to do so may result in deployment of the air bag.

2. Remove the battery top cover, if equipped.
3. Disconnect the negative battery cable. Tape the cable with insulating tape.
4. Remove the transaxle assembly.
5. Remove the clutch and flywheel assembly, if equipped with a manual transaxle.
6. Remove the flexplate-to-crankshaft bolts, the flexplate and shim plates, if equipped with an automatic transaxle.
7. Cut the oil seal lip with a knife. Install a rag to the housing and using a screwdriver, carefully pry the oil seal from the oil seal housing. Clean the gasket mounting surfaces.

To install:

8. Clean the oil seal housing. Coat the oil seal and the housing with clean engine oil.
9. Install the oil seal into the housing and tap it evenly into place with a hammer and a large diameter piece of pipe. The seal must be flush with the edge of the rear cover.
10. Install the flywheel assembly or the flexplate, as applicable, and tighten the mounting bolts to specification.
11. Install the clutch assembly, if applicable.
12. Install the transaxle.
13. Connect the negative battery cable.

ROCKER ARMS

REMOVAL & INSTALLATION

The engines covered are not equipped with rocker arms/shafts. The camshafts directly actuate the valve through a bucket type follower.

TIMING BELT FRONT COVER

REMOVAL & INSTALLATION

2.7L Engine

See Figures 124 through 128.

1. Before servicing the vehicle, refer to the Precautions Section.

➡If working near and/or around the SRS system and components, be sure to disable the SRS system. Tape the negative battery cable with insulating tape. Always disconnect the negative battery cable first.

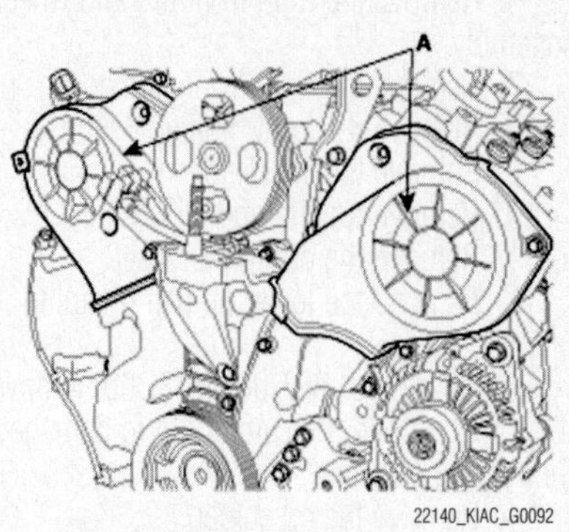

Fig. 124 Remove the timing belt upper covers (A)—2.7L engine

✲✲ CAUTION

To avoid personal injury when working on vehicles equipped with an air bag, the negative battery cable must be disconnected and at least three minutes must elapse before working on the system. Failure to do so may result in deployment of the air bag.

2. Remove the battery top cover, if equipped.
3. Disconnect the negative battery cable. Tape the cable with insulating tape.
4. Remove the engine cover
5. Remove the front right wheel and tire.
6. Remove the right side cover.
7. Remove the drive belt, the idler pulley and the tensioner pulley.

➡In removing the drive belt, fix a tool in the auto tensioner pulley bolt and turn the bolt counter clockwise.

8. Remove the timing belt upper covers (A).
9. Align the groove of the pulley with the timing mark of the timing belt cover by turning the crankshaft pulley clockwise. Check that the timing mark of the camshaft sprocket is aligned with that of the cylinder head cover. (No.1 cylinder piston at TDC)
10. Remove the engine mounting bracket.
    a. Support the engine oil pan with a jack.

✲✲ CAUTION

Put a wooden or rubber block between the jack and the engine oil pan.

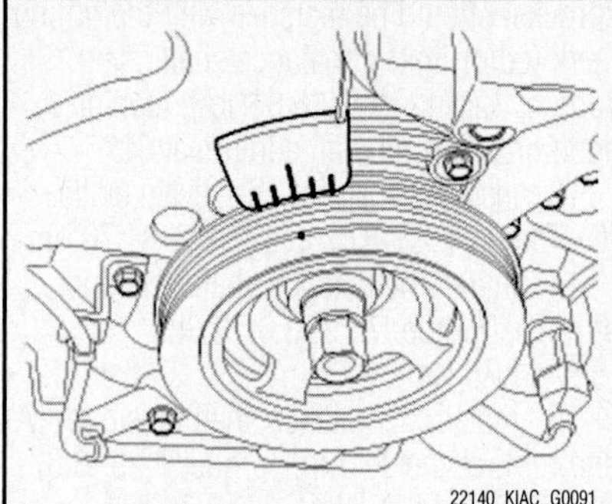

Fig. 125 Align the groove of the pulley with the timing mark of the timing belt cover—2.7L engine

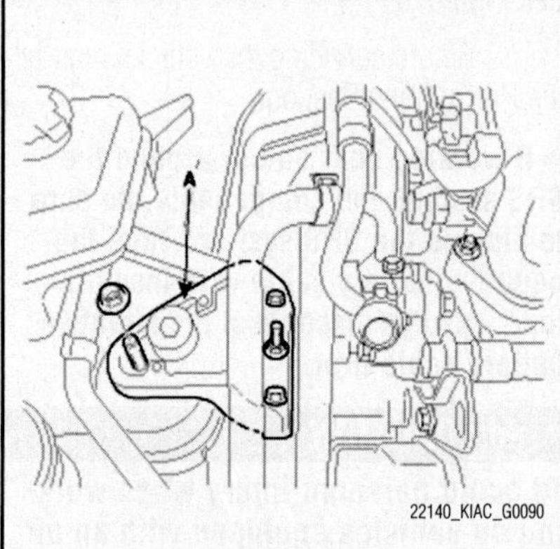

Fig. 126 Remove the engine mounting bracket (A)—2.7L engine

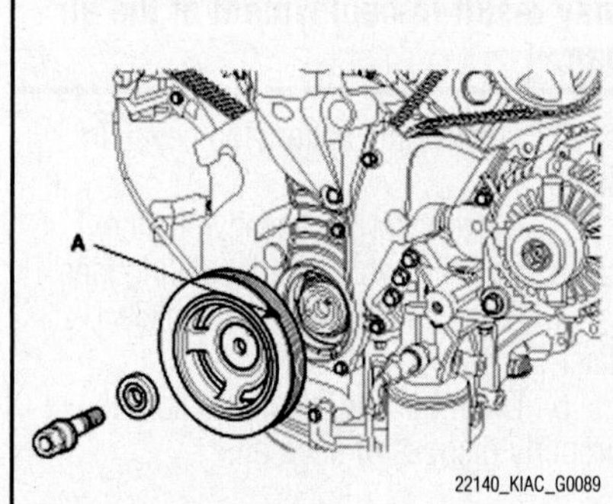

Fig. 127 Remove the crankshaft damper pulley (A)—2.7L engine

    b. Remove the engine mounting bracket (A).
11. Remove the crankshaft damper pulley (A).
12. Remove the timing belt lower cover (A).

To install:

13. Install the timing belt lower cover. Torque: 7.2–8.7 ft. lbs. (9.8–11.8 Nm).
14. Install the crankshaft damper pulley. Torque: 123.0–130.2 ft. lbs. (166.7–176.5 Nm).

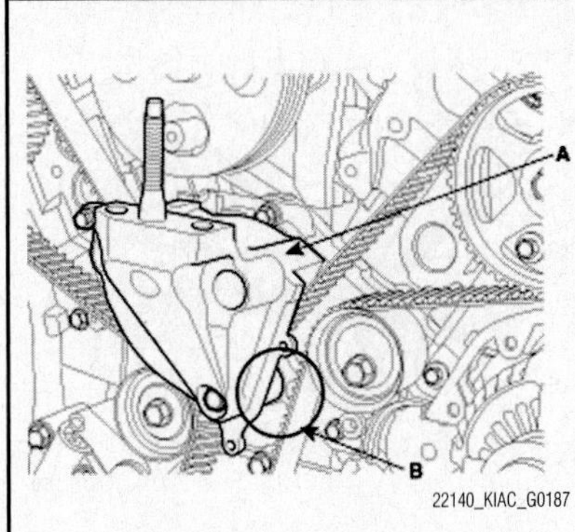

Fig. 128 Remove timing belt lower cover (A)—2.7L engine

15. Install the engine mounting bracket. Torque: 47.0–61.5 ft. lbs. (63.7–83.4 Nm).
16. Install the timing belt upper cover. Torque: 7.2–8.7 ft. lbs. (9.8–11.8 Nm).
17. Install the drive belt tensioner. Torque: 25.3–39.8 ft. lbs. (34.3–53.9 Nm).
18. Install the drive belt idler and the drive belt. Torque: 25.3–39.8 ft. lbs. (34.3–53.9 Nm).
19. Install the right side cover.
20. Install the front right wheel and tire.
21. Install the engine cover. Torque: 5.8–8.7 ft. lbs. (7.8–11.8 Nm).

TIMING BELT & SPROCKETS

REMOVAL & INSTALLATION

2.7L Engine

See Figures 126 through 133.

1. Before servicing the vehicle, refer to the Precautions Section.

➡If working near and/or around the SRS system and components, be sure to disable the SRS system. Tape the negative battery cable with insulating tape. Always disconnect the negative battery cable first.

⁂ CAUTION

To avoid personal injury when working on vehicles equipped with an air bag, the negative battery cable must be disconnected and at least three minutes must elapse before working on the system. Failure to do so may result in deployment of the air bag.

2. Remove the battery top cover, if equipped.
3. Disconnect the negative battery cable. Tape the cable with insulating tape.
4. Remove the engine cover

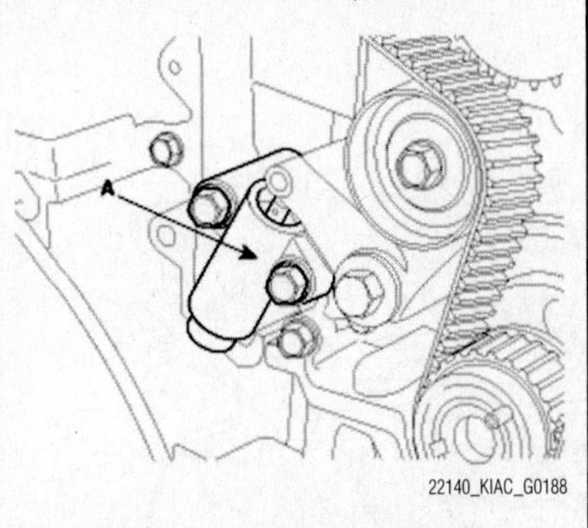

Fig. 129 Remove the timing belt auto tensioner (A)–2.7L engine

5. Remove the front right wheel and tire.
6. Remove the right side cover.
7. Remove the drive belt, the idler pulley and the tensioner pulley.

➡In removing the drive belt, fix a tool in the auto tensioner pulley bolt and turn the bolt counter clockwise.

8. Remove the timing belt upper covers (A).
9. Align the groove of the pulley with the timing mark of the timing belt cover by turning the crankshaft pulley clockwise. Check that the timing mark of the camshaft sprocket is aligned with that of the cylinder head cover. (No.1 cylinder piston at TDC)
10. Remove the engine mounting bracket.
    a. Support the engine oil pan with a jack.

⁂ CAUTION

Put a wooden or rubber block between the jack and the engine oil pan.

    b. Remove the engine mounting bracket (A).
11. Remove the crankshaft damper pulley (A).
12. Remove the timing belt lower cover (A).
13. Remove the engine support bracket.

➡After removal, a small amount of engine coolant may drain.

14. Remove the timing belt auto tensioner (A).
15. Remove the timing belt.

➡Mark the direction of rotation on the timing belt.

16. Remove the timing belt tensioner arm assembly and the idler.
17. Remove the crankshaft timing belt sprocket.

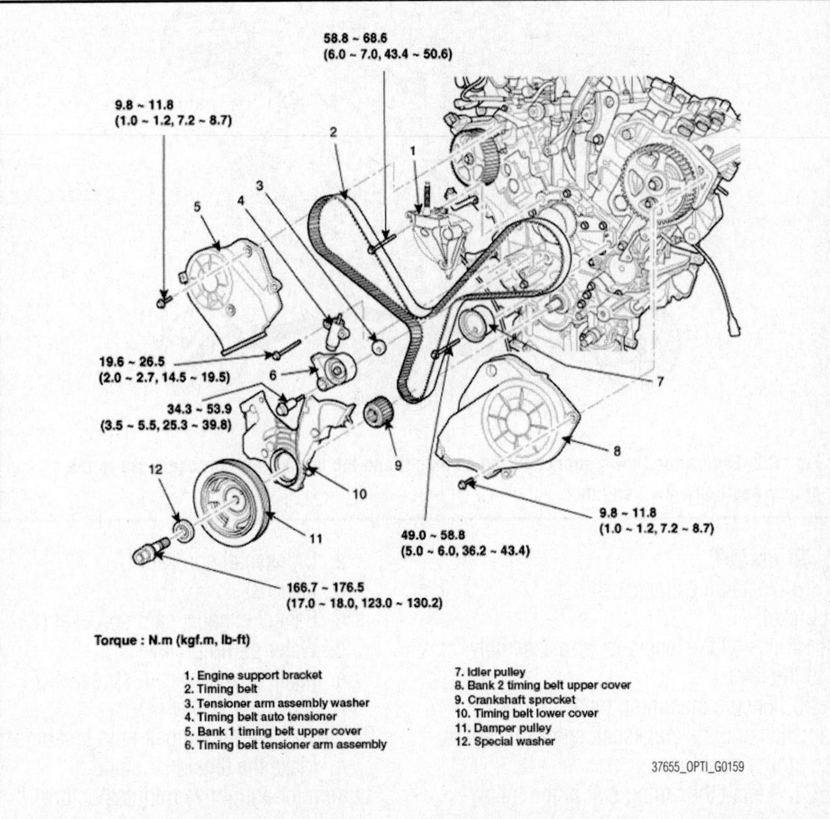

Fig. 130 Timing belt and related components with damper pulley washer—2.7L engine

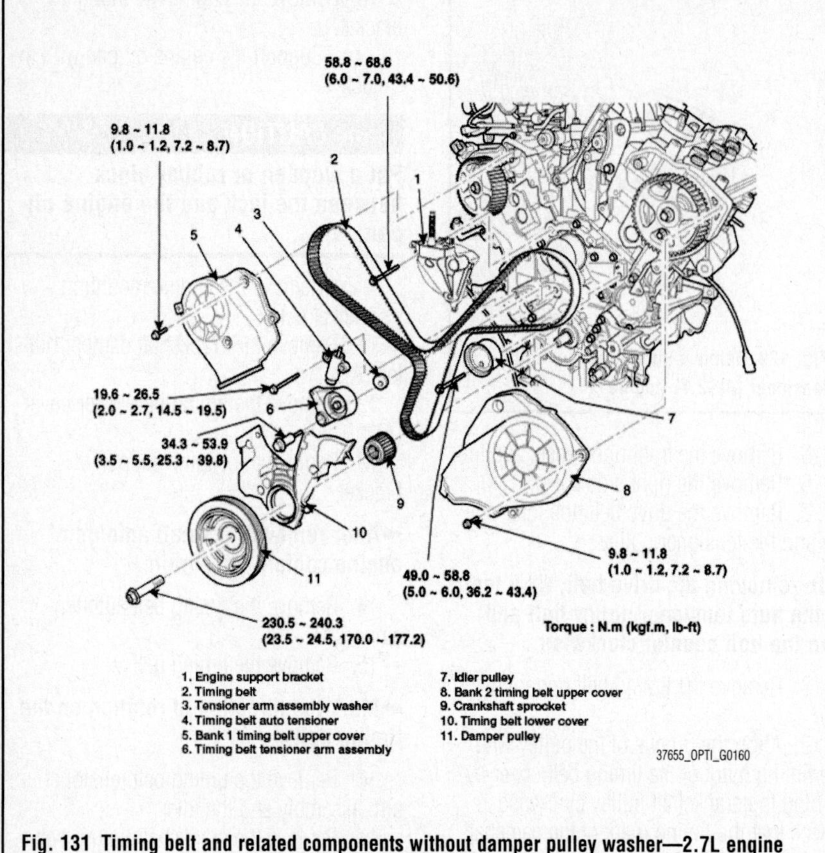

Fig. 131 Timing belt and related components without damper pulley washer—2.7L engine

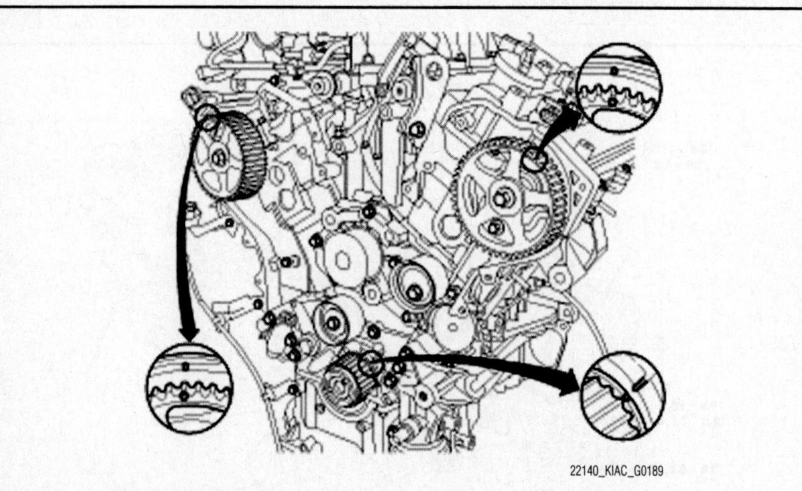

Fig. 132 Ensure the timing marks on the camshaft and the crankshaft sprockets are in the proper positions–2.7L engine

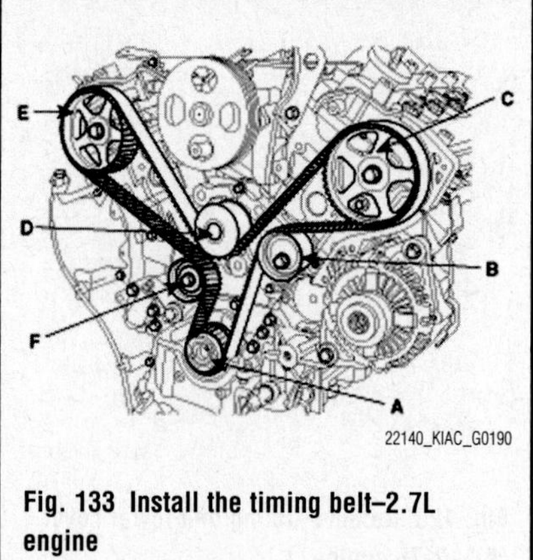

Fig. 133 Install the timing belt–2.7L engine

To install:

18. Install the crankshaft sprocket.

19. Install the tensioner arm assembly and the idler.

20. Ensure the timing marks on the camshaft and the crankshaft sprockets are in the proper positions.

21. Install the timing belt in the following order:

   a. Crankshaft sprocket (A)
   b. Idler (B)
   c. Bank 2 exhaust cam sprocket (C)
   d. Water pump pulley (D)
   e. Bank 1 exhaust cam sprocket (E)
   f. Tensioner pulley (F).

22. Install the timing belt auto tensioner.

   a. Make the tensioner stand upright for about five minutes before installing.

➡When handling the auto-tensioner observe the following:

- Do not lay down the auto tensioner.
- Do not compress the rod suddenly.
- When reinstalling the auto tensioner, ensure proper orientation.
- Do not press the rod any more when its projection from the body is 2.5mm.
- Keep the auto-tensioner upright at room temperature in winter.

   b. Install the auto tensioner to the front case with the set-pin inserted. Torque: 14.5–19.5 ft. lbs. (19.6–26.5 Nm).

23. Remove the auto tensioner set-pin.

24. Check the tension of the timing belt.

   a. Turn the crankshaft 2 revolutions clockwise, and set the number one cylinder to TDC.

➡After 5minutes, measure the length of the projected rod. Specification: 0.1969–0.2756 inches (5–7 mm).

   b. Ensure the locations of the timing marks for each sprocket.

25. Install the engine support bracket. Torque: 43.4–50.6 ft. lbs. (58.8–68.6 Nm).

26. Install the timing belt lower cover. Torque: 7.2–8.7 ft. lbs. (9.8–11.8 Nm).

27. Install the crankshaft damper pulley. Torque: 123.0–130.2 ft. lbs. (166.7–176.5 Nm).

28. Install the engine mounting bracket. Torque: 47.0–61.5 ft. lbs. (63.7–83.4 Nm).

29. Install the timing belt upper cover. Torque: 7.2–8.7 ft. lbs. (9.8–11.8 Nm).

30. Install the drive belt tensioner. Torque: 25.3–39.8 ft. lbs. (34.3–53.9 Nm).

31. Install the drive belt idler and the

drive belt. Torque: 25.3–39.8 ft. lbs. (34.3–53.9 Nm).

32. Install the right side cover.
33. Install the front right wheel and tire.
34. Install the engine cover. Torque: 5.8–8.7 ft. lbs. (7.8–11.8 Nm).

TIMING CHAIN FRONT COVER

REMOVAL & INSTALLATION

2.4L Engine

See Figures 134 through 137.

1. Before servicing the vehicle, refer to the Precautions Section.

➡If working near and/or around the SRS system and components, be sure to disable the SRS system. Tape the negative battery cable with insulating tape. Always disconnect the negative battery cable first.

⁂ CAUTION

To avoid personal injury when working on vehicles equipped with an air bag, the negative battery cable must be disconnected and at least three minutes must elapse before working on the system. Failure to do so may result in deployment of the air bag.

2. Remove the battery top cover, if equipped.
3. Disconnect the negative battery cable. Tape the cable with insulating tape.
4. Remove the engine cover.
5. Remove RH front wheel.
6. Remove RH side cover.
7. Set No.1 cylinder to TDC/compression
8. Remove the engine mount bracket.
9. Remove the accessory drive belt.
10. Remove the idler pulley.
11. Remove the accessory drive belt tensioner.

➡Tensioner pulley bolt is left - handed screw.

12. Remove the water pump pulley (A).
13. Remove the crankshaft pulley (B).
14. Remove the engine support bracket (C).
15. Disconnect the ignition coil connector.
16. Remove the ignition coils.
17. Remove the PCV hose and breather hose from the cylinder head cover.
18. Remove the cylinder head cover and gasket.
19. Remove the A/C compressor lower bolts.
20. Remove the compressor bracket.

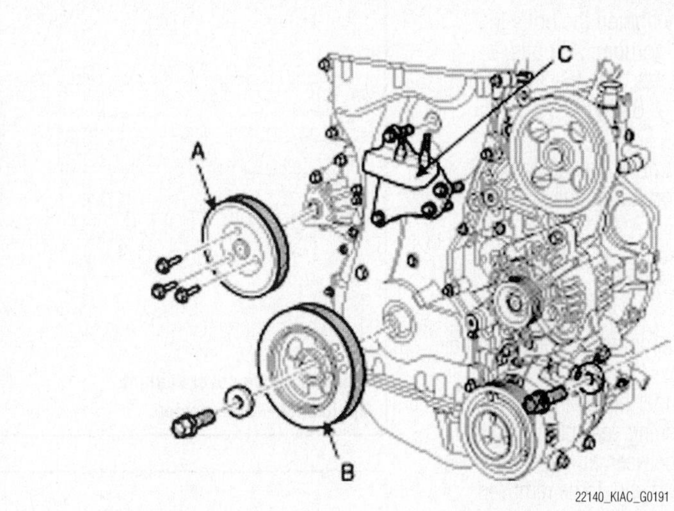

Fig. 134 Remove the water pump pulley (A), crankshaft pulley (B), and engine support bracket (C)–2.4L engine

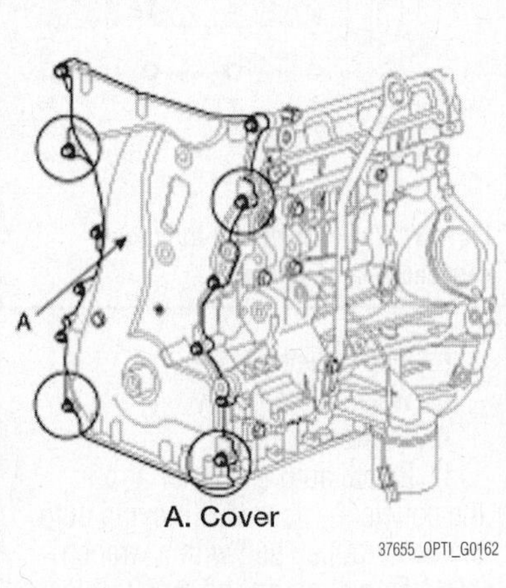

Fig. 135 Timing chain cover removal points—2.4L engine

21. Drain the engine oil.
22. Remove the oil pan.

⁂ CAUTION

Be careful not to damage the contact surfaces of cylinder block and oil pan.

23. Remove the timing chain cover by prying the portions between the cylinder head and cylinder block with a suitable tool.

⁂ CAUTION

Be careful not to damage the contact surfaces of cylinder block, cylinder head and timing chain cover.

To install:

24. Install timing chain cover.
    a. The sealant locations on chain cover and on counter parts (cylinder head, cylinder block, and ladder frame) must be free of engine oil and ETC.
    b. Before assembling the timing chain cover, the liquid sealant Loctite 5900 should be applied on the gap between cylinder head and cylinder block.
    c. After applying liquid sealant Loctite 5900 on timing chain cover, the part must be assembled within 5 minutes.
    d. The dowel pins on the cylinder block and holes on the timing chain cover should be used as a reference in order to assemble the timing chain cover to be in exact position.
    e. Torque: M6 bolts: 5.78–7.23 ft. lbs. (7.84–9.8 Nm). M8 bolts: 13.74–16.63 ft. lbs. (18.62–22.54 Nm).
    f. The firing and/or blow out test should not be performed within 30 minutes after the timing chain cover is assembled.
25. Install the oil pan.
    a. Using a gasket scraper, remove all the old packing material from the gasket surfaces.
    b. Before assembling the oil pan, the liquid sealant Loctite 5900 should be applied on oil pan. The part must be assembled within 5 minutes after the sealant was applied.

⁂ CAUTION

When applying sealant gasket, sealant must not be protruded into the inside of oil pan. To prevent leakage of oil, apply sealant gasket to the inner threads of the bolt holes.

c. Install oil pan.

d. Uniformly tighten the bolts in several passes. Torque: M8 bolts: 19.52–22.41 ft. lbs. (26.46–30.38 Nm). M6 bolts: 7.23–8.67 ft. lbs. (9.8–11.6 Nm).

e. After assembly, wait at least 30 minutes before filling the engine with oil.

26. Install air compressor bracket. Torque: 14.46–17.35 ft. lbs. (19.6–23.52 Nm).

27. Install air compressor bolts. Torque: 14.46–17.35 ft. lbs. (19.6–23.52 Nm).

28. Install cylinder head cover.

a. The hardening sealant located on the upper area between timing chain cover and cylinder head should be removed before assembling cylinder head cover.

b. After applying sealant, it should be assembled within 5 minutes.

c. The firing and/or blow out test should not be performed within 30 minutes after the cylinder head cover was assembled.

d. Torque the cylinder head cover bolts as follows:

- 1st step: 2.89–4.34 ft. lbs. (3.92–5.88 Nm)
- 2nd step: 5.78–7.23 ft. lbs. (7.84–9.8 Nm).

⁂ CAUTION

Do not reuse cylinder head cover gasket.

29. Install ignition coil.

30. Connect ignition coil connector.

31. Install engine support bracket. Torque: M10 bolts: 28.92–32.53 ft. lbs. (39.2–44.1 Nm); M8 bolts: 14.46–18.07 ft. lbs. (19.6–24.5 Nm).

32. Install crankshaft pulley. Torque: 122.9–130.13 ft. lbs. (166.6–176.4 Nm).

33. Install water pump pulley. Torque: 5.78–7.23 ft. lbs. (7.84–9.8 Nm).

34. Install drive belt tensioner and tensioner pulley. Torque: 39.7–47.0 ft. lbs. (53.9–63.7 Nm).

➡Tensioner pulley bolt is left-handed screw.

35. Install idler pulley. Torque: 39.7–47.0 ft. lbs. (53.9–63.7 Nm).

36. Install the accessory drive belt in the following order:

a. Crankshaft pulley
b. A/C pulley
c. Alternator pulley
d. Idler pulley
e. P/S pump pulley

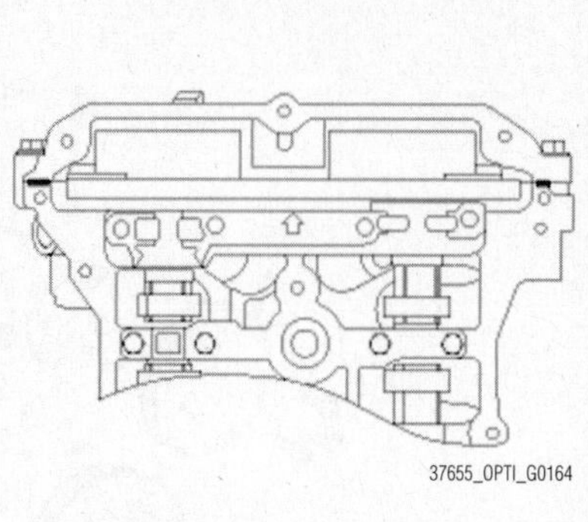

Fig. 136 Valve cover sealant application—2.4L engine

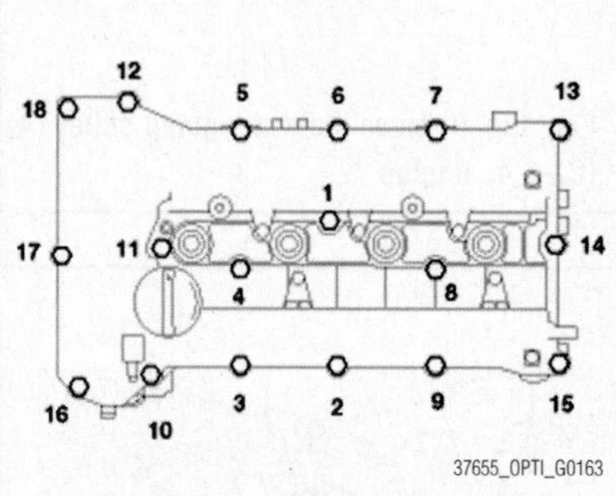

Fig. 137 Valve cover bolt tightening sequence—2.4L engine

f. Idler pulley
g. Water pump pulley
h. Tensioner pulley.
i. Rotate auto tensioner arm in the counter—clockwise moving auto tensioner pulley bolt with a wrench.

37. After putting belt on auto tensioner pulley, release the auto tensioner pulley slowly.

38. Install engine mounting bracket. Torque: 47.0–61.4 ft. lbs. (63.7–83.3 Nm).

39. Install RH side cover.

40. Install RH front wheel.

41. Install engine cover.

TIMING CHAIN & SPROCKETS

REMOVAL & INSTALLATION

2.4L Engine

See Figures 136 through 141.

1. Before servicing the vehicle, refer to the Precautions Section.

➡If working near and/or around the SRS system and components, be sure to disable the SRS system. Tape the negative battery cable with insulating tape. Always disconnect the negative battery cable first.

⁂ CAUTION

To avoid personal injury when working on vehicles equipped with an air bag, the negative battery cable must be disconnected and at least three minutes must elapse before working on the system. Failure to do so may result in deployment of the air bag.

2. Remove the battery top cover, if equipped.

3. Disconnect the negative battery cable. Tape the cable with insulating tape.

4. Remove the engine cover.

5. Remove RH front wheel.

6. Remove RH side cover.

7. Set No.1 cylinder to TDC/ compression

8. Remove the engine mount bracket.

9. Remove the accessory drive belt.

10. Remove the idler pulley.

11. Remove the accessory drive belt tensioner.

➡Tensioner pulley bolt is left - handed screw.

12. Remove the water pump pulley (A).

13. Remove the crankshaft pulley (B).

14. Remove the engine support bracket (C).

15. Disconnect the ignition coil connector.

16. Remove the ignition coils.

17. Remove the PCV hose and breather hose from the cylinder head cover.

18. Remove the cylinder head cover and gasket.

19. Remove the A/C compressor lower bolts.

20. Remove the compressor bracket.

21. Drain the engine oil.

22. Remove the oil pan.

⁂ CAUTION

Be careful not to damage the contact surfaces of cylinder block and oil pan.

23. Remove the timing chain cover by prying the portions between the cylinder head and cylinder block with a suitable tool.

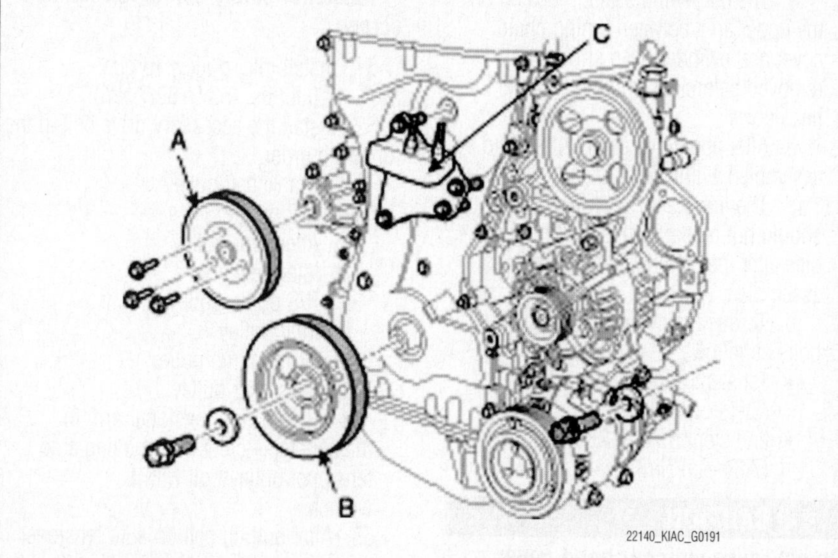

Fig. 138 Remove the water pump pulley (A), crankshaft pulley (B), and engine support bracket (C)–2.4L engine

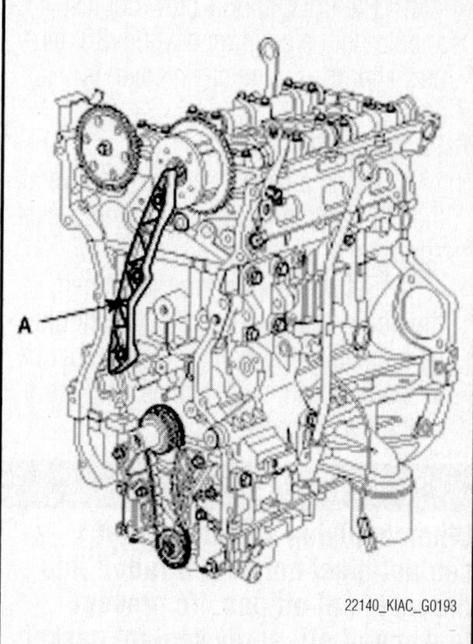

Fig. 140 Remove the timing chain guide (A)–2.4L engine

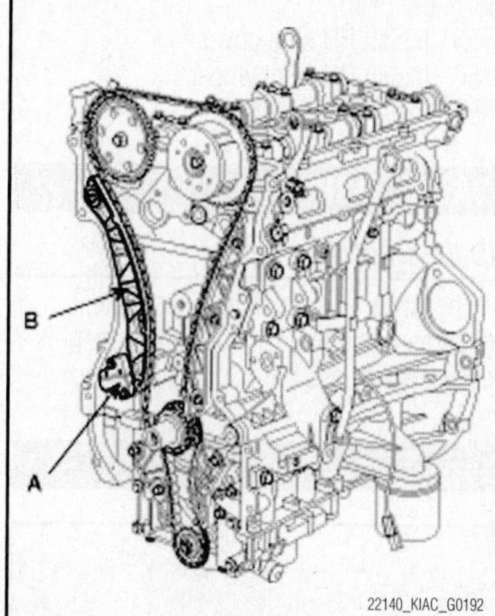

Fig. 139 Remove the timing chain tensioner (A) and timing chain tensioner arm (B)–2.4L engine

⁂ CAUTION

Be careful not to damage the contact surfaces of cylinder block, cylinder head and timing chain cover.

24. The key of crankshaft should be aligned with the mating face of main bearing cap. As a result of this, the piston of No.1 cylinder is placed at the top dead center on compression stroke.
25. Install a set pin after compressing the timing chain tensioner.
26. Remove the timing chain tensioner (A).
27. Remove the timing chain tensioner arm (B).
28. Remove the timing chain.
29. Remove the timing chain guide (A).
30. Remove the timing chain oil jet (A).
31. Remove the crankshaft chain sprocket (B).

To install:

32. Install crankshaft chain sprocket.
33. Install timing chain oil jet. Torque: 5.78–7.23 ft. lbs. (7.84–9.8 Nm).
34. The key of crankshaft should be aligned with the mating surface of main bearing cap. As a result, this places the piston on No.1 cylinder at the top dead center on compression stroke.
35. Install timing chain guide. Torque: 7.23–8.67 ft. lbs. (9.8–11.6 Nm).
36. Install timing chain.

➡To install the timing chain with no slack between each shaft (cam, crank), follow the below procedure:

- Crankshaft sprocket (A)
- Timing chain guide (B)
- Intake camshaft sprocket (C)
- Exhaust camshaft sprocket (D).

The timing mark of each sprocket should be matched with timing mark (color link) of timing chain at installing of the timing chain.

37. Install timing chain tensioner arm. Torque: 7.23–8.67 ft. lbs. (9.8–11.6 Nm).
38. Install timing chain auto tensioner and remove set pin. Torque: 7.23–8.67 ft. lbs. (9.8–11.6 Nm).
39. After rotating crankshaft 2 revolutions in regular direction (clockwise viewed from front), confirm the timing mark.
40. Install timing chain cover.

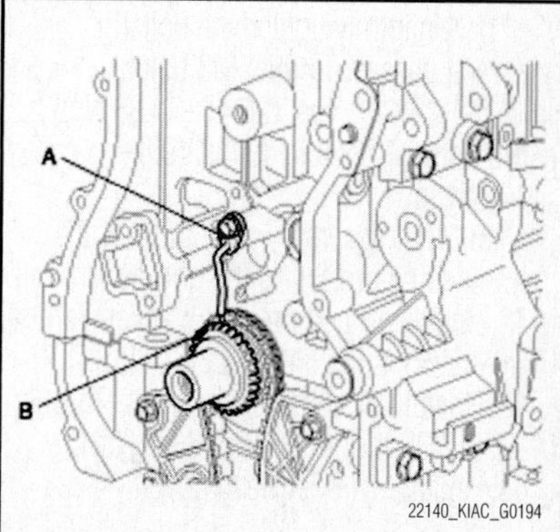

Fig. 141 Remove the timing chain oil jet (A) and crankshaft chain sprocket (B)–2.4L engine

a. The sealant locations on chain cover and on counter parts (cylinder head, cylinder block, and ladder frame) must be free of engine oil and ETC.

b. Before assembling the timing chain cover, the liquid sealant Loctite 5900 should be applied on the gap between cylinder head and cylinder block.

c. After applying liquid sealant Loctite 5900 on timing chain cover, the part must be assembled within 5 minutes.

d. The dowel pins on the cylinder block and holes on the timing chain cover should be used as a reference in order to assemble the timing chain cover to be in exact position.

e. Torque: M6 bolts: 5.78–7.23 ft. lbs. (7.84–9.8 Nm). M8 bolts: 13.74–16.63 ft. lbs. (18.62–22.54 Nm).

f. The firing and/or blow out test should not be performed within 30 minutes after the timing chain cover is assembled.

41. Install the oil pan.

a. Using a gasket scraper, remove all the old packing material from the gasket surfaces.

b. Before assembling the oil pan, the liquid sealant Loctite 5900 should be applied on oil pan. The part must be assembled within 5 minutes after the sealant was applied.

CAUTION

When applying sealant gasket, sealant must not be protruded into the inside of oil pan. To prevent leakage of oil, apply sealant gasket to the inner threads of the bolt holes.

c. Install oil pan.

d. Uniformly tighten the bolts in several passes. Torque: M8 bolts: 19.52–22.41 ft. lbs. (26.46–30.38 Nm). M6 bolts: 7.23–8.67 ft. lbs. (9.8–11.6 Nm).

e. After assembly, wait at least 30 minutes before filling the engine with oil.

42. Install air compressor bracket. Torque: 14.46–17.35 ft. lbs. (19.6–23.52 Nm).

43. Install air compressor bolts. Torque: 14.46–17.35 ft. lbs. (19.6–23.52 Nm).

44. Install cylinder head cover.

a. The hardening sealant located on the upper area between timing chain cover and cylinder head should be removed before assembling cylinder head cover.

b. After applying sealant, it should be assembled within 5 minutes.

c. The firing and/or blow out test should not be performed within 30 minutes after the cylinder head cover was assembled.

d. Torque the cylinder head cover bolts as follows:

- 1st step: 2.89–4.34 ft. lbs. (3.92–5.88 Nm)
- 2nd step: 5.78–7.23 ft. lbs. (7.84–9.8 Nm).

CAUTION

Do not reuse cylinder head cover gasket.

45. Install ignition coil.

46. Connect ignition coil connector.

47. Install engine support bracket. Torque: M10 bolts: 28.92–32.53 ft. lbs. (39.2–44.1 Nm); M8 bolts: 14.46–18.07 ft. lbs. (19.6–24.5 Nm).

48. Install crankshaft pulley. Torque: 122.9–130.13 ft. lbs. (166.6–176.4 Nm).

49. Install water pump pulley. Torque: 5.78–7.23 ft. lbs. (7.84–9.8 Nm).

50. Install drive belt tensioner and tensioner pulley. Torque: 39.7–47.0 ft. lbs. (53.9–63.7 Nm).

➡Tensioner pulley bolt is left-handed screw.

51. Install idler pulley. Torque: 39.7–47.0 ft. lbs. (53.9–63.7 Nm).

52. Install the accessory drive belt in the following order:

a. Crankshaft pulley
b. A/C pulley
c. Alternator pulley
d. Idler pulley
e. P/S pump pulley
f. Idler pulley
g. Water pump pulley
h. Tensioner pulley.
i. Rotate auto tensioner arm in the counter—clockwise moving auto tensioner pulley bolt with a wrench.

53. After putting belt on auto tensioner pulley, release the auto tensioner pulley slowly.

54. Install engine mounting bracket. Torque: 47.0–61.4 ft. lbs. (63.7–83.3 Nm).

55. Install RH side cover.

56. Install RH front wheel.

57. Install engine cover.

VALVE LASH

ADJUSTMENT

The valve lash on all engines is kept in adjustment hydraulically. No adjustment is necessary or possible.

ENGINE PERFORMANCE & EMISSION CONTROLS

CAMSHAFT POSITION (CMP) SENSOR

LOCATION

See Figures 142 through 145.

Refer to the accompanying illustrations.

REMOVAL & INSTALLATION

See Figures 142 through 145.

At this time the manufacturer does not provide removal and installation procedures for this component. The following procedure is a guideline and may differ from the vehicle you are servicing.

1. Before servicing the vehicle, refer to the Precautions Section.

➡If working near and/or around the SRS system and components, be sure to disable the SRS system. Tape the negative battery cable with insulating tape. Always disconnect the negative battery cable first.

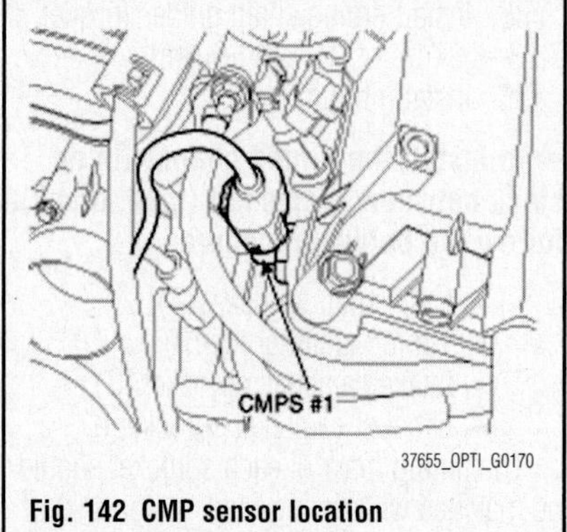

Fig. 142 CMP sensor location (sensor 1)—2.4L engine

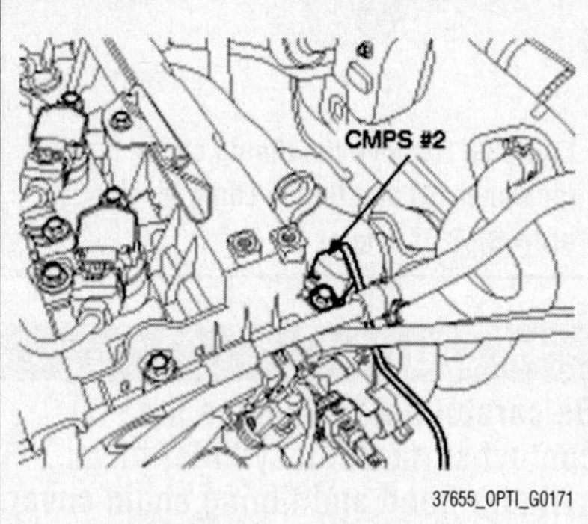

Fig. 143 CMP sensor location (sensor 2)—2.4L engine

CAUTION

To avoid personal injury when working on vehicles equipped with an air bag, the negative battery cable must be disconnected and at least three minutes must elapse before working on the system. Failure to do so may result in deployment of the air bag.

2. Remove the battery top cover, if equipped.

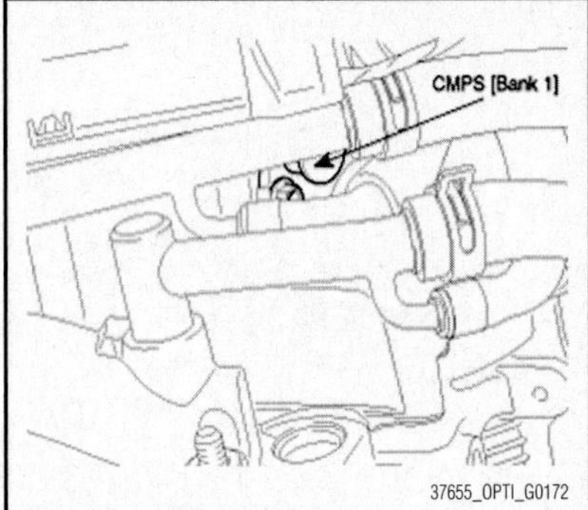

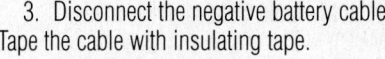

Fig. 144 CMP sensor location (bank 1)—2.7L engine

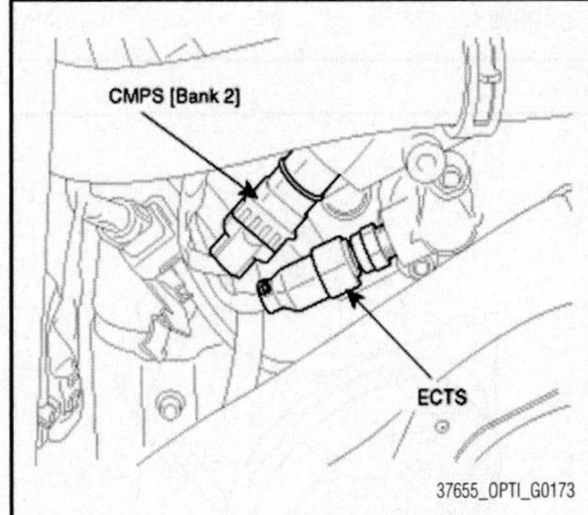

Fig. 145 CMP sensor location (bank 2)—2.7L engine

3. Disconnect the negative battery cable. Tape the cable with insulating tape.
4. Remove the necessary components in order to gain access to the sensor.
5. Disconnect the electrical connector.
6. Remove the sensor from its mounting.
7. Discard the O-ring/gasket.

To install:

➡Be sure to use new fasteners, as required.

8. Installation is the reverse of the removal procedure.

CRANKSHAFT POSITION (CKP) SENSOR

LOCATION

See Figures 146 and 147.

Refer to the accompanying illustrations.

REMOVAL & INSTALLATION

See Figures 146 and 147.

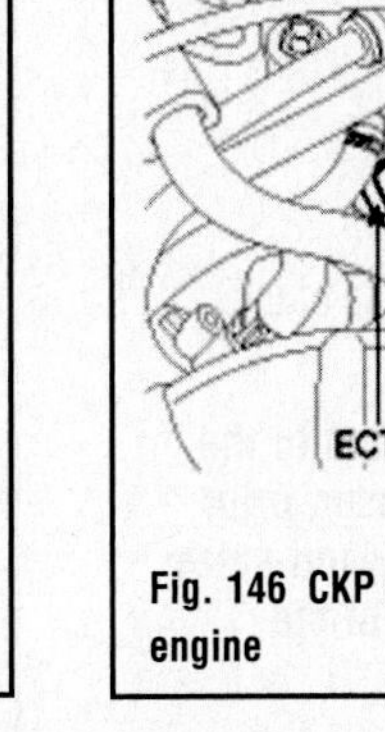

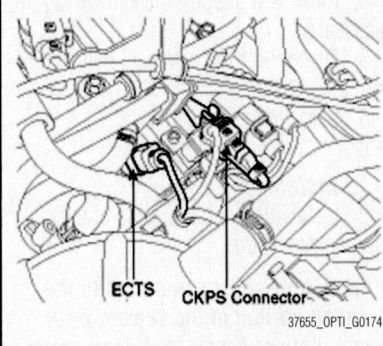

Fig. 146 CKP sensor location—2.4L engine

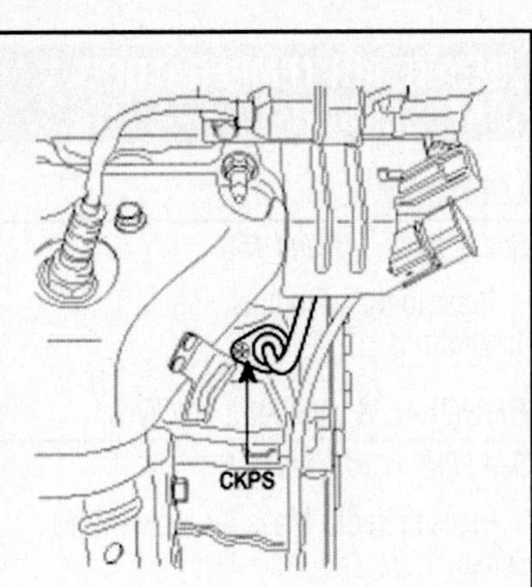

Fig. 147 CKP sensor location—2.7L engine

At this time the manufacturer does not provide removal and installation procedures for this component. The following procedure is a guideline and may differ from the vehicle you are servicing.

1. Before servicing the vehicle, refer to the Precautions Section.

➡If working near and/or around the SRS system and components, be sure to disable the SRS system. Tape the negative battery cable with insulating tape. Always disconnect the negative battery cable first.

⁂ CAUTION

To avoid personal injury when working on vehicles equipped with an air bag, the negative battery cable must be disconnected and at least three minutes must elapse before working on the system. Failure to do so may result in deployment of the air bag.

2. Remove the battery top cover, if equipped.
3. Disconnect the negative battery cable. Tape the cable with insulating tape.
4. Remove the necessary components in order to gain access to the sensor.
5. Disconnect the electrical connector.
6. Remove the sensor from its mounting.
7. Discard the O-ring/gasket.

To install:

➡Be sure to use new fasteners, as required.

8. Installation is the reverse of the removal procedure.

ENGINE COOLANT TEMPERATURE (ECT) SENSOR

LOCATION

The Engine Coolant Temperature Sensor (ECT) is located in the engine coolant passage of the cylinder head for detecting the engine coolant temperature.

REMOVAL & INSTALLATION

1. Before servicing the vehicle, refer to the Precautions Section.

➡If working near and/or around the SRS system and components, be sure to disable the SRS system. Tape the negative battery cable with insulating tape. Always disconnect the negative battery cable first.

⁂ CAUTION

To avoid personal injury when working on vehicles equipped with an air bag, the negative battery cable must be disconnected and at least three minutes must elapse before working on the system. Failure to do so may result in deployment of the air bag.

2. Remove the battery top cover, if equipped.
3. Disconnect the negative battery cable. Tape the cable with insulating tape.
4. Turn ignition switch OFF.
5. Disconnect ECT sensor connector.
6. Remove the ECT sensor.
7. Installation is the reverse of removal.

HEATED OXYGEN SENSOR (HO2S)

LOCATION

Heated Oxygen Sensors (HO2S) are installed on upstream and downstream of the Manifold Catalyst Converter (MCC).

REMOVAL & INSTALLATION

See Figure 148.

1. Before servicing the vehicle, refer to the Precautions Section.

➡If working near and/or around the SRS system and components, be sure to disable the SRS system. Tape the negative battery cable with insulating tape. Always disconnect the negative battery cable first.

⁂ CAUTION

To avoid personal injury when working on vehicles equipped with an air bag, the negative battery cable must be disconnected and at least three minutes must elapse before working on the system. Failure to do so may result in deployment of the air bag.

2. Remove the battery top cover, if equipped.
3. Disconnect the negative battery cable. Tape the cable with insulating tape.
4. Raise and support the vehicle, as necessary.
5. Disconnect the electrical connector from the sensor.
6. Remove the oxygen sensor.

To install:

7. Installation is the reverse of the removal procedure.

➡Apply anti-seize compound to the threaded portion of the sensor, prior to installation. Never apply anti-seize compound to the protector of the sensor.

INTAKE AIR TEMPERATURE (IAT) SENSOR

LOCATION

See Figures 149 and 150.

Refer to the accompanying illustrations.

REMOVAL & INSTALLATION

See Figures 149 and 150.

Please refer to Mass Air Flow (MAF) Sensor.

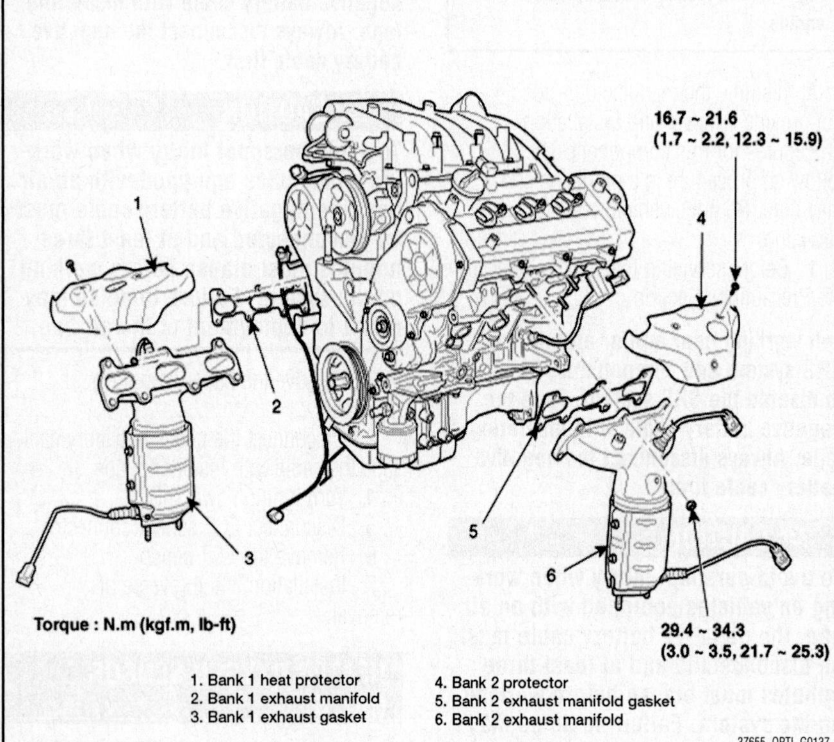

Fig. 148 Heated oxygen sensors and related components—2.7L Engine

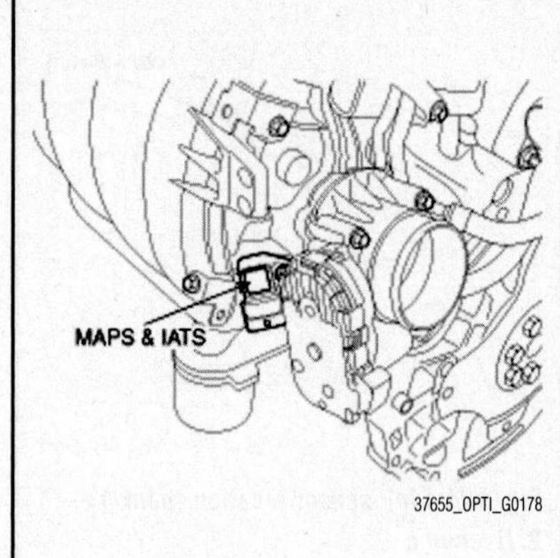

Fig. 149 IAT/MAP sensor location—2.4L engine

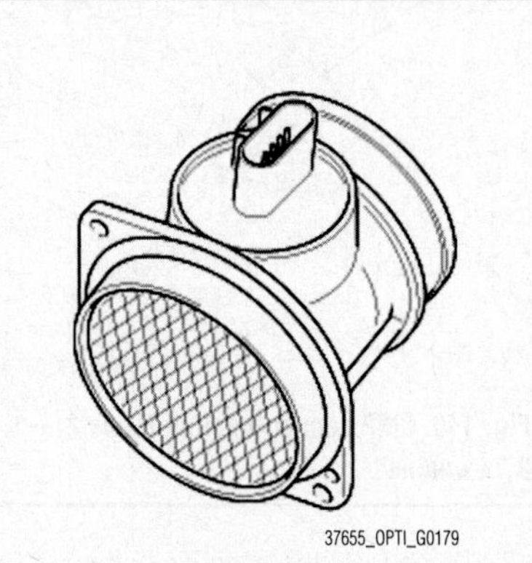

Fig. 150 IAT/MAFS sensor location—2.7L engine

KNOCK SENSOR (KS)

LOCATION

See Figures 151 and 152.

Refer to the accompanying illustrations.

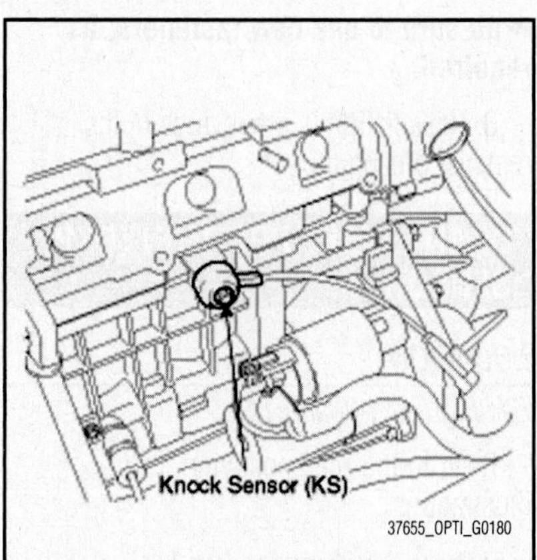

Fig. 151 Knock sensor location—2.4L engine

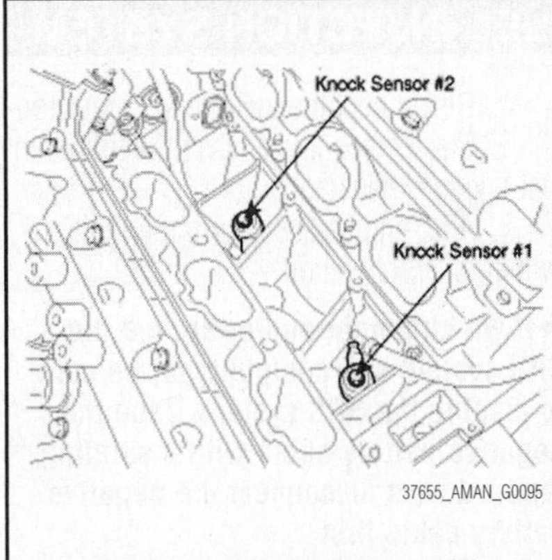

Fig. 152 Knock sensor location—2.7L engine

REMOVAL & INSTALLATION

See Figures 151 and 152.

1. Before servicing the vehicle, refer to the Precautions Section.

➡**If working near and/or around the SRS system and components, be sure to disable the SRS system. Tape the negative battery cable with insulating tape. Always disconnect the negative battery cable first.**

**** CAUTION**

To avoid personal injury when working on vehicles equipped with an air bag, the negative battery cable must be disconnected and at least three minutes must elapse before working on the system. Failure to do so may result in deployment of the air bag.

2. Remove the battery top cover, if equipped.
3. Disconnect the negative battery cable. Tape the cable with insulating tape.
4. On 2.7L engine, remove the intake manifold.
5. On 2.4L engine, remove the necessary components to gain access to the sensor.
6. Disconnect the electrical connector.
7. Remove the sensor from its mounting.

To install:

➡**Be sure to use new fasteners, as required.**

8. Installation is the reverse of the removal procedure.

MASS AIR FLOW (MAF) SENSOR

LOCATION

The 2.7L engine uses a Mass Air Flow (MAF) sensor which is located in between the air cleaner and the throttle body. This sensor is integrated with the IAT sensor.

REMOVAL & INSTALLATION

See Figures 149 and 150.

1. Before servicing the vehicle, refer to the Precautions Section.

➡**If working near and/or around the SRS system and components, be sure to disable the SRS system. Tape the negative battery cable with insulating tape. Always disconnect the negative battery cable first.**

**** CAUTION**

To avoid personal injury when working on vehicles equipped with an air bag, the negative battery cable must be disconnected and at least three minutes must elapse before working on the system. Failure to do so may result in deployment of the air bag.

2. Remove the battery top cover, if equipped.
3. Disconnect the negative battery cable. Tape the cable with insulating tape.
4. Disconnect the connector from the sensor.
5. Remove the air cleaner and air intake assembly, as required.
6. Remove the sensor from its mounting.

To install:

➡**Be sure to use new fasteners, as required.**

7. Installation is the reverse of the removal procedure.

MANIFOLD ABSOLUTE PRESSURE (MAP) SENSOR

LOCATION

See Figure 149.

The 2.4L engine uses a Manifold Absolute Pressure Sensor (MAP) sensor. This sensor is integrated with the IAT sensor.

REMOVAL & INSTALLATION

See Figure 149.

At this time the manufacturer does not provide removal and installation procedures for this component.

POWERTRAIN CONTROL MODULE (PCM)

LOCATION

See Figure 153.

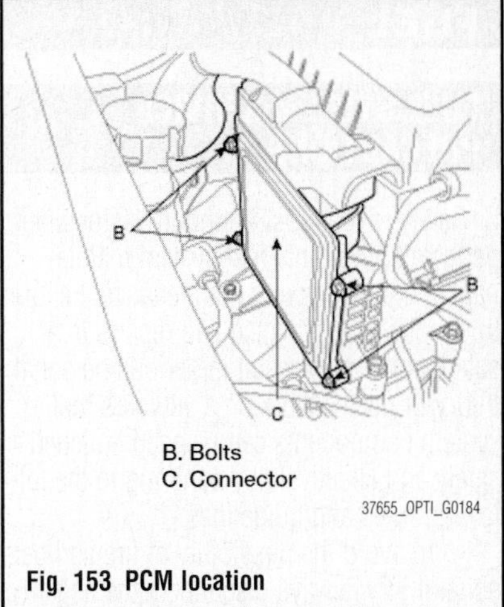

Fig. 153 PCM location

Refer to the accompanying illustration.

REMOVAL & INSTALLATION

See Figure 153.

1. Before servicing the vehicle, refer to the Precautions Section.

➡**If working near and/or around the SRS system and components, be sure to disable the SRS system. Tape the negative battery cable with insulating tape. Always disconnect the negative battery cable first.**

**** CAUTION**

To avoid personal injury when working on vehicles equipped with an air bag, the negative battery cable must be disconnected and at least three minutes must elapse before working on the system. Failure to do so may result in deployment of the air bag.

2. Remove the battery top cover, if equipped.
3. Disconnect the negative battery cable. Tape the cable with insulating tape.
4. Remove the PCM covers and disconnect the connector.
5. Remove the mounting bolts.
6. Remove the component from its mounting at the air cleaner assembly.

To install:

➡**Be sure to use new fasteners, as required.**

7. Installation is the reverse of the removal procedure.

FUEL SYSTEM SERVICE PRECAUTIONS

Safety is the most important factor when performing not only fuel system maintenance, but any type of maintenance. Failure to conduct maintenance and repairs in a safe manner may result in serious personal injury or death. Work on a vehicle's fuel system components can be accomplished safely and effectively by adhering to the following rules and guidelines.

- To avoid the possibility of fire and personal injury, always disconnect the negative battery cable unless the repair or test procedure requires that battery voltage be applied.
- Always relieve the fuel system pressure prior to disconnecting any fuel system component (injector, fuel rail, pressure regulator, etc.) fitting or fuel line connection. Exercise extreme caution whenever relieving fuel system pressure to avoid exposing skin, face and eyes to fuel spray. Please be advised that fuel under pressure may penetrate the skin or any part of the body that it contacts.
- Always place a shop towel or cloth around the fitting or connection prior to loosening to absorb any excess fuel due to spillage. Ensure that all fuel spillage is quickly removed from engine surfaces. Ensure that all fuel-soaked cloths or towels are deposited into a flame-proof waste container with a lid.
- Always keep a dry chemical (Class B) fire extinguisher near the work area.
- Do not allow fuel spray or fuel vapors to come into contact with a spark or open flame.
- Always use a second wrench when loosening or tightening fuel line connection fittings. This will prevent unnecessary stress and torsion on fuel piping. Always follow the proper torque specifications.
- Always replace worn fuel fitting O-rings with new ones. Do not substitute fuel hose where rigid pipe is installed.

FUEL SYSTEM PRESSURE

RELIEVING

1. Before servicing the vehicle, refer to the Precautions Section.
2. Disconnect the fuel pump connector.
3. Start the engine and wait until fuel in fuel line is exhausted.
4. After engine stalls, turn the ignition switch to OFF position.

➡If working near and/or around the SRS system and components, be sure to disable the SRS system. Tape the negative battery cable with insulating tape. Always disconnect the negative battery cable first.

CAUTION

To avoid personal injury when working on vehicles equipped with an air bag, the negative battery cable must be disconnected and at least three minutes must elapse before working on the system. Failure to do so may result in deployment of the air bag.

5. Remove the battery top cover, if equipped.
6. Disconnect the negative battery cable. Tape the cable with insulating tape.

FUEL FILTER

REMOVAL & INSTALLATION

See Figure 154.

The fuel filter is an integral part of the fuel pump assembly.

1. Before servicing the vehicle, refer to the Precautions Section.
2. Open the service cover, in the trunk.
3. Remove the service cover.
4. Disconnect the fuel pump connector.
5. Start the engine and wait until fuel in fuel line is exhausted.
6. After engine stalls, turn the ignition switch to OFF position.

➡If working near and/or around the SRS system and components, be sure to disable the SRS system. Tape the negative battery cable with insulating tape. Always disconnect the negative battery cable first.

CAUTION

To avoid personal injury when working on vehicles equipped with an air bag, the negative battery cable must be disconnected and at least three minutes must elapse before working on the system. Failure to do so may result in deployment of the air bag.

7. Remove the battery top cover, if equipped.
8. Disconnect the negative battery cable. Tape the cable with insulating tape.
9. Disconnect the fuel feed line and canister hoses.
10. Unscrew the fuel pump mounting bolts and remove the fuel pump assembly.
11. Service the fuel filter, as required.

To install:

➡Be sure to use new fasteners, as required.

12. Installation is the reverse of the removal procedure.

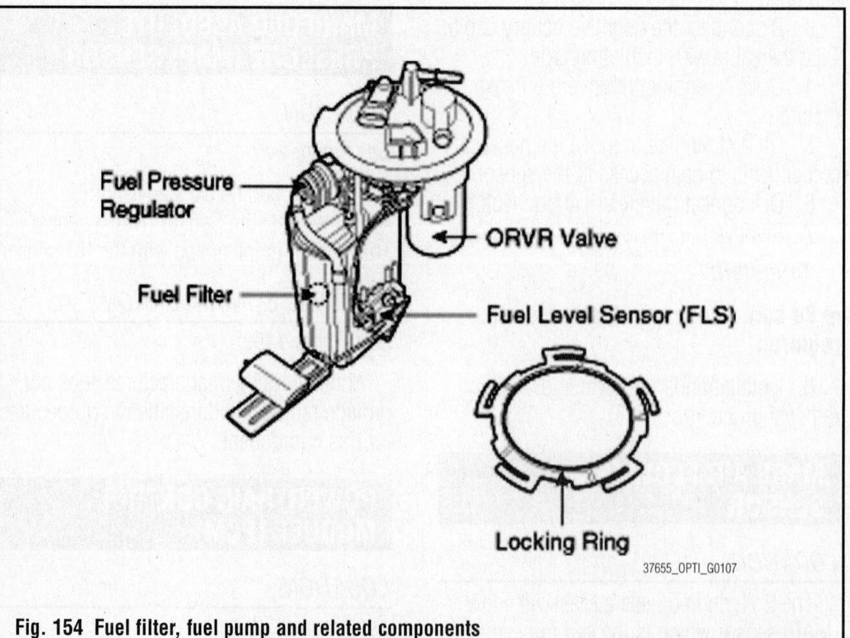

Fig. 154 Fuel filter, fuel pump and related components

FUEL PUMP

REMOVAL & INSTALLATION

See Figure 154.

1. Before servicing the vehicle, refer to the Precautions Section.
2. Open the service cover, in the trunk.
3. Remove the service cover.
4. Disconnect the fuel pump connector.
5. Start the engine and wait until fuel in fuel line is exhausted.
6. After engine stalls, turn the ignition switch to OFF position.

➡If working near and/or around the SRS system and components, be sure to disable the SRS system. Tape the negative battery cable with insulating tape. Always disconnect the negative battery cable first.

CAUTION

To avoid personal injury when working on vehicles equipped with an air bag, the negative battery cable must be disconnected and at least three minutes must elapse before working on the system. Failure to do so may result in deployment of the air bag.

7. Remove the battery top cover, if equipped.
8. Disconnect the negative battery cable. Tape the cable with insulating tape.
9. Disconnect the fuel feed line and canister hoses.
10. Unscrew the fuel pump mounting bolts and remove the fuel pump assembly.

To install:

➡Be sure to use new fasteners, as required.

11. Installation is the reverse of the removal procedure.

FUEL RAIL AND INJECTOR

REMOVAL & INSTALLATION

See Figures 155 and 156.

At this time the manufacturer does not provide removal and installation procedures for this component. Before servicing this component be sure to properly relieve the fuel system pressure.

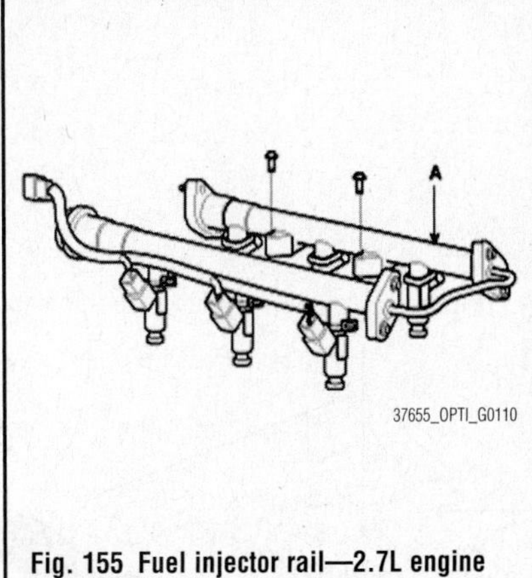

Fig. 155 Fuel injector rail—2.7L engine

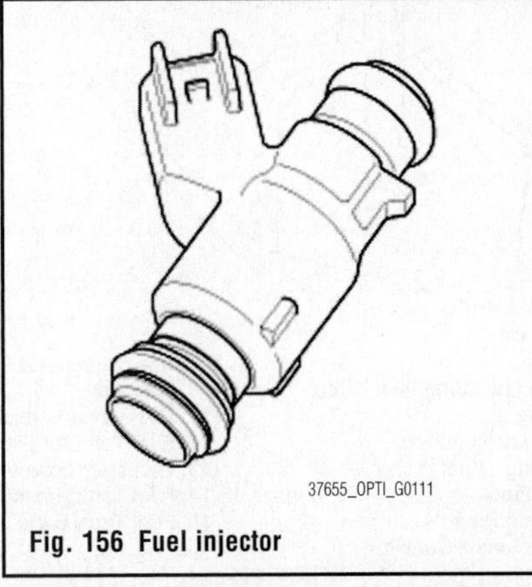

Fig. 156 Fuel injector

FUEL TANK

REMOVAL & INSTALLATION

See Figures 157 through 159.

1. Before servicing the vehicle, refer to the Precautions Section.
2. Open the service cover, in the trunk.
3. Remove the service cover.
4. Disconnect the fuel pump connector.
5. Start the engine and wait until fuel in fuel line is exhausted.
6. After engine stalls, turn the ignition switch to OFF position.

➡If working near and/or around the SRS system and components, be sure to disable the SRS system. Tape the negative battery cable with insulating tape. Always disconnect the negative battery cable first.

CAUTION

To avoid personal injury when working on vehicles equipped with an air bag, the negative battery cable must be disconnected and at least three

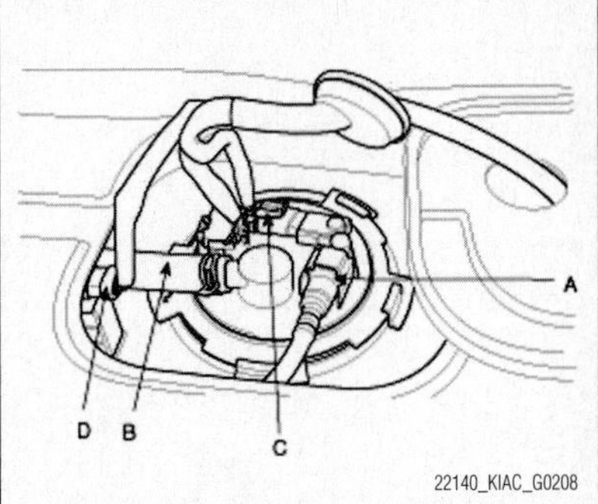

Fig. 157 Disconnect the fuel feed quick-connector (A), the vapor hose (B), the fuel tank pressure sensor connector (C) and the canister close valve connector (D)

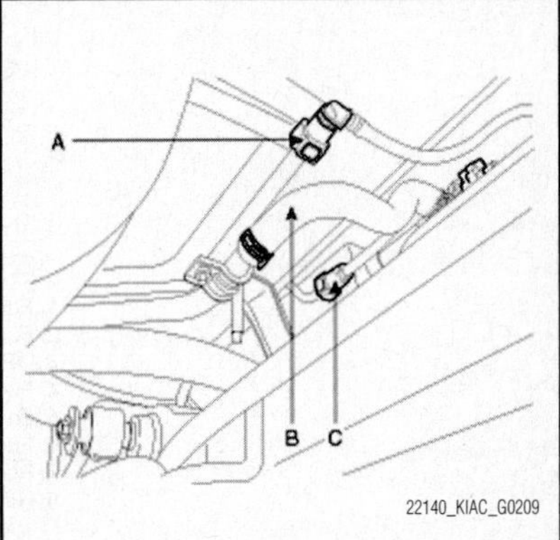

Fig. 158 Disconnect the recirculation pipe quick-connector (A), the vapor hose (B), and purge tube quick-connector (C)

minutes must elapse before working on the system. Failure to do so may result in deployment of the air bag.

7. Remove the battery top cover, if equipped.
8. Disconnect the negative battery cable. Tape the cable with insulating tape.
9. Drain the fuel tank to an acceptable level. Be sure to properly dispose of drained fuel.
10. Disconnect the fuel feed quick-connector (A), the vapor hose (B), the fuel tank pressure sensor connector (C) and the canister close valve connector (D).
11. Raise the vehicle. Support it safely.
12. Remove the main muffler.
13. Support the fuel tank with a jack and unscrew fuel tank band mounting nuts.

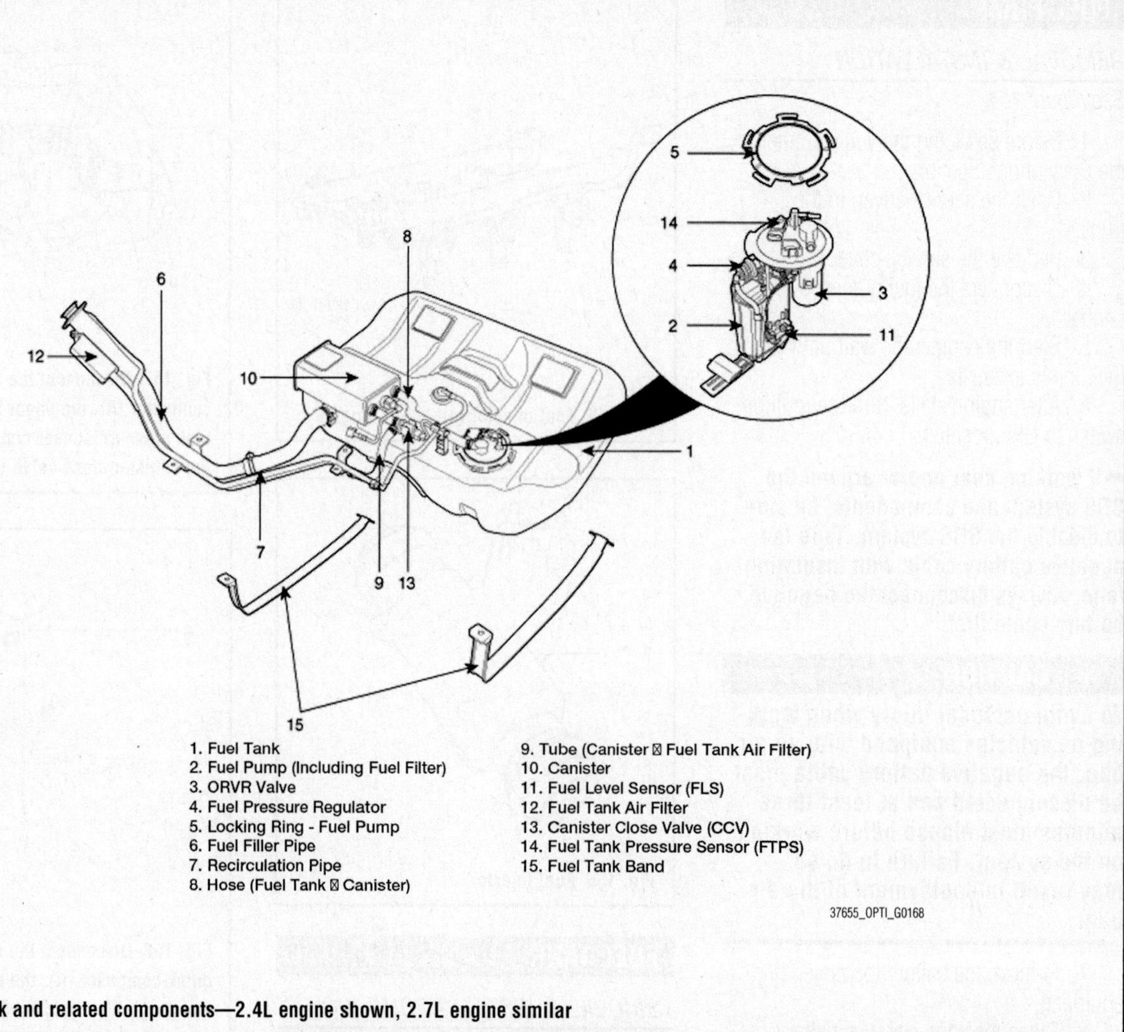

Fig. 159 Fuel tank and related components—2.4L engine shown, 2.7L engine similar

14. Disconnect the fuel filler hose.
15. Disconnect the recirculation pipe quick-connector (A), the vapor hose (B) connecting the canister with the fuel tank air filter and the purge tube quick-connector (C) connecting the canister and the intake manifold.
16. Remove the fuel tank from the vehicle with coming down the jack slowly.

To install:

➡**Be sure to use new fasteners, as required.**

17. Installation is the reverse of the removal procedure.

IDLE SPEED

ADJUSTMENT

Idle speed is controlled by the PCM and is not adjustable.

THROTTLE BODY

REMOVAL & INSTALLATION

See Figure 160.

At this time the manufacturer does not provide removal and installation procedures for this component. The following procedure is a guideline and may differ from the vehicle you are servicing.

1. Before servicing the vehicle, refer to the Precautions Section.

➡**If working near and/or around the SRS system and components, be sure to disable the SRS system. Tape the negative battery cable with insulating tape. Always disconnect the negative battery cable first.**

⁂ CAUTION

To avoid personal injury when working on vehicles equipped with an air bag, the negative battery cable must be disconnected and at least three minutes must elapse before working on the system. Failure to do so may result in deployment of the air bag.

2. Remove the battery top cover, if equipped.
3. Remove the necessary components in order to gain access to the throttle body assembly.
4. Disconnect the electrical connectors.
5. Drain the cooling system to an acceptable level. Disconnect the hose.
6. Remove the throttle body assembly from its mounting.

To install:

➡**Be sure to use new fasteners, as required.**

7. Installation is the reverse of the removal procedure.

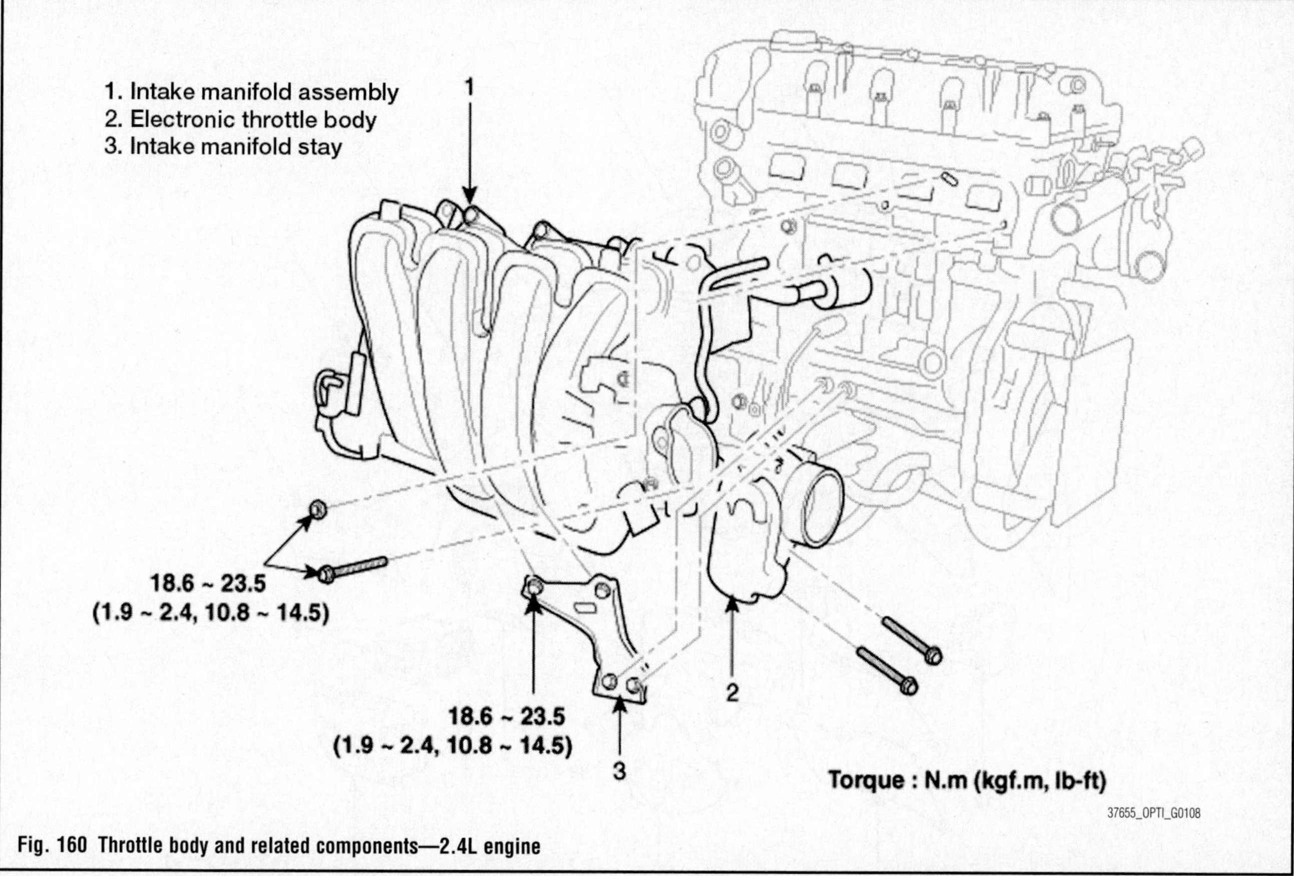

Fig. 160 Throttle body and related components—2.4L engine

HEATING & AIR CONDITIONING SYSTEM

BLOWER MOTOR

REMOVAL & INSTALLATION

1. Before servicing the vehicle, refer to the Precautions Section.

➡If working near and/or around the SRS system and components, be sure to disable the SRS system. Tape the negative battery cable with insulating tape. Always disconnect the negative battery cable first.

⁂ CAUTION

To avoid personal injury when working on vehicles equipped with an air bag, the negative battery cable must be disconnected and at least three minutes must elapse before working on the system. Failure to do so may result in deployment of the air bag.

2. Remove the battery top cover, if equipped.
3. Disconnect the negative battery cable. Tape the cable with insulating tape.
4. Disconnect the connector of the blower motor.
5. Remove the blower motor after loosening the mounting screws.

To install:

➡Be sure to use new fasteners, as required.

6. Installation is the reverse of the removal procedure.

HEATER CORE

REMOVAL & INSTALLATION

See Figures 161 through 167.

1. Before servicing the vehicle, refer to the Precautions Section.

➡If working near and/or around the SRS system and components, be sure to disable the SRS system. Tape the negative battery cable with insulating tape. Always disconnect the negative battery cable first.

⁂ CAUTION

To avoid personal injury when working on vehicles equipped with an air bag, the negative battery cable must be disconnected and at least three minutes must elapse before working on the system. Failure to do so may result in deployment of the air bag.

2. Remove the battery top cover, if equipped.
3. Disconnect the negative battery cable. Tape the cable with insulating tape.
4. Recover the refrigerant with a recovery/ recycling/ charging station.
5. When the engine is cool, drain the engine coolant from the radiator.
6. Remove the expansion valve cover.
7. Remove the bolts and the expansion valve from the evaporator core.
8. Plug or cap the lines immediately after disconnecting them to avoid moisture and dust contamination.
9. Disconnect the inlet and outlet heater hoses from the heater unit.
10. Remove the crash pad.
11. Remove the cowl cross bar assembly.
12. Remove the heater and blower unit after loosening 3 mounting bolts.

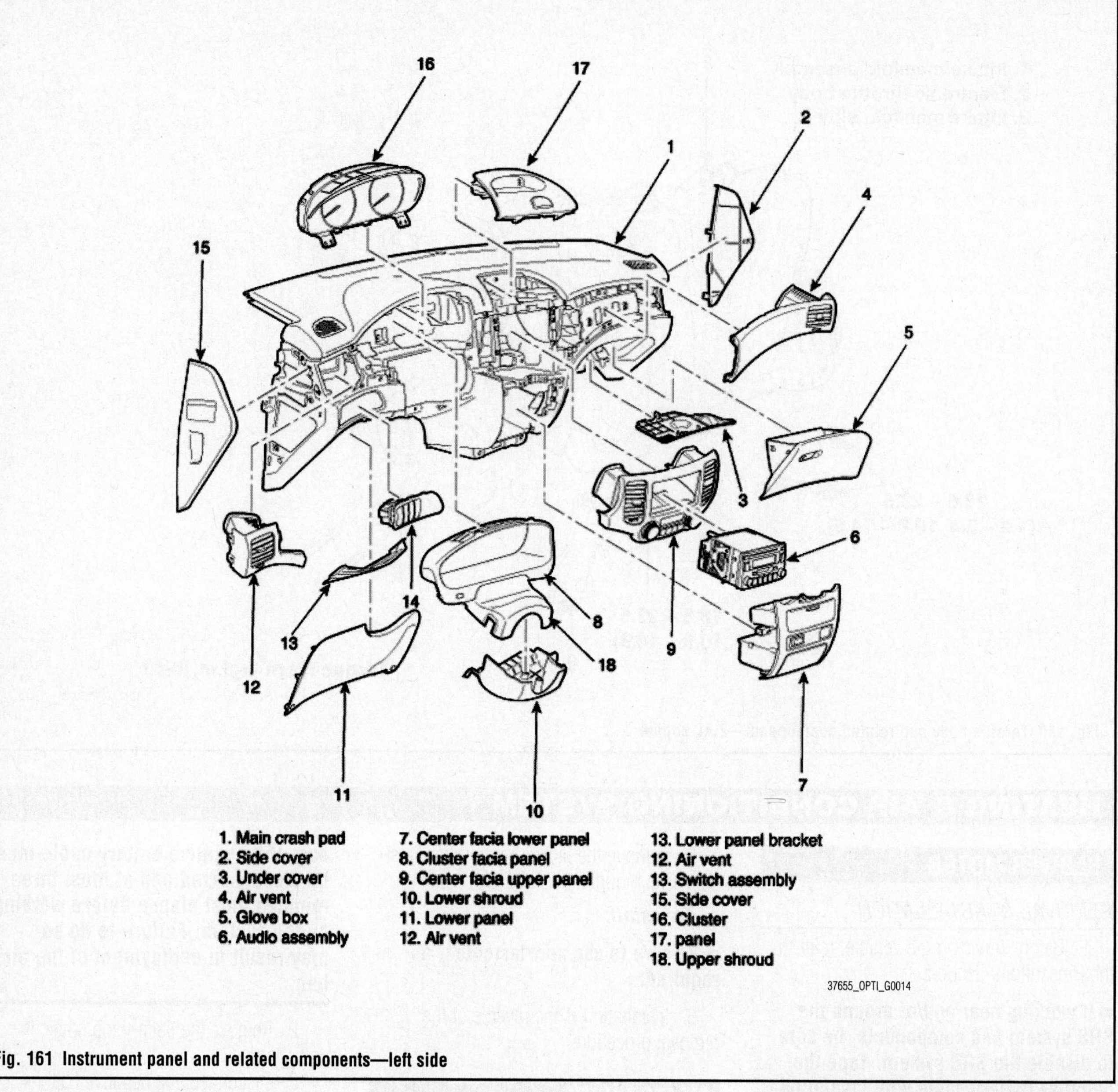

Fig. 161 Instrument panel and related components—left side

13. Remove the blower unit from heater unit after loosening 3 screws.
14. Disconnect the heater core cover and remove the heater core.
15. Disconnect the evaporator cover and remove the evaporator
16. Be careful that the inlet and outlet pipe are not bent during heater core removal, and pull out the heater core.

To install:

17. Installation is the reverse order of removal, and note these items :
   a. If you're installing a new evaporator, add refrigerant oil (ND-OIL8).
   b. Replace the O-rings with new ones at each fitting, and apply a thin coat of refrigerant oil before installing them. Be sure to use the right O-rings for R-134a to avoid leakage.
   c. Immediately after using the oil, replace the cap on the container, and seal it to avoid moisture absorption.
   d. Do not spill the refrigerant oil on the vehicle ; it may damage the paint ; if the refrigerant oil contacts the paint, wash it off immediately.
   e. Apply sealant to the grommets.
   f. Make sure that there is no air leakage.
   g. Charge the system and test its performance.
   h. Do not interchange the inlet and outlet heater hoses and install the hose clamps securely.
   i. Refill the cooling system with engine coolant.

HEATER/AC UNIT

REMOVAL & INSTALLATION

See Figures 161 through 164, 167.

1. Before servicing the vehicle, refer to the Precautions Section.

➡If working near and/or around the SRS system and components, be sure to disable the SRS system. Tape the negative battery cable with insulating tape. Always disconnect the negative battery cable first.

⁂ CAUTION

To avoid personal injury when working on vehicles equipped with an air bag, the negative battery cable must

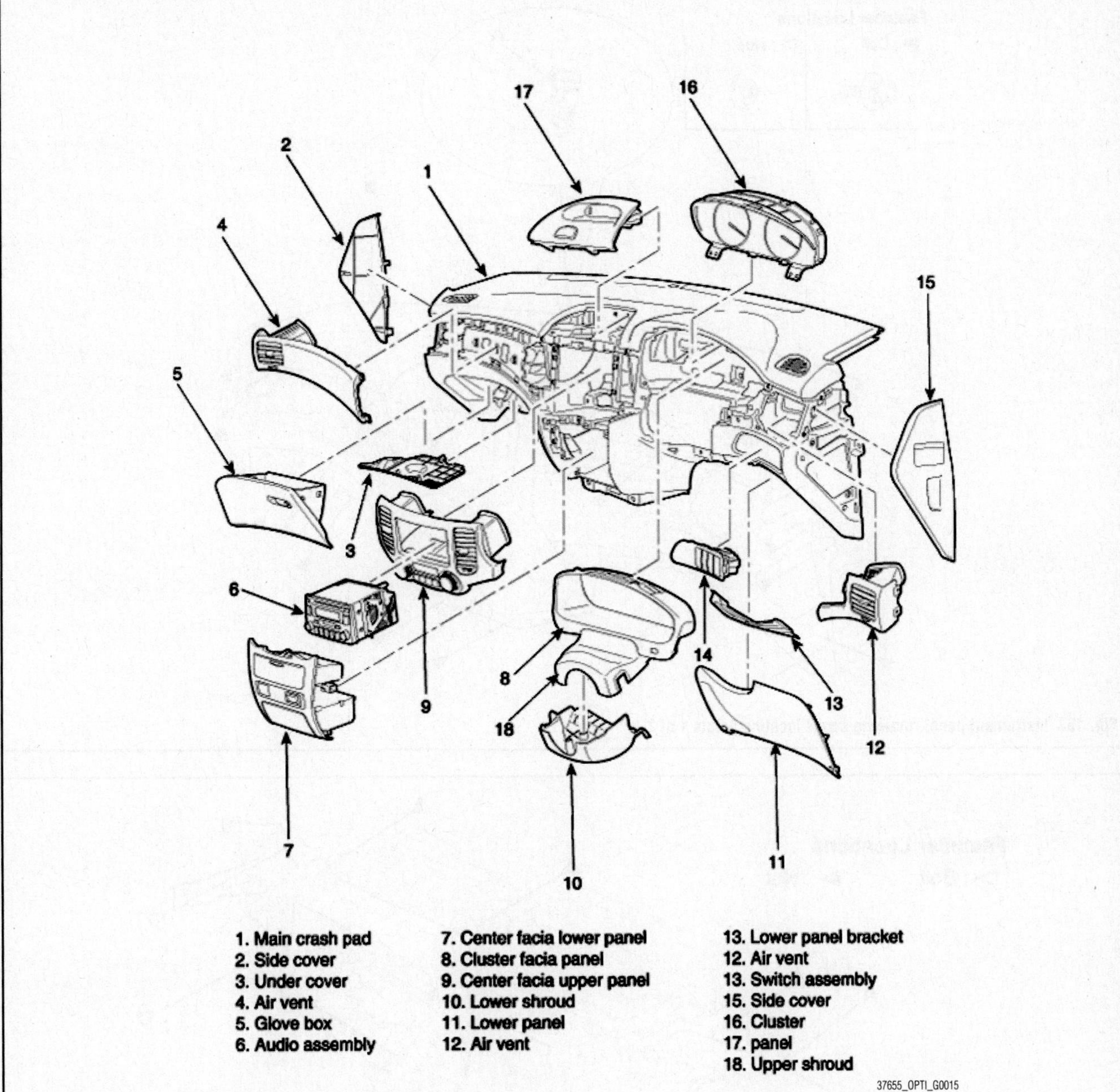

Fig. 162 Instrument panel and related components—right side

be disconnected and at least three minutes must elapse before working on the system. Failure to do so may result in deployment of the air bag.

2. Remove the battery top cover, if equipped.
3. Disconnect the negative battery cable. Tape the cable with insulating tape.
4. Recover the refrigerant with a recovery/ recycling/ charging station.
5. When the engine is cool, drain the engine coolant from the radiator.
6. Remove the expansion valve cover.
7. Remove the bolts and the expansion valve from the evaporator core.
8. Plug or cap the lines immediately after disconnecting them to avoid moisture and dust contamination.
9. Disconnect the inlet and outlet heater hoses from the heater unit.
10. Remove the crash pad.
11. Remove the cowl cross bar assembly.
12. Remove the heater and blower unit after loosening 3 mounting bolts.
13. Remove the blower unit from heater unit after loosening 3 screws.

To install:

14. Installation is the reverse order of removal, and note these items :
    a. Do not spill the refrigerant oil on

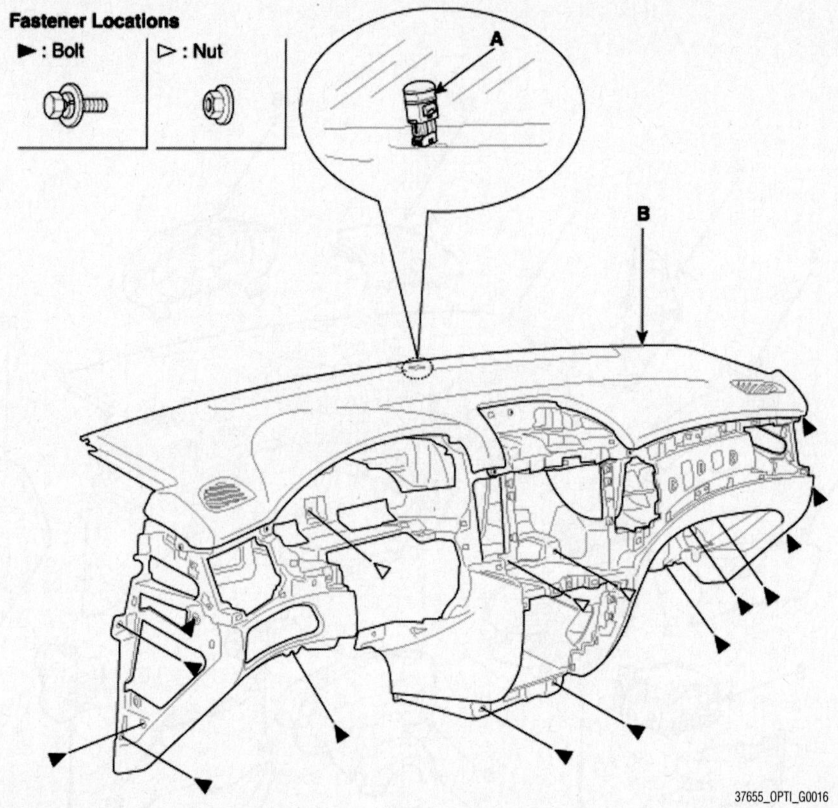

Fig. 163 Instrument panel retaining screw locating points 1 of 2

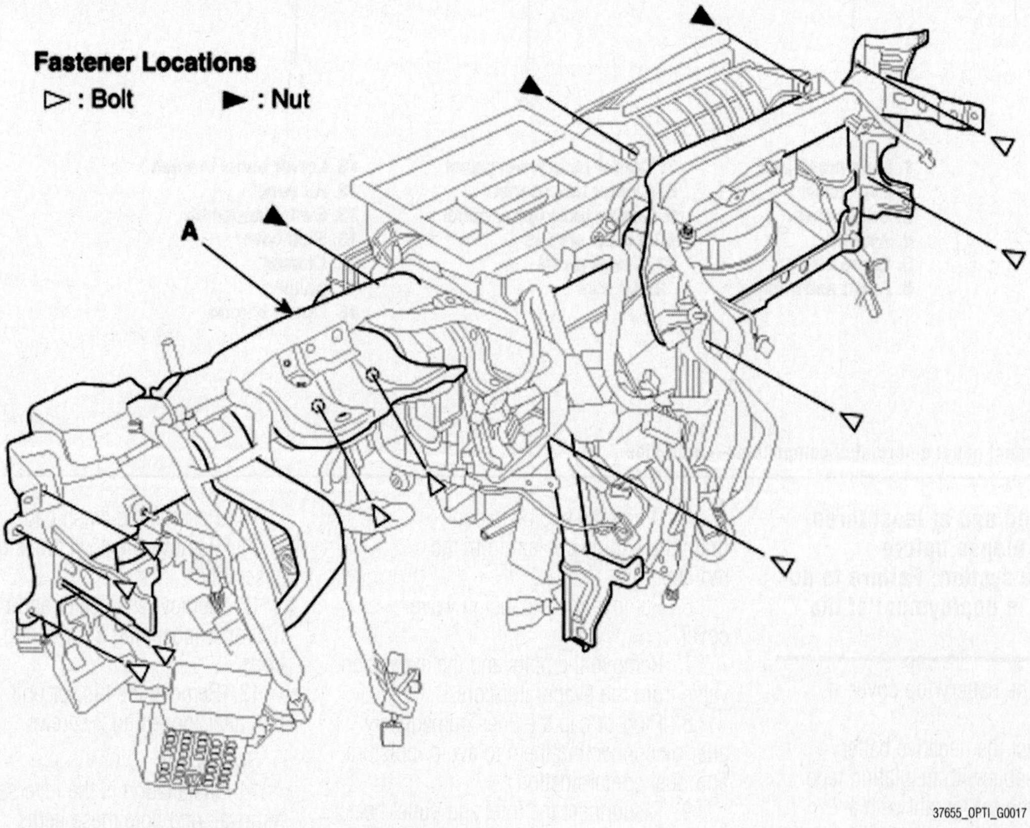

Fig. 164 Instrument panel retaining screw locating points 2 of 2

A
B
A. Cover
B. Core
37655_OPTI_G0081

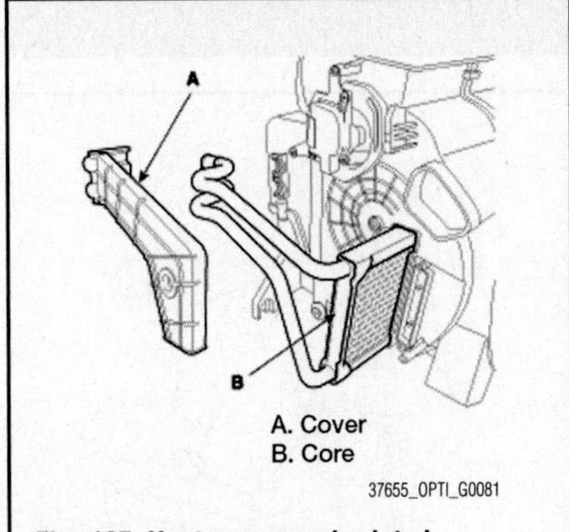

Fig. 165 Heater core and related components

B
A
A. Cover
B. Core
37655_OPTI_G0080

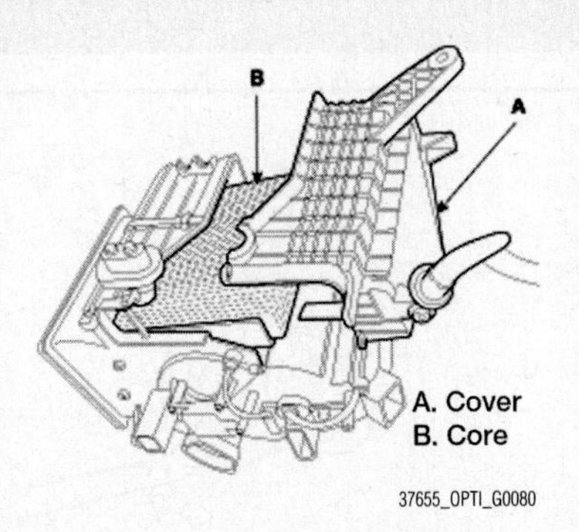

Fig. 166 Evaporator core and related components

the vehicle ; it may damage the paint ; if the refrigerant oil contacts the paint, wash it off immediately.

b. Apply sealant to the grommets.

c. Make sure that there is no air leakage.

d. Charge the system and test its performance.

e. Do not interchange the inlet and outlet heater hoses and install the hose clamps securely.

f. Refill the cooling system with engine coolant.

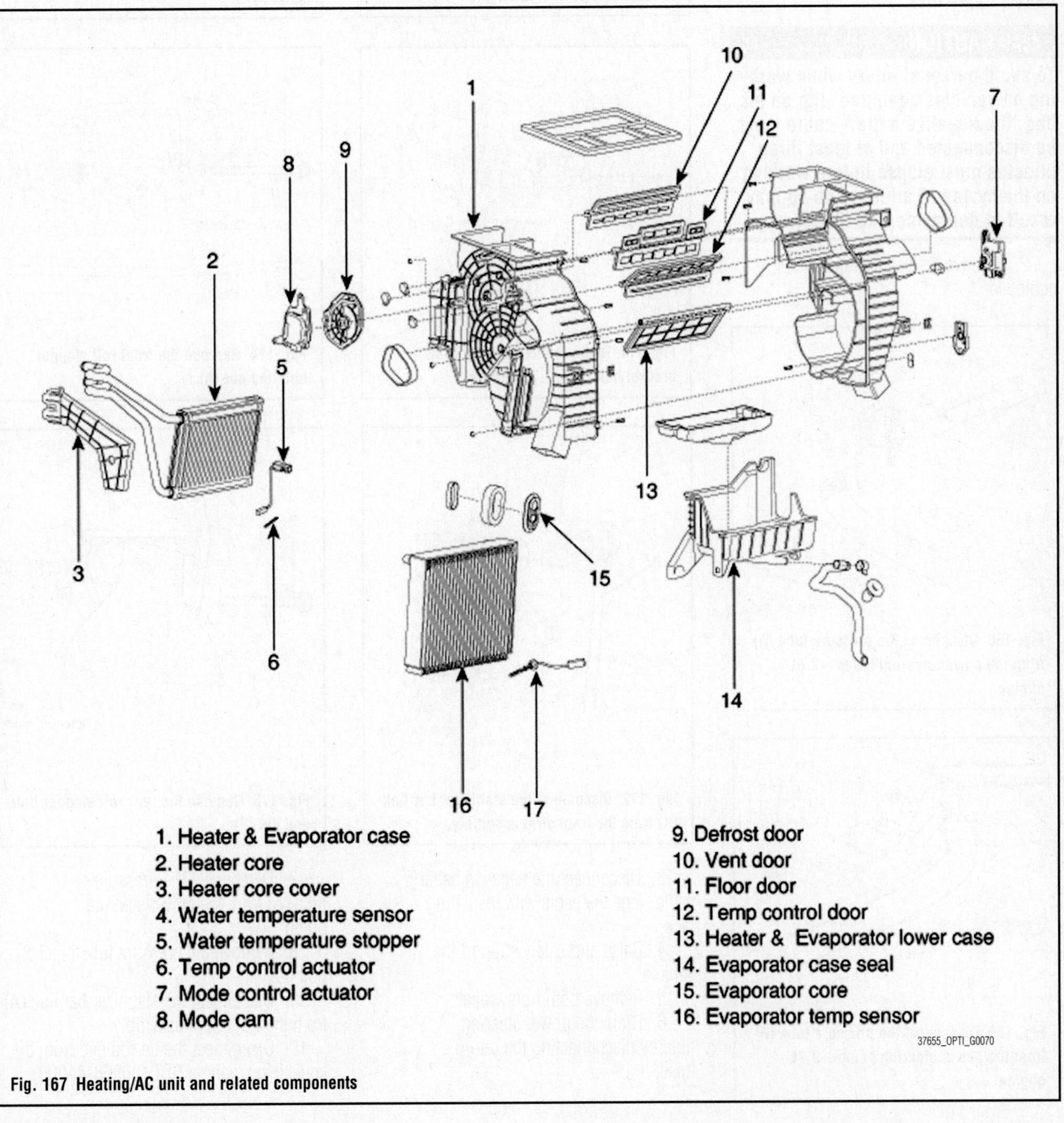

1
2
3
5
6
7
8
9
10
11
12
13
14
15
16
17
1. Heater & Evaporator case
2. Heater core
3. Heater core cover
4. Water temperature sensor
5. Water temperature stopper
6. Temp control actuator
7. Mode control actuator
8. Mode cam
9. Defrost door
10. Vent door
11. Floor door
12. Temp control door
13. Heater & Evaporator lower case
14. Evaporator case seal
15. Evaporator core
16. Evaporator temp sensor
37655_OPTI_G0070

Fig. 167 Heating/AC unit and related components

STEERING

POWER RACK & PINION STEERING GEAR

REMOVAL & INSTALLATION

See Figures 168 through 178.

1. Before servicing the vehicle, refer to the Precautions Section.

➡If working near and/or around the SRS system and components, be sure to disable the SRS system. Tape the negative battery cable with insulating tape. Always disconnect the negative battery cable first.

✻✻ CAUTION

To avoid personal injury when working on vehicles equipped with an air bag, the negative battery cable must be disconnected and at least three minutes must elapse before working on the system. Failure to do so may result in deployment of the air bag.

2. Remove the battery top cover, if equipped.

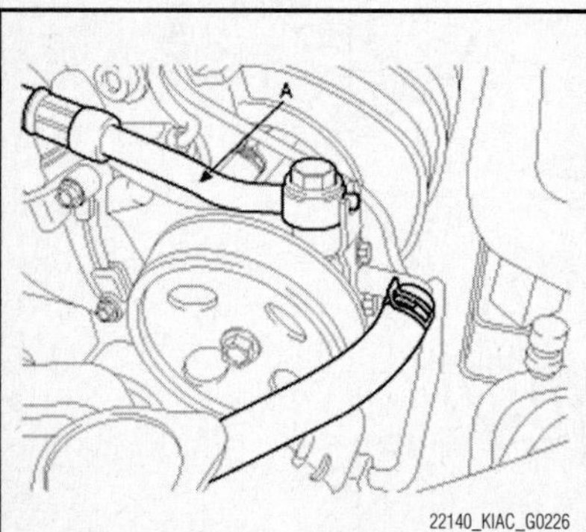

Fig. 168 Disconnect the pressure tube (A) from the power steering pump—2.4L engine

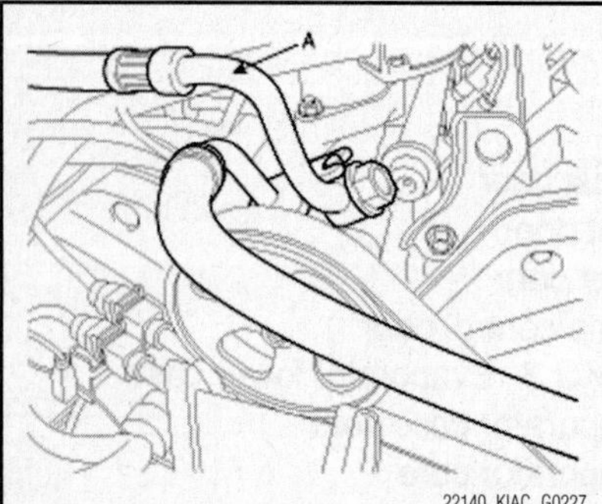

Fig. 169 Disconnect the pressure tube (A) from the power steering pump—2.7L engine

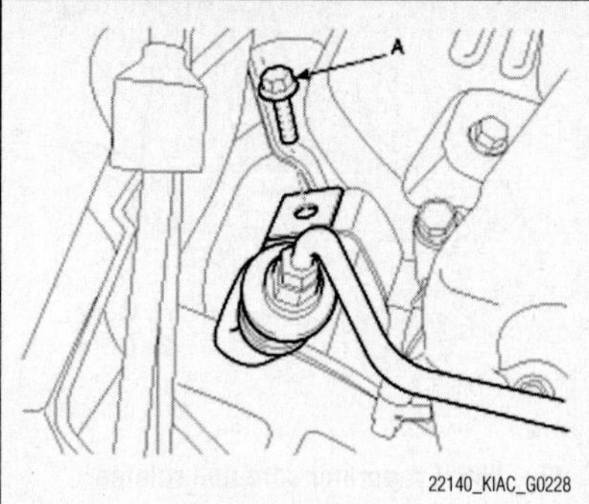

Fig. 170 Remove the pressure tube bracket bolt (A)—2.4L engine

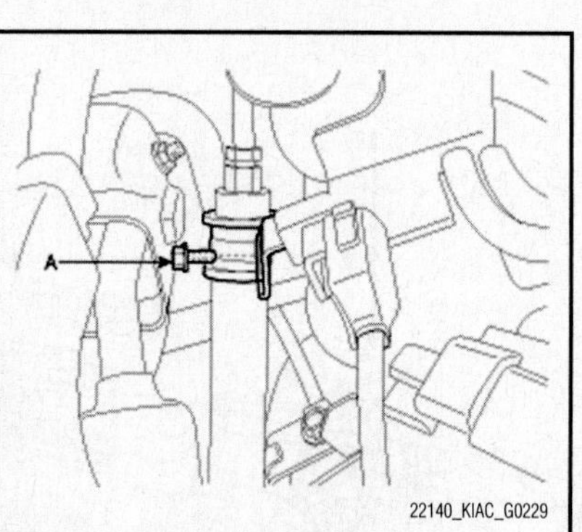

Fig. 171 Remove the pressure tube bracket bolt (A)—2.7L engine

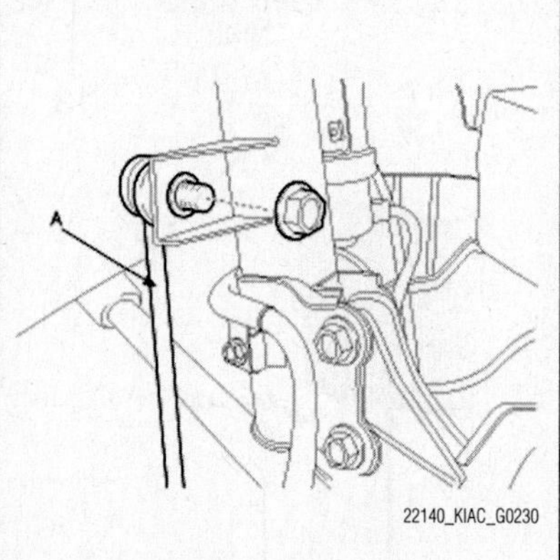

Fig. 172 Disconnect the stabilizer bar link (A) from the front strut assembly.

3. Disconnect the negative battery cable. Tape the cable with insulating tape.
4. Raise and safely support the vehicle.
5. Remove both front wheels.
6. Drain the power steering fluid by disconnecting the return hose.

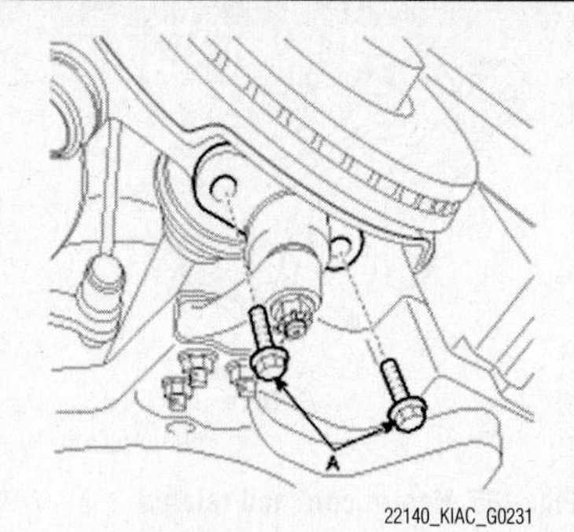

Fig. 173 Remove the lower arm ball joint bolts (A)

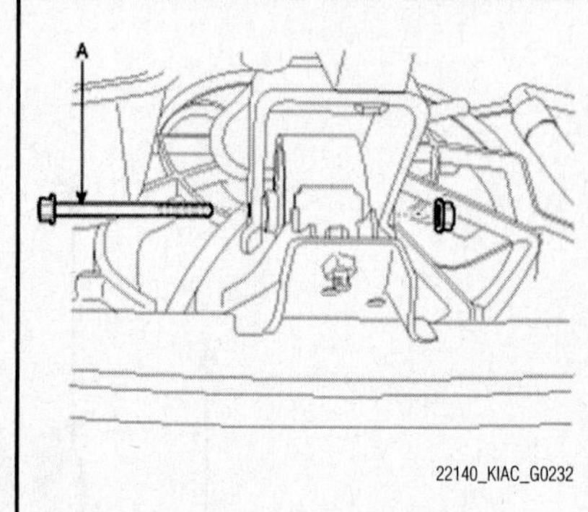

Fig. 174 Remove the front roll stopper bolt and nut (A)

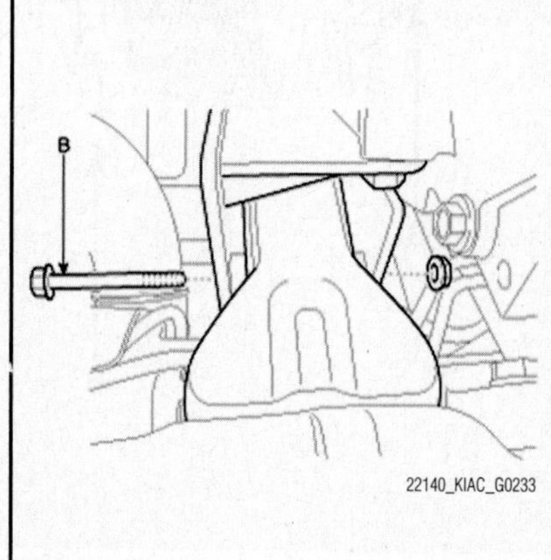

Fig. 175 Remove the rear roll stopper bolt and nut (B)

7. Disconnect the pressure tube (A) from the power steering pump.
8. Remove the pressure tube bracket bolt (A).
9. Disconnect the stabilizer bar link (A) from the front strut assembly.
10. Disconnect the tie rod end from the knuckle by using a SST (09568-4A000).

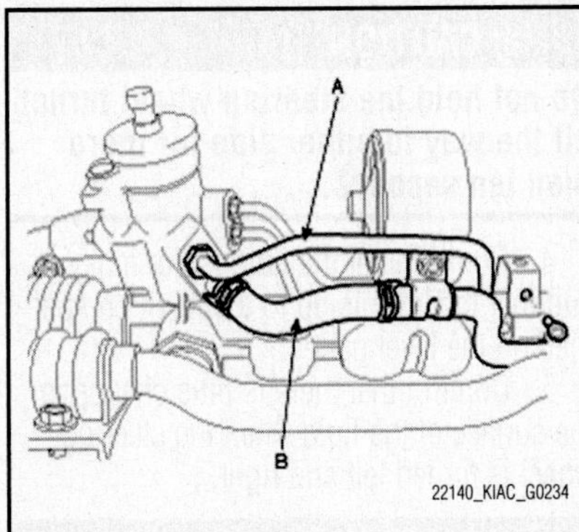

Fig. 176 Disconnect the pressure tube (A) and return (B) from the valve body housing

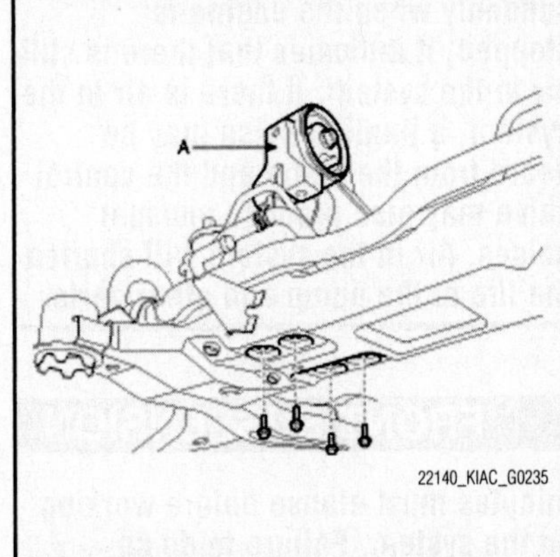

Fig. 177 Remove the rear roll stopper (A) from the sub-frame

11. Remove the lower arm ball joint bolts (A) on both sides.
12. Remove the bolt connecting steering gear pinion shaft to universal joint.
13. Remove the front and rear roll stopper bolts and nuts (A, B).
14. Remove the sub-frame.
15. Remove the heat protector.
16. Disconnect the pressure tube (A) and return (B) from the valve body housing.
17. Remove the tube bracket bolt.
18. Remove the rear roll stopper (A) from the sub-frame.
19. Remove the power steering gear box from the sub-frame.

To install:

20. Installation is the reverse of removal.
21. Use the following torque specifications:
    - Steering gear box to subframe: 43–58 ft. lbs. (60–80 Nm)
    - Rear roll stopper to subframe: 40–47 ft. lbs. (55–65 Nm)
    - Pressure and return tubes to valve body housing: 9–13 ft. lbs. (12–18 Nm)
    - Subframe bolts and nuts: 116–130 ft. lbs. (160–180 Nm)
    - Front and rear roll stopper bolts and nuts: 36–47 ft. lbs. (50–65 Nm)
    - Steering gear pinion shaft bolt: 22–25 ft. lbs. (30–35 Nm)
    - Lower arm ball joint bolts: 72–87 ft. lbs. (100–120 Nm)
    - Tie rod end castle nut: 17–25 ft. lbs. (24–34 Nm)
    - Stabilizer link to front strut assembly: 72–87 ft. lbs. (100–120 Nm)
    - Pressure tube to power steering pump: 40–47 ft. lbs. (55–65 Nm)
22. Add the power steering oil.
23. After installation, bleed the power steering system.
24. Adjust the wheel alignment.

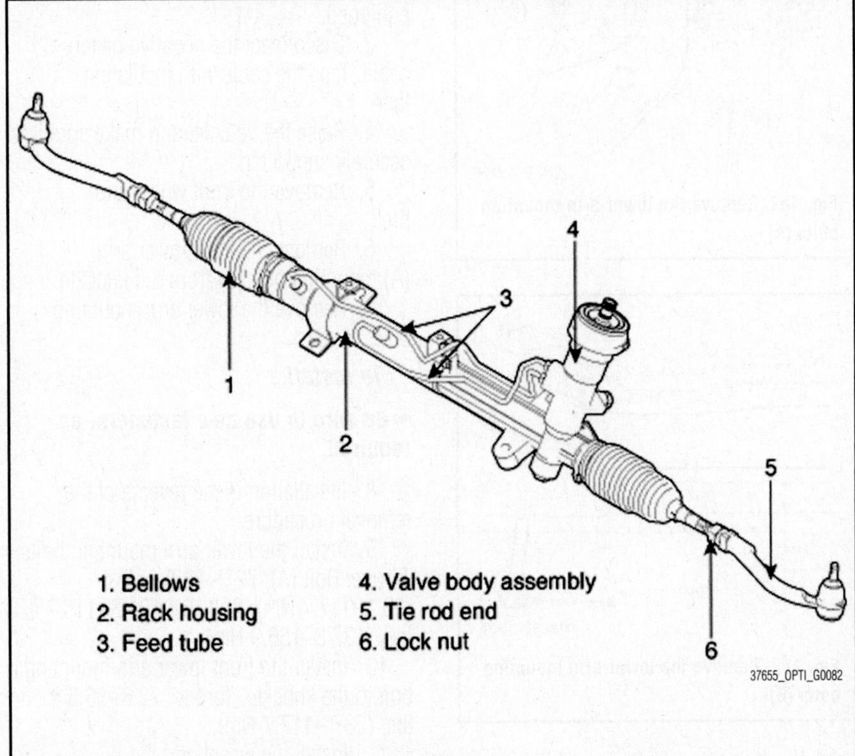

Fig. 178 Power steering unit and related components

POWER STEERING PUMP

REMOVAL & INSTALLATION

See Figure 179.

1. Before servicing the vehicle, refer to the Precautions Section.

➡If working near and/or around the SRS system and components, be sure to disable the SRS system. Tape the negative battery cable with insulating tape. Always disconnect the negative battery cable first.

⁂ CAUTION

To avoid personal injury when working on vehicles equipped with an air bag, the negative battery cable must be disconnected and at least three minutes must elapse before working on the system. Failure to do so may result in deployment of the air bag.

2. Remove the battery top cover, if equipped.
3. Disconnect the negative battery cable. Tape the cable with insulating tape.
4. Disconnect the oil pressure switch (A).
5. Disconnect the pressure tube and return hose from the power steering pump assembly.
6. Remove the drive belt.
7. Remove the power steering pump assembly from the pump bracket.

To install:

8. Installation is the reverse of removal.

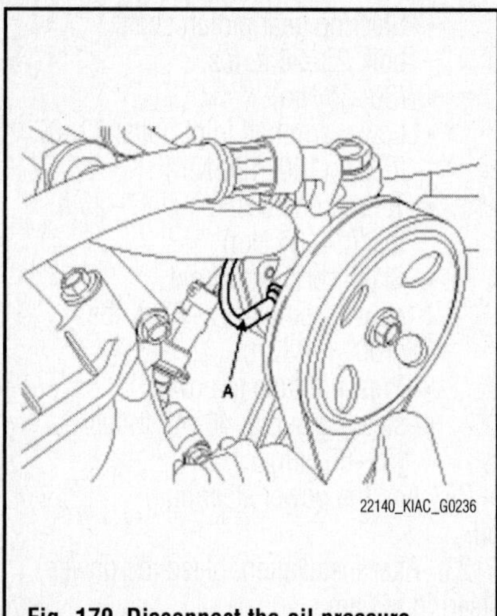

Fig. 179 Disconnect the oil pressure switch (A).

CAUTION

Ensure the pressure hose does not twist and come in contact with other components.

9. Add the power steering fluid.
10. Bleed the power steering system.

BLEEDING

1. Remove the fuel pump fuse, then start the engine and wait for the engine to stall. Next, while operating the starting motor intermittently (for 15–20 seconds), turn the steering wheel all the way to the left and then to the right five or six times.

➡During air bleeding, replenish the fluid supply so that the level never falls below the lower position of the filter. If air bleeding is done while the vehicle is idling, the air will be broken up and absorbed into the fluid. Be sure to do the bleeding only while cranking.

2. Reinstall the fuel pump fuse, and start the engine (idling).
3. Turn the steering wheel to the left and the right until there are no air bubbles in the oil reservoir.

CAUTION

Do not hold the steering wheel turned all the way to either side for more than ten seconds.

4. Confirm that the fluid is not milky, and that the level is up to the position specified on the level gauge.
5. Confirm that there is little change in the surface of the fluid when the steering wheel is turned left and right.

CAUTION

If the surface of the fluid changes considerably, air bleeding should be done again. If the fluid level rises suddenly when the engine is stopped, it indicates that there is still air in the system. If there is air in the system, a jingling noise may be heard from the pump and the control valve may also produce unusual noises. Air in the system will shorten the life of the pump and other parts.

SUSPENSION

FRONT SUSPENSION

LOWER CONTROL ARM

REMOVAL & INSTALLATION

See Figures 180 through 182.

1. Before servicing the vehicle, refer to the Precautions Section.

➡If working near and/or around the SRS system and components, be sure to disable the SRS system. Tape the negative battery cable with insulating tape. Always disconnect the negative battery cable first.

CAUTION

To avoid personal injury when working on vehicles equipped with an air bag, the negative battery cable must be disconnected and at least three minutes must elapse before working on the system. Failure to do so may result in deployment of the air bag.

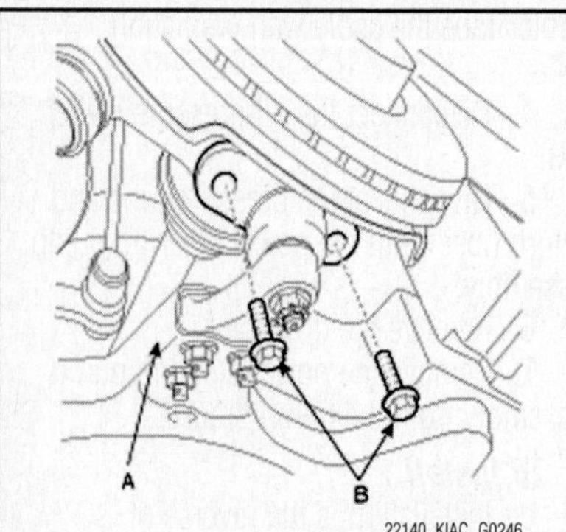

Fig. 180 Remove the front lower arm (A) mounting bolt (B) from the knuckle

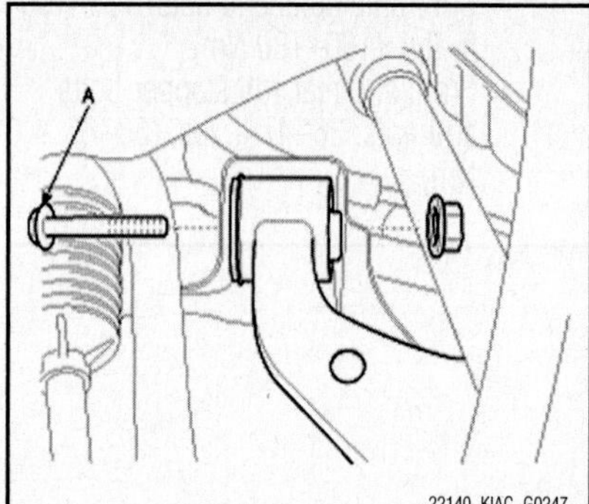

Fig. 181 Remove the lower arm mounting bolts (A)

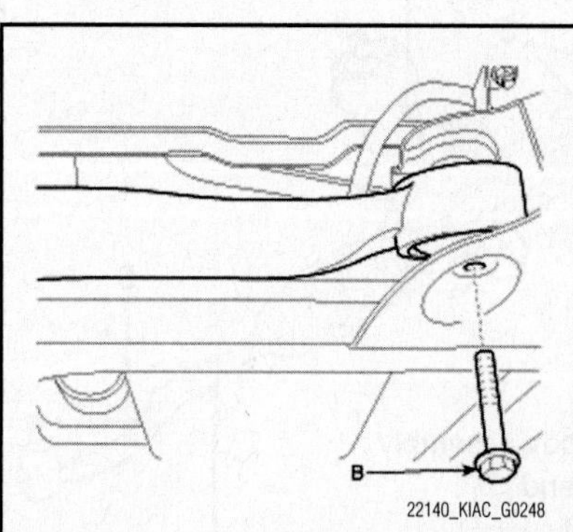

Fig. 182 Remove the lower arm mounting bolts (B)

2. Remove the battery top cover, if equipped.
3. Disconnect the negative battery cable. Tape the cable with insulating tape.
4. Raise the vehicle, and make sure it is securely supported.
5. Remove the front wheel and tire.
6. Remove the front lower arm (A) mounting bolt (B) from the knuckle.
7. Remove the lower arm mounting bolts (A, B)

To install:

➡Be sure to use new fasteners, as required.

8. Installation is the reverse of the removal procedure.
9. Install the lower arm mounting bolts. Torque: Bolt (A): 72.3–86.8 ft. lbs. (98.1–117.7 Nm); Bolt (B): 101.3–115.7 ft. lbs. (137.3–156.9 Nm).
10. Install the front lower arm mounting bolt to the knuckle. Torque: 72.3–86.8 ft. lbs. (98.1–117.7 Nm).
11. Install the wheel and the tire.

MACPHERSON STRUT

REMOVAL & INSTALLATION

See Figures 183 through 186

1. Before servicing the vehicle, refer to the Precautions Section.

➡If working near and/or around the SRS system and components, be sure to disable the SRS system. Tape the negative battery cable with insulating tape. Always disconnect the negative battery cable first.

⁂ CAUTION

To avoid personal injury when working on vehicles equipped with an air bag, the negative battery cable must be disconnected and at least three minutes must elapse before working on the system. Failure to do so may result in deployment of the air bag.

2. Remove the battery top cover, if equipped.
3. Disconnect the negative battery cable. Tape the cable with insulating tape.
4. Raise the vehicle, and make sure it is securely supported.
5. Remove the front wheel and tire.
6. Remove the brake hose bracket bolt (A, B) from the front strut assembly.
7. Remove the speed sensor (A) and wire (B) bolts from the front knuckle.
8. Remove the front stabilizer link (A) nut (B) from the strut assembly.
9. Remove the strut upper mounting nuts.
10. Remove the front strut assembly from the front knuckle.

To install:

11. Install the strut upper mounting nuts. Torque: 32.5–43.4 ft. lbs. (44.1–58.8 Nm).

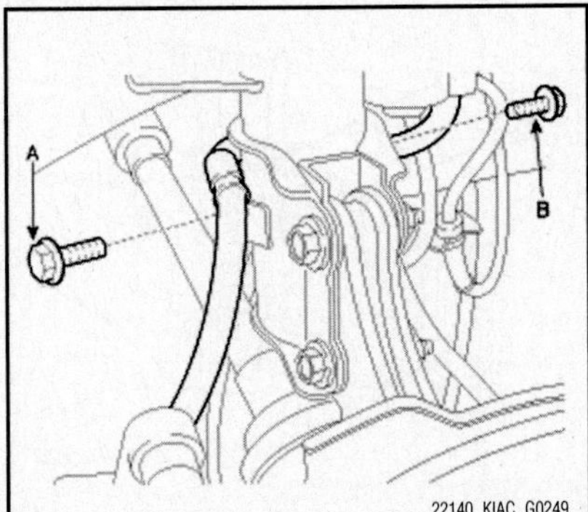

Fig. 183 Remove the brake hose bracket bolt (A, B) from the front strut assembly

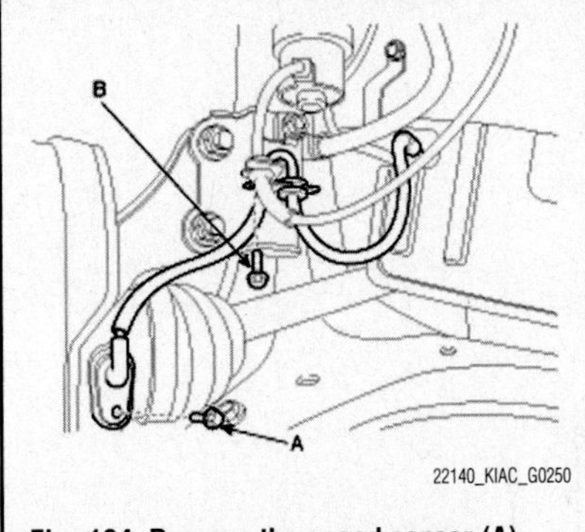

Fig. 184 Remove the speed sensor (A) and wire (B) bolts from the front knuckle

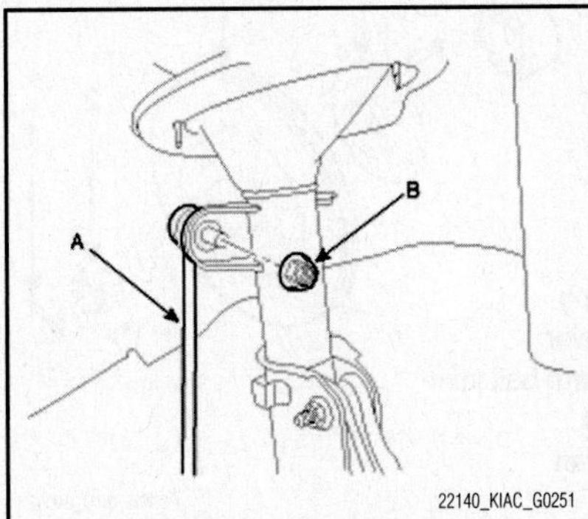

Fig. 185 Remove the front stabilizer link (A) nut (B) from the strut assembly

12. Install the front strut assembly bolts to the front knuckle. Torque: 101.3–115.7 ft. lbs. (137.3–156.9 Nm).
13. Install the front stabilizer link nut to the strut assembly. Torque: 72.3–86.8 ft. lbs. (98.1–117.7 Nm).
14. Install the speed sensor and wire bolts. Torque: 5.1–8.0 ft. lbs. (6.9–10.8 Nm).
15. Install the brake hose bracket bolt to the axle assembly. Torque: 5.1–8.0 ft. lbs. (6.9–10.8 Nm).
16. Install the wheel and the tire.

STEERING KNUCKLE

REMOVAL & INSTALLATION

See Figure 187.

1. Before servicing the vehicle, refer to the Precautions Section.

➡If working near and/or around the SRS system and components, be sure to disable the SRS system. Tape the negative battery cable with insulating tape. Always disconnect the negative battery cable first.

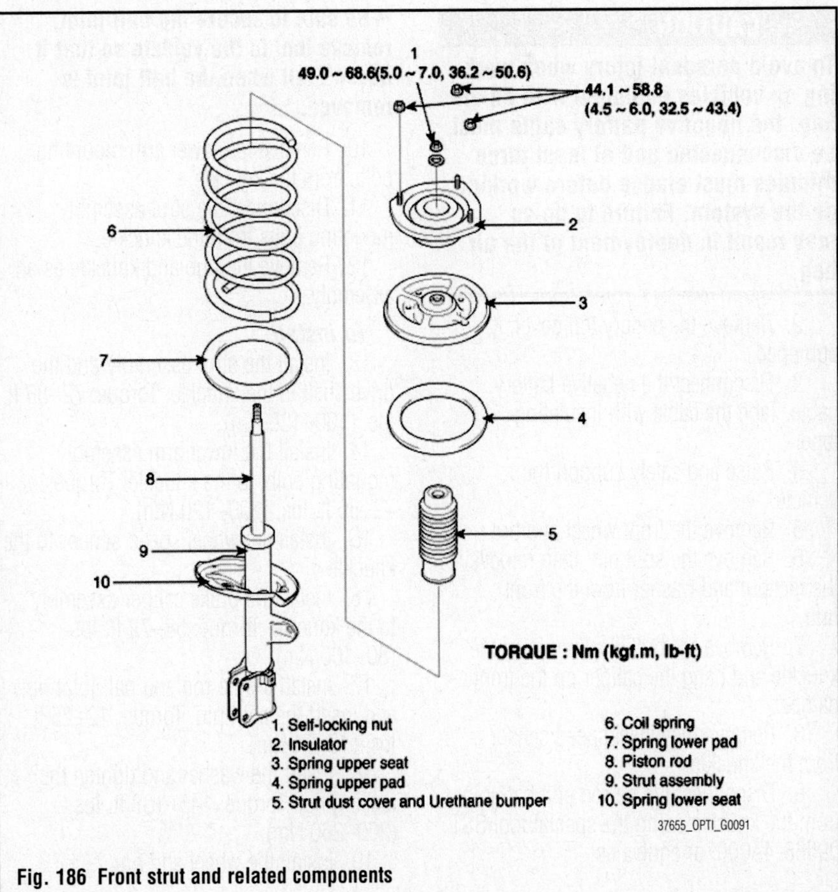

Fig. 186 Front strut and related components

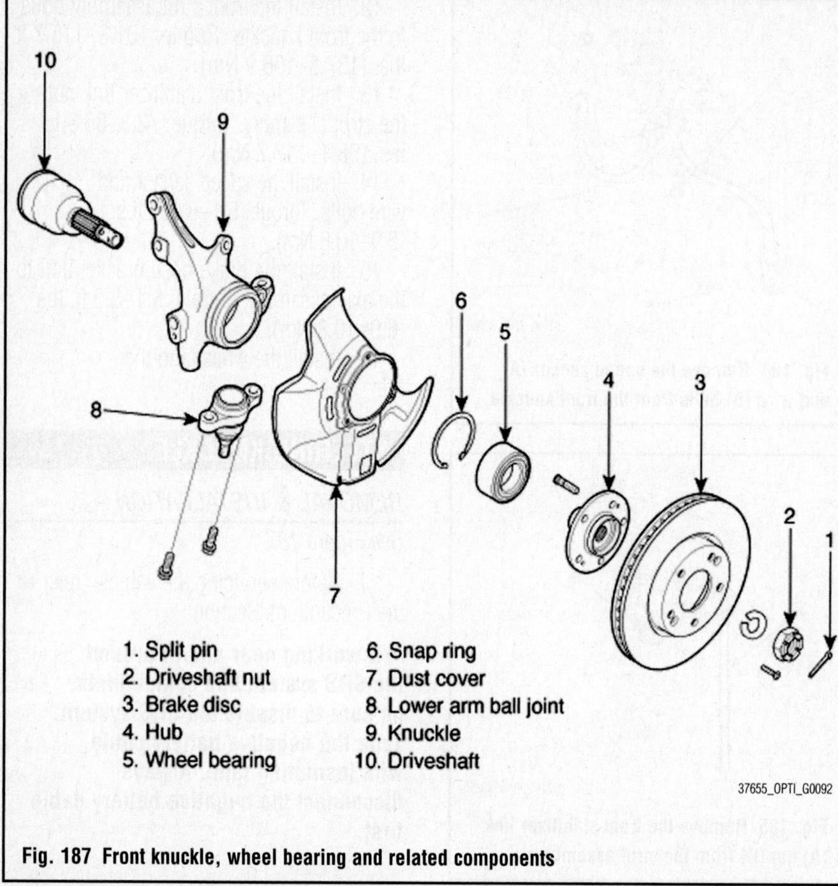

Fig. 187 Front knuckle, wheel bearing and related components

CAUTION

To avoid personal injury when working on vehicles equipped with an air bag, the negative battery cable must be disconnected and at least three minutes must elapse before working on the system. Failure to do so may result in deployment of the air bag.

2. Remove the battery top cover, if equipped.
3. Disconnect the negative battery cable. Tape the cable with insulating tape.
4. Raise and safely support the vehicle.
5. Remove the front wheel and tire.
6. Remove the split pin, then remove the locknut and washer from the front hub.
7. Remove the brake caliper from the knuckle and hang the caliper on the front damper.
8. Remove the wheel speed sensor from the knuckle.
9. Disconnect the tie rod end ball joint from the knuckle using the special tool SST 09568-4A000, or equivalent.

➡**Be sure to secure the ball joint, remove tool to the vehicle so that it doesn't fall when the ball joint is removed.**

10. Remove the lower arm mounting bolts from the knuckle.
11. Disconnect the strut assembly mounting bolts from the knuckle.
12. Remove the hub and knuckle as an assembly.

To install:

13. Install the strut assembly and the drive shaft in the knuckle. Torque: 72–86 ft. lbs. (100–120 Nm).
14. Install the lower arm assembly mounting bolts to the knuckle. Torque: 72–86 ft. lbs. (100–120 Nm).
15. Install the wheel speed sensor to the knuckle.
16. Install the brake caliper assembly to the knuckle. Torque: 58–72 ft. lbs. (80–100 Nm).
17. Install the tie rod end ball joint nut and insert the split pin. Torque: 12–25 ft. lbs. (16–34 Nm).
18. Insert the washer and tighten the locking nut. Torque: 145–188 ft. lbs. (200–260 Nm).
19. Install the wheel and tire.

STABILIZER BAR

REMOVAL & INSTALLATION

See Figures 188 through 192.

1. Before servicing the vehicle, refer to the Precautions Section.

➡**If working near and/or around the SRS system and components, be sure to disable the SRS system. Tape the negative battery cable with insulating tape. Always disconnect the negative battery cable first.**

CAUTION

To avoid personal injury when working on vehicles equipped with an air bag, the negative battery cable must be disconnected and at least three minutes must elapse before working on the system. Failure to do so may result in deployment of the air bag.

2. Remove the battery top cover, if equipped.
3. Disconnect the negative battery cable. Tape the cable with insulating tape.
4. Raise and support the vehicle safely.
5. Remove the tire and wheel assemblies, as required.
6. Remove the connecting bolt (A) between the steering universal joint assembly (B) and the pinion assembly.
7. Raise the vehicle, and make sure it is securely supported.
8. Remove the front wheel and tire.
9. Remove the front stabilizer link from the strut assembly.
10. Remove the brake caliper.
11. Remove the ball joint by using the special tool (09568-4A000).
12. Remove the lower arm (A) mounting bolts (B).

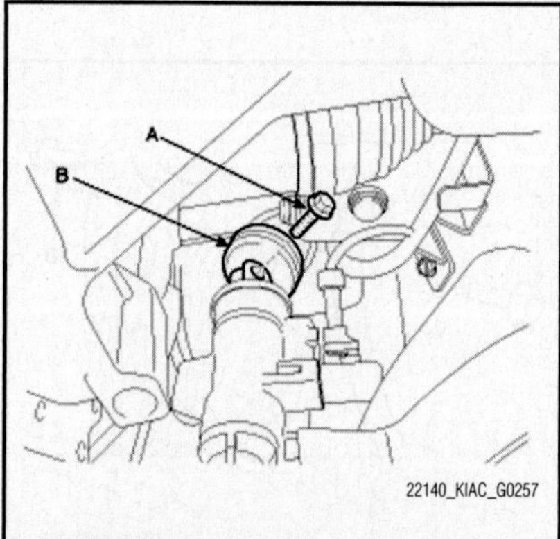

Fig. 188 Remove the connecting bolt (A) between the steering universal joint assembly (B) and the pinion assembly

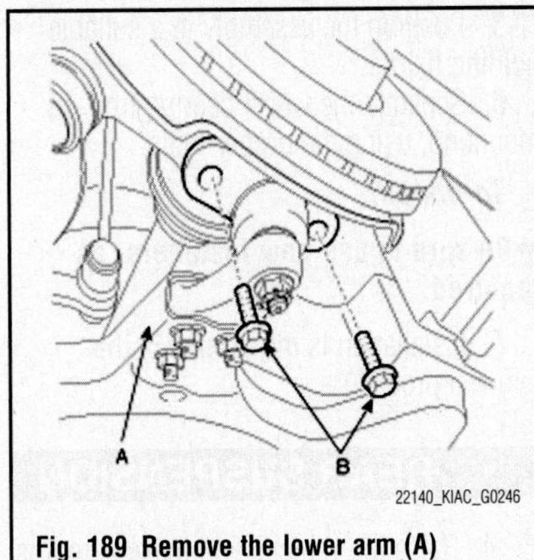

Fig. 189 Remove the lower arm (A) mounting bolt (B)

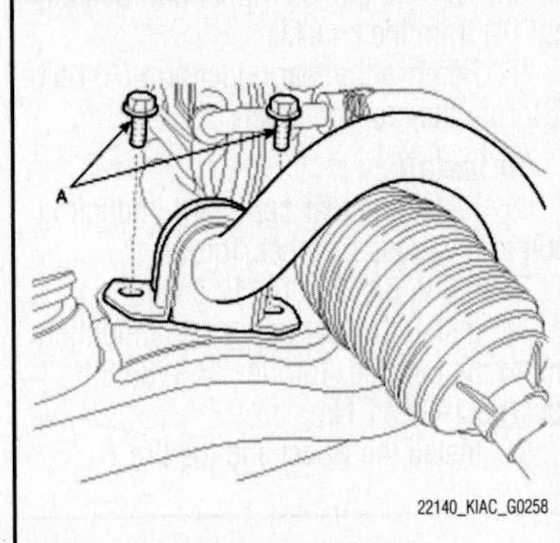

Fig. 190 Remove the stabilizer bar assembly mounting bolts (A)

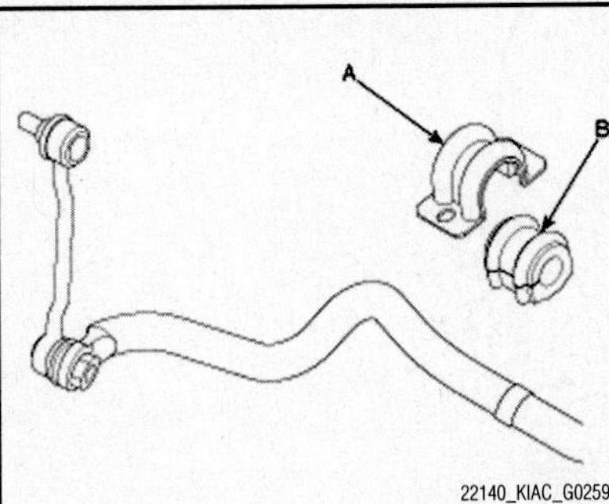

Fig. 191 Remove the brackets (A) and the bushings (B)

13. Remove the engine mounting bolts.
14. Remove the front muffler rubber hanger from the sub-frame.
15. Support the subframe with a jack, and remove the bolts and nuts.
16. After lowering the jack which supports the Sub-frame, remove both sides of the stabilizer bar assembly mounting bolts (A).
17. Remove the stabilizer bar assembly through the gap between the body and the rear side of the sub-frame.

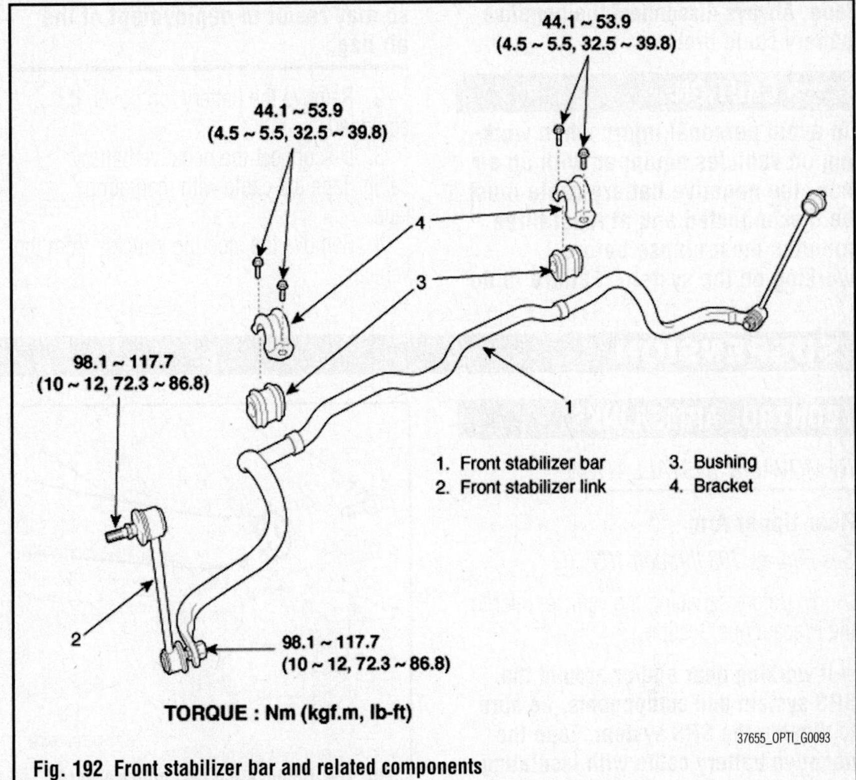

Fig. 192 Front stabilizer bar and related components

CAUTION

Be careful not to damage to the power steering related tubes.

18. Remove the brackets (A) and the bushings (B).

To install:

19. Install the bushing on the stabilizer bar.

➡Bring clamp of stabilizer bar into contact with bushing.

20. Install the bracket on the bushing.
21. After tightening the bolts of the bushing bracket temporarily, install the bushing bracket on the opposite side.
22. Install the stabilizer bar bracket mounting bolts to the subframe. Torque: 32.5–39.8 ft. lbs. (44.1–53.9 Nm).
23. After lifting the jack which supports the Sub-frame, install the four bolts of the Sub-frame and the eight bolts of the guide bracket. Torque: Bolt (A): 115.7–130.2 ft. lbs. (156.9–176.5 Nm); Bolt (B): 32.5–39.8 ft. lbs. (44.1–53.9 Nm).
24. Install the front muffler rubber hanger to the sub-frame.
25. Install the engine mounting bolts. Torque: 36.2–47.0 ft. lbs. (49.0–63.7 Nm).
26. Install both sides of the lower arm mounting bolts. Torque: 72.3–86.8 ft. lbs. (98.1–117.7 Nm).
27. Install both sides of the tie rod ends.
28. Install the brake caliper.
29. Install the front stabilizer link to the strut assembly. Torque: 72.3–86.8 ft. lbs. (98.1–117.7 Nm).
30. Install the wheel and tire.
31. Install the connecting bolt between the steering universal joint assembly and the pinion assembly. Torque: 21.7–25.3 ft. lbs. (29.4–34.3 Nm).

CAUTION

After installation, if necessary, adjust the alignment of the steering wheel and front tires.

WHEEL HUB & BEARING

REMOVAL & INSTALLATION

See Figure 187.

At this time the manufacturer does not provide removal and installation procedures for this component. The following procedure is a guideline and may differ from the vehicle you are servicing.

1. Before servicing the vehicle, refer to the Precautions Section.

➡If working near and/or around the SRS system and components, be sure to disable the SRS system. Tape the negative battery cable with insulating

tape. Always disconnect the negative battery cable first.

⁂ CAUTION

To avoid personal injury when working on vehicles equipped with an air bag, the negative battery cable must be disconnected and at least three minutes must elapse before working on the system. Failure to do so may result in deployment of the air bag.

2. Remove the battery top cover, if equipped.
3. Disconnect the negative battery cable. Tape the cable with insulating tape.
4. Remove the steering knuckle from the vehicle.
5. Position the assembly in a suitable holding fixture.
6. Separate the wheel bearing from its mounting, using the proper tools.

To install:

➡Be sure to use new fasteners, as required.

7. Installation is the reverse of the removal procedure.

CONTROL ARMS/LINKS

REMOVAL & INSTALLATION

Rear Upper Arm

See Figures 193 through 195.

1. Before servicing the vehicle, refer to the Precautions Section.

➡If working near and/or around the SRS system and components, be sure to disable the SRS system. Tape the negative battery cable with insulating tape. Always disconnect the negative battery cable first.

⁂ CAUTION

To avoid personal injury when working on vehicles equipped with an air bag, the negative battery cable must be disconnected and at least three minutes must elapse before working on the system. Failure to do so may result in deployment of the air bag.

2. Remove the battery top cover, if equipped.
3. Disconnect the negative battery cable. Tape the cable with insulating tape.

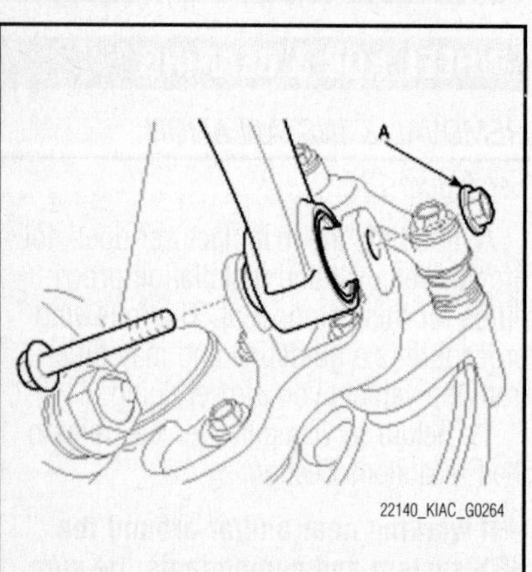

Fig. 193 Remove the rear upper arm bolt and nut (A) from the knuckle.

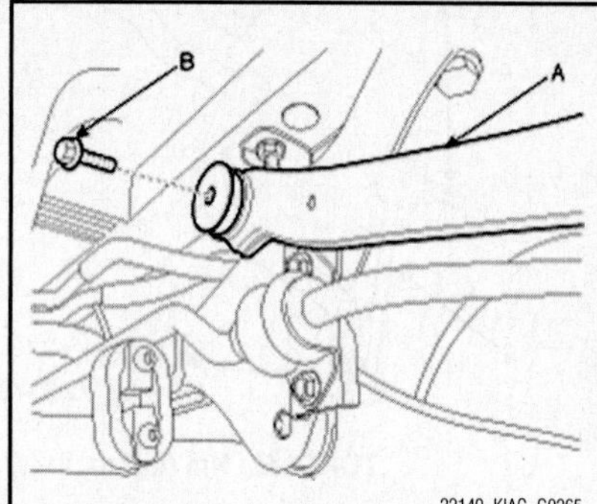

Fig. 194 Remove the rear upper arm (A) bolt (B) from the crossmember

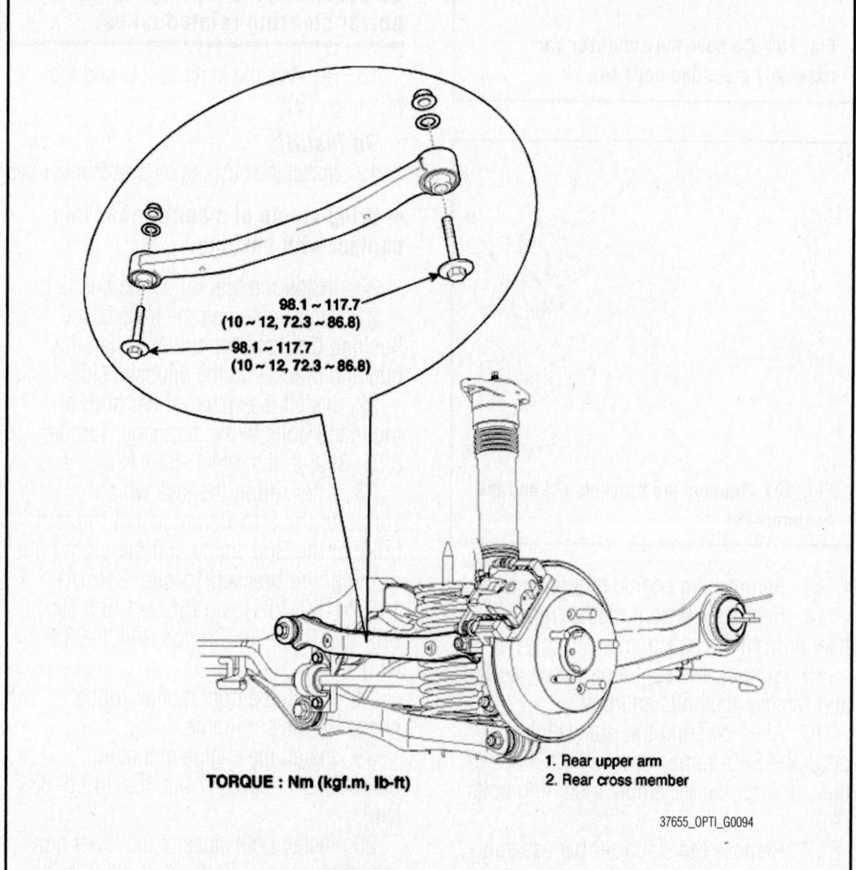

Fig. 195 Rear upper arm and related components

4. Raise the vehicle, and make sure it is securely supported.
5. Remove the rear wheel and tire.
6. Remove the rear upper arm bolt and nut (A) from the knuckle.
7. Remove the rear upper arm (A) bolt (B) from the crossmember

To install:

8. Install the rear upper arm mounting bolt to the cross member. Torque: 72.3–86.8 ft. lbs. (98.1–117.7 Nm).
9. Install the rear upper arm mounting nut to the knuckle. Torque: 72.3–86.8 ft. lbs. (98.1–117.7 Nm).
10. Install the wheel and the tire.

Rear Lower Arm

See Figures 196 through 199.

1. Before servicing the vehicle, refer to the Precautions Section.

➡If working near and/or around the SRS system and components, be sure to disable the SRS system. Tape the negative battery cable with insulating tape. Always disconnect the negative battery cable first.

⁂ CAUTION

To avoid personal injury when working on vehicles equipped with an air bag, the negative battery cable must be disconnected and at least three minutes must elapse before working on the system. Failure to do so may result in deployment of the air bag.

2. Remove the battery top cover, if equipped.
3. Disconnect the negative battery cable. Tape the cable with insulating tape.
4. Raise the vehicle, and make sure it is securely supported.

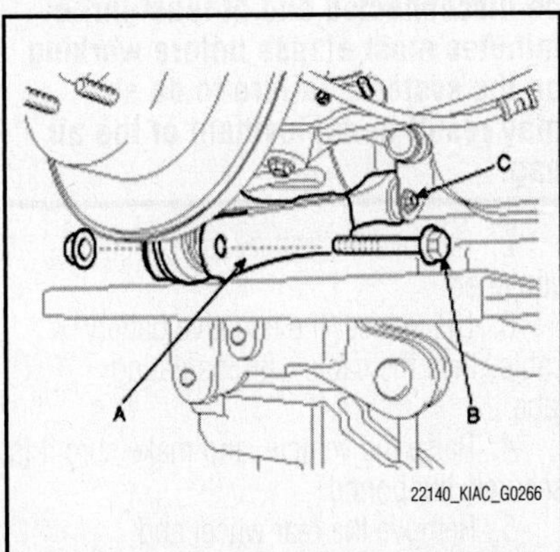

Fig. 196 Remove the lower arm bolt (B)

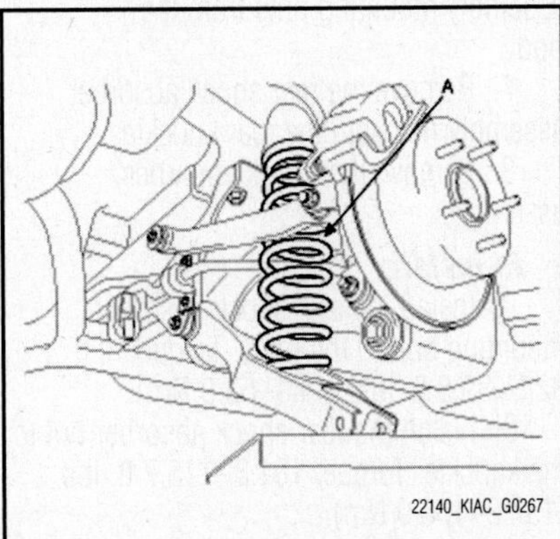

Fig. 197 Remove the coil spring (A), the lower seat, and the upper pad

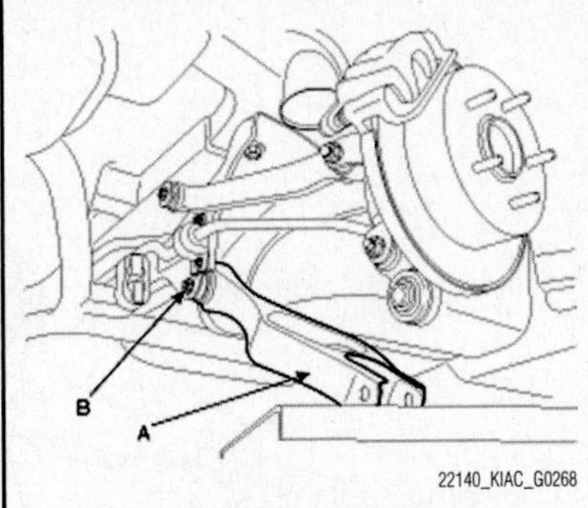

Fig. 198 Remove the lower arm (A) mounting bolts (B) from the cross member

5. Remove the rear wheel and tire.
6. Remove the lower arm bolt (B) from the rear knuckle, while supporting the lower arm (A) with a jack.
7. Loosen the lower arm bolt (C) from the cross member.
8. Remove the coil spring (A), the lower seat, and the upper pad.
9. Remove the lower arm (A) mounting bolts (B) from the cross member.

To install:

10. Pre-tighten the lower arm mounting bolts to the cross member.
11. Install the coil spring, the lower seat, and the upper pad.
12. Install the lower arm bolt to the rear knuckle and the lower arm bolt to the cross member, while supporting the lower arm with a jack. Torque: 101.3–115.7 ft. lbs. 137.3–156.9).
13. Install the wheel and the tire.

Rear Assist Arm

See Figures 200 through 203.

1. Before servicing the vehicle, refer to the Precautions Section.

➡If working near and/or around the SRS system and components, be sure to disable the SRS system. Tape the negative battery cable with insulating tape. Always disconnect the negative battery cable first.

⁂ CAUTION

To avoid personal injury when working on vehicles equipped with an air bag, the negative battery cable must be disconnected and at least three minutes must elapse before working on the system. Failure to do so may result in deployment of the air bag.

2. Remove the battery top cover, if equipped.
3. Disconnect the negative battery cable. Tape the cable with insulating tape.

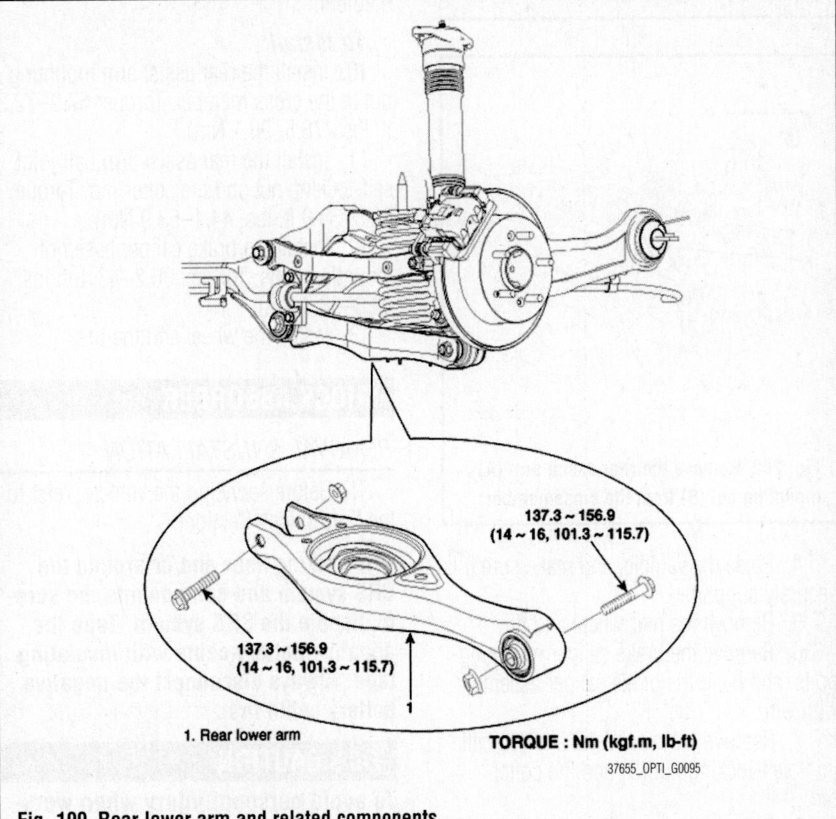

Fig. 199 Rear lower arm and related components

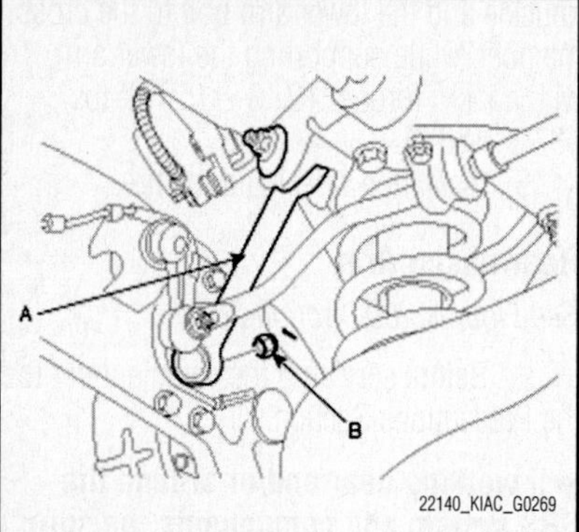

Fig. 200 Remove the rear assist arm (A) ball joint self-locking nut (B) and the cotter pin

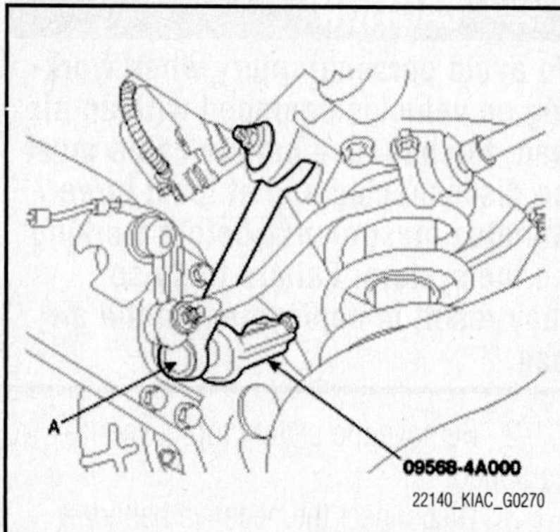

Fig. 201 Remove the rear assist arm ball joint (A) by using the special tool

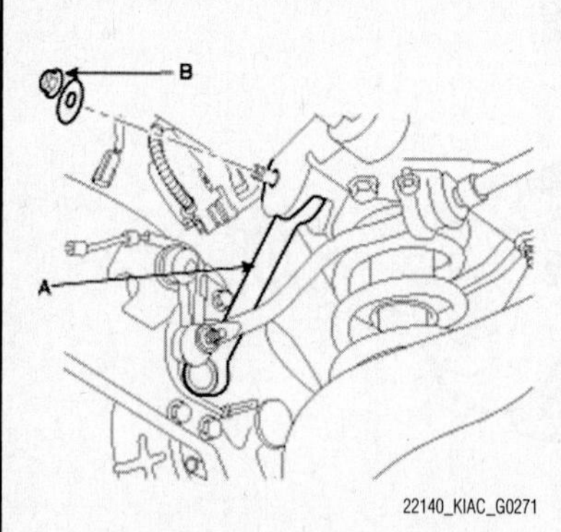

Fig. 202 Remove the rear assist arm (A) mounting nut (B) from the crossmember

4. Raise the vehicle, and make sure it is securely supported.
5. Remove the rear wheel and tire.
6. Remove the brake caliper mounting bolts, and hang the brake caliper assembly with wire.
7. Remove the rear assist arm (A) ball joint self-locking nut (B) and the cotter pin.
8. Remove the rear assist arm ball joint (A) by using the special tool (09568-4A000).
9. Remove the rear assist arm (A) mounting nut (B) from the cross member.

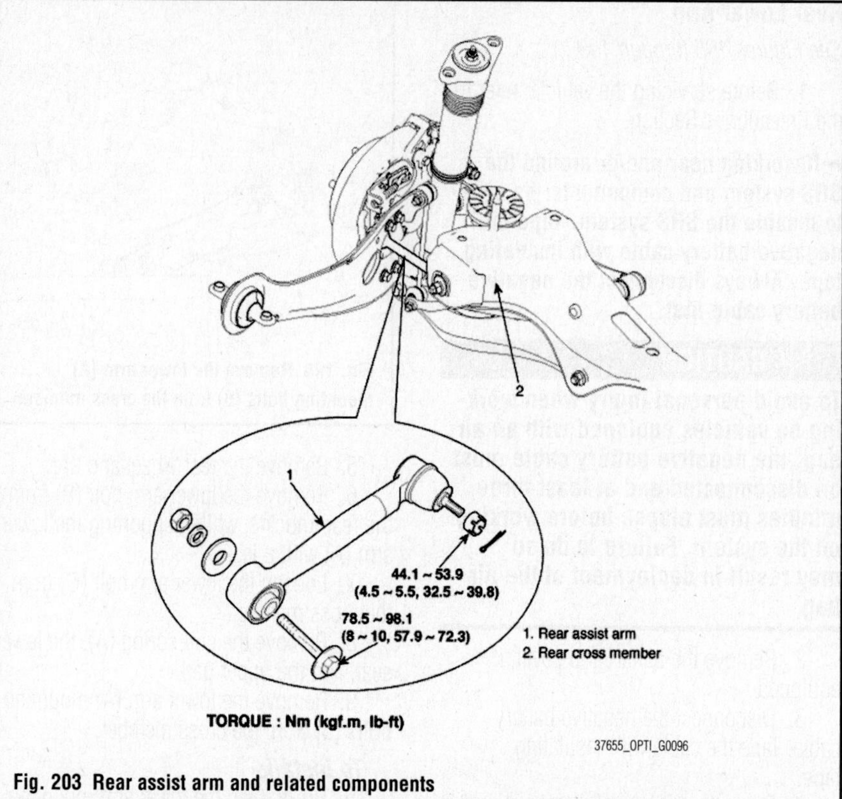

Fig. 203 Rear assist arm and related components

To install:

10. Install the rear assist arm mounting nut to the cross member. Torque: 57.9–72.3 ft. lbs. (78.5–98.1 Nm).
11. Install the rear assist arm ball joint self-locking nut and the cotter pin. Torque: 32.5–39.8 ft. lbs. 44.1–53.9 Nm).
12. Install the brake caliper assembly mounting bolts. Torque: 36.2–43.4 ft. lbs. 49.0–58.8 Nm).
13. Install the wheel and the tire.

SHOCK ABSORBER

REMOVAL & INSTALLATION

1. Before servicing the vehicle, refer to the Precautions Section.

➡If working near and/or around the SRS system and components, be sure to disable the SRS system. Tape the negative battery cable with insulating tape. Always disconnect the negative battery cable first.

⁂ CAUTION

To avoid personal injury when working on vehicles equipped with an air bag, the negative battery cable must be disconnected and at least three minutes must elapse before working on the system. Failure to do so may result in deployment of the air bag.

2. Remove the battery top cover, if equipped.
3. Disconnect the negative battery cable. Tape the cable with insulating tape.
4. Raise the vehicle, and make sure it is securely supported.
5. Remove the rear wheel and tire.
6. Remove the rear shock absorber assembly mounting nuts from the body.
7. Remove the rear shock absorber assembly nut from the rear knuckle.
8. Remove the shock absorber assembly.

To install:

9. Install the rear shock absorber mounting bolt to the body. Torque: 32.5–39.8 ft. lbs. (44.1–53.9 Nm).
10. Install the rear shock absorber nut to the knuckle. Torque: 101.3–115.7 ft. lbs. (137.3–156.9 Nm).
11. Install the wheel and the tire.

STABILIZER BAR

REMOVAL & INSTALLATION

See Figure 204.

1. Before servicing the vehicle, refer to the Precautions Section.

➡**If working near and/or around the SRS system and components, be sure to disable the SRS system. Tape the negative battery cable with insulating tape. Always disconnect the negative battery cable first.**

⁂ CAUTION

To avoid personal injury when working on vehicles equipped with an air bag, the negative battery cable must be disconnected and at least three minutes must elapse before working on the system. Failure to do so may result in deployment of the air bag.

2. Remove the battery top cover, if equipped.
3. Disconnect the negative battery cable. Tape the cable with insulating tape.
4. Raise and safely support the vehicle.
5. Remove the tire and wheel assembly.
6. Remove the left and right nuts of the stabilizer links from the trailing arms.
7. Remove the left and right mounting bolts of the stabilizer bar brackets.
8. Remove the rear stabilizer link nut from the stabilizer bar assembly.

To install:

➡**Be sure to use new fasteners, as required.**

9. Installation is the reverse of the removal procedure.

TRAILING ARM

REMOVAL & INSTALLATION

See Figure 205.

1. Before servicing the vehicle, refer to the Precautions Section.

➡**If working near and/or around the SRS system and components, be sure to disable the SRS system. Tape the negative battery cable with insulating**

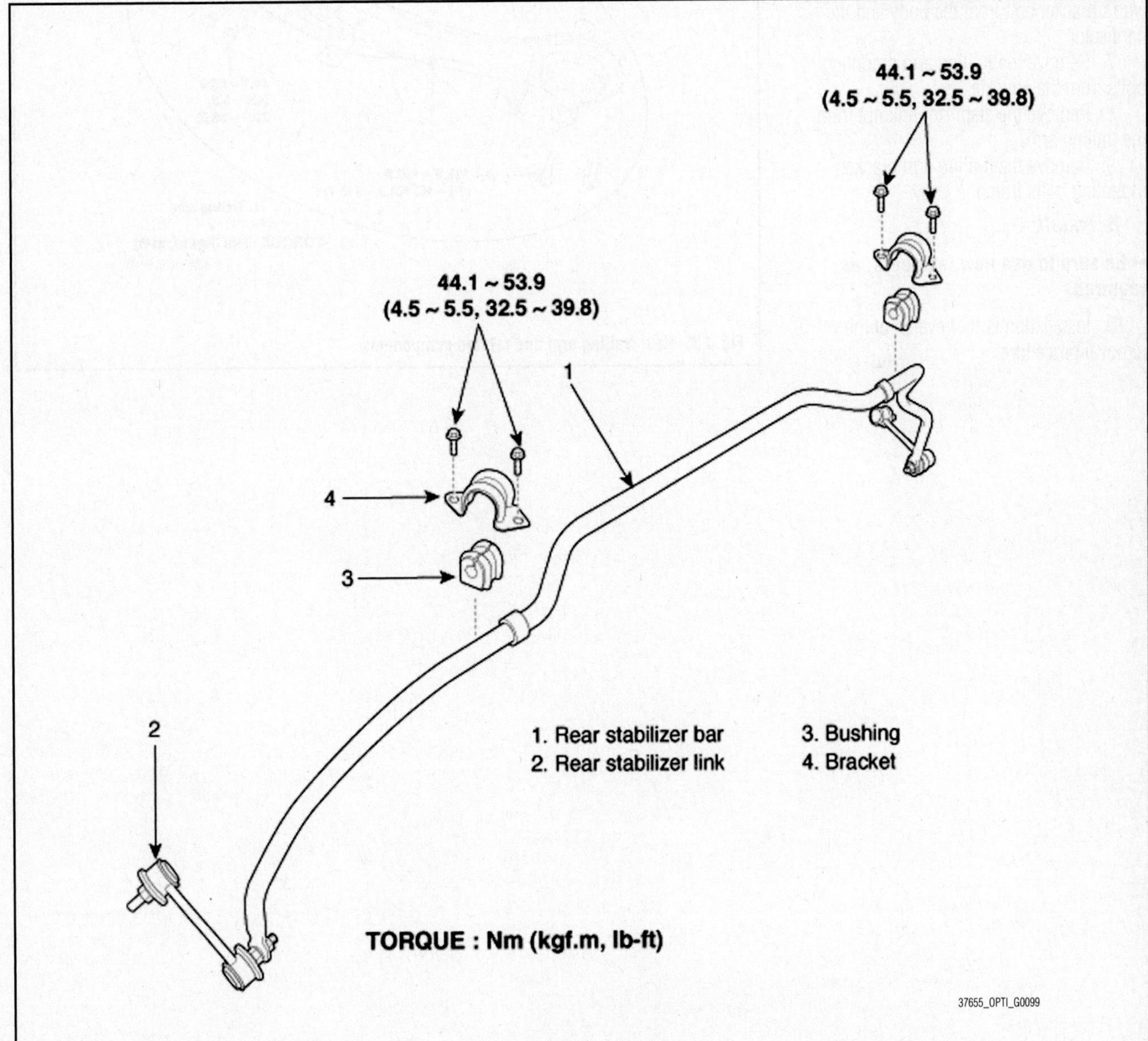

Fig. 204 Rear stabilizer bar and related components

tape. Always disconnect the negative battery cable first.

⁂ CAUTION

To avoid personal injury when working on vehicles equipped with an air bag, the negative battery cable must be disconnected and at least three minutes must elapse before working on the system. Failure to do so may result in deployment of the air bag.

2. Remove the battery top cover, if equipped.
3. Disconnect the negative battery cable. Tape the cable with insulating tape.
4. Raise and safely support the vehicle.
5. Remove the tire and wheel assembly.
6. Remove the wheel speed sensor wire's bracket bolt from the body and the connector.
7. Remove the trailing arm mounting bolts from the knuckle.
8. Remove the stabilizer link nut from the trailing arm.
9. Remove the trailing arm bracket mounting bolts from the body.

To install:

➡Be sure to use new fasteners, as required.

10. Installation is the reverse of the removal procedure.

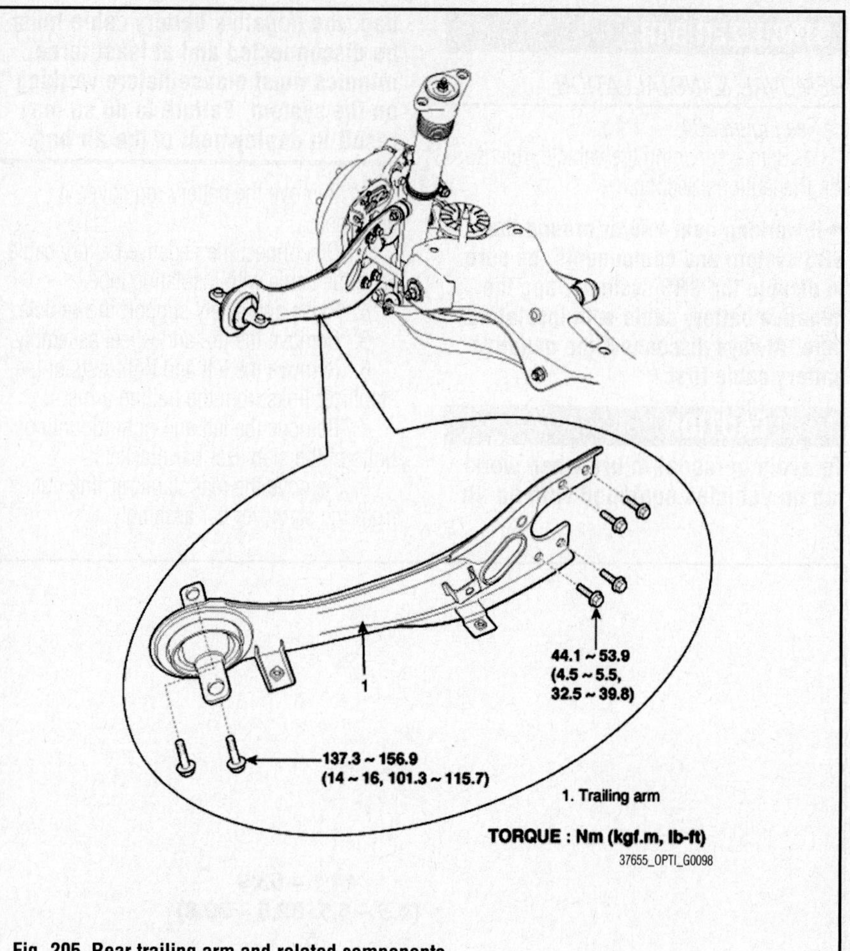

Fig. 205 Rear trailing arm and related components

KIA

Rio • Rio5

15

BRAKES 15-9

CHASSIS ELECTRICAL 15-15

DRIVE TRAIN 15-15

ENGINE COOLING 15-19

ENGINE ELECTRICAL.......15-19

ENGINE MECHANICAL......15-21

ENGINE PERFORMANCE & EMISSION CONTROLS 15-28

SPECIFICATIONS AND MAINTENANCE CHARTS

ENGINE AND VEHICLE IDENTIFICATION

Engine						
Code ①	Liters (cc)	Cu. In.	Cyl.	Fuel Sys.	Engine Type	Eng. Mfg.
3	1.6 (1599)	97.6	4	EGI	DOHC	KIA

Model Year	
Code ②	Year
9	2009
A	2010

EGI: Electronic Gasoline Injection

DOHC: Double Overhead Camshafts

① 8th digit of VIN

② 10th digit of VIN

37655_KRIO_C0001

GENERAL ENGINE SPECIFICATIONS

Year	Model	Engine Displacement Liters	Engine VIN	Net Horsepower @ rpm	Net Torque @ rpm (ft. lbs.)	Bore x Stroke (in.)	Com-pression Ratio	Oil Pressure @ rpm
2009	Rio	1.6	3	110@6000	107@4500	3.01x3.43	10.0:1	15.6@710-730
2010	Rio	1.6	3	110@6000	107@4500	3.01x3.43	10.0:1	15.6@710-730

37655_KRIO_C0002

ENGINE TUNE-UP SPECIFICATIONS

Year	Engine Displacement Liters	Spark Plug Gap (in.)	Ignition Timing (deg.)		Fuel Pump (psi)	Idle Speed (rpm)		Valve Clearance	
			MT	AT		MT	AT	Intake	Exhaust
2009	1.6	0.039-0.043	1-10B	1-10B	49.8	710-730	710-730	HYD	HYD
2010	1.6	0.039-0.043	1-10B	1-10B	49.8	710-730	710-730	HYD	HYD

NOTE: The Vehicle Emission Control Information label often reflects specification changes made during production.

The label figures must be used if they differ from those in this chart.

B: Before top dead center

HYD: Hydraulic

37655_KRIO_C0003

CAPACITIES

Year	Model	Engine Displacement Liters	Engine VIN	Engine Oil with Filter	Transaxle (pts.) Manual	Transaxle (pts.) Auto.	Fuel Tank (gal.)	Cooling System (qts.)
2009	Rio	1.6	3	3.49	4.2	12.9	11.9	6.1
2010	Rio	1.6	3	3.49	4.2	12.9	11.9	6.1

37655_KRIO_C0004

FLUID SPECIFICATIONS

Year	Model	Engine Displ. Liters (VIN)	Engine Oil	Man. Trans.	Auto. Trans.	Drive Axle Front	Drive Axle Rear	Transfer Case	Power Steering Fluid	Brake Master Cylinder	Cooling System
2009	Rio	1.6 (3)	①	②	③	NA	NA	NA	PSF-3	DOT 3	④
2010	Rio	1.6 (3)	①	②	③	NA	NA	NA	PSF-3	DOT 3	④

NA: Not Available

DOT: Department Of Transpotation

① SL (SJ) or above, 5W-20

② SAE 75/85 API GL-4 TGO-7 (MS517-14)

③ Diamond ATF SP-III or SK ATF SP-III

④ Ethylene glycol base for aluminum radiator

37655_KRIO_C0014

VALVE SPECIFICATIONS

Year	Engine Displacement Liters	Engine VIN	Seat Angle (deg.)	Face Angle (deg.)	Maximum out of Square (in.)	Spring Free Length (in.)	Stem-to-Guide Clearance (in.) Intake	Stem-to-Guide Clearance (in.) Exhaust	Stem Diameter (in.) Intake	Stem Diameter (in.) Exhaust
2009	1.6	3	45-45.5	45-45.5	①	1.7323	0.0008-0.0020	0.0014-0.0026	0.2348-0.2354	0.2343-0.2348
2010	1.6	3	45-45.5	45-45.5	①	1.7323	0.0008-0.0020	0.0014-0.0026	0.2348-0.2354	0.2343-0.2348

NA: Not Available

① Less than 1.5 degrees

② Standard range: 0.0010-0.0023 in.

Maximum value: 0.0080 in.

37655_KRIO_C0005

CAMSHAFT AND BEARING SPECIFICATIONS

All measurements are given in inches.

Year	Engine Displacement Liters	Engine VIN	Journal Diameter	Brg. Oil Clearance	Shaft End-play	Journal Runout	Journal Bore	Lobe Lift Intake	Lobe Lift Exhaust
2009	1.6	3	①	②	③	NA	NA	④	⑤
2010	1.6	3	①	②	③	NA	NA	④	⑤

NA: Not Available

① No. 1: 1.0616-1.0622 inches

No. 2, 3, 4: 0.9430-0.9437 inches

② No. 1: 0.0008-0.0024 inches

No. 2, 3, 4: 0.0012-0.0026 inches

③ 0.0039-0.0079 inches

④ 1.72241-1.73028 inches

⑤ 1.73816-1.74604 inches

37655_KRIO_C0013

CRANKSHAFT AND CONNECTING ROD SPECIFICATIONS

All measurements are given in inches.

Year	Engine Displacement Liters	Engine VIN	Crankshaft Main Brg. Journal Dia.	Crankshaft Main Brg. Oil Clearance	Crankshaft Shaft End-play	Crankshaft Thrust on No.	Connecting Rod Journal Diameter	Connecting Rod Oil Clearance	Connecting Rod Side Clearance
2009	1.6	3	1.9665-1.9672	①	0.0020-0.0069	3	1.8898-1.8905	0.0007-0.0014	0.0039-0.0098
2010	1.6	3	1.9665-1.9672	①	0.0020-0.0069	3	1.8898-1.8905	0.0007-0.0014	0.0039-0.0098

① Journal Nos. 1, 2, 4 & 5: 0.0009-0.0016 in.

Journal Nos. 3: 0.0011-0.0018 in.

37655_KRIO_C0007

PISTON AND RING SPECIFICATIONS

All measurements are given in inches.

Year	Engine Displacement Liters	Engine VIN	Piston Clearance	Ring Gap Top Compression	Ring Gap Bottom Compression	Ring Gap Oil Control	Ring Side Clearance Top Compression	Ring Side Clearance Bottom Compression	Ring Side Clearance Oil Control
2009	1.6	3	0.0008-0.0016	0.0059-0.0118	0.0138-0.0197	0.0079-0.0276	0.0016-0.0033	0.0016-0.0033	0.0031-0.0069
2010	1.6	3	0.0008-0.0016	0.0059-0.0118	0.0138-0.0197	0.0079-0.0276	0.0016-0.0033	0.0016-0.0033	0.0031-0.0069

37655_KRIO_C0006

TORQUE SPECIFICATIONS

All readings in ft. lbs.

Year	Engine Displacement Liters	Cylinder Head Bolts	Main Bearing Bolts	Rod Bearing Bolts	Crankshaft Damper Bolts	Flywheel Bolts	Manifold Intake	Manifold Exhaust	Spark Plugs	Oil Pan Drain Plug
2009	1.6	①	40-43	23-25	101-109	87-94	11-15	22-25	18-22	29-33
2010	1.6	①	40-43	23-25	101-109	87-94	11-15	22-29	18-22	29-33

① Step 1: 22 ft. lbs.
Step 2: plus 90 degrees
Step 3: Loosen fully
Step 4: 22 ft. lbs.
Step 5: Tighten 90 degrees

37655_KRIO_C0008

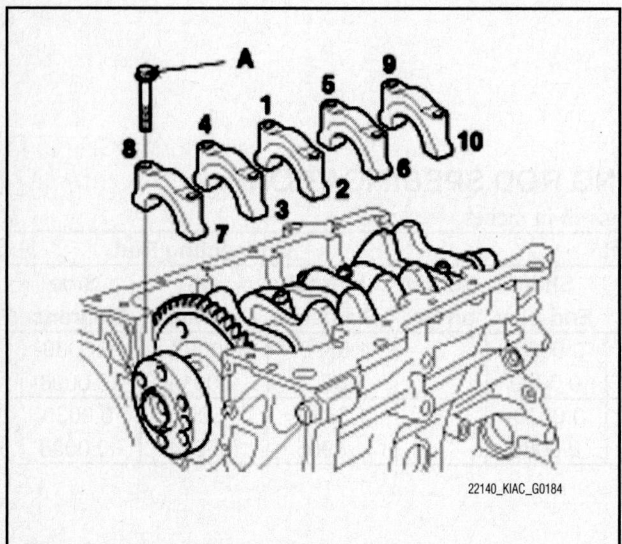

Fig. 1 Main bearing torque sequence—1.6L engine

WHEEL ALIGNMENT

Year	Model		Caster Range (+/-Deg.)	Caster Preferred Setting (Deg.)	Camber Range (+/-Deg.)	Camber Preferred Setting (Deg.)	Toe-in (in.)
2009	Rio	F	0.50	0	4.00	+0.5	0.08 +/- 0.08
		R	—	—	1.00	-0.50	0.08 +/- 0.24
2010	Rio	F	0.50	0	4.00	+0.5	0.08 +/- 0.08
		R	—	—	1.00	-0.50	0.08 +/- 0.24

37655_KRIO_C0009

TIRE, WHEEL AND BALL JOINT SPECIFICATIONS

Year	Model	OEM Tires		Tire Pressures (psi)		Wheel	Ball Joint	Lug Nut Torque
		Standard	Optional	Front	Rear	Size	Inspection	(ft. lbs.)
2009	Rio	185/65R14	195/55R15	30	30	Std: 5.0-J Opt: 5.5-J	①	65-79
2010	Rio	185/65R14	195/55R14 175/70R14	30	30	Std: 5.0-J Opt: 5.5-J	①	65-79

OEM: Original Equipment Manufacturer

PSI: Pounds Per Square Inch

STD: Standard

OPT: Optional

① Replace if any measureable movent is found.

37655_KRIO_C0010

BRAKE SPECIFICATIONS

All measurements in inches unless noted

Year	Model		Brake Disc			Brake Drum		Minimum	Brake Caliper	
			Original Thickness	Minimum Thickness	Maximum Run-out	Original Inside Diameter	Maximum Machine Diameter	Lining Thickness	Bracket Bolts (ft. lbs.)	Mounting Bolts (ft. lbs.)
2009	Rio	F	0.870	0.790	0.0012	—	—	0.079	58-72	16-23
		R	0.390	0.315	0.0012	7.87	7.91	0.079	62-69	16-23
2010	Rio	F	0.870	0.790	0.0012	—	—	0.079	62-69	16-23
		R	0.390	0.315	0.0012	7.87	7.91	0.079	62-69	16-23

F: Front

R: Rear

37655_KRIO_C0011

SCHEDULED MAINTENANCE INTERVALS

Kia - Rio

TO BE SERVICED	TYPE OF SERVICE	VEHICLE MILEAGE INTERVAL (x1000)												
		7.5	15	22.5	30	37.5	45	52.5	60	67.5	75	82.5	90	97.5
Accessory drive belts	S/I	✓	✓	✓	✓	✓	✓	✓	✓	✓	✓	✓	✓	✓
Air cleaner filter	R			✓			✓			✓			✓	
Air conditioner system	S/I													
Brake lines, hoses and connections	S/I		✓		✓		✓		✓		✓		✓	
Chassis and body fasteners	T												✓	
Cooling system hoses and coolant level	S/I		✓		✓		✓		✓		✓		✓	
CV-joint boots	S/I				✓				✓				✓	
Engine coolant	R				✓				✓				✓	
Engine oil and filter	R	✓	✓	✓	✓	✓	✓	✓	✓	✓	✓	✓	✓	✓
Exhaust system heat shields	S/I				✓				✓				✓	
Front and rear brakes	S/I	✓	✓	✓	✓	✓	✓	✓	✓	✓	✓	✓	✓	✓
Front ball joints	S/I				✓				✓				✓	
Fuel filter	R					✓					✓			
Fuel lines and hoses	S/I	✓	✓	✓	✓	✓	✓	✓	✓	✓	✓	✓	✓	✓
Locks and hinges	L	✓	✓	✓	✓	✓	✓	✓	✓	✓	✓	✓	✓	✓
Spark plugs	R								✓					
Steering operation and linkage	S/I	✓	✓	✓	✓	✓	✓	✓	✓	✓	✓	✓	✓	✓
Timing belt	R								✓					

R: Replace S/I: Service or Inspect L: Lubricate T: Tighten

FREQUENT OPERATION MAINTENANCE (SEVERE SERVICE)

If a vehicle is operated under any of the following conditions it is considered severe service

- Towing a trailer or using a camper or car-top carrier

- Repeated short trips of less than 5 miles in temperatures below freezing, or trips of less than 10 miles in any temperature

- Prolonged idling (vehicle operation in stop and go traffic).

- Operating on rough, muddy, unpaved, dusty or salt-covered roads.

- Police, taxi, delivery usage or trailer towing usage.

- Driving in extremely hot (over 90°F) conditions

Oil & oil filter: change every 5000 miles or 5 months, whichever occurs first.

Air cleaner filter: inspect every 15,000 miles or 15 months and replace everything 30,000 miles or 30 months, whichever occurs fir:

Fuel system hoses (California models only): replace every 105,000 miles

Emission system hoses (non-CA models): inspect every 55,000 or 55 months, whichever occurs first

Emission system hoses (CA models): inspect every 60,000 miles or 60 months, which occurs first

Front and rear brakes: inspect every 15,000 miles or 15 months, whichever occurs first

Chassis and body fasteners: tighten every 15,000 miles or 15 months, whichever occurs first

Locks and hinges: lubricate every 5000 miles or 5 months, whichever occurs first

37655_KRIO_C0012

BRAKES — INFORMATION AND PRECAUTIONS

ANTI-LOCK SYSTEMS

• Certain components within the ABS system are not intended to be serviced or repaired individually.

• Do not use rubber hoses or other parts not specifically specified for and ABS system. When using repair kits, replace all parts included in the kit. Partial or incorrect repair may lead to functional problems and require the replacement of components.

• Lubricate rubber parts with clean, fresh brake fluid to ease assembly. Do not use shop air to clean parts; damage to rubber components may result.

• Use only DOT 3 brake fluid from an unopened container.

• If any hydraulic component or line is removed or replaced, it may be necessary to bleed the entire system.

• A clean repair area is essential. Always clean the reservoir and cap thoroughly before removing the cap. The slightest amount of dirt in the fluid may plug an orifice and impair the system function. Perform repairs after components have been thoroughly cleaned; use only denatured alcohol to clean components. Do not allow ABS components to come into contact with any substance containing mineral oil; this includes used shop rags.

• The Anti-Lock control unit is a microprocessor similar to other computer units in the vehicle. Ensure that the ignition switch is **OFF** before removing or installing controller harnesses. Avoid static electricity discharge at or near the controller.

• If any arc welding is to be done on the vehicle, the control unit should be unplugged before welding operations begin.

DISC AND DRUM SYSTEMS

CAUTION

Dust and dirt accumulating on brake parts during normal use may contain asbestos fibers from production or aftermarket brake linings. Breathing excessive concentrations of asbestos fibers can cause serious bodily harm. Exercise care when servicing brake parts. Do not sand or grind brake lining unless equipment used is designed to contain the dust residue. Do not clean brake parts with compressed air or by dry brushing. Cleaning should be done by dampening the brake components with a fine mist of water, then wiping the brake components clean with a dampened cloth. Dispose of cloth and all residue containing asbestos fibers in an impermeable container with the appropriate label. Follow practices prescribed by the Occupational Safety and Health Administration (OSHA) and the Environmental Protection Agency (EPA) for the handling, processing, and disposing of dust or debris that may contain asbestos fibers.

BRAKES — BLEEDING THE BRAKE SYSTEM

BLEEDING PROCEDURE

BLEEDING PROCEDURE

WARNING

When bleeding the brakes, note the following:

- Do not reuse the drained fluid
- Always use Genuine DOT 3 or DOT 4 Brake Fluid. Using a non-Genuine DOT3 or DOT 4 brake fluid can cause corrosion and decrease the life of the system
- Make sure no dirt of other foreign matter is allowed to contaminate the brake fluid
- Do not spill brake fluid on the vehicle, it may damage the paint; if brake fluid does contact the paint, wash it off immediately with water
- The reservoir on the master cylinder must be at the MAX (upper) level mark at the start of bleeding procedure and checked after bleeding each brake caliper. Add fluid as required

1. Make sure the brake fluid level in the master cylinder fluid reservoir is at the MAX (upper) level line.
2. Have someone slowly pump the brake pedal several times, and then apply steady pressure
3. Loosen the right-rear brake bleed screw to allow air to escape from the system. Then tighten the bleed screw securely
4. Repeat the procedure for each wheel in the sequence shown below until air bubbles no longer appear in the fluid
5. Refill the master cylinder reservoir to the MAX (upper) level line

BLEEDING THE ABS SYSTEM

The ABS brake system is bled in the usual fashion with no special procedures required. Refer to the bleeding procedure described earlier. Make certain the master cylinder reservoir is filled before the bleeding is begun and check the level frequently.

WHEEL SPEED SENSORS

REMOVAL & INSTALLATION

Front Wheel

See Figures 2 through 5.

1. Remove the front wheel speed sensor mounting bolt.
2. Remove the front wheel speed sensor bracket.
3. Remove the front wheel guard .
4. Disconnect the front wheel speed sensor connector.
5. Remove the front wheel speed sensor.

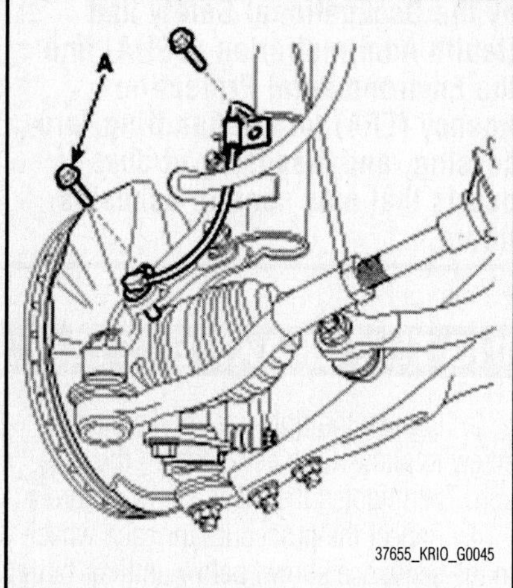

Fig. 2 Front wheel speed sensor mounting bolts

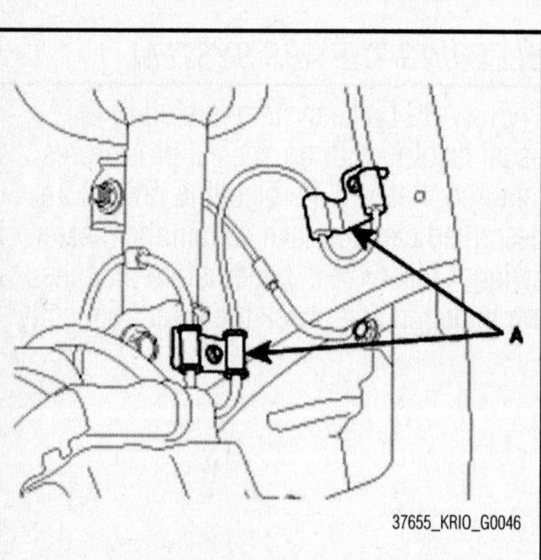

Fig. 3 Front wheel speed sensor bracket

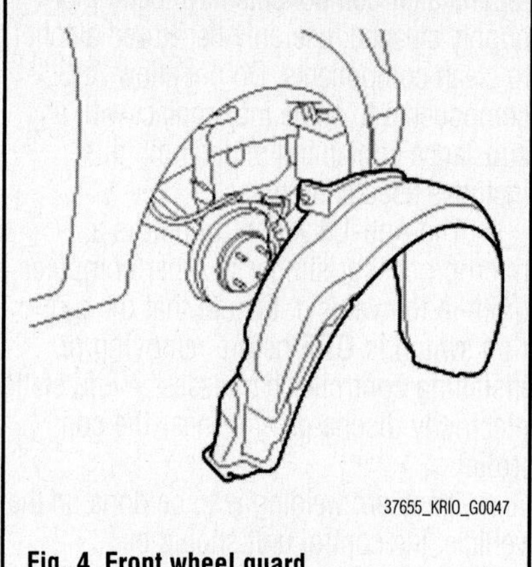

Fig. 4 Front wheel guard

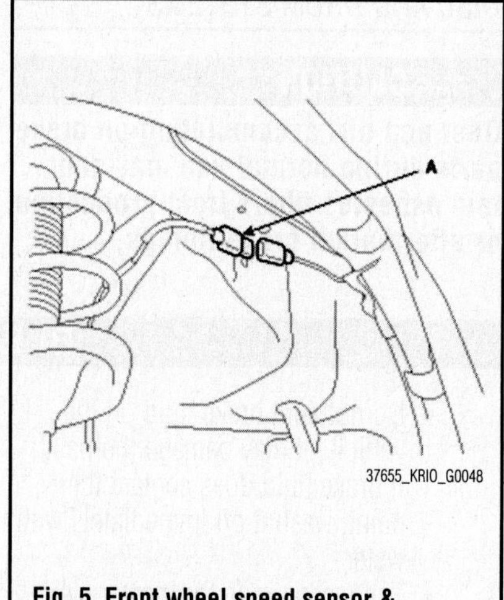

Fig. 5 Front wheel speed sensor & connector

To install:

6. Install the front wheel speed sensor.
7. Connect the front wheel speed sensor connector.
8. Install the front wheel guard.
9. Install the front wheel speed sensor bracket.
10. Install the front wheel speed sensor mounting bolt.

Rear Wheel

See Figures 6 and 7.

1. Remove the rear wheel speed sensor connector.

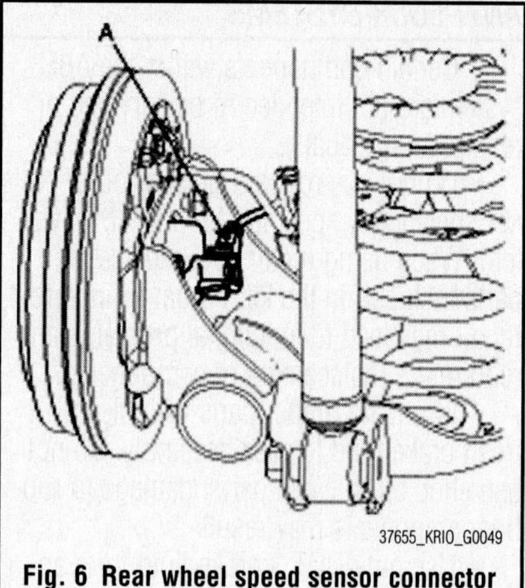

Fig. 6 Rear wheel speed sensor connector

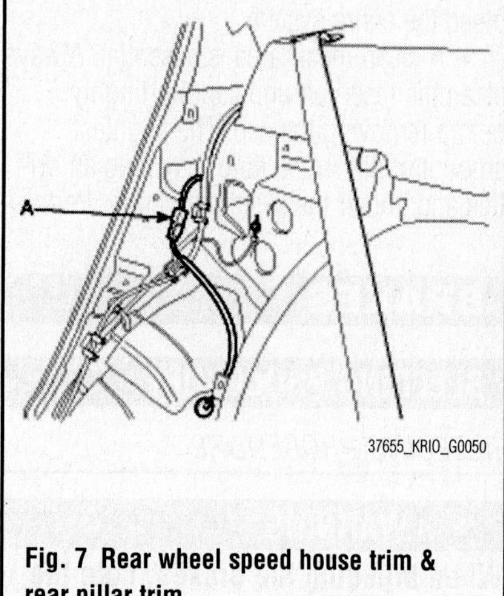

Fig. 7 Rear wheel speed house trim & rear pillar trim

2. Remove the rear seat assembly.
3. Remove the rear wheel speed house trim and rear pillar trim.
4. Disconnect the rear wheel speed sensor connector.

To install:

5. Connect the rear wheel speed sensor connector.
6. Install the rear wheel speed house trim and rear pillar trim.
7. Install the rear seat assembly.
8. Install the rear wheel speed sensor connector.

BRAKE CALIPER

REMOVAL & INSTALLATION

See Figures 8 and 9.

1. Remove guide rod and the caliper up out of the way.
2. Check the hoses and pin boots for damage and deterioration.
3. Remove the pad shims, pad retainers and pads.

To install:

4. Install the pad retainers.
5. Check the foreign material at the pad shims and the back of the pads.

CAUTION

Contaminated brake discs or pads reduce stopping ability. Keep grease off the discs and pads.

6. Install the brake pads and pad shims correctly. Install the pad with the wear indicator on the inside.

➡If you are reusing the pads, always reinstall the brake pads in their original positions to prevent a momentary loss of braking efficiency.

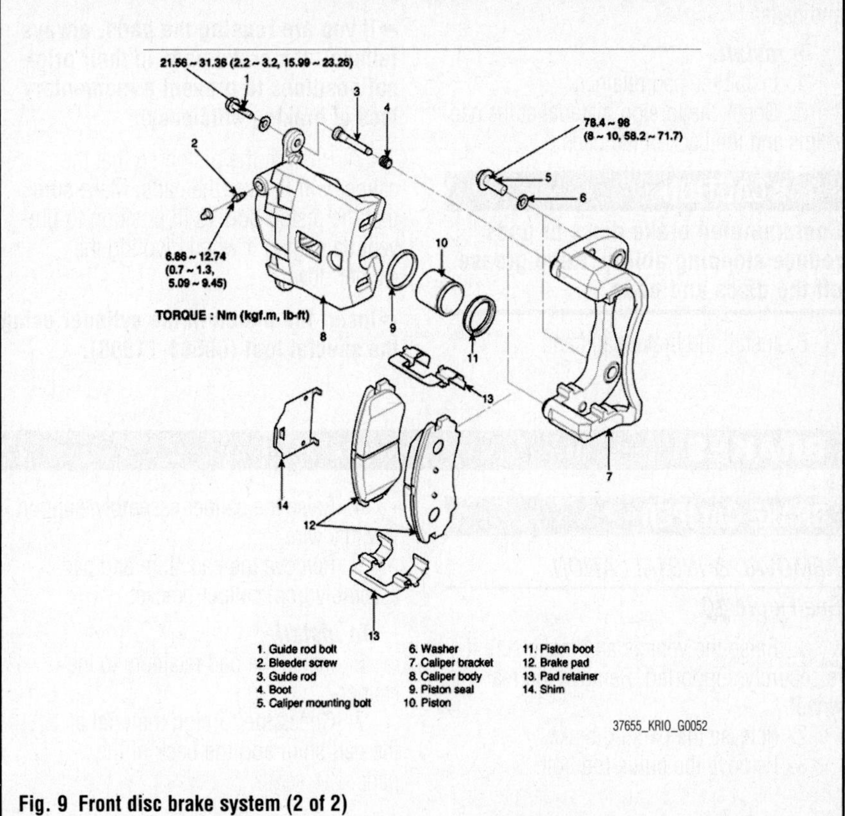

Fig. 9 Front disc brake system (2 of 2)

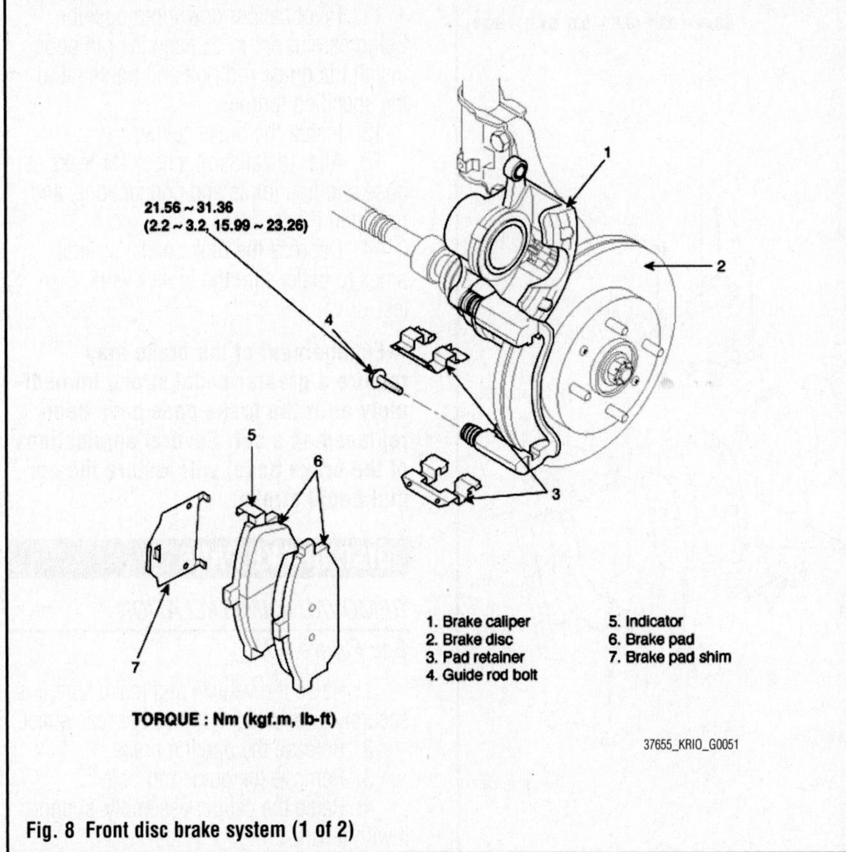

Fig. 8 Front disc brake system (1 of 2)

7. Push in the piston so that the caliper will fit over the pads. Make sure that the piston boot is in position to prevent damaging it when pivoting the caliper down.

➡Insert the piston in the cylinder using the special tool (09581-11000).

8. Pivot the caliper down into position. Being careful not to damage the pin boot, install the guide rod bolt.
9. Depress the brake pedal several times to make sure the brakes work, then test-drive.

➡Engagement of the brake may require a greater pedal stroke immediately after the brake pads have been replaced as a set. Several applications of the brake will restore the normal pedal stroke.

10. After installation, check for leaks at hose and line joints or connections, and retighten if necessary

DISC BRAKE PADS

REMOVAL & INSTALLATION

See Figures 8 and 9.

1. Remove guide rod and the caliper up out of the way.

2. Check the hoses and pin boots for damage and deterioration.

3. Remove the pad shims, pad retainers and pads.

To install:

4. Install the pad retainers.

5. Check the foreign material at the pad shims and the back of the pads.

⁂ CAUTION

Contaminated brake discs or pads reduce stopping ability. Keep grease off the discs and pads.

6. Install the brake pads and pad shims correctly. Install the pad with the wear indicator on the inside.

➡If you are reusing the pads, always reinstall the brake pads in their original positions to prevent a momentary loss of braking efficiency.

7. Push in the piston so that the caliper will fit over the pads. Make sure that the piston boot is in position to prevent damaging it when pivoting the caliper down.

➡Insert the piston in the cylinder using the special tool (09581-11000).

8. Pivot the caliper down into position. Being careful not to damage the pin boot, install the guide rod bolt.

9. Depress the brake pedal several times to make sure the brakes work, then test-drive.

➡Engagement of the brake may require a greater pedal stroke immediately after the brake pads have been replaced as a set. Several applications of the brake will restore the normal pedal stroke.

10. After installation, check for leaks at hose and line joints or connections, and retighten if necessary

BRAKES — REAR DISC BRAKES

BRAKE CALIPER

REMOVAL & INSTALLATION

See Figure 10.

1. Raise the vehicle and make sure it is securely supported. Remove the rear wheel.

2. Release the parking brake.

3. Remove the guide rod bolt.

4. Raise the caliper assembly, support it with a wire.

5. Remove the pad shim and pad assembly from caliper bracket.

To install:

6. Install the pad retainers to the caliper.

7. Check the foreign material at the pad shim and the back of the pads.

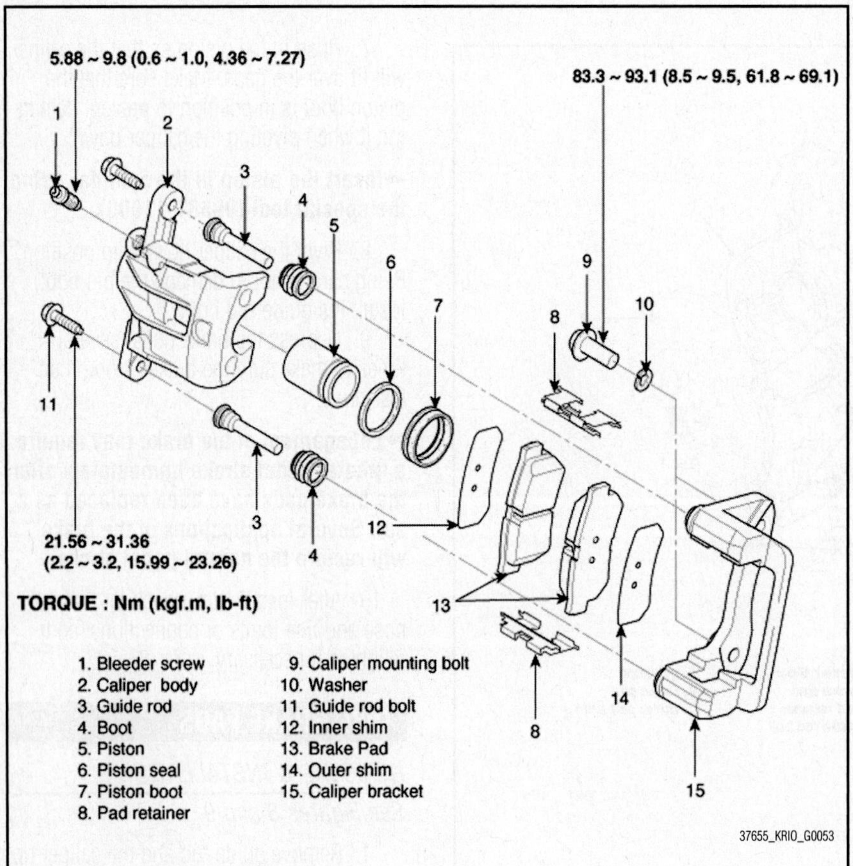

8. Contaminated brake discs or pads reduce stopping ability. Keep grease off the discs and pads.

9. Install the brake pads and pad shim on the caliper bracket.

10. If you are reusing the pads, always reinstall the brake pads in their original positions to prevent a momentary loss of braking efficiency. Push in the piston using SST (09581-11000) so that the caliper will fit over the pads. Make sure that the piston boot is in position to prevent damaging it when pivoting the caliper down.

11. Pivot caliper down into position. Being careful not to damage the pin boot, install the guide rod bolt and tighten it to the specified torque.

12. Install the brake caliper .

13. After installation, check for leaks at hose and line joints and connections, and retighten if necessary.

14. Depress the brake pedal several times to make sure the brakes work, then test-drive.

➡Engagement of the brake may require a greater pedal stroke immediately after the brake pads have been replaced as a set. Several applications of the brake pedal will restore the normal pedal stroke.

5.88 ~ 9.8 (0.6 ~ 1.0, 4.36 ~ 7.27)

83.3 ~ 93.1 (8.5 ~ 9.5, 61.8 ~ 69.1)

21.56 ~ 31.36 (2.2 ~ 3.2, 15.99 ~ 23.26)

TORQUE : Nm (kgf.m, lb-ft)

1. Bleeder screw
2. Caliper body
3. Guide rod
4. Boot
5. Piston
6. Piston seal
7. Piston boot
8. Pad retainer
9. Caliper mounting bolt
10. Washer
11. Guide rod bolt
12. Inner shim
13. Brake Pad
14. Outer shim
15. Caliper bracket

37655_KRIO_G0053

Fig. 10 Rear disc brake system

DISC BRAKE PADS

REMOVAL & INSTALLATION

See Figure 10.

1. Raise the vehicle and make sure it is securely supported. Remove the rear wheel.

2. Release the parking brake.

3. Remove the guide rod bolt.

4. Raise the caliper assembly, support it with a wire.

5. Remove the pad shim and pad assembly from caliper bracket.

To install:

6. Install the pad retainers to the caliper.

7. Check the foreign material at the pad shim and the back of the pads.

8. Contaminated brake discs or pads reduce stopping ability. Keep grease off the discs and pads.

9. Install the brake pads and pad shim on the caliper bracket.

10. If you are reusing the pads, always reinstall the brake pads in their original positions to prevent a momentary loss of braking efficiency. Push in the piston using SST (09581-11000) so that the caliper will fit over the pads. Make sure that the piston boot is in position to prevent damaging it when pivoting the caliper down.

11. Pivot caliper down into position. Being careful not to damage the pin boot, install the guide rod bolt and tighten it to the specified torque.

12. Install the brake caliper .

13. After installation, check for leaks at hose and line joints and connections, and retighten if necessary.

14. Depress the brake pedal several times to make sure the brakes work, then test-drive.

➡Engagement of the brake may require a greater pedal stroke immediately after the brake pads have been replaced as a set. Several applications of the brake pedal will restore the normal pedal stroke.

BRAKES — REAR DRUM BRAKES

BRAKE DRUM

REMOVAL & INSTALLATION

See Figure 11.

1. Remove the shoe hole down pins by pushing the shoe hole down washer and turning them.

2. Disengage the upper return spring.

3. Lower the brake shoe assembly, and remove the lower return spring. Make sure not to damage the dust cover on the wheel cylinder.

4. Disconnect the parking brake cable from the parking brake lever.

5. Remove the brake shoe assembly.

6. Remove the upper return spring, self-adjuster lever and self-adjuster spring, and separate the brake shoes.

7. Disconnect the brake line from the wheel cylinder.

8. Remove the bolt and the wheel cylinder from the backing plate.

To install:

➡During installation, note the following:

- Do not spill brake fluid on the vehicle: it may damage the paint; if brake fluid does contact the paint. Wash it off immediately with water.
- To prevent spills, cover the hose joints with rags or shop towels.
- Use only a genuine wheel cylinder special bolt.

9. Apply sealant between the wheel cylinder and backing plate, and install the wheel cylinder.

10. Connect the brake tubes to the wheel cylinder.

11. Connect the parking brake cable to the parking brake lever.

12. Clean the threaded portions of adjuster sleeve and push rod female. Coat the threads of the adjuster assembly with grease. To shorten the clevises, turn the adjuster bolt.

13. Hook the self-adjuster spring to the adjuster lever first, then to the brake shoe.

14. Install the adjuster assembly and upper return spring, noting the installation direction. Be careful not to damage the wheel cylinder dust covers.

15. Install the lower return spring.

16. Apply brake cylinder grease or equivalent rubber grease to the sliding surfaces shown. Wipe off any excess. Don't get grease on the brake linings.

17. Apply brake cylinder grease or equivalent rubber grease to the brake shoe ends and opposite edges of the shoes shown. Wipe off any excess. Don't get grease on the brake linings.

18. Install the brake shoes onto the backing plate. Be careful not to damage the wheel cylinder dust covers.

19. Install the shoe hole down pins and the shoe hole down washers.

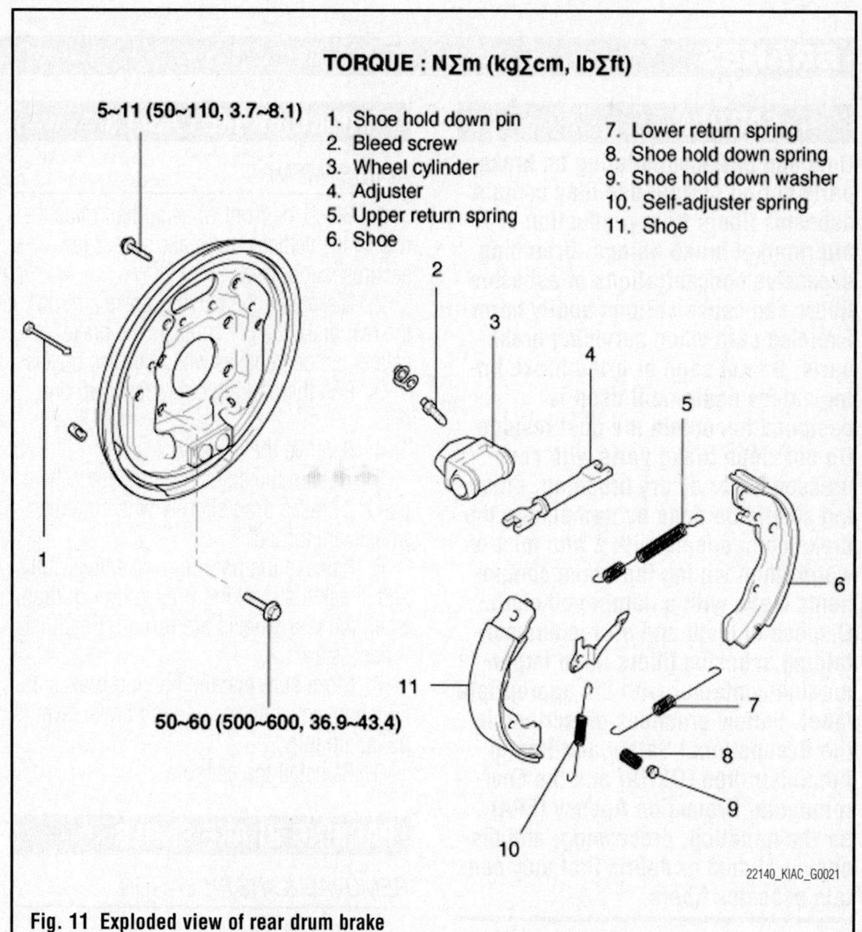

Fig. 11 Exploded view of rear drum brake

20. Hook the upper return spring.
21. Install the brake drum.
22. If the wheel cylinder has been removed, bleed the brake system.
23. Depress the brake pedal several times to set the self-adjusting brake.
24. Adjust the parking brake.

BRAKE SHOES

REMOVAL & INSTALLATION

See Figure 11.

1. Remove the shoe hole down pins by pushing the shoe hole down washer and turning them.
2. Disengage the upper return spring.
3. Lower the brake shoe assembly, and remove the lower return spring. Make sure not to damage the dust cover on the wheel cylinder.
4. Disconnect the parking brake cable from the parking brake lever.
5. Remove the brake shoe assembly.
6. Remove the upper return spring, self-adjuster lever and self-adjuster spring, and separate the brake shoes.
7. Disconnect the brake line from the wheel cylinder.
8. Remove the bolt and the wheel cylinder from the backing plate.

To install:

➡During installation, note the following:

- Do not spill brake fluid on the vehicle: it may damage the paint; if brake fluid does contact the paint. Wash it off immediately with water.
- To prevent spills, cover the hose joints with rags or shop towels.
- Use only a genuine wheel cylinder special bolt.

9. Apply sealant between the wheel cylinder and backing plate, and install the wheel cylinder.
10. Connect the brake tubes to the wheel cylinder.
11. Connect the parking brake cable to the parking brake lever.
12. Clean the threaded portions of adjuster sleeve and push rod female. Coat the threads of the adjuster assembly with grease. To shorten the clevises, turn the adjuster bolt.
13. Hook the self-adjuster spring to the adjuster lever first, then to the brake shoe.
14. Install the adjuster assembly and upper return spring, noting the installation direction. Be careful not to damage the wheel cylinder dust covers.
15. Install the lower return spring.
16. Apply brake cylinder grease or equivalent rubber grease to the sliding surfaces shown. Wipe off any excess. Don't get grease on the brake linings.
17. Apply brake cylinder grease or equivalent rubber grease to the brake shoe ends and opposite edges of the shoes shown. Wipe off any excess. Don't get grease on the brake linings.
18. Install the brake shoes onto the backing plate. Be careful not to damage the wheel cylinder dust covers.
19. Install the shoe hole down pins and the shoe hole down washers.
20. Hook the upper return spring.
21. Install the brake drum.
22. If the wheel cylinder has been removed, bleed the brake system.
23. Depress the brake pedal several times to set the self-adjusting brake.
24. Adjust the parking brake.

ADJUSTMENT

1. Depress the brake pedal several times to set the self-adjusting brake.

BRAKES — PARKING BRAKE

⁂ CAUTION

Dust and dirt accumulating on brake parts during normal use may contain asbestos fibers from production or aftermarket brake linings. Breathing excessive concentrations of asbestos fibers can cause serious bodily harm. Exercise care when servicing brake parts. Do not sand or grind brake lining unless equipment used is designed to contain the dust residue. Do not clean brake parts with compressed air or by dry brushing. Cleaning should be done by dampening the brake components with a fine mist of water, then wiping the brake components clean with a dampened cloth. Dispose of cloth and all residue containing asbestos fibers in an impermeable container with the appropriate label. Follow practices prescribed by the Occupational Safety and Health Administration (OSHA) and the Environmental Protection Agency (EPA) for the handling, processing, and disposing of dust or debris that may contain asbestos fibers.

PARKING BRAKE CABLES

ADJUSTMENT

1. Block the front wheels, then raise the rear of the vehicle and make sure it is securely supported.
2. Make sure the parking brake arm on the rear brake caliper contacts the brake caliper pin on vehicles with rear disc brakes.
3. Pull the parking brake lever up one click.
4. Remove the console.
5. Tighten the adjusting nuts until the parking brakes drag slightly when the rear wheels are turned.
6. Release the parking brake lever fully, and check that parking brakes do not drag when the rear wheels are turned. Readjust if necessary.
7. Make sure that the parking brakes are fully applied when the parking brake lever is pulled up fully.
8. Reinstall the console.

PARKING BRAKE SHOES

REMOVAL & INSTALLATION

1. Remove the console.
2. Loosen the adjusting nut and the parking brake cables.
3. Disconnect the electrical connector of parking brake switch.
4. Remove the parking brake lever assembly by loosening the bolts.
5. Remove the wheel and tire.
6. Remove the brake shoe.

To install:

7. Install the removed parts in the reverse order of removal.
8. Apply a coating of the specified grease to each sliding parts of the ratchet plate or the ratchet pawl.
9. After installing the cable adjuster, adjust the parking brake lever stroke.

ADJUSTMENT

1. Block the front wheels, then raise the rear of the vehicle and make sure it is securely supported.
2. Make sure the parking brake arm on the rear brake caliper contacts the brake caliper pin on vehicles with rear disc brakes.

3. Pull the parking brake lever up one click.
4. Remove the console.
5. Tighten the adjusting nuts until the parking brakes drag slightly when the rear wheels are turned.
6. Release the parking brake lever fully, and check that parking brakes do not drag when the rear wheels are turned. Readjust if necessary.
7. Make sure that the parking brakes are fully applied when the parking brake lever is pulled up fully.
8. Reinstall the console.

CHASSIS ELECTRICAL

AIR BAG (SUPPLEMENTAL RESTRAINT SYSTEM)

GENERAL INFORMATION

CAUTION

These vehicles are equipped with an air bag system. The system must be disarmed before performing service on, or around, system components, the steering column, instrument panel components, wiring and sensors. Failure to follow the safety precautions and the disarming procedure could result in accidental air bag deployment, possible injury and unnecessary system repairs.

PRECAUTIONS

CAUTION

Disconnect and isolate the battery negative cable before beginning any airbag system component diagnosis, testing, removal, or installation procedures. Allow system capacitor to discharge for two minutes before beginning any component service. This will disable the airbag system. Failure to disable the airbag system may result in accidental airbag deployment, personal injury, or death.

DISARMING THE SYSTEM

1. Before servicing the vehicle, refer to the Precautions Section.
2. Turn the ignition switch to the **LOCK** position.
3. Disconnect the negative battery cable.
4. Wait 10 minutes for the battery back-up power to discharge.

ARMING THE SYSTEM

1. Before servicing the vehicle, refer to the Precautions Section.
2. Connect the negative battery cable.
3. Turn the ignition switch **ON**.
4. Verify that the air bag indicator illuminates for 4–8 seconds, then goes off.

CLOCKSPRING CENTERING

Prior to installing the clock spring, align the mating mark and "NEUTRAL" position indicator of the clock spring, and, after turning the front wheels to the straight-ahead position, install the clock spring to the column switch. If the mating mark of the clock spring is not properly aligned, the steering wheel may not completely rotate during a turn, or the flat cable within the clock spring may be severed, obstructing normal operation of the SRS and possibly leading to serious injury to the vehicle's driver.

DRIVE TRAIN

CLUTCH

REMOVAL & INSTALLATION

See Figure 12.

1. Remove a transaxle assembly.
2. Remove the clutch cover bolts.
3. Remove the clutch cover and disc.
4. Using the SST (09411-25000), install the disc.
5. Tighten the clutch cover. Torque to 10–16 ft. lbs. (15–22 Nm).

To install:

6. Using the SST (09411-25000), install the disc.
7. Install the clutch cover and disc.
8. Install the clutch cover bolts.
9. Tighten the clutch cover. Torque to 10–16 ft. lbs. (15–22 Nm).
10. Install a transaxle assembly.

CLUTCH DRIVEN DISC & PRESSURE PLATE

REMOVAL & INSTALLATION

See Figure 12.

1. Remove a transaxle assembly.
2. Remove the clutch cover bolts.
3. Remove the clutch cover and disc.
4. Using the SST (09411-25000), install the disc.

To install:

5. Using the SST (09411-25000), install the disc.
6. Install the clutch cover and disc.

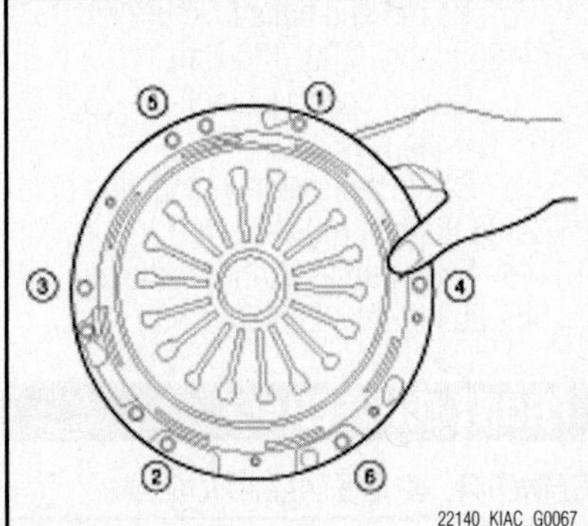

Fig. 12 Tighten clutch cover bolts in sequence

7. Install the clutch cover bolts.
8. Tighten the clutch cover. Torque to 10–16 ft. lbs. (15–22 Nm).
9. Install transaxle assembly.

ADJUSTMENTS

1. Measure the clutch pedal height (from the face of the pedal pad to the floorboard) and the clutch pedal free play (measured at the face of the pedal pad). Standard value is 0.25–0.5 inches (6–13 mm).
2. If the clutch pedal free play and height is not within the standard value range, adjust as follows:
    a. Turn and adjust the bolt within the standard value, then secure by tightening the lock nut.
    b. Turn the push rod to agree with the standard value and then secure the push rod with the lock nut.
3. If the clutch pedal free play and the distance between the clutch pedal and the floor board when the clutch is disengaged, do not meet with the standard values, it may be the result of either air in the hydraulic system or a faulty clutch master cylinder.

Bleed the air or disassemble and inspect the master cylinder or clutch.

FRONT AXLE SHAFT, BEARING & SEAL

REMOVAL & INSTALLATION

1. Raise the vehicle and remove the front wheel.
2. Remove the wheel speed sensor from the knuckle.
3. Remove the caliper assembly by loosening the bolts, and then suspend it with wire.
4. Remove the split pin, castle nut and washer from the front hub assembly.
5. Remove the ball joint assembly mounting bolt from the knuckle.
6. Remove the tie rod end split pin and castle nut.
7. Disconnect the tie rod end from the knuckle using the SST (09568 - 4A000).
8. Loosen the screws and remove the brake disc from the front hub assembly.
9. Remove the strut assembly mounting bolts and nuts.
10. Remove the hub and knuckle as an assembly.

To install:

11. Install the hub and knuckle as an assembly.
12. Install the strut assembly mounting bolts and nuts.
13. Install the brake disc from the front hub assembly, then tighten the screws.
14. Connect the tie rod end from the knuckle using the SST (09568 - 4A000).
15. Install the tie rod end split pin and castle nut.
16. Install the ball joint assembly mounting bolt from the knuckle.
17. Install the split pin, castle nut and washer from the front hub assembly.
18. Install the caliper assembly by tightening the bolts.
19. Install the wheel speed sensor from the knuckle.
20. Lower the vehicle and install the front wheel.

Disassembly

1. Remove the snap ring.
2. Remove the hub from the knuckle assembly using the SST (09545-34100).
3. Remove the wheel bearing inner race from the hub using the SST (09432-11000, 09545-34100).
4. Remove the wheel bearing outer race from the hub using the SST (09216-21600, 09216-21100).

Reassembly

1. Apply the multi-purpose grease to the contacting surface of the knuckle and wheel bearing thinly.
2. Press-in the wheel bearing to the knuckle using the SST (09216-21100).

✲✲ CAUTION

Do not strike the inner race of the wheel bearing because that can cause damage to the bearing assembly.

➡Always use a new wheel bearing assembly.

3. Install the snap ring into the groove on the knuckle.
4. Press-in the hub to the knuckle using the SST (09454-21100).

✲✲ CAUTION

Do not strike the outer race of the wheel bearing because that can cause damage to the bearing assembly.

5. Tighten the hub to the knuckle with the SST (09517-21500).
6. Rotate the hub to seat the wheel bearing.
7. Measure the wheel bearing starting torque.

➡Wheel bearing starting torque: 1.3 ft. lbs. (1.8 Nm) or below.

8. Measure the hub run-out by using a dial gauge. Runout should be 0.00157inch (0.04mm).
9. Remove the SST (09517-21500).

To install:

10. Installation is the reverse of removal.
11. Use the following torque specifications during installation:
    a. Strut assembly mounting bolts: 101–116 ft. lbs. (140–160 Nm).
    b. Tie rod end bolts to knuckle: 72–86 ft. lbs. (100–120 Nm).
    c. Tie rod end castle nut: 17–25 ft. lbs. (24–34 Nm).
    d. Ball joint assembly mounting bolt: 72–86 ft. lbs. (100–120 Nm).
    e. Front hub assembly castle nut: 145–203 ft. lbs. (200–280 Nm).

FRONT DRIVESHAFT

REMOVAL & INSTALLATION

See Figures 13 through 24.

1. Loosen the wheel nuts slightly.
2. Raise the front of the vehicle and support it with safety stands in a proper location.
3. Remove the front wheel and tire.
4. Remove the drain plug. Drain the transaxle oil.
    a. Lay a bottle keeping the gear oil under transaxle.
    b. Remove drain plug and washer in the lower part of transaxle.
5. Remove the split pin, the lock nut and the washer from the front hub under applying the brake.
6. Disconnect the tie rod end ball joint from the knuckle using the Special Tool (09568-34A000) after removing the split pin and lock nut.
7. Remove the wheel speed sensor from the knuckle.
8. Remove the lower arm mounting bolts from the knuckle.
9. Using a plastic hammer, disconnect the drive shaft from the axle hub.
10. Push the axle hub outward and separate the drive shaft from the axle hub.
11. Insert a pry bar between the transaxle case and joint case, and separate the drive shaft from the transaxle case.

✲✲ CAUTION

- Use a pry bar being careful not to damage the transaxle and joint.
- Do not insert the pry bar too deep, as this may cause damage to the oil seal. [max. depth : 0.28 in (7mm)]
- Do not pull the drive shaft by excessive force it may cause components inside the BJ or TJ joint kit to dislodge resulting in a torn boot or a damaged bearing.

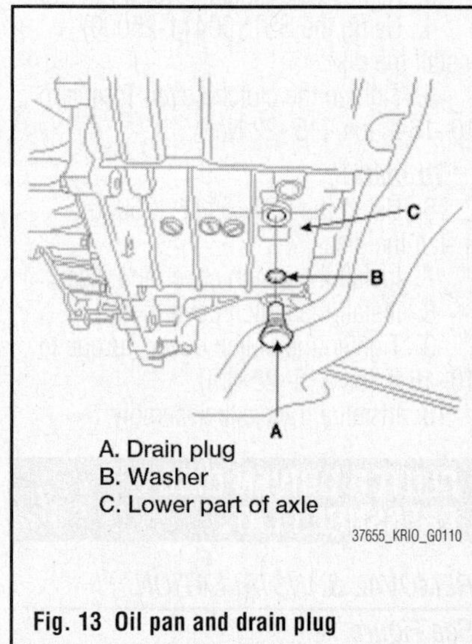

Fig. 13 Oil pan and drain plug

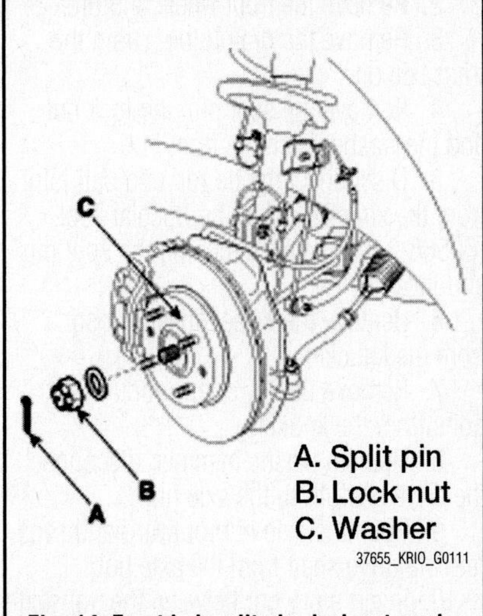

Fig. 14 Front hub split pin, lock nut, and washer

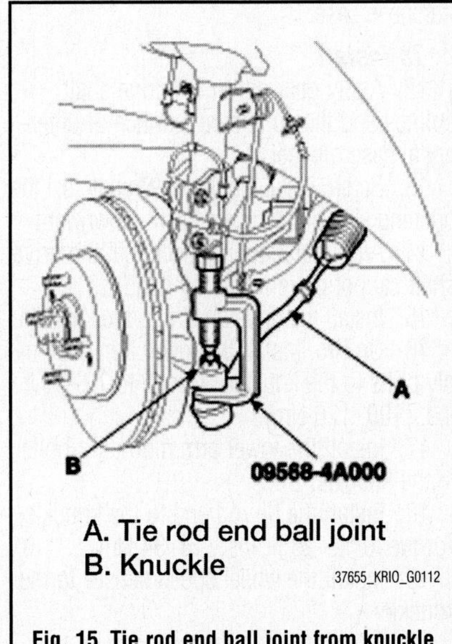

Fig. 15 Tie rod end ball joint from knuckle

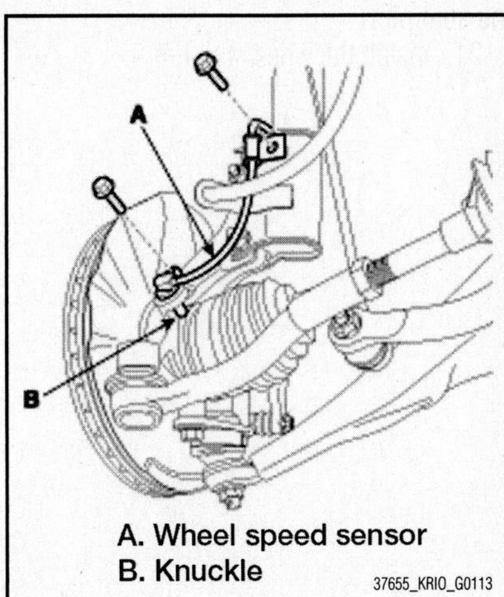

Fig. 16 Removing the wheel speed sensor from the knuckle

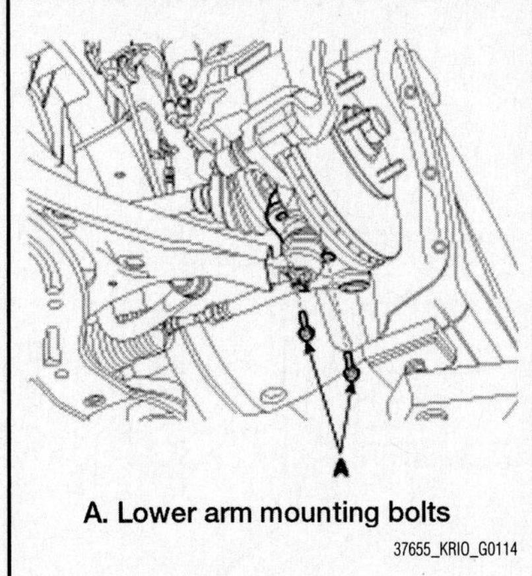

Fig. 17 Removing the lower arm mounting bolts from the knuckle

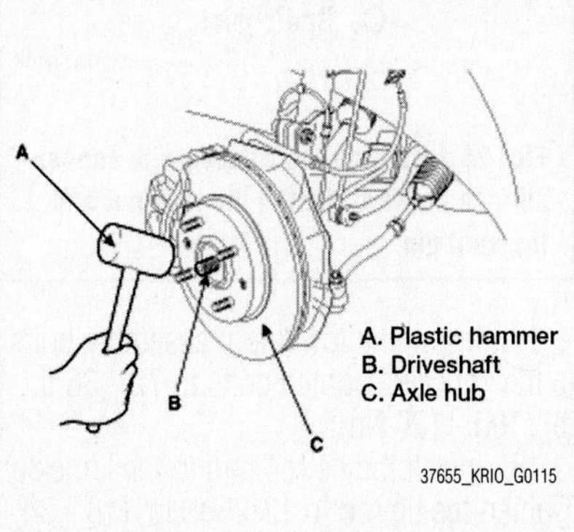

Fig. 18 Disconnecting the driveshaft from the axle hub

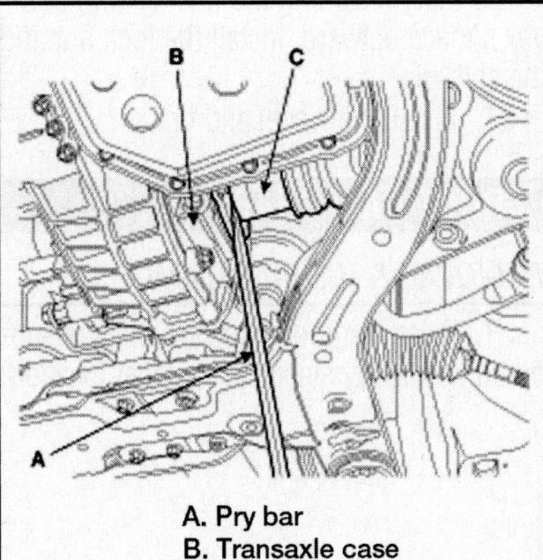

Fig. 19 Separate the driveshaft from the transaxle case

12. Pull out the drive shaft from the transaxle case.

CAUTION

- Plug the hole of the transaxle case with the oil seal cap to prevent contamination.
- Support the drive shaft properly.
- Replace the retainer ring whenever the drive shaft is removed from the transaxle case.

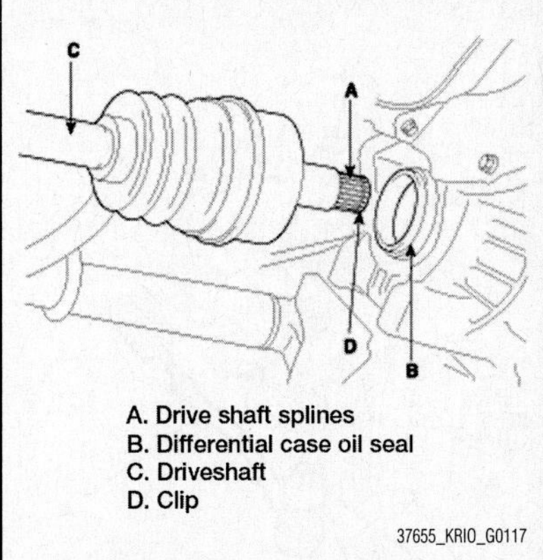

Fig. 20 Before installing the drive shaft, set the opening side of the clip facing downward

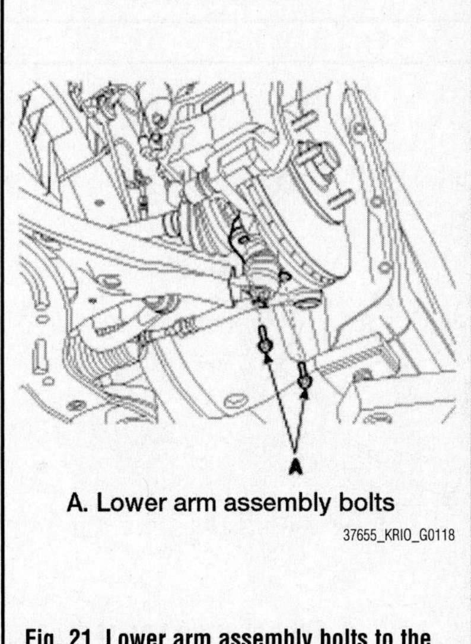

Fig. 21 Lower arm assembly bolts to the knuckle

To install:

13. Apply gear oil on the drive shaft splines and the contacting surface of differential case oil seal.

14. Before installing the drive shaft, set the opening side of the clip facing downward.

15. After installation, check that the drive shaft cannot be removed by hand.

16. Install the drive shaft into the knuckle.

CAUTION

Be careful not to damage the boot.

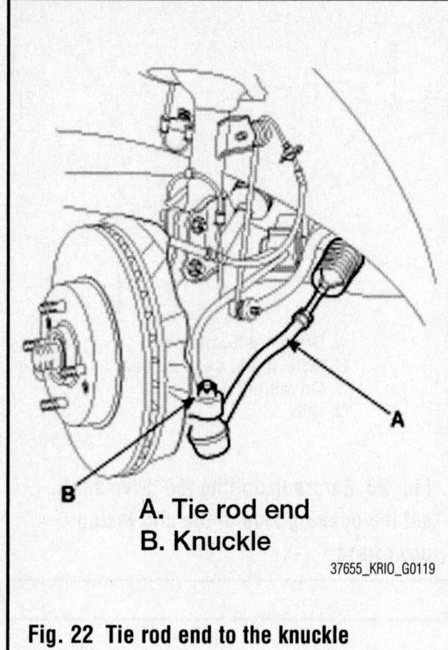

Fig. 22 Tie rod end to the knuckle

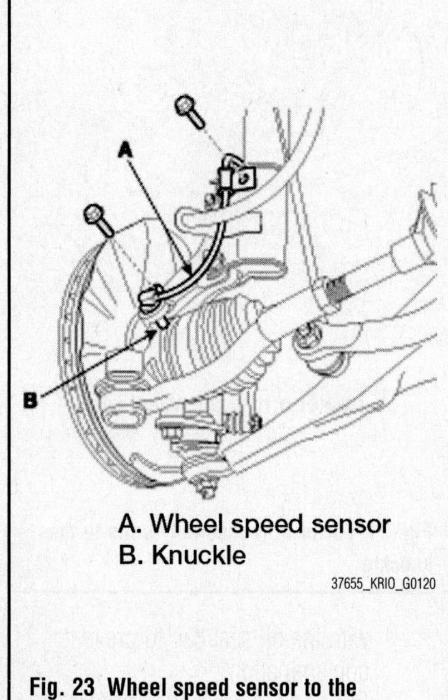

Fig. 23 Wheel speed sensor to the knuckle installation

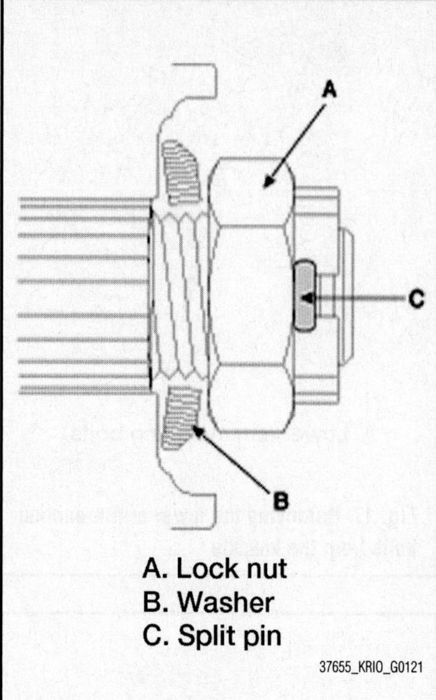

Fig. 24 Installing the washer with convex surface outward, install the lock nut and the split pin

17. Install the lower arm assembly bolts to the knuckle. Tighten bolts to 72 - 86 ft. lb. (100 - 120 Nm).
18. Install the tie rod end to the knuckle. Tighten the tie rod to 12-25 ft. lb. (16 - 34 Nm).
19. Install the wheel speed sensor to the knuckle.
20. After installing the washer with convex surface outward, install the lock nut and the split pin.
21. Install the wheel and tire.

FRONT HALFSHAFT

REMOVAL & INSTALLATION

1. Raise the front of the vehicle and support it with safety stands in a proper location.
2. Remove the front wheel and tire.
3. Remove the drain plug. Drain the transaxle oil.
4. Remove the split pin, the lock nut and the washer from the front hub.
5. Disconnect the tie rod end ball joint from the knuckle using the Special Tool (09568-34000) after removing the split pin and lock nut.
6. Remove the wheel speed sensor from the knuckle.
7. Remove the lower arm mounting bolts from the knuckle.
8. Using a plastic hammer, disconnect the drive shaft from the axle hub.
9. Push the axle hub outward and separate the drive shaft from the axle hub.
10. Insert a pry bar between the transaxle case and joint case, and separate the drive shaft from the transaxle case.
11. Pull out the drive shaft from the transaxle case.

To install:

12. Apply gear oil on the drive shaft splines and the contacting surface of differential case oil seal.
13. Before installing the drive shaft, set the opening side of the circlip facing downward.
14. After installation, check that the drive shaft cannot be removed by hand.
15. Install the drive shaft into the knuckle.
16. On Rio, install the lower arm assembly bolts to the knuckle. Torque to 72–86 ft. lbs. (100–120 Nm).
17. Install the lower arm mounting bolts to the knuckle.
18. Install the tie rod end to the knuckle. Torque to 12–25 ft. lbs. (16–34 Nm).
19. Install the wheel speed sensor to the knuckle.
20. After installing the washer with convex surface outward, install the lock nut and the split pin.
21. Install the wheel and tire

ENGINE COOLING

ENGINE FAN

REMOVAL & INSTALLATION

1. Drain the engine coolant.
2. Remove the upper radiator hose and lower radiator hose.
3. Disconnect the Automatic Transaxle Fluid (ATF) oil cooler hoses, if vehicle equipped with automatic transaxle.
4. Disconnect the fan motor connector (s).

➡**Some vehicles are equipped with two side-by-side fans.**

5. Remove the cooling fan mounting bolts and remove cooling fan.

To install:

6. Install the cooling fan and mounting bolts.
7. Connect the fan motor connector (s).
8. Install the upper radiator hose and lower radiator hose.
9. Install the ATF oil cooler hoses.
10. Fill with engine coolant.
11. Start engine and check for leaks.

RADIATOR

REMOVAL & INSTALLATION

1. Drain the engine coolant.
2. Remove the upper radiator hose and lower radiator hose.
3. Remove the ATF oil cooler hoses if vehicle is equipped with automatic transaxle.
4. Disconnect the fan motor connector.
5. Remove the cooling fan.
6. Remove the radiator upper bracket.
7. Remove the radiator.

To install:

8. Install the radiator.
9. Install the radiator upper bracket.
10. Install the fan (s) and fan shroud.
11. Install the upper radiator hose and lower radiator hose.
12. Install the ATF oil cooler hoses if vehicle is equipped with automatic transaxle.
13. Fill with engine coolant.
14. Start engine and check for leaks.

THERMOSTAT

REMOVAL & INSTALLATION

1. Drain the engine coolant so its level is below thermostat.
2. Remove the water inlet fitting, gasket and thermostat.

To install:

3. Install the water inlet fitting, gasket and thermostat.
4. Fill the engine coolant so its level is below thermostat.

WATER PUMP

REMOVAL & INSTALLATION

1. Drain the engine coolant.
2. Loosen the water pump pulley bolts.
3. Remove the drive belts.
4. Remove the water pump pulley.
5. Remove the timing belt.
6. Remove the timing belt idler.
7. Remove the water pump.
   a. Remove the 2 bolts and alternator brace.
   b. Remove the 3 bolts and remove the water pump and gasket.

To install:

8. Install the water pump.
   a. Install the water pump and a new gasket with the 3 bolts. Torque: 8.7–10.8 ft. lbs. (11.8–14.7 Nm).
   b. Install the alternator brace with the 2 bolts. Torque: 14.5–19.5 ft. lbs. (19.6–26.5 Nm.
9. Install the timing belt idler.
10. Install the timing belt.
11. Install the water pump pulley.
12. Install the drive belts.
13. Tighten the water pump pulley bolts. Torque: 5.8–7.2 ft. lbs. (7.8–9.8 Nm).
14. Fill with engine coolant.
15. Start engine and check for leaks.
16. Recheck engine coolant level.

ENGINE ELECTRICAL — CHARGING SYSTEM

ALTERNATOR

REMOVAL & INSTALLATION

1. Disconnect the battery negative terminal first, then the positive terminal.
2. Temporarily loosen the water pump pulley bolts.
3. Remove the alternator drive belt, after loosening the adjusting bolt and mounting bolt.
4. Remove the power steering pump belt.
5. Remove the water pump pulley.
6. Remove the power steering pump.
7. Remove the power steering pump bracket.
8. Disconnect the alternator connector, and remove the cable from alternator "B" terminal.
9. Remove the adjusting bolt and mounting bolt.
10. Remove the alternator brace.
11. Pull out the through bolt and then remove the alternator.

To install:

12. Push in the through bolt and then install the alternator.
13. Install the alternator brace.
14. Install the adjusting bolt and mounting bolt.
15. Connect the alternator connector, and install the cable from alternator "B" terminal.
16. Install the power steering pump bracket.
17. Install the power steering pump.
18. Install the water pump pulley.
19. Install the power steering pump belt.
20. Install the alternator drive belt, after tightening the adjusting bolt and mounting bolt.
21. Tighten the water pump pulley bolts.
22. Connect the battery negative terminal first, then the positive terminal.
23. Hand tighten the water pump pulley bolts.
24. Adjust the power steering pump belt tension.
25. Adjust the alternator belt tension after installation.
26. Tighten the water pump pulley bolts.

VOLTAGE REGULATOR

ADJUSTMENT

There is no adjustment possible for the voltage regulator. If the voltage regulator is defective, it must be replaced.

REMOVAL & INSTALLATION

The charging system included a battery, an alternator with a built-in regulator, and the charging indicator light and wire.

The Alternator has built-in diodes, each rectifying AC current to DC current. Therefore, DC current appears at alternator "B" terminal.

In addition, the charging voltage of this alternator is regulated by the battery voltage detection system. The alternator is regulated by the battery voltage detection system. The main components of the alternator are the rotor, stator, rectifier, capacitor brushes, bearings and V-ribbed belt pulley. The brush holder contains a built-in electronic voltage regulator.

ENGINE ELECTRICAL — IGNITION SYSTEM

FIRING ORDER

Firing order: 1–3–4–2

IGNITION COIL

REMOVAL & INSTALLATION

1. Remove the engine cover.
2. Remove the ignition coil.

➡When removing the ignition coil connector, pull the lock pin and push the clip.

To install:

3. Install the ignition coil.
4. Install the engine cover.

IGNITION TIMING

ADJUSTMENT

These vehicles are equipped with a Distributorless Ignition System (DIS). No adjustment is necessary or possible.

SPARK PLUGS

REMOVAL & INSTALLATION

1. Disconnect the negative battery cable.

⁂ WARNING

When disconnecting the cable, only pull on the plug cable boot, never on the wire itself!

2. Detach the spark plug cable and remove the ignition coil.
3. Use spark plug wrench to remove the spark plug (s) from the cylinder head.

⁂ WARNING

Do not let any dirt or debris get into the engine through the spark plugs holes while when are removed.

To install:

4. Install the spark plug and tighten, to 18–22 ft. lbs. (25–30 Nm)
5. Install the ignition coil (s).
6. Attach the spark plug wire (s) to the ignition coil.
7. Connect the negative battery cable.

INSPECTION

Check spark plugs for the following:

1. Broken insulator
2. Worn electrode
3. Carbon deposits
4. Damaged or broken gasket
5. Condition of the porcelain insulator at the tip of the spark plug.
6. Check the electrode gap.

Standard: 0.0394–0.0433 inches (1.0–1.1 mm).

ENGINE ELECTRICAL — STARTING SYSTEM

STARTER

REMOVAL & INSTALLATION

1. Disconnect the battery negative cable.
2. Remove the air cleaner assembly
3. Remove the shift cable and bracket, if equipped with manual transaxle.
4. Disconnect the starter cable from the B terminal on the solenoid.
5. Disconnect the connector from the S terminal.
6. Remove the 2 bolts holding the starter, then remove the starter.

To install:

7. Install the starter, then install the 2 bolts holding the starter.
8. Connect the connector from the S terminal.
9. Connect the starter cable from the B terminal on the solenoid.
10. Install the shift cable and bracket, if equipped with manual transaxle.
11. Install the air cleaner assembly
12. Connect the battery negative cable.

ENGINE MECHANICAL

ACCESSORY DRIVE BELTS

ACCESSORY BELT ROUTING

See Figure 25.

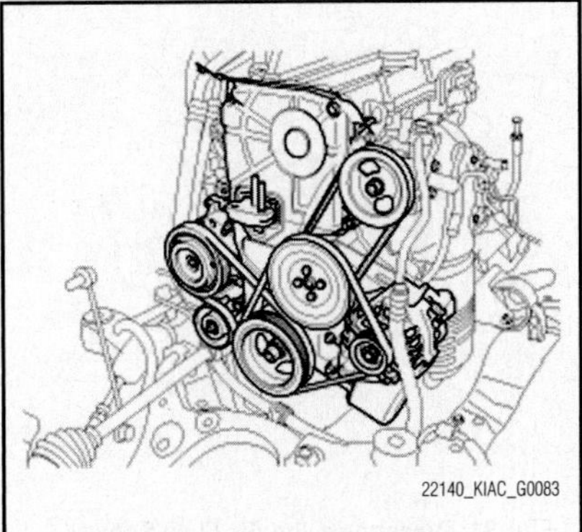

Fig. 25 Accessory drive belt routing

Refer to the accompanying illustration for belt routing.

INSPECTION

Inspect the drive belt for signs of glazing or cracking. A glazed belt will be perfectly smooth from slippage, while a good belt will have a slight texture of fabric visible. Cracks will usually start at the inner edge of the belt and run outward. All worn or damaged drive belts should be replaced immediately.

ADJUSTMENT

See Figure 26.

➡Refer to the accompanying illustration for bolt and nut locations.

1. Loosen bolt A and nuts B and C.
2. Turn adjusting bolt D and adjust the belt deflection to within the following specifications (when applying about 22 lbs. of pressure):
   a. New belt: 0.31–0.35 in. (8–9mm)
   b. Used belt: 0.35–0.39 in. (9–10mm)
3. Tighten bolt A and nuts B and C, as follows:
   a. Bolt A: 27–39 ft. lbs. (37–53 Nm)
   b. Nut B: 14–19 ft. lbs. (19–25 Nm)
   c. Nut C: 24–34 ft. lbs. (32–46 Nm)

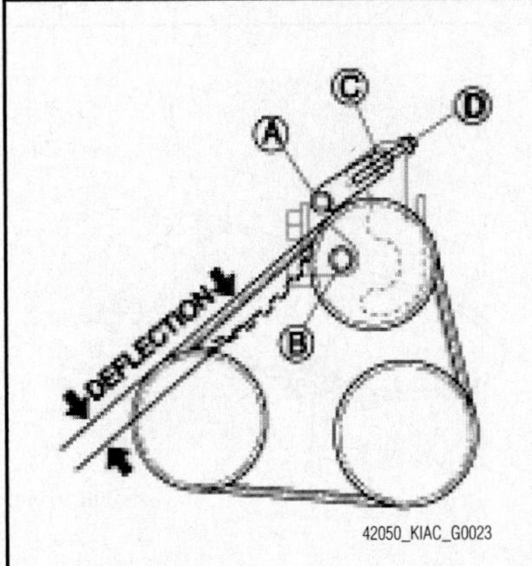

Fig. 26 Belt adjustment engines with power steering

CRANKSHAFT DAMPER (BALANCER)

REMOVAL & INSTALLATION

See Figures 27 through 30.

Engine removal is not required for this procedure.

1. Remove the engine cover.
2. Remove RH front wheel.
3. Remove 2bolts and RH side cover.
4. Temporarily loosen the water pump pulley bolts.
5. Remove the alternator drive belt
6. Remove the air conditioner compressor drive belt.
7. Remove the power steering pump drive belt.
8. Remove the 4 bolts and water pump pulley.
9. Remove the 4 bolts and timing belt upper cover.
10. Turn the crankshaft pulley, and align its groove with timing mark "T" of the timing belt cover. Check that the timing mark of camshaft sprocket is aligned with the timing mark of cylinder head cover. (No.1 cylinder compression TDC position)
11. Remove the crankshaft pulley bolt and crankshaft pulley.

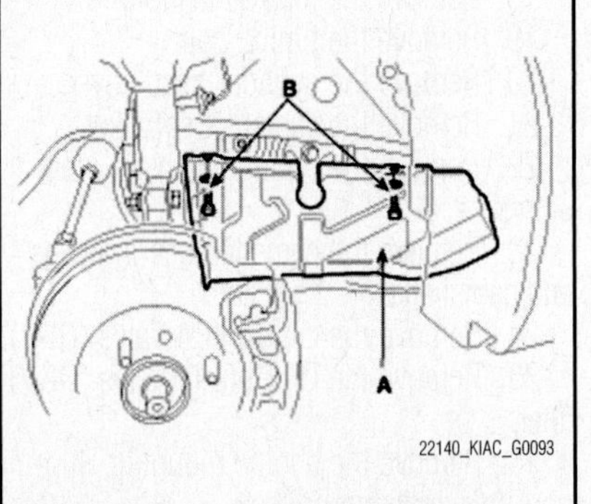

Fig. 27 Remove 2bolts (B) and RH side cover (A).

To install:

12. Install the crankshaft pulley. Make sure that crankshaft sprocket pin fits the small hole in the pulley. Torque: 101.3–108.5 ft. lbs. (137.3–147.1 Nm).

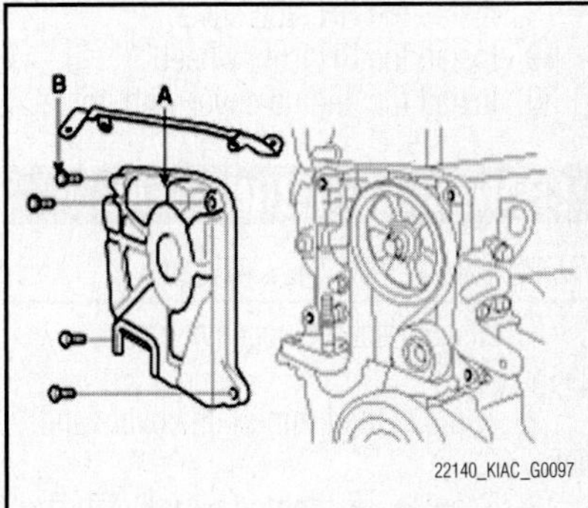

Fig. 28 Remove the 4 bolts (B) and timing belt upper cover (A).

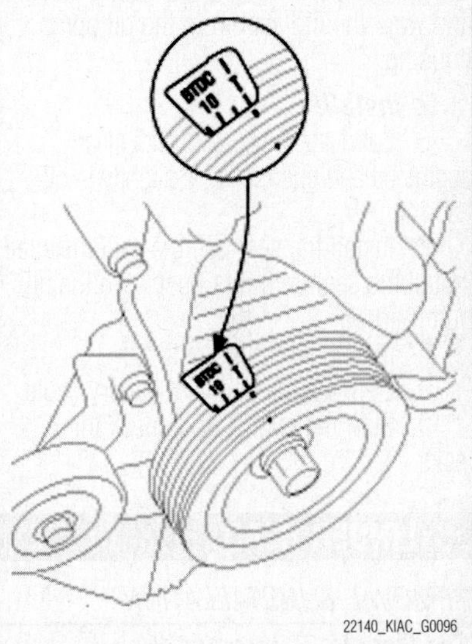

Fig. 29 Align the crankshaft pulley groove with timing mark "T" of the timing belt cover.

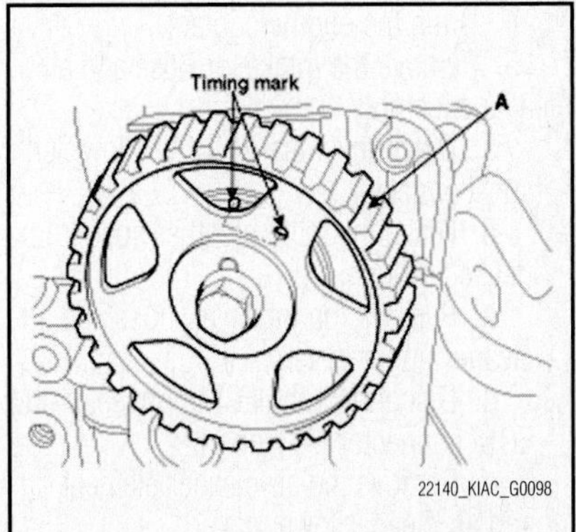

Fig. 30 Camshaft sprocket timing mark.

13. Install the timing belt upper cover. Torque: 5.8–7.2 ft. lbs. (7.8–9.8 Nm).
14. Install the water pump pulley.
15. Install the power steering pump drive belt.
16. Install the air conditioner compressor drive belt.
17. Install the alternator drive belt.
18. Install the RH side cover.
19. Install the RH front wheel.
20. Install the engine cover with bolts.

CRANKSHAFT FRONT SEAL

REMOVAL & INSTALLATION

1. Disconnect the negative battery cable.
2. Remove the timing belt covers and belt.
3. Remove the timing belt pulley using a puller.
4. Remove the oil pump bolts and the pump.
5. Wrap a suitable prytool with a rag and work the old seal from the oil pump housing.

To install:

6. Lubricate the seal lip with clean engine oil and push the seal slightly in by hand.
7. Install the seal using a seal installer. Install the seal until it is flush with the oil pump body.
8. Install the timing belt.
9. Connect the negative battery cable.
10. Start the engine and check for leaks.

CYLINDER HEAD

REMOVAL & INSTALLATION

See Figures 31 through 46.

1. Disconnect the terminals from battery and remove the battery.
2. Remove the engine cover.
3. Remove the lower cover.
4. Drain the engine coolant.
5. Remove the intake air hose and air cleaner assembly.
   a. Disconnect the AFS (Air Flow Sensor) connector.
   b. Disconnect the breather hose from intake air hose.
   c. Remove the intake air hose and air cleaner upper cover.
   d. Disconnect the ECM connector and ECM connector (A/T only).
   e. Remove the air cleaner element and air cleaner lower cover.
6. Remove the battery tray.
7. Remove the upper radiator hose and lower radiator hose.
8. Remove the heater hoses.
9. Remove the fuel hose.
10. Remove the accelerator cable by loosening the lock-nut, then slip the cable end out of the throttle linkage.
11. Disconnect the Throttle Position Sensor (TPS) connector.
12. Remove the engine wire harness connectors and wire harness clamps from cylinder head and the intake manifold.
    a. Disconnect the rear oxygen sensor connector.
    b. Disconnect the air conditioner compressor switch connector.
    c. Disconnect the knock sensor connector.
    d. Disconnect the injector connectors (No.3,4).
    e. Disconnect the injector connectors (No.1,2)
    f. Remove the wire harness bracket.
    g. Disconnect the Idle Speed Actuator (ISA) connector.
    h. Disconnect the front oxygen sensor connector.
    i. Disconnect the Crankshaft Position Sensor (CKP) connector.
    j. Disconnect the Oil Control Valve (OCV) connector.
    k. Disconnect the ignition coil connector.
    l. Disconnect the ignition coil condenser connector.
    m. Disconnect the Camshaft Position Sensor (CMP) connector.
    n. Disconnect the ground cable.
    o. Remove the wire harness bracket.
13. Disconnect the hose of the Purge Control Solenoid Valve (PCSV) side.
14. Remove the brake booster vacuum hose.
15. Remove the power steering pump and suspend the pump with a wire.
16. Remove the ignition coil.
17. Remove the exhaust manifold.
18. Remove the intake manifold.
19. Remove the timing belt.
20. Remove the cylinder head cover.
21. Remove the camshaft sprocket.
22. Remove the timing chain auto tensioner.
23. Remove the camshaft bearing caps and camshafts.
24. Remove the Oil Control Valve (OCV).
25. Remove the Oil Control Valve (OCV) filter.
26. Remove the engine mounting support bracket fixing bolts.
27. Remove the cylinder head bolts, then remove the cylinder head.
    a. Using 8mm hexagon wrench, uniformly loosen and remove the 10 cylinder head bolts, in several passes, in the sequence shown.

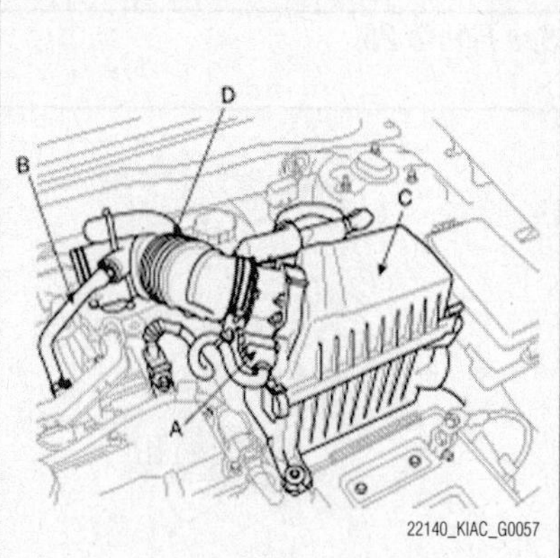

Fig. 31 Disconnect the Air Flow Sensor (AFS) connector (A), breather hose (B) from intake air hose (D), and air cleaner upper cover (C).

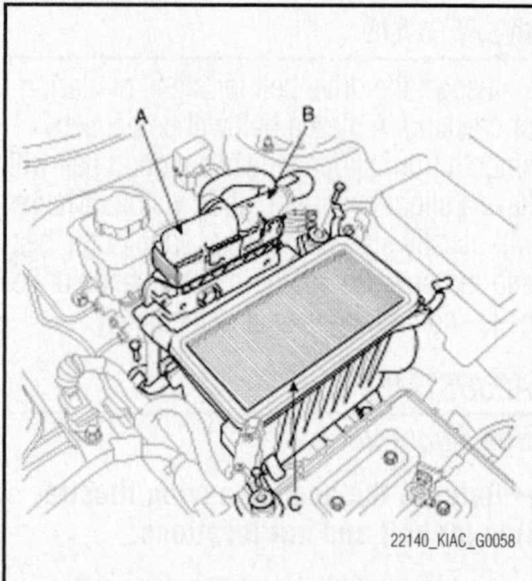

Fig. 32 Disconnect the ECM connector (A) and ECM connector (B) (A/T only).

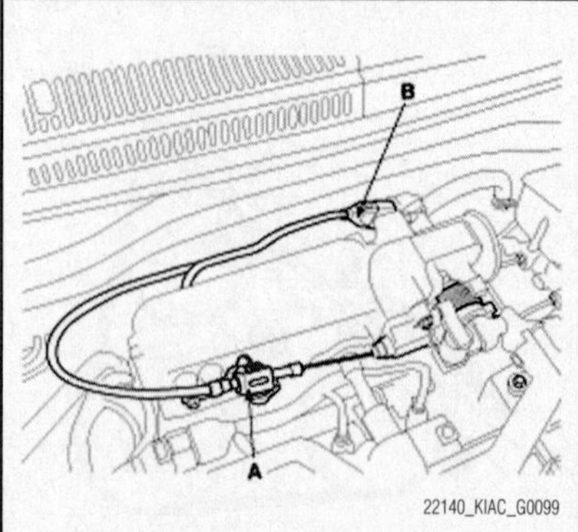

Fig. 33 Remove the accelerator cable (A). Disconnect the Throttle Position Sensor (TPS) connector (B).

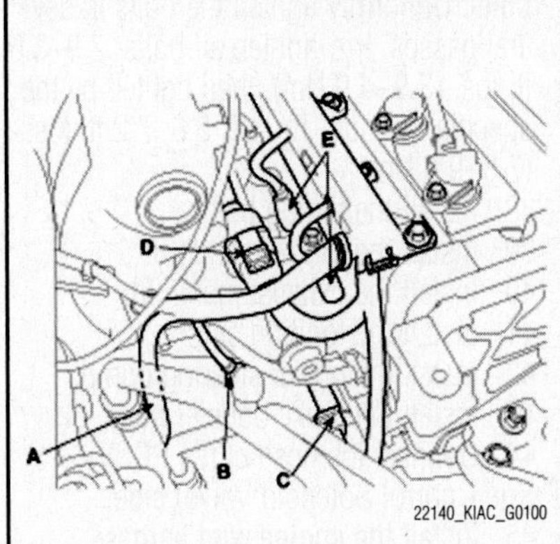

Fig. 34 Disconnect rear oxygen sensor connector (A), A/C compressor switch connector (B), injector connectors (No.3,4) (D), and injector connectors (No.1,2) (E)

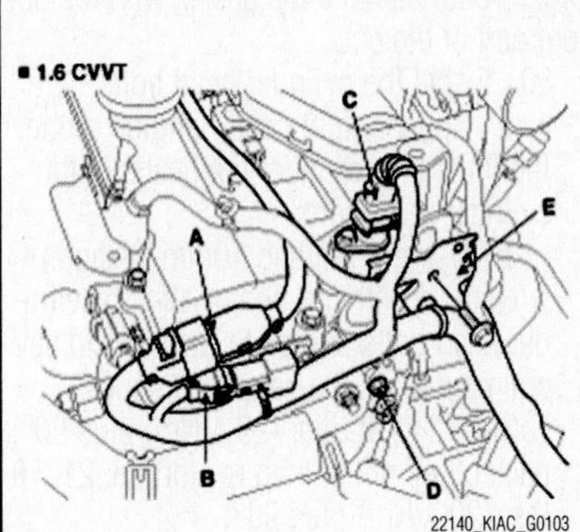

Fig. 37 Disconnect ignition coil connector (A), ignition coil condenser connector (B), Camshaft Position Sensor (CMP) connector (C), ground cable (D). and wire harness bracket (E).

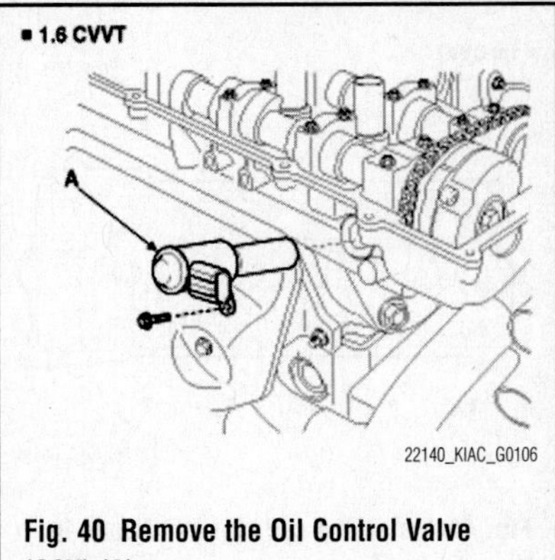

Fig. 40 Remove the Oil Control Valve (OCV) (A).

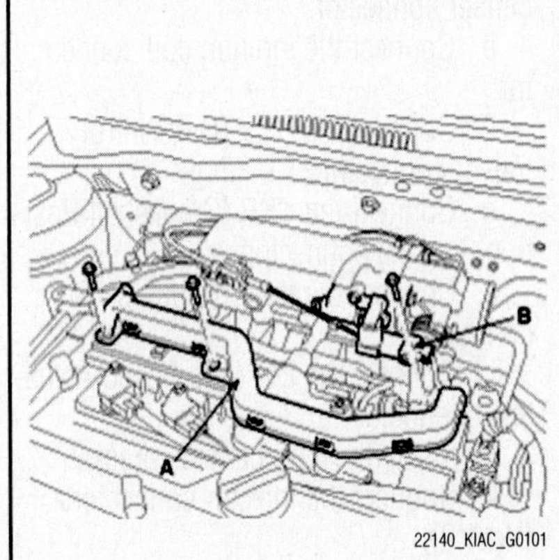

Fig. 35 Remove the wire harness bracket (A). Disconnect the Idle Speed Actuator (ISA) connector (B).

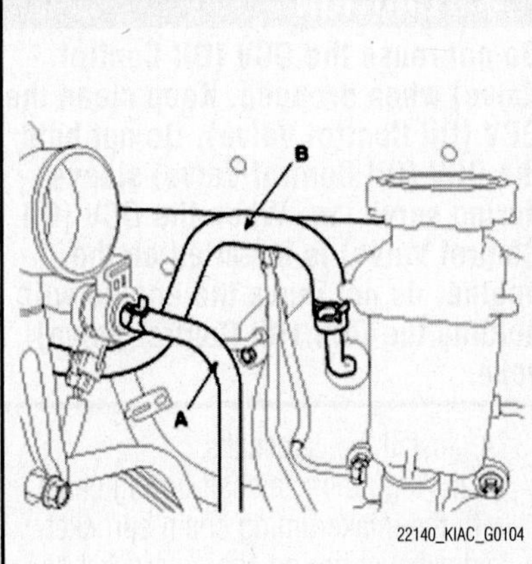

Fig. 38 Disconnect the hose (A) of Purge Control Solenoid Valve (PCVS) and brake booster vacuum hose (B).

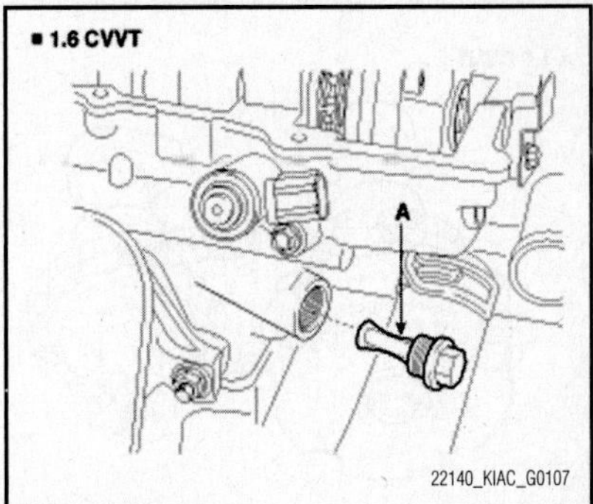

Fig. 41 Remove the Oil Control Valve (OCV) filter (A).

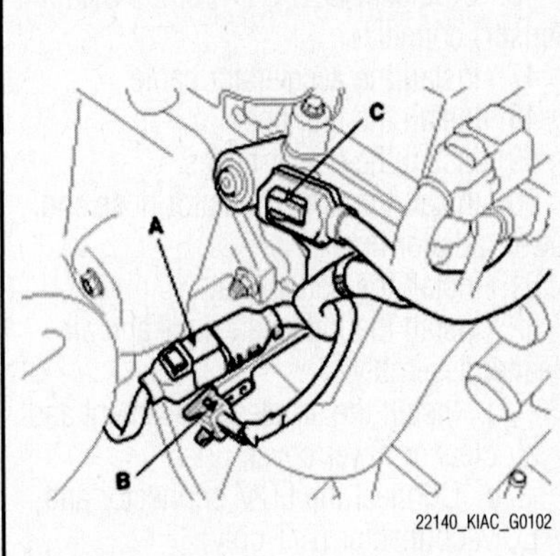

Fig. 36 Disconnect the front oxygen sensor connector (A), Crankshaft Position Sensor (CKP) connector (B), and Oil Control Valve (OCV) connector (C).

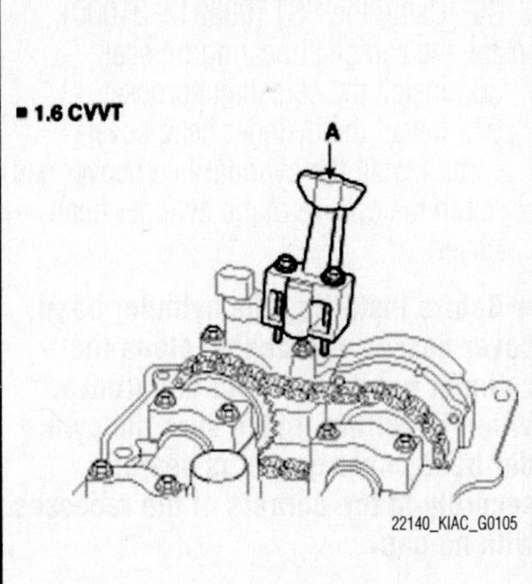

Fig. 39 Remove the timing chain auto tensioner (A).

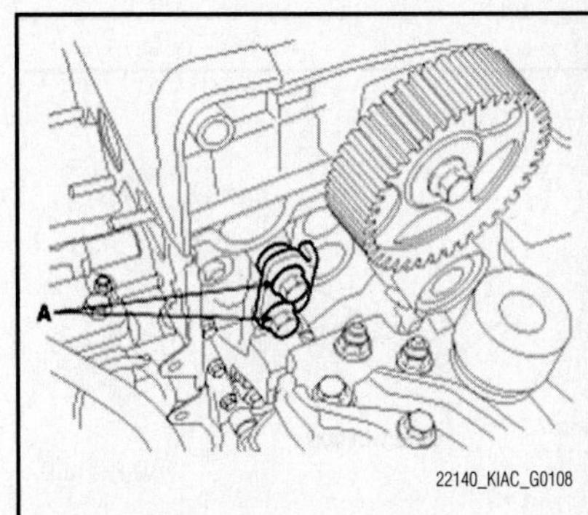

Fig. 42 Remove the engine mounting support bracket fixing bolts (A).

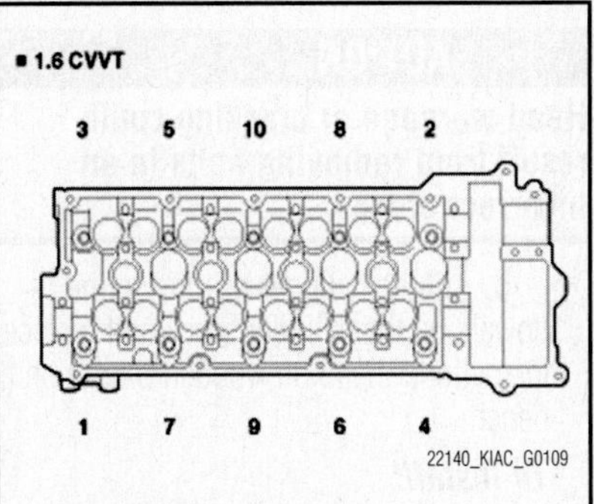

Fig. 43 Remove the cylinder head bolts in the sequence shown.

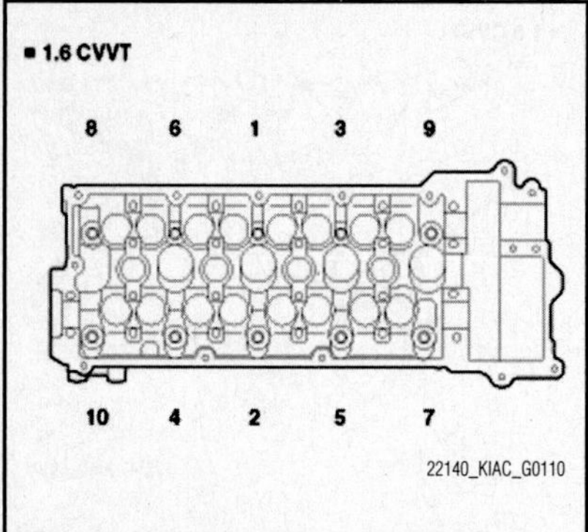

Fig. 44 Install the cylinder head bolts in the sequence shown.

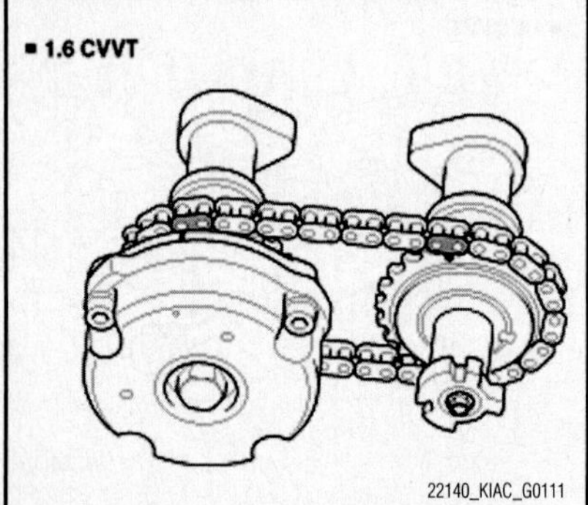

Fig. 45 Align the camshaft timing chain with the intake timing chain sprocket and exhaust timing chain sprocket as shown.

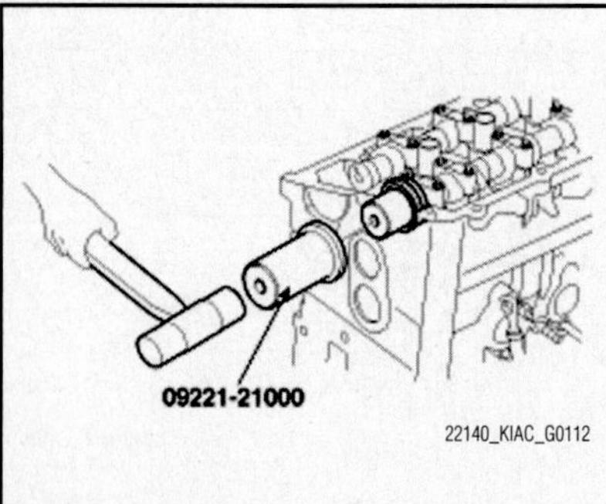

Fig. 46 Using the SST (09221 - 21000), install the camshaft bearing oil seal.

CAUTION

Head warpage or cracking could result from removing bolts in an incorrect order.

b. Lift the cylinder head from the dowels on the cylinder block and replace the cylinder head on wooden blocks on a bench.

To install:

28. Install the cylinder head gasket on the cylinder block.
29. Place the cylinder head carefully in order not to damage the gasket with the bottom part of the end.
30. Install the cylinder head bolts.
    a. Apply a light coat if engine oil on the threads and under the heads of the cylinder head bolts.
    b. Using 8mm and 10mm hexagon wrench, install and tighten the 10 cylinder head bolts and plate washers, in several passes, in the sequence shown. Torque: 21.7 ft. lbs. (29.4 Nm) plus 90°; back off all bolts, then re- torque: 21.7 ft. lbs. (29.4 Nm) plus 90°.
31. Install the engine mounting support bracket fixing bolts.
32. Install the OCV (Oil Control Valve) filter. Torque: 29.7–36.9 ft. lbs. (40.2–50.0 Nm).
33. Install the OCV (Oil Control Valve). Torque: 7.2–8.7 ft. lbs. (9.8–11.8 Nm).

CAUTION

Do not reuse the OCV (Oil Control Valve) when dropped. Keep clean the OCV (Oil Control Valve). Do not hold the OCV (Oil Control Valve) sleeve during servicing. When the OCV (Oil Control Valve) is installed on the engine, do not move the engine with holding the OCV (Oil Control Valve) yoke.

34. Install the camshafts.
    a. Align the camshaft timing chain with the intake timing chain sprocket and exhaust timing chain sprocket as shown.
    b. Install the camshaft and bearing caps. Torque: 8.7–10.1 ft. lbs. (11.8–13.7 Nm).
    c. Install the timing chain auto tensioner. Torque: 5.8–7.2 ft. lbs. (7.8–9.8 Nm)
35. Using the SST (09221 - 21000), install the camshaft bearing oil seal.
36. Install the camshaft sprocket.
37. Install the cylinder head cover.
    a. Install the cylinder head cover gasket in the groove of the cylinder head cover.

➡Before installing the cylinder head cover gasket, thoroughly clean the cylinder head cover and the groove. When installing, make sure the cylinder head cover gasket is seated securely in the corners of the recesses with no gap.

    b. Apply liquid gasket to the head cover gasket at the corners of the recess.
    c. Install the cylinder head cover with bolts. Uniformly tighten the bolts in several passes. Pre-tighten all bolts: 2.9–3.6 ft. lbs. (3.9–4.9 Nm); then tighten by the specified torque. Torque: 5.8–7.2 ft. lbs. (7.8–9.8 Nm)
38. Install the timing belt.
39. Install the intake manifold.
40. Install the exhaust manifold.
41. Install the ignition coil.
42. Install the power steering pump.
43. Install the brake booster hose.
44. Connect the hose of the PCSV (Purge Control Solenoid Valve) side.
45. Install the engine wire harness connectors and wire harness clamps to the cylinder head and the intake manifold.
    a. Install the wire harness bracket.
    b. Connect the ground cable.
    c. Connect the CMP (Camshaft position sensor) connector.
    d. Connect the ignition coil condenser connector.
    e. Connect the ignition coil connector.
    f. Connect the OCV (Oil Control Valve) connector.
    g. Connect the CKP (Crankshaft Position Sensor) connector.
    h. Connect the front oxygen sensor connector.
    i. Connect the ISA (Idle Speed Actuator) connector.
    j. Install the wire harness bracket.
    k. Connect the injector connectors (No.1,2).
    l. Connect the injector connectors (No.3,4).
    m. Connect the knock sensor connector.
    n. Connect the air conditioner compressor switch connector.
    o. Connect the rear oxygen sensor connector.
46. Connect the TPS (Throttle Position Sensor) connector.
47. Install the accelerator cable.
48. Install the fuel hose.
49. Install the heater hoses.
50. Install the upper radiator hose and lower radiator hose.
51. Install the battery tray.
52. Install the intake air hose and air cleaner assembly.
    a. Install the air cleaner element and air cleaner lower cover.
    b. Connect the ECM connector and ECM connector (A/T only).
    c. Install the intake air hose and air cleaner upper cover.
    d. Connect the breather hose to intake air hose.
    e. Connect the AFS (Air Flow Sensor) connector.

53. Install the lower cover.
54. Install the engine cover.
55. Install the battery and connect the battery terminals.
56. Fill with engine coolant.
57. Start the engine and check for leaks.
58. Recheck engine coolant level and oil level.

EXHAUST MANIFOLD

REMOVAL & INSTALLATION

1. Remove the engine cover.
2. Disconnect the front oxygen sensor connector.
3. Remove the front muffler.
4. Remove the heat protector.
5. Remove the exhaust manifold and catalytic converter assembly.
6. To install, reverse the removal procedure with new gaskets.

FLYWHEEL

REMOVAL & INSTALLATION

See Figures 47 through 49.

1. Remove the transaxle assembly.
2. Insert the special tool (09411-25000) in the clutch disc to prevent the disc from shifting.
3. Loosen the bolts which attach the clutch cover to the flywheel in a star pattern. Loosen the bolts in succession, one or two turns at a time, to avoid bending the cover.

➡Do not clean the clutch disc or the release bearing with cleaning solvent.

To install:

4. Apply multipurpose grease to the spline of the disc. Use grease CASMOLY L 9508.

⁂ CAUTION

When installing the clutch, apply grease to each part, but be careful not to apply excessive grease. It can cause clutch slippage and vibration (shudder).

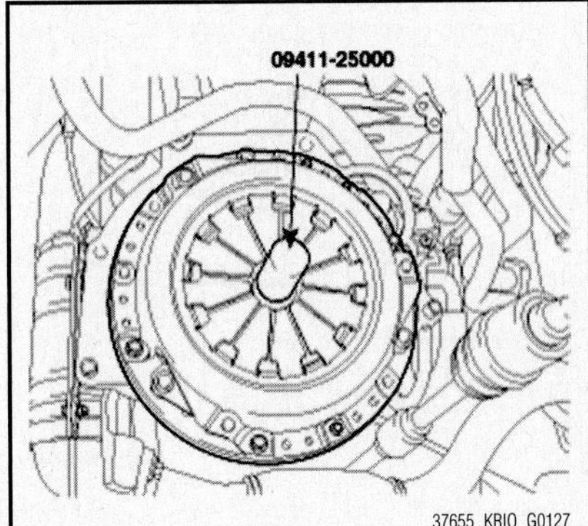

Fig. 47 Removing the transaxle assembly

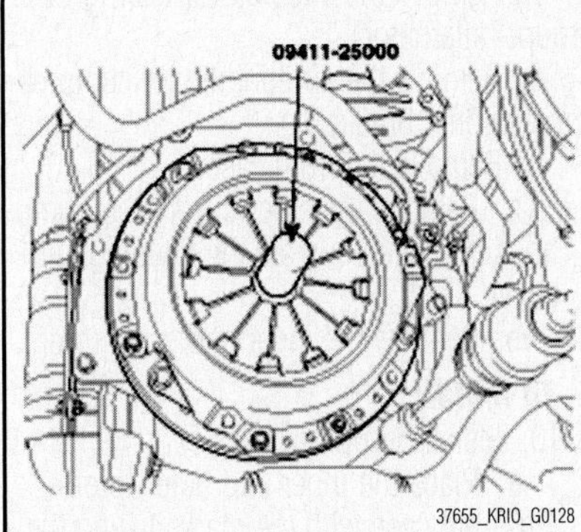

Fig. 48 Installing the clutch disc assembly to the flywheel using special tool (09411-25000).

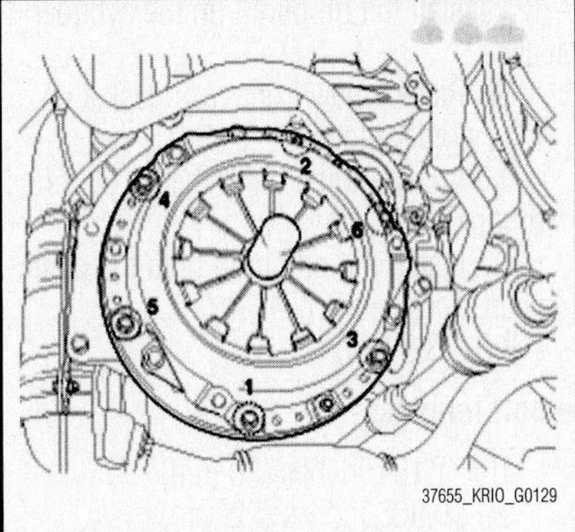

Fig. 49 Clutch cover bolt installation

5. Install the clutch disc assembly to the flywheel using the special tool (09411-25000).
6. Install the clutch cover assembly to the flywheel and temporarily tighten the bolts one or two steps at a time in a star pattern. Tighten the clutch cover bolt to 11 - 16 ft. lb. (15 - 22 Nm).
7. Install the transaxle assembly to the engine.

INTAKE MANIFOLD

REMOVAL & INSTALLATION

1. Remove the engine cover.
2. Remove the accelerator cable.
3. Disconnect the Throttle Position Sensor (TPS) connector.
4. Disconnect the Idle Speed Actuator (ISA).connector.
5. Disconnect the Positive Crankcase Ventilation (PCV) hose and breather hose.
6. Disconnect the injector connector (No. 3, 4).
7. Disconnect the injector connector (No. 1, 2).
8. Remove the heater hose, Purge Control Solenoid Valve (PCSV) hose, and the brake vacuum hose from throttle body and intake manifold.
9. Disconnect the Purge Control Solenoid Valve (PCSV) and water temperature sensor connector.
10. Remove the fuel delivery pipe.
11. Remove the intake manifold bracket.
12. Remove the intake manifold.
13. Installation is in the reverse order of removal with new gasket.

OIL PAN

REMOVAL & INSTALLATION

1. Drain the engine oil.
2. Disconnect the rear oxygen sensor connector.
3. Remove the front muffler heat protector
4. Remove the front muffler.
5. Remove the exhaust manifold and catalytic converter assembly.
6. Using the SST (09215-3C000) and remove the oil pan.

To install:

7. Install the oil pan.
   a. Using a razor blade and gasket scraper, remove all gasket material from the mating surfaces.
   b. Apply liquid gasket as an even bead, centered between the edges of the mating surface. Liquid gasket: MS 721–40A or equivalent

➡To prevent leakage of oil, apply liquid gasket to the inner threads of the bolt holes. Do not install the parts if five minutes or more have elapsed since applying the liquid gasket. After assembly, wait at least 30 minutes before filling the engine with oil.

   c. Install the oil pan with the bolts. Uniformly tighten the bolts in several passes. Torque: 7.2–8.7 ft. lbs. (9.8–11.8 Nm).
8. Install the front muffler.
9. Install the front muffler heat protector
10. Connect the rear oxygen sensor connector.
11. Fill with engine oil.

OIL PUMP

REMOVAL & INSTALLATION

See Figures 50 through 52.

1. Drain the engine oil.
2. Remove the drive belts.
3. Turn the crankshaft pulley, and align its groove with timing mark "T" of the timing belt cover.
4. Remove the timing belt. (Refer to EM - 21)
5. Remove the timing belt tensioner.
6. Remove the oil pan and oil screen.

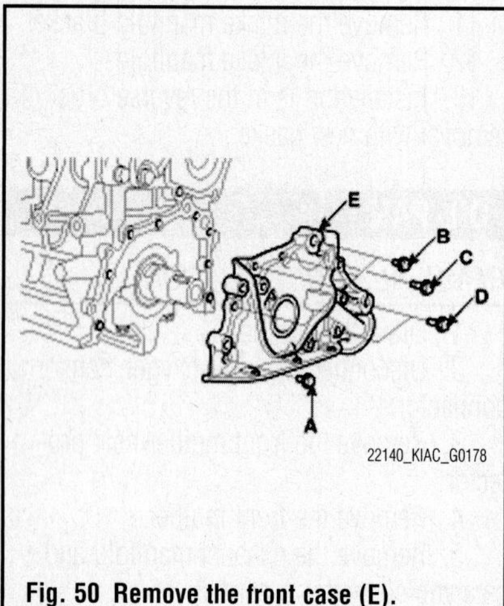

Fig. 50 Remove the front case (E).

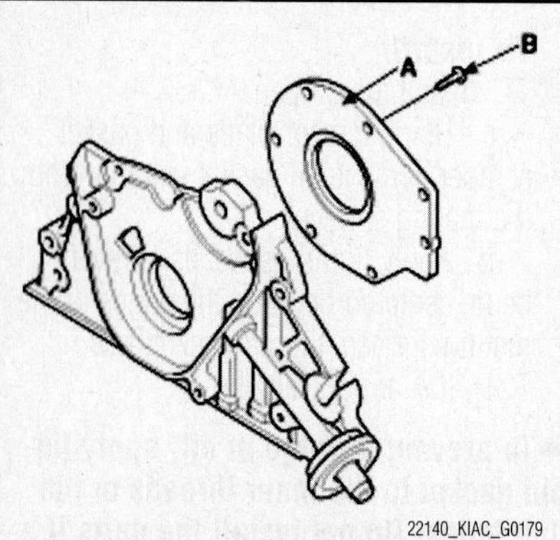

Fig. 51 Remove the screws (B) from the pump housing, then separate the housing and cover (A).

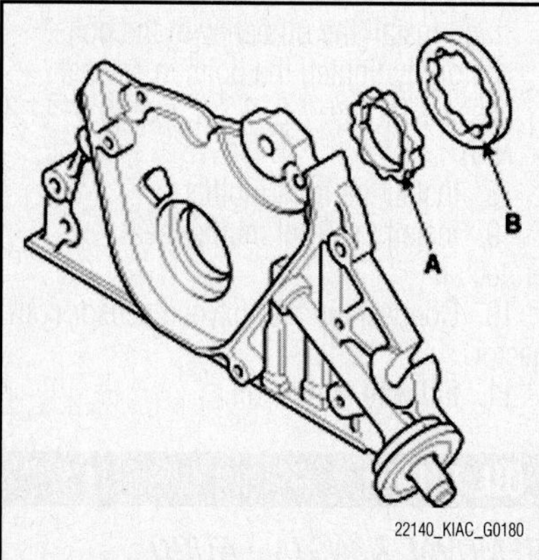

Fig. 52 Remove the inner (A) and outer (B) rotors.

7. Remove the alternator. (Refer to EE group - alternator)
8. Remove the air conditioner compressor tensioner bracket.
9. Remove the front case.
 a. Remove the screws from the pump housing, then separate the housing and cover.
 b. Remove the inner and outer rotors.

To install:

10. Install oil pump.
 a. Place the inner and outer rotors into front case with the marks facing the oil pump cover side.
 b. Install the oil pump cover to front case with the 7 screws. Torque: 4.3–5.1 ft. lbs. (5.9–6.9 Nm).
11. Check that the oil pump turns freely.
12. Install the oil pump on the cylinder block.
 a. Place a new front case gasket on the cylinder block.
 b. Apply engine oil to the lip of the oil pump seal.
 c. Install the oil pump onto the crankshaft. Torque: 13.7–17.4 ft. lbs. (18.6–23.5 Nm).

➡**Bolt lengths:**

 d. : 1.181 inches (30 mm)
 e. : 0.866 inches (22 mm)
 f. : 1.771 inches (45 mm)
 g. : 2.362 inches (60 mm)

13. Apply a light coat of oil to the seal lip.
14. Using the SST (09214-32000), install the oil seal.
15. Install the air compressor tensioner bracket.
16. Install the alternator.
17. Install the oil screen.
18. Install the oil pan. Torque: 7.2–8.7 ft. lbs. (9.8–11.8 Nm).

➡**Clean the oil pan gasket mating surfaces.**

19. Install the timing tensioner.
20. Install the timing belt.
21. Install the accessory drive belts.
22. Fill the engine oil.

REAR MAIN SEAL

REMOVAL & INSTALLATION

1. Before servicing the vehicle, refer to the Precautions Section.
2. Disconnect the negative battery cable.
3. Remove the transaxle assembly.
4. Remove the clutch and flywheel assembly, if equipped with a manual transaxle.
5. Remove the flexplate-to-crankshaft bolts, the flexplate and shim plates, if equipped with an automatic transaxle.
6. Cut the oil seal lip with a knife. Install a rag to the housing and using a screwdriver, carefully pry the oil seal from the oil seal housing. Clean the gasket mounting surfaces.

To install:

7. Clean the oil seal housing. Coat the oil seal and the housing with clean engine oil.
8. Install the oil seal into the housing and tap it evenly into place with a hammer and a large diameter piece of pipe. The seal must be flush with the edge of the rear cover.
9. Install the flywheel assembly or the flexplate, as applicable, and tighten the mounting bolts to 71–76 ft. lbs. (97–102 Nm).
10. Install the clutch assembly, if applicable.
11. Install the transaxle.
12. Connect the negative battery cable.

ROCKER ARMS & SHAFTS

REMOVAL & INSTALLATION

The engines covered are not equipped with rocker arms/shafts. The camshafts directly actuate the valve through a bucket type follower.

TIMING BELT FRONT COVER

REMOVAL & INSTALLATION

See Figures 53 through 57.

Engine removal is not required for this procedure.

1. Remove the engine cover.
2. Remove RH front wheel.
3. Remove 2bolts and RH side cover.
4. Temporarily loosen the water pump pulley bolts.
5. Remove the alternator drive belt

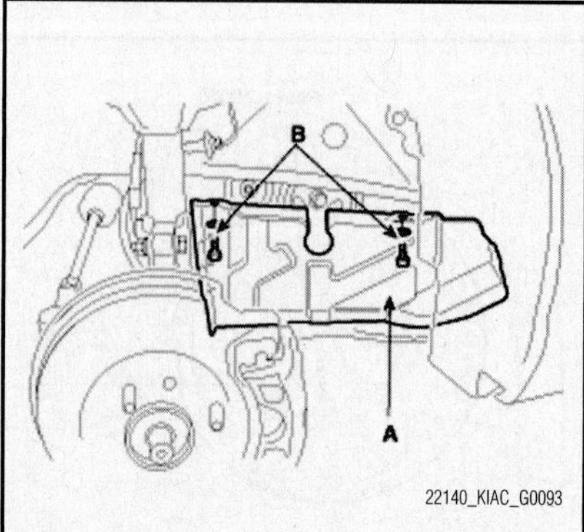

Fig. 53 Remove 2bolts (B) and RH side cover (A).

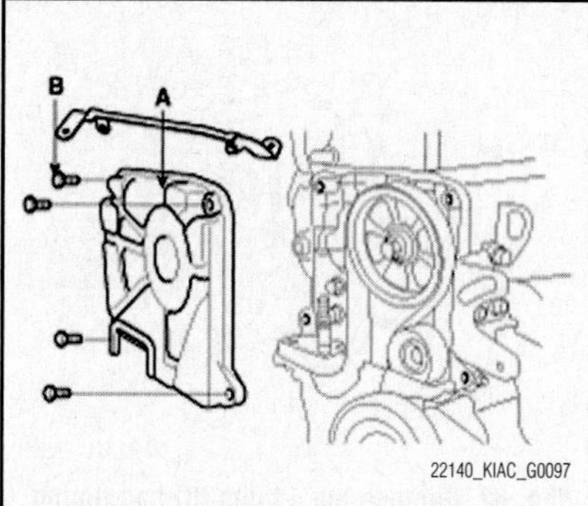

Fig. 54 Remove the 4 bolts (B) and timing belt upper cover (A).

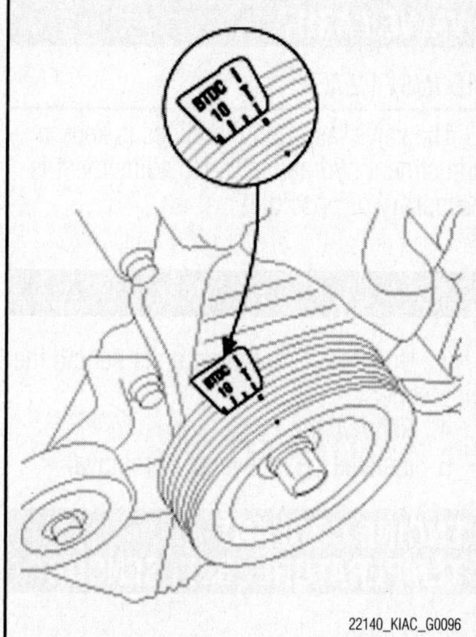

Fig. 55 Align the crankshaft pulley groove with timing mark "T" of the timing belt cover.

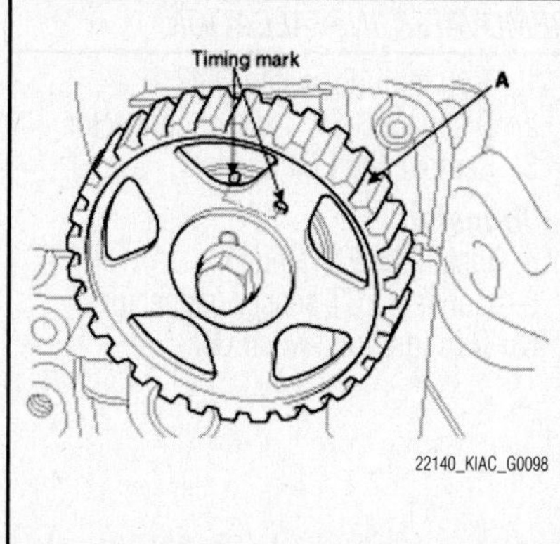

Fig. 56 Camshaft sprocket timing mark.

6. Remove the air conditioner compressor drive belt.
7. Remove the power steering pump drive belt.
8. Remove the 4 bolts and water pump pulley.

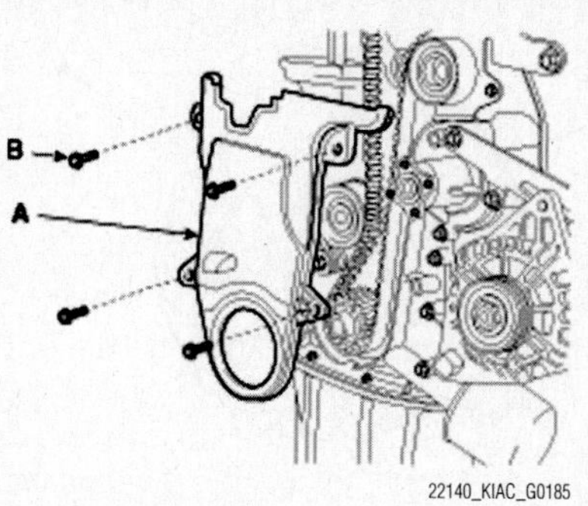

Fig. 57 Remove the 4 bolts (B) and timing belt lower cover (A).

9. Remove the 4 bolts and timing belt upper cover.
10. Turn the crankshaft pulley, and align its groove with timing mark "T" of the timing belt cover. Check that the timing mark of camshaft sprocket is aligned with the timing mark of cylinder head cover. (No.1 cylinder compression TDC position)
11. Remove the crankshaft pulley bolt and crankshaft pulley.
12. Remove the crankshaft flange.
13. Remove the 4 bolts and timing belt lower cover.

To install:

14. Install the timing belt lower cover with 4 bolts. Torque: 5.8–7.2 ft. lbs. (7.8–9.8 Nm).
15. Install the crankshaft flange.
16. Install the crankshaft pulley. Make sure that crankshaft sprocket pin fits the small hole in the pulley. Torque: 101.3–108.5 ft. lbs. (137.3–147.1 Nm).
17. Install the timing belt upper cover. Torque: 5.8–7.2 ft. lbs. (7.8–9.8 Nm).
18. Install the water pump pulley.
19. Install the power steering pump drive belt.
20. Install the air conditioner compressor drive belt.
21. Install the alternator drive belt.
22. Install the RH side cover.
23. Install the RH front wheel.
24. Install the engine cover with bolts.

TIMING BELT & SPROCKETS

REMOVAL & INSTALLATION

See Figures 58 through 62.

Engine removal is not required for this procedure.

1. Remove the engine cover.
2. Remove RH front wheel.
3. Remove 2bolts and RH side cover.
4. Temporarily loosen the water pump pulley bolts.
5. Remove the alternator drive belt
6. Remove the air conditioner compressor drive belt.
7. Remove the power steering pump drive belt.
8. Remove the 4 bolts and water pump pulley.
9. Remove the 4 bolts and timing belt upper cover.
10. Turn the crankshaft pulley, and align its groove with timing mark "T" of the timing belt cover. Check that the timing mark of camshaft sprocket is aligned with the timing mark of cylinder head cover. (No.1 cylinder compression TDC position)
11. Remove the crankshaft pulley bolt and crankshaft pulley.
12. Remove the crankshaft flange.
13. Remove the 4 bolts and timing belt lower cover.

To install:

14. Install the timing belt lower cover with 4 bolts. Torque: 5.8–7.2 ft. lbs. (7.8–9.8 Nm).
15. Install the crankshaft flange.
16. Install the crankshaft pulley. Make sure that crankshaft sprocket pin fits the small hole in the pulley. Torque: 101.3–108.5 ft. lbs. (137.3–147.1 Nm).

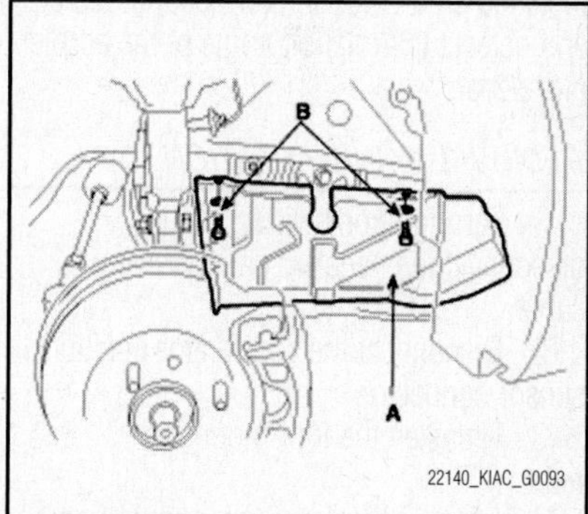

Fig. 58 Remove 2bolts (B) and RH side cover (A).

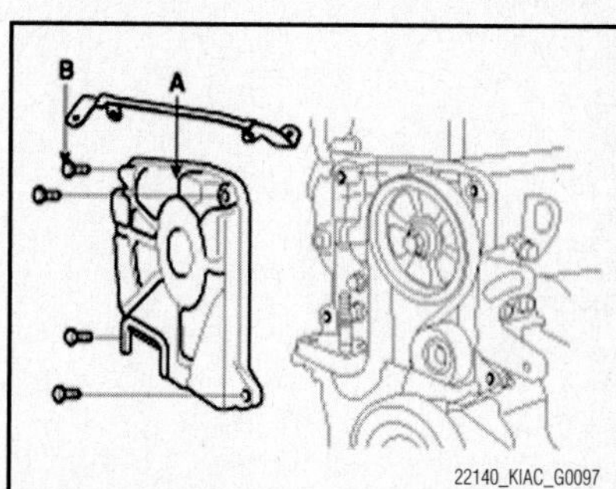

Fig. 59 Remove the 4 bolts (B) and timing belt upper cover (A).

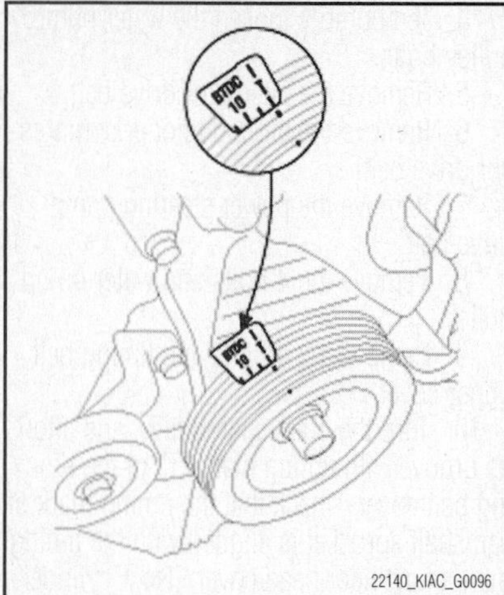

Fig. 60 Align the crankshaft pulley groove with timing mark "T" of the timing belt cover.

17. Install the timing belt upper cover. Torque: 5.8–7.2 ft. lbs. (7.8–9.8 Nm).

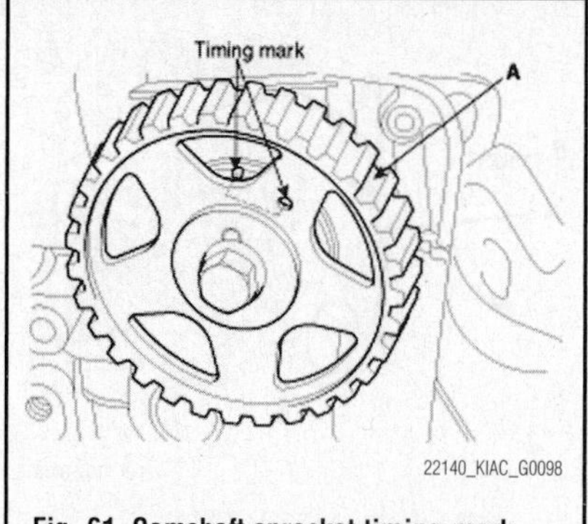

Fig. 61 Camshaft sprocket timing mark.

18. Install the water pump pulley.
19. Install the power steering pump drive belt.
20. Install the air conditioner compressor drive belt.
21. Install the alternator drive belt.
22. Install the RH side cover.
23. Install the RH front wheel.

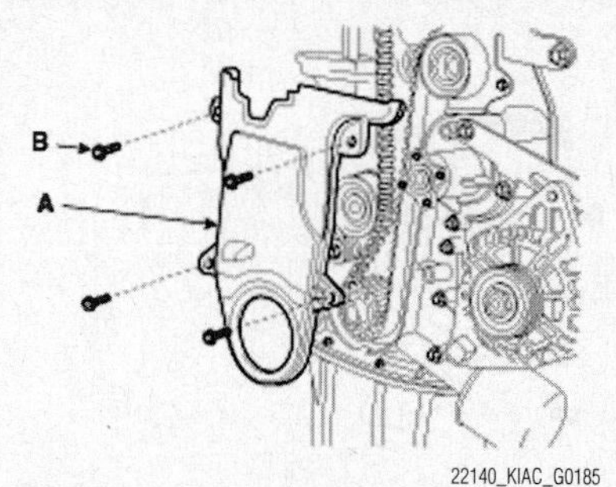

Fig. 62 Remove the 4 bolts (B) and timing belt lower cover (A).

24. Install the engine cover with bolts.

VALVE LASH

ADJUSTMENT

The valve lash on all engines is kept in adjustment hydraulically. No adjustment is necessary or possible.

ENGINE PERFORMANCE & EMISSION CONTROLS

ACCELERATOR PEDAL POSITION (APP) SENSOR

LOCATION

The Accelerator Position Sensor (APS) is installed on the accelerator pedal module and detects the rotation angle of the accelerator pedal.

REMOVAL & INSTALLATION

1. Turn ignition switch off and disconnect the negative battery cable.
2. Disconnect the accelerator position sensor connector.
3. Unfasten the four mounting nuts
4. Remove the accelerator pedal from the vehicle.
5. Installation is the reverse of removal.

CAMSHAFT POSITION (CMP) SENSOR

LOCATION

The two Camshaft Position (CMP) sensors are installed on the bank 1 and 2 engine head covers.

CRANKSHAFT POSITION (CKP) SENSOR

LOCATION

The Crankshaft Position (CKP) sensor is installed on transaxle housing

ELECTRONIC CONTROL MODULE (ECM)

REMOVAL & INSTALLATION

1. Turn the ignition switch off and disconnect the negative battery terminal.
2. Disconnect the ECM connector.
3. Unscrew 4 mounting bolts behind the air cleaner.
4. Remove the ECM.
5. Installation is reverse of removal.

ENGINE COOLANT TEMPERATURE (ECT) SENSOR

LOCATION

The Engine Coolant Temperature Sensor (ECT) is located in the engine coolant passage of the cylinder head for detecting the engine coolant temperature.

REMOVAL & INSTALLATION

1. Turn ignition switch OFF.
2. Disconnect ECT sensor connector.
3. Remove the ECT sensor.

To install:

4. Install the ECT sensor.
5. Connect ECT sensor connector.
6. Turn ignition switch ON.

FUEL SYSTEM SERVICE PRECAUTIONS

Safety is the most important factor when performing not only fuel system maintenance, but any type of maintenance. Failure to conduct maintenance and repairs in a safe manner may result in serious personal injury or death. Work on a vehicle's fuel system components can be accomplished safely and effectively by adhering to the following rules and guidelines.

- To avoid the possibility of fire and personal injury, always disconnect the negative battery cable unless the repair or test procedure requires that battery voltage be applied.
- Always relieve the fuel system pressure prior to disconnecting any fuel system component (injector, fuel rail, pressure regulator, etc.) fitting or fuel line connection. Exercise extreme caution whenever relieving fuel system pressure to avoid exposing skin, face and eyes to fuel spray. Please be advised that fuel under pressure may penetrate the skin or any part of the body that it contacts.
- Always place a shop towel or cloth around the fitting or connection prior to loosening to absorb any excess fuel due to spillage. Ensure that all fuel spillage is quickly removed from engine surfaces. Ensure that all fuel-soaked cloths or towels are deposited into a flame-proof waste container with a lid.
- Always keep a dry chemical (Class B) fire extinguisher near the work area.
- Do not allow fuel spray or fuel vapors to come into contact with a spark or open flame.
- Always use a second wrench when loosening or tightening fuel line connection fittings. This will prevent unnecessary stress and torsion on fuel piping. Always follow the proper torque specifications.
- Always replace worn fuel fitting O-rings with new ones. Do not substitute fuel hose where rigid pipe is installed.

FUEL SYSTEM PRESSURE

RELIEVING

1. Remove the rear seat cushion
2. Remove the rear seat cushion.
3. Disconnect the fuel pump connector.
4. Start the engine and wait until fuel in fuel line is exhausted.
5. After engine stalls, turn the ignition switch to OFF position.

FUEL FILTER

REMOVAL & INSTALLATION

The fuel filter is an integral part of the fuel pump assembly.

1. Remove the rear seat cushion.
2. Remove the service cover
3. Disconnect the fuel pump connector.
4. Start the engine and wait until fuel in fuel line is exhausted.
5. After engine stalls, turn the ignition switch to OFF position.
6. Disconnect the fuel feed line and canister hoses.
7. Unscrew the fuel pump mounting bolts and remove the fuel pump assembly.
8. Install the Fuel Pump according to the reverse order of REMOVAL procedure.

FUEL LEVEL SENDING UNIT

REMOVAL & INSTALLATION

1. Disconnect the fuel pump connector.
2. Start the engine and wait until fuel in fuel line is exhausted.
3. After engine stalls, turn the ignition switch to OFF position.
4. Disconnect the fuel feed quick-connector and the vapor hose .
5. Unscrew the fuel pump locking ring with the special service tool (SST No.: 09310-3K000) and remove the fuel pump assembly.
6. Installation is reverse of removal.

FUEL PUMP

REMOVAL & INSTALLATION

1. Remove the rear seat cushion.
2. Remove the service cover
3. Disconnect the fuel pump connector.
4. Start the engine and wait until fuel in fuel line is exhausted.
5. After engine stalls, turn the ignition switch to OFF position.
6. Disconnect the fuel feed line and canister hoses.
7. Unscrew the fuel pump mounting bolts and remove the fuel pump assembly.
8. Install the Fuel Pump according to the reverse order of REMOVAL procedure.

FUEL PRESSURE REGULATOR

REMOVAL & INSTALLATION

The fuel pressure regulator is an integral part of the fuel pump assembly.

1. Remove the rear seat cushion.
2. Remove the service cover
3. Disconnect the fuel pump connector.
4. Start the engine and wait until fuel in fuel line is exhausted.
5. After engine stalls, turn the ignition switch to OFF position.
6. Disconnect the fuel feed line and canister hoses.
7. Unscrew the fuel pump mounting bolts and remove the fuel pump assembly.
8. Install the Fuel Pump according to the reverse order of REMOVAL procedure.

FUEL TANK

REMOVAL & INSTALLATION

1. Remove the rear seat cushion.
2. Remove the service cover
3. Disconnect the fuel pump connector.
4. Start the engine and wait until fuel in fuel line is exhausted.
5. After engine stalls, turn the ignition switch to OFF position.
6. Disconnect the negative terminal from the battery.
7. Disconnect the fuel feed line and canister hose.
8. Raise the vehicle.
9. Remove the center muffler.
10. Support the fuel tank with a jack.
11. Remove the brake hose mounting bolts.
12. Disconnect the fuel filler pipe, the leveling hose and canister hose.
13. Remove the fuel tank mounting bolts and nuts, and then remove the fuel tank.
14. Install the Fuel Tank according to the reverse order to REMOVAL procedure.

HEATING & AIR CONDITIONING

BLOWER MOTOR

REMOVAL & INSTALLATION

1. Disconnect the negative battery terminal.
2. Disconnect the connector of the blower motor.
3. Remove the blower motor after loosening the mounting screws.
4. Installation is the reverse order of removal.

HEATER CORE

REMOVAL & INSTALLATION

See Figure 63.

1. Disconnect the negative battery terminal.
2. Recover the refrigerant with a recovery/ recycling/ charging station.
3. When the engine is cool, drain the engine coolant from the radiator.
4. Remove the bolts and the expansion valve from the evaporator core.
5. Plug or cap the lines immediately after disconnecting them to avoid moisture and dust contamination.
6. Disconnect the inlet and outlet heater hoses from the heater unit.
7. Remove the crash pad.
8. Remove the cowl cross bar assembly.
9. Disconnect the electrical connectors from the temperature control actuator, the mode control actuator and the evaporator temperature sensor.
10. Remove the heater & blower unit after loosening 2 mounting bolts.
11. Remove the blower unit from heater unit after loosening 3 screws.

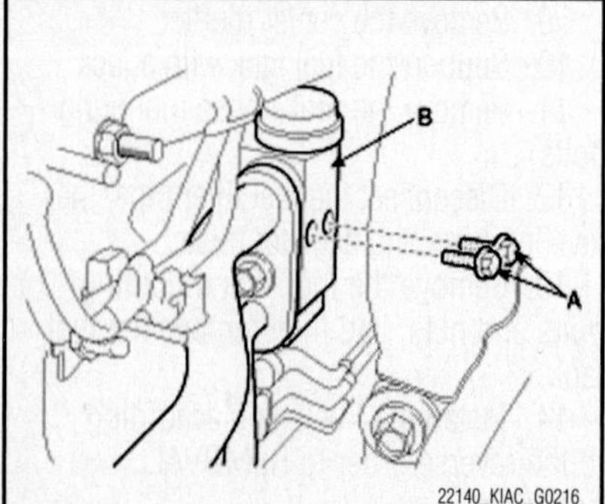

Fig. 63 Remove the bolts (A) and the expansion valve (B) from the evaporator core.

12. Remove the heater core after removing the cover.

➡Be careful that the inlet and outlet pipe are not bent during heater core removal, and pull out the heater core.

To install:

13. Install the heater core in the reverse order of removal.
14. Installation is the reverse order of removal, and note these items :
    a. If you're installing a new evaporator, add refrigerant oil (ND-OIL8).
    b. Replace the O-rings with new ones at each fitting, and apply a thin coat of refrigerant oil before installing them. Be sure to use the right O-rings for R-134a to avoid leakage.
    c. Immediately after using the oil, replace the cap on the container, and seal it to avoid moisture absorption.
    d. Do not spill the refrigerant oil on the vehicle ; it may damage the paint ; if the refrigerant oil contacts the paint, wash it off immediately.
    e. Apply sealant to the grommets.
    f. Make sure that there is no air leakage.
    g. Charge the system and test its performance.
    h. Do not interchange the inlet and outlet heater hoses and install the hose clamps securely.
    i. Refill the cooling system with engine coolant.

HEATER UNIT

REMOVAL & INSTALLATION

See Figures 64 through 68.

1. Disconnect the negative (-) battery terminal.
2. Recover the refrigerant with a recovery/ recycling/ charging station. (Refer to HA-10)
3. When the engine is cool, drain the engine coolant from the radiator.
4. Remove the bolts and the expansion valve from the evaporator core.

Plug or cap the lines immediately after disconnecting them to avoid moisture and dust contamination.

5. Disconnect the inlet and outlet heater hoses from the heater unit.

⁂ CAUTION

Engine coolant will run out when the hoses are disconnected; drain it into a clean drip pan. Be sure not to let coolant spill on electrical parts or painted surfaces. If any coolant spills, rinse it off immediately.

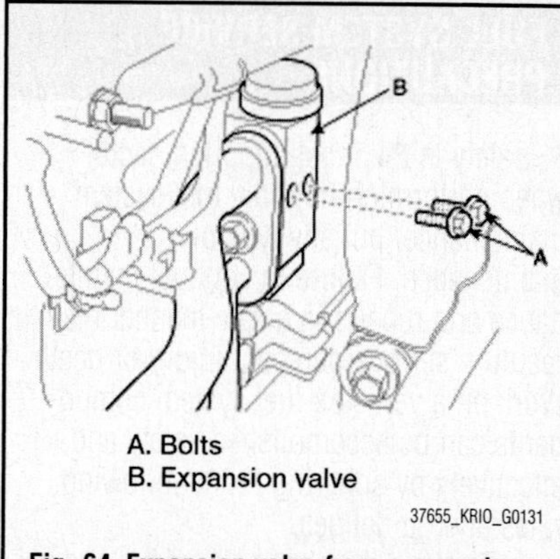

Fig. 64 Expansion valve from evaporator core removal

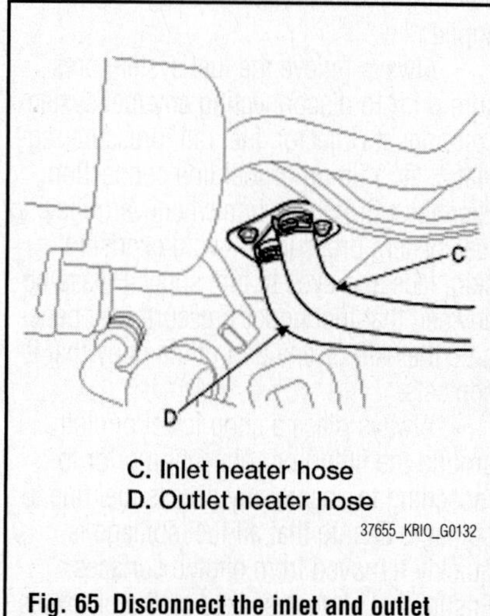

Fig. 65 Disconnect the inlet and outlet hoses from the heater unit

6. Remove the crash pad (Refer to BD group - Crash pad).
7. Remove the cowl cross bar assembly. (Refer to BD group - Crash pad)
8. Disconnect the connectors from the temperature control actuator, the mode control actuator and the evaporator temperature sensor.
9. Remove the heater & blower unit after loosening 2 mounting bolts.
10. Remove the blower unit from heater unit after loosening 3 screws.
11. Remove the heater core after removing the cover.
12. Be careful that the inlet and outlet pipe are not bent during heater core removal, and pull out the heater core.

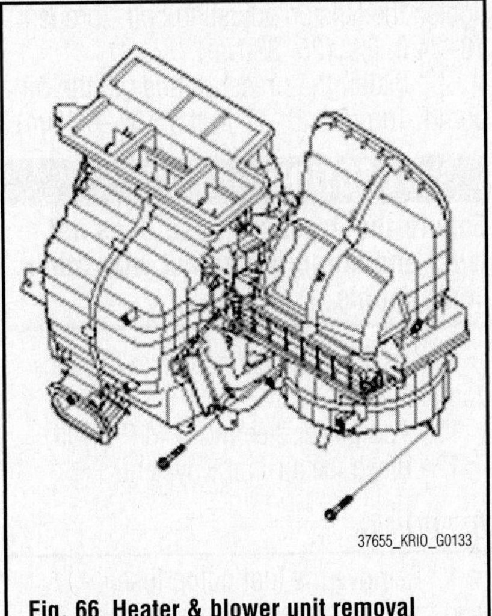

Fig. 66 Heater & blower unit removal

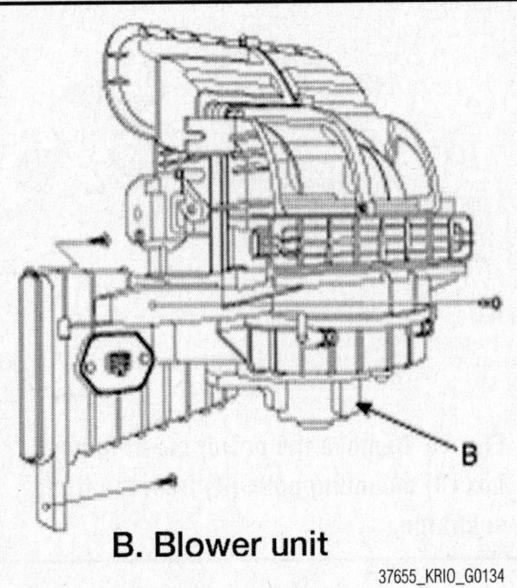

Fig. 67 Blower unit from heater unit removal

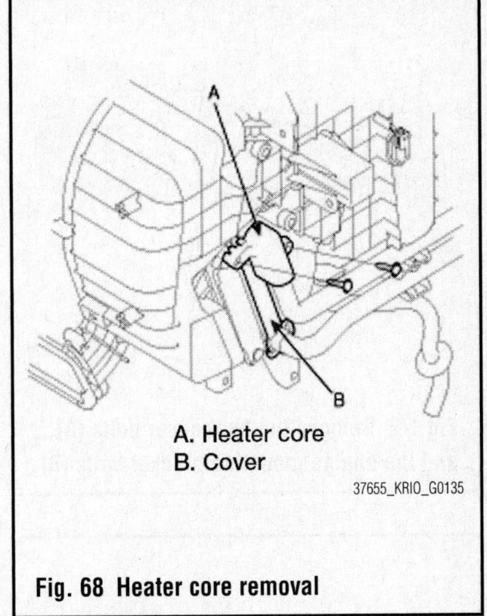

Fig. 68 Heater core removal

To install:

13. Install the heater core in the reverse order of removal.
14. Installation is the reverse order of removal, and note these items :
    a. If you're installing a new evaporator, add refrigerant oil (ND-OIL8).
    b. Replace the O-rings with new ones at each fitting, and apply a thin coat of refrigerant oil before installing them. Be sure to use the right O-rings for R-134a to avoid leakage.
    c. Immediately after using the oil, replace the cap on the container, and seal it to avoid moisture absorption.
    d. Do not spill the refrigerant oil on the vehicle ; it may damage the paint ; if the refrigerant oil contacts the paint, wash it off immediately.
    e. Apply sealant to the grommets.
    f. Make sure that there is no air leakage.
    g. Charge the system and test its performance.
    h. Do not interchange the inlet and outlet heater hoses and install the hose clamps securely.
    i. Refill the cooling system with engine coolant.

STEERING

POWER RACK & PINION STEERING GEAR

REMOVAL & INSTALLATION

See Figures 69 through 75.

1. Drain the power steering fluid by disconnecting the return hose.
2. Remove the pressure pipe from the power steering oil pump.
3. Remove the steering shaft universal joint assembly mounting bolt.
4. Raise and safely support the front of the vehicle.
5. Remove the both front tires.
6. Remove the lower arm mounting bolts and the tie rod end from the knuckle.
7. Remove the stabilizer link from the strut assembly.

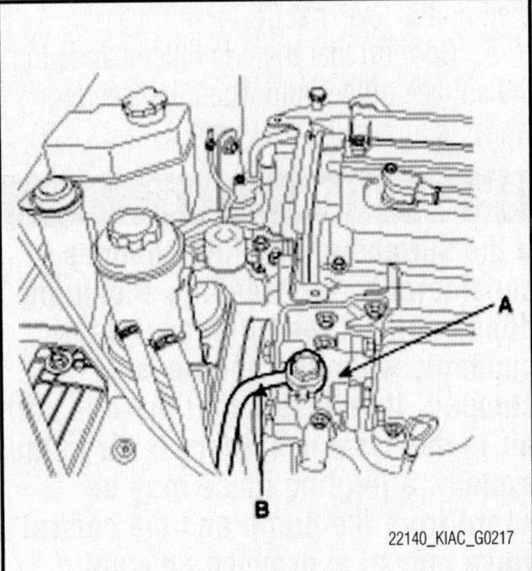

Fig. 69 Remove the pressure pipe (B) from the power steering oil pump (A).

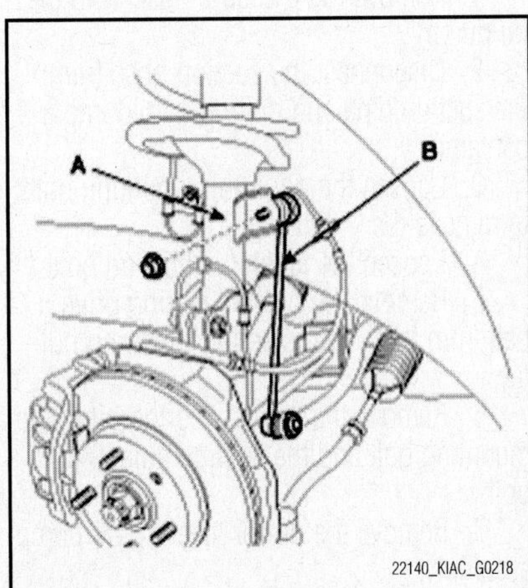

Fig. 70 Remove the stabilizer link (B) from the strut assembly (A).

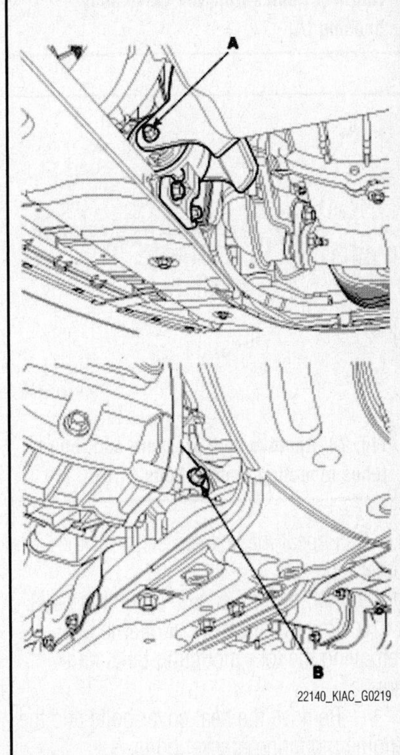

Fig. 71 Remove the engine mounting bolts (A,B).

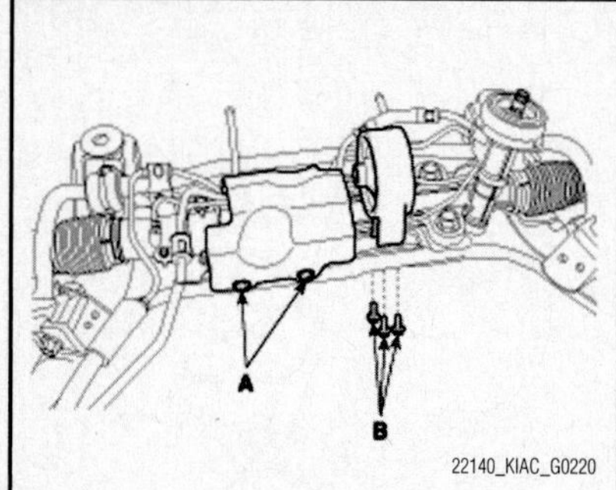

Fig. 72 Remove the heat cover bolts (A) and the engine mounting bracket bolts (B).

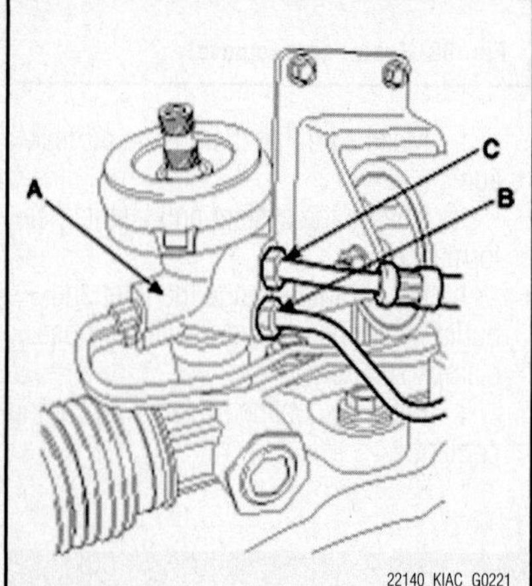

Fig. 73 Remove the both pressure (B) and return (C) tubes from the valve body housing (A).

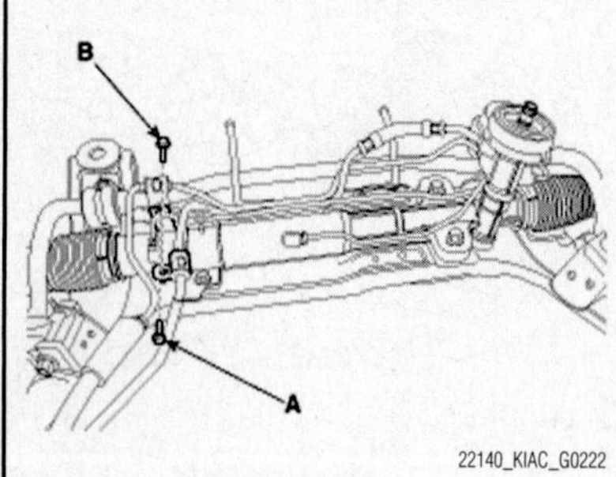

Fig. 74 Remove both pressure and return tubes mounting bracket bolts (A,B).

8. Repeat the last two steps on the other side.

9. Remove the engine mounting bolts.

10. Remove the front subframe by removing the four mounting bolts and nuts.

11. Remove the heat cover bolts and the engine mounting bracket bolts.

12. Remove the both pressure and return tubes from the valve body housing.

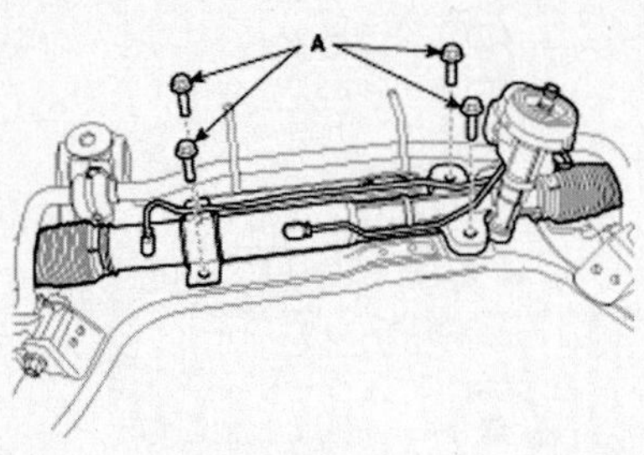

Fig. 75 Remove the power steering gear box (B) mounting bolts (A) from the front subframe.

13. Remove both pressure and return tubes mounting bracket bolts.

14. Remove the power steering gear box mounting bolts from the front subframe.

To install:

15. Installation is reverse of removal.

16. Use the following torque specifications:

- Engine mounting bracket to subframe: 36–47 ft. lbs. (50–65 Nm).
- Heat cover mounting: 5.8–8.6 ft. lbs. (8–12 Nm).
- Gear box mounting: 65–79 ft. lbs. (90–110 Nm).
- Subframe mounting: 69–86 ft. lbs. (95–120 Nm).
- Engine mounting: 36–47 ft. lbs. (50–65 Nm).
- Pressure pipe to oil pump: 40–47 ft. lbs. (55–65 Nm).

17. Bleed the air in the power steering system.

18. Check wheel alignment

POWER STEERING PUMP

REMOVAL & INSTALLATION

1. Remove the pressure hose from the oil pump.

2. Disconnect the suction hose from the suction pipe and drain the fluid into a container.

3. Loosen the oil pump mounting bolts to remove the V belt.

4. Loosen the tension adjusting bolt.

5. Remove the power steering drive belt from the power steering oil pump pulley.

6. Remove the power steering oil pump mounting bolt and the tension adjusting bolt.

7. Remove the power steering oil pump assembly.

To install:

8. After installing the oil pump to the oil pump bracket, install the V belt and tighten the tension adjusting bolt. Torque: 18–24 ft. lbs. (25–33 Nm).

9. Install the pressure hose to the oil pump. Torque: 40–47 ft. lbs. (55–65 Nm).

CAUTION

Ensure the pressure hose does not twist and come in contact with other components.

10. Install the suction hose to the suction pipe.

11. Add power steering fluid (PSF-III).

12. Bleed the air in the system.

BLEEDING

1. Remove the fuel pump fuse, then start the engine and wait for the engine to stall. Next, while operating the starting motor intermittently (for 15–20 seconds), turn the steering wheel all the way to the left and then to the right five or six times.

➡ During air bleeding, replenish the fluid supply so that the level never falls below the lower position of the filter. If air bleeding is done while the vehicle is idling, the air will be broken up and absorbed into the fluid. Be sure to do the bleeding only while cranking.

2. Reinstall the fuel pump fuse, and start the engine (idling).

3. Turn the steering wheel to the left and the right until there are no air bubbles in the oil reservoir.

CAUTION

Do not hold the steering wheel turned all the way to either side for more than ten seconds.

4. Confirm that the fluid is not milky, and that the level is up to the position specified on the level gauge.

5. Confirm that there is little change in the surface of the fluid when the steering wheel is turned left and right.

CAUTION

If the surface of the fluid changes considerably, air bleeding should be done again. If the fluid level rises suddenly when the engine is stopped, it indicates that there is still air in the system. If there is air in the system, a jingling noise may be heard from the pump and the control valve may also produce unusual noises. Air in the system will shorten the life of the pump and other parts.

LOWER BALL JOINTS

REMOVAL & INSTALLATION

See Figure 76.

1. Raise the front of the vehicle, and make sure it is securely supported.
2. Remove the front wheel and tire.
3. Remove the lower arm ball joint mounting bolts.

To install:

4. Install the lower arm ball joint mounting bolts. Torque: 72–86 ft. lbs. (100–120 Nm).
5. Install the front wheel and tire.

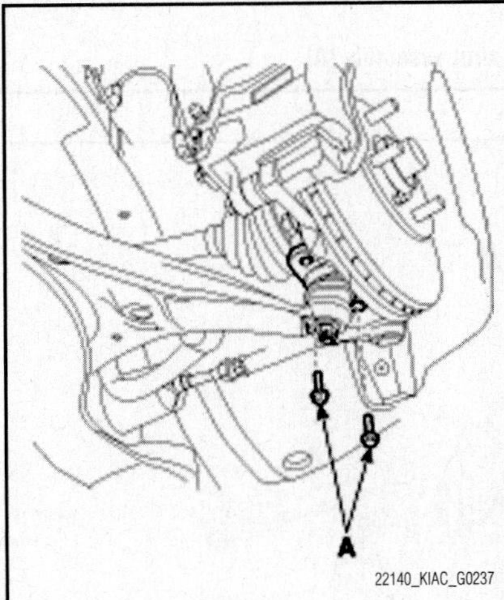

Fig. 76 Remove the lower arm ball joint mounting bolts (A).

LOWER CONTROL ARMS

REMOVAL & INSTALLATION

See Figures 76 and 77.

1. Raise the front of the vehicle, and make sure it is securely supported.
2. Remove the front wheel and tire.
3. Remove the lower arm ball joint mounting bolts.
4. Remove the lower arm mounting bolts.
5. Remove lower control arm.

To install:

6. Install the lower arm mounting bolts. Torque: A bushing: 72–86 ft. lbs. (100–120 Nm); G bushing: 72–101 ft. lbs. (100–140 Nm).
7. Install the lower arm ball joint mounting bolts. Torque: 72–86 ft. lbs. (100–120 Nm).

22140_KIAC_G0238

Fig. 77 Remove the lower arm mounting bolts (A).

8. Install the front wheel and tire.

MACPHERSON STRUT

REMOVAL & INSTALLATION

See Figures 78 through 80.

1. Raise the vehicle, and make sure it is securely supported.
2. Remove the front wheel and tire.
3. Remove the brake hose bracket and speed sensor wire mounting bolt from the strut assembly.
4. Remove the speed sensor wire mounting bolt and speed sensor.
5. Remove the nut from the stabilizer bar link.
6. Remove the strut upper mounting nuts.
7. Remove the strut lower mounting bolts and then remove the strut assembly.

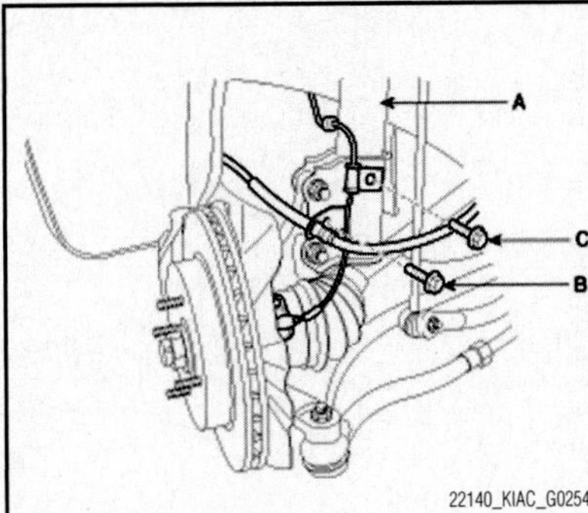

Fig. 78 Remove the brake hose bracket (B) and speed sensor wire mounting bolt (C) from the strut assembly (A).

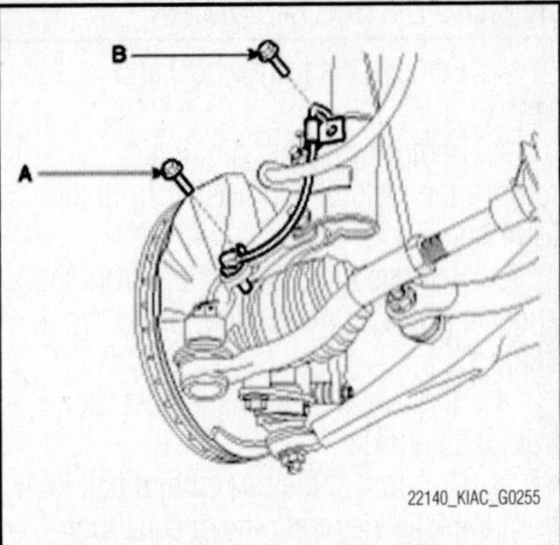

Fig. 79 Remove the speed sensor wire mounting bolt (B) and speed sensor (A).

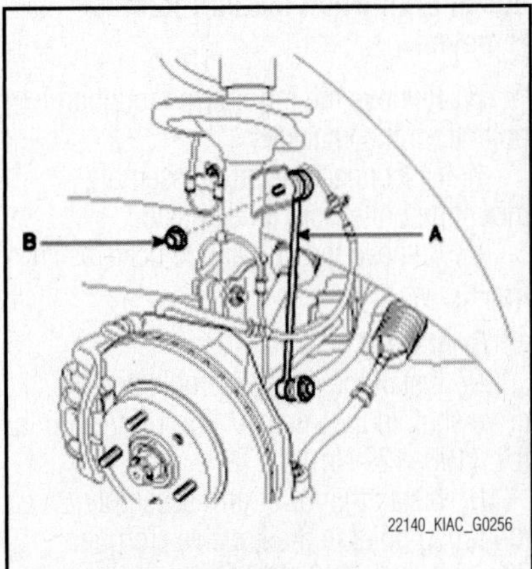

Fig. 80 Remove the nut (B) from the stabilizer bar link (A).

To install:

8. Install the strut assembly and lower mounting bolts. Torque: 72–86 ft. lbs. (100–120 Nm)
9. Install the strut upper mounting nuts. Torque: 14.4–21.6 ft. lbs. (20–30 Nm).
10. Install the nut on the stabilizer bar link. Torque: 25–32 ft. lbs. (35–45 Nm).
11. Install the speed sensor wire mounting bolt and speed sensor.
12. Install the brake hose bracket and speed sensor wire mounting bolt on the strut assembly.
13. Install the wheel and the tire.

OVERHAUL

Coil spring removal and installation and any other strut disassembly should go here

STEERING KNUCKLE

REMOVAL & INSTALLATION

1. Remove the front wheel and tire.
2. Remove the split pin, then remove the locknut and washer from the front hub.
3. Remove the brake caliper from the knuckle and hang the caliper on the front damper.
4. Remove the wheel speed sensor from the knuckle.
5. Disconnect the tie rod end ball joint from the knuckle using the special tool (09568-34000).

➡Be sure to secure the ball joint, remove tool to the vehicle so that it doesn't fall when the ball joint is removed.

6. Remove the lower arm mounting bolts from the knuckle.
7. Disconnect the strut assembly mounting bolts from the knuckle.
8. Remove the hub and knuckle as an assembly.

To install:

9. Install the strut assembly and the drive shaft in the knuckle. Torque: 72–86 ft. lbs. (100–120 Nm).
10. Install the lower arm assembly mounting bolts to the knuckle. Torque: 72–86 ft. lbs. (100–120 Nm).
11. Install the wheel speed sensor to the knuckle.
12. Install the brake caliper assembly to the knuckle. Torque: 58–72 ft. lbs. (80–100 Nm).
13. Install the tie rod end ball joint nut and insert the split pin. Torque: 12–25 ft. lbs. (16–34 Nm).
14. Insert the washer and tighten the locking nut. Torque: 145–188 ft. lbs. (200–260 Nm).
15. Install the wheel and tire.

STABILIZER BAR & LINKS

REMOVAL & INSTALLATION

See Figures 81 through 83.

1. Raise the front of the vehicle, and make sure it is securely supported.
2. Remove the front wheel and tire.
3. Remove the stabilizer bar link from the strut assembly.
4. Remove the tie rod end from the knuckle by using the special tool (09568-4A000).

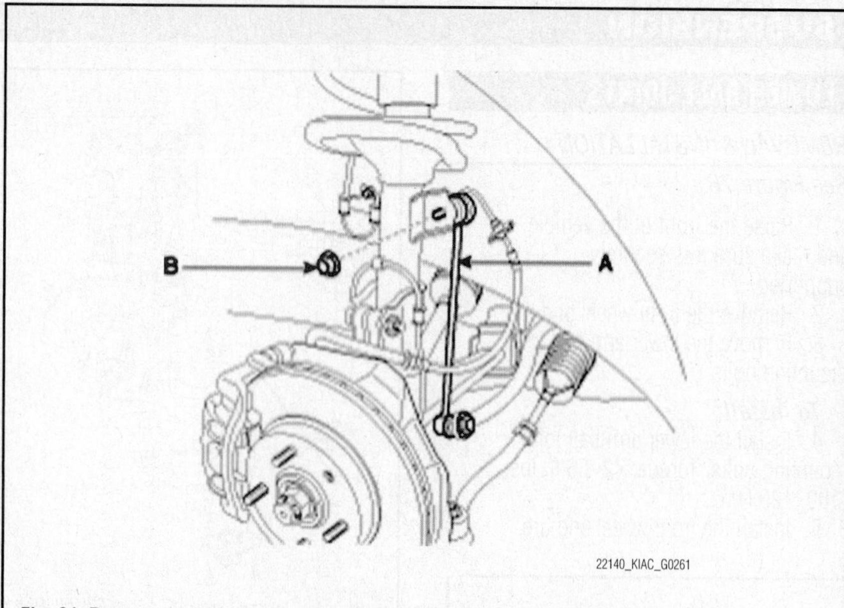

Fig. 81 Remove the stabilizer bar link (B) from the strut assembly (A).

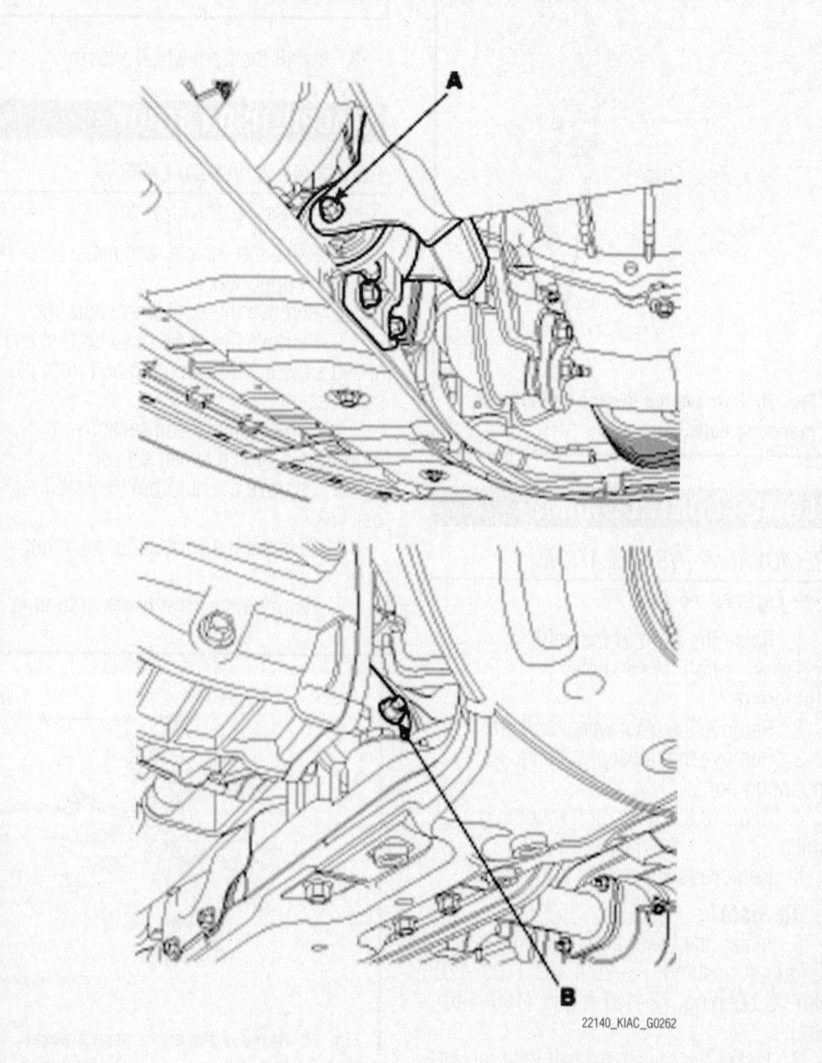

Fig. 82 Remove two engine mounting bolts (A,B).

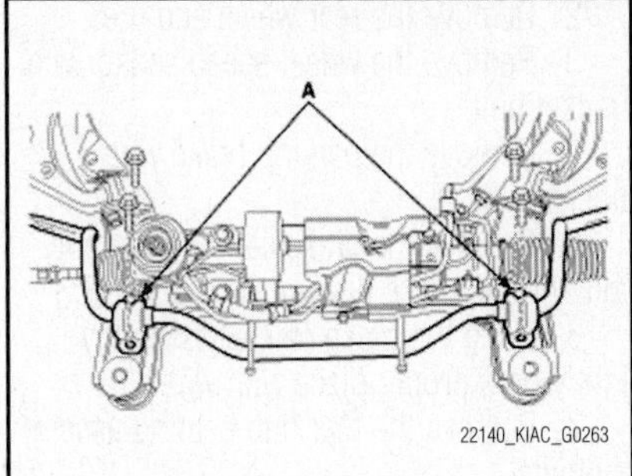

Fig. 83 Remove both stabilizer brackets (A) and two bushings.

5. Remove the two bolts for lower arm ball joint.
6. Drain power steering oil.
7. Remove the pressure pipe mounting bolt.
8. Disconnect between the return hose and tube.
9. Remove two engine mounting bolts and six subframe mounting bolts in order to remove the subframe.
10. Remove both stabilizer brackets and two bushings.
11. Remove the stabilizer bar.

To install:

12. Install the bushing on the stabilizer bar.

➡Bring clamp of stabilizer bar into contact with bushing.

13. Install the bracket on the bushing.
14. After tightening the bolts of the bushing bracket temporarily, install the bushing bracket on the opposite side.
15. Install the six subframe mounting bolts. Torque: 68–86 ft. lbs. (95–120 Nm).
16. Install the two engine mounting bolts. Torque: 36–47 ft. lbs. (50–65 Nm).
17. Install the power steering pressure pipe mounting bolt.
18. Connect the power steering return tube and hose.
19. Install the two bolts for the lower arm ball joint. Torque: 72–86 ft. lbs. (100–120 Nm).
20. Install the nut on the stabilizer bar link. Torque: 25–32 ft. lbs. (35–45 Nm).
21. Install the tie rod end on the knuckle.
22. Install the front wheel and tire.
23. Refill the power steering fluid (PSF-3).

⁂ CAUTION

After installation, bleed the air in the power steering system.

WHEEL BEARINGS

REMOVAL & INSTALLATION

1. Remove the front wheel and tire.
2. Remove the split pin, then remove the locknut and washer from the front hub.
3. Remove the brake caliper from the knuckle and hang the caliper on the front damper.
4. Remove the wheel speed sensor from the knuckle.
5. Disconnect the tie rod end ball joint from the knuckle using the special tool (09568-34000).

➡Be sure to secure the ball joint, remove tool to the vehicle so that it doesn't fall when the ball joint is removed.

6. Remove the lower arm mounting bolts from the knuckle.
7. Disconnect the strut assembly mounting bolts from the knuckle.
8. Remove the hub and knuckle as an assembly.

To install:

9. Install the strut assembly and the drive shaft in the knuckle. Torque: 72–86 ft. lbs. (100–120 Nm).
10. Install the lower arm assembly mounting bolts to the knuckle. Torque: 72–86 ft. lbs. (100–120 Nm).
11. Install the wheel speed sensor to the knuckle.
12. Install the brake caliper assembly to the knuckle. Torque: 58–72 ft. lbs. (80–100 Nm).
13. Install the tie rod end ball joint nut and insert the split pin. Torque: 12–25 ft. lbs. (16–34 Nm).
14. Insert the washer and tighten the locking nut. Torque: 145–188 ft. lbs. (200–260 Nm).
15. Install the wheel and tire.

SUSPENSION

COIL SPRING

REMOVAL & INSTALLATION

1. Remove the wheel and tire.
2. Remove the brake hose bracket.
3. Remove the wheel speed sensor wire bracket.
4. Use a jack at the bottom of the rear torsion axle beam to raise and support, then remove the rear shock absorber lower mounting bolt.
5. Remove the rear coil spring.

To install:

6. Install the upper and lower pads on the coil spring by aligning the grooves on the pads.
7. Place the coil spring with the pads on the torsion axle beam and support it with a jack.
8. Install the rear shock absorber mounting bolt by lifting the rear torsion axle beam. Torque: 72–86 ft. lbs. (100–120 Nm).
9. Install the wheel speed sensor wire bracket bolt.
10. Install the brake pressure hose bracket bolt.
11. Install the wheel and tire.

SHOCK ABSORBER

REMOVAL & INSTALLATION

1. Remove the wheel and tire.
2. After supporting the rear torsion axle beam with a jack, remove the rear shock absorber lower mounting bolt. Remove the rear shock absorber.
3. Remove the rear shock absorber mounting bolts.

To install:

4. Tighten the rear shock absorber upper mounting bolt. Torque: 28–43 ft. lbs. (40–60 Nm).
5. Placing a jack at the bottom of the rear torsion axle beam and raise the vehicle to the proper location.
6. Tighten the rear shock absorber lower mounting bolts. Torque: 72–86 ft. lbs. (100–120 Nm).

⁂ CAUTION

Check that the rear coil spring is located in the proper position.

WHEEL HUB & BEARING

REMOVAL & INSTALLATION

See Figure 84.

1. Raise the rear of the vehicle and support it with safety stand.

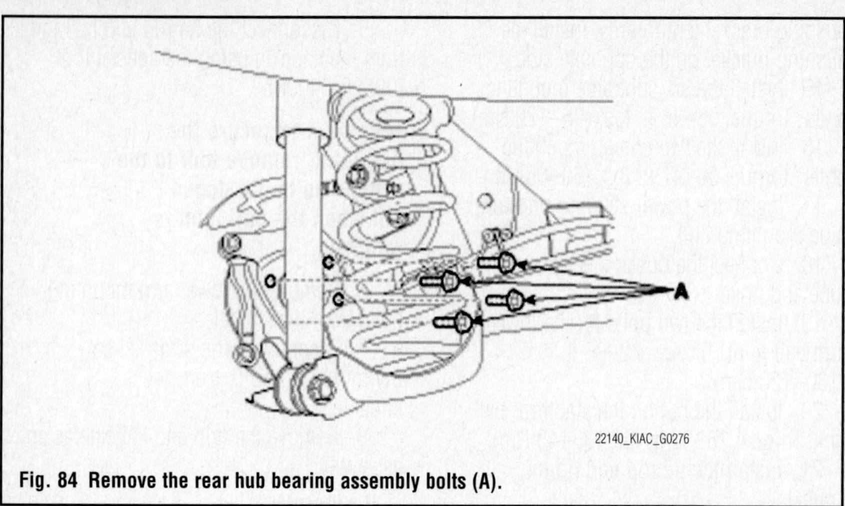

Fig. 84 Remove the rear hub bearing assembly bolts (A).

2. Remove the rear wheel and tire.
3. Remove the wheel speed sensor wire bracket bolt.
4. Remove the parking brake wire bracket bolt.
5. Remove the brake caliper assembly bolt and the brake disk mounting screw.
6. Hang the brake caliper assembly tightly on a proper place with wire.
7. Remove the rear hub bearing assembly bolts.
8. Remove the rear hub bearing assembly.
9. The installation is reverse of the removal.
10. Hub bearing assembly: Torque: 36–43 ft. lbs. (50–60 Nm).

KIA

Rondo

16

FUEL 16-46

HEATING, VENTILATION & AIR CONDITIONING.........16-48

SPECIFICATIONS AND MAINTENANCE CHARTS.....16-3

STEERING 16-49

SUSPENSION 16-51

SPECIFICATIONS AND MAINTENANCE CHARTS

ENGINE AND VEHICLE IDENTIFICATION

Engine						
Code ①	Liters (cc)	Cu. In.	Cyl.	Fuel Sys.	Engine Type	Eng. Mfg.
5	2.4 (2359)	144	4	EGI	DOHC	KIA
6	2.7 (2656)	163	6	EGI	DOHC	KIA

Model Year	
Code ②	Year
9	2009
A	2010

EGI: Electronic Gasoline Injection

DOHC: Double Overhead Camshafts

37655_ROND_C0001

GENERAL ENGINE SPECIFICATIONS

Year	Model	Engine Displacement Liters	Engine VIN	Net Horsepower @ rpm	Net Torque @ rpm (ft. lbs.)	Bore x Stroke (in.)	Com-pression Ratio	Oil Pressure psi @ rpm
2009	Rondo	2.4	5	175@6000	169@4000	3.46x3.82	10.5:1	21@1000
		2.7	6	192@6000	184@4000	3.41x2.95	10.4:1	18@700
2010	Rondo	2.4	5	175@6000	169@4000	3.46x3.82	10.5:1	21@1000
		2.7	6	192@6000	184@4000	3.41x2.95	10.4:1	18@1000

37655_ROND_C0002

ENGINE TUNE-UP SPECIFICATIONS

Year	Engine Displacement Liters	Engine VIN	Spark Plug Gap (in.)	Ignition Timing (deg.)	Fuel Pump (psi)	Idle Speed (rpm)	Valve Clearance	
							Intake	Exhaust
2009	2.4	5	0.039-0.043	①	46-49	700-900	②	③
	2.7	6	0.039-0.043	7B +/- 10	49-51	550-750	④	⑤
2010	2.4	5	0.039-0.043	8B +/- 5	49-51	520-720	②	③
	2.7	6	0.039-0.043	7B +/- 10	55	550-750	④	⑤

NA: Not available

B: Before Top Dead Center

NOTE: The Vehicle Emission Control Information label often reflects specification changes made during production. The label figures must be used if they differ from those in this chart.

① Ignition timing is controlled by electronic ignition timing system.

② 0.0039-0.0118 in

③ 0.0097-0.0157 in

④ 0.0067-0.0090 in

⑤ 0.0106-0.0129 in

37665_ROND_C0003

CAPACITIES

Year	Model	Engine Displacement Liters	Engine VIN	Engine Oil with Filter	Automatic Transaxle (pts.)	Fuel Tank (gal.)	Cooling System (qts.)
2009	Rondo	2.4	5	4.3	16.4	15.9	7.08
		2.7	6	5.2	20.1	15.9	8.7 - 8.8
2010	Rondo	2.4	5	4.5	16.4	15.8	7.08
		2.7	6	4.5	20.1	15.8	8.7 - 8.8

NA: Not available

* The quantity is for reference; the actual quantity must be set according to the Inspection and Replacement procedures.

37665_ROND_C0004

FLUID SPECIFICATIONS

Year	Model	Engine Displ. Liters	Engine Oil	Automatic Transaxle	Drive Axle Joint	Power Steering Fluid	Brake Master Cylinder
2009	Rondo	2.4	①	Diamond ATF SP-III, SK ATF SP-III	RBA/ CW - 13TJα	PSF-4	DOT 3 or DOT 4
		2.7	②	Diamond ATF SP-III, SK ATF SP-III	RBA/ CW - 13TJα	PSF-4	DOT 3 or DOT 4
2010	Rondo	2.4	③	Diamond ATF SP-III	RBA/ CW - 13TJα	PSF-4	DOT 3 or DOT 4
		2.7	③	Diamond ATF SP-III, SK ATF SP-III	RBA/ CW - 13TJα	PSF-4	DOT 3 or DOT 4

① 10W-30 API classification: SL / SM or above. Recommeded ILSAC classification: GF4 or above.

② 10W-30 API classification: SJ or SL.

③ 10W-30 API classification: SL, SM or above. Recommeded ILSAC classification: GF3, GF4, or above.

37655_ROND_C0014

VALVE SPECIFICATIONS

Year	Engine Displacement Liters	Engine VIN	Seat Angle (deg.)	Face Angle (deg.)	Maximum out of Square (in.)	Spring Free Length (in.)	Stem-to-Guide Clearance (in.) Intake	Stem-to-Guide Clearance (in.) Exhaust	Stem Diameter (in.) Intake	Stem Diameter (in.) Exhaust
2009	2.4	5	44.75-45.1	45.25-45.75	①	1.8677	0.00078-0.0019	0.00118-0.0021	0.2151-0.2157	0.2149-0.2153
	2.7	6	NA	45-45.5	①	1.8425	0.0008-0.0020	0.0014-0.0026	0.2348-0.2354	0.2343-0.2348
2010	2.4	5	44.75-45.1	45.25-45.75	①	1.8677	0.00078-0.0019	0.00118-0.0021	0.2151-0.2157	0.2149-0.2153
	2.7	6	NA	45-45.5	①	1.8425	0.0008-0.0020	0.0014-0.0026	0.2348-0.2354	0.2343-0.2348

NA: Not Available

① Less than 1.5 degrees.

37655_ROND_C0005

CAMSHAFT AND BEARING SPECIFICATIONS

All measurements are given in inches.

Year	Engine Displacement Liters	Engine VIN	Journal Diameter	Brg. Oil Clearance	Shaft End-play	Runout	Journal Bore	Lobe Lift (inches) Intake	Exhaust
2009	2.4	5	①	②	0.0015-0.0062	NA	NA	1.7362-1.7440	1.7677-1.7756
	2.7	6	1.1009-1.1016	0.0012-0.0022	0.0039-0.0079	NA	NA	1.7520	1.7520
2010	2.4	5	①	②	0.0015-0.0062	NA	NA	1.7401	1.7716
	2.7	6	1.1009-1.1016	0.0012-0.0022	0.0039-0.0079	NA	NA	1.7520	1.7520

NA: Information not available

① Intake No. 1: 1.1811 in.

Intake and Exhaust: No. 2, 3, 4, 5: 0.9449 in.

Exhaust No. 1: 1.4173 in.

② Intake No. 1: 0.0008 - 0.0022 in.

Intake No. 2, 3, 4, 5: 0.0018 - 0.0032 in.

Exhaust No. 2, 3, 4, 5: 0.0017 - 0.0032 in.

Exhaust No. 1: 0 - 0.0012 in.

37655_ROND_C0013

CRANKSHAFT AND CONNECTING ROD SPECIFICATIONS

All measurements are given in inches.

Year	Engine Displacement Liters	Engine VIN	Crankshaft Main Brg. Journal Dia.	Main Brg. Oil Clearance	Shaft End-play	Thrust on No.	Connecting Rod Journal Diameter	Oil Clearance	Side Clearance
2009	2.4	5	2.0449-2.0456	0.0010-0.0014	0.0027 0.0098	3	①	0.0011-0.0018	0.0039-0.0098
	2.7	6	2.4402-2.4409	0.0002-0.0009	0.0028-0.0098	3	①	0.0007-0.0014	0.0039-0.0098
2010	2.4	5	2.0449-2.0456	0.0007-0.0014	0.0027-0.0098	3	②	0.0009-0.0016	0.0039-0.0100
	2.7	6	2.4402-2.4409	0.0002-0.0009	0.0028-0.0098	3	①	0.0007-0.0014	0.0039-0.0098

① There are three classes of the connecting rod. It is marked: A, B, or C. Inside diameters are as follows:

A: 2.0079 - 2.0081

B: 2.0081 - 2.0083

C: 2.0083 - 2.0086

② There are three classes of the connecting rod. It is marked: A, B, or C. Inside diameters are as follows:

A: 2.0079 - 2.0081

B: 2.0081 - 2.0083

C: 2.0083 - 2.0085

37655_ROND_C0007

PISTON AND RING SPECIFICATIONS

All measurements are given in inches.

Year	Engine Displacement Liters	Engine VIN	Piston Clearance	Ring Gap Top Compression	Ring Gap Bottom Compression	Ring Gap Oil Control	Ring Side Clearance Top Compression	Ring Side Clearance Bottom Compression	Ring Side Clearance Oil Control
2009	2.4	5	0.0006-0.0014	0.0059-0.0118	0.0145-0.0204	0.0079-0.0275	0.0019 0.0031	0.0015-0.0031	0.0023-0.0059
	2.7	6	0.0008-0.0020	0.0059-0.0118	0.0118-0.0177	0.0078-0.0275	0.0016-0.0031	0.0012-0.0027	0.0024-0.0059
2010	2.4	5	0.0005-0.0013	0.0059-0.0118	0.0118-0.0204	0.0078-0.0275	0.0019-0.0031	0.0015-0.0031	0.0023-0.0051
	2.7	6	0.0008-0.0016	0.0059-0.0118	0.0118-0.0177	0.0079-0.0275	0.0016-0.0031	0.0012-0.0027	0.0024-0.0059

37655_ROND_C0006

TORQUE SPECIFICATIONS

All readings in ft. lbs.

Year	Engine Displaceme Liters	Engine VIN	Cylinder Head Bolts	Main Bearing Bolts	Rod Bearing Cap Bolts	Crankshaft Pulley Bolts	Flywheel/ Driveplate Bolts	Manifold Intake	Manifold Exhaust	Spark Plugs	Oil Pan Drain Plug
2009	2.4	5	①	②	③	123-130	87-94	14-17	④	NA	26-32
	2.7	6	⑤	⑥	⑦	123-130	53-56	14-17	22-25	NA	29-36
2010	2.4	5	⑦	⑧	⑨	123-130	87-94	14-17	⑧	15-21	26-32
	2.7	6	⑤	⑥	⑦	123-130	53-56	14-17	22-25	15-21	29-36

NA: Information not available

① 23.9-26.8 ft. lbs., plus 90-95 degrees, plus 90-95 degrees

② 10.8 ft. lbs., plus (20.3-23.1), plus 120-125 degrees

③ 13.0-15.9 ft. lbs., plus 88-92 degrees

④ Heat protector bolt 5.8 - 8.7
Nut 36.2 - 39.8
Stay bolt (M8) 13.7 - 20.3
Stay bolt (M10) ULEV 31.1 - 39.8
Stay bolt (M10) ULEV 31.1 - 32.5

⑤ 16.6 - 19.5 ft. lbs., plus 58 - 62 degrees, plus 43 - 47 degrees

⑥ M10 bolts: 19.5 - 23.9ft. lbs., plus 90 - 95 degrees
M8 bolts: 9.4 - 13.7 ft. lbs., plus 90 - 95 degrees

⑦ 13.0 - 15.9 ft. lbs., plus 90 - 94 degrees

⑧ Heat protector bolt: 5.8 - 8.7
Nut: 36.2-39.8
Stay bolt (M8): 13.7 -20.3
Stay bolt (M10), ULEV: 31.1 - 39.8
Stay bolt (M10), SULEV: 31.1 - 32.5

37655_ROND_C0008

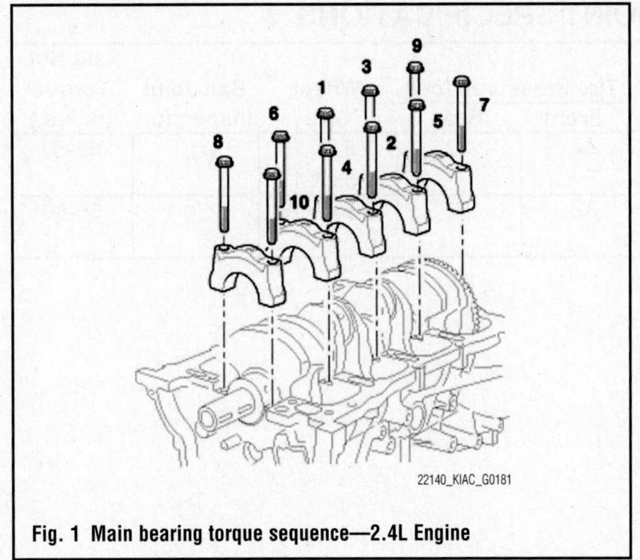

Fig. 1 Main bearing torque sequence—2.4L Engine

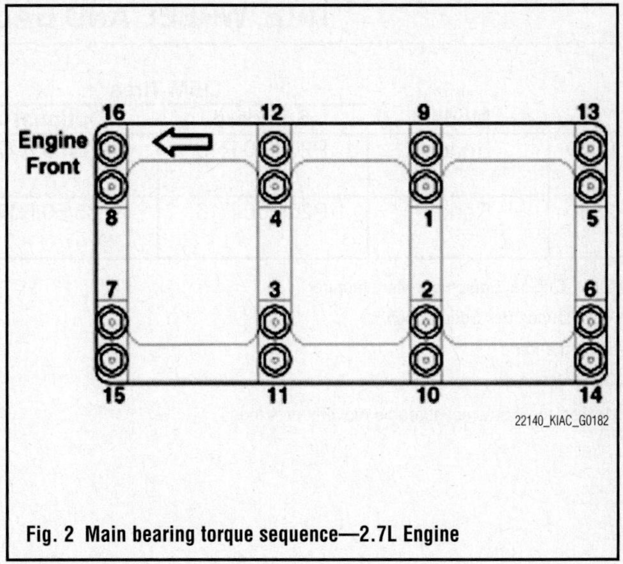

Fig. 2 Main bearing torque sequence—2.7L Engine

WHEEL ALIGNMENT

Year	Model		Caster Range (+/-Deg.)	Caster Preferred Setting (Deg.)	Camber Range (+/-Deg.)	Camber Preferred Setting (Deg.)	Toe Toe-in (in.)
2009	Rondo	F	0.5	4.74	0.5	-0.5	0 +/- 0.2
		R	—	—	0.5	-1.0	0.2 +/- 0.2
2010	Rondo	F	0.4	4.74	0.5	-0.5	0 +/- 0.2
		R	—	—	0.5	-1.0	0.2 +/- 0.2

37655_ROND_C0009

TIRE, WHEEL AND BALL JOINT SPECIFICATIONS

Year	Model	OEM Tires Standard	OEM Tires Optional	Tire Pressures (psi) Front	Tire Pressures (psi) Rear	Wheel Size	Ball Joint Inspection	Lug Nut Torque (ft. lbs.)
2009	Rondo	P205/60 R16	P225/50 R17	32	32	6.5J X 16 6.5J X 17	①	65-80
2010	Rondo	P205/60 R16	P225/50 R17	32	32	6.5J X 16 6.5J X 17	①	65-80

OEM: Original Equipment Manufacturer

PSI: Pounds Per Square Inch

STD: Standard

OPT: Optional

① Replace if any measureable movement is found.

37655_ROND_C0010

BRAKE SPECIFICATIONS

All measurements in inches unless noted

Year	Model		Brake Disc Original Thickness	Brake Disc Minimum Thickness	Brake Disc Maximum Run-out	Minimum Lining Thickness	Brake Caliper Bracket Bolts (ft. lbs.)	Brake Caliper Mounting Bolts (ft. lbs.)
2009	Rondo	F	1.020	0.940	0.0020	0.079	3.6-4.3	58-72
		R	0.390	0.310	0.0020	0.079	3.6-4.3	36-43
2010	Rondo	F	1.020	0.940	0.0020	0.079	3.6-4.3	58-72
		R	0.390	0.310	0.0020	0.079	3.6-4.3	36-43

NA: Not available/not applicable

F: Front

R: Rear

37655_ROND_C0011

SCHEDULED MAINTENANCE INTERVALS

KIA - RONDO

TO BE SERVICED	TYPE OF SERVICE	VEHICLE MILEAGE INTERVAL (x1000)												
		7.5	15	22.5	30	37.5	45	52.5	60	67.5	75	82.5	90	97.5
Accessory drive belts	S/I	✓	✓	✓	✓	✓	✓	✓	✓	✓	✓	✓	✓	✓
Air cleaner filter	S/I	✓	✓	✓	R	✓	✓	✓	R	✓	✓	✓	R	✓
Air conditioning refrigerant	S/I		✓		✓		✓		✓		✓		✓	
Automatic transaxle fluid	S/I		✓	✓	✓		✓		✓		✓		✓	
Brake fluid	S/I				✓				✓				✓	
Brake lines, hoses and connections	S/I		S/I		✓		✓		✓		✓		✓	
Chassis and body fasteners	T												✓	
Climate control air filter (for evaporator and blower) replace every 10,000 miles	R		✓	✓	✓	✓	✓	✓	✓	✓	✓	✓	✓	✓
Cooling system hoses and coolant level	S/I		✓		✓		✓		R		✓		R	
Crankcase ventilation hose	S/I				✓				✓				✓	
Driveshafts and boots	S/I		✓		✓		✓		✓		✓		✓	
Electronic throttle control	S/I		✓		✓		✓		✓		✓		✓	
Engine coolant	R				✓				✓				✓	
Engine oil and filter	R	✓	✓	✓	✓	✓	✓	✓	✓	✓	✓	✓	✓	✓
Exhaust system/ heat shields	S/I		✓		✓		✓		✓		✓		✓	
Front and rear brakes, parking brake	S/I	✓	✓	✓	✓	✓	✓	✓	✓	✓	✓	✓	✓	✓
Front ball joints	S/I				✓				✓				✓	
Fuel filter	R					✓					✓			
Fuel lines, fuel hoses, and connections	S/I	✓	✓	✓	✓	✓	✓	✓	✓	✓	✓	✓	✓	✓
Fuel tank air filter	S/I		✓		R		✓		R		✓		R	
Fuel vapor hose and filler cap	S/I				✓				✓				✓	
Locks and hinges	L	✓	✓	✓	✓	✓	✓	✓	✓	✓	✓	✓	✓	✓
Power steering pump, belt, and hoses	S/I		✓		✓				✓		✓		✓	
Spark plugs	R								✓					
Suspension mounting bolts	S/I		✓		✓		✓		✓		✓		✓	
Steering gear box, linkage,and boots/ upper and lower arm ball joints	S/I	✓	✓	✓	✓	✓	✓	✓	✓	✓	✓	✓	✓	✓
Timing belt	R				S/I				✓				✓	
Vacuum hose	S/I	✓	✓	✓	✓	✓	✓	✓	✓	✓	✓	✓	✓	✓
Valve clearance	S/I								✓					

R: Replace S/I: Service or Inspect L: Lubricate T: Tighten

FREQUENT OPERATION MAINTENANCE (SEVERE SERVICE)

If a vehicle is operated under any of the following conditions it is considered severe service

- Towing a trailer or using a camper or car-top carrier

- Repeated short trips of less than 5 miles in temperatures below freezing, or trips of less than 10 miles in any temperature

- Prolonged idling (vehicle operation in stop and go traffic).

- Operating on rough, muddy, unpaved, dusty or salt-covered roads.

- Police, taxi, delivery usage or trailer towing usage.

- Driving in extremely hot (over 90°F) conditions

Oil & oil filter: change every 5000 miles or 5 months, whichever occurs first.

Air cleaner filter: inspect every 15,000 miles or 15 months and replace everything 30,000 miles or 30 months, whichever occurs fir

Fuel system hoses (California models only): replace every 105,000 miles

Emission system hoses (non-CA models): inspect every 55,000 or 55 months, whichever occurs first

Emission system hoses (CA models): inspect every 60,000 miles or 60 months, which occurs first

Front and rear brakes: inspect every 15,000 miles or 15 months, whichever occurs first

Chassis and body fasteners: tighten every 15,000 miles or 15 months, whichever occurs first

Locks and hinges: lubricate every 5000 miles or 5 months, whichever occurs first

37665_ROND_C0012

BRAKES — INFORMATION AND PRECAUTIONS

ANTI-LOCK SYSTEMS

- Certain components within the ABS system are not intended to be serviced or repaired individually.
- Do not use rubber hoses or other parts not specifically specified for and ABS system. When using repair kits, replace all parts included in the kit. Partial or incorrect repair may lead to functional problems and require the replacement of components.
- Lubricate rubber parts with clean, fresh brake fluid to ease assembly. Do not use shop air to clean parts; damage to rubber components may result.
- Use only DOT 3 brake fluid from an unopened container.
- If any hydraulic component or line is removed or replaced, it may be necessary to bleed the entire system.
- A clean repair area is essential. Always clean the reservoir and cap thoroughly before removing the cap. The slightest amount of dirt in the fluid may plug an orifice and impair the system function. Perform repairs after components have been thoroughly cleaned; use only denatured alcohol to clean components. Do not allow ABS components to come into contact with any substance containing mineral oil; this includes used shop rags.
- The Anti-Lock control unit is a microprocessor similar to other computer units in the vehicle. Ensure that the ignition switch is **OFF** before removing or installing controller harnesses. Avoid static electricity discharge at or near the controller.
- If any arc welding is to be done on the vehicle, the control unit should be unplugged before welding operations begin.

DISC AND DRUM SYSTEMS

✲✲ CAUTION

Dust and dirt accumulating on brake parts during normal use may contain asbestos fibers from production or aftermarket brake linings. Breathing excessive concentrations of asbestos fibers can cause serious bodily harm. Exercise care when servicing brake parts. Do not sand or grind brake lining unless equipment used is designed to contain the dust residue. Do not clean brake parts with compressed air or by dry brushing. Cleaning should be done by dampening the brake components with a fine mist of water, then wiping the brake components clean with a dampened cloth. Dispose of cloth and all residue containing asbestos fibers in an impermeable container with the appropriate label. Follow practices prescribed by the Occupational Safety and Health Administration (OSHA) and the Environmental Protection Agency (EPA) for the handling, processing, and disposing of dust or debris that may contain asbestos fibers.

BRAKES — BLEEDING THE BRAKE SYSTEM

BLEEDING THE BRAKE SYSTEM

BLEEDING PROCEDURE

See Figure 3.

➡When bleeding the brakes, note the following:

- Do not reuse the drained fluid
- Always use Genuine DOT 3 or DOT 4 Brake Fluid. Using a non-Genuine DOT3 or DOT 4 brake fluid can cause corrosion and decrease the life of the system
- Make sure no dirt of other foreign matter is allowed to contaminate the brake fluid
- Do not spill brake fluid on the vehicle, it may damage the paint; if brake fluid does contact the paint, wash it off immediately with water
- The reservoir on the master cylinder must be at the MAX (upper) level mark at the start of bleeding procedure and checked after bleeding each brake caliper. Add fluid as required

1. Make sure the brake fluid level in the master cylinder fluid reservoir is at the MAX (upper) level line.
2. Have someone slowly pump the brake pedal several times, and then apply steady pressure

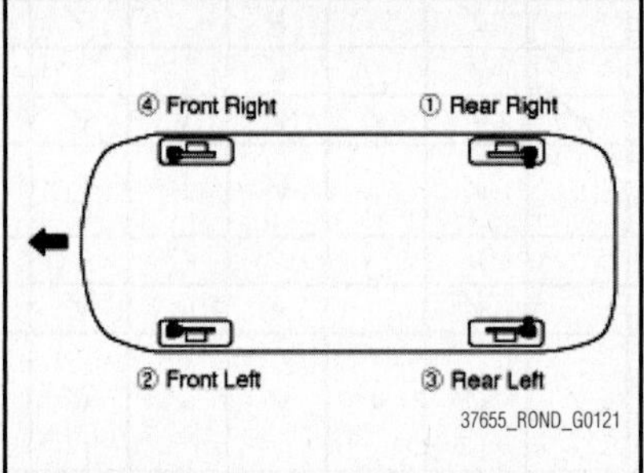

Fig. 3 Repeat the procedure for each wheel in the sequence shown until air bubbles no longer appear in the fluid

3. Loosen the right-rear brake bleed screw to allow air to escape from the system. Then tighten the bleed screw securely
4. Repeat the procedure for each wheel in the sequence shown below until air bubbles no longer appear in the fluid.

BLEEDING THE ABS SYSTEM

This procedure should be followed to ensure adequate bleeding of air and filling of the ABS unit, brake lines and master cylinder with brake fluid:

1. Remove the reservoir cap and fill the brake reservoir with brake fluid.

✲✲ CAUTION

If there is any brake fluid on any painted surface, wash it off immediately.

➡Note the following:

- When pressure bleeding, do not depress the brake pedal until instructed to do so.
- Recommended fluid: DOT3 or DOT4.

2. Connect a clear plastic tube to the wheel cylinder bleeder plug and insert the other end of the tube into a half-filled clear plastic bottle.
3. Connect the scan tool to the data link connector located underneath the dash panel.

✲✲ CAUTION

You must obey the maximum operating time of the ABS motor with the scan tool to prevent the motor pump from burning.

4. Select and operate the scan tool according to the instructions on the scan tool screen:
   a. Select KIA vehicle diagnosis.
   b. Select vehicle name.
   c. Select Anti-Lock Brake system.

d. Select air bleeding mode.
e. Press "YES" to operate the motor pump and solenoid valve.
f. Wait 120 sec. before operating the air bleeding. (If not, you may damage the motor.)
g. Perform the air bleeding.
5. Pump the brake pedal several times, and then loosen the bleeder screw until fluid starts to run out without bubbles. Then close the bleeder screw.
6. Repeat the procedure until there are no more bubbles in the fluid for each wheel.
7. Tighten the bleeder screw to 5–9 ft. lbs. (7–12 Nm).

BRAKES — ANTI-LOCK BRAKE SYSTEM (ABS)

WHEEL SPEED SENSOR

REMOVAL & INSTALLATION

Front

1. Remove the front wheel speed sensor mounting bolt and the cable mounting bolt.
2. Remove the front wheel speed sensor bracket.
3. Remove the front wheel guard .
4. Disconnect the front wheel speed sensor connector.
5. Remove the front wheel speed sensor.
6. Installation is the reverse of removal.

Rear

1. Remove the rear wheel speed sensor mounting bolt.
2. Remove the sensor cable bracket mounting bolt.
3. Disconnect the rear wheel speed sensor connector.
4. Remove the rear wheel speed sensor.
5. Installation is the reverse of removal.

BRAKES — FRONT DISC BRAKES

BRAKE CALIPER

REMOVAL & INSTALLATION

See Figures 4 and 5.

1. Raise and safely support the front of the vehicle.
2. Remove the front wheel and tire.
3. Remove the brake hose bolt and the guide rod bolts from the caliper assembly.
4. Remove the caliper assembly.

To install:

5. Install the brake disc (A).
6. Tighten the screw (B) to 3.6–4.3 ft. lbs. (4.9–5.9 Nm).
7. Install the caliper bracket and tighten the caliper mounting bolts to 57.9–72.3 ft. lbs. (78.5–98.1 Nm)
8. Install the pad retainers to the caliper bracket.
9. Install the brake pads and pad shims on the pad retainer correctly. Install the pad with the wear indicator on the inside. If you are reusing the pads, always reinstall the brake pads in their original positions to prevent a momentary loss of braking efficiency.

➡Check for foreign material at the pad shims and the back of the pads. Contaminated brake discs or pads reduce stopping ability. Keep grease off the discs and pads.

10. Push in the piston so that the caliper will fit over the pads.
11. Make sure that the piston boot is in position to prevent damaging it when installing the caliper.
12. Install the caliper assembly.

➡Be careful not to damage the piston pin boot.

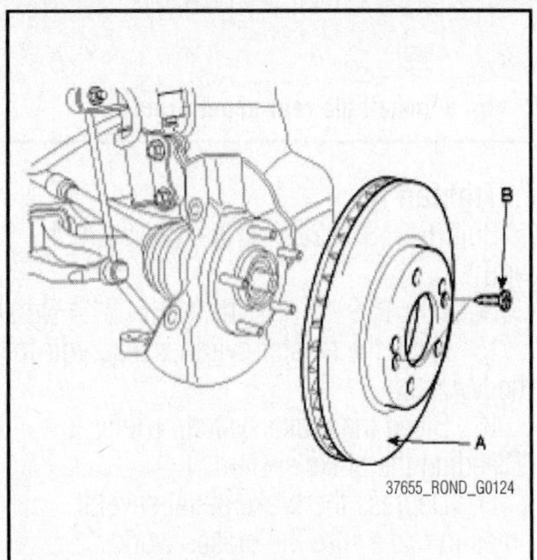

Fig. 4 Install the brake disc (A) and tighten the screw (B)

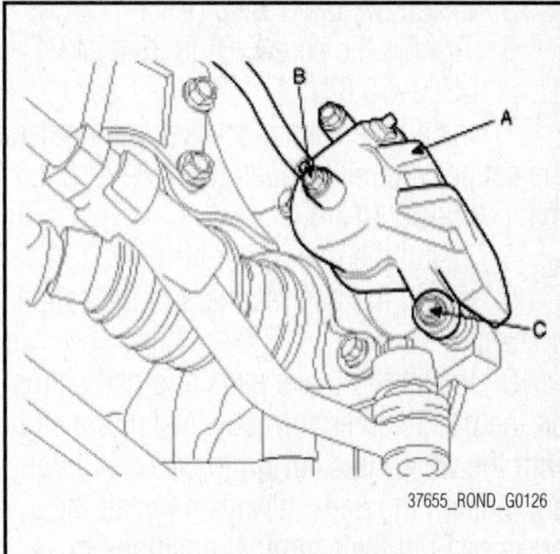

Fig. 5 Install the brake hose bolt (B) and the guide rod bolts (C) to the caliper assembly (A)

13. Install the brake hose bolt (B) and the guide rod bolts (C) to the caliper assembly (A). Tighten as follows:
    - Bolt (B): 18.1–21.7 ft. lbs. (24.5–29.4 Nm)
    - Bolt (C): 15.9–23.1 ft. lbs. (21.6–31.4 Nm)
14. Refill the master cylinder reservoir to the MAX line.
15. Bleed the brake system. (Refer to "Bleeding the brake system.")
16. Depress the brake pedal several times to make sure the brakes work.
17. Test-drive.

➡Engagement of the brake may require a greater pedal stroke immediately after the brake pads have been replaced as a set. Several applications of the brake will restore the normal pedal stroke.

18. After installation, check for leaks at hose and line joints or connections, and retighten if necessary.

DISC BRAKE PADS

REMOVAL & INSTALLATION

See Figure 4.

1. Raise and safely support the front of the vehicle.
2. Remove the front wheel and tire.
3. Remove the brake hose bolt and the guide rod bolts from the caliper assembly.
4. Remove the caliper assembly.
5. Remove the pads, the pad shims and the pad retainers from the caliper bracket.

To install:

6. Install the brake disc (A).
7. Tighten the screw (B) to 3.6–4.3 ft. lbs. (4.9–5.9 Nm).
8. Install the caliper bracket and tighten

the caliper mounting bolts to 57.9–72.3 ft. lbs. (78.5–98.1 Nm).

9. Install the pad retainers to the caliper bracket.

10. Install the brake pads and pad shims on the pad retainer correctly. Install the pad with the wear indicator on the inside. If you are reusing the pads, always reinstall the brake pads in their original positions to prevent a momentary loss of braking efficiency.

➡Check for foreign material at the pad shims and the back of the pads. Contaminated brake discs or pads reduce stopping ability. Keep grease off the discs and pads.

BRAKES

REAR DISC BRAKES

BRAKE CALIPER

REMOVAL & INSTALLATION

See Figures 4 through 6.

1. Raise and safely support the rear of the vehicle.
2. Remove the rear wheel and tire.
3. Remove the brake hose bolt and the guide rod bolts from the caliper assembly.
4. Remove the caliper assembly.

To install:

5. Install the brake disc (A).
6. Tighten the screw (B) to 36.2–43.4 ft. lbs. (49.0–58.8 Nm).
7. Install the caliper bracket and tighten the caliper mounting bolts to 57.9–72.3 ft. lbs. (78.5–98.1 Nm)
8. Install the rear upper arm (A).
9. Install the pad retainers to the caliper bracket.
10. Install the brake pads and pad shims on the pad retainer correctly. Install the pad with the wear indicator on the inside. If you are reusing the pads, always reinstall the brake pads in their original positions to prevent a momentary loss of braking efficiency.

➡Check for foreign material at the pad shims and the back of the pads. Contaminated brake discs or pads reduce stopping ability. Keep grease off the discs and pads.

11. Push in the piston so that the caliper will fit over the pads.
12. Make sure that the piston boot is in position to prevent damaging it when installing the caliper.
13. Install the caliper assembly.

➡Be careful not to damage the piston pin boot.

14. Install the brake hose bolt (B) and the guide rod bolts (C) to the caliper assembly (A).

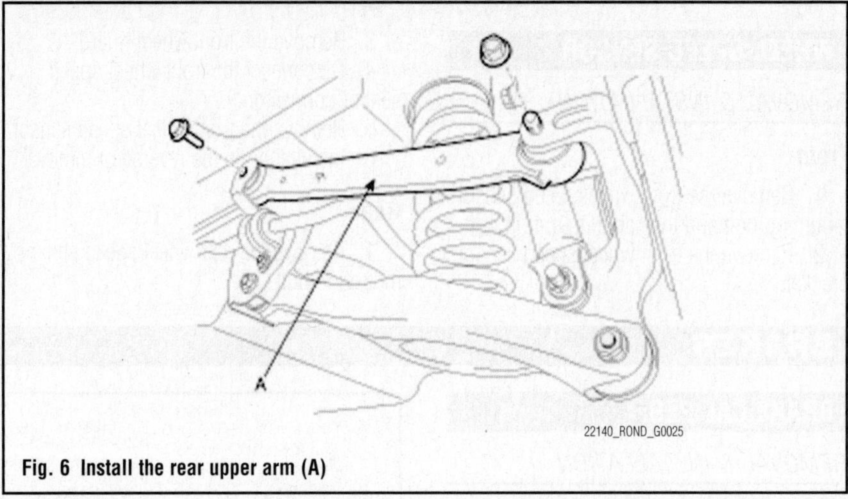

Fig. 6 Install the rear upper arm (A)

Tighten to:

Bolt (B): 18.1–21.7 ft. lbs. (24.5–29.4 Nm)

Bolt (C): 15.9–23.1 ft. lbs. (21.6–31.4 Nm)

15. Refill the master cylinder reservoir to the MAX line.
16. Bleed the brake system. (Refer to “Bleeding the brake system.”)
17. Depress the brake pedal several times to make sure the brakes work.
18. Test-drive.

➡Engagement of the brake may require a greater pedal stroke immediately after the brake pads have been replaced as a set. Several applications of the brake will restore the normal pedal stroke.

19. After installation, check for leaks at hose and line joints or connections, and retighten if necessary.

DISC BRAKE PADS

REMOVAL & INSTALLATION

See Figures 4 and 6.

1. Raise and safely support the rear of the vehicle.
2. Remove the rear wheel and tire.
3. Remove the brake hose bolt and the guide rod bolts from the caliper assembly.
4. Remove the caliper assembly.
5. Remove the pads, the pad shims and the pad retainers from the caliper bracket.

To install:

6. Install the brake disc and tighten the screw to 3.6–4.3 ft. lbs. (4.9–5.9 Nm).
7. Install the caliper bracket and tighten the caliper mounting bolts to 57.9–72.3 ft. lbs. (78.5–98.1 Nm).
8. Install the rear upper arm (A).
9. Install the pad retainers to the caliper bracket.
10. Install the brake pads and pad shims on the pad retainer correctly. Install the pad with the wear indicator on the inside. If you are reusing the pads, always reinstall the brake pads in their original positions to prevent a momentary loss of braking efficiency.

➡Check for foreign material at the pad shims and the back of the pads. Contaminated brake discs or pads reduce stopping ability. Keep grease off the discs and pads.

PARKING BRAKE PEDAL

ADJUSTMENT

Parking Brake Pedal Stroke Adjustment

See Figure 7.

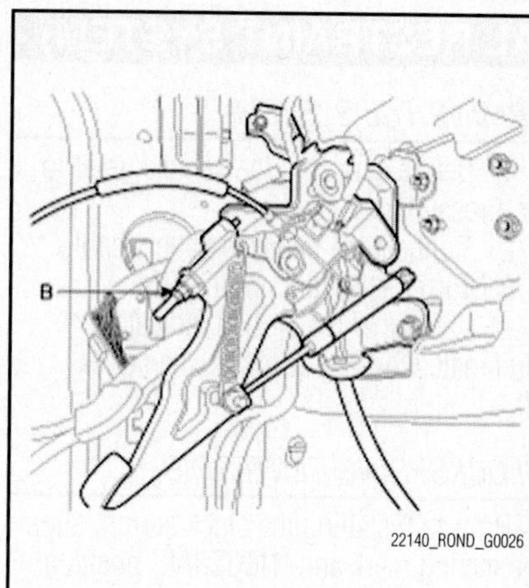

Fig. 7 Parking brake pedal stroke adjustment

1. Push the parking brake pedal to its maximum point of travel.
2. Release and repeat 2 additional times.
3. Adjust the nut (B) so that the parking brake pedal stroke is 3.46–3.86in (88–98mm) at an operating effort of 44 ft. lbs. (196N).

➡The parking brake adjustment must be carried out after adjusting the rear shoe.

4. After adjusting parking brake, verify the following:
   a. It is free from interference between the adjusting nut and pin.
   b. The brake is not dragging.

PARKING BRAKE SHOES

REMOVAL & INSTALLATION

See Figures 8 through 11.

1. Raise the vehicle, and make sure it is securely supported.
2. Remove the rear tire and wheel.
3. Remove the brake caliper.
4. Remove the rear brake disc (rotor).
5. Remove the parking brake cable (B), after removing the clip (A).
6. Remove the shoe hold down pin (A) and spring (B) by pressing and rotating the spring.

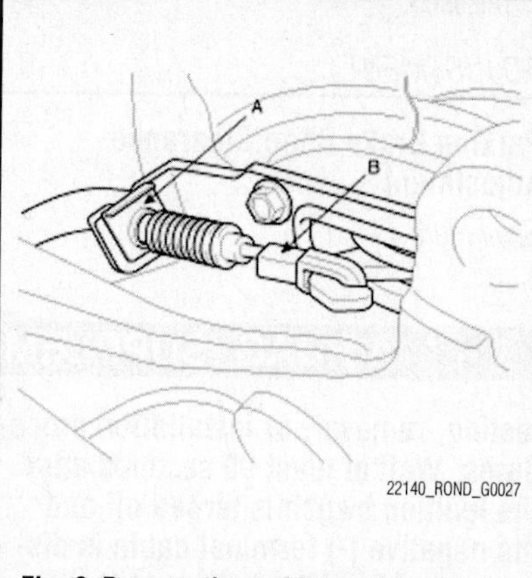

Fig. 8 Remove the parking brake cable (B), after removing the clip (A)

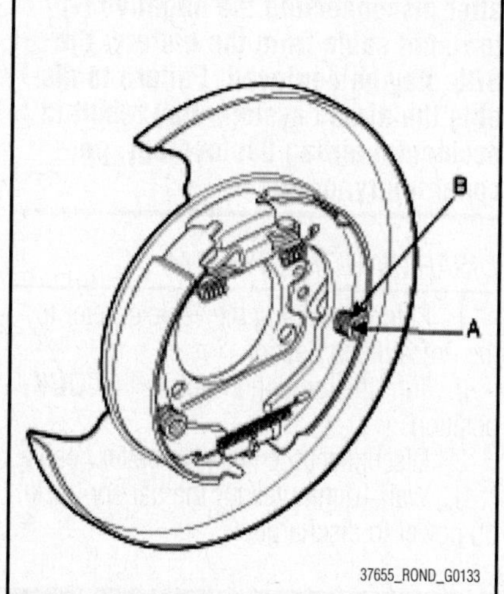

Fig. 9 Remove the shoe hold down pin and spring by pressing and rotating the spring

7. Remove the adjuster assembly (B) and the lower return spring (A).
8. Remove the upper return spring (C) and the brake shoes (D).
9. Remove the operating lever assembly (E).

To install:

10. Install the operating lever assembly (E).
11. Install the upper return spring (C) and the brake shoes (D).
12. Install the adjuster assembly(B) and the lower return spring (A).
13. Install the shoe hold down pin and spring by pressing and rotating the spring.
14. Install the parking brake cable (B), then install the clip (A).
15. Install the rear brake disc.

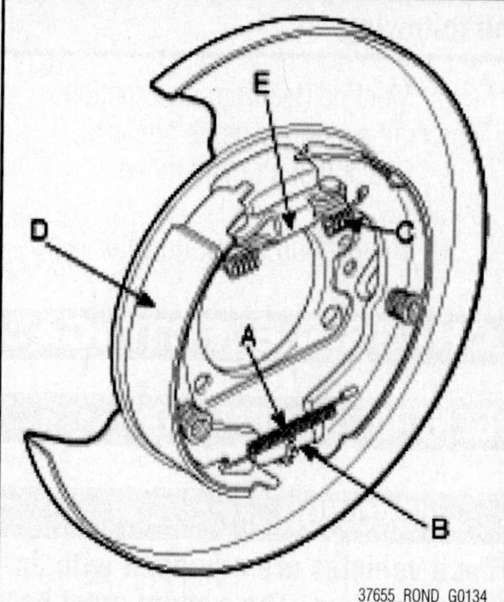

Fig. 10 Install the operating lever assembly (E)

16. Adjust the rear brake shoe clearance.
    a. Remove the plug from the disc.
    b. Rotate the toothed wheel of adjuster using a screwdriver until the disc is not moving.
    c. Return it by 5 notches in the opposite direction.
17. Install the brake caliper.
18. Install the tire and wheel.
19. Adjust the parking brake pedal.
20. If the parking brake shoe or the brake disc is replaced with a new one, perform the brake shoe bed-in procedure.
    a. While operating the parking brake pedal, drive the vehicle 0.3 miles (500 meters) at the speed of 37 mph (60 kph).
    b. Repeat the above procedure more than two times.
    c. Parking brake must hold on at 30% grade.

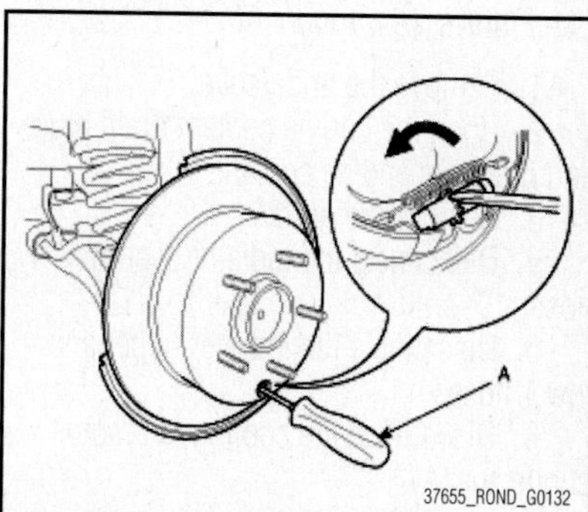

Fig. 11 Rotate the toothed wheel of adjuster using a screwdriver until the disc is not moving, then return it by 5 notches in the opposite direction

⁂ CAUTION

After adjusting parking brake, verify the following:

d. Must be free from malfunction when the parking pedal is operated.

e. Check that all parts move smoothly.

f. The parking brake indicator lamp must be on after the parking pedal is applied and must be off after the pedal is released.

ADJUSTMENT

Parking Brake Shoe Clearance Adjustment

See Figure 11.

1. Raise the vehicle, and make sure it is securely supported.
2. Remove the rear tire and wheel.
3. Remove the plug from the disc.
4. Rotate the toothed wheel of the adjuster using a screwdriver until the disc is not moving.
5. Then return it by 5 notches in the opposite direction.

CHASSIS ELECTRICAL

AIRBAG (SUPPLEMENTAL RESTRAINT SYSTEM)

GENERAL INFORMATION

⁂ CAUTION

These vehicles are equipped with an air bag system. The system must be disarmed before performing service on, or around, system components, the steering column, instrument panel components, wiring and sensors. Failure to follow the safety precautions and the disarming procedure could result in accidental air bag deployment, possible injury and unnecessary system repairs.

SERVICE PRECAUTIONS

⁂ CAUTION

Disconnect and isolate the battery negative cable before beginning any airbag system component diagnosis, testing, removal, or installation procedures. Wait at least 90 seconds after the ignition switch is turned off and the negative (-) terminal cable is disconnected from the battery before starting the operation. The SRS is equipped with a backup power source, so if work is started within 90 seconds after disconnecting the negative (-) terminal cable from the battery, the SRS may be deployed. Failure to disable the airbag system may result in accidental airbag deployment, personal injury, or death.

DISARMING THE SYSTEM

1. Before servicing the vehicle, refer to the Precautions Section.
2. Turn the ignition switch to the**LOCK**-position.
3. Disconnect the negative battery cable.
4. Wait 10 minutes for the battery back-up power to discharge.

ARMING THE SYSTEM

1. Before servicing the vehicle, refer to the Precautions Section.
2. Connect the negative battery cable.
3. Turn the ignition switch **ON**.
4. Verify that the air bag indicator illuminates for 4–8 seconds, then goes off.

CLOCKSPRING CENTERING

Prior to installing the clock spring, align the mating mark and "NEUTRAL" position indicator of the clock spring, and, after turning the front wheels to the straight-ahead position, install the clock spring to the column switch. If the mating mark of the clock spring is not properly aligned, the steering wheel may not completely rotate during a turn, or the flat cable within the clock spring may be severed, obstructing normal operation of the SRS and possibly leading to serious injury to the vehicle's driver.

ENGINE COOLING

RADIATOR

REMOVAL & INSTALLATION

2.4L Engine

See Figures 12 through 14.

1. Remove the undercover.
2. Open the engine coolant drain plug and drain the engine coolant.
3. Remove the air duct.
4. Disconnect the radiator overflow hose.
5. Disconnect the radiator upper and lower hoses.
6. Disconnect the cooling fan motor connector (A).
7. Remove the radiator grill upper cover.
8. Remove the radiator brackets (A) and mounting bolts (B).
9. Push down the radiator assembly and pull it up for removal because there is a clip (A).

To install:

10. Install the radiator assembly with the condenser.
11. Install the radiator brackets by mounting bolts.
12. Install the radiator grill upper cover.
13. Connect the cooling fan motor connector.

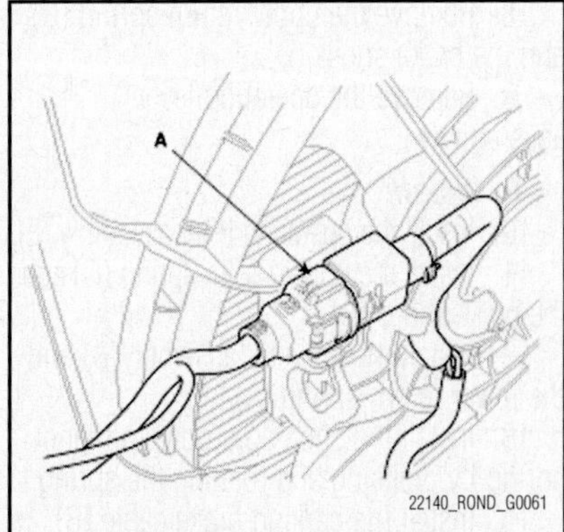

Fig. 12 Disconnect the cooling fan motor connector (A)

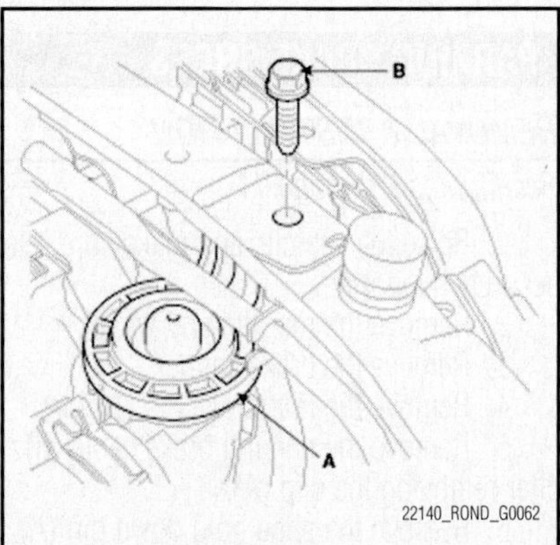

Fig. 13 Remove the radiator brackets (A) and mounting bolts (B)

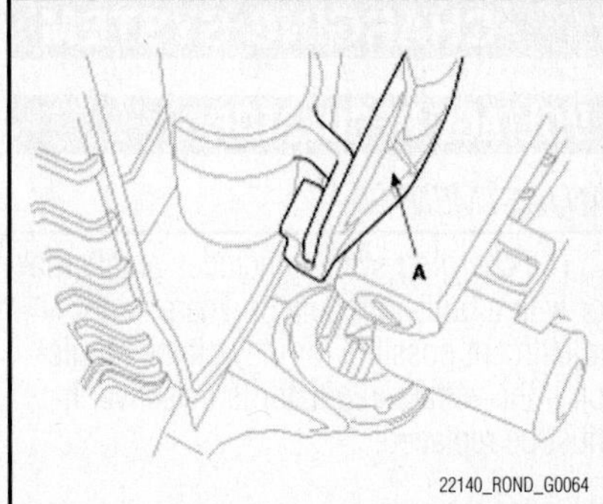

Fig. 14 Push down the radiator assembly and pull it up for removal because there is a clip (A)

14. Connect the radiator upper and lower hoses.
15. Connect the radiator overflow hose.
16. Install the air duct.
17. Tighten the engine coolant drain plug and refill engine coolant.
18. Install the undercover (A).

2.7L Engine

See Figures 15 and 16.

1. Drain the engine coolant.
2. Remove the undercover.
3. Remove the air duct.
4. Remove the upper and lower radiator hoses.
5. Disconnect the radiator fan connector.
6. Remove the front bumper upper cover.
7. Disconnect the Automatic Transaxle Fluid (ATF) cooler hoses from the radiator.
8. Remove the radiator bracket A) mounting bolt and coolant bleed hose (B).
9. Remove the condenser mounting bolts.
10. Remove the radiator from engine compartment.

To install:

11. Install the radiator assembly.
12. Install the condenser mounting bolts.
13. Install the radiator bracket mounting bolt and coolant bleed hose.
14. Install the ATF cooler hoses to the radiator.
15. Install the front bumper upper cover.
16. Reconnect the radiator fan connector.
17. Install the upper and lower radiator hoses.
18. Install the air duct.
19. Install the undercover.
20. Refill with engine coolant.
21. Start engine and check for leaks.

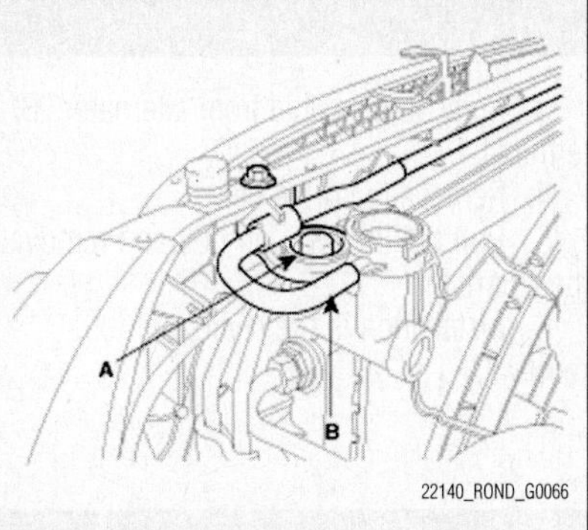

Fig. 15 Remove the left-hand radiator bracket (A) mounting bolt and coolant bleed hose (B)

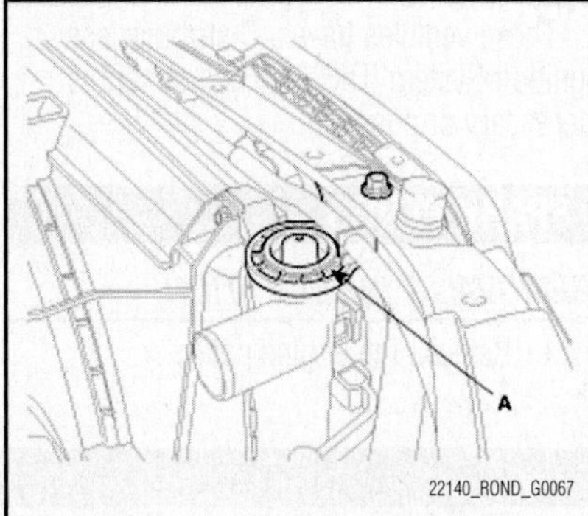

Fig. 16 Remove the right-hand radiator bracket (A)

THERMOSTAT

REMOVAL & INSTALLATION

1. Drain engine coolant so its level is below thermostat.
2. Remove water inlet and thermostat.
3. Installation is the reverse of removal.

WATER PUMP

REMOVAL & INSTALLATION

2.4L Engine

See Figure 17.

⁂ WARNING

System is under high pressure when the engine is hot. To avoid danger of releasing scalding engine coolant, remove the cap only when the engine is cool.

1. Drain the engine coolant.
2. Remove the drive belt.
3. Remove the exhaust manifold heat protector.
4. Remove the water inlet pipe nut.
5. Remove the water pump (A) and gasket.

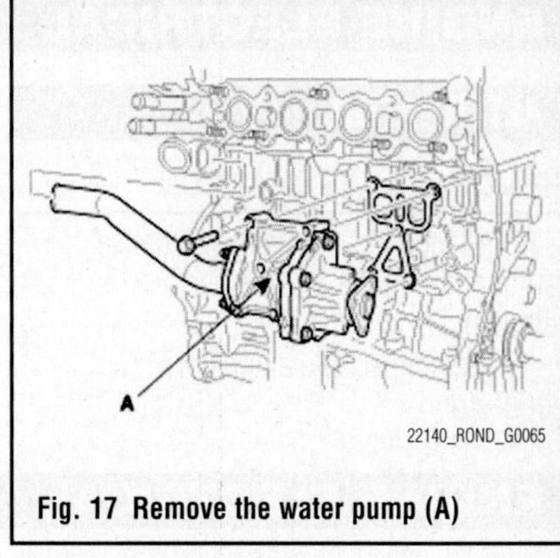

Fig. 17 Remove the water pump (A)

6. Installation is the reverse of removal. Torque as follows:
   - Water pump bolt: 13.7–17.4 ft. lbs. 18.6–23.5 Nm)
   - Water inlet pipe nut: 13.7–17.4 ft. lbs. 18.6–23.5 Nm)

2.7L Engine

See Figure 18.

⁂ WARNING

System is under high pressure when the engine is hot. To avoid danger of releasing scalding engine coolant, remove the cap only when the engine is cool.

1. Drain the engine coolant.
2. Remove accessory drive belt (A).
3. Remove the timing belt.
4. Remove the water pump (A) and gasket (B).
5. Installation is the reverse of removal. Torque as follows:
   - Water pump bolt (8_20): 10.8–15.9 ft. lbs. 14.7–21.6 Nm)
   - Water pump bolt (8_25): 10.8–15.9 ft. lbs. 14.7–21.6 Nm)
   - Water inlet pipe bolt: 12.3–14.5 ft. lbs. 16.7–19.6 Nm)

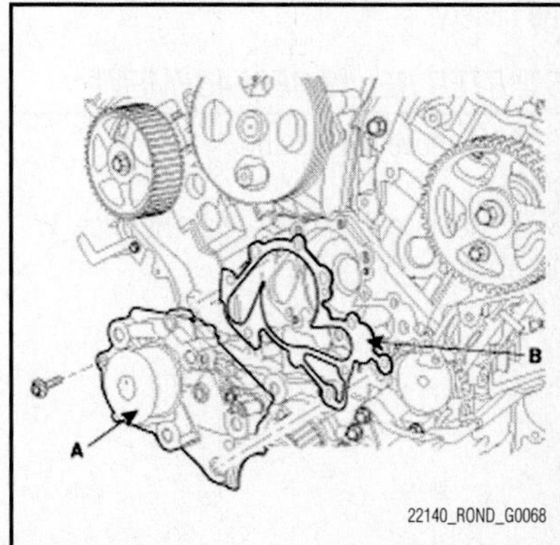

Fig. 18 Remove the water pump (A) and gasket (B)

ENGINE ELECTRICAL — CHARGING SYSTEM

ALTERNATOR

REMOVAL & INSTALLATION

1. Disconnect the battery negative terminal first, then the positive terminal.
2. Disconnect the alternator connector.
3. Remove the cable from alternator "B" terminal.
4. Remove the drive belt.
5. Pull out the through bolt and remove the alternator
6. Installation is the reverse of removal.

VOLTAGE REGULATOR

ADJUSTMENT

The charging system includes an alternator with a built-in regulator. There is no adjustment possible for the voltage regulator. If the voltage regulator is defective, it must be replaced.

ENGINE ELECTRICAL — IGNITION SYSTEM

FIRING ORDER

2.4L engine firing order: 1–3–4–2
2.7L engine firing order: 1–2–3–4–5–6

IGNITION COIL

REMOVAL & INSTALLATION

1. Remove the engine cover.
2. Remove the ignition coil connector(s).
3. Remove the ignition coil(s).
4. Installation is the reverse of removal.

IGNITION TIMING

ADJUSTMENT

These vehicles have a Distributorless Ignition System (DIS). No adjustment is necessary or possible.

SPARK PLUGS

REMOVAL & INSTALLATION

1. Remove the engine cover.
2. Remove the ignition coil connector(s).
3. Remove the ignition coil(s).
4. Using a spark plug socket, remove the spark plug (s).

CAUTION

Be careful that no contaminates enter through the spark plug holes.

5. Installation is the reverse of removal.

ENGINE ELECTRICAL — STARTING SYSTEM

STARTER

REMOVAL & INSTALLATION

See Figure 19.

1. Disconnect the battery negative cable.
2. Disconnect the starter cable (A) from the B terminal (B) on the solenoid (C), then disconnect the connector (D) from the S terminal (E).
3. Remove the 2 bolts holding the starter, then remove the starter.
4. Installation is the reverse of removal.
5. Connect the battery negative cable to the battery.

STARTER RELAY REPLACEMENT

1. Remove the fuse box cover.
2. Remove the starter relay. Replace if necessary.

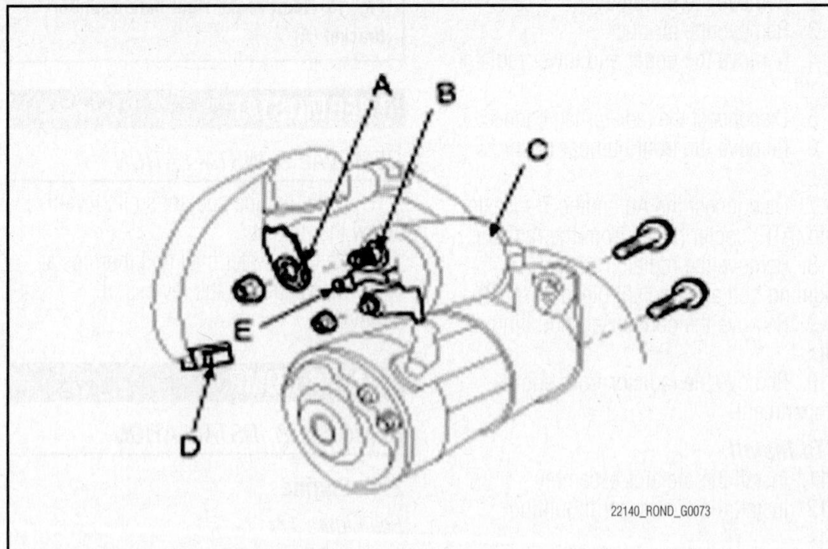

Fig. 19 Disconnect the starter cable (A) from the B terminal (B) on the solenoid (C), then disconnect the connector (D) from the S terminal (E)

ENGINE MECHANICAL

ACCESSORY DRIVE BELTS

ACCESSORY BELT ROUTING

See Figures 20 and 21.

Refer to the accompanying illustrations.

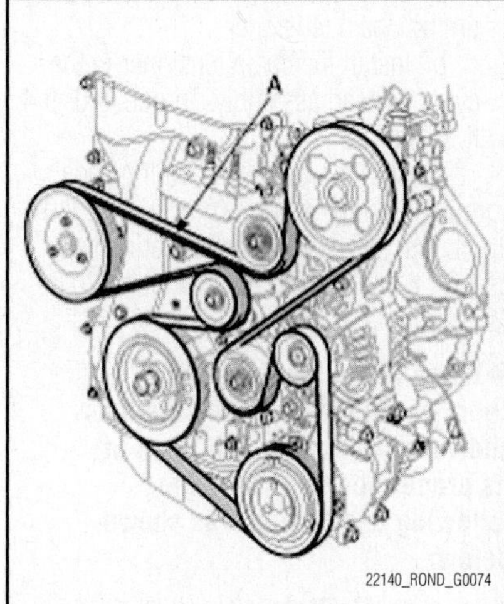

Fig. 20 Accessory belt routing—2.4L engine

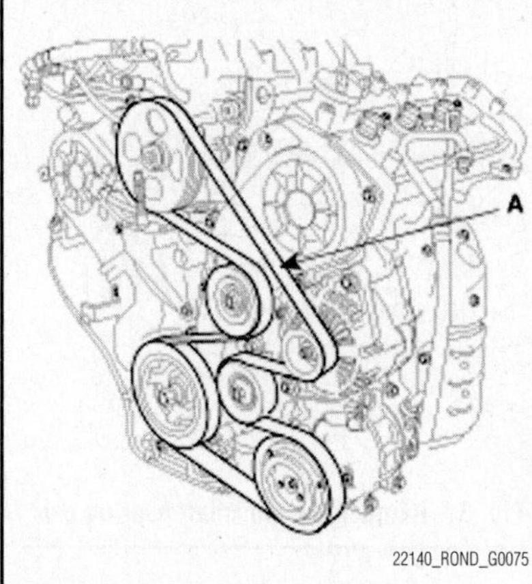

Fig. 21 Accessory belt routing—2.7L engine

INSPECTION

Visually check the belt for excessive wear, frayed cords, and signs of glazing or cracking. A glazed belt will be perfectly smooth from slippage, while a good belt will show slight fabric texture. Cracks will usually start at the inner edge of the belt and run outward. Replace worn or damaged drive belts immediately.

ADJUSTMENT

An automatic tensioner maintains belt tension. No adjustment is necessary or possible.

REMOVAL & INSTALLATION

2.4L Engine

See Figure 22.

1. Relieve tension on the tensioner pulley.
2. Remove the accessory drive belt.

CAUTION

The tensioner pulley bolt is a left-handed screw.

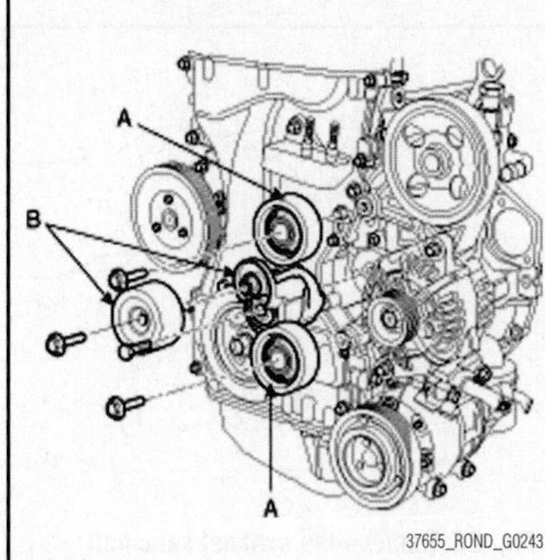

Fig. 22 Accessory belt—2.4L engine. (A) idler; (B) tensioner

To install:

3. Install accessory drive belt in the following sequence:
   a. Crankshaft pulley
   b. A/C pulley
   c. Alternator pulley
   d. Idler pulley
   e. P/C pump pulley
   f. Idler pulley
   g. Water pump pulley
   h. Tensioner pulley.
4. Rotate auto tensioner arm in the counterclockwise moving auto tensioner pulley bolt with wrench.
5. After putting belt on auto tensioner pulley, release the auto tensioner pulley slowly.
6. Note the following torque specifications:
   a. Tensioner & idler bracket bolt: 28.9–32.5 ft. lbs. (39.2–44.1 Nm)
   b. Drive belt tensioner bolt: 25.3–39.8 ft. lbs. (34.3–53.9 Nm)

2.7L Engine

See Figure 23.

1. Release tension on the accessory belt tensioner.
2. Remove the accessory drive belt.

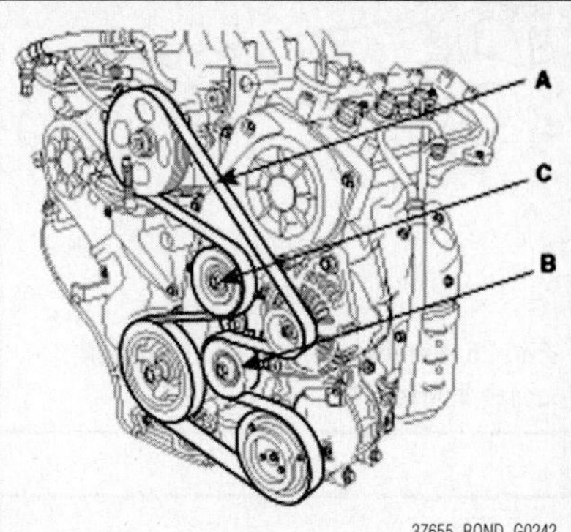

Fig. 23 Accessory belt routing—2.7L engine. (A) drive belt; (B) idler; (C) tensioner

➡In removing the drive belt, fix a tool in the auto tensioner pulley bolt and turn the bolt counter clockwise.

To install:

3. Increase the load on the accessory belt tensioner.
4. Install the accessory drive belt.
5. Release the belt tensioner to take up slack in the belt.
6. Torque specifications:
   a. Drive belt idler bolt: 25.3–39.8 ft. lbs. (34.3–53.9 Nm)
   b. Drive belt tensioner bolt: 25.3–39.8 ft. lbs. (34.3–53.9 Nm)

CAMSHAFTS & VALVE LIFTERS

REMOVAL & INSTALLATION

2.4L Engine

See Figures 24 through 29.

See also Cylinder Head Removal & Installation and Valve Lash Inspection & Adjustment

Fig. 24 Remove the front camshaft bearing cap (A)

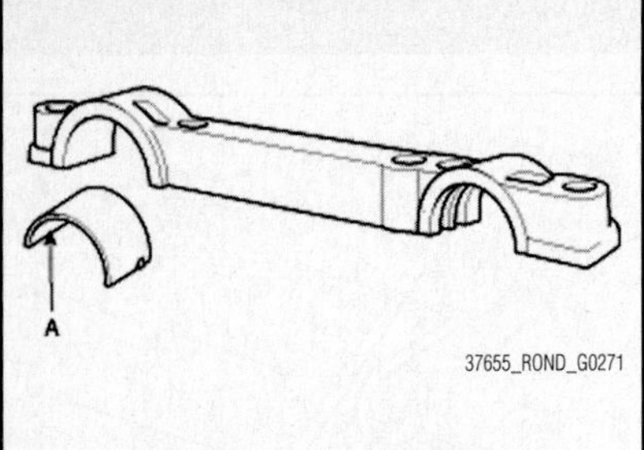

Fig. 25 Remove the exhaust camshaft upper bearing (A).

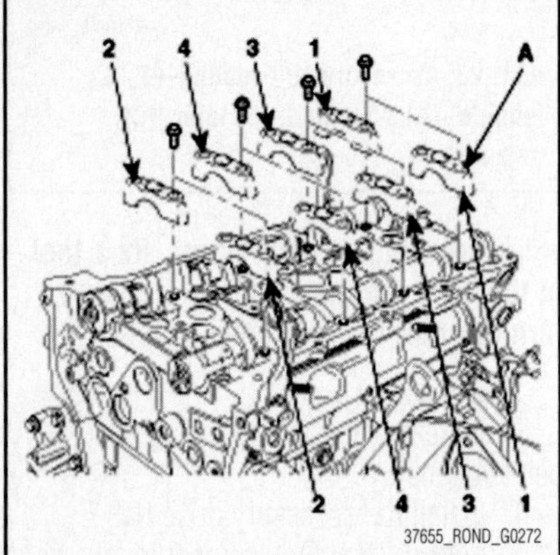

Fig. 26 Remove the camshaft bearing cap (A), in the sequence shown.

1. Remove the front camshaft bearing cap (A).
2. Remove the exhaust camshaft upper bearing (A).
3. Remove the camshaft bearing cap (A), in the sequence shown.
4. Remove the camshaft (A).
5. Remove the exhaust camshaft lower bearing (A).

To install:

➡Apply a light coat of engine oil on the camshaft journals.

6. Install the exhaust camshaft lower bearing.
7. Install the camshafts.
8. Install the exhaust camshaft upper bearing (A).
9. Install the camshaft bearing caps in their proper locations.
10. Note the following tightening order: Group A → Group B → Group C.
11. Note the following Tighten to:
    a. Step 1:
    - M6: 4.3 ft. lbs. (5.9 Nm)
    - M8: 10.8 ft. lbs. (14.7 Nm)

    b. Step 2L
    - M6: 7.9–9.4 ft. lbs. (10.8–12.7 Nm)
    - M8: 20.3–23.11 ft. lbs. (27.5–31.4 Nm)

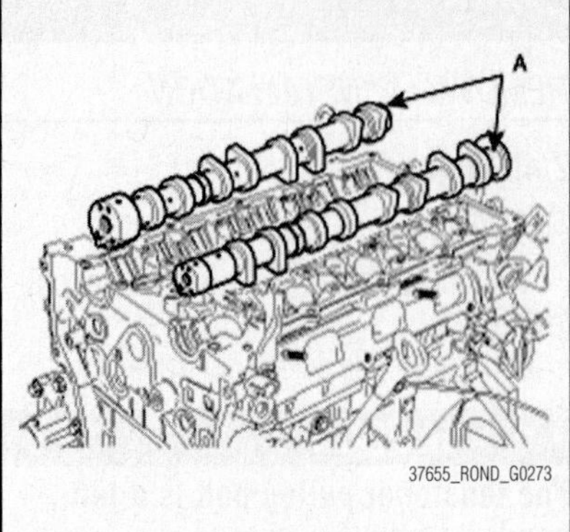

Fig. 27 Remove the camshaft.

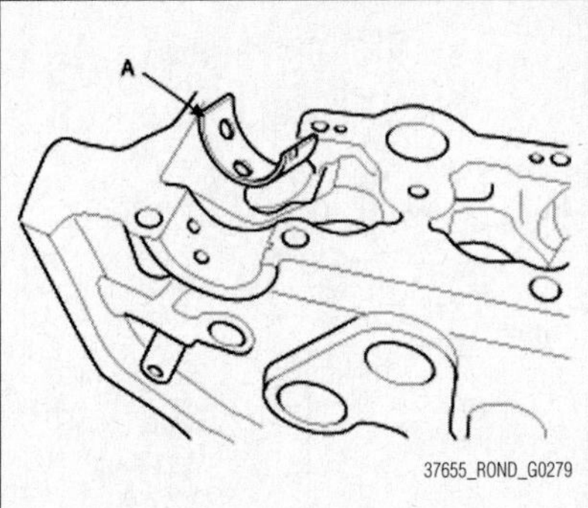

Fig. 28 Remove the exhaust camshaft lower bearing (A)

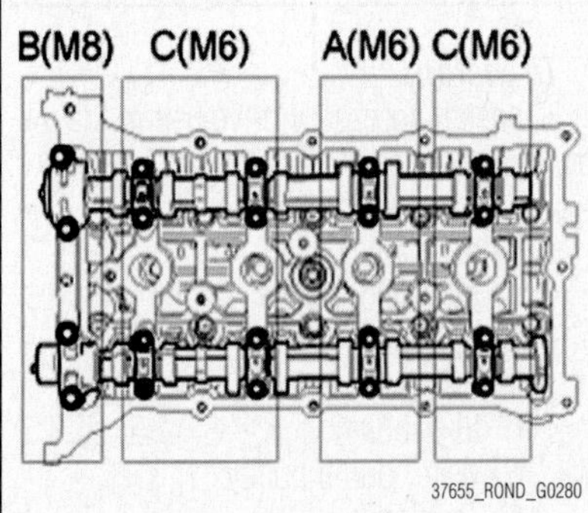

Fig. 29 Install the camshaft bearing caps in their proper locations.

2.7L Engine

See Figures 30 through 37.

1. Remove the cylinder head cover.
2. Remove the camshaft bearing cap.
3. Remove the timing chain tensioner (A).
4. Remove the camshaft.

To install:

5. Install the camshafts in the cylinder head assembly.
    a. Align the timing marks of the intake and exhaust camshaft chain sprockets and the timing chain.

⁂ CAUTION

Both timing marks should face upward in reassembly.

b. Install the intake and exhaust camshafts on the cylinder head with the timing marks aligned.

6. Install the timing chain tensioner:
    a. Insert the set pin by pressing the timing chain tensioner.
    b. Install the chain tensioner in the cylinder head assembly. Torque: 8.0–9.4 ft. lbs. (10.8–12.7 Nm).
7. Install the camshaft bearing caps. Torque: Bearing cap bolt (A: 6 × 38): 8.0–9.4 ft. lbs. (10.8–12.7 Nm). Bearing cap bolt (B: 8 × 38): 15.2–18.8 ft. lbs. (20.6–22.5 Nm).

➡When installing the bearing caps, check the marks on them as shown below and install them in its proper position. Note the following designations as shown below:

- A: (LH/RH HEAD) L (LH), R (RH)

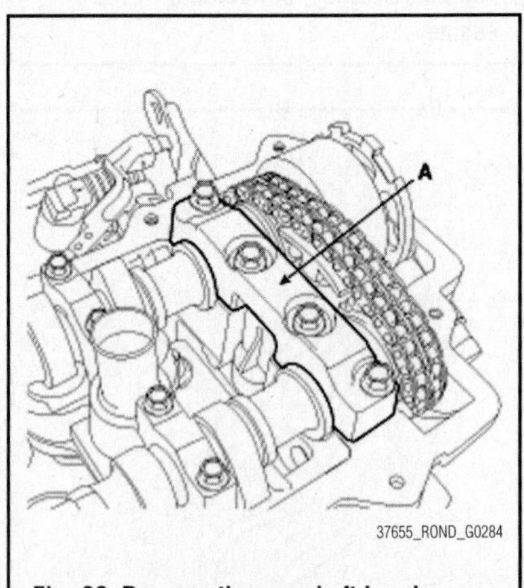

Fig. 30 Remove the camshaft bearing cap

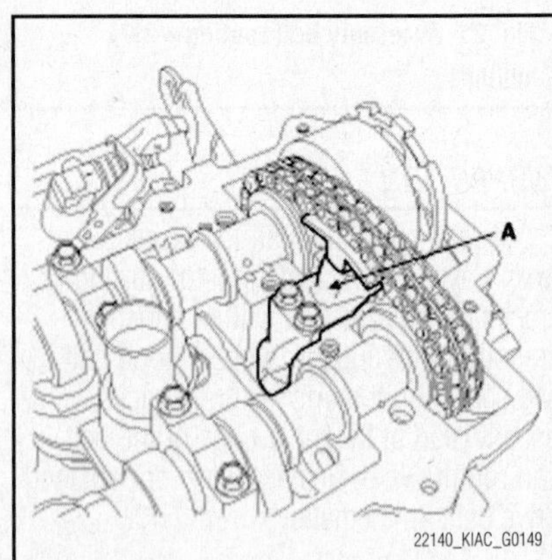

Fig. 31 Remove the timing chain tensioner (A)

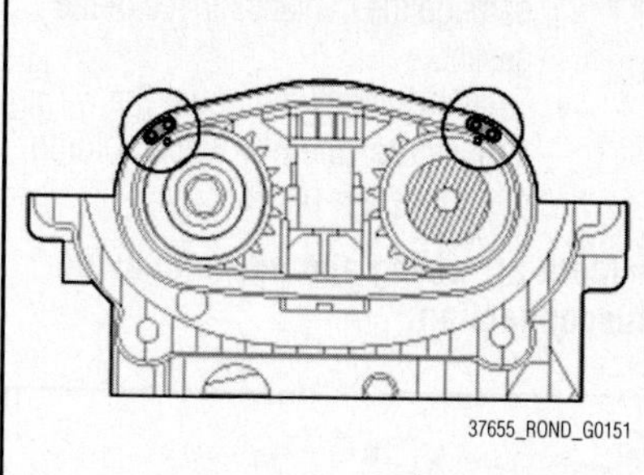
Fig. 32 Left hand camshaft chain timing mark

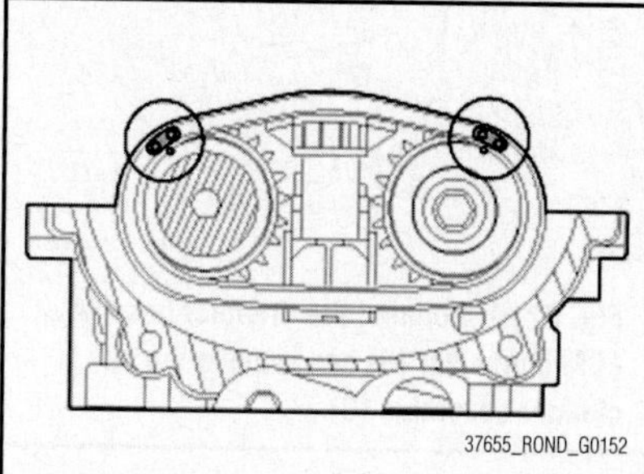
Fig. 33 Right hand camshaft chain timing mark

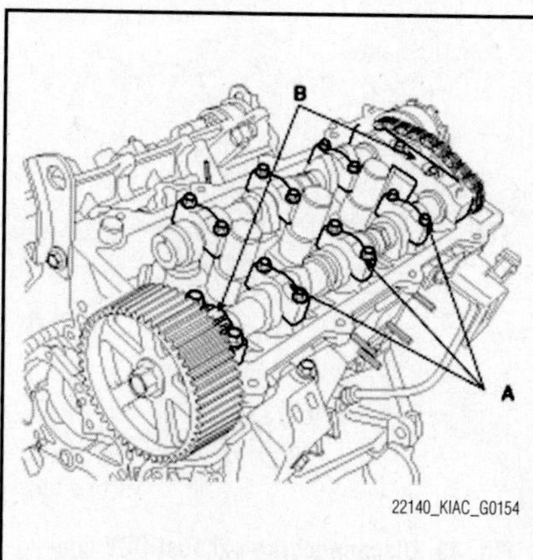

Fig. 34 Install the camshaft bearing caps

- B: (Intake/Exhaust) I (Intake), E (Exhaust)
- C: (Cap no.): 1, 2, 3

CAUTION

When installing the bearing caps, turn the crankshaft to place a piston in the middle of the block because interference between valves and pistons can occur.

8. Using the SST (09214-21000), install the camshaft oil seal.

➡Before installing, apply engine oil. The camshaft cap surface should adhere to the cylinder head assembly. Do not press an eccentric load.

9. Install the camshaft sprocket.
   a. Hold the hexagonal head wrench portion of the camshaft with a vise, and install the bolt. Torque: 65.1–79.6 ft. lbs. (88.3–107.9 Nm).
   b. When the camshaft is replaced with new one, inspect the valve clearances and then install the appropriate MLA tappet. (Refer to Valve lash inspection and adjustment.)

CAUTION

To prevent the valve from interfering with the piston, rotate the crankshaft sprocket at 3 pitches counterclockwise from the No.1 cylinder piston at TDC position before inspecting the valve clearances.

10. Install the CKP sensor connector bracket.
11. Install the bank 1 timing belt rear cover.
12. Install the bank 2 timing belt rear cover.

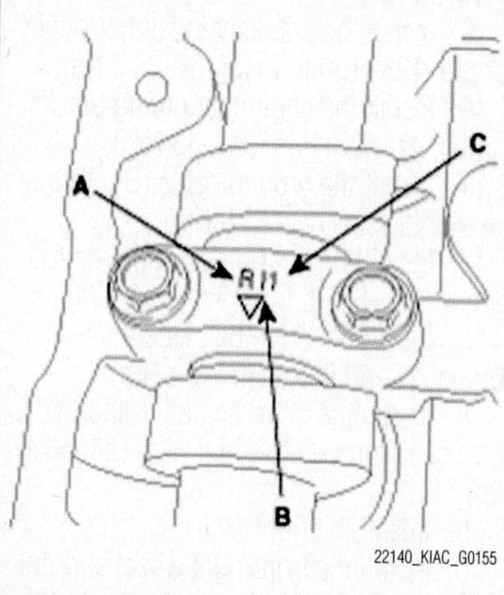

Fig. 35 When installing the bearing caps, check the marks

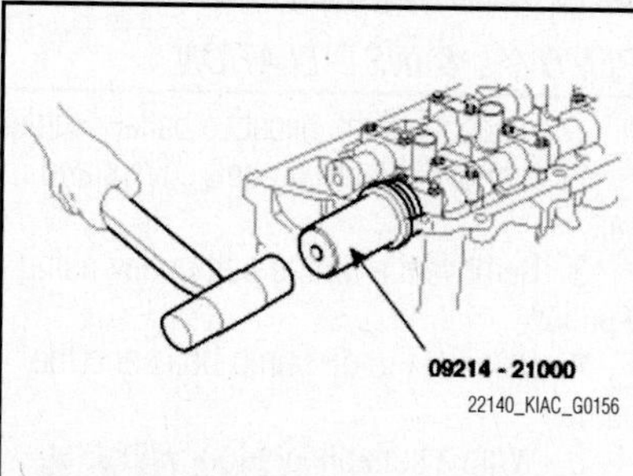

Fig. 36 Using the SST (09214-21000), install the camshaft oil seal

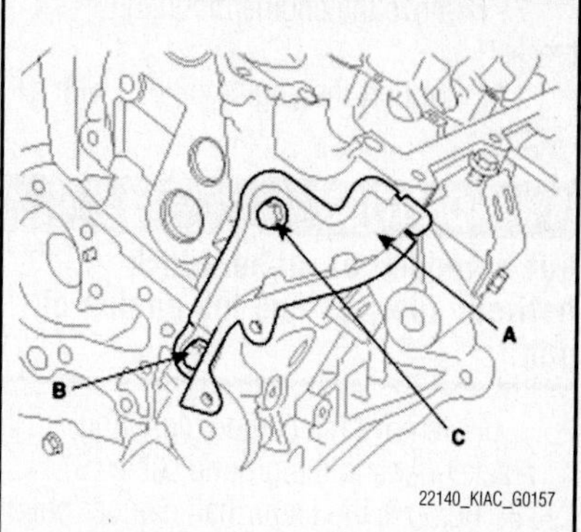

Fig. 37 Install the bank 1 timing belt rear cover

➡The length of the bolt B is longer than that of the bolt C.

13. Install the cylinder head cover.

CRANKSHAFT DAMPER

REMOVAL & INSTALLATION

2.7L Engine

See Figures 38 through 41.

1. Remove the engine cover
2. Remove the front right wheel and tire.
3. Remove the right side cover.
4. Remove the drive belt, the idler pulley and the tensioner pulley.

➡In removing the drive belt, fix a tool in the auto tensioner pulley bolt and turn the bolt counter clockwise.

5. Remove the timing belt upper covers (A).
6. Align the groove of the pulley with the timing mark of the timing belt cover by turning the crankshaft pulley clockwise. Check that the timing mark of the camshaft sprocket is aligned with that of the cylinder head cover. (No.1 cylinder piston at TDC)

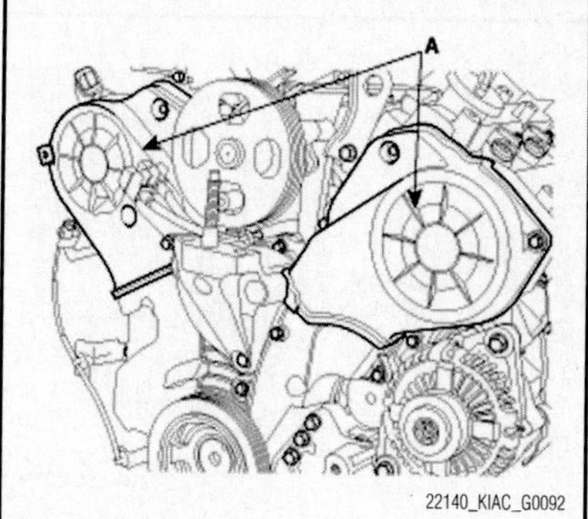

Fig. 38 Remove the timing belt upper covers (A)

7. Remove the engine mounting bracket.

a. Support the engine oil pan with a jack.

⁂ CAUTION

Put a wooden or rubber block between the jack and the engine oil pan.

b. Remove the engine mounting bracket (A) and the ground cable (B).

8. Remove the crankshaft damper pulley (A).

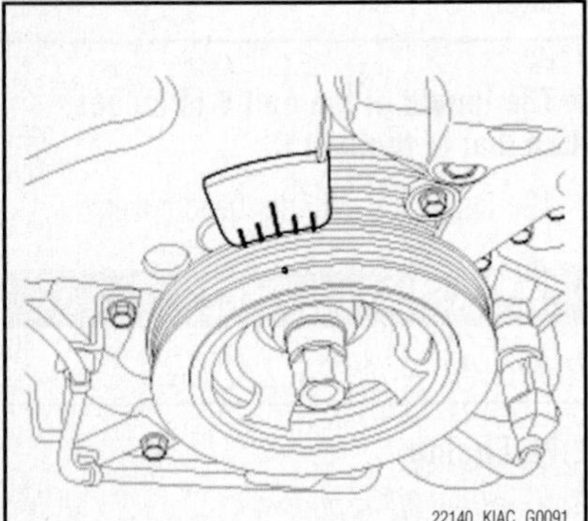

Fig. 39 Align the groove of the pulley with the timing mark of the timing belt cover

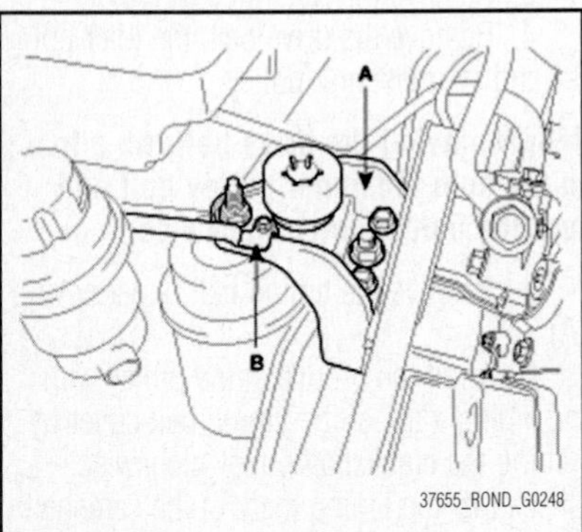

Fig. 40 Remove the engine mounting bracket (A) and the ground cable (B)

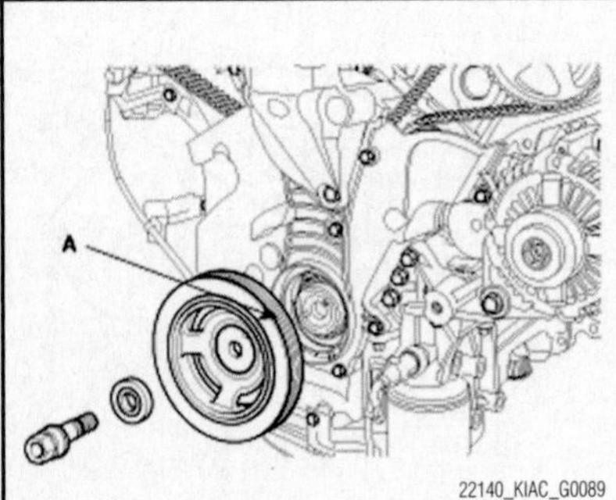

Fig. 41 Remove the crankshaft damper pulley (A)

To install:

9. Install the crankshaft damper pulley. Torque: 123–130 ft. lbs. (167–177 Nm).

10. Install the engine mounting bracket. Torque: 47–62 ft. lbs. (64–83 Nm).

11. Install the ground cable (B). Torque: 5.8–8.7 ft. lbs. (7.8–11.8 Nm).

12. Install the timing belt upper cover. Torque: 7.2–8.7 ft. lbs. (9.8–11.8 Nm).

13. Install the drive belt tensioner. Torque: 25–40 ft. lbs. (34–54 Nm).

14. Install the drive belt idler and the drive belt. Torque: 25–40 ft. lbs. (34–54 Nm).

15. Install the right side cover.

16. Install the front right wheel and tire.

17. Install the engine cover. Torque to: 5.8–8.7 ft. lbs. (7.8–11.8 Nm).

CRANKSHAFT FRONT SEAL

REMOVAL & INSTALLATION

1. Disconnect the negative battery cable.
2. Remove the timing belt covers and belt.
3. Remove the timing belt pulley using a puller.
4. Remove the oil pump bolts and the pump.
5. Wrap a suitable pry tool with a rag and work the old seal from the oil pump housing.

To install:

6. Lubricate the seal lip with clean engine oil and push the seal slightly in by hand.
7. Install the seal using a seal installer. Install the seal until it is flush with the oil pump body.
8. Install the timing belt.
9. Connect the negative battery cable.
10. Start the engine and check for leaks.

CYLINDER HEAD

REMOVAL & INSTALLATION

2.4L Engine

See Figures 42 through 53.

⁂ CAUTION

Note the following:

- Use fender covers to avoid damaging painted surfaces.
- To avoid damaging the cylinder head, wait until the engine coolant temperature drops below normal temperature before removing it.
- When handling a metal gasket, take care not to fold the gasket or damage the contact surface of the gasket.
- To avoid damage, unplug the wiring connectors carefully while holding the connector portion.

➡Mark all wiring and hoses to avoid misconnection.

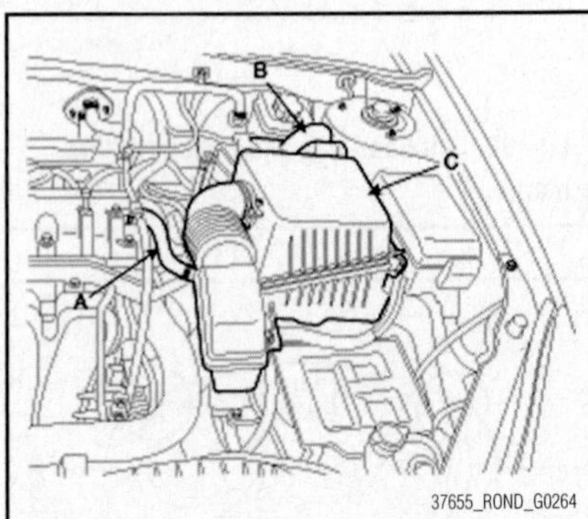

Fig. 42 Disconnect the breather hose (A), ECM connector (B) and remove the air cleaner assembly (C).

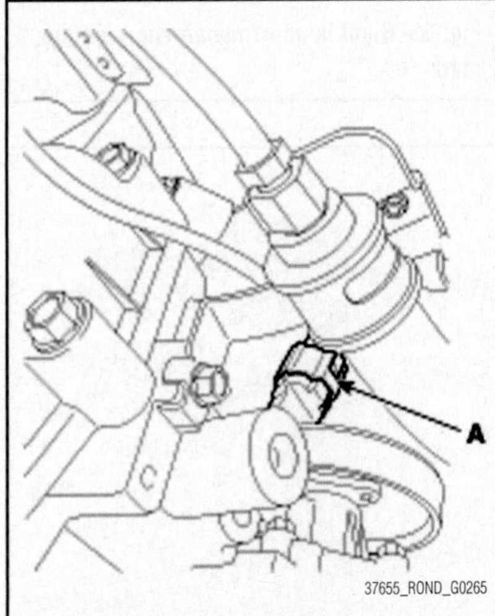

Fig. 43 Disconnect the exhaust OCV connector (A) and remove the wiring from the cylinder head cover.

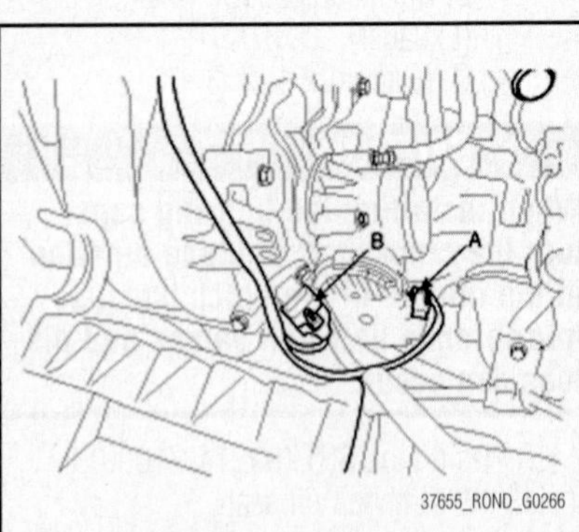

Fig. 44 Disconnect the ETC connector (A) and MAP sensor connector (B).

1. Remove the battery.
2. Remove the engine cover.
3. Remove the air duct.
4. Disconnect the breather hose (A), ECM connector (B) and remove the air cleaner assembly (C).
   a. Remove the undercover
   b. Remove the drain plug and drain the engine coolant.
5. Remove the upper and lower radiator hoses.
6. Remove the heater hoses.
7. Disconnect the exhaust OCV connector (A) and remove the wiring from the cylinder head cover.
8. Disconnect the injector connectors and ignition coil connectors.
9. Remove the PCV hose.
10. Remove the ignition coils.
11. Remove the cylinder head cover.
12. Disconnect the ETC connector (A).

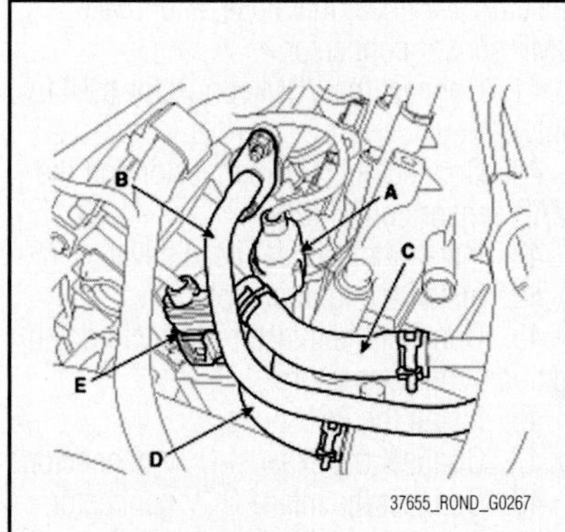

Fig. 45 Disconnect the ETC connector (A) and MAP sensor connector (B).

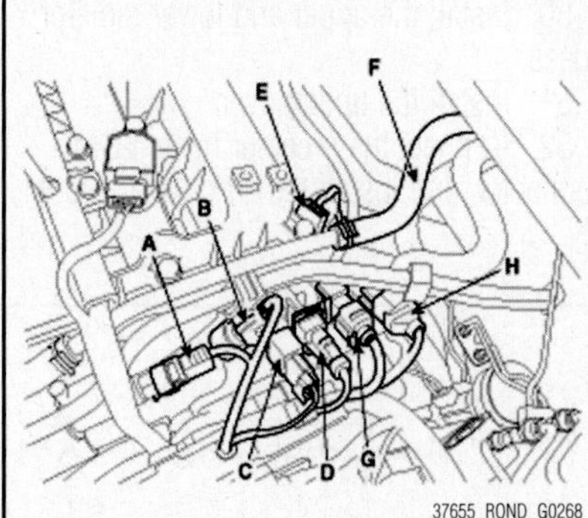

Fig. 46 Disconnect the PCSV connector (A), ECTS connector (B), condenser connector (C), CKP sensor connector (D), exhaust CMP sensor connector (E) and brake booster vacuum hose (F). Disconnect the HO2S (Bank 1 / Sensor 1) connector (G), HO2S (Bank 1 / Sensor 2) connector (H; SULEV only).

13. Disconnect the MAP sensor connector (B).
14. Disconnect the intake CMP sensor connector (A), fuel hose (B), brake booster vacuum hose (C) and PCSV hose (D).
15. Disconnect the VCM connector (E; SULEV only)
16. Disconnect the PCSV connector (A), ECTS connector (B), condenser connector (C), CKP sensor connector (D), exhaust CMP sensor connector (E) and brake booster vacuum hose (F).
17. Disconnect the HO2S (Bank 1 / Sensor 1) connector (G), HO2S (Bank 1 / Sensor 2) connector (H; SULEV only).
18. Remove the water temp control assembly (A).
19. Remove the timing chain. (Refer to Timing Chain in this group.)
20. Remove the intake & exhaust manifold. (Refer to Intake and Exhaust Manifolds in this group)
21. Remove the intake & exhaust CVVT assembly
22. Remove the camshaft. (See Camshaft section.)
23. Remove the intake Oil Control Valve (OCV; A).
24. Remove the exhaust OCV (A).
25. Remove the cylinder head bolts, then remove the cylinder head:
    a. Using a triple square wrench, uniformly loosen and remove the 10 cylinder head bolts, in several passes, in the sequence shown.

CAUTION

Head warpage or cracking could result from removing bolts in an incorrect order.

    b. Lift the cylinder head from the dowels on the cylinder block and place the cylinder head on wooden blocks on a bench.

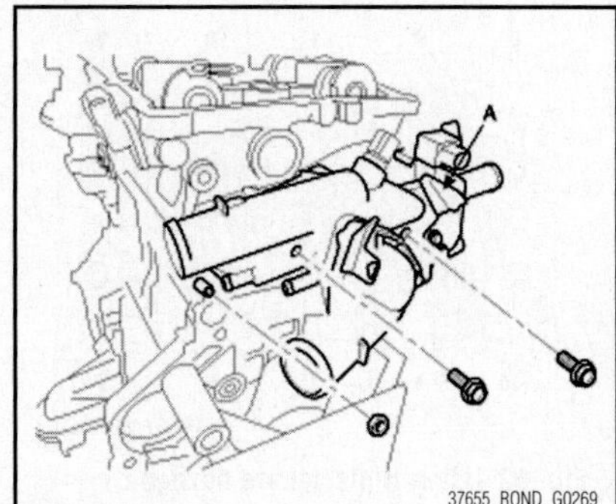

Fig. 47 Remove the water temp control assembly (A).

CAUTION

Be careful not to damage the contact surfaces of the cylinder head and cylinder block.

26. Remove the cylinder head gasket.

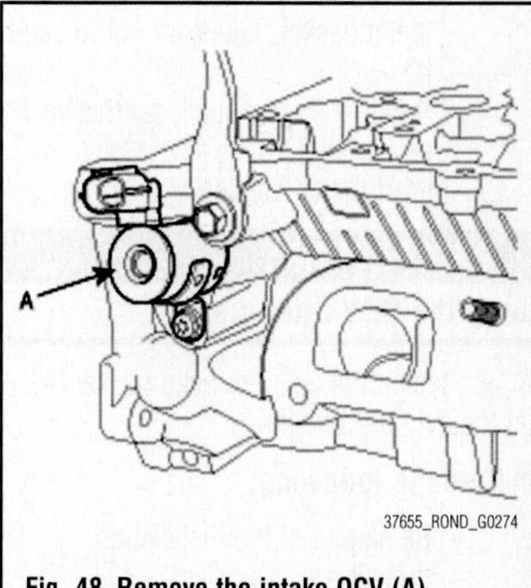

Fig. 48 Remove the intake OCV (A).

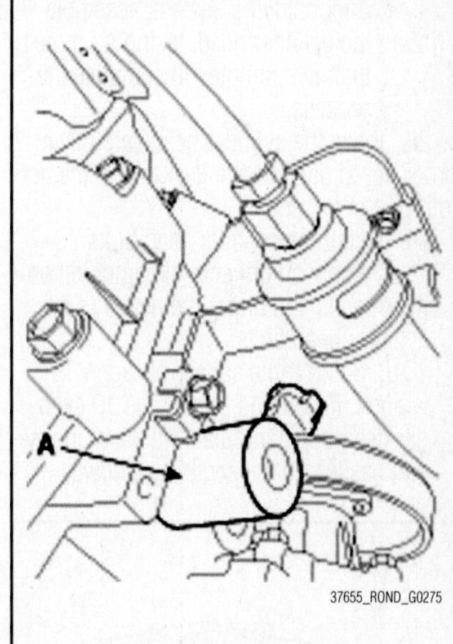

Fig. 49 Remove the exhaust OCV (A).

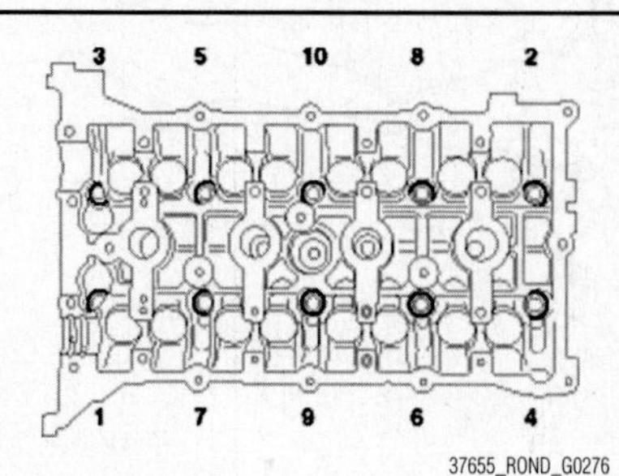

Fig. 50 Using a triple square wrench, uniformly loosen and remove the 10 cylinder head bolts, in several passes, in the sequence shown.

To install:

➡**Note the following:**

- Thoroughly clean all parts to be assembled.
- Always use a new head and manifold gasket.
- The cylinder head gasket is a metal gasket. Take care not to bend it.
- Rotate the crankshaft, set the No.1 piston at TDC.

27. Install the OCV filter.

⁂ CAUTION

Keep the OCV filter clean.

28. Install the cylinder head gasket (A) on the cylinder block.

➡**Note the following:**

- Be careful of the installation direction.
- Apply liquid gasket (LOCTITE® 5900H) on the mark (B).
- After applying sealant, assemble the cylinder head, taking no more than five minutes to complete the process.

29. Place the cylinder head carefully in order not to damage the gasket with the bottom part of the end.

30. Install the cylinder head bolts.

a. Apply a light coat of engine oil on the threads and under the heads of the cylinder head bolts.

b. Using triple square hexagon wrench, install and tighten the 10 cylinder head bolts and plate washers, in several passes, in the sequence shown.

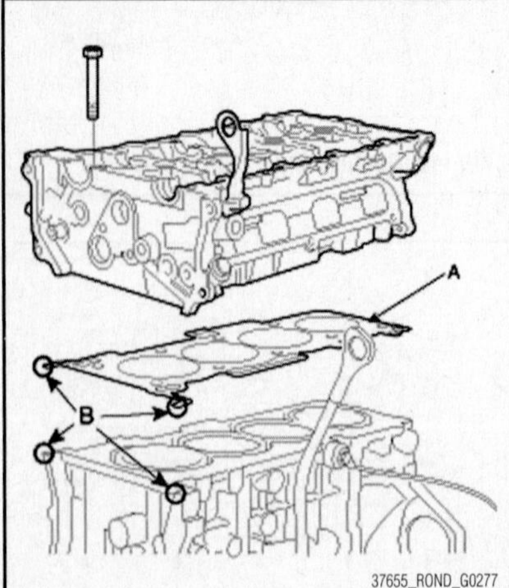

Fig. 51 Install the cylinder head gasket (A) on the cylinder block. Be careful of the installation direction. Apply liquid gasket (LOCTITE®5900H) on the mark (B).

Tighten to: 23.9–26.8 ft. lbs. (32.4–36.3 Nm) + (90–95°) + (90–95°)

➡**Always use new cylinder head bolt.**

31. Install the intake and exhaust OCV.

Tightening torque is the same for both intake and exhaust OCV:

7.2–8.7 ft. lbs. (9.8–11.8 Nm)

⁂ CAUTION

If you drop the OCV do not reuse it. Keep the OCV clean. Do not hold the OCV sleeve during servicing.

- When the OCV is installed on the engine, do not move the engine while holding the OCV yoke.

32. Install the camshafts. See Camshaft section.

33. Install the water temp control assembly (A). Assemble the water temp control assembly and water inlet pipe to the water pump assembly before tightening the water inlet pipe nuts.

Always use a new O-ring.

- Tightening torque :
- Bolt: 10.8–14.5 ft. lbs. 14.7–19.6 Nm)
- Nut: 13.7–17.4 ft. lbs. 18.6–23.5 Nm)

34. Install the timing chain. See Timing Chain section.

35. Check and adjust the valve clearance. See Valve Lash section.

36. Install the cylinder head cover:

a. Remove the hardening sealant located on the upper area between timing chain cover and cylinder head before assembling the cylinder head cover.

- After applying the sealant, assemble it within 5 minutes.
- Bead width: 0.1 in. (2.5mm)
- Sealant: LOCTITE® 5900H

b. The firing and/or blow out test should not be performed within 30 minutes after assembling the cylinder head cover.

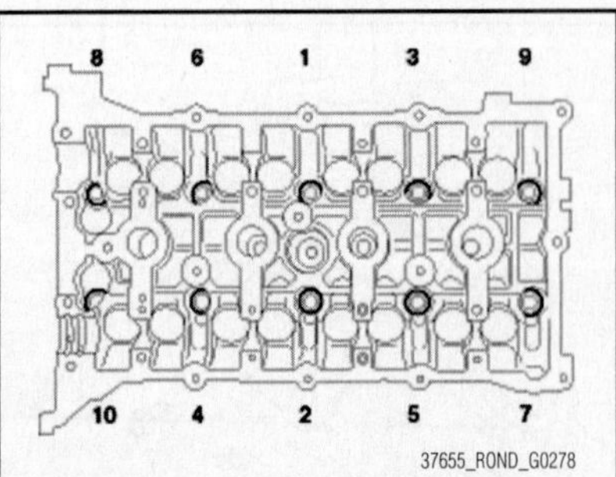

Fig. 52 Using triple square hexagon wrench, install and tighten the 10 cylinder head bolts and plate washers, in several passes, in the sequence shown.

c. Install the cylinder head cover bolts in the proper sequence, and tighten as follows:

- Step 1: 2.9–4.3 ft. lbs. 3.9–5.9 Nm)
- Step 2: 5.8–7.2 ft. lbs. 7.8–9.8 Nm)

⁂ CAUTION

Do not reuse the cylinder head cover gasket.

37. Install the intake and exhaust manifolds. (Refer to Intake and Exhaust Manifolds.)

38. Connect the PCSV connector, ECTS connector, condenser connector, CKP sensor connector, exhaust CMP sensor connector, and the brake booster vacuum hose.

39. Connect the HO2S (Bank 1 / Sensor 1) connector and the HO2S (Bank 1 / Sensor 2) connector (SULEV only).

40. Connect the brake booster vacuum hose, PCSV hose, fuel hose, and intake CMP sensor connector.

41. Connect the VCM connector (SULEV only).

42. Connect the ETC connector and the MAP sensor connector.

43. Install the cylinder head cover.

44. Install the ignition coils.

45. Connect the injector connectors and ignition coil connectors.

46. Install the PCV hose.

47. Connect the exhaust OCV connector.

48. Connect the intake OCV connector, power steering switch connector, VIS connector, OPS connector, knock sensor connector, alternator connector, and A/C switch connector.

49. Install the heater hoses.

50. Install the upper and lower radiator hoses.

51. Install the undercover.

52. Connect the breather hose, ECM connector, and install the air cleaner assembly.

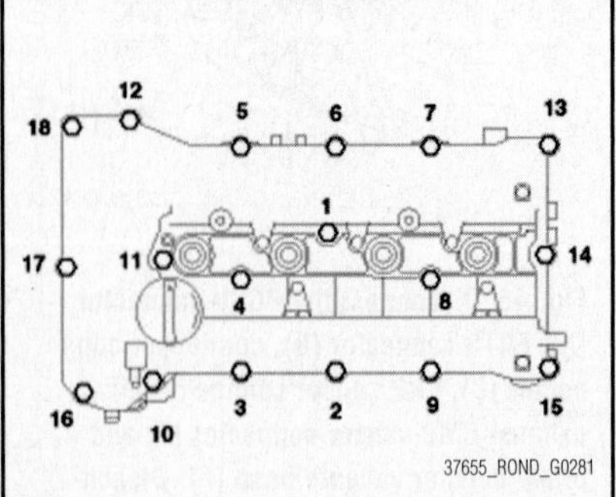

Fig. 53 Cylinder head cover bolt installation order

53. Install the air duct.

54. Install the engine cover. Torque to: 5.8–8.7 ft. lbs. (7.8–11.8 Nm).

55. Install the negative terminal to the battery.

56. Fill with engine coolant.

57. Start the engine and check for leaks.

58. Recheck the engine coolant level and the oil level.

2.7L Engine

See Figures 54 through 72.

⁂ CAUTION

Note the following:

- Use fender covers to avoid damaging painted surfaces.
- To avoid damaging the cylinder head, wait until the engine coolant temperature drops below normal temperature before removing it.
- When handling a metal gasket, take care not to fold the gasket or damage the contact surface of the gasket.
- To avoid damage, unplug the wiring connectors carefully while holding the connector portion.

➡Note the following:

- Mark all wiring and hoses to avoid misconnection.
- Turn the crankshaft pulley so that the No. 1 piston is at top dead center.

1. Remove the battery.
2. Remove the air duct.
3. Remove the engine cover.
4. Remove the air cleaner assembly. (Refer to Air Cleaner section.)
5. Remove the upper and lower radiator hoses.
6. Remove the fuel inlet hose from the delivery pipe.
7. Disconnect the engine wiring harness connectors.

 a. Disconnect the No.1/No.2 knock sensor connectors (A, B), the oil pressure switch connector (C), the ignition coil harness (D) and the No.1 VIS (Variable Induction System) connector (E).

 b. Disconnect the bank 1 front/rear O2 sensor connectors (A).

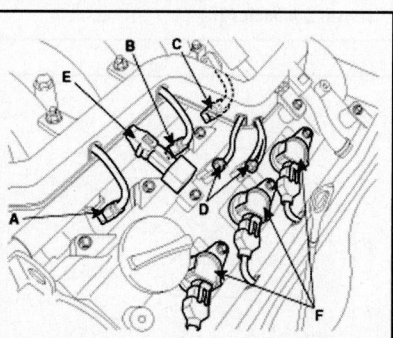

 c. Disconnect the injection connectors (A,B,C), the ground lines (D), the condenser connector (E) and the ignition coil connectors (F).

 d. Disconnect the injection harness connector (A), the No.2 VIS (Variable Induction System) connector (B), the No.1/No.2 OCV (Oil Control Valve) connectors (C, D) and the OTS (Oil Temperature Sensor) connector (E).

 e. Disconnect the MAPS (Manifold Absolute Pressure Sensor) connector (A), the ETC (Electronic Throttle Control) connector (B) and the PCSV (Purge Control Solenoid Valve) connector (C).

8. Disconnect the alternator connector and the air conditioning compressor connector.
9. Disconnect the bank 2 CMP sensor connector (A) and the ECT (Engine Coolant Temperature) sensor connector (B).

 a. Disconnect the bank 2 front/rear O2 sensor connectors (A,B) and the CKP sensor connector (C).

 b. Disconnect the bank 1 CMP sensor connector (A).

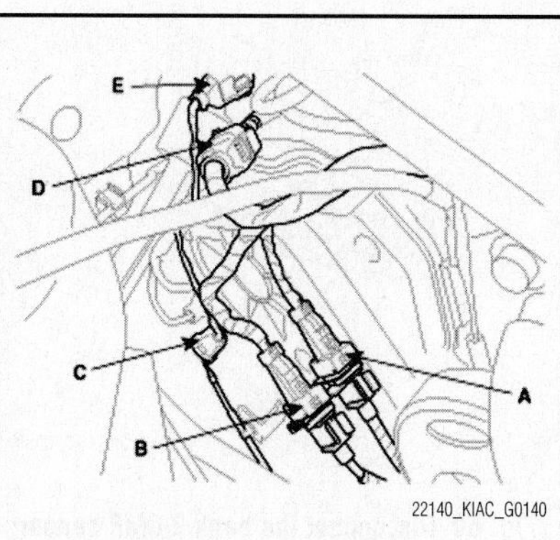

Fig. 54 Disconnect the engine wiring harness connectors

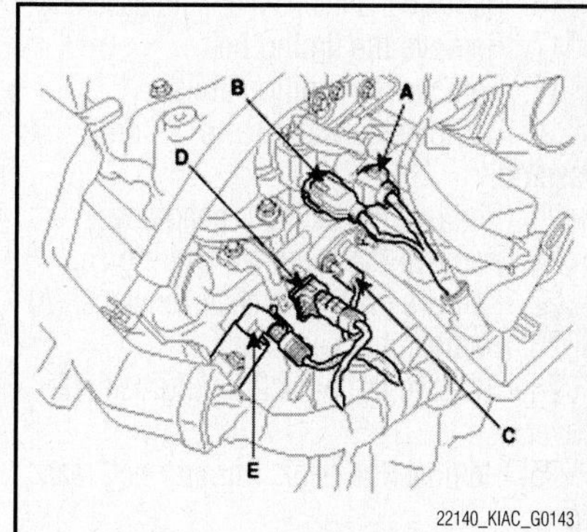

Fig. 57 Disconnect the injection harness connector (A), the No.2 VIS (Variable Induction System) connector (B), the No.1/No.2 OCV (Oil Control Valve) connectors (C,D) and the OTS (Oil Temperature Sensor) connector (E)

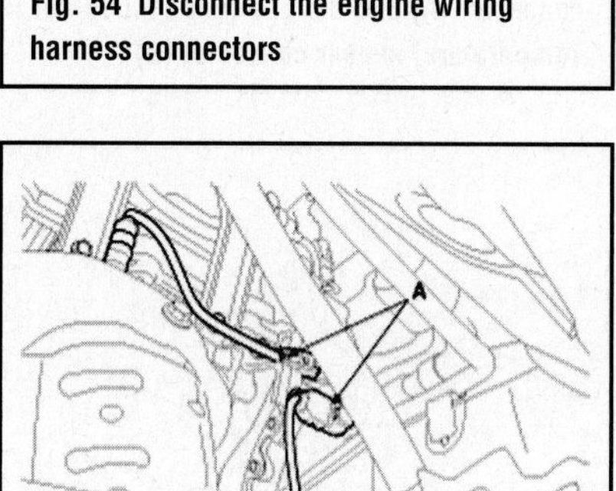

Fig. 55 Disconnect the bank 1 front/rear O2 sensor connectors (A)

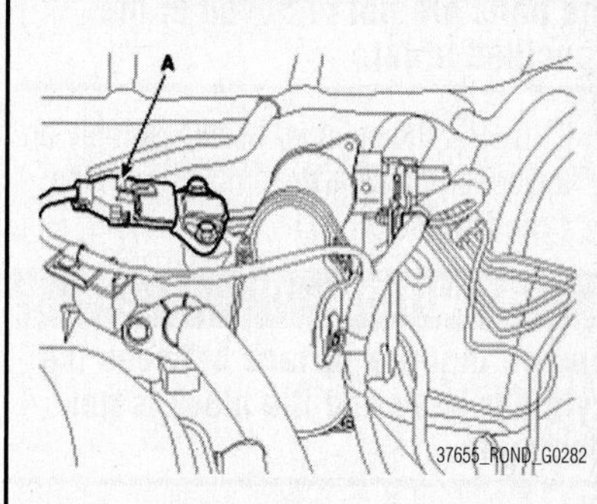

Fig. 58 Disconnect the MAPS (Manifold Absolute Pressure Sensor) connector (A)

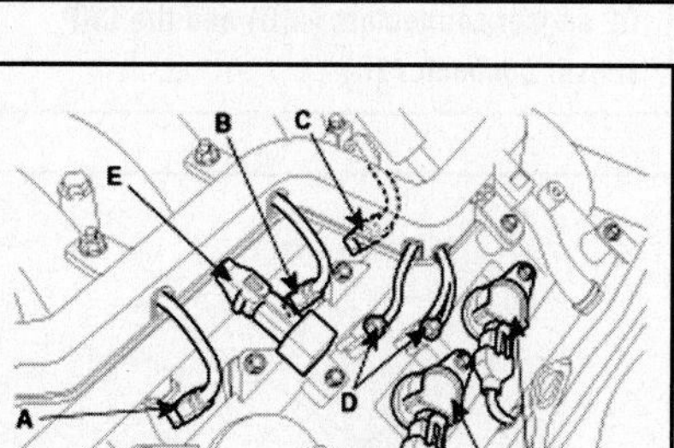

Fig. 56 Disconnect the injection connectors (A,B,C), the ground lines (D), the condenser connector (E) and the ignition coil connectors (F)

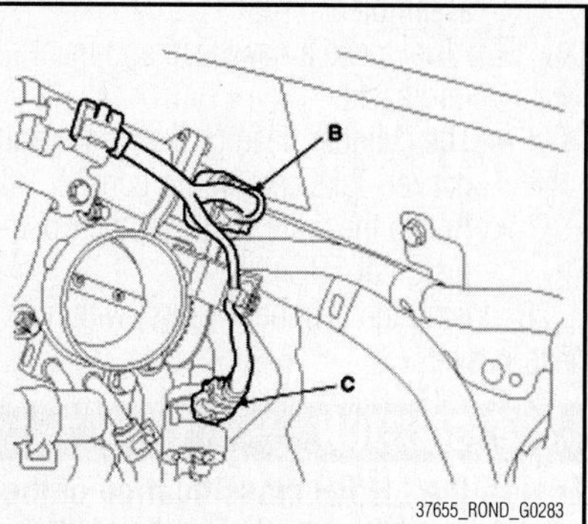

Fig. 59 Disconnect the ETC (Electronic Throttle Control) connector (B) and the PCSV (Purge Control Solenoid Valve) connector (C)

10. Remove the PCV (Purge Control Valve) hose.
11. Disconnect the brake vacuum hose.
12. Remove the heater hoses.
13. Remove the accessory drive belt.
14. Remove the power steering pump.
15. Remove the exhaust manifold assembly.
16. Remove the intake manifold assembly.
17. Remove the timing belt.
18. Remove the ignition coils.
19. Remove the water temp. control assembly.
20. Remove the cylinder head cover.
21. Remove the camshaft bearing cap.
22. Remove the timing chain tensioner (A).
23. Remove the camshaft.
24. Remove the bank 2 timing belt rear cover.
25. Remove the bank 1 timing belt rear cover.
26. Remove the CKP sensor connector bracket.
27. Remove the cylinder head assembly.
    a. Remove the bolts in 2–3 steps in the sequence shown.

CAUTION

Cylinder head damage may result if the bolts are not removed in the specified order.

    b. Put the cylinder head assembly on a wooden block after removal from the cylinder block.

CAUTION

Ensure that the surface between the cylinder head and the block is not damaged.

To install:

Note the following:

- Thoroughly clean all parts to be assembled.
- Always use a new head and manifold gasket.
- The cylinder head gasket is a metal gasket. Take care not to bend it.
- Rotate the crankshaft, set the No.1 piston at TDC.

28. Install the cylinder head(s) with new head gaskets.

CAUTION

Ensure the LH/RH classification of the cylinder head gasket when installing.

29. Tighten the cylinder head bolts with the plain washers in several steps in the sequence shown.

In assembling washers, the marked surface should face upward.

CAUTION

Always use new cylinder head bolts.

30. Before installing the cylinder head bolts, apply engine oil on the thread of the bolts and the surface of the washers.

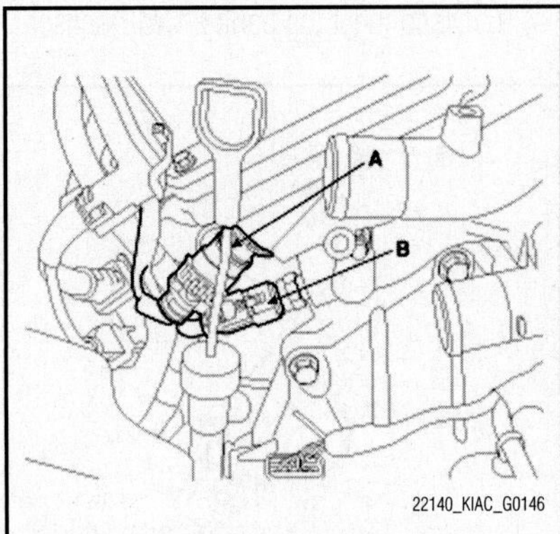

Fig. 60 Disconnect the bank 2 CMP sensor connector (A) and the ECT (Engine Coolant Temperature) sensor connector (B)

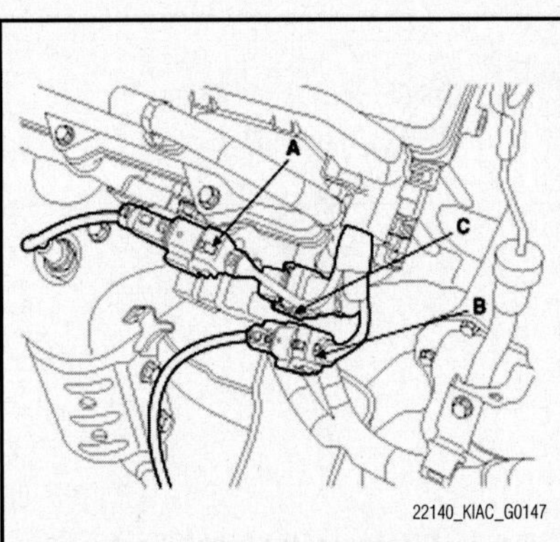

Fig. 61 Disconnect the bank 2 front/rear 02 sensor connectors (A,B) and the CKP sensor connector (C)

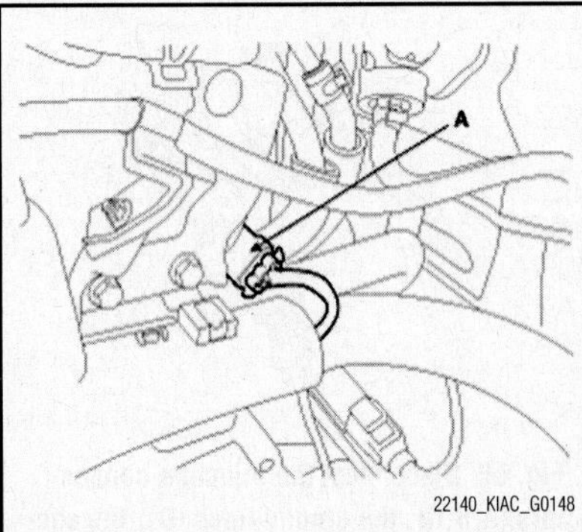

Fig. 62 Disconnect the bank 1 CMP sensor connector (A)

Torque: 18.1 ft. lbs. (24.5 Nm) plus 60°plus an additional 45°.

Using the SST (09221-4A000), tighten the bolts which need to be tightened with the angular tightening method.

31. Install the CVVT assembly and

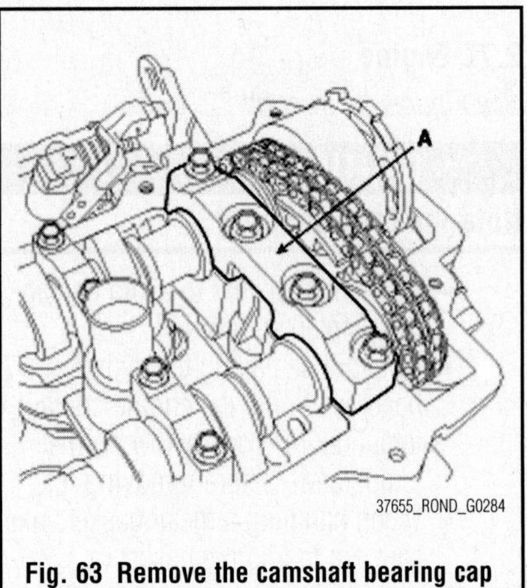

Fig. 63 Remove the camshaft bearing cap

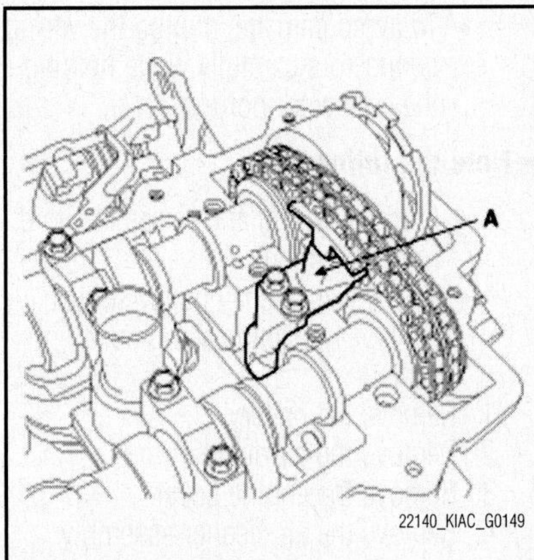

Fig. 64 Remove the timing chain tensioner (A)

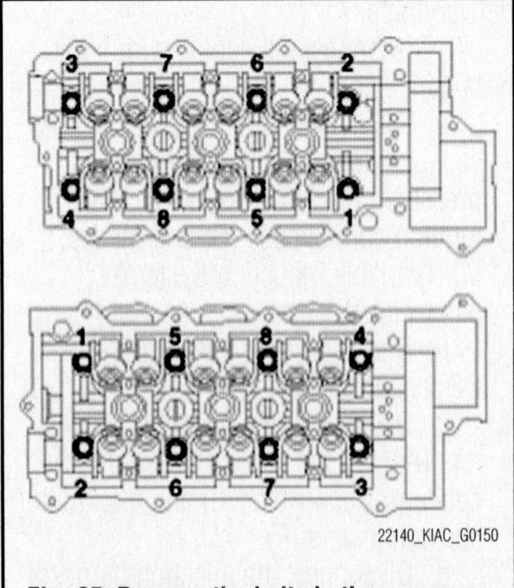

Fig. 65 Remove the bolts in the sequence shown

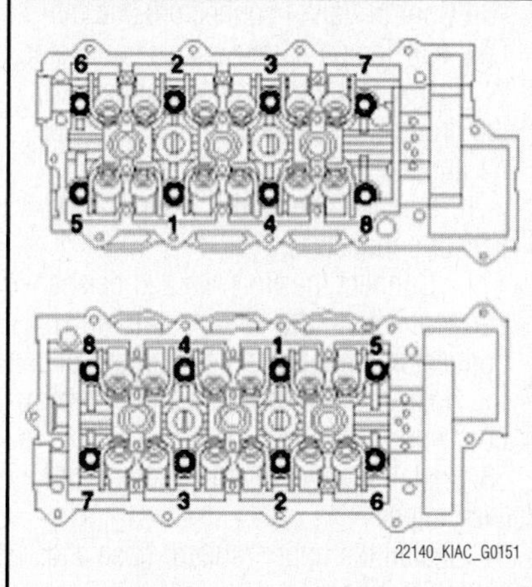

Fig. 66 Install cylinder head bolts in the sequence shown

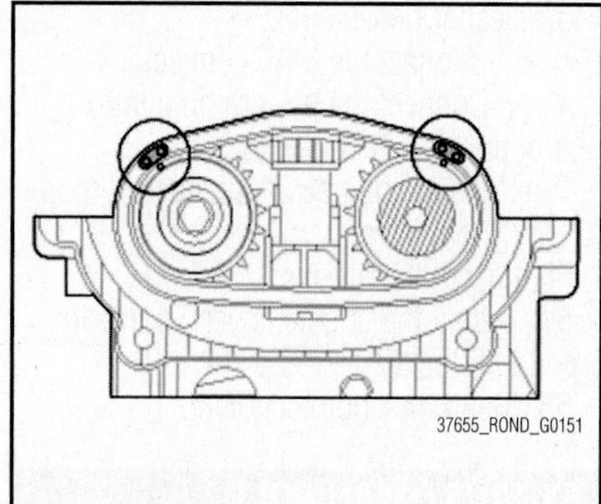

Fig. 67 Left hand camshaft chain timing mark; right side similar

camshaft chain sprocket with the dowel pin in the CVVT installed to the intake camshaft. Ensure that the pin will not be installed in the hole for oil feeding. Torque: 49.2–57.9 ft. lbs. (66.7–78.5 Nm).

➡After tightening the CVVT bolts, rotate the CVVT assembly housing counterclockwise by hand to seat the lock pin in the CVVT assembly in good position.

CAUTION

Fix the hexagonal part of the camshaft in a vise when tightening the CVVT bolts. Do not fix the CVVT housing or sprocket in a vise.

32. Install the camshafts in the cylinder head assembly.

a. Align the timing marks of the intake and exhaust camshaft chain sprockets and the timing chain.

CAUTION

Both timing marks should face upward in reassembly.

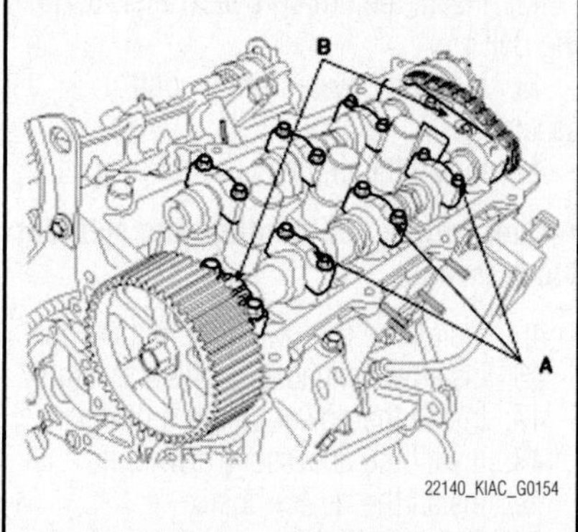

Fig. 68 Install the camshaft bearing caps

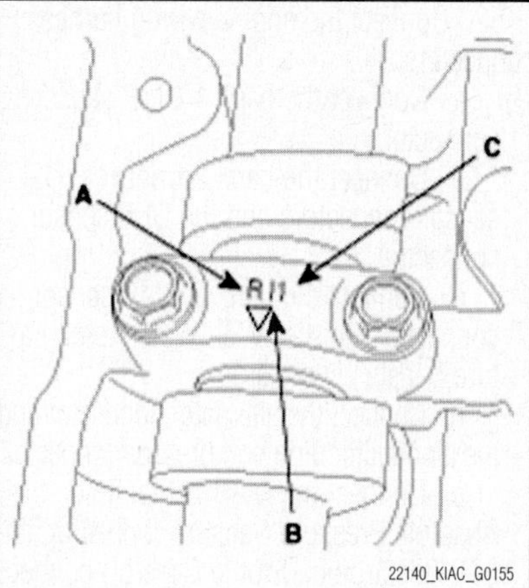

Fig. 69 When installing the bearing caps, check the marks

b. Install the intake and exhaust camshafts on the cylinder head with the timing marks aligned.

33. Install the timing chain tensioner.

a. Insert the set pin by pressing the timing chain tensioner.

b. Install the chain tensioner in the cylinder head assembly. Torque: 8.0–9.4 ft. lbs. (10.8–12.7 Nm).

34. Install the camshaft bearing caps. Torque: Bearing cap bolt (A: 6 _ 38): 8.0–9.4 ft. lbs. (10.8–12.7 Nm). Bearing cap bolt (B: 8 _ 38): 15.2–18.8 ft. lbs. (20.6–22.5 Nm).

➡When installing the bearing caps, check the marks on them as shown below and install them in its proper position. Note the following designations as shown below:

- A: (LH/RH HEAD) L (LH), R (RH)
- B: (Intake/Exhaust) I (Intake), E (Exhaust)
- C: (Cap no.): 1, 2, 3

CAUTION

When installing the bearing caps, turn the crankshaft to place a piston

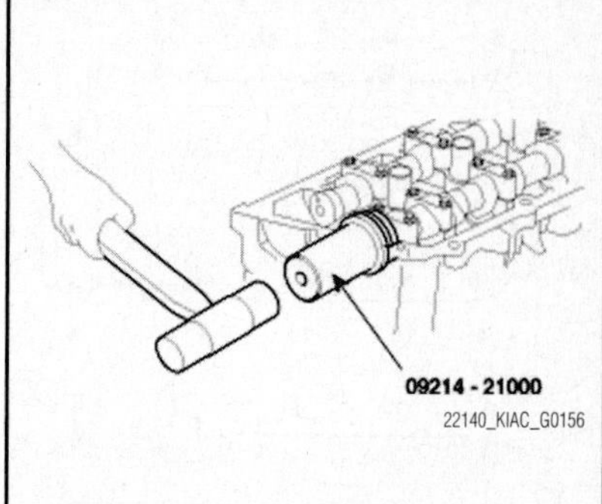

Fig. 70 Using the SST (09214-21000), install the camshaft oil seal

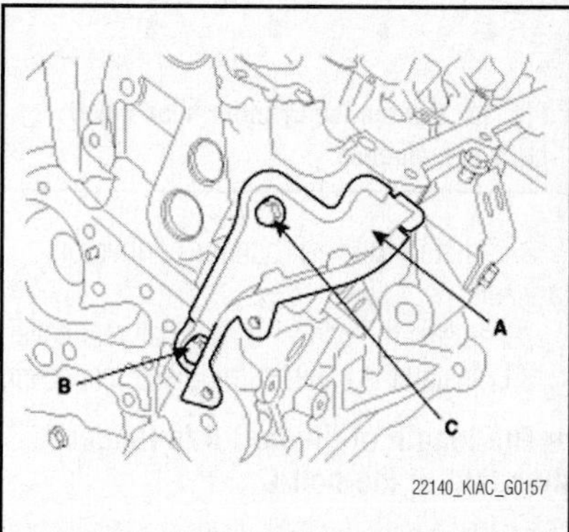

Fig. 71 Install the bank 1 timing belt rear cover

in the middle of the block because interference between valves and pistons can occur.

35. Using the SST (09214-21000), install the camshaft oil seal.

➡Before installing, apply engine oil. The camshaft cap surface should adhere to the cylinder head assembly. Do not press an eccentric load.

36. Install the camshaft sprocket.

a. Hold the hexagonal head wrench portion of the camshaft with a vise, and install the bolt. Torque: 65.1–79.6 ft. lbs. (88.3–107.9 Nm).

b. When the camshaft is replaced with new one, inspect the valve clearances and then install the appropriate MLA tappet. (Refer to Valve lash inspection and adjustment.)

CAUTION

To prevent the valve from interfering with the piston, rotate the crankshaft sprocket at 3 pitches counterclockwise from the No.1 cylinder piston at TDC position before inspecting the valve clearances.

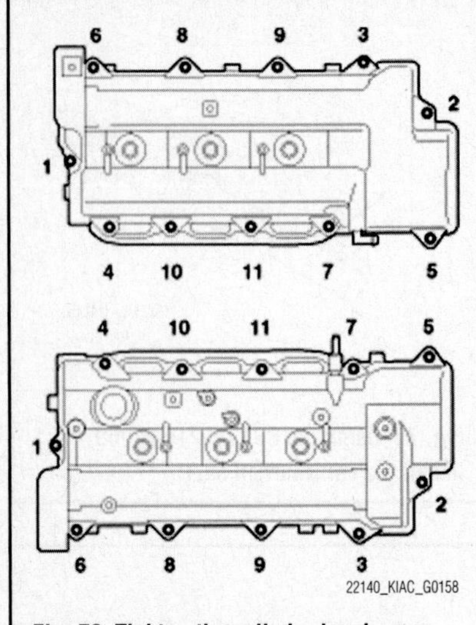

Fig. 72 Tighten the cylinder head cover bolts in sequence

37. Install the CKP sensor connector bracket.
38. Install the bank 1 timing belt rear cover.
39. Install the bank 2 timing belt rear cover.

➡The length of the bolt B is longer than that of the bolt C.

40. Install the cylinder head cover.
    a. Remove oil, dust or sealant on the upper surface of the cylinder before assembling cylinder head cover.
    b. Assemble the cylinder head cover within 5 minutes after applying liquid gasket (LOCTITE® 5900) on the camshaft cap and packing part.
    c. Tighten the cylinder head cover bolts in the sequence shown. Torque: 5.8–7.2 ft. lbs. (7.8–9.8 Nm).

➡Do not start the engine for thirty minutes after assembling the cylinder head cover. Do not reuse the cylinder head cover gasket.

41. Install the timing belt.
    a. Align the timing marks of the LH/RH camshaft sprockets.

⁂ CAUTION

To prevent the valve from interfering with the piston, rotate the crankshaft sprocket at 3 pitches counterclockwise from the No.1 cylinder piston at TDC position before aligning the timing marks of the camshaft sprockets.

    b. After aligning the timing marks of the camshaft sprockets, rotate the crankshaft sprocket at 3 pitches clockwise then align the timing mark of the crankshaft sprocket to set the crankshaft to the No.1 cylinder piston at TDC position.

42. Install the timing belt. (Refer to Timing Belt.)
43. Install the water temp. control assembly.
44. Install the engine support bracket.

➡Over torque can damage the cylinder block threads.

45. Install the intake manifold assembly.
46. Install the exhaust manifold assembly.
47. Install the power steering pump.
48. Install the accessory drive belt.
49. Install the heater hose.
50. Connect the brake vacuum hose.
51. Install the PCV (Positive Crankcase Ventilation) hose.
52. Connect the engine wiring harness connectors.
    a. Connect the bank 1 CMP sensor connector.
    b. Connect the bank 2 front/rear O2 sensor connectors and the CKP sensor connector.
    c. Connect the bank 2 CMP sensor connector and the WTS (Water Temperature Sensor) connector.
    d. Connect the alternator connector and the air conditioning compressor connector.
    e. Connect the MAPS (Manifold Absolute Pressure Sensor) connector, the ETC (Electronic Throttle Control) connector and the PCSV (Purge Control Solenoid Valve) connector.
    f. Connect the injection harness connector, the No.2 VIS (Variable Induction System) connector, the No.1/No.2 OCV (Oil Control Valve) connectors and the OTS (Oil Temperature Sensor) connector.
    g. Connect the injection connectors, the ground lines, the condenser connector and the ignition coil connectors.
    h. Connect the bank 1 front/rear O2 sensor connectors.
    i. Connect the No.1/No.2 knock sensor connectors, the oil pressure switch connector, the ignition coil harness and the No.1 VIS (Variable Induction System) connector.
53. Install the fuel inlet hose from the delivery pipe.
54. Install the upper radiator hose and lower radiator hose.
55. Install the air cleaner assembly:
    a. Install the air cleaner assembly.
    b. Connect the breather hose (B) from air cleaner hose.
    c. Connect the MAF connector.
    d. Connect the powertrain control module (PCM) connector.
    e. Install the powertrain control module (PCM) cover.
56. Install the air duct.
57. Install the engine cover. Torque to: 5.8–8.7 ft. lbs. (7.8–11.8 Nm).
58. Refill the engine coolant.

EXHAUST MANIFOLD

REMOVAL & INSTALLATION

2.4L Engines

See Figures 73 and 74.

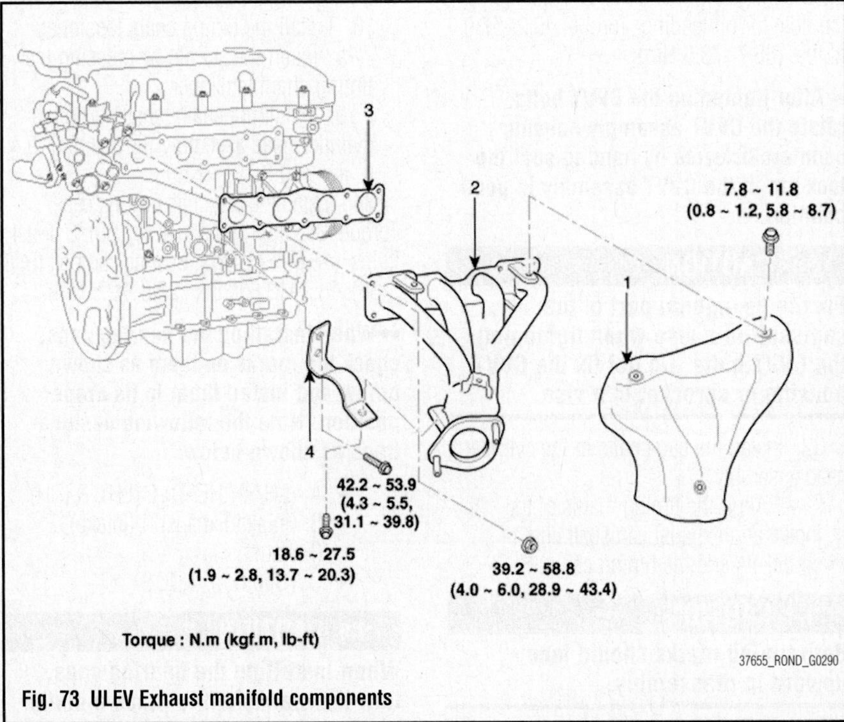

Fig. 73 ULEV Exhaust manifold components

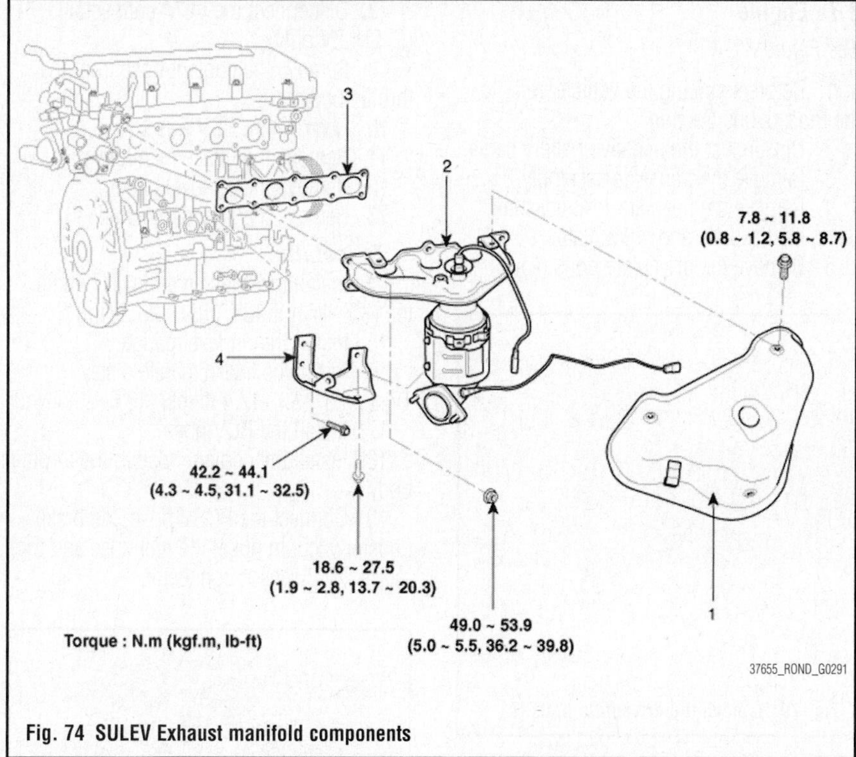

Fig. 74 SULEV Exhaust manifold components

1. Disconnect the HO2S (Bank 1 / Sensor 1) connector and the HO2S (Bank 1 / Sensor 2) connector (SULEV only).
2. Remove the front muffler.
3. Remove the exhaust manifold stay bolt.
4. Remove the heat protector.
5. Remove the exhaust manifold and gasket.

To install:

6. Install the exhaust manifold with a new gasket. Torque: 28.92–32.53 ft. lbs. (39.2–44.1 Nm).
7. Install the heat protector. Torque: 5.8–8.7 ft. lbs. (7.8–11.8 Nm).
8. Install the exhaust manifold stay bolt. Torque: 31.1–39.8 ft. lbs. (42.2–53.9 Nm).
9. Install the front muffler. Torque: 28.9–43.4 ft. lbs. (39.2–58.8 Nm).
10. Connect the HO2S (Bank 1 / Sensor 1) connector and the HO2S (Bank 1 / Sensor 2) connector (SULEV only).

2.7L Engine

See Figure 75.

1. Remove the undercover.
2. Remove the front muffler.
3. Disconnect the oxygen sensor connectors.
4. Remove the oil level gauge.
5. Remove the heat protector.
6. Remove the exhaust manifold assembly.

To install:

7. Install the exhaust manifold assembly with a new gasket. Torque: 21.7–25.3 ft. lbs. (29.4–34.3 Nm).
8. Install the heat protector. Torque: 12.3–15.9 ft. lbs. (16.7–21.6 Nm).
9. Install the front muffler assembly. Torque: 28.9–43.4 ft. lbs. (39.2–58.8 Nm).
10. Connect the oxygen sensor connector.
11. Install the lower cover.

FLYWHEEL/DRIVEPLATE

REMOVAL & INSTALLATION

2.4L Engine

F4A42 Transaxle

See Figures 76 and 77.

1. Before servicing the vehicle, refer to the Precautions Section.
2. Disconnect the negative battery cable.
3. Remove the transaxle assembly.
4. Remove the driveplate:
   a. Remove the driveplate bolts (A).

➡Remove the 6 bolts, rotating the crankshaft clockwise.

To install:

5. Install the driveplate bolts (A) by rotating the timing gear.
   Torque: 33.3–38.3 ft. lbs. (46–53 Nm).

➡Install the 6 bolts, rotating the crankshaft clockwise.

A5GF1 Transaxle

See Figure 78.

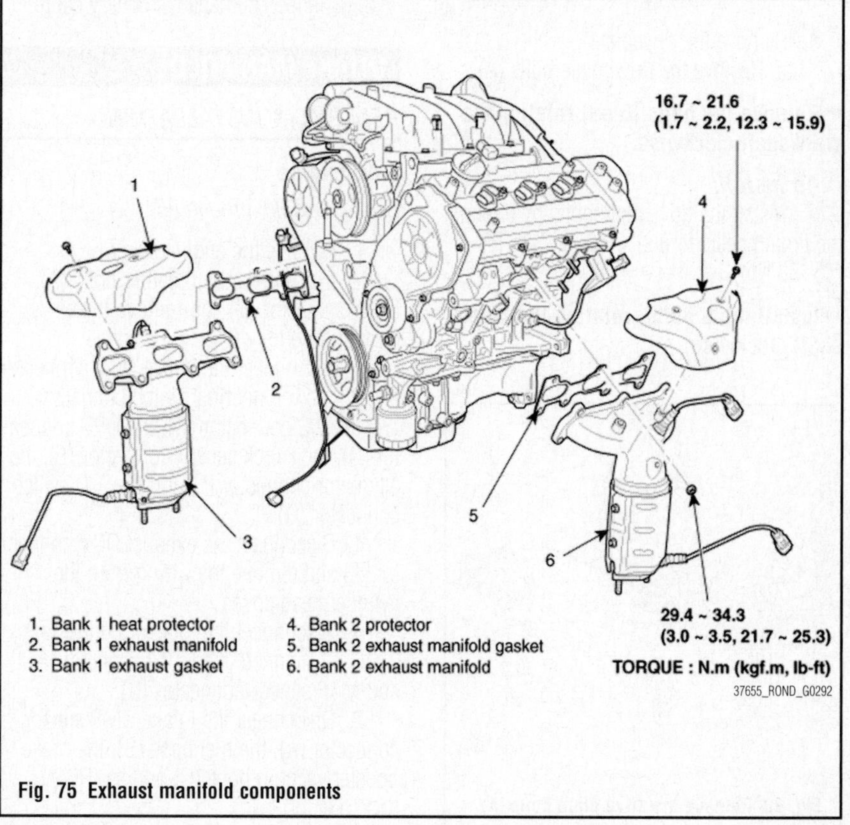

Fig. 75 Exhaust manifold components

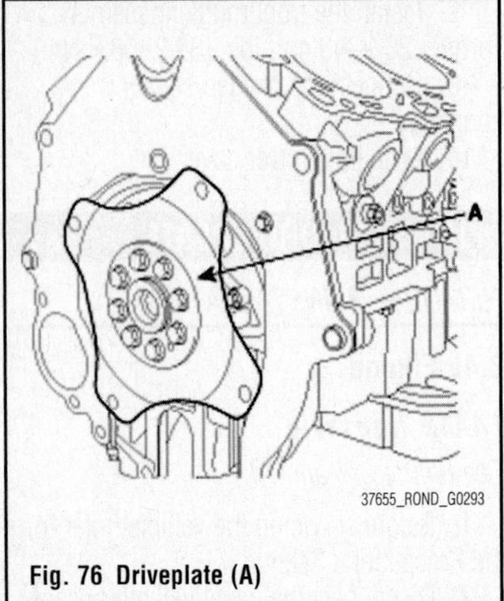

Fig. 76 Driveplate (A)

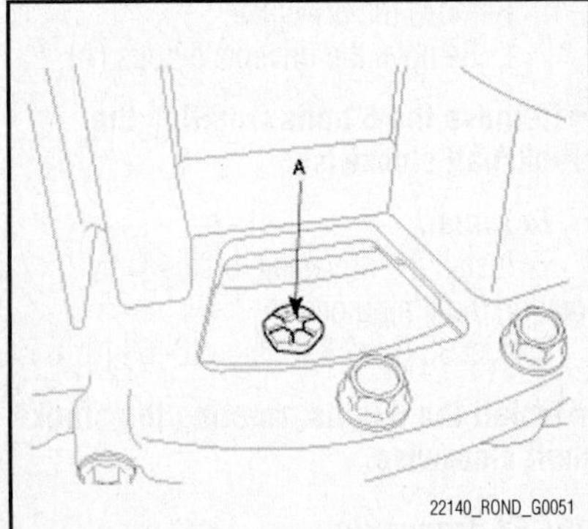

Fig. 77 Install the driveplate bolts (A) by rotating the timing gear.

1. Remove the driveplate:
   a. Remove the drive plate bolts (A).

➡**Remove the bolts (6 ea) rotating the crankshaft clockwise.**

To install:

2. Install the drive plate bolts by rotating the timing gear. Torque: 33.3–38.3 ft. lbs. (46–53 Nm).

➡**Install the 6 bolts, rotating the crankshaft clockwise.**

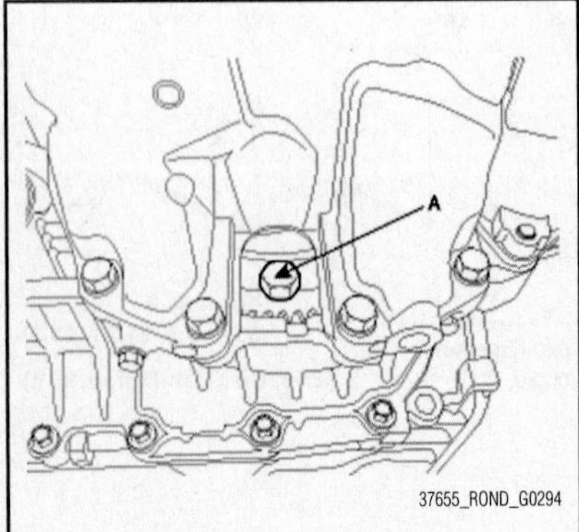

Fig. 78 Remove the drive plate bolts (A).

2.7L Engine

See Figure 79.

1. Before servicing the vehicle, refer to the Precautions Section.
2. Disconnect the negative battery cable.
3. Remove the transaxle assembly.
4. Remove the flexplate-to-crankshaft bolts, the flexplate and shim plates.
5. Remove the driveplate bolts (A).

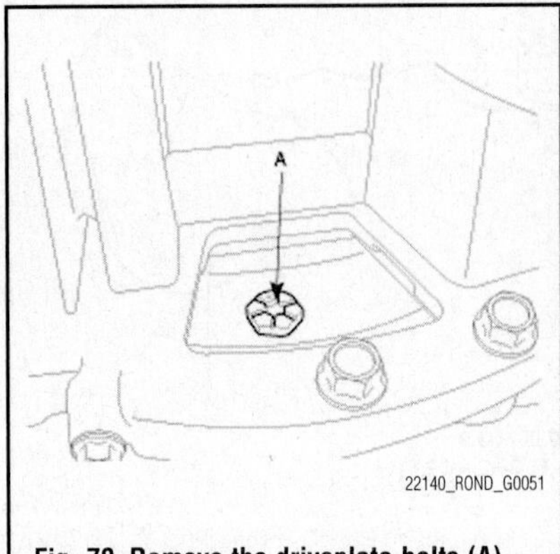

Fig. 79 Remove the driveplate bolts (A)

➡**Remove the bolts (6 ea) rotating the crankshaft clockwise.**

To install:

6. Install the flexplate and tighten the mounting bolts to 52.8–55.7 ft. lbs. (71.6–75.5 Nm).
7. Install the transaxle.
8. Connect the negative battery cable.

INTAKE MANIFOLD

REMOVAL & INSTALLATION

2.4L Engine

See Figures 80 through 86.

1. Remove the engine cover.
2. Disconnect the breather hose (A), ECM connector (B) and remove the air cleaner assembly (C).
3. Disconnect the intake OCV connector (A), the power steering switch connector (B), the VIS connector (C), the OPS connector (D), the knock sensor connector (E), the alternator connector (F) and the A/C switch connector (G).
4. Disconnect the exhaust OCV connector (A) and remove the wiring from the cylinder head cover.
5. Disconnect the injector connectors.
6. Disconnect the ETC connector (A) and MAP sensor connector (B).
7. Disconnect the intake CMP sensor connector (A), the fuel hose (B), the brake booster vacuum hose (C) and the PCSV hose (D).
8. Disconnect the VCM connector (E; SULEV only).
9. Remove the coolant hose from the throttle body.
10. Remove the PCV hose.
11. Remove the intake manifold stay (A).
12. Remove the oil level gauge.
13. Remove the intake manifold (A).

To install:

14. Install the intake manifold. Tighten to: 13.7–17.4 ft. lbs. (18.6–23.5 Nm)
15. Install the oil level gauge.
16. Install the intake manifold stay. Tighten to: 13.7–17.4 ft. lbs. (18.6–23.5 Nm)
17. Install the PCV hose.
18. Install the coolant hose to the throttle body.
19. Connect the PCSV hose, the brake booster vacuum hose, the fuel hose and the intake CMP sensor connector.

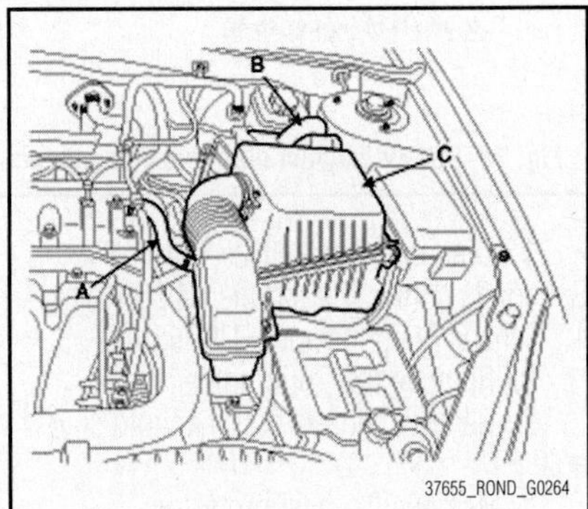

Fig. 80 Disconnect the breather hose (A), ECM connector (B) and remove the air cleaner assembly (C).

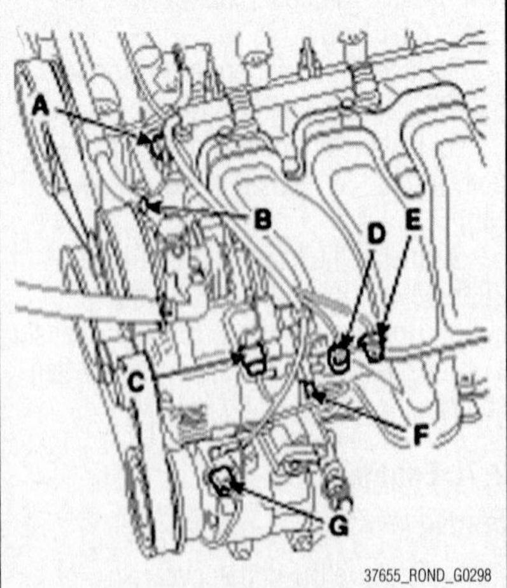

Fig. 81 Disconnect the intake OCV connector (A), the power steering switch connector (B), the VIS connector (C), the OPS connector (D), the knock sensor connector (E), the connector (F) and the A/C switch connector (G)

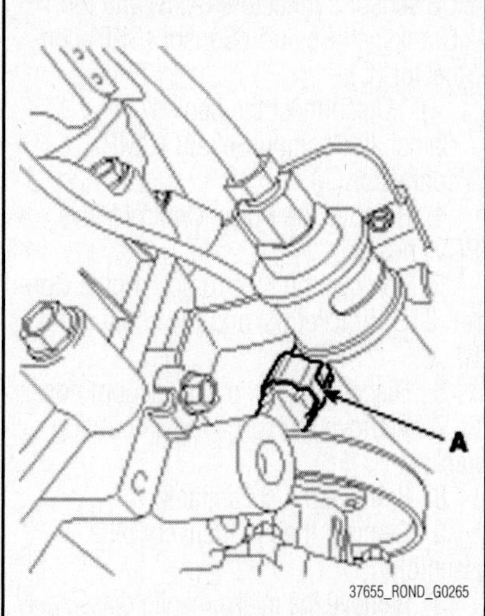

Fig. 82 Disconnect the intake OCV connector (A), the power steering switch connector (B), the VIS connector (C),

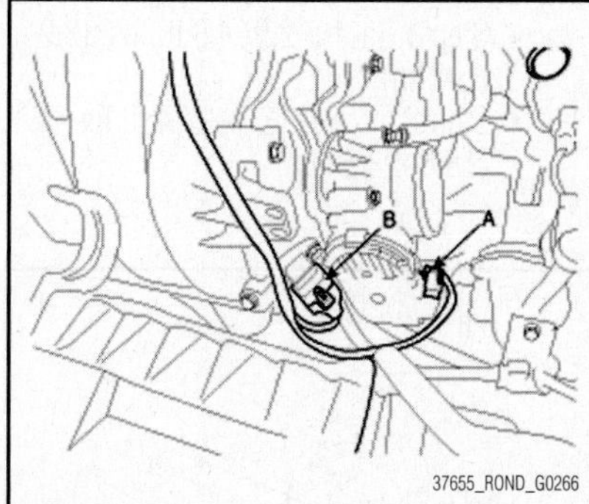

Fig. 83 Disconnect the ETC connector (A) and MAP sensor connector (B)

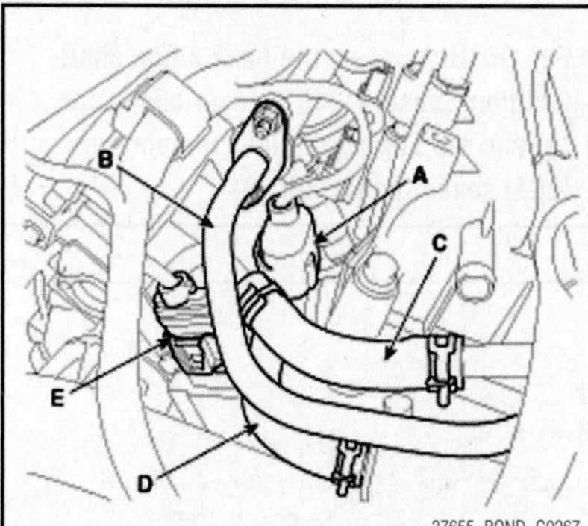

Fig. 84 Disconnect the intake CMP sensor connector (A), the fuel hose (B), the brake booster vacuum hose (C) and the PCSV hose (D). Disconnect the VCM connector (E; SULEV only)

20. Connect the VCM connector (SULEV only).

21. Connect the ETC connector and the MAP sensor connector.

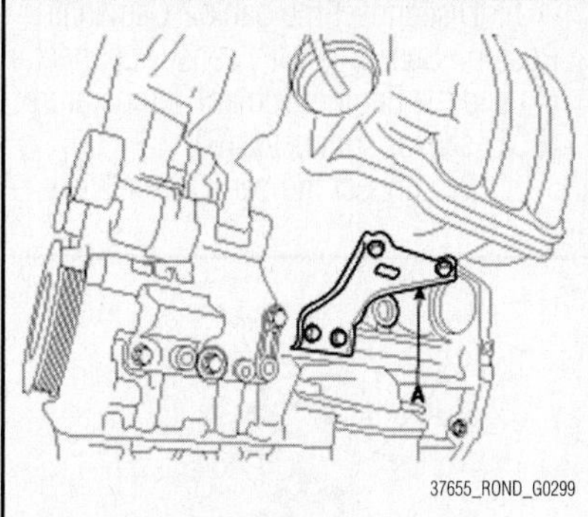

Fig. 85 Remove the intake manifold stay (A)

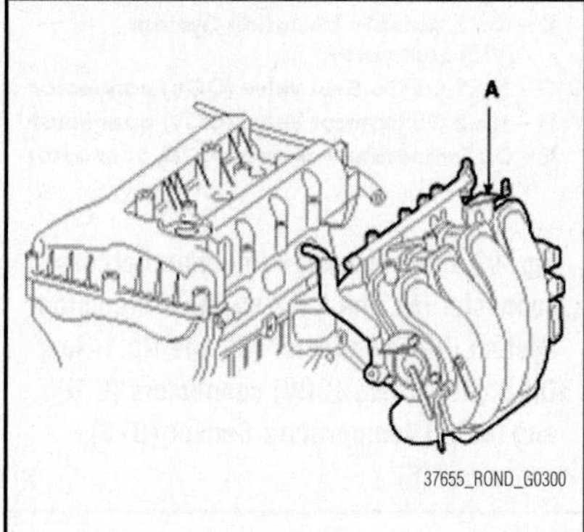

Fig. 86 Remove the intake manifold (A)

22. Connect the injector connectors.

23. Connect the exhaust OCV connector and replace the wiring from the cylinder head cover.

24. Connect the intake OCV connector, the power steering switch connector, the VIS connector, the OPS connector, the knock sensor connector, the alternator connector and the A/C switch connector.

25. Install the air cleaner assembly, then connect the breather hose and the ECM connector. Tighten to: 5.8–8.7 ft. lbs. (7.8–11,8 Nm.

26. Install the engine cover. Torque to: 5.8–8.7 ft. lbs. (7.8–11.8 Nm).

2.7L Engine

See Figures 87 through 99.

1. Remove the engine cover.
2. Remove the air cleaner assembly:
   a. Remove the Powertrain Control Module (PCM) cover (D).
   b. Disconnect the Powertrain Control Module (PCM) connector (F).
   c. Disconnect the MAF connector (A).
   d. Disconnect the breather hose (B) from the air cleaner hose.
   e. Remove the air cleaner assembly.
3. Disconnect the engine wiring harness connectors.
   a. Disconnect the No.1/No.2 knock sensor connectors (A,B), the oil pressure switch connector (C), the ignition coil harness (D) and the No.1 Variable Induction System (VIS) connector (E).
   b. Disconnect the bank 1 front/rear O2 sensor connectors (A).
   c. Disconnect the injection connectors (A ,B, C), the ground lines (D), the condenser connector (E) and the ignition coil connectors (F).
   d. Disconnect the injection harness connector (A), the No.2 Variable Induction System (VIS) connector (B), the No.1/No.2 Oil Control Valve (OCV) connectors (C,D) and the Oil Temperature Sensor (OTS) connector (E).
   e. Disconnect the Manifold Absolute Pressure Sensor (MAP) connector (A).
   f. Disconnect the Electronic Throttle Control (ETC) connector (B) and the Purge Control Solenoid Valve (PCSV) connector (C).
   g. Disconnect the alternator connector and the air conditioning compressor connector.

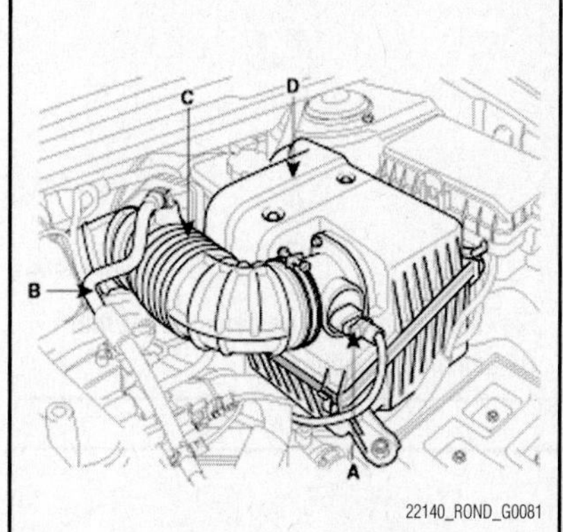

Fig. 87 Remove the Powertrain Control Module (PCM) cover (D), MAF connector (A) and breather hose (B)

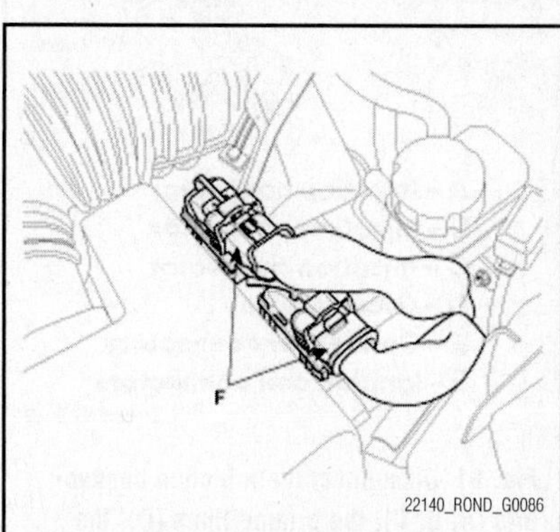

Fig. 88 Disconnect the Power Train Control Module (PCM) connector (F)

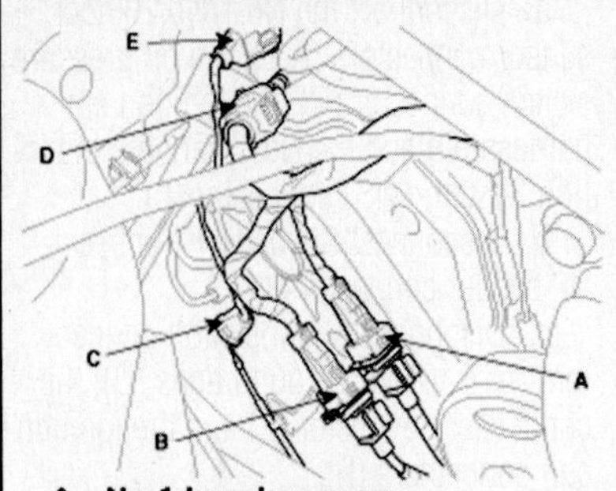

A – No.1 knock sensor
B – No.2 knock sensor
C – Oil pressure Switch connector
D – Ignition coil harness
E – No.1 Variable Induction System (VIS) connector

22140_ROND_G0087

Fig. 89 Disconnect the No.1/No.2 knock sensor connectors (A,B), the oil pressure switch connector (C), the ignition coil harness (D) and the No.1 Variable Induction System (VIS) connector (E)

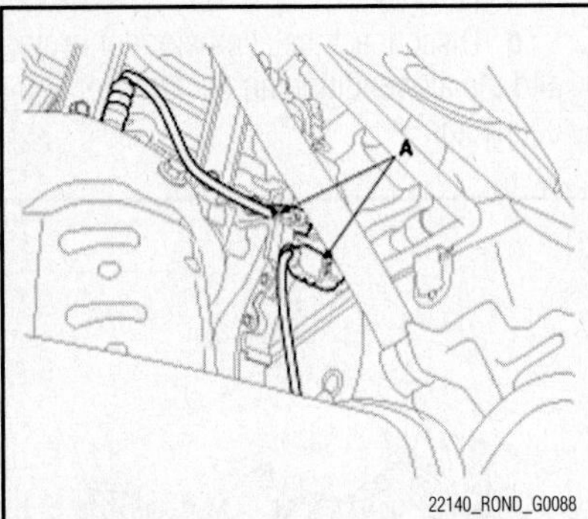

22140_ROND_G0088

Fig. 90 Disconnect the bank 1 front/rear O2 sensor connectors (A)

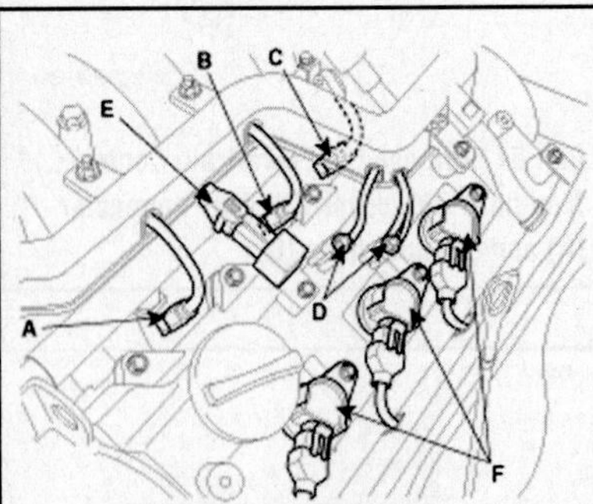

A – Injection connector
B – Injection connector
C – Injection connector
D – Ground lines
E – Condenser connectors
F – Ignition coil connectors

22140_ROND_G0089

Fig. 91 Disconnect the injection connectors (A, B, C), the ground lines (D), the condenser connector (E) and the ignition coil connectors (F)

h. Disconnect the bank 2 Camshaft Position Sensor (CMP) sensor connector (A) and the Engine Coolant Temperature (ECT) sensor connector (B).

i. Disconnect the bank 2 front/rear O2 sensor connectors (A, B) and the Crankshaft Position Sensor (CKP) connector (C).

j. Disconnect the bank 1 Camshaft Position Sensor (CMP) connector (A).

4. Remove the Purge Control Valve (PCV) hose.
5. Remove the Electronic Throttle Control (ETC) bracket (A) and the cooling hoses (B).
6. Disconnect the brake vacuum hose.
7. Remove the surge tank mounting bracket.
8. Remove the surge tank.
9. Remove the fuel delivery pipe assembly.
10. Remove the intake manifold assembly.

To install:

11. Install the intake manifold assembly with a new gasket to a cylinder head assembly. Tighten the bolts in two steps. Torque as follows:
    - Step 1: (a–h): 2.9–4.3 ft. lbs. (3.9–5.9 Nm)
    - Step 2: (1–8): 13.7–17.4 ft. lbs. (18.6–23.5 Nm)

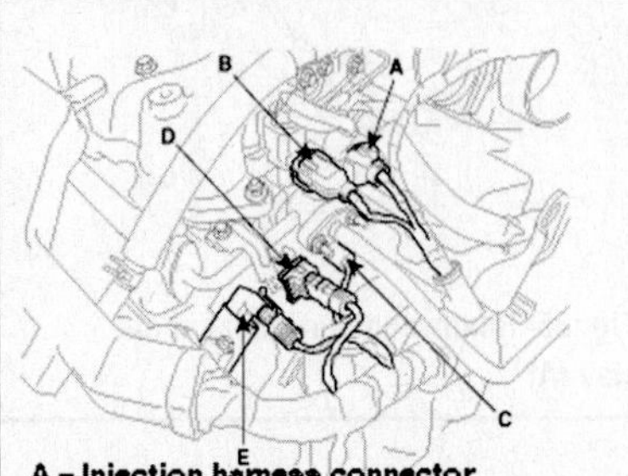

A – Injection harness connector
B – No.2 Variable Induction System (VIS) connector
C – No.1 Oil Control Valve (OCV) connector
D – No.2 Oil Control Valve (OCV) connector
E – Oil Temperature Sensor (OTS) connector

22140_ROND_G0090

Fig. 92 Disconnect the injection harness connector (A), the No.2 Variable Induction System (VIS) connector (B), the No.1/No.2 Oil Control Valve (OCV) connectors (C,D) and the Oil Temperature Sensor (OTS) connector (E)

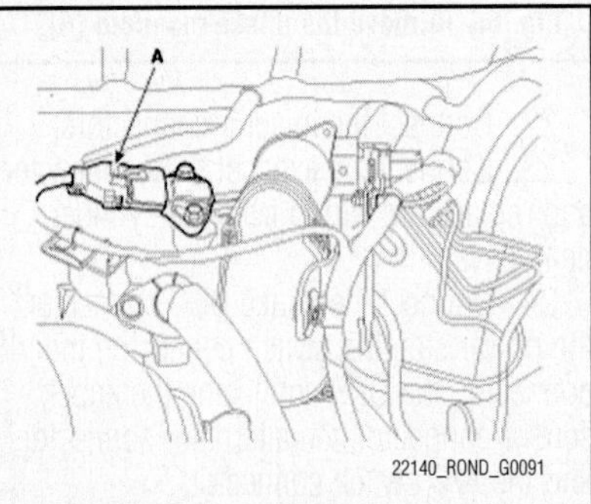

22140_ROND_G0091

Fig. 93 Disconnect the Manifold Absolute Pressure Sensor (MAP) connector (A)

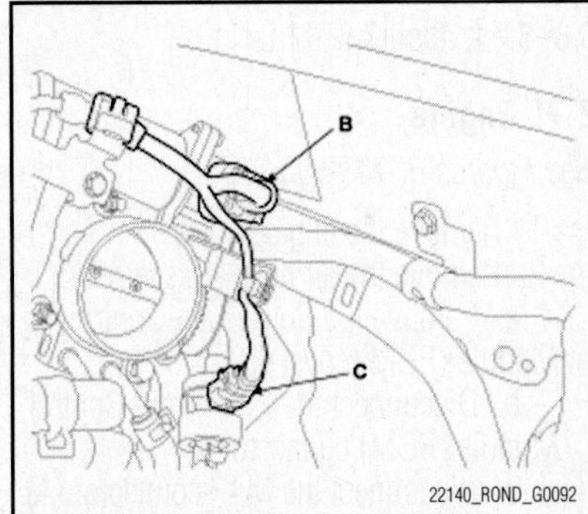

22140_ROND_G0092

Fig. 94 Disconnect the Electronic Throttle Control (ETC) connector (B) and the Purge Control Solenoid Valve (PCSV) connector (C)

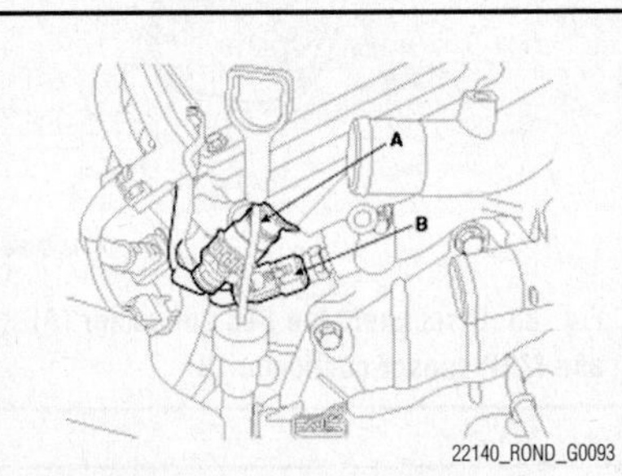

22140_ROND_G0093

Fig. 95 Disconnect the bank 2 Camshaft Position Sensor (CMP) sensor connector (A) and the Engine Coolant Temperature (ECT) sensor connector (B)

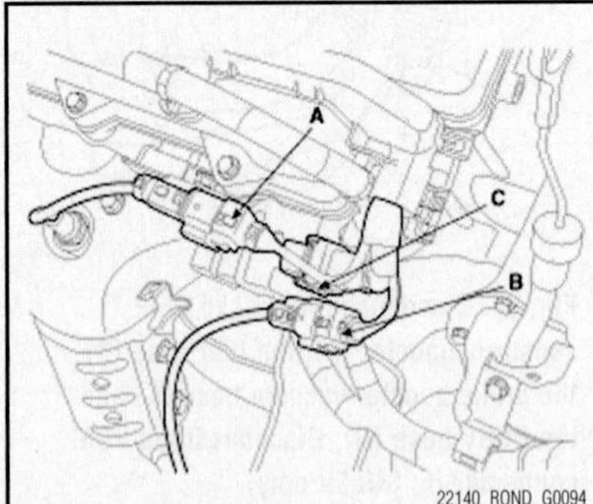

22140_ROND_G0094

Fig. 96 Disconnect the bank 2 front/rear O2 sensor connectors (A, B) and the Crankshaft Position Sensor (CKP) connector (C)

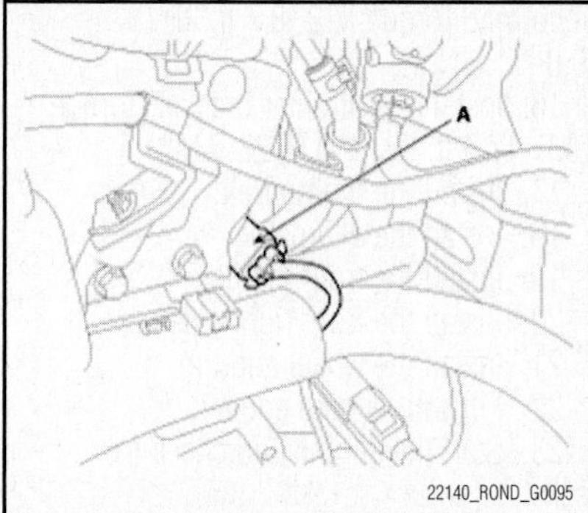

Fig. 97 Disconnect the bank 1 Camshaft Position Sensor (CMP) connector (A)

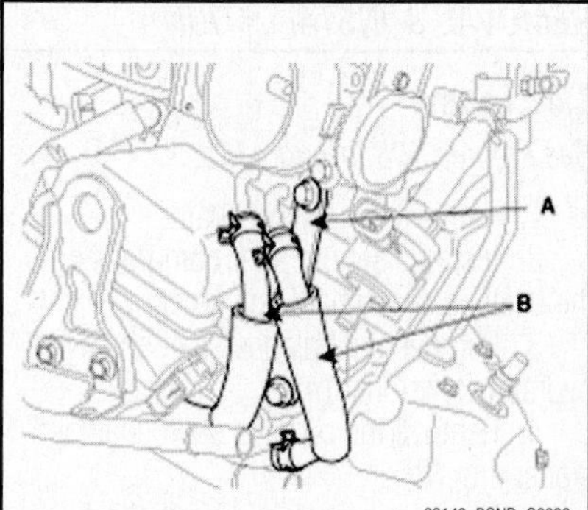

Fig. 98 Remove the Electronic Throttle Control (ETC) bracket (A) and the cooling hoses (B)

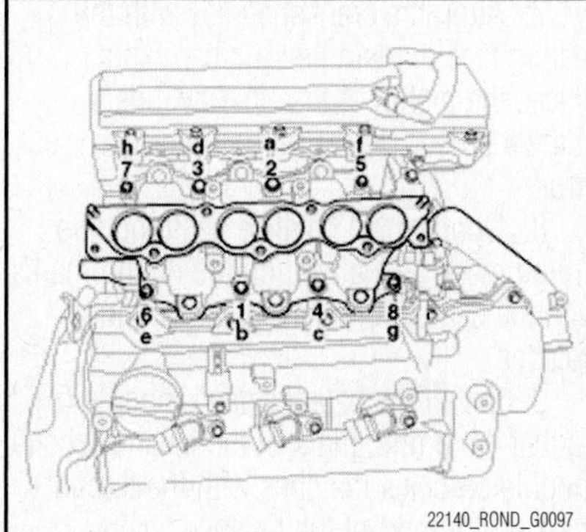

Fig. 99 Install the intake manifold assembly in two steps

⁂ CAUTION

When installing the gasket on the cylinder head, check the identification marks (LH/RH) to ensure proper installation.

12. Install the delivery pipe.
13. Connect the LH injector connector.
14. Install the surge tank. Tighten to: 13.7–17.4 ft. lbs. (18.6–23.5 Nm)
15. Install the surge tank mounting bracket. Tighten to: 13.7–17.4 ft. lbs. (18.6–23.5 Nm)
16. Install the Electronic Throttle Control (ETC) system fixing bracket.
17. Connect all hoses and connectors.
18. Install the air cleaner assembly.
19. Install the engine cover. Torque to: 5.8–8.7 ft. lbs. (7.8–11.8 Nm).

OIL PAN

REMOVAL & INSTALLATION

2.4L Engine

See Figures 100 through 105.

1. Remove the power steering pump bracket (A) and the knock sensor (B).
2. Remove the air conditioning compressor bracket(A).
3. Remove the lower oil pan (A).
    - Insert the special service tool (SST 09215-3C000) between the oil pan and the ladder frame by tapping it with a plastic hammer in the direction of arrow #1.
    - After tapping the SST with a plastic hammer along the direction of arrows #2 around more than 2/3 of the edge of the oil pan, remove the SST from the ladder frame.
    - Do not turn over the SST abruptly without tapping. It can damage the SST and the oil pan.
4. Remove the oil screen (A).
5. Remove the upper oil pan (A).
    - Insert the special service tool (SST) between the oil pan and the ladder frame by tapping it with a plastic hammer in the direction of arrow.
    - After tapping the SST with a plastic hammer along the direction of arrows around more than 2/3 of the edge of the oil pan, remove the SST from the ladder frame.
    - Do not turn over the SST abruptly without tapping. It can damage the SST and the oil pan.

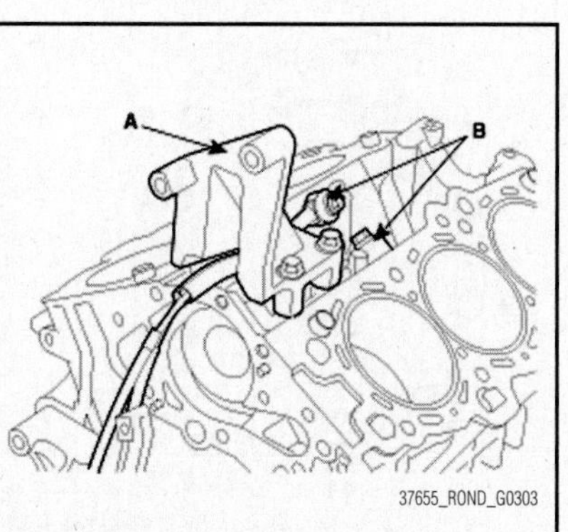

Fig. 100 Remove the power steering pump bracket (A) and the knock sensor (B)

To install:

6. Install the oil pan:

    a. Using a razor blade and gasket scraper, remove all the old packing material from the gasket surfaces.

➡Check that the mating surfaces are clean and dry before applying liquid gasket.

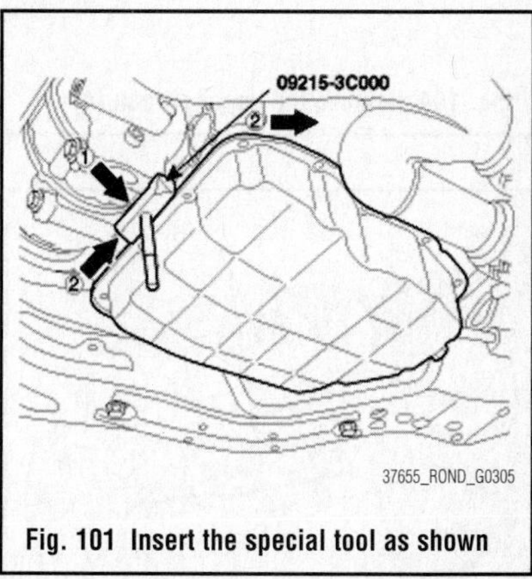

Fig. 101 Insert the special tool as shown

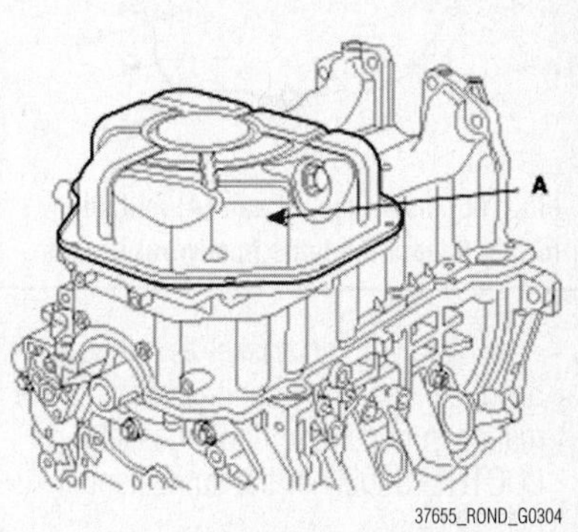

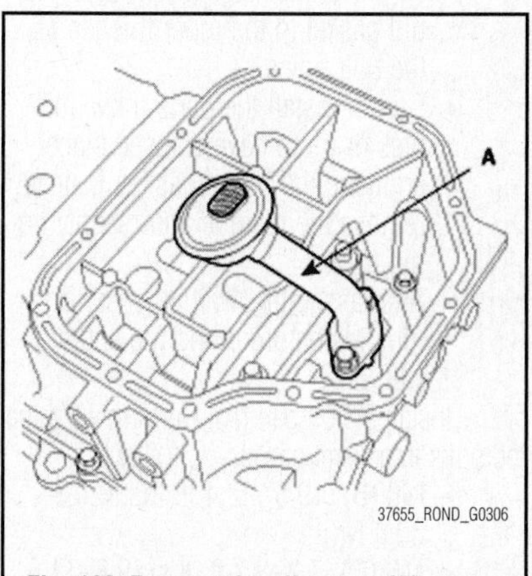

Fig. 103 Remove the oil screen (A)

Fig. 102 Remove the lower oil pan (A)

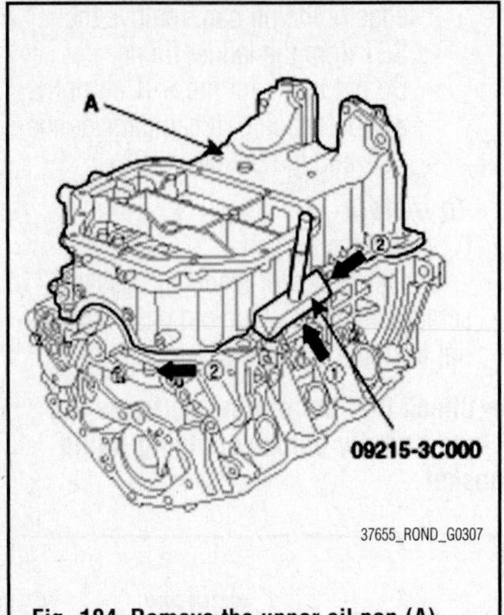

Fig. 104 Remove the upper oil pan (A)

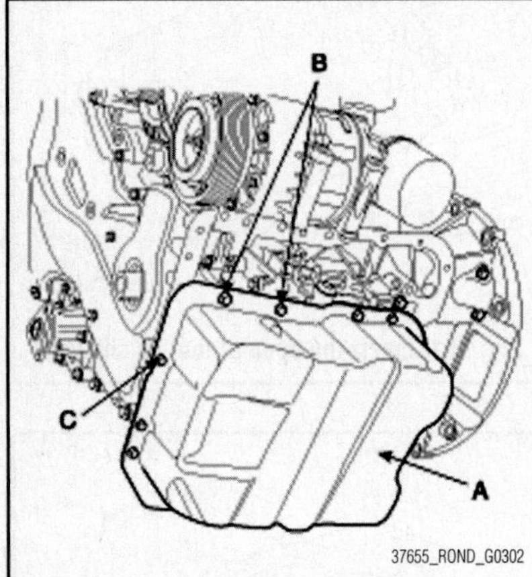

Fig. 105 Install the oil pan (A) and uniformly tighten the bolts in several passes

b. Apply liquid gasket as an even bead, centered between the edges of the mating surface. Use liquid gasket LOCTITE®5900H or the equivalent (MS721-40A).

➡Note the following:

- To prevent leakage of oil, apply liquid gasket to the inner threads of the bolt holes.
- Do not install the parts if five minutes or more have elapsed since applying the liquid gasket. Instead, remove the residue and reapply liquid gasket.
- After assembly, wait at least 30 minutes before filling the engine with oil.

7. Install the oil pan (A). Uniformly tighten the bolts in several passes, as follows:

- M8 (B): 19.5–22.4 ft. lbs. (26.5–30.4 Nm)
- M6 (C): 7.2–8.7 ft. lbs. (9.8–11.8 Nm)

2.7L Engine

See Figures 106 and 107.

1. Drain engine oil.
2. Remove the front right wheel and tire.
3. Remove the front right side cover.
4. Remove the front muffler.
5. Remove the alternator.
6. Remove the timing belt.
7. Remove the oil filter bracket.
8. Using SST (09215-3C000), remove the lower oil pan.
9. Remove the oil screen.
10. Remove the upper oil pan, using the SST (09215-3C000).

To install:

11. Install the upper oil pan.
12. Tighten the bolts in several steps uniformly. Torque: Bolts 1–15: 13.7–17.4 ft. lbs. (18.6–23.5 Nm). Bolts 16 and 17: 3.6–5.1 ft. lbs. (4.9–6.9 Nm).
13. Install the oil screen. Torque: 10.8–15.9 ft. lbs. (14.7–21.6 Nm).
14. Install the lower oil pan.
15. Tighten the bolts in several steps uniformly. Torque: 7.2–8.7 ft. lbs. (9.8–11.8 Nm).
16. Install the oil filter bracket. Torque: 13.7–17.4 ft. lbs. (18.6–23.5 Nm).
17. Install the timing belt.
18. Install the alternator.
19. Install the front muffler.
20. Install the front right side cover.
21. Install the wheel and tire.
22. Fill with engine coolant.
23. Start the engine and check for leaks.
24. Recheck the engine coolant level.

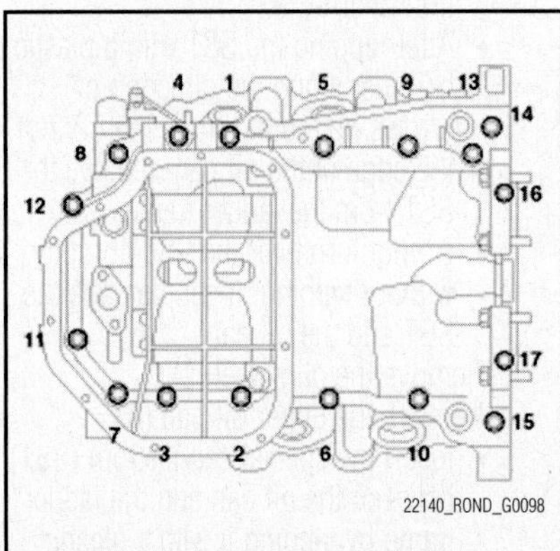

Fig. 106 Tighten the bolts in several steps uniformly

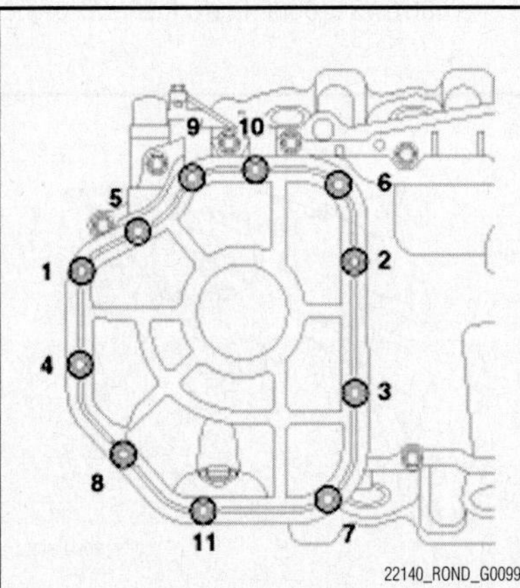

Fig. 107 Tighten the bolts in several steps uniformly

OIL PUMP

REMOVAL & INSTALLATION

2.4L Engine

See Figures 108 through 113.

1. Remove the timing chain.
2. Install a set pin after compressing the balance shaft chain tensioner.
3. Remove the balance shaft chain hydraulic tensioner (A).
4. Remove the balance shaft chain tensioner arm (B).
5. Remove the balance shaft chain guide (C).
6. Remove the oil pump & balance shaft module (A) and balance shaft chain (B).

To install:

7. Align the crankshaft key with the mating face of main bearing cap. This places the piston of the No.1 cylinder at the top dead center on compression stroke.
8. Confirm the balance shaft module timing mark. Visually align the timing marks with the centers of adjacent cast timing notches.
9. Install the balance shaft module (A) so that the timing mark of the balance shaft module sprocket matches with the timing mark (color link) of the balance shaft chain.

Torque: 12.3 ft. lbs. (16.7 Nm), then an additional 60°plus an additional 60°.

10. Bolting order:
    a. Assemble the bolts in order number as shown with seating torque 18.8 ft. lbs. (25.5 Nm).
    b. Unfasten the bolts in reverse bolting order. (4-3-2-1)
    c. Assemble the bolts in the specified bolting order as follows:
11. Install the balance shaft chain guide (C). Torque: 7.2–8.7 ft. lbs. (9.8–11.8 Nm).

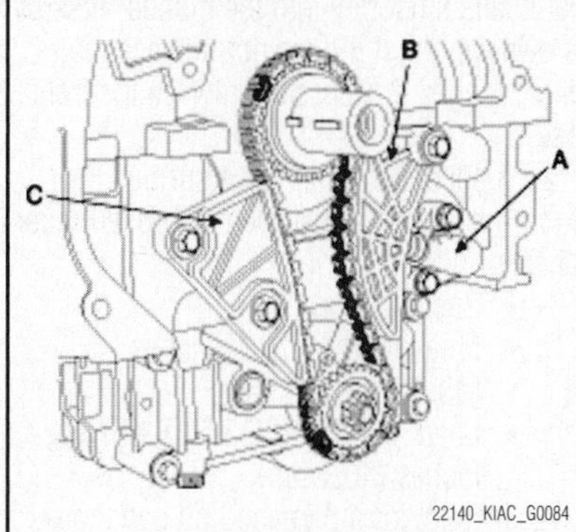

Fig. 108 Remove the balance shaft chain tensioner (A), balance shaft chain tensioner arm (B), and balance shaft chain guide (C).

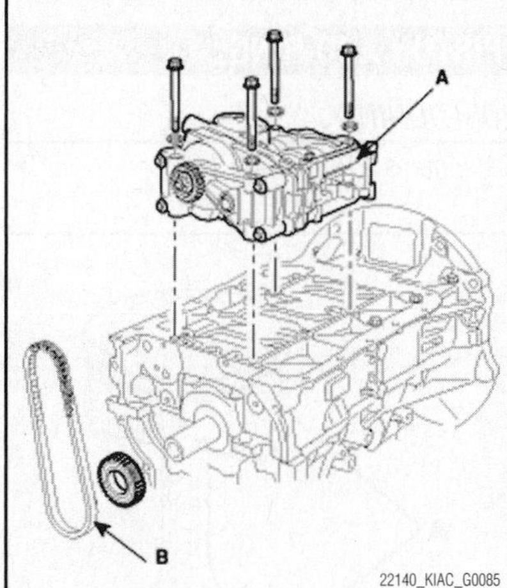

Fig. 109 Remove the oil pump & balance shaft module (A) and balance shaft chain (B).

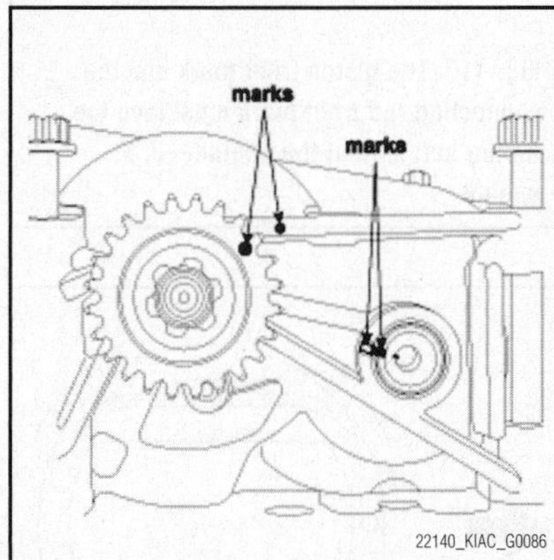

Fig. 110 Align the timing marks with the centers of adjacent cast timing notches.

12. Install the balance shaft chain tensioner arm (B). Torque: 7.2–8.7 ft. lbs. (9.8–11.8 Nm).

13. Install the balance shaft hydraulic

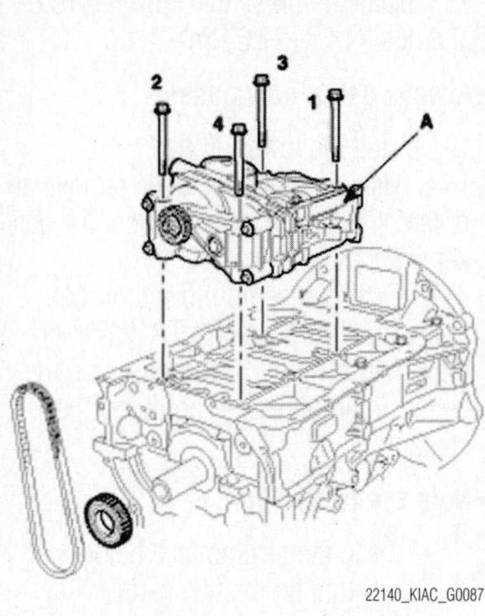

Fig. 111 Install the balance shaft module (A) so that the timing mark of the balance shaft module sprocket matches with the timing mark (color link) of the balance shaft chain.

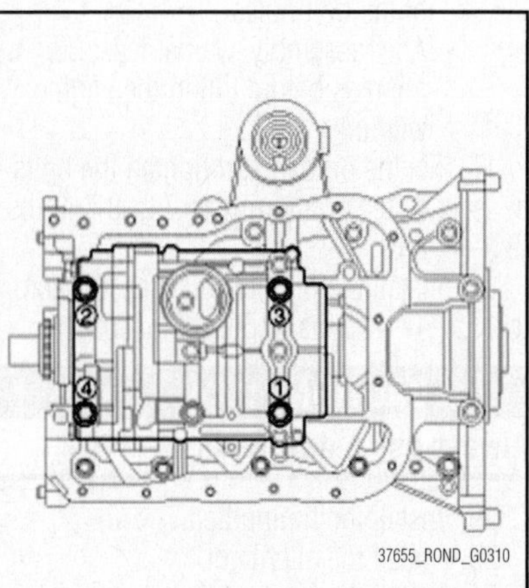
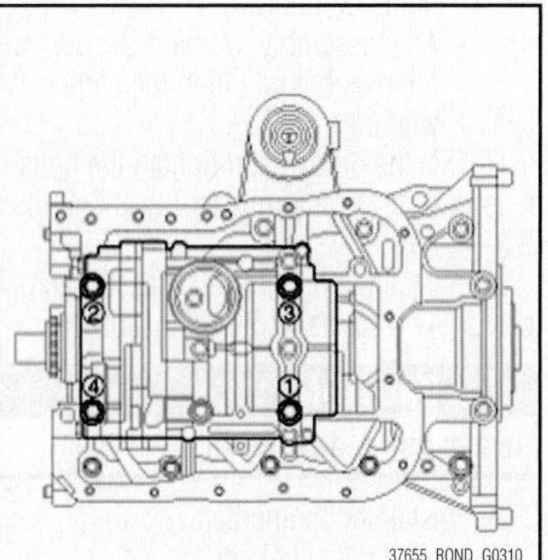

Fig. 112 Assemble the bolts in the specified bolting order

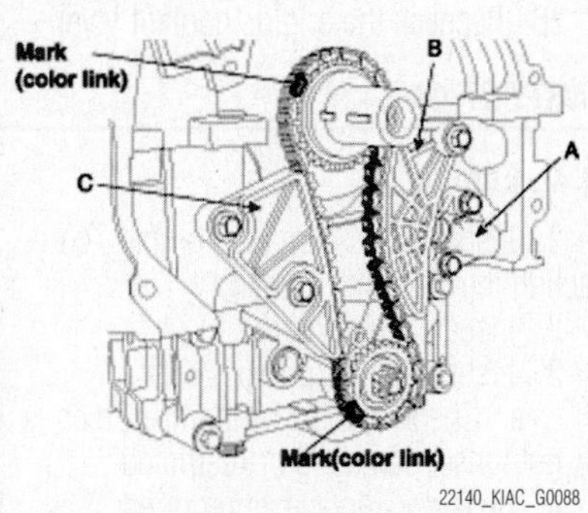

Fig. 113 Install the balance shaft chain guide (C) and the balance shaft tensioner arm (B),

tensioner (A) and remove the set pin. Torque: 7.2–8.7 ft. lbs. (9.8–11.8 Nm).

14. Confirm the timing marks.

15. Install the timing chain.

2.7L Engine

See Figures 114 and 115.

1. Drain the engine oil.
2. Remove the front right wheel and tire.
3. Remove the front right side cover.
4. Remove the front muffler.
5. Remove the alternator.
6. Remove the timing belt.
7. Remove the oil filter bracket.
8. Using SST (09215-3C000), remove the lower oil pan.
9. Remove the oil screen.
10. Remove the upper oil pan, using the SST (09215-3C000).
11. Remove the oil pump case.

To install:

12. Install the oil pump case:
   a. Using a gasket scraper, remove all of the old packing material from the gasket surfaces.
   b. Before assembling the oil pan, apply the liquid sealant TB1217H on the upper oil pan. Assemble the part within 5 minutes after applying the sealant.
   Bead width: .0984 in. (2.5mm)

➡Note the following:

- Clean the sealing face before assembling the two parts.
- Remove harmful foreign materials on the sealing face before applying sealant.
- When applying sealant gasket, the sealant must not protrude into the inside of the oil pan.
- To prevent oil leakage, apply sealant gasket to the inner threads of the bolt holes.
- After assembly, wait at least 30 minutes before filling the engine with oil.

   a. Install the oil pump case. Torque: 13.7–17.4 ft. lbs. (18.6–23.5 Nm).

⁂ CAUTION

Always use a new O-ring when installing the oil pump.

13. Using the SST (09214~33000), install the oil pump case oil seal.
14. Install the upper oil pan:
   a. Using a gasket scraper, remove all of the old packing material from the gasket surfaces.
   b. Before assembling the oil pan, apply the liquid sealant TB1217H on

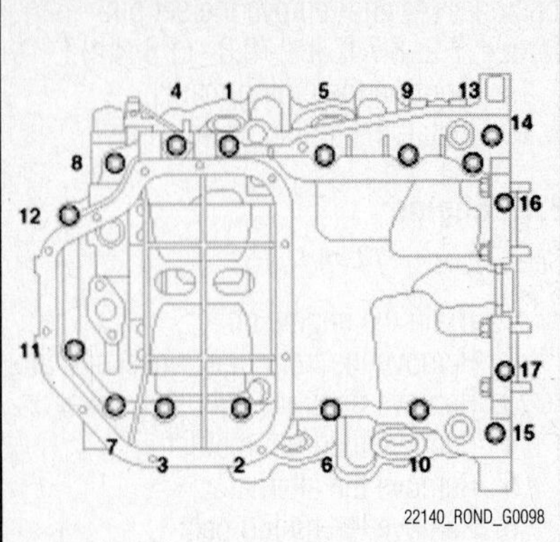

Fig. 114 Tighten the bolts in several steps uniformly

the upper oil pan. Assemble the part within 5 minutes after applying the sealant.

➡Note the following:

- Clean the sealing face before assembling the two parts.
- Remove harmful foreign materials on the sealing face before applying sealant.
- When applying sealant gasket, the sealant must not protrude into the inside of the oil pan.
- To prevent oil leakage, apply sealant gasket to the inner threads of the bolt holes.
- After assembly, wait at least 30 minutes before filling the engine with oil.

c. Fix the upper oil pan and tighten the bolts in several steps uniformly. Torque: Bolts 1–15: 13.7–17.4 ft. lbs. (18.6–23.5 Nm). Bolts 16 and 17: 3.6–5.1 ft. lbs. (4.9–6.9 Nm).

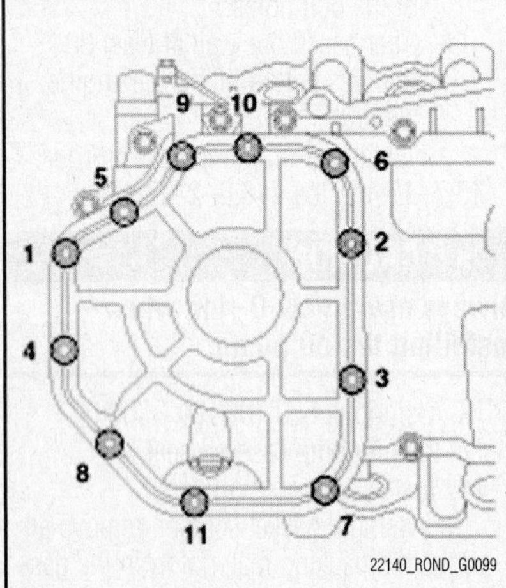

Fig. 115 Tighten the bolts in several steps uniformly

15. Install the oil screen. Torque: 10.8–15.9 ft. lbs. (14.7–21.6 Nm).

➡Always use a new gasket.

16. Install the lower oil pan:

a. Using a gasket scraper, remove all of the old packing material from the gasket surfaces.

b. Before assembling the oil pan, apply the liquid sealant TB1217H on the upper oil pan. Assemble the part within 5 minutes after applying the sealant.

➡Note the following:

- Clean the sealing face before assembling the two parts.
- Remove harmful foreign materials on the sealing face before applying sealant.
- When applying sealant gasket, the sealant must not protrude into the inside of the oil pan.
- To prevent oil leakage, apply sealant gasket to the inner threads of the bolt holes.
- After assembly, wait at least 30 minutes before filling the engine with oil.

17. Fix the oil pan and tighten the bolts in several steps uniformly to 7.2–8.7 ft. lbs. (9.8–11.8 Nm).
18. Install the oil filter bracket. Tighten to 13.7–17.4 ft. lbs. (18.6–23.5 Nm).

CAUTION

Always use a new O-ring.

19. Install the timing belt.
20. Install the alternator.
21. Install the front muffler.
22. Install the front right side cover.
23. Install the wheel and tire.
24. Fill with engine coolant.
25. Start the engine and check for leaks.
26. Recheck the engine coolant level.

INSPECTION

2.4L Engine

1. Inspect for a damaged oil pan, contacting the oil pump screen.
2. Inspect the oil pan.
3. Inspect the oil pump screen.

a. Check that the oil pump screen is not loose, damaged or restricted.

4. Repair or replace as required.

2.7L Engine

1. Inspect the relief plunger.

Coat the valve with engine oil and check that it falls smoothly into the plunger hole by its own weight. If it does not, replace the relief plunger. If necessary, replace the front case.

2. Inspect the relief valve spring.
3. Inspect for a distorted or broken relief valve spring.

- Specification:
- Free length: 1.7244 inches (43.8mm)
- Load: 0.9 lbs. (36.3 N) / 1.5787 inches (40.1mm).

4. Inspect for a damaged oil pan, contacting the oil pump screen.
5. Inspect the oil pan.
6. Inspect the oil pump screen.

a. Check that the oil pump screen is not loose, damaged or restricted.

7. Repair or replace as required.

PISTON AND RING

POSITIONING

See Figures 116 through 119.

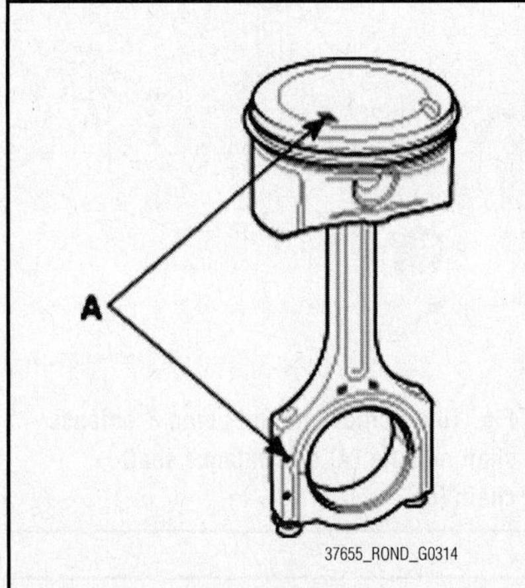

Fig. 116 The piston front mark and the connecting rod front mark must face the timing belt side of the engine—2.4L engine

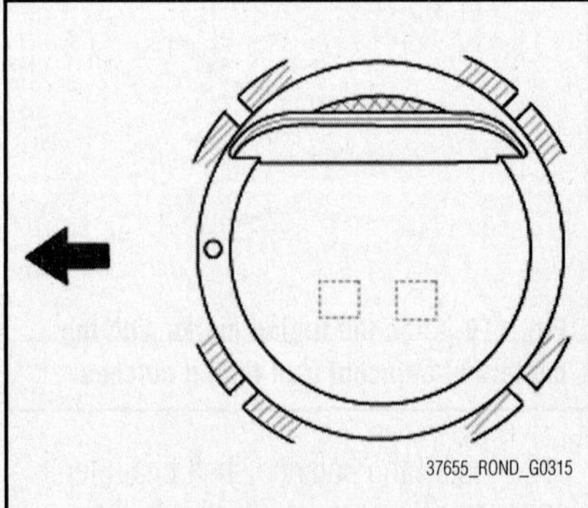

Fig. 117 Position the piston rings so that the ring ends are as shown—2.4L engine

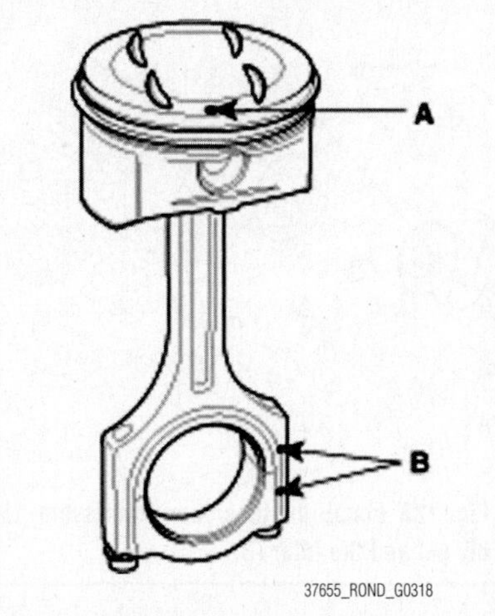

Fig. 118 The piston front mark (A) and the connecting rod front mark must face the timing belt side of the engine—2.7L engine

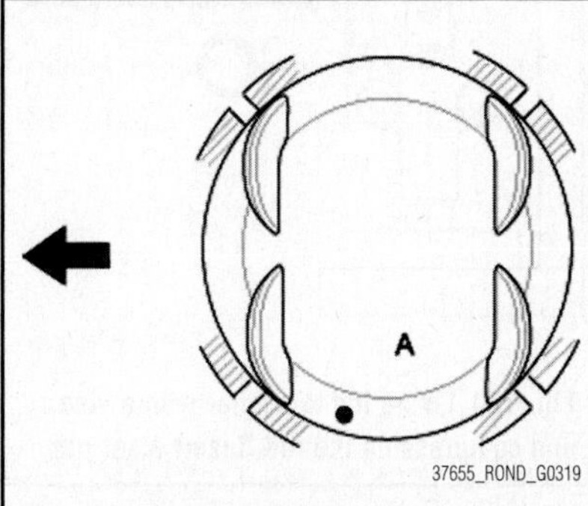

Fig. 119 Position the piston rings so that the ring ends are as shown—2.7L engine

REAR MAIN SEAL

REMOVAL & INSTALLATION

2.4L Engine

1. Cut the oil seal lip with a knife. Install a rag to the housing and using a screwdriver, carefully pry the oil seal from the oil seal housing. Clean the gasket mounting surfaces.

To install:

2. Clean the oil seal housing. Coat the oil seal and the housing with clean engine oil.
3. Install the oil seal into the housing and tap it evenly into place with a hammer and a large diameter piece of pipe. The seal must be flush with the rear oil seal retainer edge.

2.7L Engine

See Figure 120.

1. Remove the oil seal case (A).
2. Cut the oil seal lip with a knife. Install a rag to the housing and using a screwdriver, carefully pry the oil seal from the oil seal housing. Clean the gasket mounting surfaces.

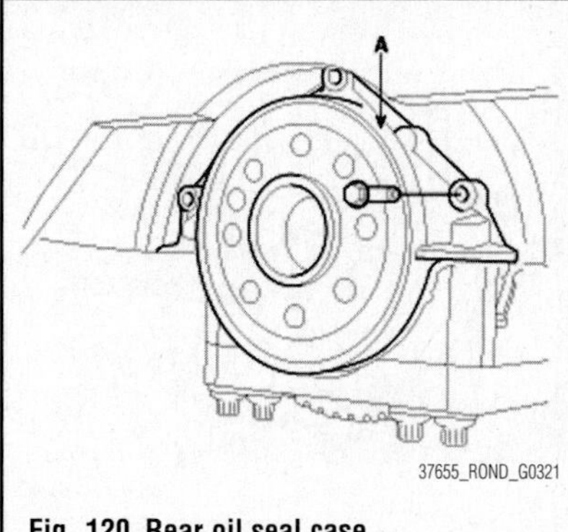

Fig. 120 Rear oil seal case

To install:

3. Install the rear oil seal case, as follows:
   a. Torque to 7.2–8.7 ft. lbs. (9.8–11.8 Nm).
   b. Clean the sealing surface face before assembling the two parts.

➡Note the following:

- Remove harmful foreign materials on the sealing face before applying sealant
- Apply sealant to the inner threads of the bolt holes.

   c. When assembling the rear oil seal case, apply the liquid sealant TB1217H to the rear oil seal case. Assemble the part within 5 minutes after applying the sealant.
   d. Install the oil seal into the housing and tap it evenly into place with a hammer and a large diameter piece of pipe. The seal must be flush with the edge of the rear cover.
   e. You can also use SST (09231-33000) to install the rear oil seal after applying engine oil on the lip of the oil seal.
4. Install the oil pump case.

ROCKER ARMS

REMOVAL & INSTALLATION

The engines covered are not equipped with rocker arms/shafts. The camshafts directly actuate the valve through a bucket type follower.

TIMING BELT FRONT COVER

REMOVAL & INSTALLATION

2.7L Engine

See Figures 121 through 125.

1. Remove the engine cover
2. Remove the front right wheel and tire.
3. Remove the right side cover.
4. Remove the drive belt, the idler pulley and the tensioner pulley.

➡In removing the drive belt, fix a tool in the auto tensioner pulley bolt and turn the bolt counter clockwise.

5. Remove the timing belt upper covers (A).

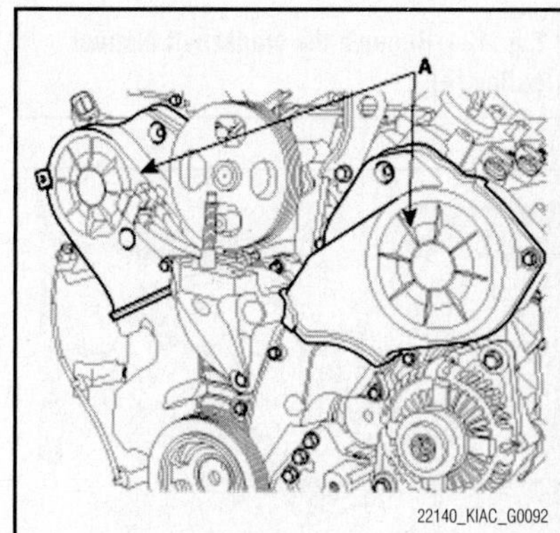

Fig. 121 Remove the timing belt upper covers (A)

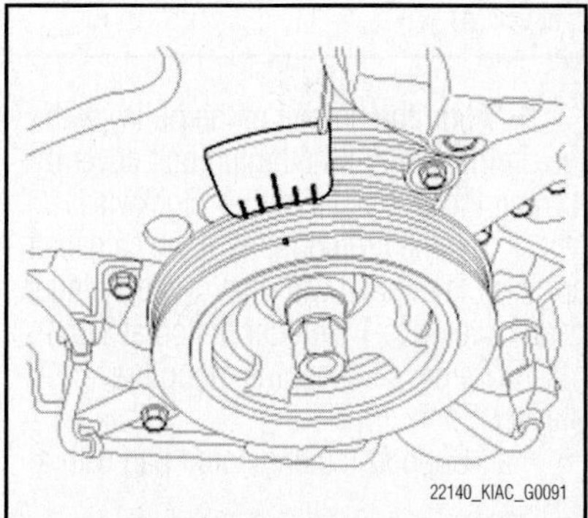

Fig. 122 Align the groove of the pulley with the timing mark of the timing belt cover

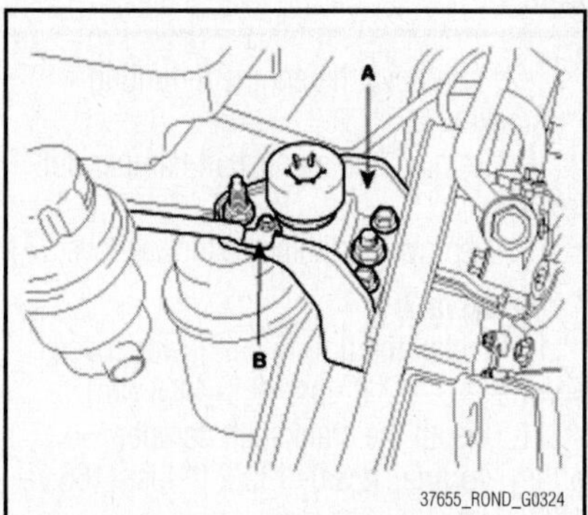

Fig. 123 Remove the engine mounting bracket (A) and the ground cable (B).

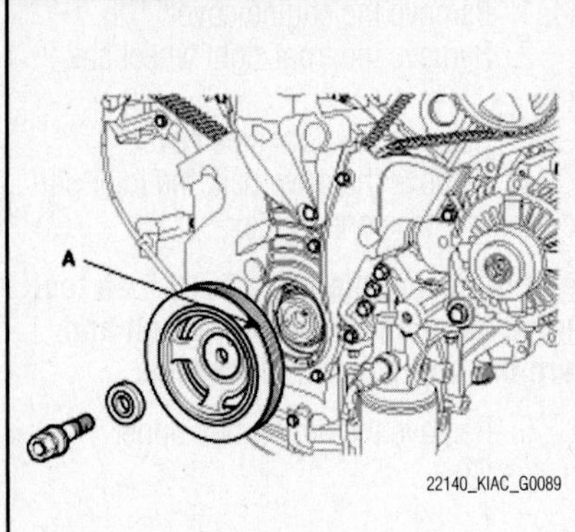

Fig. 124 Remove the crankshaft damper pulley (A)

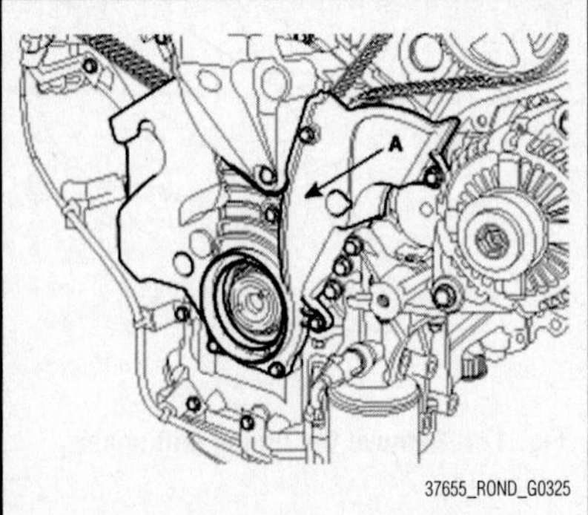

Fig. 125 Remove the timing belt lower cover (A)

6. Align the groove of the pulley with the timing mark of the timing belt cover by turning the crankshaft pulley clockwise. Check that the timing mark of the camshaft sprocket is aligned with that of the cylinder head cover. (No.1 cylinder piston at TDC)

7. Remove the engine mounting bracket.

a. Support the engine oil pan with a jack.

CAUTION

Put a wooden or rubber block between the jack and the engine oil pan.

b. Remove the engine mounting bracket (A).

8. Remove the crankshaft damper pulley (A).

9. Remove the timing belt lower cover (A)

To install:

10. Install the timing belt lower cover. Torque: 7.8–8.7 ft. lbs. (9.8–11.8 Nm).

11. Install the crankshaft damper pulley. Torque: 123.0–130.2 ft. lbs. (166.7–176.5 Nm).

12. Install the engine mounting bracket. Torque: 47.0–61.5 ft. lbs. (63.7–83.4 Nm).

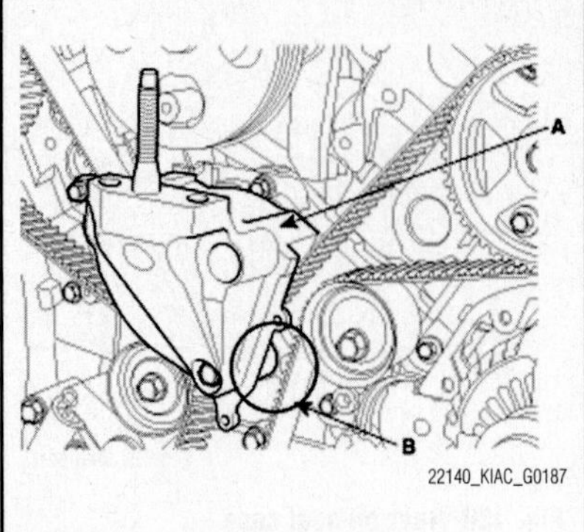

Fig. 126 Remove the engine support bracket (A)

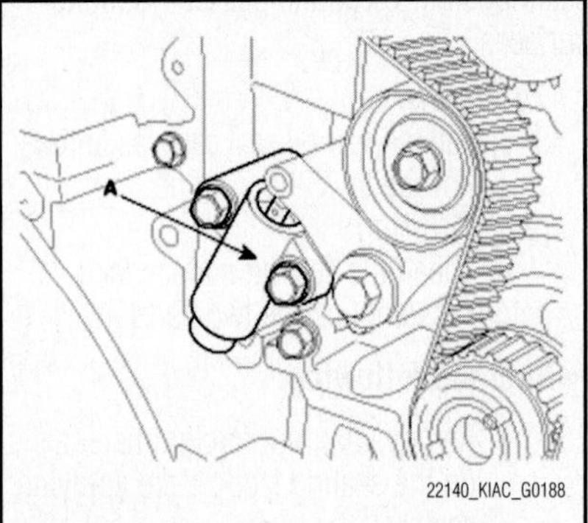

Fig. 127 Remove the timing belt auto tensioner (A)

13. Install the ground cable. Torque: 5.8–8.7 ft. lbs. (7.8–11.8 Nm).

14. Install the timing belt upper cover. Torque: 7.2–8.7 ft. lbs. (9.8–11.8 Nm).

15. Install the drive belt tensioner. Torque: 25.3–39.8 ft. lbs. (34.3–53.9 Nm).

16. Install the drive belt idler and the drive belt. Torque: 25.3–39.8 ft. lbs. (34.3–53.9 Nm).

17. Install the right side cover.

18. Install the front right wheel and tire.

19. Install the engine cover. Torque to: 5.8–8.7 ft. lbs. (7.8–11.8 Nm).

TIMING BELT & SPROCKETS

REMOVAL & INSTALLATION

2.7L Engine

See Figures 121 through 134.

1. Remove the engine cover
2. Remove the front right wheel and tire.
3. Remove the right side cover.
4. Remove the drive belt, the idler pulley and the tensioner pulley.

➡In removing the drive belt, fix a tool in the auto tensioner pulley bolt and turn the bolt counter clockwise.

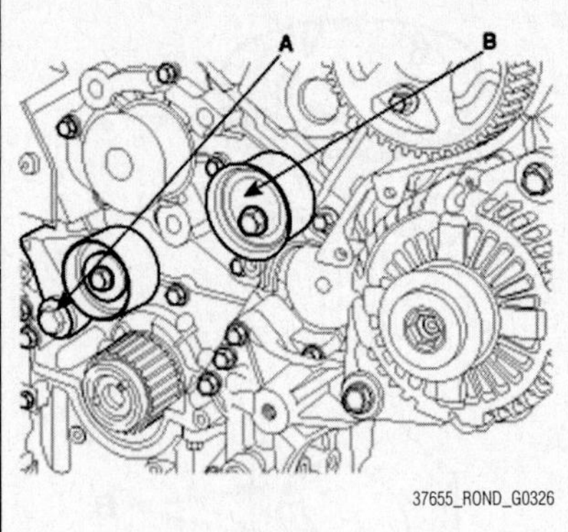

Fig. 128 Install the tensioner arm assembly (A) and the idler (B

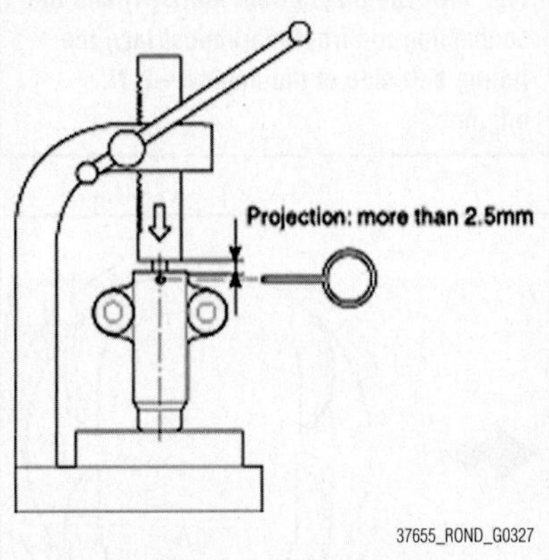

Fig. 129 Fixing the tensioner with a vise and compressing the rod, insert a set-pin

5. Remove the timing belt upper covers (A).

6. Align the groove of the pulley with the timing mark of the timing belt cover by turning the crankshaft pulley clockwise. Check that the timing mark of the camshaft sprocket is aligned with that of the cylinder head cover. (No.1 cylinder piston at TDC)

7. Remove the engine mounting bracket.

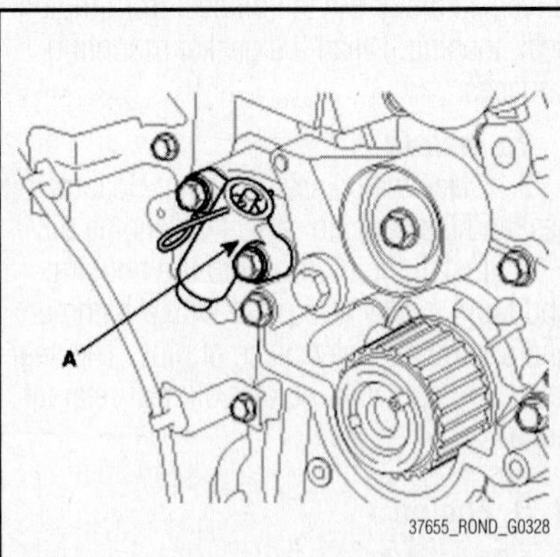

Fig. 130 Install the auto tensioner to the front case with the set-pin inserted

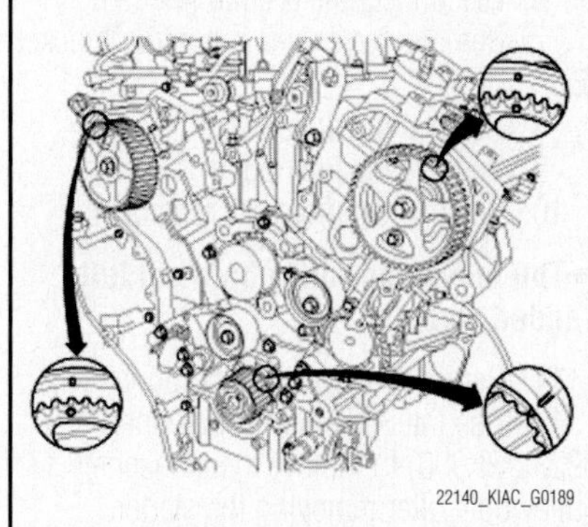

Fig. 131 Verify the alignment of the timing marks on the camshaft and the crankshaft sprockets

a. Support the engine oil pan with a jack.

⁂ CAUTION

Put a wooden or rubber block between the jack and the engine oil pan.

b. Remove the engine mounting bracket (A).

8. Remove the crankshaft damper pulley (A).
9. Remove the timing belt lower cover (A)
10. Remove the engine support bracket (A).

➡After removal, a small amount of engine coolant may drain from point (B).

11. Remove the timing belt auto tensioner (A).
12. Remove the timing belt.

➡Mark the direction of rotation on the timing belt.

13. Remove the timing belt tensioner arm assembly and the idler.
14. Remove the crankshaft timing belt sprocket.

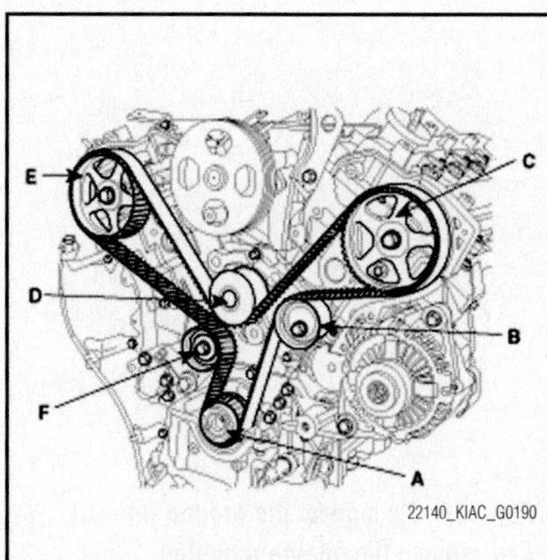

Fig. 132 Install the timing belt

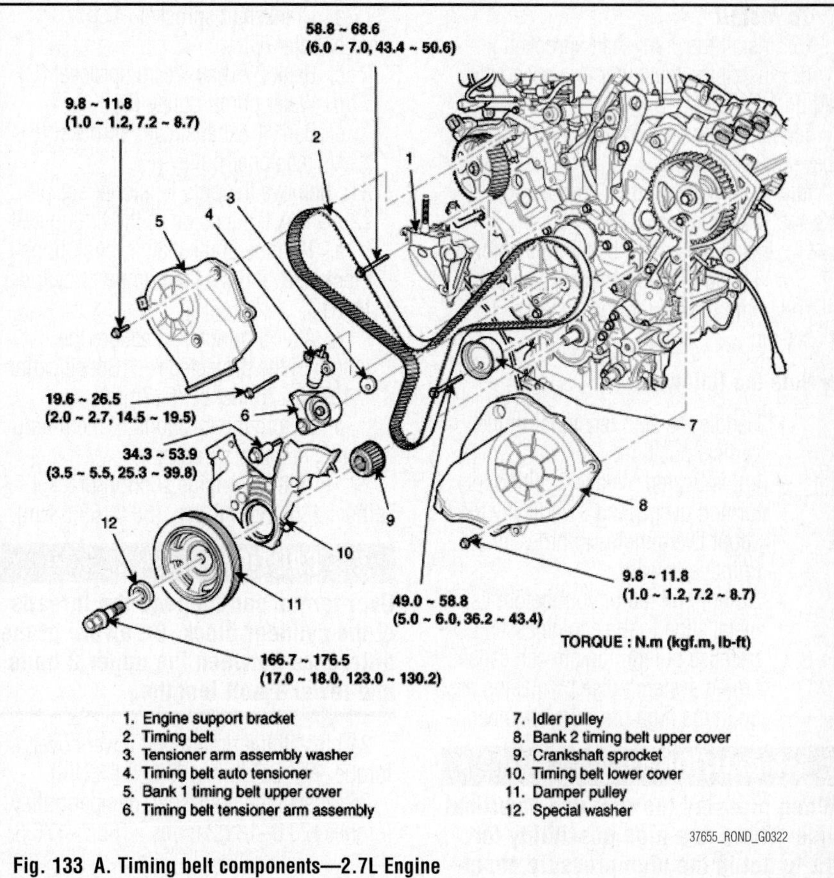

Fig. 133 A. Timing belt components—2.7L Engine

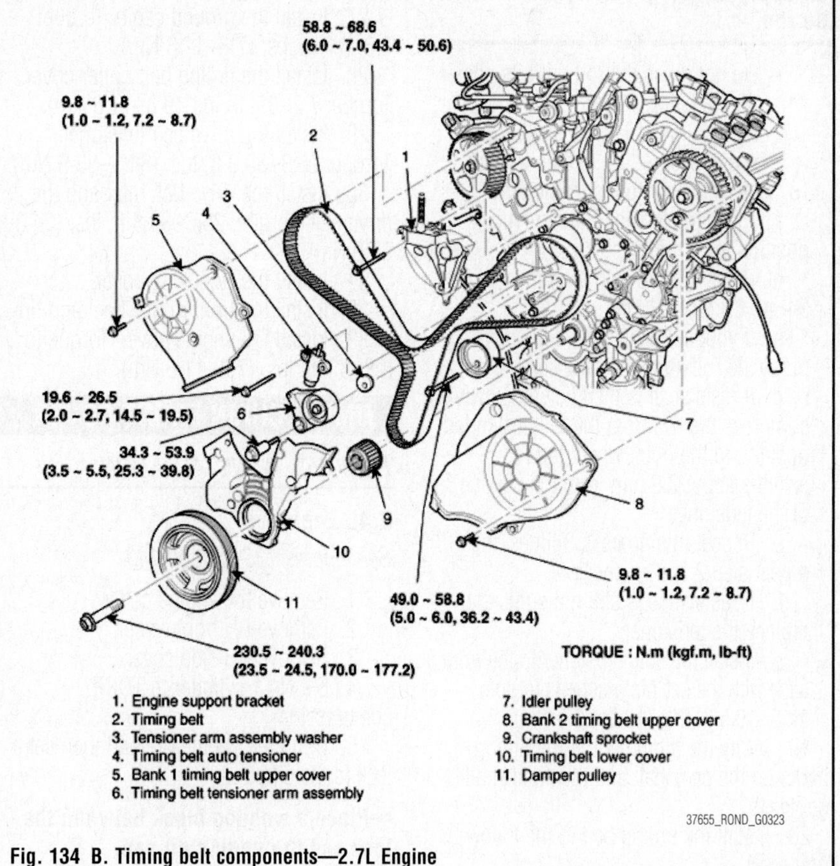

Fig. 134 B. Timing belt components—2.7L Engine

To install:

15. Install the crankshaft sprocket.

16. Install the tensioner arm assembly (A) and the idler (B).

Tensioner arm bolt torque: 25.3–39.8 ft. lbs. (34.3–53.9 Nm).

Idler pulley bolt torque: Torque: 36.2–43.4 ft. lbs. (49.0–58.8 Nm).

17. Install the timing belt auto tensioner:

a. Fixing the tensioner with a vise and compressing the rod, insert a set-pin.

➡Note the following:

- Handle the auto tensioner in the vertical position.
- If it is treated by laying, tilting, or turning over, stand it vertically for about five minutes in order for air to move upward.
- Do not pull the pin out before its installation to the engine assembly because pulling the pin can cause timing system noise by making air go in the high pressure chamber.

CAUTION

When pressing the rod, use a vertical vise. There is a high possibility for air to get in the high pressure chamber, when using a horizontal vise. Do not apply a load greater than 400N on the rod.

- Do not press the rod until its position is below 2.5 mm from the body surface. The projection of the rod is secured more than 2.5mm.

18. Before installing the tensioner:

a. Place the tensioner in a vertical position for 5 minutes.

b. Press the rod with a force of about 34–45 LB (150–200N).

c. If you feel resistance, fully compress the rod and insert the set pin.

d. If resistance is insufficient, slowly compress the rod from the fully extended position to the point where the rod extends about 2.9 mm from the surface of the tensioner.

e. Repeat this process 3 times, then repeat step 2 and re-check.

f. If resistance is still not sufficient, replace the tensioner.

g. Install the auto tensioner to the front case with the set-pin inserted. Torque: 14.5–19.5 ft. lbs. (19.6–26.5 Nm).

19. Verify the alignment of the timing marks on the camshaft and the crankshaft sprockets.

20. Install the timing belt in the following order:

a. Crankshaft sprocket (A)

b. Idler (B)

c. Bank 2 exhaust cam sprocket (C)

d. Water pump pulley (D)

e. Bank 1 exhaust cam sprocket (E)

f. Tensioner pulley (F).

21. Remove the auto tensioner set-pin.

22. Check the tension of the timing belt.

a. Turn the crankshaft 2 revolutions clockwise, and set the number 1 cylinder to TDC.

b. After 5 minutes, measure the length of the projected rod. Specification: 0.197–0.276 inches (5–7mm).

c. Ensure the locations of the timing marks for each sprocket.

23. Install the engine support bracket. Torque: 43.4–50.6 ft. lbs. (58.8–68.6 Nm).

CAUTION

Over torque can damage the threads of the cylinder block. Be aware of the difference between the upper 2 bolts and lower 1 bolt lengths.

24. Install the timing belt lower cover. Torque: 7.2–8.7 ft. lbs. (9.8–11.8 Nm).

25. Install the crankshaft damper pulley. Torque: 123.0–130.2 ft. lbs. (166.7–176.5 Nm).

26. Install the engine mounting bracket. Torque: 47.0–61.5 ft. lbs. (63.7–83.4 Nm).

27. Install the ground cable. Torque: 5.8–8.7 ft. lbs. (7.8–11.8 Nm).

28. Install the timing belt upper cover. Torque: 7.2–8.7 ft. lbs. (9.8–11.8 Nm).

29. Install the drive belt tensioner. Torque: 25.3–39.8 ft. lbs. (34.3–53.9 Nm).

30. Install the drive belt idler and the drive belt. Torque: 25.3–39.8 ft. lbs. (34.3–53.9 Nm).

31. Install the right side cover.

32. Install the front right wheel and tire.

33. Install the engine cover. Torque to: 5.8–8.7 ft. lbs. (7.8–11.8 Nm).

TIMING CHAIN FRONT COVER

REMOVAL & INSTALLATION

2.4L Engine

See Figures 135 through 140.

1. Remove the engine cover.
2. Remove RH front wheel.
3. Remove RH side cover.
4. Set No.1 cylinder to TDC/ compression.
5. Drain the engine oil, and then set a jack to the oil pan.

➡Place a wooden block between the jack and the engine oil pan.

6. Disconnect the ground line (A).
7. Remove the engine mounting bracket (B).
8. Remove the drive belt.
9. Remove the idler pulley.
10. Remove the drive belt tensioner.

➡The tensioner pulley bolt is a left-handed screw.

11. Remove the water pump pulley (A).
12. Use the SST (flywheel stopper, 09231-3K000) to remove the crankshaft pulley bolt, after removing the starter.
13. Remove the crankshaft pulley (B).
14. Remove the engine support bracket (C).
15. Disconnect the exhaust OCV connector (A) and remove the wiring from the cylinder head cover.
16. Remove the breather hose (A).
17. Remove the PCV hose.
18. Disconnect the ignition coil connectors.
19. Remove the ignition coils.
20. Remove the cylinder head cover.
21. Remove the air compressor lower bolts.
22. Remove the compressor bracket.
23. Using the SST(09215-3C000), remove the oil pan (A):
    - Insert the SST between the oil pan and the ladder frame by tapping it with a plastic hammer in the direction of the arrow.
    - After tapping the SST with a plastic hammer along the direction of the arrow around more than 2/3 of the edge of the oil pan, remove it from the ladder frame.
    - Do not turn over the SST abruptly without tapping. It damage the SST.
24. Remove the timing chain cover by prying the portions between the cylinder

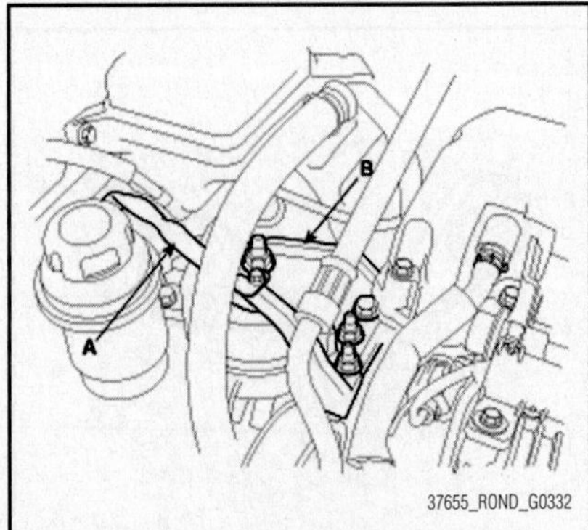

Fig. 135 Disconnect the ground line (A) and remove the engine mounting bracket (B)

head and cylinder block with a screwdriver.

⁂ CAUTION

Be careful not to damage the contact surfaces of cylinder block, cylinder head and timing chain cover.

To install:

25. Install the timing chain cover:
    a. The sealant locations on the chain cover and on counter parts (cylinder head, cylinder block, and ladder frame) must be free of engine oil, etc.
    b. Before assembling the timing chain cover, apply the liquid sealant LOCTITE®5900 on the gap between the cylinder head and cylinder block.
26. Apply the sealant without discontinuity in a bead width of 0.12 in. (3mm)
    a. After applying liquid sealant LOCTITE®5900 on timing chain cover, assemble the part within 5 minutes.
    b. Use the dowel pins on the cylinder block and the holes on the timing chain cover as a reference in order to assemble the timing chain cover to be in exact position.
    c. Torque: M6 bolts: 5.8–7.2 ft. lbs. (7.8–9.8 Nm). M8 bolts: 13.7–16.6 ft. lbs. (18.6–22.5 Nm).
    d. The firing and/or blow out test should not be performed within 30 minutes after the timing chain cover is assembled.
27. Install the oil pan:
    a. Using a gasket scraper, remove all the old packing material from the gasket surfaces.
    b. Before assembling the oil pan, apply the liquid sealant LOCTITE®5900 on the oil pan.
    c. Assemble the part within 5 minutes after applying the sealant.

⁂ CAUTION

When applying sealant gasket, do not allow sealant to protrude into the inside of oil pan. To prevent oil leakage, apply sealant gasket to the inner threads of the bolt holes.

    d. Install the oil pan.

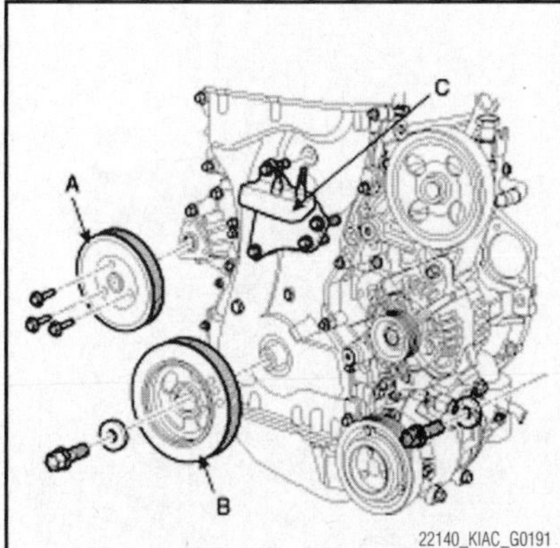

Fig. 136 Remove the water pump pulley (A), crankshaft pulley (B), and engine support bracket (C)

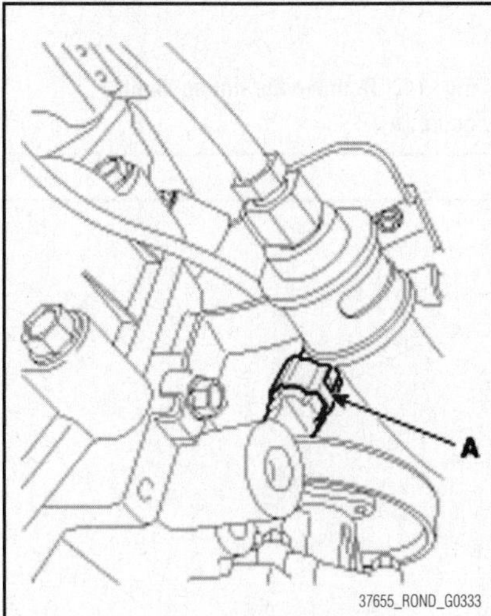

Fig. 137 Disconnect the exhaust OCV connector (A) and remove the wiring from the cylinder head cover

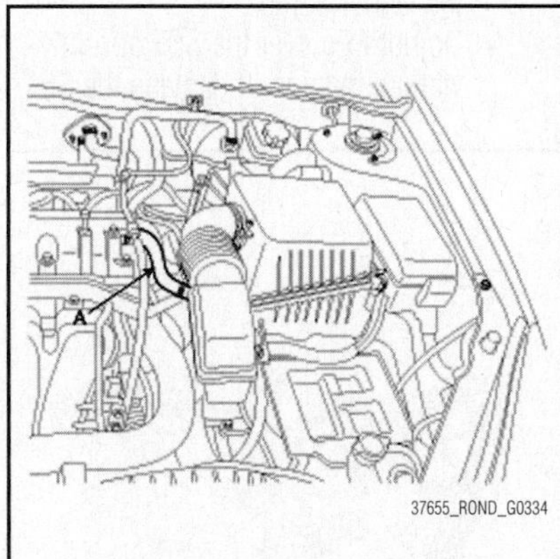

Fig. 138 Remove the breather hose (A)

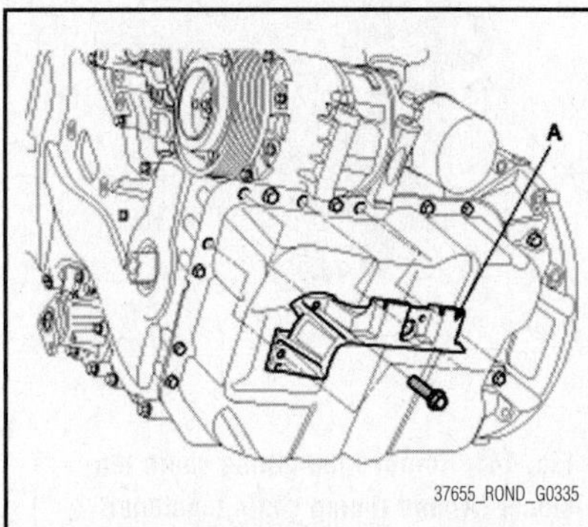

Fig. 139 Remove the breather hose (A)

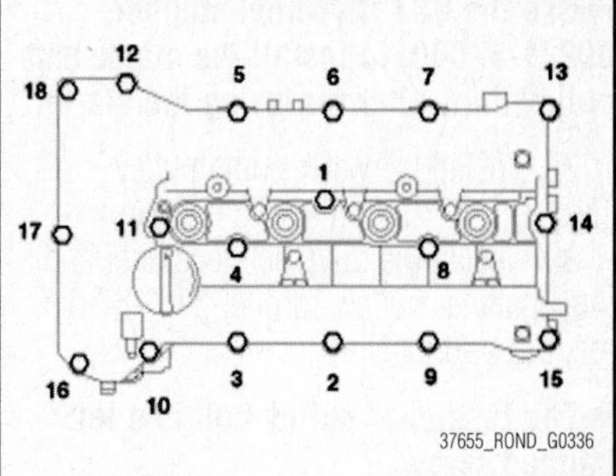

Fig. 140 Cylinder head cover bolt torque sequence

    e. Uniformly tighten the bolts in several passes. Torque: M8 bolts: 19.5–22.4 ft. lbs. (26.5–30.4 Nm). M6 bolts: 7.2–8.7 ft. lbs. (9.8–11.8 Nm).
    f. After assembly, wait at least 30 minutes before filling the engine with oil.
28. Install the air compressor bracket. Torque: 14.5–17.4 ft. lbs. (19.6–23.5 Nm).
29. Install the air compressor lower bolts. Torque: 13.7–18.1 ft. lbs. (19.6–24.5 Nm).
30. Install the cylinder head cover.
    a. The hardening sealant located on the upper area between timing chain cover and cylinder head should be removed before assembling the cylinder head cover.
    b. After applying sealant, assemble it within 5 minutes. Bead width: 0.1 in. (2.5mm)
    c. The firing and/or blow out test should not be performed within 30 minutes after the cylinder head cover was assembled.
    d. Torque the cylinder head cover bolts as follows:
       - 1st step: 2.9–4.3 ft. lbs. (3.9–5.9 Nm)
       - 2st step: 5.8–7.2 ft. lbs. (7.8–9.8 Nm)

⁂ CAUTION

Do not reuse the cylinder head cover gasket.

31. Install the ignition coils. Torque: 2.9–4.3 ft. lbs. (3.9–5.9 Nm)
32. Connect the ignition coil connectors.
33. Install the PCV hose.
34. Install the breather hose.
35. Connect the exhaust OCV connector (A) and replace the wiring on the cylinder head cover.
36. Install the engine support bracket. Torque: M10 bolts: 28.9–32.5 ft. lbs. (39.2–44.1 Nm); M8 bolts: 14.5–18.1 ft. lbs. (19.6–24.5 Nm).
37. Install the crankshaft pulley. Torque: 122.9–130.1 ft. lbs. (166.6–176.4 Nm).

➡Use the SST (flywheel stopper, 09231-3K000) to install the crankshaft pulley bolt, after removing the starter.

38. Install the water pump pulley. Torque: 5.8–7.2 ft. lbs. (7.8–9.8 Nm).
39. Install the drive belt tensioner and the tensioner pulley. Torque: 39.7–47.0 ft. lbs. (53.9–63.7 Nm).

➡The tensioner pulley bolt is a left-handed screw.

40. Install the idler. Torque: 39.7–47.0 ft. lbs. (53.9–63.7 Nm).
41. Install the accessory drive belt in the following order:
    a. Crankshaft pulley
    b. A/C pulley
    c. Alternator pulley
    d. Idler pulley
    e. P/S pump pulley
    f. Idler pulley
    g. Water pump pulley
    h. Tensioner pulley
    i. Rotate the auto tensioner arm in the counter-clockwise moving auto tensioner pulley bolt with a wrench.
42. After putting the belt on the auto tensioner pulley, release the auto tensioner pulley slowly.
43. Install engine mounting bracket and ground line. Torque: 47.0–61.5 ft. lbs. (63.7–83.4 Nm).
44. Install the RH side cover. Torque: 6.5–7.9 ft. lbs. (8.8–10.8 Nm).
45. Install the RH front wheel. Torque: 65.1–79.6 ft. lbs. (88.3–107.9 Nm).
46. Install the engine cover.

TIMING CHAIN & SPROCKETS

REMOVAL & INSTALLATION

2.4L Engine

See Figures 141 through 145.

1. Remove the engine cover.
2. Remove RH front wheel.
3. Remove RH side cover.
4. Set No.1 cylinder to TDC/compression.
5. Drain the engine oil, and then set a jack to the oil pan.

➡Place a wooden block between the jack and the engine oil pan.

6. Disconnect the ground line (A).
7. Remove the engine mounting bracket (B).
8. Remove the drive belt.
9. Remove the idler pulley.
10. Remove the drive belt tensioner.

➡The tensioner pulley bolt is a left-handed screw.

11. Remove the water pump pulley (A).
12. Use the SST (flywheel stopper, 09231-3K000) to remove the crankshaft pulley bolt, after removing the starter.
13. Remove the crankshaft pulley (B).
14. Remove the engine support bracket (C).
15. Disconnect the exhaust OCV connector (A) and remove the wiring from the cylinder head cover.
16. Remove the breather hose (A).
17. Remove the PCV hose.
18. Disconnect the ignition coil connectors.
19. Remove the ignition coils.
20. Remove the cylinder head cover.
21. Remove the air compressor lower bolts.
22. Remove the compressor bracket.
23. Using the SST(09215-3C000), remove the oil pan (A).
    - Insert the SST between the oil pan and the ladder frame by tapping it with a plastic hammer in the direction of the arrow.
    - After tapping the SST with a plastic hammer along the direction of the arrow around more than 2/3 of the edge of the oil pan, remove it from the ladder frame.
    - Do not turn over the SST abruptly without tapping. It damage the SST.

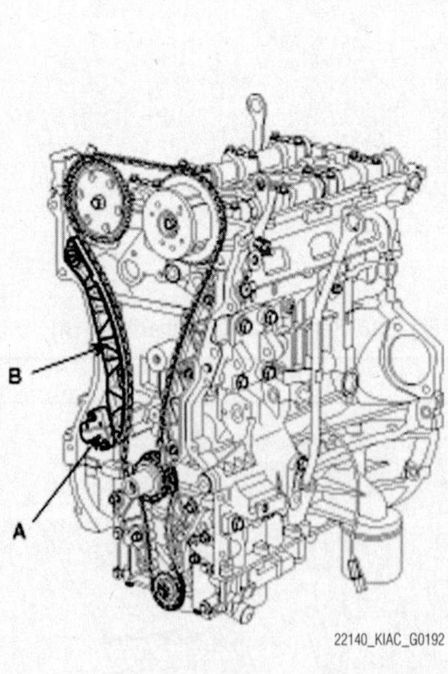

Fig. 141 Remove the timing chain tensioner (A) and timing chain tensioner arm (B)

24. Remove the timing chain cover by prying the portions between the cylinder head and cylinder block with a screwdriver.

CAUTION

Be careful not to damage the contact surfaces of cylinder block, cylinder head and timing chain cover.

25. Align the key of crankshaft with the mating face of main bearing cap. This places the piston of the No.1 cylinder at the top dead center on the compression stroke.
26. Install a set pin after compressing the timing chain tensioner.
27. Remove the timing chain tensioner (A).
28. Remove the timing chain tensioner arm (B).

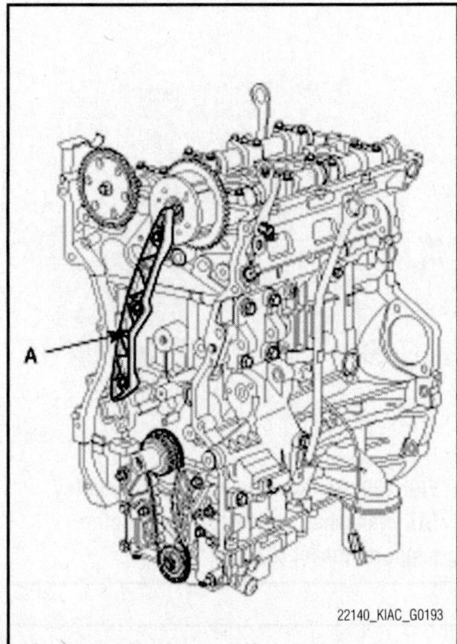

Fig. 142 Remove the timing chain guide (A)

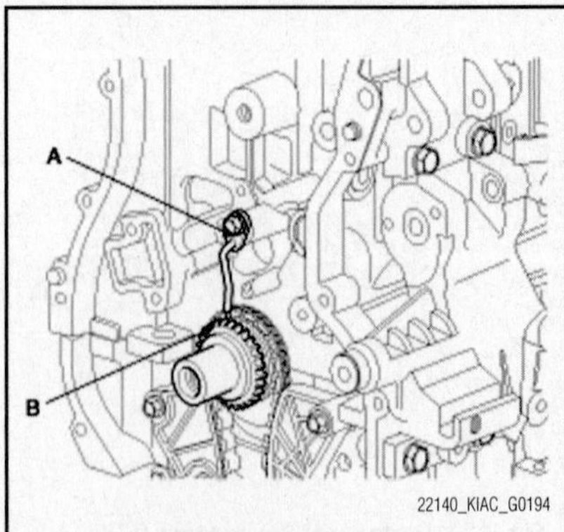

Fig. 143 Remove the timing chain oil jet (A) and crankshaft chain sprocket (B)

29. Remove the timing chain.
30. Remove the timing chain guide (A).
31. Remove the timing chain oil jet (A).
32. Remove the crankshaft chain sprocket (B).
33. Remove the balance shaft chain.

To install:

34. Install the crankshaft chain sprocket.
35. Install the timing chain oil jet. Torque: 5.8–7.2 ft. lbs. (7.8–9.8 Nm).
36. Set the crankshaft so that the key of crankshaft is aligned with the mating surface of main bearing cap. Set the intake and exhaust camshaft assembly so that the TDC mark of the intake sprocket and the exhaust sprocket is aligned with the top surface of cylinder head. This places the piston on the No.1 cylinder at top dead center on the compression stroke.
37. Install the timing chain guide. Torque: 7.2–8.7 ft. lbs. (9.8–11.6 Nm).
38. Install the timing chain. To install the timing chain with no slack between each shaft (cam, crank), follow these steps:
    a. Crankshaft sprocket (A)
    b. Timing chain guide (B)
    c. Intake camshaft sprocket (C)
    d. Exhaust camshaft sprocket (D).

Match the timing mark of each sprocket with the timing mark (color link) of the timing chain.

39. Install the timing chain tensioner arm. Torque: 7.2–8.7 ft. lbs. (9.8–11.8 Nm).
40. Install the timing chain auto tensioner and remove the set pin. Torque: 7.2–8.7 ft. lbs. (9.8–11.8 Nm).
41. After rotating the crankshaft 2 revolutions clockwise (viewed from the front), confirm the timing mark.
42. Install the timing chain cover:
    a. The sealant locations on the chain cover and on counter parts (cylinder head, cylinder block, and ladder frame) must be free of engine oil, etc.
    b. Before assembling the timing chain cover, apply the liquid sealant LOCTITE®5900 on the gap between the cylinder head and cylinder block.
43. Apply the sealant without discontinuity in a bead width of 0.12 in. (3mm).
    a. After applying liquid sealant LOCTITE®5900 on timing chain cover, assemble the part within 5 minutes.
    b. Use the dowel pins on the cylinder block and the holes on the timing chain cover as a reference in order to assemble the timing chain cover to be in exact position.
    c. Torque: M6 bolts: 5.8–7.2 ft. lbs. (7.8–9.8 Nm). M8 bolts: 13.7–16.6 ft. lbs. (18.6–22.5 Nm).
    d. The firing and/or blow out test should not be performed within 30 minutes after the timing chain cover is assembled.
44. Install the oil pan:
    a. Using a gasket scraper, remove all the old packing material from the gasket surfaces.
    b. Before assembling the oil pan, apply the liquid sealant LOCTITE®5900 on the oil pan.
    c. Assemble the part within 5 minutes after applying the sealant.

✲✲ CAUTION

When applying sealant gasket, do not allow sealant to protrude into the inside of oil pan. To prevent oil leakage, apply sealant gasket to the inner threads of the bolt holes.

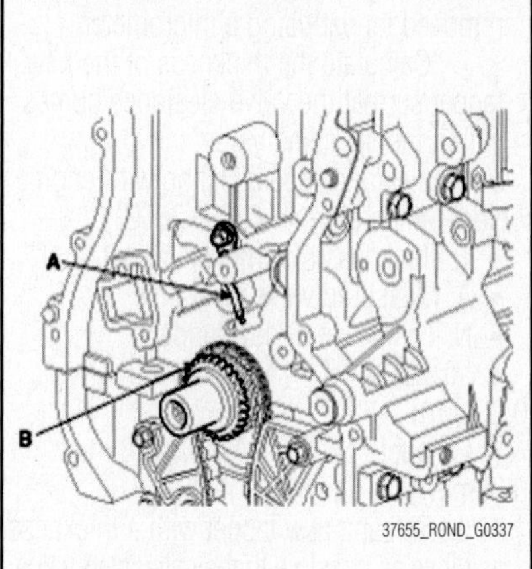

Fig. 144 Remove the crankshaft chain sprocket (B)

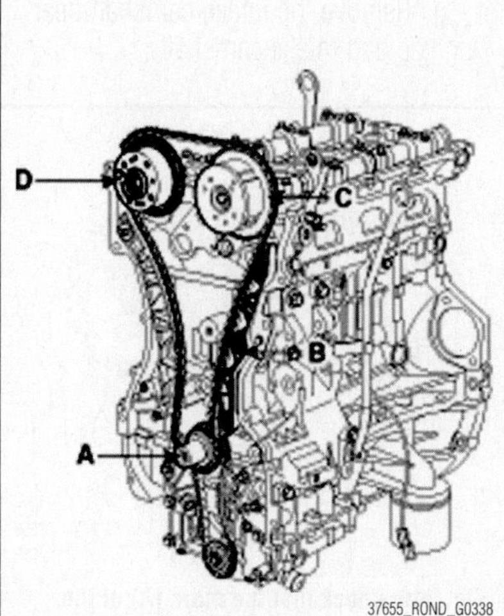

Fig. 145 Match the timing mark of each sprocket with the timing mark (color link) of the timing chain

    d. Install the oil pan.
    e. Uniformly tighten the bolts in several passes. Torque: M8 bolts: 19.5–22.4 ft. lbs. (26.5–30.4 Nm). M6 bolts: 7.2–8.7 ft. lbs. (9.8–11.8 Nm).
    f. After assembly, wait at least 30 minutes before filling the engine with oil.
45. Install the air compressor bracket. Torque: 14.5–17.4 ft. lbs. (19.6–23.5 Nm).
46. Install the air compressor lower bolts. Torque: 13.7–18.1 ft. lbs. (19.6–24.5 Nm).
47. Install the cylinder head cover.
    a. The hardening sealant located on the upper area between timing chain cover and cylinder head should be removed before assembling the cylinder head cover.
    b. After applying sealant, assemble it within 5 minutes. Bead width: 0.1 in. (2.5mm)
    c. The firing and/or blow out test should not be performed within 30 minutes after the cylinder head cover was assembled.
    d. Torque the cylinder head cover bolts as follows:
        - 1st step: 2.9–4.3 ft. lbs. (3.9–5.9 Nm)
        - 2st step: 5.8–7.2 ft. lbs. (7.8–9.8 Nm)

✲✲ CAUTION

Do not reuse the cylinder head cover gasket.

48. Install the ignition coils. Torque: 2.9–4.3 ft. lbs. (3.9–5.9 Nm)
49. Connect the ignition coil connectors.
50. Install the PCV hose.
51. Install the breather hose.
52. Connect the exhaust OCV connector (A) and replace the wiring on the cylinder head cover.
53. Install the engine support bracket. Torque: M10 bolts: 28.9–32.5 ft. lbs. (39.2–44.1 Nm); M8 bolts: 14.5–18.1 ft. lbs. (19.6–24.5 Nm).
54. Install the crankshaft pulley. Torque: 122.9–130.1 ft. lbs. (166.6–176.4 Nm).

➡Use the SST (flywheel stopper, 09231-3K000) to install the crankshaft pulley bolt, after removing the starter.

55. Install the water pump pulley. Torque: 5.8–7.2 ft. lbs. (7.8–9.8 Nm).
56. Install the drive belt tensioner and the tensioner pulley. Torque: 39.7–47.0 ft. lbs. (53.9–63.7 Nm).

➡The tensioner pulley bolt is a left-handed screw.

57. Install the idler. Torque: 39.7–47.0 ft. lbs. (53.9–63.7 Nm).

58. Install the accessory drive belt in the following order:

a. Crankshaft pulley
b. A/C pulley
c. Alternator pulley
d. Idler pulley
e. P/S pump pulley
f. Idler pulley
g. Water pump pulley
h. Tensioner pulley.
i. Rotate the auto tensioner arm in the counter-clockwise moving auto tensioner pulley bolt with a wrench.

59. After putting the belt on the auto tensioner pulley, release the auto tensioner pulley slowly.

60. Install engine mounting bracket and ground line. Torque: 47.0–61.5 ft. lbs. (63.7–83.4 Nm).

61. Install the RH side cover. Torque: 6.5–7.9 ft. lbs. (8.8–10.8 Nm).

62. Install the RH front wheel. Torque: 65.1–79.6 ft. lbs. (88.3–107.9 Nm).

63. Install the engine cover.

VALVE LASH

INSPECTION & ADJUSTMENT

2.4L Engine

See Figures 146 through 148.

➡Inspect and adjust the valve clearance when:

- The engine is cold [Engine coolant temperature: 68°F (20°C)]
- The cylinder head is installed on the cylinder block.

1. Remove the engine cover.
2. Remove the cylinder head cover:

a. Disconnect the ignition coil connectors.
b. Remove the ignition coils.
c. Remove the PCV hose and breather hose from the cylinder head cover.

3. Remove the cylinder head cover and gasket.
4. Set No.1 cylinder to TDC/compression.

a. Turn the crankshaft pulley and align its groove with the timing mark "T" of the lower timing chain cover.
b. Check that the mark of the camshaft timing sprockets are in straight line on the cylinder head surface. If not, turn the crankshaft one revolution (360°)

5. Inspect the valve clearance:

a. Check only the No. 1 cylinder: TDC/Compression. Measure the valve clearance. Using a thickness gauge, measure the clearance between the tappet and the base circle of camshaft.
b. Record the out-of-specification valve clearance measurements. They will be used later to determine required replacement of the adjusting tappet(s).

Valve clearance specification with engine coolant temperature of 68°F (20°C) Limit:

- Intake: 0.0039–0.0118 in, (0.10–0.30mm)
- Exhaust: 0.0079–0.0157 in. (0.20–0.40mm)

a. Turn the crankshaft pulley one revolution (360°) and align the groove with timing mark "T" of the lower timing chain cover.
b. Check only valves at TDC/compression. Measure the valve clearance.

6. Adjust the intake and exhaust valve clearance:

a. Set the No.1 cylinder to the TDC/compression.
b. Check the marks on the timing chain and camshaft timing sprockets.
c. Remove the service hole bolt of the timing chain cover.

⁂ CAUTION

The bolt must not be reused once it has been assembled.

d. Insert the tool SST (09242-2G000) in the service hole of the timing chain cover and release the ratchet.
e. Remove the front camshaft bearing cap (A).
f. Remove the exhaust camshaft bearing cap and exhaust camshaft.
g. Remove the intake camshaft bearing cap and intake camshaft.

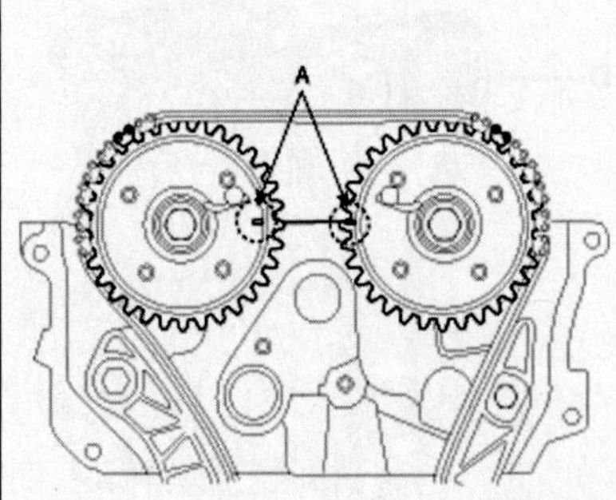

Fig. 146 Check that the mark (A) of the camshaft timing sprockets are in a straight line on the cylinder head surface as shown in the illustration. If not, turn the crankshaft one revolution (360°)

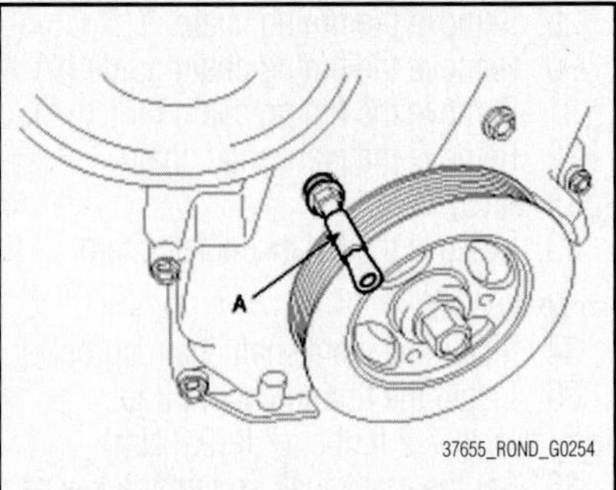

Fig. 147 Insert the tool (A) in the service hole of the timing chain cover and release the ratchet.

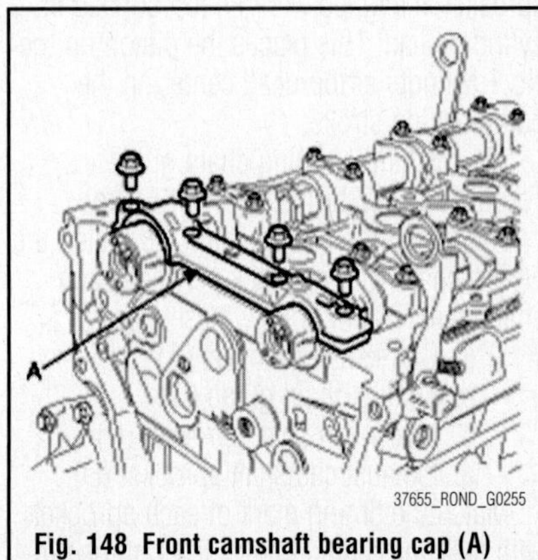

Fig. 148 Front camshaft bearing cap (A)

⁂ CAUTION

When disconnecting the timing chain from the camshaft timing sprocket, hold the timing chain.

h. Tie the timing chain with a string.

⁂ CAUTION

Be careful not to drop anything inside the timing chain cover.

i. Measure the thickness of the removed tappet using a micrometer.
j. Calculate the thickness of the new tappet so that the valve clearance comes within the specified value.

Valve clearance specification with engine coolant temperature of 68°F (20°C):

- T: Thickness of removed tappet
- A: Measured valve clearance
- N: Thickness of new tappet
- Intake: N = T + [A - 0.0079 in. (0.20mm)]
- Exhaust : N = T + [A - 0.0118 in. (0.30mm)]

a. Select a new tappet with a thickness as close as possible to the calculated value.

➡Shims are available in 47 size increments of 0.0006 in. (0.015mm) from

0.118 in. (3.00mm) to 0.1452 in. (.3.690mm).

b. Place a new tappet on the cylinder head.

c. Hold the timing chain, and place the intake camshaft and timing sprocket assembly.

d. Align the match marks on the timing chain and camshaft timing sprocket.

e. Install the intake and exhaust camshaft.

f. Install the front bearing cap.

g. Install the service hole bolt. Tighten to: 8.7–10.8 ft. lbs. (11.8–14.7Nm)

h. Turn the crankshaft two turns in the operating direction (clockwise) and realign the crankshaft sprocket and the camshaft sprocket timing marks (A).

i. Recheck the valve clearance.

Valve clearance specification with engine coolant temperature of 68°F (20°C):

- Intake: 0.0067–0.0090 in, (0.17–0.23mm)
- Exhaust: 0.0106–0.0129 in. (0.27–0.33mm)

2.7L Engine

See Figures 149 through 151.

➡Inspect and adjust the valve clearance when the engine is cold [engine coolant temperature: 20°C ± 5°C (59–77°F)] and the cylinder head is installed on the cylinder block.

1. Remove the engine cover.
2. Remove air cleaner assembly.
3. Remove the surge tank.
4. Disconnect the ignition coil connector and remove the ignition coil.
5. Remove the cylinder head cover.
6. Set the piston of the No.1 cylinder to Top Dead Center (TDC) position:

a. Turn the crankshaft pulley clockwise and align its groove with the timing mark "T" of the timing chain cover.

b. Check that the timing marks on the camshaft sprocket are in a straight line with the rocker cover mark for No. 1 cylinder TDC as shown in the illustration.

➡If not, turn the crankshaft one revolution clockwise.

7. Inspect the intake and the exhaust valve clearance:

a. With the No. 1 cylinder at TDC the valve clearance can be measured as shown below.

8. Measurement method:

a. Using a thickness gauge, measure the clearance between the tappet and the base circle of camshaft.

22140_ROND_G0100

Fig. 149 Check that the timing marks on the camshaft sprocket are in a straight line with the rocker cover mark

b. Record the out-of-specification valve clearance measurements. They will be used later to determine the required adjusting tappet(s) for replacement.

Valve clearance specification with engine coolant temperature of 68°F (20°C) Limit:

- Intake: 0.0039–0.0118 in. (0.10–0.30mm)
- Exhaust: 0.0079–0.0157 in. (0.20–0.40mm)

a. Turn the crankshaft pulley one revolution (360°) clockwise and align the groove with the timing mark "T" of the timing chain cover.

b. With the No. 4 cylinder at TDC the valve clearance can be measured as shown below.

9. Adjust the intake and the exhaust valve clearances:

a. Set the piston of the No.1 cylinder to the TDC/position.

b. Remove the timing belt.

c. Remove the camshaft bearing caps.

d. Remove the camshaft assembly.

e. Remove the Mechanical Lash Adjusters (MLA).

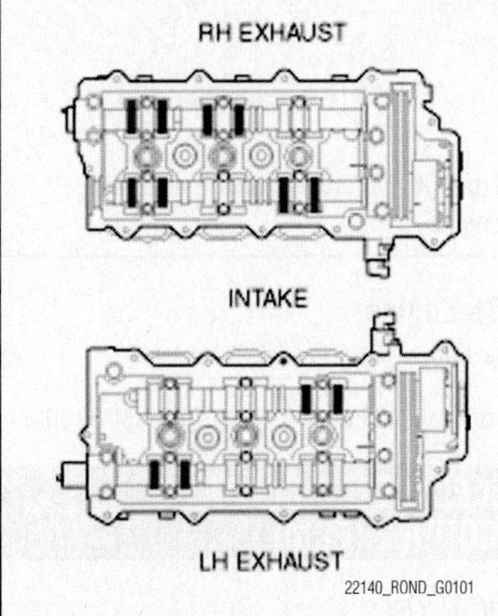

22140_ROND_G0101

Fig. 150 With the No. 1 cylinder at TDC the valve clearance can be measured

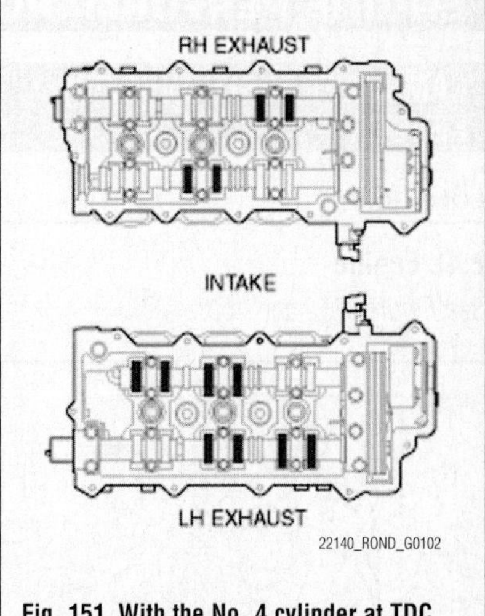

22140_ROND_G0102

Fig. 151 With the No. 4 cylinder at TDC the valve clearance can be measured

f. Measure the thickness of the removed tappet using a micrometer.

g. Calculate the thickness of the new tappet so that the valve clearance comes within the specified value.

Valve clearance specification with engine coolant temperature of 68°F (20°C):

- T: Thickness of removed tappet
- A: Measured valve clearance
- N: Thickness of new tappet
- Intake: N = T + [A - 0.0079 in. (0.20mm)]
- Exhaust : N = T + [A - 0.0118 in. (0.30mm)]

a. Select a new tappet with a thickness as close as possible to the calculated value.

➡Shims are available in 41 size increments of 0.0006 in. (0.015mm) from 0.118 in. (3.00mm) to 0.1417 in. (.3.600mm).

b. Place a new tappet on the cylinder head.

c. Apply engine oil on the periphery surface of the selected tappet.

d. Install the intake and exhaust camshafts.

e. Install the bearing caps.

f. Install the timing belt.

g. Turn the crankshaft two revolutions in the operating direction(clockwise) and realign crankshaft sprocket and camshaft sprocket timing marks(A).

h. Recheck the valve clearance.

Valve clearance specification with engine coolant temperature of 68°F (20°C):

- Intake: 0.0067–0.0090 in, (0.17–0.23mm)
- Exhaust: 0.0106–0.0129 in. (0.27–0.33mm)

ENGINE PERFORMANCE & EMISSION CONTROLS

CAMSHAFT POSITION (CMP) SENSOR

LOCATION

2.4L Engine

See Figures 152 and 153.

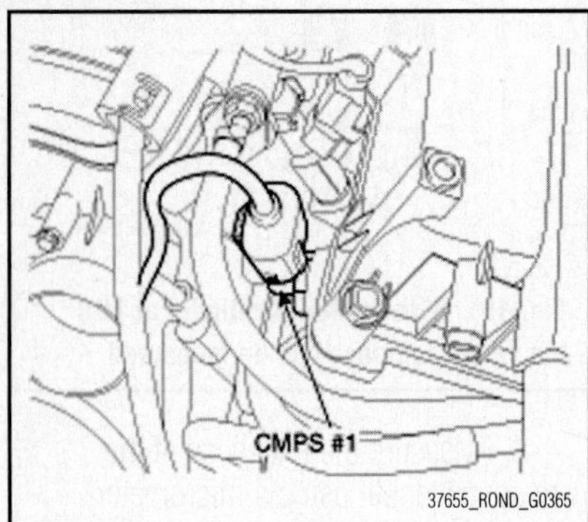

Fig. 152 CMP sensor, No. 1 Intake—2.4L Engine

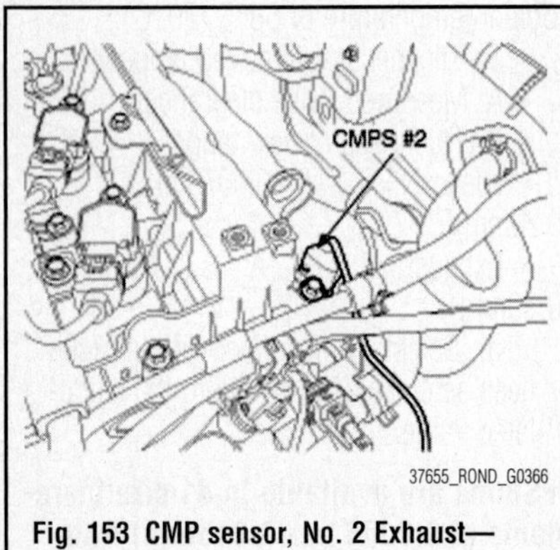

Fig. 153 CMP sensor, No. 2 Exhaust—2.4L Engine

The CMP sensors are installed on the engine head cover and use a target wheel installed on the camshaft.

2.7L Engine

See Figures 154 and 155.

The CMP sensors are installed on the engine head cover and use a target wheel installed on the camshaft.

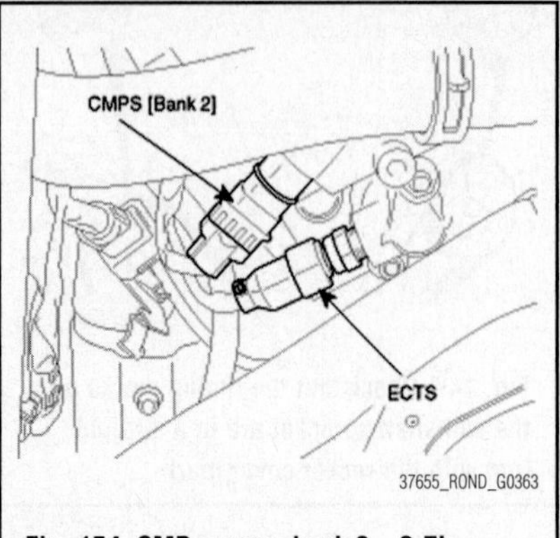

Fig. 154 CMP sensor, bank 2—2.7L Engine

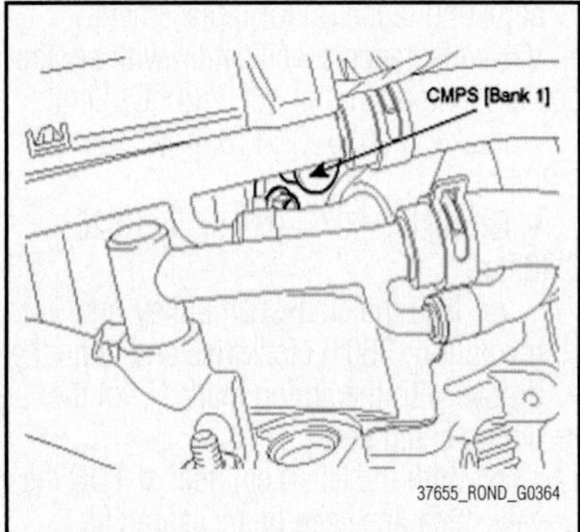

Fig. 155 CMP sensor, bank 1—2.7L Engine

CRANKSHAFT POSITION (CKP) SENSOR

LOCATION

2.4L Engine

See Figure 156.

Refer to the accompanying illustrations.

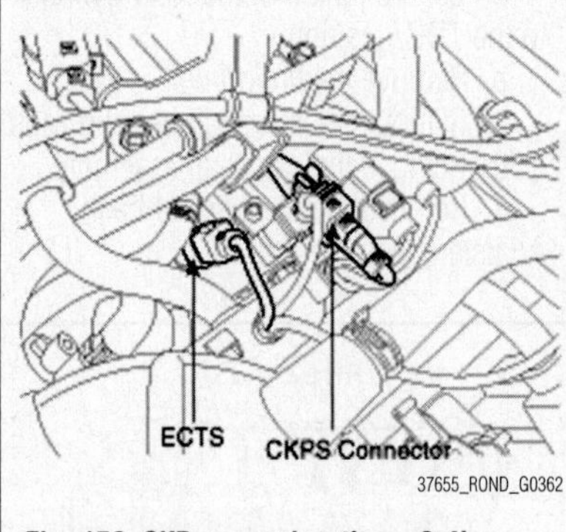

Fig. 156 CKP sensor location—2.4L Engine

2.7L Engine

See Figure 157.

Refer to the accompanying illustrations.

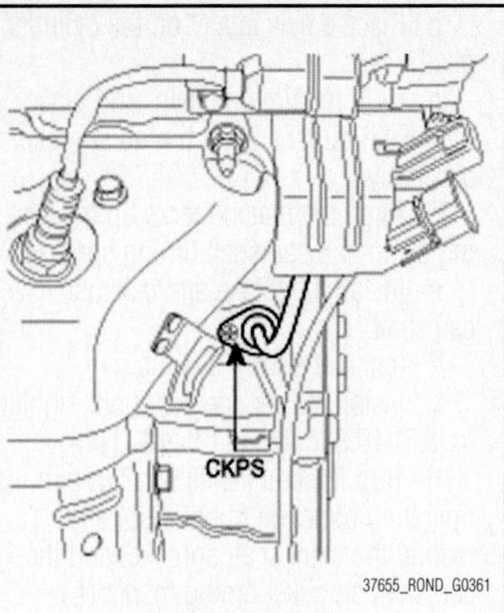

Fig. 157 CKP sensor location—2.7L Engine

ELECTRONIC CONTROL MODULE (ECM)

LOCATION

See Figures 158 and 159.

Refer to the accompanying illustrations.

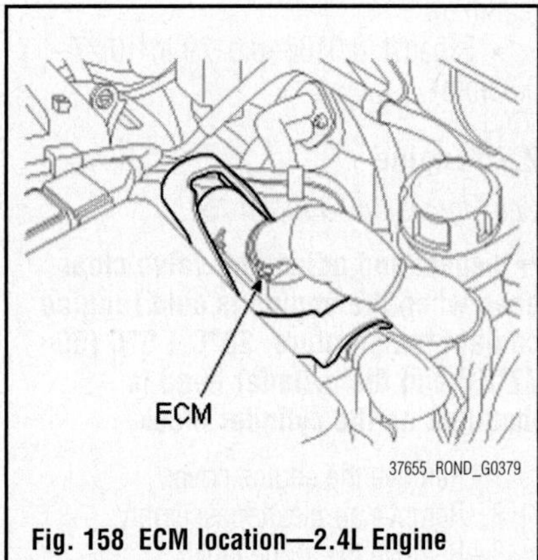

Fig. 158 ECM location—2.4L Engine

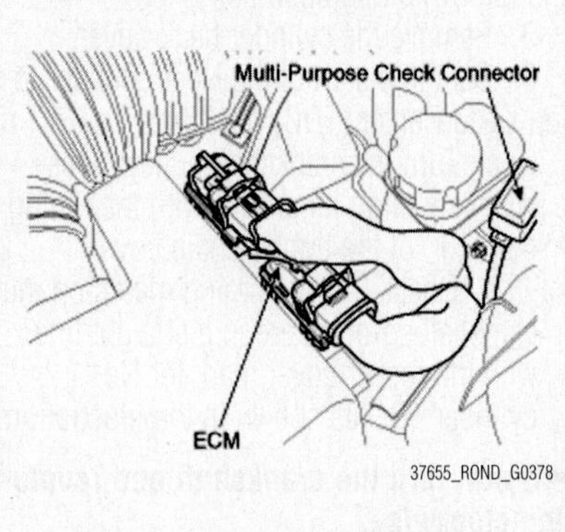

Fig. 159 ECM location—2.7L Engine

REMOVAL & INSTALLATION

See Figures 160 and 161.

1. Turn the ignition switch **OFF**.
2. Disconnect the battery negative (-) cable from the battery.
3. Remove the ECM cover and disconnect the ECM connector(s) (A).

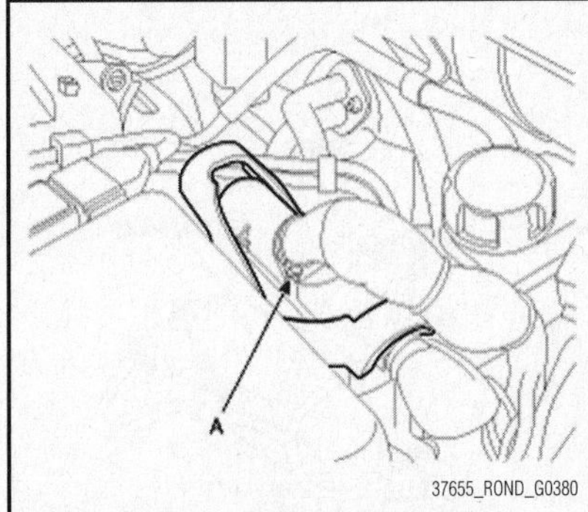

Fig. 160 Remove the ECM cover and disconnect the ECM connector(s) (A)

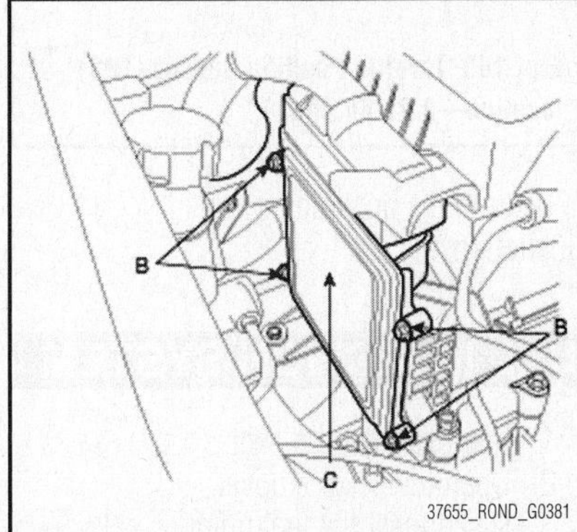

Fig. 161 Unscrew the ECM mounting bolts (B) and remove the ECM (C) from the air cleaner assembly

4. Unscrew the ECM mounting bolts (B) and remove the ECM (C) from the air cleaner assembly.
5. Install a new ECM. Tighten the mounting bolts to 7.2–8.7 ft. lbs. (9.8–11.8 Nm).

VIN REPROGRAMMING

The VIN (Vehicle Identification Number) is a number that has the vehicle's information (Maker, Vehicle Type, Vehicle Line/Series, Body Type, Engine Type, Transmission Type, Model Year, Plant Location, and so forth.

When replacing an ECM, the VIN must be programmed in the ECM. If there is no VIN in the ECM memory, the fault code, DTC P0630, is set.

⁂ WARNING

The programmed VIN cannot be changed. When writing the VIN, confirm the VIN carefully.

1. Select "Vehicle" and "Engine."
2. Select "VIN WRITING."
3. Check the ECM status:
   - VIRGIN: VIN is not programmed
   - LEARNT: VIN has been already programmed
4. Is the ECM status "VIRGIN"?
   - If yes, go to the next step.
   - If no, stop here.
5. Write the VIN with the cursor, function and number keys.

⁂ WARNING

Before pressing the "ENTER" key, confirm the VIN again because the programmed VIN cannot be changed.

6. After verifying the written VIN, press the "ENTER" key.
7. Turn the ignition switch OFF, then back ON.
8. Verify the programmed VIN in the ECM memory.

ENGINE COOLANT TEMPERATURE (ECT) SENSOR

LOCATION

See Figures 162 and 163.

Refer to the accompanying illustrations.

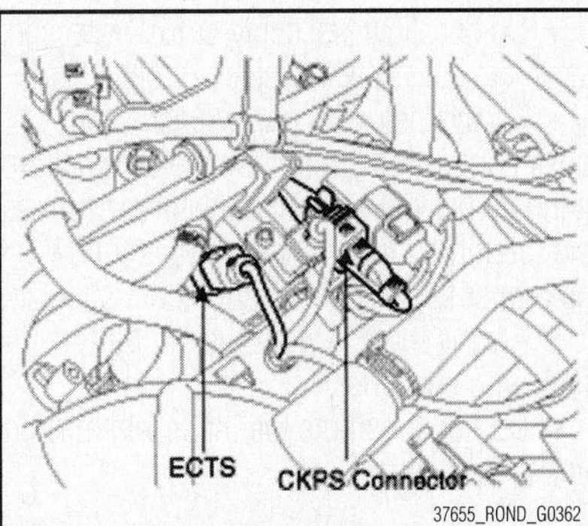

Fig. 162 ECT sensor location—2.4L Engine

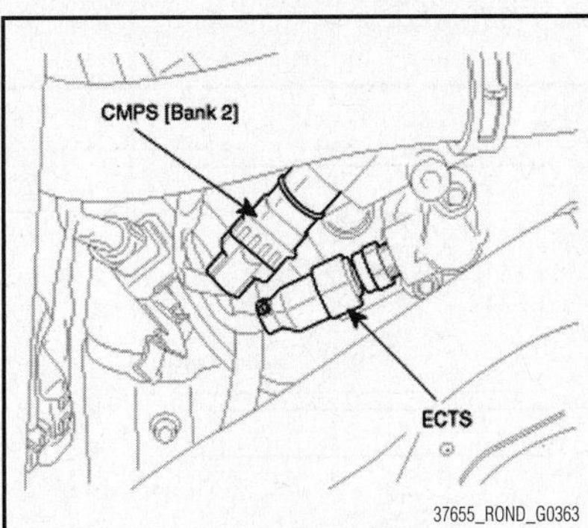

Fig. 163 ECT sensor location—2.7L Engine

REMOVAL & INSTALLATION

2.4L Engine

1. Turn the ignition switch **OFF**.
2. Disconnect the ECT sensor connector.
3. Remove the ECT sensor.
4. Installation is the reverse of the removal procedure. Tighten the ECT sensor to 21.7–28.9 ft. lbs. (29.4–39.2 Nm).

2.7L Engine

1. Turn the ignition switch **OFF**.
2. Disconnect the ECT sensor connector.
3. Remove the ECT sensor.
4. Installation is the reverse of the removal procedure. Tighten the ECT sensor to 14.5–28.9 ft. lbs. (19.6–39.2 Nm).

HEATED OXYGEN SENSOR (HO2S)

LOCATION

The Heated Oxygen Sensor (HO2S) is installed upstream and downstream of the Manifold Catalyst Converter (MCC).

REMOVAL & INSTALLATION

1. Turn the ignition switch **OFF**.
2. Disconnect the battery negative (-) cable from the battery.
3. Disconnect the HO2S connector.
4. Remove the heated oxygen sensor.
5. Installation is the reverse of the removal procedure. Tighten the sensor to 28.9–36.2 ft. lbs. (39.2–49.1 Nm).

INTAKE AIR TEMPERATURE SENSOR (IATS)

LOCATION

2.4L Engine

The Intake Air Temperature Sensor (IATS) is installed inside the Manifold Absolute Pressure (MAP) Sensor.

2.7L Engine

The Intake Air Temperature Sensor (IATS) is installed inside the Mass Air Flow (MAF) Sensor.

REMOVAL & INSTALLATION

1. Turn the ignition switch **OFF**.
2. Disconnect the IATS connector.
3. Remove the IATS.
4. Installation is the reverse of the removal procedure.

KNOCK SENSOR (KS)

LOCATION

The two Knock Sensors (KS) are installed inside the V-valley of the cylinder block.

MASS AIR FLOW (MAF) SENSOR

LOCATION

2.7L Engine

The Mass Air Flow (MAF) Sensor is located in between the air cleaner and the throttle body. It consists of a tube, a sensor assembly and a honey cell.

The MAF sensor is integrated with the Intake Air Temperature (IAT) sensor as a physical component.

MANIFOLD ABSOLUTE PRESSURE (MAP) SENSOR

LOCATION

The Manifold Absolute Pressure (MAP) sensor is installed on the surge tank.

THROTTLE POSITION SENSOR (TPS)

LOCATION

See Figures 164 and 165.

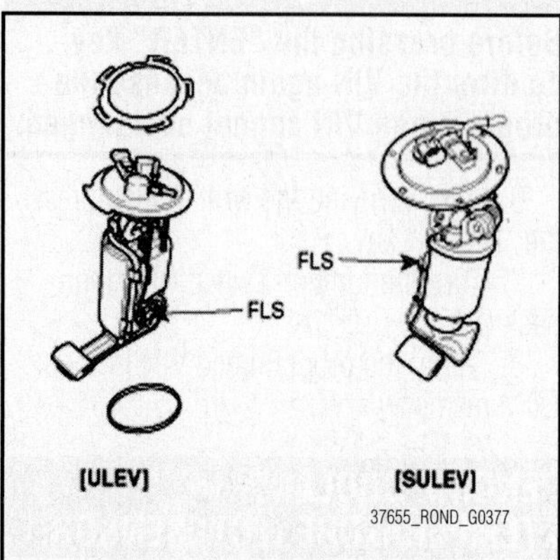

Fig. 164 ETC module (including TPS & ETC motor)—2.4L Engine

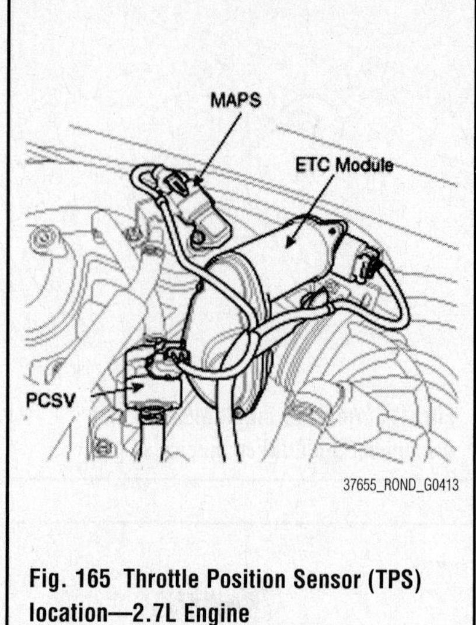

Fig. 165 Throttle Position Sensor (TPS) location—2.7L Engine

Refer to the accompanying illustrations.

FUEL

GASOLINE FUEL INJECTION SYSTEM

FUEL SYSTEM SERVICE PRECAUTIONS

Safety is the most important factor when performing not only fuel system maintenance, but any type of maintenance. Failure to conduct maintenance and repairs in a safe manner may result in serious personal injury or death. Accomplish work on a vehicle's fuel system components safely and effectively by adhering to the following rules and guidelines:

- Avoid the possibility of fire and personal injury: always disconnect the negative battery cable unless the repair or test procedure requires that battery voltage be applied.
- Always relieve the fuel system pressure prior to disconnecting any fuel system component (injector, fuel rail, pressure regulator, etc.) fitting or fuel line connection. Exercise extreme caution whenever relieving fuel system pressure to avoid exposing skin, face and eyes to fuel spray. Fuel under pressure may penetrate the skin or any part of the body that it contacts.
- Always place a shop towel or cloth around the fitting or connection prior to loosening to absorb any excess fuel due to spillage. Ensure that all fuel spillage is quickly removed from engine surfaces. Ensure that all fuel-soaked cloths or towels are deposited into a flame-proof waste container with a lid.
- Always keep a dry chemical (Class B) fire extinguisher near the work area.
- Do not allow fuel spray or fuel vapors to come into contact with a spark or open flame.
- Always use a second wrench when loosening or tightening fuel line connection fittings. This will prevent unnecessary stress and torsion on fuel piping. Always follow the proper torque specifications.
- Always replace worn fuel fitting O-rings with new ones.
- Do not substitute fuel hose where rigid pipe is installed.

FUEL SYSTEM PRESSURE

RELIEVING

See Figure 166.

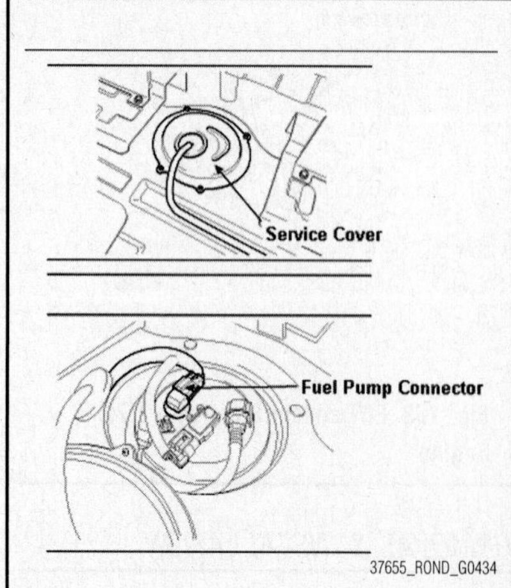

Fig. 166 Open the service cover and disconnect the fuel pump connector

1. Remove the third seat.
2. Open the service cover.
3. Disconnect the fuel pump connector.
4. Start the engine and wait until fuel in fuel line is exhausted.
5. After engine stalls, turn the ignition switch to the OFF position.
6. Disconnect the negative battery cable.

➡Reduce the fuel pressure before disconnecting the fuel feed hose, otherwise fuel will spill out.

FUEL PUMP ASSEMBLY

REMOVAL & INSTALLATION

See Figures 167 through 170.

1. Remove the third seat.
2. Open the service cover.
3. Disconnect the fuel pump connector.
4. Start the engine and wait until fuel in fuel line is exhausted.
5. After the engine stalls, turn the ignition switch to OFF position.
6. Disconnect the fuel feed tube quick-connector (A), the fuel tank pressure sensor connector (B), and the canister close valve connector (C).
7. Remove the rubber cover (D).
8. Unscrew the fuel pump plate cover (A) with the special service tool (SST No: 09310-2B200) and remove the fuel pump assembly.

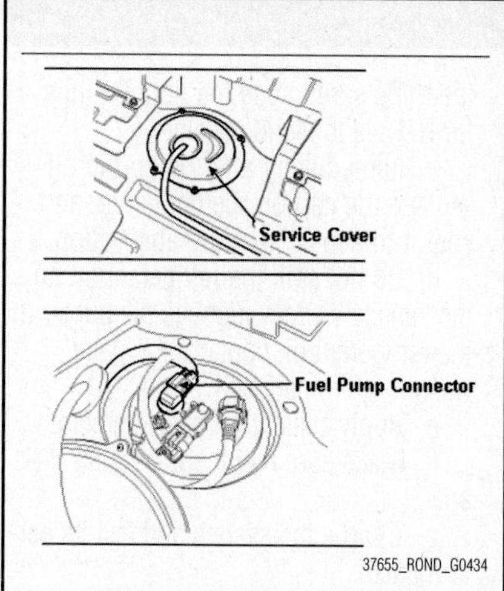

Fig. 167 Open the service cover and disconnect the fuel pump connector

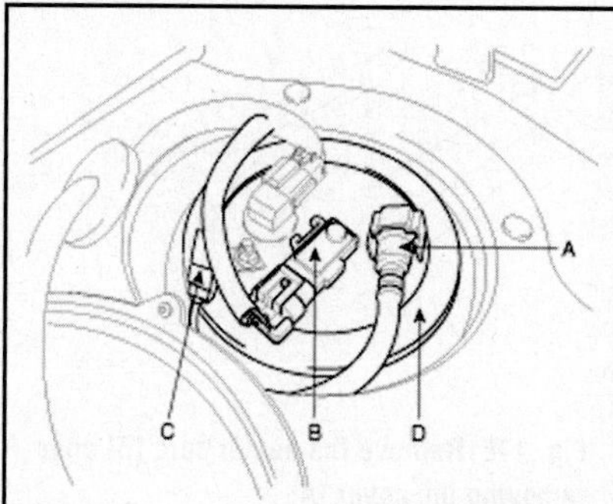

Fig. 168 Disconnect the fuel feed tube quick-connector (A), the fuel tank pressure sensor connector (B), and the canister close valve connector (C)

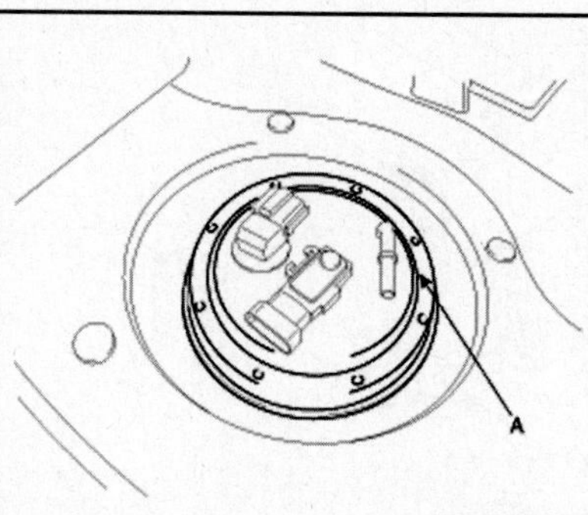

Fig. 169 Unscrew the fuel pump plate cover (A) with the special service tool (SST No.: 09310-2B200) and remove the fuel pump assembly

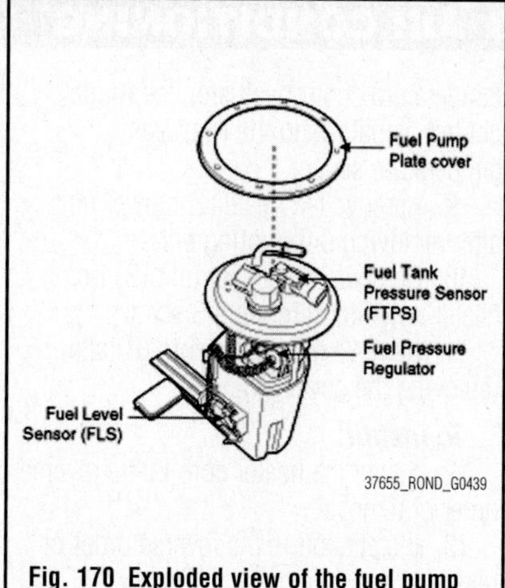

Fig. 170 Exploded view of the fuel pump

9. Installation is the reverse of the removal procedure, noting the following:
   - Fuel Pump Plate Cover Tightening Specifications: 58–72 ft. lbs. (78.5–98.1 Nm).

FUEL TANK

REMOVAL & INSTALLATION

See Figures 167, 168, 171 and 172.

1. Remove the third seat.
2. Open the service cover.
3. Disconnect the fuel pump connector.
4. Start the engine and wait until fuel in fuel line is exhausted.
5. After the engine stalls, turn the ignition switch to the OFF position.
6. Disconnect the fuel feed tube quick-connector (A), the fuel tank pressure sensor connector (B), and the canister close valve connector (C).
7. Raise and safely support the vehicle.
8. Remove the muffler assembly.
9. Support the fuel tank with a jack.
10. 2.7L Engine only: Disconnect the fuel filler hose, the ventilation hose, and the vacuum hose.
11. 2.4L Engine only: Disconnect the fuel filler hose and the vent hose.
12. Disconnect the vacuum hoses (A) from the canister (B).
13. Remove the upper arm (A), the lower arm (B), the assist arm (C), and the stabilizer bar (D) from the fuel tank & suspension member assembly.
14. Remove the mounting bolts (E) and nuts (F), and then remove the fuel tank & suspension member assembly from the vehicle.
15. Installation is reverse of removal. Fuel tank mounting bolts/nuts torque specification: 115.8–130.2 ft. lbs. (157–176.6 Nm)

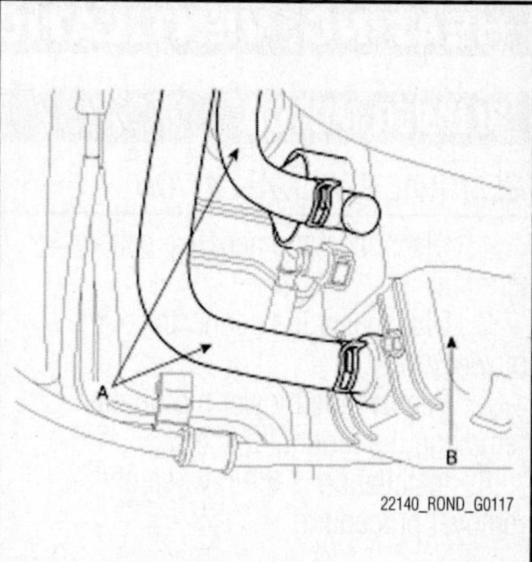

Fig. 171 Disconnect the vacuum hoses (A) from the canister (B)

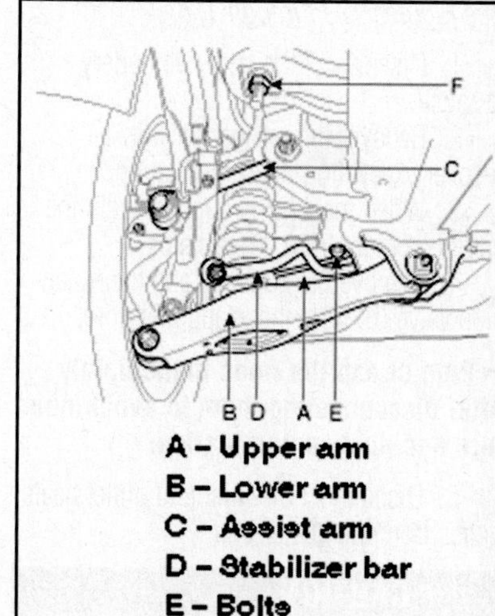

Fig. 172 Remove the upper arm (A), the lower arm (B), the assist arm (C), and the stabilizer bar (D)

HEATING, VENTILATION & AIR CONDITIONING

BLOWER MOTOR

REMOVAL & INSTALLATION

1. Disconnect the negative battery terminal.
2. Disconnect the connector of the blower motor.
3. Remove the blower motor after removing the mounting screws.
4. Installation is the reverse of the removal procedure.

HEATER CORE

REMOVAL & INSTALLATION

See Figures 173 through 176.

1. Disconnect the negative battery terminal.
2. Recover the refrigerant with a recovery/recycling/charging station.
3. When the engine is cool, drain the engine coolant from the radiator.
4. Remove the bolts (A) and the expansion valve (B) from the evaporator core.

➡Plug or cap the lines immediately after disconnecting them to avoid moisture and dust contamination.

5. Disconnect the inlet and outlet heater hoses from the heater unit.

⁂ CAUTION

Engine coolant will run out when the hoses are disconnected; drain it into a clean drip pan. Be sure not to let coolant spill on electrical parts or painted surfaces. If any coolant spills, rinse it off immediately.

6. Remove the crash pad (instrument panel).
7. Remove the cowl cross bar assembly.
8. Disconnect the connectors from the temperature control actuator, the mode control actuator and the evaporator temperature sensor.
9. Remove the heater & blower unit after removing 3 mounting bolts.
10. Remove the blower unit (B) from heater unit after removing 3 screws.
11. Remove the heater core (B) after removing the cover (A).

To install:

12. Install the heater core in the reverse order of removal.
13. Installation is the reverse order of removal, and note these items :

a. If you're installing a new evaporator, add refrigerant oil (ND-OIL8).
b. Replace the O-rings with new ones at each fitting, and apply a thin coat of refrigerant oil before installing them. Be sure to use the right O-rings for R-134a to avoid leakage.
c. Immediately after using the oil, replace the cap on the container, and seal it to avoid moisture absorption.
d. Do not spill the refrigerant oil on the vehicle ; it may damage the paint ; if the refrigerant oil contacts the paint, wash it off immediately.
e. Apply sealant to the grommets.
f. Make sure that there is no air leakage.
g. Charge the system and test its performance.
h. Do not interchange the inlet and outlet heater hoses and install the hose clamps securely.
i. Refill the cooling system with engine coolant.

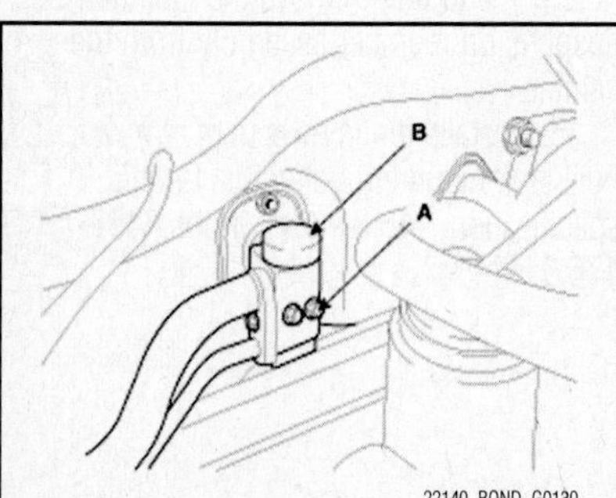

Fig. 173 Remove the bolts (A) and the expansion valve (B) from the evaporator core

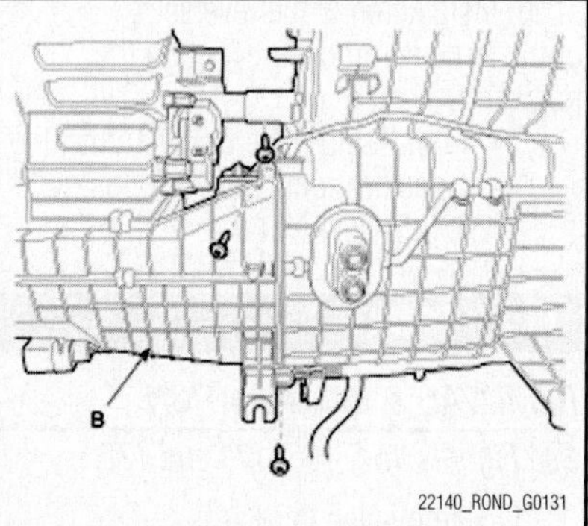

Fig. 174 Remove the blower unit (B) from heater unit

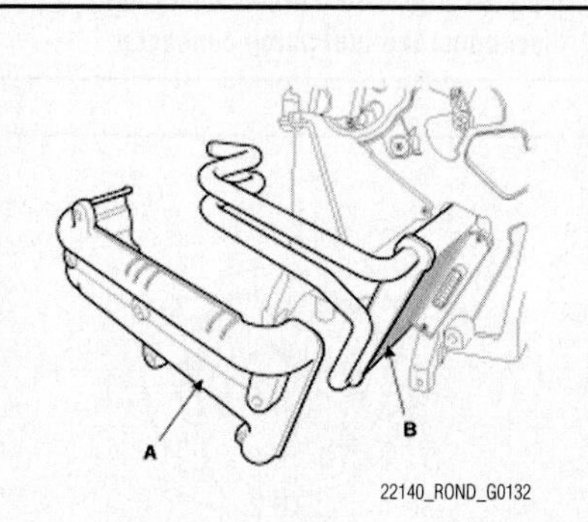

Fig. 175 Remove the heater core (B) after removing the cover (A)

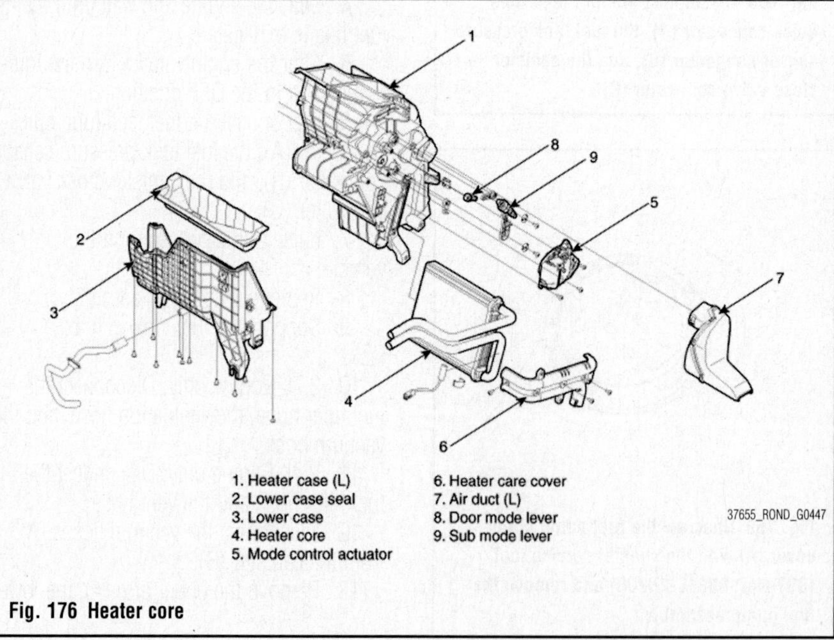

Fig. 176 Heater core

STEERING

POWER RECIRCULATING BALL STEERING GEAR

REMOVAL & INSTALLATION

See Figures 177 through 188.

1. Safely raise and support the vehicle.
2. Remove both front wheels.
3. Drain the power steering fluid by disconnecting the return hose (A).
4. Disconnect the pressure tube (A) from the power steering pump by removing the eye bolt.
5. Disconnect the stabilizer link (A) with the strut assembly by removing the nut.
6. Remove the split pin and castle nut and then disconnect the tie-rod end with the knuckle using a SST (09568-34000).
7. Separate the knuckle with the lower arm by removing the mounting bolts (A).
8. Loosen the bolt and disconnect the steering universal joint assembly with the pinion of the steering gear.

CAUTION

When you handle the steering wheel, keep it in the neutral-range to prevent damage of the clockspring inner cable.

9. Remove the muffler rubber hanger.
10. Remove the front and rear roll stopper through bolts & nuts (A, B).
11. Remove the sub-frame (A) and sub-frame stay by loosening the mounting bolts and nuts.
12. Remove the return hose mounting bracket bolt (A).
13. Remove the rear roll stopper (A) from the sub-frame by removing the mounting bolts.
14. Remove the heat protector (A).

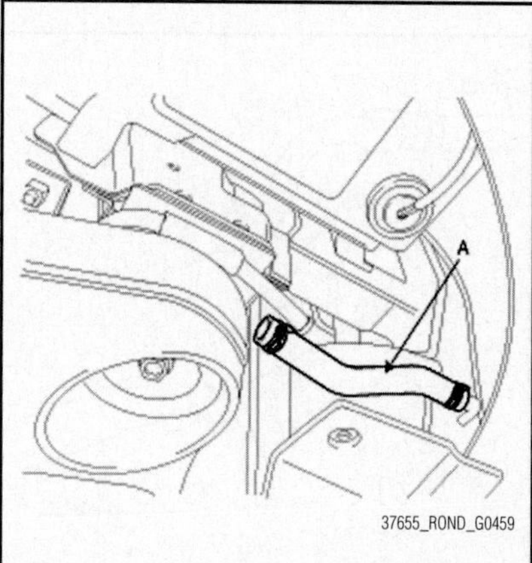

Fig. 177 Drain the power steering fluid by disconnecting the return hose (A)

15. Remove the pressure (A) and return tube (B) from the steering gear box.
16. Remove the steering gear box from the sub-frame by removing the mounting bolts.

To install:

17. Install the steering gear box to the sub frame by tightening the mounting bolts. Torque to 43–58 ft. lbs. (60–80 Nm).

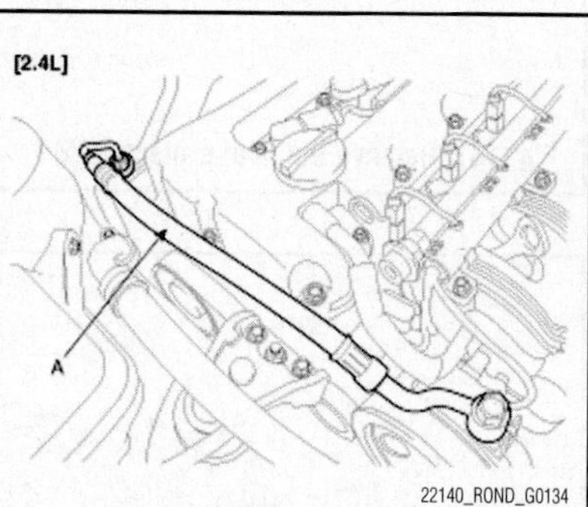

Fig. 178 Disconnect the pressure tube (A) from the power steering pump–2.4L engine

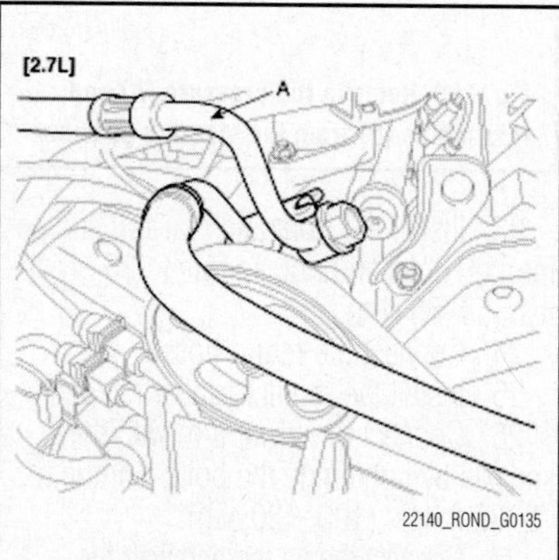

Fig. 179 Disconnect the pressure tube (A) from the power steering pump–2.7L engine

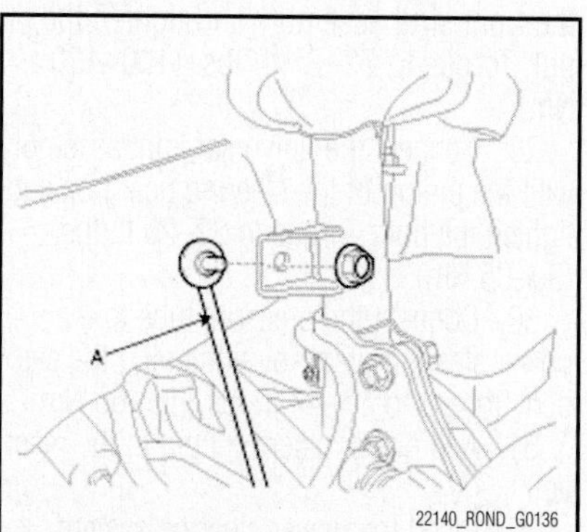

Fig. 180 Disconnect the stabilizer link (A) with the strut assembly

18. Install the pressure tube and return tube to the steering gear box and tighten the flare nut. Torque to 9–13 ft. lbs. (12–18 Nm).
19. Install the heat protector.
20. Install the rear roll stopper to the sub-frame by tightening the mounting bolts. Torque to 36–47 ft. lbs. (50–65 Nm).

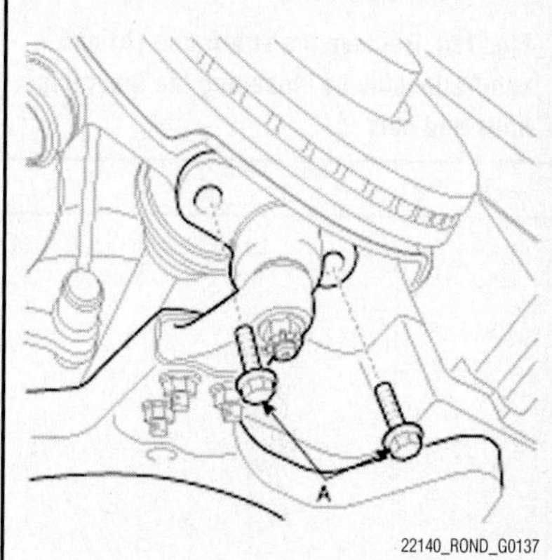

Fig. 181 Separate the knuckle with the lower arm by removing the mounting bolts (A)

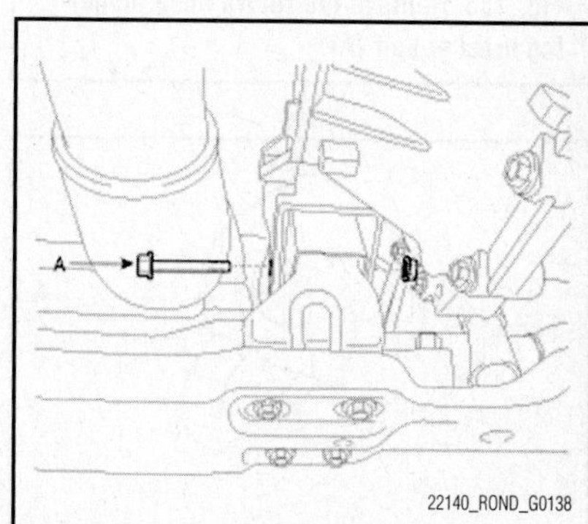

Fig. 182 Remove the front roll stopper through bolts & nuts (A)

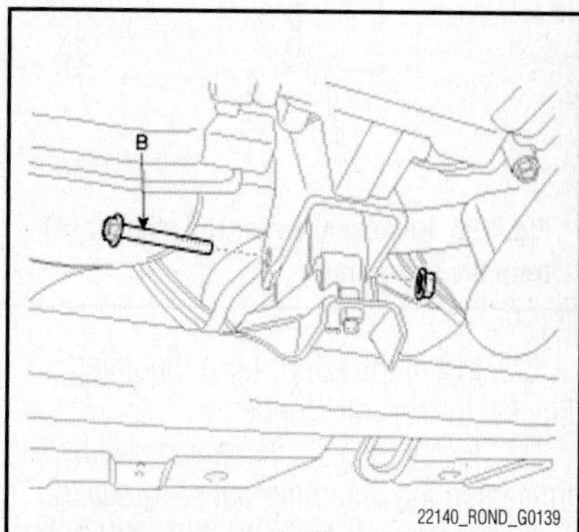

Fig. 183 Remove the rear roll stopper through bolts & nuts (B)

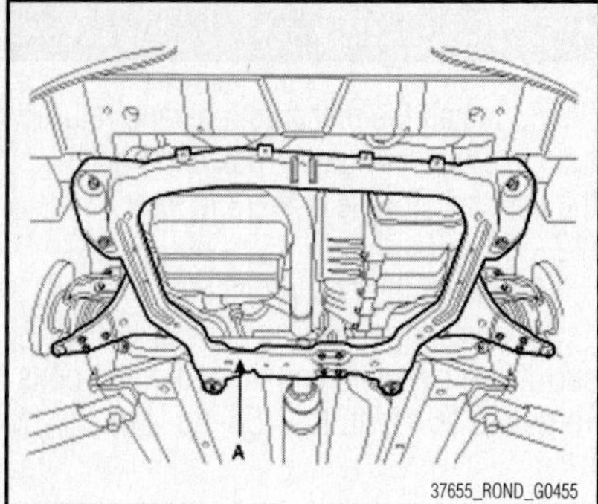

Fig. 184 Remove the sub-frame (A) and sub-frame stay by loosening the mounting bolts and nuts

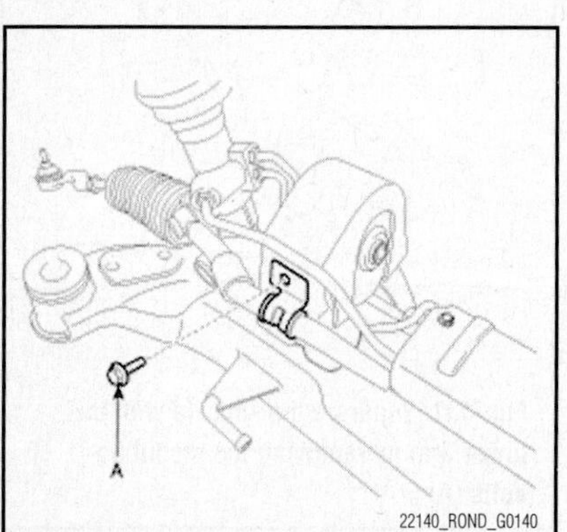

Fig. 185 Remove the return hose mounting bracket bolt (A)

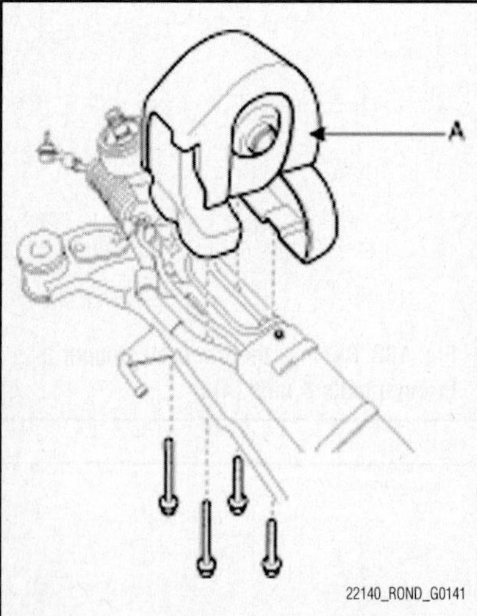

Fig. 186 Remove the rear roll stopper (A) from the sub-frame

21. Install the return hose mounting bracket to rear roll stopper.

22. Install the sub-frame and stay by tightening the mounting bolts. Torque to Sub-frame mounting bolts: 116–130 ft. Lbs. (160–180 Nm); Sub-frame stay mounting bolts: 33–40 ft. lbs. (45–55 Nm).

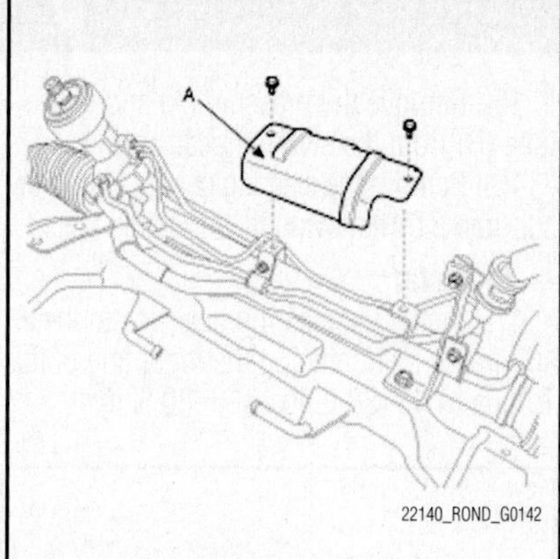

Fig. 187 Remove the heat protector (A)

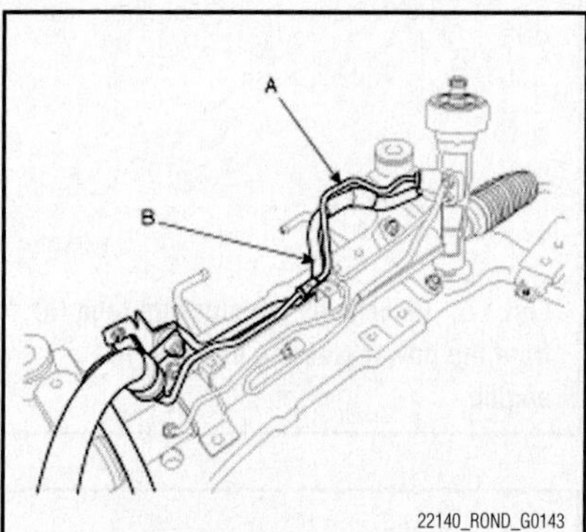

Fig. 188 Remove the pressure (A) and return tube (B) from the steering gear box

23. Install the front and rear roll stopper through bolts and nut. Torque to 36–47 ft. lbs. (50–65 Nm).

24. Connect the return hose.

25. Install the muffler rubber hanger.

26. Connect the lower arm with the knuckle by tightening the bolts. Torque to 72–87 ft. lbs. (100–120 Nm).

27. Connect the tie rod end with the knuckle and then install the castle nut and split pin. Torque to 17–25 ft. lbs. (24–34 Nm).

28. Connect the stabilizer link with the front strut assembly and tighten the nut. Torque to 72–87 ft. lbs. (100–120 Nm).

29. Connect the universal joint assembly with the pinion of the steering gear box and tighten the bolt. Torque to 22–25 ft. lbs. (30–35 Nm).

30. Connect the pressure tube to the power steering pump by tightening the eye bolt. Torque to 40–47 ft. lbs. (55–65 Nm).

31. Add power steering fluid to the reservoir.

32. Bleed the power steering system.

33. Check and adjust the front wheel alignment.

POWER STEERING PUMP

REMOVAL & INSTALLATION

See Figures 189 through 192.

1. Remove the accessory drive belt.
2. Disconnect the pressure tube (A) and return hose (B) from the power steering pump.
3. Loosen the mounting bolts and remove the power steering pump.

To install:

4. Install the power steering pump and tighten the mounting bolts. Tighten as follows:
   a. 2.4L engine: Torque to 12–19 ft. lbs. (17–26 Nm).
   b. 2.7L engine: Torque to 25–36 ft. lbs. (35–50 Nm).
5. Connect the suction hose with the suction pipe.
6. Connect the pressure tube to the power steering pump by installing the eye bolt and tighten to 36–47 ft. lbs. (55–65 Nm).
7. Install the accessory drive belt.

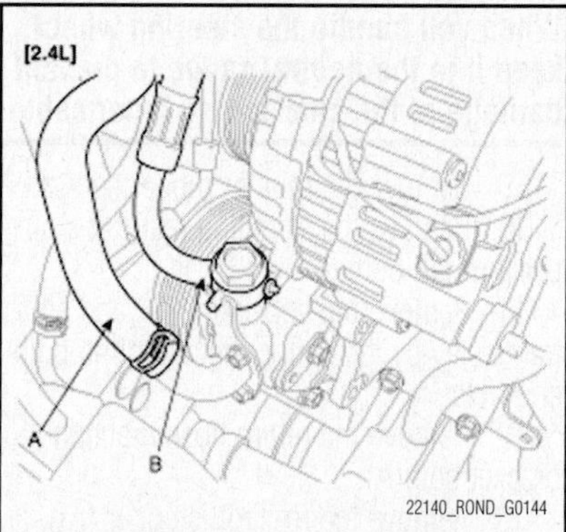

Fig. 189 Disconnect the pressure tube (A) and return hose (B) from the power steering pump—2.4L engine

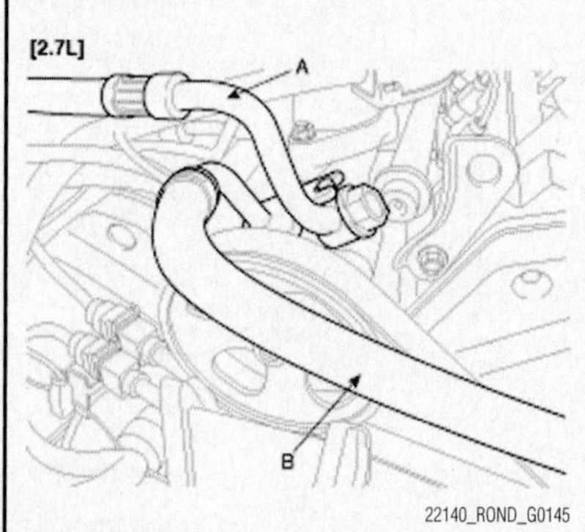

Fig. 190 Disconnect the pressure tube (A) and return hose (B) from the power steering pump—2.7L engine

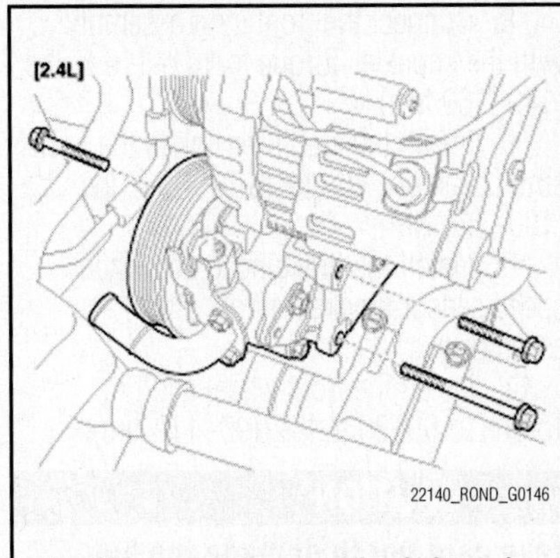

Fig. 191 Loosen the mounting bolts and remove the power steering pump—2.4L engine

8. Add power steering fluid to the reservoir.
9. Bleed the power steering system.
10. Check the power steering pump relief pressure.

BLEEDING

1. Jack up the front wheels.
2. Remove the fuel pump fuse from the fuse box.
3. Start the engine and wait for the engine to stall.

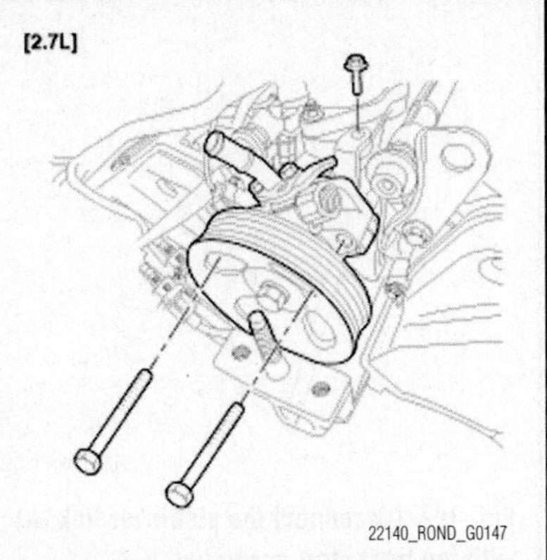

Fig. 192 Loosen the mounting bolts and remove the power steering pump—2.7L engine

4. Fill the reservoir with power steering fluid up to the upper position of the filter.

**** CAUTION**

While performing following steps, replenish the fluid so that the level never falls below the lower position of the filter.

5. Turn the steering wheel all the way to the left and then to the right 5 to 6 times while cranking (for 15 to 20 seconds).
6. Reinstall the fuel pump fuse and then start the engine.
7. Turn the steering wheel to the left and to the right until there are no air bubbles in the reservoir.

**** CAUTION**

Do not hold the steering wheel turned all the way to either side for more than ten seconds.

8. Confirm that the fluid is not milky, and that the level is up to the position specified on the level gauge.
9. Confirm that there is little change in the surface of the fluid when the steering wheel is turned left and right.

**** CAUTION**

If the surface of the fluid changes considerably, air bleeding should be done again. If the fluid level rises suddenly when the engine is stopped, it indicates that there is still air in the system. If there is air in the system, a jingling noise may be heard from the pump and the control valve may also produce unusual noises. Air in the system will shorten the life of the pump and other parts.

SUSPENSION

FRONT SUSPENSION

LOWER BALL JOINT

REMOVAL & INSTALLATION

1. Raise and safely support the front of the vehicle.
2. Remove the front wheel & tire.

**** CAUTION**

Be careful not to damage the hub bolts when removing the front wheel & tire.

3. Remove the split pin and the castle nut from the lower arm ball joint.
4. Separate the lower arm from the lower arm ball joint by using SST (09568-34000).
5. Remove the lower arm from the subframe.

To install:

6. Install the front lower arm to the subframe. Tighten the bolts to 101–116 ft. lbs. (140–160 Nm); and the bolt & nut to 72–87 ft. lbs. (100–120 Nm).
7. Connect the lower arm with the ball joint. Torque to 58–65 ft. lbs. (80–90 Nm).
8. Install the front wheel & tire. Torque to 65–80 ft. lbs. (90–110 Nm).

LOWER ARM

REMOVAL & INSTALLATION

See Figures 193 through 195.

1. Raise and safely support the front of the vehicle.
2. Remove the front wheel & tire.

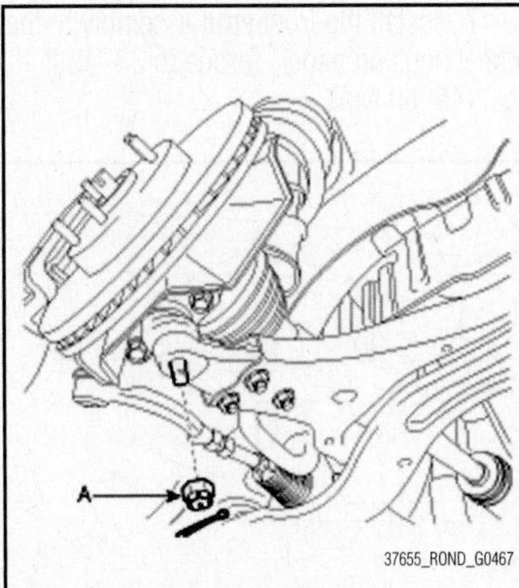

Fig. 193 Remove the split pin and the castle nut (A) from the lower arm ball joint

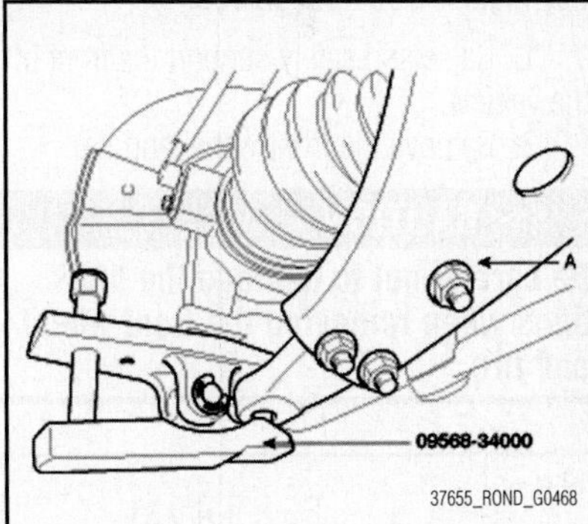

Fig. 194 Separate the lower arm from the lower arm (A) ball joint using the special service tool (SST; 09568-34000)

**** CAUTION**

Be careful not to damage the hub bolts when removing the front wheel and tire.

3. Remove the split pin and the castle nut (A) from the lower arm ball joint.
4. Separate the lower arm from the lower arm (A) ball joint using the special service tool (SST; 09568-34000).

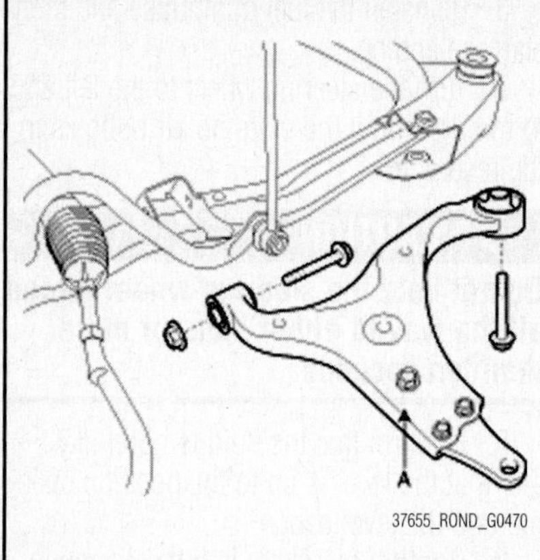

Fig. 195 Remove the lower arm (A) from the sub-frame

5. Remove the lower arm (A) from the sub-frame.

To install:

6. Install the front lower arm to the sub-frame. Tighten the bolts to 101–116 ft. lbs. (140–160 Nm); and the bolt & nut to 72–87 ft. lbs. (100–120 Nm).

7. Connect the lower arm with the ball joint. Torque to 58–65 ft. lbs. (80–90 Nm).

8. Install the front wheel & tire. Torque to 65–80 ft. lbs. (90–110 Nm).

MACPHERSON STRUT

REMOVAL & INSTALLATION

See Figures 196 through 199.

1. Raise and safely support the front of the vehicle.

2. Remove the front wheel and tire.

CAUTION

Be careful not to damage the hub bolts when removing the front wheel and tire.

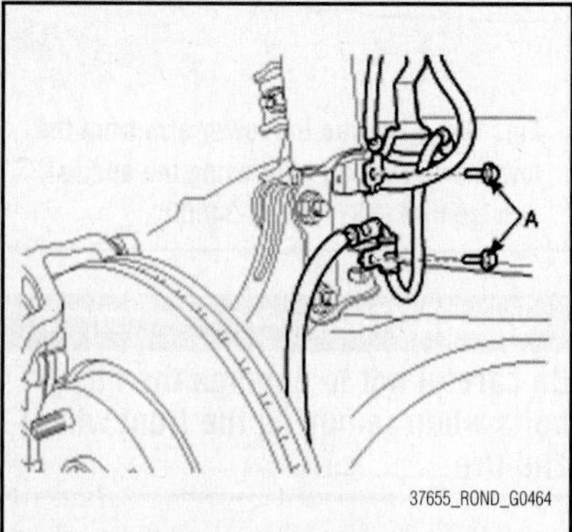

Fig. 196 Remove the brake hose and the wheel speed sensor bracket from the front strut assembly by loosening the mounting bolts (A)

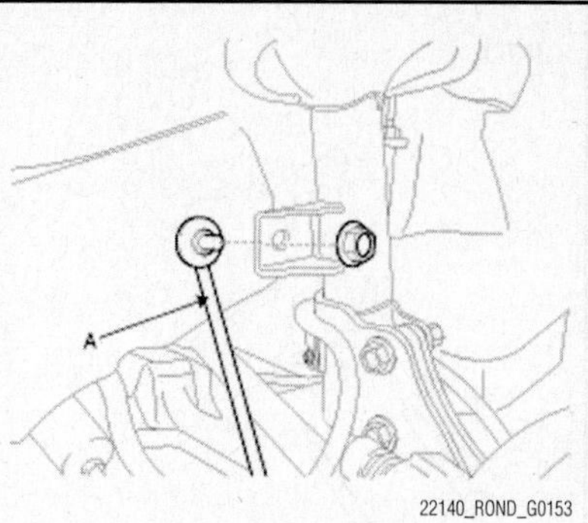

Fig. 197 Disconnect the stabilizer link (A) with the front strut assembly

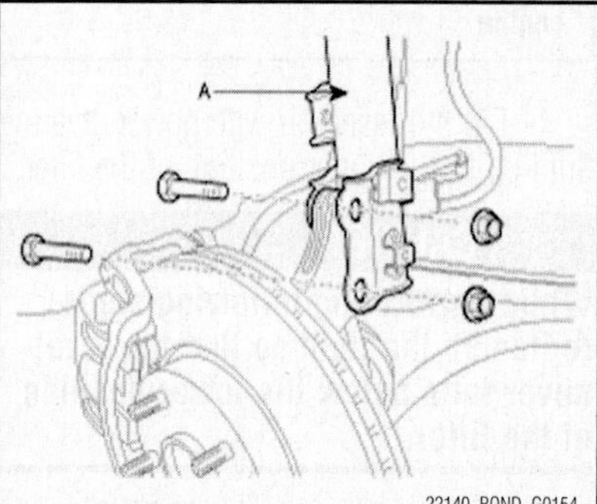

Fig. 198 Disconnect the front strut assembly (A) from the knuckle

3. Remove the brake hose and the wheel speed sensor bracket from the front strut assembly by loosening the mounting bolts (A).

4. Disconnect the stabilizer link (A) with the front strut assembly.

5. Disconnect the front strut assembly (A) from the knuckle.

6. Remove the front strut assembly from the wheel housing panel by loosening the mounting nuts (A).

To install:

7. Install the front strut assembly to the wheel housing panel. Torque to 33–43 ft. lbs. (45–60 Nm).

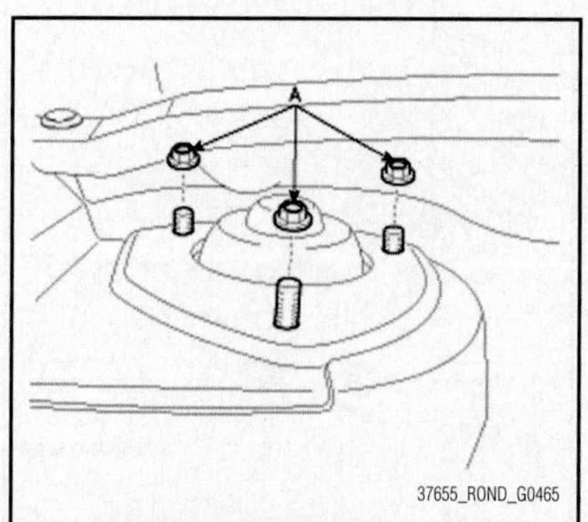

Fig. 199 Remove the front strut assembly from the wheel housing panel by loosening the mounting nuts (A)

8. Connect the front strut assembly with the knuckle. Torque to 101–116 ft. lbs. (140–160 Nm).

9. Install the stabilizer link to the front strut assembly. Torque to 72–87 ft. lbs. (100–120 Nm).

10. Install the brake hose and wheel speed sensor bracket to front strut assembly.

11. Install the front wheel and tire. Torque to 65–80 ft. lbs. (90–110 Nm).

CAUTION

Take care not to damage the hub bolts when installing the front wheel and tire.

STABILIZER BAR

REMOVAL & INSTALLATION

See Figures 200 through 209.

1. Raise and safely support the front of the vehicle.

2. Remove the front wheel and tire.

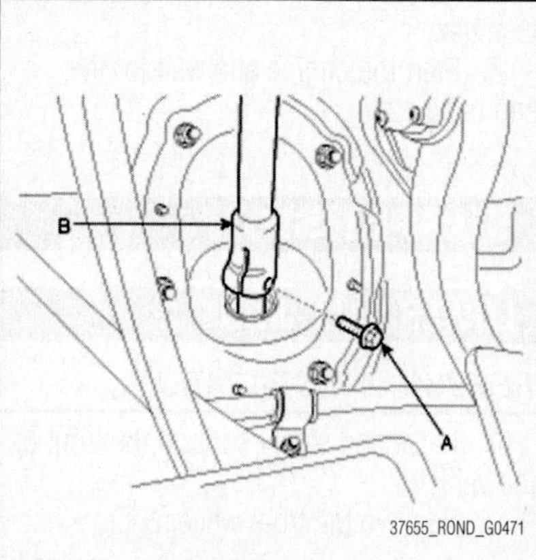

Fig. 200 Loosen the bolt (A) and then disconnect the universal joint assembly (B) from the pinion of the steering gear box

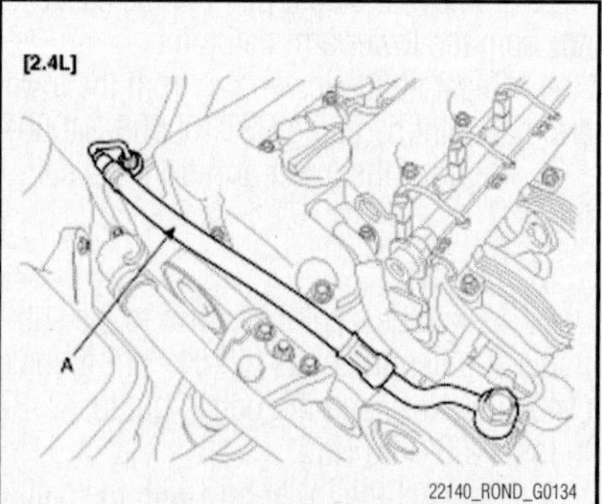

Fig. 201 Disconnect the pressure tube (A) from the power steering pump–2.4L engine

⁂ CAUTION

Take care not to damage the hub bolts when removing the front wheel and tire.

3. Loosen the bolt (A) and then disconnect the universal joint assembly (B) from the pinion of the steering gear box.

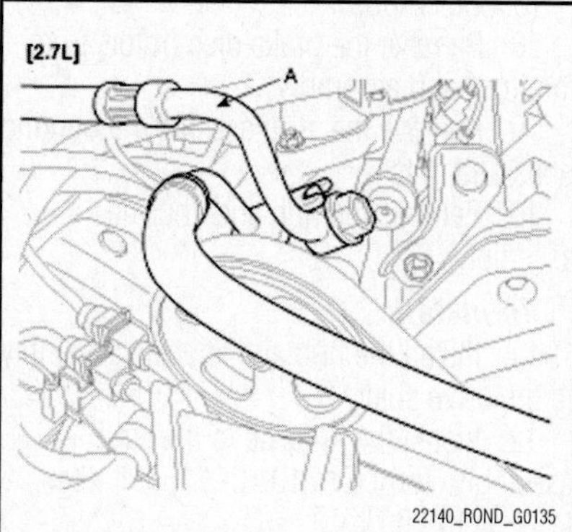

Fig. 202 Disconnect the pressure tube (A) from the power steering pump–2.7L engine

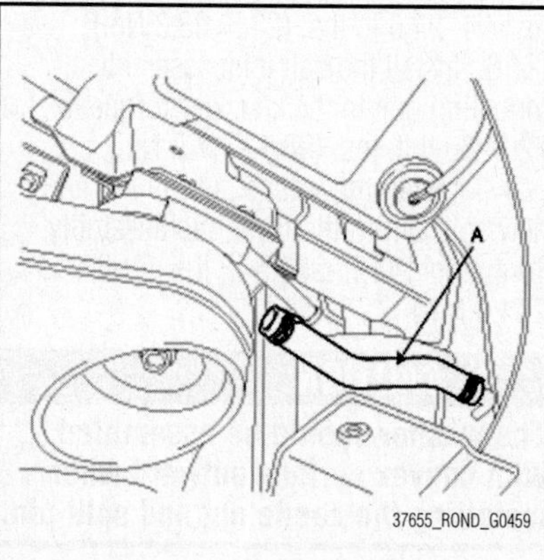

Fig. 203 Drain the power steering fluid by disconnecting the return hose (A)

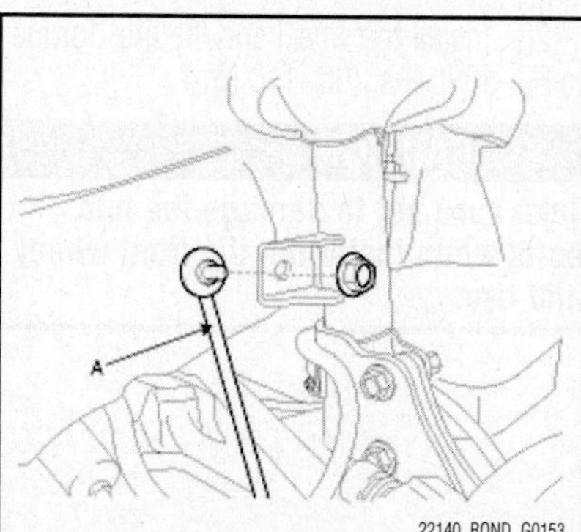

Fig. 204 Disconnect the stabilizer link (A) from the front strut assembly

⁂ CAUTION

Prevent damage to the clockspring inner cable when you handle the steering wheel: Keep it in the neutral range.

4. Disconnect the pressure tube (A) from the power steering pump by removing the eye bolt.

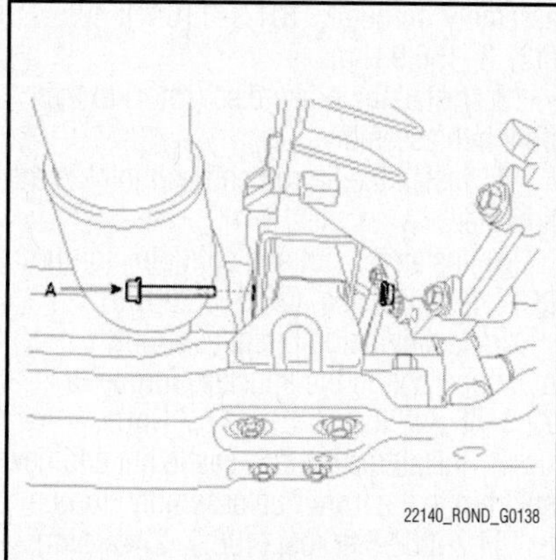

Fig. 205 Remove the front roll stopper through bolts and nuts (A)

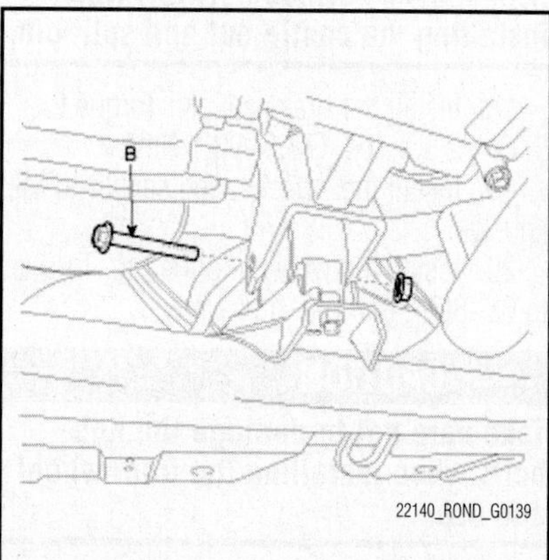

Fig. 206 Remove the rear roll stopper through bolts and nuts (B)

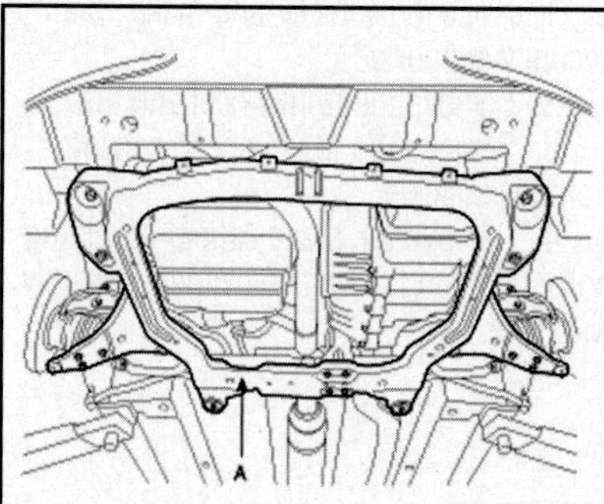

Fig. 207 Remove the sub-frame (A) and sub-frame stay by loosening the mounting bolts and nuts

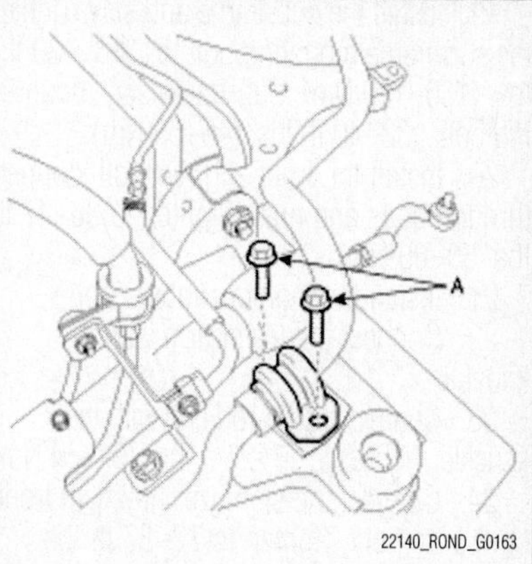

Fig. 208 Remove the stabilizer bar from the sub-frame

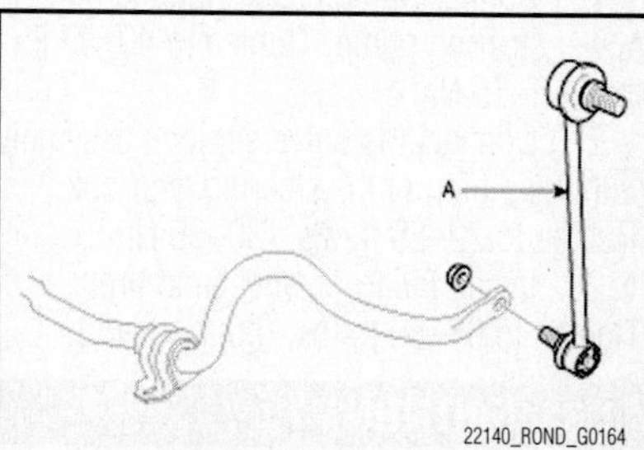

Fig. 209 Disconnect the stabilizer link (A) with the stabilizer bar

5. Drain the power steering fluid by disconnecting the return hose (A).
6. Disconnect the stabilizer link (A) from the front strut assembly
7. Disconnect the tie-rod end from the front knuckle using the special service tool (SST; 09568-34000).
8. Disconnect the front lower arm from the knuckle.
9. Remove the muffler rubber hanger.
10. Remove the front and rear roll stopper bolts and nuts (A, B).
11. Remove the sub-frame and sub-frame stay.
12. Remove the sub-frame (A) and sub-frame stay by loosening the mounting bolts and nuts.
13. Remove the stabilizer bar from the sub-frame.
14. Disconnect the stabilizer link (A) with the stabilizer bar.
15. Remove the bushing and the bracket from the stabilizer bar.

To install:

16. Install the bushing and the bracket to the stabilizer bar.
17. Connect the stabilizer link with the stabilizer bar. Tighten to 72–87 ft. lbs. (100–120 Nm).
18. Install the stabilizer to the sub-frame. Tighten to 33–40 ft. lbs. (45–55 Nm).

19. Install the sub-frame and stay. Tighten the sub-frame mounting bolt to: 116–130 ft. lbs. (160–180 Nm); Sub-frame stay mounting bolts: 33–40 ft. lbs. (45–55 Nm).
20. Install the front and rear roll stopper through bolts and nuts. Tighten to 36–47 ft. lbs. 50–65 Nm).
21. Install the muffler rubber hanger.
22. Connect the lower arm with the knuckle
23. Connect the tie-rod end with the knuckle. Tighten to 17–25 ft. lbs. (24–34 Nm).
24. Connect the stabilizer link with front strut assembly. Tighten to 72–87 ft. lbs. (100–120 Nm).
25. Connect the power steering return hose.
26. Connect the pressure tube to the power steering pump. Tighten to 40–47 ft. lbs. (55–65 Nm).
27. Connect the universal joint assembly with the pinion of the steering gear box. Tighten to 22–25 ft. lbs. (30–35 Nm).
28. Install the front wheel and tire. Tighten to 65–80 ft. lbs. (90–110 Nm).

CAUTION

Take care not to damage the hub bolts when installing the front wheel and tire.

29. Bleed the power steering system.

STEERING KNUCKLE

REMOVAL & INSTALLATION

1. Raise the vehicle, and make sure it is securely supported.
2. Remove the front wheel and tire.
3. Remove the wheel speed sensor from the knuckle.
4. Remove the brake caliper mounting bolts, and hang the brake caliper assembly with wire.
5. Remove the split pin, castle nut and washer from the front hub
6. Remove the ball joint assembly mounting bolt from the knuckle.
7. Remove the tie rod end ball joint from the knuckle.
   a. Remove the split pin and castle nut.
   b. Disconnect the ball joint from knuckle by using the special tool (09568-4A000).
8. Remove the brake disc (rotor) from the front hub assembly
9. Remove the strut assembly mounting bolts and nuts.
10. Remove the hub and knuckle assembly.

To install:

11. Install the hub and knuckle assembly to the drive shaft.
12. Install the knuckle to the strut assembly. Torque to 101.3–115.7 ft. lbs. (137.3–156.9 Nm).
13. Install the brake disc (rotor) to the front hub assembly.
14. Install the tie rod end ball joint to the knuckle.
15. Install the nut and split pin. Torque to 17.4–24.6 ft. lbs. (23.5–33.3 Nm).
16. Install the ball joint assembly mounting bolt to the knuckle. Torque to 72.3–86.8 ft. lbs. (98.1–117.7 Nm).
17. Install the washer, castle nut and new split pin to the front hub assembly. Torque to 144.7–202.5 ft. lbs. (196.1–274.6 Nm).

CAUTION

The washer should be assembled with convex surface outward when installing the castle nut and split pin.

18. Install the brake caliper. Torque to 57.9–72.3 ft. lbs. (78.5–98.1 Nm).
19. Install the wheel speed sensor to the knuckle.
20. Install the wheel and the tire. Torque to 65–80 ft. lbs. (90–110 Nm).

CAUTION

Take care not to damage the hub bolts when installing the front wheel and tire.

WHEEL HUB & BEARING

REMOVAL & INSTALLATION

1. Raise the vehicle, and make sure it is securely supported.
2. Remove the front wheel and tire.
3. Remove the wheel speed sensor from the knuckle.
4. Remove the brake caliper mounting bolts, and hang the brake caliper assembly with wire.
5. Remove the split pin, castle nut and washer from the front hub
6. Remove the ball joint assembly mounting bolt from the knuckle.
7. Remove the tie rod end ball joint from the knuckle.
   a. Remove the split pin and castle nut.
   b. Disconnect the ball joint from knuckle by using the special tool (09568-4A000).
8. Remove the brake disc (rotor) from the front hub assembly
9. Remove the strut assembly mounting bolts and nuts.
10. Remove the hub and knuckle assembly.

To install:

11. Install the hub and knuckle assembly to the drive shaft.
12. Install the knuckle to the strut assembly. Torque to 101.3–115.7 ft. lbs. (137.3–156.9 Nm).
13. Install the brake disc (rotor) to the front hub assembly.
14. Install the tie rod end ball joint to the knuckle.
15. Install the nut and split pin. Torque to 17.4–24.6 ft. lbs. (23.5–33.3 Nm).
16. Install the ball joint assembly mounting bolt to the knuckle. Torque to 72.3–86.8 ft. lbs. (98.1–117.7 Nm).
17. Install the washer, castle nut and new split pin to the front hub assembly. Torque to 144.7–202.5 ft. lbs. (196.1–274.6 Nm).

CAUTION

The washer should be assembled with convex surface outward when installing the castle nut and split pin.

18. Install the brake caliper. Torque to 57.9–72.3 ft. lbs. (78.5–98.1 Nm).
19. Install the wheel speed sensor to the knuckle.
20. Install the wheel and the tire. Torque to 65–80 ft. lbs. (90–110 Nm).

CAUTION

Take care not to damage the hub bolts when installing the front wheel and tire.

ASSIST ARM

REMOVAL & INSTALLATION

See Figures 210 and 211.

1. Raise the vehicle, and make sure it is securely supported.
2. Remove the rear wheel and tire.
3. Remove the cam bolt (A), split pin, and castle nut (B).
4. Disconnect the ball joint of the rear assist arm (A) with the carrier assembly with the SST (09568-34000).

To install:

5. Install the rear assist arm. Torque: 58–72 ft. lbs. (80–100 Nm).
6. Install the castle nut and split pin to the rear assist arm ball joint. Torque: 33–40 ft. lbs. (45–55 Nm).
7. Install the wheel and the tire. Torque: 65–80 ft. lbs. (90–110 Nm).

⁂ CAUTION

Take care not to damage the hub bolts when installing the front wheel and tire.

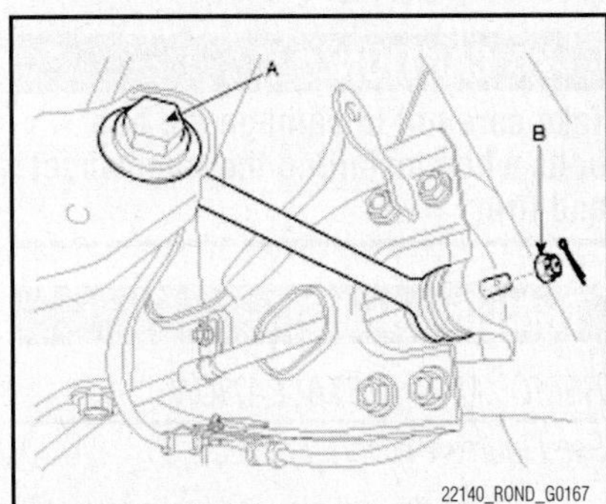

Fig. 210 Remove the cam bolt (A), split pin, and castle nut (B)

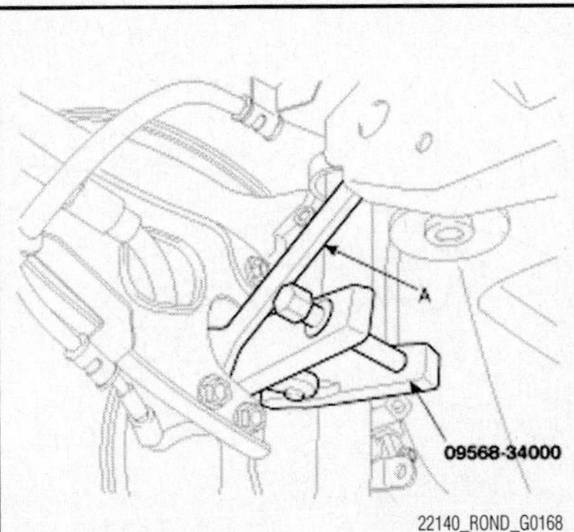

Fig. 211 Disconnect the ball joint of the rear assist arm (A) with the carrier assembly with the SST (09568-34000)

LOWER CONTROL ARM

REMOVAL & INSTALLATION

See Figures 212 and 213.

1. Raise the vehicle, and make sure it is securely supported.
2. Remove the rear wheel and tire.
3. Support the lower portion of the rear lower arm (A) with a jack.
4. Temporarily loosen the bolt (B) holding the cross member to the rear lower arm. Do not remove it.
5. Remove the bolt & nut (C) holding the rear lower arm to the carrier assembly.
6. Lower the jack and remove the coil spring (A) and the spring pad.
7. Remove the rear lower arm (C) from the cross member

To install:

8. Connect the rear lower arm with the cross member and then temporarily tighten the bolt.
9. Install the coil spring and support the lower portion of the rear lower arm with a jack.
10. Adjust height of the jack to place the bolt holding rear lower arm and carrier assembly through the mating holes.
11. Tighten the bolt and nut. Torque: 101–116 ft. lbs. (140–160 Nm).
12. Install the wheel and the tire. Torque: 65–80 ft. lbs. (90–110 Nm).

⁂ CAUTION

Take care not to damage the hub bolts when installing the front wheel and tire.

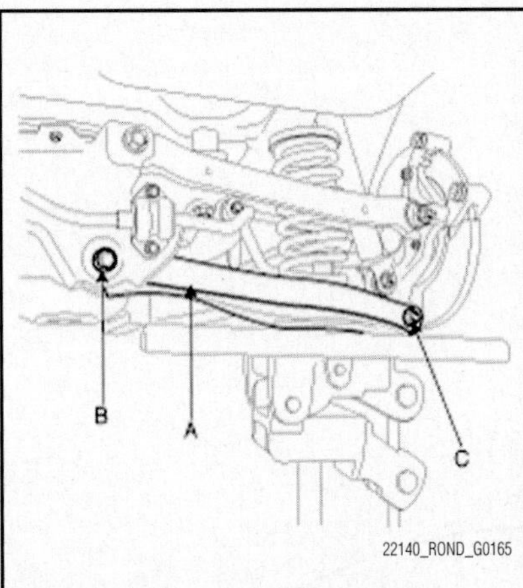

Fig. 212 Support the lower portion of the rear lower arm (A) with a jack

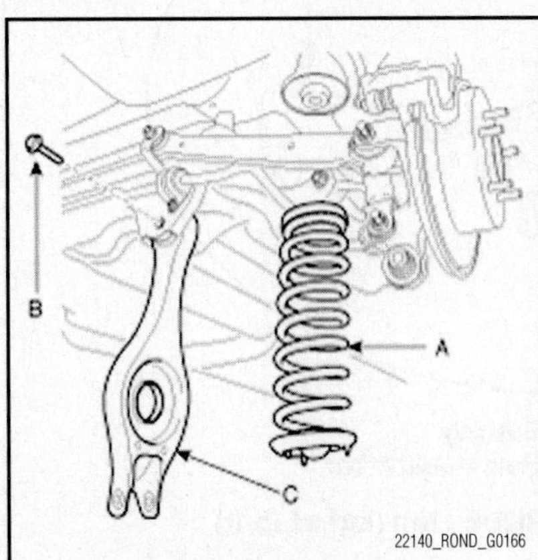

Fig. 213 Lower the jack and remove the coil spring (A) and the spring pad

SHOCK ABSORBER

REMOVAL & INSTALLATION

See Figures 212 and 214.

1. Raise the vehicle, and make sure it is securely supported.
2. Remove the rear wheel & tire.
3. Support the lower portion of the rear lower arm (A) with a jack.
4. Remove the bolt & nut holding the rear shock absorber to the carrier assembly.

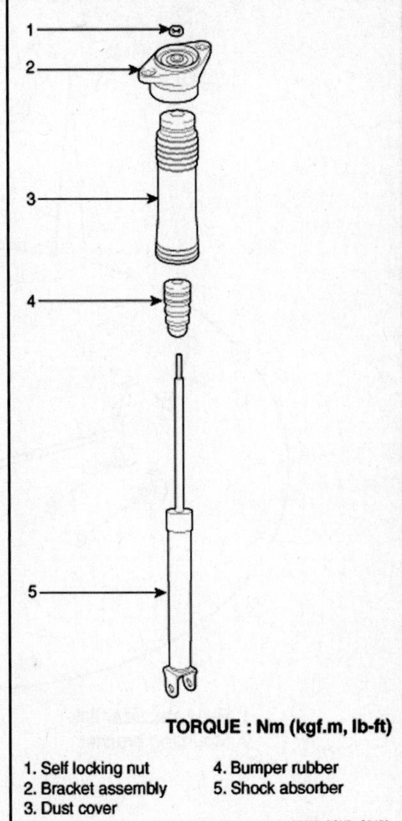

Fig. 214 Shock absorber

5. Remove the rear shock absorber from the wheel housing by removing the mounting bolts.

To install:

6. Install the rear shock absorber to the wheel housing. Torque to 33–40 ft. lbs. (45–55 Nm).

7. Connect the rear shock absorber to the carrier assembly. Torque to 101–116 ft. lbs. (140–160 Nm).

8. Install the wheel and the tire. Torque to 65–80 ft. lbs. (90–110 Nm).

**** CAUTION**

Take care not to damage the hub bolts when installing the front wheel and tire.

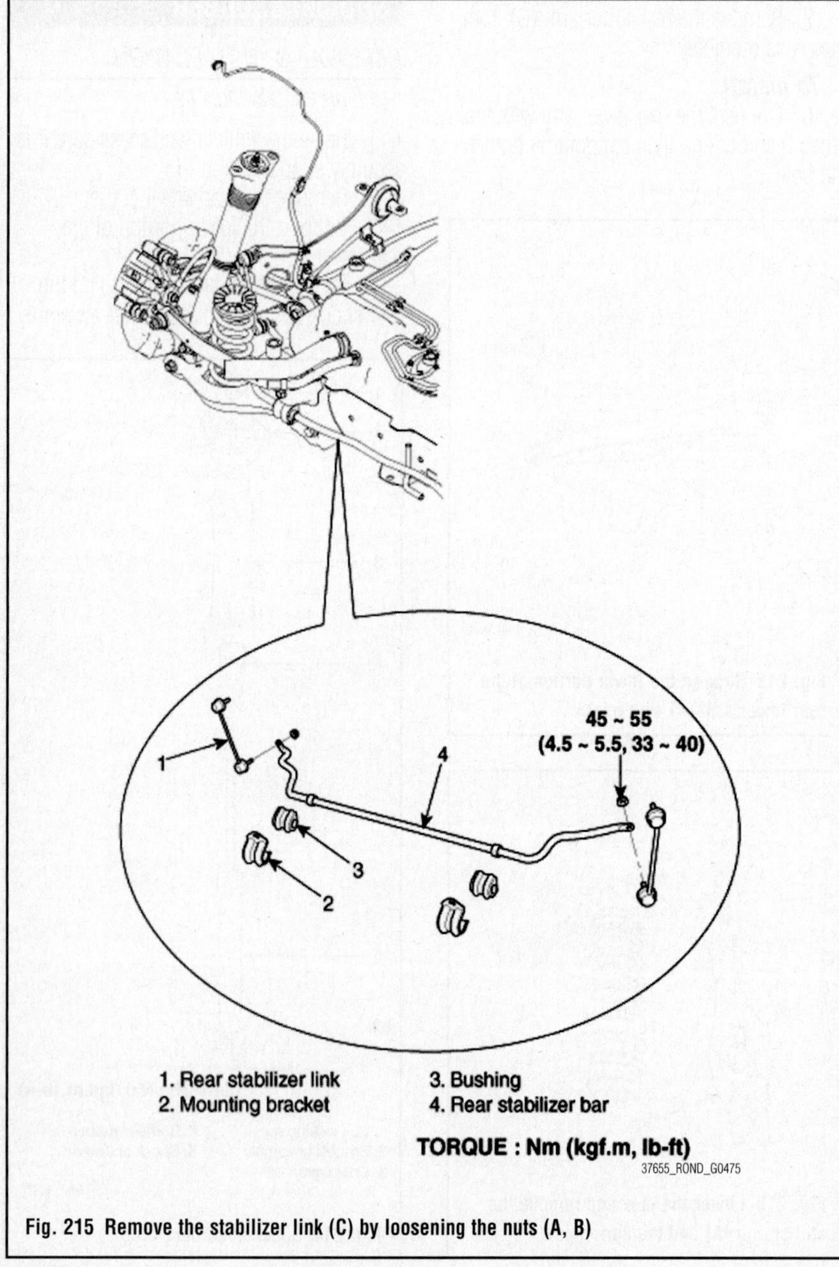

Fig. 215 Remove the stabilizer link (C) by loosening the nuts (A, B)

STABILIZER BAR

REMOVAL & INSTALLATION

See Figures 215 and 216.

1. Raise the vehicle, and make sure it is securely supported.

2. Remove the rear wheel and tire.

3. Remove the stabilizer link (C) by loosening the nuts (A, B).

4. Remove the rear stabilizer bar (B) from the crossmember by loosening the bracket mounting bolts (A).

5. Remove the mounting bracket (A) and bushing (B) from the rear stabilizer bar.

To install:

6. Install the mounting bracket (A) and bushing (B) to the rear stabilizer bar.

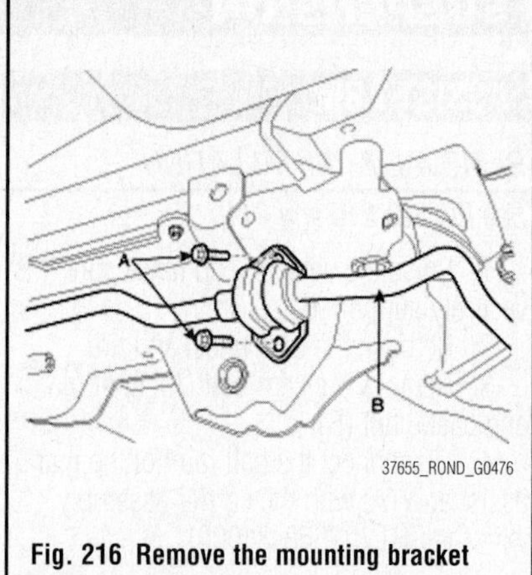

Fig. 216 Remove the mounting bracket (A) and bushing (B) from the rear stabilizer bar

7. Install the rear stabilizer bar (B) to the crossmember by tightening the bracket mounting bolts (A). Torque to 33–40 ft. lbs. (45–55 Nm).

8. Connect the rear stabilizer link (C) between the rear stabilizer bar and the trailing arm and then tighten the nut (A, B). Torque to 33–40 ft. lbs. (45–55 Nm).

9. Install the wheel and the tire. Torque to 65–80 ft. lbs. (90–110 Nm).

**** CAUTION**

Take care not to damage the hub bolts when installing the front wheel and tire.

TRAILING ARMS

REMOVAL & INSTALLATION

See Figures 217 through 221.

1. Raise the vehicle, and make sure it is securely supported.

2. Remove the rear wheel and tire.

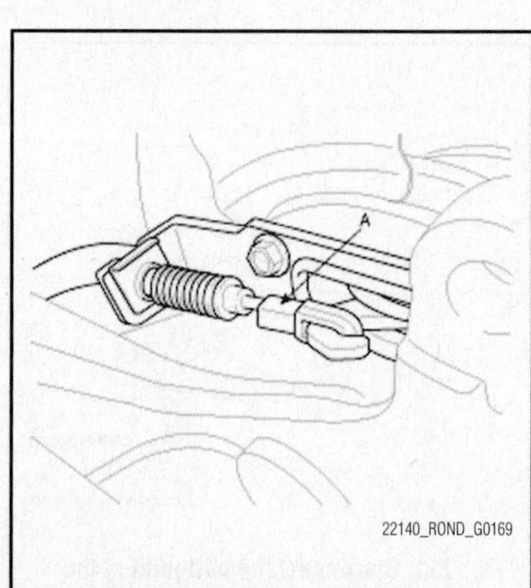

Fig. 217 Disconnect the parking brake cable (A) from the rear brake assembly

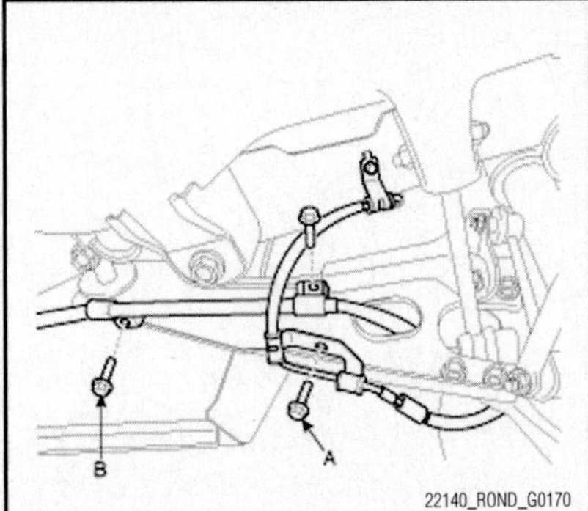

Fig. 218 Remove the wheel speed sensor and parking brake cable bracket bolts (A, B)

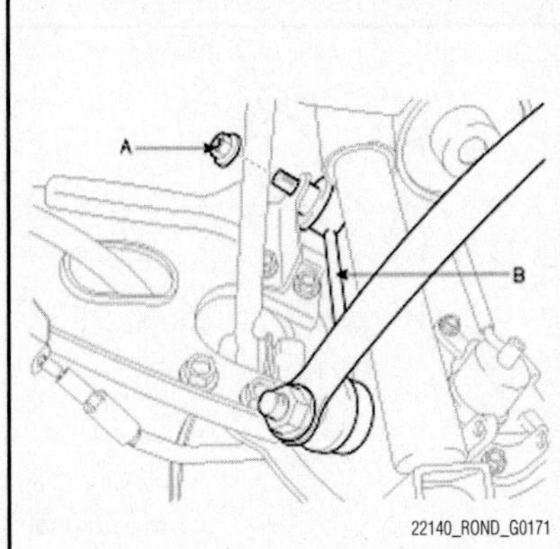

Fig. 219 Disconnect the rear stabilizer link (B) with the trailing arm by removing the nut (A)

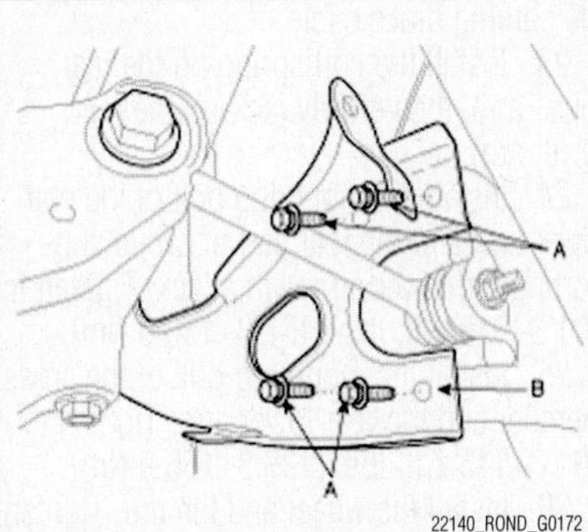

Fig. 220 Disconnect the trailing arm (B) with the carrier assembly by removing the mounting bolts (A)

3. Disconnect the parking brake cable (A) from the rear brake assembly.

4. Remove the wheel speed sensor and parking brake cable bracket bolts (A, B).

5. Disconnect the rear stabilizer link (B) with the trailing arm by removing the nut (A).

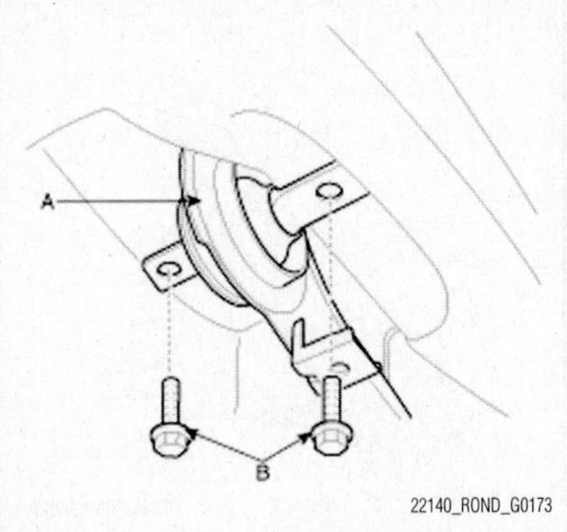

Fig. 221 Remove the rear trailing arm (A) by removing the mounting bolt (B)

6. Disconnect the trailing arm (B) with the carrier assembly by removing the mounting bolts (A).

7. Remove the rear trailing arm (A) by removing the mounting bolt (B).

To install:

8. Install the rear trailing arm to the body. Torque to 101–116 ft. lbs. (140–160 Nm).

9. Connect the trailing arm with the carrier assembly. Torque to 33–40 ft. lbs. (45–55 Nm).

10. Connect the rear stabilizer link with the trailing arm. Torque to 33–40 ft. lbs. (45–55 Nm).

11. Connect the parking brake cable with the rear brake assembly.

12. Install the wheel speed sensor and the parking brake cable bracket.

13. Install the wheel and the tire. Torque to 65–80 ft. lbs. (90–110 Nm).

**** CAUTION**

Take care not to damage the hub bolts when installing the front wheel and tire.

UPPER CONTROL ARMS

REMOVAL & INSTALLLATION

See Figure 222.

1. Raise the vehicle, and make sure it is securely supported.

2. Remove the rear wheel and tire.

3. Remove the rear upper arm (C) by removing the mounting bolts & nut (A, B).

To install:

4. Install the rear upper arm between the rear cross member and the carrier assembly. Torque: 80–94 ft. lbs. (110–130 Nm).

**** CAUTION**

Install the rear upper arm so that the letter "R" faces the rear of vehicle.

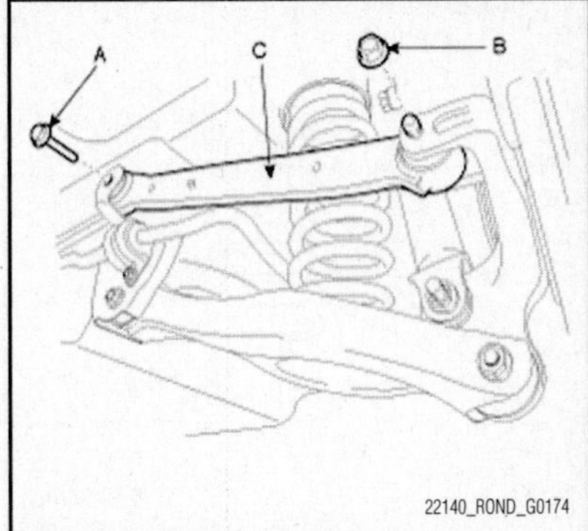

Fig. 222 Remove the rear upper arm (C) by removing the mounting bolts & nut (A, B)

5. Install the wheel and the tire. Torque to 65–80 ft. lbs. (90–110 Nm).

**** CAUTION**

Take care not to damage the hub bolts when installing the front wheel and tire.

WHEEL HUB & BEARING

REMOVAL & INSTALLATION

See Figures 223 through 229.

1. Raise the vehicle, and make sure it is securely supported.

2. Remove the rear wheel & tire.

3. Remove the mounting bolt (B) of the rear lower arm (A) and the rear carrier, while supporting the lower arm (A) with a jack.

4. Remove the mounting nut (A) of the cross member and the rear lower arm, then remove the coil spring (B) by taking down the jack.

5. Remove the wheel speed sensor.

6. Disconnect the parking brake cable.

7. Disconnect the upper arm (B) from the carrier assembly.

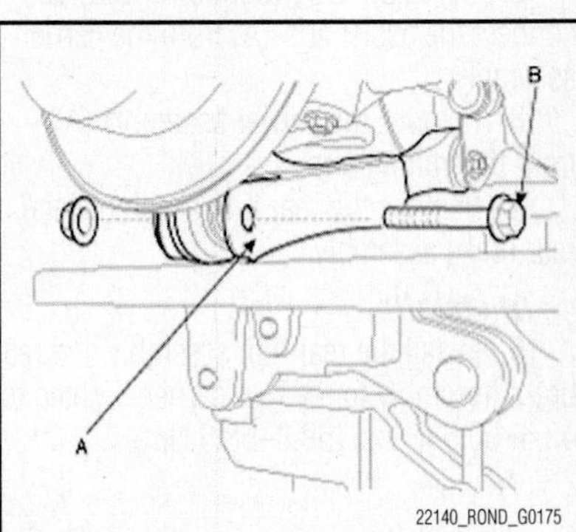

Fig. 223 Remove the mounting bolt (B) of the rear lower arm (A) and the rear carrier, while supporting the lower arm (A) with a jack

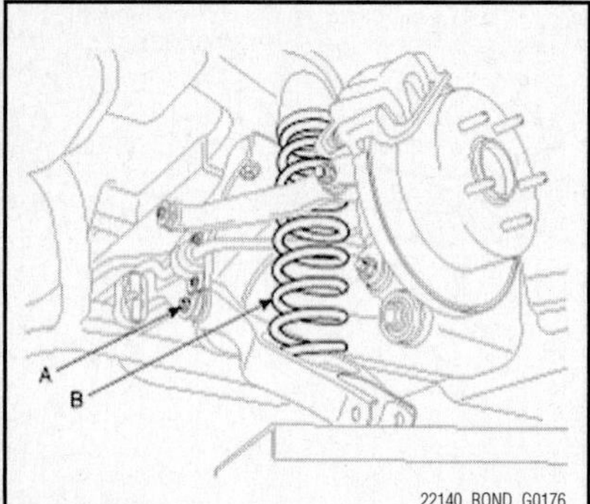

Fig. 224 Remove the mounting nut (A) of the cross member and the rear lower arm, then remove the coil spring (B) by taking down the jack

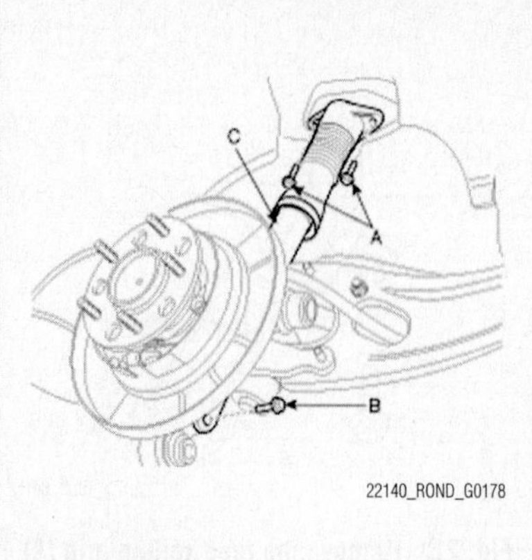

Fig. 226 Remove the rear strut assembly (C)

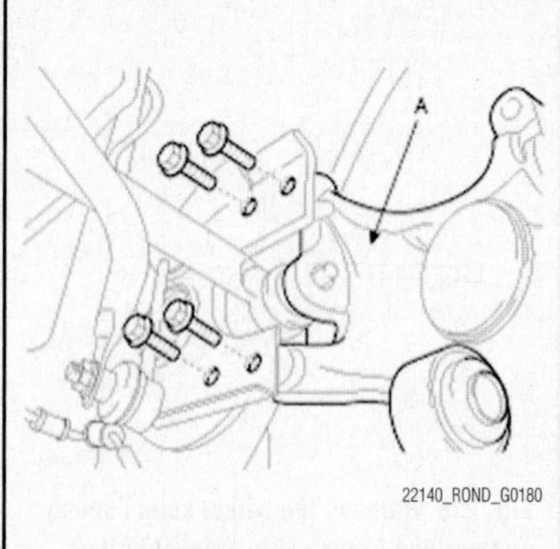

Fig. 228 Remove the carrier assembly (A) from the trailing arm

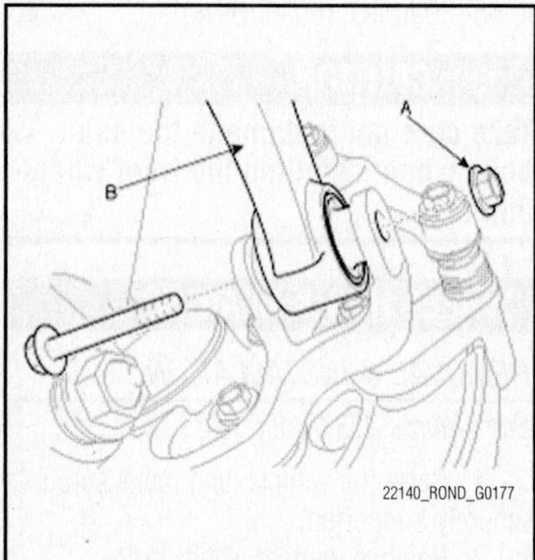

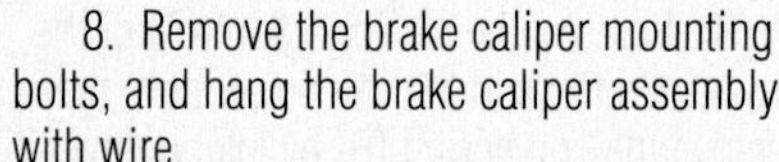

Fig. 225 Disconnect the upper arm (B) from the carrier assembly

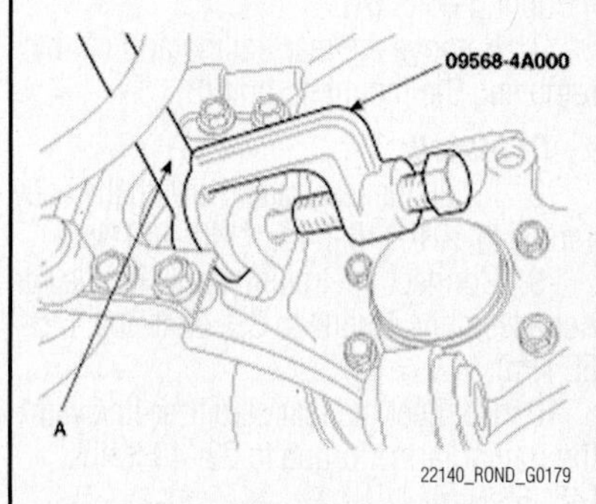

Fig. 227 Using the SST (09568-4A000), disconnect the assist arm (A) from the carrier assembly

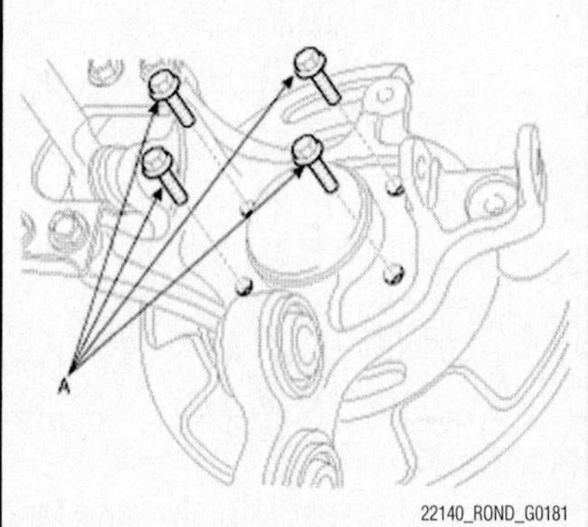

Fig. 229 Remove the rear hub assembly and rear brake assembly (A, mounting bolts)

8. Remove the brake caliper mounting bolts, and hang the brake caliper assembly with wire.
9. Remove the rear brake disc (rotor) assembly.
10. Remove the rear strut assembly (C).
11. Remove the split pin and castle nut from the assist arm.
12. Using the SST (09568-4A000), disconnect the assist arm (A) from the carrier assembly.
13. Remove the carrier assembly (A) from the trailing arm.
14. Remove the rear hub assembly and rear brake assembly.

To install:

15. Install the rear hub assembly and rear brake assembly to the rear carrier. Tighten to 43.4–50.6 ft. lbs. (58.8–68.6 Nm).
16. Install the carrier assembly to the trailing arm. Tighten to 32.5–39.8 ft. lbs. (44.1–53.9 Nm).
17. Install the assist arm to the carrier assembly. Tighten to 32.5–39.8 ft. lbs. (44.1–53.9 Nm).
18. Install the rear strut assembly. Tighten the upper bolts to 32.5–39.8 ft. lbs. (44.1–53.9 Nm); the lower bolt to 101.3–115.7 ft. lbs. (137.3–156.9 Nm).
19. Install the rear brake disc (rotor) assembly.
20. Install the brake caliper. Tighten to 36.2–43.4 ft. lbs. (49.0–58.8 Nm).
21. Install the upper arm to the carrier assembly. Tighten to 79.6–94.0 ft. lbs. (107.9–127.5 Nm).
22. Install the wheel speed sensor and the parking brake cable.
23. Install the coil spring on the rear lower arm, then slowly jack-up the rear lower arm.
24. Install the mounting bolt of the rear lower arm and the rear carrier, while supporting the lower arm with a jack. Tighten to 101.3–115.7 ft. lbs. (137.3–156.9 Nm).
25. Install the mounting bolt of the cross member and the rear lower arm. Tighten to 101.3–115.7 ft. lbs. (137.3–156.9 Nm).
26. Install the wheel and the tire. Tighten to 65–80 ft. lbs. (90–110 Nm).

CAUTION

Take care not to damage the hub bolts when installing the front wheel and tire.

KIA

Sedona

17

SPECIFICATIONS AND MAINTENANCE CHARTS

ENGINE AND VEHICLE IDENTIFICATION

Engine						
Code ①	Liters (cc)	Cu. In.	Cyl.	Fuel Sys.	Engine Type	Eng. Mfg.
3	3.8 (3778)	231	6	EGI	DOHC	KIA

Model Year	
Code ②	Year
9	2009
A	2010

EGI: Electronic Gasoline Injection

DOHC: Double Overhead Camshafts

37655_SEDO_C0001

GENERAL ENGINE SPECIFICATIONS

Year	Model	Engine Displacement Liters	Engine VIN	Net Horsepower @ rpm	Net Torque @ rpm (ft. lbs.)	Bore x Stroke (in.)	Com-pression Ratio	Oil Pressure @ rpm
2009	Sedona	3.8	3	250@6000	253@3500	3.78x3.43	10.4:01	18.77@1000
2010	Sedona	3.8	3	250@6000	253@3500	3.78x3.43	10.4:01	18.77@1000

37655_SEDO_C0002

ENGINE TUNE-UP SPECIFICATIONS

Year	Engine Displacement Liters	Engine VIN	Spark Plug Gap (in.)	Ignition Timing (deg.) ①		Fuel Pump (psi)	Idle Speed (rpm)		Valve Clearance	
				MT	AT		MT	AT	Intake	Exhaust
2009	3.8	3	0.0394-0.0433	NA	②	③	NA	550-750	0.0067-0.0090	0.0106-0.0129
2010	3.8	3	0.0394-0.0433	NA	②	③	NA	550-750	0.0067-0.0090	0.0106-0.0129

NOTE: The Vehicle Emission Control Information label often reflects specification changes made during production.
The label figures must be used if they differ from those in this chart

NA: Not Available

① Computer controled, no adjustment possible

② 5-15 degrees

③ 54.3-55.8 at idle

37655_SEDO_C0003

CAPACITIES

Year	Model	Engine Displacement Liters	Engine VIN	Engine Oil with Filter (qts.)	Transaxle (pts.) Manual	Auto.	Fuel Tank (gal.)	Cooling System (qts.)
2009	Sedona	3.8	3	5.5	—	22.0	21.1	9.1
2010	Sedona	3.8	3	5.5	—	22.0	21.1	9.1

NOTE: All capacities are approximate. Add fluid gradually and ensure a proper level is obtained.

37655_SEDO_C0004

FLUID SPECIFICATIONS

Year	Model	Engine Displ. Liters (VIN)	Engine Oil	Man. Trans.	Auto. Trans.	Drive Axle Front	Drive Axle Rear	Transfer Case	Power Steering Fluid	Brake Master Cylinder	Cooling System
2009	Sedona	3.8 (3)	①	NA	②	NA	NA	NA	PSF-3	DOT 3	③
2010	Sedona	3.8 (3)	①	NA	②	NA	NA	NA	PSF-3	DOT 3	③

NA: Not Available

DOT: Department Of Transpotation

① SL (SJ) or above, 5W-20

② Diamond ATF SP-III or SK ATF SP-III

③ Ethylene glycol base for aluminum radiator

37655_SEDO_C0014

VALVE SPECIFICATIONS

Year	Engine Displacement Liters	Engine VIN	Seat Angle (deg.)	Face Angle (deg.)	Maximum out of Square (degrees)	Spring Free Length (in.)	Stem-to-Guide Clearance (in.) Intake	Stem-to-Guide Clearance (in.) Exhaust	Stem Diameter (in.) Intake	Stem Diameter (in.) Exhaust
2009	3.8	3	44.75-45.20	45.25 45.75	NA	1.7267	0.0008-0.0019	0.0012-0.0021	0.215-0.216	0.215-0.216
2010	3.8	3	44.75-45.20	45.25 45.75	NA	1.7267	0.0008-0.0019	0.0012-0.0021	0.215-0.216	0.215-0.216

NA: Not Available

37655_SEDO_C0005

CAMSHAFT AND BEARING SPECIFICATIONS

All measurements are given in inches.

Year	Engine Displacement Liters	Engine VIN	Journal Diameter	Brg. Oil Clearance	Shaft End-play	Runout	Journal Bore	Lobe Lift Intake	Lobe Lift Exhaust
2009	3.8	3	①	②	③	NA	NA	NA	NA
2010	3.8	3	①	②	③	NA	NA	NA	NA

NA: Not Available

① No. 1: 1.1009-1.1016 inches

No. 2, 3, 4: 0.9430-0.9437 inches

② No. 1: 0.0008-0.0022 inches

No. 2, 3, 4: 0.0012-0.0026 inches

③ 0.0008-0.0071 inches

37655_SEDO_C0013

CRANKSHAFT AND CONNECTING ROD SPECIFICATIONS

All measurements are given in inches.

	Engine		Crankshaft				Connecting Rod		
Year	Displacement Liters	Engine VIN	Main Brg. Journal Dia.	Main Brg. Oil Clearance	Shaft End-play	Thrust on No.	Journal Diameter	Oil Clearance	Side Clearance
2009	3.8	3	2.7142-2.7149	0.0008-0.0016	0.0039-0.0110	3	2.2834-2.2842	0.0015-0.0022	0.0039-0.0098
2010	3.8	3	2.7142-2.7149	0.0008-0.0016	0.0039-0.0110	3	2.2834-2.2842	0.0015-0.0022	0.0039-0.0098

37655_SEDO_C0007

PISTON AND RING SPECIFICATIONS

All measurements are given in inches.

	Engine			Ring Gap			Ring Side Clearance		
Year	Displacement Liters	Engine VIN	Piston Clearance	Top Compression	Bottom Compression	Oil Control	Top Compression	Bottom Compression	Oil Control
2009	3.8	3	0.0012-0.0020	0.0067-0.0126	0.0126-0.0185	0.0078-0.0275	0.0012-0.0027	0.0012-0.0027	0.0024-0.0059
2010	3.8	3	0.0012-0.0020	0.0067-0.0126	0.0126-0.0185	0.0078-0.0275	0.0012-0.0027	0.0012-0.0027	0.0024-0.0059

37655_SEDO_C0006

TORQUE SPECIFICATIONS

All readings in ft. lbs.

Year	Engine Displacement Liters	Engine VIN	Cylinder Head Bolts	Main Bearing Bolts	Rod Bearing Bolts	Crankshaft Damper Bolts	Flywheel Bolts	Manifold Intake	Manifold Exhaust	Spark Plugs	Oil Pan Drain Plug
2009	3.8	3	①	②	③	210-224	53-57	14-17	29-33	15-22	25-32
2010	3.8	3	①	②	③	210-224	53-57	14-17	29-33	15-22	25-32

① Step 1: 28.93 ft. lbs.
Step 2: Plus 120 degrees
Step 3: Plus 90 degrees

② Bolts 1 thru 8 (inside cap bolts): 36.16 ft. lbs. plus 90 degrees
Bolts 9 thru 16 (outside cap bolts) 14.46 ft. lbs. plus 120 degrees
Bolts 17 thru 22 (side cap bolts) 21.7-23.2 ft. lbs.
See illustration in text for location information

③ Step 1: 14.46 ft. lbs.
Step 2: Plus 90 degrees

37655_SEDO_C0008

WHEEL ALIGNMENT

Year	Model		Caster Range (+/-Deg.)	Caster Preferred Setting (Deg.)	Camber Range (+/-Deg.)	Camber Preferred Setting (Deg.)	Toe-in (in.)
2009	Sedona	F	—	4° 05' +/- 30'	—	0 +/- 30'	0 +/- .0787
		R	—	—	—	-20 +/- 30'	.1378 +/- .0787
2010	Sedona	F	—	4° 05' +/- 30'	—	0 +/- 30'	0 +/- .0787
		R	—	—	—	-20 +/- 30'	.1378 +/- .0787

37655_SEDO_C0009

TIRE, WHEEL AND BALL JOINT SPECIFICATIONS

Year	Model	OEM Tires		Tire Pressure (psi)		Wheel	Ball Joint	Lug Nut Torque
		Standard	Optional	Front	Rear	Size	Inspection	(ft. lbs.)
2009	Sedona	P225/70R16	P235/60R17	35	35	②	①	65-79
2010	Sedona	P225/70R16	P235/60R17	35	35	②	①	65-79

OEM: Original Equipment Manufacturer

PSI: Pounds Per Square Inch

① Replace if any measurable movement is found.

② STD: 6.5Jx16

OPT: 6.5Jx17

37655_SEDO_C0010

BRAKE SPECIFICATIONS

All measurements in inches unless noted

Year	Model		Brake Disc			Brake Drum				Brake Caliper	
			Original Thickness	Minimum Thickness	Maximum Run-out	Original Inside Diameter	Max. Wear Limit	Maximum Machine Diameter	Minimum Lining Thickness	Bracket Bolts (ft. lbs.)	Mounting Bolts (ft. lbs.)
2009	Sedona	F	1.180	1.100	0.0012	—	—	—	—	62-72	16-23
		R	0.470	NA	0.0020	—	—	—	—	36-43	16-23
2010	Sedona	F	1.180	1.100	0.0012	—	—	—	—	62-72	16-23
		R	0.470	NA	0.0020	—	—	—	—	36-43	16-23

NA: Not Available

F: Front

R: Rear

37655_SEDO_C0011

SCHEDULED MAINTENANCE INTERVALS

Kia—Sedona

TO BE SERVICED	TYPE OF SERVIC	VEHICLE MILEAGE INTERVAL (x1000) 7.5	15	22.5	30	37.5	45	52.5	60	67.5	75	82.5	90	97.5	100	105	112.5	120
Accessory drive belts	S/I	✓	✓	✓	✓	✓	✓	✓	✓	✓	✓	✓	✓	✓	✓	✓	✓	✓
Air cleaner element	I/R		✓		✓		✓		✓		✓		✓			✓		✓
Air conditioner system	S/I	Inspect the system operation and refrigerant amount annually.																
Brake lines, hoses and connections	S/I		✓		✓		✓		✓		✓		✓			✓		✓
Chassis and body fasteners	T	✓	✓	✓	✓	✓	✓	✓	✓	✓	✓	✓	✓	✓	✓	✓	✓	✓
Cooling system hoses and coolant level	S/I		✓		✓		✓		✓		✓		✓			✓		✓
CV-joint boots					✓				✓				✓					✓
Engine coolant	R								✓				✓					✓
Engine oil and filter	R	✓	✓	✓	✓	✓	✓	✓	✓	✓	✓	✓	✓	✓	✓	✓	✓	✓
Exhaust system heat shields	S/I				✓				✓				✓					✓
Front and rear brakes	S/I				✓				✓				✓					✓
Front ball joints	S/I				✓				✓				✓					✓
Fuel filter	R					✓					✓						✓	
Fuel tank air filter	R				✓				✓				✓					
Fuel lines and hoses	S/I				✓				✓				✓					✓
Idle speed	A				✓				✓				✓					✓
Locks and hinges	L	✓	✓	✓	✓	✓	✓	✓	✓	✓	✓	✓	✓	✓	✓	✓	✓	✓
Spark plugs	R														✓			
Steering operation and linkage	S/I				✓				✓				✓					✓
Timing belt	R								✓									✓
Valve Clearance	A								✓									✓

R: Replace S/I: Inspect and service, if needed L: Lubricate A: Adjust T: Tighten

FREQUENT OPERATION MAINTENANCE (SEVERE SERVICE)

If a vehicle is operated under any of the following conditions it is considered severe service

- Towing a trailer or using a camper or car-top carrier.

- Repeated short trips of less than 5 miles in temperatures below freezing, or trips of less than 10 miles in any temperature.

- Extensive idling or low-speed driving for long distances as in heavy commercial use, such as delivery, taxi or police cars.

- Operating on rough, muddy or salt-covered roads.

- Operating on unpaved or dusty roads.

- Driving in extremely hot (over 90°F) conditions.

Engine oil and filter: replace every 5000 miles or 5 months, whichever occurs first.

Air cleaner element: inspect ever 15,000 miles or 15 months and replace every 30,000 miles or 30 months, whichever occurs first.

Fuel system hoses (California models only): replace every 105,000 miles.

Emission system hoses (non-California models): inspect every 55,000 miles or 55 months, whichever occurs first.

Emission system hoses (California models): inspect every 60,000 miles or 60 months, whichever occurs first.

Front and rear disc brakes: inspect every 15,000 miles or 15 months, whichever occurs first.

Chassis and body fasteners: tighten every 15,000 miles or 15 months, whichever occurs first.

Locks and hinges: lubricate every 5000 miles or 5 months, whichever occurs first.

37655_SEDO_C0012

BRAKES — INFORMATION AND PRECAUTIONS

ANTI-LOCK SYSTEMS

- Certain components within the ABS system are not intended to be serviced or repaired individually.
- Do not use rubber hoses or other parts not specifically specified for and ABS system. When using repair kits, replace all parts included in the kit. Partial or incorrect repair may lead to functional problems and require the replacement of components.
- Lubricate rubber parts with clean, fresh brake fluid to ease assembly. Do not use shop air to clean parts; damage to rubber components may result.
- Use only DOT 3 brake fluid from an unopened container.
- If any hydraulic component or line is removed or replaced, it may be necessary to bleed the entire system.
- A clean repair area is essential. Always clean the reservoir and cap thoroughly before removing the cap. The slightest amount of dirt in the fluid may plug an orifice and impair the system function. Perform repairs after components have been thoroughly cleaned; use only denatured alcohol to clean components. Do not allow ABS components to come into contact with any substance containing mineral oil; this includes used shop rags.
- The Anti-Lock control unit is a microprocessor similar to other computer units in the vehicle. Ensure that the ignition switch is **OFF** before removing or installing controller harnesses. Avoid static electricity discharge at or near the controller.
- If any arc welding is to be done on the vehicle, the control unit should be unplugged before welding operations begin.

DISC AND DRUM SYSTEMS

CAUTION

Dust and dirt accumulating on brake parts during normal use may contain asbestos fibers from production or aftermarket brake linings. Breathing excessive concentrations of asbestos fibers can cause serious bodily harm. Exercise care when servicing brake parts. Do not sand or grind brake lining unless equipment used is designed to contain the dust residue. Do not clean brake parts with compressed air or by dry brushing. Cleaning should be done by dampening the brake components with a fine mist of water, then wiping the brake components clean with a dampened cloth. Dispose of cloth and all residue containing asbestos fibers in an impermeable container with the appropriate label. Follow practices prescribed by the Occupational Safety and Health Administration (OSHA) and the Environmental Protection Agency (EPA) for the handling, processing, and disposing of dust or debris that may contain asbestos fibers.

BRAKES — BLEEDING THE BRAKE SYSTEM

BLEEDING PROCEDURE

BLEEDING PROCEDURE

WARNING

When bleeding the brakes, note the following:

- Do not reuse the drained fluid
- Always use Genuine DOT 3 or DOT 4 Brake Fluid. Using a non-Genuine DOT3 or DOT 4 brake fluid can cause corrosion and decrease the life of the system
- Make sure no dirt of other foreign matter is allowed to contaminate the brake fluid
- Do not spill brake fluid on the vehicle, it may damage the paint; if brake fluid does contact the paint, wash it off immediately with water
- The reservoir on the master cylinder must be at the MAX (upper) level mark at the start of bleeding procedure and checked after bleeding each brake caliper. Add fluid as required

1. Make sure the brake fluid level in the master cylinder fluid reservoir is at the MAX (upper) level line.
2. Have someone slowly pump the brake pedal several times, and then apply steady pressure
3. Loosen the right-rear brake bleed screw to allow air to escape from the system. Tighten the bleed screw securely
4. Repeat the procedure for each wheel in the sequence shown below until air bubbles no longer appear in the fluid
5. Refill the master cylinder reservoir to the MAX (upper) level line

BLEEDING THE ABS SYSTEM

The ABS brake system is bled in the usual fashion with no special procedures required. Refer to the bleeding procedure described in this section. Make certain the master cylinder reservoir is filled before the bleeding is begun and check the level frequently.

BRAKES — ANTI-LOCK BRAKE SYSTEM (ABS)

WHEEL SPEED SENSORS

REMOVAL & INSTALLATION

Front

See Figures 1 through 3.

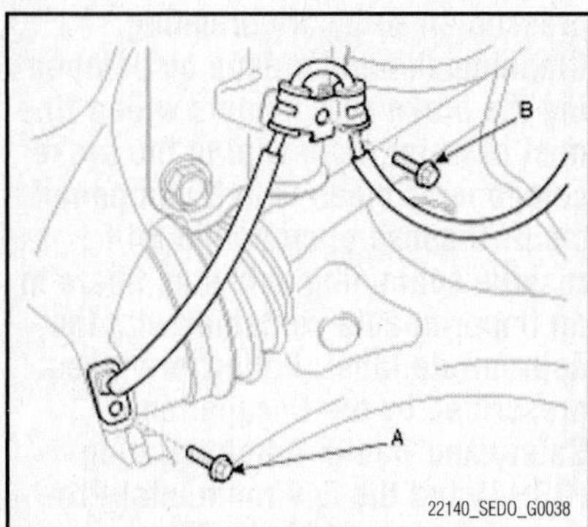

Fig. 1 Remove the front wheel speed sensor mounting bolt (A) and the wire bracket mounting bolt (B).

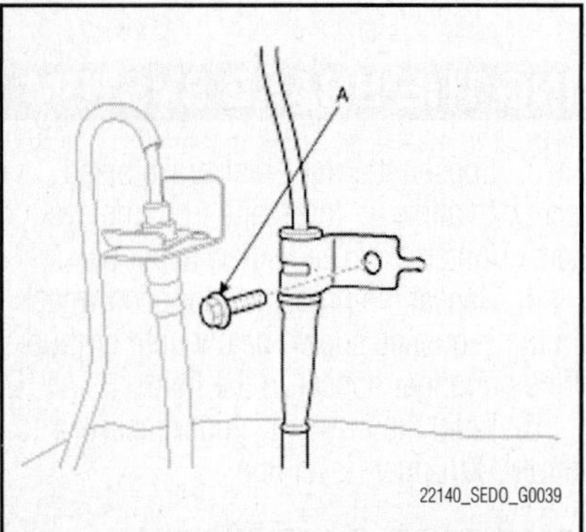

Fig. 2 Remove the wire bracket bolt (A)

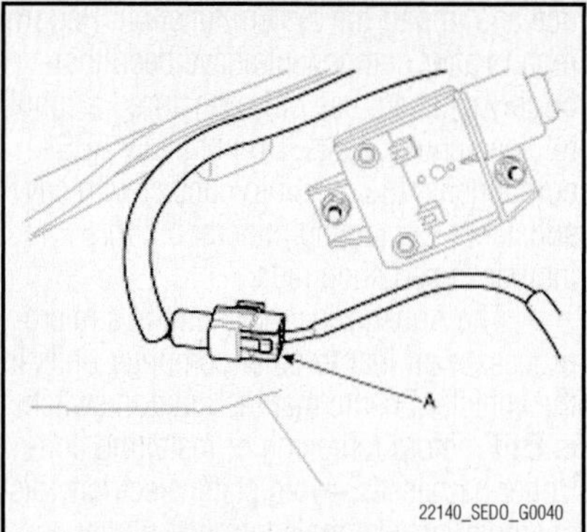

Fig. 3 Remove the front wheel speed sensor after disconnecting the wheel speed sensor connector (A).

1. Remove the front wheel speed sensor mounting bolt and the wire bracket mounting bolt.
2. Remove the wire bracket bolt.
3. Remove the front wheel guard.
4. Remove the front wheel speed sensor after disconnecting the wheel speed sensor connector.
5. Installation is the reverse of removal.

Rear

See Figures 4 and 5.

1. Remove the rear wheel speed sensor mounting bolt.
2. Remove the rear seat side pad and disconnect the rear wheel speed sensor connector.
3. Installation is the reverse of removal.

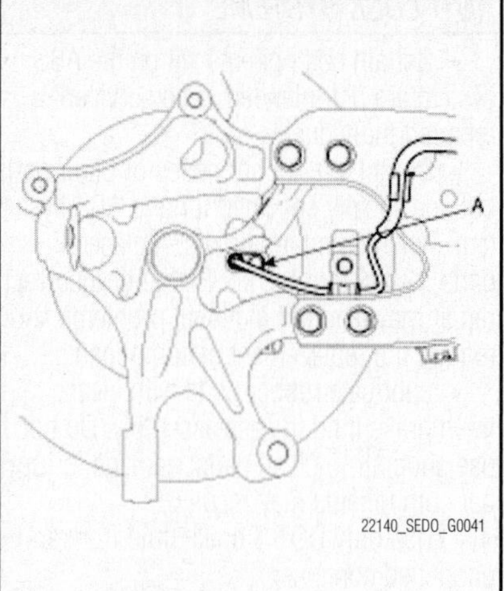

Fig. 4 Remove the rear wheel speed sensor mounting bolt (A)

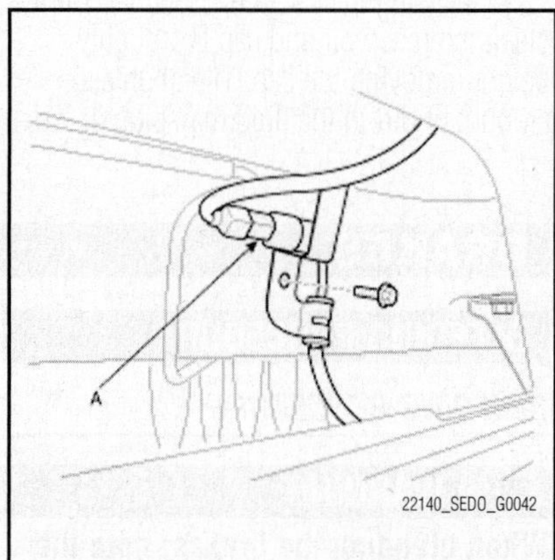

Fig. 5 Remove the rear seat side pad and disconnect the rear wheel speed sensor connector (A)

BRAKES — FRONT DISC BRAKES

BRAKE CALIPER

REMOVAL & INSTALLATION

1. Raise and safely support the front of the vehicle.
2. Remove the front wheel and tire.
3. Remove the guide rod bolts from the caliper assembly.
4. Remove the caliper assembly.
5. Installation is the reverse of removal.

BRAKE PADS

REMOVAL & INSTALLATION

1. Raise and safely support the front of the vehicle.
2. Remove the front wheel and tire.
3. Remove the guide rod bolts from the caliper assembly.
4. Remove the caliper assembly.
5. Remove the pads, the pad shims and the pad retainers from the caliper bracket.
6. Installation is the reverse of removal.

BRAKES — REAR DISC BRAKES

BRAKE CALIPER

REMOVAL & INSTALLATION

1. Raise and safely support the rear of the vehicle.
2. Remove the rear wheel and tire.
3. Remove the guide rod bolts from the caliper assembly.
4. Remove the caliper assembly.
5. Installation is the reverse of removal.

BRAKE PADS

REMOVAL & INSTALLATION

1. Raise and safely support the rear of the vehicle.
2. Remove the rear wheel and tire.
3. Remove the guide rod bolts from the caliper assembly.
4. Remove the caliper assembly.
5. Remove the pads, the pad shims and the pad retainers from the caliper bracket.
6. Installation is the reverse of removal.

BRAKES — PARKING BRAKE

PARKING BRAKE CABLES

ADJUSTMENT

See Figure 6.

1. Adjust the adjusting nut so that the parking pedal stoke is 3.46–3.86 inches (88–98 mm) after full stroke operation of parking pedal more than 3 times.
2. The parking brake adjustment must be completed after adjusting the rear shoe.

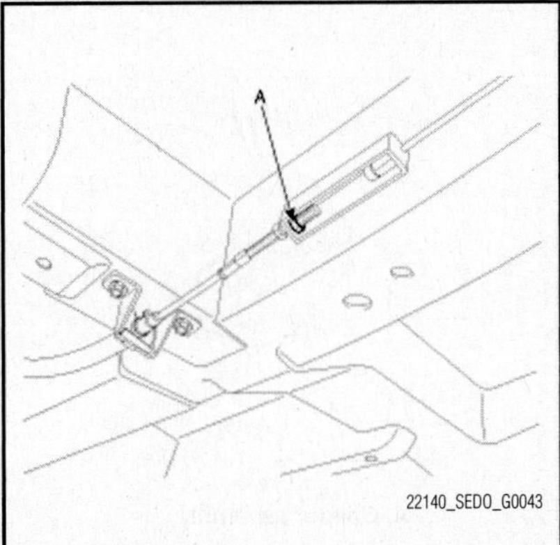

Fig. 6 Adjust the adjusting nut (A)

3. After adjusting parking brake, notice following:
   a. Must be free clearance between adjusting nut and pin.
   b. Check securely that the brake is not dragging.

PARKING BRAKE SHOES

REMOVAL & INSTALLATION

1. Raise and safely support the rear of the vehicle.
2. Remove the rear wheel and tire.
3. Remove the dust cap and the rotor.
4. Remove the hub nut and washer and remove the rear hub.
5. Remove the shoe hold down pin and spring by pressing and rotating the spring.
6. Remove the adjuster assembly and the lower return spring.
7. Remove the strut and the upper return spring.
8. Remove the retaining clip from the parking brake cable at the back of the backing plate.
9. Disconnect the parking brake cable from the brake shoe.

To install:

10. Connect the parking brake cable to the parking brake shoe.
11. Install the shoe hold down pin and spring to hold the brake shoe.
12. Install the adjuster assembly and the lower return spring.
13. Install the upper return spring and strut.
14. Grease where is necessary.
15. Install the rear hubs.
16. Install the hub nut.
17. Install the dust cap and the rotor.
18. Install the rear wheels and tires.
19. Tighten the parking brake adjusting nut.

ADJUSTMENT

1. Raise the rear of the vehicle, and make sure it is securely supported.
2. Remove the rear wheel and tire.
3. After removing the plug from the disc, rotate the toothed wheel with a screwdriver until the disc does not rotate.
4. Back off the toothed wheel by 5 notches.

CHASSIS ELECTRICAL

AIR BAG (SUPPLEMENTAL RESTRAINT SYSTEM)

GENERAL INFORMATION

CAUTION

These vehicles are equipped with an air bag system. The system must be disarmed before performing service on, or around, system components, the steering column, instrument panel components, wiring and sensors. Failure to follow the safety precautions and the disarming procedure could result in accidental air bag deployment, possible injury and unnecessary system repairs.

PRECAUTIONS

Disconnect and isolate the battery negative cable before beginning any airbag system component diagnosis, testing, removal, or installation procedures. Allow system capacitor to discharge for three minutes before beginning any component service. This will disable the airbag system. Failure to disable the airbag system may result in accidental airbag deployment, personal injury, or death.

DISARMING THE SYSTEM

1. Before servicing the vehicle, refer to the Precautions Section.
2. Turn the ignition switch to the **LOCK** position.
3. Disconnect the negative battery cable.
4. Wait 10 minutes for the battery back-up power to discharge.

ARMING THE SYSTEM

1. Before servicing the vehicle, refer to the Precautions Section.
2. Connect the negative battery cable.
3. Turn the ignition switch **ON**.
4. Verify that the air bag indicator illuminates for 4–8 seconds, then goes off.

CLOCKSPRING CENTERING

Prior to installing the clock spring, align the mating mark and "NEUTRAL" position indicator of the clock spring, and, after turning the front wheels to the straight-ahead position, install the clock spring to the column switch. If the mating mark of the clock spring is not properly aligned, the steering wheel may not completely rotate during a turn, or the flat cable within the clock spring may be severed, obstructing normal operation of the SRS and possibly leading to serious injury to the vehicle's driver.

DRIVE TRAIN

FRONT AXLE SHAFT

REMOVAL & INSTALLATION

See Figures 7 through 11.

1. Raise the vehicle and remove the wheel & tire assembly from the front hub assembly. Tighten assembly to 65–80 ft. lb. (90–110 Nm)
2. Remove the wheel speed sensor & wire by loosening the bolts.
3. Disconnect the brake hose from the strut assembly.
4. Unstake the driveshaft lock nut using a chisel and hammer.
5. Remove the driveshaft lock nut. Tighten lock nut to 177–199 ft. lb. (245–275 Nm).

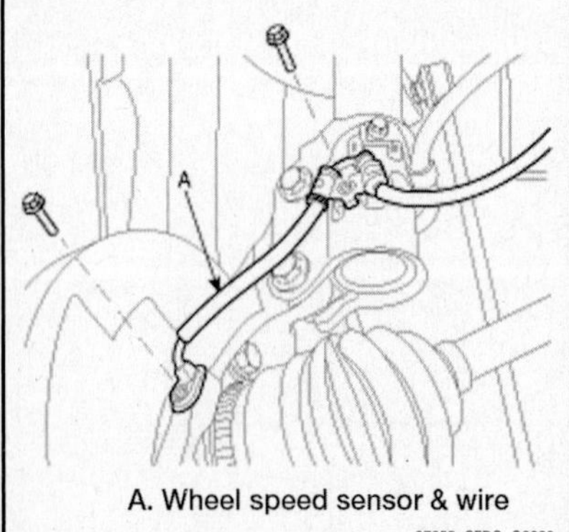

Fig. 7 Wheel speed sensor and wire removal

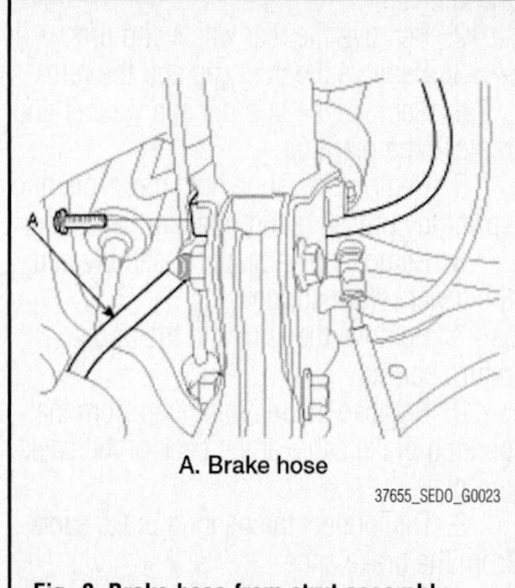

Fig. 8 Brake hose from strut assembly disconnection

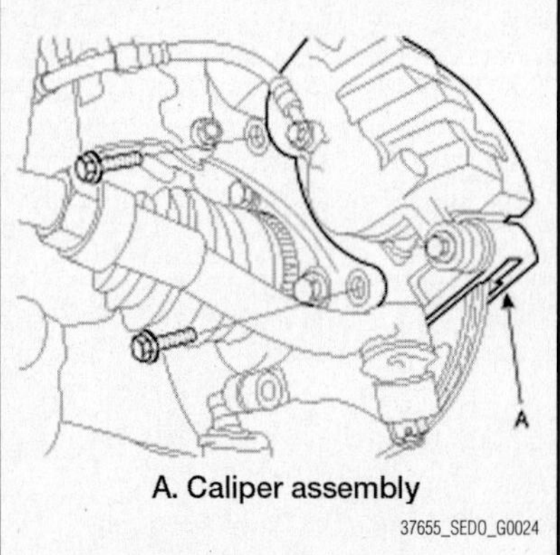

Fig. 9 Brake caliper assembly from the knuckle removal (1 of 2)

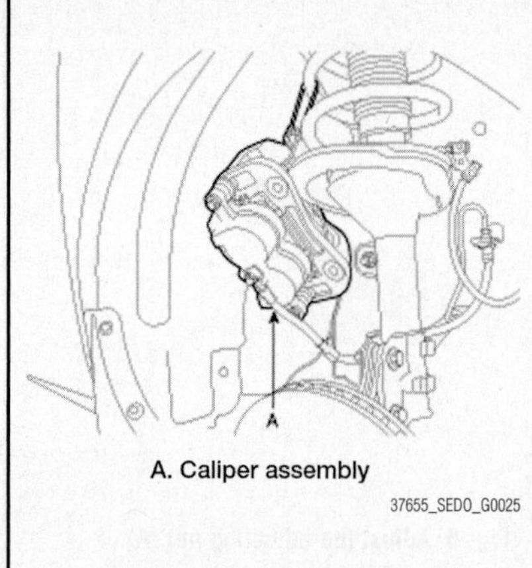

Fig. 10 Brake caliper assembly from the knuckle removal (2 of 2)

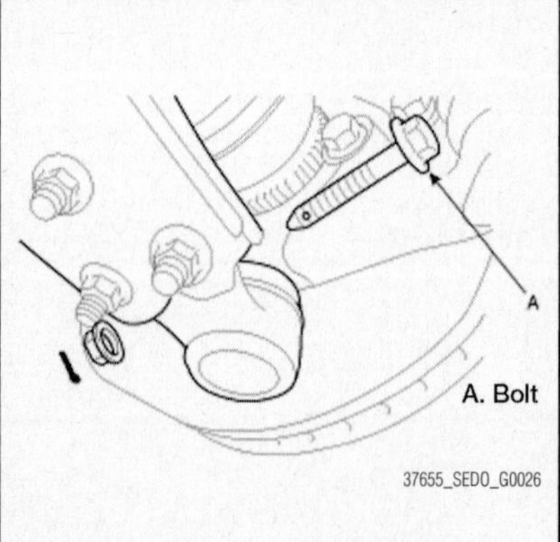

Fig. 11 Remove the split pin and lower arm bolt and nut

6. Remove the caliper assembly from the knuckle and suspend it with wire. Tighten caliper assembly to 61–72 ft. lb. (85–100 Nm).

7. Remove the split pin and castle nut from the tie rod end ball joint. Tighten nut to 43–58 ft. lb. (60–80 Nm).

8. Disconnect the tie rod end from the knuckle using a SST (09568-4A000).

9. Remove the split pin and lower arm bolt and nut. Tighten to 65–87 ft. lb. (90–120 Nm).

10. Using a plastic hammer, disconnect the driveshaft from the front hub assembly.

11. Remove the brake disc from the front hub assembly.

12. Remove the knuckle from the strut assembly by loosening the bolts. Tighten bolts to 72–87 ft. lb. (100–120 Nm).

To install:

- The driveshaft lock nut should be replaced with new ones.
- After installation driveshaft lock nut, stake the lock nut using a chisel and hammer.

13. Install the knuckle from the strut assembly by loosening the bolts. Tighten bolts to 72–87 ft. lb. (100–120 Nm).

14. Install the brake disc from the front hub assembly.

15. Using a plastic hammer, connect the driveshaft from the front hub assembly.

16. Install the split pin and lower arm bolt and nut. Tighten to 65–87 ft. lb. (90–120 Nm).

17. Connect the tie rod end from the knuckle using a SST (09568-4A000).

18. Install the split pin and castle nut from the tie rod end ball joint. Tighten nut to 43–58 ft. lb. (60–80 Nm).

19. Install the caliper assembly from the knuckle and suspend it with wire. Tighten caliper assembly to 61–72 ft. lb. (85–100 Nm).

20. Install the driveshaft lock nut. Tighten lock nut to 177–199 ft. lb. (245–275 Nm).

21. Stake the driveshaft lock nut using a chisel and hammer.

22. Connect the brake hose from the strut assembly.

23. Install the wheel speed sensor & wire by loosening the bolts.

24. Lower the vehicle and remove the wheel & tire assembly from the front hub assembly. Tighten assembly to 65–80 ft. lb. (90–110 Nm)

FRONT DRIVESHAFT

REMOVAL & INSTALLATION

See Figures 12 through 21.

1. Raise the vehicle and remove the wheel & tire assembly from the front hub assembly.

2. Loosen the drain plug and drain the ATF.

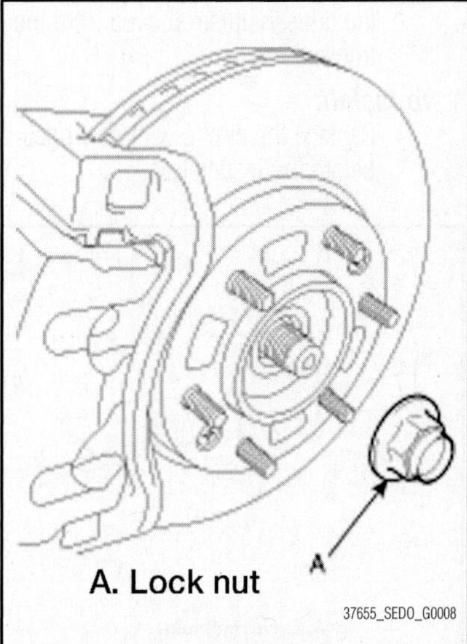

Fig. 12 Driveshaft lock nut removal

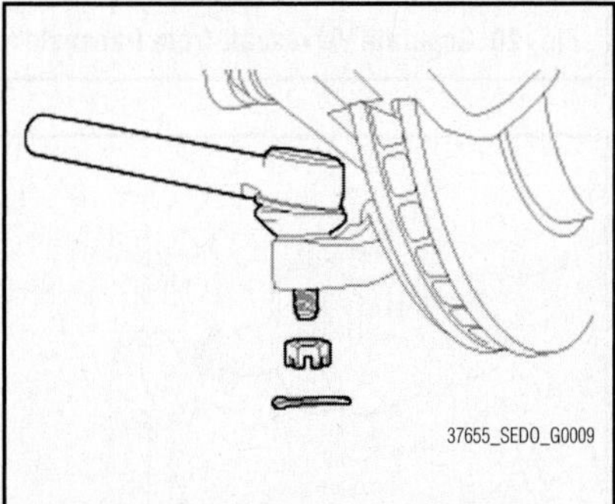

Fig. 13 Split pin and castle nut from tie rod end ball joint removal

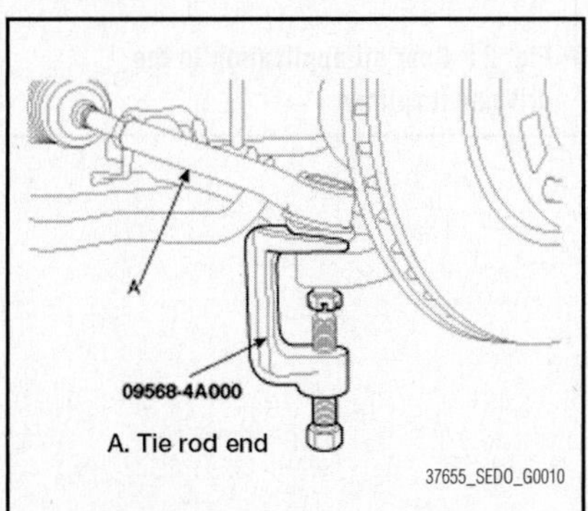

Fig. 14 Disconnect the tie rod end from the knuckle

3. Unstake the driveshaft lock nut using a chisel and hammer.

4. Remove the driveshaft lock nut. Tighten nut to 177–199 ft. lb. (245–275 Nm).

5. Remove the split pin and castle nut form the tie rod end ball joint. Tighten nut to 43–58 ft. lb. (60–80 Nm).

6. Disconnect the tie rod end from the knuckle using a SST (09568-4A000).

7. Remove the split pin and lower arm bolt and nut. Tighten to 65–87 ft. lb. (90–120 Nm).

8. Using a plastic hammer, disconnect the driveshaft from the front hub assembly.

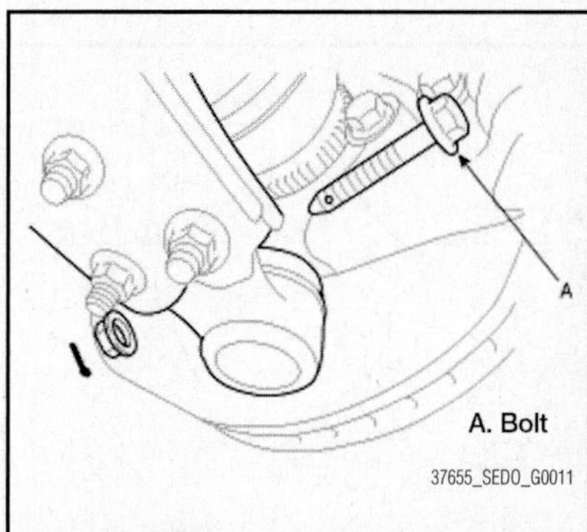

Fig. 15 Split pin and lower arm bolt and nut removal

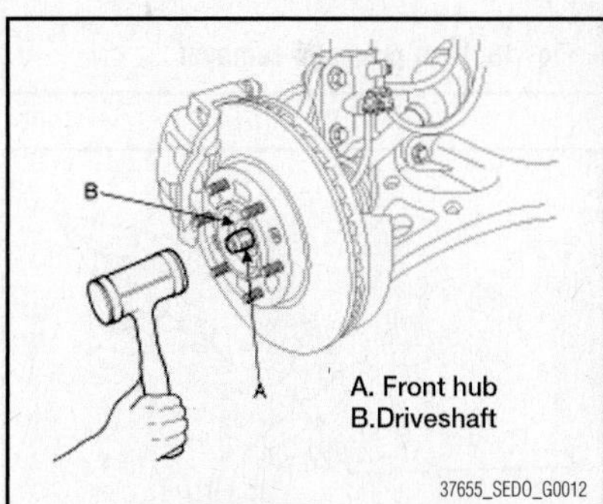

Fig. 16 Disconnect the driveshaft from the front hub assembly

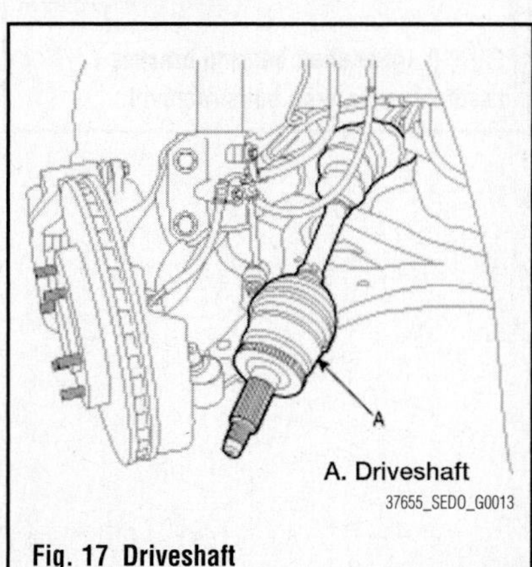

Fig. 17 Driveshaft

9. Removal of the driveshaft [RH].

a. Remove the heat protector.

b. Remove the inner shaft bearing bracket assembly mounting bolts. Tighten to 36–47 ft. lb. (50–65 Nm).

10. Insert pry bar between the transaxle case and driveshaft joint, separate driveshaft from the transaxle.

- Use a pry bar being careful not to damage the transaxle and joint.
- Do not insert a pry bar too deep, as this may cause damage to the oil seal.
- Do not pry on the driveshaft by excessive force it may cause components inside the joint kit to dislodge resulting in a torn boot or a damaged bearing.

11. Pull out the driveshaft from the transaxle case.

- Plug the hole of the transaxle case with the oil seal cap to prevent contamination.
- Replace the retainer ring whenever the driveshaft is removed from the transaxle case.

To install:

- Replace the circlip with new ones before the installation.
- Before the installation, apply the gear oil on the driveshaft splines and contacting surface of differential case oil seal.
- Plug the hole of the transaxle case with the oil seal cap to prevent contamination.
- Replace the retainer ring whenever the driveshaft is removed from the transaxle case.

12. Push in the driveshaft to the transaxle case.

13. Push the transaxle case and driveshaft joint together.

14. Install of the driveshaft [RH].

a. Install the inner shaft bearing bracket assembly mounting bolts. Tighten to 36–47 ft. lb. (50–65 Nm).

b. Install the heat protector.

15. Using a plastic hammer, connect the driveshaft to the front hub assembly.

16. Install the split pin and lower arm bolt and nut. Tighten to 65–87 ft. lb. (90–120 Nm).

17. Connect the tie rod end to the knuckle using a SST (09568-4A000).

18. Install the split pin and castle nut to the tie rod end ball joint. Tighten nut to 43–58 ft. lb. (60–80 Nm).

19. Install the driveshaft lock nut. Tighten nut to 177–199 ft. lb. (245–275 Nm).

20. Stake the driveshaft lock nut using a chisel and hammer.

21. Tighten the drain plug and drain the ATF.

22. Lower the vehicle and remove the wheel & tire assembly to the front hub assembly. Tighten assemblies to 65–80 ft. lb. (90–110 Nm).

- After installing the driveshaft joint to the transaxle case, be sure the driveshaft joint does not come out.
- The driveshaft lock nut should be replaced with new ones.
- After installation driveshaft lock nut, stake the lock nut using a chisel and hammer.

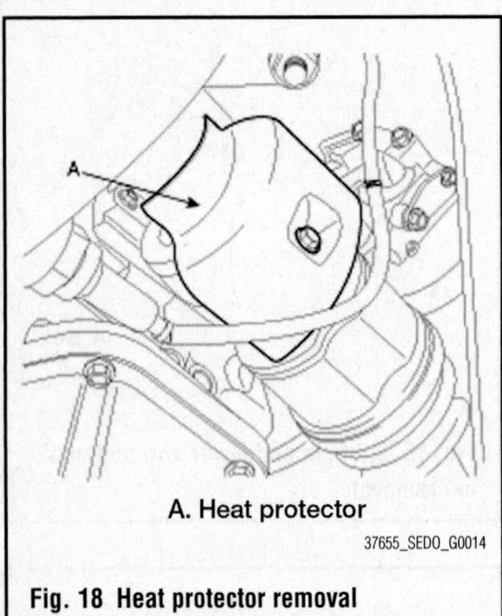

Fig. 18 Heat protector removal

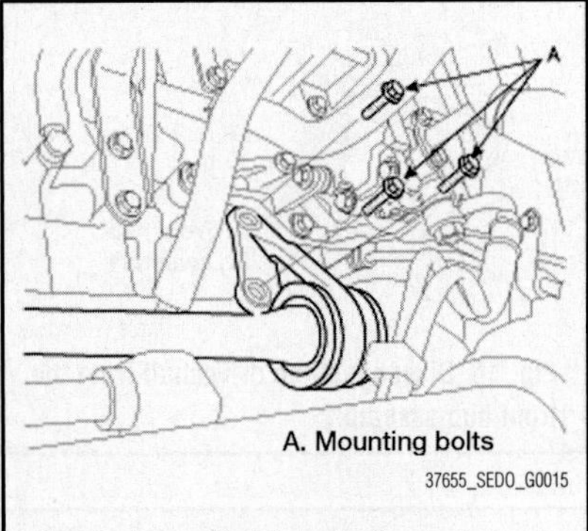

Fig. 19 Inner shaft bearing bracket assembly mounting bolts removal

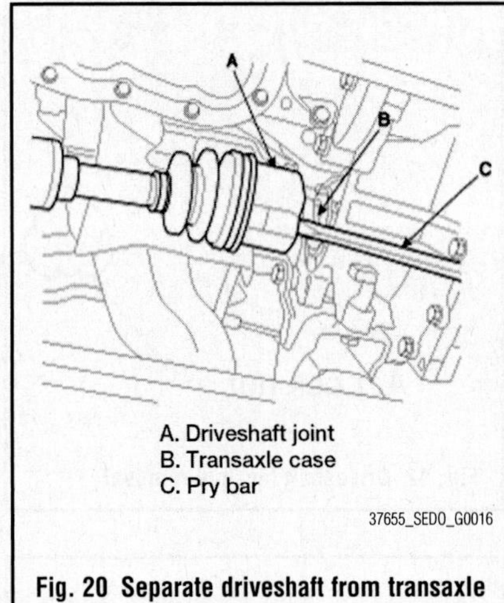

Fig. 20 Separate driveshaft from transaxle

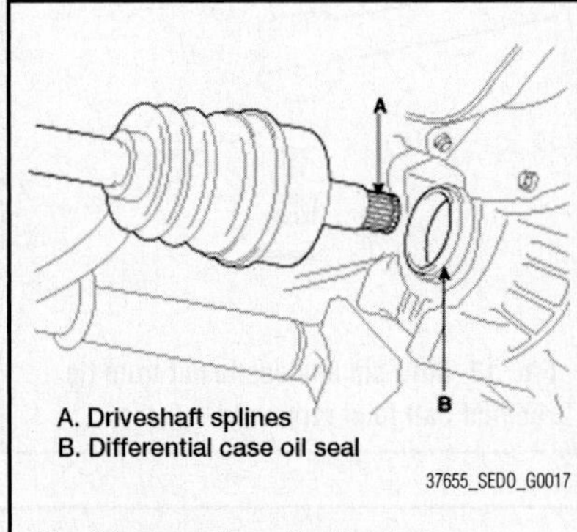

Fig. 21 Gear oil application to the driveshaft splines

ENGINE COOLING

ENGINE FAN

REMOVAL & INSTALLATION

See Figures 22 and 23.

1. Drain the engine coolant.
2. Remove the radiator grille upper cover.
3. Remove the radiator support upper member assembly.

➡The bottom side bolt which can be seen after removing the undercover should be loosened for removal of the radiator support upper member assembly.

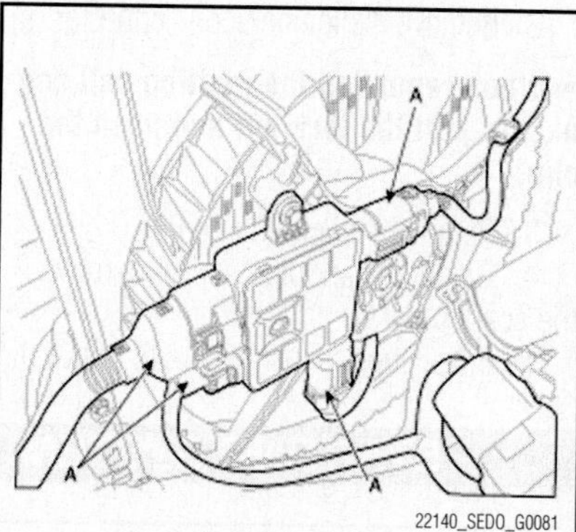

Fig. 22 Disconnect the radiator fan connectors (A).

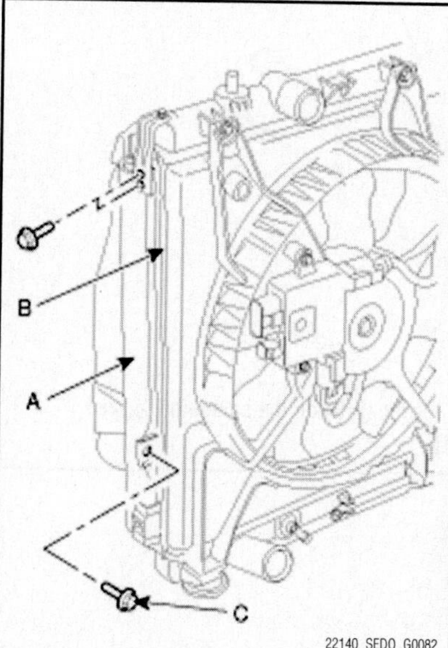

Fig. 23 Separate the condenser (A) from the radiator assembly (B) by removing the bolts (C).

4. Disconnect radiator upper and lower hoses.
5. Disconnect transaxle oil cooler hoses.
6. Disconnect the radiator fan connectors.
7. Disconnect the power steering pressure lines from the radiator assembly
8. Separate the condenser from the radiator assembly by removing the bolts.
9. Remove the radiator bracket.
10. Remove the radiator assembly.
11. Remove the radiator cooling fan

To install:

12. Install the radiator fan to the radiator.
13. Install the radiator assembly to the vehicle.
14. Install the radiator bracket.
15. Attach the condenser to the radiator assembly.
16. Connect the radiator fan connectors.
17. Connect the power steering pressure lines to the radiator assembly.
18. Connect transaxle oil cooler hoses.
19. Connect radiator upper and lower hoses.
20. Install the radiator support upper member assembly.
21. Install the radiator grille upper cover.
22. Fill with engine coolant.
23. Start engine and check for leaks.
24. Recheck engine coolant level.

RADIATOR

REMOVAL & INSTALLATION

See Figures 22 and 23.

1. Drain the engine coolant.
2. Remove the radiator grille upper cover.
3. Remove the radiator support upper member assembly.

➡The bottom side bolt which can be seen after removing the undercover should be loosened for removal of the radiator support upper member assembly.

4. Disconnect radiator upper and lower hoses.
5. Disconnect transaxle oil cooler hoses.
6. Disconnect the radiator fan connectors.
7. Disconnect the power steering pressure lines from the radiator assembly
8. Separate the condenser from the radiator assembly by removing the bolts.
9. Remove the radiator bracket.
10. Remove the radiator assembly.

To install:

11. Install the radiator assembly to the vehicle.
12. Install the radiator bracket.
13. Attach the condenser to the radiator assembly.
14. Connect the radiator fan connectors.
15. Connect the power steering pressure lines to the radiator assembly.
16. Connect transaxle oil cooler hoses.
17. Connect radiator upper and lower hoses.
18. Install the radiator support upper member assembly.
19. Install the radiator grille upper cover.
20. Fill with engine coolant.
21. Start engine and check for leaks.
22. Recheck engine coolant level.

THERMOSTAT

REMOVAL & INSTALLATION

1. Drain engine coolant so its level is below thermostat.
2. Remove water inlet and thermostat.

To install:

3. Place thermostat in thermostat housing.
    a. Install a new thermostat and gasket.
    b. Install the thermostat with the jiggle valve upward.
4. Install water inlet. Torque: 12.3—14.5 ft. lbs. (16.7–19.6 Nm).
5. Fill with engine coolant.
6. Start engine and check for leaks.

WATER PUMP

REMOVAL & INSTALLATION

1. Drain the engine coolant.
2. Remove the serpentine drive belt
3. Remove the water pump pulley
4. Remove the water pump and gasket.

To install:

5. Install the water pump and a new gasket.
6. Install the water pump pulley.
7. Install the serpentine drive belt
8. Fill with engine coolant.
9. Start engine and check for leaks.
10. Recheck engine coolant level.

ENGINE ELECTRICAL — CHARGING SYSTEM

ALTERNATOR

REMOVAL & INSTALLATION

1. Disconnect the battery negative terminal first, then the positive terminal.
2. Disconnect the alternator connector.
3. Remove the cable from alternator "B" terminal.
4. Remove the accessory drive belt.
5. Pull out the through bolt and remove the alternator
6. Installation is the reverse of removal.

VOLTAGE REGULATOR

ADJUSTMENT

The charging system includes an alternator with a built-in regulator. There is no adjustment possible for the voltage regulator. If the voltage regulator is defective, it must be replaced.

ENGINE ELECTRICAL — IGNITION SYSTEM

FIRING ORDER

3.8L engine firing order: 1–2–3–4–5–6

IGNITION COIL

REMOVAL & INSTALLATION

1. Remove the engine cover.
2. Remove the ignition coil connector (s).

➡When removing the ignition coil connector, pull the lock pin and push the clip.

3. Remove the ignition coil(s).
4. Installation is the reverse of removal.

IGNITION TIMING

ADJUSTMENT

These vehicles are equipped with a Distributorless Ignition System (DIS). No adjustment is necessary or possible.

SPARK PLUGS

REMOVAL & INSTALLATION

1. Remove the engine cover.
2. Remove the ignition coil connector (s).

➡When removing the ignition coil connector, pull the lock pin and push the clip.

3. Remove the ignition coil(s).
4. Using a spark plug socket, remove the spark plug(s).
5. Installation is the reverse of removal.

ENGINE ELECTRICAL — STARTING SYSTEM

STARTER

REMOVAL & INSTALLATION

See Figure 24.

1. Disconnect the battery negative cable.
2. Disconnect the starter cable from the B terminal on the solenoid, and disconnect the connector from the S terminal.
3. Remove the 2 bolts holding the starter, and remove the starter.
4. Installation is the reverse of removal.
5. Connect the battery negative cable to the battery.

SOLENOID OR RELAY REPLACEMENT

1. Remove the fuse box cover.
2. Remove the starter relay. Replace is necessary.

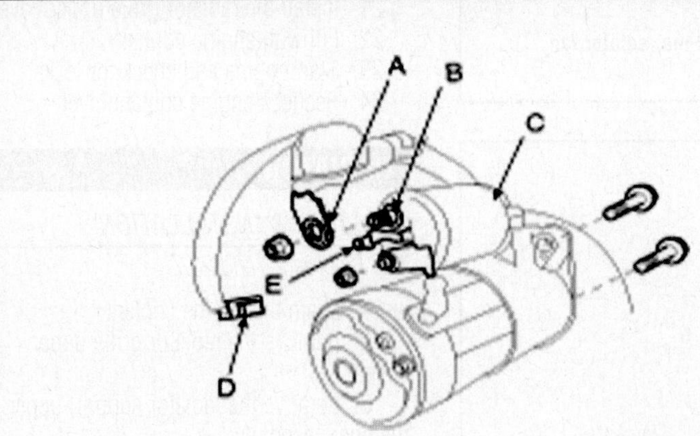

Fig. 24 Disconnect the starter cable (A) from the B terminal (B) on the solenoid (C), and disconnect the connector (D) from the S terminal (E).

ENGINE MECHANICAL

ACCESSORY DRIVE BELTS

DRIVE BELT ROUTING

See Figure 25.

Refer to the accompanying illustration.

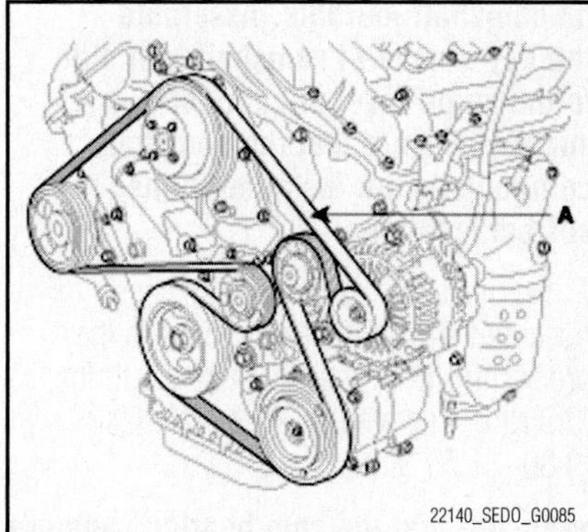

Fig. 25 Accessory belt routing

INSPECTION

Inspect the drive belt for signs of glazing or cracking. A glazed belt will be perfectly smooth from slippage, while a good belt will have a slight texture of fabric visible. Cracks will usually start at the inner edge of the belt and run outward. All worn or damaged drive belts should be replaced immediately.

ADJUSTMENT

The belt tension is maintained by an automatic tensioner. No adjustment is necessary or possible.

REMOVAL & INSTALLATION

1. Relieve tension on the tensioner pulley.
2. Remove accessory drive belt

To install:

3. Install accessory drive belt in the following sequence:
    a. Crankshaft pulley
    b. A/C pulley
    c. Idler pulley
    d. alternator pulley
    e. Water pump pulley
    f. P/S pump pulley
    g. Tensioner pulley.
4. Rotate auto tensioner arm in the counterclockwise direction moving auto tensioner pulley bolt with wrench.
5. After putting belt on auto tensioner pulley, release the auto tensioner pulley slowly.

CRANKSHAFT DAMPER (BALANCER)

REMOVAL & INSTALLATION

1. Remove the accessory drive belt.
2. Remove the crankshaft damper pulley.
3. Installation is the reverse of removal.
4. Torque crankshaft damper pulley bolt to 209.76–224.22 ft. lbs. (284.2–303.8 Nm).

CYLINDER HEAD

REMOVAL & INSTALLATION

See Figures 26 through 33.

➡**Engine removal is required for this procedure.**

1. Remove exhaust manifold.
2. Remove intake manifold.

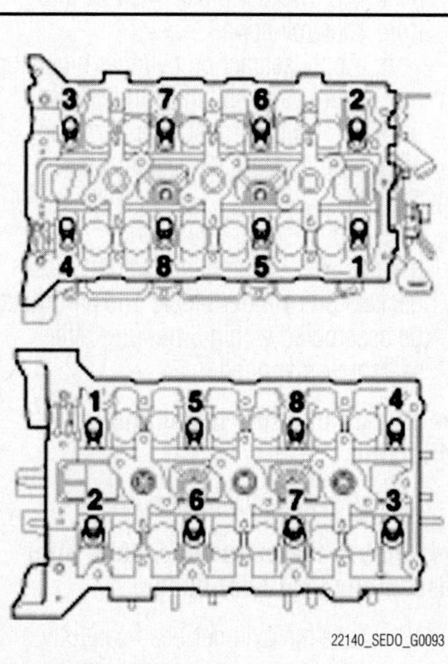

Fig. 26 Uniformly loosen and remove the 16 cylinder head bolts.

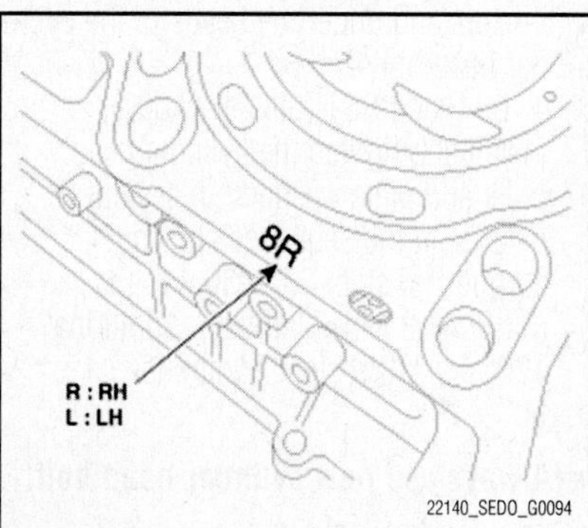

Fig. 27 Be careful of the installation direction.

3. Remove timing chain.
4. Remove water temperature control assembly.
5. Remove camshaft bearing caps.
6. Remove camshaft assembly.
7. Remove cylinder head bolts, then remove cylinder head.
    a. Uniformly loosen and remove the 16 cylinder head bolts, in several passes, in the sequence shown.

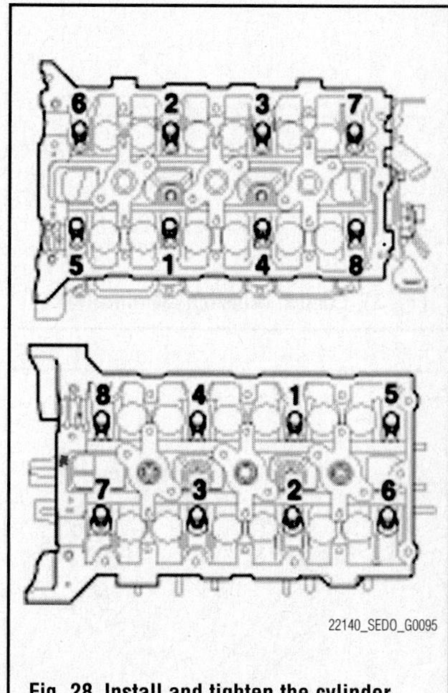

Fig. 28 Install and tighten the cylinder head bolts and plate washers, in several passes, in the sequence shown.

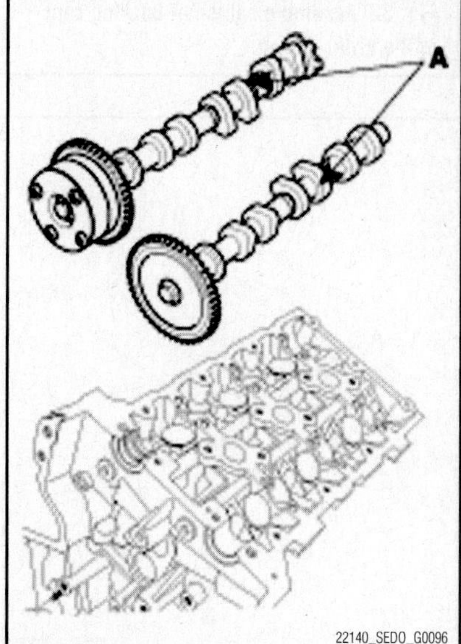

Fig. 29 Camshaft positions—left bank shown.

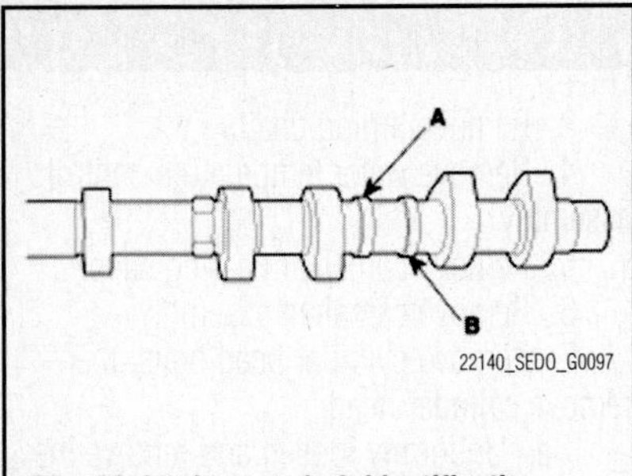

Fig. 30 Intake camshaft identification

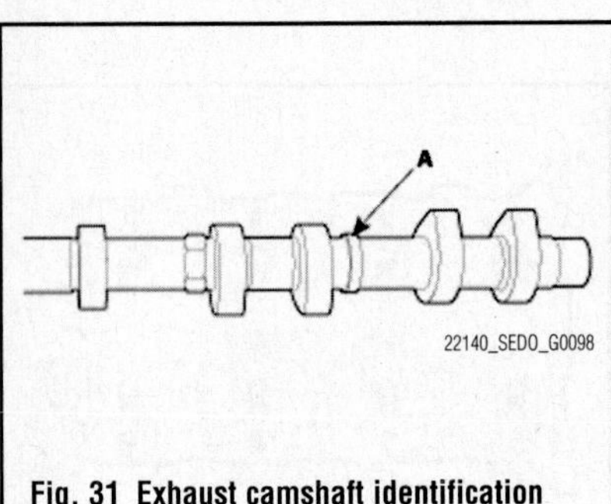

Fig. 31 Exhaust camshaft identification

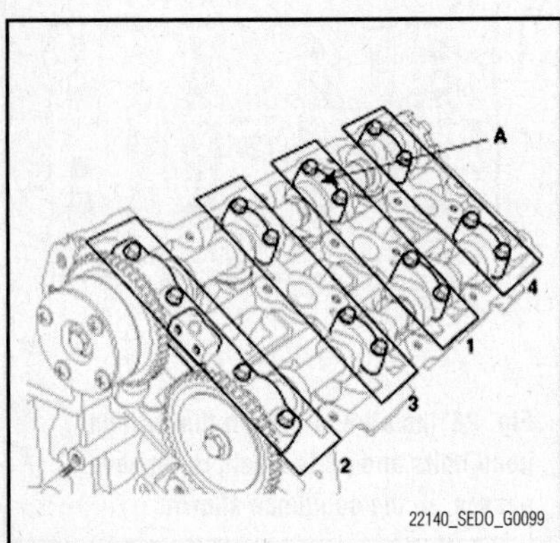

Fig. 32 Assemble camshaft bearing caps in the order shown.

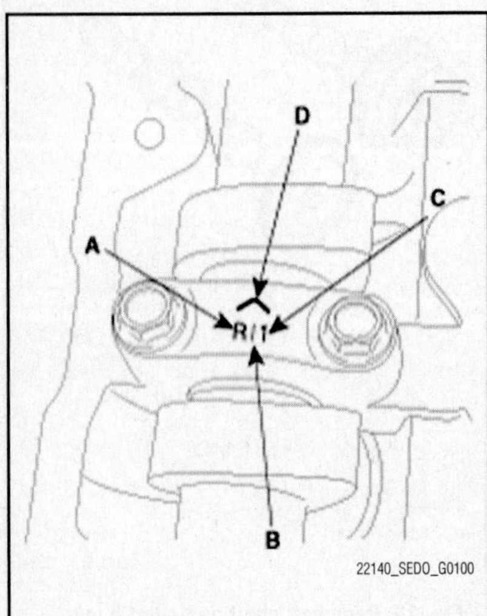

Fig. 33 Bearing cap location markers

b. Remove the 16 cylinder head bolts and plate washers.

CAUTION

Head warpage or cracking could result from removing bolts in an incorrect order.

c. Lift the cylinder head from the dowels on the cylinder block and place the cylinder head on wooden blocks on a bench.

To install:

➡Thoroughly clean all parts to be assembled. Always use a new head and manifold gasket. The cylinder head gasket is a metal gasket. Take care not to bend it.

8. If necessary, rotate the crankshaft to set theNo.1 piston at TDC.
9. Install the cylinder head.
   a. The sealant locations on cylinder head and cylinder block must be free from contamination.
   b. Apply sealant on cylinder block top face before assembling cylinder head gaskets. The part must be assembled within 5 minutes after sealant was applied.
   c. Apply sealant on cylinder head gaskets after assembling cylinder head gaskets on cylinder block. The part must be assembled within 5 minutes after sealant was applied.

➡Be careful of the installation direction.

   d. Install the cylinder head.

➡Remove the extruded sealant after assembling cylinder heads.

10. Place the cylinder head carefully in order not to damage the gasket with the bottom part of the end.
11. Install cylinder head bolts.
    a. Do not apply engine oil on the threads and under the heads of the cylinder head bolts.
    b. Using SST (09221-4A000), install and tighten the cylinder head bolts and plate washers, in several passes, in the sequence shown. Torque: 1st step: 27.5–30.4 ft. lbs. (37.3–41.2 Nm); 2nd step: Additional 120° ± 2°; 3rd step: Additional 90° ± 2°.

➡Always use new cylinder head bolt.

12. Install the CVVT and camshaft sprocket. Torque: 47.74–56.4 ft. lbs. (64.68–76.44 Nm).

➡Install camshaft-inlet to dowel pin of CVVT assembly. Ensure that the camshaft oil inlet is not obstructed. Do not rotate CVVT assembly when camshaft is installed to dowel pin of CVVT assembly.

13. Install camshafts.

➡Apply a light coat of engine oil on camshaft journals. Assemble the key groove of camshaft rear side to the same level of head top surface. Ensure that the camshaft components are installed in the correct locations.

14. Install camshaft bearing caps. Assemble camshaft bearing caps in the order shown. Torque: 1st step: 4.3 ft. lbs. (5.9 Nm); 2nd step: 7.23–8.68 ft. lbs. (9.80–11.76 Nm).

➡Ensure that the cam bearing caps are installed in the correct location and direction.

A: L(LH), R (RH) B: I(Intake), None (Exhaust) C: Journal number D: Front mark.

15. Install water temperature control assembly.
16. Install timing chain.
17. Check and adjust valve clearance.
18. Install the exhaust manifold.
19. Install the intake manifold.

EXHAUST MANIFOLD

REMOVAL & INSTALLATION

1. Remove undercover
2. Disconnect LH,RH rear oxygen sensor connector from bracket.
3. Remove front muffler
4. Remove oil level gauge.
5. Disconnect LH front oxygen sensor connector from bracket.
6. Remove LH heat protector.
7. Remove LH exhaust manifold.
8. Disconnect RH front oxygen sensor connector from bracket.
9. Remove RH heat protector.
10. Remove RH exhaust manifold.

To install:

11. Install new gasket and exhaust manifold. Torque: 28.9–32.5 ft. lbs. (39.2–44.1 Nm).
12. Install heat protector.
13. Install front muffler. Torque: 28.9–32.5 ft. lbs. (39.2–44.1 Nm).
14. Connect oxygen sensor connector.
15. Install undercover.

FLEXPLATE

REMOVAL & INSTALLATION

1. Disconnect the negative battery cable.
2. Remove the transaxle assembly.
3. Remove the flexplate-to-crankshaft bolts, the flexplate and shim plates.

To install:

4. Install the flexplate, and tighten the mounting bolts to 71–76 ft. lbs. (97–102 Nm).
5. Install the transaxle.
6. Connect the negative battery cable.

INTAKE MANIFOLD

REMOVAL & INSTALLATION

See Figures 34 through 42.

1. Remove the engine cover
2. Remove the intake air hose and air cleaner assembly.
    a. Disconnect the MAF connector.
    b. Disconnect the breather hose from air cleaner hose.
    c. Remove the intake air hose and air cleaner assembly.

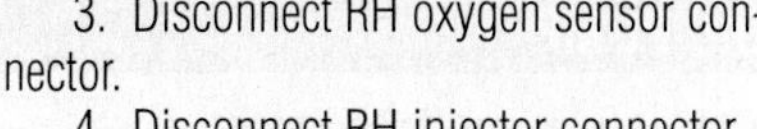

3. Disconnect RH oxygen sensor connector.
4. Disconnect RH injector connector and ignition coil connector.
5. Disconnect PCSV connector, MAP sensor connector and PCSV hose.
6. Disconnect the Electric Throttle Control (ETC) connector and knock sensor connector.
7. Remove the ETC bracket.
8. Disconnect the water hoses from ETC.

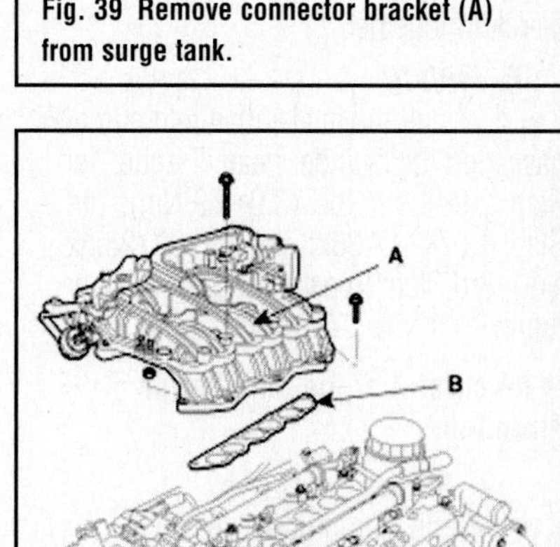

Fig. 39 Remove connector bracket (A) from surge tank.

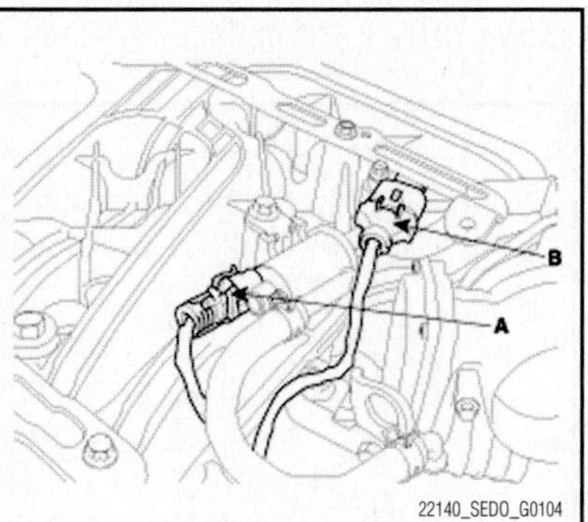

Fig. 36 Disconnect PCSV connector (A), MAP sensor connector (B) and PCSV hose.

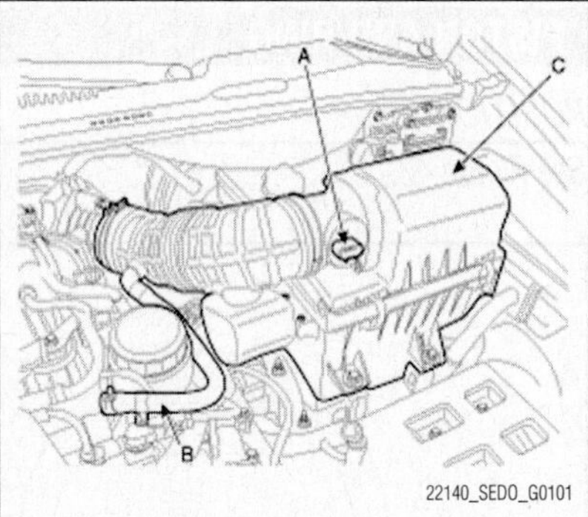

Fig. 34 Remove the intake air hose and air cleaner assembly.

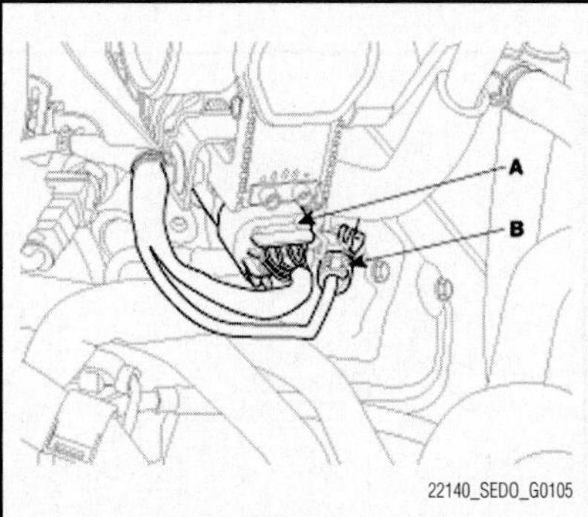

Fig. 37 Disconnect ETC connector (A) and knock sensor connector (B).

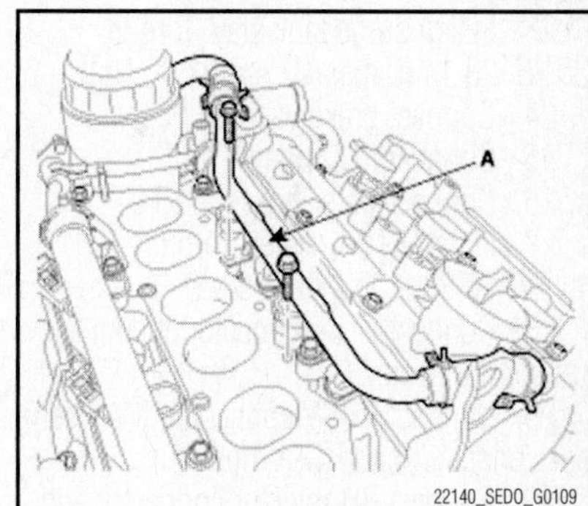

Fig. 41 Disconnect breather Pipe assembly (A).

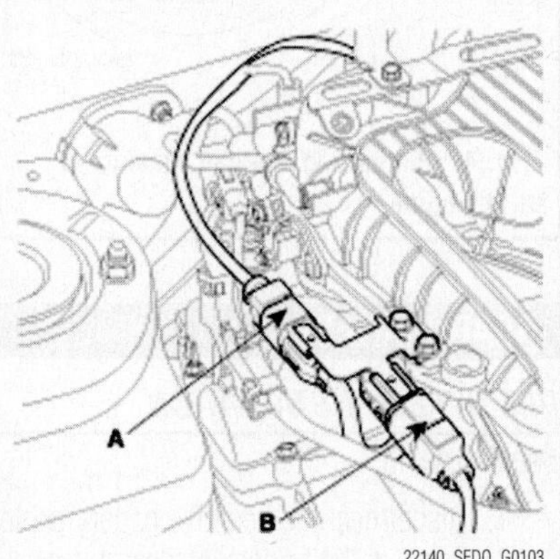

Fig. 35 Disconnect RH injector connector (A) and ignition coil connector (B).

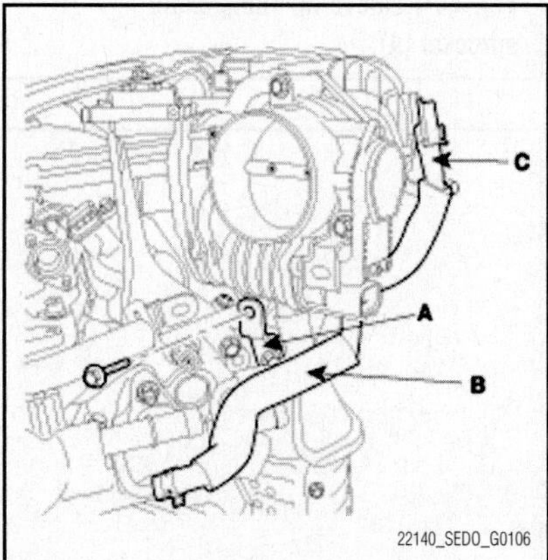

Fig. 38 Remove ETC bracket (A). Disconnect water hose (B) and PCV (C) hose.

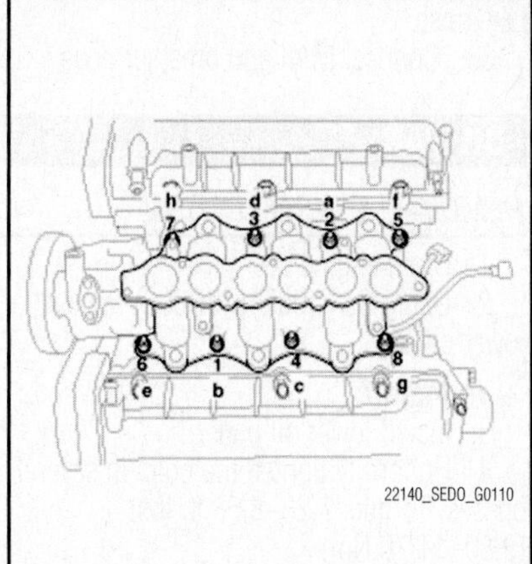

Fig. 42 Tightening sequence

9. Disconnect the PCV hose.
10. Disconnect the brake vacuum hose.
11. Remove the surge tank stay.
12. Remove the connector bracket from surge tank.
13. Remove surge tank.
14. Disconnect the breather Pipe assembly.
15. Disconnect the LH injector connector.
16. Remove the delivery pipe and intake manifold together.

To install:

17. Install the intake manifold and new gasket on the cylinder head. Torque: 1st Step: 2.9–4.3 ft. lbs. (3.9–5.9 Nm); 2nd Step: 13.74–17.36 ft. lbs. (18.62–23.52 Nm); 3rd Step: Repeat 2nd step twice or more.

➡Be careful of the installation direction.

- 1st step order: a–h.
- 2nd step order: 1–8.

18. Install delivery pipe.
19. Connect LH injector connector.
20. Connect breather pipe assembly.
21. Install surge tank.
22. Install connector bracket on the surge tank.
23. Install surge tank stay. Torque: 20.25–23.14 ft. lbs. (27.44–31.36 Nm).
24. Connect brake vacuum hose.
25. Connect PCV hose.
26. Connect water hoses to ETC.
27. Install ETC bracket. Torque: 11.57–18.80 ft. lbs. (15.68–25.48 Nm).
28. Connect ETC connector and knock sensor connector.
29. Connect PCSV connector, MAP sensor connector and PCSV hose.
30. Connect RH injector connector and ignition coil connector.
31. Connect RH oxygen sensor connector.
32. Install air cleaner upper cover and in take hose.
33. Connect MAF and breather hose.

OIL PAN

REMOVAL & INSTALLATION

1. Drain engine oil.
2. Using SST (09215-3C000) remove lower oil pan.

To install:

3. Install lower oil pan.
4. Uniformly tighten the bolts in several passes. Torque: 7.23–8.68 ft. lbs. (9.80–11.76 Nm).
5. After assembly, wait at least 30 minutes before filling the engine with oil.

OIL PUMP

REMOVAL & INSTALLATION

See Figures 43 through 46.

1. Drain engine oil.
2. Using SST (09215-3C000) remove lower oil pan.
3. Remove oil pump chain cover.
4. Remove oil pump chain sprocket.
5. Remove oil pump.

To install:

6. Install oil pump. Torque19.6 ~ 23.5Nm (2.0 ~ 2.4kgf.m, 14.5 ~ 17.4lb-ft)

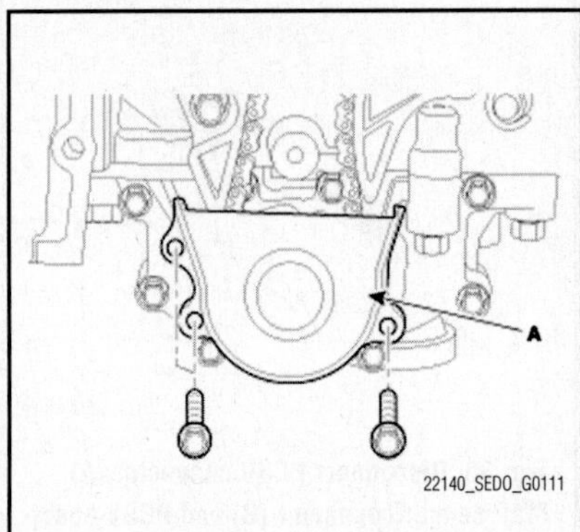

Fig. 43 Remove oil pump chain cover (A).

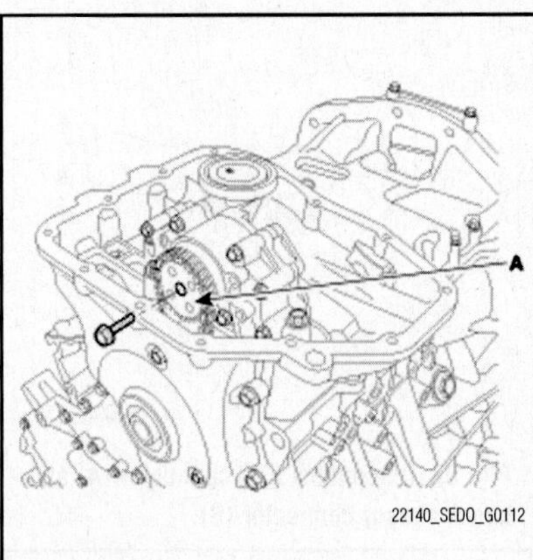

Fig. 44 Remove oil pump chain sprocket (A).

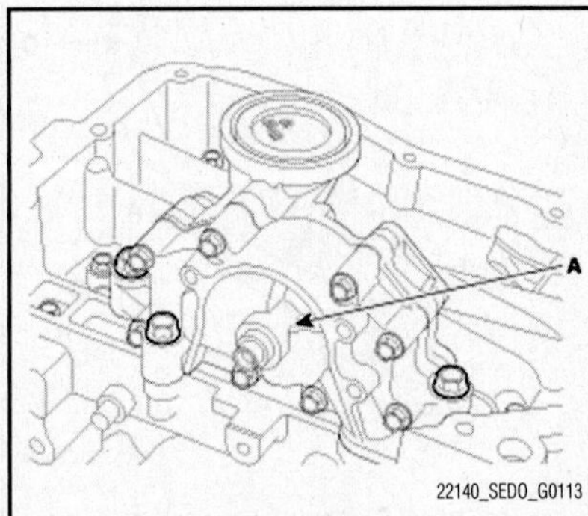

Fig. 45 Remove oil pump(A).

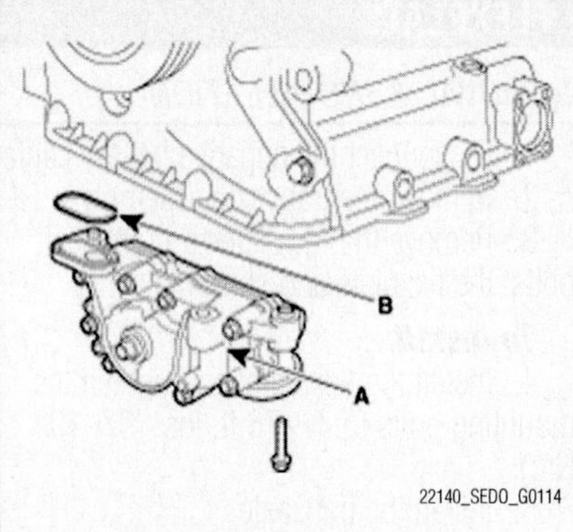

Fig. 46 Install oil pump (A). Always use a new O-ring(B).

➡Always use a new O-ring.

7. Install oil pump sprocket and oil pump chain on the oil pump. Torque: 13.74–15.91 ft. lbs. (18.62–21.56 Nm).
8. Install oil pump chain cover. Torque: 7.23–8.68 ft. lbs. (9.80–11.76 Nm).
9. Install lower oil pan.
10. Uniformly tighten the bolts in several passes. Torque: 7.23–8.68 ft. lbs. (9.80–11.76 Nm).
11. After assembly, wait at least 30 minutes before filling the engine with oil.

PISTON AND RING

POSITIONING

See Figure 47.

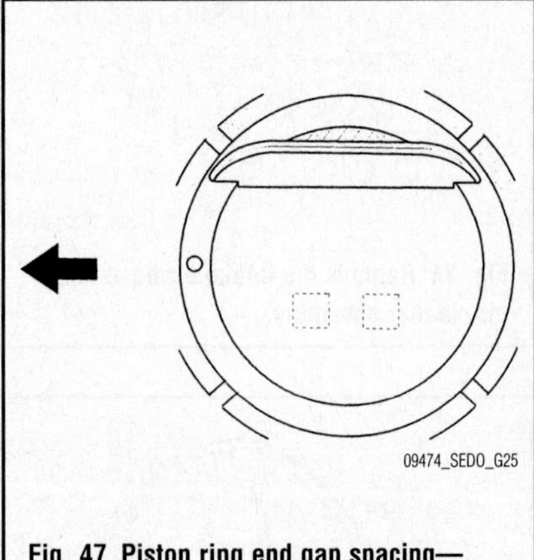

Fig. 47 Piston ring end gap spacing—3.8L Engine

REAR MAIN SEAL

REMOVAL & INSTALLATION

See Figure 48.

1. Disconnect the negative battery cable.
2. Remove the transaxle assembly.
3. Remove the flexplate-to-crankshaft bolts, the flexplate and shim plates.

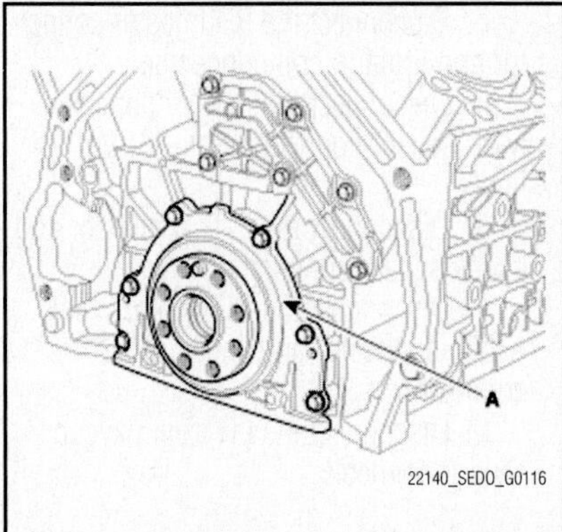

Fig. 48 Remove the rear oil seal case (A).

4. Remove the rear oil seal case.
5. Remove the oil seal.

To install:

6. Install a new oil seal.
7. Install rear oil seal case. Torque: 7.23–8.68 ft. lbs. (9.80–11.76 Nm).
8. Install the flexplate, and tighten the mounting bolts to 71–76 ft. lbs. (97–102 Nm).
9. Install the transaxle.
10. Connect the negative battery cable.

ROCKER ARMS & SHAFTS

REMOVAL & INSTALLATION

The engine covered is not equipped with rocker arms/shafts. The camshafts directly actuate the valve through a bucket type follower.

TIMING CHAIN COVER & SEAL

REMOVAL & INSTALLATION

See Figures 49 through 69.

1. Disconnect the negative battery cable.
2. Remove the engine cover
3. Remove the intake air hose and air cleaner assembly.

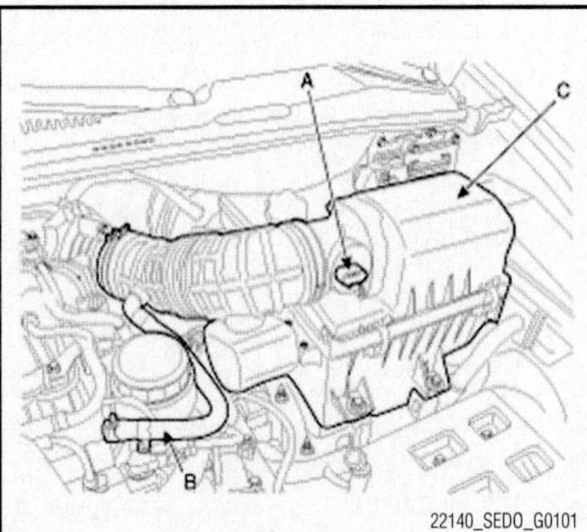

Fig. 49 Remove the intake air hose and air cleaner assembly.

a. Disconnect the Mass Air Flow (MAF) connector.

b. Disconnect the breather hose from air cleaner hose.

c. Remove the intake air hose and air cleaner assembly.

4. Remove the RH front wheel.

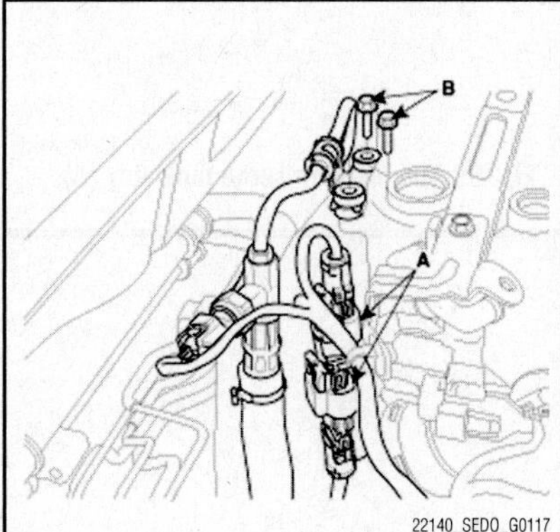

Fig. 50 Disconnect the RH oxygen sensor connector (A) and loosen the power steering hose mounting bolts (B).

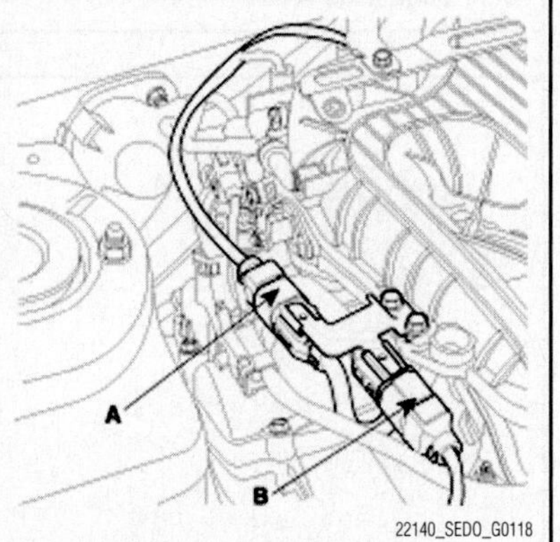

Fig. 51 Disconnect the RH injector connector (A) and ignition coil connector (B).

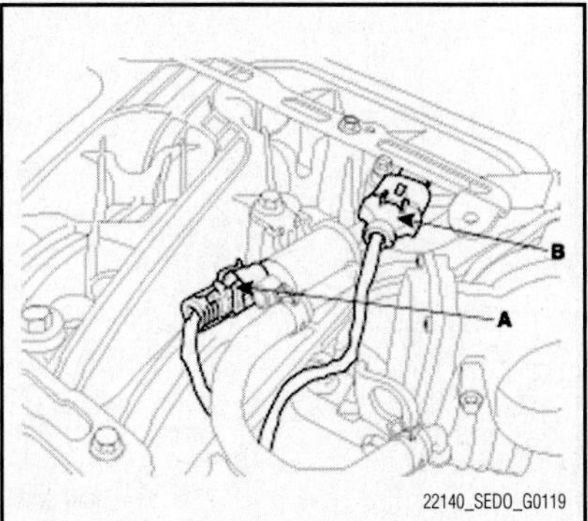

Fig. 52 Disconnect the PCSV connector (A), MAP sensor connector (B) and PCSV hose.

5. Remove the undercover
6. Remove the side cover.
7. Loosen the drain plug and drain the engine coolant.
8. Drain the engine oil.
9. Remove the surge tank.

a. Disconnect the RH oxygen sensor connector and loosen the power steering hose mounting bolts.

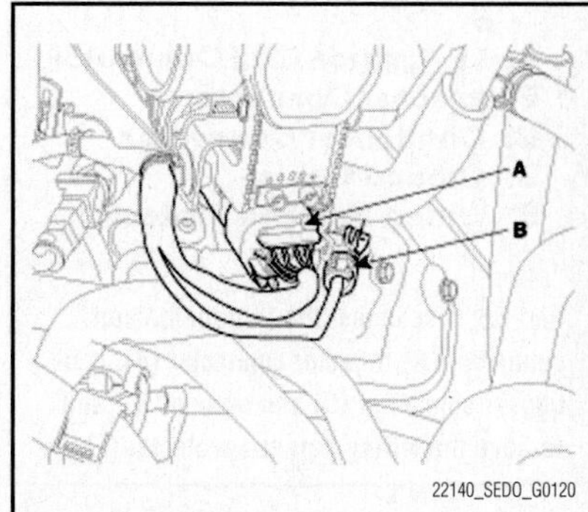

Fig. 53 Disconnect the ETC connector (A) and knock sensor connector (B).

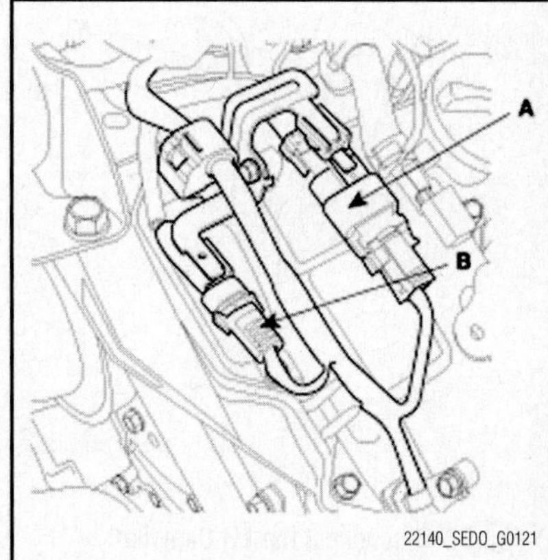

Fig. 54 Disconnect the OCV connector (A) and knock sensor connector (B).

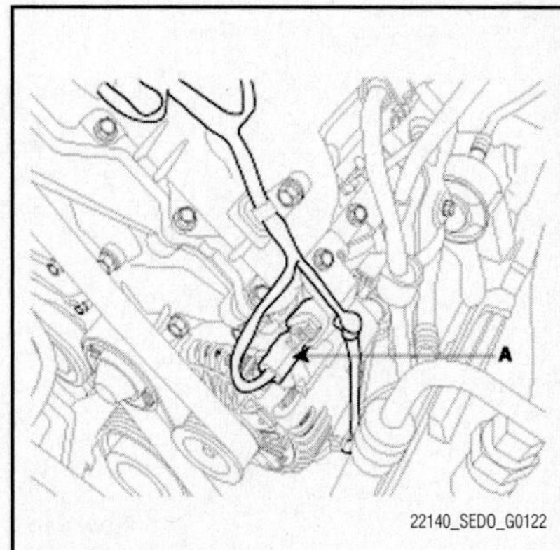

Fig. 55 Disconnect the LH front oxygen sensor connector (A).

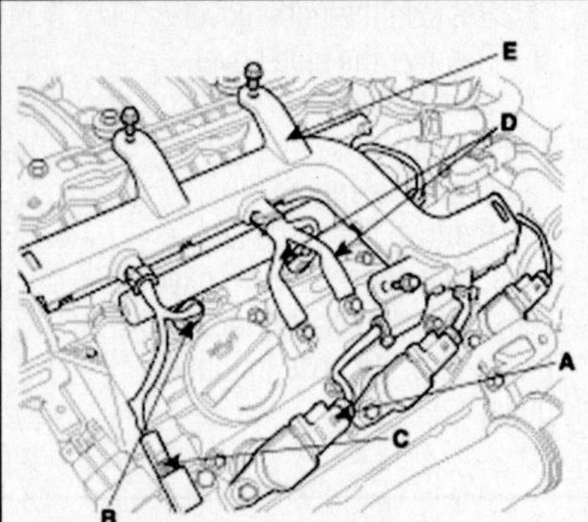

A: LH Ignition Coil Connector
B: Injector Connector
C: Condenser Connector
D: Ground Cable
E: Wiring Harness Protector

22140_SEDO_G0123

Fig. 56 Disconnect the LH ignition coil connector (A), injector connector (B), condenser connector (C) and ground (D), and remove the wiring harness protector (E).

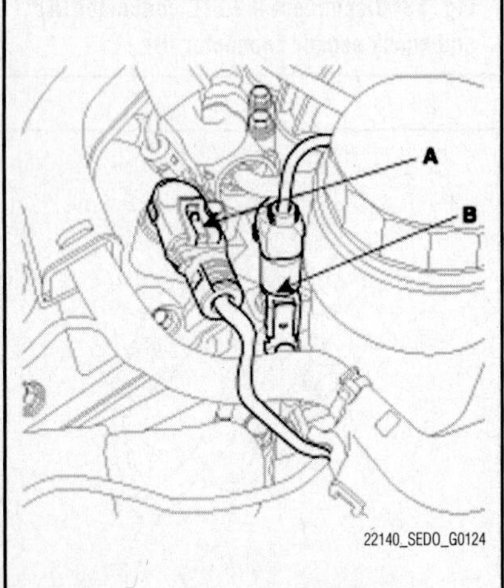

22140_SEDO_G0124

Fig. 57 Disconnect the LH Camshaft Position (CMP) Sensor (A) and oil pressure switch connector (B).

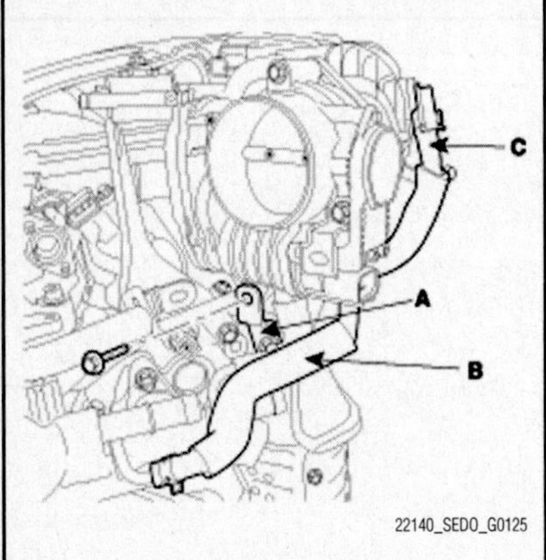

22140_SEDO_G0125

Fig. 58 Remove the Electric Throttle Control (ETC) bracket (A).

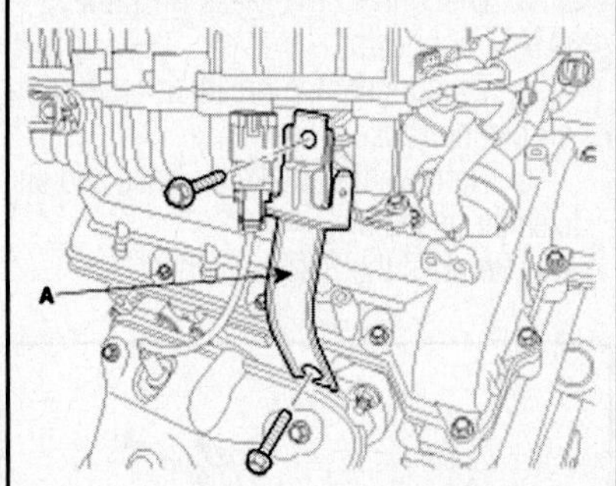

22140_SEDO_G0126

Fig. 59 Remove the surge tank stay (A).

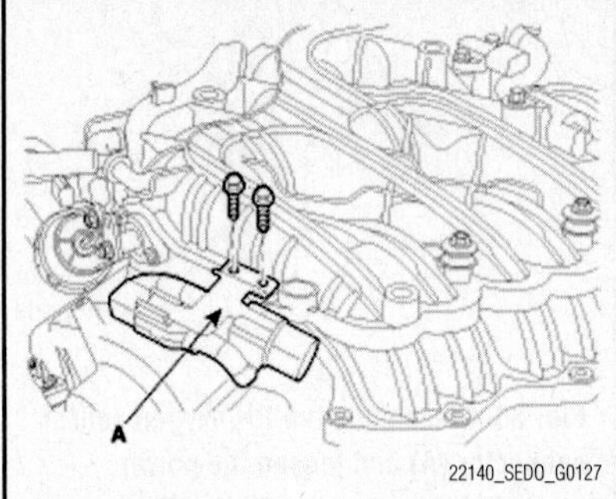

22140_SEDO_G0127

Fig. 60 Remove the connector bracket (A) from surge tank.

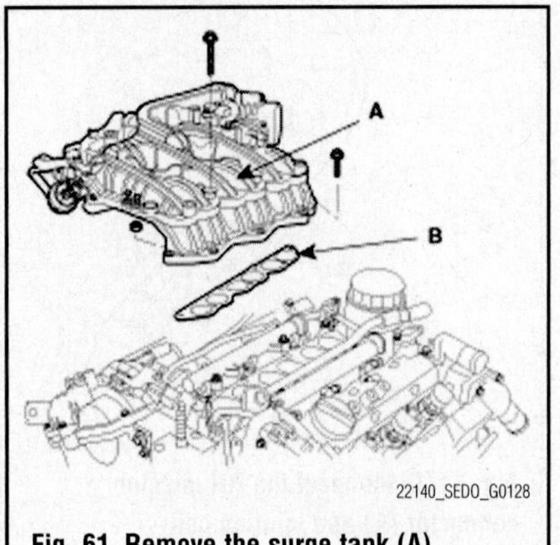

22140_SEDO_G0128

Fig. 61 Remove the surge tank (A).

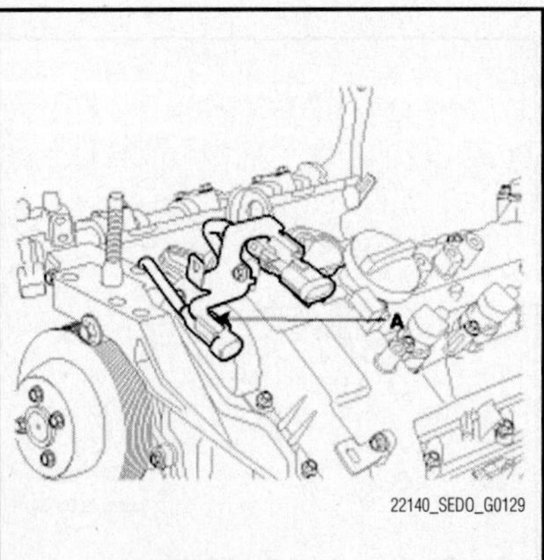

22140_SEDO_G0129

Fig. 62 Remove the connector bracket (A) from LH cylinder head cover.

b. Disconnect the RH injector connector and ignition coil connector.

c. Disconnect the PCSV connector, MAP sensor connector and PCSV hose.

d. Disconnect the ETC connector and knock sensor connector.

e. Disconnect the Oil Control Valve (OCV) connector and knock sensor connector.

f. Disconnect the LH front oxygen sensor connector.

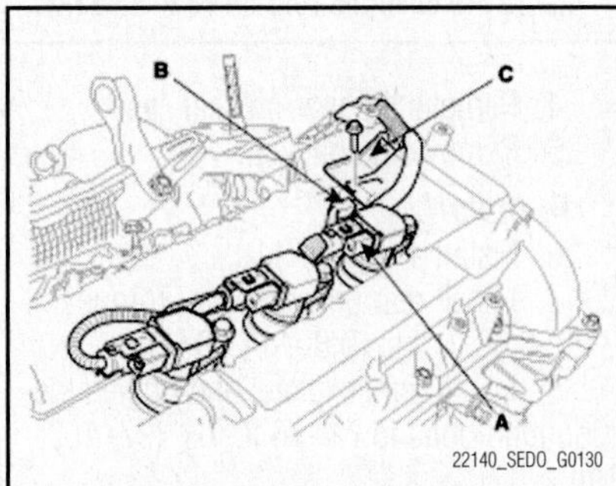

22140_SEDO_G0130

Fig. 63 Disconnect the RH ignition coil connector (A), condenser connector (B) and remove the wiring bracket (C).

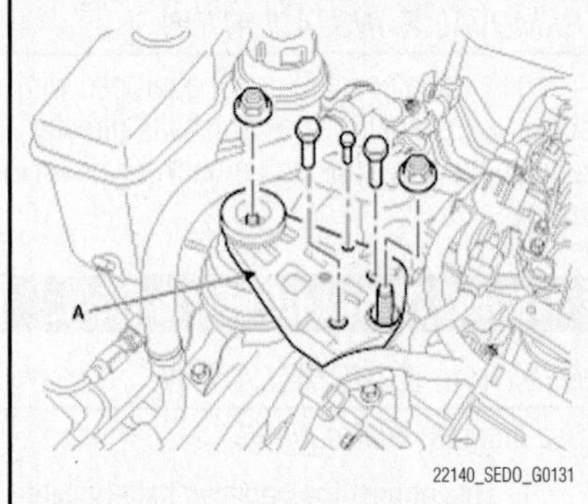

22140_SEDO_G0131

Fig. 64 Remove the engine mounting bracket (A).

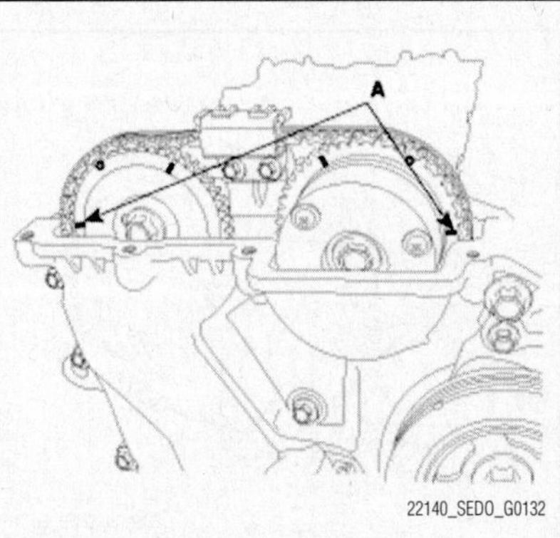

22140_SEDO_G0132

Fig. 65 Check that the mark (A) of the camshaft timing sprockets are in straight line on the cylinder head

g. Disconnect the LH ignition coil connector, injector connector, condenser connector and ground, and remove the wiring harness protector.

h. Disconnect the LH Camshaft Position (CMP) Sensor and oil pressure switch connector.

i. Remove the Electric Throttle Control (ETC) bracket.

j. Disconnect the water hoses from ETC.

k. Disconnect the PCV hose.

l. Disconnect the brake vacuum hose.

m. Remove the surge tank stay.

n. Remove the connector bracket from surge tank.

o. Remove the surge tank.

10. Remove the cylinder head cover.

a. Remove the connector bracket from LH cylinder head cover.

b. Disconnect the RH ignition coil connector, condenser connector and remove the wiring bracket.

c. Remove the LH, RH ignition coil.

d. Remove the LH, RH cylinder head covers.

11. Using SST (09215-3C000) remove lower oil pan.

12. Set a jack to the upper oil pan.

13. Remove the coolant reservoir tank.

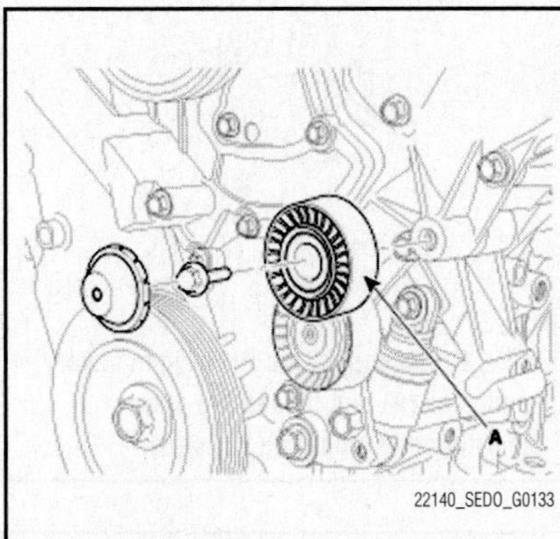

Fig. 66 Remove the drive belt idler (A).

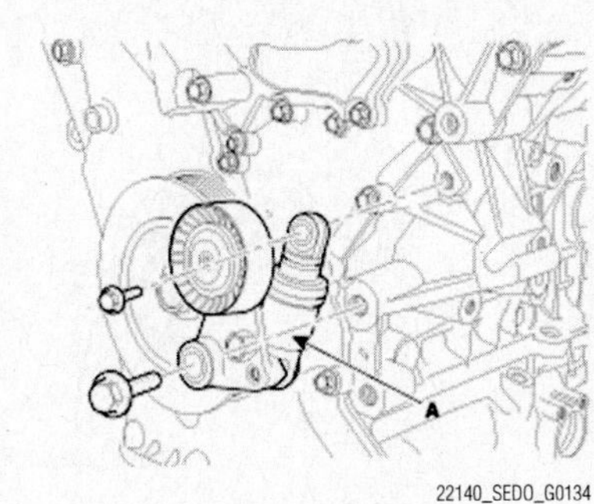

Fig. 67 Remove the drive belt auto tensioner (A).

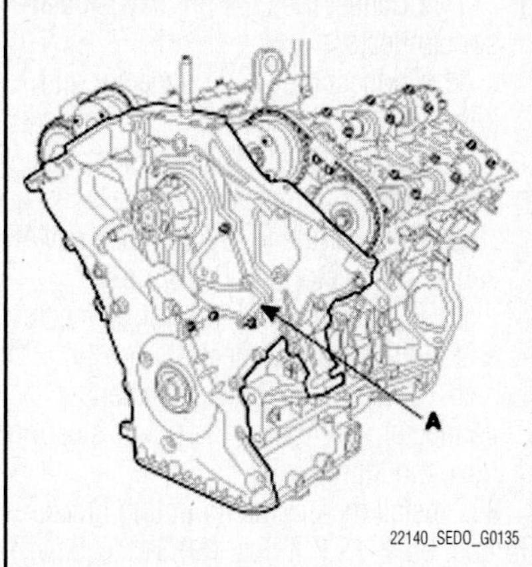

Fig. 68 Remove the timing chain cover (A).

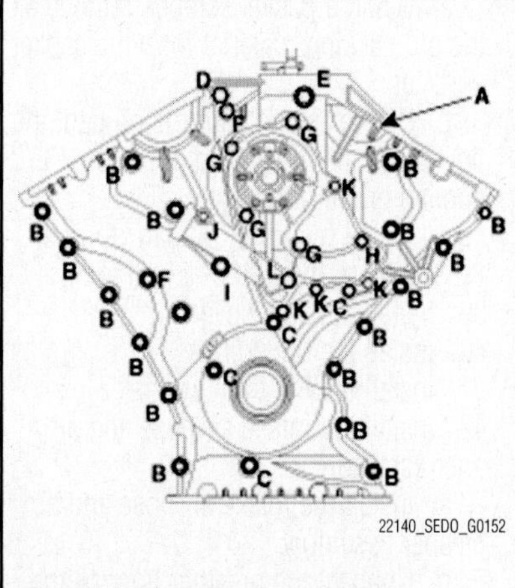

Fig. 69 Various torque locations

14. Remove the engine mounting bracket.

15. Set No.1 cylinder to TDC/compression.

a. Turn the crankshaft pulley and align its groove with the timing mark "T" of the lower timing chain cover.

➡Do not rotate the engine counterclockwise.

b. Check that the mark of the camshaft timing sprockets are in straight line on the cylinder head surface as shown. If not, turn the crankshaft one revolution (360°).

16. Remove the accessory drive belt.

17. Using SST (09231-3C300) remove the crankshaft damper pulley.

18. Lift up the engine assembly to using the jack.

19. Remove the drive belt idler.

20. Remove the drive belt auto tensioner.

21. Remove the water pump pulley.

22. Remove the timing chain cover.

To install:

23. Install the timing chain cover.

a. The sealant locations on chain cover and on counter parts (cylinder head, cylinder block, and lower oil pan) must be free of engine oil.

b. Before assembling the timing chain cover, the liquid sealant TB1217H should be applied on the gap between cylinder head and cylinder block. The part must be assembled within 5 minutes after sealant was applied.

c. Install the new gasket to the timing chain cover.

d. Install the timing chain cover.

e. The sealant locations on chain cover and on counter parts (cylinder head, cylinder block, and lower oil pan) must be free of engine oil.

➡During timing cover installation, care not to take off applied sealant on the timing cover by contact with other parts.

f. The dowel pins on the cylinder block and holes on the timing chain cover should be used as a reference in order to assemble the timing chain cover to the exact position.

Torque as follows:

- B (17ea): 13.74–15.91 ft. lbs. (18.62–21.56 Nm).
- C (4ea): 7.23–8.68 ft. lbs. (9.80–11.76 Nm).
- D (1ea): 43.40–50.63 ft. lbs. (58.80–68.80 Nm).
- E (1ea): 43.40–50.63 ft. lbs. (58.80–68.80 Nm).
- F (2ea): 18.08–19.53 ft. lbs. (24.50–26.46 Nm).
- G (4ea): 15.91–17.36 ft. lbs. (21.56–23.52 Nm).
- H (1ea): 7.23–8.68 ft. lbs. (9.80–11.76 Nm).
- I (1ea): 7.23–8.68 ft. lbs. (9.80–11.76 Nm).
- J (1ea): 7.23–8.68 ft. lbs. (9.80–11.76 Nm).
- K (4ea): 7.23–8.68 ft. lbs. (9.80–11.76 Nm).
- L (1ea): 15.91–19.53 ft. lbs. (21.56–26.46 Nm).

g. The firing and/or blow out test should not be performed within 30 minutes after the timing chain cover was assembled.

24. Install the water pump pulley.

Torque: 5.78–7.23 ft. lbs. (7.84–9.80 Nm).

25. Install the drive belt auto tensioner.

Torque: Large Bolt: 60.03–62.93 ft. lbs. (81.39–85.32 Nm); Small Bolt: 13.02–15.91 ft. lbs. (17.64–21.56 Nm).

26. Install the drive belt idler. Torque: 39.06–42.67 ft. lbs. (52.92–57.82 Nm).

27. Lower the engine assembly by using the jack.

28. Using SST (09231-3C100), install timing chain cover oil seal.

29. Using SST (09231-3C300) install the crankshaft damper pulley. Torque: 209.76–224.22 ft. lbs. (284.2–303.8 Nm).

30. Install the accessory drive belt in the following sequence:

- Crankshaft pulley
- A/C pulley
- Idler pulley
- Alternator pulley
- Water pump pulley
- P/S pump pulley
- Tensioner pulley.

31. Rotate auto tensioner arm in the counterclockwise moving auto tensioner pulley bolt with wrench.

32. After putting belt on auto tensioner pulley, release the auto tensioner pulley slowly.

33. Install the cylinder head cover.

a. The hardening sealant located on the upper area between timing chain cover and cylinder head should be removed before assembling cylinder head cover.

b. After applying sealant (TB1217H), it should be assembled within 5 minutes.

c. The firing and/or blow out test should not be performed within 30 minutes after the cylinder head cover was assembled.

d. Install the cylinder head cover bolts in sequence. Torque: 7.23–8.68 ft. lbs. (9.80–11.76 Nm).

e. Install the ignition coil.

f. Connect the RH ignition coil connector, the condenser connector and install the wiring bracket.

34. Install the surge tank and wiring connectors.

a. Install the surge tank. Torque: 7.23–8.68 ft. lbs. (9.80–11.76 Nm).

b. Install the connector bracket to the surge tank. Torque: 5.06–7.96 ft. lbs. (6.86–10.78 Nm).

c. Install the surge tank stay. Torque: 20.25–23.14 ft. lbs. (27.44–31.36 Nm).

d. Connect the brake vacuum hose.

e. Connect the PCV hose.

f. Connect the water hoses to the ETC.

g. Install the ETC bracket.

h. Connect the LH CMP sensor connector and oil pressure switch connector.

i. Install the wiring harness protector and connect the LH ignition coil connector, injector connector, condenser connector and ground.

j. Connect the LH front oxygen sensor connector.

k. Connect the OCV connector and knock sensor connector.

l. Connect the ETC connector and knock sensor connector.

m. Connect the PCSV connector, MAP sensor connector and PCSV hose.

n. Connect the RH injector connector and ignition coil connector.

o. Connect the RH oxygen sensor connector and tighten the power steering hose mounting bolts.

35. Install the engine mounting bracket. Torque: 65.1–79.6 ft. lbs. (88.3–107.9 Nm).

36. Install the coolant reservoir tank.

37. Remove the jack from the upper oil pan.

38. Install lower oil pan.

a. Using a gasket scraper, remove all the old packing material from the gasket surfaces.

b. Before assembling the oil pan, the liquid sealant TB1217H should be applied on oil pan.

c. Install the lower oil pan. Torque: 7.23–8.68 ft. lbs. (9.80–11.76 Nm).

39. Install the side cover.

40. Install the undercover.

41. Install the RH front wheel.

42. Install the intake air hose and air cleaner assembly.

a. Install the intake air hose and air cleaner assembly.

b. Connect the breather hose to the air intake hose.

c. Connect the MAF sensor connector.

43. Install the engine cover.

44. Connect the battery negative cable.

➡After installation, complete the following:

- Refill engine with engine oil.
- Refill radiator and reservoir tank with engine coolant.
- Bleed air from the cooling system.
- Start engine and let it run until it warms up. (until the radiator fan operates 3 or 4 times.)
- Turn Off the engine. Check the level in the radiator, add coolant if needed. This will allow trapped air to be removed from the cooling system.
- Put radiator cap on tightly, then run the engine again and check for leaks.

TIMING CHAIN & SPROCKETS

REMOVAL & INSTALLATION

See Figures 70 through 106.

1. Disconnect the negative battery cable.

2. Remove the engine cover

3. Remove the intake air hose and air cleaner assembly.

a. Disconnect the Mass Air Flow (MAF) connector.

b. Disconnect the breather hose from air cleaner hose.

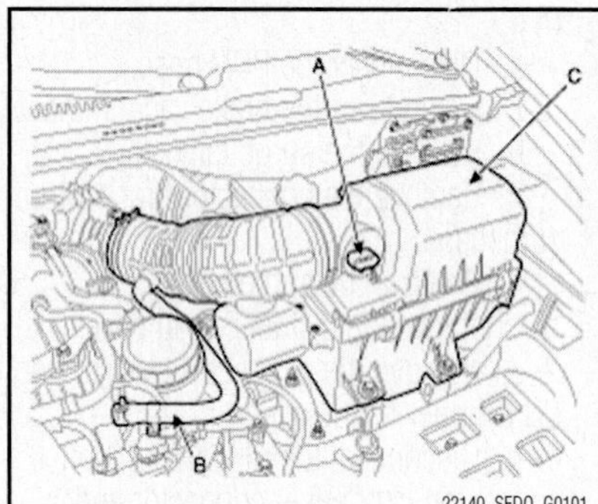

Fig. 70 Remove the intake air hose and air cleaner assembly.

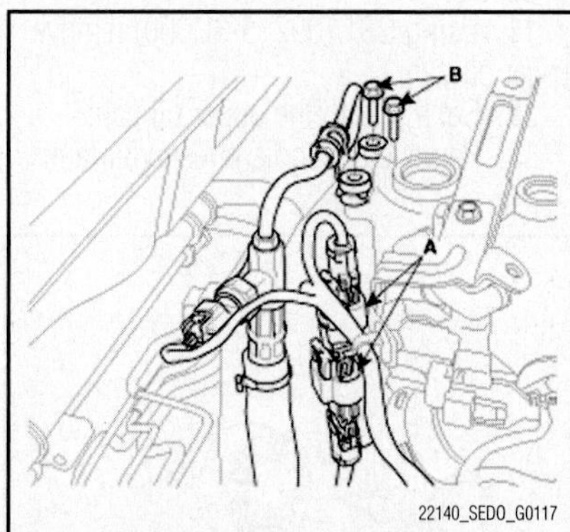

Fig. 71 Disconnect the RH oxygen sensor connector (A) and loosen the power steering hose mounting bolts (B).

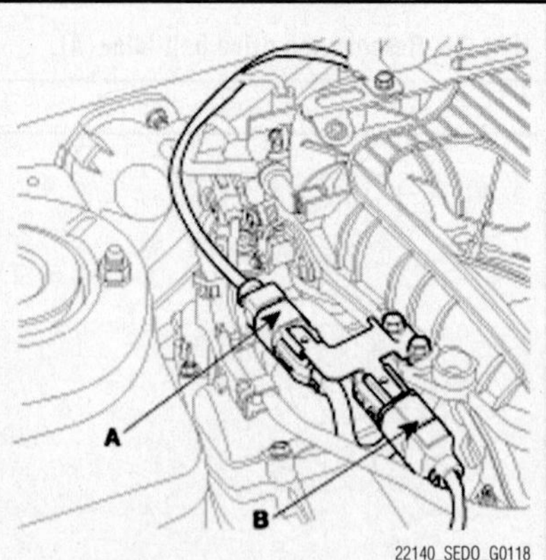

Fig. 72 Disconnect the RH injector connector (A) and ignition coil connector (B).

c. Remove the intake air hose and air cleaner assembly.

4. Remove the RH front wheel.
5. Remove the undercover
6. Remove the side cover.
7. Loosen the drain plug and drain the engine coolant.
8. Drain the engine oil.
9. Remove the surge tank.

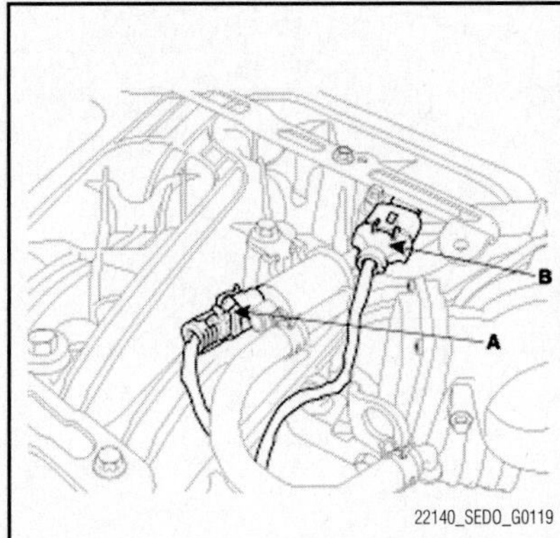

Fig. 73 Disconnect the PCSV connector (A), MAP sensor connector (B) and PCSV hose.

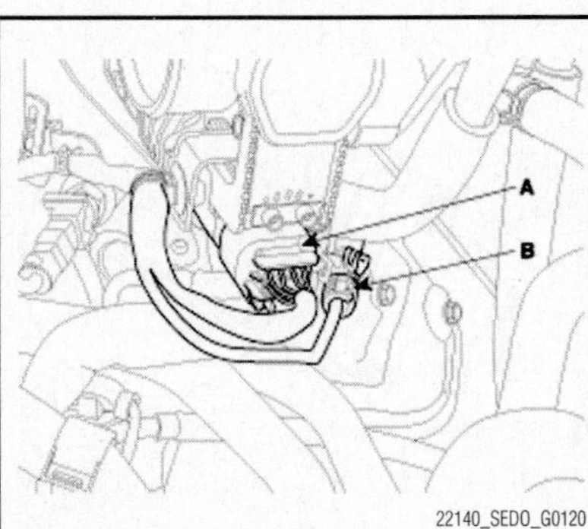

Fig. 74 Disconnect the ETC connector (A) and knock sensor connector (B).

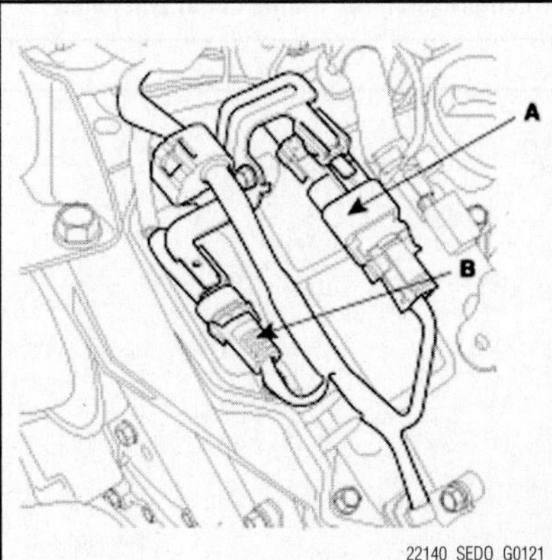

Fig. 75 Disconnect the OCV connector (A) and knock sensor connector (B).

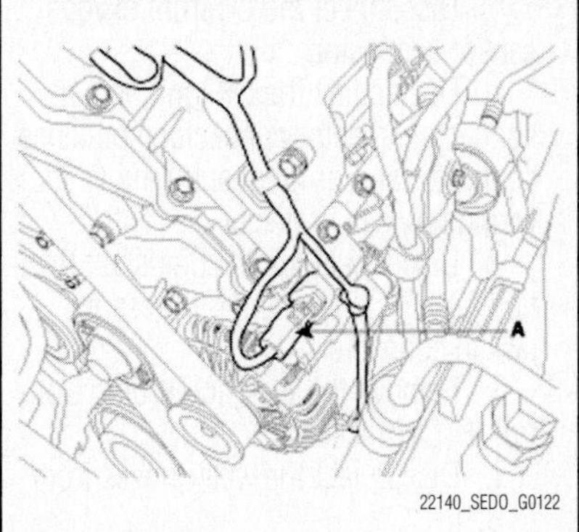

Fig. 76 Disconnect the LH front oxygen sensor connector (A).

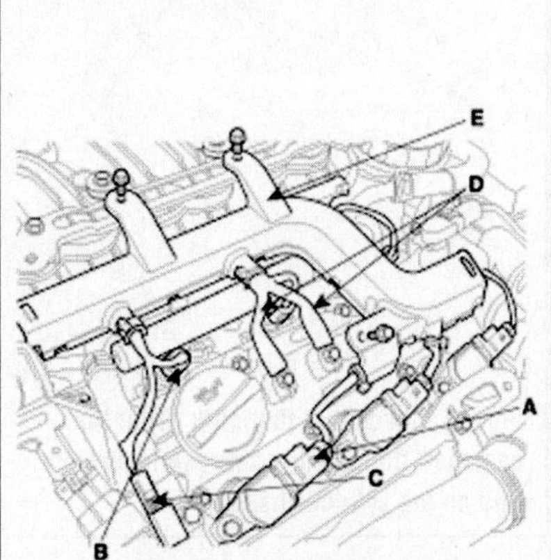

Fig. 77 Disconnect the LH ignition coil connector (A), injector connector (B), condenser connector (C) and ground (D), and remove the wiring harness protector (E).

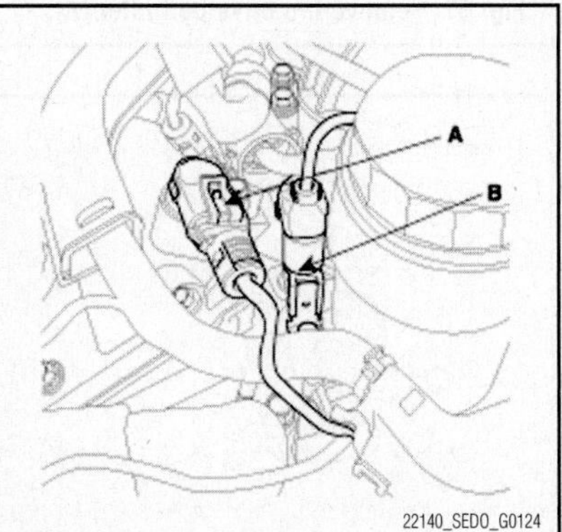

Fig. 78 Disconnect the LH Camshaft Position (CMP) Sensor and oil pressure switch connector.

a. Disconnect the RH oxygen sensor connector and loosen the power steering hose mounting bolts.

b. Disconnect the RH injector connector and ignition coil connector.

c. Disconnect the PCSV connector, MAP sensor connector and PCSV hose.

d. Disconnect the ETC connector and knock sensor connector.

e. Disconnect the Oil Control Valve (OCV) connector and knock sensor connector.

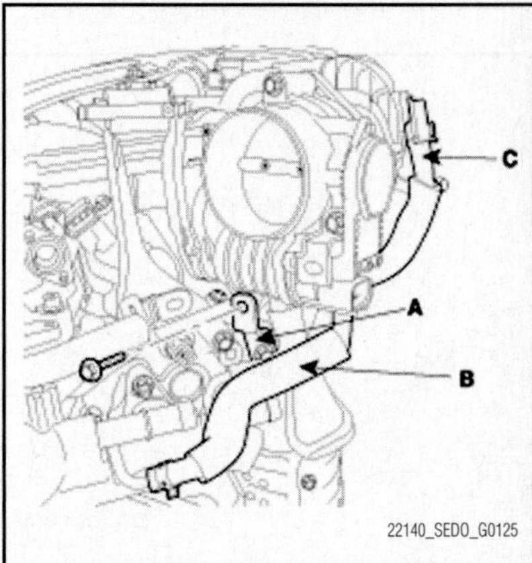

Fig. 79 Remove the Electric Throttle Control (ETC) bracket (A).

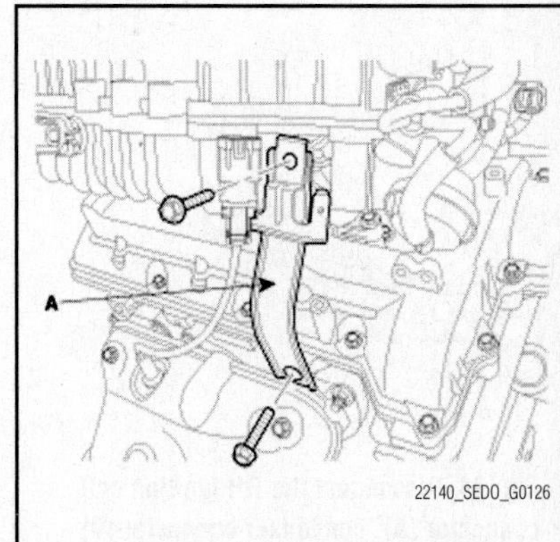

Fig. 80 Remove the surge tank stay (A).

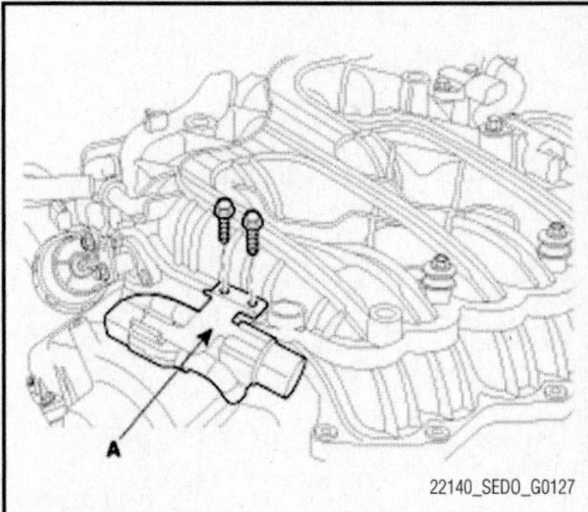

Fig. 81 Remove the connector bracket (A) from surge tank.

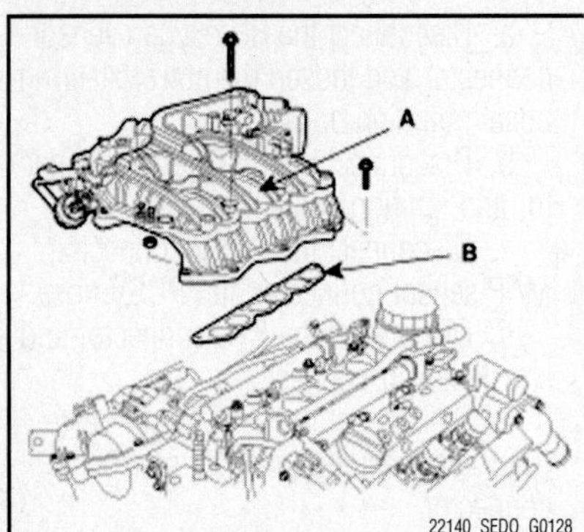

Fig. 82 Remove the surge tank (A).

f. Disconnect the LH front oxygen sensor connector.

g. Disconnect the LH ignition coil connector, injector connector, condenser connector and ground, and remove the wiring harness protector.

h. Disconnect the LH Camshaft Position (CMP) Sensor and oil pressure switch connector.

i. Remove the Electric Throttle Control (ETC) bracket.

j. Disconnect the water hoses from ETC.

k. Disconnect the PCV hose.

l. Disconnect the brake vacuum hose.

m. Remove the surge tank stay.

n. Remove the connector bracket from surge tank.

o. Remove the surge tank.

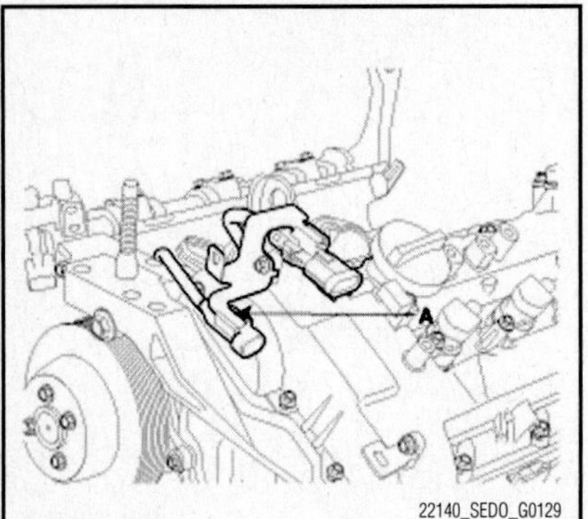

Fig. 83 Remove the connector bracket (A) from LH cylinder head cover.

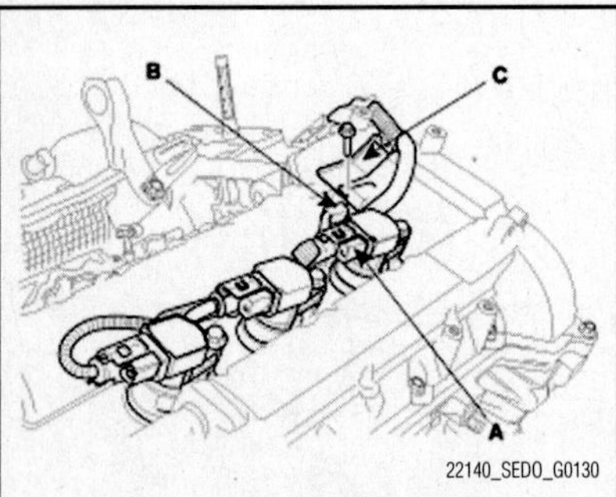

Fig. 84 Disconnect the RH ignition coil connector (A), condenser connector (B) and remove the wiring bracket (C).

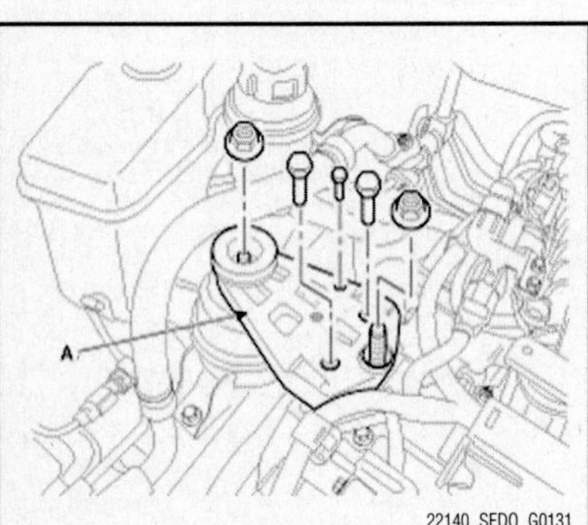

Fig. 85 Remove the engine mounting bracket (A).

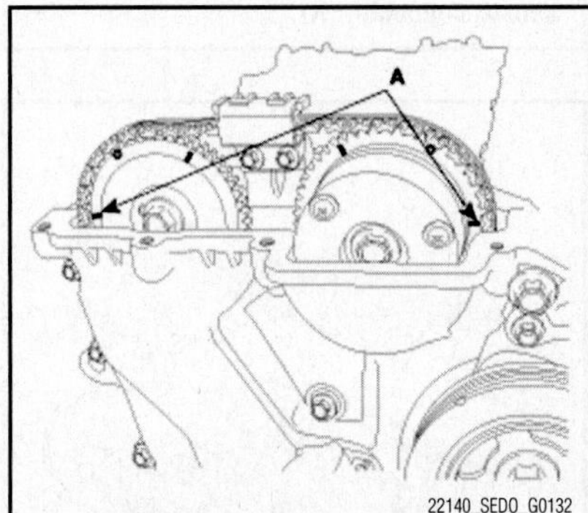

Fig. 86 Check that the mark (A) of the camshaft timing sprockets are in straight line on the cylinder head

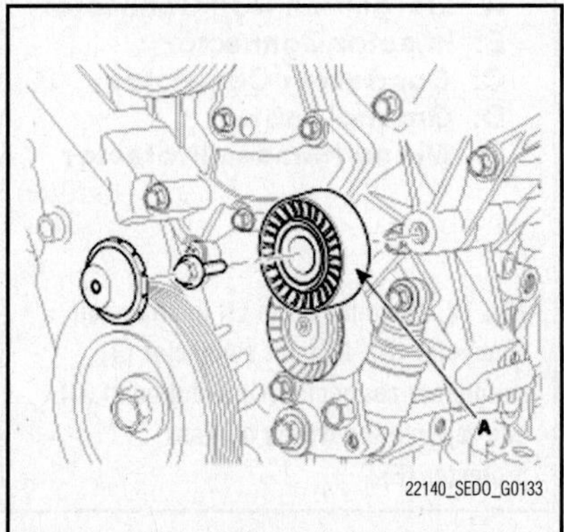

Fig. 87 Remove the drive belt idler (A).

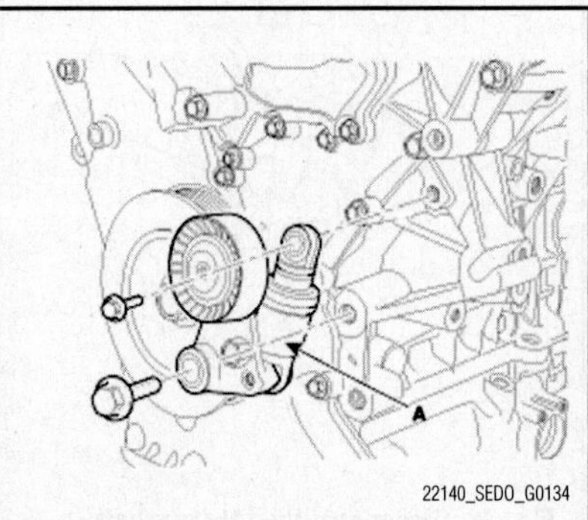

Fig. 88 Remove the drive belt auto tensioner (A).

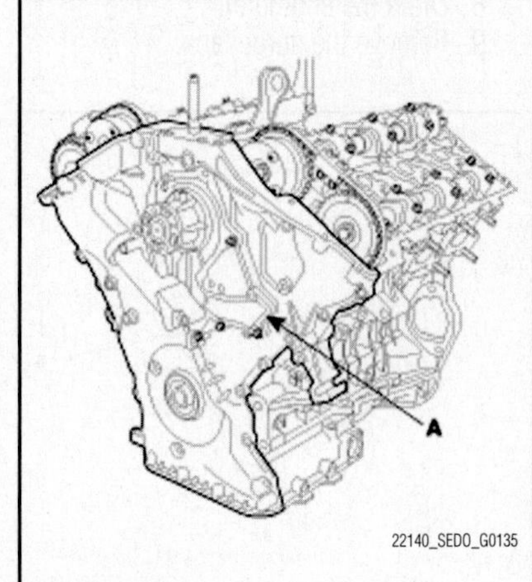

Fig. 89 Remove the timing chain cover (A).

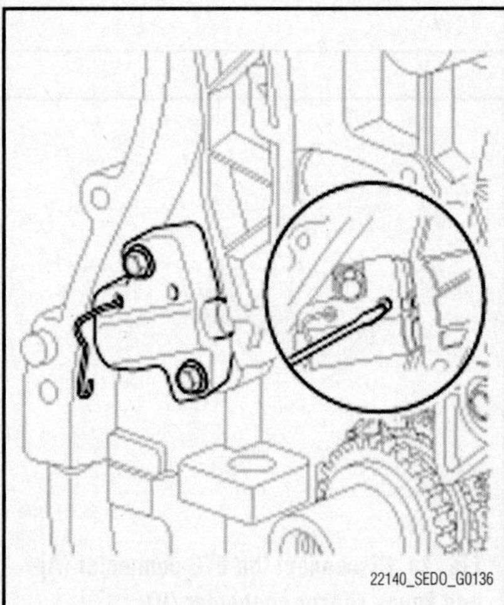

Fig. 90 Install a set pin after compressing the timing chain tensioner.

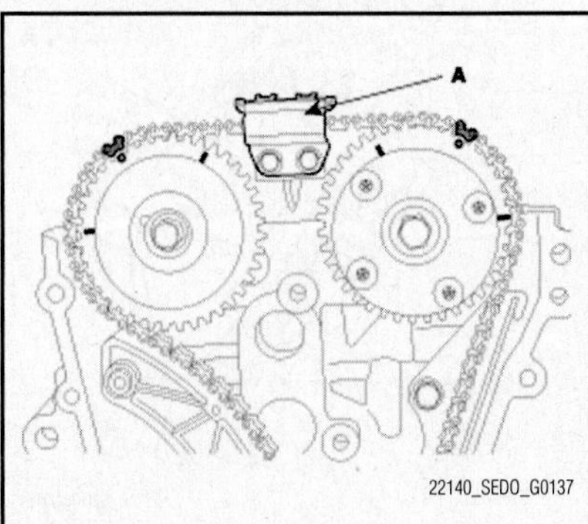

Fig. 91 Remove the RH cam-to-cam guide (A).

10. Remove the cylinder head cover.
    a. Remove the connector bracket from LH cylinder head cover.
    b. Disconnect the RH ignition coil connector, condenser connector and remove the wiring bracket.
    c. Remove the LH, RH ignition coil.
    d. Remove the LH, RH cylinder head covers.

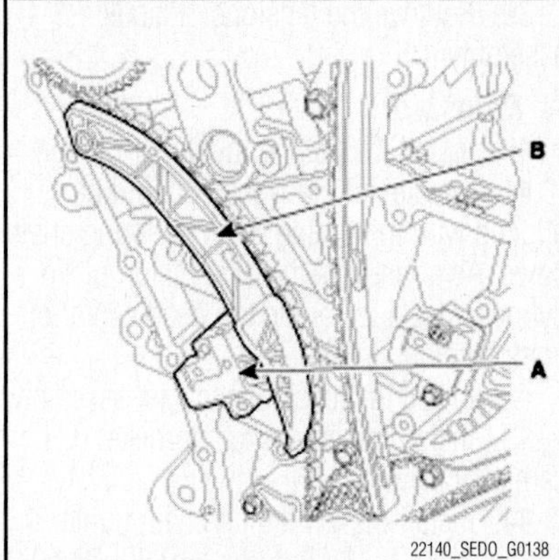

Fig. 92 Remove the RH timing chain auto tensioner (A) and RH timing chain tensioner arm.

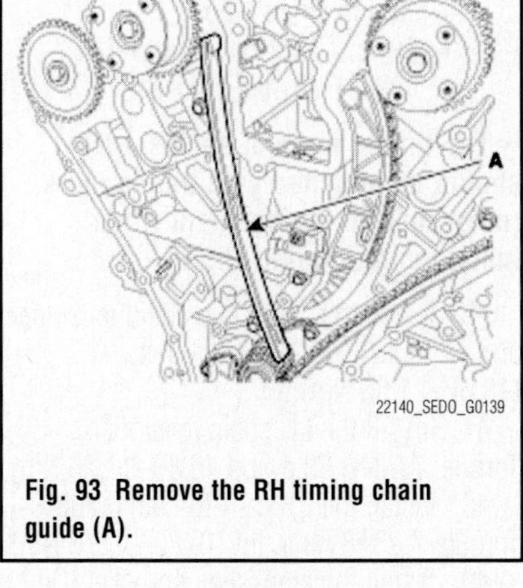

Fig. 93 Remove the RH timing chain guide (A).

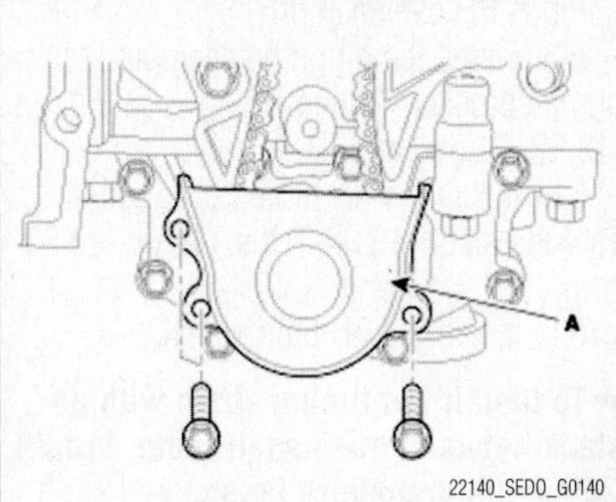

Fig. 94 Remove the oil pump chain cover (A).

11. Using SST (09215-3C000) remove lower oil pan.
12. Set a jack to the upper oil pan.
13. Remove the coolant reservoir tank.
14. Remove the engine mounting bracket.
15. Set No.1 cylinder to TDC/ compression.
    a. Turn the crankshaft pulley and align its groove with the timing mark "T" of the lower timing chain cover.

➡Do not rotate the engine counter-clockwise.

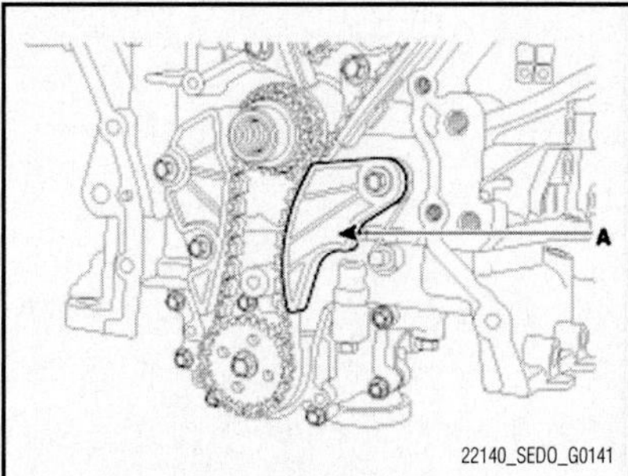

Fig. 95 Remove the oil pump chain tensioner assembly (A).

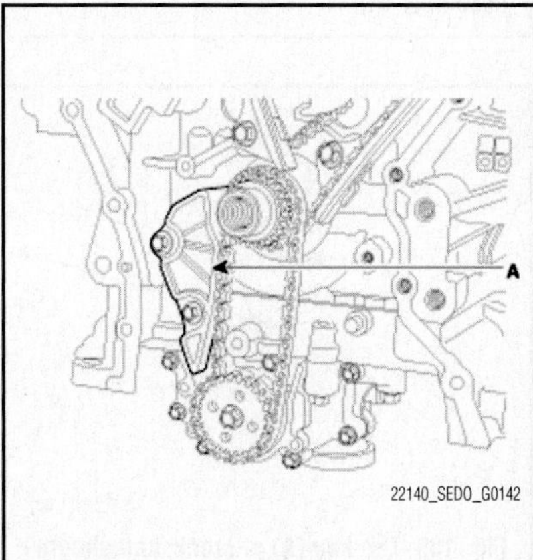

Fig. 96 Remove the oil pump chain guide (A).

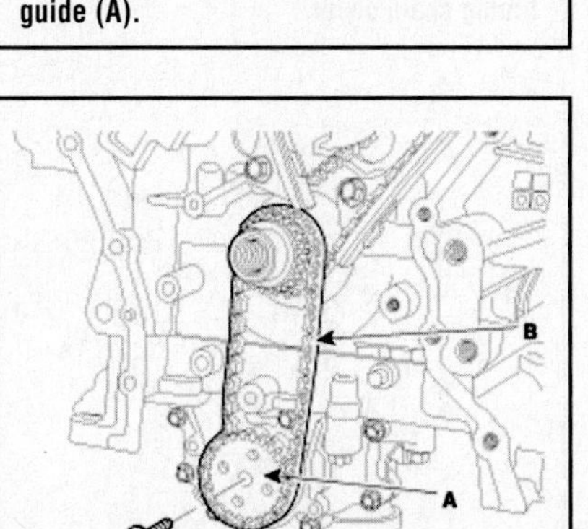

Fig. 97 Remove the oil pump chain sprocket (A) and oil pump chain (B).

    b. Check that the mark of the camshaft timing sprockets are in straight line on the cylinder head surface as shown. If not, turn the crankshaft one revolution (360°).
16. Remove the accessory drive belt.
17. Using SST (09231-3C300) remove the crankshaft damper pulley.
18. Lift up the engine assembly to using the jack.
19. Remove the drive belt idler.
20. Remove the drive belt auto tensioner.
21. Remove the water pump pulley.
22. Remove the timing chain cover.

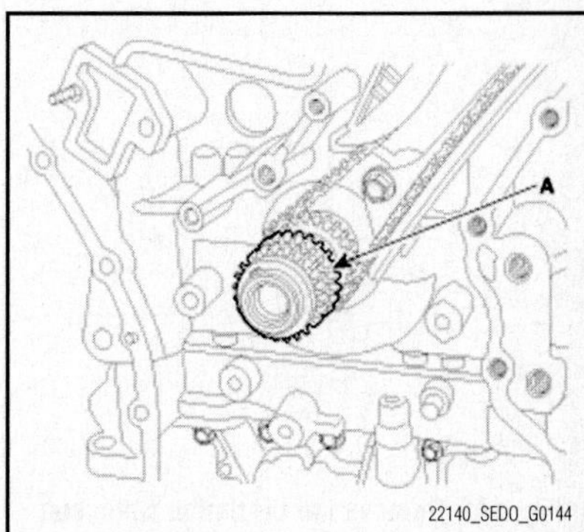

Fig. 98 Remove the crankshaft sprocket (A) (Oil pump & RH camshaft drive).

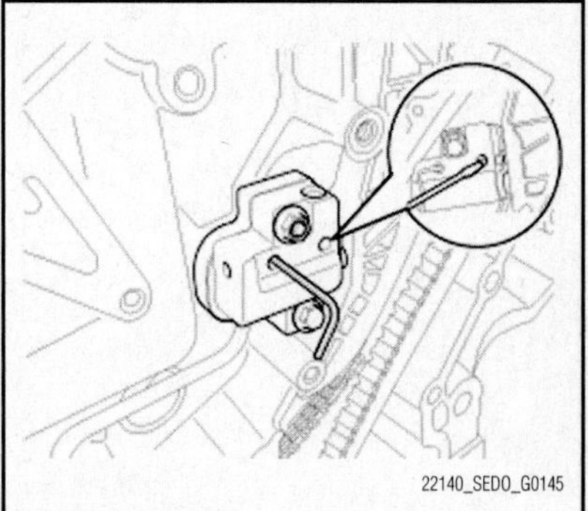

Fig. 99 Install a set pin after compressing the LH timing chain tensioner.

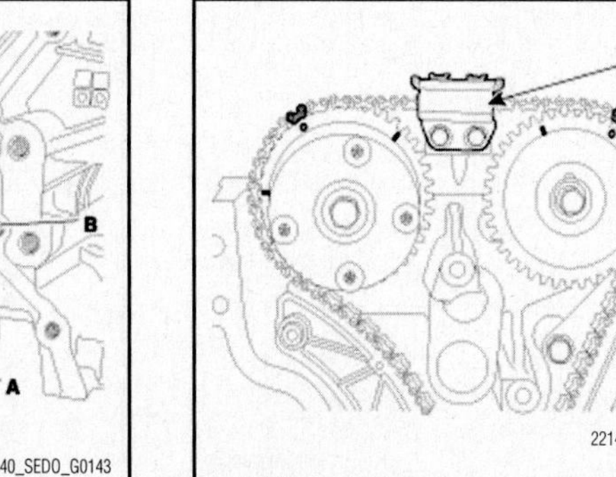

Fig. 100 Remove the LH cam-to-cam guide (A).

➡Be careful not to damage the contact surfaces of cylinder block, cylinder head and timing chain cover. Before removing the timing chain, mark the RH/LH timing chain with an identification based on the location of the sprocket because the identification mark on the chain for TDC (Top Dead Center) can be erased.

23. Install a set pin after compressing the timing chain tensioner.

24. Remove the RH cam-to-cam guide.

25. Remove the RH timing chain auto tensioner and RH timing chain tensioner arm.

26. Remove the RH timing chain.

27. Remove the RH timing chain guide.

28. Remove the oil pump chain cover.

29. Remove the oil pump chain tensioner assembly.

30. Remove the oil pump chain guide.

31. Remove the oil pump chain sprocket and oil pump chain.

32. Remove the crankshaft sprocket (Oil pump & RH camshaft drive).

33. Install a set pin after compressing the LH timing chain tensioner.

34. Remove the LH cam-to-cam guide.

35. Remove the LH timing chain auto tensioner and LH timing chain tensioner arm.

36. Remove the LH timing chain.

37. Remove the LH timing chain guide.

38. Remove the crankshaft sprocket (LH camshaft drive).

39. Remove the tensioner adapter assembly.

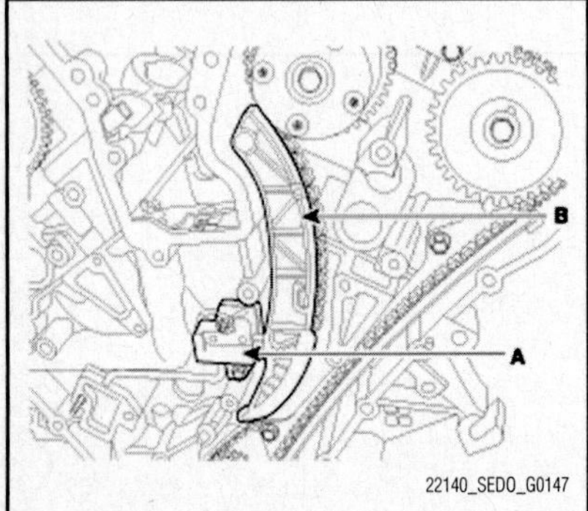

Fig. 101 Remove the LH timing chain auto tensioner (A) and LH timing chain tensioner arm (B).

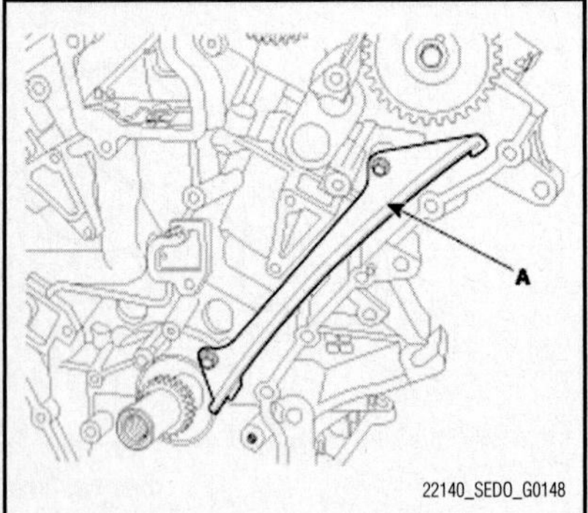

Fig. 102 Remove the LH timing chain guide (A).

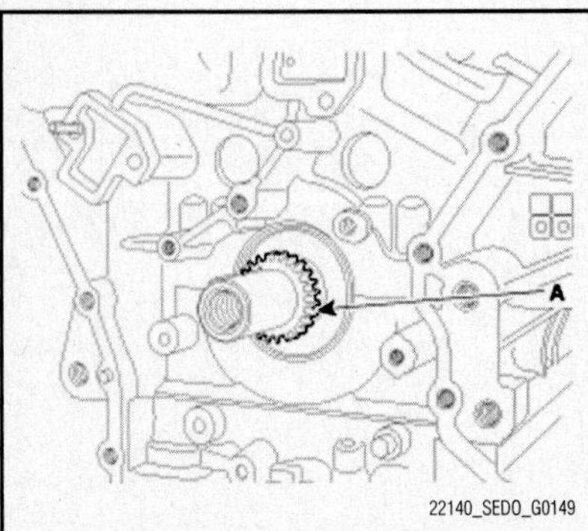

Fig. 103 Remove the crankshaft sprocket (A) (LH camshaft drive).

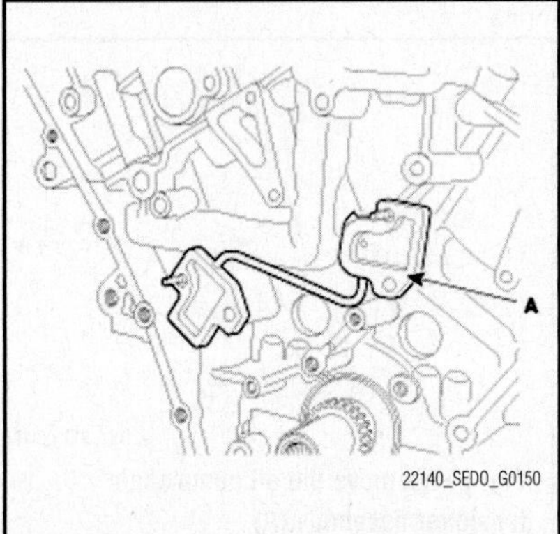

Fig. 104 Remove the tensioner adapter assembly (A).

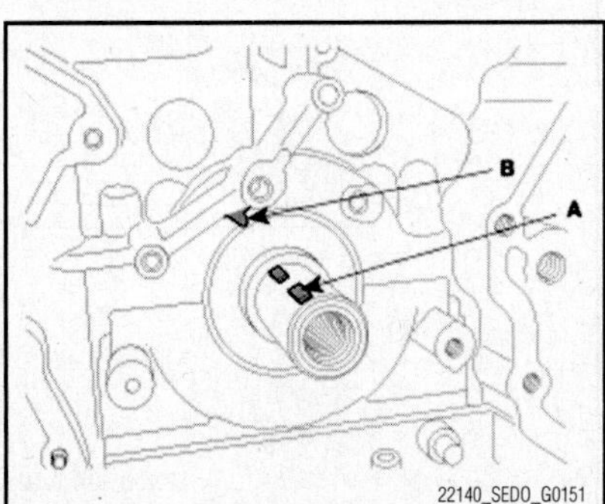

Fig. 105 The key (A) of crankshaft should be aligned with the timing mark (B) of timing chain cover.

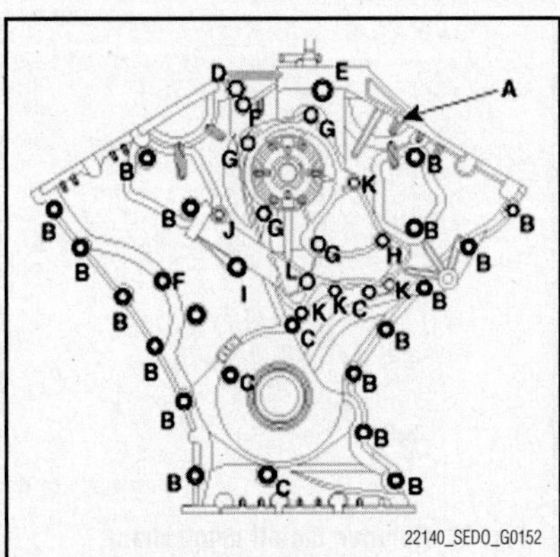

Fig. 106 Various torque locations

To install:

40. Install the jack to the upper oil pan.

41. The key of crankshaft should be aligned with the timing mark of timing chain cover. As a result of this, the piston of No.1 cylinder is placed at the top dead center on compression stroke.

42. Install the tensioner adapter assembly.

43. Install the crankshaft sprocket (LH camshaft drive).

44. Install the LH timing chain guide. Torque: 14.17–18.08 ft. lbs. (19.60–24.50 Nm).

45. Install LH timing chain.

➡To install the timing chain with no slack between each shaft (cam, crank), follow the procedure below:

- Crankshaft sprocket
- Timing chain guide
- Exhaust camshaft sprocket
- Intake camshaft sprocket.

➡The timing mark of each sprocket should be matched with timing mark (color link) of timing chain when installing the timing chain.

46. Install the LH timing chain tensioner arm. Torque: 13.74–15.91 ft. lbs. (18.62–21.56 Nm).

47. Install the LH chain tensioner. Torque: 7.23–8.68 ft. lbs. (9.80–11.76 Nm).

48. Install the LH cam-to-cam guide. Torque: 7.23–8.68 ft. lbs. (9.80–11.76 Nm).

49. Install the crankshaft sprocket (Oil pump & RH camshaft drive).

50. Install the oil pump chain and oil pump sprocket. Torque: 13.74–15.91 ft. lbs. (18.62–21.56 Nm).

51. Install the RH timing chain guide. Torque: 14.17–18.08 ft. lbs. (19.60–24.50 Nm).

52. Install the RH timing chain.

➡To install the timing chain with no slack between each shaft (cam, crank), follow the procedure below:

- Crankshaft sprocket
- Intake camshaft sprocket
- Exhaust camshaft sprocket

➡The timing mark of each sprocket should be matched with timing mark (color link) of timing chain when installing the timing chain.

53. Install the RH timing chain tensioner arm. Torque: 13.74–15.91 ft. lbs. (18.62–21.56 Nm).
54. Install the RH timing chain auto tensioner. Torque: 7.23–8.68 ft. lbs. (9.80–11.76 Nm).
55. Install the RH cam-to-cam guide. Torque: 7.23–8.68 ft. lbs. (9.80–11.76 Nm).
56. Install the oil pump chain guide. Torque: 7.23–8.68 ft. lbs. (9.80–11.76 Nm).
57. Install the oil pump chain tensioner assembly. Torque: 7.23–8.68 ft. lbs. (9.80–11.76 Nm).
58. Pull out the pins of hydraulic tensioners (LH & RH).
59. Install the oil pump chain cover. Torque: 7.23–8.68 ft. lbs. (9.80–11.76 Nm).
60. After rotating crankshaft 2 revolutions in regular direction (clockwise viewed from front), confirm the timing mark.

➡Always turn the crankshaft clockwise.

61. Install the timing chain cover.
    a. The sealant locations on chain cover and on counter parts (cylinder head, cylinder block, and lower oil pan) must be free of engine oil.
    b. Before assembling the timing chain cover, the liquid sealant TB1217H should be applied on the gap between cylinder head and cylinder block. The part must be assembled within 5 minutes after sealant was applied.
    c. Install the new gasket to the timing chain cover.
    d. Install the timing chain cover.
    e. The sealant locations on chain cover and on counter parts (cylinder head, cylinder block, and lower oil pan) must be free of engine oil.

➡During timing cover installation, care not to take off applied sealant on the timing cover by contact with other parts.

    f. The dowel pins on the cylinder block and holes on the timing chain cover should be used as a reference in order to assemble the timing chain cover to the exact position. Torque as follows:
    - B (17ea): 13.74–15.91 ft. lbs. (18.62–21.56 Nm).
    - C (4ea): 7.23–8.68 ft. lbs. (9.80–11.76 Nm).
    - D (1ea): 43.40–50.63 ft. lbs. (58.80–68.80 Nm).
    - E (1ea): 43.40–50.63 ft. lbs. (58.80–68.80 Nm).
    - F (2ea): 18.08–19.53 ft. lbs. (24.50–26.46 Nm).
    - G (4ea): 15.91–17.36 ft. lbs. (21.56–23.52 Nm).
    - H (1ea): 7.23–8.68 ft. lbs. (9.80–11.76 Nm).
    - I (1ea): 7.23–8.68 ft. lbs. (9.80–11.76 Nm).
    - J (1ea): 7.23–8.68 ft. lbs. (9.80–11.76 Nm).
    - K (4ea): 7.23–8.68 ft. lbs. (9.80–11.76 Nm).
    - L (1ea): 15.91–19.53 ft. lbs. (21.56–26.46 Nm).

    g. The firing and/or blow out test should not be performed within 30 minutes after the timing chain cover was assembled.
62. Install the water pump pulley. Torque: 5.78–7.23 ft. lbs. (7.84–9.80 Nm).
63. Install the drive belt auto tensioner. Torque: Large Bolt: 60.03–62.93 ft. lbs. (81.39–85.32 Nm); Small Bolt: 13.02–15.91 ft. lbs. (17.64–21.56 Nm).
64. Install the drive belt idler. Torque: 39.06–42.67 ft. lbs. (52.92–57.82 Nm).
65. Lower the engine assembly by using the jack.
66. Using SST (09231-3C100), install timing chain cover oil seal.
67. Using SST (09231-3C300) install the crankshaft damper pulley. Torque: 209.76–224.22 ft. lbs. (284.2–303.8 Nm).
68. Install the accessory drive belt in the following sequence:
    - Crankshaft pulley
    - A/C pulley
    - Idler pulley
    - Alternator pulley
    - Water pump pulley
    - P/S pump pulley
    - Tensioner pulley.
69. Rotate auto tensioner arm in the counterclockwise moving auto tensioner pulley bolt with wrench.
70. After putting belt on auto tensioner pulley, release the auto tensioner pulley slowly.
71. Install the cylinder head cover.
    a. The hardening sealant located on the upper area between timing chain cover and cylinder head should be removed before assembling cylinder head cover.
    b. After applying sealant (TB1217H), it should be assembled within 5 minutes.
    c. The firing and/or blow out test should not be performed within 30 minutes after the cylinder head cover was assembled.
    d. Install the cylinder head cover bolts in sequence. Torque: 7.23–8.68 ft. lbs. (9.80–11.76 Nm).
    e. Install the ignition coil.
    f. Connect the RH ignition coil connector, the condenser connector and install the wiring bracket.
72. Install the surge tank and wiring connectors.
    a. Install the surge tank. Torque: 7.23–8.68 ft. lbs. (9.80–11.76 Nm).
    b. Install the connector bracket to the surge tank. Torque: 5.06–7.96 ft. lbs. (6.86–10.78 Nm).
    c. Install the surge tank stay. Torque: 20.25–23.14 ft. lbs. (27.44–31.36 Nm).
    d. Connect the brake vacuum hose.
    e. Connect the PCV hose.
    f. Connect the water hoses to the ETC.
    g. Install the ETC bracket.
    h. Connect the LH CMP sensor connector and oil pressure switch connector.
    i. Install the wiring harness protector and connect the LH ignition coil connector, injector connector, condenser connector and ground.
    j. Connect the LH front oxygen sensor connector.
    k. Connect the OCV connector and knock sensor connector.
    l. Connect the ETC connector and knock sensor connector.
    m. Connect the PCSV connector, MAP sensor connector and PCSV hose.
    n. Connect the RH injector connector and ignition coil connector.
    o. Connect the RH oxygen sensor connector and tighten the power steering hose mounting bolts.
73. Install the engine mounting bracket. Torque: 65.1–79.6 ft. lbs. (88.3–107.9 Nm).
74. Install the coolant reservoir tank.
75. Remove the jack from the upper oil pan.
76. Install lower oil pan.
    a. Using a gasket scraper, remove all the old packing material from the gasket surfaces.
    b. Before assembling the oil pan, the liquid sealant TB1217H should be applied on oil pan.
    c. Install the lower oil pan. Torque: 7.23–8.68 ft. lbs. (9.80–11.76 Nm).
77. Install the side cover.
78. Install the undercover.
79. Install the RH front wheel.
80. Install the intake air hose and air cleaner assembly.
    a. Install the intake air hose and air cleaner assembly.

b. Connect the breather hose to the air intake hose.

c. Connect the MAF sensor connector.

81. Install the engine cover.
82. Connect the battery negative cable.

➡After installation, complete the following:

- Refill engine with engine oil.
- Refill radiator and reservoir tank with engine coolant.
- Bleed air from the cooling system.
- Start engine and let it run until it warms up. (until the radiator fan operates 3 or 4 times.)
- Turn Off the engine. Check the level in the radiator, add coolant if needed. This will allow trapped air to be removed from the cooling system.
- Put radiator cap on tightly, then run the engine again and check for leaks.

VALVE LASH

ADJUSTMENT

See Figures 107 and 108.

1. Disconnect the negative battery cable.
2. Remove the engine cover.
3. Remove the air cleaner assembly.
4. Remove the surge tank.
5. Disconnect the ignition coil connector and remove the ignition coil.
6. Disconnect the breather pipe assembly from the cylinder head cover.
7. Remove the cylinder head covers from the engine.
8. Set the No. 1 cylinder to TDC on the compression stroke.

a. Turn the crankshaft pulley and align its groove with the timing mark "T" of the lower timing chain cover.

b. Check that the mark of the camshaft timing sprockets are in straight line positioning on the cylinder head surface. If not rotate the crankshaft 360 degrees.

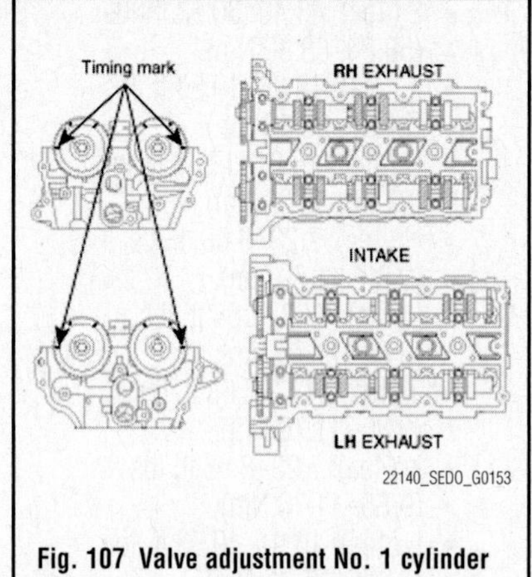

Fig. 107 Valve adjustment No. 1 cylinder

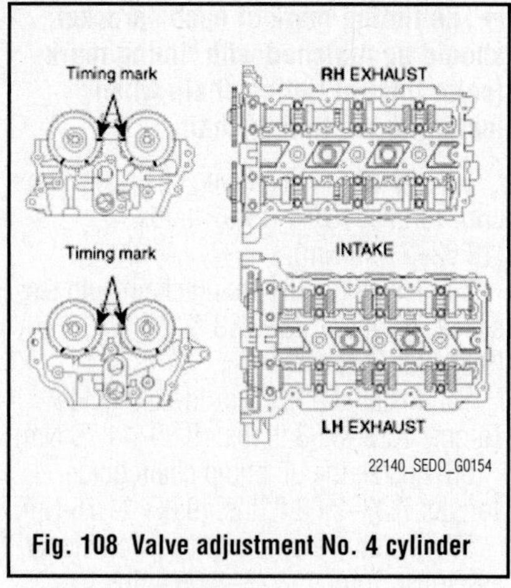

Fig. 108 Valve adjustment No. 4 cylinder

➡Do not rotate the engine counterclockwise.

9. Check the valve clearance on No. 1 cylinder by measuring the clearance between the tappet and the base circle of the camshaft.
10. Turn the crankshaft pulley one revolution and align the groove with the timing mark "T" on the lower timing chain cover.

➡Do not rotate the engine counterclockwise.

11. Check the valve clearance on No. 4 cylinder by measuring the clearance between the tappet and the base circle of the camshaft.
12. Adjust the intake and exhaust valve clearance.

a. Set the No. 1 cylinder to TDC on the compression stroke.

b. Remove the timing chain.

➡Before removing the timing chain mark the RH and LH timing chain with an identification based on the location of the sprocket. You must do this because the identification mark on the chain for TDC can be erased.

c. Remove the camshaft bearing caps.

d. Remove the camshafts.

e. Remove the MLA's.

f. Measure the thickness of the removed tappet using a micrometer.

g. Calculate the thickness of the new tappet so that the valve clearance comes within the specified value.

Valve clearance: Engine coolant temperature: 68°F (20°C

- T: Thickness of removed tappet
- A: Measured valve clearance
- N: Thickness of new tappet
- Intake: N = T + [A - 0.0079 inches (0.20 mm)]
- Exhaust: N = T + [A - 0.0118inches (0.30 mm)]

a. Select a new tappet with a thickness as close as possible to the calculated value.

➡Shims are available in 41 size increments ranging of 0.0006 inches from 0.118 inches to 0.1417 inches.

13. Place a new tappet on the cylinder head.

➡Apply clean engine oil at the selected tappet on the periphery and top surface.

14. Install the camshafts.
15. Install the bearing caps.
16. Install the timing chain.
17. Turn the crankshaft two turns in the clockwise direction and realign the crankshaft sprocket and camshaft sprocket timing marks.
18. Recheck the valve clearance.

ENGINE PERFORMANCE & EMISSION CONTROLS

ACCELERATOR PEDAL POSITION (APP) SENSOR

LOCATION

Accelerator Pedal Position (APP) sensor is installed on the accelerator pedal module.

REMOVAL & INSTALLATION

1. Turn ignition switch off and disconnect the negative battery cable.
2. Disconnect the Accelerator Pedal Position (APP) sensor connector.
3. Unfasten the four mounting nuts
4. Remove the accelerator pedal from the vehicle.
5. Installation is the reverse of removal.

CAMSHAFT POSITION (CMP) SENSOR

LOCATION

The two Camshaft Position (CMP) sensors are installed on engine head covers of banks 1 and 2.

REMOVAL & INSTALLATION

1. Turn ignition switch OFF.
2. Disconnect the CMP sensor connector.
3. Remove the bolt and the CMP sensor.
4. Installation is the reverse of removal.

CRANKSHAFT POSITION (CKP) SENSOR

LOCATION

The Crankshaft Position (CKP) Sensor is installed on transaxle housing.

REMOVAL & INSTALLATION

1. Turn ignition switch OFF.
2. Disconnect the CKP sensor connector.
3. Remove the CKP sensor.
4. Installation is the reverse of removal.

ENGINE COOLANT TEMPERATURE (ECT) SENSOR

LOCATION

The Engine Coolant Temperature (ECT) sensor is located in the engine coolant passage of the cylinder head for detecting the engine coolant temperature.

REMOVAL & INSTALLATION

1. Turn ignition switch OFF.
2. Disconnect the ECT sensor connector.
3. Remove the ECTS.
4. Installation is the reverse of removal.

HEATED OXYGEN (HO2S) SENSOR

LOCATION

The Heated Oxygen Sensor (HO2S) is installed on upstream and downstream of the Manifold Catalyst Converter (MCC).

REMOVAL & INSTALLATION

1. Disconnect the negative battery cable.
2. Disconnect the HO2S connector
3. Remove the HO2S sensor.
4. Installation is the reverse of removal.

INTAKE AIR TEMPERATURE (IAT) SENSOR

REMOVAL & INSTALLATION

1. Disconnect the negative battery cable.
2. Disconnect the IAT sensor connector
3. Remove the IAT sensor.
4. Installation is the reverse of removal.

MANIFOLD ABSOLUTE PRESSURE (MAP) SENSOR

REMOVAL & INSTALLATION

1. Disconnect the negative battery cable.
2. Disconnect the MAP sensor connector
3. Remove the MAP sensor.
4. Installation is the reverse of removal.

MASS AIR FLOW SENSOR

REMOVAL & INSTALLATION

1. Disconnect the negative battery cable.
2. Disconnect and remove the MAF sensor from the air cleaner assembly.
3. Installation is the reverse of removal.

POWERTRAIN CONTROL MODULE (PCM)

REMOVAL & INSTALLATION

1. Turn the ignition switch off and disconnect the negative battery terminal.
2. Disconnect the PCM connector.
3. Unscrew the mounting bolts behind the air cleaner.
4. Remove the PCM.
5. Installation is reverse of removal.

VARIABLE CAMSHAFT TIMING OIL CONTROL SOLENOID

LOCATION

The Continuously Variable Valve Timing (CVVT) Oil Control Valve (OCV) is located at the front of the engine block.

REMOVAL & INSTALLATION

1. Disconnect the negative battery cable.
2. Disconnect the oil control valve connector
3. Remove the oil control valve.
4. Installation is the reverse of removal.

FUEL SYSTEM SERVICE PRECAUTIONS

Safety is the most important factor when performing not only fuel system maintenance, but any type of maintenance. Failure to conduct maintenance and repairs in a safe manner may result in serious personal injury or death. Work on a vehicle's fuel system components can be accomplished safely and effectively by adhering to the following rules and guidelines.

- To avoid the possibility of fire and personal injury, always disconnect the negative battery cable unless the repair or test procedure requires that battery voltage be applied.
- Always relieve the fuel system pressure prior to disconnecting any fuel system component (injector, fuel rail, pressure regulator, etc.) fitting or fuel line connection. Exercise extreme caution whenever relieving fuel system pressure to avoid exposing skin, face and eyes to fuel spray. Please be advised that fuel under pressure may penetrate the skin or any part of the body that it contacts.
- Always place a shop towel or cloth around the fitting or connection prior to loosening to absorb any excess fuel due to spillage. Ensure that all fuel spillage is quickly removed from engine surfaces. Ensure that all fuel-soaked cloths or towels are deposited into a flame-proof waste container with a lid.
- Always keep a dry chemical (Class B) fire extinguisher near the work area.
- Do not allow fuel spray or fuel vapors to come into contact with a spark or open flame.
- Always use a second wrench when loosening or tightening fuel line connection fittings. This will prevent unnecessary stress and torsion on fuel piping. Always follow the proper torque specifications.
- Always replace worn fuel fitting O-rings with new ones. Do not substitute fuel hose where rigid pipe is installed.

FUEL SYSTEM PRESSURE

RELIEVING

1. Remove the third seat.
2. Open the service cover.
3. Disconnect the fuel pump connector.
4. Start the engine and wait until fuel in fuel line is exhausted.
5. After engine stalls, turn the ignition switch to OFF position.
6. Disconnect the negative battery cable.

FUEL FILTER

REMOVAL & INSTALLATION

See Figures 109 through 111.

1. Remove the 2nd seat (s).
2. Remove the service cover.

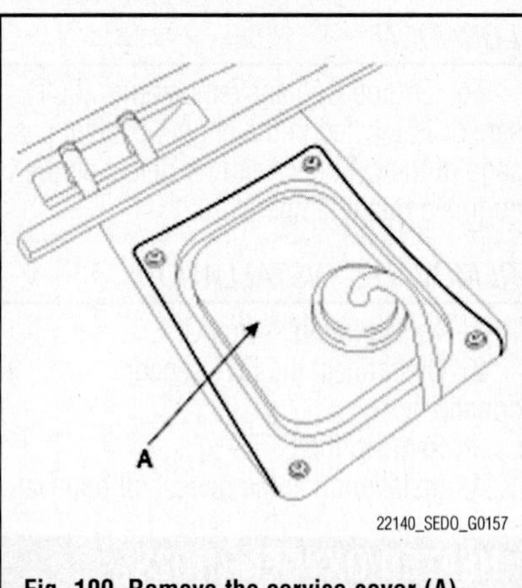

Fig. 109 Remove the service cover (A).

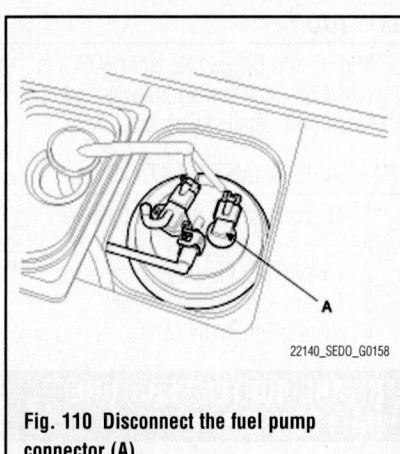

Fig. 110 Disconnect the fuel pump connector (A).

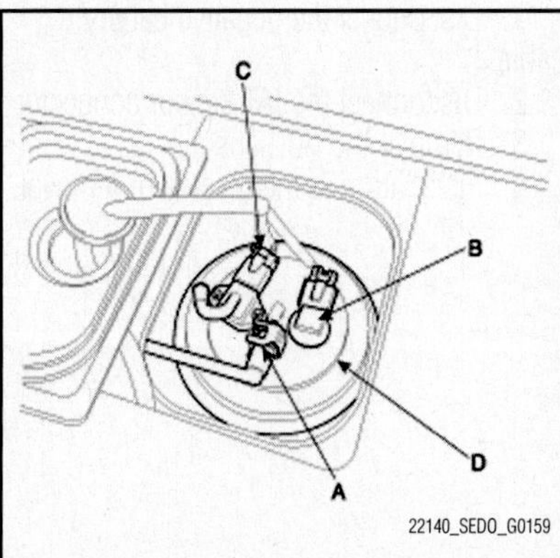

Fig. 111 Disconnect the fuel feed quick connector (A), the fuel pump connector (B) and fuel tank pressure sensor connector (C), then remove the fuel pump cover (D).

3. Disconnect the fuel pump connector.
4. Start the engine and wait until fuel in fuel line is exhausted.
5. After the engine stalls, turn the ignition switch OFF.
6. Disconnect the fuel feed quick connector, the fuel pump connector and fuel tank pressure sensor connector, then remove the fuel pump cover.
7. Remove the fuel pump from the fuel tank.

To install:

8. Installation is the reverse of removal.

FUEL LEVEL SENDING UNIT

LOCATION

The fuel level sending unit is located on the fuel pump in the fuel tank.

REMOVAL & INSTALLATION

See Figures 109 through 111.

1. Remove the 2nd seat (s).
2. Remove the service cover.
3. Disconnect the fuel pump connector.
4. Start the engine and wait until fuel in fuel line is exhausted.
5. After the engine stalls, turn the ignition switch OFF.
6. Disconnect the fuel feed quick connector, the fuel pump connector and fuel tank pressure sensor connector, then remove the fuel pump cover.
7. Remove the fuel pump from the fuel tank.

To install:

8. Installation is the reverse of removal.

FUEL PUMP

REMOVAL & INSTALLATION

See Figures 109 through 111.

1. Remove the 2nd seat(s).
2. Remove the service cover.
3. Disconnect the fuel pump connector.
4. Start the engine and wait until fuel in fuel line is exhausted.
5. After the engine stalls, turn the ignition switch OFF.
6. Disconnect the fuel feed quick connector, the fuel pump connector and fuel tank pressure sensor connector, then remove the fuel pump cover.
7. Remove the fuel pump from the fuel tank.

To install:

8. Installation is the reverse of removal.

FUEL PRESSURE REGULATOR

REMOVAL & INSTALLATION

See Figures 109 through 111.

1. Remove the rear seat(s).
2. Remove the service cover.
3. Disconnect the fuel pump connector.
4. Start the engine and wait until fuel in fuel line is exhausted.
5. After the engine stalls, turn the ignition switch OFF.
6. Disconnect the fuel feed quick connector, the fuel pump connector and fuel tank pressure sensor connector, then remove the fuel pump cover.
7. Remove the fuel pump from the fuel tank.

To install:

8. Installation is the reverse of removal.

FUEL TANK

REMOVAL & INSTALLATION

See Figures 109, 110 and 112 through 114.

1. Remove the rear seat (s).
2. Remove the service cover.
3. Disconnect the fuel pump connector.
4. Start the engine and wait until fuel in fuel line is exhausted.

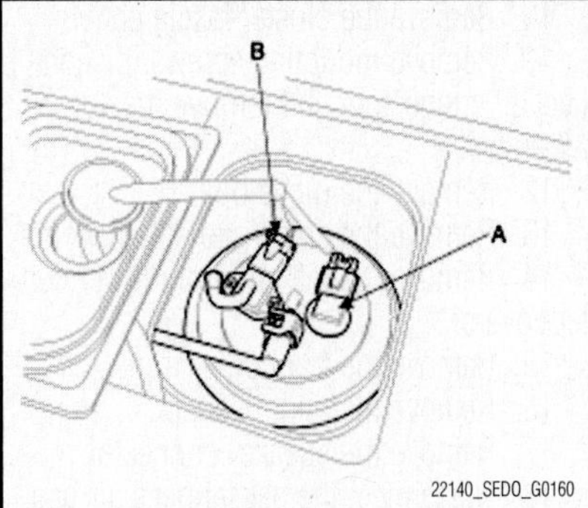

Fig. 112 Disconnect the fuel pump connector (A) and the fuel tank pressure sensor connector (B).

5. After the engine stalls, turn the ignition switch OFF.
6. Disconnect the fuel pump connector and the fuel tank pressure sensor connector.
7. Lift the vehicle and support the fuel tank with a jack.
8. Disconnect the fuel filler hose and the leveling hose.
9. Disconnect the fuel feed quick connector near canister.
10. Remove the fuel tank bands and remove the fuel tank.

To install:

11. Installation is the reverse of removal.

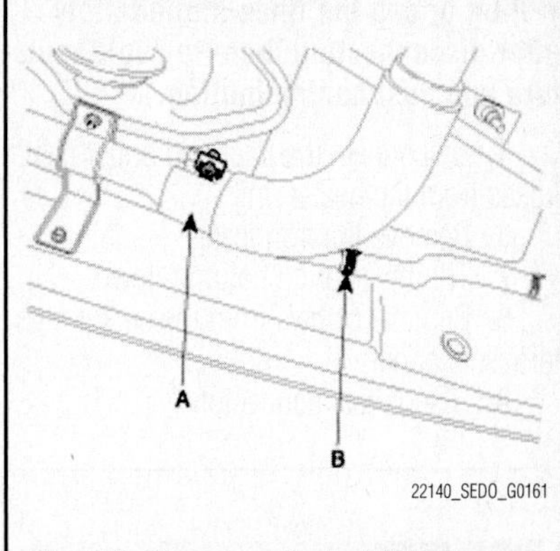

Fig. 113 Disconnect the fuel filler hose (A) and the leveling hose (B).

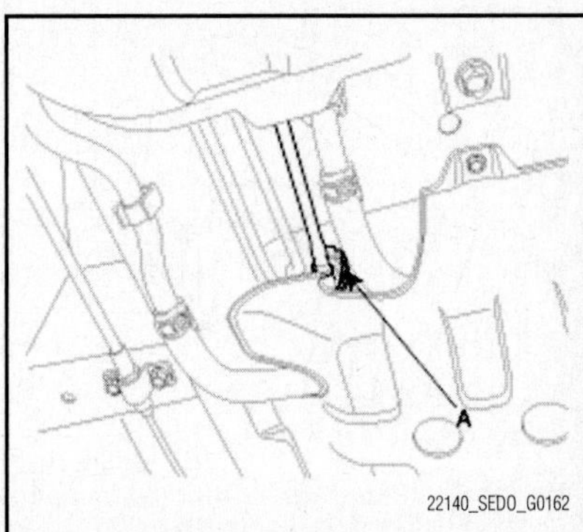

Fig. 114 Disconnect the fuel feed quick connector (A) near canister.

HEATING & AIR CONDITIONING SYSTEM

BLOWER MOTOR

REMOVAL & INSTALLATION

See Figures 115 and 116.

1. Disconnect the negative battery terminal.
2. Disconnect the connector of the blower motor.
3. Remove the blower motor after removing the mounting screws.

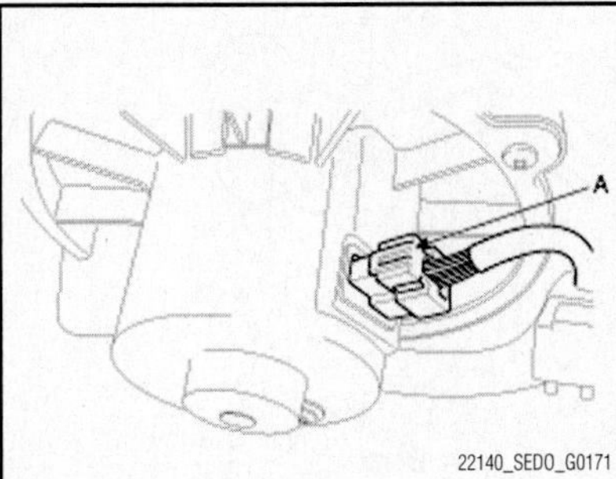

Fig. 115 Disconnect the connector (A) of the blower motor.

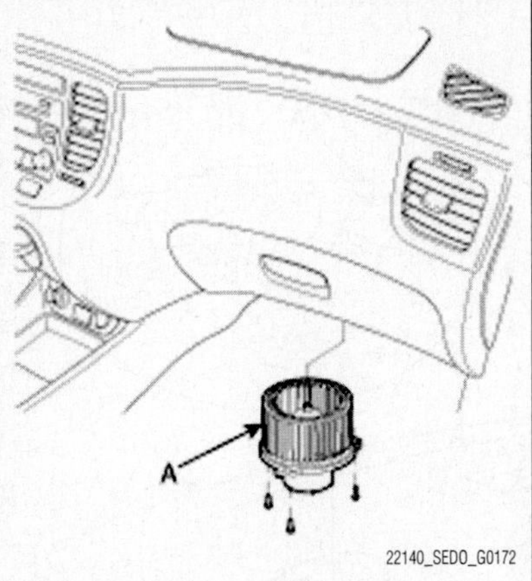

Fig. 116 Remove the blower motor (A) after removing the mounting screws.

To install:

4. Installation is the reverse of the removal procedure.

HEATER CORE

REMOVAL & INSTALLATION

See Figures 117 through 120.

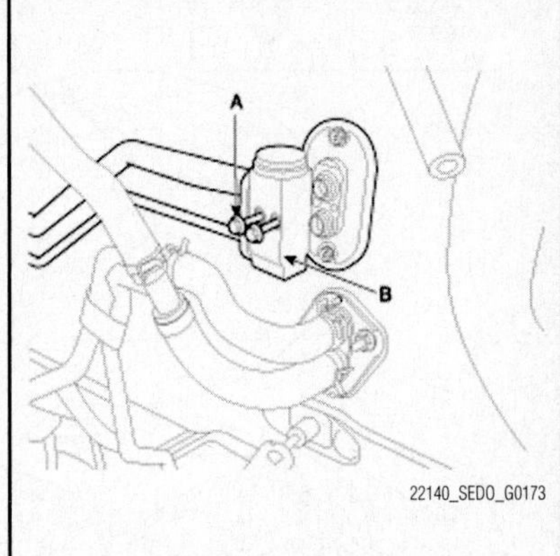

Fig. 117 Remove the bolts (A) and the expansion valve (B) from the evaporator core.

1. Disconnect the negative battery terminal.
2. Recover the refrigerant with a recovery/recycling/charging station.
3. Drain the engine coolant from the radiator.
4. Remove the bolts and the expansion valve from the evaporator core.

➡Plug or cap the lines immediately after disconnecting them to avoid moisture and dust contamination.

5. Disconnect the inlet and outlet heater hoses from the heater unit.
6. Remove the front seat.
7. Tilt the steering column down.
8. Remove screws from the center cluster fascia panel.
9. Disconnect connector.

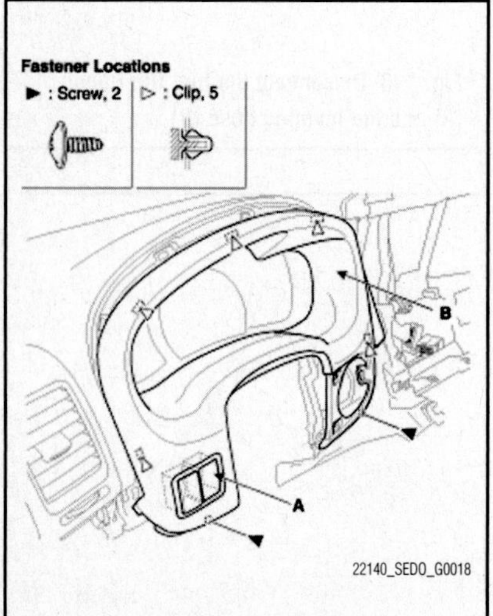

Fig. 118 Disconnect connector (A). Remove the cluster fascia panel (B).

10. Remove the cluster fascia panel.
11. Remove mounting screws, disconnect all connectors and remove the audio assembly.
12. Remove the glove box.
13. Remove the crash pad side covers.
14. Remove the left and right lower center covers.
15. Remove the front pillar trim.
16. Remove the photo sensor.
17. Remove the speaker connector.
18. Disconnect the passenger's air bag connector.
19. Remove the crash pad.
20. Remove the cowl cross bar assembly.

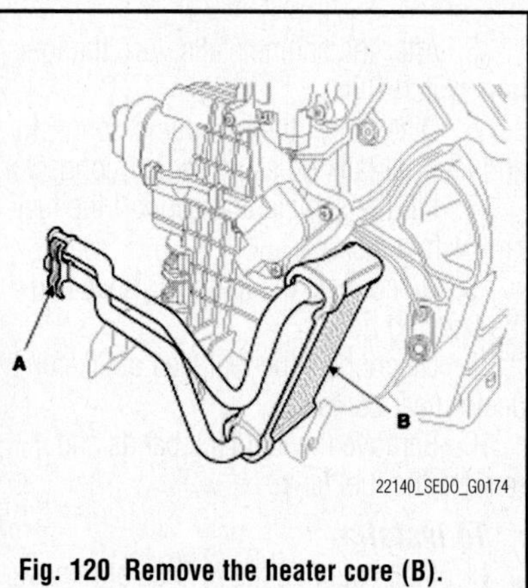

Fig. 120 Remove the heater core (B).

21. Disconnect the connectors from the temperature control actuator, the mode control actuator and the evaporator temperature sensor.
22. Remove the heater & blower unit.
23. Remove the blower unit from heater unit.
24. Remove the heater core.

➡Be careful that the inlet and outlet pipe are not bent during heater core removal.

To install:

25. Installation is the reverse order of removal, and note these items :
    a. If you're installing a new evaporator, add refrigerant oil (ND-OIL8).
    b. Replace the O-rings with new ones at each fitting, and apply a thin coat of refrigerant oil before installing them. Be sure to use the right O-rings for R-134a to avoid leakage.
    c. Apply sealant to the grommets.
    d. Make sure that there is no air leakage.
    e. Charge the system and test its performance.
    f. Do not interchange the inlet and outlet heater hoses and install the hose clamps securely.
    g. Refill the cooling system with engine coolant.

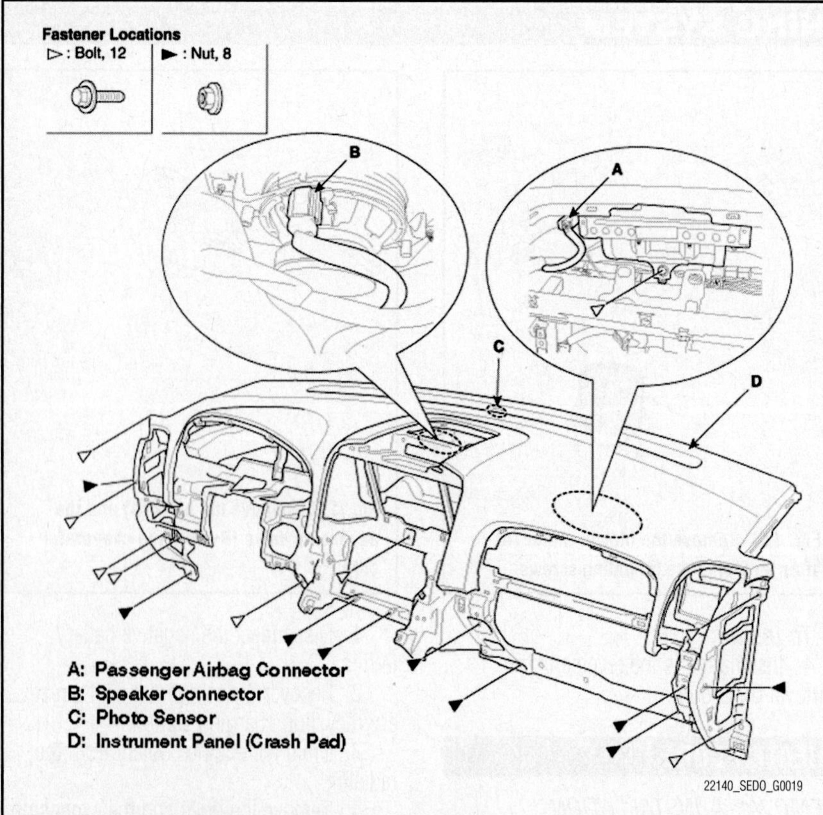
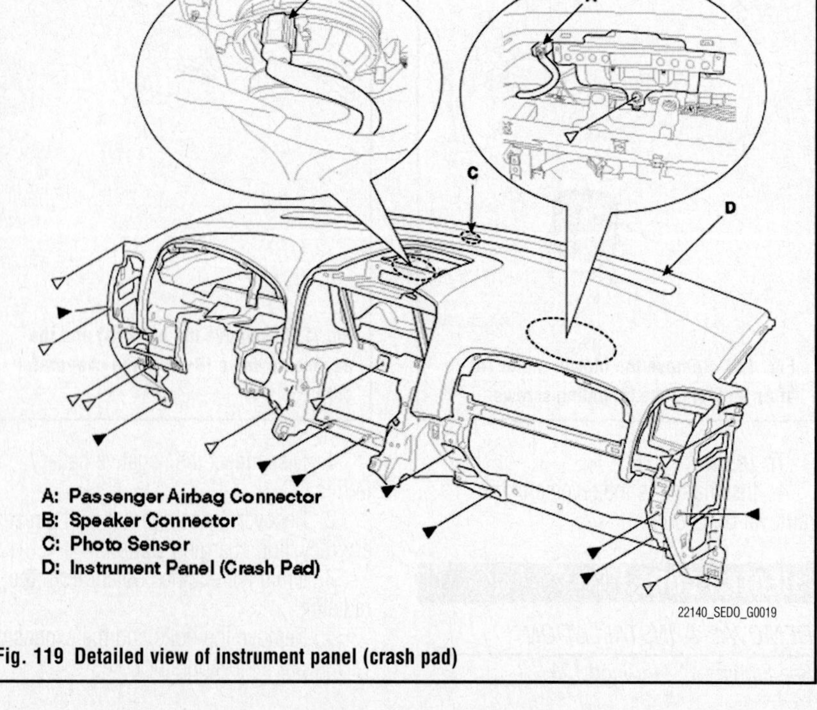

Fig. 119 Detailed view of instrument panel (crash pad)

HEATER UNIT

REMOVAL & INSTALLATION

See Figures 121 through 127.

1. Disconnect the negative (-) battery terminal.
2. Recover the refrigerant with a recovery/ recycling/ charging station. (Refer to HA-8)
3. When the engine is cool, drain the engine coolant from the radiator.

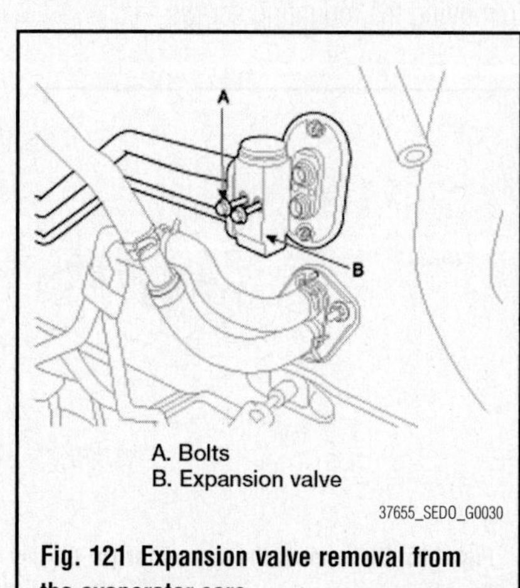

Fig. 121 Expansion valve removal from the evaporator core

4. Remove the bolts and the expansion valve from the evaporator core. Plug or cap the lines immediately after disconnecting them to avoid moisture and dust contamination.

5. Disconnect the inlet and outlet heater hoses from the heater unit.

CAUTION

Engine coolant will run out when the hoses are disconnected; drain it into a clean drip pan. Be sure not to let coolant spill on electrical parts or painted surfaces. If any coolant spills, rinse it off immediately.

6. Remove the crash pad.

7. Remove the cowl cross bar assembly.

8. Disconnect the connectors from the temperature control actuator, the mode control actuator and the evaporator temperature sensor.

9. Remove the heater & blower unit after loosening 3 mounting bolts.

10. Remove the blower unit from heater unit after loosening 2 screws.

11. Remove the heater core after remove the cover.

12. Loosen the heater unit mounting screw and clip and then remove the heater unit cover.

13. Remove the evaporator core.

14. Be careful that the inlet and outlet pipe are not bent during heater core removal, and pull out the heater core.

To install:

15. Be careful that the inlet and outlet pipe are not bent during heater core installation, and push in the heater core.

16. Install the evaporator core.

17. Tighten the heater unit mounting screw and clip and then install the heater unit cover.

18. Install the heater core after install the cover.

19. Install the blower unit to heater unit then tighten 2 screws.

20. Install the heater & blower unit then tighten 3 mounting bolts.

21. Connect the connectors from the temperature control actuator, the mode control actuator and the evaporator temperature sensor.

22. Install the cowl cross bar assembly.

23. Install the crash pad.

CAUTION

Engine coolant will run out when the hoses are disconnected; drain it into a clean drip pan. Be sure not to let coolant spill on electrical parts or painted surfaces. If any coolant spills, rinse it off immediately.

24. Disconnect the inlet and outlet heater hoses from the heater unit.

25. Remove the bolts and the expansion valve from the evaporator core. Plug or cap the lines immediately after disconnecting them to avoid moisture and dust contamination.

26. When the engine is cool, drain the engine coolant from the radiator.

27. Recover the refrigerant with a recovery/ recycling/ charging station.

28. Disconnect the negative (-) battery terminal.

a. If you're installing a new evaporator, add refrigerant oil (ND-OIL8).

b. Replace the O-rings with new ones at each fitting, and apply a thin coat of refrigerant oil before installing them. Be sure to use the right O-rings for R-134a to avoid leakage.

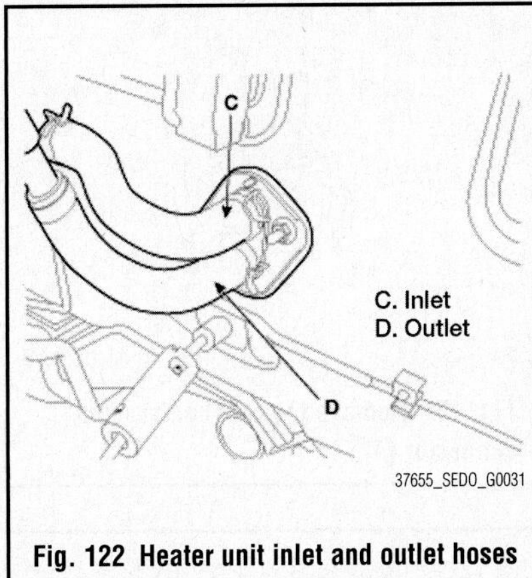

Fig. 122 Heater unit inlet and outlet hoses

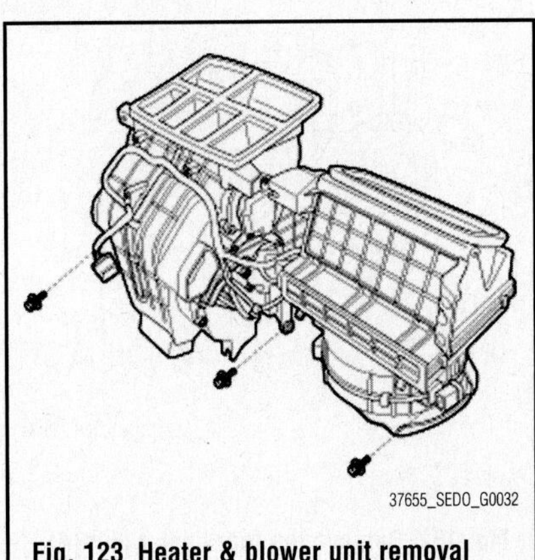

Fig. 123 Heater & blower unit removal

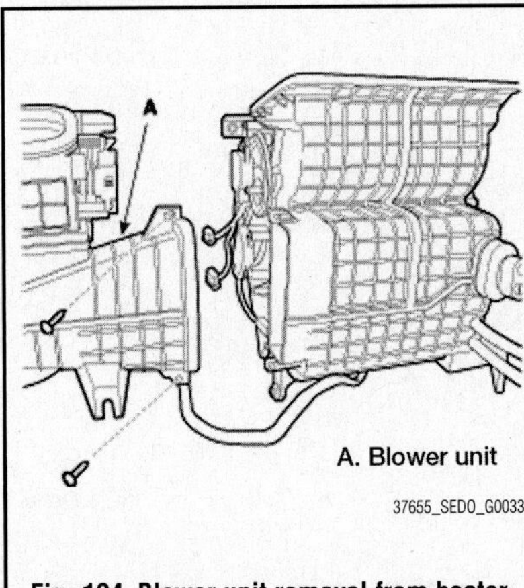

Fig. 124 Blower unit removal from heater unit removal

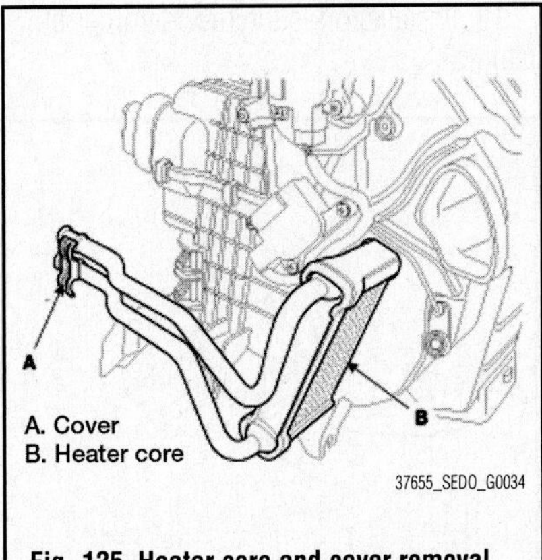

Fig. 125 Heater core and cover removal

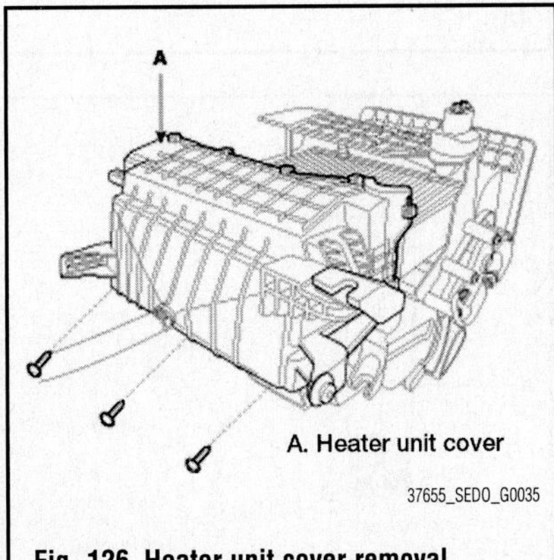

Fig. 126 Heater unit cover removal

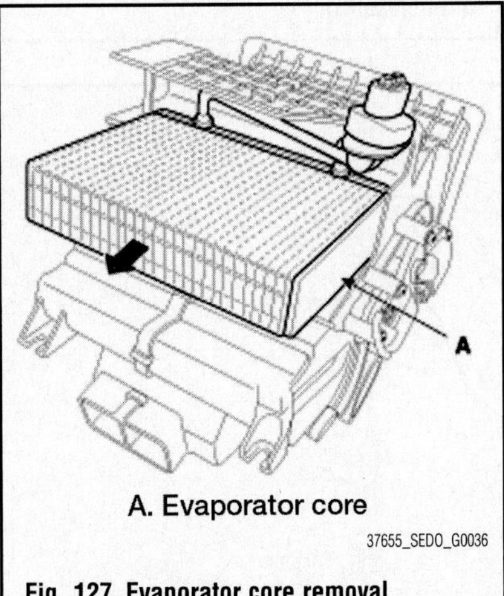

Fig. 127 Evaporator core removal

c. Immediately after using the oil, replace the cap on the container, and seal it to avoid moisture absorption.

d. Do not spill the refrigerant oil on the vehicle ; it may damage the paint ; if the refrigerant oil contacts the paint, wash it off immediately

e. Apply sealant to the grommets.

f. Make sure that there is no air leakage.

g. Charge the system and test its performance.

h. Do not interchange the inlet and outlet heater hoses and install the hose clamps securely.

i. Refill the cooling system with engine coolant.

AUXILIARY HEATING & AIR CONDITIONING SYSTEM

BLOWER MOTOR

REMOVAL & INSTALLATION

1. Disconnect the negative battery terminal.
2. Disconnect the connector of the blower motor.
3. Remove the blower motor.

To install:

4. Installation is the reverse order of removal.

HEATER CORE

REMOVAL & INSTALLATION

See Figures 128 through 133.

1. Disconnect the negative battery terminal.
2. Recover the refrigerant with a recover/recycling/charging station.
3. Drain the engine coolant from the radiator.
4. Remove luggage side trim.
5. Remove the rear speaker.
6. Remove the rear air duct.
7. Remove the rear refrigerant line.
8. Remove the heater hose, drain hose.
9. Remove the rear heater main connector.
10. Remove the rear heater unit.
11. Remove the rear temp actuator.
12. Remove the rear temp actuator lever.
13. Remove the heater core.

To install:

14. Installation is the reverse order of removal.

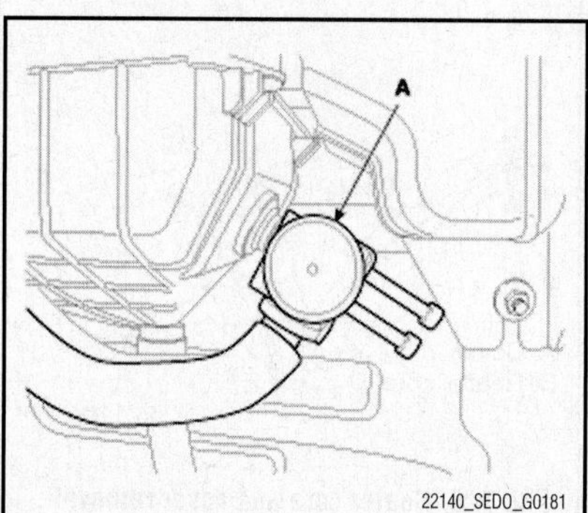

Fig. 129 Remove the rear refrigerant line (A).

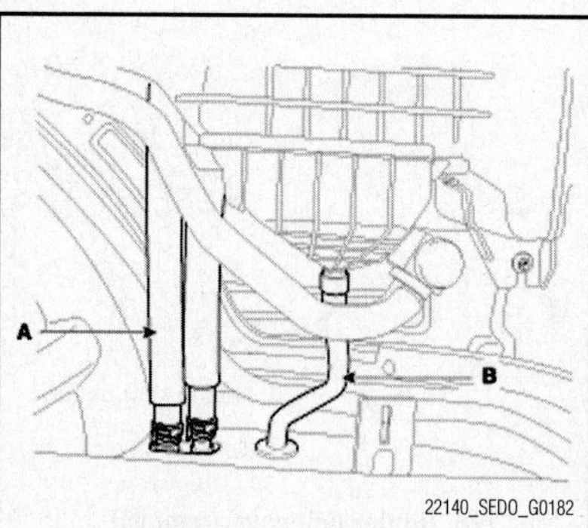

Fig. 130 Remove the heater hose (A), drain hose (B).

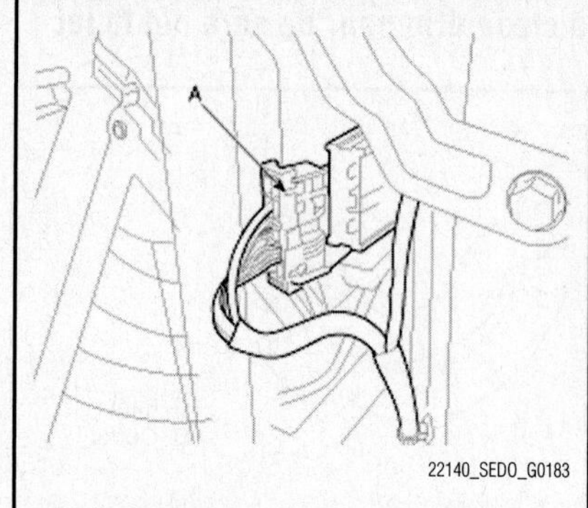

Fig. 131 Remove the rear heater main connector (A).

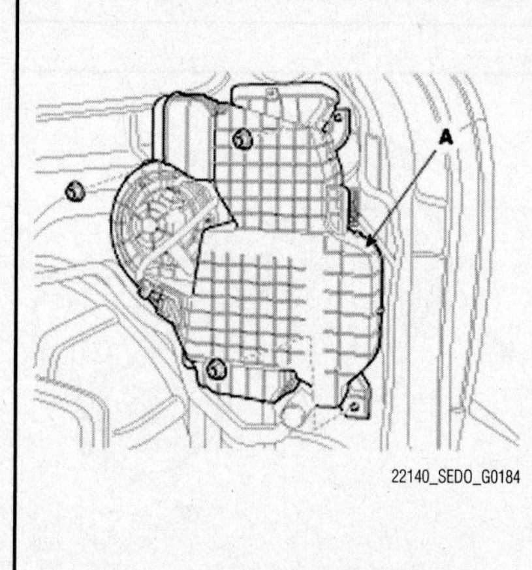

Fig. 132 Remove the rear heater unit (A).

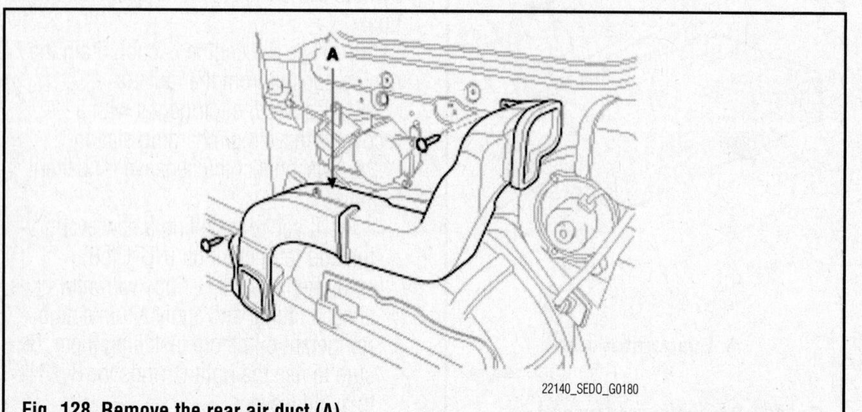

Fig. 128 Remove the rear air duct (A)

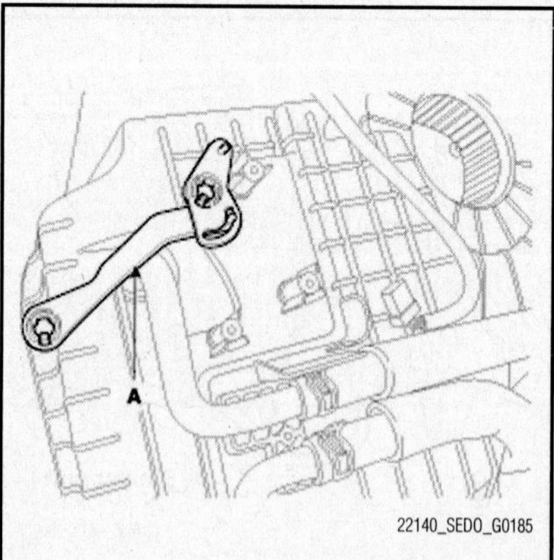

Fig. 133 Remove the rear temp actuator lever (A).

STEERING

POWER RACK & PINION STEERING GEAR

REMOVAL & INSTALLATION

See Figures 134 through 137.

1. Remove the both front wheels.
2. Drain the power steering fluid.
3. Remove the bolt connecting steering column to universal joint.
4. Disconnect the pressure tube from the power steering oil pump.
5. Disconnect the return hose.
6. Loosen the split pin and castle nut, and remove the tie rod end from the knuckle by using a SST (09568-4A000).

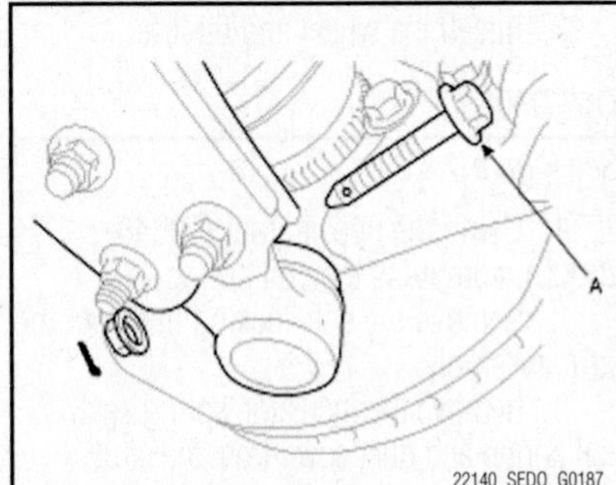

Fig. 134 Remove the split pin and lower arm bolts and nut (A).

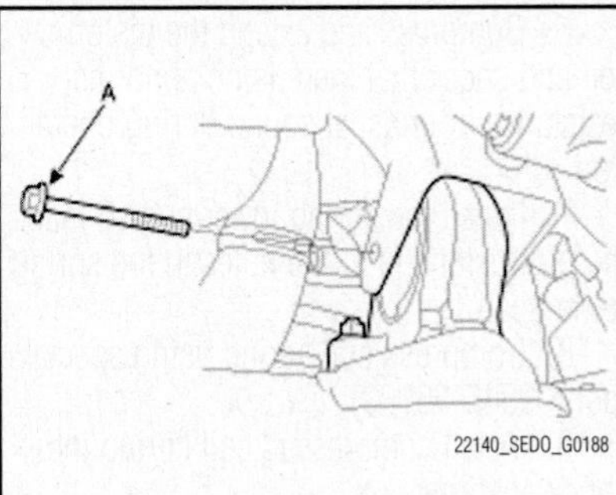

Fig. 135 Remove front and rear roll stopper bolt and nut (A).

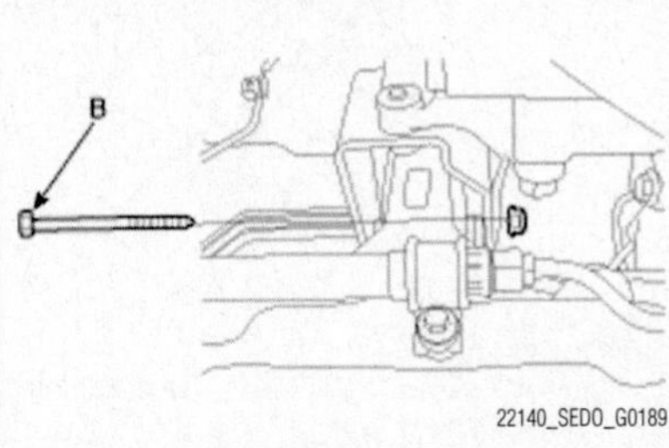

Fig. 136 Remove front and rear roll stopper bolt and nut (B).

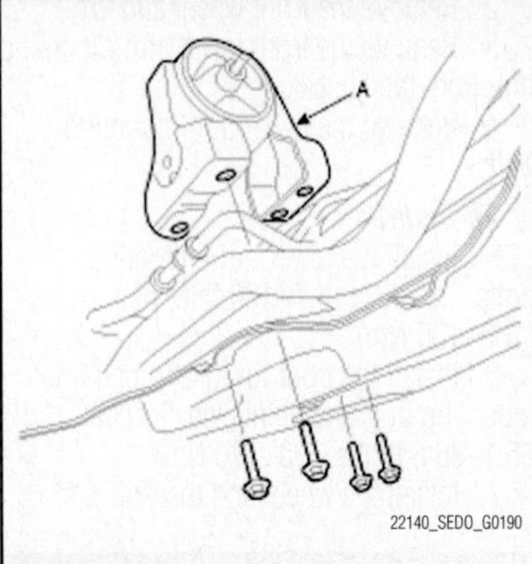

Fig. 137 Remove the rear roll stopper (A) from the sub-frame.

7. Remove the split pin and lower arm bolts and nut.
8. Disconnect the stabilizer link from the strut assembly.
9. Repeat on the other side.
10. Remove the front and rear roll stopper bolts and nuts.
11. Remove the sub-frame.
12. Remove the rear roll stopper from the sub-frame.
13. Disconnect the pressure and return lines from the valve body housing.
14. Remove the power steering gear box from the sub-frame.

To install:

15. Installation is reverse of the removal.
16. During installation, use the following torque values:
    a. Power steering gear box to sub-frame: 65–80 ft. lbs. (90–110 Nm).
    b. Stabilizer link to strut assembly: 72–87 ft. lbs. (100–120 Nm).
    c. Lower arm bolt and nut: 65–87 ft. lbs. (90–120 Nm).
    d. Tie rod end to knuckle: 36–40 ft. lbs. (50–55 Nm).
    e. Steering column to universal joint: 9.4–13.0 ft. lbs. (13–18 Nm).
17. After installation, bleed the power steering system.
18. Adjust the wheel alignment.

POWER STEERING PUMP

REMOVAL & INSTALLATION

1. Disconnect the pressure tube from the power steering pump.
2. Disconnect the suction hose from the suction pipe.
3. Remove the accessory drive belt.
4. Remove the power steering pump assembly from the pump bracket.

To install:

5. Installation is the reverse of removal.

CAUTION

The pressure tube does not twist and come in contact with other components.

6. Add power steering fluid
7. Bleed the power steering system.
8. Check the oil pump pressure.

BLEEDING

1. Remove the fuel pump fuse, then start the engine and wait for the engine to stall. Next, while operating the starting motor intermittently (for 15–20 seconds), turn the steering wheel all the way to the left and then to the right five or six times.

➡During air bleeding, replenish the fluid supply so that the level never falls below the lower position of the filter. If air bleeding is done while the vehicle is idling, the air will be broken up and absorbed into the fluid. Be sure to do the bleeding only while cranking.

2. Reinstall the fuel pump fuse, and start the engine (idling).
3. Turn the steering wheel to the left and the right until there are no air bubbles in the oil reservoir.

CAUTION

Do not hold the steering wheel turned all the way to either side for more than ten seconds.

4. Confirm that the fluid is not milky, and that the level is up to the position specified on the level gauge.
5. Confirm that there is little change in the surface of the fluid when the steering wheel is turned left and right.

CAUTION

If the surface of the fluid changes considerably, air bleeding should be done again. If the fluid level rises suddenly when the engine is stopped, it indicates that there is still air in the system. If there is air in the system, a jingling noise may be heard from the pump and the control valve may also produce unusual noises. Air in the system will shorten the life of the pump and other parts.

LOWER BALL JOINT

REMOVAL & INSTALLATION

See Figures 138 and 139.

1. Raise the front of the vehicle, and make sure it is securely supported.
2. Remove the front wheel and tire
3. Remove the front lower arm mounting bolt from the knuckle.
4. Remove the lower arm mounting bolts.

To install:

5. Install the lower arm mounting bolts. Torque: 115.7–130.2 ft. lbs. (160–180 Nm).
6. Install the front lower arm ball joint mounting bolt to the knuckle. Torque: 65.1–86.8 ft. lbs. (90–120 Nm).
7. Install the wheel and the tire.

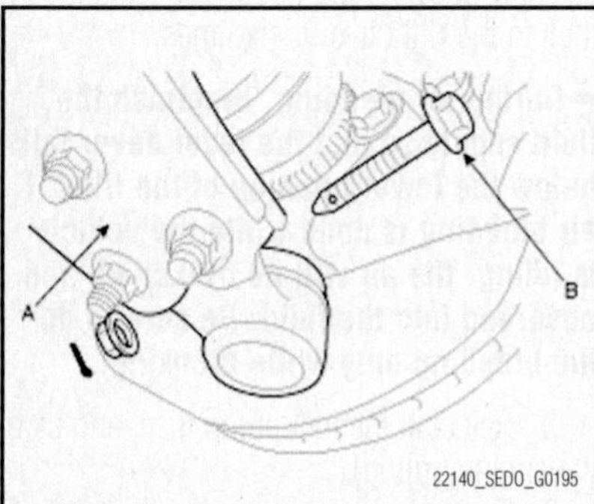

Fig. 138 Remove the front lower arm (A) mounting bolt (B) from the knuckle.

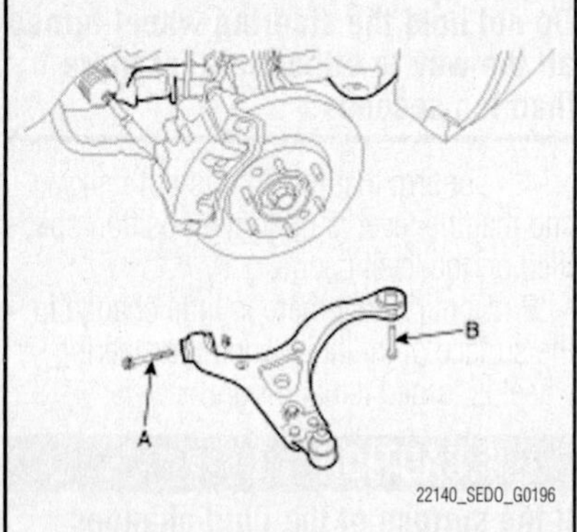

Fig. 139 Remove the lower arm mounting bolts (A, B).

LOWER CONTROL ARM

REMOVAL & INSTALLATION

See Figures 138 and 139.

1. Raise the front of the vehicle, and make sure it is securely supported.
2. Remove the front wheel and tire
3. Remove the front lower arm mounting bolt from the knuckle.
4. Remove the lower arm mounting bolts.

To install:

5. Install the lower arm mounting bolts. Torque: 115.7–130.2 ft. lbs. (160–180 Nm).
6. Install the front lower arm ball joint mounting bolt to the knuckle. Torque: 65.1–86.8 ft. lbs. (90–120 Nm).
7. Install the wheel and the tire.

MACPHERSON STRUT

REMOVAL & INSTALLATION

See Figures 140 and 141.

1. Raise the front of the vehicle, and make sure it is securely supported.
2. Remove the front wheel and tire.
3. Remove the brake hose bracket bolts from the front strut assembly.
4. Remove the speed sensor and wire bolts from the front knuckle.
5. Remove the front stabilizer link nut from the strut assembly.

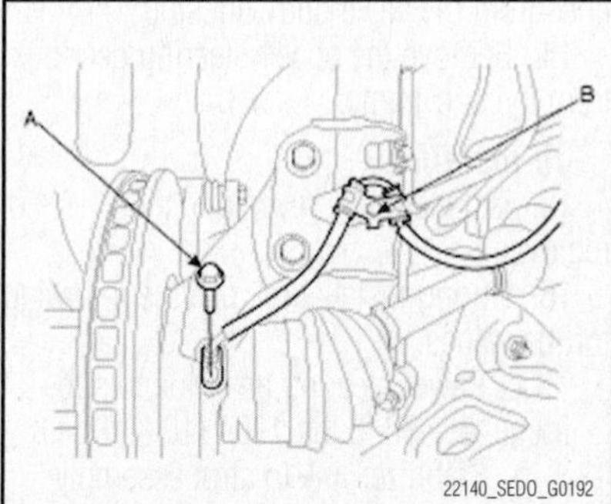

Fig. 140 Remove the speed sensor (A) and wire (B) bolts from the front knuckle.

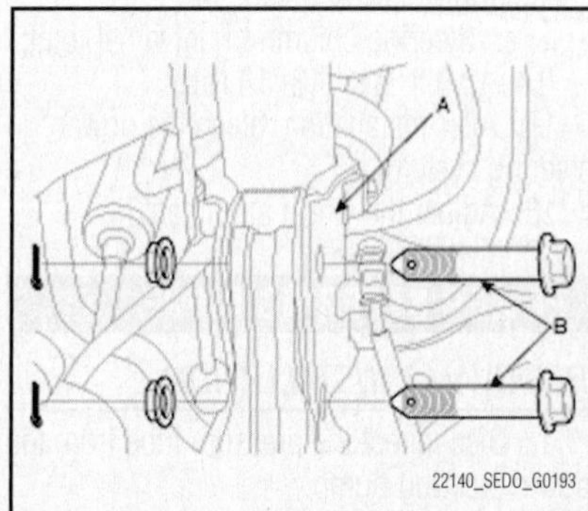

Fig. 141 Remove the front strut assembly (A) bolts (B) from the front knuckle.

6. Remove the strut upper mounting nuts.
7. Remove the front strut assembly bolts from the front knuckle.

To install:

8. Install the strut upper mounting nuts. Torque: 32.5–43.4 ft. lbs. (45–60 Nm).
9. Install the front strut assembly bolts to the front knuckle. Torque: 72.3–86.8 ft. lbs. (100–120 Nm).
10. Install the front stabilizer link nut to the strut assembly. Torque: 72.3–86.8 ft. lbs. (100–120 Nm).
11. Install the speed sensor and wire bolts.
12. Install the brake hose bracket bolt to the axle assembly.
13. Install the wheel and the tire.

OVERHAUL

See Figure 142.

1. Using the special tool (09546-26000), compress the coil spring.
2. Remove the self-locking nut from the strut assembly.
3. Remove the insulator, spring seat, coil spring and dust cover from the strut assembly.
4. Check the strut insulator for wear or damage.
5. Check rubber parts for damage or deterioration.
6. Compress and extend the piston rod and check that there is no abnormal resistance or unusual sound during operation.
7. Install the spring lower pad so that the protrusions fit in the holes in the spring lower seat.
8. Compress coil spring using special tool (09546-26000).
9. Install compressed coil spring into shock absorber.

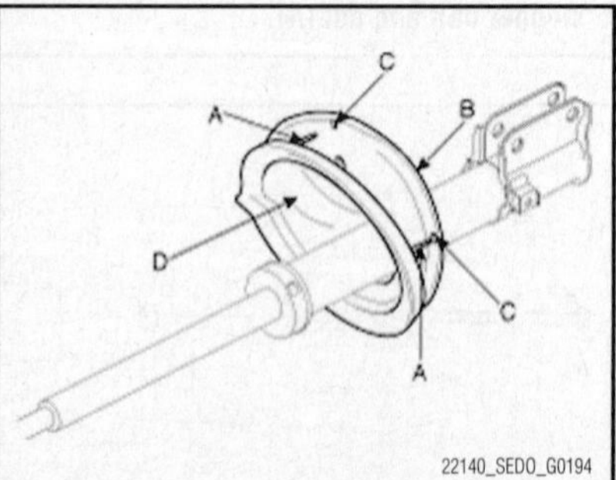

Fig. 142 Install the spring lower pad (D) so that the protrusions (A) fit in the holes (C) in the spring lower seat (B).

➡Install the coil spring with the identification mark directed toward the knuckle.

10. After fully extending the piston rod, install the spring upper seat and insulator assembly.

11. After seating the upper and lower ends of the coil spring in the upper and lower spring seat grooves correctly, tighten new self-locking nut temporarily.

12. Remove the special tool (09546-26000).

13. Tighten the self-locking nut to the specified torque. Torque: 43.4–50.6 ft. lbs. (60–70 Nm).

STEERING KNUCKLE

REMOVAL & INSTALLATION

See Figures 143 through 145.

1. Raise the vehicle and remove the wheel & tire.
2. Remove the wheel speed sensor & wire.
3. Disconnect the brake hose from the strut assembly.
4. Unstake the driveshaft lock nut using a chisel and hammer.
5. Remove the driveshaft lock nut.

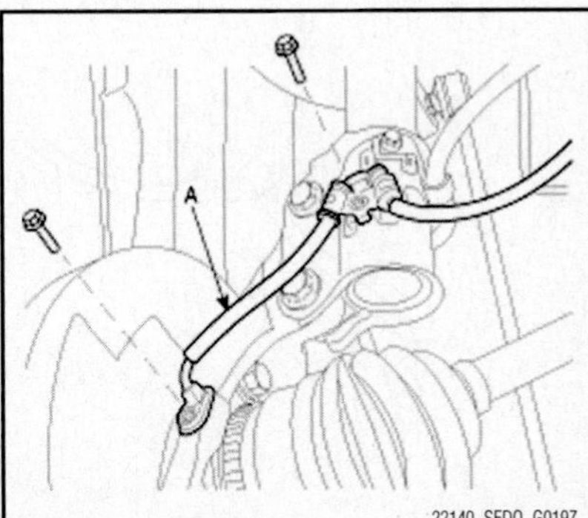

Fig. 143 Remove the wheel speed sensor & wire (A).

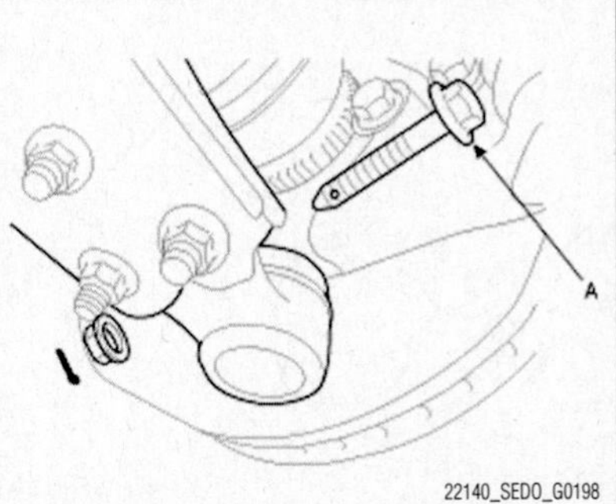

Fig. 144 Remove the split pin and lower arm bolt and nut (A).

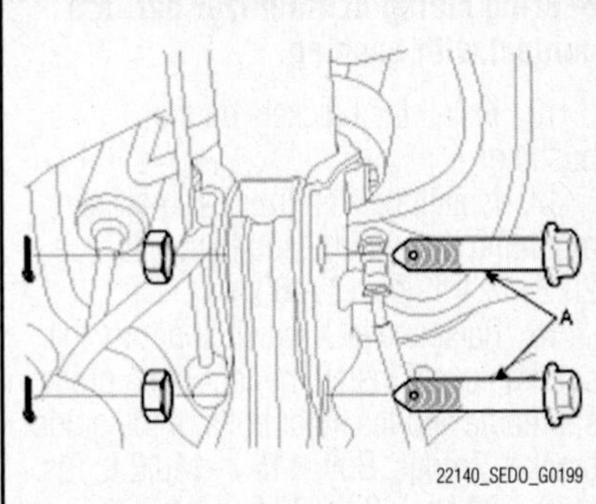

Fig. 145 Remove the knuckle from the strut assembly.

6. Remove the caliper assembly from the knuckle and suspend it with wire.
7. Remove the split pin and castle nut from the tie rod end ball joint.
8. Disconnect the tie rod end from the knuckle using a SST (09568-4A000).
9. Remove the split pin and lower arm bolt and nut.
10. Using a plastic hammer, disconnect the driveshaft from the front hub assembly.
11. Remove the disc brake rotor from the front hub assembly.
12. Remove the knuckle from the strut assembly.

To install:

13. Installation is the reverse of removal.
14. During installation, use the following torque values:
    a. Knuckle to strut assembly: 72–87 ft. lbs. (100–120 Nm).
    b. Lower arm bolt and nut: 65–87 ft. lbs. (90–120 Nm).
    c. Castle nut for tie rod end ball joint: 43–58 ft. lbs. (60–80 Nm).
    d. Brake caliper assembly to knuckle: 61–72 ft. lbs. (85–100 Nm).
    e. Driveshaft lock nut: 177–199 ft. lbs. (245–275 Nm).

➡Be sure to stake the driveshaft lock nut.

STABILIZER BAR & LINKS

REMOVAL & INSTALLATION

See Figures 146 through 153.

1. Remove the connecting bolt between the steering universal joint assembly and the pinion assembly.
2. Raise the front of the vehicle, and make sure it is securely supported.
3. Remove the front wheel and tire.
4. Remove the front stabilizer link nut from the strut assembly.
5. After removing both sides of the tie rod end self-locking nuts and cotter pins, remove the ball joints by using the special tool(09568-4A000).
6. Remove both sides of the lower arm mounting bolts.
7. Remove the engine mounting bolts.
8. Remove the twelve bolts and nuts of the sub frame by supporting it with a jack.
9. After lowering the jack which supports the sub frame, remove both sides of the stabilizer bar assembly mounting bolts.

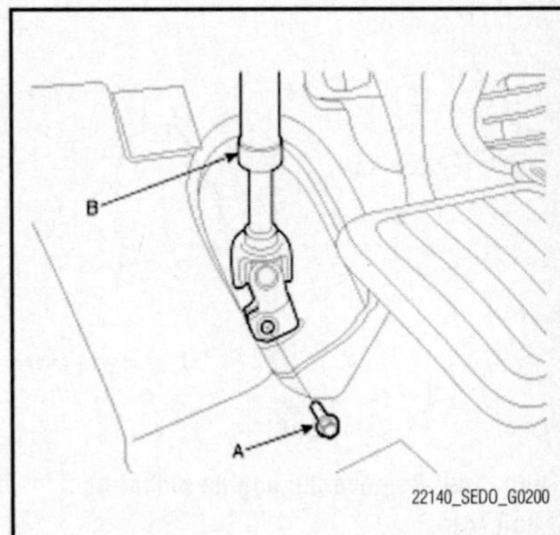

Fig. 146 Remove the connecting bolt (A) between the steering universal joint assembly (B) and the pinion assembly.

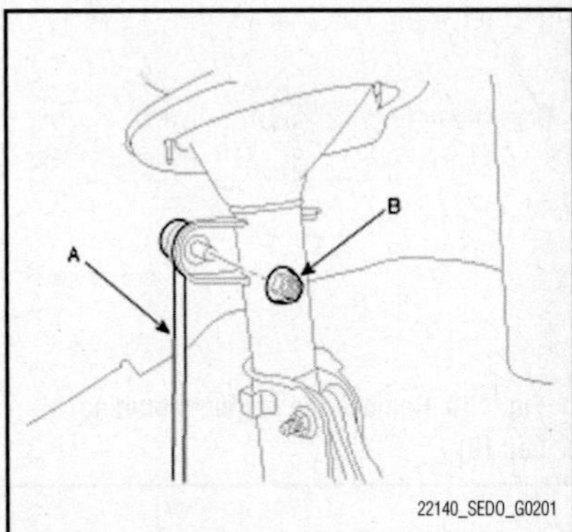

Fig. 147 Remove the front stabilizer link (A) nut (B) from the strut assembly.

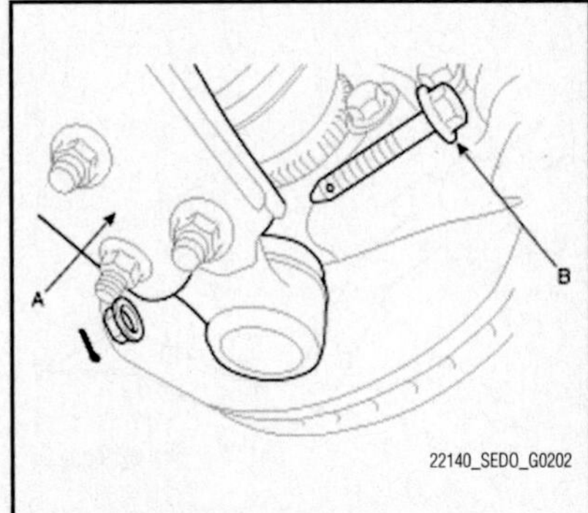

Fig. 148 Remove both sides of the lower arm (A) mounting bolts (B).

10. Remove the stabilizer bar assembly through the gap between the body and the rear side of the sub frame.

11. Remove the brackets and the bushings.

To install:

12. Install the bushings on the stabilizer bar.

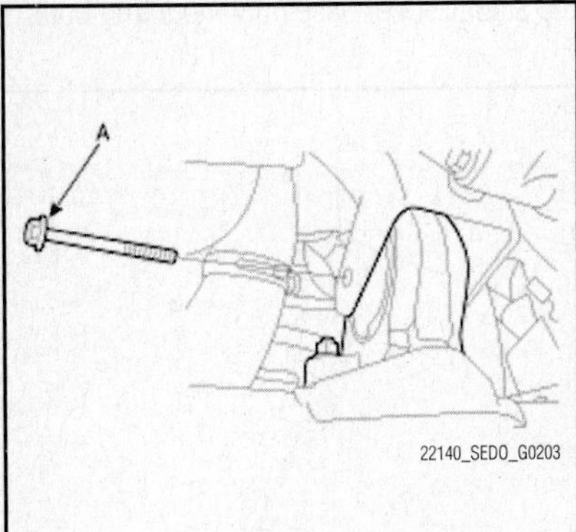

Fig. 149 Remove the engine mounting bolt (A)

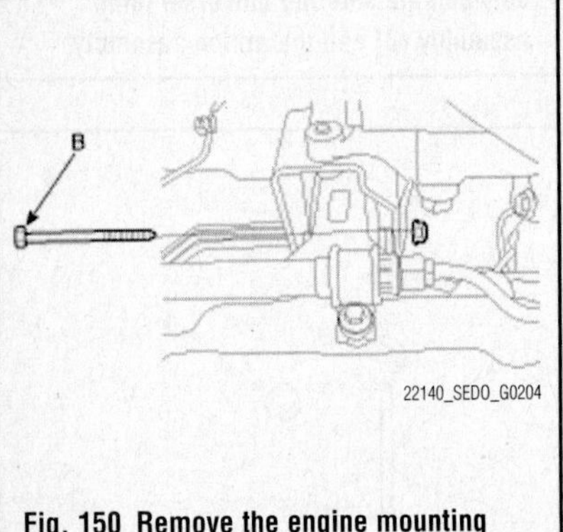

Fig. 150 Remove the engine mounting bolt (B)

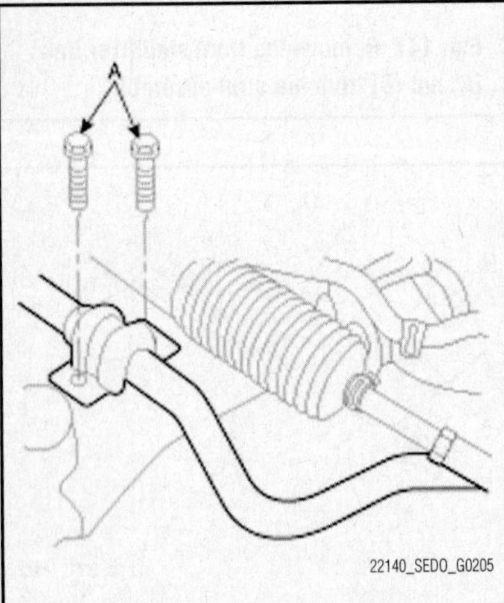

Fig. 151 Remove both sides of the stabilizer bar assembly mounting bolts (A).

➡Bring clamp of stabilizer bar into contact with bushing.

13. Install the brackets on the bushings.

14. Install the stabilizer bar bracket mounting bolts to the sub frame. Torque: 28.2–43.4 ft. lbs. (39–60 Nm).

15. Raise the jack which supports the sub frame and install the four bolts of the sub frame and the eight bolts of the guide bracket. Torque: Bolt: 115.7–130.2 ft. lbs. (160–180 Nm); Bolt: 32.5–43.4 ft. lbs. (45–60 Nm).

16. Install the engine mounting bolts. Torque: 47.0–61.5 ft. lbs. (65–85 Nm).

17. Install both sides of the lower arm mounting bolts. Torque: 65.1–86.8 ft. lbs. (90–120 Nm).

18. Install both sides of the tie rod end self-locking nuts. Insert cotter pin after nut is torqued to specifications. Torque: 43.4–57.8 ft. lbs. (65–80 Nm).

19. Install the front stabilizer link nut to the strut assembly. Torque: 72.3–86.8 ft. lbs. (100–120 Nm).

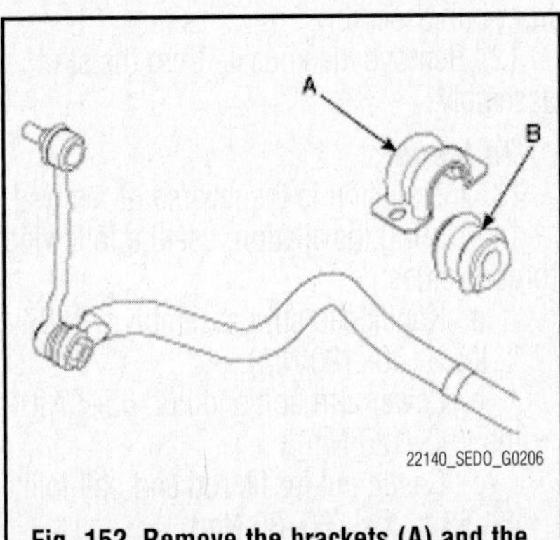

Fig. 152 Remove the brackets (A) and the bushings (B).

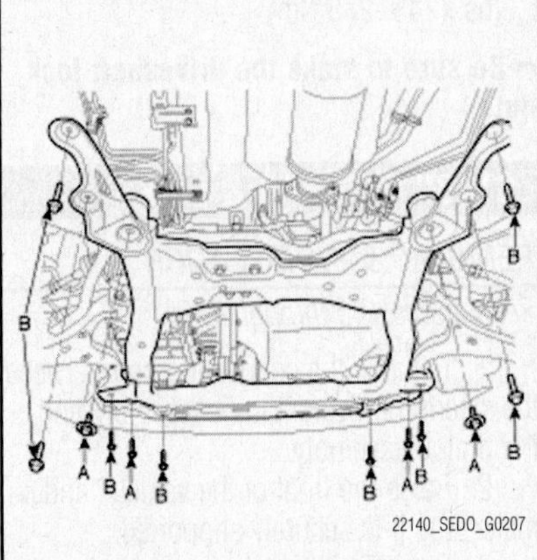

Fig. 153 Install the four bolts (A) of the sub frame and the eight bolts (B) of the guide bracket.

20. Install the wheel and the tire.

21. Install the connecting bolt between the steering universal joint assembly and the pinion assembly. Torque: 9.4–13.0 ft. lbs. (13–18 Nm).

⁂ CAUTION

After installation, if necessary, adjust the alignment of the steering wheel and front tires.

WHEEL BEARINGS

INSPECTION

1. Measure the wheel bearing starting torque: 1.45 ft. lbs. (1.97 Nm).

2. Measure the hub assembly axial play using a dial gauge. Should measure 0.008 mm or less.

REMOVAL & INSTALLATION

See Figures 154 through 157.

1. Raise the vehicle and remove the wheel & tire.

2. Remove the wheel speed sensor & wire.

3. Disconnect the brake hose from the strut assembly.

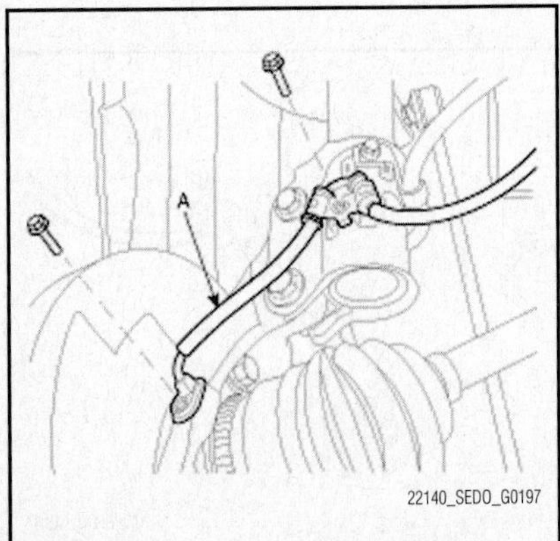

Fig. 154 Remove the wheel speed sensor & wire (A).

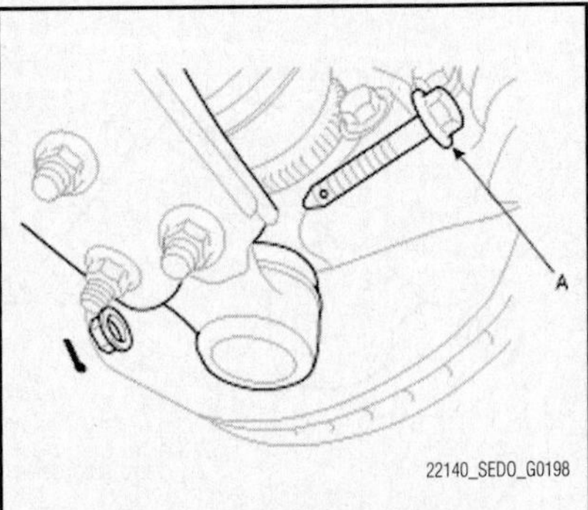

Fig. 155 Remove the split pin and lower arm bolt and nut (A).

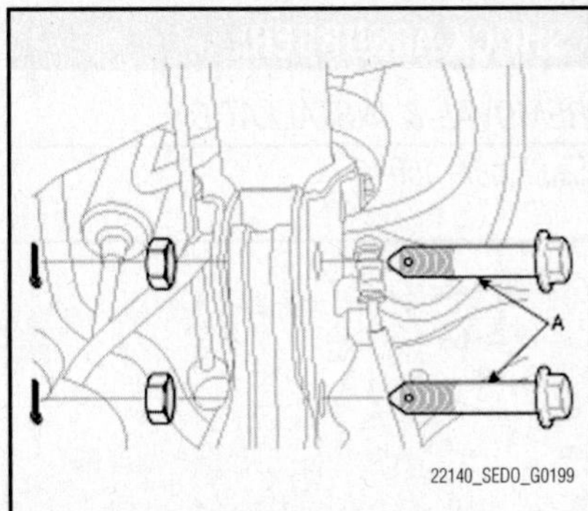

Fig. 156 Remove the knuckle from the strut assembly.

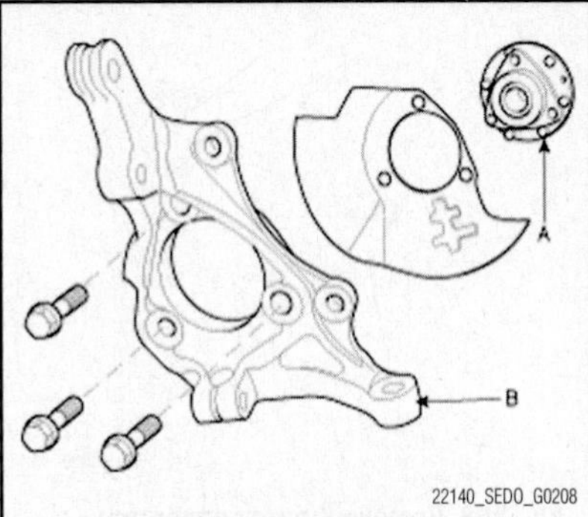

Fig. 157 Separate the hub & bearing assembly (A) from the knuckle (B).

4. Unstake the driveshaft lock nut using a chisel and hammer.
5. Remove the driveshaft lock nut.
6. Remove the caliper assembly from the knuckle and suspend it with wire.
7. Remove the split pin and castle nut from the tie rod end ball joint.
8. Disconnect the tie rod end from the knuckle using a SST (09568-4A000).
9. Remove the split pin and lower arm bolt and nut.
10. Using a plastic hammer, disconnect the driveshaft from the front hub assembly.
11. Remove the disc brake rotor from the front hub assembly.
12. Remove the knuckle from the strut assembly.
13. Separate the hub & bearing assembly from the knuckle.

To install:

14. Installation is the reverse of removal.
15. During installation, use the following torque values:
   a. Hub & bearing assembly to knuckle:116–130 ft. lbs. (160–180 Nm).
   b. Knuckle to strut assembly: 72–87 ft. lbs. (100–120 Nm).
   c. Lower arm bolt and nut: 65–87 ft. lbs. (90–120 Nm).
   d. Castle nut for tie rod end ball joint: 43–58 ft. lbs. (60–80 Nm).
   e. Brake caliper assembly to knuckle: 61–72 ft. lbs. (85–100 Nm).
   f. Driveshaft lock nut: 177–199 ft. lbs. (245–275 Nm).

➡**Be sure to stake the driveshaft lock nut.**

SUSPENSION — REAR SUSPENSION

REAR ASSIST ARM

REMOVAL & INSTALLATION

See Figures 158 through 160.

1. Raise the rear of the vehicle, and make sure it is securely supported.
2. Remove the rear wheel and tire.
3. Remove the brake caliper mounting bolts and suspend the brake caliper assembly with wire.
4. Remove the rear assist arm ball joint self-locking nut and the cotter pin.
5. Remove the rear assist arm ball joint by using the special tool(09568-4A000).
6. Remove the rear assist arm mounting nut from the cross member.

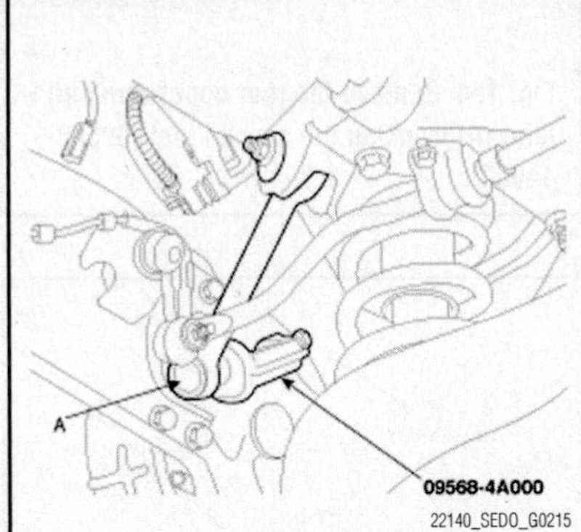

Fig. 159 Remove the rear assist arm ball joint (A) by using the special tool(09568-4A000).

To install:

7. Install the rear assist arm mounting nut to the cross member. Torque: 65.1–86.8 ft. lbs. (90–120 Nm).
8. Install the rear assist arm ball joint self-locking nut and the cotter pin. Torque: 115.7–130.2 ft. lbs. (160–180 Nm).
9. Install the brake caliper mounting bolts. Torque: 36.2–43.4 ft. lbs. (50–60 Nm).
10. Install the wheel and the tire.

REAR LOWER ARM

REMOVAL & INSTALLATION

See Figures 161 and 162.

1. Raise the rear of the vehicle, and make sure it is securely supported.

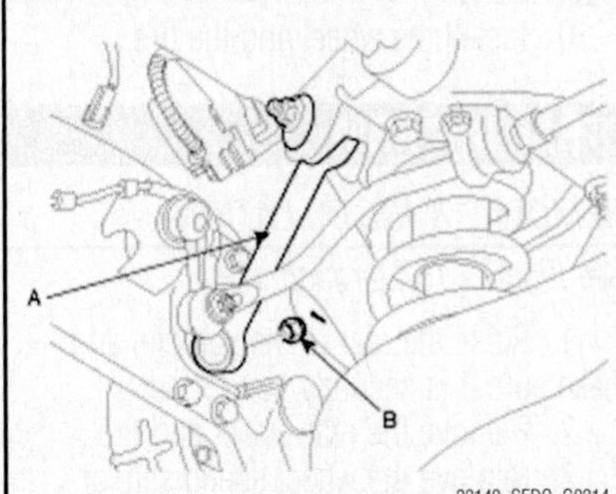

Fig. 158 Remove the rear assist arm (A) ball joint self-locking nut (B) and the cotter pin.

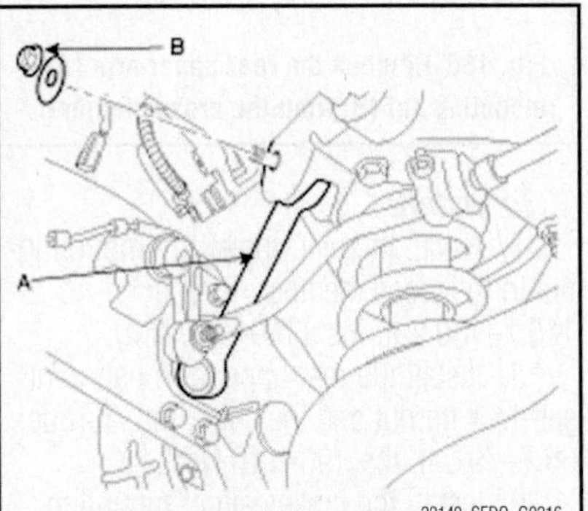

Fig. 160 Remove the rear assist arm (A) mounting nut (B) from the cross member.

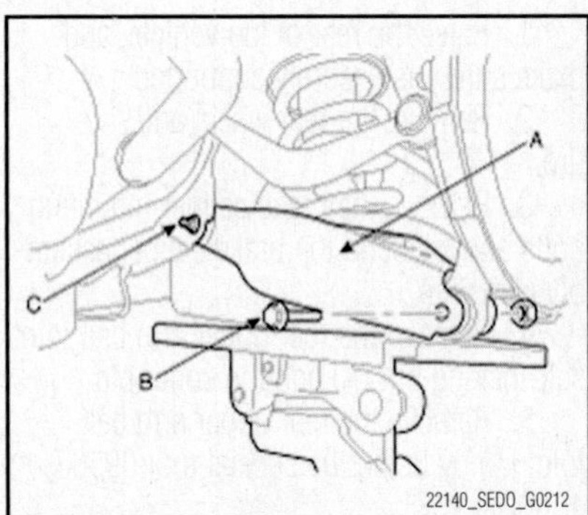

Fig. 161 Remove the lower arm bolt (B) from the rear knuckle, while supporting the lower arm (A) with a jack.

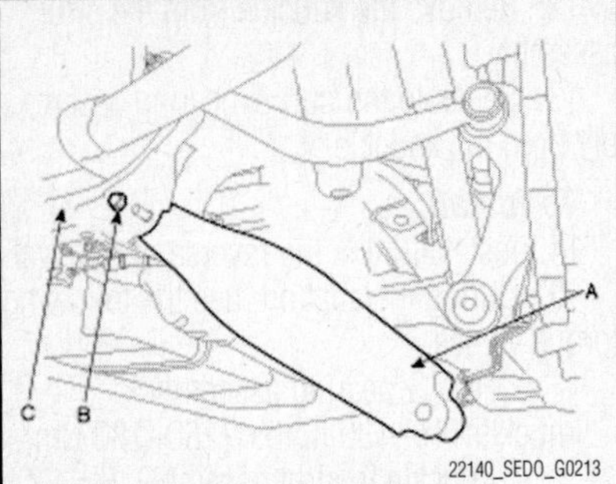

Fig. 162 Remove the lower arm (A) mounting bolts (B) from the cross member (C).

2. Remove the rear wheel and tire.
3. Remove the lower arm bolt from the rear knuckle, while supporting the lower arm with a jack.
4. Loosen the lower arm bolt from the cross member.
5. Remove the spring, the lower seat, and the upper pad.
6. Remove the lower arm mounting bolts from the cross member.

To install:

7. Install the lower arm mounting bolts to the cross member. Torque: 144.7–195.3 ft. lbs. (200–270 Nm).
8. Install the spring, the lower seat, and the upper pad.
9. Install the lower arm bolt to the rear knuckle and the lower arm bolt to the cross member with a specified torque, while supporting the lower arm with a jack. Torque: Lower arm bolt to rear knuckle: 86.8–115.7 ft. lbs. (120–160 Nm); Lower arm bolt to crossmember: 144.7–195.3 ft. lbs. (200–270 Nm).
10. Install the wheel and the tire.

REAR UPPER ARM

REMOVAL & INSTALLATION

See Figures 163 through 165.

1. Raise the rear of the vehicle, and make sure it is securely supported.
2. Remove the rear wheel and tire.
3. Remove the brake caliper mounting bolts and suspend the brake caliper assembly with wire.
4. Remove the rear upper arm ball joint self-locking nut (A) and the cotter pin.
5. Remove the rear upper arm ball joint (A) by using the special tool(09568-4A000).
6. Remove the rear upper arm (A) mounting nut (B) from the cross member.

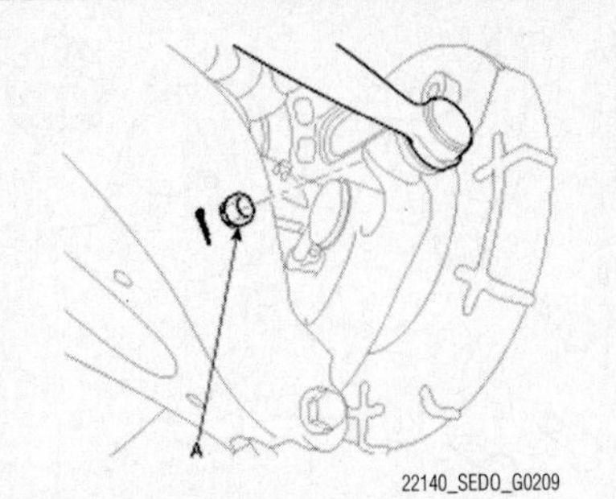

Fig. 163 Remove the rear upper arm ball joint self-locking nut (A) and the cotter pin.

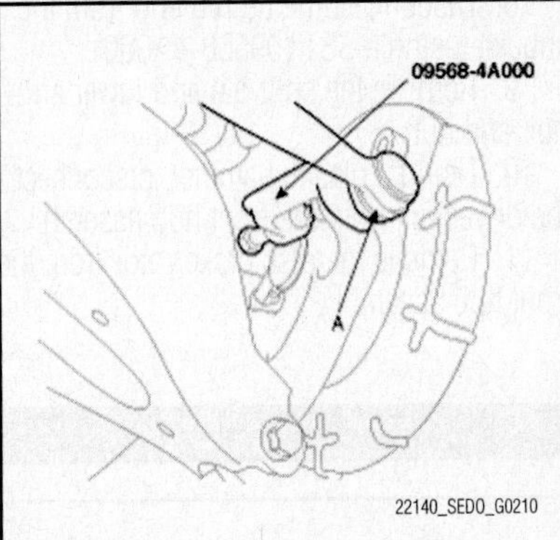

Fig. 164 Remove the rear upper arm ball joint (A) by using the special tool (09568-4A000).

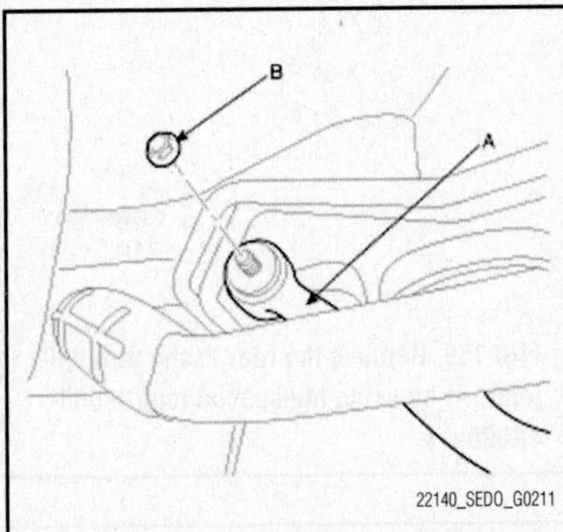

Fig. 165 Remove the rear upper arm (A) mounting nut (B) from the cross member

To install:

7. Install the rear upper arm mounting nut to the cross member. Torque: 115.7–130.2 ft. lbs. (160–180 Nm).
8. Install the rear upper arm ball joint self-locking nut and the cotter pin. Torque: 65.1–79.5 ft. lbs. (90–110 Nm).
9. Install the brake caliper mounting bolts. Torque: 36.2–43.4 ft. lbs. (50–60 Nm).
10. Install the wheel and the tire.

SHOCK ABSORBER

REMOVAL & INSTALLATION

See Figure 166.

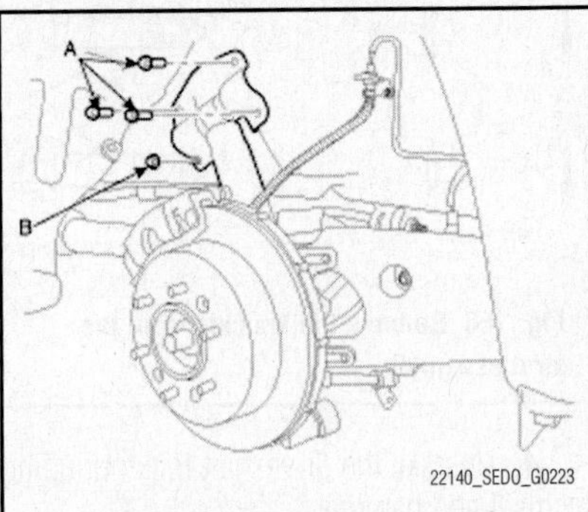

Fig. 166 Remove the rear shock absorber assembly mounting bolts (A) and nut (B) from the body.

1. Raise the rear of the vehicle, and make sure it is securely supported.
2. Remove the rear wheel and tire.
3. Remove the rear shock absorber assembly mounting bolts (A) and nut (B) from the body.
4. Remove the rear shock absorber assembly nut from the rear knuckle, and remove the shock absorber assembly.
5. Remove the rear shock absorber bracket bolt.

To install:

6. Install the connecting bolt between the rear shock absorber and the bracket. Torque: 115.7–130.2 ft. lbs. (160–180 Nm).
7. Install the rear shock absorber to the knuckle temporarily.
8. Install the rear shock absorber bracket mounting bolts and nut. Torque: Bracket bolts: 59.7–79.5 ft. lbs. (80–110 Nm); Bracket nut: 65.1–86.8 ft. lbs. (90–120 Nm).
9. Install the rear shock absorber nut to the knuckle. Torque: 115.7–130.2 ft. lbs. (160–180 Nm).
10. Install the wheel and the tire.

TRAILING ARM

REMOVAL & INSTALLATION

See Figures 167 through 172.

1. Raise the rear of the vehicle, and make sure it is securely supported.
2. Remove the rear wheel and tire.
3. Remove the wheel speed sensor wire bracket bolt from the body and the connector.
4. Remove the wheel speed sensor wire bracket bolt.

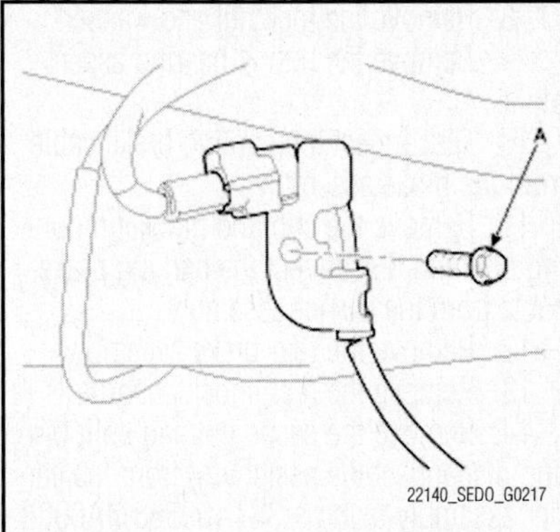

Fig. 167 Remove the wheel speed sensor wire bracket bolt (A) from the body and the connector.

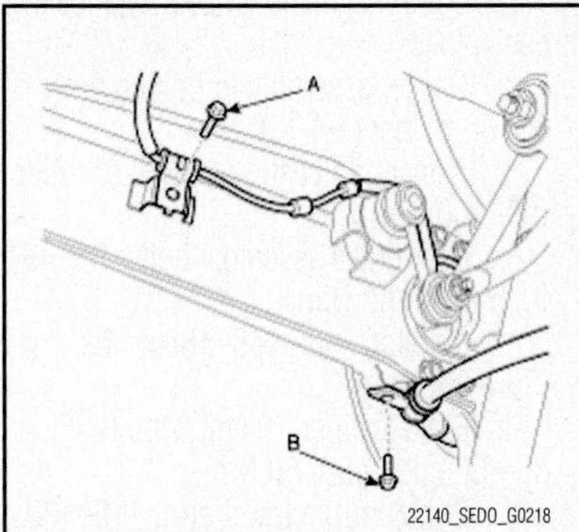

Fig. 168 Remove the wheel speed sensor wire bracket bolt (A).

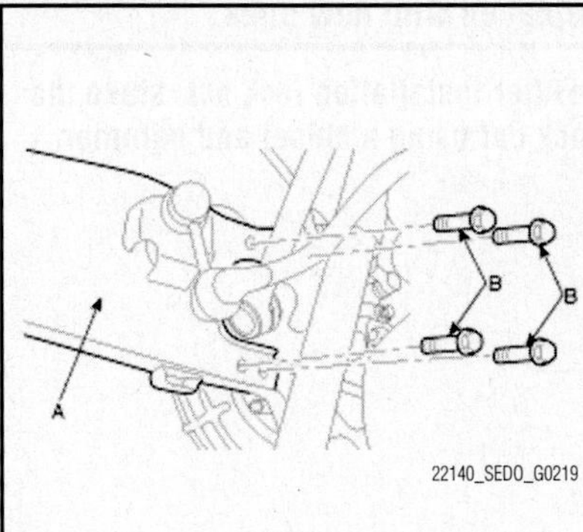

Fig. 169 Remove the trailing arm (A) mounting nuts (B) from the knuckle.

5. Remove the parking brake wire bracket bolt.
6. Remove the trailing arm mounting nuts from the knuckle.
7. Remove the stabilizer link nut from the trailing arm.
8. Remove the trailing arm bracket mounting bolts from the body.
9. Remove the connecting bolt between the trailing arm and the bracket.

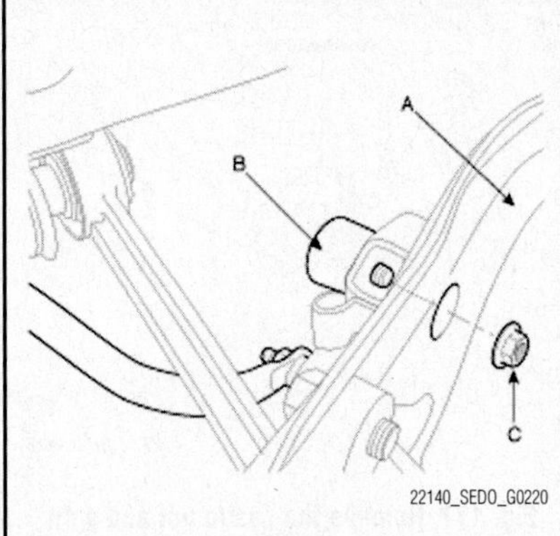

Fig. 170 Remove the stabilizer link (B) nut (C) from the trailing arm (A).

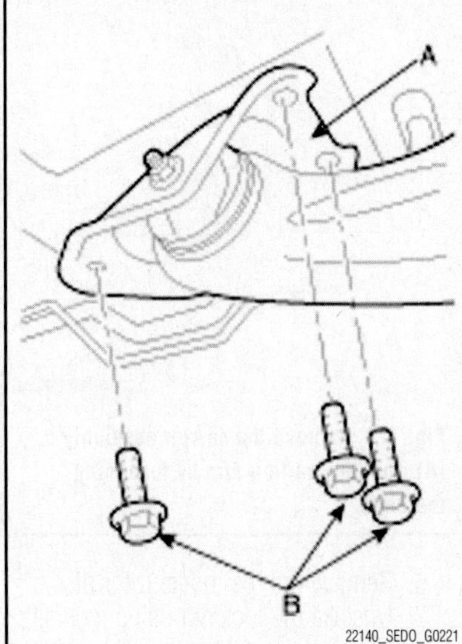

Fig. 171 Remove the trailing arm bracket (A) mounting bolts (B) from the body.

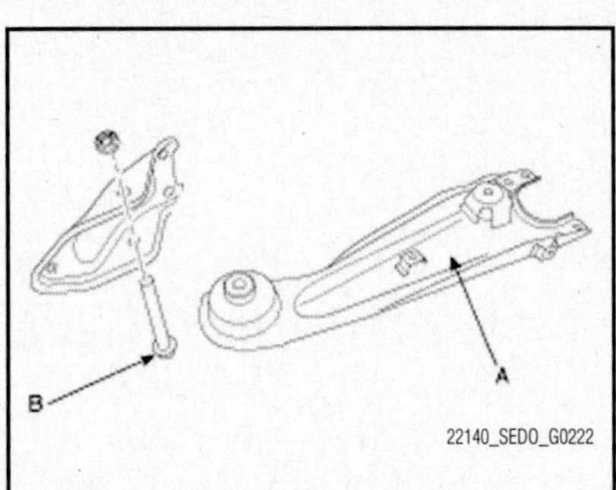

Fig. 172 Remove the connecting bolt (B) between the trailing arm (A) and the bracket.

To install:

10. Install the connecting bolt between the trailing arm and the bracket. Torque: 115.7–130.2 ft. lbs. (160–180 Nm).

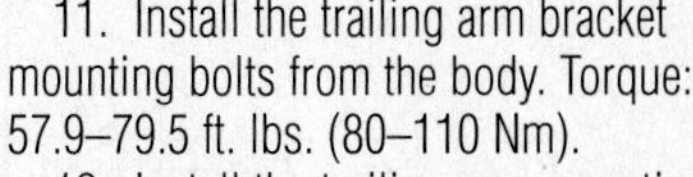
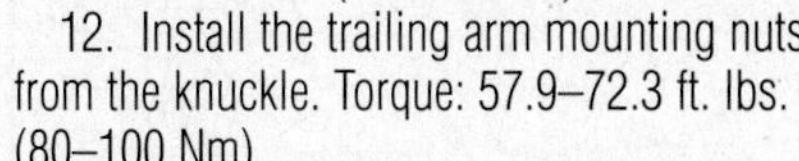

11. Install the trailing arm bracket mounting bolts from the body. Torque: 57.9–79.5 ft. lbs. (80–110 Nm).
12. Install the trailing arm mounting nuts from the knuckle. Torque: 57.9–72.3 ft. lbs. (80–100 Nm).
13. Install the stabilizer link nut to the trailing arm. Torque: 36.2–47.0 ft. lbs. (50–65 Nm).
14. Install the wheel speed sensor wire bracket bolt and the parking brake wire bracket bolt.
15. Install the wheel speed sensor wire bracket bolt from the body and the connector.
16. Install the wheel and the tire

WHEEL BEARINGS

REMOVAL & INSTALLATION

See Figures 173 through 178.

1. Remove the wheel & tire assembly.
2. Support lower part of the lower arm using a jack and remove the bolt & nut.
3. Remove the coil spring and upper pad.
4. Remove the wheel speed sensor.

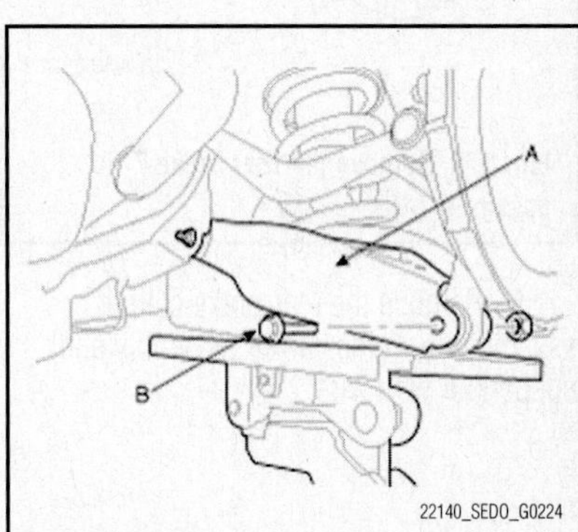

Fig. 173 Support lower part of the lower arm (A) using a jack and remove the bolt & nut (B).

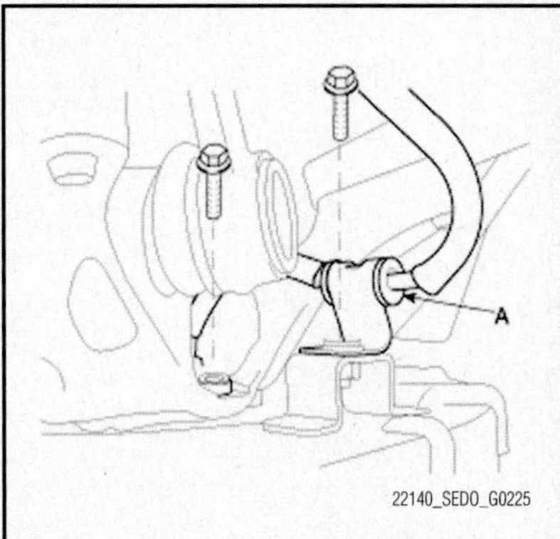

Fig. 174 Remove the wheel speed sensor (A).

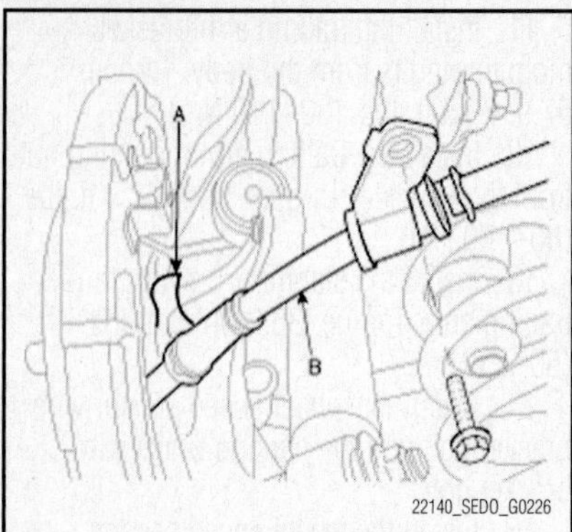

Fig. 175 Remove the clip (A) and bracket mounting bolt and disconnect the parking brake cable (B) from the carrier assembly.

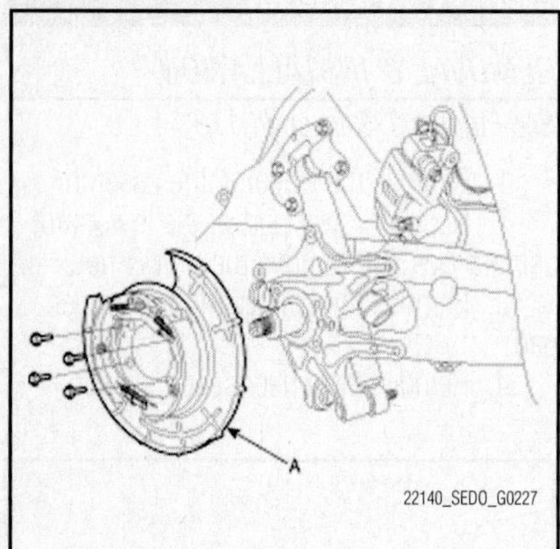

Fig. 176 Remove the rear brake assembly (A).

5. Remove the rear brake caliper assembly from the carrier assembly and suspend it with wire.

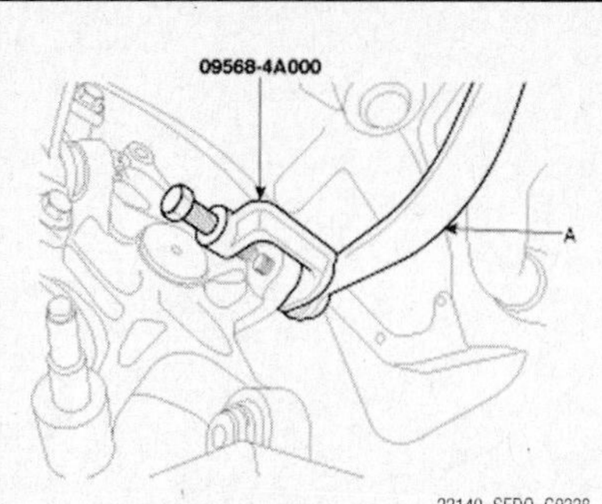

Fig. 177 Remove the castle nut and split pin and disconnect the assist arm (A) from the carrier assembly using a SST (09568-4A000).

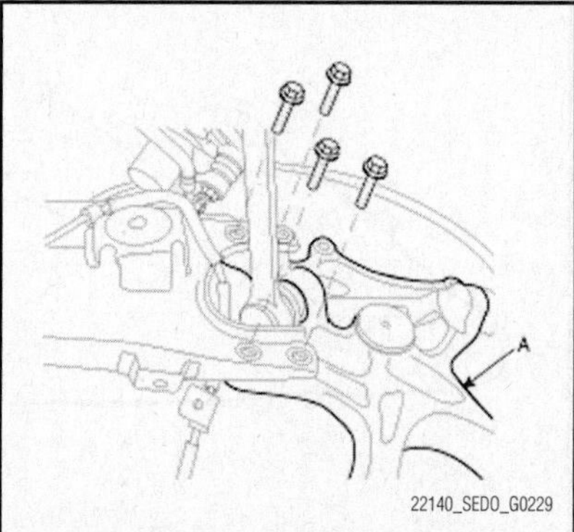

Fig. 178 Remove the carrier assembly (A) from the trailing arm by loosening the bolts.

6. Remove the rear brake rotor disc.
7. Unstake the lock nut using a chisel and hammer.
8. Remove the lock nut and washer.
9. Remove the hub & bearing assembly.
10. Disconnect the parking brake cable from the brake assembly.
11. Remove the clip and bracket mounting bolt and disconnect the parking brake cable from the carrier assembly.
12. Remove the rear brake assembly.
13. Remove the rear shock absorber.
14. Remove the castle nut and split pin and disconnect the assist arm from the carrier assembly using a SST (09568-4A000).
15. Remove the carrier assembly from the trailing arm by loosening the bolts.

To install:

16. Installation is the reverse of removal.
17. During installation, use the following torque values:
    a. Carrier assembly to trailing arm: 58–80 ft. lbs. (80–110 Nm).
    b. Assist arm castle nut: 65–72 ft. lbs (90–100 Nm).
    c. Rear brake assembly bolts: 36–43 ft. lbs. (50–60 Nm).
    d. Axle lock nut: 145–188 ft. lbs. (200–260 Nm).
    e. Brake caliper assembly bolts: 36–43 ft. lbs. (50–60 Nm).
    f. Bolt & nut to lower arm: 116–130 ft. lbs. (160–180 Nm).

CAUTION

The rear hub lock nut should be replaced with new ones.

➡After installation lock nut, stake the lock nut using a chisel and hammer.

KIA

Sorento

18

SPECIFICATIONS AND MAINTENANCE CHARTS

ENGINE AND VEHICLE IDENTIFICATION

Engine						
Code ①	Liters (cc)	Cu. In.	Cyl.	Fuel Sys.	Engine Type	Eng. Mfg.
5	3.3 (3342)	204	6	EGI	DOHC	KIA
6	3.8 (3778)	231	6	EGI	DOHC	KIA

Model Year	
Code ②	Year
9	2009

EGI: Electronic Gasoline Injection

DOHC: Double Overhead Camshafts

① 8th digit of VIN

② 10th digit of VIN

37655_SORE_C0001

GENERAL ENGINE SPECIFICATIONS

Year	Model	Engine Displacement Liters	Engine VIN	Net Horsepower @ rpm	Net Torque @ rpm (ft. lbs.)	Bore x Stroke (in.)	Com-pression Ratio	Oil Pressure @ rpm
2009	Sorento	3.3	5	242 @ 6000	228 @ 4500	3.62x3.30	10.4:01	18.77@1000
		3.8	6	262 @ 6000	260 @ 4500	3.78x3.43	10.4:01	18.77@1000

37655_SORE_C0002

ENGINE TUNE-UP SPECIFICATIONS

Year	Engine Displacement Liters	Engine VIN	Spark Plug Gap (in.)	Ignition Timing (deg.) MT	Ignition Timing (deg.) AT	Fuel Pump (psi)	Idle Speed (rpm) MT	Idle Speed (rpm) AT	Valve Clearance Intake	Valve Clearance Exhaust
2009	3.3	5	0.039-0.043	NA	①	54-56	NA	②	③	③
	3.8	6	0.039-0.043	NA	①	54-56	NA	④	③	③

NOTE: The Vehicle Emission Control Information label often reflects specification changes made during production. The label figures must be used if they differ from those in this chart

HYD: Hydraulic

① Timing is computer controlled.

② 700-900 rpm

③ Intake: 0.0039-0.0118 inches, Exhaust: 0.0078-0.0157

④ 550-750 rpm

37655_SORE_C0003

CAPACITIES

Year	Model	Engine Displacement Liters	Engine VIN	Engine Oil with Filter	Transaxle (pts.) Manual	Transaxle (pts.) Auto.	Fuel Tank (gal.)	Cooling System (qts.)
2009	Sorento	3.3	5	5.9	NA	23	21	9.4
		3.8	6	5.9	NA	23	21	9.4

NOTE: All capacities are approximate. Add fluid gradually and ensure a proper level is obtained.

37655_SORE_C0004

FLUID SPECIFICATIONS

Year	Model	Engine Displacement Liters (VIN)	Engine Oil	Auto. Trans.	Drive Axle Front	Drive Axle Rear	Transfer Case	Power Steering Fluid	Brake Master Cylinder	Cooling System
2009	Sorento	3.3 (5)	5W20	①	②	②	DEX III	PSF-3	DOT 3	③
	Sorento	3.8 (6)	5W20	①	②	②	DEX III	PSF-3	DOT 3	③

NA: Not Available

DOT: Department Of Transpotation

① APOLLOIL ATF Red-1K

② API GL5 (SAE 90) For LSD 85W-90 plus INFILREX 33 additive.

③ Ethylene Glycol base for aluminum radiators.

37655_SORE_C0014

VALVE SPECIFICATIONS

Year	Engine Displacement Liters	Engine VIN	Seat Angle (deg.)	Face Angle (deg.)	Spring out of Square (degrees)	Spring Free Length (in.)	Stem-to-Guide Clearance (in.) Intake	Stem-to-Guide Clearance (in.) Exhaust	Stem Diameter (in.) Intake	Stem Diameter (in.) Exhaust
2009	3.3	5	44.75-45.2	45.25-45.75	Less than 1.5 Degrees	1.7267	0.00078-0.0019	0.00118-0.0021	0.2151-0.216	0.2149-0.215
	3.8	6	44.75-45.2	45.25-45.75	Less than 1.5 Degrees	1.7267	0.00078-0.0019	0.00118-0.0021	0.2151-0.216	0.2149-0.215

37655_SORE_C0005

CAMSHAFT AND BEARING SPECIFICATIONS

All measurements are given in inches.

Year	Engine Displacement Liters	Engine VIN	Journal Diameter	Brg. Oil Clearance	Shaft End-play	Runout	Journal Bore	Lobe Lift Intake	Lobe Lift Exhaust
2009	3.3	5	①	②	0.0008-0.0071	NA	NA	NA	NA
	3.8	6	①	②	0.0008-0.0071	NA	NA	NA	NA

NA: Not Available

① Intake & Exhaust no. 1: 1.1009-1.1015
Intake & Exhaust nos. 2, 3, 4: 0.9430-0.9437

② Intake & Exhaust no. 1: 0.0011-0.0022
Intake & Exhaust nos. 2, 3, 4: 0.0012-0.0026

37655_SORE_C0013

CRANKSHAFT AND CONNECTING ROD SPECIFICATIONS

All measurements are given in inches.

Year	Engine Displacement Liters	Engine VIN	Crankshaft Main Brg. Journal Dia.	Crankshaft Main Brg. Oil Clearance	Crankshaft Shaft End-play	Crankshaft Thrust on No.	Connecting Rod Journal Diameter	Connecting Rod Oil Clearance	Connecting Rod Side Clearance
2009	3.3	5	2.7142-2.7149	0.0008-0.0016	0.0039-0.0110	3	2.1635-2.1642	0.0015-0.0022	0.0039-0.0098
	3.8	6	2.7142-2.7149	0.0008-0.0016	0.0039-0.0110	3	2.1635-2.1642	0.0015-0.0022	0.0039-0.0098

37655_SORE_C0008

PISTON AND RING SPECIFICATIONS

All measurements are given in inches.

Year	Engine Displacement Liters	Engine VIN	Piston Clearance	Ring Gap Top Compression	Ring Gap Bottom Compression	Ring Gap Oil Control	Ring Side Clearance Top Compression	Ring Side Clearance Bottom Compression	Ring Side Clearance Oil Control
2009	3.3	5	0.0012-0.0020	0.0067-0.0126	0.0126-0.0185	0.0078-0.0275	0.0012-0.0027	0.0012-0.0027	0.0024-0.0059
	3.8	6	0.0012-0.0020	0.0067-0.0126	0.0126-0.0185	0.0078-0.0275	0.0012-0.0027	0.0012-0.0027	0.0024-0.0059

37655_SORE_C0007

TORQUE SPECIFICATIONS

All readings in ft. lbs.

Year	Engine Displacement Liters	Engine VIN	Cylinder Head Bolts	Main Bearing Bolts	Rod Bearing Bolts	Crankshaft Damper Bolts	Flywheel Bolts	Manifold Intake	Manifold Exhaust	Spark Plugs	Oil Pan Drain Plug
2009	3.3	5	①	②	③	210-224	53-56	④	29-33	15-22	26-33
	3.8	6	①	②	③	210-224	53-56	④	29-33	15-22	26-33

① Step 1: 29 ft. lbs.
Step 2: plus 120 degrees
Step 3: plus 90 degrees
② M11 inner bolts:
Step 1: 36 ft. lbs.
Step 2: plus 90 degrees
M8 outer bolts:
Step 1: 15 ft. lbs.
Step 2: plus 120 degrees
M8 side bolts:
Step 1: 22-23 ft. lbs.
③ Step 1: 15 ft. lbs.
Step 2: plus 90 degrees
④ Bolts: 20-23 ft. lbs.
Nuts: 14-17 ft. lbs.

37655_SORE_C0006

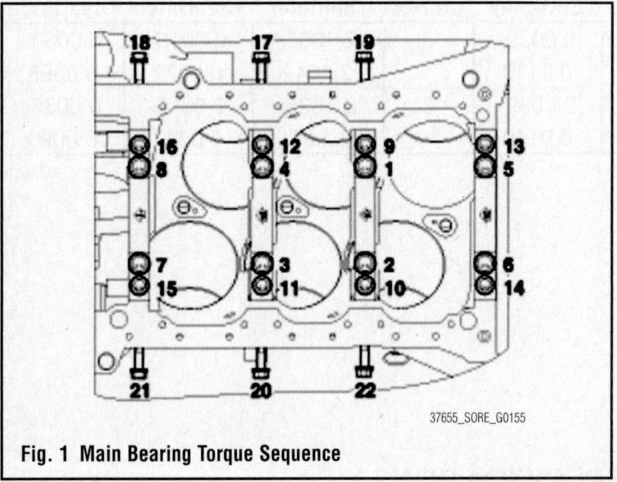

Fig. 1 Main Bearing Torque Sequence

WHEEL ALIGNMENT

Year	Model		Caster Range (+/-Deg.)	Caster Preferred Setting (Deg.)	Camber Range (+/-Deg.)	Camber Preferred Setting (Deg.)	Toe-in (Deg.)
2009	Sorento	F	0.50	3.89	0.50	0.00	0.00 +/- 0.20
		R	—	—	0.50	0.50	0.00 +/- 0.20

37655_SORE_C0009

TIRE, WHEEL AND BALL JOINT SPECIFICATIONS

Year	Model	OEM Tires		Tire Pressures (psi)		Wheel	Ball Joint	Lug Nut Torque
		Standard	Optional	Front	Rear	Size	Inspection	(ft. lbs.)
2009	Sorento	P245/70R16	P245/65R17	30	30	7JJ x16,17	①	65-86

OEM: Original Equipment Manufacturer

PSI: Pounds Per Square Inch

STD: Standard

OPT: Optional

① Replace if any measurable movement is found.

37655_SORE_C0010

BRAKE SPECIFICATIONS

All measurements in inches unless noted

Year	Model		Brake Disc			Brake Drum			Minimum	Brake Caliper	
			Original Thickness	Minimum Thickness	Maximum Run-out	Original Inside Diameter	Max. Wear Limit	Maximum Machine Diameter	Lining Thickness	Bracket Bolts (ft. lbs.)	Mounting Bolts (ft. lbs.)
2009	Sorento	F	1.100	1.020	0.0012	NA	NA	NA	0.079	16-24	47-54
		R	0.752	0.724	NA	NA	NA	NA	0.079	16-24	47-54

NA: Not Applicable

F: Front

R: Rear

37655_SORE_C0011

SCHEDULED MAINTENANCE INTERVALS
KIA—SORENTO

TO BE SERVICED	OF SERVIC	VEHICLE MILEAGE INTERVAL (x1000)													
		7.5	15	22.5	30	37.5	45	52.5	60	67.5	75	82.5	90	97.5	105
Accessory drive belts	S/I				✓				✓				✓		
Air cleaner element ①	S/I/R														
Air conditioner system	S/I	Inspect compressor operation & refrigerant every 10,000 miles, or 8 months													
Battery service	S/I		✓		✓		✓		✓		✓		✓		✓
Brake lines, hoses and connections	S/I		✓		✓		✓		✓		✓		✓		✓
Brake fluid	S/I		✓		✓		✓		✓		✓		✓		✓
Brake pedal	S/I				✓		✓		✓		✓		✓		✓
Chassis and body fasteners	T				✓				✓				✓		
Climate control filter	R	Replace every 10,000 miles, or 8 months													
Cooling system hoses and coolant level	S/I	✓	✓	✓	✓	✓	✓	✓	✓	✓	✓	✓	✓	✓	✓
Disc brakes and pads	S/I	✓	✓	✓	✓	✓	✓	✓	✓	✓	✓	✓	✓	✓	✓
Drive shafts and CV boot	S/I		✓		✓		✓		✓		✓		✓		✓
Engine coolant	R								✓				✓		
Engine oil and filter	R	✓	✓	✓	✓	✓	✓	✓	✓	✓	✓	✓	✓	✓	✓
Front ball joints	S/I				✓				✓				✓		
Fuel filter	R								✓						
Fuel lines and hoses	S/I	✓	✓	✓	✓	✓	✓	✓	✓	✓	✓	✓	✓	✓	✓
Fuel tank filter ②	S/I		✓		✓		✓		✓		✓		✓		✓
Rear differential fluid ③	S/I	✓	✓	✓	✓	✓		✓	✓		✓	✓		✓	
Front differential fluid (if equipped) ③	S/I	✓	✓	✓	✓	✓		✓	✓		✓	✓		✓	✓
Automatic transmission fluid	S/I		✓		✓		✓		✓		✓	✓	✓		✓
Transfer case oil (if equipped)	S/I	✓	✓	✓	✓	✓	✓	✓	✓	✓	✓	✓	✓	✓	✓
Emission hoses	S/I				✓				✓				✓		
Locks and hinges	L	✓	✓	✓	✓	✓	✓		✓		✓		✓		✓
Parking brake	S/I		✓		✓		✓		✓		✓		✓		✓
Power steering fluid	S/I	✓	✓	✓	✓	✓	✓	✓	✓	✓	✓	✓	✓	✓	✓
Propeller shaft	L		✓		✓		✓		✓		✓		✓		✓
Spark plugs	R	Replace every 100,000 miles, or 100 months													
Steering operation and linkage	S/I	✓	✓	✓	✓	✓	✓	✓	✓	✓	✓	✓	✓	✓	✓
Tire pressure and tread wear	S/I		✓		✓		✓		✓		✓		✓		✓
Valve clearance	A								✓						

R: Replace S/I: Inspect and service, if needed L: Lubricate A: Adjust T: Tighten

① Air cleaner element: inspect every 7,500 miles or 6 months and replace every 30,000 miles or 18 months, whichever occurs first

② Fuel tank filter replace every 30,000 miles.

③ Differential fluids replace every 22,500 miles.

37655_SORE_C0012

BRAKES — INFORMATION AND PRECAUTIONS

ANTI-LOCK SYSTEMS

- Certain components within the ABS system are not intended to be serviced or repaired individually.
- Do not use rubber hoses or other parts not specifically specified for and ABS system. When using repair kits, replace all parts included in the kit. Partial or incorrect repair may lead to functional problems and require the replacement of components.
- Lubricate rubber parts with clean, fresh brake fluid to ease assembly. Do not use shop air to clean parts; damage to rubber components may result.
- Use only DOT 3 brake fluid from an unopened container.
- If any hydraulic component or line is removed or replaced, it may be necessary to bleed the entire system.
- A clean repair area is essential. Always clean the reservoir and cap thoroughly before removing the cap. The slightest amount of dirt in the fluid may plug an orifice and impair the system function. Perform repairs after components have been thoroughly cleaned; use only denatured alcohol to clean components. Do not allow ABS components to come into contact with any substance containing mineral oil; this includes used shop rags.
- The Anti-Lock control unit is a microprocessor similar to other computer units in the vehicle. Ensure that the ignition switch is **OFF** before removing or installing controller harnesses. Avoid static electricity discharge at or near the controller.
- If any arc welding is to be done on the vehicle, the control unit should be unplugged before welding operations begin.

DISC AND DRUM SYSTEMS

**** CAUTION**

Dust and dirt accumulating on brake parts during normal use may contain asbestos fibers from production or aftermarket brake linings. Breathing excessive concentrations of asbestos fibers can cause serious bodily harm. Exercise care when servicing brake parts. Do not sand or grind brake lining unless equipment used is designed to contain the dust residue. Do not clean brake parts with compressed air or by dry brushing. Cleaning should be done by dampening the brake components with a fine mist of water, then wiping the brake components clean with a dampened cloth. Dispose of cloth and all residue containing asbestos fibers in an impermeable container with the appropriate label. Follow practices prescribed by the Occupational Safety and Health Administration (OSHA) and the Environmental Protection Agency (EPA) for the handling, processing, and disposing of dust or debris that may contain asbestos fibers.

BRAKES — BLEEDING THE BRAKE SYSTEM

BLEEDING PROCEDURE

BLEEDING PROCEDURE

See Figure 2.

1. Before servicing the vehicle, refer to the Precautions Section.
2. Remove the reservoir cap and fill the brake reservoir with brake fluid.
3. Connect a vinyl tube to the wheel cylinder bleeder screw and insert the other end of the tube in a container of brake fluid which is half full.
4. Start the engine.
5. Slowly depress the brake pedal several times.

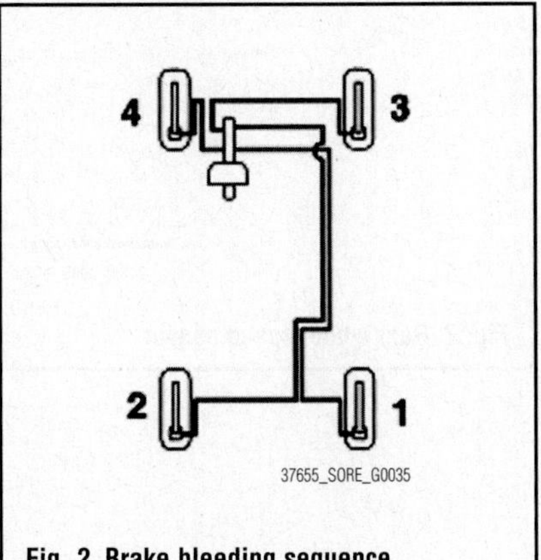

Fig. 2 Brake bleeding sequence

6. While depressing the brake pedal fully, loosen the bleeder screw until fluid runs out. Then close the bleeder screw and release the brake pedal.
7. Repeat steps 5 and 6 until there are no more bubbles in the fluid.
8. Tighten the bleeder screw to 60–79 inch lbs. (7–9 Nm).
9. Repeat the above procedure for each wheel in the sequence shown in the illustration.

BLEEDING THE ABS SYSTEM

See Figure 3.

This procedure should be followed to ensure adequate bleeding of air and the filling of the ABS unit, the brake lines, and the master cylinder with brake fluid.

1. Before servicing the vehicle, refer to the Precautions Section.
2. Remove the reservoir cap and fill the brake reservoir with brake fluid.

**** WARNING**

If there is any brake fluid on any painted surface, wash it off immediately.

➡When pressure bleeding, do not depress the brake pedal.

Recommended brake fluid: DOT3 or DOT4.

3. Connect a clear plastic tube to the wheel cylinder bleeder plug and insert the other end of the tube into a clear plastic bottle that is half filled with clean brake fluid.
4. Connect the scan tool to the data link connector located underneath the dash panel.
5. Select and operate according to the instructions on the scan tool screen.

**** CAUTION**

You must obey the maximum operating time of the ABS motor with the scan tool to prevent the motor pump from burning.

6. Select Kia vehicle diagnosis.
7. Select vehicle name.
8. Select Anti-Lock Brake system.
9. Select air bleeding mode.
10. Press "YES" to operate motor pump and solenoid valve.

**** WARNING**

Wait 60 seconds before operating the air bleeding or damage to the motor may occur.

11. Wait 60 seconds before operating the air bleeding.

12. Pump the brake pedal several times, and then loosen the bleeder screw until fluid starts to run out without bubbles. Then, close the bleeder screw.

13. Repeat until there are no more bubbles in the fluid for each wheel.

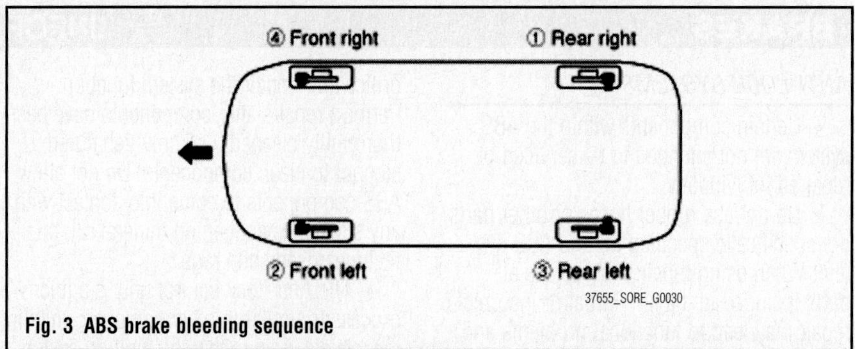

Fig. 3 ABS brake bleeding sequence

WHEEL SPEED SENSORS

REMOVAL & INSTALLATION

Front

See Figures 4 and 5.

1. Before servicing the vehicle, refer to the Precautions Section.
2. Disconnect the negative battery cable.
3. Remove the front wheel.
4. Remove the front wheel speed sensor mounting bolt.
5. Disconnect the wiring harness connector.
6. Remove the front wheel speed sensor.

To install:

7. Installation is the reverse of the removal procedure.

Rear

See Figures 4 and 6.

1. Before servicing the vehicle, refer to the Precautions Section.
2. Disconnect the negative battery cable.
3. Remove the rear wheel.
4. Disconnect the wiring harness connector.
5. Remove the bolt and the wheel speed sensor.

To install:

6. Installation is the reverse of the removal procedure.

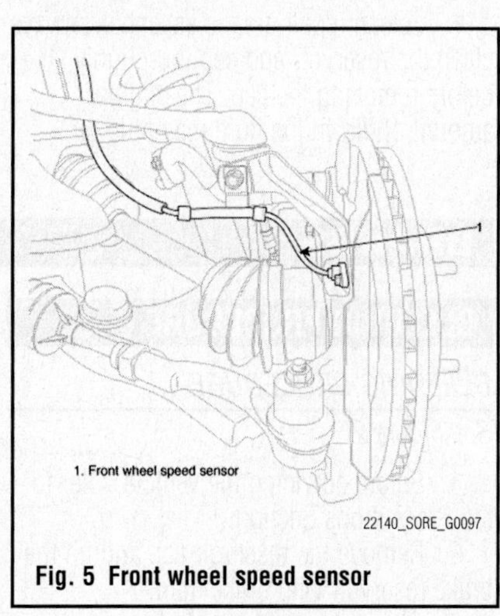

Fig. 5 Front wheel speed sensor

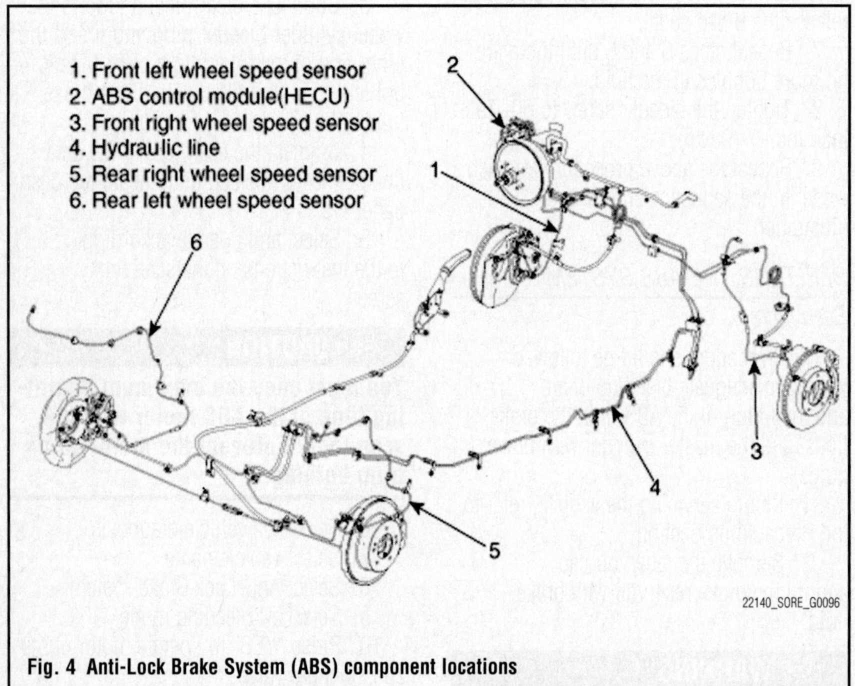

Fig. 4 Anti-Lock Brake System (ABS) component locations

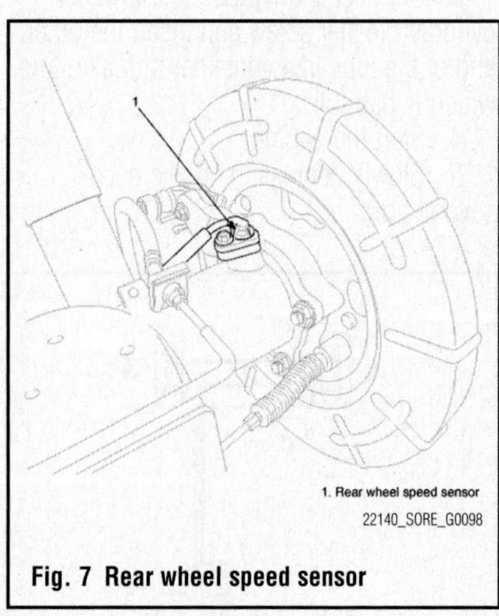

Fig. 7 Rear wheel speed sensor

BRAKES

FRONT DISC BRAKES

BRAKE CALIPER

REMOVAL & INSTALLATION

See Figure 7.

1. Remove the front wheel.
2. Disconnect the brake fluid hose.
3. Remove the caliper mounting bolts.
4. Remove the brake caliper.

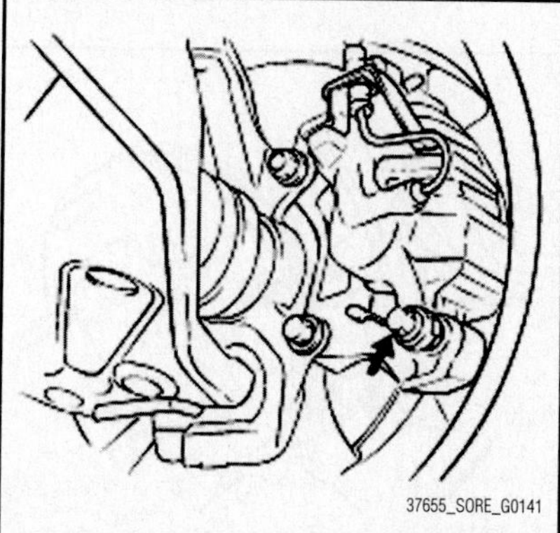

Fig. 7 Remove the caliper mounting bolts

To install:

5. Install the brake caliper. Tighten the mounting bolts to 16–24 ft. lbs. (22–32 Nm).
6. Install the brake fluid hose. Tighten the hose fitting to 12–14 ft. lbs. (17–20 Nm).
7. Bleed the brake system.
8. Install the front wheel.
9. Before attempting to move the vehicle, pump the brake pedal to seat the pads against the rotors. Make sure the vehicle has a firm brake pedal. Check the level of the brake fluid and add DOT 3 or 4 brake fluid if necessary.

DISC BRAKE PADS

REMOVAL & INSTALLATION

See Figure 8.

1. Remove the front wheel.
2. Remove the guide pin, lift the caliper assembly up and suspend it with a wire.
3. Remove the following parts from the caliper support:
   - Pad and wear sensor assembly
   - Pad spring
   - Outer shim

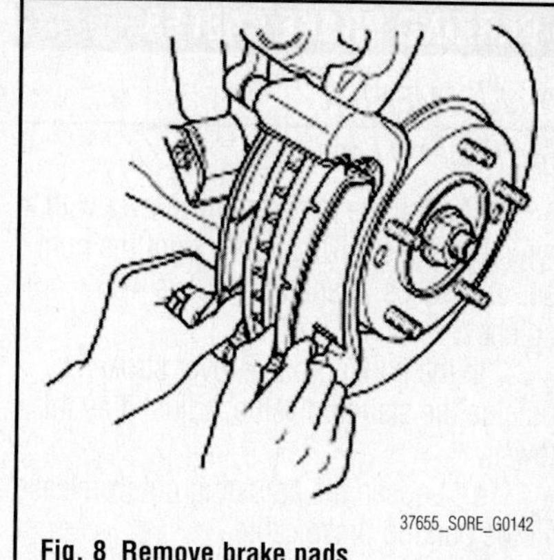

Fig. 8 Remove brake pads

To install:

4. Compress the caliper piston into the caliper bore.
5. Install the pad clips.
6. Install the inner and outer pads on each pad clip.
7. Lower the brake caliper carefully so as not to damage the boot.
8. Tighten the guide pin bolt to 16–24 ft. lbs. (22–32 Nm).
9. Install the front wheel.

BRAKES

REAR DISC BRAKES

BRAKE CALIPER

REMOVAL & INSTALLATION

See Figure 9.

1. Remove the rear wheel.
2. Disconnect the brake fluid hose.
3. Remove the caliper guide bolts.
4. Remove the brake caliper.

To install:

5. Install the brake caliper. Tighten the guide bolts to 16–23 ft. lbs. (22–32 Nm).

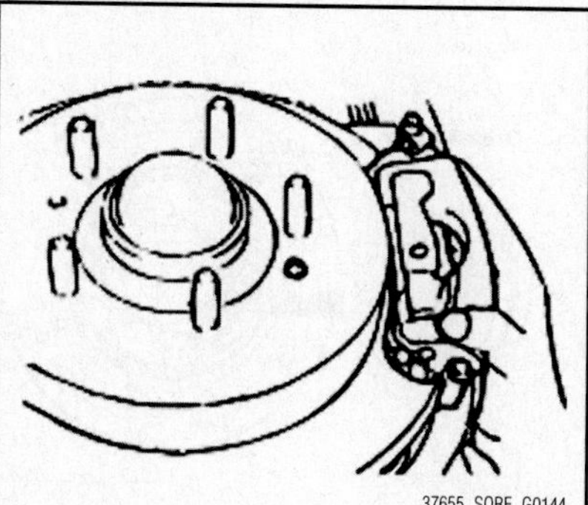

Fig. 9 Remove the rear brake caliper

6. Install the brake fluid hose. Tighten the hose fitting to 12–14 ft. lbs. (17–20 Nm).
7. Bleed the brake system.
8. Install the rear wheel.
9. Before attempting to move the vehicle, pump the brake pedal to seat the pads against the rotors. Make sure the vehicle has a firm brake pedal. Check the level of the brake fluid and add DOT 3 or 4 brake fluid if necessary.

DISC BRAKE PADS

REMOVAL & INSTALLATION

See Figure 10.

1. Remove the rear wheel.
2. Remove the guide pin bolts, lift the caliper assembly up and suspend it with a wire.
3. Before replacing the brake pads, drain brake fluid from the master cylinder reservoir until it remains half full.
4. Remove the brake pads by turning the piston in the housing assembly using special tool 09581-11000 to compress the piston.
5. Remove the inner and outer pads from the caliper.

To install:

6. Install the inner and outer brake pads, engaging the clips securely onto the caliper assembly.
7. Lower the brake caliper assembly into proper position.
8. Tighten the guide pin bolts to 16–23 ft. lbs. (22–32 Nm).
9. Install the rear wheel.
10. Check the brake fluid level and top off, if necessary.

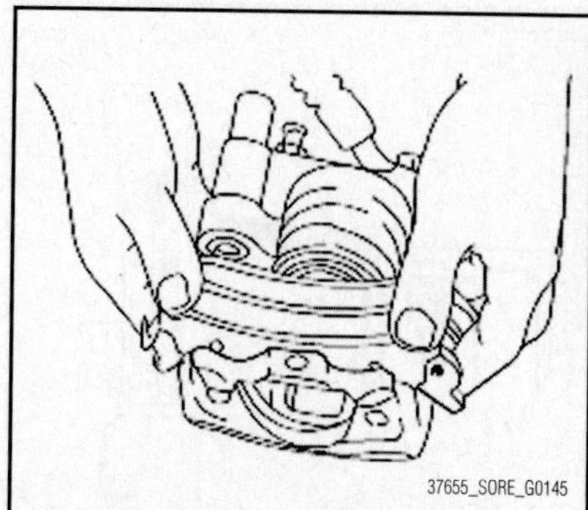

Fig. 10 Remove the inner and outer pads from the caliper

PARKING BRAKE CABLES

ADJUSTMENT

See Figures 11 and 12.

1. Pull on the parking brake lever with a force of 22 lbs. (10 kg) and count the number of notches. Standard value is 4–6 notches.
2. If the parking brake lever stroke is outside the standard value, adjust it as follows:
    a. Loosen the adjusting nut to release the parking brake cable.
    b. Remove the adjusting hole plug.
    c. Turn the adjuster in the direction of the arrow. To prevent the disc from rotating, use a screwdriver.
    d. Turn the adjuster 5 notches in the opposite direction of arrow.
    e. Turn the adjuster nut to adjust the parking brake lever stroke to the specification.

➡If the number of parking brake notches is less than the specification, loosen the adjusting nut and readjust.

3. After adjustment, raise the rear of the vehicle and check that the parking brake does not drag.

PARKING BRAKE SHOES

REMOVAL & INSTALLATION

See Figures 13 through 18.

1. Remove rear disc caliper assembly.
2. Before removing the brake disc, chalk both sides of the screw.

➡Reduce the shoe gap by turning the adjuster with appropriate tool.

3. After turning the pin to coincide with hole of spring cap, remove the shoe hold spring.
4. Remove the lower return spring.
5. Remove the parking brake cable mounting nuts.
6. Remove the parking brake shoes.

To install:

7. Install the upper return spring and brake shoes.
8. Turn the adjuster in a clockwise direction and install.
9. Install the lower return spring.
10. Install the shoe hold spring with pliers.
11. Install the disc brake and align the mark while tightening the screw.

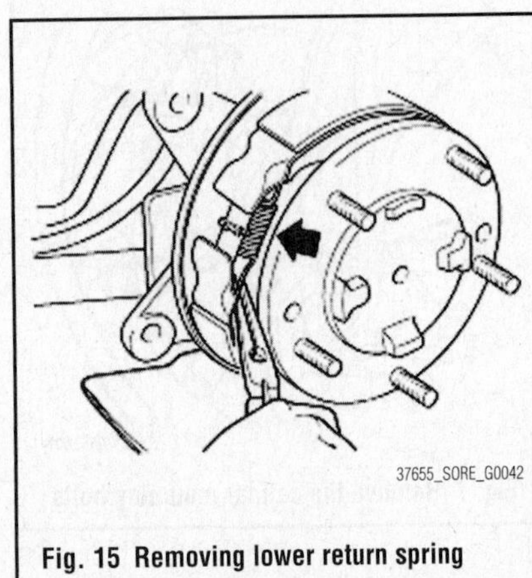

Fig. 15 Removing lower return spring

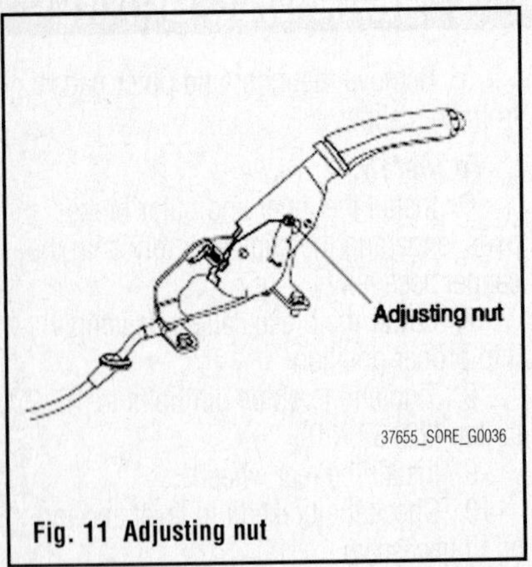

Fig. 11 Adjusting nut

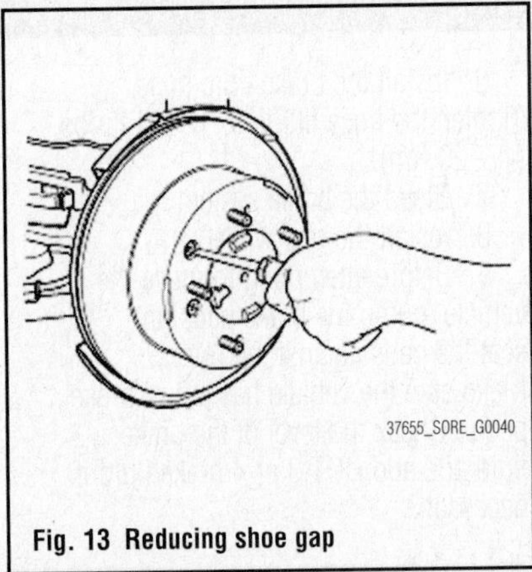

Fig. 13 Reducing shoe gap

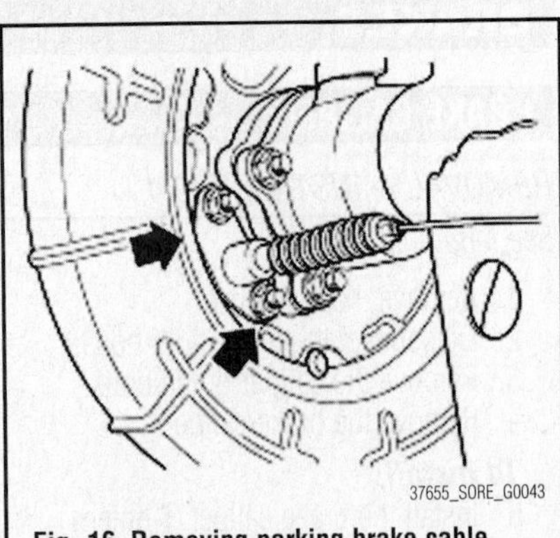

Fig. 16 Removing parking brake cable mounting nuts

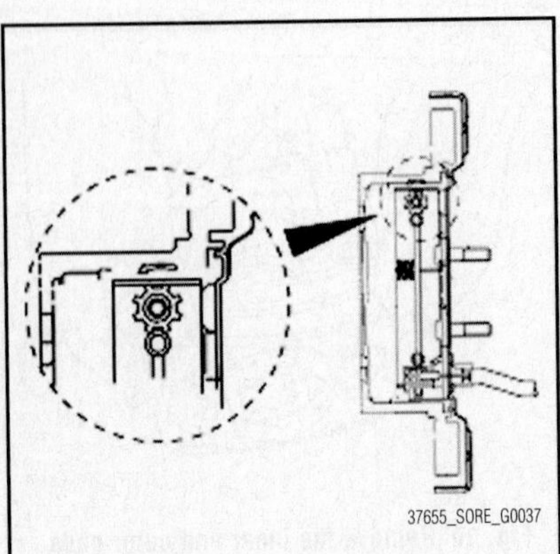

Fig. 12 Parking brake adjustment

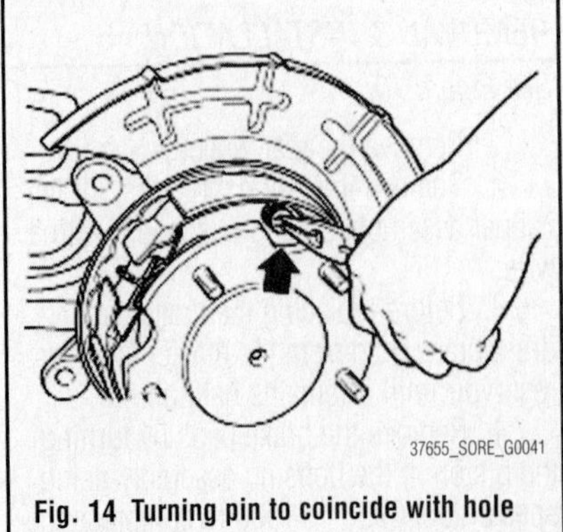

Fig. 14 Turning pin to coincide with hole of spring cap

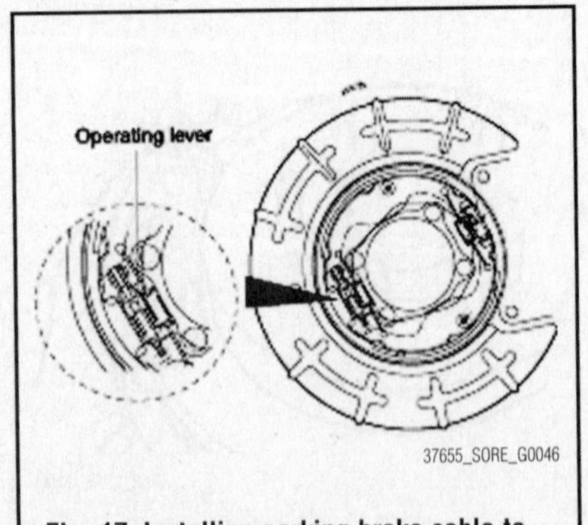

Fig. 17 Installing parking brake cable to operating lever

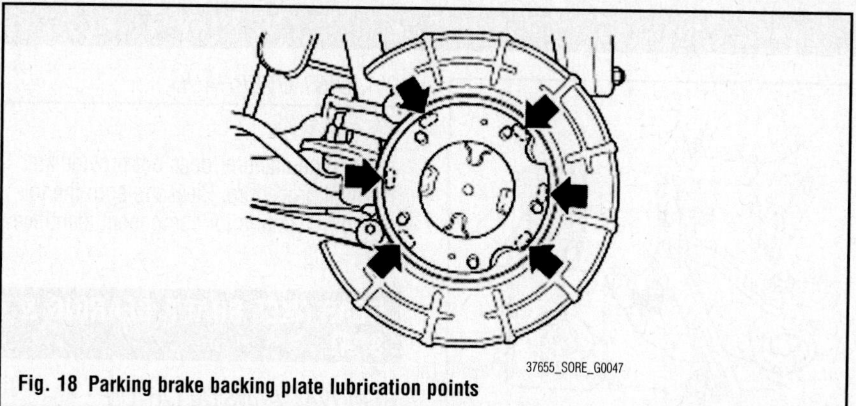
Fig. 18 Parking brake backing plate lubrication points

ADJUSTMENT

1. Raise the rear of the vehicle, and make sure it is securely supported.
2. Remove the rear wheel and tire.
3. After removing the plug from the disc, rotate the toothed wheel with a screwdriver until the disc does not rotate.
4. Back off the toothed wheel by 5 notches.

CHASSIS ELECTRICAL

AIR BAG (SUPPLEMENTAL RESTRAINT SYSTEM)

GENERAL INFORMATION

CAUTION

These vehicles are equipped with an air bag system. The system must be disarmed before performing service on, or around, system components, the steering column, instrument panel components, wiring and sensors. Failure to follow the safety precautions and the disarming procedure could result in accidental air bag deployment, possible injury and unnecessary system repairs.

SERVICE PRECAUTIONS

CAUTION

Disconnect and isolate the battery negative cable before beginning any airbag system component diagnosis, testing, removal, or installation procedures. Wait at least 90 seconds after the ignition switch is turned off and the negative (-) terminal cable is disconnected from the battery before starting the operation. The SRS is equipped with a backup power source, so if work is started within 90 seconds after disconnecting the negative (-) terminal cable from the battery, the SRS may be deployed. Failure to disable the airbag system may result in accidental airbag deployment, personal injury, or death.

DISARMING THE SYSTEM

1. Before servicing the vehicle, refer to the Precautions Section.
2. Turn the ignition switch to the **LOCK** position.
3. Disconnect the negative battery cable.
4. Wait 3minutes for the battery back-up power to discharge.

ARMING THE SYSTEM

1. Before servicing the vehicle, refer to the Precautions Section.
2. Connect the negative battery cable.
3. Turn the ignition switch **ON**.
4. Verify that the air bag indicator illuminates for 4–8 seconds, then goes off.

CLOCKSPRING CENTERING

See Figure 19.

1. Prior to installing the clockspring, confirm that the front wheels are pointed straight ahead.
2. Turn clockspring fully counter-clockwise.
3. Turn clockspring clockwise 1.2 revolutions.
4. Matchmark the clockspring housing.
5. Confirm clockspring centering in each direction from matchmark.

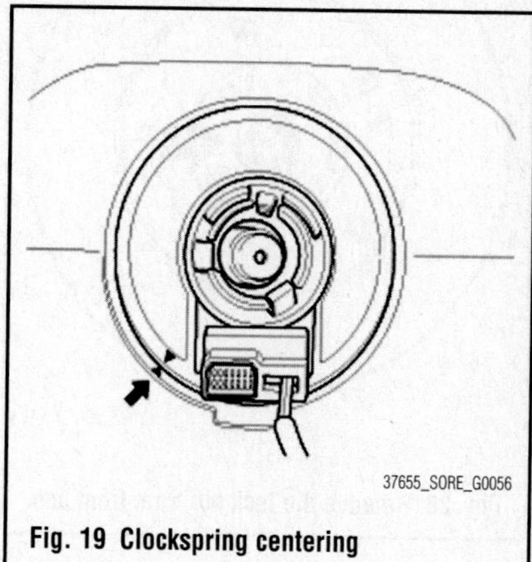
Fig. 19 Clockspring centering

DRIVE TRAIN

FRONT HALFSHAFT

REMOVAL & INSTALLATION

See Figures 20 through 22.

1. Before servicing the vehicle, refer to the Precautions Section.
2. Raise and safely support the vehicle.
3. Remove the front wheels and tires.
4. Remove the lock nut from front hub.
5. Remove the upper control arm link lock bolt, spring washer and nut.
6. Remove tie rod end cotter pin and using a ball joint puller, remove tie rod end from steering knuckle.
7. Matchmark the halfshaft for identical installation position.
8. Using a pry bar, pry the halfshaft from the differential housing.

**** WARNING**

Do not pull on the drive shaft. Doing so will damage the boots. Be sure to use the pry bar.

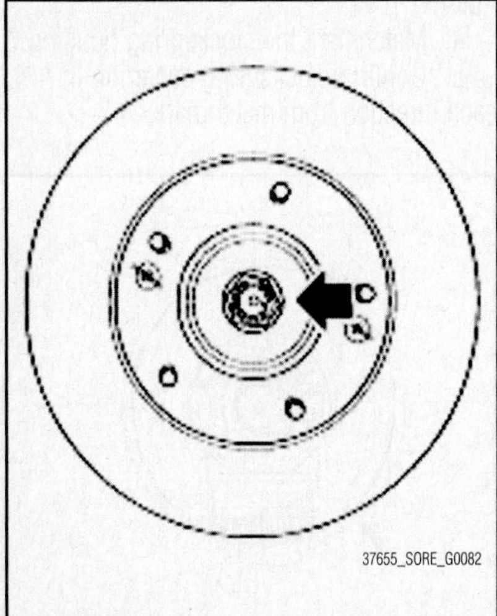

Fig. 20 Remove the lock nut from front hub

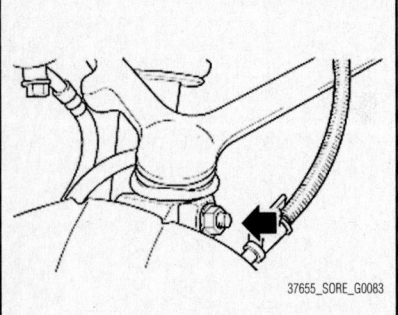

Fig. 21 Remove the upper control arm link lock bolt, spring washer and nut

37655_SORE_G0083

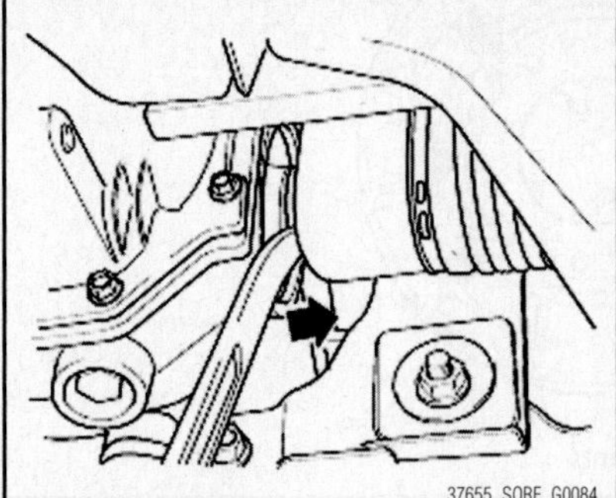

Fig. 22 Pry the halfshaft from the differential housing

9. Remove the halfshaft from the knuckle.
10. Temporarily install the knuckle to the upper arm.

To install:

11. Align the matchmark between the drive shaft and the differential and insert the shaft. Carefully insert the right-hand side of the halfshaft into the oil seal to avoid any damage.
12. Install the knuckle assembly. Tighten the tie rod ball joint to 51–57 ft. lbs. (70–80 Nm). Tighten the upper arm link lock bolt to 32–39 ft. lbs. (44–55 Nm).
13. Tighten the lock nut to 177–198 ft. lbs. (245–275 Nm) and then peen the lock nut on the end of drive shaft.
14. Install the front wheels and tires.

CV-JOINT OVERHAUL

See Figure 23.

The manufacturer does not provide an overhaul procedure. Please refer to the following illustration for component identification.

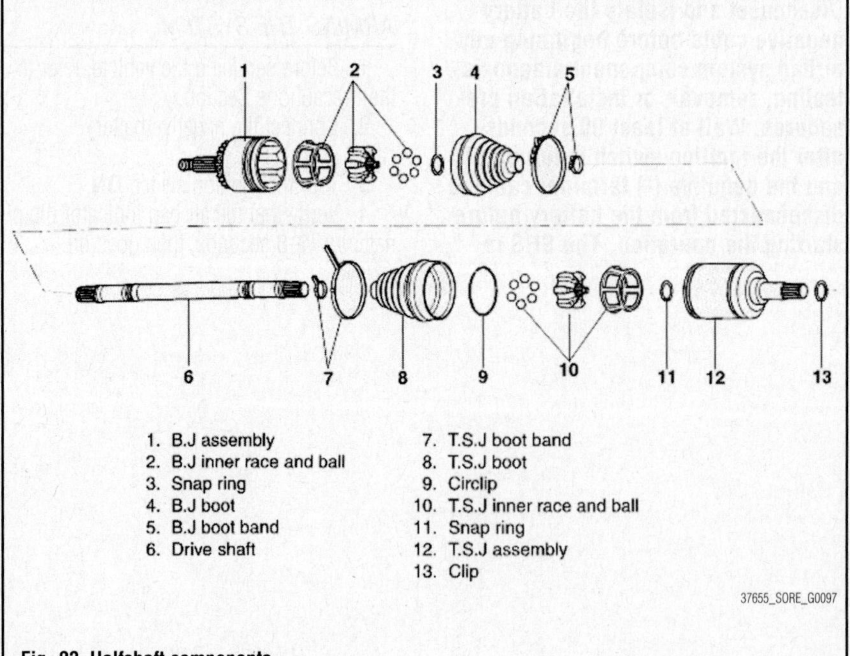

Fig. 23 Halfshaft components

REAR AXLE SHAFT, BEARING & SEAL

REMOVAL & INSTALLATION

See Figures 24 through 26.

1. Before servicing the vehicle, refer to the Precautions Section.
2. Disconnect the negative battery cable.
3. Raise and safely support the vehicle.
4. Remove the rear wheels.
5. Remove the disc brake, parking brake assembly. Refer to Brake section.
6. Remove the parking brake cable and wheel speed sensor cable.
7. Remove the rear axle shaft mounting bolts.
8. Remove the rear axle shaft.
9. Remove the bearing collar and bearing from the axle.
10. Using a slide hammer, remove the oil seal.

To install:

11. Apply grease to the oil seal lip and using the appropriate seal driver, install the new axle seal into the differential.

12. Install the new wheel bearing and retainer collar to the rear axle shaft.
13. Install the rear axle shaft. Torque the axle shaft mounting bolts to 32–44 ft. lbs. (43–60 Nm).
14. Install the wheel speed sensor and parking brake cables.
15. Install the disc brake and parking brake assembly and the rear wheels.
16. Adjust the parking brake lever.
17. Install the rear wheels.
18. Connect the negative battery cable.

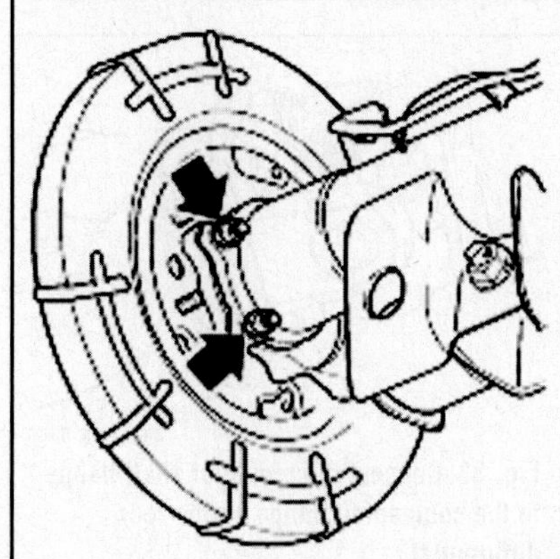

Fig. 24 Remove the rear axle shaft mounting bolts

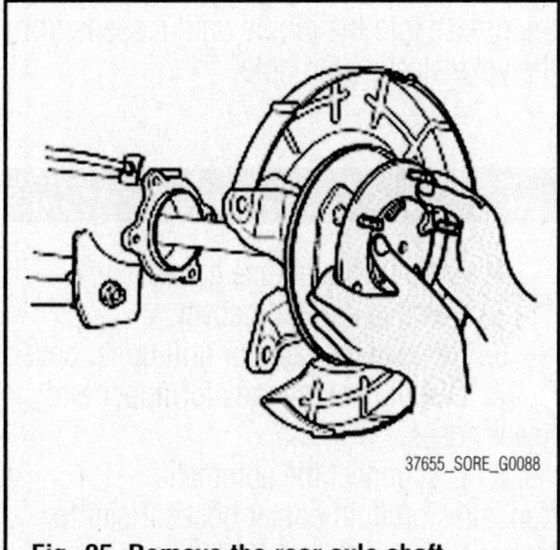

Fig. 25 Remove the rear axle shaft

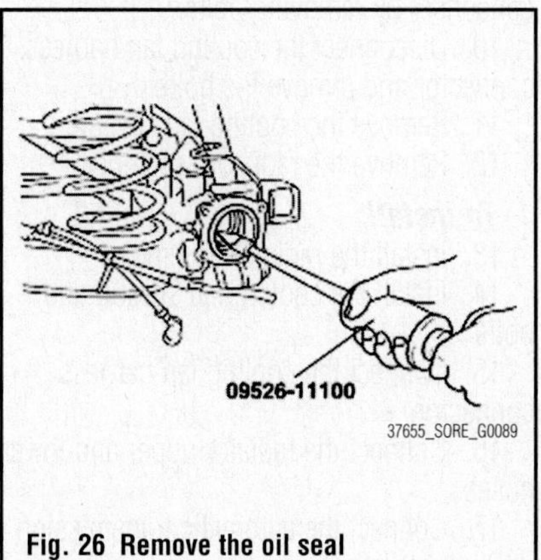

Fig. 26 Remove the oil seal

REAR DRIVESHAFT

REMOVAL & INSTALLATION

See Figures 27 through 33.

1. Before servicing the vehicle, refer to the Precautions Section.

➡On this vehicle, the rear driveshaft is referred to as propeller shaft.

2. Disconnect the negative battery cable.
3. Raise and safely support the vehicle.
4. Place matchmarks (reference marks) on the propeller shaft and the matching transfer case and differential input shafts.
5. Remove the four bolts holding universal flange to transfer case (4WD).
6. Remove the bolts holding center bearing bracket (2WD).
7. Remove the four bolts holding universal flange to differential.
8. Remove the propeller shaft.

⁂ WARNING

When removing the propeller shaft, be careful not to damage the dust cover or spline.

To install:

9. Connect the propeller shaft flange to the companion flange on the front differential (4WD):
   a. Align the matchmarks on the flange and connect the flanges with four bolts and nuts.
   b. Tighten the bolts and nuts, as follows:
      - Part-time 4WD: 36–43 ft. lbs. (50–60 Nm)
      - Full-time 4WD: 19–21 ft. lbs. (26–30 Nm)
10. Connect the front propeller shaft flange to the companion flange on the transfer case(4WD):
    a. Align the index marks on the flange

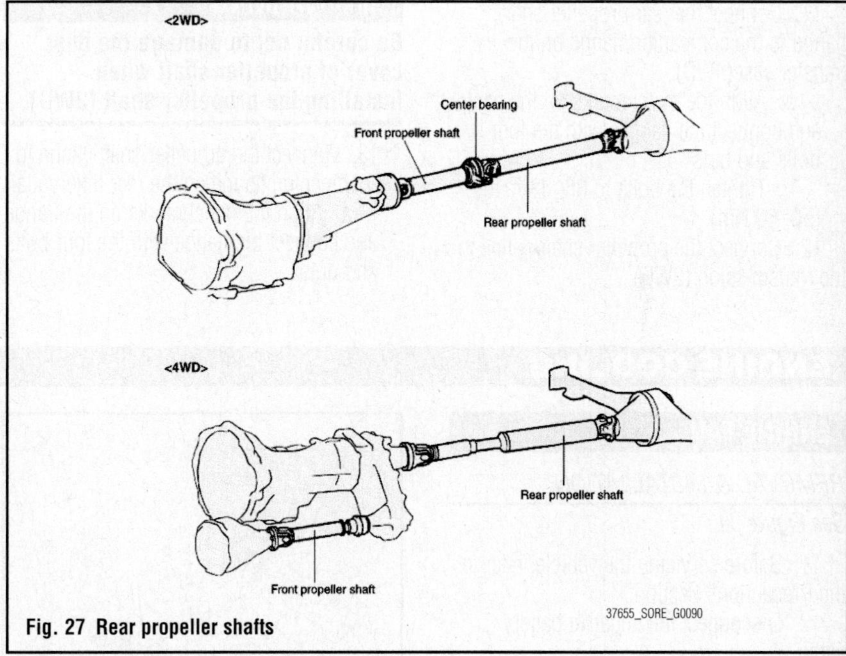

Fig. 27 Rear propeller shafts

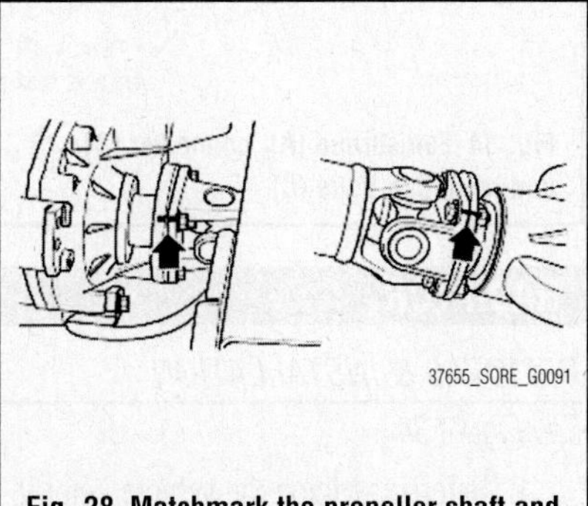

Fig. 28 Matchmark the propeller shaft and the matching transfer case and differential input shafts

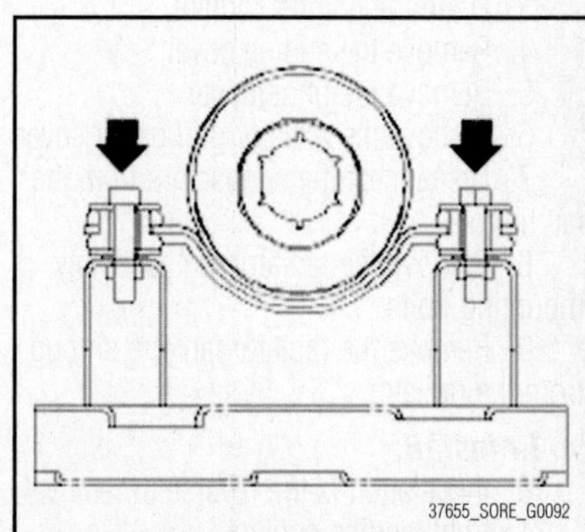

Fig. 29 Remove the four bolts holding universal flange to transfer case (4WD)

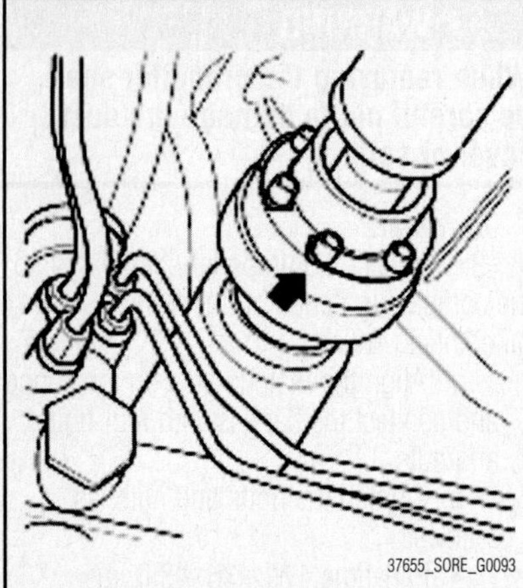

Fig. 30 Connect the propeller shaft flange to the companion flange on the front differential (4WD)

and connect the flanges with four bolts and nuts.

b. Tighten the bolts to 36–43 ft. lbs. (50–60 Nm).

11. Connect the rear propeller shaft flange to the companion flange on the transfer case(4WD):

a. Align the matchmarks on the flange and connect the flanges with the four bolts and nuts.

b. Tighten the bolts to 36–43 ft. lbs. (50–60 Nm).

12. Connect the propeller shaft spline to the transmission (2WD):

Fig. 31 Connect the front propeller shaft flange to the companion flange on transfer case(4WD)

a. Align the matchmark on the spline and then install the propeller shaft.

b. Tighten the bolts holding the center bearing bracket to 27–39 ft. lbs. (37–54 Nm).

✲✲ WARNING

Be careful not to damage the dust cover of propeller shaft when installing the propeller shaft (2WD).

13. Connect the propeller shaft flange to the companion flange on the rear differential.

a. Align the matchmarks on the flange and connect the flange with the four bolts and nuts.

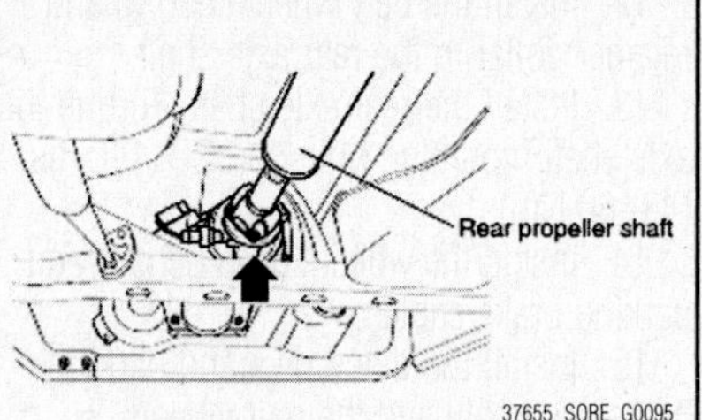

Fig. 32 Connect the rear propeller shaft flange to the companion flange on transfer case(4WD)

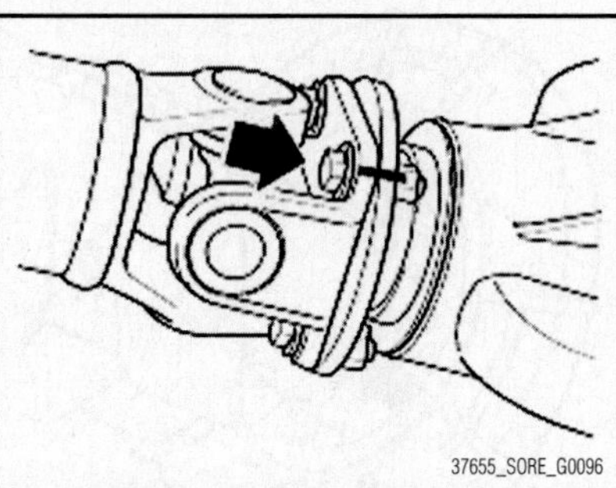

Fig. 33 Connect the propeller shaft flange to the companion flange on the rear differential

b. Tighten the bolts and nuts to 36–43 ft. lbs. (50–60 Nm).

14. After installing the propeller shaft, fill the grease into the nipple until it seeps from the yoke sleeve plug hole.

ENGINE COOLING

ENGINE FAN

REMOVAL & INSTALLATION

See Figure 34.

1. Before servicing the vehicle, refer to the Precautions Section.
2. Disconnect the negative battery cable.
3. Drain the engine coolant.
4. Remove the engine cover.
5. Remove the undercover.
6. Remove the radiator grill upper cover.
7. Disconnect the connectors from the fan motor.
8. Remove the radiator fan assembly mounting bolts.
9. Remove the radiator fan and shroud from the radiator.

To install:

10. Installation is the reverse of removal.
11. Refill engine coolant.
12. Start the engine and check for leaks.
13. Recheck engine coolant level.

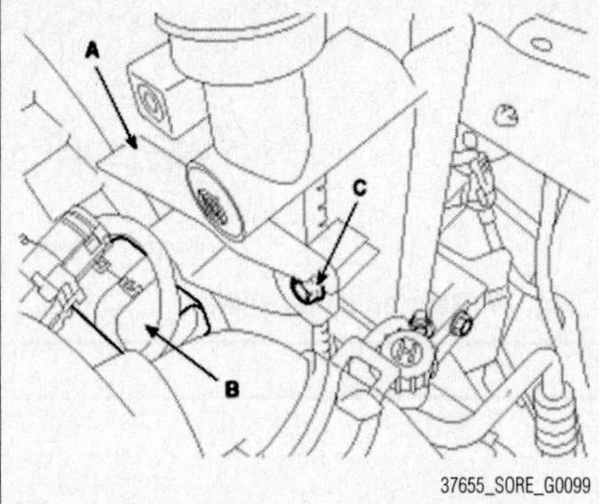

Fig. 34 Fan shroud (A), connector (B), and mounting bolts (C)

RADIATOR

REMOVAL & INSTALLATION

See Figure 34.

1. Before servicing the vehicle, refer to the Precautions Section.
2. Disconnect the negative battery cable.
3. Drain engine coolant.
4. Remove the engine cover.
5. Remove the undercover.
6. Remove the radiator grill upper cover.
7. Disconnect the radiator upper and lower hoses.
8. Disconnect the automatic transmission fluid cooler hoses. Refer to Hoses Removal & Installation.
9. Remove the radiator from the condenser by removing bolts.
10. Disconnect the cooling fan harness connector and remove the bolts.
11. Remove the cooling fan shroud.
12. Remove the radiator assembly.

To install:

13. Install the radiator assembly.
14. Install the cooling fan shroud and bolts.
15. Connect the cooling fan harness connector.
16. Connect the radiator upper and lower hoses.
17. Connect the automatic transmission fluid cooler hoses.

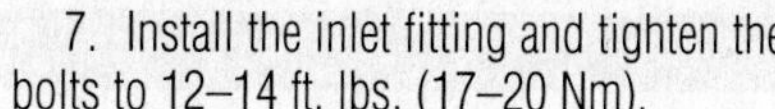

18. Install the radiator grill upper cover.
19. Install the undercover.
20. Install the engine cover.
21. Refill engine coolant.
22. Start the engine and check for leaks.
23. Recheck engine coolant level.

THERMOSTAT

REMOVAL & INSTALLATION

See Figure 35.

1. Before servicing the vehicle, refer to the Precautions Section.
2. Disconnect the negative battery cable.
3. Drain the engine coolant.
4. Remove the inlet fitting and gasket.
5. Remove the thermostat.

To install:

6. Install the thermostat with a new gasket, and with the jiggle valve on top.
7. Install the inlet fitting and tighten the bolts to 12–14 ft. lbs. (17–20 Nm).
8. Fill the cooling system.
9. Start the engine and check for leaks.

WATER PUMP

REMOVAL & INSTALLATION

See Figures 36 and 37.

1. Before servicing the vehicle, refer to the Precautions Section.
2. Disconnect the negative battery cable.
3. Drain the engine coolant.
4. Remove the drive belt. Refer to Drive Belt Removal & Installation in the Engine Mechanical section.
5. Remove the bolts and the water pump pulley.
6. Remove the cooling fan shroud.
7. Remove the water pump and gasket.

To install:

8. Install the water pump and new gaskets. Torque: Large bolt:16–17 ft. lbs. (22–24 Nm); Small bolt: 87–104 inch lbs. (10–12 Nm).

➡Make clean the contact face before assembly. When replacing a water pump, always use new gasket. When reassembling a water pump, replace the bolt with a new one.

9. Install the water pump pulley and tighten the bolts to 69–87 inch lbs. (8–10 Nm).
10. Install the drive belt.
11. Fill with engine coolant.
12. Start engine and check for leaks.
13. Recheck engine coolant level.

37655_SORE_G0102

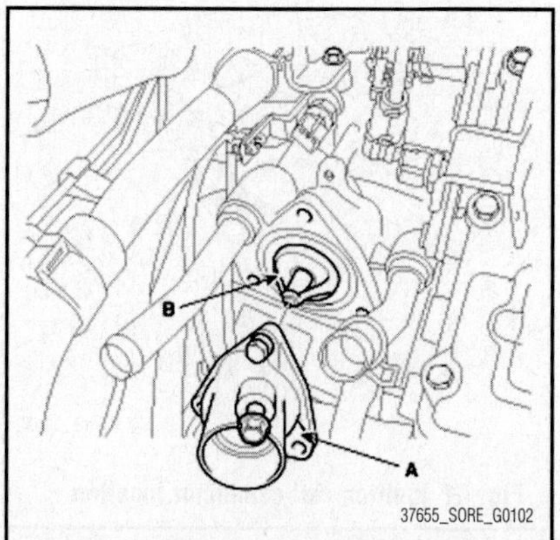

Fig. 35 Water inlet (A) and thermostat (B)

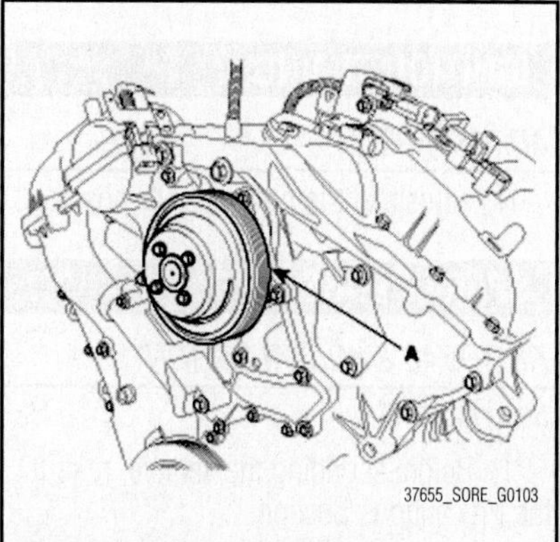

Fig. 36 Water pump pulley (A)

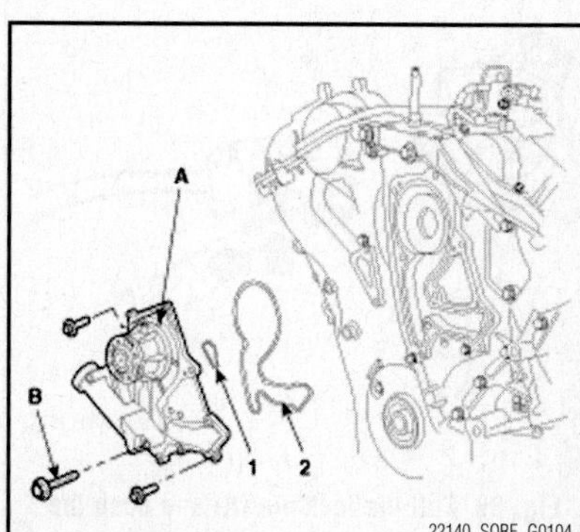

Fig. 37 Install the water pump (A) and new gaskets (1, 2).

ENGINE ELECTRICAL

CHARGING SYSTEM

ALTERNATOR

REMOVAL & INSTALLATION

See Figure 38.

1. Before servicing the vehicle, refer to the Precautions Section.
2. Disconnect the battery negative terminal first, then the positive terminal.
3. Disconnect the alternator connector, and remove the cable from alternator "B" terminal.
4. Remove the drive belt.

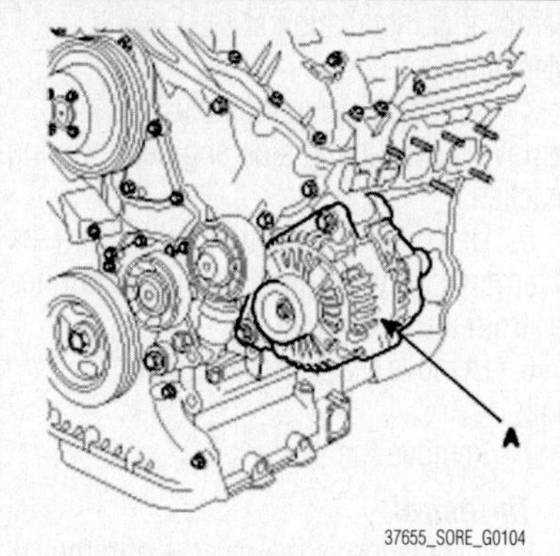

Fig. 38 Remove the alternator (A)

5. Pull out the through bolt and then remove the alternator.

To install:

6. Installation is the reverse of removal.

VOLTAGE REGULATOR

ADJUSTMENT

The charging system includes an alternator with a built-in regulator. There is no adjustment possible for the voltage regulator. If the voltage regulator is defective, it must be replaced.

ENGINE ELECTRICAL

IGNITION SYSTEM

FIRING ORDER

Firing order: 1–2–3–4–5–6

IGNITION COIL

REMOVAL & INSTALLATION

See Figures 39 and 40.

1. Before servicing the vehicle, refer to the Precautions Section.
2. Disconnect the negative battery cable.
3. Remove the engine cover.

➡**When removing the ignition coil connector, pull the lock pin and push the clip.**

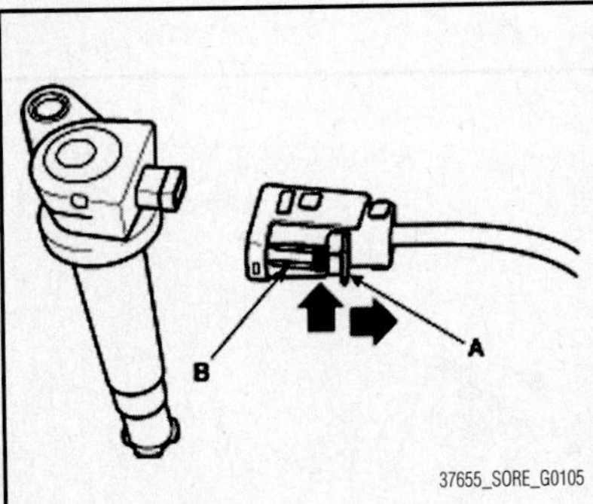

Fig. 39 Pull the lock pin (A) and push the clip (B)

37655_SORE_G0106

Fig. 40 Ignition coil connector (A) and ignition coil (B)

4. Remove the ignition coil connector.
5. Remove the ignition coil.

To install:

6. Installation is the reverse of removal.

IGNITION TIMING

ADJUSTMENT

No adjustment is necessary or possible.

SPARK PLUGS

REMOVAL & INSTALLATION

See Figure 41.

1. Before servicing the vehicle, refer to the Precautions Section.
2. Disconnect the negative battery cable.
3. Remove the engine cover.
4. Remove the ignition coil connectors.

➡**When removing the ignition coil connector, pull the lock pin and push the clip.**

5. Remove the ignition coils.
6. Using a spark plug socket, remove the spark plug.

To install:

7. Installation is the reverse of removal.

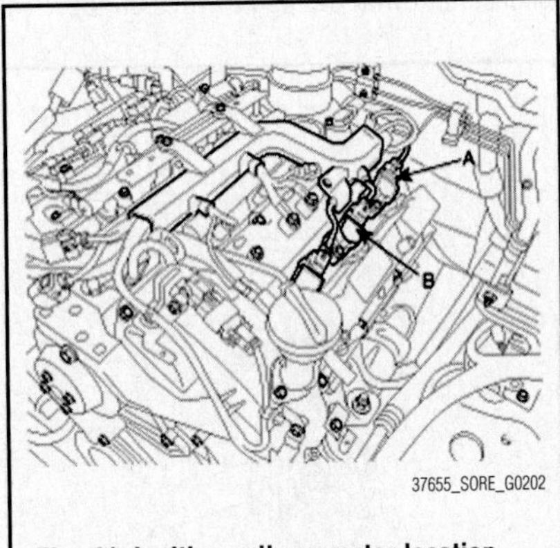

Fig. 41 Ignition coil connector location

ENGINE ELECTRICAL

STARTING SYSTEM

STARTER

REMOVAL & INSTALLATION

See Figure 42.

1. Before servicing the vehicle, refer to the Precautions Section.
2. Disconnect the negative battery cable.

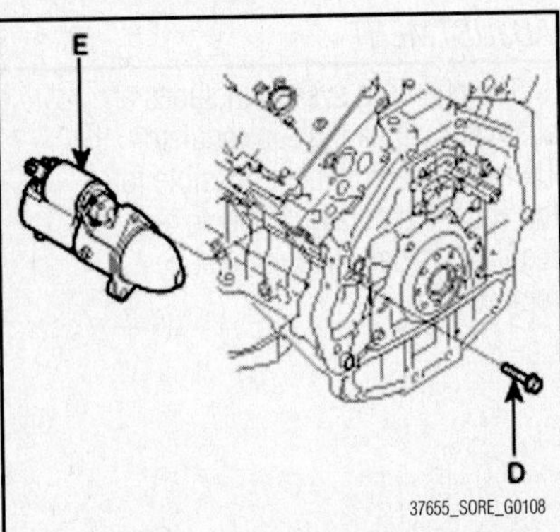

Fig. 42 Starter (E) and mounting bolts (D)

3. Remove the left-hand exhaust manifold assembly. Refer to Exhaust Manifold Removal & Installation in the Engine Mechanical section.
4. Supporting the engine with a jack, remove the left-hand side engine mounting bracket.
5. Disconnect the starter cable from the B terminal on the solenoid, and the connector from the S terminal.
6. Remove the starter mounting bolts.
7. Remove the starter.

To install:

8. Installation is the reverse of removal. Tighten the starter mounting bolts to 36–47 ft. lbs. (49–64 Nm).

SOLENOID OR RELAY REPLACEMENT

See Figure 43.

1. Before servicing the vehicle, refer to the Precautions Section.
2. Remove starter.
3. Disconnect the M-terminal on the solenoid.
4. After loosening the 3 screws, detach the solenoid.

To install:

5. Installation is reverse of removal.

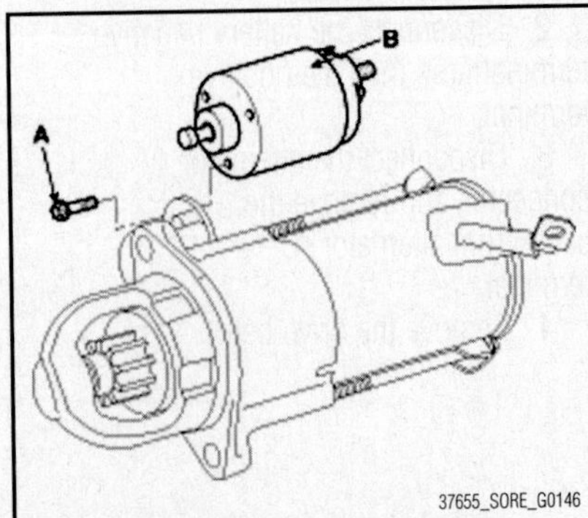

Fig. 43 Starter solenoid screw (A) and solenoid (B)

ENGINE MECHANICAL

ACCESSORY DRIVE BELTS

ACCESSORY BELT ROUTING

See Figure 44.

Refer to the accompanying illustrations.

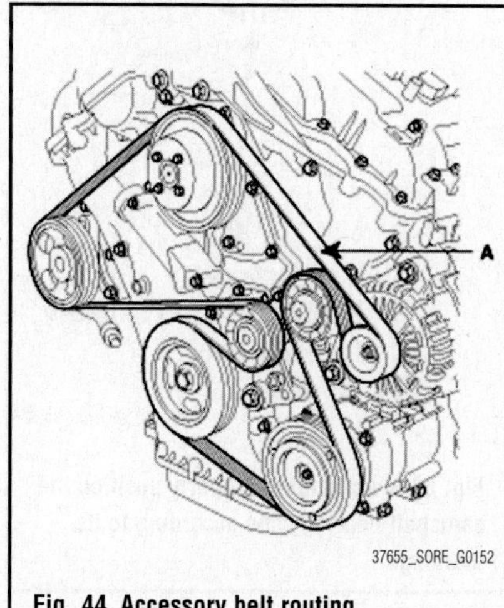

Fig. 44 Accessory belt routing

INSPECTION

See Figure 45.

Visually check the belt for excessive wear, frayed cords etc. If any defect has been found, replace the drive belt. Cracks on the rib side of a belt are considered acceptable. If the belt has chunks missing from the ribs, it should be replaced.

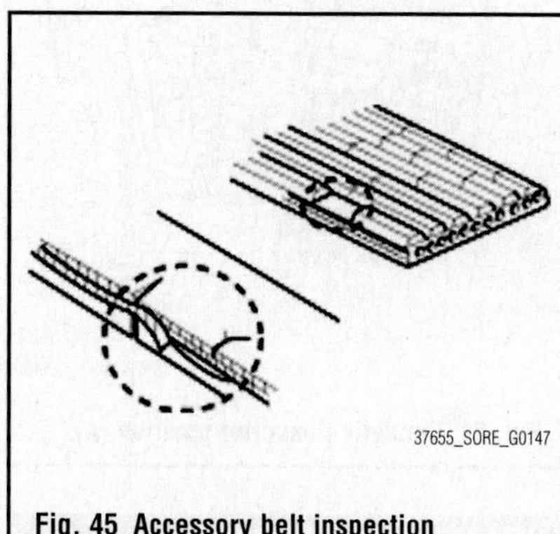

Fig. 45 Accessory belt inspection

ADJUSTMENT

The belt tension is maintained by an automatic tensioner. No adjustment is necessary or possible.

REMOVAL & INSTALLATION

1. Using the appropriate tool, remove tension of the belt tensioner.
2. Remove the serpentine belt.

To install:

3. Hang the belt on the tensioner pulley.
4. Install drive belt on remaining pulleys.

➡When installing, ensure belt centering on all pulleys.

5. Using the appropriate tool, remove tension of the belt tensioner.
6. Position belt around tensioner pulley.
7. Release tensioner to automatically position for appropriate tension.

CAMSHAFT AND VALVE LIFTERS

INSPECTION

See Figure 46.

1. Inspect the cam lobes.
   a. Using a micrometer, measure the cam lobe height.
   b. If the cam lobe height is less than specified, replace the camshaft:
   - Intake: 1.8425 in. (46.8mm)
   - Exhaust: 1.8031 in. (45.8mm)
2. Inspect the camshaft journal clearance.
   a. Clean the bearing caps and camshaft journals.
   b. Place the camshafts on the cylinder head.
   c. Lay a strip of Plastigage® across each of the camshaft journal.
   d. Install the bearing caps and tighten the bolts with specified torque.

➡Do not turn the camshaft.

   e. Remove the bearing caps.
   f. Measure the Plastigage® at its widest point.
   g. If the oil clearance is greater than specified, replace the camshaft. If necessary, replace the bearing caps and cylinder head as a set.
   h. Intake:
   - No. 1 Journal: 0.0008–0.0022 in. (0.020–0.057mm)
   - No. 2, 3, 4 Journal: 0.0012–0.0026 in. (0.030–0.067mm)

   i. Exhaust:
   - No. 1 Journal: 0.0008–0.0022 in. (0.020–0.057mm)
   - No. 2, 3, 4 Journal: 0.0012–0.0026 in. (0.030–0.067mm)

   j. Completely remove the Plastigage®.
   k. Remove the camshafts.
3. Inspect the camshaft end play.
   a. Install the camshafts.
   b. Using a dial indicator, measure the end play while moving the camshaft back and forth.

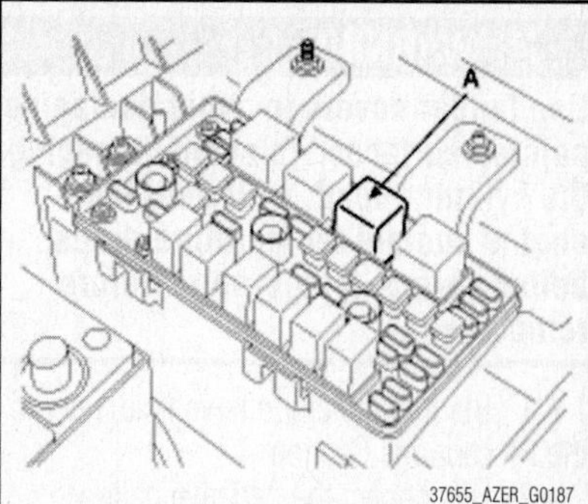

Fig. 46 Apply vinyl tape to the CVVT on all parts except the one indicated by the arrow

   c. If the end play is greater than specified, replace the camshaft. If necessary, replace the bearing caps and cylinder head as a set.
   - Camshaft End Play: 0.0008–0.0071 in. (0.02–0.18mm)

   d. Remove the camshafts.
4. Inspect the Continuous Variable Valve Timing (CVVT) assembly.
   a. Check that the CVVT assembly will not turn.
   b. Apply vinyl tape to all the parts except the one indicated by the arrow in the illustration.
   c. Wrap tape around the tip of the air gun and apply air of approx. 21 psi to the port of the camshaft. Perform this in order to release the lock pin for the maximum delay angle locking.

➡Wrap a shop rag around the CVVT as the oil may spray out when the air pressure is applied.

   d. Under the condition of air pressure being applied, turn the CVVT assembly to the advance angle side with your hand.
   - Depending on the air pressure, the CVVT assembly will turn to the advance side
   - If air is leaking from the port and air pressure cannot be maintained, the locking pin will not release
5. Except the position where the lock pin meets at the maximum delay angle, let the CVVT assembly turn back and forth and check the movable range and that there is no disturbance.
   a. The CVVT should move smoothly in the range of about 22.5°.
   b. Turn the CVVT assembly with your hand and lock it at the maximum delay angle position (counter-clockwise).

REMOVAL & INSTALLATION

See Figures 47 through 50.

WARNING

Use fender covers to avoid damaging painted surfaces. To avoid damaging the cylinder head, wait until the engine coolant temperature drops below normal temperature before removing it.

1. Before servicing the vehicle, refer to the Precautions Section.
2. Disconnect the negative battery cable.
3. Remove the engine.
4. Remove exhaust manifold. Refer to Exhaust Manifold Removal & Installation.
5. Remove the intake manifold. Refer to Intake Manifold Removal & Installation.
6. Remove the timing chain. Refer to Timing Chain Removal & Installation.
7. Remove the water temperature control assembly.
8. Remove the camshaft bearing caps.
9. Remove the camshaft assembly.

To install:

10. Apply a light coat of engine oil on camshaft journals.
11. Assemble the key groove of camshaft rear side to the same level of head top surface.
12. Install the camshaft bearing caps in the order shown.
13. Install the camshaft bearing caps in the sequence shown:
    a. Step 1—Tightening torque: 52 inch lbs. (6 Nm).
    b. Step 2—Tightening torque: 87–104 inch lbs. (10–12 Nm).
14. Be careful to get the right bank, left bank, intake side, and exhaust side in the correct position before assembling.

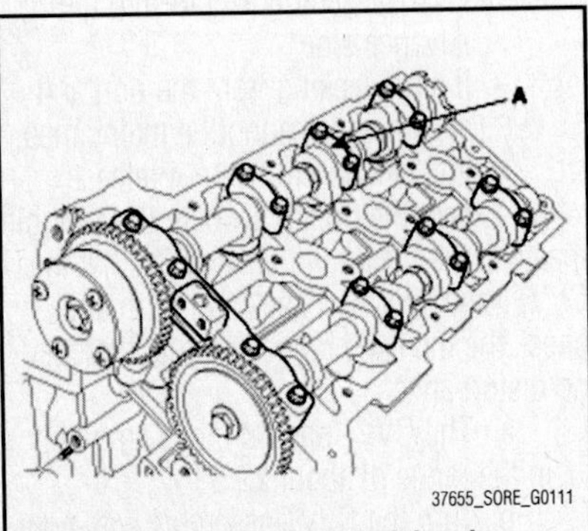

Fig. 47 Remove the camshaft bearing caps (A)

Fig. 48 Remove the camshaft assembly (A)

WARNING

Rotate the crankshaft so as not to contact the valves to the pistons by positioning the pistons 0.3937 in. (10mm) below the top of the cylinder block.

15. Install the water temperature control assembly.
16. Install the timing chain.
17. Check and adjust the valve clearance, as necessary.
18. Install the exhaust manifold.
19. Install the intake manifold.
20. Connect the negative battery cable.
21. Fill with engine coolant.
22. Start the engine and check for leaks.
23. Recheck the engine coolant level and oil level.

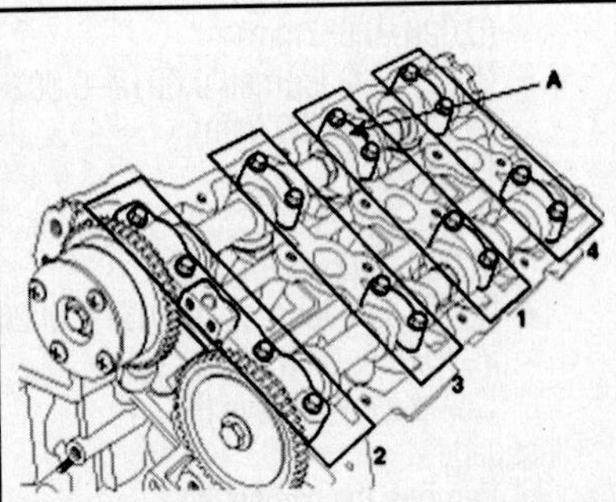
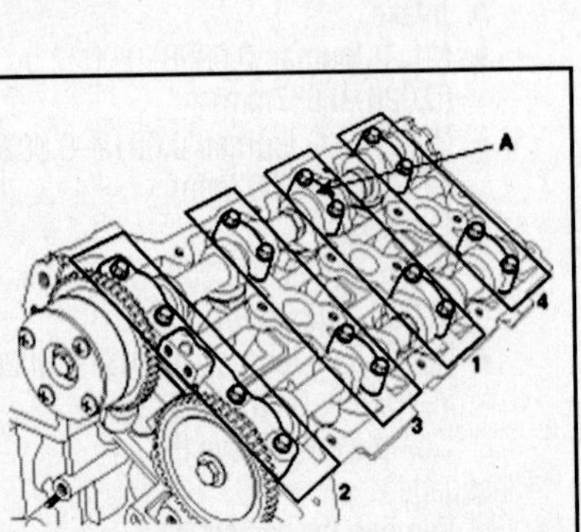

Fig. 49 Install the camshaft bearing caps in the sequence shown

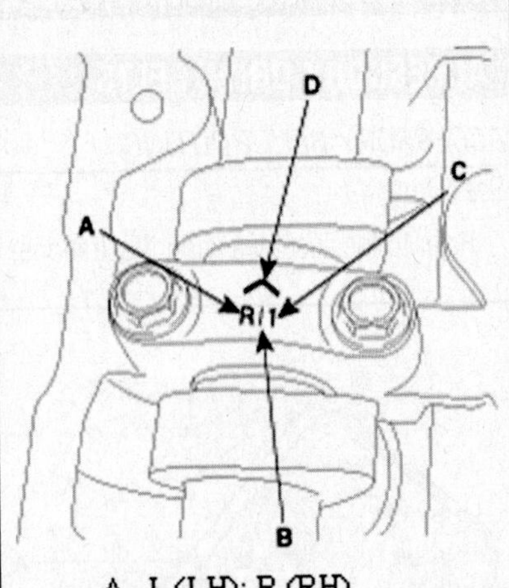

Fig. 50 Be careful to properly position the camshaft bearing caps according to its markings

CATALYTIC CONVERTER

REMOVAL & INSTALLATION

See Figure 51.

The manufacturer does not provide the removal and installation information. Refer to the illustration for location.

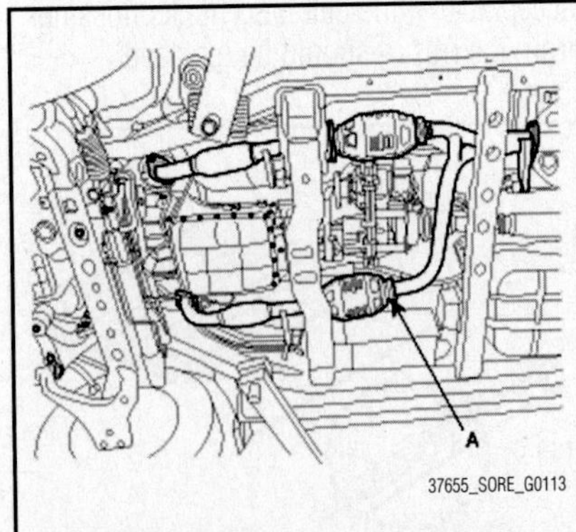

Fig. 51 Catalytic Converter position (A)

CRANKSHAFT DAMPER

REMOVAL & INSTALLATION

See Figures 52.

1. Remove accessory drive belt. Refer to Accessory Drive belt in this section.
2. Remove the crankshaft damper pulley bolt.
3. Remove the crankshaft damper pulley.

To install:

4. Installation is reverse of removal.

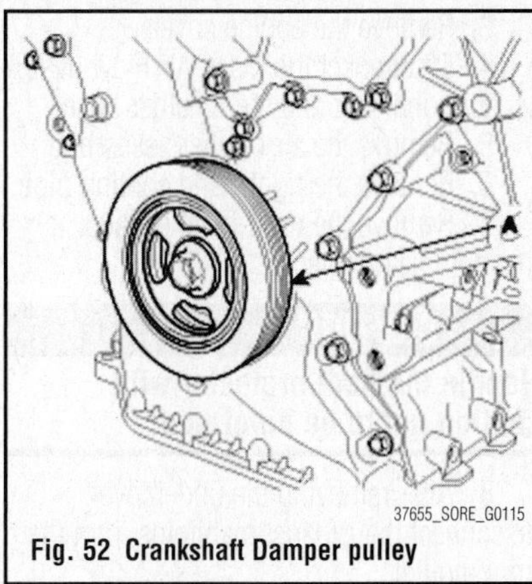

Fig. 52 Crankshaft Damper pulley

a. Tighten damper bolt to 209–224 ft. lbs. (284–304 Nm).

CRANKSHAFT FRONT SEAL

REMOVAL & INSTALLATION

See Figure 53.

1. Remove crankshaft damper. Refer to Crankshaft Damper in this section.
2. Using seal remover, remove front seal.

To install:

3. Check crankshaft for damage and repair as necessary.
4. Lubricate front seal with grease.
5. Using seal installer, install front seal ensuring seal is fully seated in front cover.
6. Install crankshaft damper.

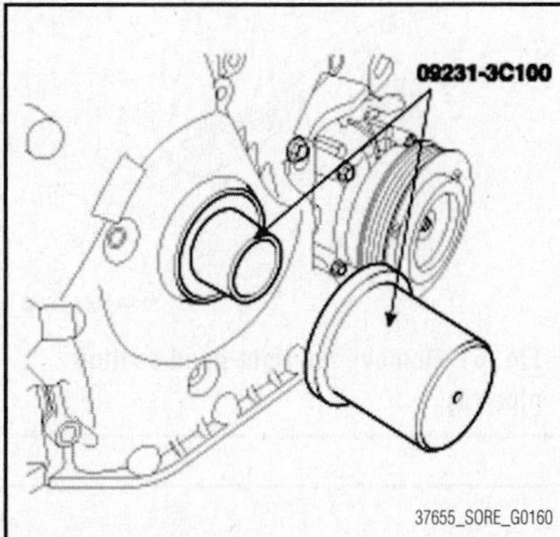

Fig. 53 Front cover oil seal installation

CYLINDER HEAD

REMOVAL & INSTALLATION

See Figures 54 through 58.

➡Engine removal is required for this procedure.

1. Remove exhaust manifold. Refer to Exhaust Manifold in this section.
2. Remove intake manifold. Refer to Intake Manifold in this section.
3. Remove timing chain. Refer to Timing Chain and Sprockets in this section.
4. Remove water temperature control assembly.
5. Remove camshaft assemblies. Refer to Camshaft and Lifters in this section.

CAUTION

Head warpage or cracking could result from removing bolts in an incorrect order.

6. Uniformly loosen and remove the 16 cylinder head bolts, in several passes, in the sequence shown.
7. Remove the 16 cylinder head bolts and plate washers.
8. Lift the cylinder head from the dowels on the cylinder block and place the cylinder head on wooden blocks on a bench.

CAUTION

Be careful not to damage the contact surfaces of the cylinder head and cylinder block.

To install:

➡Note the following:

- Thoroughly clean all parts to be assembled.
- Always use a new head and manifold gasket.
- The cylinder head gasket is a metal gasket. Take care not to bend it.

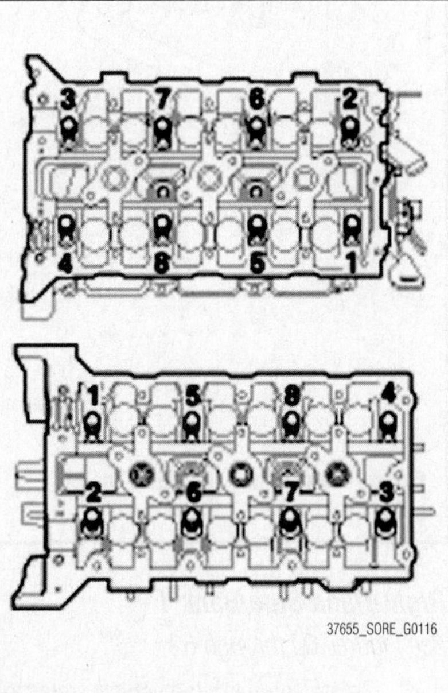

Fig. 54 Cylinder head bolt removal sequence

- Rotate the crankshaft, set the No.1 piston at TDC.

9. Install the cylinder head.

➡The sealant locations on cylinder head and cylinder block must be free of engine oil and ETC.

10. Apply sealant on cylinder block top face before assembling cylinder head gaskets.

➡The part must be assembled within 5 minutes after sealant was applied.

➡Bead width : 2.0–3.0 mm Sealant locations : 1.0–1.5mm from block surface Recommended sealant :Liquid sealant TB1217H

11. Apply sealant on cylinder head gaskets after assembling cylinder head gaskets on cylinder block.

➡The part must be assembled within 5 minutes after sealant was applied.

12. Place the cylinder head on engine block being careful not to damage the gasket.

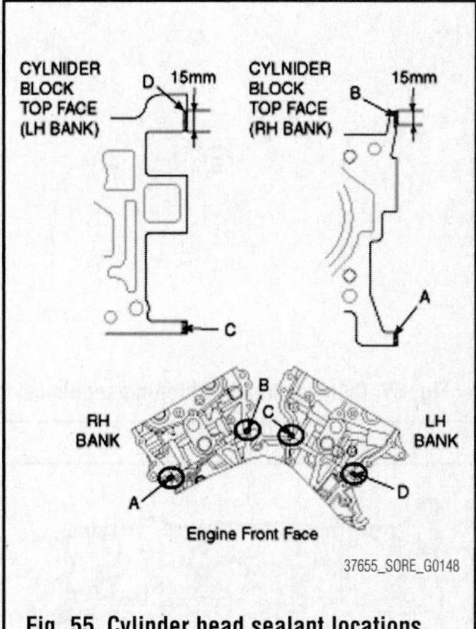

Fig. 55 Cylinder head sealant locations

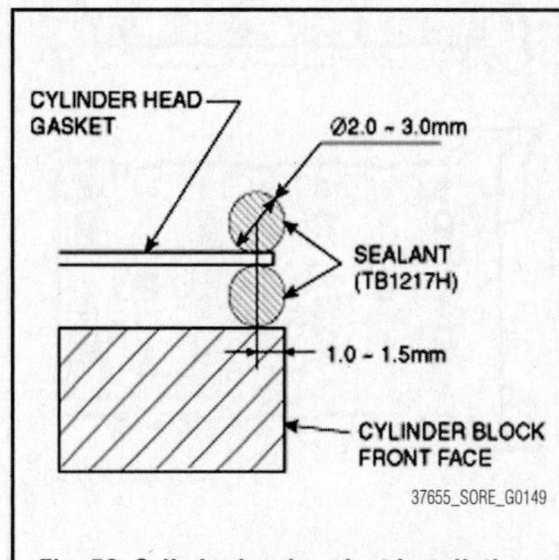

Fig. 56 Cylinder head sealant installation

13. Remove the extruded sealant after assembling cylinder heads.
14. Install cylinder head bolts.

➡Do not apply engine oil on the threads and under the heads of the cylinder head bolts. Always use new cylinder head bolts.

15. Using SST (09221-4A000), or equivalent torque angle tool, install and tighten the cylinder head bolts and plate washers, in several passes, in the sequence shown.
16. Tightening torque:
    a. Head bolt: 27–30. ft. lbs. (37–41. Nm) plus 118–122 degrees, plus an additional 88–92 degrees
    b. Bolt (A): 14–17 ft. lbs. (19–24 Nm)
17. Install water temperature control assembly.
18. Install timing chain.
19. Install intake manifold.
20. Install exhaust manifold.

➡Install engine.

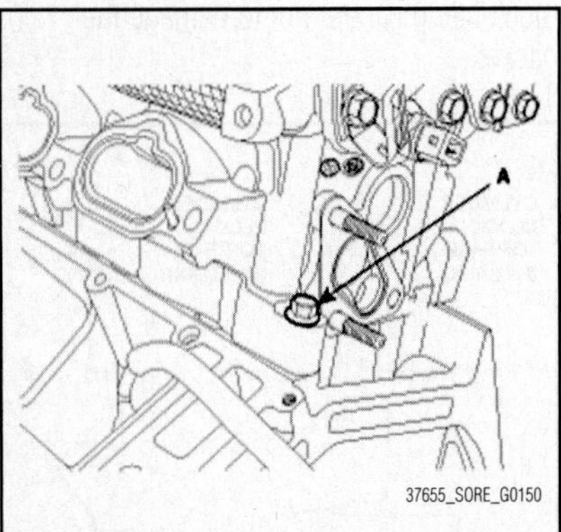

Fig. 57 Cylinder head tightening sequence

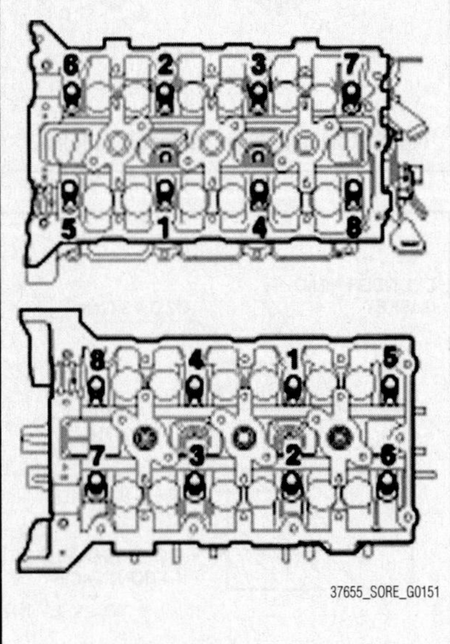

Fig. 58 Cylinder head bolt (A) location

EXHAUST MANIFOLD

REMOVAL & INSTALLATION

Left-Hand Side/Bank 2

See Figure 59.

1. Before servicing the vehicle, refer to the Precautions Section.
2. Disconnect the negative battery cable.
3. Remove the engine oil level gauge.
4. Remove the battery.
5. Remove the left-hand exhaust manifold heat protector.

CAUTION

Handle the heat protector with caution not to be deformed.

6. Remove the left-hand cooling pipe.
7. Remove the automatic transmission fluid level gauge.
8. Disconnect the oil pressure switch harness connector and the battery ground line.
9. After removing the undercover, disconnect the exhaust manifolds from the front muffler.
10. Remove the left-hand exhaust manifold.

To install:

11. To install, reverse the removal procedure.

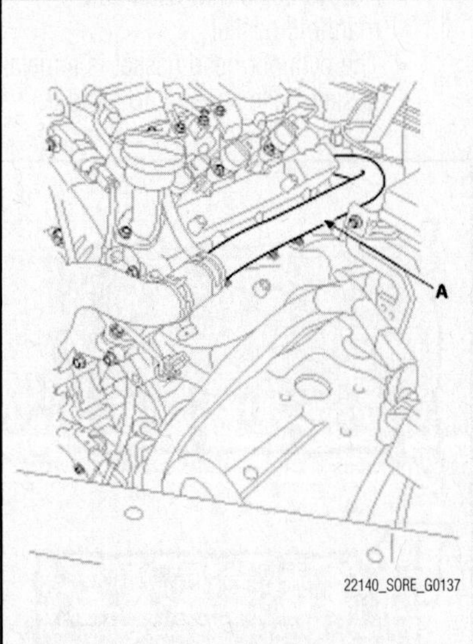

Fig. 59 Remove the left-hand cooling pipe (A)

Right-Hand Side/Bank 1

See Figures 60 through 62.

1. Before servicing the vehicle, refer to the Precautions Section.
2. Disconnect the negative battery cable.
3. Remove the engine cover.
4. Disconnect the Mass Air Flow (MAF) sensor connector and the breather hose.
5. Remove the air cleaner assembly.
6. Remove the right-hand cooling pipe.
7. Remove the right-hand exhaust manifold heat protector.

CAUTION

Handle the heat protector with caution not to be deformed.

8. After removing the undercover, disconnect the exhaust manifolds from the front muffler.

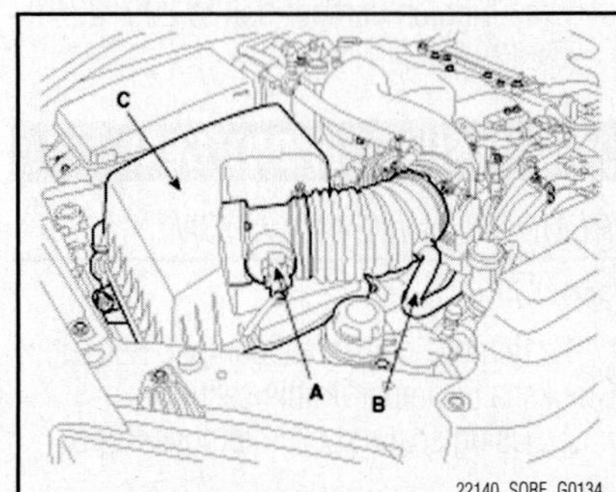

Fig. 60 Disconnect the Mass Air Flow (MAF) sensor connector (A) and the breather hose (B)

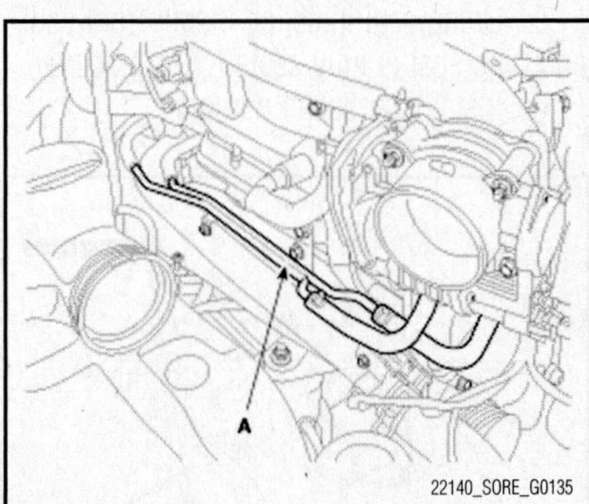

Fig. 61 Remove the right-hand cooling pipe (A)

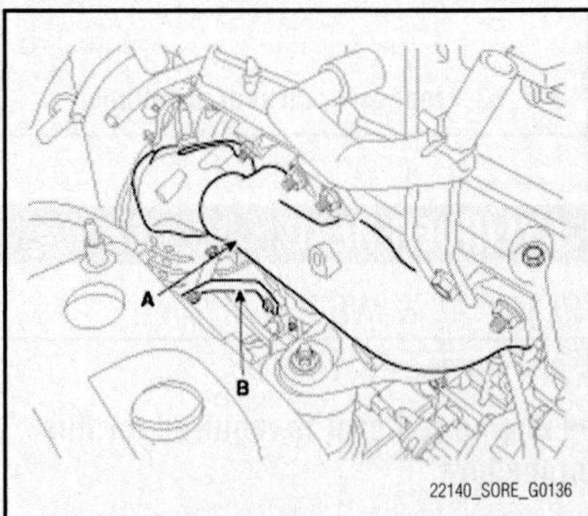

Fig. 62 Remove the right-hand exhaust manifold (A) and the stay (B)

9. Remove the right-hand exhaust manifold and the stay.

To install:

10. To install, reverse the removal procedure.

FLEXPLATE

REMOVAL & INSTALLATION

1. Remove the transmission. Refer to Transmission Removal and Installation.
2. Remove the flexplate bolts and flexplate.

To install:

3. Installation is reverse of removal.
    a. Tighten bolts in a criss-cross pattern to 53–56 ft. lbs. (30–34 Nm)

INTAKE MANIFOLD

REMOVAL & INSTALLATION

See Figures 63 through 70.

1. Before servicing the vehicle, refer to the Precautions Section.
2. Disconnect the negative battery cable.
3. Remove the engine cover.
4. Remove the engine room resonator.
5. Disconnect the MAF sensor connector and the breather hose.
6. Remove the air cleaner assembly.
7. Disconnect the other breather hose, the Purge Control Solenoid Valve (PCSV) hose, the Positive Crankcase Ventilation (PCV) hose and the Electronic Throttle Control (ETC) cooling hoses.
8. Remove the wiring over the surge tank.
    a. Disconnect the injection harness connector.
    b. Disconnect the Camshaft Position Sensor (CMP) harness connector.
    c. Disconnect the ground lines.
    d. Disconnect the ignition coil harness connector.
    e. Disconnect the condenser connector.
    f. Disconnect the Oil Control Valve (OCV) harness connector.
    g. Disconnect the Variable Induction System (VIS) solenoid valve connector.
    h. Disconnect the injector wiring (H) and ignition coil wiring.
9. Disconnect the fuel hose tube.
10. Remove heater hose.
11. Disconnect the brake vacuum hose.
12. Disconnect the surge tank stay.
13. Remove the surge tank assembly.
14. Disconnect the injector connectors.
15. Disconnect the water hose on intake manifold from the nipple on the chain cover.
16. Remove the delivery pipe and intake manifold as an assembly.

➡Except such cases as defects of injectors or pipe, do not disassemble a delivery pipe from an intake manifold because it is one of the fuel system parts, or you may have some problems in fuel system.

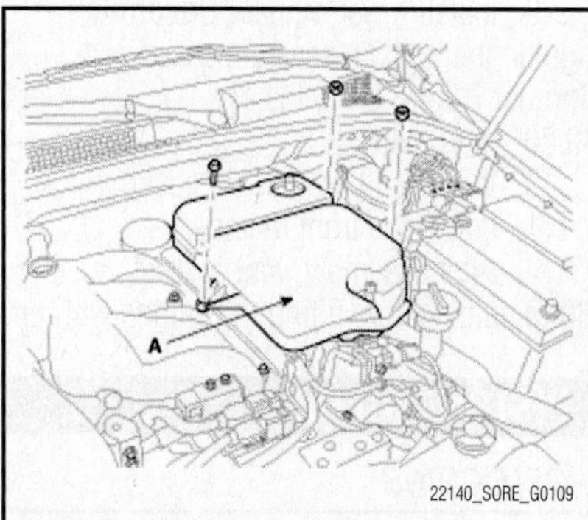

Fig. 63 Remove the engine room resonator (A)

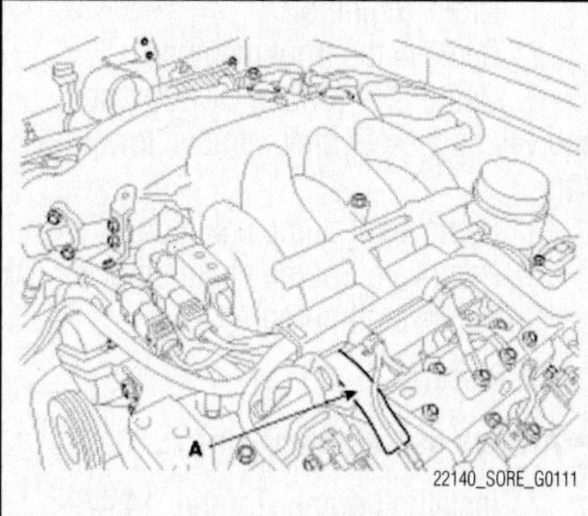

Fig. 65 Disconnect the other breather hose (A)

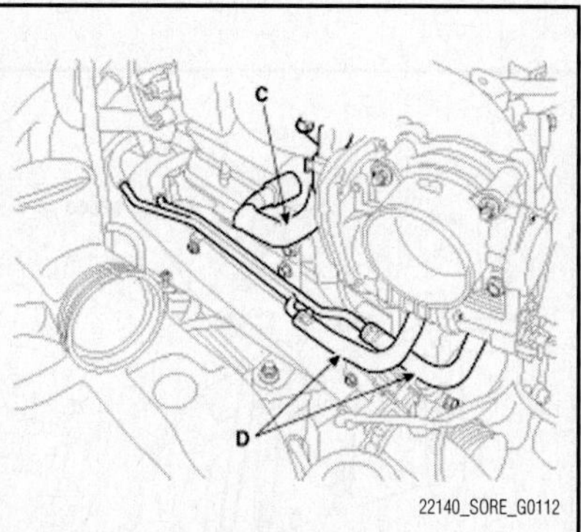
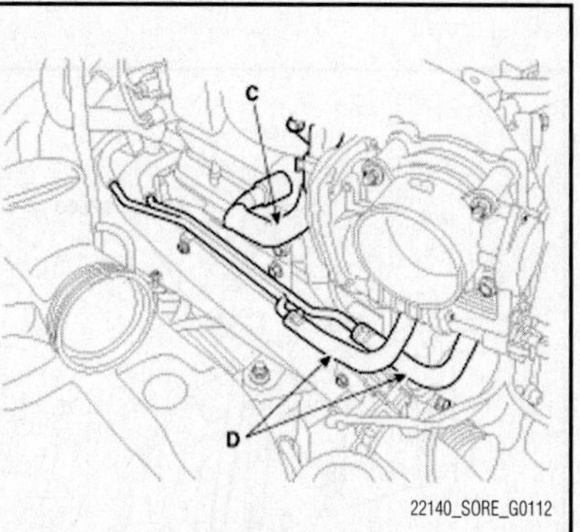

Fig. 66 Disconnect the Positive Crankcase Ventilation (PCV) hose (C) and the Electronic Throttle Control (ETC) cooling hoses (D)

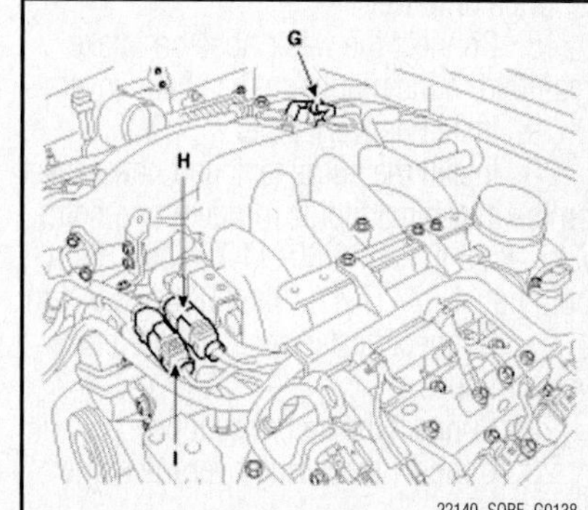

Fig. 68 Disconnect the Variable Induction System (VIS) solenoid valve connector (G), the injector wiring (H) and ignition coil wiring (I)

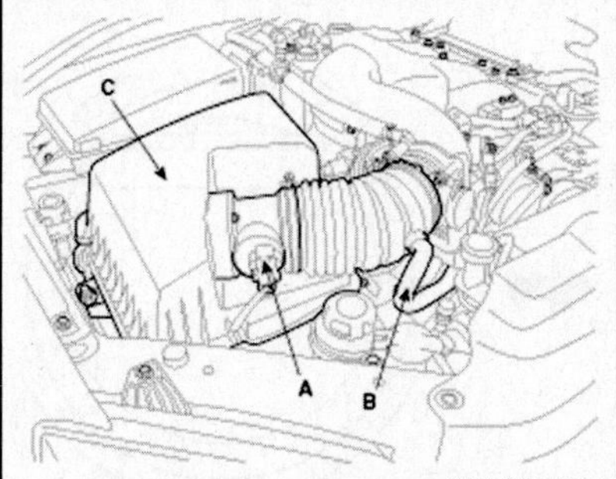

Fig. 64 Disconnect the MAF sensor connector (A) and the breather hose (B). Remove the air cleaner assembly (C)

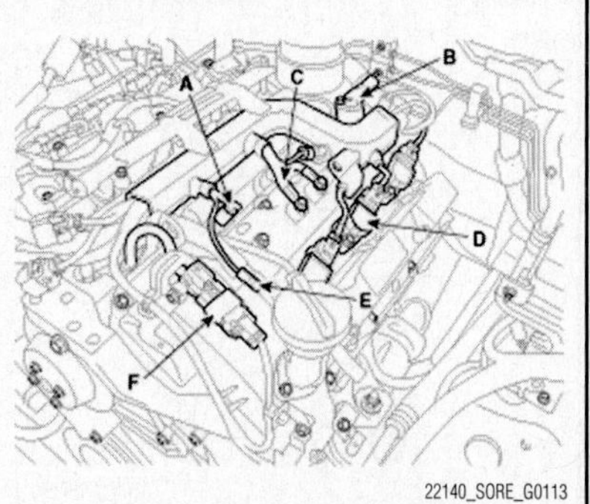

Fig. 67 Remove the wiring over the surge tank.

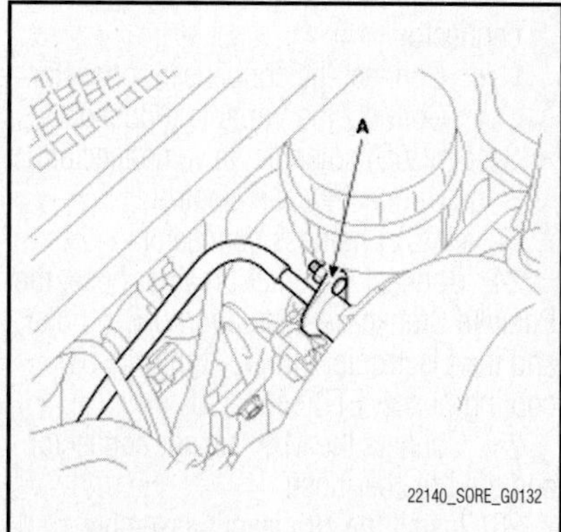

Fig. 69 Disconnect the fuel hose (A)

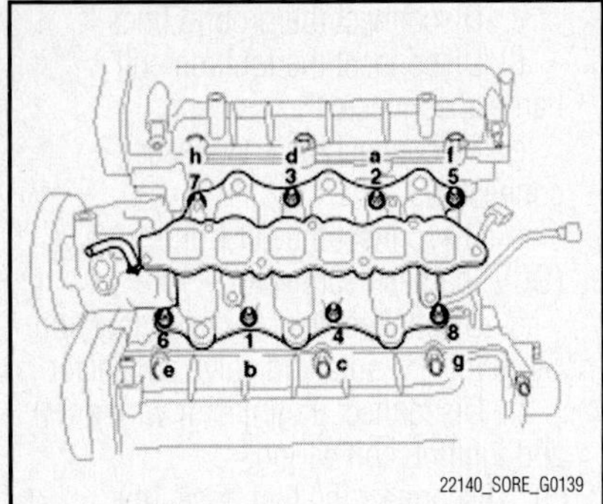

Fig. 70 Intake manifold tightening sequence

To install:

Install intake manifold and new gasket on the cylinder head using the following tightening sequence:

- 1st Step order: a—h
- 2nd Step order: 1–8.

17. Torque to:

a. 1st Step: 2.9–4.3 ft. lbs. (3.9–5.9 Nm)

b. 2nd Step: Bolts: 19.5–23.1 ft. lbs. (26.5–31.4 Nm); Nut: 13.7–17.4 ft. lbs. (18.6–23.5 Nm)

c. 3rd Step: Repeat the 2nd Step twice or more.

18. Connect the water hose on intake manifold to the nipple on the chain cover.
19. Install delivery pipe.
20. Install the surge tank and new gasket on the intake manifold. Torque: Long bolt: 7.23–8.68 ft. lbs. (9.80–11.76 Nm); Short bolt, nut: 13.7–17.4 ft. lbs. (18.6–23.5 Nm).
21. Connect heater hose and the brake vacuum hose.
22. Connect the fuel hose tube.
23. Connect the wiring over the surge tank.

a. Connect the injection harness connector.

b. Connect the Camshaft Position Sensor (CMP) harness connector.

c. Connect the ground lines.

d. Connect the ignition coil harness connector.

e. Connect the condenser connector.

f. Connect the Variable Induction System(VIS) solenoid valve connector.

g. Connect the Oil Control Valve(OCV) harness connector.

24. Connect the other breather hose, the Positive Crankcase Ventilation (PCV) hose and the Electronic Throttle Control(ETC) cooling hoses, ETC connector.
25. Connect the MAF sensor connector and the breather hose.
26. Install the air cleaner assembly.
27. Install the engine room resonator.
28. Install the engine cover.
29. Fill with engine coolant.

OIL PAN

REMOVAL & INSTALLATION

Refer to Oil Pump in this section.

OIL PUMP

REMOVAL & INSTALLATION

See Figures 71 through 74.

1. Drain engine oil.
2. Remove the front member.
3. Using SST (09215-3C000), or equivalent gasket cutter, remove lower oil pan.
4. Remove oil pump chain cover (A).
5. Remove oil pump chain sprocket (A).
6. Remove oil pump(A).

To install:

➡**Always use a new O-ring.**

7. Install oil pump. Torque: 14.47–17.36 ft. lbs. (19.60–23.52 Nm).
8. Install oil pump sprocket and oil pump chain on the oil pump. Torque: 13.74–15.91 ft. lbs. (18.62–21.56 Nm).

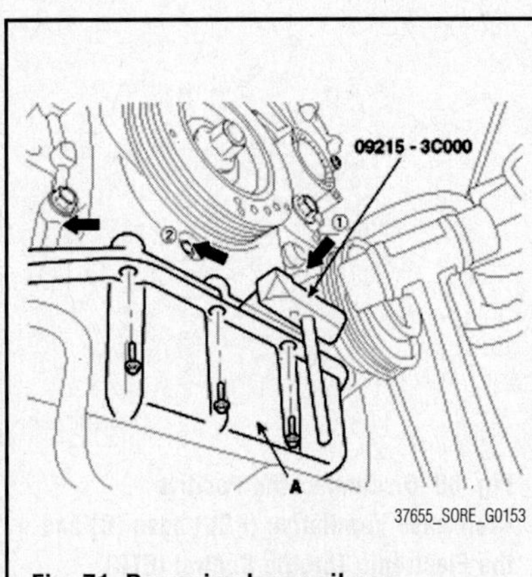

Fig. 71 Removing lower oil pan

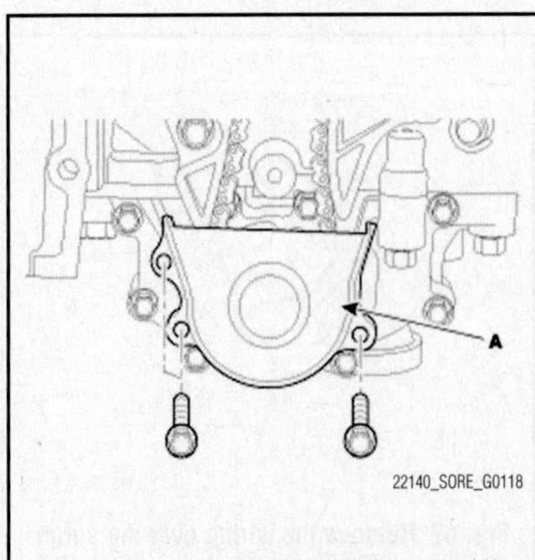

Fig. 72 Remove oil pump chain cover (A)

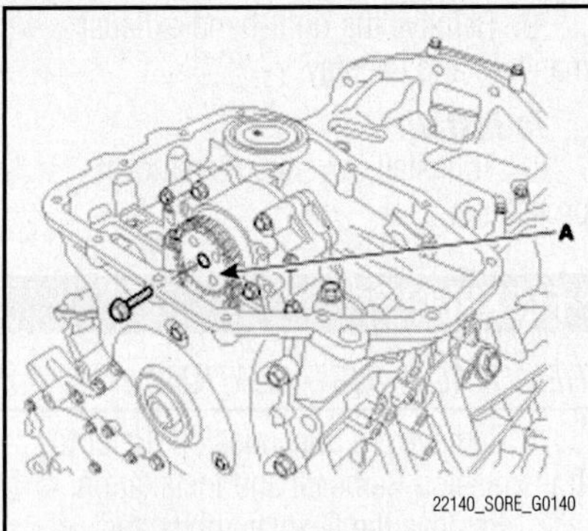

Fig. 73 Remove oil pump chain sprocket (A)

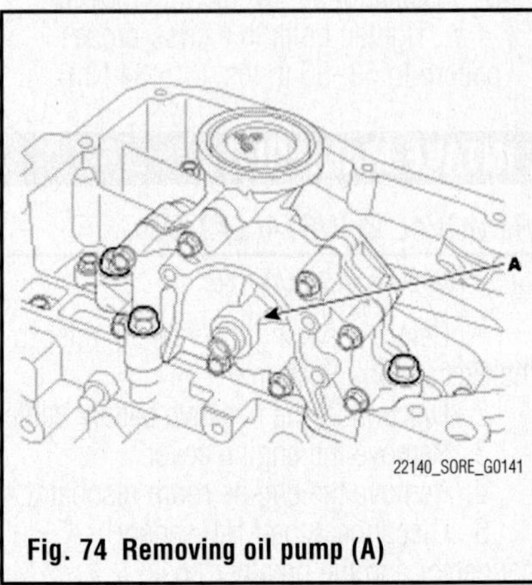

Fig. 74 Removing oil pump (A)

9. Install lower oil pan. Uniformly tighten the bolts in several passes. Torque: 7.23–8.68 ft. lbs. (9.80–11.76 Nm).
10. Install oil pump chain cover.
11. Install the front member.
12. After assembly, wait at least 30 minutes before filling the engine with oil.

PISTON AND RING

POSITIONING

See Figure 77.

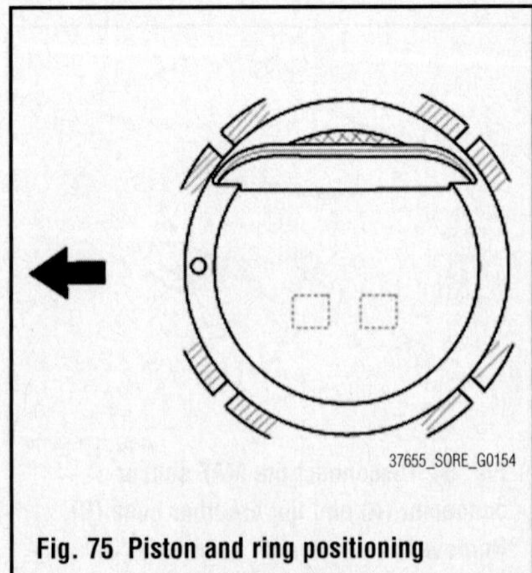

Fig. 75 Piston and ring positioning

REAR MAIN SEAL

REMOVAL & INSTALLATION

See Figure 76.

1. Disconnect the negative battery cable.
2. Remove the starter motor.
3. Remove the transmission.
4. Remove the flexplate.
5. Remove the rear oil seal case.
6. Remove and replace the rear oil seal.

To install:

7. Installation is the reverse of removal.

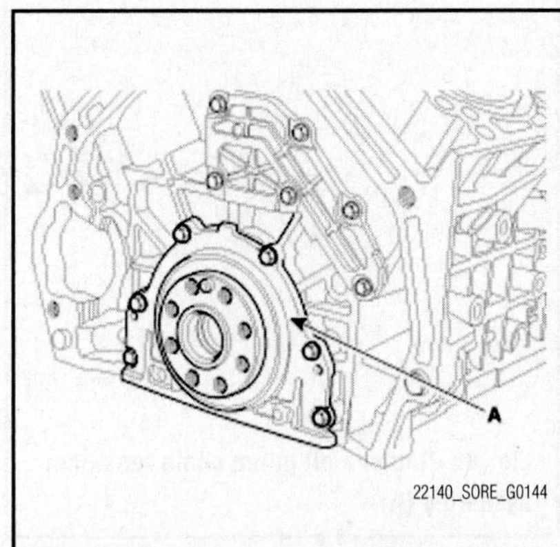

Fig. 76 Remove the rear oil seal case (A)

TIMING CHAIN FRONT COVER

REMOVAL & INSTALLATION

See Timing Chain & Sprockets for removal and installation.

TIMING CHAIN & SPROCKETS

REMOVAL & INSTALLATION

See Figures 77 through 98.

1. Remove the engine cover.
2. Remove the engine room resonator.
3. Disconnect the MAF sensor connector (A) and the breather hose (B).
4. Remove the air cleaner assembly (C).
5. Disconnect the MAF sensor connector and the breather hose.
6. Remove the air cleaner assembly.
7. Disconnect the other breather hose, the Purge Control Solenoid Valve (PCSV) hose, the Positive Crankcase Ventilation (PCV) hose and the Electronic Throttle Control (ETC) cooling hoses.
8. Remove the wiring over the surge tank:
    a. Disconnect the injection harness connector.
    b. Disconnect the camshaft position sensor (CMP) harness connector.
    c. Disconnect the ground lines.
    d. Disconnect the ignition coil harness connector.
    e. Disconnect the condenser connector.
    f. Disconnect the oil control valve (OCV) harness connector.
9. Remove the surge tank assembly.
10. Remove the cylinder head covers and gaskets.
11. Set the No. 1 cylinder to TDC/compression.
    a. Turn the crankshaft pulley and align its groove with the timing mark "T" of the lower timing chain cover.

➡Do not rotate engine counterclockwise.

    b. Check that the mark of the camshaft timing sprockets are in straight line on the cylinder head surface as shown in the illustration. If not, turn the crankshaft one revolution (360°).

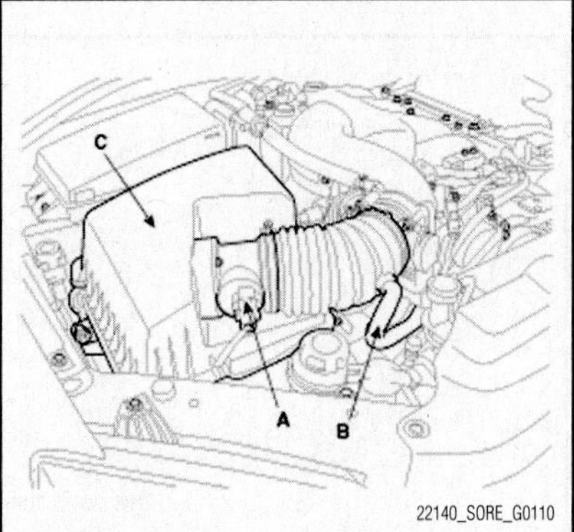

Fig. 78 Disconnect the MAF sensor connector (A) and the breather hose (B). Remove the air cleaner assembly (C)

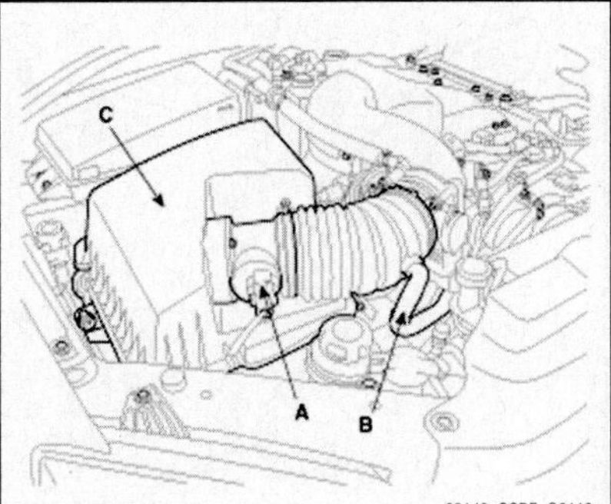

Fig. 79 MAF sensor connector (A), breather hose (B), and air cleaner assembly (C)

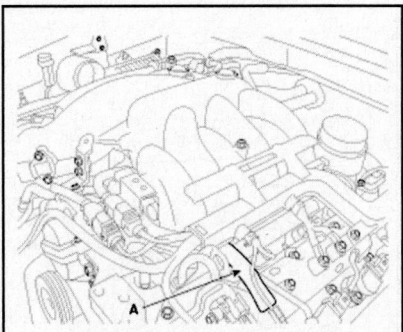

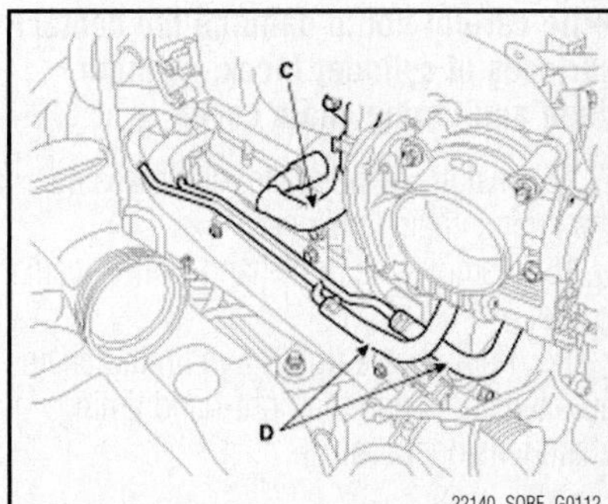

Fig. 81 Disconnect the Positive Crankcase Ventilation (PCV) hose (C) and the Electronic Throttle Control (ETC) cooling hoses (D)

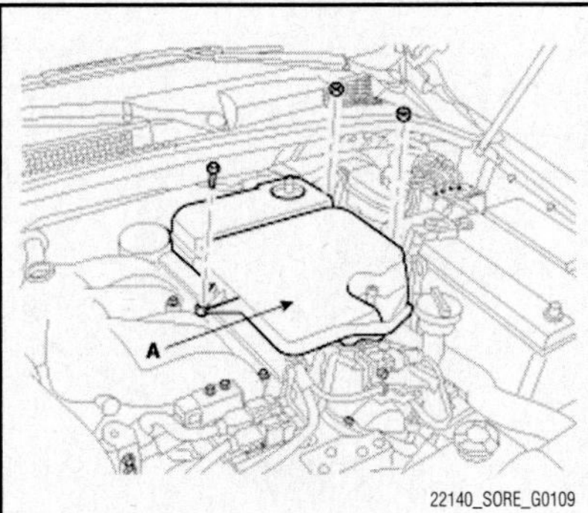

Fig. 77 Remove the engine room resonator (A)

Fig. 80 Disconnect the other breather hose (A)

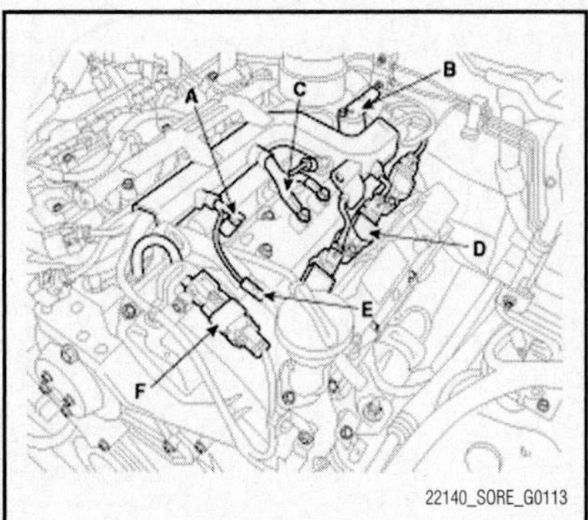

Fig. 82 Remove the wiring over the surge tank

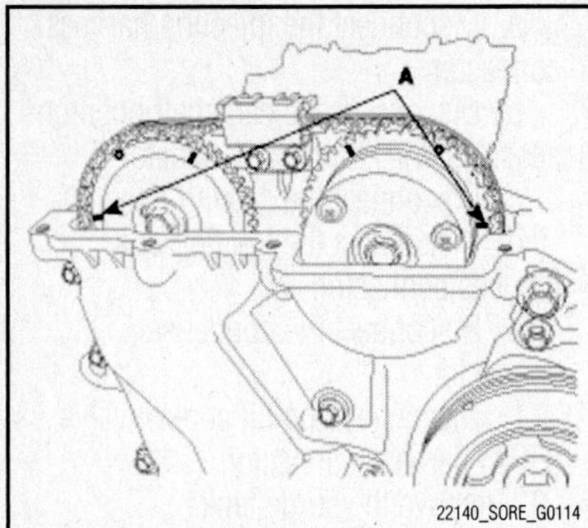

Fig. 83 Check that the mark (A) of the camshaft timing sprockets are in straight line

12. Remove the lower oil pan. Refer to Oil Pan in this section.

CAUTION

Insert the SST between the oil pan and the ladder frame by tapping it with a plastic hammer in the direction of arrow.

13. Remove the crankshaft damper pulley. Refer to Crankshaft Damper in this section.
14. Remove the timing chain cover.

➡Be careful not to damage the contact surfaces of cylinder block, cylinder head and timing chain cover.

15. Install a set pin after compressing the timing chain tensioner.
16. Remove the right-hand cam-to-cam guide.
17. Remove the right-hand timing chain auto tensioner and the right-hand timing chain tensioner arm.

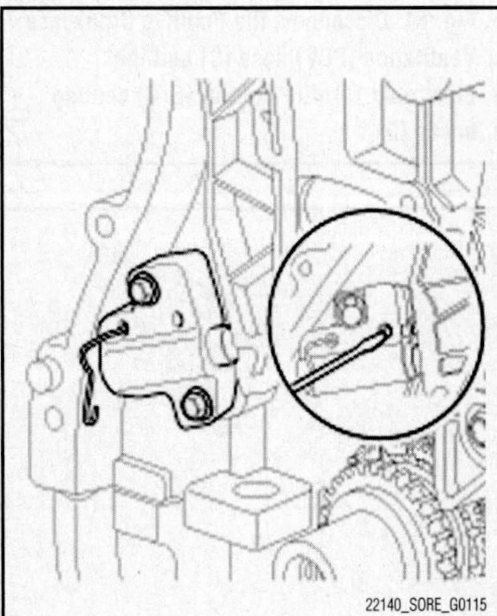

Fig. 84 Install a set pin after compressing the timing chain tensioner

18. Remove oil pump chain cover (A).
19. Remove oil pump chain tensioner assembly (A).
20. Remove oil pump chain guide.
21. Remove the right-hand timing chain.
22. Remove RH timing chain guide.
23. Remove oil pump chain sprocket and oil pump chain.
24. Remove crankshaft sprocket.
25. Remove the left-hand cam-to-cam guide.
26. Remove the left-hand timing chain auto tensioner and the left-hand timing chain tensioner arm.
27. Remove the left-hand timing chain.

To install:

28. Timing chain installation is the reverse of removal.

Install timing chain cover as follows:

➡The sealant locations on chain cover and on counter parts (cylinder head, cylinder block, and lower oil

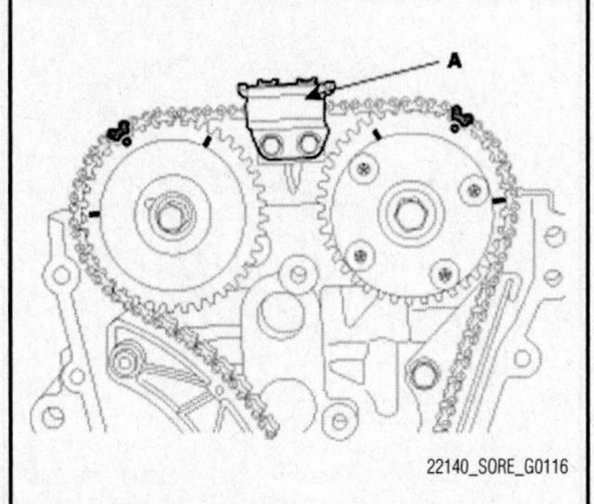

Fig. 85 Remove the right-hand cam-to-cam guide (A)

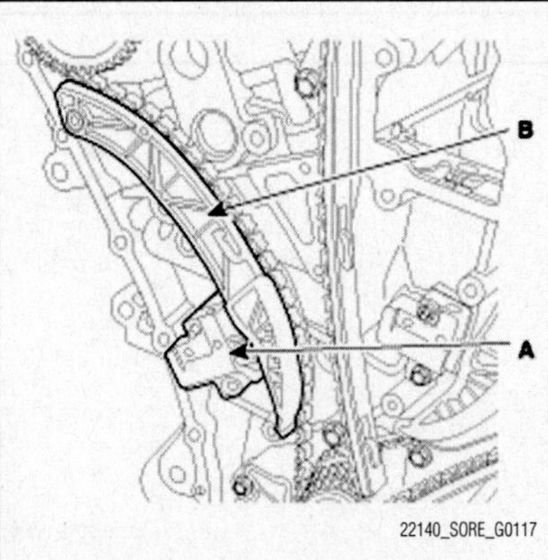

Fig. 86 Remove the right-hand timing chain auto tensioner (A) and the right-hand timing chain tensioner arm (B)

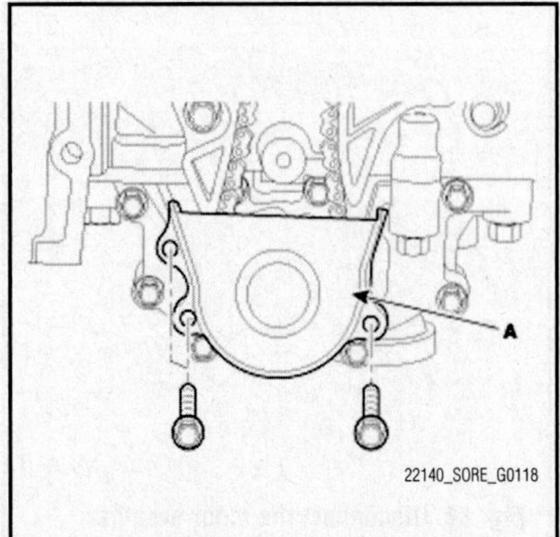

Fig. 87 Remove oil pump chain cover (A)

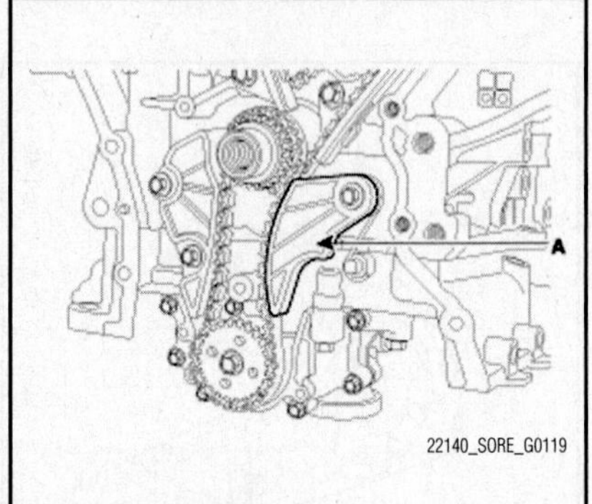

Fig. 88 Remove oil pump chain tensioner assembly (A)

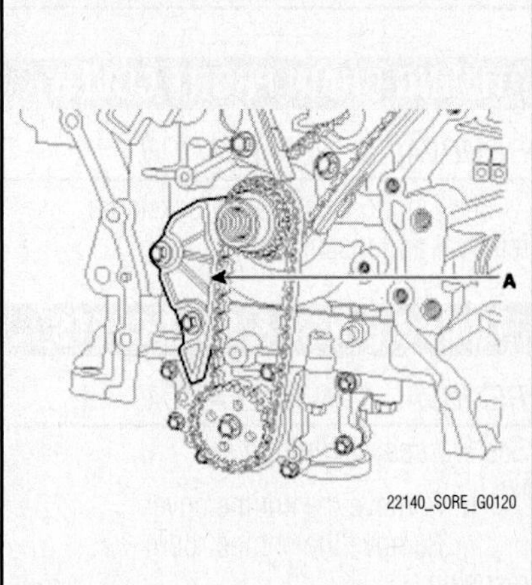

Fig. 89 Remove oil pump chain guide (A)

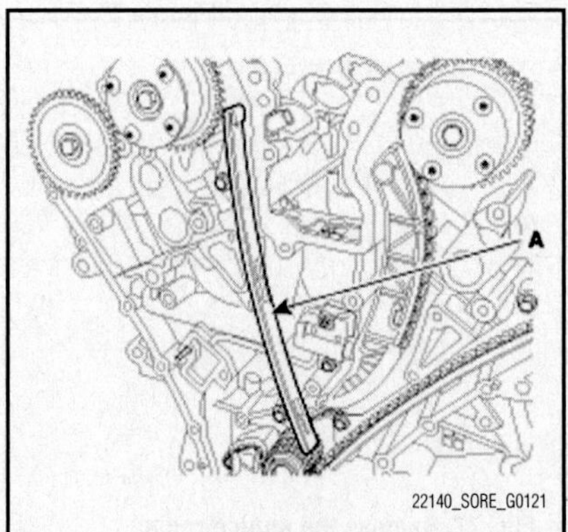

Fig. 90 Remove RH timing chain guide (A)

pan) must be free of engine oil and ETC.

29. Before assembling the timing chain cover, the liquid sealant TB1217H should be applied on the gap between cylinder head and cylinder block The part must be assembled within 5 minutes after sealant was applied. Bead width : 2.5mm (0.1in.)

30. After applying liquid sealant TB1217H on timing chain cover. The part must be assembled within 5 minutes after sealant was applied. Sealant should be applied without discontinuity. Sealant should also be applied all around the two holes of the dowel pins.

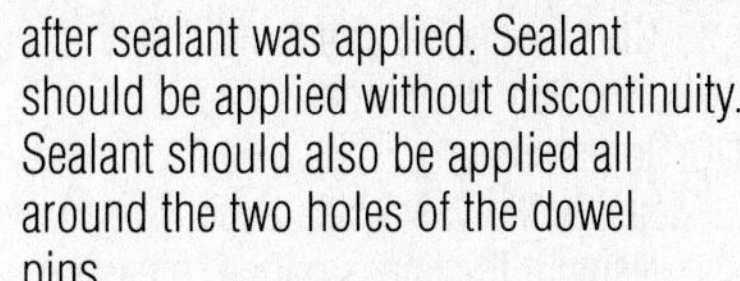

31. Install the new gasket to the timing chain cover.

➡The dowel pins on the cylinder block and holes on the timing chain cover should be used as a reference in order to assemble the timing chain cover to be in exact position.

32. Tighten front cover to the following torques in sequence shown:

- B(17): 13.74–15.91 ft. lbs. (18.62–21.56 Nm)
- C(4): 7.23–8.68 ft. lbs. (9.80–11.76 Nm)
- D(1): 43.40–50.63 ft. lbs. (58.80–68.80 Nm)
- E(1): 43.40–50.63 ft. lbs. (58.80–68.80 Nm)
- F(2): 18.08–19.53 ft .lbs. (24.50–26.46 Nm)
- G(4): 15.91–17.36 ft. lbs. (21.56–23.52 Nm)
- H(1): 7.23–8.68 ft. lbs. (9.80–11.76 Nm)
- I(1): 7.23–8.68 ft. lbs. (9.80–11.76 Nm)
- J(1): 7.23–8.68 ft. lbs. (9.80–11.76 Nm)
- K(4): 7.23–8.68 ft. lbs. (9.80–11.76 Nm)
- L(1): 15.91–19.53 ft. lbs. (21.56–26.46 Nm) - New bolt

33. Using SST(09231-3C100), or equivalent, install timing chain cover oil seal.

34. Install lower oil pan.

35. Install crankshaft damper pulley.

36. Install cylinder head cover.

➡Do not reuse cylinder head cover gasket.

37. Install ignition coil

38. Connect RH ignition coil connector, condenser connector and install wiring bracket.

39. Install connector bracket from LH cylinder head cover.

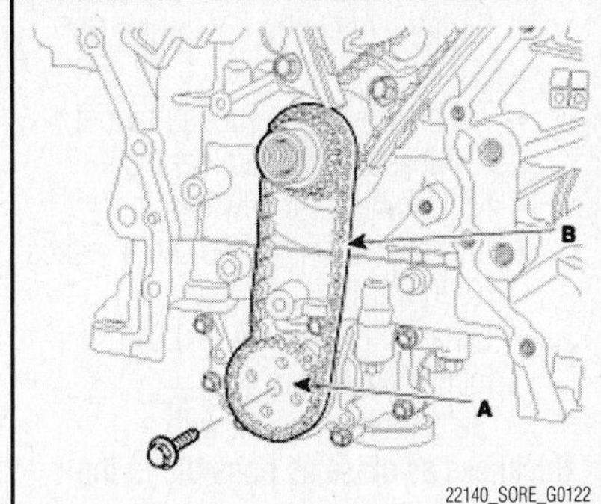

Fig. 91 Remove oil pump chain sprocket (A) and oil pump chain (B)

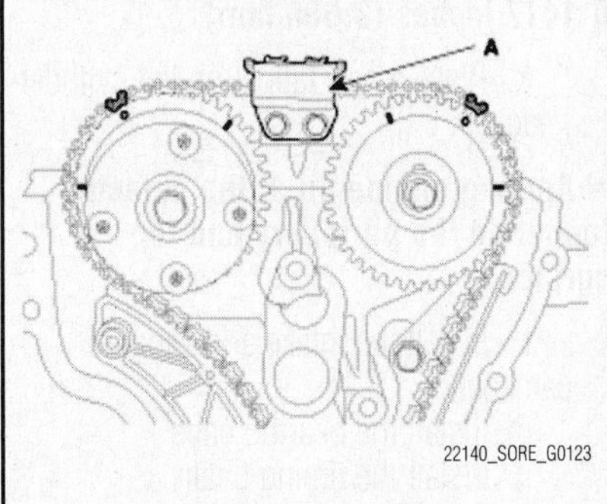

Fig. 92 Remove the left-hand cam-to-cam guide (A)

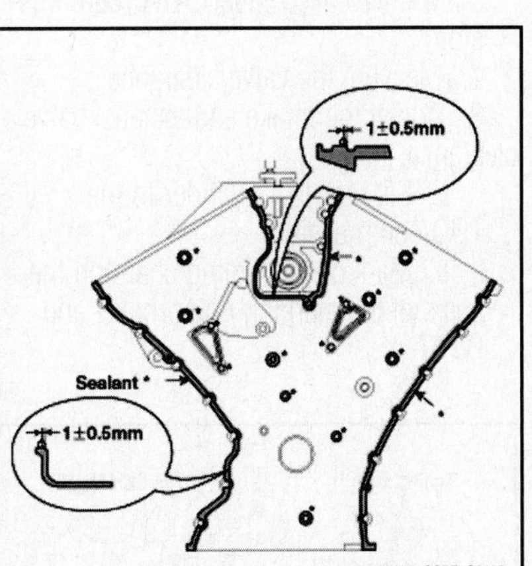

Fig. 94 Front cover sealant locations

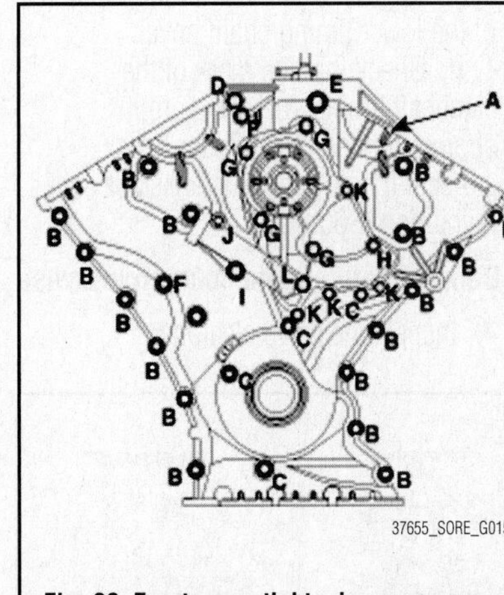

Fig. 96 Front cover tightening sequence

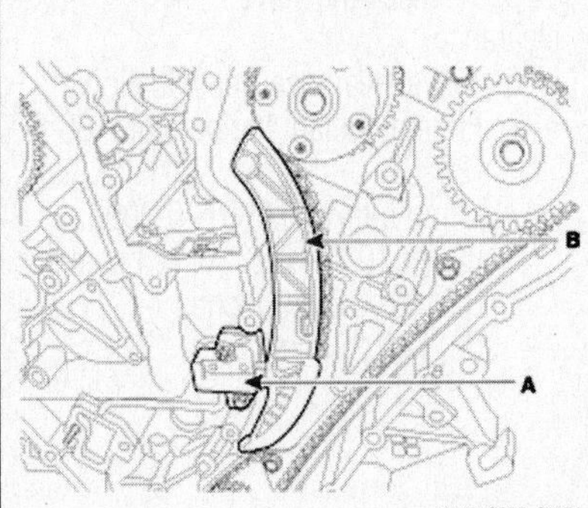

Fig. 93 Remove the left-hand timing chain auto tensioner (A) and the left-hand timing chain tensioner arm (B)

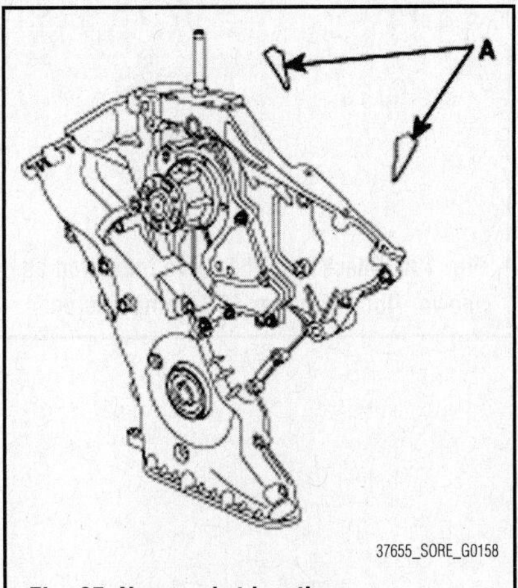

Fig. 95 New gasket locations

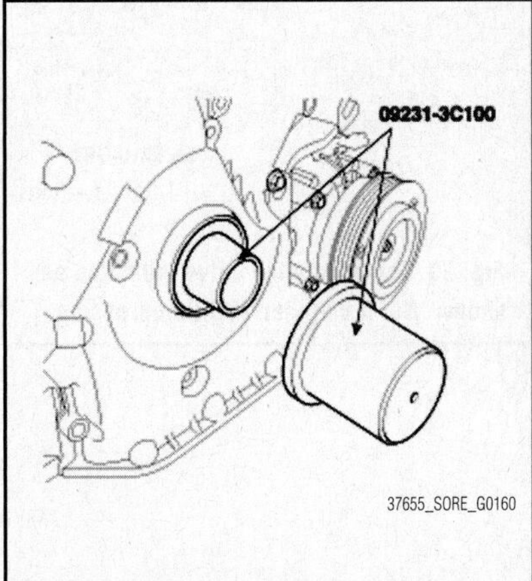

Fig. 97 Front cover oil seal installation

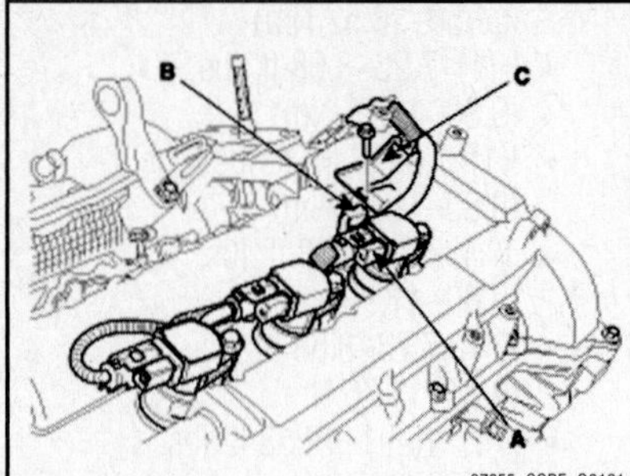

Fig. 98 Connect RH ignition coil connector (A), condenser (B) connector and install wiring bracket (C)

VALVE LASH

ADJUSTMENT

See Figures 99 and 100.

1. Remove the engine cover.
2. Remove the cylinder head covers.
3. Set No.1 cylinder to TDC/compression.
   a. Turn the crankshaft pulley and align its groove with the timing mark "T" of the lower timing chain cover.
   b. Check that the mark of the camshaft timing sprockets are in straight line on the cylinder head surface. If not, turn the crankshaft one revolution (360°)

➡Do not rotate engine counterclockwise

4. Inspect the valve clearance.

Fig. 99 Check only the valve indicated as shown. No. 1 cylinder TDC/Compression

   a. Check only the valve indicated as shown. (No. 1 cylinder : TDC/Compression).
5. Measure the valve clearance.
   a. Using a thickness gauge, measure the clearance between the tappet and the base circle of camshaft.
   b. Record the out-of-specification valve clearance measurements. They will be used later to determine the required replacement adjusting tappet.
   c. Valve clearance specification:
      - Engine coolant temperature: 68°F (20°C)
      - Limit: Intake: 0.0067–0.0090 inches (0.17–0.23 mm); Exhaust: 0.0106–0.0129 inches (0.27–0.33 mm)
6. Turn the crankshaft pulley one revolution (360°) and align the groove with timing mark "T" of the lower timing chain cover.

➡Do not rotate engine counterclockwise

   a. Check only valves indicated as shown. [NO. 4 cylinder : TDC/compression].
7. Measure the valve clearance.
8. Adjust the intake and exhaust valve clearance.
   a. Set the No.1 cylinder to the TDC/compression.
   b. Mark on the timing chain on the basis of the marking on sprocket and CVVT.

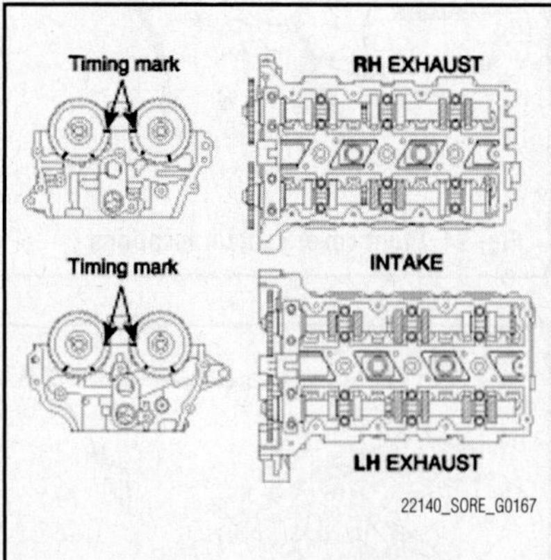

Fig. 100 Check only the valve indicated as shown. No. 4 cylinder TDC/Compression

   c. Remove the timing chain.
   d. Remove the camshaft bearing caps.
   e. Remove the camshaft assembly.
   f. Remove the Mechanical Lash Adjusters (MLA).
   g. Measure the thickness of the removed tappet using a micrometer.
   h. Calculate the thickness of a new tappet so that the valve clearance comes within the specified value.
   i. Valve clearance:
      - Engine coolant temperature: 68° (20°C)
      - T: Thickness of removed tappet
      - A: Measured valve clearance
      - N: Thickness of new tappet
      - Intake: N = T + [A – 0.0079 inches (0.20 mm)]
      - Exhaust: N = T + [A – 0.0118 inches (0.30 mm)]
   j. Select a new tappet with a thickness as close as possible to the calculated value.

➡Tappets are available in 41size increments of 0.0006 inches (0.015 mm) from 0.118 inches (3.00 mm) to 0.1417 inches (3.600 mm)

   k. Place a new tappet on the cylinder head.

➡Appling engine oil at the selected tappet on the periphery and top surface.

   l. Install the intake and exhaust camshaft.
   m. Install the bearing caps.
   n. Install the timing chain.
   o. Turn the crankshaft two turns in the operating direction(clockwise) and realign crankshaft sprocket and camshaft sprocket timing marks.
   p. Recheck the valve clearance.
   q. Valve clearance Specification:
      - Engine coolant temperature: 68°F (20°C)
      - Limit: Intake: 0.0067–0.0090 inches (0.17–0.23 mm); Exhaust: 0.0106–0.0129 inches (0.27–0.33 mm)

ENGINE PERFORMANCE & EMISSION CONTROLS

ACCELERATOR PEDAL POSITION (APP) SENSOR

LOCATION

The Accelerator Pedal Position (APP) Sensor is installed on the accelerator pedal module and detects the rotation angle of the accelerator pedal.

REMOVAL & INSTALLATION

See Figure 101.

1. Turn ignition switch off and disconnect the battery cable from the battery.
2. Disconnect the accelerator position sensor connector.
3. Unfasten the mounting bolt/nuts and remove the accelerator pedal from the vehicle.

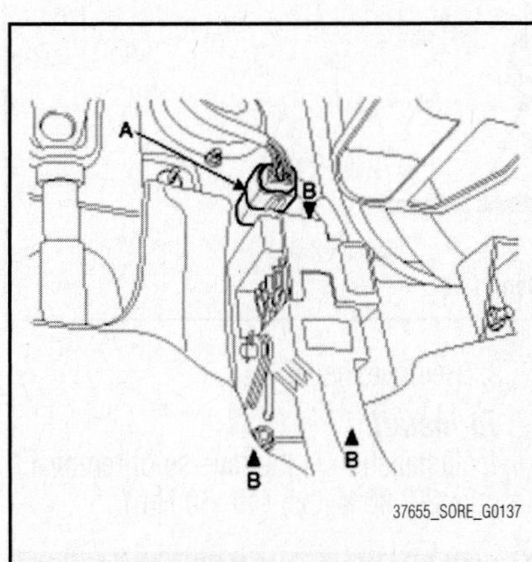

Fig. 101 Accelerator pedal removal

To install:

4. Install the accelerator pedal in according to the reverse order of removal. Tighten nuts to 6–9 ft. lbs. (8–12 Nm)

CAMSHAFT POSITION (CMP) SENSOR

LOCATION

See Figure 102.

The two Camshaft Position (CMP) sensors are installed on engine head cover of bank 1 and 2 and uses a target wheel installed on the camshaft.

REMOVAL & INSTALLATION

1. Turn ignition switch OFF.
2. Disconnect the CMP sensor connector.
3. Remove the bolt and the CMP sensor.

To install:

4. Installation is the reverse of removal.

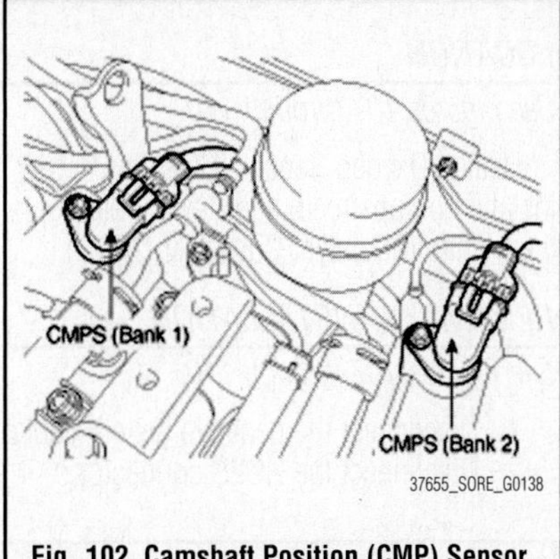

Fig. 102 Camshaft Position (CMP) Sensor

CRANKSHAFT POSITION (CKP) SENSOR

LOCATION

See Figure 103.

The Crankshaft Position (CKP) Sensor is mounted on the on transaxle housing.

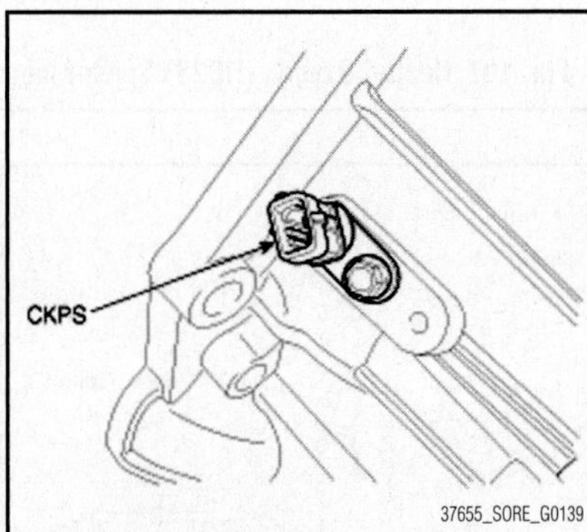

Fig. 103 Crankshaft Position (CMP) Sensor location

REMOVAL & INSTALLATION

1. Turn ignition switch OFF.
2. Disconnect the CKP sensor connector.
3. Remove the CKP sensor.

To install:

4. Installation is the reverse of removal.

ELECTRONIC CONTROL MODULE (ECM)

LOCATION

See Figure 104.

Refer to the accompanying illustration.

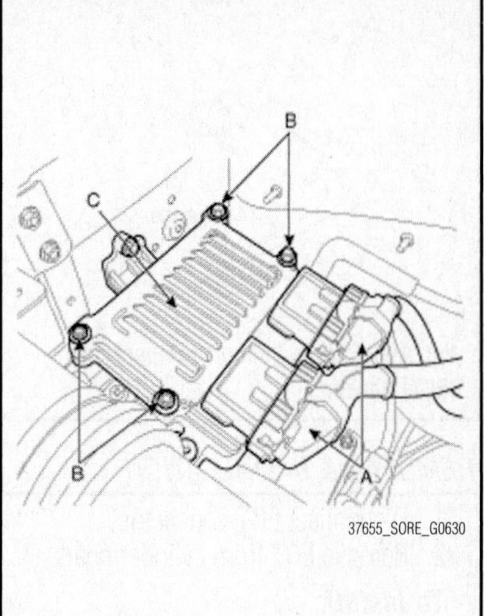

Fig. 104 Electronic Control Module (ECM) location with connectors (A) retaining bolts (B) and bracket (C).

REMOVAL & INSTALLATION

1. Turn ignition switch off.
2. Disconnect the battery (-) cable from the battery.
3. Remove the resonator.
4. Disconnect the ECM connectors.
5. Unscrew the ECM mounting bolts and remove the ECM from the bracket.

To install:

6. Install new ECM

RESET

Refer to operating handbook for scan tool to program ECM.

VIN (Vehicle Identification Number) is a number that has the vehicle's information (Maker, Vehicle Type, Vehicle Line/Series, Body Type, Engine Type, Transmission Type, Model Year, Plant Location and so forth. When replacing an ECM, the VIN must be programmed in the ECM. If there is no VIN in ECM memory, the fault code (DTC P0630) is set.

ENGINE COOLANT TEMPERATURE (ECT) SENSOR

LOCATION

See Figure 105.

Engine Coolant Temperature Sensor (ECTS) is located in the engine coolant passage of the cylinder head.

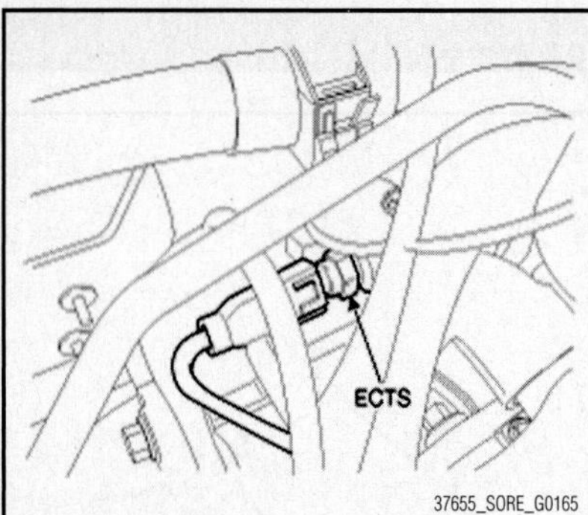

Fig. 105 Engine Coolant Temperature (ECT) Sensor

REMOVAL & INSTALLATION

1. Disconnect ECT connector.
2. Remove ECT from cylinder head.

To install:

3. Install ECT.
4. Install connector.

EVAPORATIVE EMISSIONS (EVAP) CANISTER

LOCATION

See Figure 106.

Refer to the accompanying illustration.

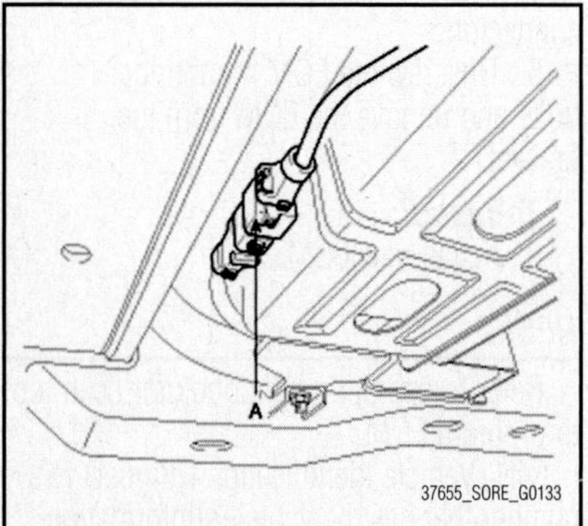

Fig. 106 Evaporative Emissions (EVAP) Canister location

REMOVAL & INSTALLATION

1. Disconnect the canister close valve connector.
2. Remove the cover and disconnect the vacuum hose.
3. Disconnect the vacuum hoses.
4. Unscrew the mounting nuts and remove the canister assembly.

To install:

5. Installation is in reverse order of removal.

HEATED OXYGEN SENSOR (HO2S)

LOCATION

See Figures 107 through 110.

Heated Oxygen Sensor (HO2S) is installed on upstream and downstream of the Manifold Catalyst Converter (MCC).

REMOVAL & INSTALLATION

See Figures 107 through 110.

1. Disconnect the negative battery cable.
2. Disconnect the HO2S connector.

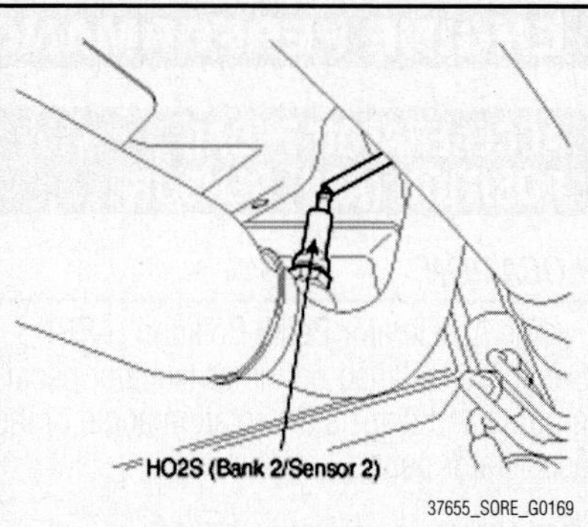

Fig. 110 Heated Oxygen (HO2S) Sensor bank 2 sensor 2

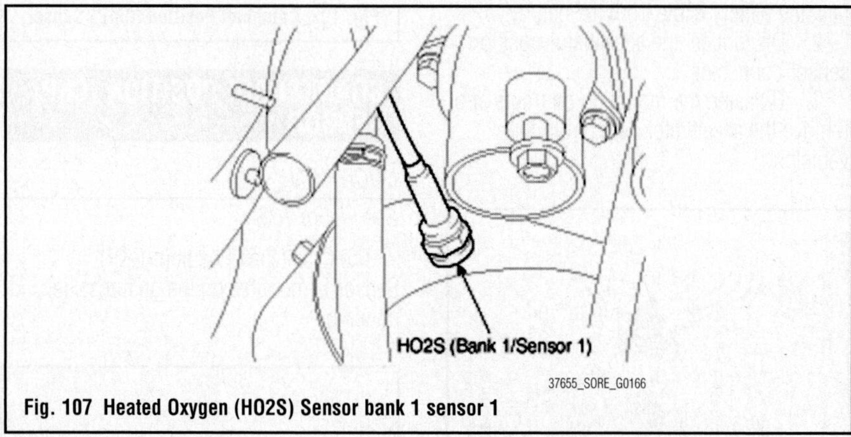

Fig. 107 Heated Oxygen (HO2S) Sensor bank 1 sensor 1

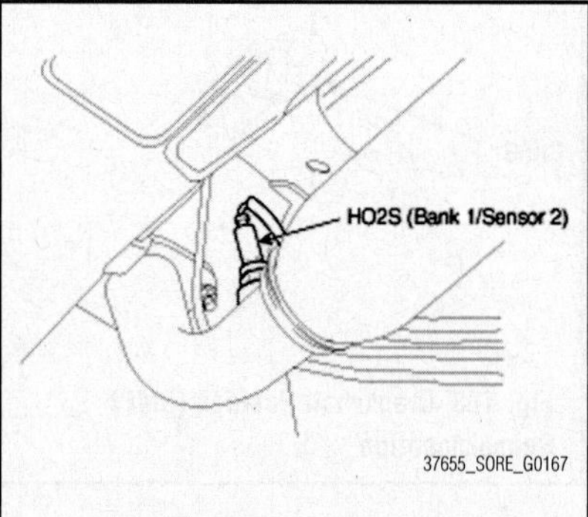

Fig. 108 Heated Oxygen (HO2S) Sensor bank 1 sensor 2

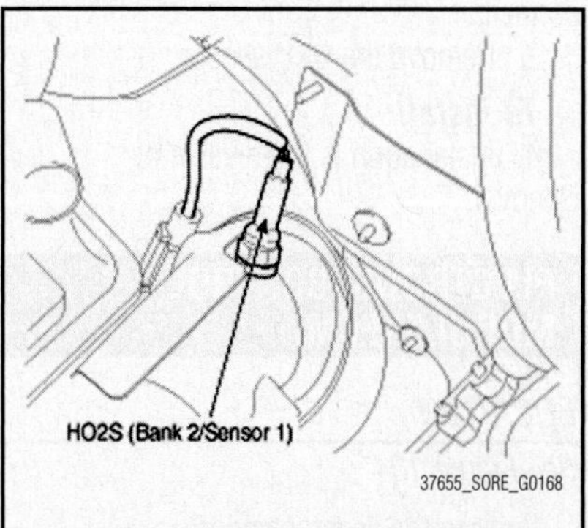

Fig. 109 Heated Oxygen (HO2S) Sensor bank 2 sensor 1

3. Remove the HO2S.

To install:

4. Installation is the reverse of removal. Torque: 29–36 ft. lbs. (40–50 Nm).

INTAKE AIR TEMPERATURE (IAT) SENSOR

LOCATION

See Figure 112.

The Intake Air Temperature (IAT) sensor is installed inside the Mass Air Flow (MAF) sensor and detects the intake air temperature.

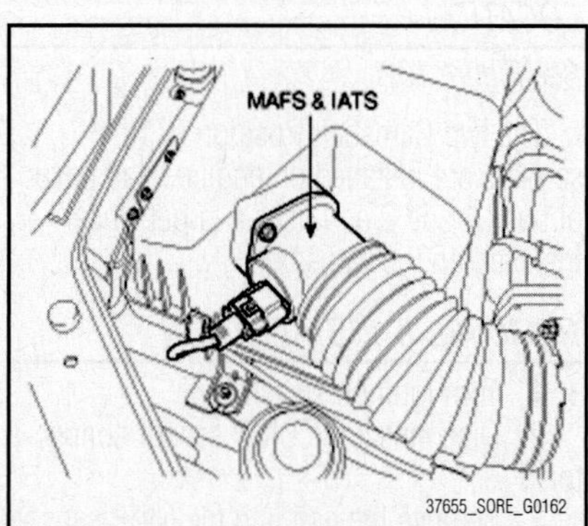

Fig. 112 Intake Air Temperature (IAT) Sensor & Mass Air Flow (MAF) sensor location

REMOVAL & INSTALLATION

1. Disconnect the negative battery cable.
2. Disconnect the IAT sensor connector.
3. Remove the IAT sensor from intake tube.

To install:

4. Installation is the reverse of removal.

KNOCK SENSOR (KS)

LOCATION

See Figures 113 and 114.

The Knock Sensor (KS) senses engine knocking and the two sensors are installed inside the V-valley of the cylinder block.

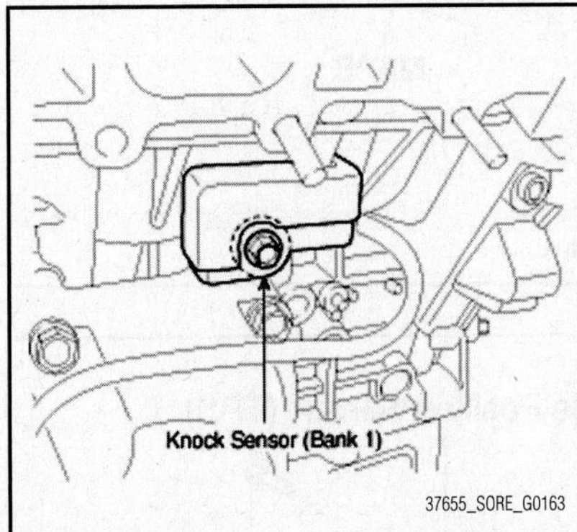

Fig. 113 Knock Sensor (KS) location bank 1

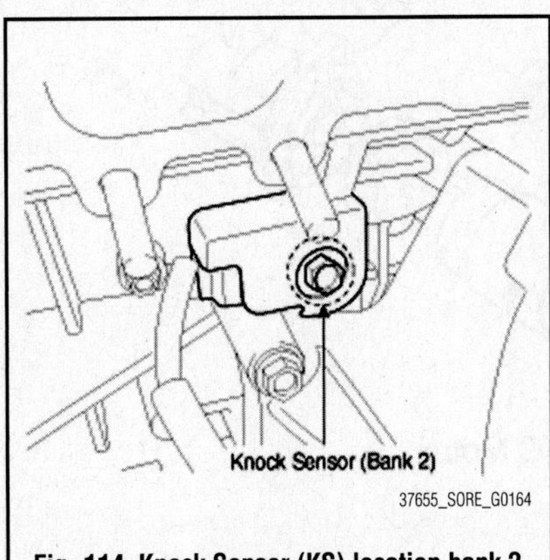

Fig. 114 Knock Sensor (KS) location bank 2

REMOVAL & INSTALLATION

1. Turn the ignition off.
2. Remove intake manifold. Refer to Intake Manifold in Engine Mechanical.
3. Disconnect the Knock Sensor (KS) connectors.
4. Remove the KS.

To install:

5. Installation is the reverse of removal.

MALFUNCTION INDICATOR LIGHT (MIL)

RESET PROCEDURE

Clearing codes resets MIL.

MASS AIR FLOW (MAF) SENSOR

LOCATION

See Figure 112.

The Mass Air Flow (MAF) sensor is a hot-film type sensor and is located in between the air cleaner and the throttle body.

REMOVAL & INSTALLATION

1. Disconnect the negative battery cable.
2. Disconnect and remove the MAF sensor from the air cleaner tube.

To install:

3. Installation is the reverse of removal.

MANIFOLD ABSOLUTE PRESSURE (MAP) SENSOR

LOCATION

See Figure 115.

The Manifold Absolute Pressure (MAP) sensor is speed-density type sensor and is installed on the surge tank.

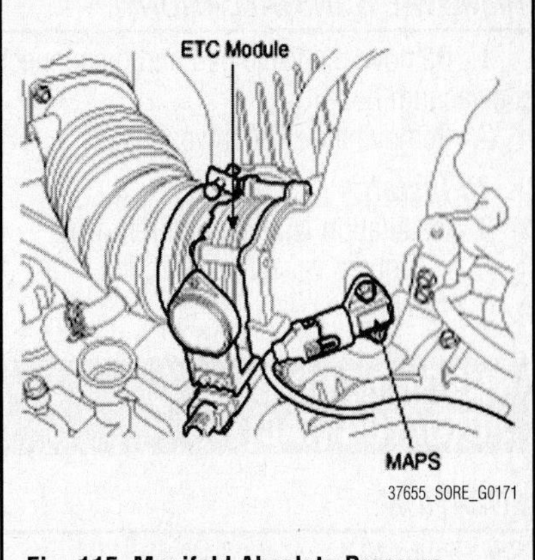

Fig. 115 Manifold Absolute Pressure (MAP) Sensor location

REMOVAL & INSTALLATION

1. Disconnect the negative battery cable.
2. Disconnect the MAP sensor connector
3. Remove the MAP sensor.

To install:

4. Installation is the reverse of removal.

POSITIVE CRANKCASE VENTILATION (PCV) VALVE

LOCATION

See Figure 116.

Refer to the accompanying illustration.

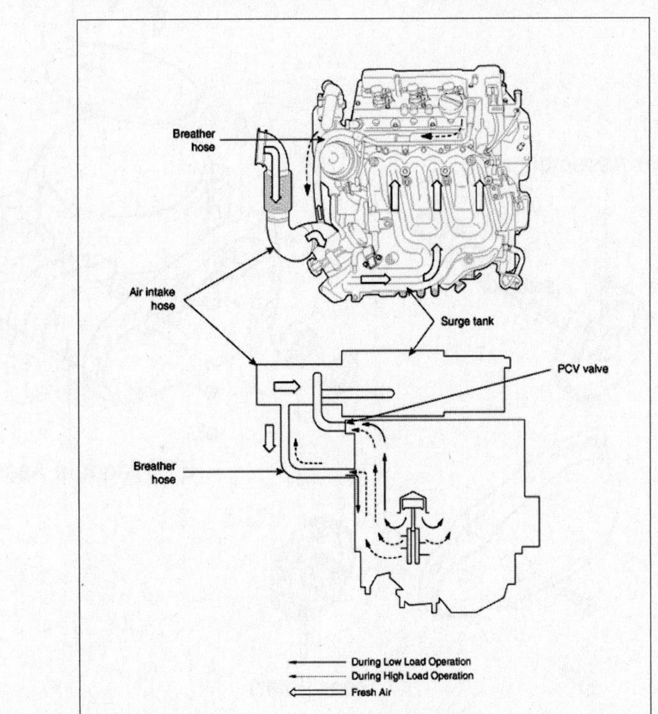

Fig. 116 Positive Crankcase Ventilation (PCV) Valve location

REMOVAL & INSTALLATION

1. Remove the valve pad and disconnect the vacuum hose.
2. Remove the PCV valve.

To install:

3. Installation is reverse of removal.
   a. Tighten valve to 6–9 ft. lbs. (8–12 Nm).

THROTTLE CONTROL ACTUATOR (TAC)

LOCATION

See Figure 117.

Refer to the accompanying illustration.

REMOVAL & INSTALLATION

See Figure 118.

1. Disconnect electrical connectors.
2. Remove cooling hoses.
3. Remove retaining nuts and bolts.
4. Remove throttle body assembly.

To install:

5. Installation is reverse of removal.

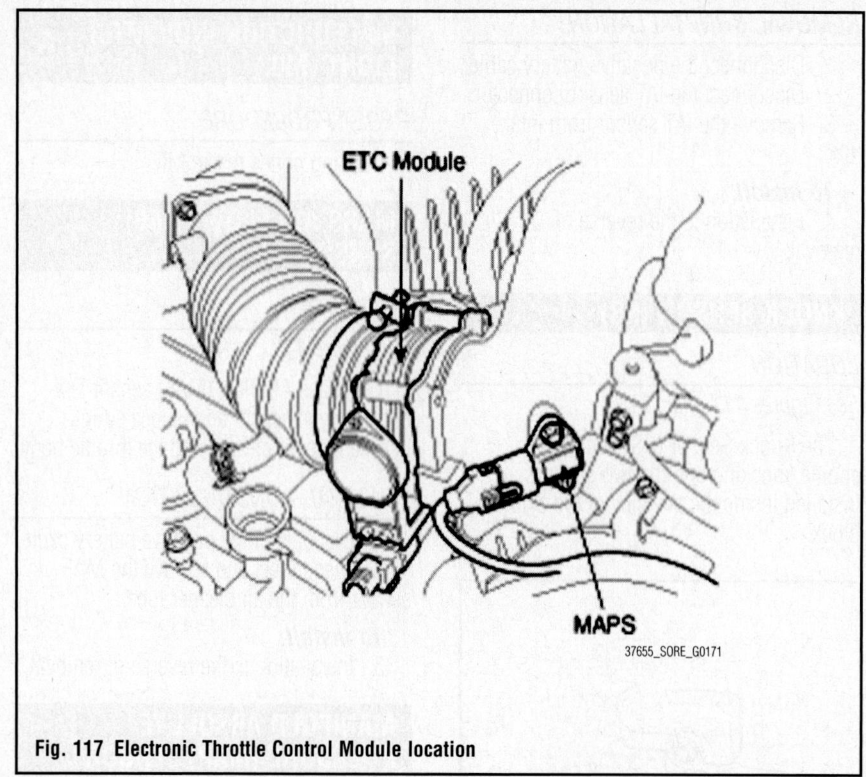

Fig. 117 Electronic Throttle Control Module location

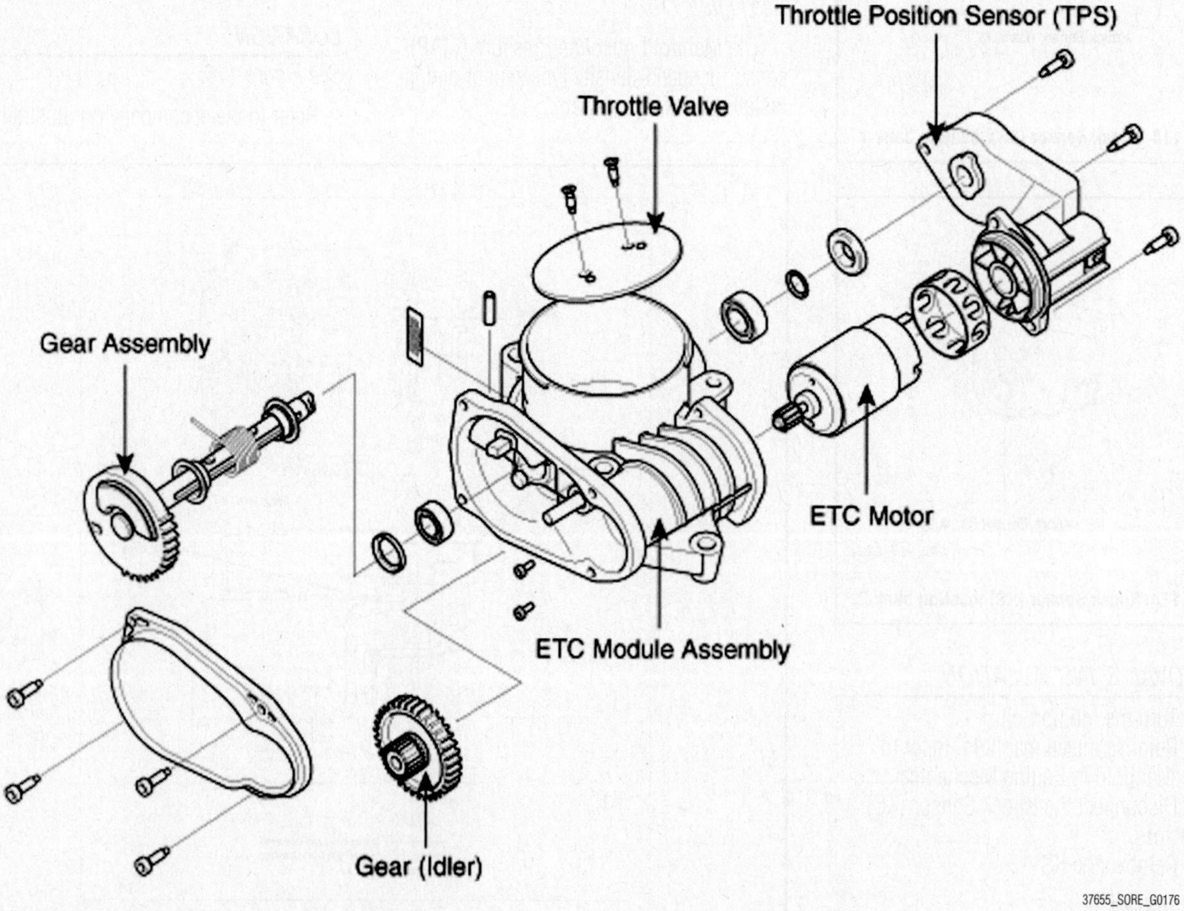

Fig. 118 Electronic Throttle Control Module components

THROTTLE POSITION SENSOR (TPS)

LOCATION

See Figure 119.

Refer to the accompanying illustration.

REMOVAL & INSTALLATION

See Figure 120.

1. Disconnect electrical connectors.
2. Remove cooling hoses.
3. Remove retaining nuts and bolts.
4. Remove throttle body assembly.

To install:

5. Installation is reverse of removal.

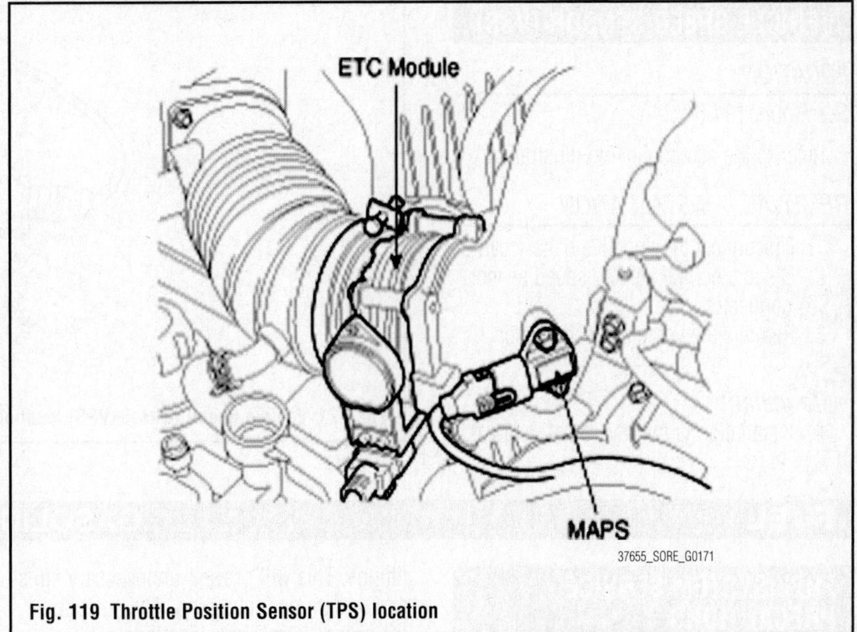

Fig. 119 Throttle Position Sensor (TPS) location

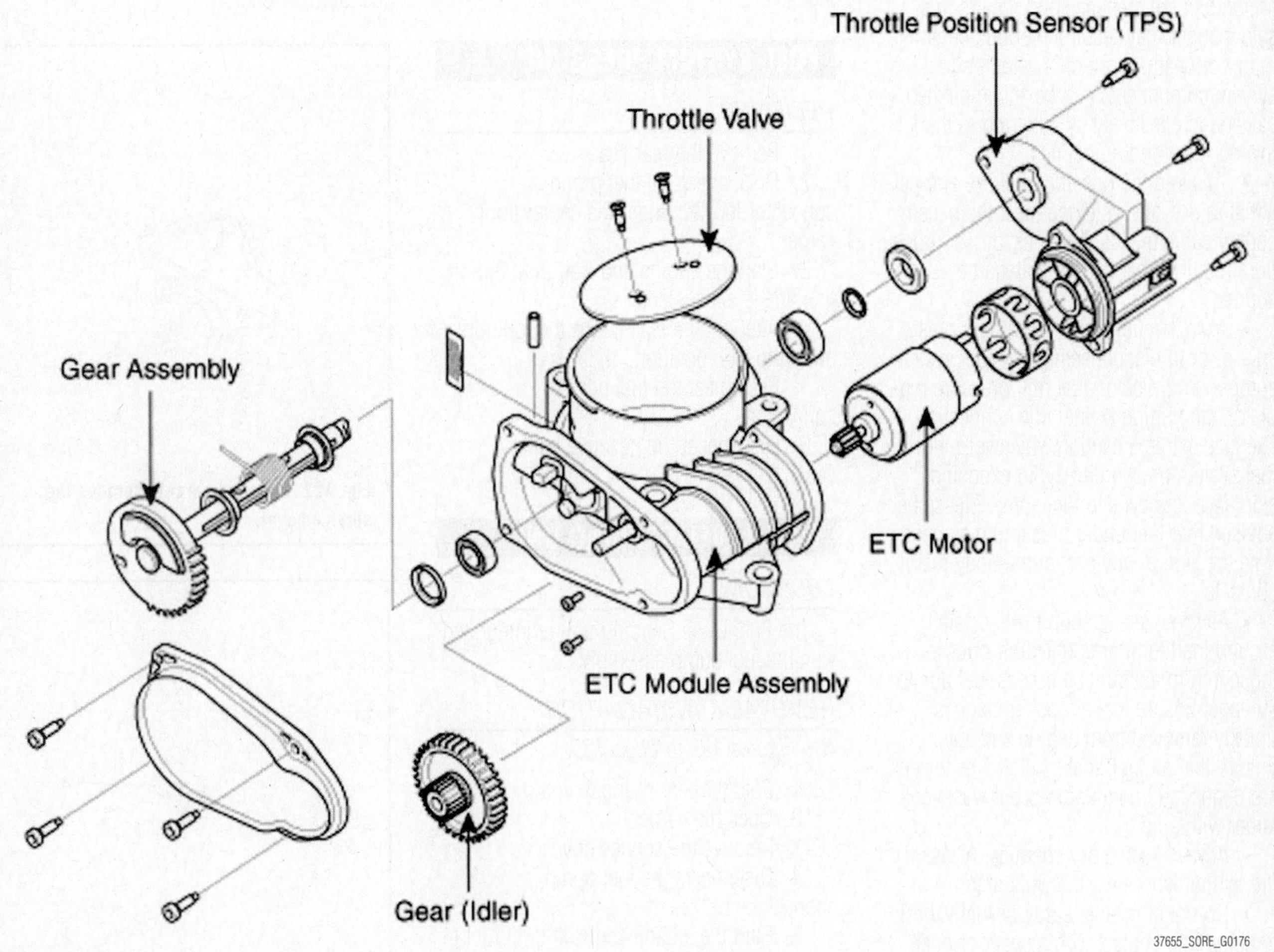

Fig. 120 Throttle Position Sensor (TPS) components

VEHICLE SPEED SENSOR (VSS)

LOCATION

See Figure 121.

Refer to the accompanying illustration.

REMOVAL & INSTALLATION

1. Disconnect the negative battery cable.
2. Disconnect the vehicle speed sensor (VSS) connector
3. Remove the vehicle speed sensor (VSS).

To install:

4. Installation is the reverse of removal.

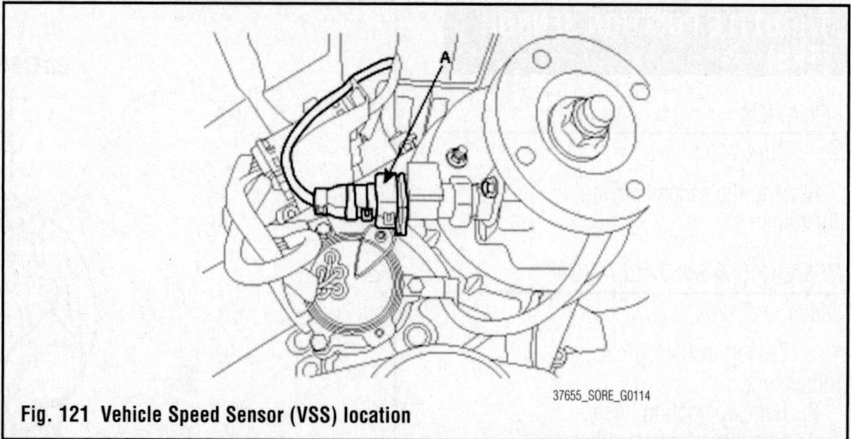

Fig. 121 Vehicle Speed Sensor (VSS) location

FUEL

GASOLINE FUEL INJECTION SYSTEM

FUEL SYSTEM SERVICE PRECAUTIONS

Safety is the most important factor when performing not only fuel system maintenance, but any type of maintenance. Failure to conduct maintenance and repairs in a safe manner may result in serious personal injury or death. Work on a vehicle's fuel system components can be accomplished safely and effectively by adhering to the following rules and guidelines.

- To avoid the possibility of fire and personal injury, always disconnect the negative battery cable unless the repair or test procedure requires that battery voltage be applied.
- Always relieve the fuel system pressure prior to disconnecting any fuel system component (injector, fuel rail, pressure regulator, etc.) fitting or fuel line connection. Exercise extreme caution whenever relieving fuel system pressure to avoid exposing skin, face and eyes to fuel spray. Please be advised that fuel under pressure may penetrate the skin or any part of the body that it contacts.
- Always place a shop towel or cloth around the fitting or connection prior to loosening to absorb any excess fuel due to spillage. Ensure that all fuel spillage is quickly removed from engine surfaces. Ensure that all fuel-soaked cloths or towels are deposited into a flame-proof waste container with a lid.
- Always keep a dry chemical (Class B) fire extinguisher near the work area.
- Do not allow fuel spray or fuel vapors to come into contact with a spark or open flame.
- Always use a second wrench when loosening or tightening fuel line connection fittings. This will prevent unnecessary stress and torsion on fuel piping. Always follow the proper torque specifications.
- Always replace worn fuel fitting O-rings with new ones. Do not substitute fuel hose where rigid pipe is installed.

FUEL SYSTEM PRESSURE

RELIEVING

1. Remove the rear seat.
2. Disconnect the fuel pump connector located under the center floor carpet..
3. Start the engine and wait until fuel in fuel line is exhausted.
4. After engine stalls, turn the ignition switch to OFF position.
5. Reconnect the fuel pump connector.
6. Disconnect the negative battery cable.

FUEL LEVEL SENDING UNIT

LOCATION

The Fuel Level Sending unit is integrated with the fuel pump assembly.

REMOVAL & INSTALLATION

See Figures 122 through 125.

1. Fold the rear seat cushion.
2. Open the carpet.
3. Remove the service cover.
4. Disconnect the fuel pump connector.
5. Start the engine and wait until fuel in fuel line is exhausted.
6. After engine stops, turn the ignition switch off.
7. Disconnect the fuel feed quick-connector and the fuel tank pressure sensor connector.
8. Remove the rubber cover.
9. Unscrew the fuel pump plate cover with the special service tool and remove the fuel pump assembly.

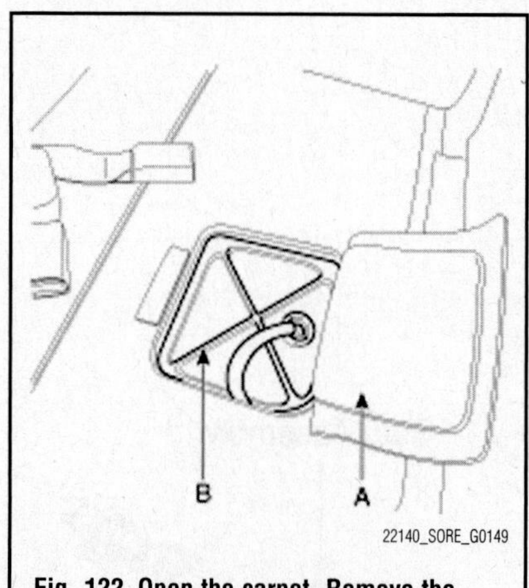

Fig. 122 Open the carpet. Remove the service cover.

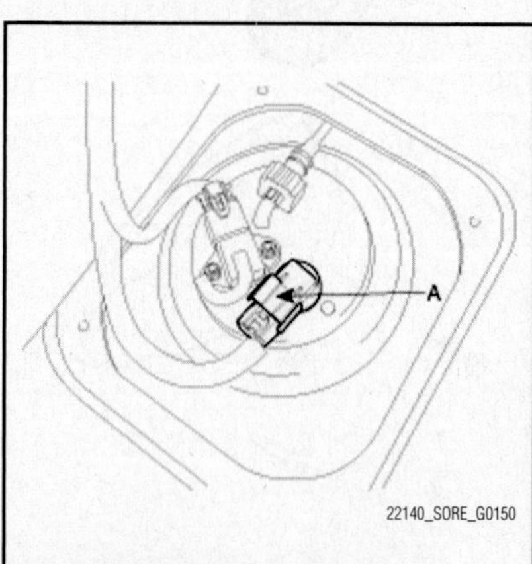

Fig. 123 Disconnect the fuel pump connector (A)

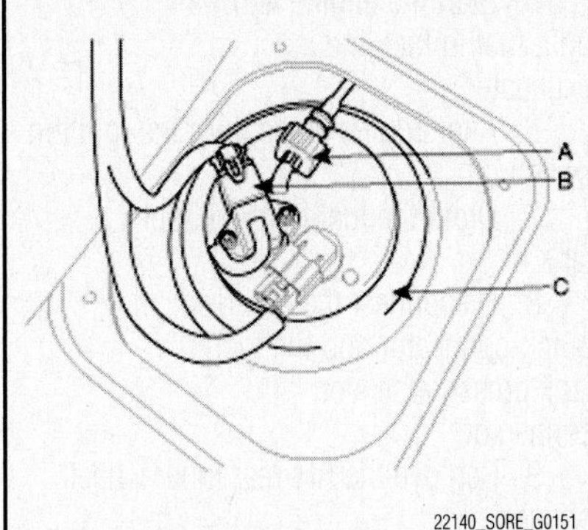

Fig. 124 Disconnect the fuel feed quick-connector (A) and the fuel tank pressure sensor connector (B). Remove the rubber cover (C).

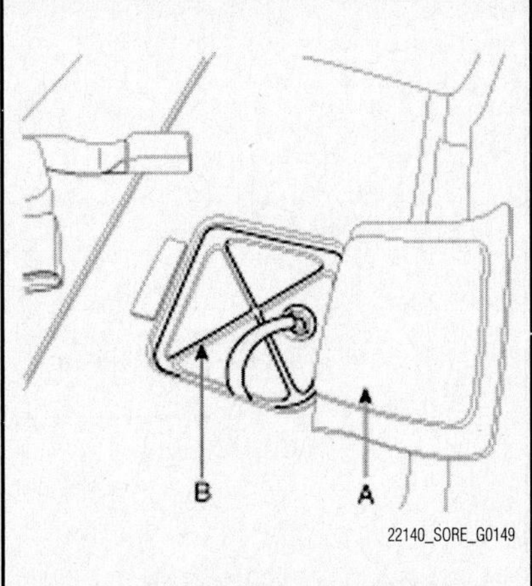

Fig. 126 Open the carpet (A). Remove the service cover (B).

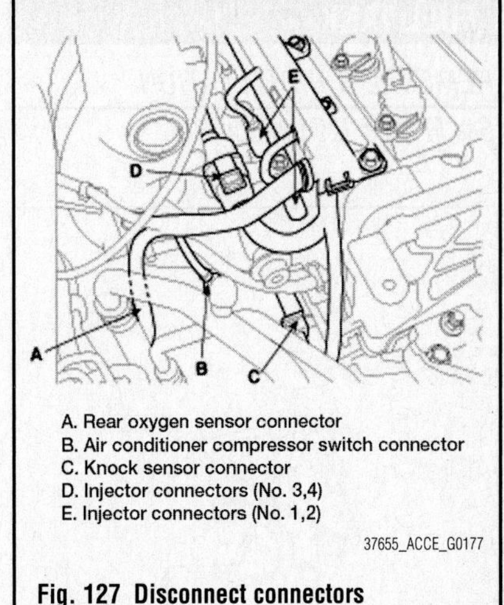

Fig. 127 Disconnect connectors

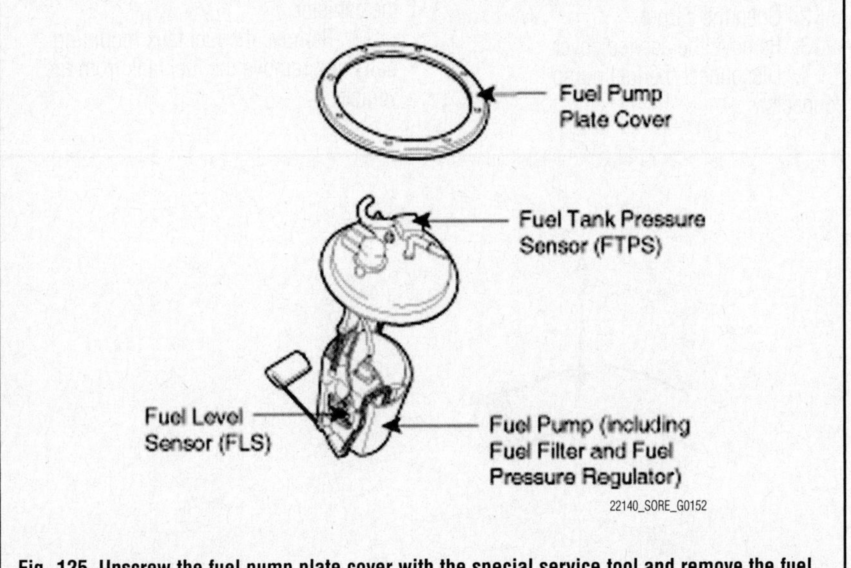

Fig. 125 Unscrew the fuel pump plate cover with the special service tool and remove the fuel pump assembly.

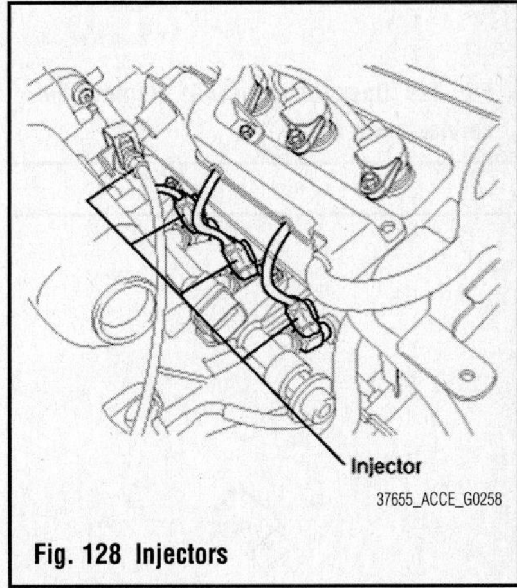

Fig. 128 Injectors

To install:

10. Installation is reverse of removal.

FUEL PUMP

REMOVAL & INSTALLATION

See Figures 123 through 126.

1. Fold the rear seat cushion.
2. Open the carpet.
3. Remove the service cover.
4. Disconnect the fuel pump connector.
5. Start the engine and wait until fuel in fuel line is exhausted.
6. After engine stops, turn the ignition switch off.
7. Disconnect the fuel feed quick-connector and the fuel tank pressure sensor connector.
8. Remove the rubber cover.
9. Unscrew the fuel pump plate cover with the special service tool and remove the fuel pump assembly.

To install:

10. Installation is reverse of removal.

FUEL PRESSURE REGULATOR

REMOVAL & INSTALLATION

See Fuel Pump Removal & Installation. The fuel pressure regulator is located inside the fuel pump.

FUEL RAIL AND INJECTOR

REMOVAL & INSTALLATION

See Figures 127 and 128.

1. Before servicing the vehicle, refer to the Precautions Section.
2. Disconnect the negative battery cable.
3. Relieve the fuel system pressure. Refer to Relieving Fuel System Pressure.
4. Remove the air intake surge tank, if necessary.
5. Remove the fuel lines.
6. Disconnect the fuel injector connectors.
7. Remove the fuel rail.
8. Separate the injectors from the supply manifold.

To install:

9. Install the injectors to the fuel supply manifold using new O-rings.
10. Install the fuel rail.
11. Install the fuel injector connectors.
12. Install the fuel lines.
13. Install the air intake surge tank, if removed.
14. Connect the negative battery cable.
15. Start the engine and check for leaks.

FUEL TANK

REMOVAL & INSTALLATION

See Figures 129 through 135.

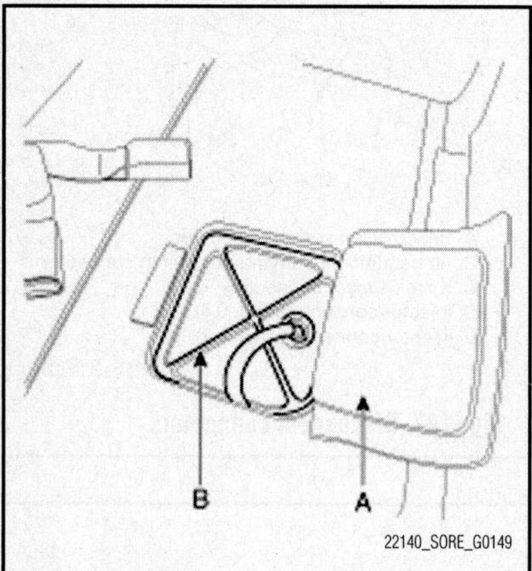

Fig. 129 Open the carpet (A). Remove the service cover (B).

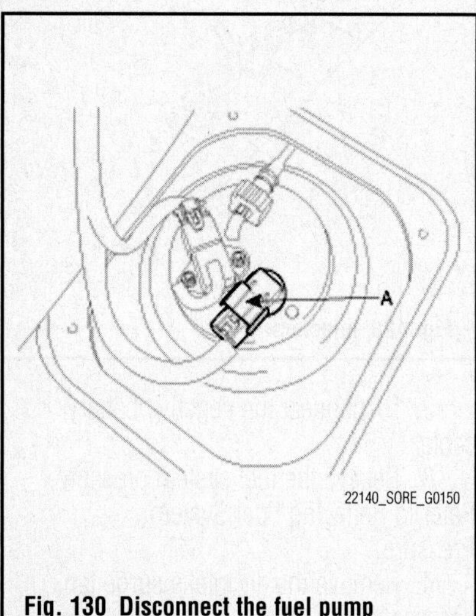

Fig. 130 Disconnect the fuel pump connector (A).

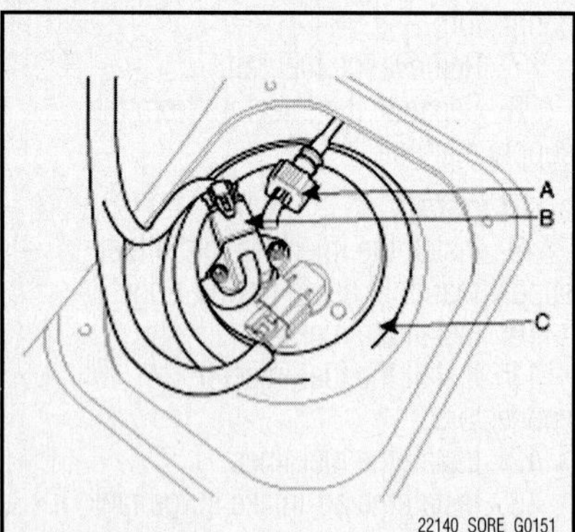

Fig. 131 Disconnect the fuel feed quick-connector (A) and the fuel tank pressure sensor connector (B). Remove the rubber cover (C).

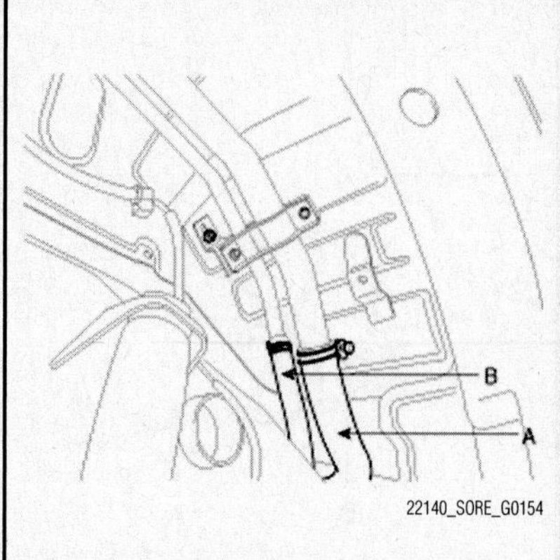

Fig. 132 Disconnect the fuel filler hose (A) and the vacuum hose (B).

1. Fold the rear seat cushion.
2. Open the carpet.
3. Remove the service cover.
4. Disconnect the fuel pump connector.
5. Start the engine and wait until fuel in fuel line is exhausted.
6. After engine stops, turn the ignition switch off.
7. Drain the fuel from the fuel tank.
8. Disconnect the fuel feed quick-connector and the fuel tank pressure sensor connector.
9. Remove the RH-rear inner wheel house.
10. Disconnect the fuel filler hose and the vacuum hose connected with the canister air filter.
11. Lift the vehicle and support the fuel tank with a jack.
12. Disconnect the vacuum hoses from the canister.
13. Remove the fuel tank mounting bolts and remove the fuel tank from the vehicle.

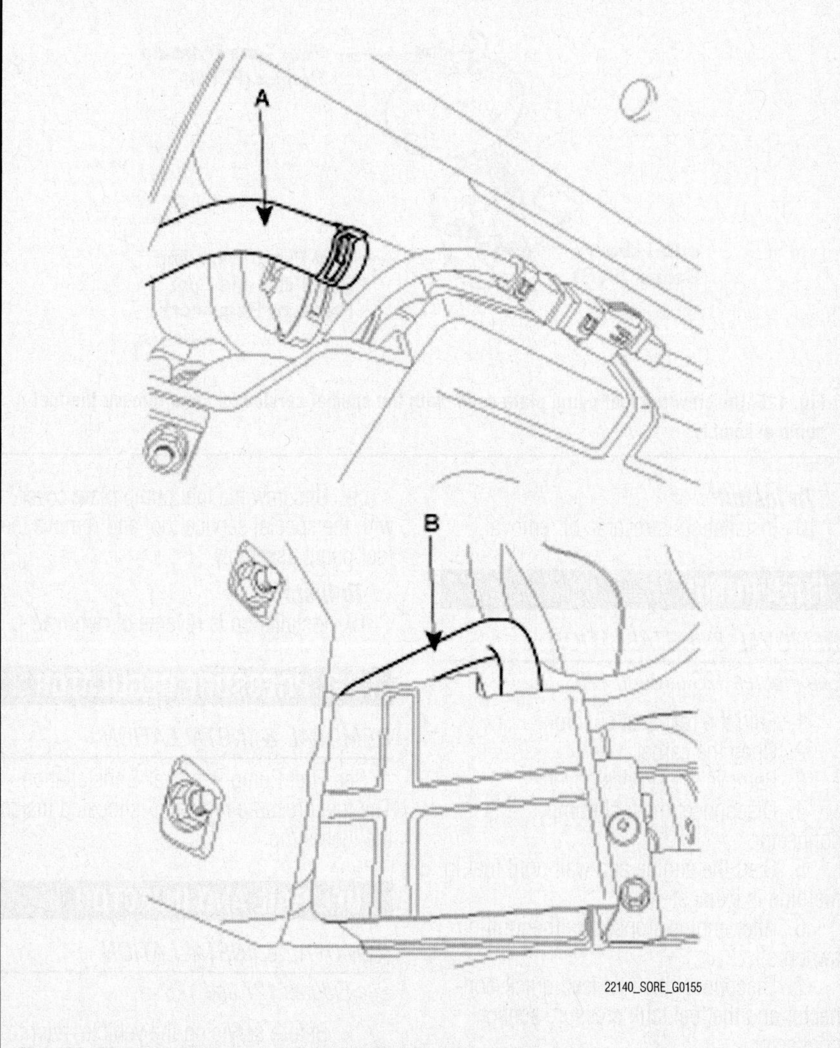

Fig. 133 Disconnect the vacuum hoses (A,B) from the canister.

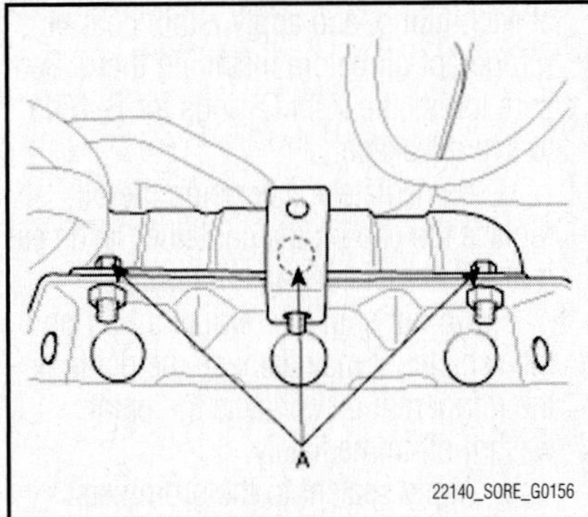

Fig. 134 Remove the fuel tank mounting bolts (A).

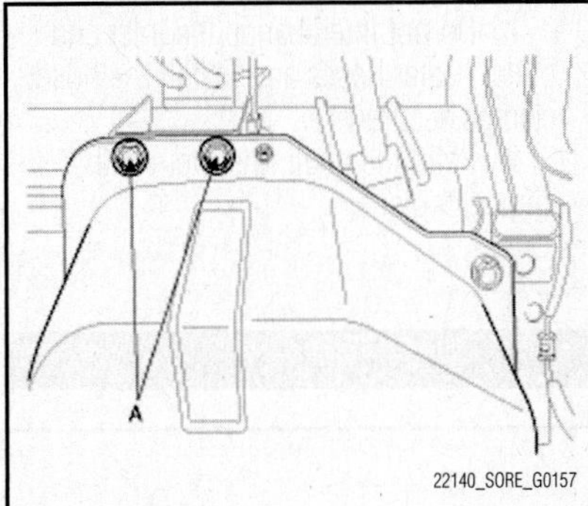

Fig. 135 Remove the fuel tank mounting bolts (A).

To install:

14. Installation is reverse of removal.
15. Torque: Fuel tank installation bolts: 36.2–43.4 ft. lbs. (49.1–58.9 Nm).

IDLE SPEED

ADJUSTMENT

Idle speed adjustment is not necessary or possible.

THROTTLE BODY

REMOVAL & INSTALLATION

See Figure 136.

1. Before servicing the vehicle, refer to the Precautions Section.
2. Turn the ignition **OFF**.
3. Remove the engine cover.
4. Remove the throttle body electrical connector.
5. Remove the throttle body bolts.
6. Remove the throttle body and gasket.

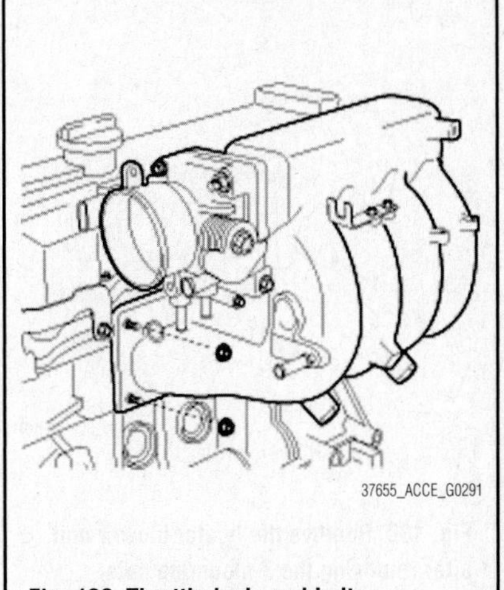

Fig. 136 Throttle body and bolts

To install:

7. Clean the throttle body gasket mating surfaces.
8. Install the throttle body and NEW gasket.
9. Install the throttle body bolts and tighten to 21 ft. lbs. (28 Nm).
10. Install the throttle body electrical connector.
11. Install the engine cover.

HEATING & AIR CONDITIONING SYSTEM

BLOWER MOTOR

REMOVAL & INSTALLATION

See Figure 137.

1. Disconnect the negative battery terminal.
2. Disconnect the connector of the blower motor.
3. Remove the blower motor after removing the mounting screws.

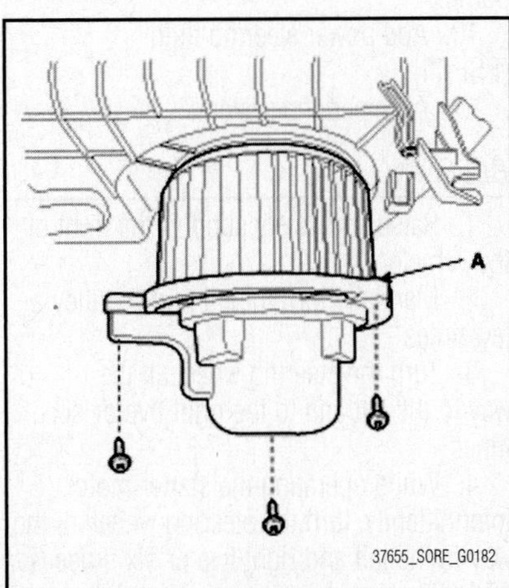

Fig. 137 Blower Motor location

To install:

4. Installation is the reverse order of removal.

HEATER CORE

REMOVAL & INSTALLATION

See Figures 138 through 140.

1. Disconnect the negative battery terminal.
2. Recover the refrigerant with a recovery/ recycling/ charging station.
3. Drain the engine coolant from the radiator.
4. Remove the bolts and the expansion valve from the evaporator core.

➡Plug or cap the lines immediately after disconnecting them to avoid moisture and dust contamination.

5. Disconnect the inlet and outlet heater hoses from the heater unit.
6. Remove the crash pad (instrument panel).
7. Remove the cowl cross bar assembly.
8. Disconnect the connectors from the temperature control actuator, the mode control actuator and the evaporator temperature sensor.
9. Remove the heater blower unit after removing the 7 mounting nuts.
10. Remove the heater core after remove the cover.

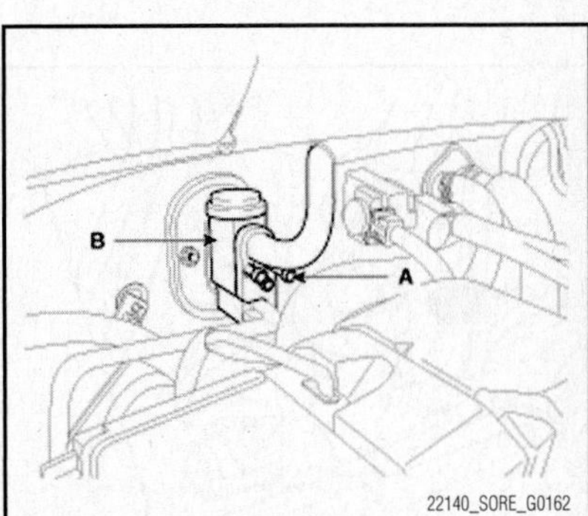

Fig. 138 Remove the bolts (A) and the expansion valve (B) from the evaporator core.

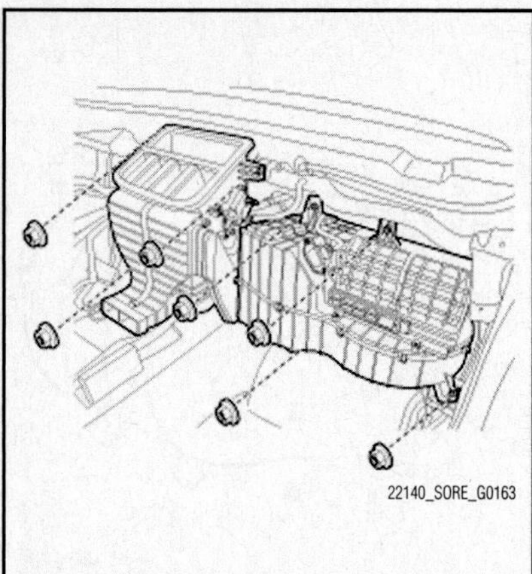

Fig. 139 Remove the heater blower unit after removing the 7 mounting nuts.

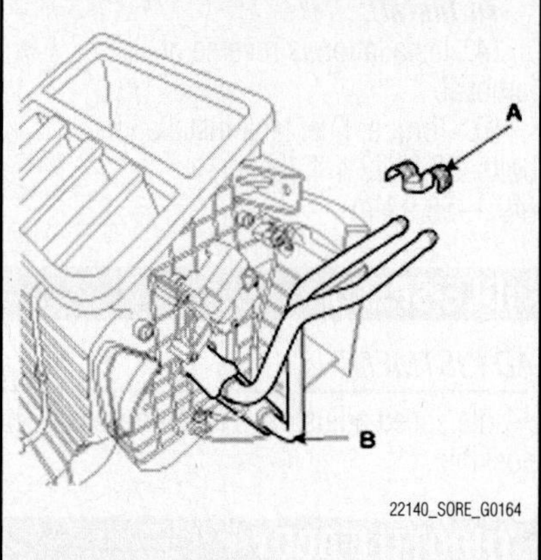

Fig. 140 Remove the heater core (B) after remove the cover (A).

11. Installation is the reverse order of removal, and note these items :

a. If you're installing a new evaporator, add refrigerant oil (ND-OIL8).

b. Replace the O-rings with new ones at each fitting, and apply a thin coat of refrigerant oil before installing them. Be sure to use the right O-rings for R-134a to avoid leakage.

c. Immediately after using the oil, replace the cap on the container, and seal it to avoid moisture absorption.

d. Do not spill the refrigerant oil on the vehicle ; it may damage the paint; if the refrigerant oil contacts the paint, wash it off immediately

e. Apply sealant to the grommets.

f. Make sure that there is no air leakage.

g. Charge the system and test its performance.

h. Do not interchange the inlet and outlet heater hoses and install the hose clamps securely.

i. Refill the cooling system with engine coolant.

STEERING

POWER RACK & PINION STEERING GEAR

REMOVAL & INSTALLATION

See Figures 141 and 142.

1. Drain the power steering fluid.
2. Disconnect the pressure tube and return tube.
3. Remove the joint assembly connecting bolt.
4. Using the special tool (0K670-321-019), disconnect the tie rod end from the knuckle arm.
5. Remove the steering gear box mounting bolts and remove the steering gear box assembly together with mounting rubber.

To install:

6. Installation is the reverse of removal.

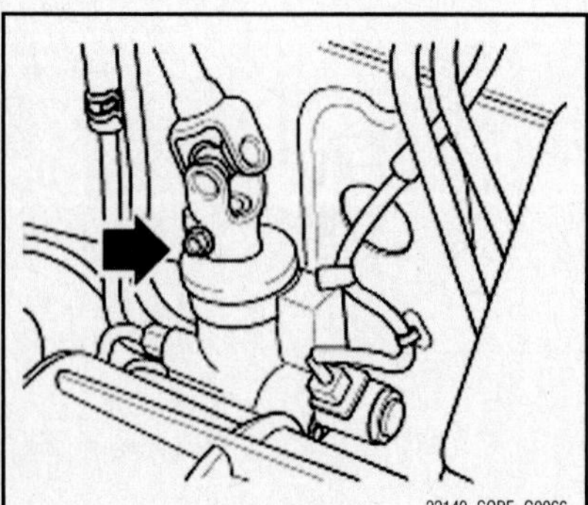

Fig. 141 Remove the joint assembly connecting bolt.

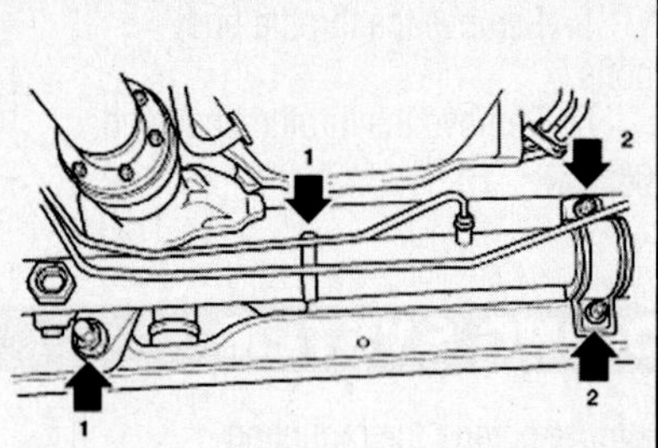

Fig. 142 Remove the steering gear box mounting bolts and remove the steering gear box assembly.

POWER STEERING PUMP

REMOVAL & INSTALLATION

See Figure 143.

1. Disconnect the pressure hose from the oil pump.
2. Disconnect the suction hose from the suction pipe, and drain the oil.
3. Remove the power steering tension adjusting bolt or flange nut.
4. Separate the belt from the power steering oil pump pulley.
5. Remove the power steering oil pump assembly.

To install:

6. Install the oil pump to the oil pump bracket. Torque: 13–16 ft. lbs. (18–23 Nm).
7. Install the belt and tighten the bolt adjusting tension.

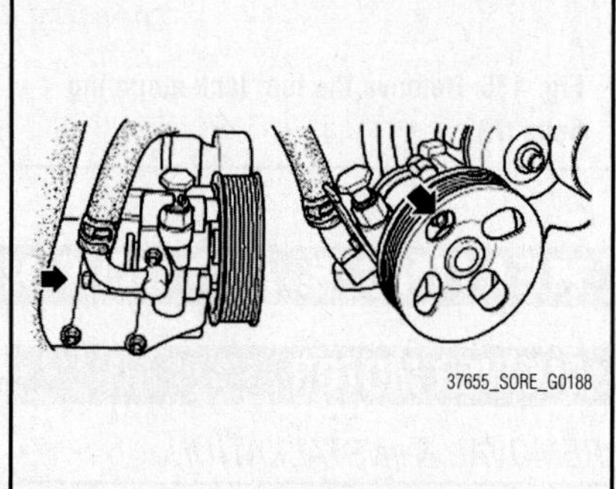

Fig. 143 Power Steering Pump bolt location

8. Install the suction hose to the oil pump.
9. Install the pressure hose to the oil pump.
10. Add power steering fluid (PSF-III).
11. Air bleed the system.

BLEEDING

1. Raise and safely support the front of the vehicle.
2. Manually turn the oil pump pulley a few times.
3. Turn the steering wheel all the way to the left and to the right five or six times.
4. While operating the starter motor intermittently, turn the steering wheel all the way to the left and right five or six times (for 15 to 20 seconds).

⁂ CAUTION

During air bleeding, replenish the fluid supply so that the level never falls below the lower position of the filter. If air bleeding is done while engine is running, the air will be broken up and absorbed into the fluid; be sure to do the bleeding only while cranking.

5. Start the engine (idling).
6. Turn the steering wheel to the left and right until there are no air bubbles in the oil reservoir.
7. Confirm that the fluid is not milky, and that the level is up to the specified position on the level gauge.
8. Confirm that there is very little change in the fluid level when the steering wheel is turned left and right.
9. Check whether or not the change in the fluid level is within 0.20 inches (5 mm) when the engine is stopped and when it is running.

⁂ CAUTION

If the change of the fluid level is 0.20 inches (5 mm) or more, the air has not been completely bled from the system, and thus must be bled completely. If the fluid level rises suddenly after the engine is stopped, the air has not been completely bled. If air bleeding is not complete, there will be abnormal noises from the pump and the flow-control valve, and this condition could cause a lessening of the life of the pump, etc.

SUSPENSION

LOWER CONTROL ARM

REMOVAL & INSTALLATION

See Figures 144 through 147.

1. Raise the front of the vehicle and support it with safety stands.
2. Remove the front wheels.
3. Remove the lower nut of control link of stabilizer bar.
4. Remove the lower nut of shock absorber.
5. Remove the bolts and nuts that joins lower arm and lower arm ball joint.
6. Remove the cotter pin and castle nut from the lower arm ball joint.
7. Remove the lower arm ball joint from the steering knuckle.
8. Remove the steering gear mounting bolts and nuts.
9. Remove the spindle from the front frame crossmember brackets during raising the steering gear box by using suitable bar.

➡Before removing the nuts of the spindles, make note of the numerical setting and mark the location on the frame bracket and plate so it can be re-installed to the same setting and location.

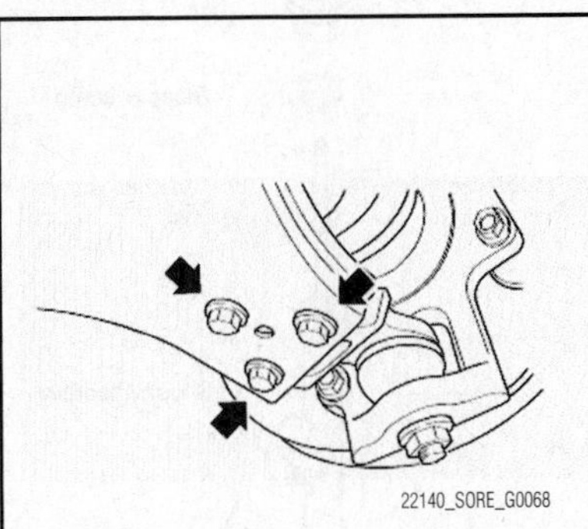

Fig. 144 Remove the bolts and nuts that joins lower arm and lower arm ball joint.

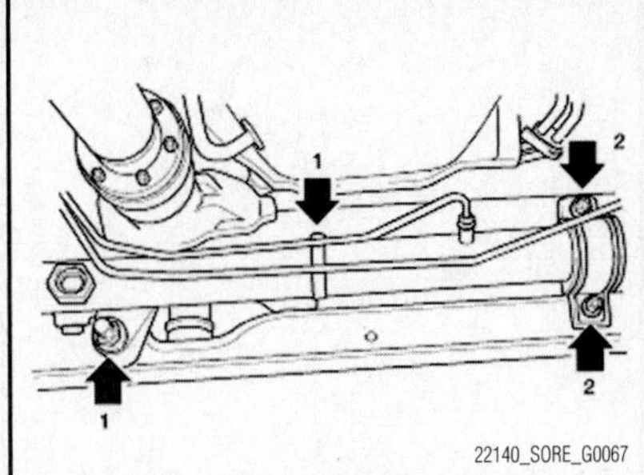

Fig. 145 Remove the steering gear box mounting bolts and nuts.

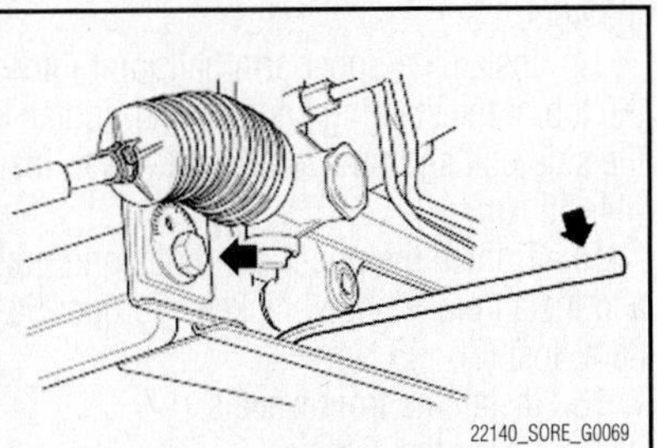

Fig. 146 Remove the spindle from the front frame crossmember brackets during raising the steering gear box by using suitable bar.

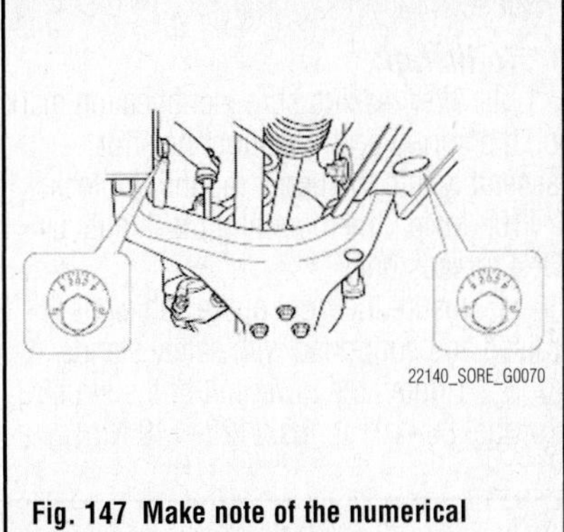

Fig. 147 Make note of the numerical setting and mark the location.

10. Remove the lower arm.
11. Remove and replace ball joints as necessary.

To install:

12. Install the lower arm ball joint to the steering knuckle. Torque: 116–145 ft. lbs. (157–196 Nm).
13. Install a new cotter pin through the castle nut.
14. Position the lower arm to the front frame crossmember brackets.
15. Position the spindle while lifting up the steering gear box by using suitable pry bar.
16. Install the lower arm spindles. Torque: 159–181 ft. lbs. (216–245 Nm).

➡Align the spindle to the numerical setting and marked location on the frame bracket and plate so the same setting and location is maintained.

17. Install the lower nut of the shock absorber. Torque: 88–101 ft. lbs. (122–140 Nm).
18. Install the lower nut of control link of stabilizer bar. Torque: 68–84 ft. lbs. (95–117 Nm).
19. Install the wheels.
20. Remove the safety stands and lower the vehicle.

➡After installation, measure the wheel alignment and adjust if necessary.

MACPHERSON STRUT

REMOVAL & INSTALLATION

See Figures 148 through 151.

1. Remove the battery.
2. Remove three strut mounting block nuts from the mounting block.
3. Raise the front of the vehicle and support it with safety stands.
4. Remove the front wheels.
5. Remove the bolt on the steering knuckle side that secures the upper arm ball joint.
6. Remove the brake hose bracket and remove the upper arm bolts and nuts.
7. Remove the strut lower nut.
8. Remove the strut assembly from the vehicle.

To install:

9. After making sure identification mark on the spring seat. Position the strut assembly into the upper mounting block.
10. Install the mounting block nuts by 3–4 threads only.
11. Insure the front of the vehicle is raised and supported with safety stands.
12. Tighten the lower nut of the strut. Torque: 88–101 ft. lbs. (122–140 Nm).

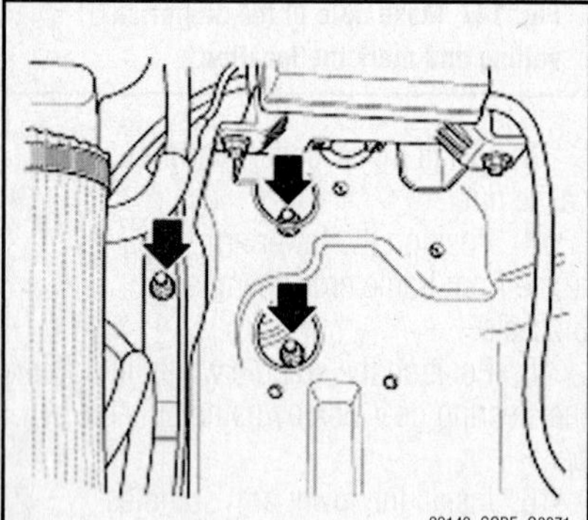

Fig. 148 Remove three strut mounting block nuts from the mounting block.

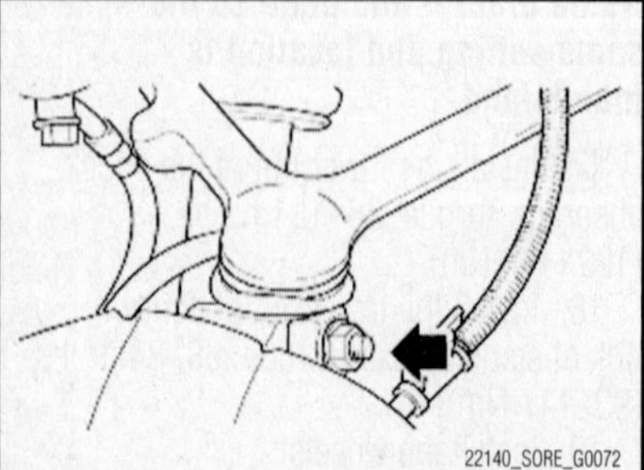

Fig. 149 Remove the bolt on the steering knuckle side that secures the upper arm ball joint.

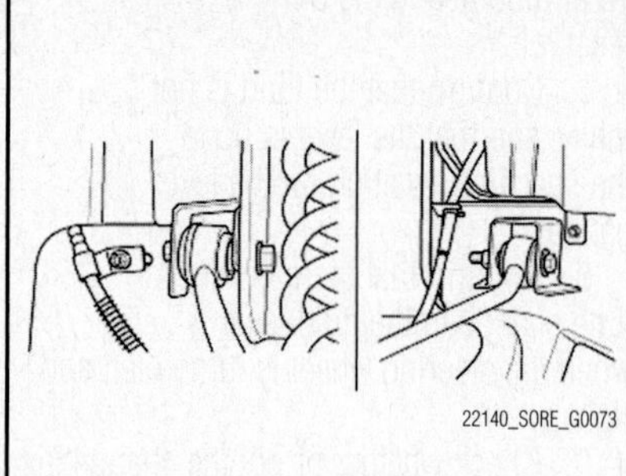

Fig. 150 Remove the brake hose bracket and remove the upper arm bolts and nuts.

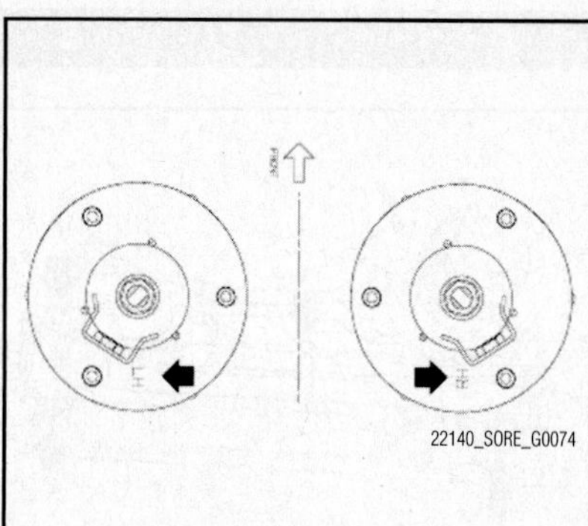

Fig. 151 Identification mark on the spring seat.

13. Position the upper arm to the frame brackets, insert the bolts and hand tighten the nuts.
14. Install the upper arm ball joint into the top of the steering knuckle and tighten the side bolt and nut. Torque: 31–39 ft. lbs. (44–55 Nm).
15. Tighten the upper arm bolts and nuts and install brake hose brackets. Torque: 54–68 ft. lbs. (76–95 Nm).
16. Install the front wheels.
17. Lower the vehicle.
18. Tighten the mounting block nuts. Torque: 31–39 ft. lbs. (44–55 Nm).
19. Install the battery mounting bracket and the battery.
20. After installing the front strut assembly, measure the wheel alignment and adjust if necessary.

OVERHAUL

See Figures 152 and 153.

1. Secure the strut in a suitable vise.
2. Loosen the piston rod nut several turns.

➡Use copperplate in the jaws of the vise to protect the shock absorber bottom bracket.

✲✲ CAUTION

Do not remove the piston rod nut until coil spring is compressed and secured.

3. While still secured in a vise, compress the coil spring with SST 0K2A1-341-001A.

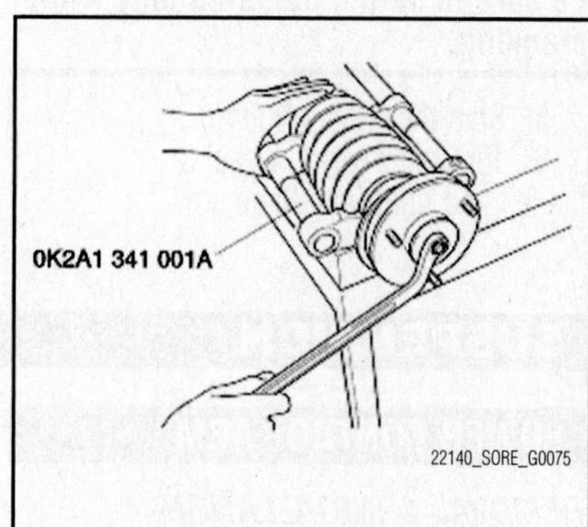

Fig. 152 Compress the coil spring with appropriate spring compressor

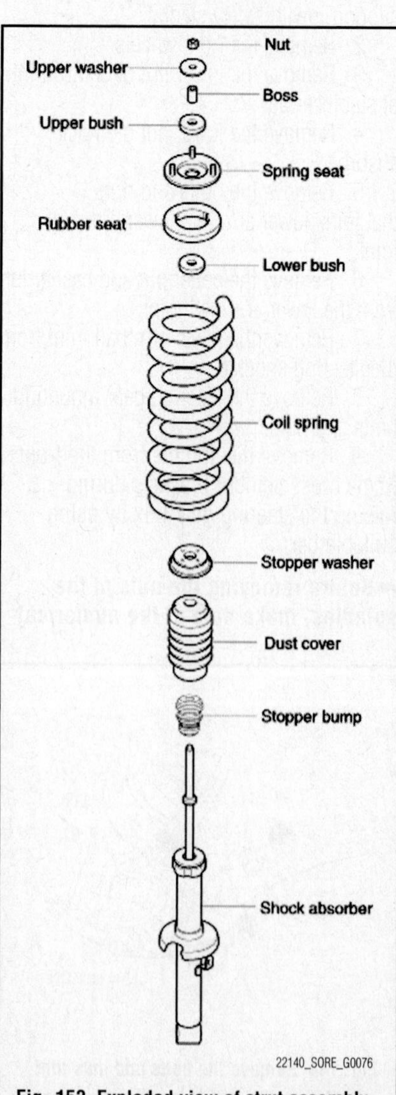

Fig. 153 Exploded view of strut assembly.

4. Remove the piston rod nut and each part as below.
5. Set the end of the coil spring to the rubber seat and install the coil spring.
6. Assemble stopper bump, dust cover, stopper washer, lower bushing, rubber seat, spring seat, boss, upper bushing and upper washer in sequence.
7. Hand tighten the piston rod nut.
8. Carefully loosen the coil spring compressor and remove it.
9. With the bottom bracket of the shock absorber still in the vice, tighten the piston rod nut. Torque: 54–68 ft. lbs. (76–95 Nm).

STEERING KNUCKLE

REMOVAL & INSTALLATION

See Figure 154.

1. Remove the vehicle speed sensor.
2. Remove brake caliper from brake rotor. Temporarily tie caliper to vehicle frame with wire. Refer to Brake Caliper in Brakes.
3. Remove the brake rotor.
4. Using a lock nut wrench (or equivalent), remove lock nut and plain washer (2WD).
5. Remove the upper arm link lock bolt, spring washer and nut. Refer to Upper Control Arm in this section.
6. Remove tie rod end from steering knuckle. Refer to Tie Rod End in Steering.
7. Remove lower arm from steering knuckle. Refer to Lower Control Arm in this section.
8. Remove steering knuckle from vehicle.

To install:

9. Put steering knuckle on the drive shaft end with upper and lower ball joints in mounting holes.
10. Attach lower arm, tighten lock nut, and install cotter pin. Torque: 116–130 ft. lbs. (160–180 Nm).
11. Attach tie rod end to knuckle, tighten nut, and install cotter pin. Torque: 51–57 ft. lbs. (70–80 Nm).
12. Insert upper arm link lock bolt with spring washer and tighten nut. Torque: 32–39 ft. lbs. (44–55 Nm).
13. Install the chamfer of plain washer toward the bearing (2WD).
14. Screw lock nut up against wheel hub assembly and using a lock nut wrench, tighten nut. Torque: 178–198 ft. lbs. (245–275 Nm).
15. To set bearing preload, use spring scale to measure. Bearing preload: 10 inch lbs.
16. Stake the flange of lock nut on the end of drive shaft.

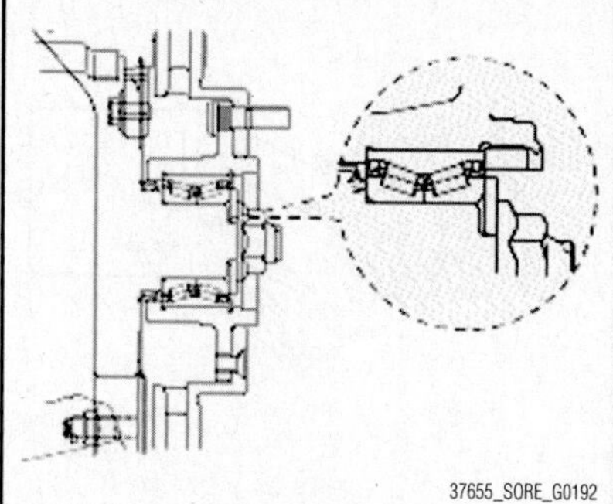

Fig. 154 Install the chamfer of plain washer toward the bearing

17. Put brake rotor on wheel bearing hub bolts and install the two retaining screws.
18. Install brake caliper and tighten two bolts. Torque: 57–75 ft. lbs. (80–104 Nm).
19. Install wheel and tire.

STABILIZER BAR

REMOVAL & INSTALLATION

See Figures 155 through 157.

1. Raise up the front of the vehicle and support it with safety stands.
2. Remove the wheels.
3. Remove the undercover.
4. Remove the nuts and damper rubbers of control link.
5. Remove the stabilizer bar bushing brackets and remove the stabilizer bar.
6. Remove the control link from the lower arm.

To install:

7. Position the control links to the lower arm.
8. Loosely tighten the control link nuts.

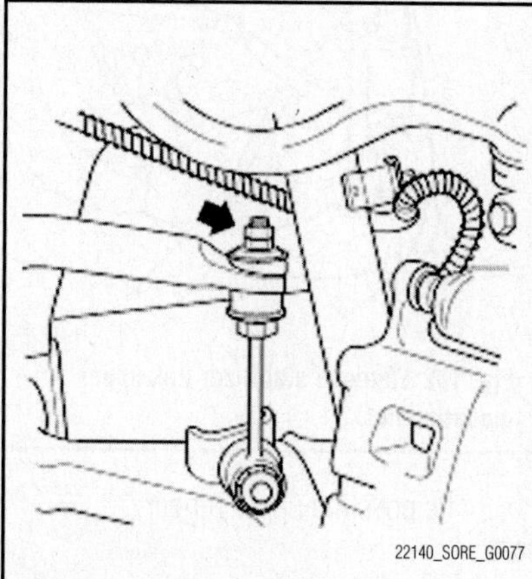

Fig. 155 Remove the nuts and damper rubbers of control link.

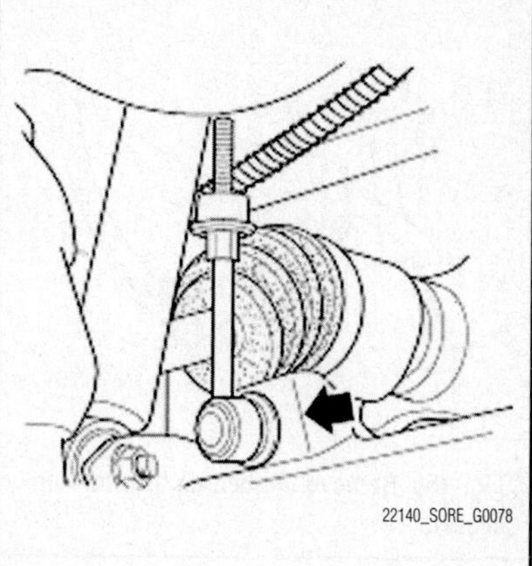

Fig. 156 Remove the control link from the lower arm.

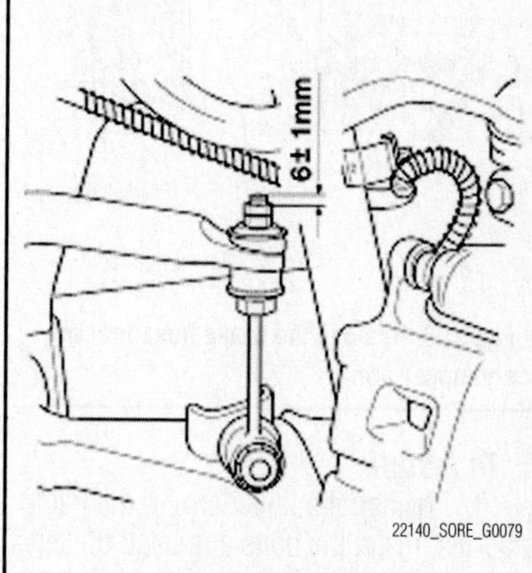

Fig. 157 Install the damper rubber and nut, and tighten to the specified length.

9. Install the stabilizer bar on the control link.
10. Align the clamp bushing inside of stabilizer bushing and install bracket. Torque: 31–39 ft. lbs. (44–55 Nm).
11. Install the damper rubber and nut, and tighten to the specified length.
12. Tighten the lower nut of control link. Torque: 68–84 ft. lbs. (95–117 Nm).

UPPER CONTROL ARM

REMOVAL & INSTALLATION

See Figures 158 through 160.

1. Raise the front of the vehicle and support it with safety stands.
2. Remove the front wheels.
3. Remove the bolt on the steering knuckle side that secures the upper arm ball link.
4. Remove the brake hose bracket.
5. Remove the upper arm.

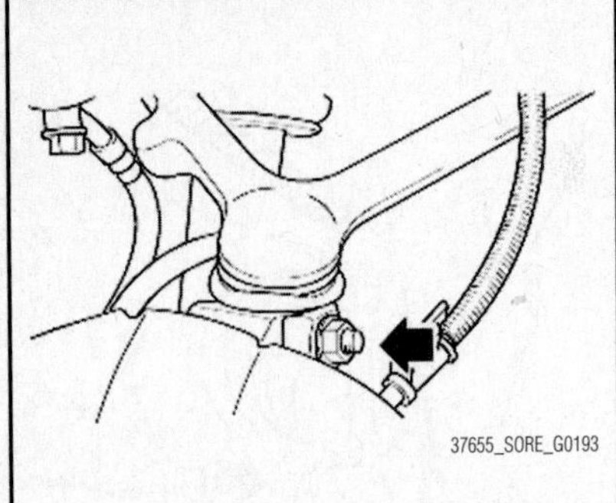

Fig. 158 Remove the bolt on the steering knuckle

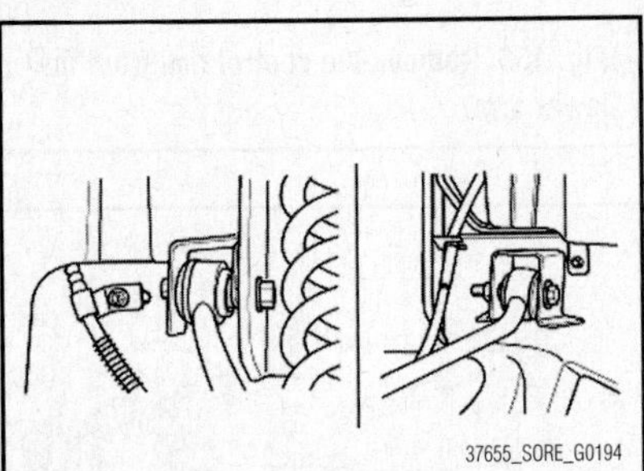

Fig. 159 Remove the brake hose bracket and upper arm

To install:

6. Position the upper arm to the frame brackets, insert the bolts and hand tighten the nuts.
7. Install the upper arm ball joint into the top of the steering knuckle. Torque: 31–39 ft. lbs. (44–55 Nm).

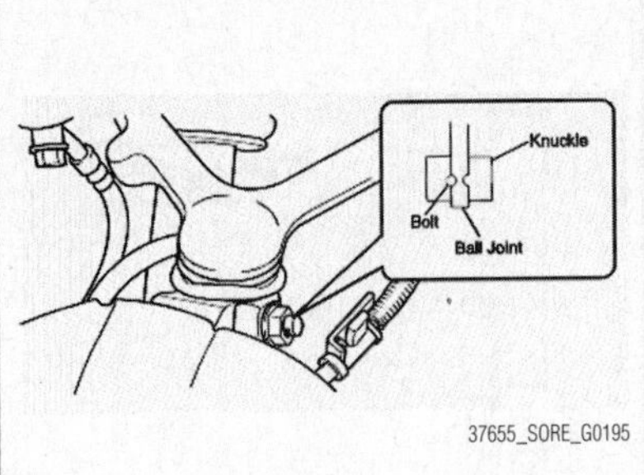

Fig. 160 Install the upper arm ball joint into the top of the steering knuckle

8. Tighten the upper arm bolts and nuts. Torque: 54–68 ft. lbs. (76–95 Nm).
9. Install brake hose brackets.
10. Install the wheels.

➡After installation, measure the wheel alignment and adjust if necessary.

WHEEL HUB & BEARING

REMOVAL & INSTALLATION

See Figure 161.

1. Remove the front knuckle. Refer to Knuckle in this section.
2. Using a screwdriver, pry out oil seal from knuckle (4WD).
3. Press the wheel hub from the knuckle (4WD).

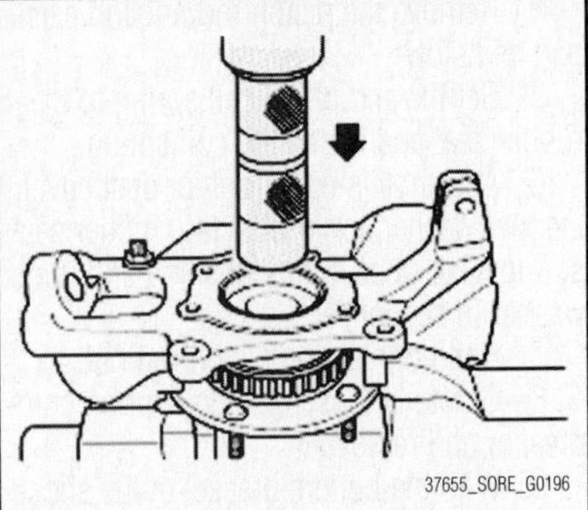

Fig. 161 Press the knuckle and remove wheel hub

4. Press the knuckle and remove wheel hub (2WD).
5. Inspect bearing for wear or damage.
6. Inspect steering knuckle for wear or damage.

To install:

7. Install the dust cover to the knuckle. Torque: 12–16 ft. lbs. (16–23 Nm).
8. Install new oil seal and press the wheel hub into the knuckle.
9. Apply grease to the wheel bearing and seal lip.
10. Install knuckle.
11. To set bearing preload, use spring scale to measure. Bearing preload: 10 inch lbs.
12. Stake the flange of lock nut on the end of drive shaft.
13. Install wheel and tire.

SUSPENSION — REAR SUSPENSION

COIL SPRING

REMOVAL & INSTALLATION

See Figures 162 through 164.

1. Raise the rear of the vehicle and support it with safety stands.
2. Remove the rear wheels.
3. Raise the rear axle housing with a jack to facilitate removal of the shock absorbers.
4. Remove stabilizer link upper mounting nut.
5. Remove the rear shock absorber lower nut and washer.
6. Remove the shock absorber upper bolt, and remove the shock absorber.
7. Lower the rear axle housing slowly to facilitate removal of the coil spring.

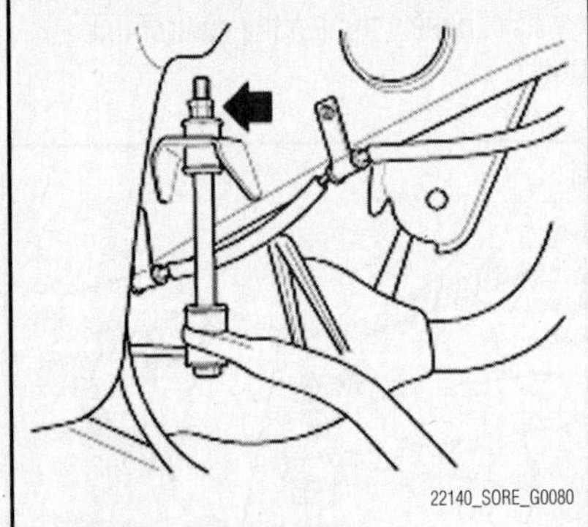

Fig. 162 Remove stabilizer link upper mounting nut

8. Remove the upper rubber seat.

To install:

9. Position the upper rubber seat to the coil spring.

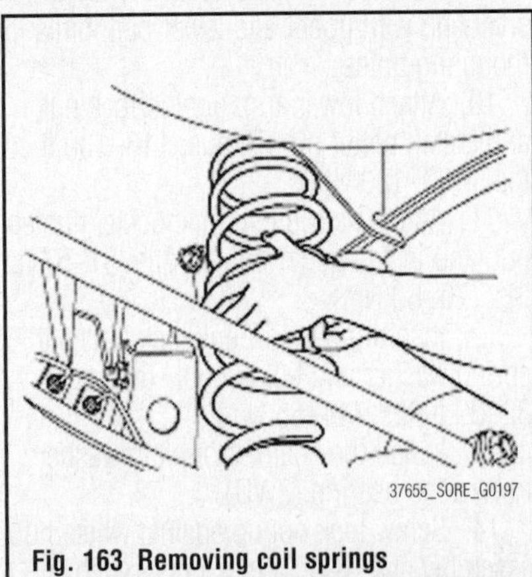

Fig. 163 Removing coil springs

**** CAUTION**

Align the spring end with the groove of the spring pad and fix the spring and the spring pad by adhering the parts with tape.

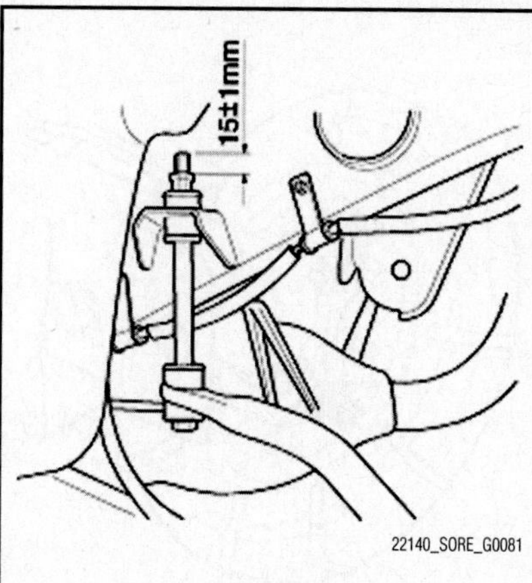

Fig. 164 Install the stabilizer link upper mounting nut to the specified length.

10. Slowly raise the rear axle housing while installing the coil spring.
11. Install the shock absorber upper nut. Torque: 88–101 ft. lbs. (122–140 Nm).
12. Install the shock absorber lower bolt. Torque: 88–101 ft. lbs. (122–140 Nm).
13. Install the stabilizer link upper mounting nut to the specified length.
14. Install the wheels
15. Remove the safety stands and lower the vehicle.

REAR LOWER ARM

REMOVAL & INSTALLATION

See Figure 165.

1. Raise the rear of the vehicle and support it with safety stands.
2. Remove the rear wheels.
3. Raise the rear axle housing to facilitate removal of the upper arm.
4. Remove shock absorber lower bolt.
5. Remove wheel speed sensor cable from rear lower arm.
6. Remove the lower arm bolts and remove the lower arm.

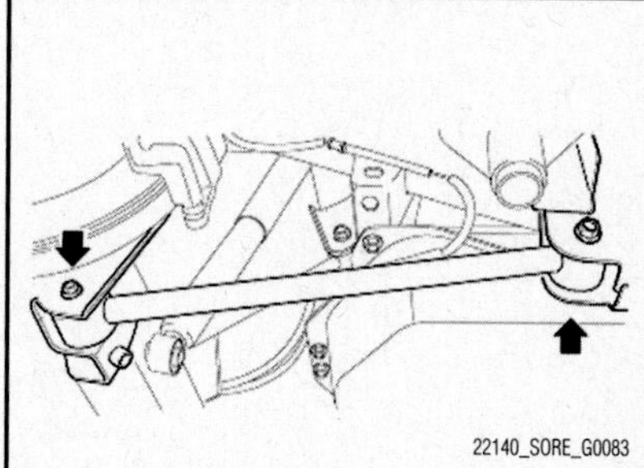

Fig. 165 Remove the lower arm bolts and remove the lower arm.

To install:

7. Install the lower arm and the bolts. Torque: 101–116 ft. lbs. (137–157 Nm).
8. Install wheel speed sensor cable to the rear lower arm.
9. Install shock absorber lower bolt. Torque: 88–101 ft. lbs. (122–140 Nm).
10. Lower the rear axle housing.
11. Install the wheels.
12. Remove the safety stands and lower the vehicle.

REAR UPPER ARM

REMOVAL & INSTALLATION

See Figure 166.

1. Raise the rear of the vehicle and support it with safety stands.
2. Remove the rear wheels.
3. Raise the rear axle housing to facilitate removal of the upper arm.
4. Remove shock absorber lower bolt.
5. Remove the upper arm bolts and remove the upper arm.

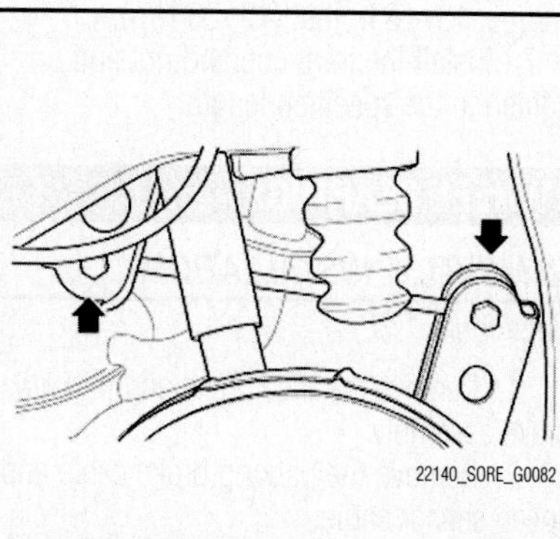

Fig. 166 Remove the upper arm bolts and remove the upper arm.

To install:

6. Install the upper arm and the bolts. Torque: 88–101 ft. lbs. (122–140 Nm).
7. Install shock absorber lower bolt. Torque: 88–101 ft. lbs. (122–140 Nm).
8. Lower the rear axle housing.
9. Install the wheels.
10. Remove the safety stands and lower the vehicle.

LATERAL ROD

REMOVAL, REPACKING & INSTALLATION

See Figure 167.

1. Raise the rear of the vehicle and support it with safety stands.
2. Remove the rear wheels.

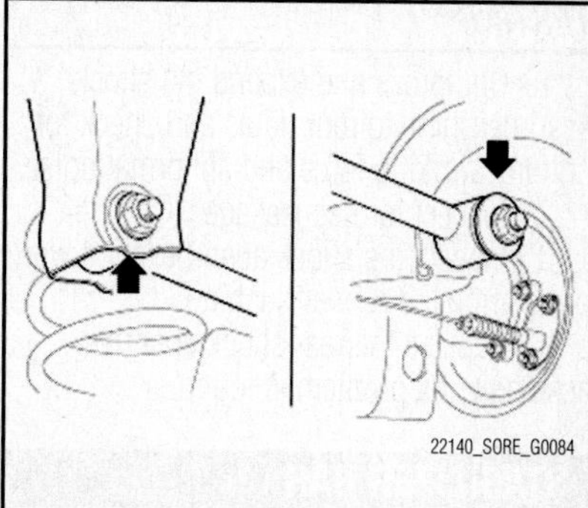

Fig. 167 Remove the lateral rod bolts and remove the lateral rod.

3. Raise the rear axle housing with a jack to facilitate removal of the lateral rod.
4. Remove shock absorber lower bolt.
5. Remove the lateral rod bolts and remove the lateral rod.

To install:

6. Install the lateral rod and the bolts. Torque: 135–155 ft. lbs. (187–215 Nm).
7. Install shock absorber lower bolt. Torque: 88–101 ft. lbs. (122–140 Nm).
8. Lower the rear axle housing.
9. Install the wheels.
10. Remove the safety stands and lower the vehicle.

SHOCK ABSORBER

REMOVAL & INSTALLATION

See Figures 162 and 164.

1. Raise the rear of the vehicle and support it with safety stands.
2. Remove the rear wheels.
3. Raise the rear axle housing with a jack to facilitate removal of the shock absorbers.
4. Remove stabilizer link upper mounting nut.
5. Remove the rear shock absorber lower nut and washer.
6. Remove the shock absorber upper bolt, and remove the shock absorber.

To install:

7. Install the shock absorber upper nut. Torque: 88–101 ft. lbs. (122–140 Nm).
8. Install the shock absorber lower bolt. Torque: 88–101 ft. lbs. (122–140 Nm).
9. Install the stabilizer link upper mounting nut to the specified length.
10. Install the wheels
11. Remove the safety stands and lower the vehicle.

TESTING

1. Compress and expand the shock absorber three to four times and check for uniform working force and abnormal noise.
2. Inspect for gas leakage.
3. Inspect the shock absorber for a worn or deteriorated rubber bushing.
4. Replace the rear shock absorber assembly if a problem is found.

STABILIZER BAR

REMOVAL & INSTALLATION

See Figures 168 and 169.

1. Support the bottom of the rear differential carrier with a jack.
2. Remove the stabilizer link mounting nut.
3. Remove the stabilizer bar bushing bracket.

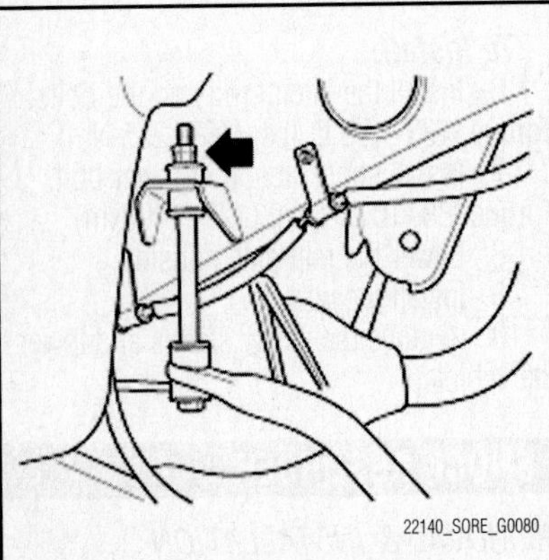

Fig. 168 Remove stabilizer link upper mounting nut.

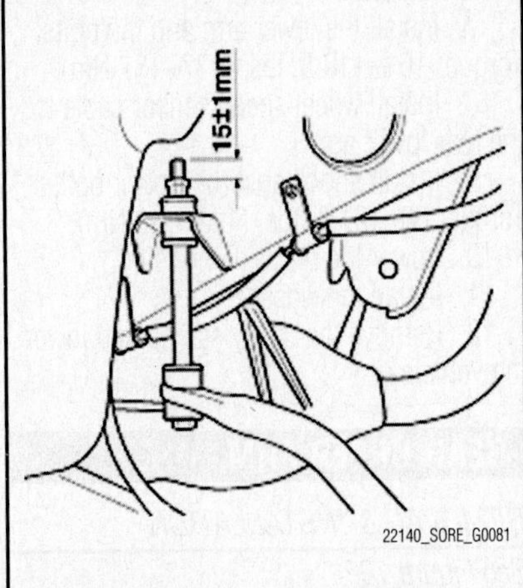

Fig. 169 Install the stabilizer link upper mounting nut to the specified length.

4. Remove the stabilizer bar.

To install:

5. Install the stabilizer bar.
6. Align the identification mark white paint on stabilizer bar with bushing and install the stabilizer bar bushing bracket. Torque: 13–16 ft. lbs. (19–23 Nm).
7. Install the joint cup and nut and tighten to the specified length.

WHEEL BEARINGS

REMOVAL & INSTALLATION

See Figure 170.

1. Remove the disc brake and parking brake assembly.
2. Remove the parking brake cable and speed sensor cable.

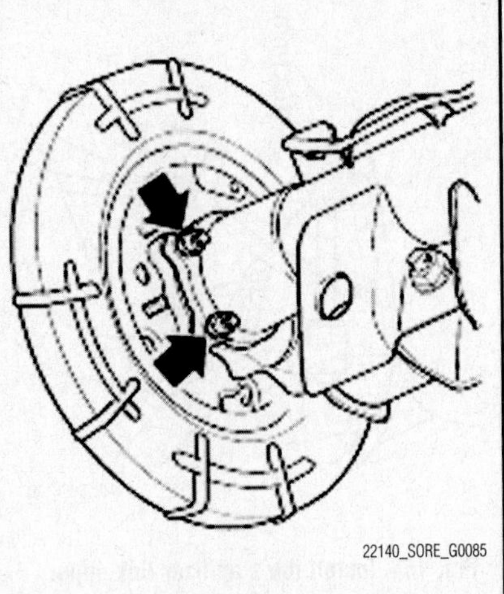

Fig. 170 Remove the rear axle shaft mounting nuts.

3. Remove the rear axle shaft mounting nuts.
4. Remove the rear axle shaft.
5. Using appropriate tool, press the bearing from the axle shaft.
6. Using the special tool (09526-11100), remove the oil seal.

To install:

7. Installation is the reverse of removal.
8. Apply grease to the oil seal lip.
9. Using the special tools (09500-11000, 09532-11500), install the oil seal.
10. Using the appropriate tool, press the new bearing to the axle shaft.
11. After installing the axle shaft, tighten the nuts. Torque: 32–44 ft. lbs. (43–60 Nm).
12. Adjust the parking brake lever stroke.

KIA

Soul

19

SPECIFICATIONS AND MAINTENANCE CHARTS

ENGINE AND VEHICLE IDENTIFICATION

Engine						
Code ①	Liters (cc)	Cu. In.	Cyl.	Fuel Sys.	Engine Type	Eng. Mfg.
1	1.6 (1591)	97.1	4	EGI	DOHC	KIA
2	2.0 (1975)	120.5	4	EGI	DOHC	KIA

Model Year	
Code ②	Year
A	2010

EGI: Electronic Gasoline Injection

DOHC: Double Overhead Camshafts

① 8th digit of VIN

② 10th digit of VIN

37655_SOUL_C0001

GENERAL ENGINE SPECIFICATIONS

Year	Model	Engine Displacement Liters	Engine VIN	Net Horsepower @ rpm	Net Torque @ rpm (ft. lbs.)	Bore x Stroke (in.)	Com-pression Ratio	Oil Pressure @ rpm
2010	Soul	1.6	1	122@6300	115@4200	3.03x3.36	10. 5:1	14.5@1000
		2.0	2	142@6000	137@4600	3.23x3.68	10.1:1	14@800

37655_SOUL_C0002

ENGINE TUNE-UP SPECIFICATIONS

Year	Engine Displacement Liters	Engine VIN	Spark Plug Gap (in.)	Ignition Timing (deg.)		Fuel Pump (psi)	Idle Speed (rpm)		Valve Clearance	
				MT	AT		MT	AT	Intake	Exhaust
2010	1.6	1	0.039-0.043	NA	NA	49.8	NA	NA	HYD	HYD
	2.0	2	0.039-0.043	NA	NA	49.8	750-850	750-850	0.008	0.011

NOTE: The Vehicle Emission Control Information label often reflects specification changes made during production.
The label figures must be used if they differ from those in this chart.

NA: Not Available

B: Before top dead center

HYD: Hydraulic

37655_SOUL_C0003

CAPACITIES

Year	Model	Engine Displacement Liters	Engine VIN	Engine Oil with Filter	Transaxle (pts.) Manual	Transaxle (pts.) Auto.	Fuel Tank (gal.)	Cooling System (qts.)
2010	Soul	1.6	1	3.49	4.0	NA	12.7	6.9
		2.0	2	4.23	4.2	13.8	12.7	7.6

⑭①▯ ⑭▯▯ ①▯㉗㉟㊳㉗㉘㊳㉛

37655_SOUL_C0004

FLUID SPECIFICATIONS

Year	Model	Engine Displ. Liters (VIN)	Engine Oil	Man. Trans.	Auto. Trans.	Power Steering Fluid	Brake Master Cylinder	Cooling System
2010	Soul	1.6 (1)	5W-20	75W/85	NA	PSF - 3	DOT3 or 4	①
		2.0 (2)	5W-30	75W/85	②	PSF - 3	DOT3 or 4	①

DOT: Department Of Transpotation

NA: Not Available

① Ethylene glycol

② GENUINE DIAMOND ATF SP-III

37655_SOUL_C0005

VALVE SPECIFICATIONS

Year	Engine Displacement Liters	Engine VIN	Seat Angle (deg.)	Face Angle (deg.)	Maximum out of Square (in.)	Spring Free Length (in.)	Stem-to-Guide Clearance (in.) Intake	Stem-to-Guide Clearance (in.) Exhaust	Stem Diameter (in.) Intake	Stem Diameter (in.) Exhaust
2010	1.6	1	45-45.5	45.25-45.75	①	1.732	0.0008-0.0019	0.0012-0.0021	0.2152-0.2157	0.2343-0.2348
	2.0	2	45-45.5	45-45.3	①	1.923	0.0008-0.0019	0.0014-0.0026	0.2348-0.2354	0.2343-0.2348

NA: Not Available

① Less than 1.5 degrees

37655_SOUL_C0006

CAMSHAFT AND BEARING SPECIFICATIONS

All measurements are given in inches.

Year	Engine Displacement Liters	Engine VIN	Journal Diameter	Brg. Oil Clearance	Shaft End-play	Runout	Journal Bore	Lobe Lift Intake	Lobe Lift Exhaust
2010	1.6	1	0.9041-0.9047	0.0011-0.0023	0.0039-0.0079	NA	NA	1.7260	1.6870
	2.0	2	1.1023	0.0011-0.0024	0.0040-0.0079	NA	NA	1.7566	1.7527

NA: Not Available

37655_SOUL_C0007

CRANKSHAFT AND CONNECTING ROD SPECIFICATIONS

All measurements are given in inches.

Year	Engine Displacement Liters	Engine VIN	Crankshaft Main Brg. Journal Dia.	Crankshaft Main Brg. Oil Clearance	Crankshaft Shaft End-play	Crankshaft Thrust on No.	Connecting Rod Journal Diameter	Connecting Rod Oil Clearance	Connecting Rod Side Clearance
2010	1.6	1	1.8875-1.8882	0.0002-0.0009	0.0020-0.0098	3	1.6517-1.6524	0.0007-0.0014	0.0039-0.0098
	2.0	2	2.2418-2.2426	0.0011-0.0018	0.0023-0.0100	3	1.7695-1.7703	0.0011-0.0018	0.004-0.0100

37655_SOUL_C0009

PISTON AND RING SPECIFICATIONS

All measurements are given in inches.

Year	Engine Displacement Liters	Engine VIN	Piston Clearance	Ring Gap Top Compression	Ring Gap Bottom Compression	Ring Gap Oil Control	Ring Side Clearance Top Compression	Ring Side Clearance Bottom Compression	Ring Side Clearance Oil Control
2010	1.6	1	0.0008-0.0016	0.0055-0.0110	0.0118-0.0177	0.0079-0.0276	0.0012-0.0028	0.0012-0.0028	0.0024-0.0059
	2.0	2	0.0008-0.0016	0.0079-0.0138	0.0146-0.0205	0.0079-0.0236	0.0015-0.0031	0.0012-0.0027	NA

NA: Not Available

37655_SOUL_C0008

TORQUE SPECIFICATIONS

All readings in ft. lbs.

Year	Engine Displacement Liters	Engine VIN	Cylinder Head Bolts	Main Bearing Bolts	Rod Bearing Bolts	Crankshaft Damper Bolts	Flywheel Bolts	Manifold Intake	Manifold Exhaust	Spark Plugs	Oil Pan Drain Plug
2010	1.6	1	①	②	②	94-101	53-56	14-17	22-25	18-22	29-33
	2.0	2	③	④	36-38	116-123	87-94	12-17	31-40	18-22	29-33

① Step 1: 13-16 ft. lbs.

Step 2: plus 90-95 degrees

Step 3: Plus 100-105 degrees

② Step 1: 13-16 ft. lbs.

Step 2: plus88-92 degrees

③ M10 bolts; 17-20 ft. lbs. plus 60-65 degrees; plus an additional 6-65 degrees.

M12 bolts: 20.3-23.1 ft. lbs., plus 60-65 degrees, plus additional 60-65 degrees

④ 20-23 ft. lbs. plus 60 degrees

37655_SOUL_C0010

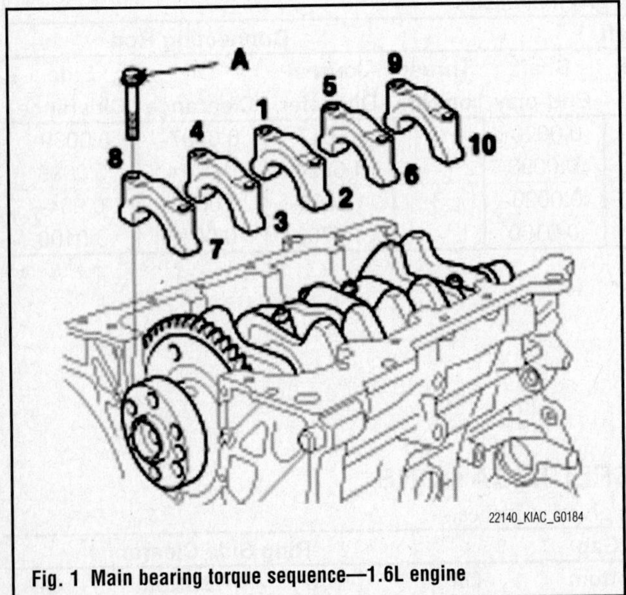

Fig. 1 Main bearing torque sequence—1.6L engine

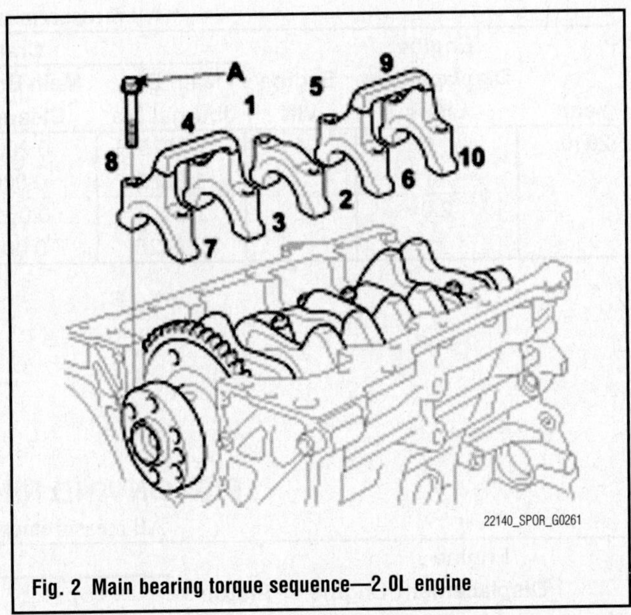

Fig. 2 Main bearing torque sequence—2.0L engine

WHEEL ALIGNMENT

Year	Model		Caster Range (+/-Deg.)	Caster Preferred Setting (Deg.)	Camber Range (+/-Deg.)	Camber Preferred Setting (Deg.)	Toe-in (Deg.)
2010	Soul	F	0.5	+3.4	0.50	0	0 +/- 0.2
		R	—	—	0.50	-1.00	0.4 +/- 0.2

37655_SOUL_C0011

TIRE, WHEEL AND BALL JOINT SPECIFICATIONS

Year	Model	OEM Tires Standard	Tire Pressures (psi) Front	Tire Pressures (psi) Rear	Wheel Size	Ball Joint Inspection	Lug Nut Torque (ft. lbs.)
2010	Soul	205/55R16	30	30	16X7	①	65-80
	Soul+	205/55R16	30	30	16X7	①	65-80
	Soul!	P225/45R18	30	30	18X7	①	65-80
	Soul sport	P225/45R18	30	30	18X7	①	65-80

OEM: Original Equipment Manufacturer

PSI: Pounds Per Square Inch

STD: Standard

OPT: Optional

① Replace if any measureable movent is found.

37655_SOUL_C0012

BRAKE SPECIFICATIONS

All measurements in inches unless noted

Year	Model		Brake Disc Original Thickness	Brake Disc Minimum Thickness	Brake Disc Maximum Run-out	Brake Drum Original Inside Diameter	Brake Drum Max. Wear Limit	Brake Drum Maximum Machine Diameter	Minimum Lining Thickness	Brake Caliper Bracket Bolts (ft. lbs.)	Brake Caliper Mounting Bolts (ft. lbs.)
2010	Soul	F	1.020	0.960	0.0016	—	—	—	0.43	58-72	16-23
		R	0.390	0.330	0.0012	6.61 ①	—	—	0.100	47-54	16-23

F: Front

R: Rear

① Drum in hat parking b rake

37655_SOUL_C0013

SCHEDULED MAINTENANCE INTERVALS

KIA - Soul

TO BE SERVICED	TYPE OF SERVICE	VEHICLE MILEAGE INTERVAL (x1000)												
		7.5	15	22.5	30	37.5	45	52.5	60	67.5	75	82.5	90	97.5
A/C system components and performance	S/I	✓	✓					✓	✓		✓	✓	✓	
Air fliter element	S/I	✓	✓					✓	✓		✓	✓	✓	
Cooling system	I	✓	✓		✓			✓	✓		✓	✓	✓	
Axle boots	I	✓	✓		✓			✓	✓		✓	✓	✓	
Battery condition	I	✓	✓		✓			✓	✓		✓	✓	✓	
Drive belts	I	✓	✓		✓			✓	✓		✓	✓	✓	
Brake/clutch fluid	I	✓	✓		✓			✓	✓		✓	✓	✓	
Brake operation and components	I	✓	✓		✓			✓	✓		✓	✓	✓	
Chassis and suspension fasteners	I	✓	✓		✓			✓	✓		✓	✓	✓	
Power steering components	I	✓	✓		✓			✓	✓		✓	✓	✓	
Locks and hinges	L	✓	✓		✓			✓	✓		✓	✓	✓	
Engine Oil and filter	R	✓	✓	✓	✓	✓	✓	✓	✓	✓	✓	✓	✓	✓
Rotate tires	S	✓	✓	✓	✓	✓	✓	✓	✓	✓	✓	✓	✓	✓
Cabin air filter	R		✓		✓			✓		✓		✓		
Air cleaner filter	R				✓				✓			✓		
Spark plugs (1.6L)	R				✓				✓			✓		
Fuel filter	I				✓									
Fuel tank (EVAP) filter	I				✓									
Automatic transmission fluid	I	Every 40,000 miles												
Manual transaxle fluid	I				✓				✓				✓	
Engine Coolant	R								✓				✓	
Timing belt (2.0L)													✓	

R: Replace S/I: Service or Inspect L: Lubricate T: Tighten

FREQUENT OPERATION MAINTENANCE (SEVERE SERVICE)

If a vehicle is operated under any of the following conditions it is considered severe service

- Towing a trailer or using a camper or car-top carrier

- Repeated short trips of less than 5 miles in temperatures below freezing, or trips of less than 10 miles in any temperature

- Prolonged idling (vehicle operation in stop and go traffic).

- Operating on rough, muddy, unpaved, dusty or salt-covered roads.

- Police, taxi, delivery usage or trailer towing usage.

- Driving in extremely hot (over 90°F) conditions

Oil & oil filter: change every 5000 miles or 5 months, whichever occurs first.

Air cleaner filter: inspect every 15,000 miles or 15 months and replace everything 30,000 miles or 30 months, whichever occurs fir:

Front and rear brakes: inspect every 15,000 miles or 15 months, whichever occurs first

37655_SOUL_C0014

BRAKES — INFORMATION AND PRECAUTIONS

ANTI-LOCK SYSTEMS

- Certain components within the ABS system are not intended to be serviced or repaired individually.
- Do not use rubber hoses or other parts not specifically specified for and ABS system. When using repair kits, replace all parts included in the kit. Partial or incorrect repair may lead to functional problems and require the replacement of components.
- Lubricate rubber parts with clean, fresh brake fluid to ease assembly. Do not use shop air to clean parts; damage to rubber components may result.
- Use only DOT 3 brake fluid from an unopened container.
- If any hydraulic component or line is removed or replaced, it may be necessary to bleed the entire system.
- A clean repair area is essential. Always clean the reservoir and cap thoroughly before removing the cap. The slightest amount of dirt in the fluid may plug an orifice and impair the system function. Perform repairs after components have been thoroughly cleaned; use only denatured alcohol to clean components. Do not allow ABS components to come into contact with any substance containing mineral oil; this includes used shop rags.
- The Anti-Lock control unit is a microprocessor similar to other computer units in the vehicle. Ensure that the ignition switch is **OFF** before removing or installing controller harnesses. Avoid static electricity discharge at or near the controller.
- If any arc welding is to be done on the vehicle, the control unit should be unplugged before welding operations begin.

DISC AND DRUM SYSTEMS

CAUTION

Dust and dirt accumulating on brake parts during normal use may contain asbestos fibers from production or aftermarket brake linings. Breathing excessive concentrations of asbestos fibers can cause serious bodily harm. Exercise care when servicing brake parts. Do not sand or grind brake lining unless equipment used is designed to contain the dust residue. Do not clean brake parts with compressed air or by dry brushing. Cleaning should be done by dampening the brake components with a fine mist of water, then wiping the brake components clean with a dampened cloth. Dispose of cloth and all residue containing asbestos fibers in an impermeable container with the appropriate label. Follow practices prescribed by the Occupational Safety and Health Administration (OSHA) and the Environmental Protection Agency (EPA) for the handling, processing, and disposing of dust or debris that may contain asbestos fibers.

BRAKES — BLEEDING THE BRAKE SYSTEM

BLEEDING PROCEDURE

BLEEDING PROCEDURE

1. Make sure the brake fluid in the reservoir is at the MAX (upper) level line.
2. Have someone slowly pump the brake pedal several times, and then apply pressure.
3. Loosen the right–rear brake bleed screw to allow air to escape from the system. Then tighten the bleed screw securely.
4. Repeat the procedure for wheel in the following sequence: right rear, left front, left rear, right front, until air bubbles no longer appear in the fluid.
5. Refill the master cylinder reservoir to MAX (upper) level line.

BLEEDING THE ABS SYSTEM

➡This procedure should be followed to ensure adequate bleeding of air and filling of the ABS unit, brake lines and master cylinder with brake fluid.

CAUTION

When pressure bleeding, do not depress the brake pedal.

1. Remove the reservoir cap and fill the brake reservoir with brake fluid.
2. Connect a clear plastic tube to the wheel cylinder bleeder plug and insert the other end of the tube into a half filled clear plastic bottle.
3. Connect the GDS to the data link connector located underneath the dash panel.
4. Select and operate according to the instructions on the GDS screen.

CAUTION

You must obey the maximum operating time of the ABS motor with the GDS to prevent the motor pump from burning.

a. Select vehicle name.
b. Select Anti–Lock Brake system.
c. Select HCU air bleeding mode.
d. Press "OK" to operate motor pump and solenoid valve.
e. Wait 60 sec. before operating the air bleeding. (If not, you may damage the motor.)
f. Perform the air bleeding.

5. Pump the brake pedal several times, and then loosen the bleeder screw until fluid starts to run out without bubbles. Then close the bleeder screw.
6. Repeat until there are no more bubbles in the fluid for each wheel.
7. Tighten bleeder screw.

WHEEL SPEED SENSORS

REMOVAL & INSTALLATION

Front

See Figure 3.

1. Remove the front wheel speed sensor mounting bolt (A).
2. Remove the front wheel guard.
3. Disconnect the front wheel speed sensor connector and remove the front wheel speed sensor.
4. Installation is the reverse of removal.
5. Tighten the front wheel speed sensor mounting bolt to 5–8 ft. lbs. (7–11 Nm).

Rear

See Figure 4.

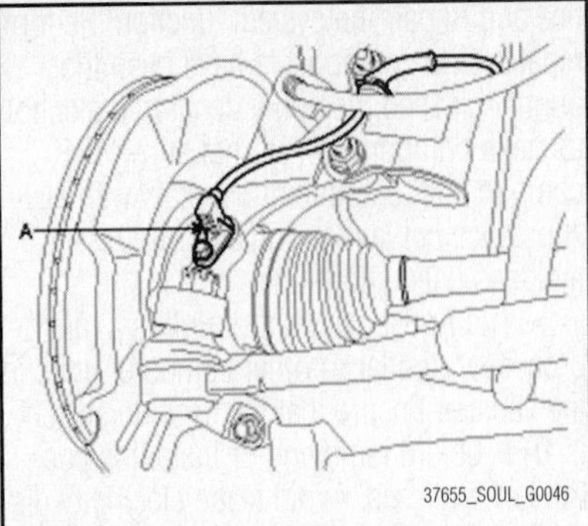

Fig. 3 Front wheel speed sensor mounting bolt (A)

1. Remove the connector after (A) removing the rear wheel speed sensor clip.

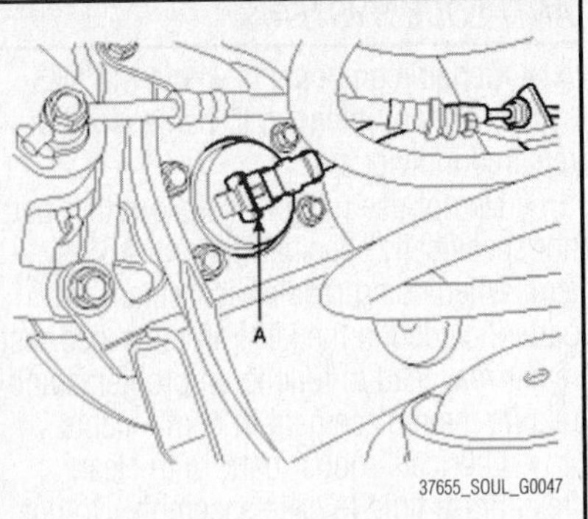

Fig. 4 Rear wheel speed sensor connector (A)

2. Remove the rear wheel speed sensor. See Rear Hub.
3. Installation is the reverse of removal.

BRAKE CALIPER

REMOVAL & INSTALLATION

See Figure 5.

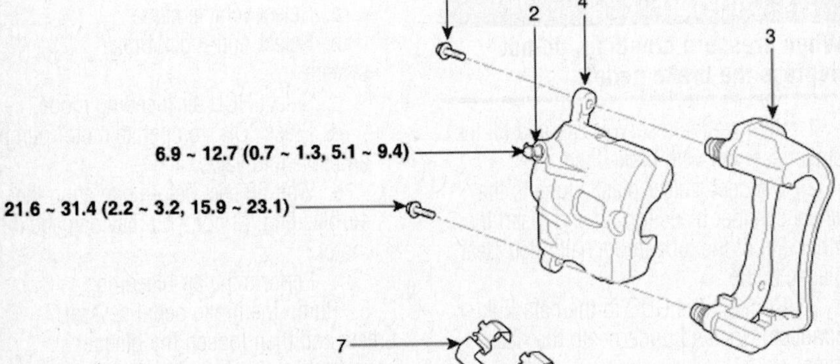

Fig. 5 Front brake components

1. Remove the front wheel & tire.
2. Loosen caliper hose eye bolt and mounting bolts, then remove the front caliper assembly.

To install:

3. Installation is the reverse of removal.
4. Use a SST (09581–11000) to push back the pads when installing the brake caliper assembly.
5. Tighten the caliper bracket bolts to 58–72 ft. lbs. (79–98 Nm).
6. Tighten the caliper hose bolt to 18–22 ft. lbs. (25–29 Nm).
7. Tighten the wheel lug nuts to 65–80 ft. lbs. (88–108 Nm).
8. After installation, bleed the brake system. See Bleeding The Brake System.

DISC BRAKE PADS

REMOVAL AND INSTALLATION

See Figure 5.

1. Remove the front wheel & tire.
2. Remove the brake hose mounting bracket knuckle mounting part.
3. Loosen the guide rod bolt and pivot the caliper up out of the way.
4. Use a SST (09581–11000) to push back the pads.
5. Replace pad shim, pad retainers and brake pads in the caliper bracket.

To install:

6. Installation is the reverse of removal.
7. Tighten the caliper–to–bracket bolts to 16–23 ft. lbs. (22–31 Nm).
8. Tighten the wheel lug nuts to 65–80 ft. lbs. (88–108 Nm).
9. After installation, bleed the brake system. See Bleeding The Brake System.

BRAKES

REAR DISC BRAKES

BRAKE CALIPER

REMOVAL & INSTALLATION

See Figure 6.

1. Remove the rear wheel & tire.
2. Loosen caliper hose eye bolt and mounting bolts, then remove the front caliper assembly.

To install:

3. Installation is the reverse of removal.
4. Use a SST (09581–11000) to push back the pads when installing the brake caliper assembly.
5. Tighten the caliper bracket bolts to 47–54 ft. lbs. (64–74 Nm).
6. Tighten the caliper hose bolt to 18–22 ft. lbs. (25–29 Nm).
7. Tighten the wheel lug nuts to 65–80 ft. lbs. (88–108 Nm).
8. After installation, bleed the brake system. See Bleeding The Brake System.

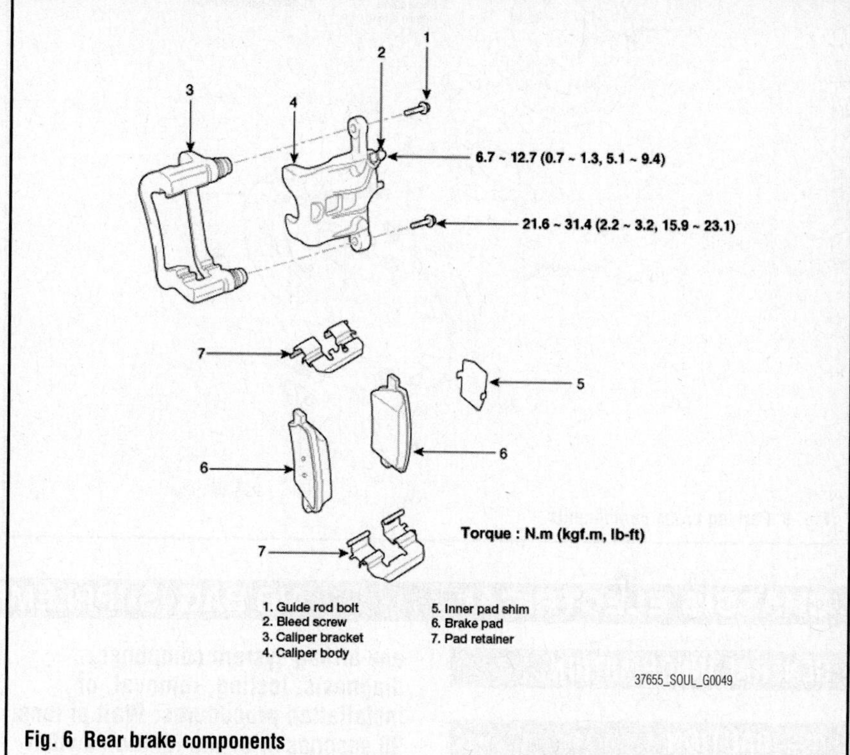

Fig. 6 Rear brake components

DISC BRAKE PADS

REMOVAL INSTALLATION

See Figure 6.

1. Remove the rear wheel & tire.
2. Remove the brake hose mounting bracket knuckle mounting part.
3. Loosen the guide rod bolt and pivot the caliper up out of the way.
4. Use a SST (09581–11000) to push back the pads.
5. Replace pad shim, pad retainers and brake pads in the caliper bracket.

To install:

6. Installation is the reverse of removal.
7. Tighten the caliper–to–bracket bolts to 16–23 ft. lbs. (22–31 Nm).
8. Tighten the wheel lug nuts to 65–80 ft. lbs. (88–108 Nm).
9. After installation, bleed the brake system. See Bleeding The Brake System.

BRAKES

PARKING BRAKE

PARKING BRAKE CABLES

ADJUSTMENT

See Figure 7.

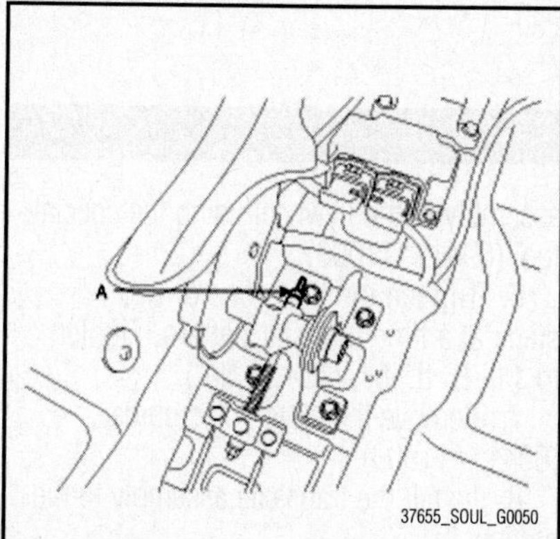

Fig. 7 Parking brake lever adjusting nut (A).

➡After parking brake repairs, adjust the brake shoe clearance, then adjust the parking brake lever stroke.

1. Adjust the parking brake lever stroke by turning adjusting nut (A).
2. Adjustment is correct when handle pulls up 5–7 clicks with 20 kg pressure.
3. Release the parking brake lever fully, and check that parking brakes do not drag when the rear wheels are turned. Readjust if necessary.

PARKING BRAKE SHOES

REMOVAL & INSTALLATION

See Figure 8.

1. Raise the vehicle, and make sure it is securely supported.
2. Remove the rear tire and wheel.
3. Remove the brake Disc. See Rear Brake Rotor.
4. Remove the rear hub unit bearing. See Rear Hub.
5. Remove the shoe hold down pin (11) and the spring (10) by pushing the retainer spring and turning the pin.
6. Remove the adjuster assembly (7) and the return spring (8).
7. Remove the parking brake cable from the brake shoe.
8. Remove the strut (3) and the strut spring (2).
9. Remove the brake shoe.
10. Install the brake shoe (4) to the back plate (1).
11. Install the shoe hold down pin (11) and the spring (10) by pushing the cup washer (9) and turning the pins.
12. After installing the strut (3) and upper return springs (6), install the adjuster assembly (7) and the lower return spring (8).

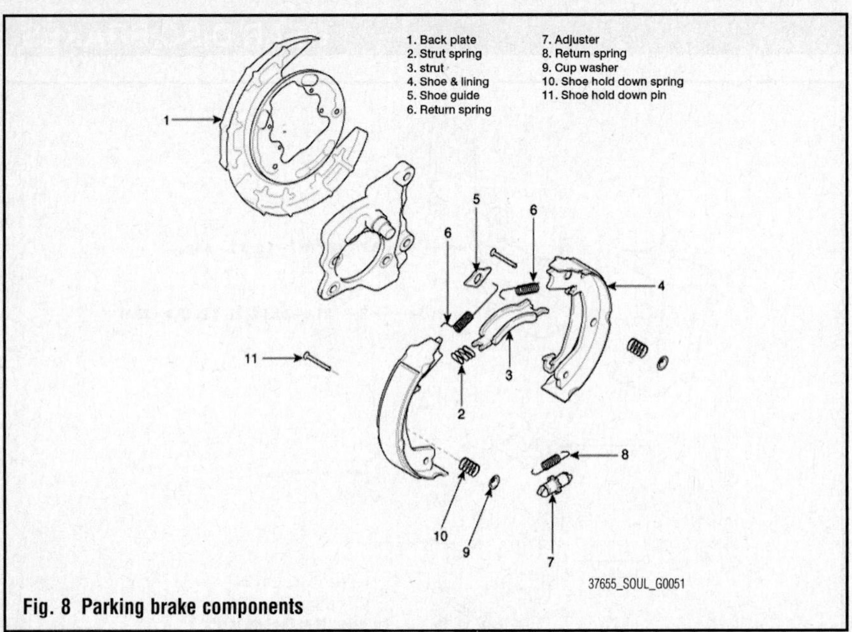

Fig. 8 Parking brake components

13. Install the parking brake cable to shoe.
14. Apply a coating of Multipurpose grease SAE J310, NLGI No.2 to each sliding point of parking brake and adjuster.
15. Install the rear brake disc, then adjust the rear brake shoe clearance. See Adjustment.
16. Install the brake caliper.
17. Install the tire and wheel.
18. Adjust the parking brake lever. See Adjustment under Parking Brake Cables.

ADJUSTMENT

1. Remove the plug from the disc.
2. Rotate the toothed wheel of adjuster by a screw driver until the disc is not moving, and then return it by 3 to 5 notches in the opposite direction.

CHASSIS ELECTRICAL

AIR BAG (SUPPLEMENTAL RESTRAINT SYSTEM)

GENERAL INFORMATION

CAUTION

These vehicles are equipped with an air bag system. The system must be disarmed before performing service on, or around, system components, the steering column, instrument panel components, wiring and sensors. Failure to follow the safety precautions and the disarming procedure could result in accidental air bag deployment, possible injury and unnecessary system repairs.

SERVICE PRECAUTIONS

CAUTION

Disconnect and isolate the battery negative cable before beginning any airbag system component diagnosis, testing, removal, or installation procedures. Wait at least 90 seconds after the ignition switch is turned off and the negative (-) terminal cable is disconnected from the battery before starting the operation. The SRS is equipped with a backup power source, so if work is started within 90 seconds after disconnecting the negative (-) terminal cable from the battery, the SRS may be deployed. Failure to disable the airbag system may result in accidental airbag deployment, personal injury, or death.

DISARMING THE SYSTEM

Disconnect the battery negative cable and wait for at least three minutes before beginning work.

ARMING THE SYSTEM

1. Connect the battery negative cable.
2. To confirm proper system operation, turn the ignition switch ON; the SRS indicator light should be turned on for about six seconds and then go off.

CLOCKSPRING CENTERING

Set the center position by getting marks between the clock spring and the cover in line. If clock spring is not centered, align the marks by turning the clock spring clockwise to the stop and then 2 revolutions counterclockwise.

DRIVE TRAIN

CLUTCH DRIVEN DISC & PRESSURE PLATE

REMOVAL & INSTALLATION

See Figures 9 and 10.

1. Remove the transaxle assembly.
2. Insert the special tool (09411–11000) in the clutch disc to prevent the disc from shifting.
3. Loosen the bolts which attach the clutch cover to the flywheel in a star pattern. Loosen the bolts in succession, one or two turns at a time, to avoid bending the cover.

To install:

4. Apply CASMOLY L 9508 multipurpose grease to the spline of the disc.

➡When installing the clutch, apply grease to each part, but be careful not to apply excessive grease. It can cause clutch slippage and vibration (shudder).

5. Temporarily install the clutch disc assembly to the flywheel using the special tool (09411–11000).
6. Tighten the bolts one or two steps at a time in a star pattern. Tighten to 11–16 ft. lbs. (15–22 Nm).
7. Remove the clutch disc guide (09411–11000).
8. Install the transaxle assembly to the engine.

ADJUSTMENTS

See Figure 11.

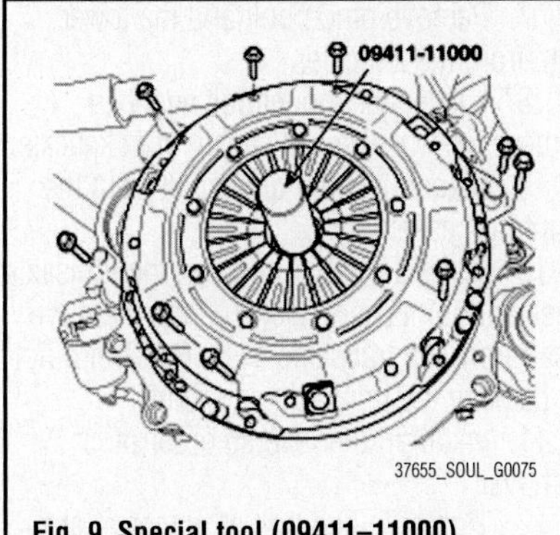

Fig. 9 Special tool (09411-11000)

1. Measure the clutch pedal height (from the face of the pedal pad to the floorboard) and the clutch pedal clevis pin play (measured at the face of the pedal pad.)

➡**If the clutch pedal height is lower than the standard value, loosen the bolt and adjust the push rod. After adjustment, adjust stopper bolt so that the clearance with pedal stopper becomes 0.5mm (0.02 in) to 1.0mm (0.04 in) and secure with lock nut.**

2. If the clutch pedal freeplay and height is not within the standard value range, adjust as follows:
   a. Turn and adjust the bolt within the standard value, then secure by tightening the lock nut.
   b. Turn the push rod to agree with the standard value and then secure the push rod with the lock nut.

➡**When adjusting the clutch pedal height or the clutch pedal play, be careful not to push the push rod toward the master cylinder.**

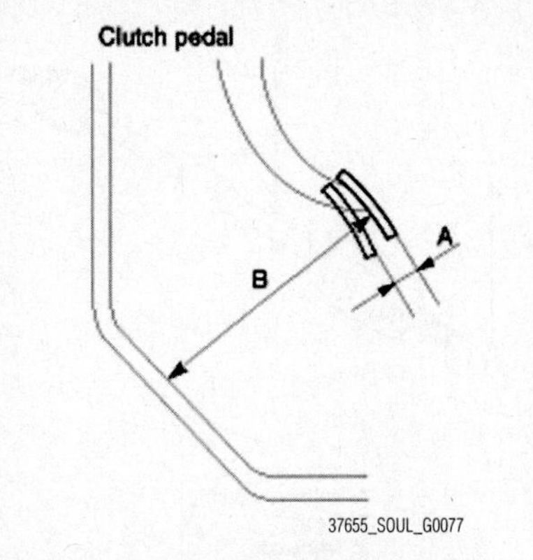

Fig. 11 Clutch pedal play height: Standard value (A) : 0.24–0.51 in. (6–13 mm) (B) : 6.64 in. (168.7 mm)

   c. If the clutch pedal free play and the distance between the clutch pedal and the floor board, do not meet with the standard values when the clutch is disengaged, it may be the result of either air in the hydraulic system or a faulty clutch master cylinder.

HYDRAULIC SYSTEM BLEEDING

BLEEDING PROCEDURE

Whenever the clutch tube, the clutch hose, and/or the clutch master cylinder have been removed, or if the clutch pedal is spongy, bleed the system.

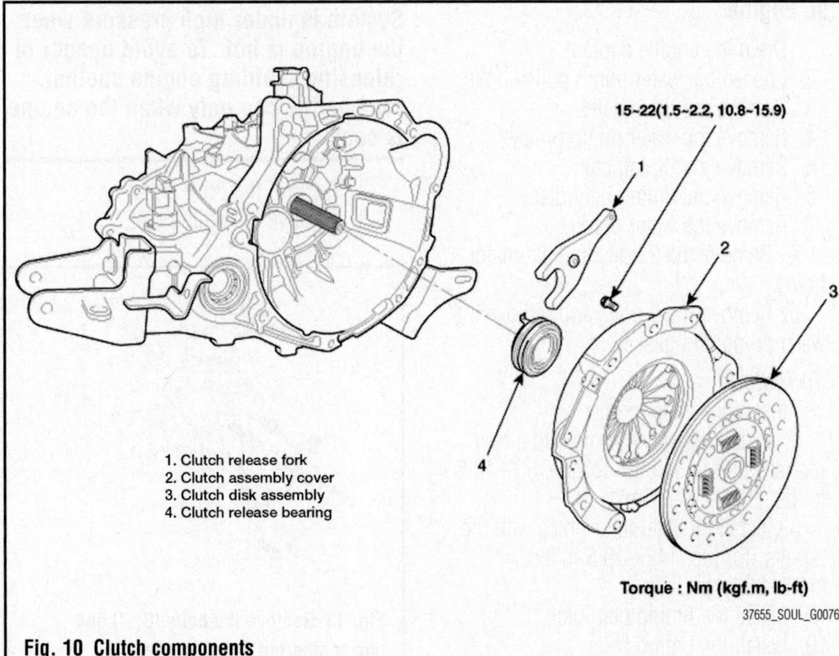

Fig. 10 Clutch components

✲✲ CAUTION

Use SAE J1703 (DOT 3 or DOT 4) fluid. Avoid mixing different brands of fluid.

1. Loosen the bleeder screw at the clutch release cylinder.
2. Depress the clutch pedal slowly until all air is expelled.
3. Hold the clutch pedal down until the bleeder is retightened.
4. Refill the clutch master cylinder with the specified fluid.

➡**Do not rapidly operate clutch while bleeding. Only operate clutch pedal through full sweep during bleeding.**

FRONT HALFSHAFT

REMOVAL & INSTALLATION

See Figure 12.

✲✲ CAUTION

Note the following:

- Use a pry bar being careful not to damage the transaxle and joint.
- Do not insert the pry bar too deep, as this may cause damage to the oil seal.
- Do not pull the driveshaft by excessive force it may cause components inside the joint kit to dislodge resulting in a torn boot or a damaged bearing.
- Plug the hole of the transaxle case with the oil seal cap to prevent contamination.
- Support the driveshaft properly.
- Replace the retainer ring whenever the driveshaft is removed from the transaxle case
- The driveshaft lock nut should always be replaced with new one.

1. Raise and support the vehicle.
2. Remove the front wheel and tire from front hub.
3. While applying brakes, remove driveshaft nut from the front hub.
4. Remove the brake hose mounting bracket.
5. Remove the brake rotor. See Rotor under Brakes.
6. Remove the tie rod end from the knuckle.
   a. Remove the split pin.
   b. Remove the castle nut.
   c. Disconnect the tie rod end from knuckle using special tool (09568–4A000 or 09568–2J100).

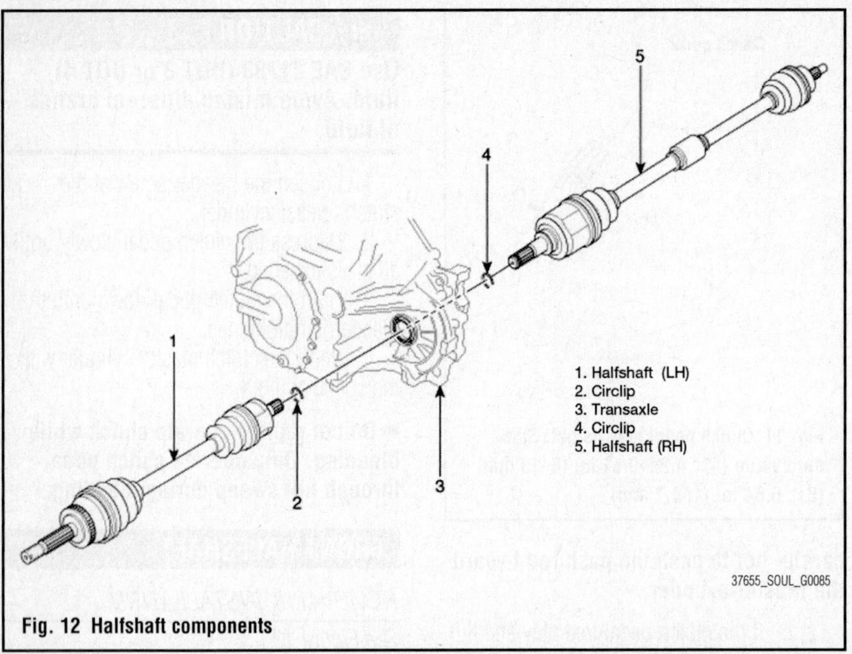

Fig. 12 Halfshaft components

7. Remove pinch bolt and the lower arm from the knuckle.
8. Loosen the mount bolt and then remove the wheel speed sensor from knuckle.
9. Disconnect the driveshaft from the front hub assembly..
10. Insert a pry bar between the transaxle case and joint case, and separate the drive shaft from the transaxle case. Be careful not to damage the transaxle and joint.
11. Install in the reverse order of removal.
12. Tighten pinch bolt at the lower arm and the knuckle to 43–52 ft. lbs. (60–72 Nm).
13. Tighten NEW driveshaft nut o 177–198 ft. lbs. (25–28 Nm). Stake nut after tightening.
14. Tighten tie rod nut to 12–25 ft. lbs. (16–348 Nm). Install cotter pin.
15. Tighten lug nuts to 65–80 ft. lbs. (90–110 Nm).

ENGINE COOLING

RADIATOR

REMOVAL & INSTALLATION

⁂ WARNING

System is under high pressure when the engine is hot. To avoid danger of releasing scalding engine coolant, remove the cap only when the engine is cool.

1. Loosen the engine coolant drain plug for draining.
2. Disconnect the radiator upper hose and the lower hose from engine.
3. Disconnect the radiator upper hose and reservoir hose from cap assembly.
4. Remove the front bumper. See Front Bumper.
5. Remove the horn and the bracket cover.
6. Remove the upper radiator mounting brackets.
7. Remove the radiator assembly after pulling back the condenser fixing bracket.
8. Installation is reverse order of removal.
9. Tighten upper radiator mounting brackets to 5–8 ft. lbs. (6.9–10.8 Nm).
10. Bleed the cooling system. See Drain & Refill Procedure.

THERMOSTAT

REMOVAL & INSTALLATION

1. Drain the engine coolant so its level is below thermostat.
2. Remove the water inlet fitting, gasket and thermostat.
3. Installation is the reverse of removal.

WATER PUMP

REMOVAL & INSTALLATION

1.6L Engine

1. Drain the engine coolant.
2. Loosen the water pump pulley bolts.
3. Remove the drive belts.
4. Remove the water pump pulley.
5. Remove the timing belt.
6. Remove the timing belt idler.
7. Remove the water pump.
   a. Remove the 2 bolts and alternator brace.
   b. Remove the 3 bolts and remove the water pump and gasket.

To install:

8. Install the water pump.
   a. Install the water pump and a new gasket with the 3 bolts. Torque: 8.7–10.8 ft. lbs. (11.8–14.7 Nm).
   b. Install the alternator brace with the 2 bolts. Torque: 14.5–19.5 ft. lbs. (19.6–26.5 Nm.
9. Install the timing belt idler.
10. Install the timing belt.
11. Install the water pump pulley.
12. Install the drive belts.
13. Tighten the water pump pulley bolts. Torque: 5.8–7.2 ft. lbs. (7.8–9.8 Nm).
14. Fill with engine coolant.
15. Start engine and check for leaks.
16. Recheck engine coolant level.

2.0L Engine

See Figures 13 through 16.

1. Drain the engine coolant.

⁂ WARNING

System is under high pressure when the engine is hot. To avoid danger of releasing scalding engine coolant, remove the cap only when the engine is cool.

2. Remove the drive belts.
3. Remove the timing belt.

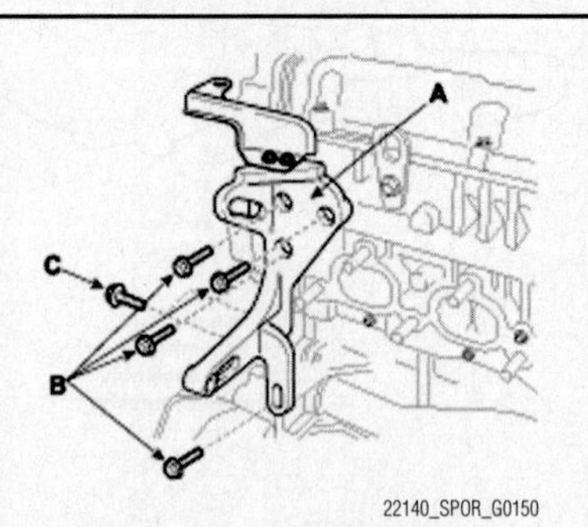

Fig. 13 Remove the bolts (B, C) and power steering pump bracket (A)

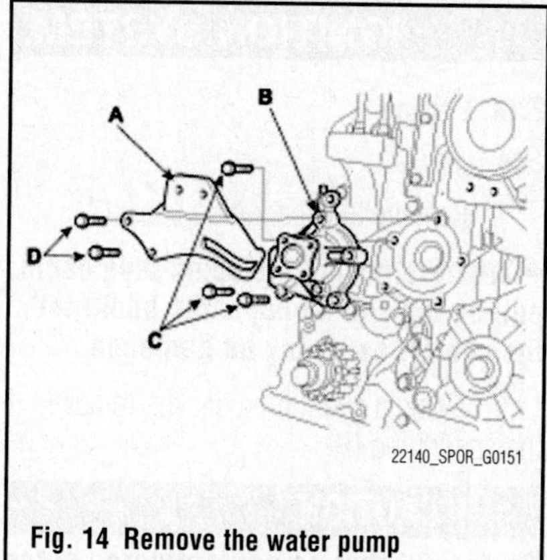

Fig. 14 Remove the water pump

4. Remove the timing belt idler.
5. Remove the power steering pump and use a wire to secure the pump to the vehicle so that it is out of the way.
6. Remove the bolts (B, C) and power steering pump bracket (A).
7. Remove the alternator.
8. Remove the water pump.
   a. Remove the 2 bolts (D) and alternator brace (A).
   b. Remove the 3 bolts (C) and remove the water pump (B) and gasket.

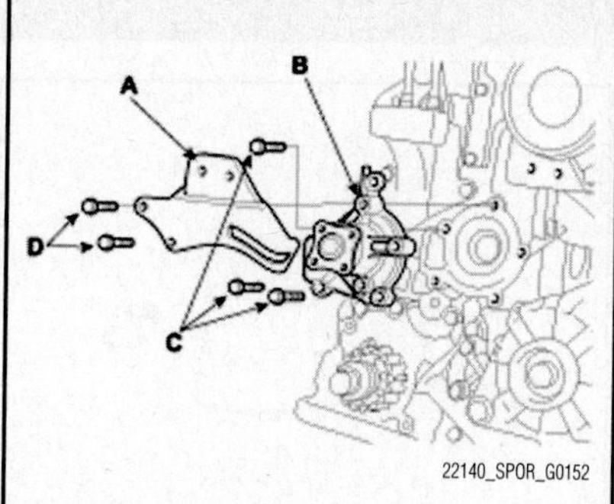

Fig. 15 Torque water pump bolts (A, B, C, D)

To install:

9. Install the water pump.
   a. Install the water pump (B) and a new gasket with the 3 bolts (C). Torque to 9–11 ft. lbs. (12–15 Nm).
   b. Install the alternator brace (A) with the 2 bolts (D). Tighten to 15–20 ft. lbs. (20–27 Nm).
10. Install the power steering pump bracket (A) and bolts (B, C). Tighten to Bolts (B): 25–36 ft. lbs. (34–49 Nm); Bolts (C): 11–15 ft. lbs. (15–20 Nm).

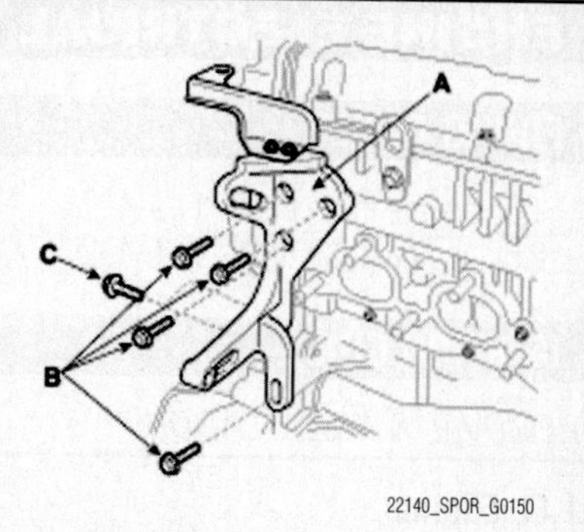

Fig. 16 Install the power steering pump bracket (A) and bolts (B, C)

11. Install the alternator.
12. Install the power steering pump.
13. Install the timing belt idler.
14. Install the timing belt.
15. Install the water pump pulley.
16. Install the drive belts.
17. Tighten the water pump pulley bolts. Tighten to 7 ft. lbs. (9 Nm).
18. Fill with engine coolant.
19. Start engine and check for leaks.
20. Recheck engine coolant level.

ENGINE ELECTRICAL

CHARGING SYSTEM

ALTERNATOR

REMOVAL & INSTALLATION

1.6L Engine

1. Disconnect the battery negative terminal first, then the positive terminal.
2. Loosen the drive belt tension adjusting bolt.
3. Remove the drive belt.
4. Disconnect the A/C compressor connector and the alternator connector, and remove the cable from alternator "B" terminal.
5. Pull out the through bolt (A) and then remove the alternator.

To install:

6. Installation is the reverse order of removal.
7. Adjust the alternator belt tension after installation. See Adjustment.

2.0L Engine

See Figures 17 and 18.

1. Disconnect the battery negative terminal first, then the positive terminal.

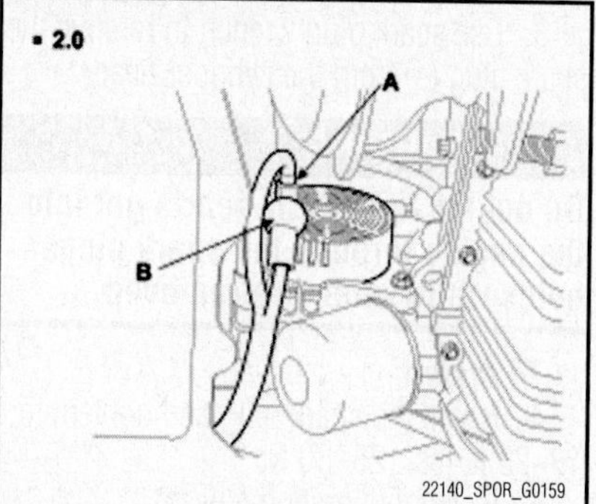

Fig. 17 Disconnect the alternator connector (A) and "B" terminal cable (B) from the alternator (C)—2.0L engine

2. Disconnect the alternator connector (A) and "B" terminal cable (B) from the alternator (C).
3. Remove the adjusting bolt (A) and mounting bolt (B), and remove the alternator belt.
4. Pull out the through bolt (C), and remove the alternator (D).
5. Installation is the reverse of removal.
6. Adjust the alternator belt tension after installation.

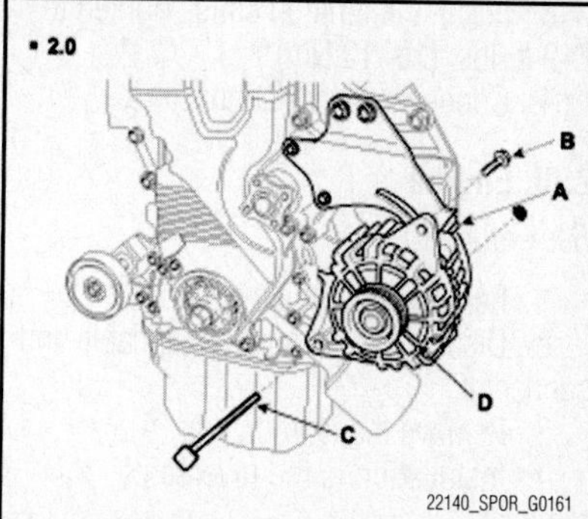

Fig. 18 Remove the adjusting bolt (A), mounting bolt (B) and through bolt (C)—2.0L engine

VOLTAGE REGULATOR

ADJUSTMENT

The charging system includes an alternator with a built–in regulator. There is no adjustment possible for the voltage regulator. If the voltage regulator is defective, it must be replaced.

REMOVAL & INSTALLATION

See Alternator Removal & Installation.

ENGINE ELECTRICAL

IGNITION SYSTEM

FIRING ORDER

1.6L engine firing order: 1–3–4–2
2.0L engine firing order: 1–3–4–2

IGNITION COIL

REMOVAL & INSTALLATION

1.6L Engine

See Figure 19.

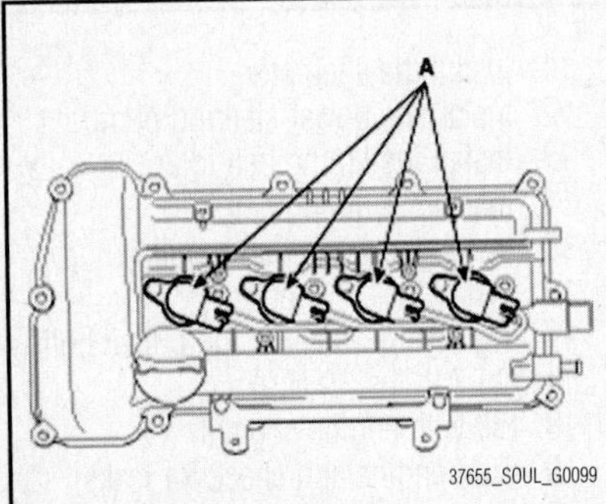

Fig. 19 Ignition Coils (A)—1.6L Engine

1. Disconnect harness connectors.
2. Remove bolts and remove coils.
3. Install the ignition coils. Tighten to 7–9 ft. lbs. (10–12 Nm).
4. Connect harness connectors.

2.0L Engine

See Figure 20.

1. Remove the engine cover.
2. Disconnect the spark plug cable and connector.
3. Remove the ignition coil (A).
4. Installation is the reverse of removal.

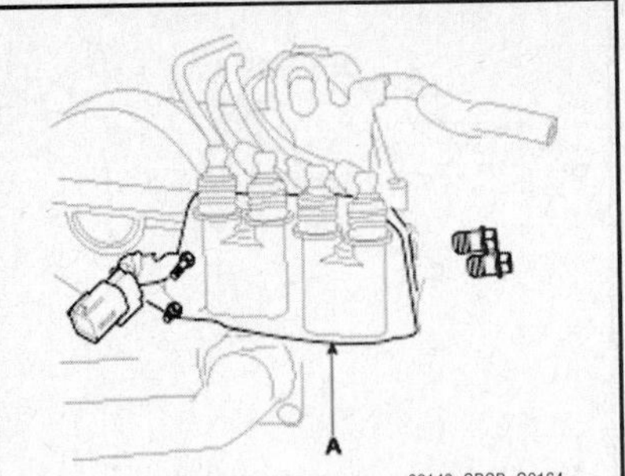

Fig. 20 Remove the ignition coil (A)—2.0L engine

IGNITION TIMING

ADJUSTMENT

These vehicles are equipped with a Distributorless Ignition System (DIS). No adjustment is necessary or possible.

SPARK PLUGS

REMOVAL & INSTALLATION

1.6L Engine

1. Disconnect the negative battery cable.
2. Remove the ignition coil (s).
3. Use spark plug wrench to remove the spark plug (s) from the cylinder head.

⁂ WARNING

Do not let any dirt or debris get into the engine through the spark plugs holes while when are removed.

To install:

4. Install the spark plug and tighten, to 18–22 ft. lbs. (25–30 Nm)
5. Install the ignition coil (s).
6. Connect the negative battery cable.

2.0L Engine

See Figure 21.

1. Remove the spark plug cable (A).

➡When removing the spark plug cable, pull on the spark plug cable boot (not the cable), as it may be damaged.

2. Using a spark plug socket, remove the spark plug (B).

⁂ CAUTION

Be careful that no contaminates enter through the spark plug holes.

3. Installation is the reverse of removal.

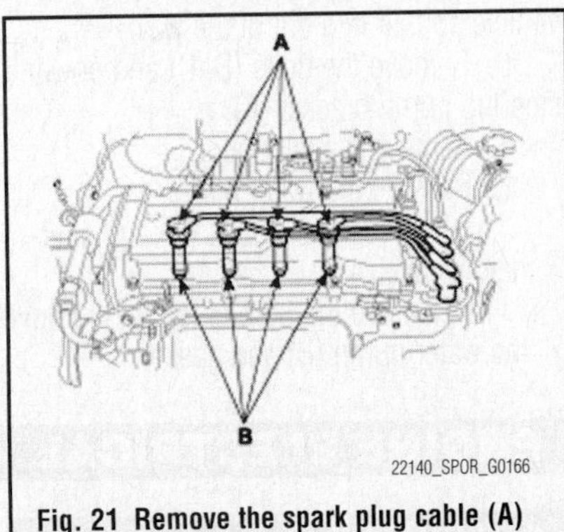

Fig. 21 Remove the spark plug cable (A)

SPARK PLUG WIRES

REMOVAL & INSTALLATION

2.0L Engine

1. Remove the spark plug cable.

➡When removing the spark plug cable, pull on the spark plug cable boot (not the cable), as it may be damaged.

2. Installation is the reverse of removal.

ENGINE ELECTRICAL

STARTING SYSTEM

STARTER

REMOVAL & INSTALLATION

See Figure 22.

1. Disconnect the battery negative cable.
2. On 1.6L engine, remove air cleaner. See Air Cleaner.
3. On all engines, disconnect the starter cable (A) from the B terminal (B) on the solenoid (C), then disconnect the connector (D) from the "S" terminal (E).
4. Remove the 2 bolts holding the starter, and remove the starter.

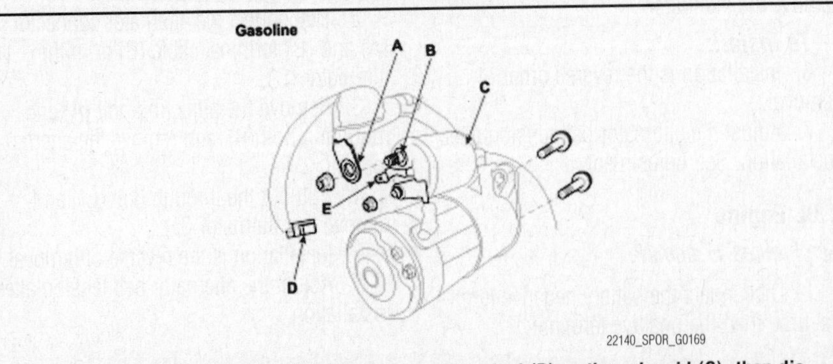

Fig. 22 Disconnect the starter cable (A) from the B terminal (B) on the solenoid (C), then disconnect the connector (D) from the "S" terminal (E)

5. Installation is the reverse of removal. Torque the nuts and bolts to 20 ft. lbs. (27 Nm).
6. Connect the battery positive cable and negative cable to the battery.

SOLENOID OR RELAY REPLACEMENT

1. Disconnect the negative battery cable.
2. Remove the starter assembly from the vehicle.
3. Remove the nut from the solenoid M–terminal.
4. Remove the solenoid retaining screws and the magnetic switch.

To install:

5. Install the magnetic switch and secure to the drive housing with the two screws.
6. Re–install the nut onto solenoid M–terminal.
7. Reinstall the starter assembly.
8. Reconnect the negative battery cable.

ENGINE MECHANICAL

ACCESSORY DRIVE BELTS

ACCESSORY BELT ROUTING

1.6L Engine

See Figures 23 and 24.

Refer to the accompanying illustrations.

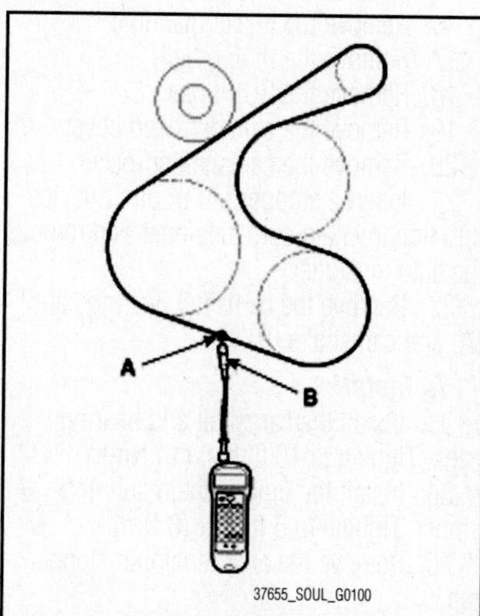

Fig. 23 Drive belt routing with A/C—1.6L Engine

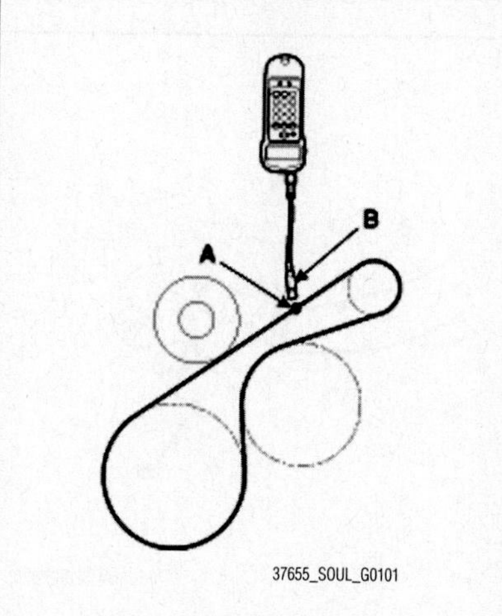

Fig. 24 Drive belt routing without A/C—1.6L Engine

2.0L Engine

See Figures 25 through 27.

Refer to the accompanying illustrations.

INSPECTION

Inspect the drive belt for signs of glazing or cracking. A glazed belt will be perfectly smooth from slippage, while a good belt will have a slight texture of fabric visible. Cracks will usually start at the inner edge of the belt and run outward. All worn or damaged drive belts should be replaced immediately.

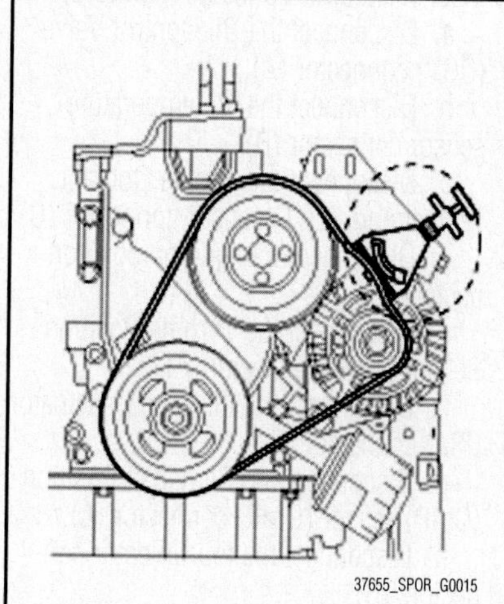

Fig. 25 Alternator belt routing—2.0L engine

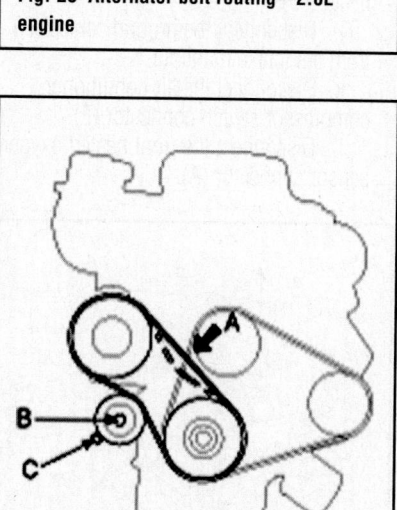

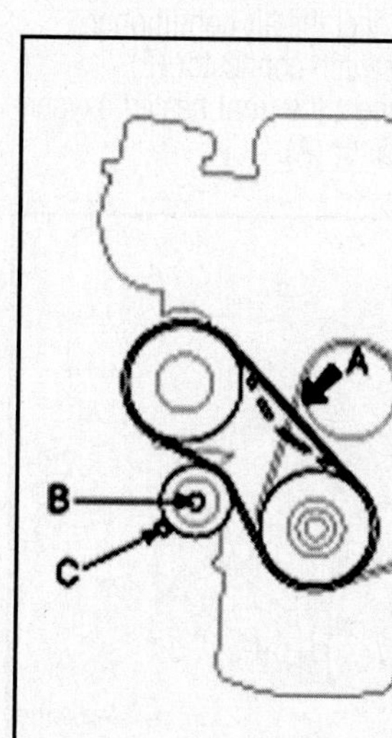

Fig. 26 A/C belt routing—2.0L engine

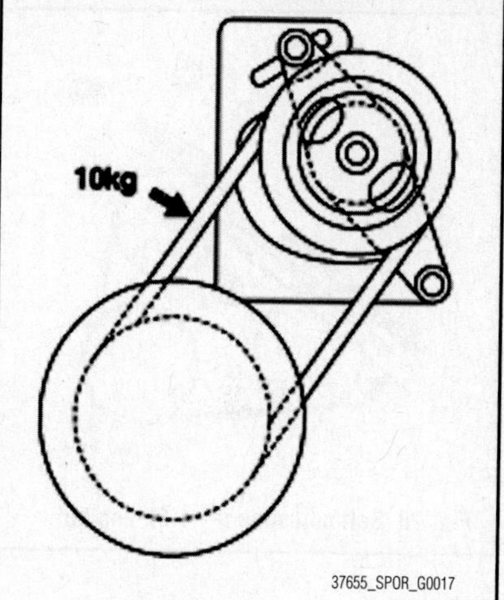

Fig. 27 P/S belt routing—2.0L engine

ADJUSTMENT

1.6L Engine

See Figure 28.

➡Refer to the accompanying illustration for bolt and nut locations.

1. Loosen bolt A and nuts B and C.
2. Turn adjusting bolt D and adjust the belt deflection to within the following specifications (when applying about 22 lbs. of pressure):
    a. New belt: 0.31–0.35 in. (8–9mm)
    b. Used belt: 0.35–0.39 in. (9–10mm)
3. Tighten bolt A and nuts B and C, as follows:
    a. Bolt A: 27–39 ft. lbs. (37–53 Nm)
    b. Nut B: 14–19 ft. lbs. (19–25 Nm)
    c. Nut C: 24–34 ft. lbs. (32–46 Nm)

2.0L Engine

The belt tension is maintained by an automatic tensioner. No adjustment is necessary or possible.

REMOVAL & INSTALLATION

1.6L Engine

See Figure 28.

➡Refer to the accompanying illustration for bolt and nut locations.

1. Loosen bolt A and nuts B and C.
2. Turn adjusting bolt D and remove belt.
3. Installation is the reverse of removal. Tension belt properly. See Adjustment.

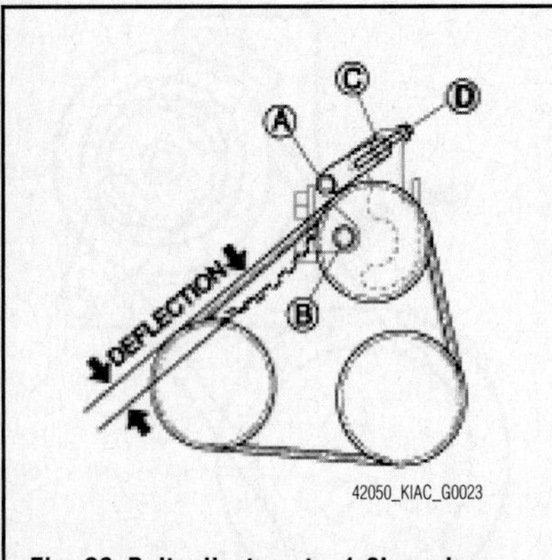

Fig. 28 Belt adjustment—1.6L engine

2.0L Engine

1. Release tension of the belt tensioner.
2. Remove the serpentine belt.

To install:

3. Hang the belt on the pulley of the tensioner and install the tensioner.

(If the tensioner is already installed, loosen its mounting bolts to allow belt installation.) Tighten the tensioner assembly bolt to 33–36 ft. lbs. (45–50 Nm).

4. Install drive belt.
5. When installing the belt on the pulley, make sure it is centered on the pulley.

CAMSHAFT AND VALVE LIFTERS

REMOVAL & INSTALLATION

1.6L Engine

No separate procedure is provided. Refer to figure and cylinder head procedure. See Cylinder Head.

2.0L Engine

See Figures 29 through 47.

➡Identify MLA (Mechanical lash adjuster), valves, valve springs as they are removed so that each item can be reinstalled in its original position.

2.0L Engine

See Figures 29 through 47.

1. Disconnect the terminals from battery.
2. Drain the engine coolant.
3. Remove the intake air hose and air cleaner assembly.
   a. Disconnect the Air Flow Sensor (AFS) connector (A).
   b. Disconnect the breather hose (B) from intake air hose.
   c. Remove the intake air hose and air cleaner assembly (C).
4. Remove the upper radiator hose (A) and lower radiator hose (B).
5. Remove the heater hoses (A).
6. Remove the accelerator cable (A) by loosening the lock–nut, then slip the cable end out of the throttle linkage.
7. Remove the engine wire harness connectors and wire harness clamps from cylinder head and the intake manifold.
   a. Disconnect the Oil Control Valve (OCV) connector (A).
   b. Disconnect the oil temperature sensor connector (B).
   c. Disconnect the Engine Coolant Temperature (ECT) sensor connector (C).
   d. Disconnect the ignition coil connector (D).
   e. Disconnect the Throttle Position Sensor (TPS) connector (A).
   f. Disconnect the Idle Speed Actuator (ISA) connector (B).
   g. Disconnect the Camshaft Position (CMP) sensor (CMP) connector (A).
   h. Disconnect the four injector connectors (B).
   i. Disconnect the knock sensor connector (C).
   j. Disconnect the ground cables (D) from the intake manifold.
   k. Disconnect the air conditioner compressor switch connector (E).
   l. Disconnect the front heated oxygen sensor connector (A).
   m. Disconnect the Crankshaft Position Sensor (CKP) connector (B).
   n. Disconnect the oil pressure switch connector (C).
   o. Disconnect the Purge Control Solenoid Valve (PCSV) connector (A).
8. Disconnect the fuel inlet hose (A) of the delivery pipe side.
9. Disconnect the hose (A) of the PCSV side.
10. Remove the brake booster vacuum hose (A).
11. Remove the power steering pump drive belt.
12. Remove the power steering pump and use a wire to secure the pump to the vehicle so that it is out of the way.
13. Remove the bolts (B, C) and power steering pump bracket (A).
14. Remove the spark plug cables.
15. Remove the exhaust manifold.
16. Remove the intake manifold.
17. Remove the timing belt.
18. Remove the PCV hose.
19. Remove the cylinder head cover.
20. Remove the camshaft sprocket.
21. Insert a stopper pin or other device into timing chain auto tensioner and remove the auto tensioner.
22. Remove the camshaft bearing caps (A) and camshafts (B).

To install:

23. Install the camshaft and bearing caps. Tighten to 10 ft. lbs. (14 Nm).
24. Install the timing chain auto tensioner. Tighten to 6 ft. lbs. (8 Nm).
25. Remove the auto tensioner stopper pin.
26. Check and adjust valve clearance.
27. Using the SST (09221–21000), install the camshaft bearing oil seal.
28. Install the camshaft sprocket.

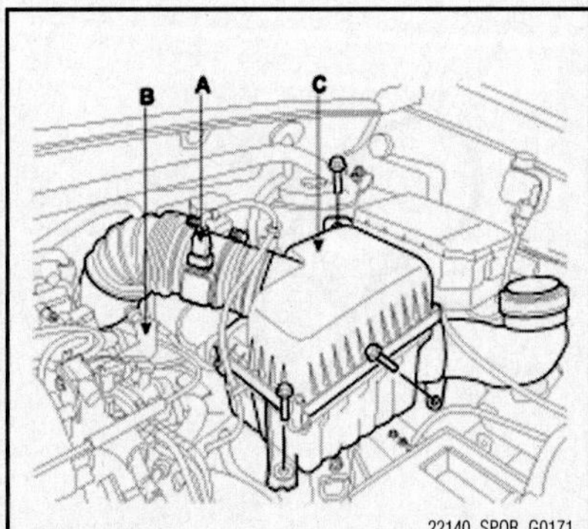

Fig. 29 Disconnect the Air Flow Sensor (AFS) connector (A)

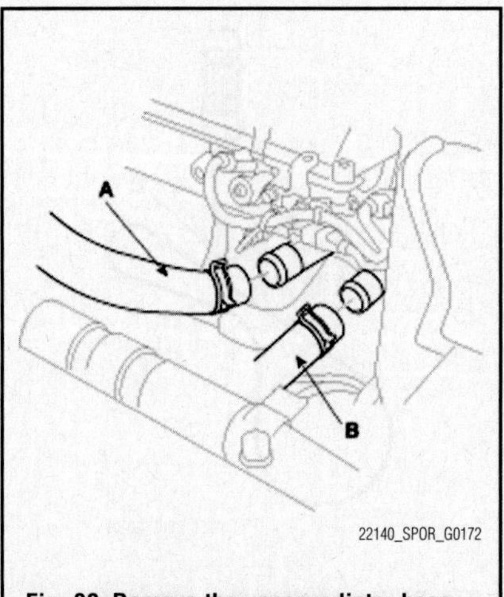

Fig. 30 Remove the upper radiator hose (A) and lower radiator hose (B)

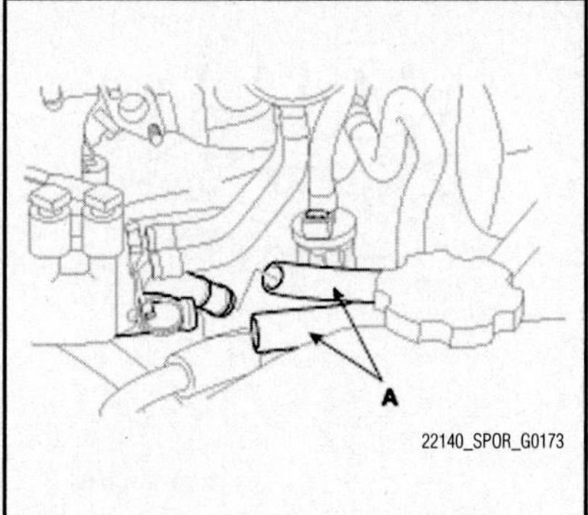

Fig. 31 Remove the heater hoses (A)

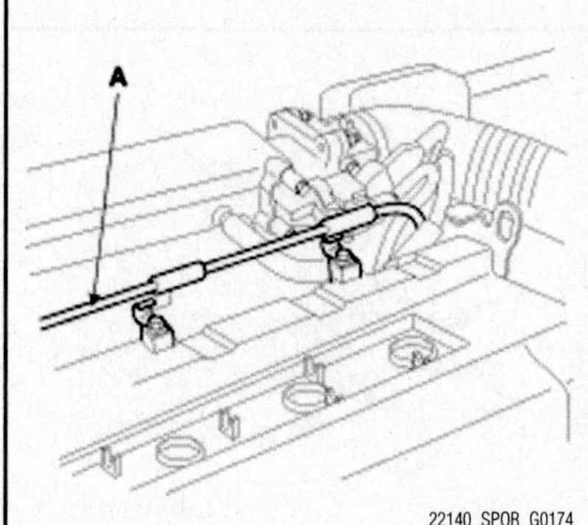

Fig. 32 Remove the accelerator cable (A)

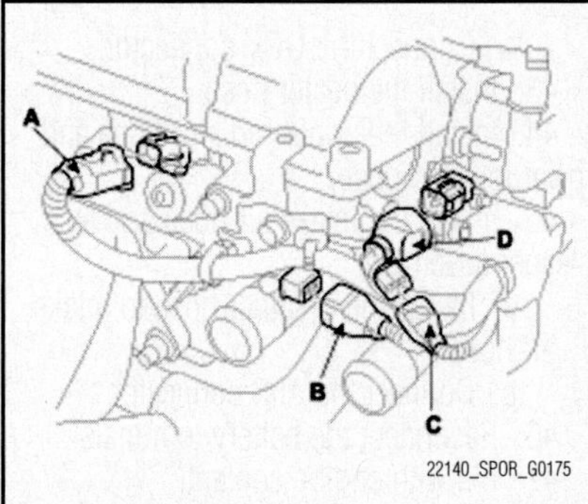

Fig. 33 Remove the engine wire harness connectors

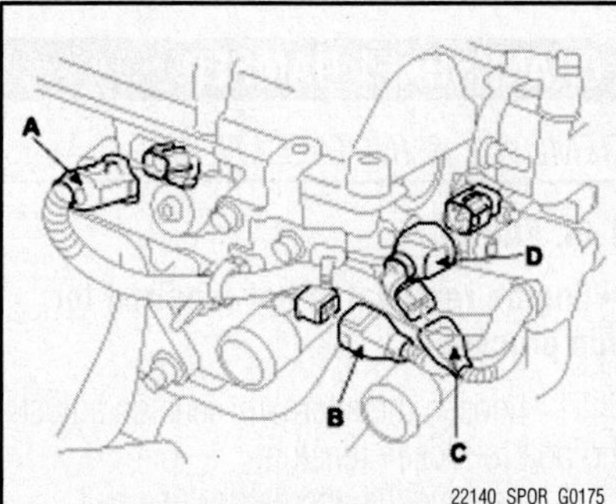

Fig. 34 Disconnect the Throttle Position Sensor (TPS) connector (A) and the Idle Speed Actuator (ISA) connector (B)

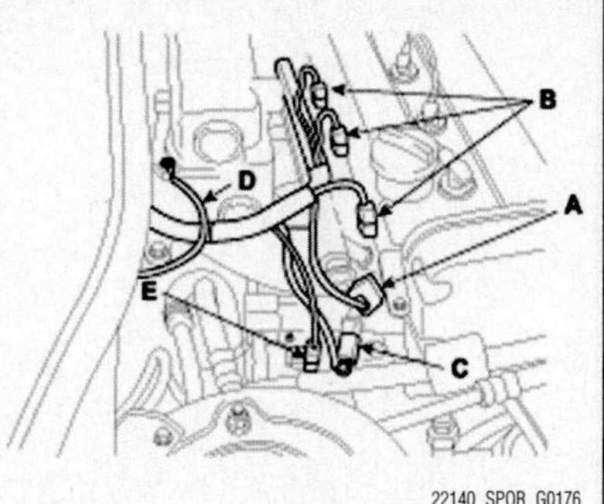

Fig. 35 Disconnect the Camshaft Position (CMP) sensor (CMP) connector (A), the four injector connectors (B), the knock sensor connector (C), the ground cables (D) and the air conditioner compressor switch connector (E)

29. Install the cylinder head cover.
    a. Install the cylinder head cover gasket in the groove of the cylinder head cover.

➡Before installing the cylinder head cover gasket, thoroughly clean the cylinder head cover and the groove. When installing, make sure the cylinder head cover gasket is seated securely in the corners of the recesses with no gap.

    b. Apply liquid gasket to the head cover gasket at the corners of the recess.

➡Use liquid gasket, Loctite® No. 5999. Check that the mating surfaces are clean and dry before applying liquid gasket. After assembly, wait at least 30 minutes before filling the engine with oil.

30. Install the cylinder head cover (A) with the 12 bolts (B). Uniformly tighten the bolts in several passes. Tighten to 6–7 ft. lbs. (8–10 Nm).
31. Install the PCV hose.
32. Install the timing belt.
33. Install the intake manifold.
34. Install the exhaust manifold.
35. Install the spark plug cables.
36. Install the power steering pump bracket and bolts. Tighten to Bolt (B): 25–36 ft. lbs. (34–49 Nm); Bolt (C): 9–11 ft. lbs. (12–15 Nm).

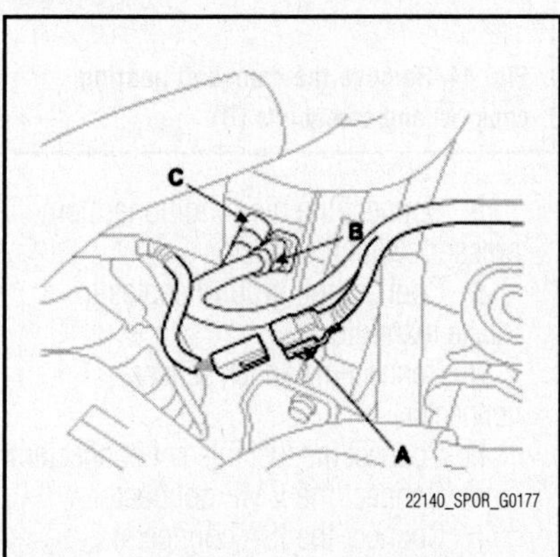

Fig. 36 Disconnect the front heated oxygen sensor connector (A), the Crankshaft Position Sensor (CKP) connector (B) and the oil pressure switch connector (C)

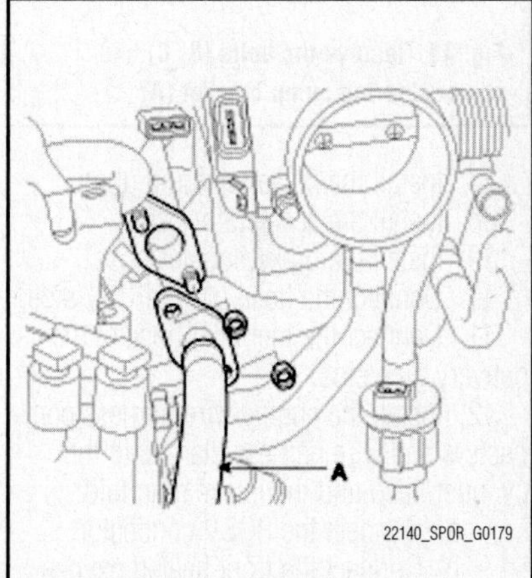

Fig. 38 Disconnect the fuel inlet hose (A) of the delivery pipe side

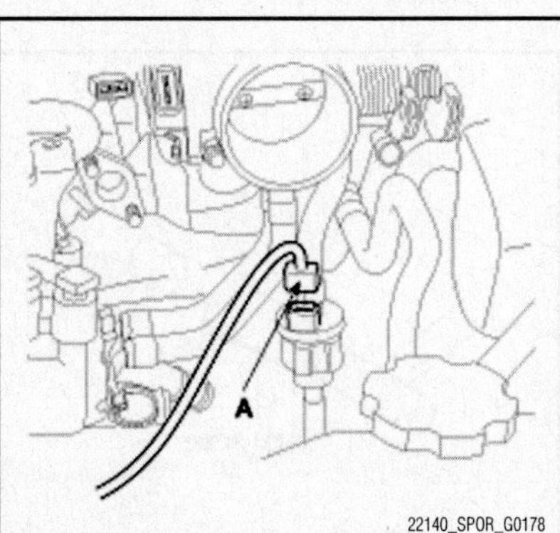

Fig. 37 Disconnect the Purge Control Solenoid Valve (PCSV) connector (A)

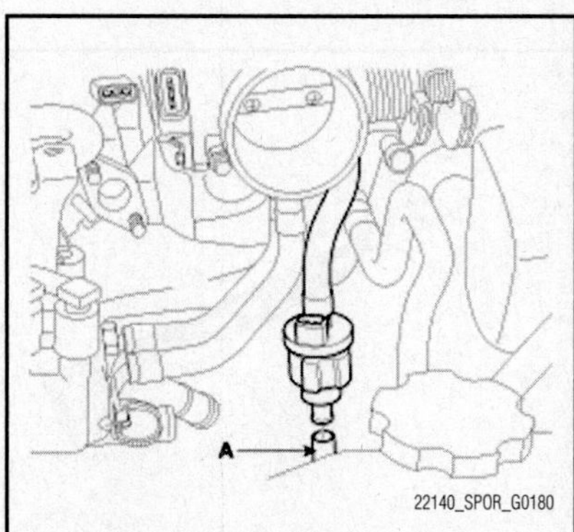

Fig. 39 Disconnect the hose (A) of the PCSV side

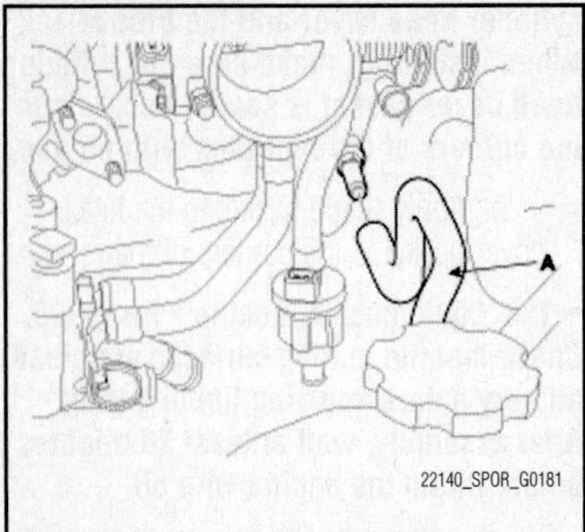

Fig. 40 Remove the brake booster vacuum hose (A)

Fig. 43 Remove the auto tensioner

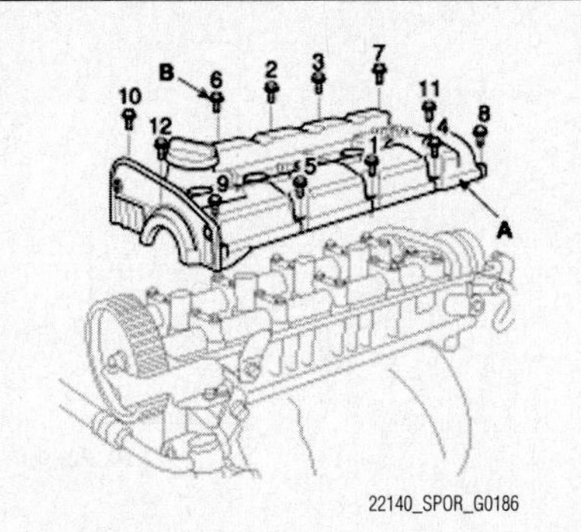

Fig. 46 Install the cylinder head cover (A) in sequence

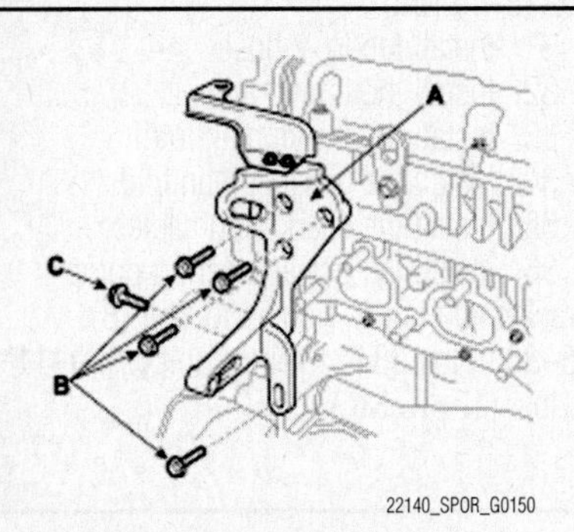

Fig. 41 Remove the bolts (B, C) and power steering pump bracket (A)

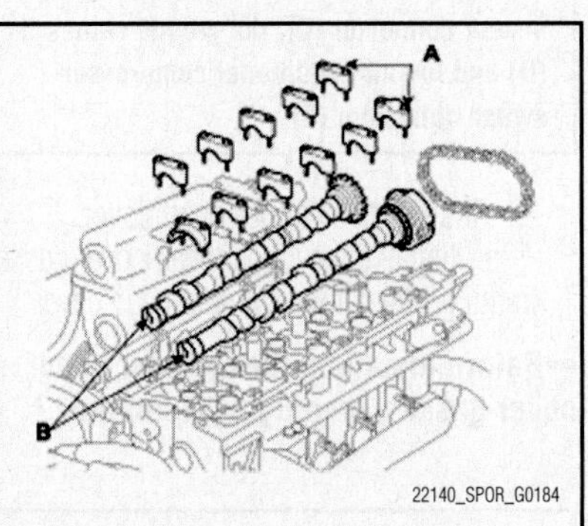

Fig. 44 Remove the camshaft bearing caps (A) and camshafts (B)

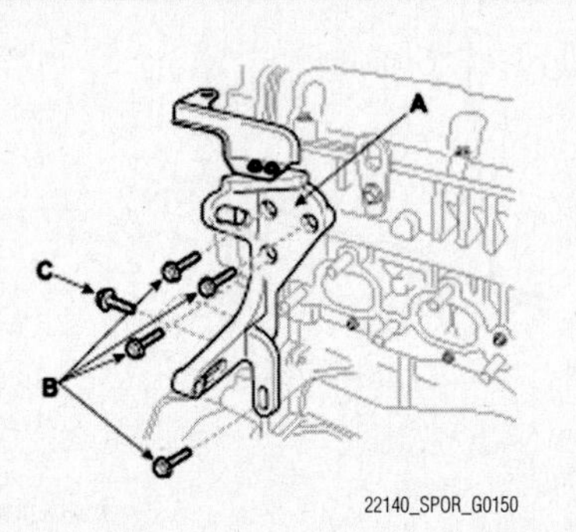

Fig. 47 Torque the power steering bracket bolts

37. Install the power steering pump.
38. Install the accelerator cable.
39. Install the brake booster hose.
40. Connect the hose of the PCSV side.
41. Connect the fuel inlet hose of the delivery pipe side.
42. Install the engine wire harness connectors and wire harness clamps to the cylinder head and the intake manifold.
    a. Connect the PCSV connector.
    b. Connect the front heated oxygen sensor connector.
    c. Connect the CKP connector.
    d. Connect the oil pressure switch connector.
    e. Connect the air conditioner compressor switch connector.
    f. Connect the ground cables to intake manifold.
    g. Connect the knock sensor connector.
    h. Connect the fuel injector connectors.
    i. Connect the CMP connector.
    j. Connect the ISA connector.
    k. Connect the TPS connector.
    l. Connect the ignition coil connector.
    m. Connect the ECT connector.
    n. Connect the oil temperature sensor connector.
    o. Connect the OCV connector.
43. Install the heater hose.
44. Install the upper radiator hose and lower radiator hose.
45. Install the intake air hose and air cleaner assembly.
    a. Install the breather hose to intake air hose.
    b. Connect the AFS connector.
46. Reconnect the battery terminals.
47. Fill with engine coolant.
48. Start the engine and check for leaks.
49. Recheck engine coolant level and oil level.

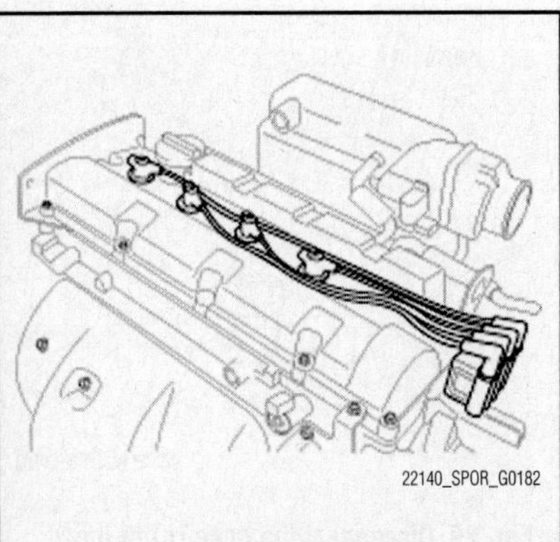

Fig. 42 Remove the spark plug cables

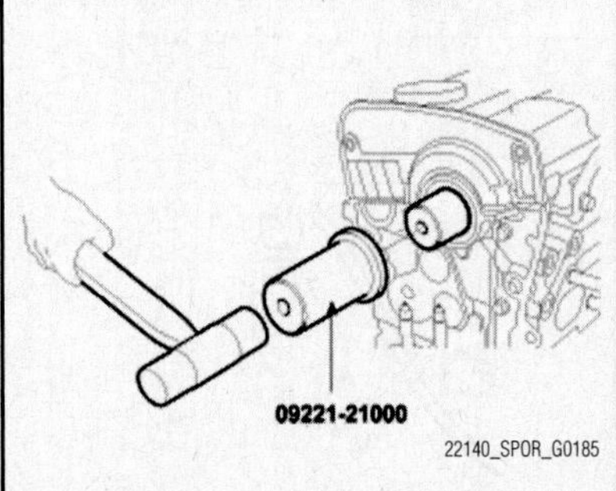

Fig. 45 Using the SST (09221–21000), install the camshaft bearing oil seal

CRANKSHAFT DAMPER

REMOVAL & INSTALLATION

1.6L Engine

➡Engine removal is not required for this procedure.

1. Loosen the alternator tension adjusting bolt to loosen tension.
2. Remove the alternator drive belt.
3. Remove the RH front wheel.
4. Remove side cover in front of crank pulley.
5. Turn the crankshaft pulley clockwise,

and align its groove with the timing mark of the timing chain cover.

6. Remove the crankshaft bolt and crankshaft pulley.

To install:

➡**When installing the pulley, the groove on the pulley should be positioned outside, not next to pointer.**

7. Install the crankshaft pulley and bolt. Tighten to 94–101 ft. lbs. (128–137 Nm).
8. Install side cover. Tighten to 6.5–8 ft. lbs. (9–11 Nm).
9. Install the front right wheel and tire.
10. Install the drive belt.
11. Tighten the alternator 12 mm mounting bolt to 15–20 ft. lbs. (20–27 Nm).
12. Tighten the alternator 14 mm mounting bolt to 22–30 ft. lbs. (30–41 Nm).

2.0L Engine

1. Disconnect the negative battery cable.
2. Remove the accessory drive belts.
3. Remove the crankshaft damper.

To install:

4. Install the crankshaft damper and tighten the bolt to 123–130 ft. lbs. (167–176 Nm).
5. Install and tension the accessory drive belts.

CRANKSHAFT FRONT SEAL

REMOVAL & INSTALLATION

1.6L Engine

See Figure 48.

1. Disconnect the negative battery cable.
2. Remove the timing belt covers and belt.
3. Remove the timing belt pulley using a puller.
4. Remove the oil pump bolts and the pump.
5. Wrap a suitable pry tool with a rag and work the old seal from the oil pump housing.

To install:

6. Lubricate the seal lip with clean engine oil and push the seal slightly in by hand.
7. Install the seal using a SST (09455–21200) seal installer. Install the seal until it is flush with the oil pump body.

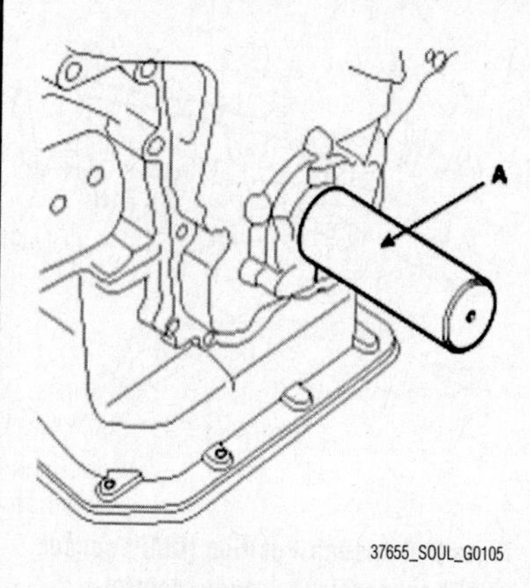

Fig. 48 Installing the seal using a SST (09455–21200) seal installer—1.6L engine

8. Install the timing belt.
9. Connect the negative battery cable.
10. Start the engine and check for leaks.

2.0L Engine

See Figures 49 and 50.

1. Drain the engine oil.
2. Remove the drive belts.
3. Turn the crankshaft pulley, and align its groove with timing mark "T" of the timing belt cover.
4. Remove the timing belt. See Timing Belt.
5. Remove the bolt and timing belt idler.
6. Remove the oil pan and oil screen.
7. Remove the alternator. See Alternator.
8. Remove the air conditioner compressor tensioner bracket.
9. Remove the bolts and front case.

To install:

10. Place a new front case gasket on the cylinder block.

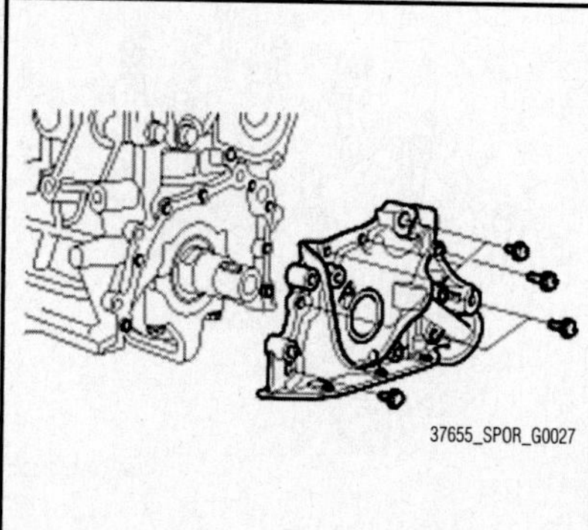

Fig. 49 Front case and bolts—2.0L engine

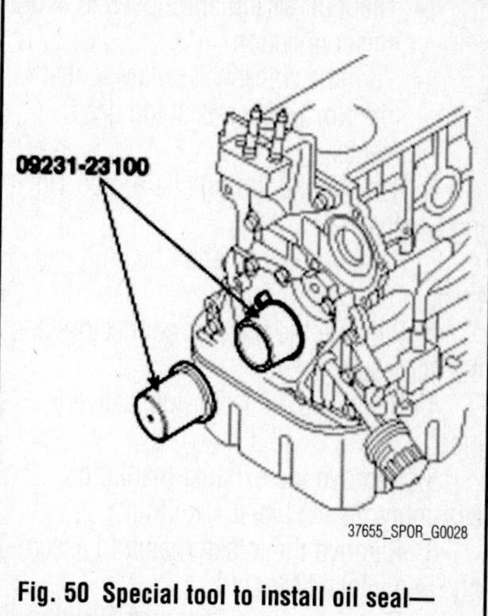

Fig. 50 Special tool to install oil seal—2.0L engine

11. Apply engine oil to the lip of the oil pump seal. Then, install the front case onto the crankshaft.
12. Tighten front case bolts to 14.5 – 19.5 ft. lbs. (19.6 – 26.5 Nm).
13. When the pump is in place, clean any excess grease off the crankshaft and check that the oil seal lip is not distorted.
14. Apply a light coat of oil to the front case oil seal lip.
15. Using special service tools (09231–23100), install the front case oil seal.
16. Install remaining components in the reverse order of removal

CYLINDER HEAD

REMOVAL & INSTALLATION

1.6L Engine

See Figures 51 through 56.

➡**Engine removal is not required for this procedure.**

**** CAUTION**

- Use Fender cover to avoid damaging painted surfaces.
- To avoid damaging the cylinder head, wait until the engine coolant temperature drops below normal temperature before removing it.
- When handling a metal gasket, take care not to fold the gasket or damage the contact surface of the gasket.
- To avoid damage, unplug the wiring connectors carefully while holding the connector portion.

- Mark all wiring and hoses to avoid misconnection.
- Turn the crankshaft pulley so that the No. 1 piston is at top dead center.

1. Remove the timing chain. See Timing Chain.
2. Remove the camshaft bearing cap bolts in the order shown.
3. Remove the injector connectors and the harness bracket.
4. Remove the injector fuel delivery pipe.
5. Remove the exhaust manifold assembly. See Exhaust Manifold.
6. Remove the intake manifold assembly. See Intake Manifold.
7. Disconnect the Camshaft Position (CMP) sensor (CMP) connector (A) and remove the purge control solenoid valve (PCSV) bracket (B) and the module hanger bracket (C).
8. Remove the water temperature control assembly and the oil control valve (OCV).
9. Remove the cylinder head bolts, then remove the cylinder head.
   a. Uniformly loosen and remove the 10 cylinder head bolts, in several passes, in the sequence shown.

CAUTION

Head warpage or cracking could result from removing bolts in an incorrect order.

10. Lift the cylinder head from the cylinder block and put the cylinder head on wooden blocks.

CAUTION

Be careful not to damage the contact surfaces of the cylinder head and cylinder block.

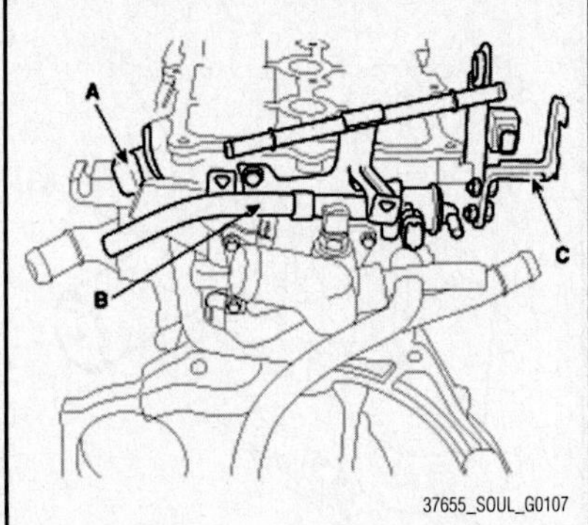

Fig. 52 Camshaft Position (CMP) sensor (CMP) connector (A), purge control solenoid valve (PCSV) bracket (B) and module hanger bracket (C)—1.6L engine

To install:

- Thoroughly clean all parts to be assembled.
- Always use a new cylinder head and manifold gasket.
- Always use a new cylinder head bolt.
- The cylinder head gasket is a metal gasket. Take care not to bend it.

11. Rotate the crankshaft, set the No.1 piston at TDC.
12. Install the cylinder head assembly.
    a. Before installing, remove the hardened sealant from the cylinder block and cylinder head surface.
    b. Before installing the cylinder head gasket, apply sealant on the upper surface of the cylinder block and reassemble the gasket within five minutes.
    c. After installing the cylinder head gasket on the cylinder block, apply sealant on the upper surface of the cylinder head gasket and reassemble in five minutes.
13. Place the cylinder head carefully not to damage the gasket.
14. Install the cylinder head bolts with washers. Always use new cylinder head bolts.
15. Tighten the 10 cylinder head bolts, in several passes, in the sequence shown.
    a. 1st step – 13.0–15.9 ft. lbs. (17.7–21.6 Nm).
    b. 2nd step – 90°–95°
    c. 3rd step – 100°–105°
16. Install the oil control valve (OCV). Tighten to 7.2–8.7 ft. lbs. (9.8–11.8 Nm).
17. Install the water temperature control assembly. Tighten to 7.2–8.7 ft. lbs. (9.8–11.8 Nm).
18. Connect the Camshaft Position (CMP) sensor (CMP) connector and install the purge control solenoid valve (PCSV) bracket and the module hanger bracket.
19. Install the intake and exhaust manifold. See Intake Manifold. See Exhaust Manifold.
20. Install the delivery pipe assembly. Tighten to 14.4–18 ft. lbs. (20–25 Nm).
21. Install the injector connector and harness bracket.
22. Install the camshafts.
    a. Before installing, apply engine oil on journals.

➡Do not make oil flow down to the front side of the cylinder head.

23. Install the camshaft bearing cap bolts in the sequence shown.
    a. Tighten M6 bolts to 8.7–10 ft. lbs. (11.8–13.7 Nm).
    b. Tighten M8 bolts to 13.7–17 ft. lbs. (18.6–22.6 Nm).
    c. After installing, check the valve clearance. See Valve Lash.

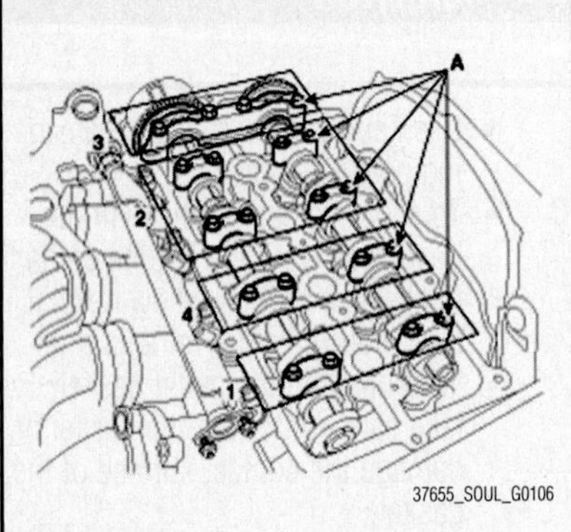

Fig. 51 Removing camshaft bearing cap bolts sequence—1.6L engine

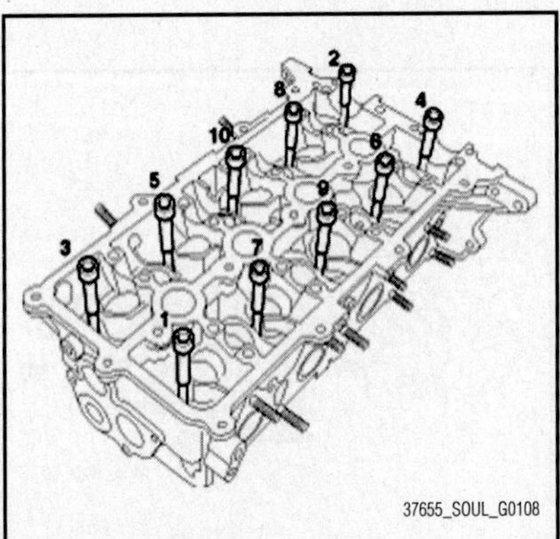

Fig. 53 Removing cylinder head bolts sequence—1.6L engine

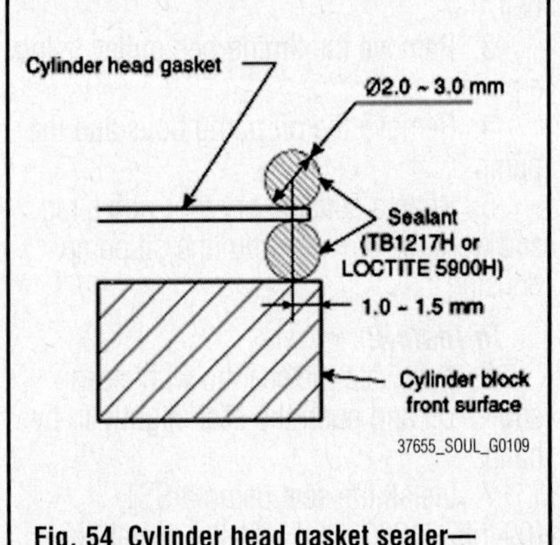

Fig. 54 Cylinder head gasket sealer—1.6L engine

Fig. 55 Cylinder head bolts torque sequence—1.6L engine

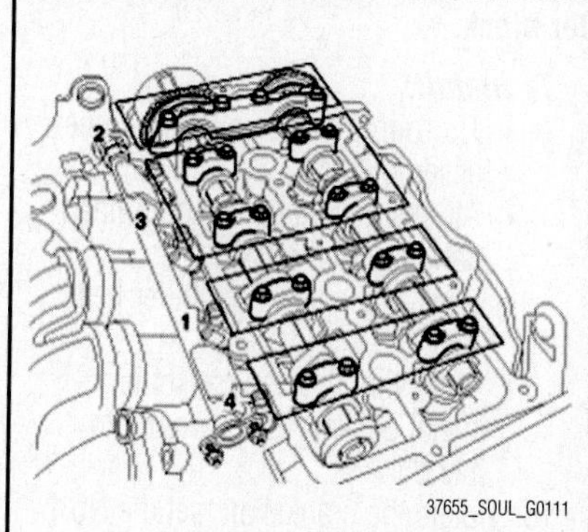

Fig. 56 Camshaft bearing cap bolts torque sequence—1.6L engine

24. Install the intake and exhaust manifold. See Intake Manifold. See Exhaust Manifold.

25. Install the timing chain. See Timing Chain.

2.0L Engine

See Figures 57 through 68.

➡Engine removal is not required for this procedure.

- Use Fender cover to avoid damaging painted surfaces.
- To avoid damaging the cylinder head, wait until the engine coolant temperature drops below normal temperature before removing it.
- When handling a metal gasket, take care not to fold the gasket or damage the contact surface of the gasket.
- To avoid damage, unplug the wiring connectors carefully while holding the connector portion.

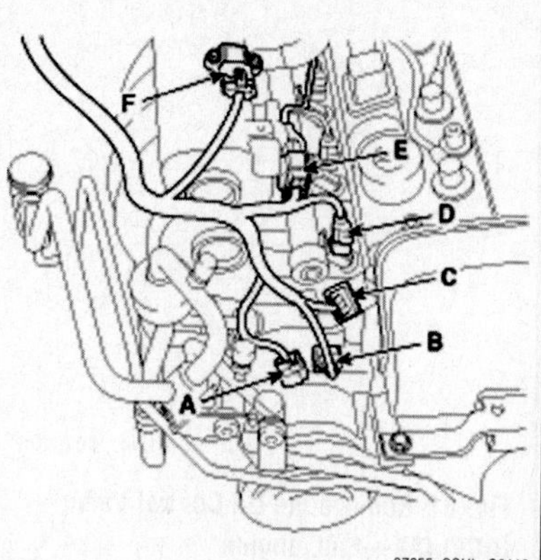

Fig. 57 A/C compressor switch connector (A), knock sensor connector (B), TDC sensor connector (C), fuel injector (No. 1) connector (D), fuel injector (No. 2, 3, 4) connector (E) and MAP sensor connector (F)—2.0L engine

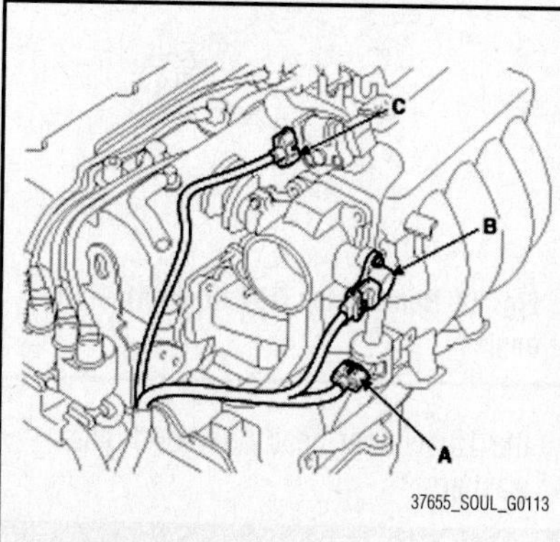

Fig. 58 PCSV connector (A), TPS connector (B) and ISCA connector (C)—2.0L engine

- Mark all wiring and hoses to avoid misconnection.

1. Disconnect the terminals from battery.
2. Remove the intake air hose and air cleaner assembly.
3. Drain the engine coolant.
4. Remove the upper radiator hose and lower radiator hose.
5. Disconnect the engine wiring harness connectors, clamps and hoses.
   a. Disconnect the A/C compressor switch connector (A).
   b. Disconnect the knock sensor connector (B).
   c. Disconnect the TDC sensor connector (C).
   d. Disconnect the fuel injector (No. 1) connector (D).
   e. Disconnect the fuel injector (No. 2, 3, 4) connector (E).
   f. Disconnect the MAP sensor connector (F).
   g. Disconnect the engine ground line.
   h. Disconnect the PCSV connector (A).
   i. Disconnect the TPS connector (B).
   j. Disconnect the ISCA connector (C).
   k. Disconnect the front heated oxygen sensor connector (A).
   l. Disconnect the rear heated oxygen sensor connector (B).
   m. Disconnect the oil pressure switch connector (C).
   n. Disconnect the OPS connector (D).
   o. Disconnect the CKPS connector (E).
   p. Disconnect the OTS connector (F).
   q. Disconnect the ECTS connector (D).
   r. Disconnect the ignition coil connector (E).
   s. Disconnect the inhibitor switch connector (F).
   t. Disconnect the brake booster vacuum hose.
   u. Disconnect the heater hoses.
   v. Disconnect the fuel inlet hose.
   w. Disconnect the PCSV hose.
6. Remove the cylinder head cover.
   a. Disconnect the spark plug cables and do not pull on the spark plug by force.

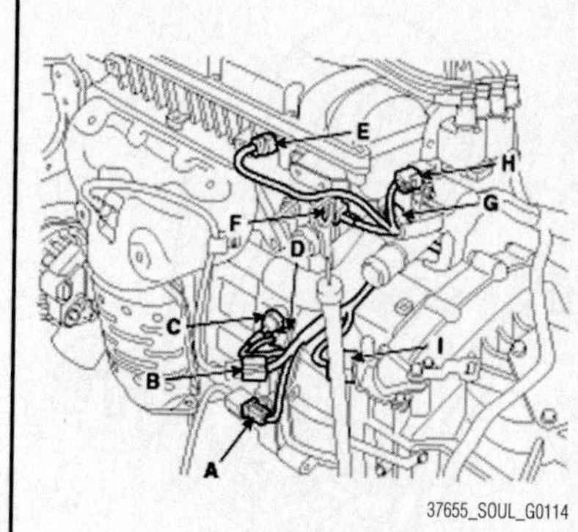

Fig. 59 Front heated oxygen sensor connector (A), rear heated oxygen sensor connector (B), oil pressure switch connector (C), OPS connector (D), CKPS connector (E), OTS connector (F), ECTS connector (G), ignition coil connector (H) and inhibitor switch connector (I)—2.0L engine

CAUTION

Pulling on or bending the cables may damage the conductor inside.

b. Disconnect the positive crankcase ventilation (PCV) hose.

c. Disconnect the accelerator cable and the auto–cruise cable from the cylinder head cover.

7. Remove the timing belt. See Timing Belt.

8. Remove the exhaust manifold and intake manifold. See Intake Manifold an d Exhaust Manifold.

9. Remove camshaft sprocket.

a. Hold the hexagonal head wrench (A) portion of the camshaft with a wrench (B), and remove the bolt and camshaft sprocket (C).

10. Remove the timing chain auto tensioner after installing the auto tensioner stopper pin.

11. Remove the camshaft bearing caps and camshafts.

12. Remove the OCV (oil control valve) (A).

13. Remove the OCV (oil control valve) filter (A).

14. Disconnect the water pipe and water hose.

15. Remove the cylinder head bolts, then remove the cylinder head.

a. Using 8mm and 10mm hexagon wrench, uniformly loosen and remove the 10 cylinder head bolts, in several passes, in the sequence shown. Remove the 10 cylinder head bolts and plate washers.

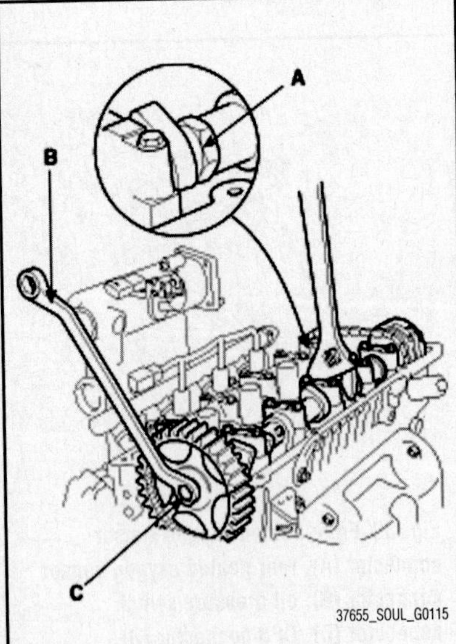

Fig. 60 Hexagonal head wrench (A) wrench (B), and bolt and camshaft sprocket (C)—2.0L engine

Fig. 61 Remove the Oil Control Valve (OCV) (A) —2.0L engine

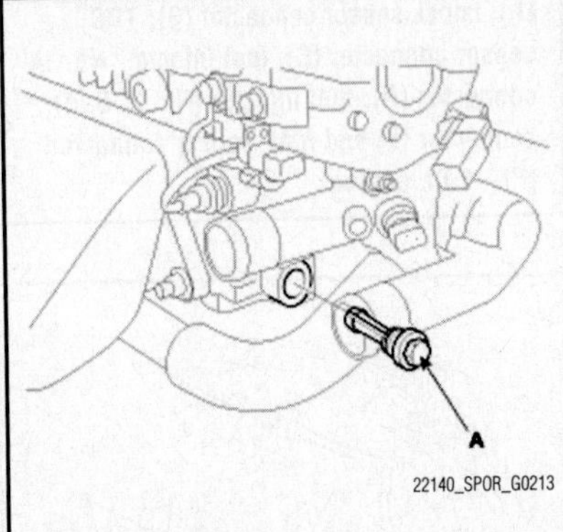

Fig. 62 Remove the OCV filter (A) —2.0L engine

CAUTION

Head warpage or cracking could result from removing bolts in an incorrect order.

b. Lift the cylinder head from the dowels on the cylinder block and place the cylinder head on wooden blocks on a bench.

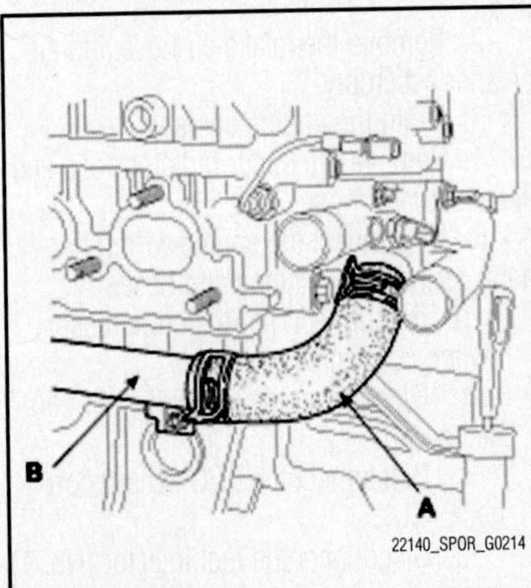

Fig. 63 Water hose (A) and water pipe (B) —2.0L engine

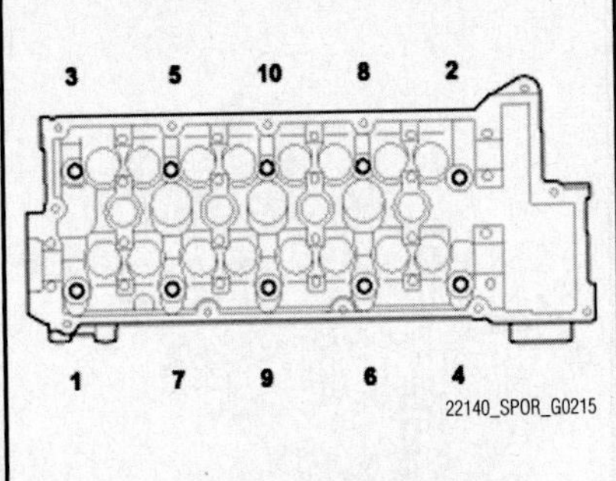

Fig. 64 Remove the 10 cylinder head bolts and plate washers, in several passes, in sequence—2.0L engine

➡Be careful not to damage the contact surfaces of the cylinder head and cylinder block.

To install:

- Thoroughly clean all parts to be assembled.
- Always use a new cylinder head and manifold gaskets.
- Always use a new cylinder head bolt.
- The cylinder head gasket is a metal gasket. Take care not to bend it.

16. Rotate the crankshaft, set the No.1 piston at TDC.

17. Install the cylinder head gasket on the cylinder block.

➡Be careful of the installation direction.

18. Install the cylinder head assembly.

19. Install the cylinder head bolts.

a. Apply a light coat if engine oil on the threads and under the heads of the cylinder head bolts.

b. Install and tighten the 10 cylinder head bolts and plate washers, in several passes, in sequence. Tighten to M10: 16.6–20 ft. lbs. (23–27 Nm+60°–65° + 60°–65°; M12: 20–23 ft. lbs. (28–31 Nm)+60°–65° +60°–65°

20. Install the OCV filter. Tighten to 30–37 ft. lbs. (40–50 Nm)

➡Always use a new OCV filter gasket. Keep the OCV filter clean.

21. Install the OCV. Tighten to 7–9 ft. lbs. (10–12 Nm)

CAUTION

Do not reuse the OCV when dropped. Keep the OCV clean. Do not hold the OCV sleeve during servicing. When

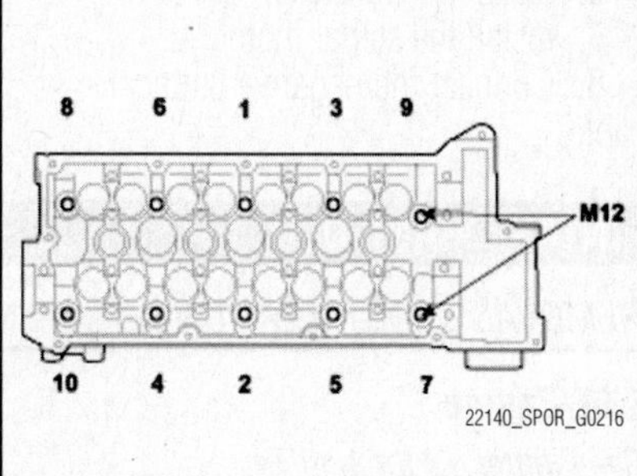

Fig. 65 Install and tighten the 10 cylinder head bolts and plate washers, in several passes, in sequence—2.0L engine

the OCV is installed on the engine, be careful not to rotate the engine while holding the yoke.

22. Align the camshaft timing chain with the intake timing chain sprocket and exhaust timing chain sprocket as shown.
23. Install the camshaft and bearing caps. Tighten to 10–11 ft. lbs. (14–15 Nm).
24. Install the timing chain auto tensioner. Tighten to 6–7 ft. lbs. (8–10 Nm).
25. Remove the auto tensioner stopper pin.
26. Check and adjust valve clearance. See Valve Lash.
27. Using the SST (09221–21000), install the camshaft bearing oil seal.
28. Install the camshaft sprocket. Tighten to 72–87 ft. lbs. (98–118 Nm).
29. Install the timing belt.
30. Install the cylinder head cover.

➡Before installing the cylinder head cover gasket, thoroughly clean the cylinder head cover and the groove. When installing, make sure the cylinder head cover gasket is seated securely in the corners of the recesses with no gap.

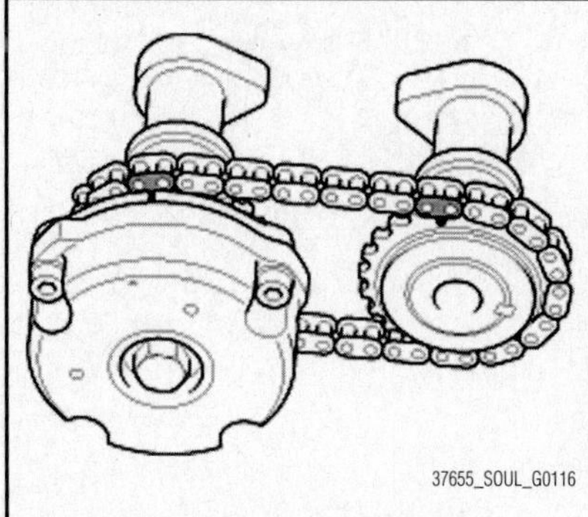

Fig. 66 Aligning the camshaft timing chain with the intake timing chain sprocket and exhaust timing chain sprocket—2.0L engine

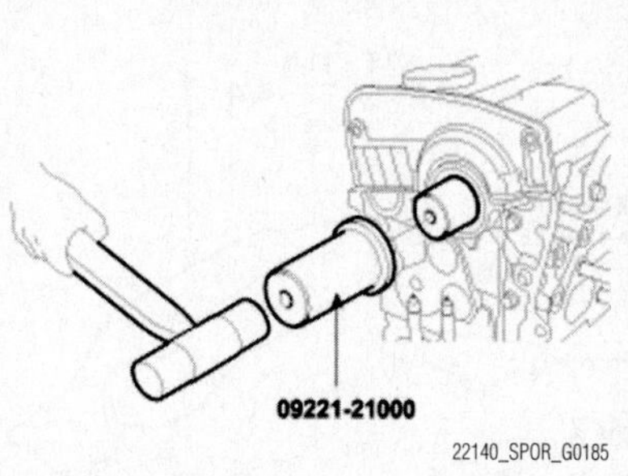

Fig. 67 Using the SST (09221–21000), install the camshaft bearing oil seal—2.0L engine

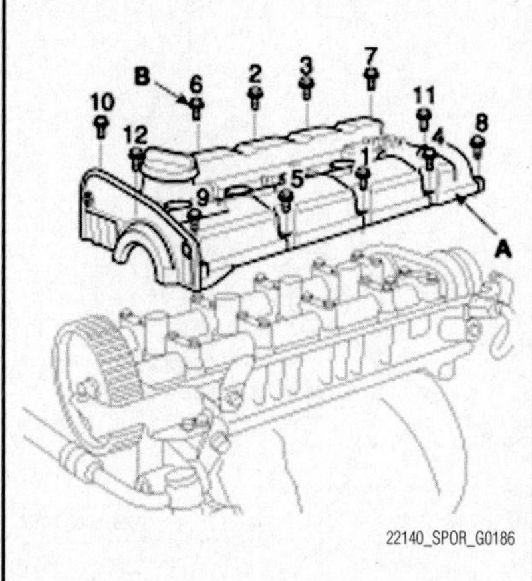

Fig. 68 Install the cylinder head cover (A) in sequence—2.0L engine

a. Install the cylinder head cover gasket in the groove of the cylinder head cover.
b. Apply liquid gasket to the head cover gasket at the corners of the recess.

➡Use liquid gasket, Loctite No. 5999. Check that the mating surfaces are clean and dry before applying liquid gasket. After assembly, wait at least 30 minutes before filling the engine with oil.

31. Install the cylinder head cover (A) with the 12 bolts (B). Uniformly tighten the bolts in several passes. Tighten to 6–7 ft. lbs. (8–10 Nm).
32. Install the PCV hose.
 a. Connect the accelerator cable and the auto–cruise cable to the cylinder head cover.
33. Install the spark plug cables.
34. Install the intake manifold.
35. Install the exhaust manifold.
36. Connect the engine wiring harness connectors, clamps and hoses.
 a. Connect the brake booster vacuum hose.
 b. Connect the heater hoses.
 c. Connect the fuel inlet hose.
 d. Connect the PCSV hose.
 e. Connect the front heated oxygen sensor connector.
 f. Connect the rear heated oxygen sensor connector.
 g. Connect the oil pressure switch connector.
 h. Connect the CKPS connector.
 i. Connect the OCV connector.
 j. Connect the OTS connector.
 k. Connect the ECTS connector.
 l. Connect the ignition coil connector.
 m. Connect the inhibitor switch connector.
 n. Connect the PCSV connector.
 o. Connect the TPS connector.
 p. Connect the ISCA connector.
 q. Connect the engine ground line.
 r. Connect the A/C compressor switch connector.
 s. Connect the knock sensor connector.
 t. Connect the TDC sensor connector.
 u. Connect the fuel injector (No. 1) connector.
 v. Connect the fuel injector (No. 2,3,4) connector.
 w. Connect the MAP sensor connector.
37. Install the upper radiator hose and lower radiator hose.
38. Install the air intake hose and air cleaner assembly.
39. Connect the battery terminals.
40. Fill with engine coolant.
41. Start the engine and check for leaks.
42. Recheck engine coolant level and oil level.

EXHAUST MANIFOLD

REMOVAL & INSTALLATION

1.6L Engine

See Figure 69.

1. Disconnect the oxygen sensor connectors and then, remove the heat protector.
2. Remove the catalytic converter.
3. Remove the exhaust manifold stay and the exhaust manifold with its gasket.
4. Installation is in the reverse order of removal

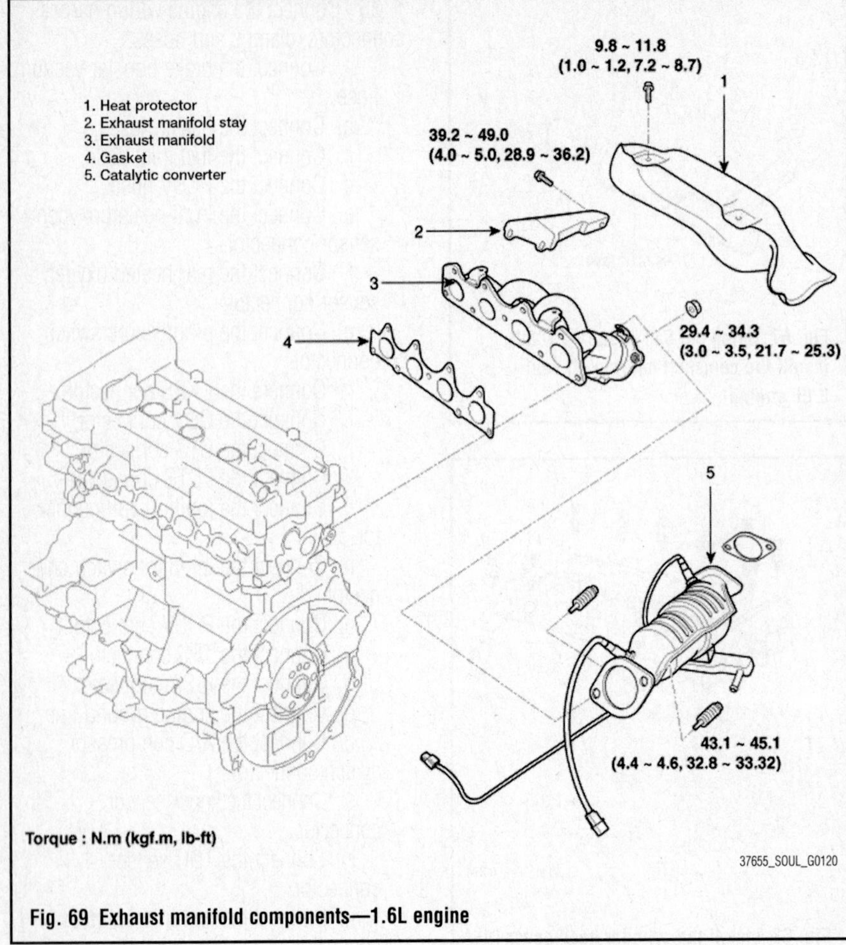

Fig. 69 Exhaust manifold components—1.6L engine

2.0L Engine

See Figures 70 and 71.

1. Remove the engine cover.
2. Disconnect the front oxygen sensor connector.
3. Remove the front muffler.
4. Remove the heat protector.
5. Remove the exhaust manifold and catalytic converter assembly.
6. Installation is in the reverse order of removal

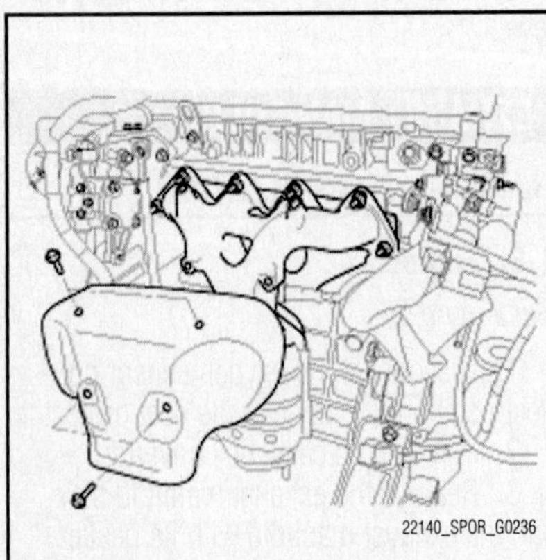

Fig. 70 Remove the heat protector

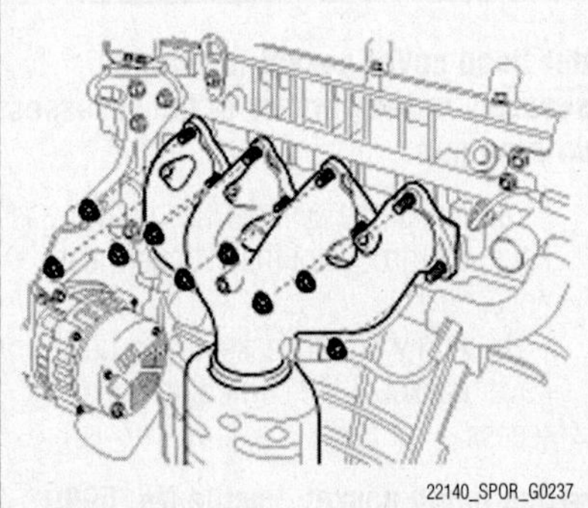

Fig. 71 Remove the exhaust manifold and catalytic converter assembly

FLYWHEEL/FLEXPLATE

REMOVAL & INSTALLATION

1. Disconnect the negative battery cable.
2. Remove the starter motor.
3. Remove the transaxle.
4. Remove the flywheel/flexplate.

To install:

5. Install the flywheel/flexplate. Tighten the bolts to 87–94 ft. lbs. (118–128 Nm).
6. Install the transaxle.
7. Install the starter motor.
8. Connect the negative battery cable.

INTAKE MANIFOLD

REMOVAL & INSTALLATION

1.6L Engine

See Figures 72 through 74.

1. Disconnect the battery terminal.
2. Disconnect the breather hose and remove the air cleaner assembly.
3. Disconnect the wiring connectors and harness clamps:
   - (1) Disconnect the TPS (Throttle Position Sensor) connector (A).
   - (2) Disconnect the ISA (Idle Speed Actuator) connector (B).
   - (3) Disconnect the CMP (Camshaft Position (CMP) sensor) connector (C).
   - (4) Disconnect the PCSV hose (D).
   - (5) Disconnect the OCV connector (A), the alternator connector (B).
   - (6) Disconnect the compressor connector (C).
   - (7) Remove the ignition coil harness mounting bolts (D).
   - (8) Remove the knock sensor bracket (E) and disconnect _ MAP sensor connector (F).
4. Remove the oil level gauge.
5. Remove the intake manifold assembly.
6. Installation is in the reverse order of removal with new gasket.

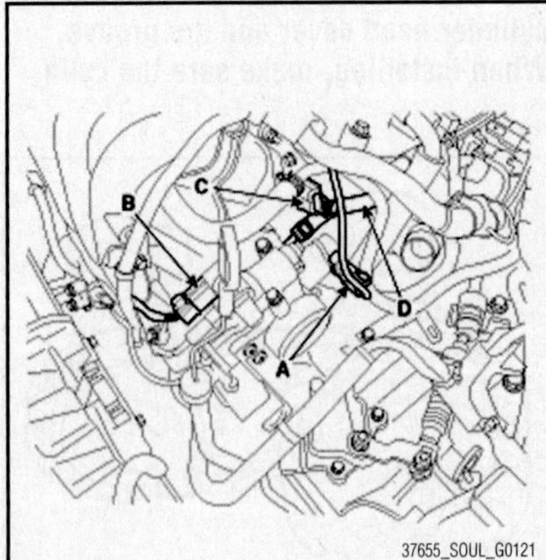

Fig. 72 TPS (Throttle Position Sensor) connector (A), ISA (Idle Speed Actuator) connector (B), CMP (Camshaft Position (CMP) sensor) connector (C) and PCSV hose (D)—1.6L engine

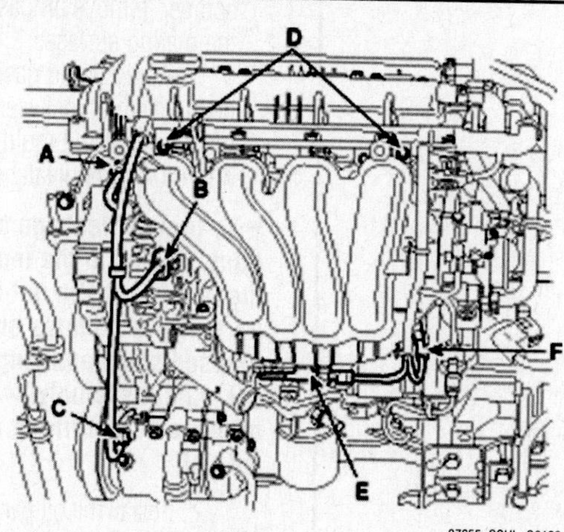

Fig. 73 OCV connector (A), alternator connector (B), compressor connector (C), ignition coil harness mounting bolts (D), knock sensor bracket (E) and MAP sensor connector (F)—1.6L engine

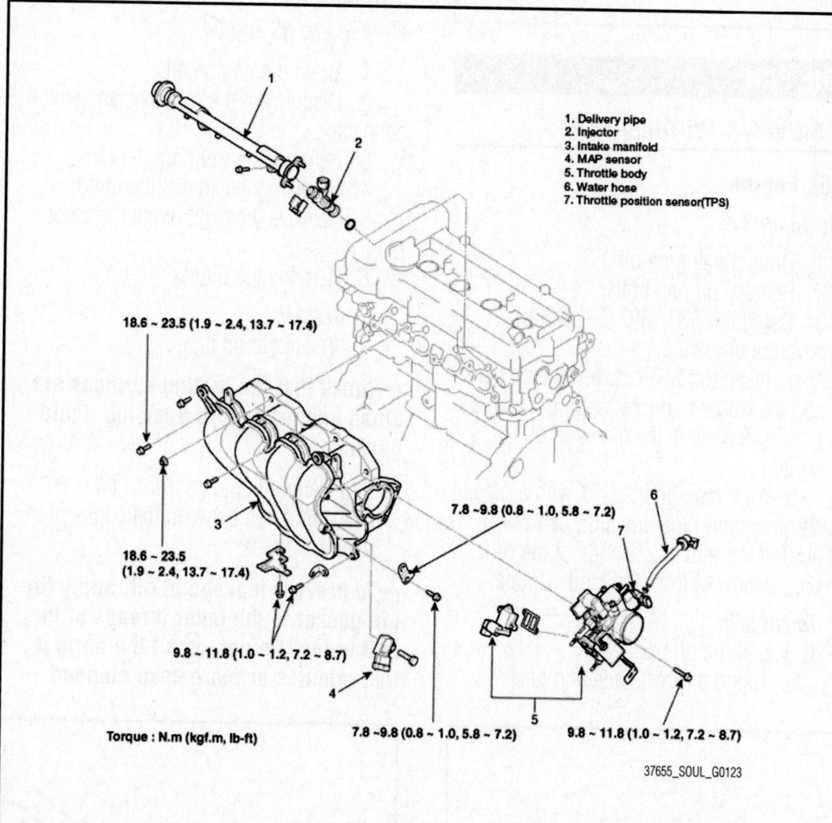

Fig. 74 Intake manifold components—1.6L engine

2.0L Engine

See Figures 75 through 81.

1. Remove the engine cover.
2. Disconnect the Throttle Position Sensor (TPS) connector (A) and Idle Speed Actuator (ISA) connector (B).
3. Disconnect the Positive Crankcase Ventilation (PCV) hose (A) and breather hose (B).
4. Remove the accelerator cable.
5. Remove the fuel delivery pipe (A).
6. Remove the heater hose (A), Purge Control Solenoid Valve (PCSV) (B) and the brake vacuum hose (C) from throttle body and intake manifold.
7. Remove the air conditioner compressor.

Fig. 75 Remove the engine cover

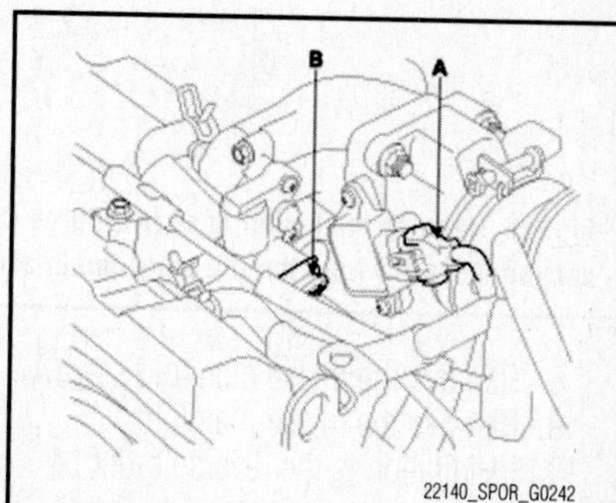

Fig. 76 Disconnect the Throttle Position Sensor (TPS) connector (A) and Idle Speed Actuator (ISA) connector (B)

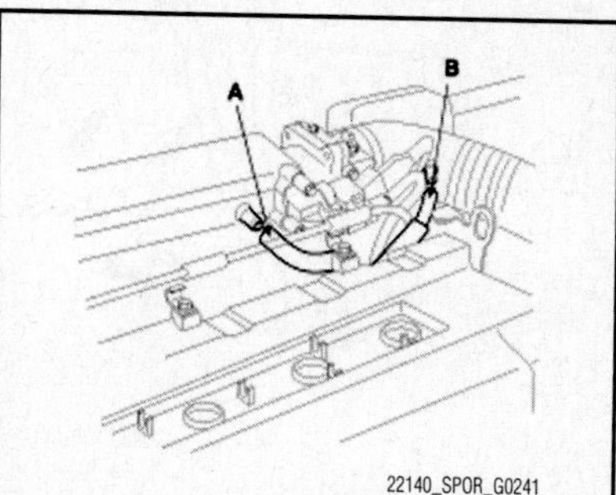

Fig. 77 Disconnect the Positive Crankcase Ventilation (PCV) hose (A) and breather hose (B)

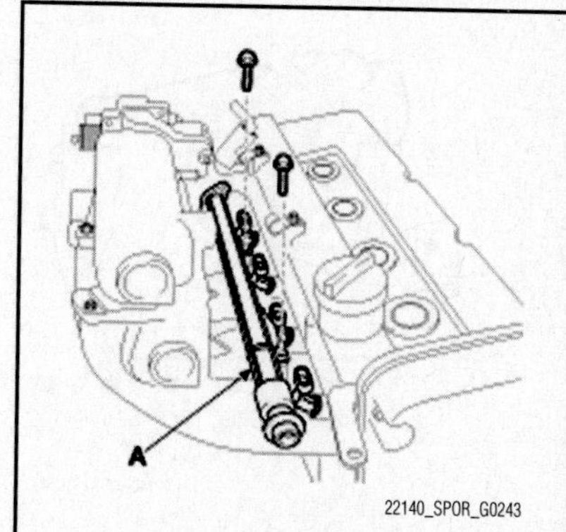

Fig. 78 Remove the fuel delivery pipe (A)

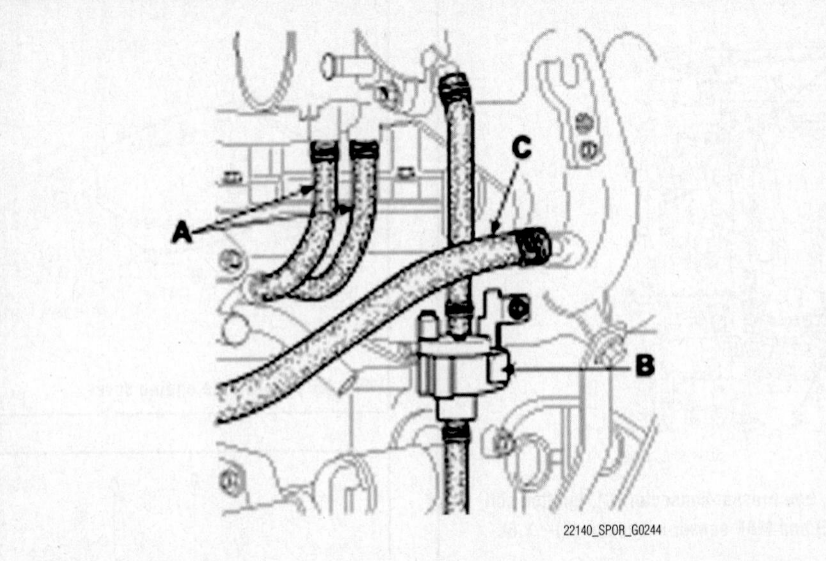

Fig. 79 Remove the heater hose (A), Purge Control Solenoid Valve (PCSV) (B) and the brake vacuum hose (C) from throttle body and intake manifold

8. Remove the intake manifold stay (A).
9. Remove the intake manifold.
10. Installation is the reverse order of removal with a new gasket.

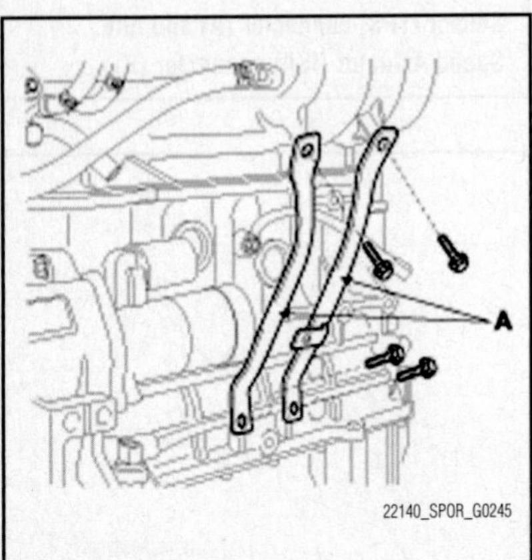

Fig. 80 Remove the intake manifold stay (A)

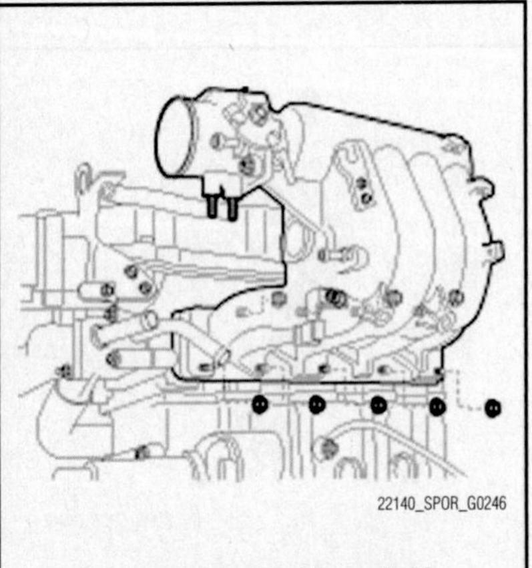

Fig. 81 Remove the intake manifold

OIL PAN

REMOVAL & INSTALLATION

1.6L Engine

See Figure 82.

1. Drain the engine oil.
2. Remove oil pan bolts.
3. Using the SST (09215–3C000), remove the oil pan.
   a. Insert the SST between the oil pan and the ladder frame by tapping it with a plastic hammer in the direction of arrow 1.
   b. After tapping the SST with a plastic hammer along the direction of arrow 2 around more than 2/3 edge of the oil pan, remove it from the ladder frame.

To install:

4. Install the oil pan.
   a. Using a razor blade and gasket scraper, remove all gasket material from the mating surfaces.
   b. Apply liquid gasket as an even bead, centered between the edges of the mating surface. Liquid gasket: MS 721–40A or equivalent

➡To prevent leakage of oil, apply liquid gasket to the inner threads of the bolt holes. Do not install the parts if five minutes or more have elapsed since applying the liquid gasket. After assembly, wait at least 30 minutes before filling the engine with oil.

   c. Install the oil pan with the bolts. Uniformly tighten the bolts in several passes. Torque: 7.2–8.7 ft. lbs. (9.8–11.8 Nm).
5. Fill with engine oil.

2.0L Engine

See Figures 83 and 84.

1. Drain the engine oil.
2. Disconnect the rear oxygen sensor connector.
3. Remove the front muffler (A).
4. Remove the exhaust manifold.
5. Remove the front muffler bracket (A).
6. Remove the oil pan.

To install:

7. Install the oil pan.

➡Check that the mating surfaces are clean and dry before applying liquid gasket.

   a. Apply liquid gasket as an even bead, centered between the edges of the mating surface.

➡To prevent leakage of oil, apply liquid gasket to the inner threads of the bolt holes. Do not install the parts if five minutes or more have elapsed

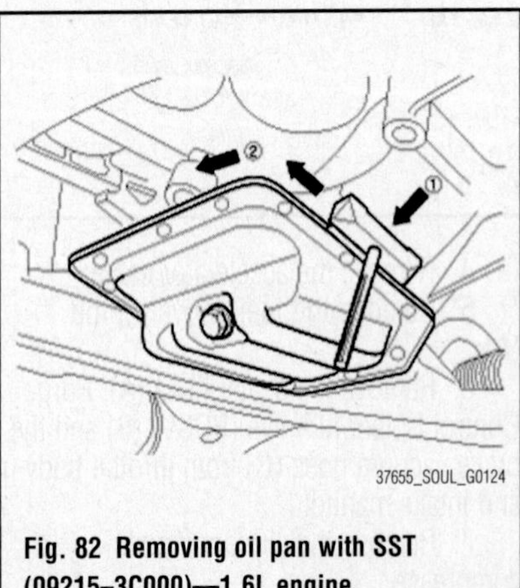

Fig. 82 Removing oil pan with SST (09215–3C000)—1.6L engine

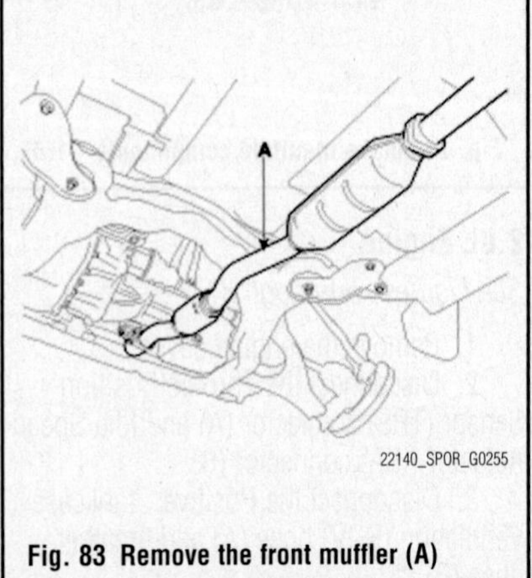

Fig. 83 Remove the front muffler (A)

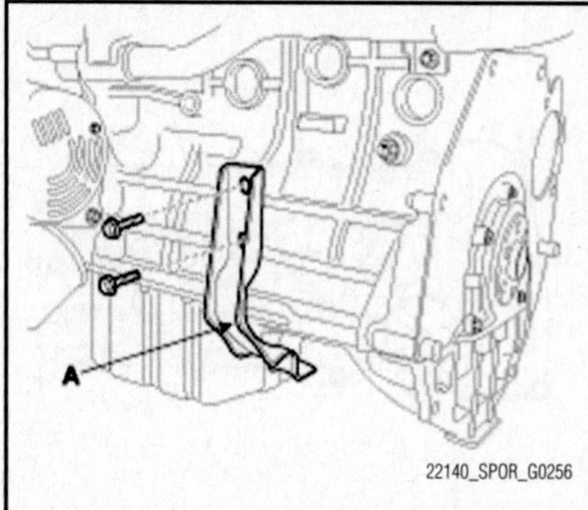

Fig. 84 Remove the front muffler bracket (A)

since applying the liquid gasket. Instead, reapply liquid gasket after removing the residue. After assembly, wait at least 30 minutes before filling the engine with oil.

b. Install the oil pan with the bolts. Uniformly tighten the bolts in several passes. Tighten to 7–9 ft. lbs. (10–12 Nm)

8. Install the front muffler bracket.
9. Install the exhaust manifold.
10. Install the front muffler.
11. Connect the rear oxygen sensor connector.
12. Fill with engine oil

OIL PUMP

REMOVAL & INSTALLATION

1.6L Engine

See Figures 85 through 91.

➡Engine removal is not required for this procedure.

1. Loosen the water pump pulley bolt and the drive idler mounting bolt.

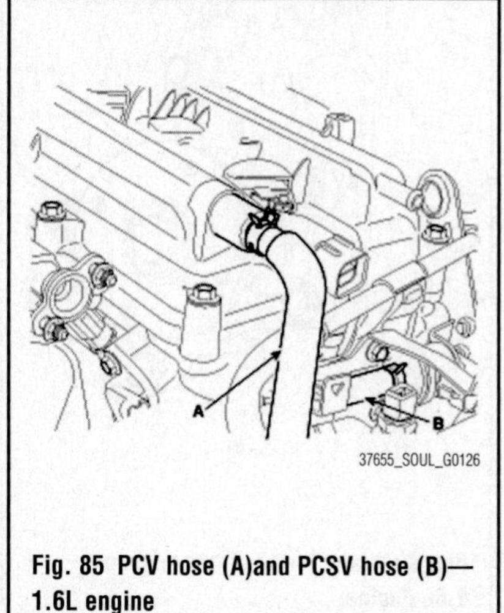

Fig. 85 PCV hose (A)and PCSV hose (B)—1.6L engine

2. Loosen the alternator tension adjusting bolt to loosen tension.
3. Remove the alternator drive belt.
4. Remove the alternator. See Alternator.
5. Remove the RH front wheel.

➡Support the engine with a jack.

6. Remove the engine mounting bracket.
7. Remove the alternator bracket.
8. Remove the engine support bracket.
9. Remove the water pump pulley.
10. Remove the water pump.
11. Remove the drive belt idler.
12. Disconnect the ignition coil connector and the breather hose from valve cover.
13. Disconnect the positive crankcase ventilation (PCV) hose and PCSV hose.
14. Remove the ignition coils.
15. Remove the cylinder head cover with its gasket.
16. Remove side cover in front of crank pulley.
17. Turn the crankshaft pulley clockwise, and align its groove with the timing mark of the timing chain cover.
18. Remove the crankshaft bolt and crankshaft pulley.
19. Remove the timing chain cover (A).
20. Remove the oil pump from timing chain cover.

To install:

⁂ CAUTION

Recheck the top dead center (TDC) marks on the crankshaft and camshaft.

21. Install the timing chain cover (A).

a. Before installing, remove the hardened sealant from the timing cover and engine surfaces.

b. Apply the liquid gasket (TB 1217H or LOCTITE 5900H) on the surface between the cylinder head and the cylinder block.

c. Apply the liquid gasket, THREE BOND 1282B or THREE BOND 1216E on the water pump contact parts of the timing chain cover and THREE BOND 1217H or LOCTITE 5900H on the rest parts. Reassemble the cover (A) within 5 minutes.

d. Align the dowel pin of the cylinder block and the holes of the oil pump. Tighten 10 mm bolts to 7–9 ft. lbs. (10–12 Nm). Tighten 12 mm bolts to 14–17 ft. lbs. (19–24 Nm).

➡After installation, do not crank engine or apply pressure on the cover for half an hour.

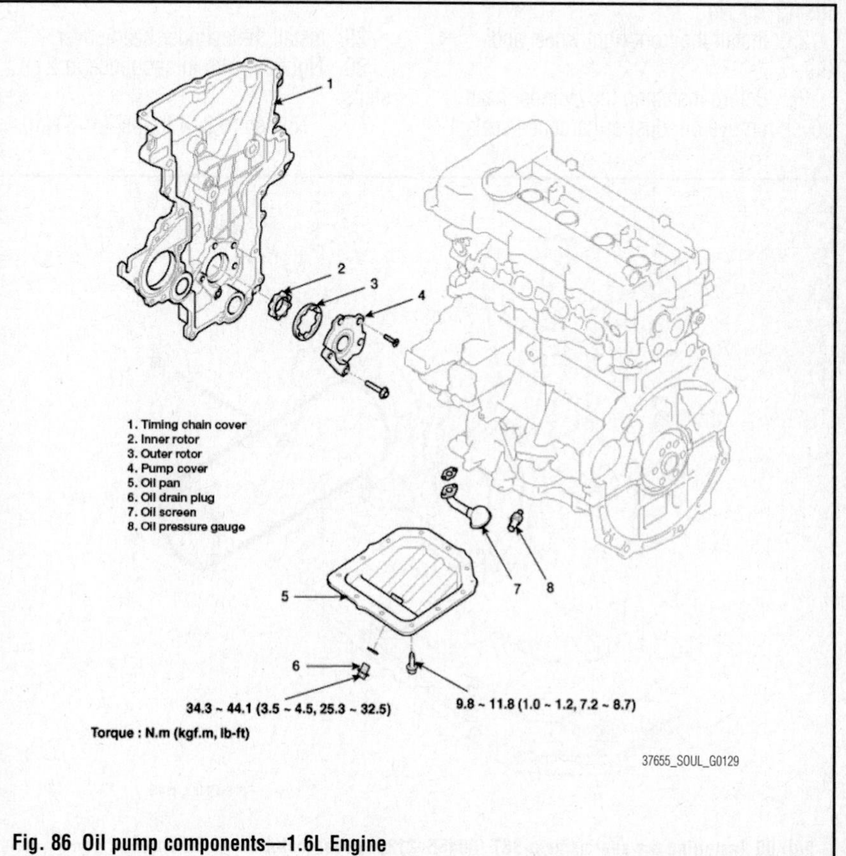

Fig. 86 Oil pump components—1.6L Engine

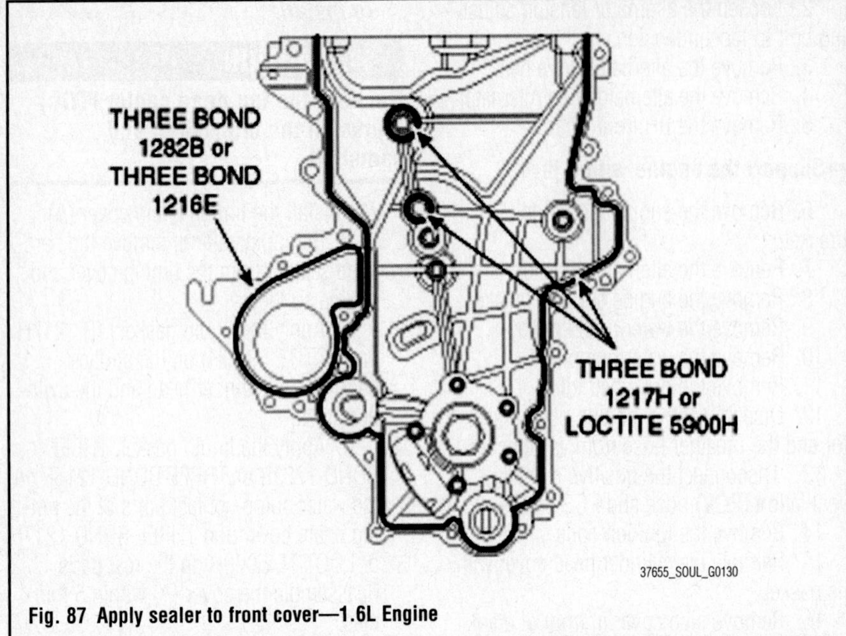

Fig. 87 Apply sealer to front cover—1.6L Engine

22. Using the SST (09455–21200), reassemble the timing chain cover oil seal.

➡**When installing the pulley, the groove on the pulley should be positioned outside, not next to pointer.**

23. Install the crankshaft pulley and bolt. Tighten to 94–101 ft. lbs. (128–137 Nm).
24. Install side cover. Tighten to 6.5–8 ft. lbs. (9–11 Nm).
25. Install the front right wheel and tire.
26. Before installing the cylinder head cover, remove oil, dust or hardened sealant from the timing chain cover and the cylinder head upper surface.
27. After applying the liquid gasket, THREE BOND 1217H or LOCTITE 5900H on the cylinder head cover, reassemble the cover within five minutes.
28. Install the cylinder head cover with a new gasket.

➡**Do not reuse the disassembled gasket.**

29. Install the cylinder head cover.
30. Tighten bolts in sequence in 2 steps:
    a. Tighten to 3–4 ft. lbs. (4–6 Nm).

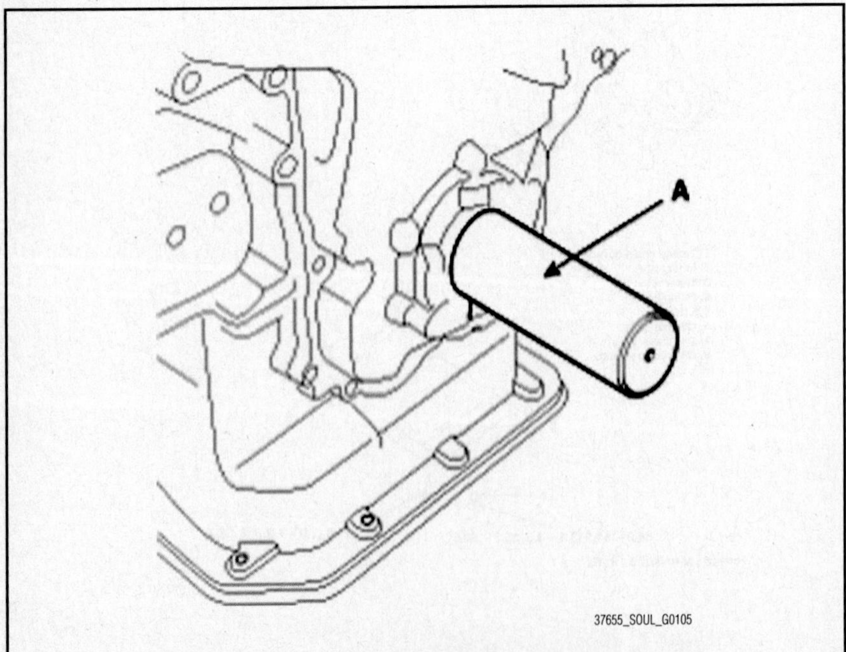

Fig. 88 Installing the seal using a SST (09455-21200) seal installer—1.6L engine

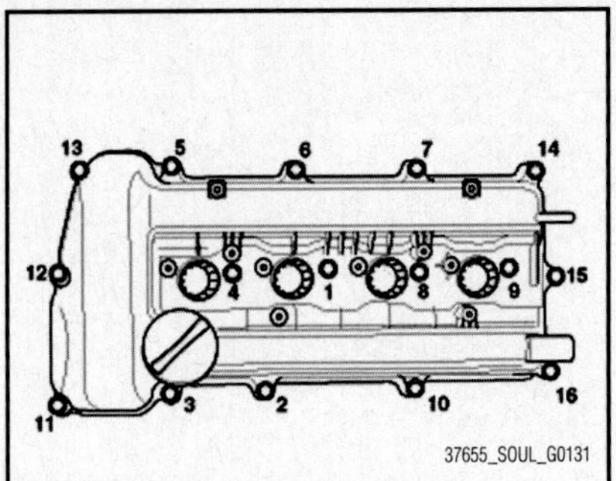

Fig. 89 Valve cover torque sequence—1.6L Engine

    b. Tighten to 6–7.2 ft. lbs. (8–10 Nm).
31. Install the ignition coils. Tighten to 7.2–9 ft. lbs. (10–12 Nm).
32. Install the PCV hose and the PCSV hose.
33. Connect the ignition coil connector and the breather hose.
34. Install the drive belt idler. Tighten to 31–40 ft. lbs. (42–54 Nm).
35. Install the water pump with a gasket. Tighten bolts in sequence to 7.2–9 ft. lbs. (10–12 Nm).
36. Install the water pump pulley. Tighten bolts in a star pattern to 7.2–9 ft. lbs. (10–12 Nm).
37. Install the engine support bracket. Tighten to 22–30 ft. lbs. (30–41 Nm).
38. Install the alternator bracket. Tighten to 15–20 ft. lbs. (20–27 Nm).
39. Install the engine mounting bracket (B).
    a. Tighten nut (C) to 47–62 ft. lbs. (64–83 Nm).

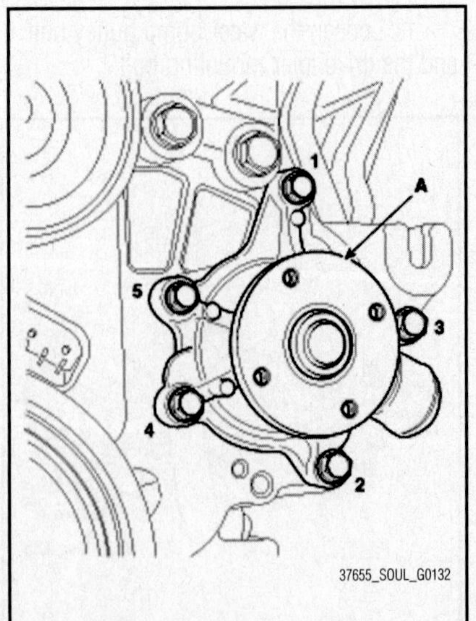

Fig. 90 Water pump torque sequence—1.6L Engine

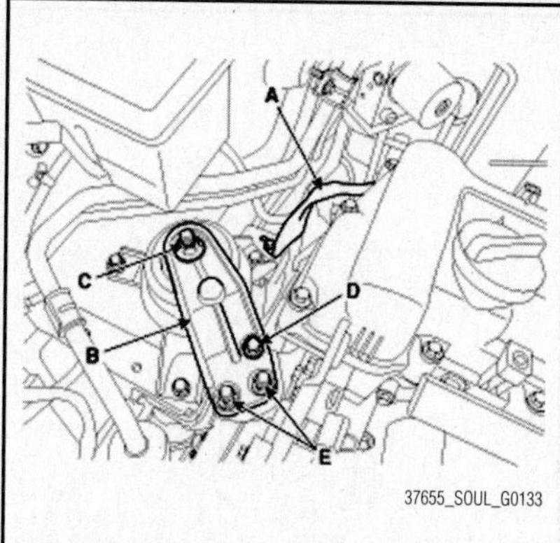

Fig. 91 Strap (A), bracket (B), nut (C) and bolt (D)—1.6L Engine

b. Tighten bolt (D) and nuts (E) to 36–47 ft. lbs. (49–64 Nm).

40. Install the alternator and snug the mounting bolts.
41. Install the drive belt.
42. Tighten the alternator 12 mm mounting bolt to 15–20 ft. lbs. (20–27 Nm).
43. Tighten the alternator 14 mm mounting bolt to 22–30 ft. lbs. (30–41 Nm).

2.0L Engine

See Figures 92 through 94.

1. Drain the engine oil.
2. Remove the drive belts.
3. Turn the crankshaft pulley, and align its groove with timing mark "T" of the timing belt cover.
4. Remove the timing belt.
5. Remove the bolt (B) and timing belt idler (A).
6. Remove the oil pan and oil screen.
7. Remove the alternator.
8. Remove the air conditioner compressor tensioner bracket (A).
9. Remove the bolts (A, B, C, D) and oil pump.

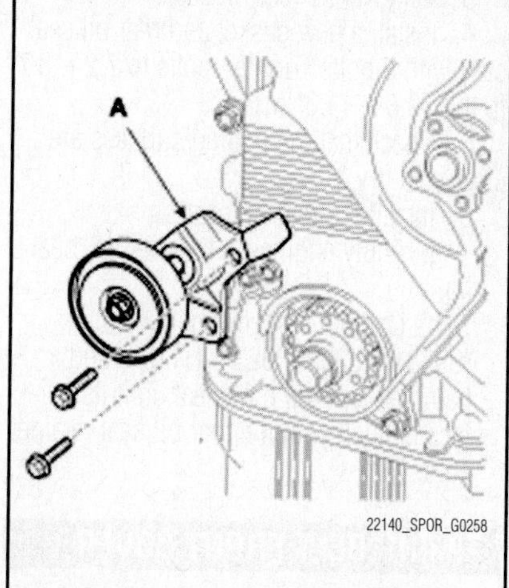

Fig. 93 Remove the air conditioner compressor tensioner bracket (A)

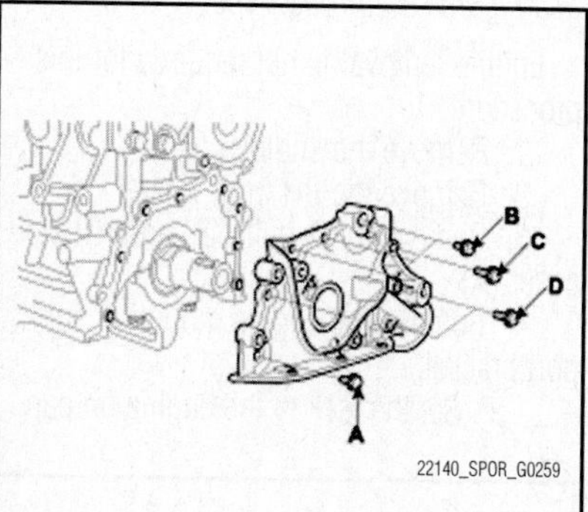

Fig. 94 Remove the bolts (A, B, C, D) and oil pump

To install:

10. Installation is the reverse of removal.
11. Fill with engine oil.
12. Start the engine and check for leaks.

PISTON AND RING

POSITIONING

See Figure 95.

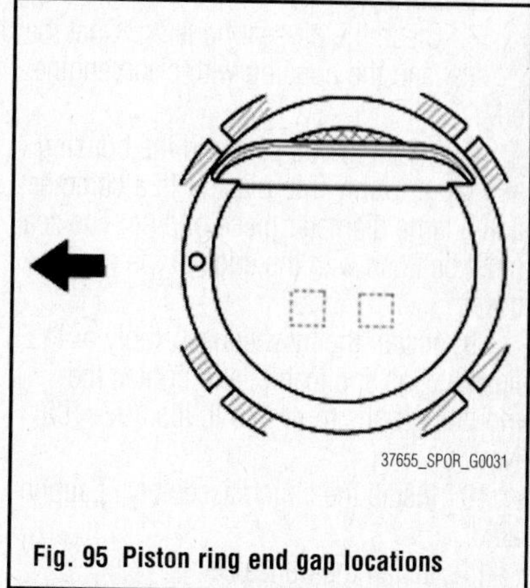

Fig. 95 Piston ring end gap locations

REAR MAIN SEAL

REMOVAL & INSTALLATION

1.6L Engine

See Figure 96.

1. Before servicing the vehicle, refer to the Precautions Section.
2. Disconnect the negative battery cable.
3. Remove the transaxle assembly.
4. Remove the clutch and flywheel assembly, if equipped with a manual transaxle.
5. Remove the flexplate–to–crankshaft bolts, the flexplate and shim plates, if equipped with an automatic transaxle.
6. Cut the oil seal lip with a

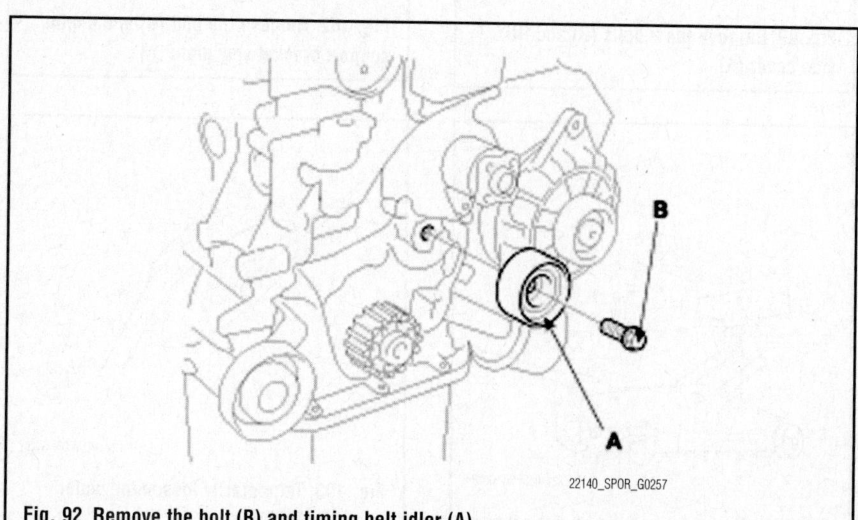

Fig. 92 Remove the bolt (B) and timing belt idler (A)

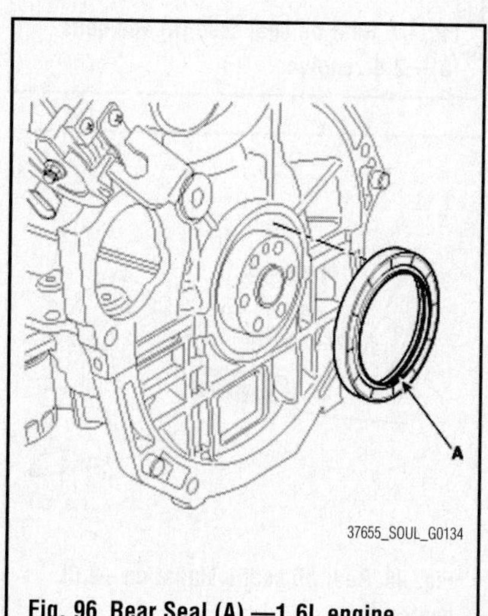

Fig. 96 Rear Seal (A) —1.6L engine

knife. Install a rag to the housing and using a screwdriver, carefully pry the oil seal from the oil seal housing. Clean the gasket mounting surfaces.

To install:

7. Clean the oil seal housing. Coat the oil seal and the housing with clean engine oil.
8. Install the oil seal into the housing and tap it evenly into place with a hammer and a large diameter piece of pipe. The seal must be flush with the edge of the rear cover.
9. Install the flywheel assembly or the flexplate, as applicable, and tighten the mounting bolts to 71–76 ft. lbs. (97–102 Nm).
10. Install the clutch assembly, if applicable.
11. Install the transaxle.
12. Connect the negative battery cable.

2.0L Engine

See Figures 97 and 98.

1. Remove the transaxle.
2. Remove the bolts and rear oil seal case.

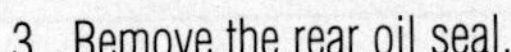

3. Remove the rear oil seal.
4. Install a new gasket and rear oil seal case with 5 bolts. Tighten bolts to 7.2 – 8.7 ft. lbs. (9.8 – 11.8Nm).
5. Check that the mating surfaces are clean and dry.
6. Install the rear oil seal.
   a. Apply engine oil to a new oil seal lip.
   b. Using the special service tools (09231–23200, 09231–H1100) and a hammer, tap in the oil seal until its surface is flush with the rear oil seal retainer edge.

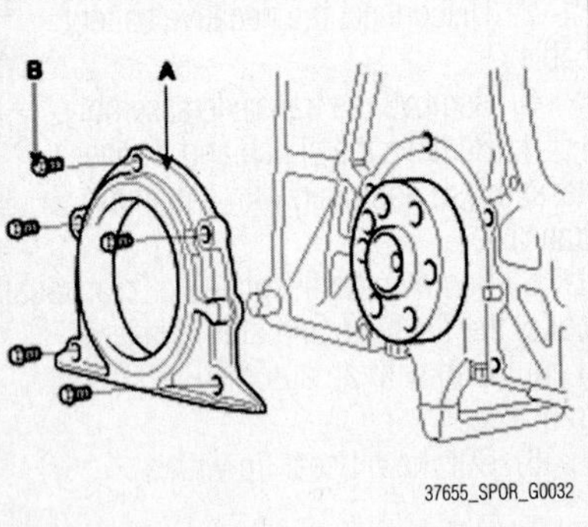

TIMING BELT FRONT COVER

REMOVAL & INSTALLATION

2.0L Engine

See Figures 99 through 109.

Engine removal is not required for this procedure.

1. Remove the engine cover.
2. Remove the RH front wheel.
3. Remove the 2 bolts (B) and RH side cover (A).
4. Remove the engine mounting support bracket.
   a. Set the jack to the engine oil pan.

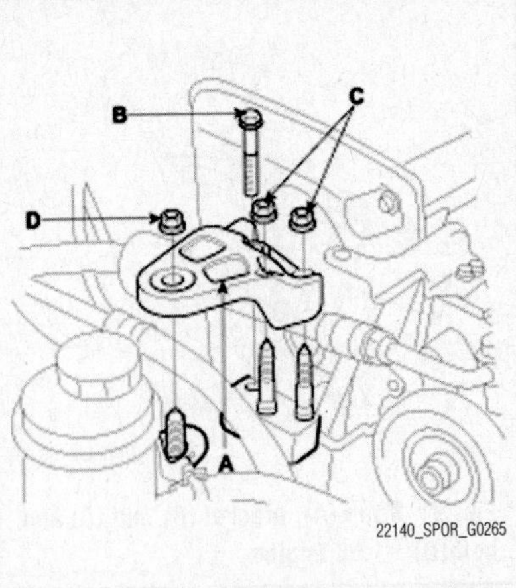

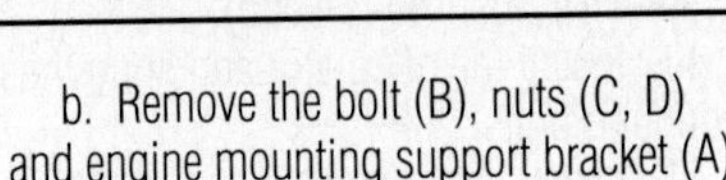

Fig. 101 Remove the bolt (B), nuts (C, D) and engine mounting support bracket (A)

   b. Remove the bolt (B), nuts (C, D) and engine mounting support bracket (A).
   c. Remove the bolt (B) and engine support bracket stay plate (A).
5. Temporarily loosen the water pump pulley bolts.
6. Remove the alternator drive belt.
7. Remove the air conditioner compressor drive belt.
8. Remove the power steering pump drive belt.

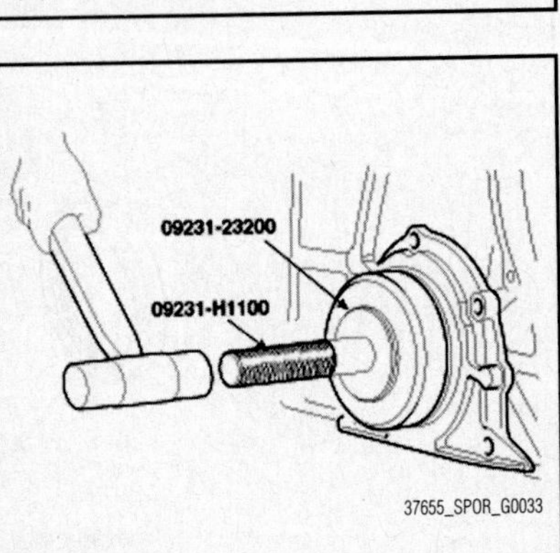

Fig. 97 Rear oil seal case (A) and bolts (B)—2.0L engine

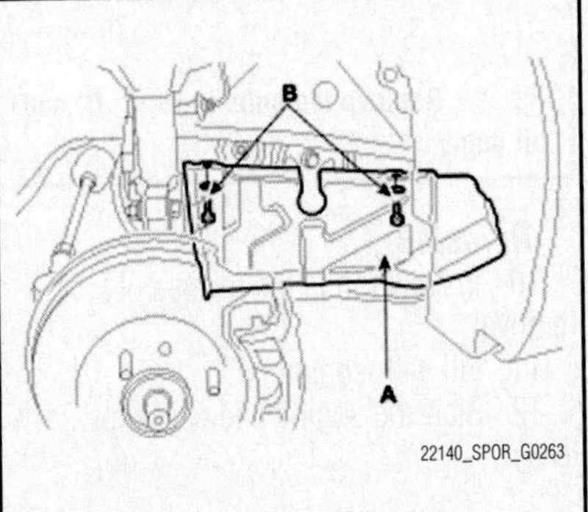

Fig. 99 Remove the 2 bolts (B) and RH side cover (A)

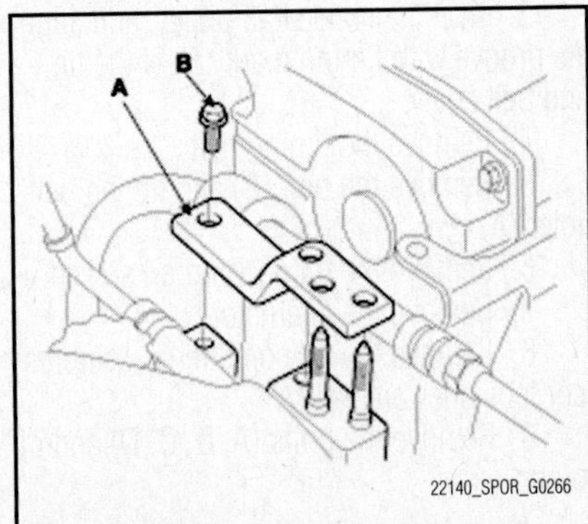

Fig. 102 Remove the bolt (B) and engine support bracket stay plate (A)

Fig. 100 Set the jack to the engine oil pan

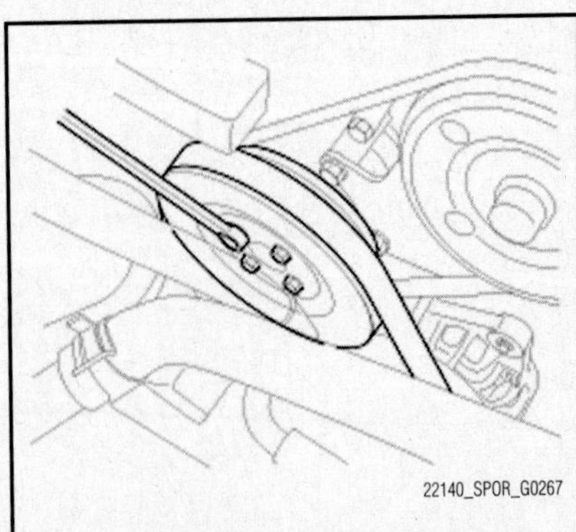

Fig. 103 Temporarily loosen the water pump pulley bolts

Fig. 98 Rear oil seal installation—2.0L engine

Fig. 104 Remove the 4 bolts (B) and timing belt upper cover (A)

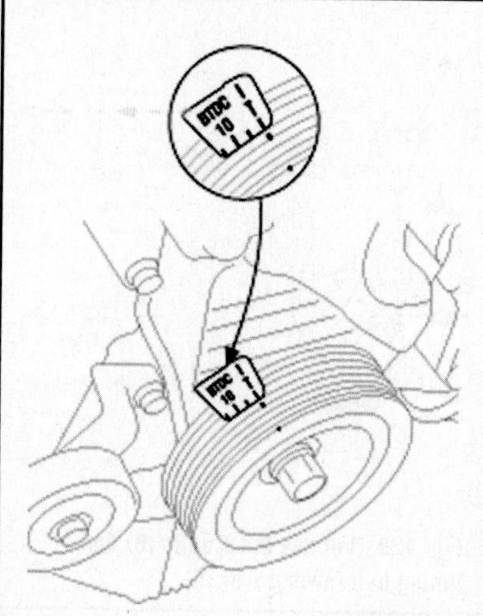

Fig. 105 Turn the crankshaft pulley, and align its groove with timing mark "T" of the timing belt cover

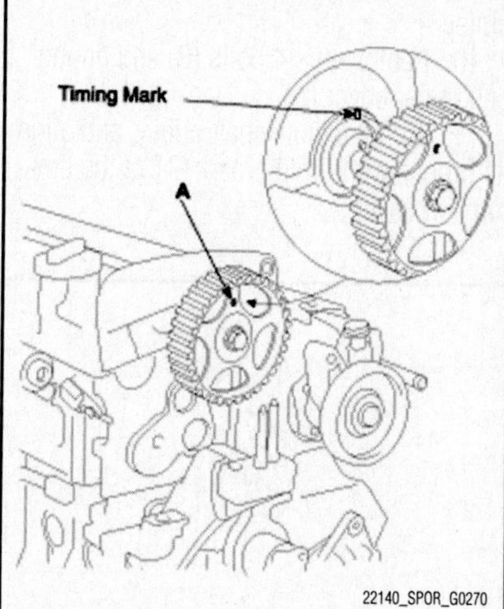

Fig. 106 Check that the timing mark of camshaft sprocket is aligned with the timing mark of cylinder head cover (No.1 cylinder compression TDC position)

9. Remove the 4 bolts and water pump pulley.
10. Remove the 4 bolts (B) and timing belt upper cover (A).
11. Turn the crankshaft pulley, and align its groove with timing mark "T" of the timing belt cover.
12. Check that the timing mark of camshaft sprocket is aligned with the timing mark of cylinder head cover. (No.1 cylinder compression TDC position)

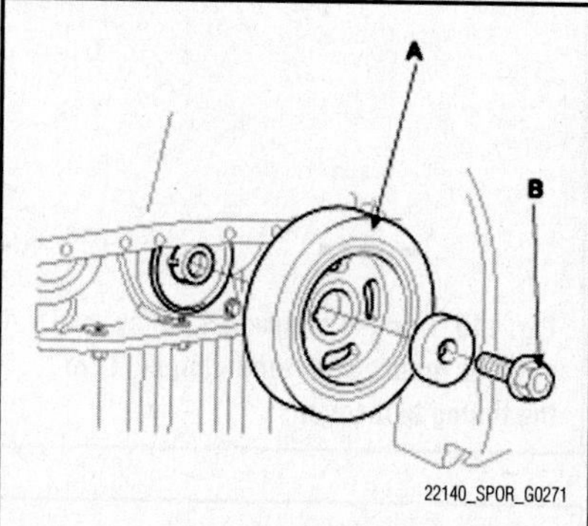

Fig. 107 Remove the crankshaft pulley bolt (B) and crankshaft pulley (A)

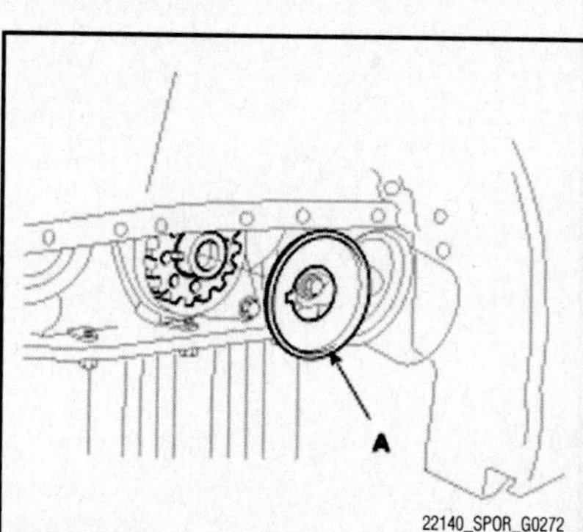

Fig. 108 Remove the crankshaft flange (A).

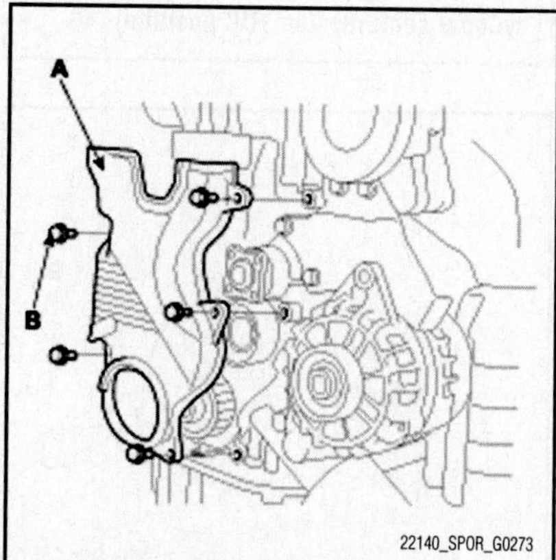

Fig. 109 Remove the 5 bolts (B) and timing belt lower cover (A)

13. Remove the crankshaft pulley bolt (B) and crankshaft pulley (A).
14. Remove the crankshaft flange (A).
15. Remove the 5 bolts (B) and timing belt lower cover (A)
16. Installation is the reverse of removal.

TIMING BELT & SPROCKETS

REMOVAL & INSTALLATION

2.0L Engine

See Figures 110 through 128.

Engine removal is not required for this procedure.

1. Remove the engine cover.
2. Remove the RH front wheel.
3. Remove the 2 bolts (B) and RH side cover (A).
4. Remove the engine mounting support bracket.
   a. Set the jack to the engine oil pan.
   b. Remove the bolt (B), nuts (C, D) and engine mounting support bracket (A).
   c. Remove the bolt (B) and engine support bracket stay plate (A).
5. Temporarily loosen the water pump pulley bolts.

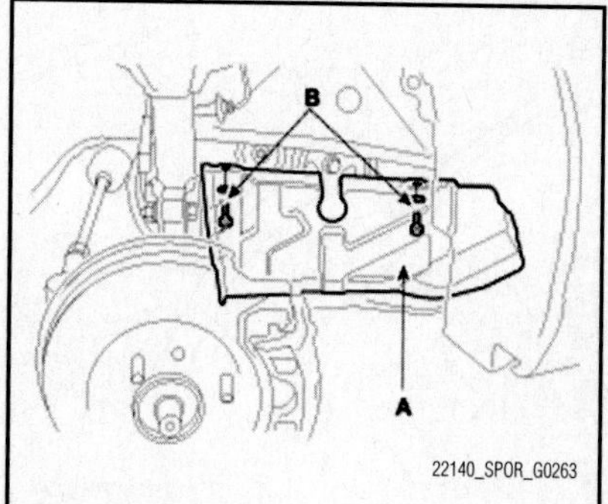

Fig. 110 Remove the 2 bolts (B) and RH side cover (A)

Fig. 111 Set the jack to the engine oil pan

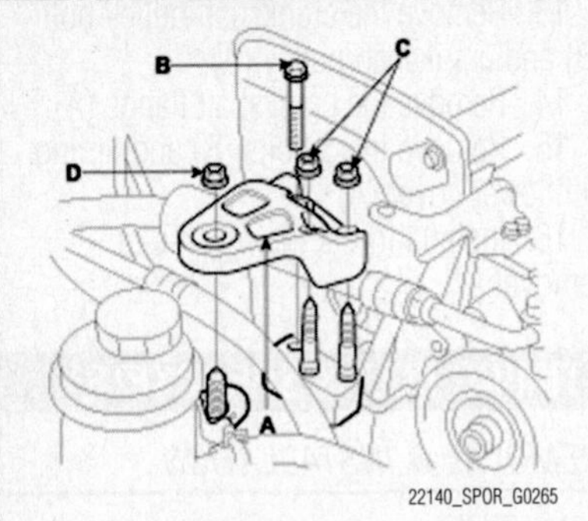

Fig. 112 Remove the bolt (B), nuts (C, D) and engine mounting support bracket (A)

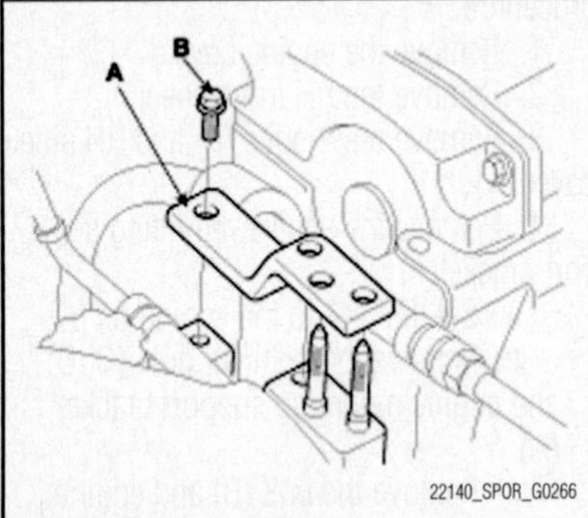

Fig. 113 Remove the bolt (B) and engine support bracket stay plate (A)

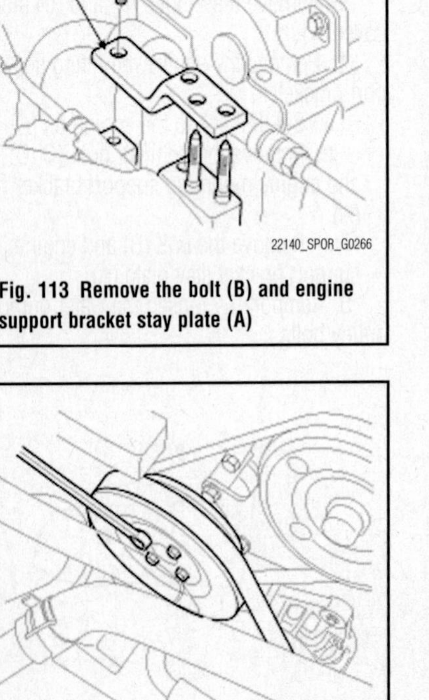

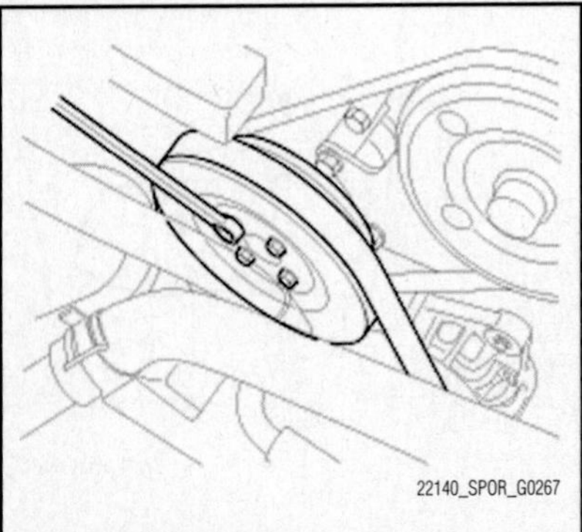

Fig. 114 Temporarily loosen the water pump pulley bolts

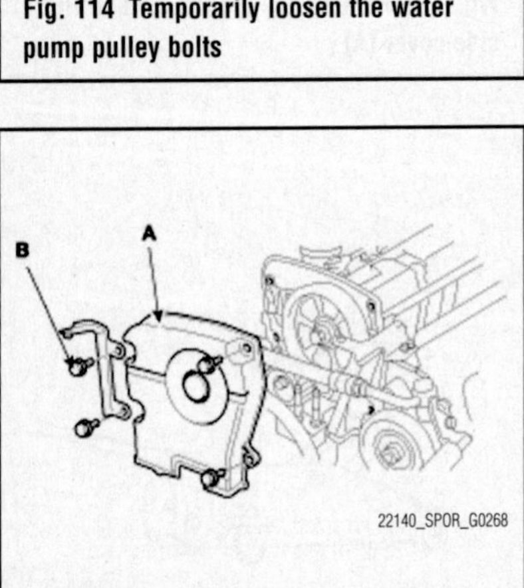

Fig. 115 Remove the 4 bolts (B) and timing belt upper cover (A)

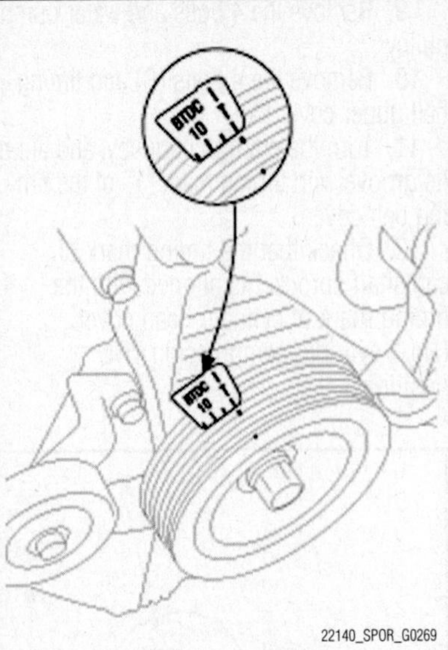

Fig. 116 Turn the crankshaft pulley, and align its groove with timing mark "T" of the timing belt cover

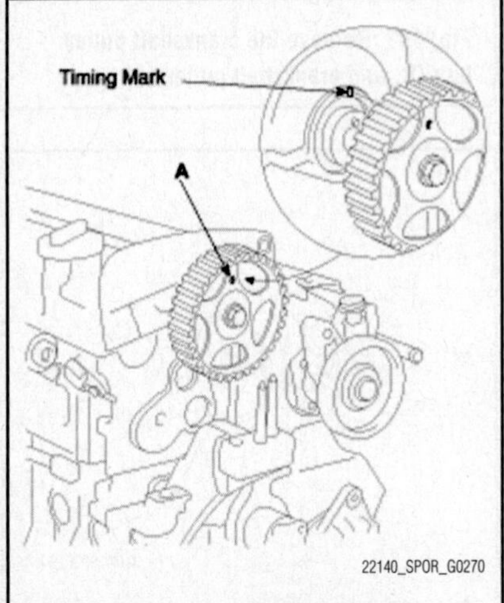

Fig. 117 Check that the timing mark of camshaft sprocket is aligned with the timing mark of cylinder head cover (No.1 cylinder compression TDC position)

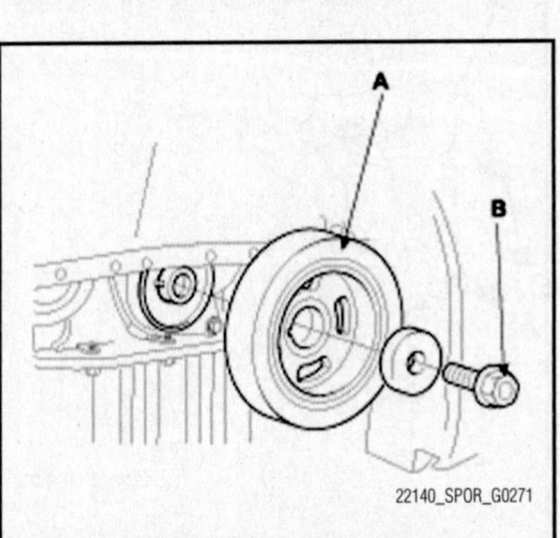

Fig. 118 Remove the crankshaft pulley bolt (B) and crankshaft pulley (A)

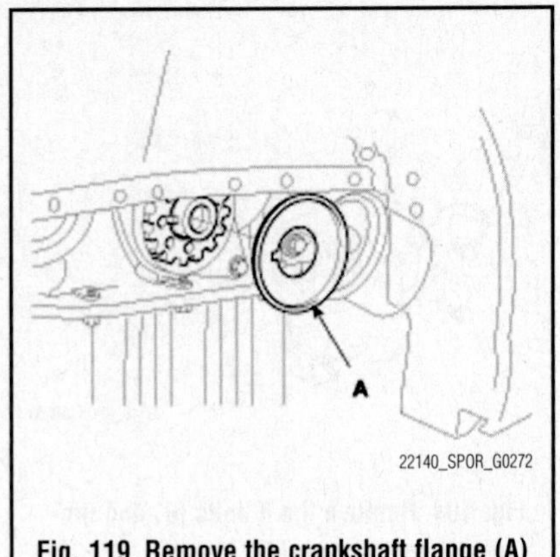

Fig. 119 Remove the crankshaft flange (A)

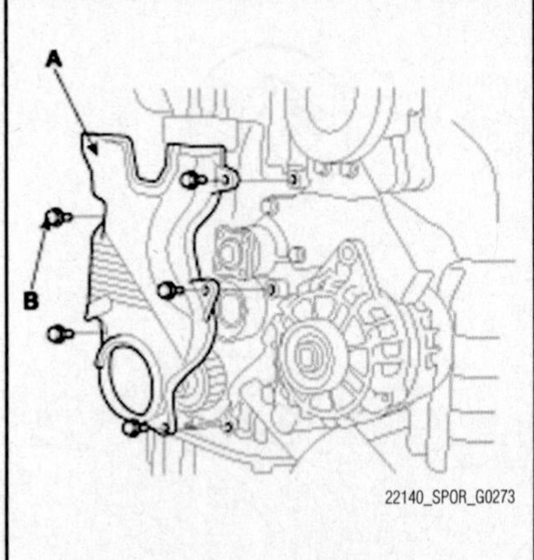

Fig. 120 Remove the 5 bolts (B) and timing belt lower cover (A)

6. Remove the alternator drive belt.
7. Remove the air conditioner compressor drive belt.
8. Remove the power steering pump drive belt.
9. Remove the 4 bolts and water pump pulley.
10. Remove the 4 bolts (B) and timing belt upper cover (A).
11. Turn the crankshaft pulley, and align its groove with timing mark "T" of the timing belt cover.

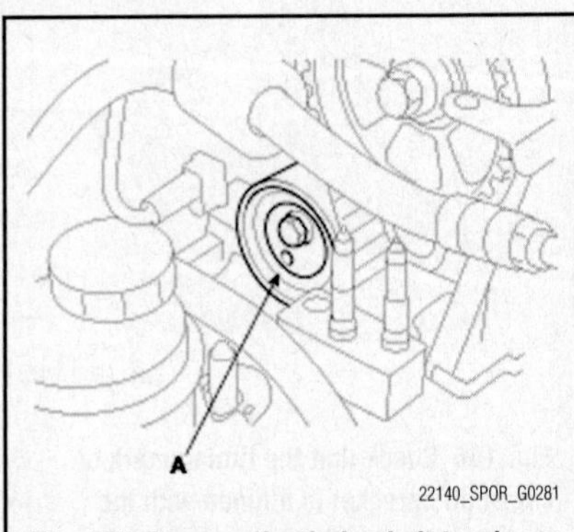

Fig. 121 Remove the timing belt tensioner (A)

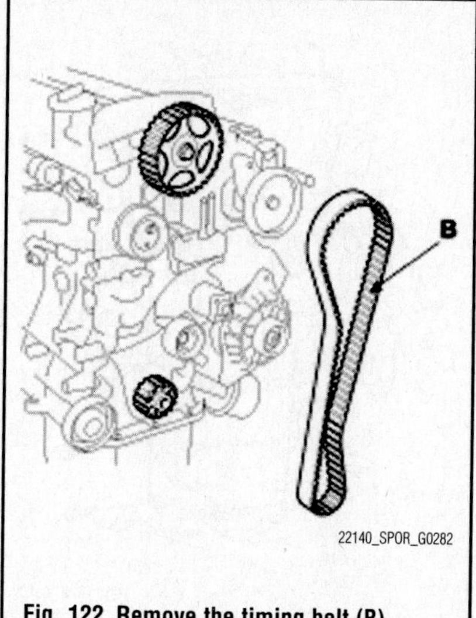

Fig. 122 Remove the timing belt (B)

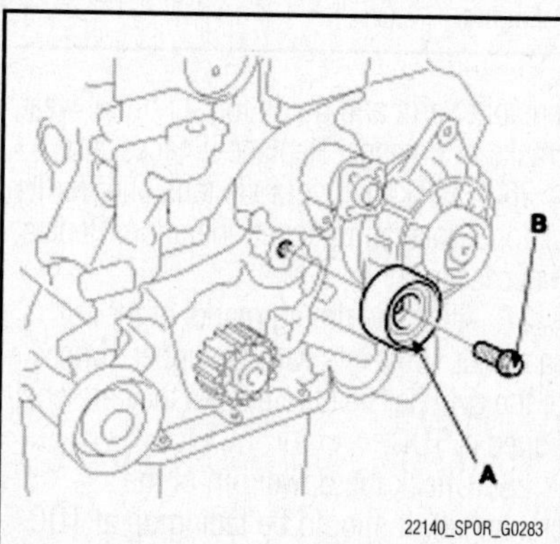

Fig. 123 Remove the bolt (B) and timing belt idler (A)

12. Check that the timing mark of camshaft sprocket is aligned with the timing mark of cylinder head cover. (No.1 cylinder compression TDC position)

13. Remove the crankshaft pulley bolt (B) and crankshaft pulley (A).

14. Remove the crankshaft flange (A).

15. Remove the 5 bolts (B) and timing belt lower cover (A)

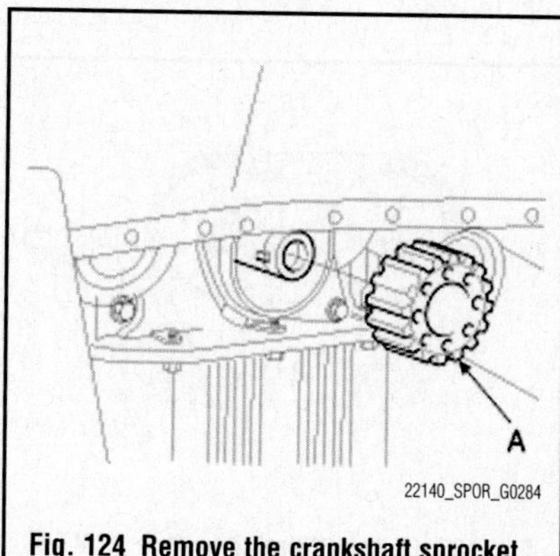

Fig. 124 Remove the crankshaft sprocket (A)

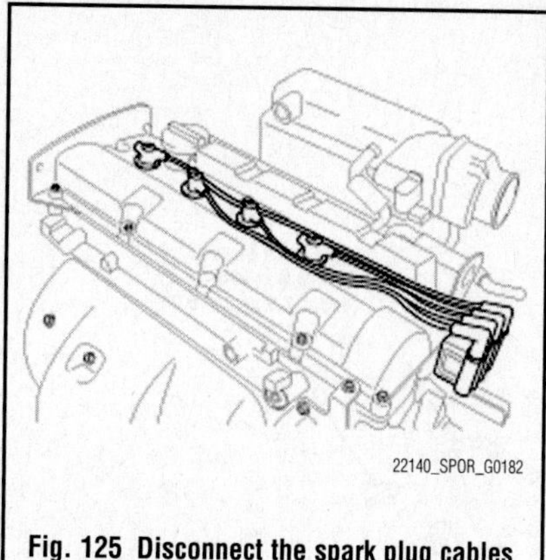

Fig. 125 Disconnect the spark plug cables

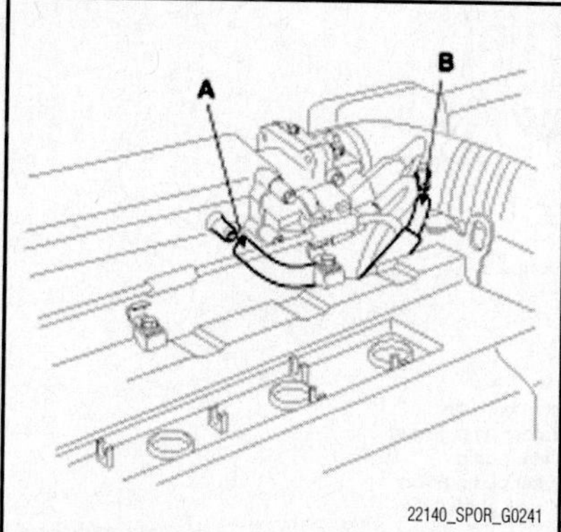

Fig. 126 Remove the Positive Crankcase Ventilation (PCV) hose (A) and breather hose (B)

16. Remove the timing belt tensioner (A) and timing belt (B)

➡If the timing belt is going to be reused, make an arrow indicating the turning direction to make sure that the belt is reinstalled in the same direction as before.

17. Remove the bolt (B) and timing belt idler (A).

18. Remove the crankshaft sprocket (A).

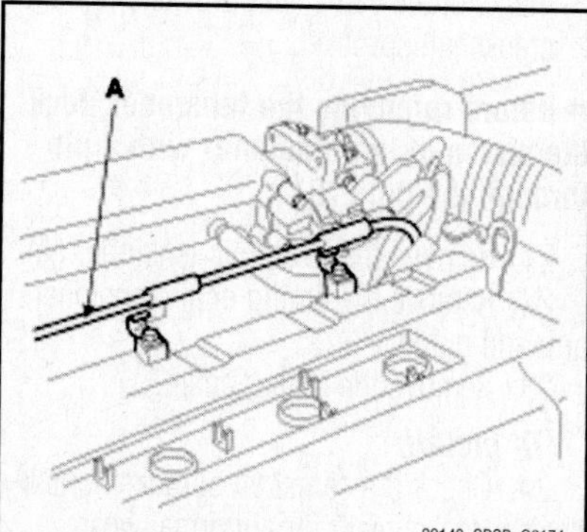

Fig. 127 Remove the accelerator cable (A)

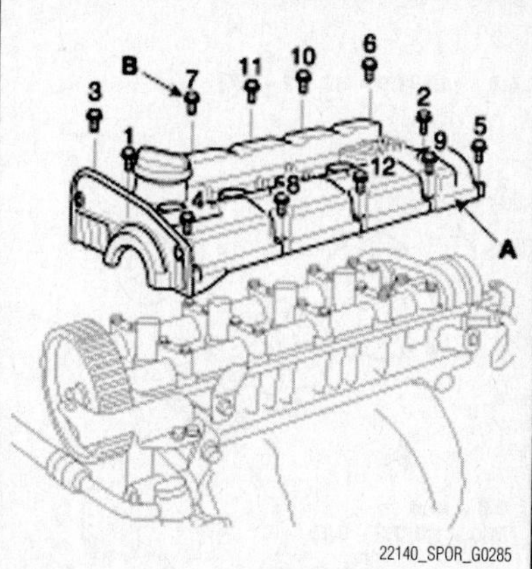

Fig. 128 Remove the cylinder head cover bolts (B) and remove the cover (A) and gasket

19. Remove the cylinder head cover.

a. Disconnect the spark plug cables and do not pull on the cable by force.

➡Pulling on or bending the cables may damage the conductor inside.

b. Remove the Positive Crankcase Ventilation (PCV) hose (A) and the breather hose (B) from the cylinder head cover.

c. Remove the accelerator cable (A) from the cylinder head cover.

d. Remove the cylinder head cover bolts (B) and remove the cover (A) and gasket.

20. Remove the camshaft sprocket.

21. Installation is the reverse of removal.

TIMING CHAIN FRONT COVER

REMOVAL & INSTALLATION

1.6L Engine

See Figures 129 through 138.

➡Engine removal is not required for this procedure.

1. Loosen the water pump pulley bolt and the drive idler mounting bolt.
2. Loosen the alternator tension adjusting bolt to loosen tension.
3. Remove the alternator drive belt.
4. Remove the alternator. See Alternator.
5. Remove the RH front wheel.

➡Support the engine with a jack.

6. Remove the engine mounting bracket.
7. Remove the alternator bracket.
8. Remove the engine support bracket.

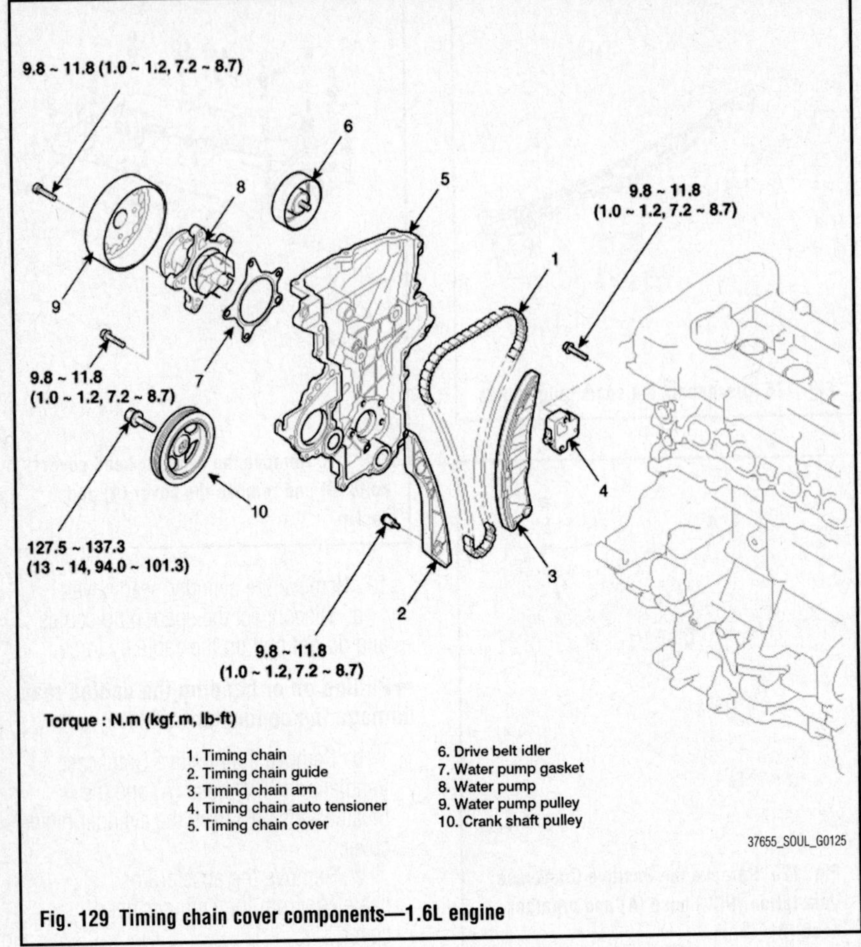

Fig. 129 Timing chain cover components—1.6L engine

9. Remove the water pump pulley.
10. Remove the water pump.
11. Remove the drive belt idler.
12. Disconnect the ignition coil connector and the breather hose from valve cover.
13. Disconnect the positive crankcase ventilation (PCV) hose and PCSV hose.
14. Remove the ignition coils.
15. Remove the cylinder head cover with its gasket.
16. Remove side cover in front of crank pulley.
17. Turn the crankshaft pulley clockwise, and align its groove with the timing mark of the timing chain cover.

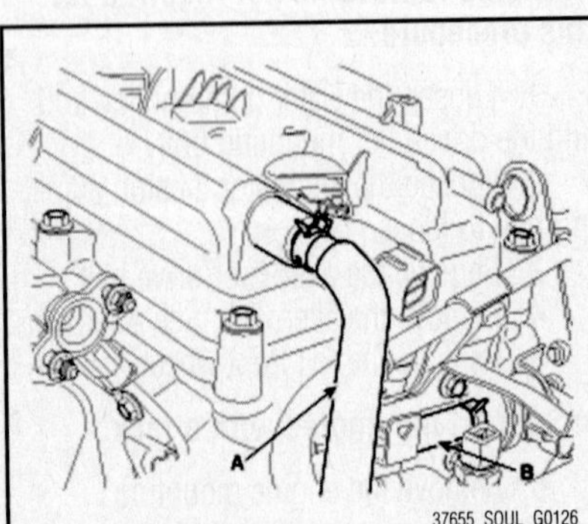

Fig. 130 PCV hose (A)and PCSV hose (B)—1.6L Engine

18. Remove the crankshaft bolt and crankshaft pulley.
19. Remove the timing chain cover (A).
20. Align the timing marks of the camshaft sprocket with the upper surface of the cylinder head to make No.1 cylinder be positioned at TDC.
  a. Check the dowel pin of the crankshaft. It should be facing up at TDC.
  b. Put paint marks on the timing chain links that meet with the timing marks of the camshaft sprockets and the crankshaft sprocket.

➡Before removing the tensioner, lock the piston of the tensioner with a pin through the hole (B).

21. Remove the hydraulic tensioner (A).
22. Remove the timing chain tensioner arm and guide.
23. Remove the timing chain.

To install:

24. Check the camshaft sprocket, crankshaft sprocket teeth for abnormal wear, cracks or damage. Replace if necessary.
25. Check a contact surface of the chain tensioner arm and guide for abnormal wear, cracks or damage. Replace if necessary.

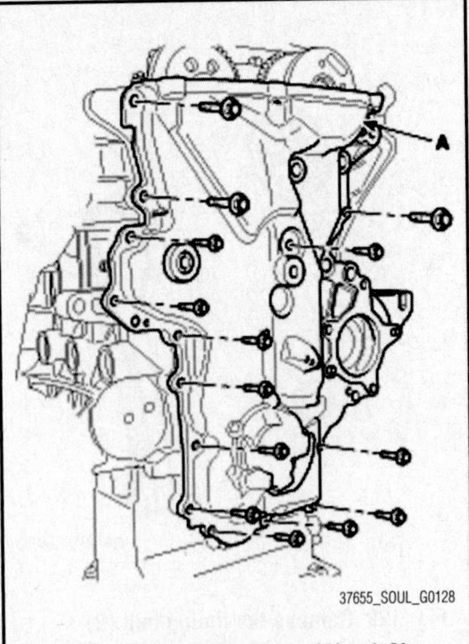

Fig. 131 Timing chain cover (A)—1.6L Engine

26. Check the hydraulic tensioner for its piston stroke and ratchet operation. Replace if necessary.
27. Align the timing marks of the camshaft sprockets with the upper surface of the cylinder head to ensure they are positioned at TDC.
28. Check the dowel pin of the crankshaft. It should be facing up at TDC.
29. Install the timing chain guide. Tighten to 7–9 ft. lbs. (10–12 Nm).
30. When installing timing chain, align the timing marks on the sprockets with paint marks of the chain.
31. Install in chain on components in this order : Crankshaft sprocket, Timing chain guide, Intake camshaft sprocket, Exhaust camshaft sprocket.
32. Install the chain tensioner arm. Tighten to 7–9 ft. lbs. (10–12 Nm).
33. Install the hydraulic tensioner and remove the pin. Tighten to 7–9 ft. lbs. (10–12 Nm).

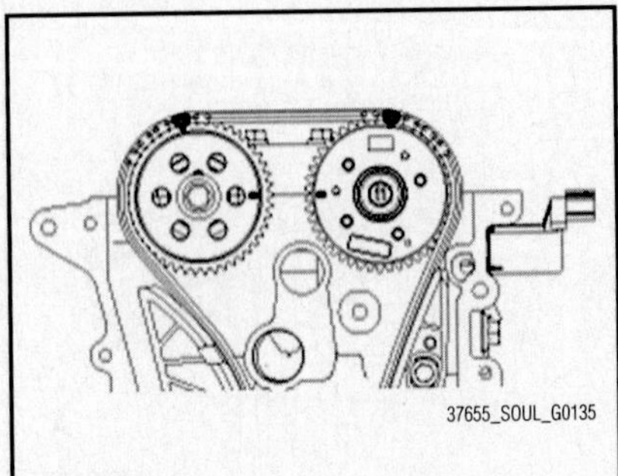

Fig. 132 Camshaft timing marks—1.6L engine

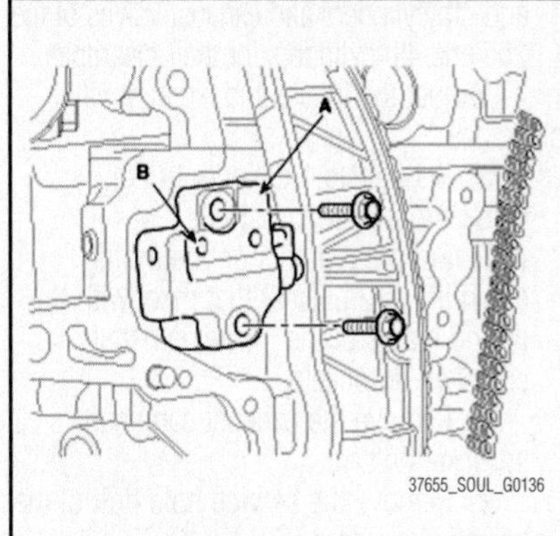

Fig. 133 Timing chain tensioner (A) and pin (B)—1.6L engine

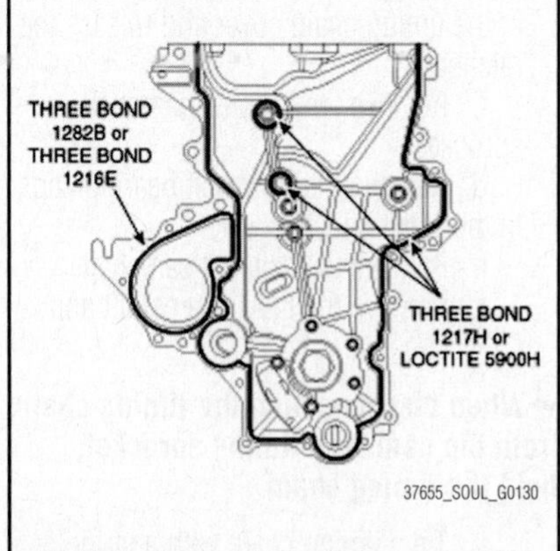

Fig. 134 Apply sealer to front cover—1.6L Engine

CAUTION

Recheck the top dead center (TDC) marks on the crankshaft and camshaft.

34. Install the timing chain cover (A).

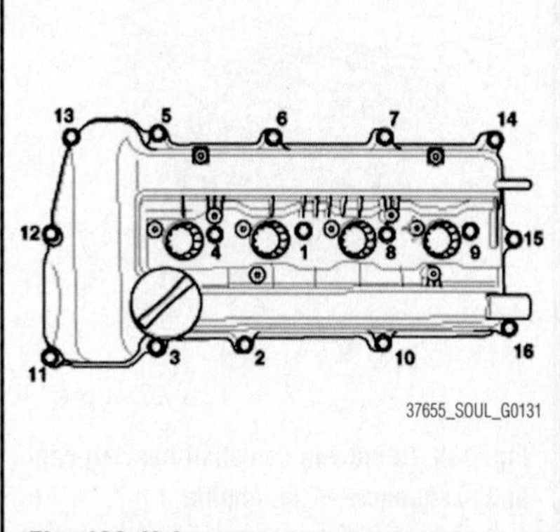

Fig. 136 Valve cover torque sequence—1.6L Engine

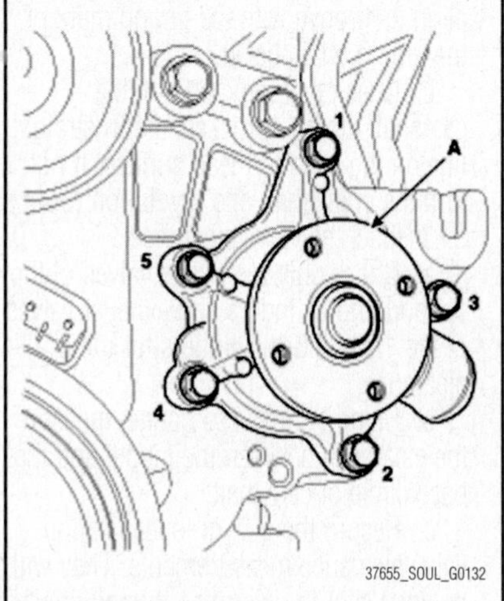

Fig. 137 Water pump torque sequence—1.6L Engine

a. Before installing, remove the hardened sealant from the timing cover and engine surfaces.

b. Apply the liquid gasket (TB 1217H or LOCTITE 5900H) on the surface between the cylinder head and the cylinder block.

c. Apply the liquid gasket, THREE BOND 1282B or THREE BOND 1216E on the water pump contact parts of the timing chain cover and THREE BOND 1217H or LOCTITE 5900H on the rest parts. Reassemble the cover (A) within 5 minutes.

d. Align the dowel pin of the cylinder block and the holes of the oil pump. Tighten 10 mm bolts to 7–9 ft. lbs. (10–12 Nm). Tighten 12 mm bolts to 14–17 ft. lbs. (19–24 Nm).

➡After installation, do not crank engine or apply pressure on the cover for half an hour.

35. Using the SST (09455–21200), reassemble the timing chain cover oil seal.

➡When installing the pulley, the groove on the pulley should be positioned outside, not next to pointer.

36. Install the crankshaft pulley and bolt. Tighten to 94–101 ft. lbs. (128–137 Nm).

37. Install side cover. Tighten to 6.5–8 ft. lbs. (9–11 Nm).

38. Install the front right wheel and tire.

39. Before installing the cylinder head cover, remove oil, dust or hardened sealant from the timing chain cover and the cylinder head upper surface.

40. After applying the liquid gasket, THREE BOND 1217H or LOCTITE 5900H on the cylinder head cover, reassemble the cover within five minutes.

41. Install the cylinder head cover with a new gasket.

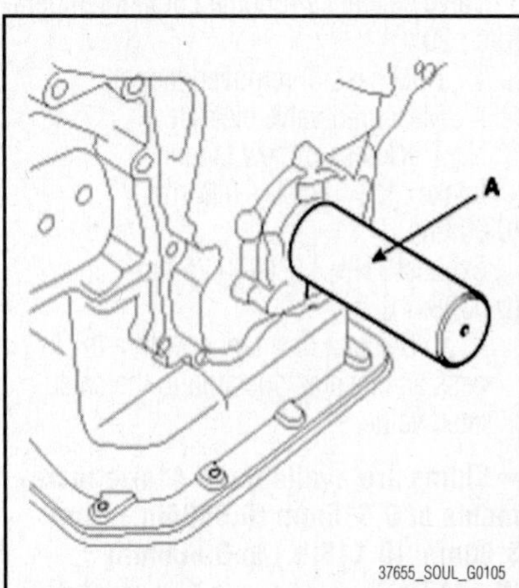

Fig. 135 Installing the seal using a SST (09455–21200) seal installer—1.6L engine

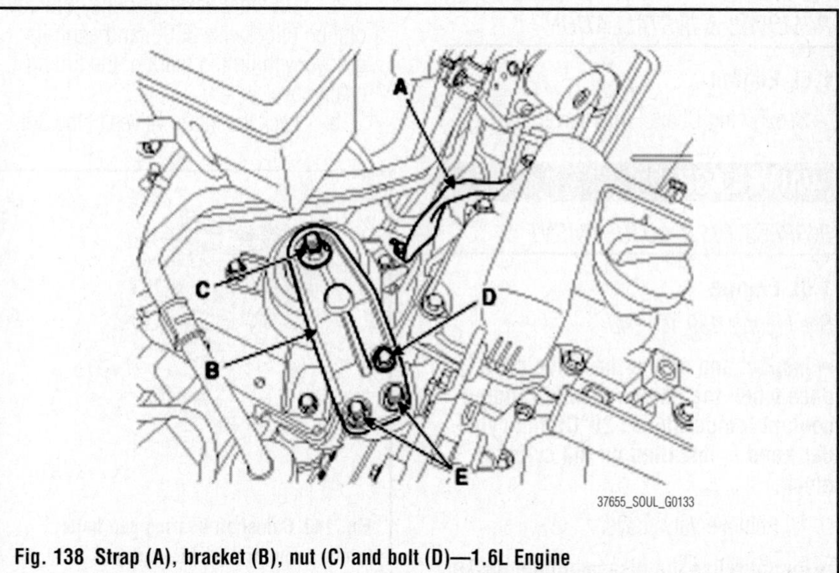

Fig. 138 Strap (A), bracket (B), nut (C) and bolt (D)—1.6L Engine

➡Do not reuse the disassembled gasket.

42. Install the cylinder head cover.
43. Tighten bolts in sequence in 2 steps:
    a. Tighten to 3–4 ft. lbs. (4–6 Nm).
    b. Tighten to 6–7.2 ft. lbs. (8–10 Nm).
44. Install the ignition coils. Tighten to 7.2–9 ft. lbs. (10–12 Nm).
45. Install the PCV hose and the PCSV hose.
46. Connect the ignition coil connector and the breather hose.
47. Install the drive belt idler. Tighten to 31–40 ft. lbs. (42–54 Nm).
48. Install the water pump with a gasket. Tighten bolts in sequence to 7.2–9 ft. lbs. (10–12 Nm).
49. Install the water pump pulley. Tighten bolts in a star pattern to 7.2–9 ft. lbs. (10–12 Nm).
50. Install the engine support bracket. Tighten to 22–30 ft. lbs. (30–41 Nm).
51. Install the alternator bracket. Tighten to 15–20 ft. lbs. (20–27 Nm).
52. Install the engine mounting bracket (B).
    a. Tighten nut (C) to 47–62 ft. lbs. (64–83 Nm).
    b. Tighten bolt (D) and nuts (E) to 36–47 ft. lbs. (49–64 Nm).
53. Install the alternator and snug the mounting bolts.
54. Install the drive belt.
55. Tighten the alternator 12 mm mounting bolt to 15–20 ft. lbs. (20–27 Nm).
56. Tighten the alternator 14 mm mounting bolt to 22–30 ft. lbs. (30–41 Nm).

TIMING CHAIN & SPROCKETS

REMOVAL & INSTALLATION

1.6L Engine

See Timing Chain Front Cover.

VALVE LASH

INSPECTION & ADJUSTMENT

1.6L Engine

See Figures 139 and 140.

➡Inspect and adjust the valve clearance when the engine is cold (Engine coolant temperature : 20°C) and cylinder head is installed on the cylinder block.

1. Remove Valve cover.

➡Do not reuse the disassembled gasket.

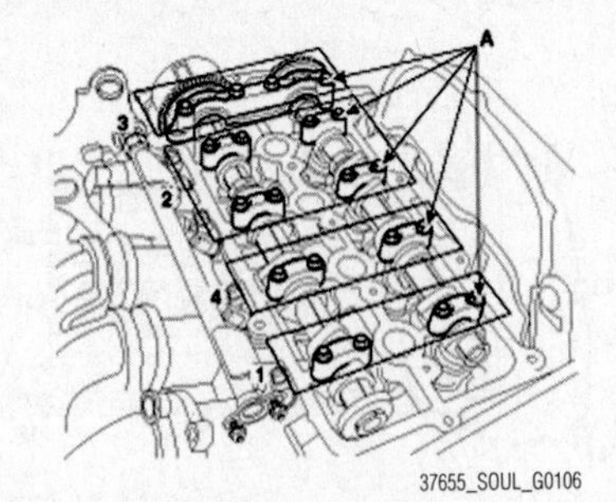

Fig. 139 Removing camshaft bearing cap bolts sequence—1.6L engine

2. Set No.1 cylinder to TDC/compression.
    a. Turn the crankshaft pulley and align its groove with the timing mark of the timing chain cover.
    b. Check that the marks of the camshaft timing sprockets are in straight line on the cylinder head surface. If not, turn the crankshaft one revolution (360°).
3. Inspect the valve clearance.
    a. Check only the intake valves of the 1st and 2nd cylinders and exhaust valves of the 1st and 3rd cylinders for their clearance.
    b. Using a thickness gauge, measure the clearance between the tappet and the base circle of camshaft.
    c. Record the out-of-specification valve clearance measurements. They will be used later to determine the required tappet for adjusting.

Valve clearance specification (Engine coolant temperature : 20°C [68°F])

Intake : 0.0067 ~ 0.0091in. (0.17 ~ 0.23mm)

Exhaust : 0.0087 ~ 0.0110in. (0.22 ~ 0.28mm)

    a. Turn the crankshaft pulley one revolution (clockwise 360°) and align its groove with timing mark of the timing chain cover.
    b. Check the intake valves of the 3rd and 4th cylinders and exhaust valves of the 2nd and 4th cylinders for their clearance.
4. Adjust the intake and exhaust valve clearance.
    a. Set the No.1 cylinder to the TDC/compression position.
    b. Put paint marks on the timing chain links (2 places) that meet with the timing marks of the intake, exhaust camshaft sprockets.
    c. Remove the exhaust camshaft sprocket bolt.
    d. Remove the service hole bolt of the timing chain cover.

➡The bolt must not be reused once it has been assembled.

    e. Insert a thin rod in the service hole of the timing chain cover and release the ratchet.
    f. Remove the exhaust camshaft sprocket.
    g. Remove the camshaft bearing caps in the order shown.
    h. Remove the exhaust camshaft.
    i. Remove the intake camshaft and CVVT module.

➡When disconnecting the timing chain from the camshaft timing sprocket, hold the timing chain.

    j. Tie a timing chain with a string.

CAUTION

Be careful not to drop anything inside timing chain cover.

    k. Measure the thickness of the removed tappet using a micrometer.
    l. Calculate the thickness of a new tappet so that the valve clearance comes within the specified value.

Valve clearance (Engine coolant temperature : 20°C)

T : Thickness of removed tappet

A : Measured valve clearance

N : Thickness of new tappet

Intake : N = T + [A – 0.20mm (0.0079in.)]

Exhaust : N = T + [A–0.25mm (0.0098in.)]

    a. Select a new tappet with a thickness as close as possible to the calculated value.

➡Shims are available in 41size increments of 0.015mm (0.0006in.) from 3.00mm (0.118in.) to 3.600mm (0.1417in.)

    b. Place a new tappet on the cylinder head.
    c. Hold the timing chain, and place

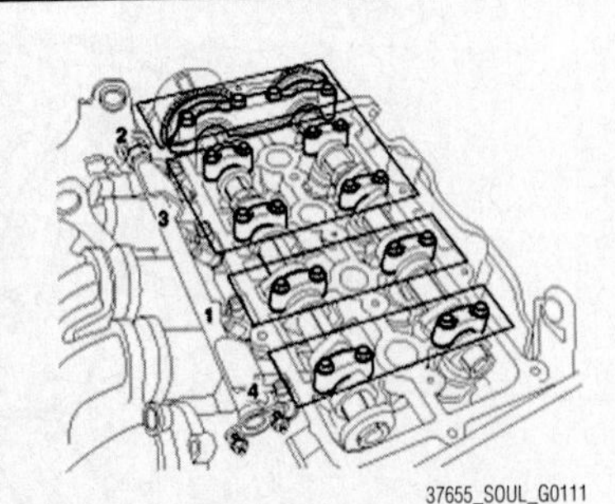

Fig. 140 Camshaft bearing cap bolts torque sequence—1.6L engine

the intake camshaft and CVVT module assembly.

d. Align the matchmarks on the timing chain and camshaft timing sprocket.

e. Install the exhaust camshaft.

f. Install the exhaust camshaft sprocket. Tighten bolt to 47–52 ft. lbs. (68–74 Nm).

g. Install the camshaft bearing caps in the sequence shown. Tighten M6 bolts to 8.7–10 ft. lbs. (11.8–13.7 Nm). Tighten M8 bolts to 13.7–17 ft. lbs. (18.6–22.6 Nm).

h. Install the service hole bolt. Tighten to 8.7–11 ft. lbs. (11.8–14.7 Nm).

i. Turn the crankshaft two turns in the operating direction (clockwise) and realign crankshaft sprocket and camshaft sprocket timing marks.

j. Recheck the valve clearance.

2.0L Engine

See Figure 141.

➡Inspect and adjust the valve clearance when the engine is cold and cylinder head is installed on the cylinder block.

1. Remove the engine cover.
2. Remove the upper timing belt cover (A).
3. Remove the valve cover.
4. Set No. 1 cylinder to TDC/compression.

a. Turn the crankshaft pulley and align its groove with the timing mark "T" of the lower timing belt cover.

b. Check that the hole of the camshaft timing pulley is aligned with the timing mark on the top of the bearing cap. If not, turn the crankshaft one revolution (360°)

5. Inspect the valve clearance

a. With engine at No. 1 cylinder TDC/Compression, check only the intake valves of the 1st and 2nd cylinders and exhaust valves of the 1st and 3rd cylinders for their clearance.

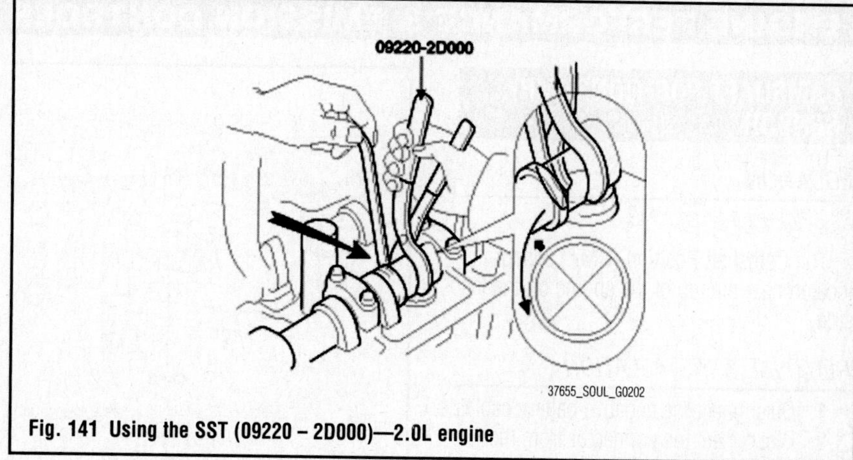

Fig. 141 Using the SST (09220 – 2D000)—2.0L engine

b. Using a thickness gauge, measure the clearance between the tappet shim and the base circle of camshaft.

- Valve clearance
- Engine coolant temperature : 68°F (20°C)
- Intake : 0.0079in. (0.20mm)
- Exhaust : 0.0110in. (0.28mm)
- Limit
- Intake : 0.0067–0.091in. (0.17–0.23mm)
- Exhaust : 0.0098–0.0122in. (0.25–0.31mm)

c. Record the out–of–specification valve clearance measurements. They will be used later to determine the required replacement adjusting shim.

d. Turn the crankshaft pulley one revolution (360°) and align the groove with timing mark "T" of the lower timing belt cover.

e. Check the intake valves of the 3rd and 4th cylinders and exhaust valves of the 2nd and 4th cylinders for their clearance.

6. Adjust the intake and exhaust valve clearance.

a. Turn the crankshaft so that the cam lobe of the camshaft on the adjusting valve is upward.

b. Using the SST (09220 – 2D000), press down the valve lifter and place the stopper between the camshaft and valve lifter and remove the special tool.

c. Remove the adjusting shim with a small screw driver and magnet.

d. Measure the thickness of the removed shim using a micrometer.

e. Calculate the thickness of a new shim so that the valve clearance comes within the specified value.

- Valve clearance (Engine coolant temperature : 68°F)
- T : Thickness of removed shim
- A : Measured valve clearance
- N : Thickness of new shim
- Intake : N = T + [A – 0.20mm (0.0079in.)]
- Exhaust : N = T + [A–0.28mm (0.0110in.)]

f. Select a new shim with a thickness as close as possible to the calculated value.

➡Shims are available in 20size increments of 0.0016in. (0.04mm) from 0.079in. (2.00mm) to 0.1087in. (2.76mm).

g. Place a new adjusting shim on the valve lifter.

h. Using the SST (09220 – 2D000), press down the valve lifter and remove the stopper.

i. Recheck the valve clearance.

ENGINE PERFORMANCE & EMISSION CONTROLS

CAMSHAFT POSITION (CMP) SENSOR

LOCATION

See Figures 142 and 143.

The Camshaft Position (CMP) sensor is located near the top of the engine on the left side.

REMOVAL & INSTALLATION

1. Disconnect the negative battery cable.
2. Disconnect the connector from the sensor.
3. Remove the mounting bolts.
4. Remove the sensor from the cylinder head.
5. Installation is the reverse or removal.

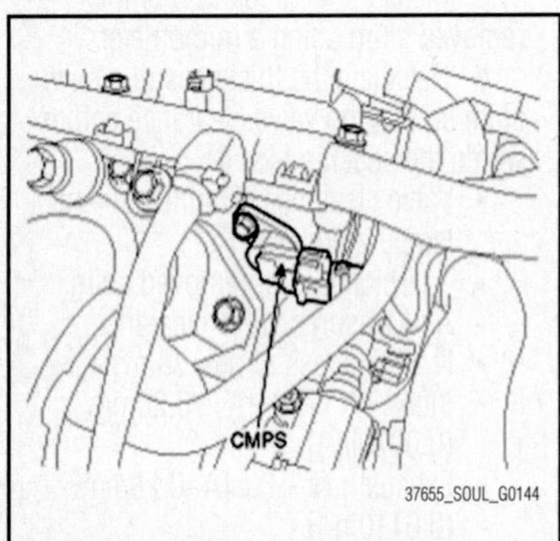

Fig. 142 CMP sensor—1.6L Engine

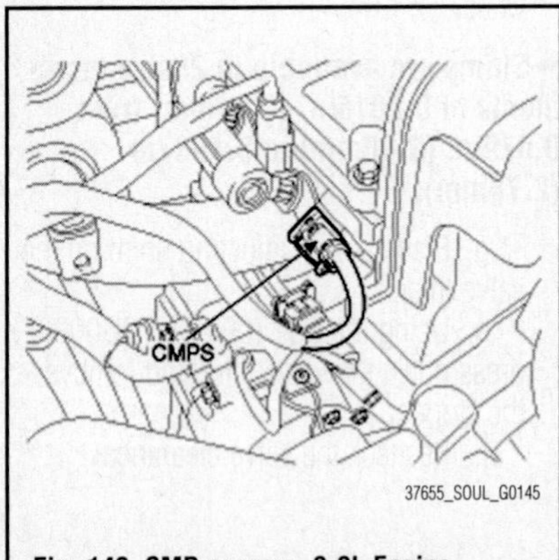

Fig. 143 CMP sensor—2.0L Engine

CRANKSHAFT POSITION (CKP) SENSOR

LOCATION

See Figures 144 and 145.

The CKP sensor is installed on the cylinder block or the transaxle housing. See Component Locations.

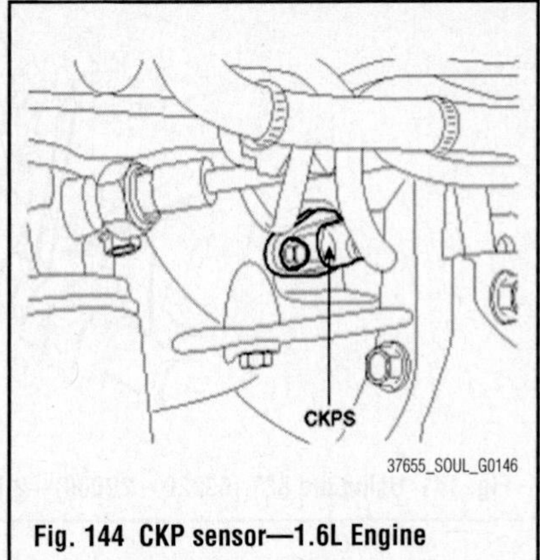

Fig. 144 CKP sensor—1.6L Engine

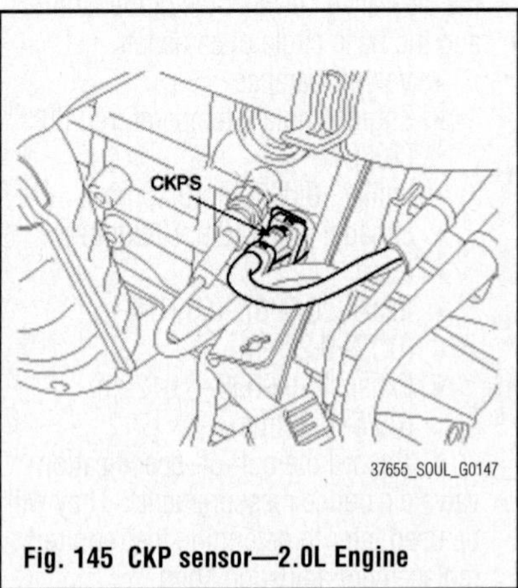

Fig. 145 CKP sensor—2.0L Engine

REMOVAL & INSTALLATION

1. Disconnect the negative battery cable.
2. Disconnect the connector from the sensor.
3. Remove the mounting bolts.
4. Remove the sensor.
5. Installation is the reverse or removal.

ELECTRONIC CONTROL MODULE (ECM)

LOCATION

See Figures 146 and 147.

Refer to the accompanying illustration.

REMOVAL & INSTALLATION

1. Turn the ignition switch off and disconnect the negative battery terminal.
2. Disconnect the ECM connector.
3. Unscrew 4 mounting bolts behind the air cleaner.
4. Remove the ECM.
5. Installation is reverse of removal.

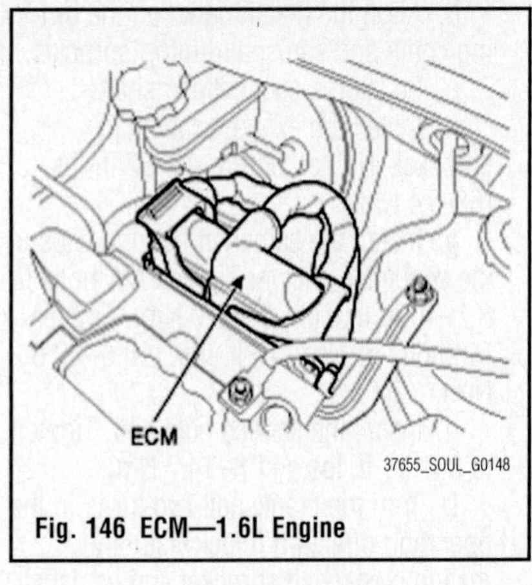

Fig. 146 ECM—1.6L Engine

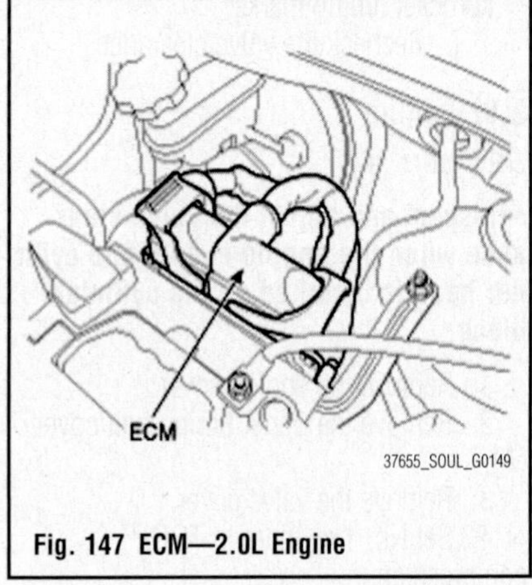

Fig. 147 ECM—2.0L Engine

VIN PROGRAMMING PROCEDURE

VIN (Vehicle Identification Number) is a number that has the vehicle's information (Maker, Vehicle Type, Vehicle Line/Series, Body Type, Engine Type, Transmission Type, Model Year, Plant Location and so forth. When replacing an ECM, the VIN must be programmed in the ECM. If there is no VIN in ECM memory, the fault code (DTC P0630) is set.

CAUTION

The programmed VIN cannot be changed. When writing the VIN, confirm the VIN carefully

1. Using scan tool, select "Vehicle" and "Engine" (For example, SEDONA 3.5L V6).
2. Select "VIN WRITING".
3. Check the ECM status.
   - VIRGIN: VIN is not programmed
   - LEARNT: VIN has been already programmed

   a. Is the ECM status "VIRGIN"? If yes, go to next step. If no, VIN has been programmed and cannot be reprogrammed.

4. Write the VIN with cursor, function and number keys.

✲✲ WARNING

Before pressing the "ENTER" key, confirm the VIN again because the programmed VIN cannot be changed.

5. After verifying the written VIN, press the "ENTER" key.
6. Turn the ignition switch OFF, then back ON.
7. Verify the programmed VIN in the ECM memory.

ENGINE COOLANT TEMPERATURE (ECT) SENSOR

LOCATION

See Figures 148 and 149.

Engine Coolant Temperature Sensor (ECTS) is located in the engine coolant passage of the cylinder head. See Component Locations.

REMOVAL & INSTALLATION

1. Disconnect the negative battery cable.

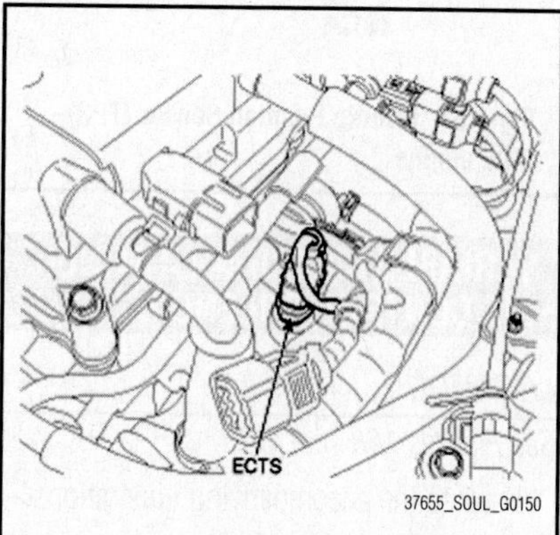

Fig. 148 ECT sensor—1.6L Engine

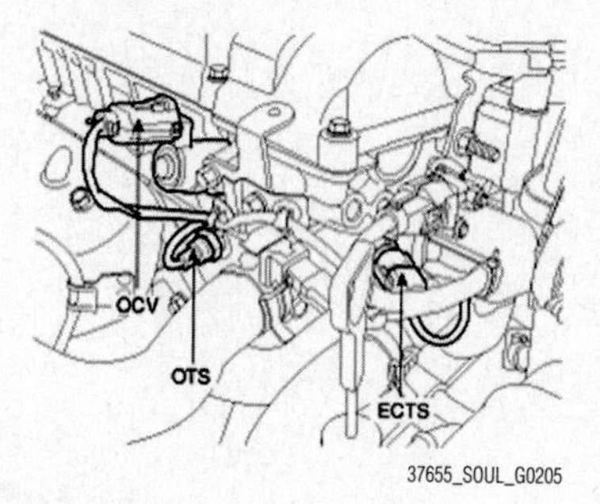

Fig. 149 Engine Coolant Temperature Sensor (ECTS), CVVT Oil Temperature Sensor (OTS) and CVVT Oil Control Valve (OCV)—2.0L Engine

2. Disconnect the connector from the sensor.
3. Drain the cooling system as required.
4. remove the sensor from its mounting.
5. Installation is the reverse of removal.
6. Refill the cooling system.

HEATED OXYGEN SENSOR (HO2S)

LOCATION

The Heated Oxygen Sensor (HO2S) is positioned in the exhaust pipe in front of and behind the TWC. See Component Locations.

REMOVAL & INSTALLATION

1. Disconnect the electrical connector from the sensor.
2. Remove the oxygen sensor.
3. Installation is the reverse of the removal procedure.

➡Apply anti-seize compound to the threaded portion of the sensor, prior to installation. Never apply anti-seize compound to the protector of the sensor.

INTAKE AIR TEMPERATURE (IAT) SENSOR

LOCATION

See Figures 150 and 151.

The Intake Air Temperature (IAT) sensor is part of MAP sensor.

REMOVAL & INSTALLATION

1. Disconnect the negative battery cable.
2. Disconnect the connector from the sensor.
3. Remove the sensor retaining screws.

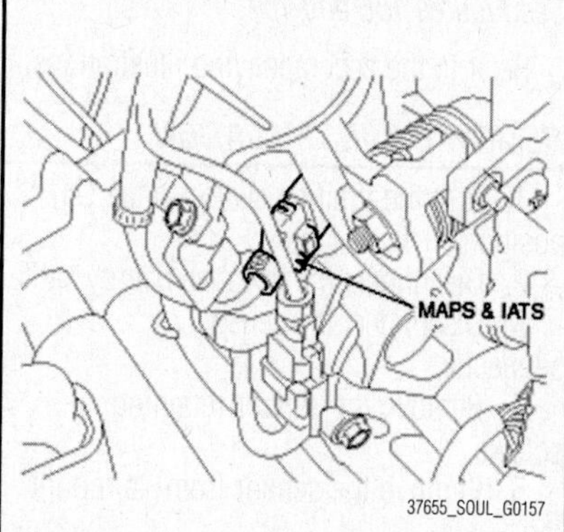

Fig. 150 IAT/MAP sensor location—1.6L engine

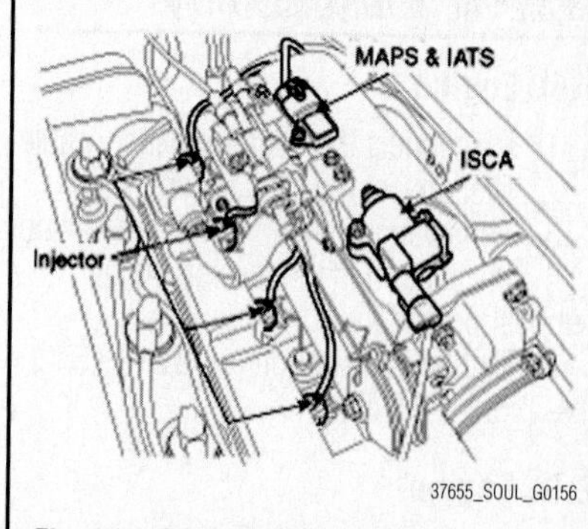

Fig. 151 ISCA , MAP/IAT and injector locations—2.0L engine

4. Remove the air cleaner and air intake assembly, as required.
5. Remove the sensor from its mounting.
6. Installation is the reverse of removal.

KNOCK SENSOR (KS)

LOCATION

See Figures 152 and 153.

Refer to the accompanying illustrations.

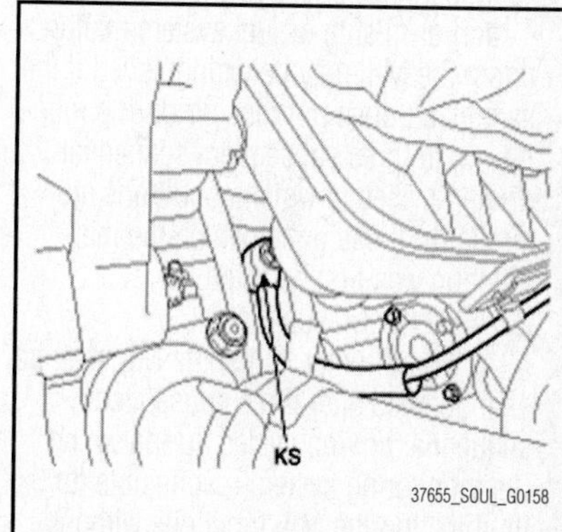

Fig. 152 Knock sensor location—1.6L engine

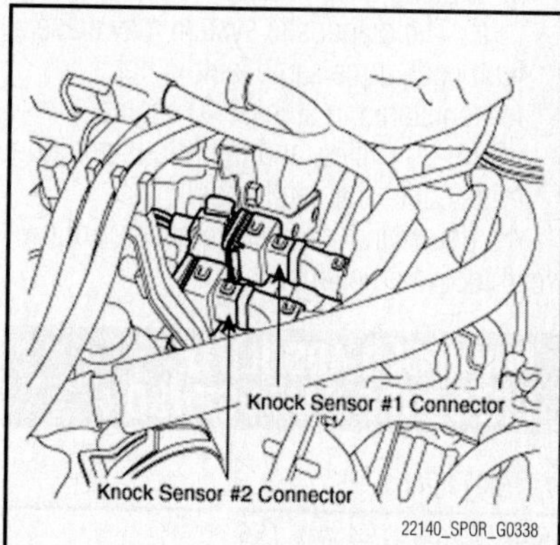

Fig. 153 Knock sensor location—2.0L engine

REMOVAL & INSTALLATION

1.6L Engine

1. Disconnect the negative battery cable.
2. Disconnect the sensor connector.
3. Remove the sensor from its mounting.
4. Installation is the reverse of the removal procedure.
5. Tighten sensor bolt to 12.3–19.5 ft. lbs. (16.7–26.5 Nm).

2.0L Engine

1. Disconnect the negative battery cable.
2. Remove the intake manifold support bracket.
3. Disconnect the sensor connector.
4. Remove the sensor from its mounting.
5. Installation is the reverse of the removal procedure.
6. Tighten sensor bolt to 12.3–19.5 ft. lbs. (16.7–26.5 Nm).

MALFUNCTION INDICATOR LIGHT (MIL)

RESET PROCEDURE

1. Misfire and Fuel System Malfunctions:
   a. For misfire or fuel system malfunctions, the MIL may be extinguished if the same fault does not reoccur during monitoring in three subsequent sequential driving cycles in which conditions are similar to those under which the malfunction was first detected.
2. All Other Malfunctions:
   a. For all other faults, the MIL may be extinguished after three subsequent sequential driving cycles during which the monitoring system responsible for illuminating the MIL functions without detecting the malfunction and if no other malfunction has been identified that would independently illuminate the MIL.
3. Erasing a fault code
   a. The diagnostic system may erase a fault code if the same fault is not re–registered in at least 40 engine warm–up cycles, and the MIL is not illuminated for that fault code.
4. Use scan tool to erase codes. Follow scan tool instructions.

MANIFOLD ABSOLUTE PRESSURE (MAP) SENSOR

LOCATION

See Figures 154 and 155.

On intake manifold, combined with IAT sensor.

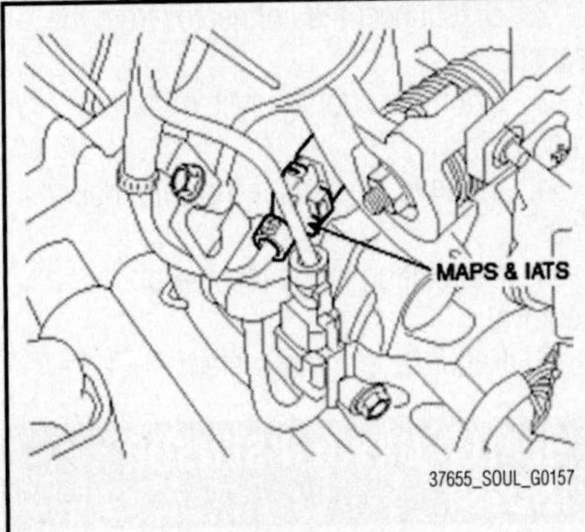

Fig. 154 IAT/MAP sensor location—1.6L engine

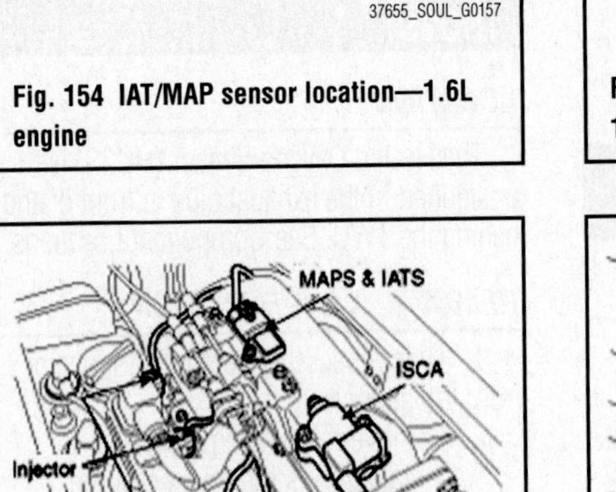

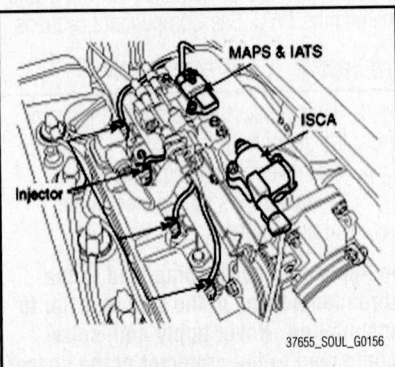

Fig. 155 ISCA , MAP/IAT and injector locations—2.0L engine

REMOVAL & INSTALLATION

1. Disconnect the negative battery cable.
2. Disconnect the connector from the sensor.
3. Remove the sensor retaining screws.
4. Remove the sensor from its mounting.
5. Installation is the reverse of the removal procedure.

THROTTLE POSITION SENSOR (TPS)

LOCATION

See Figures 156 and 157.

Refer to the accompanying illustrations.

REMOVAL & INSTALLATION

1. Turn the ignition switch to the OFF position.
2. Disconnect the negative battery cable.
3. Disconnect the sensor connector.
4. Remove the sensor retaining screws.
5. Remove the sensor from its mounting.
6. Installation is the reverse of the removal procedure.

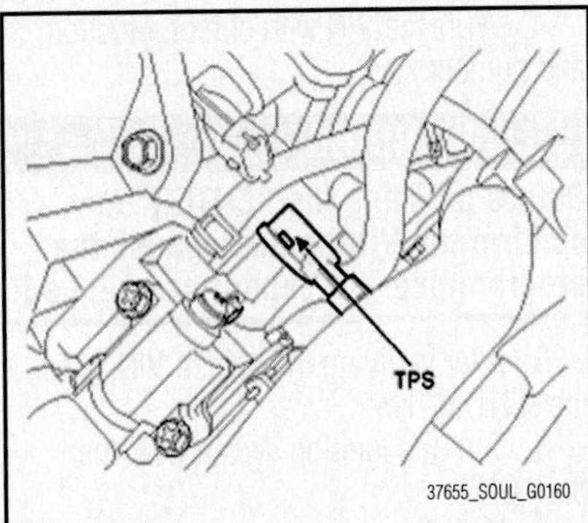

Fig. 156 Throttle Position Sensor (TPS)—1.6L engine

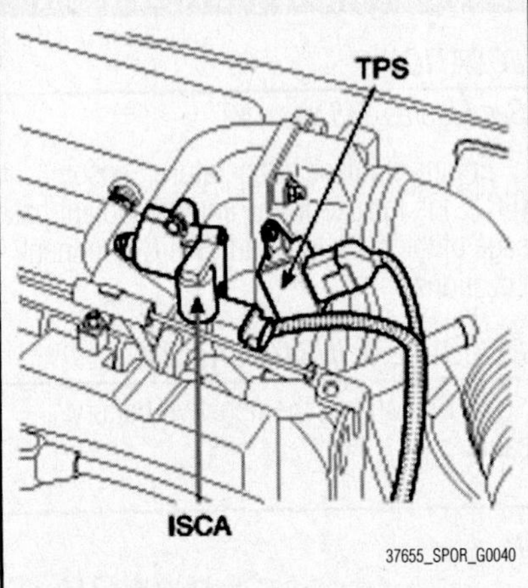

Fig. 157 Throttle Position Sensor (TPS)—2.0L engine

VARIABLE CAMSHAFT TIMING OIL CONTROL SOLENOID

LOCATION

See Figures 158 and 159.

Refer to the accompanying illustrations.

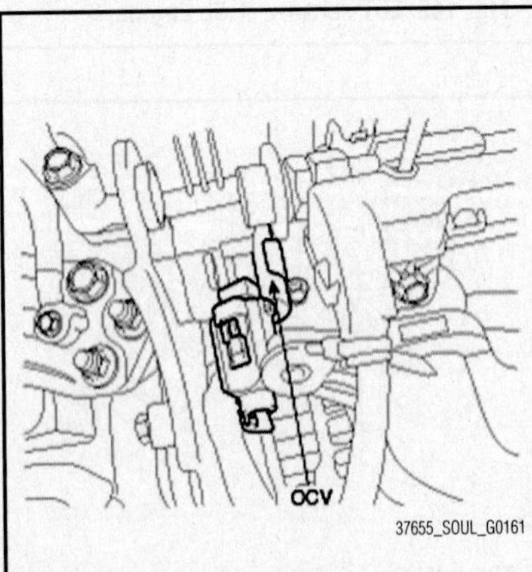

Fig. 158 CVVT Oil Control Valve (OCV)—1.6L engine

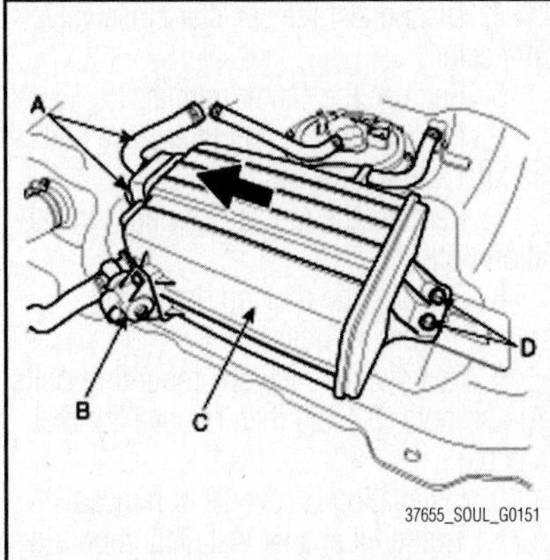

Fig. 159 Engine Coolant Temperature Sensor (ECTS), CVVT Oil Temperature Sensor (OTS) and CVVT Oil Control Valve (OCV)—2.0L Engine

REMOVAL & INSTALLATION

1. Turn the ignition switch to the OFF position.
2. Disconnect the solenoid connector.
3. Remove the solenoid from the head.
4. Installation is the reverse of the removal procedure.

VEHICLE SPEED SENSOR (VSS)

REMOVAL & INSTALLATION

See Figure 160.

1. Raise and support vehicle.
2. Remove front wheel.
3. Remove the speed sensor cable mounting bolt and speed sensor.
4. Installation is the reverse of the removal procedure.

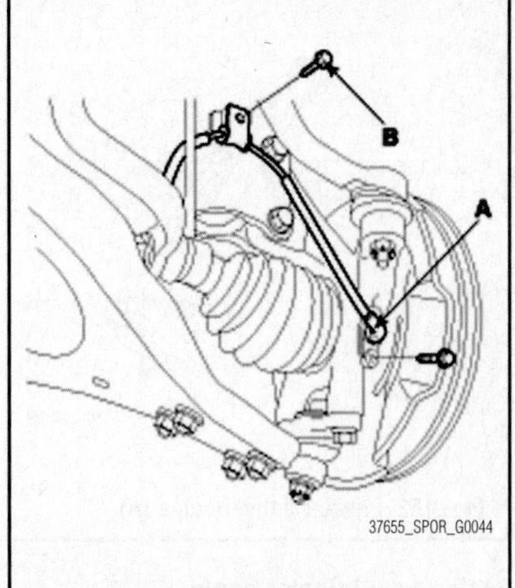

Fig. 160 Speed sensor (A) and speed sensor cable mounting bolt (B)

FUEL — GASOLINE FUEL INJECTION SYSTEM

FUEL SYSTEM SERVICE PRECAUTIONS

Safety is the most important factor when performing not only fuel system maintenance, but any type of maintenance. Failure to conduct maintenance and repairs in a safe manner may result in serious personal injury or death. Work on a vehicle's fuel system components can be accomplished safely and effectively by adhering to the following rules and guidelines.

- To avoid the possibility of fire and personal injury, always disconnect the negative battery cable unless the repair or test procedure requires that battery voltage be applied.
- Always relieve the fuel system pressure prior to disconnecting any fuel system component (injector, fuel rail, pressure regulator, etc.) fitting or fuel line connection. Exercise extreme caution whenever relieving fuel system pressure to avoid exposing skin, face and eyes to fuel spray. Please be advised that fuel under pressure may penetrate the skin or any part of the body that it contacts.
- Always place a shop towel or cloth around the fitting or connection prior to loosening to absorb any excess fuel due to spillage. Ensure that all fuel spillage is quickly removed from engine surfaces. Ensure that all fuel–soaked cloths or towels are deposited into a flame–proof waste container with a lid.
- Always keep a dry chemical (Class B) fire extinguisher near the work area.
- Do not allow fuel spray or fuel vapors to come into contact with a spark or open flame.
- Always use a second wrench when loosening or tightening fuel line connection fittings. This will prevent unnecessary stress and torsion on fuel piping. Always follow the proper torque specifications.
- Always replace worn fuel fitting O–rings with new ones. Do not substitute fuel hose where rigid pipe is installed.

FUEL SYSTEM PRESSURE

RELIEVING

1. Remove the rear seat cushion
2. Disconnect the fuel pump connector.
3. Start the engine and wait until fuel in fuel line is exhausted.
4. After engine stalls, turn the ignition switch to OFF position.

FUEL FILTER

REMOVAL & INSTALLATION

The fuel filter is an integral part of the fuel pump assembly. See Fuel Pump.

FUEL PUMP

REMOVAL & INSTALLATION

1. Remove the rear seat cushion.
2. Remove the service cover
3. Disconnect the fuel pump connector and the FTPS sensor connector.
4. Start the engine and wait until fuel in fuel line is exhausted.
5. After engine stalls, turn the ignition switch to OFF position.
6. Disconnect the fuel feed line and vapor hoses.
7. Unscrew the fuel pump mounting bolts and remove the fuel pump assembly.
8. Install the Fuel Pump according to the reverse order of REMOVAL procedure.

FUEL PRESSURE REGULATOR

REMOVAL & INSTALLATION

The fuel pressure regulator is an integral part of the fuel pump assembly. See Fuel Pump.

FUEL RAIL AND INJECTOR

REMOVAL & INSTALLATION

See Figures 161 and 162.

1. Relieve fuel system pressure. See Relieving.
2. Remove the injector connectors (A) and the harness bracket (B).
3. Remove the delivery pipe (A).
4. Remove the injectors.
5. Installation is reverse of removal.

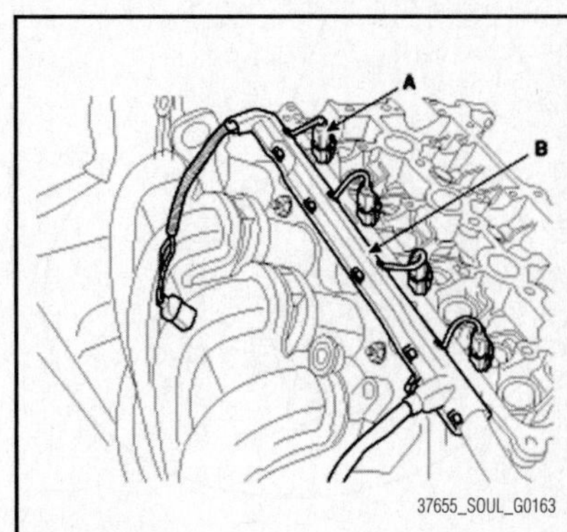

Fig. 161 injector connectors (A) and the harness bracket (B)

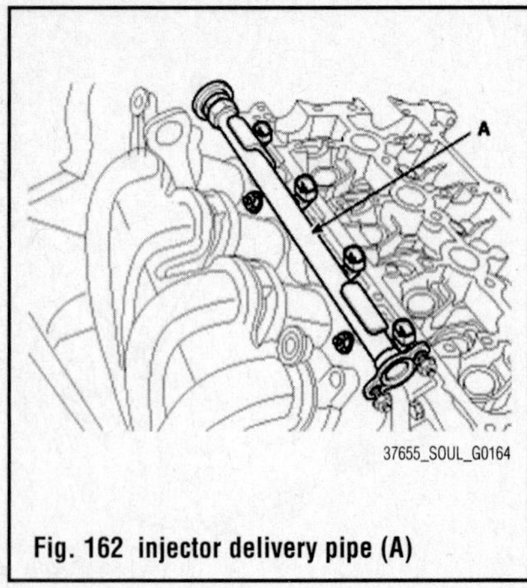

Fig. 162 injector delivery pipe (A)

➡**Use new injector seals.**

6. Tighten the delivery pipe to 14.4–18 ft. lbs. (20–25 Nm).

FUEL TANK

REMOVAL & INSTALLATION

See Figure 163.

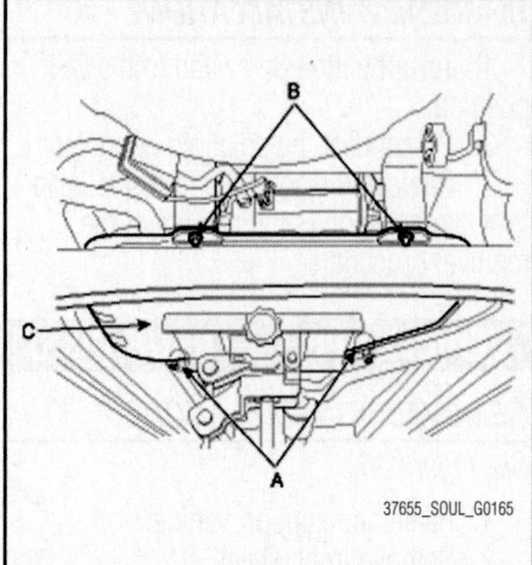

Fig. 163 Fuel tank mounting bolts (A), nuts (B) and fuel tank (C)

1. Relieve fuel system pressure. See Relieving.
2. Disconnect the fuel feed quick–connector and the vapor tube quick–connector.
3. Lift the vehicle and support the fuel tank with a jack.
4. Disconnect the canister close valve connector.
5. Remove the center muffler.
6. Disconnect the fuel filler hose and the filler ventilation hose.
7. Remove the fuel tank air filter installation bolt.
8. Remove the parking brake line bracket mounting bolts.
9. Remove the fuel tank mounting bolts (A) and nuts (B), and then remove the fuel tank (C).
10. Installation is reverse of removal.
11. Tighten fuel tank installation bolt to 29–40 ft. lbs. (39–54 Nm).
12. Tighten fuel tank installation nut to 29–40 ft. lbs. (39–54 Nm).
13. Tighten fuel tank air filter installation bolt: to 3–4 ft. lbs. (4–6 Nm).

IDLE SPEED

ADJUSTMENT

Idle speed is controlled by ECM and is not adjustable.

HEATING & AIR CONDITIONING SYSTEM

BLOWER MOTOR

REMOVAL & INSTALLATION

See Figure 164.

1. Disconnect the negative (–) battery terminal.

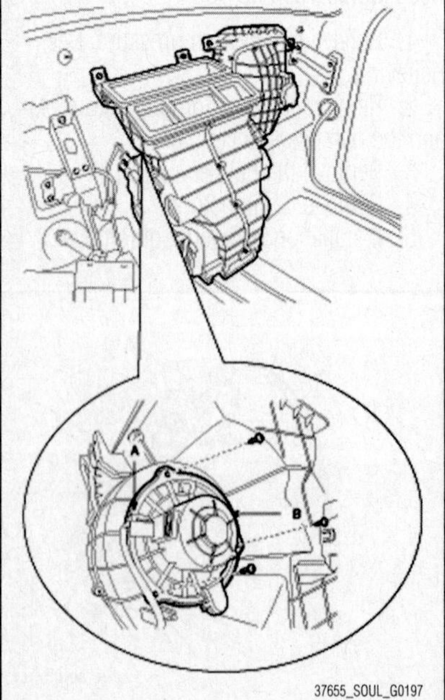

Fig. 164 Blower motor (B) and connector (A)

2. Remove the instrument panel.
3. Disconnect the connector (A) of the blower motor.
4. Remove the blower motor (B) after loosening the mounting screws.
5. Installation is the reverse order of removal.

HEATER CORE

REMOVAL & INSTALLATION

See Heater Unit.

HEATER UNIT

REMOVAL & INSTALLATION

See Figures 165 through 167.

1. Disconnect the negative (–) battery terminal.
2. Recover the refrigerant with a recovery/ recycling/ charging station.
3. When the engine is cool, drain the engine coolant from the radiator.
4. Remove the bolts and the expansion valve from the evaporator core.

➡**Plug or cap the lines immediately after disconnecting them to avoid moisture and dust contamination.**

5. Disconnect the inlet (lower) and outlet (upper) heater hoses from the heater unit.

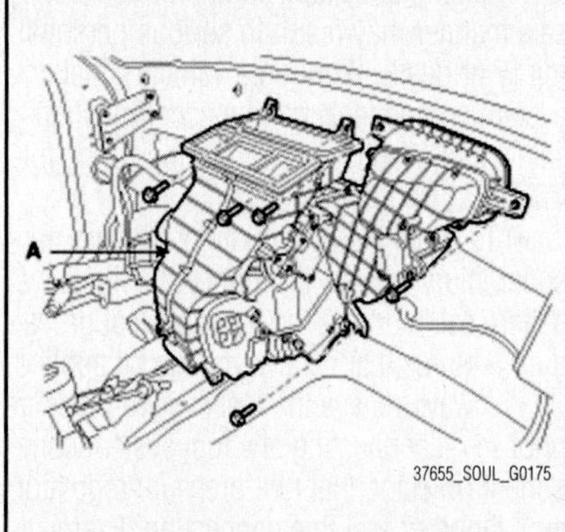

Fig. 165 Heater and blower unit (A)

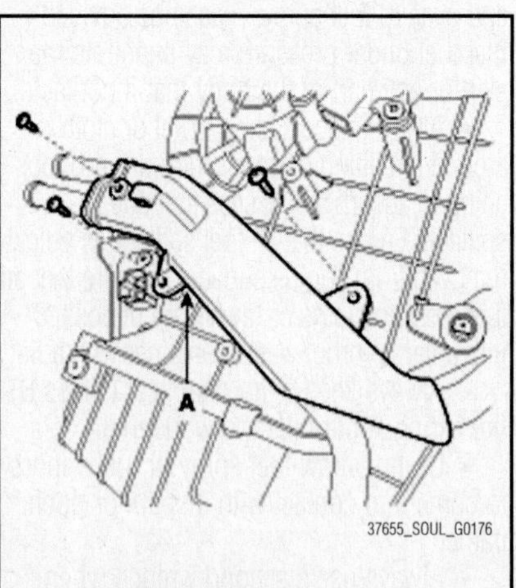

Fig. 166 Heater cover (A)

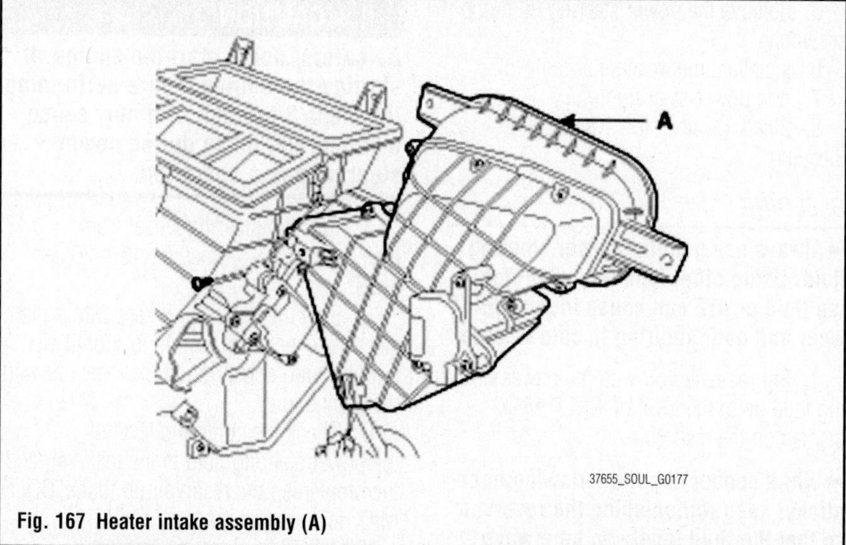

Fig. 167 Heater intake assembly (A)

➡Engine coolant will spill when the hoses are disconnected; drain it into a clean drip pan. Be sure not to let coolant spill on electrical parts or painted surfaces. If any coolant spills, rinse it off immediately.

6. Remove the instrument panel. See Instrument Panel.
7. Remove the cowl cross bar assembly.
8. Remove the heater & blower unit (A) after loosening mounting bolts.
9. Remove the cover (A).
10. Be careful that the inlet and outlet pipe are not bent during heater core removal, and pull out the heater core.
11. Remove the intake assembly (A).
12. Remove the heater unit lower case.
13. Remove the evaporator core.
14. Installation is the reverse order of removal.
15. Note these items :
    a. If you're installing a new evaporator, add refrigerant oil (ND–OIL8).
    b. Replace the O–rings with new ones at each fitting, and apply a thin coat of refrigerant oil before installing. Be sure to use the right O–rings for R-134a to avoid leakage.
    c. Immediately after using the oil, replace the cap on the container, and seal it to avoid moisture absorption.
    d. Do not spill the refrigerant oil on the vehicle ; it may damage paint ; if the refrigerant oil contacts the paint, wash off immediately.
    e. Apply sealant to the grommets.
    f. Make sure that there is no air leakage.
    g. Charge the system and test its performance.
    h. Do not interchange the inlet and outlet heater hoses and install the hose clamps securely.
    i. Refill the cooling system with engine coolant.

STEERING

POWER RACK & PINION STEERING GEAR

REMOVAL & INSTALLATION

See Figures 168 through 170.

1. Drain the power steering fluid by disconnecting the return hose.
2. Remove the front wheel & tire.
3. Disconnect the stabilizer link with the front strut assembly after loosening the nut.
4. Disconnect the tie–rod end with the knuckle using a SST (09568–2J100 or 09568–4A000).
5. Remove the two bolts for lower arm ball joint.
6. Loosen the bolt and then disconnect the universal joint assembly from the pinion of the steering gear box inside car.

⁂ CAUTION

Keep the steering wheel in straight ahead position and do not allow it to move while disassembled to prevent damage to the clock spring inner cable.

7. Remove the cross member from the body by loosening the mounting bolts and nuts. This includes the roll stopper and the crossmember to body bolts.
8. Remove the power steering gear box (B) mounting bolts (A) from the front subframe.

To install:

9. Installation is reverse of removal.
10. Install the universal joint assembly to the pinion of the steering gear box. Torque: 22–25 ft. lbs. (30–35 Nm).
11. Install the steering gear box mounting bolts. Torque: 43–58 ft. lbs. (60–80 Nm).
12. Install the lower arm assembly mounting bolt to the knuckle. Torque: 43–52 ft. lbs. (60–72 Nm).
13. Install the stabilizer link to front strut assembly nut. Torque: 72–86 ft. lbs. (100–120 Nm).
14. Install the wheel and tire. Torque: 65–80 ft. lbs. (90–110 Nm).
15. Bleed the air in the power steering system.
16. Check wheel alignment.

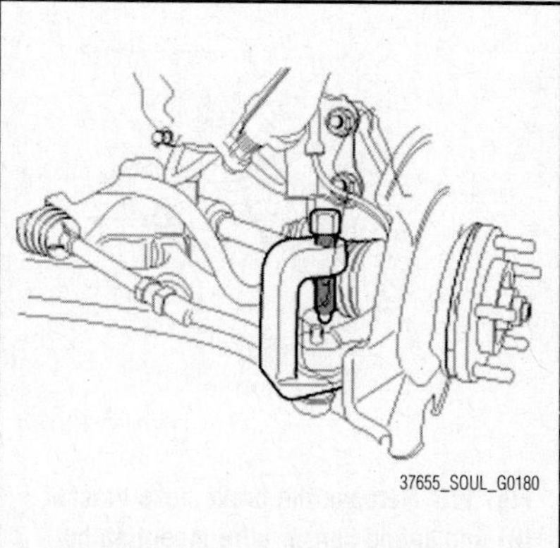

Fig. 168 Special tool 09568–4A000

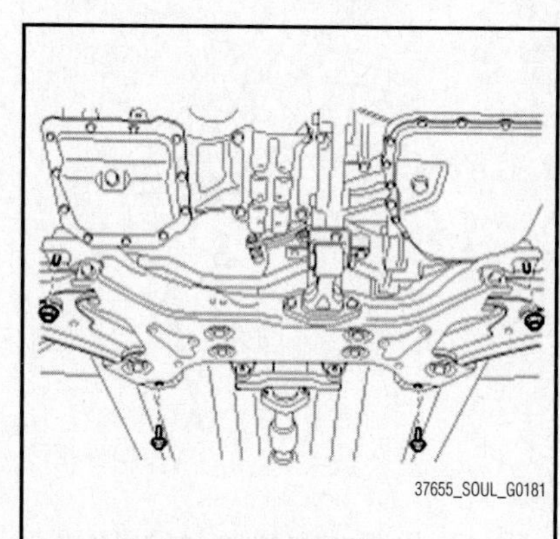

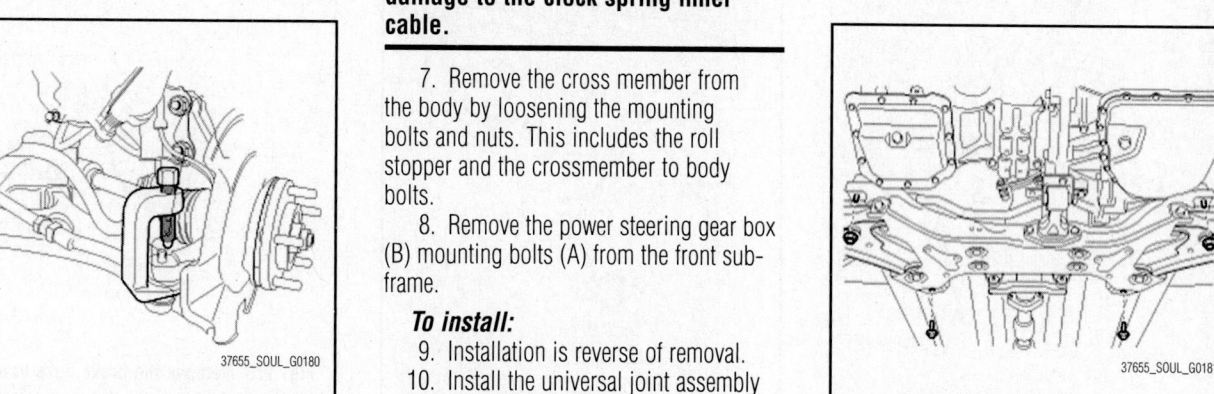

Fig. 169 Crossmember bolts

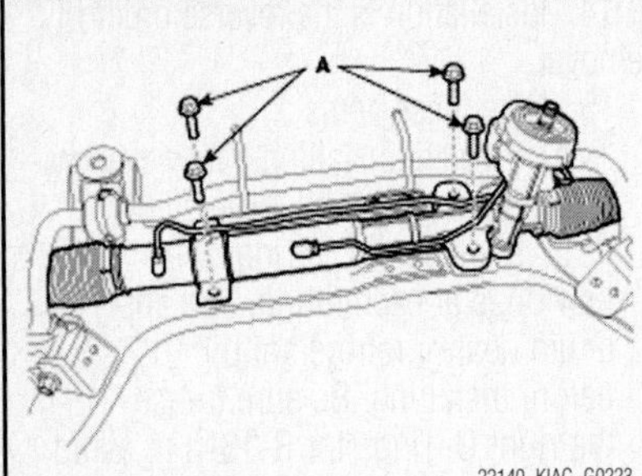

Fig. 170 Remove the power steering gear box (B) mounting bolts (A) from the front subframe.

POWER STEERING PUMP

REMOVAL & INSTALLATION

1. Disconnect the suction hose from the suction pipe and drain the fluid into a container.
2. Remove the belt.
3. Remove the pressure hose from the oil pump.
4. Remove the power steering oil pump mounting bolt.
5. Remove the power steering oil pump assembly.
6. Installation is reverse of removal.
7. Add power steering fluid.
8. Bleed the air in the system. See Bleeding.

BLEEDING

➡Always use genuine power steering fluid. Using other type of power steering fluid or ATF can cause increased wear and poor steering in cold weather.

1. Fill the reservoir with the power steering fluid up to the level of 'COLD MAX' marked on the reservoir.

➡While conducting the following operations, keep replenishing the reservoir so that the fluid level can be always between the 'COLD MAX' and the 'COLD MIN' marked on the reservoir.

2. Jack up the front wheels.
3. Crank the engine 1 ~ 2times by turning the ignition key very quickly from the 'On' position to the 'Start' position, but do not start the engine.

⁂ CAUTION

Be careful not to start the engine. If starting the engine before performing the steps 3 through 4, it may cause an abnormal noise during power steering pump operation.

4. Turn the steering wheel from lock to lock 5 ~ 6 times for 15 ~ 20 seconds.
5. Start the engine and keep turning the steering wheel from lock to lock until air bubbles stop appearing in the reservoir with the engine idle.
6. Check the color and level of the power steering fluid in the reservoir and then replenish the reservoir up to the 'COLD MAX' level as required.

If the fluid level moves up and down when turning the steering wheel, the fluid overflows out of the reservoir when the turning off the engine or the fluid has white color, it indicates that air bubbles have not been removed sufficiently from the power steering system. Therefore, repeat the steps 5 through 6 as required.

SUSPENSION

FRONT SUSPENSION

LOWER BALL JOINT

REMOVAL & INSTALLATION

Information is not available.

LOWER CONTROL ARM

REMOVAL & INSTALLATION

See Figures 171 and 172.

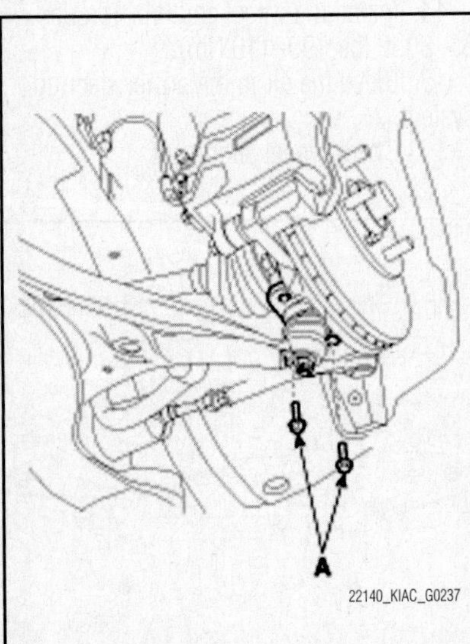

Fig. 171 Remove the lower arm ball joint mounting bolts (A).

1. Raise the front of the vehicle, and make sure it is securely supported.
2. Remove the front wheel and tire.
3. Remove the lower arm ball joint mounting bolts (A).
4. Remove the lower arm mounting bolts (A).
5. Remove lower control arm.

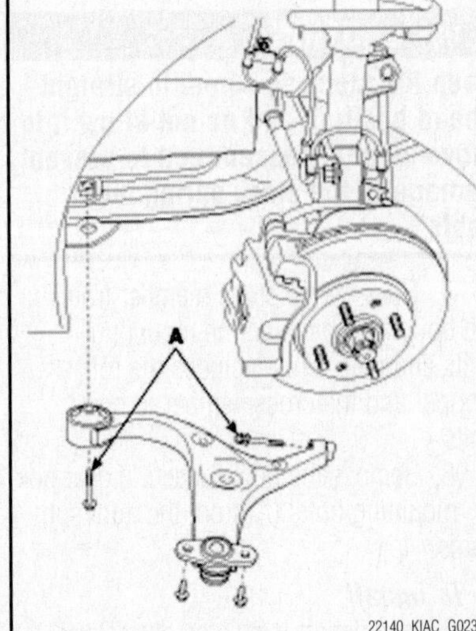

Fig. 172 Remove the lower arm mounting bolts (A).

To install:

6. Install the lower arm mounting bolts. Torque: A bushing: 72–86 ft. Lbs. (100–120 Nm); G bushing: 72–101 ft. lbs. (100–140 Nm).
7. Install the lower arm assembly mounting bolt to the knuckle. Torque: 43–52 ft. lbs. (60–72 Nm).
8. Install the wheel and tire. Torque: 65–80 ft. lbs. (90–110 Nm).

MACPHERSON STRUT

REMOVAL & INSTALLATION

See Figures 173 through 175.

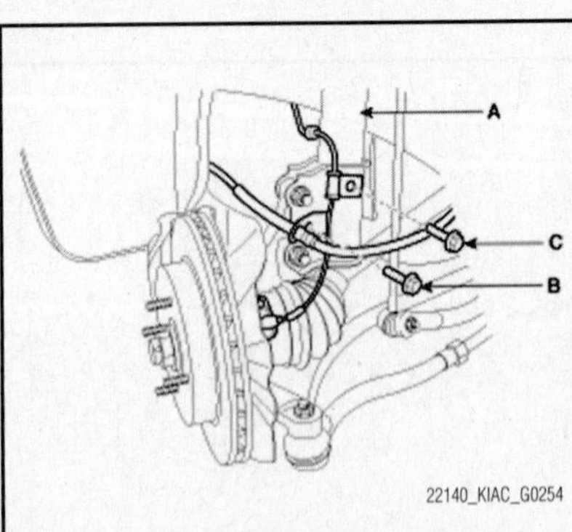

Fig. 173 Remove the brake hose bracket (B) and speed sensor wire mounting bolt (C) from the strut assembly (A).

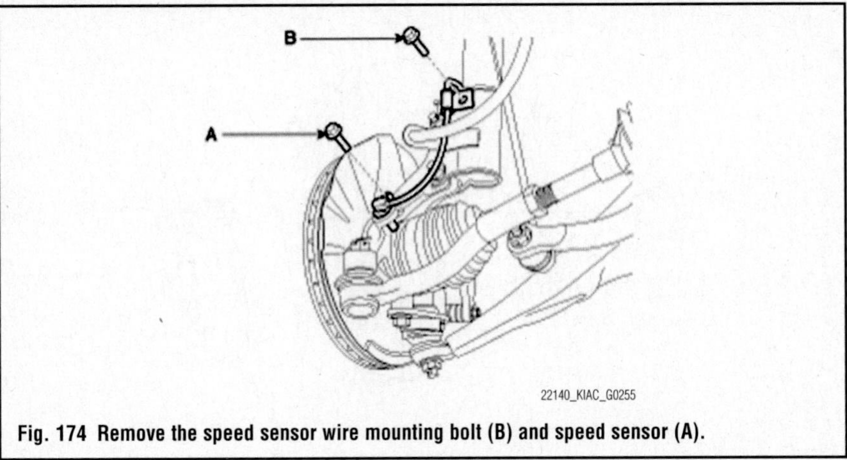

B
A
22140_KIAC_G0255
Fig. 174 Remove the speed sensor wire mounting bolt (B) and speed sensor (A).

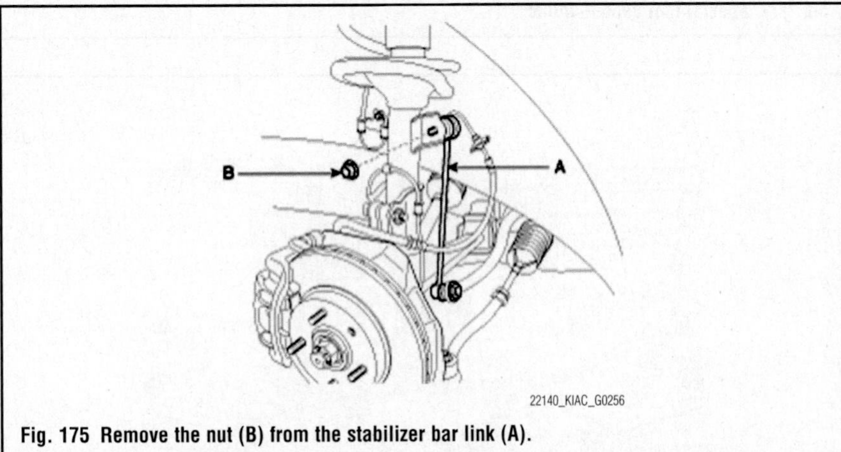

B
A
22140_KIAC_G0256
Fig. 175 Remove the nut (B) from the stabilizer bar link (A).

1. Raise the vehicle, and make sure it is securely supported.
2. Remove the front wheel and tire.
3. Remove the brake hose bracket (B) and speed sensor wire mounting bolt (C) from the strut assembly (A).
4. Remove the speed sensor wire mounting bolt (B) and speed sensor (A).
5. Remove the nut (B) from the stabilizer bar link (A).
6. Remove the strut upper mounting nuts.
7. Remove the strut lower mounting bolts and then remove the strut assembly.

To install:

8. Install the strut assembly and lower mounting bolts. Torque: 101–115 ft. lbs. (140–160 Nm)
9. Install the strut upper mounting nuts. Torque: 3.3–4.4 ft. lbs. (4.6–6 Nm).
10. Install the nut on the stabilizer bar link. Torque: 72–87 ft. lbs. (100–120 Nm).
11. Install the speed sensor wire mounting bolt and speed sensor.
12. Install the brake hose bracket and speed sensor wire mounting bolt on the strut assembly.
13. Install the wheel and the tire.

STEERING KNUCKLE

REMOVAL & INSTALLATION

See Figures 176 and 177.

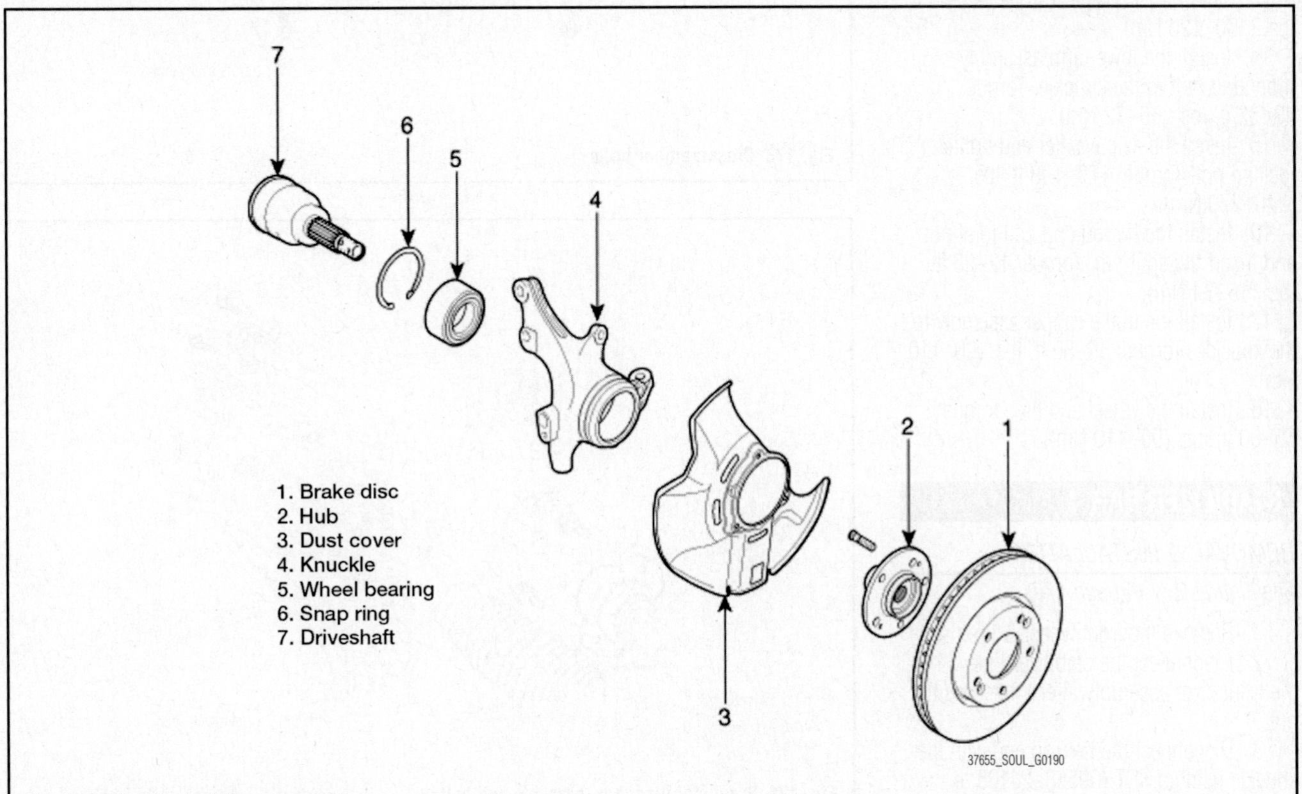

7
6
5
4
2
1
3
1. Brake disc
2. Hub
3. Dust cover
4. Knuckle
5. Wheel bearing
6. Snap ring
7. Driveshaft
37655_SOUL_G0190
Fig. 176 Steering knuckle components

1. Raise and support the vehicle.
2. Remove the front wheel and tire.
3. Remove the brake hose mounting bracket.
4. Remove the brake rotor. See Brake Rotor.
5. Remove the tie rod end ball joint from the knuckle.
   a. Remove the split pin.
   b. Remove the castle nut.
   c. Disconnect the ball joint from knuckle using the special tool (09568–4A000 or 09568–2J100).
6. Remove the driveshaft nut from the front hub.

➡The driveshaft lock nut should be replaced with new one. After installation driveshaft lock nut, stake the lock nut using a chisel and hammer.

7. Remove the lower arm from the knuckle.
8. Loosen the mount bolt and then remove the wheel speed sensor from knuckle.
9. Disconnect the driveshaft from the front hub assembly..
10. Loosen the strut mount bolts and then remove the knuckle assembly.
11. Remove the snap ring.
12. Install in the reverse order of removal.
13. Install the strut assembly and the drive shaft in the knuckle. Torque: 72–86 ft. lbs. (100–120 Nm).
14. Install the lower arm assembly mounting bolt to the knuckle. Torque: 43–52 ft. lbs. (60–72 Nm).
15. Insert the axle washer and tighten the locking nut. Torque: 177–198 ft. lbs. (240–270 Nm).
16. Install the tie rod end ball joint nut and insert the split pin. Torque: 12–25 ft. lbs. (16–34 Nm).
17. Install the brake caliper assembly to the knuckle. Torque: 65–80 ft. lbs. (90–110 Nm).
18. Install the wheel and tire. Torque: 65–80 ft. lbs. (90–110 Nm).

STABILIZER BAR

REMOVAL & INSTALLATION

See Figures 177 through 179.

1. Remove the front wheel & tire.
2. Disconnect the stabilizer link with the front strut assembly after loosening the nut.
3. Disconnect the tie–rod end with the knuckle using a SST (09568–2J100 or 09568–4A000).

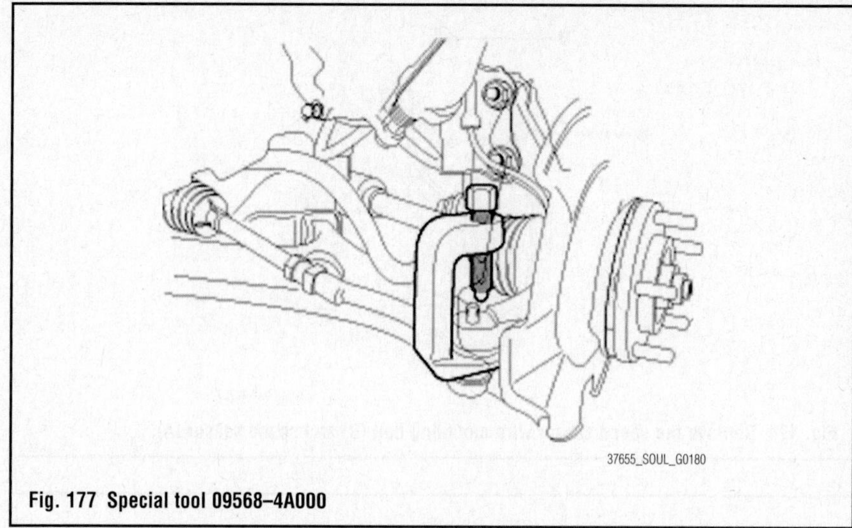

Fig. 177 Special tool 09568–4A000

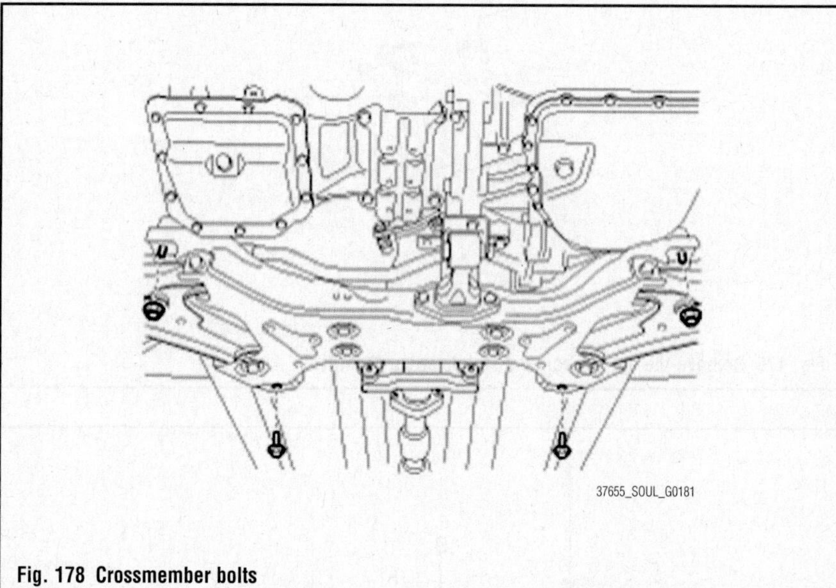

Fig. 178 Crossmember bolts

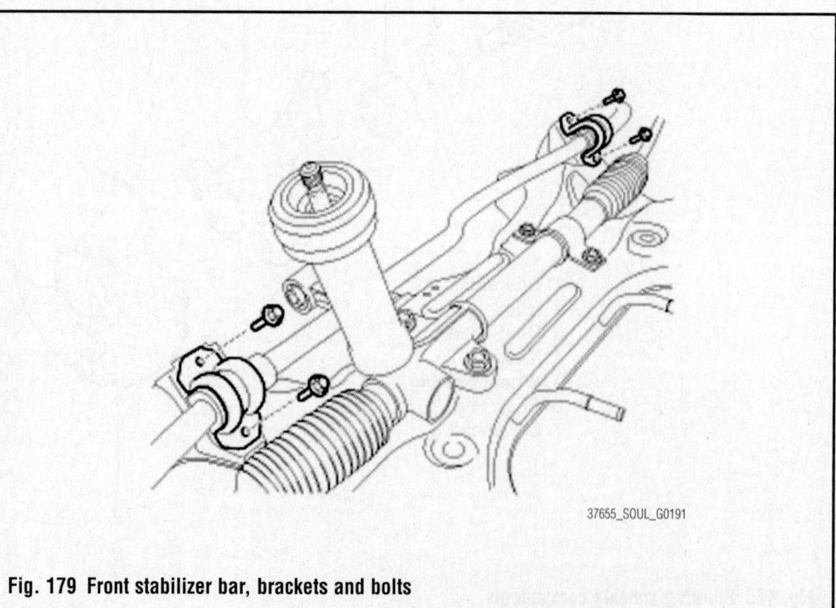

Fig. 179 Front stabilizer bar, brackets and bolts

4. Remove the two bolts for lower arm ball joint.

5. Loosen the bolt and then disconnect the universal joint assembly from the pinion of the steering gear box inside car.

⁂ CAUTION

Keep the steering wheel in straight ahead position and do not allow it to move while disassembled to prevent damage to the clock spring inner cable.

6. Remove the cross member from the body by loosening the mounting bolts and nuts. This includes the roll stopper and the crossmember to body bolts.

7. Remove the stabilizer from the cross member by loosening the bracket mounting bolts.

8. Disconnect the stabilizer link from the stabilizer bar.

9. Remove the bushing and the bracket from the stabilizer bar.

10. Installation is the reverse of removal.

11. Install the stabilizer bar brackets mounting bolts. Torque: 32–40 ft. lbs. (45–55 Nm).

12. Install the lower arm assembly mounting bolt to the knuckle. Torque: 43–52 ft. lbs. (60–72 Nm).

13. Install the stabilizer link to front strut assembly nut. Torque: 72–86 ft. lbs. (100–120 Nm).

14. Install the wheel and tire. Torque: 65–80 ft. lbs. (90–110 Nm).

WHEEL HUB & BEARING

REMOVAL & INSTALLATION

1. See Steering Knuckle.

SUSPENSION

REAR SUSPENSION

COIL SPRING

REMOVAL & INSTALLATION

1. Remove the wheel and tire.
2. Remove the brake hose bracket.
3. Remove the wheel speed sensor wire bracket.
4. Use a jack at the bottom of the rear torsion axle beam to raise and support, then remove the rear shock absorber lower mounting bolt.
5. Remove the rear coil spring.

To install:

6. Install the upper and lower pads on the coil spring by aligning the grooves on the pads.
7. Place the coil spring with the pads on the torsion axle beam and support it with a jack.
8. Install the rear shock absorber mounting bolt by lifting the rear torsion axle beam. Torque: 72–86 ft. lbs. (100–120 Nm).
9. Install the wheel speed sensor wire bracket bolt.
10. Install the brake pressure hose bracket bolt.
11. Install the wheel and tire.

SHOCK ABSORBER

REMOVAL & INSTALLATION

1. Remove the wheel and tire.
2. After supporting the rear torsion axle beam with a jack, remove the rear shock absorber lower mounting bolt. Remove the rear shock absorber.
3. Remove the rear shock absorber mounting bolts.

To install:

4. Tighten the rear shock absorber upper mounting bolt. Torque: 72–86 ft. lbs. (100–120 Nm).
5. Placing a jack at the bottom of the rear torsion axle beam and raise the vehicle to the proper location.
6. Tighten the rear shock absorber lower mounting bolts. Torque: 72–86 ft. lbs. (100–120 Nm).

⁂ CAUTION

Check that the rear coil spring is located in the proper position.

REAR TORSION BEAM AXLE

REMOVAL & INSTALLATION

See Figure 180.

1. Remove the wheel and tire.
2. Remove the rear wheel hub and bearing. See Wheel Hub & Bearing (sealed unit).
3. Remove the parking brake cable.
4. Remove the shock absorber and coil spring. See Coil Spring.
5. Remove the torsion axle to body bolts.
6. Remove the torsion axle.
7. Installation is the reverse of removal.
8. Install the axle mounting bolts. Torque: 72–86 ft. lbs. (100–120 Nm).

WHEEL HUB & BEARING

REMOVAL & INSTALLATION

See Figure 181.

1. Raise the rear of the vehicle and support it with safety stand.
2. Remove the rear wheel and tire.
3. Remove the wheel speed sensor wire bracket bolt.
4. Remove the parking brake wire bracket bolt.

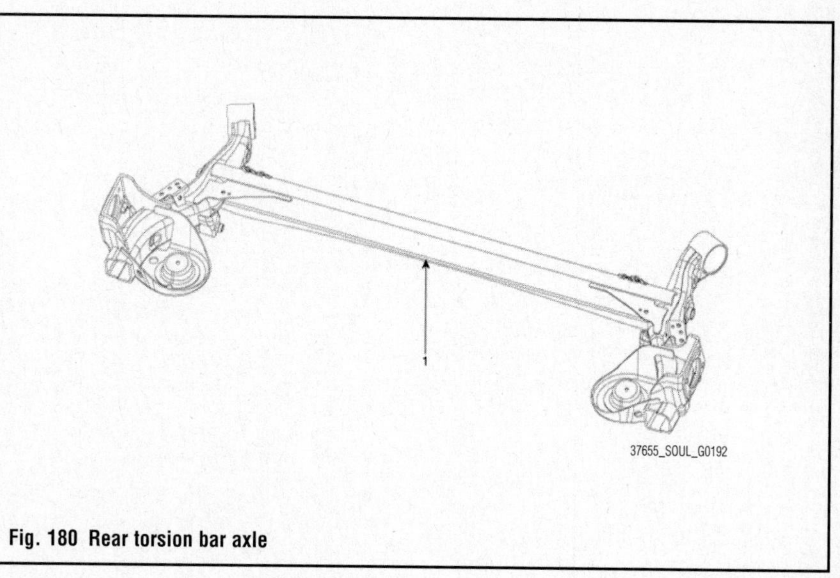

Fig. 180 Rear torsion bar axle

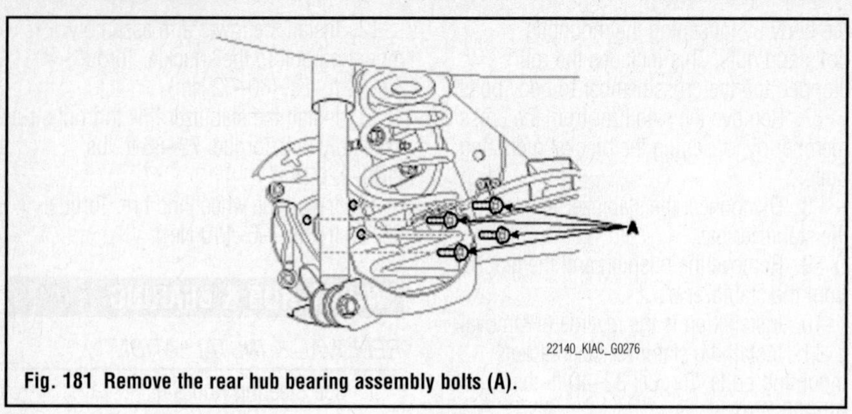

Fig. 181 Remove the rear hub bearing assembly bolts (A).

5. Remove the brake caliper assembly bolt and the brake disk mounting screw.
6. Hang the brake caliper assembly tightly on a proper place with wire.
7. Remove the rear hub bearing assembly bolts (A).
8. Remove the rear hub bearing assembly.
9. The installation is reverse of the removal.
10. Hub bearing assembly: Torque: 36–43 ft. lbs. (50–60 Nm).

KIA

Spectra

20

SPECIFICATIONS AND MAINTENANCE CHARTS

ENGINE AND VEHICLE IDENTIFICATION

Engine							Model Year	
Code ①	Liters (cc)	Cu. In.	Cyl.	Fuel Sys.	Engine Type	Eng. Mfg.	Code ②	Year
2	2.0 (2656)	163	6	EGI	DOHC	KIA	9	2009

EGI: Electronic Gasoline Injection

DOHC: Double Overhead Camshafts

① 8th digit of VIN

② 10th digit of VIN

37655_SPEC_C0001

GENERAL ENGINE SPECIFICATIONS

Year	Model	Engine Displacemen Liters	Engine VIN	Net Horsepower @ rpm	Net Torque @ rpm (ft. lbs.)	Bore x Stroke (in.)	Com-pression Ratio	Oil Pressure @ rpm
2009	Spectra	2.0	2	138@6000	136@4500	3.23x8.68	10.1:1	43-57@3000

37655_SPEC_C0002

ENGINE TUNE-UP SPECIFICATIONS

Year	Engine Displacement Liters	Engine VIN	Spark Plug Gap (in.)	Ignition Timing (deg.)		Fuel Pump (psi)	Idle Speed (rpm)		Valve Clearance	
				MT	AT		MT	AT	Intake	Exhaust
2009	2.0	2	0.039-0.043	3-7B	3-7B	49.8	750-850	750-850	HYD	HYD

NOTE: The Vehicle Emission Control Information label often reflects specification changes made during production
The label figures must be used if they differ from those in this chart.

B: Before top dead center

HYD: Hydraulic

37655_SPEC_C0003

CAPACITIES

Year	Model	Engine Displacement Liters	Engine VIN	Engine Oil with Filter	Transaxle (pts.) Manual	Transaxle (pts.) Auto.	Fuel Tank (gal.)	Cooling System (qts.)
2009	Spectra	2.0	2	4.23	4.5	13.0	14.5	6.9

37655_SPEC_C0004

FLUID SPECIFICATIONS

Year	Model	Engine Displ. Liters (VIN)	Engine Oil	Man. Trans.	Auto. Trans.	Drive Axle Front	Drive Axle Rear	Transfer Case	Power Steering Fluid	Brake Master Cylinder	Cooling System
2009	Spectra	2.0 (2)	①	NA	②	NA	NA	NA	PSF-3	DOT 3	③

NA: Not Available

DOT: Department Of Transpotation

① SL (SJ) or above, 5W-20

② Diamond ATF SP-III or SK ATF SP-III

③ Ethylene glycol base for aluminum radiator

37655_SPEC_C0014

VALVE SPECIFICATIONS

Year	Engine Displacement Liters	Engine VIN	Seat Angle (deg.)	Face Angle (deg.)	Maximum out of Square (in.)	Spring Free Length (in.)	Stem-to-Guide Clearance (in.) Intake	Stem-to-Guide Clearance (in.) Exhaust	Stem Diameter (in.) Intake	Stem Diameter (in.) Exhaust
2009	2.0	2	NA	NA	②	1.9236	0.0008-0.0019	0.0014-0.0026	0.2348-0.2354	0.2343-0.2348

NA: Not Available

① Less than 1.5 degrees

② Standard range: 0.0010-0.0023 in.

Maximum value: 0.0080 in.

37655_SPEC_C0005

CAMSHAFT AND BEARING SPECIFICATIONS

All measurements are given in inches.

Year	Engine Displacement Liters	Engine VIN	Journal Diameter	Brg. Oil Clearance	Shaft End-play	Runout	Journal Bore	Lobe Lift Intake	Lobe Lift Exhaust
2009	2.0	2	①	②	③	NA	NA	④	⑤

NA: Not Available

① No. 1: 1.1023 inches

No. 2, 3, 4: 0.9430-0.9437 inches

② No. 1: 0.0008-0.0024 inches

No. 2, 3, 4: 0.0012-0.0026 inches

③ 0.0040-0.0079 inches

④ 1.7566-1.7527 inches

⑤ 1.7527-1.7487 inches

37655_SPEC_C0013

CRANKSHAFT AND CONNECTING ROD SPECIFICATIONS

All measurements are given in inches.

Year	Engine Displacement Liters	Engine VIN	Crankshaft Main Brg. Journal Dia.	Crankshaft Main Brg. Oil Clearance	Crankshaft Shaft End-play	Crankshaft Thrust on No.	Connecting Rod Journal Diameter	Connecting Rod Oil Clearance	Connecting Rod Side Clearance
2009	2.0	2	2.2418-2.2426	0.0011-0.0018	0.0023-0.0102	3	1.8898-1.8905	0.0009-0.0017	0.0039-0.0098

① Journal Nos. 1, 2, 4 & 5: 0.0009-0.0016 in.

Journal Nos. 3: 0.0011-0.0018 in.

37655_SPEC_C0007

PISTON AND RING SPECIFICATIONS

All measurements are given in inches.

Year	Engine Displacement Liters	Engine VIN	Piston Clearance	Ring Gap Top Compression	Ring Gap Bottom Compression	Ring Gap Oil Control	Ring Side Clearance Top Compression	Ring Side Clearance Bottom Compression	Ring Side Clearance Oil Control
2009	2.0	2	0.0008-0.0016	0.0079-0.0138	0.0146-0.0205	0.0078-0.0236	0.0015-0.0031	0.0012-0.0027	0.0024-0.0059

37655_SPEC_C0006

TORQUE SPECIFICATIONS

All readings in ft. lbs.

Year	Engine Displacement Liters	Engine VIN	Cylinder Head Bolts	Main Bearing Bolts	Rod Bearing Bolts	Crankshaft Damper Bolts	Flywheel Bolts	Manifold Intake	Manifold Exhaust	Spark Plugs	Oil Pan Drain Plug
2009	2.0	2	①	②	36-38	116-123	87-94	12-17	31-40	18-22	29-33

① M10 bolts; 17-20 ft. lbs. plus 60-65 degrees; plus an additional 6-65 degrees.
M12 bolts: 20.5-33 ft. lbs., plus 60-65 degrees, plus additional 60-65 degrees

② 20-23 ft. lbs. plus 90 degrees

37655_SPEC_C0008

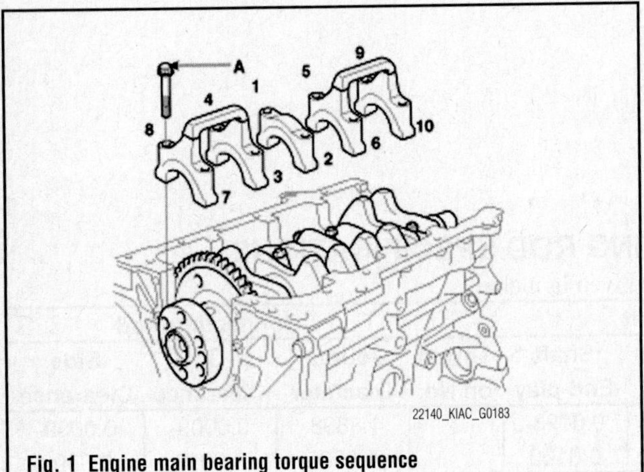

Fig. 1 Engine main bearing torque sequence

WHEEL ALIGNMENT

Year	Model		Caster Range (+/-Deg.)	Caster Preferred Setting (Deg.)	Camber Range (+/-Deg.)	Camber Preferred Setting (Deg.)	Toe-in (in.)
2009	Spectra	F	0.50	+2.60	0.50	0	0 +/- 0.08
		R	—	—	0.50	-0.92	0.16 +/- 0.08

37655_SPEC_C0009

TIRE, WHEEL AND BALL JOINT SPECIFICATIONS

Year	Model	OEM Tires		Tire Pressures (psi)		Wheel	Ball Joint	Lug Nut Torque
		Standard	Optional	Front	Rear	Size	Inspection	(ft. lbs.)
2009	Spectra	P195/60R15	P205/50R16	30	30	6.0-J	①	67-82

OEM: Original Equipment Manufacturer

PSI: Pounds Per Square Inch

STD: Standard

OPT: Optional

① Replace if any measureable movent is found.

37655_SPEC_C0010

BRAKE SPECIFICATIONS

All measurements in inches unless noted

Year	Model		Brake Disc			Brake Drum				Brake Caliper	
			Original Thickness	Minimum Thickness	Maximum Run-out	Original Inside Diameter	Max. Wear Limit	Maximum Machine Diameter	Minimum Lining Thickness	Bracket Bolts (ft. lbs.)	Mounting Bolts (ft. lbs.)
2009	Spectra	F	1.020	0.945	0.0012	—	—	—	0.079	51-63	16-24
		R	0.390	0.315	0.0012	8.00	—	8.08	0.079	51-63	16-24

F: Front

R: Rear

37655_SPEC_C0011

SCHEDULED MAINTENANCE INTERVALS

Kia - Spectra

TO BE SERVICED	TYPE OF SERVICE	VEHICLE MILEAGE INTERVAL (x1000)												
		7.5	15	22.5	30	37.5	45	52.5	60	67.5	75	82.5	90	97.5
Accessory drive belts	S/I	✓	✓	✓	✓	✓	✓	✓	✓	✓	✓	✓	✓	✓
Air cleaner filter	R			✓			✓			✓			✓	
Air conditioner system	S/I													
Brake lines, hoses and connections	S/I		✓		✓		✓		✓		✓		✓	
Chassis and body fasteners	T												✓	
Cooling system hoses and coolant level	S/I		✓		✓		✓		✓		✓		✓	
CV-joint boots	S/I				✓				✓				✓	
Engine coolant	R				✓				✓				✓	
Engine oil and filter	R	✓	✓	✓	✓	✓	✓	✓	✓	✓	✓	✓	✓	✓
Exhaust system heat shields	S/I				✓				✓				✓	
Front and rear brakes	S/I	✓	✓	✓	✓	✓	✓	✓	✓	✓	✓	✓	✓	✓
Front ball joints	S/I				✓				✓				✓	
Fuel filter	R					✓					✓			
Fuel lines and hoses	S/I	✓	✓	✓	✓	✓	✓	✓	✓	✓	✓	✓	✓	✓
Locks and hinges	L	✓	✓	✓	✓	✓	✓	✓	✓	✓	✓	✓	✓	✓
Spark plugs	R								✓					
Steering operation and linkage	S/I	✓	✓	✓	✓	✓	✓	✓	✓	✓	✓	✓	✓	✓
Timing belt	R								✓					

R: Replace S/I: Service or Inspect L: Lubricate T: Tighten

FREQUENT OPERATION MAINTENANCE (SEVERE SERVICE)

If a vehicle is operated under any of the following conditions it is considered severe service

- Towing a trailer or using a camper or car-top carrier

- Repeated short trips of less than 5 miles in temperatures below freezing, or trips of less than 10 miles in any temperature

- Prolonged idling (vehicle operation in stop and go traffic).

- Operating on rough, muddy, unpaved, dusty or salt-covered roads.

- Police, taxi, delivery usage or trailer towing usage.

- Driving in extremely hot (over 90°F) conditions

Oil & oil filter: change every 5000 miles or 5 months, whichever occurs first.

Air cleaner filter: inspect every 15,000 miles or 15 months and replace everything 30,000 miles or 30 months, whichever occurs fir

Fuel system hoses (California models only): replace every 105,000 miles

Emission system hoses (non-CA models): inspect every 55,000 or 55 months, whichever occurs first

Emission system hoses (CA models): inspect every 60,000 miles or 60 months, which occurs first

Front and rear brakes: inspect every 15,000 miles or 15 months, whichever occurs first

Chassis and body fasteners: tighten every 15,000 miles or 15 months, whichever occurs first

Locks and hinges: lubricate every 5000 miles or 5 months, whichever occurs first

37655_SPEC_C0012

BRAKES — INFORMATION AND PRECAUTIONS

ANTI-LOCK SYSTEMS

- Certain components within the ABS system are not intended to be serviced or repaired individually.
- Do not use rubber hoses or other parts not specifically specified for and ABS system. When using repair kits, replace all parts included in the kit. Partial or incorrect repair may lead to functional problems and require the replacement of components.
- Lubricate rubber parts with clean, fresh brake fluid to ease assembly. Do not use shop air to clean parts; damage to rubber components may result.
- Use only DOT 3 brake fluid from an unopened container.
- If any hydraulic component or line is removed or replaced, it may be necessary to bleed the entire system.
- A clean repair area is essential. Always clean the reservoir and cap thoroughly before removing the cap. The slightest amount of dirt in the fluid may plug an orifice and impair the system function. Perform repairs after components have been thoroughly cleaned; use only denatured alcohol to clean components. Do not allow ABS components to come into contact with any substance containing mineral oil; this includes used shop rags.
- The Anti-Lock control unit is a microprocessor similar to other computer units in the vehicle. Ensure that the ignition switch is **OFF** before removing or installing controller harnesses. Avoid static electricity discharge at or near the controller.
- If any arc welding is to be done on the vehicle, the control unit should be unplugged before welding operations begin.

DISC AND DRUM SYSTEMS

⁂ CAUTION

Dust and dirt accumulating on brake parts during normal use may contain asbestos fibers from production or aftermarket brake linings. Breathing excessive concentrations of asbestos fibers can cause serious bodily harm. Exercise care when servicing brake parts. Do not sand or grind brake lining unless equipment used is designed to contain the dust residue. Do not clean brake parts with compressed air or by dry brushing. Cleaning should be done by dampening the brake components with a fine mist of water, then wiping the brake components clean with a dampened cloth. Dispose of cloth and all residue containing asbestos fibers in an impermeable container with the appropriate label. Follow practices prescribed by the Occupational Safety and Health Administration (OSHA) and the Environmental Protection Agency (EPA) for the handling, processing, and disposing of dust or debris that may contain asbestos fibers.

BRAKES — BLEEDING THE BRAKE SYSTEM

BLEEDING PROCEDURE

BLEEDING PROCEDURE

⁂ WARNING

When bleeding the brakes, note the following:

- Do not reuse the drained fluid
- Always use Genuine DOT 3 or DOT 4 Brake Fluid. Using a non-Genuine DOT3 or DOT 4 brake fluid can cause corrosion and decrease the life of the system
- Make sure no dirt of other foreign matter is allowed to contaminate the brake fluid
- Do not spill brake fluid on the vehicle, it may damage the paint; if brake fluid does contact the paint, wash it off immediately with water
- The reservoir on the master cylinder must be at the MAX (upper) level mark at the start of bleeding procedure and checked after bleeding each brake caliper. Add fluid as required

1. Make sure the brake fluid level in the master cylinder fluid reservoir is at the MAX (upper) level line.
2. Have someone slowly pump the brake pedal several times, and then apply steady pressure
3. Loosen the right-rear brake bleed screw to allow air to escape from the system. Then tighten the bleed screw securely
4. Repeat the procedure for each wheel in the sequence shown below until air bubbles no longer appear in the fluid
5. Refill the master cylinder reservoir to the MAX (upper) level line

BLEEDING THE ABS SYSTEM

The ABS brake system is bled in the usual fashion with no special procedures required. Refer to the bleeding procedure described earlier. Make certain the master cylinder reservoir is filled before the bleeding is begun and check the level frequently.

BRAKES — ANTI-LOCK BRAKE SYSTEM (ABS)

WHEEL SPEED SENSORS

REMOVAL & INSTALLATION

Front

1. Remove the front wheel speed sensor mounting bolt.
2. Remove the mounting bolt fixed on the strut.
3. Remove the front wheel guard.
4. Remove the front wheel speed sensor after disconnecting the wheel speed sensor connector.
5. Installation is the reverse of removal.

Rear

1. Remove the rear wheel speed sensor mounting bolt.
2. Remove the mounting bolt fixed on the strut.
3. Remove rear cushion and rear bag.
4. Remove the rear seat side pad then disconnect the rear wheel speed sensor connector.
5. Installation is the reverse of removal.

BRAKES — FRONT DISC BRAKES

BRAKE CALIPER

REMOVAL & INSTALLATION

1. Remove guide rod and the caliper up out of the way.
2. Check the hoses and pin boots for damage and deterioration.
3. Remove the pad shims, pad retainers and pads.

To install:

4. Install the pad retainers.
5. Check the foreign material at the pad shims and the back of the pads.

⁂ CAUTION

Contaminated brake discs or pads reduce stopping ability. Keep grease off the discs and pads.

6. Install the brake pads and pad shims correctly. Install the pad with the wear indicator on the inside.

➡If you are reusing the pads, always reinstall the brake pads in their original positions to prevent a momentary loss of braking efficiency.

7. Push in the piston so that the caliper will fit over the pads. Make sure that the piston boot is in position to prevent damaging it when pivoting the caliper down.

➡Insert the piston in the cylinder using the special tool (09581-11000).

8. Pivot the caliper down into position. Being careful not to damage the pin boot, install the guide rod bolt.
9. Depress the brake pedal several times to make sure the brakes work, then test-drive.

➡Engagement of the brake may require a greater pedal stroke immediately after the brake pads have been replaced as a set. Several applications of the brake will restore the normal pedal stroke.

10. After installation, check for leaks at hose and line joints or connections, and retighten if necessary

DISC BRAKE PADS

REMOVAL & INSTALLATION

1. Remove guide rod and the caliper up out of the way.
2. Check the hoses and pin boots for damage and deterioration.
3. Remove the pad shims, pad retainers and pads.

To install:

4. Install the pad retainers.
5. Check the foreign material at the pad shims and the back of the pads.

⁂ CAUTION

Contaminated brake discs or pads reduce stopping ability. Keep grease off the discs and pads.

6. Install the brake pads and pad shims correctly. Install the pad with the wear indicator on the inside.

➡If you are reusing the pads, always reinstall the brake pads in their original positions to prevent a momentary loss of braking efficiency.

7. Push in the piston so that the caliper will fit over the pads. Make sure that the piston boot is in position to prevent damaging it when pivoting the caliper down.

➡Insert the piston in the cylinder using the special tool (09581-11000).

8. Pivot the caliper down into position. Being careful not to damage the pin boot, install the guide rod bolt.
9. Depress the brake pedal several times to make sure the brakes work, then test-drive.

➡Engagement of the brake may require a greater pedal stroke immediately after the brake pads have been replaced as a set. Several applications of the brake will restore the normal pedal stroke.

10. After installation, check for leaks at hose and line joints or connections, and retighten if necessary

BRAKE CALIPER

REMOVAL & INSTALLATION

1. Raise the vehicle and make sure it is securely supported. Remove the rear wheel.
2. Release the parking brake.
3. Remove the guide rod bolt.
4. Raise the caliper assembly, support it with a wire.
5. Remove the pad shim and pad assembly from caliper bracket.

To install:

6. Install the pad retainers to the caliper.
7. Check the foreign material at the pad shim and the back of the pads.
8. Contaminated brake discs or pads reduce stopping ability. Keep grease off the discs and pads.
9. Install the brake pads and pad shim on the caliper bracket.
10. If you are reusing the pads, always reinstall the brake pads in their original positions to prevent a momentary loss of braking efficiency. Push in the piston using SST (09581-11000) so that the caliper will fit over the pads. Make sure that the piston boot is in position to prevent damaging it when pivoting the caliper down.
11. Pivot caliper down into position. Being careful not to damage the pin boot, install the guide rod bolt and tighten it to the specified torque.
12. Install the brake caliper .
13. After installation, check for leaks at hose and line joints and connections, and retighten if necessary.
14. Depress the brake pedal several times to make sure the brakes work, then test-drive.

➡Engagement of the brake may require a greater pedal stroke immediately after the brake pads have been replaced as a set. Several applications of the brake pedal will restore the normal pedal stroke.

DISC BRAKE PADS

REMOVAL & INSTALLATION

1. Raise the vehicle and make sure it is securely supported. Remove the rear wheel.
2. Release the parking brake.
3. Remove the guide rod bolt.
4. Raise the caliper assembly, support it with a wire.
5. Remove the pad shim and pad assembly from caliper bracket.

To install:

6. Install the pad retainers to the caliper.
7. Check the foreign material at the pad shim and the back of the pads.
8. Contaminated brake discs or pads reduce stopping ability. Keep grease off the discs and pads.
9. Install the brake pads and pad shim on the caliper bracket.
10. If you are reusing the pads, always reinstall the brake pads in their original positions to prevent a momentary loss of braking efficiency. Push in the piston using SST (09581-11000) so that the caliper will fit over the pads. Make sure that the piston boot is in position to prevent damaging it when pivoting the caliper down.
11. Pivot caliper down into position. Being careful not to damage the pin boot, install the guide rod bolt and tighten it to the specified torque.
12. Install the brake caliper .
13. After installation, check for leaks at hose and line joints and connections, and retighten if necessary.
14. Depress the brake pedal several times to make sure the brakes work, then test-drive.

➡Engagement of the brake may require a greater pedal stroke immediately after the brake pads have been replaced as a set. Several applications of the brake pedal will restore the normal pedal stroke.

BRAKE DRUM

REMOVAL & INSTALLATION

See Figure 2.

1. Remove the shoe hole down pins by pushing the shoe hole down washer and turning them.
2. Disengage the upper return spring.
3. Lower the brake shoe assembly, and remove the lower return spring. Make sure not to damage the dust cover on the wheel cylinder.
4. Disconnect the parking brake cable from the parking brake lever.
5. Remove the brake shoe assembly.
6. Remove the upper return spring, self-adjuster lever and self-adjuster spring, and separate the brake shoes.
7. Disconnect the brake line from the wheel cylinder.
8. Remove the bolt and the wheel cylinder from the backing plate.

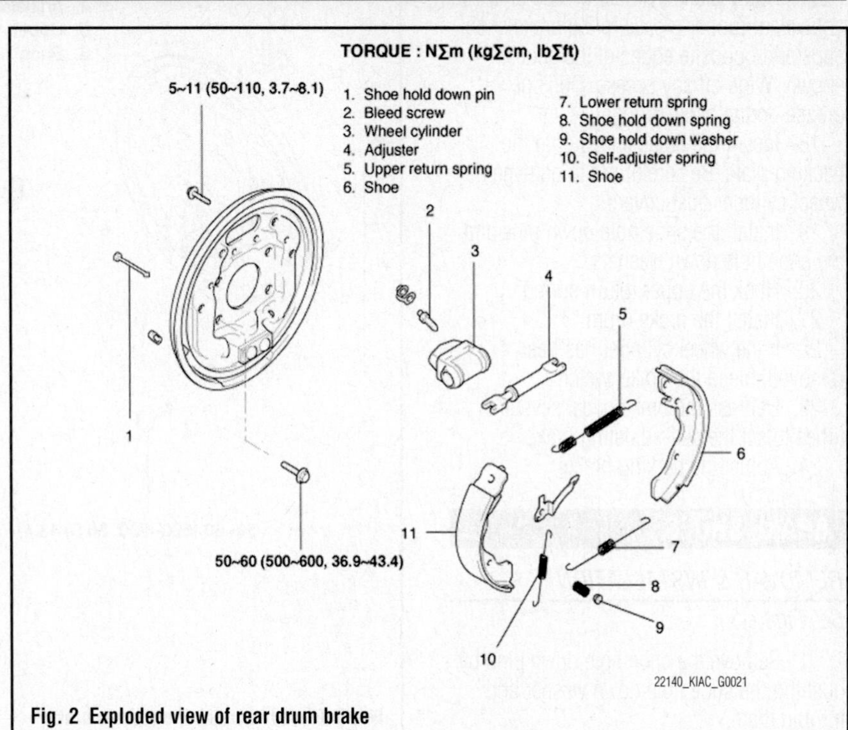

Fig. 2 Exploded view of rear drum brake

To install:

➡During installation, note the following:

- Do not spill brake fluid on the vehicle: it may damage the paint; if brake fluid does contact the paint. Wash it off immediately with water.
- To prevent spills, cover the hose joints with rags or shop towels.
- Use only a genuine wheel cylinder special bolt.

9. Apply sealant between the wheel cylinder and backing plate, and install the wheel cylinder.
10. Connect the brake tubes to the wheel cylinder.
11. Connect the parking brake cable to the parking brake lever.
12. Clean the threaded portions of adjuster sleeve and push rod female. Coat the threads of the adjuster assembly with grease. To shorten the clevises, turn the adjuster bolt.
13. Hook the self-adjuster spring to the adjuster lever first, then to the brake shoe.
14. Install the adjuster assembly and upper return spring, noting the installation direction. Be careful not to damage the wheel cylinder dust covers.
15. Install the lower return spring.
16. Apply brake cylinder grease or equivalent rubber grease to the sliding surfaces shown. Wipe off any excess. Don't get grease on the brake linings.
17. Apply brake cylinder grease or equivalent rubber grease to the brake shoe ends and opposite edges of the shoes shown. Wipe off any excess. Don't get grease on the brake linings.
18. Install the brake shoes onto the backing plate. Be careful not to damage the wheel cylinder dust covers.
19. Install the shoe hole down pins and the shoe hole down washers.
20. Hook the upper return spring.
21. Install the brake drum.
22. If the wheel cylinder has been removed, bleed the brake system.
23. Depress the brake pedal several times to set the self-adjusting brake.
24. Adjust the parking brake.

BRAKE SHOES

REMOVAL & INSTALLATION

See Figure 3.

1. Remove the shoe hole down pins by pushing the shoe hole down washer and turning them.
2. Disengage the upper return spring.
3. Lower the brake shoe assembly, and remove the lower return spring. Make sure not to damage the dust cover on the wheel cylinder.
4. Disconnect the parking brake cable from the parking brake lever.
5. Remove the brake shoe assembly.
6. Remove the upper return spring, self-adjuster lever and self-adjuster spring, and separate the brake shoes.
7. Disconnect the brake line from the wheel cylinder.
8. Remove the bolt and the wheel cylinder from the backing plate.

To install:

➡During installation, note the following:

- Do not spill brake fluid on the vehicle: it may damage the paint; if brake fluid does contact the paint. Wash it off immediately with water.
- To prevent spills, cover the hose joints with rags or shop towels.
- Use only a genuine wheel cylinder special bolt.

9. Apply sealant between the wheel cylinder and backing plate, and install the wheel cylinder.
10. Connect the brake tubes to the wheel cylinder.
11. Connect the parking brake cable to the parking brake lever.
12. Clean the threaded portions of adjuster sleeve and push rod female. Coat the threads of the adjuster assembly with grease. To shorten the clevises, turn the adjuster bolt.
13. Hook the self-adjuster spring to the adjuster lever first, then to the brake shoe.
14. Install the adjuster assembly and upper return spring, noting the installation direction. Be careful not to damage the wheel cylinder dust covers.
15. Install the lower return spring.
16. Apply brake cylinder grease or equivalent rubber grease to the sliding surfaces shown. Wipe off any excess. Don't get grease on the brake linings.
17. Apply brake cylinder grease or equivalent rubber grease to the brake shoe ends and opposite edges of the shoes shown. Wipe off any excess. Don't get grease on the brake linings.
18. Install the brake shoes onto the

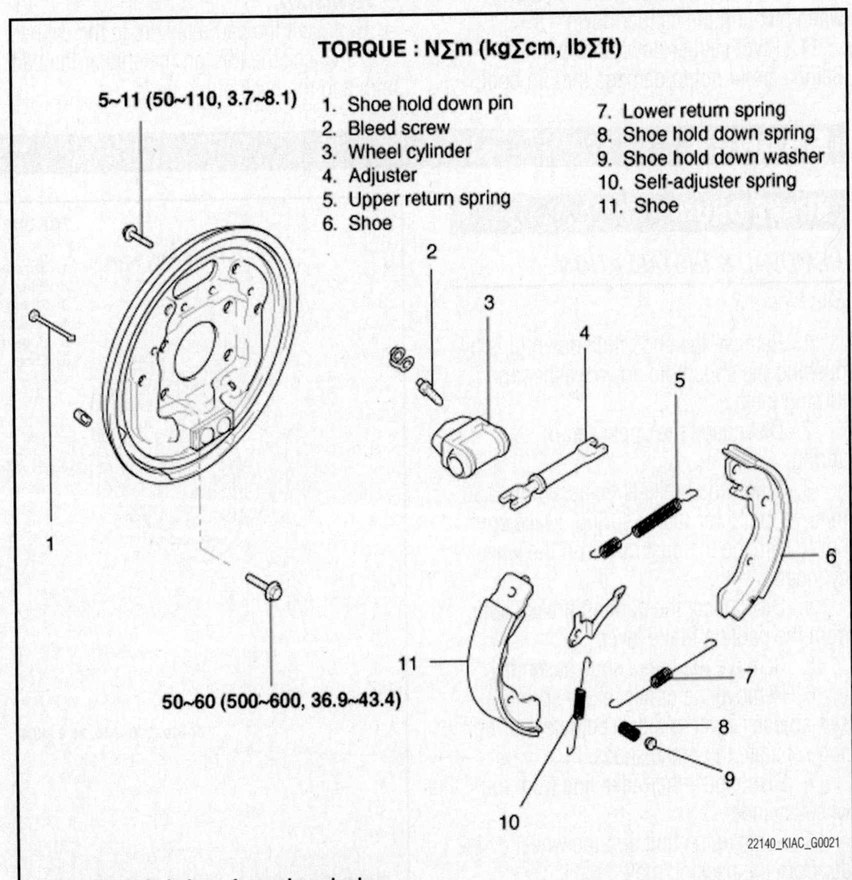

Fig. 3 Exploded view of rear drum brake

backing plate. Be careful not to damage the wheel cylinder dust covers.

19. Install the shoe hole down pins and the shoe hole down washers.

20. Hook the upper return spring.

21. Install the brake drum.

22. If the wheel cylinder has been removed, bleed the brake system.

23. Depress the brake pedal several times to set the self-adjusting brake.

24. Adjust the parking brake.

ADJUSTMENT

1. Depress the brake pedal several times to set the self-adjusting brake.

✲✲ CAUTION

Dust and dirt accumulating on brake parts during normal use may contain asbestos fibers from production or aftermarket brake linings. Breathing excessive concentrations of asbestos fibers can cause serious bodily harm. Exercise care when servicing brake parts. Do not sand or grind brake lining unless equipment used is designed to contain the dust residue. Do not clean brake parts with compressed air or by dry brushing. Cleaning should be done by dampening the brake components with a fine mist of water, then wiping the brake components clean with a dampened cloth. Dispose of cloth and all residue containing asbestos fibers in an impermeable container with the appropriate label. Follow practices prescribed by the Occupational Safety and Health Administration (OSHA) and the Environmental Protection Agency (EPA) for the handling, processing, and disposing of dust or debris that may contain asbestos fibers.

PARKING BRAKE CABLES

ADJUSTMENT

1. Block the front wheels, then raise the rear of the vehicle and make sure it is securely supported.

2. Make sure the parking brake arm on the rear brake caliper contacts the brake caliper pin on vehicles with rear disc brakes.

3. Pull the parking brake lever up one click.

4. Remove the console.

5. Tighten the adjusting nuts until the parking brakes drag slightly when the rear wheels are turned.

6. Release the parking brake lever fully, and check that parking brakes do not drag when the rear wheels are turned. Readjust if necessary.

7. Make sure that the parking brakes are fully applied when the parking brake lever is pulled up fully.

8. Reinstall the console.

PARKING BRAKE SHOES

REMOVAL & INSTALLATION

1. Remove the console.

2. Loosen the adjusting nut and the parking brake cables.

3. Disconnect the electrical connector of parking brake switch.

4. Remove the parking brake lever assembly by loosening the bolts.

5. Remove the wheel and tire.

6. Remove the brake shoe.

To install:

7. Install the removed parts in the reverse order of removal.

8. Apply a coating of the specified grease to each sliding parts of the ratchet plate or the ratchet pawl.

9. After installing the cable adjuster, adjust the parking brake lever stroke.

ADJUSTMENT

1. Block the front wheels, then raise the rear of the vehicle and make sure it is securely supported.

2. Make sure the parking brake arm on the rear brake caliper contacts the brake caliper pin on vehicles with rear disc brakes.

3. Pull the parking brake lever up one click.

4. Remove the console.

5. Tighten the adjusting nuts until the parking brakes drag slightly when the rear wheels are turned.

6. Release the parking brake lever fully, and check that parking brakes do not drag when the rear wheels are turned. Readjust if necessary.

7. Make sure that the parking brakes are fully applied when the parking brake lever is pulled up fully.

8. Reinstall the console.

CHASSIS ELECTRICAL

AIR BAG (SUPPLEMENTAL RESTRAINT SYSTEM)

GENERAL INFORMATION

**** CAUTION**

These vehicles are equipped with an air bag system. The system must be disarmed before performing service on, or around, system components, the steering column, instrument panel components, wiring and sensors. Failure to follow the safety precautions and the disarming procedure could result in accidental air bag deployment, possible injury and unnecessary system repairs.

PRECAUTIONS

**** CAUTION**

Disconnect and isolate the battery negative cable before beginning any airbag system component diagnosis, testing, removal, or installation procedures. Allow system capacitor to discharge for two minutes before beginning any component service. This will disable the airbag system. Failure to disable the airbag system may result in accidental airbag deployment, personal injury, or death.

DISARMING THE SYSTEM

1. Before servicing the vehicle, refer to the Precautions Section.
2. Turn the ignition switch to the **LOCK** position.
3. Disconnect the negative battery cable.
4. Wait 10 minutes for the battery back-up power to discharge.

ARMING THE SYSTEM

1. Before servicing the vehicle, refer to the Precautions Section.
2. Connect the negative battery cable.
3. Turn the ignition switch **ON**.
4. Verify that the air bag indicator illuminates for 4–8 seconds, then goes off.

CLOCKSPRING CENTERING

Prior to installing the clock spring, align the mating mark and "NEUTRAL" position indicator of the clock spring, and, after turning the front wheels to the straight-ahead position, install the clock spring to the column switch. If the mating mark of the clock spring is not properly aligned, the steering wheel may not completely rotate during a turn, or the flat cable within the clock spring may be severed, obstructing normal operation of the SRS and possibly leading to serious injury to the vehicle's driver.

DRIVE TRAIN

CLUTCH DRIVEN DISC & PRESSURE PLATE

REMOVAL & INSTALLATION

See Figure 4.

1. Remove a transaxle assembly.
2. Remove the clutch cover bolts.
3. Remove the clutch cover and disc.
4. Using the SST (09411-11000) (09411-25000), install the disc.
5. Tighten the clutch cover to 10–16 ft. lbs. (15–22 Nm).
6. Install transaxle assembly.

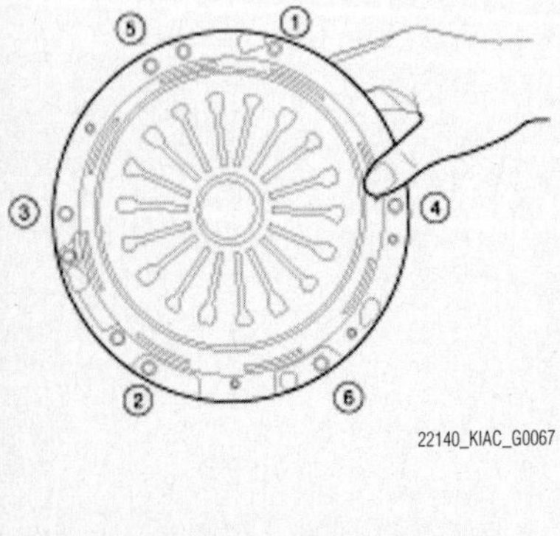

22140_KIAC_G0067

Fig. 4 Tighten clutch cover bolts in sequence—Spectra

ADJUSTMENTS

1. Measure the clutch pedal height (from the face of the pedal pad to the floorboard) and the clutch pedal free play (measured at the face of the pedal pad). Standard value is 0.25–0.5 inches (6–13 mm).
2. If the clutch pedal free play and height is not within the standard value range, adjust as follows:
    a. Turn and adjust the bolt within the standard value, then secure by tightening the lock nut.
    b. Turn the push rod to agree with the standard value and then secure the push rod with the lock nut.
3. If the clutch pedal free play and the distance between the clutch pedal and the floor board when the clutch is disengaged, do not meet with the standard values, it may be the result of either air in the hydraulic system or a faulty clutch master cylinder. Bleed the air or disassemble and inspect the master cylinder or clutch.

HYDRAULIC SYSTEM BLEEDING

BLEEDING PROCEDURE

Whenever the clutch tube, the clutch hose, and/or the clutch master cylinder have been removed, or if the clutch pedal is spongy, bleed the system.

**** CAUTION**

Avoid mixing different brands of brake fluid. Use only SAE J1703 (DOT 3 or DOT 4).

1. Loosen the bleeder screw at the clutch release cylinder.
2. Depress the clutch pedal slowly until all air is expelled.
3. Hold the clutch pedal down until the bleeder is retightened.
4. Refill the clutch master cylinder with the specified fluid.

**** CAUTION**

The rapidly-repeated operation of the clutch pedal may disrupt the release cylinder's position. During the bleeding operation, press the clutch pedal to the floor after it returns to the top of the stroke.

FRONT AXLE SHAFT, BEARING & SEAL

REMOVAL & INSTALLATION

1. Raise the vehicle and remove the front wheel.
2. Remove the wheel speed sensor from the knuckle.
3. Remove the caliper assembly by loosening the bolts, and then suspend it with wire.

4. Remove the split pin, castle nut and washer from the front hub assembly.
5. Remove the ball joint assembly mounting bolt from the knuckle.
6. Remove the tie rod end split pin and castle nut.
7. Disconnect the tie rod end from the knuckle using the SST (09568 - 4A000).
8. Loosen the screws and remove the brake disc from the front hub assembly.
9. Remove the strut assembly mounting bolts and nuts.
10. Remove the hub and knuckle as an assembly.

Disassembly

1. Remove the snap ring.
2. Remove the hub from the knuckle assembly using the SST (09545-34100).
3. Remove the wheel bearing inner race from the hub using the SST (09432-11000, 09545-34100).
4. Remove the wheel bearing outer race from the hub using the SST (09216-21600, 09216-21100).

Reassembly

1. Apply the multi-purpose grease to the contacting surface of the knuckle and wheel bearing thinly.
2. Press-in the wheel bearing to the knuckle using the SST (09216-21100).

⁂ CAUTION

Do not strike the inner race of the wheel bearing because that can cause damage to the bearing assembly.

➡Always use a new wheel bearing assembly.

3. Install the snap ring into the groove the knuckle.
4. Press-in the hub to the knuckle using the SST (09454-21100).

⁂ CAUTION

Do not strike the outer race of the wheel bearing because that can cause damage to the bearing assembly.

5. Tighten the hub to the knuckle with the SST (09517-21500).
6. Rotate the hub to seat the wheel bearing.
7. Measure the wheel bearing starting torque.

➡Wheel bearing starting torque: 1.3 ft. lbs. (1.8 Nm) or below.

8. Measure the hub run-out by using a dial gauge. Runout should be 0.00157inch (0.04mm).
9. Remove the SST (09517-21500).

To install:

10. Installation is the reverse of removal.

Use the following torque specifications during installation:

a. Strut assembly mounting bolts: 101–116 ft. lbs. (140–160 Nm).
b. Tie rod end bolts to knuckle: 72–86 ft. lbs. (100–120 Nm).
c. Tie rod end castle nut: 17–25 ft. lbs. (24–34 Nm).
d. Ball joint assembly mounting bolt: 72–86 ft. lbs. (100–120 Nm).
e. Front hub assembly castle nut: 145–203 ft. lbs. (200–280 Nm).

FRONT HALFSHAFT

REMOVAL & INSTALLATION

1. Raise the front of the vehicle and support it with safety stands in a proper location.
2. Remove the front wheel and tire.
3. Remove the drain plug. Drain the transaxle oil.
4. Remove the split pin, the lock nut and the washer from the front hub.
5. Disconnect the tie rod end ball joint from the knuckle using the Special Tool (09568-34000) after removing the split pin and lock nut.
6. Remove the wheel speed sensor from the knuckle.
7. On Rio, remove the lower arm mounting bolts from the knuckle.
8. On Spectra, remove the strut upper mounting bolts.
9. Using a plastic hammer, disconnect the drive shaft from the axle hub.
10. Push the axle hub outward and separate the drive shaft from the axle hub.
11. Insert a pry bar between the transaxle case and joint case, and separate the drive shaft from the transaxle case.
12. Pull out the drive shaft from the transaxle case.

To install:

13. Apply gear oil on the drive shaft splines and the contacting surface of differential case oil seal.
14. Before installing the drive shaft, set the opening side of the circlip facing downward.
15. After installation, check that the drive shaft cannot be removed by hand.
16. Install the drive shaft into the knuckle.
17. Install the knuckle in the strut assembly. Torque to 94–108 ft. lbs. (130–150 Nm).
18. Install the strut upper mounting bolts.
19. Install the tie rod end to the knuckle. Torque to 12–25 ft. Lbs. (16–34 Nm).
20. Install the wheel speed sensor to the knuckle.
21. After installing the washer with convex surface outward, install the lock nut and the split pin.
22. Install the wheel and tire

ENGINE COOLING

ENGINE FAN

REMOVAL & INSTALLATION

See Figure 5.

1. Drain the engine coolant.
2. Remove the upper radiator hose and lower radiator hose.
3. Disconnect the Automatic Transaxle Fluid (ATF) oil cooler hoses, if vehicle equipped with automatic transaxle.
4. Disconnect the fan motor connector (s).

➡**Some vehicles are equipped with two side-by-side fans.**

5. Remove the cooling fan mounting bolts and remove cooling fan.

To install:

6. Install the cooling fan and mounting bolts.
7. Connect the fan motor connector (s).
8. Install the upper radiator hose and lower radiator hose.
9. Install the ATF oil cooler hoses.
10. Fill with engine coolant.
11. Start engine and check for leaks.

RADIATOR

REMOVAL & INSTALLATION

See Figure 5.

1. Drain the engine coolant.
2. Remove the upper radiator hose and lower radiator hose.
3. Remove the ATF oil cooler hoses if vehicle is equipped with automatic transaxle.
4. Disconnect the fan motor connector.
5. Remove the cooling fan.
6. Remove the radiator upper bracket.
7. Remove the radiator.

To install:

8. Install the radiator.
9. Install the radiator upper bracket.
10. Install the fan (s) and fan shroud.
11. Install the upper radiator hose and lower radiator hose.
12. Install the ATF oil cooler hoses if vehicle is equipped with automatic transaxle.
13. Fill with engine coolant.
14. Start engine and check for leaks.

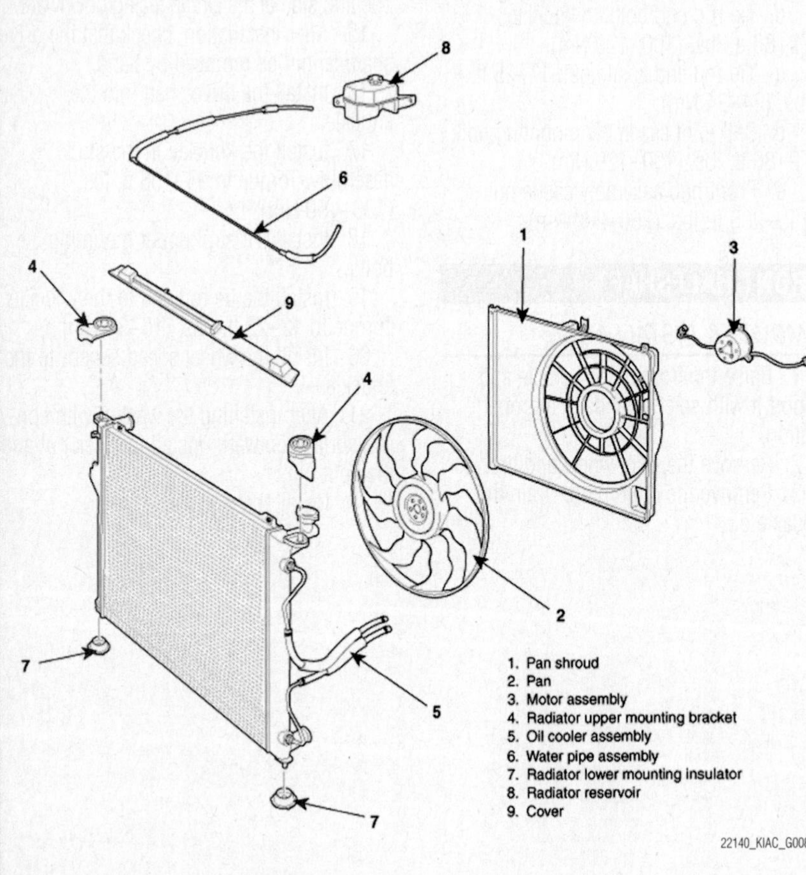

Fig. 5 Engine cooling component locations

THERMOSTAT

REMOVAL & INSTALLATION

1. Drain the engine coolant so its level is below thermostat.
2. Remove the water inlet fitting, gasket and thermostat.

To install:

3. Installation is the reverse of removal.

WATER PUMP

REMOVAL & INSTALLATION

1. Drain the engine coolant.
2. Remove drive belts.
3. Remove the timing belt.
4. Remove the timing belt idler.
5. Remove the power steering pump and the power steering pump bracket.
6. Remove the water pump.
   a. Remove the 4 bolts and pump pulley.
   b. Remove the 2 bolts, then remove the alternator brace.
   c. Remove the water pump and gasket.

To install:

7. Install the water pump.
   a. Install the water pump and a new gasket with the 3 bolts. Torque: 8.7–10.8 ft. lbs. (11.8–14.7 Nm).
   b. Install the alternator brace with the 2 bolts. Torque: 14.5–19.5 ft. lbs. (19.6–26.5 Nm).
   c. Install the pump pulley.
8. Install the power steering pump and the power steering bracket.
9. Install the timing belt idler.
10. Install the timing belt.
11. Install drive belts.
12. Fill with engine coolant.
13. Start engine and check for leaks.
14. Recheck engine coolant level.

ENGINE ELECTRICAL — CHARGING SYSTEM

ALTERNATOR

REMOVAL & INSTALLATION

See Figure 6.

1. Disconnect the battery negative terminal first, then the positive terminal.
2. Disconnect the alternator connector, and remove the cable from alternator "B" terminal.
3. Remove the accessory drive belt.
4. Pull out the through bolt and then remove the alternator.
5. Installation is the reverse of removal.

1. Nut
2. Pulley
3. Spacer
4. Front cover assembly
5. Front bearing
6. Bearing cover
7. Bearing cover bolts
8. Rotor coil
9. Rear bearing
10. Bearing cover
11. Rear cover
12. Bolts
13. Seal
14. Rectifier assembly
15. Stud bolts
16. Brush holder assembly
17. Brush holder bolts
18. Slip ring guide
19. Cover
22140_KIAC_G0294

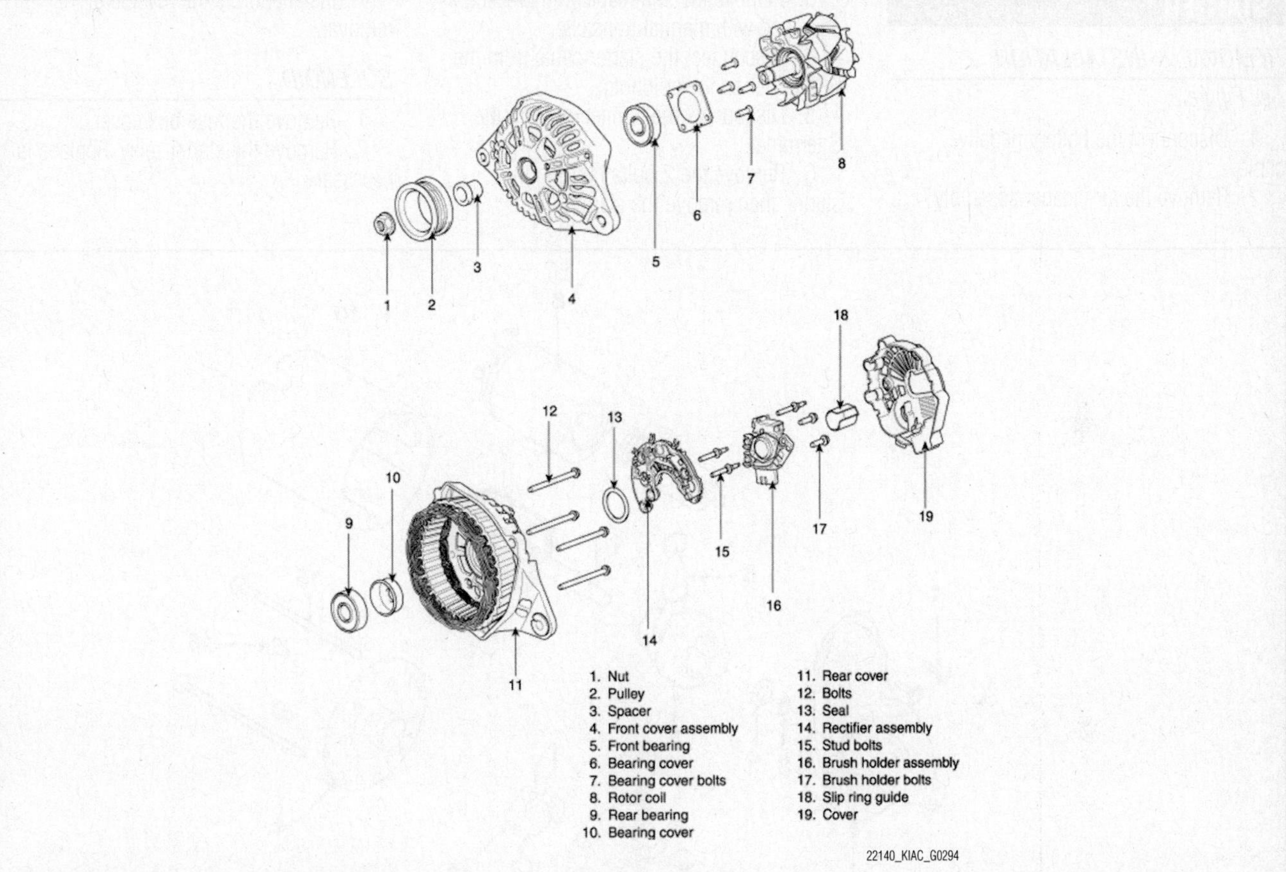

Fig. 6 Charging system component locations

ENGINE ELECTRICAL — IGNITION SYSTEM

FIRING ORDER

See Figure 7.

Firing order: 1–3–4–2

Front of Engine
Ignition Coils
1 2 3 4
67162-KIAC-G26

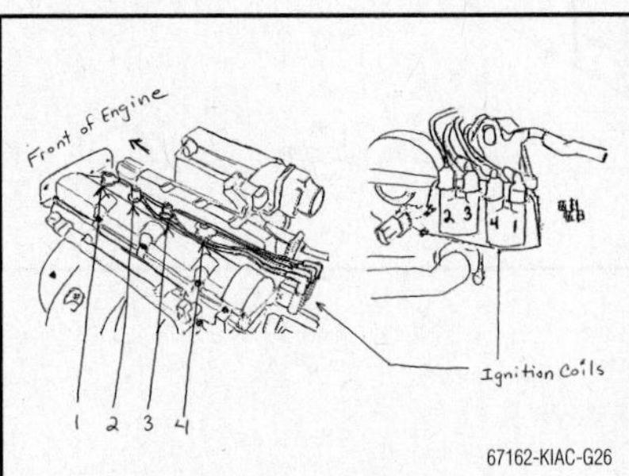

Fig. 7 Firing Order: 1–3–4–2 Distributorless ignition system

IGNITION COIL

REMOVAL & INSTALLATION

1. Remove the engine cover.
2. Remove the ignition coil.

➡When removing the ignition coil connector, pull the lock pin and push the clip.

To install:

3. Installation is the reverse of removal.

IGNITION TIMING

ADJUSTMENT

These vehicles are equipped with a Distributorless Ignition System (DIS).

No adjustment is necessary or possible.

SPARK PLUGS

REMOVAL & INSTALLATION

1. Disconnect the negative battery cable.

⁂ WARNING

When disconnecting the cable, only pull on the plug cable boot, never on the wire itself!

2. Detach the spark plug cable and remove the ignition coil.
3. Use spark plug wrench to remove the spark plug (s) from the cylinder head.

⁂ WARNING

Do not let any dirt or debris get into the engine through the spark plugs holes while when are removed.

To install:

4. Install the spark plug and tighten, to 18–22 ft. lbs. (25–30 Nm)
5. Install the ignition coil (s).
6. Attach the spark plug wire (s) to the ignition coil.
7. Connect the negative battery cable.

ENGINE ELECTRICAL

STARTING SYSTEM

STARTER

REMOVAL & INSTALLATION

See Figure 8.

1. Disconnect the battery negative cable.
2. Remove the air cleaner assembly
3. Remove the shift cable and bracket, if equipped with manual transaxle.
4. Disconnect the starter cable from the B terminal on the solenoid.
5. Disconnect the connector from the S terminal.
6. Remove the 2 bolts holding the starter, then remove the starter.
7. Installation is the reverse of removal.

SOLENOID

1. Remove the fuse box cover.
2. Remove the starter relay. Replace is necessary.

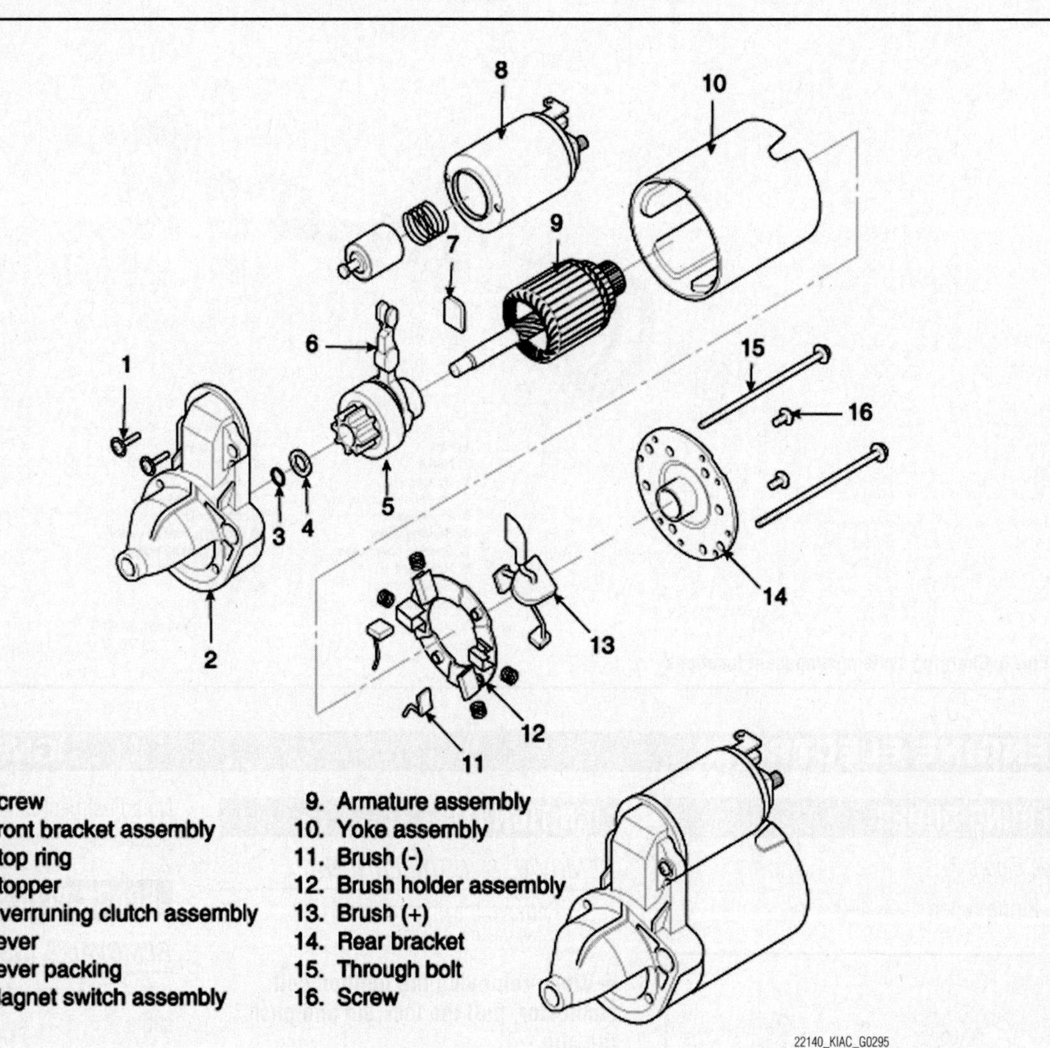

Fig. 8 Starting system component locations

ENGINE MECHANICAL

ACCESSORY DRIVE BELTS

ACCESSORY BELT ROUTING

See Figure 9.

Refer to the accompanying illustration for belt routing.

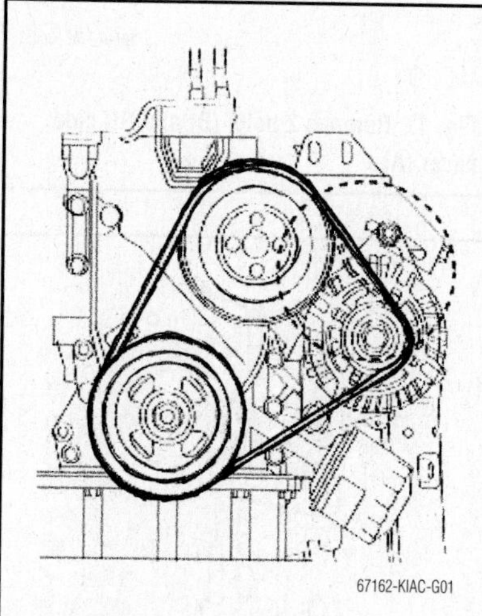

Fig. 9 Accessory drive belt routing

INSPECTION

Inspect the drive belt for signs of glazing or cracking. A glazed belt will be perfectly smooth from slippage, while a good belt will have a slight texture of fabric visible. Cracks will usually start at the inner edge of the belt and run outward. All worn or damaged drive belts should be replaced immediately.

ADJUSTMENT

A/C Compressor Belt

See Figure 10.

1. Operate the A/C one or two times a month, year round and adjust the compressor belt tension from time to time.
2. Using the deflection method, apply moderate pressure (about 22 lbs) halfway between the between the A/C compressor and crankshaft pulley.
   a. New belt: 0.197–0.217 in. (5–5.5mm)
   b. Used belt: 0.236–0.276 in. (6–7mm)
   c. Check after operation: 0.315 in. (8mm)
3. If tension is not within, adjust the belt, as follows:
   a. Loosen the tension mounting bolt.

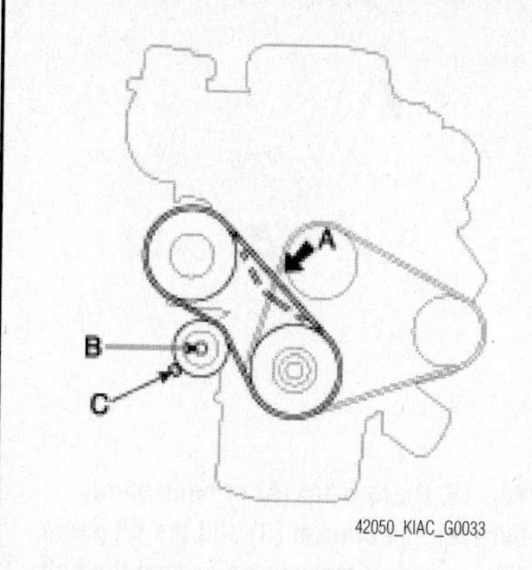

Fig. 10 Checking and adjusting the A/C compressor belt tension—2.0L engine

   b. Turn the adjusting bolt to get the proper belt tension, then retighten the mounting bolt.
   c. Recheck belt tension.

Alternator-Water Pump Belt

See Figures 11 through 13.

➡When using a new belt, first adjust the deflection or tension to the values for the new belt, then readjust the deflection or tension to the value for the used belt after the engine has run for 5 minutes.

1. Using the deflection method, apply moderate pressure (about 22 lbs) halfway between the alternator and water pump and compare with the following specifications.
   a. Used belt: 0.1969–0.2362 in. (5–6mm)

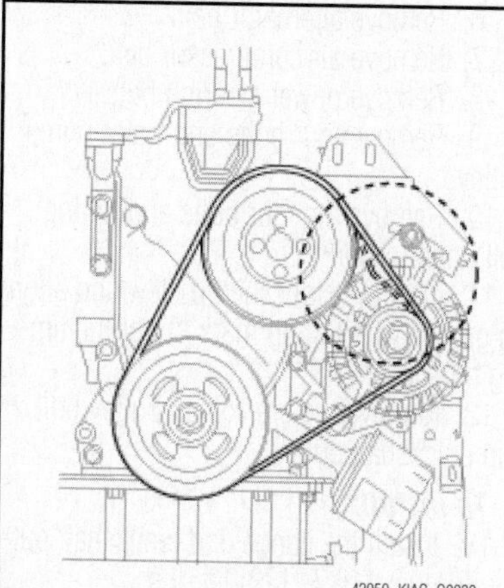

Fig. 11 Measuring the deflection between the alternator and water pump pulleys

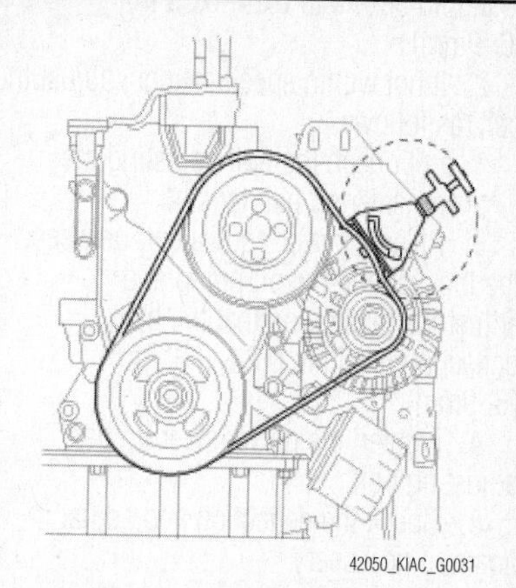

Fig. 12 Using a tension gauge to measure the alternator-water pump belt tension

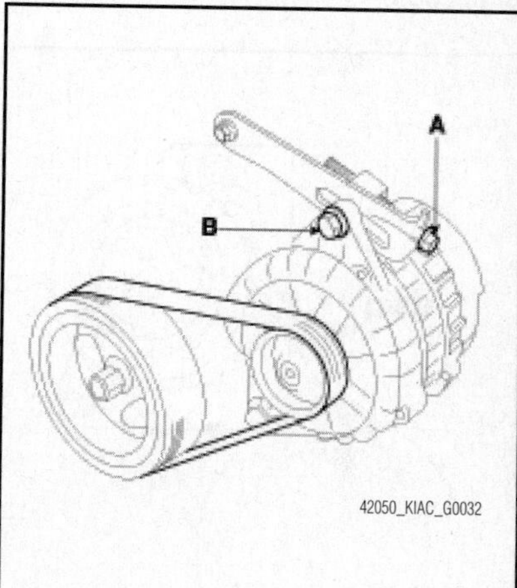

Fig. 13 Adjusting the alternator-water pump belt

   b. New belt: 0.1575–0.1969 in. (4–5mm)
2. Using a belt tension gauge, measure the drive belt tension and compare with the following specifications:
   a. Alternator belt (new): 86–103 lbs. (383–461 N)
   b. Alternator belt (used): 68–86 lbs. (304–383 N)
   c. If adjustment is necessary, loosen the adjusting bolt and the lock bolt, then move the alternator to get the proper tension and retighten the nuts.
3. Recheck the deflection or tension of the belt.

Power Steering Belt

See Figures 14 through 16.

1. Using the deflection method, apply moderate pressure (about 22 lbs)

at the point shown in the accompanying illustration and measure the deflection. It should be within 0.24–0.35 in. (6–9mm).

2. If not within specifications, adjust the belt, as follows:

a. Loosen the bolt adjusting the power steering belt tension.

3. Place a bar or equivalent, between the bracket and the oil pump and adjust the tension so that the belt deflection is within 0.24–0.35 in. (6–9mm).

4. Tighten the power steering belt adjusting bolt.

5. Check the deflection and adjust it again, if necessary

➡After turning the belt in the normal rotation direction more than once, recheck the belt deflection.

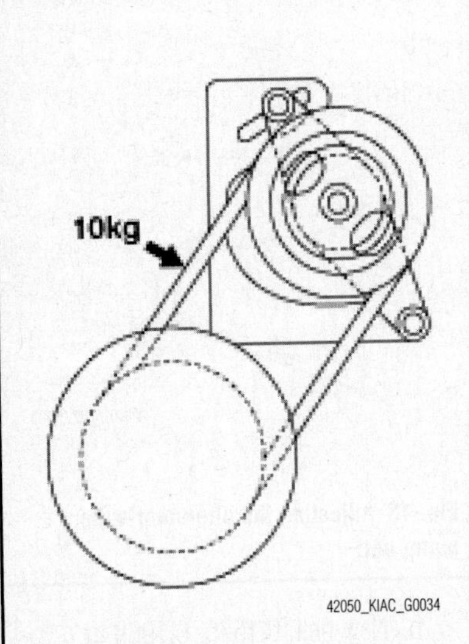

Fig. 14 Checking the power steering belt tension

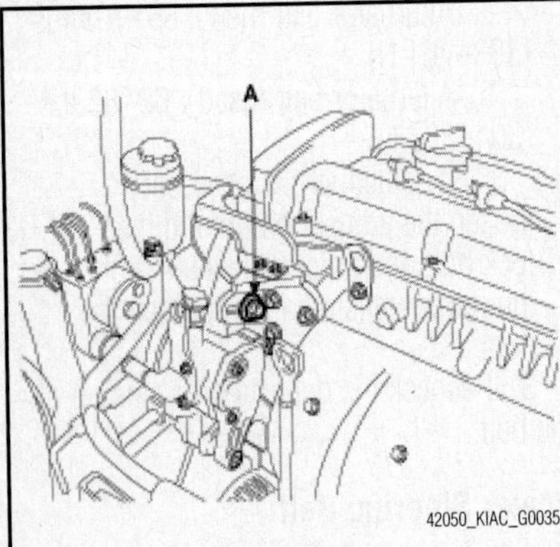

Fig. 15 Loosen the bolt adjusting (A) the power steering belt tension

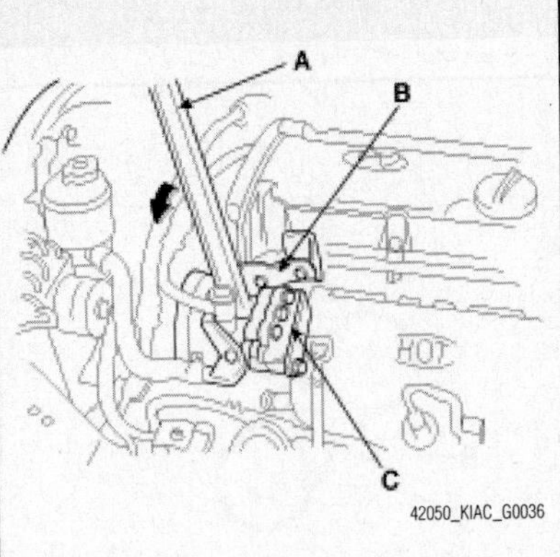

Fig. 16 Place a bar (A) or equivalent, between the bracket (B) and the oil pump (C) and adjust the tension so that the belt deflection is within specifications

CRANKSHAFT DAMPER (BALANCER)

REMOVAL & INSTALLATION

See Figures 17 through 20.

Engine removal is not required for this procedure.

1. Remove the engine cover.
2. Remove RH front wheel.
3. Remove 2bolts and RH side cover.
4. Remove the engine mount bracket.

a. Set the jack to the engine oil pan.

⁂ CAUTION

Place wooden block between the jack and engine oil pan.

b. Remove the bolt, three nuts and engine mount bracket.

c. Remove the bolt and stay plate.

5. Temporarily loosen the water pump pulley bolts.
6. Remove alternator belt.
7. Remove air compressor belt.
8. Remove power steering belt.
9. Remove four bolts and water pump pulley.
10. Remove the four bolts and timing belt upper cover.
11. Turn the crankshaft pulley, and align its groove with timing mark "T" of the timing belt cover.
12. Remove the crankshaft pulley bolt and crankshaft pulley.

To install:

13. Install the flange and crankshaft pulley. Torque: 115.7–123.0 ft. lbs. (156.9–166.7 Nm).

➡Make sure that crankshaft sprocket pin fits the small hole in the pulley.

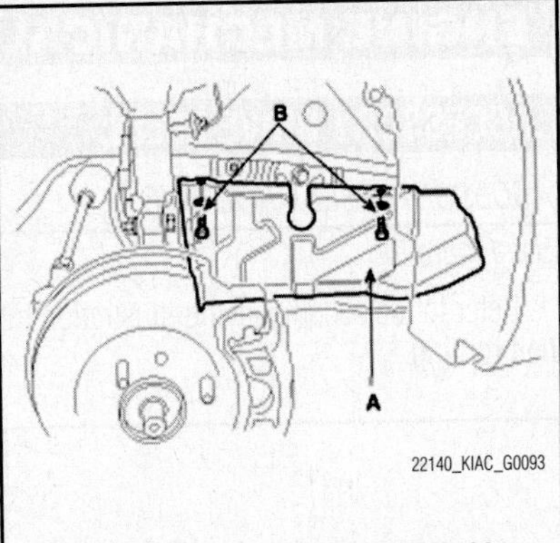

Fig. 17 Remove 2 bolts (B) and RH side cover (A).

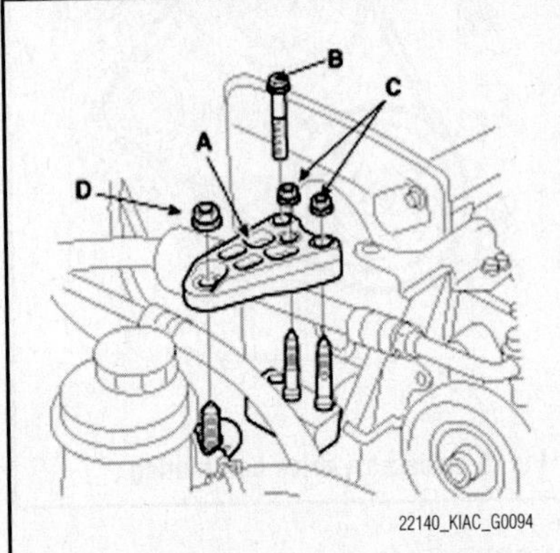

Fig. 18 Remove the bolt (B), three nuts (C, D) and engine mount bracket (A).

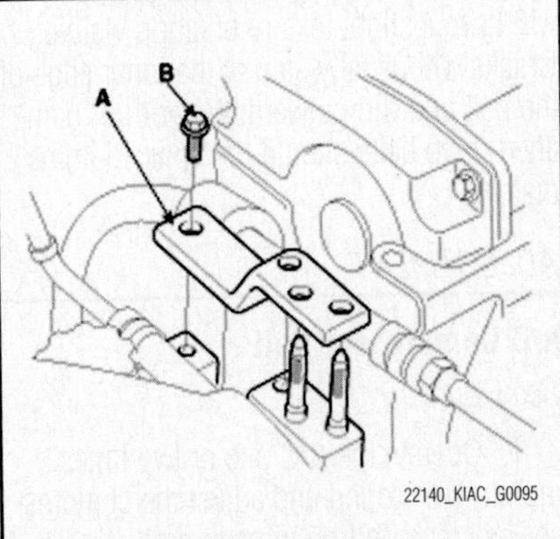

Fig. 19 Remove the bolt (B) and stay plate (A).

14. Install the timing belt upper cover. Torque: 5.87.2 ft. lbs. (7.8–9.8 Nm).
15. Install the coolant pump pulley.
16. Install power steering belt.
17. Install air compressor bolt.
18. Install alternator belt.
19. Install the engine mount bracket

a. Install the stay plate. Torque: 31.1–39.8 ft. lbs. (42.2–53.9 Nm).

b. Install engine mount bracket. Torque: 17mm nut: 50.6–68.7 ft. lbs.

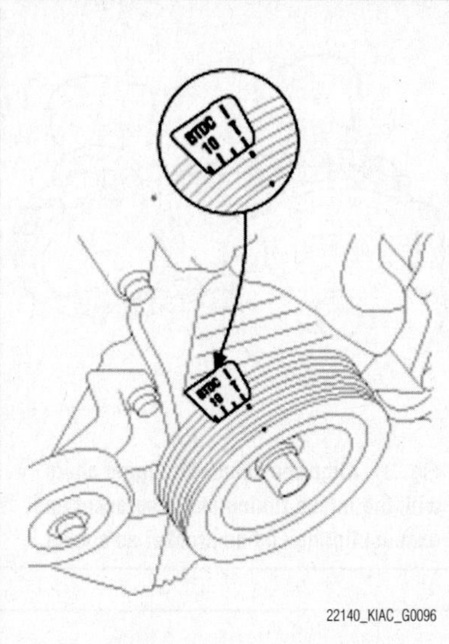

Fig. 20 Align the crankshaft pulley groove with timing mark "T" of the timing belt cover.

(68.6–3.2 Nm); 14mm nuts and bolt: 36.2–47.0 ft. lbs. (49.0–63.7 Nm).

20. Install RH side cover.
21. Install RH front wheel.
22. Install engine cover.

CRANKSHAFT FRONT SEAL

REMOVAL & INSTALLATION

1. Disconnect the negative battery cable.
2. Remove the timing belt covers and belt.
3. Remove the timing belt pulley using a puller.
4. Remove the oil pump bolts and the pump.
5. Wrap a suitable prytool with a rag and work the old seal from the oil pump housing.

To install:

6. Lubricate the seal lip with clean engine oil and push the seal slightly in by hand.
7. Install the seal using a seal installer. Install the seal until it is flush with the oil pump body.
8. Install the timing belt.
9. Connect the negative battery cable.
10. Start the engine and check for leaks.

CYLINDER HEAD

REMOVAL & INSTALLATION

See Figures 21 through 32.

1. Disconnect the battery terminals.
2. Remove the heat shield and the battery
3. Remove the engine cover.
4. Loosen the radiator drain plug and drain engine coolant.
5. Remove the intake air hose and air cleaner assembly.
   a. Disconnect the MAF connector.
   b. Disconnect the breather hose from air cleaner hose.
   c. Remove the intake air hose and air cleaner assembly.
6. Remove the upper and lower radiator hoses.
7. Remove the heater hoses.
8. Remove the accelerator cable.
9. Remove the engine wire harness connectors and wire harness clamps from the cylinder head and the intake manifold.
   a. Disconnect Oil control Valve (OCV) connector.
   b. Disconnect Oil Temperature Sensor (OTS) connector.
   c. Disconnect Engine Coolant Temperature (ECT) sensor connector.
   d. Disconnect ignition coil connector.
   e. Disconnect Throttle Position Sensor (TPS) connector.
   f. Disconnect Idle Speed Actuator (ISA) connector.
   g. Disconnect Camshaft Position Sensor (CMP) connector.
   h. Disconnect four fuel injector connectors.
   i. Disconnect Knock Sensor connector and the ground cable.
   j. Disconnect Purge Control Solenoid Valve (PCSV) connector.
   k. Disconnect front heated oxygen sensor connector.
10. Remove the fuel inlet hose from delivery pipe.
11. Remove the PCSV hose.
12. Remove the brake booster vacuum hose.
13. Remove the spark plug cable.
14. Remove the Positive Crankcase Ventilation (PCV) hose.
15. Remove the cylinder head cover.
16. Remove the timing belt.
17. Remove the exhaust manifold.
18. Remove the intake manifold.
19. Remove the camshaft sprocket.
20. Remove the timing chain auto tensioner.
21. Remove the camshaft bearing caps and camshafts.
22. Remove the Oil Control Valve (OCV).
23. Remove the Oil Control Valve (OCV) filter.
24. Remove the cylinder head bolts, then remove the cylinder head.
   a. Using 8mm and 10mm hexagon wrench, uniformly loosen and remove the 10 cylinder head bolts, in several passes, in the sequence shown. Remove the 10 cylinder head bolts and plate washers.

✲✲ CAUTION

Head warpage or cracking could result from removing bolts in an incorrect order.

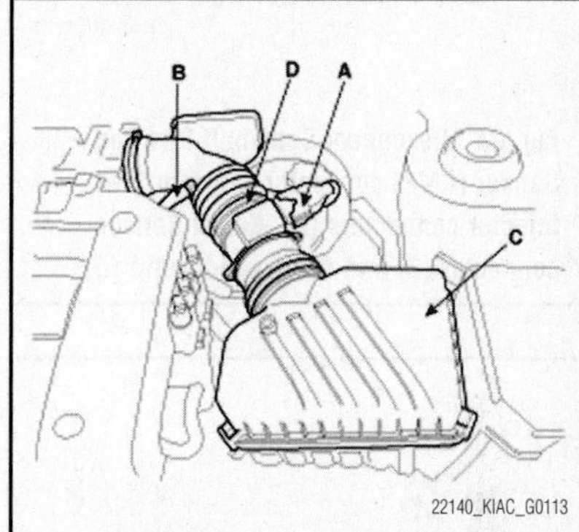

Fig. 21 Disconnect the MAF connector, (A) breather hose (B), remove air cleaner assembly (C) and air cleaner hose (D).

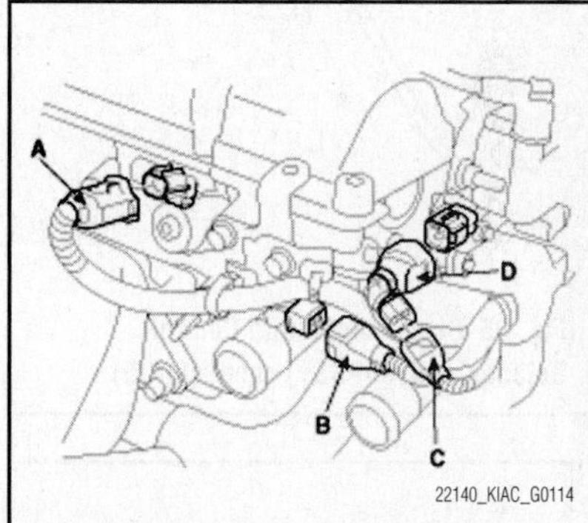

Fig. 22 Remove the engine wire harness connectors and wire harness clamps from the cylinder head and the intake manifold.

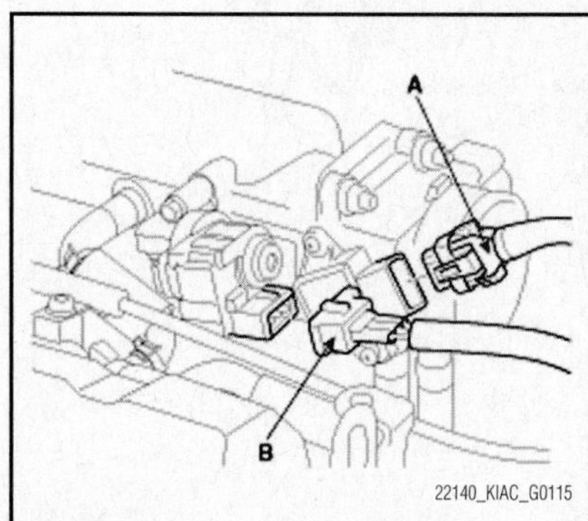

Fig. 23 Disconnect Throttle Position Sensor (TPS) connector (A) and ISA (Idle Speed Actuator) connector (B).

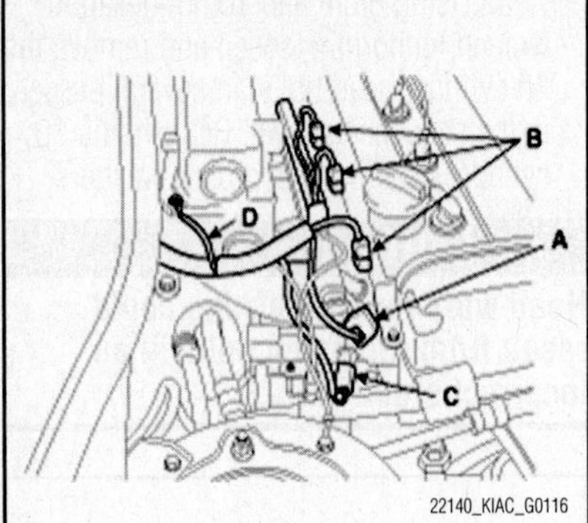

Fig. 24 Disconnect Camshaft Position Sensor (CMP) connector (A), four fuel injector connectors (B), Knock Sensor connector (C) and the ground cable (D).

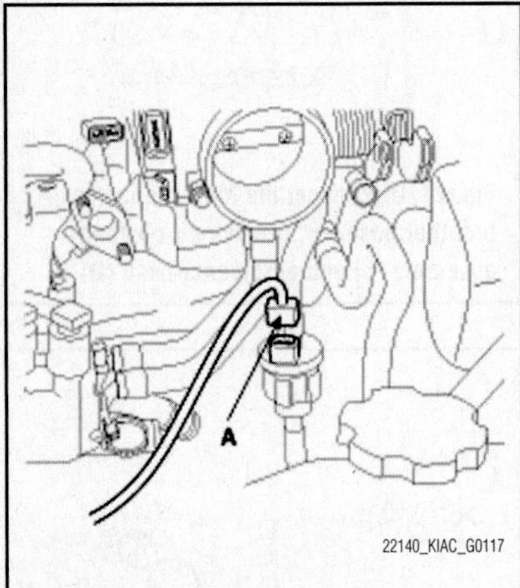

Fig. 25 Disconnect Purge Control Solenoid Valve (PCSV) connector (E)

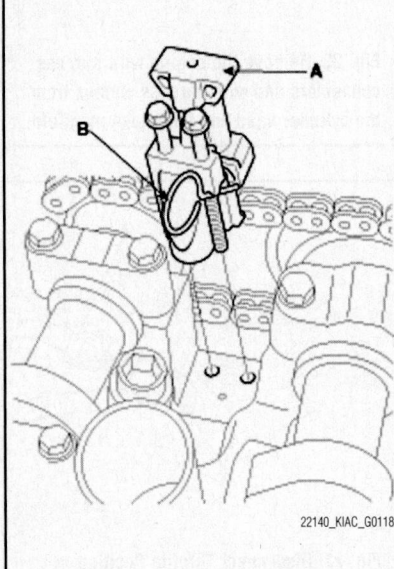

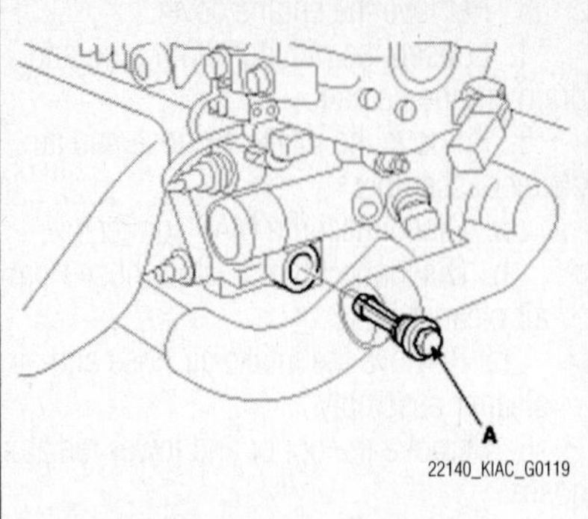

Fig. 27 Remove the OCV (oil control valve) filter (A).

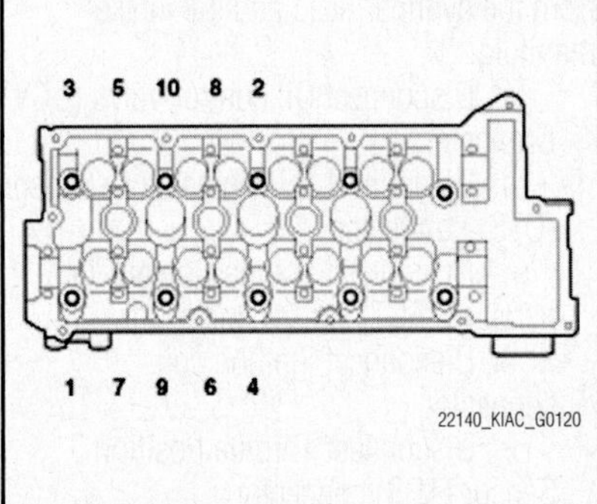

Fig. 28 Remove the cylinder head bolts in the sequence shown.

Fig. 29 Remove the Oil Control Valve (OCV) (A).

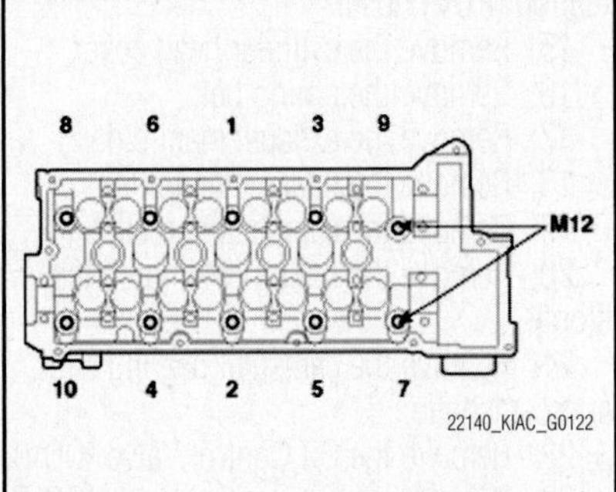

Fig. 30 Install the cylinder head bolts in the sequence shown.

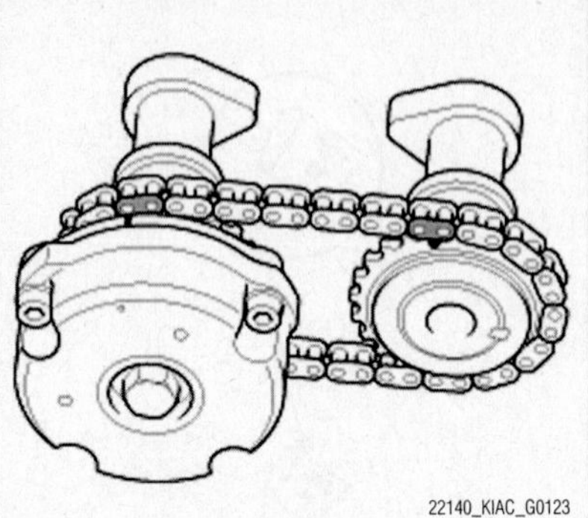

Fig. 31 Align the camshaft timing chain with the intake timing chain sprocket and exhaust timing chain sprocket as shown.

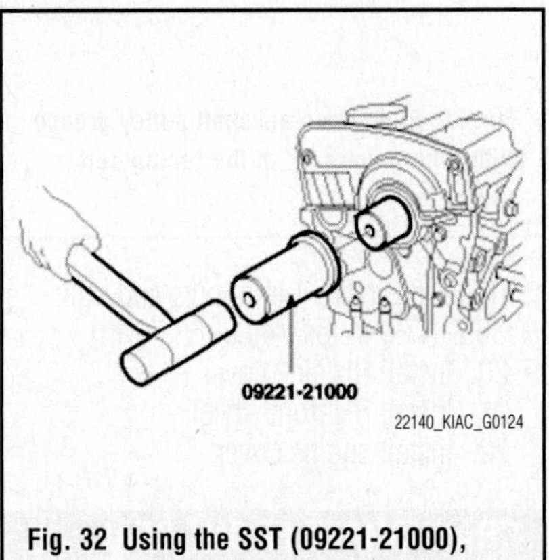

Fig. 32 Using the SST (09221-21000), install the camshaft bearing oil seal.

b. Lift the cylinder head from the dowels on the cylinder block and replace the cylinder head on wooden blocks on a bench.

To install:

25. Install the cylinder head gasket on the cylinder block.

26. Place the cylinder head carefully in order not to damage the gasket with the bottom part of the end.

27. Install the cylinder head bolts.

a. Apply a light coat if engine oil on the threads and under the heads of the cylinder head bolts.

b. Using 8mm and 10mm hexagon wrench, install and tighten the 10 cylinder head bolts and plate washers, in several passes, in the sequence shown.

Torque: M10 bolts: 16.6–19.5 ft. lbs. (22.6–26.5 Nm) plus an additional 60°–65° plus another 60°–65°. M12 bolts: 20.3–23.1 ft. lbs. (27.5–31.4 Nm) plus an additional 60°–65° plus another 60°–65°.

28. Install the Oil Control Valve (OCV) filter. Torque: 29.7–36.9 ft. lbs. (40.2–50.0 Nm).

29. Install the Oil Control Valve (OCV). Torque: 7.2–8.7 ft. lbs. (9.8–11.8 Nm).

CAUTION

Do not reuse the OCV when dropped. Keep clean the OCV. Do not hold the OCV sleeve during servicing. When the OCV is installed on the engine, do not move the engine with holding the OCV yoke.

30. Install the camshafts.
    a. Align the camshaft timing chain with the intake timing chain sprocket and exhaust timing chain sprocket as shown.
    b. Install the camshafts and bearing caps. Torque: 10.1–10.8 ft. lbs. (13.7–14.7 Nm).
    c. Install the timing chain auto tensioner. Torque: 5.8–7.2 ft. lbs. (7.8–9.8 Nm).
    d. Remove the auto tensioner stopper pin.
31. Check and adjust valve clearance.
32. Using the SST (09221-21000), install the camshaft bearing oil seal.
33. Install the camshaft sprocket.
34. Install the timing belt.
35. Install the cylinder head cover.
    a. Install the cylinder head cover gasket in the groove of the cylinder head cover.
    b. Apply liquid gasket to the head cover gasket at the corners of the recess.
    c. Install the cylinder head cover with the 12bolts. Uniformly tighten the bolts in several passes. Torque: 5.8–7.2 ft. lbs. (7.8–9.8 Nm).
36. Install the intake manifold.
37. Install the exhaust manifold.
38. Install the Positive Crankcase Ventilation (PCV).
39. Install the spark plug cable.
40. Install the accelerator cable and the auto-cruse cables.
41. Install the bake booster hose.
42. Install the Purge Control Solenoid Valve (PCSV) hose.
43. Install the fuel inlet hose.
44. Install the engine wire harness connectors and wire harness clamps to the cylinder head and the intake manifold.
    a. Front heated oxygen sensor connector.
    b. Knock sensor connector and the ground cable.
    c. Four fuel injector connectors.
    d. CMP connector.
    e. Purge Control Solenoid Valve (PCSV) connector.
    f. Idle Speed Control Actuator (ISCA) connector.
    g. Throttle Position Sensor (TPS) connector.
    h. Ignition coil connector.
    i. Engine Coolant Temperature Sensor (ECTS) sensor connector.
    j. Oil temperature sensor connector.
    k. Oil Control Valve (OCV) connector.
45. Install the heater hoses.
46. Install the upper and lower radiator hoses.
47. Install the intake air hose and air cleaner assembly.
48. Install the engine cover.
49. Connect the negative terminal to the battery.
50. Fill with engine coolant.
51. Start the engine and check for leaks.
52. Recheck engine coolant level and oil level.

EXHAUST MANIFOLD

REMOVAL & INSTALLATION

1. Remove the engine cover.
2. Disconnect the front oxygen sensor connector.
3. Remove the front muffler.
4. Remove the heat protector.
5. Remove the exhaust manifold and catalytic converter assembly.
6. To install, reverse the removal procedure with new gaskets.

INTAKE MANIFOLD

REMOVAL & INSTALLATION

1. Remove the engine cover.
2. Disconnect the Throttle Position Sensor (TPS) connector and the Idle Speed Actuator (ISA) connector.
3. Disconnect the Positive Crankcase Ventilation (PCV) hose and the breather hose.
4. Disconnect the accelerator cable.
5. Remove the fuel delivery pipe.
6. Disconnect the heater hose, Purge Control Solenoid Valve (PCSV) hose and the brake booster hose from the intake manifold and throttle body assembly.
7. Remove the intake manifold bracket.
8. Remove the intake manifold assembly.
9. To install, reverse the removal procedure with new gaskets.
10. Intake manifold assembly: Torque: 11.6–16.6 ft. lbs. (15.7–22.6 Nm).

OIL PAN

REMOVAL & INSTALLATION

1. Drain engine oil.
2. Disconnect the rear heated oxygen sensor connector.
3. Remove the front muffler.
4. Remove the front muffler mounting bracket.
5. Remove the oil pan.

To install:

6. Install the oil pan.
    a. Using a razor blade and gasket scraper, remove all gasket material from the mating surfaces.
    b. Apply liquid gasket as an even bead, centered between the edges of the mating surface. Liquid gasket: Loctite NO.5900 or equivalent

➡To prevent leakage of oil, apply liquid gasket to the inner threads of the bolt holes. Do not install the parts if five minutes or more have elapsed since applying the liquid gasket. After assembly, wait at least 30 minutes before filling the engine with oil.

    c. Install the oil pan with the bolts. Uniformly tighten the bolts in several passes. Torque: 7.2–8.7 ft. lbs. (9.8–11.8 Nm).
7. Install the front muffler bracket.
8. Install the front muffler.
9. Connect the rear oxygen sensor connector.
10. Fill with engine oil.

OIL PUMP

REMOVAL & INSTALLATION

See Figures 33 through 36.

1. Drain engine oil.
2. Remove the accessory drive belts.
3. Turn the crankshaft and align the white groove on the crankshaft pulley with the pointer on the lower cover.
4. Remove the timing belt.
5. Remove the timing belt idler.
6. Remove the oil pan and oil screen.
7. Remove the alternator.
8. Remove the air compressor tension bracket.
9. Remove the front case.
    a. Remove the screws from the pump housing, then separate the housing and cover.
    b. Remove the inner and outer rotors.

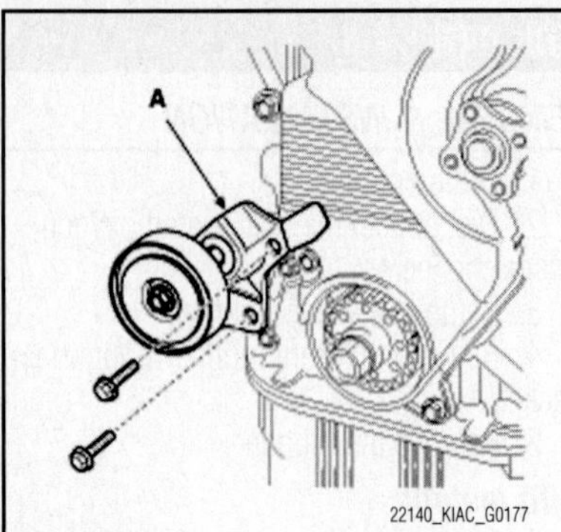

Fig. 33 Remove the air compressor tension bracket (A).

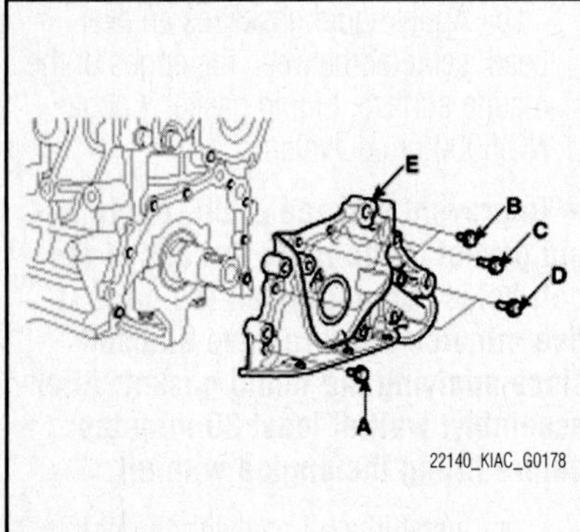

Fig. 34 Remove the front case (E).

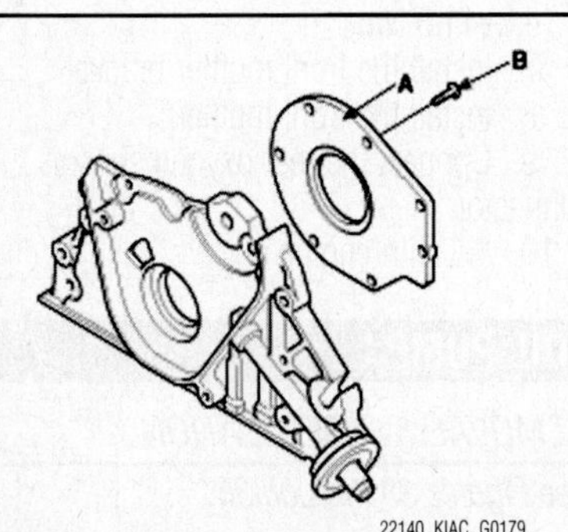

Fig. 35 Remove the screws (B) from the pump housing, then separate the housing and cover (A).

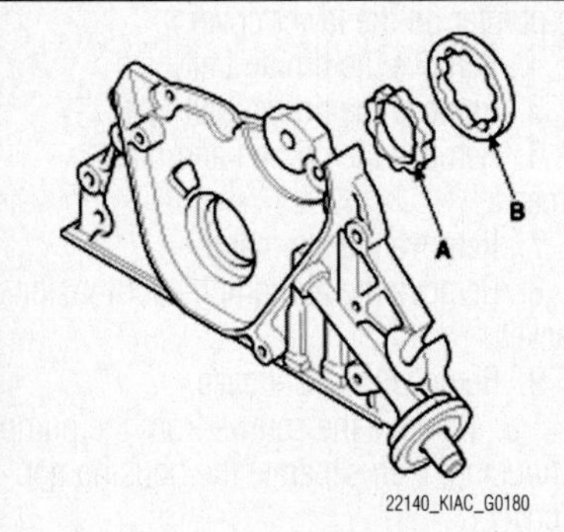

Fig. 36 Remove the inner (A) and outer (B) rotors.

To install:

10. Install oil pump.
    a. Place the inner and outer rotors into front case with the marks facing the oil pump cover side.
    b. Install the oil pump cover to front case with the 7 screws. Torque: 4.3–6.5 ft. lbs. (5.9–8.8 Nm).
11. Check that the oil pump turns freely.
12. Install the oil pump on the cylinder block.
    a. Place a new front case gasket on the cylinder block.
    b. Apply engine oil to the lip of the oil pump seal.
    c. Install the oil pump onto the crankshaft. Torque: 14.5–19.5 ft. lbs. (19.6–26.5 Nm).

➡Bolt lengths:

A : 0.98 inches (25 mm)
B : 0.787 inches (20 mm)
C : 1.496 inches (38 mm)
D : 1.771 inches (45 mm)

13. Apply a light coat of oil to the seal lip.
14. Using the SST (09214-33000), install the oil seal.
15. Install the air compressor tension bracket.
16. Install the alternator.
17. Install the oil screen.
18. Install the oil pan. Torque: 7.2–8.7 ft. lbs. (9.8–11.8 Nm).

➡Clean the oil pan gasket mating surfaces.

19. Install the timing belt idler. Torque: 31.1–39.8 ft. lbs. (42.2–53.9 Nm).
20. Install the timing belt.
21. Install the accessory drive belt.
22. Fill the engine oil.

INSPECTION

1. Inspect relief plunger.

Coat the valve with engine oil and check that it falls smoothly into the plunger hole by its own weight. If it does not, replace the relief plunger. If necessary, replace the front case.

2. Inspect relief valve spring.
3. Inspect for distorted or broken relief valve spring. Standard value:
    a. 1.6L engine:
    b. Free height: 1.835 inches (46.6 mm). Load: 13.5 lbs./1.579 inches (6.1 kg/40.1 mm).
    c. 2.0 L engine:
    d. Free height: 1.724 inches (43.8 mm). Load: 8.14 lbs./1.579 inches (3.7 kg/40.1 mm).
4. Inspect rotor side clearance.
    a. 1.6L engine:
    b. Inner rotor: 0.0016–0.0033 Inches (0.04–0.085 mm)
    c. Outer rotor: 0.0016–0.0035 inches (0.04–0.09 mm)
    d. 2.0L engine:
    e. Inner rotor: 0.0016–0.0033 Inches (0.04–0.085 mm)
    f. Outer rotor: 0.0016–0.0035 inches (0.04–0.09 mm)
5. Using a feeler gauge and precision straight edge, measure the clearance between the rotors and precision straight edge. If the side clearance is greater than maximum, replace the rotors as a set. If necessary, replace the front case.
6. Inspect rotor tip clearance.
7. Using a feeler gauge, measure the tip clearance between the inner and outer rotor tips.
    a. Tip clearance:
    b. 1.6L engine:
    c. 0.0010–0.0027 inches (0.025–0.069 mm).
    d. 2.0L engine:
    e. 0.0010–0.0027 inches (0.025–0.069 mm).

If the tip clearance is greater than maximum, replace the rotor as a set.

8. Inspect rotor body clearance.
9. Using a feeler gauge, measure the clearance between the outer rotor and body.
    a. Body clearance:
    b. 1.6L engine:
    c. 0.0024–0.0035 inches (0.60–0.090 mm).
    d. 2.0L engine:
    e. 0.0047–0.0073 inches (0.12–0.185 mm).

If the body clearance is greater than maximum, replace the rotors as a set. If necessary, replace the front case.

REAR MAIN SEAL

REMOVAL & INSTALLATION

1. Before servicing the vehicle, refer to the Precautions Section.
2. Disconnect the negative battery cable.
3. Remove the transaxle assembly.
4. Remove the clutch and flywheel assembly, if equipped with a manual transaxle.
5. Remove the flexplate-to-crankshaft bolts, the flexplate and shim plates, if equipped with an automatic transaxle.
6. Cut the oil seal lip with a knife. Install a rag to the housing and using a

screwdriver, carefully pry the oil seal from the oil seal housing. Clean the gasket mounting surfaces.

To install:

7. Clean the oil seal housing. Coat the oil seal and the housing with clean engine oil.
8. Install the oil seal into the housing and tap it evenly into place with a hammer and a large diameter piece of pipe. The seal must be flush with the edge of the rear cover.
9. Install the flywheel assembly or the flexplate, as applicable, and tighten the mounting bolts to 71–76 ft. lbs. (97–102 Nm).
10. Install the clutch assembly, if applicable.
11. Install the transaxle.
12. Connect the negative battery cable.

ROCKER ARMS & SHAFTS

REMOVAL & INSTALLATION

The engines covered are not equipped with rocker arms/shafts. The camshafts directly actuate the valve through a bucket type follower.

TIMING BELT FRONT COVER

REMOVAL & INSTALLATION

See Figures 37 through 41.

Engine removal is not required for this procedure.

1. Remove the engine cover.
2. Remove RH front wheel.
3. Remove 2 bolts and RH side cover.
4. Remove the engine mount bracket.
   a. Set the jack to the engine oil pan.

CAUTION

Place wooden block between the jack and engine oil pan.

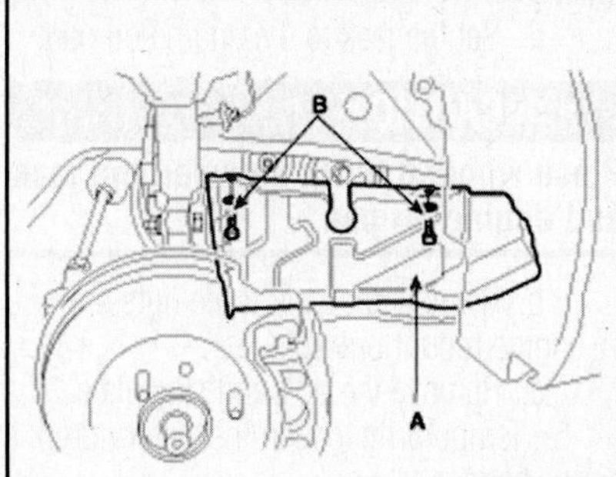

Fig. 37 Remove 2 bolts (B) and RH side cover (A).

   b. Remove the bolt, three nuts and engine mount bracket.
   c. Remove the bolt and stay plate.
5. Temporarily loosen the water pump pulley bolts.
6. Remove alternator belt.

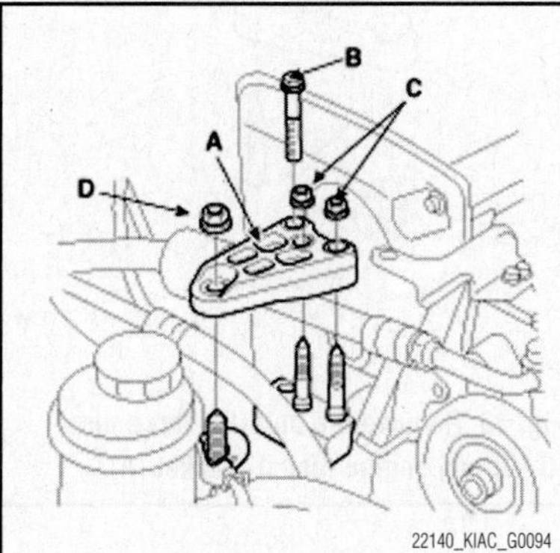

Fig. 38 Remove the bolt (B), three nuts (C, D) and engine mount bracket (A).

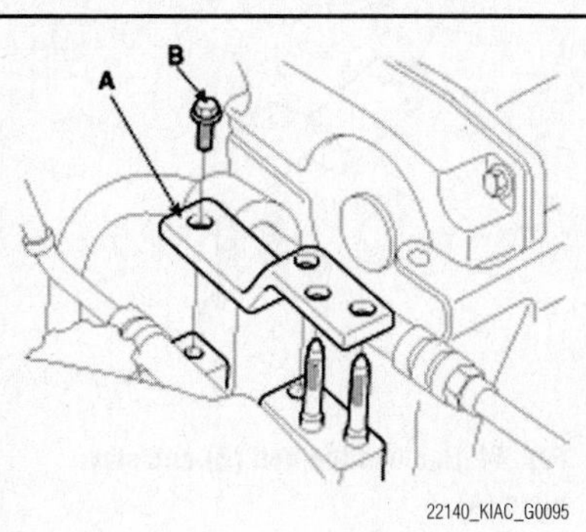

Fig. 39 Remove the bolt (B) and stay plate (A).

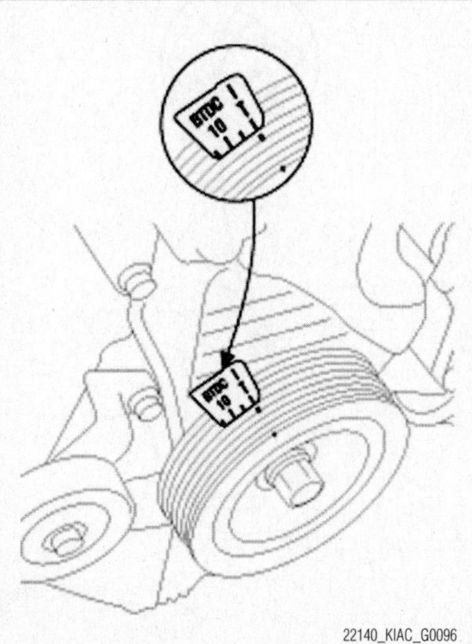

Fig. 40 Align the crankshaft pulley groove with timing mark "T" of the timing belt cover.

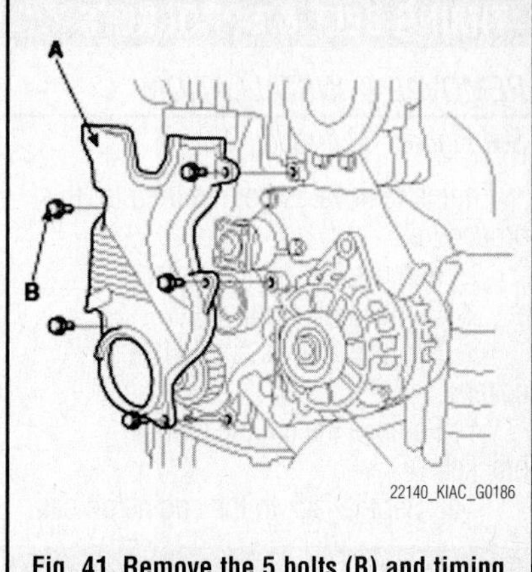

Fig. 41 Remove the 5 bolts (B) and timing belt lower cover (A).

7. Remove air compressor belt.
8. Remove power steering belt.
9. Remove four bolts and water pump pulley.
10. Remove the four bolts and timing belt upper cover.
11. Turn the crankshaft pulley, and align its groove with timing mark "T" of the timing belt cover.
12. Remove the crankshaft pulley bolt and crankshaft pulley.
13. Remove the crankshaft flange.
14. Remove the 5bolts and timing belt lower cover.

To install:

15. Install the timing belt lower cover with 5 bolts. Torque: 5.8–7.2 ft. lbs. (7.8–9.8 Nm).
16. Install the flange and crankshaft pulley. Torque: 115.7–123.0 ft. lbs. (156.9–166.7 Nm).

➡Make sure that crankshaft sprocket pin fits the small hole in the pulley.

17. Install the timing belt upper cover. Torque: 5.8–7.2 ft. lbs. (7.8–9.8 Nm).
18. Install the coolant pump pulley.
19. Install power steering belt.
20. Install air compressor bolt.
21. Install alternator belt.
22. Install the engine mount bracket
    a. Install the stay plate. Torque: 31.1–39.8 ft. lbs. (42.2–53.9 Nm).
    b. Install engine mount bracket. Torque: 17mm nut: 50.6–68.7 ft. lbs. (68.6–3.2 Nm); 14mm nuts and bolt: 36.2–47.0 ft. lbs. (49.0–63.7 Nm).
23. Install RH side cover.
24. Install RH front wheel.
25. Install engine cover.

TIMING BELT & SPROCKETS

REMOVAL & INSTALLATION

See Figures 42 through 46.

Engine removal is not required for this procedure.

1. Remove the engine cover.
2. Remove RH front wheel.
3. Remove 2 bolts and RH side cover.
4. Remove the engine mount bracket.
    a. Set the jack to the engine oil pan.

CAUTION

Place wooden block between the jack and engine oil pan.

    b. Remove the bolt, three nuts and engine mount bracket.
    c. Remove the bolt and stay plate.
5. Temporarily loosen the water pump pulley bolts.
6. Remove alternator belt.
7. Remove air compressor belt.
8. Remove power steering belt.
9. Remove four bolts and water pump pulley.
10. Remove the four bolts and timing belt upper cover.
11. Turn the crankshaft pulley, and align its groove with timing mark "T" of the timing belt cover.
12. Remove the crankshaft pulley bolt and crankshaft pulley.
13. Remove the crankshaft flange.
14. Remove the 5 bolts and timing belt lower cover.

To install:

15. Install the timing belt lower cover with 5 bolts. Torque: 5.8–7.2 ft. lbs. (7.8–9.8 Nm).
16. Install the flange and crankshaft pulley. Torque: 115.7–123.0 ft. lbs. (156.9–166.7 Nm).

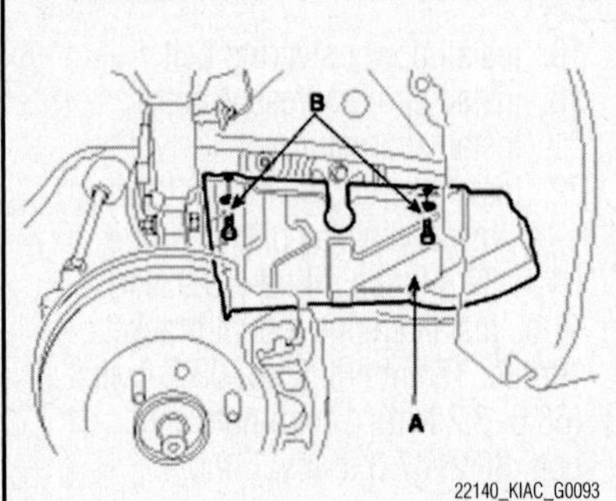

Fig. 42 Remove 2 bolts (B) and RH side cover (A).

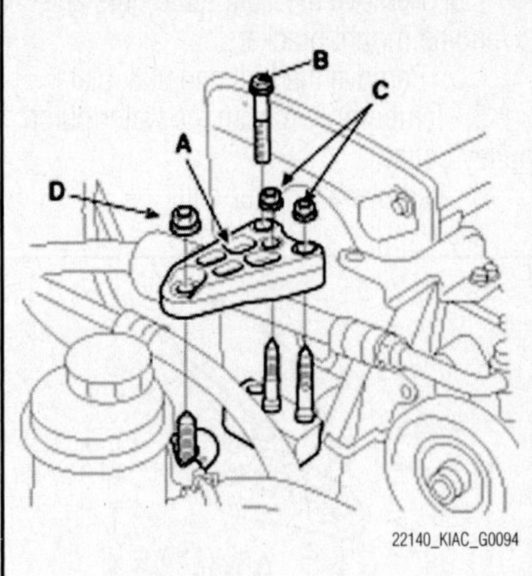

Fig. 43 Remove the bolt (B), three nuts (C, D) and engine mount bracket (A).

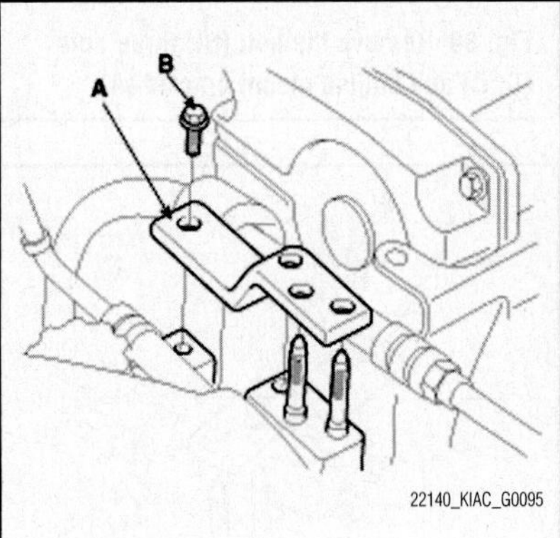

Fig. 44 Remove the bolt (B) and stay plate (A).

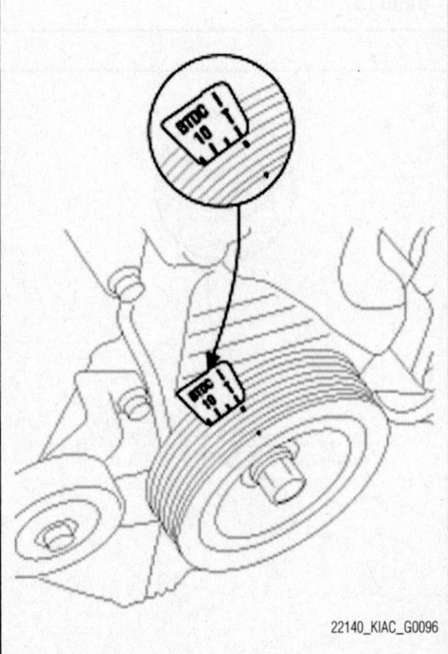
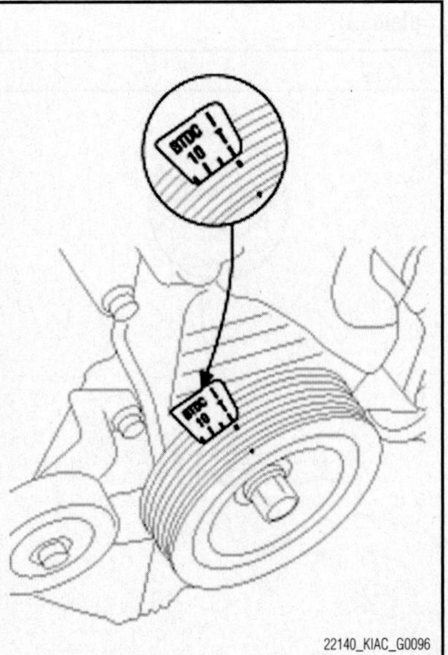

Fig. 45 Align the crankshaft pulley groove with timing mark "T" of the timing belt cover.

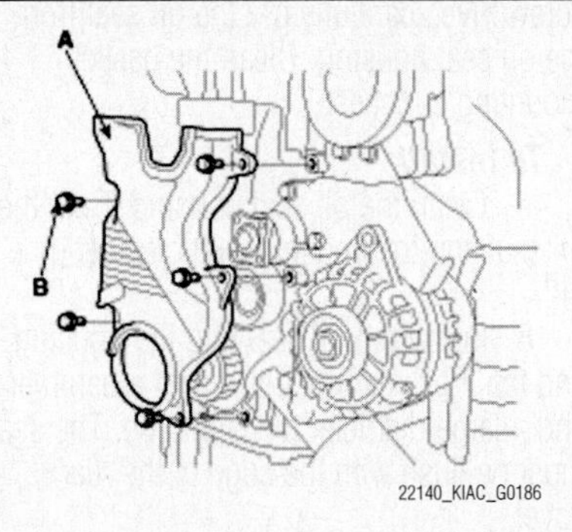

Fig. 46 Remove the 5 bolts (B) and timing belt lower cover (A).

➡Make sure that crankshaft sprocket pin fits the small hole in the pulley.

17. Install the timing belt upper cover. Torque: 5.8–7.2 ft. lbs. (7.8–9.8 Nm).
18. Install the coolant pump pulley.
19. Install power steering belt.
20. Install air compressor bolt.
21. Install alternator belt.
22. Install the engine mount bracket
    a. Install the stay plate. Torque: 31.1–39.8 ft. lbs. (42.2–53.9 Nm).
    b. Install engine mount bracket. Torque: 17mm nut: 50.6–68.7 ft. lbs. (68.6–3.2 Nm); 14mm nuts and bolt: 36.2–47.0 ft. lbs. (49.0–63.7 Nm).
23. Install RH side cover.
24. Install RH front wheel.
25. Install engine cover.

TIMING BELT LOWER COVER

REMOVAL & INSTALLATION

See Figures 42 through 46.

Engine removal is not required for this procedure.

1. Remove the engine cover.
2. Remove RH front wheel.
3. Remove 2 bolts and RH side cover.
4. Remove the engine mount bracket.
    a. Set the jack to the engine oil pan.

CAUTION

Place wooden block between the jack and engine oil pan.

    b. Remove the bolt, three nuts and engine mount bracket.
    c. Remove the bolt and stay plate.
5. Temporarily loosen the water pump pulley bolts.
6. Remove alternator belt.
7. Remove air compressor belt.
8. Remove power steering belt.

9. Remove four bolts and water pump pulley.

10. Remove the four bolts and timing belt upper cover.

11. Turn the crankshaft pulley, and align its groove with timing mark "T" of the timing belt cover.

12. Remove the crankshaft pulley bolt and crankshaft pulley.

13. Remove the crankshaft flange.

14. Remove the 5 bolts and timing belt lower cover.

To install:

15. Install the timing belt lower cover with 5 bolts. Torque: 5.8–7.2 ft. lbs. (7.8–9.8 Nm).

16. Install the flange and crankshaft pulley. Torque: 115.7–123.0 ft. lbs. (156.9–166.7 Nm).

➡Make sure that crankshaft sprocket pin fits the small hole in the pulley.

17. Install the timing belt upper cover. Torque: 5.8–7.2 ft. lbs. (7.8–9.8 Nm).

18. Install the coolant pump pulley.

19. Install power steering belt.

20. Install air compressor bolt.

21. Install alternator belt.

22. Install the engine mount bracket

a. Install the stay plate. Torque: 31.1–39.8 ft. lbs. (42.2–53.9 Nm).

b. Install engine mount bracket. Torque: 17mm nut: 50.6–68.7 ft. lbs. (68.6–3.2 Nm); 14mm nuts and bolt: 36.2–47.0 ft. lbs. (49.0–63.7 Nm).

23. Install RH side cover.

24. Install RH front wheel.

25. Install engine cover.

VALVE LASH

INSPECTION

The valve lash on all engines is kept in adjustment hydraulically. No adjustment is necessary or possible.

ENGINE PERFORMANCE & EMISSION CONTROLS

ACCELERATOR PEDAL POSITION (APP) SENSOR

LOCATION

The Accelerator Position Sensor (APS) is installed on the accelerator pedal module and detects the rotation angle of the accelerator pedal.

REMOVAL & INSTALLATION

1. Turn ignition switch off and disconnect the negative battery cable.
2. Disconnect the accelerator position sensor connector.
3. Unfasten the four mounting nuts
4. Remove the accelerator pedal from the vehicle.
5. Installation is the reverse of removal.

CAMSHAFT POSITION (CMP) SENSOR

LOCATION

The two Camshaft Position (CMP) sensors are installed on the bank 1 and 2 engine head covers.

CRANKSHAFT POSITION (CKP) SENSOR

LOCATION

The Crankshaft Position (CKP) sensor is installed on transaxle housing.

ELECTRONIC CONTROL MODULE (ECM)

REMOVAL & INSTALLATION

1. Turn the ignition switch off and disconnect the negative battery terminal.
2. Disconnect the ECM connector.
3. Unscrew 4 mounting bolts behind the air cleaner.
4. Remove the ECM.

To install:

5. Installation is reverse of removal.

ENGINE COOLANT TEMPERATURE (ECT) SENSOR

LOCATION

The Engine Coolant Temperature Sensor (ECT) is located in the engine coolant passage of the cylinder head for detecting the engine coolant temperature.

REMOVAL & INSTALLATION

1. Turn ignition switch OFF.
2. Disconnect ECT sensor connector.
3. Remove the ECT sensor.

To install:

4. Installation is the reverse of removal.

HEATED OXYGEN (HO2S) SENSOR

LOCATION

Heated Oxygen Sensor (HO2S) is installed on upstream and downstream of the Manifold Catalyst Converter (MCC).

INTAKE AIR TEMPERATURE (IAT) SENSOR

LOCATION

The Intake Air Temperature (IAT) sensor is located in the engine compartment on the air cleaner housing.

KNOCK SENSOR (KS)

LOCATION

The Knock Sensor (KS) senses engine knocking and the two sensors are installed inside the V-valley of the cylinder block.

MASS AIR FLOW (MAF) SENSOR

LOCATION

The Mass Air Flow (MAF) sensor is located in between the air cleaner and the throttle body.

MANIFOLD ABSOLUTE PRESSURE (MAP) SENSOR

LOCATION

The Manifold Absolute Pressure Sensor (MAP) sensor is installed on the surge tank.

FUEL SYSTEM SERVICE PRECAUTIONS

Safety is the most important factor when performing not only fuel system maintenance, but any type of maintenance. Failure to conduct maintenance and repairs in a safe manner may result in serious personal injury or death. Work on a vehicle's fuel system components can be accomplished safely and effectively by adhering to the following rules and guidelines.

- To avoid the possibility of fire and personal injury, always disconnect the negative battery cable unless the repair or test procedure requires that battery voltage be applied.
- Always relieve the fuel system pressure prior to disconnecting any fuel system component (injector, fuel rail, pressure regulator, etc.) fitting or fuel line connection. Exercise extreme caution whenever relieving fuel system pressure to avoid exposing skin, face and eyes to fuel spray. Please be advised that fuel under pressure may penetrate the skin or any part of the body that it contacts.
- Always place a shop towel or cloth around the fitting or connection prior to loosening to absorb any excess fuel due to spillage. Ensure that all fuel spillage is quickly removed from engine surfaces. Ensure that all fuel-soaked cloths or towels are deposited into a flame-proof waste container with a lid.
- Always keep a dry chemical (Class B) fire extinguisher near the work area.
- Do not allow fuel spray or fuel vapors to come into contact with a spark or open flame.
- Always use a second wrench when loosening or tightening fuel line connection fittings. This will prevent unnecessary stress and torsion on fuel piping. Always follow the proper torque specifications.
- Always replace worn fuel fitting O-rings with new ones. Do not substitute fuel hose where rigid pipe is installed.

FUEL SYSTEM PRESSURE

RELIEVING

1. Remove the rear seat cushion
2. Remove the rear seat cushion.
3. Disconnect the fuel pump connector.
4. Start the engine and wait until fuel in fuel line is exhausted.
5. After engine stalls, turn the ignition switch to OFF position.

FUEL FILTER

REMOVAL & INSTALLATION

The fuel filter is an integral part of the fuel pump assembly.

1. Remove the rear seat cushion.
2. Remove the service cover
3. Disconnect the fuel pump connector.
4. Start the engine and wait until fuel in fuel line is exhausted.
5. After engine stalls, turn the ignition switch to OFF position.
6. Disconnect the fuel feed line and canister hoses.
7. Unscrew the fuel pump mounting bolts and remove the fuel pump assembly.

To install:

8. Install the Fuel Pump according to the reverse order of REMOVAL procedure.

FUEL LEVEL SENDING UNIT

LOCATION

The fuel level sending unit is located on the fuel pump in the fuel tank.

REMOVAL & INSTALLATION

1. Disconnect the fuel pump connector.
2. Start the engine and wait until fuel in fuel line is exhausted.
3. After engine stalls, turn the ignition switch to OFF position.
4. Disconnect the fuel feed quick-connector and the vapor hose .
5. Unscrew the fuel pump locking ring with the special service tool (SST No.: 09310-3K000) and remove the fuel pump assembly.

To install:

6. Installation is reverse of removal.

FUEL PUMP

REMOVAL & INSTALLATION

1. Remove the rear seat cushion.
2. Remove the service cover
3. Disconnect the fuel pump connector.
4. Start the engine and wait until fuel in fuel line is exhausted.
5. After engine stalls, turn the ignition switch to OFF position.
6. Disconnect the fuel feed line and canister hoses.
7. Unscrew the fuel pump mounting bolts and remove the fuel pump assembly.

To install:

8. Install the Fuel Pump according to the reverse order of REMOVAL procedure.

FUEL PRESSURE REGULATOR

REMOVAL & INSTALLATION

The fuel pressure regulator is an integral part of the fuel pump assembly.

1. Remove the rear seat cushion.
2. Remove the service cover
3. Disconnect the fuel pump connector.
4. Start the engine and wait until fuel in fuel line is exhausted.
5. After engine stalls, turn the ignition switch to OFF position.
6. Disconnect the fuel feed line and canister hoses.
7. Unscrew the fuel pump mounting bolts and remove the fuel pump assembly.
8. Install the Fuel Pump according to the reverse order of REMOVAL procedure.

FUEL TANK

REMOVAL & INSTALLATION

See Figure 47.

1. Remove the rear seat cushion.
2. Remove the service cover
3. Disconnect the fuel pump connector.
4. Start the engine and wait until fuel in fuel line is exhausted.
5. After engine stalls, turn the ignition switch to OFF position.
6. Disconnect the negative terminal from the battery.

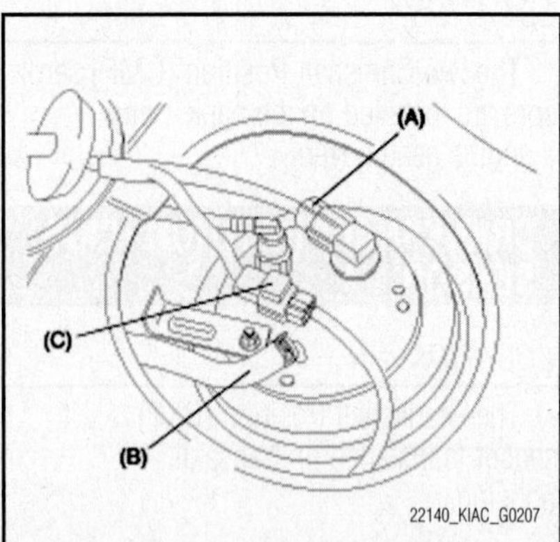

Fig. 47 Disconnect the fuel feed line (B) and the Fuel Tank Pressure Sensor (FTPS) connector (C).

7. Disconnect the fuel feed line and the Fuel Tank Pressure Sensor (FTPS) connector.

➡**Cover the hose connection with a shop towel to prevent splashing of fuel caused by residual pressure in the fuel line**

8. Lift the vehicle.
9. Remove the main muffler.
10. Disconnect the fuel filler hose, the canister drain hose and the vapor tube.
11. Unfasten two fuel tank band mounting bolts, and then remove the fuel tank from the vehicle.

To install:

12. Install the Fuel Tank in the reverse order of the removal procedure.

HEATING & AIR CONDITIONING SYSTEM

BLOWER MOTOR

REMOVAL & INSTALLATION

1. Disconnect the negative battery terminal.
2. Disconnect the connector of the blower motor.
3. Remove the blower motor after loosening the mounting screws.

To install:

4. Installation is the reverse order of removal.

HEATER CORE

REMOVAL & INSTALLATION

1. Disconnect the negative battery terminal.
2. Recover the refrigerant with a recovery/recycling/charging station
3. When the engine is cool, drain the engine coolant from the radiator.
4. Disconnect the inlet and outlet heater hoses from the heater unit.
5. Remove the bolts and the expansion valve from the evaporator core.
6. Plug or cap the lines immediately after disconnecting them to avoid moisture and dust contamination.
7. Remove the crash pad.
8. Disconnect the connectors from the temp. actuator, the mode actuator and the thermistor, then remove the mounting nuts.
9. Remove the heater and evaporator unit after loosening the mounting screws.
10. Remove the self-tapping screws and the upper bracket, the side cover.

➡**Be careful not to bend the inlet and outlet pipes during heater core removal.**

11. Pull out the heater core.

To install:

12. Install the heater core in the reverse order of removal.
13. Install in the reverse order of removal, and note these items:
    a. If you're installing a new evaporator, add refrigerant oil (ND-OIL8).
    b. Replace the O-rings with new ones at each fitting, and apply a thin coat of refrigerant oil before installing them. Be sure to use the right O-rings for R-134a to avoid leakage.
    c. Immediately after using the oil, replace the cap on the container, and seal it to avoid moisture absorption.
    d. Do not spill the refrigerant oil on the vehicle ; it may damage the paint ; if the refrigerant oil contacts the paint, wash it off immediately
    e. Apply sealant to the grommets.
    f. Make sure that there is no air leakage.
    g. Charge the system and test its performance.
    h. Do not interchange the inlet and outlet heater hoses and install the hose clamps securely.
    i. Refill the cooling system with engine coolant.

STEERING

POWER RACK & PINION STEERING GEAR

REMOVAL & INSTALLATION

See Figures 48 and 49.

1. Drain the power steering fluid.
2. Remove the air intake hose assembly.
3. Disconnect the pressure tube and the return tube fittings from the gear box.
4. Remove the steering shaft universal joint assembly mounting bolt.
5. Raise and safely support the front of the vehicle.
6. Remove the both front wheels.
7. Disconnect the tie rod from the knuckle by using the special tool (09568-34000).
8. Remove the dust cover of the stabilizer bar (LH side) mounting bracket.
9. Remove the mounting bolt and mounting clamp of power steering gear box, and the clamp holding the pressure tube and the return tube.

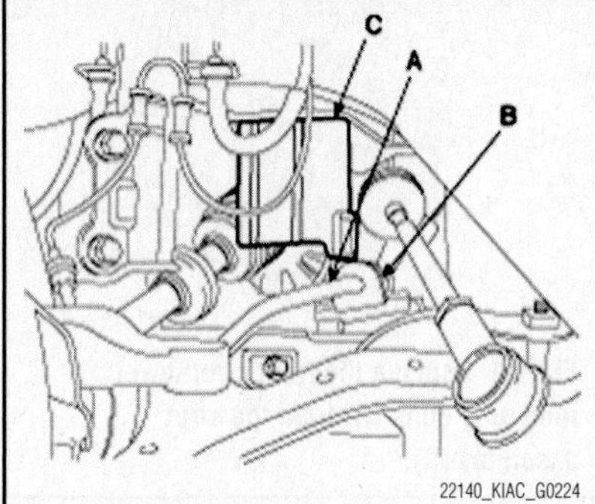

Fig. 48 Remove the dust cover (C) of the stabilizer bar (A) (LH side) mounting bracket (B).

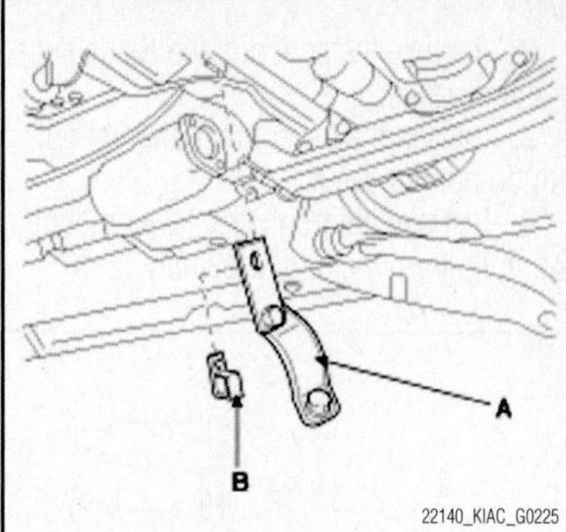

Fig. 49 Remove the mounting bolt and mounting clamp (A) of power steering gear box, and the clamp (B) holding the pressure tube and the return tube.

10. Pull the power steering gear box assembly toward the left side of the vehicle.

To install:

11. Installation is the reverse of removal.
12. Push in the power steering gear box assembly on the left side of the vehicle.

13. Bleed the air in the power steering system.

POWER STEERING PUMP

REMOVAL & INSTALLATION

1. Remove the pressure hose from the oil pump.
2. Disconnect the suction hose from the suction pipe and drain the fluid into a container.
3. Loosen the oil pump mounting bolts to remove the V belt.
4. Loosen the tension adjusting bolt.
5. Remove the power steering drive belt from the power steering oil pump pulley.
6. Remove the power steering oil pump mounting bolt and the tension adjusting bolt.
7. Remove the power steering oil pump assembly.

To install:

8. After installing the oil pump to the oil pump bracket, install the V belt and tighten the tension adjusting bolt. Torque: 18–24 ft. lbs. (25–33 Nm).
9. Install the pressure hose to the oil pump. Torque: 40–47 ft. lbs. (55–65 Nm).

CAUTION

Ensure the pressure hose does not twist and come in contact with other components.

10. Install the suction hose to the suction pipe.
11. Add power steering fluid (PSF-III).
12. Bleed the air in the system.

BLEEDING

1. Remove the fuel pump fuse, then start the engine and wait for the engine to stall. Next, while operating the starting motor intermittently (for 15–20 seconds), turn the steering wheel all the way to the left and then to the right five or six times.

➡During air bleeding, replenish the fluid supply so that the level never falls below the lower position of the filter. If air bleeding is done while the vehicle is idling, the air will be broken up and absorbed into the fluid. Be sure to do the bleeding only while cranking.

2. Reinstall the fuel pump fuse, and start the engine (idling).
3. Turn the steering wheel to the left and the right until there are no air bubbles in the oil reservoir.

CAUTION

Do not hold the steering wheel turned all the way to either side for more than ten seconds.

4. Confirm that the fluid is not milky, and that the level is up to the position specified on the level gauge.
5. Confirm that there is little change in the surface of the fluid when the steering wheel is turned left and right.

CAUTION

If the surface of the fluid changes considerably, air bleeding should be done again. If the fluid level rises suddenly when the engine is stopped, it indicates that there is still air in the system. If there is air in the system, a jingling noise may be heard from the pump and the control valve may also produce unusual noises. Air in the system will shorten the life of the pump and other parts.

SUSPENSION

FRONT SUSPENSION

LOWER BALL JOINT

REMOVAL & INSTALLATION

See Figures 50 and 51.

1. Remove the front wheel and tire.
2. Remove the split pin, the castle nut and the washer.
3. Loosen the lower arm ball joint nut, but do not remove it.
4. Remove the front strut lower mounting bolts from the strut assembly.

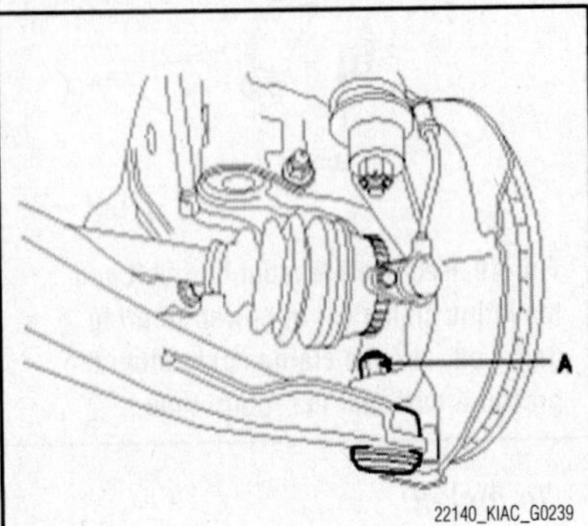

Fig. 50 Loosen the lower arm ball joint nut (A).

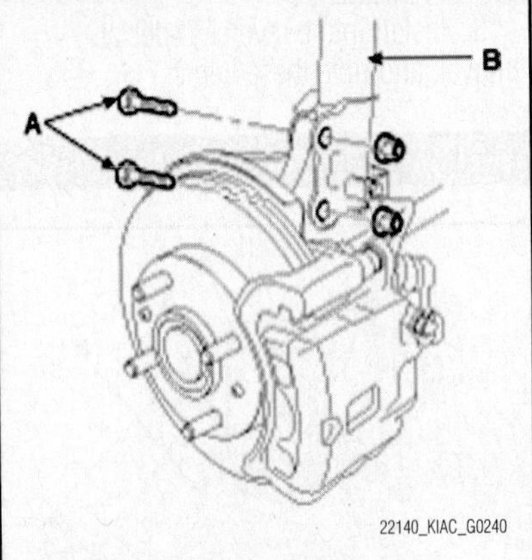

Fig. 51 Remove the front strut lower mounting bolts (A) from the strut assembly (B).

5. Push the axle hub outward to install the Special tool (09568-34000) easily.
6. Using the Special Tool (09568 - 34000), disconnect the lower arm ball joint from the lower arm.

To install:

7. Installation is the reverse of the removal procedure.

LOWER CONTROL ARMS

REMOVAL & INSTALLATION

See Figures 50 through 56.

1. Remove the front wheel and tire.
2. Remove the split pin, the castle nut and the washer.
3. Loosen the lower arm ball joint nut, but do not remove it.
4. Remove the front strut lower mounting bolts from the strut assembly.
5. Push the axle hub outward to install the Special tool (09568-34000) easily.
6. Using the Special Tool (09568 - 34000), disconnect the lower arm ball joint from the lower arm.
7. Remove the stabilizer link nut.
8. Temporarily install the strut lower mounting bolt.
9. To the lower arm mounting bolt, remove the right side cover.
10. Remove the lower arm mounting bolts.
11. Remove the lower arm assembly after completely removing the nut of lower arm ball joint.

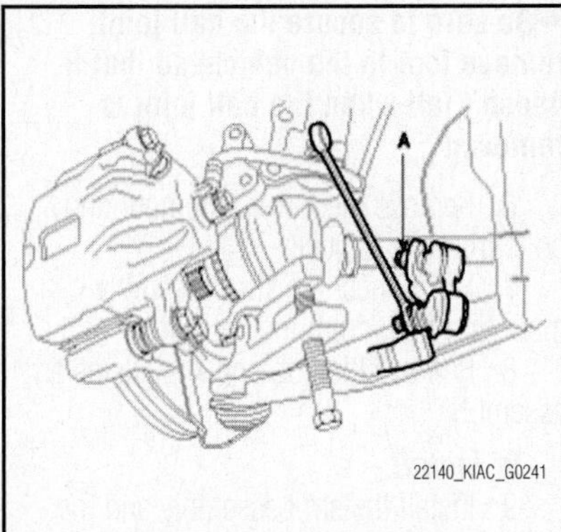

Fig. 52 Remove the stabilizer link nut (A).

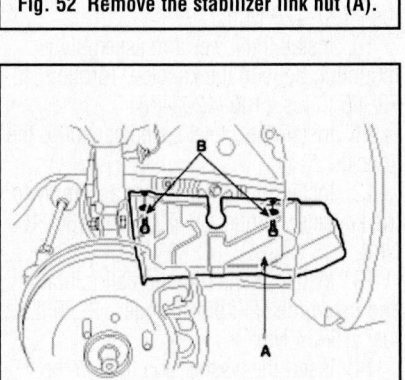

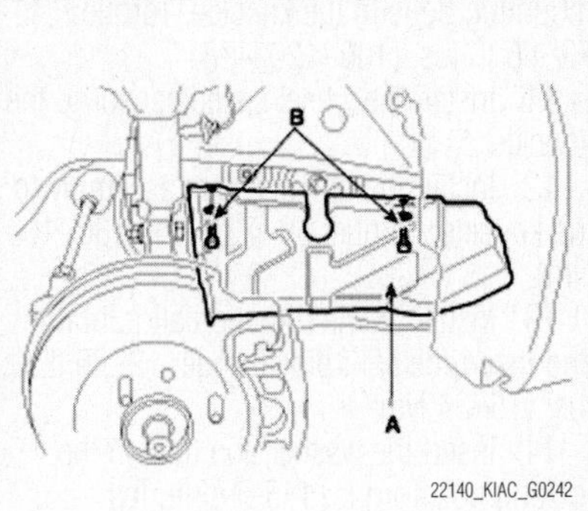

Fig. 53 Remove the right side cover (A).

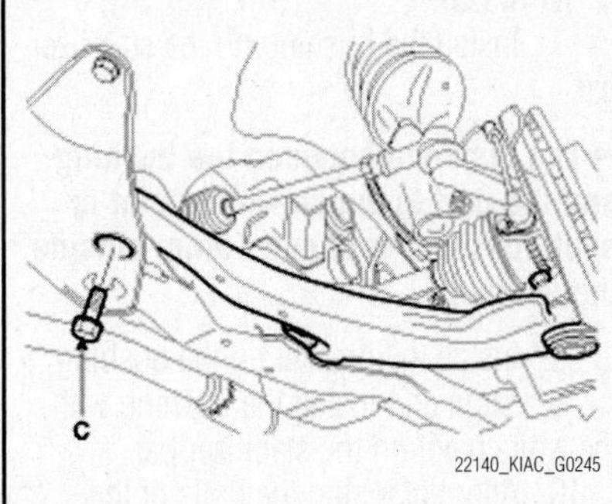

Fig. 56 Remove the lower arm mounting bolt (C).

To install:

12. Installation is the reverse of the removal procedure.

MACPHERSON STRUT

REMOVAL & INSTALLATION

See Figures 57 and 58.

1. Raise the vehicle, and make sure it is securely supported.
2. Remove the front wheel and tire.
3. Detach the brake hose bracket from the front strut assembly.
4. In case of the vehicles equipped with Anti-lock Brake system, remove the wheel speed sensor from the knuckle.
5. Remove the strut upper mounting nuts.
6. Remove the strut lower mounting bolts and then remove the strut assembly.

To install:

7. Installation is the reverse of the removal procedure.

STEERING KNUCKLE

REMOVAL & INSTALLATION

1. Remove the front wheel and tire.
2. Remove the split pin, then remove the locknut and washer from the front hub.
3. Remove the brake caliper from the knuckle and hang the caliper on the front damper.
4. Remove the wheel speed sensor from the knuckle.
5. Disconnect the tie rod end ball joint from the knuckle using the special tool (09568-34000).

➡Be sure to secure the ball joint, remove tool to the vehicle so that it doesn't fall when the ball joint is removed.

6. Remove the lower arm mounting bolts from the knuckle.
7. Disconnect the strut assembly mounting bolts from the knuckle.
8. Remove the hub and knuckle as an assembly.

To install:

9. Install the strut assembly and the drive shaft in the knuckle. Torque: 72–86 ft. lbs. (100–120 Nm).
10. Install the lower arm assembly mounting bolts to the knuckle. Torque: 72–86 ft. lbs. (100–120 Nm).
11. Install the wheel speed sensor to the knuckle.
12. Install the brake caliper assembly to the knuckle. Torque: 58–72 ft. lbs. (80–100 Nm).
13. Install the tie rod end ball joint nut and insert the split pin. Torque: 12–25 ft. lbs. (16–34 Nm).
14. Insert the washer and tighten the locking nut. Torque: 145–188 ft. lbs. (200–260 Nm).
15. Install the wheel and tire.

STABILIZER BAR & LINKS

REMOVAL & INSTALLATION

See Figure 59.

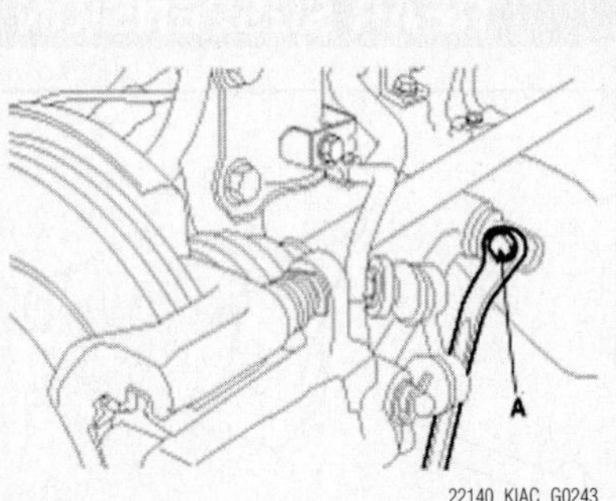

Fig. 54 Remove the lower arm mounting bolt (A).

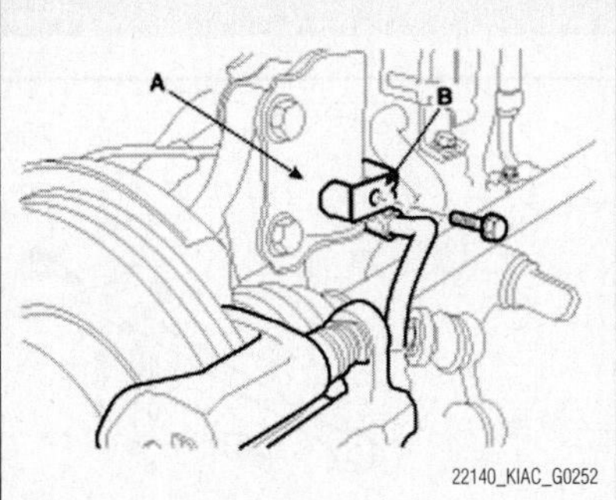

Fig. 57 Detach the brake hose bracket (B) from the front strut assembly (A).

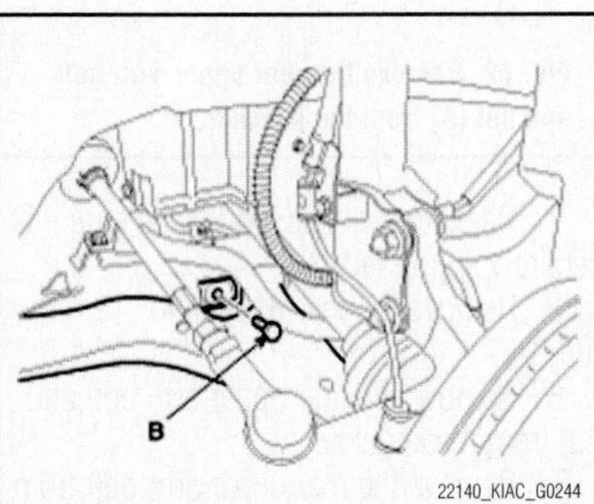

Fig. 55 Remove the lower arm mounting bolt (B).

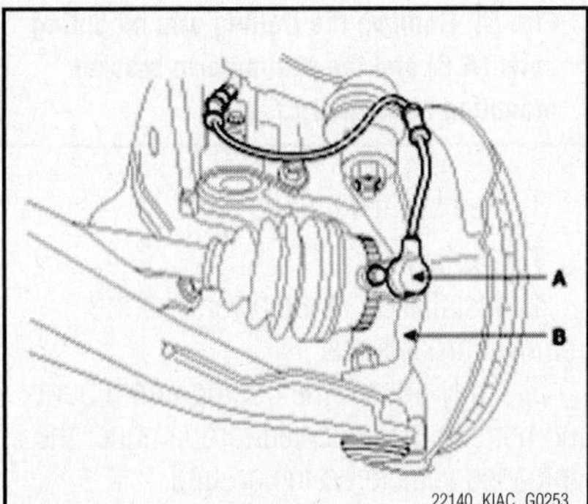

Fig. 58 R remove the wheel speed sensor (A) from the knuckle (B).

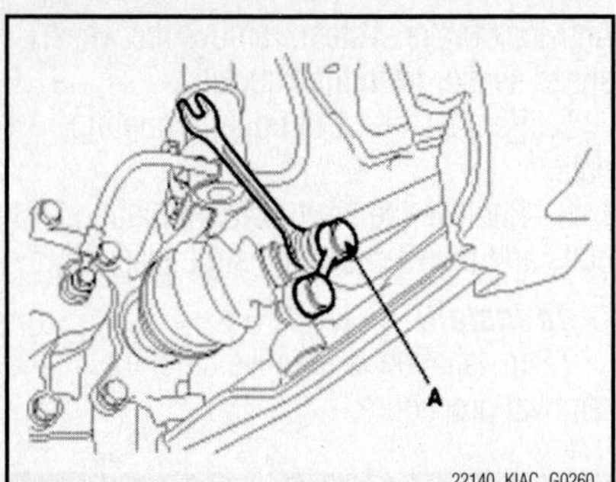

Fig. 59 Remove the stabilizer bar link assembly (A).

1. Remove the front wheel and tire.
2. Remove the stabilizer bar link assembly.
3. Remove the stabilizer bar link on the opposite side in the same way.

CAUTION

Be careful not to damage the ball joint boot.

4. Remove the stabilizer bracket and bushing.
5. Remove the stabilizer bar link on the opposite side in the same way.
6. Remove the stabilizer out of the vehicle's right side.

To install:

7. Install the bushing on the stabilizer bar.

➡The distance between the bushing, and the part to which white paint is applied, must continue 10mm outside the vehicle.

8. Install the bracket on the bushing
9. Align and install the bushing with the white paint on the stabilizer bar.
10. After tightening the bolts of the bushing bracket temporarily, install the bushing bracket on the opposite side.

WHEEL HUB & BEARING

REMOVAL & INSTALLATION

1. Remove the front wheel and tire.
2. Remove the split pin, then remove the locknut and washer from the front hub.
3. Remove the brake caliper from the knuckle and hang the caliper on the front damper.
4. Remove the wheel speed sensor from the knuckle.
5. Disconnect the tie rod end ball joint from the knuckle using the special tool (09568-34000).

➡Be sure to secure the ball joint, remove tool to the vehicle so that it doesn't fall when the ball joint is removed.

6. Remove the lower arm mounting bolts from the knuckle.
7. Disconnect the strut assembly mounting bolts from the knuckle.
8. Remove the hub and knuckle as an assembly.

To install:

9. Install the strut assembly and the drive shaft in the knuckle. Torque: 72–86 ft. lbs. (100–120 Nm).
10. Install the lower arm assembly mounting bolts to the knuckle. Torque: 72–86 ft. lbs. (100–120 Nm).
11. Install the wheel speed sensor to the knuckle.
12. Install the brake caliper assembly to the knuckle. Torque: 58–72 ft. lbs. (80–100 Nm).
13. Install the tie rod end ball joint nut and insert the split pin. Torque: 12–25 ft. lbs. (16–34 Nm).
14. Insert the washer and tighten the locking nut. Torque: 145–188 ft. lbs. (200–260 Nm).
15. Install the wheel and tire.

SUSPENSION

REAR SUSPENSION

CONTROL ARMS/LINKS

REMOVAL & INSTALLATION

Trailing Arm

See Figures 60 and 61.

1. After removing the bolt, detach the parking brake cable (B) which is fixed on the rear trailing arm bracket.
2. Remove the trailing arm mounting bolts and the trailing arm bracket mounting bolts.

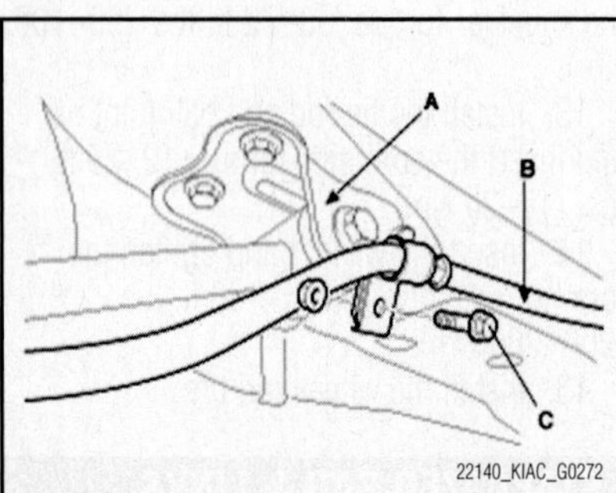

Fig. 60 Remove the bolt (C), detach the parking brake cable (B) on the rear trailing arm bracket (A).

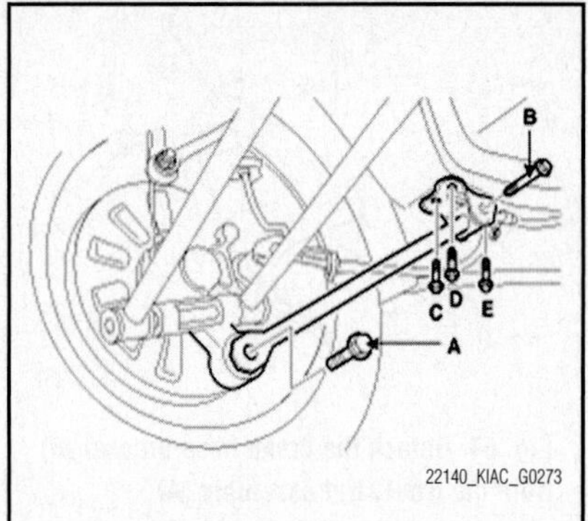

Fig. 61 Remove the trailing arm mounting bolts (A,B) and the trailing arm bracket mounting bolts (C,D,E).

3. Remove the trailing arm.

To install:

4. Installation is the reverse of the removal procedures.
5. Fully tighten the trailing arm mounting bolts to the specified torque under the unloaded vehicle on the ground.

Rear Upper Arm

See Figures 62 and 63.

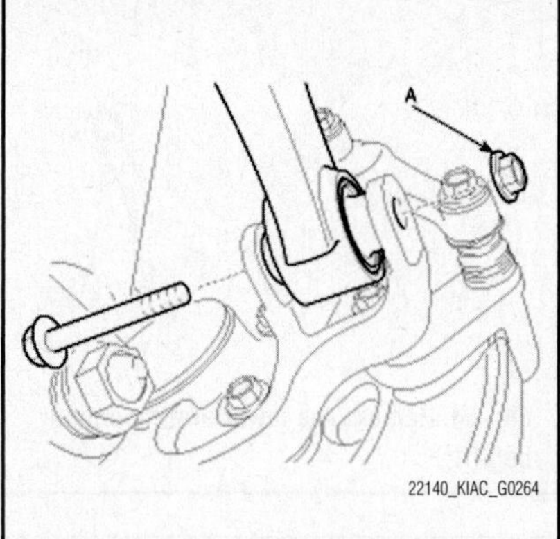

Fig. 62 Remove the rear upper arm bolt and nut (A) from the knuckle.

1. Raise the vehicle, and make sure it is securely supported.
2. Remove the rear wheel and tire.
3. Remove the rear upper arm bolt and nut from the knuckle.
4. Remove the rear upper arm bolt from the cross member.

To install:

5. Install the rear upper arm mounting

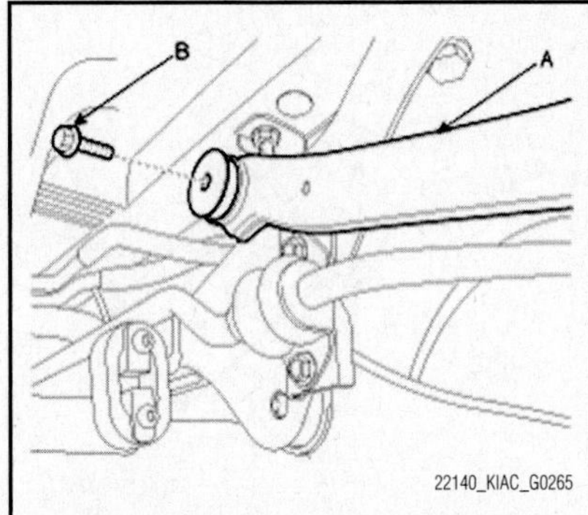

Fig. 63 Remove the rear upper arm (A) bolt (B) from the cross member.

bolt to the cross member. Torque: 72.3–86.8 ft. lbs. (98.1–117.7 Nm).

6. Install the rear upper arm mounting nut to the knuckle. Torque: 72.3–86.8 ft. lbs. (98.1–117.7 Nm).
7. Install the wheel and the tire.

Rear Lower Arm

See Figures 64 through 66.

1. Raise the vehicle, and make sure it is securely supported.
2. Remove the rear wheel and tire.
3. Remove the lower arm bolt from the rear knuckle, while supporting the lower arm with a jack.

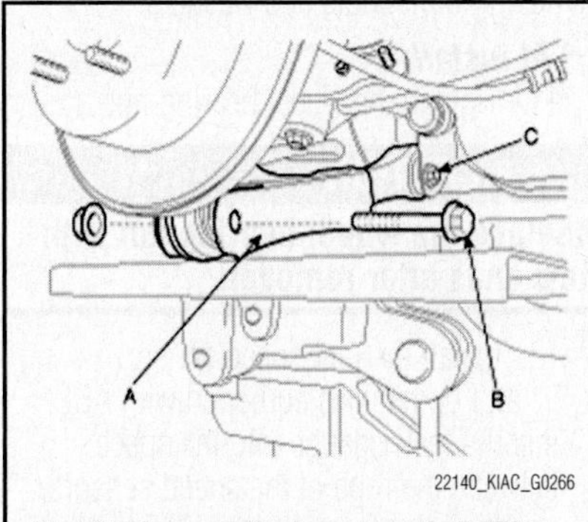

Fig. 64 Remove the lower arm bolt (B).

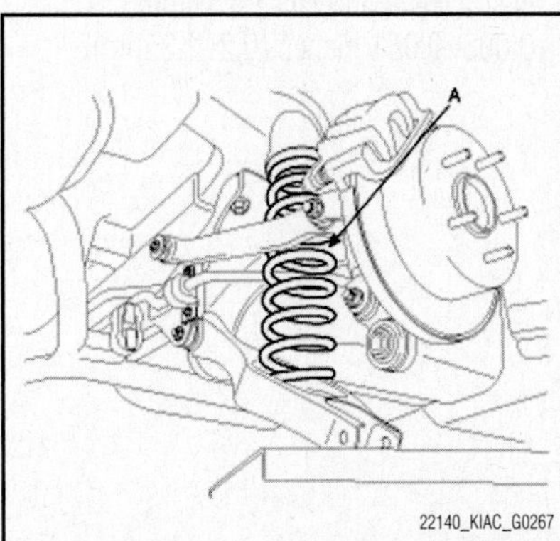

Fig. 65 Remove the coil spring (A), the lower seat, and the upper pad.

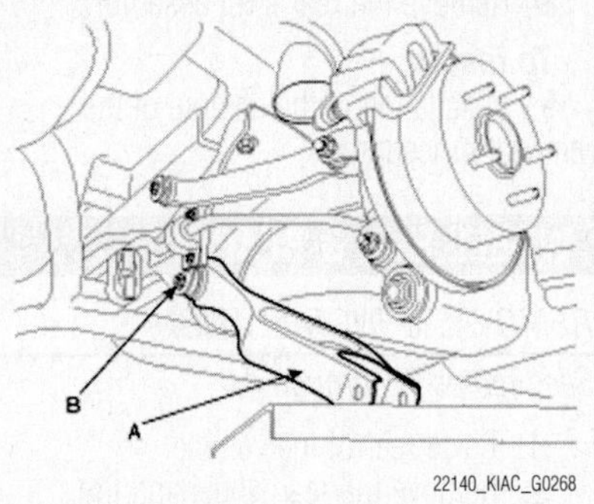

Fig. 66 Remove the lower arm (A) mounting bolts (B) from the cross member.

4. Loosen the lower arm bolt from the cross member.
5. Remove the coil spring, the lower seat, and the upper pad.
6. Remove the lower arm mounting bolts from the cross member.

To install:

7. Pre-tighten the lower arm mounting bolts to the cross member.
8. Install the coil spring, the lower seat, and the upper pad.
9. Install the lower arm bolt to the rear knuckle and the lower arm bolt to the cross member, while supporting the lower arm with a jack. Torque: 101.3–115.7 ft. lbs. 137.3–156.9).
10. Install the wheel and the tire.

Rear Assist Arm

See Figures 67 through 69.

1. Raise the vehicle, and make sure it is securely supported.
2. Remove the rear wheel and tire.
3. Remove the brake caliper mounting bolts, and hang the brake caliper assembly with wire.
4. Remove the rear assist arm ball joint self-locking nut and the cotter pin.
5. Remove the rear assist arm ball joint by using the special tool (09568-4A000).

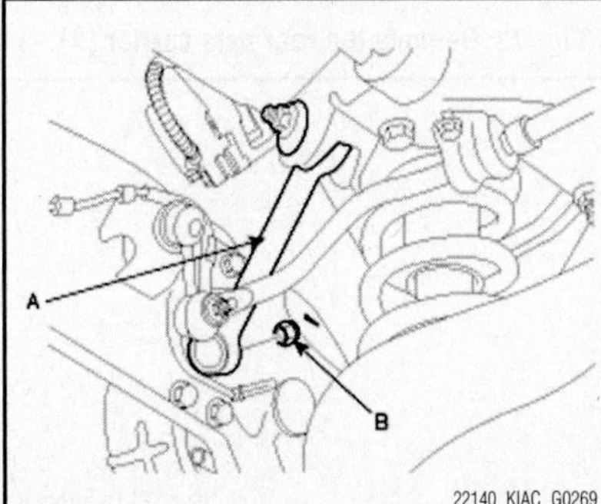

Fig. 67 Remove the rear assist arm (A) ball joint self-locking nut (B) and the cotter pin.

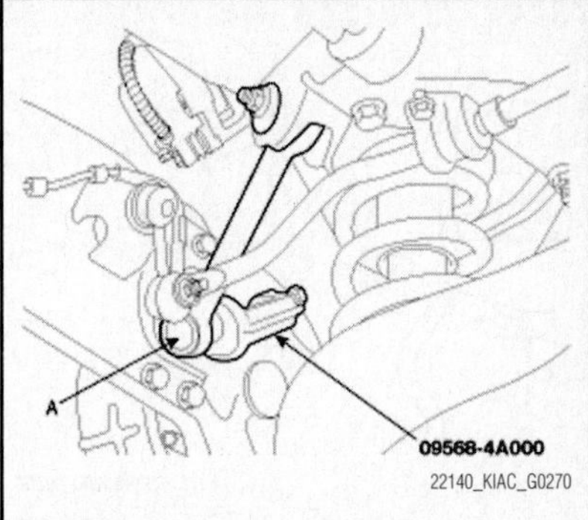

Fig. 68 Remove the rear assist arm ball joint (A) by using the special tool.

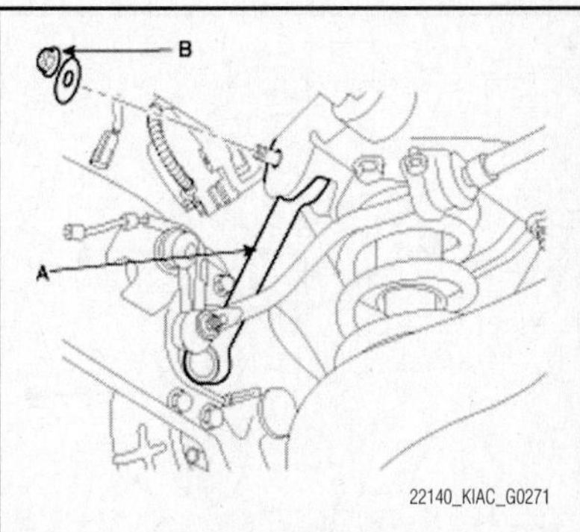

Fig. 69 Remove the rear assist arm (A) mounting nut (B) from the cross member.

6. Remove the rear assist arm mounting nut from the cross member.

To install:

7. Install the rear assist arm mounting nut to the cross member. Torque: 57.9–72.3 ft. lbs. (78.5–98.1 Nm).
8. Install the rear assist arm ball joint self-locking nut and the cotter pin. Torque: 32.5–39.8 ft. lbs. 44.1–53.9 Nm).
9. Install the brake caliper assembly mounting bolts. Torque: 36.2–43.4 ft. lbs. 49.0–58.8 Nm).
10. Install the wheel and the tire.

MACPHERSON STRUTS

REMOVAL & INSTALLATION

See Figures 70 and 71.

1. Remove the rear seat.
   a. Raise the rear cushion.
   b. Remove the mounting bolts between rear cushion and rear seatback.
   c. Remove the mounting bolts to both end parts of rear seatback.
2. Remove the rear strut upper mounting nuts.

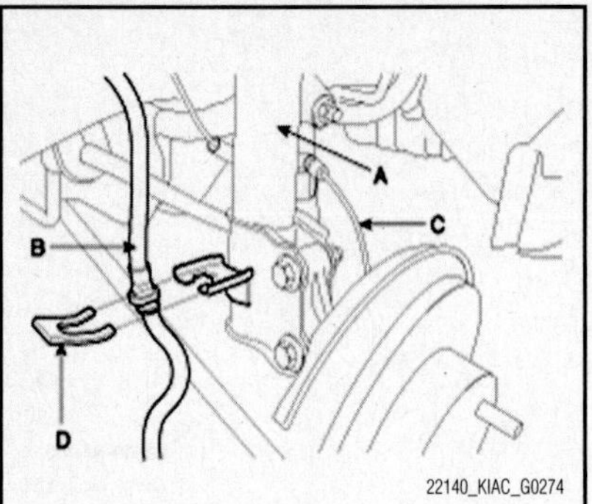

Fig. 70 Disconnect the brake hose (B) by removing the clip (D).

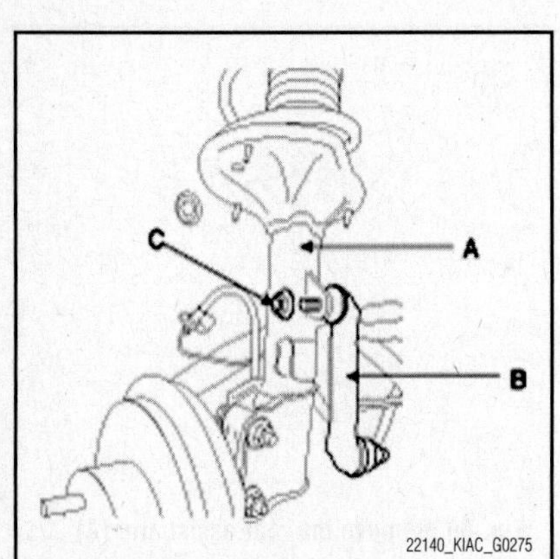
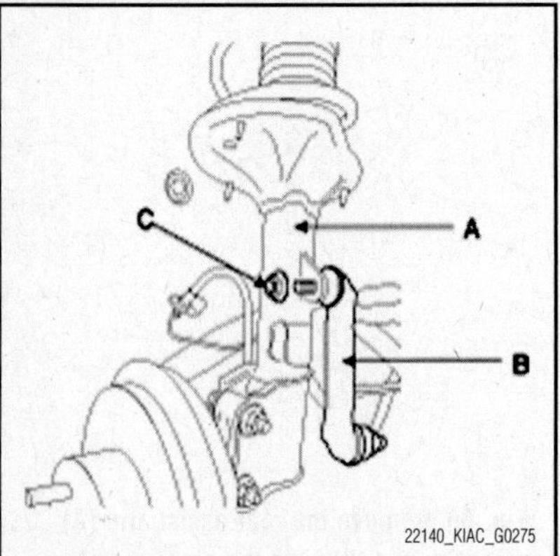

Fig. 71 Remove the stabilizer bar link (B) from the strut (A).

3. Raise the vehicle, and make sure it is securely supported.
4. Remove the rear wheel and tire.
5. Disconnect the brake hose (B) and wheel speed sensor wiring (C) from the rear strut (A).
   a. Disconnect the brake hose (B) by removing the clip (D)
   b. Disconnect the wheel speed sensor wiring.
6. After unfastening stabilizer bar link nut (C), remove the stabilizer bar link (B) from the strut (A).
7. Remove the lower strut mounting bolts.
8. Remove the rear strut assembly.

To install:

9. Installation is the reverse of the removal procedures.

WHEEL HUB & BEARING

REMOVAL & INSTALLATION

See Figures 72 through 74.

1. Raise rear of the vehicle.
2. Remove the rear wheel and tire.
3. Remove the brake caliper assembly bolt and the brake disk mounting screw.

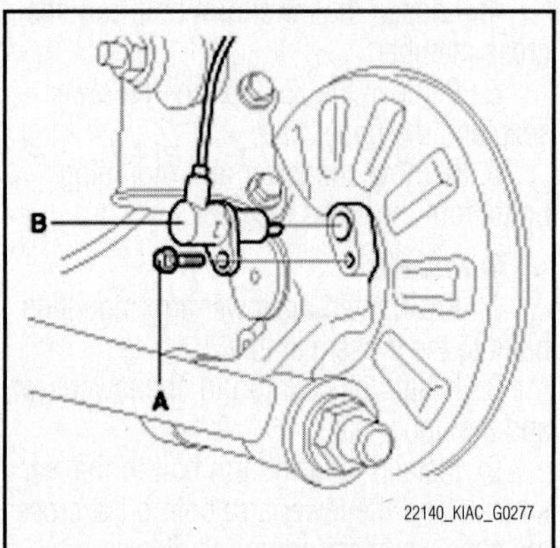

Fig. 72 Remove the bolt (A) and the remove the rear wheel speed sensor (B).

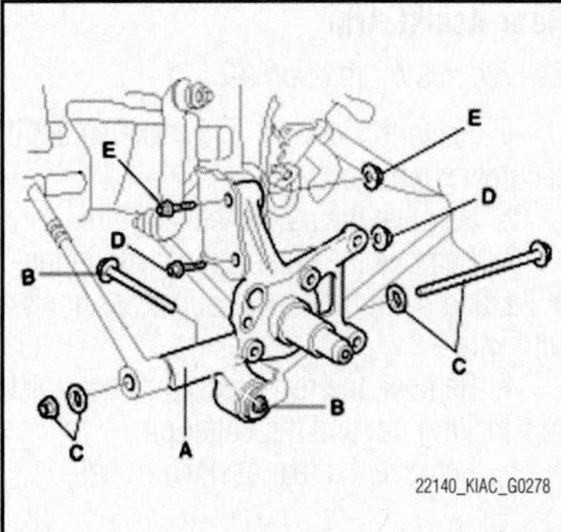

Fig. 73 Remove the rear axle carrier (A).

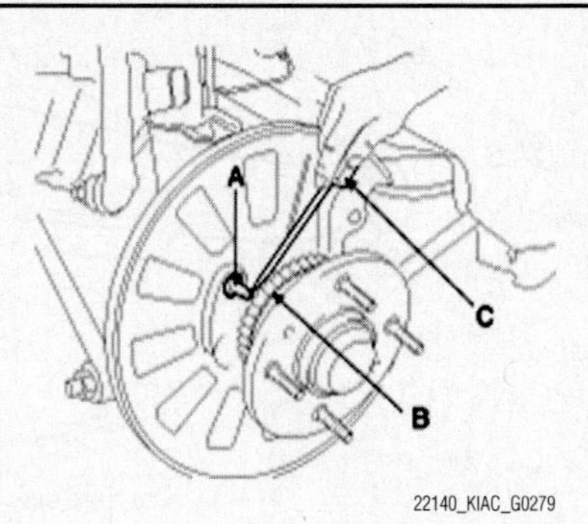

Fig. 74 For vehicles equipped with ABS, insert a feeler gauge (C).

4. Hang the brake caliper assembly tightly on a proper place with wire.
5. Remove the brake disc rotor.
6. Remove the bolt and the remove the rear wheel speed sensor.
7. Remove the wheel hub dust cap.
8. Remove the wheel bearing nut.
9. Remove the rear wheel hub assembly.
   a. Remove the dust cover.
   b. The rear hub assembly should not be disassembled.
   c. For vehicles equipped with ABS, care must be taken not to scratch or damage the teeth of the rotor.
10. Remove the rear axle carrier by removing bolts, nuts and washers.

To install:

11. Tighten the wheel bearing nut.

✱✱ CAUTION

Replace the wheel bearing nut with new ones after removal.

12. Install the rear speed sensor.
   a. For vehicles equipped with ABS, insert a feeler gauge into the space between the robe of the speed sensors and the rotor teeth surface. Tighten the speed sensors to the position where the clearance at all places is within 0.008–0.051 inches (0.2–1.3 mm).

KIA

Sportage

21

ENGINE PERFORMANCE & EMISSION CONTROLS 21-46

FUEL 21-49

GASOLINE FUEL INJECTION SYSTEM 21-49

HEATING & AIR CONDITIONING SYSTEM...21-51

SPECIFICATIONS AND MAINTENANCE CHARTS.....21-3

STEERING....21-52

SUSPENSION....21-54

FRONT SUSPENSION 21-54

REAR SUSPENSION 21-57

ENGINE AND VEHICLE IDENTIFICATION

Engine							Model Year	
Code ①	Liters (cc)	Cu. In.	Cyl.	Fuel Sys.	Engine Type	Eng. Mfg.	Code ②	Year
2	2.0 (1975)	120.5	4	MFI	DOHC	KIA	9	2009
4	2.0 (1975)	120.5	4	MFI	DOHC	KIA	A	2010
3	2.7 (2656)	162	6	MFI	DOHC	KIA		

MFI: Multi-port Fuel Injection

DOHC: Double Overhead Camshafts

① 8th digit of VIN

② 10th digit of VIN

37655_SPOR_C0001

GENERAL ENGINE SPECIFICATIONS

Year	Model	Engine Displacement Liters	Engine VIN	Net Horsepower @ rpm	Net Torque @ rpm (ft. lbs.)	Bore x Stroke (in.)	Com-pression Ratio	Oil Pressure @ rpm
2009	Sportage	2.0	2, 4	140@6000	136@4500	3.23x3.68	10:01	23 ①
		2.7	3	173@6000	178@4000	3.41x2.95	10:01	14 ①
2010	Sportage	2.0	2, 4	140@6000	136@4500	3.23x3.68	10:01	23 ①
		2.7	3	173@6000	178@4000	3.41x2.95	10:01	14 ①

① At 800 RPM idle minimum

37655_SPOR_C0002

ENGINE TUNE-UP SPECIFICATIONS

Year	Engine Displacement Liters	Engine VIN	Spark Plug Gap (in.)	Ignition Timing (deg.)		Fuel Pump (psi)	Idle Speed (rpm)		Valve Clearance (in.)	
				MT	AT		MT	AT	Intake	Exhaust
2009	2.0	2, 4	0.039-0.043	NA	NA	50	800	800	0.0047-0.0110	0.0079-0.0150
	2.7	3	0.039-0.043	NA	NA	50	800	800	HYD.	HYD.
2010	2.0	2, 4	0.039-0.043	NA	NA	50	800	800	0.0047-0.0110	0.0079-0.0150
	2.7	3	0.039-0.043	NA	NA	50	800	800	HYD.	HYD.

B: Before Top Dead Center

HYD: Hydraulic lash adjusters

NA: Information not available

37655_SPOR_C0003

CAPACITIES

Year	Model	Engine Displacement Liters	Engine VIN	Engine Oil with Filter (qts.)	Transmission (pts.) Manual	Transmission (pts.) Auto.	Transfer Case (pts.)	Drive Axle Front (pts.)	Drive Axle Rear (pts.)	Fuel Tank (gal.)	Cooling System (qts.)
2009	Sportage	2.0	2, 4	4.33	4.4	16.4	1.6	①	1.8	15.3	6.3
		2.7	3	5.07	—	16.4	1.6	①	1.8	14.3	8.2
2010	Sportage	2.0	2, 4	4.33	4.4	16.4	1.6	①	1.8	15.3	6.3
		2.7	3	5.07	—	16.4	1.6	①	1.8	14.3	8.2

NOTE: All capacities are approximate. Add fluid gradually and check to be sure a proper fluid level is obtained.

① Included in transaxle capacity

37655_SPOR_C0004

FLUID SPECIFICATIONS

Year	Model	Engine Displ. Liters	Engine Oil	Man. Trans.	Auto. Trans.	Drive Axle Front	Drive Axle Rear	Transfer Case	Power Steering Fluid	Brake Master Cylinder	Cooling System
2009	Sportage	2.0	5W20	75W/85	①	②	80W/90	80W/90	PSF-3	DOT 3	③
	Sportage	2.7	5W20	—	①	②	80W/90	80W/90	PSF-3	DOT 3	③
2010	Sportage	2.0	5W20	75W/90	①	②	80W/90	80W/90	PSF-3	DOT 3	③
	Sportage	2.7	5W20	—	①	②	80W/90	80W/90	PSF-3	DOT 3	③

DOT: Department Of Transpotation

NA: Not available

① Diamond ATF SP-III or SK ATF SP III

② Included with transaxle

③ Ethyene Glycol base

37655_SPOR_C0015

VALVE SPECIFICATIONS

Year	Engine Displacement Liters	Engine VIN	Seat Angle (deg.)	Face Angle (deg.)	Spring Test Pressure (lbs. @ in.)	Spring Installed Height (in.)	Stem-to-Guide Clearance (in.) Intake	Stem-to-Guide Clearance (in.) Exhaust	Stem Diameter (in.) Intake	Stem Diameter (in.) Exhaust
2009	2.0	2, 4	45-45.5	45-45.5	①	①	0.0008-0.0020	0.0014-0.0026	0.2348-0.2354	0.2343-0.2348
	2.7	3	45-45.5	45-45.5	48.4 @ 1.3780	1.3780	0.0008-0.0020	0.0012-0.0026	0.2350-0.2354	0.2340-0.2350
2010	2.0	2, 4	45-45.5	45-45.5	①	①	0.0008-0.0020	0.0014-0.0026	0.2348-0.2354	0.2343-0.2348
	2.7	3	45-45.5	45-45.5	48.4 @ 1.3780	1.3780	0.0008-0.0020	0.0012-0.0026	0.2350-0.2354	0.2340-0.2350

① Intake: 41.4 lbs. @ 1.5354 in.

Exhaust: 90.4 lbs. @ 1.2008 in.

Lengths given are installed height.

37655_SPOR_C0006

CAMSHAFT SPECIFICATIONS CHART

All measurements are given in inches.

Year	Engine Displ.	Engine VIN	Journal Dia.	Brg. Oil Clearance	Shaft End-play	Runout	Lobe Height Intake	Lobe Height Exhaust
2009	2.0	2, 4	1.1009-1.1016	0.0008-0.0024	0.0039-0.0079	0.0012	1.7527-1.7605	1.7487-1.7566
	2.7	3	1.0222-1.0228	0.0007-0.0024	0.0039-0.0059	0.0012	1.7303-1.7382	1.7303-1.7382
2010	2.0	2, 4	1.1009-1.1016	0.0008-0.0024	0.0039-0.0079	0.0012	1.7527-1.7605	1.7487-1.7566
	2.7	3	1.0222-1.0228	0.0007-0.0024	0.0039-0.0059	0.0012	1.7303-1.7382	1.7303-1.7382

37655_SPOR_C0007

CRANKSHAFT AND CONNECTING ROD SPECIFICATIONS

All measurements are given in inches.

Year	Engine Displacement Liters	Engine VIN	Crankshaft Main Brg. Journal Dia.	Crankshaft Main Brg. Oil Clearance	Crankshaft Shaft End-play	Crankshaft Thrust on No.	Connecting Rod Journal Diameter	Connecting Rod Oil Clearance	Connecting Rod Side Clearance
2009	2.0	2, 4	2.2418-2.2426	0.0011-0.0019	0.0024-0.0102	3	1.8898-1.8905	0.0009-0.0017	0.0039-0.0098
	2.7	3	2.4402-2.4409	0.0002-0.0009	0.0028-0.0098	3	1.8891-1.8898	0.0007-0.0014	0.0039-0.0098
2010	2.0	2, 4	2.2418-2.2426	0.0011-0.0019	0.0024-0.0102	3	1.8898-1.8905	0.0009-0.0017	0.0039-0.0098
	2.7	3	2.4402-2.4409	0.0002-0.0009	0.0028-0.0098	3	1.8891-1.8898	0.0007-0.0014	0.0039-0.0098

37655_SPOR_C0005

PISTON AND RING SPECIFICATIONS

All measurements are given in inches.

Year	Engine Displacement Liters	Engine VIN	Piston Clearance	Ring Gap Top Compression	Ring Gap Bottom Compression	Ring Gap Oil Control	Ring Side Clearance Top Compression	Ring Side Clearance Bottom Compression	Ring Side Clearance Oil Control
2009	2.0	2, 4	0.0008-0.0016	0.0091-0.0150	0.0130-0.0189	0.0079-0.0276	0.0016-0.0031	0.0012-0.0028	0.0024-0.0059
	2.7	3	0.0004-0.0012	0.0079-0.0138	0.0146-0.0205	0.0079-0.0276	0.0016-0.0031	0.0012-0.0028	SNUG
2010	2.0	2, 4	0.0008-0.0016	0.0091-0.0150	0.0130-0.0189	0.0079-0.0276	0.0016-0.0031	0.0012-0.0028	0.0024-0.0059
	2.7	3	0.0004-0.0012	0.0079-0.0138	0.0146-0.0205	0.0079-0.0276	0.0016-0.0031	0.0012-0.0028	SNUG

37655_SPOR_C0008

TORQUE SPECIFICATIONS

All readings in ft. lbs.

Year	Engine Displacement Liters	Engine VIN	Cylinder Head Bolts	Main Bearing Bolts	Rod Bearing Bolts	Crankshaft Damper Bolts	Flywheel Bolts	Manifold Intake	Manifold Exhaust	Spark Plugs	Oil Pan Drain Plug
2009	2.0	2, 4	①	②	36-38	123-130	87-94	12-17	31-40	15-22	29-32
	2.7	3	③	④	⑤	130-138	53-56	14-15	22-26	15-22	25-33
2010	2.0	2, 4	①	②	36-38	123-130	87-94	12-17	31-40	15-22	29-32
	2.7	3	③	④	⑤	130-138	53-56	14-15	22-26	15-22	25-33

① 10x99 bolts:
Step 1: 16.6-19.5 ft. lbs.
Step 2: + 60-65 degrees
Step 3: +60-65 degrees
12x151 bolts:
Step 1: 20-23 ft. lbs.
Step 2: + 60-65 degrees
Step 3: +60-65 degrees

② Step 1: 20-23 ft. lbs.
Step 2: + 60-65 degrees

③ Step 1: 18 ft. lbs.
Step 2: +58-62 degrees
Step 3: +43-47 degrees

④ M10 bolts:
Step 1: 20-24 ft. lbs
Step 2: +90-94 degrees
M8 bolts:
Step 1: 10-14 ft. lbs
Step 2: +90-94 degrees

⑤ Step 1: 12-15 ft. lbs.
Step 2: +90-94 degrees

37655_SPOR_C0009

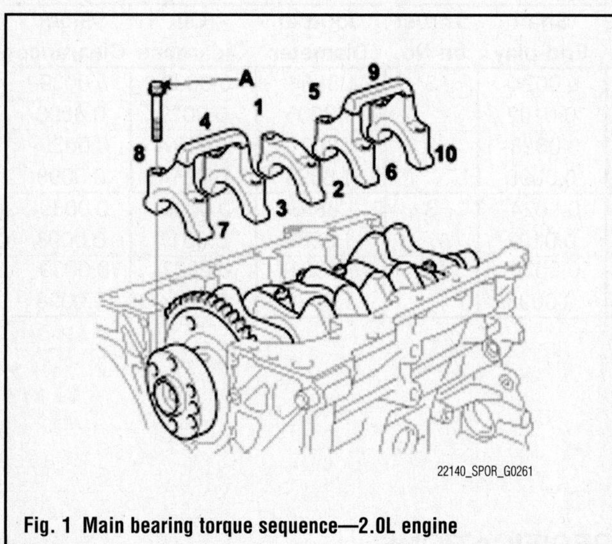

Fig. 1 Main bearing torque sequence—2.0L engine

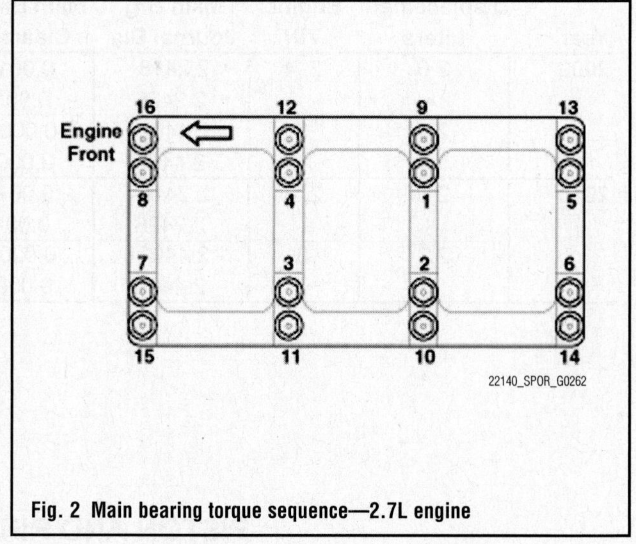

Fig. 2 Main bearing torque sequence—2.7L engine

WHEEL ALIGNMENT

Year	Model		Caster Range (+/-Deg.)	Caster Preferred Setting (Deg.)	Camber Range (+/-Deg.)	Camber Preferred Setting (Deg.)	Toe-in (in.)
2009	Sportage	F	0° 30'	3° 36'	-0.5° to 0.5°	0°	0+/-0.79
		R	—	—	0°30'	0°55'	4.6+3,-1
2010	Sportage	F	0° 30'	3° 36'	-0.5° to 0.5°	0°	0+/-0.79
		R	—	—	0°30'	0°55'	4.6+3,-1

37655_SPOR_C0010

TIRE, WHEEL AND BALL JOINT SPECIFICATIONS

Year	Model	OEM Tires Standard	OEM Tires Optional	Tire Pressures (psi) Front	Tire Pressures (psi) Rear	Wheel Size	Ball Joint Inspection	Lug Nut Torque (ft. lbs.)
2009	Sportage	P215/65R16	P235/60R16	30	30	6.5J	①	66-81
2010	Sportage	P215/65R16	P235/60R16	30	30	6.5J	①	66-81

OEM: Original Equipment Manufacturer

PSI: Pounds Per Square Inch

① Replace if any wear shown.

37655_SPOR_C0011

BRAKE SPECIFICATIONS

All measurements in inches unless otherwise noted

Year	Model		Brake Disc Original Thickness	Brake Disc Minimum Thickness	Brake Disc Maximum Run-out	Brake Drum ① Original Inside Diameter	Brake Drum ① Max. Wear Limit	Brake Drum ① Maximum Machine Diameter	Minimum Lining Thickness	Caliper Mounting Bolts (ft. lbs.)	Adaptor Plate Bolts (ft. lbs.)
2009	Sportage	F	1.020	0.961	0.001	—	—	—	0.079	16-23	58-72
		R	0.390	0.315	0.002	9.00	9.079	—	0.079	16-23	—
2010	Sportage	F	1.020	0.961	0.001	—	—	—	0.079	16-23	58-72
		R	0.390	0.315	0.002	9.00	9.079	—	0.079	16-23	—

① Parking brake

37655_SPOR_C0012

SCHEDULED MAINTENANCE INTERVALS - NORMAL SERVICE

Kia Sportage

TO BE SERVICED	TYPE OF SERVICE	VEHICLE MILEAGE INTERVAL (x1000)													
		7.5	15	22.5	30	37.5	45	52.5	60	67.5	75	82.5	90	97.5	105
Accessory drive belt 2.0L ①	I	✓	✓	✓	✓	✓	✓	✓	✓	✓	✓	✓	✓	✓	✓
Accessory drive belt 2.7L ①	I				✓				✓				✓		
Air cleaner filter	I	✓	✓	✓		✓	✓	✓		✓	✓	✓		✓	✓
Air cleaner filter	R				✓				✓				✓		
A/C filter	R	Every 10,000 miles													
A/C system	I	Inspect refrigerant amount & operation annually													
Automatic transaxle fluid	I		✓		✓		✓		✓		✓		✓		✓
Ball joints	S/I				✓				✓				✓		
Battery condition	S/I		✓		✓		✓		✓		✓		✓		✓
Brake/Clutch fluid	I		✓		✓		✓		✓		✓		✓		✓
Brake lines & connections	S/I		✓		✓		✓		✓		✓		✓		✓
Chassis/body fasteners	S/I				✓				✓				✓		
Disc brakes	S/I	✓	✓	✓	✓	✓	✓	✓	✓	✓	✓	✓	✓	✓	✓
Driveshaft, boots & U-joints	S/I		✓		✓		✓		✓		✓		✓		✓
Engine coolant	R	Replace at 60,000 miles, then every 30,000 mile thereafter													
Engine oil & filter	R	Every 7,500 miles or 12 months													
EVAP canister filter	I				✓				✓				✓		
Exhaust system & heat	S/I		✓		✓		✓		✓		✓		✓		✓
Fuel filter	R					✓					✓				
Fuel tank, cap, lines & hoses	S/I	✓	✓	✓	✓	✓	✓	✓	✓	✓	✓	✓	✓	✓	✓
Halfshafts and boots	S/I		✓		✓		✓		✓		✓		✓		✓
Locks & hinges	L	✓	✓	✓	✓	✓	✓	✓	✓	✓	✓	✓	✓	✓	✓
Manual transaxle fluid ②	I					✓					✓				
Parking brake	I		✓		✓		✓		✓		✓		✓		✓
Power steering fluid	I	✓	✓	✓	✓	✓	✓	✓	✓	✓	✓	✓	✓	✓	✓
Rear differential fluid ②	S/I					✓					✓				
Spark plugs 2.0L platinum	R								✓						
Spark plugs 2.7L iridium	R	Every 100,000 miles or 10 years													
Steering operation, linkage & hoses	S/I	✓	✓	✓	✓	✓	✓	✓	✓	✓	✓	✓	✓	✓	✓
Timing belt	S/I				✓								✓		
Timing belt	R								✓						
Tires	Rotate	✓	✓	✓	✓	✓	✓	✓	✓	✓	✓	✓	✓	✓	✓
Transfer case fluid ②	I					✓					✓				
Vacuum and crankcase hoses	S/I				✓				✓				✓		
Valve clearance - 2.0L	I	Every 60,000 miles or 4 years													
Water pump	I	Inspect when replacing drive belt or timing belt													

R: Replace S/I: Inspect and service, if needed L: Lubricate

① Replace when excessive cracks occur or tension is not maintained

② Replace fluid whenever submerged in water

37655_SPOR_C0013

SCHEDULED MAINTENANCE INTERVALS - FREQUENT/SEVERE SERVICE

Kia Sportage

TO BE SERVICED	TYPE OF SERVICE	VEHICLE MILEAGE INTERVAL (x1000)													
		3	6	9	12	15	18	21	24	27	30	33	36	39	42
Accessory drive belt 2.0L ①	I	✓	✓	✓	✓	✓	✓	✓	✓	✓	✓	✓	✓	✓	✓
Accessory drive belt 2.7L ①	I						✓						✓		
Air cleaner filter	I	Inspect more frequently depending on condition and if necessary, replace													
A/C filter	R	Every 10,000 miles													
A/C system	I	Inspect refrigerant amount & operation annually													
Automatic transaxle fluid	R	Replace every 60,000 miles													
Ball joints	S/I				✓				✓				✓		
Battery condition	S/I		✓		✓		✓		✓		✓		✓		✓
Brake/Clutch fluid	I	✓	✓	✓	✓	✓	✓	✓	✓	✓	✓	✓	✓	✓	✓
Brake lines & connections	S/I	✓	✓	✓	✓	✓	✓	✓	✓	✓	✓	✓	✓	✓	✓
Chassis/body fasteners	S/I				✓				✓				✓		
Disc brakes	S/I	✓	✓	✓	✓	✓	✓	✓	✓	✓	✓	✓	✓	✓	✓
Driveshaft, boots & U-joints	S/I		✓		✓		✓		✓		✓		✓		✓
Engine coolant	R	Replace at 60,000 miles, then every 30,000 mile thereafter													
Engine oil & filter	R	Every 3,750 miles or 6 months													
EVAP canister filter	I				✓				✓				✓		
Exhaust system & heat	S/I		✓		✓		✓		✓		✓		✓		✓
Fuel filter	R	Replace every 37,500 miles													
Fuel tank, cap, lines & hoses	S/I	✓	✓	✓	✓	✓	✓	✓	✓	✓	✓	✓	✓	✓	✓
Halfshafts and boots	S/I	Inspect every 7,500 miles or 6 months													
Locks & hinges	L	✓	✓	✓	✓	✓	✓	✓	✓	✓	✓	✓	✓	✓	✓
Manual transaxle fluid ②	R	Replace every 75,000 miles													
Parking brake	I	✓	✓	✓	✓	✓	✓	✓	✓	✓	✓	✓	✓	✓	✓
Power steering fluid	I	✓	✓	✓	✓	✓	✓	✓	✓	✓	✓	✓	✓	✓	✓
Rear differential fluid ②	R	Replace every 75,000 miles													
Spark plugs	I	Inspect more frequently depending on condition and if necessary, replace													
Steering operation, linkage & hoses	S/I	✓	✓	✓	✓	✓	✓	✓	✓	✓	✓	✓	✓	✓	✓
Timing belt	S/I				✓								✓		
Timing belt	R								✓						
Tires	S/I	✓	✓	✓	✓	✓	✓	✓	✓	✓	✓	✓	✓	✓	✓
Transfer case fluid ②	R	Replace every 75,000 miles													
Vacuum and crankcase hoses	S/I				✓				✓				✓		
Valve clearance - 2.0L	I	Every 60,000 miles or 4 years													
Water pump	I	Inspect when replacing drive belt or timing belt													

R: Replace S/I: Inspect and service, if needed L: Lubricate

① Replace when excessive cracks occur or tension is not maintained

② Replace fluid whenever submerged in water

If a vehicle is operated under any of the following conditions it is considered severe service

- **Repeated short distance driving.**
- **Extensive use of brakes or driving in mountainous areas.**
- **Driving for a prolonged period in cold temperatures and/or extremely humid conditions.**
- **Prolonged idling (vehicle operation in stop and go traffic).**
- **Operating on rough, muddy, unpaved, dusty or salt-covered roads.**
- **Police, taxi, delivery usage or trailer towing usage.**
- **Driving more than 50% heavy city traffic in extremely hot (over 90°F) conditions**

37655_SPOR_C0014

BRAKES — INFORMATION AND PRECAUTIONS

ANTI-LOCK SYSTEMS

• Certain components within the ABS system are not intended to be serviced or repaired individually.

• Do not use rubber hoses or other parts not specifically specified for and ABS system. When using repair kits, replace all parts included in the kit. Partial or incorrect repair may lead to functional problems and require the replacement of components.

• Lubricate rubber parts with clean, fresh brake fluid to ease assembly. Do not use shop air to clean parts; damage to rubber components may result.

• Use only DOT 3 brake fluid from an unopened container.

• If any hydraulic component or line is removed or replaced, it may be necessary to bleed the entire system.

• A clean repair area is essential. Always clean the reservoir and cap thoroughly before removing the cap. The slightest amount of dirt in the fluid may plug an orifice and impair the system function. Perform repairs after components have been thoroughly cleaned; use only denatured alcohol to clean components. Do not allow ABS components to come into contact with any substance containing mineral oil; this includes used shop rags.

• The Anti-Lock control unit is a microprocessor similar to other computer units in the vehicle. Ensure that the ignition switch is **OFF** before removing or installing controller harnesses. Avoid static electricity discharge at or near the controller.

• If any arc welding is to be done on the vehicle, the control unit should be unplugged before welding operations begin.

DISC AND DRUM SYSTEMS

⁂ CAUTION

Dust and dirt accumulating on brake parts during normal use may contain asbestos fibers from production or aftermarket brake linings. Breathing excessive concentrations of asbestos fibers can cause serious bodily harm. Exercise care when servicing brake parts. Do not sand or grind brake lining unless equipment used is designed to contain the dust residue. Do not clean brake parts with compressed air or by dry brushing. Cleaning should be done by dampening the brake components with a fine mist of water, then wiping the brake components clean with a dampened cloth. Dispose of cloth and all residue containing asbestos fibers in an impermeable container with the appropriate label. Follow practices prescribed by the Occupational Safety and Health Administration (OSHA) and the Environmental Protection Agency (EPA) for the handling, processing, and disposing of dust or debris that may contain asbestos fibers.

BRAKES — BLEEDING THE BRAKE SYSTEM

BLEEDING PROCEDURE

BLEEDING PROCEDURE

1. Before servicing the vehicle, refer to the Precautions Section.

⁂ WARNING

Do not reuse the drained fluid. Always use Genuine DOT3 or DOT 4 Brake Fluid. Using unapproved brake fluid can cause corrosion and decrease the life of the system. Make sure no dirt of other foreign matter is allowed to contaminate the brake fluid. Do not spill brake fluid on the vehicle, it may damage the paint; if brake fluid does contact the paint, wash it off immediately with water.

➡The reservoir on the master cylinder must be at the MAX (upper) level mark at the start of bleeding procedure and checked after bleeding each brake caliper. Add fluid as required.

2. Have someone slowly pump the brake pedal several times, then apply pressure.
3. Connect a length of clear plastic tube to the bleeder nipple and place the other end in a jar half full of clean brake fluid.
4. Loosen the right-rear brake bleed screw to allow air to escape from the system. Then tighten the bleed screw securely.
5. Repeat the procedure for each wheel in the sequence shown until air bubbles no longer appear in the fluid.
6. Refill the master cylinder reservoir to MAX (upper) level line.

BLEEDING THE ABS SYSTEM

The ABS brake system is bled in the usual fashion with no special procedures required. Refer to the bleeding procedure described in this section. Make certain the master cylinder reservoir is filled before the bleeding is begun and check the level frequently.

BRAKES — ANTI-LOCK BRAKE SYSTEM (ABS)

WHEEL SPEED SENSORS

REMOVAL & INSTALLATION

Front

See Figure 3.

1. Remove the front wheel speed sensor mounting bolt (A).
2. Remove the front wheel guard (B), after removing the mud guard (A).
3. Remove the front wheel speed sensor after disconnecting the wheel speed sensor connector (A).

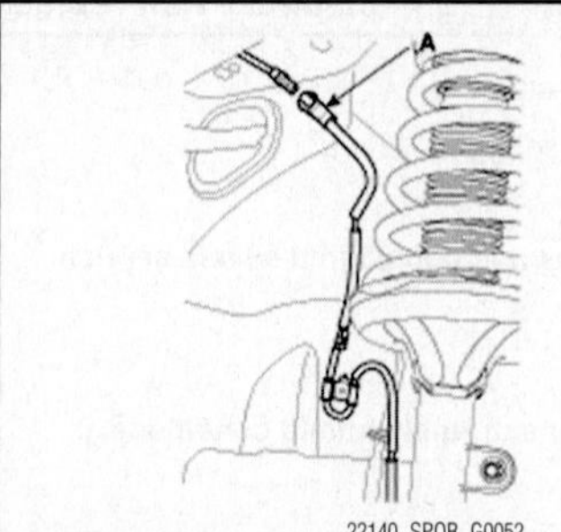

Fig. 3 Remove the front wheel speed sensor after disconnecting the wheel speed sensor connector (A)

Rear

1. Remove the rear wheel speed sensor mounting bolt (A).
2. Remove the rear seat side pad and disconnect the rear wheel speed sensor connector (A).

WHEEL SPEED SENSOR RINGS (TOOTHED RINGS)

REMOVAL & INSTALLATION

Refer to the Halfshaft procedure in Drive Train.

BRAKE CALIPER

REMOVAL & INSTALLATION

1. Raise and safely support the front of the vehicle.
2. Remove the front wheel.
3. Disconnect the brake fluid hose.
4. Remove the caliper mounting bolts.
5. Remove the brake caliper.

To install:

6. Install the brake caliper. Tighten the mounting bolts to 16–24 ft. lbs. (22–32 Nm).
7. Install the brake fluid hose. Tighten the hose fitting to 12–14 ft. lbs. (17–20 Nm).
8. Bleed the brake system.
9. Install the front wheel.
10. Before attempting to move the vehicle, pump the brake pedal to seat the pads against the rotors. Make sure the vehicle has a firm brake pedal. Check the level of the brake fluid and add DOT 3 brake fluid if necessary.

DISC BRAKE PADS

REMOVAL & INSTALLATION

See Figures 4 through 8.

1. Raise and safely support the front of the vehicle.
2. Remove the front wheel.
3. Remove the guide bolt (B), lift the caliper assembly (A) up and suspend it with a wire.
4. Remove the following parts from the caliper support:

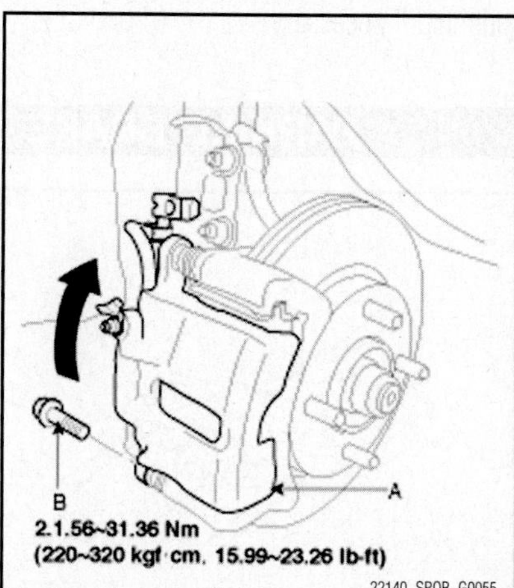

Fig. 4 Remove the guide bolt (B), lift the caliper assembly (A)

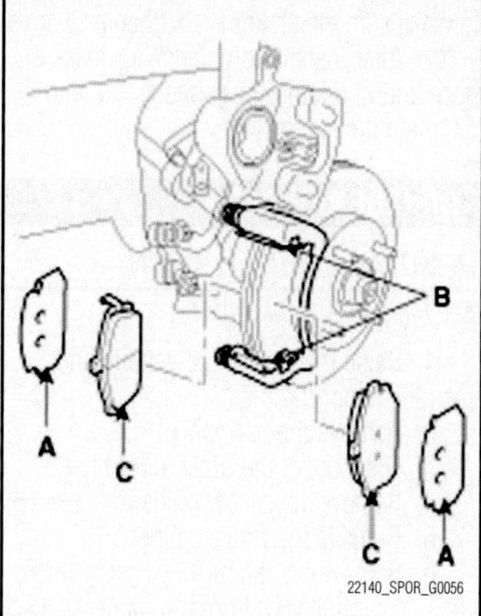

Fig. 5 Remove pad shim (A), pad retainer (B) and pad assembly (C)

- Pad shims (A)
- Pad retainer (B)
- Pad assembly (C)

To install:

5. Install the pad retainers (A) on the caliper bracket.
6. Check the foreign material at the pad shims and the back of the pads.

➡Contaminated brake discs or pads reduce stopping ability. Keep grease off the discs and pads.

7. Install the brake pads (B) and pad shims (A) correctly. Install the pad with the wear indicator (C) on the inside.

➡If you are reusing the pads, always reinstall the brake pads in their original positions to prevent a momentary loss of braking efficiency.

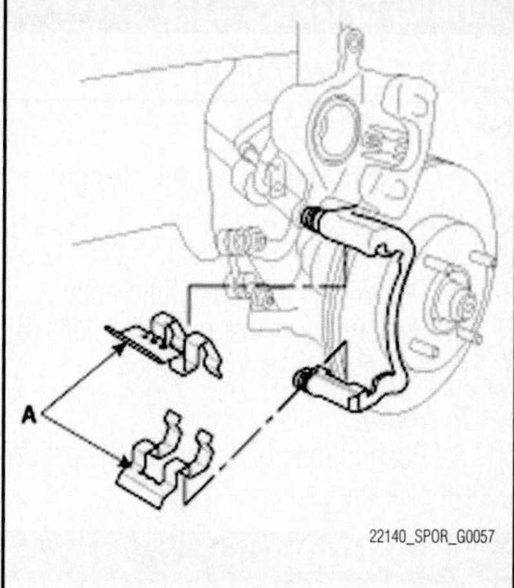

Fig. 7 Install the pad retainers (A) on the caliper bracket

8. Push in the piston (A) so that the caliper will fit over the pads. Make sure that the piston boot is in position to prevent damaging it when pivoting the caliper down.
9. Pivot the caliper down into position. Being careful not to damage the pin boot, install the guide rod bolt (B). Tighten to 15.99–23.26 ft. lbs. (21.56–31.36 Nm).
10. Depress the brake pedal several times to make sure the brakes work, then test-drive.

➡Engagement of the brake may require a greater pedal stroke immediately after the brake pads have been replaced as a set. Several applications of the brake will restore the normal pedal stroke.

11. After installation, check for leaks at hose and line joints or connections, and retighten if necessary.

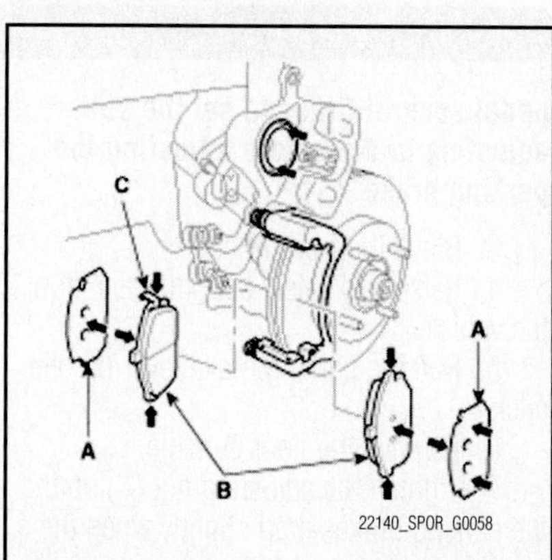

Fig. 6 Install the brake pads (B) and pad shims (A)

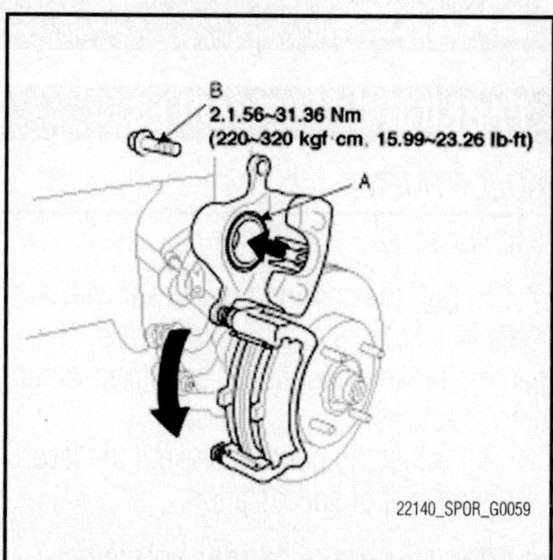

Fig. 8 Caliper piston (A) and guide rod bolt (B)

BRAKE CALIPER

REMOVAL & INSTALLATION

See Figure 9.

1. Raise the rear of the vehicle and securely support.
2. Remove the rear wheel.
3. Disconnect the brake fluid hose.
4. Remove the caliper guide rod bolts (B).
5. Remove the brake caliper.

To install:

6. Push in the piston so that the caliper will fit over the pads.

✲✲ CAUTION

Make sure that the piston boot is in position to prevent damaging it when pivoting the caliper down.

7. Pivot caliper down into position.
8. Install the guide rod bolt. Tighten to 16–23 ft. lbs. (22–31 Nm).
9. Reconnect the brake fluid line. Tighten the hose fitting to 12–14 ft. lbs. (17–20 Nm).

Fig. 9 Remove the caliper guide rod bolts (B)

10. Depress the brake pedal several time to make sure the brakes work, then test-drive.
11. After installation, check for leaks at hose and line joints or connections, and retighten if necessary.

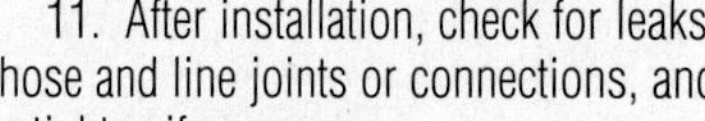

DISC BRAKE PADS

REMOVAL & INSTALLATION

See Figures 6, 9 through 11.

1. Raise the rear of the vehicle and securely support.
2. Remove the rear wheel.
3. Disconnect the brake fluid hose.
4. Remove the caliper guide rod bolts (B).
5. Remove the brake caliper.
6. Remove pad shim(A), pad retainer(B) and pad assembly(C) from the caliper bracket.

To install:

7. Install the pad retainers(A) on the caliper bracket.
8. Check the foreign material at the pad shims and the back of the pads.

➡Contaminated brake discs or pads reduce stopping ability. Keep grease off the discs and pads.

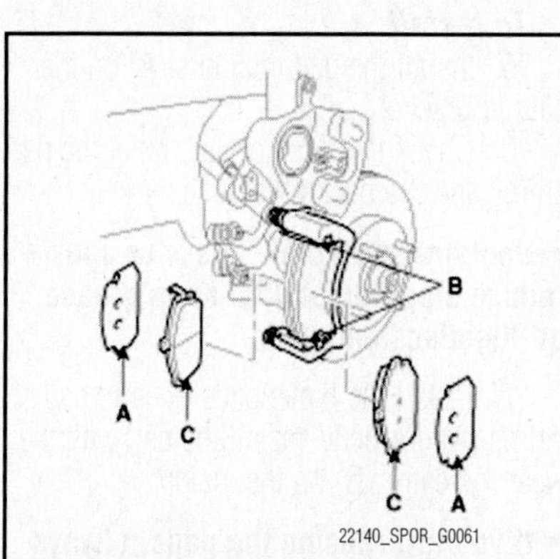

Fig. 10 Remove pad shim(A), pad retainer(B) and pad assembly(C)

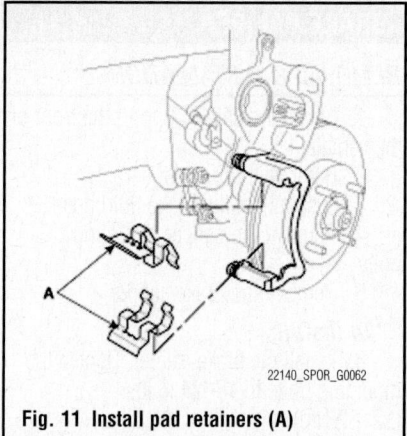

Fig. 11 Install pad retainers (A)

9. Install the brake pads. Install the pad with the wear indicator on the inside.

➡If you are reusing the pads, always reinstall the brake pads in their original positions to prevent a momentary loss of braking efficiency.

10. Push in the piston so that the caliper will fit over the pads.

✲✲ CAUTION

Make sure that the piston boot is in position to prevent damaging it when pivoting the caliper down.

11. Pivot caliper down into position.
12. Install the guide rod bolt. Tighten to 15.99–23.26 ft. lbs. (21.56–31.36 Nm).
13. Depress the brake pedal several time to make sure the brakes work, then test-drive.
14. After installation, check for leaks at hose and line joints or connections, and retighten if necessary.

PARKING BRAKE CABLES

ADJUSTMENT

See Figure 12.

1. Pull the parking brake lever with force to fully apply the parking brake. The parking brake lever should be locked within the 7–8 clicks.
2. Adjust the parking brake if the lever clicks are out of specification.

➡After rear brake caliper servicing, loosen the parking brake adjusting nut, start the engine and depress the brake pedal several times to set the self-adjusting brake before adjusting the parking brake.

3. Block the front wheels.
4. Raise and safely support the rear of the vehicle.
5. Pull the parking brake lever up one click.
6. Remove the floor console.
7. Tighten the adjusting nut (A) until the parking brakes drag slightly when the rear wheels are turned.
8. Release the parking brake lever completely, and check if parking brakes are not

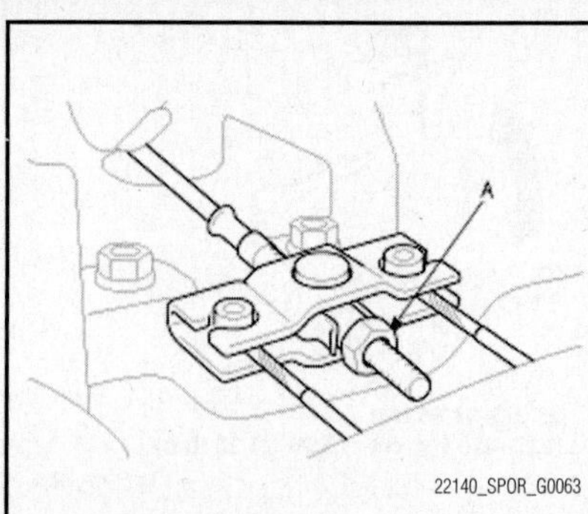

Fig. 12 Tighten the parking brake adjusting nut (A)

dragged when the rear wheels are turned. Readjust if necessary.

9. Make sure that the parking brakes are fully applied when the parking brake lever is pulled up completely.

10. Reinstall the floor console.

PARKING BRAKE SHOES

REMOVAL & INSTALLATION

1. Remove rear disc caliper assembly.

2. Before removing the brake disc, chalk both sides of the screw.

➡Reduce the shoe gap by turning the adjuster with appropriate tool.

3. After turning the pin to coincide with hole of spring cap, remove the shoe hold spring.

4. Remove the lower return spring.

5. Remove the parking brake cable mounting nuts.

6. Remove the parking brake shoes.

To install:

7. Install the upper return spring and brake shoes

8. Turn the adjuster in clockwise direction and install.

9. Install the lower return spring.

10. Install the shoe hold spring with pliers.

11. Install the disc brake and align the mark while tightening the screw.

ADJUSTMENT

1. Raise the rear of the vehicle, and make sure it is securely supported.

2. Remove the rear wheel and tire.

3. After removing the plug from the disc, rotate the toothed wheel with a screwdriver until the disc does not rotate.

4. Back off the toothed wheel by 5 notches.

CHASSIS ELECTRICAL

AIR BAG (SUPPLEMENTAL RESTRAINT SYSTEM)

GENERAL INFORMATION

⁂ CAUTION

These vehicles are equipped with an air bag system. The system must be disarmed before performing service on, or around, system components, the steering column, instrument panel components, wiring and sensors. Failure to follow the safety precautions and the disarming procedure could result in accidental air bag deployment, possible injury and unnecessary system repairs.

PRECAUTIONS

⁂ CAUTION

Disconnect and isolate the battery negative cable before beginning any airbag system component diagnosis, testing, removal, or installation procedures. Allow system capacitor to discharge for three minutes before beginning any component service. This will disable the airbag system. Failure to disable the airbag system may result in accidental airbag deployment, personal injury, or death.

DISARMING THE SYSTEM

1. Before servicing the vehicle, refer to the Precautions Section.

2. Turn the ignition switch to the **LOCK** position.

3. Disconnect the negative battery cable.

4. Wait 10 minutes for the battery back-up power to discharge.

ARMING THE SYSTEM

1. Before servicing the vehicle, refer to the Precautions Section.

2. Connect the negative battery cable.

3. Turn the ignition switch **ON**.

4. Verify that the air bag indicator illuminates for 4–8 seconds, then goes off.

CLOCKSPRING CENTERING

Prior to installing the clock spring, align the mating mark and "NEUTRAL" position indicator of the clock spring, and, after turning the front wheels to the straight-ahead position, install the clock spring to the column switch. If the mating mark of the clock spring is not properly aligned, the steering wheel may not completely rotate during a turn, or the flat cable within the clock spring may be severed, obstructing normal operation of the SRS and possibly leading to serious injury to the vehicle's driver.

DRIVE TRAIN

CLUTCH DRIVEN DISC & PRESSURE PLATE

REMOVAL & INSTALLATION

See Figures 13 through 15.

1. Remove the manual transaxle assembly.

2. Insert the special tool (09411-11000) in the clutch disc to prevent the disc from falling.

3. Remove the bolts which attach the clutch cover to the flywheel in a star pattern. Loosen the bolts in succession, one or two turns at a time, to avoid bending the cover flange.

➡Do not clean the clutch disc or the release bearing with cleaning solvent.

4. Remove the release fork shaft and bushing.

To install:

5. Apply multipurpose grease to the spline of the disc.

⁂ CAUTION

When installing the clutch, apply grease to each part, but be careful not to apply excessive grease. It can cause clutch slippage and shudder.

Fig. 13 Insert the special tool (09411-11000) in the clutch disc to prevent the disc from falling

6. Install the clutch disc assembly to the flywheel using the special tool (09411-11000).

7. Install the clutch cover assembly to the flywheel and temporarily tighten the bolts one or two steps at a time in a star pattern. Tighten to Clutch cover bolt: 11–16 ft. lbs. (15–22 Nm).

8. Align the bearing (A) to the release fork (B) and then install it to the sleeve of the housing.

CAUTION

Apply multipurpose grease to the bearing sleeve, contact point of the release fork (B) and the bushing inner surface (C).

9. Install the release lever to the release fork.

10. Install the manual transaxle assembly to the engine.

CAUTION

If the transaxle assembly is installed to the engine without performing this step, the release bearing can be separated, as the release fork rotates freely.

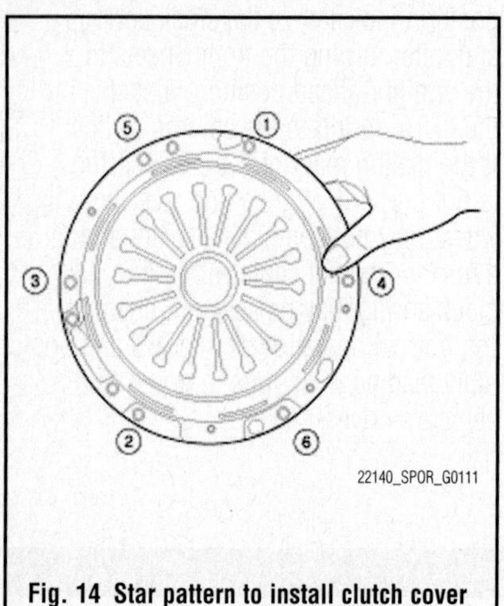

Fig. 14 Star pattern to install clutch cover

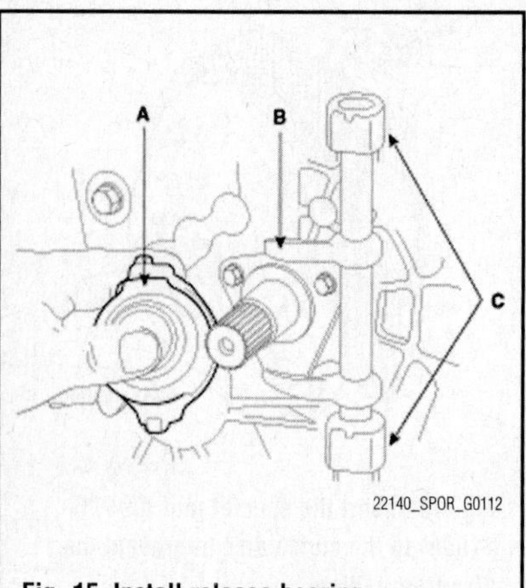

Fig. 15 Install release bearing

HYDRAULIC SYSTEM BLEEDING

BLEEDING PROCEDURE

See Figures 16 and 17.

CAUTION

Use only SAE J1703 (DOT 3 or DOT 4). Avoid mixing different brands of fluid.

1. Loosen the bleeder screw (B) at the clutch release cylinder (A).

2. Pump the clutch pedal slowly until all air is expelled.

3. Hold the clutch pedal down until the bleeder is retightened.

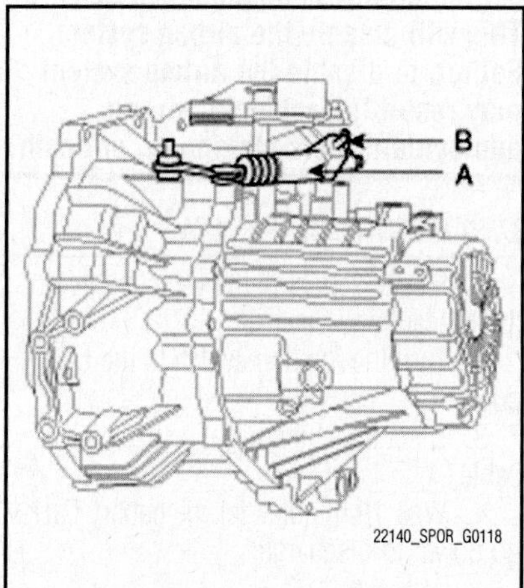

Fig. 16 Loosen the bleeder screw (B) at the clutch release cylinder (A)

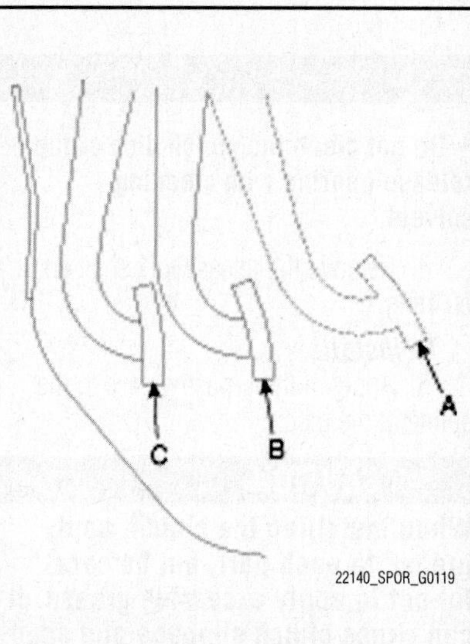

Fig. 17 Clutch pedal positioning for bleeding operation

4. Refill the clutch master cylinder with the specified fluid.

CAUTION

The rapidly-repeated operation of the clutch pedal in B-C range may disrupt the release cylinder's position. During the bleeding operation, press the clutch pedal to the floor after it returns to the "A" point.

FRONT AXLE SHAFT, BEARING & SEAL

REMOVAL & INSTALLATION

See Figures 18 through 27.

1. Raise the front of the vehicle, and make sure it is securely supported.

2. Remove the front wheels and tires.

3. Remove the split pin (A), the castle nut (B) and the washer (C) from the front hub.

4. Remove the wheel speed sensor (B) from the knuckle (A).

5. Remove the split pin (A) and the castle nut (B) from the tie rod end.

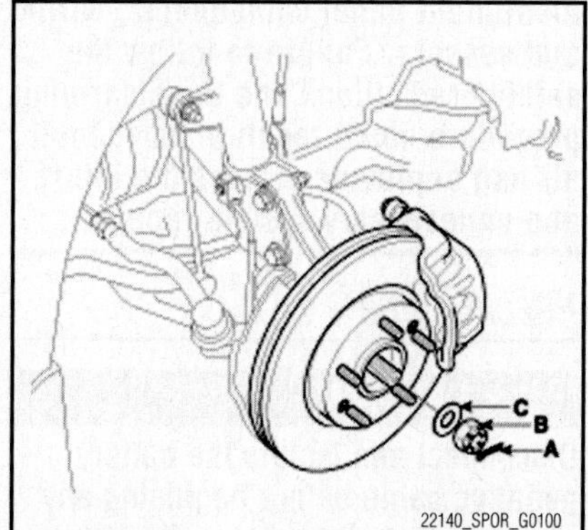

Fig. 18 Remove the split pin (A), the castle nut (B) and the washer (C) from the front hub

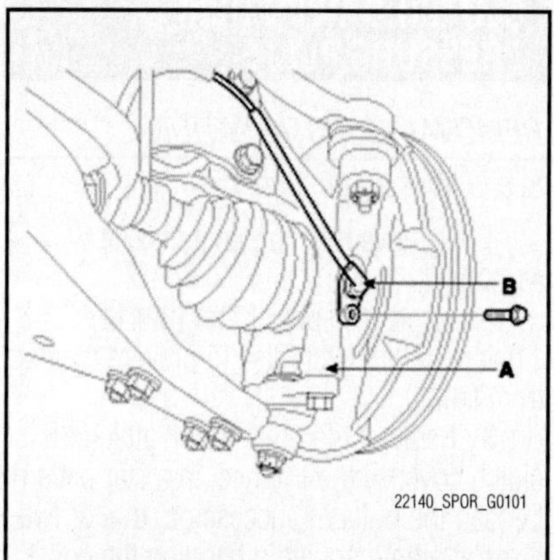

Fig. 19 Remove the wheel speed sensor (B) from the knuckle (A)

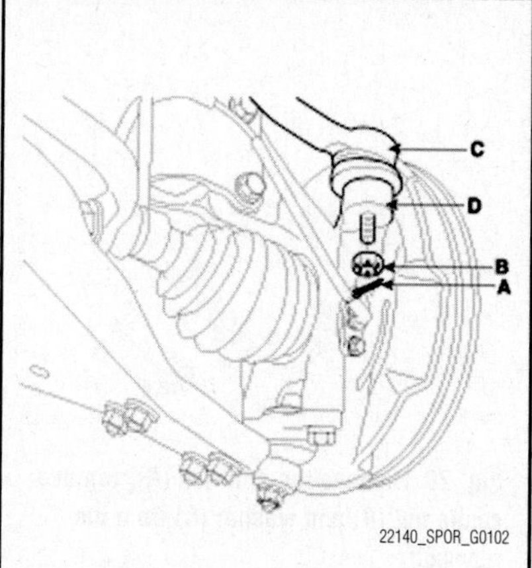

Fig. 20 Remove the split pin (A) and the castle nut (B) from the tie rod end

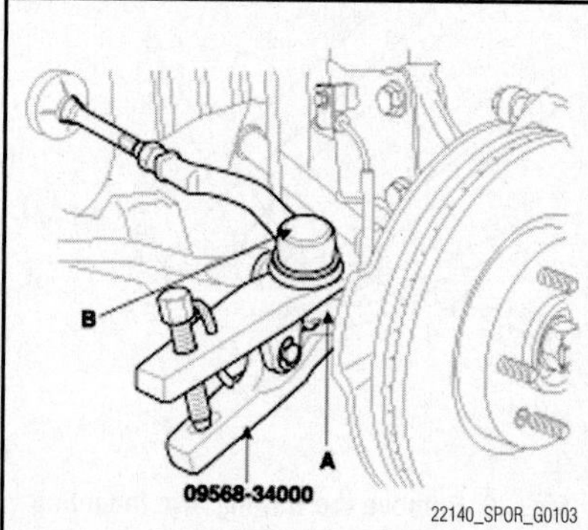

Fig. 21 Disconnect the ball joint (B) from knuckle (A) using the special tool (09568-34000)

6. Disconnect the ball joint (B) from knuckle (A) using the special tool (09568-34000).

⁂ CAUTION

Apply a few drops of oil to the special tool. (Boot contact part)

7. Remove the lower arm ball joint mounting bolts (A).
8. Using a plastic hammer (A), disconnect the driveshaft (C) from the axle hub (B).
9. Push the axle hub (B) outward and separate the driveshaft (C) from the axle hub (B).
10. Insert a pry bar (A) between the transaxle case (B) and joint case (C), and separate the driveshaft (D) from the transaxle case.

⁂ CAUTION

Use a pry bar being careful not to damage the transaxle or joint. Do not insert the pry bar too deep, as this may cause damage to the oil seal. Do not pull the driveshaft with excessive force as it may cause components inside the axle shaft joint to dislodge resulting in a torn boot or a damaged bearing. Plug the hole of the transaxle case with the oil seal cap to prevent contamination. Replace the retainer ring whenever the driveshaft is removed from the transaxle case.

To install:

11. Apply gear oil on the driveshaft oil seal case contacting surface (B) and transaxle case splines (A).
12. Before installing the driveshaft (C), set the opening side of the circlip (D) facing downward.
13. After installation, check that the driveshaft cannot be removed by hand.
14. Install the drive shaft into the knuckle.

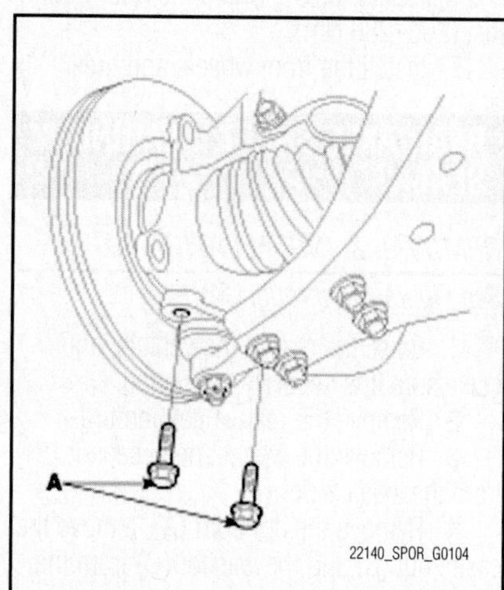

Fig. 22 Remove the lower arm ball joint mounting bolts (A)

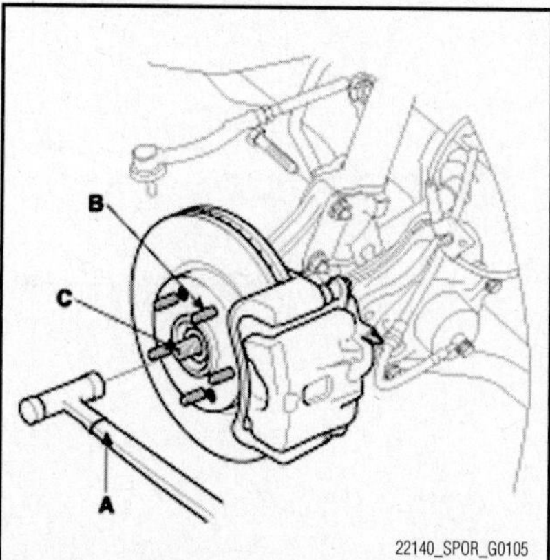

Fig. 23 Using a plastic hammer (A), disconnect the driveshaft (C) from the axle hub (B)

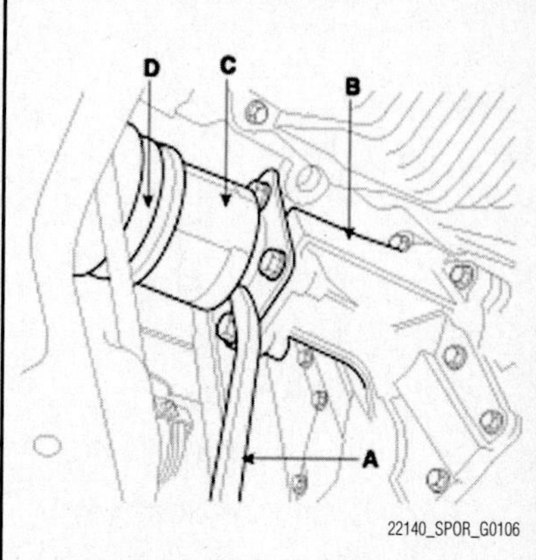

Fig. 24 Separate the driveshaft (D) from the transaxle case

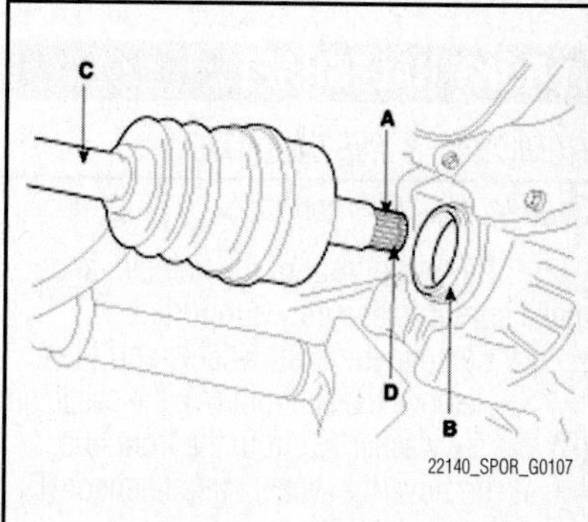

Fig. 25 Apply gear oil on the driveshaft oil seal case contacting surface (B) and transaxle case splines (A)–2WD

15. Install the lower arm mounting bolts. Tighten to 73.8–88.5 ft. lbs. (100–120 Nm).
16. Install the washer (B) with convex surface outward. Install the castle nut (A) and the split pin (C). Tighten to 147.5–206.6 ft. lbs. (200–280 Nm).
17. Install the front wheels and tires.

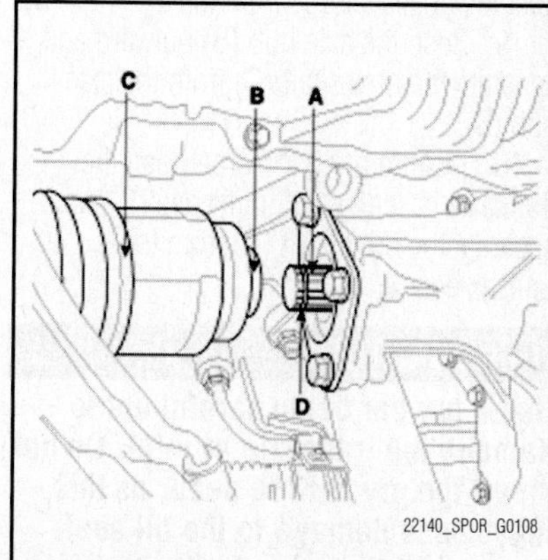

Fig. 26 Apply gear oil on the driveshaft oil seal case contacting surface (B) and transaxle case splines (A)–4WD

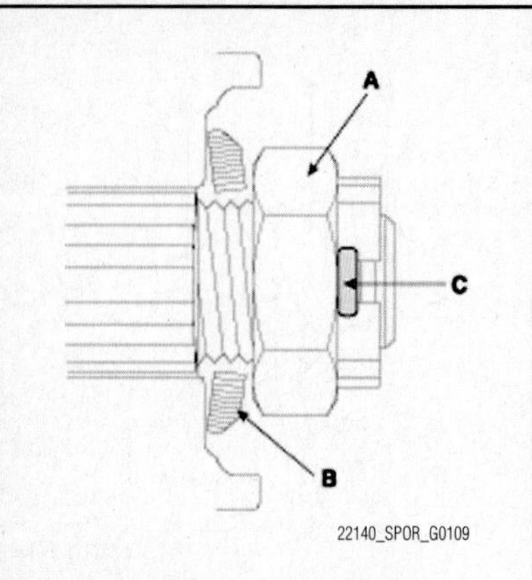

Fig. 27 Install the washer (B) with convex surface outward. Install the castle nut (A) and the split pin (C)

FRONT HALFSHAFT

REMOVAL & INSTALLATION

See Figures 18 through 27.

1. Raise the front of the vehicle, and make sure it is securely supported.
2. Remove the front wheels and tires.
3. Remove the split pin (A), the castle nut (B) and the washer (C) from the front hub.
4. Remove the wheel speed sensor (B) from the knuckle (A).
5. Remove the split pin (A) and the castle nut (B) from the tie rod end.
6. Disconnect the ball joint (B) from knuckle (A) using the special tool (09568-34000).

⁂ CAUTION

Apply a few drops of oil to the special tool. (Boot contact part)

7. Remove the lower arm ball joint mounting bolts (A).
8. Using a plastic hammer (A), disconnect the driveshaft (C) from the axle hub (B).
9. Push the axle hub (B) outward and separate the driveshaft (C) from the axle hub (B).
10. Insert a pry bar (A) between the transaxle case (B) and joint case (C), and separate the driveshaft (D) from the transaxle case.

⁂ CAUTION

Use a pry bar being careful not to damage the transaxle or joint. Do not insert the pry bar too deep, as this may cause damage to the oil seal. Do not pull the driveshaft with excessive force as it may cause components inside the axle shaft joint to dislodge resulting in a torn boot or a damaged bearing. Plug the hole of the transaxle case with the oil seal cap to prevent contamination. Replace the retainer ring whenever the driveshaft is removed from the transaxle case.

To install:

11. Apply gear oil on the driveshaft oil seal case contacting surface (B) and transaxle case splines (A).
12. Before installing the driveshaft (C), set the opening side of the circlip (D) facing downward.
13. After installation, check that the driveshaft cannot be removed by hand.
14. Install the drive shaft into the knuckle.
15. Install the lower arm mounting bolts. Tighten to 74–89 ft. lbs. (100–120 Nm).
16. Install the washer (B) with convex surface outward. Install the castle nut (A) and the split pin (C). Tighten to 148–207 ft. lbs. (200–280 Nm).
17. Install the front wheels and tires.

REAR AXLE SHAFT, BEARING & SEAL

REMOVAL & INSTALLATION

See Figures 28 through 32.

1. Raise the rear of the vehicle, and make sure it is securely supported.
2. Remove the rear wheel and tire.
3. Remove the wheel speed sensor (B) from the axle carrier (A).
4. Remove the split pin (A), remove the castle nut (B) and the washer (C) from the rear hub.
5. Remove the trailing arm mounting bolt (B) from the knuckle (A).

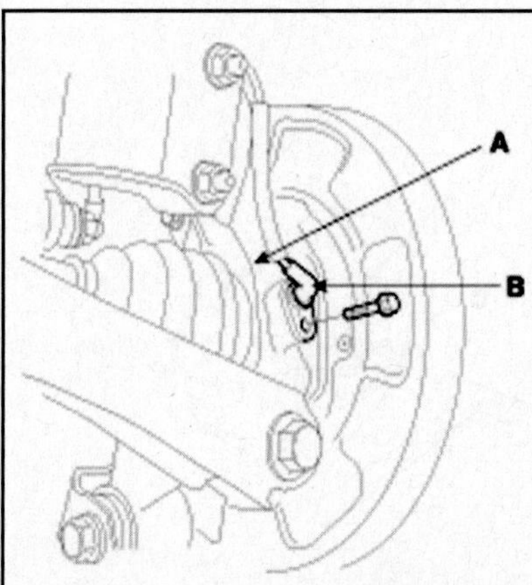

Fig. 28 Remove the wheel speed sensor (B) from the axle carrier (A)

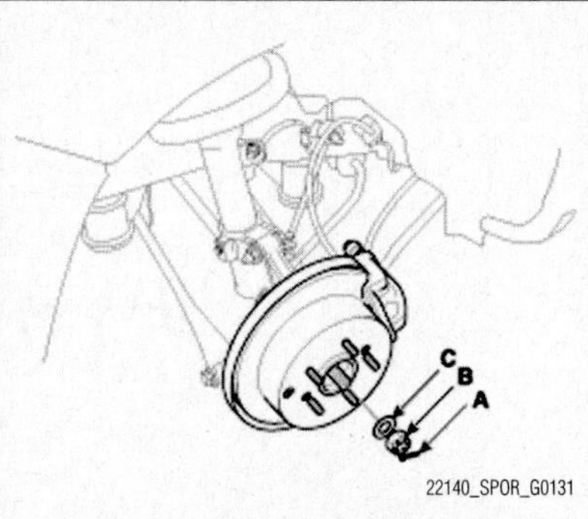

Fig. 29 Remove the split pin (A), remove castle nut (B) and washer (C) from the rear hub

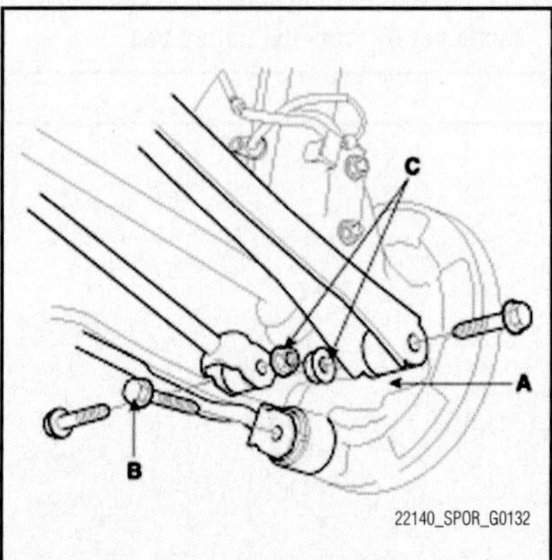

Fig. 30 Remove the trailing arm mounting bolt (B) from the knuckle (A)

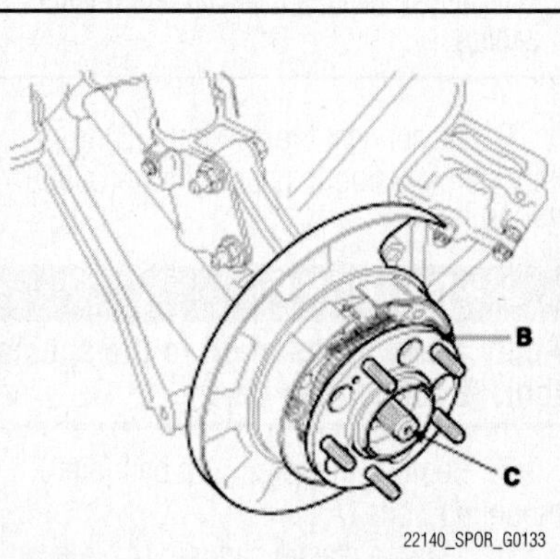

Fig. 31 Push the axle hub (B) outward and separate the driveshaft (C) from the axle hub (B)

6. Remove the suspension arm mounting nuts (C).
7. Push the axle hub (B) outward and separate the driveshaft (C) from the axle hub (B).
8. Insert a pry bar (A) between the differential case (B) and joint case (C), and separate the driveshaft(D) from the differential case.

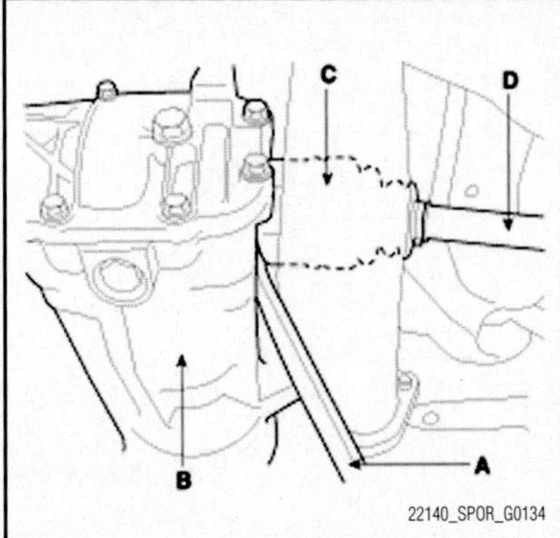

Fig. 32 Insert a pry bar (A) between the differential case (B) and joint case (C)

CAUTION

Use a pry bar(A) being careful not to damage the transaxle and joint. Do not insert the pry bar (A) too deep, as this may cause damage to the oil seal. Do not pull the driveshaft by excessive force because it may cause components inside the axle joint kit to dislodge resulting in a torn boot or a damaged bearing. Plug the hole of the transaxle case with the oil seal cap to prevent contamination. Replace the retainer ring whenever the driveshaft is removed from the transaxle case.

To install:

9. Apply gear oil on the driveshaft differential case contacting surface and driveshaft splines.
10. Before installing the driveshaft, set the opening side of the circlip facing downward.
11. After installation, check that the driveshaft cannot be removed by hand.
12. Install the axle into the knuckle.
13. Install the suspension arm mounting nuts and trailing arm mounting bolt to the knuckle. Tighten to
    - Suspension arm mounting nuts: 103.8–118.0 ft. lbs. (140–160 Nm)
    - Trailing arm mounting bolt: 73.8–88.5 ft. lbs. (100–120 Nm)
14. Install the washer, castle nut and split pin to the rear hub. Tighten to 147.5–206.6 ft. lbs. (200–280 Nm)
15. Install the wheel speed sensor to the knuckle.
16. Install the rear wheel and tire.

REAR DRIVESHAFT

REMOVAL & INSTALLATION

1. After making a match mark on the rubber coupling and rear differential companion, remove the propeller shaft mounting bolts.
2. Remove the center bearing bracket mounting bolts.
3. After making a match mark on the flange yoke and transaxle companion, remove the propeller shaft mounting bolts.

➡If a grease leak is shown around the universal joint, be sure to put grease in the universal joint through the nipple enough until grease come out of the universal joint.

To install:

4. Installation is the reverse of the removal procedures.
5. Install according to match mark of transaxle companion (or rear differential companion) and propeller shaft.
6. Tighten front propeller shaft mounting bolt to 36.9-44.3 ft. lb. (50-60 Nm)
7. Tighten center bearing bracket mounting bolt to 36.9-47 ft. lb. (50-65 Nm)
8. Tighten rear propeller shaft mounting bolt to 73.8-88.5 ft. lb. (100-120 Nm)

REAR HALFSHAFT

REMOVAL & INSTALLATION

See Figures 33 through 42.

1. Raise the front of the vehicle, and make sure it is securely supported.
2. Remove the front wheels and tires.
3. Remove the split pin (A), the castle nut (B) and the washer (C) from the front hub.
4. Remove the wheel speed sensor (B) from the knuckle (A).
5. Remove the split pin (A) and the castle nut (B) from the tie rod end.
6. Disconnect the ball joint (B) from knuckle (A) using the special tool (09568-34000).

CAUTION

Apply a few drops of oil to the special tool. (Boot contact part)

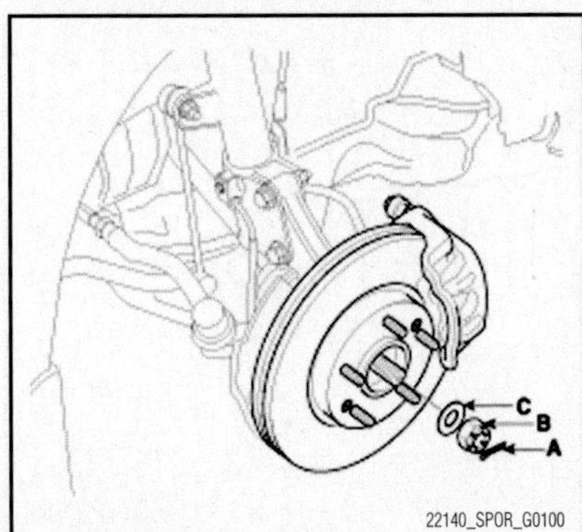

Fig. 33 Remove the split pin (A), the castle nut (B) and the washer (C) from the front hub

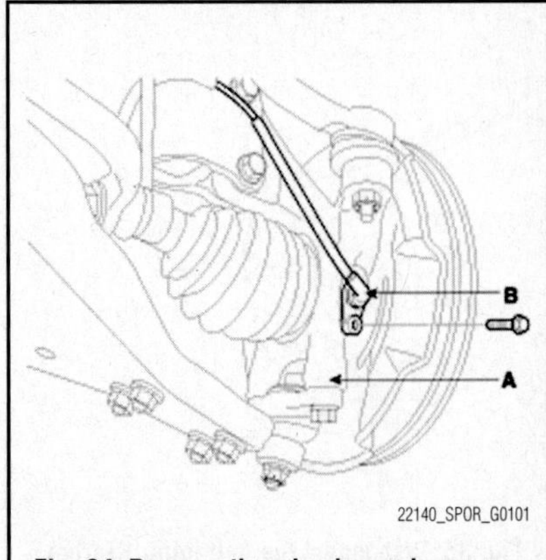

Fig. 34 Remove the wheel speed sensor (B) from the knuckle (A)

7. Remove the lower arm ball joint mounting bolts (A).
8. Using a plastic hammer (A), disconnect the driveshaft (C) from the axle hub (B).
9. Push the axle hub (B) outward and separate the driveshaft (C) from the axle hub (B).
10. Insert a pry bar (A) between the transaxle case (B) and joint case (C), and separate the driveshaft (D) from the transaxle case.

CAUTION

Use a pry bar being careful not to damage the transaxle or joint. Do not insert the pry bar too deep, as this may cause damage to the oil seal. Do not pull the driveshaft with excessive force as it may cause components inside the axle shaft joint to dislodge resulting in a torn boot or a damaged bearing. Plug the hole of the transaxle case with the oil seal cap to prevent contamination. Replace the retainer ring whenever the driveshaft is removed from the transaxle case.

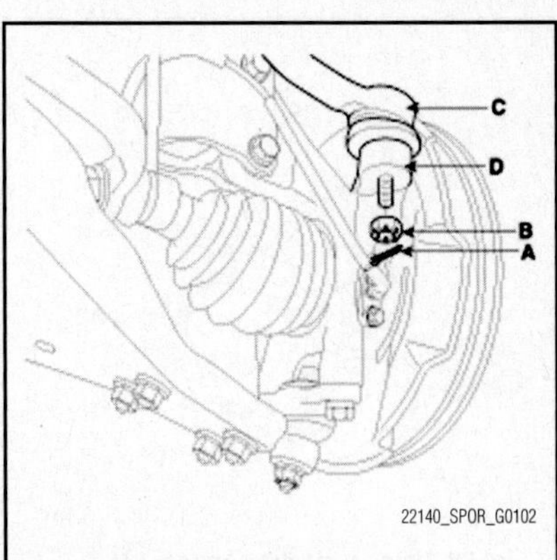

Fig. 35 Remove the split pin (A) and the castle nut (B) from the tie rod end

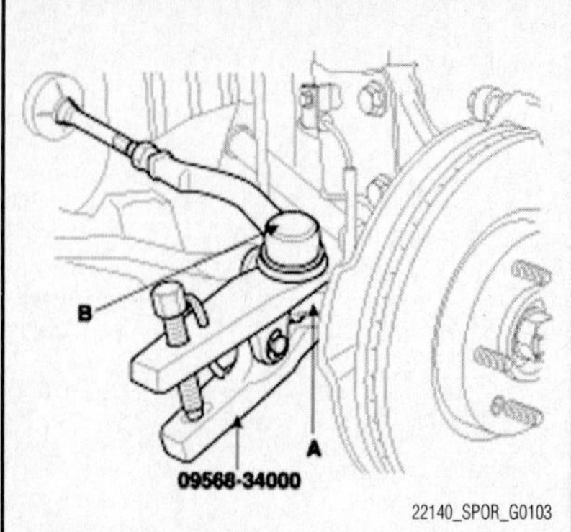

Fig. 36 Disconnect the ball joint (B) from knuckle (A) using the special tool (09568-34000)

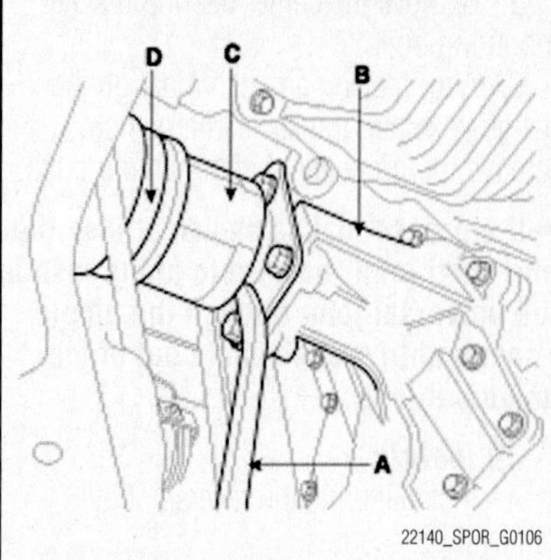

Fig. 39 Separate the driveshaft (D) from the transaxle case

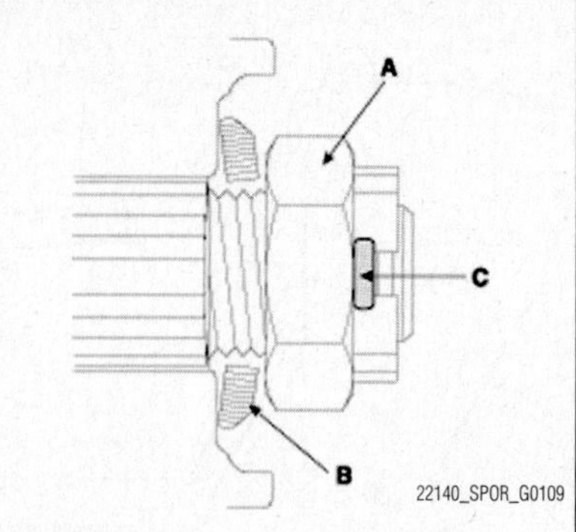

Fig. 42 Install the washer (B) with convex surface outward. Install the castle nut (A) and the split pin (C)

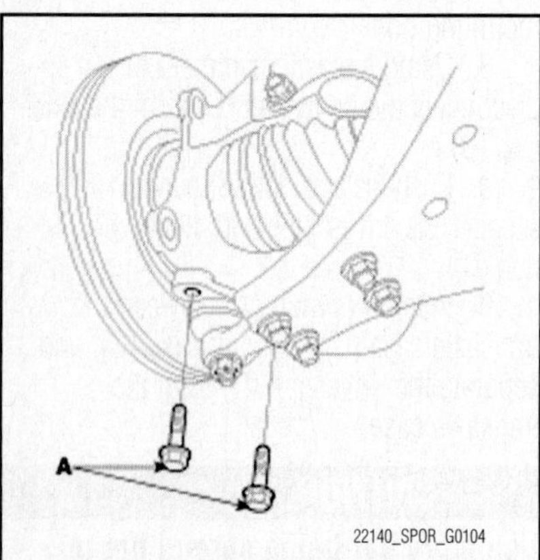

Fig. 37 Remove the lower arm ball joint mounting bolts (A)

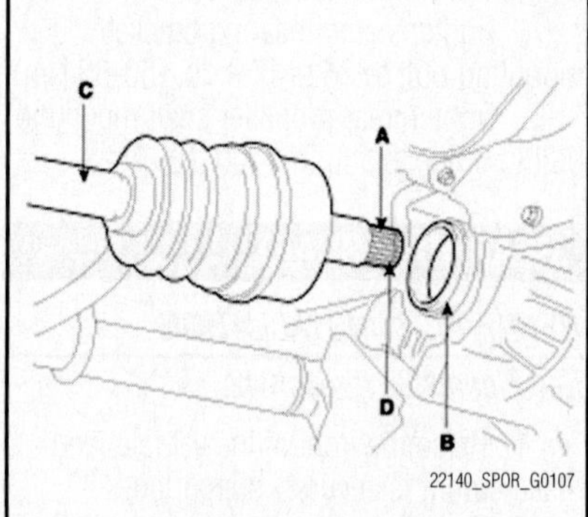

Fig. 40 Apply gear oil on the driveshaft oil seal case contacting surface (B) and transaxle case splines (A)—2WD

To install:

11. Apply gear oil on the driveshaft oil seal case contacting surface (B) and transaxle case splines (A).
12. Before installing the driveshaft (C), set the opening side of the circlip (D) facing downward.
13. After installation, check that the driveshaft cannot be removed by hand.
14. Install the drive shaft into the knuckle.
15. Install the lower arm mounting bolts. Tighten to 73.8–88.5 ft. lbs. (100–120 Nm).
16. Install the washer (B) with convex surface outward. Install the castle nut (A) and the split pin (C). Tighten to 148–207 ft. lbs. (200–280 Nm).
17. Install the front wheels and tires.

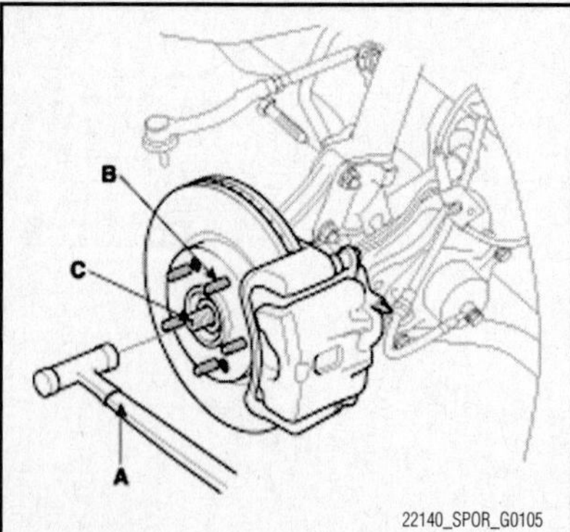

Fig. 38 Using a plastic hammer (A), disconnect the driveshaft (C) from the axle hub (B)

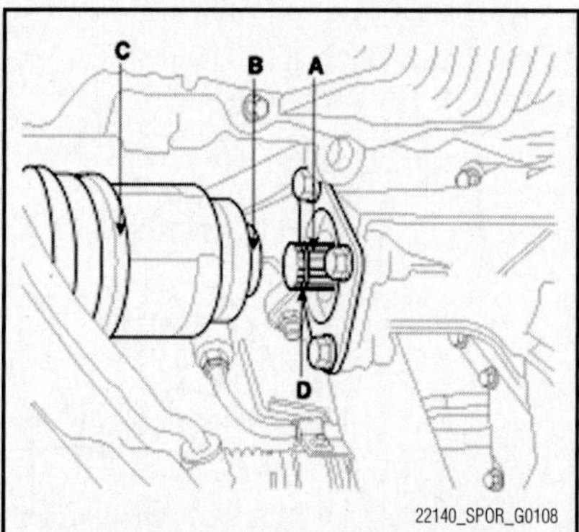

Fig. 41 Apply gear oil on the driveshaft oil seal case contacting surface (B) and transaxle case splines (A)—4WD

TRANSFER CASE ASSEMBLY

REMOVAL & INSTALLATION

See Figure 43.

1. Remove the battery terminal.
2. Lift up the vehicle.
3. Remove the propeller shaft.
4. Remove the front muffler (A).
5. Remove the RH driveshaft.
6. Remove the oil drain plug and drain the fluid.
7. After draining, re-tighten the oil drain plug. Tighten to 29–43 ft. lbs. (39–59 Nm)
8. Support the transfer assembly with a jack.
9. Remove the transfer assembly by removing the mounting bolts.

CAUTION

Remove the transfer bracket mounting bolts (2EA) together.

To install:

10. Remove the filler plug.
11. Refill with 0.8 qt (0.8L) SAE 80W90.
12. Fix it in proper position with mounting bolts.

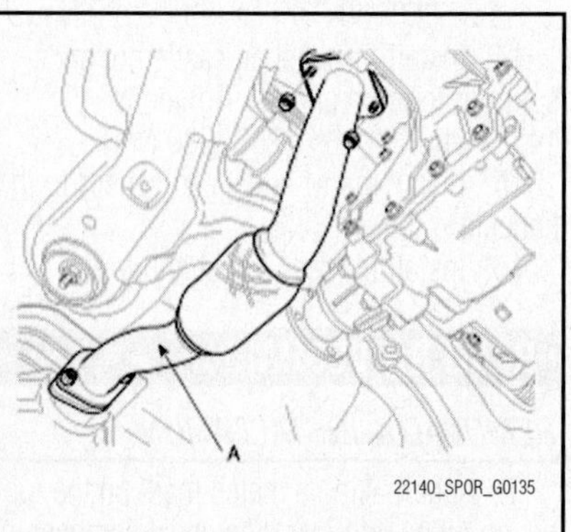

Fig. 43 Remove the front muffler (A)

ENGINE COOLING

ENGINE FAN

REMOVAL & INSTALLATION

See Figures 44 through 48.

1. Remove the undercover (A).
2. Drain the engine coolant. Remove the radiator cap to speed draining.
3. Remove the air duct (A).
4. Remove the battery and tray.
5. Remove the coolant reservoir tank (A).
6. Remove the radiator upper hose (B).

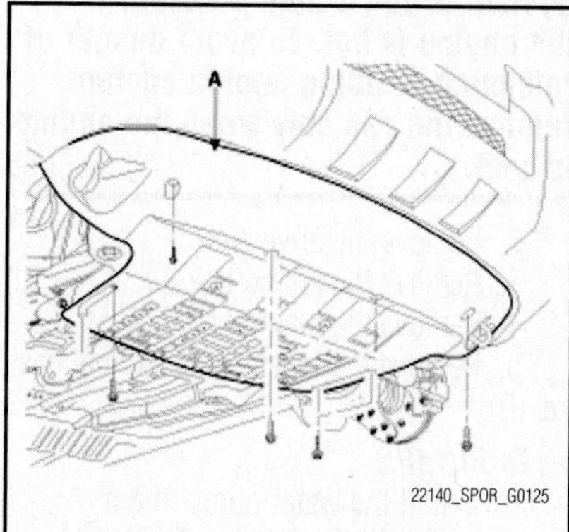

Fig. 44 Remove the undercover (A)

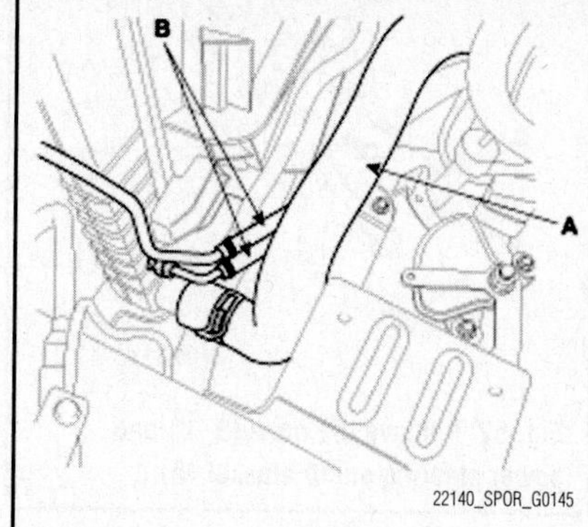

Fig. 47 Remove the radiator lower hose (A) and the ATF oil cooler hose (B)

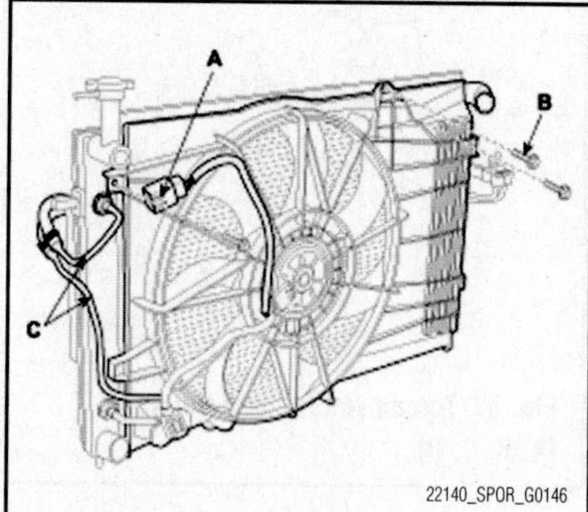

Fig. 48 Remove the cooling fan motor connector (A), the cooling fan motor assembly mounting bolt (B) and the ATF oil cooler pipe (C)

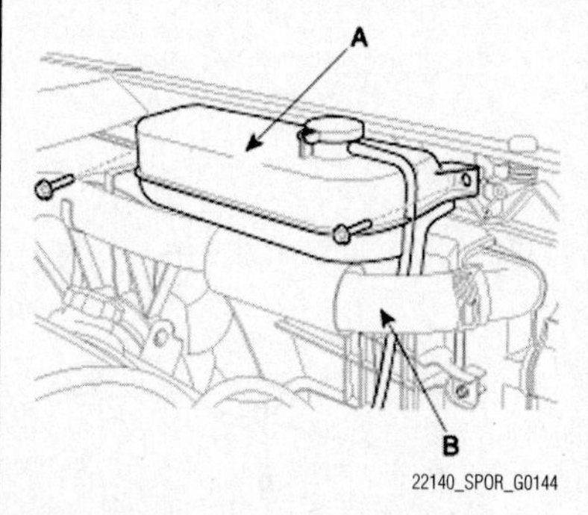

Fig. 46 Remove the coolant reservoir tank (A) and the radiator upper hose (B)

7. Remove the radiator lower hose (A).
8. Remove the ATF oil cooler hose (B).
9. Remove the cooling fan motor connector (A).
10. Remove the cooling fan motor assembly mounting bolt (B).
11. Remove the ATF oil cooler pipe (C).
12. Remove the cooling fan motor assembly by pulling it from the radiator.

To install:

13. Installation is the reverse of removal.
14. Fill with engine coolant.
15. Start engine and check for leaks.

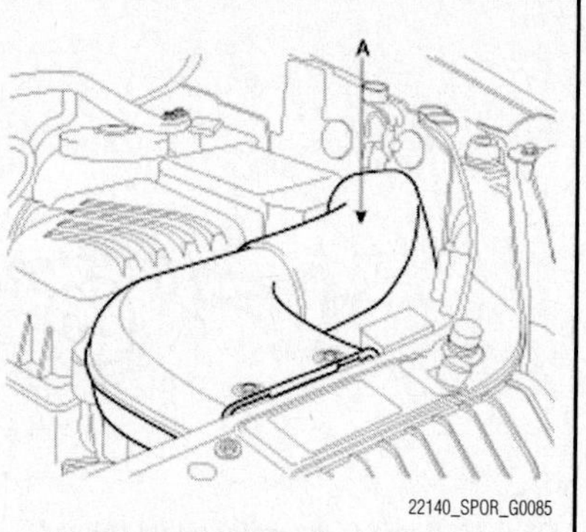

Fig. 45 Remove the air duct (A)

RADIATOR

REMOVAL & INSTALLATION

See Figures 44 through 46, 49 through 51.

1. Remove the undercover (A).
2. Drain the engine coolant. Remove the radiator cap to speed draining.
3. Remove the air duct (A).
4. Remove the battery and tray.
5. Remove the coolant reservoir tank (A).
6. Remove the radiator upper hose (B).
7. Remove the radiator lower hose (A).
8. Remove the ATF oil cooler hose (B).
9. Remove the cooling fan motor connector (A).
10. Remove the cooling fan motor assembly mounting bolt (B).
11. Remove the ATF oil cooler pipe (C).

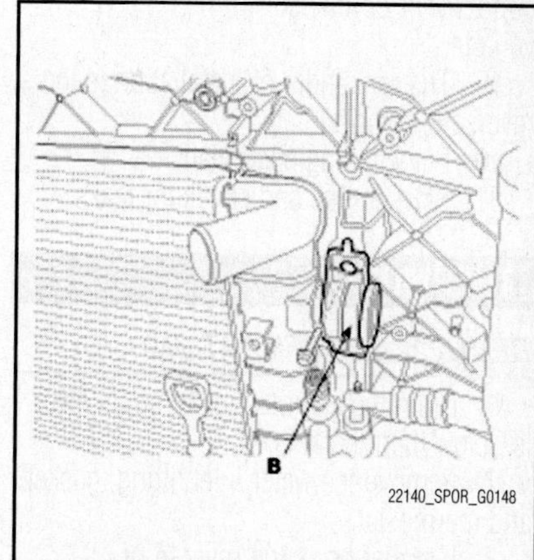

Fig. 50 Remove the radiator bracket (B)

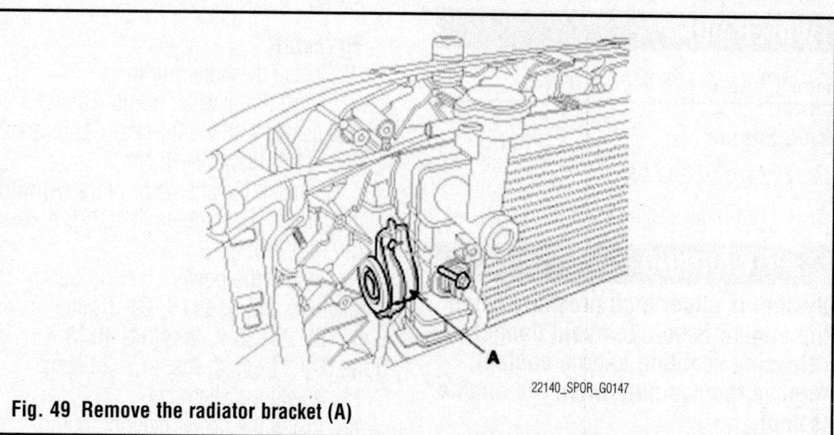

Fig. 49 Remove the radiator bracket (A)

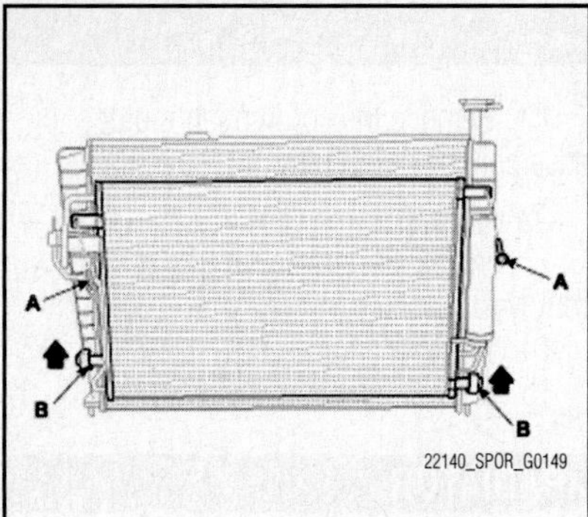

Fig. 51 Remove the condenser mounting bolt (A) and the condenser bracket (B)

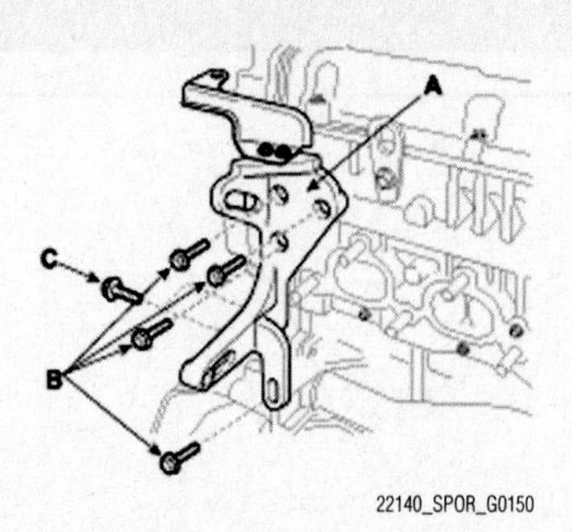

Fig. 52 Remove the bolts (B, C) and power steering pump bracket (A)

12. Remove the cooling fan motor assembly by pulling it from the radiator.
13. Remove the radiator brackets (A, B).
14. Remove the condenser mounting bolt (A).
15. Remove the condenser bracket (B) by pulling the condenser from the radiator.
16. Remove the radiator from engine room.

To install:

17. Install the cooling fan onto the radiator.
18. Install the radiator onto the air conditioner condenser in the vehicle.
19. The remainder of installation is the reverse of removal.
20. Fill with engine coolant.
21. Start engine and check for leaks.

THERMOSTAT

REMOVAL & INSTALLATION

1. Drain the engine coolant so its level is below thermostat.
2. Remove the water inlet fitting, gasket and thermostat.
3. Installation is the reverse of removal.

WATER PUMP

REMOVAL & INSTALLATION

2.0L Engine

See Figures 52 and 53.

1. Drain the engine coolant.

WARNING

System is under high pressure when the engine is hot. To avoid danger of releasing scalding engine coolant, remove the cap only when the engine is cool.

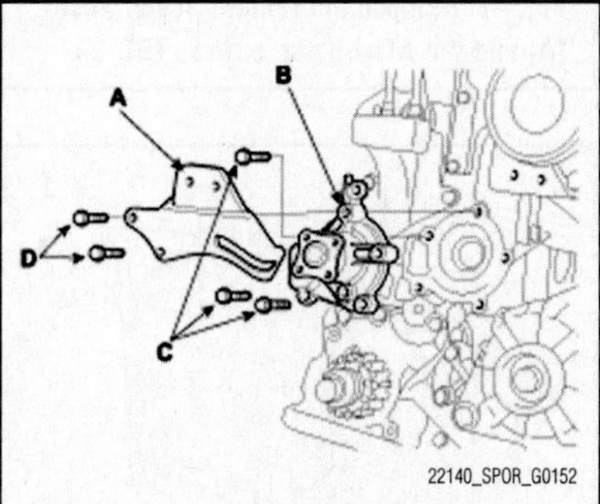

Fig. 53 Torque water pump bolts (A, B, C, D)

2. Remove the drive belts.
3. Remove the timing belt.
4. Remove the timing belt idler.
5. Remove the power steering pump and use a wire to secure the pump to the vehicle so that it is out of the way.
6. Remove the bolts (B, C) and power steering pump bracket (A).
7. Remove the alternator.
8. Remove the water pump.
   a. Remove the 2 bolts (D) and alternator brace (A).
   b. Remove the 3 bolts (C) and remove the water pump (B) and gasket.

To install:

9. Install the water pump.
   a. Install the water pump (B) and a new gasket with the 3 bolts (C). Torque to 9–11 ft. lbs. (12–15 Nm).
   b. Install the alternator brace (A) with the 2 bolts (D). Tighten to 15–20 ft. lbs. (20–27 Nm).
10. Install the power steering pump bracket (A) and bolts (B, C). Tighten to Bolts (B): 25–36 ft. lbs. (34–49 Nm); Bolts (C): 11–15 ft. lbs. (15–20 Nm).
11. Install the alternator.
12. Install the power steering pump.
13. Install the timing belt idler.
14. Install the timing belt.
15. Install the water pump pulley.
16. Install the drive belts.
17. Tighten the water pump pulley bolts. Tighten to 7 ft. lbs. (9 Nm).
18. Fill with engine coolant.
19. Start engine and check for leaks.
20. Recheck engine coolant level.

2.7L Engine

See Figures 54 and 55.

1. Drain the engine coolant.

WARNING

System is under high pressure when the engine is hot. To avoid danger of releasing scalding engine coolant, remove the cap only when the engine is cool.

2. Remove the drive belt.
3. Remove the timing belt.
4. Remove the timing belt idler.
5. Remove the water pump (A) and gasket (B).

To install:

6. Install the water pump and a new gasket with the 8 bolts. Tighten to 11–16 ft. lbs. (15–22 Nm).

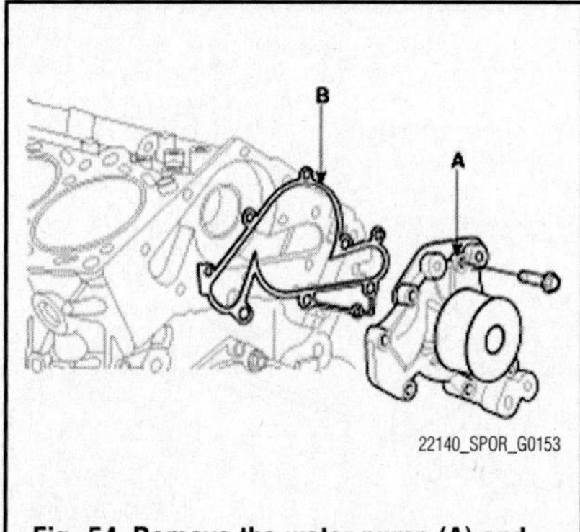

Fig. 54 Remove the water pump (A) and gasket (B)

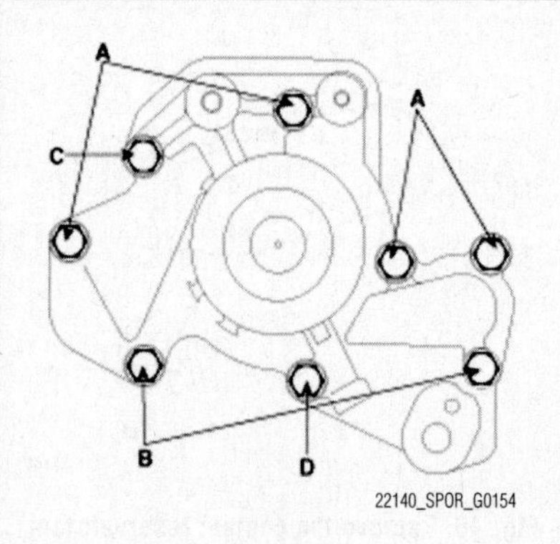

Fig. 55 Water pump bolt locations by size

a. Water pump bolt locations by size:
- A: 4 each 8x25
- B: 2 each 8x30
- C: 1 each 8x32
- D: 1 each 8x40

7. Install the timing belt idler.
8. Install the timing belt.
9. Install drive belt.
10. Fill with engine coolant.
11. Start engine and check for leaks.
12. Recheck engine coolant level.

ENGINE ELECTRICAL — CHARGING SYSTEM

ALTERNATOR

REMOVAL & INSTALLATION

See Figures 56 through 59.

1. Disconnect the battery negative terminal first, then the positive terminal.
2. Disconnect the alternator connector (A) and "B" terminal cable (B) from the alternator (C).
3. Remove the adjusting bolt (A) and mounting bolt (B), and remove the alternator belt.
4. Pull out the through bolt (C), and remove the alternator (D).
5. Installation is the reverse of removal.

Fig. 56 Disconnect the alternator connector (A) and "B" terminal cable (B) from the alternator (C)—2.0L engine

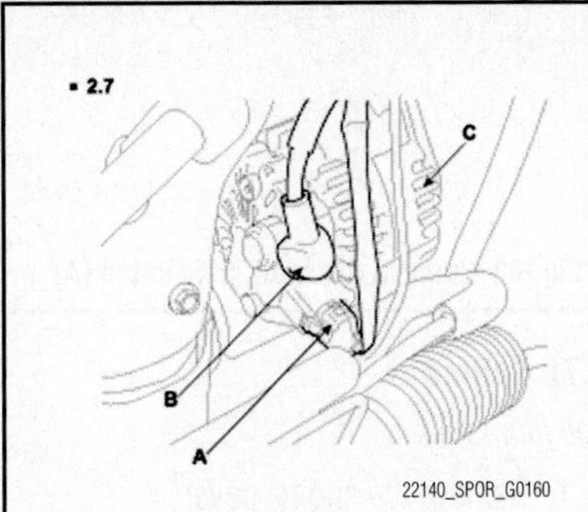

Fig. 57 Disconnect the alternator connector (A) and "B" terminal cable (B) from the alternator (C)—2.7L engine

6. Adjust the alternator belt tension after installation.

VOLTAGE REGULATOR

ADJUSTMENT

The charging system includes an alternator with a built-in regulator. There is no adjustment possible for the voltage regulator. If the voltage regulator is defective, it must be replaced.

REMOVAL & INSTALLATION

See Alternator Removal & Installation.

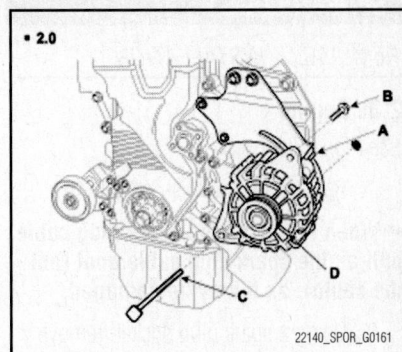

Fig. 58 Remove the adjusting bolt (A), mounting bolt (B) and through bolt (C)—2.0L engine

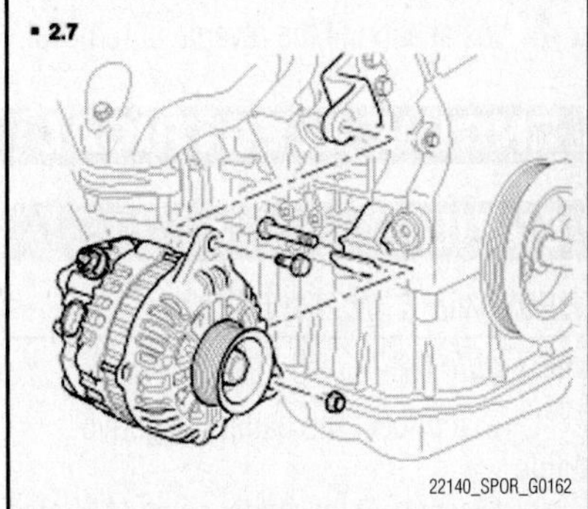

Fig. 59 Remove the mounting bolt and through bolt—2.7L engine

ENGINE ELECTRICAL — IGNITION SYSTEM

FIRING ORDER

2.0L engine firing order: 1–3–4–2

2.7L engine firing order: 1–2–3–4–5–6

IGNITION COIL

REMOVAL & INSTALLATION

See Figures 60 and 61.

1. Remove the engine cover.
2. Disconnect the spark plug cable and connector.
3. Remove the ignition coil (A).
4. Installation is the reverse of removal.

Fig. 60 Remove the ignition coil (A)—2.0L engine

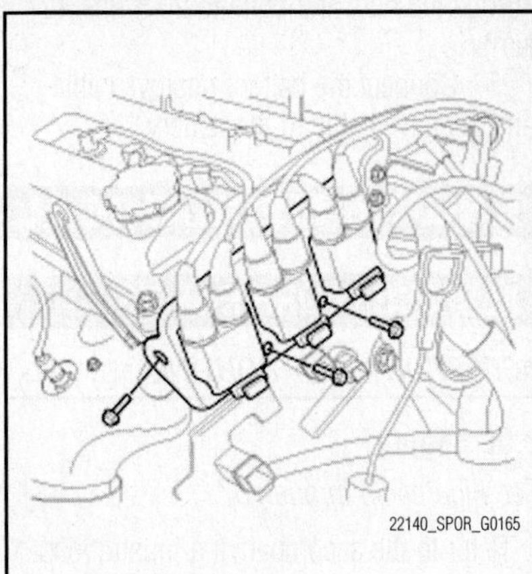

Fig. 61 Remove the ignition coil—2.7L engine

IGNITION TIMING

ADJUSTMENT

These vehicles are equipped with a Distributorless Ignition System (DIS). No adjustment is necessary or possible.

SPARK PLUGS

REMOVAL & INSTALLATION

2.0L Engine

See Figure 62.

1. Remove the spark plug cable (A).

➡When removing the spark plug cable, pull on the spark plug cable boot (not the cable), as it may be damaged.

2. Using a spark plug socket, remove the spark plug (B).

⁂ CAUTION

Be careful that no contaminates enter through the spark plug holes.

3. Installation is the reverse of removal.

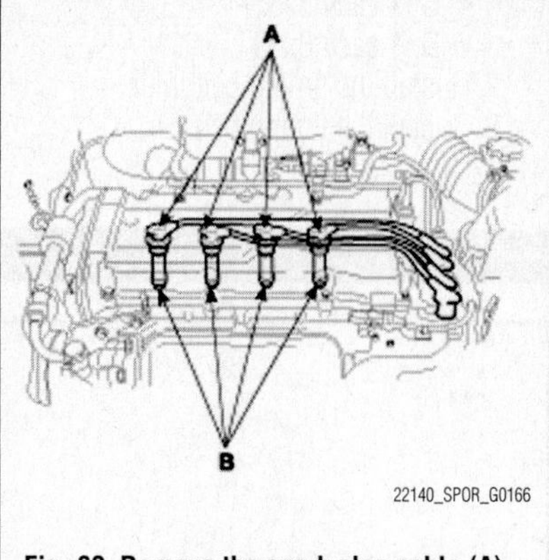

Fig. 62 Remove the spark plug cable (A)

2.7L Engine

See Figure 63.

1. Remove the engine cover.
2. Disconnect the VIS actuator connectors and injector connectors.
3. Remove the accelerator cable.
4. Remove surge tank sub assembly.
5. Remove the spark plug cable.
6. Remove the spark plug.
7. Installation is the reverse of removal.

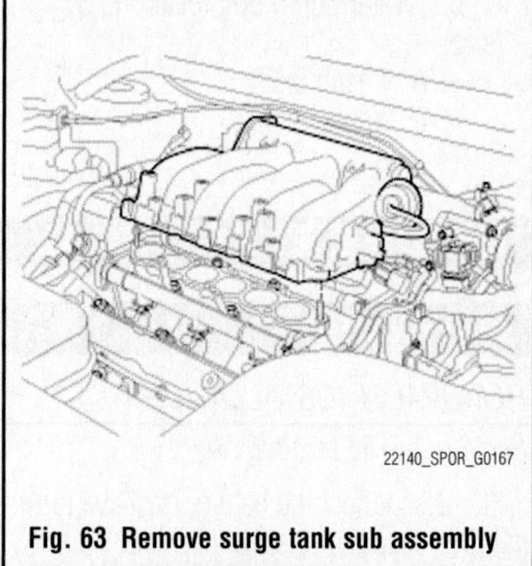

Fig. 63 Remove surge tank sub assembly

SPARK PLUG WIRES

REMOVAL & INSTALLATION

1. Remove the spark plug cable.

➡When removing the spark plug cable, pull on the spark plug cable boot (not the cable), as it may be damaged.

2. Installation is the reverse of removal.

ENGINE ELECTRICAL — STARTING SYSTEM

STARTER

REMOVAL & INSTALLATION

See Figure 64.

1. Disconnect the battery negative cable.
2. Disconnect the starter cable (A) from the B terminal (B) on the solenoid (C), then disconnect the connector (D) from the "S" terminal (E).
3. Remove the 2 bolts holding the starter, and remove the starter.
4. Installation is the reverse of removal. Torque the nuts and bolts to 20 ft. lbs. (27 Nm).
5. Connect the battery positive cable and negative cable to the battery.

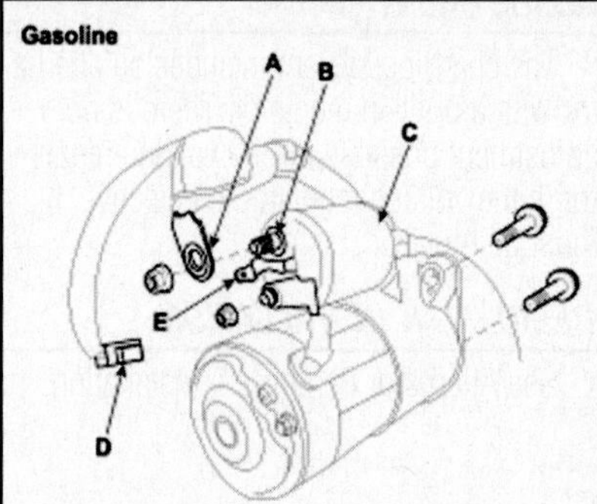

Fig. 64 Disconnect the starter cable (A) from the B terminal (B) on the solenoid (C), then disconnect the connector (D) from the "S" terminal (E)

SOLENOID OR RELAY REPLACEMENT

1. Disconnect the negative battery cable.
2. Remove the starter assembly from the vehicle.
3. Remove the nut from the solenoid M-terminal.
4. Remove the solenoid retaining screws and the magnetic switch.

To install:

5. Install the magnetic switch and secure to the drive housing with the two screws.
6. Re-install the nut onto solenoid M-terminal.
7. Reinstall the starter assembly.
8. Reconnect the negative battery cable.

ENGINE MECHANICAL

ACCESSORY DRIVE BELTS

ACCESSORY BELT ROUTING

2.0L Engine

See Figures 65 through 67.

Refer to the accompanying illustrations.

2.7L Engine

See Figure 68.

Refer to the accompanying illustration.

INSPECTION

Inspect the drive belt for signs of glazing or cracking. A glazed belt will be perfectly smooth from slippage, while a good belt will have a slight texture of fabric visible. Cracks will usually start at the inner edge of the belt and run outward. All worn or damaged drive belts should be replaced immediately.

ADJUSTMENT

The belt tension is maintained by an automatic tensioner. No adjustment is necessary or possible.

REMOVAL & INSTALLATION

1. Release tension of the belt tensioner.
2. Remove the serpentine belt.

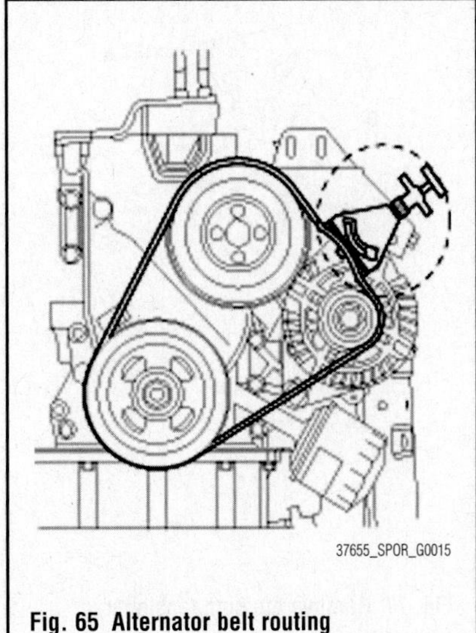

Fig. 65 Alternator belt routing

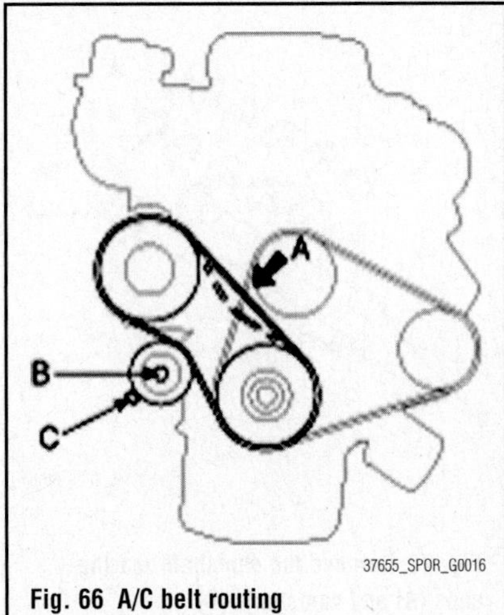

Fig. 66 A/C belt routing

To install:

3. Hang the belt on the pulley of the tensioner and install the tensioner.

(If the tensioner is already installed, loosen its mounting bolts to allow belt installation.) Tighten the tensioner assembly bolt to 33–36 ft. lbs. (45–50 Nm).

4. Install drive belt.
5. When installing the belt on the pulley, make sure it is centered on the pulley.

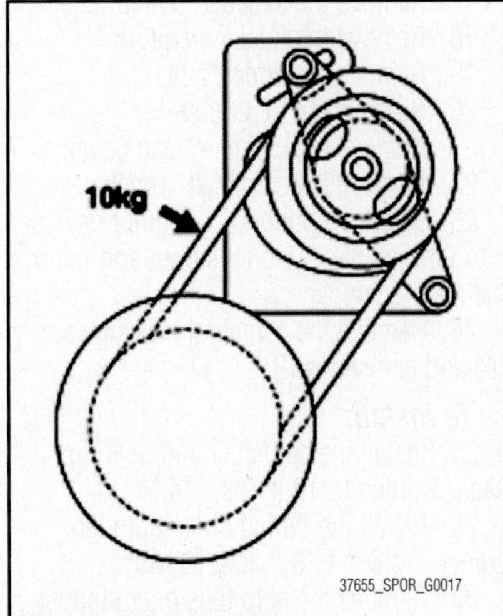

Fig. 67 P/S belt routing

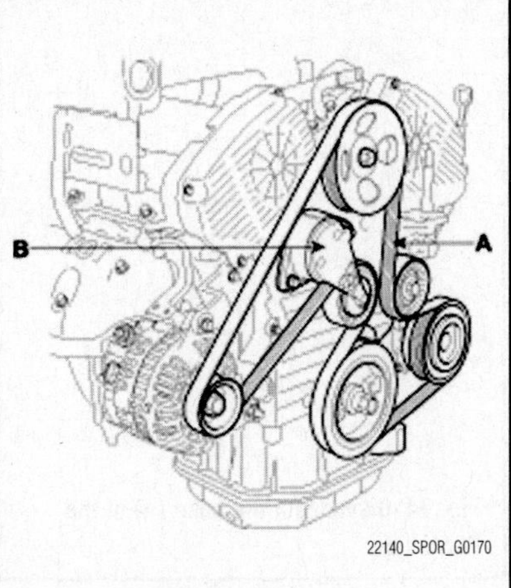

Fig. 68 Serpentine belt routing—2.7L engine

CAMSHAFT AND VALVE LIFTERS

REMOVAL & INSTALLATION

2.0L Engine

See Figures 69 through 81.

1. Remove the air duct.
2. Disconnect the terminals from battery.
3. Remove the engine cover.
4. Drain the engine coolant.
5. Remove the intake air hose and air cleaner assembly.
   a. Disconnect the Air Flow Sensor (AFS) connector.
   b. Disconnect the breather hose from intake air hose.
   c. Remove the intake air hose and air cleaner assembly.
6. Remove the upper radiator hose and lower radiator hose.
7. Remove the heater hoses.
8. Remove the accelerator cable by loosening the lock-nut, then slip the cable end out of the throttle linkage.
9. Remove the engine wire harness connectors and wire harness clamps from cylinder head and the intake manifold.
   a. Disconnect the Oil Control Valve (OCV) connector (A).
   b. Disconnect the oil temperature sensor connector (B).
   c. Disconnect the Engine Coolant Temperature (ECT) sensor connector (C).
   d. Disconnect the ignition coil connector (D).
   e. Disconnect the Throttle Position Sensor (TPS) connector (A).
   f. Disconnect the Idle Speed Actuator (ISA) connector (B).
   g. Disconnect the Camshaft Position Sensor (CMP) connector (A).
   h. Disconnect the four injector connectors (B).
   i. Disconnect the knock sensor connector (C).
   j. Disconnect the ground cables (D) from the intake manifold.
   k. Disconnect the air conditioner compressor switch connector (E).
   l. Disconnect the front heated oxygen sensor connector (A).
   m. Disconnect the Crankshaft Position Sensor (CKP) connector (B).
   n. Disconnect the oil pressure switch connector (C).

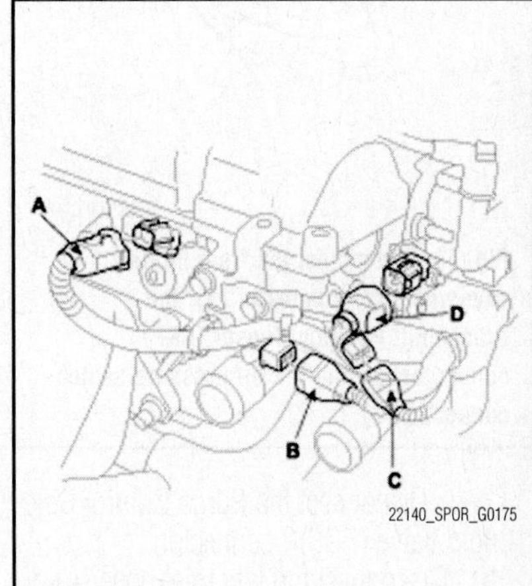

Fig. 69 Remove the engine wire harness connectors

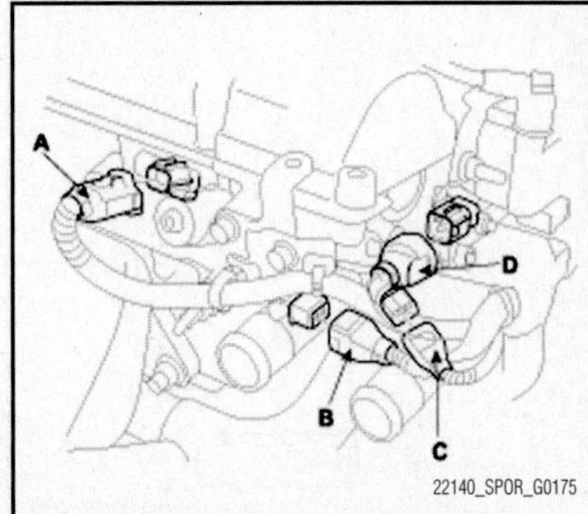

Fig. 70 Disconnect the Throttle Position Sensor (TPS) connector (A) and the Idle Speed Actuator (ISA) connector (B)

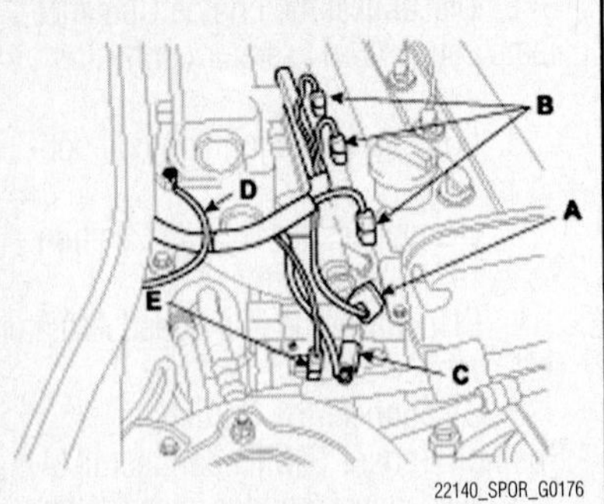

Fig. 71 Disconnect the Camshaft Position Sensor (CMP) connector (A), the four injector connectors (B), the knock sensor connector (C), the ground cables (D) and the air conditioner compressor switch connector (E)

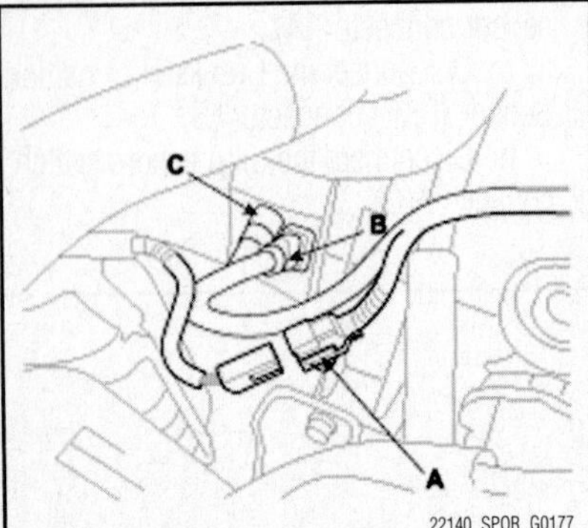

Fig. 72 Disconnect the front heated oxygen sensor connector (A), the Crankshaft Position Sensor (CKP) connector (B) and the oil pressure switch connector (C)

o. Disconnect the Purge Control Solenoid Valve (PCSV) connector.

10. Disconnect the fuel inlet hose (A) of the delivery pipe side.

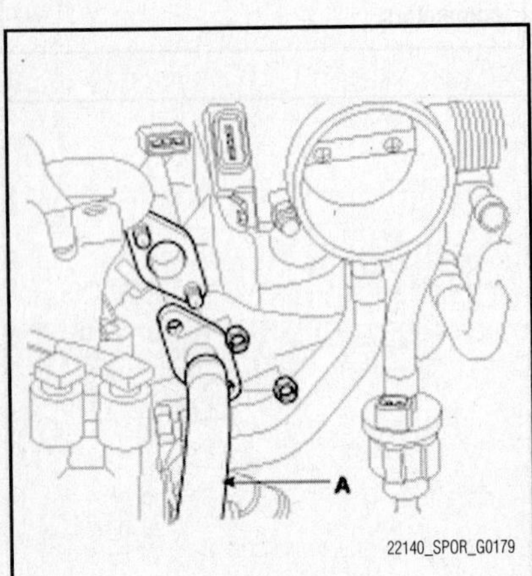

Fig. 73 Disconnect the fuel inlet hose (A) of the delivery pipe side

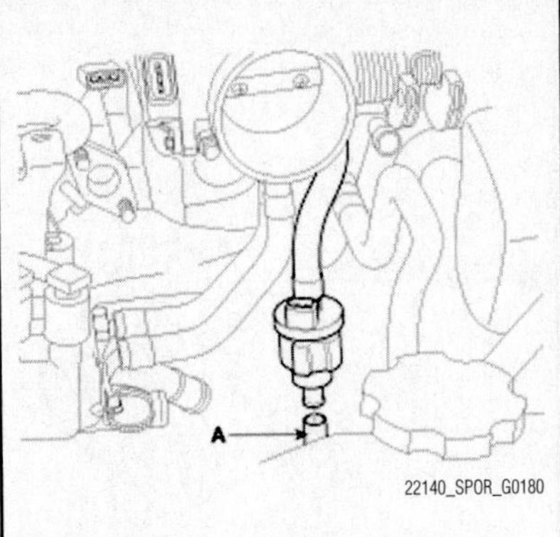

Fig. 74 Disconnect the hose (A) of the PCSV side

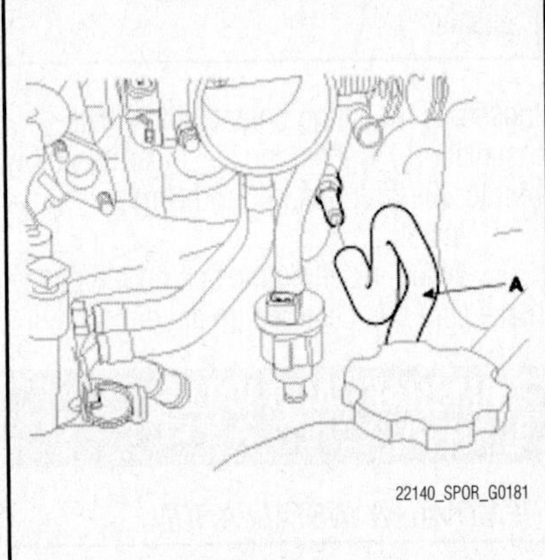

Fig. 75 Remove the brake booster vacuum hose (A)

11. Disconnect the hose (A) of the PCSV side.

12. Remove the brake booster vacuum hose (A).

13. Remove the power steering pump drive belt.

14. Remove the power steering pump and use a wire to secure the pump to the vehicle so that it is out of the way.

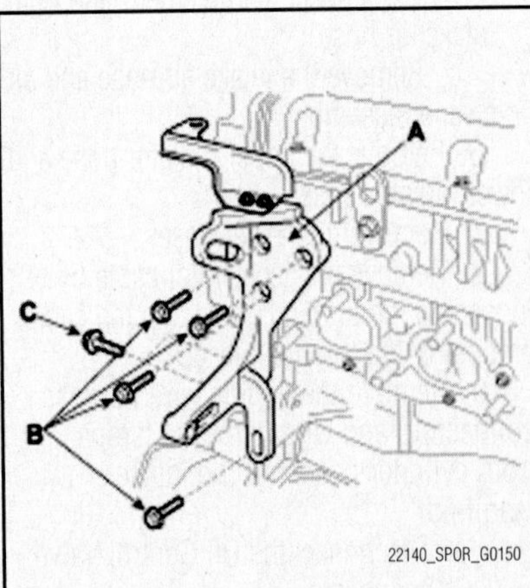

Fig. 76 Remove the bolts (B, C) and power steering pump bracket (A)

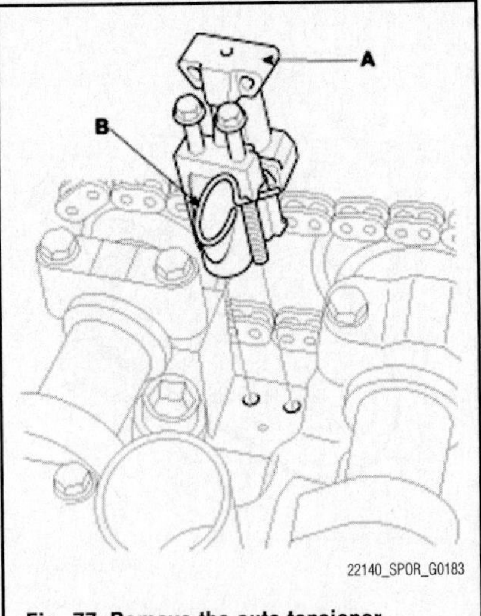

Fig. 77 Remove the auto tensioner

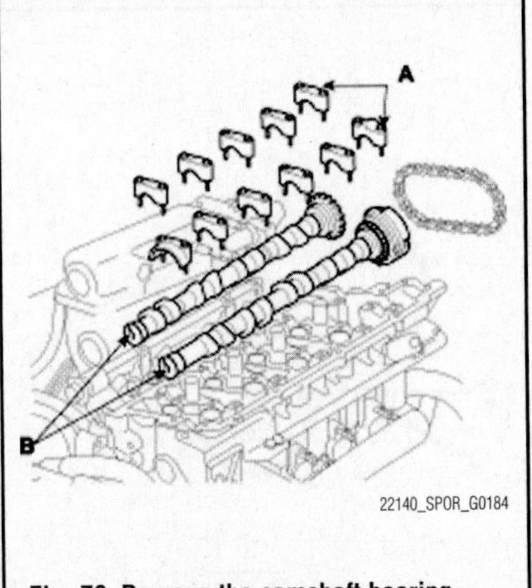

Fig. 78 Remove the camshaft bearing caps (A) and camshafts (B)

15. Remove the bolts (B, C) and power steering pump bracket (A).

16. Remove the spark plug cables.

17. Remove the exhaust manifold.

18. Remove the intake manifold.

19. Remove the timing belt.

20. Remove the PCV hose.

21. Remove the cylinder head cover.

22. Remove the camshaft sprocket.

23. Insert a stopper pin or other device into timing chain auto tensioner and remove the auto tensioner.

24. Remove the camshaft bearing caps (A) and camshafts (B).

To install:

25. Install the camshaft and bearing caps. Tighten to 10 ft. lbs. (14 Nm).

26. Install the timing chain auto tensioner. Tighten to 6 ft. lbs. (8 Nm).

27. Remove the auto tensioner stopper pin.

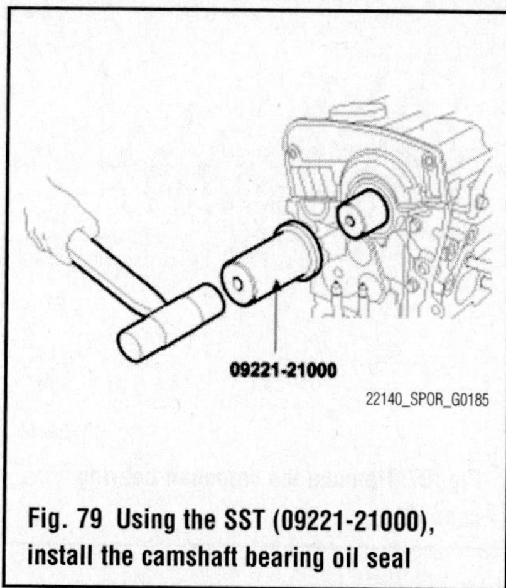

Fig. 79 Using the SST (09221-21000), install the camshaft bearing oil seal

28. Check and adjust valve clearance.
29. Using the SST (09221-21000), install the camshaft bearing oil seal.
30. Install the camshaft sprocket.
31. Install the cylinder head cover.
 a. Install the cylinder head cover gasket in the groove of the cylinder head cover.

➡Before installing the cylinder head cover gasket, thoroughly clean the cylinder head cover and the groove. When installing, make sure the cylinder head cover gasket is seated securely in the corners of the recesses with no gap.

 b. Apply liquid gasket to the head cover gasket at the corners of the recess.

➡Use liquid gasket, Loctite No. 5999. Check that the mating surfaces are clean and dry before applying liquid gasket. After assembly, wait at least 30 minutes before filling the engine with oil.

32. Install the cylinder head cover (A) with the 12 bolts(B). Uniformly tighten the bolts in several passes. Tighten to 6–7 ft. lbs. (8–10 Nm).
33. Install the PCV hose.
34. Install the timing belt.
35. Install the intake manifold.
36. Install the exhaust manifold.
37. Install the spark plug cables.
38. Install the power steering pump bracket and bolts. Tighten to Bolt (B): 25–36 ft. lbs. (34–49 Nm); Bolt (C): 9–11 ft. lbs. (12–15 Nm).
39. Install the power steering pump.
40. Install the accelerator cable.
41. Install the brake booster hose.
42. Connect the hose of the PCSV side.
43. Connect the fuel inlet hose of the delivery pipe side.
44. Install the engine wire harness connectors and wire harness clamps to the cylinder head and the intake manifold.

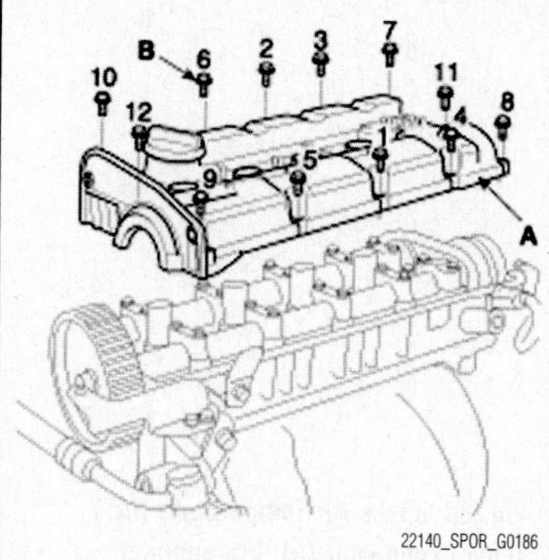

Fig. 80 Install the cylinder head cover (A) in sequence

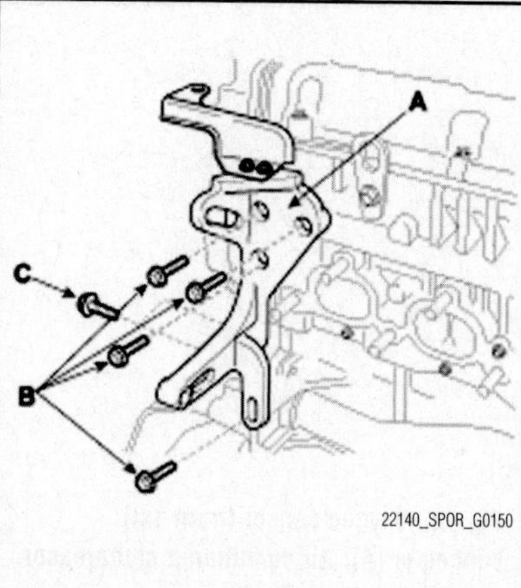

Fig. 81 Torque the power steering bracket bolts

 a. Connect the PCSV connector.
 b. Connect the front heated oxygen sensor connector.
 c. Connect the CKP connector.
 d. Connect the oil pressure switch connector.
 e. Connect the air conditioner compressor switch connector.
 f. Connect the ground cables to intake manifold.
 g. Connect the knock sensor connector.
 h. Connect the fuel injector connectors.
 i. Connect the CMP connector.
 j. Connect the ISA connector.
 k. Connect the TPS connector.
 l. Connect the ignition coil connector.
 m. Connect the ECT connector.
 n. Connect the oil temperature sensor connector.
 o. Connect the OCV connector.
45. Install the heater hose.
46. Install the upper radiator hose and lower radiator hose.
47. Install the intake air hose and air cleaner assembly.
 a. Install the intake air hose, air cleaner assembly and bolts. Tighten to 6–7 ft. lbs. (8–10 Nm).
 b. Install the breather hose to intake air hose.
 c. Connect the AFS connector.
48. Install the engine cover.
49. Reconnect the battery terminals.
50. Install the air duct.
51. Fill with engine coolant.
52. Start the engine and check for leaks.
53. Recheck engine coolant level and oil level.

2.7L Engine

See Figures 82 through 90.

1. Remove the air duct.
2. Remove the cowl grill and wiper motor.
3. Remove the strut bar.
4. Disconnect the negative terminal from the battery.
5. Drain the engine coolant. Remove the radiator cap to speed draining.
6. Remove the engine cover.
7. Remove the intake air hose and air cleaner assembly.
 a. Disconnect the Air Flow Sensor (AFS) connector.
 b. Disconnect the breather hose from air cleaner hose.

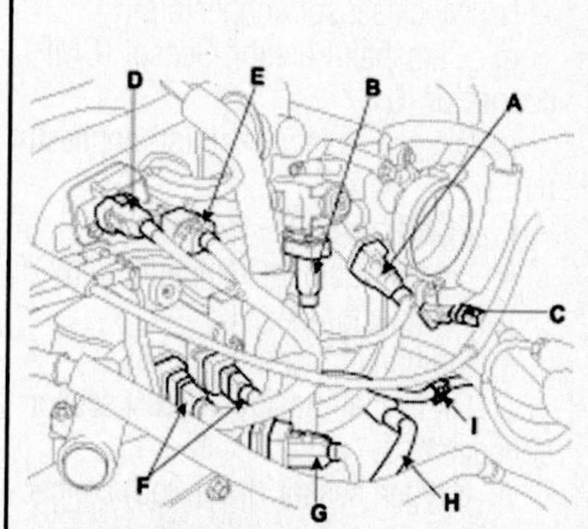

Fig. 82 Remove the engine wire harness connectors and wire harness clamps from the cylinder head and the intake manifold

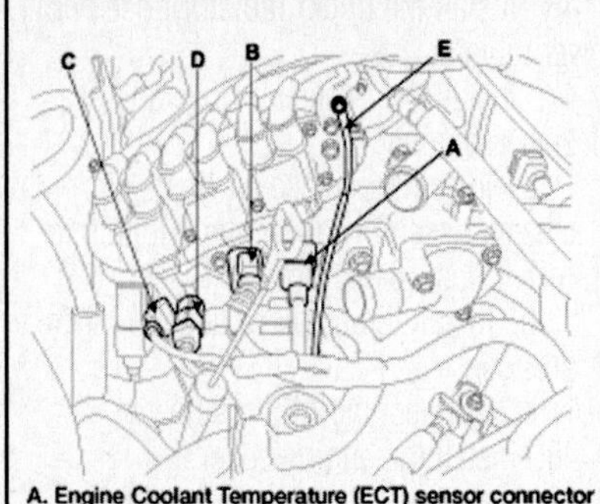

A. Engine Coolant Temperature (ECT) sensor connector
B. Ignition coil connector
C. Crankshaft Position Sensor (CKP) connector
D. Oxygen sensor (front 2nd) connector
E. Ground cable

22140_SPOR_G0192

Fig. 83 Additional engine harness connectors

c. Remove the intake air hose and air cleaner assembly.

8. Remove the coolant reservoir tank.

9. Remove the upper radiator hose and lower radiator hose.

10. Remove the heater hoses and throttle body heater hose.

11. Remove the engine wire harness connectors and wire harness clamps from the cylinder head and the intake manifold.

a. Throttle Position Sensor (TPS) connector (A).

b. Idle Speed Actuator (ISA) connector (B).

c. Purge Control Solenoid Valve (PCSV) connector (C).

d. VIS actuator connector (D).

e. Injector connector (E).

f. Knock sensor connectors (F).

g. Camshaft Position Sensor (CMP) connector (G).

h. Oxygen sensor (rear 1st) connector (H).

i. PCSV hose (I).

j. Engine Coolant Temperature (ECT) sensor connector (A).

k. Ignition coil connector (B).

l. Crankshaft position (CKP) sensor connector (C).

m. Oxygen sensor (front 2nd) connector (D).

n. Ground cable (E).

o. Three fuel injector connectors (A).

p. Disconnect ground cable from the cowl panel.

q. Intake Air Temperature (IAT) sensor connector (A).

r. VIS actuator connector (B).

s. Oxygen sensor (rear 2nd) connector (C).

t. Power steering pump switch (D).

u. Oxygen sensor (front 1st) connector (A).

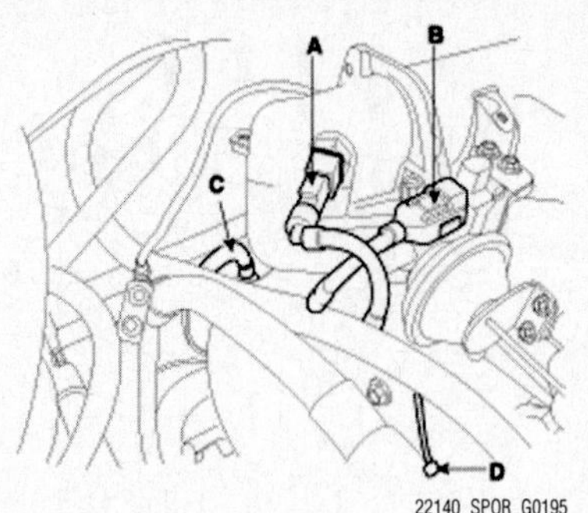

22140_SPOR_G0195

Fig. 84 Intake Air Temperature (IAT) sensor connector (A). VIS actuator connector (B). Oxygen sensor (rear 2nd) connector (C). Power steering pump switch (D).

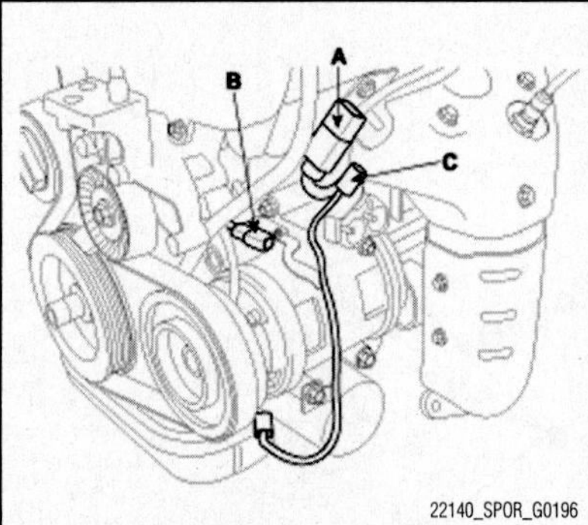

22140_SPOR_G0196

Fig. 85 Oxygen sensor (front 1st) connector (A). Air conditioner compressor switch connector (B). Oil pressure sensor connector (C).

v. Air conditioner compressor switch connector (B).

w. Oil pressure sensor connector (C).

12. Remove the fuel inlet hose (A) from delivery pipe.

13. Remove the brake booster vacuum hose.

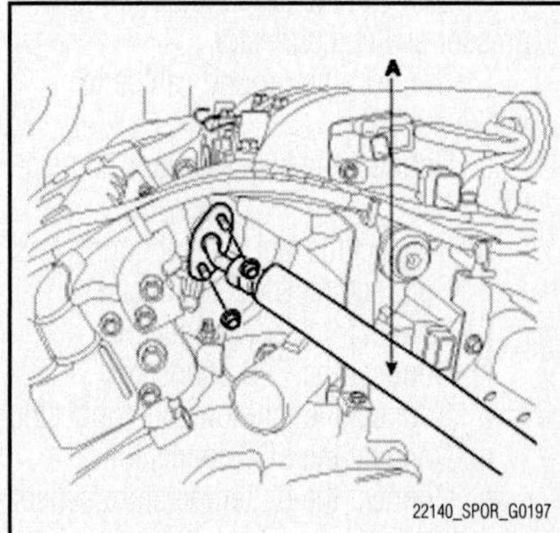

22140_SPOR_G0197

Fig. 86 Remove the fuel inlet hose (A) from delivery pipe

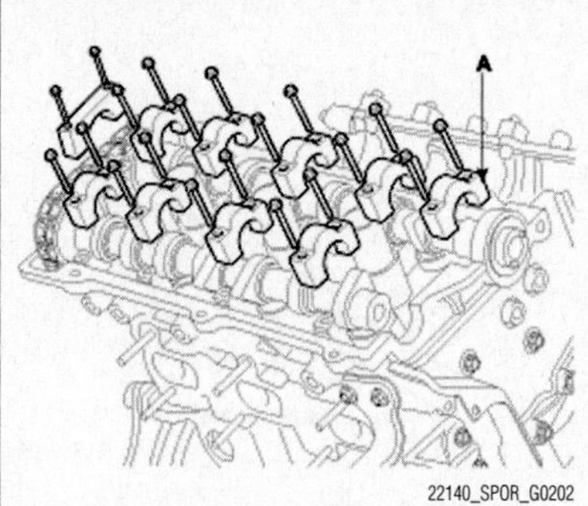

22140_SPOR_G0202

Fig. 87 Remove the camshaft bearing caps (A)

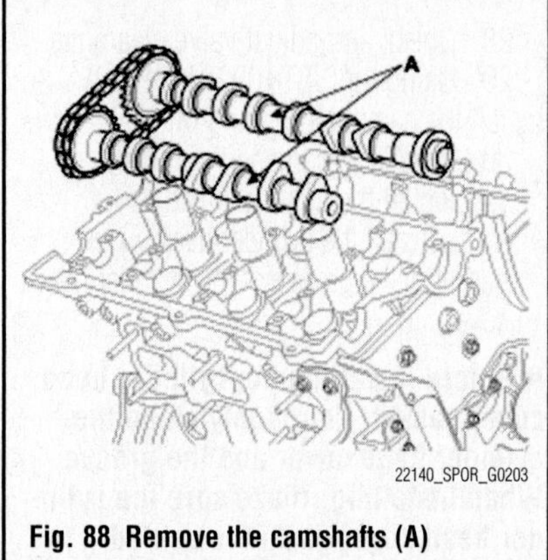

22140_SPOR_G0203

Fig. 88 Remove the camshafts (A)

14. Remove the accelerator cable by loosening the locknut, then slip the cable end out of the throttle linkage.

15. Remove the PCV hose.

16. Remove the intake manifold.

17. Remove the power steering pump.

18. Remove the exhaust manifold.

19. Remove the timing belt.

20. Remove the spark plug cable.

21. Remove the cylinder head covers.

22. Remove the camshaft sprocket.

23. Remove the camshaft bearing caps (A).

24. Remove the camshafts (A).

To install:

25. Install the camshafts.

a. Align the camshaft timing chain with the intake timing chain sprocket and exhaust timing chain sprocket.

b. Install the camshaft.

c. Install the camshaft bearing caps. Tighten to M6 (38mm): 7–9 ft. lbs. (10–12 Nm) (Mark7); M6 (50mm): 10–12 ft. lbs. (14–16 Nm) (Mark10)

➡Apply new engine oil to the thrust portion and journal of the camshafts. Apply a light coat of engine oil on the

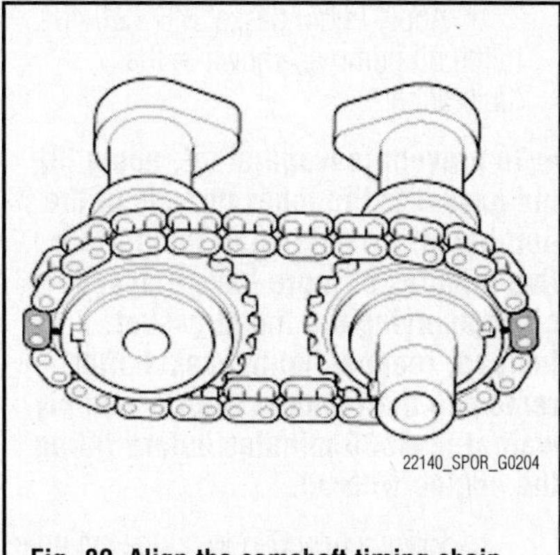

Fig. 89 Align the camshaft timing chain

threads and under the heads of the bearing cap bolts.

26. Using the SST (09214-21000), install the camshaft bearing oil seal.
27. Install the camshaft sprocket.
   a. Temporarily install the camshaft sprocket bolts.
   b. Hold the hexagonal head wrench portion of the camshaft with a wrench, and tighten the camshaft sprocket bolts. Tighten to Camshaft sprocket bolt: 65–80 ft. lbs. (90–110 Nm)
28. Install semi-circular packing.
29. Install the cylinder head cover.
   a. Install the cylinder head cover gasket in the groove of the cylinder head cover.

➡Before installing the head cover gasket, thoroughly clean the head cover gasket and the groove. When installing, make sure the head cover gasket is seated securely in the corners of the recesses with no gap.

   b. Apply liquid gasket to the head cover gasket at the corners of the recess.

➡Use liquid gasket, Loctite No.5699. Check that the mating surfaces are clean and dry before applying liquid gasket. After assembly, wait at least 30 minutes before filling the engine with oil.

   c. Install the cylinder head covers with the 16bolts. Uniformly tighten the bolts in several passes. Tighten to 6–7.4 ft. lbs. (8–10 Nm)
30. Install the spark plug cable.
31. Install the timing belt.
32. Install the exhaust manifold.
33. Install the power steering pump.
34. Install the intake manifold.
35. Install the PCV hose.
36. Install the accelerator cable.
37. Install the brake booster vacuum hose.

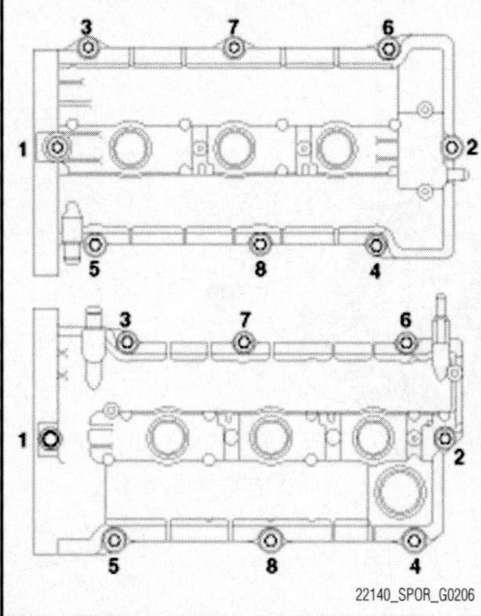

Fig. 90 Install the cylinder head cover bolts in sequence

38. Install the fuel inlet hose.
39. Install the engine wire harness connectors and wire harness clamps to the cylinder head and the intake manifold.
   a. Oil pressure sensor connector.
   b. Air conditioner compressor switch connector.
   c. Oxygen sensor (front 1st) connector.
   d. Power steering pump switch.
   e. Oxygen sensor (rear 2nd) connector.
   f. VIS actuator connector.
   g. IAT sensor connector.
   h. Connect the ground cable to the cowl panel.
   i. Three fuel injector connectors.
   j. Ground cable.
   k. Oxygen sensor connector.
   l. Crankshaft position sensor connector.
   m. Ignition coil connector.
   n. ECT sensor connector.
   o. PCSV hose.
   p. Oxygen sensor (rear 1st) connector.
   q. CMP connector.
   r. Knock sensor connector.
   s. Injector connector.
   t. VIS actuator connector.
   u. PCSV connector.
   v. ISA connector.
   w. TPS connector.
40. Install the heater hoses and throttle body heater hose.
41. Install the upper and lower radiator hose.
42. Install the coolant reservoir tank.
43. Install the intake air hose and air cleaner assembly.
   a. Install the intake air hose and air cleaner assembly.
   b. Connect the breather hose from air cleaner hose.
   c. Connect the AFS connector.
44. Install the engine cover.
45. Connect the negative terminal to the battery.
46. Install the strut bar.
47. Install the cowl grill and wiper motor.
48. Install the air duct.
49. Fill with engine coolant.
50. Start the engine and check for leaks.
51. Recheck engine coolant level and oil level.

CRANKSHAFT DAMPER

REMOVAL & INSTALLATION

1. Disconnect the negative battery cable.
2. Remove the accessory drive belts.
3. Remove the crankshaft damper.

To install:

4. Install the crankshaft damper and tighten the bolt.
   a. 2.0L engine: 123–130 ft. lbs. (167–176 Nm).
   b. 2.7L engine: 130–138 ft. lbs. (180–190 Nm).
5. Install and tension the accessory drive belts.

CRANKSHAFT FRONT SEAL

REMOVAL & INSTALLATION

2.0L Engine

See Figures 91 and 92.

1. Drain the engine oil.
2. Remove the drive belts.
3. Turn the crankshaft pulley, and align its groove with timing mark "T" of the timing belt cover.
4. Remove the timing belt. See Timing Belt.
5. Remove the bolt and timing belt idler.
6. Remove the oil pan and oil screen.
7. Remove the alternator. See Alternator.
8. Remove the air conditioner compressor tensioner bracket.
9. Remove the bolts and front case.

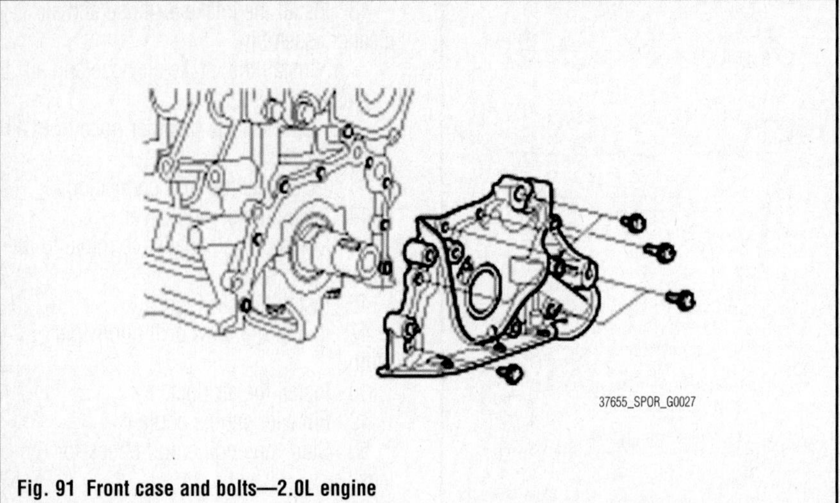

Fig. 91 Front case and bolts—2.0L engine

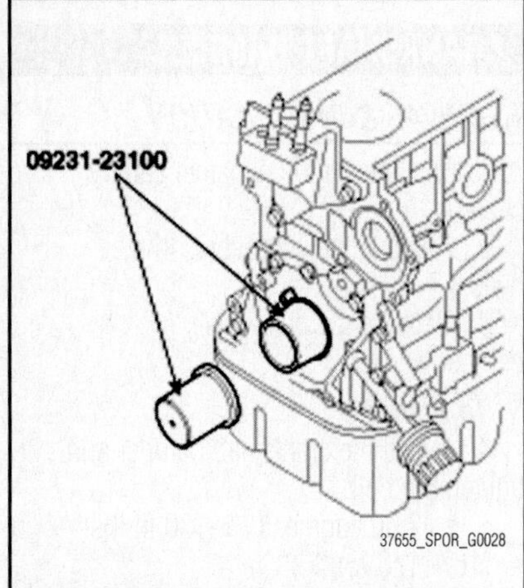

Fig. 92 Special tool to install oil seal—2.0L engine

To install:

10. Place a new front case gasket on the cylinder block.
11. Apply engine oil to the lip of the oil pump seal. Then, install the front case onto the crankshaft.
12. Tighten front case bolts to 14.5 - 19.5 ft. lb. (19.6 - 26.5 Nm).
13. When the pump is in place, clean any excess grease off the crankshaft and check that the oil seal lip is not distorted.
14. Apply a light coat of oil to the front case oil seal lip.
15. Using special service tools (09231-23100), install the front case oil seal.
16. Install remaining components in the reverse order of removal

2.7L Engine

See Figure 93.

1. Drain engine oil.
2. Remove RH front wheel.
3. Remove RH side cover.
4. Remove the front exhaust pipe. See Catalytic Converter.
5. Remove the alternator from engine. See Alternator.
6. Remove the drive belt.
7. Turn the crankshaft and align the white groove on the crankshaft pulley with the pointer on the lower cover.
8. Remove the timing belt. See Timing Belt.
9. Remove the oil pan and oil screen.
10. Remove the oil pump case.

To install:

11. Install the oil pump on the cylinder block.
    a. Remove any old liquid gasket and be careful not to drop any oil on the contact surfaces of the oil pump and cylinder block. Using a razor blade and gasket scraper, remove all the old liquid gasket from the gasket surfaces and sealing grooves. Using a non-residue solvent, clean both sealing surfaces.

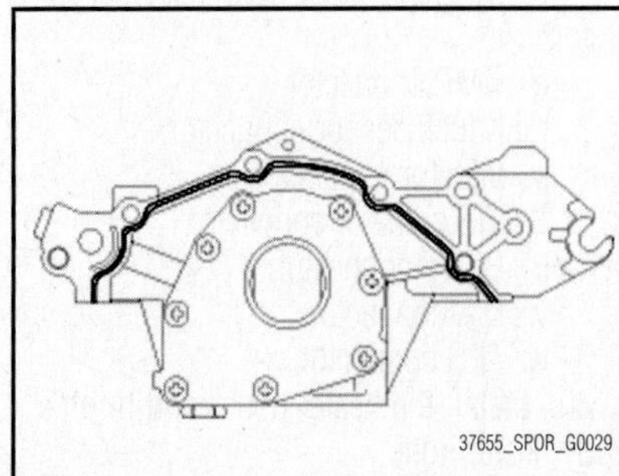

Fig. 93 Gasket application to oil pump—2.7L engine

    b. Apply liquid gasket MS 721-40A to the oil pump as shown in the illustration.

➡To prevent leakage of oil, apply liquid gasket to the inner threads of the bolt holes. Do not install the parts if five minutes or more have elapsed since applying the liquid gasket. Instead, reapply liquid gasket after removing the residue. After assembly, wait at least 30 minutes before filling the engine with oil.

    c. Place a new O-ring on the cylinder block.
    d. Engage the spline teeth of the oil pump drive gear with large teeth of the crankshaft, and slide the oil pump on the crankshaft.
    e. Install the oil pump with 5 bolts. Uniformly tighten the bolts in several passes to 8.8 - 11 ft. lb. (12 - 15Nm).
12. Apply a light coat of oil to the seal lip.
13. Using the special tool(09214-33000), install the oil seal.
14. Install the oil pan and oil screen.
15. Install the timing belt.
16. Install the drive belt
17. Install the alternator.
18. Install the front exhaust pipe.
19. Install the RH front wheel.
20. Fill engine with oil.
21. Start engine and check for leaks.
22. Recheck engine oil level.

CYLINDER HEAD

REMOVAL & INSTALLATION

2.0L Engine

See Figures 94 through 108.

1. Remove the air duct.
2. Disconnect the terminals from battery.
3. Remove the engine cover.
4. Drain the engine coolant.
5. Remove the intake air hose and air cleaner assembly.
    a. Disconnect the Air Flow Sensor (AFS) connector.
    b. Disconnect the breather hose from intake air hose.
    c. Remove the intake air hose and air cleaner assembly.
6. Remove the upper radiator hose and lower radiator hose.
7. Remove the upper radiator hose and lower radiator hose.
8. Remove the heater hoses.

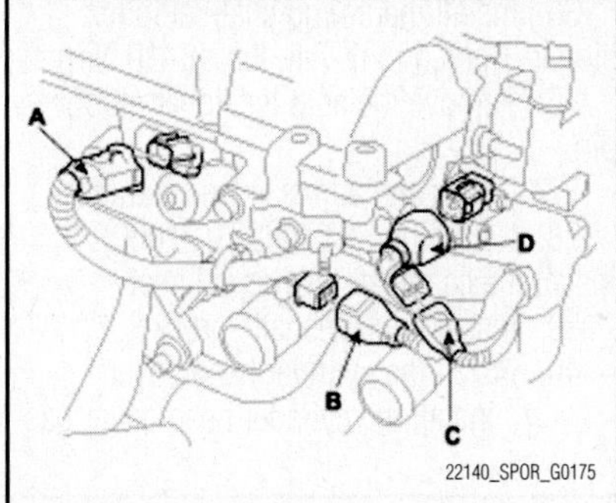

Fig. 94 Remove the engine wire harness connectors

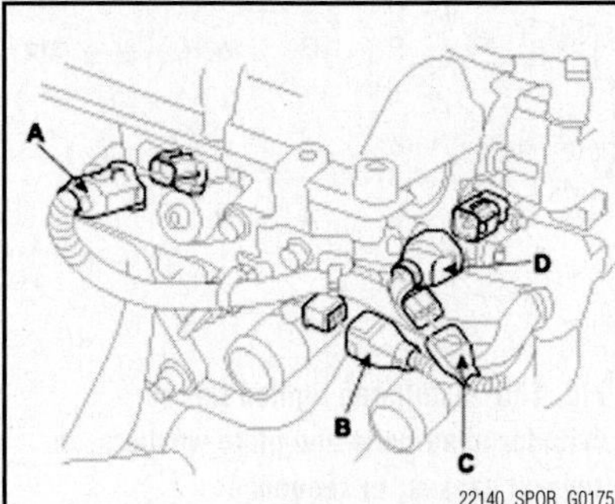

Fig. 95 Disconnect the Throttle Position Sensor (TPS) connector (A) and the Idle Speed Actuator (ISA) connector (B)

9. Remove the accelerator cable by loosening the lock-nut, then slip the cable end out of the throttle linkage.

10. Remove the engine wire harness connectors and wire harness clamps from cylinder head and the intake manifold.

a. Disconnect the Oil Control Valve (OCV) connector (A).

b. Disconnect the oil temperature sensor connector (B).

c. Disconnect the Engine Coolant Temperature (ECT) sensor connector (C).

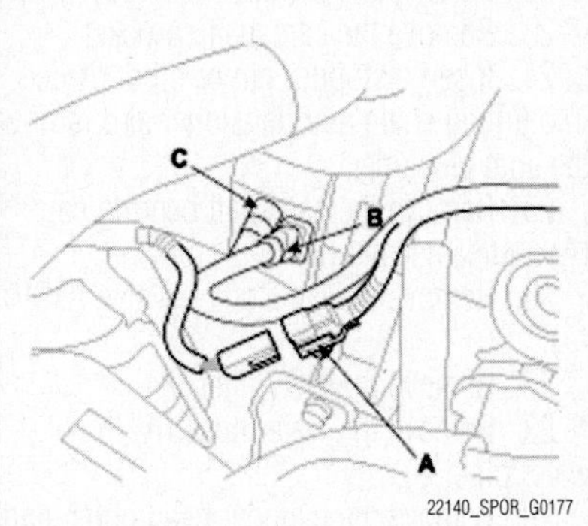

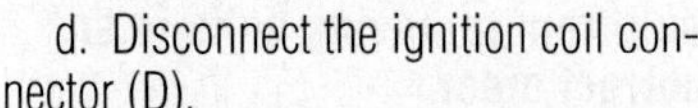

Fig. 97 Disconnect the front heated oxygen sensor connector (A), the Crankshaft Position Sensor (CKP) connector (B) and the oil pressure switch connector (C)

d. Disconnect the ignition coil connector (D).

e. Disconnect the Throttle Position Sensor (TPS) connector (A).

f. Disconnect the Idle Speed Actuator (ISA) connector (B).

g. Disconnect the Camshaft Position Sensor (CMP) connector (A).

h. Disconnect the four injector connectors (B).

i. Disconnect the knock sensor connector (C).

j. Disconnect the ground cables (D) from the intake manifold.

k. Disconnect the air conditioner compressor switch connector (E).

l. Disconnect the front heated oxygen sensor connector (A).

m. Disconnect the Crankshaft Position Sensor (CKP) connector (B).

n. Disconnect the oil pressure switch connector (C).

o. Disconnect the Purge Control Solenoid Valve (PCSV) connector (A).

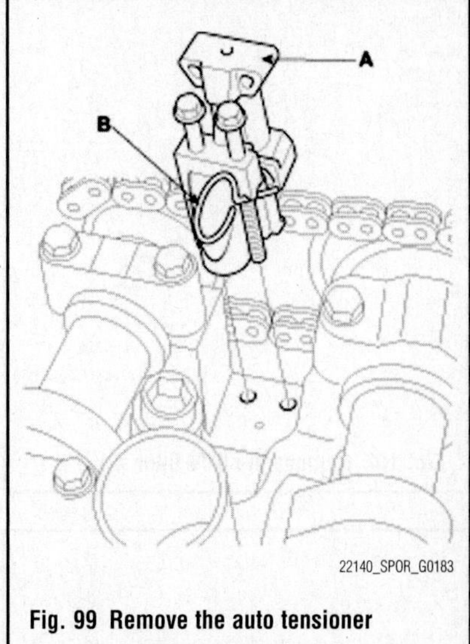

Fig. 99 Remove the auto tensioner

11. Disconnect the fuel inlet hose (A) of the delivery pipe side.

12. Disconnect the hose (A) of the PCSV side.

13. Remove the brake booster vacuum hose (A).

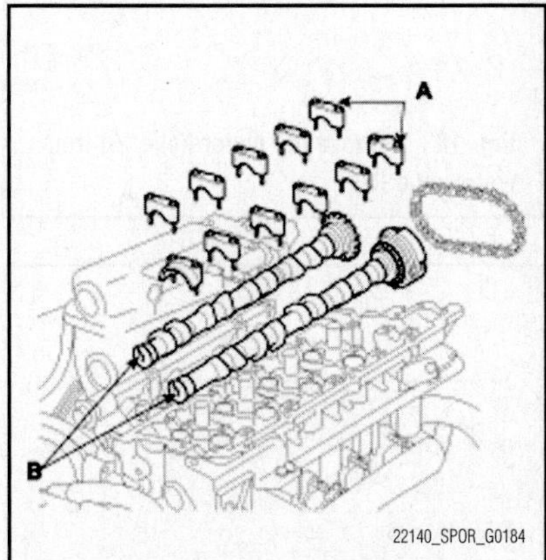

Fig. 100 Remove the camshaft bearing caps (A) and camshafts (B)

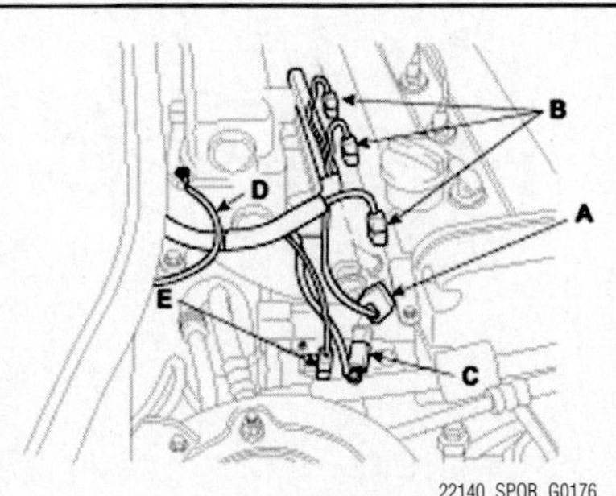

Fig. 96 Disconnect the Camshaft Position Sensor (CMP) connector (A), the four injector connectors (B), the knock sensor connector (C), the ground cables (D) and the air conditioner compressor switch connector (E)

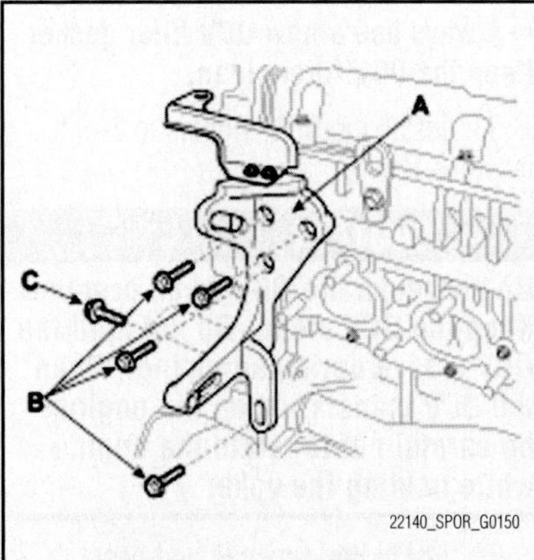

Fig. 98 Remove the bolts (B, C) and power steering pump bracket (A)

Fig. 101 Remove the Oil Control Valve (OCV) (A)

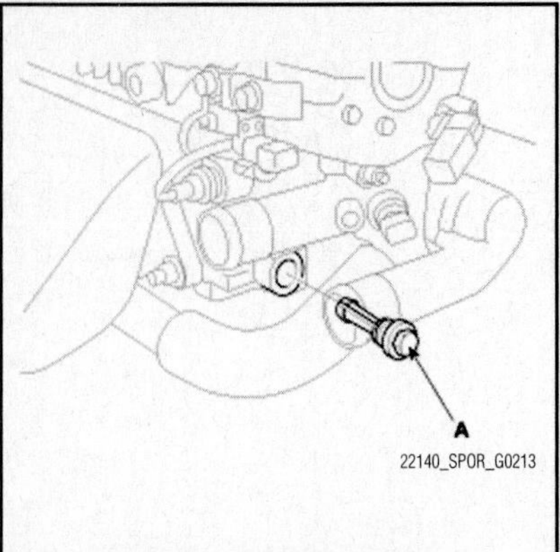

Fig. 102 Remove the OCV filter (A)

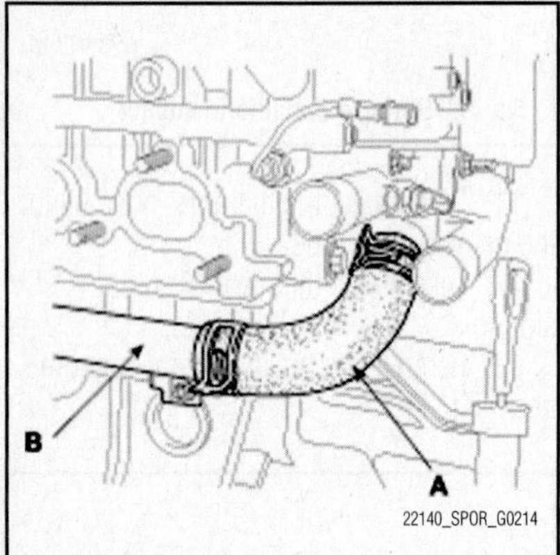

Fig. 103 Remove the water hose (A) from water pipe (B)

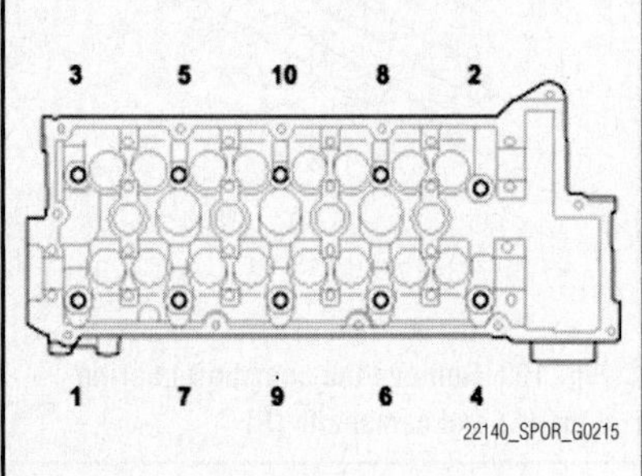

Fig. 104 Remove the 10 cylinder head bolts and plate washers, in several passes, in sequence

14. Remove the power steering pump drive belt.
15. Remove the power steering pump and use a wire to secure the pump to the vehicle so that it is out of the way.
16. Remove the bolts (B, C) and power steering pump bracket (A).
17. Remove the spark plug cables.
18. Remove the exhaust manifold.
19. Remove the intake manifold.
20. Remove the timing belt.
21. Remove the PCV hose.
22. Remove the cylinder head cover.
23. Remove the camshaft sprocket.
24. Insert a stopper pin or other device into timing chain auto tensioner and remove the auto tensioner.
25. Remove the camshaft bearing caps (A) and camshafts (B).
26. Remove the Oil Control Valve (OCV) (A).
27. Remove the OCV filter (A).
28. Remove the water hose (A) from water pipe (B).
29. Remove the cylinder head bolts, and remove the cylinder head.
    a. Uniformly loosen and remove the 10 cylinder head bolts and plate washers, in several passes, in sequence.

CAUTION

Head warpage or cracking could result from removing bolts in an incorrect order.

    b. Lift the cylinder head from the dowels on the cylinder block and replace the cylinder head on wooden blocks on a bench.

To install:

30. Install the cylinder head gasket on the cylinder block.

➡Be careful of the installation direction.

31. Install the cylinder head bolts.
    a. Apply a light coat if engine oil on the threads and under the heads of the cylinder head bolts.
    b. Install and tighten the 10 cylinder head bolts and plate washers, in several passes, in sequence. Tighten to M10: 1677–20 ft. lbs. (23–27 Nm+60°–65° + 60°–65°; M12: 20–23 ft. lbs. (28–31 Nm)+60°–65° +60°–65°
32. Install the OCV filter. Tighten to 30–37 ft. lbs. (40–50 Nm)

➡Always use a new OCV filter gasket. Keep the OCV filter clean.

33. Install the OCV. Tighten to 7–9 ft. lbs. (10–12 Nm)

CAUTION

Do not reuse the OCV when dropped. Keep the OCV clean. Do not hold the OCV sleeve during servicing. When the OCV is installed on the engine, be careful not to rotate the engine while holding the yoke.

34. Install the camshaft and bearing caps. Tighten to 10–11 ft. lbs. (14–15 Nm).
35. Install the timing chain auto tensioner. Tighten to 6–7 ft. lbs. (8–10 Nm).
36. Remove the auto tensioner stopper pin.
37. Check and adjust valve clearance.
38. Using the SST (09221-21000), install the camshaft bearing oil seal.
39. Install the camshaft sprocket.
40. Install the cylinder head cover.
    a. Install the cylinder head cover gas-

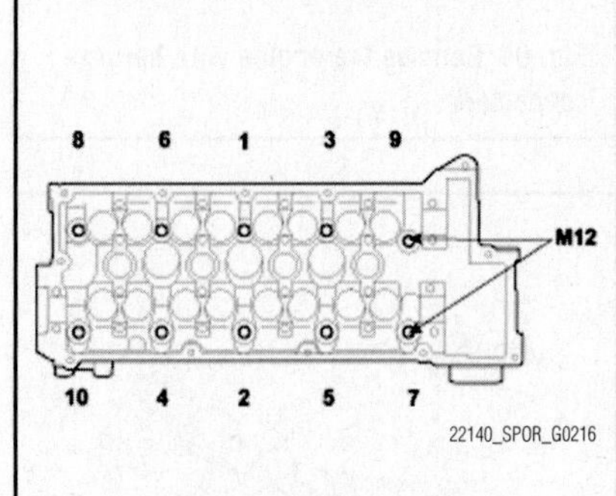

Fig. 105 Install and tighten the 10 cylinder head bolts and plate washers, in several passes, in sequence

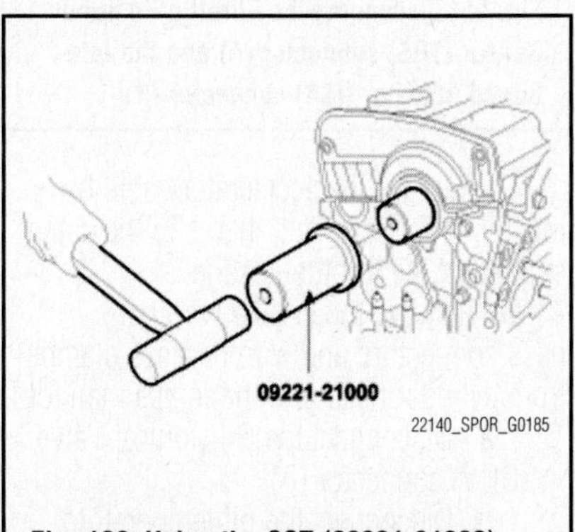

Fig. 106 Using the SST (09221-21000), install the camshaft bearing oil seal

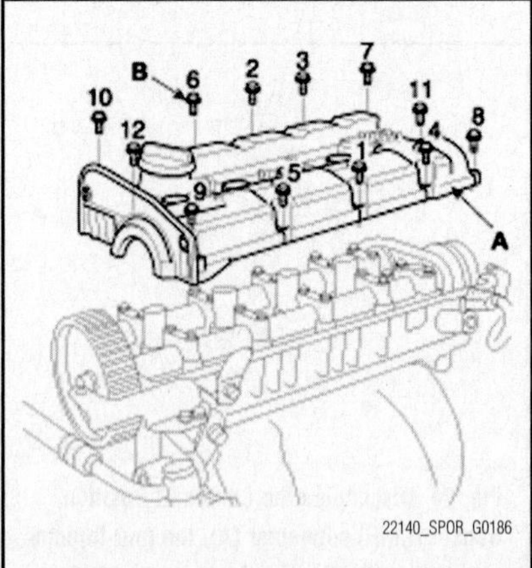

Fig. 107 Install the cylinder head cover (A) in sequence

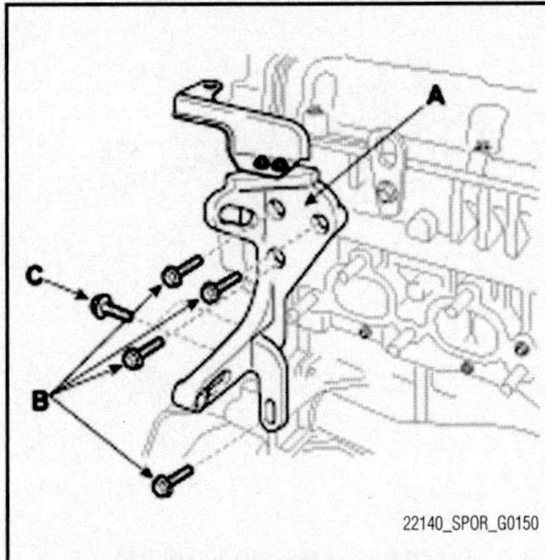

Fig. 108 Torque the power steering bracket bolts

ket in the groove of the cylinder head cover.

➡**Before installing the cylinder head cover gasket, thoroughly clean the cylinder head cover and the groove. When installing, make sure the cylinder head cover gasket is seated securely in the corners of the recesses with no gap.**

b. Apply liquid gasket to the head cover gasket at the corners of the recess.

➡**Use liquid gasket, Loctite No. 5999. Check that the mating surfaces are clean and dry before applying liquid gasket. After assembly, wait at least 30 minutes before filling the engine with oil.**

41. Install the cylinder head cover (A) with the 12 bolts (B). Uniformly tighten the bolts in several passes. Tighten to 6–7 ft. lbs. (8–10 Nm).
42. Install the PCV hose.
43. Install the timing belt.
44. Install the intake manifold.
45. Install the exhaust manifold.
46. Install the spark plug cables.
47. Install the power steering pump bracket and bolts. Tighten to Bolt (B): 25–36 ft. lbs. (34–49 Nm); Bolt (C): 9–11 ft. lbs. (12–15 Nm).
48. Install the power steering pump.
49. Install the accelerator cable.
50. Install the brake booster hose.
51. Connect the hose of the PCSV side.
52. Connect the fuel inlet hose of the delivery pipe side.
53. Install the engine wire harness connectors and wire harness clamps to the cylinder head and the intake manifold.
   a. Connect the PCSV connector.
   b. Connect the front heated oxygen sensor connector.
   c. Connect the CKP connector.
   d. Connect the oil pressure switch connector.
   e. Connect the air conditioner compressor switch connector.
   f. Connect the ground cables to intake manifold.
   g. Connect the knock sensor connector.
   h. Connect the fuel injector connectors.
   i. Connect the CMP connector.
   j. Connect the ISA connector.
   k. Connect the TPS connector.
   l. Connect the ignition coil connector.
   m. Connect the ECT connector.
   n. Connect the oil temperature sensor connector.
   o. Connect the OCV connector.
54. Install the heater hose.
55. Install the upper radiator hose and lower radiator hose.
56. Install the intake air hose and air cleaner assembly.
   a. Install the intake air hose, air cleaner assembly and bolts. Tighten to 6–7 ft. lbs. (8–10 Nm).
   b. Install the breather hose to intake air hose.
   c. Connect the AFS connector.
57. Install the engine cover.
58. Reconnect the battery terminals.
59. Install the air duct.
60. Fill with engine coolant.
61. Start the engine and check for leaks.
62. Recheck engine coolant level and oil level.

2.7L Engine

See Figures 109 through 124.

1. Remove the air duct.
2. Remove the cowl grill and wiper motor.
3. Remove the strut bar.
4. Disconnect the negative terminal from the battery.
5. Drain the engine coolant. Remove the radiator cap to speed draining.
6. Remove the engine cover.
7. Remove the intake air hose and air cleaner assembly.
   a. Disconnect the Air Flow Sensor (AFS) connector.
   b. Disconnect the breather hose from air cleaner hose.
   c. Remove the intake air hose and air cleaner assembly.
8. Remove the coolant reservoir tank.
9. Remove the upper radiator hose and lower radiator hose.

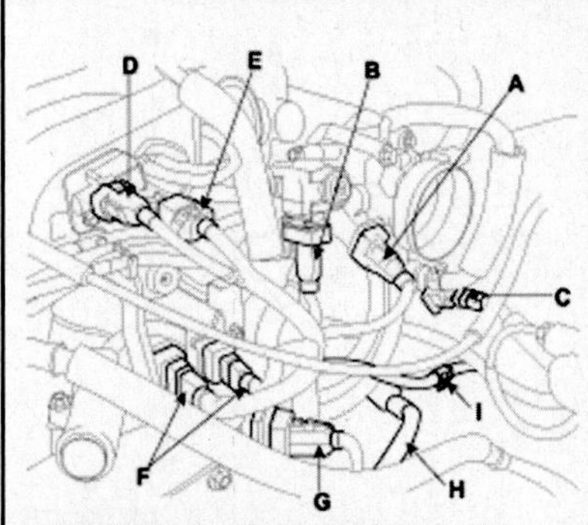

Fig. 109 Remove the engine wire harness connectors and wire harness clamps.

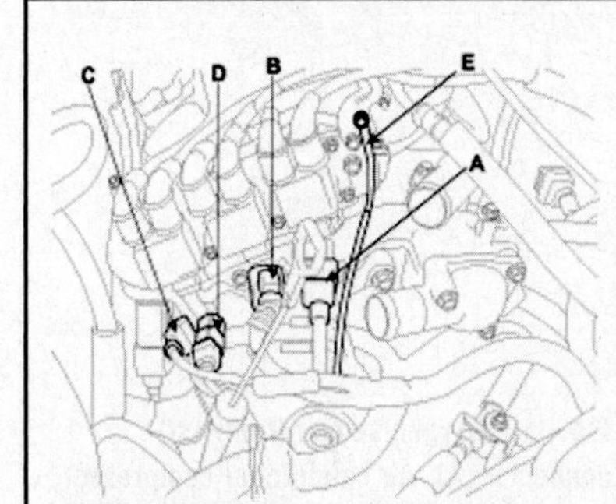

Fig. 110 Additional engine harness connectors

10. Remove the heater hoses and throttle body heater hose.
11. Remove the engine wire harness connectors and wire harness clamps from the cylinder head and the intake manifold.
   a. Throttle Position Sensor (TPS) connector (A).
   b. Idle Speed Actuator (ISA) connector (B).
   c. Purge Control Solenoid Valve (PCSV) connector (C).
   d. VIS actuator connector (D).
   e. Injector connector (E).
   f. Knock sensor connectors (F).
   g. Camshaft Position Sensor (CMP) connector (G).
   h. Oxygen sensor (rear 1st) connector (H).

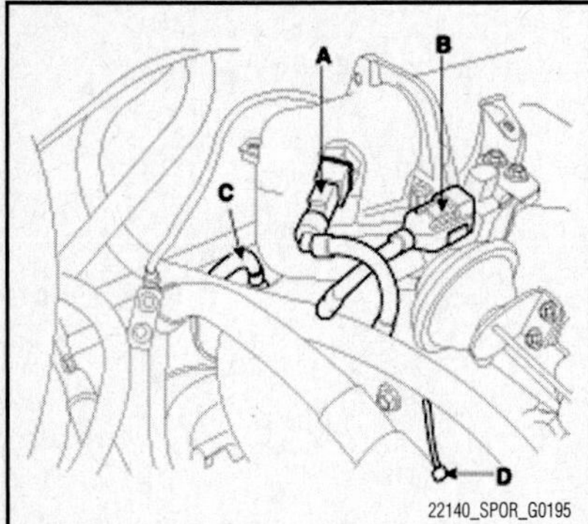

Fig. 111 Intake Air Temperature (IAT) sensor connector (A). VIS actuator connector (B). Oxygen sensor (rear 2nd) connector (C). Power steering pump switch (D).

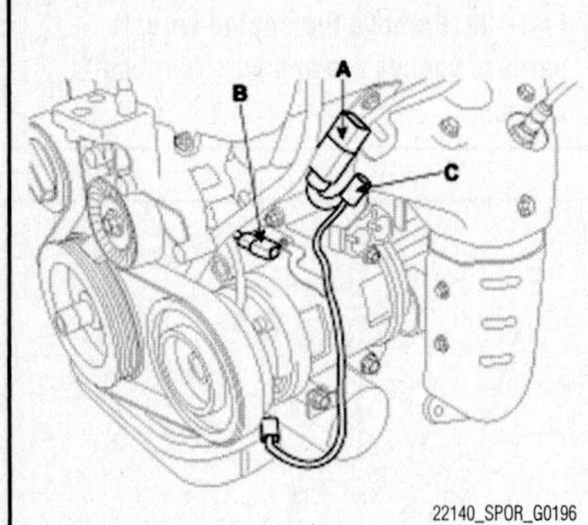

Fig. 112 Oxygen sensor (front 1st) connector (A). Air conditioner compressor switch connector (B). Oil pressure sensor connector (C).

i. PCSV hose (I).

j. Engine Coolant Temperature (ECT) sensor connector (A).

k. Ignition coil connector (B).

l. Crankshaft position (CKP) sensor connector (C).

m. Oxygen sensor (front 2nd) connector (D).

n. Ground cable (E).

o. Three fuel injector connectors (A).

p. Disconnect ground cable (A) from the cowl panel.

q. Intake Air Temperature (IAT) sensor connector (A).

r. VIS actuator connector (B).

s. Oxygen sensor (rear 2nd) connector (C).

t. Power steering pump switch (D).

u. Oxygen sensor (front 1st) connector (A).

v. Air conditioner compressor switch connector (B).

w. Oil pressure sensor connector (C).

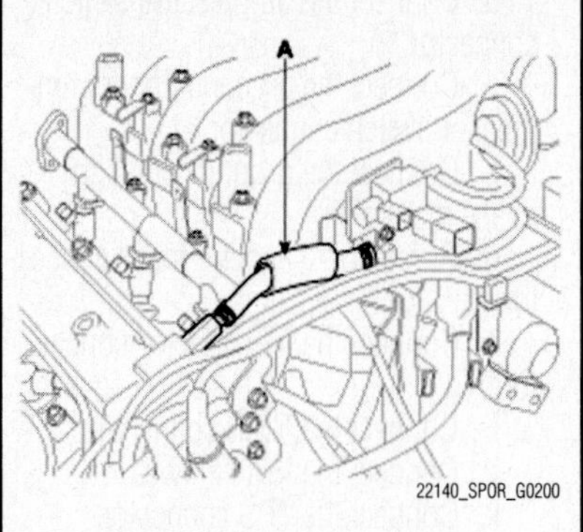

Fig. 113 Remove the PCV hose (A).

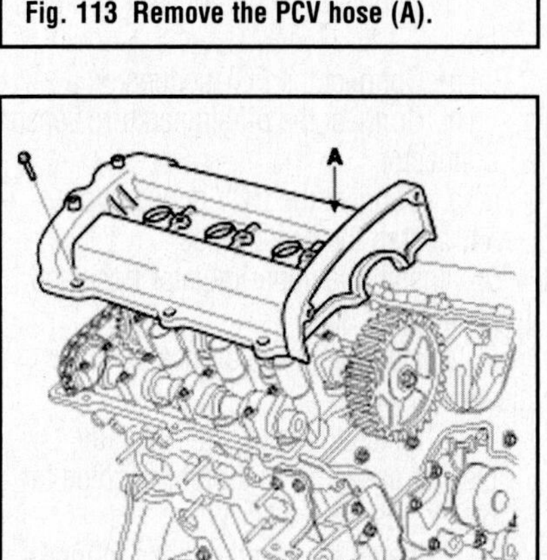

Fig. 114 Remove the cylinder head covers (A).

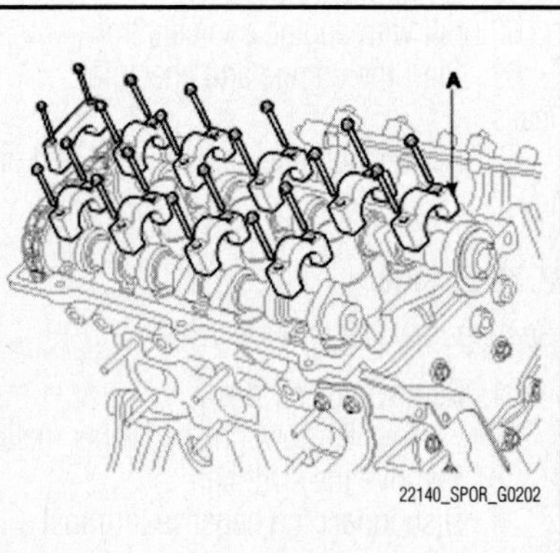

Fig. 115 Remove the camshaft bearing caps (A).

12. Remove the fuel inlet hose (A) from delivery pipe.
13. Remove the brake booster vacuum hose (A).
14. Remove the accelerator cable by loosening the locknut, then slip the cable end out of the throttle linkage.
15. Remove the PCV hose (A).
16. Remove the intake manifold.
17. Remove the power steering pump.
18. Remove the exhaust manifold.
19. Remove the timing belt.

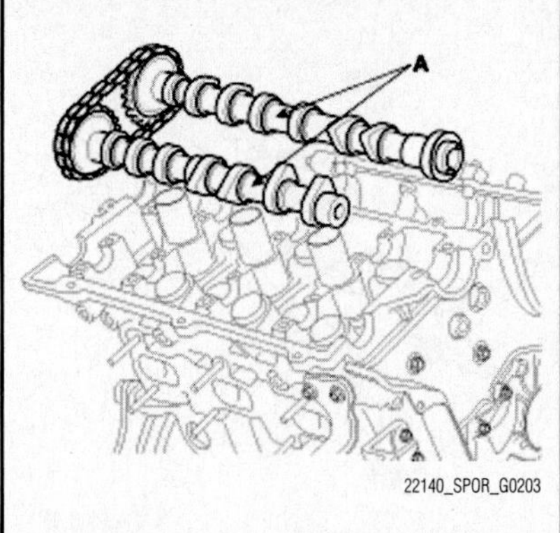

Fig. 116 Remove the camshafts (A).

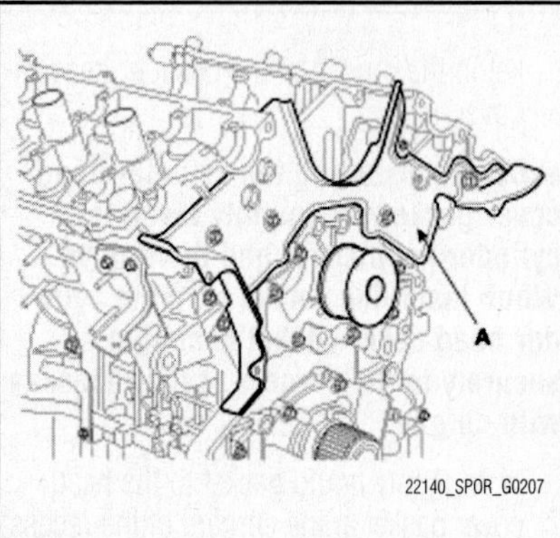

Fig. 117 Remove the timing belt rear cover (A).

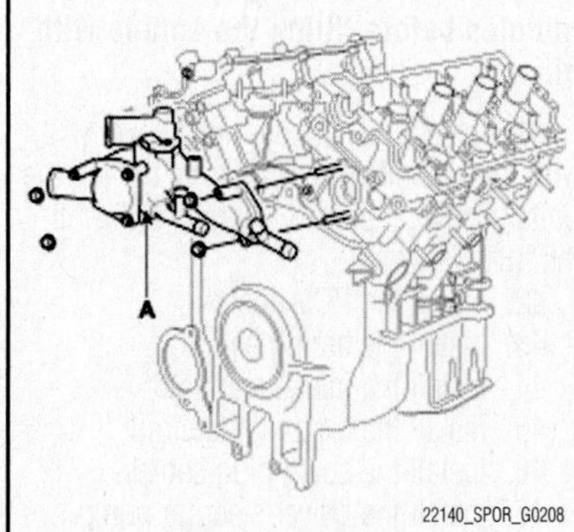

Fig. 118 Remove the water temperature control assembly (A) and water pipe.

20. Remove the spark plug cable.
21. Remove the cylinder head covers (A).
22. Remove the camshaft sprocket.
23. Remove the camshaft bearing caps (A).
24. Remove the camshafts (A).
25. Remove the timing belt rear cover (A).
26. Remove the water temperature control assembly (A) and water pipe.
27. Remove the cylinder head bolts, and remove the cylinder heads.

a. Uniformly loosen and remove the 8 cylinder head bolts and plate washers on

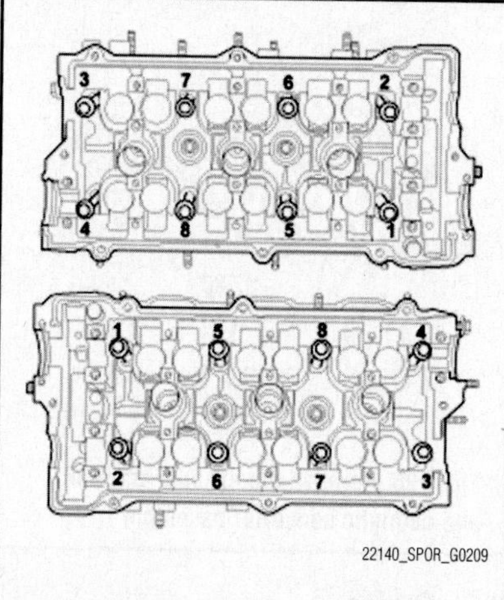

Fig. 119 Remove the 16 cylinder head bolts in several passes and in sequence.

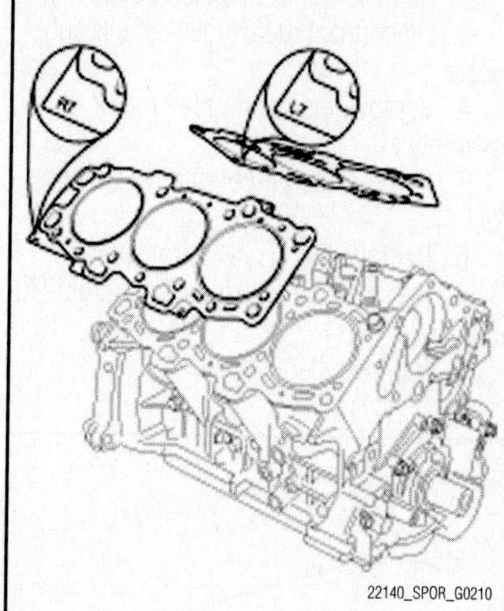

Fig. 120 Install the cylinder head gaskets on the cylinder block.

each cylinder head in several passes and in sequence.

⁂ CAUTION

Head warpage or cracking could result from removing bolts in an incorrect order.

b. Lift the cylinder head from the dowels on the cylinder block and place the cylinder head on wooden blocks on a bench.

To install:

28. Install the cylinder head gaskets on the cylinder block.

➡Be careful of the installation direction.

29. Place the cylinder head quietly in order not to damage the gasket with the bottom part of the end.

30. Install cylinder head bolts.

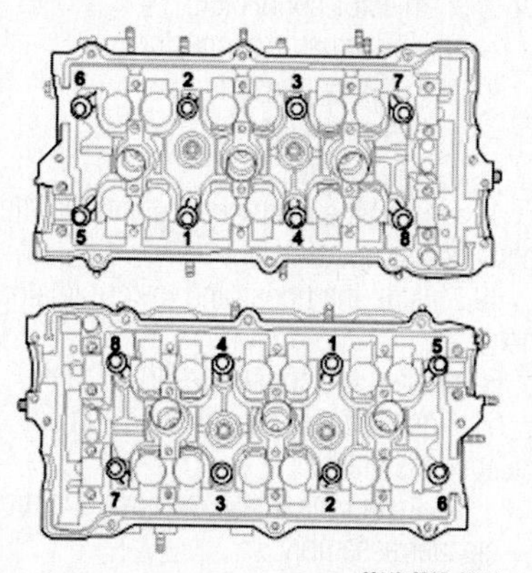

Fig. 121 Install and uniformly tighten the cylinder head bolts on each cylinder head in several passes and in sequence

a. Apply a light coat if engine oil on the threads and under the heads of the cylinder head bolts.

b. Install the plate washer to the cylinder head bolt.

c. Install and uniformly tighten the cylinder head bolts on each cylinder head in several passes and in sequence.

⁂ CAUTION

If any of the cylinder head bolts does not meet the torque specification, repeat the cylinder head bolt torque tightening sequence.

d. Tighten to 218 ft. lbs. (5 Nm)

e. Retighten the cylinder head bolts by 60° in the same sequential order.

f. Again, retighten the cylinder head bolts by 45° in the same sequential order.

31. Install the water pipe and water temperature control assembly. Tighten to 11–14 ft. lbs. (15–20 Nm)

32. Install the timing belt rear cover. Tighten to 7–9 ft. lbs. (10–12 Nm)

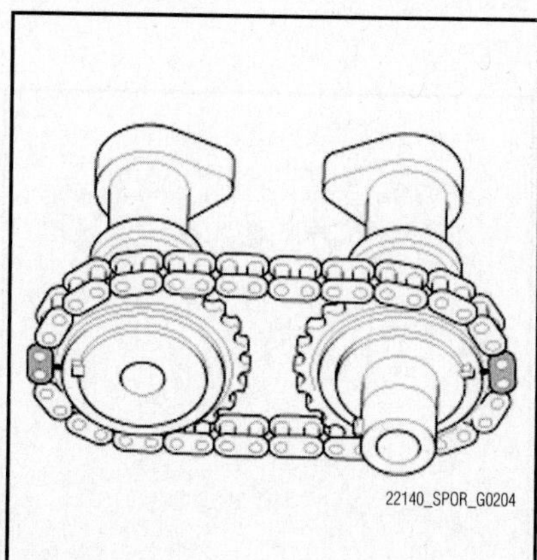

Fig. 122 Align the camshaft timing chain

33. Install the camshafts.

a. Align the camshaft timing chain with the intake timing chain sprocket and exhaust timing chain sprocket.

b. Install the camshaft.

c. Install the camshaft bearing caps. Tighten to M6 (38mm): 7–9 ft. lbs. (10–12 Nm) (Mark7); M6 (50mm): 10–12 ft. lbs. (14–16 Nm) (Mark10)

➡Apply new engine oil to the thrust portion and journal of the camshafts. Apply a light coat of engine oil on the threads and under the heads of the bearing cap bolts.

34. Using the SST (09214-21000), install the camshaft bearing oil seal.

35. Install the camshaft sprocket.

a. Temporarily install the camshaft sprocket bolts.

b. Hold the hexagonal head wrench portion of the camshaft with a wrench, and tighten the camshaft sprocket bolts. Tighten to Camshaft sprocket bolt: 65–80 ft. lbs. (90–110 Nm)

36. Install semi-circular packing.

37. Install the cylinder head cover.

a. Install the cylinder head cover gasket in the groove of the cylinder head cover.

➡Before installing the head cover gasket, thoroughly clean the head cover gasket and the groove. When installing, make sure the head cover gasket is seated securely in the corners of the recesses with no gap.

b. Apply liquid gasket to the head cover gasket at the corners of the recess.

➡Use liquid gasket, Loctite No.5699. Check that the mating surfaces are clean and dry before applying liquid gasket. After assembly, wait at least 30 minutes before filling the engine with oil.

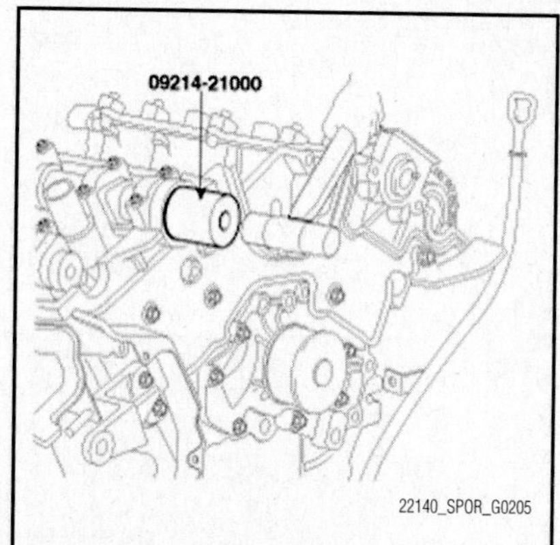

Fig. 123 Using the SST (09214-21000), install the camshaft bearing oil seal.

c. Install the cylinder head covers with the 16bolts. Uniformly tighten the bolts in several passes. Tighten to 6–7.4 ft. lbs. (8–10 Nm)

38. Install the spark plug cable.
39. Install the timing belt.
40. Install the exhaust manifold.
41. Install the power steering pump.
42. Install the intake manifold.
43. Install the PCV hose.
44. Install the accelerator cable.
45. Install the brake booster vacuum hose.
46. Install the fuel inlet hose.
47. Install the engine wire harness connectors and wire harness clamps to the cylinder head and the intake manifold.
    a. Oil pressure sensor connector.
    b. Air conditioner compressor switch connector.
    c. Oxygen sensor (front 1st) connector.
    d. Power steering pump switch.
    e. Oxygen sensor (rear 2nd) connector.
    f. VIS actuator connector.
    g. IAT sensor connector.
    h. Connect the ground cable to the cowl panel.
    i. Three fuel injector connectors.
    j. Ground cable.
    k. Oxygen sensor connector.
    l. Crankshaft position sensor connector.
    m. Ignition coil connector.
    n. ECT sensor connector.
    o. PCSV hose.
    p. Oxygen sensor (rear 1st) connector.
    q. CMP connector.
    r. Knock sensor connector.
    s. Injector connector.
    t. VIS actuator connector.
    u. PCSV connector.
    v. ISA connector.
    w. TPS connector.
48. Install the heater hoses and throttle body heater hose.
49. Install the upper and lower radiator hose.
50. Install the coolant reservoir tank.
51. Install the intake air hose and air cleaner assembly.
    a. Install the intake air hose and air cleaner assembly.
    b. Connect the breather hose from air cleaner hose.
    c. Connect the AFS connector.
52. Install the engine cover.
53. Connect the negative terminal to the battery.
54. Install the strut bar.
55. Install the cowl grill and wiper motor.
56. Install the air duct.
57. Fill with engine coolant.
58. Start the engine and check for leaks.
59. Recheck engine coolant level and oil level.

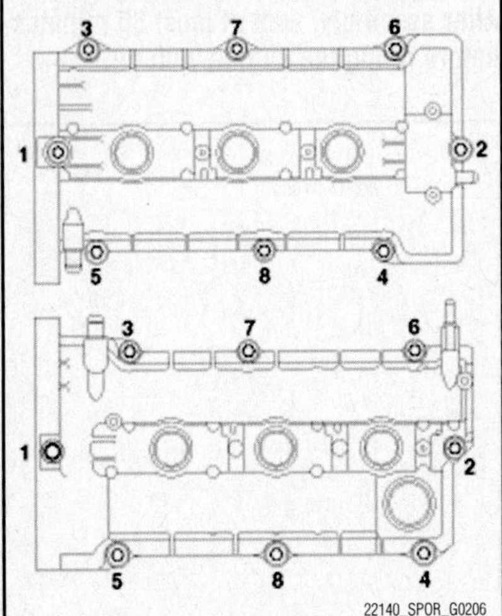

Fig. 124 Install the cylinder head cover bolts in sequence

EXHAUST MANIFOLD

REMOVAL & INSTALLATION

2.0L Engine

See Figures 125 and 126.

1. Remove the engine cover.
2. Disconnect the front oxygen sensor connector.
3. Remove the front muffler.
4. Remove the heat protector.
5. Remove the exhaust manifold and catalytic converter assembly.
6. Installation is in the reverse order of removal

Fig. 125 Remove the heat protector

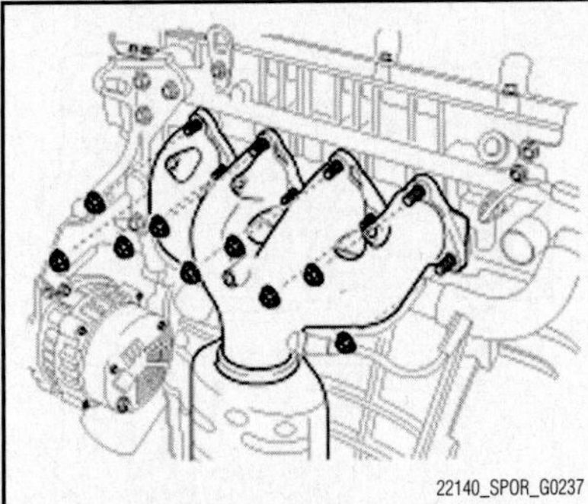

Fig. 126 Remove the exhaust manifold and catalytic converter assembly

2.7L Engine

See Figures 127 through 130.

1. Remove the undercover (A).
2. Remove the front exhaust pipe (A).
3. Disconnect the oxygen sensor connector.
4. Remove the cooling fan motor assembly.
5. Remove the air conditioner compressor.
6. Remove the LH heat protector (A).
7. Remove the LH exhaust manifold (A) and gasket (B).
8. Remove the alternator.

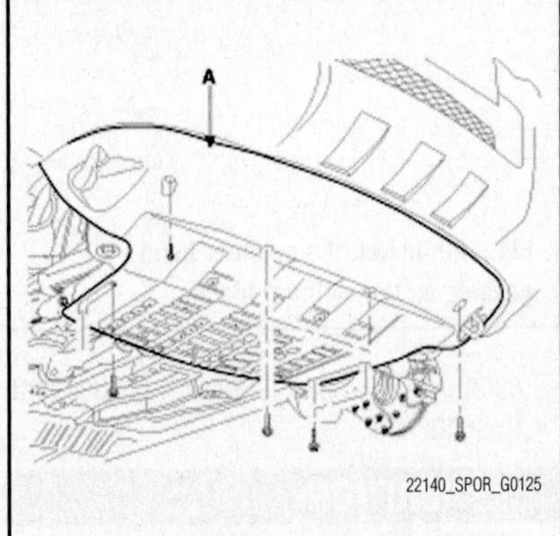

Fig. 127 Remove the undercover (A)

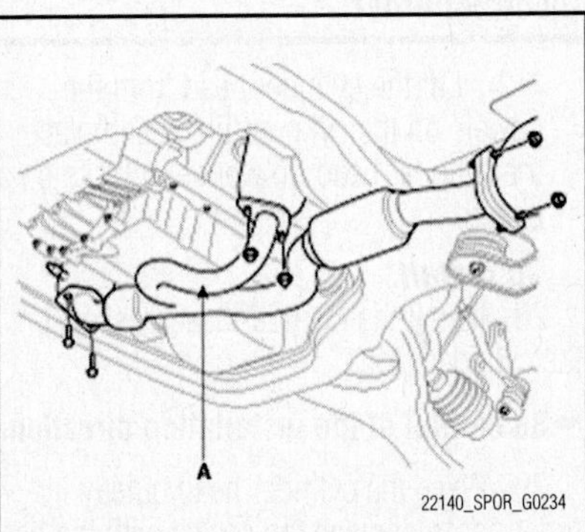

Fig. 128 Remove the front exhaust pipe (A)

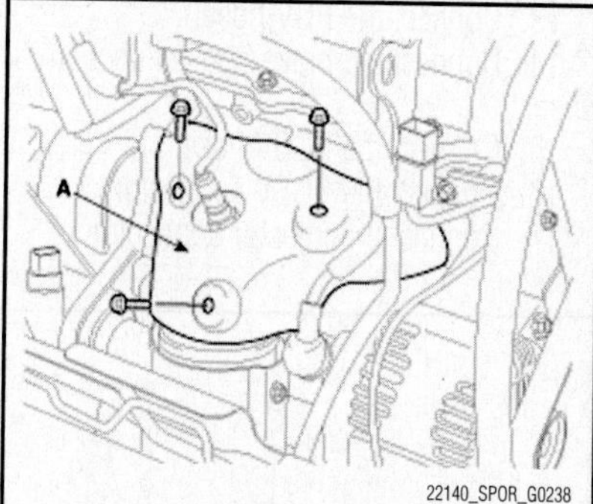

Fig. 129 Remove the LH heat protector (A)

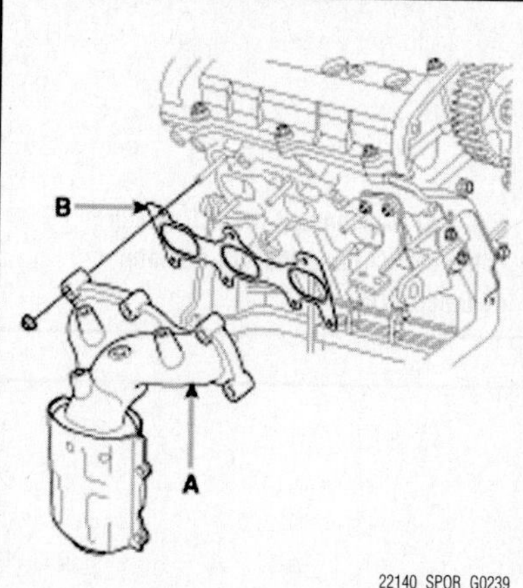

Fig. 130 Remove the LH exhaust manifold (A) and gasket (B)

9. Remove the RH drive shaft.
10. Remove the RH heat protector.
11. Remove the RH exhaust manifold and gasket.

To install:

12. Install the RH exhaust manifold and gasket. Tighten to 22–26 ft. lbs. (30–35 Nm).
13. Install the RH heat protector. Tighten to 12–16 ft. lbs. (17–22 Nm).
14. Install the RH drive shaft.
15. Install the alternator.
16. Install the exhaust manifold and gasket. Tighten to 22–26 ft. lbs. (30–35 Nm).
17. Install the heat protector. Tighten to 9–11 Ft. lbs. (12–15 Nm).
18. Install the air conditioner compressor.
19. Install the cooling fan motor assembly.
20. Connect the oxygen sensor connector.
21. Install the front exhaust pipe. Tighten to 22–30 ft. lbs. (30–40 Nm).
22. Install the undercover.

FLYWHEEL/FLEXPLATE

REMOVAL & INSTALLATION

1. Disconnect the negative battery cable.
2. Remove the starter motor.
3. Remove the transaxle.
4. Remove the flywheel/flexplate.

To install:

5. Install the flywheel/flexplate. Tighten the bolts to 87–94 ft. lbs. (118–128 Nm).
6. Install the transaxle.
7. Install the starter motor.
8. Connect the negative battery cable.

INTAKE MANIFOLD

REMOVAL & INSTALLATION

2.0L Engine

See Figures 131 through 137.

1. Remove the engine cover.
2. Disconnect the Throttle Position Sensor (TPS) connector (A) and Idle Speed Actuator (ISA) connector (B).
3. Disconnect the Positive Crankcase Ventilation (PCV) hose (A) and breather hose (B).
4. Remove the accelerator cable.
5. Remove the fuel delivery pipe (A).

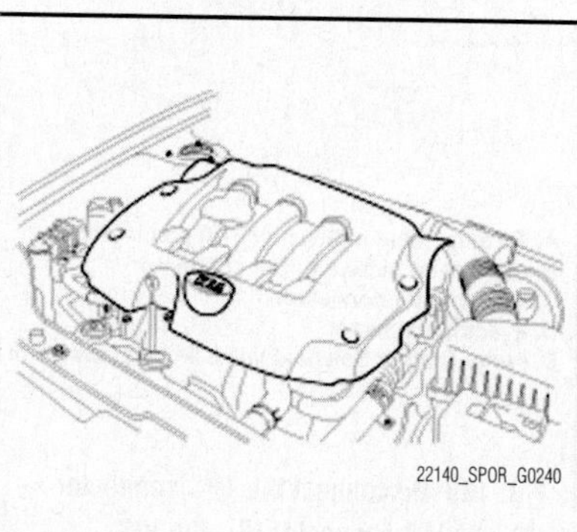

Fig. 131 Remove the engine cover

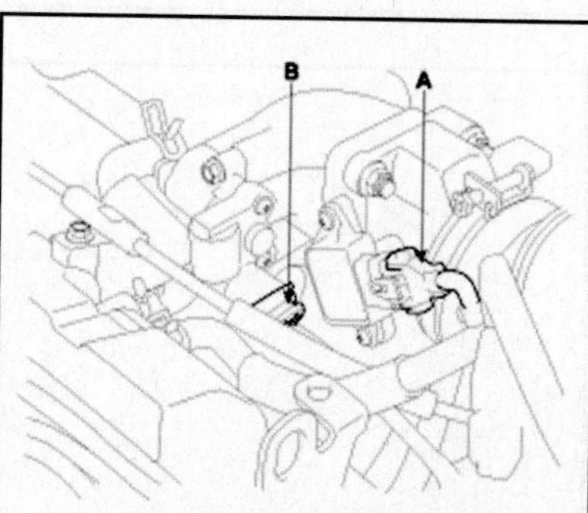

Fig. 132 Disconnect the Throttle Position Sensor (TPS) connector (A) and Idle Speed Actuator (ISA) connector (B)

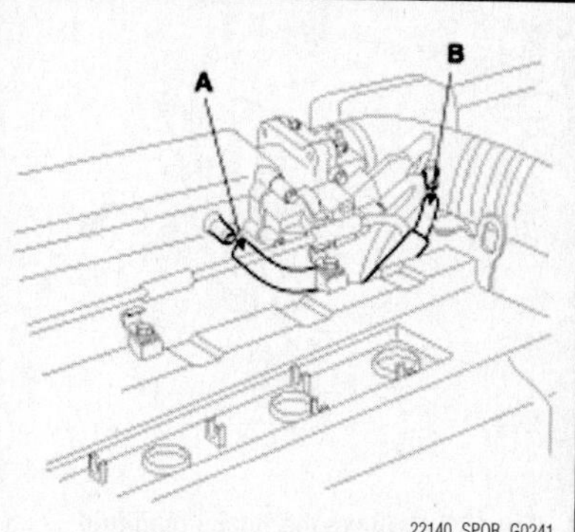

Fig. 133 Disconnect the Positive Crankcase Ventilation (PCV) hose (A) and breather hose (B)

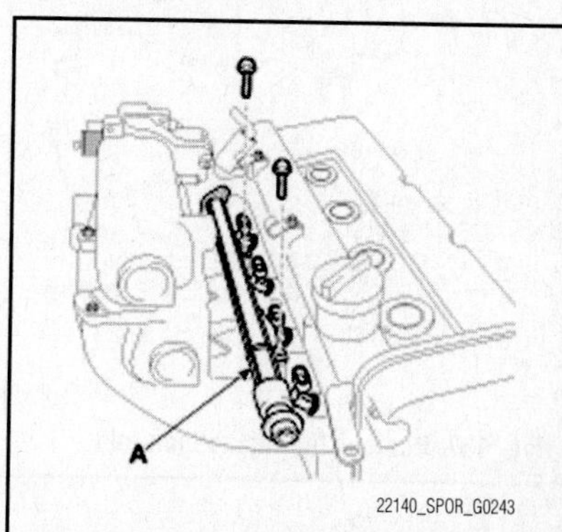

Fig. 134 Remove the fuel delivery pipe (A)

6. Remove the heater hose (A), Purge Control Solenoid Valve (PCSV) (B) and the brake vacuum hose (C) from throttle body and intake manifold.
7. Remove the air conditioner compressor.
8. Remove the intake manifold stay (A).
9. Remove the intake manifold.
10. Installation is the reverse order of removal with a new gasket.

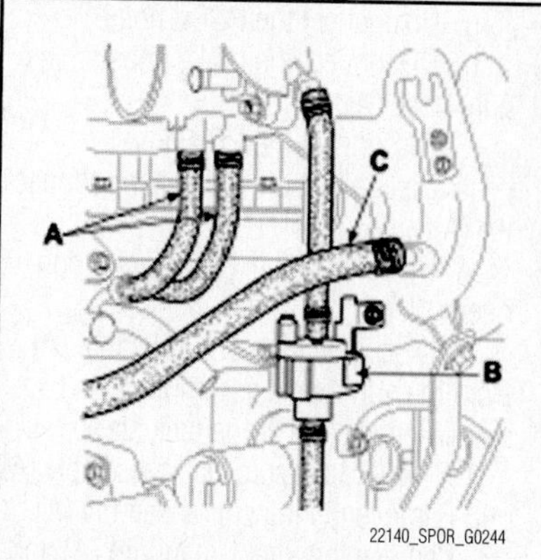

Fig. 135 Remove the heater hose (A), Purge Control Solenoid Valve (PCSV) (B) and the brake vacuum hose (C) from throttle body and intake manifold

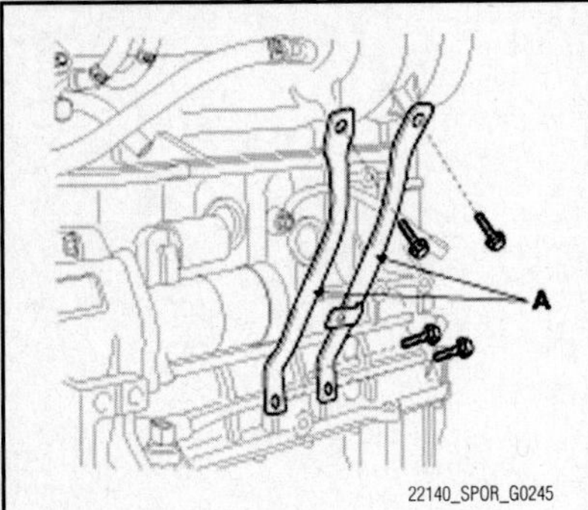

Fig. 136 Remove the intake manifold stay (A)

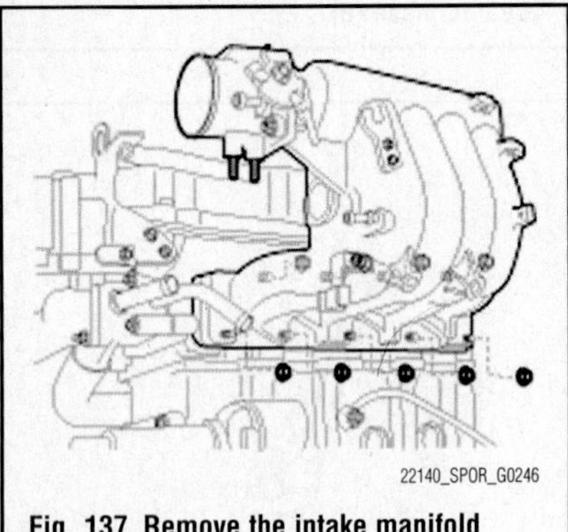

Fig. 137 Remove the intake manifold

2.7L Engine

See Figures 138 through 147.

1. Remove the engine cover.
2. Remove air cleaner hose.
3. Remove surge tank assembly.
   a. Disconnect the accelerator cable.
   b. Disconnect the TPS connector (A).
   c. Disconnect the ISA connector (B).
   d. Disconnect the VIS actuator connector (C).
   e. Disconnect the injector connector (D).
   f. Disconnect the PCSV connector (E).
   g. Disconnect the PCSV hose.
   h. Disconnect the brake booster vacuum hose (A).
   i. Disconnect the PCV hose.
   j. Disconnect the IAT sensor connector (A).
   k. Disconnect the VIS actuator connector (B).
   l. Disconnect the ground cable (A) from the surge tank assembly.
   m. Remove the surge tank stay.
   n. Remove the surge tank assembly (A).
4. Remove the injector assembly (A).
5. Remove the intake manifold (A) and gasket.

To install:

6. Install the intake manifold and gasket. Tighten to 14–15 ft. lbs. (19–21 Nm).
7. Install the injector assembly.
8. Install the surge tank assembly. Tighten to 11–15 ft. lbs. (15–20 Nm).
9. Install the surge tank stay. Tighten to 11–15 ft. lbs. (15–20 Nm).
10. Install the ground cable.
11. Connect the VIS actuator connector.
12. Connect the IAT sensor connector.
13. Connect the PCV hose.
14. Connect the brake booster vacuum hose.
15. Connect the PCSV hose.
16. Connect the PCSV connector.
17. Connect the injector connector.

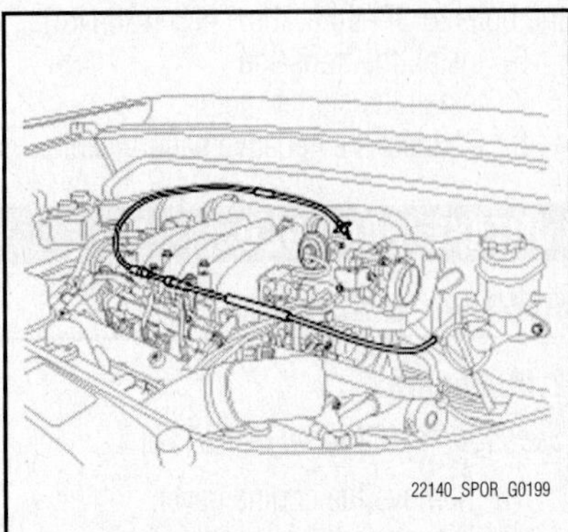

Fig. 138 Disconnect the accelerator cable

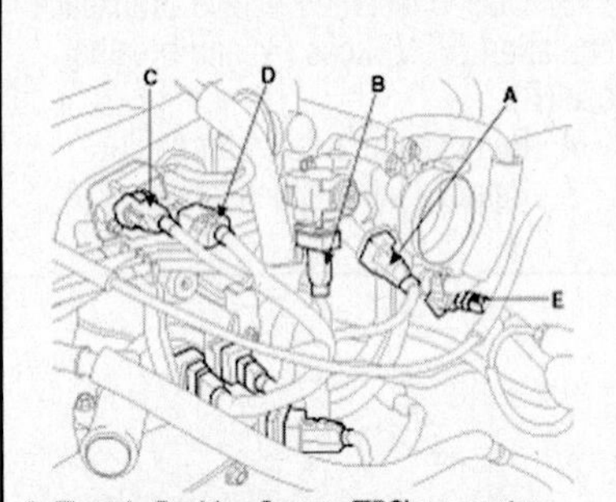

A. Throttle Position Sensor (TPS) connector
B. Idle Speed Actuator (ISA) connector
C. VIS actuator connector
D. Injector connector
E. Purge Control Solenoid Valve (PCSV) connector

Fig. 139 Disconnect the TPS connector (A), the ISA connector (B), the VIS actuator connector (C), the injector connector (D) and the PCSV connector (E)

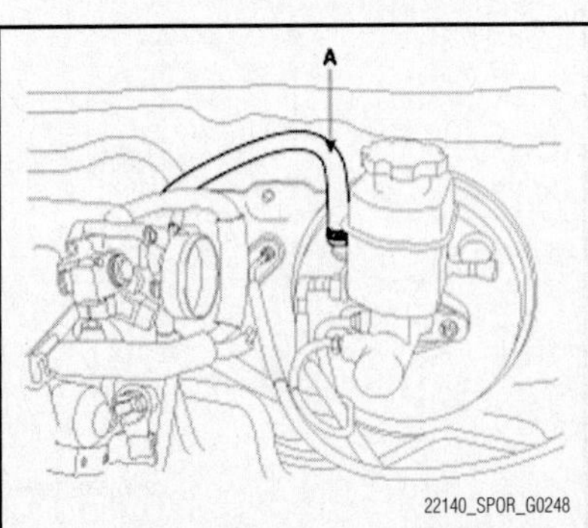

Fig. 140 Disconnect the brake booster vacuum hose (A)

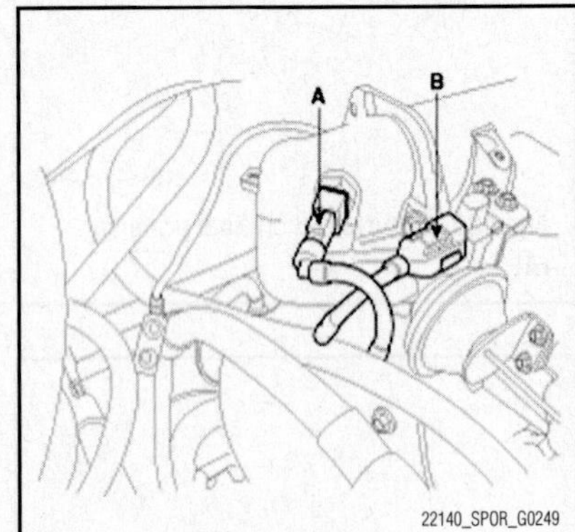

Fig. 141 Disconnect the IAT sensor connector (A) and the VIS actuator connector (B)

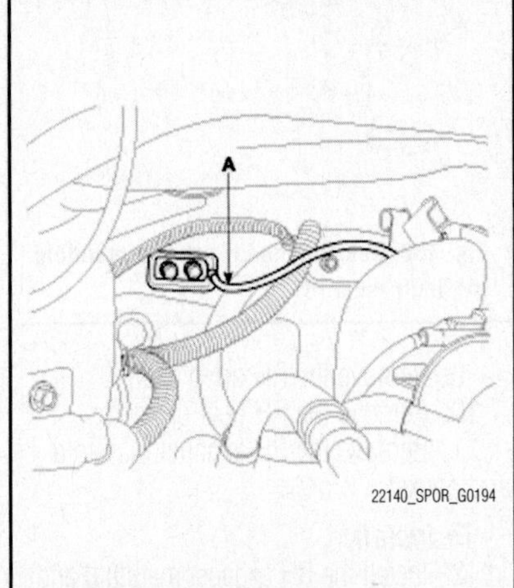

Fig. 142 Disconnect the ground cable (A) from the surge tank assembly

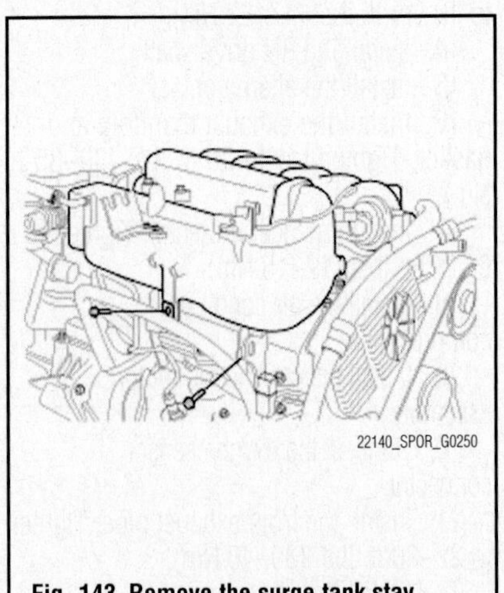

Fig. 143 Remove the surge tank stay

18. Connector the VIS actuator connector.
19. Connector the ISA connector.
20. Connector the TPS connector.
21. Connector the actuator cable.
22. Install the air cleaner hose.
23. Install the engine cover.

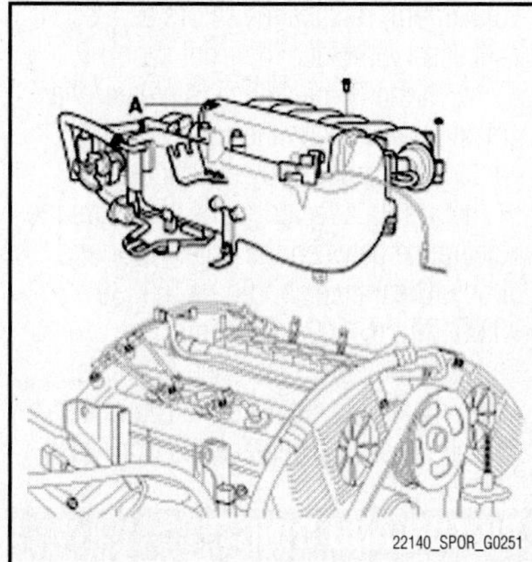

Fig. 144 Remove the surge tank assembly (A)

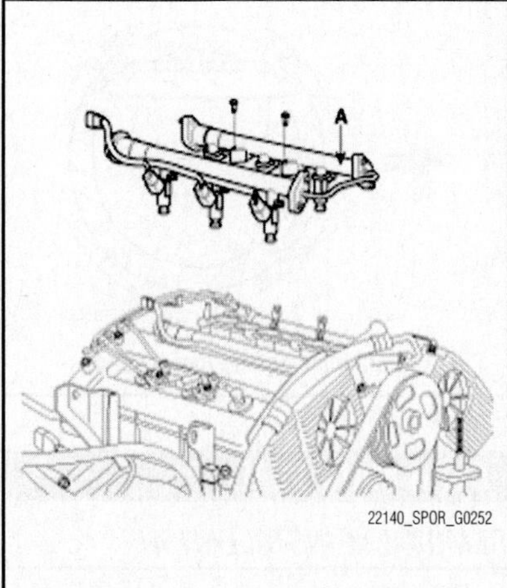

Fig. 145 Remove the injector assembly (A)

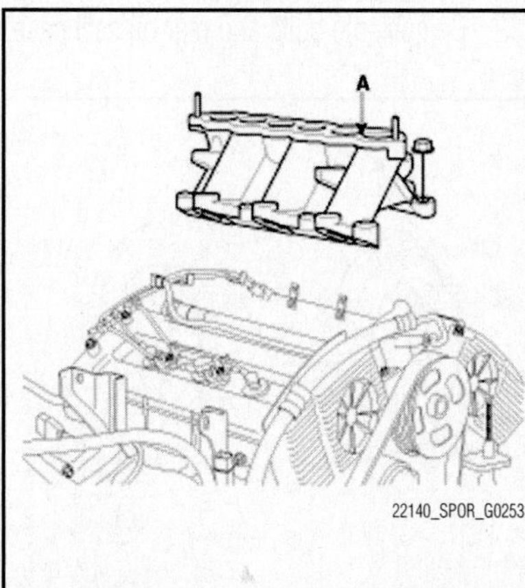

Fig. 146 Remove the intake manifold (A) and gasket

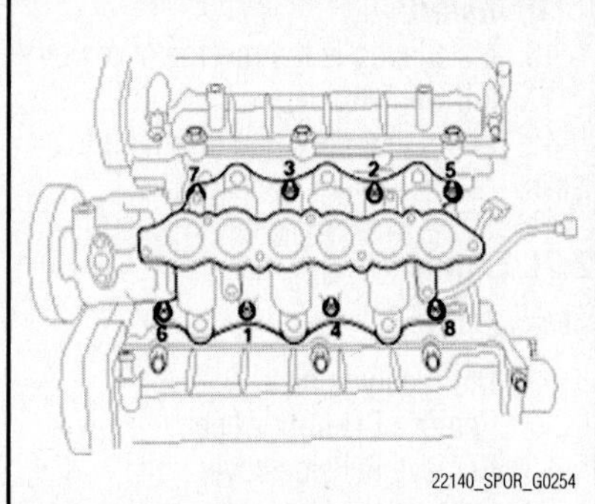

Fig. 147 Install the intake manifold and gasket

OIL PAN

REMOVAL & INSTALLATION

2..0L Engine

See Figures 148 and 149.

1. Drain the engine oil.
2. Disconnect the rear oxygen sensor connector.
3. Remove the front muffler (A).
4. Remove the exhaust manifold.
5. Remove the front muffler bracket (A).
6. Remove the oil pan.

To install:

7. Install the oil pan.

➡Check that the mating surfaces are clean and dry before applying liquid gasket.

a. Apply liquid gasket as an even bead, centered between the edges of the mating surface.

➡To prevent leakage of oil, apply liquid gasket to the inner threads of the bolt holes. Do not install the parts if five minutes or more have elapsed since applying the liquid gasket. Instead, reapply liquid gasket after removing the residue.

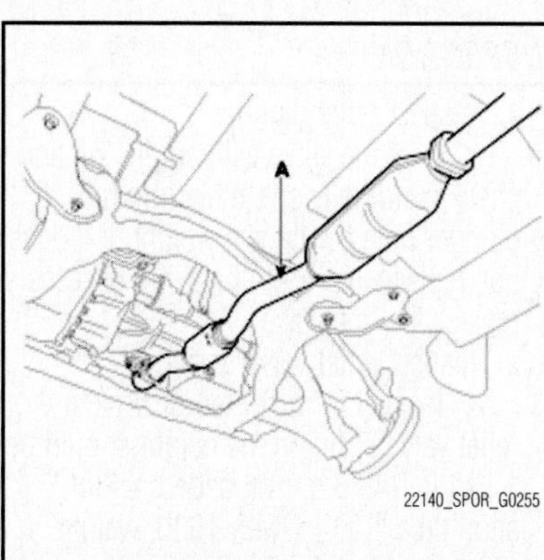

Fig. 148 Remove the front muffler (A)

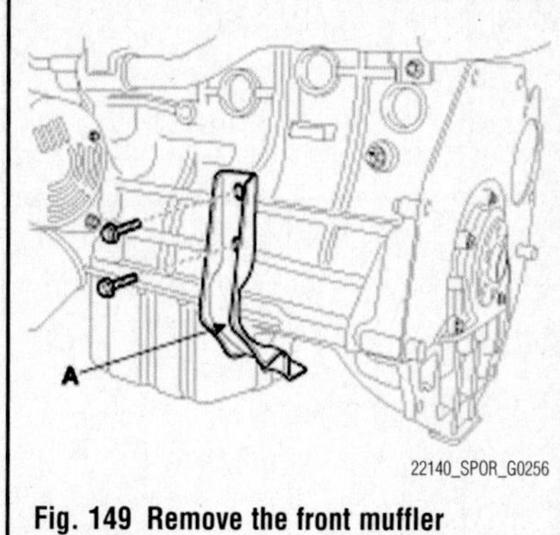

Fig. 149 Remove the front muffler bracket (A)

After assembly, wait at least 30 minutes before filling the engine with oil.

b. Install the oil pan with the bolts. Uniformly tighten the bolts in several passes. Tighten to 7–9 ft. lbs. (10–12 Nm)

8. Install the front muffler bracket.
9. Install the exhaust manifold.
10. Install the front muffler.
11. Connect the rear oxygen sensor connector.
12. Fill with engine oil

2.7L Engine

1. Drain engine oil.
2. Remove RH front wheel.
3. Remove RH side cover.
4. Remove the front exhaust pipe.
5. Remove the alternator from engine.
6. Remove the drive belt.
7. Turn the crankshaft and align the white groove on the crankshaft pulley with the pointer on the lower cover.
8. Remove the timing belt.
9. Remove the oil pan and oil screen.

To install:

10. Install the oil pan and oil screen.
11. Install the timing belt.
12. Install the drive belt.
13. Install the alternator.
14. Install the front exhaust pipe.
15. Install the RH front wheel.
16. Fill engine with oil.
17. Start engine and check for leaks.
18. Recheck engine oil level.

OIL PUMP

REMOVAL & INSTALLATION

2.0L Engine

See Figures 150 through 152.

1. Drain the engine oil.
2. Remove the drive belts.

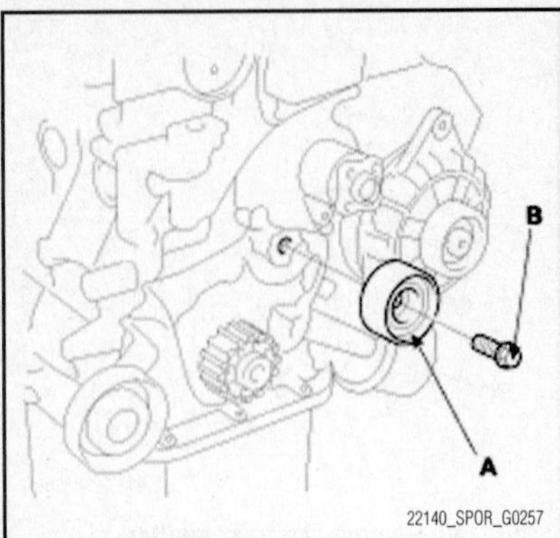

Fig. 150 Remove the bolt (B) and timing belt idler (A)

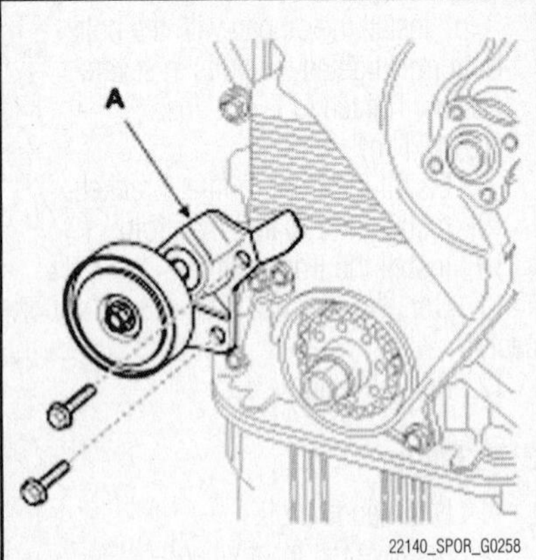

Fig. 151 Remove the air conditioner compressor tensioner bracket (A)

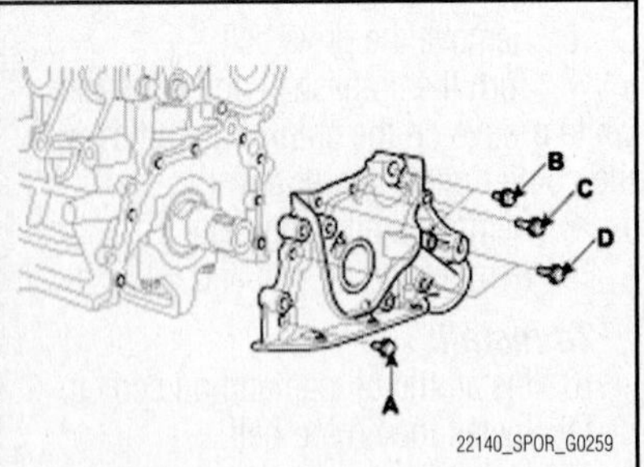

Fig. 152 Remove the bolts (A, B, C, D) and oil pump

3. Turn the crankshaft pulley, and align its groove with timing mark "T" of the timing belt cover.
4. Remove the timing belt.
5. Remove the bolt (B) and timing belt idler (A).
6. Remove the oil pan and oil screen.
7. Remove the alternator.
8. Remove the air conditioner compressor tensioner bracket (A).
9. Remove the bolts (A, B, C, D) and oil pump.

To install:

10. Installation is the reverse of removal.
11. Fill with engine oil.
12. Start the engine and check for leaks.

2.7L Engine

See Figure 153.

1. Drain engine oil.
2. Remove RH front wheel.
3. Remove RH side cover.
4. Remove the front exhaust pipe.
5. Remove the alternator from engine.
6. Remove the drive belt.
7. Turn the crankshaft and align the white groove on the crankshaft pulley with the pointer on the lower cover.
8. Remove the timing belt.
9. Remove the oil pan and oil screen.
10. Remove the oil pump case (A).

To install:

11. Install the oil pump.
12. Install the oil pan and oil screen.
13. Install the timing belt.
14. Install the drive belt.
15. Install the alternator.
16. Install the front exhaust pipe.
17. Install the RH front wheel.
18. Fill engine with oil.
19. Start engine and check for leaks.
20. Recheck engine oil level.

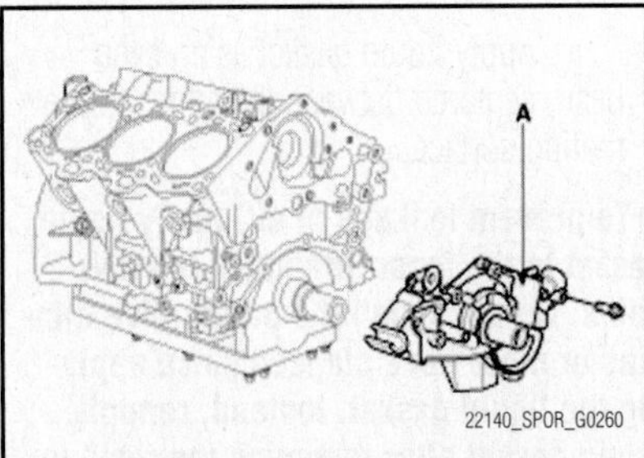

Fig. 153 Remove the oil pump case (A)

INSPECTION

1. Inspect relief plunger.
   a. Coat the valve with engine oil and check that it falls smoothly into the plunger hole by its own weight. If it does not, replace the relief plunger. If necessary, replace the front case.
2. Inspect relief valve spring.
   a. Inspect for distorted or broken relief valve spring. Free height should be 1.724 in. (43.8 mm). Loaded height should be 1.54 in. with 10 lb. weight (39.1 mm with 4.6kg weight).
3. Inspect rotor side clearance.
   a. Using a feeler gauge and precision straight edge, measure the clearance between the rotors and precision straight edge resting on pump body. Clearance should be 0.0016 - 0.0037in. (0.04 - 0.095mm).
   b. If the side clearance is greater than maximum, replace the rotors as a set. If necessary, replace the front case.
   c. If the tip clearance is greater than maximum, replace the rotor as a set.
4. Inspect rotor body clearance.
   a. Using a feeler gauge, measure the clearance between the outer rotor and body. Clearance should be 0.0039 - 0.0017in. (0.100 - 0.181mm).
   b. If the body clearance is greater than maximum, replace the rotors as a set. If necessary, replace the front case

PISTON AND RING

POSITIONING

See Figure 154.

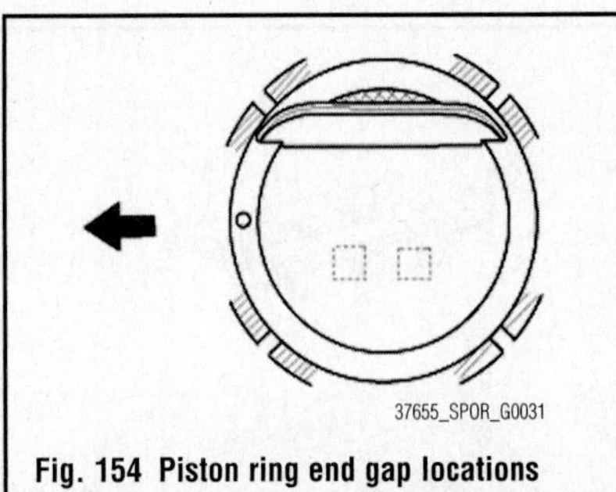

Fig. 154 Piston ring end gap locations

REAR MAIN SEAL

REMOVAL & INSTALLATION

2.0L Engine

See Figures 155 and 156.

1. Remove the transaxle.
2. Remove the bolts and rear oil seal case.

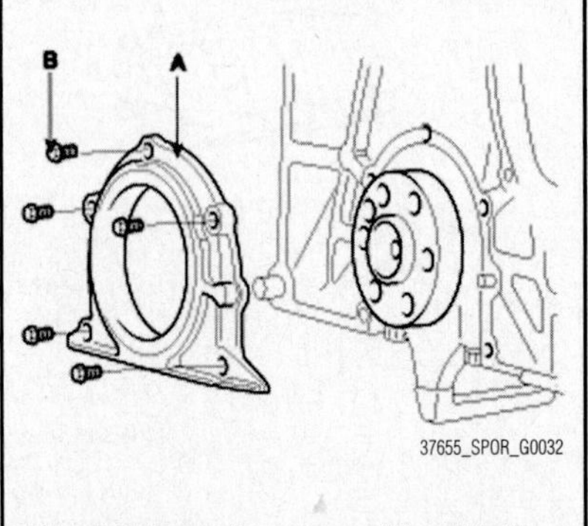

Fig. 155 Rear oil seal case (A) and bolts (B)—2.0L engine

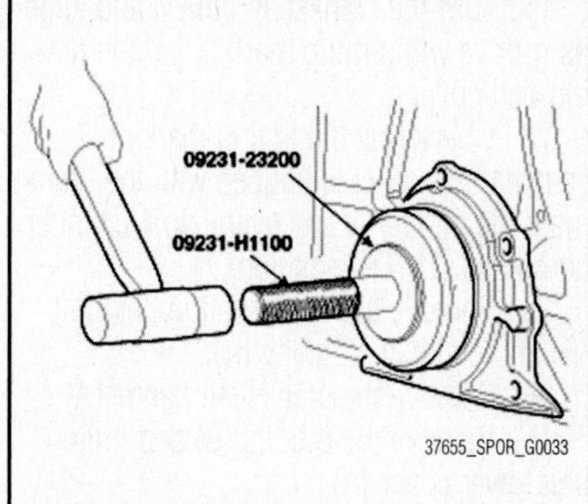

Fig. 156 Rear oil seal installation—2.0L engine

3. Remove the rear oil seal.
4. Install a new gasket and rear oil seal case with 5 bolts. Tighten bolts to 7.2 - 8.7 ft. lb. (9.8 - 11.8Nm).
5. Check that the mating surfaces are clean and dry.
6. Install the rear oil seal.
 a. Apply engine oil to a new oil seal lip.
 b. Using the special service tools (09231-23200, 09231-H1100) and a hammer, tap in the oil seal until its surface is flush with the rear oil seal retainer edge.

2.7L Engine

See Figures 157 and 158.

1. Remove the transaxle. See Transaxle.
2. Remove the lower and upper oil pan. See Oil Pan.
3. Remove the bolts and rear oil seal case.
4. Remove the rear oil seal.
5. Install a new gasket and rear oil seal case with 5 bolts. Tighten bolts to 7.2 - 8.7 ft. lb. (9.8 - 11.8Nm).
6. Check that the mating surfaces are clean and dry.
7. Install the rear oil seal.

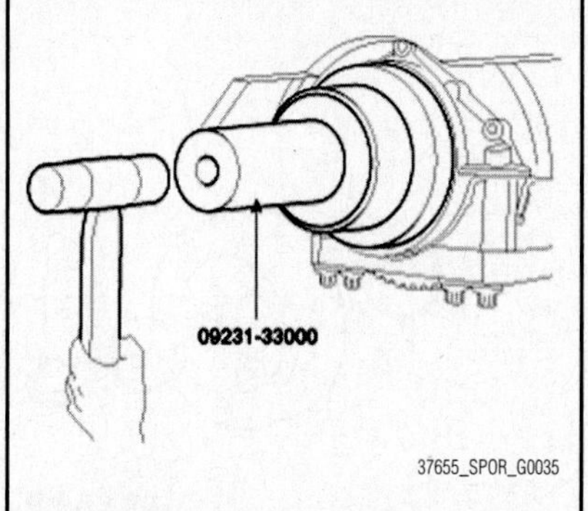

Fig. 158 Rear oil seal installation—2.7L engine

 a. Apply engine oil to a new oil seal lip.
 b. Using special service tool (09231-33000) and a hammer, tap in the oil seal until its surface is flush with the rear oil seal retainer edge.
8. Install the lower and upper oil pan.
9. Install the transaxle.

TIMING BELT FRONT COVER

REMOVAL & INSTALLATION

2.0L Engine

See Figures 159 through 169.

Engine removal is not required for this procedure.

1. Remove the engine cover.
2. Remove the RH front wheel.
3. Remove the 2 bolts (B) and RH side cover (A).
4. Remove the engine mounting support bracket.
 a. Set the jack to the engine oil pan.
 b. Remove the bolt (B), nuts (C, D) and engine mounting support bracket (A).
 c. Remove the bolt (B) and engine support bracket stay plate (A).
5. Temporarily loosen the water pump pulley bolts.

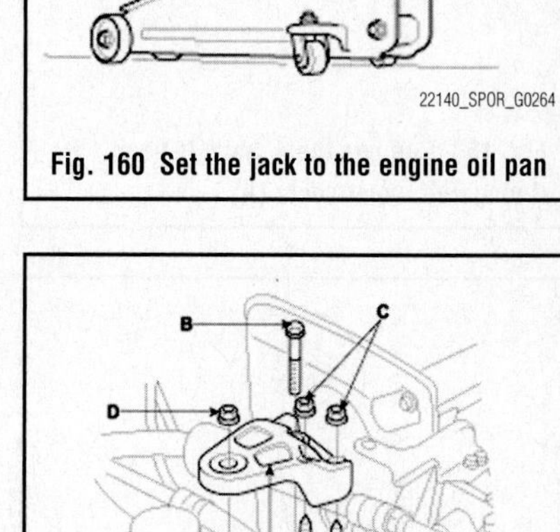

Fig. 160 Set the jack to the engine oil pan

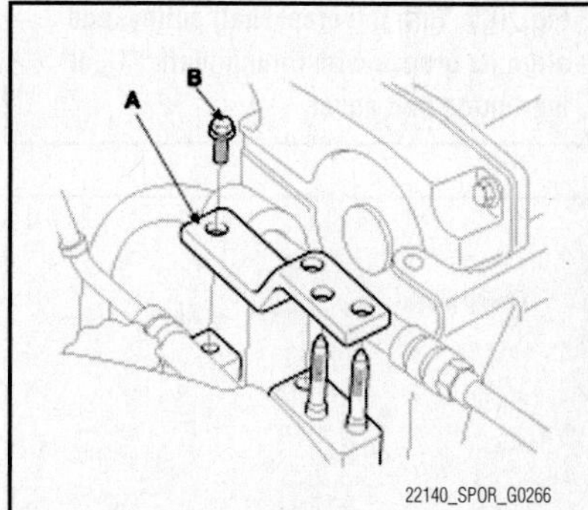

Fig. 161 Remove the bolt (B), nuts (C, D) and engine mounting support bracket (A)

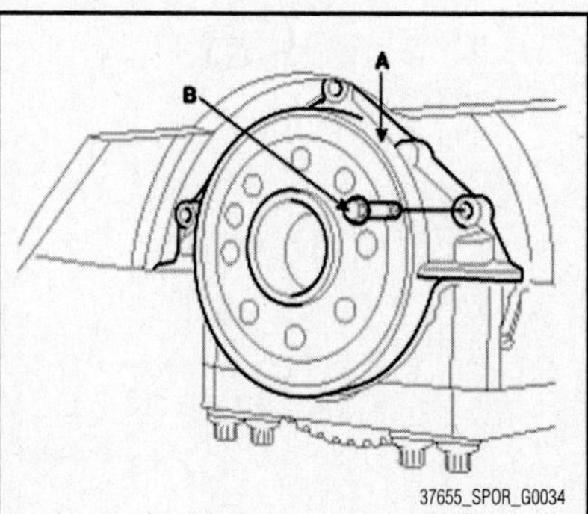

Fig. 162 Remove the bolt (B) and engine support bracket stay plate (A)

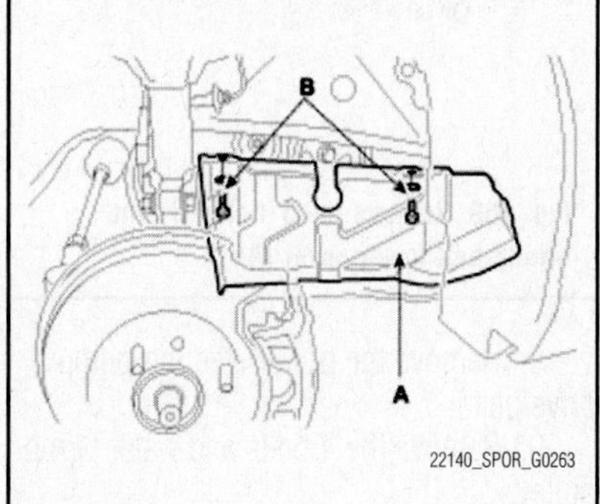

Fig. 157 Rear oil seal case (A) and bolt (B)—2.7L engine

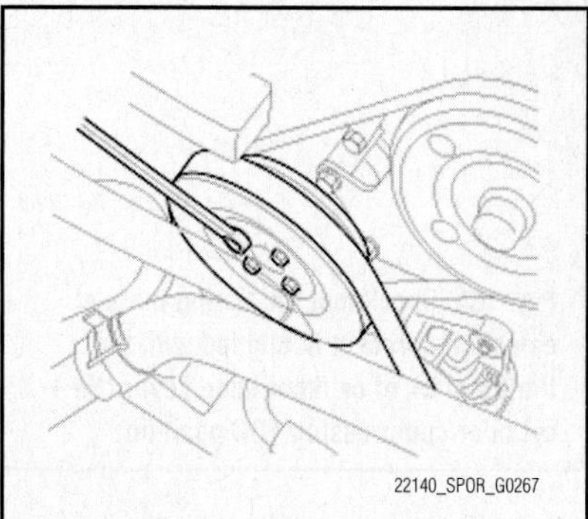

Fig. 159 Remove the 2 bolts (B) and RH side cover (A)

Fig. 163 Temporarily loosen the water pump pulley bolts

Fig. 164 Remove the 4 bolts (B) and timing belt upper cover (A)

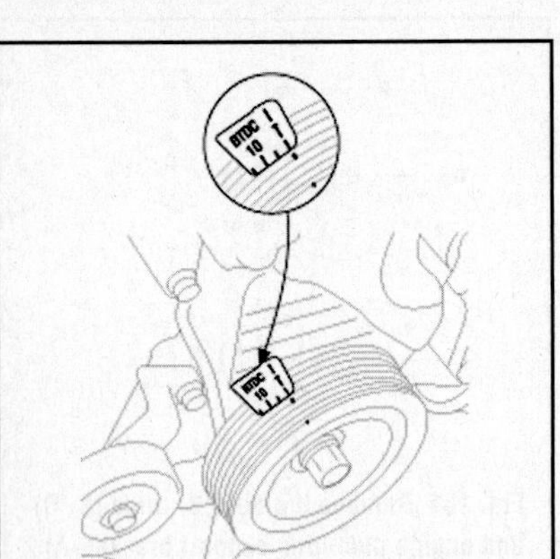

Fig. 165 Turn the crankshaft pulley, and align its groove with timing mark "T" of the timing belt cover

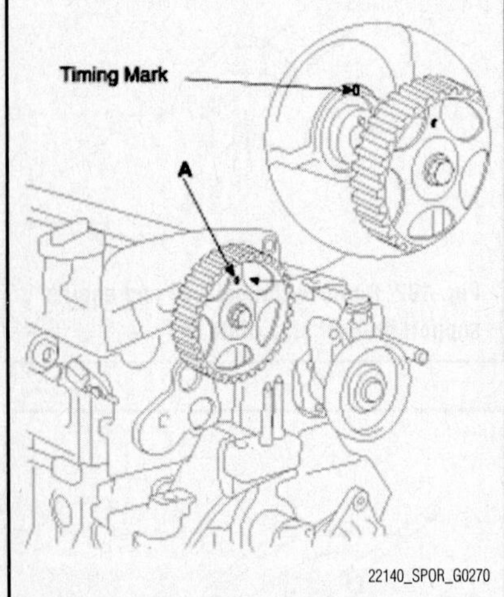

Fig. 166 Check that the timing mark of camshaft sprocket is aligned with the timing mark of cylinder head cover (No.1 cylinder compression TDC position)

6. Remove the alternator drive belt.
7. Remove the air conditioner compressor drive belt.

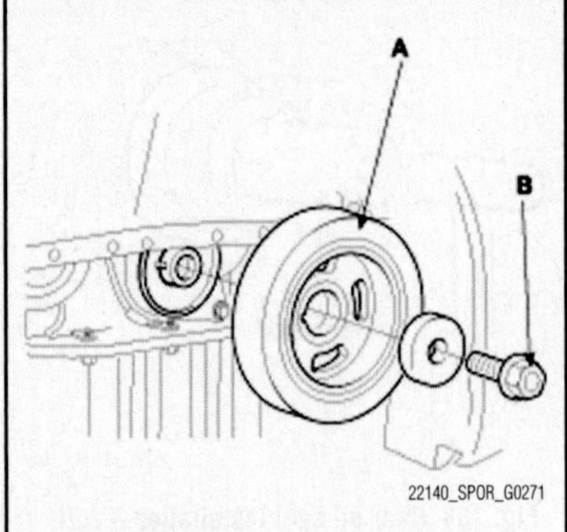

Fig. 167 Remove the crankshaft pulley bolt (B) and crankshaft pulley (A)

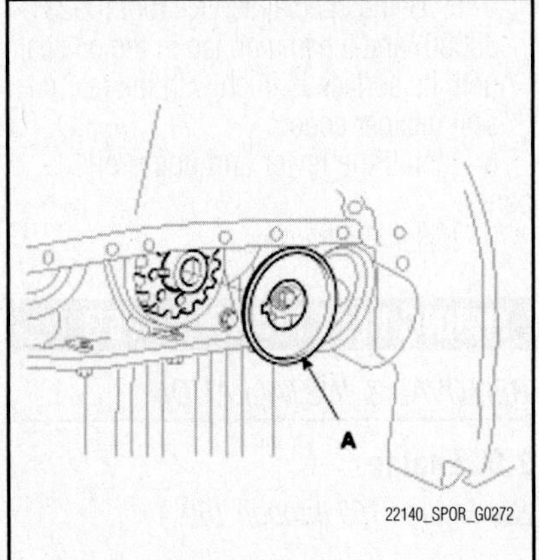

Fig. 168 Remove the crankshaft flange (A).

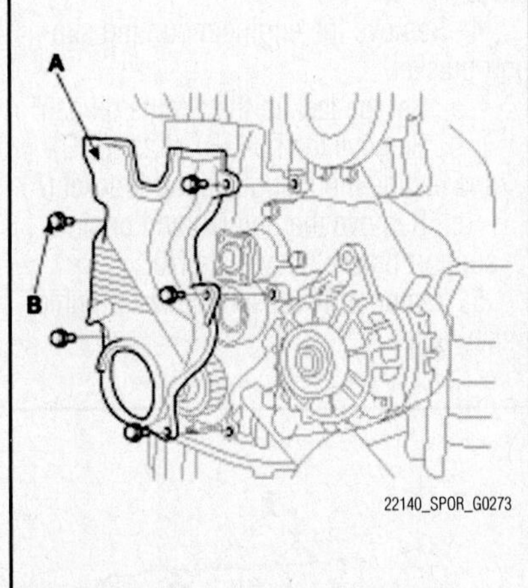

Fig. 169 Remove the 5 bolts (B) and timing belt lower cover (A)

8. Remove the power steering pump drive belt.
9. Remove the 4 bolts and water pump pulley.
10. Remove the 4 bolts (B) and timing belt upper cover (A).
11. Turn the crankshaft pulley, and align its groove with timing mark "T" of the timing belt cover.
12. Check that the timing mark of camshaft sprocket is aligned with the timing mark of cylinder head cover. (No.1 cylinder compression TDC position)
13. Remove the crankshaft pulley bolt (B) and crankshaft pulley (A).
14. Remove the crankshaft flange (A).
15. Remove the 5 bolts (B) and timing belt lower cover (A)
16. Installation is the reverse of removal.

2.7L Engine

See Figures 170 through 176.

Engine removal is not required for this procedure.

1. Remove the engine cover.
2. Remove RH front wheel.
3. Remove 2bolts (B) and RH side cover (A).
4. Turn the crankshaft pulley, and align its groove with timing mark "T" of the timing belt cover.

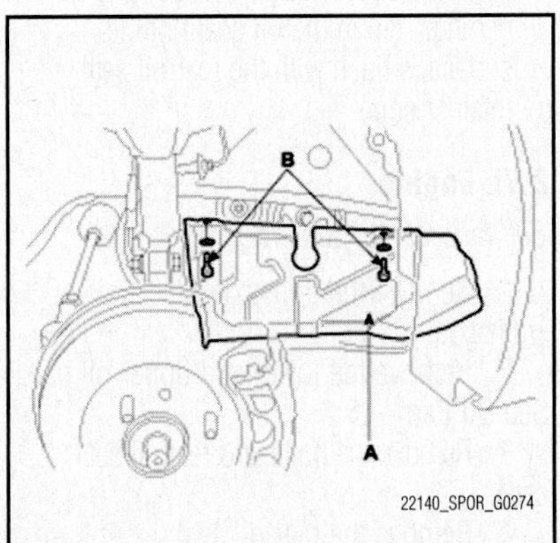

Fig. 170 Remove 2bolts (B) and RH side cover (A)

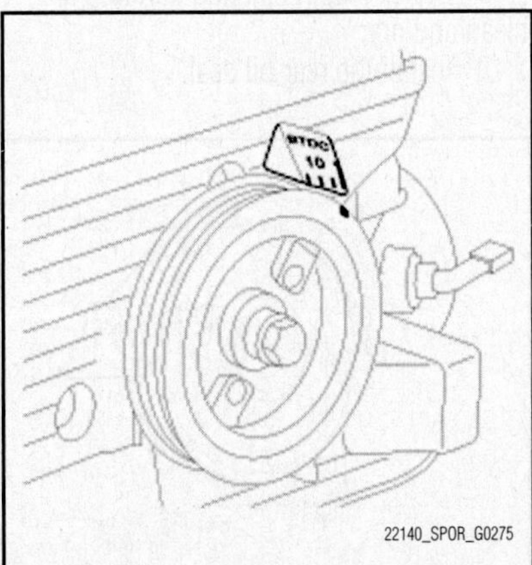

Fig. 171 Turn the crankshaft pulley, and align its groove with timing mark "T" of the timing belt cover

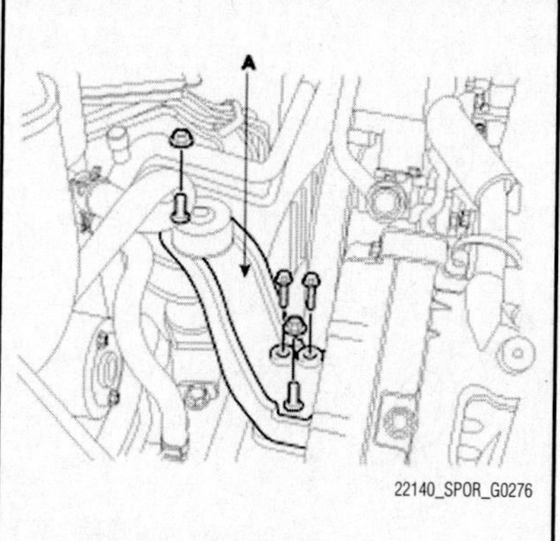

Fig. 172 Remove the engine mount bracket (A)

➡Always turn the crankshaft clockwise.

5. Remove drive belt and belt tensioner.
6. Remove the engine mount bracket.
    a. Set the jack to the engine oil pan.

➡Place wooden block between the jack and engine oil pan.

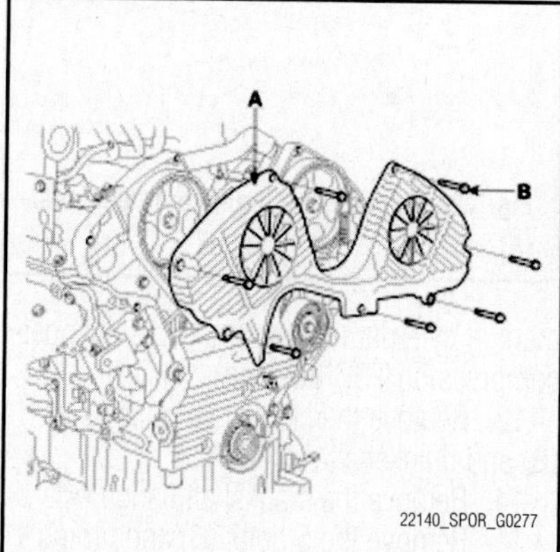

Fig. 173 Remove the 7 bolts (B) and timing belt upper cover (A)

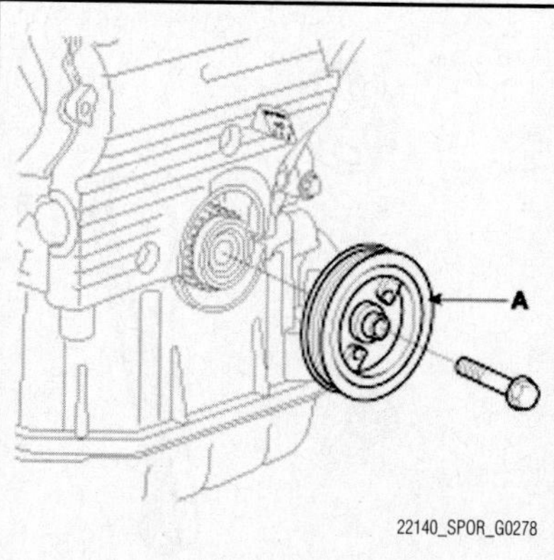

Fig. 174 Remove the crankshaft pulley bolt and crankshaft pulley (A)

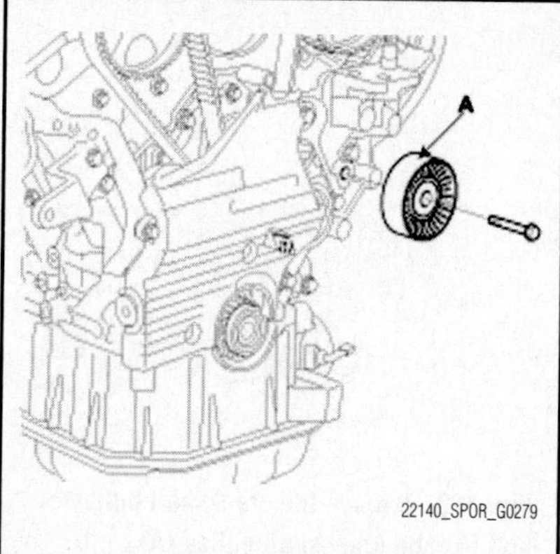

Fig. 175 Remove the drive belt idler pulley (A)

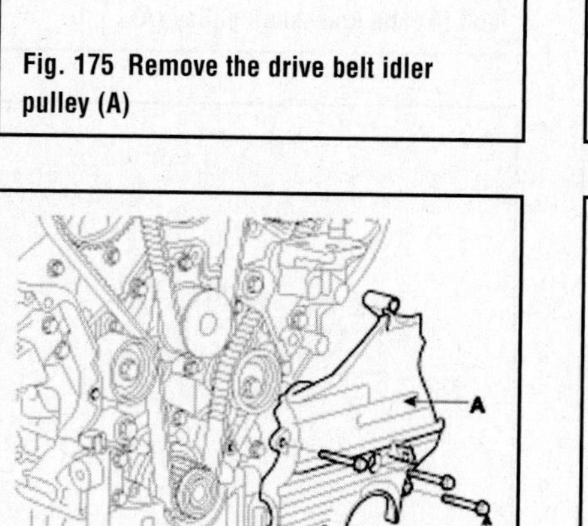

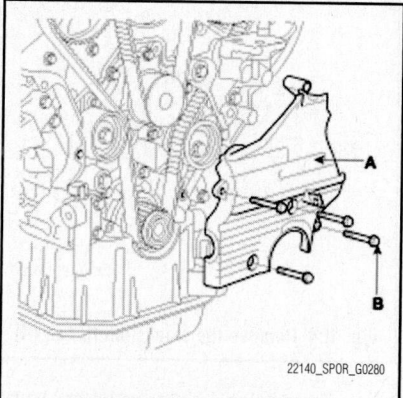

Fig. 176 Remove the 4 bolts (B) and timing belt lower cover (A)

    b. Remove the engine mount bracket (A).
7. Remove the power steering pump.
8. Remove the 7 bolts (B) and timing belt upper cover (A).
9. Remove the crankshaft pulley bolt and crankshaft pulley (A).
10. Remove the drive belt idler pulley (A).
11. Remove the 4 bolts (B) and timing belt lower cover (A).
12. Installation is the reverse of removal.

TIMING BELT & SPROCKETS

REMOVAL & INSTALLATION

2.0L Engine

See Figures 177 through 190.

Engine removal is not required for this procedure.

1. Remove the engine cover.
2. Remove the RH front wheel.
3. Remove the 2 bolts (B) and RH side cover (A).

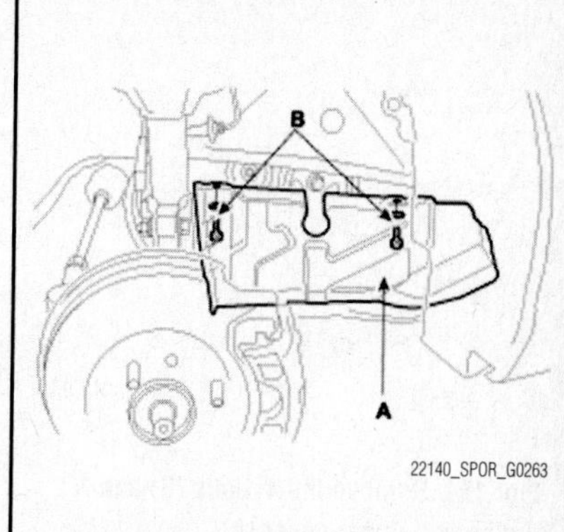

Fig. 177 Remove the 2 bolts (B) and RH side cover (A)

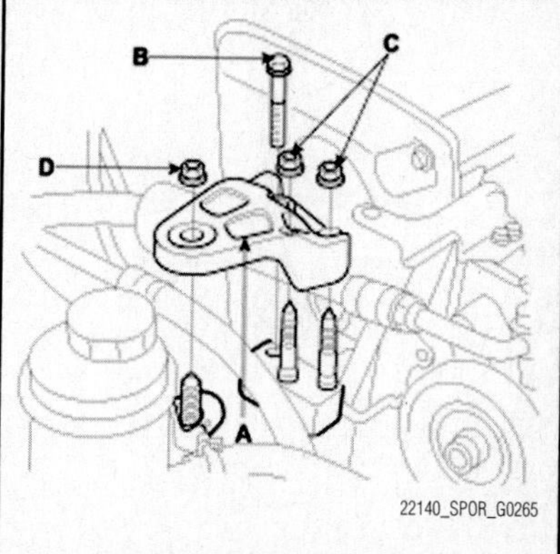

Fig. 178 Remove the bolt (B), nuts (C, D) and engine mounting support bracket (A)

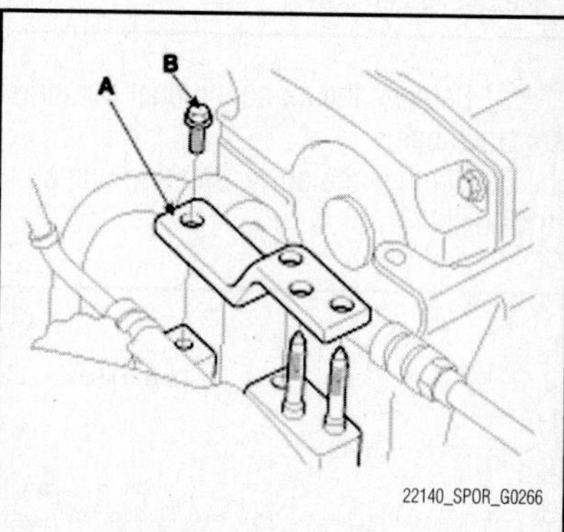

Fig. 179 Remove the bolt (B) and engine support bracket stay plate (A)

4. Remove the engine mounting support bracket.
    a. Set the jack to the engine oil pan.
    b. Remove the bolt (B), nuts (C, D) and engine mounting support bracket (A).
    c. Remove the bolt (B) and engine support bracket stay plate (A).
5. Temporarily loosen the water pump pulley bolts.
6. Remove the alternator drive belt.

Fig. 180 Remove the 4 bolts (B) and timing belt upper cover (A)

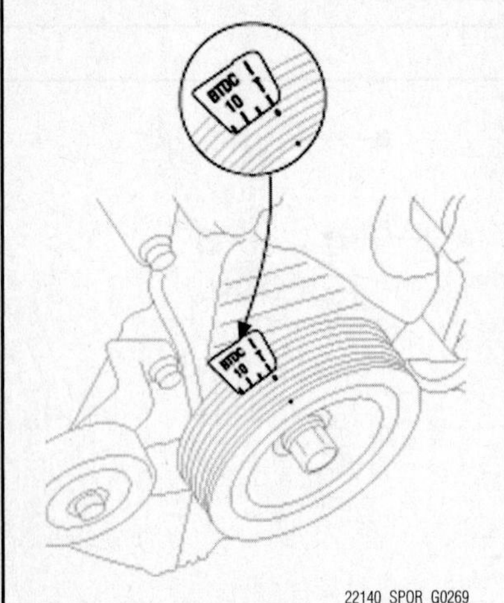

Fig. 181 Turn the crankshaft pulley, and align its groove with timing mark "T" of the timing belt cover

7. Remove the air conditioner compressor drive belt.
8. Remove the power steering pump drive belt.

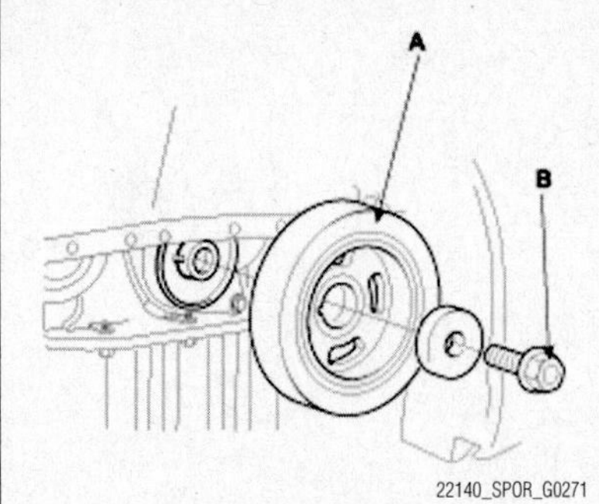
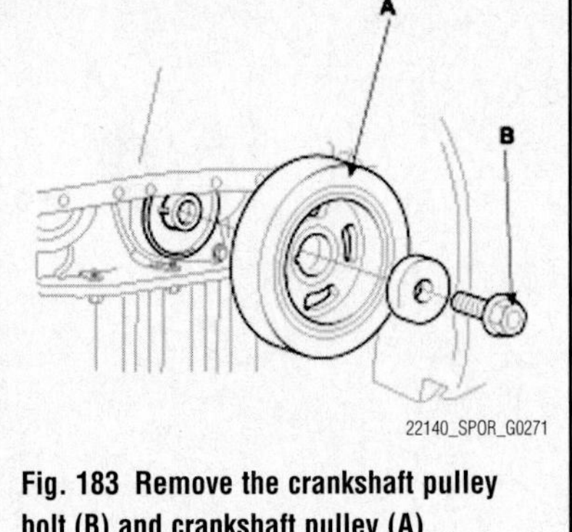

Fig. 183 Remove the crankshaft pulley bolt (B) and crankshaft pulley (A)

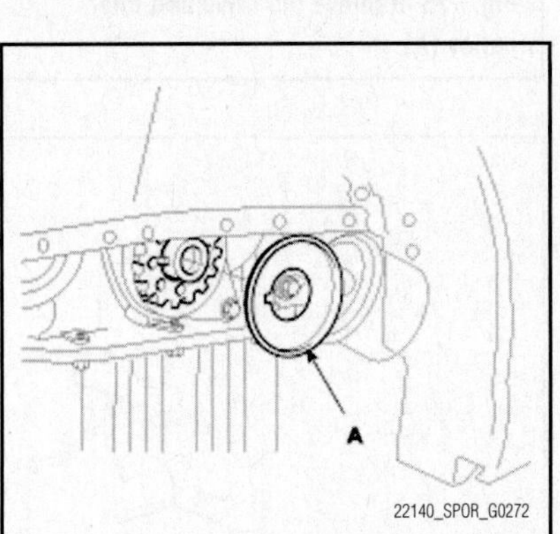

Fig. 184 Remove the crankshaft flange (A)

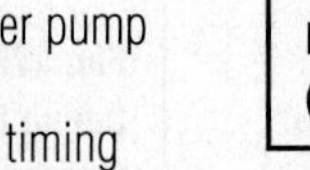

9. Remove the 4 bolts and water pump pulley.
10. Remove the 4 bolts (B) and timing belt upper cover (A).
11. Turn the crankshaft pulley, and align its groove with timing mark "T" of the timing belt cover.
12. Check that the timing mark of camshaft sprocket is aligned with the timing mark of cylinder head cover. (No.1 cylinder compression TDC position)

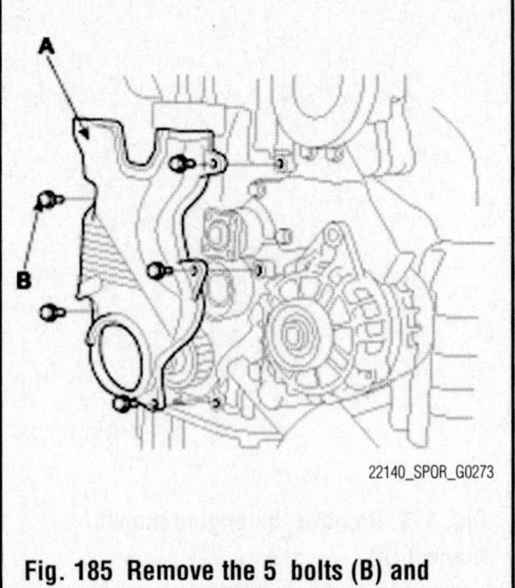

Fig. 185 Remove the 5 bolts (B) and timing belt lower cover (A)

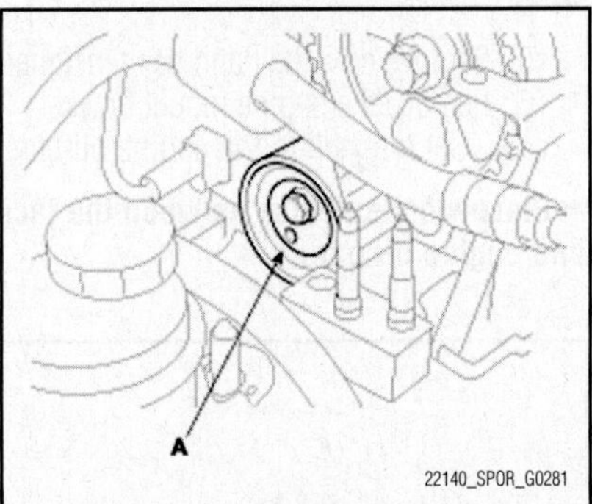

Fig. 186 Remove the timing belt tensioner (A)

13. Remove the crankshaft pulley bolt (B) and crankshaft pulley (A).
14. Remove the crankshaft flange (A).
15. Remove the 5 bolts (B) and timing belt lower cover (A)
16. Remove the timing belt tensioner (A) and timing belt (B)

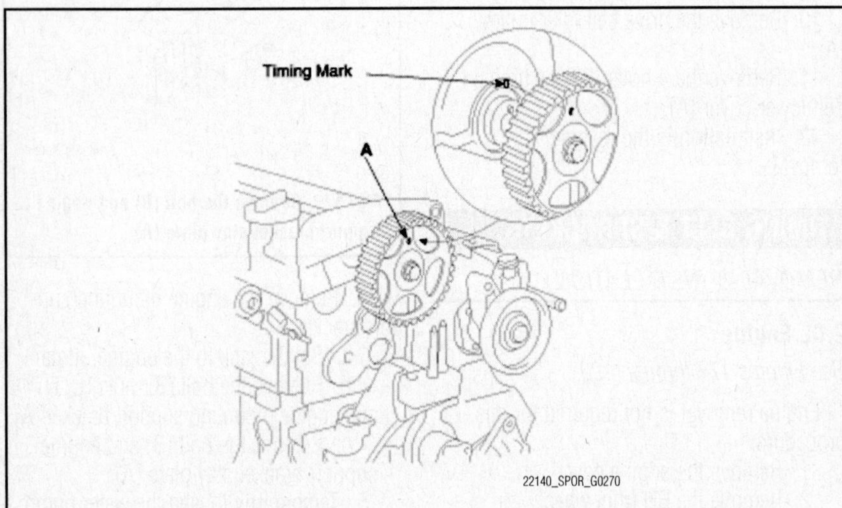

Fig. 182 Check that the timing mark of camshaft sprocket is aligned with the timing mark of cylinder head cover (No.1 cylinder compression TDC position)

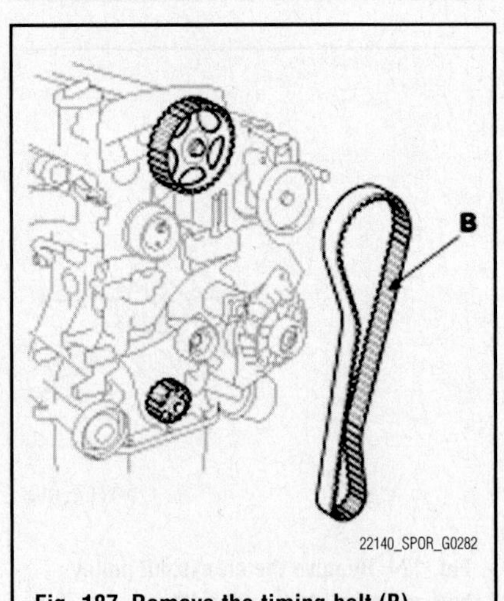

Fig. 187 Remove the timing belt (B)

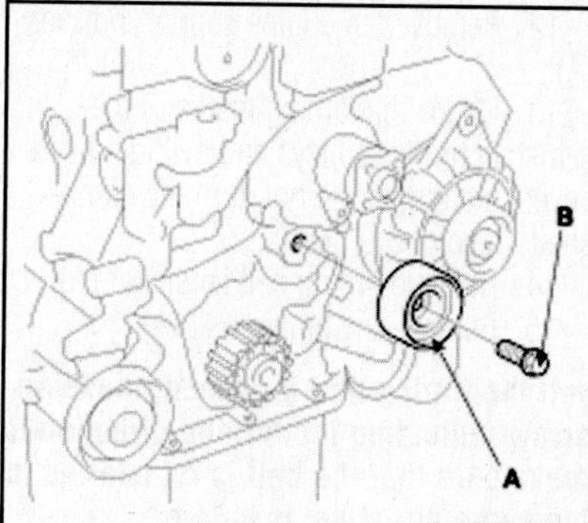

Fig. 188 Remove the bolt (B) and timing belt idler (A)

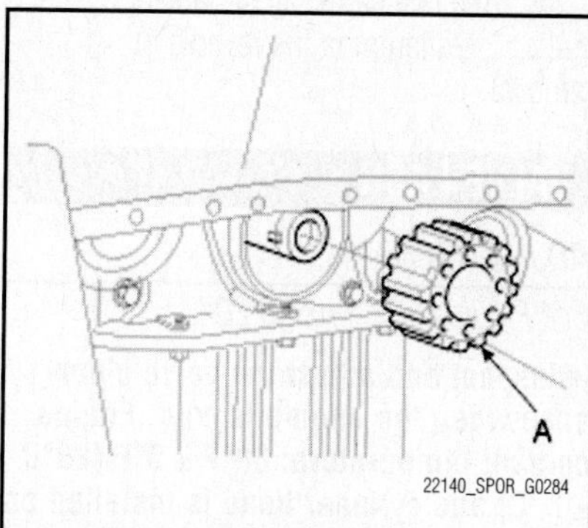

Fig. 189 Remove the crankshaft sprocket (A)

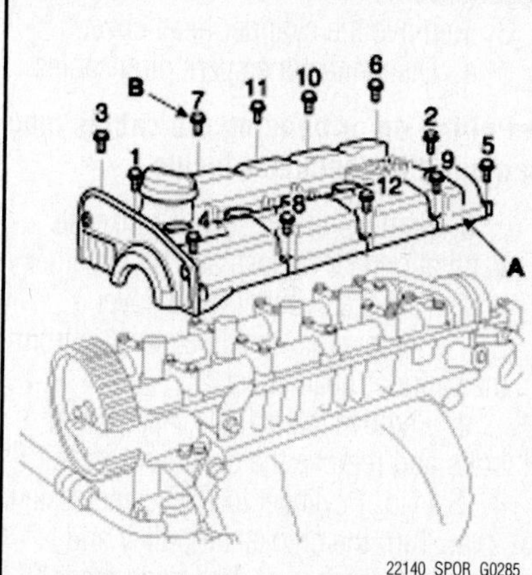

Fig. 190 Remove the cylinder head cover bolts (B) and remove the cover (A) and gasket

➡If the timing belt is going to be reused, make an arrow indicating the turning direction to make sure that the belt is reinstalled in the same direction as before.

17. Remove the bolt (B) and timing belt idler (A).
18. Remove the crankshaft sprocket (A).
19. Remove the cylinder head cover.

a. Disconnect the spark plug cables and do not pull on the cable by force.

➡Pulling on or bending the cables may damage the conductor inside.

b. Remove the Positive Crankcase Ventilation (PCV) hose (A) and the breather hose (B) from the cylinder head cover.

c. Remove the accelerator cable from the cylinder head cover.

d. Remove the cylinder head cover bolts (B) and remove the cover (A) and gasket.

20. Remove the camshaft sprocket.
21. Installation is the reverse of removal.

2.7L Engine

See Figures 191 through 201.

Engine removal is not required for this procedure.

1. Remove the engine cover.
2. Remove RH front wheel.
3. Remove 2 bolts (B) and RH side cover (A).

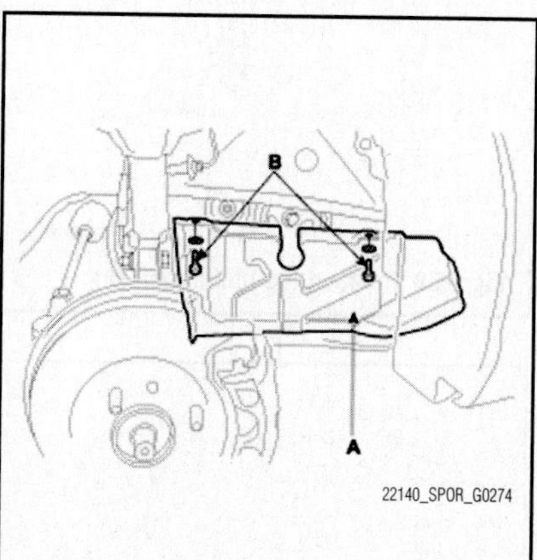

Fig. 191 Remove 2 bolts (B) and RH side cover (A)

Fig. 192 Turn the crankshaft pulley, and align its groove with timing mark "T" of the timing belt cover

4. Turn the crankshaft pulley, and align its groove with timing mark "T" of the timing belt cover.

➡Always turn the crankshaft clockwise.

5. Remove drive belt and belt tensioner.
6. Remove the engine mount bracket.

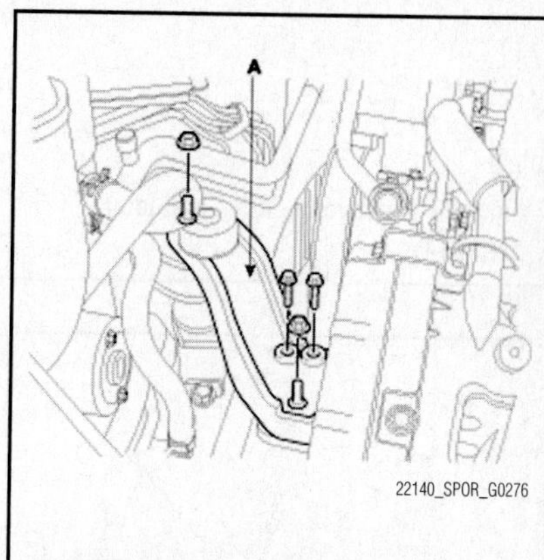

Fig. 193 Remove the engine mount bracket (A)

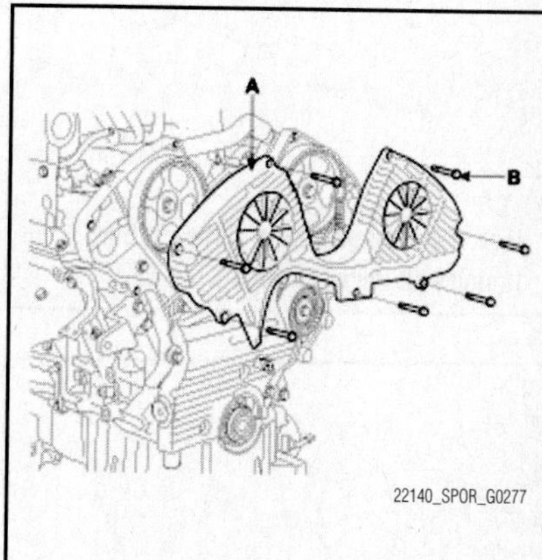

Fig. 194 Remove the 7 bolts (B) and timing belt upper cover (A)

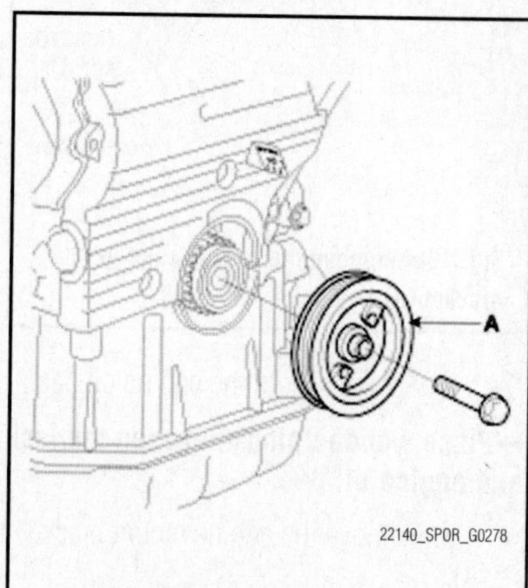

Fig. 195 Remove the crankshaft pulley bolt and crankshaft pulley (A)

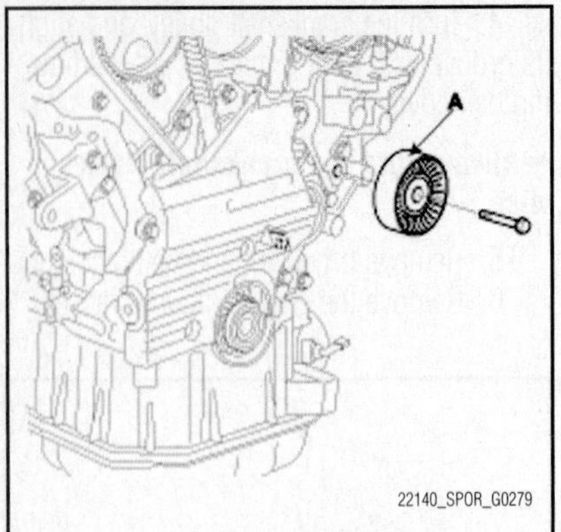

Fig. 196 Remove the drive belt idler pulley (A)

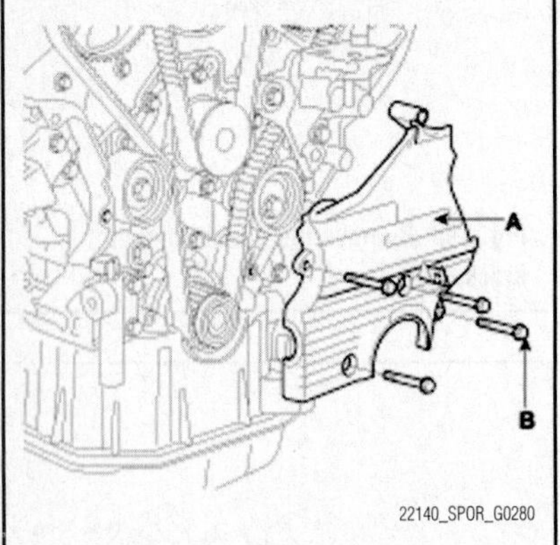

Fig. 197 Remove the 4 bolts (B) and timing belt lower cover (A).

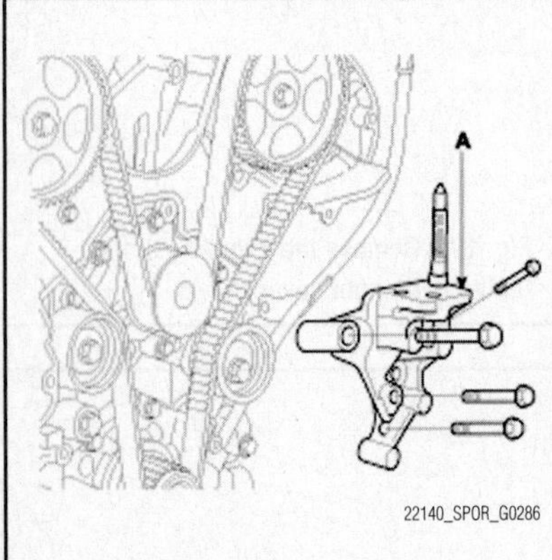

Fig. 198 Remove the engine support bracket (A)

a. Set the jack to the engine oil pan.

➡Place wooden block between the jack and engine oil pan.

b. Remove the engine mount bracket (A).

7. Remove the power steering pump.

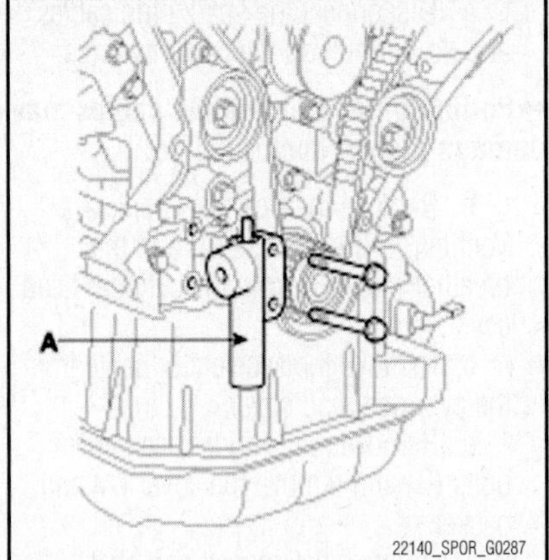

Fig. 199 Remove timing belt tensioner (A)

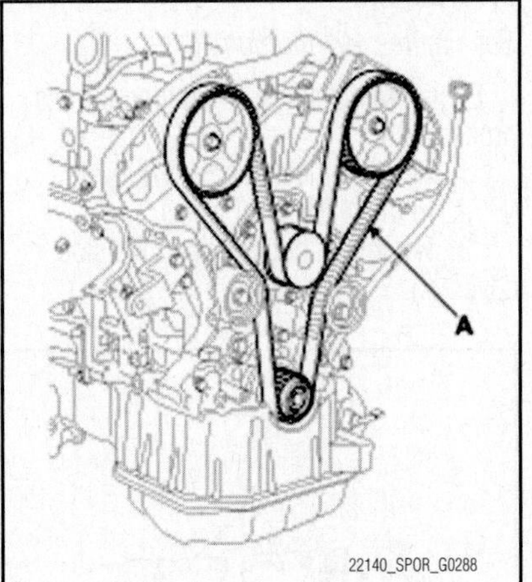

Fig. 200 Remove the timing belt (A)

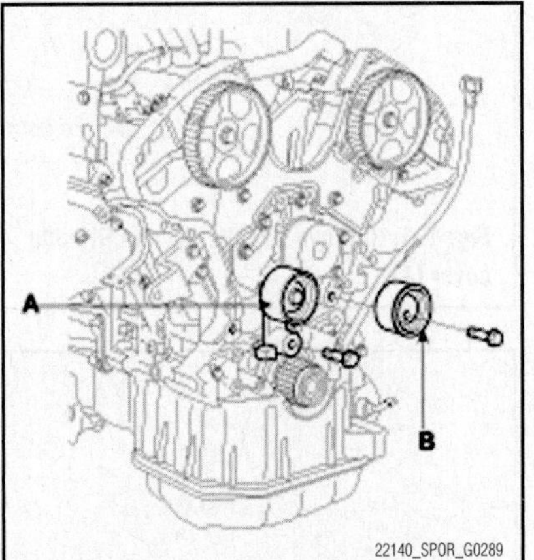

Fig. 201 Remove the tensioner pulley (A) and timing belt idler pulley (B)

8. Remove the 7 bolts (B) and timing belt upper cover (A).
9. Remove the crankshaft pulley bolt and crankshaft pulley (A).
10. Remove the drive belt idler pulley (A).
11. Remove the 4 bolts (B) and timing belt lower cover (A).
12. Remove the engine support bracket (A).
13. Check that timing marks of the camshaft timing pulleys and cylinder head covers are aligned. If not, turn the crankshaft 1revolution (360°).
14. Remove timing belt tensioner (A).
15. Remove the timing belt (A).

➡If the timing belt is reused, make an arrow indicating the turning direction to make sure that the belt is reinstalled in the same direction as before.

16. Remove the tensioner pulley (A) and timing belt idler pulley (B).
17. Remove the crankshaft sprocket.
18. Remove camshaft sprockets.
19. Installation is the reverse of removal.

VALVE LASH

ADJUSTMENT

See Figures 202 through 207.

➡Inspect and adjust the valve clearance when the engine is cold. Engine coolant temperature: 68°F ± 9°F (20°C ± 5°C) and cylinder head is installed on cylinder block.

1. Remove the engine cover.
2. Remove the bolts and timing belt upper cover.
3. Remove the cylinder head cover.
   a. Disconnect the spark plug cables.

➡Pulling on or bending the cables may damage the conductor inside.

   b. Remove the Positive Crankcase Ventilation (PCV) hose and the breather hose from the cylinder head cover.
   c. Remove the accelerator cable from the cylinder head cover.
   d. Remove the cylinder head cover bolts and remove the cover and gasket.
4. Set No. 1 cylinder to TDC/compression.
   a. Turn the crankshaft pulley and align its groove with the timing mark "T" of the lower timing belt cover.
   b. Check that the hole of the camshaft timing pulley (A) is aligned with the timing mark of the bearing cap. If not, turn the crankshaft one revolution (360°).
5. Inspect the valve clearance.
   a. Check only the valves indicated. No. 1 cylinder : TDC/compression. Measure the valve clearance.
   b. Using a thickness gauge, measure the clearance between the tappet shim and the base circle of camshaft.
   c. Record the out-of-specification valve clearance measurements. They will

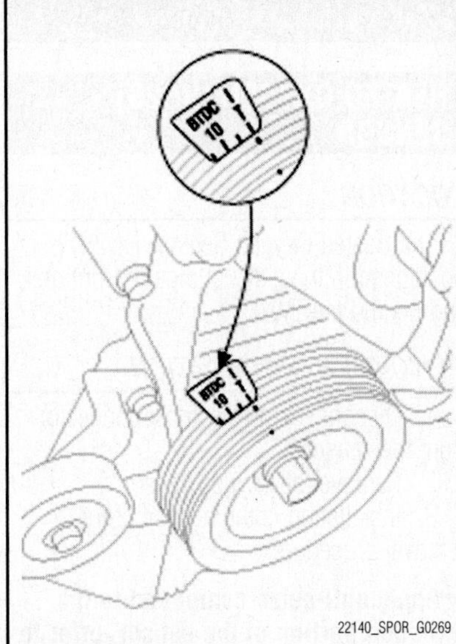

Fig. 202 Turn the crankshaft pulley, and align its groove with timing mark "T" of the timing belt cover

be used later to determine the required replacement adjusting shim.

d. Valve clearance Specification:

- Engine coolant temperature: 68°F ± 9°F (20°C ± 5°C)
- Intake: 0.0079inches (0.20 mm)
- Exhaust: 0.0110inches (0.28 mm)

e. Limit:

- Intake: 0.0047–0.0110 inches (0.12–0.28mm)
- Exhaust: 0.0079–0.0150 inches (0.20–0.38 mm)

f. Turn the crankshaft pulley one revolution (360°) and align the groove with the timing mark "T" of lower timing belt cover.

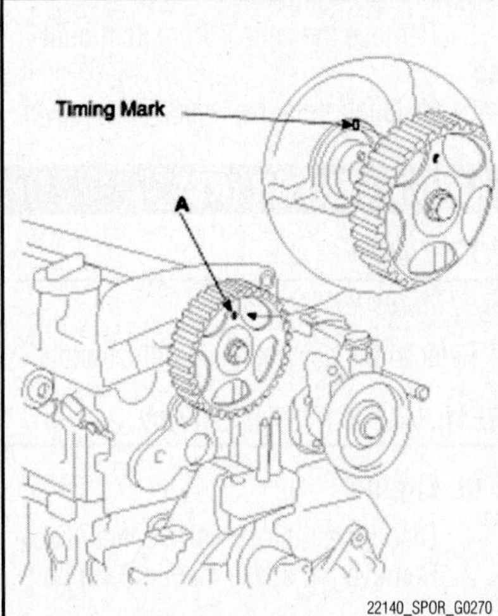

Fig. 203 Check that the timing mark of camshaft sprocket is aligned with the timing mark of cylinder head cover (No.1 cylinder compression TDC position)

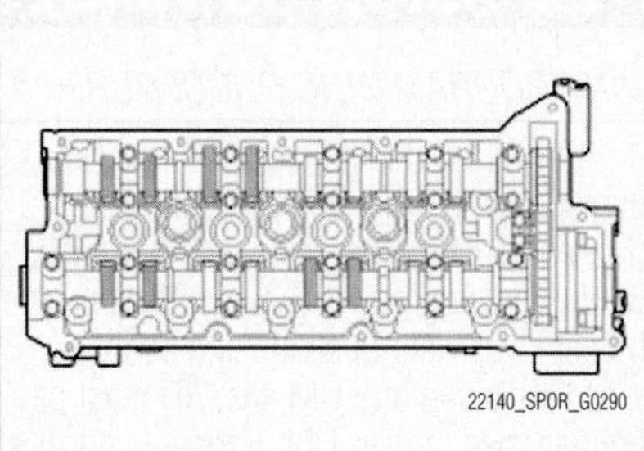
Fig. 204 Check only the valves indicated. No. 1 cylinder: TDC/compression

g. Check only valves indicated. No. 4 cylinder: TDC/compression. Measure the valve clearance.

6. Adjust the intake and exhaust valve clearance.

a. Turn the crankshaft so that the lobe of the camshaft on the adjusting valve is upward.

b. Using the SST (09220–2D000), press down the valve lifter and place the stopper between the camshaft and valve lifter and remove the special tool.

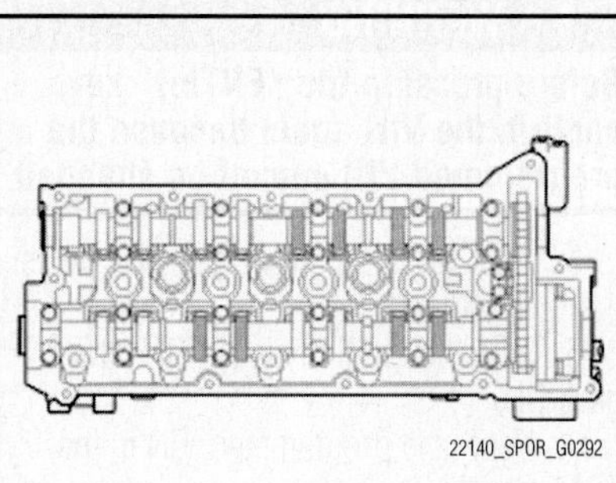
Fig. 205 Check only the valves indicated. No. 4 cylinder: TDC/compression

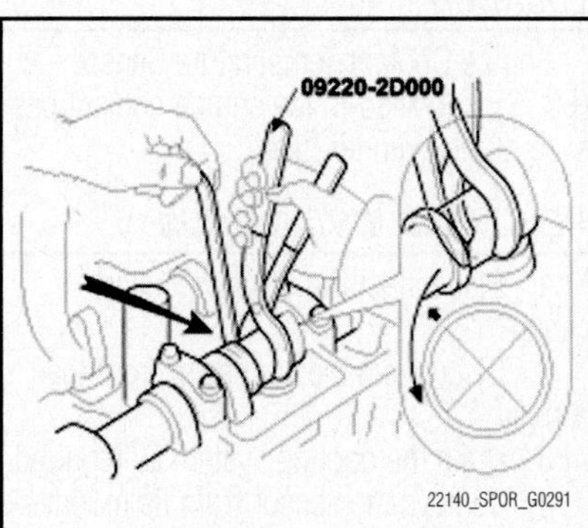

Fig. 206 Using the SST (09220–2D000)

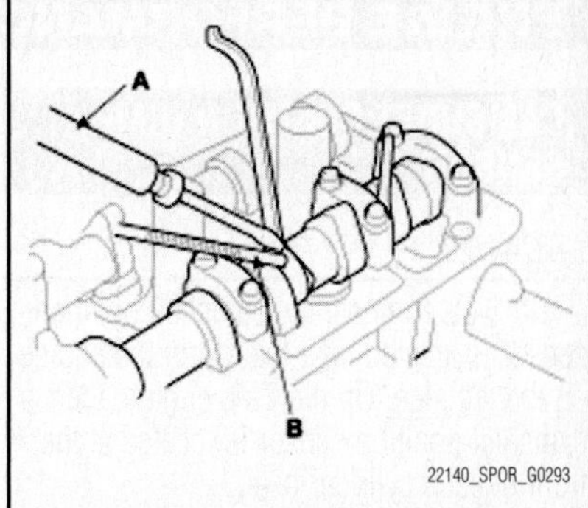

Fig. 207 Remove the adjusting shim with a small screw driver (A) and magnet (B)

c. Remove the adjusting shim with a small screw driver (A) and magnet (B).

d. Measure the thickness of the removed shim using a micrometer.

e. Calculate the thickness of a new shim so that the valve clearance comes within the specified value.

f. Valve clearance:

- Engine coolant temperature: 68°F ± 9°F (20°C ± 5°C)
- T: Thickness of removed shim
- A: Measured valve clearance
- N: Thickness of new shim
- Intake: N = T + [A - 0.0079 inches (0.20 mm)]
- Exhaust: N = T + [A - 0.0110 inches (0.28 mm)]

g. Select a new shim with a thickness as close as possible to the calculated value.

➡Shims are available in 20 size increments of 0.0016 inches (0.04 mm) from 0.0787 inches (2.00 mm) to 0.1087 inches (2.76 mm)

h. Place a new adjusting shim on the valve lifter.

i. Using SST (09220 - 2D000), press down the valve lifter and remove the stopper.

j. Recheck the valve clearance.

k. Valve clearance:

- Engine coolant temperature: 68°F ± 9°F (20°C ± 5°C)
- Specification:
- Intake: 0.0079 inches (0.20 mm)
- Exhaust: 0.0110 inches (0.28 mm)
- Limit (After adjusting valve clearance)
- Intake: 0.0067–0.0091 inches (0.17–0.23 mm)
- Exhaust: 0.0098–0.0122 inches (0.25–0.31 mm)

INSPECTION

See Adjustment.

ENGINE PERFORMANCE & EMISSION CONTROLS

CAMSHAFT POSITION (CMP) SENSOR

LOCATION

On 2.0L engines, the camshaft position sensor is located near the top of the engine on the left side. On the 2.7L engine, the camshaft position sensor is located at the front of each cylinder head.

REMOVAL & INSTALLATION

1. Disconnect the negative battery cable.
2. Disconnect the connector from the sensor.
3. Remove the mounting bolts.
4. Remove the sensor from the cylinder head.
5. Installation is the reverse or removal.

CRANKSHAFT POSITION (CKP) SENSOR

LOCATION

The CKP sensor is installed on the cylinder block or the transaxle housing.

REMOVAL & INSTALLATION

1. Disconnect the negative battery cable.
2. Disconnect the connector from the sensor.
3. Remove the mounting bolts.
4. Remove the sensor from the cylinder head.
5. Installation is the reverse or removal.

ELECTRONIC CONTROL MODULE (ECM)

LOCATION

Under left side of instrument panel.

REMOVAL & INSTALLATION

1. Disconnect the negative battery cable.
2. Remove the lower inner trim.
3. Detach the floor mat.
4. Remove the protective cover.
5. Remove the ECM bracket retaining nuts.
6. Remove the clip from the bracket.
7. Disconnect the connectors.
8. Remove the ECM from the vehicle.
9. Installation is the reverse of removal.

➡When replacing the ECM, be careful not to use the wrong part number as damage to the injection system could occur.

VIN PROGRAMMING PROCEDURE

VIN (Vehicle Identification Number) is a number that has the vehicle's information (Maker, Vehicle Type, Vehicle Line/Series, Body Type, Engine Type, Transmission Type, Model Year, Plant Location and so forth. When replacing an ECM, the VIN must be programmed in the ECM. If there is no VIN in ECM memory, the fault code (DTC P0630) is set.

⁂ CAUTION

The programmed VIN cannot be changed. When writing the VIN, confirm the VIN carefully

1. Using scan tool, select "Vehicle" and "Engine" (For example, SEDONA 3.5L V6).
2. Select "VIN WRITING".
3. Check the ECM status:
   - VIRGIN: VIN is not programmed
   - LEARNT: VIN has been already programmed

   a. Is the ECM status "VIRGIN"? If yes, go to next step. If no, VIN has been programmed and cannot be reprogrammed.
4. Write the VIN with cursor, function and number keys.

⁂ WARNING

Before pressing the "ENTER" key, confirm the VIN again because the programmed VIN cannot be changed.

5. After verifying the written VIN, press the "ENTER" key.
6. Turn the ignition switch OFF, then back ON.
7. Verify the programmed VIN in the ECM memory.

ENGINE COOLANT TEMPERATURE (ECT) SENSOR

LOCATION

Engine Coolant Temperature Sensor (ECTS) is located in the engine coolant passage of the cylinder head.

REMOVAL & INSTALLATION

1. Disconnect the negative battery cable.
2. Disconnect the connector from the sensor.
3. Drain the cooling system as required.
4. remove the sensor from its mounting.
5. Installation is the reverse of removal.
6. Refill the cooling system.

HEATED OXYGEN (HO2S) SENSOR

LOCATION

The Heated Oxygen Sensor (HO2S) is positioned in the exhaust pipe in front of and behind the TWC.

REMOVAL & INSTALLATION

1. Disconnect the electrical connector from the sensor.
2. Remove the oxygen sensor.
3. Installation is the reverse of the removal procedure.

➡Apply anti-seize compound to the threaded portion of the sensor, prior to installation. Never apply anti-seize compound to the protector of the sensor.

INTAKE AIR TEMPERATURE (IAT) SENSOR

LOCATION

The Intake Air Temperature (IAT) sensor is installed into the Mass Air Flow (MAF) Sensor.

REMOVAL & INSTALLATION

1. Disconnect the negative battery cable.
2. Disconnect the connector from the sensor.
3. Remove the sensor retaining screws.
4. Remove the air cleaner and air intake assembly, as required.
5. Remove the sensor from its mounting.
6. Installation is the reverse of removal.

KNOCK SENSOR (KS)

LOCATION

See Figures 208 and 209.

Refer to the accompanying illustration.

REMOVAL & INSTALLATION

2.0L Engine

1. Disconnect the negative battery cable.
2. Remove the intake manifold support bracket.
3. Disconnect the sensor connector.
4. Remove the sensor from its mounting.
5. Installation is the reverse of the removal procedure.
6. Tighten sensor bolt to 12.3 - 19.5 ft. lb. (16.7 - 26.5 Nm).

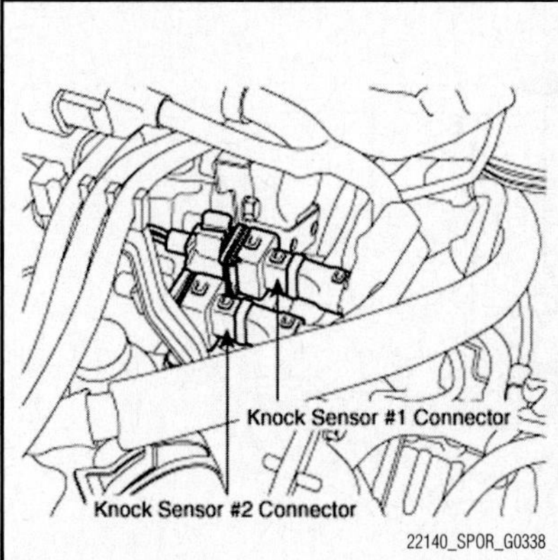

Fig. 208 Knock sensor location—2.0L engine

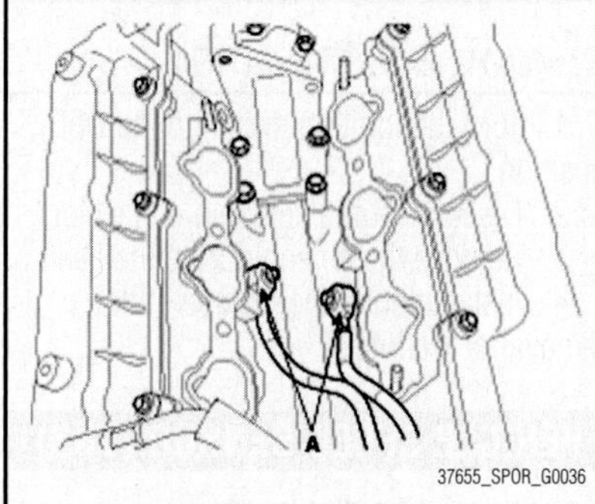

Fig. 209 Knock sensor location—2.7L engine

2.7L Engine

1. Disconnect the negative battery cable.
2. Remove the intake manifold.
3. Disconnect the sensor connector.
4. Remove the sensor from its mounting.
5. Installation is the reverse of the removal procedure.
6. Tighten sensor bolt to 12.5 - 19 ft. lb. (17 - 26 Nm).

MALFUNCTION INDICATOR LIGHT (MIL)

RESET PROCEDURE

1. Misfire and Fuel System Malfunctions:
   a. For misfire or fuel system malfunctions, the MIL may be extinguished if the same fault does not reoccur during monitoring in three subsequent sequential driving cycles in which conditions are similar to those under which the malfunction was first detected.
2. All Other Malfunctions:
   a. For all other faults, the MIL may be extinguished after three subsequent sequential driving cycles during which the monitoring system responsible for illuminating the MIL functions without detecting the malfunction and if no other malfunction has been identified that would independently illuminate the MIL.
3. Erasing a fault code
   a. The diagnostic system may erase a fault code if the same fault is not re-registered in at least 40 engine warm-up cycles, and the MIL is not illuminated for that fault code.
4. Use scan tool to erase codes. Follow scan tool instructions.

MASS AIR FLOW (MAF) SENSOR

LOCATION

2.7L Engine

The Mass Air Flow (MAF) sensor is located between the air cleaner assembly and the throttle body.

REMOVAL & INSTALLATION

2.7L Engine

1. Disconnect the negative battery cable.
2. Disconnect the connector from the sensor.
3. Remove the air cleaner and air intake assembly, as required.
4. Remove the sensor from its mounting.
5. Installation is the reverse of the removal procedure.

MANIFOLD ABSOLUTE PRESSURE (MAP) SENSOR

LOCATION

2.0L Engine

On intake manifold, combined with IAT sensor.

REMOVAL & INSTALLATION

2.0L Engine

1. Disconnect the negative battery cable.
2. Disconnect the connector from the sensor.
3. Remove the sensor retaining screws.
4. Remove the sensor from its mounting.
5. Installation is the reverse of the removal procedure.

OUTPUT SHAFT SPEED (OSS) SENSOR

LOCATION

See Figure 210.

Refer to the accompanying illustration.

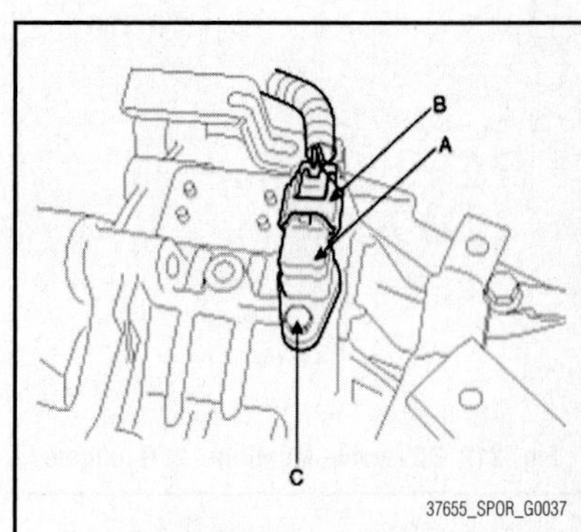

Fig. 210 Output shaft speed sensor (A), sensor connector (B) and bolt (C)

REMOVAL & INSTALLATION

1. Remove the battery and air cleaner.
2. Remove the output shaft speed sensor.
   a. Disconnect the output shaft speed sensor connector.
   b. Remove the bolt.
   c. Inspect the output shaft speed sensor bore.
3. Apply a light coat of automatic transaxle fluid to the O-ring seal before installation.
4. Installation is the reverse of removal.

POSITIVE CRANKCASE VENTILATION (PCV) VALVE

LOCATION

2.0L Engine

See Figure 211.

Refer to the accompanying illustration.

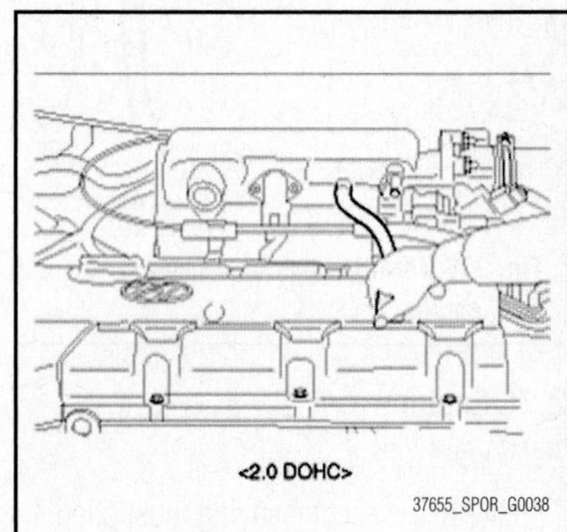

Fig. 211 PCV valve location—2.0L engine

2.7L Engine

See Figure 212.

Refer to the accompanying illustration.

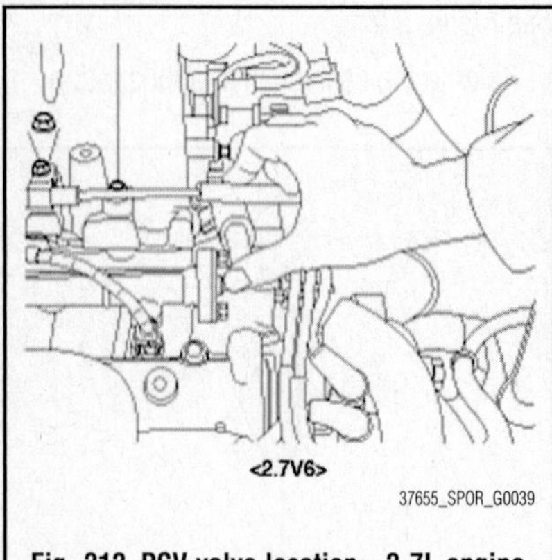

Fig. 212 PCV valve location—2.7L engine

REMOVAL & INSTALLATION

1. Remove PCV valve from valve cover.
2. Remove PCV valve from hose.
3. Installation is the reverse of the removal procedure.

THROTTLE POSITION SENSOR (TPS)

LOCATION

2.0L Engine

See Figure 213.

Refer to the accompanying illustration.

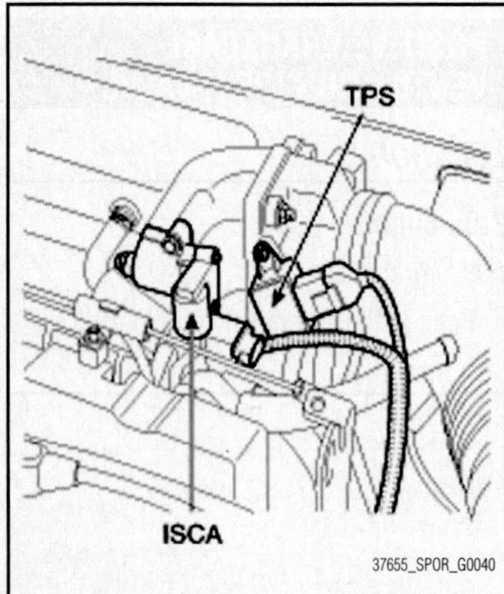

Fig. 213 Throttle Position Sensor (TPS)—2.0L engine

2.7L Engine

See Figure 214.

Refer to the accompanying illustration.

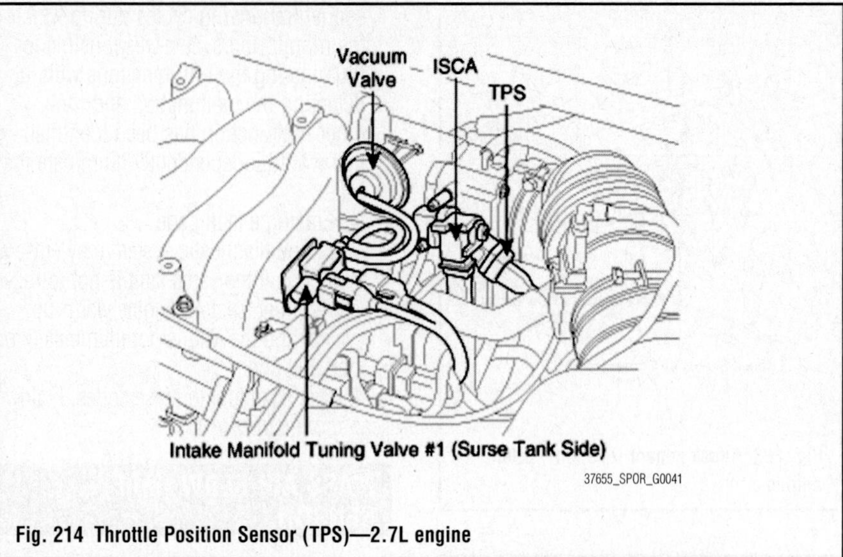

Fig. 214 Throttle Position Sensor (TPS)—2.7L engine

REMOVAL & INSTALLATION

1. Turn the ignition switch to the OFF position.
2. Disconnect the negative battery cable.
3. Disconnect the sensor connector.
4. Remove the sensor retaining screws.
5. Remove the sensor from its mounting.
6. Installation is the reverse of the removal procedure.

VARIABLE CAMSHAFT TIMING OIL CONTROL SOLENOID

LOCATION

See Figure 215.

Refer to the accompanying illustration.

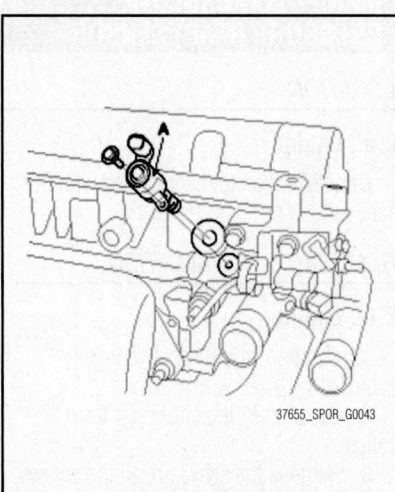

Fig. 215 OCV(Oil Control Valve)(A)—2.0L engine

REMOVAL & INSTALLATION

1. Turn the ignition switch to the OFF position.
2. Disconnect the solenoid connector.
3. Remove the solenoid from the head.
4. Installation is the reverse of the removal procedure.

VEHICLE SPEED SENSOR (VSS)

REMOVAL & INSTALLATION

See Figure 216.

1. Raise and support vehicle.
2. Remove front wheel.
3. Remove the speed sensor cable mounting bolt and speed sensor.
4. Installation is the reverse of the removal procedure.

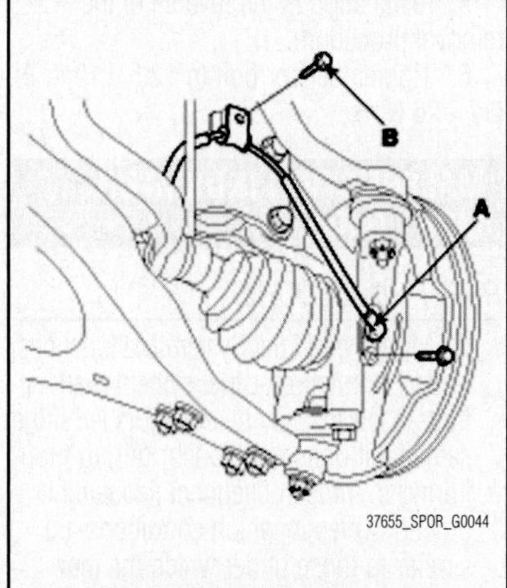

Fig. 216 Speed sensor (A) and speed sensor cable mounting bolt (B)

FUEL SYSTEM SERVICE PRECAUTIONS

Safety is the most important factor when performing not only fuel system maintenance, but any type of maintenance. Failure to conduct maintenance and repairs in a safe manner may result in serious personal injury or death. Work on a vehicle's fuel system components can be accomplished safely and effectively by adhering to the following rules and guidelines.

- To avoid the possibility of fire and personal injury, always disconnect the negative battery cable unless the repair or test procedure requires that battery voltage be applied.
- Always relieve the fuel system pressure prior to disconnecting any fuel system component (injector, fuel rail, pressure regulator, etc.) fitting or fuel line connection. Exercise extreme caution whenever relieving fuel system pressure to avoid exposing skin, face and eyes to fuel spray. Please be advised that fuel under pressure may penetrate the skin or any part of the body that it contacts.
- Always place a shop towel or cloth around the fitting or connection prior to loosening to absorb any excess fuel due to spillage. Ensure that all fuel spillage is quickly removed from engine surfaces. Ensure that all fuel-soaked cloths or towels are deposited into a flame-proof waste container with a lid.
- Always keep a dry chemical (Class B) fire extinguisher near the work area.
- Do not allow fuel spray or fuel vapors to come into contact with a spark or open flame.
- Always use a second wrench when loosening or tightening fuel line connection fittings. This will prevent unnecessary stress and torsion on fuel piping. Always follow the proper torque specifications.
- Always replace worn fuel fitting O-rings with new ones. Do not substitute fuel hose where rigid pipe is installed.

FUEL SYSTEM PRESSURE

RELIEVING

1. Remove the rear seat.
2. Disconnect the fuel pump connector located under the center floor carpet..
3. Start the engine and wait until fuel in fuel line is exhausted.
4. After engine stalls, turn the ignition switch to OFF position.
5. Reconnect the fuel pump connector.
6. Disconnect the negative battery cable.

FUEL FILTER

REMOVAL & INSTALLATION

See Fuel Pump Removal & Installation. The fuel filter is located inside the fuel pump.

FUEL LEVEL SENDING UNIT

REMOVAL & INSTALLATION

➡The fuel sending unit is separate from the fuel pump unit. It is located in top of the fuel tank.

1. Remove the rear seat cushion.
2. Open the left service cover.
3. Disconnect the fuel pump connector.
4. Start the engine and wait until fuel in fuel line is exhausted.
5. After engine stalls, turn the ignition switch to OFF position.
6. Open the right service cover.
7. Disconnect the sub fuel sender connector and the suction tube quick-connector.
8. Remove the sub fuel sender installation bolts and remove the sub fuel sender.
9. Installation is reverse of removal.
10. Tighten the sub fuel sender installation bolts to 1.4 - 2.2 ft. lb. (2.0 - 2.9 Nm).

➡When installing the Sub fuel sender, be careful not to get the seal-ring entangled.

FUEL PUMP

REMOVAL & INSTALLATION

See Figures 217 through 219.

1. Turn ignition switch to "OFF" and disconnect the battery negative terminal.
2. Remove the rear seat cushion.
3. Remove the service cover (A) under the carpet.
4. Disconnect the fuel pump wiring connector (A).
5. Remove the fuel feed hose (B) and suction hose (C).
6. Remove the fuel pump plate mounting bolts (D).

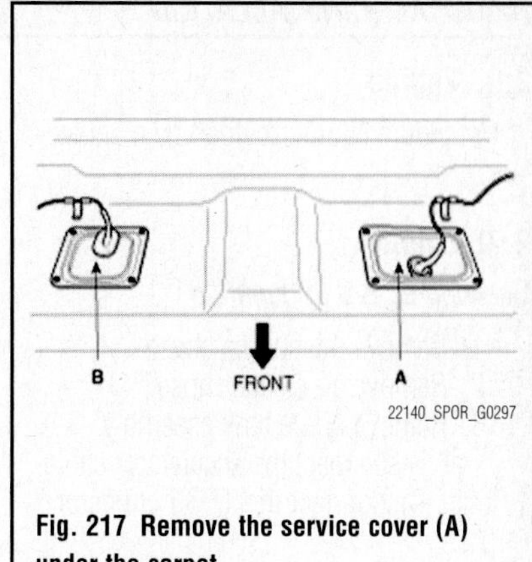

Fig. 217 Remove the service cover (A) under the carpet

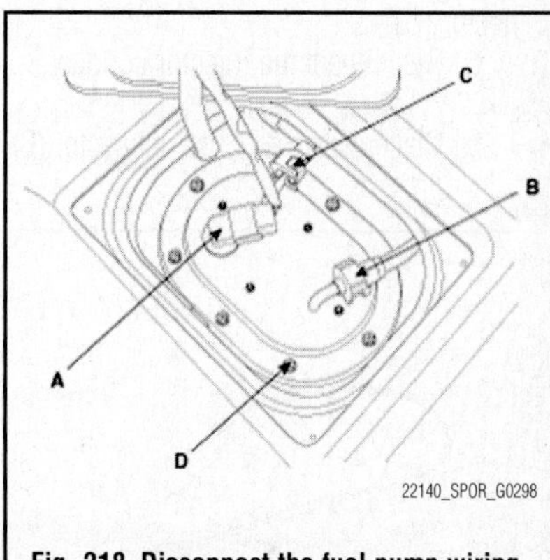

Fig. 218 Disconnect the fuel pump wiring connector (A)

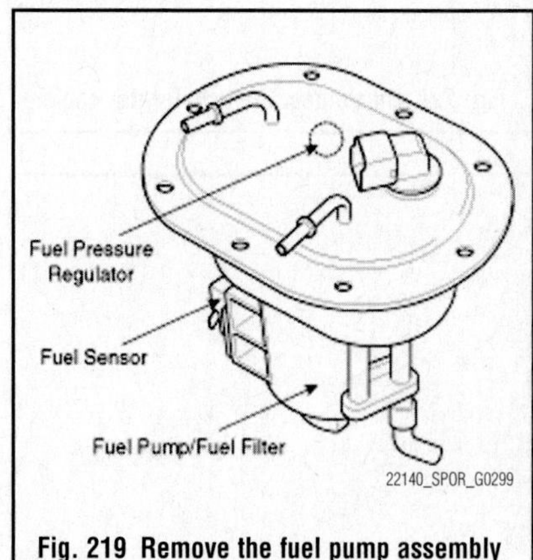

Fig. 219 Remove the fuel pump assembly

7. Remove the fuel pump assembly.
8. Installation is the reverse of removal.

FUEL PRESSURE REGULATOR

REMOVAL & INSTALLATION

See Fuel Pump Removal & Installation. The fuel tank pressure regulator is located inside the fuel pump.

FUEL RAIL AND INJECTOR

REMOVAL & INSTALLATION

2.0L Engine

Procedure is not provided by manufacturer.

2.7L Engine

See Figures 220 through 227.

1. Remove the engine cover.
2. Remove air cleaner hose.
3. Remove surge tank assembly.
   a. Disconnect the accelerator cable.
   b. Disconnect the TPS connector (A).
   c. Disconnect the ISA connector (B).
   d. Disconnect the VIS actuator connector (C).
   e. Disconnect the injector connector (D).
   f. Disconnect the PCSV connector (E).

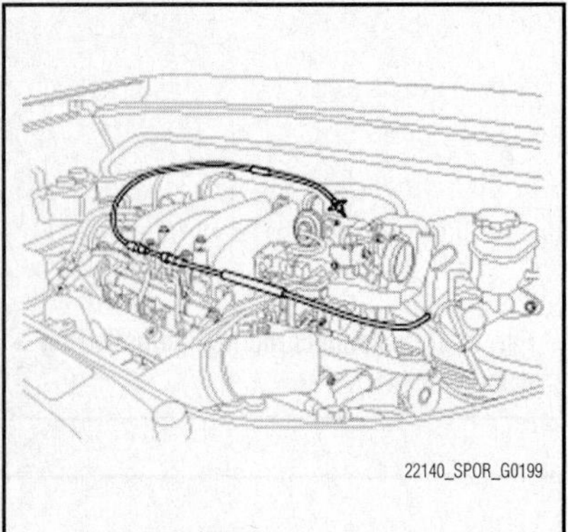

22140_SPOR_G0199

Fig. 220 Disconnect the accelerator cable

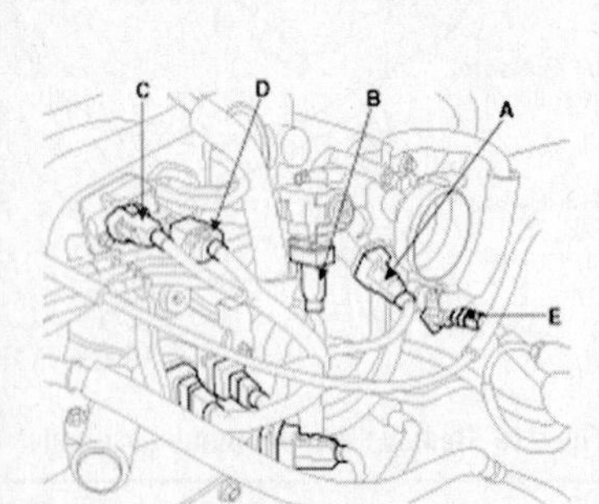

A. Throttle Position Sensor (TPS) connector
B. Idle Speed Actuator (ISA) connector
C. VIS actuator connector
D. Injector connector
E. Purge Control Solenoid Valve (PCSV) connector

22140_SPOR_G0247

Fig. 221 Disconnect the TPS connector (A), the ISA connector (B), the VIS actuator connector (C), the injector connector (D) and the PCSV connector (E)

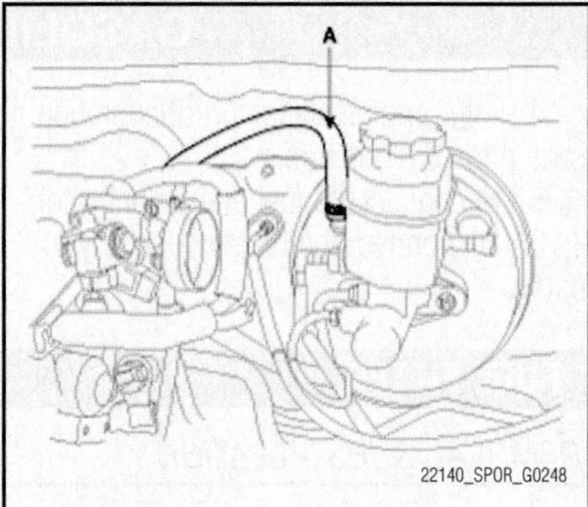

22140_SPOR_G0248

Fig. 222 Disconnect the brake booster vacuum hose (A)

   g. Disconnect the PCSV hose.
   h. Disconnect the brake booster vacuum hose (A).
   i. Disconnect the PCV hose.
   j. Disconnect the IAT sensor connector (A).
   k. Disconnect the VIS actuator connector (B).
   l. Disconnect the ground cable (A) from the surge tank assembly.
   m. Remove the surge tank stay.
   n. Remove the surge tank assembly (A).

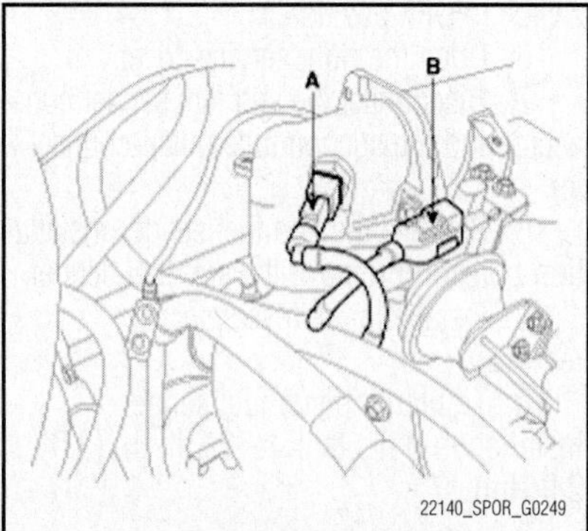

22140_SPOR_G0249

Fig. 223 Disconnect the IAT sensor connector (A) and the VIS actuator connector (B)

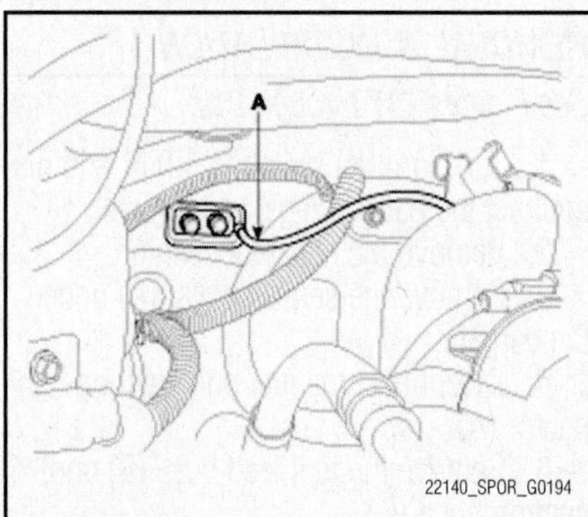

22140_SPOR_G0194

Fig. 224 Disconnect the ground cable (A) from the surge tank assembly

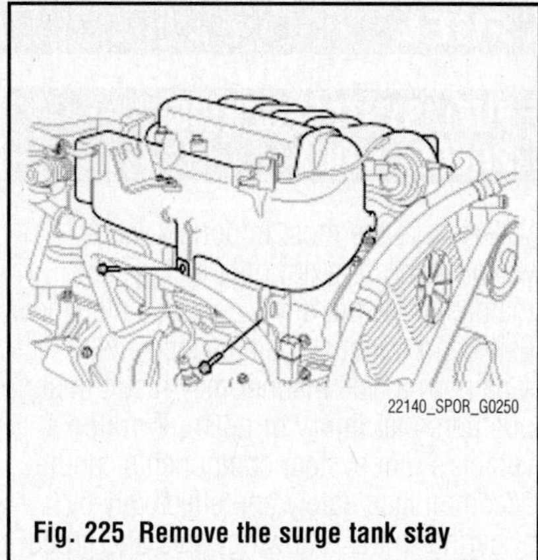

22140_SPOR_G0250

Fig. 225 Remove the surge tank stay

4. Remove the injector assembly (A).

To install:

5. Install the injector assembly.
6. Install the surge tank assembly. Tighten to 11–15 ft. lbs. (15–20 Nm).
7. Install the surge tank stay. Tighten to 11–15 ft. lbs. (15–20 Nm).

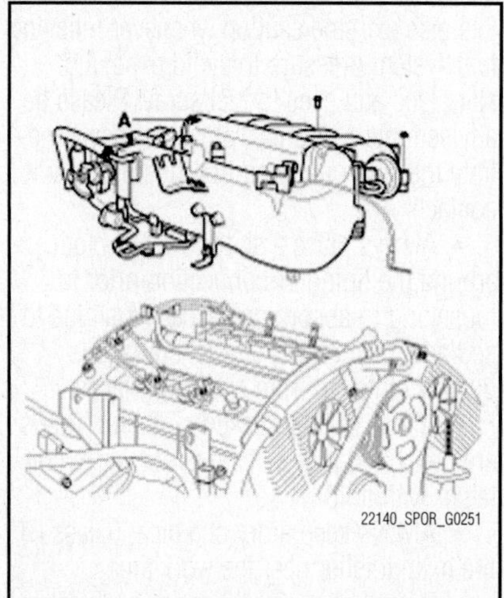

22140_SPOR_G0251

Fig. 226 Remove the surge tank assembly (A)

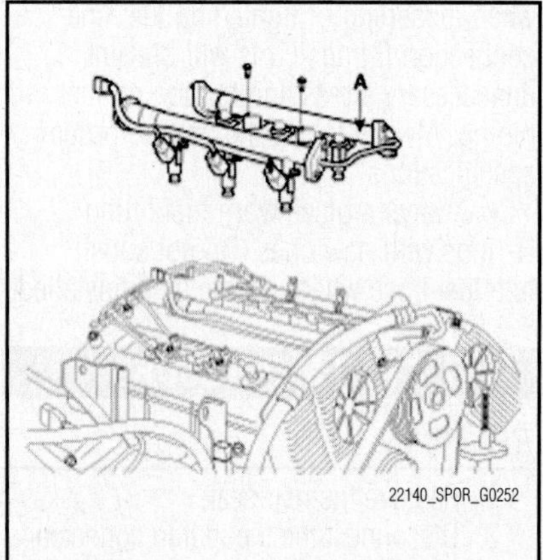

22140_SPOR_G0252

Fig. 227 Remove the injector assembly (A)

8. Install the ground cable.
9. Connect the VIS actuator connector.
10. Connect the IAT sensor connector.
11. Connect the PCV hose.
12. Connect the brake booster vacuum hose.
13. Connect the PCSV hose.
14. Connect the PCSV connector.
15. Connect the injector connector.
16. Connector the VIS actuator connector.
17. Connector the ISA connector.
18. Connector the TPS connector.
19. Connector the actuator cable.
20. Install the air cleaner hose.
21. Install the engine cover.

FUEL TANK

REMOVAL & INSTALLATION

See Figures 228 through 231.

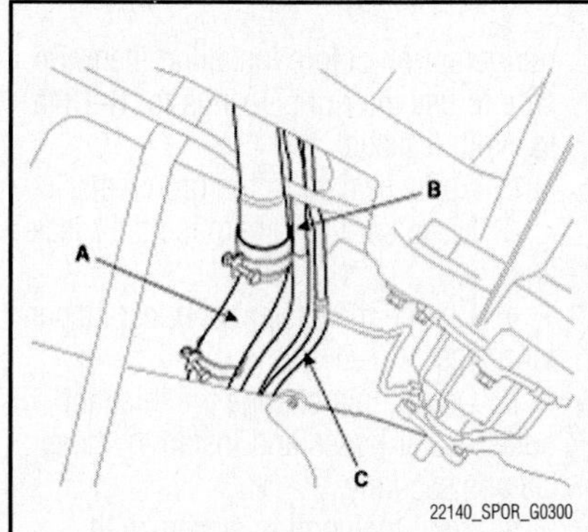

Fig. 228 Disconnect the fuel filler hose (A), fuel leveling hose (B) and ventilation hose (C)

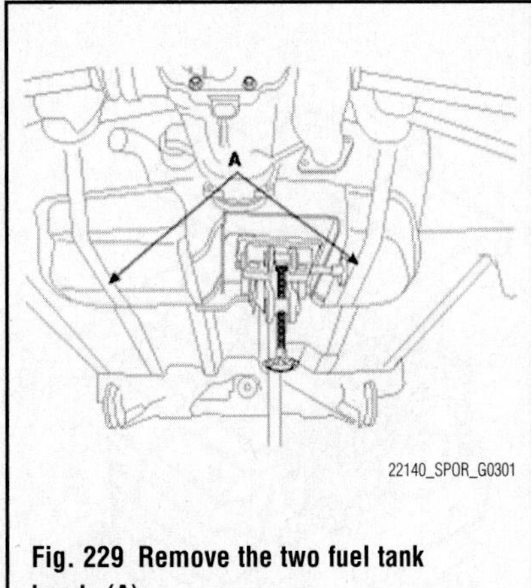

Fig. 229 Remove the two fuel tank bands (A)

1. Remove the front and main muffler assembly.
2. Remove the rear drive shaft (For 4WD only).

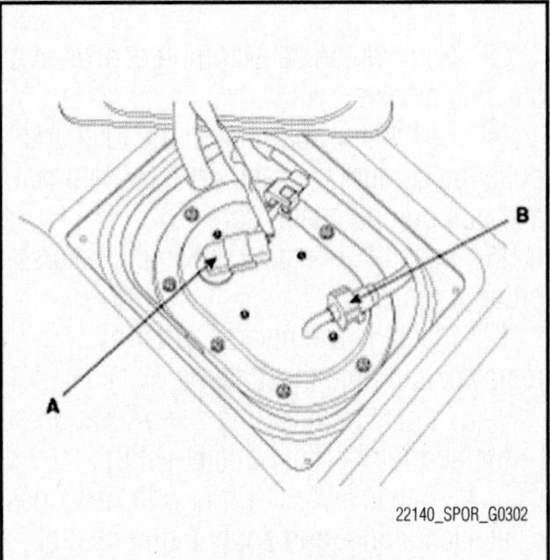

Fig. 230 disconnect the fuel pump connector (A) and fuel feed hose(B)

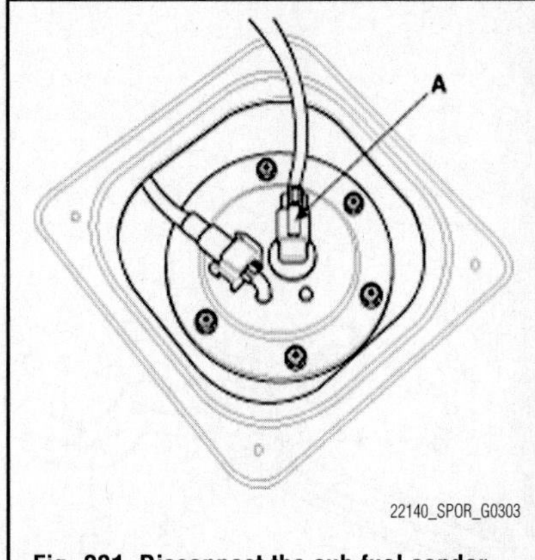

Fig. 231 Disconnect the sub fuel sender connector (A)

3. Disconnect the fuel filler hose (A), fuel leveling hose (B) and ventilation hose (C).
4. Lift the vehicle and support the fuel tank with a jack, and remove the two fuel tank bands (A).
5. Remove the four parking brake mounting bolts.
6. Lower the fuel tank slowly until work space is made, and then disconnect the fuel pump connector (A) and fuel feed hose (B).
7. Disconnect the sub fuel sender connector (A).
8. Remove the fuel tank.
9. Installation is the reverse of removal.

IDLE SPEED

Idle speed is controlled by the ECM and is not adjustable.

HEATING & AIR CONDITIONING SYSTEM

BLOWER MOTOR

REMOVAL & INSTALLATION

See Figure 232.

1. Disconnect the negative battery terminal.
2. Remove the heater unit.
3. Disconnect the connectors from the fresh and recirculation actuator, the blower relay, the blower motor and power mosfet.
4. Remove the self-tapping screws (A), the mounting nut (B), the mounting bolt (C) and the blower unit (D).

➡Make sure that there is no air leaking out of the blower and duct joints.

5. Install in the reverse order of removal.

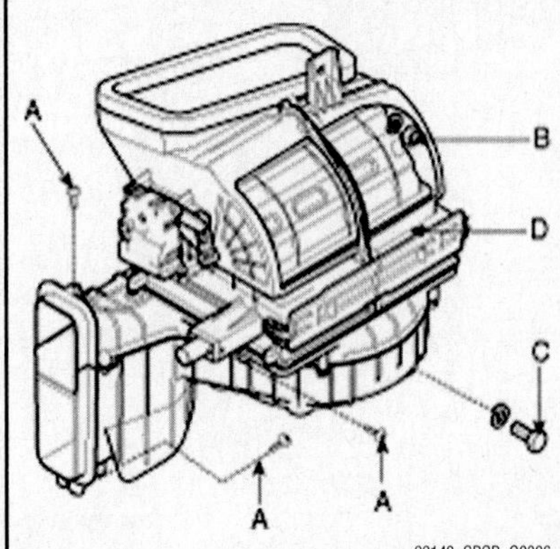

Fig. 232 Remove the self-tapping screws (A), the mounting nut (B), the mounting bolt (C) and the blower unit (D)

HEATER CORE

REMOVAL & INSTALLATION

See Figures 233 through 235.

1. Disconnect the negative battery terminal.
2. Recover the refrigerant with a recovery/recycling/charging station.
3. Remove the bolts (A) and the expansion valve (B) from the evaporator core.

CAUTION

Plug or cap the lines immediately after disconnecting them to avoid moisture and dust contamination.

4. When the engine is cool, disconnect

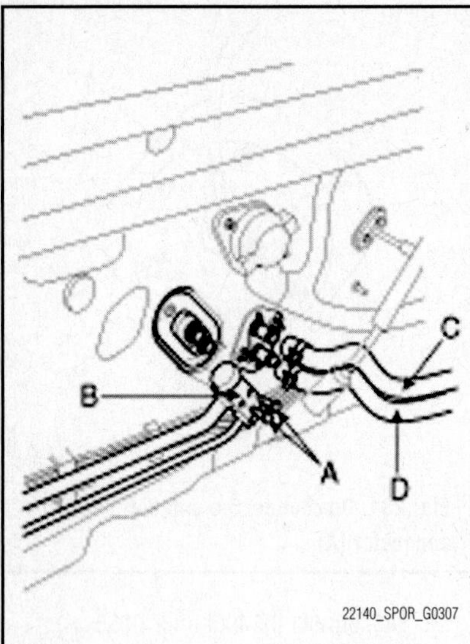

Fig. 233 Remove the bolts (A) and the expansion valve (B) from the evaporator core

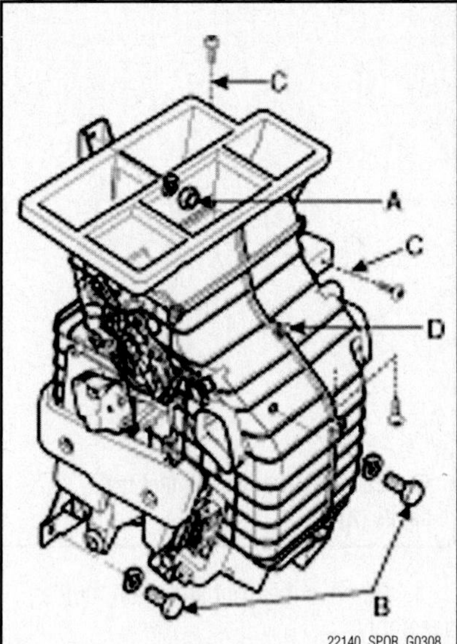

Fig. 234 Remove the mounting nut (A) and the mounting bolts (B)

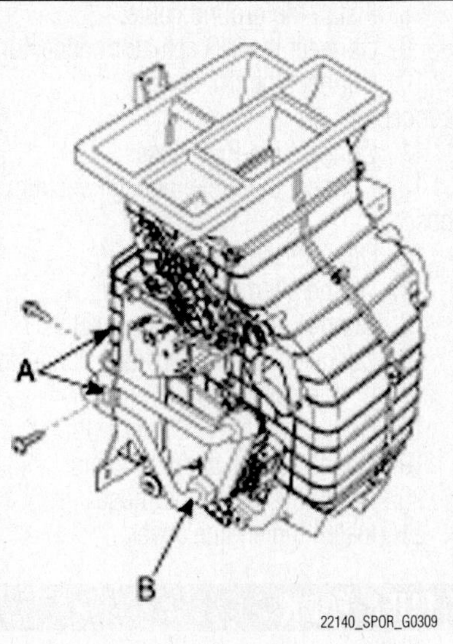

Fig. 235 Be careful not to bend the inlet and outlet pipes during heater core (A) removal, and pull out the heater core (B)

the inlet (D) and outlet (C) heater hoses from the heater unit.

5. Remove the crash pad (instrument panel). Refer to Instrument Panel Removal and Installation.

6. Remove the cross member.

7. Disconnect the connectors from the temperature control actuator, the mode control actuator and the evaporator temperature sensor, and remove the mounting nut (A) and the mounting bolts (B).

8. Remove the heater and evaporator unit (D) after removing the mounting screws (C).

9. Remove the self-tapping screws and the side bracket.

10. Be careful not to bend the inlet and outlet pipes during heater core (A) removal, and pull out the heater core (B).

11. Install the heater core in the reverse order of removal.

12. Install in the reverse order of removal, and note these items :

a. If you're installing a new evaporator, add refrigerant oil (ND-OIL8).

b. Replace the O-rings with new ones at each fitting, and apply a thin coat of refrigerant oil before installing them. Be sure to use the right O-rings for R-134a to avoid leakage.

c. Apply sealant to the grommets.

d. Make sure that there is no air leakage.

e. Charge the system and test its performance.

f. Do not interchange the inlet and outlet heater hoses and install the hose clamps securely.

g. Refill the cooling system with engine coolant.

STEERING

POWER RACK & PINION STEERING GEAR

REMOVAL & INSTALLATION

See Figures 236 through 245.

1. Disconnect the cover fixing clip (A) on the steering shaft universal joint driver side interior.

2. Remove the noise covers (B).

3. Remove the universal joint and the gear box mounting bolt (A) and disconnect the universal joint (B) from the gear box.

4. Raise the vehicle.

5. Remove the front tires.

6. Remove the engine undercover.

7. After removing the split pin, disconnect the tie rod (A) from the knuckle (B) by using the special tool (09568-34000).

8. Remove the stabilizer link (B) from the strut assembly (A).

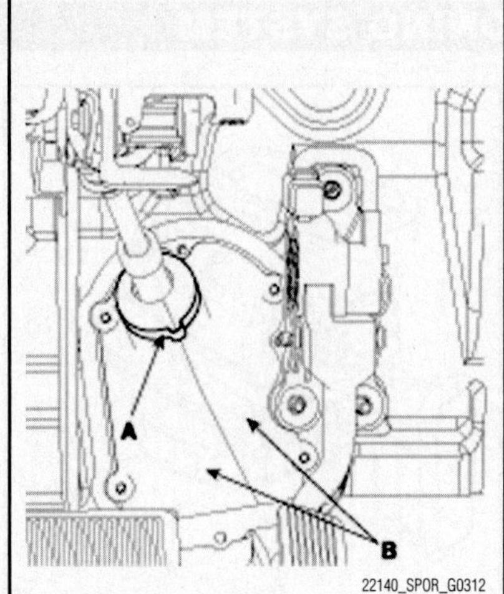

Fig. 236 Disconnect the cover fixing clip (A). Remove the noise covers (B)

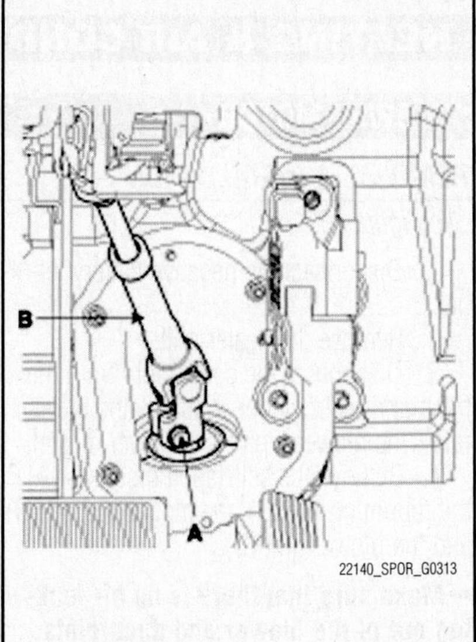

Fig. 237 Disconnect the universal joint from the gear box

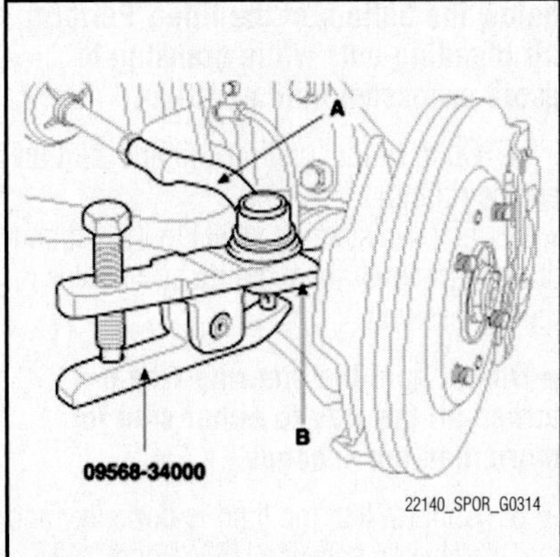

Fig. 238 Disconnect the tie rod (A) from the knuckle (B)

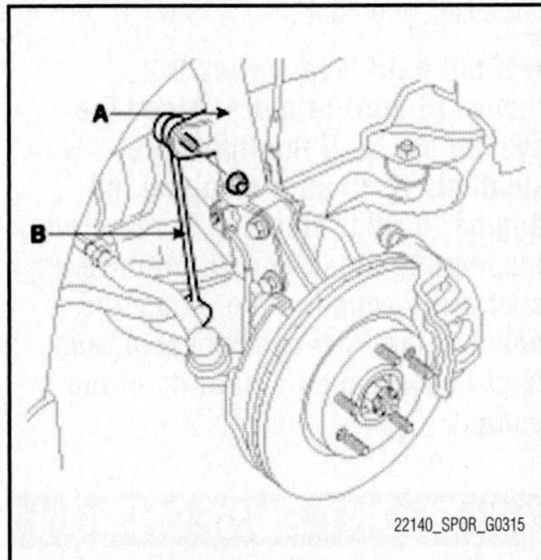

Fig. 239 Remove the stabilizer link (B) from the strut assembly (A)

9. Remove the two bolts (A) for lower arm ball joint.

10. Remove the propeller shaft (A) (4WD) and the front muffler assembly (B).

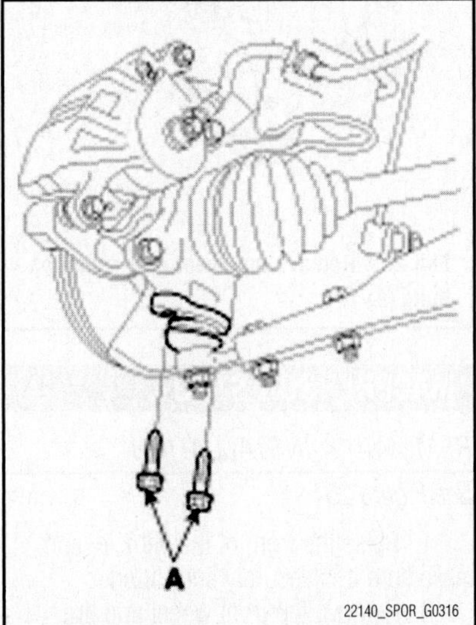

Fig. 240 Remove the two bolts (A) for lower arm ball joint

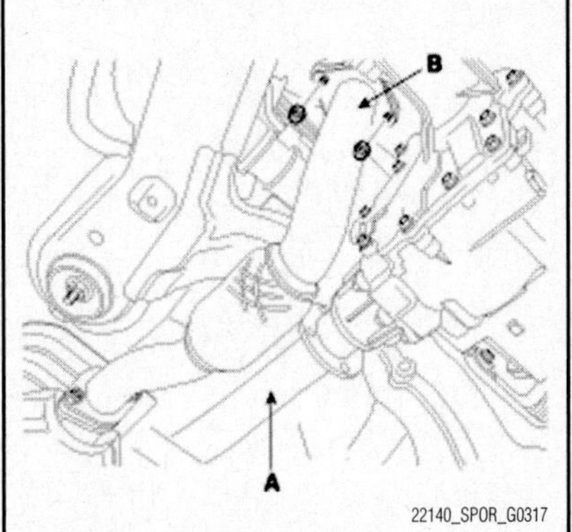

Fig. 241 Remove the propeller shaft (A) (4WD) and the front muffler assembly (B)

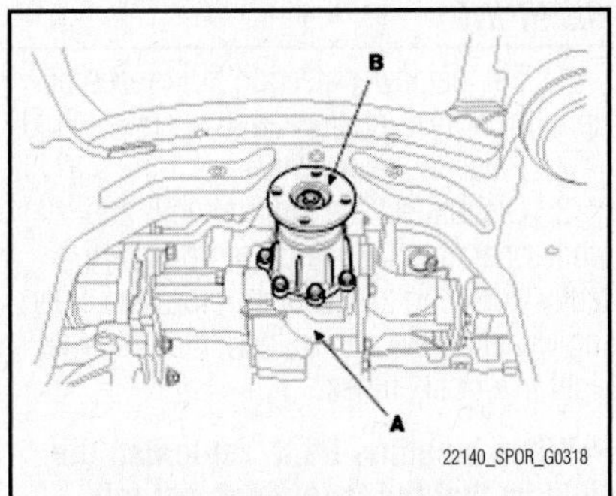

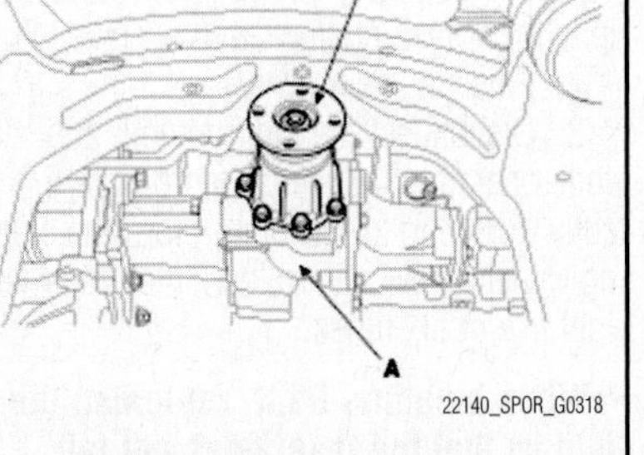

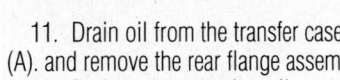

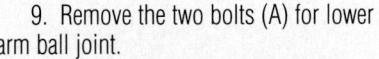

Fig. 242 Drain oil from the transfer case (A) and remove the rear flange assembly (B)

11. Drain oil from the transfer case (A). and remove the rear flange assembly (B).

12. Drain power steering oil.

13. Remove the connector (A) for the PS pressure tubes.

14. Remove two engine mounting bolts (B, C) and six subframe mounting bolts to remove the subframe (A).

15. Remove the power steering gearbox (A) after removing four mounting bolts (B) of the power steering gearbox.

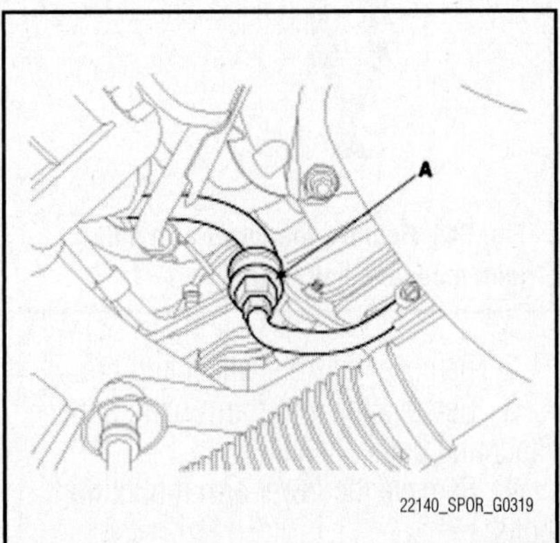

Fig. 243 Remove the connector (A) for the PS pressure tubes

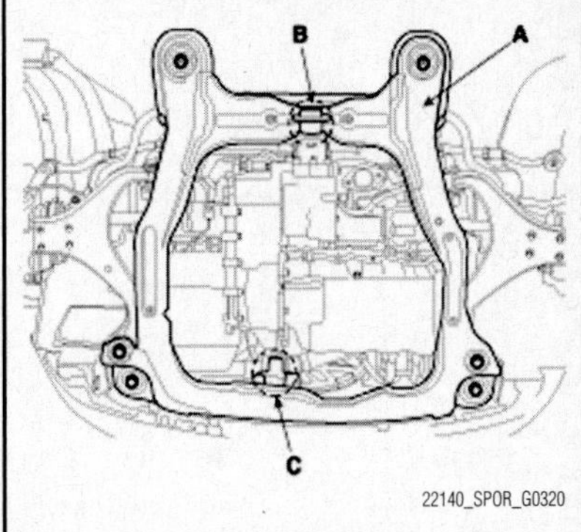

Fig. 244 Remove two engine mounting bolts (B, C) and six subframe mounting bolts to remove the subframe (A)

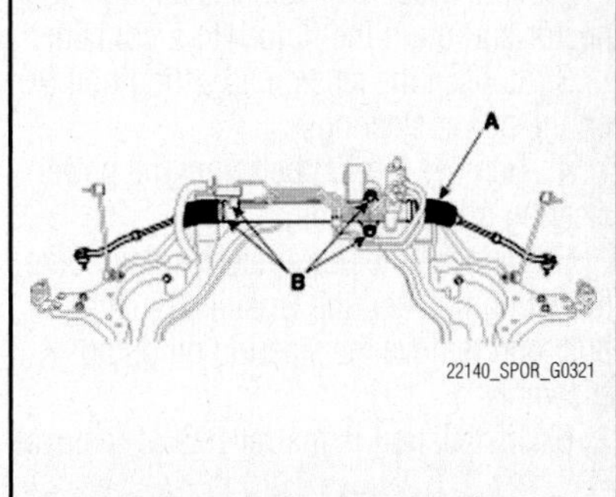

Fig. 245 Remove the power steering gearbox (A) after removing four mounting bolts (B) of the power steering gearbox.

16. Installation is the reverse of removal.

POWER STEERING PUMP

REMOVAL & INSTALLATION

See Figures 246 through 248.

1. Remove the bolt attaching the wiring bracket, and move the wiring to the side.

2. Remove the pressure hose (A) from the oil pump (B), and disconnect

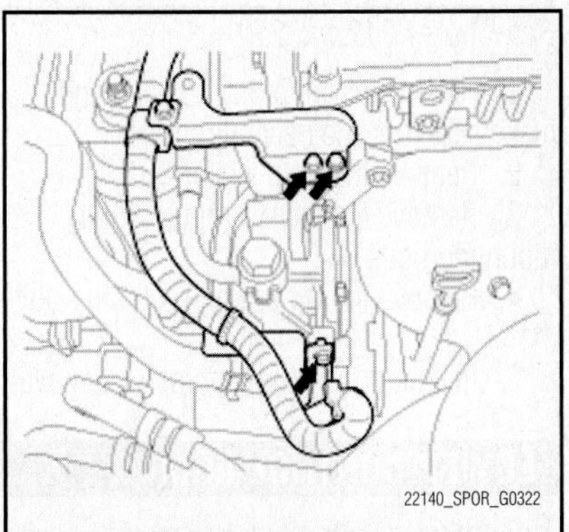

Fig. 246 Remove the bolt attaching the wiring bracket, and move the wiring to the side

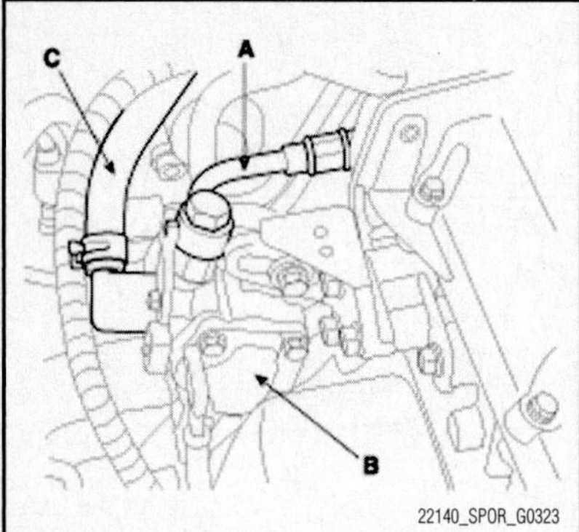

Fig. 247 Remove the pressure hose (A) from the oil pump (B), and disconnect the suction hose (C)

the suction hose (C) from the suction connector and drain the fluid into a container.

3. Loosen the tension adjusting bolt (A) on the power steering.
4. Remove the "V" belt from the power steering oil pump pulley.
5. Remove the power steering oil pump mounting bolt and the tension adjusting bolt, and remove the steering oil pump assembly.
6. Installation is the reverse of removal.

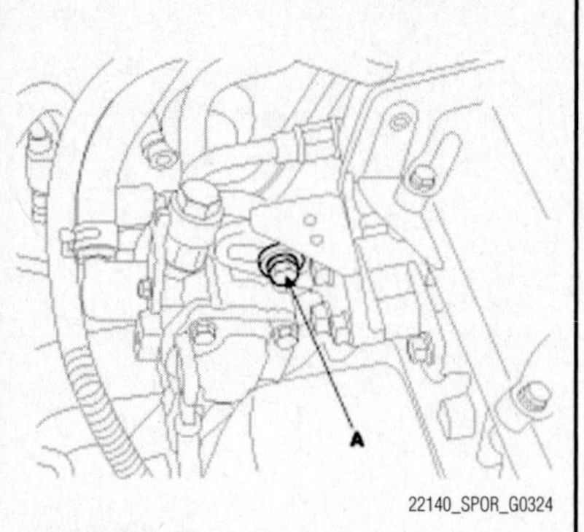

Fig. 248 Loosen the tension adjusting bolt (A) on the power steering pump

BLEEDING

1. Fill the power steering fluid reservoir up to the "MAX" position with specified fluid.
2. Raise the front wheels.
3. Disconnect the ignition coil, and while operating the starter motor intermittently (for 15 to 20 seconds), turn the steering wheel all the way to the left and to the right five or six times.

➡When bleeding fluid, replenish the fluid so that the level does not fall below the bottom of the filter. Perform air bleeding only while cranking to avoid excessive fluid aeration.

4. Connect the ignition coil and start the engine (idling).
5. Turn the steering wheel to the left and to the right, until there are no air bubbles in the oil reservoir.

➡Do not hold the steering wheel turned all the way to either side for more than ten seconds.

6. Confirm that the fluid is not milky and that the level is between "MAX" and "MIN" mark on the reservoir.
7. Check that there is a little change in the fluid level when the steering wheel is turned left and right.

➡If the fluid level varies 0.2 inches (5 mm) or more, bleed the system again. If the fluid level suddenly rises after stopping the engine, further bleeding is required. Incomplete bleeding will produce a chattering sound in the pump and noise in the flow control valve, and lead to decreased durability of the pump.

SUSPENSION

FRONT SUSPENSION

CONTROL LINKS

REMOVAL & INSTALLATION

1. Raise the front of the vehicle, and make sure it is securely supported.
2. Remove the front wheel and tire.
3. Remove the stabilizer bar link from the strut assembly.
4. Installation is the reverse of removal.

LOWER BALL JOINT

REMOVAL & INSTALLATION

See Figures 249 and 250.

1. Raise the front of the vehicle, and make sure it is securely supported.
2. Remove the front wheel and tire.
3. Remove the lower arm ball joint mounting bolts (A).
4. Remove the lower arm mounting bolts (A).
5. Installation is the reverse of removal.

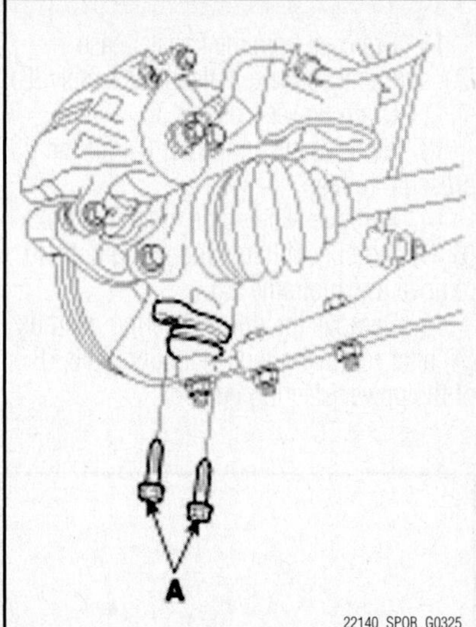

Fig. 249 Remove the lower arm ball joint mounting bolts (A)

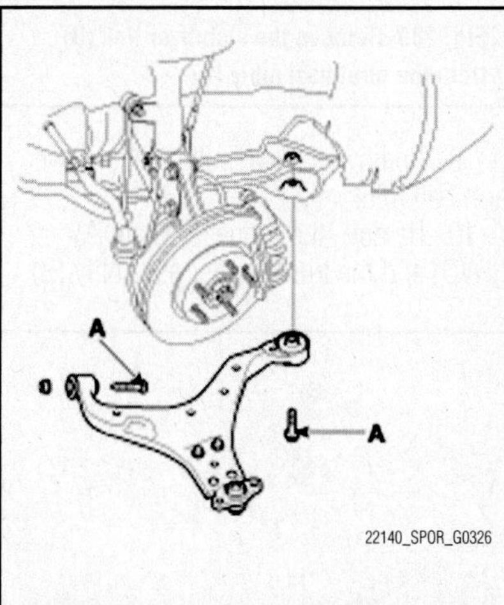

Fig. 250 Remove the lower arm mounting bolts (A)

LOWER CONTROL ARM

REMOVAL & INSTALLATION

1. Raise the front of the vehicle, and make sure it is securely supported.
2. Remove the front wheel and tire.
3. Remove the lower arm ball joint mounting bolts.
4. Remove the lower arm mounting bolts.
5. Installation is the reverse of removal.

MACPHERSON STRUT

REMOVAL & INSTALLATION

See Figure 251.

1. Raise the front of the vehicle, and make sure it is securely supported.
2. Remove the front wheel and tire.
3. Remove the brake hose bracket and speed sensor cable-mounting bolt from the strut assembly.

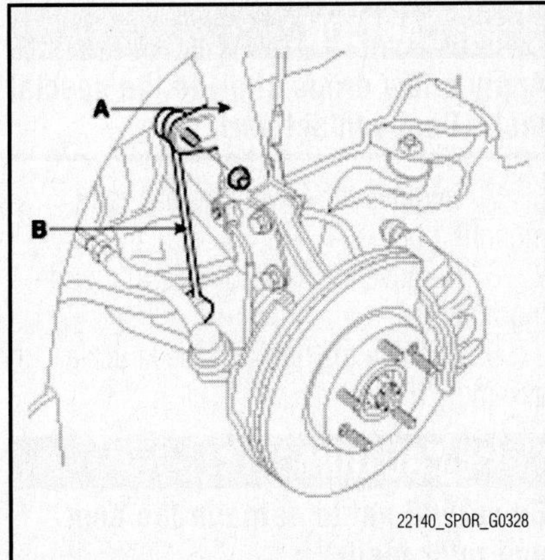

Fig. 251 Remove the stabilizer bar link (B) from the strut assembly (A)

4. Remove the speed sensor cable-mounting bolt and speed sensor.
5. Remove the nut from the stabilizer bar link.
6. Remove the strut upper mounting nuts.
7. Remove the strut lower mounting bolts and remove the strut assembly.

To install:

8. Install the strut assembly and install the strut lower mounting bolts. Tighten to 103–118 ft. lbs. (140–160 Nm).
9. Install the strut upper mounting nuts. Tighten to 33–44 ft. lbs. (45–60 Nm).
10. Install the nut on the stabilizer bar link. Tighten to 74–88 ft. lbs. (100–120 Nm).
11. Install the speed sensor cable mounting bolt and speed sensor. Tighten to 5–8 ft. lbs. (7–11 Nm).
12. Install the brake hose bracket and speed sensor cable mounting bolt on the strut assembly. Tighten to 5–8 ft. lbs. (7–11 Nm).
13. Install the front wheel and tire.

STABILIZER BAR

REMOVAL & INSTALLATION

See Figures 251, 252 through 256.

1. Raise the front of the vehicle, and make sure it is securely supported.
2. Remove the front wheel and tire.
3. Remove the stabilizer bar link (B) from the strut assembly (A).
4. Remove the two bolts (A) for lower arm ball joint.
5. Remove the propeller shaft (A) (4WD) and the front muffler assembly (B).

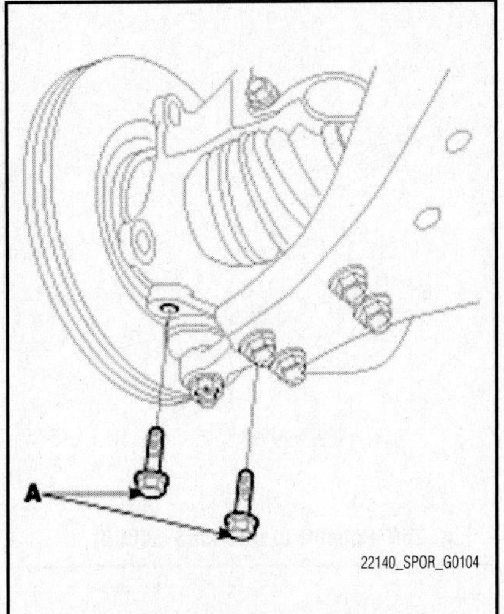

Fig. 252 Remove the lower arm ball joint mounting bolts (A)

6. Drain oil from the transfer case (A) and remove the rear flange assembly (B).
7. Drain power steering oil.
8. Remove the connector (A) for PS pressure tubes.

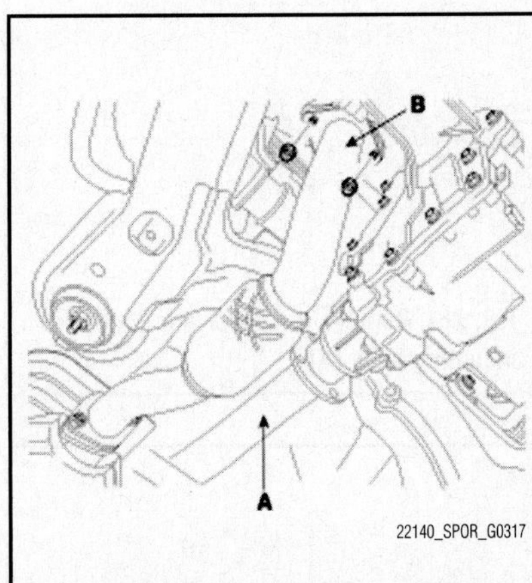

Fig. 253 Remove the propeller shaft (A) (4WD) and the front muffler assembly (B)

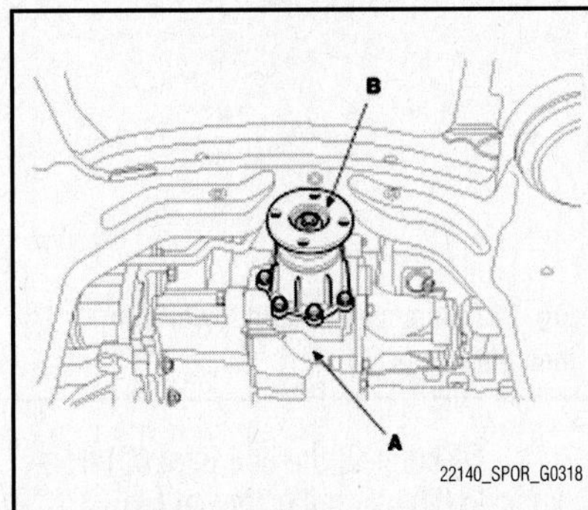

Fig. 254 Drain oil from the transfer case (A) and remove the rear flange assembly (B)

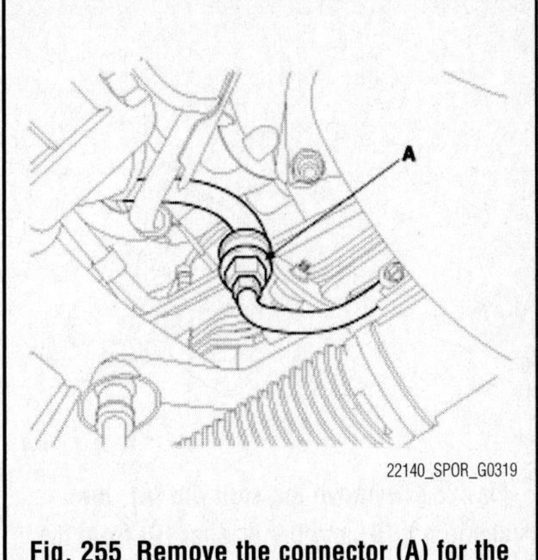

Fig. 255 Remove the connector (A) for the PS pressure tubes

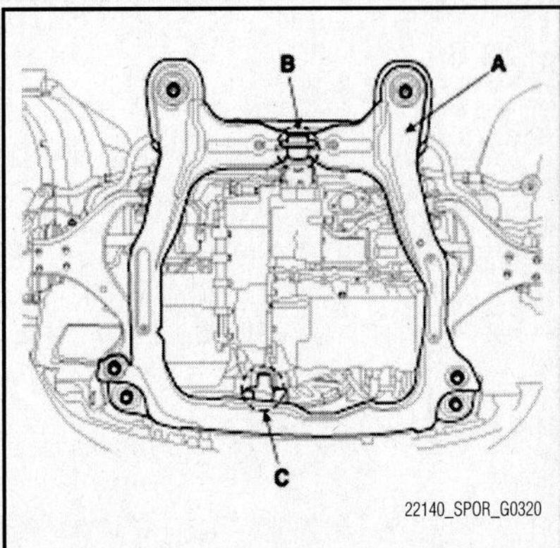

Fig. 256 Remove two engine mounting bolts(B, C) and six subframe mounting bolts to remove the subframe (A)

9. Remove two engine mounting bolts (B, C) and six subframe mounting bolts in order to remove the subframe (A).
10. Remove both stabilizer brackets and two bushings respectively.
11. Remove the stabilizer bar.

CAUTION

Be careful not to do damage to pressure tubes.

12. Installation is the reverse of removal.

STEERING KNUCKLE

REMOVAL & INSTALLATION

See Figures 257 through 262.

1. Raise the front of the vehicle, and make sure it is securely supported.
2. Remove the front wheel and tire.
3. Remove the split pin (A), the castle nut (B) and the washer (C) from the front hub.
4. Remove the caliper mounting bolts and hang the caliper assembly to one side.

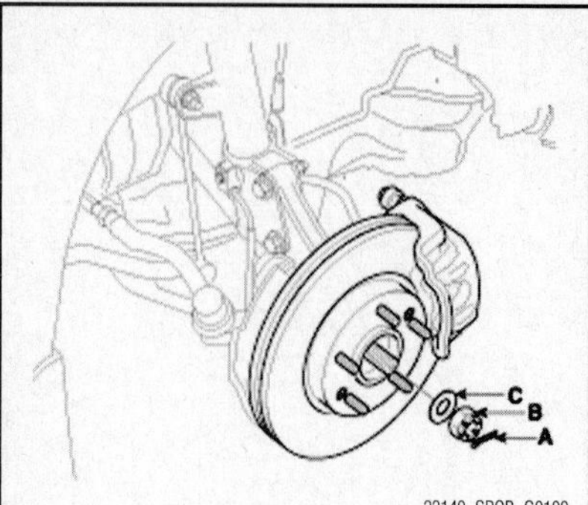

Fig. 257 Remove the split pin (A), the castle nut (B) and the washer (C) from the front hub

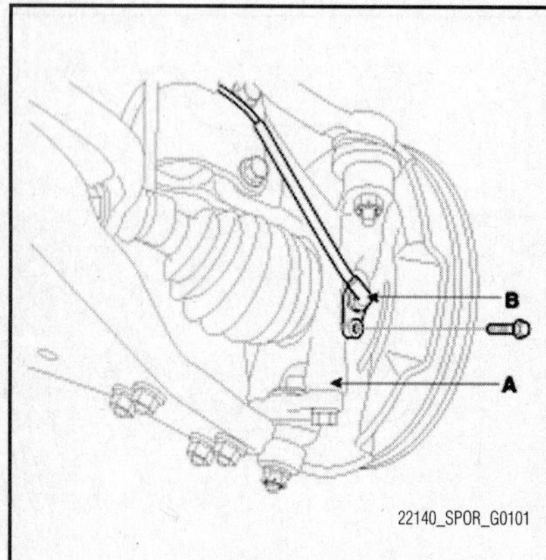

Fig. 258 Remove the wheel speed sensor (B) from the knuckle (A)

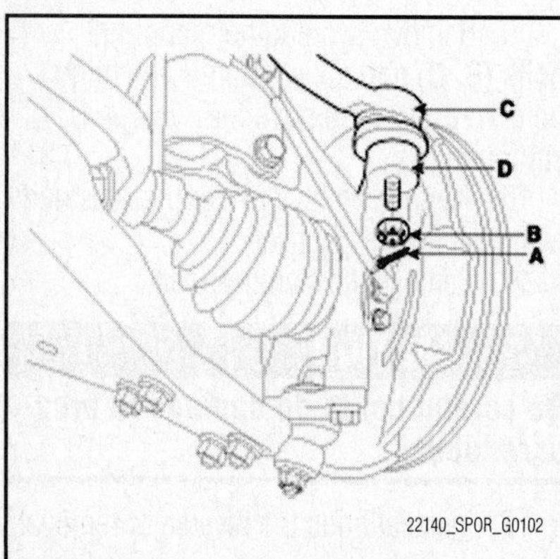

Fig. 259 Disconnect the tie rod end ball joint (C) from the knuckle (D)

5. Remove the wheel speed sensor (B) from the knuckle (A).

6. Disconnect the tie rod end ball joint (C) from the knuckle (D) using the special tool (09568-34000).
   a. Remove the split pin (A).
   b. Remove the castle nut (B).

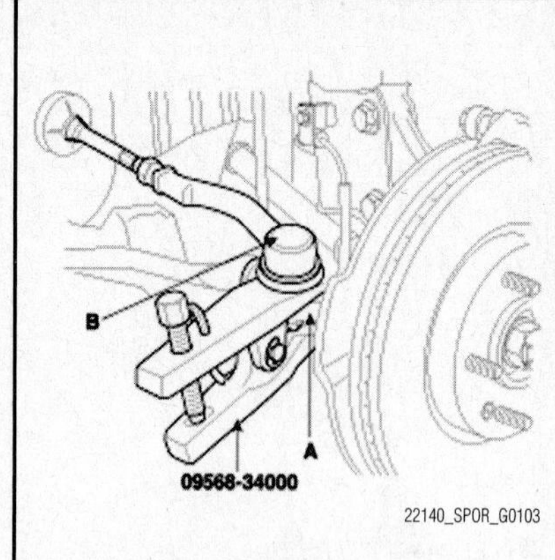

Fig. 260 Special tool (09568-34000)

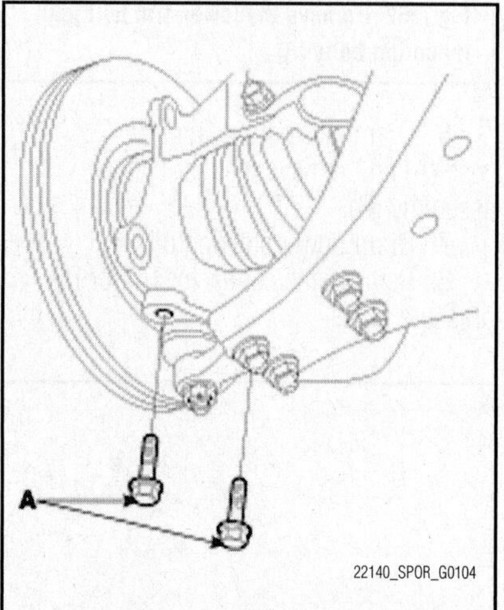

Fig. 261 Remove the lower arm ball joint mounting bolts (A)

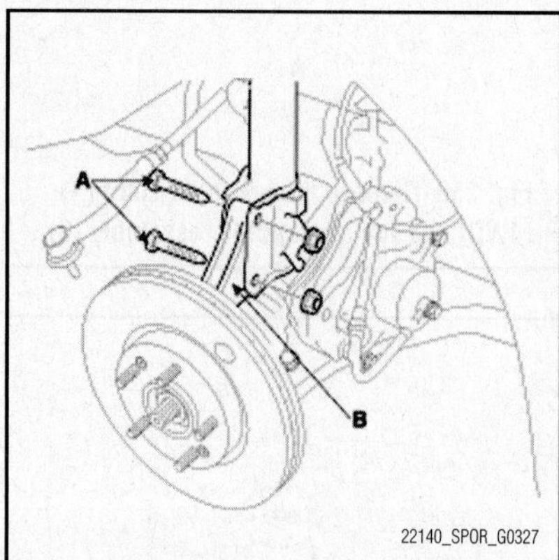

Fig. 262 Remove the strut lower arm mounting bolts (A)

   c. Disconnect the ball joint (C) from knuckle (D) using the special tool (09568-34000).

✻✻ CAUTION

Apply a few drops of oil to the special tool. (Boot contact part)

7. Remove the lower arm ball joint mounting bolts (A).

8. Remove the strut lower arm mounting bolts (A).

9. Remove the hub and the knuckle assembly (B).

✻✻ CAUTION

Be careful not to damage the boot and rotor teeth.

10. Installation is the reverse of removal.

WHEEL HUB & BEARING

REMOVAL & INSTALLATION

See Figures 257 through 262.

1. Raise the front of the vehicle, and make sure it is securely supported.

2. Remove the front wheel and tire.

3. Remove the split pin (A), the castle nut (B) and the washer (C) from the front hub.

4. Remove the caliper mounting bolts and hang the caliper assembly to one side.

5. Remove the wheel speed sensor (B) from the knuckle (A).

6. Disconnect the tie rod end ball joint (C) from the knuckle (D) using the special tool (09568-34000).
   a. Remove the split pin (A).
   b. Remove the castle nut (B).
   c. Disconnect the ball joint (C) from knuckle (D) using the special tool (09568-34000).

✻✻ CAUTION

Apply a few drops of oil to the special tool. (Boot contact part)

7. Remove the lower arm ball joint mounting bolts (A).

8. Remove the strut lower arm mounting bolts (A).

9. Remove the hub and the knuckle assembly (B).

✻✻ CAUTION

Be careful not to damage the boot and rotor teeth.

10. Installation is the reverse of removal.

MACPHERSON STRUTS

REMOVAL & INSTALLATION

See Figures 263 through 265.

1. Raise the rear of the vehicle, and make sure it is securely supported.
2. Remove the rear wheel and tire.
3. Remove the speed sensor cable mounting bolt (A).
4. Remove the stabilizer bar link nut (B).
5. Remove the strut upper mounting nuts (A).
6. Remove the strut lower mounting bolts (A) and remove the strut assembly (B).
7. Installation is the reverse of removal.

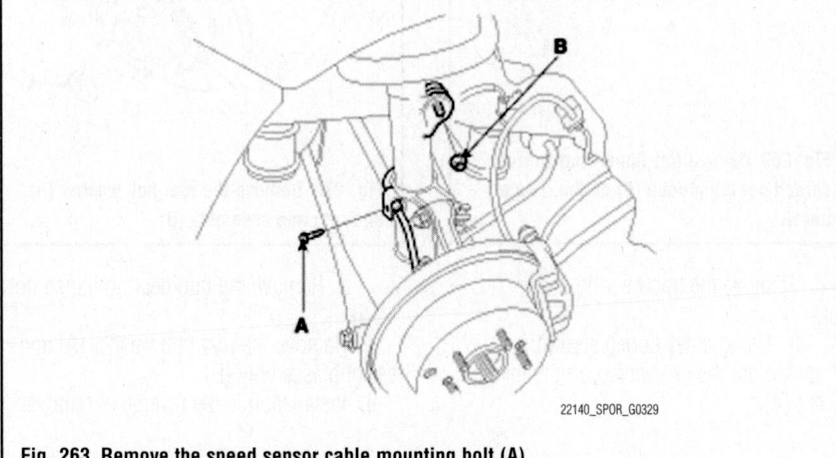

Fig. 263 Remove the speed sensor cable mounting bolt (A)

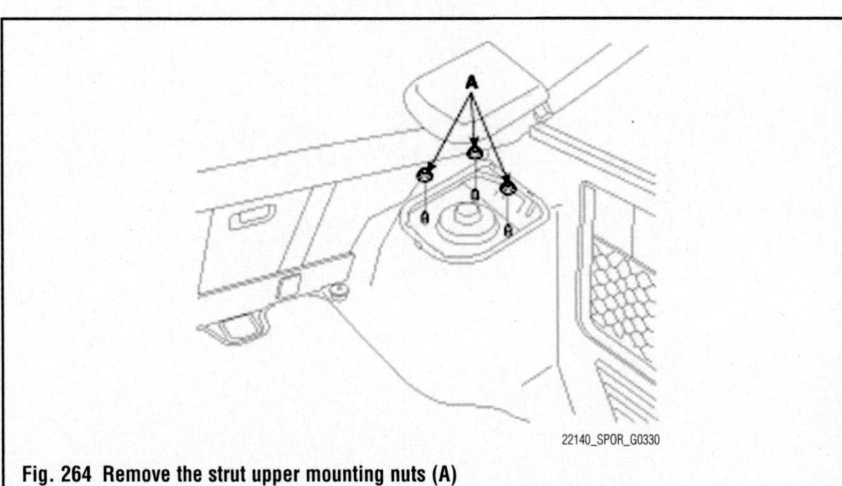

Fig. 264 Remove the strut upper mounting nuts (A)

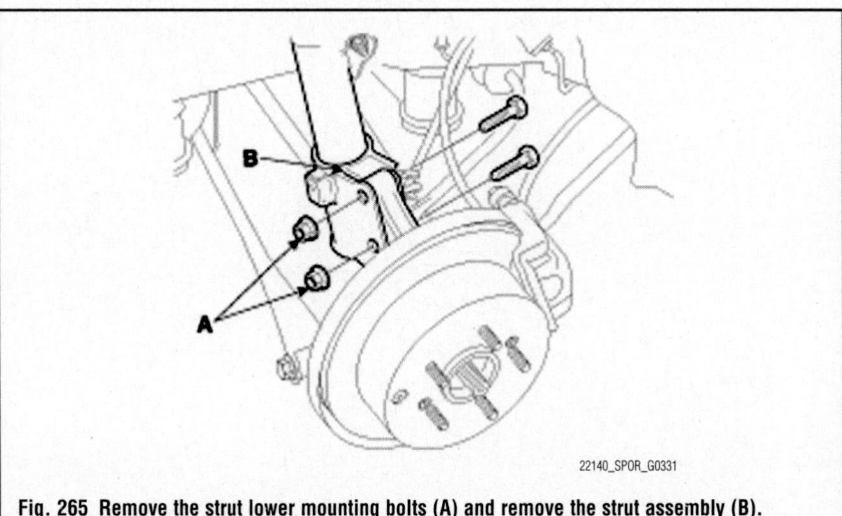

Fig. 265 Remove the strut lower mounting bolts (A) and remove the strut assembly (B).

WHEEL HUB & BEARING

REMOVAL & INSTALLATION

See Figures 266 through 270.

1. Raise the rear of the vehicle, and make sure it is securely supported.
2. Remove the rear wheel and tire.
3. Remove the brake caliper mounting bolts and hang the caliper assembly to one side.
4. Remove the wheel speed sensor (B) from the axle carrier (A).

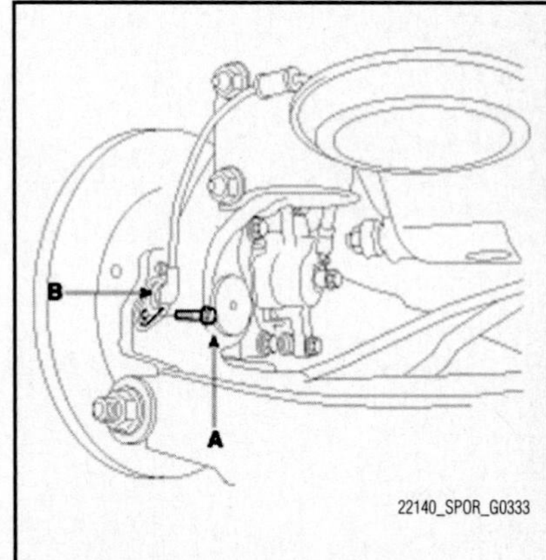

Fig. 266 Remove the wheel speed sensor (B) from the axle carrier (A)

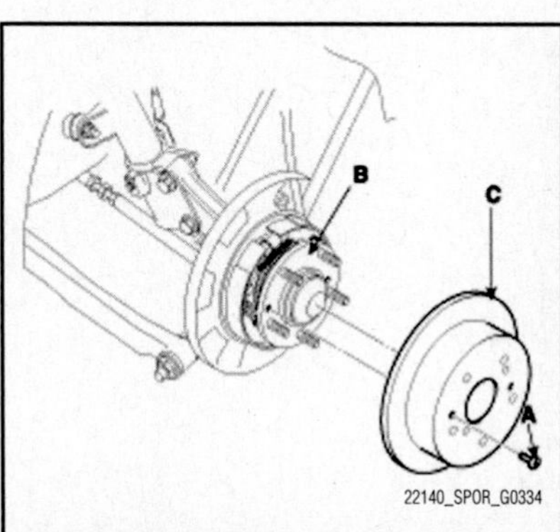

Fig. 267 Remove the brake disc mounting screw (A), and remove the brake disc (C) from the hub (B)

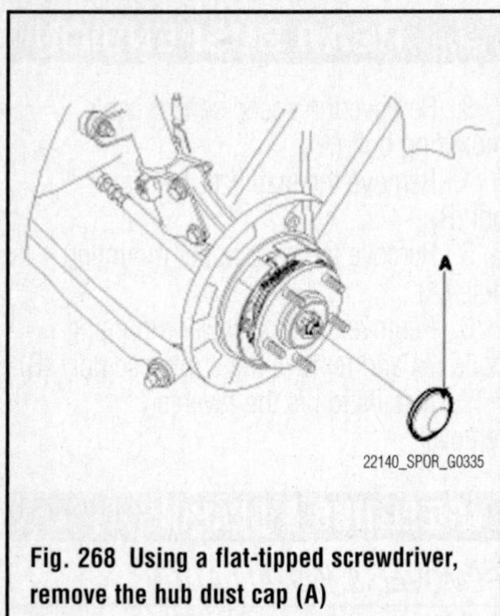

Fig. 268 Using a flat-tipped screwdriver, remove the hub dust cap (A)

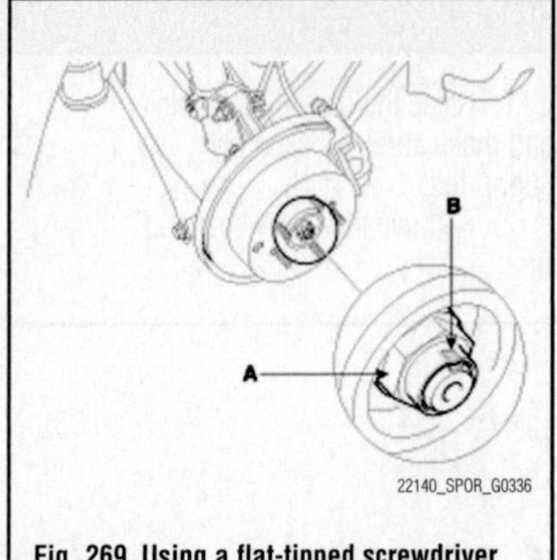

Fig. 269 Using a flat-tipped screwdriver, spread out the groove (B) on the flange nut (A)

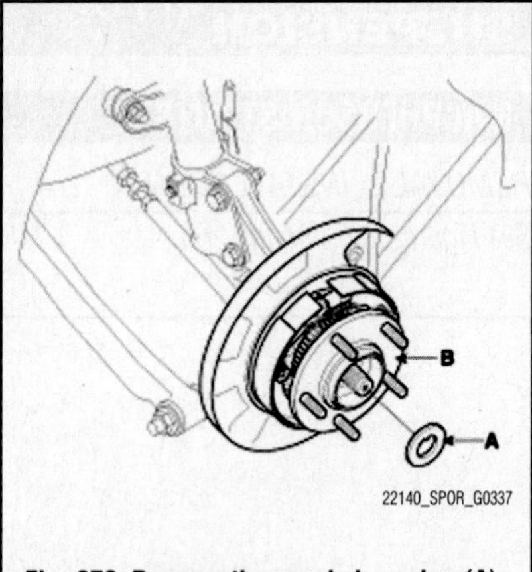

Fig. 270 Remove the rear hub washer (A) and rear hub assembly (B).

5. Remove the brake disc mounting screw (A), and remove the brake disc (C) from the hub (B).
6. Using a flat-tipped screwdriver, remove the hub dust cap (A).
7. Remove the hub bearing flange nut.
   a. Using a flat-tipped screwdriver, spread out the groove (B) on the flange nut (A)
   b. Remove the hub bearing flange nut (A).
8. Remove the rear hub washer (A) and rear hub assembly (B).
9. Installation is the reverse of removal.

KIA

22

Diagnostic Trouble Codes

DIAGNOSTIC TROUBLE CODES

OBD II VEHICLE APPLICATIONS

KIA

Amanti
2009–2010
- 3.8L .... VIN 5

Borrego
2009–2010
- 3.8L .... VIN 1
- 4.6L .... VIN 2

Forte
2010
- 2.0L .... VIN 2
- 2.4L .... VIN 3

Optima
2009–2010
- 2.4L .... VIN 3
- 2.7L .... VIN 4

Rio
2009–2010
- 1.6L .... VIN 3

Rondo
2009–2010
- 2.4L .... VIN 5
- 2.7L .... VIN 6

Sedona
2009–2010
- 3.8L .... VIN 3

Sorento
2009–2010
- 3.3L .... VIN 5
- 3.8L .... VIN 6

Soul
2010
- 1.6L .... VIN 1
- 2.0L .... VIN 2

Spectra
2009
- 2.0L .... VIN 2

Sportage
2009–2010
- 2.0L .... VIN 2, 4
- 2.7L .... VIN 3

OBD II Trouble Code List (P0XXX Codes)

DTC	Trouble Code Title, Conditions & Possible Causes
DTC: P0011 **2T CCM, MIL: Yes** **Year:** 2009, 2010 **Model:** Optima **Engine:** 2.4L L4 VIN 8, 2.7L V6 VIN 4 **Transmission:** All	**"A" Camshaft Position Timing Over Advanced or System Performance (Bank 1):** Monitor deviation between camshaft position set point and actual value. **Possible Causes:** • Oil leakage • Faulty oil pump • Faulty intake valve control solenoid
DTC: P0011 **2T PCM, MIL: Yes** **Year:** 2009, 2010 **Model:** Amanti, Borrego, Rio, Rio5, Sorento, Soul, Spectra, Spectra5, Sportage **Engine:** 1.6L L4 VIN 1, 1.6L L4 VIN 3, 2.0L L4 VIN 1, 2.0L L4 VIN 2, 3.3L V6 VIN 5, 3.8L V6 VIN 1, 3.8L V6 VIN 5, 3.8L V6 VIN 6, 4.6L V8 VIN 2 **Transmission:** All	**'A' Camshaft Position - Timing Over-Advanced or System Performance (Bank 1):** The CVVT (Continuously Variable Valve Timing) system is installed to the chain sprocket of the intake camshaft. This system controls the intake camshaft to provide the optimal valve timing. The PCM controls the Oil Control Valve(OCV), based on the signals output from mass air flow, throttle position and engine coolant temperature. The CVVT controller regulates the intake camshaft angle using oil pressure through the OCV. PCM detects CAM phasing average rate while cam signal is normally generating. PCM determines that a fault exists and a DTC is stored while vehicle is tip - in and out driving for 5 minutes. **Possible Causes:** • Oil level and condition • OCV • CVVT
DTC: P0012 **2T PCM, MIL: Yes** **Year:** 2009, 2010 **Model:** Amanti, Borrego, Rio, Rio5, Sorento, Soul **Engine:** 1.6L L4 VIN 1, 1.6L L4 VIN 3, 2.0L L4 VIN 2, 3.3L V6 VIN 5, 3.8L V6 VIN 1, 3.8L V6 VIN 5, 3.8L V6 VIN 6 **Transmission:** All	**'A' Camshaft Position - Timing Over-Retarded (Bank 1):** The CVVT (Continuously Variable Valve Timing) system is installed to the chain sprocket of the intake camshaft. This system controls the intake camshaft to provide the optimal valve timing. The PCM controls the Oil Control Valve(OCV), based on the signals output from mass air flow, throttle position and engine coolant temperature. The CVVT controller regulates the intake camshaft angle using oil pressure through the OCV. PCM monitors CAM phaser error while CMP signal is normally generating and vehicle is driving in 2000 ~ 3000rpm .If the CAM phaser does not move although PCM commands OCV duty cycle PCM determines that a fault exists and a DTC is stored. **Possible Causes:** • Oil level and condition • OCV • CVVT
DTC: P0014 **2T PCM, MIL: Yes** **Year:** 2009, 2010 **Model:** Borrego **Engine:** 3.8L V6 VIN 1, 4.6L V8 VIN 2 **Transmission:** All	**"B" Camshaft Position -Timing Over-Advanced or System Performance (Bank 1):** The CVVT (Continuously Variable Valve Timing) system is installed to the chain sprocket of the intake camshaft. This system controls the intake camshaft to provide the optimal valve timing. The PCM controls the Oil Control Valve(OCV), based on the signals output from mass air flow, throttle position and engine coolant temperature. The CVVT controller regulates the intake camshaft angle using oil pressure through the OCV. PCM detects CAM phasing average rate while cam signal is normally generating. PCM determines that a fault exists and a DTC is stored while vehicle is tip - in and out driving for 5 minutes. **Possible Causes:** • Oil level and condition • OCV • CVVT
DTC: P0014 **2T PCM, MIL: Yes** **Year:** 2009 **Model:** Borrego **Engine:** 3.8L V6 VIN 1, 4.6L V8 VIN 2 **Transmission:** All	**'B' Camshaft Position - Timing Over-Advanced or System Performance (Bank 1):** The CVVT (Continuously Variable Valve Timing) system is installed to the chain sprocket of the intake camshaft. This system controls the intake camshaft to provide the optimal valve timing. The PCM controls the Oil Control Valve(OCV), based on the signals output from mass air flow, throttle position and engine coolant temperature. The CVVT controller regulates the intake camshaft angle using oil pressure through the OCV. PCM detects CAM phasing average rate while cam signal is normally generating. PCM determines that a fault exists and a DTC is stored while vehicle is tip - in and out driving for 5 minutes. **Possible Causes:** • Oil level and condition • OCV • CVVT

DTC	Trouble Code Title, Conditions & Possible Causes
DTC: P0015 **2T PCM, MIL: Yes** **Year:** 2009 **Model:** Borrego **Engine:** 3.8L V6 VIN 1 **Transmission:** All	**'B' Camshaft Position - Timing Over-Retarded (Bank 1):** The CVVT (Continuously Variable Valve Timing) system is installed to the chain sprocket of the intake camshaft. This system controls the intake camshaft to provide the optimal valve timing. The PCM controls the Oil Control Valve(OCV), based on the signals output from mass air flow, throttle position and engine coolant temperature. The CVVT controller regulates the intake camshaft angle using oil pressure through the OCV. PCM monitors CAM phaser error while CMP signal is normally generating and vehicle is driving in 2000 ~ 3000rpm .If the CAM phaser does not move although PCM commands OCV duty cycle PCM determines that a fault exists and a DTC is stored. **Possible Causes:** • Oil level and condition • OCV • CVVT
DTC: P0015 **2T PCM, MIL: Yes** **Year:** 2009, 2010 **Model:** Borrego **Engine:** 3.8L V6 VIN 1, 4.6L V8 VIN 2 **Transmission:** All	**"B" Camshaft Position -Timing Over-Retarded (Bank 1):** Enable Conditions • CAM signal is normally generating • Vehicle is on driving (2000 ~ 3000RPM) for 5 minutes Threshold Value Case 1 • 5 CAD < Cam Actual Position < 50 CAD • Duty Cycle > 90% or Duty Cycle < 10% Case 2 • Cam Position error > 15 CAD (Difference between Actual Position and Desire Position is more than 15°) • Timing Counter > 80 Diagnosis Time • Continuous (within 5min.) **Possible Causes:** • Engine Oil • OCV stuck • CVVT stuck
DTC: P0016 **2T CCM, MIL: Yes** **Year:** 2009, 2010 **Model:** Optima **Engine:** 2.4L L4 VIN 8, 2.7L V6 VIN 4 **Transmission:** All	**Crankshaft Position- Camshaft Position Correlation (Bank 1 Sensor "A"):** Monitor camshaft position in the full retard condition or during CVVT control. Camshaft switching out of 109 to 141 degrees in full retard position, 70 to 140 degrees CRK during CVVT control. **Possible Causes:** • Abnormal installation of camshaft • Abnormal installation of crankshaft • Abnormal installation of tone wheel
DTC: P0016 **2T PCM, MIL: Yes** **Year:** 2009, 2010 **Model:** Amanti, Borrego, Rio, Rio5, Sorento, Soul, Spectra, Spectra5, Sportage **Engine:** 1.6L L4 VIN 1, 1.6L L4 VIN 3, 2.0L L4 VIN 1, 2.0L L4 VIN 2, 3.3L V6 VIN 5, 3.8L V6 VIN 1, 3.8L V6 VIN 5, 3.8L V6 VIN 6, 4.6L V8 VIN 2 **Transmission:** All	**Crankshaft Position - Camshaft Position Correlation (Bank 1 Sensor A):** The CVVT (Continuously Variable Valve Timing) system is installed to the chain sprocket of the intake camshaft. This system controls the intake camshaft to provide the optimal valve timing. The PCM controls the Oil Control Valve(OCV), based on the signals output from mass air flow, throttle position and engine coolant temperature. The CVVT controller regulates the intake camshaft angle using oil pressure through the OCV. PCM monitors timing misalignment while no active faults are present and fully warmed up engine oil at idle. If the timing is misaligned PCM determines that a fault exists and a DTC is stored. **Possible Causes:** • Oil level and condition • OCV • CVVT • Timing mark alignment
DTC: P0017 **2T PCM, MIL: Yes** **Year:** 2009 **Model:** Borrego **Engine:** 3.8L V6 VIN 1, 4.6L V8 VIN 2 **Transmission:** All	**Crankshaft Position- Camshaft Position Correlation (Bank 1 Sensor B):** Deviation between CKPS and CMPS(EX-CMPS Bank 1) is bigger than the threshold value **Possible Causes:** • Poor connection • Contamination of Oil / Clog of Oil path • CKPS, CMPS • OCV • CVVT

DTC	Trouble Code Title, Conditions & Possible Causes
DTC: P0018 **2T PCM, MIL: Yes** **Year:** 2009 **Model:** Borrego, Sorento **Engine:** 3.3L V6 VIN 5, 3.8L V6 VIN 1, 3.8L V6 VIN 6, 4.6L V8 VIN 2 **Transmission:** All	**Crankshaft Position- Camshaft Position Correlation (Bank 2 Sensor A):** Deviation between CKPS and CMPS (IN CMPS Bank 2) is bigger than the threshold value **Possible Causes:** • Poor connection • Contamination of Oil / Clog of Oil path • CKPS, CMPS • OCV • CVVT
DTC: P0019 **2T PCM, MIL: Yes** **Year:** 2009 **Model:** Borrego **Engine:** 3.8L V6 VIN 1, 4.6L V8 VIN 2 **Transmission:** All	**Crankshaft Position- Camshaft Position Correlation (Bank 2 Sensor B):** Deviation between CKPS and CMPS (EX CMPS Bank 2) is bigger than the threshold value **Possible Causes:** • Poor connection • Contamination of Oil / Clog of Oil path • CKPS, CMPS • OCV • CVVT
DTC: P0021 **2T PCM, MIL: Yes** **Year:** 2009 **Model:** Borrego, Sorento **Engine:** 3.3L V6 VIN 5, 3.8L V6 VIN 1, 3.8L V6 VIN 6, 4.6L V8 VIN 2 **Transmission:** All	**'A' Camshaft Position- Timing Over-Advanced or System Performance (Bank 2):** delay between the target angle and the real angle over 10 times • Time after | target VVT angle –VVT angle | > 6 deg(2 sec) • Time after engine start > 10 ~ 60 sec • Coolant temperature : 0~120°C • Engine oil temperature : 0~130°C • 1000 < rpm < 6600rpm **Possible Causes:** • Poor connection • Contamination of Oil / Clog of Oil path • OCV • CVVT
DTC: P0022 **2T CCM, MIL: Yes** **Year:** 2009 **Model:** Borrego, Sorento **Engine:** 3.3L V6 VIN 5, 3.8L V6 VIN 1, 3.8L V6 VIN 6 **Transmission:** All	**"A" Camshaft Position- Timing Over Retarded (Bank 2):** Determines if the phaser is stuck or has a steady state error. Off sets available. Cam velocity below threshold at 15 CAD/s. **Possible Causes:** • Engine oil • OCV • CVVT stuck • Faulty PCM
DTC: P0024 **2T PCM, MIL: Yes** **Year:** 2009 **Model:** Borrego **Engine:** 3.8L V6 VIN 1, 4.6L V8 VIN 2 **Transmission:** All	**'B' Camshaft Position- Timing Over-Advanced or System Performance (Bank 2):** Enable Conditions • CAM signal is normally generating • Accelerate and decelerate more than 10 times within 5 minutes – while driving Threshold value • Cam phasing is abnormally fast or slow **Possible Causes:** • Excessive phasing system leakage • Binding Oil pressure • (ex. Blockage in OCV filter) • Faulty OCV
DTC: P0025 **2T PCM, MIL: Yes** **Year:** 2009 **Model:** Borrego **Engine:** 3.8L V6 VIN 1 **Transmission:** All	**'B' Camshaft Position- Timing Over-Retarded (Bank 2):** CAM phaser does not move although ECM commands OCV duty cycle **Possible Causes:** • Engine Oil • OCV stuck • CVVT stuck
DTC: P0026 **2T PCM, MIL: Yes** **Year:** 2009 **Model:** Borrego, Sorento **Engine:** 3.3L V6 VIN 5, 3.8L V6 VIN 1, 3.8L V6 VIN 6 **Transmission:** All	**Intake Valve Control Solenoid Circuit Range/Performance (Bank 1):** Determines if oil control valve is stuck. Valve cleaning not in progress. CAM Actual Position is too high or low and Difference between Cam Actual Position and Desired Position is higher than 20 degrees **Possible Causes:** • Oil pressure loss • OCV seizure • Faulty PCM

DTC	Trouble Code Title, Conditions & Possible Causes
DTC: P0027 **2T PCM, MIL: Yes** **Year:** 2009 **Model:** Borrego **Engine:** 3.8L V6 VIN 1 **Transmission:** All	**Exhaust Valve Control Solenoid Circuit Range/Performance (Bank 1):** CAM Actual Position is too high or low and Difference between Cam Actual Position and Desired Position is higher than 20 degrees **Possible Causes:** • Oil Pressure Loss • OCV seizure
DTC: P0028 **2T CCM, MIL: Yes** **Year:** 2009 **Model:** Borrego, Sorento **Engine:** 3.3L V6 VIN 5, 3.8L V6 VIN 1, 3.8L V6 VIN 6 **Transmission:** All	**Intake Valve Control Solenoid Circuit Range/Performance (Bank 2):** Determines if oil control valve is stuck. Valve cleaning not in progress. CAM Actual Position is too high or low and Difference between Cam Actual Position and Desired Position is higher than 20 degrees **Possible Causes:** • Oil pressure loss • OCV seizure
DTC: P0029 **2T PCM, MIL: Yes** **Year:** 2009 **Model:** Borrego **Engine:** 3.8L V6 VIN 1 **Transmission:** All	**Exhaust Valve Control Solenoid Circuit Range/Performance (Bank 2):** Determines if oil control valve is stuck. Valve cleaning not in progress. CAM Actual Position is too high or low and Difference between Cam Actual Position and Desired Position is higher than 20 degrees **Possible Causes:** • Oil Pressure Loss • OCV seizure
DTC: P0030 **2T CCM, MIL: Yes** **Year:** 2009, 2010 **Model:** Amanti, Borrego, Optima, Rio, Rio5, Sorento, Soul, Spectra, Spectra5, Sportage **Engine:** 1.6L L4 VIN 1, 1.6L L4 VIN 3, 2.0L L4 VIN 1, 2.0L L4 VIN 2, 2.4L L4 VIN 8, 2.7L V6 VIN 3, 2.7L V6 VIN 4, 3.3L V6 VIN 5, 3.8L V6 VIN 1, 3.8L V6 VIN 5, 3.8L V6 VIN 6, 4.6L V8 VIN 2 **Transmission:** All	**HO2S-11 Heater Circuit Malfunction:** Engine started, engine runtime over 3 minutes, and the PCM determined the resistance of the HO2S heater was more than a calculated amount. **Possible Causes:** • HO2S heater control circuit is open or shorted to ground • HO2S heater control circuit is shorted to system power (B+) • HO2S heater is damaged or has failed • PCM has failed
DTC: P0031 **2T CCM, MIL: Yes** **Year:** 2009, 2010 **Model:** Amanti, Borrego, Optima, Rio, Rio5, Sorento, Soul, Spectra, Spectra5, Sportage **Engine:** 1.6L L4 VIN 1, 1.6L L4 VIN 3, 2.0L L4 VIN 1, 2.0L L4 VIN 2, 2.4L L4 VIN 8, 2.7L V6 VIN 3, 2.7L V6 VIN 4, 3.3L V6 VIN 5, 3.8L V6 VIN 1, 3.8L V6 VIN 5, 3.8L V6 VIN 6, 4.6L V8 VIN 2 **Transmission:** All	**O2 Sensor Heater Circuit Low (Bank 1/Sensor 1):** Heater check, low. Open or short circuit. **Possible Causes:** • Open in battery and control circuit • Short to ground in control circuit (pin 48 to 36) • Faulty HO2S heater • Faulty PCM
DTC: P0032 **2T CCM, MIL: Yes** **Year:** 2009, 2010 **Model:** Amanti, Borrego, Optima, Rio, Rio5, Sorento, Soul, Spectra, Spectra5, Sportage **Engine:** 1.6L L4 VIN 1, 1.6L L4 VIN 3, 2.0L L4 VIN 1, 2.0L L4 VIN 2, 2.4L L4 VIN 8, 2.7L V6 VIN 3, 2.7L V6 VIN 4, 3.3L V6 VIN 5, 3.8L V6 VIN 1, 3.8L V6 VIN 5, 3.8L V6 VIN 6, 4.6L V8 VIN 2 **Transmission:** All	**O2 Sensor Heater Circuit High (Bank 1/Sensor 1):** Heater check, high. Short circuit. **Possible Causes:** • Short to battery in control circuit • Faulty HO2S heater • Faulty PCM

DTC	Trouble Code Title, Conditions & Possible Causes
DTC: P0036 **2T CCM, MIL: Yes** **Year:** 2009, 2010 **Model:** Amanti, Borrego, Optima, Rio, Rio5, Sorento, Soul, Spectra, Spectra5, Sportage **Engine:** 1.6L L4 VIN 1, 1.6L L4 VIN 3, 2.0L L4 VIN 1, 2.0L L4 VIN 2, 2.4L L4 VIN 8, 2.7L V6 VIN 3, 2.7L V6 VIN 4, 3.3L V6 VIN 5, 3.8L V6 VIN 1, 3.8L V6 VIN 5, 3.8L V6 VIN 6, 4.6L V8 VIN 2 **Transmission:** All	**HO2S-12 Heater Circuit Malfunction:** Engine started, engine runtime over 3 minutes, and the PCM determined the resistance of the HO2S heater was more than a calculated amount. **Possible Causes:** • HO2S heater control circuit is open or shorted to ground • HO2S heater control circuit is shorted to system power (B+) • HO2S heater is damaged or has failed • PCM has failed
DTC: P0037 **2T CCM, MIL: Yes** **Year:** 2009, 2010 **Model:** Amanti, Borrego, Optima, Rio, Rio5, Sorento, Soul, Spectra, Spectra5, Sportage **Engine:** 1.6L L4 VIN 1, 1.6L L4 VIN 3, 2.0L L4 VIN 1, 2.0L L4 VIN 2, 2.4L L4 VIN 8, 2.7L V6 VIN 3, 2.7L V6 VIN 4, 3.3L V6 VIN 5, 3.8L V6 VIN 1, 3.8L V6 VIN 5, 3.8L V6 VIN 6, 4.6L V8 VIN 2 **Transmission:** All	**O2 Sensor Heater Circuit Low (Bank 1/Sensor 2):** Heater check, low. Open or short circuit. **Possible Causes:** • Open in battery and control circuit • Short to ground in control circuit • Faulty HO2S heater • Faulty PCM
DTC: P0038 **2T CCM, MIL: Yes** **Year:** 2009, 2010 **Model:** Amanti, Borrego, Optima, Rio, Rio5, Sorento, Soul, Sportage **Engine:** 1.6L L4 VIN 1, 1.6L L4 VIN 3, 2.0L L4 VIN 2, 2.4L L4 VIN 8, 2.7L V6 VIN 3, 2.7L V6 VIN 4, 3.3L V6 VIN 5, 3.8L V6 VIN 1, 3.8L V6 VIN 5, 3.8L V6 VIN 6, 4.6L V8 VIN 2 **Transmission:** All	**O2 Sensor Heater Circuit High (Bank 1/Sensor 2):** Heater check, high. Short circuit. **Possible Causes:** • Short to battery in control circuit • Faulty HO2S heater • Faulty PCM
DTC: P0050 **2T CCM, MIL: Yes** **Year:** 2009, 2010 **Model:** Amanti, Borrego, Sorento, Sportage **Engine:** 2.7L V6 VIN 3, 3.3L V6 VIN 5, 3.8L V6 VIN 1, 3.8L V6 VIN 5, 3.8L V6 VIN 6, 4.6L V8 VIN 2 **Transmission:** All	**OHO2S Heater Control Circuit High (Bank 2/Sensor 1):** Evaluate O2 sensor element temperature via measuring element resistance. Sensor preheating and full heating phases finished. **Possible Causes:** • Related fuse blown or missing • Heater control circuit open or short • Power supply circuit open or short • Contact resistance in connectors • Faulty HO2S
DTC: P0051 **2T CCM, MIL: Yes** **Year:** 2009, 2010 **Model:** Amanti, Borrego, Sorento, Sportage **Engine:** 2.7L V6 VIN 3, 3.3L V6 VIN 5, 3.8L V6 VIN 1, 3.8L V6 VIN 5, 3.8L V6 VIN 6, 4.6L V8 VIN 2 **Transmission:** All	**OHO2S Heater Circuit Low (Bank 2/Sensor 1):** Short circuit to ground on front HO2S heater line. Battery voltage above 10 volts. **Possible Causes:** • Related fuse blown or missing • Open or short to ground in power supply or control harness • Contact resistance in connectors • Faulty HO2S

DTC	Trouble Code Title, Conditions & Possible Causes
DTC: P0052 **2T CCM, MIL: Yes** **Year:** 2009, 2010 **Model:** Amanti, Borrego, Sorento, Sportage **Engine:** 2.7L V6 VIN 3, 3.3L V6 VIN 5, 3.8L V6 VIN 1, 3.8L V6 VIN 5, 3.8L V6 VIN 6, 4.6L V8 VIN 2 **Transmission:** All	**OHO2S Heater Circuit High (Bank 2/Sensor 1):** Open or short circuit to battery line on front HO2S heater line. Battery voltage above 10 volts. **Possible Causes:** • Open or short to battery in control harness • Contact resistance in connectors • Faulty HO2S
DTC: P0056 **2T CCM, MIL: Yes** **Year:** 2009, 2010 **Model:** Amanti, Borrego, Sorento, Sportage **Engine:** 2.7L V6 VIN 3, 3.3L V6 VIN 5, 3.8L V6 VIN 1, 3.8L V6 VIN 5, 3.8L V6 VIN 6, 4.6L V8 VIN 2 **Transmission:** All	**HO2S Heater Control Circuit (Bank 2/Sensor 2):** Evaluate O2 sensor element temperature via measuring element resistance. Sensor preheating and full heating phases finished. **Possible Causes:** • Related fuse blown or missing • Heater control circuit open or short • Power supply circuit open or short • Contact resistance in connections • Faulty HO2S
DTC: P0057 **2T CCM, MIL: Yes** **Year:** 2009, 2010 **Model:** Amanti, Borrego, Sorento, Sportage **Engine:** 2.7L V6 VIN 3, 3.3L V6 VIN 5, 3.8L V6 VIN 1, 3.8L V6 VIN 5, 3.8L V6 VIN 6, 4.6L V8 VIN 2 **Transmission:** All	**HO2S Heater Circuit Low (Bank 2/Sensor 2):** Check short circuit to ground on rear HO2S heater line. **Possible Causes:** • Related fuse blown or missing • Open or short to ground in power supply or control harness • Contact resistance in connections • Faulty HO2S
DTC: P0058 **2T CCM, MIL: Yes** **Year:** 2009, 2010 **Model:** Amanti, Borrego, Sorento, Sportage **Engine:** 2.7L V6 VIN 3, 3.3L V6 VIN 5, 3.8L V6 VIN 1, 3.8L V6 VIN 5, 3.8L V6 VIN 6, 4.6L V8 VIN 2 **Transmission:** All	**HO2S Heater Circuit High (Bank 2/Sensor 2):** Check short circuit to ground on rear HO2S heater line. **Possible Causes:** • Open or short to battery in control harness • Contact resistance in connections • Faulty HO2S
DTC: P0068 **2T PCM, MIL: Yes** **Year:** 2009, 2010 **Model:** Rio, Rio5, Soul **Engine:** 1.6L L4 VIN 1, 1.6L L4 VIN 3 **Transmission:** All	**MAP(MAF)-Throttle Position Sensor Correlation:** Enable Conditions • Correction factor for secondary load > 1.2 or Correction factor for secondary load < 0.8 Threshold Value • Time for secondary load adaptation > 300s Diagnostic Time • 1sec **Possible Causes:** • Poor connection • TPS • MAFS • ECM/PCM
DTC: P0075 **2T CCM, MIL: Yes** **Year:** 2009 **Model:** Borrego **Engine:** 4.6L V8 VIN 2 **Transmission:** All	**Intake Valve Control Solenoid Circuit (Bank 1):** Circuit continuity check, open. **Possible Causes:** • Poor connection • Open or short to ground in power circuit • Open in control circuit • OCV • Faulty ECM/PCM

DTC	Trouble Code Title, Conditions & Possible Causes
DTC: P0075 **2T PCM, MIL: Yes** **Year:** 2009, 2010 **Model:** Rio, Rio5, Soul **Engine:** 1.6L L4 VIN 1, 1.6L L4 VIN 3 **Transmission:** All	**Intake Valve Control Solenoid Circuit (Bank 1):** Enable Conditions Threshold Value • Disconnected Diagnostic Time • Continuous **Possible Causes:** • Poor connection • Open or Short to ground in power circuit • Open in control circuit • OCV • ECM/PCM
DTC: P0076 **2T CCM, MIL: Yes** **Year:** 2009, 2010 **Model:** Amanti, Borrego, Optima, Rio, Rio5, Sorento, Soul, Spectra, Spectra5, Sportage **Engine:** 1.6L L4 VIN 1, 1.6L L4 VIN 3, 2.0L L4 VIN 1, 2.0L L4 VIN 2, 2.4L L4 VIN 8, 2.7L V6 VIN 4, 3.3L V6 VIN 5, 3.8L V6 VIN 1, 3.8L V6 VIN 5, 3.8L V6 VIN 6, 4.6L V8 VIN 2 **Transmission:** All	**Intake Valve Control Solenoid Circuit Low (Bank 1):** PCM sets the code if it detects that the intake valve control solenoid control circuit is short to ground. Electrical check. **Possible Causes:** • Faulty ECM/PCM • Short to ground in control circuit • Contact resistance in connectors • Faulty intake valve control solenoid
DTC: P0077 **2T CCM, MIL: Yes** **Year:** 2009, 2010 **Model:** Amanti, Borrego, Optima, Rio, Rio5, Sorento, Soul, Spectra, Spectra5, Sportage **Engine:** 1.6L L4 VIN 1, 1.6L L4 VIN 3, 2.0L L4 VIN 1, 2.0L L4 VIN 2, 2.4L L4 VIN 8, 2.7L V6 VIN 4, 3.3L V6 VIN 5, 3.8L V6 VIN 1, 3.8L V6 VIN 5, 3.8L V6 VIN 6, 4.6L V8 VIN 2 **Transmission:** All	**Intake Valve Control Solenoid Circuit High (Bank 1):** PCM sets the code if it detects that the OCV control circuit is open or short to battery. Electrical check. **Possible Causes:** • Open or short to battery in control circuit • Contact resistance in connectors • Faulty intake valve control solenoid • Faulty PCM
DTC: P0078 **2T PCM, MIL: Yes** **Year:** 2009 **Model:** Borrego **Engine:** 4.6L V8 VIN 2 **Transmission:** All	**Exhaust Valve Control Solenoid Circuit (Bank 1):** Enable Conditions • - Threshold value • Open Diagnosis Time • Continuous **Possible Causes:** • Poor Connection • Open or Short to ground in power circuit • Open in control circuit • OCV

DTC	Trouble Code Title, Conditions & Possible Causes
DTC: P0079 **2T PCM, MIL: Yes** **Year:** 2009 **Model:** Borrego **Engine:** 3.8L V6 VIN 1, 4.6L V8 VIN 2 **Transmission:** All	**Exhaust Valve Control Solenoid Circuit Low (Bank 1):** Enable Conditions • No disabling Faults Present • Engine Running • $11V \leq$ Battery Voltage $\leq 16V$ Threshold value • Short to ground or open circuit Diagnosis Time • Continuous (More than 5 seconds failure for every 10 seconds test) **Possible Causes:** • Poor Connection • Open in Power circuit • Open or short to ground in Control Circuit • OCV • ECM
DTC: P0080 **2T PCM, MIL: Yes** **Year:** 2009 **Model:** Borrego **Engine:** 3.8L V6 VIN 1, 4.6L V8 VIN 2 **Transmission:** All	**Exhaust Valve Control Solenoid Circuit High(Bank 1):** Enable Conditions • No disabling Faults Present • Engine Running • $11V \leq$ Battery Voltage $\leq 16V$ Threshold value • Short to battery Diagnosis Time • Continuous (More than 5 seconds failure for every 10 seconds test) **Possible Causes:** • Poor Connection • Short to battery in Control Circuit • OCV • ECM
DTC: P0081 **2T PCM, MIL: Yes** **Year:** 2009 **Model:** Borrego **Engine:** 4.6L V8 VIN 2 **Transmission:** All	**Intake Valve Control Solenoid Circuit (Bank 2):** Enable Conditions • - Threshold value • Open Diagnosis Time • Continuous **Possible Causes:** • Poor Connection • Open or Short to ground in power circuit • Open in control circuit • OCV
DTC: P0082 **2T CCM, MIL: Yes** **Year:** 2009 **Model:** Amanti, Borrego, Sorento **Engine:** 3.3L V6 VIN 5, 3.8L V6 VIN 1, 3.8L V6 VIN 5, 3.8L V6 VIN 6, 4.6L V8 VIN 2 **Transmission:** All	**Intake Valve Control Solenoid Circuit Low (Bank 2):** Detects a short to ground or open circuit of VCPD bank 1 intake circuit output. No disabling faults present. Engine running. Enable time delay equal to or greater than 0.5 second. **Possible Causes:** • Poor connection • Open in power circuit • Open or short to ground in control circuit • OCV • Faulty PCM
DTC: P0083 **2T CCM, MIL: Yes** **Year:** 2009 **Model:** Amanti, Borrego, Sorento **Engine:** 3.3L V6 VIN 5, 3.8L V6 VIN 1, 3.8L V6 VIN 5, 3.8L V6 VIN 6, 4.6L V8 VIN 2 **Transmission:** All	**Intake Valve Control Solenoid Circuit High (Bank 2):** Detects a short to battery of VCPD bank 1 intake circuit output. No disabling faults present. Engine running. Enable time delay equal to or greater than 0.5 second. **Possible Causes:** • poor connection • Short to battery in control circuit • OCV • Faulty PCM

DTC	Trouble Code Title, Conditions & Possible Causes
DTC: P0084 **2T PCM, MIL: Yes** **Year:** 2009 **Model:** Borrego **Engine:** 4.6L V8 VIN 2 **Transmission:** All	**Exhaust Valve Control Solenoid Circuit (Bank 2):** Enable Conditions • - Threshold value • Open Diagnosis Time • Continuous **Possible Causes:** • Poor Connection • Open or Short to ground in power circuit • Open in control circuit • OCV
DTC: P0085 **2T PCM, MIL: Yes** **Year:** 2009 **Model:** Borrego **Engine:** 3.8L V6 VIN 1 **Transmission:** All	**Exhaust Valve Control Solenoid Circuit Low (Bank 2):** Enable Conditions • No disabling Faults Present • Engine Running • 11V ≤ Battery Voltage ≤ 16V Threshold value • Short to ground or open circuit Diagnosis Time • Continuous (More than 5 seconds failure for every 10 seconds test) **Possible Causes:** • Poor Connection • Open in Power circuit • Open or short to ground in Control Circuit • OCV • ECM
DTC: P0085 **2T PCM, MIL: Yes** **Year:** 2009 **Model:** Borrego **Engine:** 3.8L V6 VIN 1, 4.6L V8 VIN 2 **Transmission:** All	**Exhaust Valve Control Solenoid Circuit Low (Bank 2):** Enable Conditions • - Threshold value • Short circuit to ground Diagnosis Time • Continuous **Possible Causes:** • Poor Connection • Open or Short to ground in power circuit • Open in control circuit • OCV
DTC: P0086 **2T PCM, MIL: Yes** **Year:** 2009 **Model:** Borrego **Engine:** 3.8L V6 VIN 1, 4.6L V8 VIN 2 **Transmission:** All	**Exhaust Valve Control Solenoid Circuit High (Bank 2):** Enable Conditions • No disabling Faults Present • Engine Running • 11V ≤ Battery Voltage ≤ 16V Threshold value • Short to battery Diagnosis Time • Continuous (More than 5 seconds failure for every 10 seconds test **Possible Causes:** • Poor Connection • Short to battery in Control Circuit • OCV • ECM

DTC	Trouble Code Title, Conditions & Possible Causes
DTC: P0101 **2T CCM, MIL: Yes** **Year:** 2009, 2010 **Model:** Amanti, Borrego, Optima, Sorento, Spectra, Spectra5, Sportage **Engine:** 2.0L L4 VIN 1, 2.0L L4 VIN 2, 2.4L L4 VIN 8, 2.7L V6 VIN 3, 2.7L V6 VIN 4, 3.3L V6 VIN 5, 3.8L V6 VIN 1, 3.8L V6 VIN 5, 3.8L V6 VIN 6, 4.6L V8 VIN 2 **Transmission:** All	**MAF or Volume Airflow Sensor Performance:** Engine started, engine running, and the PCM detected too much difference between the Actual MAF sensor signal and a Threshold MAF sensor value in memory during the CCM Rationality test. **Possible Causes:** • MAF sensor signal circuit has high resistance • MAF sensor ground circuit has high resistance • MAF sensor is damaged or has failed (it may be contaminated) • PCM has failed
DTC: P0102 **2T CCM, MIL: Yes** **Year:** 2009, 2010 **Model:** Amanti, Borrego, Optima, Sorento, Spectra, Spectra5, Sportage **Engine:** 2.0L L4 VIN 1, 2.0L L4 VIN 2, 2.4L L4 VIN 8, 2.7L V6 VIN 3, 2.7L V6 VIN 4, 3.3L V6 VIN 5, 3.8L V6 VIN 1, 3.8L V6 VIN 5, 3.8L V6 VIN 6, 4.6L V8 VIN 2 **Transmission:** All	**MAF Sensor Circuit Low Input:** Key on or engine running and the PCM detected the MAF sensor signal was less than 0.39 volt for more than 1 second. **Possible Causes:** • MAF sensor signal circuit is shorted to ground • MAF sensor power circuit open between sensor and main relay • MAF sensor is damaged or has failed • PCM has failed
DTC: P0103 **2T CCM, MIL: Yes** **Year:** 2009, 2010 **Model:** Amanti, Borrego, Optima, Sorento, Spectra, Spectra5, Sportage **Engine:** 2.0L L4 VIN 1, 2.0L L4 VIN 2, 2.4L L4 VIN 8, 2.7L V6 VIN 3, 2.7L V6 VIN 4, 3.3L V6 VIN 5, 3.8L V6 VIN 1, 3.8L V6 VIN 5, 3.8L V6 VIN 6, 4.6L V8 VIN 2 **Transmission:** All	**MAF Sensor Circuit High Input:** Key on or engine running and the PCM detected the MAF sensor signal was more than 3.90 volts, condition met for 1 second. **Possible Causes:** • MAF sensor signal circuit shorted to VREF or system power • MAF sensor ground circuit is open between sensor and ground • MAF sensor is damaged or has failed • PCM has failed
DTC: P0105 **2T PCM, MIL: Yes** **Year:** 2009 **Model:** Borrego, Sorento **Engine:** 3.3L V6 VIN 5, 3.8L V6 VIN 1, 3.8L V6 VIN 6 **Transmission:** All	**Manifold Absolute Pressure/Barometric Pressure Circuit:** Enable Conditions • No Disabling Fault Present • Shutdown time > 20 minutes • Engine running Threshold value • The difference between the signal at key-on and the signal at engine start < 0.5 kPa Diagnosis Time • For 3 seconds out of 5 seconds **Possible Causes:** • Faulty MAPS
DTC: P0106 **2T CCM, MIL: Yes** **Year:** 2009, 2010 **Model:** Amanti, Borrego, Rio, Rio5, Sorento, Soul, Sportage **Engine:** 1.6L L4 VIN 1, 1.6L L4 VIN 3, 2.0L L4 VIN 2, 3.3L V6 VIN 5, 3.8L V6 VIN 1, 3.8L V6 VIN 5, 3.8L V6 VIN 6 **Transmission:** All	**Manifold Absolute Pressure Sensor Performance:** Engine running and the PCM detected the MAP sensor signal was less than 2.20 volts but more than 0.40 volt during the CCM Rationality test. **Possible Causes:** • Loss of 5-volt supply from the PCM (circuit open or grounded) • MAP sensor signal circuit is open or grounded • MAP sensor is damaged or has failed • PCM has failed

DTC	Trouble Code Title, Conditions & Possible Causes
DTC: P0107 **2T CCM, MIL: Yes** **Year:** 2009, 2010 **Model:** Amanti, Borrego, Rio, Rio5, Sorento, Soul, Sportage **Engine:** 1.6L L4 VIN 1, 1.6L L4 VIN 3, 2.0L L4 VIN 2, 3.3L V6 VIN 5, 3.8L V6 VIN 1, 3.8L V6 VIN 5, 3.8L V6 VIN 6 **Transmission:** All	**Manifold Absolute Pressure Sensor Low Input:** Key on or engine running and the PCM detected the MAP sensor signal was less than 0.20 volt, condition met for 1 second. **Possible Causes:** • Loss of 5-volt supply from the PCM (circuit open or grounded) • MAP sensor signal circuit is shorted to ground • MAP sensor is damaged or has failed • PCM has failed
DTC: P0108 **2T CCM, MIL: Yes** **Year:** 2009, 2010 **Model:** Amanti, Borrego, Rio, Rio5, Sorento, Soul, Sportage **Engine:** 1.6L L4 VIN 1, 1.6L L4 VIN 3, 2.0L L4 VIN 2, 3.3L V6 VIN 5, 3.8L V6 VIN 1, 3.8L V6 VIN 5, 3.8L V6 VIN 6 **Transmission:** All	**Manifold Absolute Pressure Sensor High Input:** Key on or engine running and the PCM detected the MAP sensor signal was more than 4.90 volts, condition met for 1 second. **Possible Causes:** • MAP sensor ground circuit open • MAP sensor signal circuit shorted to VREF or system power • MAP sensor is damaged or has failed • PCM has failed
DTC: P0109 **2T PCM** **Year:** 2009 **Model:** Borrego, Sorento **Engine:** 3.3L V6 VIN 5, 3.8L V6 VIN 1, 3.8L V6 VIN 6 **Transmission:** All	**Manifold Absolute Pressure/Barometric Pressure Circuit Intermittent:** Enable Conditions • Engine running • I ΔAPS I < 5% • Engine Speed > 800rpm Threshold value • MAP_stable –MAP_current| >10 % Diagnosis Time • - MIL On Condition • NO MIL ON(DTC only) **Possible Causes:** • Connecting condition • Open or short to ground in power circuit • Open or short to ground in signal circuit • MAPS • ECM
DTC: P0110 **2T CCM, MIL: Yes** **Year:** 2009 **Model:** Amanti, Borrego, Sorento **Engine:** 3.3L V6 VIN 5, 3.8L V6 VIN 1, 3.8L V6 VIN 5, 3.8L V6 VIN 6 **Transmission:** All	**Intake Air Temperature Sensor Circuit Malfunction:** Engine runtime over 600 seconds and the PCM detected the IAT signal indicated more than 262(F, or that it indicated less than -38°F at any time during the CCM test. **Possible Causes:** • IAT sensor signal circuit open or shorted to ground • IAT sensor signal circuit shorted to VREF or system power • IAT sensor is damaged or has failed • PCM has failed
DTC: P0111 **2T CCM, MIL: Yes** **Year:** 2009, 2010 **Model:** Amanti, Borrego, Optima, Rio, Rio5, Sorento, Soul, Spectra, Spectra5, Sportage **Engine:** 1.6L L4 VIN 1, 1.6L L4 VIN 3, 2.0L L4 VIN 1, 2.0L L4 VIN 2, 2.4L L4 VIN 8, 2.7L V6 VIN 3, 2.7L V6 VIN 4, 3.3L V6 VIN 5, 3.8L V6 VIN 1, 3.8L V6 VIN 5, 3.8L V6 VIN 6, 4.6L V8 VIN 2 **Transmission:** All	**Intake Air Temperature Sensor 1 Circuit Range/Performance:** If the sensor is out of specification, a code is set. Output voltage is monitored. Engine coolant is above 167 degrees F. Vehicle speed is above 30MPH for more than 60 seconds. Vehicle speed is below 7mph for more than 30 seconds. **Possible Causes:** • Poor connection • Faulty IATS • Faulty PCM

DTC	Trouble Code Title, Conditions & Possible Causes
DTC: P0112 **2T CCM, MIL: Yes** **Year:** 2009, 2010 **Model:** Amanti, Borrego, Optima, Rio, Rio5, Sorento, Soul, Spectra, Spectra5, Sportage **Engine:** 1.6L L4 VIN 1, 1.6L L4 VIN 3, 2.0L L4 VIN 1, 2.0L L4 VIN 2, 2.4L L4 VIN 8, 2.7L V6 VIN 3, 2.7L V6 VIN 4, 3.3L V6 VIN 5, 3.8L V6 VIN 1, 3.8L V6 VIN 5, 3.8L V6 VIN 6, 4.6L V8 VIN 2 **Transmission:** All	**Intake Air Temperature Sensor Circuit Low Input:** Key on or engine running and the PCM detected the IAT sensor signal was less than 0.2 volt during the CCM test. **Possible Causes:** • IAT sensor signal circuit shorted to ground • IAT sensor is damaged or has failed • PCM has failed
DTC: P0113 **2T CCM, MIL: Yes** **Year:** 2009, 2010 **Model:** Amanti, Borrego, Optima, Rio, Rio5, Sorento, Soul, Spectra, Spectra5, Sportage **Engine:** 1.6L L4 VIN 1, 1.6L L4 VIN 3, 2.0L L4 VIN 1, 2.0L L4 VIN 2, 2.4L L4 VIN 8, 2.7L V6 VIN 3, 2.7L V6 VIN 4, 3.3L V6 VIN 5, 3.8L V6 VIN 1, 3.8L V6 VIN 5, 3.8L V6 VIN 6, 4.6L V8 VIN 2 **Transmission:** All	**Intake Air Temperature Sensor Circuit High Input:** Key on or the engine running and the PCM detected the IAT sensor signal was more than 4.9 volts during the CCM test. **Possible Causes:** • IAT sensor signal circuit shorted to VREF or system power • IAT sensor ground circuit is open • IAT sensor is damaged or has failed • PCM has failed
DTC: P0115 **2T CCM, MIL: Yes** **Year:** 2009, 2010 **Model:** Amanti, Borrego, Rio, Rio5, Sorento, Soul **Engine:** 1.6L L4 VIN 1, 1.6L L4 VIN 3, 3.3L V6 VIN 5, 3.8L V6 VIN 1, 3.8L V6 VIN 5, 3.8L V6 VIN 6 **Transmission:** All	**Engine Coolant Temperature Sensor Circuit Malfunction:** Engine runtime over 600 seconds, and the PCM detected the IAT signal indicated more than 280°F, or that it indicated less than -38°F at any time during the CCM test. **Possible Causes:** • ECT sensor signal circuit open or shorted to ground • ECT sensor signal circuit shorted to VREF or system power • ECT sensor is damaged or has failed • PCM has failed
DTC: P0116 **2T CCM, MIL: Yes** **Year:** 2009, 2010 **Model:** Amanti, Borrego, Optima, Rio, Rio5, Sorento, Soul, Spectra, Spectra5, Sportage **Engine:** 1.6L L4 VIN 1, 1.6L L4 VIN 3, 2.0L L4 VIN 1, 2.0L L4 VIN 2, 2.4L L4 VIN 8, 2.7L V6 VIN 3, 2.7L V6 VIN 4, 3.3L V6 VIN 5, 3.8L V6 VIN 1, 3.8L V6 VIN 5, 3.8L V6 VIN 6, 4.6L V8 VIN 2 **Transmission:** All	**Engine Coolant Temperature Sensor Range/Performance:** Engine running for 20 minutes, and the PCM detected an ECT signal remained at less than a specified value during the CCM test period. **Possible Causes:** • Check for low coolant level or incorrect coolant mixture • Cooling system component failure (thermostat stuck open) • ECT sensor is out of calibration or it is "skewed" • ECT sensor is damaged or has failed
DTC: P0117 **2T CCM, MIL: Yes** **Year:** 2009, 2010 **Model:** Amanti, Borrego, Optima, Rio, Rio5, Sorento, Soul, Spectra, Spectra5, Sportage **Engine:** 1.6L L4 VIN 1, 1.6L L4 VIN 3, 2.0L L4 VIN 1, 2.0L L4 VIN 2, 2.4L L4 VIN 8, 2.7L V6 VIN 3, 2.7L V6 VIN 4, 3.3L V6 VIN 5, 3.8L V6 VIN 1, 3.8L V6 VIN 5, 3.8L V6 VIN 6, 4.6L V8 VIN 2 **Transmission:** All	**Engine Coolant Temperature Sensor Circuit Low Input:** Key on or engine running and the PCM detected the ECT sensor signal was less than 0.20 volt during the CCM test. **Possible Causes:** • ECT sensor signal circuit shorted to ground • ECT sensor is damaged or has failed • PCM has failed

DTC	Trouble Code Title, Conditions & Possible Causes
DTC: P0118 **2T CCM, MIL: Yes** **Year:** 2009, 2010 **Model:** Amanti, Borrego, Optima, Rio, Rio5, Sorento, Soul, Spectra, Spectra5, Sportage **Engine:** 1.6L L4 VIN 1, 1.6L L4 VIN 3, 2.0L L4 VIN 1, 2.0L L4 VIN 2, 2.4L L4 VIN 8, 2.7L V6 VIN 3, 2.7L V6 VIN 4, 3.3L V6 VIN 5, 3.8L V6 VIN 1, 3.8L V6 VIN 5, 3.8L V6 VIN 6, 4.6L V8 VIN 2 **Transmission:** All	**Engine Coolant Temperature Sensor Circuit High Input:** Key on or engine running and the PCM detected the ECT sensor signal was more than 4.90 volts during the CCM test. **Possible Causes:** • ECT sensor signal circuit is open between sensor and the PCM • ECT sensor ground circuit is open between sensor and ground • ECT sensor signal circuit is shorted to VREF • ECT sensor is damaged or has failed • PCM has failed
DTC: P0119 **2T PCM, MIL: Yes** **Year:** 2009, 2010 **Model:** Borrego, Optima, Sportage **Engine:** 2.4L L4 VIN 8, 2.7L V6 VIN 3, 2.7L V6 VIN 4, 4.6L V8 VIN 2 **Transmission:** All	**Engine Coolant Temperature Circuit Intermittent:** Enable Conditions • - Threshold value • Δ ECT sensor voltage > 0.35V Diagnosis Time • 6 sec **Possible Causes:** • Poor Connection • Short to ground in signal circuit • ECTS
DTC: P0120 **2T CCM, MIL: Yes** **Year:** 2009 **Model:** Amanti **Engine:** 3.8L V6 VIN 5 **Transmission:** All	**Throttle Position Sensor Circuit Malfunction:** Key on or engine running and the PCM detected the TP sensor signal was less than 0.14 volt, or that it was more than 4.96 volts. **Possible Causes:** • TP sensor signal circuit open or shorted to ground • TP sensor signal circuit shorted to VREF or system power (B+) • TP sensor is damaged or has failed • PCM has failed
DTC: P0121 **1T PCM, MIL: Yes** **Year:** 2009, 2010 **Model:** Borrego, Soul, Sportage **Engine:** 1.6L L4 VIN 1, 2.0L L4 VIN 2, 2.7L V6 VIN 3, 4.6L V8 VIN 2 **Transmission:** All	**Throttle/Pedal Position Sensor/Switch 'A' Circuit Range/Performance:** Enable Conditions • Engine speed > 480rpm • ECT > 75C • Engine load < 95% Threshold value • I TPS1 Position – TPS2 Position I > 6.3% Diagnosis Time • 0.5 sec **Possible Causes:** • Poor Connection • TPS • ECM
DTC: P0121 **2T CCM, MIL: Yes** **Year:** 2009, 2010 **Model:** Optima, Rio, Rio5, Spectra, Spectra5 **Engine:** 1.6L L4 VIN 3, 2.0L L4 VIN 1, 2.0L L4 VIN 2, 2.4L L4 VIN 8, 2.7L V6 VIN 4 **Transmission:** All	**Throttle Position Sensor Performance:** Engine at idle speed, and the PCM detected the TP sensor signal was outside of the idle speed limit (i.e., 1.20-2.1 volts at idle is the limit). **Possible Causes:** • MAF sensor or TP sensor ground circuit has high resistance • MAF sensor has drifted out of calibration • TP sensor has drifted out of calibration • TP sensor is damaged or has failed

DTC	Trouble Code Title, Conditions & Possible Causes
DTC: P0122 **2T CCM, MIL: Yes** **Year:** 2009, 2010 **Model:** Amanti, Borrego, Optima, Rio, Rio5, Sorento, Soul, Spectra, Spectra5, Sportage **Engine:** 1.6L L4 VIN 1, 1.6L L4 VIN 3, 2.0L L4 VIN 1, 2.0L L4 VIN 2, 2.4L L4 VIN 8, 2.7L V6 VIN 3, 2.7L V6 VIN 4, 3.3L V6 VIN 5, 3.8L V6 VIN 1, 3.8L V6 VIN 5, 3.8L V6 VIN 6, 4.6L V8 VIN 2 **Transmission:** All	**Throttle Position Sensor 'A' Circuit Low Input:** Key on or engine running and the PCM detected the TP sensor 'A' signal was less than 0.170-0.200 volt during the CCM test. **Possible Causes:** • TP sensor signal circuit is shorted to ground • TP sensor VREF circuit is open or shorted to ground • TP sensor is damaged or has failed • PCM has failed
DTC: P0123 **2T CCM, MIL: Yes** **Year:** 2009, 2010 **Model:** Amanti, Borrego, Optima, Rio, Rio5, Sorento, Soul, Spectra, Spectra5, Sportage **Engine:** 1.6L L4 VIN 1, 1.6L L4 VIN 3, 2.0L L4 VIN 1, 2.0L L4 VIN 2, 2.4L L4 VIN 8, 2.7L V6 VIN 3, 2.7L V6 VIN 4, 3.3L V6 VIN 5, 3.8L V6 VIN 1, 3.8L V6 VIN 5, 3.8L V6 VIN 6, 4.6L V8 VIN 2 **Transmission:** All	**Throttle Position Sensor 'A' Circuit High Input:** Key on or engine running and the PCM detected the TP sensor signal was more than 4.60-4.80 volts during the CCM test. **Possible Causes:** • TP sensor signal circuit is open • TP sensor ground circuit is open • TP sensor signal circuit is shorted to VREF or system power • TP sensor is damaged or has failed\ • PCM has failed
DTC: P0124 **2T CCM, MIL: Yes** **Year:** 2009, 2010 **Model:** Rio, Rio5, Soul **Engine:** 1.6L L4 VIN 3, 2.0L L4 VIN 2 **Transmission:** All	**Throttle/Pedal Position Sensor/Switch "A" Circuit Intermittent:** Rationality check. Rate of change in throttle angle 0.1221 percent. Engine speed 600 rpm. Coolant temperature 167 degrees F. **Possible Causes:** • Poor connection • TPS • Faulty ECM/PCM
DTC: P0125 **2T ECT, MIL: Yes** **Year:** 2009, 2010 **Model:** Amanti, Borrego, Sorento, Soul, Spectra, Spectra5, Sportage **Engine:** 1.6L L4 VIN 1, 2.0L L4 VIN 1, 2.0L L4 VIN 2, 2.7L V6 VIN 3, 3.3L V6 VIN 5, 3.8L V6 VIN 1, 3.8L V6 VIN 5, 3.8L V6 VIN 6 **Transmission:** All	**Insufficient Coolant Temperature for Closed Loop:** DTC P0116, P0117 and P0118 not set, engine run time from 6-8 minutes, and the PCM detected the ECT signal did not reach the closed loop temperature of at least 68°F. **Possible Causes:** • Check for low coolant level or incorrect coolant mixture • Cooling system component failure (thermostat stuck open) • ECT sensor is out of calibration ("skewed") or it has failed • PCM has failed
DTC: P0128 **2T ECT, MIL: Yes** **Year:** 2009, 2010 **Model:** Amanti, Borrego, Rio, Rio5, Sorento, Soul, Spectra, Spectra5, Sportage **Engine:** 1.6L L4 VIN 1, 1.6L L4 VIN 3, 2.0L L4 VIN 1, 2.0L L4 VIN 2, 2.7L V6 VIN 3, 3.3L V6 VIN 5, 3.8L V6 VIN 1, 3.8L V6 VIN 5, 3.8L V6 VIN 6, 4.6L V8 VIN 2 **Transmission:** All	**Thermostat Malfunction:** ECT sensor signal less than 140°F at startup, then with the engine running the PCM detected the ECT sensor signal did not reach 170°F with the engine "modeling" temperature (runtime) over 190°F. **Possible Causes:** • ECT sensor is out of calibration or it is "skewed" • Check the operation of the thermostat (it may be stuck open) • Inspect for low coolant level or for an incorrect coolant mixture

DTC	Trouble Code Title, Conditions & Possible Causes
DTC: P0130 **2T CCM, MIL: Yes** **Year:** 2009, 2010 **Model:** Borrego, Optima, Rio, Rio5, Soul, Sportage **Engine:** 1.6L L4 VIN 1, 1.6L L4 VIN 3, 2.0L L4 VIN 2, 2.4L L4 VIN 8, 2.7L V6 VIN 3, 2.7L V6 VIN 4, 4.6L V8 VIN 2 **Transmission:** All	**HO2S-11 (Bank 1 Sensor 1) Circuit Malfunction:** Engine started, engine running in closed loop at a speed over 3 mph for 2-3 minutes, and the PCM detected the HO2S signal was more than 1.4 volts, or it was less than 0.02 volt during the CCM test. **Possible Causes:** • HO2S signal circuit open or shorted to ground • HO2S signal circuit shorted to VREF or system power (B+) • HO2S is damaged or has failed • PCM has failed
DTC: P0131 **2T CCM, MIL: Yes** **Year:** 2009, 2010 **Model:** Amanti, Borrego, Optima, Rio, Rio5, Sorento, Soul, Spectra, Spectra5, Sportage **Engine:** 1.6L L4 VIN 1, 1.6L L4 VIN 3, 2.0L L4 VIN 1, 2.0L L4 VIN 2, 2.4L L4 VIN 8, 2.7L V6 VIN 3, 2.7L V6 VIN 4, 3.3L V6 VIN 5, 3.8L V6 VIN 1, 3.8L V6 VIN 5, 3.8L V6 VIN 6, 4.6L V8 VIN 2 **Transmission:** All	**HO2S-11 (Bank 1 Sensor 1) Circuit Low Input:** Engine started, engine running in closed loop at a speed over 3 mph for 2-3 minutes, and the PCM detected the HO2S signal remained at less than 300 mv during the CCM test. **Possible Causes:** • HO2S signal circuit open or shorted to ground • HO2S signal ground circuit is open • HO2S is contaminated or has failed • PCM has failed
DTC: P0132 **2T CCM, MIL: Yes** **Year:** 2009, 2010 **Model:** Amanti, Borrego, Optima, Rio, Rio5, Sorento, Soul, Spectra, Spectra5, Sportage **Engine:** 1.6L L4 VIN 1, 1.6L L4 VIN 3, 2.0L L4 VIN 1, 2.0L L4 VIN 2, 2.4L L4 VIN 8, 2.7L V6 VIN 3, 2.7L V6 VIN 4, 3.3L V6 VIN 5, 3.8L V6 VIN 1, 3.8L V6 VIN 5, 3.8L V6 VIN 6, 4.6L V8 VIN 2 **Transmission:** All	**HO2S-11 (Bank 1 Sensor 1) Circuit High Input:** Engine started, engine running in closed loop at a speed over 3 mph for 2-3 minutes, and the PCM detected the front HO2S signal remained fixed at more than 600 mv during the CCM test. **Possible Causes:** • HO2S signal tracking (wet/oily) in connector causing a short • HO2S signal circuit shorted to VREF or system power (B+) • HO2S signal circuit is open, or the ground circuit is open • HO2S heater supply voltage is open or the HO2S has failed • PCM has failed
DTC: P0133 **2T O2S2, MIL: Yes** **Year:** 2009, 2010 **Model:** Amanti, Borrego, Optima, Rio, Rio5, Sorento, Soul, Spectra, Spectra5, Sportage **Engine:** 1.6L L4 VIN 1, 1.6L L4 VIN 3, 2.0L L4 VIN 1, 2.0L L4 VIN 2, 2.4L L4 VIN 8, 2.7L V6 VIN 3, 2.7L V6 VIN 4, 3.3L V6 VIN 5, 3.8L V6 VIN 1, 3.8L V6 VIN 5, 3.8L V6 VIN 6, 4.6L V8 VIN 2 **Transmission:** All	**HO2S-11 (Bank 1 Sensor 1) Slow Response:** Engine started, engine running in closed loop at a speed over 3 mph for 2-3 minutes, and the PCM detected the average ratio between the HO2S Actual and maximum allowed frequency during 100 Lambda cycles was more than the Threshold value (e.g., 0.66 Hz). **Possible Causes:** • Exhaust leak present in the exhaust manifold or exhaust pipes • Front HO2S failed, or front & rear HO2S connections reversed • HO2S has deteriorated, is contaminated or has failed
DTC: P0134 **2T O2S2, MIL: Yes** **Year:** 2009, 2010 **Model:** Amanti, Borrego, Optima, Rio, Rio5, Sorento, Soul, Sportage **Engine:** 1.6L L4 VIN 1, 1.6L L4 VIN 3, 2.0L L4 VIN 2, 2.4L L4 VIN 8, 2.7L V6 VIN 3, 2.7L V6 VIN 4, 3.3L V6 VIN 5, 3.8L V6 VIN 1, 3.8L V6 VIN 5, 3.8L V6 VIN 6, 4.6L V8 VIN 2 **Transmission:** All	**HO2S-11 (Bank 1 Sensor 1) No Activity Detected:** Engine started, engine running in closed loop at a speed over 3 mph for more than 2 minutes, and the PCM detected the HO2S signal stroke rationality was more than the threshold value (0.250 volt) during the Heated Oxygen Sensor Monitor test. **Possible Causes:** • HO2S signal circuit is open or shorted to ground • HO2S has deteriorated, is contaminated or has failed • PCM has failed

DTC	Trouble Code Title, Conditions & Possible Causes
DTC: P0135 **2T , MIL: Yes** **Year:** 2009, 2010 **Model:** Rio, Rio5, Soul **Engine:** 1.6L L4 VIN 1, 1.6L L4 VIN 3 **Transmission:** All	**HO2S-11 (Bank 1 Sensor 1) Heater Circuit Malfunction:** Engine started, engine running and PCM detected the HO2S heater current was more than 2 amps, or it was less than 0.25 amps. **Possible Causes:** • HO2S power feed circuit from the Main Relay is open • HO2S heater control circuit is open • HO2S heater element has high resistance, is shorted or open • HO2S heater is damaged or has failed • PCM has failed
DTC: P0136 **2T CCM, MIL: Yes** **Year:** 2009, 2010 **Model:** Borrego, Optima, Rio, Rio5, Soul, Spectra, Spectra5, Sportage **Engine:** 1.6L L4 VIN 1, 1.6L L4 VIN 3, 2.0L L4 VIN 1, 2.0L L4 VIN 2, 2.4L L4 VIN 8, 2.7L V6 VIN 3, 2.7L V6 VIN 4, 4.6L V8 VIN 2 **Transmission:** All	**HO2S-12 (Bank 1 Sensor 2) Circuit Malfunction:** Engine started, engine running in closed loop at a speed over 3 mph for 2-3 minutes and the PCM detected the HO2S signal was more than 1.4 volts, or that it was less than 0.02 volt during the CCM test. **Possible Causes:** • HO2S signal circuit open or shorted to ground • HO2S signal circuit shorted to VREF or system power (B+) • HO2S is damaged or has failed • PCM has failed
DTC: P0137 **2T CCM, MIL: Yes** **Year:** 2009, 2010 **Model:** Amanti, Borrego, Optima, Rio, Rio5, Sorento, Soul, Spectra, Spectra5, Sportage **Engine:** 1.6L L4 VIN 1, 1.6L L4 VIN 3, 2.0L L4 VIN 1, 2.0L L4 VIN 2, 2.4L L4 VIN 8, 2.7L V6 VIN 3, 2.7L V6 VIN 4, 3.3L V6 VIN 5, 3.8L V6 VIN 1, 3.8L V6 VIN 5, 3.8L V6 VIN 6, 4.6L V8 VIN 2 **Transmission:** All	**HO2S-12 (Bank 1 Sensor 2) Circuit Low Input:** Engine started, engine running in closed loop at a speed over 3 mph, and the PCM detected the HO2S signal remained at less than 300 mv during the CCM test. **Possible Causes:** • HO2S signal circuit open or shorted to ground • HO2S signal ground circuit is open • HO2S is contaminated or has failed • PCM has failed
DTC: P0138 **2T CCM, MIL: Yes** **Year:** 2009, 2010 **Model:** Amanti, Borrego, Optima, Rio, Rio5, Sorento, Soul, Spectra, Spectra5, Sportage **Engine:** 1.6L L4 VIN 1, 1.6L L4 VIN 3, 2.0L L4 VIN 1, 2.0L L4 VIN 2, 2.4L L4 VIN 8, 2.7L V6 VIN 3, 2.7L V6 VIN 4, 3.3L V6 VIN 5, 3.8L V6 VIN 1, 3.8L V6 VIN 5, 3.8L V6 VIN 6, 4.6L V8 VIN 2 **Transmission:** All	**HO2S-12 (Bank 1 Sensor 2) Circuit High Input:** Engine started, engine running in closed loop at a speed over 3 mph, and the PCM detected the front HO2S signal remained fixed at more than 600 mv during the CCM test. **Possible Causes:** • HO2S signal tracking (wet/oily) in connector causing a short • HO2S signal circuit shorted to VREF or system power (B+) • HO2S signal circuit is open, or the ground circuit is open • HO2S heater supply voltage is open or the HO2S has failed • PCM has failed
DTC: P0139 **2T O2S2, MIL: Yes** **Year:** 2009, 2010 **Model:** Amanti, Borrego, Optima, Rio, Rio5, Sorento, Soul, Spectra, Spectra5, Sportage **Engine:** 1.6L L4 VIN 1, 1.6L L4 VIN 3, 2.0L L4 VIN 1, 2.0L L4 VIN 2, 2.4L L4 VIN 8, 2.7L V6 VIN 3, 2.7L V6 VIN 4, 3.3L V6 VIN 5, 3.8L V6 VIN 1, 3.8L V6 VIN 5, 3.8L V6 VIN 6, 4.6L V8 VIN 2 **Transmission:** All	**HO2S-12 (Bank 1 Sensor 2) Slow Response:** Engine started, engine running in closed loop at a speed over 3 mph, and the PCM detected the average ratio between the HO2S Actual and maximum allowed frequency during 100 Lambda cycles was more than the Threshold value (e.g., 0.66 Hz). **Possible Causes:** • Exhaust leak present in the exhaust manifold or exhaust pipes • Front HO2S failed, or front & rear HO2S connections reversed • HO2S has deteriorated, is contaminated or has failed

DTC	Trouble Code Title, Conditions & Possible Causes
DTC: P0140 **2T O2S2, MIL: Yes** **Year:** 2009, 2010 **Model:** Amanti, Borrego, Optima, Rio, Rio5, Sorento, Soul, Spectra, Spectra5, Sportage **Engine:** 1.6L L4 VIN 1, 1.6L L4 VIN 3, 2.0L L4 VIN 1, 2.0L L4 VIN 2, 2.4L L4 VIN 8, 2.7L V6 VIN 3, 2.7L V6 VIN 4, 3.3L V6 VIN 5, 3.8L V6 VIN 1, 3.8L V6 VIN 5, 3.8L V6 VIN 6, 4.6L V8 VIN 2 **Transmission:** All	**HO2S-12 (Bank 1 Sensor 2) No Activity Detected:** Engine started, engine running in closed loop at cruise speed at over 3 mph for more than 2 minutes, and the PCM detected the HO2S signal stroke rationality was more than the threshold value (0.250 volt) during the Oxygen Sensor Monitor test. **Possible Causes:** • HO2S signal circuit is open or shorted to ground • HO2S has deteriorated, is contaminated or has failed • PCM has failed
DTC: P0141 **2T , MIL: Yes** **Year:** 2009, 2010 **Model:** Rio, Rio5, Soul **Engine:** 1.6L L4 VIN 1, 1.6L L4 VIN 3 **Transmission:** All	**HO2S-12 (Bank 1 Sensor 2) Heater Circuit Malfunction:** Engine started, engine running and PCM detected the HO2S heater current was more than 2 amps, or it was less than 0.25 amps. **Possible Causes:** • HO2S power feed circuit from the Main Relay is open • HO2S heater control circuit is open • HO2S heater element has high resistance, is shorted or open • HO2S heater is damaged or has failed • PCM has failed
DTC: P0150 **2T O2S2, MIL: Yes** **Year:** 2009, 2010 **Model:** Borrego, Sportage **Engine:** 2.7L V6 VIN 3, 4.6L V8 VIN 2 **Transmission:** All	**HO2S-21 (Bank 2 Sensor 1) Slow Response:** Engine started, engine running in closed loop at a speed over 3 mph, and the PCM detected the average ratio between the HO2S Actual and maximum allowed frequency during 100 Lambda cycles was more than the Threshold value (e.g., 0.66 Hz) during the test. **Possible Causes:** • Exhaust leak present in the exhaust manifold or exhaust pipes • Front HO2S failed, or front & rear HO2S connections reversed • HO2S has deteriorated, is contaminated or has failed
DTC: P0151 **2T O2S2, MIL: Yes** **Year:** 2009, 2010 **Model:** Amanti, Borrego, Sorento, Sportage **Engine:** 2.7L V6 VIN 3, 3.3L V6 VIN 5, 3.8L V6 VIN 1, 3.8L V6 VIN 5, 3.8L V6 VIN 6, 4.6L V8 VIN 2 **Transmission:** All	**HO2S Circuit Low Voltage (Bank 2 Sensor 1):** Engine started, engine running in closed loop at a speed over 3 mph, and the PCM detected the average ratio between the HO2S Actual and maximum allowed frequency during 100 Lambda cycles was more than the Threshold value (e.g., 0.66 Hz) during the test. **Possible Causes:** • Poor connection • Short to ground in harness • Faulty PCM • HO2S has deteriorated, is contaminated or has failed
DTC: P0152 **2T CCM, MIL: Yes** **Year:** 2009, 2010 **Model:** Amanti, Borrego, Sorento, Sportage **Engine:** 2.7L V6 VIN 3, 3.3L V6 VIN 5, 3.8L V6 VIN 1, 3.8L V6 VIN 5, 3.8L V6 VIN 6, 4.6L V8 VIN 2 **Transmission:** All	**HO2S-21 (Bank 2 Sensor 1) Circuit Malfunction:** Engine started, engine running in closed loop at a speed over 3 mph for more than 2-3 minutes, and the PCM detected an unexpected voltage condition on the HO2S signal circuit. **Possible Causes:** • HO2S signal circuit open or shorted to ground • HO2S signal circuit is shorted to VREF or to system power (B+) • HO2S is damaged or has failed • PCM has failed
DTC: P0153 **2T , MIL: Yes** **Year:** 2009, 2010 **Model:** Amanti, Borrego, Sorento, Sportage **Engine:** 2.7L V6 VIN 3, 3.3L V6 VIN 5, 3.8L V6 VIN 1, 3.8L V6 VIN 5, 3.8L V6 VIN 6, 4.6L V8 VIN 2 **Transmission:** All	**HO2S-21 (Bank 2 Sensor 1) Slow Response:** DTC P0155 not set, engine started, engine running at idle speed in closed loop, and PCM detected the HO2S-12 response time to switch from rich-to-lean or from lean-to-rich was over one second. **Possible Causes:** • HO2S signal circuit is open or shorted to ground • HO2S element is contaminated or it has failed • HO2S heater is damaged or has failed • Intake air leaks, exhaust manifold leaks or PCV system leaks • MAF sensor out of calibration (it may be dirty or contaminated)

DTC	Trouble Code Title, Conditions & Possible Causes
DTC: P0154 **2T O2S2, MIL: Yes** **Year:** 2009, 2010 **Model:** Amanti, Borrego, Sorento, Sportage **Engine:** 2.7L V6 VIN 3, 3.3L V6 VIN 5, 3.8L V6 VIN 1, 3.8L V6 VIN 5, 3.8L V6 VIN 6, 4.6L V8 VIN 2 **Transmission:** All	**HO2S-21 (Bank 2 Sensor 1) No Activity Detected:** Engine started, engine running in closed loop at a speed over 3 mph for more than 2 minutes, and the PCM detected the HO2S signal stroke rationality was more than the threshold value (0.250 volt) during the Oxygen Sensor Monitor test. **Possible Causes:** • HO2S signal circuit is open or shorted to ground • HO2S has deteriorated, is contaminated or has failed • PCM has failed
DTC: P0156 **2T CCM, MIL: Yes** **Year:** 2009, 2010 **Model:** Borrego, Sportage **Engine:** 2.7L V6 VIN 3, 4.6L V8 VIN 2 **Transmission:** All	**HO2S-22 (Bank 2 Sensor 2) Circuit Malfunction:** Engine started, engine running in closed loop at a speed over 3 mph for more than 2 minutes, and the PCM detected an unexpected low voltage condition on the HO2S-22 circuit during the CCM test. **Possible Causes:** • HO2S signal circuit open between the HO2S and the PCM • HO2S signal circuit shorted to ground between HO2S and PCM • HO2S is damaged or has failed • PCM has failed
DTC: P0157 **2T CCM, MIL: Yes** **Year:** 2009, 2010 **Model:** Amanti, Borrego, Sorento, Sportage **Engine:** 2.7L V6 VIN 3, 3.3L V6 VIN 5, 3.8L V6 VIN 1, 3.8L V6 VIN 5, 3.8L V6 VIN 6, 4.6L V8 VIN 2 **Transmission:** All	**HO2S Circuit Low Voltage (Bank 2 Sensor 2):** The signal voltage of the front or rear sensor changes the rear circuit voltage specification when air fuel ratio is rich, a DTC is set. Out of range low failure (ground short open circuit). **Possible Causes:** • Poor connection • Short to ground in HO2S circuit • Faulty HO2S • Faulty PCM
DTC: P0158 **2T CCM, MIL: Yes** **Year:** 2009, 2010 **Model:** Amanti, Borrego, Sorento, Sportage **Engine:** 2.7L V6 VIN 3, 3.3L V6 VIN 5, 3.8L V6 VIN 1, 3.8L V6 VIN 5, 3.8L V6 VIN 6, 4.6L V8 VIN 2 **Transmission:** All	**HO2S Circuit High Voltage (Bank 2 Sensor 2):** The signal voltage is higher than 1.2 volts after open in circuit. Out of range high failure. **Possible Causes:** • Poor connection • Short to battery in HO2S circuit • Faulty HO2S • Faulty PCM
DTC: P0159 **2T O2S2, MIL: Yes** **Year:** 2009, 2010 **Model:** Amanti, Borrego, Sorento, Sportage **Engine:** 2.7L V6 VIN 3, 3.3L V6 VIN 5, 3.8L V6 VIN 1, 3.8L V6 VIN 5, 3.8L V6 VIN 6, 4.6L V8 VIN 2 **Transmission:** All	**HO2S-22 (Bank 2 Sensor 2) Slow Response:** Engine started, engine running in closed loop at a speed over 3 mph for more than 2 minutes, and the PCM detected the average ratio between the HO2S Actual and maximum allowed frequency during 100 Lambda cycles was more than the Threshold value (e.g., 0.66 Hz) during the Oxygen Sensor Monitor test. **Possible Causes:** • Exhaust leak present in the exhaust manifold or exhaust pipes • Front HO2S failed, or front & rear HO2S connections reversed • HO2S has deteriorated, is contaminated or has failed
DTC: P0160 **2T CCM, MIL: Yes** **Year:** 2009, 2010 **Model:** Amanti, Borrego, Sorento, Sportage **Engine:** 2.7L V6 VIN 3, 3.3L V6 VIN 5, 3.8L V6 VIN 1, 3.8L V6 VIN 5, 3.8L V6 VIN 6, 4.6L V8 VIN 2 **Transmission:** All	**HO2S-22 (Bank 2 Sensor 2) Circuit Malfunction:** Engine started, engine running in closed loop at a speed over 3 mph, and the PCM detected an unexpected high voltage condition on the HO2S circuit during the CCM test. **Possible Causes:** • HO2S signal circuit is shorted to VREF or to the Heater power • HO2S signal circuit is shorted to system power (B+) • HO2S is damaged or has failed • PCM has failed

DTC	Trouble Code Title, Conditions & Possible Causes
DTC: P0170 **2T Fuel, MIL: Yes** **Year:** 2009, 2010 **Model:** Borrego, Optima, Soul, Spectra, Spectra5, Sportage **Engine:** 2.0L L4 VIN 1, 2.0L L4 VIN 2, 2.4L L4 VIN 8, 2.7L V6 VIN 3, 2.7L V6 VIN 4, 4.6L V8 VIN 2 **Transmission:** All	**Fuel System Too Rich or Too Lean (Bank 1):** DTC P0171and P0172 not set, engine running in closed loop for over 2 minutes, and the PCM detected the amount of rich or lean Fuel Trim correction exceeded the Threshold maximum. **Possible Causes:** • Air leaks present in the exhaust manifold or exhaust pipes • Air being drawn in from leaks in engine gaskets or other seals • Incorrect fuel pressure, or one or more fuel injectors has failed • Front HO2S element is contaminated or has failed • A "fuel control" sensor is out of calibration (BARO, ECT or IAT)
DTC: P0171 **2T Fuel, MIL: Yes** **Year:** 2009, 2010 **Model:** Amanti, Borrego, Rio, Rio5, Sorento, Soul, Spectra, Spectra5, Sportage **Engine:** 1.6L L4 VIN 1, 1.6L L4 VIN 3, 2.0L L4 VIN 1, 2.0L L4 VIN 2, 2.7L V6 VIN 3, 3.3L V6 VIN 5, 3.8L V6 VIN 1, 3.8L V6 VIN 5, 3.8L V6 VIN 6, 4.6L V8 VIN 2 **Transmission:** All	**Fuel System Too Lean (Bank 1):** Engine running in closed loop at a speed of over 5 mph for 2-3 minutes, and the PCM detected the Lambda correction value exceeded the "high" Threshold limit, condition met for 200 seconds. **Possible Causes:** • Air leaks in intake manifold, exhaust pipes or exhaust manifold • One or more injectors restricted or pressure regulator has failed • Air is being drawn in from leaks in gaskets or other seals • O2S element is deteriorated or has failed • A "fuel control" sensor is out of calibration (ECT, IAT or MAP)
DTC: P0172 **2T Fuel, MIL: Yes** **Year:** 2009, 2010 **Model:** Amanti, Borrego, Rio, Rio5, Sorento, Soul, Spectra, Spectra5, Sportage **Engine:** 1.6L L4 VIN 1, 1.6L L4 VIN 3, 2.0L L4 VIN 1, 2.0L L4 VIN 2, 2.7L V6 VIN 3, 3.3L V6 VIN 5, 3.8L V6 VIN 1, 3.8L V6 VIN 5, 3.8L V6 VIN 6, 4.6L V8 VIN 2 **Transmission:** All	**Fuel System Too Rich (Bank 1):** Engine running in closed loop at a speed of over 5 mph for 2-3 minutes, and the PCM detected the Lambda correction value exceeded the "low" Threshold limit, condition met for 240 seconds. **Possible Causes:** • One or more injectors leaking or pressure regulator is leaking • O2S element is deteriorated or has failed • EVAP vapor recovery system has failed (canister full of fuel) • Base engine fault (i.e., cam timing incorrect, engine oil too high • A "fuel control" sensor is out of calibration (ECT, IAT or MAP)
DTC: P0173 **2T Fuel, MIL: Yes** **Year:** 2009, 2010 **Model:** Borrego, Sportage **Engine:** 2.7L V6 VIN 3, 4.6L V8 VIN 2 **Transmission:** All	**Fuel Trim Too Rich or Too Lean (Bank 2):** Engine running in closed loop, and the PCM detected the Fuel system was too rich or too lean during two or more consecutive trips. **Possible Causes:** • Base engine "mechanical" fault affecting one or more cylinders • EVAP system component has failed or canister fuel saturated • Exhaust leaks located in front of the HO2S location • Fuel control sensor is out of calibration (i.e., ECT, IAT or MAF) • Fuel delivery system supplying too much fuel during cruise or idle periods (e.g., faulty fuel pump, or faulty pressure regulator) • Fuel injector(s) is leaking or stuck partially open (one or more) • HO2S is contaminated, deteriorated or it has failed
DTC: P0174 **2T Fuel, MIL: Yes** **Year:** 2009, 2010 **Model:** Amanti, Borrego, Sorento, Sportage **Engine:** 2.7L V6 VIN 3, 3.3L V6 VIN 5, 3.8L V6 VIN 1, 3.8L V6 VIN 5, 3.8L V6 VIN 6, 4.6L V8 VIN 2 **Transmission:** All	**Fuel System Too Lean (Bank 2):** Engine running in closed loop at a speed of over 5 mph for 2-3 minutes, and the PCM detected the Lambda correction value exceeded the "high" Threshold limit, condition met for 200 seconds. **Possible Causes:** • Air leaks in intake manifold, exhaust pipes or exhaust manifold • One or more injectors restricted or pressure regulator has failed • Air is being drawn in from leaks in gaskets or other seals • O2S element is deteriorated or has failed • A "fuel control" sensor is out of calibration (ECT, IAT or MAP)

DTC	Trouble Code Title, Conditions & Possible Causes
DTC: P0175 **2T Fuel, MIL: Yes** **Year:** 2009, 2010 **Model:** Amanti, Borrego, Sorento, Sportage **Engine:** 2.7L V6 VIN 3, 3.3L V6 VIN 5, 3.8L V6 VIN 1, 3.8L V6 VIN 5, 3.8L V6 VIN 6, 4.6L V8 VIN 2 **Transmission:** All	**Fuel System Too Rich (Bank 2):** Engine running in closed loop at a speed of over 5 mph for 2-3 minutes, and the PCM detected the Lambda correction value exceeded the "low" Threshold limit, condition met for 240 seconds. **Possible Causes:** • One or more injectors leaking or pressure regulator is leaking • O2S element is deteriorated or has failed • EVAP vapor recovery system has failed (canister full of fuel) • Base engine fault (i.e., cam timing incorrect, engine oil too high • A "fuel control" sensor is out of calibration (ECT, IAT or MAP)
DTC: P0196 **2T CCM, MIL: Yes** **Year:** 2009, 2010 **Model:** Amanti, Borrego, Optima, Sorento, Soul, Spectra, Spectra5, Sportage **Engine:** 2.0L L4 VIN 1, 2.0L L4 VIN 2, 2.4L L4 VIN 8, 2.7L V6 VIN 4, 3.3L V6 VIN 5, 3.8L V6 VIN 1, 3.8L V6 VIN 5, 3.8L V6 VIN 6 **Transmission:** All	**Engine Oil Temperature Sensor Range/Performance:** Stuck oil temperature sensor signal or unusual low or high signal. Condition 1 (signal high or low), engine coolant temperature more than 158 degrees F and oil temperature less than 68 degrees F. Condition 2 (signal high or low), engine coolant temperature less than 158 degrees F and oil temperature above 212 degrees F. Condition 3 (stuck signal) engine coolant temperature less than 104 degrees F. **Possible Causes:** • Contact resistance in connectors • faulty OTS
DTC: P0197 **2T CCM, MIL: Yes** **Year:** 2009, 2010 **Model:** Amanti, Borrego, Sorento, Soul, Sportage **Engine:** 2.0L L4 VIN 2, 3.3L V6 VIN 5, 3.8L V6 VIN 1, 3.8L V6 VIN 5, 3.8L V6 VIN 6 **Transmission:** All	**Engine Oil temperature Sensor Low Input:** Signal voltage lower than the possible range of a properly operating OTS. Voltage range check. Engine coolant temperature less than 212 degrees F. Oil temperature above 309 degrees F. **Possible Causes:** • Short circuit to ground • Contact resistance in connectors • faulty OTS
DTC: P0198 **2T CCM, MIL: Yes** **Year:** 2009, 2010 **Model:** Amanti, Borrego, Optima, Sorento, Soul, Spectra, Spectra5, Sportage **Engine:** 2.0L L4 VIN 1, 2.0L L4 VIN 2, 2.4L L4 VIN 8, 2.7L V6 VIN 4, 3.3L V6 VIN 5, 3.8L V6 VIN 1, 3.8L V6 VIN 5, 3.8L V6 VIN 6 **Transmission:** All	**Engine Oil Temperature Sensor High Input:** Signal voltage higher than the possible range of a properly operating OTS. Voltage range check. Five minutes after engine start if engine coolant temperature less than 14 degrees F. Oil temperature minus 33 degrees F. **Possible Causes:** • Open circuit to battery • Contact resistance in connectors • faulty OTS
DTC: P0201 **2T CCM, MIL: Yes** **Year:** 2009, 2010 **Model:** Borrego, Rio, Rio5, Soul **Engine:** 1.6L L4 VIN 1, 1.6L L4 VIN 3, 4.6L V8 VIN 2 **Transmission:** All	**Cylinder 1 Injector Circuit Malfunction:** Engine running and the PCM detected the identified fuel injector control circuit signal was more than the upper limit, or that it was less than the lower limit, or that no control signal was present. **Possible Causes:** • Main relay power supply circuit to the injector is open • Fuel injector 1 control circuit is open or shorted to ground • Fuel injector 1 is damaged or has failed • Injector "driver" circuit in the PCM is damaged or has failed
DTC: P0202 **2T CCM, MIL: Yes** **Year:** 2009, 2010 **Model:** Borrego, Rio, Rio5, Soul **Engine:** 1.6L L4 VIN 1, 1.6L L4 VIN 3, 4.6L V8 VIN 2 **Transmission:** All	**Cylinder 2 Injector Circuit Malfunction:** Engine running and the PCM detected the identified fuel injector control circuit signal was more than the upper limit, or that it was less than the lower limit, or that no control signal was present. **Possible Causes:** • Main relay power supply circuit to the injector is open • Fuel injector 2 control circuit is open or shorted to ground • Fuel injector 2 is damaged or has failed • Injector "driver" circuit in the PCM is damaged or has failed

DTC	Trouble Code Title, Conditions & Possible Causes
DTC: P0203 **2T CCM, MIL: Yes** **Year:** 2009, 2010 **Model:** Borrego, Rio, Rio5, Soul **Engine:** 1.6L L4 VIN 1, 1.6L L4 VIN 3, 4.6L V8 VIN 2 **Transmission:** All	**Cylinder 3 Injector Circuit Malfunction:** Engine running and the PCM detected the identified fuel injector control circuit signal was more than the upper limit, or that it was less than the lower limit, or that no control signal was present. **Possible Causes:** • Main relay power supply circuit to the injector is open • Fuel injector 3 control circuit is open or shorted to ground • Fuel injector 3 is damaged or has failed • Injector "driver" circuit in the PCM is damaged or has failed
DTC: P0204 **2T CCM, MIL: Yes** **Year:** 2009, 2010 **Model:** Borrego, Rio, Rio5, Soul **Engine:** 1.6L L4 VIN 1, 1.6L L4 VIN 3, 4.6L V8 VIN 2 **Transmission:** All	**Cylinder 4 Injector Circuit Malfunction:** Engine running and the PCM detected the identified fuel injector control circuit signal was more than the upper limit, or that it was less than the lower limit, or that no control signal was present. **Possible Causes:** • Main relay power supply circuit to the injector is open • Fuel injector 4 control circuit is open or shorted to ground • Fuel injector 4 is damaged or has failed • Injector "driver" circuit in the PCM is damaged or has failed
DTC: P0205 **2T CCM, MIL: Yes** **Year:** 2009 **Model:** Borrego **Engine:** 4.6L V8 VIN 2 **Transmission:** All	**Cylinder 5 Injector Circuit Malfunction:** Engine running and the PCM detected the identified fuel injector control circuit signal was more than the upper limit, or that it was less than the lower limit, or that no control signal was present. **Possible Causes:** • Main relay power supply circuit to the injector is open • Fuel injector 5 control circuit is open or shorted to ground • Fuel injector 5 is damaged or has failed • Injector "driver" circuit in the PCM is damaged or has failed
DTC: P0206 **2T CCM, MIL: Yes** **Year:** 2009 **Model:** Borrego **Engine:** 4.6L V8 VIN 2 **Transmission:** All	**Cylinder 6 Injector Circuit Malfunction:** Engine running and the PCM detected the identified fuel injector control circuit signal was more than the upper limit, or that it was less than the lower limit, or that no control signal was present. **Possible Causes:** • Main relay power supply circuit to the injector is open • Fuel injector 6 control circuit is open or shorted to ground • Fuel injector 6 is damaged or has failed • Injector "driver" circuit in the PCM is damaged or has failed
DTC: P0207 **2T CCM, MIL: Yes** **Year:** 2009 **Model:** Borrego **Engine:** 4.6L V8 VIN 2 **Transmission:** All	**Cylinder 7 Injector Circuit Malfunction:** Engine running and the PCM detected the identified fuel injector control circuit signal was more than the upper limit, or that it was less than the lower limit, or that no control signal was present. **Possible Causes:** • Main relay power supply circuit to the injector is open • Fuel injector 6 control circuit is open or shorted to ground • Fuel injector 6 is damaged or has failed • Injector "driver" circuit in the PCM is damaged or has failed
DTC: P0208 **2T CCM, MIL: Yes** **Year:** 2009 **Model:** Borrego **Engine:** 4.6L V8 VIN 2 **Transmission:** All	**Cylinder 8 Injector Circuit Malfunction:** Engine running and the PCM detected the identified fuel injector control circuit signal was more than the upper limit, or that it was less than the lower limit, or that no control signal was present. **Possible Causes:** • Main relay power supply circuit to the injector is open • Fuel injector 6 control circuit is open or shorted to ground • Fuel injector 6 is damaged or has failed • Injector "driver" circuit in the PCM is damaged or has failed
DTC: P0217 **2T CCM, MIL: Yes** **Year:** 2009 **Model:** Amanti, Borrego, Sorento **Engine:** 3.3L V6 VIN 5, 3.8L V6 VIN 1, 3.8L V6 VIN 5, 3.8L V6 VIN 6 **Transmission:** All	**Engine Coolant Over Temperature Condition:** Engine running and no disabling faults present. Coolant sensor within range. **Possible Causes:** • Poor connection • Lack of engine coolant • Faulty water pump • ECTS • Faulty PCM

DTC	Trouble Code Title, Conditions & Possible Causes
DTC: P0219 **1T PCM** **Year:** 2009 **Model:** Borrego **Engine:** 4.6L V8 VIN 2 **Transmission:** All	**Engine Overspeed Condition:** Enable Conditions • Coolant sensor is normal • No disabling faults present(DTCs related to MAFS/MAPS, catalyst, fuel system or engine oil temperature sensor) • Coolant Temperature at startup < 45°C(113 °F) • Engine running state • Coolant temperature > 50°C(122 °F) • Intake air temperature < 35°C(95 °F) Threshold value • Coolant temperature ≥ 110°C (230 °F) (Average airflow< 30 g/sec and filtered airflow< 50 g/sec) Diagnosis Time • Once per driving cycle (about 2 minutes) **Possible Causes:** • ECTS
DTC: P0221 **2T CCM, MIL: Yes** **Year:** 2009, 2010 **Model:** Borrego, Optima **Engine:** 2.4L L4 VIN 8, 2.7L V6 VIN 4, 4.6L V8 VIN 2 **Transmission:** All	**Throttle/Pedal Position Sensor/Switch "B" Circuit Range/Performance:** The ECM compares the TPS1 and TPS2 signal, and sets a code as required. Enabling conditions are as follows; no engine stop and engine start, no TPS adaptation request, no relevant failure. **Possible Causes:** • Poor connection or damaged harness • Air leakage in intake system • Faulty TPS2
DTC: P0222 **2T CCM, MIL: Yes** **Year:** 2009, 2010 **Model:** Amanti, Borrego, Optima, Sorento **Engine:** 2.4L L4 VIN 8, 2.7L V6 VIN 4, 3.3L V6 VIN 5, 3.8L V6 VIN 1, 3.8L V6 VIN 5, 3.8L V6 VIN 6, 4.6L V8 VIN 2 **Transmission:** All	**Throttle/Pedal Position Sensor/Switch "B" Circuit Low Input:** The DTC is recorded if the output voltage of the TPS 1 is lower than threshold value (Vtps1 less than or equal to 0.2 volt). TPS 1 low input. **Possible Causes:** • Poor connection • Open or short to ground in TPS circuit • Faulty TPS • Faulty PCM
DTC: P0223 **2T CCM, MIL: Yes** **Year:** 2009, 2010 **Model:** Amanti, Borrego, Optima, Sorento **Engine:** 2.4L L4 VIN 8, 2.7L V6 VIN 4, 3.3L V6 VIN 5, 3.8L V6 VIN 1, 3.8L V6 VIN 5, 3.8L V6 VIN 6, 4.6L V8 VIN 2 **Transmission:** All	**Throttle/Pedal Position Sensor/Switch "B" Circuit High Input:** The DTC is recorded if the output voltage of the TPS 1 higher than threshold value (Vtps1 greater than or equal to 4.85 volts, load value, EV less than 70 percent) when TPS 2 (Vtps2 less than or equal to 2.5 volts) is normal. TPS 1 high input. **Possible Causes:** • Poor connection • Open or short to ground in TPS circuit • Faulty TPS • Faulty PCM
DTC: P0230 **1T CCM, MIL: Yes** **Year:** 2009, 2010 **Model:** Amanti, Borrego, Optima, Rio, Rio5, Sorento, Soul, Spectra, Spectra5, Sportage **Engine:** 1.6L L4 VIN 1, 1.6L L4 VIN 3, 2.0L L4 VIN 1, 2.0L L4 VIN 2, 2.4L L4 VIN 8, 2.7L V6 VIN 3, 2.7L V6 VIN 4, 3.3L V6 VIN 5, 3.8L V6 VIN 1, 3.8L V6 VIN 5, 3.8L V6 VIN 6, 4.6L V8 VIN 2 **Transmission:** All	**Fuel Pump Circuit Malfunction:** Key on, and then the PCM detected an unexpected voltage condition on the fuel pump circuit through the fuel pump monitoring input. **Possible Causes:** • Fuel pump control circuit is open or shorted to ground • Fuel pump relay power circuit from ignition switch is open • Fuel pump relay is damaged or has failed • PCM has failed

DTC	Trouble Code Title, Conditions & Possible Causes
DTC: P0231 **2T CCM, MIL: Yes** **Year:** 2009, 2010 **Model:** Rio, Rio5, Soul **Engine:** 1.6L L4 VIN 1, 1.6L L4 VIN 3 **Transmission:** All	**Electric Fuel Pump Relay Open Or Short Circuit:** Circuit continuity check, high. **Possible Causes:** • Poor connection • Short to power in control circuit • Fuel pump relay • Faulty ECM/PCM
DTC: P0232 **2T CCM, MIL: Yes** **Year:** 2009, 2010 **Model:** Rio, Rio5, Soul **Engine:** 1.6L L4 VIN 1, 1.6L L4 VIN 3 **Transmission:** All	**Electric Fuel Pump Relay Short Circuit:** Circuit continuity check, low. **Possible Causes:** • Poor connection • Short to ground in control circuit • Fuel pump relay • Faulty ECM/PCM
DTC: P0261 **2T CCM, MIL: Yes** **Year:** 2009, 2010 **Model:** Amanti, Borrego, Optima, Rio, Rio5, Sorento, Soul, Spectra, Spectra5, Sportage **Engine:** 1.6L L4 VIN 1, 1.6L L4 VIN 3, 2.0L L4 VIN 1, 2.0L L4 VIN 2, 2.4L L4 VIN 8, 2.7L V6 VIN 3, 2.7L V6 VIN 4, 3.3L V6 VIN 5, 3.8L V6 VIN 1, 3.8L V6 VIN 5, 3.8L V6 VIN 6, 4.6L V8 VIN 2 **Transmission:** All	**Fuel Injector 1 Circuit Low Input:** Engine running and the PCM detected the Injector 1 signal was in a low signal state (0 volt) with the injector commanded off in the test. **Possible Causes:** • Main relay power supply circuit to the injector is open • Fuel injector 1 control circuit is shorted to ground • Fuel injector 1 is damaged or has failed • Injector "driver" circuit in the PCM is damaged or has failed
DTC: P0262 **2T CCM, MIL: Yes** **Year:** 2009, 2010 **Model:** Amanti, Borrego, Optima, Rio, Rio5, Sorento, Soul, Spectra, Spectra5, Sportage **Engine:** 1.6L L4 VIN 1, 1.6L L4 VIN 3, 2.0L L4 VIN 1, 2.0L L4 VIN 2, 2.4L L4 VIN 8, 2.7L V6 VIN 3, 2.7L V6 VIN 4, 3.3L V6 VIN 5, 3.8L V6 VIN 1, 3.8L V6 VIN 5, 3.8L V6 VIN 6, 4.6L V8 VIN 2 **Transmission:** All	**Fuel Injector 1 Circuit High Input:** Engine running and the PCM detected the Injector 1 signal was in a high signal state (12 volts) with the injector commanded off in the test. **Possible Causes:** • Fuel injector 1 control circuit is shorted to system power (B+) • Fuel injector 1 is damaged or has failed • Injector 1 "driver" circuit in the PCM is damaged or has failed
DTC: P0264 **2T CCM, MIL: Yes** **Year:** 2009, 2010 **Model:** Amanti, Borrego, Optima, Rio, Rio5, Sorento, Soul, Spectra, Spectra5, Sportage **Engine:** 1.6L L4 VIN 1, 1.6L L4 VIN 3, 2.0L L4 VIN 1, 2.0L L4 VIN 2, 2.4L L4 VIN 8, 2.7L V6 VIN 3, 2.7L V6 VIN 4, 3.3L V6 VIN 5, 3.8L V6 VIN 1, 3.8L V6 VIN 5, 3.8L V6 VIN 6, 4.6L V8 VIN 2 **Transmission:** All	**Fuel Injector 2 Circuit Low Input:** Engine running and the PCM detected the Injector 2 signal was in a low signal state (0 volt) with the injector commanded off in the test. **Possible Causes:** • Main relay power supply circuit to the injector is open • Fuel injector 2 control circuit is shorted to ground • Fuel injector 2 is damaged or has failed • Injector 2 "driver" circuit in the PCM is damaged or has failed

DTC	Trouble Code Title, Conditions & Possible Causes
DTC: P0265 **2T CCM, MIL: Yes** **Year:** 2009, 2010 **Model:** Amanti, Borrego, Optima, Rio, Rio5, Sorento, Soul, Spectra, Spectra5, Sportage **Engine:** 1.6L L4 VIN 1, 1.6L L4 VIN 3, 2.0L L4 VIN 1, 2.0L L4 VIN 2, 2.4L L4 VIN 8, 2.7L V6 VIN 3, 2.7L V6 VIN 4, 3.3L V6 VIN 5, 3.8L V6 VIN 1, 3.8L V6 VIN 5, 3.8L V6 VIN 6, 4.6L V8 VIN 2 **Transmission:** All	**Fuel Injector 2 Circuit High Input:** Engine running and the PCM detected the Injector 2 signal was in a high signal state (12 volts) with the injector commanded off in the test. **Possible Causes:** • Fuel injector 2 control circuit is shorted to system power (B+) • Fuel injector 2 is damaged or has failed • Injector 2 "driver" circuit in the PCM is damaged or has failed
DTC: P0267 **2T CCM, MIL: Yes** **Year:** 2009, 2010 **Model:** Amanti, Borrego, Optima, Rio, Rio5, Sorento, Soul, Spectra, Spectra5, Sportage **Engine:** 1.6L L4 VIN 1, 1.6L L4 VIN 3, 2.0L L4 VIN 1, 2.0L L4 VIN 2, 2.4L L4 VIN 8, 2.7L V6 VIN 3, 2.7L V6 VIN 4, 3.3L V6 VIN 5, 3.8L V6 VIN 1, 3.8L V6 VIN 5, 3.8L V6 VIN 6, 4.6L V8 VIN 2 **Transmission:** All	**Fuel Injector 3 Circuit Low Input:** Engine running and the PCM detected the Injector 3 signal was in a low signal state (0 volt) with the injector commanded off in the test. **Possible Causes:** • Main relay power supply circuit to the injector is open • Fuel injector 3 control circuit is shorted to ground • Fuel injector 3 is damaged or has failed • Injector 3 "driver" circuit in the PCM is damaged or has failed
DTC: P0268 **2T CCM, MIL: Yes** **Year:** 2009, 2010 **Model:** Amanti, Borrego, Optima, Rio, Rio5, Sorento, Soul, Spectra, Spectra5, Sportage **Engine:** 1.6L L4 VIN 1, 1.6L L4 VIN 3, 2.0L L4 VIN 1, 2.0L L4 VIN 2, 2.4L L4 VIN 8, 2.7L V6 VIN 3, 2.7L V6 VIN 4, 3.3L V6 VIN 5, 3.8L V6 VIN 1, 3.8L V6 VIN 5, 3.8L V6 VIN 6, 4.6L V8 VIN 2 **Transmission:** All	**Fuel Injector 3 Circuit High Input:** Engine running and the PCM detected the Injector 3 signal was in a high signal state (12 volts) with the injector commanded off in the test. **Possible Causes:** • Fuel injector 3 control circuit is shorted to system power (B+) • Fuel injector 3 is damaged or has failed • Injector 3 "driver" circuit in the PCM is damaged or has failed
DTC: P0270 **2T CCM, MIL: Yes** **Year:** 2009, 2010 **Model:** Amanti, Borrego, Optima, Rio, Rio5, Sorento, Soul, Spectra, Spectra5, Sportage **Engine:** 1.6L L4 VIN 1, 1.6L L4 VIN 3, 2.0L L4 VIN 1, 2.0L L4 VIN 2, 2.4L L4 VIN 8, 2.7L V6 VIN 3, 2.7L V6 VIN 4, 3.3L V6 VIN 5, 3.8L V6 VIN 1, 3.8L V6 VIN 5, 3.8L V6 VIN 6, 4.6L V8 VIN 2 **Transmission:** All	**Fuel Injector 4 Circuit Low Input:** Engine running and the PCM detected the Injector 4 signal was in a low signal state (0 volt) with the injector commanded off in the test. **Possible Causes:** • Fuel injector 4 control circuit is shorted to system power (B+) • Fuel injector 4 is damaged or has failed • Injector 4 "driver" circuit in the PCM is damaged or has failed

DTC	Trouble Code Title, Conditions & Possible Causes
DTC: P0271 **2T CCM, MIL: Yes** **Year:** 2009, 2010 **Model:** Amanti, Borrego, Optima, Rio, Rio5, Sorento, Soul, Spectra, Spectra5, Sportage **Engine:** 1.6L L4 VIN 1, 1.6L L4 VIN 3, 2.0L L4 VIN 1, 2.0L L4 VIN 2, 2.4L L4 VIN 8, 2.7L V6 VIN 3, 2.7L V6 VIN 4, 3.3L V6 VIN 5, 3.8L V6 VIN 1, 3.8L V6 VIN 5, 3.8L V6 VIN 6, 4.6L V8 VIN 2 **Transmission:** All	**Fuel Injector 4 Circuit High Input:** Engine running and the PCM detected the Injector 4 signal was in a high signal state (12 volts) with the injector commanded off in the test. **Possible Causes:** • Fuel injector 4 control circuit is shorted to system power (B+) • Fuel injector 4 is damaged or has failed • Injector 4 "driver" circuit in the PCM is damaged or has failed
DTC: P0273 **2T CCM, MIL: Yes** **Year:** 2009, 2010 **Model:** Amanti, Borrego, Sorento, Sportage **Engine:** 2.7L V6 VIN 3, 3.3L V6 VIN 5, 3.8L V6 VIN 1, 3.8L V6 VIN 5, 3.8L V6 VIN 6, 4.6L V8 VIN 2 **Transmission:** All	**Cylinder 5- Injector Circuit Low:** The PCM sets the DTC if the control circuit is shorted to ground. Driver stage check. **Possible Causes:** • Open in power supply harness • Short to ground in control harness • Contact resistance in connectors • Faulty injector
DTC: P0274 **2T CCM, MIL: Yes** **Year:** 2009, 2010 **Model:** Amanti, Borrego, Sorento, Sportage **Engine:** 2.7L V6 VIN 3, 3.3L V6 VIN 5, 3.8L V6 VIN 1, 3.8L V6 VIN 5, 3.8L V6 VIN 6, 4.6L V8 VIN 2 **Transmission:** All	**Cylinder 5- Injector Circuit High:** The PCM sets the DTC if the control circuit is open or shorted to battery voltage. Driver stage check. **Possible Causes:** • Open or short to battery control harness • Contact resistance in connectors • Faulty injector
DTC: P0276 **2T CCM, MIL: Yes** **Year:** 2009, 2010 **Model:** Amanti, Borrego, Sorento, Sportage **Engine:** 2.7L V6 VIN 3, 3.3L V6 VIN 5, 3.8L V6 VIN 1, 3.8L V6 VIN 5, 3.8L V6 VIN 6, 4.6L V8 VIN 2 **Transmission:** All	**Cylinder 6- Injector Circuit Low:** The PCM sets the DTC if the control circuit is shorted to ground. Driver stage check. **Possible Causes:** • Open in power supply harness • Short to ground in control harness • Contact resistance in connectors • Faulty injector
DTC: P0277 **2T CCM, MIL: Yes** **Year:** 2009, 2010 **Model:** Amanti, Borrego, Sorento, Sportage **Engine:** 2.7L V6 VIN 3, 3.3L V6 VIN 5, 3.8L V6 VIN 1, 3.8L V6 VIN 5, 3.8L V6 VIN 6, 4.6L V8 VIN 2 **Transmission:** All	**Cylinder 6- Injector Circuit High:** The PCM sets the DTC if the control circuit is open or shorted to battery voltage. Driver stage check. **Possible Causes:** • Open or short to battery control harness • Contact resistance in connectors • Faulty injector

DTC	Trouble Code Title, Conditions & Possible Causes
DTC: P0279 **2T PCM, MIL: Yes** **Year:** 2009 **Model:** Borrego **Engine:** 4.6L V8 VIN 2 **Transmission:** All	**Cylinder 7 Injector Circuit Low:** Enable Conditions • - Threshold value • Short to ground Diagnosis Time • 5 sec **Possible Causes:** • Poor connection • Open or short to ground in power harness • Open or short to ground in control harness • Injector • ECM
DTC: P0280 **2T PCM, MIL: Yes** **Year:** 2009 **Model:** Borrego **Engine:** 4.6L V8 VIN 2 **Transmission:** All	**Cylinder 7 - Injector Circuit High:** Enable Conditions • - Threshold value • Short to battery Diagnosis Time • 5 sec **Possible Causes:** • Poor connection • Short to battery in harness • Injector • ECM
DTC: P0282 **2T PCM, MIL: Yes** **Year:** 2009 **Model:** Borrego **Engine:** 4.6L V8 VIN 2 **Transmission:** All	**Cylinder 8 - Injector Circuit Low:** Enable Conditions • - Threshold value • Short to ground Diagnosis Time • 5 sec **Possible Causes:** • Poor connection • Open or short to ground in power harness • Open or short to ground in control harness • Injector • ECM
DTC: P0283 **2T PCM, MIL: Yes** **Year:** 2009 **Model:** Borrego **Engine:** 4.6L V8 VIN 2 **Transmission:** All	**Cylinder 8 - Injector Circuit High:** Enable Conditions • - Threshold value • Short to battery Diagnosis Time • 5 sec **Possible Causes:** • Poor connection • Short to battery in harness • Injector • ECM

DTC	Trouble Code Title, Conditions & Possible Causes
DTC: P0300 **2T MISFIRE, MIL: Yes** **Year:** 2009, 2010 **Model:** Amanti, Borrego, Rio, Rio5, Sorento, Soul, Spectra, Spectra5, Sportage **Engine:** 1.6L L4 VIN 1, 1.6L L4 VIN 3, 2.0L L4 VIN 1, 2.0L L4 VIN 2, 2.7L V6 VIN 3, 3.3L V6 VIN 5, 3.8L V6 VIN 1, 3.8L V6 VIN 5, 3.8L V6 VIN 6, 4.6L V8 VIN 2 **Transmission:** All	**Random Cylinder Misfire Detected:** Engine runtime 3 seconds, engine speed change under 1200 rpm, and the PCM detected a random misfire condition in more than one cylinder during the 200 revolution or the 1000 revolution test range. **NOTE: If the misfire is severe, the MIL will flash on/off on the 1st trip!** **Possible Causes:** • Vehicle driven under low fuel condition (less than 1/8 of a tank) • CKP or CMP sensor signal erratic or out of phase • Base engine problem affecting two or more engine cylinders • Ignition system problem affecting two or more engine cylinders
DTC: P0301 **2T MISFIRE, MIL: Yes** **Year:** 2009, 2010 **Model:** Amanti, Borrego, Rio, Rio5, Sorento, Soul, Spectra, Spectra5, Sportage **Engine:** 1.6L L4 VIN 1, 1.6L L4 VIN 3, 2.0L L4 VIN 1, 2.0L L4 VIN 2, 2.7L V6 VIN 3, 3.3L V6 VIN 5, 3.8L V6 VIN 1, 3.8L V6 VIN 5, 3.8L V6 VIN 6, 4.6L V8 VIN 2 **Transmission:** All	**Cylinder 1 Misfire Detected:** Engine runtime 3 seconds, engine speed change under 1200 rpm, and the PCM detected a misfire condition present in one cylinder during the 200 (Catalyst) or 1000 revolution (Emission) test range. **NOTE: If the misfire is severe, the MIL will flash on/off on the 1st trip!** **Possible Causes:** • Fuel metering (fuel injector dirty) problem affecting Cylinder 1 • Base engine (compression) problem affecting Cylinder 1 • Ignition system (spark plug or plug wire) problem on Cylinder 1
DTC: P0302 **2T MISFIRE, MIL: Yes** **Year:** 2009, 2010 **Model:** Amanti, Borrego, Rio, Rio5, Sorento, Soul, Spectra, Spectra5, Sportage **Engine:** 1.6L L4 VIN 1, 1.6L L4 VIN 3, 2.0L L4 VIN 1, 2.0L L4 VIN 2, 2.7L V6 VIN 3, 3.3L V6 VIN 5, 3.8L V6 VIN 1, 3.8L V6 VIN 5, 3.8L V6 VIN 6, 4.6L V8 VIN 2 **Transmission:** All	**Cylinder 2 Misfire Detected:** Engine runtime 3 seconds, engine speed change under 1200 rpm, and the PCM detected a misfire condition present in one cylinder during the 200 (Catalyst) or 1000 revolution (Emission) test range. **NOTE: If the misfire is severe, the MIL will flash on/off on the 1st trip!** **Possible Causes:** • Fuel metering (fuel injector dirty) problem affecting Cylinder 2 • Base engine (compression) problem affecting Cylinder 2 • Ignition system (spark plug or plug wire) problem on Cylinder 2
DTC: P0303 **2T MISFIRE, MIL: Yes** **Year:** 2009, 2010 **Model:** Amanti, Borrego, Rio, Rio5, Sorento, Soul, Spectra, Spectra5, Sportage **Engine:** 1.6L L4 VIN 1, 1.6L L4 VIN 3, 2.0L L4 VIN 1, 2.0L L4 VIN 2, 2.7L V6 VIN 3, 3.3L V6 VIN 5, 3.8L V6 VIN 1, 3.8L V6 VIN 5, 3.8L V6 VIN 6, 4.6L V8 VIN 2 **Transmission:** All	**Cylinder 3 Misfire Detected:** Engine runtime 3 seconds, engine speed change under 1200 rpm, and the PCM detected a misfire condition present in one cylinder during the 200 (Catalyst) or 1000 revolution (Emission) test range. **NOTE: If the misfire is severe, the MIL will flash on/off on the 1st trip!** **Possible Causes:** • Fuel metering (fuel injector dirty) problem affecting Cylinder 3 • Base engine (compression) problem affecting Cylinder 3 • Ignition system (spark plug or plug wire) problem on Cylinder 3
DTC: P0304 **2T MISFIRE, MIL: Yes** **Year:** 2009, 2010 **Model:** Amanti, Borrego, Rio, Rio5, Sorento, Soul, Spectra, Spectra5, Sportage **Engine:** 1.6L L4 VIN 1, 1.6L L4 VIN 3, 2.0L L4 VIN 1, 2.0L L4 VIN 2, 2.7L V6 VIN 3, 3.3L V6 VIN 5, 3.8L V6 VIN 1, 3.8L V6 VIN 5, 3.8L V6 VIN 6, 4.6L V8 VIN 2 **Transmission:** All	**Cylinder 4 Misfire Detected:** Engine runtime 3 seconds, engine speed change under 1200 rpm, and the PCM detected a misfire condition present in one cylinder during the 200 (Catalyst) or 1000 revolution (Emission) test range. **NOTE: If the misfire is severe, the MIL will flash on/off on the 1st trip!** **Possible Causes:** • Fuel metering (fuel injector dirty) problem affecting Cylinder 4 • Base engine (compression) problem affecting Cylinder 4 • Ignition system (spark plug or plug wire) problem on Cylinder 4

DTC	Trouble Code Title, Conditions & Possible Causes
DTC: P0305 **2T MISFIRE, MIL: Yes** **Year:** 2009, 2010 **Model:** Amanti, Borrego, Sorento, Sportage **Engine:** 2.7L V6 VIN 3, 3.3L V6 VIN 5, 3.8L V6 VIN 1, 3.8L V6 VIN 5, 3.8L V6 VIN 6, 4.6L V8 VIN 2 **Transmission:** All	**Cylinder 5 Misfire Detected:** Engine runtime 3 seconds, engine speed change under 1200 rpm, and the PCM detected a misfire condition present in one cylinder during the 200 (Catalyst) or 1000 revolution (Emission) test range. **NOTE: If the misfire is severe, the MIL will flash on/off on the 1st trip!** **Possible Causes:** • Fuel metering (fuel injector dirty) problem affecting Cylinder 5 • Base engine (compression) problem affecting Cylinder 5 • Ignition system (spark plug or plug wire) problem on Cylinder 5
DTC: P0306 **2T MISFIRE, MIL: Yes** **Year:** 2009, 2010 **Model:** Amanti, Borrego, Sorento, Sportage **Engine:** 2.7L V6 VIN 3, 3.3L V6 VIN 5, 3.8L V6 VIN 1, 3.8L V6 VIN 5, 3.8L V6 VIN 6, 4.6L V8 VIN 2 **Transmission:** All	**Cylinder 6 Misfire Detected:** Engine runtime 3 seconds, engine speed change under 1200 rpm, and the PCM detected a misfire condition present in one cylinder during the 200 (Catalyst) or 1000 revolution (Emission) test range. **NOTE: If the misfire is severe, the MIL will flash on/off on the 1st trip!** **Possible Causes:** • Fuel metering (fuel injector dirty) problem affecting Cylinder 6 • Base engine (compression) problem affecting Cylinder 6 • Ignition system (spark plug or plug wire) problem on Cylinder 6
DTC: P0307 **T PCM, MIL: Yes** **Year:** 2009 **Model:** Borrego **Engine:** 4.6L V8 VIN 2 **Transmission:** All	**Cylinder 7 - Misfire Detected:** Case1 Threshold Value • Misfire rate > 3.25% Diagnostic Time • 1000 rev MIL ON Condition • 2 driving cycle Case2 Threshold Value • Misfire rate > 2.4~11.7% Diagnostic Time • 200 rev MIL ON Condition • Immediately (Blink) **Possible Causes:** • Poor connection • Ignition system • Fuel system • Intake/exhaust air system • Ignition timing • Injector

DTC	Trouble Code Title, Conditions & Possible Causes
DTC: P0308 **T PCM, MIL: Yes** **Year:** 2009 **Model:** Borrego **Engine:** 4.6L V8 VIN 2 **Transmission:** All	**Cylinder 8 - Misfire Detected:** Case1 Threshold Value • Misfire rate > 3.25% Diagnostic Time • 1000 rev MIL ON Condition • 2 driving cycle Case2 Threshold Value • Misfire rate > 2.4~11.7% Diagnostic Time • 200 rev MIL ON Condition • Immediately (Blink) **Possible Causes:** • Poor connection • Ignition system • Fuel system • Intake/exhaust air system • Ignition timing • Injector
DTC: P0315 **2T CCM, MIL: Yes** **Year:** 2009, 2010 **Model:** Optima, Sorento, Soul, Spectra, Spectra5, Sportage **Engine:** 2.0L L4 VIN 1, 2.0L L4 VIN 2, 2.4L L4 VIN 8, 2.7L V6 VIN 3, 2.7L V6 VIN 4, 3.3L V6 VIN 5, 3.8L V6 VIN 6 **Transmission:** All	**Segment Time Acquisition Incorrect:** A misfire induces a decrease in the engine speed and causes a variation in the segment period. Monitor segment time adaptation. **Possible Causes:** • Improperly installed target wheel • Contact resistance in connectors • Faulty PCM • Faulty CKPS
DTC: P0315 **2T PCM** **Year:** 2009, 2010 **Model:** Amanti, Borrego, Sportage **Engine:** 2.0L L4 VIN 2, 3.8L V6 VIN 1, 3.8L V6 VIN 5 **Transmission:** All	**Camshaft Position (CMP) Sensor Error:** • 10% ≤ Engine load < 90% • 2000 rpm ≤ engine speed ≤ 4000 rpm • TEC(Tooth Error Correction) RPM stability timer > 10sec • 0°C(32°F) < coolant temperature < 110°C(230°F) • Not active disabling faults **Possible Causes:** • Loosened CKPS • Target wheel • PCM
DTC: P0325 **2T CCM, MIL: Yes** **Year:** 2009, 2010 **Model:** Amanti, Borrego, Optima, Sorento, Soul, Spectra, Spectra5, Sportage **Engine:** 2.0L L4 VIN 1, 2.0L L4 VIN 2, 2.4L L4 VIN 8, 2.7L V6 VIN 3, 2.7L V6 VIN 4, 3.3L V6 VIN 5, 3.8L V6 VIN 1, 3.8L V6 VIN 5, 3.8L V6 VIN 6 **Transmission:** All	**Knock Sensor Circuit Malfunction (Bank 1):** Engine running at over 1000 rpm for 5 seconds, and the PCM did not detect enough variation in the KS signals (e.g., 0.049 volt). **Possible Causes:** • Knock sensor signal circuit open or shorted to ground • Knock sensor signal circuit shorted to VREF or system power • Knock sensor is damaged or has failed • PCM has failed

DTC	Trouble Code Title, Conditions & Possible Causes
DTC: P0326 **2T CCM, MIL: Yes** **Year:** 2009, 2010 **Model:** Amanti, Borrego, Rio, Rio5, Sorento, Soul **Engine:** 1.6L L4 VIN 1, 1.6L L4 VIN 3, 3.3L V6 VIN 5, 3.8L V6 VIN 1, 3.8L V6 VIN 5, 3.8L V6 VIN 6 **Transmission:** All	**Knock Sensor Circuit Malfunction (Bank 1):** Engine speed from 1000-2200 rpm, ECT sensor signal more than 104°F, engine load more than 2 ms, and the PCM detected the Knock sensor signal was out of range at a calculated engine speed. **Possible Causes:** • Knock sensor signal circuit open or shorted to ground • Knock sensor signal circuit shorted to VREF or system power • Knock sensor is damaged or has failed • PCM has failed
DTC: P0327 **2T CCM, MIL: Yes** **Year:** 2009, 2010 **Model:** Borrego, Rio, Rio5, Soul **Engine:** 1.6L L4 VIN 1, 1.6L L4 VIN 3, 4.6L V8 VIN 2 **Transmission:** All	**Knock Sensor 1 Circuit Low Input:** Engine speed greater than 2600 rpm. **Possible Causes:** • Poor connection • Open or short to ground in signal circuit • Knock sensor is damaged or has failed • PCM/ECM has failed
DTC: P0328 **2T CCM, MIL: Yes** **Year:** 2009, 2010 **Model:** Borrego, Rio, Rio5, Soul **Engine:** 1.6L L4 VIN 1, 1.6L L4 VIN 3, 4.6L V8 VIN 2 **Transmission:** All	**Knock Sensor 1 Circuit High Input:** Coolant temperature more than 104 degrees F. **Possible Causes:** • Poor connection • Short to power in signal circuit • Knock sensor is damaged or has failed • PCM/ECM has failed
DTC: P032C **2T PCM, MIL: Yes** **Year:** 2009 **Model:** Borrego **Engine:** 4.6L V8 VIN 2 **Transmission:** All	**Min error of DFP_KS3: internal failure path number: knock sensor 3:** Enable Conditions • ECT > 40C • Engine speed > 2400rpm • Engine load > 35% Threshold value • Normalized reference level < lower limit, f(rpm) Diagnosis Time • Continuous **Possible Causes:** • Poor connection • Open or short to ground in signal circuit • Knock sensor
DTC: P032D **2T PCM, MIL: Yes** **Year:** 2009 **Model:** Borrego **Engine:** 4.6L V8 VIN 2 **Transmission:** All	**Max error of DFP_KS3: internal failure path number: knock sensor 3:** Enable Conditions • ECT > 40C • Engine speed > 2400rpm • Engine load > 35% Threshold value • Normalized reference level > upper limit, f(rpm) Diagnosis Time • Continuous **Possible Causes:** • Poor connection • Short to power in signal circuit • Knock sensor
DTC: P0330 **2T CCM, MIL: Yes** **Year:** 2009, 2010 **Model:** Amanti, Borrego, Sorento, Sportage **Engine:** 2.7L V6 VIN 3, 3.3L V6 VIN 5, 3.8L V6 VIN 1, 3.8L V6 VIN 5, 3.8L V6 VIN 6 **Transmission:** All	**Knock Sensor Circuit Malfunction (Bank 2):** Engine running at over 1000 rpm for 5 seconds, and the PCM did not detect enough variation in the KS signals (e.g., 0.049 volt). **Possible Causes:** • Knock sensor signal circuit open or shorted to ground • Knock sensor signal circuit shorted to VREF or system power • Knock sensor is damaged or has failed • PCM has failed

DTC	Trouble Code Title, Conditions & Possible Causes
DTC: P0331 **2T CCM, MIL: Yes** **Year:** 2009 **Model:** Amanti, Borrego, Sorento **Engine:** 3.3L V6 VIN 5, 3.8L V6 VIN 1, 3.8L V6 VIN 5, 3.8L V6 VIN 6 **Transmission:** All	**Knock Sensor 2 Circuit Range/Performance (Bank 2):** Signal short. Pressure in intake manifold is normal. Engine speed is equal to or less than 1600 rpm. **Possible Causes:** • Poor connection • Short in harness • Faulty knock sensor • Faulty PCM
DTC: P0332 **2T PCM, MIL: Yes** **Year:** 2009 **Model:** Borrego **Engine:** 4.6L V8 VIN 2 **Transmission:** All	**Knock Sensor 2 Circuit Low (Bank 2):** Enable Conditions • ECT > 40C • Engine speed > 2400rpm • Engine load > 35% Threshold value • Normalized reference level < lower limit, f(rpm) Diagnosis Time • Continuous **Possible Causes:** • Poor connection • Open or short to ground in signal circuit • Knock sensor
DTC: P0333 **2T PCM, MIL: Yes** **Year:** 2009 **Model:** Borrego **Engine:** 4.6L V8 VIN 2 **Transmission:** All	**Knock Sensor 2 Circuit High (Bank 2):** Enable Conditions • ECT > 40C • Engine speed > 1600rpm • Engine load > 35% Threshold value • Normalized reference level > upper limit, f(rpm) Diagnosis Time • Continuous **Possible Causes:** • Poor connection • Short to power in signal circuit • Knock sensor
DTC: P0335 **1T CCM, MIL: Yes** **Year:** 2009, 2010 **Model:** Amanti, Optima, Rio, Rio5, Spectra, Spectra5 **Engine:** 1.6L L4 VIN 3, 2.0L L4 VIN 1, 2.0L L4 VIN 2, 2.4L L4 VIN 8, 2.7L V6 VIN 4, 3.8L V6 VIN 5 **Transmission:** All	**Crankshaft Position Sensor Circuit Malfunction:** Engine cranking for 5 seconds, and the PCM did not detect any CKP sensor signals, or the vehicle was driven to a speed over 15.5 mph, and the PCM did not detect any CKP sensor signals for 5 seconds. **Possible Causes:** • CKP sensor signal circuit open or shorted to ground • CKP sensor signal circuit shorted to VREF or system power • CKP sensor is damaged or has failed • PCM has failed
DTC: P0335 **2T PCM, MIL: Yes** **Year:** 2009, 2010 **Model:** Borrego, Sorento, Soul, Sportage **Engine:** 1.6L L4 VIN 1, 2.0L L4 VIN 2, 2.7L V6 VIN 3, 3.3L V6 VIN 5, 3.8L V6 VIN 1, 3.8L V6 VIN 6, 4.6L V8 VIN 2 **Transmission:** All	**Crankshaft Position Sensor 'A' Circuit:** Enable Conditions • IG "ON", Cranking or engine-off during driving • No DTC related to CAM • Camshaft position sensor state change Threshold value • No signal counter > 6 times. Diagnosis Time • Continuous **Possible Causes:** • Poor connection • Open in harness • CKP sensor • ECM

<table>
<tr><th>DTC</th><th>Trouble Code Title, Conditions & Possible Causes</th></tr>
<tr><td>DTC: P0336
2T PCM, MIL: Yes
Year: 2009, 2010
Model: Borrego, Sorento, Soul
Engine: 1.6L L4 VIN 1, 3.3L V6 VIN 5, 3.8L V6 VIN 1, 3.8L V6 VIN 6, 4.6L V8 VIN 2
Transmission: All</td><td>Crankshaft Position Sensor 'A' Circuit Range/Performance:
Enable Conditions
• IG "ON", Cranking or engine-off during driving
• No DTC related to CAM
• Camshaft position sensor state change
Threshold value
• Implausible signal counter > 32 times
Diagnosis Time
• Continuous
Possible Causes:
• Poor connection
• Noise
• Short in harness
• Target wheel
• ECM</td></tr>
<tr><td>DTC: P0336
1T CCM, MIL: Yes
Year: 2009, 2010
Model: Amanti, Optima, Rio, Rio5
Engine: 1.6L L4 VIN 3, 2.4L L4 VIN 8, 2.7L V6 VIN 4, 3.8L V6 VIN 5
Transmission: All</td><td>Crankshaft Position Sensor Performance:
Engine running and the PCM detected the number of CKP sensor signals counted (between the reference mark gap) did not equal the Actual number of available teeth (i.e., the CKP signals were out or the normal window" of operation with the CMP sensor signals okay).
Possible Causes:
• CKP sensor signal circuit connections loose (intermittent fault)
• CKP sensor wiring harness has a connection fault (intermittent)
• CKP to Target Wheel "air gap" is incorrect
• CKP sensor is damaged or has failed</td></tr>
<tr><td>DTC: P0337
2T CCM, MIL: Yes
Year: 2009, 2010
Model: Rio, Rio5, Soul
Engine: 1.6L L4 VIN 1, 1.6L L4 VIN 3
Transmission: All</td><td>Crankshaft Position Sensor "A" Circuit Low Input:
If the output voltage of the CKPS remains low for more than two seconds. When the change of the CMPS output voltage is zero, the PCM determines a fault and stores a code. Change in output voltage (delta sign Vckp) is monitored.
Possible Causes:
• Poor connection
• Open or short to ground in CKPS circuit
• Faulty CKPS
• Faulty PCM</td></tr>
<tr><td>DTC: P0338
2T CCM, MIL: Yes
Year: 2009, 2010
Model: Rio, Rio5, Soul
Engine: 1.6L L4 VIN 1, 1.6L L4 VIN 3
Transmission: All</td><td>Crankshaft Position Sensor "A" Circuit High Input:
If the output voltage of the CKPS remains high for more than two seconds. When the change of the CMPS output voltage is zero, the PCM determines a fault and stores a code. Change in output voltage (delta sign Vckp) is monitored.
Possible Causes:
• Poor connection
• Open or short to ground in CKPS circuit
• Faulty CKPS
• Faulty PCM</td></tr>
<tr><td>DTC: P0339
2T CCM, MIL: Yes
Year: 2009, 2010
Model: Rio, Rio5, Soul
Engine: 1.6L L4 VIN 1, 1.6L L4 VIN 3
Transmission: All</td><td>Crankshaft Position Sensor "A" Circuit:
Signal check. Edge counter of camshaft position sensor 8.
Possible Causes:
• Poor connection
• Open or short in signal circuit
• CKPS
• Faulty ECM/PCM</td></tr>
<tr><td>DTC: P033C
2T PCM, MIL: Yes
Year: 2009
Model: Borrego
Engine: 4.6L V8 VIN 2
Transmission: All</td><td>Min error of DFP_KS4: internal failure path number: knock sensor 4:
Enable Conditions
• ECT > 40C
• Engine speed > 2400rpm
• Engine load > 35%
Threshold value
• Normalized reference level > upper limit, f(rpm)
Diagnosis Time
• Continuous
Possible Causes:
• Poor connection
• Open or short to ground in signal circuit
• Knock sensor</td></tr>
</table>

DTC	Trouble Code Title, Conditions & Possible Causes
DTC: P033D **2T PCM, MIL: Yes** **Year:** 2009 **Model:** Borrego **Engine:** 4.6L V8 VIN 2 **Transmission:** All	**Max error of DFP_KS4: internal failure path number: knock sensor 4:** Enable Conditions • ECT > 40C • Engine speed > 2400rpm • Engine load > 35% Threshold value • Normalized reference level > upper limit, f(rpm) Diagnosis Time • Continuous **Possible Causes:** • Poor connection • Open or short to ground in signal circuit • Knock sensor
DTC: P0340 **2T CCM, MIL: Yes** **Year:** 2009, 2010 **Model:** Amanti, Optima, Rio, Rio5, Spectra, Spectra5 **Engine:** 1.6L L4 VIN 3, 2.0L L4 VIN 1, 2.0L L4 VIN 2, 2.4L L4 VIN 8, 2.7L V6 VIN 4, 3.8L V6 VIN 5 **Transmission:** All	**Camshaft Position Sensor Circuit Malfunction:** Engine speed over 600 rpm, and the PCM detected less than one CMP sensor signal was present, condition met for 1.5 seconds. **Possible Causes:** • CMP sensor signal circuit is open or shorted to ground • CMP sensor signal circuit is shorted to VREF or system power • CMP sensor is damaged or has failed • PCM has failed
DTC: P0340 **2T PCM, MIL: Yes** **Year:** 2009, 2010 **Model:** Borrego, Sorento, Soul, Sportage **Engine:** 1.6L L4 VIN 1, 2.0L L4 VIN 2, 2.7L V6 VIN 3, 3.3L V6 VIN 5, 3.8L V6 VIN 1, 3.8L V6 VIN 6, 4.6L V8 VIN 2 **Transmission:** All	**Camshaft Position Sensor 'A' Circuit (Bank 1 Or Single Sensor):** Enable Conditions • Reference mark found • No CKP sensor error • No 1 tooth off error Threshold value • Abnormal phase edges (High or Low) (No signal counter > 2 times) Diagnosis Time • Continuous **Possible Causes:** • Poor connection • Open in harness • CMPS(B1-Intake) • ECM
DTC: P0341 **2T CCM, MIL: Yes** **Year:** 2009, 2010 **Model:** Amanti, Optima, Rio, Rio5 **Engine:** 1.6L L4 VIN 3, 2.4L L4 VIN 8, 2.7L V6 VIN 4, 3.8L V6 VIN 5 **Transmission:** All	**Camshaft Position Sensor Circuit Malfunction:** No signal or no signal switching is detected. Crankshaft sensor is normal. Battery voltage is between 10 and 16 volts. **Possible Causes:** • Open or short in CMPS circuit • Faulty CMPS • Faulty PCM
DTC: P0341 **2T PCM, MIL: Yes** **Year:** 2009, 2010 **Model:** Borrego, Sorento, Soul **Engine:** 1.6L L4 VIN 1, 3.3L V6 VIN 5, 3.8L V6 VIN 1, 3.8L V6 VIN 6, 4.6L V8 VIN 2 **Transmission:** All	**Camshaft Position Sensor 'A' Circuit Range/Performance (Bank 1 Or Single Sensor):** Enable Conditions • Reference mark found • No CKP sensor error • No 1 tooth off error Threshold value • Implausible signal counter > 10 times Diagnosis Time • Continuous **Possible Causes:** • Poor connection • Short in harness • electrical noise • Target wheel • CMPS(B1-Intake) • ECM

DTC	Trouble Code Title, Conditions & Possible Causes
DTC: P0342 **2T CCM, MIL: Yes** **Year:** 2009, 2010 **Model:** Rio, Rio5, Soul **Engine:** 1.6L L4 VIN 1, 1.6L L4 VIN 3 **Transmission:** All	**Camshaft Position Sensor Low Input:** Engine speed over 600 rpm, and the PCM detected an unexpected low voltage condition on the CMP sensor signal circuit. **Possible Causes:** • CMP sensor signal circuit is open or shorted to ground • CMP sensor ground circuit is open • CMP sensor is damaged or has failed • PCM has failed
DTC: P0343 **2T CCM, MIL: Yes** **Year:** 2009, 2010 **Model:** Rio, Rio5, Soul **Engine:** 1.6L L4 VIN 1, 1.6L L4 VIN 3 **Transmission:** All	**Camshaft Position Sensor High Input:** Engine cranking and the PCM detected the CMP sensor signal was above a threshold value stored in memory during the CCM test. **Possible Causes:** • CMP sensor signal circuit is shorted to VREF or system power • CMP sensor is damaged or has failed • PCM has failed
DTC: P0345 **2T PCM, MIL: Yes** **Year:** 2009 **Model:** Borrego **Engine:** 4.6L V8 VIN 2 **Transmission:** All	**Camshaft Position Sensor 'A' Circuit (Bank 2):** Enable Conditions • Reference mark found • No CKP sensor error • No 1 tooth off error Threshold value • Abnormal phase edges (High or Low) • No signal counter > 2 times Diagnosis Time • Continuous **Possible Causes:** • Poor connection • Open in harness • CMPS(B2-Intake) • ECM
DTC: P0346 **2T CCM, MIL: Yes** **Year:** 2009 **Model:** Amanti, Borrego, Sorento **Engine:** 3.3L V6 VIN 5, 3.8L V6 VIN 1, 3.8L V6 VIN 5, 3.8L V6 VIN 6, 4.6L V8 VIN 2 **Transmission:** All	**Camshaft Position Sensor "A" Circuit Range/Performance (Bank 2):** Engine running and the PCM detected the CMP sensor signal was above a threshold value stored in memory during the CCM test. **Possible Causes:** • Poor connection • Open or Short in harness • Electrical noise • Target wheel • CMPS • PCM has failed
DTC: P0350 **1T PCM** **Year:** 2009, 2010 **Model:** Sportage **Engine:** 2.7L V6 VIN 3 **Transmission:** All	**Ignition Coil Primary / Secondary Circuit:** Threshold Value • Coolant temperature >75°C(167°F) Threshold Value • Failure on 3 cylinder or more Diagnostic Time • 255 revolutions **Possible Causes:** • Open or short in power supply circuit • Open or short in control circuit • Contact resistance in connectors • Faulty ignition coil
DTC: P0351 **2T CCM, MIL: Yes** **Year:** 2009, 2010 **Model:** Amanti, Borrego, Sorento, Sportage **Engine:** 2.7L V6 VIN 3, 3.3L V6 VIN 5, 3.8L V6 VIN 1, 3.8L V6 VIN 5, 3.8L V6 VIN 6 **Transmission:** All	**Ignition Coil 'A' Circuit Malfunction:** Engine started, engine running, and the PCM detected an unexpected voltage condition on the Ignition Coil 'A' primary circuit. **Possible Causes:** • Ignition Coil 'A' primary circuit is open or shorted to ground • Ignition Coil 'A' power circuit is open (test power from I/P fuse) • Ignition Coil 'A' is damaged or has failed • PCM has failed

DTC	Trouble Code Title, Conditions & Possible Causes
DTC: P0352 **2T CCM, MIL: Yes** **Year:** 2009, 2010 **Model:** Amanti, Borrego, Sorento, Sportage **Engine:** 2.7L V6 VIN 3, 3.3L V6 VIN 5, 3.8L V6 VIN 1, 3.8L V6 VIN 5, 3.8L V6 VIN 6 **Transmission:** All	**Ignition Coil 'B' Circuit Malfunction:** Engine started, engine running, and the PCM detected an unexpected voltage condition on the Ignition Coil 'B' primary circuit. **Possible Causes:** • Ignition Coil 'B' primary circuit is open or shorted to ground • Ignition Coil 'B' power circuit is open (test power from I/P fuse) • Ignition Coil 'B' is damaged or has failed • PCM has failed
DTC: P0353 **2T CCM, MIL: Yes** **Year:** 2009, 2010 **Model:** Amanti, Borrego, Sorento, Sportage **Engine:** 2.7L V6 VIN 3, 3.3L V6 VIN 5, 3.8L V6 VIN 1, 3.8L V6 VIN 5, 3.8L V6 VIN 6 **Transmission:** All	**Ignition Coil 'C' Circuit Malfunction:** Engine started, engine running, and the PCM detected an unexpected voltage condition on the Ignition Coil 'C' primary circuit. **Possible Causes:** • Ignition Coil 'C' primary circuit is open or shorted to ground • Ignition Coil 'C' power circuit is open (test power from I/P fuse) • Ignition Coil 'C' is damaged or has failed • PCM has failed
DTC: P0354 **2T CCM, MIL: Yes** **Year:** 2009, 2010 **Model:** Amanti, Borrego, Sorento, Sportage **Engine:** 2.7L V6 VIN 3, 3.3L V6 VIN 5, 3.8L V6 VIN 1, 3.8L V6 VIN 5, 3.8L V6 VIN 6 **Transmission:** All	**Ignition Coil 'D' Circuit Malfunction:** Engine started, engine running, and the PCM detected an unexpected voltage condition on the Ignition Coil 'D' primary circuit. **Possible Causes:** • Ignition Coil 'D' primary circuit is open or shorted to ground • Ignition Coil 'D' power circuit is open (test power from I/P fuse) • Ignition Coil 'D' is damaged or has failed • PCM has failed
DTC: P0355 **2T CCM, MIL: Yes** **Year:** 2009, 2010 **Model:** Amanti, Borrego, Sorento, Sportage **Engine:** 2.7L V6 VIN 3, 3.3L V6 VIN 5, 3.8L V6 VIN 1, 3.8L V6 VIN 5, 3.8L V6 VIN 6 **Transmission:** All	**Ignition Coil 'E' Circuit Malfunction:** Engine started, engine running, and the PCM detected an unexpected voltage condition on the Ignition Coil 'E' primary circuit. **Possible Causes:** • Ignition Coil 'E' primary circuit is open or shorted to ground • Ignition Coil 'E' power circuit is open (test power from I/P fuse) • Ignition Coil 'E' is damaged or has failed • PCM has failed
DTC: P0356 **2T CCM, MIL: Yes** **Year:** 2009, 2010 **Model:** Amanti, Borrego, Sorento, Sportage **Engine:** 2.7L V6 VIN 3, 3.3L V6 VIN 5, 3.8L V6 VIN 1, 3.8L V6 VIN 5, 3.8L V6 VIN 6 **Transmission:** All	**Ignition Coil 'F' Circuit Malfunction:** Engine started, engine running, and the PCM detected an unexpected voltage condition on the Ignition Coil 'F' primary circuit. **Possible Causes:** • Ignition Coil 'F' primary circuit is open or shorted to ground • Ignition Coil 'F' power circuit is open (test power from I/P fuse) • Ignition Coil 'F' is damaged or has failed • PCM has failed
DTC: P0365 **2T PCM, MIL: Yes** **Year:** 2009 **Model:** Borrego **Engine:** 4.6L V8 VIN 2 **Transmission:** All	**Camshaft Position Sensor B Circuit Malfunction (Bank 1):** Enable Conditions • Reference mark found • No CKP sensor error • No 1 tooth off error Threshold value • Abnormal phase edges (High or Low) (No signal counter > 2 times) Diagnosis Time • Continuous **Possible Causes:** • Poor connection • Open in harness • CMPS(B1-Exhaust) • ECM

DTC	Trouble Code Title, Conditions & Possible Causes
DTC: P0366 **2T PCM, MIL: Yes** **Year:** 2009 **Model:** Borrego **Engine:** 3.8L V6 VIN 1, 4.6L V8 VIN 2 **Transmission:** All	**Camshaft Position Sensor 'B' Circuit Range/Performance (Bank 1):** Enable Conditions • Reference mark found • No CKP sensor error • No 1 tooth off error Threshold value • Implausible signal counter > 10 times Diagnosis Time • Continuous **Possible Causes:** • Poor connection • Short in harness • electrical noise • Target wheel • CMPS(B1-Exhaust) • ECM
DTC: P0390 **2T PCM, MIL: Yes** **Year:** 2009 **Model:** Borrego **Engine:** 4.6L V8 VIN 2 **Transmission:** All	**Camshaft Position Sensor 'B' Circuit (Bank 2):** Enable Conditions • Reference mark found • No CKP sensor error • No 1 tooth off error Threshold value • Abnormal phase edges (High or Low) (No signal counter > 2 times) Diagnosis Time • Continuous **Possible Causes:** • Poor connection • Open in harness • CMPS(B2-Exhaust) • ECM
DTC: P0391 **2T PCM, MIL: Yes** **Year:** 2009 **Model:** Borrego **Engine:** 3.8L V6 VIN 1, 4.6L V8 VIN 2 **Transmission:** All	**Camshaft Position Sensor 'B' Circuit Range/Performance (Bank 2):** Enable Conditions • Reference mark found • No CKP sensor error • No 1 tooth off error Threshold value • Implausible signal counter > 10 times Diagnosis Time • Continuous **Possible Causes:** • Poor connection • Short in harness • electrical noise • Target wheel • CMPS(B2-Exhaust) • ECM
DTC: P0420 **2T CAT1, MIL: Yes** **Year:** 2009, 2010 **Model:** Amanti, Borrego, Rio, Rio5, Sorento, Soul, Spectra, Spectra5, Sportage **Engine:** 1.6L L4 VIN 1, 1.6L L4 VIN 3, 2.0L L4 VIN 1, 2.0L L4 VIN 2, 2.7L V6 VIN 3, 3.3L V6 VIN 5, 3.8L V6 VIN 1, 3.8L V6 VIN 5, 3.8L V6 VIN 6, 4.6L V8 VIN 2 **Transmission:** All	**Catalyst Efficiency Below Normal (Bank 1):** DTC P0130, P0133, P0134, P0135, P0136, P0139, P0140 and P0141 not set, engine started, engine running in closed loop at a speed of 45-60 mph for 2-3 minutes, and the PCM detected that the rear HO2S and front HO2S voltage amplitudes were too similar. **Possible Causes:** • Air leaks at the exhaust manifold or in the exhaust pipes • Catalytic converter is damaged or has failed • Front HO2S or rear HO2S is contaminated with fuel or moisture • Front HO2S or rear HO2S is contaminated with fuel or moisture

DTC	Trouble Code Title, Conditions & Possible Causes
DTC: P0430 **2T CAT1, MIL: Yes** **Year:** 2009, 2010 **Model:** Amanti, Borrego, Sorento, Sportage **Engine:** 2.7L V6 VIN 3, 3.3L V6 VIN 5, 3.8L V6 VIN 1, 3.8L V6 VIN 5, 3.8L V6 VIN 6, 4.6L V8 VIN 2 **Transmission:** All	**Catalyst Efficiency Below Normal (Bank 2):** DTC P0150, P0153, P0154, P0155, P0156, P0160 and P0161 not set, engine started, engine running in closed loop at 45-60 mph for 2-3 minutes, and the PCM detected the rear HO2S and front HO2S voltage amplitudes were too similar during the Catalyst Monitor test. **Possible Causes:** • Air leaks at the exhaust manifold or in the exhaust pipes • Catalytic converter is damaged or has failed • Front HO2S or rear HO2S is contaminated with fuel or moisture • Front HO2S or rear HO2S heater is damaged or has failed • Front HO2S or rear HO2S is contaminated with fuel or moisture
DTC: P0441 **2T EVAP, MIL: Yes** **Year:** 2009, 2010 **Model:** Amanti, Borrego, Sorento, Soul, Spectra, Spectra5, Sportage **Engine:** 2.0L L4 VIN 1, 2.0L L4 VIN 2, 2.7L V6 VIN 3, 3.3L V6 VIN 5, 3.8L V6 VIN 1, 3.8L V6 VIN 5, 3.8L V6 VIN 6 **Transmission:** All	**EVAP System Malfunction:** ECT sensor signal less than158°F at startup, IAT sensor signal more than 9.05°F, system voltage more than 10.9 volts engine runtime 15-20 minutes at cruise speed, then returned to idle speed, VSS indicating 0 mph, load value 2.2 ms, canister load factor less than 4.0, fuel tank pressure less than 0.5" Hg, then after the Idle Control system and Fuel Trim stabilized, the PCM detected a continuous purge condition in the EVAP system during the Purge flow test. **Possible Causes:** • Small hoses or cuts present in the EVAP vapor hoses/lines • EVAP canister purge solenoid is damaged or is stuck open • PCM has failed
DTC: P0442 **2T EVAP, MIL: Yes** **Year:** 2009, 2010 **Model:** Amanti, Borrego, Rio, Rio5, Sorento, Soul, Spectra, Spectra5, Sportage **Engine:** 1.6L L4 VIN 1, 1.6L L4 VIN 3, 2.0L L4 VIN 1, 2.0L L4 VIN 2, 2.7L V6 VIN 3, 3.3L V6 VIN 5, 3.8L V6 VIN 1, 3.8L V6 VIN 5, 3.8L V6 VIN 6, 4.6L V8 VIN 2 **Transmission:** All	**EVAP System Small Leak (0.040") Detected:** ECT sensor signal less than158°F at startup, IAT sensor signal more than 9.05°F, system voltage more than 10.9 volts engine runtime 15-20 minutes at cruise speed, then returned to idle speed, VSS indicating 0 mph, load value 2.2 ms, canister load factor less than 4.0, fuel tank pressure less than 0.5" Hg, then after the Idle Control system and Fuel Trim had stabilized, the PCM detected a fuel vapor leak (as small as 0.040") in the EVAP system during the EVAP Leak Test. **Possible Causes:** • Fuel filler cap damaged, cross-threaded or loosely installed • Small leaks or cuts present in the EVAP vapor hoses/lines • EVAP purge valve is damaged or has failed • PCM has failed
DTC: P0444 **2T CCM, MIL: Yes** **Year:** 2009, 2010 **Model:** Amanti, Borrego, Rio, Rio5, Sorento, Soul, Spectra, Spectra5, Sportage **Engine:** 1.6L L4 VIN 1, 1.6L L4 VIN 3, 2.0L L4 VIN 1, 2.0L L4 VIN 2, 2.7L V6 VIN 3, 3.3L V6 VIN 5, 3.8L V6 VIN 1, 3.8L V6 VIN 5, 3.8L V6 VIN 6, 4.6L V8 VIN 2 **Transmission:** All	**EVAP Emission System- Purge Control Valve Circuit Open:** Engine running. Checking output signals from PCSV every 10 seconds, under detecting condition. **Possible Causes:** • Poor connection • Open or short to ground in harness • PCVS • PCM
DTC: P0445 **2T CCM, MIL: Yes** **Year:** 2009, 2010 **Model:** Amanti, Borrego, Sorento, Soul, Spectra, Spectra5, Sportage **Engine:** 1.6L L4 VIN 1, 2.0L L4 VIN 1, 2.0L L4 VIN 2, 2.7L V6 VIN 3, 3.3L V6 VIN 5, 3.8L V6 VIN 1, 3.8L V6 VIN 5, 3.8L V6 VIN 6 **Transmission:** All	**EVAP Emission System- Purge Control Valve Circuit Shorted:** Engine running. Checking output signals from PCSV every 10 seconds, under detecting condition. **Possible Causes:** • Poor connection • Short to battery in harness • PCVS • PCM
DTC: P0446 **2T CCM, MIL: Yes** **Year:** 2009, 2010 **Model:** Rio, Rio5, Soul **Engine:** 1.6L L4 VIN 1, 1.6L L4 VIN 3 **Transmission:** All	**EVAP Emission System- Vent Control Circuit:** CCV stuck open. Time after engine start greater than 600 seconds. Idle speed controller activated. Coolant temperature less than 12 degrees F. **Possible Causes:** • Poor connection • CCV • ECM/PCM

<table>
<tr><th>DTC</th><th>Trouble Code Title, Conditions & Possible Causes</th></tr>
<tr><td>DTC: P0447
2T CCM, MIL: Yes
Year: 2009, 2010
Model: Amanti, Borrego, Sorento, Soul, Spectra, Spectra5, Sportage
Engine: 2.0L L4 VIN 1, 2.0L L4 VIN 2, 2.7L V6 VIN 3, 3.3L V6 VIN 5, 3.8L V6 VIN 1, 3.8L V6 VIN 5, 3.8L V6 VIN 6, 4.6L V8 VIN 2
Transmission: All</td><td>EVAP Emission System- Vent Control Circuit Open:
Detects a short to ground or open circuit on vent valve output circuit. No disabling faults present. Engine running.
Possible Causes:
• Poor connection
• Open or short in power circuit
• Open or short in control circuit
• CCV
• ECM/PCM</td></tr>
<tr><td>DTC: P0448
2T CCM, MIL: Yes
Year: 2009, 2010
Model: Amanti, Borrego, Sorento, Soul, Spectra, Spectra5, Sportage
Engine: 2.0L L4 VIN 1, 2.0L L4 VIN 2, 2.7L V6 VIN 3, 3.3L V6 VIN 5, 3.8L V6 VIN 1, 3.8L V6 VIN 5, 3.8L V6 VIN 6, 4.6L V8 VIN 2
Transmission: All</td><td>EVAP Emission System- Vent Control Circuit Shorted:
Detects a short to battery on vent valve output circuit. No disabling faults present. Engine running.
Possible Causes:
• Poor connection
• Short to battery in CCV circuit
• CCV
• ECM/PCM</td></tr>
<tr><td>DTC: P0449
2T CCM, MIL: Yes
Year: 2009, 2010
Model: Borrego, Rio, Rio5, Soul, Spectra, Spectra5, Sportage
Engine: 1.6L L4 VIN 1, 1.6L L4 VIN 3, 2.0L L4 VIN 1, 2.0L L4 VIN 2, 2.7L V6 VIN 3, 4.6L V8 VIN 2
Transmission: All</td><td>EVAP Emission System- Vent Valve/Solenoid Circuit:
Circuit continuity check open.
Possible Causes:
• Poor connection
• Open or short to ground in power circuit
• CCV
• ECM/PCM</td></tr>
<tr><td>DTC: P0450
2T CCM, MIL: Yes
Year: 2009, 2010
Model: Borrego, Rio, Rio5, Soul
Engine: 1.6L L4 VIN 1, 1.6L L4 VIN 3, 4.6L V8 VIN 2
Transmission: All</td><td>EVAP Pressure Sensor Circuit Malfunction:
Key on or engine running and the PCM detected the Fuel Tank Pressure sensor signal was more than 4.9 volts or less than 0.14 volt.
Possible Causes:
• FTP sensor signal circuit open or shorted to ground
• FTP sensor signal circuit shorted to VREF or system power
• FTP sensor is damaged or has failed
• PCM has failed</td></tr>
<tr><td>DTC: P0451
2T CCM, MIL: Yes
Year: 2009, 2010
Model: Amanti, Borrego, Rio, Rio5, Sorento, Soul, Spectra, Spectra5, Sportage
Engine: 1.6L L4 VIN 1, 1.6L L4 VIN 3, 2.0L L4 VIN 1, 2.0L L4 VIN 2, 2.7L V6 VIN 3, 3.3L V6 VIN 5, 3.8L V6 VIN 1, 3.8L V6 VIN 5, 3.8L V6 VIN 6, 4.6L V8 VIN 2
Transmission: All</td><td>EVAP Pressure Sensor Performance:
Engine at idle speed with the vehicle speed indicating 0 mph, then with the EVAP Vent Control solenoid commanded “on”, the PCM detected the FTP sensor signal variation was less than 15 mv.
Possible Causes:
• FTP sensor signal or ground circuit has high resistance
• FTP sensor is damaged or out of calibration
• EVAP canister close valve (CCV) is stuck closed
• PCM has failed</td></tr>
<tr><td>DTC: P0452
2T CCM, MIL: Yes
Year: 2009, 2010
Model: Amanti, Borrego, Rio, Rio5, Sorento, Soul, Spectra, Spectra5, Sportage
Engine: 1.6L L4 VIN 1, 1.6L L4 VIN 3, 2.0L L4 VIN 1, 2.0L L4 VIN 2, 2.7L V6 VIN 3, 3.3L V6 VIN 5, 3.8L V6 VIN 1, 3.8L V6 VIN 5, 3.8L V6 VIN 6, 4.6L V8 VIN 2
Transmission: All</td><td>EVAP Pressure Sensor Circuit Low Input:
Key on or engine running and the PCM detected the Fuel Tank Pressure (FTP) sensor signal was less than 0.14 volt during the test.
Possible Causes:
• FTP sensor signal circuit is shorted to ground
• FTP sensor is damaged or has failed
• PCM has failed</td></tr>
</table>

DTC	Trouble Code Title, Conditions & Possible Causes
DTC: P0453 **2T CCM, MIL: Yes** **Year:** 2009, 2010 **Model:** Amanti, Borrego, Rio, Rio5, Sorento, Soul, Spectra, Spectra5, Sportage **Engine:** 1.6L L4 VIN 1, 1.6L L4 VIN 3, 2.0L L4 VIN 1, 2.0L L4 VIN 2, 2.7L V6 VIN 3, 3.3L V6 VIN 5, 3.8L V6 VIN 1, 3.8L V6 VIN 5, 3.8L V6 VIN 6, 4.6L V8 VIN 2 **Transmission:** All	**EVAP Pressure Sensor Circuit High Input:** Key on or engine running and the PCM detected the Fuel Tank Pressure (FTP) sensor signal was more than 4.90 volts during the test. **Possible Causes:** • FTP sensor signal circuit is open • FTP sensor ground circuit is open • FTP sensor is damaged or has failed • PCM has failed
DTC: P0454 **2T CCM, MIL: Yes** **Year:** 2009, 2010 **Model:** Amanti, Borrego, Sorento, Soul, Spectra, Spectra5, Sportage **Engine:** 2.0L L4 VIN 1, 2.0L L4 VIN 2, 2.7L V6 VIN 3, 3.3L V6 VIN 5, 3.8L V6 VIN 1, 3.8L V6 VIN 5, 3.8L V6 VIN 6 **Transmission:** All	**EVAP Emission System- Pressure Sensor Intermittent:** The PCM measures pressure stability in the fuel tank, by means of a sensor for a predetermined duration. If fluctuation is larger than predetermined threshold a DTC is set. Sensor signal noise check. **Possible Causes:** • Contact resistance in connectors • Faulty FTPS • Faulty ECM/PCM • Faulty FTPS
DTC: P0455 **2T EVAP, MIL: Yes** **Year:** 2009, 2010 **Model:** Amanti, Borrego, Rio, Rio5, Sorento, Soul, Spectra, Spectra5, Sportage **Engine:** 1.6L L4 VIN 1, 1.6L L4 VIN 3, 2.0L L4 VIN 1, 2.0L L4 VIN 2, 2.7L V6 VIN 3, 3.3L V6 VIN 5, 3.8L V6 VIN 1, 3.8L V6 VIN 5, 3.8L V6 VIN 6, 4.6L V8 VIN 2 **Transmission:** All	**EVAP System Large Leak Detected:** ECT input at 38-95°F at startup, engine running at a steady throttle at over 6.2 mph for over 2 minutes, and the PCM detected the FTP sensor signal indicated more than -15 kPa in the EVAP Leak test. **Possible Causes:** • Fuel filler cap damaged, cross-threaded or loosely installed • Small leaks or cuts present in the EVAP vapor hoses/lines • EVAP purge valve is damaged or has failed • PCM has failed
DTC: P0456 **2T EVAP, MIL: Yes** **Year:** 2009, 2010 **Model:** Amanti, Borrego, Rio, Rio5, Sorento, Soul, Spectra, Spectra5, Sportage **Engine:** 1.6L L4 VIN 1, 1.6L L4 VIN 3, 2.0L L4 VIN 1, 2.0L L4 VIN 2, 2.7L V6 VIN 3, 3.3L V6 VIN 5, 3.8L V6 VIN 1, 3.8L V6 VIN 5, 3.8L V6 VIN 6, 4.6L V8 VIN 2 **Transmission:** All	**EVAP System Very Small Leak (0.020") Detected:** ECT input at 38-95°F at startup, vehicle driven at a steady speed of over 6.2 mph for 2 minutes, and then the PCM detected a very small leak (less than 0.020") in the EVAP system during the Leak test. **Possible Causes:** • Fuel filler is damaged, cross-threaded, loose or missing • Fuel filler pipe is damaged, or a fuel vapor hose is leaking • Rollover valve or ORVR (valve) had failed allowing fuel in lines • Canister close valve clogged or stuck in open or closed position • Purge solenoid valve is damaged or installed improperly • FTP sensor is damaged or has failed • Leaks in the charcoal canister, or at the fuel tank seals
DTC: P0457 **2T CCM, MIL: Yes** **Year:** 2009, 2010 **Model:** Rio, Rio5 **Engine:** 1.6L L4 VIN 3 **Transmission:** All	**Evaporative Emission System- Leak Detected (Tank Cap Loose/Off):** Large leak caused by fuel cap loosened. **Possible Causes:** • Fuel cap • Faulty ECM/PCM
DTC: P0458 **2T CCM, MIL: Yes** **Year:** 2009, 2010 **Model:** Borrego, Rio, Rio5, Soul **Engine:** 1.6L L4 VIN 1, 1.6L L4 VIN 3, 4.6L V8 VIN 2 **Transmission:** All	**Evaporative Emission System Purge Control valve Circuit Low:** Circuit continuity check, low. **Possible Causes:** • Poor connection • Short to ground in control circuit • PCSV • Faulty ECM/PCM

DTC	Trouble Code Title, Conditions & Possible Causes
DTC: P0459 **2T CCM, MIL: Yes** **Year:** 2009, 2010 **Model:** Borrego, Rio, Rio5, Soul **Engine:** 1.6L L4 VIN 1, 1.6L L4 VIN 3, 3.8L V6 VIN 1, 4.6L V8 VIN 2 **Transmission:** All	**Evaporative Emission System Purge Control valve Circuit High:** Circuit continuity check, high. **Possible Causes:** • Poor connection • Short to power in control circuit • PCSV • Faulty ECM/PCM
DTC: P0461 **2T CCM, MIL: Yes** **Year:** 2009, 2010 **Model:** Amanti, Borrego, Optima, Rio, Rio5, Sorento, Soul, Sportage **Engine:** 1.6L L4 VIN 1, 1.6L L4 VIN 3, 2.4L L4 VIN 8, 2.7L V6 VIN 3, 2.7L V6 VIN 4, 3.3L V6 VIN 5, 3.8L V6 VIN 1, 3.8L V6 VIN 5, 3.8L V6 VIN 6, 4.6L V8 VIN 2 **Transmission:** All	**Fuel Level Sensor "A" Circuit Range/Performance:** Filtered and unfiltered signal of fuel sensor are monitored. **Possible Causes:** • Poor connection • Faulty fuel level sensor • Faulty PCM
DTC: P0462 **2T CCM, MIL: Yes** **Year:** 2009, 2010 **Model:** Amanti, Borrego, Optima, Rio, Rio5, Sorento, Soul, Sportage **Engine:** 1.6L L4 VIN 1, 1.6L L4 VIN 3, 2.4L L4 VIN 8, 2.7L V6 VIN 3, 2.7L V6 VIN 4, 3.3L V6 VIN 5, 3.8L V6 VIN 1, 3.8L V6 VIN 5, 3.8L V6 VIN 6, 4.6L V8 VIN 2 **Transmission:** All	**Fuel Level Sensor Input Low (Sticking):** Key on or engine running system voltage more than 10 volts, and the PCM detected the fuel level sensing unit signal was less than 0.2 volt. **Possible Causes:** • Fuel level sending unit signal circuit shorted to VREF • Fuel level sending unit signal circuit shorted to system power • Fuel level sensing unit is damaged or the fuel tank is damaged • BCM or PCM has failed
DTC: P0463 **2T CCM, MIL: Yes** **Year:** 2009, 2010 **Model:** Amanti, Borrego, Optima, Rio, Rio5, Sorento, Sportage **Engine:** 1.6L L4 VIN 3, 2.4L L4 VIN 8, 2.7L V6 VIN 3, 2.7L V6 VIN 4, 3.3L V6 VIN 5, 3.8L V6 VIN 1, 3.8L V6 VIN 5, 3.8L V6 VIN 6, 4.6L V8 VIN 2 **Transmission:** All	**Fuel Level Sensor Input High (Sticking):** Key on or engine running system voltage more than 10 volts, and the PCM detected the fuel level sensing unit signal was more than 4.5 volts. **Possible Causes:** • Fuel level sending unit signal circuit shorted to VREF • Fuel level sending unit signal circuit shorted to system power • Fuel level sensing unit is damaged or the fuel tank is damaged • BCM or PCM has failed
DTC: P0464 **2T CCM** **Year:** 2009, 2010 **Model:** Amanti, Borrego, Optima, Sorento, Sportage **Engine:** 2.4L L4 VIN 8, 2.7L V6 VIN 3, 2.7L V6 VIN 4, 3.3L V6 VIN 5, 3.8L V6 VIN 1, 3.8L V6 VIN 5, 3.8L V6 VIN 6 **Transmission:** All	**Fuel Level Sensor "A" Circuit Intermittent:** Check signal for fluctuation. The ECM sets the DTC if the fuel level signal is higher than the threshold value (signal fluctuation greater than 50 percent). **Possible Causes:** • Contact resistance in connectors • Short to battery in fuel level (FLS) circuit • Faulty ECM
DTC: P0480 **2T CCM, MIL: Yes** **Year:** 2009 **Model:** Amanti, Borrego, Sorento **Engine:** 3.3L V6 VIN 5, 3.8L V6 VIN 1, 3.8L V6 VIN 5, 3.8L V6 VIN 6 **Transmission:** All	**Fan 1 Control Circuit Malfunction:** This will detect a short to ground, to battery or open circuit of fan relay output. Fault information provided by an output driver chip. No disabling faults present. Engine running. Enable time delay equal or greater than 0.5 seconds. **Possible Causes:** • Poor connection • Open in power circuit to cooling fan • Open or short in control circuit to PCM • Faulty fan relay • Faulty cooling fan module • Faulty PCM

DTC	Trouble Code Title, Conditions & Possible Causes
DTC: P0481 **2T CCM, MIL: Yes** **Year:** 2009 **Model:** Borrego, Sorento **Engine:** 3.3L V6 VIN 5, 3.8L V6 VIN 1, 3.8L V6 VIN 6 **Transmission:** All	**Fan 2 Control Circuit Malfunction:** This will detect a short to ground, to battery or open circuit of fan relay output. Fault information provided by an output driver chip. No disabling faults present. Engine running. Enable time delay equal or greater than 0.5 seconds. **Possible Causes:** • Poor connection • Open in power circuit to cooling fan • Open or short in control circuit to PCM • Faulty fan relay
DTC: P0496 **2T CCM, MIL: Yes** **Year:** 2009, 2010 **Model:** Borrego, Rio, Rio5, Soul **Engine:** 1.6L L4 VIN 1, 1.6L L4 VIN 3, 4.6L V8 VIN 2 **Transmission:** All	**Evaporative Emission System High Purge Flow:** Fuel tank pressure behavior (canister purge valve stuck). Time after engine start 600 seconds. Idle speed controller activated. Mixture adaptation activated. Coolant temperature at start 11.88 degrees F. Tank ventilation must be active for 10 seconds. **Possible Causes:** • Leakage at the fuel evaporative system • PCSV • Faulty ECM/PCM
DTC: P0497 **2T CCM, MIL: Yes** **Year:** 2009, 2010 **Model:** Borrego, Rio, Rio5, Soul **Engine:** 1.6L L4 VIN 1, 1.6L L4 VIN 3, 4.6L V8 VIN 2 **Transmission:** All	**Evaporative Emission System Low Purge Flow:** Fuel tank pressure behavior (canister purge valve stuck). Time after engine start 600 seconds. Idle speed controller activated. Mixture adaptation activated. Coolant temperature at start 11.88 degrees F. Tank ventilation must be active for 10 seconds. **Possible Causes:** • Clog in the fuel evaporative system • PCSV • Faulty ECM/PCM
DTC: P0498 **2T CCM, MIL: Yes** **Year:** 2009, 2010 **Model:** Rio, Rio5, Soul **Engine:** 1.6L L4 VIN 1, 1.6L L4 VIN 3 **Transmission:** All	**Evaporative Emission System Vent Valve Control Circuit Low:** Circuit continuity check, low. **Possible Causes:** • Poor connection • Short to ground in control circuit • CCV • Faulty ECM/PCM
DTC: P0499 **2T CCM, MIL: Yes** **Year:** 2009, 2010 **Model:** Rio, Rio5, Soul **Engine:** 1.6L L4 VIN 1, 1.6L L4 VIN 3 **Transmission:** All	**Evaporative Emission System Vent Valve Control Circuit High:** Circuit continuity check, high. **Possible Causes:** • Poor connection • Short to power in control circuit • CCV • Faulty ECM/PCM
DTC: P0501 **2T CCM, MIL: Yes** **Year:** 2009, 2010 **Model:** Amanti, Rio, Rio5, Spectra, Spectra5 **Engine:** 1.6L L4 VIN 3, 2.0L L4 VIN 1, 2.0L L4 VIN 2, 3.8L V6 VIN 5 **Transmission:** All	**Vehicle Speed Sensor Performance:** Engine running in gear at high speed and lover for over 1 second, and the PCM did not detect any VSS signals. **Possible Causes:** • VSS signal circuit is open or shorted to ground • VSS signal circuit is shorted to VREF or to system power (B+) • VSS is damaged or has failed • PCM has failed
DTC: P0501 **2T CCM, MIL: Yes** **Year:** 2009, 2010 **Model:** Borrego, Sorento, Soul, Sportage **Engine:** 2.0L L4 VIN 2, 2.7L V6 VIN 3, 3.3L V6 VIN 5, 3.8L V6 VIN 1, 3.8L V6 VIN 6, 4.6L V8 VIN 2 **Transmission:** All	**Vehicle Speed Sensor Range/Performance:** Engine running in gear at high speed and for over 1 second, and the PCM did not detect any VSS signals. **Possible Causes:** • VSS signal circuit is open or shorted to ground • VSS signal circuit is shorted to VREF or to system power (B+) • VSS is damaged or has failed • PCM has failed

DTC	Trouble Code Title, Conditions & Possible Causes
DTC: P0502 **2T PCM, MIL: Yes** **Year:** 2009 **Model:** Borrego **Engine:** 4.6L V8 VIN 2 **Transmission:** All	**Vehicle Speed Sensor 'A' Circuit Low Input:** Enable Conditions • Vehicle speed > 0mph Threshold Value • Old and New value are same for 10sec Diagnostic Time • 10sec **Possible Causes:** • Poor connection • Open or Short in harness • ESC(ESP) control unit • ECM
DTC: P0503 **2T PCM, MIL: Yes** **Year:** 2009 **Model:** Borrego **Engine:** 4.6L V8 VIN 2 **Transmission:** All	**Vehicle Speed Sensor 'A' Intermittent/Erratic/High:** Enable Conditions • Vehicle speed raw signal > 168mph **Possible Causes:** • Poor connection • Short in harness • ESC(ESP) control unit • ECM
DTC: P0504 **2T CCM, MIL: Yes** **Year:** 2009, 2010 **Model:** Amanti, Borrego, Optima, Sorento **Engine:** 2.4L L4 VIN 8, 2.7L V6 VIN 4, 3.3L V6 VIN 5, 3.8L V6 VIN 1, 3.8L V6 VIN 5, 3.8L V6 VIN 6, 4.6L V8 VIN 2 **Transmission:** All	**Brake Switch "A"/"B" Correlation (2):** Plausibility check between brake light switch and brake test switch. Engine running. Time between brake light switch and brake test switch do not correlate longer than 10 seconds. **Possible Causes:** • Open or short circuit in harness • Poor connection or damaged harness • Faulty brake warning lamp or brake test switch
DTC: P0505 **2T CCM, MIL: Yes** **Year:** 2009, 2010 **Model:** Rio, Rio5, Soul, Sportage **Engine:** 1.6L L4 VIN 1, 1.6L L4 VIN 3, 2.0L L4 VIN 2 **Transmission:** All	**Idle Speed System Malfunction:** Engine running at idle speed while in closed loop, and the PCM detected the Actual idle speed more than 100-200 rpm above or below the Target idle speed during the test. **Possible Causes:** • High resistance between the main relay and IAC valve • High resistance between PCM and IAC valve control circuits • IAC valve is damaged or has failed • The throttle plate is carbon fouled (it may need to be cleaned)
DTC: P0506 **2T CCM, MIL: Yes** **Year:** 2009, 2010 **Model:** Amanti, Borrego, Optima, Rio, Rio5, Sorento, Soul, Spectra, Spectra5, Sportage **Engine:** 1.6L L4 VIN 1, 1.6L L4 VIN 3, 2.0L L4 VIN 1, 2.0L L4 VIN 2, 2.4L L4 VIN 8, 2.7L V6 VIN 3, 2.7L V6 VIN 4, 3.3L V6 VIN 5, 3.8L V6 VIN 1, 3.8L V6 VIN 5, 3.8L V6 VIN 6, 4.6L V8 VIN 2 **Transmission:** All	**Idle Speed Lower Than Expected:** Engine running at idle speed while in closed loop, and the PCM detected the Actual idle speed was more than 100 rpm below the Target idle speed during the CCM test. **Possible Causes:** • High resistance between the main relay and IAC valve • High resistance between PCM and IAC valve control circuits • IAC valve is damaged or has failed • The throttle plate is carbon fouled (it may need to be cleaned)

DTC	Trouble Code Title, Conditions & Possible Causes
DTC: P0507 **2T CCM, MIL: Yes** **Year:** 2009, 2010 **Model:** Amanti, Borrego, Optima, Rio, Rio5, Sorento, Soul, Spectra, Spectra5, Sportage **Engine:** 1.6L L4 VIN 1, 1.6L L4 VIN 3, 2.0L L4 VIN 1, 2.0L L4 VIN 2, 2.4L L4 VIN 8, 2.7L V6 VIN 3, 2.7L V6 VIN 4, 3.3L V6 VIN 5, 3.8L V6 VIN 1, 3.8L V6 VIN 5, 3.8L V6 VIN 6, 4.6L V8 VIN 2 **Transmission:** All	**Idle Speed Higher Than Expected:** Engine running at idle speed while in closed loop, and the PCM detected the Actual idle speed was more than 200 rpm above the Target idle speed during the CCM test. **Possible Causes:** • High resistance between the main relay and IAC valve • High resistance between PCM and IAC valve control circuits • IAC valve is damaged or has failed • Intake air leak located below the throttle plate assembly
DTC: P050B **2T PCM, MIL: Yes** **Year:** 2009 **Model:** Amanti, Borrego, Sorento **Engine:** 3.3L V6 VIN 5, 3.8L V6 VIN 1, 3.8L V6 VIN 5, 3.8L V6 VIN 6 **Transmission:** All	**Cold Start Ignition Timing Performance:** Enable Conditions • After engine overnight soaking • Vehicle is not rapidly accelerating or decelerating. • 11V< Battery Voltage < 16V • Engine is running. • NO DTC related to CKPS, Ignition coil, and Misfire Threshold value • The actual spark timing > the commanded spark timing + 15° or < the commanded spark timing -15° Diagnosis Time • Within 1 minute after cold-starting **Possible Causes:** • Faulty Ignition Coil • Faulty PCM
DTC: P0532 **2T CCM** **Year:** 2009, 2010 **Model:** Amanti, Borrego, Rio, Rio5, Sorento, Soul **Engine:** 1.6L L4 VIN 1, 1.6L L4 VIN 3, 3.3L V6 VIN 5, 3.8L V6 VIN 1, 3.8L V6 VIN 5, 3.8L V6 VIN 6, 4.6L V8 VIN 2 **Transmission:** All	**A/C Refrigerant Pressure Sensor "A" Circuit Low Input:** Detects sensor signal short to low voltage. Engine works. Sensor output 0.05 volt. **Possible Causes:** • Poor connection • Open in power circuit • Open or short to ground in signal circuit • Faulty A/C pressure sensor • Faulty PCM
DTC: P0533 **2T CCM** **Year:** 2009, 2010 **Model:** Amanti, Borrego, Rio, Rio5, Sorento, Soul **Engine:** 1.6L L4 VIN 1, 1.6L L4 VIN 3, 3.3L V6 VIN 5, 3.8L V6 VIN 1, 3.8L V6 VIN 5, 3.8L V6 VIN 6, 4.6L V8 VIN 2 **Transmission:** All	**A/C Refrigerant Pressure Sensor "A" Circuit High Input:** Detects sensor signal short to high voltage. Engine works. Sensor output 4.65 volts. **Possible Causes:** • Poor connection • Open in signal circuit open • Open in ground circuit • Faulty A/C pressure sensor • Faulty PCM
DTC: P0551 **2T CCM, MIL: Yes** **Year:** 2009, 2010 **Model:** Optima, Sportage **Engine:** 2.4L L4 VIN 8, 2.7L V6 VIN 3, 2.7L V6 VIN 4 **Transmission:** All	**Power Steering Pressure Sensor/Switch Circuit Range/Performance:** If a power steering switch signal is ON when the engine speed is more than 2500 rpm, load value is greater than 55 percent and engine coolant temperature is above 50 degrees F, the DTC will set. Signal of power steering pressure switch is monitored. **Possible Causes:** • Poor connection • Faulty power steering switch • Open or short in power steering switch • Faulty PCM

DTC	Trouble Code Title, Conditions & Possible Causes
DTC: P0552 **2T PCM** **Year:** 2009 **Model:** Borrego **Engine:** 4.6L V8 VIN 2 **Transmission:** All	**Power Steering Pressure Sensor/Switch Circuit Low input:** Enable Conditions • IG On Threshold value • Sensor output voltage < 0.234V Diagnosis Time • 0.3 sec **Possible Causes:** • Poor connection • Open in power circuit • Open or short to ground in signal circuit • Faulty P/S pressure sensor • Faulty ECM
DTC: P0552 **T PCM** **Year:** 2009 **Model:** Borrego **Engine:** 4.6L V8 VIN 2 **Transmission:** All	**Power Steering Pressure Sensor/Switch Circuit Low input:** Enable Conditions • IG On Threshold value • Sensor output voltage < 0.234V Diagnosis Time • 0.3 sec MIL On Condition • NO MIL **Possible Causes:** • Poor connection • Open in power circuit • Open or short to ground in signal circuit • Faulty P/S pressure sensor • Faulty ECM
DTC: P0553 **T PCM** **Year:** 2009 **Model:** Borrego **Engine:** 4.6L V8 VIN 2 **Transmission:** All	**Power Steering Pressure Sensor/Switch Circuit High input:** Enable Conditions • IG On Threshold value • Sensor output voltage < 0.234V Diagnosis Time • 0.3 sec MIL On Condition • NO MIL **Possible Causes:** • Poor connection • Open in power circuit • Open or short to ground in signal circuit • Faulty P/S pressure sensor • Faulty ECM
DTC: P0560 **2T CCM, MIL: Yes** **Year:** 2009, 2010 **Model:** Rio, Rio5, Soul, Spectra, Spectra5, Sportage **Engine:** 1.6L L4 VIN 1, 1.6L L4 VIN 3, 2.0L L4 VIN 1, 2.0L L4 VIN 2, 2.7L V6 VIN 3 **Transmission:** All	**Battery Backup Line Circuit Malfunction:** Engine runtime over 4 minutes and the PCM did not detect any system voltage on the Battery Backup circuit for 5 seconds. **Possible Causes:** • Battery backup circuit to the PCM is open • Battery backup fuse to the PCM is open or missing • Battery backup circuit to the PCM has high resistance • PCM has failed
DTC: P0561 **2T CCM** **Year:** 2009, 2010 **Model:** Rio, Rio5, Soul **Engine:** 1.6L L4 VIN 1, 1.6L L4 VIN 3 **Transmission:** All	**System Voltage Unstable:** Engine runtime over 4 minutes, and the PCM detected the system voltage rapidly changed its value by more than 3 volts. **NOTE: If the Battery Backup circuit is open, the vehicle will not run.** **Possible Causes:** • Charging system problem (charging voltage interrupted) • Backup voltage circuit to the PCM open (intermittent fault) • PCM has failed

DTC	Trouble Code Title, Conditions & Possible Causes
DTC: P0562 **2T CCM** **Year:** 2009, 2010 **Model:** Amanti, Borrego, Rio, Rio5, Sorento, Soul, Spectra, Spectra5, Sportage **Engine:** 1.6L L4 VIN 1, 1.6L L4 VIN 3, 2.0L L4 VIN 1, 2.0L L4 VIN 2, 2.7L V6 VIN 3, 3.3L V6 VIN 5, 3.8L V6 VIN 1, 3.8L V6 VIN 5, 3.8L V6 VIN 6, 4.6L V8 VIN 2 **Transmission:** All	**System Voltage Low Input:** Engine runtime over 4 minutes, and the PCM detected the system voltage was less than 8.0 volts, condition met for 5 seconds. **NOTE: If the Battery Backup circuit is open, the vehicle will not run.** **Possible Causes:** • Charging system problem (charging voltage too low) • Battery backup circuit to the PCM has high resistance • Backup voltage circuit to the PCM open (intermittent fault) • PCM has failed
DTC: P0563 **2T CCM** **Year:** 2009, 2010 **Model:** Amanti, Borrego, Rio, Rio5, Sorento, Soul, Spectra, Spectra5, Sportage **Engine:** 1.6L L4 VIN 1, 1.6L L4 VIN 3, 2.0L L4 VIN 1, 2.0L L4 VIN 2, 2.7L V6 VIN 3, 3.3L V6 VIN 5, 3.8L V6 VIN 1, 3.8L V6 VIN 5, 3.8L V6 VIN 6, 4.6L V8 VIN 2 **Transmission:** All	**System Voltage High Input:** Engine runtime over 4 minutes, and the PCM detected the system voltage was more than 17.0 volts, condition met for 5 seconds. **NOTE: If the Battery Backup circuit is open, the vehicle will not run.** **Possible Causes:** • Charging system problem (charging voltage too high) • Backup voltage circuit to the PCM open (intermittent fault) • PCM has failed
DTC: P0564 **2T CCM, MIL: Yes** **Year:** 2009 **Model:** Borrego **Engine:** 4.6L V8 VIN 2 **Transmission:** All	**Cruise Control Multifunction Input "A" Circuit:** Invalid voltage range check. A DTC code is set for the following conditions. Check SET/COAST switch stuck. Check RES/ACC switch stuck. **Possible Causes:** • Open or short in harness • Poor connection or damaged harness • Faulty cruise control remote control switch
DTC: P0571 **2T CCM, MIL: Yes** **Year:** 2009 **Model:** Amanti, Borrego, Sorento **Engine:** 3.3L V6 VIN 5, 3.8L V6 VIN 1, 3.8L V6 VIN 5, 3.8L V6 VIN 6 **Transmission:** All	**Brake Switch "A" Circuit:** PCM detects brake light input signal when the vehicle stops. VSS is normal. Vehicle speed 0 mph, during one second or more. **Possible Causes:** • Poor connection • Open or short to ground in signal circuit • Faulty PCM
DTC: P0600 **2T CCM, MIL: Yes** **Year:** 2009, 2010 **Model:** Optima, Spectra, Spectra5 **Engine:** 2.0L L4 VIN 1, 2.0L L4 VIN 2, 2.4L L4 VIN 8, 2.7L V6 VIN 4 **Transmission:** All	**CAN Communication Bus:** CAN message transfer incorrect? **Possible Causes:** • Open or short in CAN line • Contact resistance in connectors • Faulty PCM
DTC: P0601 **1T PCM, MIL: Yes** **Year:** 2009 **Model:** Amanti, Borrego, Sorento **Engine:** 3.3L V6 VIN 5, 3.8L V6 VIN 1, 3.8L V6 VIN 5, 3.8L V6 VIN 6 **Transmission:** All	**PCM (Internal Controller) Checksum Error:** Key on or engine running for 1 second, and the PCM detected an internal checksum data error during the initial Self-Test. **Possible Causes:** • Clear the trouble codes and retest for this trouble code. If the same trouble code resets, the PCM has failed and must be replaced to repair this problem.
DTC: P0602 **2T CCM, MIL: Yes** **Year:** 2009 **Model:** Amanti, Borrego, Sorento **Engine:** 3.3L V6 VIN 5, 3.8L V6 VIN 1, 3.8L V6 VIN 5, 3.8L V6 VIN 6 **Transmission:** All	**EEPROM Programming Error:** Check internal CPU **Possible Causes:** • Faulty PCM

DTC	Trouble Code Title, Conditions & Possible Causes
DTC: P0604 **2T PCM, MIL: Yes** **Year:** 2009 **Model:** Amanti, Borrego, Sorento **Engine:** 3.3L V6 VIN 5, 3.8L V6 VIN 1, 3.8L V6 VIN 5, 3.8L V6 VIN 6, 4.6L V8 VIN 2 **Transmission:** All	**PCM or TCM Internal Random Access Memory Error:** Key on or engine running and the PCM or TCM detected an Internal Random Access Memory (RAM) error was present. **Possible Causes:** • Poor terminal contact at the ECM Backup Voltage circuit • PCM or TCM has an internal problem or has failed
DTC: P0605 **1T PCM, MIL: Yes** **Year:** 2009, 2010 **Model:** Borrego, Optima, Rio, Rio5, Soul, Spectra, Spectra5, Sportage **Engine:** 1.6L L4 VIN 1, 1.6L L4 VIN 3, 2.0L L4 VIN 1, 2.0L L4 VIN 2, 2.4L L4 VIN 8, 2.7L V6 VIN 3, 2.7L V6 VIN 4, 4.6L V8 VIN 2 **Transmission:** All	**PCM (Internal Controller) ROM Error:** Key on for 1 second, and the PCM detected an internal ROM error occurred during the initial Self-Test. **Possible Causes:** • Clear the trouble codes and retest for this trouble code. If the same trouble code resets, the PCM has failed and must be replaced to repair this problem.
DTC: P0606 **2T CCM, MIL: Yes** **Year:** 2009 **Model:** Amanti, Borrego, Sorento, Spectra, Spectra5 **Engine:** 2.0L L4 VIN 1, 2.0L L4 VIN 2, 3.3L V6 VIN 5, 3.8L V6 VIN 1, 3.8L V6 VIN 5, 3.8L V6 VIN 6, 4.6L V8 VIN 2 **Transmission:** All	**ECM Processor (ECU-Self Test Failed):** Controller error. No electrical fault of the front HO2S. **Possible Causes:** • Faulty PCM
DTC: P061B **2T CCM** **Year:** 2009 **Model:** Amanti, Borrego, Sorento **Engine:** 3.3L V6 VIN 5, 3.8L V6 VIN 1, 3.8L V6 VIN 5, 3.8L V6 VIN 6, 4.6L V8 VIN 2 **Transmission:** All	**Internal Control Module Torque Calculation Performance:** Desired torque error. **Possible Causes:** • Faulty PCM
DTC: P0624 **2T CCM** **Year:** 2009, 2010 **Model:** Rio, Rio5 **Engine:** 1.6L L4 VIN 3 **Transmission:** All	**Fuel Cap Lamp Control Circuit:** Circuit continuity check, (high, low, or open). **Possible Causes:** • Poor connection • Open or short • Instrument cluster • Faulty ECM/PCM
DTC: P0625 **2T CCM** **Year:** 2009 **Model:** Borrego **Engine:** 4.6L V8 VIN 2 **Transmission:** All	**Alternator Field "F" Terminal Circuit Low:** Electrical check. **Possible Causes:** • Short to battery in harness • Poor connection or damaged harness
DTC: P0626 **2T CCM** **Year:** 2009 **Model:** Borrego **Engine:** 4.6L V8 VIN 2 **Transmission:** All	**Alternator Field "F" Terminal Circuit High:** Electrical check. Time after ignition ON, 1 second. Engine speed 0. No main relay error. **Possible Causes:** • Open or short to ground in harness • Faulty charging system

DTC	Trouble Code Title, Conditions & Possible Causes
DTC: P0630 **2T CCM, MIL: Yes** **Year:** 2009, 2010 **Model:** Amanti, Borrego, Optima, Rio, Rio5, Sorento, Soul, Spectra, Spectra5, Sportage **Engine:** 1.6L L4 VIN 1, 1.6L L4 VIN 3, 2.0L L4 VIN 1, 2.0L L4 VIN 2, 2.4L L4 VIN 8, 2.7L V6 VIN 3, 2.7L V6 VIN 4, 3.3L V6 VIN 5, 3.8L V6 VIN 1, 3.8L V6 VIN 5, 3.8L V6 VIN 6, 4.6L V8 VIN 2 **Transmission:** All	**VIN Not Programmed Or Incompatible- ECM/PCMECM:** PCM internal check. Enable condition, ignition ON. VIN does not exist in boot area. **Possible Causes:** • PCM is new and has not yet been programmed • Faulty PCM
DTC: P0638 **2T CCM, MIL: Yes** **Year:** 2009, 2010 **Model:** Amanti, Borrego, Optima, Sorento **Engine:** 2.4L L4 VIN 8, 2.7L V6 VIN 4, 3.3L V6 VIN 5, 3.8L V6 VIN 1, 3.8L V6 VIN 5, 3.8L V6 VIN 6, 4.6L V8 VIN 2 **Transmission:** All	**Throttle Actuator Control Range/Performance:** ETS position control malfunction. Battery voltage more than 5 volts. **Possible Causes:** • Throttle stuck • Open in motor circuit • Faulty motor • Faulty PCM
DTC: P0641 **2T CCM, MIL: Yes** **Year:** 2009 **Model:** Amanti, Borrego, Sorento **Engine:** 3.3L V6 VIN 5, 3.8L V6 VIN 1, 3.8L V6 VIN 5, 3.8L V6 VIN 6 **Transmission:** All	**Sensor Reference Voltage "A" Circuit Open:** Sensor reference voltage check. Ignition ON. **Possible Causes:** • Short in sensor power supply line • Faulty PCM
DTC: P0642 **2T CCM, MIL: Yes** **Year:** 2009, 2010 **Model:** Borrego, Optima, Rio, Rio5, Soul **Engine:** 1.6L L4 VIN 1, 1.6L L4 VIN 3, 2.4L L4 VIN 8, 2.7L V6 VIN 4, 4.6L V8 VIN 2 **Transmission:** All	**Sensor Reference Voltage "A" Circuit Low:** Sensor reference voltage check. Battery voltage 11-16 volts. **Possible Causes:** • Spoor connection • Short to ground in 5V, voltage circuit. • ECM/PCM
DTC: P0643 **2T CCM, MIL: Yes** **Year:** 2009, 2010 **Model:** Optima, Rio, Rio5, Soul **Engine:** 1.6L L4 VIN 1, 1.6L L4 VIN 3, 2.4L L4 VIN 8, 2.7L V6 VIN 4 **Transmission:** All	**Sensor Reference Voltage "A" Circuit High:** Sensor reference voltage check. Battery voltage 11-16 volts. **Possible Causes:** • Spoor connection • Short to power in 5V, voltage circuit. • ECM/PCM
DTC: P0645 **2T CCM** **Year:** 2009, 2010 **Model:** Borrego, Rio, Rio5, Soul **Engine:** 1.6L L4 VIN 1, 1.6L L4 VIN 3, 4.6L V8 VIN 2 **Transmission:** All	**A/C Clutch Relay Control Circuit:** DTC is set if the PCM detects that the relay line is open or shorted to ground or battery line. Circuit continuity check. **Possible Causes:** • Open in battery and control circuit • Short to ground in control circuit • Short to battery in control circuit • Faulty A/C relay • Faulty PCM

DTC	Trouble Code Title, Conditions & Possible Causes
DTC: P0646 **2T CCM** **Year:** 2009, 2010 **Model:** Borrego, Rio, Rio5, Sorento, Soul **Engine:** 1.6L L4 VIN 1, 1.6L L4 VIN 3, 3.3L V6 VIN 5, 3.8L V6 VIN 1, 3.8L V6 VIN 6, 4.6L V8 VIN 2 **Transmission:** All	**A/C Clutch Relay Control Circuit Low:** Detects circuit short to low voltage. No DTC exists. Engine works. After 0.5 seconds. **Possible Causes:** • Poor connection • Open or short to ground in A/C relay circuit • Faulty A/C relay • Faulty PCM
DTC: P0647 **2T CCM** **Year:** 2009, 2010 **Model:** Amanti, Borrego, Rio, Rio5, Sorento, Soul **Engine:** 1.6L L4 VIN 1, 1.6L L4 VIN 3, 3.3L V6 VIN 5, 3.8L V6 VIN 1, 3.8L V6 VIN 5, 3.8L V6 VIN 6, 4.6L V8 VIN 2 **Transmission:** All	**A/C Clutch Relay Control Circuit High:** Detects circuit short to high voltage. No DTC exists. Engine works. After 0.5 seconds. **Possible Causes:** • Poor connection • Short to power in A/C relay circuit • Faulty A/C relay • Faulty PCM
DTC: P0650 **2T CCM** **Year:** 2009, 2010 **Model:** Amanti, Borrego, Optima, Rio, Rio5, Sorento, Soul, Spectra, Spectra5, Sportage **Engine:** 1.6L L4 VIN 1, 1.6L L4 VIN 3, 2.0L L4 VIN 1, 2.0L L4 VIN 2, 2.4L L4 VIN 8, 2.7L V6 VIN 3, 2.7L V6 VIN 4, 3.3L V6 VIN 5, 3.8L V6 VIN 1, 3.8L V6 VIN 5, 3.8L V6 VIN 6, 4.6L V8 VIN 2 **Transmission:** All	**Malfunction Indicator Lamp Circuit Malfunction:** Key on or engine running and the PCM detected an unexpected voltage condition on the Malfunction Indicator Lamp (MIL) circuit. **Possible Causes:** • MIL control circuit open • MIL control circuit shorted to ground • MIL "bulb" is damaged or missing • PCM has failed (MIL control "driver" may be open or shorted)
DTC: P0651 **2T CCM, MIL: Yes** **Year:** 2009 **Model:** Amanti, Sorento **Engine:** 3.3L V6 VIN 5, 3.8L V6 VIN 5, 3.8L V6 VIN 6 **Transmission:** All	**Sensor reference Voltage "B" Circuit Open:** Sensor reference voltage check. Key ON. **Possible Causes:** • Short in sensor power supply line • Faulty PCM
DTC: P0652 **2T CCM, MIL: Yes** **Year:** 2009, 2010 **Model:** Borrego, Optima **Engine:** 2.4L L4 VIN 8, 2.7L V6 VIN 4, 4.6L V8 VIN 2 **Transmission:** All	**Sensor Reference Voltage "B" Circuit Low:** Electrical check. Ignition ON. APS2 voltage 0.7 volt, for at least 0.04 second. **Possible Causes:** • Open or short to ground in power circuit • Poor connection or damaged harness • Faulty ECM
DTC: P0653 **2T CCM, MIL: Yes** **Year:** 2009, 2010 **Model:** Optima **Engine:** 2.4L L4 VIN 8, 2.7L V6 VIN 4 **Transmission:** All	**Sensor Reference Voltage "B" Circuit High:** Electrical check. Ignition ON. TPS voltage 5.5 volts, for at least 0.04 second. **Possible Causes:** • Open or short to ground in power circuit • Poor connection or damaged harness • Faulty ECM
DTC: P0660 **2T CCM, MIL: Yes** **Year:** 2009 **Model:** Amanti, Borrego, Sorento **Engine:** 3.3L V6 VIN 5, 3.8L V6 VIN 1, 3.8L V6 VIN 5, 3.8L V6 VIN 6, 4.6L V8 VIN 2 **Transmission:** All	**Intake Manifold Tuning Valve Control Circuit/Open (Bank 1):** Signal low, high. **Possible Causes:** • Poor connection • Open or short in VIS circuit • Faulty VIS • Faulty PCM

DTC	Trouble Code Title, Conditions & Possible Causes
DTC: P0661 **2T CCM** **Year:** 2009, 2010 **Model:** Borrego, Sportage **Engine:** 2.7L V6 VIN 3, 4.6L V8 VIN 2 **Transmission:** All	**Intake Manifold Tuning Valve Control Circuit Low (Bank 1) Solenoid Type:** DTC is set if the ECM detects that the valve control circuit is shorted to ground. Driver stage check. **Possible Causes:** • Open in power supply harness • Short to ground in control harness • Contact resistance in connectors • Faulty valve
DTC: P0662 **2T CCM** **Year:** 2009, 2010 **Model:** Borrego, Sportage **Engine:** 2.7L V6 VIN 3, 4.6L V8 VIN 2 **Transmission:** All	**Intake Manifold Tuning Valve Control Circuit High (Bank 1) Solenoid Type:** DTC is set if the ECM detects that the valve control circuit is open or shorted to battery voltage. Driver stage check. **Possible Causes:** • Open or short to battery in control harness • Contact resistance in connectors • Faulty valve
DTC: P0664 **2T CCM, MIL: Yes** **Year:** 2009, 2010 **Model:** Sportage **Engine:** 2.7L V6 VIN 3 **Transmission:** All	**Intake Manifold Tuning Valve Control Circuit Low (Bank 2) Solenoid Type:** DTC is set if the ECM detects that the valve control circuit is shorted to ground. Driver stage check. **Possible Causes:** • Open in power supply harness • Short to ground in control harness • Contact resistance in connectors • Faulty valve
DTC: P0664 **2T CCM, MIL: Yes** **Year:** 2009, 2010 **Model:** Sportage **Engine:** 2.7L V6 VIN 3 **Transmission:** All	**Intake Manifold Tuning Valve Control Circuit Low (Bank 2):** Signal low, high. **Possible Causes:** • Poor connection • Open or short in VIS #2 circuit • Faulty VIS #2 • Faulty PCM
DTC: P0685 **2T CCM** **Year:** 2009 **Model:** Amanti, Borrego, Sorento **Engine:** 3.3L V6 VIN 5, 3.8L V6 VIN 1, 3.8L V6 VIN 5, 3.8L V6 VIN 6 **Transmission:** All	**ECM/PCM Power Relay Control Circuit/Open:** Engine running. Ignition voltage less than or equal to 11 volts. **Possible Causes:** • poor connection • Open or short to in control circuit • Main relay • PCM
DTC: P0698 **2T CCM, MIL: Yes** **Year:** 2009, 2010 **Model:** Borrego, Optima **Engine:** 2.4L L4 VIN 8, 2.7L V6 VIN 4, 4.6L V8 VIN 2 **Transmission:** All	**Sensor Reference Voltage "C" Circuit Low:** Electrical check. Ignition ON. ASP1 voltage less than 0.7 volt, for at least 0.1 second. **Possible Causes:** • Open or short to ground in power circuit • Poor connection or damaged harness • Faulty ECM
DTC: P0699 **2T CCM, MIL: Yes** **Year:** 2009, 2010 **Model:** Optima **Engine:** 2.4L L4 VIN 8, 2.7L V6 VIN 4 **Transmission:** All	**Sensor Reference Voltage "C" Circuit High:** Electrical check. Ignition ON. APS1 voltage 5.5 volts, for at least 0.1 second. **Possible Causes:** • Open or short to ground in power circuit • Poor connection or damaged harness • Faulty ECM

DTC	Trouble Code Title, Conditions & Possible Causes
DTC: P0700 **2T CCM, MIL: Yes** **Year:** 2009, 2010 **Model:** Borrego, Optima, Rio, Rio5, Soul, Spectra, Spectra5, Sportage **Engine:** 1.6L L4 VIN 3, 2.0L L4 VIN 1, 2.0L L4 VIN 2, 2.4L L4 VIN 8, 2.7L V6 VIN 3, 2.7L V6 VIN 4, 3.8L V6 VIN 1, 4.6L V8 VIN 2 **Transmission:** All	**TCU Request For MIL "ON":** Engine at normal operating temperature. Check for additional DTC's. **Possible Causes:** • poor connection • TCM • PCM/ECM
DTC: P0707 **1T PCM, MIL: Yes** **Year:** 2009 **Model:** Amanti **Engine:** 3.8L V6 VIN 5 **Transmission:** All	**Transaxle Range Switch Circuit Low Input:** Enable Conditions • Engine state = "RUN" • 11V ≤ Battery Voltage ≤ 16V • TPS ≥ 3% Threshold value • No signal detected Diagnostic Time • More than 30seconds Fail Safe • Recognition as previous signal. - When P-D or R-D or D-R SHIFT is detected, it is regarded as N-D or N-R though "N" signal is not detected - When sports mode S/W is ON without P,R,N, D-RANGE signals, it is regarded sports mode. (DTC is not set) **Possible Causes:** • Open or short in circuit • Faulty Shift cable adjustment • Faulty Inhibitor switch and Manual control lever position adjustment • Faulty TRANSAXLE RANGE SWITCH • Faulty TCM(PCM)
DTC: P0708 **1T PCM, MIL: Yes** **Year:** 2009 **Model:** Amanti **Engine:** 3.8L V6 VIN 5 **Transmission:** All	**Transaxle Range Switch Circuit High Input:** Enable Conditions • Engine state = "RUN" • 11V ≤ Battery Voltage ≤ 16V • TPS ≥ 3% Threshold value • Multiple signal Diagnostic Time • More than 30sec Fail Safe • Recognition as previous signal - When signal is input "D" and "N" at the same time, TCM regards it as "N" RANGE - After PCM/TCM Reset, If the if the PCM/TCM detects multiple signal or no signal, then it holds the 3rd gear position **Possible Causes:** • Open or short in TRANSAXLE RANGE SWITCH • Faulty Shift cable adjustment • Faulty Inhibitor switch and Manual control lever position adjustment • Faulty TRANSAXLE RANGE SWITCH • Faulty PCM

DTC	Trouble Code Title, Conditions & Possible Causes
DTC: P0711 **1T PCM, MIL: Yes** **Year:** 2009 **Model:** Amanti **Engine:** 3.8L V6 VIN 5 **Transmission:** All	**Transaxle Fluid Temperature Sensor Rationality:** Enable Conditions 1) • Intake air temperature ≥ -25°C(-13°F) • Engine state = RUN • No error with relations other sensors • Engine be cooled sufficiently Enable Conditions 2) • Engine state = RUN • Average start up temperature of TM stuck diagnostic ≤ 55°C(131°F) Threshold Value 1) • ATF Temp - Coolant Temp ≥ 20°C(68°F) Threshold Value 2) • ATF Temp - TM start up Temp ≤ 0.5°C(32.9°F) Diagnostic Time 1) • more than 1 second Diagnostic Time 2) • more than 900 seconds Fail Safe • Learning control and Intelligent shift are inhibited • Fluid temperature is regarded as 80°C(176°F) **Possible Causes:** • Sensor signal circuit is short to ground • Faulty sensor • Faulty PCM
DTC: P0712 **2T CCM, MIL: Yes** **Year:** 2009 **Model:** Amanti **Engine:** 3.8L V6 VIN 5 **Transmission:** All	**Transmission Fluid Temperature Low Input:** Key on or engine running and the PCM detected the TFT sensor signal indicated less than 0.49 volt 300°F) for more than 1 second. **NOTE: The TFT sensor signal at 68°F is 4.0 volts, and at 266°F it is 1.5 volts.** **Possible Causes:** • TFT sensor signal circuit shorted to ground (sensor to TCM) • TFT sensor signal circuit shorted to ground (TCM to PCM) • TFT sensor is damaged or has failed • PCM is damaged
DTC: P0713 **2T CCM, MIL: Yes** **Year:** 2009 **Model:** Amanti **Engine:** 3.8L V6 VIN 5 **Transmission:** All	**Transmission Fluid Temperature High Input:** Key on or engine running and the PCM detected the TFT sensor signal indicated more than 4.57 volts (-40°F) during the CCM test period. **NOTE: The TFT sensor signal at 68°F is 4.0 volts, and at 266°F it is 1.5 volts.** **Possible Causes:** • TFT sensor signal circuit is open (sensor circuit to the TCM) • TFT sensor signal circuit is open (TCM circuit to the PCM) • TFT sensor is damaged or has failed • PCM is damaged
DTC: P0717 **2T CCM, MIL: Yes** **Year:** 2009 **Model:** Amanti **Engine:** 3.8L V6 VIN 5 **Transmission:** All	**Input/Turbine Speed Sensor No Signal:** Vehicle drive to a speed of over 25 mph, gear ratio indicating the vehicle is in Drive, 2nd or 1st gear, and the PCM detected that the Input/Turbine Speed sensor signal indicated less than 98 rpm. **Possible Causes:** • Input/Turbine speed sensor signal circuit open or shorted • Input/Turbine speed sensor is damaged or has failed • PCM has failed
DTC: P0722 **2T CCM, MIL: Yes** **Year:** 2009 **Model:** Amanti **Engine:** 3.8L V6 VIN 5 **Transmission:** All	**Output Shaft Speed Sensor No Signal:** VEHICLE DRIVEN IN DRIVE, 2ND OR 1ST GEAR, GEAR RATIO INDICATING THE VEHICLE IS IN DRIVE, 2ND OR 1ST GEAR, INPUT SHAFT SPEED OVER 775 RPM, AND THE PCM DID NOT DETECT ANY OSS SENSOR SIGNALS DURING THE TEST. **Possible Causes:** • OUTPUT SHAFT SPEED SENSOR SIGNAL CIRCUIT OPEN OR SHORTED • OUTPUT SHAFT SPEED SENSOR IS DAMAGED OR HAS FAILED • PCM HAS FAILED

DTC	Trouble Code Title, Conditions & Possible Causes
DTC: P0731 **2T CCM, MIL: Yes** **Year:** 2009 **Model:** Amanti **Engine:** 3.8L V6 VIN 5 **Transmission:** All	**TCM Incorrect First Gear Ratio:** Vehicle driven at 12-32 mph in 3rd gear, shift solenoids 'A', 'B' and 'C', input/turbine speed sensor and TFT sensor inputs all indicating okay, and the PCM detected the 1st gear ratio was too high. **Possible Causes:** • ATF fluid level too low or line pressure low • Control valve stuck or solenoid valve is damaged or has failed • Forward clutch, 3-4 brake band or 1-way clutch No. 1 slippage • PCM has failed • TSB 2TD007 (12/01) contains information related to this code
DTC: P0732 **2T CCM, MIL: Yes** **Year:** 2009 **Model:** Amanti **Engine:** 3.8L V6 VIN 5 **Transmission:** All	**TCM Incorrect Second Gear Ratio:** Vehicle driven at 17-60 mph in 2nd gear, shift solenoids 'A', 'B' and 'C', input/turbine speed sensor and TFT sensor inputs all indicating okay, and the PCM detected the 2nd gear ratio was too high. **Possible Causes:** • ATF fluid level too low or line pressure low • Control valve stuck or solenoid valve is damaged or has failed • Forward clutch, 2-4 brake band or 1-way clutch No. 1 slippage • PCM has failed • TSB 2TD004 (8/00) contains information related to this code • TSB 2TD007 (12/01) contains information related to this code • TSB 3TD008 (3/02) contains information related to this code
DTC: P0733 **2T CCM, MIL: Yes** **Year:** 2009 **Model:** Amanti **Engine:** 3.8L V6 VIN 5 **Transmission:** All	**TCM Third Gear Incorrect Ratio:** Vehicle driven at 19-32 mph in 3rd gear, shift solenoids 'A', 'B' and 'C', input/turbine speed sensor and TFT sensor inputs all indicating okay, and the PCM detected the 3rd gear ratio was too high. **Possible Causes:** • ATF fluid level too low or line pressure low • Control valve stuck or solenoid valve is damaged or has failed • Forward clutch, 3-4 brake band or 1-way clutch No. 1 slippage • PCM has failed • TSB 2TD004 (8/00) contains information related to this code • TSB 2TD007 (12/01) contains information related to this code • TSB 3TD008 (3/02) contains information related to this code
DTC: P0734 **2T CCM, MIL: Yes** **Year:** 2009 **Model:** Amanti **Engine:** 3.8L V6 VIN 5 **Transmission:** All	**TCM Fourth Gear Incorrect Ratio:** Vehicle driven at 44-65 mph in 4th gear, shift solenoids 'A', 'B' and 'C', input/turbine speed sensor and TFT sensor inputs all indicating okay, and the PCM detected the 4th gear ratio was too high. **Possible Causes:** • ATF fluid level too low or line pressure low • Control valve stuck or solenoid valve is damaged or has failed • 2-4 brake band or 3-4 clutch slippage • PCM has failed • TSB 2TD004 (8/00) contains information related to this code • TSB 2TD007 (12/01) contains information related to this code • TSB 3TD008 (3/02) contains information related to this code
DTC: P0735 **2T CCM, MIL: Yes** **Year:** 2009 **Model:** Amanti **Engine:** 3.8L V6 VIN 5 **Transmission:** All	**TCM Fifth Gear Incorrect Ratio:** Vehicle driven at 44-65 mph in 5th gear, shift solenoids 'A', 'B' and 'C', input/turbine speed sensor and TFT sensor inputs all indicating okay, and the PCM detected the 5th gear ratio was too high. If this code sets 4 times or more, the PCM will lock the gear into 3rd gear. **Possible Causes:** • ATF fluid level too low or line pressure low • Control valve stuck or solenoid valve is damaged or has failed • 2-4 brake band or 3-4 clutch slippage • PCM has failed

DTC	Trouble Code Title, Conditions & Possible Causes
DTC: P0741 **1T PCM, MIL: Yes** **Year:** 2009 **Model:** Amanti **Engine:** 3.8L V6 VIN 5 **Transmission:** All	**Torque Converter Clutch Stuck Off:** Enable Conditions • Always Threshold value • TCC duty > 0% or TCC abnormal slip counter ≥ 4 Diagnostic Time • 1 second Fail Safe • Damper clutch abnormal system (If diagnosis code P0741 is output four times, TORQUE CONVERTER(DAMPER) CLUTCH is not controlled by PCM/TCM) **Possible Causes:** • TORQUE CONVERTER(DAMPER) CLUTCH : TCC • Faulty TCC or oil pressure system • Faulty TCC solenoid valve • Faulty body control valve • Faulty PCM/TCM
DTC: P0742 **1T PCM, MIL: Yes** **Year:** 2009 **Model:** Amanti **Engine:** 3.8L V6 VIN 5 **Transmission:** All	**Torque Converter Clutch Stuck On:** Enable Conditions • Throttle position > 20% • Output speed > 500 rpm • Manifold air pressure >60 kPa • A/T range switch D,SP • TCC stuck on delay timer >5 secs Threshold value • {Engine rpm - Input speed sensor rpm}≤ 20 rpm Diagnostic Time • More than 1sec Fail Safe • Damper clutch abnormal system (If diagnosis code P0741 is output four times, TORQUE CONVERTER(DAMPER) CLUTCH is not controlled by PCM/TCM) **Possible Causes:** • TORQUE CONVERTER(DAMPER) CLUTCH : TCC • Faulty TCC or oil pressure system • Faulty TCC solenoid valve • Faulty body control valve • Faulty TCM(PCM)
DTC: P0743 **2T CCM, MIL: Yes** **Year:** 2009 **Model:** Amanti **Engine:** 3.8L V6 VIN 5 **Transmission:** All	**TCC Solenoid Circuit Malfunction:** Engine running in gear with VSS inputs received, and the PCM detected an unexpected voltage condition on the TCC circuit. **Possible Causes:** • TCC solenoid control circuit is open or shorted to ground • TCC solenoid control circuit is shorted to system power (B+) • TCC solenoid is damaged or has failed • PCM or TCM has failed
DTC: P0748 **1T PCM, MIL: Yes** **Year:** 2009 **Model:** Amanti **Engine:** 3.8L V6 VIN 5 **Transmission:** All	**VFS Solenoid - Open Or Short To Ground:** Enable Conditions • 16V > Voltage Battery >11V • In gear state(no gear shifting) 500msec is passed from turn on the relay • A/T Relay = ON • Engine state = RUN Threshold value • Out of available voltage range Diagnostic Time • More than 2 seconds Fail Safe • Locked in 3rd gear (Control relay off) **Possible Causes:** • Open or short in circuit • Faulty VFS SOLENOID VALVE • Faulty PCM/TCM

DTC	Trouble Code Title, Conditions & Possible Causes
DTC: P0750 **1T PCM, MIL: Yes** **Year:** 2009 **Model:** Amanti **Engine:** 3.8L V6 VIN 5 **Transmission:** All	**Low And Reverse Solenoid Valve Circuit - Open Or Short To Ground:** Enable Conditions • 16V > Voltage Battery >11V • In gear state(no gear shifting) 500msec is passed from turn on the relay • A/T Relay = ON • Engine state = RUN Threshold value • Out of available voltage range Diagnostic Time • More than 5 seconds Fail Safe • Locked in 3rd gear.(Control relay off) **Possible Causes:** • Open or short in circuit • Faulty LR SOLENOID VALVE • Faulty PCM/TCM
DTC: P0755 **1T PCM, MIL: Yes** **Year:** 2009 **Model:** Amanti **Engine:** 3.8L V6 VIN 5 **Transmission:** All	**Under Drive Solenoid Valve Circuit - Open Or Short To Ground:** Enable Conditions • 16V > Voltage Battery >11V • In gear state(no gear shifting) 500msec is passed from turn on the relay • A/T Relay = ON • Engine state = RUN Threshold value • Out of available voltage range Diagnostic Time • More than 5 seconds Fail Safe • Locked in 3rd gear.(Control relay off) **Possible Causes:** • Open or short in circuit • Faulty UD SOLENOID VALVE • Faulty PCM/TCM
DTC: P0760 **1T PCM, MIL: Yes** **Year:** 2009 **Model:** Amanti **Engine:** 3.8L V6 VIN 5 **Transmission:** All	**Second Solenoid Valve Circuit - Open Or Short To Ground:** Enable Conditions • 16V > Voltage Battery >11V • In gear state(no gear shifting) 500msec is passed from turn on the relay • A/T Relay = ON • Engine state = RUN Threshold value • Out of available voltage range Diagnostic Time • More than 5 seconds Fail Safe • Locked in 3rd gear.(Control relay off) **Possible Causes:** • Open or short in circuit • Faulty 2ND SOLENOID VALVE • Faulty PCM/TCM
DTC: P0765 **2T CCM, MIL: Yes** **Year:** 2009 **Model:** Amanti **Engine:** 3.8L V6 VIN 5 **Transmission:** All	**TCM Overdrive Solenoid Circuit Malfunction:** Engine running in gear, VSS inputs received, and PCM detected an unexpected voltage condition on the O/D valve circuit during the test. **Possible Causes:** • O/D solenoid control circuit open or shorted to ground • O/D solenoid control circuit shorted to system power (B+) • O/D solenoid valve is damaged or has failed • PCM or TCM has failed

DTC	Trouble Code Title, Conditions & Possible Causes
DTC: P0770 **2T CCM, MIL: Yes** **Year:** 2009 **Model:** Amanti **Engine:** 3.8L V6 VIN 5 **Transmission:** All	**TCM RED Solenoid Circuit Malfunction:** Engine running in gear, VSS inputs received, and PCM detected an unexpected voltage condition on the RED solenoid circuit in the test. **Possible Causes:** • RED control circuit open or shorted to ground • RED control circuit shorted to system power (B+) • RED is damaged or has failed • PCM or TCM has failed
DTC: P0885 **1T PCM, MIL: Yes** **Year:** 2009 **Model:** Amanti **Engine:** 3.8L V6 VIN 5 **Transmission:** All	**A/T Control Relay - Open Or Short To Ground:** Enable Conditions • 16V > Voltage Battery >11V • Time after TCM turns on >0.5sec Threshold value • 16V > Voltage Battery >11V Diagnostic Time • 2.375 seconds Fail Safe • Locked in 3 rd gear.(control relay off) **Possible Causes:** • Open or short in circuit • Faulty A/T control relay • Faulty PCM/TCM
DTC: P0890 **1T PCM, MIL: Yes** **Year:** 2009 **Model:** Amanti **Engine:** 3.8L V6 VIN 5 **Transmission:** All	**TCM Power Relay Sense Circuit Low:** Enable Conditions • 16V > Voltage Battery >11V • Time after TCM turns on >0.5sec Threshold value • Feedback Voltage ≤ 0.5V Diagnostic Time • 2 seconds Fail Safe • Locked in 3 rd gear.(control relay off) **Possible Causes:** • Open or short in circuit • Faulty A/T control relay • Faulty PCM/TCM
DTC: P0891 **1T PCM, MIL: Yes** **Year:** 2009 **Model:** Amanti **Engine:** 3.8L V6 VIN 5 **Transmission:** All	**TCM Power Relay Sense Circuit High:** Enable Conditions • 16V > Voltage Battery >11V • Time after TCM turns on >0.5sec Threshold value • Feedback Voltage ≥ 20V Diagnostic Time • 2 seconds Fail Safe • Locked in 3 rd gear.(control relay off) **Possible Causes:** • Open or short in circuit • Faulty A/T control relay • Faulty PCM/TCM

OBD II Trouble Code List (P1XXX Codes)

DTC	Trouble Code Title, Conditions & Possible Causes
DTC: P1106 **2T CCM** **Year:** 2009, 2010 **Model:** Amanti, Borrego, Sedona, Sorento **Engine:** 3.3L V6 VIN 5, 3.8L V6 VIN 1, 3.8L V6 VIN 3, 3.8L V6 VIN 5, 3.8L V6 VIN 6 **Transmission:** All	**Manifold Absolute Pressure Sensor Circuit Short- Intermittent High Input:** This code detects an intermittent short to high in either the signal circuit or the MAP sensor. **Possible Causes:** • Poor connection • Short to battery in signal circuit • Open in ground circuit • Faulty MAPS • Faulty PCM
DTC: P1107 **2T CCM** **Year:** 2009 **Model:** Amanti, Borrego, Sorento **Engine:** 3.3L V6 VIN 5, 3.8L V6 VIN 1, 3.8L V6 VIN 5, 3.8L V6 VIN 6 **Transmission:** All	**Manifold Absolute Pressure Sensor Circuit Short- Intermittent Low Input:** This code detects an intermittent short to high in either the signal circuit or the MAP sensor. **Possible Causes:** • Poor connection • Open or short to ground in the power circuit • Open or short to ground in the signal circuit • Faulty MAPS • Faulty PCM
DTC: P1111 **2T CCM** **Year:** 2009 **Model:** Amanti, Borrego, Sorento **Engine:** 3.3L V6 VIN 5, 3.8L V6 VIN 1, 3.8L V6 VIN 5, 3.8L V6 VIN 6 **Transmission:** All	**Intake Air Temperature Sensor Circuit Short- Intermittent High Input:** This code detects a continuous short to high in either the signal circuit or the sensor. **Possible Causes:** • Poor connection • Open or short in signal circuit • Open in ground circuit • Faulty IATS • Faulty PCM
DTC: P1112 **2T CCM** **Year:** 2009 **Model:** Amanti, Borrego, Sorento **Engine:** 3.3L V6 VIN 5, 3.8L V6 VIN 1, 3.8L V6 VIN 5, 3.8L V6 VIN 6 **Transmission:** All	**Intake Air Temperature Sensor Circuit Short- Intermittent Low Input:** This code detects a continuous short to high in either the signal circuit or the sensor. **Possible Causes:** • Poor connection • Short to ground in the signal circuit • Open in ground circuit • Faulty IATS • Faulty PCM
DTC: P1114 **2T CCM** **Year:** 2009 **Model:** Amanti, Borrego, Sorento **Engine:** 3.3L V6 VIN 5, 3.8L V6 VIN 1, 3.8L V6 VIN 5, 3.8L V6 VIN 6 **Transmission:** All	**Engine Coolant temperature Sensor Circuit- Intermittent Low Input:** This code detects an intermittent short to ground in the signal circuit or the sensor. **Possible Causes:** • Poor connection • Short to ground in signal circuit • Open in ground circuit • Faulty ECTS • Faulty PCM
DTC: P1115 **1T ECT** **Year:** 2009 **Model:** Amanti, Borrego, Sorento **Engine:** 3.3L V6 VIN 5, 3.8L V6 VIN 1, 3.8L V6 VIN 5, 3.8L V6 VIN 6 **Transmission:** All	**Engine Coolant temperature Sensor Circuit- Intermittent High Input:** Enable Conditions Case 1 • Time after start-up > 120 sec. Case 2 • Time from IG "OFF" to IG "ON" > 360 min. • Intake air temperature ≥ -10°C(14°F) • Engine running Threshold value • Intermittently engine coolant temperature sensor's voltage > 4.9V Diagnosis Time • Continuous **Possible Causes:** • Poor Connection • Open or short to battery in signal Circuit • Open in Ground Circuit. • Faulty ECTS • Faulty PCM

DTC	Trouble Code Title, Conditions & Possible Causes
DTC: P1295 **2T CCM** **Year:** 2009 **Model:** Amanti, Borrego, Sorento **Engine:** 3.3L V6 VIN 5, 3.8L V6 VIN 1, 3.8L V6 VIN 5, 3.8L V6 VIN 6 **Transmission:** All	**Electronic Throttle Control (ETC) System Malfunction- Power Management:** This code is set is there is a problem in the power management system. Ignition ON. **Possible Causes:** • TPS malfunction • TPS malfunction plus MAFS malfunction • MAP malfunction plus TPS malfunction • Faulty PCM
DTC: P1505 **2T CCM, MIL: Yes** **Year:** 2009, 2010 **Model:** Rio, Rio5, Soul, Sportage **Engine:** 1.6L L4 VIN 1, 1.6L L4 VIN 3, 2.0L L4 VIN 2, 2.7L V6 VIN 3 **Transmission:** All	**IAC Valve Opening Coil Signal Low:** Engine runtime over 5 seconds, and the PCM detected the IAC Valve Opening Coil signal remained in a low state during the test. **Possible Causes:** • IAC valve control signal is open • IAC valve control signal is shorted to ground • IAC valve is damaged or has failed • PCM has failed (IAC "driver" circuit may be open in the PCM)
DTC: P1506 **2T CCM, MIL: Yes** **Year:** 2009, 2010 **Model:** Rio, Rio5, Soul, Sportage **Engine:** 1.6L L4 VIN 1, 1.6L L4 VIN 3, 2.0L L4 VIN 2, 2.7L V6 VIN 3 **Transmission:** All	**IAC Valve Opening Coil Signal High:** Engine runtime over 5 seconds, and the PCM detected the IAC Valve Opening Coil signal remained in a high state during the test. **Possible Causes:** • IAC valve control signal is shorted to system power (B+) • IAC valve is damaged or has failed • PCM has failed (IAC "driver" circuit may be shorted in the PCM)
DTC: P1507 **2T CCM, MIL: Yes** **Year:** 2009, 2010 **Model:** Rio, Rio5, Soul, Spectra, Spectra5, Sportage **Engine:** 1.6L L4 VIN 1, 1.6L L4 VIN 3, 2.0L L4 VIN 1, 2.0L L4 VIN 2, 2.7L V6 VIN 3 **Transmission:** All	**IAC Valve Closing Coil Signal Low:** Engine runtime over 5 seconds, and the PCM detected the IAC Valve Closing Coil signal remained in a low state during the test. **Possible Causes:** • IAC valve control signal is open • IAC valve control signal is shorted to ground • IAC valve is damaged or has failed • PCM has failed (IAC "driver" circuit may be open in the PCM)
DTC: P1508 **2T CCM, MIL: Yes** **Year:** 2009, 2010 **Model:** Rio, Rio5, Spectra, Spectra5, Sportage **Engine:** 1.6L L4 VIN 3, 2.0L L4 VIN 1, 2.0L L4 VIN 2, 2.7L V6 VIN 3 **Transmission:** All	**IAC Valve Closing Coil Signal High:** Engine runtime over 5 seconds, and the PCM detected the IAC Valve Closing Coil signal remained in a high state during the test. **Possible Causes:** • IAC valve control signal is shorted to system power (B+) • IAC valve is damaged or has failed • PCM has failed (IAC "driver" circuit may be shorted in the PCM)
DTC: P1523 **1T PCM** **Year:** 2009, 2010 **Model:** Amanti, Borrego, Sedona, Sorento **Engine:** 3.3L V6 VIN 5, 3.8L V6 VIN 1, 3.8L V6 VIN 3, 3.8L V6 VIN 5, 3.8L V6 VIN 6 **Transmission:** All	**Throttle Actuator Control System - Throttle Valve Stuck:** Enable Conditions • ETC Power Control Mode • TPS 1 & 2 = normal • Sensor Supply voltage = Normal Threshold value • The throttle did not return to default range within 1 to 4 seconds of turning off. That is, (TPS1's signal > 0.9V AND TPS1's signal < 1.85V) or (TPS2's signal < 1.85V AND TPS2's signal > 0.9V) when the power to the ETC motor is turned off. Diagnosis Time • Continuous **Possible Causes:** • Carbon in throttle • Broken Throttle return spring • throttle sticky • throttle icy • PCM

DTC	Trouble Code Title, Conditions & Possible Causes
DTC: P1550 **2T CCM** **Year:** 2009, 2010 **Model:** Rio, Rio5, Soul **Engine:** 1.6L L4 VIN 1, 1.6L L4 VIN 3 **Transmission:** All	**Knock Sensor Evaluation IC:** Circuit continuity check, pulse test. **Possible Causes:** • Poor connection • Open or short in control circuit • Faulty knock sensor • Faulty PCM
DTC: P1560 **2T CCM** **Year:** 2009, 2010 **Model:** Rio, Rio5, Soul **Engine:** 1.6L L4 VIN 1, 1.6L L4 VIN 3 **Transmission:** All	**Knock Control SPI (Serial Port Interface) Check:** SPI communication check. **Possible Causes:** • Poor connection • Faulty ECM/PCM
DTC: P1610 **1T PCM, MIL: Yes** **Year:** 2009, 2010 **Model:** Amanti, Borrego, Optima, Soul **Engine:** 2.0L L4 VIN 2, 2.4L L4 VIN 8, 2.7L V6 VIN 4, 3.8L V6 VIN 5, 4.6L V8 VIN 2 **Transmission:** All	**Non-Immobilizer-EMS connected to an Immobilizer:** Non-Immobilizer-EMS connected to an Immobilizer **Possible Causes:** • EMS • PCM
DTC: P161B **2T CCM, MIL: Yes** **Year:** 2009 **Model:** Amanti, Borrego, Sorento **Engine:** 3.3L V6 VIN 5, 3.8L V6 VIN 1, 3.8L V6 VIN 5, 3.8L V6 VIN 6 **Transmission:** All	**PCM Internal Error- Torque Calculating:** This code is set if delivered torque is grossly different from the desired torque. **Possible Causes:** • Intake air leakage • Faulty ETS system • Clogged exhaust system • Faulty PCM
DTC: P1674 **1T PCM, PATS: Yes** **Year:** 2009, 2010 **Model:** Amanti, Borrego, Soul **Engine:** 2.0L L4 VIN 2, 3.8L V6 VIN 5, 4.6L V8 VIN 2 **Transmission:** All	**Transponder Status Error:** Enable Conditions • IG ON (On Registering TP Procedure) Threshold value • Key not in 'VIRGIN' Status or with invalid ID code Detecting time • Immediately **Possible Causes:** • Invalid transponder.
DTC: P1675 **1T PCM, PATS: Yes** **Year:** 2009, 2010 **Model:** Amanti, Borrego, Soul **Engine:** 2.0L L4 VIN 2, 3.8L V6 VIN 5, 4.6L V8 VIN 2 **Transmission:** All	**Transponder Programming Error:** Enable Conditions • IG ON(During the authentication) Threshold value • Invalid characteristic data • No transponder or more than two transponder is detected by coil antenna Detecting time • Immediately **Possible Causes:** • Invalid transponder.
DTC: P1676 **T , PATS: Yes** **Year:** 2009, 2010 **Model:** Amanti, Borrego, Soul **Engine:** 2.0L L4 VIN 2, 3.8L V6 VIN 5, 4.6L V8 VIN 2 **Transmission:** All	**SMARTRA Message Error:** Enable Conditions • IG ON Threshold value • SMARTRA Message error Detecting time • Immediately **Possible Causes:** • Faulty SMARTRA

DTC	Trouble Code Title, Conditions & Possible Causes
DTC: P1690 **2T CCM** **Year:** 2009, 2010 **Model:** Amanti, Borrego, Soul **Engine:** 2.0L L4 VIN 2, 3.8L V6 VIN 5, 4.6L V8 VIN 2 **Transmission:** All	**SMARTRA Error:** No answer from SMARTRA. Invalid message from SMARTRA to ECM. **Possible Causes:** • Open or short in antenna or SMARTRA circuit • Antenna • SMARTRA • Faulty transponder • Faulty ECM
DTC: P1691 **2T CCM** **Year:** 2009, 2010 **Model:** Amanti, Borrego, Soul **Engine:** 2.0L L4 VIN 2, 3.8L V6 VIN 5, 4.6L V8 VIN 2 **Transmission:** All	**Antenna Error:** Antenna error. **Possible Causes:** • Open or short in antenna or SMARTRA circuit • Antenna • SMARTRA • Faulty transponder • Faulty ECM
DTC: P1692 **1T PCM, PATS: Yes** **Year:** 2009, 2010 **Model:** Amanti, Borrego, Soul **Engine:** 2.0L L4 VIN 2, 3.8L V6 VIN 5, 4.6L V8 VIN 2 **Transmission:** All	**Immobilizer Lamp Error:** Enable Conditions • IG ON Threshold value • Short to GND, Wiring open **Possible Causes:** • Short Circuit in immobilizer lamp circuit. • Open/Short in control harness • Faulty PCM/ECM
DTC: P1693 **1T PCM, PATS: Yes** **Year:** 2009, 2010 **Model:** Amanti, Borrego, Soul **Engine:** 2.0L L4 VIN 2, 3.8L V6 VIN 5, 4.6L V8 VIN 2 **Transmission:** All	**Transponder No Response Error/Invalid Response:** Enable Conditions • IG ON Threshold value • Corrupted data from Transponder • More than one TP in the magnetic field • No TP(Key without TP) in the magnetic field **Possible Causes:** • Invalid transponder.
DTC: P1694 **2T CCM** **Year:** 2009, 2010 **Model:** Amanti, Optima, Soul **Engine:** 2.0L L4 VIN 2, 2.4L L4 VIN 8, 2.7L V6 VIN 4, 3.8L V6 VIN 5 **Transmission:** All	**ECM Signal Error:** Invalid request from ECM or corrupted data **Possible Causes:** • Open or short in antenna or SMARTRA circuit • Antenna • SMARTRA • Faulty transponder • Faulty ECM
DTC: P1695 **2T CCM** **Year:** 2009, 2010 **Model:** Amanti, Soul **Engine:** 2.0L L4 VIN 2, 3.8L V6 VIN 5 **Transmission:** All	**EEPROM Error:** Inconsistent data from EEPROM. Invalid write operation from EEPROM. Not plausible immobilizer indicator store in ECM. No valid data from SMARTRA after three attempts from the ECM. Invalid tester message or unexpected request from tester. **Possible Causes:** • Open or short in antenna or SMARTRA circuit • Antenna • SMARTRA • Faulty transponder • Faulty ECM
DTC: P1696 **1T PCM, PATS: Yes** **Year:** 2009, 2010 **Model:** Amanti, Borrego, Soul **Engine:** 2.0L L4 VIN 2, 3.8L V6 VIN 5, 4.6L V8 VIN 2 **Transmission:** All	**Authentication Fail:** Enable Conditions • IG ON Threshold value • Virgin TP at PCM/ECM status "Learnt" • Learnt(Invalid) TP at PCM/ECM status "Learnt" **Possible Causes:** • Invalid transponder.

DTC	Trouble Code Title, Conditions & Possible Causes
DTC: P1697 **1T PCM, PATS: Yes** **Year:** 2009 **Model:** Amanti, Borrego **Engine:** 3.8L V6 VIN 5, 4.6L V8 VIN 2 **Transmission:** All	**Hi-Scan Message Error:** Enable Conditions • IG ON Threshold value • Invalid request - Protocol layer violation - Check sum error **Possible Causes:** • Poor connection between scanner and diagnostic connector • Up-to-date of Scanner program
DTC: P1699 **1T PCM, PATS: Yes** **Year:** 2009, 2010 **Model:** Amanti, Borrego, Soul **Engine:** 2.0L L4 VIN 2, 3.8L V6 VIN 5, 4.6L V8 VIN 2 **Transmission:** All	**Twice IG ON over trial:** This is a special function for engine start by vehicle manufacturer. The engine can be started for moving from the production line to an area where the key teaching is proceeded. Enable Conditions • IG ON Threshold value • Twice IGN ≥ 32 times **Possible Causes:** • Over time trial of Twice IGN
DTC: P169A **1T PCM, PATS: Yes** **Year:** 2009, 2010 **Model:** Amanti, Borrego, Soul **Engine:** 2.0L L4 VIN 2, 3.8L V6 VIN 5, 4.6L V8 VIN 2 **Transmission:** All	**SMARTRA Authentication Fail:** Enable Conditions • IG ON Threshold value • Virgin SMARTRA at Learnt EMS • Neutral SMARTRA at Learnt EMS • Incorrect the Authentication of EMS and SMARTRA • Locking of SMARTRA **Possible Causes:** • Locking of SMARTRA

OBD II Trouble Code List (P2XXX Codes)

DTC	Trouble Code Title, Conditions & Possible Causes
DTC: P2096 **2T CCM, MIL: Yes** **Year:** 2009, 2010 **Model:** Amanti, Borrego, Optima, Rio, Rio5, Sorento, Soul, Spectra, Spectra5 **Engine:** 1.6L L4 VIN 1, 1.6L L4 VIN 3, 2.0L L4 VIN 1, 2.0L L4 VIN 2, 2.4L L4 VIN 8, 2.7L V6 VIN 4, 3.3L V6 VIN 5, 3.8L V6 VIN 1, 3.8L V6 VIN 5, 3.8L V6 VIN 6, 4.6L V8 VIN 2 **Transmission:** All	**Post Catalyst Fuel Trim System Too Lean (Bank 1):** Case 1: Monitoring deviation of fuel trim control (long term). No relevant failure. Long term fuel trim active. Case 2: Monitoring deviation of fuel trim control (short term). No relevant failure. Short term fuel trim active. Current engine speed less than 500 rpm. Current mass air flow less than 400mg/rev. Current lambda correction mean value less than 4 percent. **Possible Causes:** • Three way catalytic converter (TWC) • Rear HO2S
DTC: P2097 **2T CCM, MIL: Yes** **Year:** 2009, 2010 **Model:** Amanti, Borrego, Optima, Rio, Rio5, Sorento, Soul, Spectra, Spectra5 **Engine:** 1.6L L4 VIN 1, 1.6L L4 VIN 3, 2.0L L4 VIN 1, 2.0L L4 VIN 2, 2.4L L4 VIN 8, 2.7L V6 VIN 4, 3.3L V6 VIN 5, 3.8L V6 VIN 1, 3.8L V6 VIN 5, 3.8L V6 VIN 6, 4.6L V8 VIN 2 **Transmission:** All	**Post Catalyst Fuel Trim System Too Rich (Bank 1):** Case 1: Monitoring deviation of fuel trim control (long term). No relevant failure. Long term fuel trim active. Case 2: Monitoring deviation of fuel trim control (short term). No relevant failure. Short term fuel trim active. Current engine speed less than 500 rpm. Current mass air flow less than 400mg/rev. Current lambda correction mean value less than 4 percent. **Possible Causes:** • Three way catalytic converter (TWC) • Rear HO2S

DTC	Trouble Code Title, Conditions & Possible Causes
DTC: P2098 **2T CCM, MIL: Yes** **Year:** 2009 **Model:** Amanti, Borrego, Sorento **Engine:** 3.3L V6 VIN 5, 3.8L V6 VIN 1, 3.8L V6 VIN 5, 3.8L V6 VIN 6, 4.6L V8 VIN 2 **Transmission:** All	**Post Catalyst Fuel Trim System Too Lean (Bank 2):** Case 1: Monitoring deviation of fuel trim control (long term). No relevant failure. Long term fuel trim active. Case 2: Monitoring deviation of fuel trim control (short term). No relevant failure. Short term fuel trim active. Current engine speed less than 500 rpm. Current mass air flow less than 400mg/rev. Current lambda correction mean value less than 4 percent. **Possible Causes:** • Three way catalytic converter (TWC) • Rear HO2S
DTC: P2099 **2T CCM, MIL: Yes** **Year:** 2009 **Model:** Amanti, Borrego, Sorento **Engine:** 3.3L V6 VIN 5, 3.8L V6 VIN 1, 3.8L V6 VIN 5, 3.8L V6 VIN 6, 4.6L V8 VIN 2 **Transmission:** All	**Post Catalyst Fuel Trim System Too Rich (Bank 2):** Case 1: Monitoring deviation of fuel trim control (long term). No relevant failure. Long term fuel trim active. Case 2: Monitoring deviation of fuel trim control (short term). No relevant failure. Short term fuel trim active. Current engine speed less than 500 rpm. Current mass air flow less than 400mg/rev. Current lambda correction mean value less than 4 percent. **Possible Causes:** • Three way catalytic converter (TWC) • Rear HO2S
DTC: P2101 **2T CCM** **Year:** 2009, 2010 **Model:** Borrego, Optima **Engine:** 2.4L L4 VIN 8, 2.7L V6 VIN 4, 4.6L V8 VIN 2 **Transmission:** All	**Throttle Actuator Control Motor Circuit Range/Performance:** Hardware check. Battery voltage 9 volts. ECU power stage error. **Possible Causes:** • Poor connection or damaged harness • Faulty ETC motor
DTC: P2104 **2T CCM, MIL: Yes** **Year:** 2009, 2010 **Model:** Amanti, Borrego, Optima, Sorento **Engine:** 2.4L L4 VIN 8, 2.7L V6 VIN 4, 3.3L V6 VIN 5, 3.8L V6 VIN 1, 3.8L V6 VIN 5, 3.8L V6 VIN 6 **Transmission:** All	**Electronic Throttle Control (ETC) System Malfunction- Forced Idle:** This code is set if the system is in forced idle mode. Ignition ON. **Possible Causes:** • Faulty AFS • Faulty AFS plus brake • Faulty AFS plus vehicle speed sensor • Faulty AFS plus brake plus vehicle speed sensor • Faulty PCM
DTC: P2105 **2T CCM, MIL: Yes** **Year:** 2009, 2010 **Model:** Amanti, Borrego, Optima, Sorento **Engine:** 2.4L L4 VIN 8, 2.7L V6 VIN 4, 3.3L V6 VIN 5, 3.8L V6 VIN 1, 3.8L V6 VIN 5, 3.8L V6 VIN 6 **Transmission:** All	**Electronic Throttle Control (ETC) System Malfunction- Forced Engine Shutdown:** This code is set if the system is in forced engine shutdown mode. Ignition ON. **Possible Causes:** • Faulty AFS plus MAPS plus ETS • Faulty PCM
DTC: P2106 **2T CCM, MIL: Yes** **Year:** 2009, 2010 **Model:** Amanti, Borrego, Optima, Sorento **Engine:** 2.4L L4 VIN 8, 2.7L V6 VIN 4, 3.3L V6 VIN 5, 3.8L V6 VIN 1, 3.8L V6 VIN 5, 3.8L V6 VIN 6, 4.6L V8 VIN 2 **Transmission:** All	**Electronic Throttle Control (ETC) System Malfunction- Forced Limited Power:** This code is set if the system is in forced limited power mode. Ignition ON. **Possible Causes:** • Faulty APS • Faulty APS + Brake • Faulty APS + vehicle speed sensor • Faulty APS + vehicle speed sensor + brake • Faulty PCM
DTC: P2118 **2T CCM, MIL: Yes** **Year:** 2009, 2010 **Model:** Optima **Engine:** 2.4L L4 VIN 8, 2.7L V6 VIN 4 **Transmission:** All	**Throttle Actuator Control Motor Circuit Range/Performance/Throttle Actuator Control Motor Circuit High:** Motor circuit High. Ignition switch ON. **Possible Causes:** • Poor connection • Short to battery in ETS motor circuit • Faulty ETS motor • Faulty PCM

DTC	Trouble Code Title, Conditions & Possible Causes
DTC: P2118 **2T CCM, MIL: Yes** **Year:** 2009, 2010 **Model:** Optima **Engine:** 2.4L L4 VIN 8, 2.7L V6 VIN 4 **Transmission:** All	**Throttle Actuator Control Motor Circuit Range/Performance/Throttle Actuator Control Motor Circuit Low:** Motor circuit low. Ignition switch ON. **Possible Causes:** • Poor connection • Short to ground in ETS motor circuit • Faulty ETS motor • Faulty PCM
DTC: P2118 **2T CCM, MIL: Yes** **Year:** 2009, 2010 **Model:** Borrego, Optima **Engine:** 2.4L L4 VIN 8, 2.7L V6 VIN 4, 4.6L V8 VIN 2 **Transmission:** All	**Throttle Actuator Control Motor Current Range/Performance/Throttle Actuator Control Motor Circuit Open:** Vb open. Motor relay ON. Voltage to detect circuit open less than or equal to 4.0 volts. **Possible Causes:** • Poor connection • Open in ETS relay circuit • Faulty ETS relay/fuse • Faulty PCM
DTC: P2119 **2T CCM, MIL: Yes** **Year:** 2009, 2010 **Model:** Optima **Engine:** 2.4L L4 VIN 8, 2.7L V6 VIN 4 **Transmission:** All	**Throttle Actuator Control Module Performance/Throttle Actuator Control System Stuck Closed (IG OFF):** Valve stuck closed (#1). Ignition switch OFF. TPS output as throttle valve is closed less than 0.025 volt. **Possible Causes:** • Poor connector • Faulty throttle valve • Faulty ETS motor • Faulty PCM
DTC: P2122 **2T CCM, MIL: Yes** **Year:** 2009, 2010 **Model:** Amanti, Borrego, Optima, Sorento **Engine:** 2.4L L4 VIN 8, 2.7L V6 VIN 4, 3.3L V6 VIN 5, 3.8L V6 VIN 1, 3.8L V6 VIN 5, 3.8L V6 VIN 6, 4.6L V8 VIN 2 **Transmission:** All	**Throttle/Pedal Position Sensor/Switch "D" Circuit Low Input:** Accelerator position sensor (APS1) low input. ETS/PCM communication is normal. Output voltage of APS1 is less than 0.2 volt. **Possible Causes:** • Poor connector • Faulty APS1 • Open or short in APS1 circuit • Faulty PCM
DTC: P2123 **2T CCM, MIL: Yes** **Year:** 2009, 2010 **Model:** Amanti, Borrego, Optima, Sorento **Engine:** 2.4L L4 VIN 8, 2.7L V6 VIN 4, 3.3L V6 VIN 5, 3.8L V6 VIN 1, 3.8L V6 VIN 5, 3.8L V6 VIN 6, 4.6L V8 VIN 2 **Transmission:** All	**Throttle/Pedal Position Sensor/Switch "D" Circuit High Input:** Accelerator position sensor (APS1) high input. ETS/PCM communication is normal. Output voltage of APS1 is equal to or greater than 4.9 volts. Output voltage of APS2 is less than 4.1 volts. **Possible Causes:** • Poor connector • Faulty APS1 • Open or short in APS1 circuit • Faulty PCM
DTC: P2127 **2T CCM, MIL: Yes** **Year:** 2009, 2010 **Model:** Amanti, Borrego, Optima, Sorento **Engine:** 2.4L L4 VIN 8, 2.7L V6 VIN 4, 3.3L V6 VIN 5, 3.8L V6 VIN 1, 3.8L V6 VIN 5, 3.8L V6 VIN 6, 4.6L V8 VIN 2 **Transmission:** All	**Throttle/Pedal Position Sensor/Switch "E" Circuit Low Input:** Accelerator position sensor (APS2) low input. ETS/PCM communication is normal. Output voltage of APS2 is less than 0.2 volt. **Possible Causes:** • Poor connection • Faulty APS2 • Open or short in APS2 circuit • Faulty PCM
DTC: P2128 **2T CCM, MIL: Yes** **Year:** 2009, 2010 **Model:** Amanti, Borrego, Optima, Sorento **Engine:** 2.4L L4 VIN 8, 2.7L V6 VIN 4, 3.3L V6 VIN 5, 3.8L V6 VIN 1, 3.8L V6 VIN 5, 3.8L V6 VIN 6, 4.6L V8 VIN 2 **Transmission:** All	**Throttle/Pedal Position Sensor/Switch "E" Circuit High Input:** Accelerator position sensor (APS2) high input. ETS/PCM communication is normal. Output voltage of APS2 is greater than or equal to 4.9 volts. Output voltage of ASP1 is less than 4.1 volts. **Possible Causes:** • Poor connection • Faulty APS2 • Open or short in APS2 circuit • Faulty PCM

DTC	Trouble Code Title, Conditions & Possible Causes
DTC: P2135 **2T CCM, MIL: Yes** **Year:** 2009 **Model:** Amanti, Borrego, Sorento **Engine:** 3.3L V6 VIN 5, 3.8L V6 VIN 1, 3.8L V6 VIN 5, 3.8L V6 VIN 6 **Transmission:** All	**Throttle/Pedal Position Sensor/Switch "A"/"B" Voltage Correlation:** Determines if TPS #1 disagrees with TPS #2. Ignition "ON". **Possible Causes:** • Poor connection • Open or short in TPS circuit • Faulty TPS • Faulty PCM
DTC: P2138 **2T CCM, MIL: Yes** **Year:** 2009, 2010 **Model:** Amanti, Borrego, Optima, Sorento **Engine:** 2.4L L4 VIN 8, 2.7L V6 VIN 4, 3.3L V6 VIN 5, 3.8L V6 VIN 1, 3.8L V6 VIN 5, 3.8L V6 VIN 6, 4.6L V8 VIN 2 **Transmission:** All	**Throttle/Pedal Position Sensor/Switch "D/E" Voltage Correlation:** Monitoring abnormal APS. Output voltage of APS1: 0.2 to 4.9 volts. Output voltage of APS2: 0.2 to 4.9 volts. Ignition switch ON. **Possible Causes:** • Poor connection • Faulty APS • Faulty PCM
DTC: P2159 **2T CCM, MIL: Yes** **Year:** 2009, 2010 **Model:** Optima, Soul **Engine:** 1.6L L4 VIN 1, 2.4L L4 VIN 8, 2.7L V6 VIN 4 **Transmission:** All	**Vehicle Speed Sensor "B" Range/Performance:** Plausibility check. Enabling conditions are as follows: engine speed greater than 2100 rpm, air mass flow greater than 0.44 g/rev, no fuel injection shut off, coolant temperature 140 degrees F. **Possible Causes:** • Open or short in harness • Poor connection or damaged harness • VSS
DTC: P2173 **2T CCM, MIL: Yes** **Year:** 2009 **Model:** Amanti, Borrego, Sorento **Engine:** 3.3L V6 VIN 5, 3.8L V6 VIN 1, 3.8L V6 VIN 5, 3.8L V6 VIN 6 **Transmission:** All	**Electronic Throttle Control (ETC) System Malfunction- High Air Flow Detected:** The engine airflow measurements are not based on throttle position. They are compared with throttle position based on estimated air flow. If measured air flow is much higher, the throttle body may not be throttling the engine. Engine running. Throttle actuation mode is not off. MAP sensor is not failed. MAF sensor is not failed. IAT sensor is not failed. **Possible Causes:** • Air leakage between TPS and MAFS • Faulty throttle body • Faulty PCM
DTC: P2187 **2T CCM, MIL: Yes** **Year:** 2009, 2010 **Model:** Amanti, Borrego, Optima, Sorento **Engine:** 2.4L L4 VIN 8, 2.7L V6 VIN 4, 3.3L V6 VIN 5, 3.8L V6 VIN 1, 3.8L V6 VIN 5, 3.8L V6 VIN 6, 4.6L V8 VIN 2 **Transmission:** All	**System Too Lean At Idle (Additive) (Bank 1):** Engine coolant temperature 140 degrees F. Intake air temperature 140 degrees F. System voltage greater than 11 volts. Closed loop active. **Possible Causes:** • Sensors related to fuel trim • Intake system • Fuel pressure • Faulty PCM
DTC: P2188 **2T CCM, MIL: Yes** **Year:** 2009, 2010 **Model:** Amanti, Borrego, Optima, Sorento **Engine:** 2.4L L4 VIN 8, 2.7L V6 VIN 4, 3.3L V6 VIN 5, 3.8L V6 VIN 1, 3.8L V6 VIN 5, 3.8L V6 VIN 6, 4.6L V8 VIN 2 **Transmission:** All	**System Too Rich At Idle (Additive) (Bank 1):** Engine coolant temperature 140 degrees F. Intake air temperature 140 degrees F. System voltage greater than 11 volts. Closed loop active. **Possible Causes:** • Sensors related to fuel trim • Intake system • Fuel pressure • Faulty PCM
DTC: P2189 **2T CCM, MIL: Yes** **Year:** 2009 **Model:** Amanti, Borrego, Sorento **Engine:** 3.3L V6 VIN 5, 3.8L V6 VIN 1, 3.8L V6 VIN 5, 3.8L V6 VIN 6, 4.6L V8 VIN 2 **Transmission:** All	**System Too Lean At Idle (Additive) (Bank 2):** Engine coolant temperature 140 degrees F. Intake air temperature 140 degrees F. System voltage greater than 11 volts. Closed loop active. **Possible Causes:** • Sensors related to fuel trim • Intake system • Fuel pressure • Faulty PCM

DTC	Trouble Code Title, Conditions & Possible Causes
DTC: P2190 **2T CCM, MIL: Yes** **Year:** 2009 **Model:** Amanti, Borrego, Sorento **Engine:** 3.3L V6 VIN 5, 3.8L V6 VIN 1, 3.8L V6 VIN 5, 3.8L V6 VIN 6, 4.6L V8 VIN 2 **Transmission:** All	**System Too Rich At Idle (Additive) (Bank 2):** Engine coolant temperature 140 degrees F. Intake air temperature 140 degrees F. System voltage greater than 11 volts. Closed loop active. **Possible Causes:** • Sensors related to fuel trim • Intake system • Fuel pressure • Faulty PCM
DTC: P2191 **2T CCM, MIL: Yes** **Year:** 2009, 2010 **Model:** Borrego, Optima **Engine:** 2.4L L4 VIN 8, 2.7L V6 VIN 4, 4.6L V8 VIN 2 **Transmission:** All	**System Too Lean At Higher Load (Multiple) (Bank 1):** Fuel trim limit. Coolant temperature greater than 158 degrees F. Intake air temperature less than 176 degrees F. Throttle angle less than 60 percent. Integrated air mass greater than 10 grams. Closed loop control enabled. No transient control phase. No canister purge phase. Air mass1 40 to 80kg/h. Air mass2 greater than 100kg/h. **Possible Causes:** • Faulty ignition system • EVAP PCSV malfunction • Faulty fuel injectors • Leak in exhaust system • Faulty MAP, TPS, ECTS • Faulty front HO2S • Faulty PCM
DTC: P2192 **2T CCM, MIL: Yes** **Year:** 2009, 2010 **Model:** Borrego, Optima **Engine:** 2.4L L4 VIN 8, 2.7L V6 VIN 4, 4.6L V8 VIN 2 **Transmission:** All	**System Too Rich At Higher Load (Bank 1):** Fuel trim limit. Coolant temperature greater than 158 degrees F. Intake air temperature less than 176 degrees F. Throttle angle less than 60 percent. Integrated air mass greater than 10 grams. Closed loop control enabled. No transient control phase. No canister purge phase. Engine load1 30 to 55 percent. Engine load2 greater than 70 percent. **Possible Causes:** • Faulty ignition system • EVAP PCSV malfunction • Faulty fuel injectors • Leak in exhaust system • Faulty MAP, TPS, ECTS • Faulty front HO2S • Faulty PCM
DTC: P2193 **2T PCM, MIL: Yes** **Year:** 2009 **Model:** Borrego **Engine:** 4.6L V8 VIN 2 **Transmission:** All	**System Too Lean at Higher Load (Bank 2):** Enable Conditions • 12.3%< load < 65% • Engine speed > 500rpm • ECT > 70C • IAT < 80C Threshold value • Long Term FT (multiplicative) > 1.23 Diagnosis Time • 10 sec **Possible Causes:** • Air leakage • Improper fuel pressure • PCV valve stuck • Clogging of injector
DTC: P2194 **2T PCM, MIL: Yes** **Year:** 2009 **Model:** Borrego **Engine:** 4.6L V8 VIN 2 **Transmission:** All	**System Too Rich at Higher Load (Bank 2):** Enable Conditions • 12.3%< load < 65% • Engine speed > 500rpm • ECT > 70C • IAT < 80C Threshold value • Long Term FT (multiplicative) < 0.77 Diagnosis Time • 10 sec **Possible Causes:** • Blocking of intake system • Fuel leakage in injector • Improper fuel pressure

DTC	Trouble Code Title, Conditions & Possible Causes
DTC: P2195 **2T CCM, MIL: Yes** **Year:** 2009 **Model:** Amanti, Borrego, Sorento, Spectra, Spectra5 **Engine:** 2.0L L4 VIN 1, 2.0L L4 VIN 2, 3.3L V6 VIN 5, 3.8L V6 VIN 1, 3.8L V6 VIN 5, 3.8L V6 VIN 6 **Transmission:** All	**HO2S Signal Stuck Lean (Bank 1 Sensor 1):** Sensor characteristic line shifted to lean. No relevant failure. No misfire detected. Fuel trim control active. **Possible Causes:** • Contact resistance in connectors • Faulty HO2S • Faulty PCM
DTC: P2196 **2T CCM, MIL: Yes** **Year:** 2009 **Model:** Amanti, Borrego, Sorento, Spectra, Spectra5 **Engine:** 2.0L L4 VIN 1, 2.0L L4 VIN 2, 3.3L V6 VIN 5, 3.8L V6 VIN 1, 3.8L V6 VIN 5, 3.8L V6 VIN 6 **Transmission:** All	**HO2S Signal Stuck Rich (Bank 1 Sensor 1):** Sensor characteristic line shifted to lean. No relevant failure. No misfire detected. Fuel trim control active. **Possible Causes:** • Contact resistance in connectors • Faulty HO2S • Faulty PCM
DTC: P2197 **2T CCM, MIL: Yes** **Year:** 2009 **Model:** Amanti, Borrego, Sorento **Engine:** 3.3L V6 VIN 5, 3.8L V6 VIN 1, 3.8L V6 VIN 5, 3.8L V6 VIN 6 **Transmission:** All	**HO2S Signal Stuck Lean (Bank 2 Sensor 1):** Determines if O2 sensor indicates lean exhaust while in power enrichment. Sensor not in cooled status flag. Not in transient conditions status flag. Device control not active. Engine running. Minimum air flow present is equal or greater than 2 g/s. Engine coolant warm (140 degrees F. Above conditions met for at least 1.5L seconds. **Possible Causes:** • Poor connection • Faulty HO2S • Faulty PCM
DTC: P2198 **2T CCM, MIL: Yes** **Year:** 2009 **Model:** Amanti, Borrego, Sorento **Engine:** 3.3L V6 VIN 5, 3.8L V6 VIN 1, 3.8L V6 VIN 5, 3.8L V6 VIN 6 **Transmission:** All	**HO2S Signal Stuck Rich (Bank 2 Sensor 1):** Determines if O2 sensor indicates rich exhaust while in decal fuel cut off (DFCO). Sensor not in cooled status flag. Not in transient conditions status flag. Device control not active. Engine running. Minimum air flow present is equal or greater than 2 g/s. Ignition voltage equal to or greater than 10 volts. Fuel reduction not active. Engine running long enough (more than 60 seconds). Engine coolant warm (140 degrees F. Above conditions met for at least 1.5L seconds. **Possible Causes:** • Poor connection • Faulty HO2S • Faulty PCM
DTC: P2226 **2T CCM, MIL: Yes** **Year:** 2009, 2010 **Model:** Rio, Rio5, Soul **Engine:** 1.6L L4 VIN 1, 1.6L L4 VIN 3 **Transmission:** All	**Barometric Pressure Circuit:** Rationality check. **Possible Causes:** • Clog at sensing hole • Faulty ECM
DTC: P2227 **2T CCM, MIL: Yes** **Year:** 2009, 2010 **Model:** Rio, Rio5, Soul **Engine:** 1.6L L4 VIN 1, 1.6L L4 VIN 3 **Transmission:** All	**Barometric Pressure Circuit Range/Performance:** Rationality check. **Possible Causes:** • Clog at sensing hole • Faulty ECM
DTC: P2228 **2T CCM, MIL: Yes** **Year:** 2009, 2010 **Model:** Rio, Rio5, Soul **Engine:** 1.6L L4 VIN 1, 1.6L L4 VIN 3 **Transmission:** All	**Barometric Pressure Circuit Low Input:** Signal check low. **Possible Causes:** • Faulty ECM

DTC	Trouble Code Title, Conditions & Possible Causes
DTC: P2229 **2T CCM, MIL: Yes** **Year:** 2009, 2010 **Model:** Rio, Rio5, Soul **Engine:** 1.6L L4 VIN 1, 1.6L L4 VIN 3 **Transmission:** All	**Barometric Pressure Circuit High Input:** Signal check, high. **Possible Causes:** • Faulty ECM
DTC: P2231 **2T CCM, MIL: Yes** **Year:** 2009 **Model:** Spectra, Spectra5 **Engine:** 2.0L L4 VIN 1, 2.0L L4 VIN 2 **Transmission:** All	**HO2S Signal Circuit Shorted To Heater Circuit (Bank 1 Sensor 1):** Front HO2S signal monitoring. Exhaust temperature greater than 752 degrees F. No relevant failure. Amplitude of forced lambda simulation less than 0.05. Period time of forced lambda simulation less than 2.55 seconds. Current engine speed less than 500 rpm. Current mass air flow less than 400 mg/rev. **Possible Causes:** • Contact resistance in connectors • Interference in HO2S
DTC: P2232 **2T CCM, MIL: Yes** **Year:** 2009, 2010 **Model:** Rio, Rio5, Soul **Engine:** 1.6L L4 VIN 1, 1.6L L4 VIN 3 **Transmission:** All	**HO2S Signal Circuit Shorted To Heater Circuit (Bank 1 Sensor 2):** Rationality check. **Possible Causes:** • Poor connection • Short to power in signal circuit • B1S2 • Faulty ECM
DTC: P2237 **2T CCM, MIL: Yes** **Year:** 2009 **Model:** Spectra, Spectra5 **Engine:** 2.0L L4 VIN 1, 2.0L L4 VIN 2 **Transmission:** All	**HO2S Pumping Current Circuit/Open Bank 1, Sensor 1:** Open circuit of front HO2S circuit. No relevant failure. **Possible Causes:** • Contact resistance in connectors • Open or short to ground in HO2S circuit • Front HO2S sensor
DTC: P2243 **2T CCM, MIL: Yes** **Year:** 2009 **Model:** Spectra, Spectra5 **Engine:** 2.0L L4 VIN 1, 2.0L L4 VIN 2 **Transmission:** All	**HO2S Reference Voltage Circuit/Open Bank 1, Sensor 1:** Open circuit of front HO2S circuit. No relevant failure. **Possible Causes:** • Contact resistance in connectors • Open or short to ground in HO2S circuit • Front HO2S sensor
DTC: P2251 **2T CCM, MIL: Yes** **Year:** 2009 **Model:** Spectra, Spectra5 **Engine:** 2.0L L4 VIN 1, 2.0L L4 VIN 2 **Transmission:** All	**HO2S Reference Ground Circuit/Open Bank 1, Sensor 1:** Open circuit of front HO2S circuit. No pump current malfunction. **Possible Causes:** • Contact resistance in connectors • Open or short to ground in HO2S circuit • Front HO2S sensor
DTC: P2270 **2T CCM, MIL: Yes** **Year:** 2009 **Model:** Amanti, Borrego, Sorento, Spectra, Spectra5 **Engine:** 2.0L L4 VIN 1, 2.0L L4 VIN 2, 3.3L V6 VIN 5, 3.8L V6 VIN 1, 3.8L V6 VIN 5, 3.8L V6 VIN 6 **Transmission:** All	**HO2S Signal Stuck Lean (Bank 1 Sensor 2):** Plausibility check during shift of lambda set point to rich from lean. No fuel cut off. No full load phase. No fuel trim error detected. Delay time to start diagnosis: 13 to 30 seconds. No relevant failure. **Possible Causes:** • Three way catalytic converter (TWC) • Air leakage in exhaust system • Faulty rear HO2S sensor
DTC: P2271 **2T CCM, MIL: Yes** **Year:** 2009 **Model:** Amanti, Borrego, Sorento, Spectra, Spectra5 **Engine:** 2.0L L4 VIN 1, 2.0L L4 VIN 2, 3.3L V6 VIN 5, 3.8L V6 VIN 1, 3.8L V6 VIN 5, 3.8L V6 VIN 6 **Transmission:** All	**O2 Signal Stuck Rich (Bank 1/2 Sensor 1):** Plausibility check during shift of lambda set point to rich from lean. No fuel cut off. No full load phase. No fuel trim error detected. Delay time to start diagnosis: 13 to 30 seconds. No relevant failure. **Possible Causes:** • Three way catalytic converter (TWC) • Air leakage in exhaust system • Faulty rear HO2S sensor

DTC	Trouble Code Title, Conditions & Possible Causes
DTC: P2272 **2T CCM, MIL: Yes** **Year:** 2009 **Model:** Amanti, Borrego, Sorento **Engine:** 3.3L V6 VIN 5, 3.8L V6 VIN 1, 3.8L V6 VIN 5, 3.8L V6 VIN 6 **Transmission:** All	**HO2S Signal Stuck Lean (Bank 2 Sensor 2):** Determines if O2 sensor indicates lean exhaust while in power enrichment mode. Sensor not in cooled status flag. Not in transient conditions status flag. Device control not active. Engine running. Minimum air flow present is equal or greater than 2 g/s. Ignition voltage equal to or greater than 10 volts. Fuel reduction not active. Engine running long enough (more than 60 seconds). Engine coolant warm (140 degrees F. Above conditions met for at least 2.5 seconds. **Possible Causes:** • Poor connection • Faulty HO2S • Faulty PCM
DTC: P2273 **2T CCM, MIL: Yes** **Year:** 2009 **Model:** Amanti, Borrego, Sorento **Engine:** 3.3L V6 VIN 5, 3.8L V6 VIN 1, 3.8L V6 VIN 5, 3.8L V6 VIN 6 **Transmission:** All	**HO2S Signal Stuck Rich (Bank 2 Sensor 2):** Determines if O2 sensor indicates rich exhaust while in decal fuel cut off (DFCO). Sensor not in cooled status flag. Not in transient conditions status flag. Device control not active. Engine running. Minimum air flow present is equal or greater than 2 g/s. Ignition voltage equal to or greater than 10 volts. Fuel reduction not active. Engine running long enough (more than 60 seconds). Engine coolant warm (140 degrees F. Above conditions met for at least 2.0L seconds. **Possible Causes:** • Poor connection • Faulty HO2S • Faulty PCM
DTC: P2414 **2T CCM, MIL: Yes** **Year:** 2009 **Model:** Spectra, Spectra5 **Engine:** 2.0L L4 VIN 1, 2.0L L4 VIN 2 **Transmission:** All	**HO2S Exhaust Sample Error Bank 1 Sensor 1:** Sensor not mounted. Plausibility check in part load or full load conditions. Sensor tip temperature 1202 degrees F. Part load or full load. No relevant failure. **Possible Causes:** • Incorrect installation of HO2S sensor • Contact resistance in connectors
DTC: P2422 **2T CCM, MIL: Yes** **Year:** 2009 **Model:** Amanti, Sorento **Engine:** 3.3L V6 VIN 5, 3.8L V6 VIN 5, 3.8L V6 VIN 6 **Transmission:** All	**Evaporative Emission System- Canister Clogging:** Ignition voltage 10-16 volts. Barometric pressure 72kpa. Engine run time, one second. **Possible Causes:** • Faulty canister close valve • Clogging of canister air filter • Open in ground harness of FTPS • Faulty PCM
DTC: P2422 **2T PCM, MIL: Yes** **Year:** 2009 **Model:** Borrego **Engine:** 3.8L V6 VIN 1 **Transmission:** All	**Evaporative Emission System Vent Valve Stuck Closed:** Enable Conditions • 10 V < Battery voltage < 16 V • Barometric pressure > 72 kPa (0.72 bar) • Engine coolant temperature at startup - Intake air temperature at startup < 6.7°C(12 °F) • Engine coolant temperature at startup: 4.5 ~ 35°C(40 ~ 95 °F) • Intake air temperature at startup: 4.5 ~ 35°C(40 ~ 95 °F) • Fuel level: 15 ~ 85 % Threshold value • Fuel tank's vacuum at purging > a prescribed threshold Diagnosis Time • Continuous **Possible Causes:** • Faulty Canister Close Valve • Clogging of canister air filter
DTC: P2507 **1T PCM, MIL: Yes** **Year:** 2009 **Model:** Amanti, Borrego, Sorento **Engine:** 3.3L V6 VIN 5, 3.8L V6 VIN 1, 3.8L V6 VIN 5, 3.8L V6 VIN 6 **Transmission:** All	**ECM/PCM Power Input Signal Low:** DTC Strategy • Monitor the battery power input line Threshold value • Open or short to ground in line Diagnosis Time • Continuous (More than 5 sec. failure for every 10 sec. test) **Possible Causes:** • Poor connection • Open or short to ground in line • Faulty PCM

DTC	Trouble Code Title, Conditions & Possible Causes
DTC: P2610 **2T CCM, MIL: Yes** **Year:** 2009 **Model:** Amanti, Borrego, Sorento **Engine:** 3.3L V6 VIN 5, 3.8L V6 VIN 1, 3.8L V6 VIN 5, 3.8L V6 VIN 6, 4.6L V8 VIN 2 **Transmission:** All	**ECM/PCM Internal Engine Off Timer Performance:** The LPC SPI diagnostic allows the low power counter to count down and simultaneously enables a test timer to run for a calibrated length of time and then compares the lapsed time recorded by the counter to make a pass/fail determination. Engine running. Enough time (10 seconds). Battery voltage 8 volts. No memory failure. **Possible Causes:** • Faulty PCM
DTC: P2626 **2T CCM, MIL: Yes** **Year:** 2009 **Model:** Spectra, Spectra5 **Engine:** 2.0L L4 VIN 1, 2.0L L4 VIN 2 **Transmission:** All	**HO2S Pumping Current Trim Circuit Open Bank 1 Sensor 1:** Check open circuit of front HO2S sensor. **Possible Causes:** • Contact resistance in connectors • Open or short to ground in HO2S circuit • Faulty canister • Faulty front HO2S sensor
DTC: P2A00 **2T CCM, MIL: Yes** **Year:** 2009 **Model:** Amanti, Borrego, Sorento **Engine:** 3.3L V6 VIN 5, 3.8L V6 VIN 1, 3.8L V6 VIN 5, 3.8L V6 VIN 6 **Transmission:** All	**O2 Sensor Not Ready (Bank 1 Sensor 1):** Detects loss of O2 ready status, which would lead to open loop fueling operation, a default mode. Engine running. Ignition ON. DFCO not present too long (less than 15 seconds). No disabling faults present. All of the above for at least 20 seconds. **Possible Causes:** • Poor connection • Faulty HO2S • Faulty PCM
DTC: P2A01 **2T O2S2, MIL: Yes** **Year:** 2009 **Model:** Amanti, Borrego, Sorento **Engine:** 3.3L V6 VIN 5, 3.8L V6 VIN 1, 3.8L V6 VIN 5, 3.8L V6 VIN 6 **Transmission:** All	**HO2S Circuit Range/Performance (Bank 1 / Sensor 2):** Enable Conditions • After engine warming-up • Deceleration fuel cut-off state • No other disabling faults Threshold value • The average time for voltage drop > approx. 0.3 seconds Diagnosis Time • During deceleration fuel cut-off **Possible Causes:** • Poor connection • Faulty HO2S • Faulty PCM
DTC: P2A03 **2T CCM, MIL: Yes** **Year:** 2009 **Model:** Amanti, Borrego, Sorento **Engine:** 3.3L V6 VIN 5, 3.8L V6 VIN 1, 3.8L V6 VIN 5, 3.8L V6 VIN 6 **Transmission:** All	**O2 Sensor Not Ready (Bank 1 Sensor 2):** Detects loss of O2 ready status, which would lead to open loop fueling operation, a default mode. Engine running. Ignition ON. DFCO not present too long (less than 15 seconds). No disabling faults present. All of the above for at least 20 seconds. **Possible Causes:** • Poor connection • Faulty HO2S • Faulty PCM
DTC: P2A04 **2T O2S2, MIL: Yes** **Year:** 2009 **Model:** Amanti, Borrego, Sorento **Engine:** 3.3L V6 VIN 5, 3.8L V6 VIN 1, 3.8L V6 VIN 5, 3.8L V6 VIN 6 **Transmission:** All	**HO2S Circuit Range/Performance (Bank 2 / Sensor 2):** Enable Conditions • After engine warming-up • Deceleration fuel cut-off state • No other disabling faults Threshold value • The average time for voltage drop > approx. 0.3 seconds Diagnosis Time • During deceleration fuel cut-off **Possible Causes:** • Poor connection • Faulty HO2S • Faulty PCM

OBD II Trouble Code List (U0XXX Codes)

DTC	Trouble Code Title, Conditions & Possible Causes
DTC: U0001 **2T PCM, MIL: Yes** **Year:** 2009, 2010 **Model:** Amanti, Borrego, Optima, Rio, Rio5, Sedona, Sorento, Soul, Sportage **Engine:** 1.6L L4 VIN 1, 1.6L L4 VIN 3, 2.0L L4 VIN 2, 2.4L L4 VIN 8, 2.7L V6 VIN 4, 3.3L V6 VIN 5, 3.8L V6 VIN 1, 3.8L V6 VIN 3, 3.8L V6 VIN 5, 3.8L V6 VIN 6, 4.6L V8 VIN 2 **Transmission:** All	**CAN Communication Malfunction:** Enable Conditions • Engine Run Time ≥ 2sec. • Ignition Voltage ≥ 11V Threshold value • CAN communication error Diagnosis Time • Continuous **Possible Causes:** • CAN BUS • CAN communication module component
DTC: U0101 **2T PCM, MIL: Yes** **Year:** 2009, 2010 **Model:** Borrego, Rio, Rio5, Sorento, Soul, Sportage **Engine:** 1.6L L4 VIN 1, 1.6L L4 VIN 3, 2.0L L4 VIN 2, 2.7L V6 VIN 3, 3.3L V6 VIN 5, 3.8L V6 VIN 1, 3.8L V6 VIN 6, 4.6L V8 VIN 2 **Transmission:** All	**Lost Communication With TCM:** Enable Conditions • Engine Run Time ≥ 2sec. • Ignition Voltage ≥ 11V Threshold value • CAN communication error with TCM Diagnosis Time • Continuous **Possible Causes:** • TCM faulty • CAN communication • line between ECM and TCM
DTC: U0122 **T PCM** **Year:** 2009 **Model:** Borrego **Engine:** 4.6L V8 VIN 2 **Transmission:** All	**CAN TCS Communication:** Enable Conditions • B+ > 10.8V • Time after Ignition switch on = 0.5 sec Threshold value • Duration of communication lost with TCS > 2 sec Diagnosis Time • Continuous MIL On Condition • No MIL **Possible Causes:** • TCS faulty • CAN communication line between ECM and TCS

LEXUS

ES350

23

BRAKES 23-8

CHASSIS ELECTRICAL 23-13

DRIVE TRAIN 23-13

ENGINE COOLING 23-16

ENGINE ELECTRICAL 23-18

ENGINE MECHANICAL......23-20

ENGINE PERFORMANCE & EMISSION CONTROLS 23-40

SPECIFICATIONS AND MAINTENANCE CHARTS

ENGINE AND VEHICLE IDENTIFICATION

Engine							Model Year	
Code ①	Liters (cc)	Cu. In.	Cyl.	Fuel Sys.	Engine Type	Eng. Mfg.	Code ②	Year
2GR-FE	3.5 (3456)	211	6	SFI	DOHC	Toyota	9	2009
							A	2010

SFI: Sequential Fuel Injection

DOHC: Double Overhead Camshaft

NA: Information not available

① Stamped on the left side of the engine block

② 10th digit of the Vehicle Identification Number (VIN)

3768X_ES35_C0001

GENERAL ENGINE SPECIFICATIONS

Year	Model	Engine Displacement Liters	Engine Series ID	Net Horsepower @ rpm	Net Torque @ rpm (ft. lbs.)	Bore x Stroke (in.)	Com-pression Ratio	Oil Pressure @ rpm
2009	ES350	3.5	2GR-FE	272@6200	254@4700	3.70x3.27	10.8:1	36-78@3000
2010	ES350	3.5	2GR-FE	272@6200	254@4700	3.70x3.27	10.8:1	36-78@3000

3768X_ES35_C0002

ENGINE TUNE-UP SPECIFICATIONS

Year	Engine Displacement Liters	Engine ID	Spark Plug Gap (in.)	Ignition Timing (deg.)*	Fuel Pump (psi)	Idle Speed (rpm)	Valve Clearance	
							Intake	Exhaust
2009	3.5	2GR-FE	0.043	NA	44-50	650-750	NA	NA
2010	3.5	2GR-FE	0.043	NA	44-50	650-750	NA	NA

NOTE: The Vehicle Emission Control Information label often reflects specification changes made during production.

The label figures must be used if they differ from those in this chart.

NA: Not available

3768X_ES35_C0003

CAPACITIES

Year	Model	Engine Displacement Liters	Engine ID	Engine Oil with Filter (qts.)	Transmission (qts.) 5-Spd	Transmission (qts.) Auto.*	Transfer Case (pts.)	Drive Axle Front (pts.)	Drive Axle Rear (pts.)	Fuel Tank (gal.)	Cooling System (qts.)
2009	ES350	3.5	2GR-FE	6.4	—	6.8	—	—	—	18.5	9.5
2010	ES350	3.5	2GR-FE	6.4	—	6.8	—	—	—	18.5	9.5

3768X_ES35_C0004

FLUID SPECIFICATIONS

Year	Model	Engine Displacement Liters	Engine ID/VIN	Engine Oil	Auto. Trans.	Drive Axle	Power Steering Fluid	Brake Master Cylinder
2009	ES350	3.5	2GR-FE	5W-20	—	—	ATF Dexron II Or III	DOT 3
2010	ES350	3.5	2GR-FE	5W-20	—	—	ATF Dexron II Or III	DOT 3

DOT: Department Of Transpotation

3768X_ES35_C0005

VALVE SPECIFICATIONS

Year	Engine Displacement Liters	Engine ID	Seat Angle (deg.)	Face Angle (deg.)	Spring Test Pressure (lbs. @ in.)	Spring Installed Height (in.)	Stem-to-Guide Clearance (in.) Intake	Stem-to-Guide Clearance (in.) Exhaust	Stem Diameter (in.) Intake	Stem Diameter (in.) Exhaust
2009	3.5	2GR-FE	45	44.5	NA	NA	0.0010-0.0024	0.0012-0.0026	0.2154-0.2159	0.2151-0.2157
2010	3.5	2GR-FE	45	44.5	NA	NA	0.0010-0.0024	0.0012-0.0026	0.2154-0.2159	0.2151-0.2157

NA: Information not available

3768X_ES35_C0006

CRANKSHAFT AND CONNECTING ROD SPECIFICATIONS

All measurements are given in inches.

Year	Engine Displacement Liters	Engine ID	Crankshaft Main Brg. Journal Dia.	Crankshaft Main Brg. Oil Clearance	Crankshaft Shaft End-play	Crankshaft Thrust on No.	Connecting Rod Journal Diameter	Connecting Rod Oil Clearance	Connecting Rod Side Clearance
2009	3.5	2GR-FE	2.4011-2.4016	0.0010-0.0019	0.0016-0.0095	2	2.0863-2.0866	0.0018-0.0026	0.0059-0.0157
2010	3.5	2GR-FE	2.4011-2.4016	0.0010-0.0019	0.0016-0.0095	2	2.0863-2.0866	0.0018-0.0026	0.0059-0.0157

3768X_ES35_C0007

PISTON AND RING SPECIFICATIONS

All measurements are given in inches.

Year	Engine Displ. Liters	Engine ID	Piston Clearance	Ring Gap Top Comp.	Ring Gap Bottom Comp.	Ring Gap Oil Control	Ring Side Clearance Top Comp.	Ring Side Clearance Bottom Comp.	Ring Side Clearance Oil Control
2009	3.5	2GR-FE	0.0018-0.0020	0.0098-0.0138	0.0197-0.0236	0.0039-0.0157	0.0008-0.0028	0.0008-0.0024	0.0028-0.0059
2010	3.5	2GR-FE	0.0018-0.0020	0.0098-0.0138	0.0197-0.0236	0.0039-0.0157	0.0008-0.0028	0.0008-0.0024	0.0028-0.0059

3768X_ES35_C0008

TORQUE SPECIFICATIONS

All readings in ft. lbs.

Year	Engine Displacement Liters	Engine ID	Cylinder Head Bolts	Main Bearing Bolts	Rod Bearing Bolts	Crankshaft Damper Bolts	Flywheel Bolts	Manifold Intake	Manifold Exhaust	Spark Plugs	Oil Pan Drain Plug
2009	3.5	2GR-FE	①	②	③	184	61	15	15	13	30
2010	3.5	2GR-FE	①	②	③	184	61	15	15	13	30

① Step 1: 10mm bolts to 27 ft. lbs.
Step 2: 10mm point cap bolts plus 90 degrees
Step 3: 10mm point cap bolts plus 90 degrees

② Step 1: 16 cap bolts to 45 ft. lbs.
Step 2: 16 cap bolts plus 90 degrees
Step 3: 8 side bolts to 38 ft. lbs.
Step 4: Front bolts to 22 ft. lbs.

③ Step 1: 18 ft. lbs.
Step 2: Plus 90 degrees

3768X_ES35_C0009

WHEEL ALIGNMENT

Year	Model		Caster Range (+/-Deg.)	Caster Preferred Setting (Deg.)	Camber Range (+/-Deg.)	Camber Preferred Setting (Deg.)	Toe-in (in.)	Steering Axis Inclination (Deg.)
2009	ES350	Front	0.75	2.95	0.75	-0.72	0+/-0.08	12.33
		Rear	—	—	0.75	-1.35	0.16+/-0.08	—
2010	ES350	Front	0.75	2.95	0.75	-0.72	0+/-0.08	12.33
		Rear	—	—	0.75	-1.35	0.16+/-0.08	—

3768X_ES35_C0010

TIRE, WHEEL AND BALL JOINT SPECIFICATIONS

Year	Model	OEM Tires Standard	OEM Tires Optional	Tire Pressures (psi) Front	Tire Pressures (psi) Rear	Wheel Size	Ball Joint Inspection	Lug Nut Torque (ft. lbs.)
2009	ES350	P215/55R17	—	32	32	7-J	①	76
2010	ES350	P215/55R17	—	32	32	7-J	①	76

OEM: Original Equipment Manufacturer

PSI: Pounds Per Square Inch

STD: Standard

OPT: Optional

① Replace if any measurable movement is found.

3768X_ES35_C0011

BRAKE SPECIFICATIONS

All measurements in inches unless noted

Year	Model		Original Thickness	Brake Disc Minimum Thickness	Maximum Runout	Minimum Lining Thickness	Brake Caliper Bracket Bolts (ft. lbs.)	Brake Caliper Mounting Bolts (ft. lbs.)
2009	ES350	F	1.102	0.983	0.0020	0.039	79	25
		R	0.390	0.334	0.0059	0.039	46	20
2010	ES350	F	1.102	0.983	0.0020	0.039	79	25
		R	0.390	0.334	0.0059	0.039	46	20

F: Front

R: Rear

3768X_ES35_C0012

SCHEDULED MAINTENANCE INTERVALS

LEXUS—ES350

TO BE SERVICED	TYPE OF SERVICE	VEHICLE MILEAGE INTERVAL (x1000)												
		5	10	15	20	25	30	35	40	45	50	55	60	90
Engine oil & filter	R	✓	✓	✓	✓	✓	✓	✓	✓	✓	✓	✓	✓	✓
Automatic transmission fluid	S/I			✓			✓			✓			✓	✓
Ball joints & dust covers	S/I			✓			✓			✓			✓	✓
Bolts & nuts on chassis & body	S/I			✓			✓			✓			✓	✓
Brake linings & drums	S/I	✓	✓	✓	✓	✓	✓	✓	✓	✓	✓	✓	✓	✓
Brake line pipes & hoses	S/I			✓			✓			✓			✓	✓
Brake pads & discs (front & rear)	S/I	✓	✓	✓	✓	✓	✓	✓	✓	✓	✓	✓	✓	✓
Brake fluid	R						✓						✓	✓
Rack and pinion assembly	S/I			✓			✓			✓			✓	✓
Steering linkage & boots	S/I			✓			✓			✓			✓	✓
Air cleaner filter	R						✓						✓	✓
Spark plugs ①	R													
Drive belts	S/I												✓	✓
Exhaust pipes & mountings	S/I			✓			✓			✓			✓	✓
Fuel lines & connections	S/I						✓						✓	✓
Engine coolant ②	S/I			✓			✓			✓			✓	✓
Fuel tank cap gasket	S/I						✓						✓	✓
Rotate tires	S/I			✓			✓			✓			✓	
Clean air conditioning filter ③	S/I			✓			✓			✓			✓	
Axle shaft bolts	S/I			✓			✓			✓			✓	✓
Brake pad thickness and rotor runout	S/I						✓						✓	✓

R: Replace S/I: Service or Inspect

① Spark plugs are replaced at 120,000 miles

② Replace engine coolant at 100,000 miles and then inspect every 15,000 miles

③ Replace air conditioning filter every 30,000 miles

FREQUENT OPERATION MAINTENANCE (SEVERE SERVICE)

If a vehicle is operated under any of the following conditions it is considered severe service:

- Extremely dusty areas.

- 50% or more of the vehicle operation is in 32°C (90°F) or higher temperatures, or constant temperatures below 0°C (32°F).

- Prolonged idling (vehicle operation in stop and go traffic).

- Frequent short running periods (engine does not warm to normal operating temperatures).

- Police, taxi, delivery usage or trailer towing usage.

Air cleaner filter: service or inspect every 5000 miles

Ball joints & dust covers: service or inspect every 5000 miles.

Bolts & nuts on chassis & body: service or inspect every 5000 miles.

Axle shaft bolts: service or inspect every 5000 miles.

Steering linkage: service or inspect every 5000 miles.

3768X_ES35_C0013

BRAKES — INFORMATION AND PRECAUTIONS

ANTI-LOCK SYSTEMS

• Certain components within the ABS system are not intended to be serviced or repaired individually.

• Do not use rubber hoses or other parts not specifically specified for and ABS system. When using repair kits, replace all parts included in the kit. Partial or incorrect repair may lead to functional problems and require the replacement of components.

• Lubricate rubber parts with clean, fresh brake fluid to ease assembly. Do not use shop air to clean parts; damage to rubber components may result.

• Use only DOT 3 brake fluid from an unopened container.

• If any hydraulic component or line is removed or replaced, it may be necessary to bleed the entire system.

• A clean repair area is essential. Always clean the reservoir and cap thoroughly before removing the cap. The slightest amount of dirt in the fluid may plug an orifice and impair the system function. Perform repairs after components have been thoroughly cleaned; use only denatured alcohol to clean components. Do not allow ABS components to come into contact with any substance containing mineral oil; this includes used shop rags.

• The Anti-Lock control unit is a microprocessor similar to other computer units in the vehicle. Ensure that the ignition switch is **OFF** before removing or installing controller harnesses. Avoid static electricity discharge at or near the controller.

• If any arc welding is to be done on the vehicle, the control unit should be unplugged before welding operations begin.

DISC AND DRUM SYSTEMS

CAUTION

Dust and dirt accumulating on brake parts during normal use may contain asbestos fibers from production or aftermarket brake linings. Breathing excessive concentrations of asbestos fibers can cause serious bodily harm. Exercise care when servicing brake parts. Do not sand or grind brake lining unless equipment used is designed to contain the dust residue. Do not clean brake parts with compressed air or by dry brushing. Cleaning should be done by dampening the brake components with a fine mist of water, then wiping the brake components clean with a dampened cloth. Dispose of cloth and all residue containing asbestos fibers in an impermeable container with the appropriate label. Follow practices prescribed by the Occupational Safety and Health Administration (OSHA) and the Environmental Protection Agency (EPA) for the handling, processing, and disposing of dust or debris that may contain asbestos fibers.

BRAKES — BLEEDING THE BRAKE SYSTEM

BLEEDING PROCEDURE

Except ABS

See Figure 1.

If any work is done on the brake system or if air in the brake lines is suspected, bleed the system of air.

➡Do not let brake fluid remain on painted surfaces. Wash it off immediately.

1. Fill the reservoir with brake fluid.

➡If the master cylinder has been disassembled or if the reservoir becomes empty, bleed the air from the master cylinder.

2. Bleed the brake master cylinder as follows:
   a. Disconnect the brake lines from the master cylinder.
   b. Slowly depress the brake pedal and hold it.
   c. Block off the outer holes with your fingers, and release the brake pedal.
   d. Repeat the previous 2 steps 3 or 4 times.
3. Bleed the brake line as follows:
   a. Connect the vinyl tube to the brake caliper.

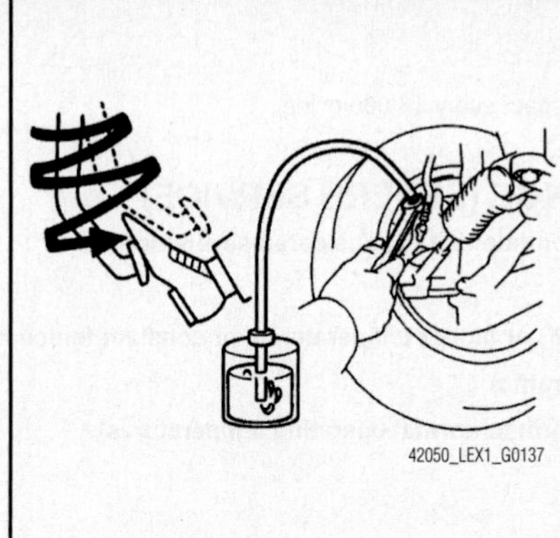

Fig. 1 Bleeding the brake line

   b. Depress the brake pedal several times, then loosen the bleeder plug with the pedal held down.
   c. At the point when fluid stops coming out, tighten the bleeder plug, then release the brake pedal.
   d. Repeat the previous 2 steps until all the air in the fluid has been bled out.
   e. Repeat the above procedure to bleed the air out of the brake line for each wheel.
   f. Tighten the bleeder plug to 8 ft. lbs. (11 Nm).
4. Bleed the brake actuator as follows:
   a. Remove the reservoir cap.
   b. Install the SST 09992-00242, 09992-00350 to the reservoir.
   c. Connect the vinyl tube to the bleeder plug of the brake actuator.
   d. Using SST, apply the 14.2 psi (98.1kpa) of pressure to the reservoir.
   e. Loosen the bleeder plug.
   f. Bleed the air out of the brake actuator, tighten the bleeder plug to 74 inch lbs. (8.3 Nm).
5. Check the fluid level and add fluid if necessary.

ABS System

➡After performing the usual air bleeding in the brake system, if the height or feel of the brake pedal cannot be obtained, perform air bleeding in the brake actuator assembly with a hand–held tester by following procedures below. Make sure that the brake fluid in the master cylinder reservoir tank does not become empty.

1. Depress the brake pedal more than 20 times with the engine off.
2. Connect the hand–held tester to the DLC3, then turn the ignition switch to the ON position.
3. Do not start the engine.
4. Select "AIR BLEEDING" on the hand–held tester. Please refer to the

Hand–Held Tester Operator's Manual for further details.

5. Bleed the air out of the regular brake line in"Step1: Increase" on the hand_held tester display. Perform the air bleeding by following the steps displayed on the hand–held tester. Make sure that the brake fluid in the master cylinder reservoir tank does not become empty.

6. Connect the vinyl tube to either one of the bleeder plugs.

7. Depress the brake pedal several times, then loosen the bleeder plug of one of the above wheels with the pedal depressed.

8. When fluid stops coming out, tighten the bleeder plug, then release the brake pedal.

9. Repeat (2) and (3) until all air in the fluid is completely bled out.

10. Tighten the bleeder plug to 73 inch lbs. (8.3 Nm).

11. Repeat the above procedure to bleed the air out of the brake line for each wheel.

12. Bleed the air out of the suction line in"Step2: Inhalation" on the hand–held tester display.

13. Connect the vinyl tube to the bleeder plug at the right front wheel or the right rear wheel and loosen the bleeder plug.

14. Operate the brake actuator assembly using the hand–held tester to bleed the air.

➡The operation stops automatically in 4 seconds. At this time, be sure to release the brake pedal.

15. Check that the operation has stopped, by referring to the hand–held tester display.

16. Repeat (2) and (3) until all the air in the fluid is completely bled out.

17. Tighten the bleeder plug.

18. For the rest of the wheels, bleed the air in the same way as stated in the above procedure.

19. Bleed the air out of the pressure reduction line in"Step3: Decrease" on the hand–held tester display.

20. Connect a vinyl tube to either one of the bleeder plugs.

21. Loosen the bleeder plug.

22. Using the hand–held tester, operate the brake actuator assembly using hand–held tester, completely depress the brake pedal and keep it.

➡The operation stops automatically in 4 seconds. When performing this procedure continuously, an interval of at least 20 seconds is required. When the operation is completed, the brake pedal slightly goes down. This is a normal phenomenon caused when the solenoid opens. During this procedure, the pedal seems heavy, but completely depress it so that the brake fluid comes out from the bleeder plug. Be sure to keep depressing the brake pedal. Never depress and release the pedal repeatedly.

23. Tighten the bleeder plug, and then release the brake pedal.

24. Repeat 3 previous steps until all the air in the fluid is completely bled out.

25. Tighten the bleeder plug.

26. Repeat the above procedure to bleed the air out of the brake line for each wheel.

27. Bleed the air out of the regular brake line again in"Step4: Increase" on the hand–held tester display.

28. Connect the vinyl tube to either one of the bleeder plug.

29. Depress the brake pedal several times, then loosen the bleeder plug of one of the above wheels with the pedal depressed.

30. When fluid stops coming out, tighten the bleeder plug, then release the brake pedal.

31. Repeat the previous 2 steps until all the air in the fluid is completely bled out.

32. Tighten the bleeder plug.

33. Repeat the above procedure to bleed the air out of the brake line for each wheel.

WHEEL SPEED SENSORS

REMOVAL & INSTALLATION

1. Before servicing the vehicle, refer to the Precautions Section.
2. Remove the front wheel.
3. Remove the 3 screws and the front wheel opening extension pad
4. Remove the front fender liner.
5. Remove the front speed sensor, as follows:
   a. Disconnect the speed sensor connector and clamp.
   b. Remove the 2 bolts and separate the speed sensor harness from the body and shock absorber assembly.
   c. Disengage the 2 claws from the steering knuckle.
   d. Remove the bolt and the front speed sensor.

➡Do not allow foreign matter to attach to the sensor tip.

➡Clean the installation hole and surface for the speed sensor every time the speed.

To install:

➡Do not twist the wire harness for the front speed sensor when installing the speed sensor.

➡The bolt B tightens the brake flexible hose and front speed sensor together. Make sure that the flexible hose is positioned over the front speed sensor.

➡Install the stopper firmly to the hole in the body.

6. To install, reverse removal procedure.

7. Tighten the front speed sensor with the bolt to 71 inch lbs. (8 Nm).

➡Do not allow foreign matter to attach to the sensor tip.

8. Tighten the sensor harness brackets with the 2 bolts to the body and shock absorber assembly and 44 inch lbs. (5 Nm), and then to 14 ft. lbs. (19 Nm).

9. Install the front wheel and tighten the lug nuts to 76 ft. lbs. (103 Nm).

10. Check ABS speed sensor signal.

BRAKE CALIPER

REMOVAL & INSTALLATION

See Figure 2.

1. Before servicing the vehicle, refer to the Precautions Section.
2. Remove the front wheel.

➡Do not let brake fluid sit on painted surfaces, as it will eat through the paint. Wash it off immediately.

3. Drain brake fluid.
4. Remove the union bolt and gasket from the disc brake caliper assembly, then disconnect the flexible hose.

➡Remove the disc brake caliper assembly while holding both of the brake pads or the anti-squeal springs may fall off the brake pads.

5. Hold the front disc brake caliper slide pin and remove the 2 bolts and disc brake caliper assembly.

To install:

6. Install the disc brake caliper assembly with the 2 bolts and tighten to 25 ft. lbs. (34 Nm).
7. Check the installation of the anti-squeal springs. Visually check for any clearance between the brake pad and front disc brake pad support plates.

➡If the anti-squeal springs are installed correctly, there will be no clearance between the brake pad and the front disc brake pad support plates. If there is a clearance, the anti-squeal springs may not be installed properly.

➡Check all 4 contact surfaces between the brake pad and the front disc brake pad support plates.

Fig. 2 Removing the union bolt and gasket

8. Connect the flexible hose with the union bolt and a new gasket and tighten to 21 ft. lbs. (29 Nm).

➡Install the front brake flexible hose lock securely in the lock hole in the disc brake caliper.

9. Fill reservoir with brake fluid.
10. Bleed master cylinder.
11. Bleed brake line.
12. Bleed brake actuator assembly.
13. Inspect for brake fluid leak.
14. Inspect brake fluid level in reservoir.
15. Install the front wheel and tighten the lug nuts to 76 ft. lbs. (103 Nm).

DISC BRAKE PADS

REMOVAL & INSTALLATION

See Figure 3.

1. Before servicing the vehicle, refer to the Precautions Section.
2. Remove the 2 front disc brake caliper slide pins (upper and lower) from the front disc brake caliper mounting.
3. Remove brake cylinder.
4. Remove the 2 anti-squeal springs.
5. Remove the 2 brake pads from the front disc brake caliper mounting.

To install:

6. Install the 2 brake pads with front anti-squeal shims to the front disc brake caliper mounting.

➡When replacing worn pads, the front anti-squeal springs must be replaced at the same time.

➡Be sure to install the anti-squeal springs into the front disc brake pad installation holes as far as they will go.

7. Install the 2 front disc brake caliper slide pins (upper and lower) from the front disc brake caliper mounting.
8. Install the brake cylinder.

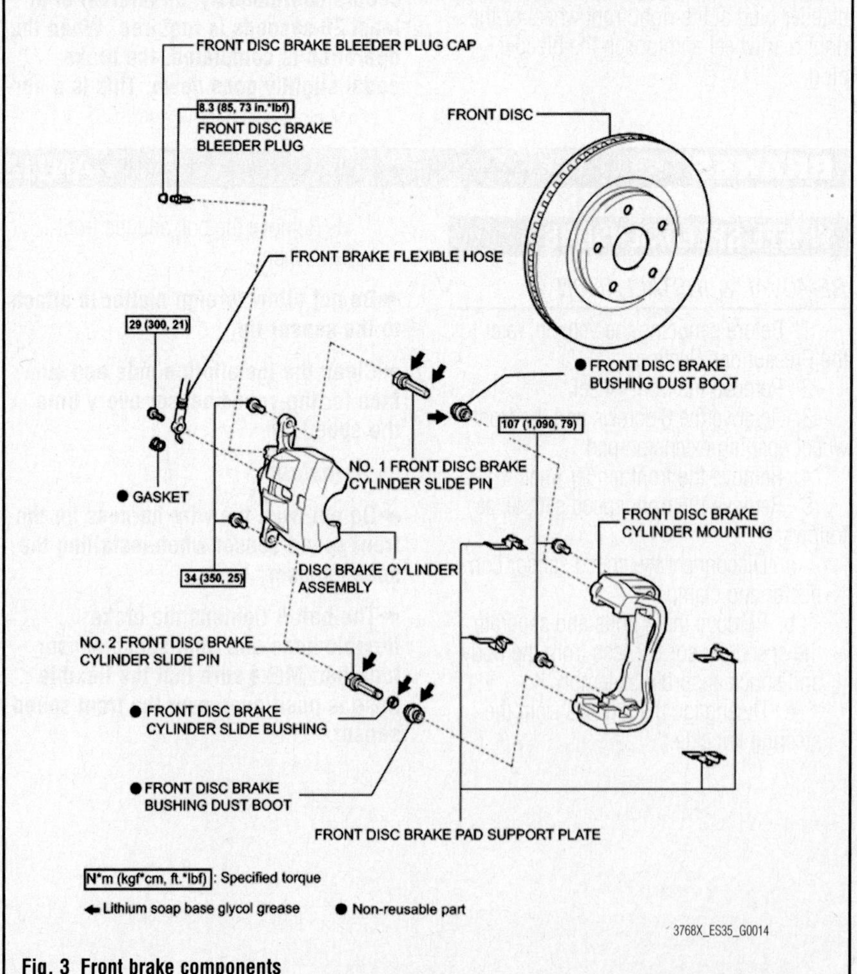

Fig. 3 Front brake components

BRAKES

REAR DISC BRAKES

BRAKE CALIPER

REMOVAL & INSTALLATION

1. Before servicing the vehicle, refer to the Precautions Section.
2. Remove the rear wheel.
3. Drain brake fluid.

WARNING

Do not let brake fluid sit on painted surfaces, as it will eat through the paint. Wash it off immediately.

4. Remove the union bolt and the gasket from the rear disc brake caliper assembly, then disconnect the rear brake flexible hose.
5. Hold the 2 rear disc brake caliper slide pins and remove the 2 bolts and rear disc brake caliper assembly.

To install:

6. Install the rear disc brake caliper assembly with the 2 bolts and tighten to 20 ft. lbs. (77 Nm).
7. Connect the rear brake flexible hose with the union bolt and a new gasket and tighten to 24 ft. lbs. (33 Nm).
8. Fill reservoir with brake fluid.
9. Bleed brake line.
10. Inspect for brake fluid leak.
11. Inspect brake fluid level in reservoir.
12. Install the rear wheel and tighten the lug nuts to 76 ft. lbs. (103 Nm).

DISC BRAKE PADS

REMOVAL & INSTALLATION

1. Before servicing the vehicle, refer to the Precautions Section.
2. Remove the 2 rear disc brake caliper slide pins (upper and lower) from the rear disc brake caliper mounting.
3. Remove brake cylinder.
4. Remove the 2 brake pads with the rear anti-squeal shims.

To install:

5. Installation is the reverse of removal procedure.

BRAKES

PARKING BRAKE

PARKING BRAKE CABLES

ADJUSTMENT

See Figure 4.

1. Slowly depress the parking brake pedal all the way, and count the number of clicks. It should be 3–6 clicks at 68.3 ft. lbs. (300 N).
2. Adjust the parking brake pedal travel as follows:
   a. Depress the parking brake pedal 3 notches to make a room for the procedure, and loosen the lock nut.
   b. Return the parking brake pedal to the original position.
   c. Turn the adjusting nut until the parking brake pedal travel is correct.
   d. Depress the parking brake pedal 3 notches to make a room for the procedure, and tighten the lock nut to 48 inch lbs. (5.4 Nm).
   e. Return the parking brake pedal to the original position.
   f. Check whether parking brake drags or not.
   g. When operating the parking brake pedal, check that the parking brake pedal indicator light lights up.

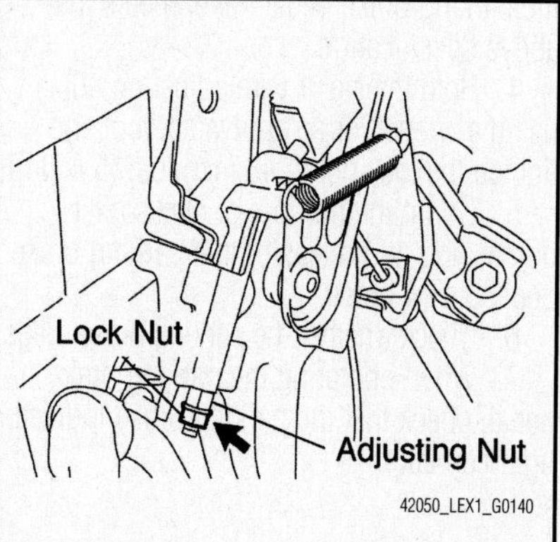

Fig. 4 Parking brake adjuster and lock nut.

PARKING BRAKE SHOES

REMOVAL & INSTALLATION

See Figures 5 and 6.

1. Remove the rear wheel.
2. Remove the 2 bolts and separate the rear disc brake caliper assembly. Do not disconnect the flexible hose from the disc brake caliper assembly.
3. Remove the parking brake shoe adjusting hole plug from the rear disc.

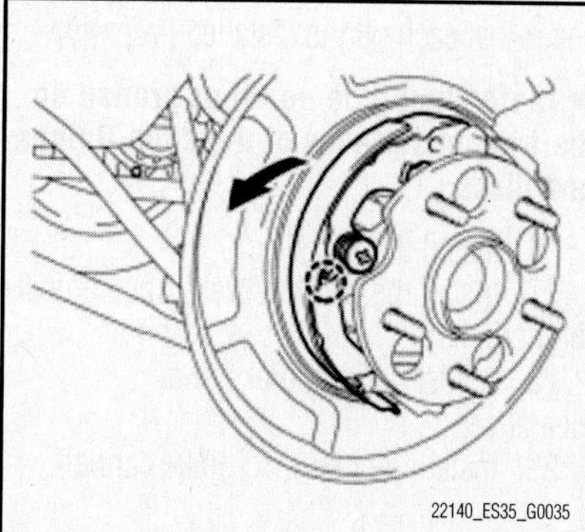

Fig. 5 Remove the No. 1 parking brake shoe assembly

4. Release the parking brake and place the matchmarks on the rear disc and the axle hub.
5. Remove the rear disc.

➡**If the disc cannot be removed easily, turn the shoe adjuster until the disc turns freely.**

6. Using needle-nose pliers, remove the 2 parking brake shoe return tension No. 1 springs.
7. Remove the parking brake shoe strut and the parking brake shoe strut compression spring.
8. Remove the No. 1 parking brake shoe assembly, as follows:
   a. Release the claw of the parking brake shoe hold down spring No. 2 cup.
   b. Remove the No. 1 parking brake shoe assembly as shown in the illustration.
   c. Remove the parking brake shoe hold down spring No. 1 cup, the parking brake shoe hold down spring, the parking brake shoe hold down spring No. 2 cup, and the parking brake shoe hold down spring No. 1 pin.
9. Remove the parking brake shoe adjusting screw set.
10. Remove the parking brake shoe return tension No. 2 spring.
11. Remove the No. 2 parking brake shoe assembly, as follows:
   a. Release the claw of the parking brake shoe hold down spring No. 2 cup.
   b. Remove the No. 2 parking brake shoe assembly as shown in the illustration.

c. Remove the parking brake shoe hold down spring No. 1 cup, the parking brake shoe hold down spring, the parking brake shoe hold down spring No. 2 cup, and the parking brake shoe hold down spring No. 2 pin.

d. Using needle-nose pliers, disconnect the No. 3 parking brake cable assembly from the parking brake shoe lever.

➡Be careful not to damage the No. 3 parking brake cable assembly.

12. Using a screwdriver, remove the C-washer, shim and the parking brake shoe lever.

13. Remove the parking brake shoe guide plate set bolt and the parking brake shoe guide plate.

To install:

14. Apply high temperature grease to the backing plate where it contacts the shoe.

15. Apply adhesive (Toyota Genuine Adhesive 1344, Three Bond 1344 or equivalent) to the threads of the parking brake shoe guide plate set bolt.

16. Install the parking brake shoe guide plate with the parking brake shoe guide plate set bolt and tighten to 13 ft. lbs. (18 Nm).

17. Install the parking brake shoe lever and shim to the No. 2 parking brake shoe assembly with a new C-washer.

18. Using a feeler gauge, measure the clearance between the No. 2 parking brake shoe assembly and parking brake shoe lever. Standard clearance: Less than 0.35 mm (0.014 in.).

19. If the clearance is not as specified, replace the shim with one of the correct size.

20. Install the No. 2 parking brake shoe assembly as follows:

a. Using needle-nose pliers, connect the No. 3 parking brake cable assembly to the parking brake shoe lever.

b. Install the No. 2 parking brake shoe assembly with the parking brake shoe hold down spring No. 2 pin, the parking brake shoe hold down spring No. 2 cup, the parking brake shoe hold down spring and the parking brake shoe hold down spring No. 1 cup.

c. Engage the claw of the parking brake shoe hold down spring No. 2 cup to the No. 2 parking brake shoe assembly.

21. Install the parking brake shoe adjusting screw set, as follows:

a. Apply high temperature grease to the parking brake shoe adjusting screw set as shown in the illustration.

b. Install the parking brake shoe return tension No. 2 spring to the No. 1 parking brake shoe assembly and the No. 2 parking brake shoe assembly.

c. Install the parking brake shoe adjusting screw set to the No. 1 parking brake shoe assembly and the No. 2 parking brake shoe assembly.

22. Install the No. 1 parking brake shoe assembly as follows:

a. Install the No. 1 parking brake shoe assembly with the parking brake shoe hold down spring No. 1 pin, parking brake shoe hold down spring No. 2 cup, parking brake shoe hold down spring and parking brake shoe hold down spring No. 1 cup.

b. Engage the claw of the parking brake shoe hold down spring No. 2 cup to the No. 1 parking brake shoe assembly.

23. Attach the parking brake shoe strut and the parking brake shoe strut compression spring to the No. 1 parking brake shoe assembly and No. 2 parking brake shoe assembly.

24. Using needle-nose pliers, install the 2 parking brake shoe return tension No. 1 springs. First install the front side spring and then the rear side spring.

25. Inspect parking brake installation and check that each part is installed properly.

➡There should be no oil or grease on the friction surfaces of the shoe linings and discs.

26. Install the rear disc.

27. Install the parking brake shoe adjusting hole plug.

28. Adjust parking brake shoe clearance.

29. Install the rear disc brake caliper assembly with the 2 bolts and tighten to 46 ft. lbs. (62 Nm).

30. Install the rear wheel.

31. Adjust the parking brake pedal travel.

32. Bed in parking brake shoes to discs, as follows:

a. Drive the vehicle at about 31 mph (50 km/h) on a safe, level and dry road.

b. Depress the parking brake pedal with 34 lbs. (150 N) of force.

33. Drive the vehicle about 0.25 miles (400 m) in this condition.

a. Repeat this procedure 3 times using 5-minute intervals between each procedure to prevent the parking brake assembly from overheating.

34. Remove the rear wheel.

35. Adjust parking brake shoe clearance.

36. For A/T vehicles, adjust the parking brake pedal travel.

37. For M/T vehicles, adjust the parking brake pedal travel.

38. Install the rear wheel and tighten the lug nuts to 76 ft. lbs. (103 Nm).

ADJUSTMENT

1. Adjust parking brake shoe clearance, as follows:

a. Temporarily install the hub nuts.

b. Remove the shoe adjusting hole plug, turn the adjuster and expand the shoes until the disc locks.

c. Contract the shoe adjuster until the disc rotates smoothly. Standard: returns 8 notches

d. Check that the disc has no brake drag.

e. Install the shoe adjusting hole plug.

Parking Brake Pedal Travel Adjustment

1. Depress the parking brake pedal. Hold the No. 1 wire adjusting nut using a wrench and loosen the lock nut.

2. Release the parking brake pedal.

3. Turn the No. 1 wire adjusting nut until the parking brake pedal travel meets the above specification.

4. Hold the No. 1 wire adjusting nut using a wrench or an equivalent tool and tighten the lock nut to 48 inch lbs. (5.4 Nm).

5. Count the number of clicks after depressing and releasing the parking brake pedal 3 or 4 times.

6. Check whether the parking brake drags.

7. When operating the parking brake pedal, check that the parking brake indicator light comes on.

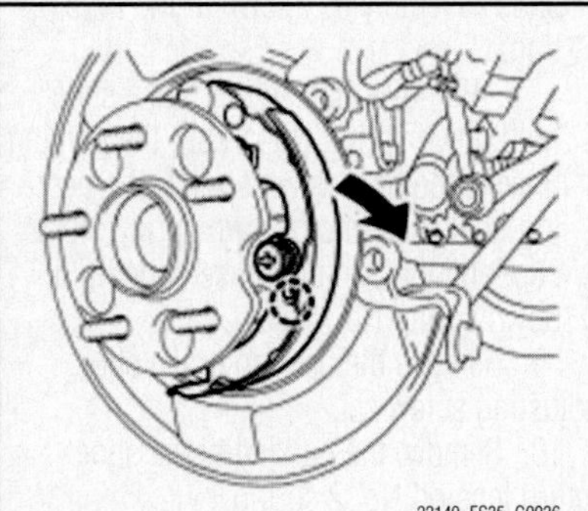

Fig. 6 Remove the No. 2 parking brake shoe assembly

GENERAL INFORMATION

These vehicles are equipped with an air bag system. The system must be disarmed before performing service on, or around, system components, the steering column, instrument panel components, wiring and sensors. Failure to follow the safety precautions and the disarming procedure could result in accidental air bag deployment, possible injury and unnecessary system repairs.

SERVICE PRECAUTIONS

CAUTION

Disconnect and isolate the battery negative cable before beginning any airbag system component diagnosis, testing, removal, or installation procedures. Allow system capacitor to discharge for two minutes before beginning any component service. This will disable the airbag system. Failure to disable the airbag system may result in accidental airbag deployment, personal injury, or death.

DISARMING THE SYSTEM

To avoid personal injury when working on vehicles equipped with an air bag, the negative battery cable must be disconnected and at least 90 seconds must elapse before working on the system. Failure to do so may result in deployment of the air bag.

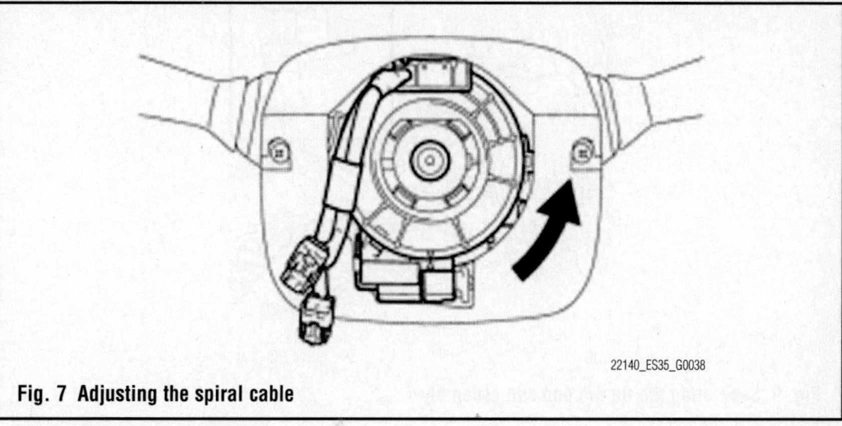

Fig. 7 Adjusting the spiral cable

ARMING THE SYSTEM

To arm the system after service is finished, connect the negative battery cable.

CLOCKSPRING CENTERING

1. Before servicing the vehicle, refer to the Precautions Section.
2. Check that the ignition switch is **OFF**.
3. Check that the battery negative (-) terminal is disconnected.

CAUTION

After removing the terminal, wait for at least 90 seconds before starting the operation.

4. Rotate the spiral cable counter clockwise slowly by hand until it feels firm.
5. Rotate the spiral cable clockwise approximately 2.5 turns to align the marks.

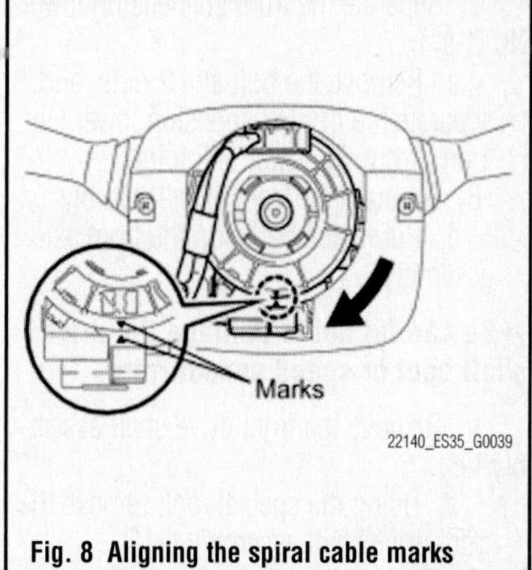

Fig. 8 Aligning the spiral cable marks

➡**Do not turn the spiral cable by the airbag wire harness.**

➡**The spiral cable will rotate approximately 2.5 turns to both the left and right from the center.**

DRIVE TRAIN

FRONT DRIVESHAFT

REMOVAL & INSTALLATION

See Figures 9 through 11.

➡**Use the same procedures for the RH side and LH side.**

➡**The procedures listed below are for the LH side.**

1. Drain the automatic transaxle fluid.
2. Remove the front wheel.
3. Remove the front axle hub nut.
   a. Using the special tool and a hammer, release the staked part of the front axle hub nut.

➡**Loosen the staked part of the nut completely, otherwise the thread of the drive shaft may be damaged.**

   b. While applying the brakes, remove the front axle hub nut.
4. Separate the front stabilizer link assembly.
   a. Remove the nut and separate the front stabilizer link assembly.

➡**If the ball joint turns together with the nut, use a hexagon wrench (6 mm) to hold the stud.**

5. Separate the front speed sensor.
   a. Remove the bolt and clamp, and separate the speed sensor wire and flexible hose from the shock absorber.
   b. Remove the bolt and separate the speed sensor from the steering knuckle.

➡**Prevent foreign matter from adhering to the speed sensor.**

➡**Be careful not to damage the speed sensor.**

6. Separate the tie rod end sub assembly.
   a. Remove the cotter pin and nut.
   b. Using a special tool, separate the tie rod end sub assembly from the steering knuckle.

➡**Make sure that the string of the SST is securely tied to the vehicle.**

➡**Be careful not to damage the ball joint dust cover.**

➡**Be careful not to damage the steering knuckle.**

➡**Be careful not to damage the front disc brake dust cover.**

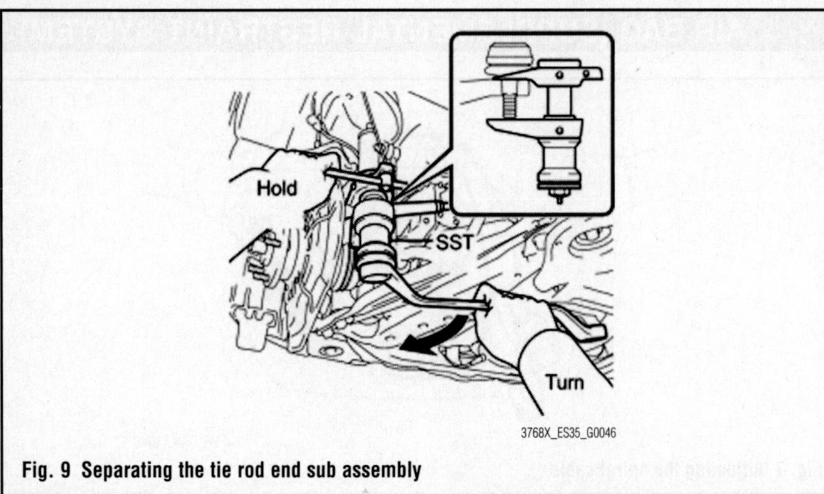

Fig. 9 Separating the tie rod end sub assembly

7. Separate the front suspension lower No. 1 arm.

a. Remove the bolt and 2 nuts, and separate the front suspension lower No. 1 arm from the lower ball joint.

8. Separate the front axle assembly.

a. Put matchmarks on the front axle assembly.

➡Be careful not to damage the drive shaft boot or speed sensor rotor.

9. Remove the front drive shaft assembly LH.

a. Using the special tool, remove the front drive shaft assembly LH.

10. Remove the front drive shaft assembly RH.

a. Using a screwdriver, remove the bearing bracket hole snap ring.

b. Remove the bolt and front drive shaft assembly RH front the drive shaft bearing bracket.

➡Do not damage the boot or oil seal.

11. Secure the front axle hub sub assembly.

a. Secure the front axle hub bearing.

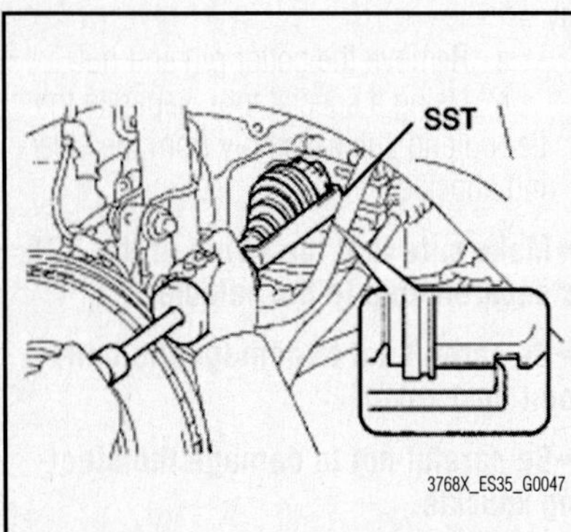

Fig. 10 Removing the front drive shaft assembly LH

➡The hub bearing may be damaged if it is subjected to the vehicle's full weight, such as moving the vehicle with the drive shaft removed. If it is necessary to place the vehicle's weight on the hub bearing, first support it with SST.

12. Inspect the front drive shaft assembly.

To install:

13. Install the front drive shaft assembly LH.

a. Coat the spline of the inboard joint shaft assembly with ATF.

b. Align the shaft splines and install the drive shaft assembly LH with a brass bar and hammer.

➡Set the shaft snap ring with the opening side facing down.

➡Be careful not to damage the drive shaft dust cover, boot, or oil seal.

➡Move the drive shaft assembly while keeping it level.

14. Install the front drive shaft assembly RH.

a. Coat the spline of the inboard joint shaft assembly with ATF.

b. Install the front drive shaft to the assembly.

c. Using a screwdriver, install a new bearing bracket hole snap ring.

➡Do not damage the boot or oil seal.

➡Move the drive shaft assembly while keeping it level.

d. Install a new bolt. Tighten to 24 ft. lbs. (32 Nm).

15. Install the front axle assembly.

a. Align the matchmarks and install the front drive shaft assembly to the front axle hub sub assembly.

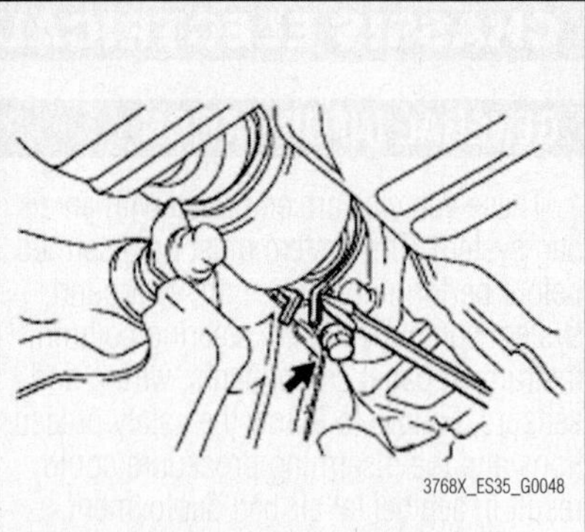

Fig. 11 Removing the front drive shaft assembly RH

➡Be careful not to damage the drive shaft boot and speed sensor.

16. Install the front suspension lower No. 1 arm with the bolt and 2 nuts. Tighten to 55 ft. lbs. (75 Nm).

17. Install the tie rod end sub assembly.

a. Install the tie rod end sub assembly to the steering knuckle with the nut. Tighten to 36 ft. lbs. (49 Nm).

b. Install a new cotter pin.

➡If the holes for the cotter pin are not aligned, tighten the nut up to 60°further.

18. Install the front speed sensor.

a. Install the front speed sensor to the steering knuckle with the bolt. Tighten to 71 inch lbs. (8 Nm).

➡Prevent foreign matter from adhering to the speed sensor.

➡Be careful not to damage the speed sensor.

b. Install the flexible hose and the speed sensor wire to the shock absorber with the bolt and set the clamp of the sensor on the knuckle. Tighten to 14 ft. lbs. (19 Nm).

➡Be careful not to damage the speed senor.

➡Prevent foreign matter from adhering to the speed sensor.

➡Do not twist the sensor wire when installing the speed sensor.

19. Install the front stabilizer link assembly with the nut. Tighten to 55 ft. lbs. (74 Nm).

➡If the ball joint turn together with the nut, use a hexagon wrench (6 mm) to hold the stud.

20. Install the front axle hub nut.

a. Clean the threaded parts on the drive shaft and axle hub nut using a non residue solvent.

➡Be sure to perform this work for a new drive shaft.

➡Keep the threaded parts free of oil and foreign objects.

b. Using a socket wrench (30 mm), install a new axle hub nut. Tighten to 217 ft. lbs. (294 Nm).

21. Install the front wheel. Tighten to 76 ft. lbs. (103 Nm).
22. Add automatic transaxle fluid.
23. Adjust the front wheel alignment.
24. Check the ABS speed sensor signal.

FRONT HALFSHAFTS

REMOVAL & INSTALLATION

See Figures 12 through 14.

1. Before servicing the vehicle, refer to the Precautions Section.
2. Remove the engine under cover.
3. Remove the drain plug and gasket, and then drain the automatic transaxle fluid.
4. Install a new gasket and drain plug and tighten to 36 ft. lbs. (49 Nm).
5. Remove front wheel.
6. Using SST (SST: 09930-00010) and hammer, release the staked part of the front axle hub nut.

➡Loosen the staked part of the nut completely, otherwise the screw of the drive shaft may be damaged.

7. While applying the brakes, remove the front axle hub nut.
8. Remove the nut and separate the front stabilizer link assembly.

➡If the ball joint turns together with the nut, use a hexagon wrench (6mm) to hold the stud.

9. Remove the bolt and clip, and separate the speed sensor wire and flexible hose from the shock absorber.
10. Remove the bolt and separate the front speed sensor from the steering knuckle.

➡Do not allow foreign matter to adhere to the speed sensor. Be careful not to damage the speed sensor.

11. Separate tie rod end sub-assembly, as follows:

a. Remove the cotter pin and nut.

b. Using SST (SST: 09628-62011) or equivalent, separate the tie rod end sub-assembly from the steering knuckle.

➡Do not damage the ball joint dust cover.

12. Remove the bolt and 2 nuts, and separate the lower No. 1 front suspension arm sub-assembly from the lower ball joint.
13. Put matchmarks on the front drive shaft assembly and the axle hub.
14. Using a plastic hammer, separate the front drive shaft assembly from the front axle hub sub-assembly.

➡Be careful not to damage the drive shaft boot and speed sensor rotor.

15. Remove the front drive shaft assembly(s), as follows:

a. For left front drive shaft, use SST (SST: 09520-01010, SST: 09520-24010) or equivalent, and remove the left front drive shaft assembly.

➡Be careful not to damage the drive shaft dust cover, boot and oil seal. Be careful not to drop the drive shaft assembly.

b. For right front drive shaft, use a screwdriver and remove the bearing bracket hole snap ring.

c. Remove the bolt and right front drive shaft assembly from the drive shaft bearing bracket.

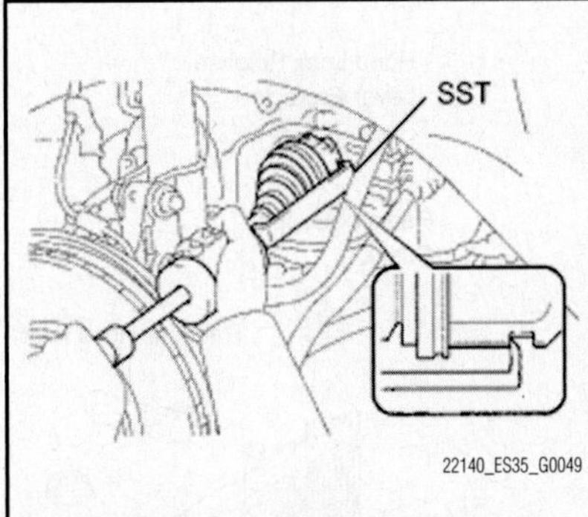

Fig. 12 Remove the left front drive shaft assembly

➡Do not damage the boot and oil seal.

16. Fix front axle hub bearing. The hub bearing could be damaged if it is subjected to the vehicle's full weight, such as moving the vehicle with the drive shaft removed. If it is necessary to place the vehicle's weight on the hub bearing, first support it with SST (SST: 09608-16042).

To install:

17. Install the front drive shaft assembly(s), as follows:

a. Coat the spline of the inboard joint shaft assembly with automatic transaxle fluid.

b. For the left front drive shaft, align the shaft splines and install the drive shaft assembly with a brass bar and hammer.

➡Set the shaft snap ring with the opening side facing down. Be careful not to damage the drive shaft dust cover, boot, and oil seal. Move the drive shaft assembly while keeping it level.

c. For the right front drive shaft, install the drive shaft and use a screwdriver to install a new bearing bracket hole snap ring, and install a new bolt tightened to 24 ft. lbs. (32 Nm).

➡Be careful not to damage the drive shaft dust cover, boot and oil seal.

➡Move the drive shaft assembly while keeping it level.

18. Align the matchmarks and install the front drive shaft assembly to the front axle hub sub-assembly.

➡Be careful not to damage the drive shaft boot and speed sensor rotor.

19. Install the lower ball joint to the lower No. 1 front suspension arm sub-assembly with the bolt and 2 nuts and tighten to 55 ft. lbs. (75 Nm).
20. Install the tie rod end sub-assembly

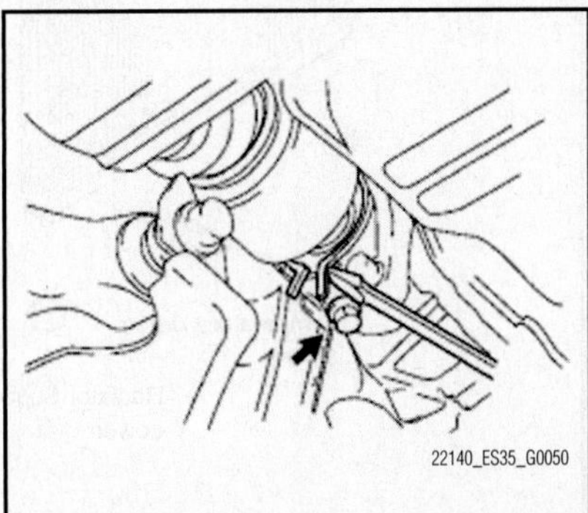

Fig. 13 Remove the right front drive shaft assembly

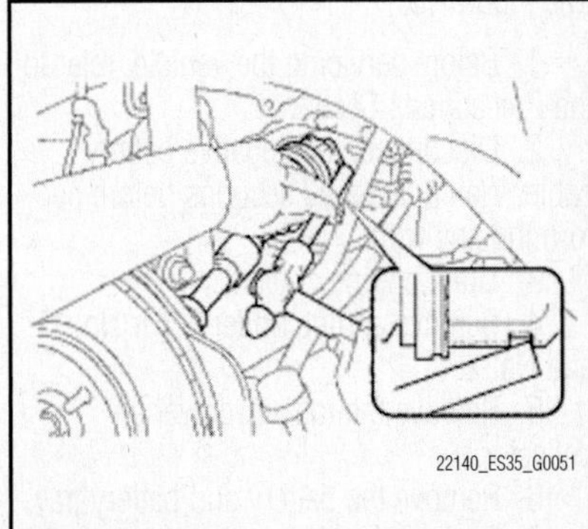

Fig. 14 Install the left front drive shaft assembly

to the steering knuckle with the nut and tighten to 36 ft. lbs. (49 Nm).

21. Install a new cotter pin. If the holes for the cotter pin are not aligned, tighten the nut up to 60°further.

22. Install the front speed sensor to the steering knuckle with the bolt and tighten to 71 inch lbs. (8 Nm).

23. Install the flexible hose and the speed sensor to the shock absorber with the bolt and set the sensor clip on the knuckle and tighten to 14 ft. lbs. (19 Nm).

➡Be careful not to damage the speed sensor. Do not allow foreign matter to adhere to the speed sensor. Do not twist the sensor wire when installing the speed sensor.

24. Install the stabilizer link assembly with the nut and tighten to 55 ft. lbs. (74 Nm).

➡If the ball joint turns together with the nut, use a hexagon wrench (6 mm) to hold the stud.

25. Clean the threaded parts on the drive shaft and axle hub nut using a non-residue solvent.

➡Be sure to perform this work for a new drive shaft. Keep the threaded parts free of oil and foreign objects.

26. Using a socket wrench (30 mm), install a new axle hub nut and tighten to 217 ft. lbs. (294 Nm).

27. Using a chisel and hammer, stake the front axle hub nut.

28. Install front wheel.

29. Add automatic transaxle fluid.

30. Inspect automatic transaxle fluid.

31. Inspect and adjust front wheel alignment.

32. Install the engine under cover.

33. Check the ABS speed sensor signal.

ENGINE COOLING

ENGINE FAN

REMOVAL & INSTALLATION

1. Before servicing the vehicle, refer to the Precautions Section.
2. Disconnect the negative battery cable. Wait at least 90 seconds before performing any work.
3. Drain the engine coolant.
4. Remove the radiator assembly (the fan assembly is attached).
5. Separate the fan assembly from the radiator.

To install:

6. Install the fan assembly to the radiator and tighten the fasteners to 44 inch lbs. (5 Nm).
7. Install the radiator and fan assembly.
8. Refill the engine cooling system.
9. Connect the negative battery cable.
10. Start the engine and check system for coolant leaks.
11. Install the engine under cover No. 1.

RADIATOR

REMOVAL & INSTALLATION

See Figure 15.

1. Before servicing the vehicle, refer to the Precautions Section.
2. Disconnect the negative battery cable. Wait at least 90 seconds before performing any work.
3. Drain engine coolant.
4. Remove engine under cover No. 1 and No. 2.
5. Remove the radiator lower air deflector.
6. Remove the battery and battery tray.
7. Remove the air cleaner inlet assembly.
8. Remove the air cleaner assembly.

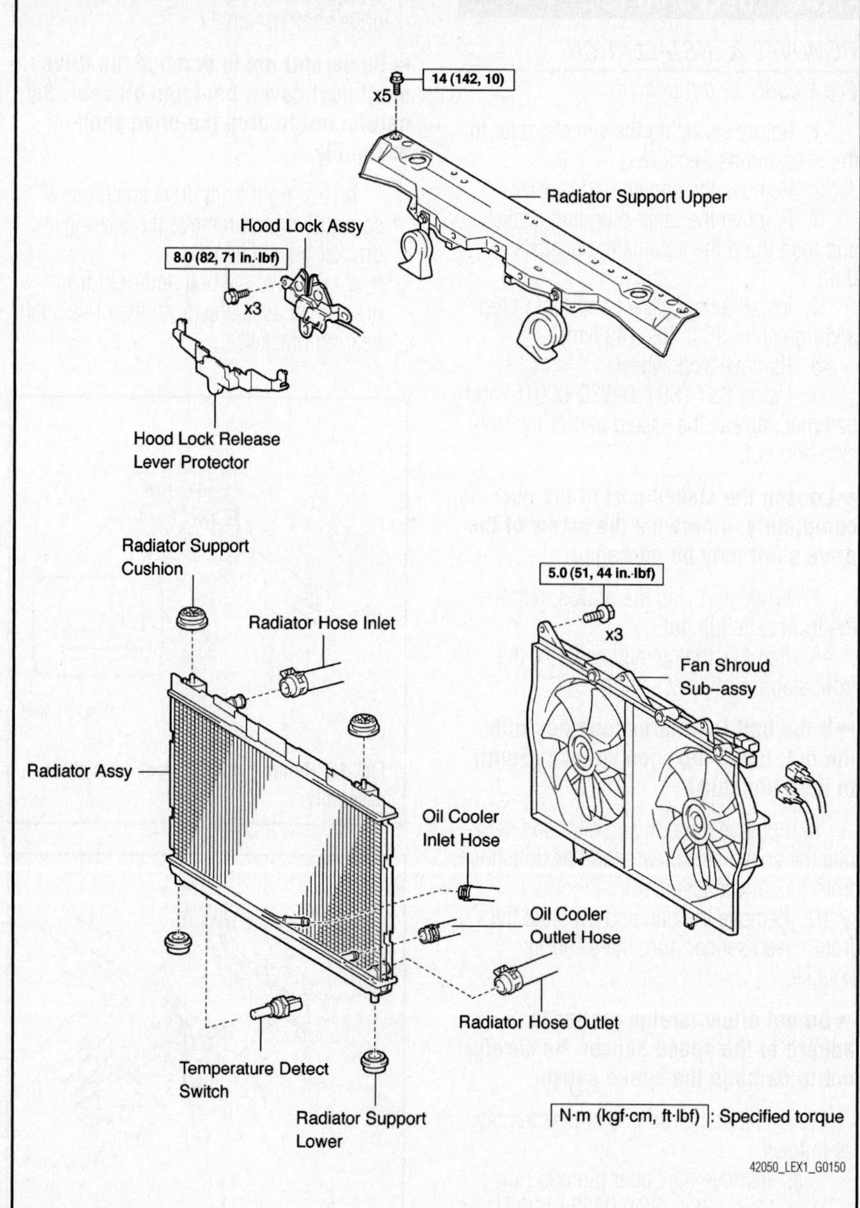

Fig. 15 Exploded view of the radiator assembly and related components.

9. Remove the air cleaner bracket.
10. Remove the air cleaner inlet No. 1.
11. Disconnect the radiator inlet and outlet hoses.
12. Disconnect the oil cooler hoses.
13. Using a screwdriver with the tip wrapped in tape, remove the hood lock protector.

➡Removing the protector damages the clips inside the protector, and therefore the use of a new protector is necessary on installation.

14. Disconnect the hood switch connector, remove the 3 bolts, then separate the hood lock.
15. Remove the radiator support upper as follows:
    a. Disconnect the horn connectors.
    b. Remove the 5 bolts and radiator support upper.
16. Disconnect the 2 wire harness connectors, and remove the radiator.
17. Remove the 2 radiator support cushions from the radiator.
18. Remove the 2 radiator support lowers from the radiator.
19. Disconnect the temperature detect switch connector, remove the 3 bolts, then remove the fan shroud with motor.
20. Remove the temperature detect switch.

To install:

21. Install the temperature detect switch.
22. Install the fan shroud with motor and tighten the bolts to 44 inch lbs. (5 Nm). Connect the temperature detect switch connector.
23. Install the 2 radiator support lowers to the radiator.
24. Install the 2 radiator support cushions to the radiator.
25. Install the radiator assembly and connect the 2 wire harness connectors.
26. Install the radiator support upper and tighten the 5 bolts to 10 ft. lbs. (14 Nm).
27. Connect the horn connectors.
28. Install the hood lock and tighten the bolts to 71 inch lbs. (8 Nm). Connect the hood switch connector.
29. Install the hood lock protector.
30. Connect the oil cooler hoses.
31. Connect the radiator inlet and outlet hoses.
32. Install the air cleaner inlet NO. 1.
33. Install the air cleaner bracket.
34. Install the air cleaner assembly.
35. Install the air cleaner inlet assembly.
36. Install the battery and battery tray.
37. Install the radiator lower air deflector.
38. Install engine under cover No. 1 and No. 2.
39. Inspect the vacuum hose connections.
40. Refill the engine cooling system.
41. Connect the negative battery cable.
42. Start the engine and check system for coolant leaks.

THERMOSTAT

REMOVAL & INSTALLATION

See Figures 16 and 17.

1. Before servicing the vehicle, refer to the Precautions Section.
2. Drain engine coolant.
3. Remove the V-bank cover sub-assembly.
4. Remove the RH front fender apron seal.
5. Remove the RH No. 2 engine mounting stay.

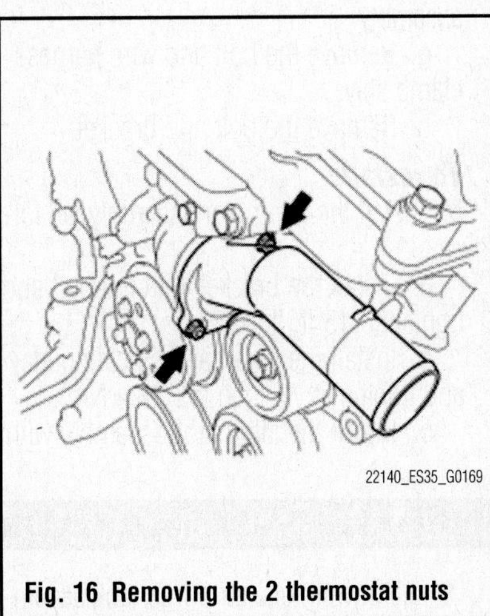

Fig. 16 Removing the 2 thermostat nuts

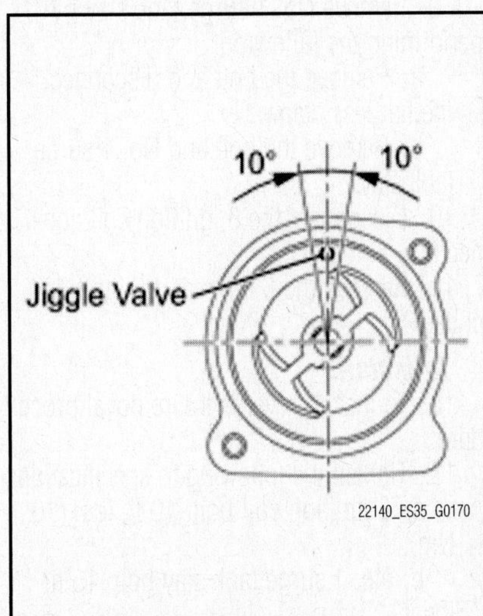

Fig. 17 Radiator jiggle valve

6. Remove the drive belt.
7. Remove No. 2 idler pulley sub-assembly.
8. Separate the radiator hose outlet.
9. Remove the 2 nuts and disconnect the water inlet from the cylinder block.
10. Remove the thermostat and the gasket from the thermostat.

To install:

11. Install a new gasket to the thermostat.
12. Install a new gasket to the thermostat.

➡The jiggle valve may be set within 10°of either side of the prescribed position.

13. Install the thermostat with the jiggle valve facing up.
14. Install the water inlet and tighten to 7 ft. lbs. (10 Nm).
15. The remainder of installation is the reverse of removal.
16. After installation, inspect for coolant leak.

WATER PUMP

REMOVAL & INSTALLATION

See Figures 18.

1. Before servicing the vehicle, refer to the Precautions Section.
2. Remove engine assembly and transaxle. secure engine stand.
3. Remove RH front No. 1 engine mounting bracket.
4. Remove the No. 2 idler pulley sub-assembly, as follows:
    a. Remove the 2 bolts, 2 idler pulley cover plates and 2 idler pulley sub-assemblies.
5. Remove the 5 bolts and V-ribbed belt tensioner assembly.
6. Using SST: 09960-10010, hold the water pump pulley. Remove the 4 bolts and water pump pulley.
7. Remove water inlet housing, as follows:
    a. Separate the water hose.
    b. Remove the 2 bolts, nut and water inlet housing.
    c. Remove the water inlet housing gasket and water outlet pipe O-ring.
8. Remove the 16 bolts, water pump assembly and water pump gasket.

To install:

➡Make sure that there is no oil on the threads of the A bolts.

➡Be sure to replace the 2 C bolts with new ones or reuse them after applying adhesive (Part No. 08833-00080, three bond 1344 or equivalent).

9. Install a new water pump gasket and the water pump assembly with the 16 bolts and tighten to:
 a. Bolt A: 15 ft. lbs. (21 Nm)
 b. Bolt B: 81 inch lbs. (9.1 Nm).
 c. Bolt C: 81 inch lbs. (9.1 Nm).
10. Install a new water inlet housing No. 1 gasket and water outlet pipe O-ring..

➡Be careful not to allow the O-ring to get caught between the parts.

11. Install water inlet housing, as follows:

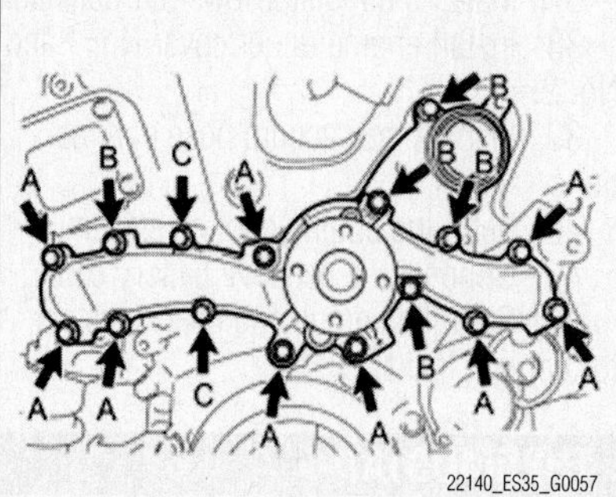

Fig. 18 Water pump bolt location and tightening sequence

 a. Install a new No. 1 water inlet housing gasket and water outlet pipe O-ring.
 b. Install the water inlet with the 2 bolts and nut and tighten to 7 ft. lbs. (10 Nm).
12. Temporarily install the water pump pulley with the 4 bolts.
 a. Using SST (SST: 09960-10010) or equivalent, hold the water pump pulley.
 b. Tighten the 4 bolts to 15 ft. lbs. (21 Nm).
13. Install the V-ribbed belt tensioner assembly with the 5 bolts and tighten to 32 ft. lbs. (43 Nm).
14. Install the 2 idler pulley cover plates and idler pulley sub-assemblies with the 2 bolts and tighten to 32 ft. lbs. (43 Nm).
15. To complete installation, reverse remaining removal procedure.
16. Add engine coolant.
17. Inspect for coolant leak.

ENGINE ELECTRICAL

CHARGING SYSTEM

ALTERNATOR

REMOVAL & INSTALLATION

1. Disconnect the negative battery cable.
2. Remove the V-bank cover sub-assembly.
3. Remove the V-ribbed belt.
4. Remove the alternator assembly, as follows:
 a. Disconnect the wire harness clamp.
 b. Remove the terminal cap.
 c. Remove the nut and disconnect the wire harness from terminal B.
 d. Disconnect the alternator connector from the alternator assembly.
 e. Remove the nut from the cylinder block.
 f. Remove the 2 bolts and alternator assembly.
 g. Remove the bolt and wire harness clamp stay.
 h. Remove the bolt and bracket.

To install:

5. Install the alternator assembly, as follows:
 a. Install the bracket with the bolt and tighten to 15 ft. lbs. (20 Nm).
 b. Install the wire harness clamp stay and tighten to 74 inch lbs. (8.4 Nm).
 c. Install the alternator assembly with the 2 bolts and tighten to 32 ft. lbs. (43 Nm).
 d. Install the nut to the cylinder block and tighten to 15 ft. lbs. (20 Nm).
 e. Connect the alternator connector to the alternator assembly.
 f. Install the alternator wire with the nut and tighten to 87 inch lbs. (9.8 Nm).
 g. Install the terminal cap.
 h. Connect the wire harness clamp.
6. Install the V-ribbed belt.
7. Install the V-bank cover sub-assembly.
8. Connect the negative battery cable.
9. Perform initialization.

ENGINE ELECTRICAL

IGNITION SYSTEM

FIRING ORDER

Firing order for 3.5L engine:
1–2–3–4–5–6

IGNITION COIL

REMOVAL & INSTALLATION

See Figure 19.

1. Before servicing the vehicle, refer to the Precautions Section.
2. Disconnect the negative battery cable.
3. Drain and recycle the engine coolant.
4. Remove windshield wiper link assembly.
5. Remove cowl top panel outer sub-assembly.
6. Remove v-bank cover sub-assembly.
7. Remove air cleaner cap sub-assembly.
8. Remove intake air surge tank assembly.
9. Remove No. 1 surge tank stay by performing the following:
 a. Remove the bolt and disconnect the harness clamp.
 b. Remove the bolt and No. 1 surge tank stay.
10. Disconnect the 6 ignition coil connectors.
11. Remove the 6 bolts and 6 ignition coils.

To install:

12. To install, reverse the removal procedure.
13. Tighten the following to specification:
 a. 6 ignition coil bolt: 10 ft. lbs. (10 Nm).
 b. No. 1 surge tank stay bolt: 15 ft. lbs. (21 Nm).
 c. No. 1 surge tank stay bolt and clamp: 62 inch lbs. (7 Nm).

IGNITION TIMING

INSPECTION

1. Warm up engine.

➡A warmed up engine should have an engine coolant temperature of over 80°C (176°F), have an engine oil temperature of 60°C (140°F), and the engine rpm should be stabilized.

2. When using the hand_held tester or OBD II scan tool, perform the following:
 a. Connect the hand–held tester or OBD II scan tool to the DLC3.
 b. Enter DATA LIST MODE on the hand–held tester or OBD II scan tool. Refer to the hand–held tester or OBD II

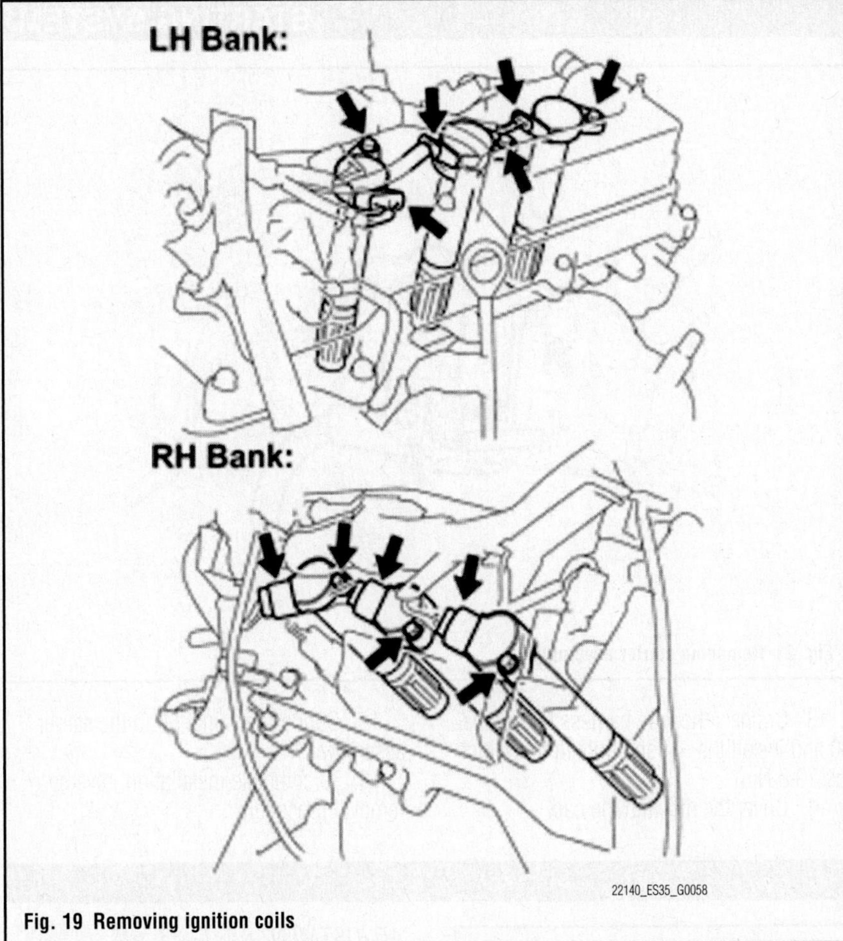

Fig. 19 Removing ignition coils

scan tool operator's manual if you need help to select DATA LIST.

c. Ignition timing should measure 8–12°BTDC.

3. When not using the hand_held tester or OBD II scan tool, perform the following:

a. Using SST 09843–18040, connect terminals 13 (TC) and 4 (CG) of the DLC3.

➡Make sure of the terminal numbers before connecting them. Connecting wrong terminals can damage the engine. Turn OFF all electrical systems before connecting the terminals. Perform this inspection after the cooling fan motor is turned OFF.

b. Remove the V–bank cover.

c. Pull out the engine wire colored by black and red as shown in the illustration.

d. Connect the tester terminal of the timing light to the engine.

➡Use a timing light which detects the first signal.

e. Inspect the ignition timing at idle. Ignition timing should measure 8–12°BTDC.

➡When checking the ignition timing, the transmission is in the neutral position.

f. Run the engine at 1,000–1,300 rpm for 5 seconds, then check that the engine speed returns to the idle speed.

g. Disconnect terminals 13 (TC) and 4 (CG) of the DLC3.

h. Inspect the ignition timing at idle. Ignition timing should measure 7–24°BTDC.

i. Confirm that the ignition timing advances when the engine speed is increased.

j. Remove the timing light.

ADJUSTMENT

The ignition timing is controlled by the Powertrain Control Module (PCM). No adjustment is necessary or possible.

SPARK PLUGS

REMOVAL & INSTALLATION

See Figure 20.

1. Before servicing the vehicle, refer to the Precautions Section.
2. Remove the V-bank cover.
3. Remove the intake air surge tank.
4. Disconnect the 6 ignition coil connectors.
5. Remove the 6 bolts and 6 ignition coils.
6. Using a 16 mm (0.63 in.) plug wrench, remove the spark plugs.

To install:

7. Installation is the reverse of removal, noting the following:

a. Torque the ignition coils to 66 inch lbs. (7.5 Nm) and the spark plugs to 13 ft. lbs (18 Nm).

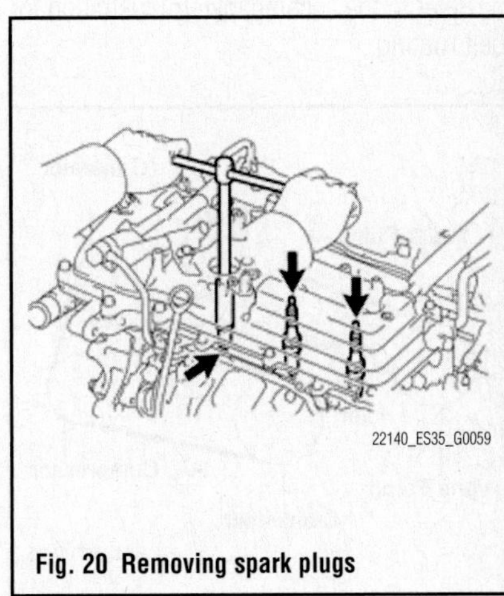

Fig. 20 Removing spark plugs

STARTER

REMOVAL & INSTALLATION

See Figure 21.

1. Before servicing the vehicle, refer to the Precautions Section.
2. Disconnect the negative battery cable.
3. Remove cool air intake duct seal.
4. Remove v-bank cover sub-assembly.
5. Remove air cleaner inlet assembly.
6. Remove air cleaner cap sub-assembly.
7. Remove air cleaner case sub-assembly.
8. Remove No. 1 air cleaner inlet.
9. Disconnect the terminal 50 connector from the starter assembly.
10. Remove the nut and disconnect the wire harness from terminal 30.
11. Remove the 2 bolts and starter assembly.

To install:

12. Install the starter assembly with the 2 bolts and tighten to 26 ft. lbs. (37 Nm).

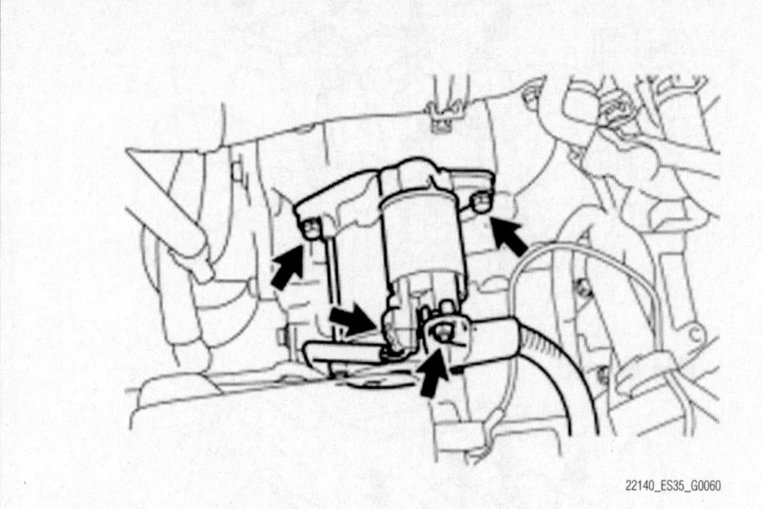

Fig. 21 Removing starter assembly

13. Connect the wire harness to terminal 30 and install the nut and tighten to 87 inch lbs. (9.8 Nm).
14. Cover the nut with the cap.
15. Connect terminal 50 to the starter assembly.
16. To complete installation, reverse removal procedure.

ENGINE MECHANICAL

ACCESSORY DRIVE BELTS

ACCESSORY BELT ROUTING

See Figure 22.

Refer to the accompanying illustration for belt routing.

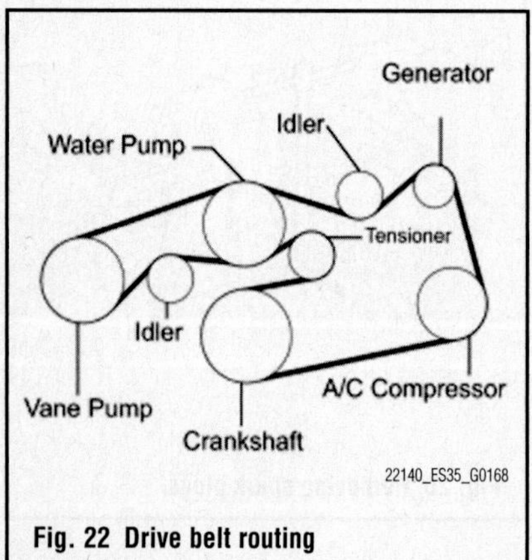

Fig. 22 Drive belt routing

INSPECTION

See Figure 23.

Visually check the V-ribbed belt for excessive wear, frayed cords, etc. If any defect has been found, replace the V-ribbed belt.

REPLACE

CORRECT INCORRECT

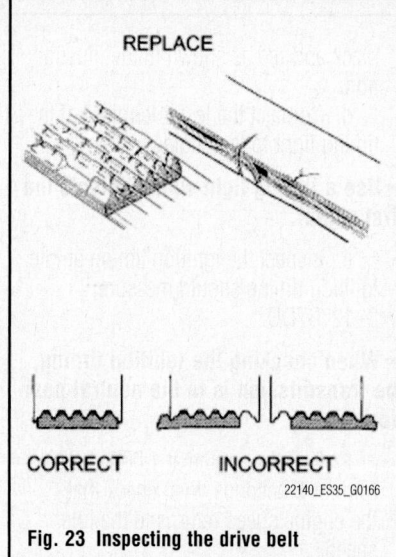

22140_ES35_G0166

Fig. 23 Inspecting the drive belt

- Cracks on the rib side of a belt is considered acceptable. If the belt has chunks missing from the ribs, it should be replaced
- A "new belt" is a belt which has been used for less than 5 minutes with the engine running
- A "used belt" is a belt which has been used for 5 minutes or more with the engine running

ADJUSTMENT

These engines are equipped with automatic belt tensioners. Adjusting the belt tension is not possible or necessary.

REMOVAL & INSTALLATION

See Figure 24.

1. Before servicing the vehicle, refer to the Precautions Section.
2. Remove the right hand front wheel.
3. Remove the right hand front fender apron seal.
4. Remove the V-bank cover sub-assembly.
5. Using Special Tool: 09249—63010, release the belt tension by turning the belt tensioner counterclockwise, and remove the V-ribbed belt from the belt tensioner.
6. While turning the belt tensioner counterclockwise, aligns with its holes and then insert the 5 mm bi-hexagon wrench into the holes to fix the V-ribbed belt tensioner.
7. Remove the v-ribbed belt.

To install:

8. To install, reverse removal procedure.
9. If it is difficult to install the V-ribbed belt, perform the following procedure:

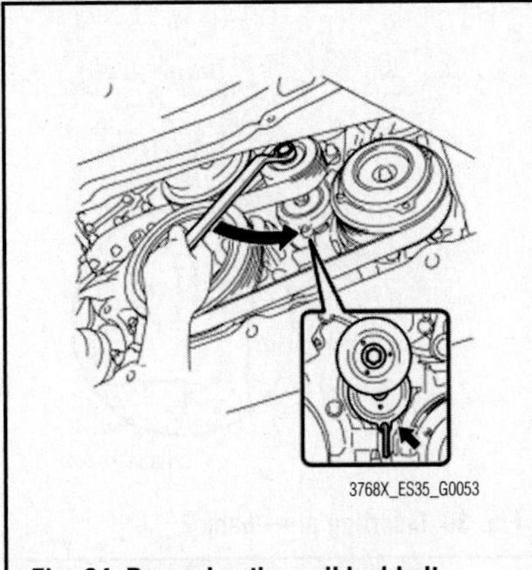

Fig. 24 Removing the v-ribbed belt

a. Put the V-ribbed belt on every pulley except the tensioner pulley.

b. While releasing the belt tension by turning the belt tensioner counterclockwise, put the V-ribbed belt on the tensioner pulley.

➡Put the backside of the V-ribbed belt on the tensioner pulley and idler pulley. Check that the V-ribbed belt is properly set to each pulley.

10. After installing the V-ribbed belt, check that it fits properly in the ribbed grooves. Check to confirm that the belt has not slipped out of the grooves on the bottom of the crank pulley by hand.

11. Tighten the right hand front wheel to: 76 ft. lbs. (103 Nm).

CAMSHAFT AND VALVE LIFTERS

REMOVAL & INSTALLATION

See Figures 25 through 45.

1. Before servicing the vehicle, refer to the Precautions Section.
2. Remove the engine assembly.
3. Install on engine stand.
4. Remove the oil filler cap and gasket.
5. Remove the spark plugs and ignition coil assembly.
6. Remove the drain plug and gasket.
7. Remove the ventilation valve.
8. Remove the 4 bolts and 4 camshaft position sensors.
9. Remove the 4 bolts and 4 camshaft timing oil control valves.
10. Remove the bolt and Crankshaft Position (CKP) sensor.
11. Remove the 2 oil pipe unions and oil pipe. Remove the LH oil control valve filter and gaskets.
12. Remove the oil pipe bolt. Remove the 2 oil pipe unions and oil pipe. Remove the RH oil control valve filter and gaskets.
13. Remove the cylinder block water drain cock sub-assembly, as follows:
    a. Remove the water drain cocks from the cylinder block.
    b. Remove the water drain cock plugs from the water drain cocks.
14. Remove the oil filter.
15. Remove the crankshaft pulley, as follows:
    a. Using SST (SST: 09213-70011, SST: 09330-00021) or equivalent, loosen the crankshaft pulley bolt.
    b. Using SST (SST: 09950-50013) or equivalent, remove the crankshaft pulley bolt and crankshaft pulley.
16. Remove the 6 bolts and the left hand No. 1 front engine mounting bracket. Using "Torx®"socket wrench E8, remove the 2 stud bolts.
17. Remove the water inlet housing, as follows:
    a. Remove the 2 nuts, water inlet and thermostat.
    b. Remove the gasket.
    c. Remove the drain cock plug.
    d. Remove the drain cock.
    e. Remove the 2 stud bolts.
    f. Remove the 2 bolts, nut, and water inlet housing.
    g. Remove the 2 O-rings.
18. Remove the water outlet, as follows:
    a. Remove the 2 bolts, 4 nuts and water outlet.
    b. Remove the 2 gaskets and O-ring.
19. Remove the 12 bolts, valve cover (for Bank 1) and gasket, then remove the 3 gaskets.
20. Remove the 12 bolts, valve cover (for Bank 2) and gasket, then remove the 3 gaskets.
21. Remove the No. 2 oil pan sub-assembly.
22. Remove the oil strainer sub-assembly.
23. Remove the oil pan sub-assembly.
24. Remove the No. 1 oil pan baffle plate.
25. Remove the engine rear oil seal, as follows:
    a. Remove the 6 bolts.
    b. Using a screwdriver with the tip taped, pry out the oil seal retainer. Be careful not to damage the oil seal retainer.

➡Be careful not to damage the engine rear oil seal retainer.

➡Tape the screwdriver tip before use.

26. Place the oil seal retainer on wooden blocks. Using a screwdriver and a hammer, tap out the oil seal.
27. Remove the water pump assembly.
28. Remove the timing chain cover sub-assembly.
29. Remove timing chain case oil seal.
30. Set the No. 1 cylinder to TDC/compression.
31. Remove the No. 1 chain tensioner assembly.
32. Remove the chain tensioner slipper.
33. Remove the chain sub-assembly.
34. Remove the idle sprocket assembly.
35. Remove the No. 1 chain vibration damper.
36. Remove crankshaft timing sprocket. Remove the 2 pulley set keys from the crankshaft.
37. Remove camshaft timing gears and No. 2 chain (for Bank 1), as follows:
    a. While raising the No. 2 chain tensioner, insert a pin of 0.039 in (1.0 mm) into the hole to fix the No. 2 chain tensioner.

➡Be careful not to damage the cylinder head with the wrench.

➡Do not disassemble the camshaft timing gear assemblies.

    b. Hold the hexagonal portion of the

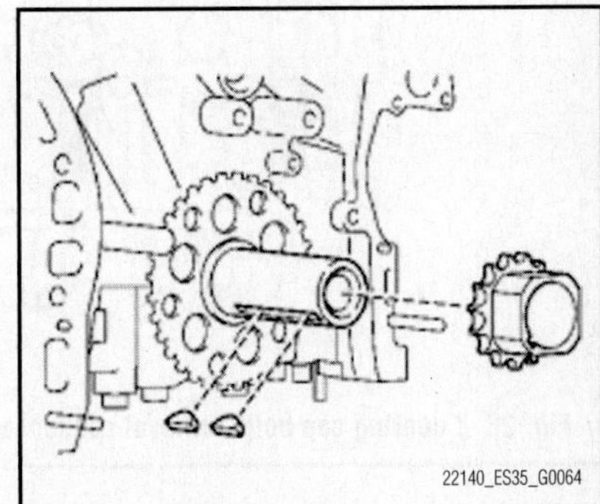

Fig. 25 Removing 2 pulley set keys from the crankshaft

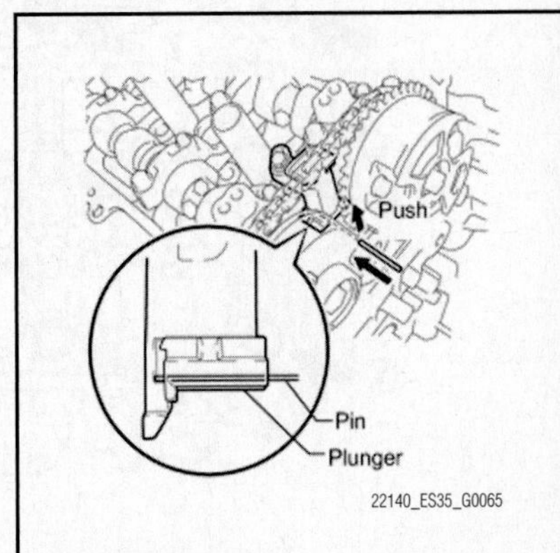

Fig. 26 Inserting pin

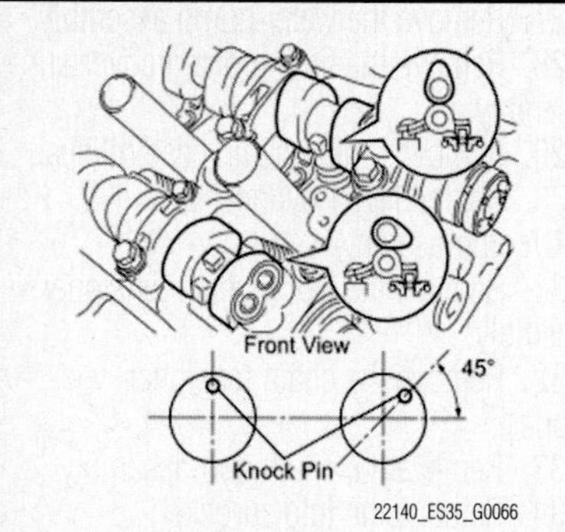

Fig. 27 Camshafts knock pin positioning—bank 1

camshaft with a wrench, and remove the 2 bolts and 2 camshaft timing gears.

c. Remove the No. 2 chain.

38. Remove the bolt and No. 2 chain tensioner assembly.

39. Remove camshaft bearing cap (for Bank 1), as follows:

a. Check that the camshafts are positioned as shown in the illustration.

b. Uniformly loosen and remove the 8 bearing cap bolts in the sequence shown in the illustration.

c. Uniformly loosen and remove the 12 bearing cap bolts in the sequence shown in the illustration. Uniformly loosen the bolts while keeping the camshaft level.

d. Remove the 5 bearing caps.

40. Remove the camshaft.

41. Remove the No. 2 camshaft.

42. If necessary, remove the right hand camshaft housing sub-assembly by prying between the cylinder head and the camshaft housing with a screwdriver with the tip taped.

➡**Be careful not to damage the contact surfaces of the cylinder head and the camshaft housing.**

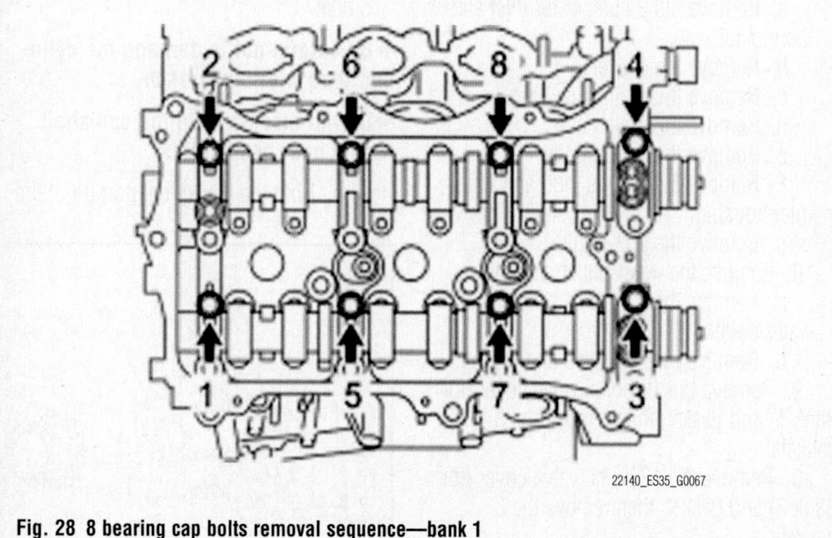

Fig. 28 8 bearing cap bolts removal sequence—bank 1

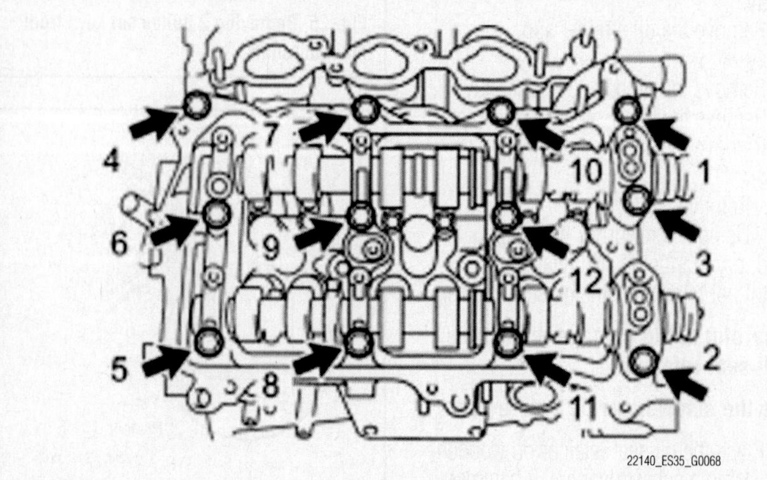

Fig. 29 12 bearing cap bolts removal sequence—bank 1

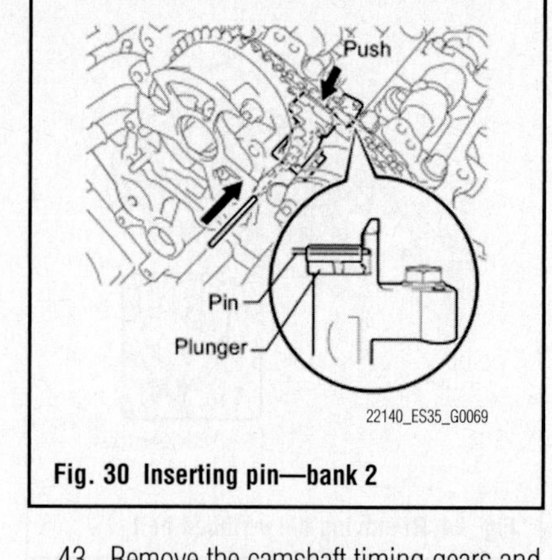

Fig. 30 Inserting pin—bank 2

43. Remove the camshaft timing gears and No. 2 chain (for Left-hand Bank), as follows:

a. While pushing down the No. 3 chain tensioner, insert a pin of 1.0 mm (0.039 in.) into the hole to fix the No. 3 chain tensioner.

b. Hold the hexagonal portion of the camshaft with a wrench, and remove the 2 bolts and 2 camshaft timing gears.

➡**Be careful not to damage the cylinder head with the wrench.**

➡**Do not disassemble the camshaft timing gear assemblies.**

c. Remove the No. 2 chain.

44. Remove the bolt and No. 3 chain tensioner.

45. Remove the camshaft bearing cap (bank 2), as follows:

a. Make sure that the knock pin of the camshaft is positioned as shown in the illustration.

b. Uniformly loosen and remove the 8 bearing cap bolts in the sequence shown in the illustration.

c. Uniformly loosen and remove the 13 bearing cap bolts in the sequence shown in the illustration. Loosen the bolts while keeping the camshaft level.

d. Remove the 5 camshaft bearing caps.

46. Remove the No. 3 camshaft.

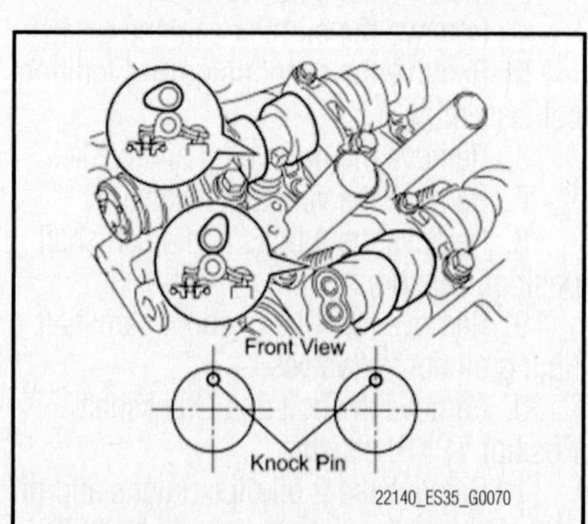

Fig. 31 Camshafts knock pin positioning—bank 2

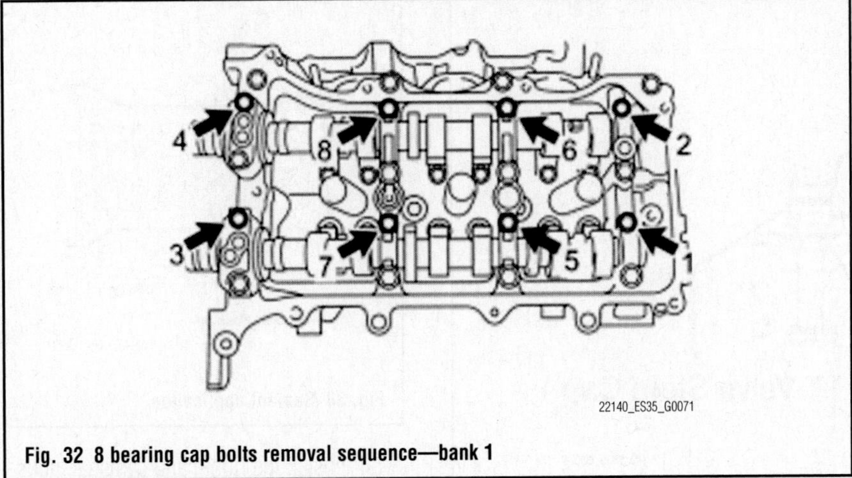

Fig. 32 8 bearing cap bolts removal sequence—bank 1

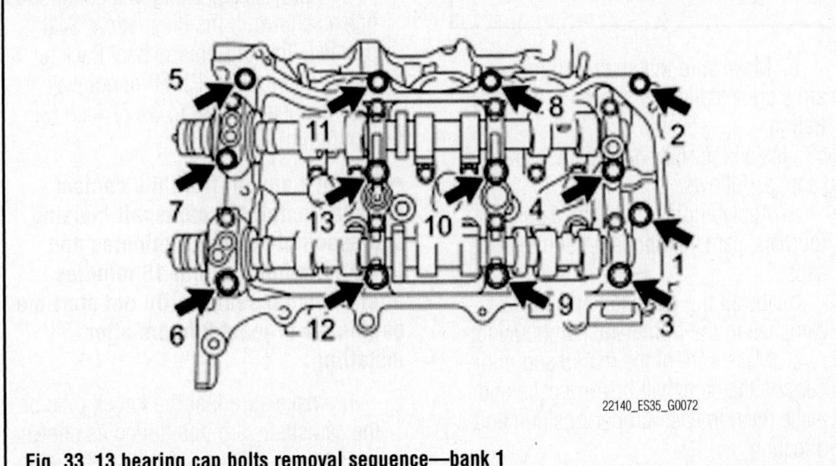

Fig. 33 13 bearing cap bolts removal sequence—bank 1

47. Remove the No. 4 camshaft.

48. If necessary, remove the left hand camshaft housing sub-assembly by prying between the cylinder head and the camshaft housing with a screwdriver with the tip taped.

➡Be careful not to damage the contact surfaces of the cylinder head and the camshaft housing.

49. Remove the No. 1 valve rocker arm sub-assembly, as follows:

a. Remove the 24 valve rocker arms.

➡Arrange the removed parts in the correct order.

50. Remove the valve lash adjuster assembly, as follows:

a. Remove the 24 valve lash adjusters from the cylinder head.

➡Arrange the removed parts in the correct order.

To install:

➡Keep the lash adjuster free of dirt and foreign objects. Only use clean engine oil.

51. Install the valve lash adjuster assembly, as follows:

a. Place the lash adjuster into a container filled with engine oil. Insert the SST 09276-75010 tip into the lash adjuster's plunger and use the tip to press down on the check ball inside the plunger.

b. Squeeze the SST and lash adjuster together to move the plunger up and down 5 to 6 times.

c. Check the movement of the plunger and bleed the air. Make sure the plunger moves up and down.

➡When bleeding air from the high-pressure chamber, make sure that the tip of the SST is actually pressing the check ball as shown in the illustration. If the check ball is not pressed, air will not bleed.

d. After bleeding the air, remove the SST. Then, try to press the plunger quickly and firmly with a finger. Make sure the plunger is very difficult to move. If the result is not as specified, replace the lash adjuster.

➡Install the lash adjuster to the same place it was removed from.

e. Install the lash adjusters.

52. Install the No. 1 valve rocker arm sub-assembly, as follows:

a. Apply engine oil to the lash adjuster tips and valve stem cap ends.

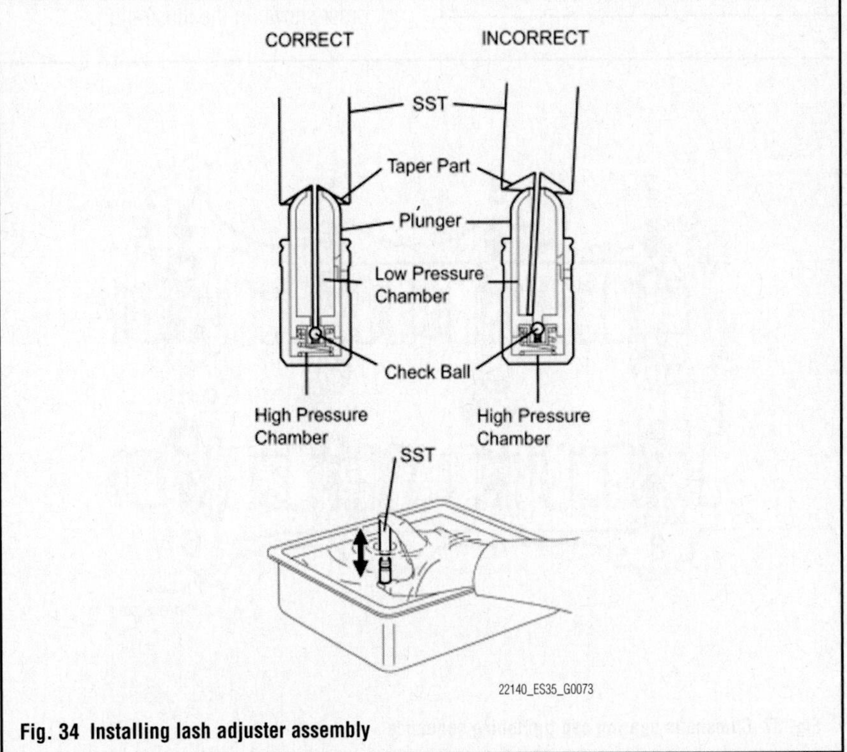

Fig. 34 Installing lash adjuster assembly

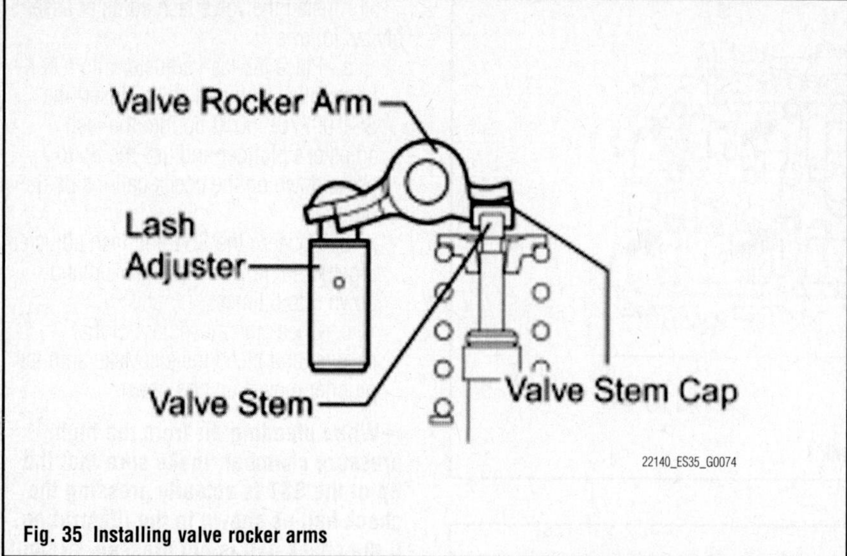

Fig. 35 Installing valve rocker arms

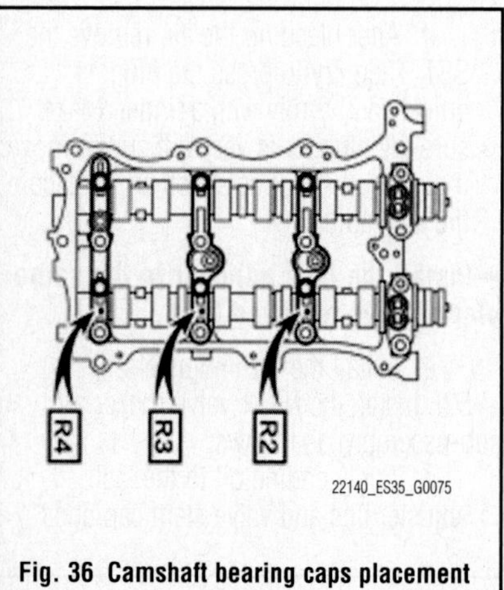

Fig. 36 Camshaft bearing caps placement

b. Make sure that the valve rocker arms are installed as shown in the illustration.

53. Install the right hand camshaft bearing cap, as follows:

a. Apply engine oil to the camshaft journals, camshaft housing and bearing caps.

b. Install the camshaft and No. 2 camshaft to the right camshaft housing.

c. Make sure of the marks and numbers on the camshaft bearing caps and place them in each proper position and direction.

d. Temporarily tighten the 8 bearing cap bolts to 7 ft. lbs. (10 Nm) in the order shown in the illustration.

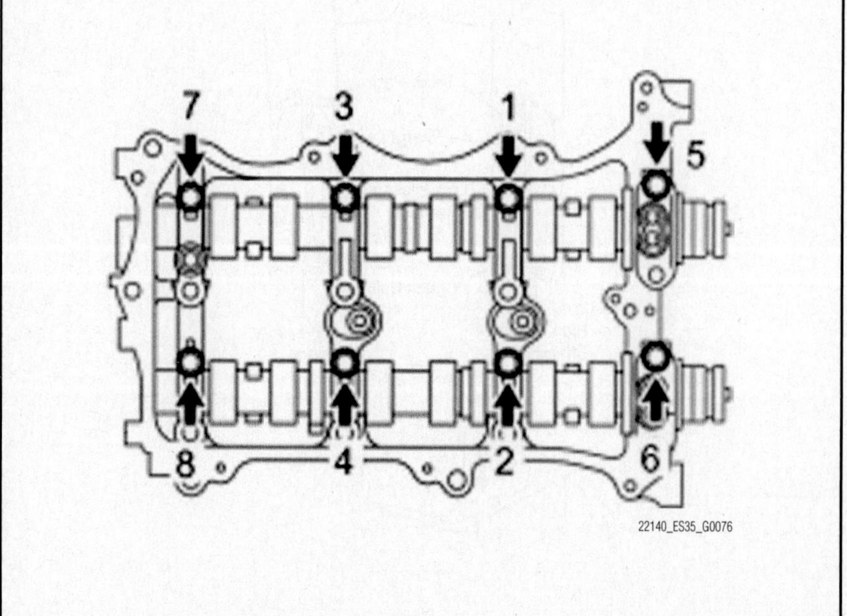

Fig. 37 Camshafts bearing cap tightening sequence

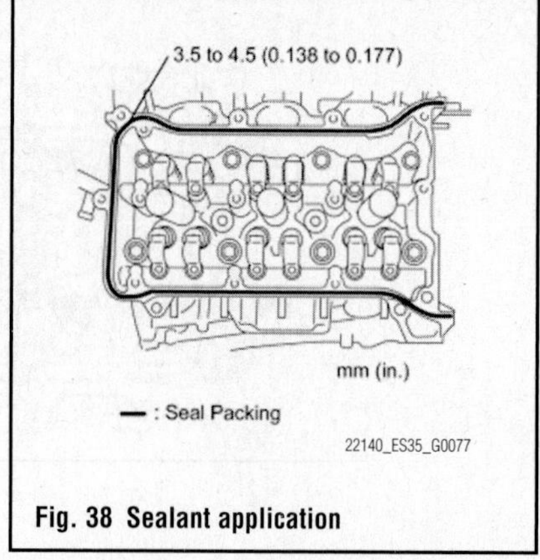

Fig. 38 Sealant application

54. Install the right hand camshaft housing sub-assembly, as follows:

a. Apply seal packing in a continuous line as shown in the illustration. Seal packing: Toyota Genuine Seal Packing Black, Three Bond 1207B or equivalent. Seal diameter: 0.138 to 0.177 inch (3.5 to 4.5 mm).

➡Remove any oil from the contact surface. Install the camshaft housing sub-assembly within 3 minutes and tighten the bolts within 15 minutes after applying sealant. Do not start the engine for at least 2 hours after installing.

b. Make sure that the knock pins of the camshafts are positioned as shown. Install the right hand camshaft housing and tighten the 12 bolts in the order shown in the illustration to 21 ft. lbs. (28 Nm).

➡When installing the camshaft housing RH, it is necessary to correctly position the camshafts as shown in the illustration. Failure to correctly position these parts may result in damage due to contact between the pistons and

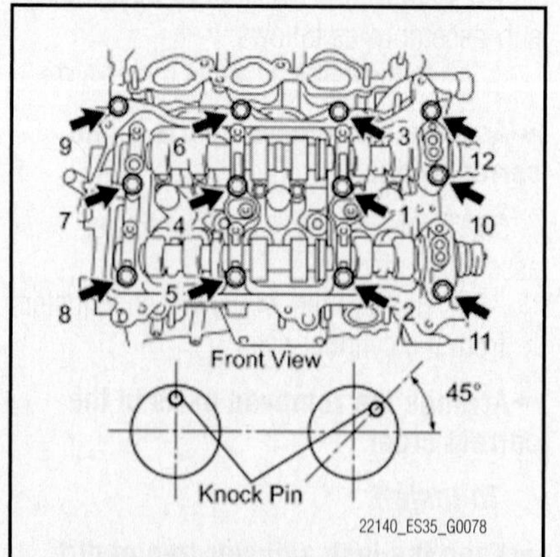

Fig. 39 Camshaft bolt tightening sequence and knock pin position

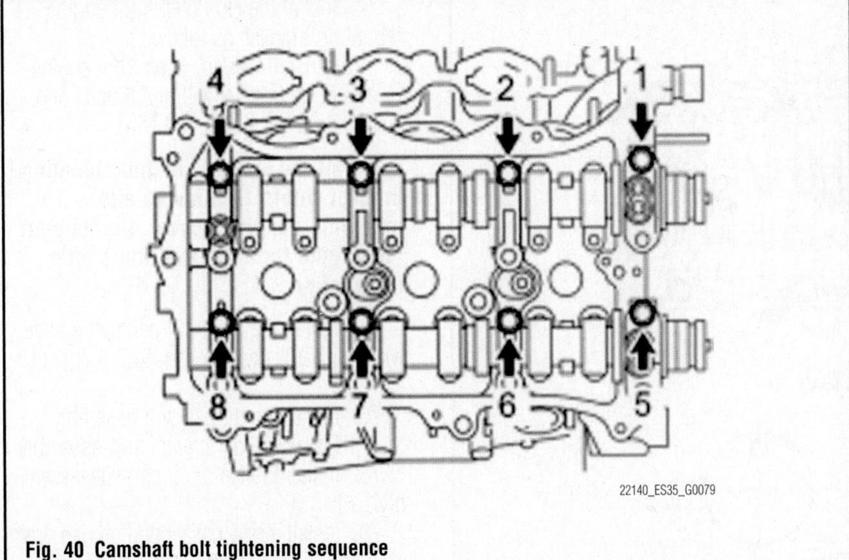

Fig. 40 Camshaft bolt tightening sequence

valves. If a camshaft is rotated with a piston at TDC, valve contact will occur.

➡If any of the bolts are loosened during installation, remove the camshaft housing, clean the installation surfaces, and reapply seal packing.

➡If the camshaft housing is removed because any of the bolts are loosened during installation, make sure that the previously applied seal packing does not enter any oil passages.

c. Tighten the 8 bolts to 12 ft. lbs. (16 Nm) in the order shown in the illustration.

55. Install the camshaft bearing cap (bank 2), as follows:

a. Apply engine oil to the camshaft journals, camshaft housing and bearing caps.

b. Install the No. 3 camshaft and No. 4 camshaft to the left hand camshaft housing.

c. Make sure of the marks and numbers on the camshaft bearing caps and place them in each proper position and direction.

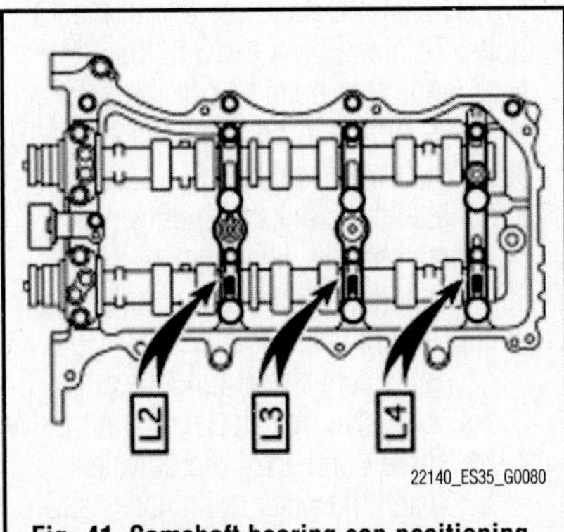

Fig. 41 Camshaft bearing cap positioning

d. Temporarily tighten the 8 bolts in the order shown in the illustration to 7 ft. lbs. (10 Nm).

56. Make sure that the valve rocker arm is installed.

57. Install the left camshaft housing sub-assembly, as follows:

a. Apply seal packing in a continuous line as shown in the illustration. Seal packing: Toyota Genuine Seal Packing Black, Three Bond 1207B or equivalent. Seal diameter: 0.138 to 0.177 inch (3.5 to 4.5 mm).

➡Remove any oil from the contact surface. Install the camshaft housing sub-assembly within 3 minutes and tighten the bolts within 15 minutes after applying sealant. Do not start the engine for at least 2 hours after installing.

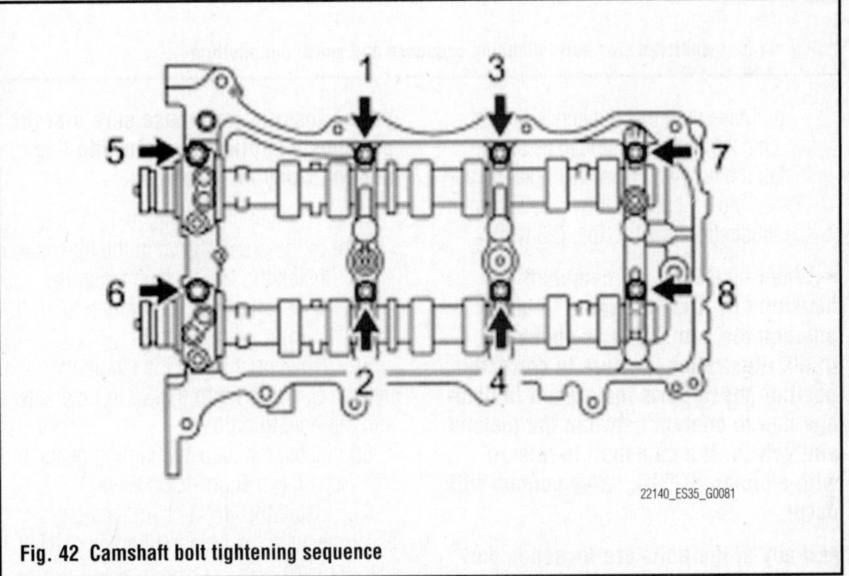

Fig. 42 Camshaft bolt tightening sequence

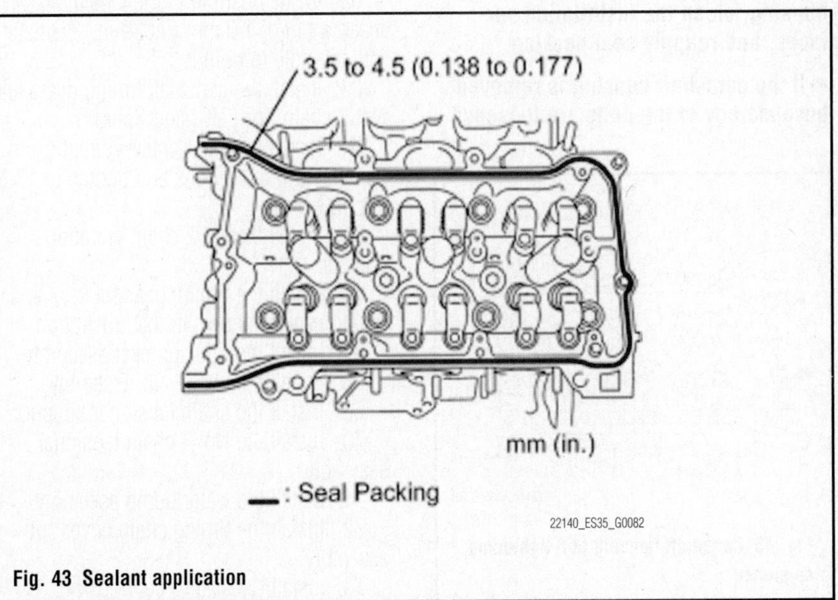

Fig. 43 Sealant application

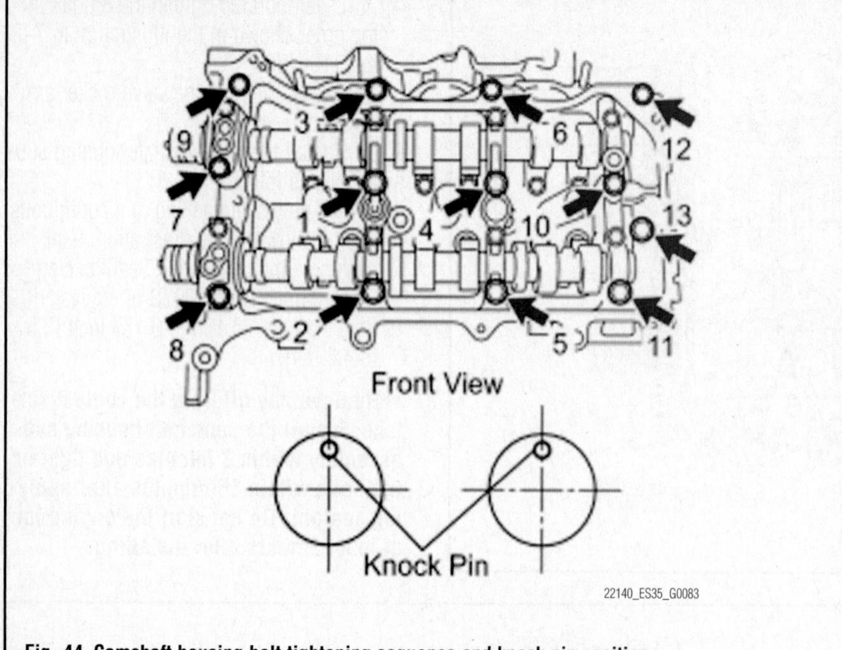

Fig. 44 Camshaft housing bolt tightening sequence and knock pin position

b. Make sure that the knock pins of the camshafts are positioned as shown. Install the left hand camshaft housing and tighten the 13 bolts in the order shown in the illustration to 21 ft. lbs. (28 Nm).

➡**When installing the camshaft housing LH, it is necessary to correctly position the camshafts as shown in the illustration. Failure to correctly position these parts may result in damage due to contact between the pistons and valves. If a camshaft is rotated with a piston at TDC, valve contact will occur.**

➡**If any of the bolts are loosened during installation, remove the camshaft housing, clean the installation surfaces, and reapply seal packing.**

➡**If the camshaft housing is removed because any of the bolts are loosened during installation, make sure that the previously applied seal packing does not enter any oil passages.**

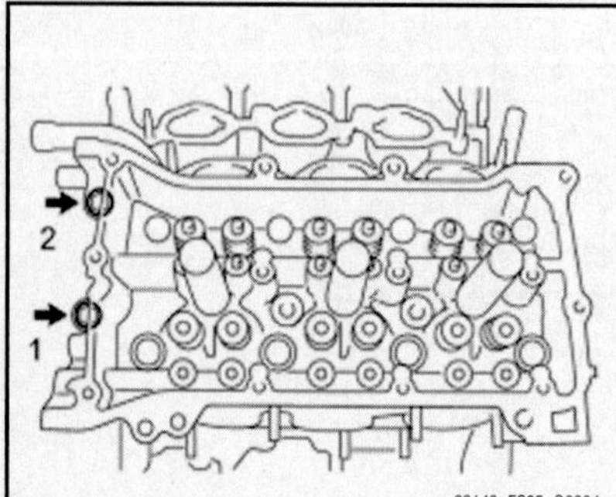

Fig. 45 Camshaft housing bolt tightening sequence

c. Tighten the 8 bolts to 12 ft. lbs. (16 Nm) in the order shown in the illustration.

58. Install the No. 2 chain tensioner assembly with the bolt and tighten to 15 ft. lbs. (21 Nm).

59. While pushing in the tensioner, insert a pin of 1.0 mm (0.039 in.) diameter into the hole to hold it.

60. Install the camshaft timing gears and No. 2 chain (for Right-hand Bank).

61. Install the No. 3 chain tensioner assembly with the bolt and tighten to 15 ft. lbs. (21 Nm).

62. While pushing in the tensioner, insert a pin of 1.0 mm (0.039 in.) diameter into the hole to hold it.

63. Install the camshaft timing gears and No. 2 chain (for Left-hand Bank).

64. Install the No 1 chain vibration damper with the 2 bolts and tighten to 17 ft. lbs. (23 Nm).

65. Install the No 2 chain vibration damper.

66. Install the timing gear set keys and timing gear as shown in the illustration.

67. Install the idle sprocket assembly.

68. Install the chain sub-assembly.

69. Install the chain tensioner slipper.

70. Install the No. 1 chain tensioner assembly.

71. Install the water pump assembly.

72. Install the timing chain cover sub-assembly.

73. Install the water inlet housing.

74. Install the No. 1 left front engine mounting bracket, as follows:

a. Install the No. 1 left front engine mounting bracket with the 6 bolts and tighten to 40 ft. lbs. (54 Nm).

➡**Install the water inlet and mounting bracket within 15 minutes after installing the chain cover. Do not start the engine for at least 2 hours after installation.**

75. Install the No. 1 oil pan baffle plate with the 7 bolts and tighten to 7 ft. lbs. (10 Nm).

76. Install the oil pan sub-assembly.

77. Install the oil strainer sub-assembly.

78. Install the No. 2 oil pan sub-assembly.

79. Install a new gasket and oil pan drain plug and tighten to 30 ft. lbs. (40 Nm).

80. Install the cylinder head cover sub-assembly, as follows:

a. Apply seal packing (Toyota Genuine Seal Packing Black, Three Bond 1207B or equivalent).

➡**Remove any oil from the contact surface. Install the crankcase within 3 minutes after applying seal packing. Do not start the engine for at least 2 hours after installation.**

b. Install the gasket to the head cover.

c. Install the head cover with the 12 bolts. Tighten bolt A to 15 ft. lbs. (21 Nm), and other bolts to 7 ft. lbs. (10 Nm). Be certain to tighten bolt 1.

81. Install the left-hand cylinder head cover sub-assembly, as follows:

a. Apply seal packing (Toyota Genuine Seal Packing Black, Three Bond 1207B or equivalent).

➡**Remove any oil from the contact surface. Install the crankcase within 3 minutes after applying seal packing. Do not start the engine for at least 2 hours after installation.**

b. Install the gasket to the head cover.

c. Install the head cover with the 14 bolts. Tighten bolt A to 15 ft. lbs. (21 Nm), and other bolts to 7 ft. lbs. (10 Nm). Be certain to tighten bolts 1 and 10.

82. Install water outlet.

83. Install the crankshaft pulley.

84. Install the oil filter element.

85. Install the cylinder block water drain cock sub-assembly, as follows:

a. Apply adhesive around the drain cocks. Adhesive: Toyota Genuine Adhesive 1324, Three Bond 1324 or Equivalent.

b. Install the water drain cocks and tighten to 18 ft. lbs. (25 Nm). Do not rotate the drain cocks more than 1 revolution (360°) after tightening the drain cocks with the specified torque. Do not loosen after setting correctly.

c. Install the water drain cock plug to the water drain cocks and tighten to 9 ft. lbs. (13 Nm).

86. Install the No. 1 oil pipe.
87. Install the oil pipe.
88. Install the Crankshaft Position (CKP) sensor with the bolt and tighten to 7 ft. lbs. (10 Nm).
89. Install the 4 camshaft timing oil control valves with the 4 bolts and tighten to 7 ft. lbs. (10 Nm).
90. Install the 4 camshaft position sensors with the 4 bolts and tighten to 7 ft. lbs. (10 Nm).
91. Install the ventilation valve sub-assembly, as follows:

a. Apply adhesive (Toyota Genuine Adhesive 1324, Three Bond 1324 or equivalent) around the ventilation valve.

b. Install the ventilation valve and tighten to 20 ft. lbs. (27 Nm).

92. Install the 6 spark plugs and the ignition coil assembly.
93. Install the oil filler cap sub-assembly.
94. Remove the engine stand.
95. Install the engine assembly.

CRANKSHAFT DAMPER

REMOVAL & INSTALLATION

See Figure 46.

1. Before servicing the vehicle, refer to the Precautions Section.
2. Disconnect the negative battery cable. Wait at least 90 seconds before performing any other work.
3. It may be necessary to remove the radiator assembly for access.
4. Remove the accessory drive belt(s).
5. Using SST 09213-54015 (91651-60855), 09330-00021, loosen the crankshaft pulley bolt.
6. Using SST 09950-50013 (09951-05010, 09952-05010, 09953-05010, and 09954-05030) and the pulley bolt, remove the crankshaft pulley.

➡Before using SST, apply lubricating oil on the threads and tip of the center bolt.

To install:

7. Align the pulley set key with the key groove of the pulley, and slide on the pulley.
8. Using SST 09213-54015 (91651-60855), 09330-00021, install the pulley bolt.
9. Tighten the crankshaft pulley bolt to 184 ft. lbs. (250 Nm).
10. Install the accessory drive belt(s).
11. Install the radiator, if necessary.
12. Connect the negative battery cable.

CRANKSHAFT FRONT SEAL

REMOVAL & INSTALLATION

See Timing Chain Cover and Seal.

CYLINDER HEAD

REMOVAL & INSTALLATION

See Figures 47 through 53.

1. Before servicing the vehicle, refer to the Precautions Section.
2. Remove the engine assembly with transaxle.
3. Secure engine.
4. Remove the oil filler cap sub-assembly.
5. Remove the spark plugs and ignition coil assembly.
6. Remove the oil pan drain plug and gasket.
7. Remove the ventilation valve sub-assembly.
8. Remove the camshaft position sensor.
9. Remove the camshaft timing oil control valve assembly.
10. Remove Crankshaft Position (CKP) sensor.
11. Remove the No. 1 oil pipe.
12. Remove the oil pipe.
13. Remove the cylinder block water drain cock sub-assembly.
14. Remove the oil filter.
15. Remove the crankshaft pulley.
16. Remove the left hand No. 1 front engine mounting bracket.
17. Remove the water inlet housing.
18. Remove the water outlet.
19. Remove the cylinder head covers and gaskets.
20. Remove the No. 2 oil pan sub-assembly.
21. Remove the oil strainer sub-assembly.
22. Remove the oil pan sub-assembly.
23. Remove the No. 1 oil pan baffle plate.
24. Remove the engine rear oil seal.
25. Remove the water pump assembly.
26. Remove the timing chain cover.
27. Set the No. 1 cylinder to TDC/compression.
28. Remove the No. 1 chain tensioner assembly.
29. Remove the chain tensioner slipper.
30. Remove the chain sub-assembly.
31. Remove the idle sprocket assembly.
32. Remove the No. 1 and 2 chain vibration damper.

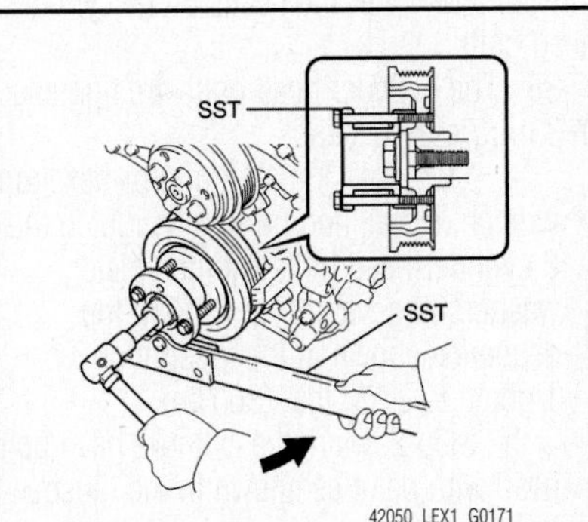

Fig. 46 Using special tools to remove the crankshaft pulley.

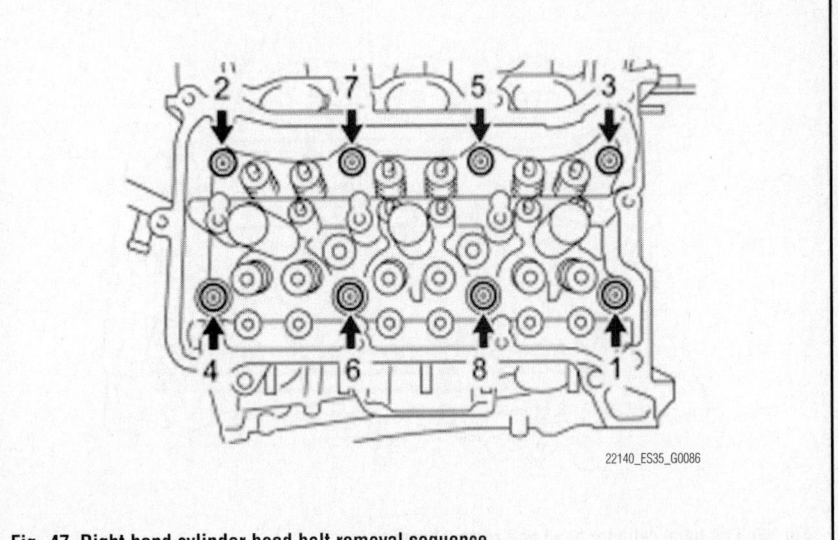

Fig. 47 Right hand cylinder head bolt removal sequence

➡Be careful not to drop washers into the cylinder head.

➡Cylinder head warpage or cracking could result from removing bolts in an incorrect order.

➡Be sure to keep separate the removed parts for each installation position.

a. Remove the cylinder head and gasket.

33. Remove the right hand cylinder head sub-assembly, as follows:

a. Using a 10 mm bi-hexagon wrench, uniformly loosen the 8 bolts in the sequence shown in the illustration. Remove the 8 cylinder head bolts and plate washers.

✲✲ WARNING

Be careful not to drop washers into the cylinder head.

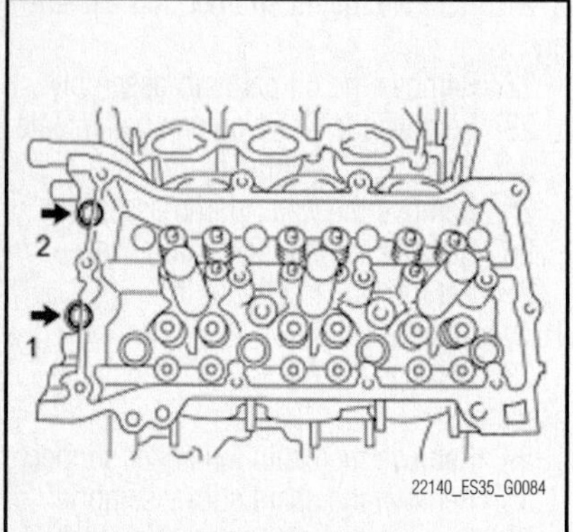

Fig. 48 Left hand cylinder head sub-assembly bolt removal sequence

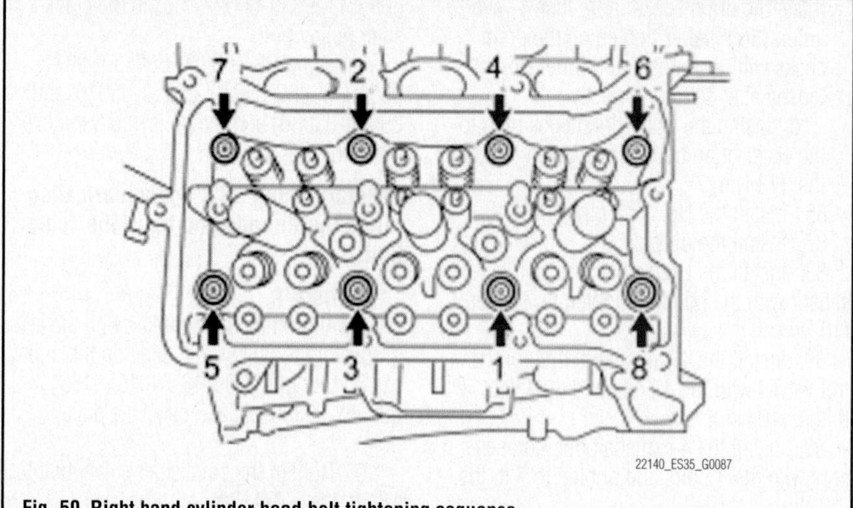

Fig. 50 Right hand cylinder head bolt tightening sequence

✲✲ WARNING

Cylinder head warpage or cracking could result from removing bolts in an incorrect order.

➡Be sure to keep separate the removed parts for each installation position.

b. Remove the cylinder head and gasket.

34. Remove the left hand cylinder head sub-assembly, as follows:

a. Uniformly loosen and remove the 2 bolts in the sequence shown in the illustration.

➡Be careful not to drop washers into the cylinder head.

➡Cylinder head warpage or cracking could result from removing bolts in an incorrect order.

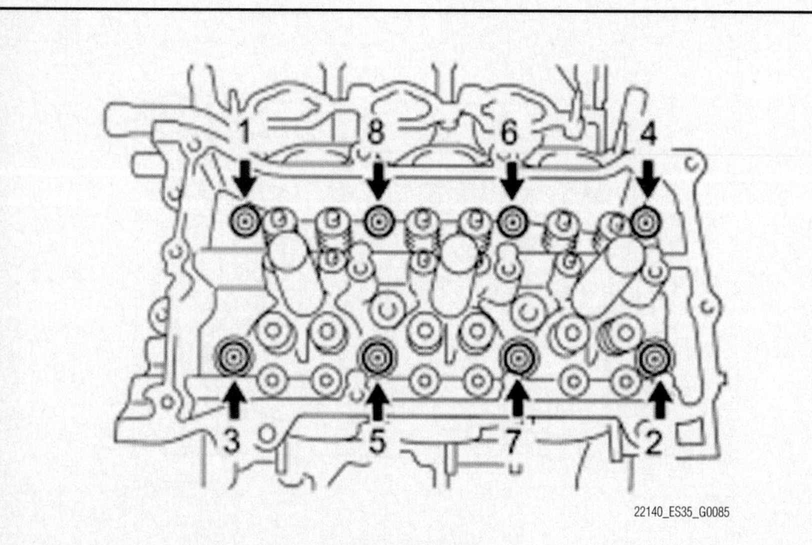

Fig. 49 Left hand cylinder head bolt removal sequence

b. Using a 10 mm bi-hexagon wrench, uniformly loosen the 8 bolts in the sequence shown in the illustration. Remove the 8 cylinder head bolts and plate washers.

c. Remove the cylinder head and gasket

To install:

35. Place the right hand cylinder head gasket on the cylinder block surface with the front face of the Lot No. stamp upward.

➡Be careful of the installation direction.

➡Gently place the cylinder head in order not to damage the gasket with the bottom part of the head.

36. Place the cylinder head on the cylinder block.

✲✲ CAUTION

Do not allow oil to adhere to the mounting surface of the cylinder head.

37. Apply a light coat of engine oil to the threads and under the heads of the cylinder head bolts.

38. The cylinder head bolts are tightened in 3 progressive steps:

a. Step 1: Using a 10 mm bi-hexagon wrench, install and uniformly tighten the 8 cylinder head bolts with the plate washers in several steps and in the sequence shown in the illustration. Tighten to 27 ft. lbs. (36 Nm).

b. Step 2: Mark the cylinder head bolt head with paint as shown in the illustration. Tighten the cylinder head bolts another 90°.

c. Step 3: Tighten the cylinder head bolts an additional 90°. Check

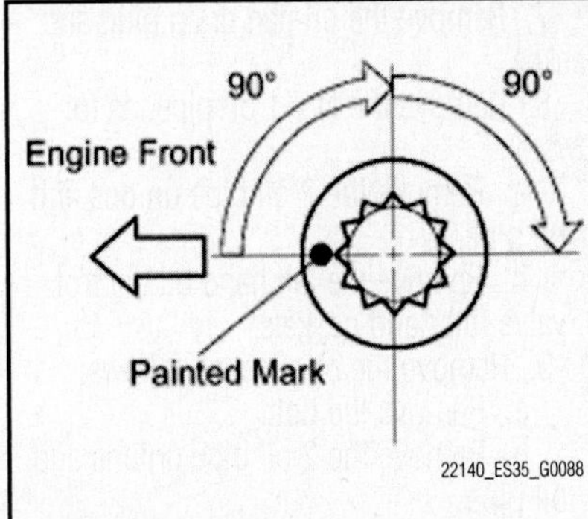

Fig. 51 **Mark the cylinder head bolt and tighten another 90°**

that the painted mark is now facing rearward.

d. Seal packing will seep out on the engine's front side. Thoroughly wipe clean any seal packing.

39. Place the left hand cylinder head gasket on the cylinder block surface with the front face of the Lot No. stamp upward.

➡**Be careful of the installation direction.**

➡**Gently place the cylinder head in order not to damage the gasket with the bottom part of the head.**

40. Place the cylinder head on the cylinder block.

➡**Do not allow oil to adhere to the mounting surface of the cylinder head.**

41. Apply a light coat of engine oil to the threads and under the heads of the cylinder head bolts.

42. The cylinder head bolts are tightened in 3 progressive steps:

a. Step 1: Using a 10 mm bi-hexagon wrench, install and uniformly tighten the 8 cylinder head bolts with the plate washers in several steps and in the sequence shown in the illustration. Tighten to 27 ft. lbs. (36 Nm).

b. Step 2: Mark the cylinder head bolt head with paint as shown in the illustration. Tighten the cylinder head bolts another 90°.

c. Step 3: Tighten the cylinder head bolts an additional 90°. Check that the painted mark is now facing rearward.

d. Tighten the 2 bolts in the order shown in the illustration to 22 ft. lbs. (30 Nm). Only use the specifications stated above when tightening the bolts 1 and 2 shown in the illustration.

e. Seal packing will seep out on the engine's front side. Thoroughly wipe clean any seal packing.

43. Install the No. 2 chain tensioner assembly.

44. Install the No. 3 chain tensioner.

45. Install the No. 1 and 2 chain vibration damper.

46. Install the idle sprocket assembly.

47. Install the chain sub-assembly.

48. Install the chain tensioner slipper.

49. Install the No. 1 chain tensioner assembly.

50. Install the water pump assembly.

51. Install the timing chain cover.

52. Install the water inlet housing.

53. Install the left hand No. 1 front engine mounting bracket.

54. Install the No. 1 oil pan baffle plate.

55. Install the oil pan sub-assembly.

56. Install the oil strainer sub-assembly.

57. Install the No. 2 oil pan sub-assembly.

58. Install the oil pan drain plug and gasket.

59. Install the cylinder head covers and gaskets

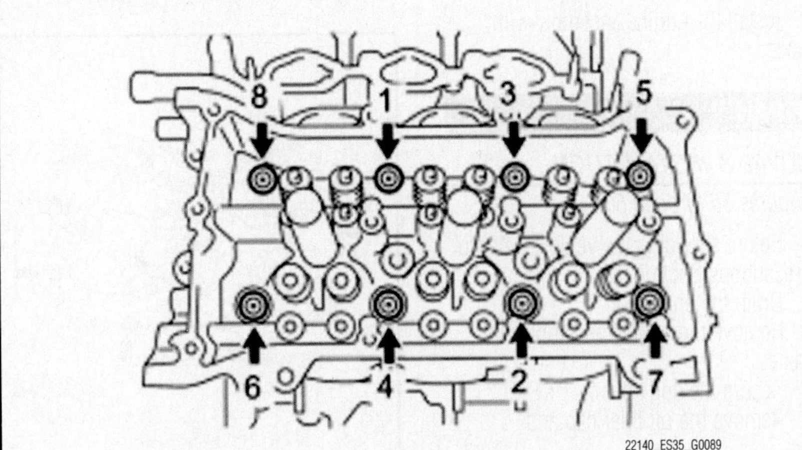

Fig. 52 **Left hand cylinder head bolt tightening sequence**

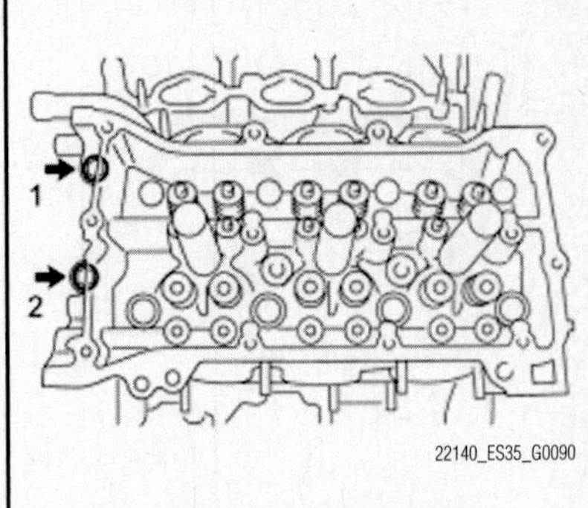

Fig. 53 **Left hand cylinder head sub-assembly bolt tightening sequence**

60. Install the water outlet.

61. Install the crankshaft pulley.

62. Install the oil filter.

63. Install the cylinder block water drain cock sub-assembly.

64. Install the No. 1 oil pipe.

65. Install the oil pipe.

66. Install the Crankshaft Position (CKP) sensor.

67. Install the camshaft timing oil control valve assembly.

68. Install the camshaft position sensor.

69. Install the ventilation valve sub-assembly.

70. Install the spark plugs and ignition coil assembly.

71. Install the oil filler cap sub-assembly.

72. Install the engine assembly with transaxle.

EXHAUST MANIFOLD

REMOVAL & INSTALLATION

See Figures 54 and 55.

1. Before servicing the vehicle, refer to the Precautions Section.

2. Remove the engine assembly with transaxle.

3. Secure the engine.

4. Remove the ignition coil assembly.

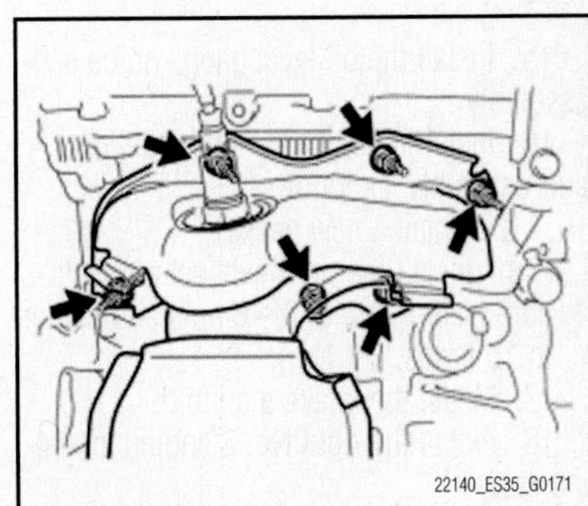

Fig. 54 **Removing the exhaust manifold 6 nuts—Right**

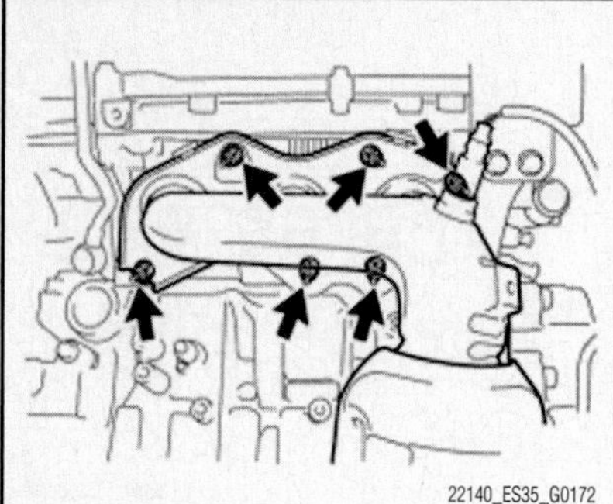

Fig. 55 Removing the exhaust manifold 6 nuts—Left

5. Remove the right No. 2 engine mounting stay.
6. Remove the intake manifold.
7. Remove the right exhaust manifold sub-assembly, as follows:
   a. Uniformly loosen and remove the 6 nuts.
   b. Remove the manifold and gasket.
8. Remove the oil level gauge guide sub-assembly.
9. Remove the bolt, nut and No. 2 manifold stay.
10. Remove the 3 bolts and No. 2 exhaust manifold heat insulator.
11. Remove the left exhaust manifold sub-assembly, as follows:
   a. Uniformly loosen and remove the 6 nuts.
   b. Remove the manifold and gasket.

To install:

12. Install the left exhaust manifold sub-assembly, as follows:
   a. Install a new gasket.
   b. Install the left exhaust manifold sub-assembly with the 6 nuts and tighten to 15 ft. lbs. (21 Nm).
13. Install the No. 2 exhaust manifold heat insulator with the 3 bolts and tighten to 75 inch lbs. (8.5 Nm).
14. Install the No. 2 manifold stay with the bolt and nut and tighten to 25 ft. lbs. (34 Nm).
15. Install the oil level gauge guide sub-assembly.
16. Install the right exhaust manifold sub-assembly, as follows:
   a. Install a new gasket.
   b. Install the right exhaust manifold sub-assembly with the 6 nuts and tighten to 15 ft. lbs. (21 Nm).
17. Install the intake manifold.
18. Install the right No. 2 engine mounting stay.
19. Install the ignition coil assembly.
20. Install the engine assembly with transaxle.

INTAKE MANIFOLD

REMOVAL & INSTALLATION

1. Before servicing the vehicle, refer to the Precautions Section.
2. Remove the engine assembly with transaxle.
3. Secure the engine.
4. Remove the ignition coil assembly.
5. Remove the right No. 2 engine mounting stay.
6. Remove the intake manifold, as follows:
   a. Uniformly loosen and remove the 6 bolts and 4 nuts.
   b. Remove the intake manifold and 2 gaskets.

To install:

7. Install the intake manifold, as follows:

⁂ WARNING

DO NOT apply oil to the intake manifold and cylinder head sub-assembly bolts.

   a. Set a new gasket on each cylinder head.

➡Align the port holes of the gasket and cylinder head.

➡Make sure that the gasket is installed in the correct direction.

   b. Set the intake manifold on the cylinder heads.
   c. Install and tighten the 6 bolts and 4 nuts uniformly in several steps to 15 ft. lbs. (21 Nm).
8. Install the right No. 2 engine mounting stay.
9. Install the ignition coil assembly.
10. Install the engine assembly with transaxle.

OIL PAN

REMOVAL & INSTALLATION

See Figures 56 through 60.

1. Before servicing the vehicle, refer to the Precautions Section.
2. Drain the engine oil.
3. Remove the engine assembly with transaxle.
4. Secure the engine.
5. Remove the oil filler cap and gasket.
6. Remove the oil pan drain plug and gasket.
7. Remove the oil pan drain plug and gasket.
8. Remove the No. 1 oil pipe, as follows:
   a. Remove the 2 oil pipe unions and oil pipe.
   b. Remove the left hand oil control valve filter and gaskets.
9. Remove the oil pipe, as follows:
   a. Remove the bolt.
   b. Remove the 2 oil pipe unions and oil pipe.
   c. Remove the right oil control valve filter and gaskets.
10. Remove the oil filter element, as follows:
   a. Remove the drain plug. Do not remove the O-ring.
   b. Connect the hose to the pipe.
   c. Insert the pipe with the hose into the oil filter cap.
   d. Make sure that the oil is completely drained and remove the pipe and O-ring.
   e. Using SST (SST: 09228-06501) or equivalent, remove the oil filter cap.
   f. Remove the oil filter element and O-ring from the oil filter cap. Do not use any tools when removing the O-ring to prevent the O-ring groove from being damaged.
11. Remove the No. 2 oil pan sub-assembly, as follows:
   a. Remove the 16 bolts and 2 nuts.
   b. Insert the blade of SST (SST: 09032-00100) or equivalent tool between the oil pans. Cut through the applied sealer and remove the No. 2 oil pan sub-assembly.

➡Be careful not to damage the contact surfaces of the oil pans.

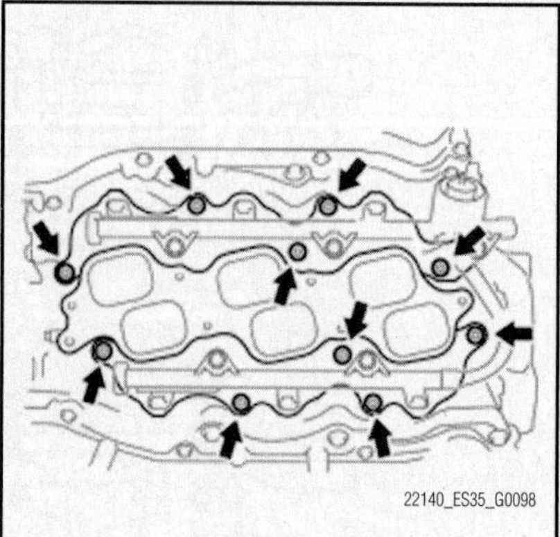

Fig. 56 No. 2 oil pan sub-assembly removal

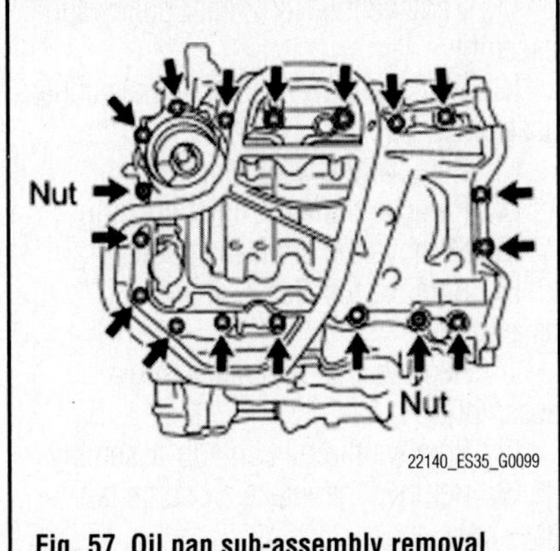

Fig. 57 Oil pan sub-assembly removal

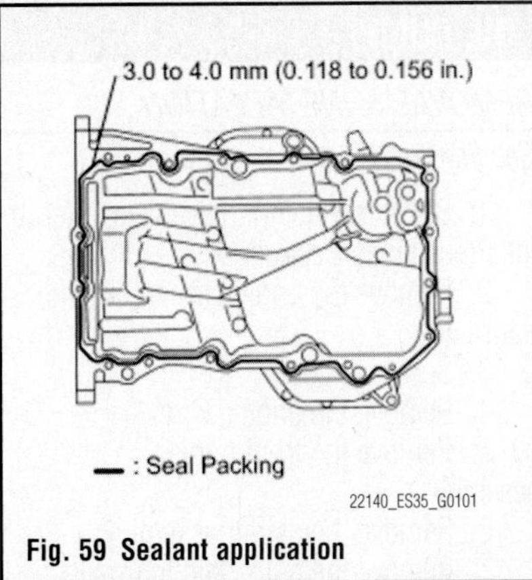

Fig. 59 Sealant application

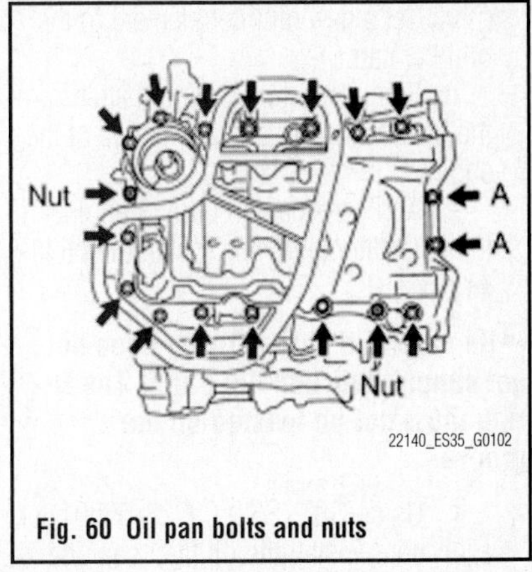

Fig. 60 Oil pan bolts and nuts

12. Remove the oil pan sub-assembly, as follows:

a. Remove the 16 bolts and 2 nuts.

➡**Be sure to clean the bolts and stud bolts and check the threads for cracks or other damage.**

b. Remove the oil pan by prying between the oil pan and cylinder block with a taped screwdriver.

➡**Be careful not to damage the contact surfaces of the cylinder block and oil pan.**

c. Remove the 2 O-rings.

To install:

13. Install the oil pan sub-assembly, as follows:

a. Using an E8 "TORX®" socket wrench, install the stud bolts as shown in the illustration. Tighten to 7 ft. lbs (10 Nm).

b. Install 2 new O-rings.

➡**Remove any oil from the contact surface.**

➡**Install the oil pan within 3 minutes after applying seal packing.**

➡**Do not start the engine for at least 2 hours after installing.**

c. Apply seal packing (Toyota Genuine Seal Packing Black, Three Bond 1207B or equivalent) in a continuous line as shown in the illustration. Seal diameter: 3.0 to 4.0 mm (0.118 to 0.156 in.).

d. Install the oil pan with the 16 bolts and 2 nuts and tighten to 7 ft. lbs (10 Nm), and 15 ft. lbs (21 Nm).

14. Install the No. 2 oil pan sub-assembly, as follows:

➡**Remove any oil from the contact surface.**

➡**Install the No. 2 oil pan within 3 minutes after applying seal packing.**

➡**Do not start the engine for at least 2 hours after installing.**

a. Using an E6 "TORX®" socket wrench, install the stud bolts as shown in the illustration and tighten to 35 inch lbs (4 Nm).

b. Apply seal packing (Toyota Genuine Seal Packing Black, Three Bond 1207B or equivalent) in a continuous line as shown in the illustration. Seal diameter: 3.0 to 4.0 mm (0.118 to 0.156 in.).

c. Install the No. 2 oil pan with the 16 bolts and 2 nuts and tighten to 7 ft. lbs (10 Nm).

15. Install the oil pan drain plug and a new gasket. Tighten to 30 ft. lbs (40 Nm).

16. Install the oil filter element, as follows:

a. Clean the inside of the oil filter cap, the threads and O-ring groove.

b. Apply a small amount of engine oil to a new O-ring and install it to the oil filter cap.

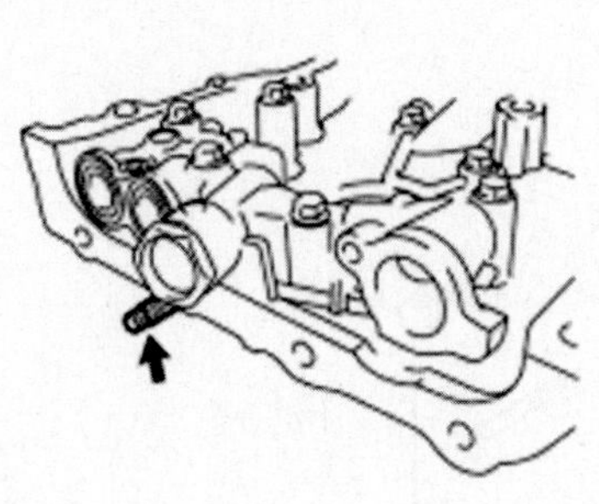

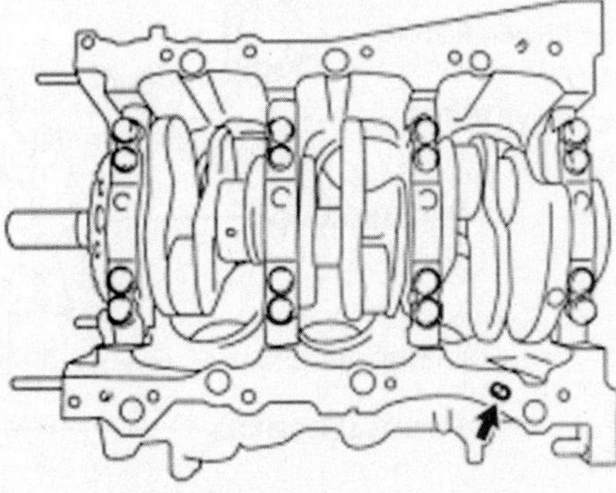

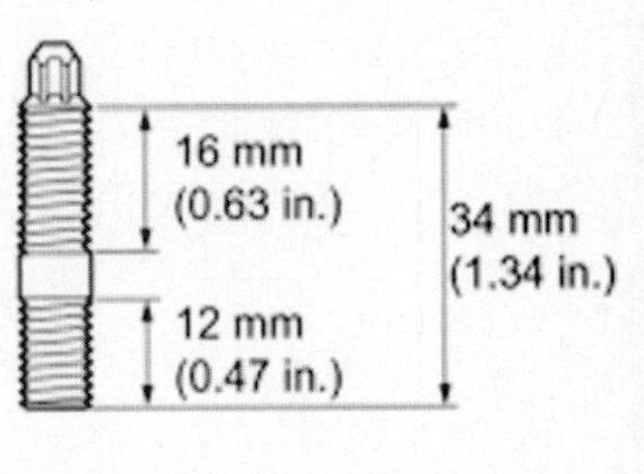

Fig. 58 Locating stud bolts

c. Set a new oil filter element to the oil filter cap.

d. Remove dirt or foreign matter from the installation surface and inside of the engine.

e. Apply a small amount of engine oil to the O-ring again and install the oil filter cap.

➡Be careful that the O-ring does not get caught between the parts. The O-ring must not be twisted on the groove.

f. Using SST (SST: 09228-06501) or equivalent, install the oil filter cap and tighten to 18 ft. lbs (25 Nm). Make sure that the oil filter is installed securely as shown in the illustration.

17. Install the oil filler cap sub-assembly.
18. Install the engine assembly with transaxle.
19. Check for oil leaks.

OIL PUMP

REMOVAL & INSTALLATION

See Figure 61.

1. Before servicing the vehicle, refer to the Precautions Section.
2. Remove the engine assembly with transaxle.
3. Secure engine.
4. Remove the engine wire.
5. Remove the front frame assembly.
6. Remove the starter assembly.
7. Remove the automatic transaxle assembly.
8. Remove the oil level gauge guide sub-assembly.
9. Remove the right and left exhaust manifold sub-assemblies.
10. Remove the drive plate and ring gear sub-assembly.
11. Remove the No. 2 idler pulley sub-assembly.
12. Remove the V-ribbed belt tensioner assembly.
13. Remove the water pump pulley.
14. Remove the water inlet housing.
15. Remove the crankshaft pulley.
16. Remove the No. 2 oil pan sub-assembly.
17. Remove the oil strainer sub-assembly.
18. Remove the oil pan sub-assembly.
19. Remove the intake air surge tank assembly.
20. Remove the ignition coil assembly.
21. Remove the No. 1 and 2 oil pipes.
22. Remove the right and left cylinder head cover sub-assemblies.
23. Remove the timing chain or belt cover sub-assembly.
24. Remove the timing gear case or timing chain case oil seal, as follows:

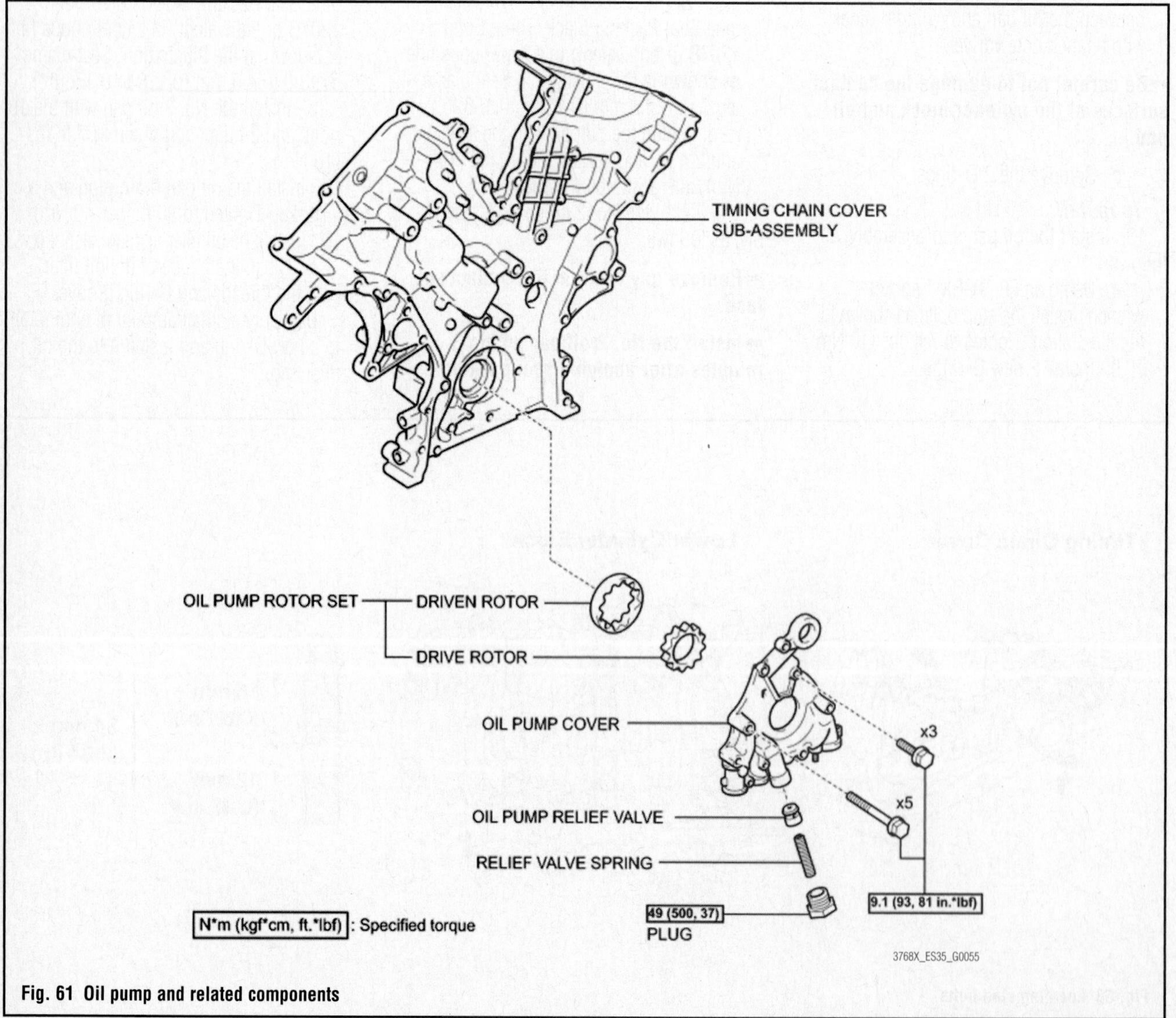

Fig. 61 Oil pump and related components

a. Using a screwdriver with the tip taped, pry out the oil seal.

To install:

25. Install timing gear case or timing chain case oil seal, as follows:

a. Using SST (SST: 09316-60011) or equivalent tool, tap in a new oil seal until its surface is flush with the timing chain case edge.

➡Keep the lip free from foreign matter.

➡Do not tap on the oil seal at an angle.

➡Make sure that the oil seal edge does not stick out of the timing chain case.

b. Apply MP grease to the oil seal lip.

26. Install timing chain or belt cover sub-assembly.
27. Install the right and left cylinder head cover sub-assemblies.
28. Install the No. 1 and 2 oil pipes.
29. Install the ignition coil assembly.
30. Install the intake air surge tank assembly.
31. Install the oil pan sub-assembly.
32. Install the oil strainer sub-assembly.
33. Install the No. 2 oil pan sub-assembly.
34. Install the crankshaft pulley.
35. Install the water inlet housing.
36. Install the water pump pulley.
37. Install the V-ribbed belt tensioner assembly.
38. Install the No. 2 idler pulley sub-assembly.
39. Install the drive plate and ring gear sub-assembly.
40. Install the right and left exhaust manifold sub-assemblies.
41. Install the oil level gauge guide sub-assembly.
42. Install the automatic transaxle assembly.
43. Install the starter assembly.
44. Install the front frame assembly.
45. Install the engine wire.
46. Install the engine assembly with transaxle.

PISTON AND RING

POSITIONING

See Figure 62.

REAR MAIN SEAL

REMOVAL & INSTALLATION

See Figures 63 and 64.

1. Before servicing the vehicle, refer to the Precautions Section.

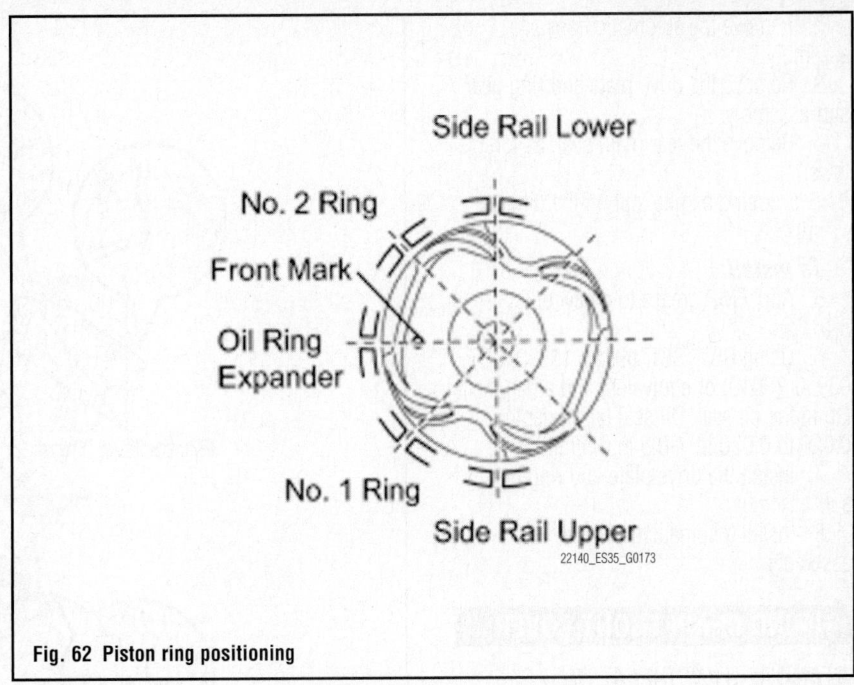

Fig. 62 Piston ring positioning

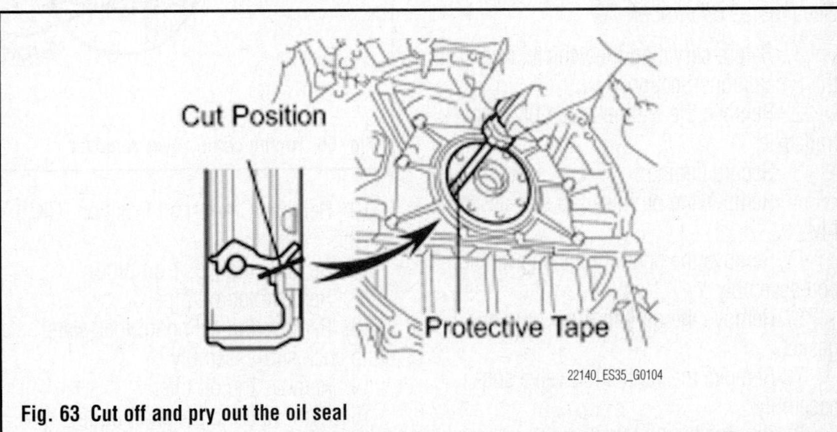

Fig. 63 Cut off and pry out the oil seal

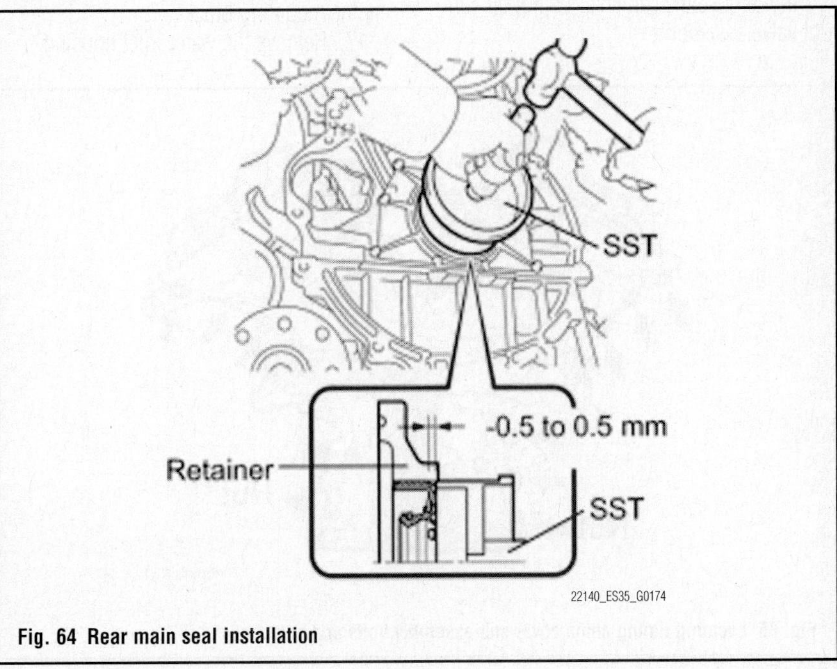

Fig. 64 Rear main seal installation

2. Remove the automatic transaxle assembly.

3. Remove the drive plate and ring gear sub-assembly.

4. Remove the rear main seal, as follows:

a. Using a knife, cut off the oil seal lip.

To install:

5. Apply MP grease to a new oil seal lip.

6. Using SST (SST: 09223-15030, SST: 09950-70010) or equivalent and a hammer, tap in the oil seal. Oil seal tap in depth: -0.020 to 0.020 in. (-0.5 to 0.5 mm)

7. Install the drive plate and ring gear sub-assembly.

8. Install automatic transaxle assembly.

TIMING CHAIN FRONT COVER

REMOVAL & INSTALLATION

See Figures 65 through 70.

1. Before servicing the vehicle, refer to the Precautions Section.

2. Remove the engine assembly with transaxle.

3. Secure engine.

4. Remove the oil filler cap sub-assembly.

5. Remove the spark plugs and ignition coil assembly.

6. Remove the oil pan drain plug and gasket.

7. Remove the ventilation valve sub-assembly.

8. Remove the camshaft position sensor.

9. Remove the camshaft timing oil control valve assembly.

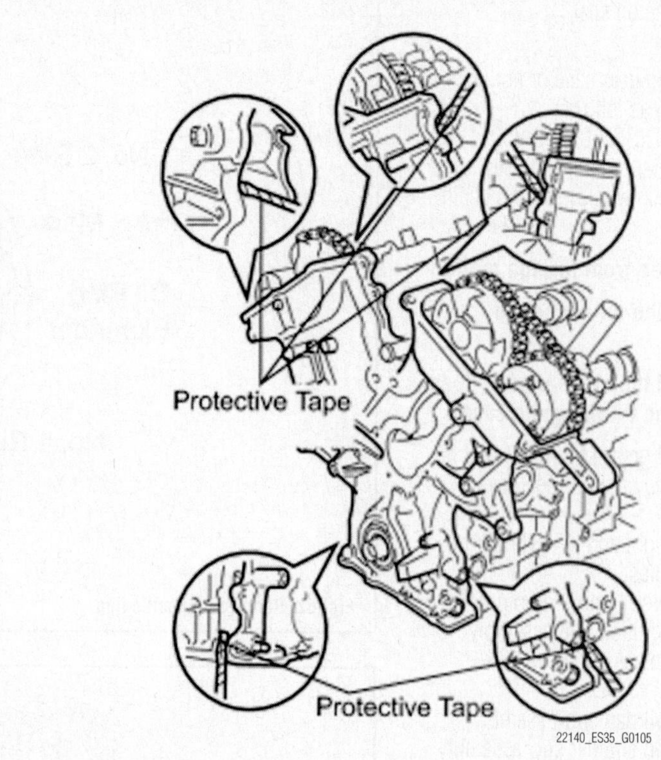

Fig. 66 Timing chain cover removal

10. Remove Crankshaft Position (CKP) sensor.

11. Remove the No. 1 oil pipe.

12. Remove the oil pipe.

13. Remove the cylinder block water drain cock sub-assembly.

14. Remove the oil filter.

15. Remove the crankshaft pulley.

16. Remove the left hand No. 1 front engine mounting bracket.

17. Remove the water inlet housing.

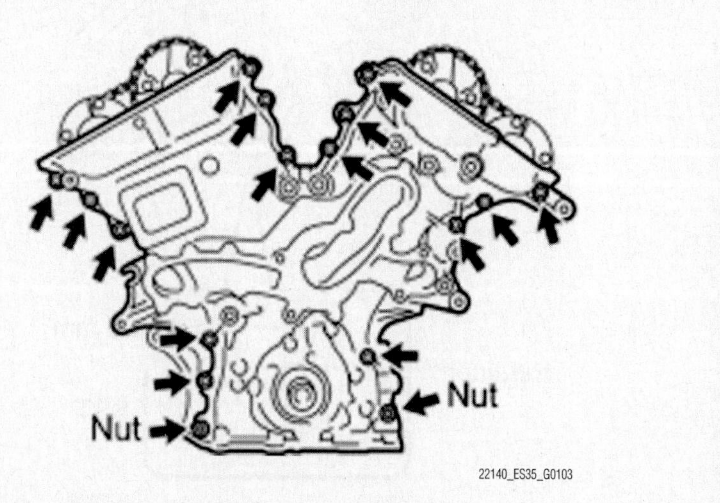

Fig. 65 Locating timing chain cover sub-assembly bolts and nuts

18. Remove the water outlet.

19. Remove the left-hand cylinder head cover sub-assembly and gasket.

20. Remove the cylinder head cover sub-assembly and gasket.

21. Remove the No. 2 oil pan sub-assembly.

22. Remove the oil strainer sub-assembly.

23. Remove the oil pan sub-assembly.

24. Remove the water pump assembly.

25. Remove the timing chain cover sub-assembly, as follows:

a. Remove the 15 bolts and 2 nuts as shown in the illustration.

➡Be careful not to damage the contact surfaces of the cylinder head, cylinder block and chain cover.

b. Remove the timing chain cover by prying between the timing chain cover and cylinder head or cylinder block with a screwdriver with the tip taped.

c. Remove the 4 bolts, chain cover plate and gasket.

d. Remove the gasket.

To install:

26. Install timing gear case or timing chain cover oil seal, as follows:

➡Keep the lip free from foreign matter.

A
A
A
A
C
C
C
C
C
: Seal Packing
3.0 mm or more
(0.118 in.)
22140_ES35_G0106

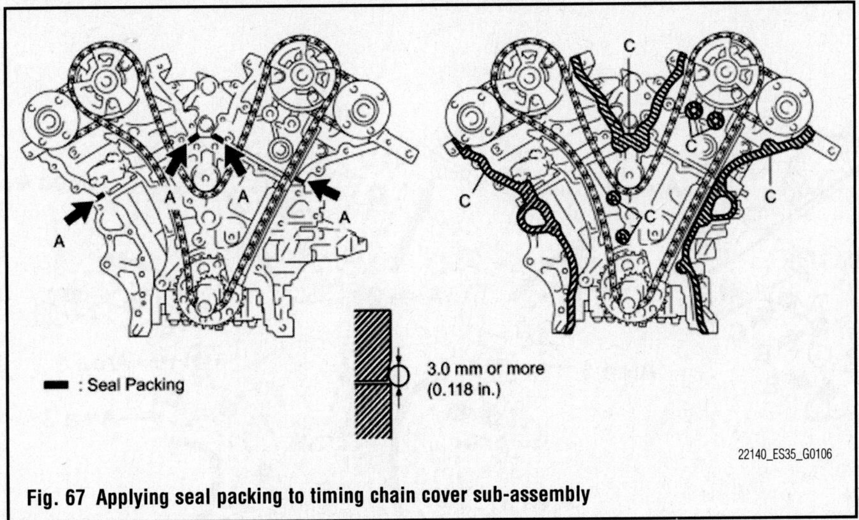

Fig. 67 Applying seal packing to timing chain cover sub-assembly

Drive Rotor
Spline
Crank-
shaft
22140_ES35_G0108

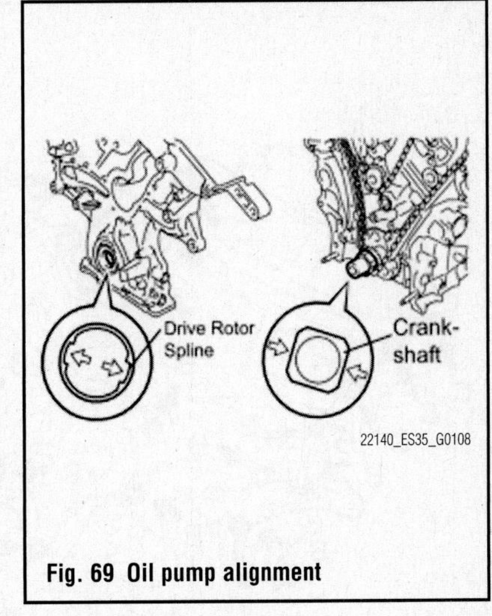

Fig. 69 Oil pump alignment

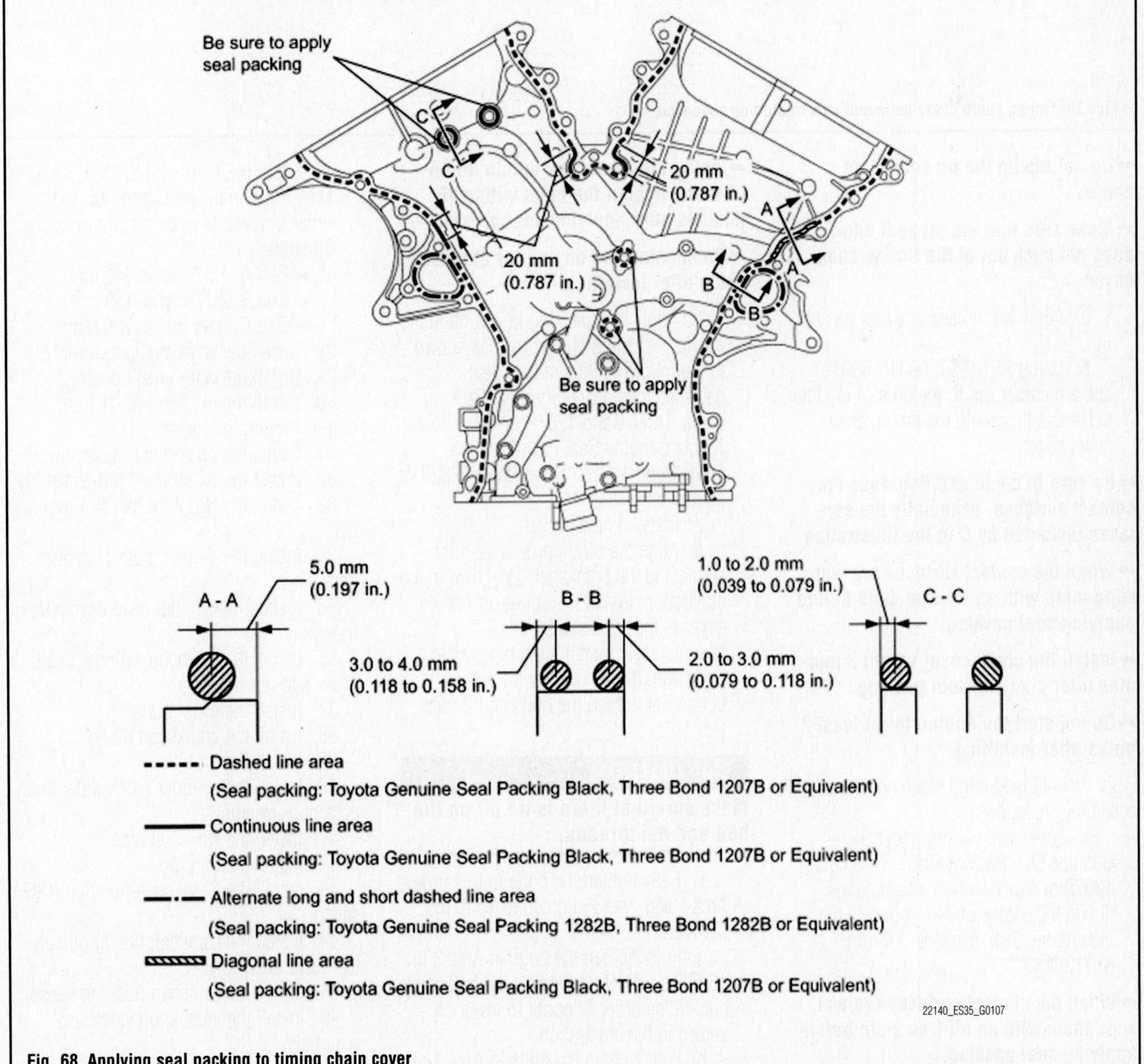

Be sure to apply
seal packing
C
C
20 mm
(0.787 in.)
20 mm
(0.787 in.)
A
A
B
B
Be sure to apply
seal packing
A - A
5.0 mm
(0.197 in.)
B - B
3.0 to 4.0 mm
(0.118 to 0.158 in.)
1.0 to 2.0 mm
(0.039 to 0.079 in.)
2.0 to 3.0 mm
(0.079 to 0.118 in.)
C - C
Dashed line area
(Seal packing: Toyota Genuine Seal Packing Black, Three Bond 1207B or Equivalent)
Continuous line area
(Seal packing: Toyota Genuine Seal Packing Black, Three Bond 1207B or Equivalent)
Alternate long and short dashed line area
(Seal packing: Toyota Genuine Seal Packing 1282B, Three Bond 1282B or Equivalent)
Diagonal line area
(Seal packing: Toyota Genuine Seal Packing Black, Three Bond 1207B or Equivalent)
22140_ES35_G0107

Fig. 68 Applying seal packing to timing chain cover

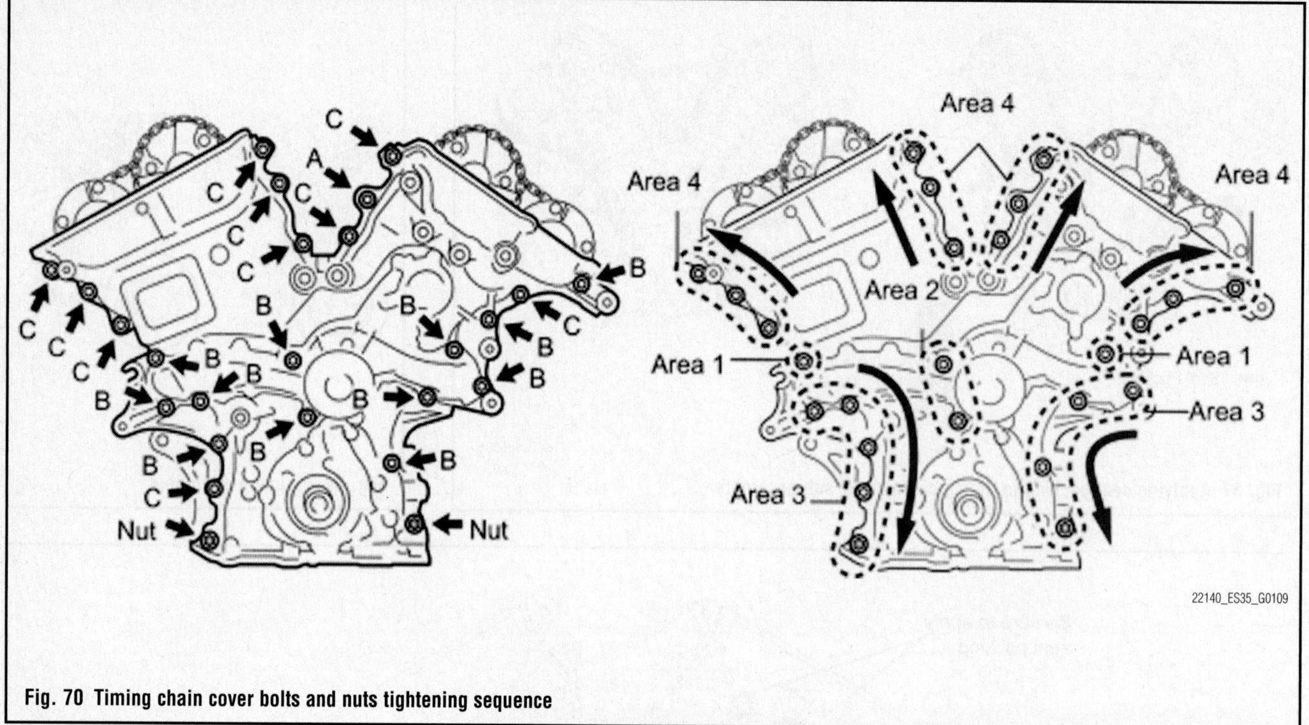

Fig. 70 Timing chain cover bolts and nuts tightening sequence

➡Do not tap on the oil seal at an angle.

➡Make sure that the oil seal edge does not stick out of the timing chain cover.

a. Apply MP grease to a new oil seal lip.

b. Using SST (SST: 09316-60011) and a hammer, tap in the oil seal until its surface is flush with the timing chain cover edge.

➡Be sure to clean and degrease the contact surfaces, especially the surfaces indicated by C in the illustration.

➡When the contact surfaces are wet, wipe them with an oil-free cloth before applying seal packing.

➡Install the chain cover within 3 minutes after applying seal packing.

➡Do not start the engine for at least 2 hours after installing.

27. Install the timing chain cover sub-assembly, as follows:

a. Apply seal packing (Toyota Genuine Seal Packing Black, Three Bond 1207B or equivalent) in a continuous line to the engine unit as shown in the illustration. Seal diameter: 3.0 mm (0.118 in.).

➡When the contact surfaces are wet, wipe them with an oil-free cloth before applying seal packing.

➡Install the crankcase within 3 minutes and tighten the bolts within 15 minutes after applying seal packing.

➡Do not start the engine for at least 2 hours after installing.

b. Apply seal packing in a continuous line to the timing chain cover as shown in the following illustration. Seal packing: Toyota Genuine Seal Packing Black, Three Bond 1207B or equivalent, Toyota Genuine Seal Packing Black, Three Bond 1282B, Three Bond 1282B or equivalent.

c. Install a new gasket.

d. Align the oil pump's drive rotor spline and the crankshaft as shown in the illustration. Install the spline and chain cover to the crankshaft.

e. Loosely install the timing chain cover with the 23 bolts and 2 nuts, but do not tighten the bolts and 2 nuts yet.

⁂ CAUTION

Make sure that there is no oil on the bolt and nut threads.

f. Fully tighten the bolts in this order: Area 1 and Area 2, tighten to 15 ft. lbs. (21 Nm).

g. Fully tighten the bolts in Area 3 to 15 ft. lbs. (21 Nm). Tighten the bolts and nuts in the order of upper to lower as shown in the illustration.

h. Fully tighten the bolts in Area 4 to 32 ft. lbs. (43 Nm), and to 15 ft. lbs. (21 Nm). Tighten the bolts and nuts in the order of lower to upper as shown in the illustration.

- Bolt A: 1.57 inches (40 mm)
- Bolt B: 2.17 inches (55 mm)
- Bolt C: 0.98 inches (25 mm)

28. Install the water pump assembly.
29. Install the water inlet housing.
30. Install the left hand No. 1 front engine mounting bracket.
31. Install the oil pan sub-assembly.
32. Install the oil strainer sub-assembly.
33. Install the No. 2 oil pan sub-assembly.
34. Install the oil pan drain plug and gasket.
35. Install the cylinder head cover sub-assembly.
36. Install the left-hand cylinder head cover sub-assembly.
37. Install the water outlet.
38. Install the crankshaft pulley.
39. Install the oil filter.
40. Install the cylinder block water drain cock sub-assembly.
41. Install the No. 1 oil pipe.
42. Install the oil pipe.
43. Install the Crankshaft Position (CKP) sensor.
44. Install the camshaft timing oil control valve assembly.
45. Install the camshaft position sensor.
46. Install the ventilation valve sub-assembly.

47. Install the spark plugs and ignition coil assembly.
48. Install the oil filler cap sub-assembly.
49. Install the water pump assembly.
50. Install the engine assembly with transaxle.

TIMING CHAIN & SPROCKETS

REMOVAL & INSTALLATION

See Figures 71 through 82.

1. Before servicing the vehicle, refer to the Precautions Section.
2. Remove the engine assembly with transaxle.
3. Secure engine.
4. Remove the oil filler cap sub-assembly.
5. Remove the spark plugs and ignition coil assembly.
6. Remove the oil pan drain plug and gasket.
7. Remove the ventilation valve sub-assembly.
8. Remove the camshaft position sensor.
9. Remove the camshaft timing oil control valve assembly.
10. Remove Crankshaft Position (CKP) sensor.
11. Remove the No. 1 oil pipe.
12. Remove the oil pipe.
13. Remove the cylinder block water drain cock sub-assembly.
14. Remove the oil filter.
15. Remove the crankshaft pulley.
16. Remove the left hand No. 1 front engine mounting bracket.
17. Remove the water inlet housing.
18. Remove the water outlet.
19. Remove the left-hand cylinder head cover sub-assembly and gasket.
20. Remove the cylinder head cover sub-assembly and gasket.

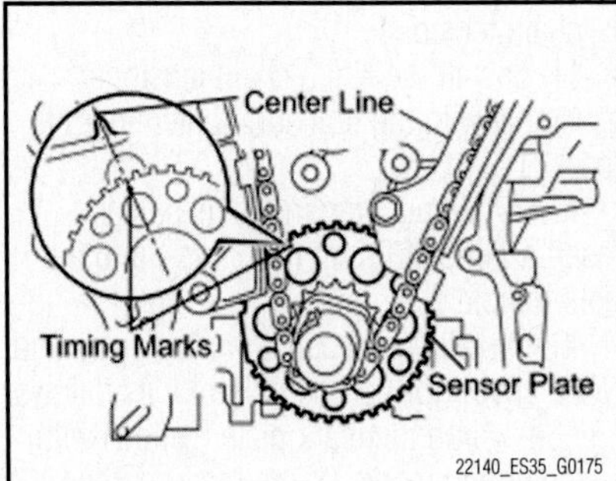

Fig. 71 Set the timing mark on the crank angle sensor plate

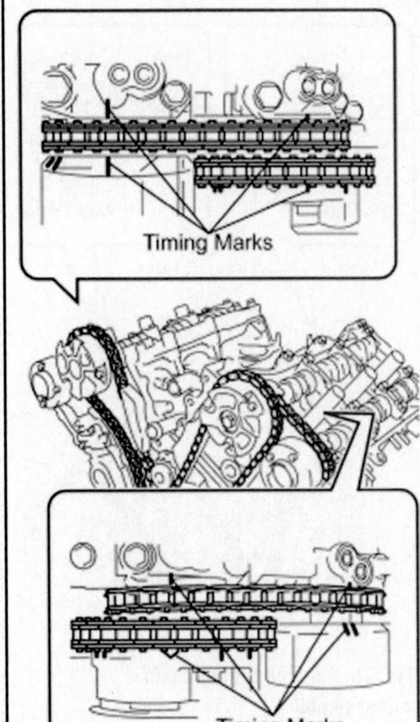

Fig. 72 Check that the timing marks of the camshaft timing gears

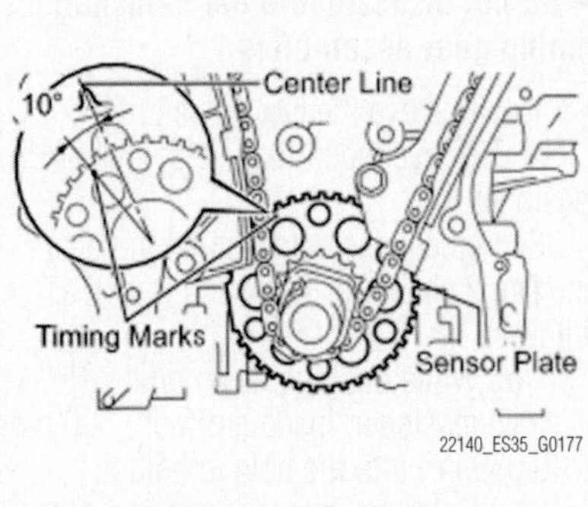

Fig. 73 Turning the crankshaft counter-clockwise 10°

21. Remove the No. 2 oil pan sub-assembly.
22. Remove the oil strainer sub-assembly.
23. Remove the oil pan sub-assembly.
24. Remove the water pump assembly.
25. Remove the timing chain cover and seal.
26. Set no. 1 cylinder to TDC/compression, as follows:
    a. Temporarily tighten the pulley set bolt. Set the timing mark on the crank angle sensor plate to the RH block bore center line (TDC/compression).

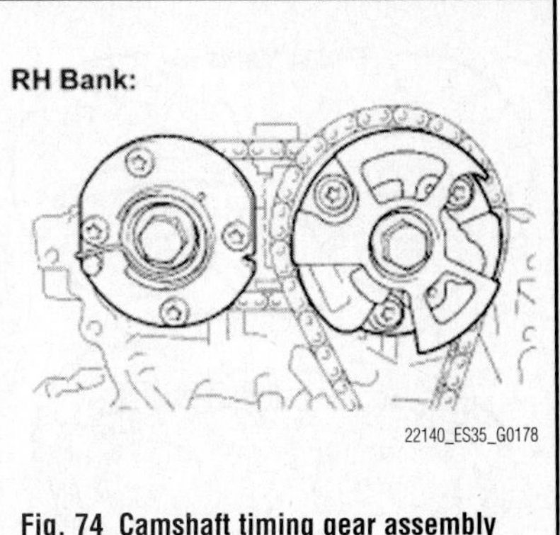

Fig. 74 Camshaft timing gear assembly positioning

    b. Check that the timing marks of the camshaft timing gears are aligned with the timing marks of the bearing cap as shown in the illustration. If not, turn the crankshaft 1 revolution (360°) and align the timing marks as above.
27. Remove the No. 1 chain tensioner assembly, as follows:
    a. Move the stopper plate upward to release the lock, and push the plunger deep into the tensioner.
    b. Move the stopper plate downward to set the lock, and insert a pin of _1.27 mm (0.05 in.) into the stopper plate's hole.
    c. Remove the 2 bolts and chain tensioner.
28. Remove the chain tensioner slipper.
29. Remove the chain sub-assembly, as follows:
    a. Turn the crankshaft counterclockwise 10° to loosen the chain of the crankshaft timing sprocket.
    b. Remove the pulley set bolt.
    c. Remove the chain from the crankshaft timing sprocket and place it on the crankshaft.

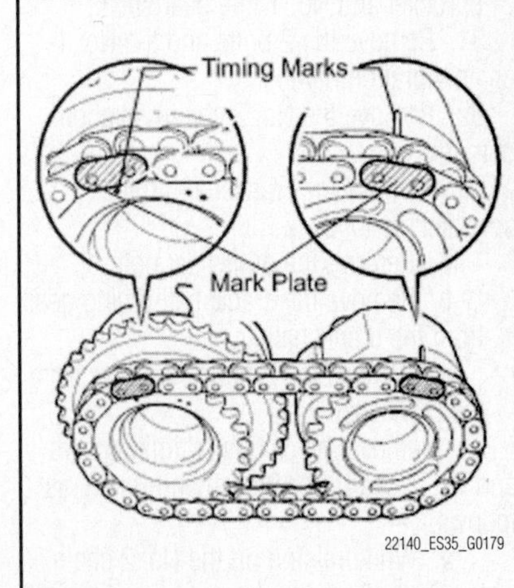

Fig. 75 Aligning No. 2 timing chain

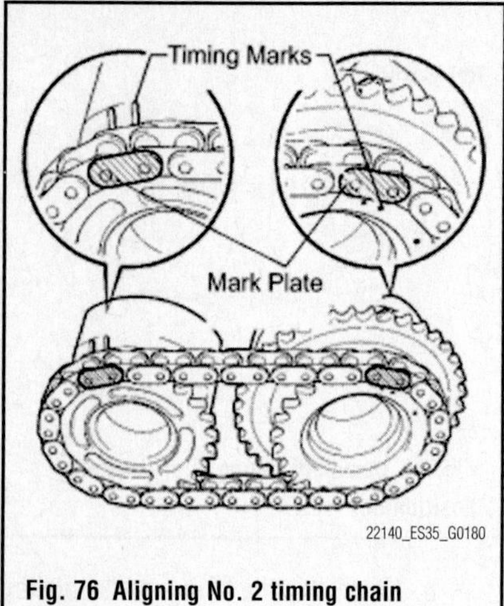

Fig. 76 Aligning No. 2 timing chain (bank 2)

Fig. 77 Crankshaft timing sprocket

d. Turn the camshaft timing gear assembly on the right hand bank clockwise (approximately 60°) and set it as shown in the illustration. Be sure to loosen the chain between the banks.

e. Remove the chain.

30. Remove the idle sprocket assembly, as follows:

a. Using a 10 mm hexagon wrench, remove the No. 2 idle gear shaft, sprocket and No. 1 idle gear shaft.

31. Remove the 2 bolts and the No. 1 chain vibration damper.

32. Remove the No. 2 chain vibration damper.

33. Remove the crankshaft timing sprocket, as follows:

a. Remove the pulley set bolt.

b. Remove the crankshaft timing gear from the crankshaft.

c. Remove the 2 pulley set keys from the crankshaft.

34. Remove the camshaft timing gears and No. 2 chain (for Right-hand Bank), as follows:

a. While raising up the No. 2 chain tensioner, insert a pin of _1.0 mm (0.039 in.) into the hole to hold it.

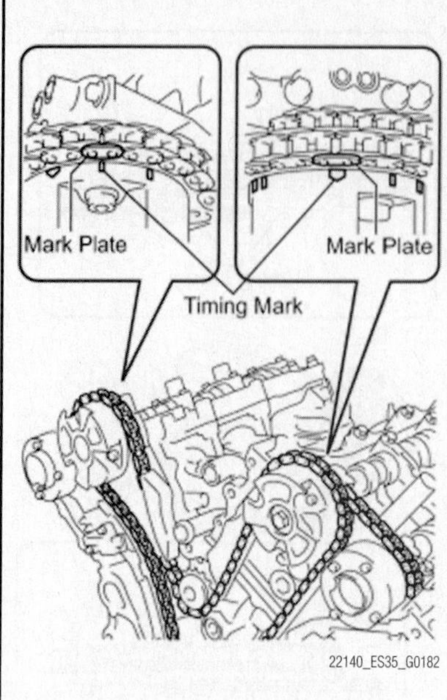

Fig. 78 Aligning timing chain sub-assembly

b. Hold the hexagonal portion of the camshaft with a wrench, and remove the 2 bolts and 2 camshaft timing gears.

➡Be careful not to damage the cylinder head with the wrench.

➡Do not disassemble the camshaft timing gear assemblies.

c. Remove the No. 2 chain.

35. Remove the bolt and No. 2 chain tensioner.

36. Remove the camshaft timing gears and No. 2 chain (for Left-hand Bank), as follows:

a. While pushing down on the No. 3 chain tensioner, insert a pin of _ 1.0 mm (0.039 in.) into the hole to hold it.

b. Hold the hexagonal portion of the camshaft with a wrench, and remove the 2 bolts and 2 camshaft timing gears.

➡Be careful not to damage the cylinder head with the wrench.

➡Do not disassemble the camshaft timing gear assemblies.

c. Remove the No. 2 chain.

37. Remove the bolt and the No. 3 chain tensioner.

To install:

38. Install the No. 2 chain tensioner assembly with the bolt and tighten to 15 ft. lbs. (21 Nm).

39. While pushing in the tensioner, insert a pin of 1.0 mm (0.039 in.) into the hole to hold it.

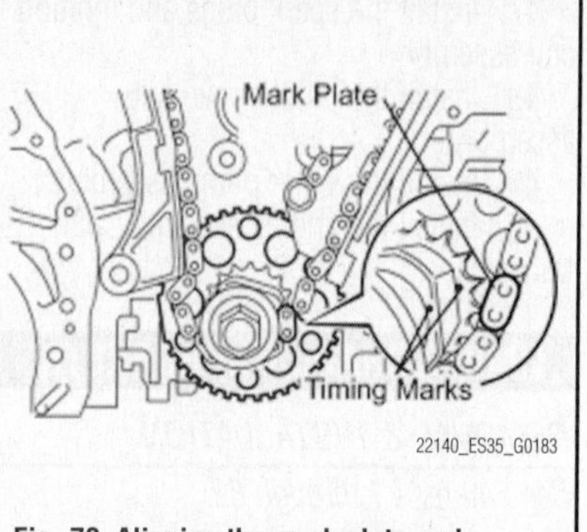

Fig. 79 Aligning the mark plate and timing mark

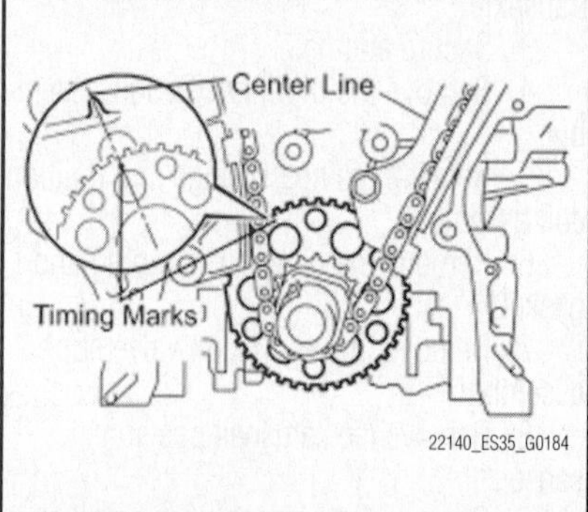

Fig. 80 Turning the crankshaft clockwise (TDC/compression)

40. Install the camshaft timing gears and No. 2 chain (for Right-hand Bank), as follows:

a. Align the mark plate with the timing marks (1-dot mark) of the camshaft timing gears as shown.

b. Apply a light coat of engine oil to the bolt threads and bolt-seating surface.

c. Align the knock pin of the camshaft with pin hole of the camshaft timing gear. Install the camshaft timing gear and the right camshaft timing exhaust gear with the No. 2 chain installed.

d. Hold the hexagonal portion of the camshaft with the wrench and tighten the two bolts to 74 ft. lbs. (100 Nm).

e. Remove the pin from the No. 2 chain tensioner.

41. Install the No. 3 chain tensioner assembly with the bolt and tighten to 15 ft. lbs. (21 Nm).

42. While pushing in the tensioner, insert a pin of 1.0 mm (0.039 in.) into the hole to hold it.

43. Install the camshaft timing gears and No. 2 chain (for Left-hand Bank), as follows:

a. Align the mark plate (yellow) with the timing marks (2-dot mark) of the camshaft timing gears as shown.

b. Apply a light coat of engine oil to the bolt threads and bolts seating surface.

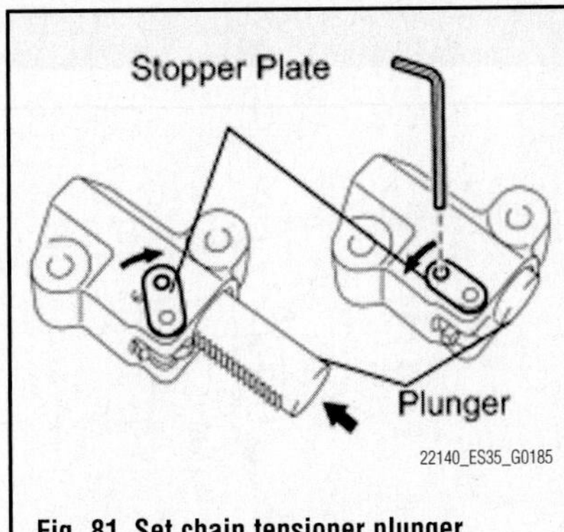

Fig. 81 Set chain tensioner plunger position

c. Align the knock pin of the camshaft with pin hole of the camshaft timing gear. Install the camshaft timing gear and the left camshaft timing exhaust gear with the No. 2 chain installed.

d. Hold the hexagonal portion of the camshaft with the wrench and tighten the two bolts to 74 ft. lbs. (100 Nm).

e. Remove the pin from the No 2 chain tensioner.

44. Install the No. 1 and 2 chain vibration dampers.

45. Install the timing gear set keys and crankshaft timing sprocket as shown in the illustration.

46. Install the idle sprocket assembly, as follows:

a. Apply a light coat of engine oil to the rotating surface of the No. 1 idle gear shaft.

b. Temporarily install the No. 1 idle gear shaft and idle sprocket with the No. 2 idle gear shaft while aligning the knock pin of the No. 1 idle gear with the knock pin groove of the cylinder block. Be careful of the idle gear direction.

c. Using a 10 mm hexagon wrench, tighten the No. 2 idle gear shaft to 44 ft. lbs. (60 Nm).

d. After installing the idle sprocket assembly, check that the idle sprocket turns smoothly.

47. Install the chain sub-assembly, as follows:

a. Align the mark plate and timing marks as shown in the illustration and install the chain. The camshaft mark plate is orange.

b. Do not pass the chain over the crankshaft, just put it on.

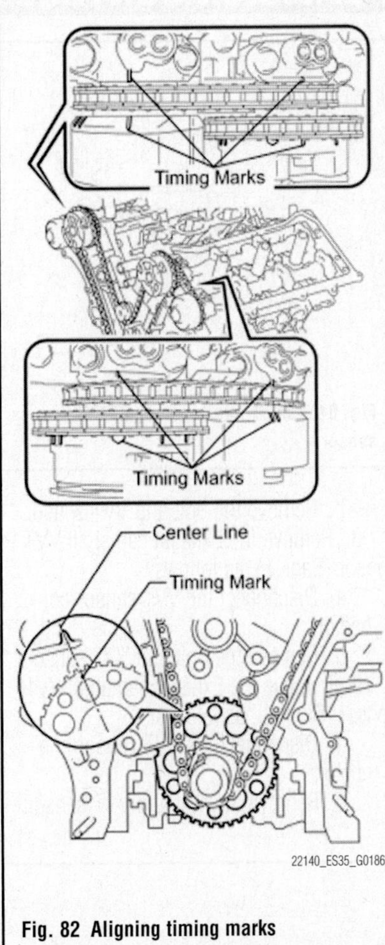

Fig. 82 Aligning timing marks

c. Turn the camshaft timing gear assembly on the right bank counterclockwise to tighten the chain between the banks.

➡When the idle sprocket assembly is reused, align the timing chain plate with the mark on the sprocket in order to tighten the chain between the banks.

d. Align the mark plate and timing marks as shown in the illustration and install the chain onto the crankshaft timing sprocket. The crankshaft to mark plate is yellow.

e. Temporarily tighten the pulley set bolt.

f. Turn the crankshaft clockwise to set it to the right-hand block bore more centerline. (TDC/compression).

48. Install the chain tensioner slipper.

49. Install the No. 1 chain tensioner assembly, as follows:

a. Move the stopper plate upward to release the lock, and push the plunger deep into the tensioner.

b. Move the stopper plate downward to set the lock, and insert a hexagon wrench into the hole of the stopper plate.

c. Install the No. 1 chain tensioner with the 2 bolts and tighten to 7 ft. lbs. (10 Nm).

d. Remove the lock pin of the No. 1 chain tensioner. Check that each timing mark is aligned with the crankshaft at TDC/compression.

e. Remove the pulley set bolt.

50. Install timing chain cover and seal.

51. Install the water pump assembly.

52. Install the water inlet housing.

53. Install the left hand No. 1 front engine mounting bracket.

54. Install the oil pan sub-assembly.

55. Install the oil strainer sub-assembly.

56. Install the No. 2 oil pan sub-assembly.

57. Install the oil pan drain plug and gasket.

58. Install the cylinder head cover sub-assembly.

59. Install the left-hand cylinder head cover sub-assembly.

60. Install the water outlet.

61. Install the crankshaft pulley.

62. Install the oil filter.

63. Install the cylinder block water drain cock sub-assembly.

64. Install the No. 1 oil pipe.

65. Install the oil pipe.

66. Install the Crankshaft Position (CKP) sensor.

67. Install the camshaft timing oil control valve assembly.

68. Install the camshaft position sensor.

69. Install the ventilation valve sub-assembly.

70. Install the spark plugs and ignition coil assembly.

71. Install the oil filler cap sub-assembly.

72. Install the engine assembly with transaxle.

VALVE LASH

ADJUSTMENT

The 3.5L (2GR-FE) engine is equipped with hydraulic valves which are not adjustable.

ENGINE PERFORMANCE & EMISSION CONTROLS

CAMSHAFT POSITION (CMP) SENSOR

LOCATION

See Figure 83.

Refer to the accompanying illustration.

REMOVAL & INSTALLATION

See Figures 84 through 87.

1. Before servicing the vehicle, refer to the Precautions Section.
2. Drain and recycle the engine coolant.
3. Disconnect the negative battery cable.
4. Remove the V-bank cover sub-assembly.
5. Remove the windshield wiper link assembly.
6. Remove the front cowl top outside panel.
7. Remove the Intake camshaft VVT sensor (Bank 1), as follows:
   a. Disconnect the VVT sensor connector.
   b. Remove the bolt and VVT sensor.
8. Remove the Exhaust camshaft VVT sensor (Bank 1), as follows:
   a. Disconnect the VVT sensor connector.
   b. Remove the bolt and VVT sensor.
9. Remove the Exhaust camshaft VVT sensor (Bank 2), as follows:
   a. Disconnect the VVT sensor connector.
   b. Remove the bolt and VVT sensor.

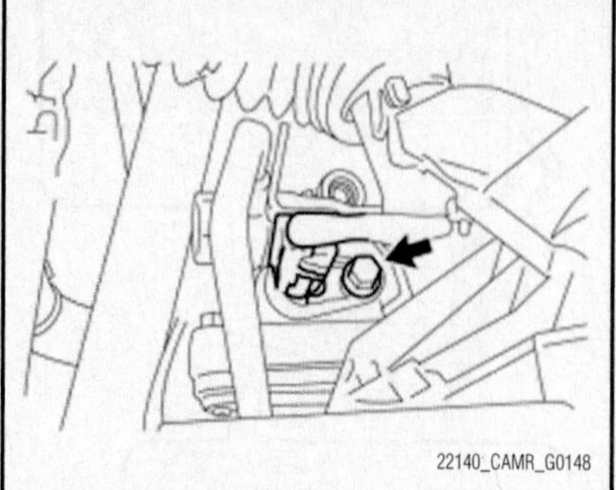

Fig. 84 Bank 1 intake camshaft VVT sensor

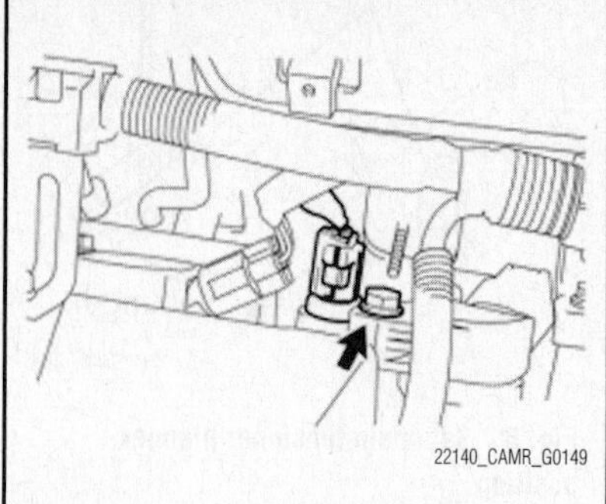

Fig. 85 Bank 1 exhaust camshaft VVT sensor

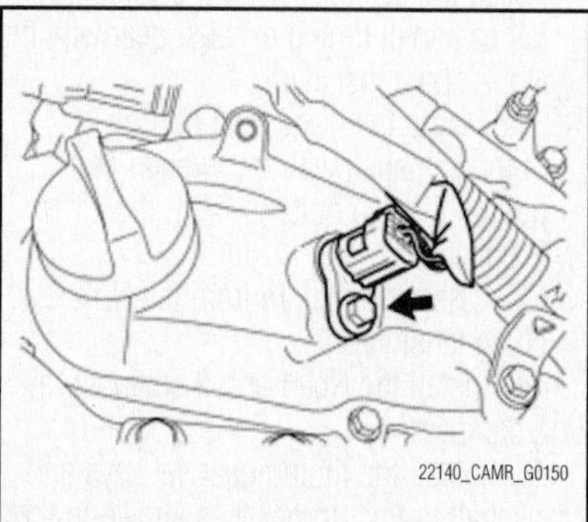

Fig. 86 Bank 2 intake camshaft VVT sensor

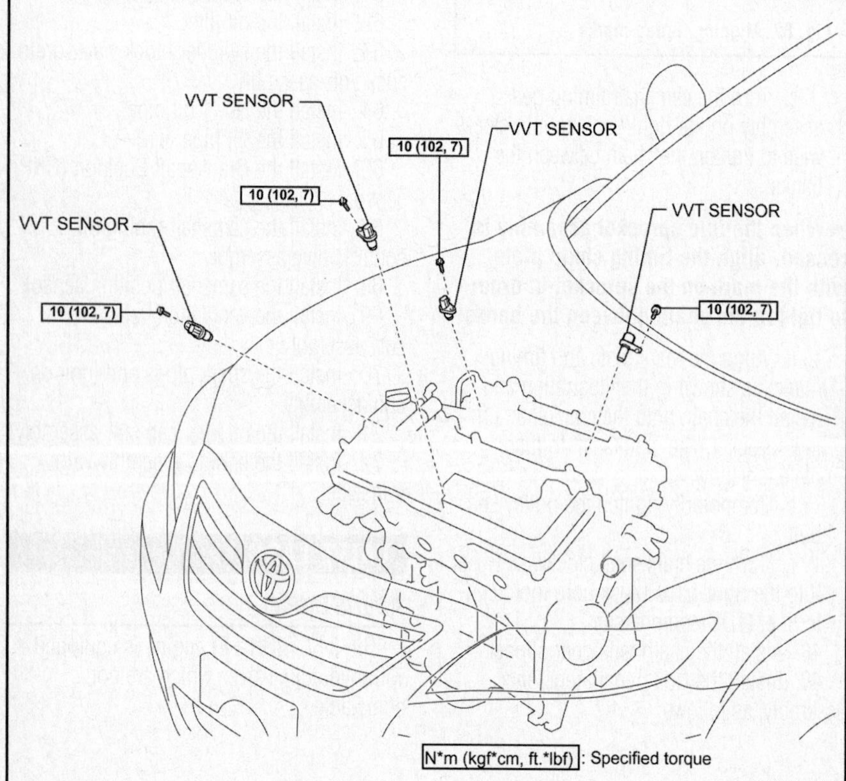

Fig. 83 VVT sensor location

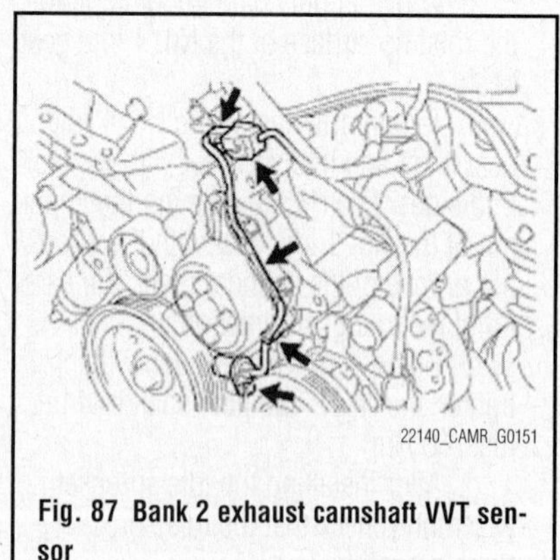

Fig. 87 Bank 2 exhaust camshaft VVT sensor

10. Remove the Intake camshaft VVT sensor (Bank 2), as follows:
    a. Disconnect the VVT sensor connector.
    b. Remove the bolt and VVT sensor.

CRANKSHAFT POSITION (CKP) SENSOR

LOCATION

See Figure 88.

Refer to the accompanying illustration.

Fig. 88 Crankshaft Position (CKP) sensor location

REMOVAL & INSTALLATION

See Figure 89.

1. Before servicing the vehicle, refer to the Precautions Section.
2. Disconnect the negative battery cable.
3. Remove alternator assembly.
4. Disconnect the cooler compressor assembly.
5. Remove the Crankshaft Position (CKP) sensor connector.
6. Remove the bolt, and then remove the Crankshaft Position (CKP) sensor.

To install:

7. Apply a light coat of engine oil to the O-ring on the Crankshaft Position (CKP) sensor.
8. Install the Crankshaft Position (CKP) sensor with the bolt and tighten to 7ft. lbs. (10 Nm).
9. Connect the Crankshaft Position (CKP) sensor connector.
10. The remainder of installation is the reverse of the removal procedure.

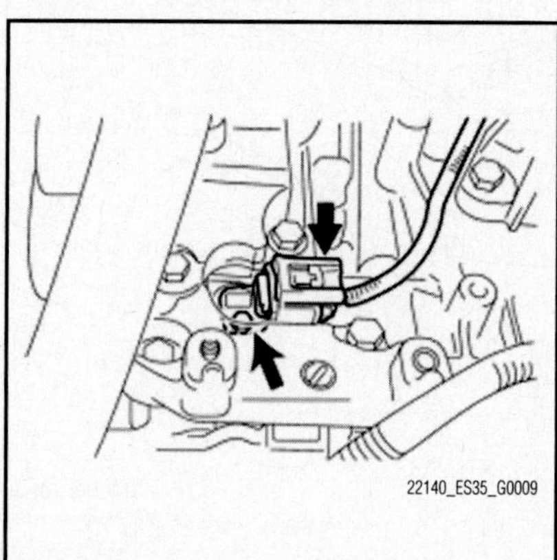

Fig. 89 Installing Crankshaft Position (CKP) sensor

ELECTRONIC CONTROL MODULE (ECM)

LOCATION

See Figure 90.

Refer to the accompanying illustration for ECM location.

REMOVAL & INSTALLATION

See Figure 91.

1. Before servicing the vehicle, refer to the Precautions Section.
2. Disconnect the negative battery cable.
3. Remove both windshield wiper arm and blade assemblies.
4. Remove the cowl top ventilator louver sub-assembly.
5. Remove the windshield wiper motor and link assembly.
6. Remove the outer cowl top panel.
7. Remove the ECM, as follows:
   a. Remove the 3 nuts.
   b. Separate the ECM from the body. When separating the ECM, do not apply excessive force to the wire harness.
   c. Raise the 2 levers while pushing the locks on the 2 levers, and disconnect the 2 ECM connectors.

➡After disconnecting the connectors, make sure that dirt, water or other foreign matter does not contact the connections of the connectors.

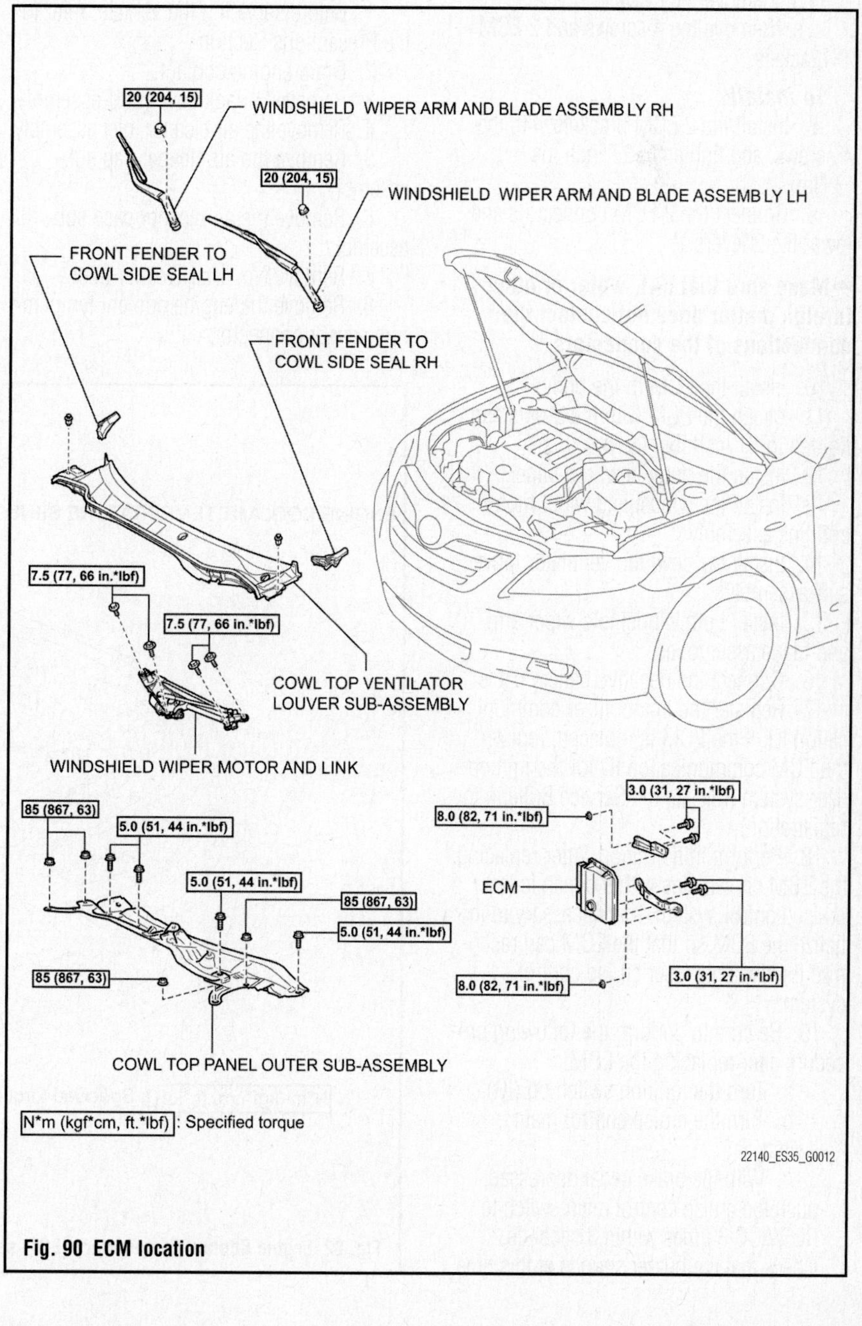

Fig. 90 ECM location

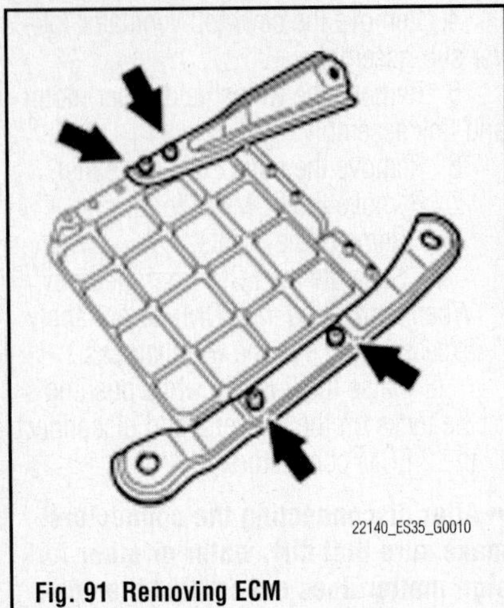

Fig. 91 Removing ECM

d. Remove the ECM.

e. Remove the 4 screws and 2 ECM brackets.

To install:

8. Install the 2 ECM brackets with the 4 screws, and tighten to 27 inch lbs. (3 Nm).

9. Connect the 2 ECM connectors and lower the 2 levers.

➡Make sure that dirt, water or other foreign matter does not contact the connections of the connectors.

10. Install the ECM to the body.
11. Attach the ECM with the 3 nuts and tighten to 71 inch lbs. (8 Nm).
12. Install the outer cowl top panel.
13. Install the windshield wiper motor and link assembly.
14. Install the cowl top ventilator louver sub-assembly.
15. Install both windshield wiper arm and blade assemblies.
16. Connect the negative battery cable.
17. Register the immobilizer communication ID. If the ECM is replaced, register the ECM communication ID for the immobilizer system (refer to the Service Bulletin for registration).
18. Perform initialization. After replacing the ECM on vehicles with a dynamic laser cruise control system, it is necessary to initialize the ECM so that the ECM can recognize the dynamic laser cruise control system.
19. Be sure to perform the following procedure after replacing the ECM:

a. Turn the ignition switch on (IG).

b. Turn the cruise control main switch on.

c. With the brake pedal depressed, push the cruise control main switch to RES/ACC 3 times within 3 seconds. Check that the buzzer sounds at this time.

➡Do not turn the headlight dimmer switch on at this time because the optical axis automatic adjustment mode has already started, which may lead to an incorrect optical axis setting. If the headlight dimmer switch is turned on by mistake, readjust the optical axis.

ENGINE COOLANT TEMPERATURE (ECT) SENSOR

LOCATION

See Figure 92.

Refer to the accompanying illustration for ECT sensor location.

REMOVAL & INSTALLATION

1. Before servicing the vehicle, refer to the Precautions Section.
2. Drain engine coolant.
3. Remove V-bank cover sub-assembly.
4. Remove the air cleaner inlet assembly.
5. Remove the air cleaner cap sub-assembly.
6. Remove the air cleaner case sub-assembly.
7. Remove No. 1 air cleaner inlet.
8. Remove the engine coolant temperature sensor connector.
9. Using a 19 mm deep socket wrench, remove the engine coolant temperature sensor and gasket.

To install:

10. Installation is the reverse of the removal procedure. Torque the engine coolant temperature sensor to 15 ft. lbs. (20 Nm).

HEATED OXYGEN SENSOR (HO2S)

LOCATION

See Figure 93.

Refer to the accompanying illustration for the location of the Heated Oxygen Sensor (HO2S).

REMOVAL & INSTALLATION

See Figures 94 and 95.

1. Before servicing the vehicle, refer to the Precautions Section.
2. Remove the front exhaust pipe assembly.
3. Disconnect the 2 oxygen sensor connectors.
4. Using Special Tool: 09224-00010 or equivalent, remove the 2 oxygen sensors from the front pipe assembly.

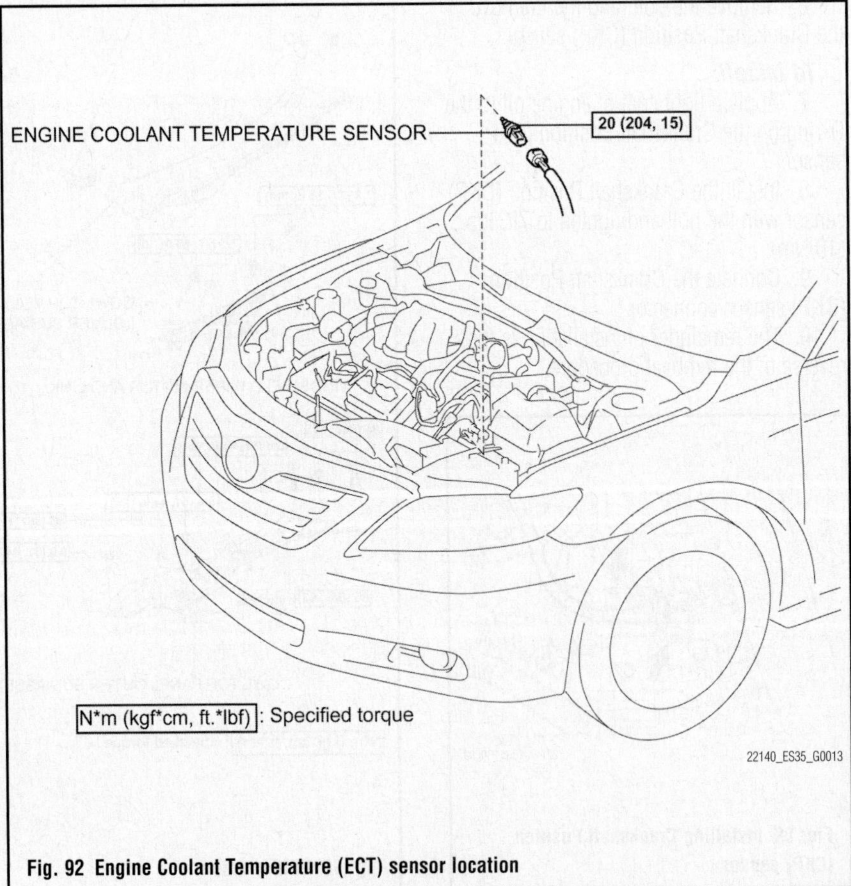

Fig. 92 Engine Coolant Temperature (ECT) sensor location

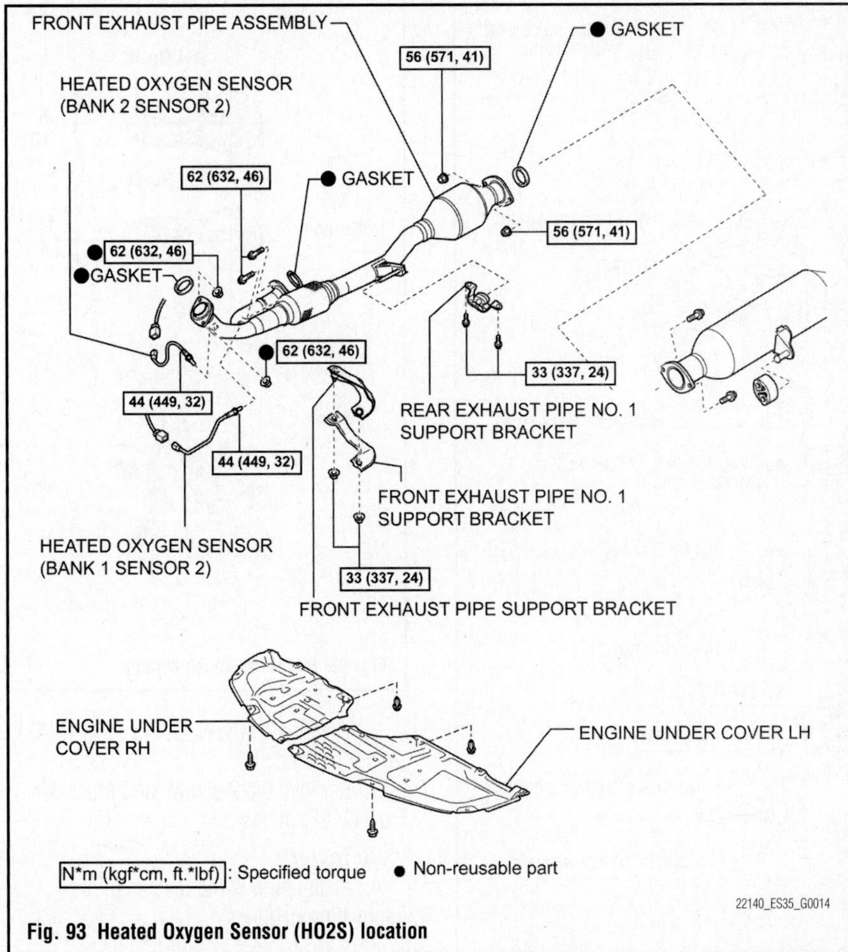

Fig. 93 Heated Oxygen Sensor (HO2S) location

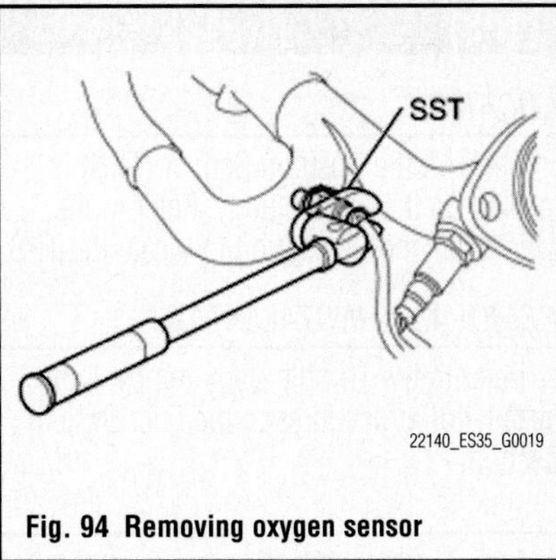

Fig. 94 Removing oxygen sensor

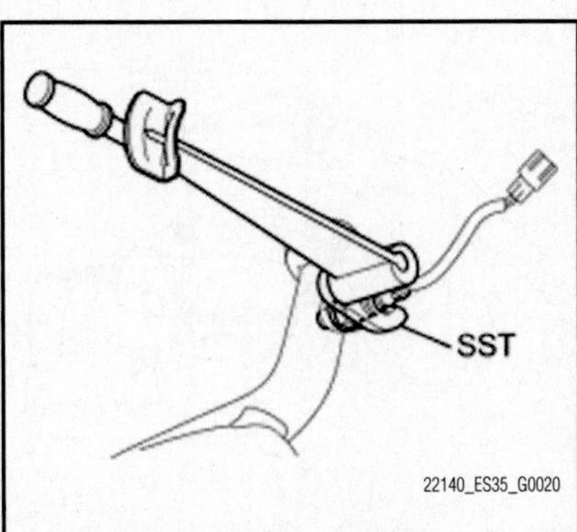

Fig. 95 Installing oxygen sensor

To install:

5. Install the 2 oxygen sensors to the front pipe assembly. Tighten to 32 ft. lbs. (44 Nm) and 30 ft. lbs. (40 Nm). Use a torque wrench with a fulcrum length of 300 mm (11.81 in.).
6. Connect the 2 oxygen sensor connectors.
7. Remove the front exhaust pipe assembly.

INTAKE AIR TEMPERATURE (IAT) SENSOR

LOCATION

The Intake Air Temperature (IAT) sensor is mounted on the Mass Air Flow (MAF) meter.

REMOVAL & INSTALLATION

See Mass Air Flow (MAF) meter.

KNOCK SENSOR (KS)

LOCATION

See Figure 96.

Refer to the accompanying illustration to locate the knock sensor.

REMOVAL & INSTALLATION

See Figures 97 and 98.

**** CAUTION**

Observe all applicable safety precautions when working around fuel. Whenever servicing the fuel system, always work in a well ventilated area. Do not allow fuel spray or vapors to come in contact with a spark or open flame. Keep a dry chemical fire extinguisher near the work area. Always keep fuel in a container specifically designed for fuel storage; also, always properly seal fuel containers to avoid the possibility of fire or explosion.

1. Before servicing the vehicle, refer to the Precautions Section.
2. Properly discharge the fuel system pressure.
3. Disconnect battery negative cable.
4. Drain and recycle the engine coolant.
5. Remove both plastic engine under covers.
6. Remove windshield wiper arms and blade assemblies.
7. Remove both of the front fender to cowl side seals.
8. Remove cowl top ventilator louver sub-assembly.
9. Remove the windshield wiper motor and link assembly.
10. Remove the outer cowl top panel.
11. Remove the cool air intake duct seal.
12. Remove the V-bank cover sub-assembly.
13. Remove air cleaner inlet assembly.
14. Remove the air cleaner cap sub-assembly.
15. Remove air cleaner case sub-assembly.
16. Remove the intake air surge tank.
17. Remove no. 1 air cleaner inlet.
18. Separate fuel tube sub-assembly.
19. Remove the intake manifold.
20. Disconnect the 2 knock control sensor connectors.
21. Remove the 2 bolts and 2 knock control sensors.

To install:

22. Install the 2 knock control sensors with the 2 bolts as shown in the illustration and tighten to 15 ft. lbs. (20 Nm).
23. Connect the 2 knock control sensor connectors.

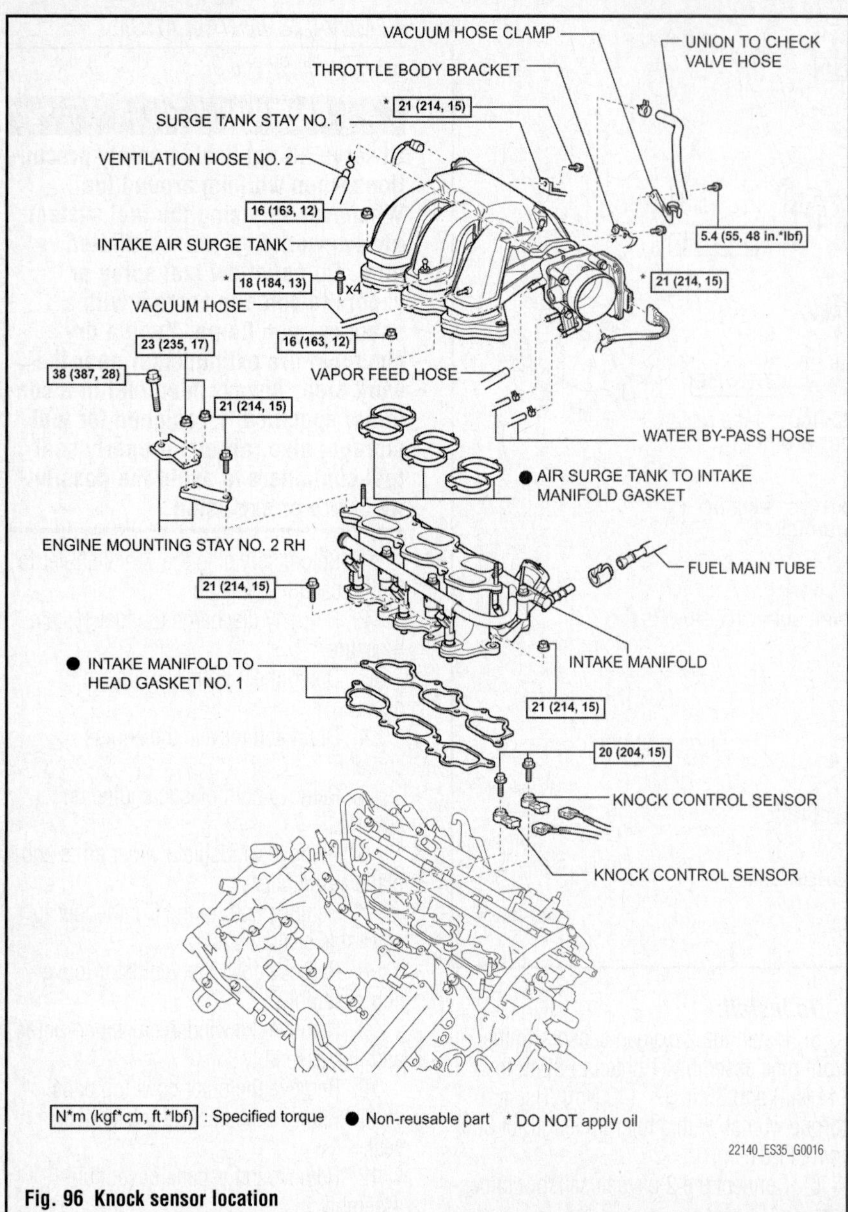

Fig. 96 Knock sensor location

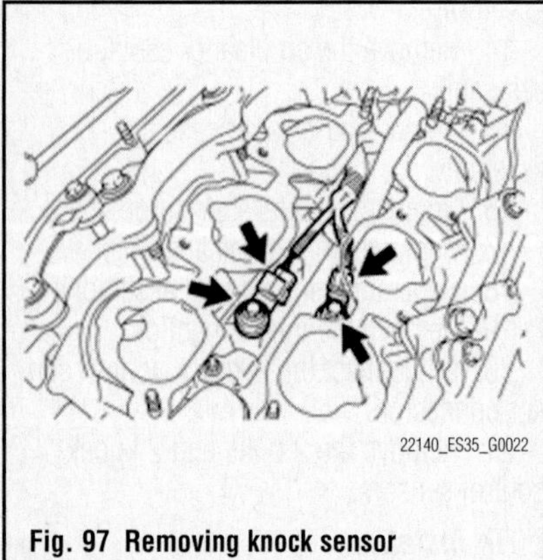

Fig. 97 Removing knock sensor

24. The remainder of installation is the reverse of the removal procedure.

25. Inspect for fuel leak and check the function of throttle body.

MALFUNCTION INDICATOR LIGHT (MIL)

RESET PROCEDURE

Clear the DTC codes.

MASS AIR FLOW (MAF) SENSOR

LOCATION

The Mass Air Flow (MAF) sensor is between the throttle body and air cleaner housing.

REMOVAL & INSTALLATION

See Figure 99.

1. Before servicing the vehicle, refer to the Precautions Section.

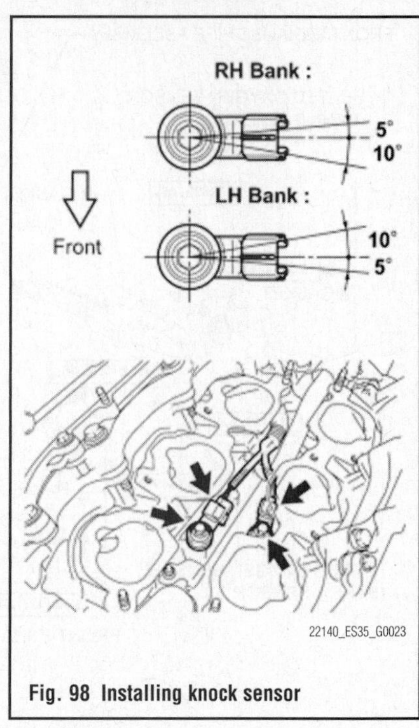

Fig. 98 Installing knock sensor

2. Disconnect the Mass Air Flow (MAF) sensor connector.

3. Remove the 2 screws and Mass Air Flow (MAF) meter.

To install:

4. Installation is the reverse of the removal procedure.

THROTTLE POSITION SENSOR (TPS)

LOCATION

The Throttle Position Sensor (TPS) is located on the throttle body. Refer to the accompanying illustration to locate the TPS.

REMOVAL & INSTALLATION

Refer to the Throttle Body removal and installation procedures in the Fuel System Section.

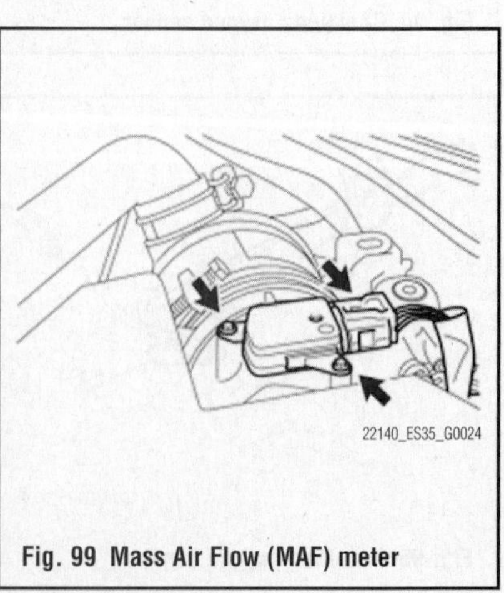

Fig. 99 Mass Air Flow (MAF) meter

FUEL SYSTEM SERVICE PRECAUTIONS

Safety is the most important factor when performing not only fuel system maintenance, but any type of maintenance. Failure to conduct maintenance and repairs in a safe manner may result in serious personal injury or death. Work on a vehicle's fuel system components can be accomplished safely and effectively by adhering to the following rules and guidelines.

- To avoid the possibility of fire and personal injury, always disconnect the negative battery cable unless the repair or test procedure requires that battery voltage be applied.
- Always relieve the fuel system pressure prior to disconnecting any fuel system component (injector, fuel rail, pressure regulator, etc.) fitting or fuel line connection. Exercise extreme caution whenever relieving fuel system pressure to avoid exposing skin, face and eyes to fuel spray. Please be advised that fuel under pressure may penetrate the skin or any part of the body that it contacts.
- Always place a shop towel or cloth around the fitting or connection prior to loosening to absorb any excess fuel due to spillage. Ensure that all fuel spillage is quickly removed from engine surfaces. Ensure that all fuel-soaked cloths or towels are deposited into a flame-proof waste container with a lid.
- Always keep a dry chemical (Class B) fire extinguisher near the work area.
- Do not allow fuel spray or fuel vapors to come into contact with a spark or open flame.
- Always use a second wrench when loosening or tightening fuel line connection fittings. This will prevent unnecessary stress and torsion on fuel piping. Always follow the proper torque specifications.
- Always replace worn fuel fitting O-rings with new ones. Do not substitute fuel hose where rigid pipe is installed.

FUEL SYSTEM PRESSURE

RELIEVING

See Figure 100.

⁂ CAUTION

Perform the following procedures to prevent fuel from spilling out before removing any fuel system parts.

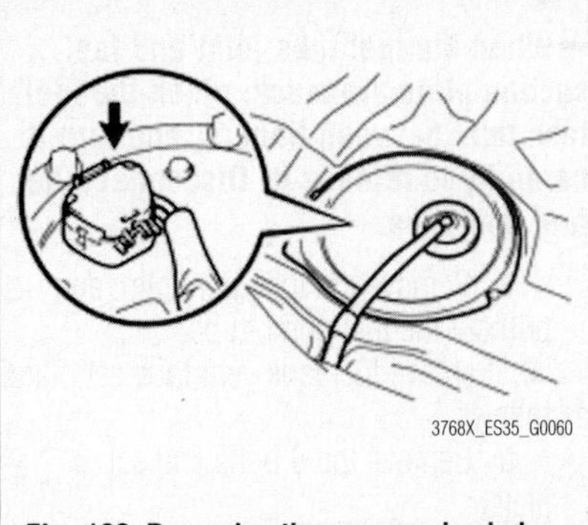

Fig. 100 Removing the rear service hole cover

⁂ CAUTION

Pressure will still remain in the fuel line even after performing the following procedures. When disconnecting the fuel line, cover it with a shop rag or a piece of cloth to prevent fuel from spraying or coming out.

1. Remove the rear seat cushion assembly.
2. Remove the rear floor service hole cover.
3. Disconnect the fuel pump connector.
4. Start the engine.
5. After the engine stops, turn the engine switch off.

➡DTC P0171/25 (fuel problem) may be detected.

6. Crank the engine again. Check that the engine does not start.
7. Remove the fuel tank cap to discharge pressure from the fuel tank.
8. Disconnect the cable from the negative (-) battery terminal.
9. Reconnect the fuel pump connector.
10. Install the rear floor service hole cover.
11. Install the rear seat.
12. Check that there are no fuel leaks from the fuel system after doing any maintenance or repairs.

FUEL FILTER

REMOVAL & INSTALLATION

See Figure 101.

1. Before servicing the vehicle, refer to the precautions section.
2. Remove the fuel pump from the vehicle.
3. Remove the fuel pump filter, as follows:

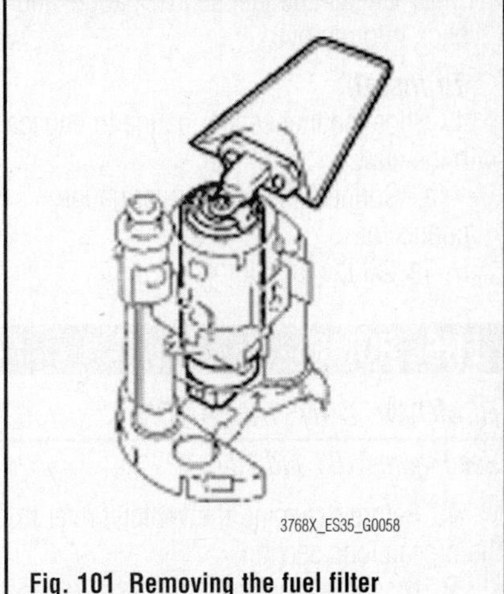

Fig. 101 Removing the fuel filter

➡Do not damage the fuel pump filter. Do not remove the suction filter.

a. Using a screwdriver, pry out the clips.
b. Pull out the fuel pump filter from the fuel pump.

To install:

4. Install the fuel pump filter with a new clip.
5. Install the fuel pump.

FUEL LEVEL SENDING UNIT

LOCATION

The fuel level sending unit is located on the fuel pump.

REMOVAL & INSTALLATION

See Figure 102.

1. Before servicing the vehicle, refer to the precautions section.
2. Remove the fuel pump from the vehicle.
3. Remove the fuel sender gauge assembly, as follows:

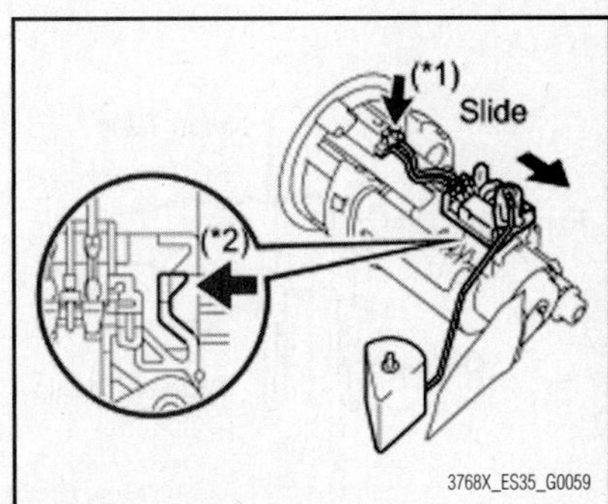

Fig. 102 Disconnecting the fuel sender gauge connector

a. Disconnect the fuel sender gauge connector.

b. Unlock the fuel sender gauge, and slide it to remove.

To install:

4. Slide the fuel sender gauge to engage with the claw.

a. Connect the fuel sender gauge connector.

5. Install the fuel pump.

FUEL PUMP

REMOVAL & INSTALLATION

See Figures 103 and 104.

1. Before servicing the vehicle, refer to the precautions section.
2. Discharge fuel system pressure.
3. Disconnect battery negative cable.
4. Remove the rear seat cushion assembly.
5. Remove the rear floor service hole cover.
6. Disconnect the fuel pump connector.
7. Separate the fuel pump tube sub-assembly, as follows:

➡Check if there is any dirt or mud around the connector before this operation and remove the dirt as necessary.

➡Be careful of mud because the quick connector has an O-ring which seals the pipe and connector that can be contaminated.

➡Do not use any tools in this operation.

➡Do not bend or twist the nylon tube. Cover the fuel tube joint with a plastic bag.

➡When the fuel tube joint and fuel suction plate are stuck, pinch the fuel tank tube between fingers, and turn it carefully to release it. Disconnect the fuel tank tube.

a. Remove the tube joint clip, and pull out the fuel pump tube.

8. Remove fuel tank vent tube set plate, as follows:

a. Remove the 8 bolts and set plate.

9. Remove fuel suction tube assembly with pump and gauge, as follows:

a. Pull out the fuel suction tube from the fuel tank.

➡Do not damage the fuel pump filter.

➡Be careful not to bend the arm of the fuel sender gauge.

b. Remove the gasket from the fuel suction tube.

To install:

10. Install a new gasket to the fuel suction tube.
11. Install the fuel suction tube.

➡Do not damage the fuel pump filter.

➡Be careful not to bend the arm of the fuel sender gauge.

12. Install the fuel tank vent tube set plate, as follows:

a. Align the mark of the set plate with the fuel suction tube.

b. Install the set plate with the 8 bolts and tighten to 52 inch lbs. (5.9 Nm).

13. Install the fuel pump tube with the tube joint clip.

➡Check that there is no scratches or foreign objects on the connecting part.

➡Check that the fuel tube joint is inserted securely.

➡Check that the tube joint clip is on the collar of the fuel tube joint.

➡After installing the tube joint clip, check that the fuel tube joint is pulled off.

14. Connect battery negative cable.
15. Inspect for fuel leak.
16. Install the rear floor service hole cover.
17. Install the rear seat cushion assembly.

FUEL RAIL AND INJECTOR

REMOVAL & INSTALLATION

See Figures 105 through 107.

1. Properly discharge the fuel system pressure.
2. Disconnect battery negative cable.
3. Drain engine coolant.
4. Remove both windshield wiper arm and blade assemblies.

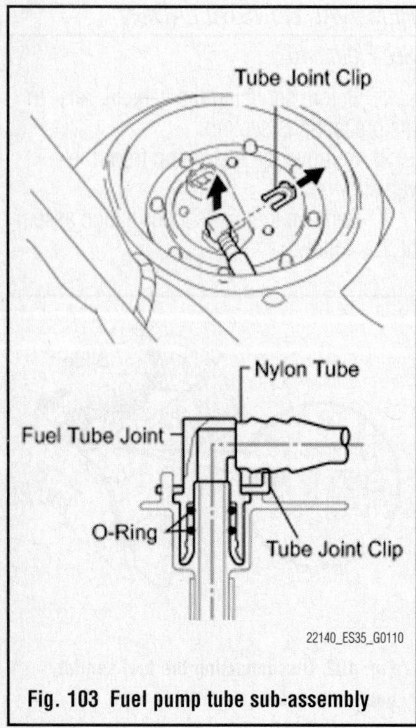

Fig. 103 Fuel pump tube sub-assembly

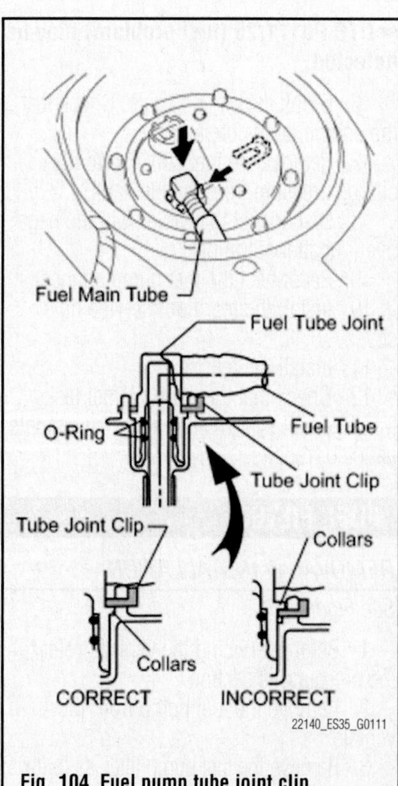

Fig. 104 Fuel pump tube joint clip

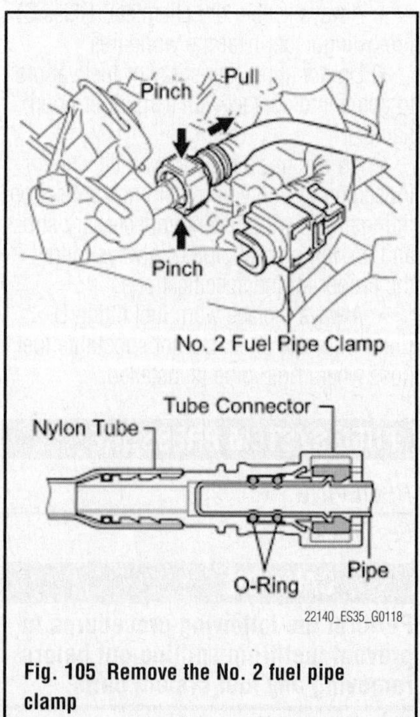

Fig. 105 Remove the No. 2 fuel pipe clamp

5. Remove the right cowl top ventilator louver.

6. Remove the windshield wiper motor and link assembly.

7. Remove front fender to cowl side seal.

8. Remove the cowl top ventilator louver sub-assembly.

9. Remove windshield wiper motor and link assembly.

10. Remove cowl top panel outer sub-assembly.

11. Remove the V-bank cover sub-assembly.

12. Remove the air cleaner cap with air cleaner hose.

13. Remove the intake air surge tank.

14. Disconnect the fuel tube sub-assembly, as follows:

a. Remove the No. 2 fuel pipe clamp.

b. Pinch the tube connector and then pull out the fuel pipe.

➡Check that there is no dirt or other foreign objects around the connector before removing fuel tube, and clean the connector as necessary.

➡It is necessary to prevent mud or dirt from entering the connector. If mud or dirt gets in the connector, the O-rings may not seal properly.

➡Do not use any tools in this operation.

➡Do not bend, kink or twist the nylon tube. Protect the connector by covering it with a plastic bag.

➡When the pipe and connector are stuck, push and pull the connector to release and pull the connector out carefully.

15. Remove the fuel injector assembly, as follows:

a. Disconnect the 6 fuel injector connectors.

b. Remove the 5 bolts and fuel delivery pipe together with the 6 fuel injectors.

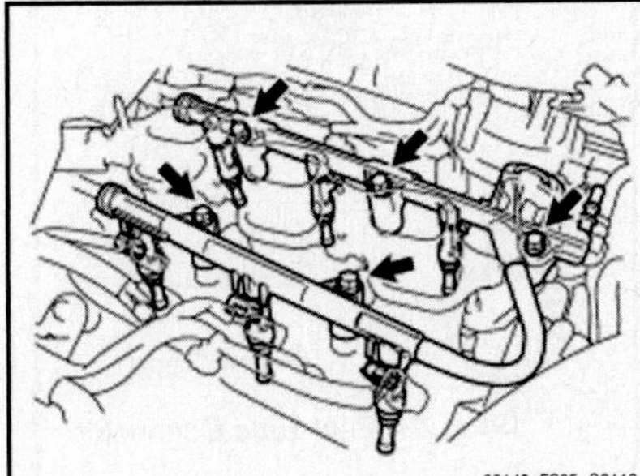

Fig. 106 Removing the 5 bolts and fuel delivery pipe together with the 6 fuel injectors

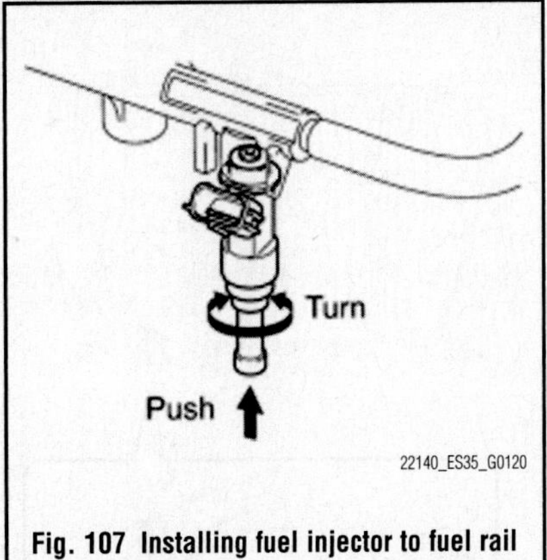

Fig. 107 Installing fuel injector to fuel rail

➡Be careful not to drop the fuel injectors when removing the fuel delivery pipe.

c. Remove the 6 insulators from the intake manifold.

d. Pull out the fuel injector from the fuel delivery pipe.

e. Remove the 6 O-rings from the injectors.

To install:

16. Install the fuel injector assembly, as follows:

a. Apply a light coat of spindle oil or gasoline to new O-rings, and install one to each injector.

b. Apply a light coat of spindle oil or gasoline where the fuel delivery pipe contacts the O-ring.

➡Be careful not to twist the O-ring.

➡After installing the fuel injector, check that it turns smoothly. If not, reinstall it with a new O-ring.

c. Push the fuel injector while twisting it back and forth to install it in the fuel delivery pipe.

d. Position the fuel injector connector outward.

e. Install 6 new insulators to the intake manifold.

f. Place the fuel delivery pipe and the 6 fuel injectors together to the intake manifold.

➡Be careful not to drop the fuel injectors when installing the fuel delivery pipe.

g. Temporarily install the 6 bolts which are used to hold the fuel delivery pipe to the intake manifold.

➡After installing the fuel injector, check that it turns smoothly. If not, reinstall it with a new O-ring.

h. Tighten the 5 bolts which are used to hold the fuel delivery pipe to the intake manifold to 15 ft. lbs. (21 Nm).

17. Push in the tube connector to the pipe until the tube connector makes a "click" sound.

➡Before connecting the tube, make sure that it is not damaged. Make sure that there is no dirt present on the connecting surfaces.

➡After connecting, check if the fuel tube connector and the pipe are securely connected by pulling on them.

18. Install the No. 2 fuel pipe clamp.

19. Te remainder of installation is the reverse of the removal procedure.

20. Check for coolant leak and fuel leak.

FUEL TANK

REMOVAL & INSTALLATION

See Figures 108 through 113.

1. Before servicing the vehicle, refer to the precautions section.

2. Discharge fuel system pressure.

3. Disconnect battery negative cable.

4. Remove rear seat cushion assembly.

5. Remove rear floor service hole cover.

6. Separate fuel pump tube sub-assembly.

7. Remove fuel tank vent tube set plate.

8. Remove fuel suction tube assembly with pump and gauge.

9. Drain fuel.

10. Remove center exhaust pipe assembly.

11. Disconnect no. 2 parking brake cable assembly.

12. Disconnect no. 3 parking brake cable assembly.

13. Remove rear stabilizer BAR No. 1 bracket.

14. Remove lower center fuel tank protector, as follows:

a. Remove the 4 bolts and the (Except SE grade).

b. Remove the 4 bolts and 2 clips (for SE Grade).

c. Remove the fuel tank protector (for SE Grade).

➡Check that there is no dirt or other foreign objects around the connector before removing fuel tubes, and clean the connector as necessary.

➡It is necessary to prevent mud or dirt from entering the connector. If mud or dirt gets in the connector, the O-rings may not seal properly.

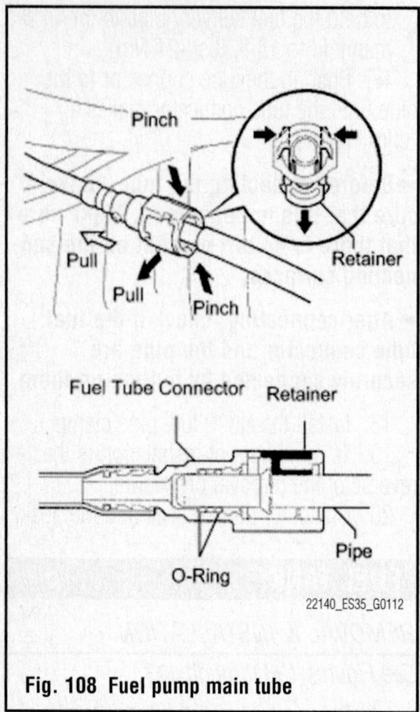

Fig. 108 Fuel pump main tube

➡Do not use any tools in these operations.

➡Do not bend, kink or twist the nylon tubes. Protect the connector by covering it with a plastic bag.

➡When the pipe and connector are stuck, push and pull the connector to release and pull the connector out carefully.

15. Disconnect the fuel pump tube, as follows:
 a. Pinch the tabs of the retainer to

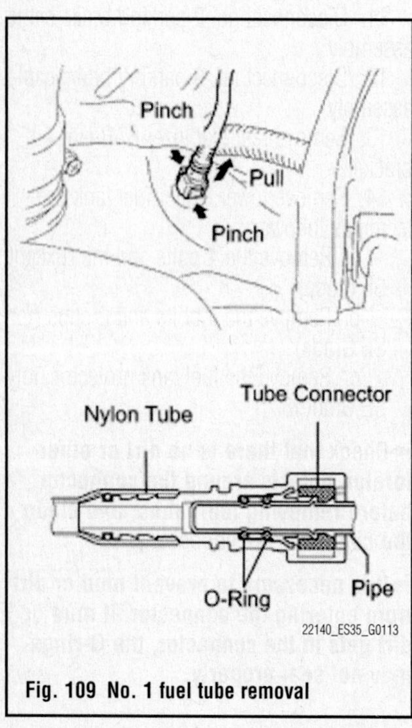

Fig. 109 No. 1 fuel tube removal

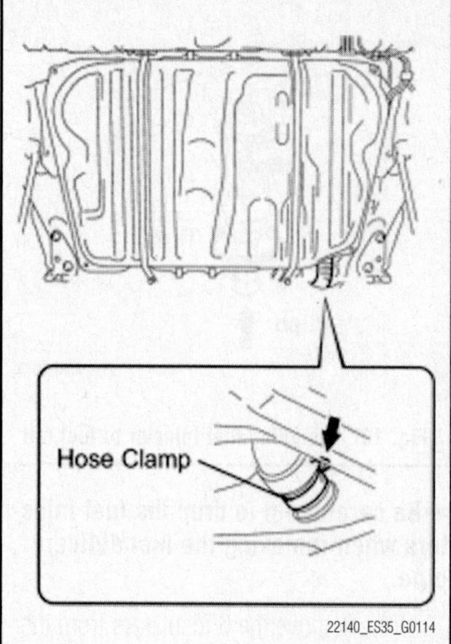

Fig. 110 Disconnecting the fuel tank to filter pipe hose—except PZEV

remove the lock claws and pull it down as shown in the illustration.
 b. Pull out the fuel tank main tube.

16. Pinch the tube connector and then pull out the No. 1 fuel tube.

17. Set up a transmission jack underneath the fuel tank.

18. Remove the 2 set bolts of the fuel tank bands.

➡Check that there is no dirt or other foreign objects around the connector before removing fuel tubes, and clean the connector as necessary.

➡It is necessary to prevent mud or dirt from entering the connector. If mud or dirt gets in the connector, the O-rings may not seal properly.

➡Do not use any tools in these operations.

➡Do not bend, kink or twist the nylon tubes. Protect the connector by covering it with a plastic bag.

➡When the pipe and connector are stuck, push and pull the connector to release and pull the connector out carefully.

19. Remove the hose clamp and disconnect the fuel tank to filter pipe hose (except PZEV).

20. Remove the clamp and disconnect the fuel tank to filter pipe hose (for PZEV).

21. Slightly lower the transmission jack.

22. Disconnect the fuel tank vent hose from the charcoal canister, as follows:
 a. Push the connector deep into the charcoal canister to release the locking pin.
 b. Pinch portion A.
 c. Pull out the connector.

23. Remove the 2 pins and 2 fuel tank bands as shown in the illustration.

24. Remove the 4 clip nuts.

To install:

25. Install the 4 clip nuts.

26. Install the 2 fuel tank bands with the 2 pins.

27. Connect the fuel tank vent hose.

28. Connect the fuel tank inlet pipe with the fuel filter pipe clamp.

29. Tighten the 2 set bolts of the fuel tank bands to 29 ft. lbs. (39 Nm).

30. Connect the No. 1 fuel tube, as follows:
 a. Push the fuel tube connector into the pipe until the fuel tube connector makes a "click" sound.

➡Check that there is no damage or foreign objects on the connected part.

➡After connecting, check if the fuel tube connector and the pipe are securely connected by trying to pull them apart.

31. Connect the fuel pump tube, as follows:
 a. Push in the fuel pump tube connector to the pipe and push up the retainer so that the claws engage.

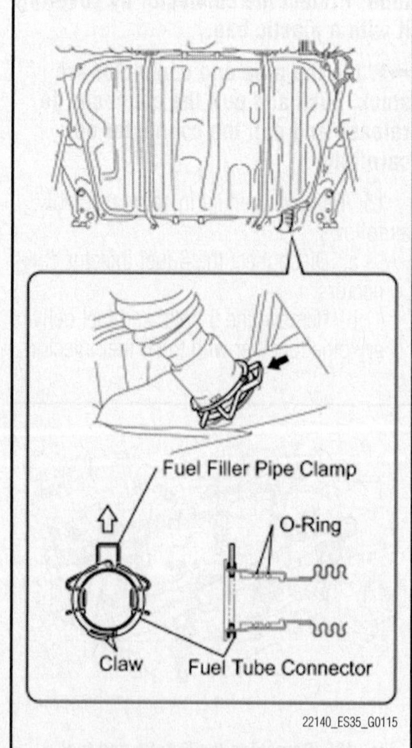

Fig. 111 Disconnecting the fuel tank to filter pipe hose—PZEV

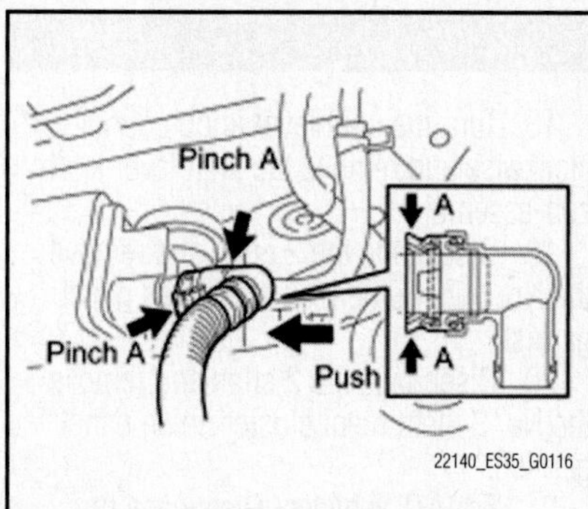

Fig. 112 Disconnect the fuel tank vent hose from the charcoal canister

➡Check that there is no damage or foreign objects on the connected part.

➡After connecting, check if the fuel tube connector and the pipe are securely connected by trying to pull them apart.

32. Install the lower center fuel tank protector and tighten to 48 inch lbs. (5.4 Nm).
33. Install the No. 3 parking brake cable assembly with the bolt and nut and tighten to 53 inch lbs. (6 Nm), and 75 inch lbs. (8.5 Nm).
34. Install the No. 2 parking brake cable assembly with the bolt and nut and tighten to 53 inch lbs. (6 Nm), and 75 inch lbs. (8.5 Nm).
35. Install the center exhaust pipe assembly.
36. Install the fuel suction tube assembly with pump and gauge.
37. Install the fuel tank vent tube set plate.
38. Connect the fuel pump tube sub-assembly.
39. Add fuel.
40. Connect battery negative cable.
41. Inspect for fuel leak and exhaust gas leak.
42. Install the rear floor service hole cover.
43. Install the rear seat cushion assembly.

IDLE SPEED

ADJUSTMENT

Idle speed is maintained by the ECM. No adjustment is necessary or possible.

THROTTLE BODY

REMOVAL & INSTALLATION

See Figures 114 through 117.

1. Before servicing the vehicle, refer to the precautions section.
2. Disconnect battery negative cable.
3. Drain engine coolant.
4. Remove cool air intake duct seal.
5. Remove the V-bank cover sub-assembly.
6. Remove air cleaner inlet assembly.
7. Remove air cleaner cap sub-assembly, as follows:
    a. Disconnect the 3 vacuum hoses.
    b. Disconnect the Mass Air Flow (MAF) sensor connector (1).
    c. Disconnect the No. 2 ventilation hose (2).
    d. Disconnect the hose band (3).
    e. Disconnect the 3 bands, and remove the air cleaner cap sub-assembly.
8. Remove air cleaner case sub-assembly.
9. Remove No. 1 air cleaner inlet.
10. Disconnect the throttle body connector and clamp.
11. Disconnect the 2 water by-pass hoses from the throttle body.
12. Remove the 4 bolts and throttle body.
13. Remove the throttle body gasket from the intake air surge tank.

To install:

14. Install a new throttle body gasket to the intake air surge tank.
15. Install the throttle w/ motor body assembly and wire harness clamp stay to the intake air surge tank with the 4 bolts and tighten to 7 ft. lbs. (10 Nm).
16. Connect the throttle w/ motor body assembly connector.
17. Te remainder of installation is the reverse of the removal procedure.
18. Check for coolant leak.
19. Check the function of the throttle body.

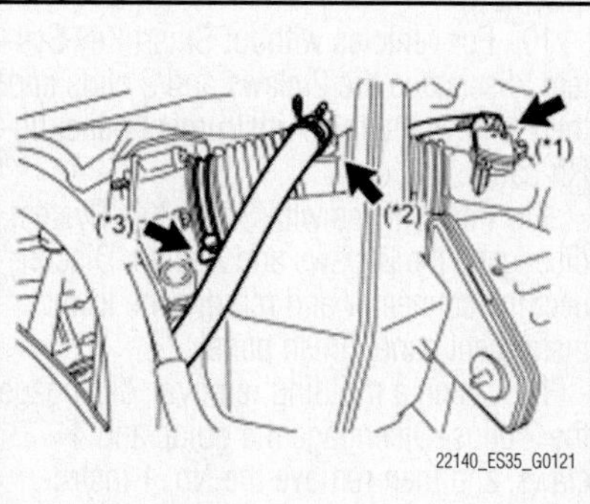

Fig. 114 Removing air cleaner cap sub-assembly

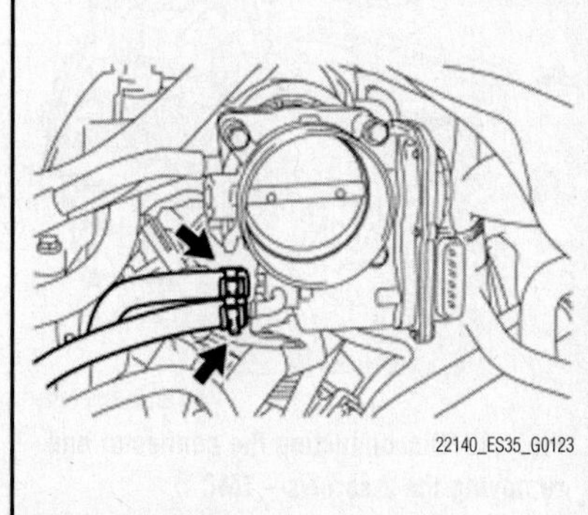

Fig. 116 Disconnecting the 2 water by-pass hoses

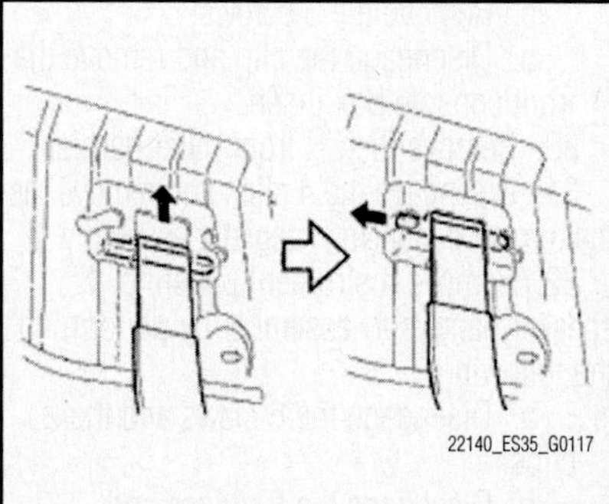

Fig. 113 Removing the 2 pins and 2 fuel tank bands

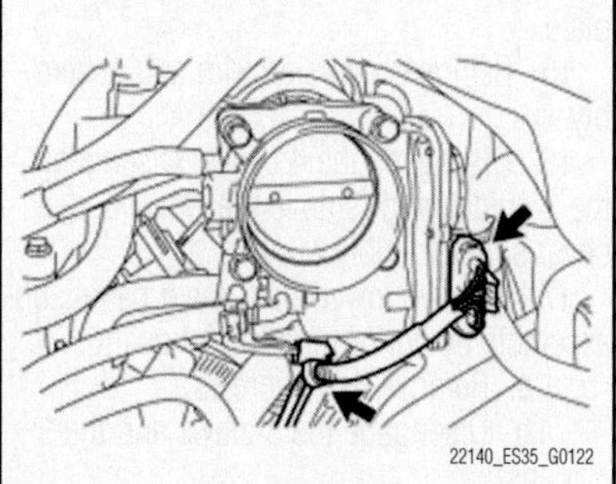

Fig. 115 Disconnecting the throttle body connector

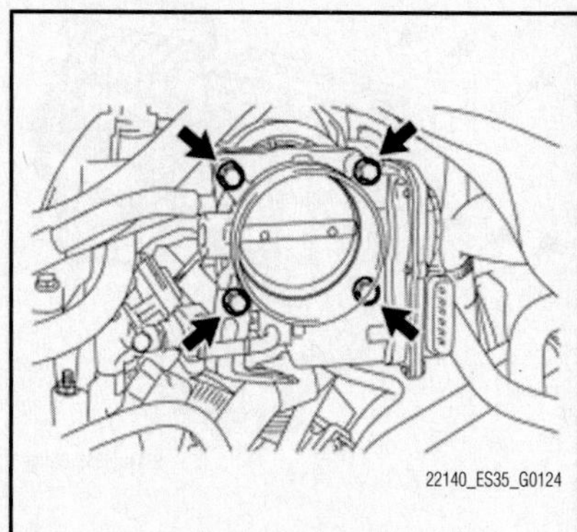

Fig. 117 Removing throttle body bolts

HEATING & AIR CONDITIONING SYSTEM

BLOWER MOTOR

REMOVAL & INSTALLATION

See Figures 118 and 119.

1. Drain and recycle the engine coolant.
2. Disconnect the negative battery cable.
3. Remove instrument panel.
4. For TMC made:
   a. Disconnect the connector.
   b. Remove the 2 screws and blower assembly.
   c. Remove the 3 screws and blower with fan motor sub-assembly.
5. For TMMK made:
   a. Remove cooler expansion valve.
   b. Remove the connector and clamp, and disconnect the wire harness.
   c. Remove the 6 screws and then the blower assembly with the cooler evaporator sub-assembly.
   d. Remove the 3 screws and blower with fan motor sub-assembly.

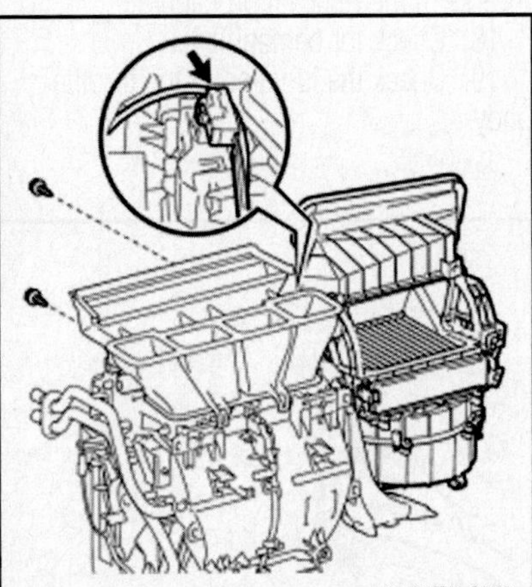

Fig. 118 Disconnecting the connector and removing the 2 screws—TMC

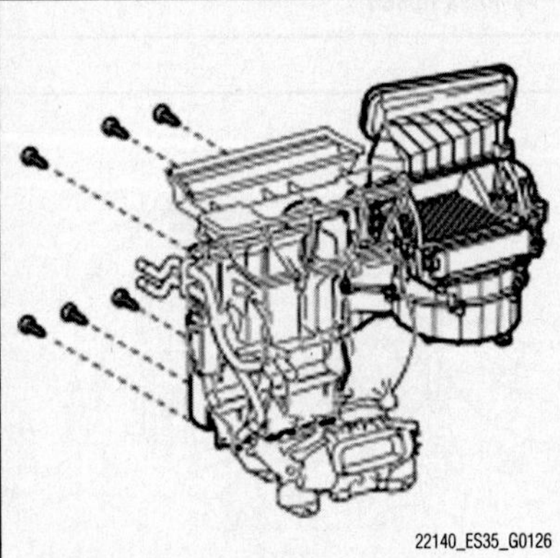

Fig. 119 Removing the 6 screws and blower assembly—TMMK

To install:

6. To install, reverse the removal procedure.

HEATER CORE

REMOVAL & INSTALLATION

See Figures 120 and 121.

1. Before servicing the vehicle, refer to the Precautions Section.

CAUTION

Wait for 90 seconds after disconnecting the cable to prevent airbag deployment.

2. Disconnect battery negative terminal.
3. Remove the lower No. 2 and No. 3 steering wheel covers.
4. Remove the steering pad.
5. Remove the steering wheel assembly.
6. Remove the LH front door scuff plate.
7. Remove the LH cowl side trim sub-assembly.
8. Remove steering column cover.
9. Remove the turn signal switch assembly.
10. For vehicles without Smart Key System, disengage the 2 claws and 2 clips and then remove the lower instrument panel finish panel.
11. For vehicles with Smart Key System, disengage the 2 claws and 2 clips. Disconnect the connector and remove the lower instrument panel finish panel.
12. Using a molding remover, disengage the 2 clips. Disengage the guide and 4 claws, and then remove the No. 1 instrument cluster finish panel.
13. Remove the 4 screws. Disconnect each connector and remove the combination meter assembly.
14. Remove the RH front door scuff plate.
15. Remove cowl side trim sub-assembly.
16. Disengage the 4 claws. Disengage the 2 guides and remove the No. 2 under cover sub-assembly.
17. Remove lower instrument panel sub-assembly by performing the following:
    a. Remove the 4 screws.
    b. Disengage the 3 claws and the 3 clips.
    c. Disconnect the connector and remove the lower instrument panel sub-assembly.
18. Turn the shift lever knob counterclockwise and remove the shift lever knob sub-assembly.
19. Disengage the 2 clips and remove the No. 1 instrument cluster finish panel garnish.
20. Disengage the 2 clips and remove the No. 2 instrument cluster finish panel garnish.
21. For A/T vehicles, Disengage the 6 claws and the 3 clips, and then remove the floor shift position indicator housing sub-assembly. If equipped with Seat Heater System, disconnect each connector.
22. For M/T vehicles, open the lid of the upper console panel. Apply protective tape to the area. Using a molding remover, disengage the 2 claws and the 5 clips, and then remove the upper console panel.
23. Disengage the 3 claws and the 5 clips. Disconnect the connector and remove the upper console rear panel sub-assembly.
24. Remove instrument panel no. 2 register assembly by performing the following:
    a. Apply protective tape to the areas.
    b. Using a molding remover, disengage the 3 clips.
    c. Using a molding remover, disengage the 4 clips.
    d. Disconnect the connector and remove the instrument panel No. 2 register assembly.
25. Remove radio receiver with heater control panel assembly.
26. Remove the console box pocket
27. Remove the console box pocket.
28. Remove the console box assembly by performing the following:
    a. Remove the 2 screws.
    b. Disengage the clamp.
    c. Remove the 2 bolts and the console box assembly.
29. Remove both of the front console box inserts by performing the following:
    a. Remove the 3 screws.
    b. Disengage the clip and remove the front console box insert.
30. Remove the LH front pillar garnish.
31. Disengage the 4 clips and remove the instrument panel No. 1 register assembly.
32. Remove instrument panel no. 1 speaker panel sub-assembly by performing the following:
    a. Disengage the 6 claws and the 2 clips.
    b. Disengage the 2 guides and remove the instrument panel No. 1 speaker panel sub-assembly.

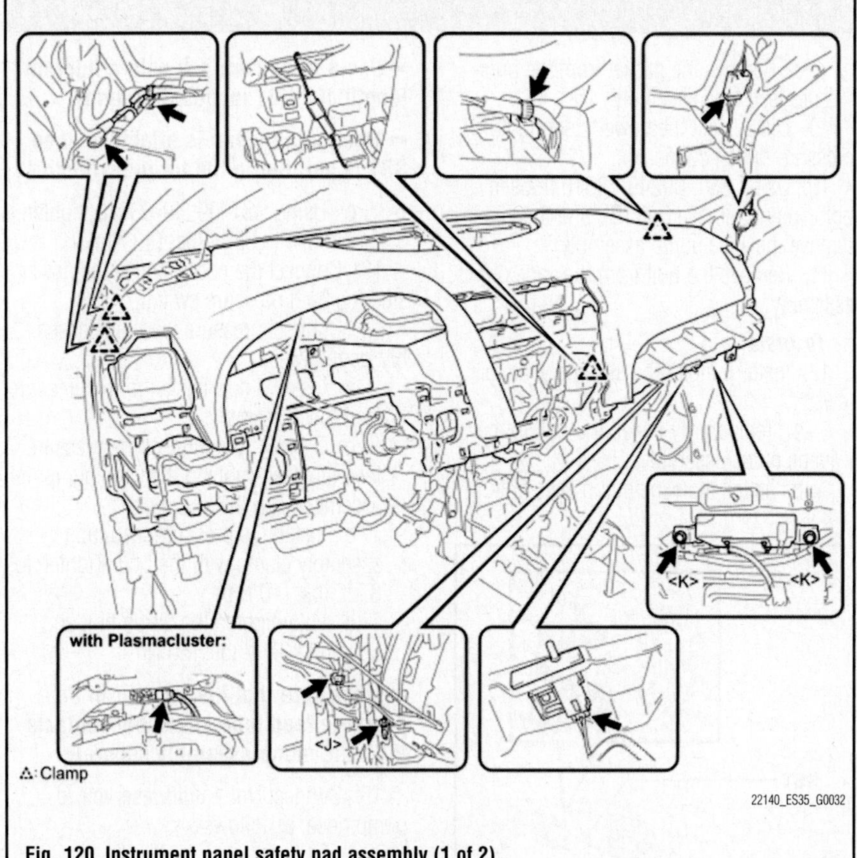

Fig. 120 Instrument panel safety pad assembly (1 of 2)

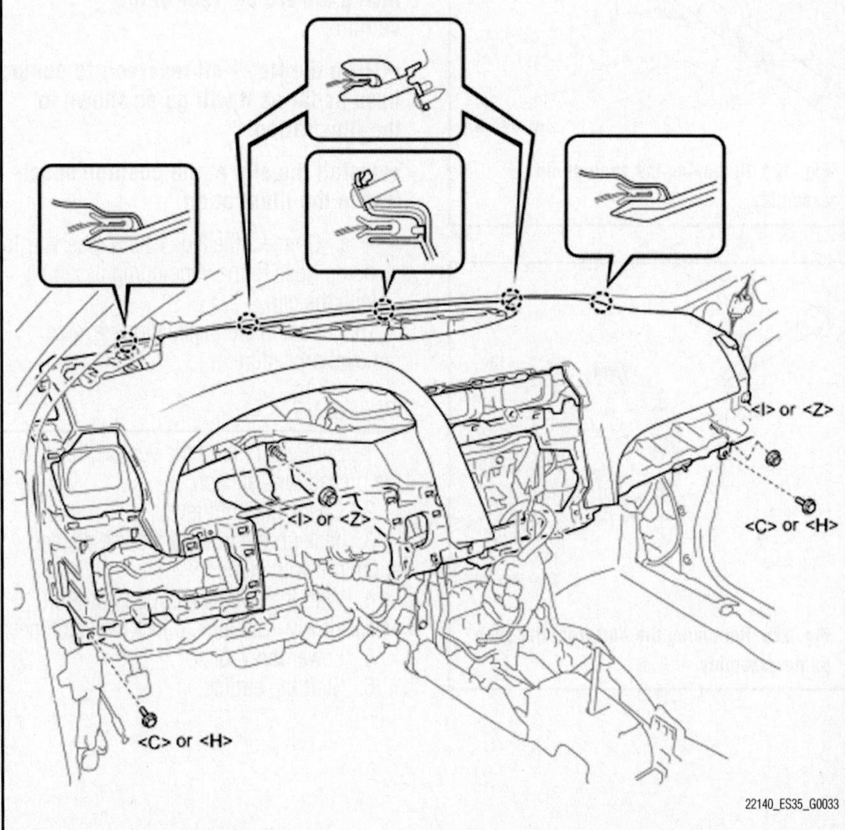

Fig. 121 Instrument panel safety pad assembly (2 of 2)

33. Remove RH front no. 2 speaker assembly.
34. Remove the RH front pillar garnish.
35. Disengage the 4 clips and remove the instrument panel No. 3 register assembly.
36. Remove instrument panel no. 2 speaker panel sub-assembly by performing the following:
    a. Disengage the 6 claws and the 2 clips.
    b. Disengage the 2 guides and remove the instrument panel No. 2 speaker panel sub-assembly.
37. Remove LH and RH front no. 2 speaker assemblies.
38. Remove no. 1 defroster nozzle garnish by performing the following:
    a. Disengage the 8 clips and the 4 guides.
    b. Disconnect each connector and remove the No. 1 defroster nozzle garnish.
39. Disconnect instrument panel wire assembly.
40. Remove instrument panel safety pad assembly by performing the following:
    a. Disengage each clamp.
    b. Disconnect each connector.
    c. Remove the bolt.

CAUTION

Some models covered by this manual may be equipped with a Supplemental Restraint System (SRS), which uses an air bag. Whenever working near any of the SRS components, such as the impact sensors, the air bag module, steering column and instrument panel, disable the SRS, as described in the Chassis Electrical Section.

    d. Remove the 2 passenger airbag bolts.
    e. If equipped with Plasmacluster, disconnect the connector
    f. Disconnect the connector
    g. Remove the 2 bolts.
    h. Disengage the 5 claws and remove the instrument panel safety pad assembly.
    i. Disengage the claw and remove the 5 instrument panel stays.
41. Remove heater core as necessary.

To install:

42. Installation is the reverse of the removal procedure.
43. Perform initialization.
44. Inspect the steering pad.
45. Inspect the SRS warning light.

STEERING

POWER STEERING PUMP

REMOVAL & INSTALLATION

See Figures 122 through 125.

1. Before servicing the vehicle, refer to the precautions section.
2. Drain power steering fluid.
3. Remove RH engine under cover.
4. Remove RH front fender apron seal.
5. Remove v-bank cover sub-assembly.
6. Remove fan and generator v belt.
7. Slide the clip and disconnect the No. 1 fluid reservoir to pump hose from the vane pump assembly.
8. Disconnect pressure feed tube assembly, as follows:
    a. Remove the union bolt and disconnect the pressure feed tube assembly from the vane pump assembly.
    b. Remove the bolt and separate the pressure feed tube clamp.

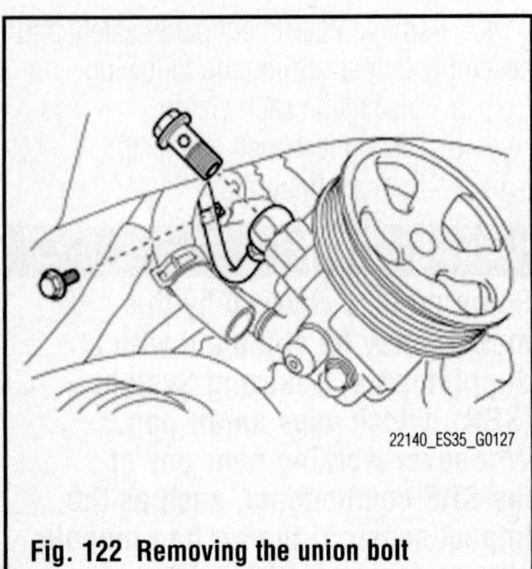

Fig. 122 Removing the union bolt

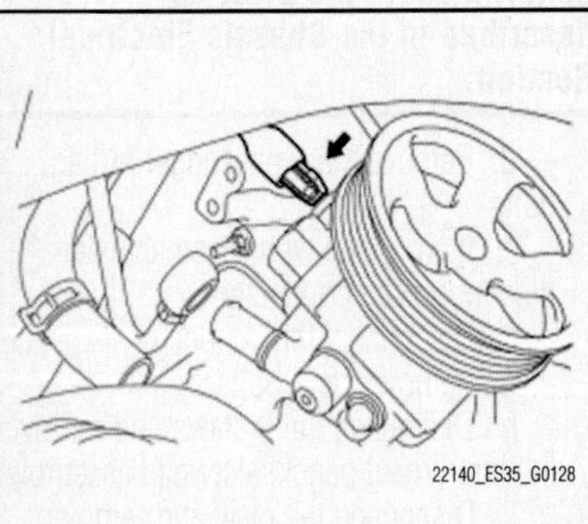

Fig. 123 Disconnecting the power steering fluid pressure switch connector

    c. Remove the gasket from the pressure feed tube assembly.
9. Disconnect the power steering fluid pressure switch connector.
10. Using SST: 09249-63010, loosen bolt (A) and remove bolt (B), and then remove the vane pump assembly.
11. Remove the bolt from the vane pump assembly.

To install:

12. Install vane pump assembly, as follows:
    a. Temporarily install the bolt to the vane pump assembly.
    b. Install the vane pump assembly.

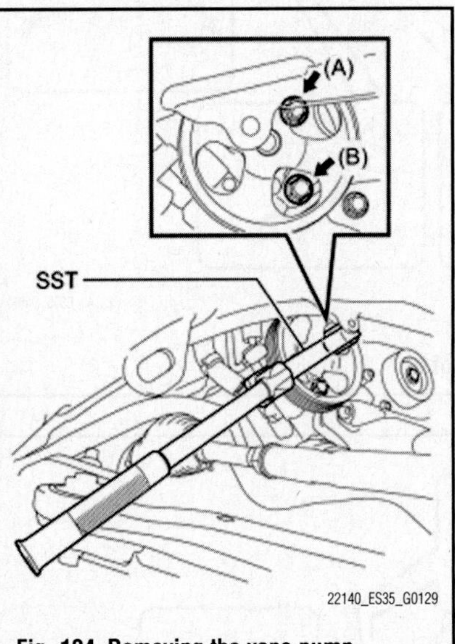

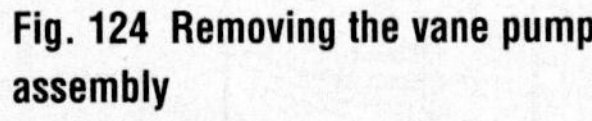

Fig. 124 Removing the vane pump assembly

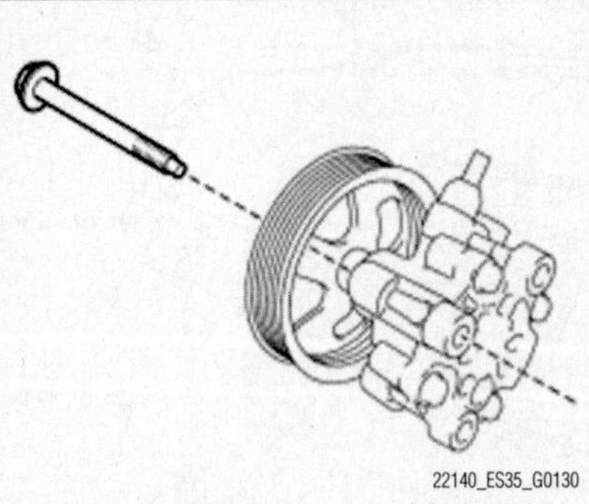

Fig. 125 Removing the bolt from the vane pump assembly

➡Use a torque wrench with a fulcrum length of 11.81 inches (300 mm).

➡This torque value is effective when SST is parallel to the torque wrench.

    c. Using SST: 09249-63010, tighten the 2 bolts to 32 ft. lbs. (43 Nm).
13. Connect the connector to the power steering fluid pressure switch.
14. Connect pressure feed tube assembly, as follows:
    a. Install a new gasket to the pressure feed tube assembly.
    b. Temporarily connect the pressure feed tube assembly to the vane pump assembly with the union bolt.
    c. Install the pressure feed tube assembly clamp with the bolt. Tighten to 87 ft. lbs. (10 Nm).
    d. Fully tighten the union bolt and tighten to 37 ft. lbs. (50 Nm).

➡Make sure that the stopper of the pressure feed tube assembly contacts the vane pump assembly securely.

15. Connect No. 1 fluid reservoir to pump hose, as follows:

➡Connect the No. 1 oil reservoir to pump hose with the paint mark facing toward the rear of the vehicle.

➡Push the No. 1 oil reservoir to pump hose as far as it will go as shown in the illustration.

➡Install the clip at the position specified in the illustration.

    a. Connect the No. 1 fluid reservoir to pump hose to the vane pump assembly with the clip.
16. To complete installation, reverse removal procedure.

BLEEDING

1. Before servicing the vehicle, refer to the precautions section.
2. Check the fluid level.
3. Jack up the front of the vehicle and support it with stands.
4. With the engine stopped, turn the wheel slowly from lock to lock several times.
5. Lower the vehicle.
6. Start the engine.

COIL SPRING

REMOVAL & INSTALLATION

See Figures 126 through 131.

1. Before servicing the vehicle, refer to the precautions section.
2. Remove the front shock absorber.
3. As shown in the illustration, secure the front shock absorber with coil spring in a vise using aluminum plates by clamping onto a double nutted bolt affixed to the bracket at the bottom of the absorber.

➡Do not use an impact wrench.

➡If the front coil spring is compressed at an angle, using 2 SST will make the work easier.

4. Using SST: 09727-30021, compress the front coil spring.
5. Remove the front suspension support sub-assembly, front suspension support bearing, front coil spring upper seat, front coil spring upper insulator, front coil spring, front spring bumper, and front coil spring lower insulator from the front shock absorber.

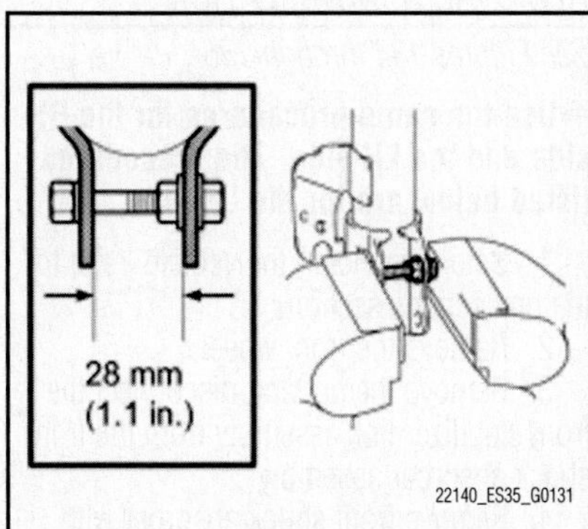

Fig. 126 Secure the front shock absorber

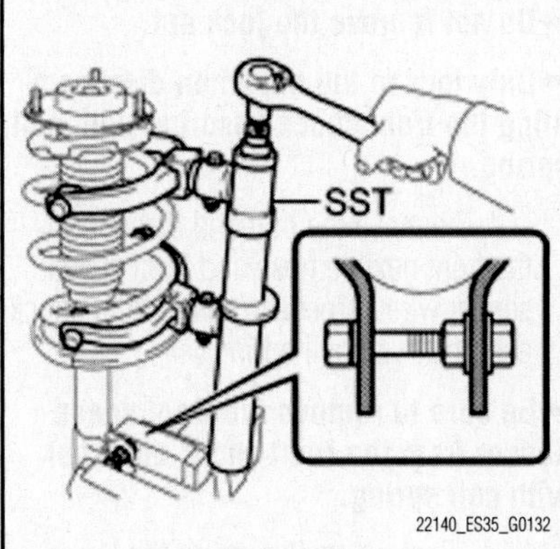

Fig. 127 Removing coil spring components

To install:

6. Install front coil spring as follows:
   a. Install the front spring bumper to the piston rod.

➡Align the 2 protrusions of the front coil spring lower insulator and the 2 holes in the front shock absorber.

➡Do not use an impact wrench.

   b. Install the front coil spring lower insulator onto the front shock absorber.
   c. Using SST: 09727-30021, compress the front coil spring.

➡The smaller diameter end of the front coil spring must face upward.

➡Fit the lower end of the front coil spring into the gap of the insulator.

   d. Install the front coil spring to the front shock absorber.

➡Any misalignment between the front shock absorber lower bracket and the matchmark must be +/-5°.

   e. Install the front coil spring upper insulator as shown in the illustration.

➡Any misalignment between the front shock absorber lower bracket and the matchmark must be +/-5°.

7. Install the front coil spring upper seat with the mark facing to the outside of the vehicle.

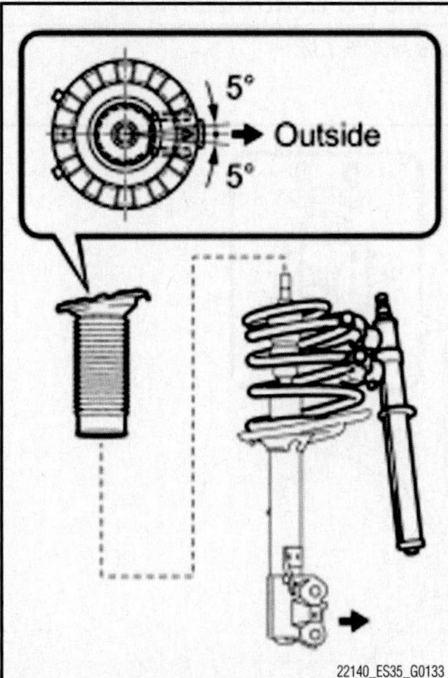

Fig. 128 Installing the front coil spring upper insulator

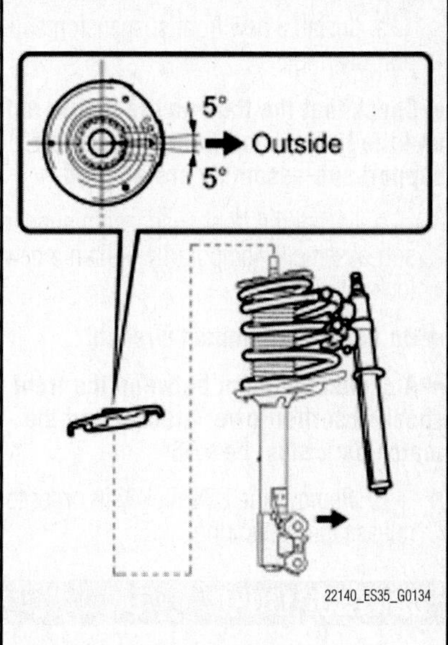

Fig. 129 Installing the front coil spring upper seat

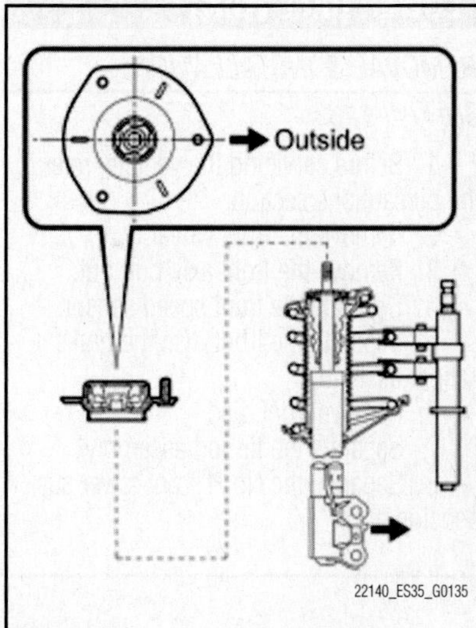

Fig. 130 the front suspension support sub-assembly

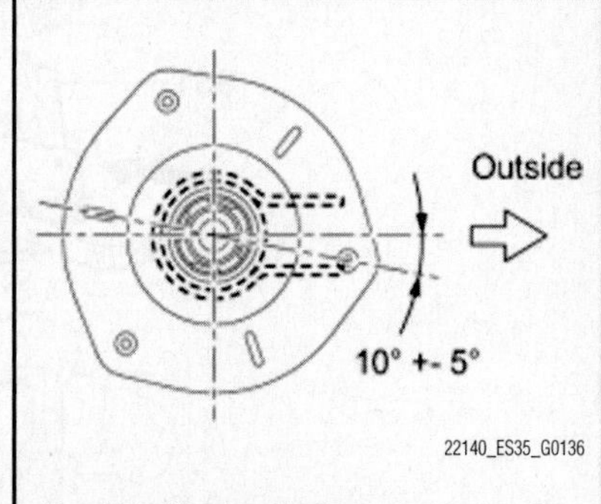

Fig. 131 Aligning the front shock absorber lower bracket and arrows

➡If there is foreign matter inside the front suspension support bearing, replace it with a new one.

a. Install a new front suspension support bearing.

➡Check that the flats on the piston rod and the flats on the front suspension support sub-assembly are aligned.

b. Install the front suspension support sub-assembly. Temporarily tighten a new lock nut.

➡Do not use an impact wrench.

➡Any misalignment between the front shock absorber lower bracket and the matchmark must be +/-5°.

c. Remove the SST slowly in order to release the coil spring.

CONTROL LINKS

REMOVAL & INSTALLATION

See Front Stabilizer Bar.

LOWER BALL JOINT

REMOVAL & INSTALLATION

See Figure 132.

1. Before servicing the vehicle, refer to the precautions section.
2. Remove the front wheel.
3. Remove the front axle hub nut.
4. Separate the front speed sensor.
5. Separate the front disc the brake caliper assembly.
6. Remove front disc.
7. Separate the tie rod assembly.
8. Separate the No. 1 front lower suspension arm.
9. Remove the front axle assembly.
10. Remove front wheel No. 1 bearing dust deflector.
11. Remove front axle hub hole snap ring.
12. Remove front axle hub.
13. Remove front disc brake dust cover.
14. Remove the front lower ball joint assembly, as follows:
   a. Secure the steering knuckle in a vise using aluminum plates.
   b. Remove the cotter pin and castle nut.

➡Do not damage the dust cover of the ball joint.

➡Do not damage the steering knuckle.

   c. Using SST (SST: 09628-62011) or equivalent, remove the front lower ball joint assembly.

To install:

15. Installation is the reverse of the removal procedure, noting the following:
   a. Install the front lower ball joint assembly to the steering knuckle with the castle nut and tighten to 91 ft. lbs. (123 Nm). Further tighten the nut up to 60° if the holes for the cotter pin are not aligned.
   b. Inspect and adjust the front wheel alignment.
   c. Inspect the ABS speed sensor signal.

LOWER CONTROL ARM

REMOVAL & INSTALLATION

See Figure 133.

➡Removal of the lower control arm requires the removal of the engine and transaxle.

1. Before servicing the vehicle, refer to the Precautions Section.
2. Remove or disconnect the following:
   - Transverse engine mounting insulator
   - Engine and transaxle assembly
   - 2 bolts on front side of suspension arm
   - Bolt and nut on rear side of suspension arm and lower arm
   - Lower bush stopper

To install:

3. Install or connect the following:
   - Lower bush arm stopper
   - 2 bolt on the front side to 148 ft. lbs. (200 Nm)
   - Rear side bolt and nut to 152 ft. lbs. (206 Nm)
   - Transverse engine mounting insulator to 64 ft. lbs. (87 Nm)
   - Engine and transaxle assembly

SHOCK ABSORBERS

REMOVAL & INSTALLATION

See Figures 134 through 136.

➡Use the same procedures for the RH side and the LH side. The procedures listed below are for the LH side.

1. Before servicing the vehicle, refer to the precautions section.
2. Remove the front wheel.
3. Remove the nut and disconnect the front stabilizer link assembly from the front shock absorber assembly.
4. Remove front shock absorber with coil spring, as follows:
   a. Loosen the lock nut of the front shock absorber with coil spring.

➡Do not remove the lock nut.

➡Only loosen the nut when disassembling the front shock absorber with coil spring.

   b. Remove the bolt and disconnect the front flexible hose and front speed sensor wire harness from the front shock absorber with coil spring.

➡Be sure to remove the front speed sensor from the front shock absorber with coil spring.

   c. Remove the 2 nuts on the lower side of the front shock absorber with coil spring.

Fig. 132 Remove the front lower ball joint assembly

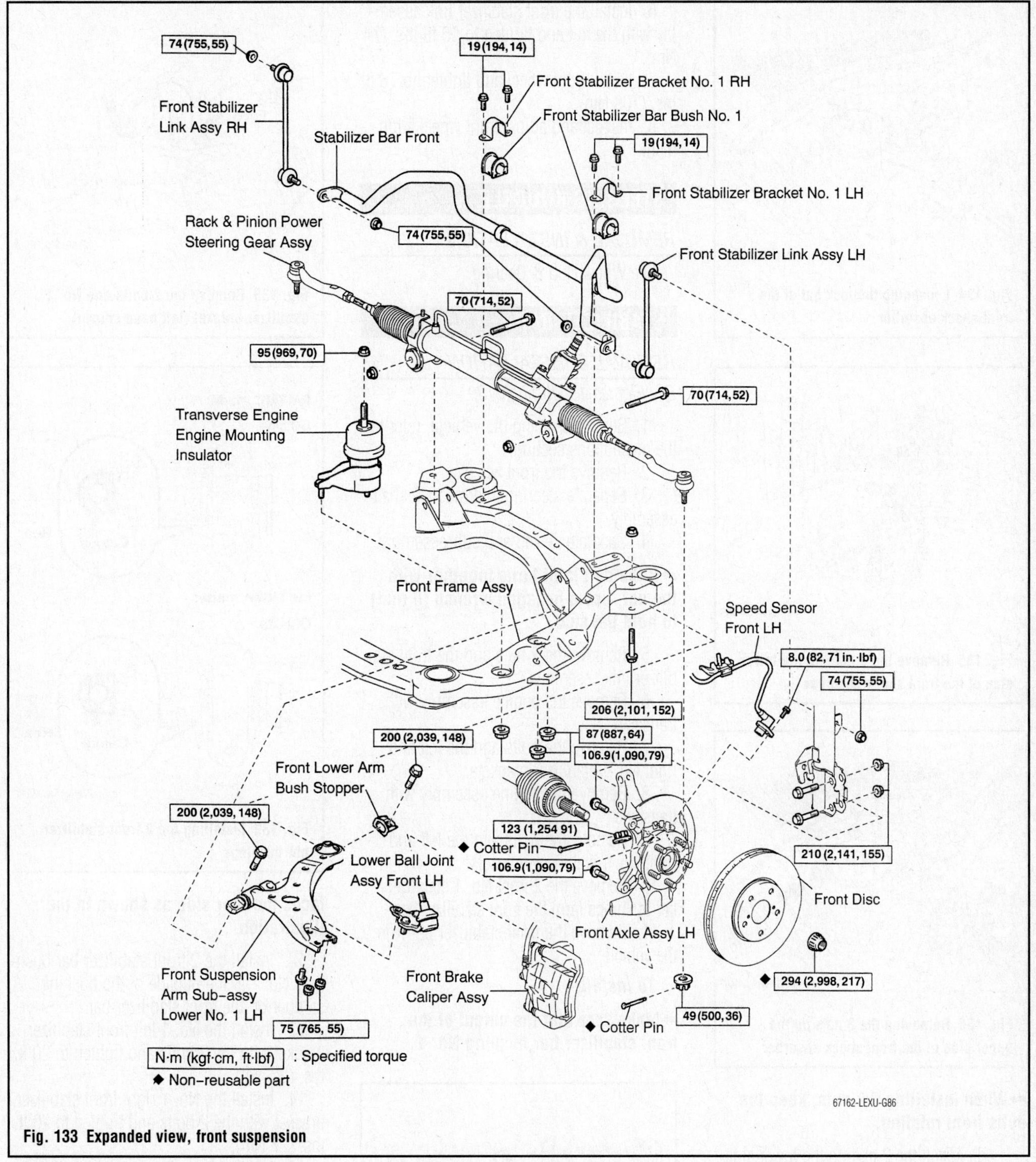

Fig. 133 Expanded view, front suspension

➡When removing the nuts, keep the bolts from rotating.

➡Keep the bolts inserted to secure the front axle assembly.

d. Remove the 3 nuts on the upper side of the front shock absorber with coil spring.

e. Lower the front axle assembly, and remove the 2 bolts on the lower side of the front shock absorber.

➡Make sure that the front speed sensor is disconnected from the front shock absorber with coil spring.

f. Remove the front shock absorber with coil spring.

To install:

5. Install front shock absorber with coil spring, as follows:

a. Install the front shock absorber with coil spring to the front axle assembly and insert the 2 bolts from the front side of the vehicle.

b. Slowly jack up the vehicle using a wooden block and install the front shock absorber with coil spring (upper side) to the vehicle.

c. Install the 3 nuts to the upper side of the front shock absorber with coil spring and tighten to 63 ft. lbs. (85 Nm).

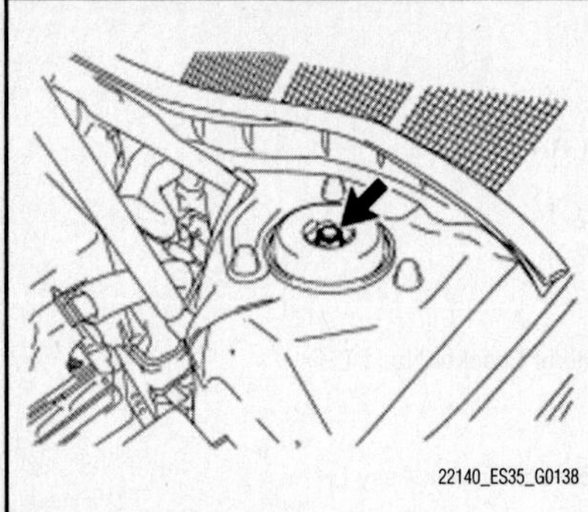

Fig. 134 Loosening the lock nut of the front shock absorber

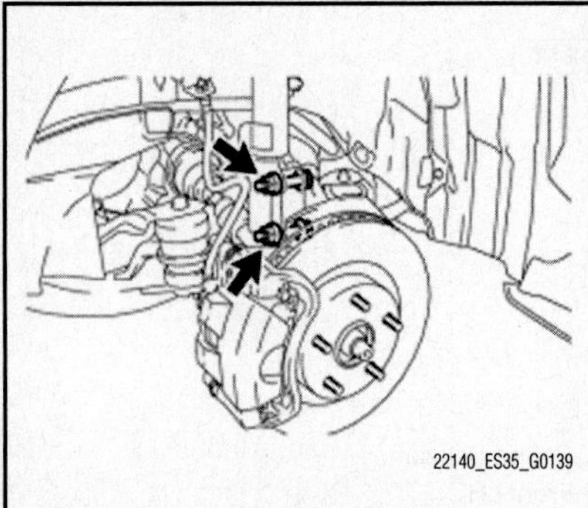

Fig. 135 Remove the 2 nuts on the lower side of the front shock absorber

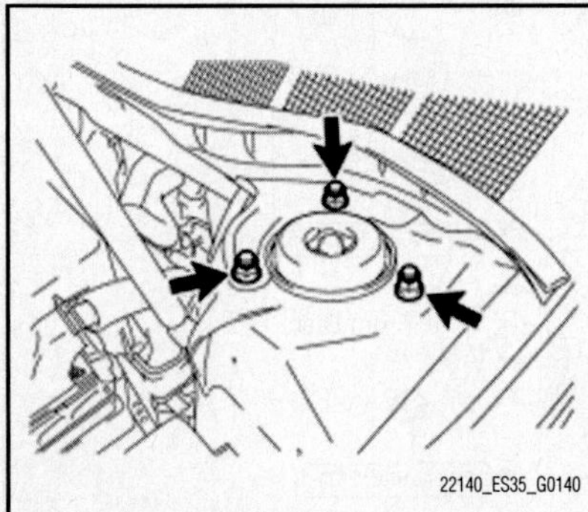

Fig. 136 Removing the 3 nuts on the upper side of the front shock absorber

➡When installing the nuts, keep the bolts from rotating.

d. Install the 2 nuts to the lower side of the front shock absorber with coil spring and tighten to 155 ft. lbs. (210 Nm).

e. Install the front flexible hose and front speed sensor wire harness with the bolt and tighten to 14 ft. lbs. (19 Nm).

f. Fully tighten the lock nut and tighten to 52 ft. lbs. (70 Nm).

➡If the ball joint turns together with the nut, use a hexagon wrench (6 mm) to hold the stud.

6. Install the front stabilizer link assembly with the nut and tighten to 55 ft. lbs. (74 Nm).
7. Install front wheel and tighten to 76 ft. lbs. (103 Nm).
8. Inspect and adjust front wheel alignment.

STEERING KNUCKLE

REMOVAL & INSTALLATION

See Wheel Hub & Bearing.

STABILIZER BAR

REMOVAL & INSTALLATION

See Figures 137 through 139.

1. Before servicing the vehicle, refer to the precautions section.
2. Remove the front wheels.
3. Separate steering intermediate shaft assembly.
4. Separate tie rod end sub-assembly.

➡If the ball joint turns together with the nut, use a hexagon wrench (6 mm) to hold the stud.

5. Remove the 2 nuts and the front stabilizer link assembly
6. Remove the engine assembly with transaxle.
7. Remove the bolts and the left and right No. 1 stabilizer brackets.
8. Remove the engine assembly with transaxle.
9. Remove the bolts and the left and right No. 1 stabilizer brackets.
10. Remove the 2 front No. 1 stabilizer bar bushings from the front stabilizer bar.
11. Remove the front stabilizer bar from the vehicle.

To install:

➡Make sure that the cutout of the front stabilizer bar bushing No. 1 faces the rear side as shown in the illustration.

Fig. 137 Remove the 2 nuts and front stabilizer link assembly (left hand shown)

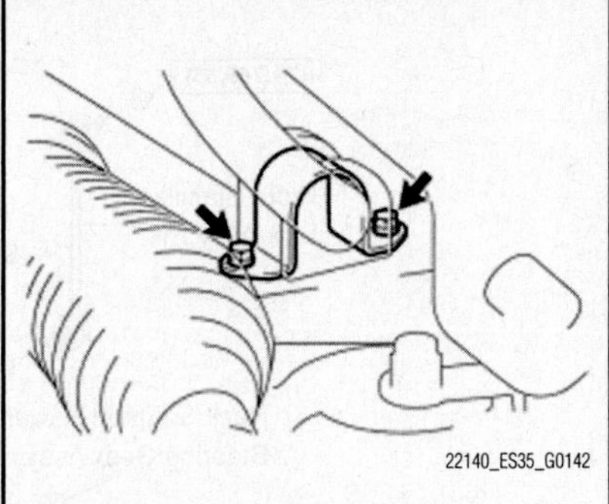

Fig. 138 Remove the 2 bolts and No. 1 stabilizer bracket (left hand shown)

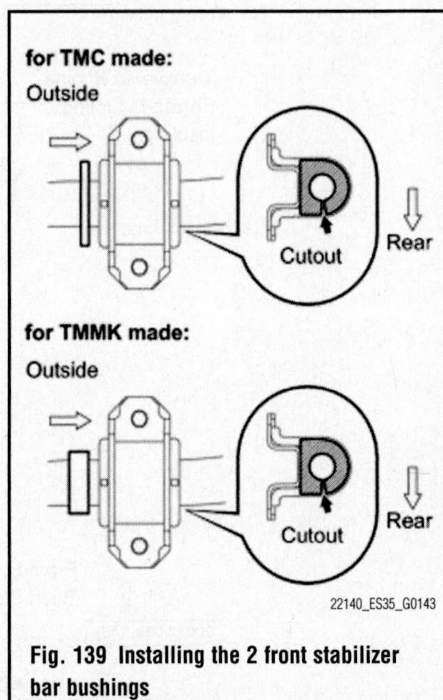

Fig. 139 Installing the 2 front stabilizer bar bushings

12. Install the 2 front stabilizer bar bushings No. 1 to the outside of the bushing stopper on the front stabilizer bar.
13. Install the No. 1 left front stabilizer bracket with the 2 bolts and tighten to 20 ft. lbs. (27 Nm).
14. Install the No. 1 right front stabilizer bracket with the 2 bolts and tighten to 20 ft. lbs. (27 Nm).
15. Install the engine assembly with transaxle.
16. Install the left front stabilizer link assembly with the 2 nuts and tighten to 55 ft. lbs. (74 Nm).
17. Install the right front stabilizer link assembly with the 2 nuts and tighten to 55 ft. lbs. (74 Nm).
18. To complete installation, reverse removal procedure.
19. Inspect and adjust the front wheel alignment.

WHEEL BEARINGS

REMOVAL & INSTALLATION

See Figures 140 through 146.

1. Before servicing the vehicle, refer to the Precautions Section.
2. Remove front wheel.
3. Remove front axle hub nut.
4. Separate front speed sensor.
5. Remove the 2 bolts and separate the front disc brake caliper assembly from the steering knuckle. Use wire or an equivalent tool to keep the brake caliper from hanging down by the flexible hose.
6. Remove front disc.
7. Separate tie rod end sub-assembly.
8. Separate front suspension lower no. 1 arm.
9. Remove front axle assembly.
10. Using a screwdriver with its tip wrapped with vinyl tape, remove the No. 1 front wheel bearing dust deflector. Be careful not to damage the steering knuckle.

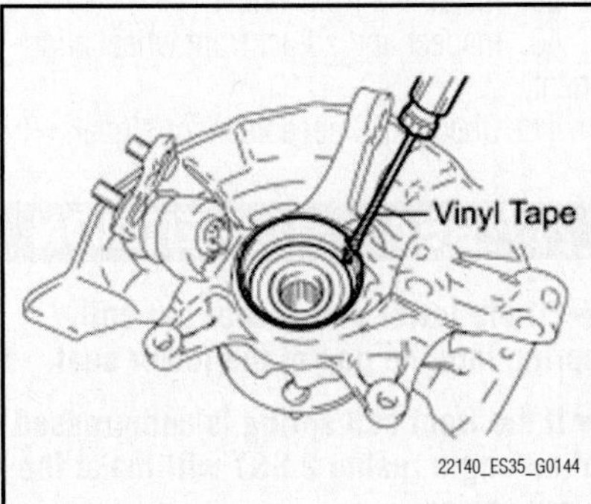

Fig. 140 remove the No. 1 front wheel bearing dust deflector

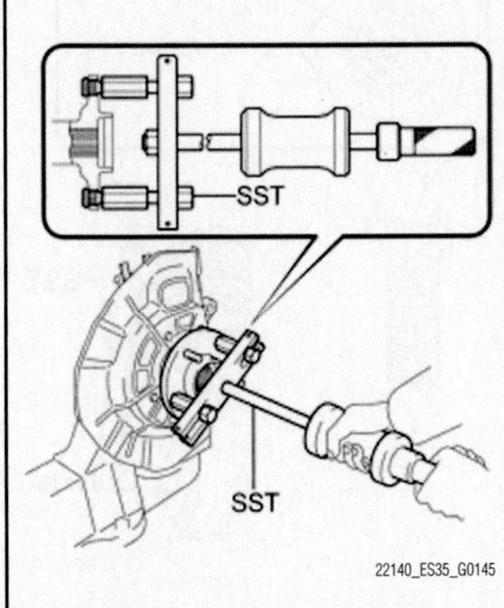

Fig. 141 Remove the front axle hub sub-assembly

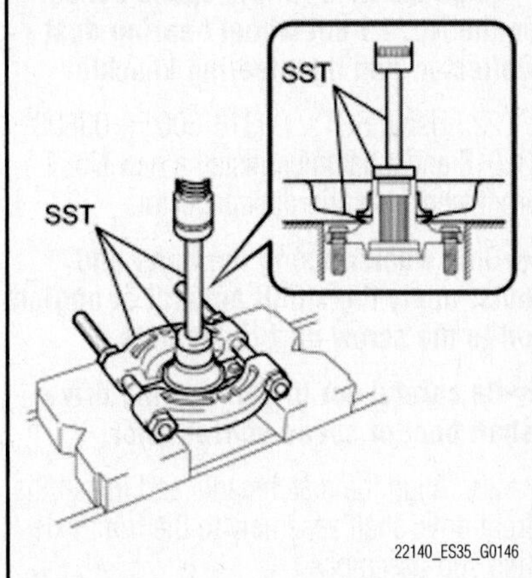

Fig. 142 Remove the bearing inner race (outside) from the front axle hub sub-assembly

11. Using snap ring pliers, remove the front axle hub hole snap ring.
12. Remove front axle hub sub-assembly by performing the following:
    a. Hold the front axle assembly between aluminum plates in a vise.

➡Do not over tighten the vise.

    b. Using SST 09520-00031, remove the front axle hub sub-assembly.

➡Be careful not to drop the front axle hub sub-assembly.

    c. Using SST 09555-55010, SST: 09950-60010 and SST: 09950-70010 and a press, remove the bearing inner race (outside) from the front axle hub sub-assembly.
13. Remove the 4 bolts and disc brake dust cover from the steering knuckle.

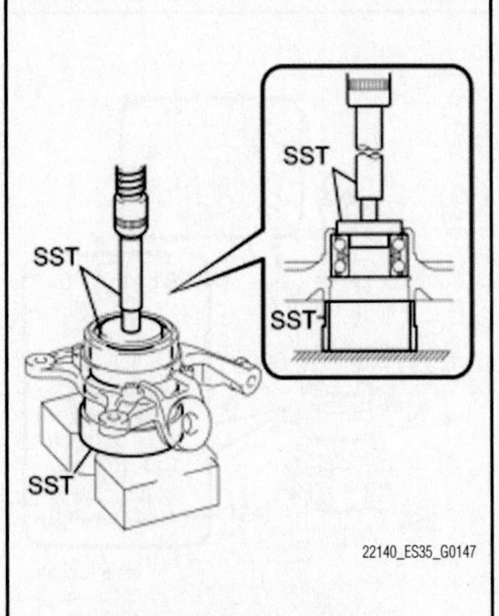

Fig. 143 Pressing the front axle hub bearing

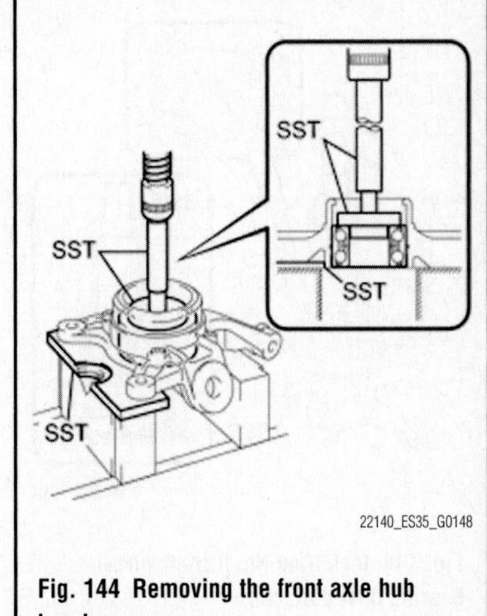

Fig. 144 Removing the front axle hub bearing

14. Remove front lower ball joint assembly.
15. Remove front axle hub bearing by performing the following:
    a. Place the bearing inner race (outside) on the front axle hub bearing.
    b. Using SST 09527-17011, SST: 09950-60010 and a press, press the front axle hub bearing until it contacts the SST: 09950-70010.
    c. Using SST: 09527-20011, SST: 09950-60010 to make the steering knuckle horizontal, fix it to the V-block.
16. Using SST : 09950-70010 and a press, remove the front axle hub bearing from the steering knuckle.

To install:

17. Using SST's: 09950-60020, 09950-70010 and a press, install a new front axle hub bearing to the steering knuckle.

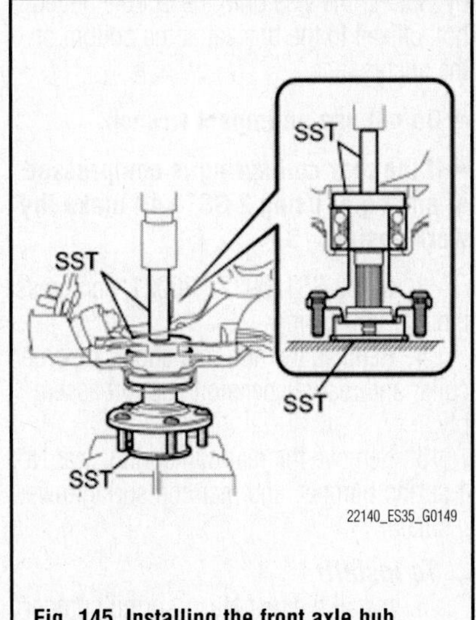

Fig. 145 Installing the front axle hub sub-assembly

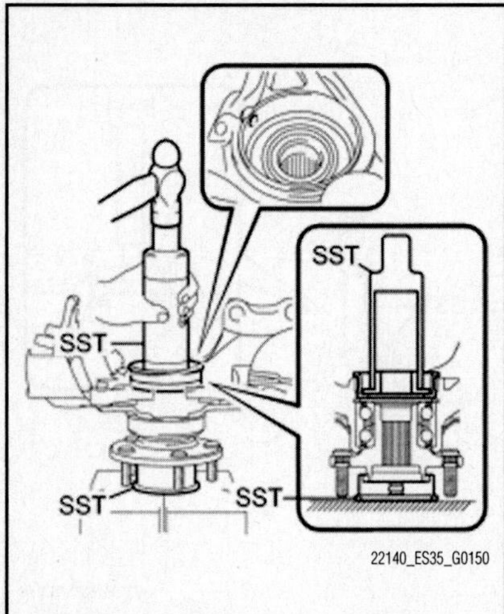

Fig. 146 Installing No. 1 front wheel bearing dust deflector

18. Install front lower ball joint assembly.

19. Install the disc brake dust cover to the steering knuckle with the 4 bolts and tighten to 73 inch lbs. (8.3 Nm).

20. Using SST's: 09608-32010, 09950-60020, 09950-70010 and a press, install the front axle hub sub-assembly.

21. Using snap ring pliers, install a new front axle hub hole snap ring.

➡Align the hole for the speed sensor in the No. 1 front wheel bearing dust deflector with the steering knuckle.

22. Using SST's: 09316-60011, 09608-32010 and a hammer, install a new No. 1 front wheel bearing dust deflector.

➡Only when reusing the bolts and nuts, apply the small amount of engine oil to the screw part of the nuts.

➡Be careful not to damage the drive shaft boot or speed sensor rotor.

23. Align the matchmarks and install the front drive shaft assembly to the front axle hub sub-assembly.

24. Install the steering knuckle with the front axle hub sub-assembly to the front shock absorber assembly with the 2 bolts and 2 nuts and tighten to 155 ft. lbs. (210 Nm).

25. Install the lower No. 1 front suspension arm sub-assembly.

26. Install the tie rod end sub-assembly.

27. Install the front disc.

28. Install the front disc brake caliper assembly with the 2 bolts to the steering knuckle and tighten to 79 ft. lbs. (107 Nm).

29. Clean the threaded parts on the drive shaft and axle hub nut using a non-residue solvent.

➡Be sure to perform this work for a new drive shaft.

➡Keep the threaded parts free of oil and foreign objects.

30. Using a 30 mm socket wrench, install the front axle hub nut and tighten to 217 ft. lbs. (294 Nm).

31. Remove the 2 bolts and separate the front disc brake caliper assembly from the steering knuckle.

32. Remove the front disc.

33. Inspect front axle hub bearing looseness.

34. Inspect front axle hub runout.

35. Install the front disc.

36. Install the front disc brake caliper assembly with the 2 bolts to the steering knuckle and tighten to 79 ft. lbs. (107 Nm).

37. Install the front speed sensor.

38. Using a chisel and hammer, stake the axle hub nut.

39. Install the front wheel.

40. Inspect and adjust front wheel alignment.

41. Check ABS speed sensor signal.

COIL SPRING

REMOVAL & INSTALLATION

See Figures 147 through 149.

1. Before servicing the vehicle, refer to the precautions section.

2. Secure the rear shock absorber with coil spring in a vise using aluminum plates by closing the vise onto the double nutted bolt affixed to the bracket at the bottom of the absorber.

➡Do not use an impact wrench.

➡If the rear coil spring is compressed at an angle, using 2 SST will make the work easier.

3. Using SST: 09727-30021, compress the rear coil spring

4. Remove the nut, rear shock absorber collar and rear suspension support assembly.

5. Remove the rear coil spring, rear No. 1 spring bumper, and rear coil spring lower insulator.

To install:

6. Install the rear No. 1 spring bumper to the piston rod.

7. Install the rear coil spring lower insulator onto the rear shock absorber.

➡Do not use an impact wrench.

8. Using SST: 09727-30021, compress the rear coil spring.

➡The smaller diameter end must face upward.

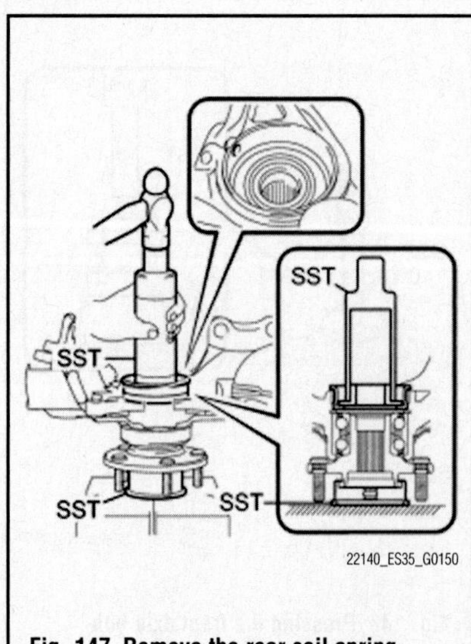

Fig. 147 Remove the rear coil spring

➡Fit the lower end of the rear coil spring into the gap of the lower seat.

➡If the front coil spring is compressed at an angle, using 2 SST will make the work easier.

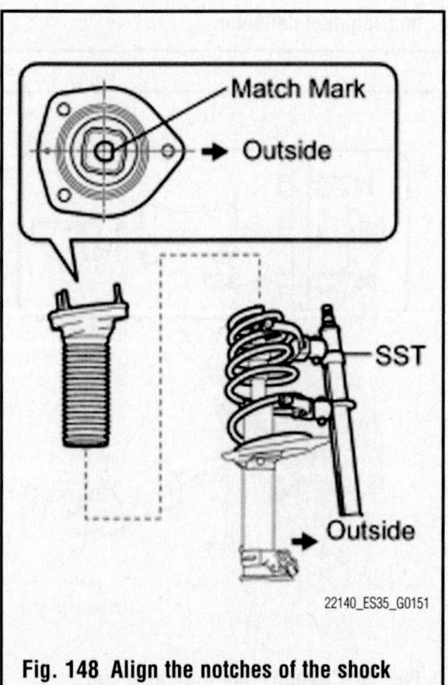

Fig. 148 Align the notches of the shock absorber

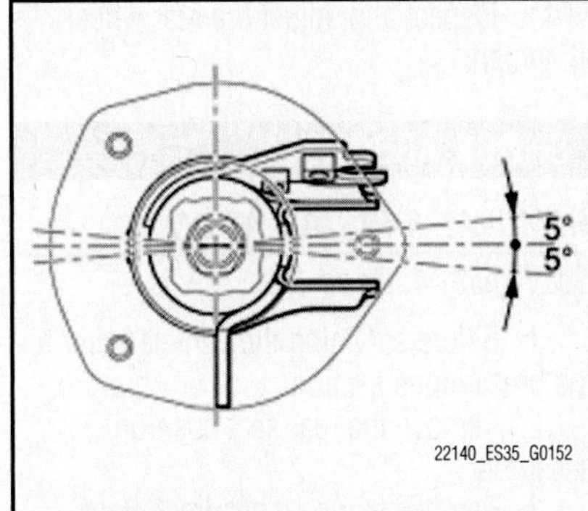

Fig. 149 Lining up the rear suspension support assembly's stud bolts

9. Install the rear coil spring to the rear shock absorber.

➡Align the notches of the piston rod and the rear suspension support assembly as shown in the illustration before installing the rear suspension support assembly.

10. Install the rear suspension support assembly.

11. Align the notches of the shock absorber with the notch of the rear suspension support assembly so that the notches face the outside of the vehicle.

12. Install the rear shock absorber collar.

13. Loosely tighten a new lock nut to the rear suspension piston rod.

➡Do not use an impact wrench.

➡When lining up the rear suspension support assembly's stud bolts at the middle point between the two sides of the bracket, the maximum permissible degree of error is plus or minus 5°.

14. Release the spring while adjusting the rear suspension support assembly to the position shown in the illustration, and remove the SST from the rear coil spring.

CONTROL ARMS/LINKS

REMOVAL & INSTALLATION

No. 1 Suspension Arm

See Figures 150 through 155.

1. Before servicing the vehicle, refer to the precautions section.

➡Check if an old gasket still remains on the pipe. If so, remove it. Also, check if any bolts or nuts are rusted. If so, replace them.

2. Remove rear wheel.
3. Remove center exhaust pipe assembly.

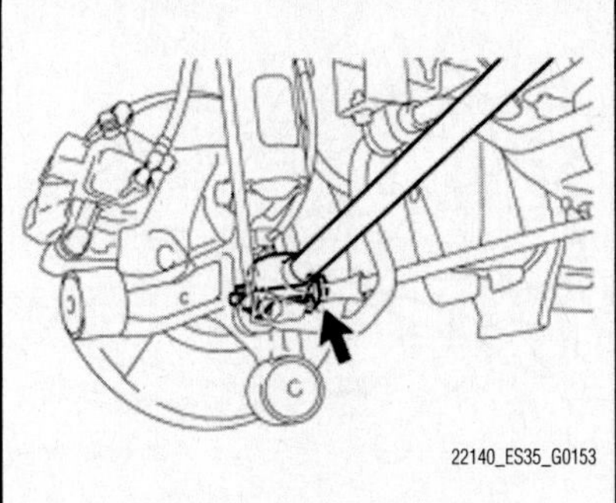

Fig. 150 Removing the bolt, nut and the rear No. 1 suspension arm—LH shown

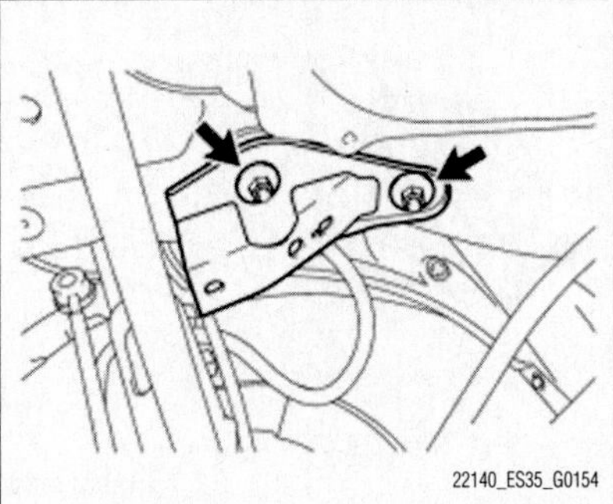

Fig. 151 Removing the 2 nuts and the LH rear suspension member lower stopper

4. Remove tail exhaust pipe assembly.
5. Separate both rear stabilizer link assemblies.
6. Remove rear stabilizer bar no. 2 and no. 1 bracket.
7. Remove rear stabilizer bar.
8. Remove rear stabilizer bushing.
9. Separate rear strut rod.

➡When removing the bolt, keep the nut from rotating.

10. Remove the bolt, nut and separate the rear suspension No. 2 arm (outer side) from the rear axle carrier.

➡When removing the bolt, keep the nut from rotating.

11. Remove the bolt, nut and the rear No. 1 suspension arm (outer side) from the rear axle carrier.
12. Remove the 2 nuts and the LH rear suspension member lower stopper.
13. Remove the 2 nuts and the RH rear suspension member lower stopper.
14. Support the rear suspension member with a jack.
15. Remove the 2 bolts, and the rear suspension member sub-assembly.
16. Remove the bolt and rear No. 1 suspension arm assembly.

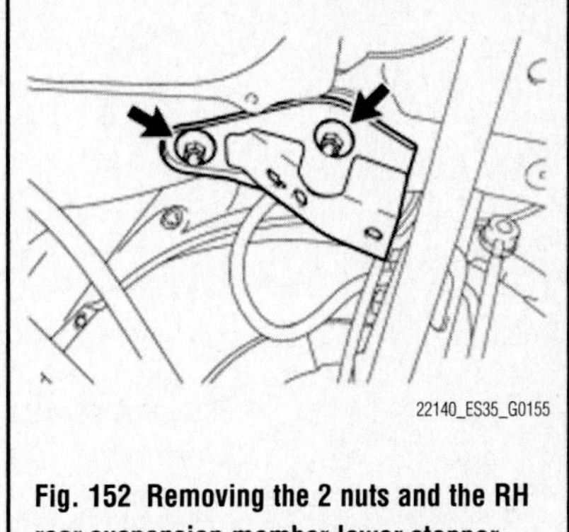

Fig. 152 Removing the 2 nuts and the RH rear suspension member lower stopper

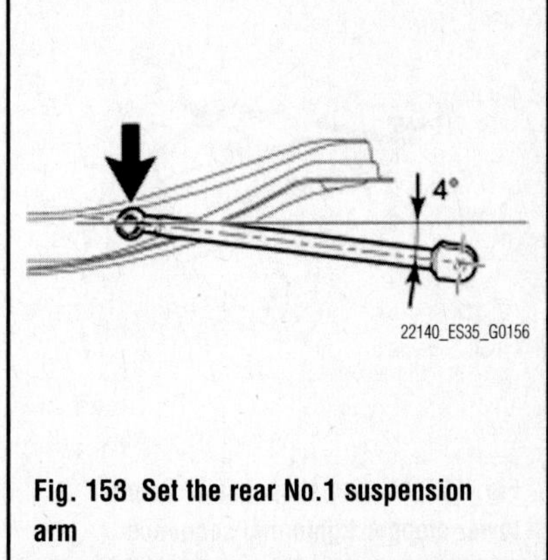

Fig. 153 Set the rear No.1 suspension arm

To install:

17. Install the No. 1 rear suspension arm (inner side) with the bolt, and temporarily tighten the bolt.
18. Install the rear No. 1 suspension arm so that the bracket leans toward the front side of the vehicle.
19. Ensure that the paint mark faces the rear side of the vehicle.
20. Set the rear No.1 suspension arm in the position shown in the illustration, and fully tighten the bolt to 74 ft. lbs. (100 Nm).
21. Raise the rear suspension member with a jack. Install the rear suspension member with the 2 bolts and tighten to 41 ft. lbs. (56 Nm).
22. Install both the rear suspension member lower stoppers with the 2 nuts and tighten to:
    a. Nut A: 41 ft. lbs. (55 Nm).
    b. Nut B: 28 ft. lbs. (38 Nm).

➡Insert the bolt from the front of the vehicle and temporarily install the bolt.

23. Connect the rear No.1 suspension arm (outer side) to the rear axle carrier with the bolt and nut and temporarily tighten the bolt and nut. When temporarily tightening the bolt, keep the nut from rotating.

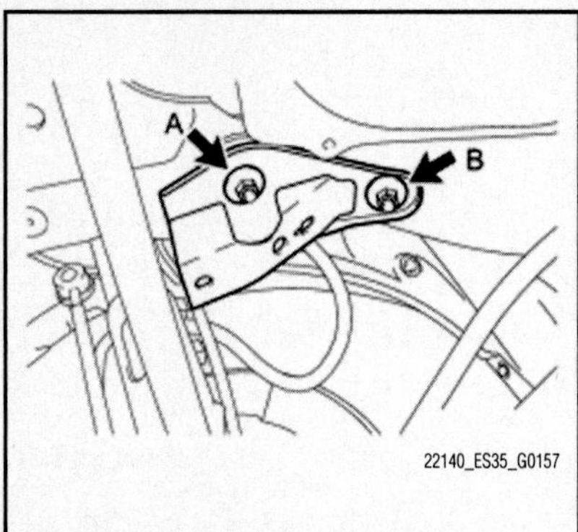

Fig. 154 LH rear suspension member lower stopper tightening sequence

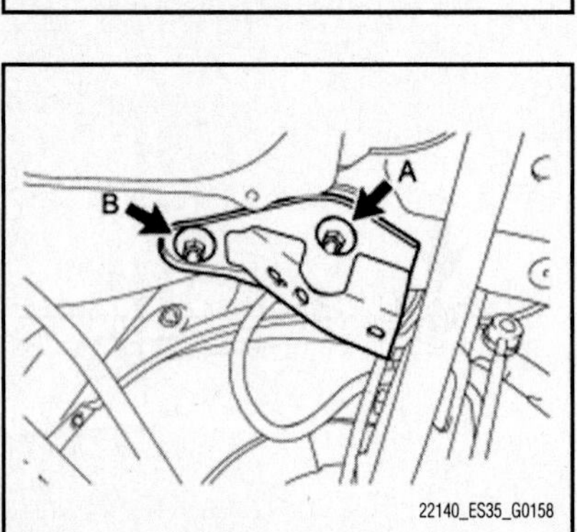

Fig. 155 LH rear suspension member lower stopper tightening sequence

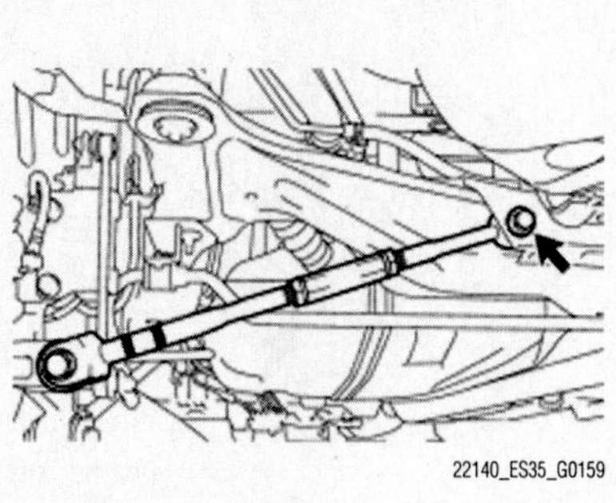

Fig. 156 Remove the bolt, and disconnect the rear No. 2 suspension arm (inner side)

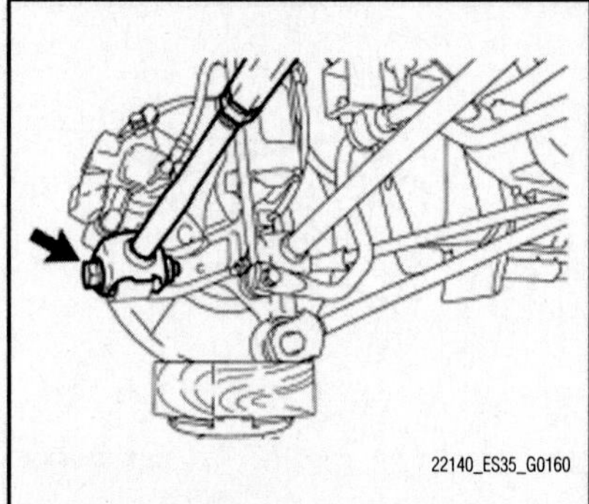

Fig. 157 Removing the bolt, nut and the rear No. 2 suspension arm (outer side)

➡Insert the bolt from the inside of the vehicle and temporarily install the bolt.

24. Connect the strut rod assembly rear to the axle carrier with the bolt and nut and temporarily tighten the bolt. When temporarily tightening the bolt, keep the nut from rotating.
25. Jack up the rear axle carrier, placing a wooden block to avoid damage. Apply load to the suspension so that the installed bolt of the rear No. 1 suspension arm (inner side) is horizontally aligned with the center of the rear axle hub.
26. Fully tighten rear No. 1 suspension arm and tighten the bolt to 74 ft. lbs. (100 Nm).
27. Fully tighten rear No. 2 suspension arm and tighten the bolt to 74 ft. lbs. (100 Nm).
28. To complete installation, reverse removal procedure.

No. 2 Suspension Arm

See Figures 156 and 157.

1. Before servicing the vehicle, refer to the precautions section.
2. Remove the rear wheel.
3. Remove the bolt, and disconnect the rear No. 2 suspension arm (inner side).

➡When removing the bolt, keep the nut from rotating.

4. Remove the bolt, nut and the rear No. 2 suspension arm (outer side) from the rear axle carrier.

To install:

➡Ensure that the paint mark faces to the rear of the vehicle.

5. Install the rear No. 2 suspension arm (inner side) with the bolt, and temporarily tighten the bolt.

➡When temporarily tightening the bolt, keep the nut from rotating.

6. Connect the rear No. 2 suspension arm (outer side) to the rear axle carrier with the bolt and nut, and temporarily tighten the bolt.
7. Stabilize suspension.
8. Fully tighten the rear No. 2 suspension arm bolt (inner side) to 74 ft. lbs. (100 Nm).
9. Fully tighten the rear No. 2 suspension arm bolt (outer side) to 74 ft. lbs. (100 Nm).
10. Install the rear wheel.
11. Inspect and adjust the rear wheel alignment.

SHOCK ABSORBER

REMOVAL & INSTALLATION

See Figures 158 and 159.

1. Before servicing the vehicle, refer to the precautions section.
2. Remove the rear seat cushion assembly.
3. Remove rear seat headrest plate cover.
4. Remove rear seat headrest assembly.
5. Remove the rear seatback assembly.
6. Remove the rear wheel.
7. Separate LH rear stabilizer link assembly.
8. Remove the 2 bolts, and disconnect the rear brake flexible hose and rear speed sensor from the rear shock absorber with coil spring and rear axle carrier.
9. Remove the 4 claws and the rear suspension support No. 1 cover.
10. Remove the rear shock absorber with coil spring, as follows:

➡Do not remove the lock nut.

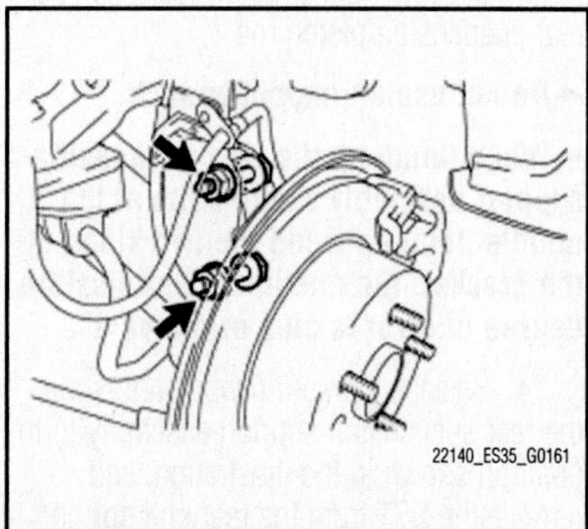

Fig. 158 Loosen the 2 nuts on the lower side of the shock absorber

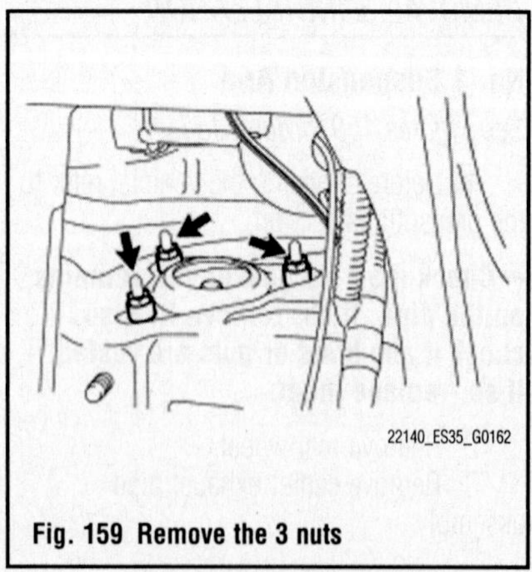

Fig. 159 Remove the 3 nuts

➡**Only loosen the nut when disassembling the rear shock absorber with coil spring.**

a. Loosen the lock nut of the rear shock absorber with coil spring.

➡**When removing the nuts, keep the bolts from rotating.**

➡**Keep one bolt inserted to secure the hub and disc rotor.**

b. Remove the 2 nuts and 2 bolts on the lower side of the rear shock absorber with coil spring.

c. Remove the 3 nuts on the upper side of the rear shock absorber with coil spring.

➡**Make sure that the rear speed sensor is disconnected from the rear shock absorber with coil spring.**

d. Lower the rear axle carrier, and remove the 2 bolts on the lower side of the rear shock absorber with coil spring.

To install:

11. Install the rear shock absorber with coil spring to the rear axle carrier assembly and insert the 2 bolts from the rear of the vehicle.
12. Slowly jack up the vehicle using a wooden block and install the rear shock absorber with coil spring (upper side) to the vehicle.
13. Install the 3 nuts to the upper side of the rear shock absorber with coil spring and tighten to 29 ft. lbs. (39 Nm).

➡**When installing the nuts, keep the bolts from rotating.**

14. Install the 2 nuts and 2 bolts to the lower side of the rear shock absorber with coil spring and tighten 133 ft. lbs. (180 Nm).
15. Fully tighten the lock nut to 41 ft. lbs. (55 Nm).
16. Connect rear speed sensor.
17. Install the LH rear stabilizer link assembly.
18. Engage the 4 claws and install the rear suspension support No. 1 cover.
19. To complete installation, reverse remaining removal.
20. Check abs speed sensor signal.
21. Inspect and adjust rear wheel alignment.

STABILIZER BAR

REMOVAL & INSTALLATION

See Figures 160 and 161.

Fig. 160 Removing the 2 bolts and No. 2 bracket

Fig. 161 Removing the 2 bolts and No. 1 bracket

1. Before servicing the vehicle, refer to the precautions section.
2. Remove rear wheels.
3. Remove tail exhaust pipe assembly.
4. Center exhaust pipe assembly.
5. Remove rear stabilizer link assembly.
6. Remove the 2 bolts and rear stabilizer bar No. 2 bracket.
7. Remove the 2 bolts and rear stabilizer bar No. 1 bracket.
8. Remove the 2 rear stabilizer bushings from the rear stabilizer bar.
9. Remove rear stabilizer bar.

To install:

10. Install the 2 rear stabilizer bushings to the outside of the stopper ring on the stabilizer bar.
11. Install the rear stabilizer bar No. 2 and No. 1 bracket.
12. Install the rear stabilizer bar with the 2 bolts and tighten to 23 ft. lbs. (31 Nm).
13. Install rear stabilizer link assembly.
14. To complete installation, reverse remaining removal.
15. Check abs speed sensor signal.
16. Inspect and adjust rear wheel alignment.

WHEEL BEARINGS

REMOVAL & INSTALLATION

See Figure 162.

➡**Use the same procedures for the RH side and LH side.**

➡**The procedures listed below are for the LH side.**

1. Before servicing the vehicle, refer to the precautions section.
2. Remove the rear wheel.
3. Separate the rear disc brake caliper assembly, as follows:
   a. Remove the bolt and separate the flexible hose from the shock absorber.
   b. Remove the 2 bolts and separate the rear disc brake caliper assembly.
4. Remove the rear disc.
5. Disconnect the skid control sensor connector.
6. Remove the 4 bolts and the rear axle hub and bearing assembly.

To install:

7. Install the hub and bearing assembly with the 4 bolts and tighten to 59 ft. lbs. (80 Nm).
8. Connect the skid control sensor connector. Do not twist the sensor wire.
9. Inspect rear axle hub bearing looseness.
10. Inspect rear axle hub runout.
11. Install the rear disc.
12. Install the rear disc brake caliper assembly, as follows:
    a. Install the rear disc brake caliper with the 2 bolts and tighten to 46 ft. lbs. (62 Nm).
    b. Install the flexible hose with the bolt and tighten to 14 ft. lbs. (19 Nm).
13. Install the rear wheel.

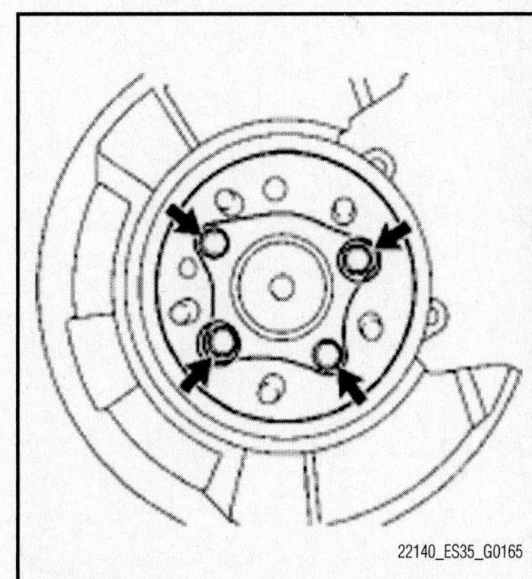

Fig. 162 Remove the 4 bolts and the rear axle hub and bearing assembly

14. Inspect and adjust the rear wheel alignment.
15. Check ABS speed sensor signal.

ADJUSTMENT

See Figure 163.

➡The rear wheel bearings are non-adjustable. If the wheel bearing is out of specifications, replace the wheel bearing.

Check the backlash in bearing shaft direction and the axle hub deviation. Maximum for backlash should be 0.0020 in. (0.05mm) and for axle hub deviation 0.020 in. (0.05mm).

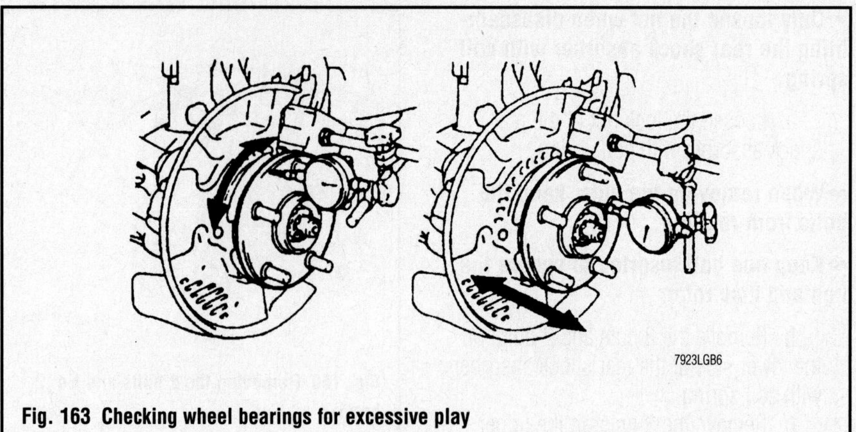

Fig. 163 Checking wheel bearings for excessive play

LEXUS

GS350 • GS460

24

SPECIFICATIONS AND MAINTENANCE CHARTS

ENGINE AND VEHICLE IDENTIFICATION

Engine							Model Year	
Code ①	Liters (cc)	Cu. In.	Cyl.	Fuel Sys.	Engine Type	Eng. Mfg.	Code ②	Year
2GR-FSE	3.5 (3456)	211	V6	SFI D4S	DOHC	Toyota	9	2009
1UR-FSE	4.6 (4608)	282	V8	SFI D4S	DOHC	Toyota	A	2010

SFI: Sequential Multi-port Fuel Injection

DOHC: Double Overhead Camshaft

① Located on the timing belt cover

② 10th digit of the VIN

3768X_GS46_C0001

GENERAL ENGINE SPECIFICATIONS

All measurements are given in inches.

Year	Model	Engine Displacement Liters (cc)	Engine ID/VIN	Fuel System Type	Net Horsepower @ rpm	Net Torque @ rpm (ft. lbs.)	Bore x Stroke (in.)	Com-pression Ratio	Oil Pressure @ rpm
2009	GS350	3.5 (3456)	2GR-FSE	SFID4S	303@6400	274@4800	3.70x3.27	11.8:1	36@2500
	GS460	4.6 (4608)	1UR-FSE	SFID4S	N/A	N/A	3.70x3.27	N/A	31@2500
2010	GS350	3.5 (3456)	2GR-FSE	SFID4S	303@6400	274@4800	3.70x3.27	11.8:1	36@2500
	GS460	4.6 (4608)	1UR-FSE	SFID4S	N/A	N/A	3.70x3.27	N/A	31@2500

SFI : Sequential Multi-port Fuel Injection

3768X_GS46_C0002

ENGINE TUNE-UP SPECIFICATIONS

Year	Engine Displacement Liters	Engine ID/VIN	Spark Plug Gap (in.)	Ignition Timing (deg.)	Fuel Pump (psi)	Idle Speed (rpm)	Valve Clearance	
							Intake	Exhaust
2009	3.5	2GR-FSE	0.039-0.043	8-18B	①	600-700	HYD	HYD
	4.6	1UR-FSE	0.039-0.043	8-12B	①	700-800	HYD	HYD
2010	3.5	2GR-FSE	0.039-0.043	8-18B	①	600-700	HYD	HYD
	4.6	1UR-FSE	0.039-0.043	8-12B	①	700-800	HYD	HYD

NOTE: The Vehicle Emission Control Information label often reflects specification changes made during production. The label figures must be used if they differ from those in this chart.

B: Before top dead center

HYD: Hydraulic Valve Lifters

① Low pressure system 28-85 psi. High pressure system 508-653 psi.

3768X_GS46_C0003

CAPACITIES

Year	Model	Engine Displacement Liters	Engine ID/VIN	Engine Oil with Filter	Transmission (qts.) Auto.	Drive Axle (pts.)	Fuel Tank (gal.)	Cooling System (qts.)
2009	GS350	3.5	2GR-FSE	①	②	③	18.7	9.6
	GS460	4.6	1UR-FSE	9.1	11.1	2.8	18.7	11.7
2010	GS350	6.5	2GR-FSE	①	②	③	18.7	9.6
	GS460	4.6	1UR-FSE	9.1	11.1	2.8	18.7	11.7

NOTE: All capacities are approximate. Add fluid gradually and check to be sure a proper fluid level is obtained.

① 2WD models: 6.6 qts.

AWD models: 6.7 qts.

② 2WD models: 8.4 qts. Overhaul

AWD models: 10.6 qts. Overhaul

③ 2WD models: 2.84 pts.

AWD models front: 1.4 pts.

AWD models rear: 2.8 pts.

3768X_GS46_C0004

FLUID SPECIFICATIONS

Year	Model	Engine Displacement Liters	Engine ID/VIN	Engine Oil	Auto. Trans. ①	Drive Axle ②	Power Steering Fluid	Brake Master Cylinder	Engine Coolant ③
2009	GS350	3.5	2GR-FSE	5W-30	ATF-WS	75W-90	NA	DOT 3	Toyota coolant
	GS460	4.6	1UR-FSE	5W-30	ATF-WS	75W-90	NA	DOT 3	Toyota coolant
2010	GS350	3.5	2GR-FSE	5W-30	ATF-WS	75W-90	NA	DOT 3	Toyota coolant
	GS460	4.6	1UR-FSE	5W-30	ATF-WS	75W-90	NA	DOT 3	Toyota coolant

NA: Not Applicable

DOT: Department Of Transpotation

① The use of genuine Toyota ATF-WS is recommended

② Synthetic GL-5 (75W-90) or equivalent

③The use of genuine Toyota engine coolant is recommended or similar ethylene glycol based non-silicate, non-amine, non- nitrite, and non- borate coolant

3768X_GS46_C0014

VALVE SPECIFICATIONS

Year	Engine Displacement Liters	Engine ID/VIN	Seat Angle (deg.)	Face Angle (deg.)	Spring Test Pressure (lbs. @ in.)	Spring Free-Length (in.)	Stem-to-Guide Clearance (in.)		Stem Diameter (in.)	
							Intake	Exhaust	Intake	Exhaust
2009	3.5	2GR-FSE	NA	NA	N/A	2.035	0.0010-0.0024	0.0012-0.0026	0.2154-0.2159	0.2151-0.2157
	4.6	1UR-FSE	45	44.5	N/A	2.035	0.0010-0.0024	0.0012-0.0026	0.2154-0.2159	0.2152-0.2157
2010	3.5	2GR-FSE	NA	NA	N/A	2.035	0.0010-0.0024	0.0012-0.0026	0.2154-0.2159	0.2151-0.2157
	4.6	1UR-FSE	45	44.5	N/A	2.035	0.0010-0.0024	0.0012-0.0026	0.2154-0.2159	0.2152-0.2157

NA: Not Available

3768X_GS46_C0005

CAMSHAFT AND BEARING SPECIFICATIONS CHART

All measurements are given in inches.

Year	Engine Displacement Liters	Engine ID/VIN	Journal Dia.	Brg. Oil Clearance	Shaft End-play	Runout	Journal Bore	Lobe Height	
								Intake	Exhaust
2009	3.5	2GR-FSE	①	②	0.0031-0.0051	0.0016	NA	1.7448-1.7487	1.7457-1.7496
	4.6	1UR-FSE	③	②	N/A	0.0016	NA	1.7438-1.7497	1.7447-1.7506
2010	3.5	2GR-FSE	①	②	0.0031-0.0051	0.0016	NA	1.7448-1.7487	1.7457-1.7496
	4.6	1UR-FSE	③	②	N/A	0.0016	NA	1.7438-1.7497	1.7447-1.7506

NA: Not Available

① Journal 1: 1.4152-1.4157
All Others: 1.0220-1.0226

② Oil clearance 1: 0.0016-0.0031
All Others: 0.0010-0.0024

③ Journal 1: 1.1794-1.1799
All Others: 1.0220-1.0226

3768X_GS46_C0013

CRANKSHAFT AND CONNECTING ROD SPECIFICATIONS

All measurements are given in inches.

	Engine		Crankshaft				Connecting Rod		
Year	Displacement Liters	Engine ID/VIN	Main Brg. Journal Dia.	Main Brg. Oil Clearance	Shaft End-play	Thrust on No.	Journal Diameter	Oil Clearance	Side Clearance
2009	3.5	2GR-FSE	2.4011-2.4016	0.0018-0.0026	0.0016-0.0094	2	2.0863-2.0866	0.0018-0.0026	0.0059-0.0157
	4.6	1UR-FSE	2.6373-2.6378	①	0.0008-0.0087	3	2.0859-2.0866	0.0010-0.0020	0.0059-0.0217
2010	3.5	2GR-FSE	2.4011-2.4016	0.0018-0.0026	0.0016-0.0094	2	2.0863-2.0866	0.0018-0.0026	0.0059-0.0157
	4.6	1UR-FSE	2.6373-2.6378	①	0.0008-0.0087	3	2.0859-2.0866	0.0010-0.0020	0.0059-0.0217

NA: Not Available

① Journal No. 1 and 5: 0.0007 - 0.0012 inch
Remaining journals: 0.0009 - 0.0015 inch

3768X_GS46_C0008

PISTON AND RING SPECIFICATIONS

All measurements are given in inches.

	Engine			Ring Gap			Ring Side Clearance		
Year	Displacement Liters	Engine ID/VIN	Piston Clearance	Top Compression	Bottom Compression	Oil Control	Top Compression	Bottom Compression	Oil Control
2009	3.5	2GR-FSE	0.0008-0.0020	0.0091-0.0130	0.0138-0.0177	0.0039-0.0157	0.0008-0.0028	0.0008-0.0024	0.0008-0.0028
	4.6	1UR-FSE	0.0014-0.0022	0.0091-0.0130	0.0138-0.0177	0.0039-0.0157	0.0008-0.0028	0.0008-0.0024	0.0008-0.0028
2010	3.5	2GR-FSE	0.0008-0.0020	0.0091-0.0130	0.0138-0.0177	0.0039-0.0157	0.0008-0.0028	0.0008-0.0024	0.0008-0.0028
	4.6	1UR-FSE	0.0014-0.0022	0.0091-0.0130	0.0138-0.0177	0.0039-0.0157	0.0008-0.0028	0.0008-0.0024	0.0008-0.0028

3768X_GS46_C0007

TORQUE SPECIFICATIONS

All readings in ft. lbs.

Year	Engine Displacement Liters	Engine ID/VIN	Cylinder Head Bolts	Main Bearing Bolts	Rod Bearing Bolts	Crankshaft Damper Bolts	Flywheel Bolts	Manifold Intake	Manifold Exhaust	Spark Plugs	Oil Pan Drain Plug
2009	3.5	2GR-FSE	①	②	③	184	61	15	30	13	30
	4.6	1UR-FSE	④	⑤	③	221	22	15	15	13	30
2010	3.5	2GR-FSE	①	②	③	184	61	15	30	13	30
	4.6	1UR-FSE	④	⑤	③	221	22	15	15	13	30

① Step 1: 27 ft. lbs.
Step 2: Tighten an additional 90 degrees
Step 3: Tighten an additional 90 degrees
14 mm bolt: 22 ft. lbs.

② Inside position: 45 ft. lbs.
Inside position Step 2: Tighten an additional 90 degrees
Cylinder block Side postion: 19 ft. lbs.

③ Step 1: 30 ft. lbs.
Step 2: Plus 90 degrees

④ Step 1: 27 ft. lbs.
Step 2: Tighten an addit
Step 3: Tighten an additional 90 degrees
12 mm bolt: 15 ft. lbs.

⑤ Inside position: 45 ft. lbs.
Outside position Step 1: 20 ft. lbs.
Outside position Step 2: Tighten an additional 90 degrees
Cylinder block Side postion: 33 ft. lbs.

3768X_GS46_C0006

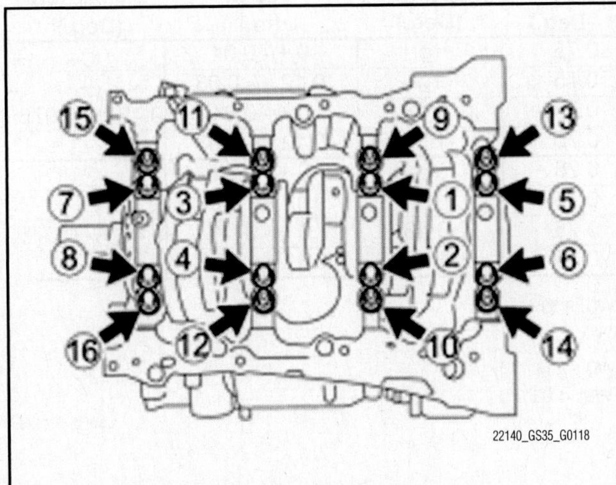

Fig. 1 Main bearing torque sequence first step—3.5L

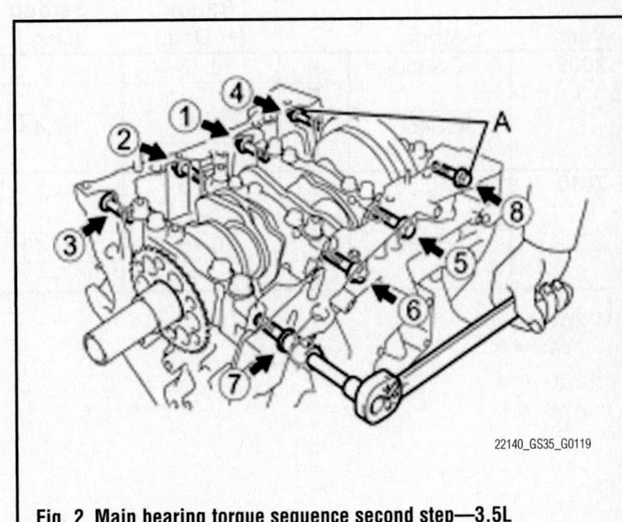

Fig. 2 Main bearing torque sequence second step—3.5L

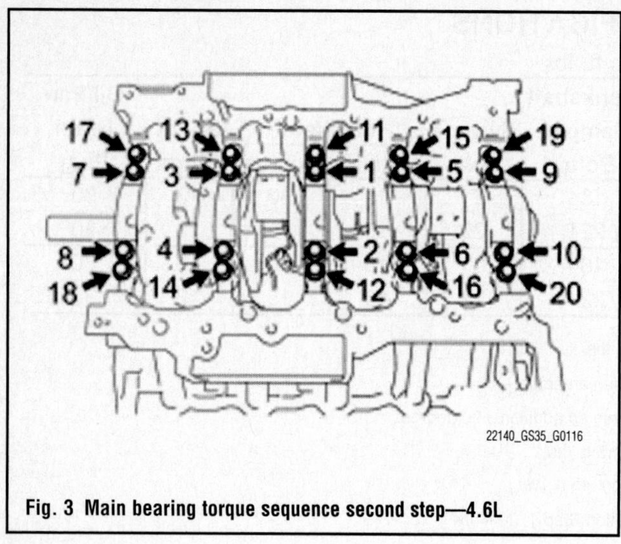

Fig. 3 Main bearing torque sequence second step—4.6L

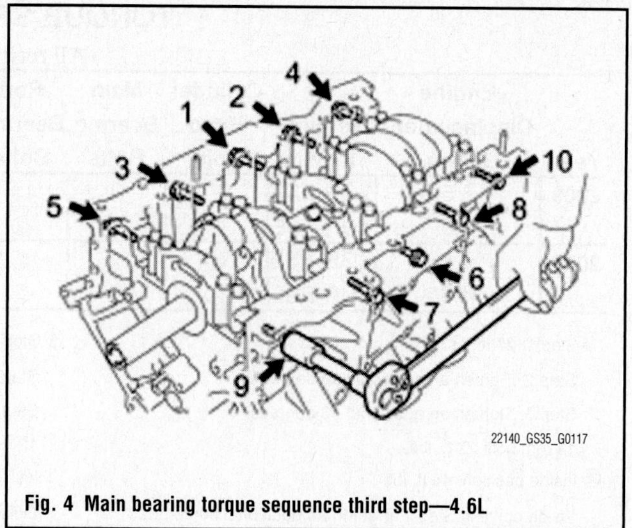

Fig. 4 Main bearing torque sequence third step—4.6L

WHEEL ALIGNMENT

Year	Model		Caster Range (+/-Deg.)	Caster Preferred Setting (Deg.)	Camber Range (+/-Deg.)	Camber Preferred Setting (Deg.)	Toe-in (in.)	Steering Axis Inclination (Deg.)
2009	GS350	F	0.75	①	0.75	②	0 +/- 0.04	③
		R	—	—	0.75	④	0.12 +/- 0.08	—
	GS460	F	0.75	+7.43	0.75	-0.43	0 +/- 0.04	9.43 +/- 0.075
		R	—	—	0.75	-1.32	0.12 +/- 0.08	—
2010	GS350	F	0.75	①	0.75	②	0 +/- 0.04	③
		R	—	—	0.75	④	0.12 +/- 0.08	—
	GS460	F	0.75	+7.43	0.75	-0.43	0 +/- 0.04	9.43 +/- 0.075
		R	—	—	0.75	-1.32	0.12 +/- 0.08	—

① 2WD: +7.38
AWD: +4.88

② 2WD: -0.38
AWD: -0.42

③ 2WD: 9.38
AWD: 11.18

④ 2WD: -1.17
AWD: -1.05

3768X_GS46_C0009

TIRE, WHEEL AND BALL JOINT SPECIFICATIONS

Year	Model	OEM Tires		Tire Pressures (psi)		Wheel	Ball Joint	Lug Nuts
		Standard	Optional	Front	Rear	Size	Inspection	(ft. lbs.)
2009	GS350/460	225/50R17 94W	245/40R18 93V	Std: 33 Opt: 33	Std: 33 Opt: 33	Std:17x 7.5-JJ Opt:18x8-JJ	U: 9-30 in. ①	76
2010	GS350/460	225/50R17 94W	245/40R18 93V	Std: 33 Opt: 33	Std: 33 Opt: 33	Std:17x 7.5-JJ Opt:18x8-JJ	U: 9-30 in. ①	76

OEM: Original Equipment Manufacturer

PSI: Pounds Per Square Inch

STD: Standard

OPT: Optional

U: Upper

① Torque required in inch lbs. to rotate ball joint when removed from the knuckle

3768X_GS46_C0010

BRAKE SPECIFICATIONS

All measurements in inches unless noted

Year	Model	Front Brake Disc			Rear Brake Disc			Minimum	Brake Caliper	
									Bracket Bolts	Mounting Bolts
		Original Thickness	Minimum Thickness	Maximum Run-out	Original Thickness	Minimum Thickness	Maximum Run-out	Lining Thickness	(ft. lbs.)	(ft. lbs.)
2009	GS350	1.181	1.063	0.0020	0.709	0.650	0.0020	0.0390	①	②
	GS460	1.181	1.063	0.0020	0.709	0.650	0.0020	0.0390	①	②
2010	GS350	1.181	1.063	0.0020	0.709	0.650	0.0020	0.0390	①	②
	GS460	1.181	1.063	0.0020	0.709	0.650	0.0020	0.0390	①	②

① Front: 58 ft. lbs.
Rear: 40 ft. lbs.

② Front: 58 ft. lbs.
Rear: 18 ft. lbs.

3768X_GS46_C0011

SCHEDULED MAINTENANCE INTERVALS

Lexus—GS350 and GS460

TO BE SERVICED	TYPE OF SERVICE	VEHICLE MILEAGE INTERVAL (x1000)													
		5	10	15	20	25	30	35	40	45	50	55	60	90	120
Engine oil & filter	R	✓	✓	✓	✓	✓	✓	✓	✓	✓	✓	✓	✓	✓	✓
Automatic transmission fluid	S/I			✓			✓			✓			✓	✓	✓
Ball joints & dust covers	S/I			✓			✓			✓			✓	✓	✓
Bolts & nuts on chassis & body	S/I			✓			✓			✓			✓	✓	✓
Brake linings & drums	S/I	✓	✓	✓	✓	✓	✓	✓	✓	✓	✓	✓	✓	✓	✓
Brake line pipes & hoses	S/I			✓			✓			✓			✓	✓	✓
Brake pads & discs (front & rear)	S/I	✓	✓	✓	✓	✓	✓	✓	✓	✓	✓	✓	✓	✓	✓
Brake fluid	R						✓						✓	✓	✓
Rack and pinion assembly	S/I			✓			✓			✓			✓	✓	✓
Steering linkage & boots	S/I			✓			✓			✓			✓	✓	✓
Air cleaner filter	R						✓						✓	✓	✓
Spark plugs ①	R														✓
Drive belts	S/I												✓	✓	✓
Exhaust pipes & mountings	S/I			✓			✓			✓			✓	✓	✓
Fuel lines & connections	S/I						✓						✓	✓	✓
Engine coolant ②	S/I			✓			✓			✓			✓	✓	
Rear differential & transfer case oil	S/I			✓			✓			✓			✓	✓	✓
Fuel tank cap gasket	S/I						✓						✓	✓	✓
Rotate tires	S/I			✓			✓			✓			✓		✓
Clean air conditioning filter ③	S/I			✓			✓			✓			✓		✓
Axle shaft bolts	S/I			✓			✓			✓			✓	✓	✓
Brake pad thickness and rotor runout	S/I						✓						✓	✓	✓

R: Replace S/I: Service or Inspect

① Spark plugs are replaced at 120,000 miles

② Replace engine coolant at 100,000 miles and then inspect every 15,000 miles

③ Replace air conditioning filter every 30,000 miles

FREQUENT OPERATION MAINTENANCE (SEVERE SERVICE)

If a vehicle is operated under any of the following conditions it is considered severe service:

- Extremely dusty areas.

- 50% or more of the vehicle operation is in 32°C (90°F) or higher temperatures, or constant temperatures below 0°C (32°F).

- Prolonged idling (vehicle operation in stop and go traffic).

- Frequent short running periods (engine does not warm to normal operating temperatures).

- Police, taxi, delivery usage or trailer towing usage.

Air cleaner filter: service or inspect every 5000 miles

Rear differential & transfer case oil: replace every 15,000 miles.

Ball joints & dust covers: service or inspect every 5000 miles.

Bolts & nuts on chassis & body: service or inspect every 5000 miles.

Axle shaft bolts: service or inspect every 5000 miles.

Steering linkage: service or inspect every 5000 miles.

Air filter: service or inspect every 15,000 miles.

3768X_GS46_C0012

BRAKES — INFORMATION AND PRECAUTIONS

ANTI-LOCK SYSTEMS

- Certain components within the ABS system are not intended to be serviced or repaired individually.
- Do not use rubber hoses or other parts not specifically specified for and ABS system. When using repair kits, replace all parts included in the kit. Partial or incorrect repair may lead to functional problems and require the replacement of components.
- Lubricate rubber parts with clean, fresh brake fluid to ease assembly. Do not use shop air to clean parts; damage to rubber components may result.
- Use only DOT 3 brake fluid from an unopened container.
- If any hydraulic component or line is removed or replaced, it may be necessary to bleed the entire system.
- A clean repair area is essential. Always clean the reservoir and cap thoroughly before removing the cap. The slightest amount of dirt in the fluid may plug an orifice and impair the system function. Perform repairs after components have been thoroughly cleaned; use only denatured alcohol to clean components. Do not allow ABS components to come into contact with any substance containing mineral oil; this includes used shop rags.
- The Anti-Lock control unit is a microprocessor similar to other computer units in the vehicle. Ensure that the ignition switch is **OFF** before removing or installing controller harnesses. Avoid static electricity discharge at or near the controller.
- If any arc welding is to be done on the vehicle, the control unit should be unplugged before welding operations begin.

DISC AND DRUM SYSTEMS

CAUTION

Dust and dirt accumulating on brake parts during normal use may contain asbestos fibers from production or aftermarket brake linings. Breathing excessive concentrations of asbestos fibers can cause serious bodily harm. Exercise care when servicing brake parts. Do not sand or grind brake lining unless equipment used is designed to contain the dust residue. Do not clean brake parts with compressed air or by dry brushing. Cleaning should be done by dampening the brake components with a fine mist of water, then wiping the brake components clean with a dampened cloth. Dispose of cloth and all residue containing asbestos fibers in an impermeable container with the appropriate label. Follow practices prescribed by the Occupational Safety and Health Administration (OSHA) and the Environmental Protection Agency (EPA) for the handling, processing, and disposing of dust or debris that may contain asbestos fibers.

BRAKES — BLEEDING THE BRAKE SYSTEM

BLEEDING PROCEDURE

BLEEDING & FILLING PROCEDURE

3.5L Engines

➡If any work is performed on the brake system or if air in the brake lines is suspected, bleed the air from the brake system.

➡Note the following:

- Move the shift lever to the P position and apply the parking brake before bleeding the brakes.
- Add brake fluid to keep the level between MIN and MAX lines of the reservoir while bleeding the brakes.
- If brake fluid leaks onto any painted surface, wash or otherwise remove it completely.

1. Top off the reservoir using fluid SAE J1703 or FMVSS No. 116 DOT3.

➡If the master cylinder is reinstall or if the reservoir becomes empty, bleed the air from the master cylinder. To avoid brake fluid from adhering, cover the painted surface with a shop rag or piece of cloth.

2. Bleed master cylinder:
   a. Disconnect the 2 brake lines from the master cylinder.
   b. Slowly depress the brake pedal and hold it.
   c. Cover the 2 outer holes with fingers, and release the brake pedal.
   d. Repeat (b) and (c) 3 or 4 times.
   e. Using SST 09023-00101, connect the brake lines to the master cylinder. Tighten 11 ft. lbs. (15 Nm).

➡Bleed the air from the wheel furthest from the master cylinder.

3. Bleed brake line:
   a. Connect the vinyl tube to the bleeder plug.
   b. Depress the brake pedal several times, then loosen the bleeder plug with the pedal depressed.
   c. When fluid stops coming out, tighten the bleeder plug, then release the brake pedal.
   d. Repeat (b) and (c) until all the air in the fluid is completely bled out.
   e. Using SST, tighten the bleeder plug completely. Tighten to 8 ft. lbs. (11 Nm).
   f. Repeat the above procedures for each wheel to bleed the air from the brake line.

➡After bleeding the air from the brake system, if the height or feel of the brake pedal cannot be obtained, bleed the air from the ABS and TRACTION actuator assembly with the Techstream by following the procedures below.

4. Bleed ABS and traction actuator assembly:
   a. Depress the brake pedal more than 20 times with the engine switch off.
   b. Connect the Techstream to the DLC3, then turn the engine switch on (IG).

➡Do not start the engine.

   c. Turn the Techstream on and select "AIR BLEEDING" on the screen.

➡Note the following:

- Refer to the Techstream operator's manual for further details.
- Bleed the air by following the steps displayed on the Techstream.

   d. Bleed the air according to "Step 1: Increase" on the Techstream display.

➡Make sure that the master cylinder reservoir tank does not become empty of brake fluid.

- Connect the vinyl tube to either one of the bleeder plugs.
- Depress the brake pedal several times, then loosen the bleeder plug connected to the vinyl tube with the pedal depressed.

- When fluid stops coming out, tighten the bleeder plug, then release the brake pedal.
- Repeat (2) and (3) until all the air in the fluid is completely bled out.
- Using SST, tighten the bleeder plug completely. Tighten to 8 ft. lbs. (11 Nm).
- For the rest of the wheels, bleed the air in the same way as stated in the above procedures.

e. Bleed the air from the pressure reduction line according to "Step 3: Decrease" on the Techstream display.

➡Note the following:

- Bleed the pressure reduction line by following the steps displayed on the Techstream.
- Make sure that the master cylinder reservoir tank does not become empty of brake fluid.
- Connect a vinyl tube to either one of the bleeder plugs.
- Loosen the bleeder plug.
- Using the Techstream, operate the ABS and TRACTION actuator assembly, completely depress the brake pedal and hold it.
- The operation stops automatically in 4 seconds. When performing this procedure continuously, an interval of at least 20 seconds is required.
- When the operation is completed, the brake pedal slightly goes down. This is a normal phenomenon when the solenoid opens.
- During this procedure, the pedal seems heavy, but completely depress it so that the brake fluid comes out from the bleeder plug.
- Be sure to keep the brake pedal depressed. Never depress and release the pedal repeatedly.
- Tighten the bleeder plug, then release the brake pedal.
- Repeat steps "2" to "4" until all the air in the fluid is completely bled out.
- Using SST, tighten the bleeder plug completely. Tighten to 8 ft. lbs. (11 Nm).
- Repeat the above procedures for the rest of the brakes to bleed the air from the brake line.
- Bleed the air by following the steps displayed on the Techstream.
- Make sure that the master cylinder reservoir tank does not become empty of brake fluid.

f. Bleed the air from the brake line again according to "Step 4: Increase" on the Techstream display.

- Connect a vinyl tube to either one of the bleeder plugs.
- Depress the brake pedal several times, then loosen the bleeder plug connected to the vinyl tube with the pedal depressed.
- When fluid stops coming out, tighten the bleeder plug, then release the brake pedal.
- Repeat (2) and (3) until all the air in the fluid is completely bled out.
- Using SST, tighten the bleeder plug completely. Tighten to 8 ft. lbs. (11 Nm).
- Repeat the above procedures for each brake to bleed the air from the brake line.

g. Finish "AIR BLEEDING" on the Techstream and turn off the power.

- Disconnect the Techstream from the DLC3.
- Turn the engine switch off.

5. Check the fluid level and add fluid if necessary. Fluid: SAE J1703 or FMVSS No. 116 DOT3.

6. If fluid leaks, tighten or replace the leaking part.

4.6L Engines

⁂ CAUTION

Bleeding without the Techstream may result in air remaining in the brake hydraulic system. As this can cause an accident, be sure to use the Techstream for air bleeding.

- Move the shift lever to the P position and apply the parking brake before bleeding the brakes.
- Add brake fluid to keep the level between MIN and MAX lines of the reservoir while bleeding the brakes.
- If the pump motor operates while air remains inside the brake reservoir tube no.1 hose, the air will enter the actuator, resulting in difficulty in bleeding.
- Keep the 2 ABS motor relays removed until instructed to reinstall them to prevent air from entering the brake reservoir tube no.1 hose.
- The actuator pump motor and solenoid can be operated by the driver even if the engine switch is off.
- Although a buzzer may sound due to declined accumulator pressure during bleeding, keep on bleeding.
- DTCs indicating a malfunction in the ABS motor relay or pressure sensor are stored after bleeding. Clear the DTCs when instructed during or after bleeding.

1. Disable brake control:

a. Move the shift lever to the P position and apply the parking brake.

b. Connect the Techstream to the DLC3 with the engine switch off.

c. Remove the 2 ABS motor relays with the engine room relays from No.3 block.

d. Turn the engine switch on (IG).

➡Do not start the engine.

e. Turn the Techstream on and enter the following menus: Chassis / ABS/VSC/TRC / Utility / Electronically Controlled Brake system Utility / Electronically Controlled Brake system Invalid.

➡If brake fluid leaks onto any painted surface, wash or otherwise remove it completely. Do not place the fluid can on the reservoir inlet because brake fluid may overflow.

2. Add reservoir with brake fluid:

a. Add brake fluid into the reservoir. Fluid: SAE J1703 or FMVSS No. 116 DOT3.

3. Bleed brake reservoir tube no.1 hose:

a. Connect SST 09992-00242, 09992-00350 to the reservoir with the brake reservoir pressure adapter.

b. Connect the vinyl tube to the bleeder plug of the actuator.

c. Loosen the bleeder plug of the actuator.

d. Use the SST to boost the pressure in the reservoir. Standard: 7.3 to 11.6 psi (50 to 80 kPa).

e. Drain approximately 100 cc of fluid.

f. Tighten the bleeder plug and boost the pressure in the reservoir again. Then loosen the bleeder plug to bleed air.

➡Repeat the above procedures 5 times or more.

g. When air is completely bled from the hose between the reservoir and the actuator, tighten the bleeder plug. Tighten to 74 inch lbs. (8.3 Nm).

➡If the master cylinder has been disassembled or if the reservoir becomes empty, bleed the air from the master cylinder.

4. Bleed master cylinder:

a. Disconnect the brake lines from the master cylinder.

b. Slowly depress and hold the brake pedal.

c. Cover the outer holes with fingers, and release the brake pedal.

d. Repeat (b) and (c) 3 or 4 times.

e. Connect the brake lines to the master cylinder. Tighten to 11 ft. lbs. (15 Nm).

➡**Bleed the air from the wheel furthest from the master cylinder.**

5. Bleed front brake system:

a. Connect the vinyl tube to the bleeder plug.

b. Depress the brake pedal several times, then loosen the bleeder plug with the pedal depressed.

c. When fluid stops coming out, tighten the bleeder plug, then release the brake pedal.

d. Repeat (b) and (c) until all the air in the fluid is completely bled out.

e. Using SST 09023-00101, tighten the bleeder plug completely. Tighten to 8 ft. lbs. (11 Nm).

f. Repeat the above procedures for other wheel to bleed the air from the brake line.

➡**Bleed the air by following the steps displayed on the Techstream.**

6. Bleed rear brake system:

a. Turn the engine switch off.

b. Install the 2 ABS motor relays to the engine room relay block No.3.

➡**Install ABS motor relays before bleeding the air from the rear brake system.**

c. Turn the engine switch on (IG) and turn the Techstream on.

d. Cancel "DISABLE BRAKE CONTROL" on the Techstream.

➡**If the brake control has been disabled brake control.**

e. Clear the DTC.

f. Turn the Techstream on and enter the following menus: Chassis / ABS/VSC/TRC / Utility / Electronically Controlled Brake system Utility / Electronically Controlled Brake system Invalid.

g. With the brake pedal depressed, bleed the bleeder plug on the rear disc brake caliper LH.

➡**Keep the fluid inside the reservoir above the LOW level by replenishing.**

➡**Depress and hold the brake pedal. After the solenoid operates for approximately 30 seconds, release the brake pedal to stop the solenoid. Repeat the procedures until air is completely bled from the rear brake system. The brake warning light comes on and the buzzer sound while bleeding, but they do not indicate a malfunction.**

h. Tighten the bleeder plug after bleeding. Tighten to 8 ft. lbs. (11 Nm).

i. With the brake pedal depressed, bleed the bleeder plug on the rear disc brake caliper RH.

➡ **Keep the fluid inside the reservoir above the MIN level by replenishing.**

➡**Depress and hold the brake pedal. After the solenoid operates for approximately 30 seconds, release the brake pedal to stop the solenoid. Repeat the procedures until air is completely bled from the rear brake system. The brake warning light comes on and the buzzer sound while bleeding, but they do not indicate a malfunction.**

j. Tighten the bleeder plug after bleeding. Tighten to 8 ft. lbs. (11 Nm).

k. Cancel "DISABLE BRAKE CONTROL" on the Techstream.

⁂ CAUTION

Be sure to perform this procedure before removal of the actuator.

➡ **Perform accumulator zero down by following the steps displayed on the Techstream.**

7. Perform accumulator zero down:

a. Drain the brake fluid in the brake fluid reservoir tank near the MIN line.

b. Connect the Techstream to the DLC3 with the engine switch off.

c. Turn the Techstream on and repeat the following steps 5 times:

- Turn the engine switch on (IG).
- Turn the Techstream on and enter the following menus: Chassis / ABS/VSC/TRC / Utility / Electronically Controlled Brake system Utility / Zero Down.
- When the buzzer sounds, turn the engine switch off.

➡ **Keep the fluid inside the reservoir above the MIN level by replenishing.**

➡ **Accumulator pressure is released and accumulated repeatedly, which circulates the fluid inside the accumulator zero down (accumulator depressurizing). The pump motor rotates and accumulator is pressurized every time the engine switch is turned from off to on (IG).**

8. Adjust the fluid level to the MAX line with the engine switch on (IG).

9. Clear the DTC.

10. When the brake actuator assembly is replaced, perform linear valve offset learning.

LINEAR VALVE OFFSET LEARNING

Perform initialization of linear solenoid valve and calibration when the skid control ECU, brake actuator or brake pedal stroke sensor is replaced. Follow the procedure to perform initialization.

- If there is a problem with battery (12 V) voltage, initialization of linear solenoid valve and calibration cannot be completed normally. Check the battery voltage before performing initialization of linear solenoid valve and calibration.
- If the actuator's temperature is high, initialization of linear solenoid valve and calibration may not be completed normally. If so, wait until the temperature drops and then perform initialization of linear solenoid valve and calibration.
- If the engine switch is turned off, the brake pedal is operated or vehicle speed is input while the linear solenoid valve offset learning is being performed, the learning will be canceled.

1. Clear stored value of initialization of linear solenoid valve and calibration (when using Techstream):

a. Connect the Techstream to the DLC3.

b. Move the shift lever to the P position.

c. Turn the engine switch on (IG) with the brake pedal released.

d. Perform the Reset Memory function under the ABS/VSC/TRAC menu.

e. Perform initialization of linear solenoid valve.

f. Perform zero point calibration of yaw rate and acceleration sensor.

2. Perform initialization of linear solenoid valve and calibration (when using Techstream):

a. Connect the Techstream to the DLC3.

b. Move the shift lever to the P position.

c. Turn the engine switch on (IG) with the brake pedal released.

➡ **If the linear solenoid valve offset learning is performed without turning the engine switch on (IG), the learning process may not be completed properly because of insufficient battery voltage. When the linear solenoid valve offset learning is interrupted, or the learning process is**

performed with the shift lever not in the P position, DTC C1345/66 will be stored.

d. Set the Techstream to Test Mode (select "Electronically Control Brake Utility").

e. Leave the vehicle stationary without depressing the brake pedal for 1 or 2 minutes.

f. Check that the interval between blinks of the brake control warning light changes from 1 second to 0.25 seconds.

➡ The time needed to complete initialization of linear solenoid valve and calibration varies depending on battery voltage. The brake control warning light blinks at 1 second intervals during the initialization of linear solenoid valve and calibration. The brake control warning light blinks at 0.25 seconds intervals if the Test Mode is normal.

g. When the brake control warning light changes to the Test Mode display, check that DTC C1345/66, which indicates trouble with stroke sensor zero point learning, is not output.

h. Enter the normal mode from the Test Mode by following the Techstream prompts.

3. Clear stored value of initialization of linear solenoid valve and calibration (when using sst check wire):

a. Move the shift lever to the P position.

b. Turn the engine switch on (IG) with the brake pedal released.

c. Using SST 09843-18040, connect and disconnect terminals TS and CG of the DLC3 4 times or more within 8 seconds.

d. Check that no codes other than ABS code 42, VSC code 45 and Electronically Controlled Brake System code 48, 66, or 95 are stored in the diagnostic system.

➡ The ABS warning, brake control warning and SLIP indicator lights do not indicate the normal system code.

e. Remove SST from the terminals of the DLC3.

4. PERFORM INITIALIZATION OF LINEAR SOLENOID VALVE AND CALIBRATION (WHEN USING SST CHECK WIRE):

a. Turn the engine switch off.

b. Using SST 09843-18040, connect terminals TS and CG of the DLC3.

c. Move the shift lever to the P position.

d. Turn the engine switch on (IG) with the brake pedal released.

➡ If the linear solenoid valve offset learning is performed without turning the engine switch on (IG), the learning process may not be completed properly because of insufficient battery voltage. When the linear solenoid valve offset learning is interrupted, or the learning process is performed with the shift lever not in the P position, DTC C1345/66 will be stored.

e. Leave the vehicle stationary without depressing the brake pedal for 1 or 2 minutes.

f. Check that the interval between blinks of the brake control warning light changes from 1 second to 0.25 seconds.

➡ The time needed to complete initialization of linear solenoid valve and calibration varies depending on the battery voltage. The brake control warning light blinks at 1 second intervals during the initialization of linear solenoid valve and calibration and changes to the Test Mode display. The brake control warning light blinks at 0.25 seconds intervals if the Test Mode is normal.

g. When the brake control warning light changes to the Test Mode display, check that DTC C1345/66, which indicates trouble with stroke sensor zero point learning is not output.

h. Turn the engine switch off and disconnect SST from the DLC3.

BLEEDING THE ABS SYSTEM

➡After performing the usual air bleeding in the brake system, if the height or feel of the brake pedal cannot be obtained, perform air bleeding in the brake actuator assembly with a hand–held tester by following procedures below. Make sure that the brake fluid in the master cylinder reservoir tank does not become empty.

1. Depress the brake pedal more than 20 times with the engine off.
2. Connect the hand–held tester to the DLC3, then turn the ignition switch to the ON position.
3. Do not start the engine.
4. Select "AIR BLEEDING" on the hand–held tester. Please refer to the Hand–Held Tester Operator's Manual for further details.
5. Bleed the air out of the regular brake line in "Step1: Increase" on the hand–held tester display. Perform the air bleeding by following the steps displayed on the hand–held tester. Make sure that the brake fluid in the master cylinder reservoir tank does not become empty.
6. Connect the vinyl tube to either one of the bleeder plugs.
7. Depress the brake pedal several times, then loosen the bleeder plug of one of the above wheels with the pedal depressed.
8. When fluid stops coming out, tighten the bleeder plug, then release the brake pedal.
9. Repeat (2) and (3) until all air in the fluid is completely bled out.
10. Tighten the bleeder plug to 73 inch lbs. (8.3 Nm).
11. Repeat the above procedure to bleed the air out of the brake line for each wheel.
12. Bleed the air out of the suction line in "Step2: Inhalation" on the hand–held tester display.
13. Connect the vinyl tube to the bleeder plug at the right front wheel or the right rear wheel and loosen the bleeder plug.
14. Operate the brake actuator assembly using the hand–held tester to bleed the air.

➡The operation stops automatically in 4 seconds. At this time, be sure to release the brake pedal.

15. Check that the operation has stopped, by referring to the hand–held tester display.
16. Repeat (2) and (3) until all the air in the fluid is completely bled out.
17. Tighten the bleeder plug.
18. For the rest of the wheels, bleed the air in the same way as stated in the above procedure.
19. Bleed the air out of the pressure reduction line in "Step 3: Decrease" on the hand–held tester display.
20. Connect a vinyl tube to either one of the bleeder plugs.
21. Loosen the bleeder plug.
22. Using the hand–held tester, operate the brake actuator assembly using hand–held tester, completely depress the brake pedal and keep it.

➡The operation stops automatically in 4 seconds. When performing this procedure continuously, an interval of at least 20 seconds is required. When the operation is completed, the brake pedal slightly goes down. This is a normal phenomenon caused when the solenoid opens. During this procedure, the pedal seems heavy, but completely depress it so that the brake fluid comes out from the bleeder plug. Be sure to keep depressing the brake pedal. Never depress and release the pedal repeatedly.

23. Tighten the bleeder plug, then release the brake pedal.
24. Repeat 3 previous steps until all the air in the fluid is completely bled out.
25. Tighten the bleeder plug.
26. Repeat the above procedure to bleed the air out of the brake line for each wheel.
27. Bleed the air out of the regular brake line again in "Step 4: Increase" on the hand–held tester display.
28. Connect the vinyl tube to either one of the bleeder plug.
29. Depress the brake pedal several times, then loosen the bleeder plug of one of the above wheels with the pedal depressed.
30. When fluid stops coming out, tighten the bleeder plug, then release the brake pedal.
31. Repeat the previous 2 steps until all the air in the fluid is completely bled out.
32. Tighten the bleeder plug.
33. Repeat the above procedure to bleed the air out of the brake line for each wheel.

BRAKES — ANTI-LOCK BRAKE SYSTEM (ABS)

WHEEL SPEED SENSORS

REMOVAL & INSTALLATION

Front Wheel

2WD Models

See Figure 5.

1. Disconnect the speed sensor connector from the speed sensor.

➡Prevent foreign matter from adhering to the speed sensor. Be careful not to damage the speed sensor.

2. Remove front disc brake caliper assembly.
3. Remove front axle hub sub-assembly.

➡Replace the hub and bearing assembly if it is dropped or receives a strong shock.

4. Remove the wheel hub and bearing assembly and speed sensor.

➡Note the following:

- Keep the sensor away from magnets.
- Pull the skid control sensor off straight, taking care not to make contact with the skid control sensor rotor.
- If the speed sensor rotor is damaged or deformed, replace the hub and bearing assembly.
- Do not scratch the area where the skid control sensor contacts the hub and bearing assembly.
- Do not attach foreign matter to the speed sensor rotor.

5. To install, reverse removal procedure.

4WD Vehicles

1. Raise and support the vehicle.
2. Remove the appropriate wheel.
3. Disconnect the speed sensor front connector.
4. Remove the bolt and speed sensor front.

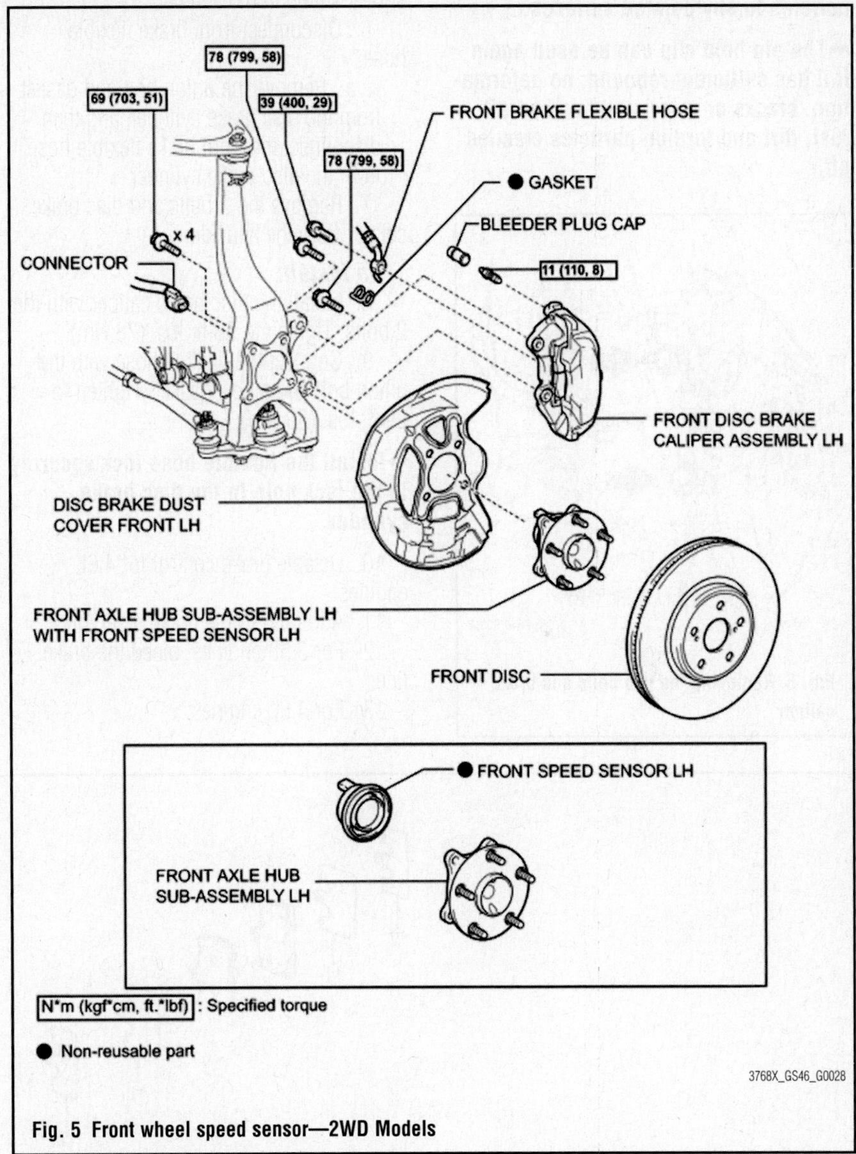

Fig. 5 Front wheel speed sensor—2WD Models

➡Prevent foreign matter from attaching to the sensor tip.

5. To install, reverse removal procedure.

Rear Wheel

1. Raise and support the vehicle.
2. Remove the appropriate wheel.
3. Disconnect the speed sensor rear connector.
4. Remove the bolt and rear speed sensor.

➡Prevent foreign matter from attaching to the sensor tip.

5. To install, reverse removal procedure.

BRAKE CALIPER

REMOVAL & INSTALLATION

See Figure 6.

1. For 4.6L engines, remove the 2 ABS motor relays with the engine switch off.
2. Remove front wheel.

➡Wash brake fluid off immediately if it adheres to any painted surfaces.

➡The pin hold clip can be used again if it has sufficient rebound; no deformation, cracks or wear; and has had all rust, dirt and foreign particles cleaned off.

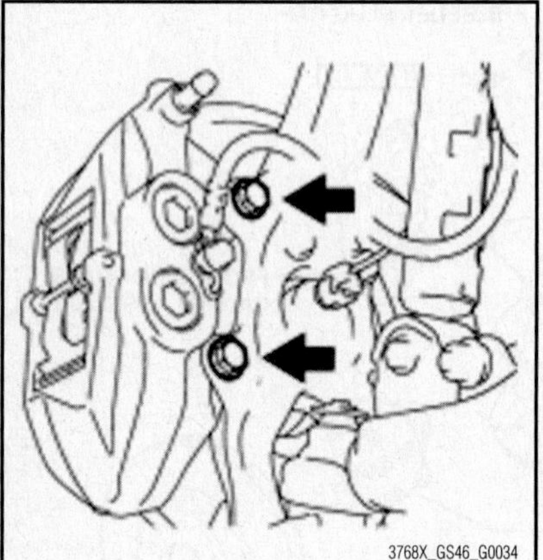
3768X_GS46_G0034

Fig. 6 Removing the two bolts and brake caliper

3. Remove front disc brake anti-rattle with hole pin:
   a. Remove the pin hold clip.
   b. Remove the front disc brake anti-rattle with hole pin while pushing on the anti-rattle spring.
4. Remove front disc brake anti-rattle spring.
5. Remove the 2 pads and the 2 anti-squeal shims from each pad.
6. Disconnect front brake flexible hose:
   a. Remove the union bolt and gasket from the disc brake cylinder, and then disconnect the front brake flexible hose from the disc brake cylinder.
7. Remove the 2 bolts and disc brake caliper from the knuckle.

To install:

8. Install the disc brake caliper with the 2 bolts. Tighten to 58 ft. lbs. (78 Nm).
9. Connect the flexible hose with the union bolt and a new gasket. Tighten to 29 ft. lbs. (39 Nm).

➡Install the flexible hose lock securely in the lock hole in the disc brake cylinder.

10. Disable brake control for 4.6L engines.
11. Add fluid to the brake reservoir.
12. For 3.5L engines, bleed the brake line.
13. For 4.6L engines:
    a. Bleed the front brake system.
    b. Install the ABS motor relay.
14. Check the brake fluid level.
15. Check and clear DTC's
16. Inspect for brake fluid leaks.

DISC BRAKE PADS

REMOVAL & INSTALLATION

See Figures 7 through 9.

1. For 4.6L engines, remove the 2 ABS motor relays with the engine switch off.
2. Remove front wheel.

➡Wash brake fluid off immediately if it adheres to any painted surfaces.

➡The pin hold clip can be used again if it has sufficient rebound; no deforma-

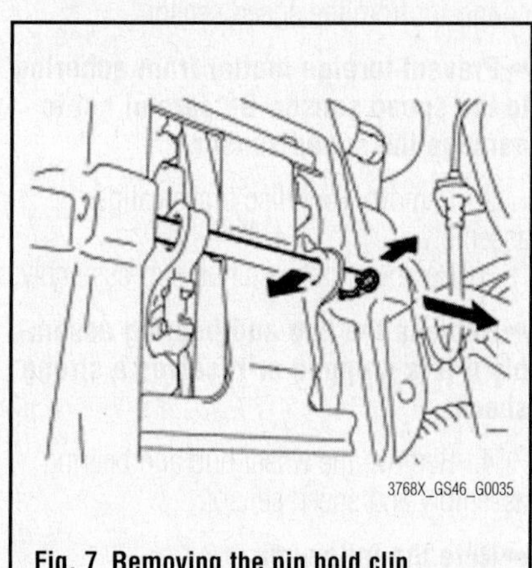
3768X_GS46_G0035

Fig. 7 Removing the pin hold clip

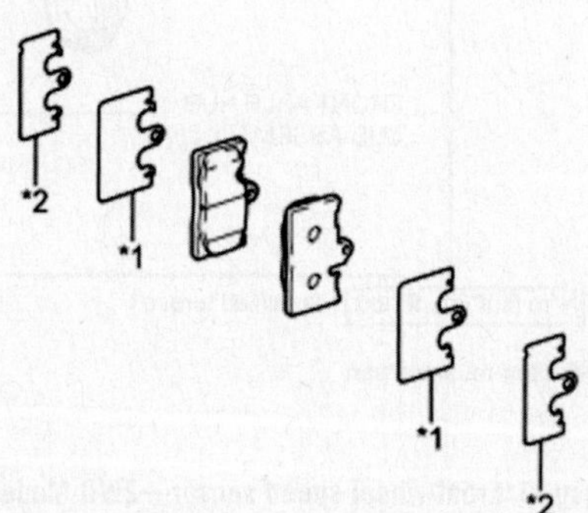

*1	No. 1 Anti-squeal Shim (Large)
*2	No. 2 Anti-squeal Shim (Small)

NOTICE:
- When replacing worn pads, the anti-squeal shims must be replaced together with the pads.
- The No. 1 anti-squeal shim is larger than the No. 2 anti-squeal shim. Make sure to install them in the correct order.

3768X_GS46_G0036

Fig. 8 Identifying the anti-squeal shims

tion, cracks or wear; and has had all rust, dirt and foreign particles cleaned off.**

3. Remove front disc brake anti-rattle with hole pin:

a. Remove the pin hold clip.

b. Remove the front disc brake anti-rattle with hole pin while pushing on the anti-rattle spring.

4. Remove front disc brake anti-rattle spring.

5. Remove the 2 pads and the 2 anti-squeal shims from each pad.

To install:

6. Install the No. 1 anti-squeal shim and No. 2 anti-squeal shim to each disc brake pad.

7. Install the disc brake pad to the disc brake caliper assembly.

8. Install front disc brake anti-rattle spring.

➡The anti-rattle spring can be used again if it has sufficient rebound; no deformation, cracks or wear; and has had all rust, dirt and foreign particles cleaned off.

➡The pin hold clip can be used again if it has sufficient rebound; no deformation, cracks or wear; and has had all rust, dirt and foreign particles cleaned off.

9. Install front disc brake anti-rattle with hole pin:

a. Install the hole pin while pushing on the front disc brake anti-rattle spring.

b. Install the pin hold clip.

10. To complete installation, reverse the remaining removal procedure.

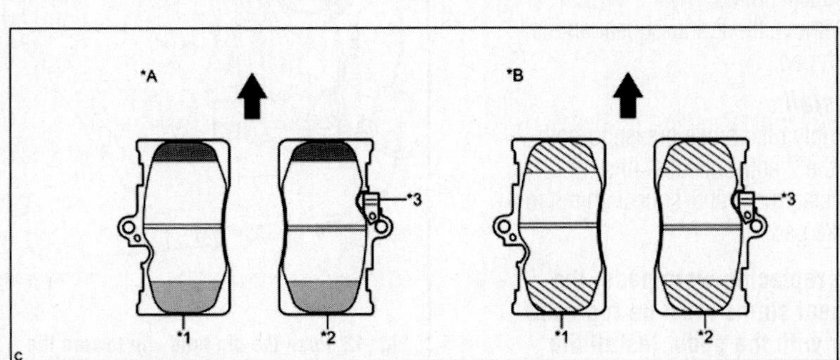

Text in Illustration

*A	for Type A	*B	for Type B
*1	Outer Pad	*2	Inner Pad
*3	Pad Wear Indicator	-	-
[arrow symbol]	Direction of Disc Rotation for Forward Movement	[symbol]	Chamfered Edge (Small)
[symbol]	Chamfered Edge (Large)	[symbol]	Chamfered Edge (Same Size)

NOTICE:

- for Type A: The upper and lower edges of the brake pads are chamfered differently. Install each brake pad to the correct side and in the correct position as shown in the illustration.
- for Type B: The upper and lower edges of the brake pads are the equally chamfered.
- The disc brake pad with the pad wear indicator is positioned on the inside of the vehicle.
- Install the disc brake pads in the correct directions.
- There should be no oil or grease on the shims and friction surfaces of the pads and disc.

3768X_GS46_G0037

Fig. 9 Brake pad installation orientation

BRAKES — REAR DISC BRAKES

BRAKE CALIPER

REMOVAL & INSTALLATION

See Figure 10.

1. Remove the 2 ABS motor relays with the engine switch off.

2. Remove rear wheel.

3. Drain brake fluid.

4. Remove disc brake pad kit:

a. Disengage the both ends of the anti-squeal spring from the 2 brake pad holes, and turn the anti-squeal spring.

b. Push the pin hold clip toward the brake rotor to disengage it from the brake pad, and then remove it from the 2 pad guide pins.

c. Holding the anti-squeal spring, remove the 2 pad guide pins from the rear disc brake caliper assembly.

d. Remove the anti-squeal spring from the rear disc brake caliper assembly.

e. Remove the 2 brake pads with the anti-squeal shims.

5. Remove the 2 anti-squeal shims from each pad.

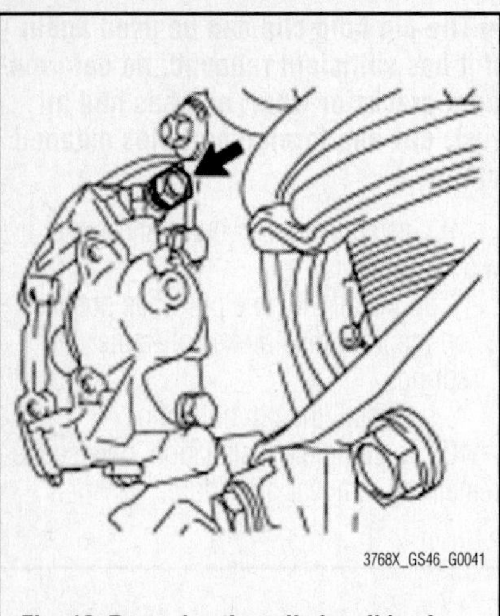

Fig. 10 Removing the cylinder slide pin

6. Remove the union bolt and gasket from the rear disc brake cylinder, and then disconnect the flexible hose.
7. Remove the cylinder slide pin and then tilt the disc brake caliper toward the rear of the vehicle.
8. Remove the 2 bolts and rear disc brake caliper together with the caliper support bracket.
9. Remove the 2 No. 1 caliper plates from the caliper support bracket.

To install:

10. Install 2 new No. 1 caliper plates to the caliper support bracket.
11. Install the rear disc brake caliper together with the caliper support bracket with the 2 bolts. Tighten to 48 ft. lbs. (65 Nm).
12. Install the rear disc brake caliper with the cylinder slide pin. Tighten to 18 ft. lbs. (25 Nm).
13. Install the brake pads.
14. Disable brake control for 4.6L engines.
15. Add fluid to the brake reservoir.
16. For 3.5L engines, bleed the brake line.
17. For 4.6L engines:
    a. Bleed the front brake system.
    b. Install the ABS motor relay.
18. Check the brake fluid level.
19. Check and clear DTC's
20. Inspect for brake fluid leaks.

DISC BRAKE PADS

REMOVAL & INSTALLATION

See Figures 11 through 16.

1. Remove the 2 ABS motor relays with the engine switch off.
2. Remove rear wheel.
3. Drain brake fluid.
4. Remove disc brake pad kit:
    a. Disengage the both ends of the anti-squeal spring from the 2 brake pad holes, and turn the anti-squeal spring.
    b. Push the pin hold clip toward the brake rotor to disengage it from the brake pad, and then remove it from the 2 pad guide pins.
    c. Holding the anti-squeal spring, remove the 2 pad guide pins from the rear disc brake caliper assembly.
    d. Remove the anti-squeal spring from the rear disc brake caliper assembly.
    e. Remove the 2 brake pads with the anti-squeal shims.
5. Remove the 2 anti-squeal shims from each pad.

To install:

6. Apply disc brake grease to both sides of the 2 anti-squeal shims.
7. Install the 2 anti-squeal shims to each brake pad.

➡When replacing worn pads, the anti-squeal shims must be replaced together with the pads. Install the anti-squeal shims in the correct positions and directions.

8. Install the 2 brake pads to the rear disc brake cylinder.
9. Install the pad guide pin (upper) to the rear disc brake cylinder.
10. Insert the end of the anti-squeal spring to the brake pad (inner).
11. Holding the anti-squeal spring, install the pad guide pin (lower) to the rear disc brake cylinder.
12. Engage the pin hold clip to the holes of both pad guide pins, and insert the pin hold clip to the brake pad.

➡Make sure that the claw of the pin hold clip is engaged to the brake pad hole.

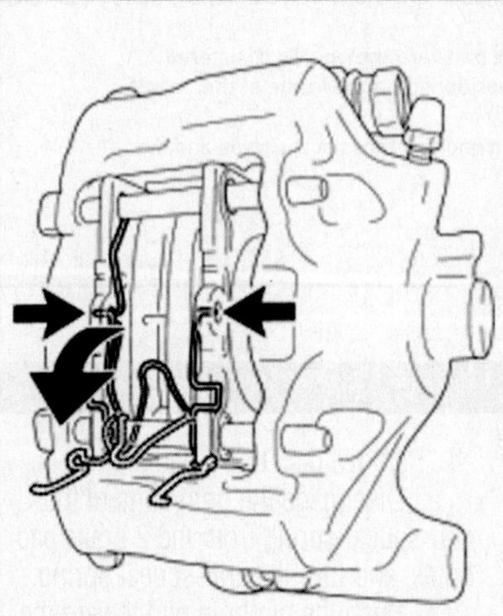

Fig. 11 Disengage the both ends of the anti-squeal spring from the 2 brake pad holes

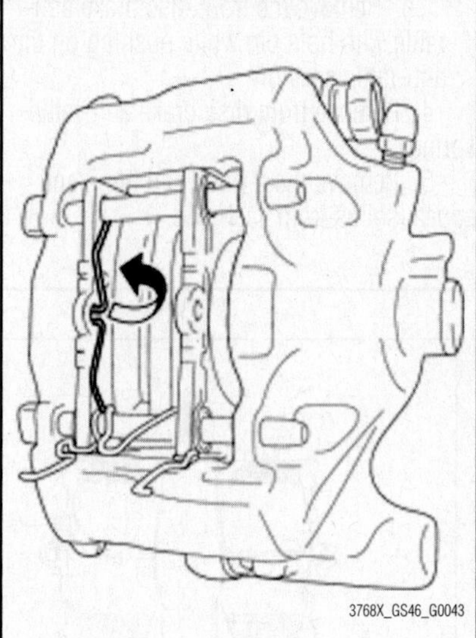

Fig. 12 Push the pin hold clip toward the brake rotor to disengage it from the brake pad

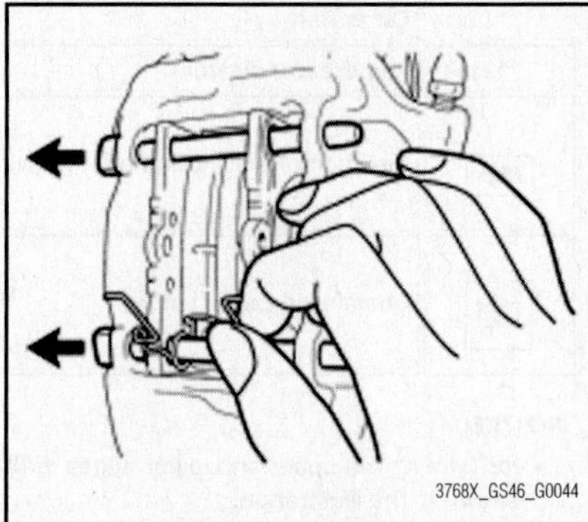

Fig. 13 Holding the anti-squeal spring, remove the 2 pad guide pins from the rear disc brake caliper assembly

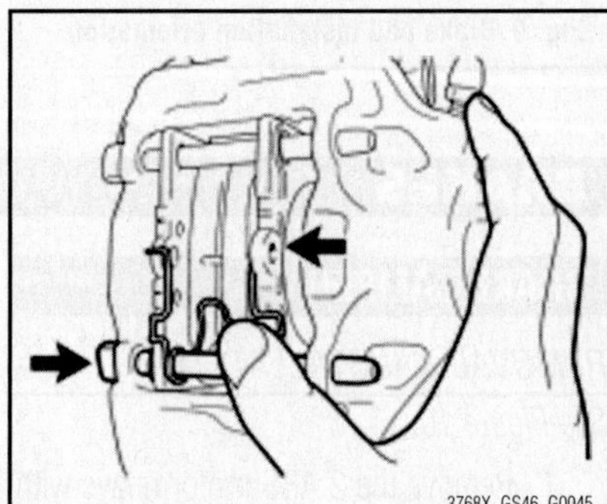

Fig. 14 Insert the end of the anti-squeal spring to the brake pad (inner).

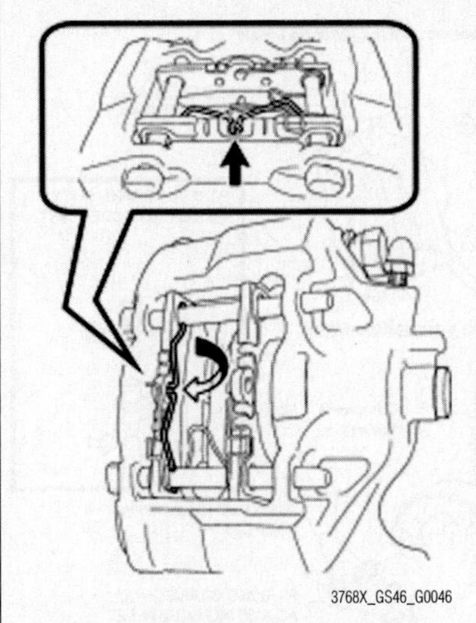

Fig. 15 Engage the pin hold clip to the holes of both pad guide pins

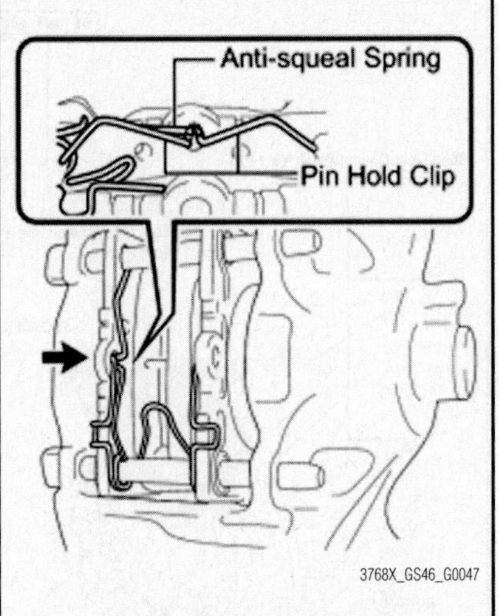

Fig. 16 Insert the end of the anti-squeal spring to the brake pad hole

13. Insert the end of the anti-squeal spring to the brake pad hole.

➡Make sure that the anti-squeal spring and pin hold clip are placed. Make sure that the pin hold clip cannot be removed easily from the pad guide pins.

14. Disable brake control for 4.6L engines.
15. Add fluid to the brake reservoir.
16. For 3.5L engines, bleed the brake line.
17. For 4.6L engines:
    a. Bleed the front brake system.
    b. Install the ABS motor relay.
18. Check the brake fluid level.
19. Check and clear DTC's
20. Inspect for brake fluid leaks.

BRAKES — PARKING BRAKE

PARKING BRAKE CABLES

ADJUSTMENT

See Figure 17.

1. Depress the parking brake pedal. Hold the No. 1 wire adjusting nut using a wrench and loosen the lock nut.
2. Release the parking brake pedal.
3. Turn the No. 1 wire adjusting nut until the parking brake pedal travel is at 7–9 notches at 67.5 lbs. (300 Nm).
4. Hold the wire adjusting No. 1 nut using a wrench or equivalent tool and tighten the lock nut to 53 inch lbs. (6 Nm).
5. Count the number of clicks after depressing and releasing the parking brake pedal 3 or 4 times.
6. Check whether the parking brake drags or not.
7. When operating the parking brake pedal, check that the parking brake indicator light comes on.

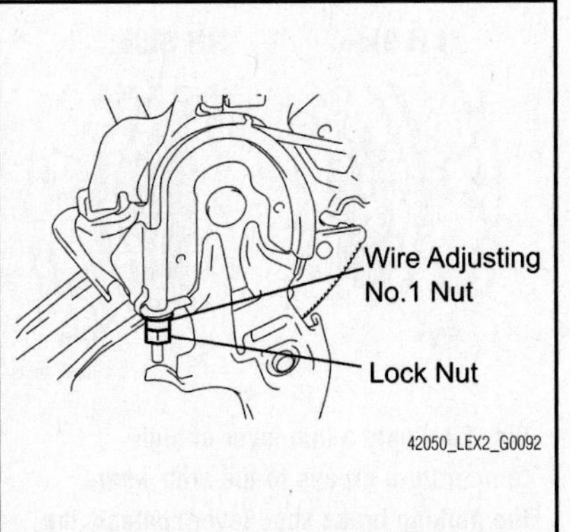

Fig. 17 Parking brake adjuster lock nuts

PARKING BRAKE SHOES

REMOVAL & INSTALLATION

See Figures 18 through 24.

1. Before servicing the vehicle, refer to the precautions in the beginning of this section.
2. Disconnect the negative battery cable.
3. Remove the rear wheel.
4. Remove the 2 bolts and separate the rear disc brake caliper assembly.

➡Hang the caliper with wire or equivalent.

5. Remove the caliper No.1 plate (2 pieces).
6. Release the parking brake.
7. Remove the rear disc.
8. Using SST 09703-30011, remove the parking brake shoe return No.2 spring.
9. Using SST 09703-30011, remove the parking brake shoe return No.1 spring.
10. Slide the parking brake shoe, and remove the parking brake shoe adjusting screw set.
11. Using SST 09718-00010, remove the shoe hold down spring No.1 cup, compression No.1 spring and shoe hold down spring No.1 pin. Remove the parking brake No.1 shoe assembly.
12. Using SST 09718-00010, remove the shoe hold down spring No.1 cup, compression No.1 spring and shoe hold down spring No.1 pin. Remove the parking brake No.2 shoe assembly.

➡Use the service hole to retain the hold down spring No.1 pin with your finger.

13. Remove parking brake shoe lever
14. Remove the rear axle hub and bearing assembly.
15. Remove the 2 nuts and parking brake anchor block.

To install:

16. Install the parking brake anchor block with 2 nuts. Tighten to 56 ft. lbs. (76 Nm).
17. Install the rear axle hub and bearing assembly.
18. Apply a thin layer of high temperature grease to the area where the parking brake plate contacts the parking brake shoe.

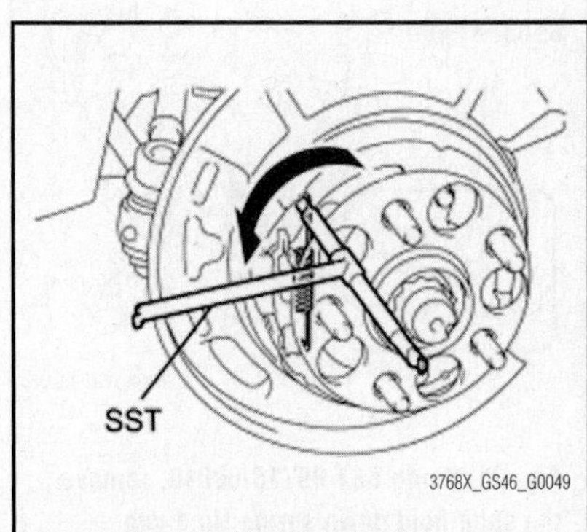

Fig. 18 Using SST 09703-30011, remove the parking brake shoe return No.2 spring

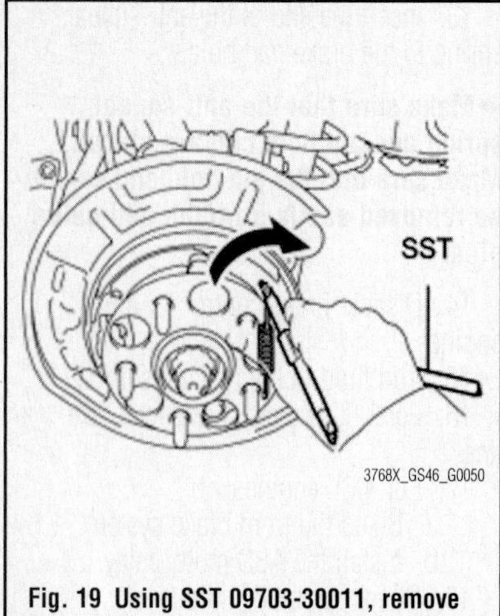

Fig. 19 Using SST 09703-30011, remove the parking brake shoe return No.1 spring

19. Apply a thin layer of high temperature grease to the area where the parking brake shoe lever contacts the parking brake anchor block.

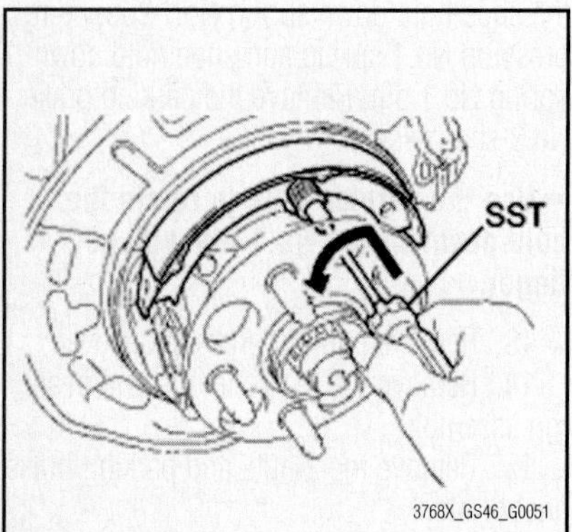

Fig. 20 Using SST 09718-00010, remove the shoe hold down spring No.1 cup, compression No.1 spring and shoe hold down spring No.1 pin

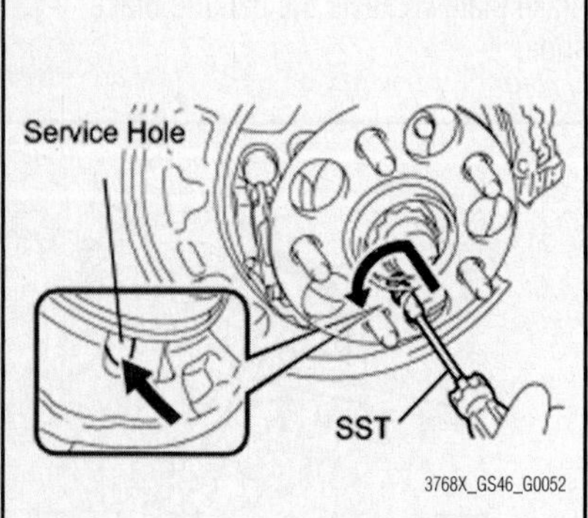

Fig. 21 Using SST 09718-00010, remove the shoe hold down spring No.1 cup, compression No.1 spring and shoe hold down spring No.1 pin

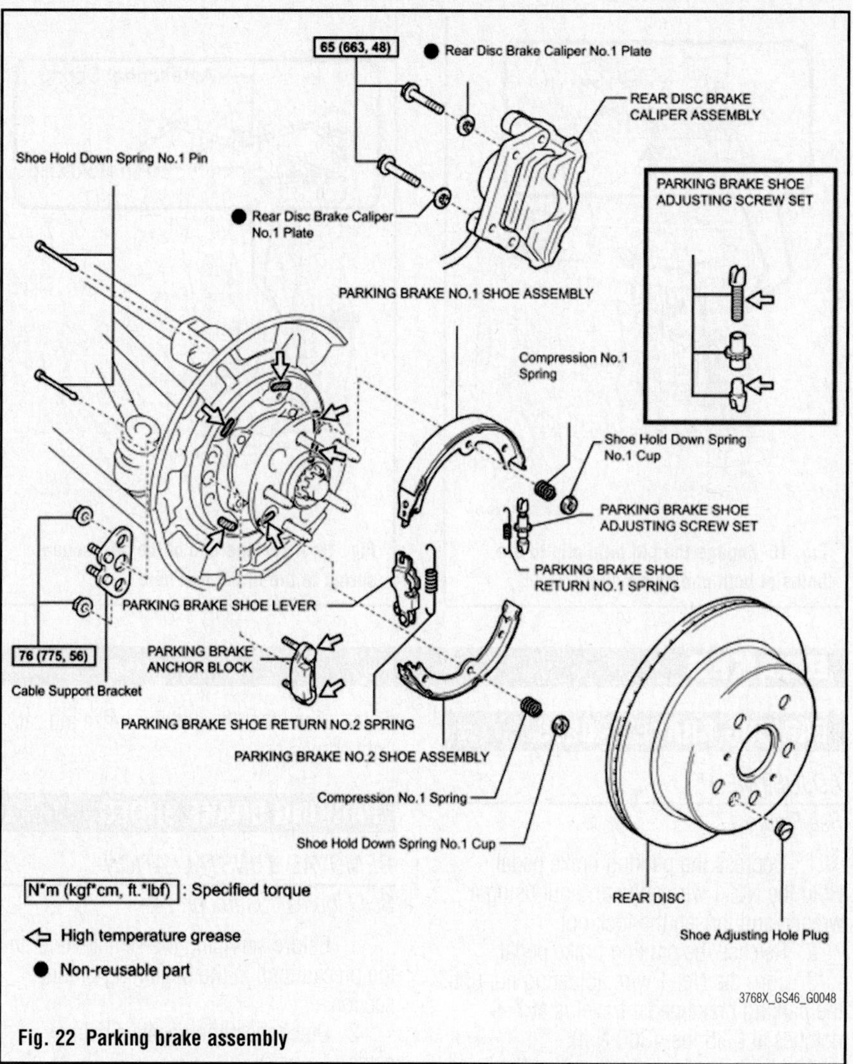

Fig. 22 Parking brake assembly

20. Install the parking brake shoe lever to the parking brake cable assembly.

21. Using SST 09718-00010, install the parking brake No.2 shoe assembly with the shoe hold down spring No.1 cup, compression No.1 spring and shoe hold down spring No.1 pin.

➡Use the service hole to retain the shoe hold down spring No.1 pin with your finger.

22. Using SST 09718-00010, install the parking brake No.1 shoe assembly with the shoe hold down spring No.1 cup, compression No.1 spring and shoe hold down spring No.1 pin.

23. Apply high temperature grease to the thread and all joining areas of the parking brake shoe adjusting screw set.

24. Install the parking brake shoe adjusting screw set.

25. Install the parking brake shoe return No.1 spring.

26. Install the parking brake shoe return No.2 spring.

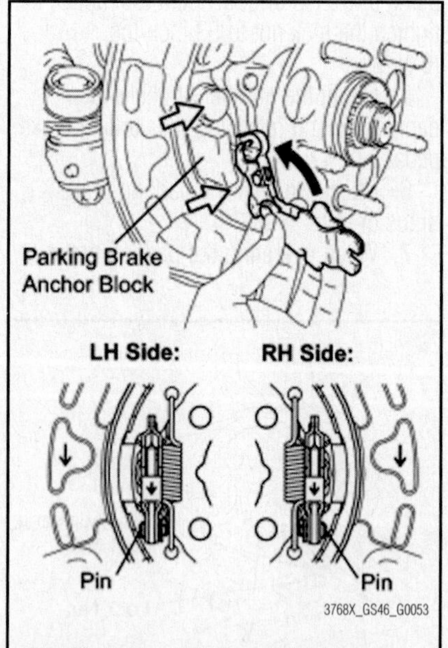

Fig. 23 Apply a thin layer of high temperature grease to the area where the parking brake shoe lever contacts the parking brake anchor block

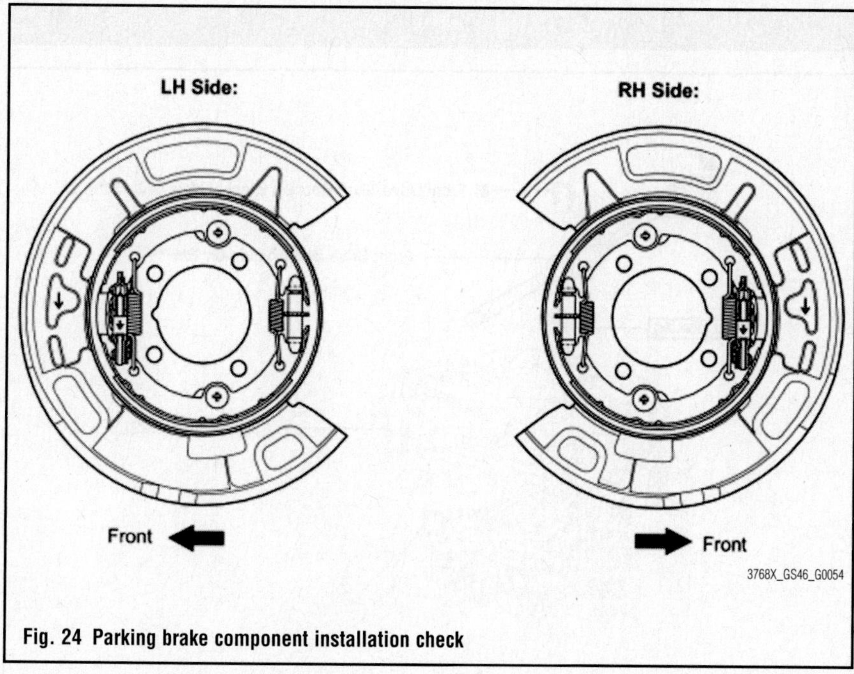

Fig. 24 Parking brake component installation check

27. Make sure that all the parts are installed properly. If necessary, reinstall properly.

28. Align the matchmarks on the rear disc and rear axle hub.

29. Adjust parking brake shoe clearance.

30. Install the rear disc brake caliper assembly with the 2 bolts and 2 new caliper No.1 plates. Tighten 48 ft. lbs. (65 Nm).

➡Do not twist the brake hose. Make sure that the bolts are free from damage and foreign matter. Do not overtighten the bolts.

31. Install rear wheel.

32. Inspect parking brake pedal travel.

33. Adjust parking brake pedal travel.

ADJUSTMENT

See Figure 25.

1. Temporarily install the hub nuts.
2. Remove the hole plug, and turn the adjuster and expand the shoes until the disc locks.
3. Contract the shoe adjuster until the disc can rotate smoothly. Standard: Return 7–8 notches
4. Check shoe is no brake drag.
5. Install the hole plug.

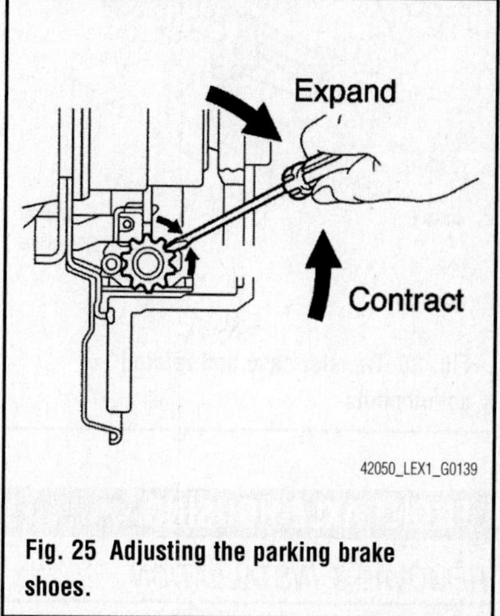

Fig. 25 Adjusting the parking brake shoes.

CHASSIS ELECTRICAL

AIR BAG (SUPPLEMENTAL RESTRAINT SYSTEM)

GENERAL INFORMATION

⁂ CAUTION

These vehicles are equipped with an air bag system. The system must be disarmed before performing service on, or around, system components, the steering column, instrument panel components, wiring and sensors. Failure to follow the safety precautions and the disarming procedure could result in accidental air bag deployment, possible injury and unnecessary system repairs.

SERVICE PRECAUTIONS

⁂ CAUTION

Disconnect and isolate the battery negative cable before beginning any airbag system component diagnosis, testing, removal, or installation procedures. Wait at least 90 seconds after the ignition switch is turned off and the negative (-) terminal cable is disconnected from the battery before starting the operation. The SRS is equipped with a backup power source, so if work is started within 90 seconds after disconnecting the negative (-) terminal cable from the battery, the SRS may be deployed. Failure to disable the airbag system may result in accidental airbag deployment, personal injury, or death.

DISARMING THE SYSTEM

To avoid personal injury when working on vehicles equipped with an air bag, the negative battery cable must be disconnected and at least 90 seconds must elapse before working on the system. Failure to do so may result in deployment of the air bag.

ARMING THE SYSTEM

To rearm the air bag system, simply reconnect the battery cable(s).

CLOCKSPRING CENTERING

1. Center the spiral cable as follows:
   a. Check that the ignition switch is OFF.
   b. Check that the battery negative terminal is disconnected. Do not start the operation for 90 seconds after removing the terminal.
   c. Turn the cable counterclockwise by hand until it becomes harder to turn.
   d. Then rotate the cable clockwise about 2.5 turns to align the marks. The cable will rotate about 2.5 turns to both left and right of the center.

DRIVE TRAIN

TRANSFER CASE ASSEMBLY

REMOVAL & INSTALLATION

See Figure 26.

➡Remove and install transfer case with automatic transmission.

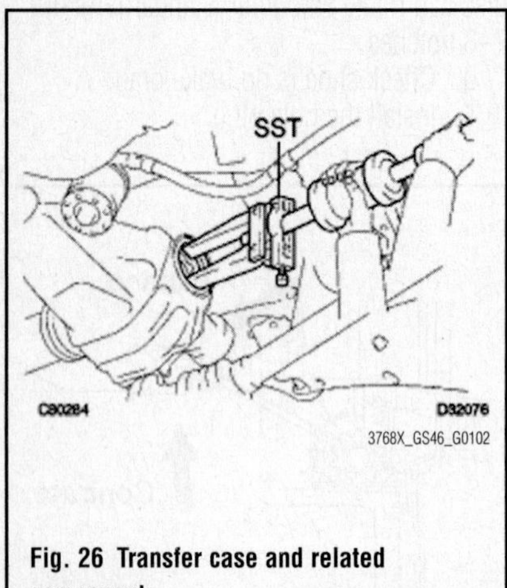

Fig. 26 Transfer case and related components

FRONT AXLE HOUSING

REMOVAL & INSTALLATION

4WD Models

See Figure 27.

➡Do not drop the differential carrier. Do not damage the installation surface when removing the differential carrier assembly. The remaining oil may leak out when removing the differential carrier assembly.

1. Remove the differential filler plug and gasket.
2. Remove the differential drain plug and gasket, and drain the oil.
3. Remove the engine assembly.
4. Support the differential carrier with a jack.
5. Remove the 3 bolts and differential carrier.

To install:

6. Support the differential carrier assembly with a jack. Temporarily install the differential carrier assembly to the engine assembly with the 3 bolts.
7. Fully tighten the 3 bolts to install the differential carrier assembly. Tighten to 64 ft. lbs. (87 Nm).
8. Install the engine.

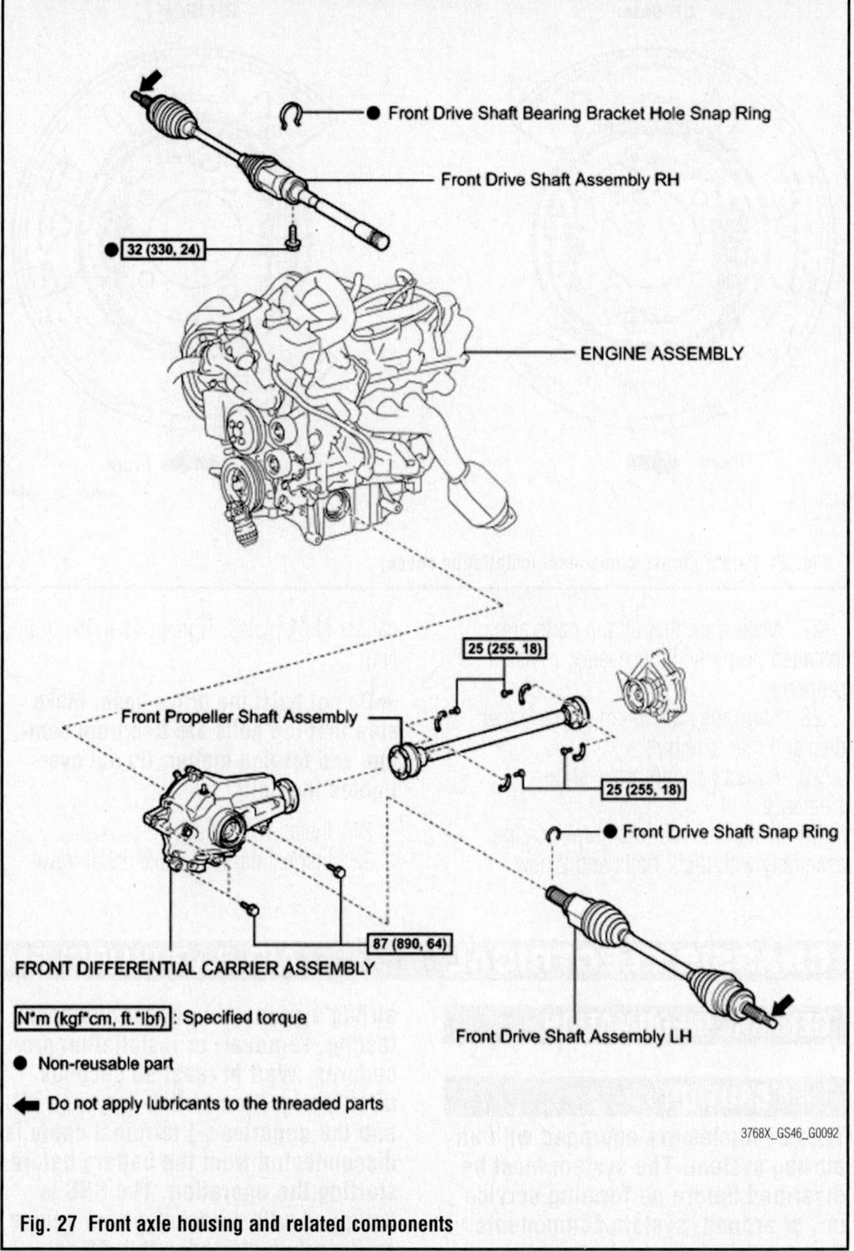

Fig. 27 Front axle housing and related components

FRONT DRIVESHAFT

REMOVAL & INSTALLATION

See Figure 28.

4WD Models

1. Remove the front exhaust pipe assembly
2. Put matchmarks on the transfer companion flange and the front propeller shaft.
3. Using a screwdriver or an equivalent, hold the transfer companion flange.
4. Using a socket hexagon wrench 6 mm, loosen the 6 bolts.

➡Be careful not to damage the front propeller shaft.

5. Using a screwdriver or an equivalent, hold the transfer companion flange.
6. Using a socket hexagon wrench 6 mm, loosen the 6 bolts.
7. Remove the 12 bolts, 4 constant velocity universal joint washers and front propeller shaft assembly.
8. Check that the propeller shaft is not cracked or damaged.
9. Check that there is no excessive play in the joint.

To install:

10. Align the matchmarks on the transfer companion flange and the front propeller shaft.

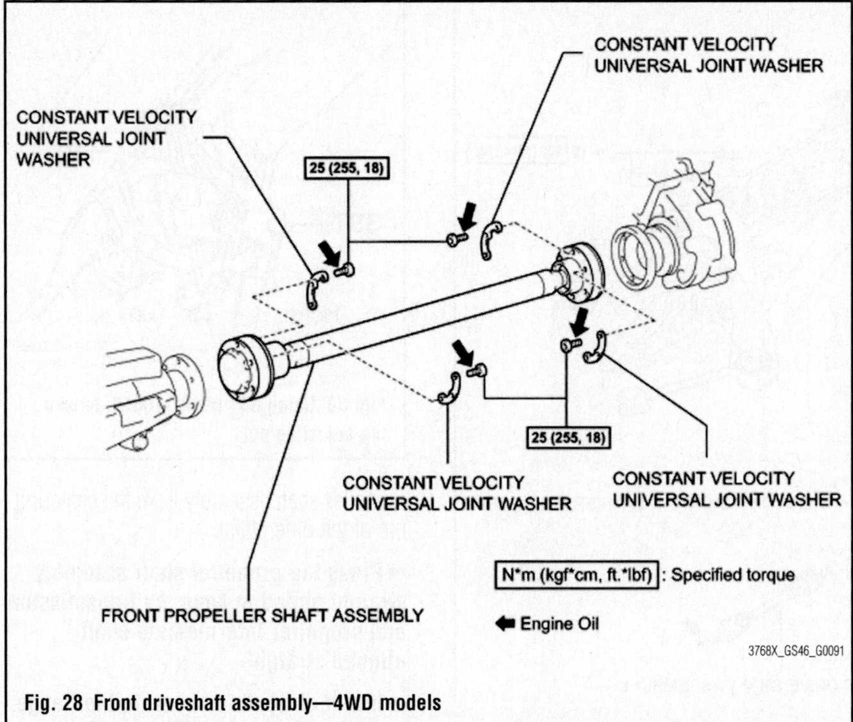

Fig. 28 Front driveshaft assembly—4WD models

11. Temporarily tighten the front propeller shaft with the 6 bolts and 2 constant velocity universal joint washers.
12. Align the matchmarks on the differential companion flange and the front propeller shaft.
13. Temporarily tighten the front propeller shaft with the 6 bolts and 2 constant velocity universal joint washers.

➡If reusing the bolts, apply engine oil to the bolt before use.

14. Using a screwdriver or an equivalent, hold the transfer companion flange.
15. Using a socket hexagon wrench 6 mm, fully tighten the front propeller shaft with the 6 bolts.

➡Be careful not to damage the front propeller shaft.

16. Using a screwdriver or an equivalent, hold the transfer companion flange.
17. Using a socket hexagon wrench 6 mm, fully tighten the front propeller shaft with the 6 bolts. Tighten to 18 ft. lbs. (25 Nm).
18. To complete installation, reverse remaining removal procedure.

FRONT HALFSHAFTS

REMOVAL & INSTALLATION

4WD Models

1. Remove front wheels.
2. Separate speed sensor.
3. Release the staked part of the front axle hub nut.
4. Remove the front axle hub nut.
5. Remove the 2 bolts and steering knuckle from the lower ball joint.
6. Separate the tie rod end from the steering knuckle.

➡Pay careful attention not to damage the steering knuckle because it is made of aluminum and may be damaged easily. If the steering knuckle spacer has come off, replace the steering knuckle with a new one.

7. Using a plastic hammer, separate the front axle assembly from the drive shaft assembly.
8. Remove the front drive shaft assemblies.

To install:

9. Coat the spline of the inboard joint shaft assembly with gear oil.
10. Set the shaft snap ring with the opening side facing down.
11. Align the shaft splines and install the drive shaft assembly with a brass bar and hammer.
12. Install the front drive shaft assembly to the front axle assembly.
13. Install the tie rod end to the steering knuckle with the nut and tighten to 48 ft. lbs. (65 Nm). Install a new clip.
14. Install the steering knuckle to the lower ball joint with the 2 bolts and tighten to 89 ft. lbs. (120 Nm).
15. Clean the threaded parts on the drive shaft and front axle hub nut using a non-residue solvent.
16. Install a new axle hub nut and tighten to 217 ft. lbs. (295 Nm).
17. Using a chisel and hammer, stake the front axle hub nut.
18. Install speed sensor front.
19. Install front wheel.
20. Inspect alignment.

REAR AXLE HOUSING

REMOVAL & INSTALLATION

See Figure 29.

1. Remove rear wheel.
2. Drain differential oil:
   a. Using a hexagon wrench (10 mm), remove the differential filler plug and gasket.
   b. Using a hexagon wrench (10 mm), remove the differential drain plug and gasket, and drain the oil.

CAUTION

As the differential oil is hot right after driving, allow some time for it to cool down sufficiently.

3. Remove the exhaust pipe assembly.
4. Remove the driveshaft.
5. Remove the halfshafts.
6. Support the rear differential carrier assembly with a jack.
7. Using a hexagon wrench (12 mm), remove the 3 hexagon bolts.
8. Remove the 2 bolts, 2 lower differential mount stoppers and rear differential carrier assembly.
9. Remove the 2 upper differential mount stoppers from the rear differential carrier assembly.

➡Note the following:

- Slowly lower the jack and remove the rear differential carrier assembly.
- Do not drop the rear differential carrier assembly.
- Do not damage the installation surface when removing the rear differential carrier assembly.
- Residual oil may leak when the rear differential carrier assembly is removed.

10. To install, reverse the removal procedure.
11. Add rear differential oil.
12. Inspect and adjust rear differential oil.
13. Inspect and adjust rear wheel alignment.

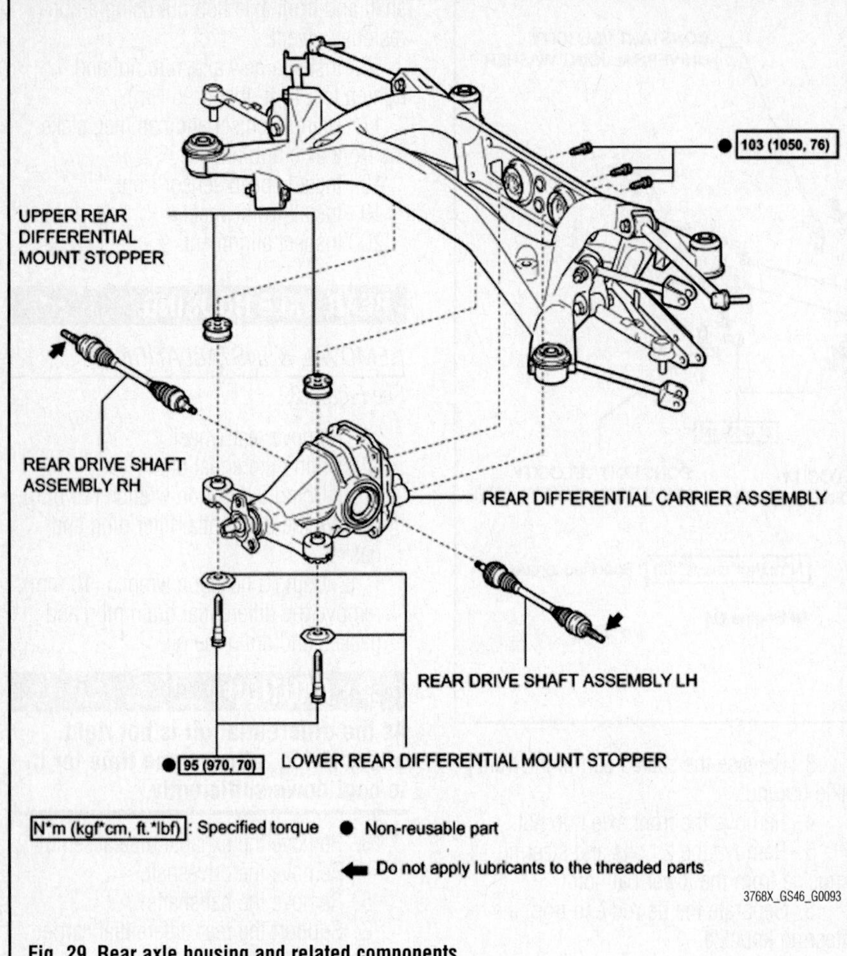

Fig. 29 Rear axle housing and related components

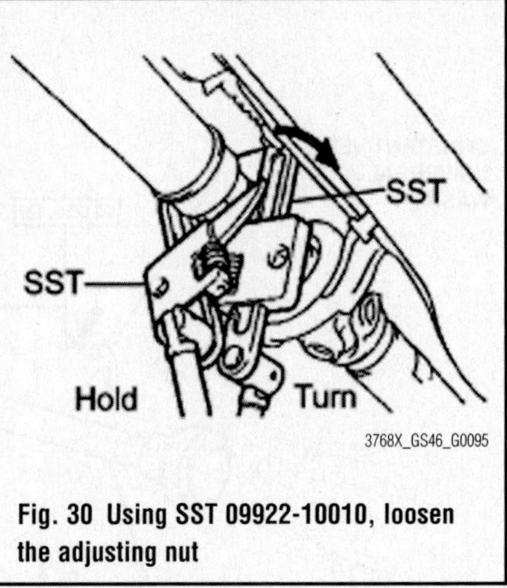

Fig. 30 Using SST 09922-10010, loosen the adjusting nut

14. Inspect for rear differential oil leak.
15. Inspect for exhaust gas leak.
16. Inspect speed sensor signal.

REAR DRIVESHAFT

REMOVAL & INSTALLATION

2WD Models

See Figures 30 through 35.

1. Remove no. 2 engine under cover (for 3.5L engines).
2. Remove front floor cover center LH and RH (for 4.6L engines).
3. Remove rear no. 1 floor panel brace (for 4.6L engines).
4. Remove front floor brace center (for 3.5L engines).
5. Remove front center floor brace (for 4.6L engines).
6. Disconnect Heated Oxygen Sensor (HO2S)(for 3.5L engines).
7. Disconnect Heated Oxygen Sensor (HO2S)(for 4.6L engines).
8. Remove front exhaust pipe assembly.
9. Remove the 4 nuts, grommet and front floor No. 1 heat insulator.
10. Remove the 4 nuts and air guide plate outside RH.
11. Using SST 09922-10010, loosen the adjusting nut until it can be turned by hand.
12. Put matchmarks on the transmission companion flange, flexible coupling and propeller intermediate shaft.
13. Remove the 3 bolts, 3 washers and 3 nuts.

➡The propeller intermediate shaft and flexible coupling should not be separated.

14. Put matchmarks on the differential companion flange, flexible coupling and propeller shaft assembly.
15. Remove the 3 bolts, 3 washers and 3 nuts.
16. Remove the 2 bolts and 2 center support bearing washers.

➡Some vehicles are not equipped with the center support bearing washers.

17. Push the propeller shaft assembly straight forward to compress the rear propeller shaft assembly and pull out the rear propeller shaft assembly from the centering pin of the differential.

➡Press the propeller shaft assembly straight ahead to keep the transmission and propeller intermediate shaft aligned straight.

➡If it is difficult to separate the flange from the flexible coupling, pry it using a screwdriver.

18. Pull the propeller shaft outward from the vehicle's rear.

➡The propeller intermediate shaft and propeller shaft assembly should not be separated.

To install:

19. Apply grease to the flexible coupling centering bushings. Grease: Molybdenum disulphide lithium base NLGI No. 2.
20. Align the matchmarks on the transmission companion flange and flexible coupling.
21. Install and tighten the 3 bolts, 3 washers and 3 nuts. Tighten to 58 ft. lbs. (79 Nm).

➡ The bolts should be installed from the propeller shaft side.

22. Align the matchmarks on the differential companion flange and flexible coupling.
23. Install and tighten the 3 bolts, 3 washers and 3 nuts. Tighten to 58 ft. lbs. (79 Nm).
24. Temporarily install the 2 center support bearing set bolts with the 2 center support bearing washers.
25. Adjust the dimension between the edge surface of the center support bearing and the edge surface of the cushion to 0.4528 to 0.5315 inches (11.5 to 13.5 mm) respectively as shown in the illustration.
26. Check that the center line of the bracket is at right angles to the shaft axial direction.

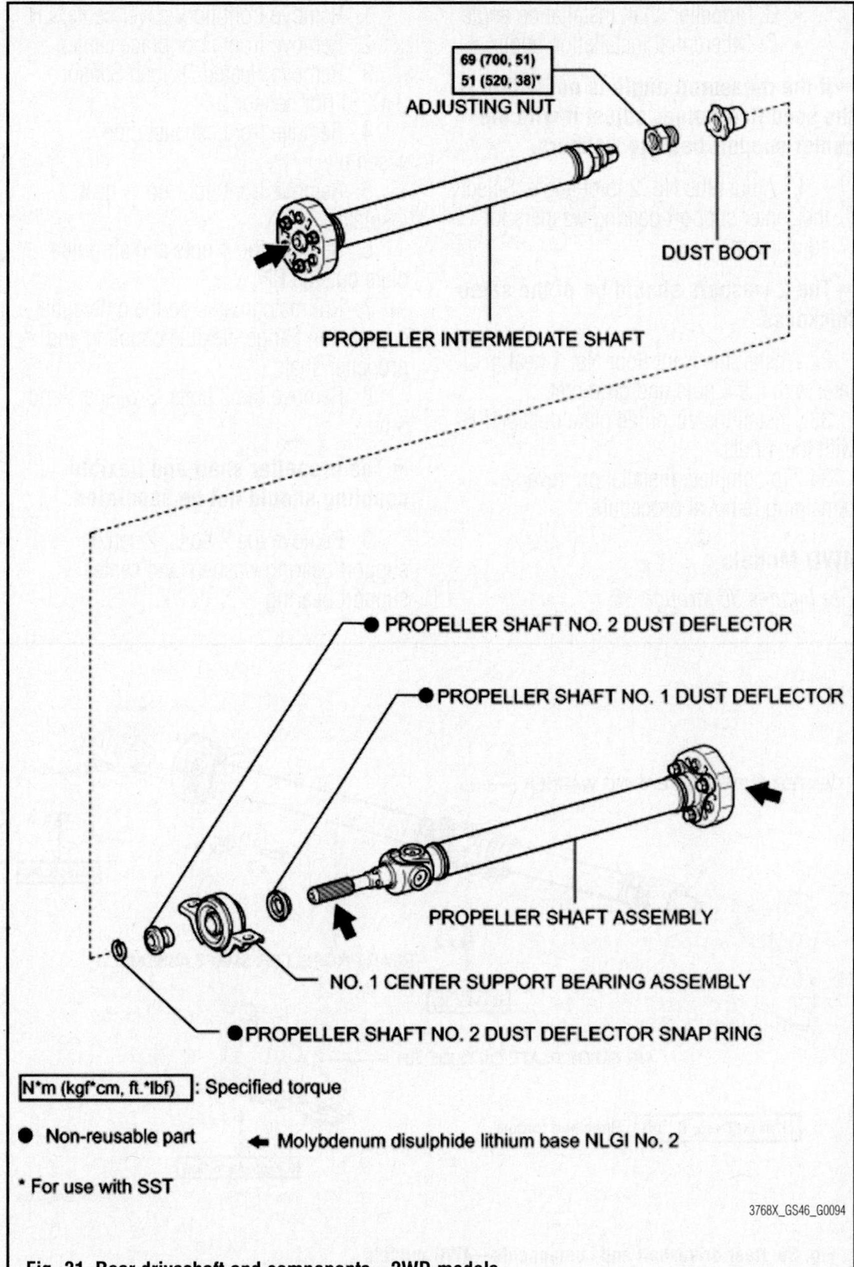

Fig. 31 Rear driveshaft and components—2WD models

27. Tighten the 2 bolts to 36 ft. lbs. (49 Nm).
28. Using SST, tighten the adjusting nut to 38 ft. lbs. (51 Nm).
29. If not using the SST, tighten the adjusting nut to 51 ft. lbs. (69 Nm).
30. For 4.6L Engines, inspect and adjust no. 2 and no. 3 joint angle:
    a. Stabilize the propeller shaft and differential. Turn the propeller shaft several times by hand to stabilize the center support bearing.
    b. Check the No. 2 and No. 3 joint angles. Using SST, measure the installation angle of the propeller intermediate shaft and propeller shaft assembly.

➡The SST should be set directly underneath the shaft.

➡Measure the installation angle by placing SST in the positions shown in the illustration.

c. Using SST: 09370-50010, measure the installation angle of the differential.
d. Calculate the No. 2 joint angle:
- No. 2 joint angle: A - B = -0°20' to -1°20'
- A: Propeller intermediate shaft installation angle
- B: Propeller shaft assembly installation angle

e. Calculate the No. 3 joint angle:
- B - C = 0°53' to 1°53'
- B: Propeller shaft assembly installation angle
- C: Differential installation angle

➡If the measured angle is not within the specified range, adjust it with the center support bearing washers.

f. Adjust the No. 2 joint angle. Select the center support bearing washers for adjustment.

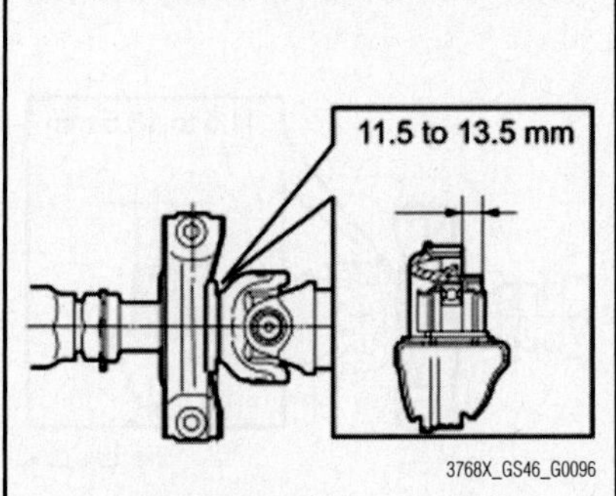

Fig. 32 Adjust the dimension between the edge surface of the center support bearing and the edge surface of the cushion

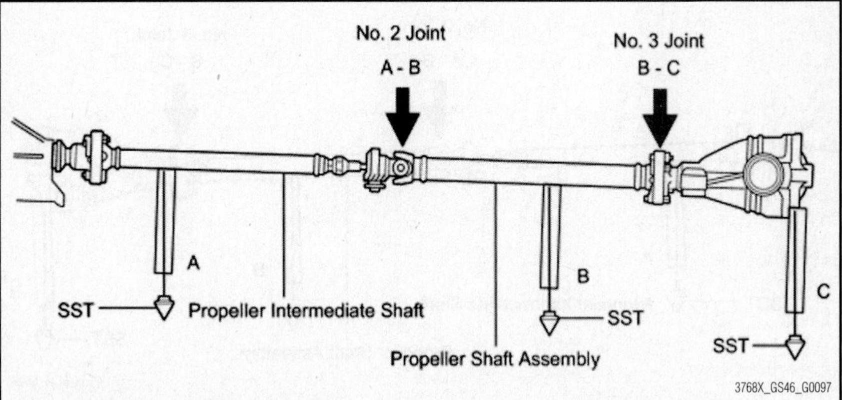

Fig. 33 Measuring the installation angle of the propeller intermediate shaft and propeller shaft assembly

31. For 3.5L engines, inspect and adjust no. 2 and no. 3 joint angles:

a. Stabilize the propeller shaft and differential. Turn the propeller shaft several times by hand to stabilize the center support bearing.

b. Check the No. 2 and No. 3 joint angles. Using SST, measure the installation angle of the propeller intermediate shaft and propeller shaft assembly.

➡**The SST should be set directly underneath the shaft.**

➡**Measure the installation angle by placing SST in the positions shown in the illustration.**

c. Using SST: 09370-50010, measure the installation angle of the differential.

d. Calculate the No. 2 joint angle:

- No. 2 joint angle: A - B = 0°11' to 1°11'
- A: Intermediate shaft installation angle
- B: Propeller shaft installation angle

e. Calculate the No. 3 joint angle:

- No. 3 joint angle: B - C = 0°53' to 1°53'
- B: Propeller shaft installation angle
- C: Differential installation angle

➡**If the measured angle is not within the specified range, adjust it with the center support bearing washers.**

f. Adjust the No. 2 joint angle. Select the center support bearing washers for adjustment.

➡**The 2 washers should be of the same thickness.**

32. Install the front floor No. 1 heat insulator with the 4 nuts and grommet.

33. Install the air guide plate outside RH with the 4 nuts.

34. To complete installation, reverse remaining removal procedure.

4WD Models

See Figures 36 through 38.

1. Remove front floor cover center LH.
2. Remove front floor brace center.
3. Remove Heated Oxygen Sensor (HO2S) (for sensor 2).
4. Remove front exhaust pipe assembly.
5. Remove front floor no. 1 heat insulator.
6. Remove the 4 nuts and air guide plate outside RH.
7. Put matchmarks on the differential companion flange, flexible coupling and propeller shaft.
8. Remove the 3 bolts, 3 washers and 3 nuts.

➡**The propeller shaft and flexible coupling should not be separated.**

9. Remove the 2 bolts, 2 center support bearing washers and center support bearing.

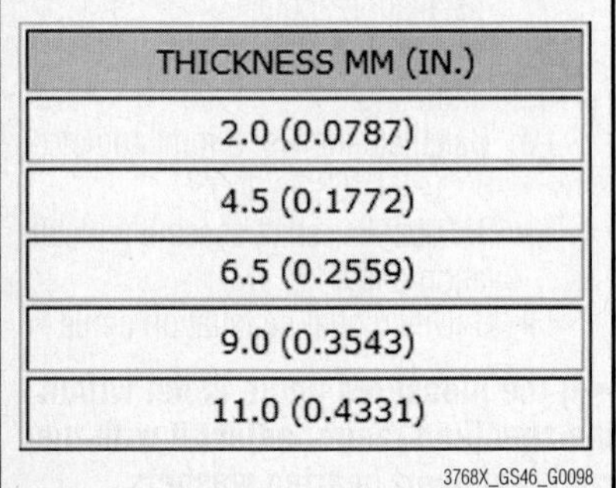

THICKNESS MM (IN.)
2.0 (0.0787)
4.5 (0.1772)
6.5 (0.2559)
9.0 (0.3543)
11.0 (0.4331)

3768X_GS46_G0098

Fig. 34 Center support bearing washer thickness

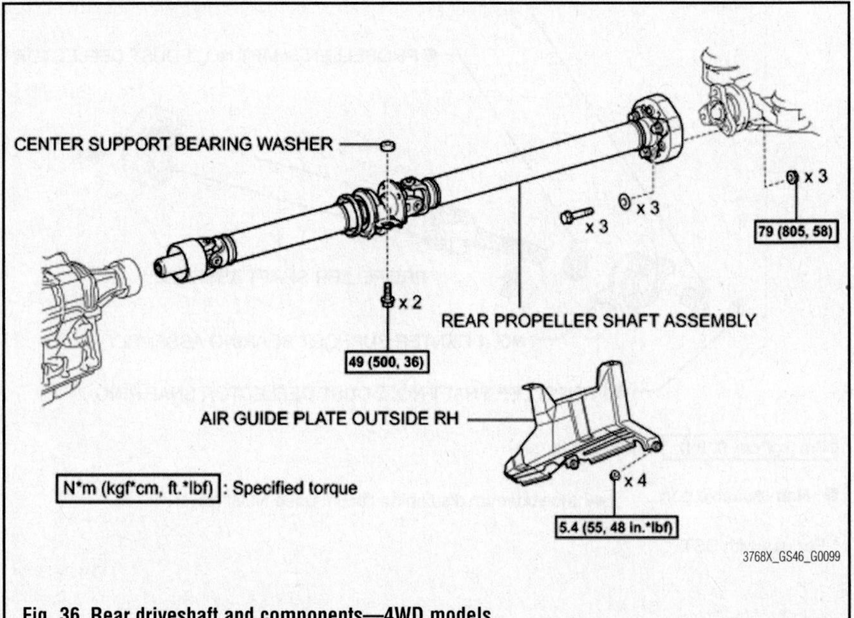

Fig. 36 Rear driveshaft and components—4WD models

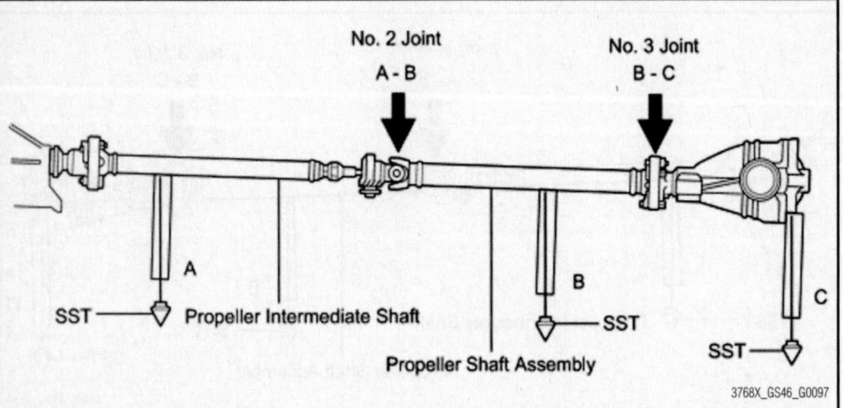

Fig. 35 Measuring the installation angle of the propeller intermediate shaft and propeller shaft assembly

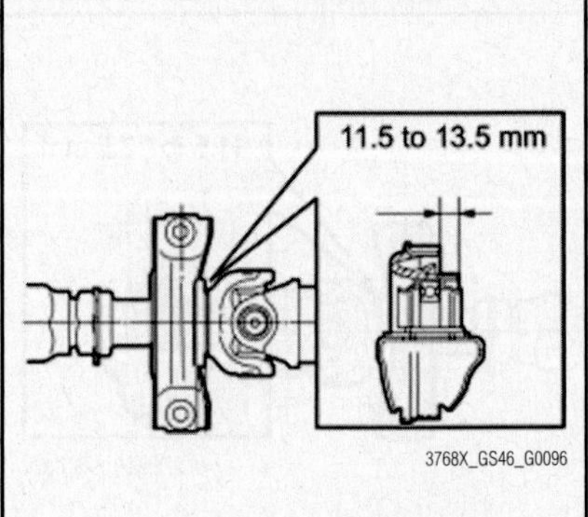

Fig. 37 Adjust the dimension between the edge surface of the center support bearing and the edge surface of the cushion

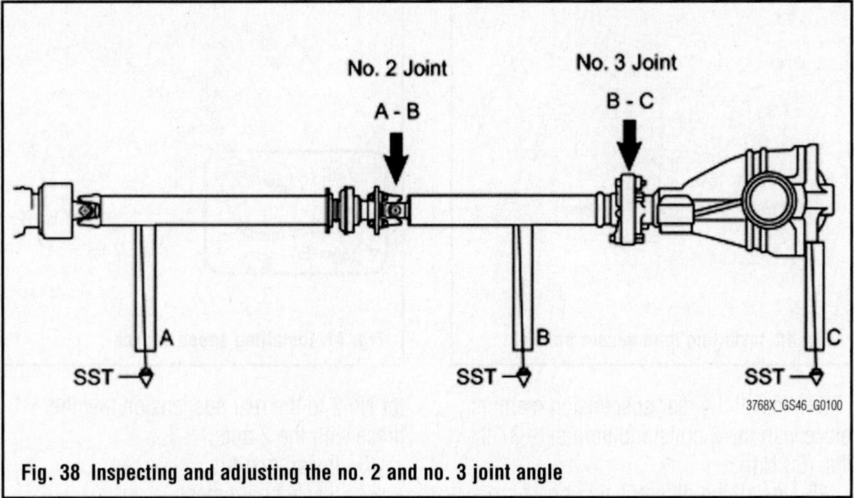

Fig. 38 Inspecting and adjusting the no. 2 and no. 3 joint angle

10. Insert SST 09325-40010 in the transmission to prevent oil leakage.

➡**Be careful not to damage the oil seal.**

To install:

11. Remove the SST: 09325-40010 from the transmission.
12. Insert the yoke of the intermediate shaft into the transmission.

➡**Be careful not to damage the oil seal.**

13. Install the 2 center support bearing washers and center support bearing, and temporarily tighten the 2 bolts.
14. Align the matchmarks on the differential companion flange and flexible coupling.
15. Install and torque the 6 bolts, washers and nuts. Tighten to 58 ft. lbs. (79 Nm).

➡**Be careful not to damage the flexible coupling centering bushing.**

➡**The bolts should be installed from the propeller shaft side.**

Adjust the dimension between the edge surface of the center support bearing and the edge surface of the cushion to 0.4528 to 0.5315 inches (11.5 to 13.5 mm) respectively as shown in the illustration.

16. Check that the center line of the bracket is at right angles to the shaft axial direction.
17. Tighten the 2 bolts to 36 ft. lbs. (49 Nm).
18. Inspect and adjust no. 2 and no. 3 joint angle:
    a. Stabilize the propeller shaft and differential. Turn the propeller shaft several times by hand to stabilize the center support bearing.
    b. Check the No. 2 and No. 3 joint angles. Using SST 09370-50010, measure the installation angle of the intermediate shaft and propeller shaft.

➡ **The SST should be set directly underneath the shaft.**

    c. Using SST, measure the installation angle of the differential.

➡ **Measure the installation angle by placing the SST in the position shown in the illustration.**

    d. Calculate the No. 2 joint angle.
    - No. 2 joint angle: A - B = 0°16' to 1°16'
    - A: Intermediate shaft installation angle
    - B: Propeller shaft installation angle
    e. Calculate the No. 3 joint angle.
    - No. 3 joint angle:
    - B - C = 0°55' to 1°55'
    - B: Propeller shaft installation angle
    - C: Differential installation angle
    f. Adjust the No. 2 joint angle. Select the center support bearing washers for adjustment.
19. To complete installation, reverse remaining removal procedure.

REAR HALFSHAFTS

REMOVAL & INSTALLATION

See Figures 39 through 41.

1. Remove rear tire.
2. Remove the bolt and nut, and separate the load sensing valve sensor bracket and stabilizer link assembly.
3. Remove the 2 nuts and differential support protector No.2 from the suspension member brace.
4. Remove the 2 bolts and suspension member brace.
5. Remove the 2 bolts, and separate the parking brake cable No.3.
6. Release the staked part of the axle shaft nut.
7. Remove the axle shaft nut.
8. Remove the 2 bolts, and separate the speed sensor from the axle carrier.
9. Remove the 2 bolts, and disconnect the rear disc brake caliper assembly.
10. Remove the caliper plates No.1 from the brake caliper.
11. Remove rear disc.
12. Separate upper control arm assembly rear.
13. Separate rear suspension arm assembly.
14. Push the rear axle carrier toward the outside of the vehicle. Using a plastic hammer, separate the rear drive shaft assembly from the rear axle carrier.
15. Remove the rear drive shaft assembly.

To install:

16. Coat the spline of the inboard joint shaft assembly with gear oil.
17. Set the shaft snap ring with the opening side facing down.
18. Align the shaft splines and install the drive shaft assembly with a brass bar and hammer.
19. Install the rear drive shaft assembly to the rear axle carrier.
20. Install upper control arm assembly.
21. Temporarily tighten upper control arm assembly.
22. Temporarily tighten rear suspension arm assemblies.
23. Install the stabilizer link assembly and the load sensing valve sensor bracket to the rear suspension arm assembly No.2 with the bolt and nut and tighten to 20 ft. lbs. (27 Nm).
24. Install rear disc.
25. Install the rear disc brake caliper assembly and caliper plates No.1 with the 2 bolts and tighten to 40 ft. lbs. (54 Nm).
26. Install the speed sensor to the rear axle carrier with the 2 bolts and tighten to 75 inch lbs. (8.5 Nm) and 53 inch lbs. (6.0 Nm).
27. Clean the threaded parts on the drive shaft and rear axle shaft nut using a non-residue solvent.
28. Install a new axle shaft nut and tighten to 214 ft. lbs. (290 Nm).
29. Using a chisel and a hammer, stake the axle shaft nut.
30. Install the parking brake cable assembly No.3 with the 2 bolts and tighten to 14 ft. lbs. (19 Nm).
31. Stabilize suspension.

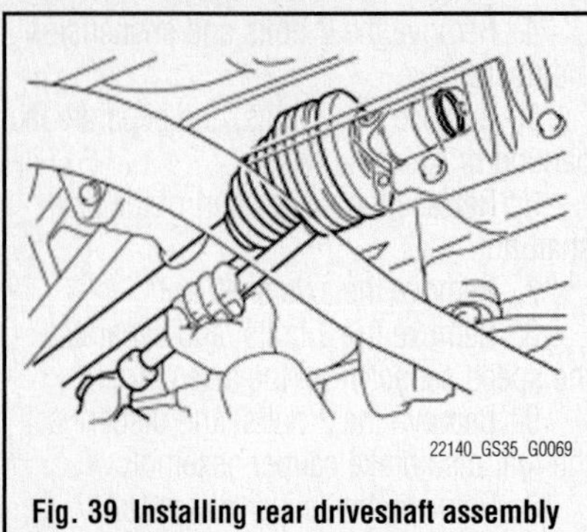

Fig. 39 Installing rear driveshaft assembly

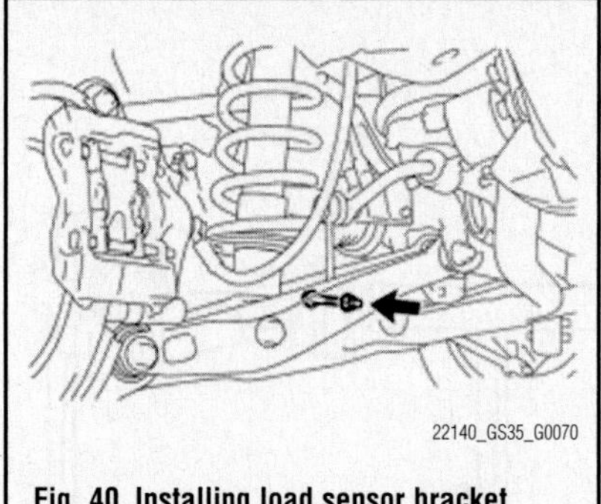

Fig. 40 Installing load sensor bracket

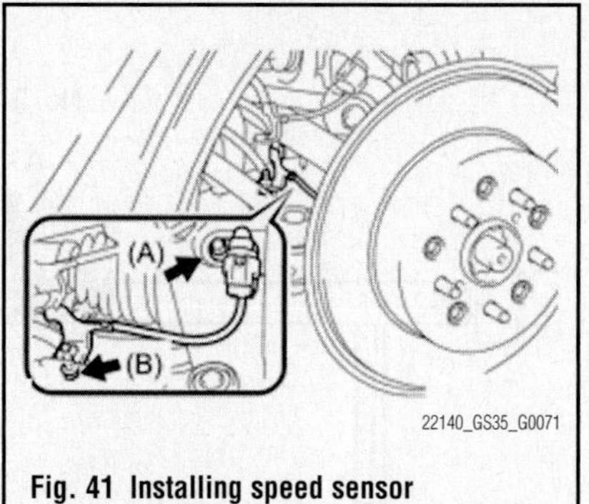

Fig. 41 Installing speed sensor

32. Fully tighten upper control arm assembly.
33. Fully tighten rear suspension arm assemblies.
34. Install the rear suspension member brace with the 2 bolts and tighten to 37 ft. lbs. (50 Nm).
35. Install the differential support protector No.2 to the rear suspension member brace with the 2 nuts.
36. Install rear tire.
37. Inspect alignment.

ENGINE COOLING

ENGINE FAN

REMOVAL & INSTALLATION

1. Before servicing the vehicle, refer to the precautions in the beginning of this section.
2. Disconnect the negative battery cable. Wait at least 90 seconds before performing any work.
3. Drain the engine coolant.
4. Remove the radiator assembly (the fan assembly is attached).
5. Separate the fan assembly from the radiator.

To install:

6. Install the fan assembly to the radiator and tighten the fasteners to 44 inch lbs. (5 Nm).
7. Install the radiator and fan assembly.
8. Refill the engine cooling system.
9. Connect the negative battery cable.
10. Start the engine and check system for coolant leaks.
11. Install the engine under cover No. 1.

RADIATOR

REMOVAL & INSTALLATION

3.5L Engines

1. Before servicing the vehicle, refer to the precautions in the beginning of this section.
2. Disconnect the negative battery cable. Wait at least 90 seconds before performing any work.
3. Remove the engine under cover.
4. Drain engine coolant.
5. Remove the 2 nuts and V-bank cover.
6. Using a clip remover, remove the 7 clips and cool air intake duct seal.
7. Remove the left and right engine room side covers.
8. Remove the bolt and No. 1 air cleaner inlet.
9. Remove the air cleaner cap with air cleaner hose.
10. Remove the front bumper cover as follows:
    a. Using a clip remover, remove the 2 clips. Put protective tape under the front fender.
    b. Remove the 6 screws and 5 bolts.
    c. Pull the bumper cover to detach the 3 claws on the LH side.
    d. Pull the bumper cover to detach the 3 claws on the RH side and remove the bumper cover.
    e. Remove millimeter wave radar sensor assembly.
11. Remove the 4 clips and the radiator support opening cover.
12. Remove the hood lock control cable cover.
13. Remove the hood lock assembly.
14. Remove the upper radiator support sub-assembly.
15. Disconnect the hose from the water inlet with thermostat and radiator tank upper.
16. Disconnect the hose from the water inlet and radiator tank lower.
17. Remove the radiator assembly as follows:
    a. Disconnect the harness from the cooling fan ECU.
    b. Remove the 4 bolts and radiator.

➡Do not allow the cooler condenser and radiator to come into contact with each other.

    c. Detach the 3 claws and remove the fan.
    d. Remove the 2 radiator support cushions and 2 radiator lower supports from the radiator.

To install:

18. Install the radiator assembly as follows:
    a. Install the 2 radiator support cushions and 2 radiator lower supports to the radiator.
    b. Install the fan to the radiator, and attach the 3 claws.
    c. Align the radiator with the condenser. Then install the radiator.
19. Connect the radiator hose to the water inlet with thermostat and radiator tank upper.
20. Connect the radiator hose to the water inlet and radiator tank lower.
21. Install the upper radiator support sub-assembly.
22. Install the hood lock assembly.
23. Install the hood lock control cable cover.
24. Install the radiator support opening cover.
25. Refill the engine cooling system.
26. Check for engine coolant leaks.
27. Install millimeter wave radar sensor assembly
28. Install the front bumper cover.
29. Install the air cleaner cap with air cleaner hose.
30. Install the No. 1 air cleaner inlet with the bolt and tighten to 44 inch lbs. (5 Nm).

31. Install the V-bank cover and tighten the 2 nuts to 44 inch lbs. (5 Nm).
32. Install the left and right engine room side covers.
33. Install the cool air intake duct seal.
34. Install the engine under cover.
35. Connect the negative battery cable.
36. Perform the system initialization procedure.

4.6L Engines

1. Remove cool air intake duct seal.
2. Remove engine room side cover LH and RH.
3. Remove engine room side cover RH.
4. Remove v-bank cover sub-assembly.
5. Remove no. 1 air cleaner inlet.
6. Remove air cleaner cap sub-assembly.
7. Remove air cleaner case sub-assembly.
8. Remove engine under cover.
9. Drain engine coolant.
10. Remove front bumper assembly.
11. Remove front bumper energy absorber.
12. Remove radiator support opening cover. Remove the 6 clips and radiator support opening cover.
13. Remove millimeter wave radar sensor assembly (w/ Dynamic Radar Cruise Control System)
14. Remove hood lock control cable cover. Remove the 3 screws, claw and the hood lock control cable cover.
15. Remove hood lock assembly.
16. Remove radiator reserve tank assembly. Disconnect 2 reserve tank hoses, then remove the 2 bolts and radiator reserve tank assembly.
17. Remove engine room ECU outlet duct. Remove the engine room ECU outlet duct from the engine room ECU box.
18. Remove upper radiator support sub-assembly:
    a. Disconnect the smog ventilation sensor connector.
    b. Disconnect the 2 horn connectors. Separate the 5 clamps and the wire harness from the upper radiator support.
    c. Separate the 2 clamps and the wire harness from the upper radiator support. Separate the 3 clamps and the wire harness from the fan shroud.
    d. Disconnect the 2 cooling fan motor connectors.
    e. Remove the 5 bolts and upper radiator support.
19. Disconnect oil cooler hose. Disconnect the 2 oil cooler hoses from the oil cooler pipe
20. Disconnect no. 1 radiator hose. Disconnect the No. 1 radiator hose from the radiator assembly.
21. Disconnect no. 2 radiator hose. Disconnect the No. 2 radiator hose from the radiator assembly.
22. Remove radiator and fan assembly:
    a. Remove the 4 bolts, and separate the cooler condenser assembly from the radiator assembly.
    b. Remove the radiator assembly from the vehicle together with the fan assembly.

➡ Make sure that the cooler condenser assembly and radiator assembly do not come into contact with each other.

23. Make sure that the cooler condenser assembly and radiator assembly do not come into contact with each other. Remove the 2 radiator support cushions.
24. Remove lower radiator support. Remove the 2 lower radiator supports
25. Remove radiator assembly.
    a. Disconnect the 2 oil cooler hoses from the radiator assembly.
    b. Remove the 2 bolts and oil cooler pipe.
    c. Remove the 2 bolts.(d) Release the snap fit and lift the fan assembly with motor from the radiator assembly.
26. To install, reverse the removal procedure.
27. Add engine coolant. Inspect for coolant leak.

THERMOSTAT

REMOVAL & INSTALLATION

3.5L Engines

See Figure 42.

1. Before servicing the vehicle, refer to the precautions in the beginning of this section.
2. Disconnect the negative battery cable. Wait at least 90 seconds before performing any other work.

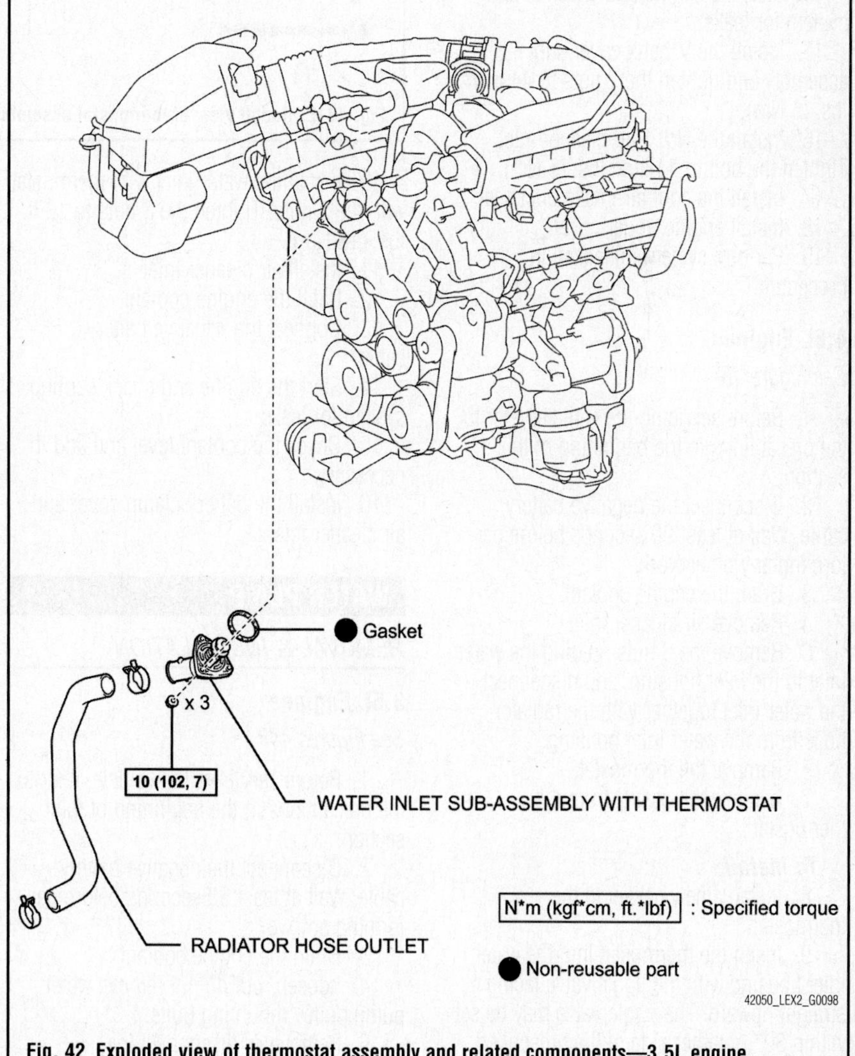

Fig. 42 Exploded view of thermostat assembly and related components—3.5L engine

3. Remove engine under cover.

4. Remove the cool air intake duct seal.

5. Remove the bolt and No. 1 air cleaner inlet.

6. Remove the 2 nuts and the V-bank cover sub-assembly.

7. Drain the engine coolant.

8. Disconnect the hose from the water inlet with thermostat.

9. Remove the 3 nuts, water inlet with thermostat and gasket.

To install:

10. Install a new gasket and the water inlet with thermostat with the 3 nuts. Tighten the 3 nuts to 7 ft. lbs. (10 Nm).

11. Connect the radiator outlet hose to the radiator assembly and secure it with the clip. Make sure that the claws on the clip are positioned outward with space between the hose end and the clip end should be less than 1–5mm. Hose contact with the stopper should be less than 2.5mm.

12. Refill the engine coolant.

13. Connect the negative battery cable.

14. Start the engine and check cooling system for leaks.

15. Install the V-bank cover sub-assembly and tighten the 2 nuts to 44 inch lbs. (5 Nm).

16. Install the No. 1 air cleaner inlet. Tighten the bolt to 44 inch lbs. (5 Nm).

17. Install the cool air intake duct seal.

18. Install engine under cover.

19. Perform system initialization procedure.

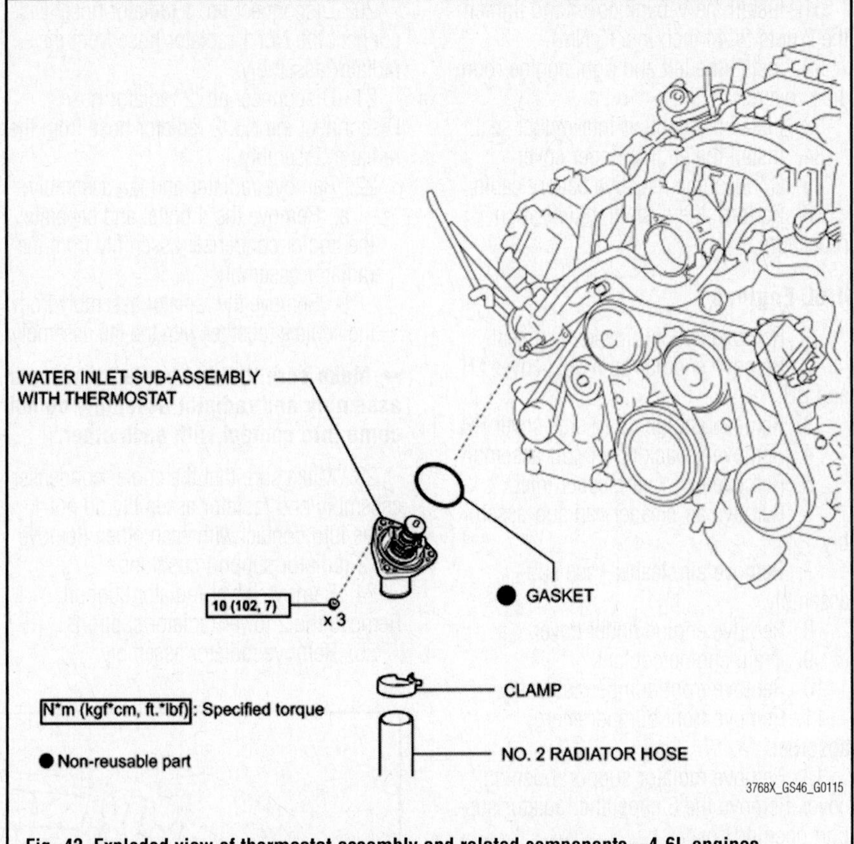

Fig. 43 Exploded view of thermostat assembly and related components—4.6L engines

4.6L Engines

See Figure 43.

1. Before servicing the vehicle, refer to the precautions in the beginning of this section.

2. Disconnect the negative battery cable. Wait at least 90 seconds before performing any other work.

3. Drain the engine coolant.

4. Remove air cleaner inlet.

5. Remove the 3 nuts holding the water inlet to the inlet housing, and disconnect the water inlet together with the radiator hose from the water inlet housing.

6. Remove the thermostat.

7. Remove the gasket from the thermostat.

To install:

8. Install a new gasket to the thermostat.

9. Insert the thermostat into the water inlet housing with the jiggle valve facing straight upward. The jiggle valve may be set within 30° of either side of the prescribed position.

10. Install the water inlet with thermostat with the 3 nuts. Tighten the 3 nuts to 13 ft. lbs. (18 Nm).

11. Install air cleaner inlet.

12. Refill the engine coolant.

13. Connect the negative battery cable.

14. Start the engine and check cooling system for leaks.

15. Check the coolant level and add if necessary.

16. Install the battery clamp cover and air cleaner inlet.

WATER PUMP

REMOVAL & INSTALLATION

3.5L Engines

See Figures 44.

1. Before servicing the vehicle, refer to the precautions in the beginning of this section.

2. Disconnect the negative battery cable. Wait at least 90 seconds before performing any work.

3. Drain the engine coolant.

4. Loosen, but do not remove water pump pulley mounting bolts.

5. Remove or disconnect the following:

- Engine under cover
- Cool air intake duct seal
- No. 1 inlet air cleaner
- V-bank engine cover
- Serpentine drive belt
- Injector driver
- Water pump pulley
- No. 2 engine cover
- No. 1 engine cover
- Mounting fasteners and injector driver unit
- Radiator inlet and outlet hoses
- 5 hoses from the water inlet housing
- Water inlet housing, gasket and O-ring
- Bolt, cover plate and No. 2 idler pulley
- Bolt, cover plate and belt tensioner pulley (marking: L)

➡Do not turn the bolt "L" counter-clockwise.

- 16 bolts, water pump assembly and water pump gasket

To install:

6. Install a new water pump gasket and water pump assembly with the 16 mounting bolts and tighten as follows:

- Bolt A: 15 ft. lbs. (21 Nm)
- Bolt B: 81 inch lbs. (9 Nm)
- Bolt C: 81 inch lbs. (9 Nm)

➡Be sure to replace 2 bolts C with new ones or reuse them after applying adhesive 1344. Make sure that there is no oil on the threads of the A bolts.

7. Install or connect the following:
   - Belt tensioner pulley and cover plate with bolt. Torque the bolt to 32 ft. lbs. (43 Nm).

WARNING

Be careful when tightening the bolt because it is left-hand threaded.

   - No. 2 idler pulley, cover plate and bolt. Torque the bolt to 32 ft. lbs. (43 Nm).
   - Water inlet housing, new gasket and O-ring. Torque the 4 mounting bolts and nut to 7 ft. lbs. (10 Nm).

➡Be careful not to allow the O-ring to get caught between parts.

   - 5 hoses to the water inlet housing
   - Inlet and outlet radiator hoses and new clamps
   - Water pump pulley with 4 bolts finger tight only
   - Injector driver unit
   - No. 1 engine cover
   - No. 2 engine cover
   - Serpentine drive belt
   - V-bank engine cover
   - No. 1 inlet air cleaner
   - Cool air intake duct seal
   - Engine under cover
   - Negative battery cable

8. Torque the 4 water pump pulley bolts to 15 ft. lbs. (21 Nm).
9. Add engine coolant.
10. Start the engine, check for leaks and bleed the cooling system.
11. Recheck all fluid levels and add if necessary.
12. Perform system initialization (which includes power window control system, sliding roof system, clearance sonar system and variable gear ratio steering system) procedure as follows:

   - Power window control system

a. Turn the ignition switch on.
b. Open power window halfway by pressing power window switch.
c. Fully pull up the switch until the power window is fully closed and continue to hold the switch for at least 1 second.
d. Check that the AUTO UP / DOWN function operates normally.

➡If the remote UP / DOWN function does not operate after the conditions 1), 2), or 3) is satisfied, the power window regulator master switch may have a malfunction.

   - Sliding roof system

e. Turn the ignition switch on.
f. If the sliding roof is opened, close it fully.
g. Push the open switch of the slide switch, or the up switch of the tilt switch on the personal light, making the sliding roof tilt up approximately 1 second, tilt down, slide open, slide close.
h. Sliding roof stops at the fully closed position.
i. Finish the initialization.
j. Check that the operation works normally with AUTO operation.

   - Clearance sonar system

k. Turn the ignition switch on.
l. Turn the clearance sonar main switch ON.
m. Turn the steering wheel to the full left and right lock position.

➡Make sure to completely turn the steering wheel to the left and right full lock position.

n. Confirm that the learning operation has been completed by checking the multi-information display.
o. At an area with few turns and curves, and minimal traffic, drive at 20 km/h or more for 5 minutes or more.

   - Variable gear ratio steering system

p. Turn the ignition switch on, and check that the master warning light and VSC/ABS warning lights illuminate for a few seconds.

➡If the warning lights remain on or blink, repair the applicable system.

q. Drive the vehicle on a straight road at 35 km/h (22 mph) or more for 5 seconds or longer.
r. Confirm that steering angle sensor initialization is completed by doing the following:

   - Drive the vehicle on a straight road at 60 km/h (37 mph) or more for 30 seconds or longer.
   - Stop the vehicle (engine running).
   - Slowly turn the steering wheel from lock to lock.
   - If it turns approximately 2.7 turns, steering angle sensor initialization is completed. If it turns approximately 3.2 turns, steering angle sensor initialization is not completed.

13. Road test the vehicle.

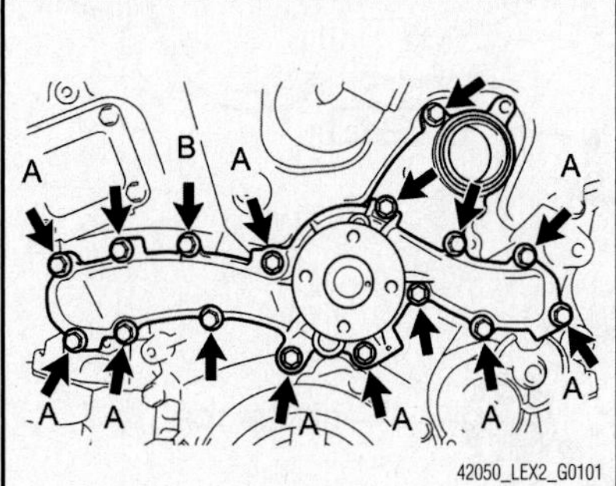

Fig. 44 Water pump mounting bolts—3.5L (2GR-FSE) engine

4.6L Engines

See Figures 45 and 46.

1. Remove engine under cover.
2. Drain engine coolant.
3. Remove v-bank cover sub-assembly.
4. Remove cool air intake duct seal.
5. Remove no. 1 air cleaner inlet.
6. Remove engine room side cover RH.
7. Remove air cleaner cap sub-assembly.
8. Remove radiator reserve tank assembly.
9. Remove v-ribbed belt.
10. Disconnect no. 2 radiator hose.

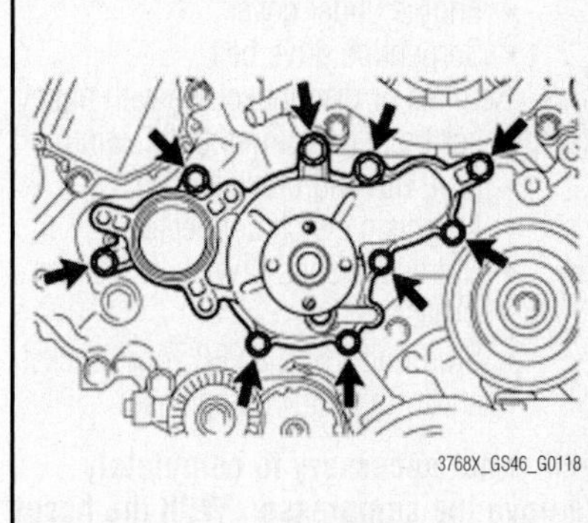

Fig. 45 Remove the 9 bolts, water pump and gasket

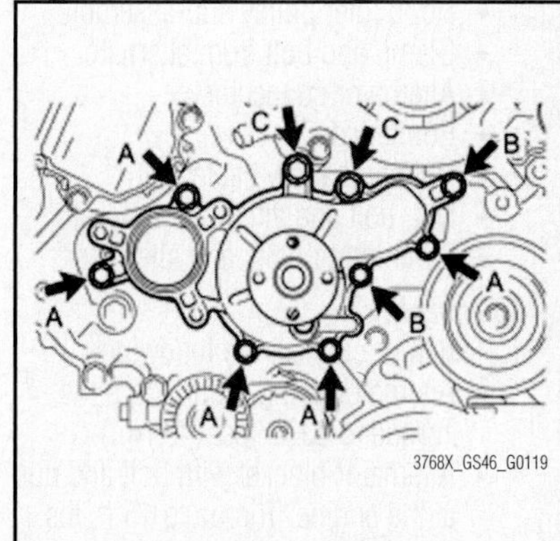

Fig. 46 Water pump bolt identification

11. Disconnect the No. 2 radiator hose from the water inlet sub-assembly with thermostat.

12. Disconnect the No. 5 water by-pass hose from the water inlet housing.

13. Disconnect the water inlet hose from the water inlet housing.

14. Disconnect the No. 3 water by-pass hose from the water inlet housing.

15. Remove the 3 bolts, water inlet housing and gasket.

16. Using a screwdriver or an equivalent, hold the water pump pulley.

17. Remove the 4 bolts and water pump pulley.

18. Remove the 9 bolts, water pump and gasket.

To install:

19. Install the water pump and a new gasket with the 9 bolts as shown in the illustration.

a. Bolt A: tighten to 15 ft. lbs. (20 Nm).

b. Bolt B: tighten to 17 ft. lbs. (23 Nm).

c. Bolt C: tighten to 35 ft. lbs. (47 Nm).

20. Install water pump pulley:

a. Temporarily install the pulley with the 4 bolts.

b. Using a screwdriver or an equivalent, hold the pulley and tighten the 4 bolts. Tighten to 15 ft. lbs. (21 Nm).

21. Install water inlet housing:

a. Install a new gasket to the water pump.

b. Install the water inlet housing with the 3 bolts. Tighten to 15 ft. lbs. (21 Nm).

22. To complete the installation, reverse the remaining removal procedure.

ENGINE ELECTRICAL — CHARGING SYSTEM

ALTERNATOR

REMOVAL & INSTALLATION

3.5L Engine

1. Before servicing the vehicle, refer to the precautions in the beginning of this section.
2. Remove or disconnect the following:
   - Negative battery cable. Wait at least 90 seconds before performing any other work.
   - V-bank cover sub-assembly
   - Cool air intake duct seal
   - Left engine room side cover
   - Engine under cover
   - Serpentine drive belt
3. Remove or disconnect the with pulley compressor by performing the following:
   - Bolt, nut and bracket (AWD)
   - Magnetic clutch connector
   - Nut and 3 bolts (2WD); or 2 bolts (AWD)
   - Stud bolt using an E8 Torx® socket and with pulley compressor

➡It is not necessary to completely remove the compressor. With the hoses connected to the compressor, hang the compressor on the vehicle body with a rope.

   - No. 2 idler pulley sub-assembly
   - Clamp and bolt from alternator
   - Alternator connector
   - Rubber cap
   - Nut and battery cable.
   - Nut, bolt and alternator bracket
   - 2 mounting bolts and alternator

To install:

4. Install or connect the following:
   - Alternator and 2 mounting bolts. Torque to 32 ft. lbs. (43 Nm).
   - Alternator bracket with bolt and nut at the engine. Torque to 15 ft. lbs. (20 Nm).
   - Battery cable. Tighten the nut to 87 inch lbs. (10 Nm).
   - Wire harness bracket and clamp to the alternator
   - Alternator wire to terminal.
   - Terminal cap
   - No. 2 idler pulley and cover plate with the bolt. Tighten bolt to 32 ft. lbs. (43 Nm).
   - With pulley compressor assembly
   - Serpentine drive belt
   - Engine under cover
   - Left engine room side cover
   - Cool air intake duct seal
   - V-bank cover sub-assembly
   - Negative battery cable

4.6L Engines

See Figure 47.

1. Before servicing the vehicle, refer to the precautions in the beginning of this section.
2. Remove or disconnect the following:
   - Negative battery cable. Wait at least 90 seconds before performing any other work.
   - Air cleaner inlet
   - Accessory drive belt
   - Oil pan protector
   - Engine under cover
   - Power steering pump
   - Alternator harness connectors

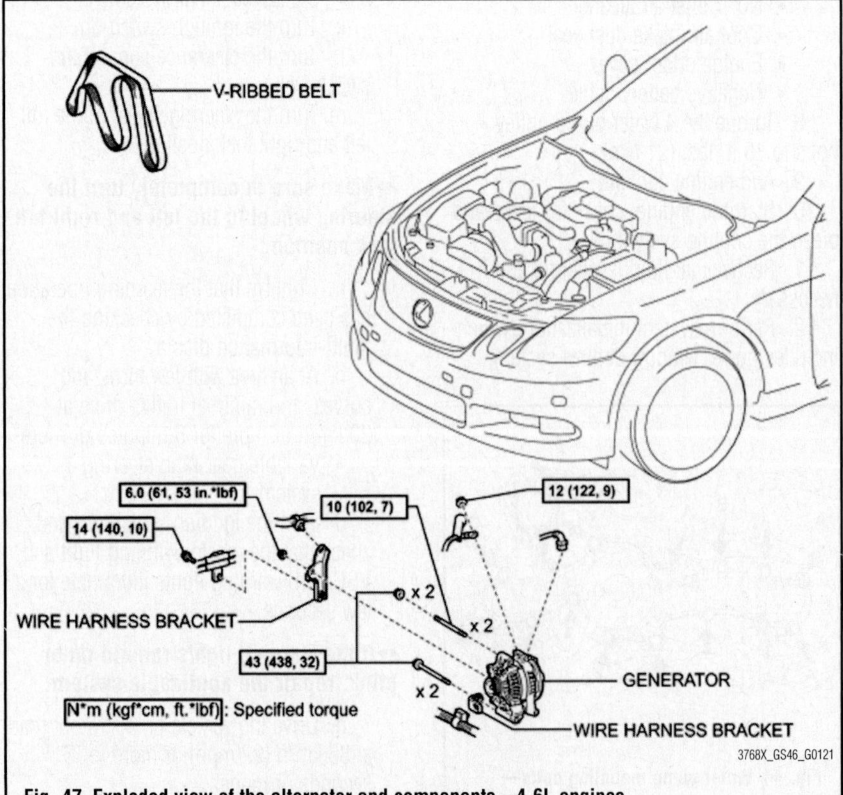

Fig. 47 Exploded view of the alternator and components—4.6L engines

- Heated Oxygen (HO_2S) sensor wiring
- Alternator

To install:

3. Install or connect the following:

- Alternator. Tighten the fasteners to 29 ft. lbs. (39 Nm).
- HO_2S sensor wiring
- Alternator harness connectors
- Power steering pump
- Engine under cover
- Oil pan protector
- Accessory drive belt
- Air cleaner inlet
- Negative battery cable

ENGINE ELECTRICAL

IGNITION SYSTEM

FIRING ORDER

See Figures 48 and 49.

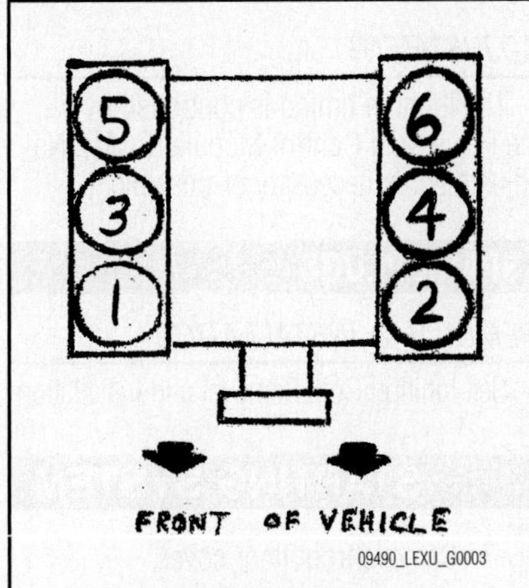

Fig. 48 3.5L (2GR-FSE) Engines Firing order: 1–2–3–4–5–6 Distributorless ignition system (one coil per cylinder)

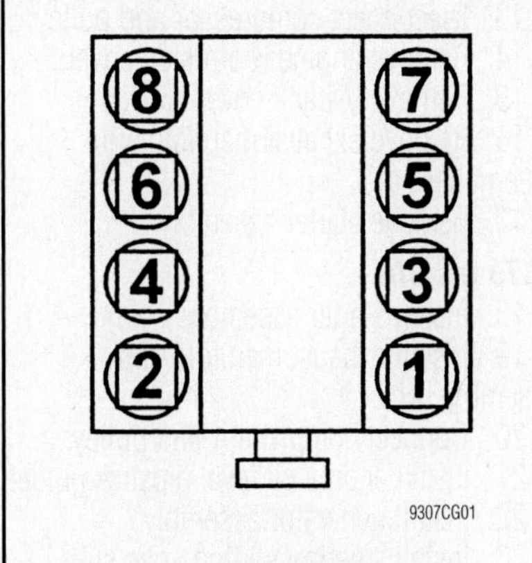

Fig. 49 4.6L (1UR-FSE) Engines Firing order: 1–8–4–3–6–5–7–2 Distributorless ignition system (one coil on each cylinder)

IGNITION COIL

REMOVAL & INSTALLATION

GS350

1. Before servicing the vehicle, refer to the precautions in the beginning of this section.
2. Disconnect the negative battery cable. Wait at least 90 seconds before performing any other work.
3. Remove cool air intake duct seal.
4. Remove the right and left engine room side covers.
5. Remove V-bank cover.
6. Disconnect the ventilation hose from the cylinder head.
7. Disconnect the Mass Air Flow (MAF) meter connector.
8. Disconnect the clamp from the air cleaner.
9. Disconnect the VSV from the EVAP.
10. Disconnect the 4 clamps.
11. Remove the hose clamp and air cleaner cap with air cleaner hose.
12. Disconnect the 2 wire harness clamps.
13. Disconnect the 6 ignition coil connectors.
14. Remove the 6 bolts and 6 ignition coils.

To install:

15. Install the 6 ignition coils with the 6 bolts and torque to 7 ft. lbs. (10 Nm).
16. Connect the 6 ignition coil connectors.
17. Connect the 2 wire harness clamps.
18. Install the air cleaner cap with the air cleaner hose assembly with the 4 clamps and hose clamp.

➡Be sure to install the air cleaner assembly so that the screw part of the hose clamp is as shown in the illustration.

19. Install the VSV for EVAP to the air cleaner hose.
20. Connect the MAF meter connector and clamp to the air cleaner.
21. Connect the ventilation hose to the cylinder head cover with the clamp.
22. Install V-bank cover.
23. Install the right and left engine room side covers.
24. Install cool air intake duct seal.
25. Connect the negative battery cable.
26. Perform system initialization procedure.

GS460 Models

1. Disconnect cable from negative battery terminal.
2. Remove V-bank cover sub-assembly.
3. Remove cool air intake duct seal.
4. Remove engine room side covers.
5. Remove air cleaner assembly.
6. Remove skid control ECU.
7. Remove battery.
8. Remove no. 1 battery tray support.
9. Remove power steering ECU assembly.
10. Remove the 2 bolts and battery tray.
11. Separate No. 1 engine room relay block assembly.
12. Remove the 2 wire harness clamps.
13. Disconnect the 8 ignition coil connectors and remove ignitions coils along with tube gaskets.

To install:

14. To install, reverse removal procedure

IGNITION TIMING

INSPECTION

GS350

See Figures 50 and 51.

1. Warm up the engine and stop the engine.

➡A warmed up engine should have an engine coolant temperature of over 80°C (176°F), have an engine oil temperature of 60°C (140°F), and the engine rpm should be stabilized.

2. When using the intelligent tester, perform the following:
   a. Connect the intelligent tester to the DLC3.

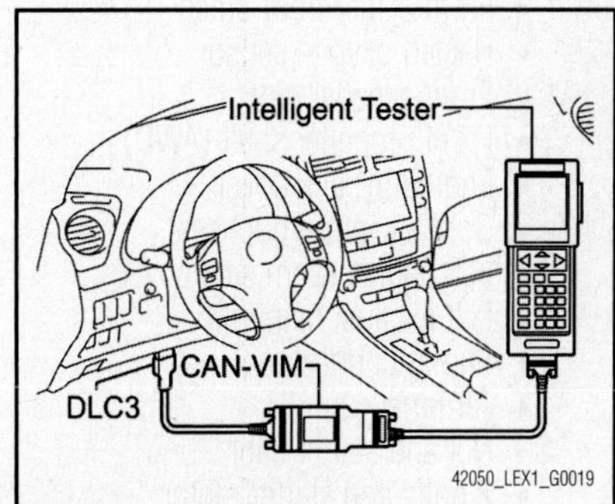

Fig. 50 Using the intelligent tester to check the ignition timing.

b. Start the engine and idle it.

c. Push the intelligent tester main switch ON.

d. Enter the following items: Powertrain / Engine and ECT / Data list / IGN Advance.

e. Ignition timing should measure 5–15°BTDC at idle.

➡Refer to the intelligent tester operator's manual for further details.

3. When not using the intelligent tester, perform the following:

a. Remove the V-bank cover sub-assembly.

b. Connect the tester probe of a timing light to the wire of the ignition connector for No. 1 cylinder. Use a timing light which can detect the first signal.

c. Using SST 09843-18040, connect terminals TC and CG of the DLC3.

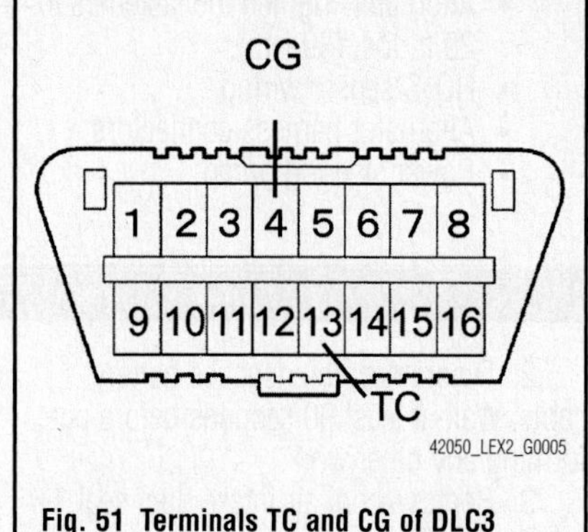

Fig. 51 Terminals TC and CG of DLC3 GS350

⁂ CAUTION

Confirm the terminal numbers before connecting them. Connecting the wrong terminals can damage the engine. When checking the ignition timing, the transmission should be in neutral.

d. Using a timing light, check the ignition timing. Ignition timing should measure 8–12°BTDC at idle.

e. Remove the SST from the DLC3.

f. Check the ignition timing. Ignition timing should measure 5–15°BTDC at idle.

g. Check that the ignition timing advances immediately when the engine speed is increased.

h. Disconnect the timing light from the engine.

i. Install the V-bank cover sub-assembly.

ADJUSTMENT

The ignition timing is controlled by the Powertrain Control Module (PCM). No adjustment is necessary or possible.

SPARK PLUGS

REMOVAL & INSTALLATION

See ignition Coil, Removal and Installation.

ENGINE ELECTRICAL — STARTING SYSTEM

STARTER

REMOVAL & INSTALLATION

3.5L (2GR-FSE) Engine

1. Before servicing the vehicle, refer to the precautions in the beginning of this section.
2. Remove or disconnect the following:
   - Negative battery cable. Wait at least 90 seconds before performing any other work.
   - Front console upper panel garnish
   - Front console upper panel assembly
   - Left and right instrument panel finish panel ends
   - Console box plate
   - Console box register assembly
   - Console box
   - Left front seat
   - Engine under cover
   - Front center floor brace
   - Heated oxygen sensor
   - Front exhaust pipe
   - Front propeller shaft (AWD)
   - Engine V-bank cover
   - Cool air intake duct seal
   - Left engine room side cover
   - Left exhaust manifold
   - Starter connector
   - Terminal cap
   - Nut and starter cable
   - 2 bolts and starter motor

To install:

3. Install or connect the following:
   - Starter assembly with 2 bolts. Torque to 43 ft. lbs. (58 Nm).
   - Starter wires
   - Left exhaust manifold
   - Front exhaust pipe assembly
   - Heated oxygen sensor
   - Front center floor brace. Torque to 65 inch lbs. (7 Nm).
   - Negative battery cable
   - Front propeller shaft assembly (AWD)
   - Engine under cover
   - Left engine room side cover
   - Cool air intake duct seal
   - Engine V-bank cover
   - Left front seat
   - Console box
   - Console box register assembly
   - Console box plate
   - Left and right instrument panel finish panel ends
   - Front console upper panel assembly
   - Front console upper panel garnish
   - System initialization
4. Start the vehicle and check for exhaust leaks.

4.6L (1UR-FSE) Engine

1. Disconnect cable from negative battery terminal.
2. Remove engine under covers.
3. Remove cool air intake duct seal.
4. Remove air cleaner.
5. Remove V-ribbed belt.
6. Remove the front stabilizer bar.
7. Remove front floor cover.
8. Remove front center floor brace.
9. Remove front exhaust pipe assembly.
10. Remove steering sliding yoke sub-assembly.
11. Remove generator assembly.
12. Remove engine oil level dipstick guide.
13. Disconnect compressor and pulley.
14. Remove engine room side covers.
15. Remove V-bank cover.
16. Remove exhaust manifold sub-assemblies.
17. Remove starter assembly.

To install:

18. Install starter assembly.
19. Install exhaust manifold sub-assembly LH.
20. Connect compressor and pulley.
21. Install engine oil level dipstick guide.
22. Install generator assembly.
23. Install steering sliding yoke sub-assembly.
24. Install front exhaust pipe assembly.
25. Install front floor covers.
26. Install front center floor brace.
27. Install the front stabilizer bar.
28. Install V-ribbed belt.
29. Install air cleaner.
30. Connect cable to negative battery terminal.
31. Install engine under covers.
32. Install V-bank cover.
33. Install engine room side covers.
34. Install cool air intake duct seal.

ENGINE MECHANICAL

ACCESSORY DRIVE BELTS

ACCESSORY BELT ROUTING

See Figures 52 and 53.

Refer to the accompanying illustrations.

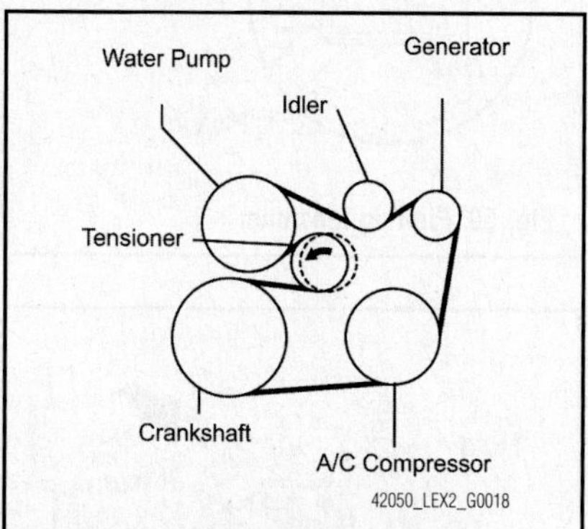

Fig. 52 Serpentine drive belt routing—3.5L (2GR-FSE) engine

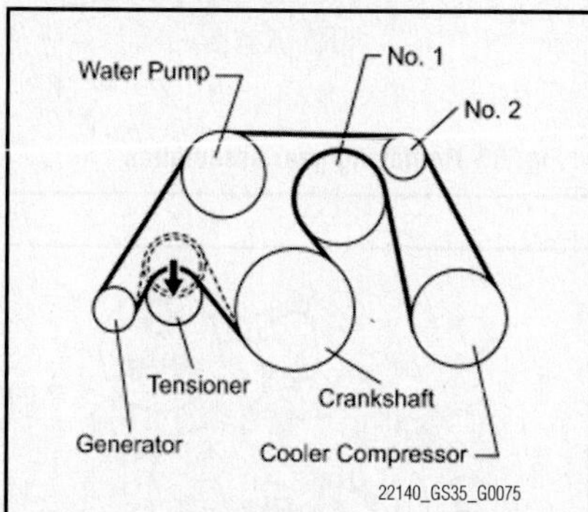

Fig. 53 Serpentine drive belt routing—4.6L (1UR-FSE) engine

INSPECTION

Inspect the drive belt for signs of glazing or cracking. A glazed belt will be perfectly smooth from slippage, while a good belt will have a slight texture of fabric visible. Cracks will usually start at the inner edge of the belt and run outward. All worn or damaged drive belts should be replaced immediately.

ADJUSTMENT

These engines are equipped with automatic belt tensioners. Adjusting the belt tension is not possible or necessary.

REMOVAL & INSTALLATION

3.5L (2GR-FSE) Engine

See Figures 54 and 55.

1. Before servicing the vehicle, refer to the precautions in the beginning of this section.
2. Disconnect the negative battery cable. Wait at least 90 seconds before performing any other work.
3. Remove the cool air intake duct seal.
4. Remove the air cleaner inlet.
5. Remove the V-bank cover.
6. While releasing the belt tension by turning the belt tensioner counterclockwise, and remove the drive belt from the belt tensioner.
7. While turning the belt tensioner counterclockwise, align with its holes, and then insert a 5mm bi-hexagon wrench into the holes to fix the belt tensioner.
8. Visually check the drive belt for excessive wear, frayed cords, chunks missing from its ribs, etc. If any defect has been found, replace the belt.
9. Check that nothing gets caught in the tensioner by turning it clockwise and counterclockwise. If a malfunction exists, replace the tensioner.

To install:

10. Install the drive belt.
11. While turning the belt tensioner counterclockwise, remove the bar.
12. Put the backside of the drive belt on the tensioner pulley and idler pulley. Check that the belt is properly set to each pulley.

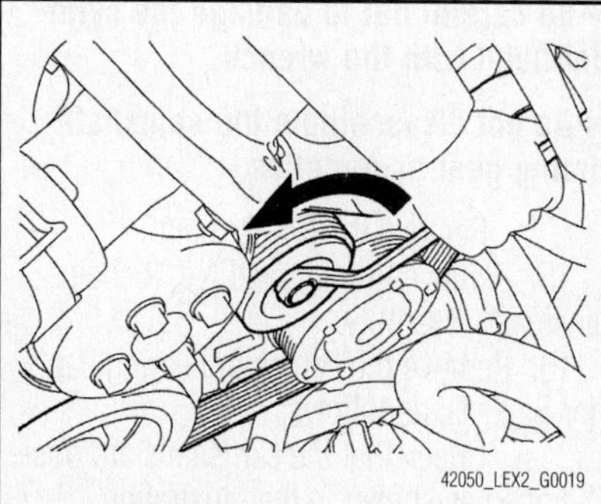

Fig. 54 Loosening the drive belt tension—2GR-FSE engine

Fig. 55 Fix the belt tensioner in place by inserting a 5mm bi-hexagon wrench into the holes—2GR-FSE engine

13. If it is difficult to install the drive belt, perform the following procedure:
    a. Put the belt on every part except the tensioner pulley as shown in the routing illustration.
    b. While releasing the belt tension by turning the belt tensioner counterclockwise, put the belt on the tensioner pulley.
14. Install the V-bank cover.
15. Install the air cleaner inlet with the bolt.
16. Install the cool air intake duct seal.
17. Connect the negative battery cable.

4.6L (1UR-FSE) Engines

See Figures 56 and 57.

1. Before servicing the vehicle, refer to the precautions in the beginning of this section.
2. Disconnect the negative battery cable. Wait at least 90 seconds before performing any other work.
3. Remove the air cleaner inlet.
4. Loosen the belt tension by turning the belt tensioner counterclockwise.

➡The pulley bolt for the belt tensioner has a left-hand thread.

5. Remove the drive belt.
6. Visually check the drive belt for excessive wear, frayed cords, chunks missing from its ribs, etc. If any defect has been found, replace the belt.

To install:

7. Set the drive belt to everything except the idler pulley No. 2, as shown in the routing illustration.
8. Loosen the belt by turning the belt tensioner counterclockwise.
9. Then set the belt to the idler pulley.
10. After a new belt has been installed, check that the mark is within range B as shown in the illustration.
11. Install the air cleaner inlet.
12. Connect the negative battery cable.

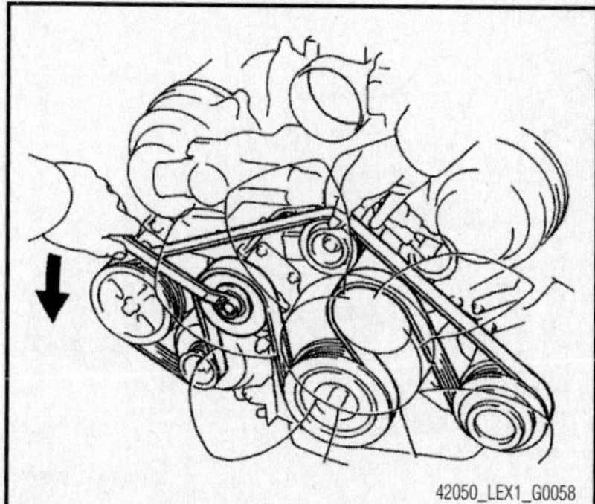

Fig. 56 Loosening the drive belt tension—1UZ-FE engines

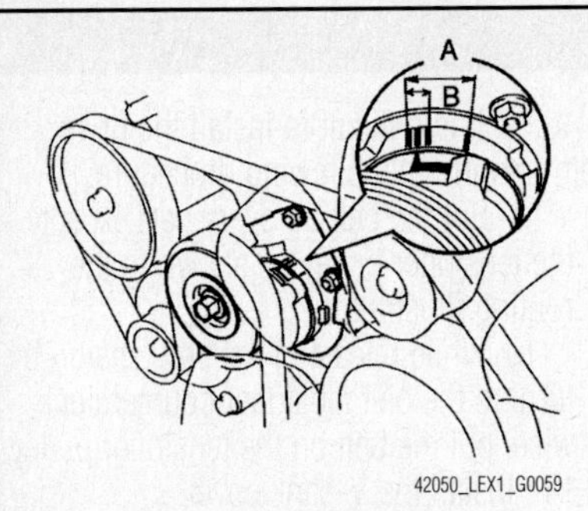

Fig. 57 Check that the mark is within range B as shown on the tensioner pulley—1UR-FSE engines

CAMSHAFT AND VALVE LIFTERS

REMOVAL & INSTALLATION

3.5L (2GR-FSE) Engine

See Figures 58 through 82.

1. Before servicing the vehicle, refer to the precautions in the beginning of this section.
2. Drain the cooling system.
3. Drain the engine oil.
4. Relieve the fuel system pressure.
5. Remove or disconnect the following:
   - Negative battery cable. Wait at least 90 seconds before performing any other work
   - Oil filler cap and gasket
   - Radiator cap
   - No. 1 and 2 engine hangers
   - Spark plugs
   - Ventilation valve
   - 4 camshaft position sensors
   - 4 camshaft timing oil control valves
   - Crankshaft Position (CKP) sensor
   - Left and right side oil check valve bolt, oil pipe union and oil pipe
   - Left and right side oil control valve filter and gaskets

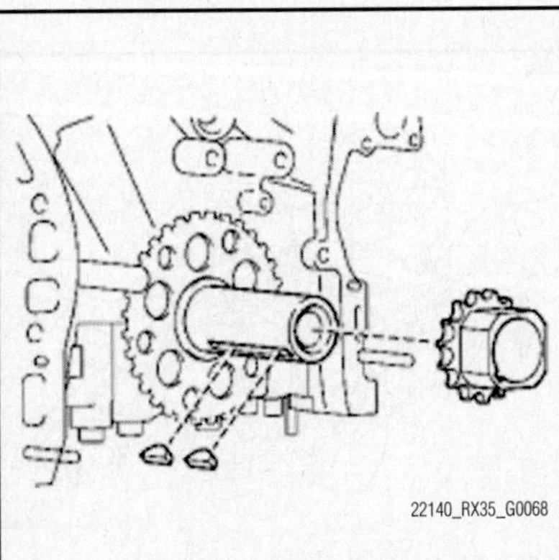

Fig. 58 Crankshaft timing sprocket

   - Cylinder block water drain cocks
   - Oil filter element
   - Water inlet and thermostat assembly
   - Rear water by-pass joint
   - Cylinder head cover and gaskets
6. Set the No. 1 cylinder to TDC/compression.
7. Remove the No. 1 chain tensioner assembly.
8. Remove the chain tensioner slipper.
9. Remove the chain sub-assembly.
10. Remove the idle sprocket assembly.
11. Remove the No. 1 chain vibration damper.
12. Remove the No. 2 chain vibration damper.
13. Remove the crankshaft timing sprocket, as follows:
    a. Remove the crankshaft timing sprocket from the crankshaft.
    b. Remove the 2 pulley set keys from the crankshaft.
14. Remove the camshaft timing gears and No. 2 chain (for Bank 1), as follows:
    a. While raising the No. 2 chain tensioner, insert a pin of 1.0 mm (0.039 in.) into the hole to fix the No. 2 chain tensioner.
    b. Hold the hexagonal portion of the camshaft with a wrench, and remove the 2 bolts and 2 camshaft timing gear assemblies.

➡Be careful not to damage the cylinder head with the wrench.

➡Do not disassemble the camshaft timing gear assemblies.

    c. Remove the No. 2 chain.
15. Remove the bolt and No. 2 chain tensioner assembly.
16. Remove the camshaft bearing cap (for Bank 1), as follows:
    a. Check that the camshafts are positioned as shown in the illustration.
    b. Uniformly loosen and remove the 8 bearing cap bolts in several steps and in the sequence shown in the illustration.
    c. Uniformly loosen and remove the 12 bearing cap bolts in several steps and in the sequence shown in the illustration.

➡Uniformly loosen the bolts while keeping the camshaft level.

    d. Remove the 5 camshaft bearing caps.
17. Remove the camshaft.
18. Remove the No. 2 camshaft.
19. Remove the right hand camshaft housing sub-assembly by prying between the cylinder head and the camshaft housing with a screwdriver with the tip taped.

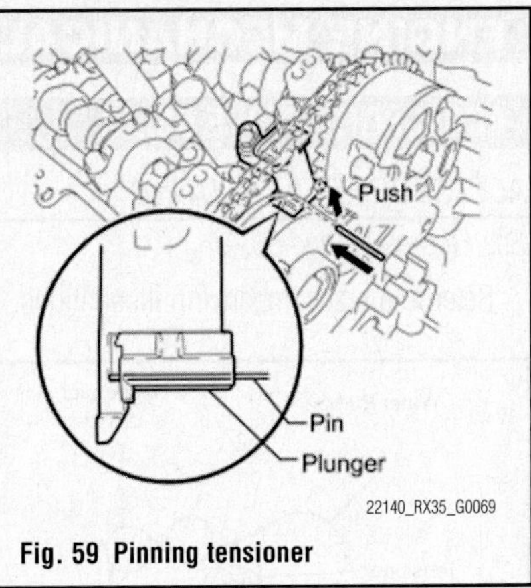

Fig. 59 Pinning tensioner

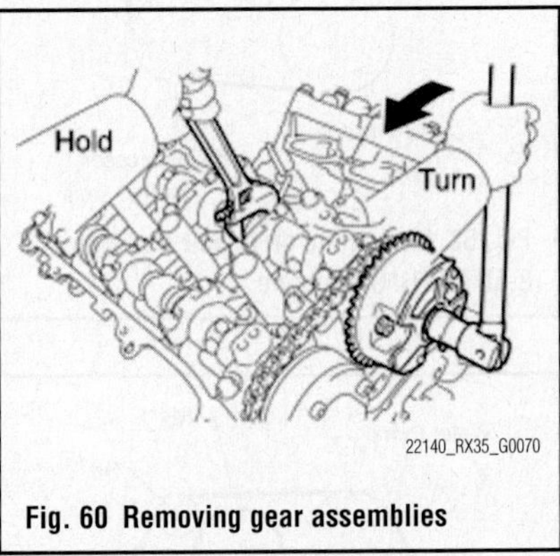

Fig. 60 Removing gear assemblies

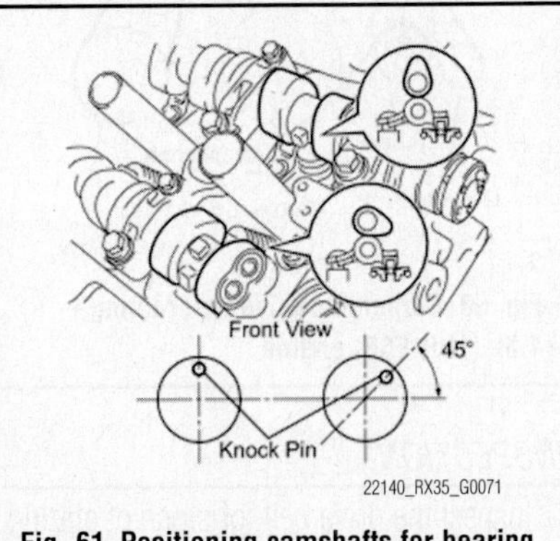

Fig. 61 Positioning camshafts for bearing cap removal bank 1

➡Be careful not to damage the contact surfaces of the cylinder head and the camshaft housing.

20. Remove the camshaft timing gears and No. 2 chain (for Bank 2), as follows:
    a. While pushing down the No. 3 chain tensioner, insert a pin of 1.0 mm (0.039 in.) into the hole to fix the No. 3 chain tensioner.
    b. Hold the hexagonal portion of the camshaft with a wrench, and remove the 2 bolts and 2 camshaft timing gear assemblies.

➡Be careful not to damage the cylinder head with the wrench.

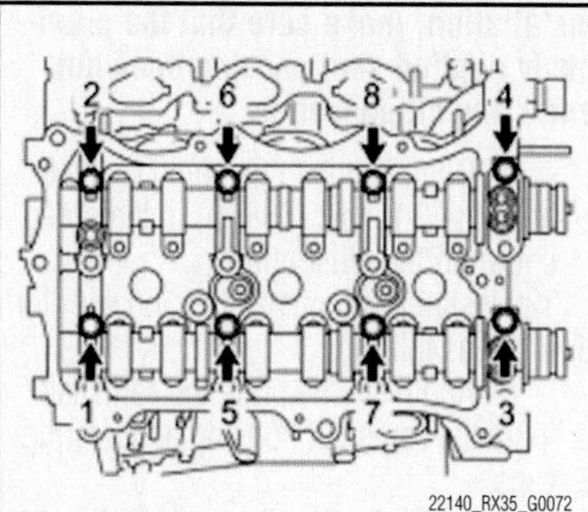
Fig. 62 Camshafts bearing cap removal sequence bank 1

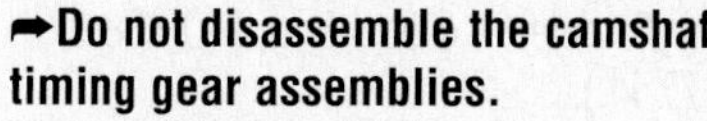
➡Do not disassemble the camshaft timing gear assemblies.

c. Remove the No. 2 chain.

21. Remove the bolt and No. 3 chain tensioner.

22. Remove the camshaft bearing cap (for Bank 2), as follows:

a. Check that the camshafts are positioned as shown in the illustration.

b. Uniformly loosen and remove the 8 bearing cap bolts in the sequence shown in the illustration.

c. Uniformly loosen and remove the 13 bearing cap bolts in the sequence shown in the illustration.

➡Uniformly loosen the bolts while keeping the camshaft level.

d. Remove the 5 camshaft bearing caps.

23. Remove the No. 3 camshaft.

24. Remove the No. 4 camshaft.

25. Remove the left hand camshaft housing sub-assembly by prying between the cylinder head and the camshaft housing with a screwdriver with the tip taped.

➡Be careful not to damage the contact surfaces of the cylinder head and the camshaft housing.

26. Remove the No. 1 valve rocker arm sub-assembly, as follows:

a. Remove the 24 valve rocker arms.

➡Arrange the removed parts in the correct order.

27. Remove the valve lash adjuster assembly, as follows:

a. Remove the 24 valve lash adjusters from the cylinder head.

➡Arrange the removed parts in the correct order.

To install:

28. Install the valve lash adjuster assembly.

29. Install the No. 1 valve rocker arm sub-assembly, as follows:

a. Apply engine oil to the lash adjuster tip and valve stem cap end.

b. Make sure that the valve rocker arms are installed as shown in the illustration.

30. Install the camshaft bearing cap (for Bank 1), as follows:

a. Apply engine oil to the camshaft journals, camshaft housing and bearing caps.

b. Install the camshaft and No. 2 camshaft to the right camshaft housing.

c. Make sure of the marks and numbers on the camshaft bearing caps and

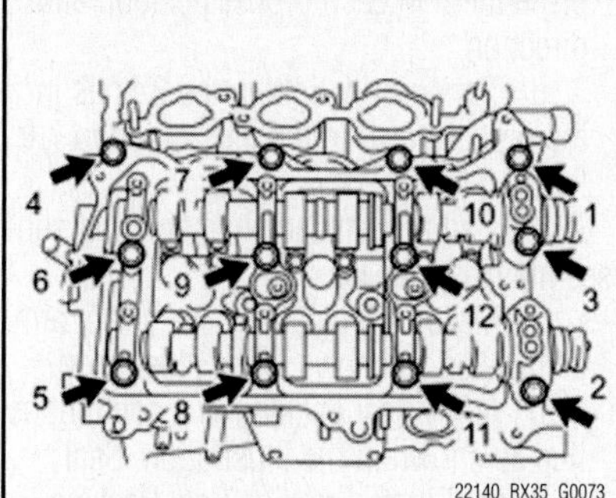
Fig. 63 Bearing cap loosening sequence bank 1

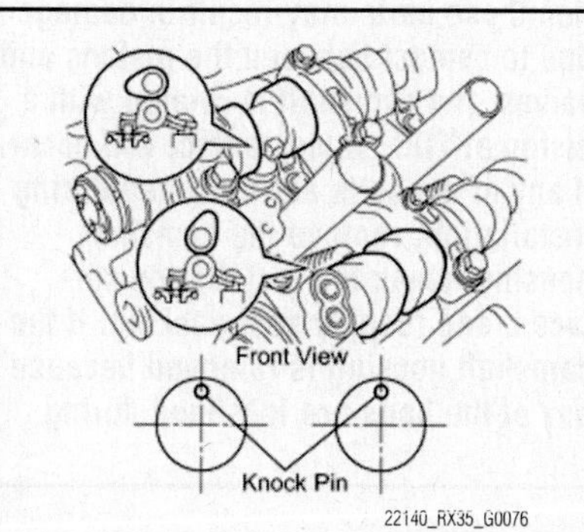

Fig. 66 Positioning camshafts for bearing cap removal bank 2

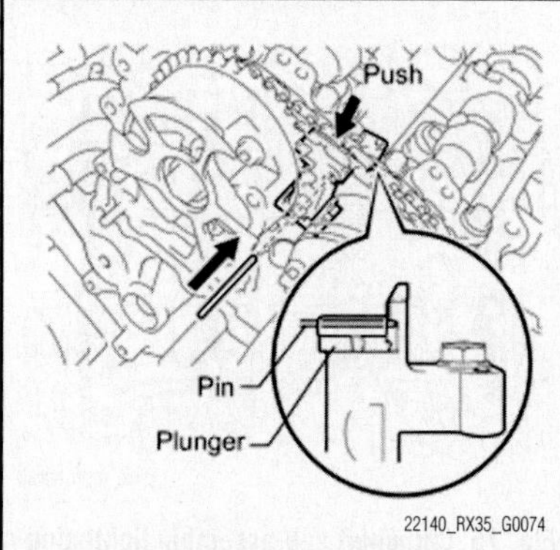

Fig. 64 Pinning No. 3 tensioner

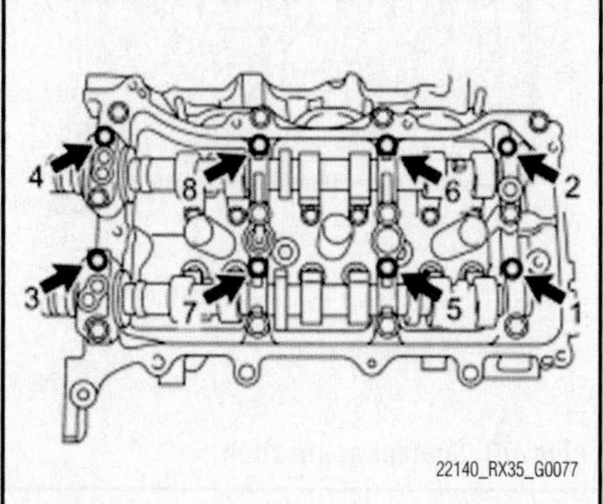
Fig. 67 Camshafts bearing cap removal sequence bank 2

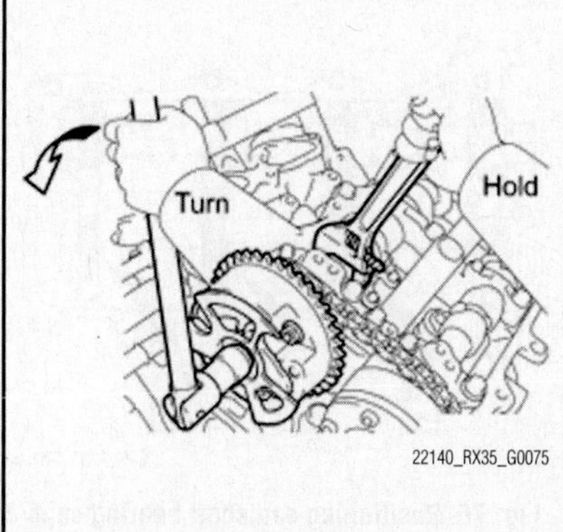

Fig. 65 Removing gear assemblies

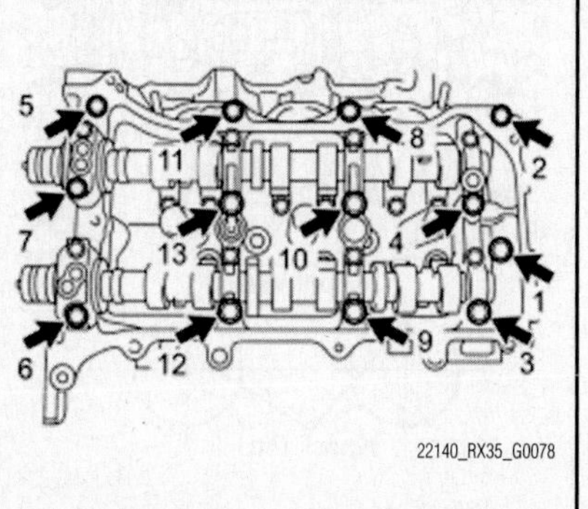
Fig. 68 Bearing cap loosening sequence bank 2

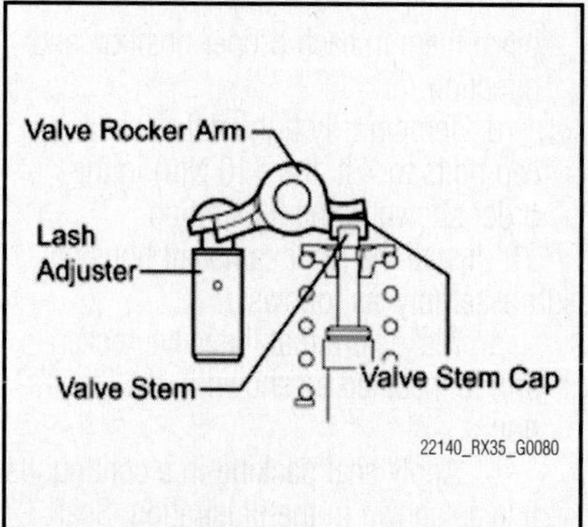

Fig. 69 Installing valve rocker arms bank 1

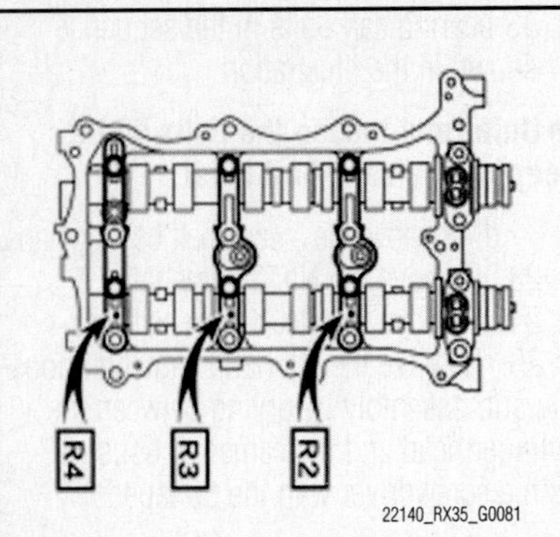

Fig. 70 Camshaft bearing caps placement bank 1

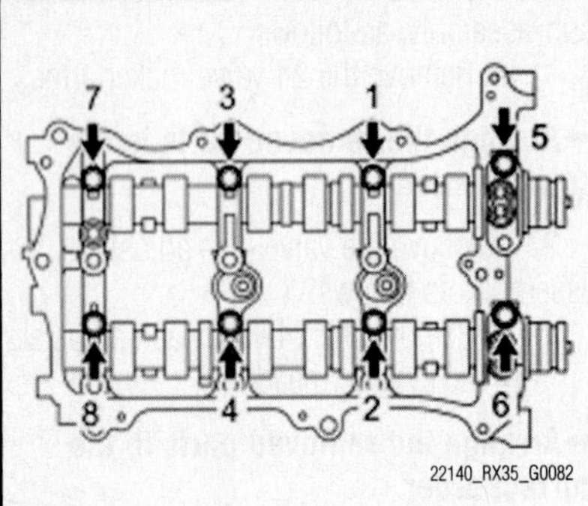

Fig. 71 Camshafts bearing cap tightening sequence bank 1

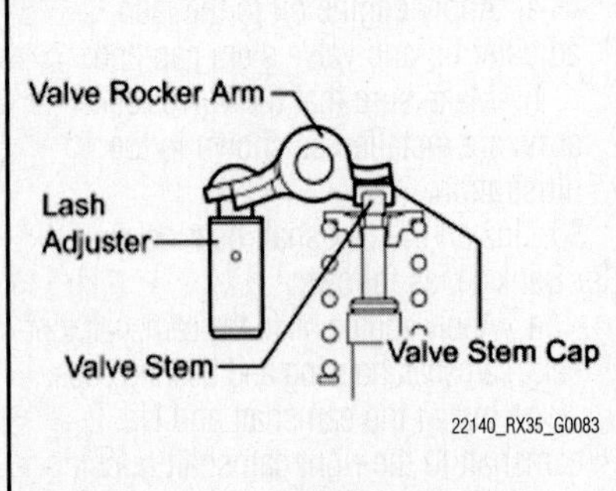

Fig. 72 Valve rocker arm installation bank 1

place them in each proper position and direction.

d. Temporarily tighten the 8 bearing cap bolts to 7 ft. lbs. (10 Nm) in the order shown in the illustration.

31. Install the right camshaft housing sub-assembly, as follows:

a. Make sure that the valve rocker arm is installed as shown in the illustration.

b. Apply seal packing in a continuous line as shown in the illustration. Seal packing: Toyota Genuine Seal Packing Black, Three Bond 1207B or equivalent. Seal diameter: 3.5 to 4.5 mm (0.138 to 0.177 in.).

➡Remove any oil from the contact surface. Install the camshaft housing sub-assembly within 3 minutes. Do not start the engine for at least 2 hours after installing.

c. Install the camshaft housing and tighten the 12 bolts in the order shown in the illustration to 21 ft. lbs. (28 Nm).

➡When installing the camshaft housing, it is necessary to correctly position the camshafts as shown in the removal illustration. Failure to correctly position these parts may result in damage due to contact between the pistons and valves. If a camshaft is rotated with a piston at TDC, valve contact will occur. If any of the bolts are loosened during installation, remove the camshaft housing, clean the installation surfaces, and reapply seal packing. If the camshaft housing is removed because any of the bolts are loosened during installation, make sure that the previously applied seal packing does not enter any oil passages.

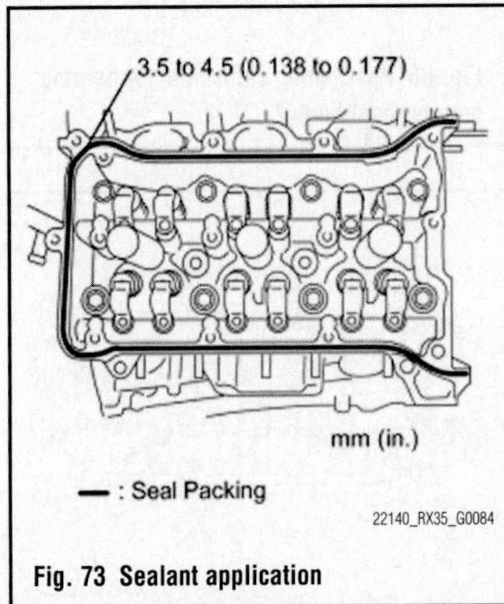

Fig. 73 Sealant application

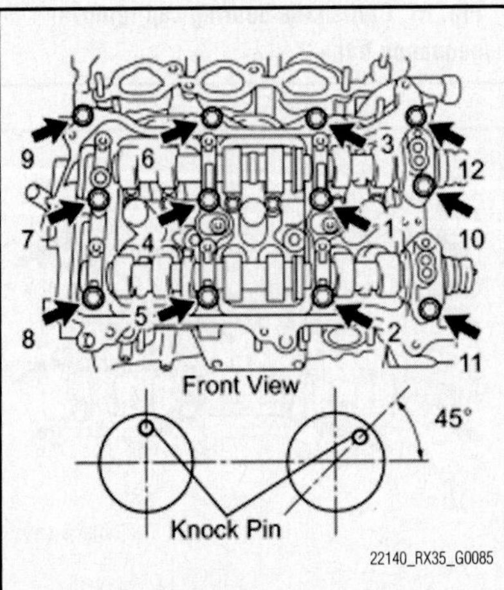

Fig. 74 Camshaft sub-assembly tightening sequence 12 bolt bank 1

d. Complete the tightening of the 8 bolts to 12 ft. lbs. (16 Nm) in the order shown in the illustration.

32. Install the camshaft bearing cap (for Bank 2), as follows:

a. Apply engine oil to the camshaft journals, camshaft housing and bearing caps.

b. Install the No. 3 camshaft and No. 4 camshaft to the left camshaft housing.

c. Make sure of the marks and numbers on the camshaft bearing caps and place them in each proper position and direction.

d. Temporarily tighten the 8 bolts in the order shown in the illustration to 7 ft. lbs. (10 Nm).

33. Install the left camshaft housing sub-assembly, as follows:

a. Make sure that the valve rocker arm is installed as shown in the illustration.

b. Apply seal packing in a continuous line as shown in the illustration. Seal packing: Toyota Genuine Seal Packing Black, Three Bond 1207B or equivalent. Seal diameter: 3.5 to 4.5 mm (0.138 to 0.177 in.).

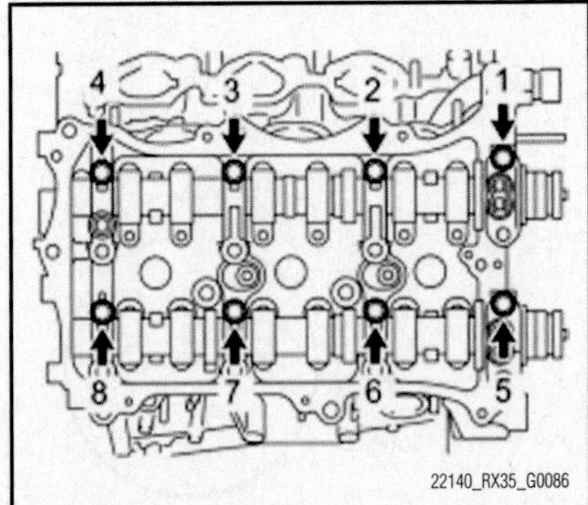

Fig. 75 Camshaft sub-assembly tightening sequence 8 bolt bank 1

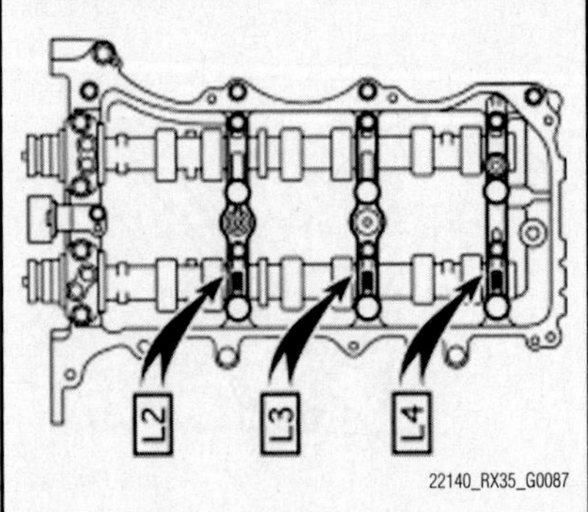

Fig. 76 Positioning camshaft bearing caps bank 1

➡**Remove any oil from the contact surface. Install the camshaft housing sub-assembly within 3 minutes. Do not start the engine for at least 2 hours after installing.**

c. Install the camshaft housing and tighten the 13 bolts in the order shown in the illustration to 21 ft. lbs. (28 Nm).

➡**When installing the camshaft housing, it is necessary to correctly position the camshafts as shown in the removal illustration. Failure to correctly position these parts may result in damage due to contact between the pistons and valves. If a camshaft is rotated with a piston at TDC, valve contact will occur. If any of the bolts are loosened during installation, remove the camshaft housing, clean the installation surfaces, and reapply seal packing. If the camshaft housing is removed because any of the bolts are loosened during installation, make sure that the previously applied seal packing does not enter any oil passages.**

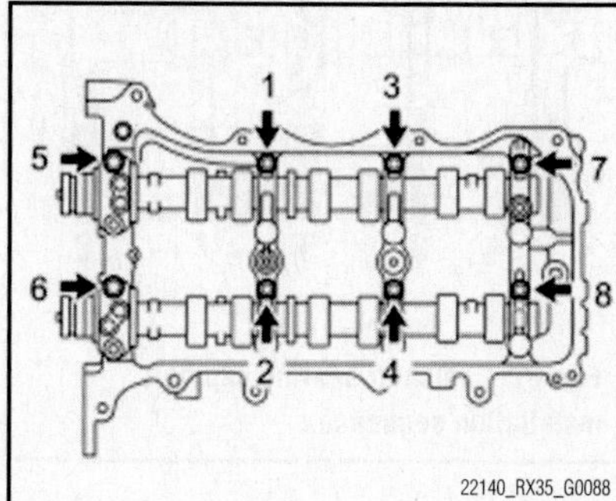

Fig. 77 Camshaft bolt tightening sequence bank 1

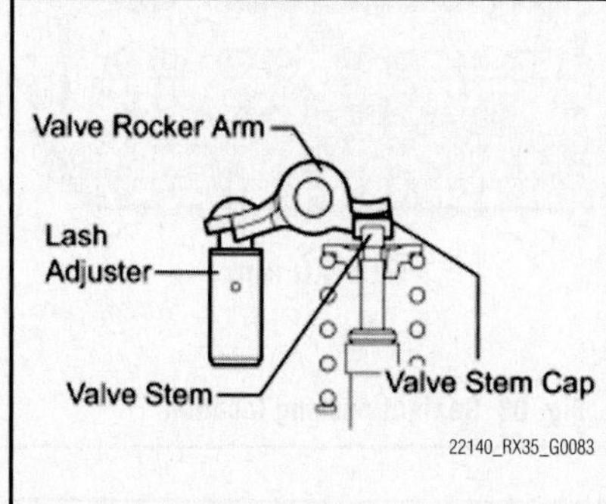

Fig. 78 Valve rocker arm installation bank 2

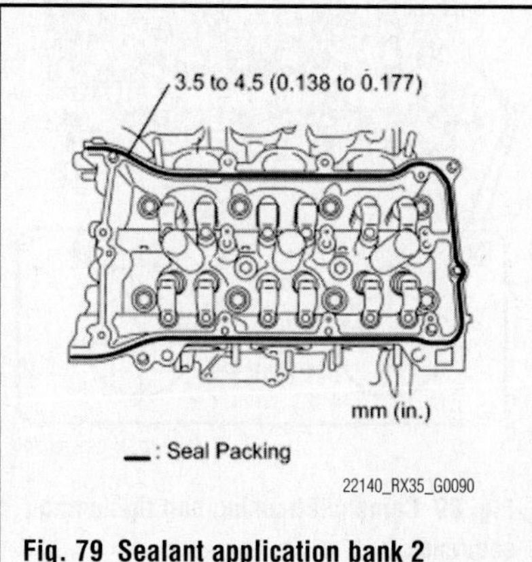

Fig. 79 Sealant application bank 2

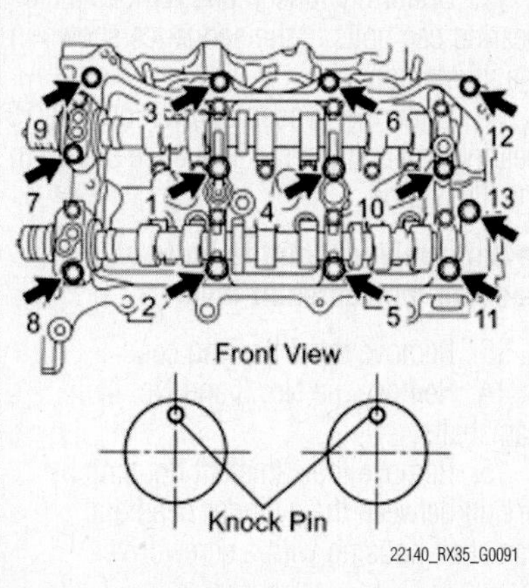

Fig. 80 Camshaft sub-assembly tightening sequence bank 2

d. Complete the tightening of the 8 bolts to 12 ft. lbs. (16 Nm) in the order shown in the illustration.

34. Install the No. 2 chain tensioner assembly with the bolt and tighten to 15 ft. lbs. (21 Nm).

35. While pushing in the tensioner, insert a pin of 1.0 mm (0.039 in.) diameter into the hole to fix it.

36. Install the camshaft timing gears and No. 2 chain (for Bank 1).

37. Install the No. 3 chain tensioner assembly with the bolt and tighten to 15 ft. lbs. (21 Nm).

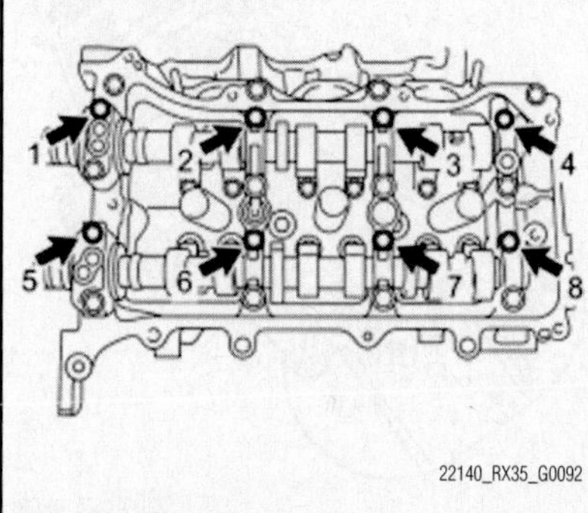

Fig. 81 Camshaft sub-assembly tightening sequence bank 2

Fig. 82 Crankshaft timing sprocket bank 2

38. While pushing in the tensioner, insert a pin of 1.0 mm (0.039 in.) diameter into the hole to hold it.

39. Install the camshaft timing gears and No. 2 chain (for Bank 2).

40. Install the chain vibration damper with the 2 bolts and tighten to 17 ft. lbs. (23 Nm).

41. Install the 2 chain vibration dampers.

42. Install the 2 timing gear set keys and crankshaft timing sprocket as shown in the illustration.

43. Install the idle sprocket assembly.

44. Install chain sub-assembly.

45. Install the chain tensioner slipper.

46. Install the No. 1 chain tensioner assembly.

47. Install the following:

- Timing chain cover,
- Cylinder head cover
- New gaskets, O-ring and rear water by-pass joint. Torque to 7 ft. lbs. (10 Nm).
- New gasket, water inlet and thermostat assembly. Torque to 7 ft. lbs. (10 Nm).
- Oil filter element
- Adhesive sealer, left and right side cylinder block water drain cocks. Torque to 22 ft. lbs. (30 Nm).
- Left and right side oil control valve filter and gaskets
- Left and right side oil check valve bolt, oil pipe union and oil pipe. Torque to 44 ft. lbs. (60 Nm).
- Crankshaft Position (CKP) sensor. Torque to 7 ft. lbs. (10 Nm).
- 4 camshaft timing oil control valves
- 4 camshaft position sensors
- Adhesive and ventilation valve
- Spark plugs. Torque to 13 ft. lbs. (18 Nm).
- Engine hangers. Torque to 24 ft. lbs. (33 Nm).
- Radiator cap
- Oil filler cap and gasket
- Negative battery cable

48. Refill the coolant and engine oil. Start the engine and check for leaks or abnormal conditions. Perform and road test. Then, recheck for leaks and recheck fluid levels.

4.6L (1UR-FSE) Engine

See Figures 83 through 92.

1. Discharge fuel system pressure.
2. Disconnect cable from negative battery terminal.
3. Remove timing chain cover sub-assembly.
4. Set No. 1 cylinder to TDC/compression.
5. Remove No. 1 chain tensioner assemblies.
6. Remove No. 1 chain vibration damper.
7. Remove chain sub-assembly.
8. Remove No. 3 chain tensioner assembly.
9. Remove camshaft bearing cap.
10. Make sure that the knock pin of the camshaft is positioned as shown in the illustration.
11. Uniformly loosen and remove the 8 bearing cap bolts in the sequence shown in the illustration
12. Uniformly loosen and remove the 18 bearing cap bolts in the sequence shown in the illustration

➡Uniformly loosen the bolts while keeping the camshaft level

13. Remove the 7 bearing caps.
14. Remove the No. 3 and No. 4 camshafts.
15. Remove the camshaft housing by prying between the cylinder head and camshaft housing with a screwdriver.

To install:

16. Apply a light coat of engine oil to the camshaft journals, camshaft housing and bearing caps.
17. Install the No. 3 and No. 4 camshafts to the camshaft housing.
18. Confirm the marks and numbers on the camshaft bearing caps and place them in their proper positions and directions.
19. Temporarily install the 8 bolts in the order shown in the illustration.
20. Make sure that the valve rocker arms are installed correctly
21. Apply seal packing in a continuous line as shown

➡Remove any oil from the contact surface. Install the camshaft housing within 3 minutes and tighten the bolts within 15 minutes after applying seal packing. Do not start the engine for at least 2 hours after the installation.

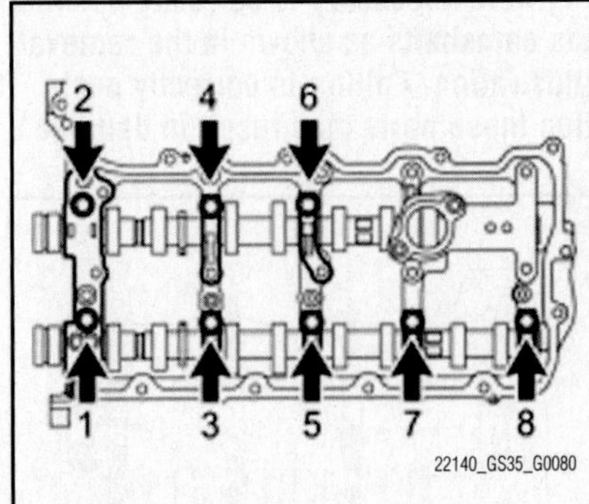

Fig. 87 Camshaft bearing cap bolt installation sequence

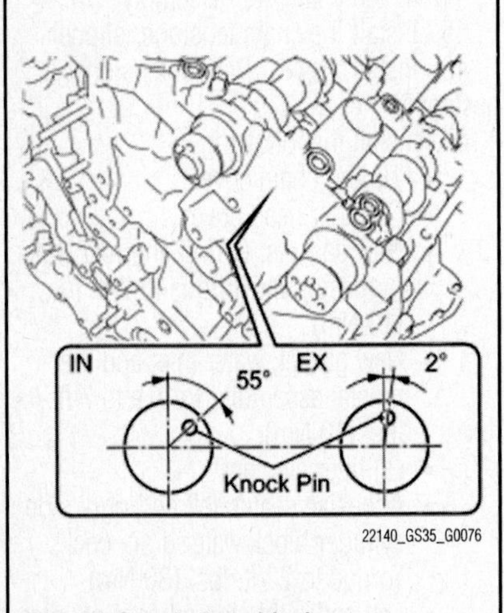

Fig. 83 Camshaft positioning

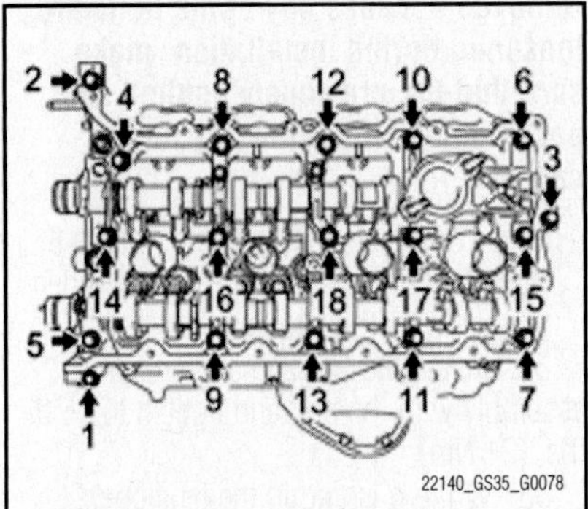

Fig. 85 Bearing cap bolt removal sequence second

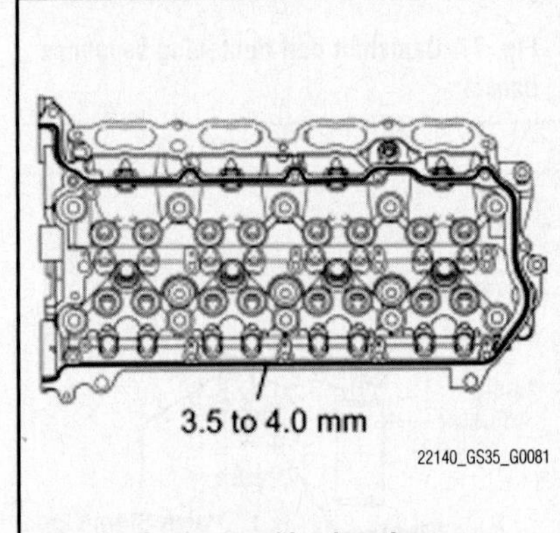

Fig. 88 Sealant packing location

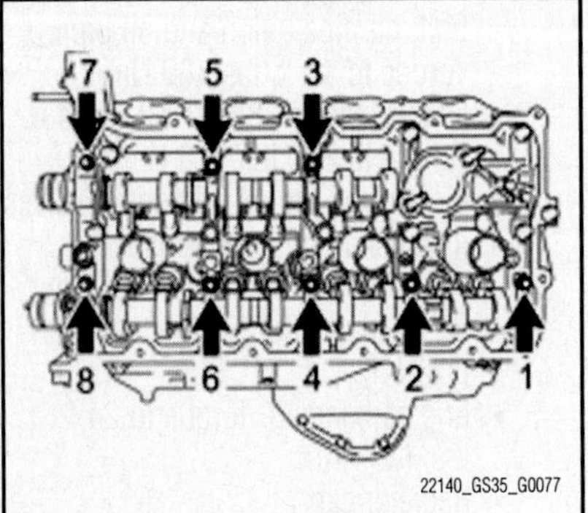

Fig. 84 Bearing cap bolt removal sequence first

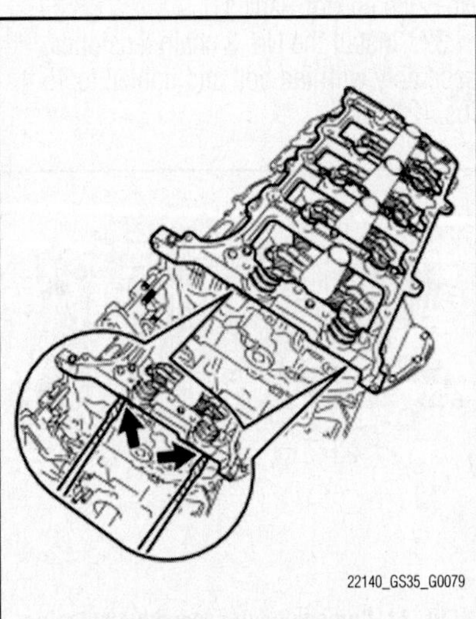

Fig. 86 Camshaft housing removal

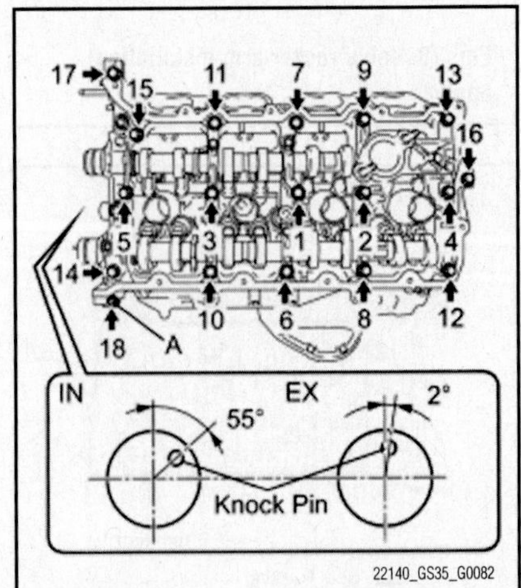

Fig. 89 Camshaft housing bolt tightening sequence

22. Install the camshaft housing, and install the 12 bolts and tighten to 7 ft. lbs. (10 nm) and 22 Ft. lbs. (30 Nm), in the order shown.

➡Make sure that each knock pin of the camshafts is positioned as shown in the illustration before installing the camshaft housing.

23. Tighten the 8 bolts to 12 ft. lbs. (16 Nm) in the order shown in the illustration
24. Install No. 3 chain tensioner assembly.
25. Install chain sub-assembly.
26. Install chain tensioner slipper.
27. Install No. 1 chain tensioner assembly.
28. Install No. 1 chain vibration damper.
29. Using a wrench, hold the hexagonal portion of the No. 3 camshaft.
30. Using a 12 mm socket hexagon wrench, tighten the camshaft timing gear assembly with a new bolt to 58 ft. lbs. (79 Nm).
31. Using a wrench to hold the hexagonal portion of the No. 4 camshaft, tighten the camshaft timing exhaust gear assembly with the bolt to 74 ft. lbs. (100 Nm).

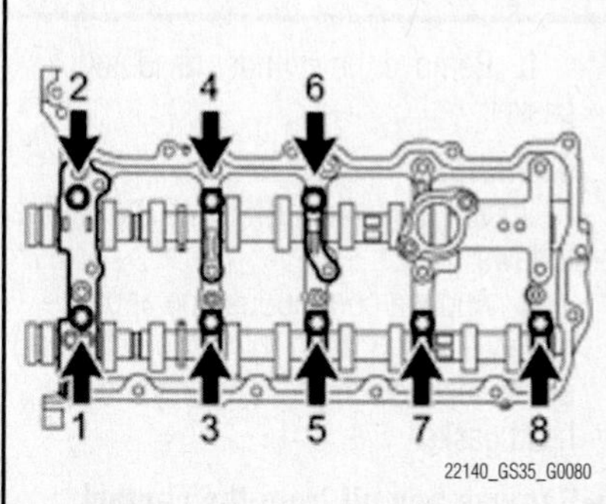

Fig. 90 Camshaft bearing cap bolt tightening sequence

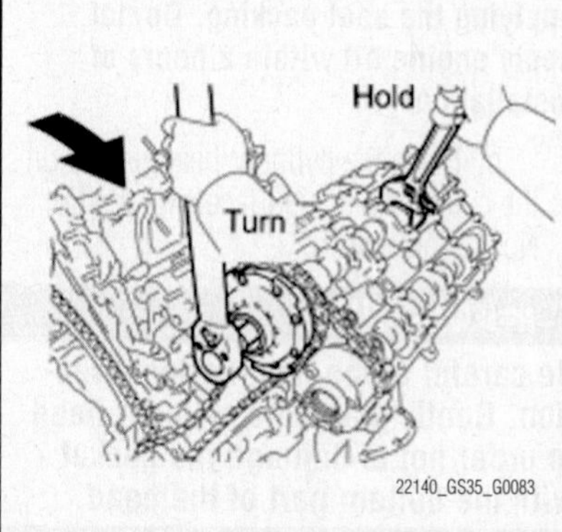

Fig. 91 No. 3 camshaft timing gear tightening

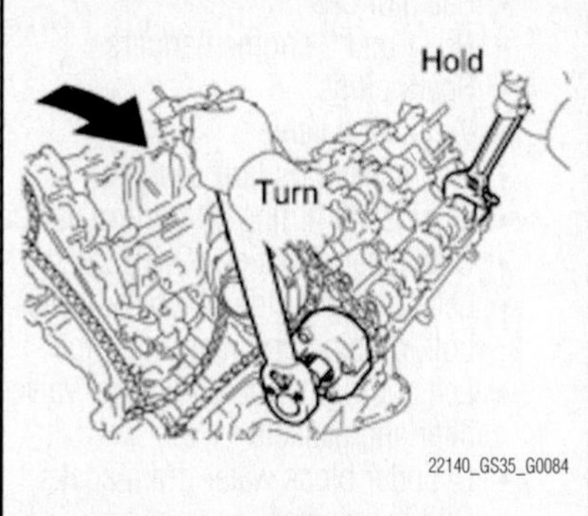

Fig. 92 No. 4 camshaft timing gear tightening

32. Check No. 1 cylinder to TDC/compression.
33. Install timing chain and gears.
34. Install timing chain cover sub-assembly.
35. Connect cable from negative battery terminal.
36. Perform similar procedure for opposite bank.

Perform initialization , if necessary.

CATALYTIC CONVERTER

REMOVAL & INSTALLATION

The catalytic converter is integrated with the exhaust manifold and has to be removed as a unit.

CRANKSHAFT DAMPER

REMOVAL & INSTALLATION

See Figure 93.

1. Before servicing the vehicle, refer to the precautions in the beginning of this section.

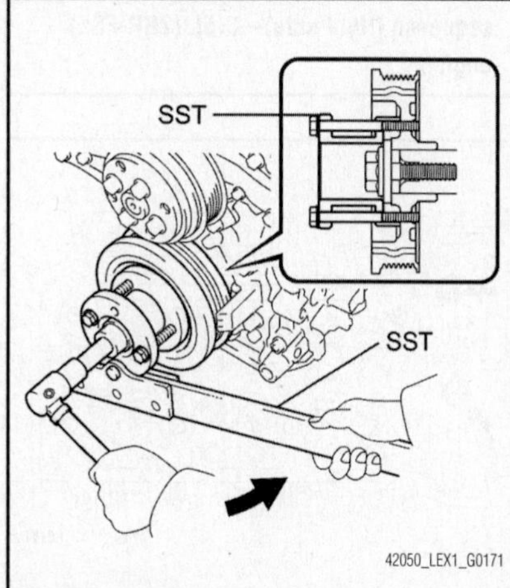

Fig. 93 Using special tools to remove the crankshaft pulley.

2. Disconnect the negative battery cable. Wait at least 90 seconds before performing any other work.
3. It may be necessary to remove the radiator assembly for access.
4. Remove the accessory drive belt(s).
5. Using SST 09213-54015 (91651-60855), 09330-00021, loosen the crankshaft pulley bolt.
6. Using SST 09950-50013 (09951-05010, 09952-05010, 09953-05010, 09954-05030)and the pulley bolt, remove the crankshaft pulley.

➡Before using SST, apply lubricating oil on the threads and tip of the center bolt.

To install:

7. Align the pulley set key with the key groove of the pulley, and slide on the pulley.
8. Using SST 09213-54015 (91651-60855), 09330-00021, install the pulley bolt.
9. Tighten the crankshaft pulley bolt (depending on the engine) to the following torque values:
   - 3.5L (2GR-FSE) engine: 184 ft. lbs. (250 Nm)
   - 4.6L (1UR-FSE) and 4.3L (3UZ-FE) engines: 181 ft. lbs. (245 Nm)
10. Install the accessory drive belt(s).
11. Install the radiator, if necessary.
12. Connect the negative battery cable.

CRANKSHAFT FRONT SEAL

REMOVAL & INSTALLATION

See Figures 94 and 95.

1. Before servicing the vehicle, refer to the precautions in the beginning of this section.
2. Disconnect the negative battery cable. Wait at least 90 seconds before performing any other work.
3. Air cleaner assembly.

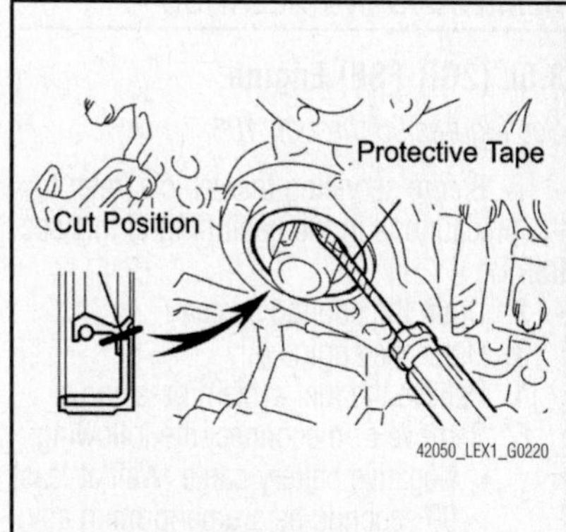

Fig. 94 Removal of the front oil pump seal—4.6L (1UR-FSE) engines

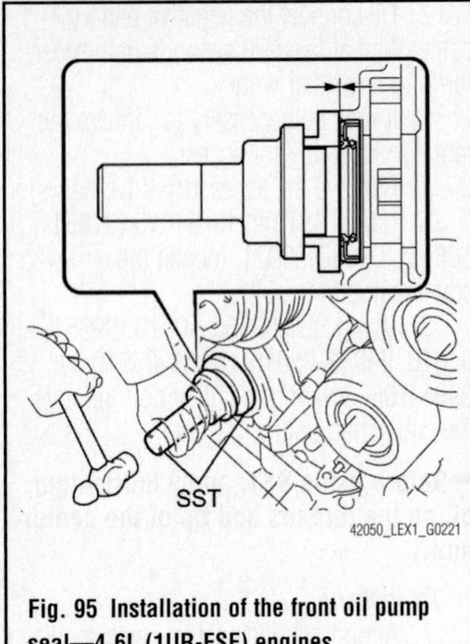

Fig. 95 Installation of the front oil pump seal—4.6L (1UR-FSE) engines

4. Remove the radiator reserve assembly.
5. Remove the accessory drive belt.
6. Remove the crankshaft pulley.
7. Using a screwdriver with its tip taped, pry out the timing chain case oil seal.

➡**After the removal, check the crankshaft for damage. If it is damaged, smooth the surface with 400-grit sandpaper.**

To install:

8. Using SST 09223-22010, 09506-35010 and a hammer, tap in the oil seal until its surface is flush with the rear oil seal retainer edge. Keep the lip free of foreign matter. Do not tap the oil seal at an angle.
9. Install the crankshaft pulley.
10. Install the accessory drive belt.
11. Install the radiator reserve assembly.
12. Air cleaner assembly.
13. Inspect for oil leaks.
14. Connect the negative battery cable.

CYLINDER HEAD

REMOVAL & INSTALLATION

3.5L (2GR-FSE) Engine

See Figures 96 through 101.

1. Before servicing the vehicle, refer to the precautions in the beginning of this section.
2. Drain the cooling system.
3. Drain the engine oil.
4. Relieve the fuel system pressure.
5. Remove or disconnect the following:
   - Negative battery cable. Wait at least 90 seconds before performing any other work
   - Oil filler cap and gasket
   - Radiator cap
   - No. 1 and 2 engine hangers
   - Spark plugs
   - Ventilation valve
   - 4 camshaft position sensors
   - 4 camshaft timing oil control valves
   - Crankshaft Position (CKP) sensor
   - Left and right side oil check valve bolt, oil pipe union and oil pipe
   - Left and right side oil control valve filter and gaskets
   - Cylinder block water drain cocks
   - Oil filter element
   - Water inlet and thermostat assembly
   - Rear water by-pass joint
   - Cylinder head cover and gaskets
   - Timing chain cover, timing chain and timing chain sprockets
   - Camshaft and camshaft housing assembly
   - Valve rocker arms. Arrange the removed rocker arms in the correct order.
   - Valve lash adjusters. Arrange the removed valve lash adjusters in the correct order.
6. Remove the cylinder head (left or right) as follows:

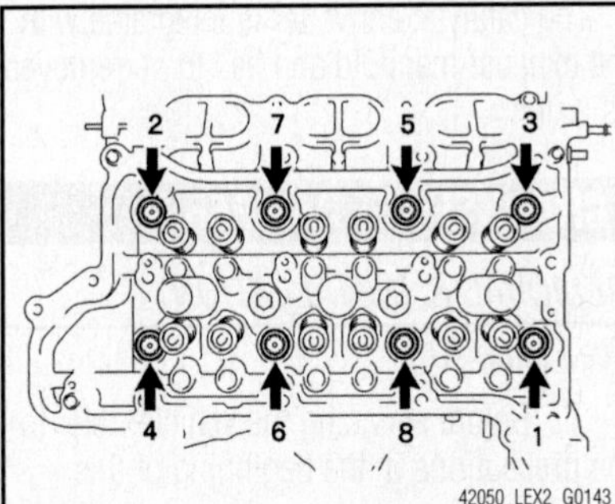

Fig. 96 Cylinder head bolt loosening sequence (right side)—3.5L (2GR-FSE) engine

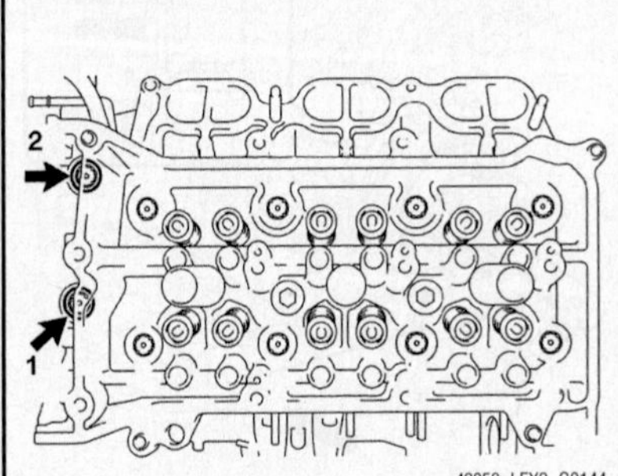

Fig. 97 Cylinder head 14mm bolt loosening sequence (left side)—3.5L (2GR-FSE) engine

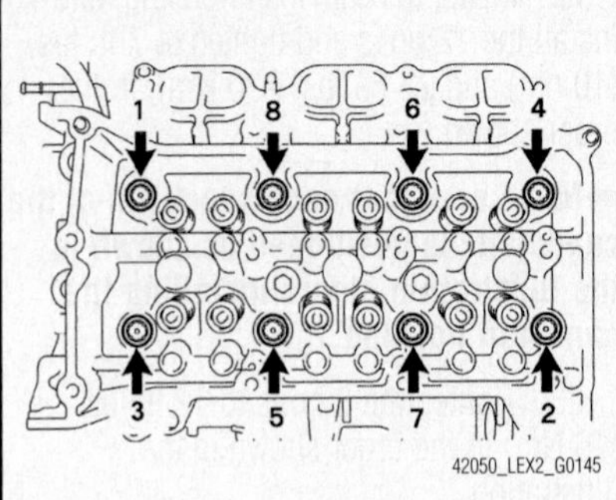

Fig. 98 Cylinder head bolt loosening sequence (left side)—3.5L (2GR-FSE) engine

a. Using a 10mm bi-hexagon wrench, uniformly loosen the 8 bolts in the sequence shown in the illustration. Remove the 8 cylinder head bolts and plate washers.

WARNING

Be careful not to drop washers into the cylinder head. Cylinder head warpage or cracking could result from removing bolts in an incorrect order. Be sure to keep separate the removed parts for each installation position.

b. Remove the cylinder head and gasket.

To install:

7. Install the cylinder head to the engine as follows:

a. Apply a continuous line approximately 2.5 to 3.0mm (0.098 to 0.118 in.) of the seal packing to a new cylinder head gasket.

➡**Remove any oil from the contact surface. Install the cylinder head gasket within 3 minutes after applying the seal packing. Install the cylinder head bolt within 15 minutes after applying the seal packing. Do not apply engine oil within 2 hours of installation.**

b. Place the cylinder head gasket on the cylinder block surface with the Lot No. stamp upward.

WARNING

Be careful of the installation direction. Gently place the cylinder head in order not to damage the gasket with the bottom part of the head.

c. Place the cylinder head on the cylinder block.

⁂ WARNING

Be careful not to allow oil to adhere to the bottom part of the cylinder head.

d. Apply a light coat of engine oil to the threads and under the heads of the cylinder head bolts.

e. Using a 10mm bi-hexagon wrench, install and uniformly tighten the 8 cylinder head bolts with the plate washers to 27 ft. lbs. (36 Nm) in the sequence shown in the illustration. If any of the bolts does not meet the torque, replace it.

f. Mark the forward edge of each bolt with paint, then retighten each bolt, in proper sequence, an additional 90 degrees. Check that each painted mark is now at a 90 degrees angle to the front. The paint mark should have been applied to the bolt in the 9 o'clock position and should now be in the 12 o'clock position.

g. Tighten each bolt again, in proper sequence, an additional 90 degrees. Check that each painted mark is now facing rearward.

h. Tighten the 2 bolts on the left cylinder head in the order shown in the illustration. Torque to 22 ft. lbs. (30 Nm).

➡Do not use the tightening procedure for a plastic region bolt (if equipped) when tightening bolts 1 and 2 shown in the illustration.

i. Seal packing will seep out on the engine's front side. Thoroughly wipe off seeped out seal packing.

8. Install or connect the following:
- Valve lash adjusters
- Valve rocker arms
- Camshaft and camshaft housing assembly
- Timing chain cover, timing chain and timing chain sprockets
- Cylinder head cover

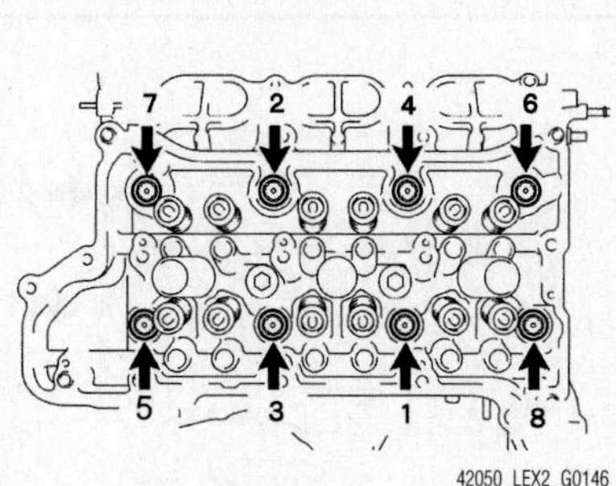

Fig. 99 Cylinder head bolt tightening sequence (right side)—3.5L (2GR-FSE) engine

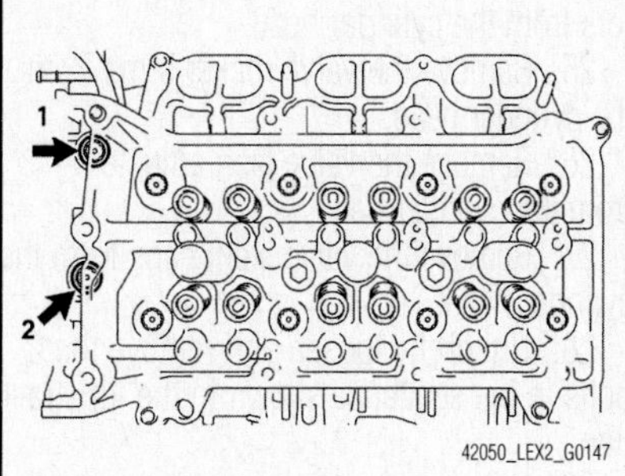

Fig. 100 Cylinder head 14mm bolt tightening sequence (left side)—3.5L (2GR-FSE) engine

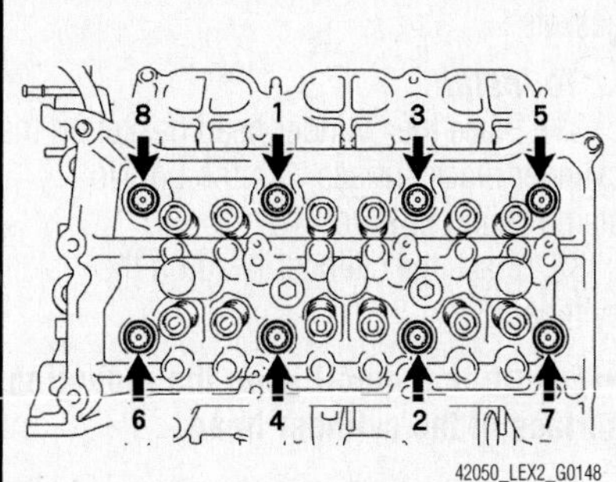

Fig. 101 Cylinder head bolt tightening sequence (left side)—3.5L (2GR-FSE) engine

- New gaskets, O-ring and rear water by-pass joint. Torque to 7 ft. lbs. (10 Nm).
- New gasket, water inlet and thermostat assembly. Torque to 7 ft. lbs. (10 Nm).
- Oil filter element
- Adhesive sealer, left and right side cylinder block water drain cocks. Torque to 22 ft. lbs. (30 Nm).
- Left and right side oil control valve filter and gaskets
- Left and right side oil check valve bolt, oil pipe union and oil pipe. Torque to 44 ft. lbs. (60 Nm).
- Crankshaft Position (CKP) sensor. Torque to 7 ft. lbs. (10 Nm).
- 4 camshaft timing oil control valves
- 4 camshaft position sensors
- Adhesive and ventilation valve
- Spark plugs. Torque to 13 ft. lbs. (18 Nm).
- Engine hangers. Torque to 24 ft. lbs. (33 Nm).
- Radiator cap
- Oil filler cap and gasket
- Negative battery cable

9. Refill the coolant and engine oil. Start the engine and check for leaks or abnormal conditions. Perform and road test. Then, recheck for leaks and recheck fluid levels.

4.6L (1UR-FSE) Engine

See Figures 102 through 118.

1. Discharge fuel system pressure.
2. Disconnect cable from negative battery terminal.
3. Remove timing chain cover sub-assembly.
4. Set No. 1 cylinder to TDC/compression.
5. Remove No. 1 chain tensioner assemblies.
6. Remove No. 1 chain vibration damper.
7. Remove chain sub-assembly.
8. Remove No. 3 chain tensioner assembly.
9. Remove camshaft bearing cap.
10. Make sure that the knock pin of the camshaft is positioned as shown in the illustration.
11. Uniformly loosen and remove the 8 bearing cap bolts in the sequence shown in the illustration

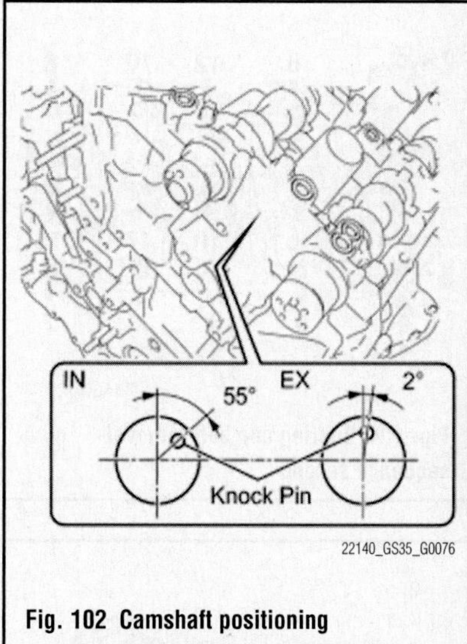

Fig. 102 Camshaft positioning

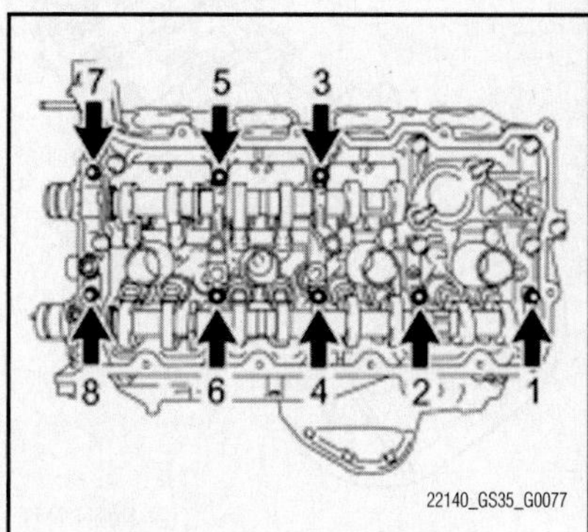

Fig. 103 Bearing cap bolt removal sequence first

12. Uniformly loosen and remove the 18 bearing cap bolts in the sequence shown in the illustration

➡**Uniformly loosen the bolts while keeping the camshaft level**

13. Remove the 7 bearing caps
14. Remove the No. 3 and No. 4 camshafts.
15. Remove the camshaft housing by prying between the cylinder head and camshaft housing with a screwdriver.
16. Remove the intake manifold.
17. Remove the fuel pressure pulsation damper assembly and No. 1 fuel pipe sub-assembly.
18. Remove fuel pipe sub-assemblies.
19. Remove engine covers.
20. Remove the ventilation hose from the ventilation valve.
21. Disconnect the fuel pressure sensor connector.
22. Disconnect the injector connectors.
23. Remove the bolts and delivery pipe from the cylinder head.

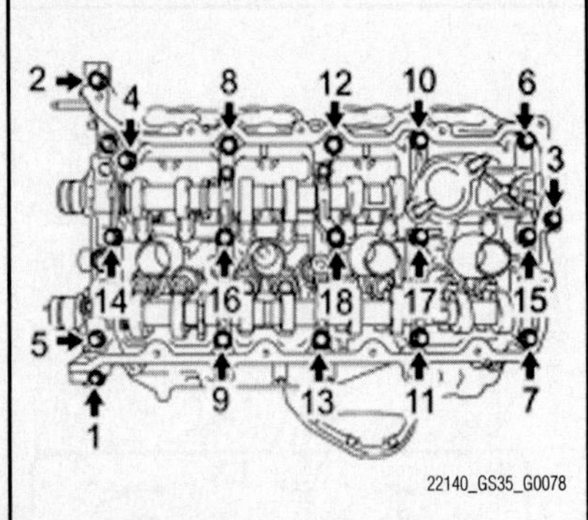

Fig. 104 Bearing cap bolt removal sequence second

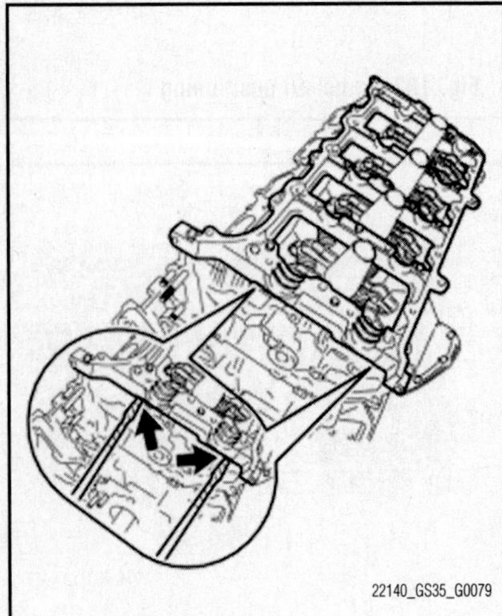

Fig. 105 Camshaft housing removal

24. Remove the injector vibration insulators from the cylinder head.
25. Remove the valve rocker arms from the cylinder head.
26. Remove the valve lash adjusters from the cylinder head.
27. Remove the valve stem caps from the cylinder head.
28. Uniformly loosen and remove the 2 bolts in the sequence shown in the illustration.
29. Using a 10 mm bi-hexagon wrench, uniformly loosen the 10 bolts in the sequence shown in the illustration. Remove the 10 cylinder head bolts and plate washers
30. Remove the cylinder heads and gaskets

To install:

31. Place the cylinder head gasket on the cylinder block surface with the Lot No. stamp facing upward.
32. Place the cylinder head on the cylinder block.

➡**Ensure that no oil is on the mounting surface of the cylinder head.**

33. Apply a light coat of engine oil to the threads and under the heads of the cylinder head bolts.

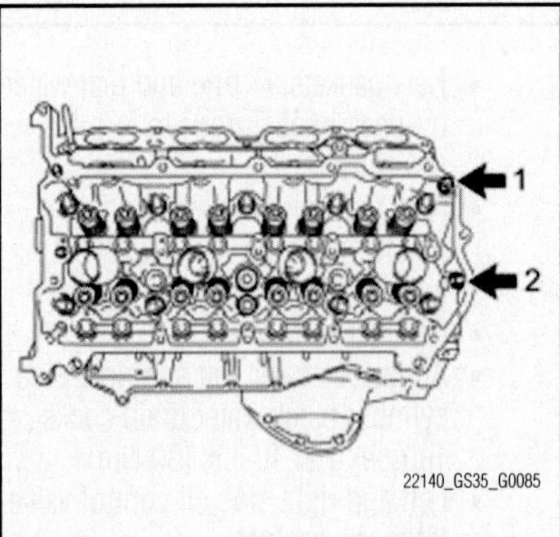

Fig. 106 Bolt loosening sequence Bank 1

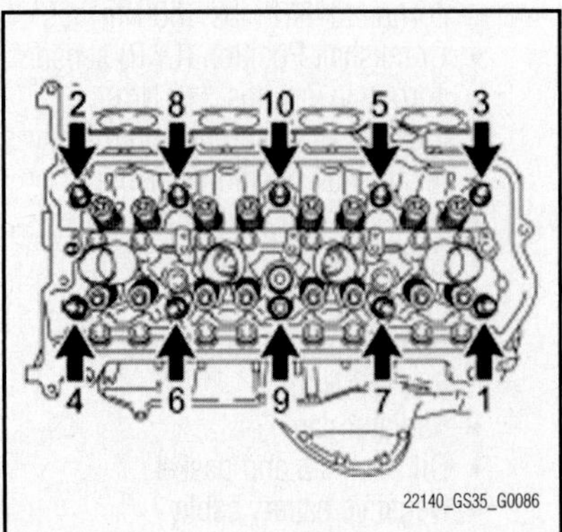

Fig. 107 Bolt loosening sequence Bank 2

34. Using a 10 mm bi-hexagon wrench, install and uniformly tighten the 10 cylinder head bolts to 27 ft. lbs. (36 Nm) with the plate washers in several steps, in the sequence shown in the illustration.
35. Mark each cylinder head bolt head with paint as shown in the illustration.
36. Tighten the cylinder head bolts another 90° in the sequence shown in step 1.
37. Tighten the cylinder head bolts by an additional 90° in the sequence shown in step 1.
38. Check that the painted marks are now facing rearward.

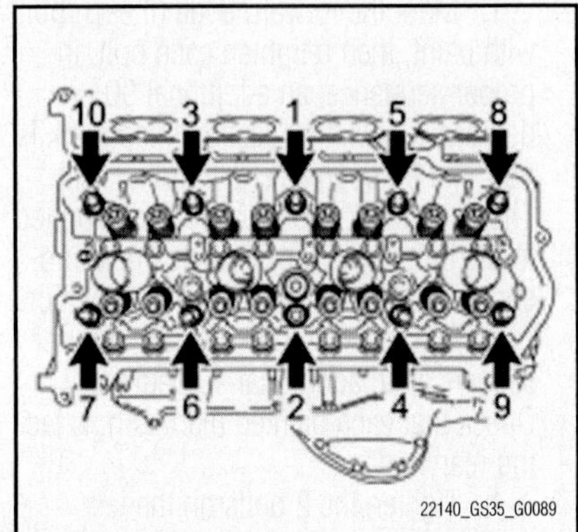

Fig. 108 Bolt tightening sequence Bank 1 first

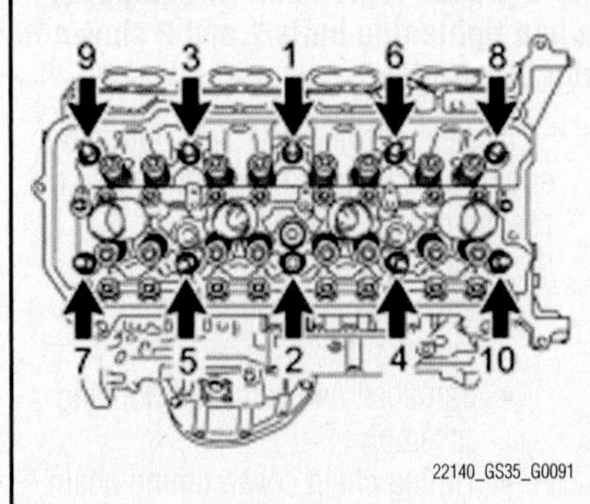

Fig. 109 Bolt tightening sequence Bank 2 first

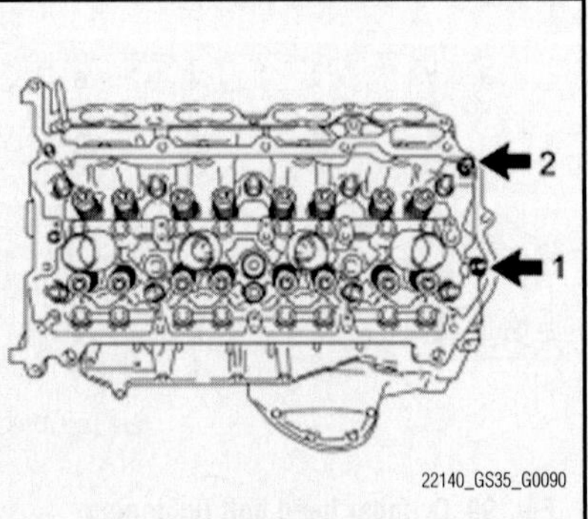

Fig. 110 Bolt tightening sequence Bank 1 second

39. Uniformly install the 2 cylinder head bolts to 15 ft. lbs. (21 Nm).

40. Apply a light coat of engine oil to the valve stem caps.

41. Install the valve stem caps to the cylinder head.

42. Install the valve lash adjusters to the cylinder head.

43. Install the ventilation hose from the ventilation valve.

44. Install fuel pipe sub-assemblies.

45. Install the fuel pressure pulsation damper assembly and No. 1 fuel pipe sub-assembly.

46. Install the intake manifold.

47. Apply engine oil to the lash adjuster tips and valve stem cap ends.

48. Make sure that the valve rocker arms are installed as shown in the illustration.

49. Install the bolts and delivery pipe to the cylinder head.

50. Connect the injector connectors.

51. Connect the fuel pressure sensor connector.

52. Install the injector vibration insulators to the cylinder head.

53. Apply a light coat of engine oil to the camshaft journals, camshaft housing and bearing caps.

54. Install the No. 3 and No. 4 camshafts to the camshaft housing.

55. Confirm the marks and numbers on the camshaft bearing caps and place them in their proper positions and directions.

56. Temporarily install the 8 bolts in the order shown in the illustration.

57. Make sure that the valve rocker arms are installed correctly

58. Apply seal packing in a continuous line as shown

➡Remove any oil from the contact surface. Install the camshaft housing within 3 minutes and tighten the bolts within 15 minutes after applying seal packing. Do not start the engine for at least 2 hours after the installation.

59. Install the camshaft housing, and install the 12 bolts and tighten to 7 ft. lbs. (10 nm) and 22 Ft. lbs. (30 Nm), in the order shown.

➡Make sure that each knock pin of the camshafts is positioned as shown in the illustration before installing the camshaft housing.

60. Tighten the 8 bolts to 12 ft. lbs. (16 Nm) in the order shown in the illustration

61. Install No. 3 chain tensioner assembly.

62. Install chain sub-assembly.

63. Install chain tensioner slipper.

64. Install No. 1 chain tensioner assembly.

65. Install No. 1 chain vibration damper.

66. Using a wrench, hold the hexagonal portion of the No. 3 camshaft.

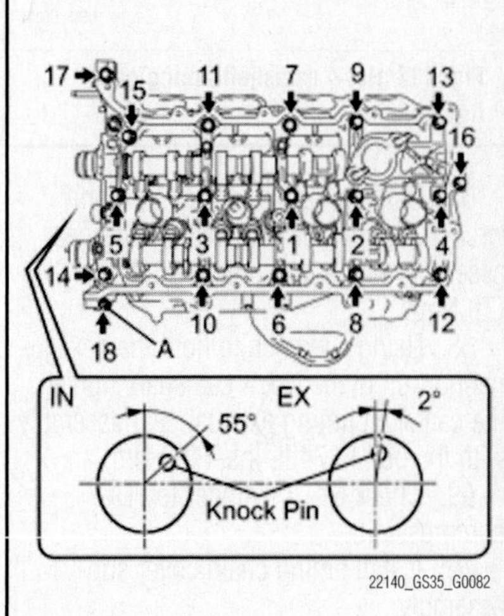

Fig. 115 Camshaft housing bolt tightening sequence

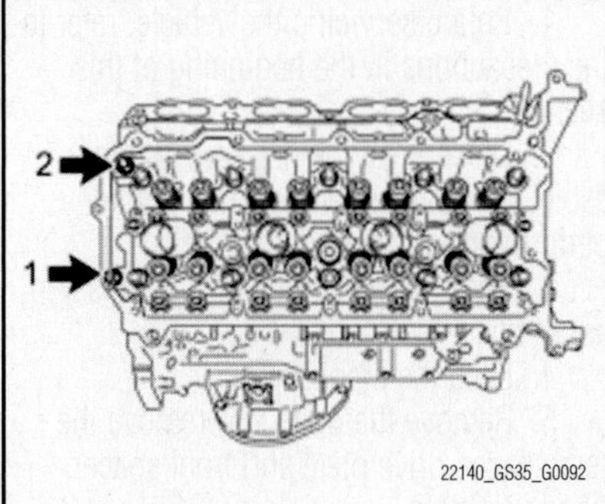

Fig. 111 Bolt tightening sequence Bank 2 second

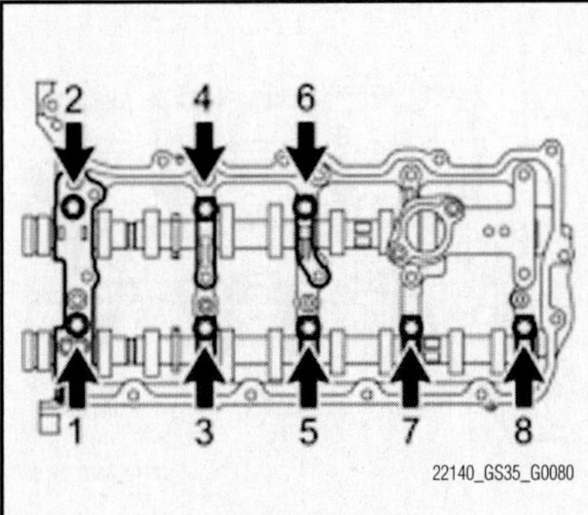

Fig. 113 Camshaft bearing cap bolt installation sequence

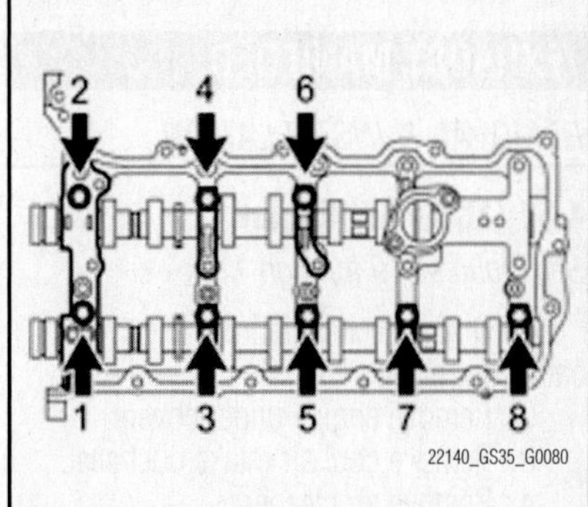

Fig. 116 Camshaft bearing cap bolt tightening sequence

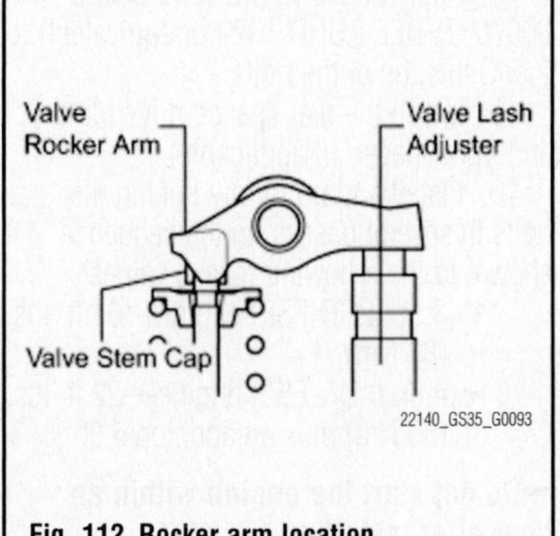

Fig. 112 Rocker arm location

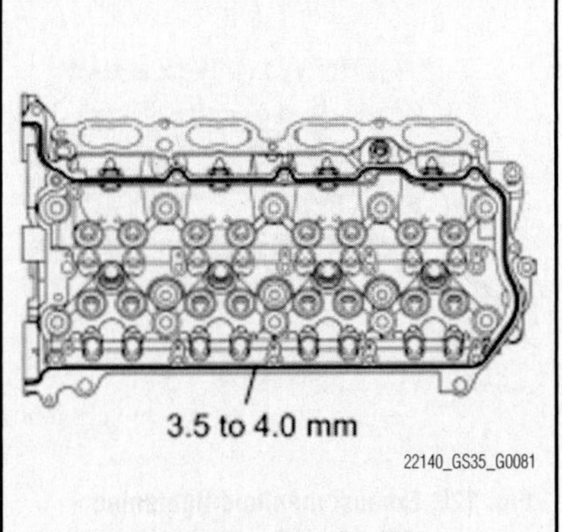

Fig. 114 Sealant packing location

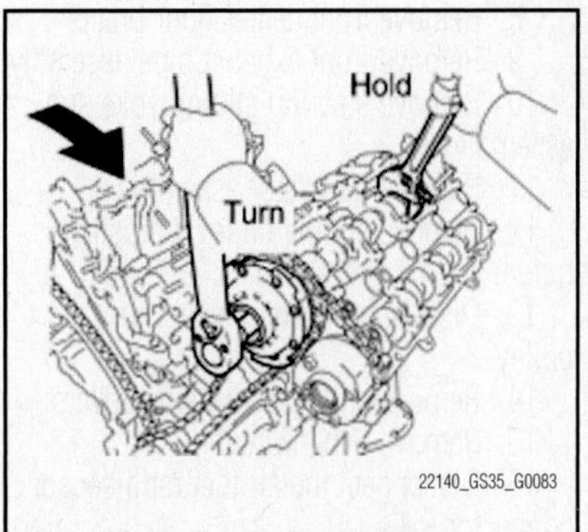

Fig. 117 No. 3 camshaft timing gear tightening

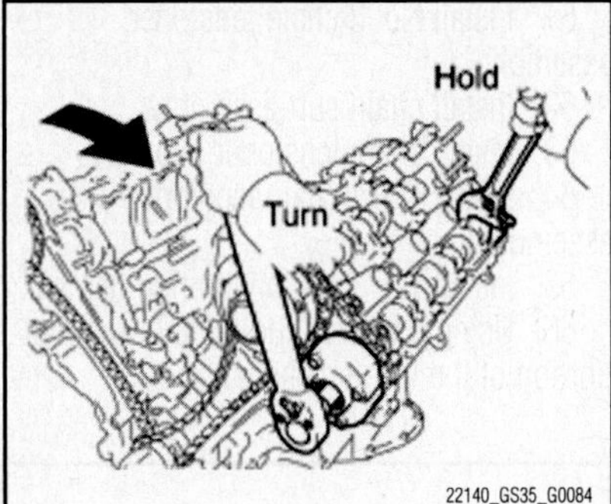

Fig. 118 No. 4 camshaft timing gear tightening

67. Using a 12 mm socket hexagon wrench, tighten the camshaft timing gear assembly with a new bolt to 58 ft. lbs. (79 Nm).

68. Using a wrench to hold the hexagonal portion of the No. 4 camshaft, tighten the camshaft timing exhaust gear assembly with the bolt to 74 ft. lbs. (100 Nm).

69. Check No. 1 cylinder to TDC/ compression.

70. Install timing chain cover sub-assembly.

71. Install engine covers.

72. Connect cable from negative battery terminal.

Perform initialization , if necessary.

EXHAUST MANIFOLD

REMOVAL & INSTALLATION

4.6L (1UR-FSE) Engine

See Figures 119 through 120.

1. Disconnect cable from negative battery terminal.
2. Remove engine under covers.
3. Remove cool air intake duct seal.
4. Remove air cleaner.
5. Remove V-ribbed belt.
6. Remove the front stabilizer bar.
7. Remove front floor cover.
8. Remove front center floor brace.
9. Remove front exhaust pipe assembly.
10. Remove steering sliding yoke sub-assembly.
11. Remove generator assembly.
12. Remove engine oil level dipstick guide.
13. Disconnect compressor and pulley.
14. Remove engine room side covers.
15. Remove V-bank cover.
16. Disconnect the air fuel ratio sensor connector RH.
17. Remove the 3 bolts and No. 1 exhaust manifold heat insulator.
18. Remove the 8 nuts and exhaust manifold RH.
19. Remove the gasket.
20. Disconnect the air fuel ratio sensor connector LH.
21. Remove the bolt and wire harness bracket LH.
22. Remove the 3 bolts and No. 2 exhaust manifold heat insulator.
23. Remove the 8 nuts and exhaust manifold LH.
24. Remove the gasket.

To install:

25. Install a new gasket LH.
26. Install the exhaust manifold LH, and install 8 new nuts and tighten to 15 ft. lbs. (21 Nm) in the order shown in the illustration.
27. Install the exhaust manifold heat insulator with the 3 bolts.
28. Install the wire harness bracket with the bolt.
29. Connect the air fuel ratio sensor connector.
30. Install air fuel ratio sensor.
31. Install a new gasket RH.
32. Install the exhaust manifold LH, and install 8 new nuts and tighten to 15 ft. lbs. (21 Nm) in the order shown in the illustration.

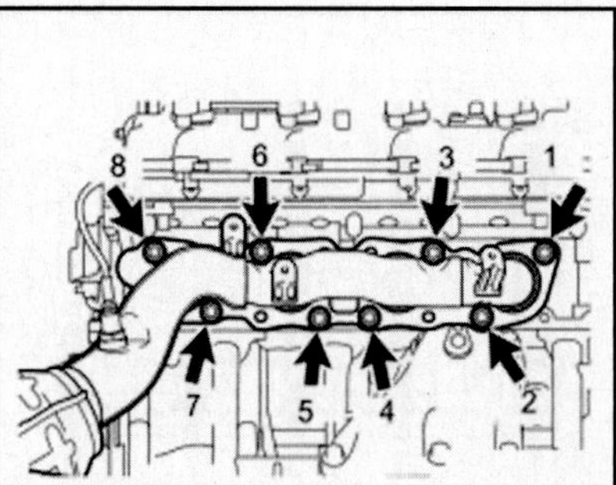

Fig. 119 Exhaust manifold tightening sequence LH

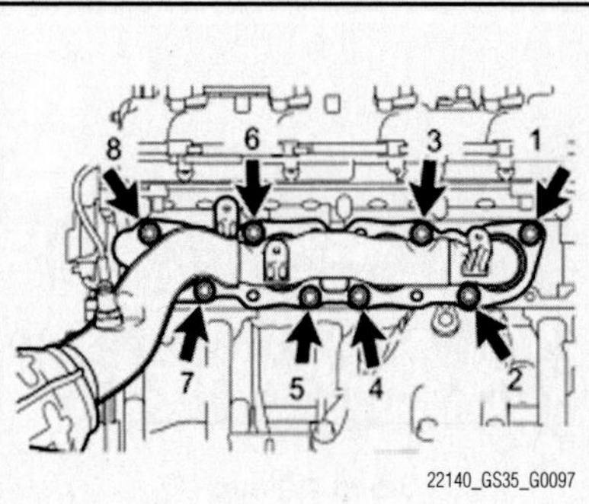

Fig. 120 Exhaust manifold tightening sequence RH

33. Install the exhaust manifold heat insulator with the 3 bolts.
34. Install the wire harness bracket with the bolt.
35. Connect the air fuel ratio sensor connector.
36. Install air fuel ratio sensor.
37. Connect compressor and pulley.
38. Install engine oil level dipstick guide.
39. Install generator assembly.
40. Install steering sliding yoke sub-assembly.
41. Install front exhaust pipe assembly.
42. Install front floor covers.
43. Install front center floor brace.
44. Install the front stabilizer bar.
45. Install V-ribbed belt.
46. Install air cleaner.
47. Connect cable to negative battery terminal.
48. Install engine under covers.
49. Install V-bank cover.
50. Install engine room side covers.
51. Install cool air intake duct seal.

FLEXPLATE

REMOVAL & INSTALLATION

See Figures 121 and 122.

1. Before servicing the vehicle, refer to the precautions in the beginning of this section.
2. Disconnect the negative battery cable. Wait at least 90 seconds before performing any other work
3. Remove the engine or transmission assembly.
4. Hold the crankshaft.
5. Remove the bolts and remove the rear spacer, drive plate and front spacer. (if applicable)

To install:

6. Hold the crankshaft.
7. Clean the bolts and bolt holes.
8. Apply adhesive (Part No. 08833-00070, THREE BOND 1324 or equivalent) to 2 or 3 threads of the bolts.
9. Install the rear spacer, drive plate and front spacer. (if applicable)
10. Install and uniformly tighten the bolts in several passes, in the sequence shown, to the following torque values:
    - 3.5L (2GR-FSE) engine—61 ft. lbs. (83 Nm)
    - 4.6L (1UR-FSE) engine—22 ft. lbs. (30 Nm) plus an additional 90°

➡Do not start the engine within an hour after installing.

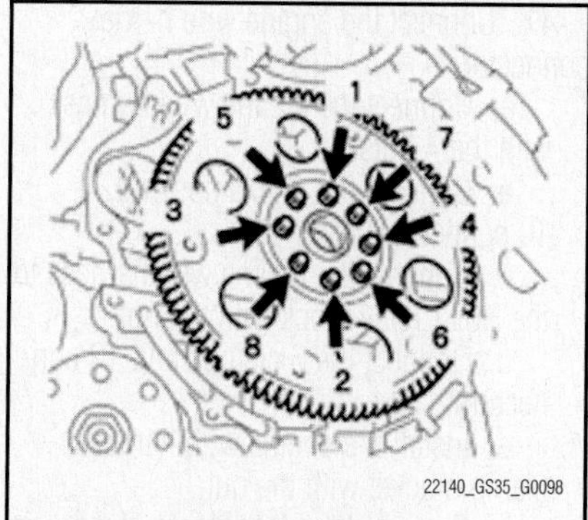

Fig. 121 Drive plate tightening sequence—3.5L

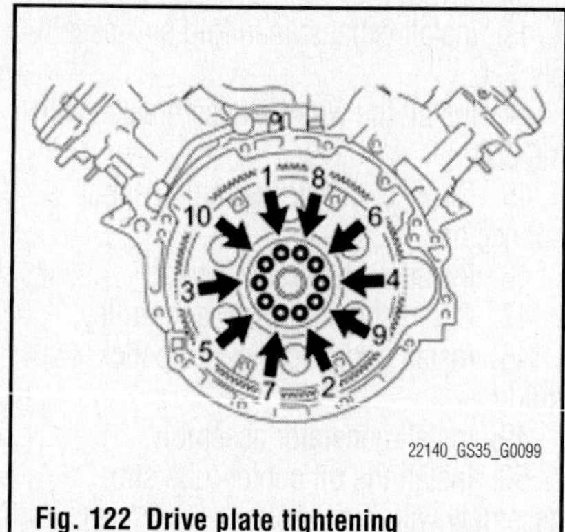

Fig. 122 Drive plate tightening sequence—4.6L

11. Install the engine or transmission assembly.
12. Connect the negative battery cable.

INTAKE MANIFOLD

REMOVAL & INSTALLATION

4.6L (1UR-FSE) Engine

See Figures 123 through 127.

1. Disconnect cable from negative battery terminal.
2. Remove engine under covers.
3. Drain coolant.
4. Remove cool air intake duct seal.
5. Remove air cleaner.
6. Remove throttle body.
7. Remove V-ribbed belt.
8. Remove the front stabilizer bar.
9. Remove front floor cover.
10. Remove front center floor brace.
11. Remove front exhaust pipe assembly.
12. Remove steering sliding yoke sub-assembly.
13. Remove generator assembly.
14. Remove engine oil level dipstick guide.
15. Disconnect compressor and pulley.
16. Remove engine room side covers.
17. Remove V-bank cover.

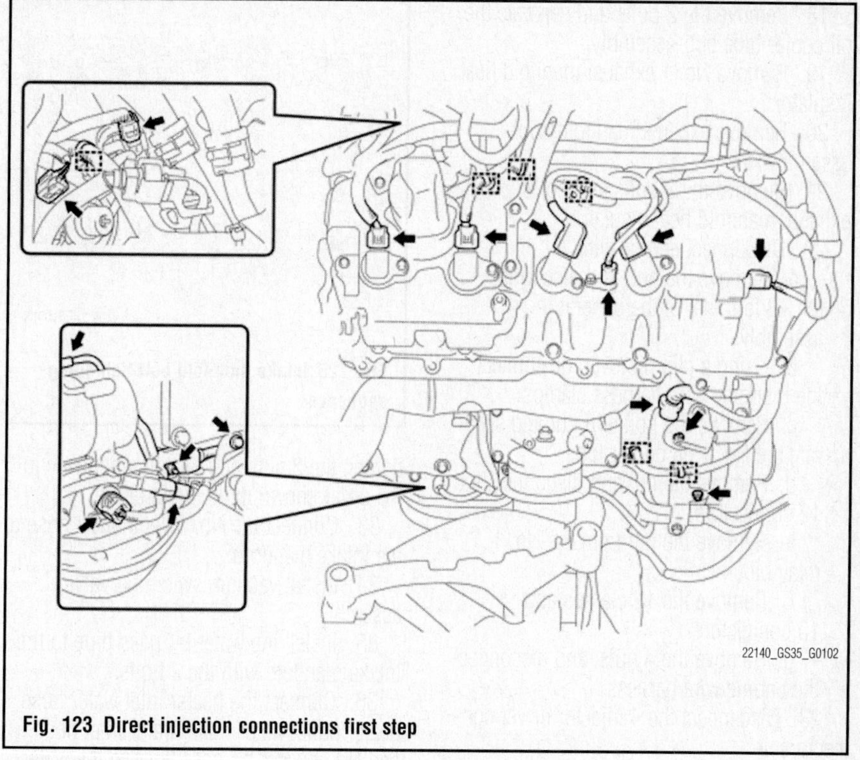

Fig. 123 Direct injection connections first step

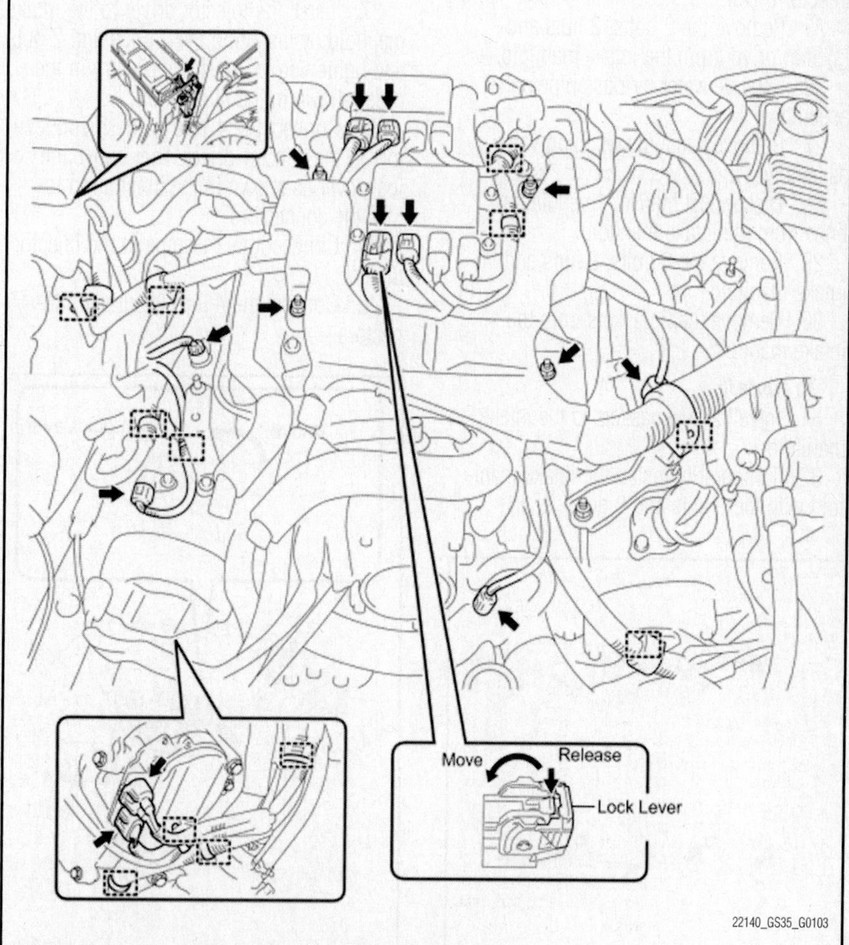

Fig. 124 Direct injection connections second step

18. Remove the 2 bolts and separate the oil cooler tube sub-assembly.

19. Remove No. 1 exhaust manifold heat insulator.

20. Remove exhaust manifold sub-assembly RH.

21. Remove the 3 bolts and no. 3 exhaust manifold heat insulator.

22. Disconnect engine wire.

a. Remove the nut, and disconnect the +B terminal of the generator assembly.

b. Using a clip remover, disconnect the 6 engine wire harness clamps.

c. Remove the bolt and nut and separate the No. 1 oil cooler tube.

d. Remove the nut and disconnect the 12 connectors.

e. Remove the nut from the No. 1 relay block.

f. Remove the 12 clamps and 10 connectors.

g. Remove the 4 nuts, and disconnect the engine wire harness.

23. Disconnect the 4 injector driver connectors

24. Disconnect the 2 clamps from the injector driver.

25. Remove the 2 bolts, 2 nuts and injector driver from the intake manifold.

26. Remove water by-pass pipe sub-assembly.

27. Remove vacuum switching valve assembly.

28. Disconnect the No. 1 ventilation hose from the intake manifold.

29. Remove the 8 bolts, 2 nuts and intake manifold.

30. Remove the 2 gaskets from the intake manifold.

To install:

31. Install 2 new gaskets to the intake manifold.

32. Temporarily install the intake manifold with the 2 nuts and 8 bolts. Then

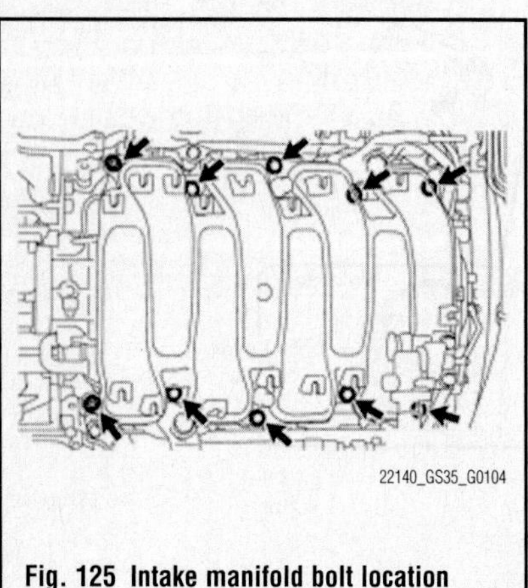

Fig. 125 Intake manifold bolt location

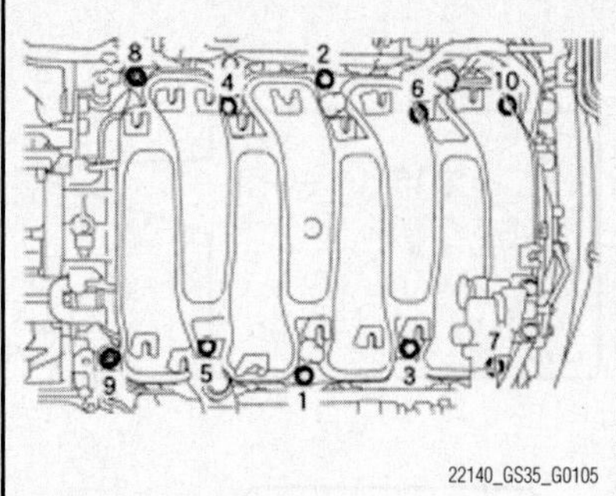

Fig. 126 Intake manifold bolt tightening sequence

tighten the 2 nuts and 8 bolts uniformly in the order shown in the illustration

33. Connect the No. 1 ventilation hose to the intake manifold.

34. Install vacuum switching valve assembly.

35. Install the water by-pass pipe to the intake manifold with the 2 bolts.

36. Connect the heater inlet water hose, heater outlet water hose, water inlet hose, and No. 3 water by-pass hose to the water by-pass pipe with the 4 clamps.

37. Install the injector driver to the intake manifold by installing the 2 bolts and 2 nuts and tightening to 7 ft. lbs. (10 Nm) in the order shown in the illustration.

38. Connect the 4 wire harness connectors to the injector driver. Then move the lock lever as shown in the illustration to lock the connectors.

39. Connect the 2 clamps to the injector driver.

40. Connect the 4 injector driver connectors.

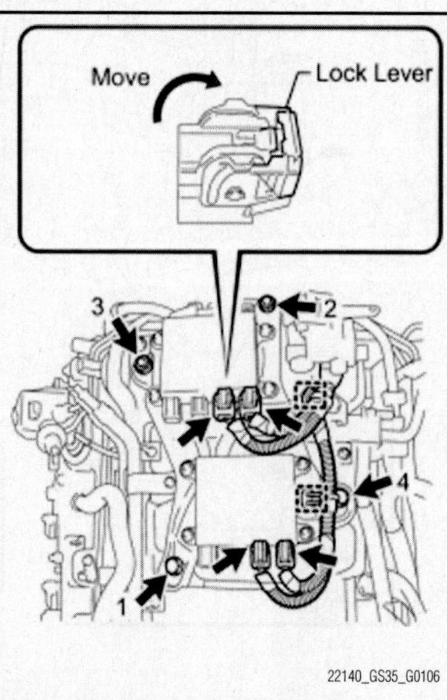

Fig. 127 Injector driver installation

41. Connect the engine wire harness connector.

a. Connect the engine wire harness with the 4 nuts.

b. Connect the 12 clamps and 10 connectors.

c. Connect the engine wire harness to the No. 1 relay block with the nut.

d. Connect the 6 clamps and 12 connectors.

e. Install the engine wire harness clamp bracket with the nut.

f. Connect the +B terminal of the generator assembly with the nut.

42. Install the exhaust manifold heat insulator with the 3 bolts.

43. Install exhaust manifold sub-assembly RH.

44. Install the wire harness bracket with the bolt.

45. Connect the air fuel ratio sensor connector.

46. Install air fuel ratio sensor.

47. Connect compressor and pulley.

48. Install engine oil level dipstick guide.

49. Install generator assembly.

50. Install the oil cooler tube sub-assembly with the 2 bolts.

51. Install steering sliding yoke sub-assembly.

52. Install front exhaust pipe assembly.

53. Install front floor covers.

54. Install front center floor brace.

55. Install the front stabilizer bar.

56. Install V-ribbed belt.

57. Install throttle body.

58. Install air cleaner.

59. Connect cable to negative battery terminal.

60. Fill coolant.

61. Install engine under covers.

62. Install V-bank cover.

63. Install engine room side covers.

64. Install cool air intake duct seal.

65. Connect cable to negative battery terminal.

66. Check for leaks.

67. Perform initialization, if necessary.

OIL PAN

REMOVAL & INSTALLATION

3.5L (2GR-FSE) Engine

See Figures 128 and 129.

➡The No. 1 oil pan cannot be removed with the engine in the vehicle. The engine and transmission must be removed as a unit, then separated. See

ENGINE ASSEMBLY section. It may be possible to remove the No. 2 oil pan from the vehicle while the engine is still in the vehicle.

1. Before servicing the vehicle, refer to the precautions in the beginning of this section.
2. Drain the engine oil.
3. Remove or disconnect the following:
   - Negative battery cable from the battery
   - Engine/transmission assembly
   - Oil filter element
   - Oil filter bracket (AWD)
4. Remove the No. 2 oil pan sub-assembly as follows:
   a. Remove the 15 bolts and 2 nuts (2WD).
   b. Remove the 14 bolts and 2 nuts (AWD).
   c. Insert the blade of a prying tool between the oil pans. Cut through the applied sealer and remove the No. 2 oil pan sub-assembly.

➡Be careful not to damage the contact surfaces of the oil pans.

5. Remove the oil with strainer pipe sub-assembly (AWD) as follows:
   a. Remove the 3 mounting nuts.

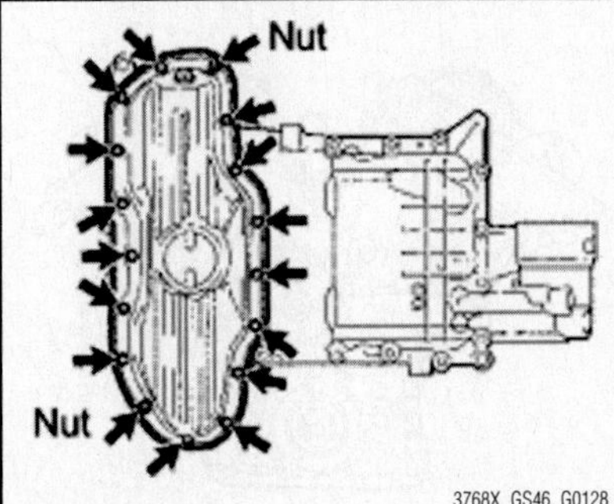

Fig. 128 Remove the 15 bolts and 2 nuts (2WD)

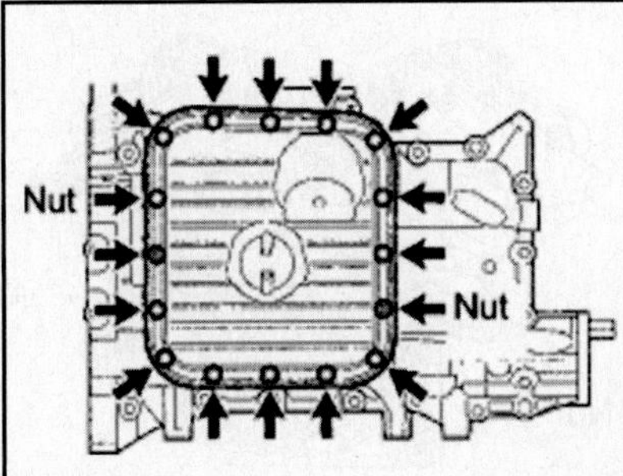

Fig. 129 Remove the 14 bolts and 2 nuts (AWD)

   b. Remove the oil with strainer pipe and gasket.
6. Remove the oil pan sub-assembly as follows:
   a. Remove the 16 bolts and 2 nuts.

➡Be sure to clean the bolts and stud bolts and check the threads for cracks or other damage.

   b. Remove the oil pan by prying between the oil pan and cylinder block with a screwdriver.

⁂ WARNING

Be careful not to damage the contact surfaces of the cylinder block and oil pan.

   c. Remove the 2 O-rings.

To install:

7. Install the oil pan sub-assembly as follows:
   a. When replacing a stud bolt, install it by using an E6 Torx® socket wrench. Torque the stud bolt to 35 inch lbs. (4 Nm).
   b. Apply seal packing in a continuous line of 0.118–0.156 inches (3.0–4.0mm) in diameter.

➡Remove any oil from the contact surface. Install the oil pan within 3 minutes after applying seal packing. Do not start the engine for at least 2 hours after installing.

   c. Install the oil pan with the 16 bolts and 2 nuts. Torque the A bolts to 7 ft. lbs. (10 Nm). Torque the remaining bolts to 15 ft. lbs. (21 Nm).
8. Install the oil with strainer pipe sub-assembly (AWD) as follows:
   a. Install a new gasket.
   b. Install the oil with strainer pipe with the 3 nuts. Torque to 7 ft. lbs. (10 Nm).
9. Install the No. 2 oil pan sub-assembly as follows:
   a. Apply seal packing in a continuous line of 0.118–0.156 inches (3.0–4.0mm) in diameter.

➡Remove any oil from the contact surface. Install the oil pan No. 2 within 3 minutes after applying seal packing. Do not start the engine for at least 2 hours after installing.

   b. Install the oil pan with the 16 bolts and 2 nuts (2WD). Torque the bolts to 7 ft. lbs. (10 Nm).
   c. Install the oil pan with the 14 bolts and 2 nuts (AWD). Torque the bolts to 7 ft. lbs. (10 Nm).
10. Install or connect the following:
   - New gasket and oil filter bracket (AWD). Torque the bolt and 2 nuts to 15 ft. lbs. (21 Nm).
   - Oil filter element
   - Engine/transmission assembly
   - Negative battery cable from the battery
   - Engine with oil

4.6L (1UR-FSE) Engines

See Figure 130.

➡The oil pan cannot be removed with the engine in the vehicle. The engine and transmission must be removed as a unit, then separated. It may be possible to remove the oil pan from the vehicle while the engine is still in the vehicle.

1. Before servicing the vehicle, refer to the precautions in the beginning of this section.
2. Remove or disconnect the following:
   - Engine/transmission assembly
3. Remove the 15 bolts and 2 nuts and remove the No. 2 oil pan.
4. Insert the blade of an oil pan cutter between the oil pans. Cut through the applied sealer and remove the No. 2 oil pan sub-assembly.
5. Remove the 14 bolts and 2 nuts and remove the No. 1 oil pan.
6. Remove the oil pan sub-assembly by prying between the oil pan and cylinder block with a screwdriver.

To install:

7. Apply the seal packing to oil pan No. 1.
8. Install the oil pan sub-assembly with the 14 bolts and 2 nuts.
9. Uniformly tighten the bolts and nuts in several passes to the following:
   a. Bolts A to 7 ft. lbs. (10 Nm).
   b. Bolts B 26 ft. lbs, (35 Nm).
   c. Nut to 26 ft. lbs, (35 Nm).
10. Apply seal packing to oil pan No. 2.

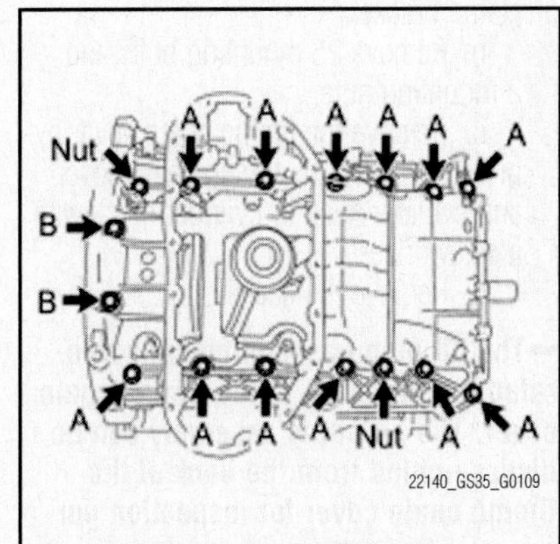

Fig. 130 Oil pan No. 1 bolt location

11. Install the oil pan with the 15 bolts and 2 nuts to 7 ft. lbs. (10 Nm). Uniformly tighten the bolts and nuts in several passes.

12. Install or connect the following:

- Engine/transaxle assembly
- All fluids

OIL PUMP

REMOVAL & INSTALLATION

3.5L (2GR-FSE) Engine

See Figures 131 through 136.

➡The oil pump cannot be removed with the engine in the vehicle. The engine and transmission must be removed as a unit, then separated.

1. Before servicing the vehicle, refer to the precautions in the beginning of this section.
2. Remove or disconnect the following:
   - Engine/transmission assembly
   - Front differential assembly (AWD)
   - Serpentine drive belt
   - No. 2 idler pulley
   - Alternator
   - A/C compressor unit, if necessary
   - Left and right engine mounting brackets
   - Serpentine belt tensioner
   - Water pump pulley
   - Fuel injector driver
   - Intake air surge tank assembly and No. 2 surge tank stay
   - Water hose joint
   - Crankshaft pulley
   - Water inlet
   - Oil pan assembly
   - Oil strainer
   - No. 1 and 2 fuel pipes
   - High pressure side fuel pump
   - Ignition coil assembly
   - No. 1 and 2 oil pipes
   - Left and right cylinder head covers
3. Remove the timing chain cover assembly as follows:
   a. Remove bolt and wiring harness clamp bracket.
   b. Remove 25 mounting bolts and 2 mounting nuts.
   c. Remove the timing chain cover by prying between the timing chain cover and cylinder head or cylinder block with a screwdriver.
   d. Remove the gasket.

➡The oil pump assembly is incorporated into the back of the timing chain cover. The oil pump assembly can be disassembled from the back of the timing chain cover for inspection purposes.

Fig. 131 Location of mounting bolts for oil pump cover behind the timing chain cover—3.5L (2GR-FSE) engine

To install:

4. Install the timing chain cover assembly as follows:
   a. Apply seal packing in a continuous line of 0.197–0.217 inches (5.0–5.5mm) in diameter to the engine at the seam where the cylinder head meets the camshaft bearing cap assembly and the cylinder head meets the cylinder block.

➡Be sure to clean, degrease and dry the contact surfaces before applying the seal packing. Install the component within 3 minutes after applying seal packing. Do not start the engine for at least 2 hours after installing.

   b. Apply seal packing in a continuous line of 0.138–0.158 inches (3.5–4.0mm) in diameter to the timing chain cover.
   c. Install a new gasket.
   d. Align the oil pump drive rotor spline and the crankshaft. Install the spline and chain cover to the crankshaft.
   e. Temporarily tighten the timing chain cover with the 25 bolts and nuts.
   f. Fully tighten the 3 bolts shown in the illustration. Torque bolt A to 32 ft. lbs. (43 Nm). Torque the 2 remaining bolts to 15 ft. lbs (21 Nm).
   g. Fully tighten the 3 bolts shown in the illustration. Torque the bolts to 15 ft. lbs (21 Nm).
   h. Fully tighten the 7 bolts and 2 nuts shown in the illustration. Torque the bolts to 15 ft. lbs (21 Nm).

➡Be sure to tighten the bolts and nuts in order of upper to lower.

   i. Fully tighten the 12 bolts shown in the illustration. Torque the bolts to 15 ft. lbs (21 Nm).

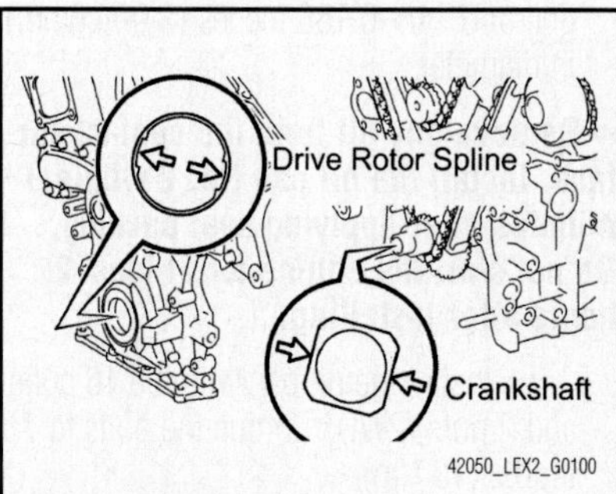

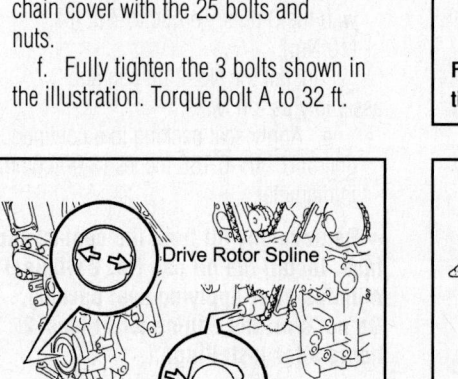

Fig. 132 Align the oil pump drive rotor spline and the crankshaft—3.5L (2GR-FSE) engine

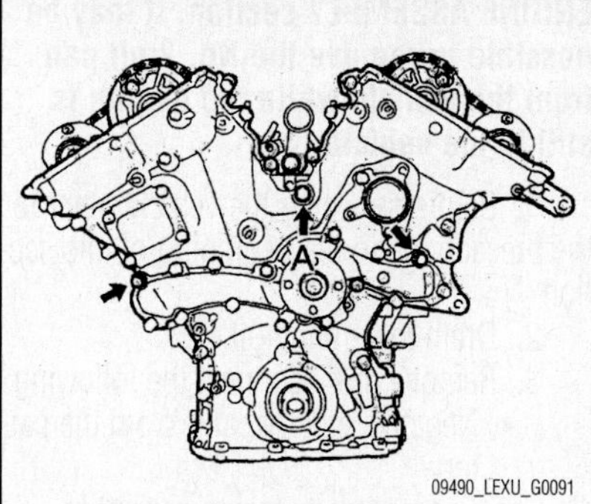

Fig. 133 Location of 3 bolts to be tightened first (location of bolt A shown)—3.5L (2GR-FSE) engine

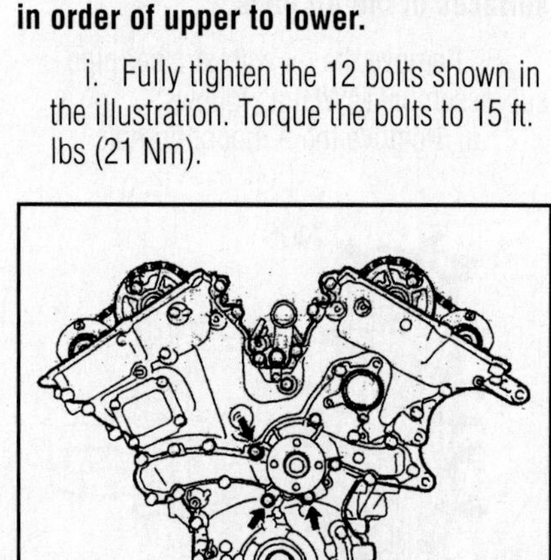

Fig. 134 Location of 3 bolts to be tightened second—3.5L (2GR-FSE) engine

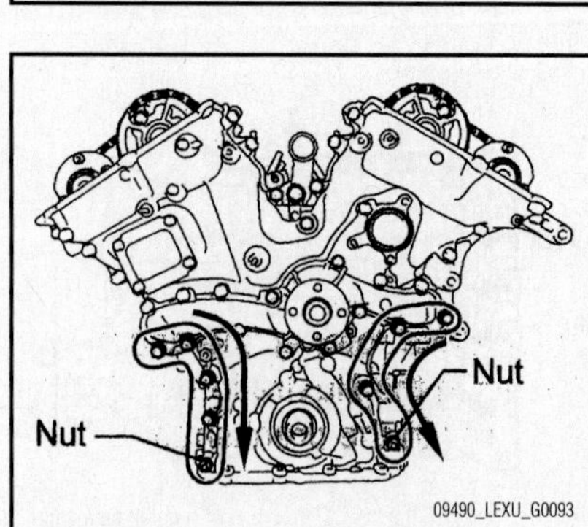

Fig. 135 Location of 7 bolts and 2 nuts to be tightened third—3.5L (2GR-FSE) engine

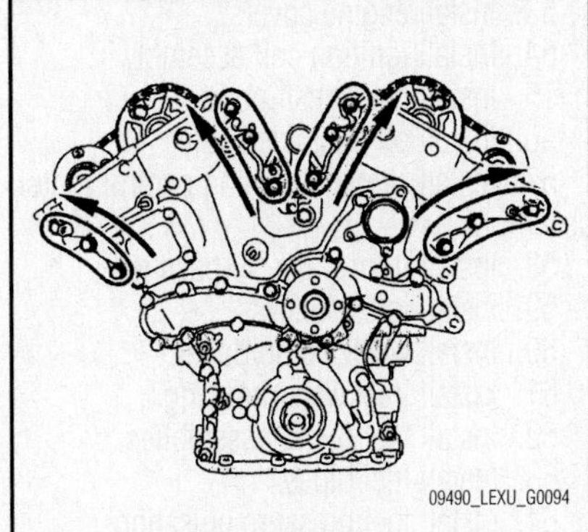

Fig. 136 Location of 12 bolts to be tightened fourth—3.5L (2GR-FSE) engine

➡Be sure to tighten the bolts in order of lower to upper.

j. Install the bolt and wiring harness bracket. Torque the bolts to 7 ft. lbs (10 Nm).

5. Install or connect the following:
   - Left and right cylinder head covers
   - No. 1 and 2 oil pipes
   - Ignition coil assembly
   - High pressure side fuel pump
   - No. 1 and 2 fuel pipes
   - Oil strainer
   - Oil pan assembly
   - Water inlet
   - Crankshaft pulley. Torque the bolt to 192 ft. lbs (260 Nm).
   - Water hose joint
   - Intake air surge tank assembly and No. 2 surge tank stay
   - Fuel injector driver
   - Water pump pulley. Torque the bolt to 15 ft. lbs (21 Nm).
   - Serpentine belt tensioner
   - Left and right engine mounting brackets
   - A/C compressor unit, if necessary
   - Alternator
   - No. 2 idler pulley
   - Serpentine drive belt
   - Front differential assembly (AWD)
   - Engine/transmission assembly

4.6L (1UR-FSE) Engine

See Figures 137 through 142.

1. Discharge fuel system pressure.
2. Disconnect cable from negative battery terminal.
3. Remove intake manifold.
4. Remove battery.
5. Remove battery tray.
6. Remove power steering ECU assembly.
7. Remove wire harness clamp bracket.
8. Drain engine oil.
9. Remove front bumper assembly.
10. Remove radiator support opening cover.

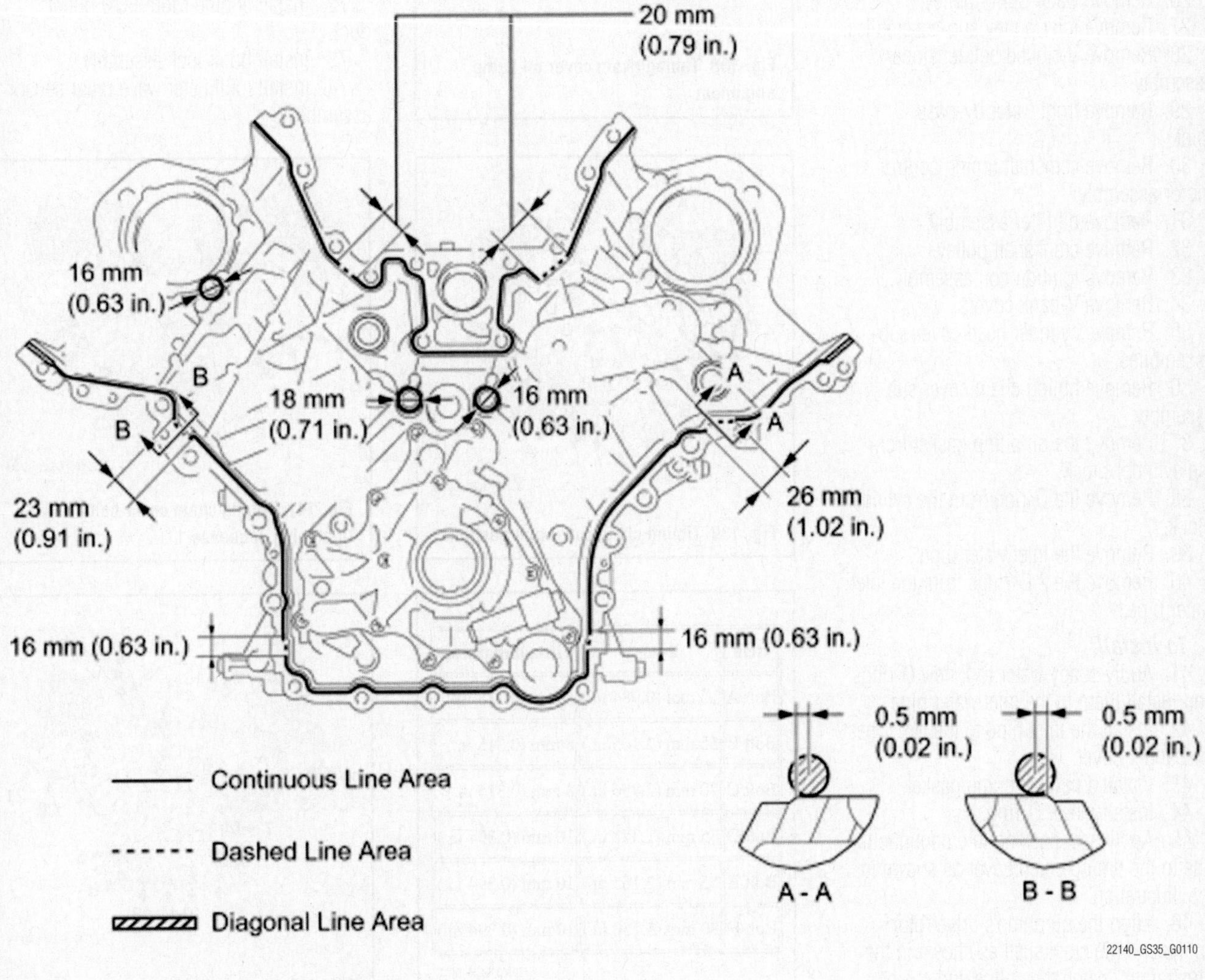

Fig. 137 Timing chain cover sealant position

11. Remove millimeter wave radar sensor assembly.
12. Remove hood lock assembly.
13. Remove engine room ECU outlet duct.
14. Remove upper radiator support sub-assembly.
15. Disconnect oil cooler hose.
16. Disconnect radiator hoses and radiator assembly.
17. Separate compressor and pulley.
18. Remove exhaust manifold sub-assembly lh.
19. Disconnect wire harness.
20. Remove the bolt, and disconnect the clamp and ground cables.
21. Remove engine cover sub-assemblies.
22. Remove fuel pressure pulsation damper assembly.
23. Remove fuel pipe sub-assemblies.
24. Remove fuel pump assembly.
25. Remove inlet water housing.
26. Remove water pump pulley.
27. Remove idler pulley sub-assemblies.
28. Remove V-ribbed belt tensioner assembly.
29. Remove front water by-pass joint.
30. Remove camshaft timing control motor assembly.
31. Remove oil filter assembly.
32. Remove crankshaft pulley.
33. Remove ignition coil assembly.
34. Remove V-bank covers.
35. Remove cylinder head cover sub-assemblies.
36. Remove timing chain cover sub-assembly.
37. Remove the oil pump gasket from the cylinder block.
38. Remove the O-ring from the cylinder block.
39. Remove the inlet water pipe.
40. Remove the 2 O-rings from the inlet water pipe.

To install:

41. Apply soapy water to 2 new O-rings and install them to the inlet water pipe.
42. Install the inlet pipe to the No. 1 heat exchanger cover.
43. Install a new oil pump gasket.
44. Install a new O-ring.
45. Apply seal packing in a continuous line to the timing chain cover as shown in the illustration.
46. Align the oil pump's drive rotor spline and the crankshaft as shown in the illustration. Install the spline and chain cover to the crankshaft.
47. Temporarily tighten the timing chain cover with the 30 bolts and nut.
48. Tighten the 11 bolts in several steps, in the sequence shown in the illustration.
49. Temporarily tighten the belt tensioner with the standard bolt and 6 mm hexagon wrench bolt.
50. Tighten the 21 bolts and nut in several steps, in the sequence shown in the illustration.
51. Install 2 new gaskets and 2 plugs.
52. Install cylinder head cover sub-assemblies.

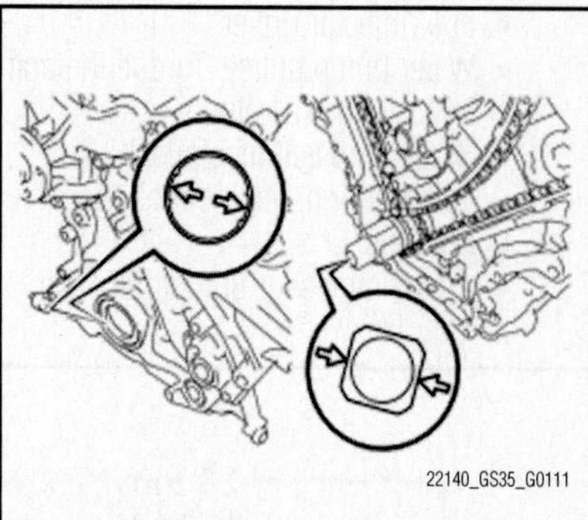

Fig. 138 Timing chain cover oil pump alignment

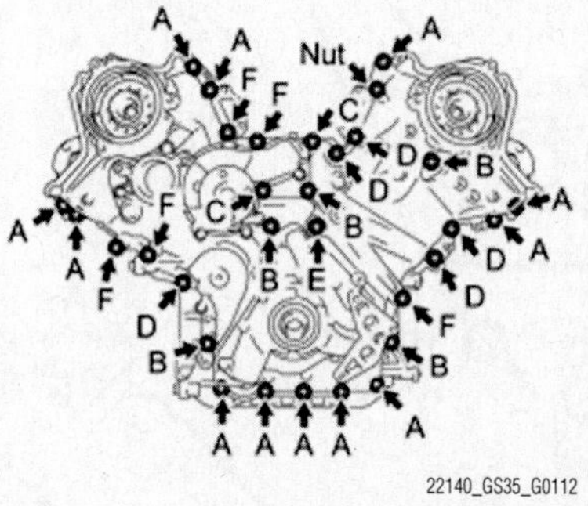

22140_GS35_G0112

Fig. 139 Timing chain cover bolt position

Item	Length	Thread diameter
Bolt A	25 mm (0.984 in.)	8 mm (0.315 in.)
Bolt B	55 mm (2.165 in.)	8 mm (0.315 in.)
Bolt C	70 mm (2.756 in.)	8 mm (0.315 in.)
Bolt D	35 mm (1.378 in.)	10 mm (0.394 in.)
Bolt E	55 mm (2.165 in.)	10 mm (0.394 in.)
Bolt F	80 mm (3.150 in.)	10 mm (0.394 in.)

22140_GS35_G0113

Fig. 140 Timing chain cover bolt legend

53. Install engine covers.
54. Install ignition coil assembly.
55. Install crankshaft pulley.
56. Install oil filter assembly.
57. Install camshaft timing control motor assembly.
58. Install front water by-pass joint.
59. Install idler pulleys.
60. Install water pump pulley.
61. Install inlet water housing.
62. Install fuel pump assemblies.
63. Install fuel pipes.
64. Install fuel pressure pulsation damper assembly.
65. Install engine covers.
66. Connect wire harness.
67. Connect ground wire.
68. Install exhaust manifold sub-assembly lh.
69. Install compressor and pulley.
70. Install radiator assembly and hoses.
71. Install upper radiator support sub-assembly.
72. Install engine room ECU outlet duct.
73. Install hood lock assembly.
74. Install millimeter wave radar sensor assembly.

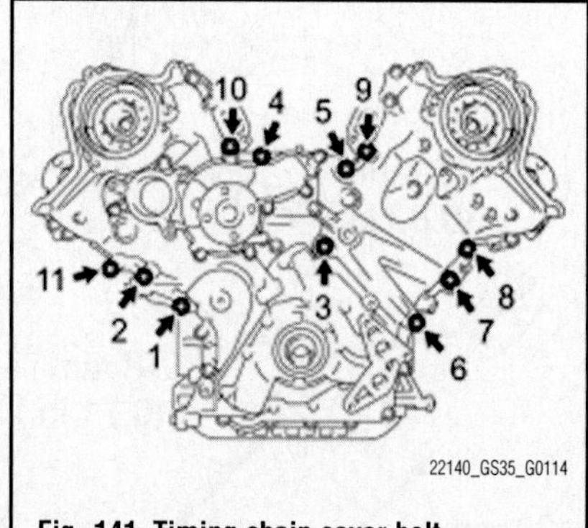

Fig. 141 Timing chain cover bolt tightening sequence 1

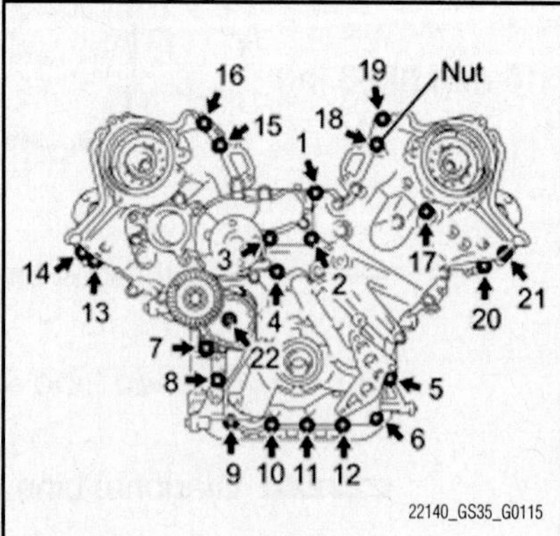

Fig. 142 Timing chain cover bolt tightening sequence 2

75. Install radiator support opening cover.
76. Install front bumper assembly.
77. Install wire harness clamp bracket.
78. Install engine room relay block assembly.
79. Install battery tray.
80. Install power steering ECU assembly.
81. Install battery.
82. Install intake manifold.
83. Add engine oil.
84. Inspect for oil leak.
85. Inspect for oil leak.
86. Perform initialization.

PISTON AND RING

POSITIONING

See Figure 143.

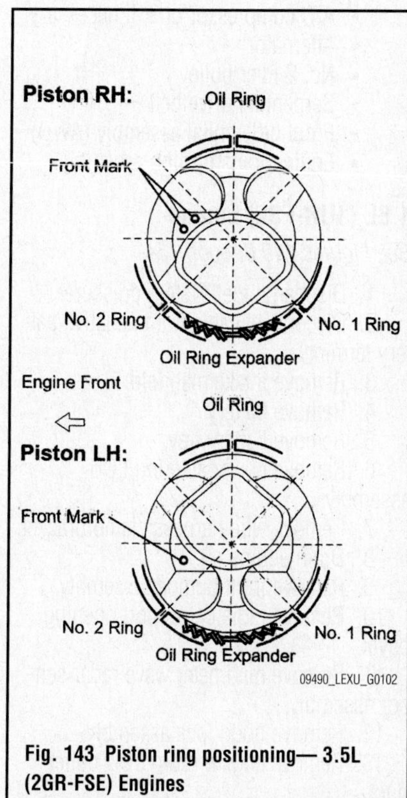

Fig. 143 Piston ring positioning— 3.5L (2GR-FSE) Engines

REAR MAIN SEAL

REMOVAL & INSTALLATION

See Figure 144.

1. If the rear oil seal retainer is removed from the cylinder block, perform the following:
    a. Using a screwdriver and hammer, tap out the oil seal.
    b. Using SST 09223-15030, 09950-70010 (09951-07100) and a hammer, tap in a new oil seal until its surface is flush with the rear oil seal retainer edge.

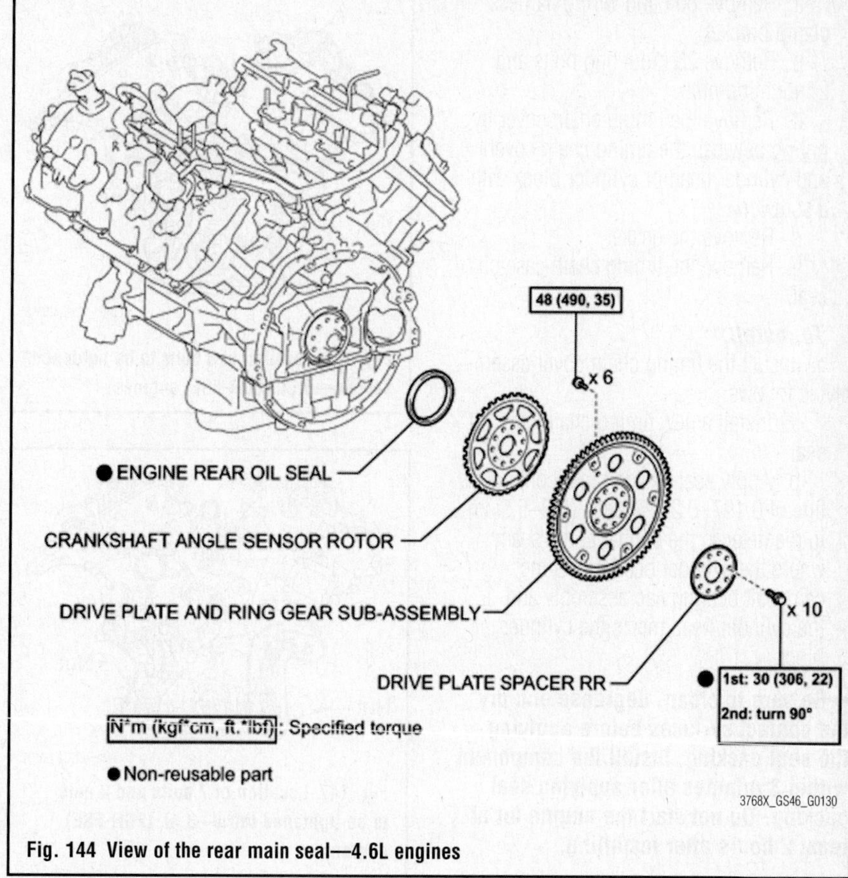

Fig. 144 View of the rear main seal—4.6L engines

    c. Apply MP grease to the oil seal lip.
2. If the rear seal retainer is installed on the cylinder block, perform the following:
    a. Using a knife, cut off the oil seal lip.
    b. Using a screwdriver, pry out the oil seal.

⁂ CAUTION

Be careful not to damage the crankshaft. Tape the screwdriver tip.

    c. Apply MP grease to a new oil seal lip.
    d. Using SST 09223-15030, 09950-70010 (09951-07100) and a hammer, tap in the oil seal until its surface is flush with the rear oil seal retainer edge.

ROCKER ARMS

REMOVAL & INSTALLATION

See Cylinder Head.

TIMING CHAIN FRONT COVER

REMOVAL & INSTALLATION

3.5L (2GR-FSE) Engine

See Figures 145 through 148.

1. Before servicing the vehicle, refer to the precautions in the beginning of this section.
2. Remove or disconnect the following:
    - Engine/transmission assembly
    - Front differential assembly (AWD)
    - Serpentine drive belt
    - No. 2 idler pulley
    - Alternator
    - A/C compressor unit, if necessary
    - Left and right engine mounting brackets
    - Serpentine belt tensioner
    - Water pump pulley
    - Fuel injector driver
    - Intake air surge tank assembly and No. 2 surge tank stay
    - Water hose joint
    - Crankshaft pulley
    - Water inlet
    - Oil pan assembly
    - Oil strainer
    - No. 1 and 2 fuel pipes
    - High pressure side fuel pump
    - Ignition coil assembly
    - No. 1 and 2 oil pipes
    - Left and right cylinder head covers
3. Remove the timing chain cover assembly as follows:

a. Remove bolt and wiring harness clamp bracket.

b. Remove 25 mounting bolts and 2 mounting nuts.

c. Remove the timing chain cover by prying between the timing chain cover and cylinder head or cylinder block with a screwdriver.

d. Remove the gasket.

e. Remove the timing chain case oil seal.

To install:

4. Install the timing chain cover assembly as follows:

a. Install a new timing chain case oil seal.

b. Apply seal packing in a continuous line of 0.197–0.217 inches (5.0–5.5mm) in diameter to the engine at the seam where the cylinder head meets the camshaft bearing cap assembly and the cylinder head meets the cylinder block

➡Be sure to clean, degrease and dry the contact surfaces before applying the seal packing. Install the component within 3 minutes after applying seal packing. Do not start the engine for at least 2 hours after installing.

c. Apply seal packing in a continuous line of 0.138–0.158 inches (3.5–4.0mm) in diameter to the timing chain cover

d. Install a new gasket

e. Align the oil pump drive rotor spline and the crankshaft. Install the spline and chain cover to the crankshaft.

f. Temporarily tighten the timing chain cover with the 25 bolts and nuts.

g. Fully tighten the 3 bolts shown in the illustration. Torque bolt A to 32 ft. lbs. (43 Nm). Torque the 2 remaining bolts to 15 ft. lbs (21 Nm).

h. Fully tighten the 3 bolts shown in the illustration. Torque the bolts to 15 ft. lbs (21 Nm).

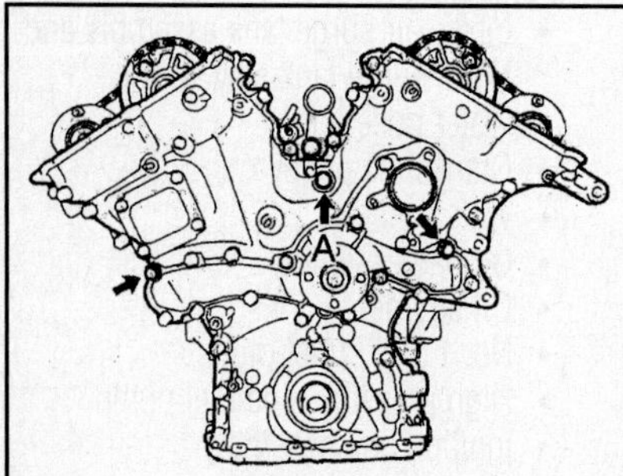

Fig. 145 Location of 3 bolts to be tightened first (location of bolt A shown)—3.5L (2GR-FSE) engines

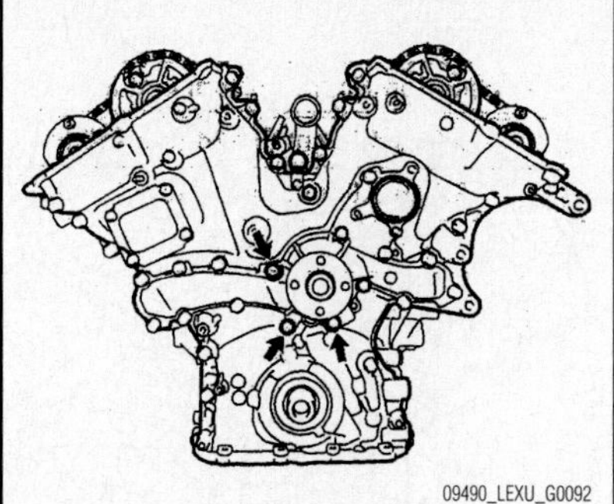

Fig. 146 Location of 3 bolts to be tightened second—3.5L (2GR-FSE) engines

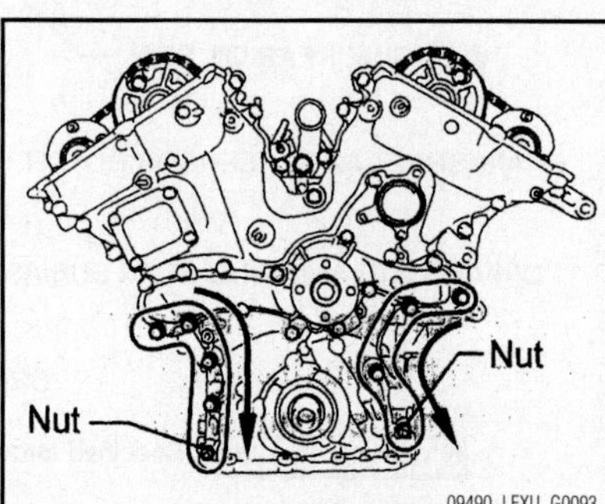

Fig. 147 Location of 7 bolts and 2 nuts to be tightened third—3.5L (2GR-FSE) engines

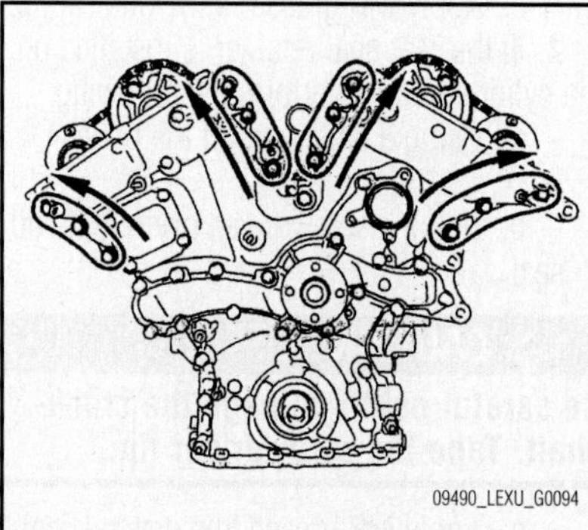

Fig. 148 Location of 12 bolts to be tightened fourth—3.5L (2GR-FSE) engine

i. Fully tighten the 7 bolts and 2 nuts shown in the illustration. Torque the bolts to 15 ft. lbs (21 Nm).

➡Be sure to tighten the bolts and nuts in order of upper to lower.

j. Fully tighten the 12 bolts shown in the illustration. Torque the bolts to 15 ft. lbs (21 Nm).

➡Be sure to tighten the bolts in order of lower to upper.

k. Install the bolt and wiring harness bracket. Torque the bolts to 7 ft. lbs (10 Nm).

5. Install or connect the following:

- Left and right cylinder head covers
- No. 1 and 2 oil pipes
- Ignition coil assembly
- High pressure side fuel pump
- No. 1 and 2 fuel pipes
- Oil strainer
- Oil pan assembly
- Water inlet
- Crankshaft pulley. Torque the bolt to 192 ft. lbs (260 Nm).
- Water hose joint
- Intake air surge tank assembly and No. 2 surge tank stay
- Fuel injector driver
- Water pump pulley. Torque the bolt to 15 ft. lbs (21 Nm).
- Serpentine belt tensioner
- Left and right engine mounting brackets
- A/C compressor unit, if necessary
- Alternator
- No. 2 idler pulley
- Serpentine drive belt
- Front differential assembly (AWD)
- Engine/transmission assembly

4.6L (1UR-FSE) Engine

See Figures 149 through 154.

1. Discharge fuel system pressure.
2. Disconnect cable from negative battery terminal.
3. Remove intake manifold.
4. Remove battery.
5. Remove battery tray.
6. Remove power steering ECU assembly.
7. Remove wire harness clamp bracket.
8. Drain engine oil.
9. Remove front bumper assembly.
10. Remove radiator support opening cover.
11. Remove millimeter wave radar sensor assembly.
12. Remove hood lock assembly.
13. Remove engine room ECU outlet duct.
14. Remove upper radiator support sub-assembly.
15. Disconnect oil cooler hose.
16. Disconnect radiator hoses and radiator assembly.
17. Separate compressor and pulley.
18. Remove exhaust manifold sub-assembly lh.
19. Disconnect wire harness.
20. Remove the bolt, and disconnect the clamp and ground cables.
21. Remove engine cover sub-assemblies.
22. Remove fuel pressure pulsation damper assembly.

23. Remove fuel pipe sub-assemblies.
24. Remove fuel pump assembly.
25. Remove inlet water housing.
26. Remove water pump pulley.
27. Remove idler pulley sub-assemblies.
28. Remove V-ribbed belt tensioner assembly.
29. Remove front water by-pass joint.
30. Remove camshaft timing control motor assembly.
31. Remove oil filter assembly.
32. Remove crankshaft pulley.
33. Remove ignition coil assembly.
34. Remove V-bank covers.
35. Remove cylinder head cover sub-assemblies.
36. Remove timing chain cover sub-assembly.
37. Remove the oil pump gasket from the cylinder block.
38. Remove the O-ring from the cylinder block.
39. Remove the inlet water pipe.
40. Remove the 2 O-rings from the inlet water pipe.

To install:

41. Apply soapy water to 2 new O-rings and install them to the inlet water pipe.
42. Install the inlet pipe to the No. 1 heat exchanger cover.
43. Install a new oil pump gasket.
44. Install a new O-ring.
45. Apply seal packing in a continuous line to the timing chain cover as shown in the illustration.
46. Align the oil pump's drive rotor spline and the crankshaft as shown in the illustration. Install the spline and chain cover to the crankshaft
47. Temporarily tighten the timing chain cover with the 30 bolts and nut
48. Tighten the 11 bolts in several steps, in the sequence shown in the illustration
49. Temporarily tighten the belt tensioner with the standard bolt and 6 mm hexagon wrench bolt.
50. Tighten the 21 bolts and nut in several steps, in the sequence shown in the illustration.
51. Install 2 new gaskets and 2 plugs.
52. Install cylinder head cover sub-assemblies.
53. Install engine covers.
54. Install ignition coil assembly.
55. Install crankshaft pulley.
56. Install oil filter assembly.
57. Install camshaft timing control motor assembly.
58. Install front water by-pass joint.
59. Install idler pulleys.
60. Install water pump pulley.
61. Install inlet water housing.
62. Install fuel pump assemblies.
63. Install fuel pipes.
64. Install fuel pressure pulsation damper assembly.
65. Install engine covers.
66. Connect wire harness.
67. Connect ground wire.
68. Install exhaust manifold sub-assembly lh.

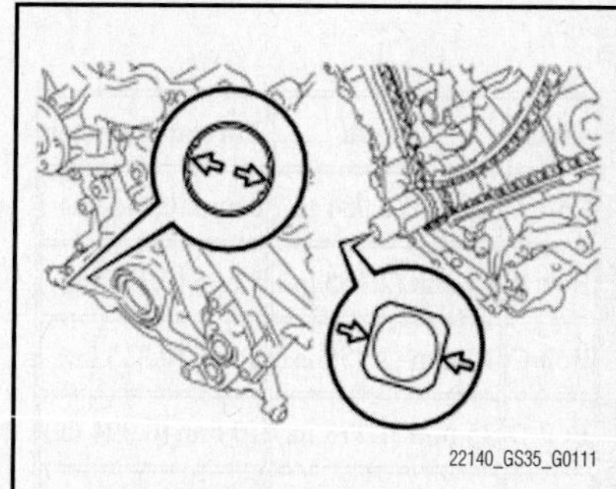

Fig. 150 Timing chain cover oil pump alignment

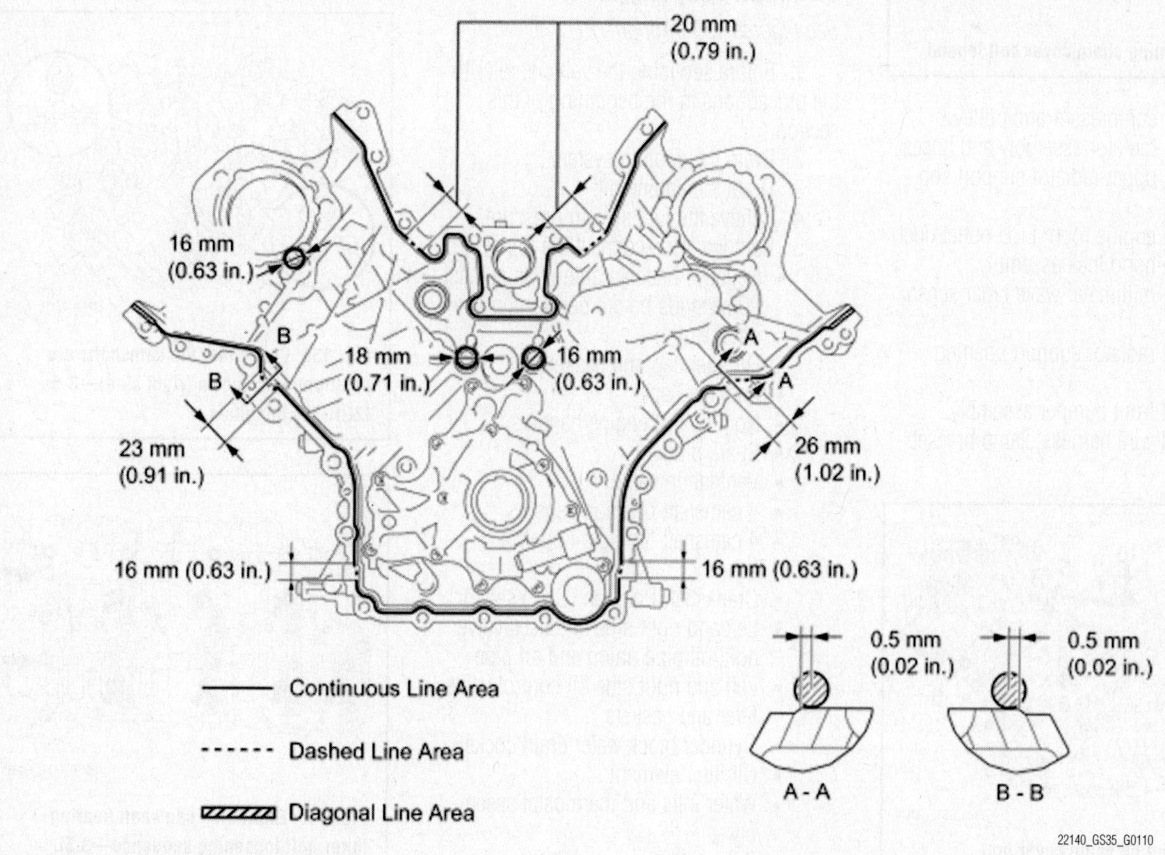

Fig. 149 Timing chain cover sealant position

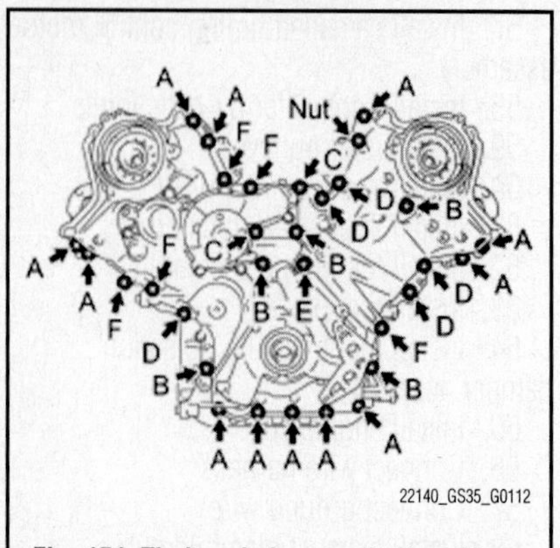

Fig. 151 Timing chain cover bolt position

Item	Length	Thread diameter
Bolt A	25 mm (0.984 in.)	8 mm (0.315 in.)
Bolt B	55 mm (2.165 in.)	8 mm (0.315 in.)
Bolt C	70 mm (2.756 in.)	8 mm (0.315 in.)
Bolt D	35 mm (1.378 in.)	10 mm (0.394 in.)
Bolt E	55 mm (2.165 in.)	10 mm (0.394 in.)
Bolt F	80 mm (3.150 in.)	10 mm (0.394 in.)

22140_GS35_G0113

Fig. 152 Timing chain cover bolt legend

69. Install compressor and pulley.
70. Install radiator assembly and hoses.
71. Install upper radiator support sub-assembly.
72. Install engine room ECU outlet duct.
73. Install hood lock assembly.
74. Install millimeter wave radar sensor assembly.
75. Install radiator support opening cover.
76. Install front bumper assembly.
77. Install wire harness clamp bracket.

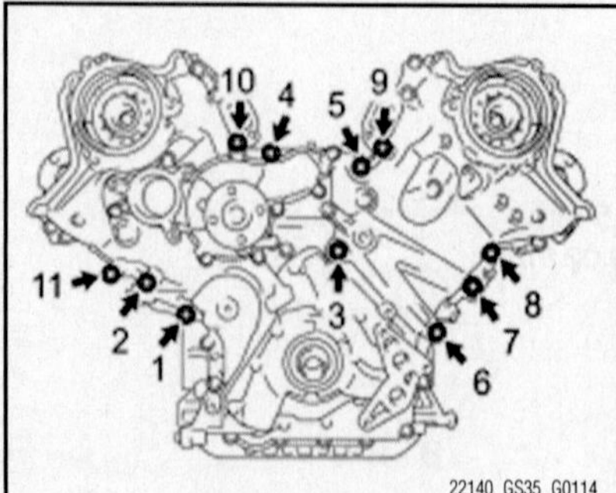

Fig. 153 Timing chain cover bolt tightening sequence 1

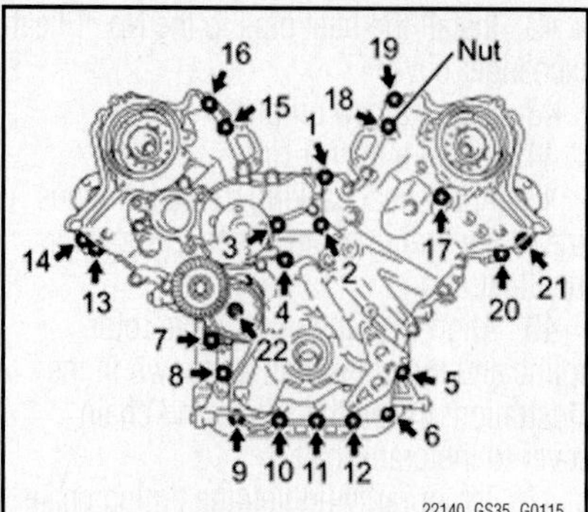

Fig. 154 Timing chain cover bolt tightening sequence 2

78. Install engine room relay block assembly LH.
79. Install battery tray.
80. Install power steering ECU assembly.
81. Install battery.
82. Install intake manifold.
83. Add engine oil.
84. Inspect for oil leak.
85. Inspect for oil leak.
86. Perform initialization.

TIMING CHAIN & SPROCKETS

REMOVAL & INSTALLATION

3.5L (2GR-FSE) Engine

See Figures 155 through 171.

1. Before servicing the vehicle, refer to the precautions in the beginning of this section.
2. Drain the cooling system.
3. Drain the engine oil.
4. Relieve the fuel system pressure.
5. Remove or disconnect the following:
   - Negative battery cable. Wait at least 90 seconds before performing any other work
   - Oil filler cap and gasket
   - Radiator cap
   - No. 1 and 2 engine hangers
   - Spark plugs
   - Ventilation valve
   - 4 camshaft position sensors
   - 4 camshaft timing oil control valves
   - Crankshaft Position (CKP) sensor
   - Left and right side oil check valve bolt, oil pipe union and oil pipe
   - Left and right side oil control valve filter and gaskets
   - Cylinder block water drain cocks
   - Oil filter element
   - Water inlet and thermostat assembly
   - Rear water by-pass joint
   - Cylinder head cover and gaskets
   - Timing chain cover, timing chain and timing chain sprockets
   - No. 2 chain tensioner assembly
6. Remove the camshaft timing gears and No. 2 chain (right) as follows:
   a. While raising the No. 2 chain tensioner, insert a pin of 1.0mm (0.039 in.) into the hole to fix the No. 2 chain tensioner.
   b. Hold the hexagonal portion of the camshaft with a wrench, and remove the 2 bolts and 2 camshaft timing gear assemblies.
   c. Remove the No. 2 chain.
   - No. 2 chain tensioner
7. Remove the camshaft bearing cap (right) as follows:
   a. Check that the camshafts are positioned as shown in the illustration.
   b. Uniformly loosen and remove the 9 bearing cap bolts in the sequence shown in the illustration.
   c. Uniformly loosen and remove the 14 bearing cap bolts in the sequence shown in the illustration.
   d. Remove the 6 bearing caps.
   - Camshaft
   - No. 2 camshaft
   - Camshaft housing sub-assembly (right) by prying between the cylin-

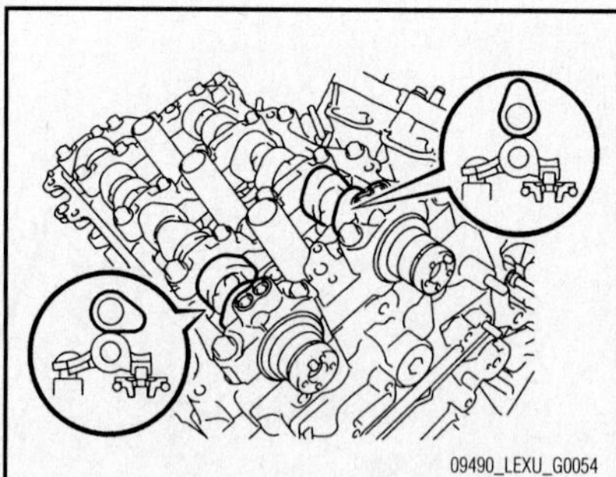

Fig. 155 Check that the camshafts are positioned as shown (right side)—3.5L (2GR-FSE) engines

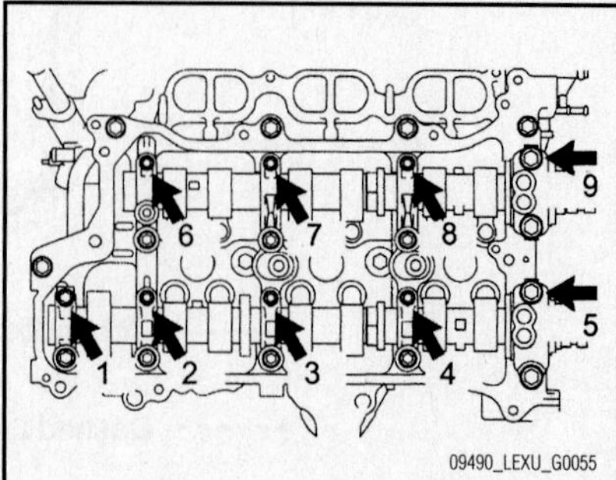

Fig. 156 Right side camshaft bearing inner bolt loosening sequence—3.5L (2GR-FSE) engines

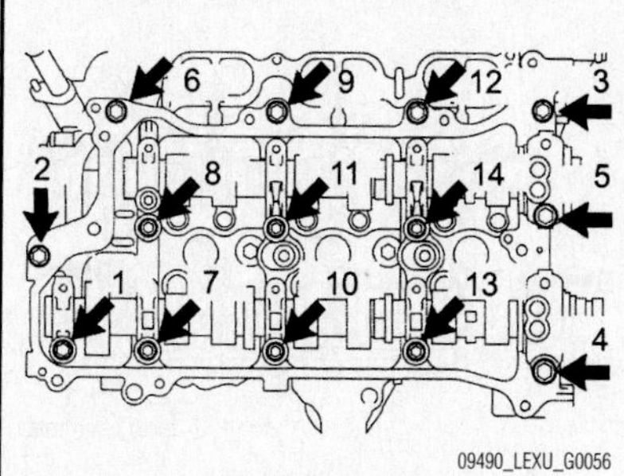

Fig. 157 Right side camshaft bearing outer bolt loosening sequence—3.5L (2GR-FSE) engines

der head and camshaft housing sub-assembly (right) with a screwdriver.

⁂ WARNING

Be careful not to damage the contact surfaces of the cylinder head and camshaft housing.

8. Remove the camshaft timing gears and No. 2 chain (left) as follows:
   a. While pushing down the No. 3 chain tensioner, insert a pin of 1.0mm (0.039 in.) into the hole to fix the No. 3 chain tensioner.
   b. Hold the hexagonal portion of the camshaft with a wrench, and remove the 2 bolts and 2 camshaft timing gear assemblies.
   c. Remove the No. 2 chain.
   - No. 3 chain tensioner
9. Remove the camshaft bearing cap (left) as follows:
   a. Check that the camshafts are positioned as shown in the illustration.
   b. Uniformly loosen and remove the 8 bearing cap bolts in the sequence shown in the illustration.
   c. Uniformly loosen and remove the 13 bearing cap bolts in the sequence shown in the illustration.
   d. Remove the 5 bearing caps.
   - No. 3 camshaft
   - No. 4 camshaft
   - Camshaft housing sub-assembly (left) by prying between the cylinder head and camshaft housing sub-assembly (left) with a screwdriver.

⁂ WARNING

Be careful not to damage the contact surfaces of the cylinder head and camshaft housing.

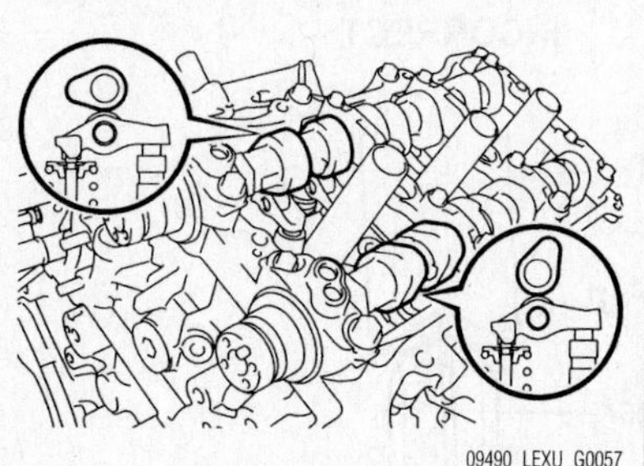

Fig. 158 Check that the camshafts are positioned as shown (left side)—3.5L (2GR-FSE) Engines

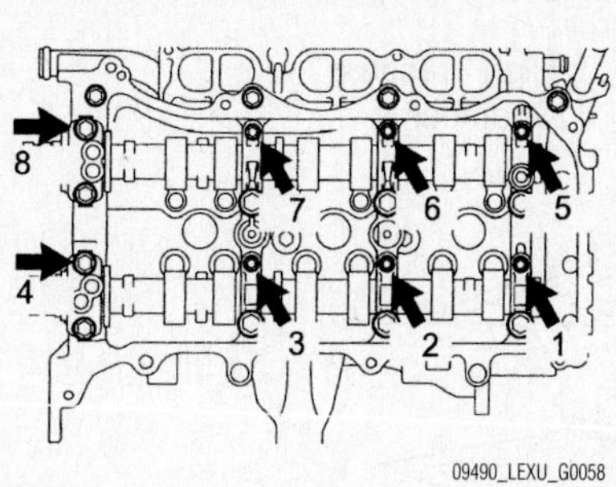

Fig. 159 Left side camshaft bearing inner bolt loosening sequence—3.5L (2GR-FSE) Engines

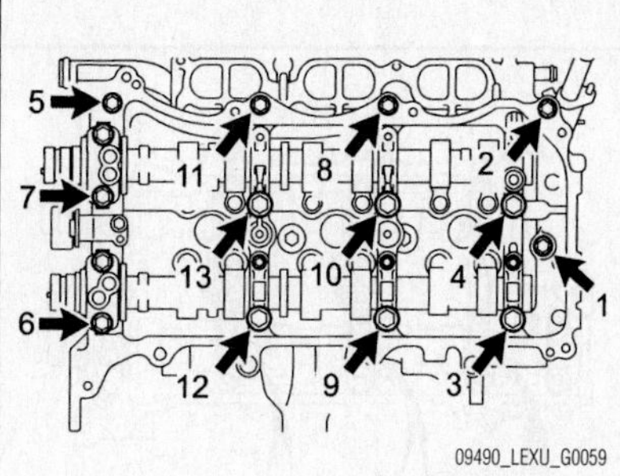

Fig. 160 Left side camshaft bearing outer bolt loosening sequence—3.5L (2GR-FSE) Engines

- Valve rocker arms. Arrange the removed rocker arms in the correct order.
- Valve lash adjusters. Arrange the removed valve lash adjusters in the correct order.

To install:

10. Install valve lash adjusters as follows:

⁂ WARNING

Keep the lash adjuster free of dirt and foreign objects. Only use clean engine oil.

   a. Place the lash adjuster into a container filled with engine oil.
   b. Insert the SST's tip (09276-75010) into the lash adjuster's plunger and use the tip to press down on the check ball inside the plunger as is shown.
   c. Squeeze the SST and lash adjuster together to move the plunger up and down 5 to 6 times.
   d. Check the movement of the plunger and bleed the air.
   e. After bleeding the air, remove SST. Then, try to quickly and firmly press the plunger with a finger. If the result is not as specified, replace the lash adjuster.
   f. Install the lash adjusters.

➡Install the lash adjuster to the same place it was removed from.

11. Install No. 1 valve rocker arm assembly as follows:
   a. Apply engine oil to the lash adjuster tip and valve stem cap end.
   b. Make sure that the valve rocker arms are installed as shown in the illustration.
12. Install right side camshaft bearing cap as follows:
   a. Apply engine oil to the camshaft journals, camshaft housing and bearing caps.
   b. Install the camshaft and camshaft No. 2 to the right camshaft housing.
   c. Make sure of the marks and numbers on the camshaft bearing caps and place them in each proper position and direction.
   d. Temporarily tighten the 9 bolts to 7 ft. lbs. (10 Nm) in the order shown in the illustration.
13. Install the right side camshaft housing assembly as follows:
   a. Make sure that the valve rocker arm is installed as shown in the illustration.
   b. Apply seal packing in a continuous line approximately 0.138–0.158 inches (3.5–4.0mm) wide.

➡Remove any oil from the contact surface. Install the camshaft housing assembly within 3 minutes. Do not start the engine for at least 2 hours after installing.

   c. Install the right camshaft housing and tighten the 14 bolts to 21 ft. lbs. (28 Nm) in the order shown in the illustration.

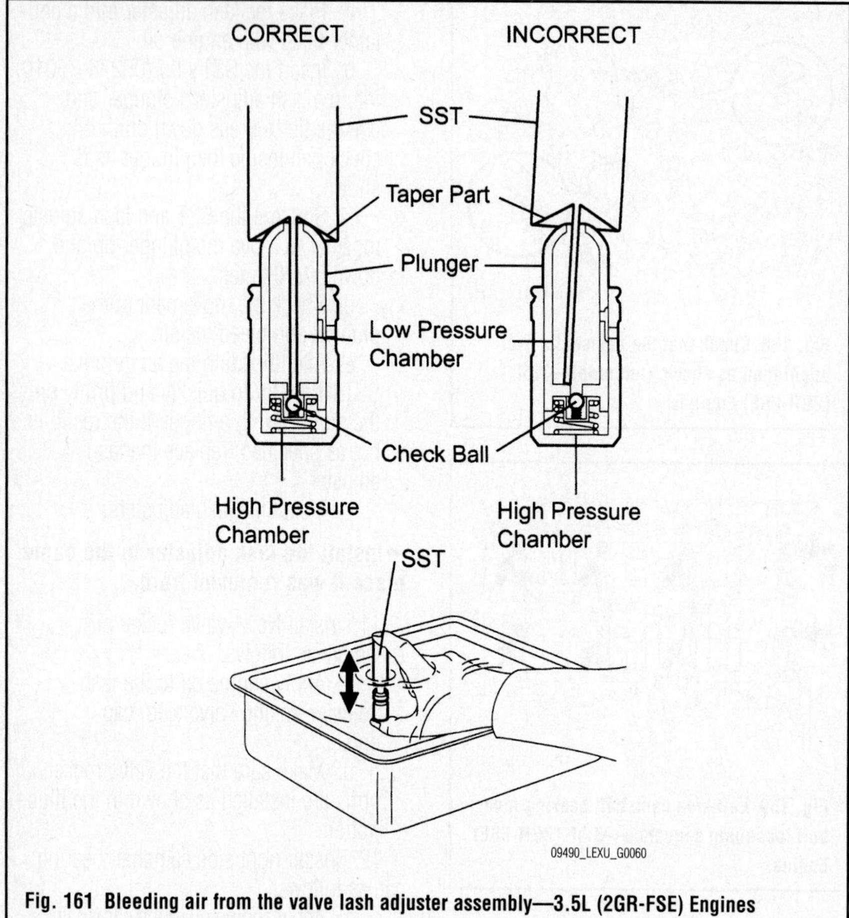

Fig. 161 Bleeding air from the valve lash adjuster assembly—3.5L (2GR-FSE) Engines

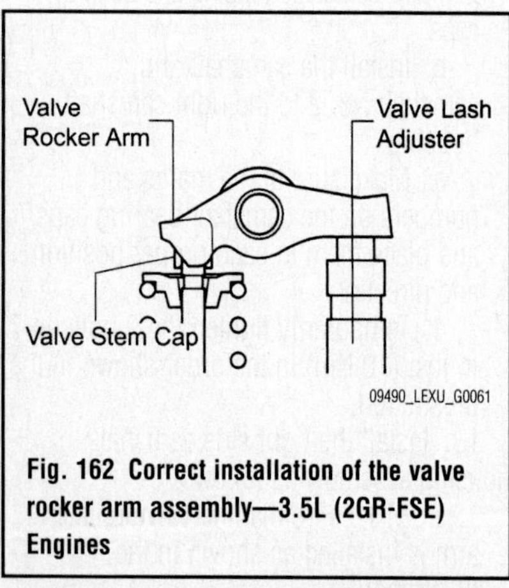

Fig. 162 Correct installation of the valve rocker arm assembly—3.5L (2GR-FSE) Engines

⁂ WARNING

When installing the camshaft housing, it is necessary to correctly position the camshafts as shown in the illustration. Failure to correctly position these parts may result in damage due to contact between the pistons and valves. If a camshaft is rotated with a piston at TDC, valve contact will occur. If any of the bolts are loosened during installation, remove the camshaft housing, clean the installation surfaces, and reapply seal packing. If the camshaft housing is removed because any of the bolts are loosened during installation, make sure that the previously applied seal packing does not enter any oil passages.

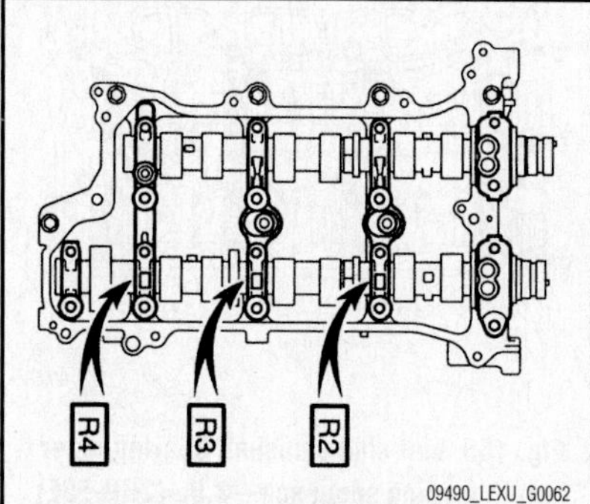

Fig. 163 Make sure of the marks and numbers on the right camshaft bearing caps and place them in each proper position and direction—3.5L (2GR-FSE) Engines

d. Tighten the 9 bolts to 12 ft. lbs. (16 Nm) in the order shown in the illustration.

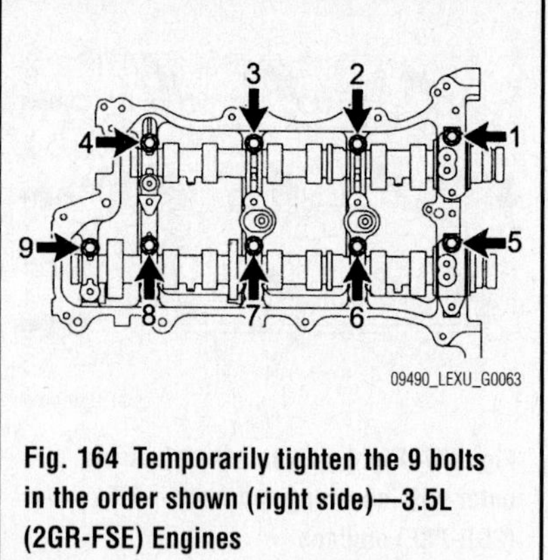

Fig. 164 Temporarily tighten the 9 bolts in the order shown (right side)—3.5L (2GR-FSE) Engines

14. Install left side camshaft bearing cap as follows:
    a. Apply engine oil to the camshaft journals, camshaft housing and bearing caps.
    b. Install camshaft No. 3 and camshaft No. 4 to the left camshaft housing.
    c. Make sure the marks and numbers on the camshaft bearing caps and place them in each proper position and direction.
    d. Temporarily tighten the 8 bolts to 7 ft. lbs. (10 Nm) in the order shown in the illustration.
15. Install the left side camshaft housing assembly as follows:
    a. Make sure that the valve rocker arm is installed as shown in the illustration.
    b. Apply seal packing in a continuous line approximately 0.197–0.217 inches (5.0–5.5mm) wide.

➡Remove any oil from the contact surface. Install the camshaft housing assembly within 3 minutes. Do not start the engine for at least 2 hours after installing.

    c. Install the left camshaft housing and tighten the 13 bolts to 21 ft. lbs. (28 Nm) in the order shown in the illustration.

⁂ WARNING

When installing the camshaft housing, it is necessary to correctly position the camshafts as shown in the illustration. Failure to correctly position these parts may result in damage due to contact between the pistons and valves. If a camshaft is rotated with a piston at TDC, valve contact will occur. If any of the bolts are loosened during installation, remove the camshaft housing, clean the installation surfaces, and reapply seal packing. If the camshaft housing is removed because any of the bolts are loosened during installation,

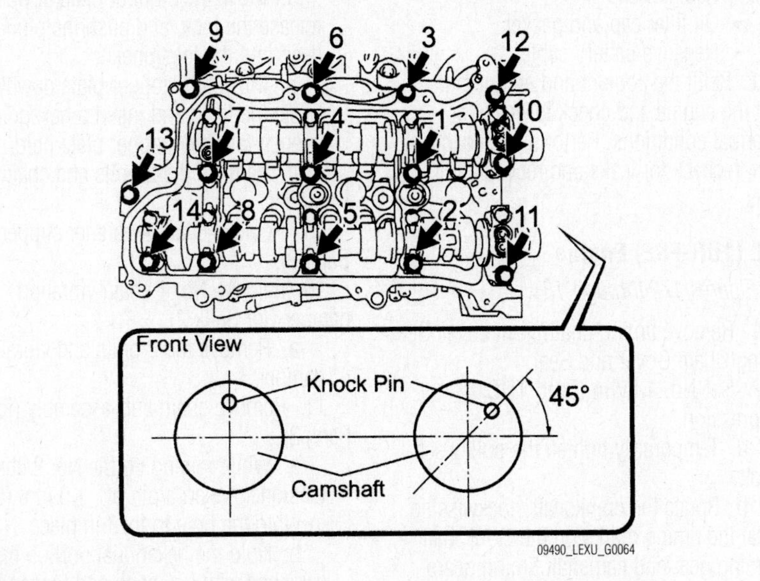

Fig. 165 Right camshaft housing assembly bolt tightening sequence and camshaft positioning—3.5L (2GR-FSE) Engines

make sure that the previously applied seal packing does not enter any oil passages.

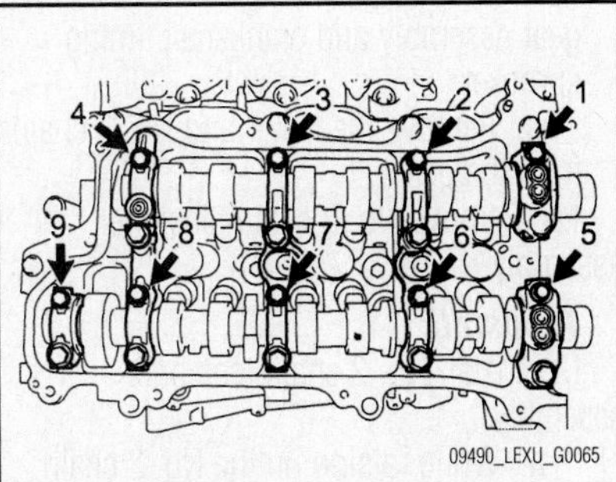
Fig. 166 Right side camshaft bearing inner bolt tightening sequence—3.5L (2GR-FSE) Engines

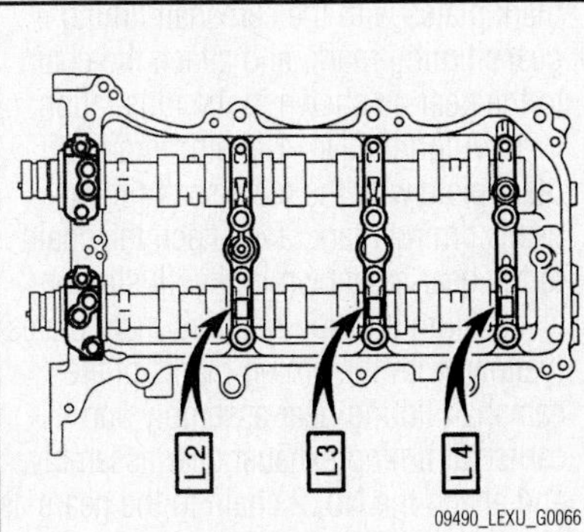

Fig. 167 Make sure of the marks and numbers on the left camshaft bearing caps and place them in each proper position and direction—3.5L (2GR-FSE) Engines

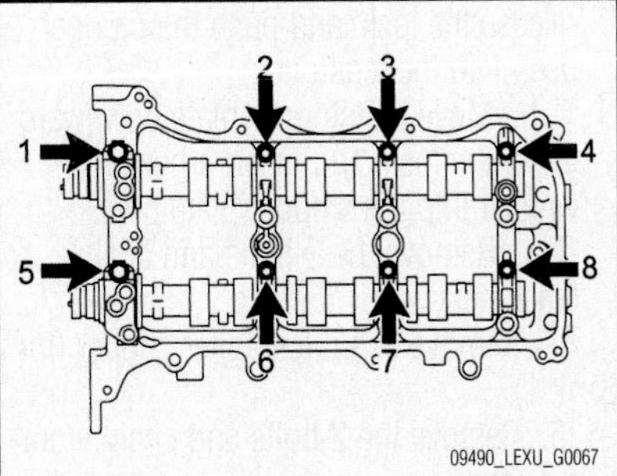
Fig. 168 Temporarily tighten the 8 bolts in the order shown (left side)—3.5L (2GR-FSE) Engines

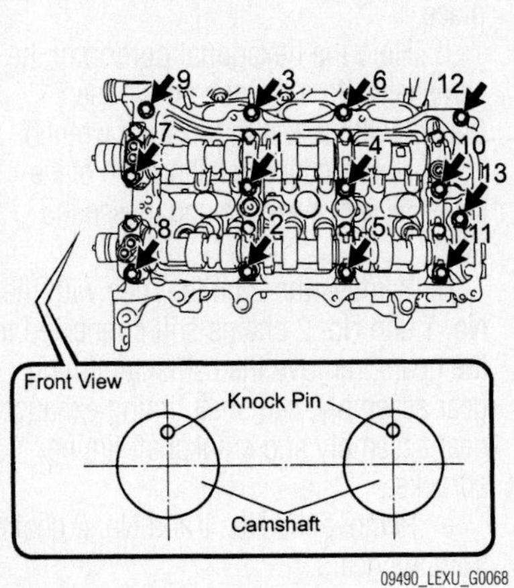

Fig. 169 Left camshaft housing assembly bolt tightening sequence and camshaft positioning—3.5L (2GR-FSE) Engines

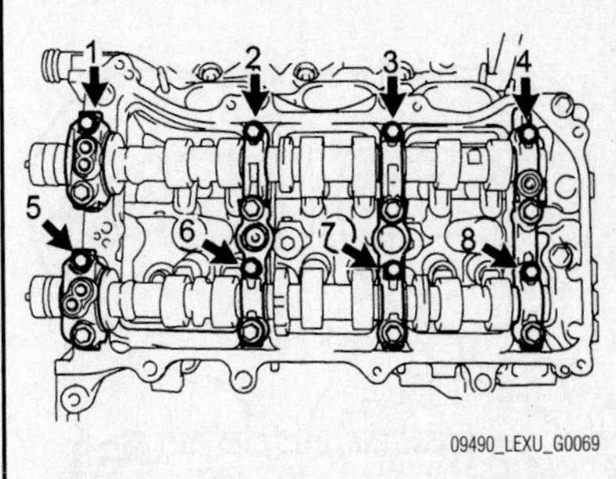
Fig. 170 Left side camshaft bearing inner bolt tightening sequence—3.5L (2GR-FSE) Engines

d. Tighten the 8 bolts to 12 ft. lbs. (16 Nm) in the order shown in the illustration.

16. Install No. 2 chain tensioner assembly as follows:

a. Install the No. 2 chain tensioner with the bolt. Torque the bolt to 15 ft. lbs. (21 Nm).

b. While pushing in the tensioner, insert a pin of 0.039 in. (1.0mm) into the hole to fix it.

17. Install right side camshaft timing gears and No. 2 chain.

a. Align the mark plate (yellow) with the timing marks (1-dot mark) of the camshaft timing gears as shown in the illustration.

b. Apply a light coat of engine oil to the bolt threads and bolt-seating surface.

c. Align the knock pin of the camshaft with the pin hole of the camshaft timing gear. Install the camshaft timing gear and camshaft timing exhaust gear (right) with the No. 2 chain installed.

d. Hold the hexagonal portion of the camshaft with a wrench, and tighten the 2 bolts to 74 ft. lbs. (100 Nm)

e. Remove the pin from the chain tensioner.

18. Install No. 3 chain tensioner assembly as follows:

a. Install the chain tensioner with the bolt. Torque the bolt to 15 ft. lbs. (21 Nm).

b. While pushing in the tensioner, insert a pin of 0.039 in. (1.0mm) into the hole to hold it.

19. Install left side camshaft timing gears and No. 2 chain.

a. Align the mark plate (yellow) with the timing marks (2-dot mark) of the camshaft timing gears as shown in the illustration.

b. Apply a light coat of engine oil to the bolt threads and bolt-seating surface.

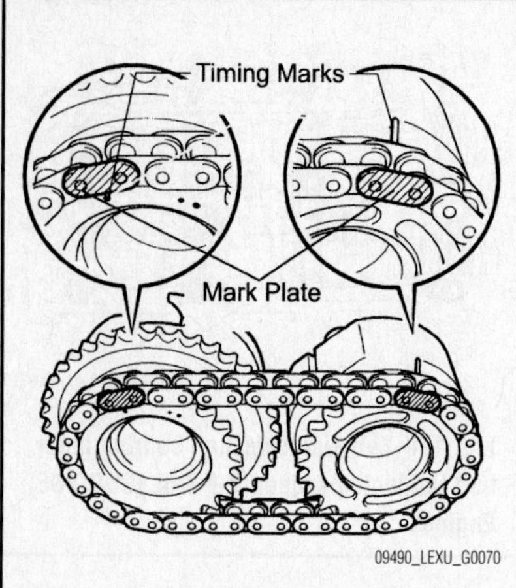

Fig. 171 Align the mark plate with the timing marks of the camshaft timing gears (right side shown, left side similar)

c. Align the knock pin of the camshaft with the pin hole of the camshaft timing gear. Install the camshaft timing gear and camshaft timing exhaust gear (left) with the No. 2 chain installed.

d. Hold the hexagonal portion of the camshaft with a wrench, and tighten the 2 bolts to 74 ft. lbs. (100 Nm)

e. Remove the pin from the chain tensioner.

20. Install or connect the following:

- Timing chain cover, timing chain and timing chain sprockets
- Cylinder head cover and gaskets
- New gaskets, O-ring and rear water by-pass joint. Torque to 7 ft. lbs. (10 Nm).
- New gasket, water inlet and thermostat assembly. Torque to 7 ft. lbs. (10 Nm).
- Oil filter element
- Adhesive sealer, left and right side cylinder block water drain cocks. Torque to 22 ft. lbs. (30 Nm).
- Left and right side oil control valve filter and new gaskets
- Left and right side oil check valve bolt, oil pipe union and oil pipe. Torque to 44 ft. lbs. (60 Nm).
- Crankshaft Position (CKP) sensor. Torque to 7 ft. lbs. (10 Nm).
- 4 camshaft timing oil control valves. Torque to 7 ft. lbs. (10 Nm).
- 4 camshaft position sensors. Torque to 7 ft. lbs. (10 Nm).
- Adhesive and ventilation valve. Torque to 20 ft. lbs. (27 Nm).
- Spark plugs. Torque to 13 ft. lbs. (18 Nm).
- Engine hangers. Torque to 24 ft. lbs. (33 Nm).
- Radiator cap
- Oil filler cap and gasket
- Negative battery cable

21. Refill the coolant and engine oil. Start the engine and check for leaks or abnormal conditions. Perform and road test. Then, recheck for leaks and recheck fluid levels.

4.6L (1UR-FSE) Engine

See Figures 172 through 181.

1. Remove timing chain front cover. See Timing Chain Cover and Seal.

2. Set No. 1 cylinder to TDC/compression.

a. Temporarily tighten the pulley set bolt.

b. Rotate the crankshaft clockwise so that the timing marks on the crankshaft timing gear and camshaft timing gears are as shown in the illustration.

3. Remove No. 1 chain tensioner assembly (for Bank 1).

a. Move the stopper plate upward to release the lock, and push the plunger deep into the tensioner.

b. Move the stopper plate downward to set the lock, and insert a hexagon wrench into the stopper plate hole.

c. Remove the 2 bolts and chain tensioner.

4. Remove chain tensioner slipper (for Bank 1).

5. Remove the 2 bolts and chain vibration damper.

6. Remove chain sub-assembly (for Bank 1).

a. While pushing down the No. 3 chain tensioner, insert a pin of φ1.0 mm (0.039 in.) into the hole to fix it in place.

b. Hold the hexagonal portion of the camshaft with a wrench and loosen the bolt with a 12mm hexagon wrench.

c. Hold the hexagonal portion of the camshaft with a wrench and loosen the bolt.

d. Remove the 2 bolts. Then with the No. 1 and No. 2 chains still connected to the gears, remove the camshaft timing gear assembly, camshaft timing exhaust gear assembly and crankshaft timing sprocket.

e. Remove the No. 1 and No. 2 chains from the gears.

7. Remove No. 3 chain tensioner assembly.

a. Remove the 2 bolts and chain tensioner.

8. Remove No. 1 chain tensioner assembly (for Bank 2).

a. Move the stopper plate upward to release the lock, and push the plunger deep into the tensioner.

b. Move the stopper plate downward to set the lock, and insert a hexagon wrench into the stopper plate hole.

c. Remove the 2 bolts and chain tensioner.

9. Remove chain tensioner slipper (for Bank 2).

10. Remove No. 1 chain vibration damper (for Bank 2).

a. Remove the 2 bolts and vibration damper.

11. Remove chain sub-assembly (for Bank 2).

a. While raising up the No. 2 chain tensioner, insert a pin of _1.0 mm (0.039 in.) into the hole to fix it in place.

b. Hold the hexagonal portion of the camshaft with a wrench and loosen the bolt with a 12 mm hexagon wrench.

c. Hold the hexagonal portion of the camshaft with a wrench and loosen the bolt.

d. Remove the 2 bolts. Then with the No. 1 and No. 2 chains still connected to the gears, remove the camshaft timing gear assembly, camshaft timing exhaust gear assembly and crankshaft timing sprocket.

e. Remove the No. 1 and No. 2 chains from the gears.

12. Remove No. 2 chain tensioner assembly.

To install:

13. Install No. 2 chain tensioner assembly

a. While raising up the No. 2 chain tensioner, insert a pin of _1.0 mm (0.039 in.) into the hole to fix it in place.

14. Install chain sub-assembly (for Bank 2).

a. Align the No. 1 chain's orange mark plates with the camshaft timing gear's timing mark, and attach the chain to the gear as shown in the illustration.

b. Align the No. 1 chain's orange mark plate with the crankshaft timing gear's timing mark, and attach the chain to the gear as shown in the illustration

c. Align the No. 2 chain's mark plates (yellow) with the timing marks of the camshaft timing gear assembly and camshaft timing exhaust gear assembly, and attach the No. 2 chain to the gears as shown in the illustration.

➡The crankshaft timing gear and camshaft exhaust gear assembly will be installed with the No. 1 and No. 2 chains connected to the gears.

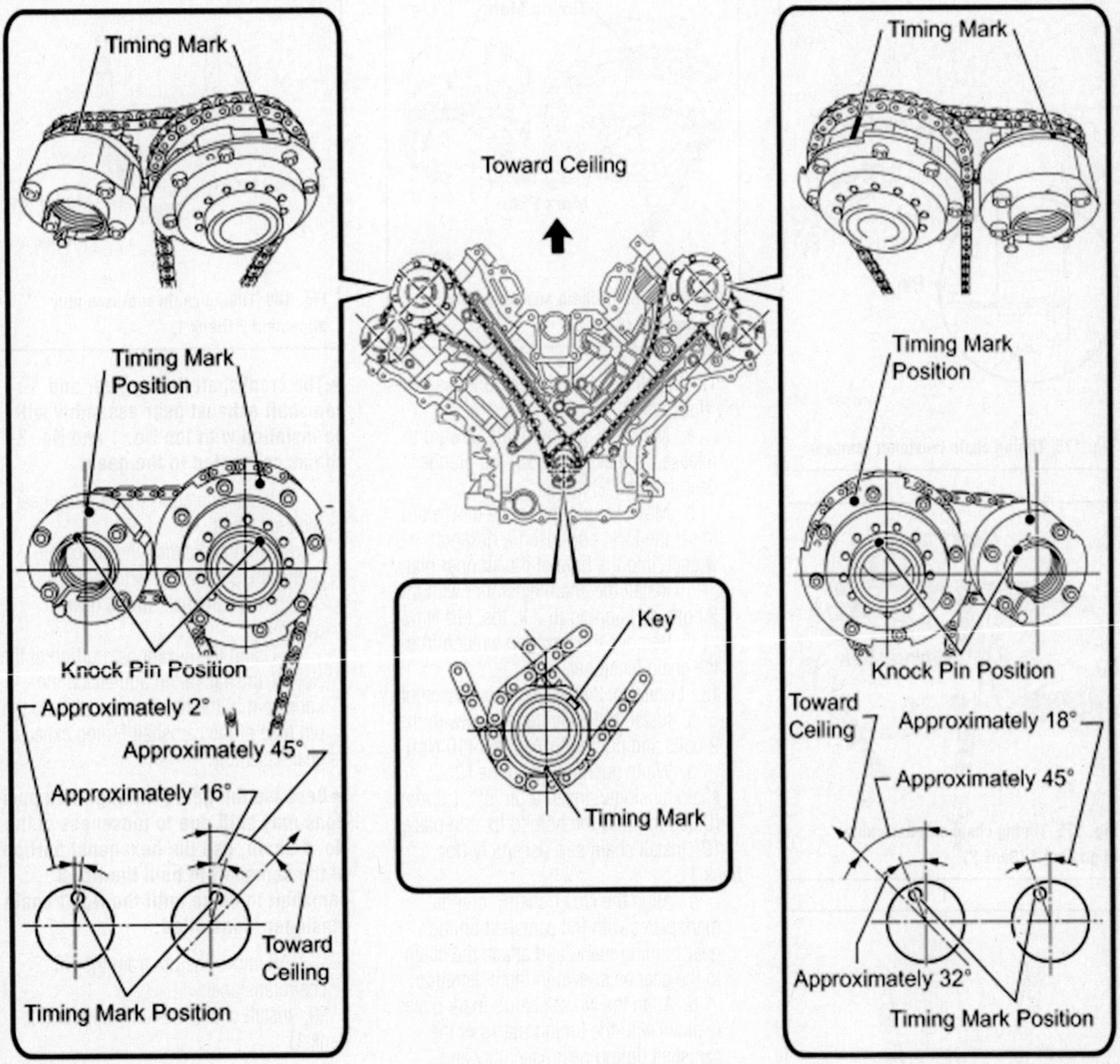

Fig. 172 Timing chain alignment

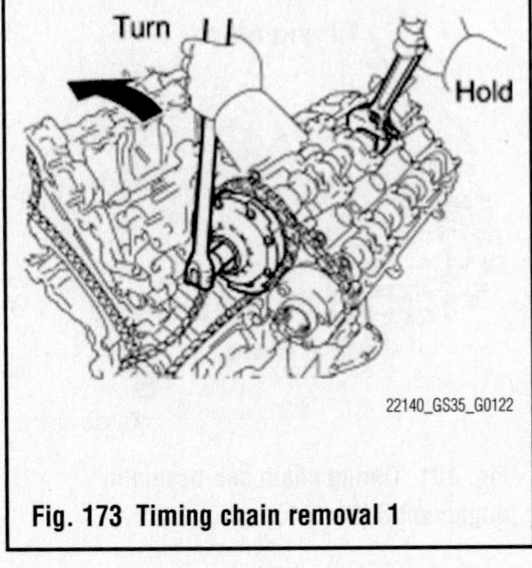

Fig. 173 Timing chain removal 1

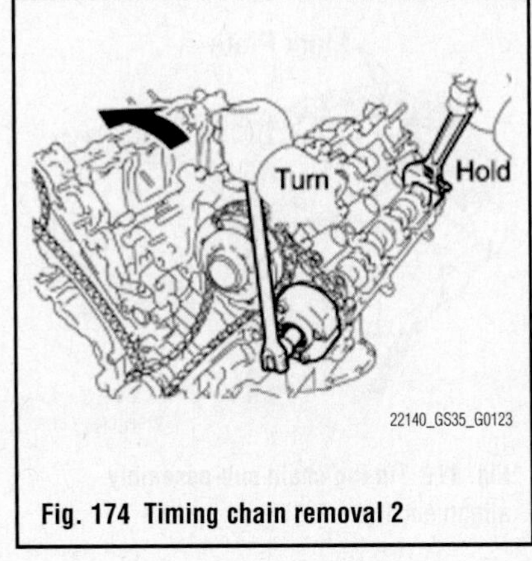

Fig. 174 Timing chain removal 2

d. Install the crankshaft timing gear to the crankshaft.

e. Align and attach the knock pin of the No. 1 camshaft with the pin hole of the camshaft timing gear assembly.

f. Using the hexagonal portion of the No. 2 camshaft, align and attach the knock pin of the No. 2 camshaft with the pin hole of the camshaft timing exhaust gear assembly.

g. Remove the pin from the No. 2 chain tensioner.

15. Install No. 1 chain vibration damper (for Bank 2).

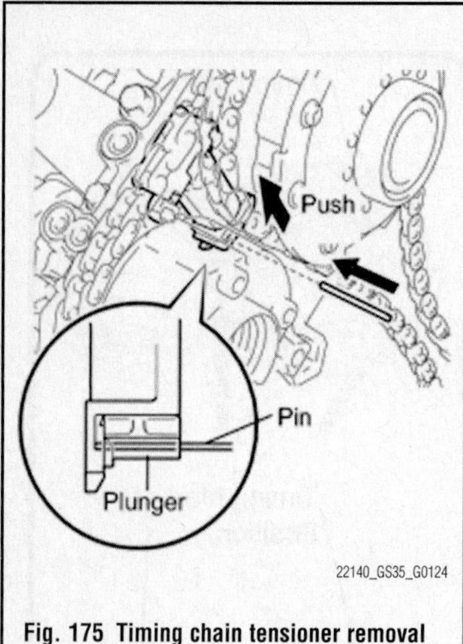

Fig. 175 Timing chain tensioner removal

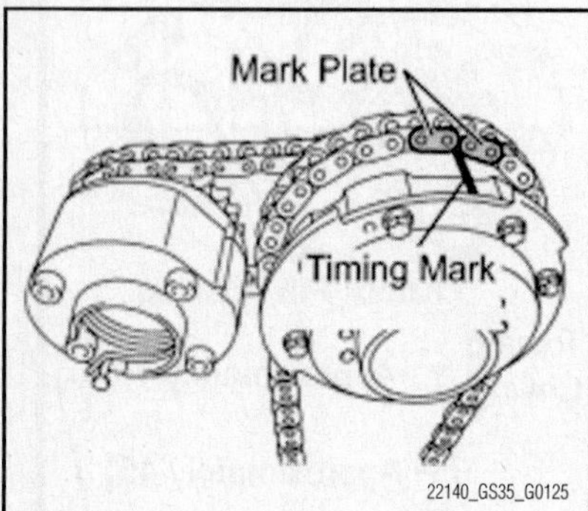

Fig. 176 Timing chain sub-assembly alignment 1 (Bank 2)

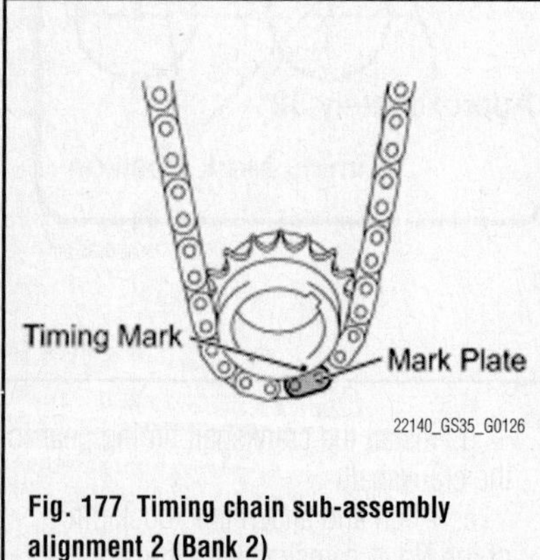

Fig. 177 Timing chain sub-assembly alignment 2 (Bank 2)

a. Install the vibration damper with the 2 bolts and tighten to 15 ft. lbs. (21 Nm).

16. Install chain tensioner slipper (for Bank 2).

➡**If you cannot install the chain tensioner slipper due to the tension of the chain, use the hexagonal portion of the camshaft to loosen the chain, and then install the chain tensioner slipper.**

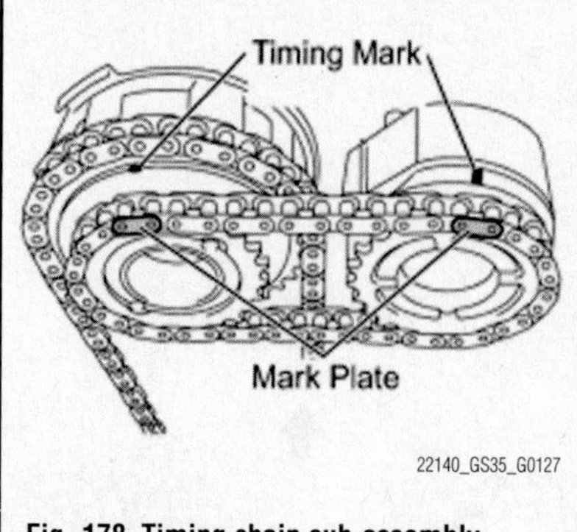

Fig. 178 Timing chain sub-assembly alignment 3 (Bank 2)

17. Install No. 1 chain tensioner assembly (for Bank 2).

a. Move the stopper plate upward to release the lock, and push the plunger deep into the tensioner.

b. Move the stopper plate downward to set the lock, and insert a hexagon wrench into the hole of the stopper plate.

c. Install the chain tensioner with the 2 bolts and tighten to 7 ft. lbs. (10 Nm).

d. Remove the hexagon wrench from the chain tensioner.

18. Install No. 3 chain tensioner assembly.

a. Install the chain tensioner with the 2 bolts and tighten to 7 ft. lbs. (10 Nm) .

b. While pushing down the No. 2 chain tensioner, insert a pin of _1.0 mm (0.039 in.) into the hole to fix it in place.

19. Install chain sub-assembly (for Bank 1).

a. Align the No. 1 chain's orange mark plates with the camshaft timing gear's timing mark, and attach the chain to the gear as shown in the illustration.

b. Align the No. 2 chain's mark plates (yellow) with the timing marks of the camshaft timing gear assembly and camshaft timing exhaust gear assembly, and attach the No. 2 chain to the gears as shown in the illustration

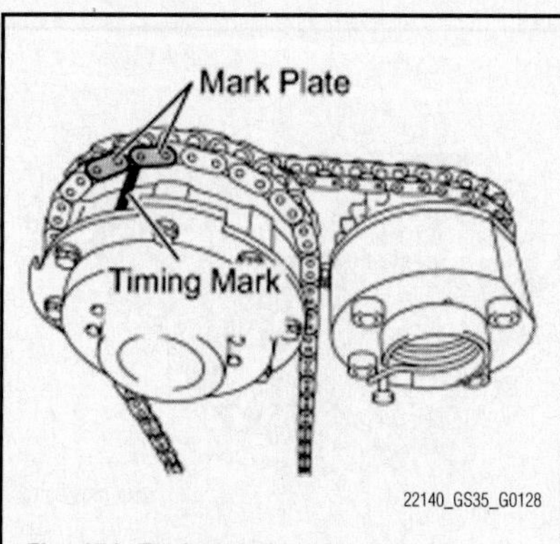

Fig. 179 Timing chain sub-assembly alignment 1 (Bank 1)

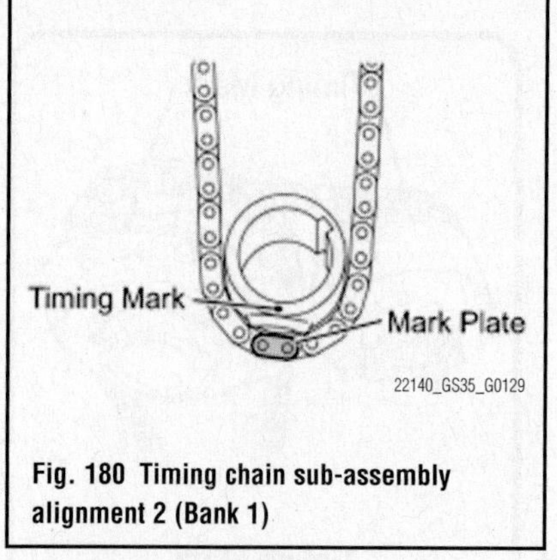

Fig. 180 Timing chain sub-assembly alignment 2 (Bank 1)

➡**The crankshaft timing gear and camshaft exhaust gear assembly will be installed with the No. 1 and No. 2 chains connected to the gears**

c. Install the crankshaft timing gear to the crankshaft.

d. Align and attach the knock pin of the No. 3 camshaft with the pin hole of the camshaft timing gear assembly.

e. Using the hexagonal portion of the No. 4 camshaft, align and attach the knock pin of the No. 4 camshaft with the pin hole of the camshaft timing exhaust gear assembly.

➡**Because the gears' timing mark positions may shift due to looseness of the No. 1 chain, use the hexagonal portion of the camshaft to hold the No. 3 camshaft in place until the No. 1 chain tensioner is installed.**

f. Remove the pin from the No. 2 chain tensioner.

20. Install chain tensioner slipper (for Bank 1).

➡**If you cannot install the chain tensioner slipper due to the tension of the chain, use the hexagonal portion of the**

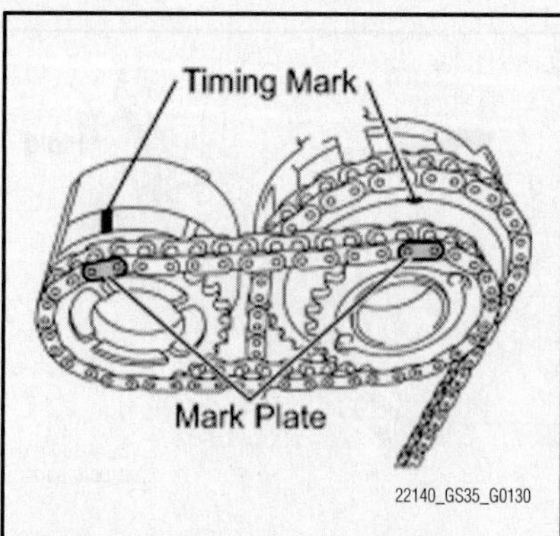

Fig. 181 Timing chain sub-assembly alignment 3 (Bank 1)

camshaft to loosen the chain and install the chain tensioner.

21. Install No. 1 chain tensioner assembly (for Bank 1).
    a. Move the stopper plate upward to release the lock, and push the plunger deep into the tensioner.
    b. Move the stopper plate downward to set the lock, and insert a hexagon wrench into the hole of the stopper plate.
    c. Install the chain tensioner and gasket with the 2 bolts and tighten to 7 ft. lbs. (10 Nm).
22. Install No. 1 chain vibration damper (for Bank 1).
    a. Install the vibration damper with the 2 bolts.
    b. Remove the hexagon wrench from the No. 1 chain tensioner.
23. Tighten camshaft timing gear assembly for Bank 1.
    a. Using a wrench, hold the hexagonal portion of the No. 3 camshaft.
    b. Using a 12 mm socket hexagon wrench, tighten the camshaft timing gear assembly with a new bolt and tighten to 58 ft. lbs. (79 Nm).
    c. Using a wrench to hold the hexagonal portion of the No. 4 camshaft, tighten the camshaft timing exhaust gear assembly with the bolt and tighten to 74 ft. lbs. (100 Nm).
24. Tighten camshaft timing gear assembly for Bank 2.
    a. Using a wrench, hold the hexagonal portion of the No. 1 camshaft.
    b. Using a 12 mm socket hexagon wrench, tighten the camshaft timing gear assembly with a new bolt and tighten to 58 ft. lbs. (79 Nm).
    c. Using a wrench to hold the hexagonal portion of the No. 2 camshaft, tighten the camshaft timing exhaust gear assembly with the bolt and tighten to 74 ft. lbs. (100 Nm).
25. Check No. 1 cylinder to TDC/compression.
    a. Temporarily tighten the pulley set bolt.
    b. Rotate the crankshaft clockwise so that the timing marks on the crankshaft timing gear and camshaft timing gears are as shown in the illustration.
26. Install timing chain cover.

VALVE LASH

ADJUSTMENT

No adjustment is necessary.

ENGINE PERFORMANCE & EMISSION CONTROLS

CAMSHAFT POSITION (CMP) SENSOR

LOCATION

See Figure 182.

Refer to the accompanying illustration.

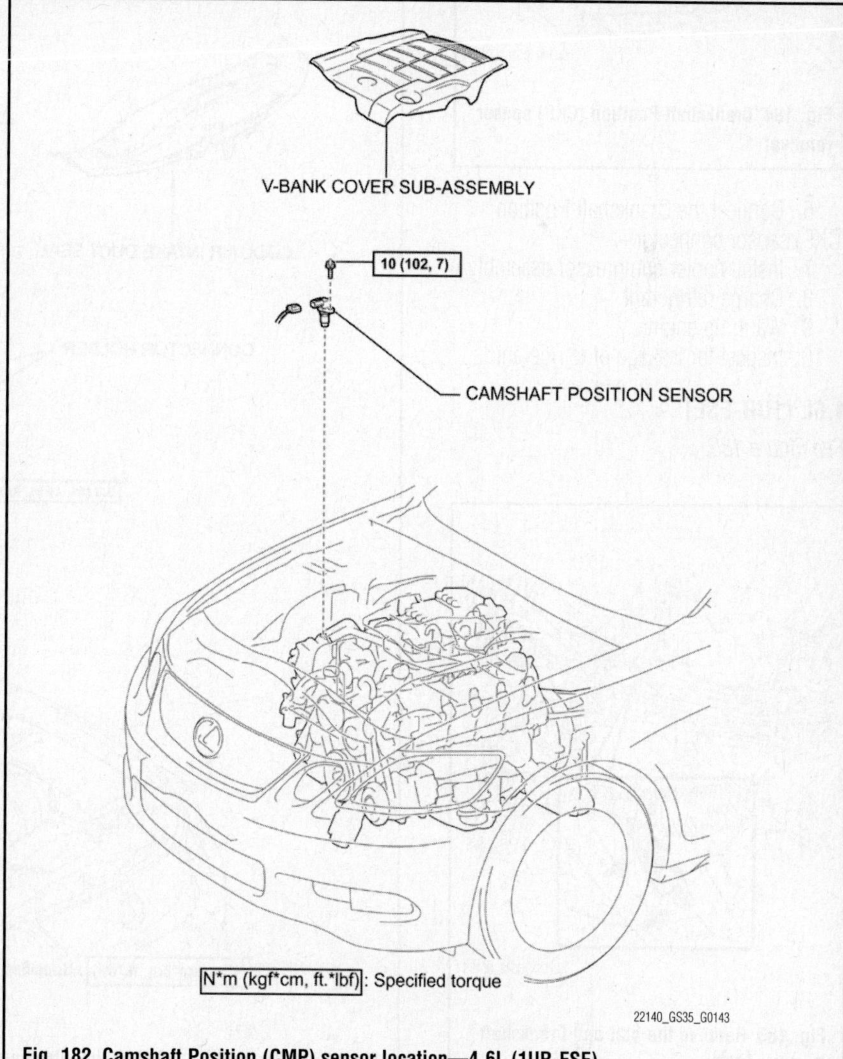

Fig. 182 Camshaft Position (CMP) sensor location—4.6L (1UR-FSE)

REMOVAL & INSTALLATION

4.6L (1UR-FSE) Engines

See Figure 183.

1. Remove V-bank cover sub-assembly.
2. Disconnect the Camshaft Position (CMP) sensor connector.
3. Remove the bolt and camshaft position sensor.

To install:

4. Reverse removal procedure

CRANKSHAFT POSITION (CKP) SENSOR

REMOVAL & INSTALLATION

3.5L (2GR-FSE)

See Figure 184.

1. Remove air conditioning cooler assembly.
2. Disconnect the sensor connector.
3. Remove the bolt and sensor.

To install:

4. Apply a coat of engine oil to the O-ring of the sensor.
5. Install the sensor with the bolt and tighten to 7 ft. lbs. (10 Nm).

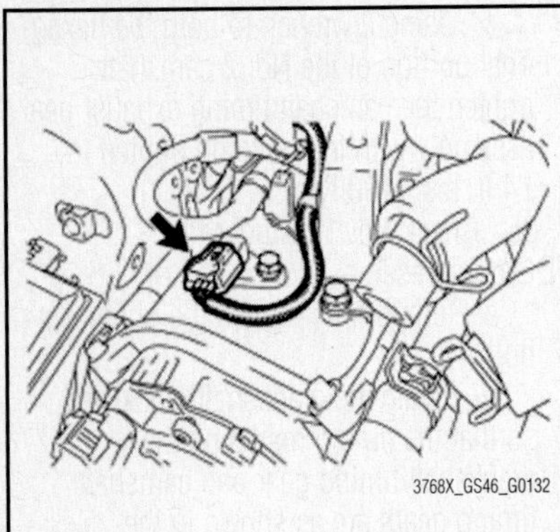
Fig. 183 Camshaft Position (CMP) sensor

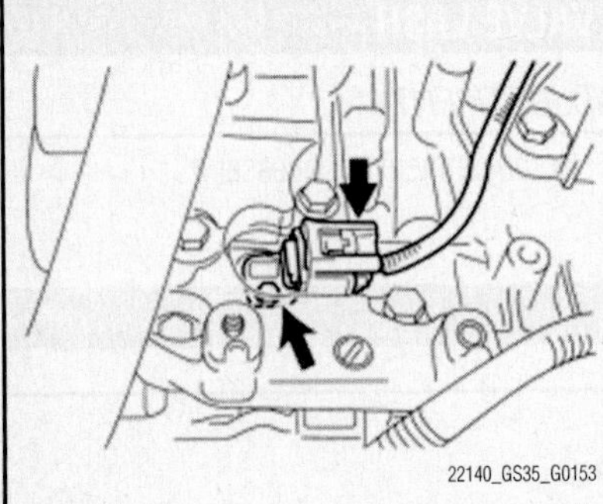
Fig. 184 Crankshaft Position (CKP) sensor removal

6. Connect the Crankshaft Position (CKP) sensor connector.
7. Install cooler compressor assembly.
8. Charge refrigerant.
9. Warm up engine.
10. Inspect for leakage of refrigerant.

4.6L (1UR-FSE)

See Figure 185.

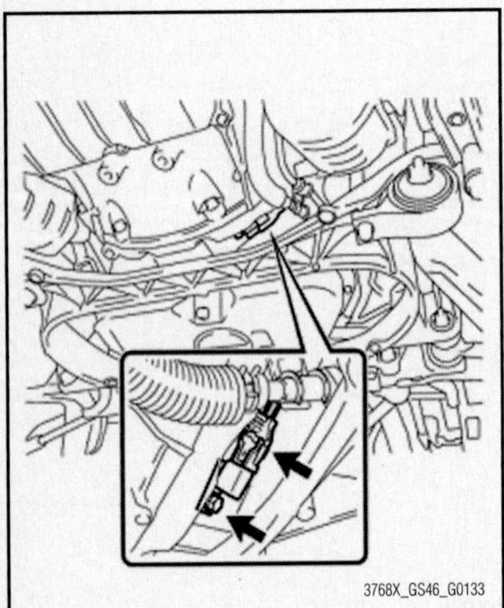
Fig. 185 Remove the bolt and Crankshaft Position (CKP) sensor

1. Remove No. 2 engine under cover.
2. Disconnect the Crankshaft Position (CKP) sensor connector.
3. Remove the bolt and Crankshaft Position (CKP) sensor.

To install:

4. Install the Crankshaft Position (CKP) sensor with the bolt and tighten to 7 ft. lbs. (10 Nm).
5. Connect the Crankshaft Position (CKP) sensor connector.
6. Inspect for oil leak.
7. Install No. 2 engine under cover.

ELECTRONIC CONTROL MODULE (ECM)

LOCATION

See Figure 186.

Refer to the accompanying illustrations.

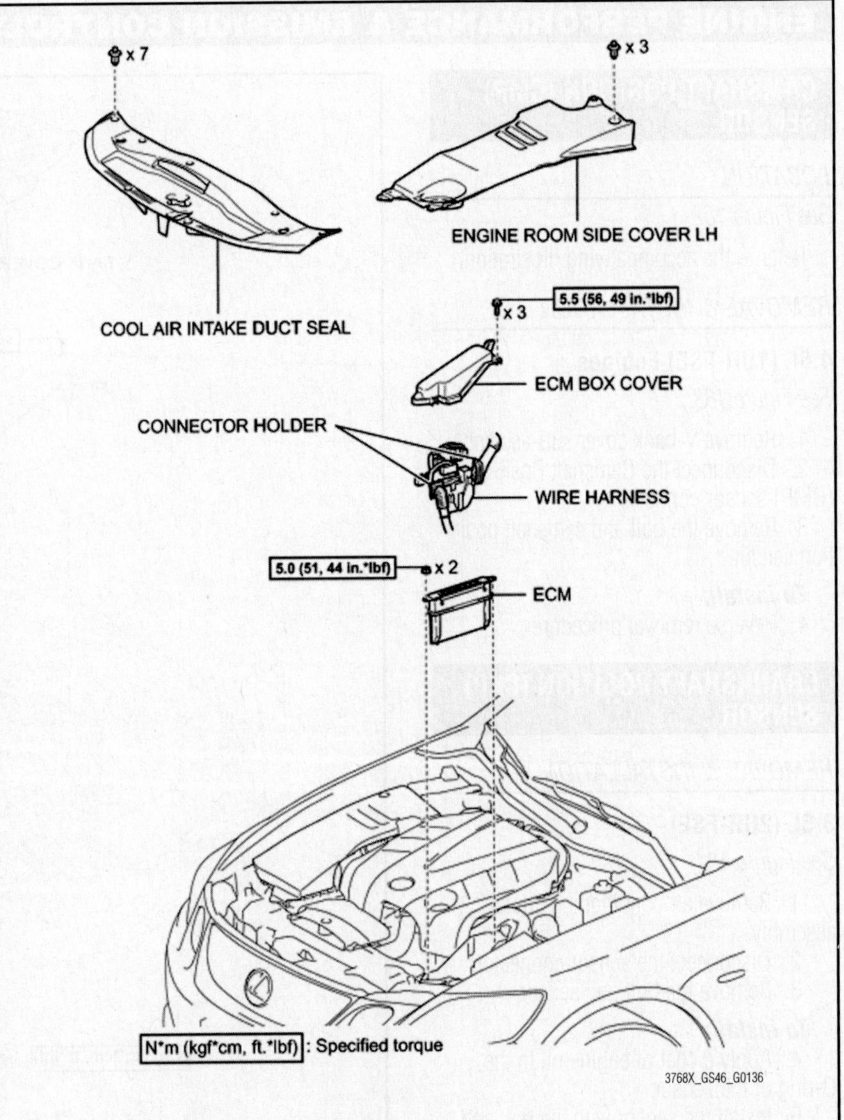

Fig. 186 ECM location—GS350 models shown, GS460 similar

REMOVAL & INSTALLATION

See Figures 187 and 188.

1. Disconnect cable from negative battery terminal.

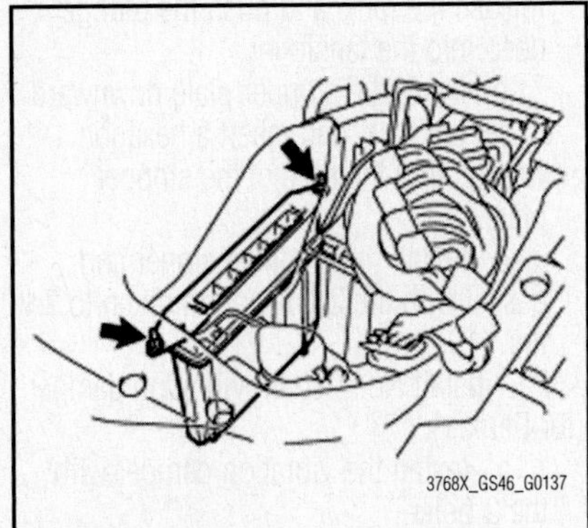
Fig. 187 Removing the ECM—GS350 models

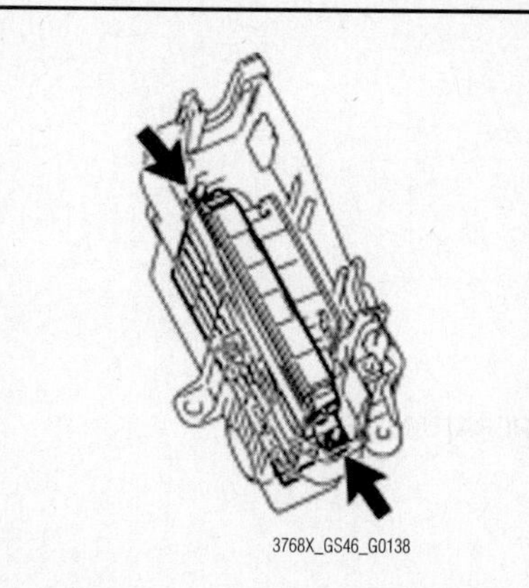

Fig. 188 Removing the ECM—GS350 models

2. Remove cool air intake duct seal.
3. Remove engine room side cover LH.
4. Remove V-bank cover sub-assembly.
5. Remove the 3 bolts and engine room ECU cover.
6. Detach the claw and disconnect the No. 4 connector holder.
7. Disconnect the 6 ECM connectors.
8. Remove the 2 nuts and ECM.

To install:

9. Install the ECM with the 2 nuts.
10. Connect the 6 ECM connectors.
11. Connect the No. 4 connector holder.
12. Install the engine room ECU cover with the 3 bolts.
13. Install V-bank cover sub-assembly.
14. Install engine room side cover LH.
15. Install cool air intake duct seal.
16. Connect cable to negative battery terminal.
17. Perform reset memory.

ENGINE COOLANT TEMPERATURE (ECT) SENSOR

LOCATION

See Figures 189 and 190.

Refer to the accompanying illustrations.

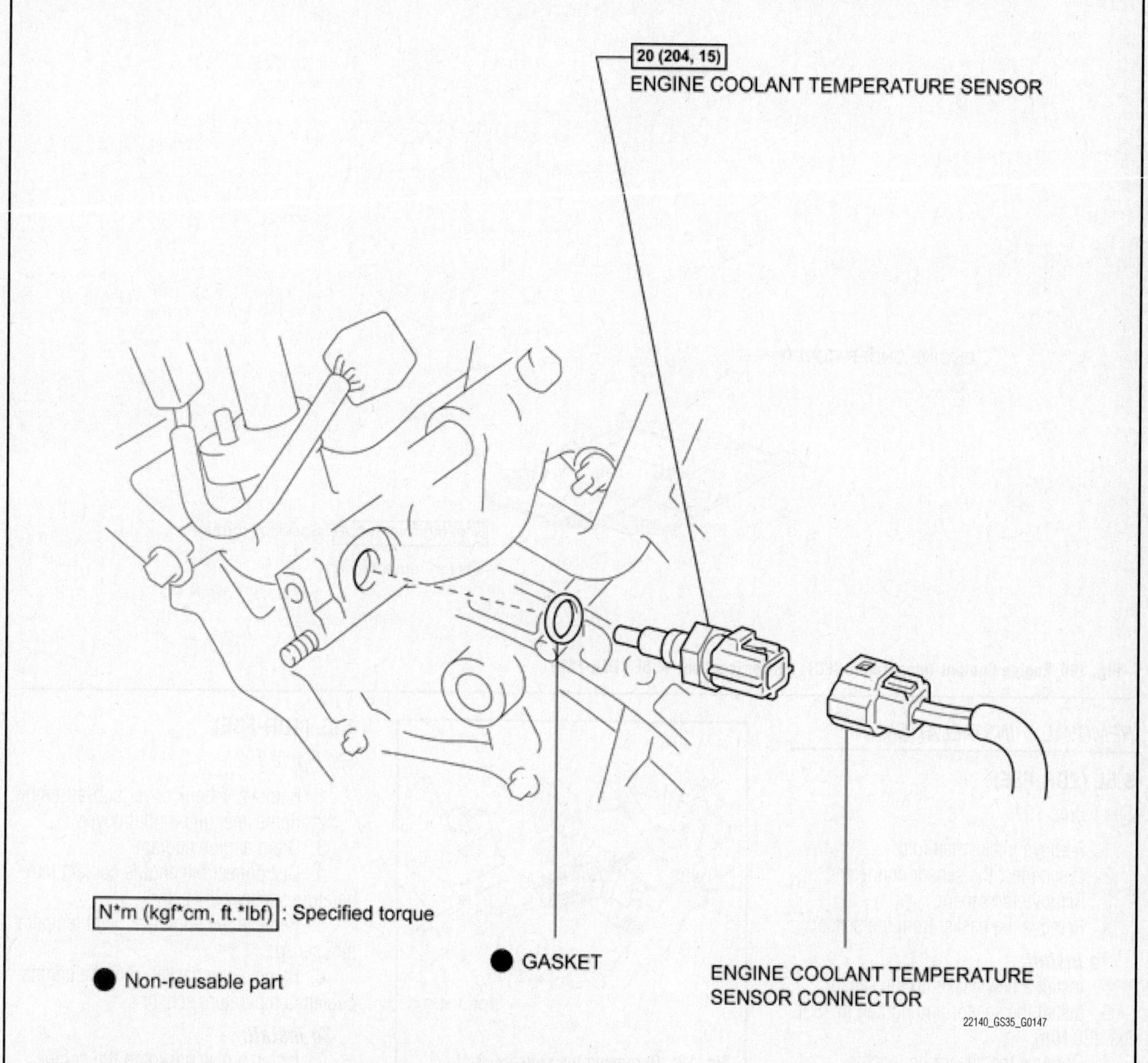

Fig. 189 Engine Coolant Temperature (ECT) sensor location—3.5L (2GR-FSE)

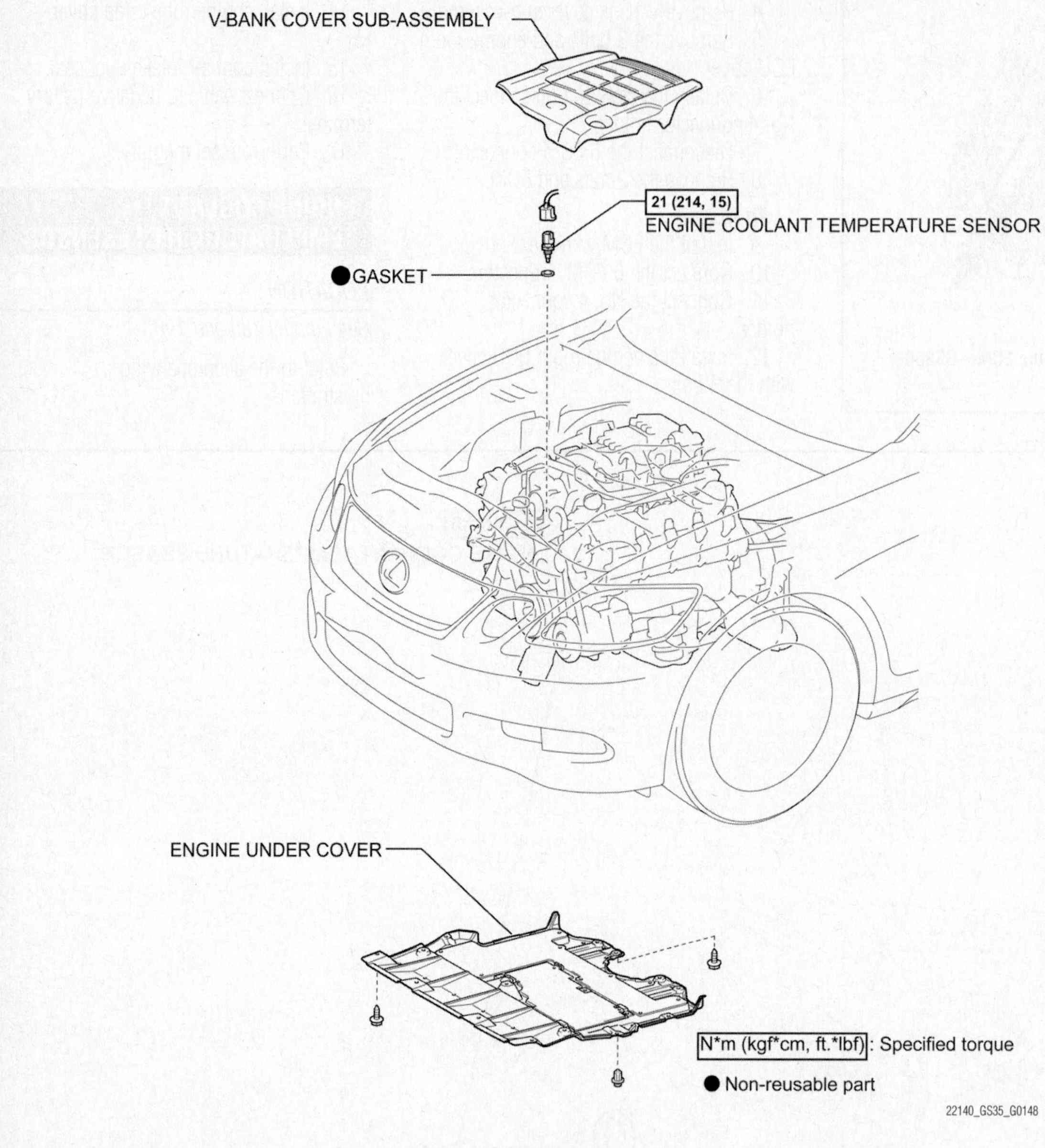

Fig. 190 Engine Coolant Temperature (ECT) sensor location—4.6L (1UR-FSE)

REMOVAL & INSTALLATION

3.5L (2GR-FSE)

See Figure 191.

1. Remove intake manifold.
2. Disconnect the sensor connector.
3. Remove the sensor.
4. Remove the gasket from the sensor.

To install:

5. Install a new gasket to the sensor.
6. Install the sensor and tighten to 15 ft. lbs. (20 Nm).
7. Connect the sensor connector.
8. Install intake manifold.

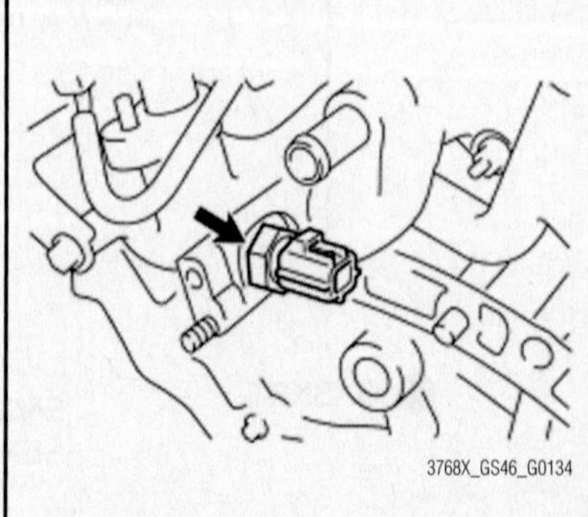

Fig. 191 Removing the sensor—3.5L engines

4.6L (1UR-FSE)

See Figure 192.

1. Remove V-bank cover sub-assembly.
2. Remove engine under cover.
3. Drain engine coolant.
4. Disconnect the engine coolant temperature sensor connector.
5. Remove the engine coolant temperature sensor.
6. Remove the gasket from the engine coolant temperature sensor.

To install:

7. Install a new gasket to the engine coolant temperature sensor.

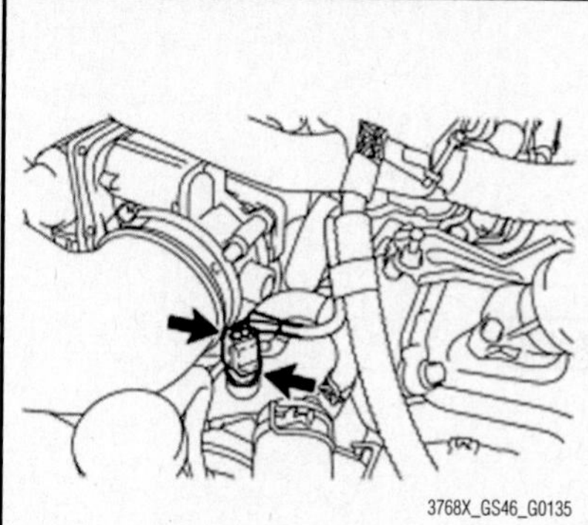

Fig. 192 Removing the sensor—4.6L engines

8. Install the engine coolant temperature sensor and tighten to 15 ft. lbs. (20 Nm).
9. Connect the engine coolant temperature sensor connector.
10. Add engine coolant.
11. Inspect for coolant leak.
12. Install V-bank cover sub-assembly.
13. Install engine under cover.

HEATED OXYGEN SENSOR (HO2S)

LOCATION

See Figure 193.

Refer to the accompanying illustration.

REMOVAL & INSTALLATION

1. Remove front seat assembly RH.
2. Disconnect the Heated Oxygen Sensor (HO2S) connector.
3. Remove the grommet.
4. Remove the Heated Oxygen Sensor (HO2S) from the front exhaust pipe assembly.
5. Remove front seat assembly LH.
6. Disconnect the Heated Oxygen Sensor (HO2S) connector.
7. Remove the grommet.
8. Remove the Heated Oxygen Sensor (HO2S) from the front exhaust pipe assembly.

To install:

9. Reverse removal procedure, and tighten to 32 ft. lbs. (44 Nm).

KNOCK SENSOR (KS)

LOCATION

See Figures 194 and 195.

Refer to the accompanying illustrations.

REMOVAL & INSTALLATION

3.5L Engines

See Figure 196.

1. Remove intake manifold and fuel injection assembly.
2. Disconnect the 2 knock sensor connectors.
3. Remove the 2 bolts and 2 knock sensors.
4. To install, reverse the removal procedure.

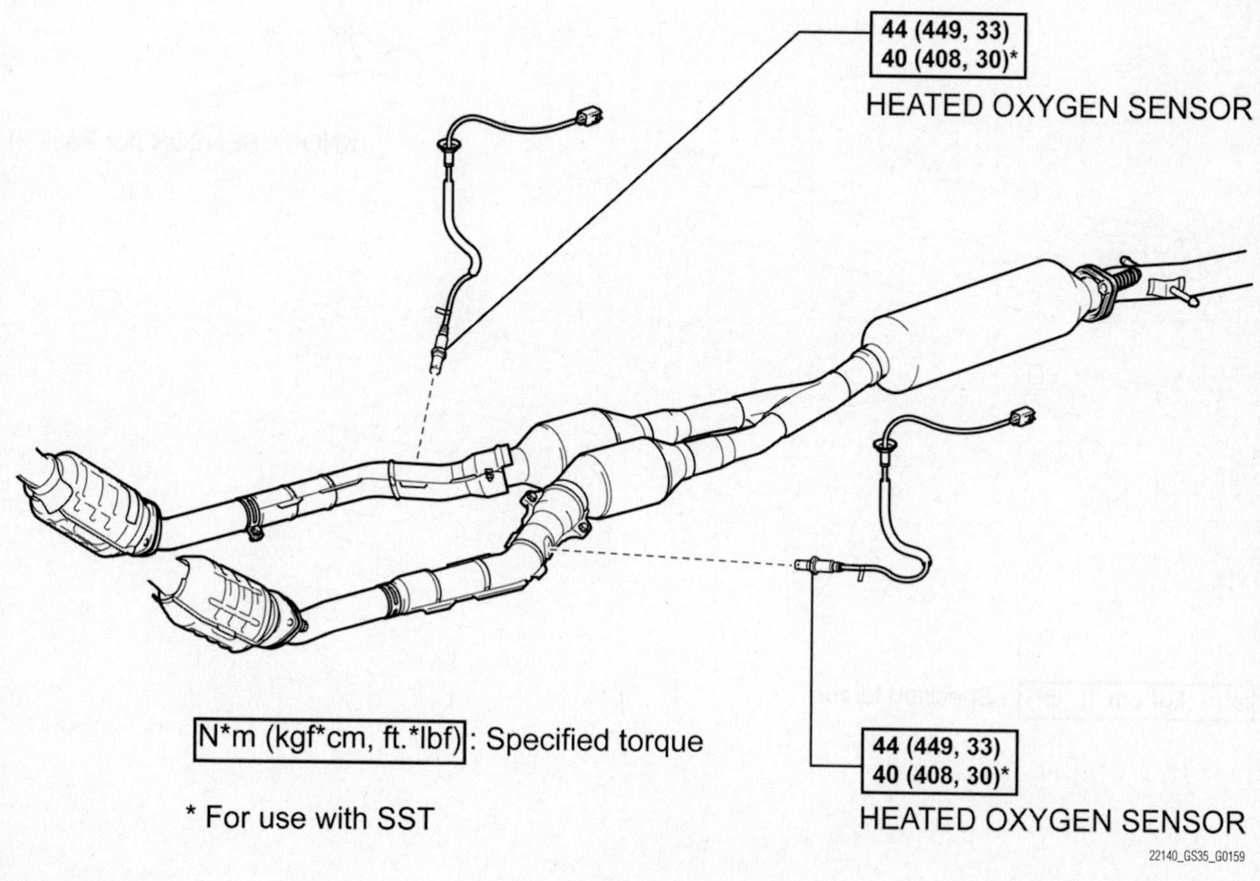

Fig. 193 Heated Oxygen Sensor (HO2S) location

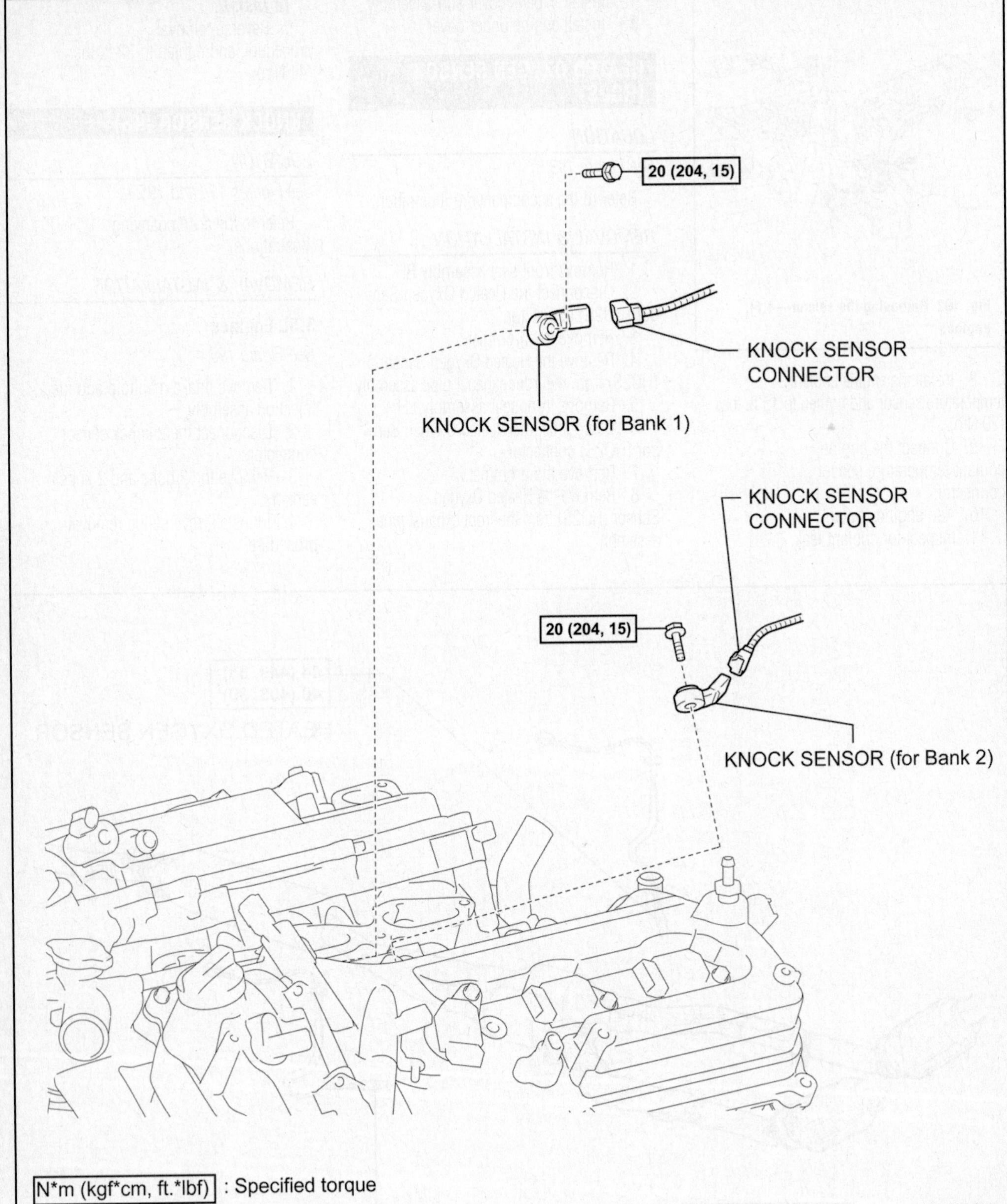

Fig. 194 Knock sensor location—3.5L (2GR- FSE)

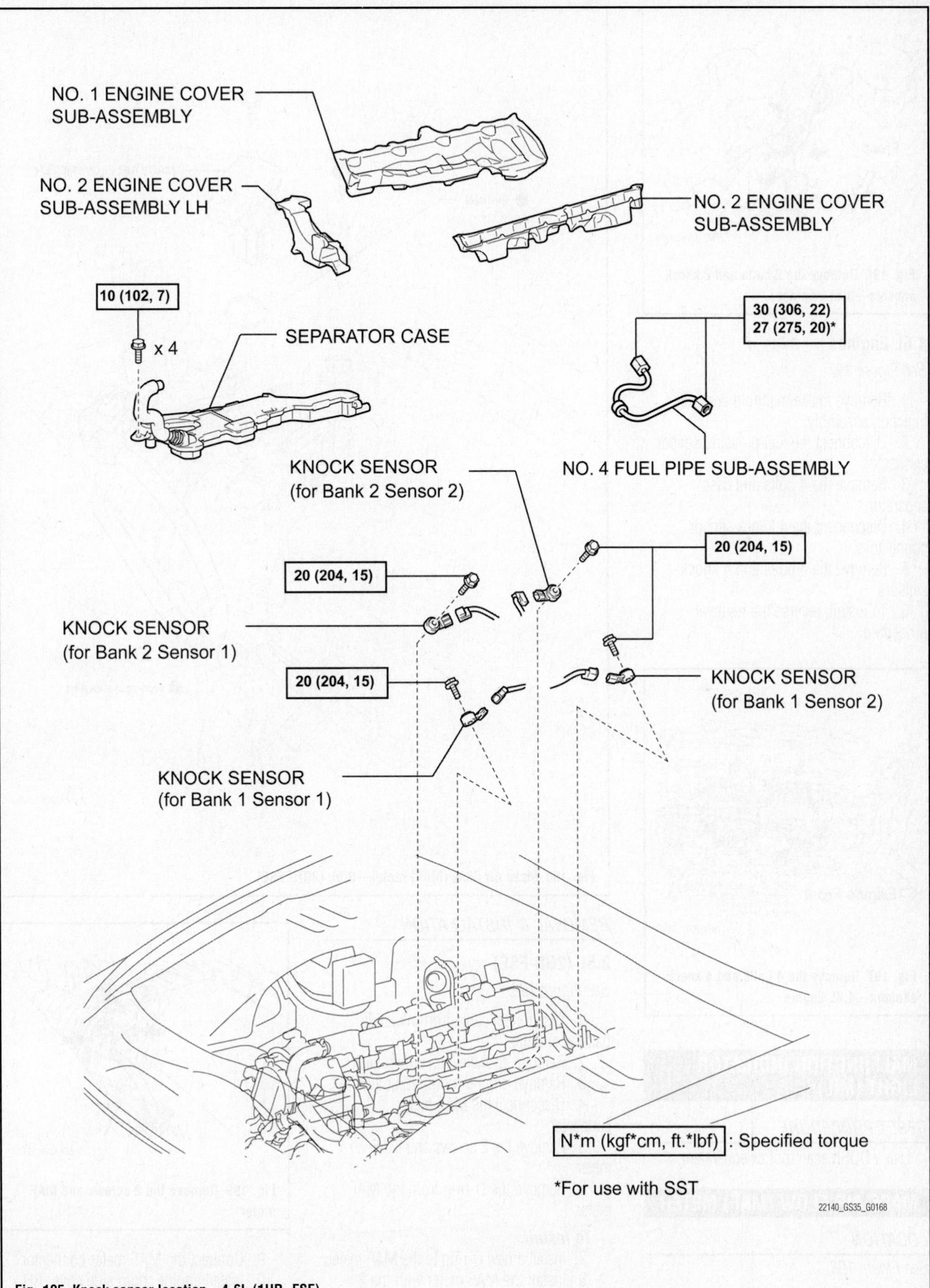

Fig. 195 Knock sensor location—4.6L (1UR- FSE)

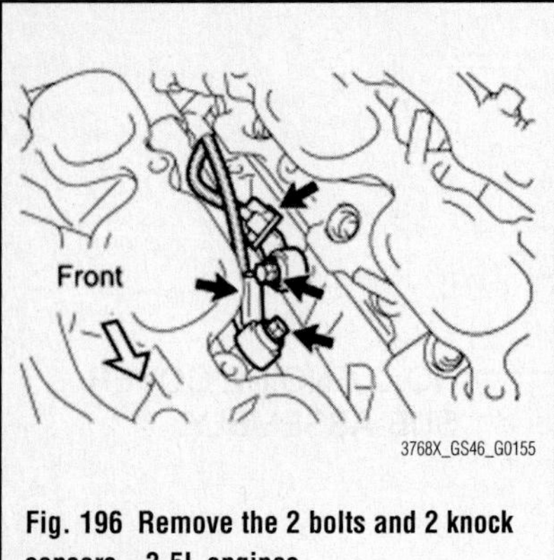

Fig. 196 Remove the 2 bolts and 2 knock sensors—3.5L engines

4.6L Engines

See Figure 197.

1. Remove intake manifold and fuel injection assembly.
2. Disconnect the fuel pressure sensor connector.
3. Remove the 4 bolts and case separator.
4. Disconnect the 4 knock sensor connectors.
5. Remove the 4 bolts and 4 knock sensors.
6. To install, reverse the removal procedure.

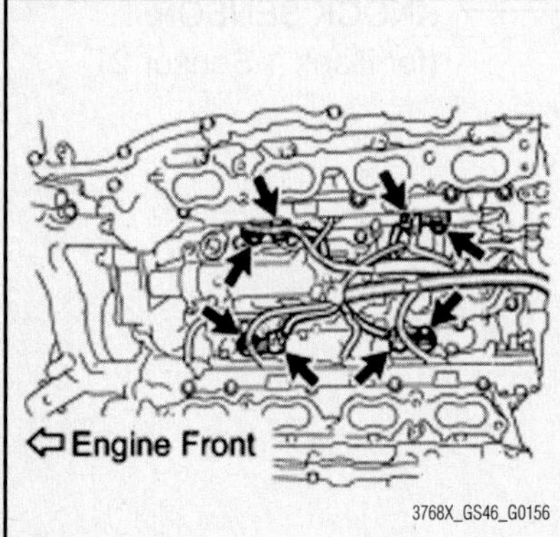

Fig. 197 Remove the 4 bolts and 4 knock sensors—4.6L Engine

MALFUNCTION INDICATOR LIGHT (MIL)

RESET PROCEDURE

Use a ODBII scan tool or equivalent.

MASS AIR FLOW (MAF) METER

LOCATION

See Figure 198.

Refer to the accompanying illustration.

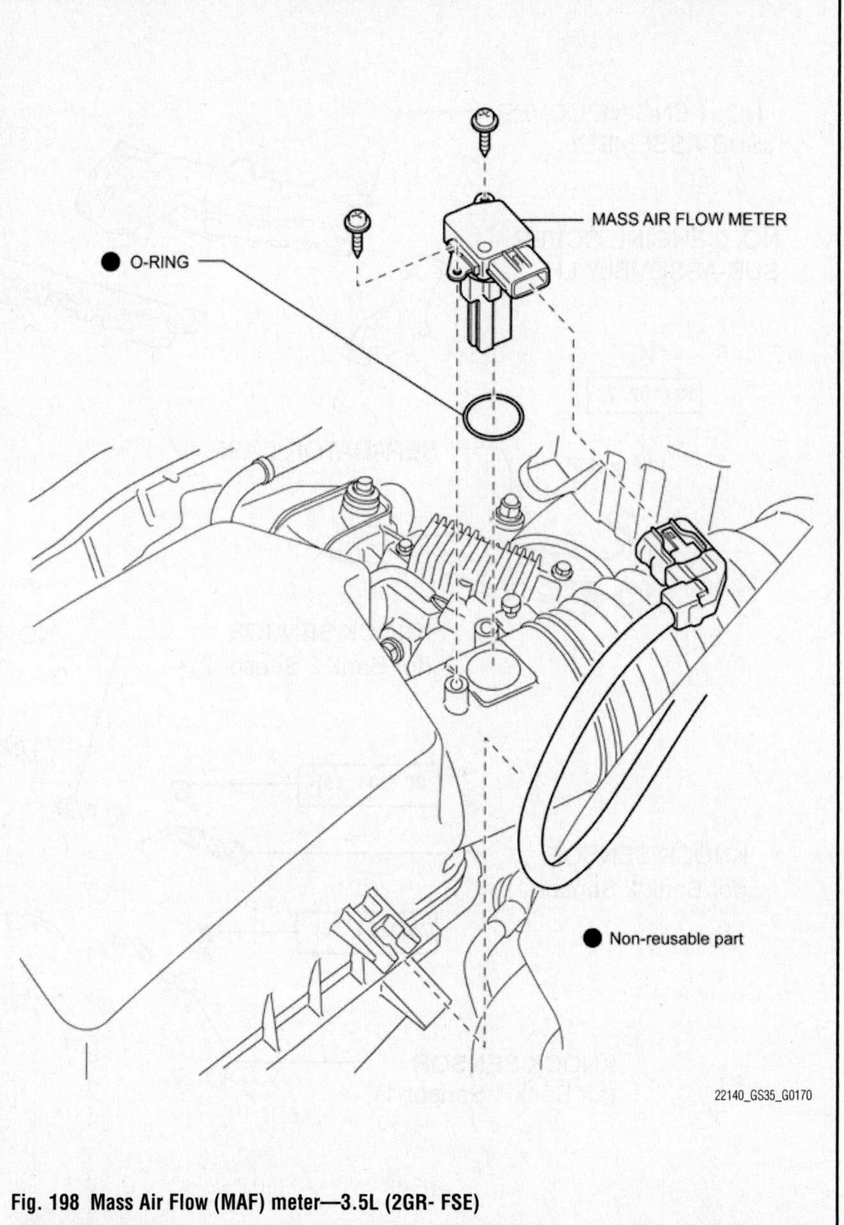

Fig. 198 Mass Air Flow (MAF) meter—3.5L (2GR- FSE)

REMOVAL & INSTALLATION

3.5L (2GR-FSE)

See Figure 199.

1. Disconnect cable from negative battery terminal.
2. Remove cool air intake duct seal.
3. Remove engine room side cover RH.
4. Disconnect the MAF meter connector.
5. Remove the 2 screws and MAF meter.
6. Remove the O-ring from the MAF meter.

To install:

7. Install a new O-ring to the MAF meter.
8. Install the MAF meter with the 2 screws.

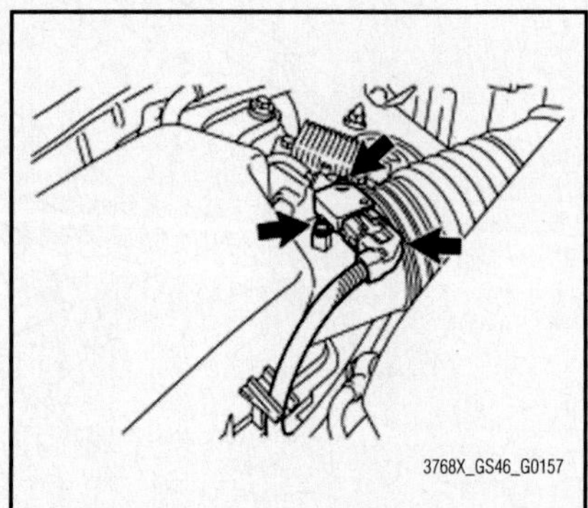

Fig. 199 Remove the 2 screws and MAF meter

9. Connect the MAF meter connector.
10. Install engine room side cover RH.
11. Install cool air intake duct seal.

12. Connect cable to negative battery terminal.
13. Perform initialization.

4.6L (1UR-FSE)

See Figure 200.

1. Disconnect cable from negative battery terminal.
2. Remove cool air intake duct seal.
3. Remove engine room side cover RH.
4. Remove V-bank cover.
5. Disconnect the MAF meter connector.
6. Remove the 2 screws and MAF meter.
7. Remove the O-ring from the MAF meter.

To install:

8. Install a new O-ring to the MAF meter.
9. Install the MAF meter with the 2 screws.

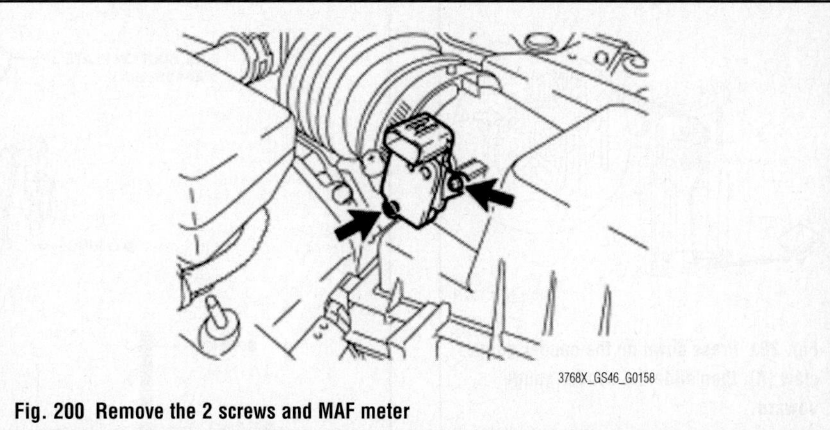

Fig. 200 Remove the 2 screws and MAF meter

10. Connect the MAF meter connector.
11. Install engine room side cover RH.
12. Install V-bank cover.
13. Install cool air intake duct seal.
14. Connect cable to negative battery terminal.
15. Perform initialization.

FUEL — GASOLINE FUEL INJECTION SYSTEM

FUEL SYSTEM SERVICE PRECAUTIONS

Safety is the most important factor when performing not only fuel system maintenance, but any type of maintenance. Failure to conduct maintenance and repairs in a safe manner may result in serious personal injury or death. Work on a vehicle's fuel system components can be accomplished safely and effectively by adhering to the following rules and guidelines.

- To avoid the possibility of fire and personal injury, always disconnect the negative battery cable unless the repair or test procedure requires that battery voltage be applied.
- Always relieve the fuel system pressure prior to disconnecting any fuel system component (injector, fuel rail, pressure regulator, etc.) fitting or fuel line connection. Exercise extreme caution whenever relieving fuel system pressure to avoid exposing skin, face and eyes to fuel spray. Please be advised that fuel under pressure may penetrate the skin or any part of the body that it contacts.
- Always place a shop towel or cloth around the fitting or connection prior to loosening to absorb any excess fuel due to spillage. Ensure that all fuel spillage is quickly removed from engine surfaces. Ensure that all fuel-soaked cloths or towels are deposited into a flame-proof waste container with a lid.
- Always keep a dry chemical (Class B) fire extinguisher near the work area.
- Do not allow fuel spray or fuel vapors to come into contact with a spark or open flame.
- Always use a second wrench when loosening or tightening fuel line connection fittings. This will prevent unnecessary stress and torsion on fuel piping. Always follow the proper torque specifications.
- Always replace worn fuel fitting O-rings with new ones. Do not substitute fuel hose where rigid pipe is installed.

FUEL SYSTEM PRESSURE

RELIEVING

1. Before servicing the vehicle, refer to the precautions in the beginning of this section.
2. Remove the fuse for the electronic fuel pump.
3. Start the engine until the engine stalls.
4. Disconnect the negative battery terminal.
5. Place a catch-pan under the joint to be disconnected. A large quantity of fuel may be released when the joint is opened.
6. Wear eye or full-face protection.
7. Place a shop towel over the area and slowly release the joint using a wrench of the correct size.
8. Allow any fuel left in the line to bleed off slowly before fully disconnecting the joint.
9. Plug the opened lines immediately to prevent fuel spillage or the entry of dirt.
10. Dispose of the released fuel properly.
11. After connecting fuel lines, install the fuse for the fuel pump and start the engine.
12. Check for leaks and repair as needed.

FUEL FILTER

REMOVAL & INSTALLATION

Fuel filters are in the fuel tank along with the fuel pump.

FUEL LEVEL SENDING UNIT

OPERATION

Uses a variable resistor to regulate voltage return to gauge.

REMOVAL & INSTALLATION

See Figure 201.

1. Before servicing the vehicle, refer to the precautions in the beginning of this section.
2. Relieve the fuel system pressure.
3. Disconnect the negative battery cable. Wait at least 90 seconds before performing any other work.
4. Remove rear seat cushion assembly.
5. Remove room No. 3 partition pad.
6. Remove rear floor No. 2 service hole cover.
7. Remove the fuel pump and sending gauge assembly.
8. Remove the fuel sending gauge from the fuel pump assembly as follows:

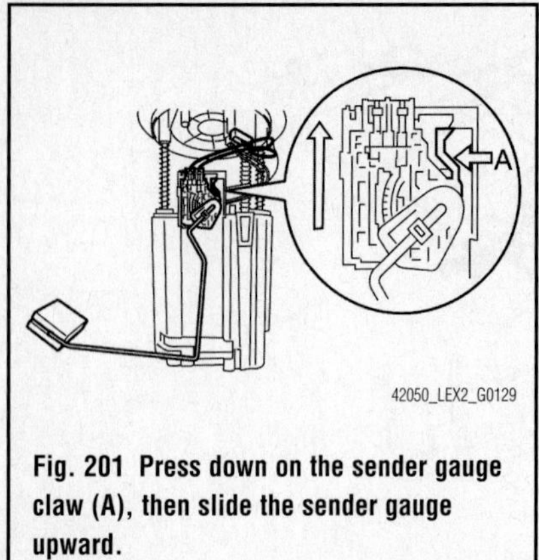

Fig. 201 Press down on the sender gauge claw (A), then slide the sender gauge upward.

a. Disconnect the fuel sender gauge connector.

b. Press down on the sender gauge claw. Then slide the sender gauge upward.

To install:

9. Install the fuel sending gauge to the fuel pump assembly as follows:

a. Set the fuel sender gauge to the No. 1 fuel sub-tank. Then slide the sender gauge downward to install.

b. Connect the fuel sender gauge connector.

10. Install the fuel pump and sending gauge assembly.
11. Install rear floor No. 2 service hole cover.
12. Install room No. 3 partition pad.
13. Install rear seat cushion assembly.
14. Connect the negative battery cable.

FUEL PUMP

REMOVAL & INSTALLATION

See Figure 202.

1. Before servicing the vehicle, refer to the precautions in the beginning of this section.
2. Relieve the fuel system pressure.
3. Remove or disconnect the following:
   - Negative battery cable. Wait at least 90 seconds before performing any other work.
   - Rear seat bottom
   - Partition cover
   - Floor service hole cover
   - Fuel pump electrical connector

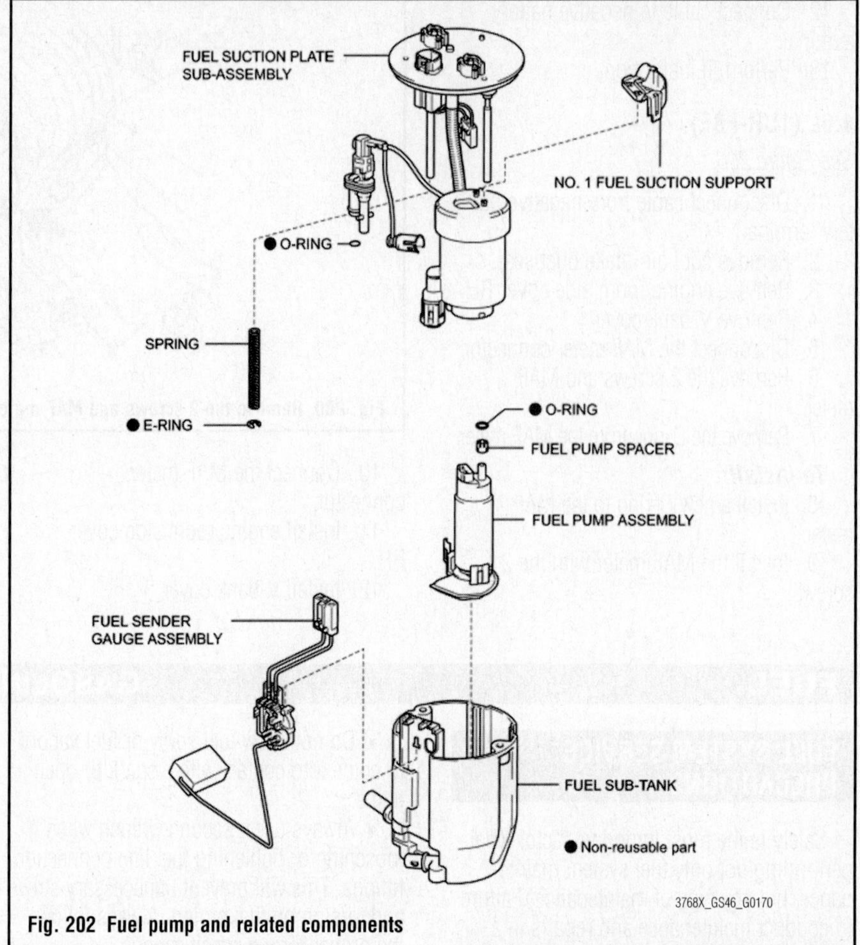

Fig. 202 Fuel pump and related components

   - Fuel main tube and fuel pump tube from the top of the fuel pump
   - Mounting bolts, or retaining ring using SST 09808-14020
   - Pump, bracket and set plate as an assembly

To install:

4. Install or connect the following:
   - A new gasket on the set plate
   - Fuel pump and bracket assembly. Torque the mounting bolts to 31 inch lbs. (3.5 Nm). If equipped with a retaining ring, use SST 09808-14020 to tighten the retainer 2 full turns so that the mark on the ring lines up within the 2 marks indicated next to the ring on the fuel tank.
   - Fuel main tube and fuel pump tube to the top of the fuel pump
   - Fuel pump electrical connector
   - Floor service hole cover
   - Partition cover
   - Rear seat bottom
   - Negative battery cable
5. Start the engine; check the fuel system for leaks

FUEL RAIL AND INJECTOR

REMOVAL & INSTALLATION

3.5L (2GR-FSE) Engine

See Figures 203 through 205.

1. Before servicing the vehicle, refer to the Precautions section.
2. Relieve fuel system pressure.
3. Remove or disconnect the following:
   - Negative battery cable. Wait at least 90 seconds before performing any other work.
   - Coolant
   - Cool air intake duct seal
   - Engine under cover
   - Right engine room side cover
   - V-bank cover
   - Intake air surge tank
   - Intake manifold
   - Fuel pressure pulsation damper
   - No. 1 and 2 fuel pipes
   - High pressure side fuel pump
   - No. 2 and 3 fuel pipes
   - No. 1 and 2 fuel delivery pipes
   - Fuel injectors, O-rings and seals

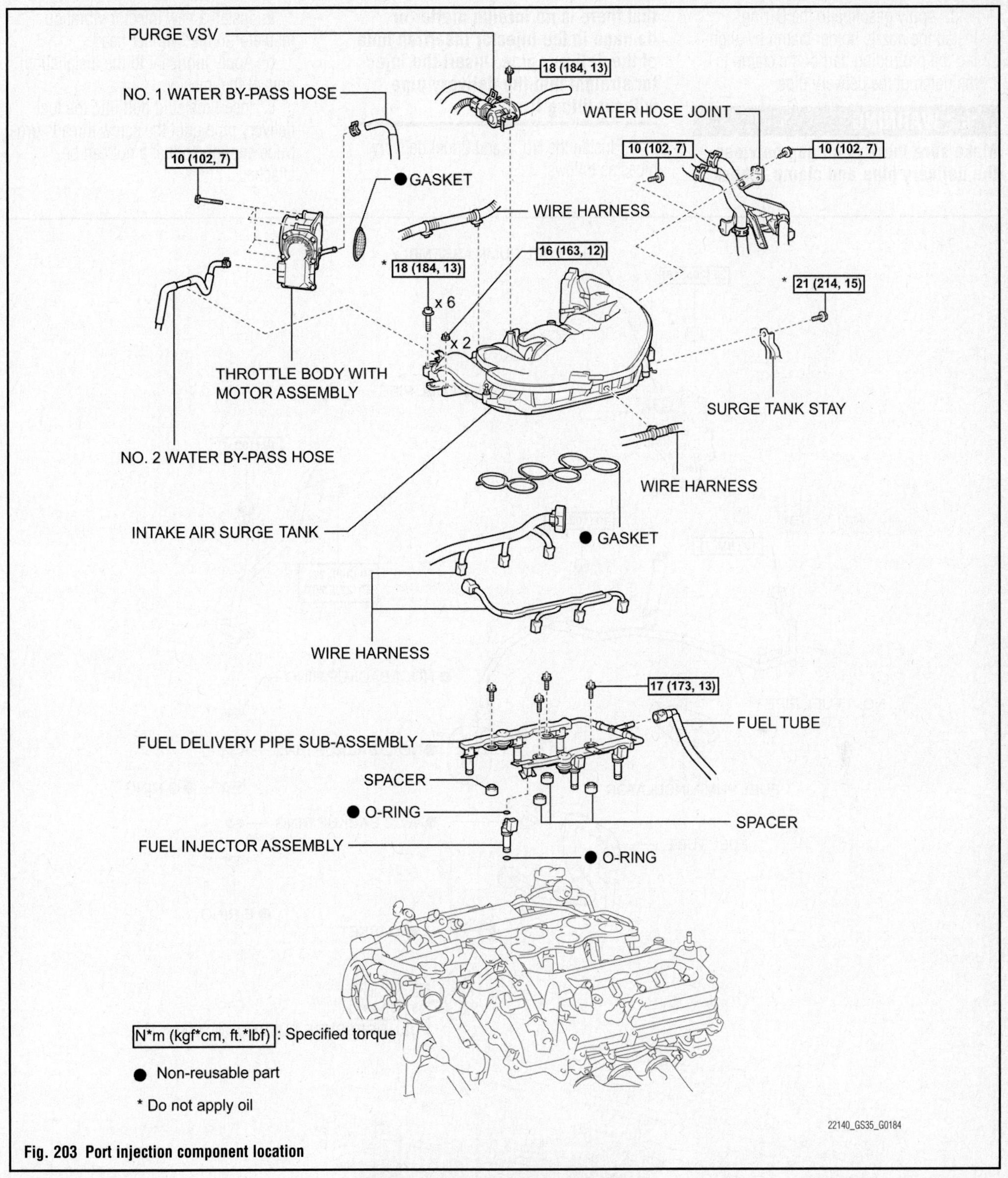

Fig. 203 Port injection component location

To install:

4. Install or connect the following:
   - 2 new seals to each injector
5. Install the fuel injectors as follows:
   a. Install a new O-ring, new backup rings (No. 1, No. 2, No. 3) and new E-ring to the fuel injector.

WARNING

Check that there is no foreign matter or damaged areas in the injector's O-ring groove. Check that the installation direction of the No. 1 and No. 2 backup ring are correct. Make sure the backup rings and O-ring are installed in the correct order. Check that the alignment openings of the backup rings are not overlapped or stretched. After installing the O-ring, check that it is not contaminated with foreign matter and is not damaged.

b. Install the injector nozzle holder clamp.

c. Apply gasoline to the O-ring. Install the nozzle holder clamp by aligning the protruding part of the clamp to the notch of the delivery pipe.

WARNING

Make sure there is no gap between the delivery pipe and clamp. Check that there is no foreign matter or damage in the injector insertion hole of the delivery pipe. Insert the injector straight into the delivery pipe without tilting it.

6. Install the No. 1 and 2 fuel delivery pipes as follows:

a. Install a new injector vibration insulator to the cylinder head.

b. Apply lubricant to the installation hole of the injector.

c. Insert the stud bolt into the fuel delivery pipe until the screw threads protrude enough so that a nut can be attached.

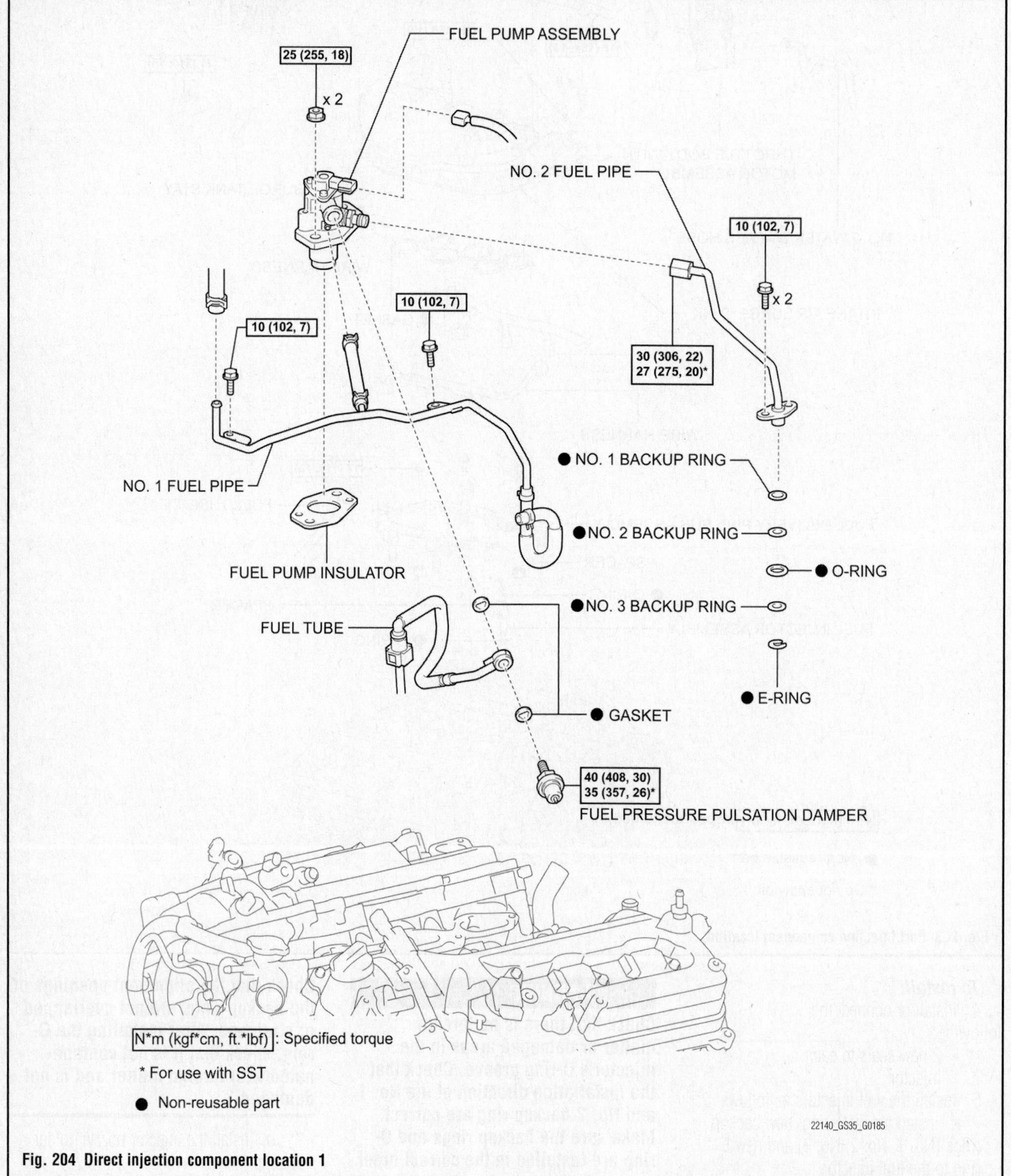

Fig. 204 Direct injection component location 1

➡If an injector is dropped, replace it with a new one. Check that there is no foreign matter or damage in the injector insertion hole of the delivery pipe. Be extremely careful not to touch or strike the tips of the injectors. When inserting the fuel delivery pipe, push it in evenly without tilting it.

d. Install the fuel delivery pipe by uniformly tightening the 2 bolts and 2 nuts in several passes to 15 ft. lbs. (21 Nm).

e. Connect the 3 connectors and 2 clamps.

- No. 2 and 3 fuel pipes and all new rings and seals. Torque No. 3 fuel pipe fastener to 7 ft. lbs. (10 Nm).
- High pressure side fuel pump
- No. 1 and 2 fuel pipes. Torque fasteners to 7 ft. lbs. (10 Nm).
- Fuel pressure pulsation damper and new gasket. Torque to 28 ft. lbs. (40 Nm).
- Intake manifold
- Intake air surge tank

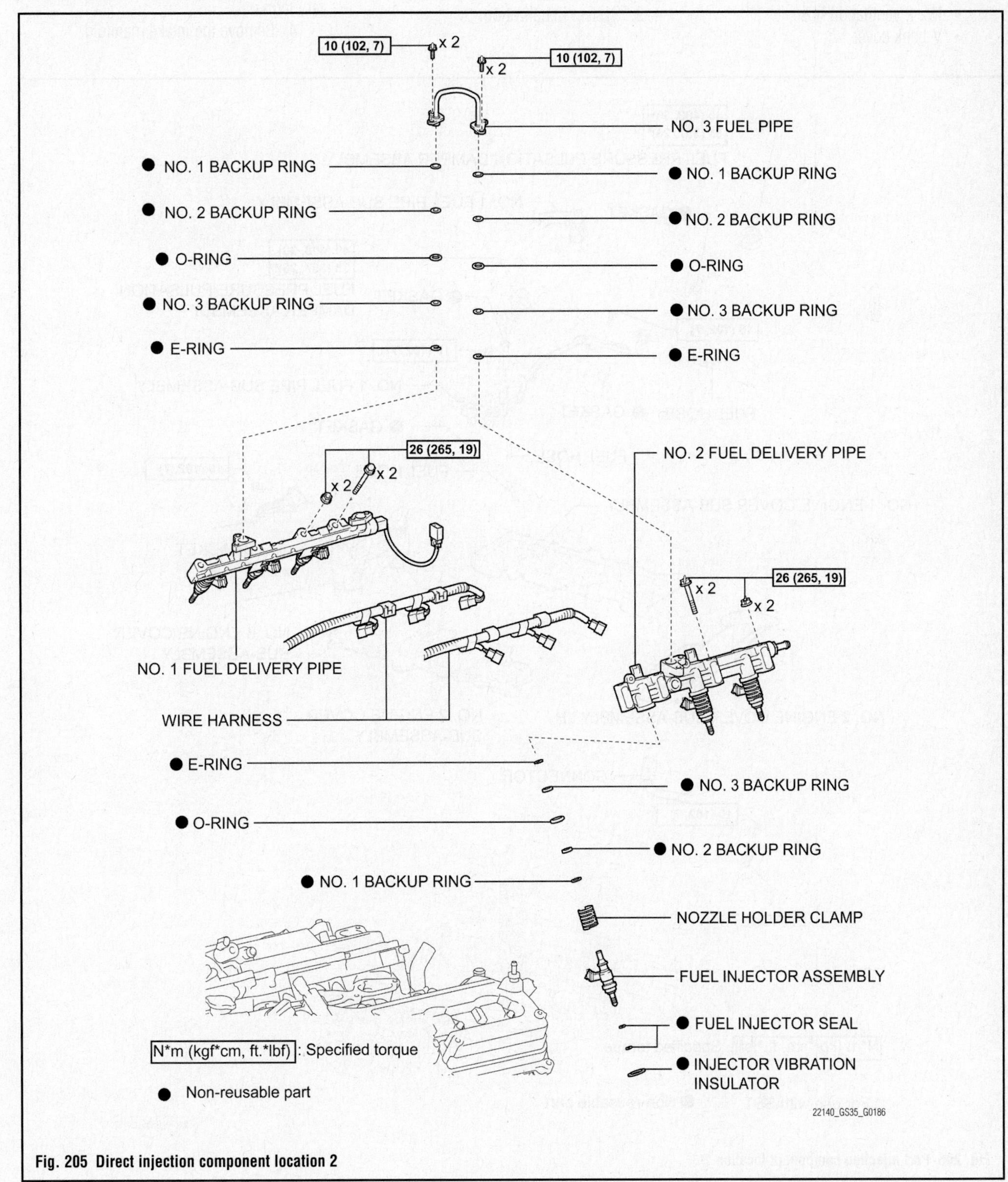

Fig. 205 Direct injection component location 2

- Fuel main tube
- Ventilation hose
- Union to check valve hose
- Engine rear cover
- Inlet and outlet heater water hoses
- Water by-pass hose
- Cold start injector. Torque the bolts to 7 ft. lbs. (10 Nm).
- Air cleaner cap with air cleaner hose
- No. 2 ventilation hose
- V-bank cover
- Right engine room side cover
- Cool air intake duct seal
- Engine under cover
- Negative battery cable

7. Refill the cooling system. Start the engine and check for coolant and fuel leaks and proper operation.
8. Check the function of the throttle body unit
9. System initialization.

4.6L (1UR-FSE)

See Figures 206 through 210.

1. Before servicing the vehicle, refer to the precautions in the beginning of this section.
2. Discharge fuel system pressure.
3. Disconnect cable from negative battery terminal.
4. Remove the intake manifold.

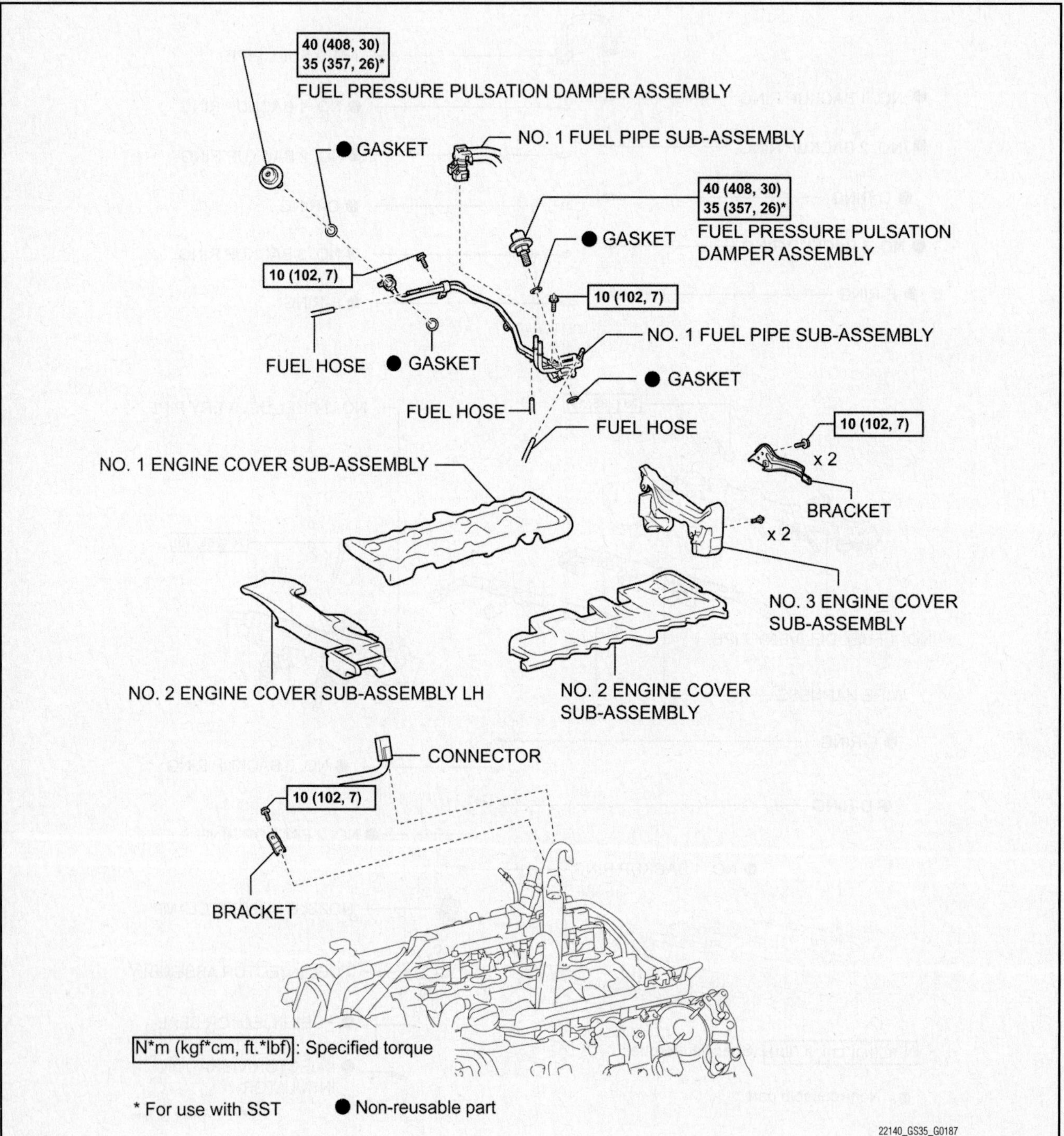

Fig. 206 Port injection component location 1

5. Remove the fuel pressure pulsation damper assembly and fuel pipe sub-assembly.

6. Remove additional fuel pipe sub-assemblies.

7. Remove the engine cover sub-assembly.

8. Remove the ventilation hose from the ventilation valve.

9. Disconnect the fuel pressure sensor connector.

10. Disconnect the injector connectors.

11. Remove the bolts and delivery pipes from the cylinder heads.

➡Make sure that the fuel delivery pipe is disconnected from the delivery pipe. Be extremely careful not to touch or strike the tips of the injectors. Pull and remove the fuel delivery pipe in a straight line without tilting it.

12. Remove the injector vibration insulators from the cylinder heads.

To install:

13. Install a new injector seals.

14. Install new injector vibration insulators to the cylinder heads.

15. Install the delivery pipe spacers for direct injection.

16. Apply lubricant to the installation injector seal and holes of the injectors.

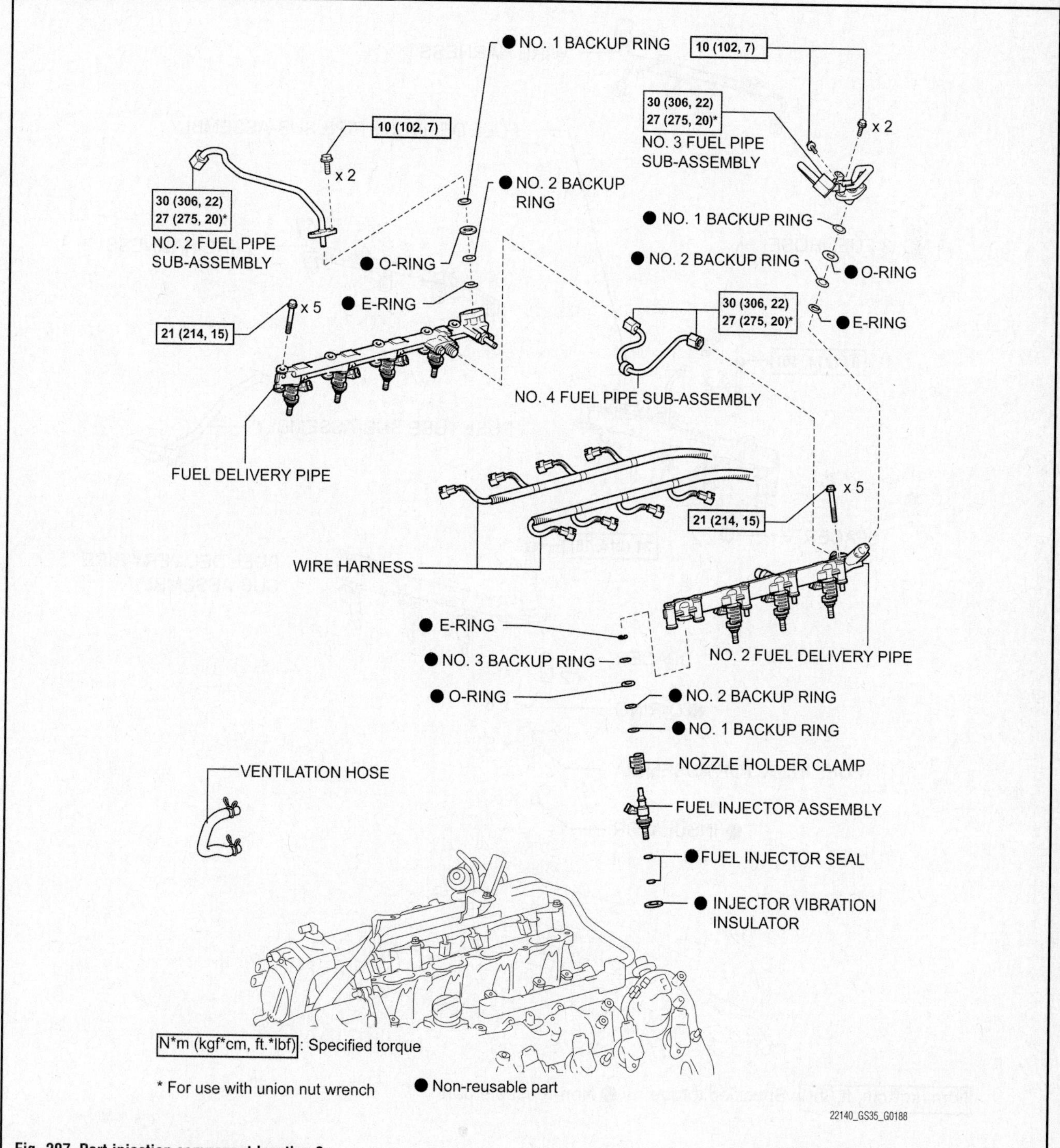

Fig. 207 Port injection component location 2

17. Install the delivery pipe (with injector) to the intake manifold and tighten to 15 ft. lbs. (21 Nm).

➡If an injector is dropped or the tips of the injectors are struck, replace it with a new one. Check that there is no foreign matter or damage to the injector insertion hole of the cylinder head. When inserting the fuel delivery pipe, push it in evenly without tilting it.

18. Connect the connectors.
19. Connect the fuel pressure sensor connector.
20. Connect the fuel hoses.
21. Install the ventilation hose to the ventilation valve.
22. Temporarily install the fuel pipe.
23. Using a 19mm union nut wrench, tighten the fuel pipe to 22 ft. lbs. (30 Nm), in the order shown.

➡After installing the fuel pipe, check that the fuel pipe protector contacts with the separator case.

24. Install the engine cover sub-assemblies.
25. Install fuel pipe sub-assemblies.
26. Install the fuel pressure pulsation damper assembly and No. 1 fuel pipe sub-assembly
27. Install the intake manifold.

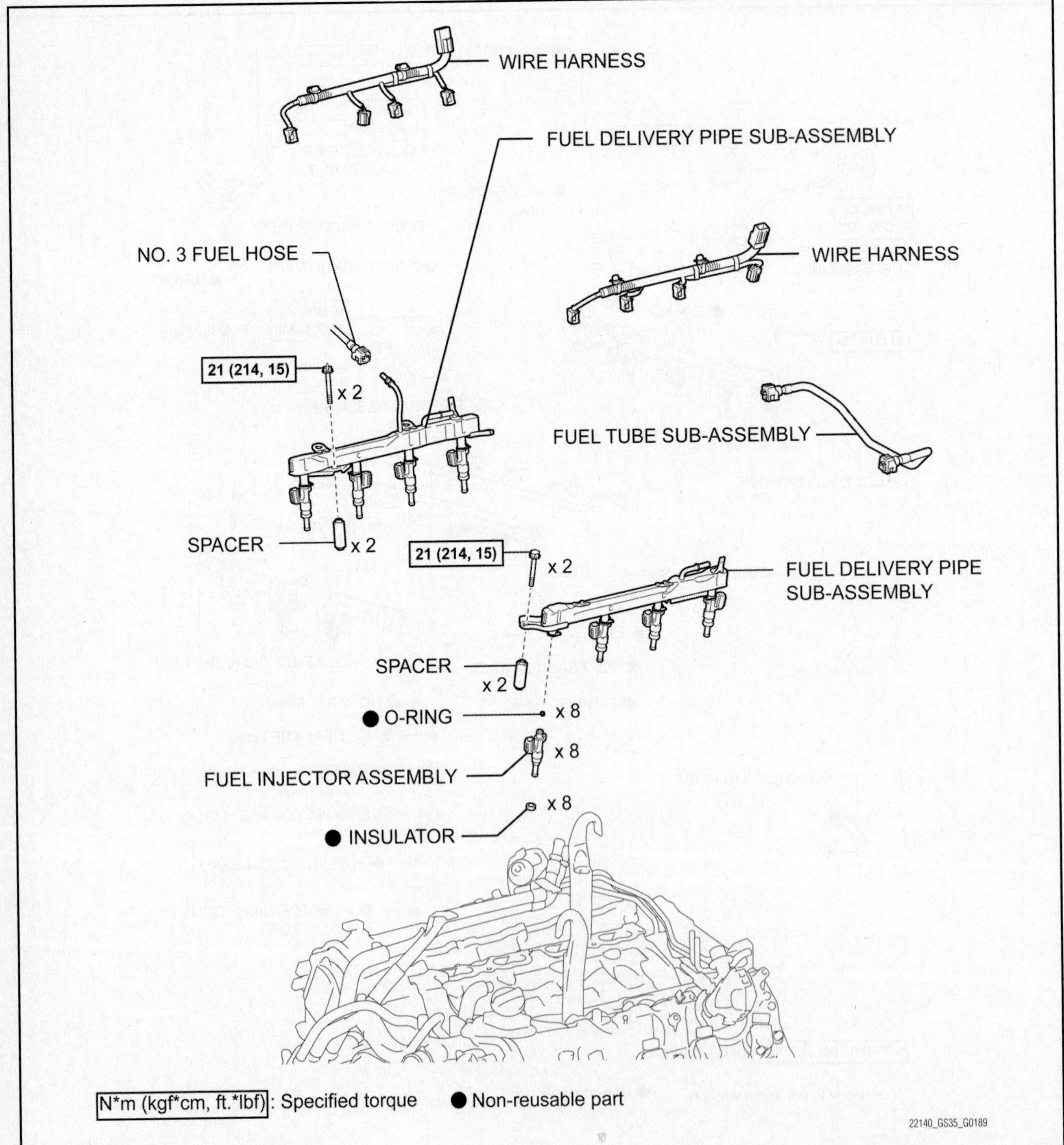

Fig. 208 Direct injection component location

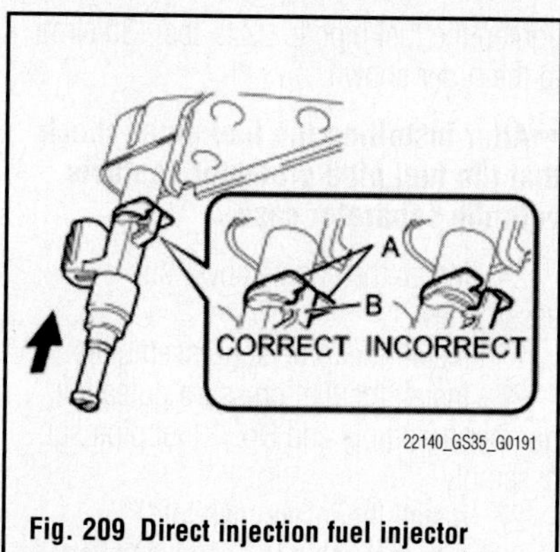

Fig. 209 Direct injection fuel injector installation

- Negative battery cable. Wait at least 90 seconds before performing any other work.
- Coolant
- Cool air intake duct seal
- Engine under cover
- Right engine room side cover
- V-bank cover
- Intake air surge tank
- Intake manifold
- Fuel pressure pulsation damper
- No. 1 and 2 fuel pipes
- High pressure side fuel pump
- No. 2 and 3 fuel pipes

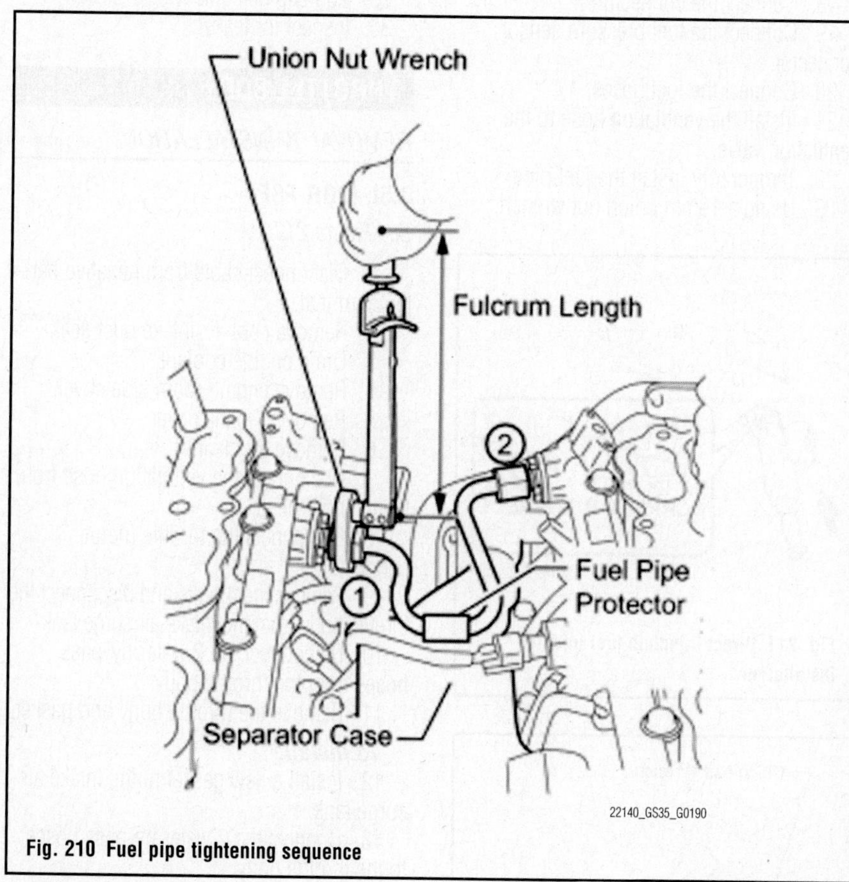

Fig. 210 Fuel pipe tightening sequence

28. Connect cable to negative battery terminal.
29. Add engine coolant.
30. Inspect for leaks.

FUEL TANK

REMOVAL & INSTALLATION

3.5L (2GR-FSE) Engine

See Figures 203 through 205.

1. Before servicing the vehicle, refer to the precautions in the beginning of this section.
2. Relieve fuel system pressure.
3. Remove or disconnect the following:

- No. 1 and 2 fuel delivery pipes
- Fuel injectors, O-rings and seals

To install:

4. Install or connect the following:
   - 2 new seals to each injector
5. Install the fuel injectors as follows:

a. Install a new O-ring, new backup rings (No. 1, No. 2, No. 3) and new E-ring to the fuel injector.

⁂ WARNING

Check that there is no foreign matter or damaged areas in the injector's O-ring groove. Check that the installation direction of the No. 1 and No. 2 backup ring are correct. Make sure the backup rings and O-ring are installed in the correct order. Check that the alignment openings of the backup rings are not overlapped or stretched. After installing the O-ring, check that it is not contaminated with foreign matter and is not damaged.

b. Install the injector nozzle holder clamp.

c. Apply gasoline to the O-ring. Install the nozzle holder clamp by aligning the protruding part of the clamp to the notch of the delivery pipe.

⁂ WARNING

Make sure there is no gap between the delivery pipe and clamp. Check that there is no foreign matter or damage in the injector insertion hole of the delivery pipe. Insert the injector straight into the delivery pipe without tilting it.

6. Install the No. 1 and 2 fuel delivery pipes as follows:

a. Install a new injector vibration insulator to the cylinder head.

b. Apply lubricant to the installation hole of the injector.

c. Insert the stud bolt into the fuel delivery pipe until the screw threads protrude enough so that a nut can be attached.

➡If an injector is dropped, replace it with a new one. Check that there is no foreign matter or damage in the injector insertion hole of the delivery pipe. Be extremely careful not to touch or strike the tips of the injectors. When inserting the fuel delivery pipe, push it in evenly without tilting it.

d. Install the fuel delivery pipe by uniformly tightening the 2 bolts and 2 nuts in several passes to 15 ft. lbs. (21 Nm).

e. Connect the 3 connectors and 2 clamps.

- No. 2 and 3 fuel pipes and all new rings and seals. Torque No. 3 fuel pipe fastener to 7 ft. lbs. (10 Nm).
- High pressure side fuel pump
- No. 1 and 2 fuel pipes. Torque fasteners to 7 ft. lbs. (10 Nm).
- Fuel pressure pulsation damper and new gasket. Torque to 28 ft. lbs. (40 Nm).
- Intake manifold

- Intake air surge tank
- Fuel main tube
- Ventilation hose
- Union to check valve hose
- Engine rear cover
- Inlet and outlet heater water hoses
- Water by-pass hose
- Cold start injector. Torque the bolts to 7 ft. lbs. (10 Nm).
- Air cleaner cap with air cleaner hose
- No. 2 ventilation hose
- V-bank cover
- Right engine room side cover
- Cool air intake duct seal
- Engine under cover
- Negative battery cable

7. Refill the cooling system. Start the engine and check for coolant and fuel leaks and proper operation.
8. Check the function of the throttle body unit
9. System initialization.

4.6L (1UR-FSE)

See Figures 206 through 208, 211 and 212.

1. Before servicing the vehicle, refer to the precautions in the beginning of this section.
2. Discharge fuel system pressure.
3. Disconnect cable from negative battery terminal.
4. Remove the intake manifold.
5. Remove the fuel pressure pulsation damper assembly and fuel pipe sub-assembly.
6. Remove additional fuel pipe sub-assemblies.
7. Remove the engine cover sub-assembly.
8. Remove the ventilation hose from the ventilation valve.
9. Disconnect the fuel pressure sensor connector.
10. Disconnect the injector connectors.
11. Remove the bolts and delivery pipes from the cylinder heads.

➡Make sure that the fuel delivery pipe is disconnected from the delivery pipe. Be extremely careful not to touch or strike the tips of the injectors. Pull and remove the fuel delivery pipe in a straight line without tilting it.

12. Remove the injector vibration insulators from the cylinder heads.

To install:

13. Install a new injector seals.
14. Install new injector vibration insulators to the cylinder heads.
15. Install the delivery pipe spacers for direct injection.
16. Apply lubricant to the installation injector seal and holes of the injectors.
17. Install the delivery pipe (with injector) to the intake manifold and tighten to 15 ft. lbs. (21 Nm).

➡If an injector is dropped or the tips of the injectors are struck, replace it with a new one. Check that there is no foreign matter or damage to the injector insertion hole of the cylinder head. When inserting the fuel delivery pipe, push it in evenly without tilting it.

18. Connect the connectors.
19. Connect the fuel pressure sensor connector.
20. Connect the fuel hoses.
21. Install the ventilation hose to the ventilation valve.
22. Temporarily install the fuel pipe.
23. Using a 19mm union nut wrench, tighten the fuel pipe to 22 ft. lbs. (30 Nm), in the order shown.

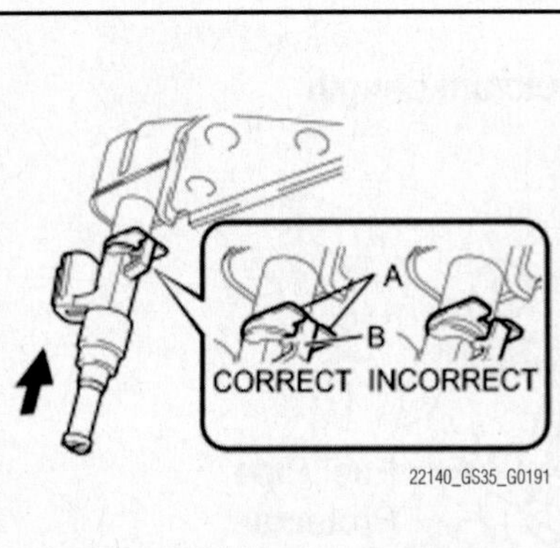

Fig. 211 Direct injection fuel injector installation

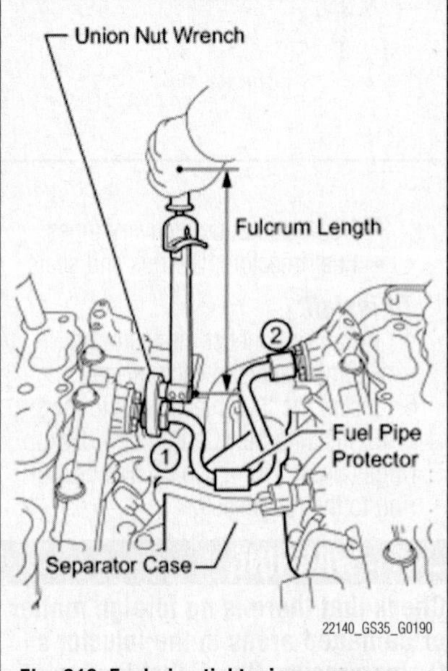

Fig. 212 Fuel pipe tightening sequence

➡After installing the fuel pipe, check that the fuel pipe protector contacts with the separator case.

24. Install the engine cover sub-assemblies.
25. Install fuel pipe sub-assemblies.
26. Install the fuel pressure pulsation damper assembly and No. 1 fuel pipe sub-assembly
27. Install the intake manifold.
28. Connect cable to negative battery terminal.
29. Add engine coolant.
30. Inspect for leaks.

THROTTLE BODY

REMOVAL & INSTALLATION

3.5L (2GR-FSE)

See Figure 213.

1. Disconnect cable from negative battery terminal.
2. Remove cool air intake duct seal.
3. Drain engine coolant.
4. Remove engine room side cover.
5. Remove V-bank cover.
6. Remove air cleaner.
7. Disconnect the ventilation hose from the cylinder head.
8. Disconnect the throttle motor connector.
9. Remove the 4 bolts and disconnect the throttle body from the intake air surge tank.
10. Disconnect the 2 water by-pass hoses from the throttle body.
11. Remove the throttle body and gasket.

To install:

12. Install a new gasket to the intake air surge tank.
13. Connect the 2 water by-pass hoses to the throttle body.

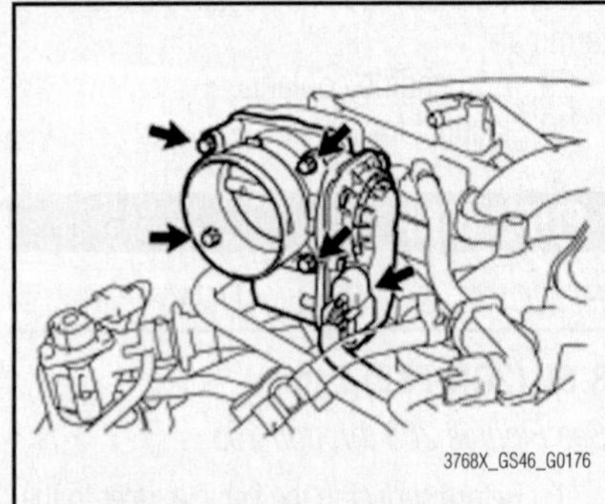

Fig. 213 Remove the 4 bolts and disconnect the throttle body from the intake air surge tank

14. Install the throttle body with the 4 bolts an tighten to 7 ft. lbs. (10 Nm).
15. Connect the throttle motor connector.
16. Install the air cleaner.
17. Connect the ventilation hose to the cylinder head cover with the clamp.
18. Connect cable to negative battery terminal.
19. Add engine coolant.
20. Inspect for coolant leak.
21. Inspect function of throttle body.
22. Install V-bank cover.
23. Install engine room side cover.
24. Install cool air intake duct seal.
25. Perform initialization.

4.6L (1UR-FSE) Engine

See Figure 214.

1. Before servicing the vehicle, refer to the precautions in the beginning of this section.
2. Disconnect the negative battery cable. Wait at least 90 seconds before performing any other work.
3. Remove the V-bank cover.
4. Remove engine under covers
5. Drain the engine coolant.
6. Remove the air cleaner inlet.
7. Remove the intake air connector.
8. Disconnect the throttle motor connector.
9. Remove the 4 bolts and throttle body
10. Slide the clamps, and disconnect the No. 4 and No. 5 water by-pass hoses from the throttle body

To install:

11. Connect the No. 4 water by-pass hose and No. 5 water by-pass hose to the throttle body
12. Install the throttle body with the 4 bolts and tighten to 7 ft. lbs. (10 Nm)
13. Connect the throttle motor connector.
14. Install intake air connector.

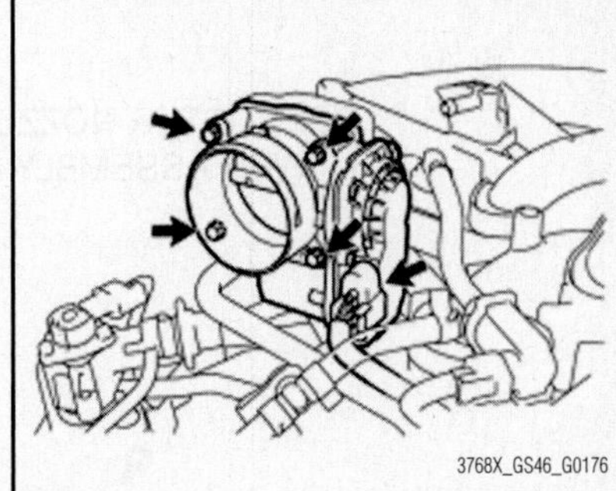

Fig. 214 Remove the 4 bolts and throttle body

15. Install air cleaner inlet.
16. Refill the engine cooling system.
17. Connect the negative battery cable.
18. Check the cooling system for leaks.
19. Install engine under covers
20. Install V-bank cover.

HEATING & AIR CONDITIONING SYSTEM

BLOWER MOTOR

REMOVAL & INSTALLATION

See Figure 215.

1. Before servicing the vehicle, refer to the precautions in the beginning of this section.
2. Disconnect the negative battery cable. Wait 90 seconds before doing any further work while the airbag system de-energizes.
3. Remove the instrument panel assembly.
4. Remove the air conditioner unit assembly. See HEATER CORE Removal & Installation procedure in this section.
5. Remove the blower assembly as follows:
   a. Disconnect the connector.
   b. Remove the screw and nut.
   c. Release the 2 claws and remove the blower assembly.
6. Remove the damper servo sub-assembly as follows:
   a. Detach the claw and remove the lever.
   b. Remove the 2 screws and damper servo.
7. Remove the 3 screws and the blower motor.

To install:

8. Install the blower motor to the blower unit assembly and tighten the 3 screws.
9. Install the damper servo sub-assembly as follows:
   a. Install the damper servo with the 2 screws.
   b. Attach the claw and install the lever.
10. Install the blower assembly as follows:
   a. Install the blower assembly with the 2 claws, screw and nut. Tighten the screw to 27 inch lbs. (3 Nm); tighten the nut to 7 ft. lbs. (10 Nm).
   b. Connect the wiring connector.
11. Install the air conditioner unit assembly. See HEATER CORE Removal & Installation procedure in this section.
12. Install the instrument panel assembly.
13. Connect the negative battery cable.

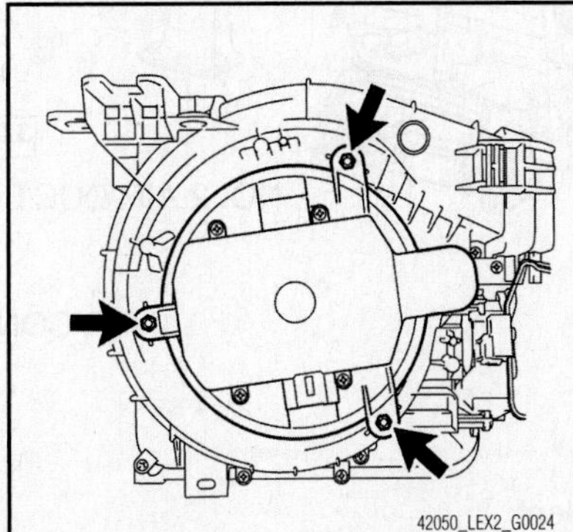

Fig. 215 Blower motor screw locations

HEATER CORE

REMOVAL & INSTALLATION

See Figures 216 and 217.

1. Before servicing the vehicle, refer to the precautions in the beginning of this section.
2. Discharge the A/C system.
3. Set radio receiver assembly to shipment mode as follows:
   a. Be sure that all discs and tapes have been removed from the unit.
   b. Be sure that the engine switch off.
   c. While simultaneously pressing the "SEEK UP" and "DISC" switches, turn the engine switch on (ACC).

➡The CD loading door indicator light blinks during mode setting and it remains lit after the setting is completed.

   d. Turn the engine switch off.
4. Disconnect the negative battery cable. Wait 90 seconds before doing any further work while the airbag system de-energizes.
5. Drain the cooling system into a clean container for reuse.
6. Align the front wheels facing straight ahead.
7. Remove or disconnect the following:
   - Cool air intake duct seal
   - Left and right engine room side covers

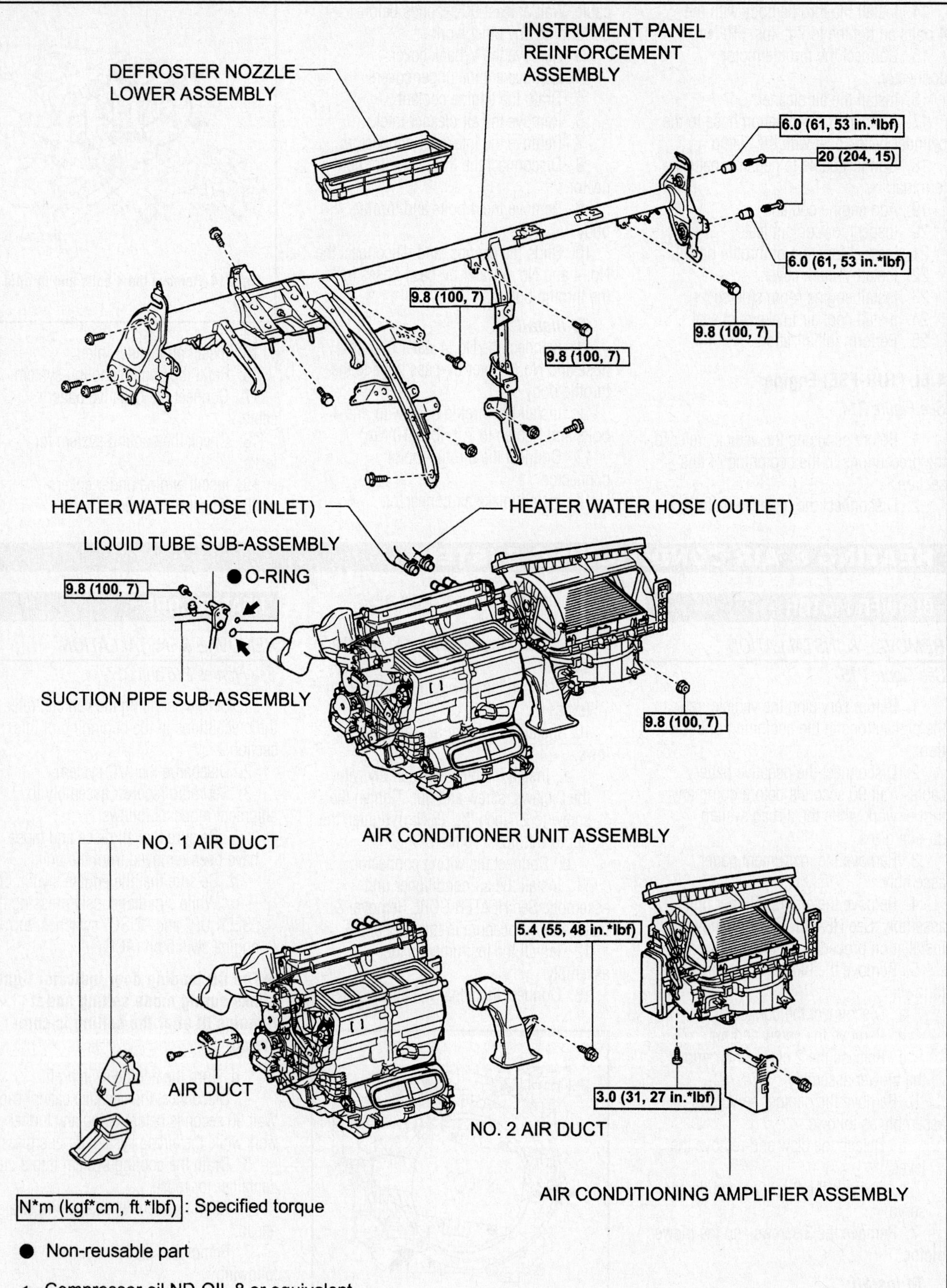

Fig. 216 Exploded view of the A/C unit assembly, instrument panel reinforcement and related components

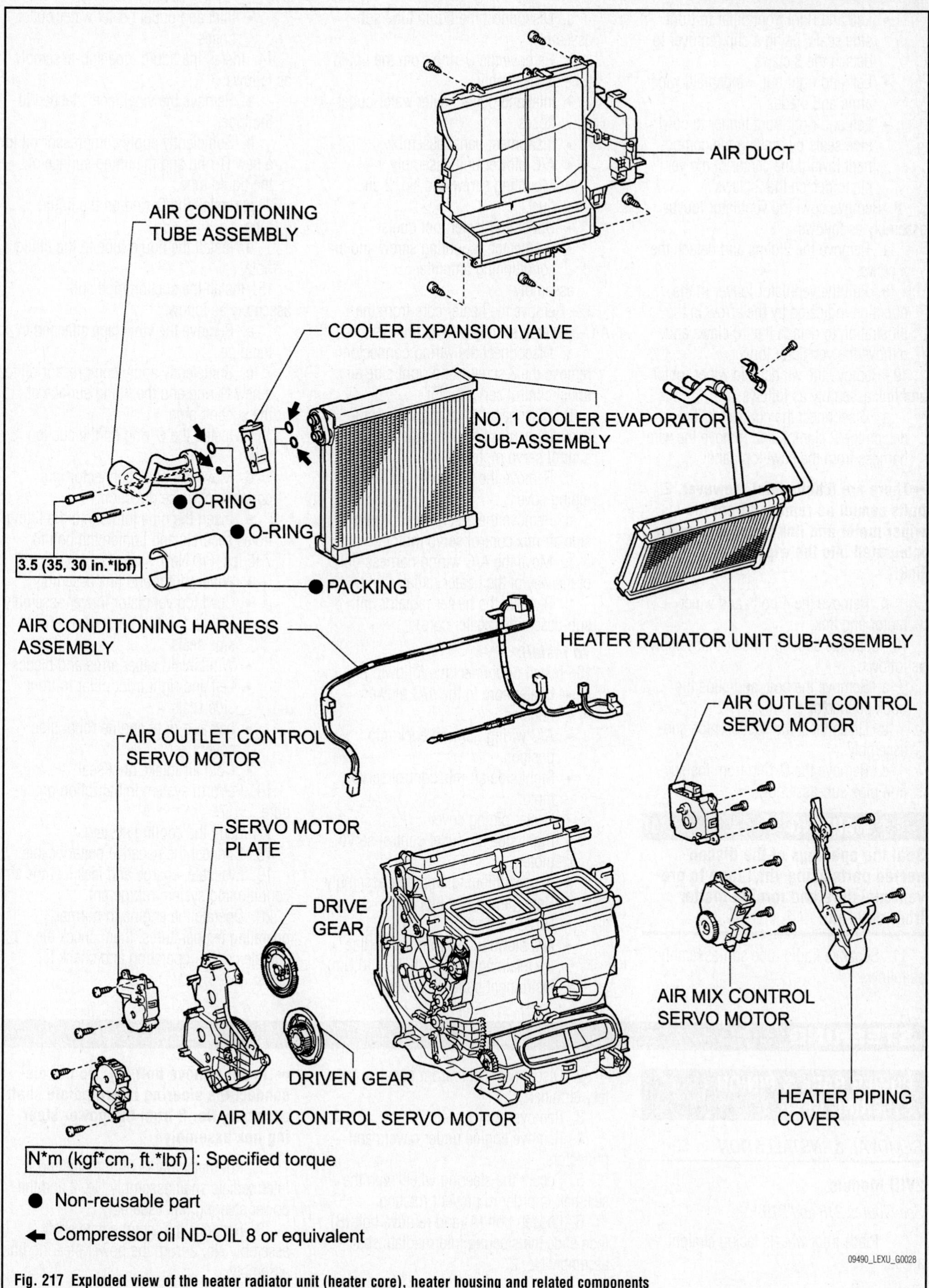

Fig. 217 Exploded view of the heater radiator unit (heater core), heater housing and related components

- Left and right front pillar to front side seals, using a clip remover to detach the 3 claws
- Left and right nut, windshield wiper arms and blades
- Left and right front fender to cowl side seals by moving the component toward the center of the vehicle to detach the 2 claws

8. Remove cowl top ventilator louver assembly as follows:

a. Remove the 2 clips and detach the 5 claws.

b. Pull the ventilator louver in the direction indicated by the arrow in the illustration to detach the 10 claws and remove the ventilator louver.

9. Remove the windshield wiper motor and link assembly as follows:

a. Disconnect the connector. Then detach the 2 clamps and remove the wire harness from the cowl top panel.

➡There are 6 bolts total, however, 2 bolts cannot be removed from the wiper motor and link because they are integrated into the wiper motor and link.

b. Remove the 4 bolts and wiper motor and link.

10. Separate suction pipe sub-assembly as follows:

a. Remove the bolt, and slide the hook connector.

b. Disconnect the suction pipe sub-assembly.

c. Remove the O-ring from the suction pipe sub-assembly.

WARNING

Seal the openings of the disconnected parts using vinyl tape to prevent moisture and foreign matter from entering.

11. Separate liquid tube sub-assembly as follows:

a. Disconnect the liquid tube sub-assembly.

b. Remove the O-ring from the liquid tube subassembly.

- Inlet and outlet heater water outlet hoses
- Instrument panel assembly.
- A/C blower unit assembly
- Mounting screw and No. 2 air duct
- Upper and lower foot ducts
- Connector, mounting screw and air conditioning amplifier assembly

12. Remove the heater core from the A/C-blower assembly unit as follows:

a. Disconnect the wiring connector, remove the 3 screws and right side air outlet control servo motor.

b. Disengage 3 clamps, connector and wire harness to right side air mix control servo motor.

c. Remove the 2 screws and heater piping cover.

d. Remove the 3 screws and right side air mix control servo motor.

e. Move the A/C wiring harness out of the way of the heater radiator unit.

f. Remove the heater radiator unit sub-assembly (heater core)

To install:

13. Install or connect the following:

- Heater core to the A/C blower housing
- A/C wiring harness back into position
- Right side air mix control servo motor
- Heater piping cover
- Right side air outlet control servo motor
- Air conditioning amplifier assembly
- Upper and lower foot ducts
- No. 2 air duct
- A/C blower unit assembly. Torque the retaining nut to 7 ft. lbs. (10 Nm).
- Instrument panel assembly.
- Inlet and outlet heater water outlet hoses

14. Install the liquid tube sub-assembly as follows:

a. Remove the vinyl tape attached to the tube.

b. Sufficiently apply compressor oil to a new O-ring and the fitting surface of the liquid tube.

c. Install the O-ring on the liquid tube.

d. Install the liquid tube to the fitting hole.

15. Install the suction pipe sub-assembly as follows:

a. Remove the vinyl tape attached to the pipe.

b. Sufficiently apply compressor oil to a new O-ring and the fitting surface of the suction pipe.

c. Install the O-ring on the suction pipe.

d. Move the hook connector in a counterclockwise direction.

e. Insert the pipe joints into the fitting holes securely and tighten the bolt to 7 ft. lbs. (10 Nm).

- Wiper motor and link assembly
- Cowl top ventilator louver assembly
- Left and right front fender to cowl side seals
- Windshield wiper arms and blades
- Left and right front pillar to front side seals
- Left and right engine room side covers
- Cool air intake duct seal

16. Perform system initialization procedure.

17. Refill the cooling system.

18. Connect the negative battery cable.

19. Evacuate, charge and leak test the air conditioning system refrigerant.

20. Operate the engine to normal operating temperatures; then, check the climate control operation and check for leaks.

STEERING

POWER RACK & PINION STEERING GEAR

REMOVAL & INSTALLATION

2WD Models

See Figures 218 and 219.

1. Place front wheels facing straight ahead.

2. Disconnect cable from negative battery terminal.

3. Remove front wheels.

4. Remove engine under covers and protectors.

5. Secure the steering wheel with the seat belt in order to prevent rotation.

6. Loosen bolt (A) and remove bolt (B), then slide the steering intermediate shaft assembly No. 2.

➡Do not remove bolt (A). Do not disconnect the steering intermediate shaft assembly No. 2 from the power steering link assembly.

7. Put matchmarks on the steering intermediate shaft assembly No. 2 and the power steering link assembly.

8. Separate the intermediate shaft assembly No. 2 from the power steering link assembly.

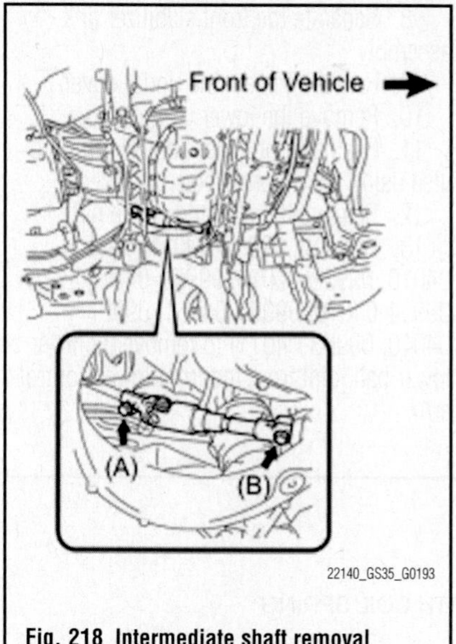

Fig. 218 Intermediate shaft removal

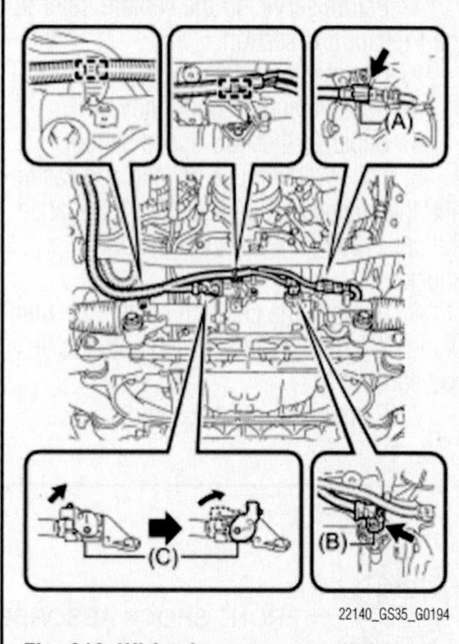

Fig. 219 Wiring harness connections

9. Separate the tie rod ends from the steering knuckles.
10. Remove the 2 clamps to disconnect the wire harness from the bracket.
11. Disconnect 2 connectors (A) and (B) from the power steering link assembly.
12. Release the lock of connector (C) and disconnect connector (C) from the power steering link assembly.
13. Remove the 2 bolts, 2 washers, 2 nuts, and the power steering link assembly from the front suspension cross member.

To install:

14. Install the power steering link assembly with the 2 bolts, 2 washers and 2 nuts and tighten to 87 ft. lbs. (118 Nm).
15. Connect wire harness connector (C) to the power steering link assembly and securely lock the connector.
16. Connect 2 wire harness connectors (A) and (B) to the power steering link assembly.
17. Install the 2 wire harness clamps to the power steering link assembly.
18. Connect the tie rod end LH to the steering knuckle with the nut and tighten to 50 ft. lbs. (65 Nm).
19. Install a new clip.
20. Align the matchmarks on the intermediate shaft assembly No. 2 and the power steering link assembly.
21. Install bolt (A) and tighten the 2 bolts and tighten to 26 ft. lbs. (35 Nm).
22. Install engine under covers and protectors.
23. Install front wheels.
24. Connect cable from negative battery terminal.
25. Inspect and adjust front wheel alignment.
26. Initialize rotation angle sensor and calibrate torque sensor zero point.
27. Perform variable gear ratio steering system calibration (for 1UR-FSE).
28. Perform initialization.

4WD Models

1. Place front wheels facing straight ahead.
2. Disconnect cable from negative battery terminal.
3. Remove front wheels.
4. Remove engine under covers and protectors.
5. Secure the steering wheel with the seat belt in order to prevent rotation.
6. Loosen bolt (A) and remove bolt (B), then slide the steering intermediate shaft assembly No. 2.

➡Do not remove bolt (A). Do not disconnect the steering intermediate shaft assembly No. 2 from the power steering link assembly.

7. Put matchmarks on the steering intermediate shaft assembly No. 2 and the power steering link assembly.
8. Separate the intermediate shaft assembly No. 2 from the power steering link assembly.
9. Separate the tie rod ends from the steering knuckles.
10. Remove engine assembly with transmission.
11. Remove the 2 bolts, 2 washers, 2 nuts, and the power steering link assembly from the front suspension cross member.

To install:

12. Install the power steering link assembly with the 2 bolts and 2 nuts and tighten to 75 ft. lbs. (102 Nm).
13. Install engine assembly with transmission.
14. Connect the tie rod end LH to the steering knuckle with the nut and tighten to 50 ft. lbs. (65 Nm).
15. Install a new clip.
16. Align the matchmarks on the intermediate shaft assembly No. 2 and the power steering link assembly.
17. Install bolt (A) and tighten the 2 bolts and tighten to 26 ft. lbs. (35 Nm).
18. Install engine under covers and protectors.
19. Install front wheels.
20. Connect cable from negative battery terminal.
21. Inspect and adjust front wheel alignment.
22. Initialize rotation angle sensor and calibrate torque sensor zero point.
23. Perform initialization.

POWER STEERING PUMP

The LEXUS GS350 and GS460 models are equipped with electronic power steering. These models do not utilize a power steering (vane) pump assembly or power steering fluid.

SUSPENSION

FRONT SUSPENSION

CONTROL LINKS

REMOVAL & INSTALLATION

See Figure 220.

1. Remove front wheel.
2. Remove the 2 nuts and the front stabilizer link assemblies.

To install:

3. Install the front stabilizer link assembly with the nuts and tighten to 62 ft. lbs. (84 Nm).
4. Install front wheel.

LOWER BALL JOINT

REMOVAL & INSTALLATION

See Figure 221.

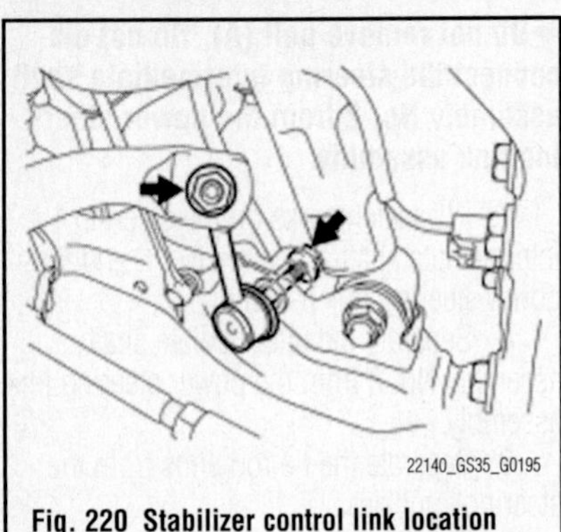

Fig. 220 Stabilizer control link location

1. Before servicing the vehicle, refer to the Precautions section.
2. Remove the front wheel.
3. Remove engine under cover.
4. Separate the ABS speed sensor.
5. For AWD, remove the clip and castle nut, then using SST 09628-00011, separate the upper ball joint from the steering knuckle.
6. Remove the clip and castle nut, then using SST 09628_62011, separate the tie rod end.
7. Remove the strut assembly.
8. Separate the front stabilizer link assembly.
9. Remove the engine under cover.
10. Remove the lower control arm.
11. Fix the front lower control arm in a vise using aluminum plates.
12. Remove the clip and castle nut.
13. Use SST 09950-40011 (09951-04010, 09952-04010, 09953-04020, 09954-04010, 09955-04051, 09957-04010, 09958-04011) to remove the front lower ball joint from the front lower control arm.

FRONT SHOCK ABSORBER WITH COIL SPRING
FRONT SUSPENSION LOWER ARM
LOWER ARM NO.2 BRACKET SUB-ASSEMBLY
135 (1,380, 100)
113 (1,150, 83)
● Clip
86 (877, 63)
84 (857, 62)
50 (510, 37)
162 (1,650, 120)
84 (857, 62)
FRONT STABILIZERLINK ASSEMBLY
157 (1,600, 116)
204 (2,080, 150)
● Clip
65 (663, 48)
TIE ROD ASSEMBLY
FRONT LOWER BALL JOINT
120 (1,220, 89)

N*m (kgf*cm, ft.*lbf) : Specified torque ● Non-reusable part

42050_LEX2_G0140

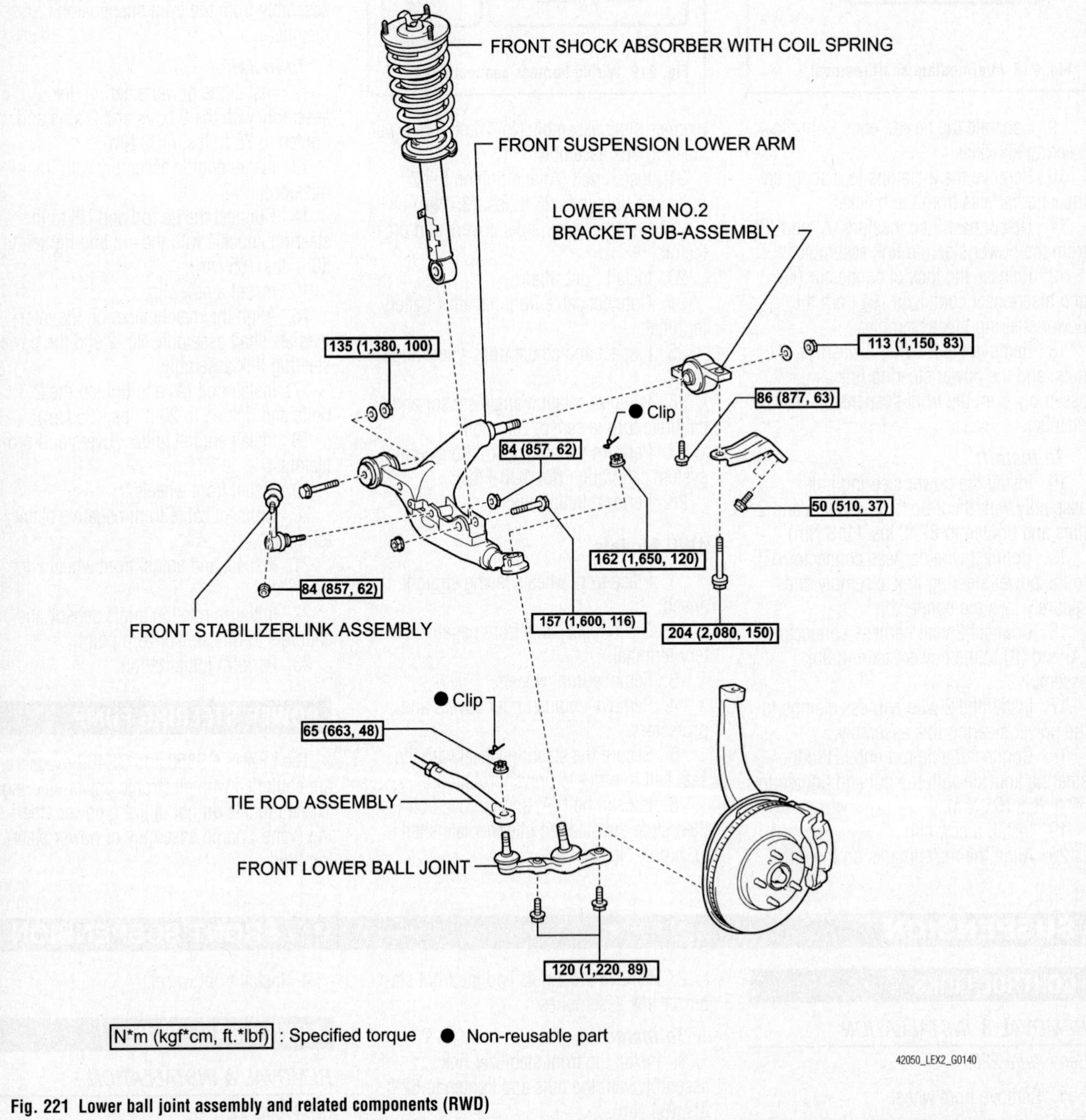

Fig. 221 Lower ball joint assembly and related components (RWD)

14. Inspect the lower control arm ball joint as follows:

a. Flip the ball joint stud back and forth 5 times.

b. Temporarily install the nut, and use a torque wrench to turn the nut continuously at a rate of 3 to 5 seconds per turn. Take the torque reading on the 5th turn.

- Turning torque: 53 inch lbs (6 Nm)

c. Check the dust boots for cracks or grease leakage. If the value is not within the specified range, replace the front lower ball joint with a new one.

To install:

15. Install the front lower ball joint to the front lower control arm with the nut. Ensure that the thread and taper are free of oil or other foreign matter. Tighten the nut to 120 ft. lbs. (162 Nm) for RWD, and 92 ft. lbs. (125 Nm) for AWD.

16. Install a new clip to the front lower ball joint. Further tighten the nut up to 60° if the holes for the cotter pin are not aligned.

17. Install the lower control arm.

18. Connect the front stabilizer link assembly.

19. Install the strut assembly, but temporarily tighten the bolts at the lower control arm.

20. Connect the tie rod end to the steering knuckle Tighten the nut to 50 ft. lbs. (65 Nm). Install a new cotter pin. If the holes for the clip are not aligned, tighten the nut up to 60° further.

21. For AWD, install the steering knuckle to the front suspension upper control arm, and tighten it with the castle nut to 64 ft. lbs. (87 Nm). Install a new clip to the steering knuckle. Further tighten the nut up to 60° if the holes for the cotter pin are not aligned.

22. Connect the ABS speed sensor.

23. Stabilize the suspension as follows:

a. Install the front wheels. Tighten the lug nuts to 76 ft. lbs. (103 Nm).

b. Lower the vehicle and bounce it up and down several times to stabilize the front suspension.

c. Remove the front wheels.

d. Jack up the front suspension lower arm placing a wooden block in between. Apply a load to the front suspension so that the front suspension lower arm is placed in a horizontal position.

24. Fully tighten the strut assembly bolts at the lower control arm.

25. Fully tighten the lower control arm.

26. Install engine under cover.

27. Install the front wheel.

28. Inspect and adjust wheel alignment.

LOWER CONTROL ARM

REMOVAL & INSTALLATION

See Figure 222.

1. Before servicing the vehicle, refer to the Precautions section.

2. Raise the vehicle on a hoist, so that front suspension components are hanging and accessible.

3. Remove or disconnect the following:
- Front wheels.
- Engine under covers.
- Brake caliper(s); Do NOT disconnect the brake hose; hang the caliper without stress on the hose
- Tie rod end from steering knuckle
- Stabilizer bar link from stabilizer bar
- Height control sensor link, if equipped, from shock absorber bracket
- Shock absorber lower mount
- Lower control arm set bolts (loosen only)
- Lower ball joint from No. 2 lower control arm (lower suspension arm)
- Steering gear assembly
- Strut bar bracket
- No. 1 lower control arm (lower suspension arm); matchmark adjusting cam to crossmember

To install:

4. To install, reverse the removal procedure, noting the following torque settings:
- No. 1 lower control arm (lower suspension arm) bolt to shock absorber bracket: 44 ft. lbs. (59 Nm)
- No. 1 lower control arm (lower suspension arm) adjusting cam bolt and nut to crossmember: 127 ft. lbs. (172 Nm)
- Strut bar bracket bolts: 43 ft. lbs. (58 Nm)
- Strut bar bracket nut: 112 ft. lbs. (152 Nm)
- No. 2 lower control arm (lower suspension arm) nuts: 122 ft. lbs. (164 Nm)
- Lower shock absorber mounting bolt and nut: 116 ft. lbs. (157 Nm)
- Stabilizer bar link nut: 83 ft. lbs. (113 Nm)
- Stabilizer bar link-to-stabilizer bar bolt and nut: 43 ft. lbs. (55 Nm)
- Tie rod end nut: 64 ft. lbs. (87 Nm)
- Brake caliper bolts: 87 ft. lbs. (118 Nm)
- Front wheel nuts: 76 ft. lbs. (103 Nm)

STEERING KNUCKLE

REMOVAL & INSTALLATION

See Figure 223.

1. Before servicing the vehicle, refer to the Precautions section.

2. Remove the front wheel.

3. Disconnect the front brake caliper assembly. Support the brake caliper securely.

4. Remove the front brake disc

5. Remove the clip and nut, then using SST 09610-20012, disconnect the tie rod end.

6. Remove the cotter pin and nut, then using SST 09628-6201 1, remove the lower ball joint from the lower control arm.

7. Remove the cotter pin and castle nut, then using SST09628_62011, separate the upper ball joint from the steering knuckle.

8. Remove the steering knuckle from the vehicle.

9. If necessary, remove the lower ball joint and hub and bearing assembly from the steering knuckle.

To install:

10. If necessary, install the hub and bearing assembly and lower ball joint to the steering knuckle.

11. Install the steering knuckle into the vehicle.

12. Connect the upper ball joint to the steering knuckle. Install castle nut and new cotter pin. Tighten to 64 ft. lbs. (87 Nm).

13. Connect the lower ball joint to lower control arm. Install castle nut and new cotter pin. Tighten to 64 ft. lbs. (87 Nm).

14. Connect the tie rod end to the steering knuckle. Install castle nut and new cotter pin. Tighten to 48 ft. lbs. (65 Nm).

15. Install the front brake disc.

16. Install the front brake caliper assembly.

17. Install the front wheel.

18. Inspect and adjust wheel alignment.

STRUT

REMOVAL & INSTALLATION

See Figure 224.

The strut removal procedure also includes the separation of the coil spring.

1. Before servicing the vehicle, refer to the Precautions section.

2. Remove or disconnect the following:
- Negative battery cable.
- Front wheel

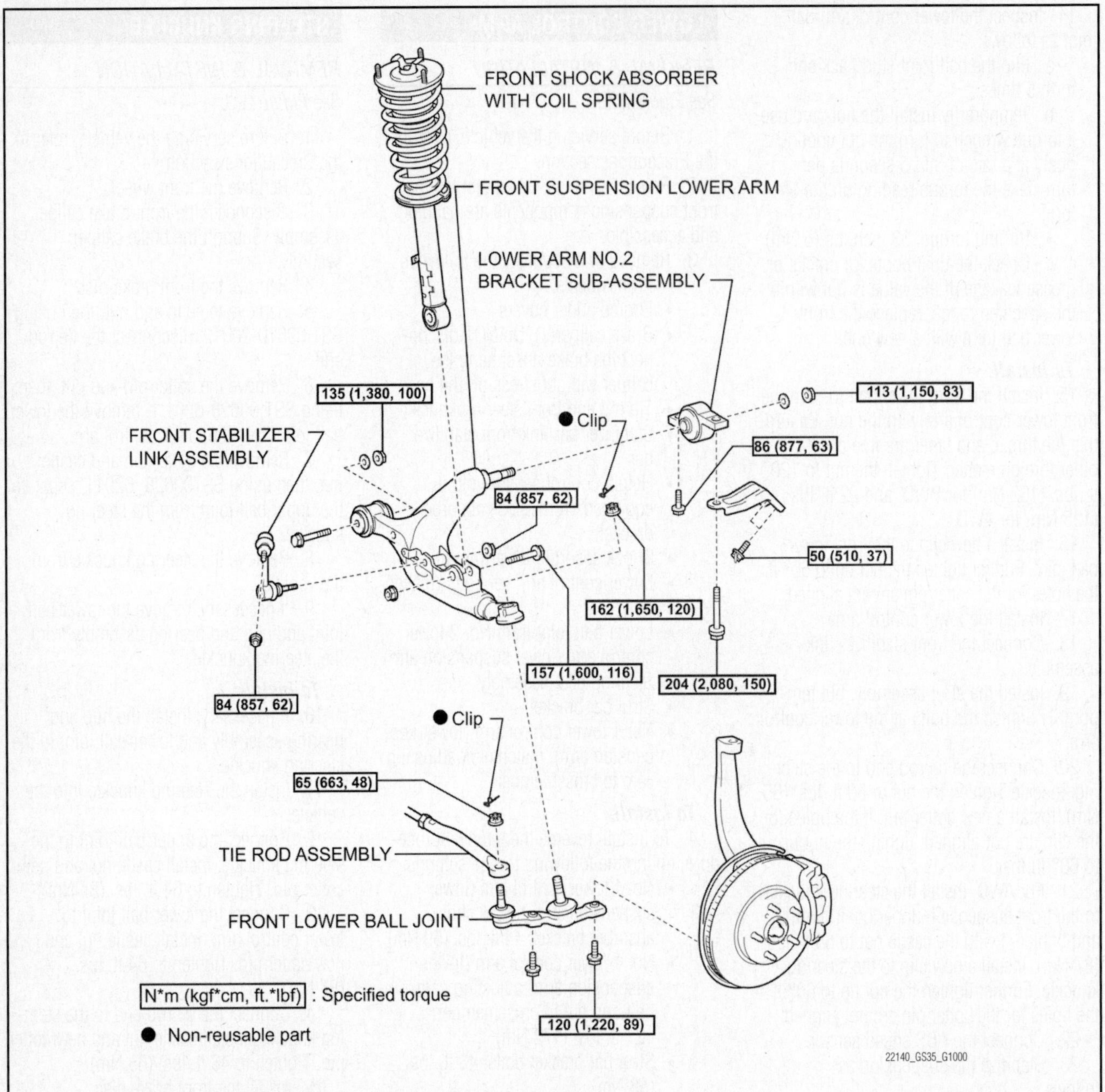

FRONT SHOCK ABSORBER WITH COIL SPRING

FRONT SUSPENSION LOWER ARM

LOWER ARM NO.2 BRACKET SUB-ASSEMBLY

135 (1,380, 100)

113 (1,150, 83)

FRONT STABILIZER LINK ASSEMBLY

● Clip

86 (877, 63)

84 (857, 62)

50 (510, 37)

162 (1,650, 120)

157 (1,600, 116)

204 (2,080, 150)

84 (857, 62)

● Clip

65 (663, 48)

TIE ROD ASSEMBLY

FRONT LOWER BALL JOINT

N*m (kgf*cm, ft.*lbf) : Specified torque

● Non-reusable part

120 (1,220, 89)

22140_GS35_G1000

Fig. 222 Exploded view of the front lower control arm and related components

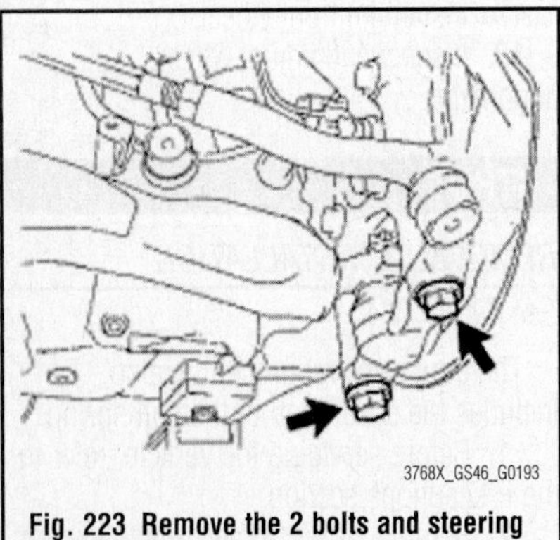

Fig. 223 Remove the 2 bolts and steering knuckle from the lower ball joint—2WD models

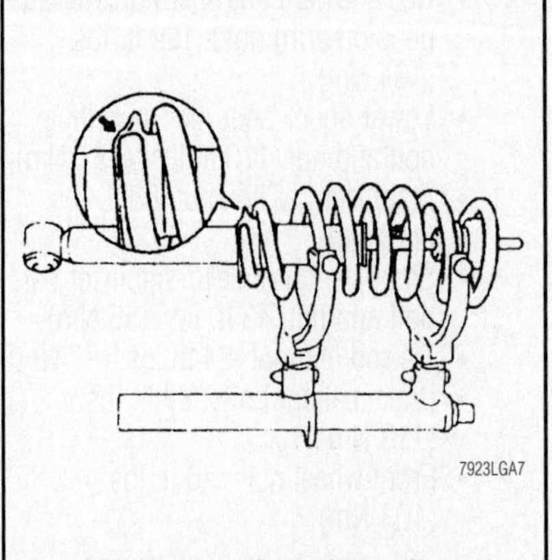

Fig. 224 Matching the spring to the seat

- Brake caliper, leaving the line attached

3. Loosen the 3 upper strut mounting nuts.
4. Loosen, but do not remove, the upper strut rod nut.

⁂ CAUTION

Do NOT remove the upper strut nut at this time.

5. Remove or disconnect the following:
   - Anti-lock Brake System (ABS) speed sensor and harness
   - Upper suspension arm from the steering knuckle

- Stabilizer bar from the link and remove the bracket
- Strut from the lower suspension arm.
- 3 upper strut mounting nuts and remove the strut

6. Compress the coil spring.
7. Remove or disconnect the following:
    - Piston rod locknut
    - Suspension support, coil spring and bumper
8. If disposing the strut, perform the following procedure:
    a. Fully extend the strut rod.
    b. Drill a hole near the bottom of the shock to remove the gas inside.

CAUTION

The gas is harmless, but be careful of chips that may fly up when the gas is released.

To install:

9. Install or connect the following:
    - Spring bumper
    - Coil spring
    - Suspension support to the rod and temporarily install a new nut
10. Turn the suspension support so one of the bolts on the support faces the same direction as shown in the illustration.

➡Align the bolt so a line drawn between the rod and bolt would be at 90°to the direction of the lower bushing.

11. Install or connect the following:
    - Spring compressor
    - Strut and tighten the upper retaining nuts to 41 ft. lbs. (56 Nm)
    - New upper strut rod nut to 20 ft. lbs. (27 Nm)
    - Strut to the lower arm and temporarily tighten the nut and bolt
    - Stabilizer bar bracket and tighten the bolts to 21 ft. lbs. (28 Nm)
    - The stabilizer bar to the link and tighten the bolts to 29 ft. lbs. (39 Nm)
    - Upper suspension arm to the steering knuckle. Tighten the nut to 64 ft. lbs. (87 Nm) and install a new cotter pin.
    - ABS speed sensor and tighten the bolt to 69 inch lbs. (8 Nm)
    - Caliper
    - Wheel
12. Bounce the vehicle several times to stabilize the suspension.
13. Tighten the lower strut bolt and nut to 116 ft. lbs. (157 Nm).
14. Check the front wheel alignment.

STABILIZER BAR

REMOVAL & INSTALLATION

All Wheel Drive

1. Before servicing the vehicle, refer to the Precautions section.
2. Disconnect the negative battery cable. Wait at least 90 seconds before performing any other work.
3. Remove the engine assembly with transmission.
4. Remove the 2 nuts and the left front stabilizer link assembly. Repeat for the right side. If the ball joint turns together with the nut, use a hexagon (6mm) wrench to hold the stud.
5. Remove the 2 bolts and the left front No. 2 stabilizer bracket from the front suspension crossmember. Repeat for the right side.
6. Remove the 2 front No. 1 stabilizer bar bushings from the front stabilizer bar.
7. Remove the front stabilizer bar from the vehicle.

To install:

8. Install the front stabilizer bar to the vehicle. The identification mark must be on the right side of the vehicle when installing the front stabilizer bar.
9. Install the 2 front No. 1 stabilizer bar bushings outside the bush stoppers on the front stabilizer bar as shown in the illustration. Be sure to install the front No. 1 stabilizer bar bushings so that the cutouts face the front of the vehicle.
10. Install the left front No. 2 stabilizer bracket on the vehicle with the 2 bolts. Repeat for the right side. Tighten the bolts to 57 ft. lbs. (78 Nm).
11. Install the left front stabilizer link assembly with the 2 nuts and tighten to 62 ft. lbs. (84 Nm). Repeat for the right side.
12. Install the engine assembly with transmission
13. Connect the negative battery cable.

With Active Stability Control

See Figures 225 through 227.

1. Disconnect cable from negative battery terminal.
2. Remove front wheel.
3. Remove engine under cover.
4. Remove front fender wheel opening moulding RH.
5. Remove front fender liner RH.
6. Disconnect the 2 wire harness clamps from the bracket.
7. Using the procedures below, disconnect the ECU connector.
    a. Release the lever's lock. (*1).
    b. Press the claw and move the lever in the direction of the arrow in the illustration. (*2).
    c. Disconnect the ECU connector. (*3).

➡When disconnecting the connector, do not apply excessive force to the wire harness.

8. Remove the 2 nuts and links from the stabilizer bar.
9. Remove rear engine under covers.
10. Remove the EPS wire harness.
11. Remove the 2 bolts and bracket from the crossmember.
12. Remove front No. 2 stabilizer bracket RH.
    a. Remove the 2 bolts, bracket and control actuator RH.

➡The front active stabilizer control actuator is very heavy. Be careful not to drop it.

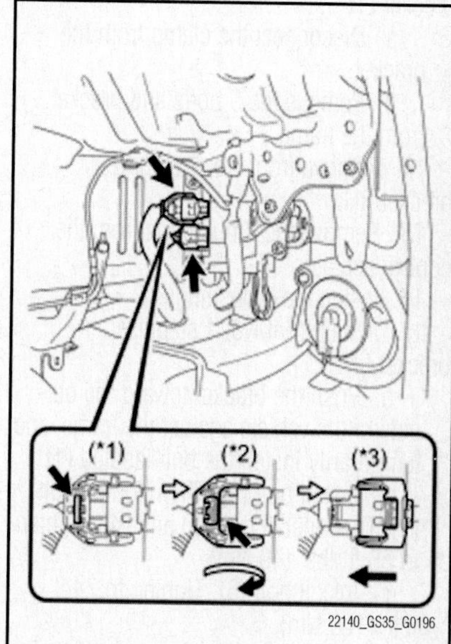

Fig. 225 Stabilizer control ECU connections

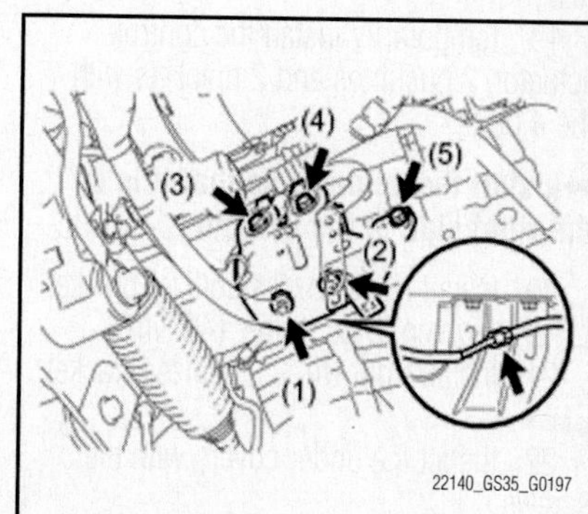

Fig. 226 Stabilizer bracket LH bolt location

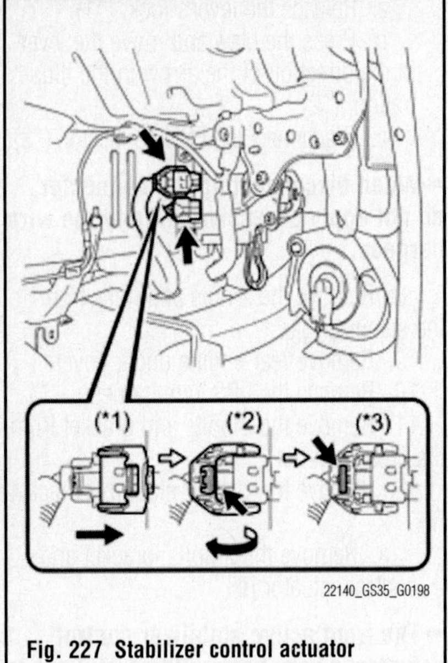

Fig. 227 Stabilizer control actuator

13. Remove front No. 1 stabilizer bracket LH.
  a. Disconnect the clamp from the bracket.
  b. Remove the 5 bolts and bracket from the frame.
14. Remove front No. 1 stabilizer bracket RH.
15. Remove the 2 bushings from the control actuator.
16. Install the 2 bushings.
17. Install front No. 1 stabilizer bracket LH.
  a. Push the bracket toward the outside of the vehicle against the frame, and temporarily install the bolt labeled (1).
  b. Install bolt (2). Then tighten bolt (1). Then install bolt (3) and (4). Tighten to 36 ft. lbs. (49 Nm).
  c. Install bolt (5). Tighten to 74 ft. lbs. (100 Nm).
  d. Install the wire clamp to the bracket.
18. Install front No. 1 stabilizer bracket RH.
19. Temporarily install the control actuator, 2 bushings and 2 brackets with the 4 bolts.

➡Install the actuator so that it is on the right side of the vehicle.

20. Install front No. 2 stabilizer bracket LH and tighten to 36 ft. lbs. (49 Nm).
21. Install front No. 2 stabilizer bracket RH.
22. Install the under covers with the screws.
23. Install the link to the lower arm side and stabilizer bar side with the 2 nuts and tighten to 62 ft. lbs. (84 Nm).
24. Install front stabilizer link assembly RH.
25. Connect front active stabilizer control actuator assembly.
  a. Connect the connector to the front active stabilizer control ECU. (*1).
  b. Rotate the lever in the direction of the arrow until a "click" sound is heard. (*2).
  c. Lock the lever's lock. (*3).
26. Install the wire clamp to the bracket.
27. Install front fender liner RH.
28. Install front fender wheel opening moulding RH.
29. Install engine under cover.
30. Install front wheel.
31. Connect cable to negative battery terminal.
32. Perform initialization.

With RWD Non-Active Stability Control

See Figures 228 through 230.

1. Before servicing the vehicle, refer to the Precautions section.
2. Remove the front wheels.
3. Remove engine under cover.
4. Remove the left and right rear engine under covers.
5. Remove the 2 nuts and the left front stabilizer link assembly. Repeat for the right side. If the ball joint turns together with the nut, use a hexagon (6mm) wrench to hold the stud.
6. Remove the 2 bolts and the left front No. 2 stabilizer bracket from the front suspension crossmember. Repeat for the right side.
7. Remove the 4 bolts and the left front No. 1 stabilizer bracket from the frame. Repeat for the right side.
8. Remove the 2 front No. 1 stabilizer bar bushings from the front stabilizer bar.
9. Remove the front stabilizer bar from the vehicle.

To install:

10. Install the front stabilizer bar to the vehicle. The identification mark must be on the right side of the vehicle when installing the front stabilizer bar.
11. Install the 2 front No. 1 stabilizer bar bushings as shown in the illustration. Be sure to install the front No. 1 stabilizer bar bushings so that the cutouts face the front of the vehicle.
12. Install the left front No. 1 stabilizer bracket as follows:
  a. Temporarily install the left front stabilizer No.1 bracket to the side member with the 4 bolts.
  b. Tighten the bolt (A) while pressing the left front stabilizer No.1 bracket.

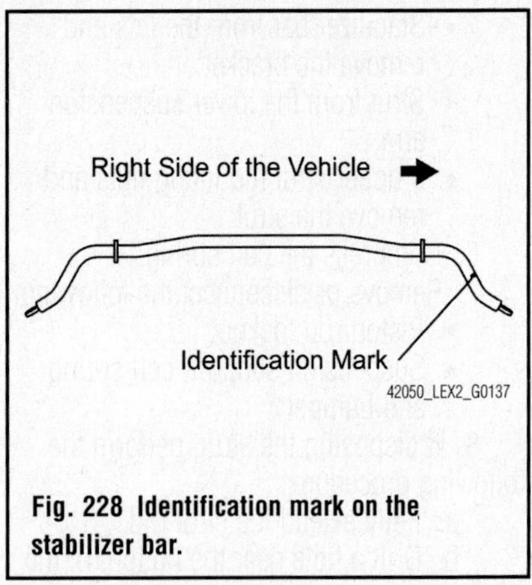

Fig. 228 Identification mark on the stabilizer bar.

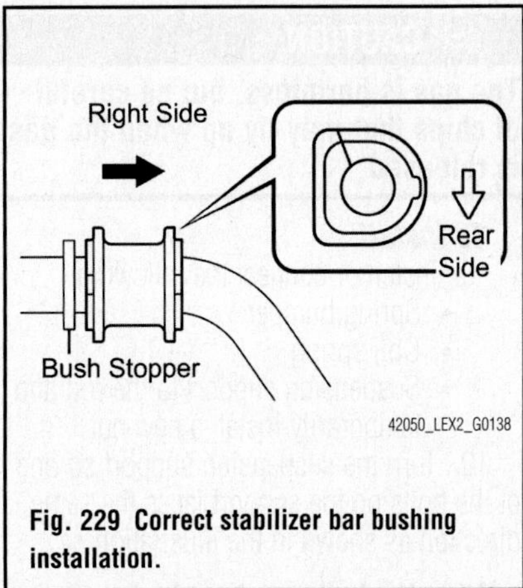

Fig. 229 Correct stabilizer bar bushing installation.

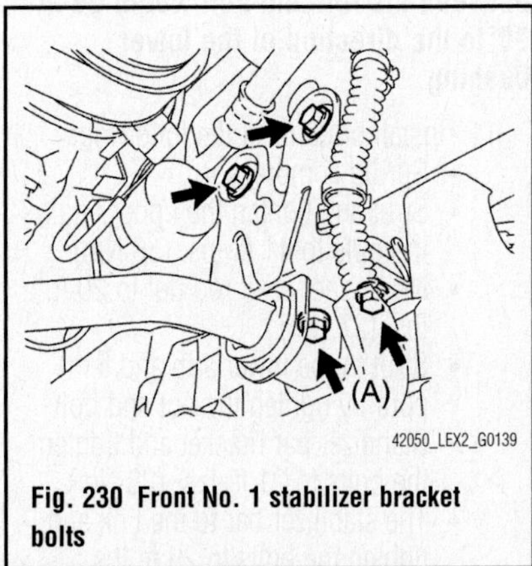

Fig. 230 Front No. 1 stabilizer bracket bolts

  c. Tighten the bolts to 36 ft. lbs. (49 Nm).
  d. Install the right side following the same procedures as for the left side.
13. Install the left front No. 2 stabilizer bracket on the vehicle with the 2 bolts. Repeat for the right side. Tighten the bolts to 36 ft. lbs. (49 Nm).
14. Install the left front stabilizer link assembly with the 2 nuts and tighten to 62 ft. lbs. (84 Nm). Repeat for the right side. If the ball joint turns together with the nut, use a hexagon (6mm) wrench to hold the stud.

15. Install the left and right rear engine under covers.

16. Install engine under cover.

17. Install the front wheels.

18. Inspect and adjust the front wheel alignment.

UPPER BALL JOINT

REMOVAL & INSTALLATION

The upper ball joint is an integral part of the upper arm and is not replaced separately. The upper ball joint replacement is accomplished by replacing the upper arm.

UPPER CONTROL ARM

REMOVAL & INSTALLATION

See Figure 231.

1. Before servicing the vehicle, refer to the Precautions section.
2. Remove or disconnect the following:
   - Negative battery cable
   - Wheel
3. Loosen the 3 upper strut mounting nuts.
4. Loosen, but do not remove, the upper strut rod nut.

⁂ CAUTION

Do NOT completely remove the upper strut nut at this time.

5. Remove or disconnect the following:
   - Brake caliper, leaving the line attached and secure it out of the way
   - Anti-lock Brake System (ABS) speed sensor and harness
   - Cotter pin and nut from the upper control arm
   - Upper control arm from the steering knuckle
   - Stabilizer bar from the link and remove the bracket

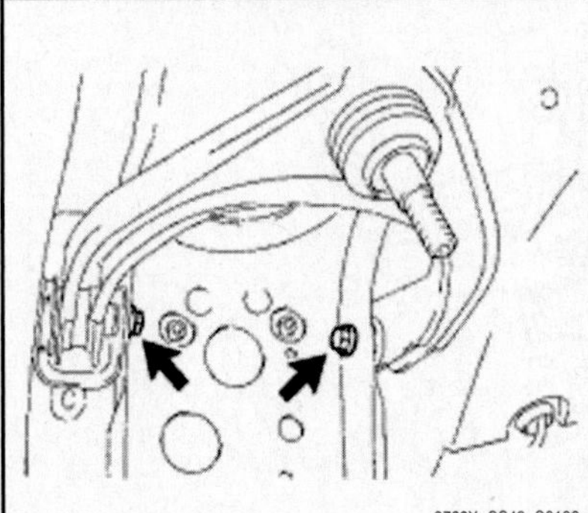

Fig. 231 Remove the 2 bolts and front suspension upper arm

   - Cotter pin and nut from the lower control arm
   - Strut from the lower suspension arm
   - 3 upper strut mounting nuts and remove the strut
   - Mounting bolts holding the upper control arm to the frame
   - Upper control arm from the vehicle

To install:

6. Install or connect the following:
   - Upper suspension arm and tighten the mounting bolts to 39 ft. lbs. (53 Nm)
   - Strut and tighten the upper retaining nuts to 41 ft. lbs. (56 Nm). Tighten the new upper strut rod nut to 20 ft. lbs. (27 Nm).
   - Strut to the lower arm and temporarily tighten the nut and bolt
   - Stabilizer bar bracket and tighten the bolts to 21 ft. lbs. (28 Nm)
   - Stabilizer bar to the link and tighten the bolts to 29 ft. lbs. (39 Nm)
   - Upper suspension arm to the steering knuckle. Tighten the nut to 64 ft. lbs. (87 Nm) and install a new cotter pin.
   - ABS speed sensor and tighten the bolt to 69 inch lbs. (8 Nm)
   - Caliper
   - Front wheel
7. Lower the vehicle.
8. Bounce the vehicle several times to stabilize the suspension.
9. Tighten the lower strut bolt and nut to 116 ft. lbs. (157 Nm).
10. Check the front wheel alignment.

WHEEL HUB & BEARING

REMOVAL & INSTALLATION

See Figure 232.

1. Before servicing the vehicle, refer to the Precautions section.
2. Remove or disconnect the following:
   - Negative battery cable
   - Front wheel
   - Caliper, leaving the brake line connected and suspend it out of the way

⁂ WARNING

Never allow the brake caliper to hang freely from the brake hose.

   - Rotor
   - Anti-lock Brake System (ABS) speed sensor and harness
   - Tie rod from the arm on the lower ball joint

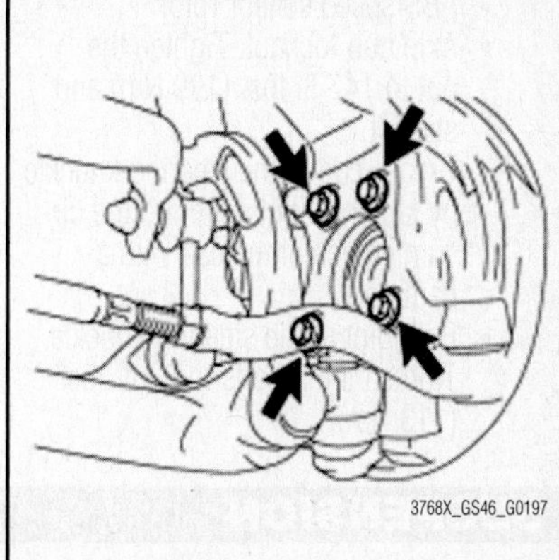

Fig. 232 Remove the 4 bolts, front axle hub sub-assembly and front disc brake dust cover

   - Upper suspension arm from the steering knuckle
   - Steering knuckle from the lower control arm
   - Ball joint from the steering knuckle
   - Front hub grease cap
3. Clamp the hub in a soft jaw vise.
4. Using a hammer and chisel, loosen the staked part of the locknut.
5. Remove or disconnect the following:
   - Locknut
   - ABS speed sensor rotor

➡Do NOT scratch the serrations of the sensor rotor.

   - Brake dust cover bolts and shift the cover toward the outside.
   - Hub from the steering knuckle
   - Inner bearing race from the hub shaft
   - Oil seal from the knuckle
   - Bearing snapring from the steering knuckle
   - Bearing from the steering knuckle

To install:

6. Install or connect the following:
   - New bearing into the steering knuckle

➡If the inner race and balls come loose from the bearing outer race, be sure to install them on the same side as before.

   - Snapring
   - New outside inner race and tap in the new seal. Tap the seal until it is flush with the end surface of the steering knuckle.
   - Brake dust cover to the knuckle and tighten the bolts to 74 inch lbs. (8 Nm)
   - Hub into the steering knuckle

- ABS speed sensor rotor
- Axle hub locknut. Tighten the nut to 147 ft. lbs. (199 Nm) and stake it.
- Grease cap to the steering knuckle by tapping lightly around the circumference of the cap with a hammer
- Ball joint to the steering knuckle. Tighten the 2 bolts to 83 ft. lbs. (113 Nm).
- Steering knuckle to the upper and lower suspension arms. Tighten the upper nut to 64 ft. lbs. (87 Nm) and the lower nut to 95 ft. lbs. (127 Nm). Install a new cotter pin on the lower nut. Install the clip on the upper suspension arm nut.
- Tie rod end to the steering knuckle. Tighten the nut to 64 ft. lbs. (87 Nm) and install a new cotter pin.
- Rotor, disc brake pads and the brake caliper
- ABS speed sensor and harness. Tighten the sensor retaining bolt to 69 inch lbs. (8 Nm).
- Wheel

7. Lower the vehicle and connect the negative battery cable.
8. Check the front wheel alignment.

CONTROL ARMS/LINKS

REMOVAL & INSTALLATION

Upper No. 1 Control Arm

See Figure 233.

1. Remove rear wheel.
2. Remove differential No.2 support protector.
3. Remove the bolt and nut, and separate the rear stabilizer link assembly and the load sensing valve sensor bracket from the rear No.2 suspension arm.
4. Remove the bolt and nut, and separate the rear shock absorber with coil spring from the rear No.2 suspension arm assembly.
5. Remove the bolt, nut washer, and separate the rear upper No.1 control arm from the rear axle carrier.
6. Remove the bolt, nut washer, and the rear upper No.1 control arm from the rear suspension member.

To install:

7. Temporarily install the rear upper No.1 control arm with the bolt, nut and washer.
8. Temporarily install the rear shock absorber with coil spring with the bolt and nut.
9. Temporarily install the rear stabilizer link assembly and the load sensing valve sensor bracket to the rear No.2 suspension arm assembly with the bolt and nut.
10. Jack up the axle carrier, with a wooden block placed between the jack and axle carrier, to apply load to the suspension so that the rear drive shaft assembly becomes level.
11. Fully tighten the nut on the rear No.1 upper control arm assembly to 119 ft. lbs. (161 Nm).
12. Fully tighten the bolt holding the rear shock absorber with coil spring to 81 ft. lbs. (110 Nm).
13. Fully tighten the nut holding the rear stabilizer link assembly to 20 ft. lbs. (27 Nm).
14. Install differential no.2 support protector.
15. Install rear wheel.
16. Inspect and adjust rear wheel alignment.

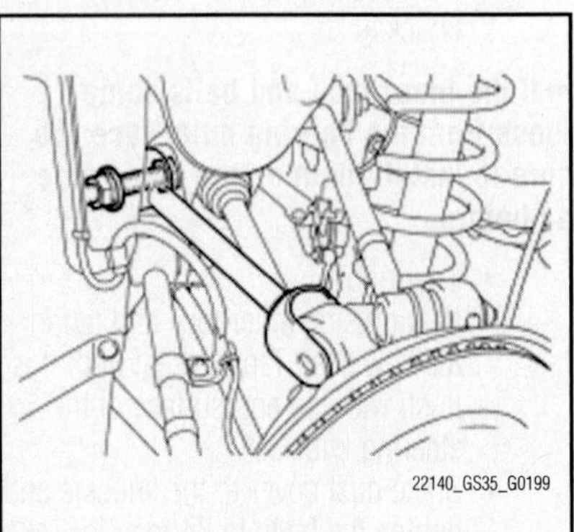

Fig. 233 Installing No. 1 upper control arm

Upper No. 2 Control Arm

See Figure 234.

1. Remove rear wheel.
2. Remove differential No.2 support protector.
3. Separate rear shock absorber with coil spring.
4. Remove rear upper No.2 control arm.
   a. Remove the nut on the axle carrier side.
   b. Using Service Tool, separate the rear upper No.2 control arm.
5. Remove the bolt, nut, washer, and rear upper No.2 control arm.

➡Push the axle carrier downward.

To install:

6. Install the stud of the rear upper No.2 control arm, and temporarily tighten a new nut.
7. Install the rear upper control arm, and temporarily tighten the bolt, nut and washer.
8. Fully tighten the nut of the rear upper No.2 control arm to 52 ft. lbs. (70 Nm).
9. Temporarily tighten rear shock absorber with coil spring.
10. Jack up the axle carrier, with a wooden block placed between the jack and axle carrier, to apply load to the suspension so that the rear drive shaft assembly becomes level.

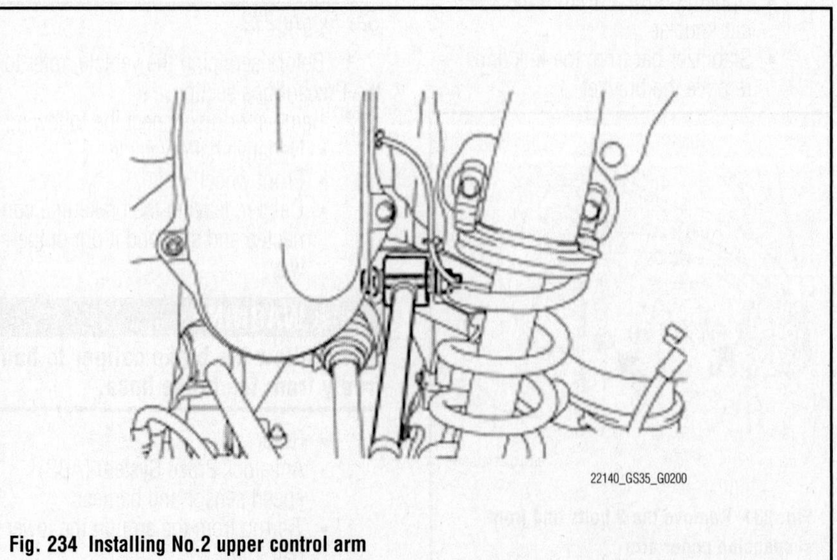

Fig. 234 Installing No.2 upper control arm

11. Fully tighten the nut on the rear suspension member side of the rear upper No.2 control arm to 107 ft. lbs. (145 Nm).

12. Fully tighten rear shock absorber with coil spring.

13. Install differential no.2 support protector.

14. Install rear wheel.

15. Inspect and adjust rear wheel alignment.

SHOCK ABSORBER

REMOVAL & INSTALLATION

1. Before servicing the vehicle, refer to the Precautions section.

2. Remove the corresponding rear wheel(s).

3. Remove the luggage compartment trim front cover.

4. Remove the rear fender apron seal.

5. Remove the rear lower suspension arm:

   a. Remove 2 bolts, nuts and the No. 1 lower suspension arm (trailing arm).

   b. Remove the bolt and nuts and disconnect the stabilizer bar link (and height control link, if equipped) from the No. 2 lower suspension arm.

   c. Remove the bolt and nut and disconnect the strut from the No. 2 lower suspension arm.

   d. Place matchmarks on the adjusting cam and on the No. 2 lower suspension arm.

   e. Remove the nut and adjusting cams.

   f. Remove the bolt, nut and washer and the No. 2 lower suspension arm.

6. Remove the 3 upper retaining nuts. Loosen, but Do NOT remove the center stud nut.

7. Remove the lower strut mounting bolt.

8. Remove the strut assembly, with the coil spring.

9. Compress the coil spring in a suitable spring compressor.

➡Do NOT remove these items with an impact wrench; it will damage the spring compressor.

10. Remove or disconnect the following:

- Suspension support nut
- Washer
- 2 cushions
- Collar
- Suspension support
- Upper insulator
- Lower cup
- Spring bumper

11. Carefully release the spring compressor and remove the coil spring.

12. If the shock absorber is being replaced, fully extend the shock absorber rod and drill a hole to discharge the gas from the cylinder. Drill this hole about 5-8 in. (130-185mm) from lower mounting bolt hole.

To install:

13. With a new shock absorber in place, install the suspension support and compress the coil spring.

14. With a non-impact wrench, install the coil spring to the shock absorber, fitting the lower end of the spring into the recent of the spring seat on the shock absorber.

15. Install the spring bumper, lower cup, cushion, collar, upper insulator, suspension support, cushion and washer onto the shock absorber. Temporarily tighten a new nut.

16. Rotate the suspension support so the rod and one of the bolts on the suspension support are aligned with the lower shock mounting hole such that while the lower mounting bolt hole is in proper position, the 3 studs on the top of the strut assembly will align with the holes in the body.

17. Carefully remove the spring compressor.

18. Install the strut assembly into the vehicle. Torque the 2 bolts on top of the coil spring to 13 ft. lbs. (18 Nm), the 3 nuts on the top mounting studs to 47 ft. lbs. (64 Nm), and the nut in the center of the upper strut mounting to 20 ft. lbs. (27 Nm).

19. Install the bolt, nut and washer and the No. 2 lower suspension arm. Torque bolt to 81 ft. lbs. (110 Nm).

20. Position both adjusting cams, referencing the matchmarks made during removal. Install and torque the retaining nuts to 81 ft. lbs. (110 Nm).

21. Install the strut assembly lower mounting bolt to the No. 2 lower suspension arm. Torque the nut to 81 ft. lbs. (110 Nm).

22. Install the stabilizer bar link (and height control link, if equipped) to the No. 2 lower suspension arm. Torque the bolt and nut to 22 ft. lbs. (30 Nm).

23. Install the No. 1 lower suspension arm (trailing arm) and torque the bolts and nuts to 55 ft. lbs. (75 Nm).

24. Install the rear fender apron seal.

25. Install the luggage compartment trim front cover.

26. Install the rear wheel(s). Torque the wheel nuts to 76 ft. lbs. (103 Nm).

STABILIZER BAR

REMOVAL & INSTALLATION

See Figure 235.

1. Remove rear wheel.

2. Remove the nut, and separate the rear stabilizer link assembly LH and RH from the rear stabilizer bar.

3. Remove the 4 bolts and the rear stabilizer bar.

➡The stabilizer bracket and bush are built onto the stabilizer bar. If the bracket and/or bush detach from the bar, replace the bar.

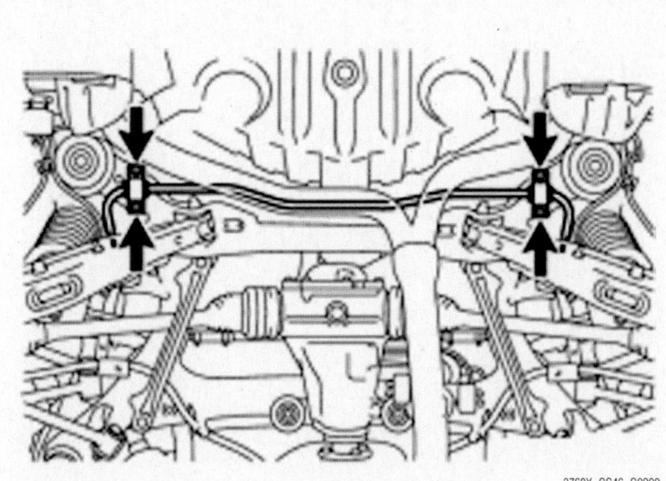

Fig. 235 Remove the 4 bolts and the rear stabilizer bar

4. To install, reverse the removal procedure.

5. Install the rear stabilizer bar with the 4 bolts in a criss cross pattern. Tighten to 24 ft. lbs. (32 Nm).

WHEEL BEARINGS

REMOVAL & INSTALLATION

See Figure 236.

1. Remove rear tire.
2. Separate load sensing valve sensor bracket.
3. Remove rear axle shaft nut.
4. Separate rear disc brake caliper assembly.
5. Remove rear disc.
6. Separate speed sensor rear.
7. Remove parking brake shoe no. 2 return spring.
8. Remove parking brake shoe no. 1 return spring.
9. Remove parking brake shoe adjusting screw set.
10. Remove parking brake shoe assembly no.1.

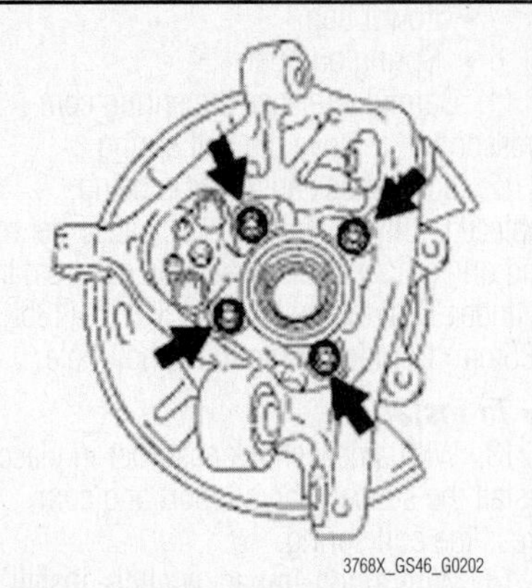

Fig. 236 Remove the 4 bolts and axle hub and bearing assembly

11. Remove parking brake shoe assembly no.2.
12. Install parking brake shoe lever.
13. Remove the 2 nuts and the parking brake anchor block sub-assembly from the rear axle carrier sub-assembly.
14. Separate upper control arm assembly rear no.2.
15. Separate upper control arm assembly rear no.1.
16. Separate rear suspension arm assembly no.1.
17. Separate rear suspension arm assembly no.2.
18. Separate toe control link sub-assembly.
19. Remove rear axle assembly:
    a. Using a plastic hammer, separate the drive shaft from the rear axle carrier sub-assembly.
    b. Remove the rear axle assembly.
20. Using a screwdriver, remove the bearing dust deflector No.2 from the rear axle carrier sub-assembly.
21. Hold the axle hub and bearing assembly in a vise between aluminum plates.
22. Remove the 4 bolts and axle hub and bearing assembly from the rear axle carrier sub-assembly.
23. To install, reverse the removal procedure.
24. Install the axle hub and bearing assembly to the rear axle carrier sub-assembly with the 4 bolts. Tighten to 52 ft. lbs. (70 Nm).

LEXUS

GS450h

25

SPECIFICATIONS AND MAINTENANCE CHARTS

ENGINE AND VEHICLE IDENTIFICATION

Engine							Model Year	
Code ①	Liters (cc)	Cu. In.	Cyl.	Fuel Sys.	Engine Type	Eng. Mfg.	Code ②	Year
2GR-FSE	3.5 (3456)	210.8	6	SFI	DOHC	Lexus	9	2009
							A	2010

SFI: Sequential Fuel Injection

DOHC: Double Overhead Camshaft

① Stamped on the left side of the engine block

② 10th digit of the Vehicle Identification Number (VIN)

3768X_GS45_C0001

GENERAL ENGINE SPECIFICATIONS

Year	Model	Engine Displacement Liters	Engine Series ID	Net Horsepower @ rpm	Net Torque @ rpm (ft. lbs.)	Bore x Stroke (in.)	Com-pression Ratio	Oil Pressure @ rpm
2009	GS450h	3.5	2GR-FSE	292@6400	267@4800	3.70x3.27	11.8:1	36.4@2500
2010	GS450h	3.5	2GR-FSE	292@6400	267@4800	3.70x3.27	11.8:1	36.4@2500

3768X_GS45_C0002

ENGINE TUNE-UP SPECIFICATIONS

Year	Engine Displacement Liters	Engine ID	Spark Plug Gap (in.)	Ignition Timing (deg.)	Fuel Pump (psi)	Idle Speed (rpm)	Valve Clearance	
							Intake	Exhaust
2009	3.5	2GR-FSE	0.039-0.043	8-12B ①	28-85 ②	950-1050	0.0010-0.0024	0.0012-0.0026
2010	3.5	2GR-FSE	0.039-0.043	8-12B ①	28-85 ②	950-1050	0.0010-0.0024	0.0012-0.0026

NOTE: The Vehicle Emission Control Information label often reflects specification changes made during production. The label figures must be used if they differ from those in this chart.

B: Before top dead center

① With terminals TC and CG of DLC3 connected

② For high pressure pump the readings should be 508-653.

3768X_GS45_C0003

CAPACITIES

Year	Model	Engine Displacement Liters	Engine ID	Engine Oil with Filter (qts.)	Automatic Transmission (qts.)	Rear Drive Axle (pts.) ①	Fuel Tank (gal.)	Cooling System (qts.)
2009	GS450h	3.5	2GR-FSE	6.7	5.7	2.7-3.0	17.2	9.8 ②
2010	GS450h	3.5	2GR-FSE	6.7	5.7	2.7-3.0	17.2	9.8 ②

① Synthetic GL-5 (75W-90) or equivalent

②The use of genuine Toyota engine coolant is recommended or similar

ethylene glycol based non-silicate, non-amine, non- nitrite, and non- borate coolant

3768X_GS45_C0005

FLUID SPECIFICATIONS

Year	Model	Engine Displacement Liters	Engine ID/VIN	Engine Oil	Auto. Trans. ①	Drive Axle ②	Power Steering Fluid	Brake Master Cylinder	Engine Coolant ③
2009	GS450h	3.5	2GR-FSE	5W-30	ATF-WS	75W-90	NA	DOT 3	Toyota coolant
2010	GS450h	3.5	2GR-FSE	5W-30	ATF-WS	75W-90	NA	DOT 3	Toyota coolant

NA: Not Applicable

DOT: Department Of Transpotation

① The use of genuine Toyota ATF-WS is recommended

② Synthetic GL-5 (75W-90) or equivalent

③The use of genuine Toyota engine coolant is recommended or similar

ethylene glycol based non-silicate, non-amine, non- nitrite, and non- borate coolant

3768X_GS45_C0004

VALVE SPECIFICATIONS

Year	Engine Displacement Liters	Engine ID	Seat Angle (deg.)	Face Angle (deg.)	Inner Spring free length (in.) ①	Spring Installed Height (in.)	Stem-to-Guide Clearance (in.)		Stem Diameter (in.)	
							Intake	Exhaust	Intake	Exhaust
2009	3.5	2GR-FSE	45	40.5	2.0354	NA	0.0010-0.0024	0.0012-0.0026	0.2154-0.2159	0.2152 0.2157
2010	3.5	2GR-FSE	45	40.5	2.0354	NA	0.0010-0.0024	0.0012-0.0026	0.2154-0.2159	0.2152 0.2157

NA: Not Available

① If the free length is not as specified , replace the inner spring

3768X_GS45_C0006

CAMSHAFT AND BEARING SPECIFICATIONS CHART

All measurements are given in inches.

Year	Engine Displ. Liters	Engine ID/VIN	Journal Dia.	Brg. Oil Clearance	Shaft End-play	Runout	Journal Bore	Lobe Height	
								Intake	Exhaust
2009	3.5	2GR-FSE	①	②	0.0031-0.0051	0.0016	NA	1.7448-1.7487	1.7457-1.7496
2010	3.5	2GR-FSE	①	②	0.0031-0.0051	0.0016	NA	1.7448-1.7487	1.7457-1.7496

N/A: Not Available

① Journal 1: 1.4152-1.4157

All Others: 1.0220-1.0226

② Oil clearance 1: 0.0016-0.0031

All Others: 0.0010-0.0024

3768X_GS45_C0008

CRANKSHAFT AND CONNECTING ROD SPECIFICATIONS

All measurements are given in inches.

Year	Engine Displacement Liters	Engine ID	Crankshaft				Connecting Rod		
			Main Brg. Journal Dia.	Main Brg. Oil Clearance	Shaft End-play	Thrust on No.	Journal Diameter	Oil Clearance	Side Clearance
2009	3.5	2GR-FSE	2.4011-2.4016	①	0.0016-0.0094	2	2.0863-2.0866	0.0015-0.0026	0.0059-0.0118
2010	3.5	2GR-FSE	2.4011-2.4016	①	0.0016-0.0094	2	2.0863-2.0866	0.0015-0.0026	0.0059-0.0118

① Journals 1 and 4: 0.0006 - 0.0013 in.

Journals 2 and 3: 0.0010 - 0.0018 in.

3768X_GS45_C0007

PISTON AND RING SPECIFICATIONS

All measurements are given in inches.

Year	Engine Displ. Liters	Engine ID	Piston Clearance	Ring Gap			Ring Side Clearance		
				Top Comp.	Bottom Comp.	Oil Control	Top Comp.	Bottom Comp.	Oil Control
2009	3.5	2GR-FSE	0.0017-0.0020	0.0091-0.0130	0.01138-0.0177	0.0039-0.0157	0.0008-0.0028	0.0008-0.0024	0.0008-0.0028
2010	3.5	2GR-FSE	0.0017-0.0020	0.0091-0.0130	0.01138-0.0177	0.0039-0.0157	0.0008-0.0028	0.0008-0.0024	0.0008-0.0028

3768X_GS45_C0009

TORQUE SPECIFICATIONS

All readings in ft. lbs.

Year	Engine Displacement Liters	Engine ID	Cylinder Head Bolts	Main Bearing Bolts	Rod Bearing Bolts	Crankshaft Damper Bolts	Flywheel Bolts	Manifold Intake	Manifold Exhaust	Spark Plugs	Oil Pan Drain Plug
2009	3.5	2GR-FSE	①	②	③	184	61	15	15	13	30
2010	3.5	2GR-FSE	①	②	③	184	61	15	15	13	30

① Step 1: 12 point bolts to 27 ft. lbs.
Step 2: 12 point bolts plus 90 degrees
Step 3: 12 point bolts plus 90 degrees
Step 4: Tighten the two remaining bolts to 22 ft. lbs.
② Step 1: 12 point cap bolts to 45 ft. lbs.
Step 2: 12 point cap bolts plus 90 degrees
Step 3: Hex head side bolts to 19 ft. lbs.
③ Step 1: 30 ft. lbs.
Step 2: Plus 90 degrees

3768X_GS45_C0010

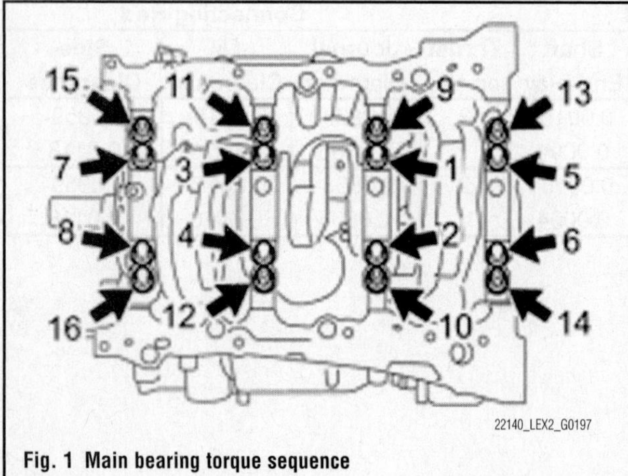

Fig. 1 Main bearing torque sequence

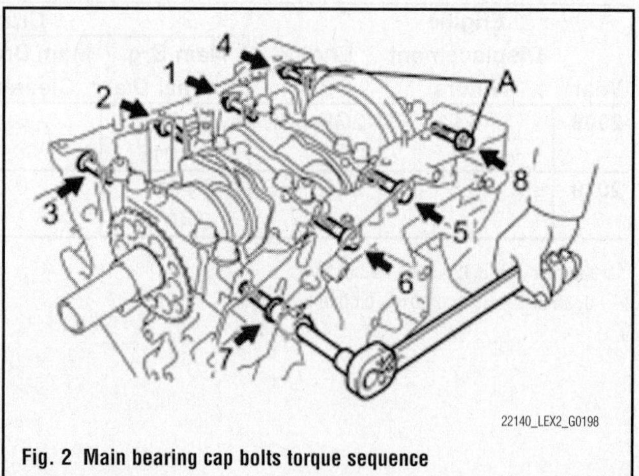

Fig. 2 Main bearing cap bolts torque sequence

WHEEL ALIGNMENT

Year	Model		Caster Range (+/-Deg.)	Caster Preferred Setting (Deg.)	Camber Range (+/-Deg.)	Camber Preferred Setting (Deg.)	Toe-in (in.)
2009	GS450h	F	0.75	+7.43	0.75	-0.43	0.04+/-0.08
		R	N/A	N/A	0.75	-1.32	0.12+/-0.08
2010	GS450h	F	0.75	+7.43	0.75	-0.43	0.04+/-0.08
		R	N/A	N/A	0.75	-1.32	0.12+/-0.08

N/A: Not Applicable

F: Front

R: Rear

3768X_GS45_C0011

TIRE, WHEEL AND BALL JOINT SPECIFICATIONS

Year	Model	OEM Tires Standard	OEM Tires Optional	Tire Pressures (psi) Front	Tire Pressures (psi) Rear	Wheel Size	Ball Joint Inspection	Lug Nut Torque (ft. lbs.)
2009	GS450h	P245/40R18	245/40ZR18	35	35	8-JJ	①	76
2010	GS450h	P245/40R18	245/40ZR18	35	35	8-JJ	①	76

OEM: Original Equipment Manufacturer

PSI: Pounds Per Square Inch

STD: Standard

OPT: Optional

① Replace if any measurable movement is found.

3768X_GS45_C0012

BRAKE SPECIFICATIONS

All measurements in inches unless noted

Year	Model		Brake Disc Original Thickness	Brake Disc Minimum Thickness	Brake Disc Maximum Runout	Minimum Lining Thickness	Brake Caliper Bracket Bolts (ft. lbs.)	Brake Caliper Mounting Bolts (ft. lbs.)
2009	GS450h	F	1.181	1.063	0.0020	0.039	NA	57
		R	0.709	0.650	0.0020	0.039	40	18
2010	GS450h	F	1.181	1.063	0.0020	0.039	NA	57
		R	0.709	0.650	0.0020	0.039	40	18

NA: Not Applicable:

F: Front

R: Rear

3768X_GS45_C0013

SCHEDULED MAINTENANCE INTERVALS

2009-10 Lexus Hybrid GS450h

TO BE SERVICED	TYPE OF SERVICE	VEHICLE MILEAGE INTERVAL (x1000)																	
		5	10	15	20	25	30	35	40	45	50	55	60	65	70	75	80	85	90
Engine oil & filter	R	✓	✓	✓	✓	✓	✓	✓	✓	✓	✓	✓	✓	✓	✓	✓	✓	✓	✓
Auto transmission fluid	S/I												✓						✓
Ball joints & dust covers	S/I	✓	✓	✓	✓	✓	✓	✓	✓	✓	✓	✓	✓	✓	✓	✓	✓	✓	✓
Bolts & nuts on chassis & body	S/I	✓	✓		✓	✓		✓	✓		✓	✓		✓	✓	✓	✓	✓	✓
Brake line pipes & hoses	S/I			✓			✓			✓			✓			✓			✓
Brake fluid	R						✓			✓			✓			✓			✓
Brake pads & discs (front & rear)	S/I	✓	✓	✓	✓	✓	✓	✓	✓	✓	✓	✓	✓	✓	✓	✓	✓	✓	✓
Drive belts	R												✓			✓			✓
Propeller shaft grease	S/I	✓	✓	✓	✓	✓	✓	✓	✓	✓	✓	✓	✓	✓	✓	✓	✓	✓	✓
Steering knuckle & chassis grease	S/I	✓	✓	✓	✓	✓	✓	✓	✓	✓	✓	✓	✓	✓	✓	✓	✓	✓	✓
Steering linkage	S/I	✓	✓	✓	✓	✓	✓	✓	✓	✓	✓	✓	✓	✓	✓	✓	✓	✓	✓
Air cleaner filter	R						✓						✓		✓	✓			✓
Air conditioner filter	R						✓						✓						✓
Spark plugs	R	Replace at 120,000 miles																	
Exhaust pipes & mountings	S/I			✓			✓			✓			✓			✓			✓
Fuel lines & connections	S/I												✓						✓
Engine coolant	R	Replace at 120,000 miles																	
Engine/Inverter coolant	R																		
Rear differential fluid	S/I												✓						✓
Rotate Tires	S/I	✓	✓	✓	✓	✓	✓	✓	✓	✓	✓	✓	✓	✓	✓	✓	✓	✓	✓

R: Replace S/I: Service or Inspect

FREQUENT OPERATION MAINTENANCE (SEVERE SERVICE)

If a vehicle is operated under any of the following conditions it is considered severe service:

- Extremely dusty areas.

- 50% or more of the constant operation is in 32°C (90°F) or higher temperatures, or in temperatures below 0°C (32°F).

- Prolonged idling (vehicle operation in stop and go traffic).

- Frequent short running periods (engine does not warm to normal operating temperatures).

- Police, taxi, delivery usage or trailer towing usage.

Air cleaner filter: service or inspect every 3750 miles

Engine oil & filter: replace every 3750 miles.

Ball joints & dust covers: service or inspect every 7500 miles.

Bolts & nuts on chassis & body: service or inspect every 7500 miles.

Brake pads & discs (front & rear): service or inspect every 7500 miles.

Steering knuckle & chassis grease: service or inspect every 7500 miles.

Steering linkage: service or inspect every 7500 miles.

Exhaust pipes & mountings: service or inspect every 15,000 miles.

3768X_GS45_C0014

BRAKES — INFORMATION AND PRECAUTIONS

ANTI-LOCK SYSTEMS

- Certain components within the ABS system are not intended to be serviced or repaired individually.
- Do not use rubber hoses or other parts not specifically specified for and ABS system. When using repair kits, replace all parts included in the kit. Partial or incorrect repair may lead to functional problems and require the replacement of components.
- Lubricate rubber parts with clean, fresh brake fluid to ease assembly. Do not use shop air to clean parts; damage to rubber components may result.
- Use only DOT 3 brake fluid from an unopened container.
- If any hydraulic component or line is removed or replaced, it may be necessary to bleed the entire system.
- A clean repair area is essential. Always clean the reservoir and cap thoroughly before removing the cap. The slightest amount of dirt in the fluid may plug an orifice and impair the system function. Perform repairs after components have been thoroughly cleaned; use only denatured alcohol to clean components. Do not allow ABS components to come into contact with any substance containing mineral oil; this includes used shop rags.
- The Anti-Lock control unit is a microprocessor similar to other computer units in the vehicle. Ensure that the ignition switch is **OFF** before removing or installing controller harnesses. Avoid static electricity discharge at or near the controller.
- If any arc welding is to be done on the vehicle, the control unit should be unplugged before welding operations begin.

DISC AND DRUM SYSTEMS

CAUTION

Dust and dirt accumulating on brake parts during normal use may contain asbestos fibers from production or aftermarket brake linings. Breathing excessive concentrations of asbestos fibers can cause serious bodily harm. Exercise care when servicing brake parts. Do not sand or grind brake lining unless equipment used is designed to contain the dust residue. Do not clean brake parts with compressed air or by dry brushing. Cleaning should be done by dampening the brake components with a fine mist of water, then wiping the brake components clean with a dampened cloth. Dispose of cloth and all residue containing asbestos fibers in an impermeable container with the appropriate label. Follow practices prescribed by the Occupational Safety and Health Administration (OSHA) and the Environmental Protection Agency (EPA) for the handling, processing, and disposing of dust or debris that may contain asbestos fibers.

BRAKES — BLEEDING THE BRAKE SYSTEM

BLEEDING PROCEDURE

BLEEDING PROCEDURE

See Figure 3.

1. Before servicing the vehicle, refer to the Precautions section.

   If any work is done on the brake system or if air in the brake lines is suspected, bleed the system of air.

➡Do not let brake fluid remain on painted surfaces. Wash it off immediately.

2. Fill the reservoir with brake fluid.

➡If the master cylinder has been disassembled or if the reservoir becomes empty, bleed the air from the master cylinder.

3. Bleed the brake master cylinder as follows:
   - Disconnect the brake lines from the master cylinder.
   - Slowly depress the brake pedal and hold it.
   - Block off the outer holes with your fingers, and release the brake pedal.
   - Repeat the previous 2 steps 3 or 4 times.

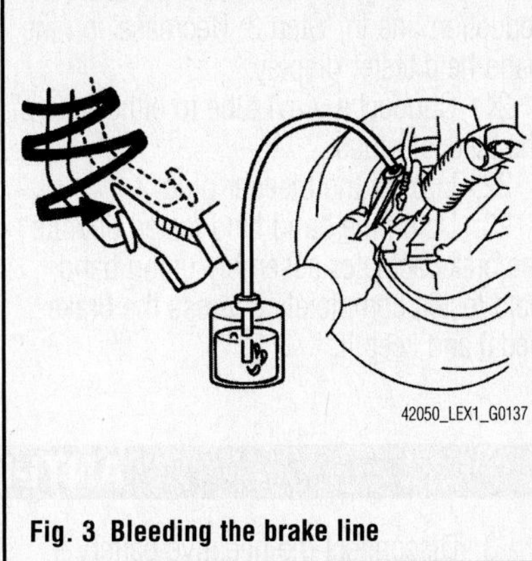

Fig. 3 Bleeding the brake line

4. Bleed the brake line as follows:
   - Connect the vinyl tube to the brake caliper.
   - Depress the brake pedal several times, then loosen the bleeder plug with the pedal held down.
   - At the point when fluid stops coming out, tighten the bleeder plug, then release the brake pedal.
   - Repeat the previous 2 steps until all the air in the fluid has been bled out.
   - Repeat the above procedure to bleed the air out of the brake line for each wheel.
   - Tighten the bleeder plug to 8 ft. lbs. (11 Nm).
5. Bleed the brake actuator as follows:
   - Remove the reservoir cap.
   - Install the SST 09992-00242, 09992-00350 to the reservoir.
   - Connect the vinyl tube to the bleeder plug of the brake actuator.
   - Using SST, apply the 14.2 psi (98.1kpa) of pressure to the reservoir.
   - Loosen the bleeder plug.
   - Bleed the air out of the brake actuator, tighten the bleeder plug to 74 inch lbs. (8.3 Nm).
6. Check the fluid level and add fluid if necessary.

WARNING

Clean, high quality brake fluid is essential to the safe and proper operation of the brake system. You should always buy the highest quality brake fluid that is available. If the brake fluid becomes contaminated, drain and flush the system, then refill the master cylinder with new fluid. Never reuse any brake fluid. Any brake fluid that is removed from the system should be discarded. Also, do not allow any brake fluid to come in contact with a painted surface; it will damage the paint.

BLEEDING THE ABS SYSTEM

➡After performing the usual air bleeding in the brake system, if the height or feel of the brake pedal cannot be obtained, perform air bleeding in the brake actuator assembly with a hand held tester by following procedures below. Make sure that the brake fluid in the master cylinder reservoir tank does not become empty.

1. Before servicing the vehicle, refer to the Precautions section.
2. Depress the brake pedal more than 20 times with the engine off.
3. Connect the hand held tester to the DLC3, then turn the ignition switch to the ON position.
4. Do not start the engine.
5. Select "AIR BLEEDING" on the hand held tester. Please refer to the Hand Held Tester Operator's Manual for further details.
6. Bleed the air out of the regular brake line in "Step1: Increase" on the hand held tester display. Perform the air bleeding by following the steps displayed on the hand held tester. Make sure that the brake fluid in the master cylinder reservoir tank does not become empty.
7. Connect the vinyl tube to either one of the bleeder plugs.
8. Depress the brake pedal several times, then loosen the bleeder plug of one of the above wheels with the pedal depressed.
9. When fluid stops coming out, tighten the bleeder plug, then release the brake pedal.
10. Repeat (2) and (3) until all air in the fluid is completely bled out.
11. Tighten the bleeder plug to 73 inch lbs. (8.3 Nm).
12. Repeat the above procedure to bleed the air out of the brake line for each wheel.
13. Bleed the air out of the suction line in "Step 2: Inhalation" on the hand held tester display.
14. Connect the vinyl tube to the bleeder plug at the right front wheel or the right rear wheel and loosen the bleeder plug.
15. Operate the brake actuator assembly using the hand held tester to bleed the air.

➡The operation stops automatically in 4 seconds. At this time, be sure to release the brake pedal.

16. Check that the operation has stopped, by referring to the hand held tester display.
17. Repeat (2) and (3) until all the air in the fluid is completely bled out.
18. Tighten the bleeder plug.
19. For the rest of the wheels, bleed the air in the same way as stated in the above procedure.
20. Bleed the air out of the pressure reduction line in "Step 3: Decrease" on the hand held tester display.
21. Connect a vinyl tube to either one of the bleeder plugs.
22. Loosen the bleeder plug.
23. Using the hand held tester, operate the brake actuator assembly using hand held tester, completely depress the brake pedal and keep it.

➡The operation stops automatically in 4 seconds. When performing this procedure continuously, an interval of at least 20 seconds is required. When the operation is completed, the brake pedal slightly goes down. This is a normal phenomenon caused when the solenoid opens. During this procedure, the pedal seems heavy, but completely depress it so that the brake fluid comes out from the bleeder plug. Be sure to keep depressing the brake pedal. Never depress and release the pedal repeatedly.

24. Tighten the bleeder plug, then release the brake pedal.
25. Repeat 3 previous steps until all the air in the fluid is completely bled out.
26. Tighten the bleeder plug.
27. Repeat the above procedure to bleed the air out of the brake line for each wheel.
28. Bleed the air out of the regular brake line again in "Step 4: Increase" on the hand-held tester display.
29. Connect the vinyl tube to either one of the bleeder plug.
30. Depress the brake pedal several times, then loosen the bleeder plug of one of the above wheels with the pedal depressed.
31. When fluid stops coming out, tighten the bleeder plug, then release the brake pedal.
32. Repeat the previous 2 steps until all the air in the fluid is completely bled out.
33. Tighten the bleeder plug.
34. Repeat the above procedure to bleed the air out of the brake line for each wheel.

WHEEL SPEED SENSORS

REMOVAL & INSTALLATION

Front Wheel

See Figures 4 through 6.

⁂ WARNING

When the brake caliper is removed from the brake disc, do not start the hybrid system. If the hybrid system is started, the brake fluid pressure may increase.

1. Turn the power switch OFF, and check that the "READY" indicator turns off.
2. Before servicing the vehicle, refer to the Precautions section.
3. Disconnect the negative battery cable.
4. Remove the front wheel.
5. Disconnect the connector from the front speed sensor LH.

➡Be careful not to damage the speed sensor. Prevent foreign matter from adhering to the speed sensor

6. Disconnect the brake caliper assembly.
7. Remove the front brake rotor.
8. Remove the front speed sensor as follows:
   - Install the 3 hub bolts. Placing an aluminum plate below the hub and bearing, hold them in a vise.

➡Replace the hub and bearing if it is dropped or receives a strong shock.

9. Using a pin punch (3 mm) or hammer, remove the 2 pins from the SST (09520-00031) and separate the attachment (09521-00010).

To install:

10. Wipe off sealant attached to the sensor's fitting surface with non-residue solvent.
11. Install a new sensor to the hub and bearing. The sensor connector should be placed in the upper position.
12. Using SST and a press, press in the speed sensor to the hub and bearing.
13. Install the front axle hub sub-assembly

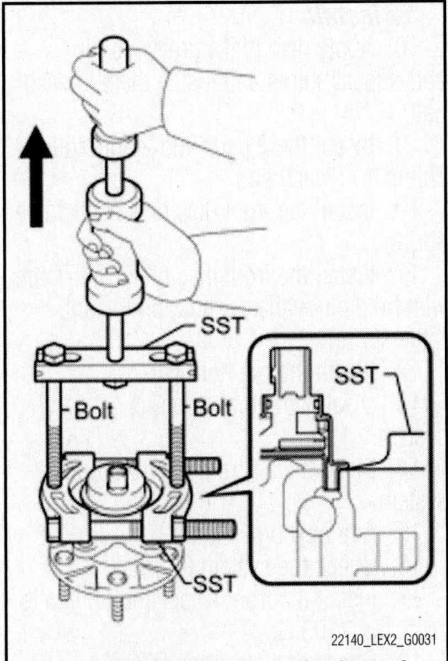

Fig. 4 Removal of the front wheel speed sensor—GS450h

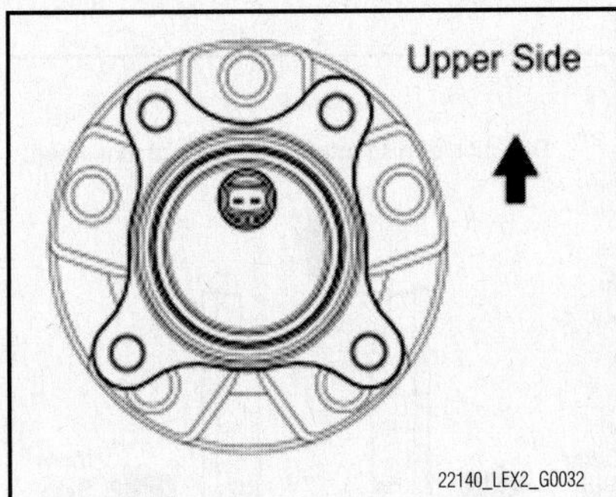

Fig. 5 Installation position of the front brake sensor—GS450h

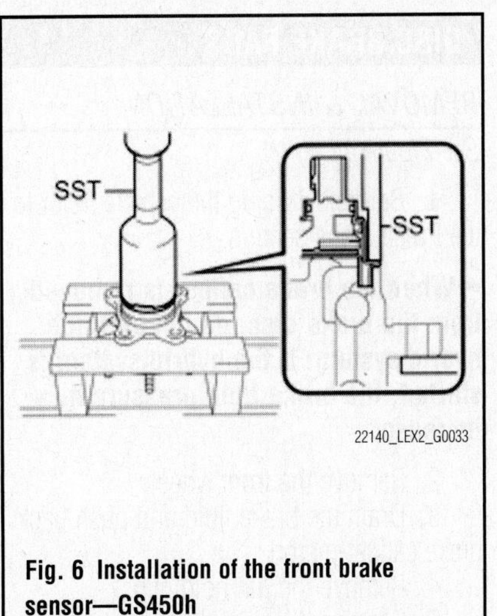

Fig. 6 Installation of the front brake sensor—GS450h

14. Check the front axle hub for looseness. If the run out exceeds 0.0020 inch (0.05 mm) maximum, replace the axle hub.
15. Install the front brake rotor.
16. Install the disc brake caliper with the 2 bolts. Tighten to 29 ft. lbs. (39 Nm).
17. Connect the sensor connector to the front speed sensor
18. Install the front wheel and tighten to 76 ft. lbs. (103 Nm).

Rear

See Figure 7.

WARNING

When the brake caliper is removed from the brake disc, do not start the hybrid system. If the hybrid system is started, the brake fluid pressure may increase.

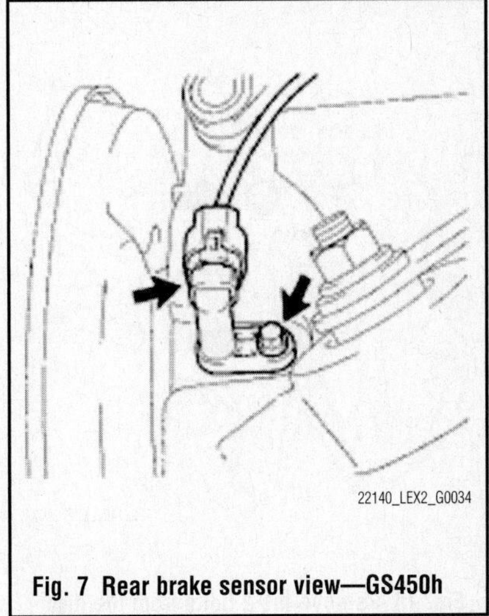
Fig. 7 Rear brake sensor view—GS450h

1. Before servicing the vehicle, refer to the Precautions section.
2. Remove the rear wheel.
3. Disconnect the rear speed sensor connector.
4. Remove the mounting bolt and remove rear speed sensor.

To install:

5. Install the sensor with the bolt and tighten to 87 inch lbs. (8.5 Nm).
6. Reconnect the rear speed sensor connector.
7. Check the speed sensor signal.

BRAKES — FRONT DISC BRAKES

BRAKE CALIPER

REMOVAL & INSTALLATION

See Figures 8 through 10.

1. Before servicing the vehicle, refer to the Precautions section.

➡When the brake caliper is removed from the brake disc, do not start the hybrid system. If the hybrid system is started, the brake fluid pressure may increase.

2. Remove the front wheel.
3. Drain the brake fluid.
4. Remove the pin hold clip.
5. Remove the front disc brake anti-rattle with hole pin while pushing on the anti-rattle spring.
6. Remove the front disc brake anti-rattle spring.
7. Remove the 2 pads and 2 anti-squeal shims from each pad.
8. Remove the union bolt and gasket from the disc brake caliper, then disconnect the flexible hose from the disc brake caliper.
9. Remove the 2 bolts and disc brake caliper from the knuckle.

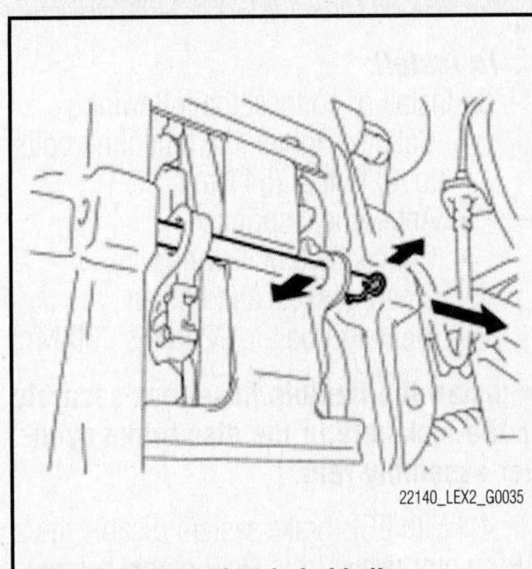
Fig. 8 Remove the pin hold clip

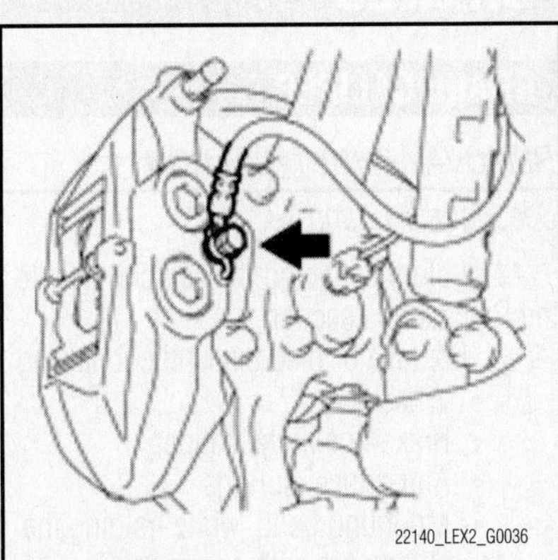
Fig. 9 Remove the flexible hose from the disc brake caliper

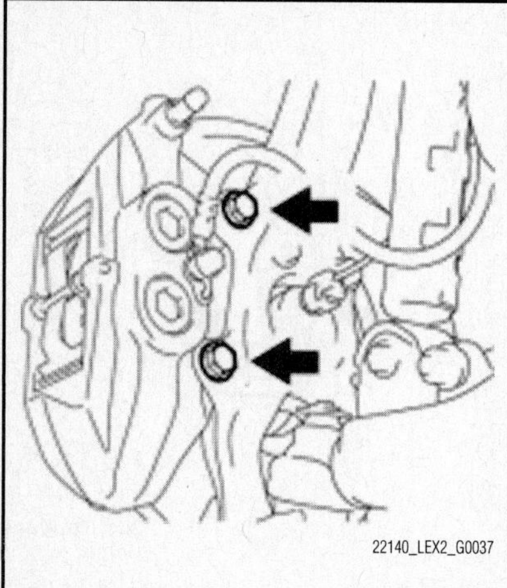

Fig. 10 Remove the 2 bolts from the disc brake caliper

To install:

10. Install the disc brake caliper with the 2 bolts and tighten to 57 ft. lbs. (78 Nm).
11. Connect the flexible hose with the union bolt and a new gasket. Tighten to 29 ft. lbs. (39 Nm).
12. Apply disc brake grease to the anti-squeal shims and install them to each pad.
13. Install the brake anti-rattle spring.
14. Install the hole pin while pushing on the anti-rattle spring.
15. Install the clip.
16. Disable the brake control system.
17. Bleed the air from the front brake system.
18. Add new brake fluid.
19. Check the system for leaks.
20. Install the front wheel and tighten to 76 ft. lbs. (103 Nm).
21. Check and clear any DTC.
22. Test drive the vehicle and burnish in the pads and rotor.

DISC BRAKE PADS

REMOVAL & INSTALLATION

See Figure 11.

1. Before servicing the vehicle, refer to the Precautions section.

➡When the brake caliper is removed from the brake disc, do not start the hybrid system. If the hybrid system is started, the brake fluid pressure may increase.

2. Remove the front wheel.
3. Drain the brake fluid and push back brake caliper piston
4. Remove the pin hold clip.
5. Remove the front disc brake anti-rattle with hole pin while pushing on the anti-rattle spring.
6. Remove the front disc brake anti-rattle spring.
7. Remove the 2 pads and 2 anti-squeal shims from each pad.

To install:

8. Apply disc brake grease to the anti-squeal shims and install them to each pad.
9. Install the 2 pads and 2 anti-squeal shims from each pad.
10. Install the front disc brake anti-rattle spring.
11. Install the front disc brake anti-rattle with hole pin while pushing on the anti-rattle spring.
12. Install the pin hold clip.
13. Disable the brake control system.
14. Bleed the air from the front brake system.
15. Add new brake fluid.
16. Check the system for leaks.
17. Install the front wheel and tighten to 76 ft. lbs. (103 Nm).
18. Check and clear any DTC.
19. Test drive the vehicle and burnish in the pads and rotor.

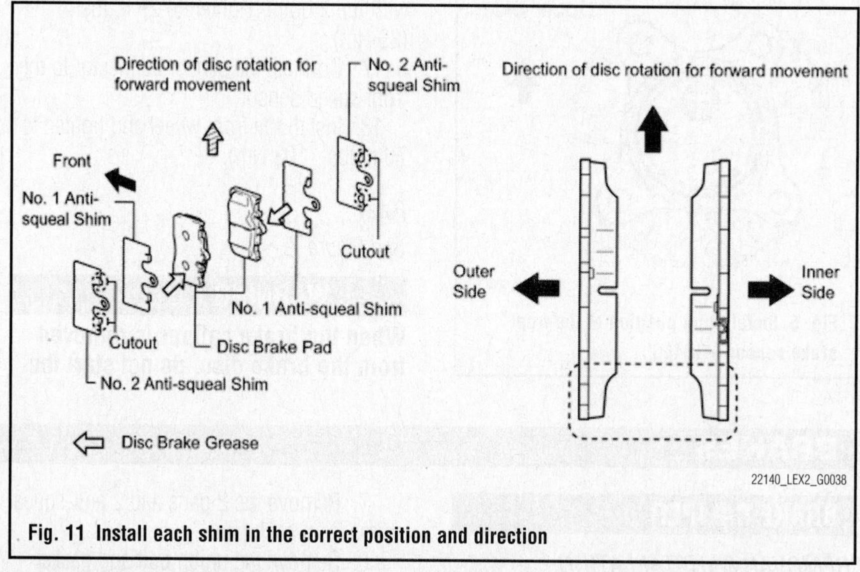

Fig. 11 Install each shim in the correct position and direction

BRAKE CALIPER

REMOVAL & INSTALLATION

See Figures 12 and 13.

1. Before servicing the vehicle, refer to the Precautions section.
2. Remove or disconnect the following:
   - Wheels
   - Brake line at the caliper
   - Anti-squeal springs
   - Mounting bolts, while holding the sliding pin with a wrench
   - Caliper assembly

To install:

3. Install or connect the following:
   - Caliper. Tighten the mounting bolts to 40 ft. lbs. (54 Nm).
   - Anti-squeal springs
   - Connect the brake line with 2 new gaskets and tighten the union bolt to 22 ft. lbs. (30 Nm)

➡Install the flexible hose lock securely in the lock hole in the disc brake cylinder assembly rear.

4. With ECB brake system disable the 2 ABS motor relays from No.3 block. before bleeding the system. Install scanner to verify system is off.
5. Bleed the brake system of any air present.
6. Install relays if previously removed.
7. Check and clear DTC with scanner.
8. Check for fluid leaks.
9. Install the rear wheels and tighten to 76 ft. lbs. (103 Nm).
10. Road test the vehicle and burnish in brakes.

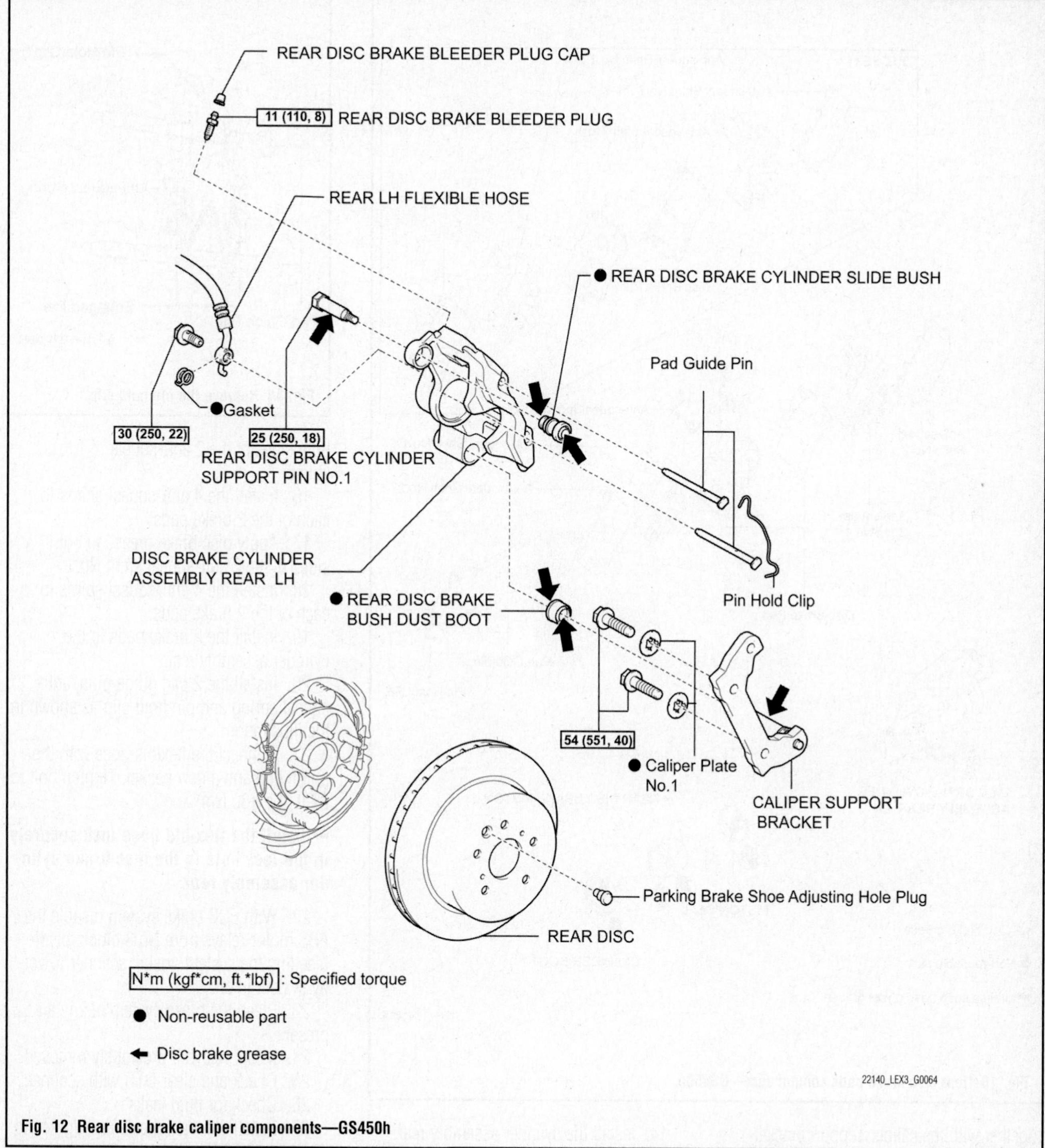

Fig. 12 Rear disc brake caliper components—GS450h

DISC BRAKE PADS

REMOVAL & INSTALLATION

See Figure 14.

1. Before servicing the vehicle, refer to the Precautions section.
2. Rear wheels.
3. Remove the pin hold clip and disengage the engaged parts of anti-squeal spring.
4. Remove the 2 pad guide pins and anti-squeal spring.
5. Remove the 2 brake pads with the anti-squeal shims.
6. Remove the 2 anti-squeal shims from each pad.
7. Remove the union bolt and gasket from the rear disc brake cylinder, and then disconnect the flexible hose.
8. Remove the cylinder slide pin and then tilt the disc brake cylinder toward the rear of the vehicle.
9. Remove the 2 bolts and rear disc brake cylinder together with the caliper support bracket.
10. Remove the 2 No. 1 caliper plate from the caliper support bracket.

To install:

11. Install 2 new caliper plate No.1 to the caliper support bracket.
12. Install the cylinder assembly rear

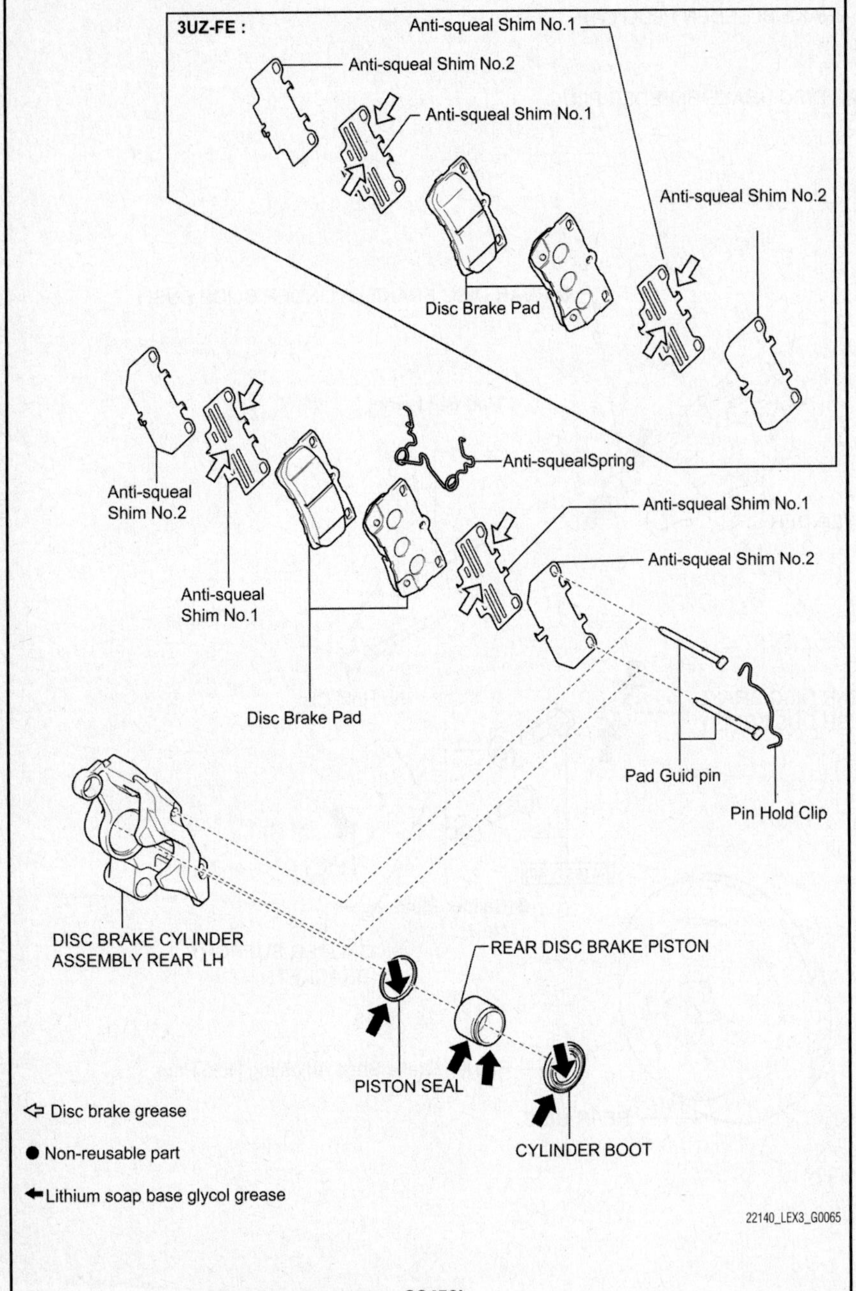

Fig. 13 Rear disc brake pads components—GS450h

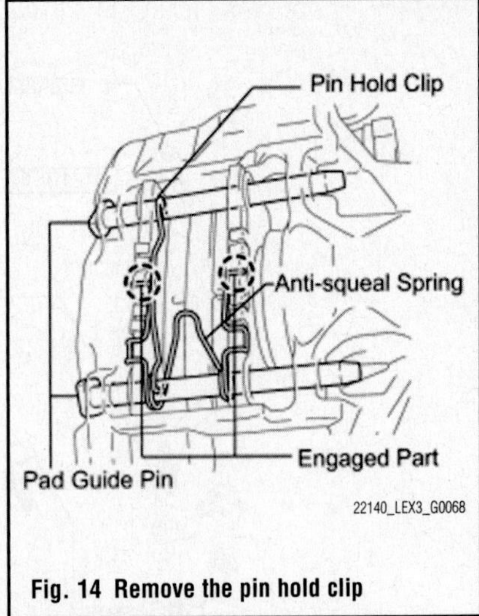

Fig. 14 Remove the pin hold clip

together with the caliper support bracket with the 2 bolts.

13. Tighten the support bolts to 40 ft. lbs. (54 Nm).

14. Install the cylinder assembly rear with the cylinder slide pin. Tighten to 18 ft. lbs. (25 Nm).

15. Apply disc brake grease to both sides of the 2 anti-squeal shims

16. Install the 4 anti-squeal shims to each of the 2 brake pads.

17. Apply disc brake grease to both sides of the 2 anti-squeal shim No.1.

18. Install the 4 anti-squeal shims to each of the 2 brake pads.

19. Install the 2 brake pads to the cylinder assembly rear.

20. Install the 2 pad guide pins, anti-squeal spring and pin hold clip as shown in the illustration.

21. Connect the flexible hose with the union bolt and a new gasket. Tighten bolt to 22 ft. lbs. (30 Nm).

➡Install the flexible hose lock securely in the lock hole in the disc brake cylinder assembly rear.

22. With ECB brake system disable the 2 ABS motor relays from No.3 block. before bleeding the system. Install scanner to verify system is off.

23. Bleed the brake system of any air present.

24. Install relays if previously removed.

25. Check and clear DTC with scanner.

26. Check for fluid leaks.

27. Install the rear wheels and tighten to 76 ft. lbs. (103 Nm).

28. Test drive the vehicle and burnish in the pads and rotor.

BRAKES — PARKING BRAKE

PARKING BRAKE CABLES

ADJUSTMENT

See Figure 15.

1. Depress the parking brake pedal. Hold the No. 1 wire adjusting nut using a wrench and loosen the lock nut.

2. Release the parking brake pedal.

3. Turn the No. 1 wire adjusting nut until the parking brake pedal travel is at 7–9 notches at 67.5 ft, lbs. (300 Nm).

4. Hold the wire adjusting No. 1 nut using a wrench or equivalent tool and tighten the lock nut to 53 inch lbs. (6 Nm).

5. Count the number of clicks after depressing and releasing the parking brake pedal 3 or 4 times.

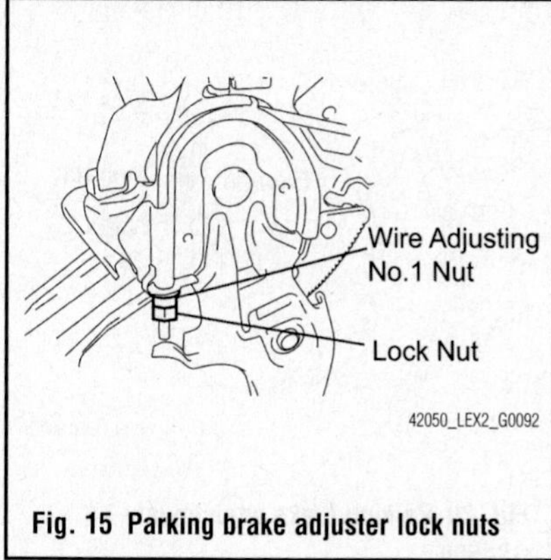

Fig. 15 Parking brake adjuster lock nuts

6. Check whether the parking brake drags or not.
7. When operating the parking brake pedal, check that the parking brake indicator light comes on.

PARKING BRAKE SHOES

REMOVAL & INSTALLATION

See Figures 16 through 19.

1. Remove the rear wheels.
2. Remove the 2 bolts and separate the rear disc brake caliper assembly.

➡Hang the caliper with wire or equivalent.

3. Remove the caliper mounting bolts.
4. Release the parking brake.
5. Put matchmarks on the rear disc and rear axle hub.
6. Remove the rear disc.

➡If the disc cannot be removed easily, turn the shoe adjuster until the wheel turns smoothly.

7. Using SST 09703-30011, remove both parking brake shoe return springs.
8. Slide the parking brake shoe, and remove the parking brake shoe adjusting screw set.
9. Using SST 09718-00010, remove

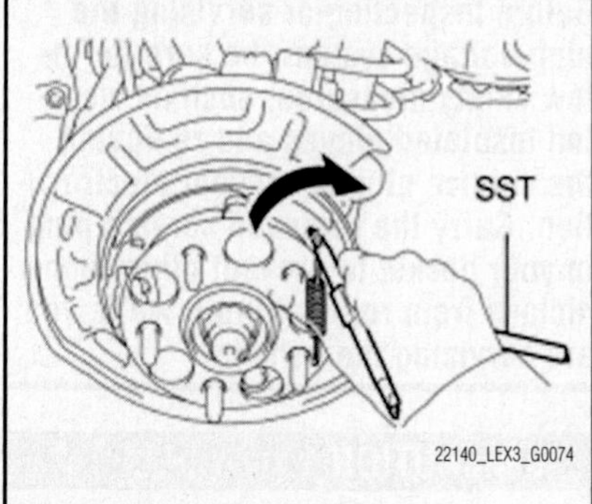

Fig. 16 Parking brake return spring removal

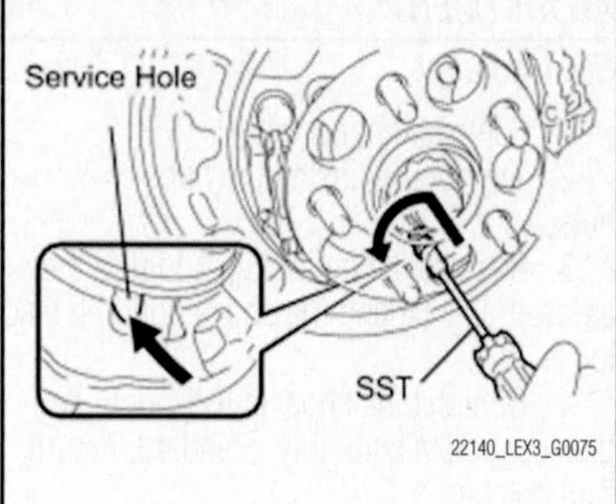

Fig. 17 Removal of parking brake hold down spring

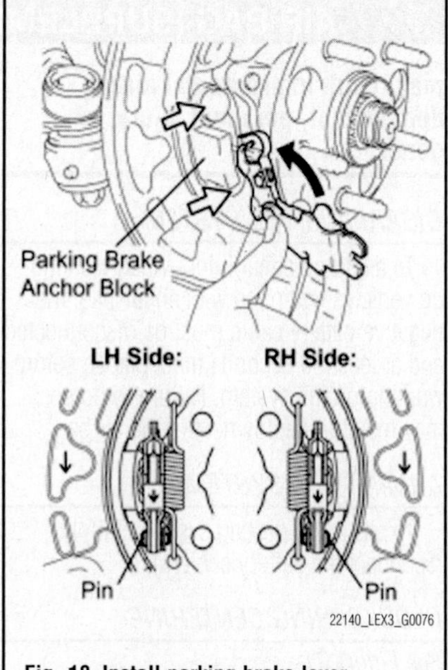

Fig. 18 Install parking brake lever

the shoe hold down spring No.1 cup, compression No.1 spring and shoe hold down spring No.1 pin.

10. Remove the parking brake No.1 shoe assembly.
11. Repeat for No 2 shoe.
12. Remove the parking brake shoe lever.

To install:

13. Apply a thin layer of high temperature grease to the area where the parking brake plate contacts the parking brake shoe
14. Apply a thin layer of high temperature grease to the area where the parking brake shoe lever contacts the parking brake anchor block.
15. Install the parking brake shoe lever to the parking brake cable assembly.

➡Take care to install the correct parking brake shoe lever because the direction of the pin is different between left and right.

16. Using SST, install the parking brake No.2 shoe assembly with the shoe hold down spring No.1 cup, compression No.1 spring and shoe hold down spring No.1 pin.
17. Repeat for No. 2 shoe.
18. Apply high temperature grease to the thread and all joining areas of the parking brake shoe adjusting screw set.
19. Install the parking brake shoe adjusting screw set.

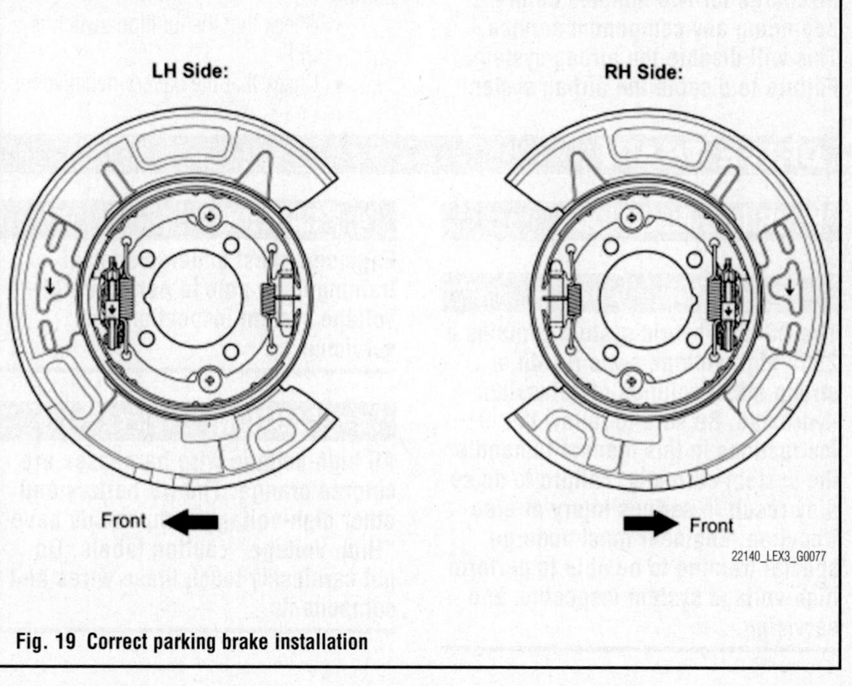

Fig. 19 Correct parking brake installation

20. Install the both parking brake shoe return springs.

21. Make sure that all the parts are installed properly.

22. Align the matchmarks on the rear disc and rear axle hub.

23. Adjust parking brake clearance.

24. Install rear brake caliper and tighten mounting bolts to 40 ft. lbs. (54 Nm).

25. Install the rear wheel. Tighten to 76 ft. lbs. (103 Nm).

26. Inspect parking brake travel and adjust as needed.

ADJUSTMENT

See Figure 20.

1. Remove the rear wheel.
2. Temporarily install the hub nuts.
3. Remove the hole plug, and turn the adjuster and expand the shoes until the disc locks.
4. Contract the shoe adjuster until the disc can rotate smoothly. Standard: Return 7–8 notches
5. Check that shoe is no brake drag.
6. Install the hole plug.

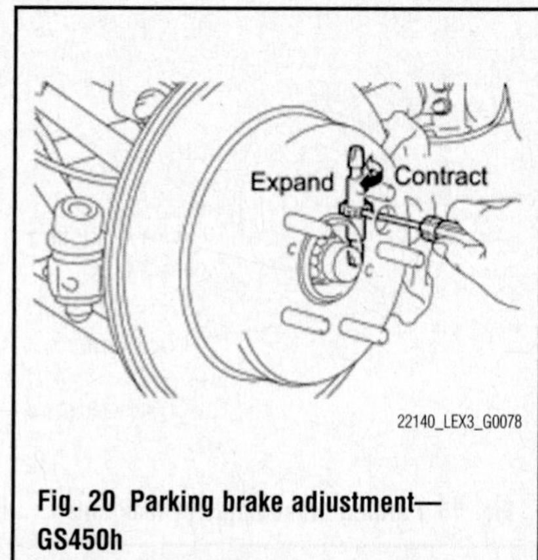

Fig. 20 Parking brake adjustment—GS450h

CHASSIS ELECTRICAL

AIR BAG (SUPPLEMENTAL RESTRAINT SYSTEM)

GENERAL INFORMATION

These vehicles are equipped with an air bag system. The system must be disarmed before performing service on, or around, system components, the steering column, instrument panel components, wiring and sensors. Failure to follow the safety precautions and the disarming procedure could result in accidental air bag deployment, possible injury and unnecessary system repairs.

SERVICE PRECAUTIONS

CAUTION

Disconnect and isolate the battery negative cable before beginning any airbag system component diagnosis, testing, removal, or installation procedures. Allow system capacitor to discharge for two minutes before beginning any component service. This will disable the airbag system. Failure to disable the airbag system may result in accidental airbag deployment, personal injury, or death.

DISARMING THE SYSTEM

To avoid personal injury when working on vehicles equipped with an air bag, the negative battery cable must be disconnected and at least 90 seconds must elapse before working on the system. Failure to do so may result in deployment of the air bag.

ARMING THE SYSTEM

To rearm the air bag system, simply reconnect the battery cable(s).

CLOCKSPRING CENTERING

See Figure 21.

1. Center the clockspring spiral cable as follows:
   - Check that the ignition switch is OFF.
   - Check that the battery negative terminal is disconnected. Do not start the operation for 90 seconds after removing the terminal.
   - Turn the cable counterclockwise by hand until it becomes harder to turn.
   - Then rotate the cable clockwise about 2.5 turns to align the marks. The cable will rotate about 2.5 turns to both left and right of the center.

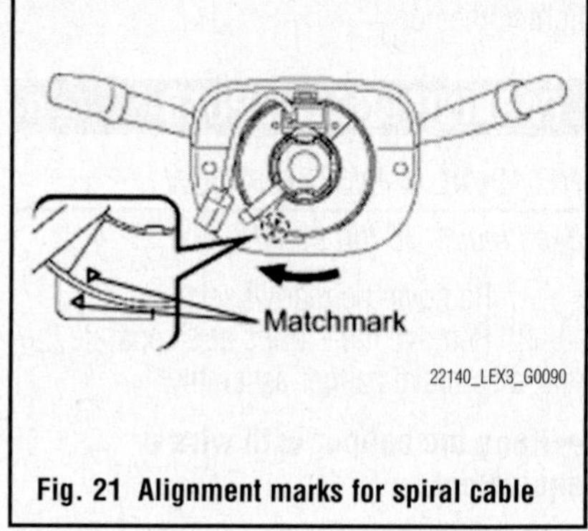

Fig. 21 Alignment marks for spiral cable

DRIVE TRAIN

HYBRID PRECAUTIONS

CAUTION

The GS450h hybrid system contains a 288V high-voltage system with a strong alkali solution of potassium hydroxide. Be sure to follow the instructions in this manual to handle the system correctly. Failure to do so may result in serious injury or electrocution. Engineer must undergo special training to be able to perform high-voltage system inspection and servicing.

CAUTION

Engineers must undergo special training to be able to perform high-voltage system inspection and servicing.

CAUTION

All high-voltage wire harnesses are colored orange. The HV battery and other high-voltage components have "High Voltage" caution labels. Do not carelessly touch these wires and components.

CAUTION

Before inspecting or servicing the high-voltage system, be sure to follow safety measures, such as wearing insulated gloves and removing the service plug to prevent electrocution. Carry the removed service plug in your pocket to prevent other technicians from reinstalling it while you are servicing the vehicle.

CAUTION

After removing the service plug, wait 10 minutes before touching any of

the high-voltage connectors and terminals.

⁂ CAUTION

Be sure to install the service plug before starting the hybrid system. Starting the hybrid system with the service plug removed may damage the vehicle.

⁂ CAUTION

Before wearing insulated gloves, make sure that they are not cracked, ruptured, torn, or damaged in any way. Do not wear wet insulated gloves.

⁂ CAUTION

When servicing the vehicle, do not carry metal objects like mechanical pencils or scales that can be dropped accidentally and cause a short circuit.

⁂ CAUTION

Before touching a bare high-voltage terminal, wear insulated gloves and use an electrical tester to ensure that the terminal is not charged with electricity (approximately 0 V).

1. After disconnecting or exposing a high-voltage connector or terminal, insulate it immediately using insulation tape.

⁂ CAUTION

The screw of a high-voltage terminal should be tightened firmly to the specified torque. Both insufficient and excessive torque can cause failure.

⁂ CAUTION

Use a "CAUTION: HIGH VOLTAGE. DO NOT TOUCH DURING OPERATION." sign to notify other engineers that a high-voltage system is being inspected and/or repaired.

⁂ CAUTION

Do not place the battery upside down while removing and installing it.

⁂ CAUTION

After servicing the high-voltage system and before reinstalling the service plug, check again that you have not left a part or tool inside, that the high-voltage terminal screws are firmly tightened, and that the connectors are correctly connected.

2. The LEXUS GS450h automatically turns the engine ON and OFF when the READY light on the instrument panel is ON. To avoid injury, remove the key from the vehicle before inspecting or servicing the engine compartment.

REAR DIFFERENTIAL CARRIER

REMOVAL & INSTALLATION

See Figure 22.

1. Before servicing the vehicle, refer to the Precautions section.
2. Disconnect the negative battery cable.
3. Remove the rear wheels.
4. Using a 10 mm hexagon wrench, remove the differential drain plug and gasket, and drain the oil.
5. Remove the rear exhaust pipe.
6. Remove the LH and RH drive axles.
7. Remove the propeller with center bearing shaft.
8. Support the differential carrier with a jack
9. Using a 12 mm hexagon wrench, remove the 3 hexagon bolts.
10. Remove the 2 bolts, 2 lower differential mount stoppers and rear differential carrier.

To install:

11. Support the rear differential carrier with a jack.
12. Using a 12 mm hexagon wrench, temporarily install the differential carrier to the suspension member with 3 new bolts.
13. Install the mount stopper upper and mount stopper lower. Temporarily install the differential carrier rear to the suspension member with 2 new bolts.
14. Using a 12 mm hexagon wrench and torque wrench, tighten the 3 bolts to 105 ft. lbs. (102 Nm).
15. Using a torque wrench, tighten 2 bolts to the 70 ft. lbs. (95 Nm).
16. Install the rear drive axles and tighten the new axle shaft nuts to 214 ft. lbs. (290 Nm). Using a chisel and a hammer, stake the axle shaft nut.
17. Install the rear propeller shaft.
18. Install 75W-90 synthetic gear oil.
19. Install the rear wheels.
20. Install the rear exhaust pipe assembly.
21. Inspect and adjust the rear wheel alignment
22. Check for exhaust leaks.
23. Check the ABS speed sensor operation.

REAR DRIVESHAFT

REMOVAL & INSTALLATION

See Figures 23 through 26.

1. Before servicing the vehicle, refer to the Precautions section.
2. Remove the front exhaust pipe.
3. Remove the 4 nuts and front heat insulator.
4. Using SST, loosen the adjusting nut until it can be turned by hand.
5. Put matchmarks on the transmission companion flange, flexible coupling and intermediate shaft.
6. Remove the 3 bolts, 3 washers and 3 nuts.
7. Put matchmarks on the differential companion flange, flexible coupling and propeller shaft.
8. Remove the 2 center support bearing dampers and 2 center support bearing washers. Some vehicles are not equipped with the center support bearing washers.

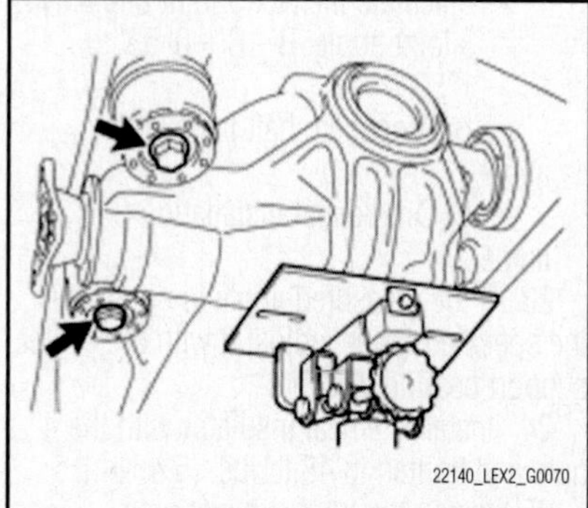

Fig. 22 Rear differential supported for removal

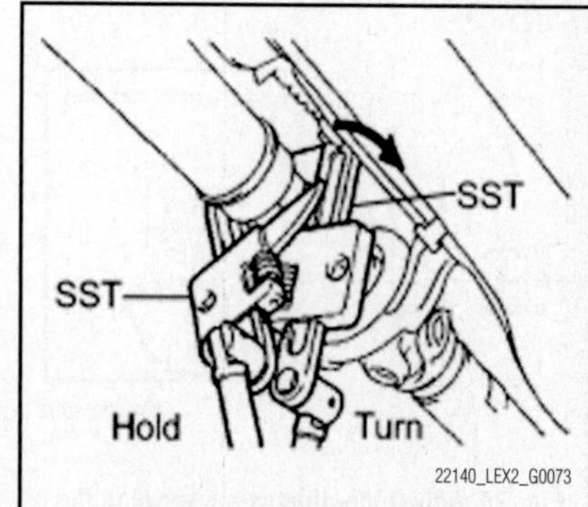

Fig. 23 Using SST, loosen the adjusting nut until it can be turned by hand

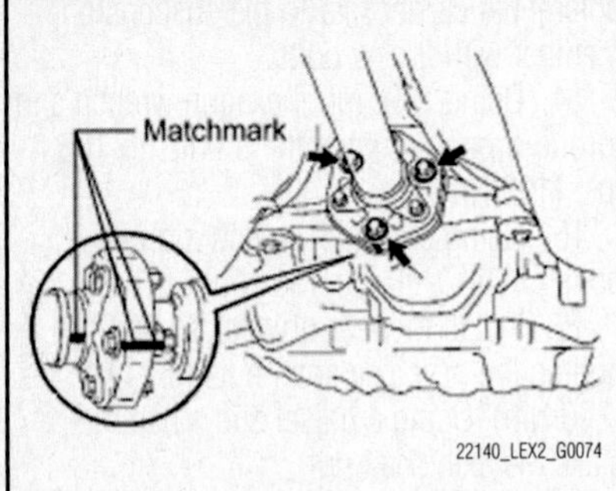

Fig. 24 Put matchmarks on the transmission companion flange, flexible coupling and intermediate shaft.

9. Push the rear propeller shaft straight forward to compress the propeller shaft and pull out the propeller shaft from the centering pin of the differential.

10. If it is difficult to separate the flange from the flexible coupling, pry it using a screwdriver

11. Pull the propeller shaft outward from the vehicle's rear. The intermediate shaft and propeller shaft should not be separated.

To install:

12. Apply grease to the flexible coupling centering bushings

13. Align the matchmarks on the transmission companion flange and flexible coupling.

14. Install and tighten the 6 bolts, washers and nuts to 58 ft. lbs. (79 Nm).

15. Align the matchmarks on the differential companion flange and flexible coupling.

16. Install and tighten the 6 bolts, washers and nuts to 58 ft. lbs. (79 Nm).

17. Temporarily install the 2 center support bearing dampers with the adjusting washers.

18. Adjust the dimension between the edge surface of the center support bearing and the edge surface of the cushion to 0.452 to 0.531 inch (11.5 to 13.5 mm) as shown in illustration.

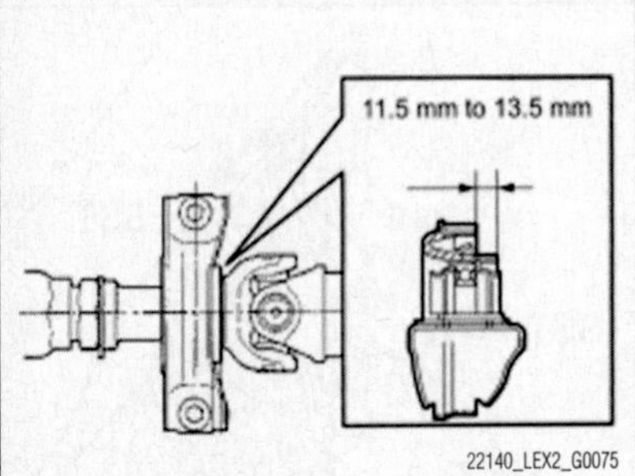

Fig. 25 Adjust the dimension between the edge surface of the center support bearing and the edge surface of the cushion

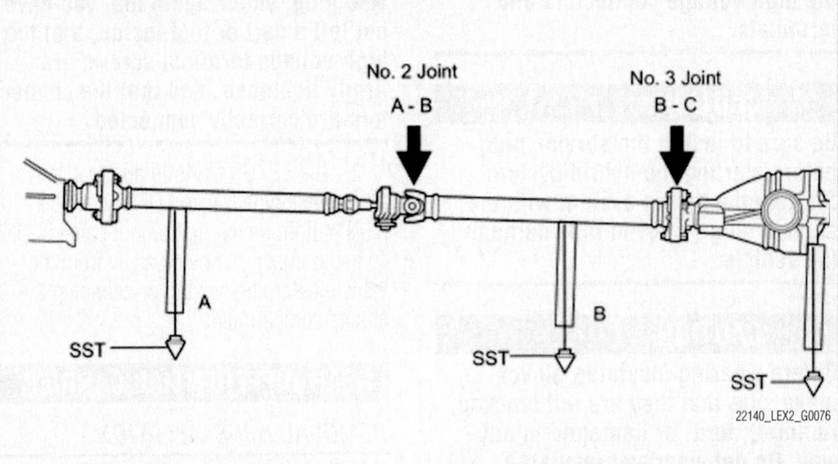

Fig. 26 Measure the installation angle by placing SST in the positions shown in the illustration.

19. Check that the center line of the bracket is at right angles to the shaft axial direction.

20. Tighten the 2 bolts to 36 ft. lbs. (49 Nm).

21. Using SST, tighten the adjusting nut to 51 ft. lbs. (69 Nm).

22. Inspect and adjust No. 2 and No 3 joint angle as follows:

- Stabilize the propeller shaft and differential.
- Turn the propeller shaft several times by hand to stabilize the center support bearing.
- Check the No. 2 and No. 3 joint angles.
- Using SST, measure the installation angle of the intermediate shaft and propeller shaft.
- Using SST, measure the installation angle of the differential.
- (3) Calculate the No. 2 joint angle. No. 2 joint angle: A - B = 0°11' to 1°11'

a. A: Intermediate shaft installation angle

b. B: Propeller shaft installation angle

- Calculate the No. 3 joint angle: No. 3 joint angle: B - C = 0°53' to 1°53'

c. B: Propeller shaft installation angle

d. C: Differential installation angle

23. If the measured angle is not within the specified range, adjust it with the center support bearing washers.

24. Install the heat insulator with the 4 nuts and tighten to 48 ft. lbs. (5.4 Nm)

25. Install the front exhaust pipe assembly.

26. Check for exhaust leaks.

REAR HALFSHAFTS

REMOVAL & INSTALLATION

1. Before servicing the vehicle, refer to the Precautions section.

2. Turn the power switch OFF, and check that the "READY" indicator turns off.

3. Remove the rear wheel.

4. Remove the bolt, nut, load sensing valve sensor bracket and stabilizer link.

5. Remove the 2 nuts and differential support protector from the suspension member brace

6. Remove the 2 bolts and suspension member brace.

7. Remove the 2 bolts and disconnect the parking brake cable.

8. Using SST and a hammer, release the staked part of the axle shaft nut.

9. While depressing the brake pedal, remove the axle shaft nut.

10. Remove the 2 bolts and speed sensor from the axle carrier.

11. Remove the 2 bolts and disconnect the rear disc brake caliper.

12. Remove the caliper plates from the brake caliper.

13. Make a match mark and remove the brake rotor.

14. Remove the rear upper control arm assembly.

15. Remove the rear upper control arm assembly.

16. Remove the rear upper suspension arm assembly.

17. Remove the rear upper suspension arm assembly.

18. Push the rear axle carrier toward the outside of the vehicle. Using a plastic-faced hammer, disconnect the rear drive shaft from the rear axle carrier.

To install:

19. Turn the power switch OFF, and check that the "READY" indicator turns off.
20. Coat the spline of the inboard joint shaft with gear oil.
21. Set the shaft snap ring with the opening side facing down.
22. Align the shaft splines and install the drive shaft with a brass bar and hammer. Be careful not to damage the drive shaft dust cover, boot and oil seal.
23. Install the rear drive shaft to the rear axle carrier.
24. Install the rear upper control arm assembly.
25. Install the rear upper control arm assembly.
26. Install the rear upper suspension arm assembly.
27. Install the rear upper suspension arm assembly.
28. Install the stabilizer link and the load sensing valve sensor bracket to the rear suspension arm with the bolt and nut. Tighten to 20 ft. lbs. (54 Nm).
29. Install the brake rotor.
30. Install the rear brake caliper assembly and tighten the bolts to 40 ft. lbs. (54 Nm).
31. Install the speed sensor to the rear axle carrier. Tighten to 75 inch lbs. (8.5 Nm).
32. Install the rear drive axles and tighten the new axle shaft nuts to 214 ft. lbs. (290 Nm). Using a chisel and a hammer, stake the axle shaft nut.
33. Check and add fluid to differential if needed.
34. Install the parking brake cable with the 2 bolts and tighten to 14 ft. lbs. (19 Nm).
35. Stabilize the rear suspension.
36. Tighten the rear upper control arm assembly to 119 ft. lbs. (161 Nm).
37. Tighten the rear upper control arm assembly to 119 ft. lbs. (161 Nm).
38. Tighten the rear upper suspension arm assembly to 74 ft. lbs. (105 Nm).
39. Tighten the rear upper suspension arm assembly to 74 ft. lbs. (105 Nm).
40. Install the member cover with the 2 screws.
41. Install the No2 support protector.
42. Install the rear wheel and tighten to 76 ft. lbs. (103 Nm).
43. Inspect and adjust the rear wheel alignment.
44. Inspect and adjust the headlights.

REAR PINION SEAL

REMOVAL & INSTALLATION

See Figures 27 and 28.

1. Before servicing the vehicle, refer to the Precautions section.
2. Drain the gear oil.
3. Remove the rear propeller shaft.
4. Using a torque wrench, measure the total preload and record.
5. Using a punch and a hammer, un-stake the staked part of the drive pinion nut.
6. Using SST 09330-00021 to hold the flange, remove the drive pinion nut.
7. With a puller remove the rear pinion flange.
8. Remove and replace rear dust deflector if it is damaged.

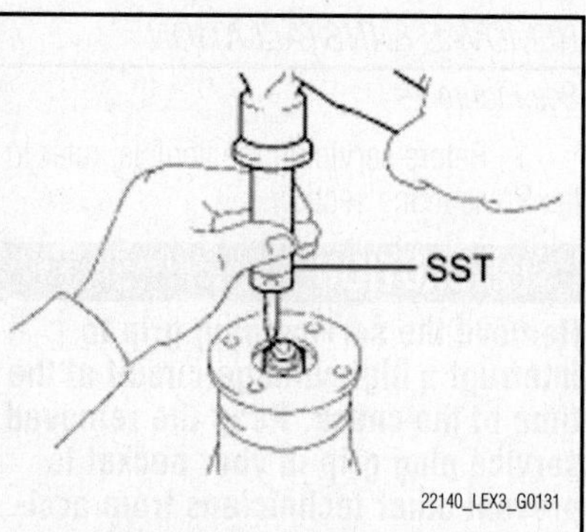

Fig. 27 Un-stake the staked part of the drive pinion nut

9. Carefully remove the oil seal from the differential carrier.

To install:

10. Install the companion flange with a flange installer.
11. Coat the threads of the drive pinion nut with hypoid gear oil LSD.
12. Using SST 09330-00021 to hold the flange, tighten the drive pinion nut.
13. Tighten the nut approximately 1,000 kgf cm, 72 ft. lbs. (98 Nm), and tighten it further while checking the preload.
14. Turn the bearing clockwise and counterclockwise several times to stabilize it
15. Using a torque wrench, measure the preload.
16. Drive pinion preload (at starting):
    a. New bearing 10 to 15.0 inch lbs. (1.21 to 1.70 Nm)
    b. Reused bearing 3.5 to 6.2 inch lbs. (0.40 to 0.70 Nm)

➡**Record the preload for total preload measurement.**

17. If the preload is not within the specified range, adjust the differential drive pinion preload or repair as necessary.
18. Fill the differential with synthetic gear oil GL-5 75W-90 or equivalent.

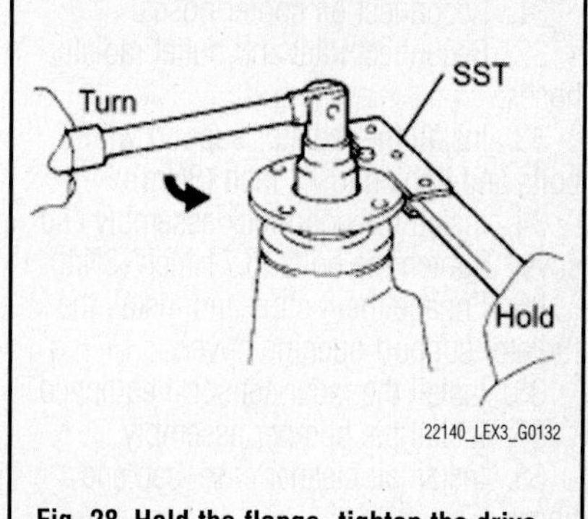

Fig. 28 Hold the flange, tighten the drive pinion nut

ENGINE COOLING

ENGINE FAN

REMOVAL & INSTALLATION

Refer to radiator removal procedure.

RADIATOR

REMOVAL & INSTALLATION

1. Before servicing the vehicle, refer to the Precautions section.
2. Disconnect the negative battery cable.
3. Remove the engine under cover.
4. Drain the engine coolant.
5. Remove the engine V-Bank cover
6. Remove the cool air intake duct seal.
7. Remove the RH and LH engine room side covers.
8. Remove the No 1 air cleaner inlet.
9. Disconnect the inlet and outlet radiator hoses.
10. Disconnect the 2 oil cooler hoses and hose clamp.
11. Remove air cleaner cap with hose.
12. Remove the air cleaner case.
13. Remove the hybrid oil pump motor controller.
14. Remove the hybrid vehicle ECU.
15. Remove the front bumper cover.

16. Remove the radar sensor if equipped.

17. Remove the hood lock control cable cover.

18. Remove the hood lock assembly.

19. Remove the radiator upper support assembly.

20. Disconnect the harness from the cooling fan ECU.

21. Remove the 4 bolts and radiator.

22. Detach the 3 claws and remove the fan.

23. Remove the 2 radiator support cushions and 2 radiator lower supports from the radiator.

To install:

24. Install the 2 radiator support cushions and 2 radiator lower supports to the radiator.

25. Install the fan to the radiator, and attach the 3 claws.

26. Install the radiator with the 4 bolts and tighten to 44 inch lbs. (5 Nm).

27. Connect the cooling fan ECU connector.

28. Install the oil pump motor controller with the 3 bolts and tighten to 53 inch lbs. (6 Nm).

29. Install the oil pump motor controller with bracket and tighten the 2 bolts to 44 inch lbs. (5 Nm).

30. Install the hybrid vehicle ECU.

31. Reconnect oil cooler hoses.

32. Reconnect inlet and outlet radiator hoses.

33. Install the radiator support with 5 bolts and tighten to 71 inch (8Nm).

34. Install the hood lock assembly and cover. Tighten the bolts to 71 inch (8Nm).

35. Engage the 4 clips and install the radiator support opening cover.

36. Install the radar sensor if equipped.

37. Install the bumper assembly.

38. Install air cleaner case, cap and hoses.

39. Install engine coolant and bleed the cooling system.

40. Inspect for engine coolant leaks.

41. Install the engine cover and side covers.

THERMOSTAT

REMOVAL & INSTALLATION

1. Before servicing the vehicle, refer to the Precautions section.

2. Disconnect the negative battery cable.

3. Remove the engine under cover.

4. Remove the cool air intake duct seal.

5. Remove the bolts and air cleaner inlet.

6. Drain the engine coolant.

7. Remove the engine V-bank cover.

8. Disconnect radiator hose outlet.

9. Remove the 3 nuts, water inlet with thermostat and gasket.

To install:

10. Install a new gasket and the water inlet with thermostat with the 3 nuts.

11. Tighten the nuts to 7 ft. lbs. (10 Nm).

12. Reconnect the radiator hose outlet.

13. Connect the negative battery cable.

14. Install engine coolant and bleed the cooling system.

15. Inspect for engine coolant leaks.

16. Install the engine cover and side covers.

17. Install air cleaner inlet.

18. Install the air cleaner inlet with the bolts and tighten to 44 inch lbs. (4 Nm).

19. Install engine under cover.

WATER PUMP

REMOVAL & INSTALLATION

See Figure 29.

1. Before servicing the vehicle, refer to the Precautions section.

⁂ CAUTION

Remove the service plug grip to interrupt a high voltage circuit at the time of the check. Keep the removed service plug grip in your pocket to prevent other technicians from accidentally reconnecting it while you are servicing the vehicle. All the high voltage wiring connectors are colored in orange.

2. Disconnect the negative battery cable.

3. Wear insulated gloves. Remove the service plug grip after sliding the lever of the service plug grip.

4. Remove the engine under cover.

5. Remove the cool air intake duct seal.

6. Remove the bolts and air cleaner inlet.

7. Drain the engine coolant.

8. Disconnect the radiator inlet and outlet hoses.

9. Rotate the tensioner pulley counterclockwise to loosen the belt tension. Then remove the V belt.

10. Remove the 3 nuts and engine harness cover (LH bank side).

11. Remove the 3 nuts and cover.

12. Disconnect the 4 injector driver connectors.

13. Move the lock lever in the direction indicated by the arrow to release the connector lock. Disconnect the 3 connectors with wire harness locks and the connector from the injector driver.

14. Remove the bolt, 2 nuts and injector driver.

15. Disconnect the 5 hoses.

16. Remove the 4 bolts, nut and water inlet.

17. Remove the water inlet housing gasket and water outlet pipe's O-ring.

18. Remove the compressor with motor assembly. Carefully set aside.

19. Remove the V-belt tensioner.

20. Remove the 4 bolts and water pump pulley.

21. Remove the 16 bolts, water pump and gasket.

To install:

22. Install a new gasket and the water pump with the 16 bolts.

23. Tighten the bolts as follows:

- Tighten bolt A to 15 ft. lbs. (21 Nm).
- Tighten bolt B and C to 81 inch lbs. (9 Nm).

24. Be sure to replace the 2 bolts labeled C with new ones or reuse them after applying adhesive.

25. Temporarily install the pulley with the 4 bolts.

26. Using SST, hold the pulley and tighten the 4 bolts. Tighten to 15 ft. lbs. (21 Nm).

27. Install a new gasket to the water inlet housing.

28. Install a new O-ring to the water outlet pipe.

29. Install the water inlet with the 4 bolts and nut. Tighten to 15 ft. lbs. (21 Nm).

30. Connect the 5 hoses.

31. Install the compressor with motor assembly. Tighten mounting bolts to 32 ft. lbs. (43 Nm).

32. Install the belt tensioner assembly.

33. Install the injector driver and tighten mounting nuts and bolt to 7 ft. lbs. (10 Nm).

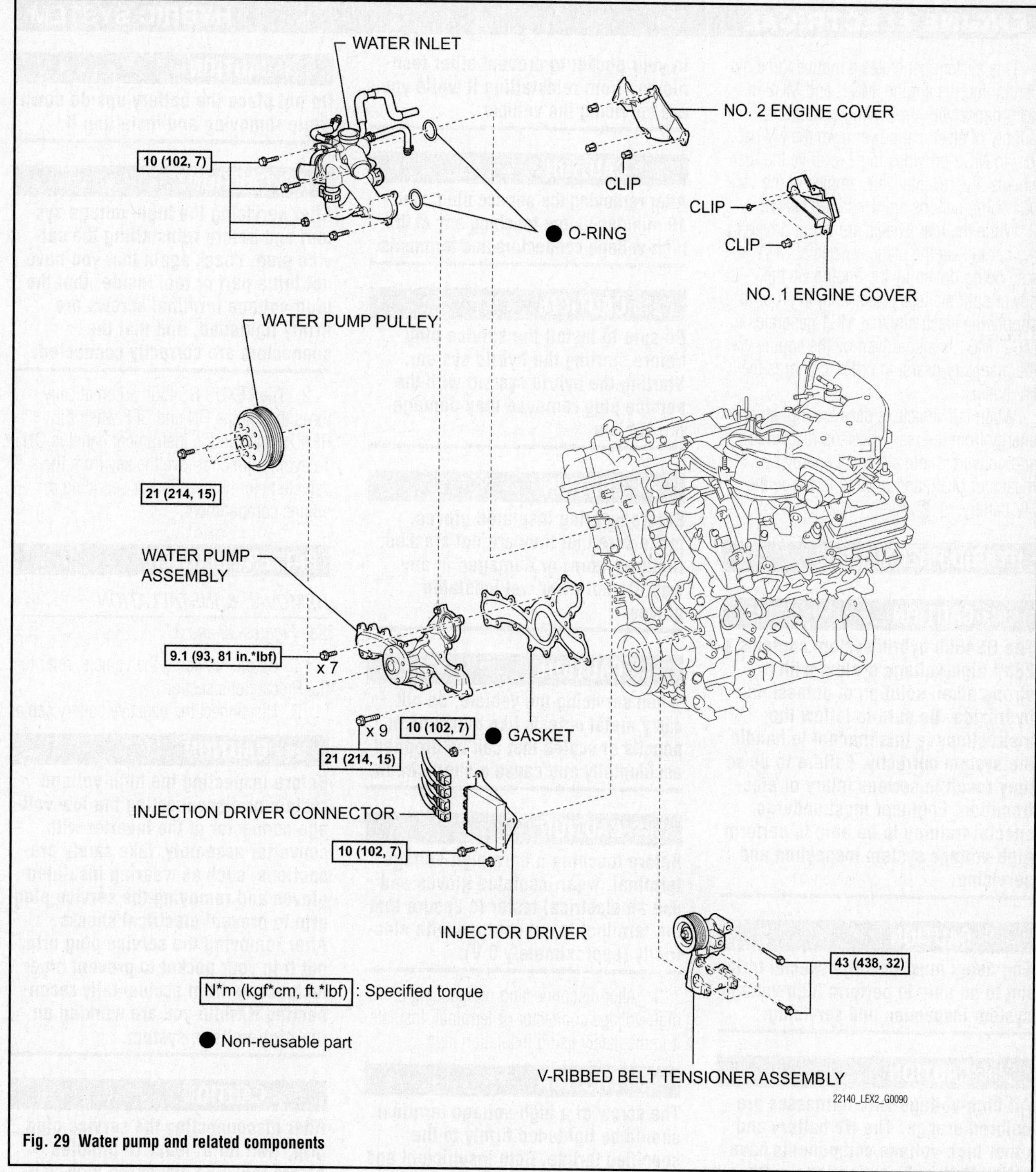

Fig. 29 Water pump and related components

34. Connect the 4 injector driver connectors.
35. Install the No 1 and 2 engine covers.
36. Install the engine wire harness cover with the 3 nuts and tighten to 7 ft. lbs. (10 Nm).
37. Rotate the tensioner pulley counterclockwise, and then install the V-belt.
38. Reconnect radiator inlet and outlet hoses.
39. Install the No 1 air cleaner inlet.
40. Connect the negative battery cable.
41. Install the service plug.
42. Install engine coolant and bleed the cooling system.
43. Inspect for engine coolant leaks.
44. Install the engine cover and side covers.
45. Install air cleaner inlet.
46. Install the air cleaner inlet with the bolts and tighten to 44 inch lbs. (4 Nm).
47. Install engine under cover.

This system generates a motive force by combining the engine, MG1, and MG2 in accordance with the driving conditions. Supply of electrical power from the HV battery to MG2 provides force to drive the rear wheels. Representative examples of the various combinations are described below.

While the rear wheels are being driven by the engine via the planetary gears, MG1 is also being driven by the engine via the power split planetary gear unit, in order to supply the electricity that MG1 generates to MG2. MG1 is also driven by the engine via the planetary gears, in order to charge the HV battery.

When the vehicle is decelerating, kinetic energy from the rear wheels is recovered and converted into electrical energy by means of MG2, and used to recharge the HV battery.

HYBRID PRECAUTIONS

**** CAUTION**

The GS450h hybrid system contains a 288V high-voltage system with a strong alkali solution of potassium hydroxide. Be sure to follow the instructions in this manual to handle the system correctly. Failure to do so may result in serious injury or electrocution. Engineer must undergo special training to be able to perform high-voltage system inspection and servicing.

**** CAUTION**

Engineers must undergo special training to be able to perform high-voltage system inspection and servicing.

**** CAUTION**

All high-voltage wire harnesses are colored orange. The HV battery and other high-voltage components have "High Voltage" caution labels. Do not carelessly touch these wires and components.

**** CAUTION**

Before inspecting or servicing the high-voltage system, be sure to follow safety measures, such as wearing insulated gloves and removing the service plug to prevent electrocution. Carry the removed service plug in your pocket to prevent other technicians from reinstalling it while you are servicing the vehicle.

**** CAUTION**

After removing the service plug, wait 10 minutes before touching any of the high-voltage connectors and terminals.

**** CAUTION**

Be sure to install the service plug before starting the hybrid system. Starting the hybrid system with the service plug removed may damage the vehicle.

**** CAUTION**

Before wearing insulated gloves, make sure that they are not cracked, ruptured, torn, or damaged in any way. Do not wear wet insulated gloves.

**** CAUTION**

When servicing the vehicle, do not carry metal objects like mechanical pencils or scales that can be dropped accidentally and cause a short circuit.

**** CAUTION**

Before touching a bare high-voltage terminal, wear insulated gloves and use an electrical tester to ensure that the terminal is not charged with electricity (approximately 0 V).

1. After disconnecting or exposing a high-voltage connector or terminal, insulate it immediately using insulation tape.

**** CAUTION**

The screw of a high-voltage terminal should be tightened firmly to the specified torque. Both insufficient and excessive torque can cause failure.

**** CAUTION**

Use a "CAUTION: HIGH VOLTAGE. DO NOT TOUCH DURING OPERATION." sign to notify other engineers that a high-voltage system is being inspected and/or repaired.

**** CAUTION**

Do not place the battery upside down while removing and installing it.

**** CAUTION**

After servicing the high-voltage system and before reinstalling the service plug, check again that you have not left a part or tool inside, that the high-voltage terminal screws are firmly tightened, and that the connectors are correctly connected.

2. The LEXUS GS450h automatically turns the engine ON and OFF when the READY light on the instrument panel is ON. To avoid injury, remove the key from the vehicle before inspecting or servicing the engine compartment.

BATTERY BLOWER

REMOVAL & INSTALLATION

See Figures 30 and 31.

1. Before servicing the vehicle, refer to the Precautions section.
2. Disconnect the negative battery cable.

**** CAUTION**

Before inspecting the high-voltage system or disconnecting the low voltage connector of the inverter with converter assembly, take safety precautions, such as wearing insulated gloves and removing the service plug grip to prevent electrical shocks. After removing the service plug grip, put it in your pocket to prevent other technicians from accidentally reconnecting it while you are working on the high-voltage system.

**** CAUTION**

After disconnecting the service plug grip, wait for at least 10 minutes before touching any of the high-voltage connectors or terminals. After waiting, check the voltage at the inspection point in the inverter with converter assembly. The voltage should be 0 V before beginning work.

3. Remove the luggage compartment floor mat.
4. Remove the battery service hole cover LH.

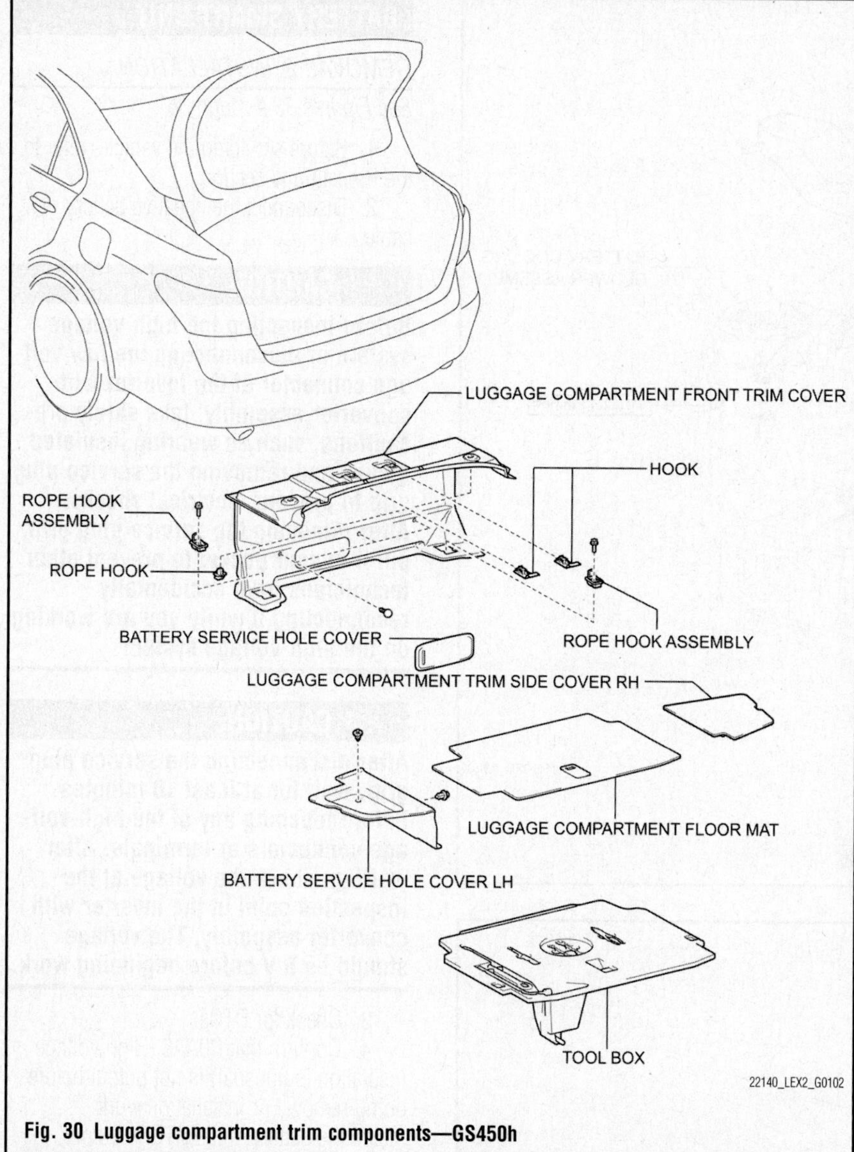

Fig. 30 Luggage compartment trim components—GS450h

5. Remove the hybrid service plug grip.
6. Remove the tool box.
7. Remove the rope hook assembly.
8. Remove the luggage compartment front trim cover.
9. Remove the No. 2 HV battery intake duct. Wear insulated gloves.
10. Remove the 2 nuts and battery cooling blower connector.
11. Separate the No. 3 HV battery intake duct and No. 4 HV battery intake duct.
12. Remove the battery cooling blower assembly together with the No. 3 HV battery intake duct from the HV battery.
13. Remove the No. 3 HV battery intake duct from the battery cooling blower assembly.

To install:

14. Install the No. 3 HV battery intake duct to the battery cooling blower assembly.
15. Install the battery cooling blower assembly together with the No. 3 HV battery intake duct to the HV battery. Wear insulated gloves.
16. Connect the No. 3 HV battery intake duct to the No. 4 HV battery intake duct.
17. Install the 2 nuts and connector. Tighten the HV blower mounting bolts to 71 inch lbs. (8 Nm).
18. Install the No. 2 HV battery intake duct with the 2 clips.
19. Install the luggage compartment front trim cover.
20. Install the rope trim assembly.
21. Install the tool box.
22. Install luggage compartment inner trim cover RH.
23. Install the service plug grip.
24. Install the battery service hole cover.
25. Install the luggage compartment floor mat.
26. Connect the negative battery cable.

BATTERY FUSE

REMOVAL & INSTALLATION

See Figure 32.

1. Before servicing the vehicle, refer to the Precautions section.
2. Disconnect the negative battery cable.

CAUTION

Before inspecting the high-voltage system or disconnecting the low voltage connector of the inverter with converter assembly, take safety precautions, such as wearing insulated gloves and removing the service plug grip to prevent electrical shocks. After removing the service plug grip, put it in your pocket to prevent other technicians from accidentally reconnecting it while you are working on the high-voltage system.

CAUTION

After disconnecting the service plug grip, wait for at least 10 minutes before touching any of the high-voltage connectors or terminals. After waiting, check the voltage at the inspection point in the inverter with converter assembly. The voltage should be 0 V before beginning work.

3. Remove the electric battery fuse as follows:
   - Release the 2 claws and remove the service plug grip cover.
   - Remove the 2 bolts and electric vehicle fuse.

To install:

4. Install the electric vehicle fuse with the 2 bolts. Tighten the bolts to 48 inch lbs. (5.4 Nm).
5. Install the service plug grip cover with the 2 craws.
6. Install the service plug grip.
7. Connect the negative battery cable.
8. Install the battery service hole cover.

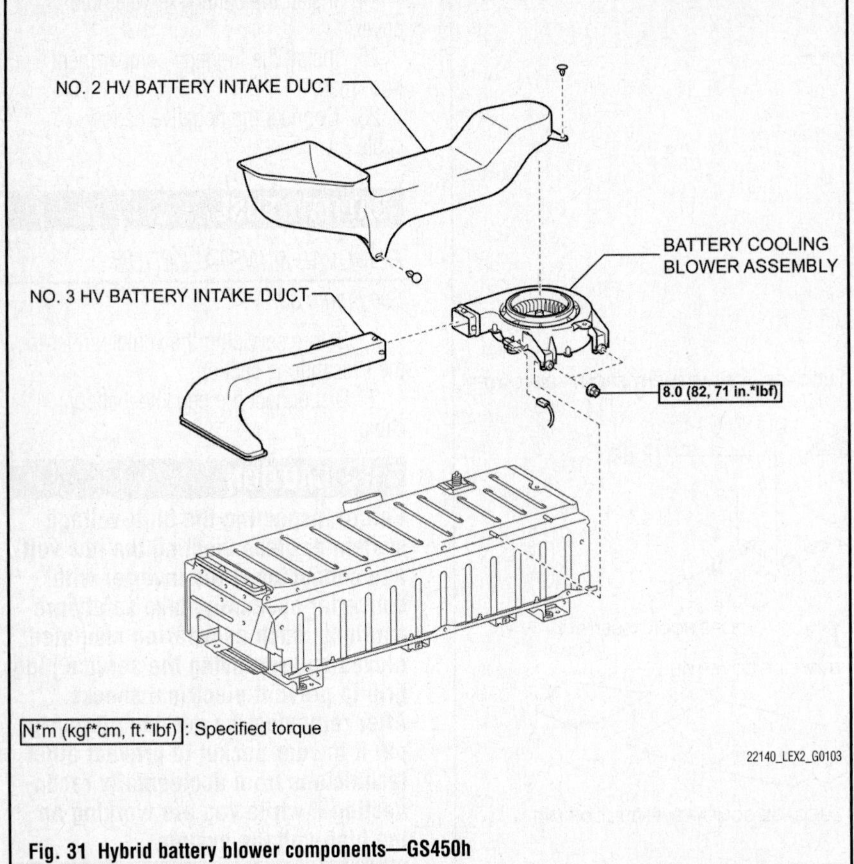

Fig. 31 Hybrid battery blower components—GS450h

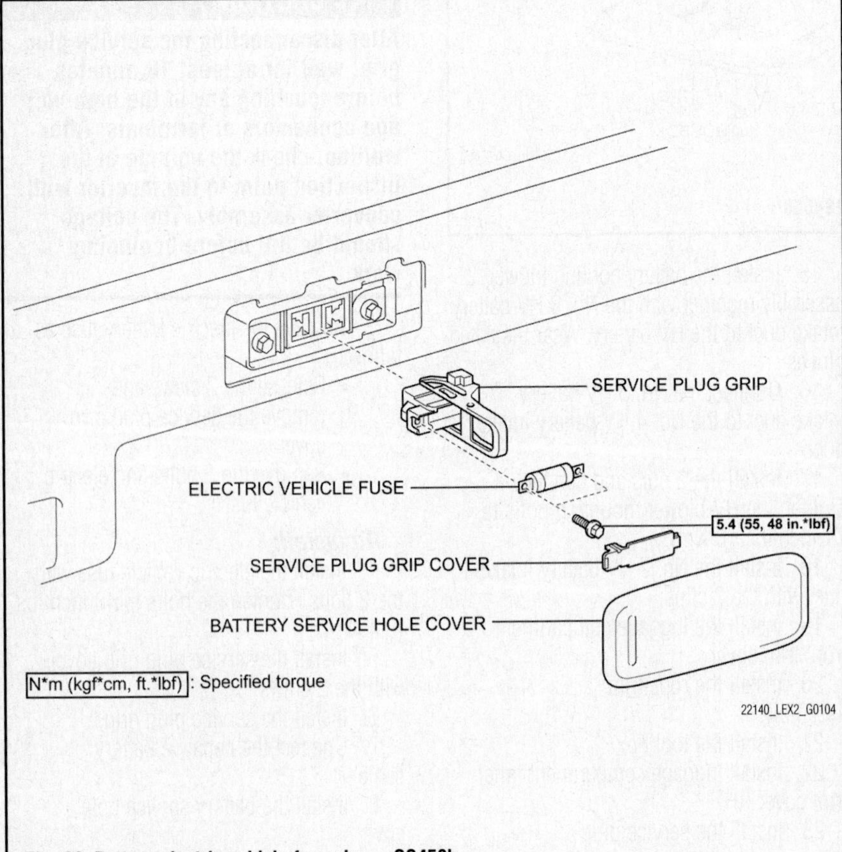

Fig. 32 Battery electric vehicle fuse view—GS450h

BATTERY SMART UNIT

REMOVAL & INSTALLATION

See Figures 33 through 35.

1. Before servicing the vehicle, refer to the Precautions section.
2. Disconnect the negative battery cable.

✲✲ CAUTION

Before inspecting the high-voltage system or disconnecting the low voltage connector of the inverter with converter assembly, take safety precautions, such as wearing insulated gloves and removing the service plug grip to prevent electrical shocks. After removing the service plug grip, put it in your pocket to prevent other technicians from accidentally reconnecting it while you are working on the high-voltage system.

✲✲ CAUTION

After disconnecting the service plug grip, wait for at least 10 minutes before touching any of the high-voltage connectors or terminals. After waiting, check the voltage at the inspection point in the inverter with converter assembly. The voltage should be 0 V before beginning work.

3. Check for DTCs
4. Confirm that P0AA6 (High voltage insulation is unusual) is not output before doing removal or installation work inside the battery. If the DTC is output, perform troubleshooting procedures first.

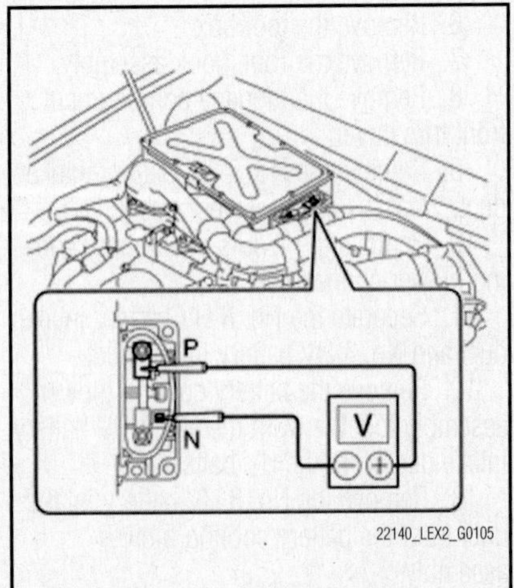

Fig. 33 Measure the voltage between the terminals of the 2 phase connectors

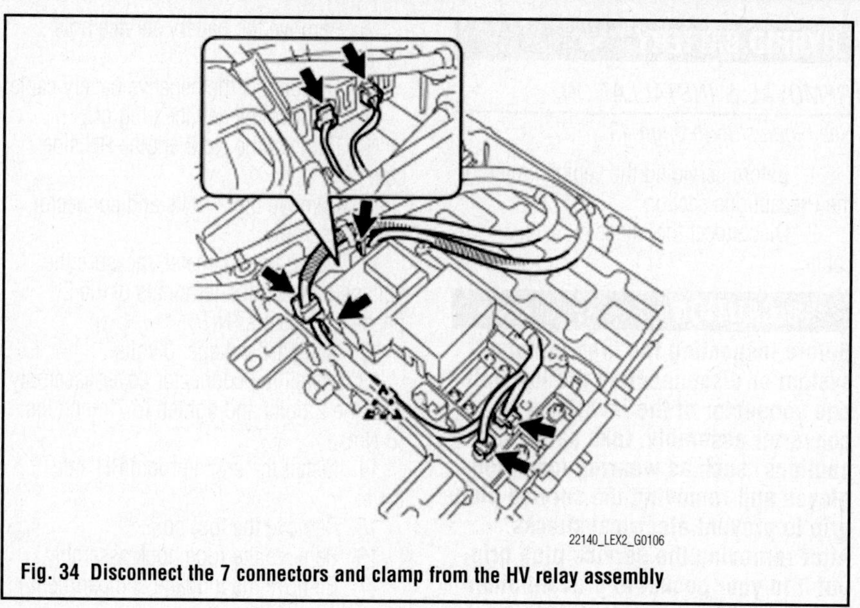

Fig. 34 Disconnect the 7 connectors and clamp from the HV relay assembly

5. Remove the luggage compartment floor mat.
6. Remove the battery service hole cover.
7. Disconnect the negative battery cable.
8. Remove the service plug grip.
9. Remove the No.2 engine RH side cover.
10. Remove the 2 bolts and connector cover assembly.
11. Using the voltmeter, measure the voltage between the terminals of the 2 phase connectors (N-P).
12. Standard voltage: 0 volts
13. Install the connector cover assembly with the 2 bolts and tighten to 72 inch lbs. (8 Nm).
14. Install the engine room RH side cover.

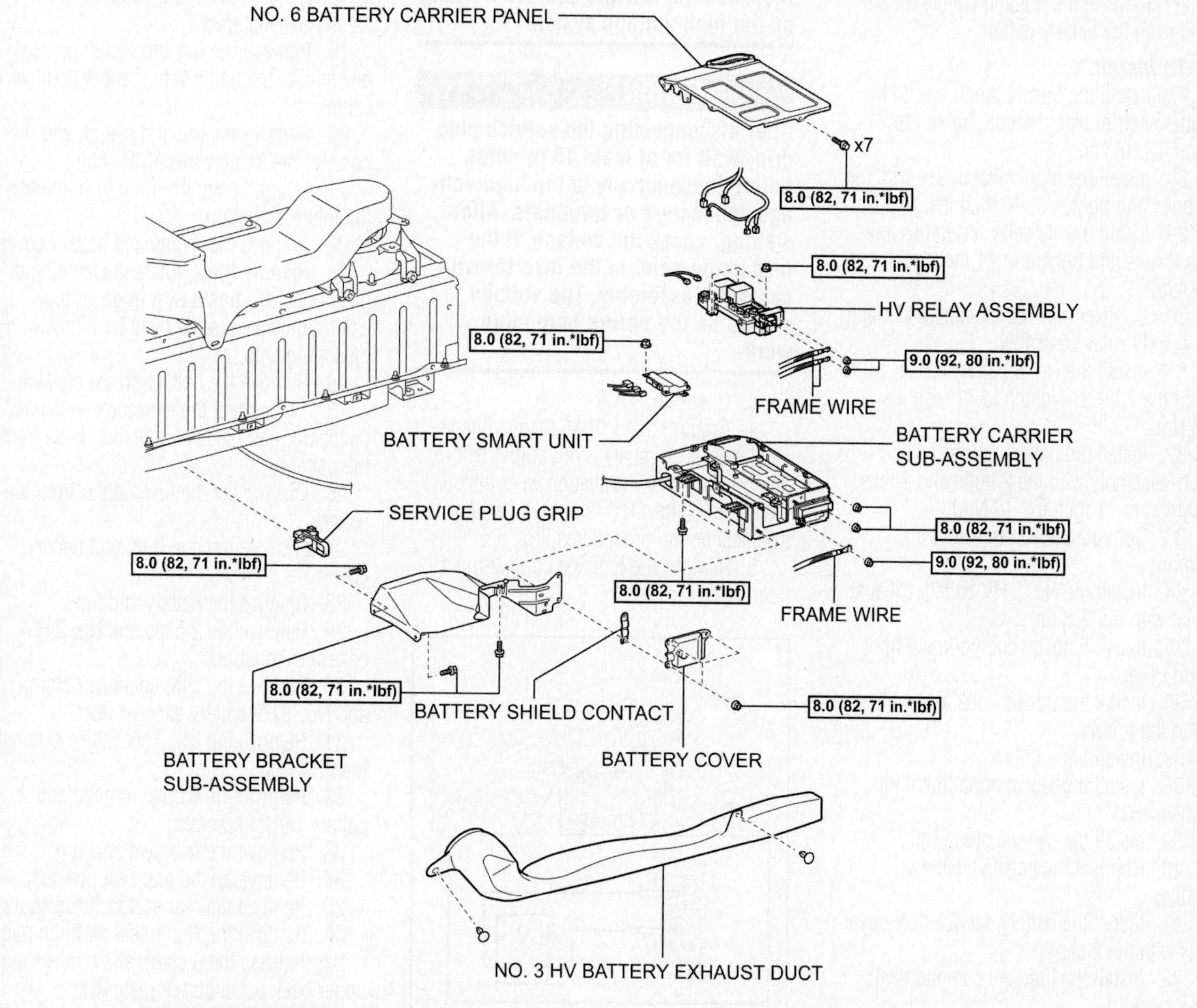

NO. 6 BATTERY CARRIER PANEL
x7
8.0 (82, 71 in.*lbf)
8.0 (82, 71 in.*lbf)
HV RELAY ASSEMBLY
8.0 (82, 71 in.*lbf)
9.0 (92, 80 in.*lbf)
FRAME WIRE
BATTERY SMART UNIT
BATTERY CARRIER SUB-ASSEMBLY
SERVICE PLUG GRIP
8.0 (82, 71 in.*lbf)
8.0 (82, 71 in.*lbf)
9.0 (92, 80 in.*lbf)
8.0 (82, 71 in.*lbf)
FRAME WIRE
8.0 (82, 71 in.*lbf)
BATTERY SHIELD CONTACT
8.0 (82, 71 in.*lbf)
BATTERY BRACKET SUB-ASSEMBLY
BATTERY COVER
NO. 3 HV BATTERY EXHAUST DUCT
N*m (kgf*cm, ft.*lbf) : Specified torque
22140_LEX2_G0107

Fig. 35 Smart battery unit and related components

15. Remove the tool box.
16. Remove the rope hook assembly.
17. Remove the luggage compartment front trim cover.
18. Remove the 2 clips and No. 3 HV battery exhaust duct.
19. Remove the nut and 2 clamps, and disconnect the frame wire (battery positive cable).
20. Remove the battery bracket sub-assembly.
21. Remove the battery carrier sub-assembly.
22. Remove the No.6 battery carrier panel.
23. Disconnect the 7 connectors and clamp from the HV relay assembly.
24. Remove the 3 nuts and HV relay assembly from the battery carrier.
25. Remove the 2 bolts and high voltage fuse.
26. Remove the nut and battery smart unit from the battery carrier.

To install:

27. Install the battery smart unit to the battery carrier with the nut. Tighten to 71 inch lbs. (8 Nm).
28. Install the high voltage fuse with the 2 bolts and tighten to 40 inch lbs. (4.5 Nm).
29. Install the HV relay assembly with the 3 nuts and tighten to 71 inch lbs. (8 Nm).
30. Connect the 7 connectors and clamp to the HV relay assembly.
31. Install the No. 6 battery carrier panel with the 7 bolts. Tighten to 71 inch lbs. (8 Nm).
32. Install the battery carrier sub-assembly with the 2 bolts and 2 nuts. Tighten to 71 inch lbs. (8 Nm).
33. Connect the 3 connectors and 2 clamps
34. Install the No. 3 HV battery exhaust duct with the 2 clips.
35. Install luggage compartment front trim cover.
36. Install the 2 rope hook assemblies with the 2 bolts.
37. Install the tool box.
38. Install luggage compartment trim cover RH.
39. Install the service plug grip.
40. Connect the negative battery cable.
41. Install the battery service hole cover LH with the 2 clips.
42. Install the luggage compartment floor mat.
43. Some vehicle systems require initialization after reconnecting the cable to the negative battery terminal.

HYBRID BATTERY

REMOVAL & INSTALLATION

See Figures 36 through 43.

1. Before servicing the vehicle, refer to the Precautions section.
2. Disconnect the negative battery cable.

CAUTION

Before inspecting the high-voltage system or disconnecting the low voltage connector of the inverter with converter assembly, take safety precautions, such as wearing insulated gloves and removing the service plug grip to prevent electrical shocks. After removing the service plug grip, put it in your pocket to prevent other technicians from accidentally reconnecting it while you are working on the high-voltage system.

CAUTION

After disconnecting the service plug grip, wait for at least 10 minutes before touching any of the high-voltage connectors or terminals. After waiting, check the voltage at the inspection point in the inverter with converter assembly. The voltage should be 0 V before beginning work.

3. Check for DTCs
4. Confirm that P0AA6 (High voltage insulation is unusual) is not output before doing removal or installation work inside the battery. If the DTC is output, perform troubleshooting procedures first.
5. Remove the luggage compartment floor mat.

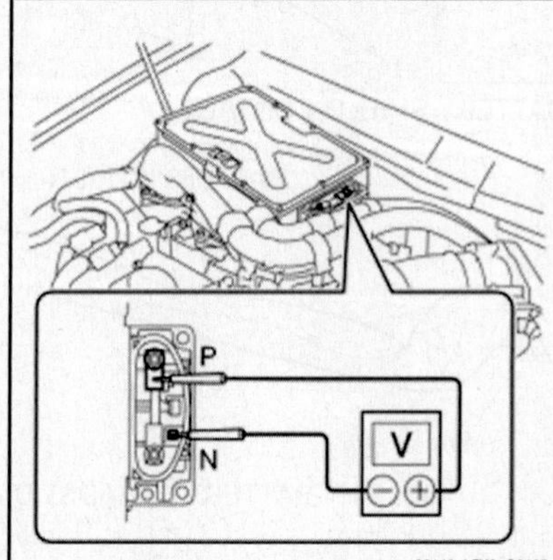

Fig. 36 Measure the voltage between the terminals of the 2 phase connectors

6. Remove the battery service hole cover.
7. Disconnect the negative battery cable.
8. Remove the service plug grip.
9. Remove the No.2 engine RH side cover.
10. Remove the 2 bolts and connector cover assembly.
11. Using the voltmeter, measure the voltage between the terminals of the 2 phase connectors (N-P).
12. Standard voltage: 0 volts
13. Install the connector cover assembly with the 2 bolts and tighten to 72 inch lbs. (8 Nm).
14. Install the engine room RH side cover.
15. Remove the tool box.
16. Remove the rope hook assembly.
17. Remove the luggage compartment front trim cover.
18. Remove the 2 clips and No. 3 HV battery exhaust duct.
19. Remove the nut and 2 clamps, and disconnect the frame wire (battery positive cable).
20. Remove the nut and clamp, and disconnect the frame wire (AMD cable).
21. Using the service plug grip, release the interlock button.
22. Remove the 3 nuts and battery cover.
23. Remove the 2 nuts and clamp, and disconnect the frame wire (high voltage cable). Insulate the removed terminals with insulating tape.
24. Remove the battery shield contact.
25. Disconnect the thermometer sensor connector and battery ventilation hose from the battery.
26. Disconnect the fuse holder from the battery.
27. Remove the nut, bolt, and battery clamp.
28. Remove the battery and tray.
29. Remove the 2 clips and No. 2 HV battery exhaust duct.
30. Remove the clip, connector clamp and No. 1 HV battery exhaust duct.
31. Remove the No. 2 HV intake exhaust duct.
32. Remove the 6 bolts, 4 nuts, and 2 battery carrier brackets.
33. Remove the rear seat cushion.
34. Remove all the rear seat headrest.
35. Remove the rear seatback assembly.
36. Remove the No. 1 seat partition pad.
37. Release the 6 claws and remove the upper back panel corner plate RH.
38. Remove the 2 upper roof side inner covers.
39. Install the tool box.

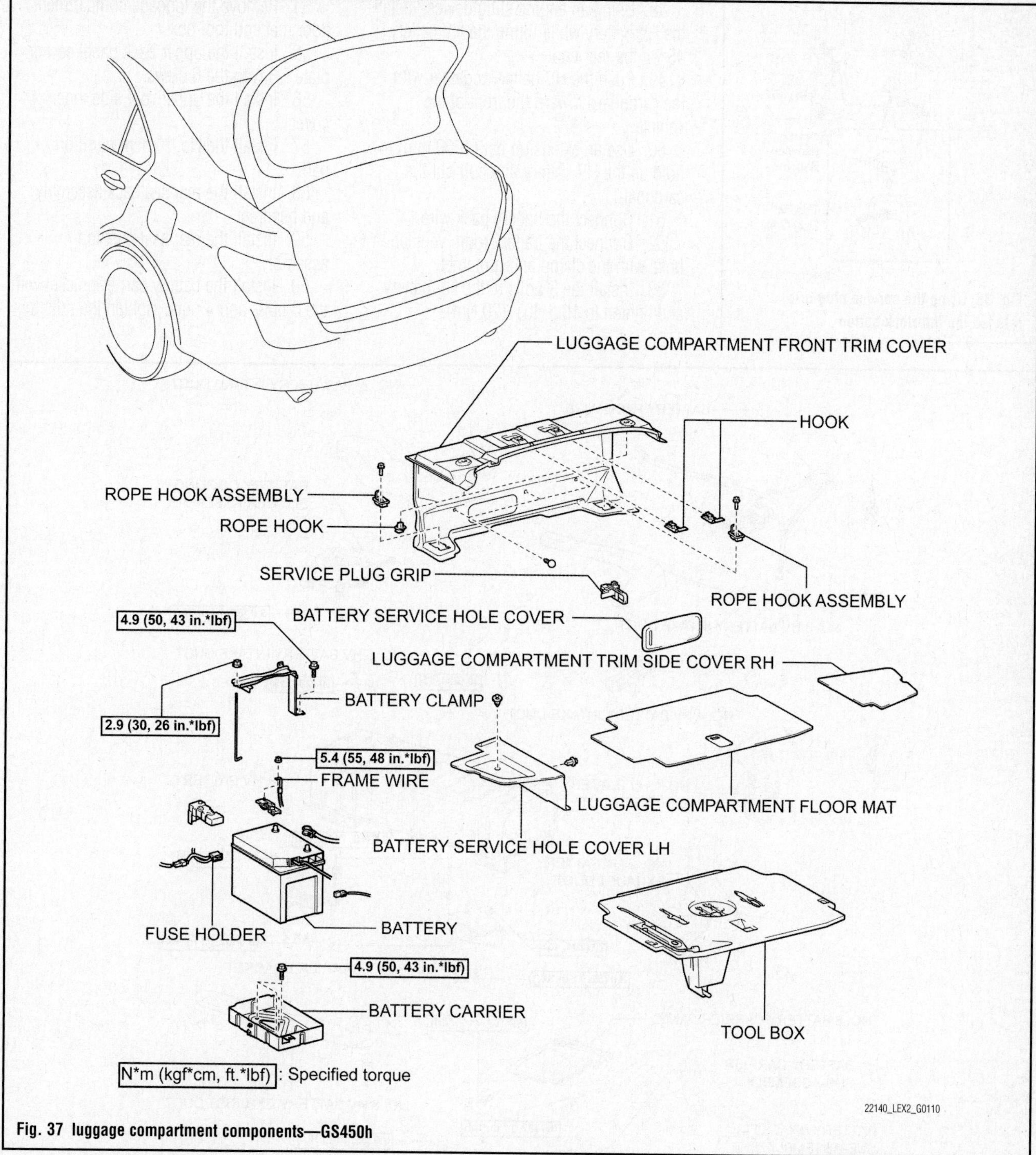

Fig. 37 luggage compartment components—GS450h

40. Install the floor mat. Reverse the luggage compartment floor mats to install them to the vehicle.
41. Remove the 3 bolts from the HV battery.
42. Remove the clamp grommet and battery room ventilation hose.
43. Disconnect the battery pack wire.
44. Prepare a piece of cardboard of 780 mm (30.71 in.) X 700 mm (27.56 in.) or larger.
45. Using an extension bar (1000 mm) to hold up the HV battery, insert the cardboard until it cannot be inserted any further.

➡Bind the frame wire with electrical tape to prevent it from getting caught with the parts. Before holding up the HV battery, check that the extension bar goes through to the front side of the battery from the cabin side.

46. Pull the HV battery together with the cardboard toward the rear of the vehicle.
47. Using an engine sling device, remove the HV battery while tilting the HV battery 45° at the rear end.

To install:

➡Place a piece of cardboard or other similar object to protect the HV battery and vehicle body from damage.

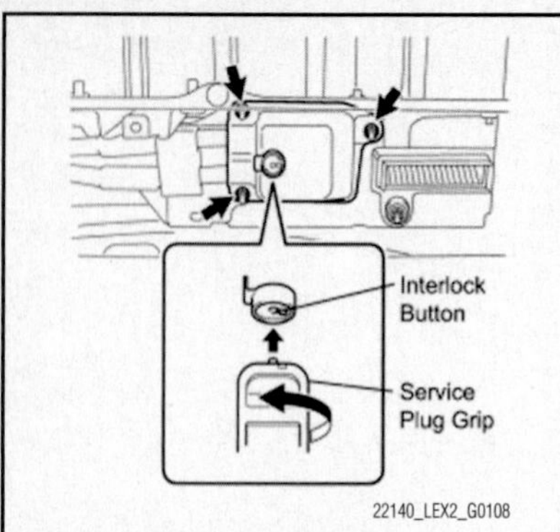

Fig. 38 Using the service plug grip, release the interlock button

48. Using an engine sling device, install the HV battery while tilting the HV battery 45° at the rear end.

49. Push the HV battery together with the cardboard toward the front of the vehicle.

50. Use an extension bar (1000 mm) to hold up the HV battery and pull out the cardboard.

51. Connect the battery pack wire.

52. Connect the battery room ventilation hose with the clamp and grommet.

53. Install the 3 bolts to the HV battery and tighten to 15 ft. lbs. (20 Nm).

54. Remove the luggage compartment floor mat and tool box.

55. Install the upper back panel corner plate RH with the 6 claws.

56. Install the upper roof side inner cover.

57. Install the No. 1 room partition pad.

58. Install the rear seatback assembly and headrest.

59. Install the rear seat cushion assembly.

60. Install the battery carrier bracket with the 6 bolts and 4 nuts. Tighten the nuts to

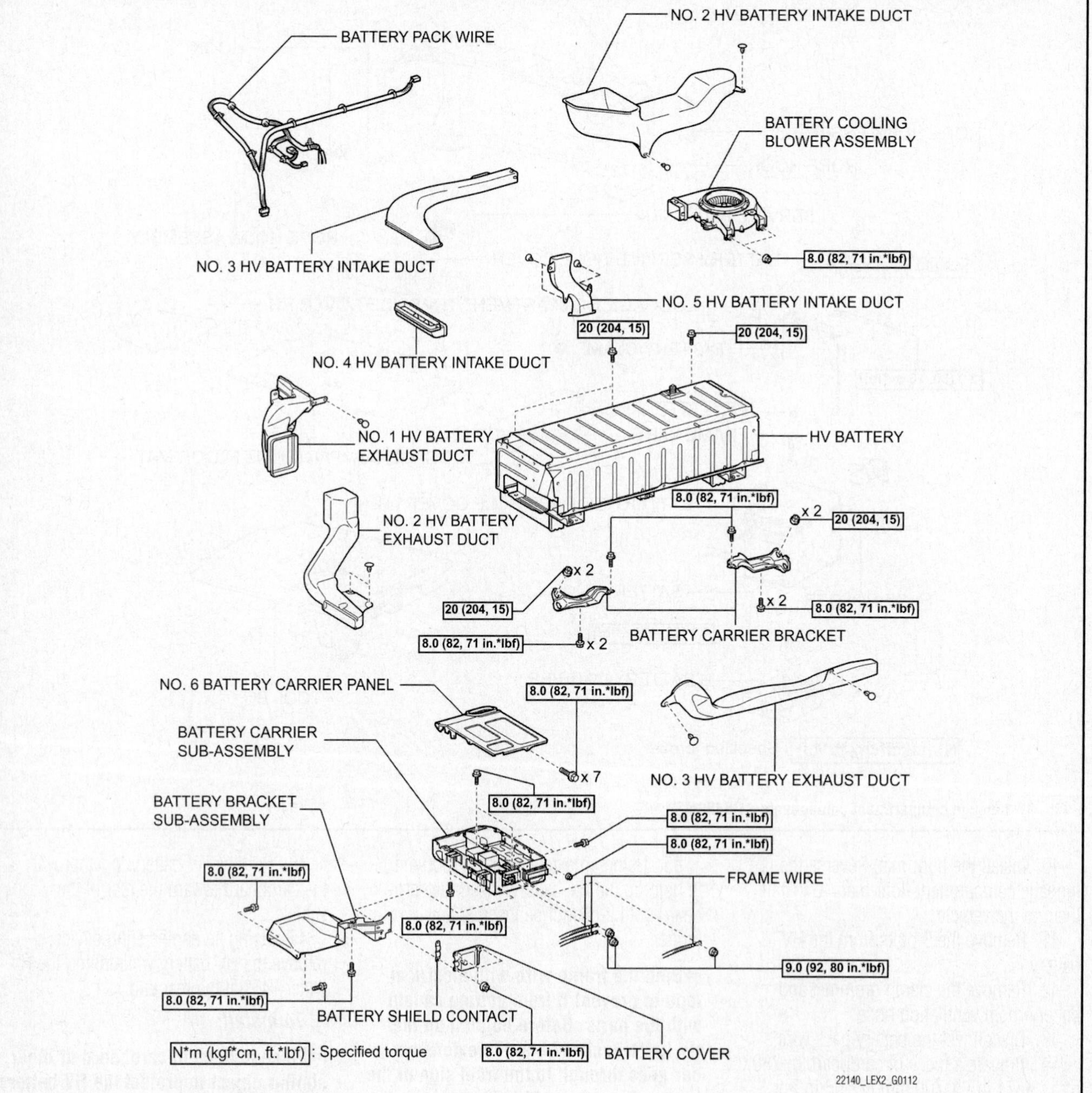

Fig. 39 Hybrid battery and related components

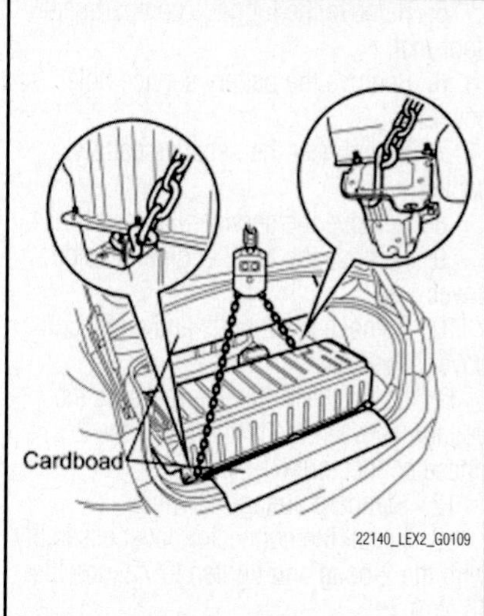

Fig. 40 Hybrid battery removal shown—GS450h

15 ft. lbs. (20 Nm). Tighten the bolts to 71 inch lbs. (8 Nm).

61. Install the battery cooling blower motor assembly.
62. Install the battery intake and exhaust ducts.
63. Install the battery tray and battery.
64. Install the battery clamp.
65. Connect the fuse holder to the battery clamp.
66. Connect the battery thermometer sensor connector and battery ventilation hose to the battery.
67. Install the battery shield contact.
68. Connect the 2 frame wires (high voltage cable) with the 2 nuts and clamp. Tighten to 80 inch lbs. (9 Nm). Be sure to match the red marks on the frame wires to the red marks on the HV relay assembly and install the frame wires.
69. Install the battery cover and press the interlock button.
70. Secure the battery cover with the 3 nuts and tighten to 71 inch lbs. (8 Nm).

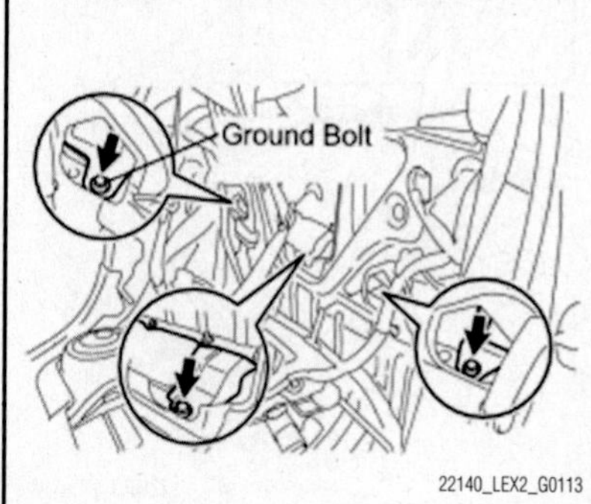

Fig. 41 Install the 3 bolts to the HV battery

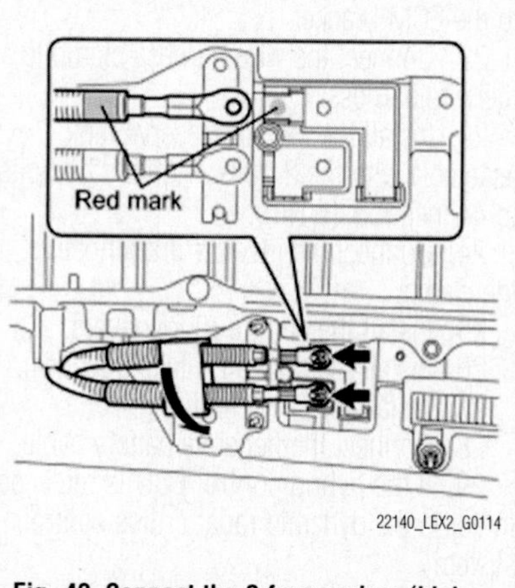

Fig. 42 Connect the 2 frame wires (high voltage cable)

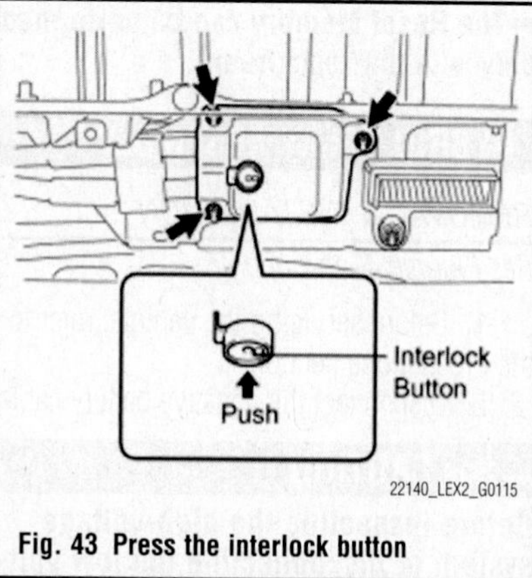

Fig. 43 Press the interlock button

71. Connect the frame wire (AMD cable) with the nut and clamp and tighten to 80 inch lbs. (9 Nm).
72. Connect the frame wire (battery positive cable) with the nut and 2 clamps.
73. Install the No. 3 HV battery exhaust duct with the 2 clips.
74. Install the front luggage compartment front trim cover.
75. Install the 2 rope hook assemblies with the 2 bolts.
76. Install the tool box.
77. Install the luggage compartment side trim cover.
78. Install the service plug grip.
79. Connect the negative battery cable.
80. Install the battery service hole cover LH with the 2 clips.
81. Install the luggage compartment floor mat.
82. Some vehicle systems require initialization after reconnecting the cable to the negative battery terminal.

HYBRID CONTROL ECU

REMOVAL & INSTALLATION

See Figures 44 and 45.

1. Before servicing the vehicle, refer to the Precautions section.
2. Disconnect the negative battery cable.

CAUTION

Before inspecting the high-voltage system or disconnecting the low voltage connector of the inverter with converter assembly, take safety precautions, such as wearing insulated gloves and removing the service plug grip to prevent electrical shocks. After removing the service plug grip, put it in your pocket to prevent other technicians from accidentally reconnecting it while you are working on the high-voltage system.

CAUTION

After disconnecting the service plug grip, wait for at least 10 minutes before touching any of the high-voltage connectors or terminals. After waiting, check the voltage at the inspection point in the inverter with converter assembly. The voltage should be 0 V before beginning work.

3. Check for DTCs
4. Confirm that P0AA6 (High voltage insulation is unusual) is not output before doing removal or installation work inside the battery. If the DTC is output, perform troubleshooting procedures first.
5. Remove the luggage compartment floor mat.
6. Remove the battery service hole cover.
7. Remove the service plug grip.
8. Disconnect the negative battery cable.
9. Remove the cool air intake seal.
10. Remove the LH engine room side cover.
11. Remove the radiator reserve tank as follows:
    a. Disconnect the inverter drain hose from the clamp.
    b. Remove the 2 bolts and radiator reserve tank assembly.
12. Remove the Hybrid ECU bracket as follows:
    a. Disconnect the wire harness clamp from the wire harness.

b. Disconnect the 3 wire harness clamps from the ECM bracket.

c. Remove the 6 bolts, nut and ECM bracket.

13. Raise the lock lever and disconnect the 2 connectors.

14. Make sure that the lock lever is raised 90° as shown in the illustration before disconnecting the connectors. Failure to do this may cause the connectors to break.

15. Remove the hybrid vehicle control ECU.

To install:

16. Install the Hybrid ECU as follows:

17. Connect the 2 connectors to the hybrid vehicle control ECU and push each lock lever down to lock the connectors.

18. Set the ECM bracket to the hybrid vehicle control ECU and ECM

19. Install the hybrid vehicle control ECU with the 2 bolts (A). Tighten the bolts to 49 inch lbs. (5.6 Nm). Tighten the bolts to 27 inch lbs. (3 Nm).

20. Install the ECM bracket with the 4 bolts (B) and nut. Tighten the nuts to 27 inch lbs. (3 Nm).

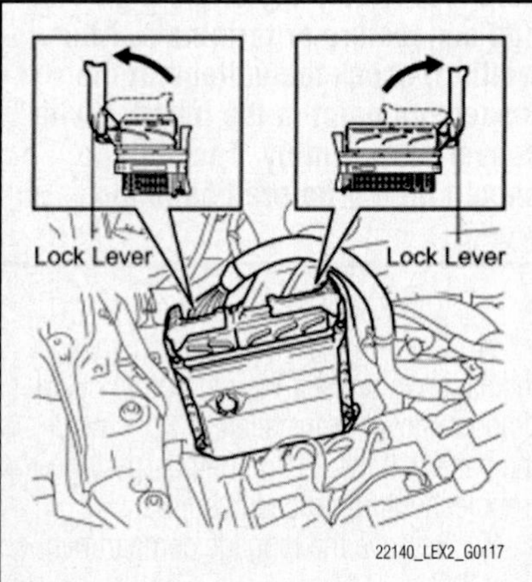

Fig. 44 Hybrid vehicle control ECU removal—GS450h

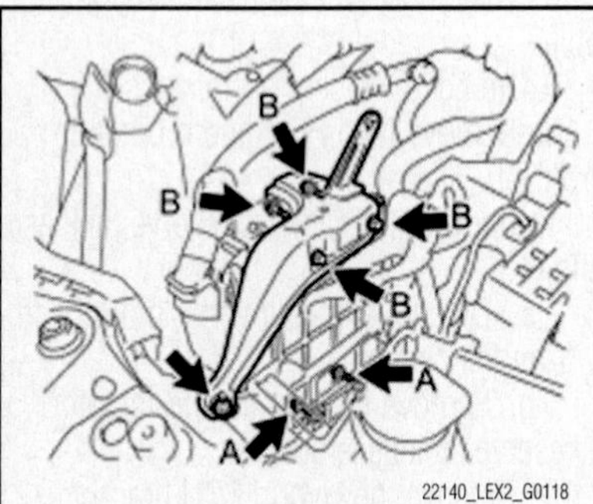

Fig. 45 Hybrid vehicle control ECU mounting bolts (A) and (B)

21. Connect the 3 wire harness clamps to the ECM bracket.

22. Connect the wire harness clamp to the wire harness.

23. Install the radiator reserve tank assembly with the 2 bolts. Tighten the bolts to 44 inch lbs. (5 Nm).

24. Connect the inverter drain hose to the clamp.

25. Install the engine room covers.

26. Install the cool air intake duct seal.

27. Install the service plug grip.

28. Connect the negative battery cable.

29. If the hybrid control ECU is replaced, initialize the dynamic radar cruise control system

30. If the hybrid vehicle control ECU is replaced, initialize the learning value of the automatic transmission.

➡The Reset Memory can be performed only with the Techstream.

HYBRID RELAY ASSEMBLY

REMOVAL & INSTALLATION

See Figures 46 through 48.

1. Before servicing the vehicle, refer to the Precautions section.

2. Disconnect the negative battery cable.

⁂ CAUTION

Before inspecting the high-voltage system or disconnecting the low voltage connector of the inverter with converter assembly, take safety precautions, such as wearing insulated gloves and removing the service plug grip to prevent electrical shocks. After removing the service plug grip, put it in your pocket to prevent other technicians from accidentally reconnecting it while you are working on the high-voltage system.

⁂ CAUTION

After disconnecting the service plug grip, wait for at least 10 minutes before touching any of the high-voltage connectors or terminals. After waiting, check the voltage at the inspection point in the inverter with converter assembly. The voltage should be 0 V before beginning work.

3. Check for DTCs

4. Confirm that P0AA6 (High voltage insulation is unusual) is not output before doing removal or installation work inside the battery. If the DTC is output, perform troubleshooting procedures first.

5. Remove the luggage compartment floor mat.

6. Remove the battery service hole cover.

7. Disconnect the negative battery cable.

8. Remove the service plug grip.

9. Remove the No.2 engine RH side cover.

10. Remove the 2 bolts and connector cover assembly.

11. Using the voltmeter, measure the voltage between the terminals of the 2 phase connectors (N-P).

12. Standard voltage: 0 volts

13. Install the connector cover assembly with the 2 bolts and tighten to 72 inch lbs. (8 Nm).

14. Install the engine room RH side cover.

15. Remove the tool box.

16. Remove the rope hook assembly.

17. Remove the luggage compartment front trim cover.

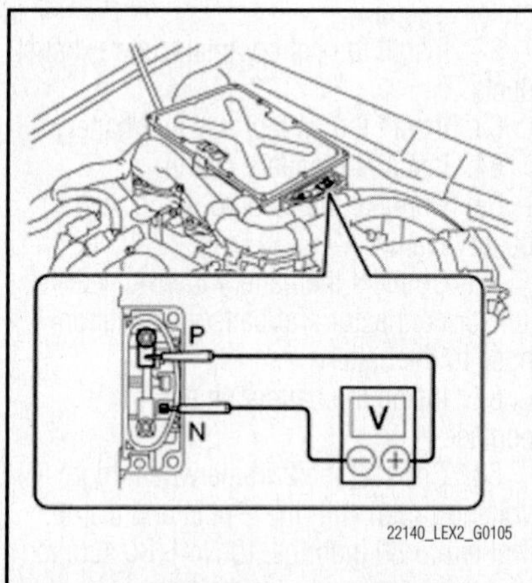

Fig. 46 Measure the voltage between the terminals of the 2 phase connectors

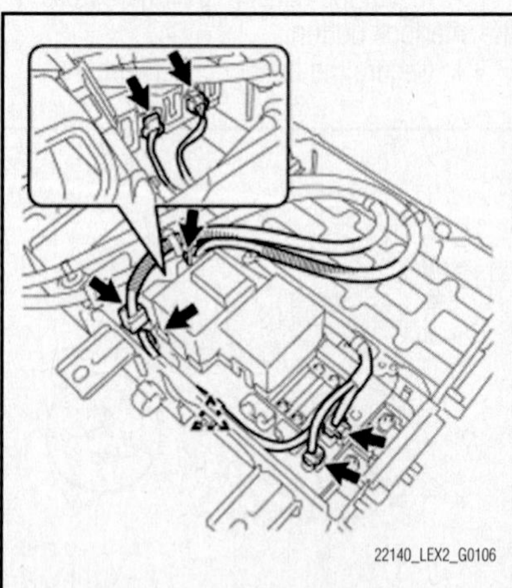

Fig. 47 Disconnect the 7 connectors and clamp from the HV relay assembly

18. Remove the 2 clips and No. 3 HV battery exhaust duct.

19. Remove the nut and 2 clamps, and disconnect the frame wire (battery positive cable).

20. Remove the battery bracket sub-assembly.

21. Remove the battery carrier sub-assembly.

22. Remove the No.6 battery carrier panel.

23. Disconnect the 7 connectors and clamp from the HV relay assembly.

24. Remove the 3 nuts and HV relay assembly from the battery carrier.

To install:

25. Install the high voltage fuse with the 2 bolts and tighten to 40 inch lbs. (4.5 Nm).

26. Install the HV relay assembly with the 3 nuts and tighten to 71 inch lbs. (8 Nm).

27. Connect the 7 connectors and clamp to the HV relay assembly.

28. Install the No. 6 battery carrier panel with the 7 bolts. Tighten to 71 inch lbs. (8 Nm).

29. Install the battery carrier sub-assembly with the 2 bolts and 2 nuts. Tighten to 71 inch lbs. (8 Nm).

30. Connect the 3 connectors and 2 clamps

31. Install the No. 3 HV battery exhaust duct with the 2 clips.

32. Install luggage compartment front trim cover.

33. Install the 2 rope hook assemblies with the 2 bolts.

34. Install the tool box.

35. Install luggage compartment trim cover RH.

36. Install the service plug grip.

37. Connect the negative battery cable.

38. Install the battery service hole cover LH with the 2 clips.

39. Install the luggage compartment floor mat.

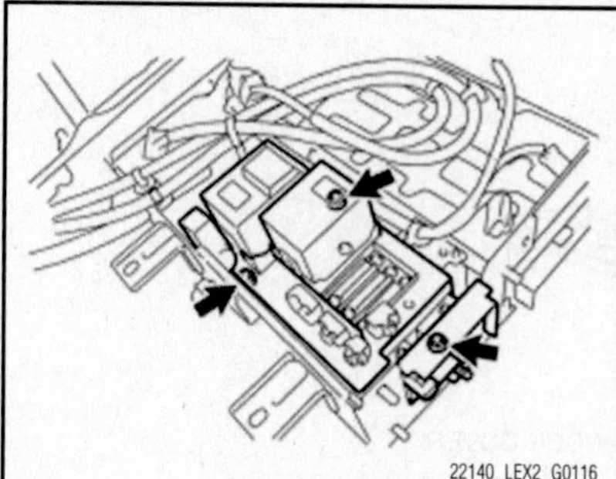

Fig. 48 Hybrid relay assembly view—GS450h

40. Some vehicle systems require initialization after reconnecting the cable to the negative battery terminal.

HYBRID VEHICLE CONVERTER

REMOVAL & INSTALLATION

See Figure 49.

1. Before servicing the vehicle, refer to the Precautions section.

2. Disconnect the negative battery cable.

CAUTION

Before inspecting the high-voltage system or disconnecting the low voltage connector of the inverter with converter assembly, take safety precautions, such as wearing insulated gloves and removing the service plug grip to prevent electrical shocks. After removing the service plug grip, put it in your pocket to prevent other technicians from accidentally reconnecting it while you are working on the high-voltage system.

CAUTION

After disconnecting the service plug grip, wait for at least 10 minutes before touching any of the high-voltage connectors or terminals. After waiting, check the voltage at the inspection point in the inverter with converter assembly. The voltage should be 0 V before beginning work.

3. Check for DTCs

4. Confirm that P0AA6 (High voltage insulation is unusual) is not output before doing removal or installation work inside the battery. If the DTC is output, perform troubleshooting procedures first.

5. Remove the luggage compartment floor mat.

6. Remove the battery service hole cover.

7. Remove the service plug grip.

8. Disconnect the negative battery cable.

9. Remove the engine room side covers.

10. Remove the 2 bolts and connector cover assembly.

11. Using the voltmeter, measure the voltage between the terminals of the 2 phase connectors (N-P).

12. Standard voltage: 0 volts.

13. Install the connector cover assembly and side covers.

14. Remove the luggage compartment trim RH side cover.

15. Remove the tool box.

16. Remove the rope hook assembly.

17. Remove the luggage compartment front trim cover.

18. Install the tool box.

19. Remove the No. 3 HV battery exhaust duct.

20. Disconnect the frame wire.

21. Remove the battery bracket assembly.

22. Remove the carrier assembly.

23. Remove the 7 bolts and No. 6 battery carrier panel from the battery carrier. Do not drop the bolts in the cooling air inlet.

24. Remove the Hybrid Vehicle (HV) converter as follows:

- Disconnect the 3 connectors and 3 clamps.
- Remove the 4 bolts and hold up the hybrid vehicle converter
- Remove the connector and hybrid vehicle converter from the battery carrier.

To install:

25. Install the HV converter as follows:

26. Connect the connector to the hybrid vehicle converter.

27. Install the hybrid vehicle converter to the battery carrier with the 4 bolts. Tighten the bolt to 71 inch lbs. (8 Nm).

28. Connect the 3 connectors and 3 clamps.

29. Install the No. 6 battery carrier panel with the 7 bolts. Tighten to 71 inch lbs. (8 Nm).

30. Install the battery carrier sub-assembly with the 2 bolts and 2 nuts. Tighten to 71 inch lbs. (8 Nm).

31. Connect the 3 connectors and 2 clamps

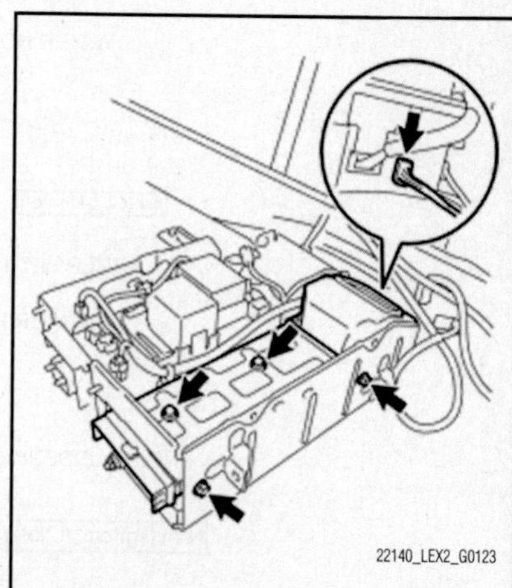

Fig. 49 Remove the connector and hybrid vehicle converter

32. Install the No. 3 HV battery exhaust duct with the 2 clips.
33. Install luggage compartment front trim cover.
34. Install the 2 rope hook assemblies with the 2 bolts.
35. Install the tool box.
36. Install luggage compartment trim cover RH.
37. Install the service plug grip.
38. Connect the negative battery cable.
39. Install the battery service hole cover LH with the 2 clips.
40. Install the luggage compartment floor mat.
41. Some vehicle systems require initialization after reconnecting the cable to the negative battery terminal.

HYBRID WATER PUMP WITH MOTOR

REMOVAL & INSTALLATION

See Figure 50.

1. Before servicing the vehicle, refer to the Precautions section.
2. Disconnect the negative battery cable.

CAUTION

Before inspecting the high-voltage system or disconnecting the low voltage connector of the inverter with converter assembly, take safety precautions, such as wearing insulated gloves and removing the service plug grip to prevent electrical shocks. After removing the service plug grip, put it in your pocket to prevent other technicians from accidentally reconnecting it while you are working on the high-voltage system.

CAUTION

After disconnecting the service plug grip, wait for at least 10 minutes before touching any of the high-voltage connectors or terminals. After waiting, check the voltage at the inspection point in the inverter with converter assembly. The voltage should be 0 V before beginning work.

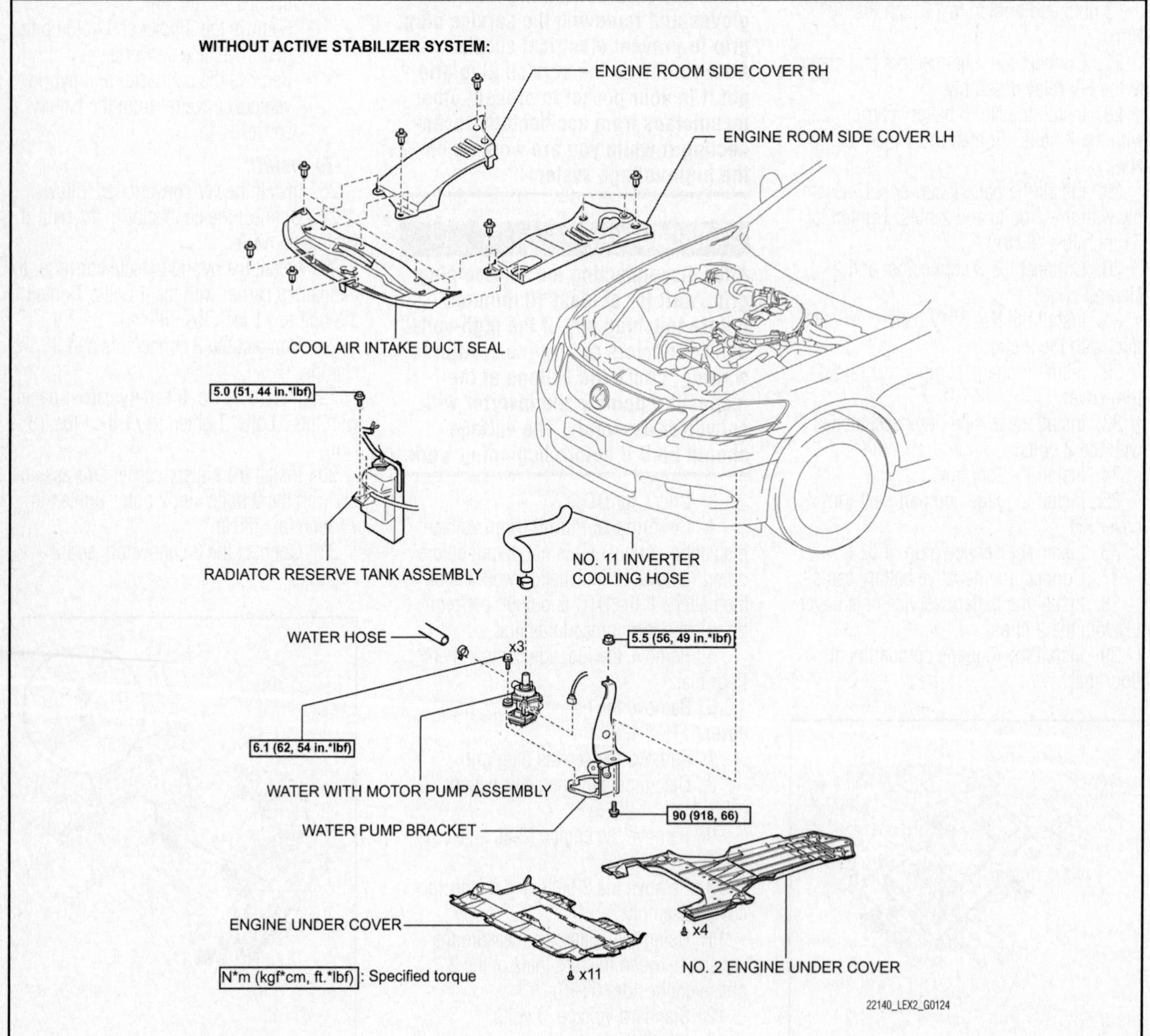

Fig. 50 Hybrid water pump removal—GS450h

3. Remove the air intake duct seal.
4. Remove the engine room side covers.
5. Remove the engine under cover.
6. Remove the No. 2 engine under cover.
7. Drain the engine coolant from the inverter.
8. Remove the radiator reserve tank.
9. Remove the water pump with motor as follows:

- Remove the 2 clips and disconnect the No. 11 inverter cooling hose and water hose from the water with motor and bracket pump assembly.
- Disconnect the connector and 2 clamps from the water with motor and bracket pump assembly.

➡Wrap the connectors of the vehicle wire harness and the water with motor and bracket pump assembly with electrical tape to prevent LLC from contaminating the connectors.

- Remove the nut from the water with motor and bracket pump assembly.
- Remove the bolt and water pump with motor and bracket pump assembly.

To install:

10. Install the water pump with motor and bracket pump assembly with the bolt. Tighten to 66 ft. lbs. (90 Nm).
11. Install the nut to the water pump with motor and bracket pump assembly. Tighten to 49 inch lbs. (5.5 Nm).
12. Connect the connector and 2 clamps to the water with motor and bracket pump assembly.
13. Connect the No. 11 inverter cooling hose and water hose to the water with motor and bracket pump assembly with the 2 clips.
14. Install the radiator reserve tank assembly.
15. Add fresh coolant to the inverter, water pump system. Check the system for leaks.
16. Install the engine under covers.
17. Install the cool air intake duct seal.

INVERTER WITH CONVERTER

REMOVAL & INSTALLATION

See Figures 51 through 54.

1. Before servicing the vehicle, refer to the Precautions section.
2. Disconnect the negative battery cable.

CAUTION

Before inspecting the high-voltage system or disconnecting the low voltage connector of the inverter with converter assembly, take safety precautions, such as wearing insulated gloves and removing the service plug grip to prevent electrical shocks. After removing the service plug grip, put it in your pocket to prevent other technicians from accidentally reconnecting it while you are working on the high-voltage system.

CAUTION

After disconnecting the service plug grip, wait for at least 10 minutes before touching any of the high-voltage connectors or terminals. After waiting, check the voltage at the inspection point in the inverter with converter assembly. The voltage should be 0 V before beginning work.

3. Check for DTCs
4. Confirm that P0AA6 (High voltage insulation is unusual) is not output before doing removal or installation work inside the battery. If the DTC is output, perform troubleshooting procedures first.
5. Remove the luggage compartment floor mat.
6. Remove the battery service hole cover.
7. Remove the service plug grip.
8. Disconnect the negative battery cable.
9. Remove the cool air intake seal.
10. Remove the LH and RH engine room side cover.
11. Remove the LH and RH upper fender protector.
12. Remove the LH and RH wiper arm assembly.
13. Remove the front fender cowl seals.
14. Pull the ventilator louver in the direction indicated by the arrow in the illustration to detach the 10 claws and remove the ventilator louver.
15. Remove the engine under covers.
16. Drain the coolant for the inverter.
17. Remove the connector cover assembly.
18. Remove the 2 bolts and connector cover assembly.
19. Using the voltmeter, measure the voltage between the terminals of the 2 phase connectors (N-P).
20. Standard voltage: 0 volts.
21. Remove the 2 bolts (A).
22. While loosening bolt B, disconnect the generator cable. Removing bolts A and B in a wrong order may cause the cable and the inverter with converter to break.

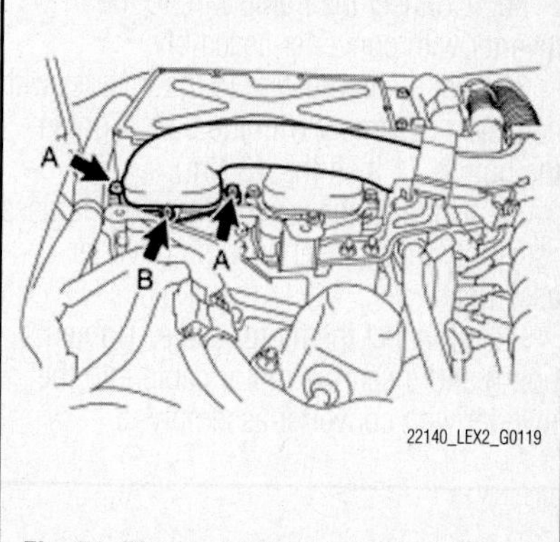

Fig. 51 Disconnect the generator cable

23. Separate the motor cable.
24. Disconnect the air conditioning harness assembly. Insulate the removed terminals with insulating tape.
25. Remove the bolt and disconnect the frame wire
26. Remove the 2 bolts and No. 6 inverter bracket
27. Remove the inverter with converter assembly as follows:

- Raise the lock lever and disconnect the inverter with converter connector. Cover the hole where the connector was connected with tape or equivalent (non-residue type) to prevent entry of foreign matter.
- Disconnect the wire harness clamp.
- Remove the 2 clips, and disconnect the No. 8 inverter cooling hose and No. 13 inverter cooling hose.
- Cover the disconnected water hose and pipe with a plastic bag and tape to prevent coolant from spilling.
- Remove the 4 bolts and inverter with converter assembly.

➡2 people are needed to remove the inverter with converter assembly.

To install:

28. Install the inverter with converter assembly as follows:

- Install the inverter with converter assembly with 4 new bolts. Tighten the bolts to 71 inch lbs. (8 Nm).
- Connect the No. 8 inverter cooling hose and No. 13 inverter cooling hose with the 2 clips.
- Connect the inverter with converter connector and lock the connector with the lock lever.
- Connect the wire harness clamp.

29. Install the No. 6 inverter bracket with 2 new and tighten the bolts to 71 inch lbs. (8 Nm).

30. Connect the frame wire to the inverter with converter assembly.
31. Fix the frame wire to the inverter with converter assembly with the bolt. Tighten the bolt to 71 inch lbs. (8 Nm).
32. Connect the air conditioning harness assembly to the inverter with converter assembly.
33. Connect the motor cable. Tighten bolt B and insert the motor cable into the inverter with converter assembly.
34. Fix the motor cable to the inverter with converter assembly with the 2 bolts (A). Tighten the bolt to 71 inch lbs. (8 Nm).
35. Install the inverter motor cable bracket with the 2 nuts (C).). Tighten the bolt to 71 inch lbs. (8 Nm).
36. Connect the generator cable. Tighten bolt B and insert the generator cable into the inverter with converter.
37. Fix the generator cable to the inverter with converter assembly with the 2 bolts (A).
38. Install the connector cover assembly with the 2 bolts. Tighten the bolt to 71 inch lbs. (8 Nm).
39. Install the service plug grip.
40. Connect the negative battery cable.
41. Install the battery service hole cover.
42. Install the luggage compartment floor mat.
43. Install coolant and to the inverter and bleed the system.

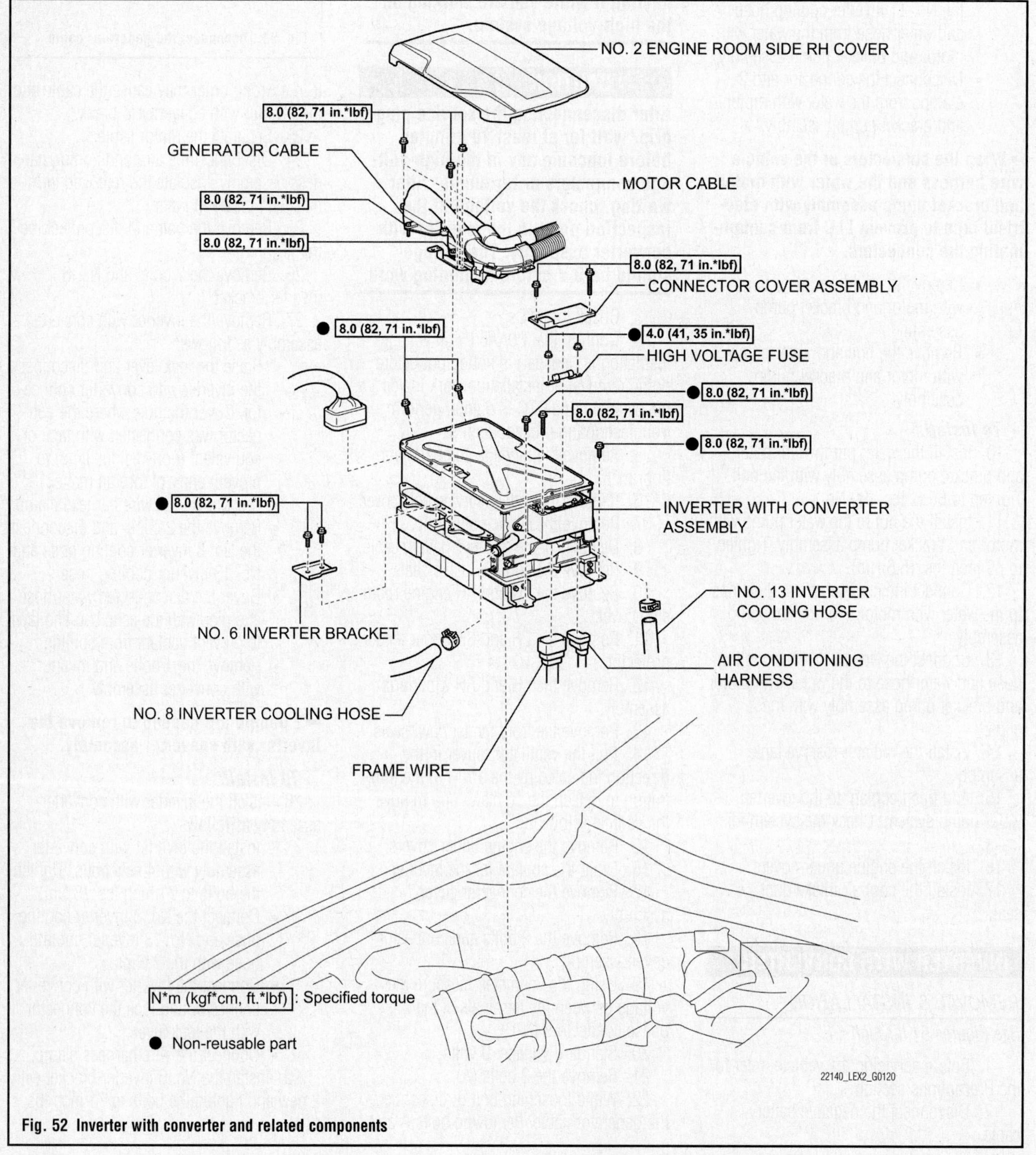

Fig. 52 Inverter with converter and related components

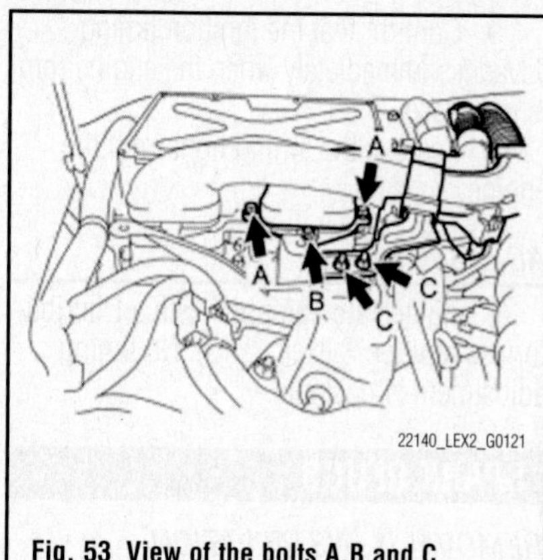

Fig. 53 View of the bolts A,B and C

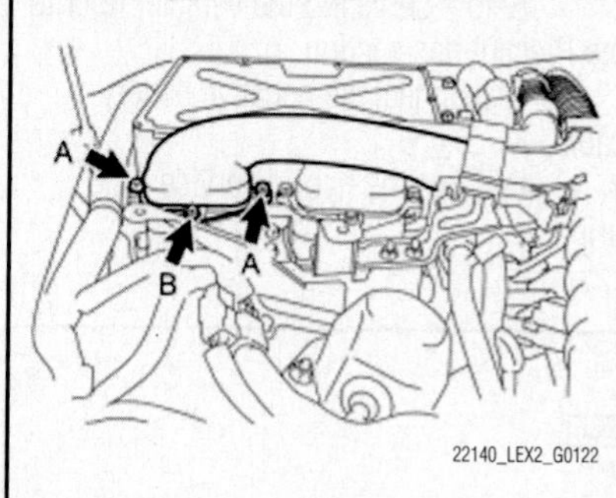

Fig. 54 View of the generator cable bolts A, and B

44. Check for coolant leaks.
45. Install the engine under covers.
46. Install the cowl top ventilator sub-assembly
47. Install the LH and RH fender to cowl seal.
48. Install LH and RH wiper arm assembly.
49. Install both upper front fender protectors.
50. Engage the 3 clips and install the No. 2 engine room side RH cover.
51. Install the engine room covers.
52. Install the cool air intake duct seal.
53. Some vehicle systems require initialization after reconnecting the cable to the negative battery terminal.

SERVICE PLUG GRIP

REMOVAL & INSTALLATION

See Figure 55.

1. Before servicing the vehicle, refer to the Precautions section.
2. Remove the luggage compartment floor mat.
3. Disconnect the negative battery cable. Wait at least 90 seconds after disconnecting the cable from the negative (-) battery terminal to prevent airbag and seat belt pretensioner activation.
4. Remove the service plug grip as follows:
   - Remove the battery service hole cover.
   - Wear insulated gloves. Remove the service plug grip after sliding the lever of the service plug grip.

⁂ CAUTION

Keep the removed service plug grip in your pocket to prevent other technicians from accidentally reconnecting it while you are servicing the vehicle. After removing the service plug grip, wait for at least 10 minutes before touching any of the high-voltage connectors or terminals.

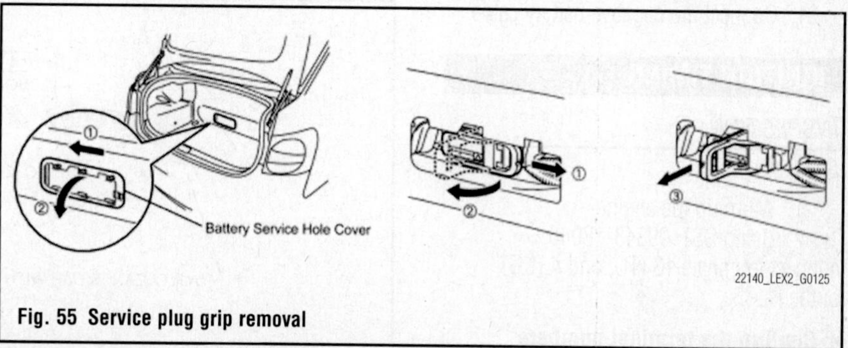

Fig. 55 Service plug grip removal

ENGINE ELECTRICAL

IGNITION SYSTEM

FIRING ORDER

See Figure 56.

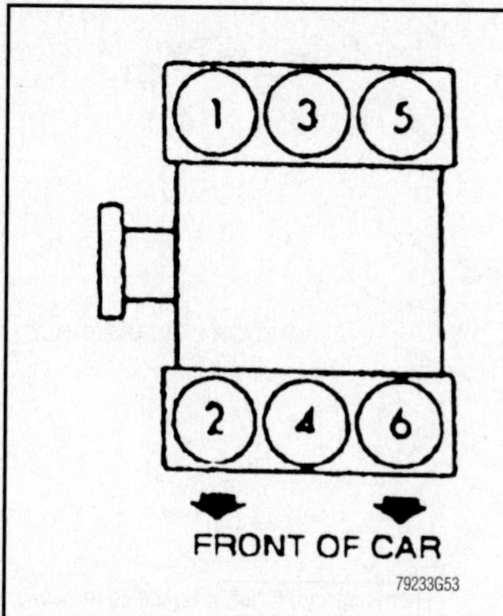

Fig. 56 3.5L (2GR-FSE) Engine Firing Order

IGNITION COIL

REMOVAL & INSTALLATION

See Figure 58.

1. Before servicing the vehicle, refer to the Precautions section.
2. Disconnect the negative battery cable.
3. Remove the hybrid service plug grip.

⁂ CAUTION

Before inspecting the high-voltage system or disconnecting the low voltage connector of the inverter with converter assembly, take safety precautions, such as wearing insulated gloves and removing the service plug grip to prevent electrical shocks. After removing the service plug grip, put it in your pocket to prevent other technicians from accidentally reconnecting it while you are working on the high-voltage system.

⁂ CAUTION

After disconnecting the service plug grip, wait for at least 10 minutes before touching any of the high-voltage connectors or terminals. After waiting, check the voltage at the inspection point in the inverter with converter assembly. The voltage should be 0 V before beginning work

4. Remove the cool air intake duct seal.
5. Remove the engine room side cover.
6. Remove the V-bank cover.
7. Disconnect the No 2 ventilation hose.

8. Remove the air cleaner cap and hose.
9. Detach the 3 wire harness clamps and disconnect the noise filter connector.
10. Disconnect the 6 ignition coil connectors.
11. Remove the 6 bolts and the 6 ignition coils.

To install:

12. Install the 6 ignition coils with the 6 bolts and tighten to 7 ft. lbs. (10 Nm).
13. Connect the 6 ignition coil connectors.
14. Attach the 3 wire harness clamps and connect the noise filter connector.
15. Install the air cleaner cap and hose.
16. Reconnect the No 2 ventilation hose.
17. Install the V-bank cover.
18. Install the engine room side cover.
19. Install the cool air intake duct seal.
20. Install the service plug grip.
21. Connect the negative battery cable.

IGNITION TIMING

INSPECTION

See Figure 57.

1. Warm up the engine.
2. Using SST: 09843-18040, connect terminals 13 (TC) and 4 (CG) of the DLC3.

➡Confirm the terminal numbers before connecting them. Connecting the wrong terminals can damage the engine.

➡Turn off all electrical systems before connecting the terminals.

➡Perform this inspection after the cooling fan motor is turned off.

3. Remove the V-bank cover sub-assembly.
4. Pull out the red lead wire harness.
5. Connect the tester terminal of the timing light to the red lead wire as shown in the illustration.

➡Use a timing light which detects the No. 1 cylinder ignition signal.

6. Check the ignition timing at idle. Standard ignition timing: 8 to 12° BTDC at idle.

➡When checking the ignition timing, the transmission should be in the neutral position.

➡Run the engine at 1000 to 1300 rpm for 5 seconds, and then check that the engine rpm returns to idle speed.

7. Disconnect terminals 13 (TC) and 4 (CG) of the DLC3.
8. Check the ignition timing at idle. Standard ignition timing: 7 to 24° BTDC at idle.
9. Confirm that the ignition timing advances immediately when the engine rpm is increased.
10. Remove the timing light from the engine.

ADJUSTMENT

All engines are equipped with a Distributorless Ignition System (DIS). No timing adjustment is possible.

SPARK PLUGS

REMOVAL & INSTALLATION

See Figure 58.

1. Before servicing the vehicle, refer to the Precautions section.
2. Disconnect the negative battery cable.
3. Remove the hybrid service plug grip.

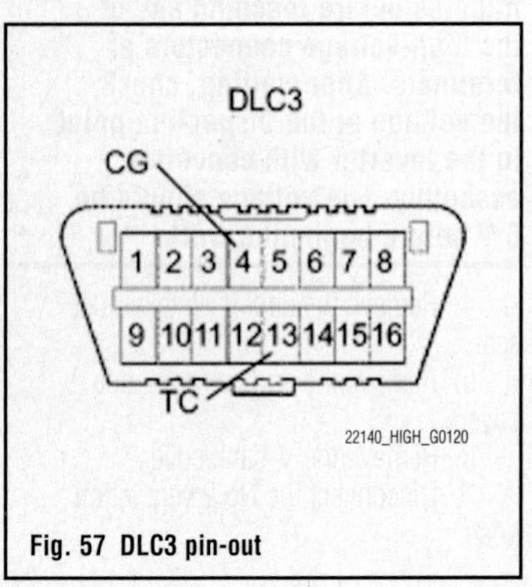

Fig. 57 DLC3 pin-out

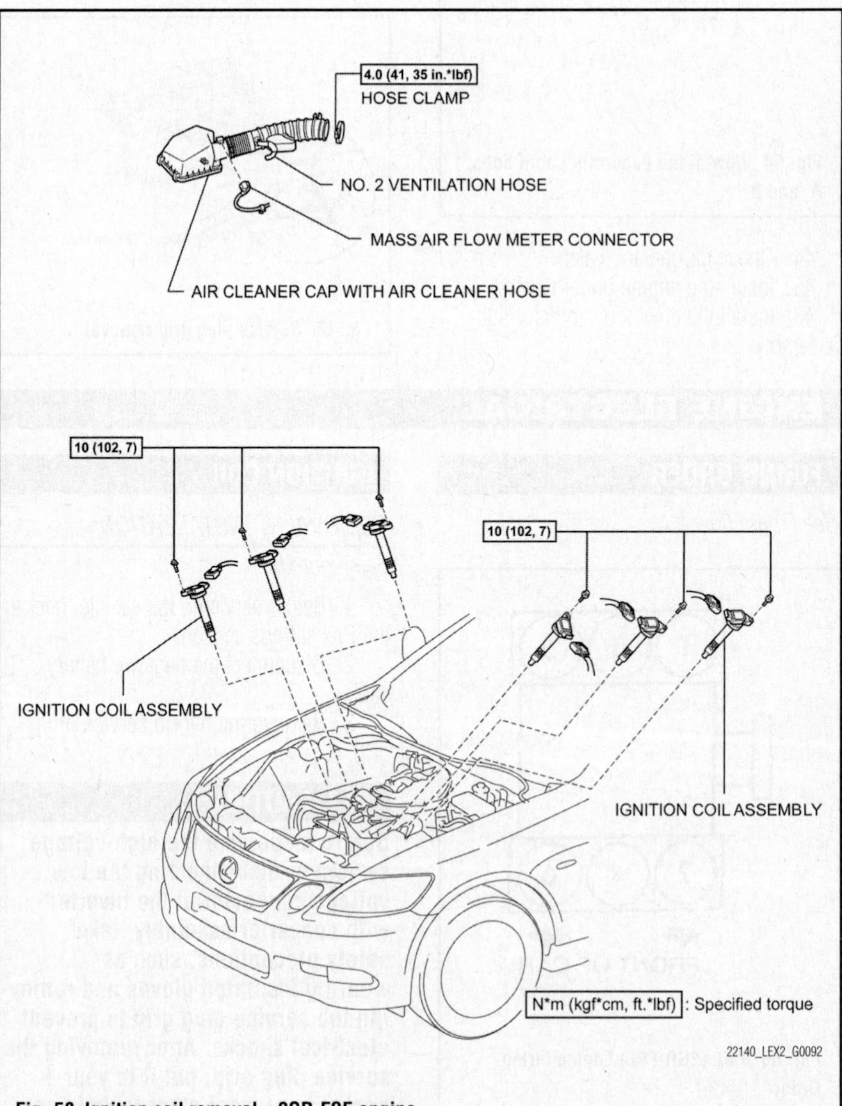

Fig. 58 Ignition coil removal—2GR-FSE engine

✲✲ CAUTION

Before inspecting the high-voltage system or disconnecting the low voltage connector of the inverter with converter assembly, take safety precautions, such as wearing insulated gloves and removing the service plug grip to prevent electrical shocks. After removing the service plug grip, put it in your pocket to prevent other technicians from accidentally reconnecting it while you are working on the high-voltage system.

✲✲ CAUTION

After disconnecting the service plug grip, wait for at least 10 minutes before touching any of the high-voltage connectors or terminals. After waiting, check the voltage at the inspection point in the inverter with converter assembly. The voltage should be 0 V before beginning work

4. Remove the cool air intake duct seal.
5. Remove the engine room side cover.
6. Remove the V-bank cover.
7. Disconnect the No 2 ventilation hose.
8. Remove the air cleaner cap and hose.
9. Detach the 3 wire harness clamps and disconnect the noise filter connector.
10. Disconnect the 6 ignition coil connectors.
11. Remove the 6 bolts and the 6 ignition coils.
12. Remove the 6 spark plugs.

To install:

13. Check that the spark plug gap is between: 0.039–0.043 inch.
14. Install the 6 spark plugs and tighten to 13 ft. lbs. (16 Nm).
15. Install the 6 ignition coils with the 6 bolts and tighten to 7 ft. lbs. (10 Nm).
16. Connect the 6 ignition coil connectors.
17. Attach the 3 wire harness clamps and connect the noise filter connector.
18. Install the air cleaner cap and hose.
19. Reconnect the No 2 ventilation hose.
20. Install the V-bank cover.
21. Install the engine room side cover.
22. Install the cool air intake duct seal.
23. Install the service plug grip.
24. Connect the negative battery cable.

ENGINE MECHANICAL

ACCESSORY DRIVE BELTS

ACCESSORY BELT ROUTING

See Figure 59.

Refer to the accompanying illustration for belt routing.

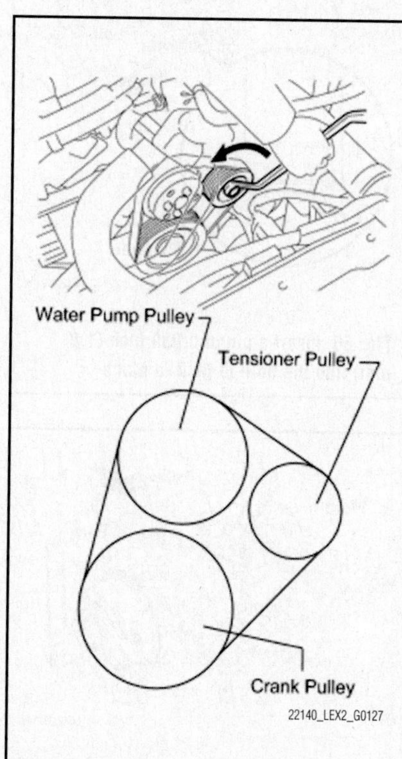

Fig. 59 2GR-FSE drive belt routing—GS450h

INSPECTION

See Figure 60.

Visually check the drive belt for excessive wear, frayed cords, etc. If any defect has been found, replace the drive belt.

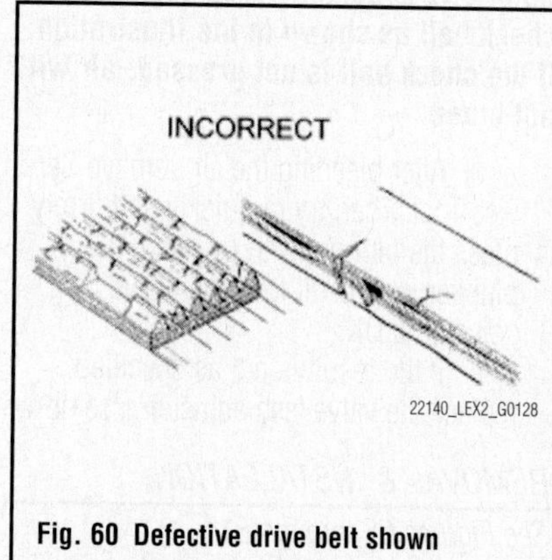

Fig. 60 Defective drive belt shown

ADJUSTMENT

The drive belt system is equipped with an automatic belt tensioner.

REMOVAL & INSTALLATION

See Figure 61.

1. Before servicing the vehicle, refer to the Precautions section.
2. Remove the cool air duct seal.
3. Remove the No. 1 air cleaner inlet.
4. Rotate the tensioner pulley counterclockwise to loosen the belt tension. Then remove the V belt.

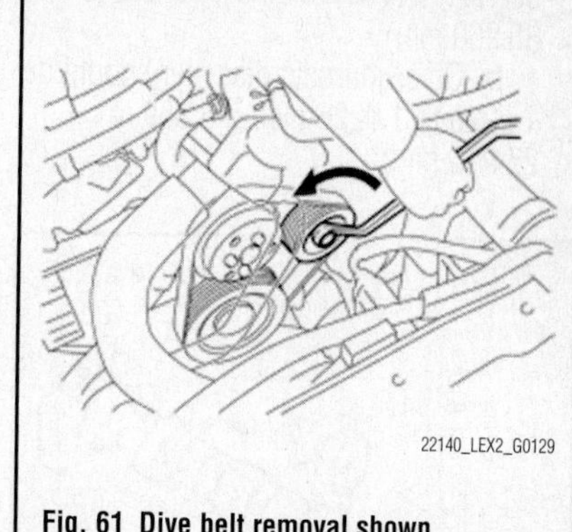

Fig. 61 Dive belt removal shown

➡The pulley bolt for the belt tensioner has a left-handed thread. Do not perform this procedure from underneath the vehicle, as the tensioner may be damaged.

To install:

5. Rotate the tensioner pulley counterclockwise, and then install the V belt.

➡After installing a new belt, run the engine for approximately 5 minutes and then recheck the tension.

6. Install the air cleaner inlet with the bolt. And tighten to 44 inch lbs. (5.0 Nm).
7. Install the cool air duct seal.

CAMSHAFT AND VALVE LIFTERS

INSPECTION

See Figures 62 through 64.

1. To inspect the camshaft, place the camshaft on V-blocks.
   a. Using a dial indicator, measure the circle run out at the center journal.
2. Maximum run out 0.0016 inches (0.04 mm).
   a. If the run out is greater than the maximum, replace the camshaft.

➡**Check the oil clearance after replacing the camshaft.**

3. Using a micrometer, measure the cam lobe height.
4. Using a micrometer, measure the journal diameter.
   a. 3.5L No.1 journal diameter should be 1.4152 to 1.4157 inch (35.946 to 35.960 mm)
   b. Other journals diameter should be 1.0220 to 1.0226 inch (25.959 to 25.975 mm)

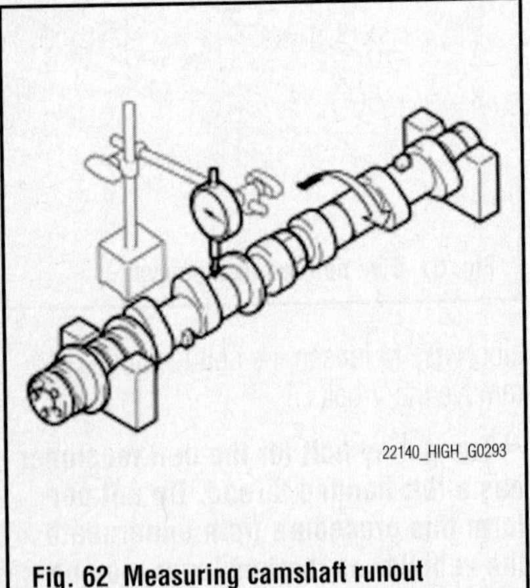

22140_HIGH_G0293

Fig. 62 Measuring camshaft runout

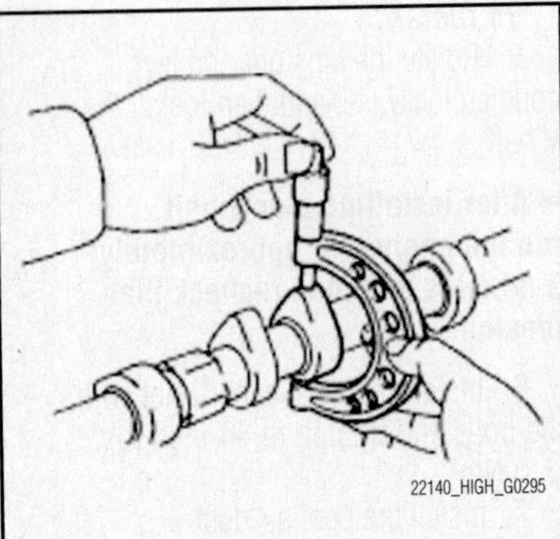

22140_HIGH_G0295

Fig. 63 Measuring camshaft lobe height

ITEM	SPECIFIED CONDITION
Intake	44.318 to 44.418 mm (1.7448 to 1.7487 in.)
Exhaust	44.341 to 44.441 mm (1.7457 to 1.7496 in.)

22140_LEX2_G0145

Fig. 64 Standard camshaft lobe height—3.5L engine

   c. If the journal diameter is not as specified, check the oil clearance.
5. To inspect the lifters, place the valve lash adjuster assembly into a container filled with engine oil.
   a. Insert Service Tool tip into the valve lash adjuster assembly's plunger and use the tip to press down on the check ball inside the plunger.
   b. Squeeze Service Tool and valve lash adjuster assembly together to move the plunger up and down 5 to 6 times.
   c. Check the movement of the plunger and bleed the air. If plunger moves up and down freely, valve lash adjuster is OK.

➡**When bleeding air from the high-pressure chamber, make sure that the tip of SST is actually pressing the check ball as shown in the illustration. If the check ball is not pressed, air will not bleed.**

   d. After bleeding the air, remove Service Tool. Then try to quickly and firmly press the plunger with by hand. If plunger is difficult to move, valve lash adjuster is OK.
   e. If the result is not as specified, replace the valve lash adjuster assembly.

REMOVAL & INSTALLATION

See Figures 65 through 81.

Before servicing any vehicle, please be sure to read all of the following precautions, which deal with personal safety, prevention of component damage, and important points to take into consideration when servicing a motor vehicle:

CAUTION

The GS450h hybrid system contains a 288V high-voltage system with a strong alkali solution of potassium hydroxide. Be sure to follow the instructions in this manual to handle the system correctly. Failure to do so may result in serious injury or electrocution.

Engineer must undergo special training to be able to perform high-voltage system inspection and servicing.

CAUTION

All high-voltage wire harness connectors are colored orange. The HV battery and other high-voltage components have "High Voltage" caution labels. Do not carelessly touch these wires and components.

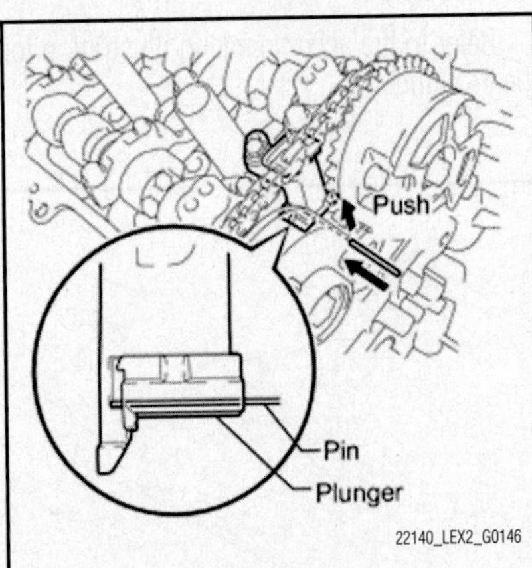

22140_LEX2_G0146

Fig. 65 Insert a pin of 0.039 inch (1.0 mm) into the hole to fix it in place

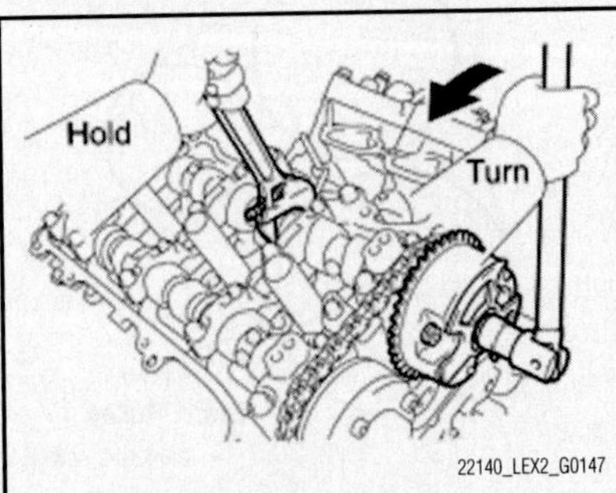

22140_LEX2_G0147

Fig. 66 Hold the hexagonal portion of the camshaft with a wrench, and remove the 2 bolts and 2 camshaft timing gears

⁂ CAUTION

Before inspecting or servicing the high-voltage system, be sure to follow safety measures, such as wearing insulated gloves and removing the service plug to prevent electrocution. Carry the removed service plug in your pocket to prevent other technicians from reinstalling it while you are servicing the vehicle.

⁂ CAUTION

After removing the service plug, wait 10 minutes before touching any of the high-voltage connectors and terminals.

Before servicing any vehicle, please be sure to read all of the following precautions, which deal with personal safety, prevention of component damage, and important points to take into consideration when servicing a motor vehicle:

1. Discharge the fuel system.
2. Recover the refrigerant from the A/C system using a refrigerant recovery unit.
3. Place the front wheels in the straight ahead position.

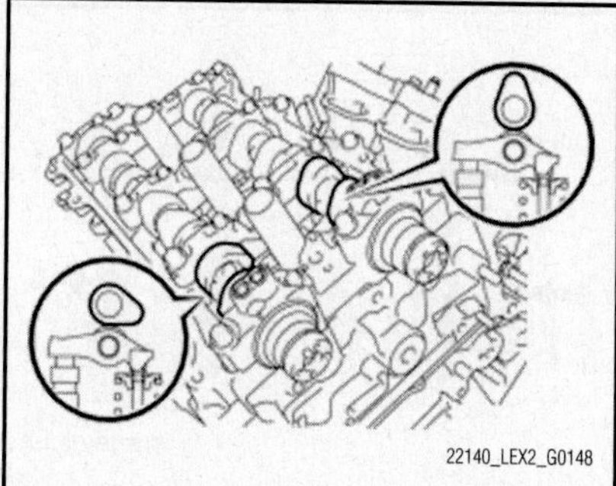

Fig. 67 Make sure that the knock pin of the camshaft is positioned as shown

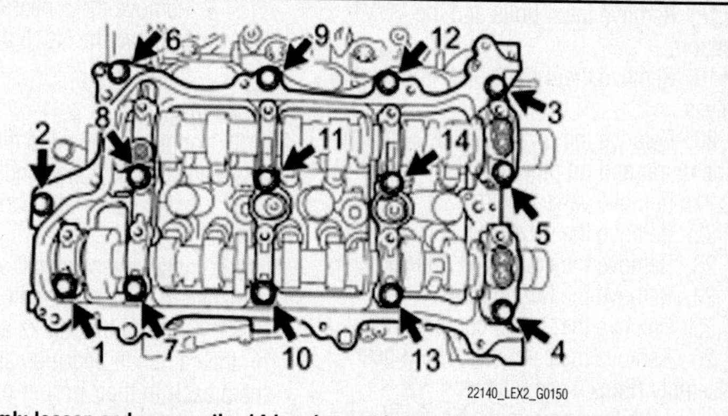

Fig. 69 Uniformly loosen and remove the 14 bearing cap bolts in the sequence shown

4. Disconnect the negative battery cable.
5. Remove the hybrid service plug grip.

⁂ CAUTION

Before inspecting the high-voltage system or disconnecting the low voltage connector of the inverter with converter assembly, take safety precautions, such as wearing insulated gloves and removing the service plug grip to prevent electrical shocks. After removing the service plug grip, put it in your pocket to prevent other technicians from accidentally reconnecting it while you are working on the high-voltage system.

⁂ CAUTION

After disconnecting the service plug grip, wait for at least 10 minutes before touching any of the high-voltage connectors or terminals. After waiting, check the voltage at the inspection point in the inverter with converter assembly. The voltage should be 0 V before beginning work.

6. Remove the luggage floor mat.
7. Remove the service plug cover.
8. Remove the service plug.
9. Disconnect the negative battery cable.
10. Drain the engine oil and coolant.
11. Remove the cool air intake duct seal.
12. Remove the engine room side cover.
13. Remove the V-bank cover.
14. Remove hoses and electrical connectors.
15. Remove the air intake surge tank sub-assembly.
16. Remove the ignition coil connectors and coils.
17. Remove the ventilation valve.

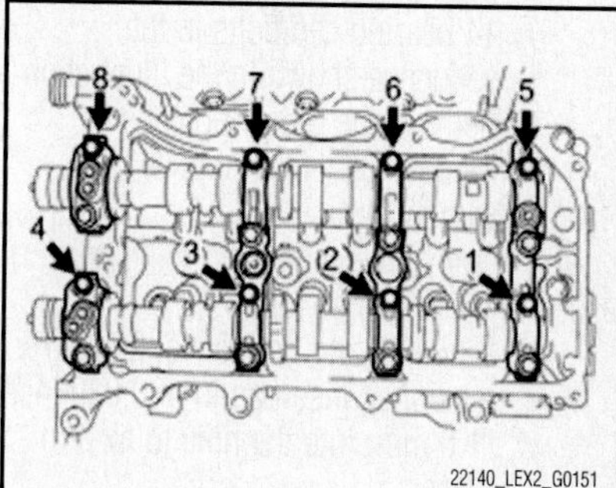

Fig. 70 Uniformly loosen and remove the 8 bearing cap bolts in the sequence shown

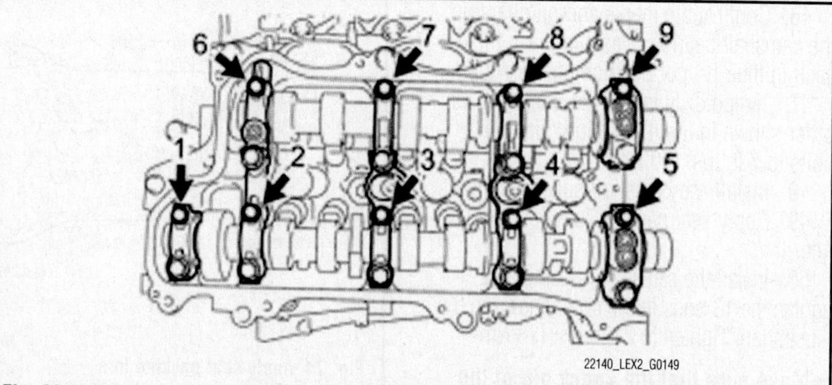

Fig. 68 Uniformly loosen and remove the 9 bearing cap bolts in the sequence shown

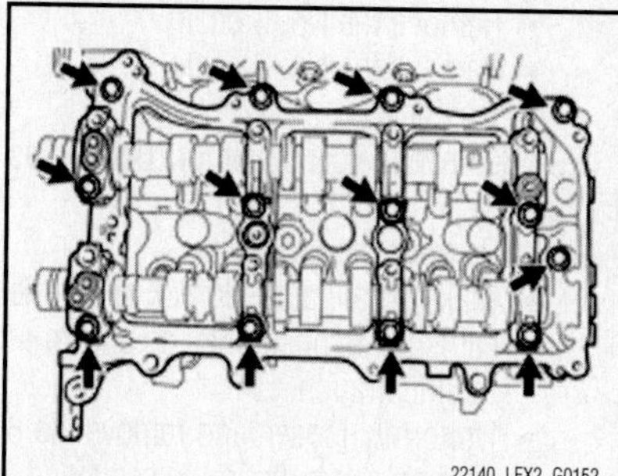

Fig. 71 Uniformly loosen and remove the 13 bearing cap bolts in the sequence shown

18. Remove the 4 bolts and the 4 VVT sensors.

19. Remove the 4 bolts and 4 oil control valves.

20. Remove the oil check valve bolt, oil pipe union and oil pipe.

21. Remove valve covers.

22. Remove the drive belt.

23. Remove the crankshaft pulley.

24. Remove the timing cover.

25. Remove the timing components.

26. Remove the camshaft timing gear assembly (Bank 1) as follows:

- While raising up the No. 2 chain tensioner, insert a pin of 0.039 inch (1.0 mm) into the hole to fix it in place.
- Hold the hexagonal portion of the camshaft with a wrench, and remove the 2 bolts and 2 camshaft timing gears.
- Remove the No. 2 chain.
- Remove the No. 2 Chain tensioner

27. Remove the camshaft for (Bank 1) as follows:

- Remove the 3 gaskets.
- Make sure that the knock pin of the camshaft is positioned as shown in the illustration.
- Uniformly loosen and remove the 9 bearing cap bolts in the sequence shown in the illustration.
- Uniformly loosen and remove the 14 bearing cap bolts in the sequence shown in the illustration.
- Remove the 6 bearing caps.
- Remove the No. 1 and No. 2 camshafts.

28. Remove the camshaft timing gear assembly (Bank 2) as follows:

- While raising up the No. 2 chain tensioner, insert a pin of 0.039 inch (1.0 mm) into the hole to fix it in place.
- Hold the hexagonal portion of the camshaft with a wrench, and remove the 2 bolts and 2 camshaft timing gears.
- Remove the No. 3 chain.
- Remove the No. 3 chain tensioner.

29. Remove the camshaft for (Bank 2) as follows:

- Remove the 3 gaskets.
- Make sure that the knock pin of the camshaft is positioned as shown in the illustration.
- Uniformly loosen and remove the 8 bearing cap bolts in the sequence shown in the illustration.
- Uniformly loosen and remove the 13 bearing cap bolts in the sequence shown in the illustration.
- Remove the 5 bearing caps.
- Remove the No. 3 and No. 4 camshafts.

To install:

30. Apply engine oil to the camshaft journals, camshaft housings and bearing caps.

31. Install camshaft bearing cap for (Bank 1).

32. Install the camshaft and No. 2 camshaft to the camshaft housing.

33. Confirm the marks and numbers on the camshaft bearing caps and place them each in their proper position and direction.

34. Temporarily install the 8 bolts in the order shown in the illustration.

35. Tighten bolts to 7 ft. lbs. (10 Nm).

36. Install camshaft assembly sub-assembly.

37. Apply seal packing in a continuous line as shown in the illustration.

38. Install the camshaft housing, and install the 12 bolts in the order shown in the illustration. Tighten to 21 ft. lbs. (28 Nm).

➡Make sure that the knock pin of the camshaft is positioned as shown in the illustration before installing the camshaft housing.

39. Tighten the 9 bolts in the order shown in the illustration. Tighten bolts to 12 ft. lbs. (16 Nm).

40. Install 3 new gaskets as shown in the illustration.

41. Apply engine oil to the camshaft journals, camshaft housings and bearing caps.

42. Install camshaft bearing cap for (Bank 1).

43. Install the camshaft and No. 2 camshaft to the camshaft housing.

44. Confirm the marks and numbers on the camshaft bearing caps and place them each in their proper position and direction.

45. Install the camshaft No. 3 and camshaft No. 4 to the camshaft housing.

46. Confirm the marks and numbers on the camshaft bearing caps and place them each in their proper position and direction.

47. Temporarily install the 8 bolts in the order shown in the illustration. Tighten bolts to 7 ft. lbs. (10 Nm).

48. Install the camshaft sub-assembly.

49. Apply seal packing in a continuous line.

50. Install the camshaft housing and tighten the 13 bolts in the order shown in the illustration. Tighten to 21 ft. lbs. (28 Nm).

➡Make sure that the knock pin of the camshaft is positioned as shown in the illustration before installing the camshaft housing.

51. Tighten the 8 bolts in the order shown in the illustration. Tighten to 12 ft. lbs. (16 Nm).

52. Install the chain tensioner and tighten the mounting the bolt to 15 ft. lbs. (21 Nm).

53. While pushing in the tensioner, insert a pin of 1.0 mm (0.039 in.) into the hole to fix it in place.

54. Install the camshaft timing gears and No.2 chain for (Bank 1) as follows:

55. Align the mark plate (yellow) with the

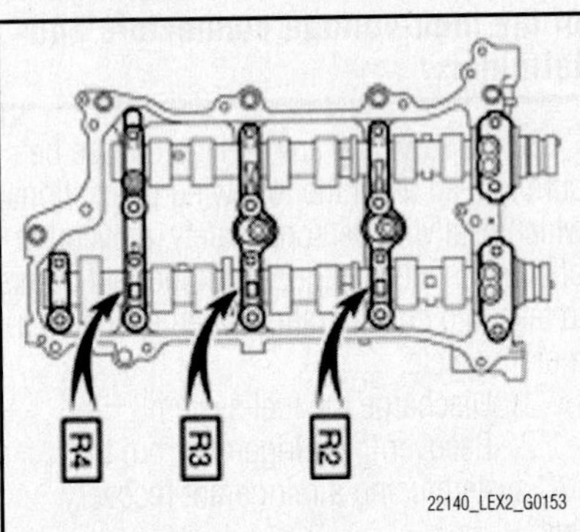

Fig. 72 Confirm the marks and numbers on the camshaft bearing caps

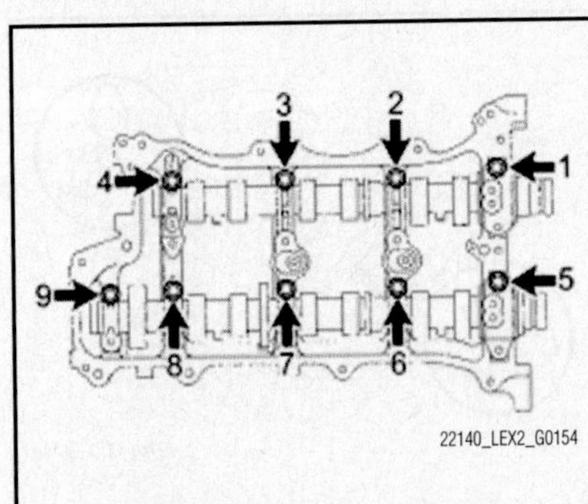

Fig. 73 Temporarily install the 8 bolts in the order shown

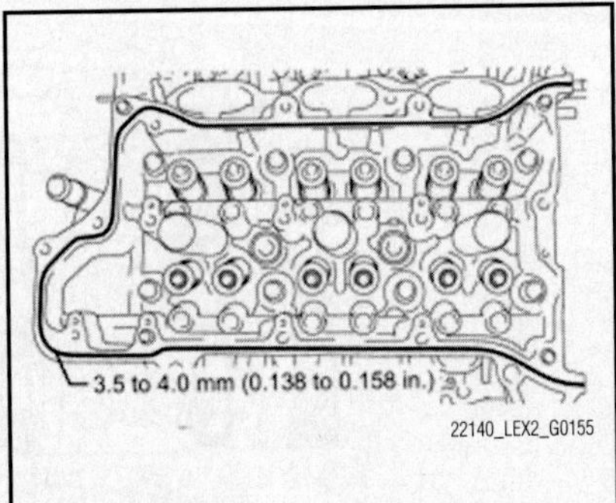

Fig. 74 Apply seal packing in a continuous line

timing marks (1 dot mark) of the camshaft timing gears as shown in the illustration.

56. Apply a small amount of engine oil to the bolt threads and bolt-seating surface.

57. Align the knock pin of the camshaft with the pin hole of the camshaft timing gear. Install the camshaft timing gear and camshaft timing exhaust gear with the No. 2 chain installed

58. Hold the hexagonal portion of the camshaft with a wrench, and tighten the 2 bolts to 74 ft. lbs. (100 Nm).

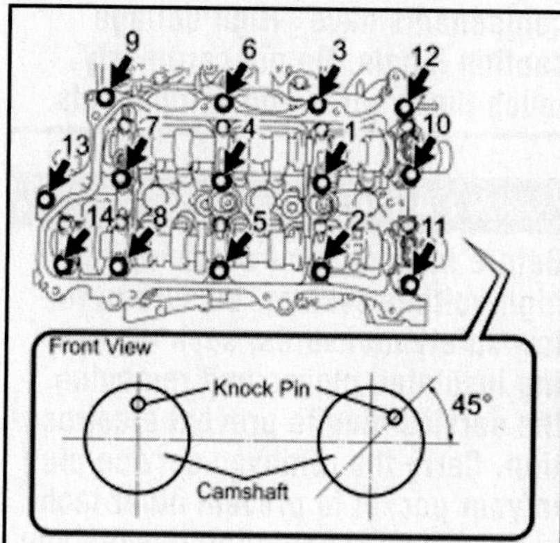

Fig. 75 Install the 12 bolts in the order shown

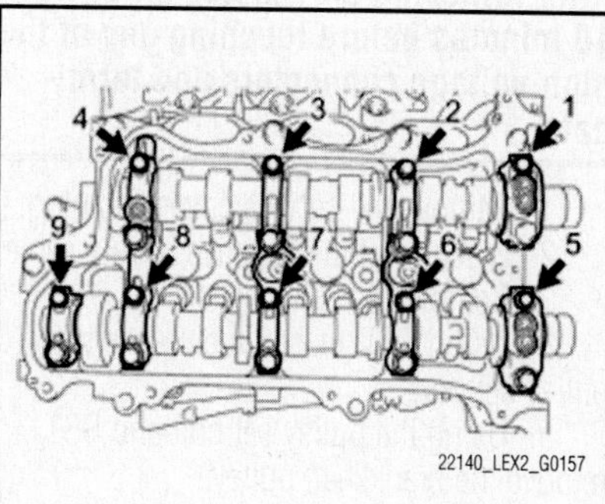

Fig. 76 Tighten the 9 bolts in the order shown

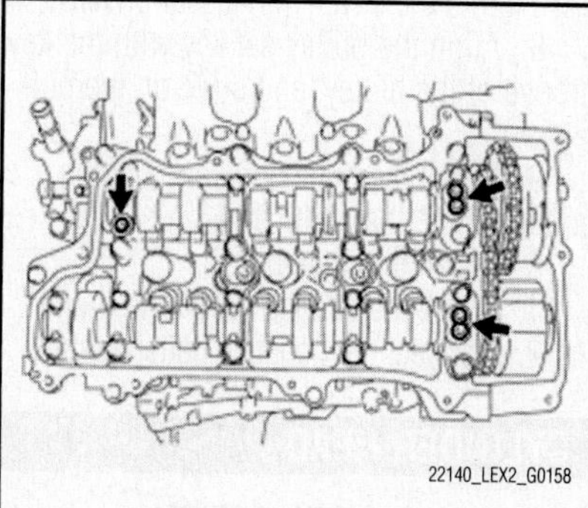

Fig. 77 Install 3 new gaskets as shown

59. Remove the pin from the No. 2 chain tensioner.

60. Install the chain tensioner and tighten the mounting the bolt to 15 ft. lbs. (21 Nm).

61. While pushing in the tensioner, insert a pin of 1.0 mm (0.039 in.) into the hole to fix it in place.

62. Install the camshaft timing gears No. 2 for (Bank 2) as follows:

63. Align the mark plate (yellow) with the timing marks (1 dot mark) of the camshaft timing gears as shown in the illustration.

64. Apply a small amount of engine oil to the bolt threads and bolt-seating surface.

65. Align the knock pin of the camshaft with the pin hole of the camshaft timing gear. Install the camshaft timing gear and camshaft timing exhaust gear with the No. 2 chain installed

66. Hold the hexagonal portion of the camshaft with a wrench, and tighten the 2 bolts to 74 ft. lbs. (100 Nm).

67. Remove the pin from the No. 2 chain tensioner.

68. Install the timing components.

69. Install the timing cover.

70. Install the crankshaft pulley.

71. Install the drive belt.

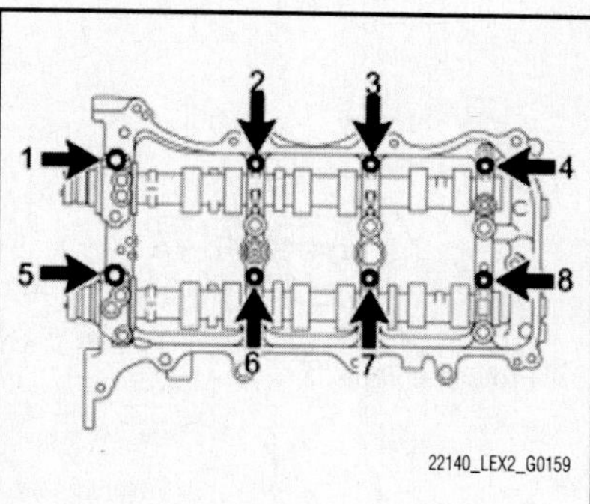

Fig. 78 Temporarily install the 8 bolts in the order shown

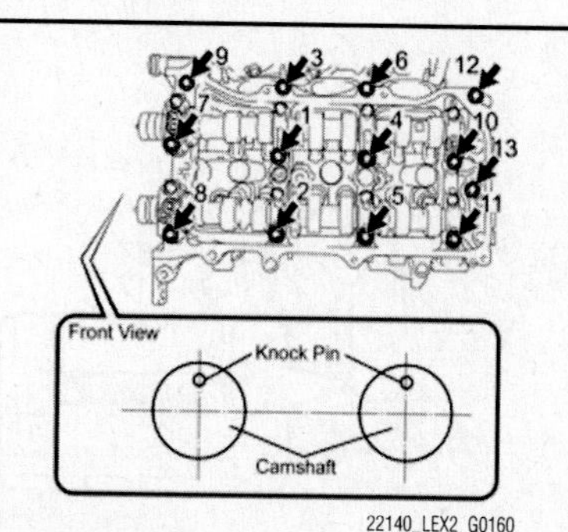

Fig. 79 Install the camshaft housing and tighten the 13 bolts in the order shown

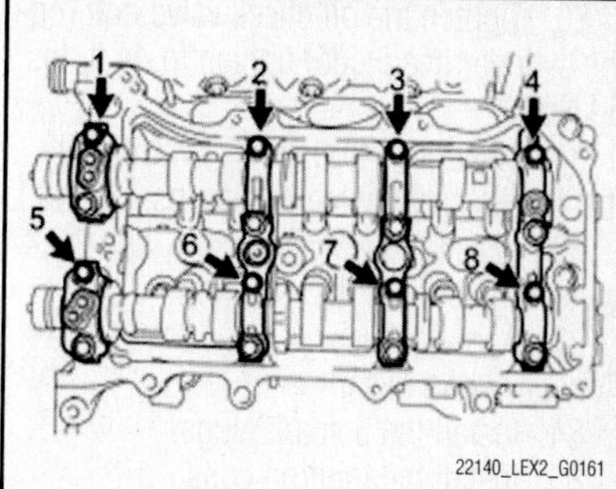

Fig. 80 Tighten the 8 bolts in the order shown

72. Install the 4 bolts and 4 oil control valves.

73. Install the 4 bolts and the 4 VVT sensors.

74. Install the oil check valve bolt, oil pipe union and oil pipe.

75. Install the ventilation valve.

76. Install the head cover with the 14 bolts as follows:

- Tighten bolt A to 15 ft. lbs. (21 Nm)
- Tighten the remainder of the bolts to 7ft. lbs. (10 Nm).

➡Do not start the engine for at least 2 hours after the installation.

77. Install the oil control valve filter RH and LH to the oil pipe union. Install new gaskets and temporarily install the oil pipe (on the head cover side).

78. Install a new gasket and temporarily install the oil pipe (on the cylinder head side) with the oil pipe check valve bolt.

79. Tighten the oil pipe union (on the head cover side) tighten to 44 ft. lbs. 60 Nm).

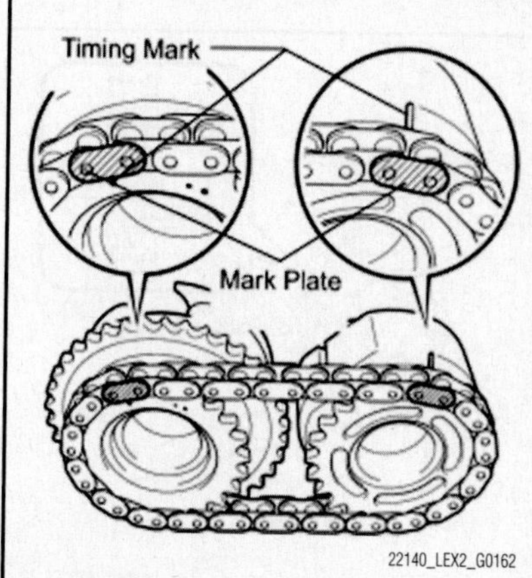

Fig. 81 Align the mark plate (yellow) with the timing marks (1 dot mark) of the camshaft timing gears

80. Tighten the oil check valve bolt (on the cylinder head side) tighten to 44 ft. lbs. 60 Nm).

81. Install the 4 oil control valves with the 4 bolts and tighten to 7 ft. lbs. (10 Nm).

82. Install the 4 VVT sensors with the 4 bolts. Tighten 7 ft. lbs. (10 Nm).

83. Apply adhesive around the ventilation valve and install.

84. Install the 6 spark plugs.

85. Install the ignition coils.

86. Install the air intake surge tank sub-assembly.

87. Install all hoses and electrical connectors.

88. Install the service plug and cover.

89. Connect the negative battery cable.

90. Install the luggage floor mat.

CATALYTIC CONVERTER

REMOVAL & INSTALLATION

The catalytic converter is integrated with the exhaust manifold and exhaust pipe and must be removed as an assembly. Refer to the appropriate section.

CRANKSHAFT DAMPER

REMOVAL & INSTALLATION

See Figures 82 and 83.

1. Using SST, loosen the crankshaft pulley set bolt.
2. Using the pulley set bolt and SST, remove the crankshaft pulley.

To install:

3. Align the pulley set key with the key groove of the pulley, and slide on the pulley.
4. Install the pulley bolt and tighten bolt to 184 ft. lbs. (240 Nm).

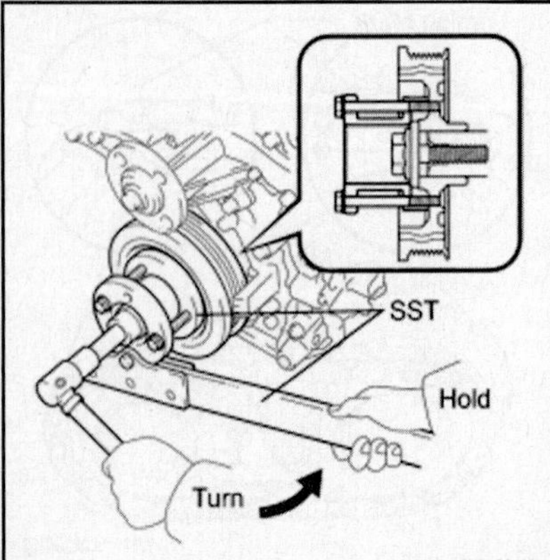

Fig. 82 Loosen the crankshaft pulley set bolt using SST: 09213-70011 and 09330-00021

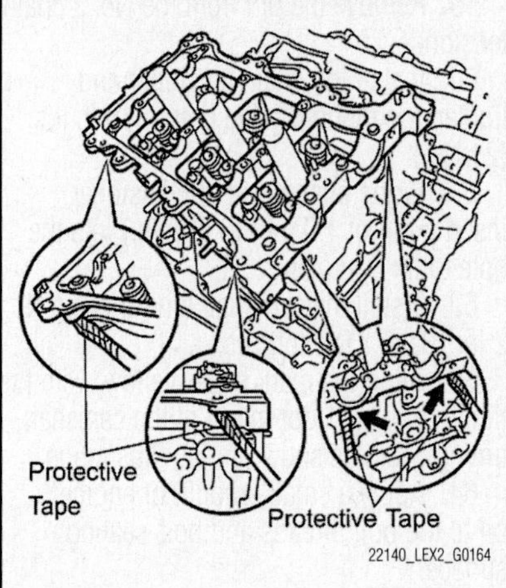

Fig. 83 Using the pulley set bolt and SST, remove the crankshaft pulley

CRANKSHAFT FRONT SEAL

REMOVAL & INSTALLATION

See Figures 84 and 85.

⁂ CAUTION

The GS450h hybrid system contains a 288V high-voltage system with a strong alkali solution of potassium hydroxide. Be sure to follow the instructions in this manual to handle the system correctly. Failure to do so may result in serious injury or electrocution. Engineer must undergo special training to be able to perform high-voltage system inspection and servicing.

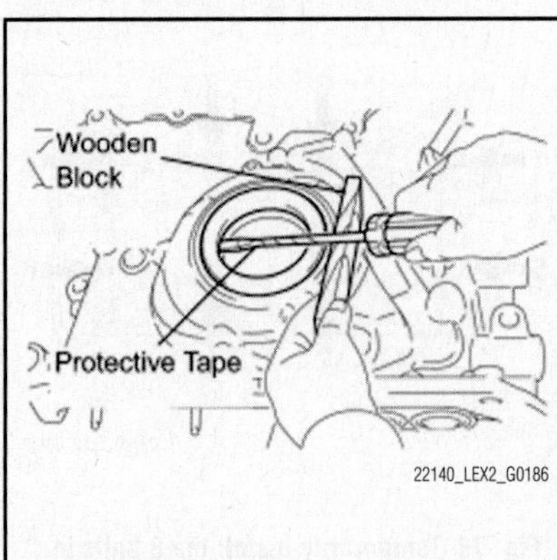

Fig. 84 Front crank seal removal

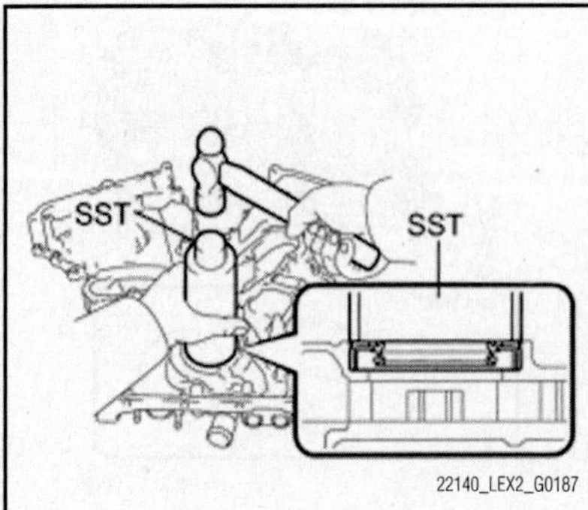

Fig. 85 Front crank seal installation

⁂ CAUTION

All high-voltage wire harness connectors are colored orange. The HV battery and other high-voltage components have "High Voltage" caution labels. Do not carelessly touch these wires and components.

⁂ CAUTION

Before inspecting or servicing the high-voltage system, be sure to follow safety measures, such as wearing insulated gloves and removing the service plug to prevent electrocution. Carry the removed service plug in your pocket to prevent other technicians from reinstalling it while you are servicing the vehicle.

⁂ CAUTION

After removing the service plug, wait 10 minutes before touching any of the high-voltage connectors and terminals.

1. Remove the service cover and plug.
2. Disconnect the negative battery cable.
3. Remove the drive belt.
4. Using SST, loosen the crankshaft pulley set bolt.
5. Using the pulley set bolt and SST, remove the crankshaft pulley.
6. Using a screwdriver and wooden block, pry out the oil seal.

To install:

7. Install the front crankshaft seal and drive it into the cover with a seal driver.
8. Align the pulley set key with the key groove of the pulley, and slide on the pulley.
9. Install the pulley bolt and tighten bolt to 184 ft. lbs. (240 Nm).
10. Install the drive belt.
11. Install the service cover and plug.
12. Connect the negative battery cable.

CYLINDER HEAD

REMOVAL & INSTALLATION

See Figures 86 through 93.

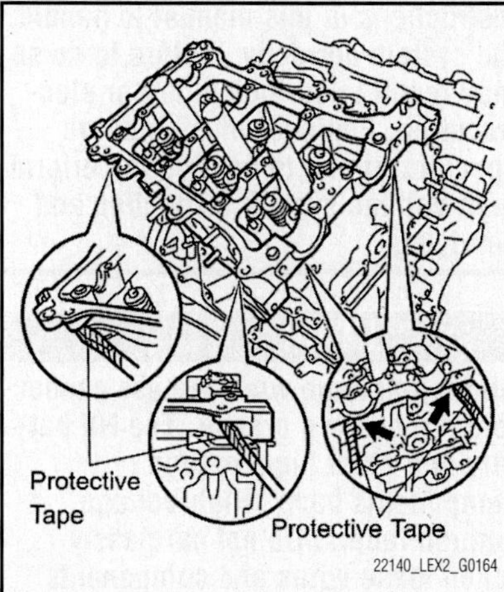

Fig. 86 Location of the prying points when removing the camshaft housing—3.5L (2GR-FSE) engine

CAUTION

The GS450h hybrid system contains a 288V high-voltage system with a strong alkali solution of potassium hydroxide. Be sure to follow the instructions in this manual to handle the system correctly. Failure to do so may result in serious injury or electrocution. Engineer must undergo special training to be able to perform high-voltage system inspection and servicing.

CAUTION

All high-voltage wire harness connectors are colored orange. The HV battery and other high-voltage components have "High Voltage" caution labels. Do not carelessly touch these wires and components.

CAUTION

Before inspecting or servicing the high-voltage system, be sure to follow safety measures, such as wearing insulated gloves and removing the service plug to prevent electrocution. Carry the removed service plug in your pocket to prevent other technicians from reinstalling it while you are servicing the vehicle.

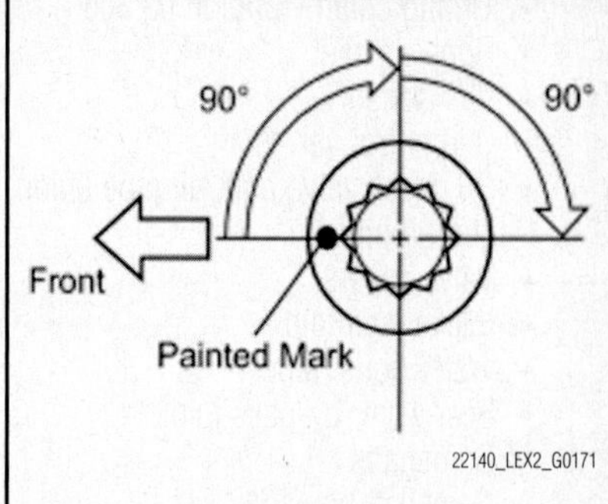

Fig. 91 Mark the cylinder head bolt heads with paint as shown in the illustration

CAUTION

After removing the service plug, wait 10 minutes before touching any of the high-voltage connectors and terminals.

1. Before servicing the vehicle, refer to the Precautions Section.
2. Remove the service grip.
3. Disconnect the negative battery cable.
4. Drain the cooling system.
5. Drain the engine oil.
6. Relieve the fuel system pressure.
7. Remove or disconnect the following:
   - Timing chain

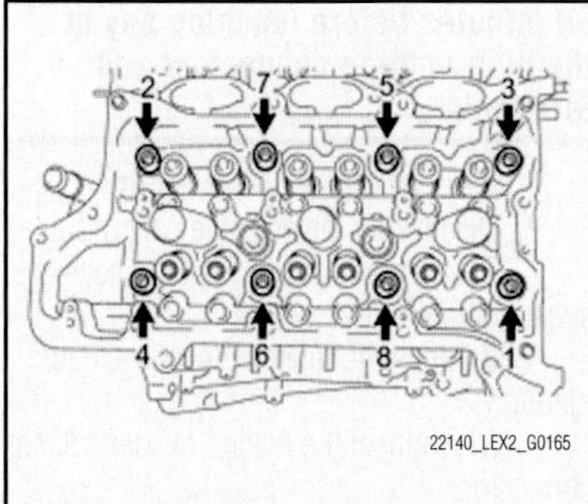

Fig. 87 Loosen the cylinder head mounting bolts in several steps (Bank 1)

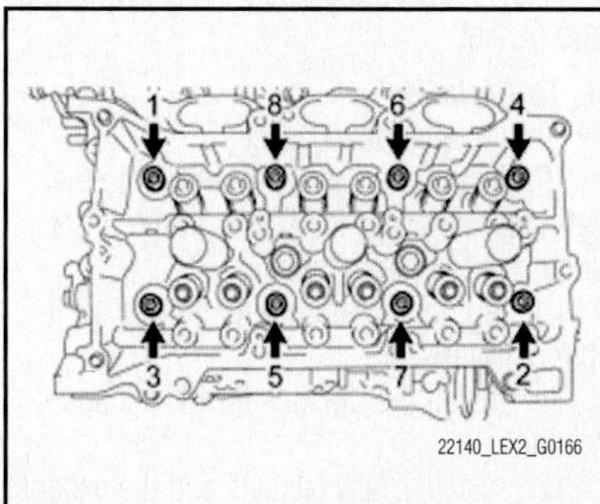

Fig. 88 Loosen the cylinder head mounting bolts in several steps (Bank 2)

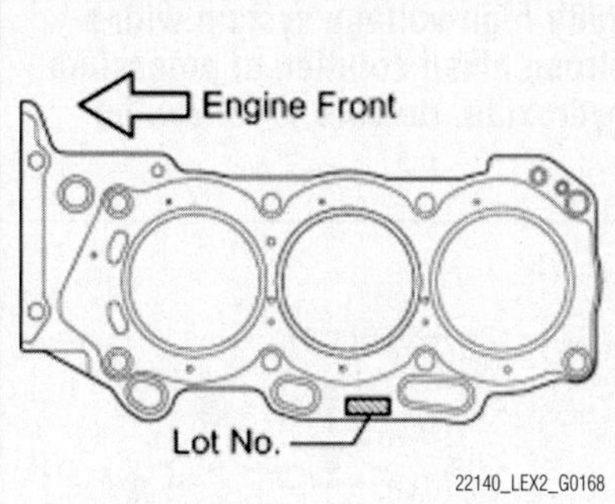

Fig. 89 Place the cylinder head gasket on the cylinder block surface with the front face of the Lot No. stamp upward.

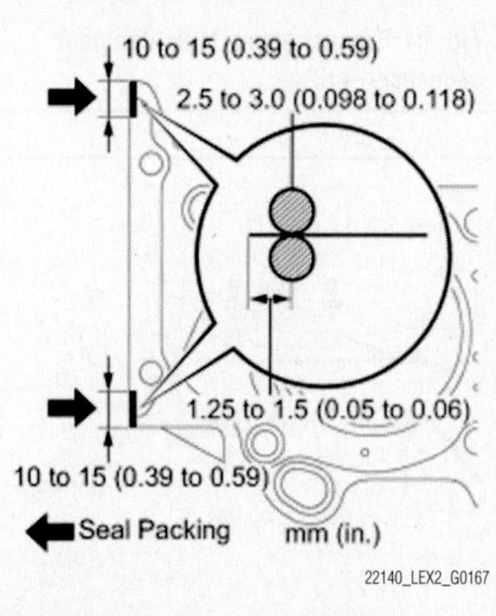

Fig. 90 Apply a continuous line of the seal packing to a new cylinder head

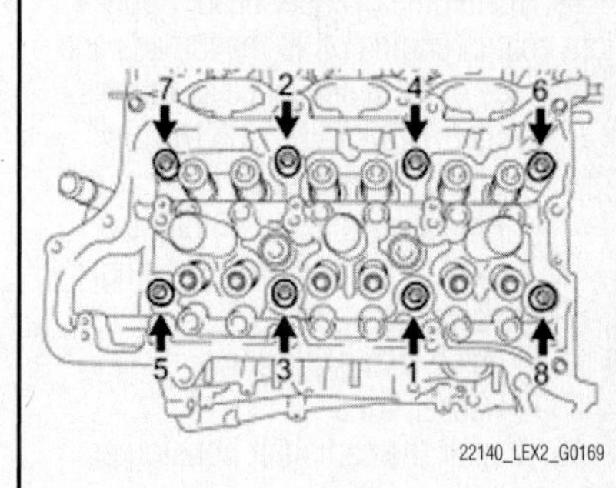

Fig. 92 Cylinder head tightening sequence—3.5L (2GR-FSE) engine (Bank 1)

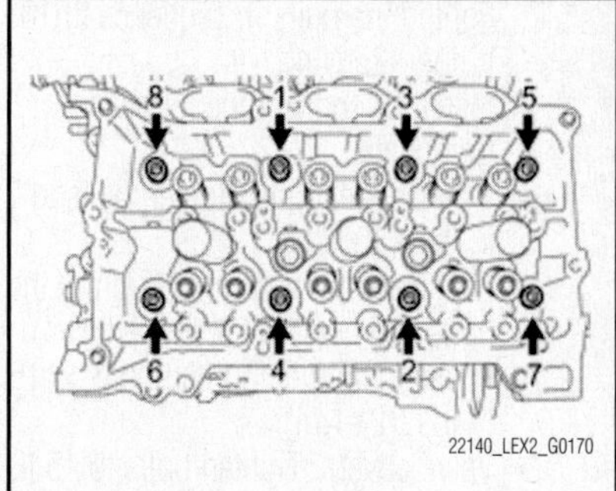

Fig. 93 Cylinder head tightening sequence—3.5L (2GR-FSE) engine (Bank 2)

- Timing chain vibration damper
- Upper intake
- VVT sensor
- Oil control valves
- Oil check valve bolt, oil pipe union and oil pipe
- Valve covers
- Intake manifold
- Water outlet pipe
- Rear water by-pass joint
- Camshafts
- Camshaft housing
- Rocker arms
- Valve lash adjusters
- Valve stem cap

8. Loosen the cylinder head mounting bolts in several steps in the sequence shown.

➡Head warpage or cracking could result from removing bolts in an incorrect order.

9. Remove the head bolts and plate washers.
10. Remove the cylinder head and gasket.

To install:

11. Install a new cylinder head gasket with the Lot number stamp upper side facing upward.
12. Apply a continuous line of the seal packing to a new cylinder head gasket as shown in the illustration.
13. Install the cylinder head. Apply a light coat of engine oil to the threads and tighten the bolts in sequence as follows:
    - Step 1: 27 ft. lbs. (36 Nm)
    - Step 2: Plus 90 degrees
    - Plus an additional 90 degrees
14. Install or connect the following:
    - Valve stem cap
    - Valve lash adjusters
    - Rocker arms
15. Install the camshaft housing as follows:
    - Apply 0.138–0.177 in. (3.5–4.5mm) wide bead of sealant to the contact surface
    - Install the camshaft housing and tighten the bolts in sequence to 18 ft. lbs. (25 Nm).
    - Install or connect the following:
    - Camshafts
    - Water outlet pipe. Tighten bolts to 7 ft. lbs. (10 Nm).
    - Rear water by-pass joint. Tighten the bolts and nuts to 7 ft. lbs. (10 Nm).
    - Intake manifold. Tighten bolts to 15 ft. lbs. (21 Nm).
    - Valve covers. Tighten bolts to 15 ft. lbs. (21 Nm).
    - Upper intake
    - VVT sensor
    - Oil control valves
    - Oil check valve bolt, oil pipe union and oil pipe
    - Timing chain vibration damper
    - Timing chain
16. Install the service plug grip
17. Connect the negative battery cable.
18. Refill the cooling system to the correct level.
19. Refill the engine with oil to the correct level.
20. Start then engine and check for leaks.

EXHAUST MANIFOLD

REMOVAL & INSTALLATION

See Figures 94 and 95.

Before servicing any vehicle, please be sure to read all of the following precautions, which deal with personal safety, prevention of component damage, and important points to take into consideration when servicing a motor vehicle:

⁂ CAUTION

The GS450h hybrid system contains a 288V high-voltage system with a strong alkali solution of potassium hydroxide. Be sure to follow the instructions in this manual to handle the system correctly. Failure to do so may result in serious injury or electrocution. Engineer must undergo special training to be able to perform high-voltage system inspection and servicing.

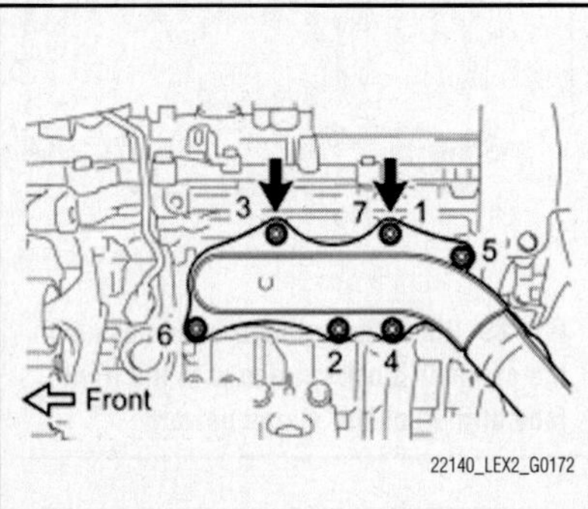

Fig. 94 Exhaust manifold tightening sequence—LH

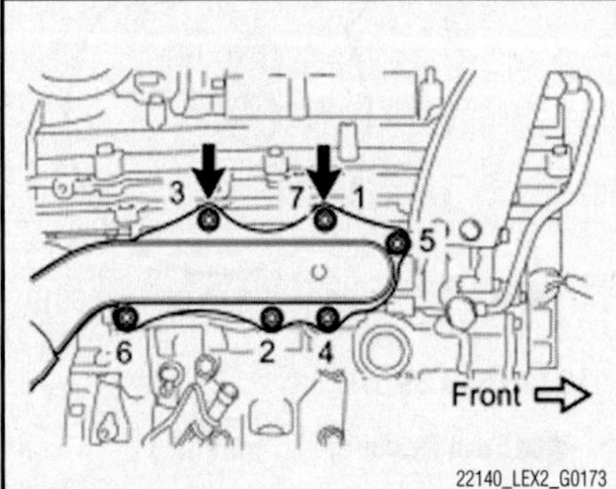

Fig. 95 Exhaust manifold tightening sequence—RH

⁂ CAUTION

All high-voltage wire harness connectors are colored orange. The HV battery and other high-voltage components have "High Voltage" caution labels. Do not carelessly touch these wires and components.

⁂ CAUTION

Before inspecting or servicing the high-voltage system, be sure to follow safety measures, such as wearing insulated gloves and removing the service plug to prevent electrocution. Carry the removed service plug in your pocket to prevent other technicians from reinstalling it while you are servicing the vehicle.

⁂ CAUTION

After removing the service plug, wait 10 minutes before touching any of the high-voltage connectors and terminals.

1. Remove the service plug grip.
2. Disconnect the negative battery cable.
3. Remove any components to access exhaust manifold.
4. Remove the dipstick for RH exhaust manifold.
5. Disconnect the heated oxygen sensor connector.
6. Disconnect the exhaust pipe front ends from the exhaust manifold and remove the 2 gaskets.
7. Remove the 6 nuts, exhaust manifold and gasket.

To install:

8. Install a new gasket.
9. Install the exhaust manifold to the cylinder head with the 6 nuts in the order shown in the illustration.
10. Tighten the mounting nuts to 15 ft. lbs. (21 Nm).
11. Install the dipstick for RH exhaust manifold.
12. Install 2 new gaskets and the exhaust pipe front ends to the exhaust manifolds with the 4 bolts and 2 nuts. Tighten to 29 ft. lbs. (39 Nm).

13. Install any components that may have been removed to access exhaust manifold.
14. Remove the service plug grip.
15. Disconnect the negative battery cable.
16. Check for exhaust leaks.

FLYWHEEL

REMOVAL & INSTALLATION

See Figures 96 through 99.

Before servicing any vehicle, please be sure to read all of the following precautions, which deal with personal safety, prevention of component damage, and important points to take into consideration when servicing a motor vehicle:

CAUTION

The GS450h hybrid system contains a 288V high-voltage system with a strong alkali solution of potassium hydroxide. Be sure to follow the instructions in this manual to handle the system correctly. Failure to do so may result in serious injury or electrocution. Engineer must undergo special training to be able to perform high-voltage system inspection and servicing.

CAUTION

All high-voltage wire harness connectors are colored orange. The HV battery and other high-voltage components have "High Voltage" caution labels. Do not carelessly touch these wires and components.

CAUTION

Before inspecting or servicing the high-voltage system, be sure to follow safety measures, such as wearing insulated gloves and removing the service plug to prevent electrocution. Carry the removed service plug in your pocket to prevent other technicians from reinstalling it while you are servicing the vehicle.

Fig. 96 Crankshaft holding tool

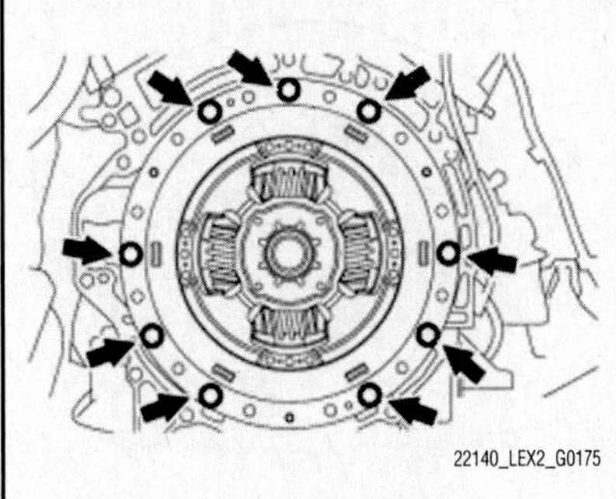

Fig. 97 Remove the 9 bolts and transmission input damper cover

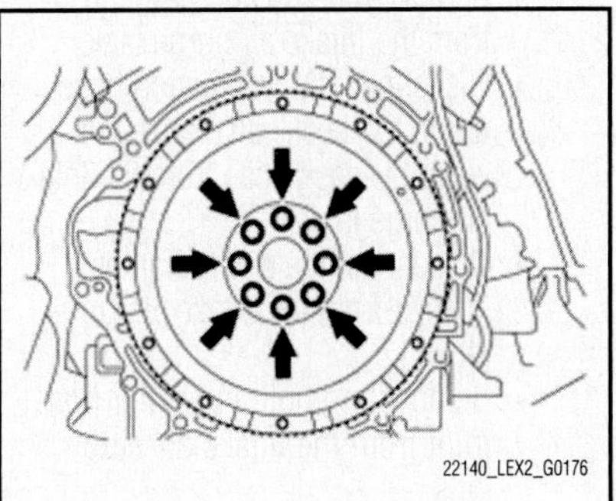

Fig. 98 Remove the 9 bolts and transmission input damper cover

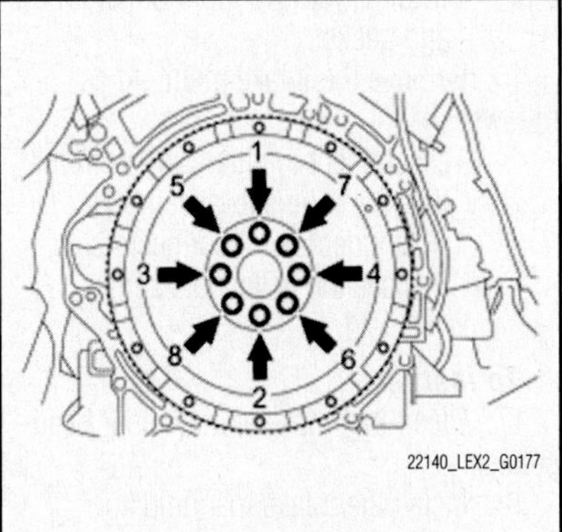

Fig. 99 Flywheel tightening sequence

CAUTION

After removing the service plug, wait 10 minutes before touching any of the high-voltage connectors and terminals.

1. Remove the service plug grip.
2. Disconnect the negative battery cable.
3. Remove the hybrid vehicle transmission assembly.
4. Remove the flywheel sub-assembly as follows:
   - Using SST, hold the crankshaft.
   - Remove the 9 bolts and transmission input damper cover.
   - Remove the 8 bolts and flywheel.

To install:

5. Install the flywheel sub-assembly as follows:
   - Using SST, hold the crankshaft.
   - Apply adhesive to 2 or 3 threads of the mounting bolts end.
   - Install and tighten the 8 mounting bolts uniformly in several steps to 61 ft. lbs. (83 Nm).

➡Do not start the engine for at least 1 hour after installing.

   - Install the transmission input damper cover with the 9 bolts. Tighten to 36 ft. lbs. (49 Nm).
   - Install the hybrid vehicle transmission assembly.
6. Install the service plug grip.
7. Connect the negative battery cable.

INTAKE MANIFOLD

REMOVAL & INSTALLATION

See Figures 100 and 101.

CAUTION

The GS450h hybrid system contains a 288V high-voltage system with a strong alkali solution of potassium hydroxide. Be sure to follow the instructions in this manual to handle the system correctly. Failure to do so may result in serious injury or electrocution. Engineer must undergo special training to be able to perform high-voltage system inspection and servicing.

CAUTION

All high-voltage wire harness connectors are colored orange. The HV battery and other high-voltage components have "High Voltage" caution labels. Do not carelessly touch these wires and components.

CAUTION

Before inspecting or servicing the high-voltage system, be sure to follow safety measures, such as wearing insulated gloves and removing

the service plug to prevent electrocution. Carry the removed service plug in your pocket to prevent other technicians from reinstalling it while you are servicing the vehicle.

⁂ CAUTION

After removing the service plug, wait 10 minutes before touching any of the high-voltage connectors and terminals.

1. Discharge the fuel system.
2. Remove the hybrid service plug grip.
3. Disconnect the negative battery cable.

⁂ CAUTION

Before inspecting the high-voltage system or disconnecting the low voltage connector of the inverter with converter assembly, take safety precautions, such as wearing insulated gloves and removing the service plug grip to prevent electrical shocks. After removing the service plug grip, put it in your pocket to prevent other technicians from accidentally reconnecting it while you are working on the high-voltage system.

⁂ CAUTION

After disconnecting the service plug grip, wait for at least 10 minutes before touching any of the high-voltage connectors or terminals. After waiting, check the voltage at the inspection point in the inverter with converter assembly. The voltage should be 0 V before beginning work.

4. Drain the engine coolant.
5. Remove the cool air intake duct seal.
6. Remove the engine room side cover.
7. Remove the V-bank cover.
8. Remove the air cleaner cap and hose.
9. Remove the wiper arm assembly LH and RH
10. Remove the cowl top ventilator louver sub-assembly.
11. Drain the engine coolant.
12. Disconnect the No. 2 ventilator hose.
13. Remove the air cleaner cap with air cleaner hose.
14. Remove the throttle body with motor assembly.
15. Remove the air intake surge tank as follows:
    - Disconnect the purge line hose from the intake air surge tank.

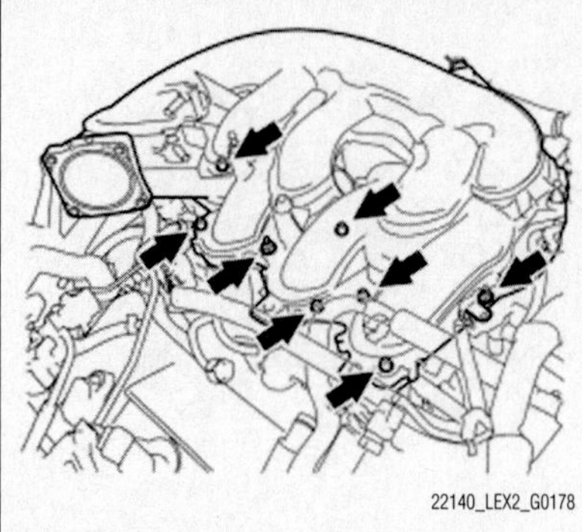

Fig. 100 Air intake surge tank removal

- Remove the bolt and purge VSV from the intake air surge tank.
- Disconnect the ventilation hose, union to check valve hose and water by-pass hose from the intake air surge tank.
- Disconnect the 4 wire harness clamps from the intake air surge tank.
- Remove the bolt and water hose joint from the intake air surge tank.
- Remove the bolt and disconnect the surge tank stay from the intake air surge tank.
- Using a 5 mm hexagon socket wrench, remove the 6 bolts, 2 nuts and gasket.

16. Remove the intake manifold as follows:
    - Disconnect the fuel tube from the delivery pipe sub-assembly.
    - Disconnect the 4 connectors.
    - Remove the bolts, nuts and manifold.

To install:

17. Clean the intake and cylinder head surface.
18. Install the intake manifold as follows:

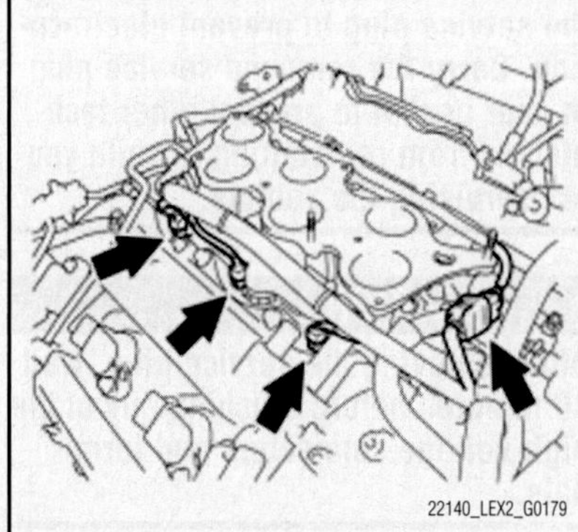

Fig. 101 Disconnect the 4 connectors

- Install a new gasket and the intake manifold with the 4 bolts and 4 nuts. Tighten to 15 ft. lbs. (21 Nm).
- Connect the 4 connectors.
- Push in the fuel tube connector to the delivery pipe sub-assembly and push up the retainer to engage the claws.
- After connecting the fuel tube, align the paint marks on the delivery pipe.

19. Install the air intake surge tank as follows:
    - Install a new gasket to the intake air surge tank.
    - Install the intake air surge tank with the 2 nuts. Tighten to 12 ft. lbs. (16 Nm).
    - Using a 5 mm hexagon socket wrench, install the 6 bolts. Tighten to 13 ft. lbs. (18 Nm).
    - Install the surge tank stay to the intake air surge tank with the bolt. Tighten to 15 ft. lbs. (21 Nm).
    - Install the water hose joint to the intake air surge tank with the bolt. Tighten to 15 ft. lbs. (21 Nm).
    - Tighten to 15 ft. lbs. (21 Nm).
    - Connect the 4 wire harness clamps to the intake air surge tank.
    - Connect the ventilation hose, union to check valve hose and water by-pass hose to the intake air surge tank.
    - Install the purge VSV to the intake air surge tank with the bolt. Tighten to 13 ft. lbs. (18 Nm).
    - Connect the purge line hose to the intake air surge tank.
20. Install the throttle body with motor assembly. Tighten the mounting bolts to 7 ft. lbs. (10 Nm).
21. Connect the throttle motor connector.
22. Install the air cleaner cap with air cleaner hose.
23. Reconnect the No. 2 ventilator hose.
24. Install the cowl top ventilator louver sub-assembly.
25. Install the wiper arm assembly LH and RH
26. Install the air cleaner cap and hose.
27. Install the V-bank cover.
28. Install the engine room side cover.
29. Install the cool air intake duct seal.
30. Install and bleed engine coolant.

OIL PAN

REMOVAL & INSTALLATION

See Figures 102 through 105.

⁂ CAUTION

The GS450h hybrid system contains a 288V high-voltage system with a strong alkali solution of potassium hydroxide. Be sure to follow the instructions in this manual to handle the system correctly. Failure to do so may result in serious injury or electrocution. Engineer must undergo special training to be able to perform high-voltage system inspection and servicing.

⁂ CAUTION

All high-voltage wire harness connectors are colored orange. The HV battery and other high-voltage components have "High Voltage" caution labels. Do not carelessly touch these wires and components.

⁂ CAUTION

Before inspecting or servicing the high-voltage system, be sure to follow safety measures, such as wearing insulated gloves and removing the service plug to prevent electrocution. Carry the removed service plug in your pocket to prevent other technicians from reinstalling it while you are servicing the vehicle.

⁂ CAUTION

After removing the service plug, wait 10 minutes before touching any of the high-voltage connectors and terminals.

1. Discharge the fuel system.
2. Remove the hybrid service plug grip.
3. Disconnect the negative battery cable.

⁂ CAUTION

Before inspecting the high-voltage system or disconnecting the low voltage connector of the inverter with converter assembly, take safety precautions, such as wearing insulated gloves and removing the service plug grip to prevent electrical shocks. After removing the service plug grip, put it in your pocket to prevent other technicians from accidentally reconnecting it while you are working on the high-voltage system.

⁂ CAUTION

After disconnecting the service plug grip, wait for at least 10 minutes before touching any of the high-voltage connectors or terminals. After waiting, check the voltage at the inspection point in the inverter with converter assembly. The voltage should be 0 V before beginning work.

4. Drain the engine coolant.
5. Remove the engine assembly.
6. Remove the No. 2 oil pan sub-assembly as follows:
   - Remove the 15 bolts and 2 nuts.
   - Insert the blade of oil pan seal cutter between the oil pans. Cut through the applied sealer and remove the No. 2 oil pan sub-assembly.

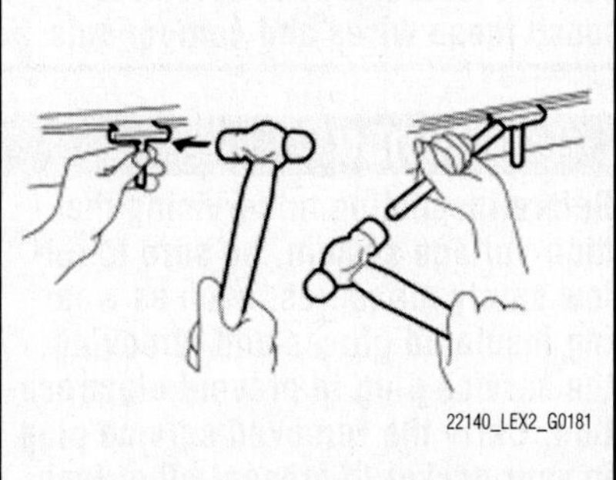

Fig. 102 Cut through the applied sealer and remove the No. 2 oil pan sub-assembly

for Bank 2 Side:

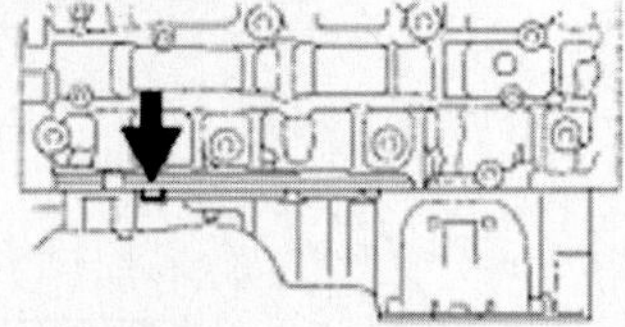

for Bank 1 Side:

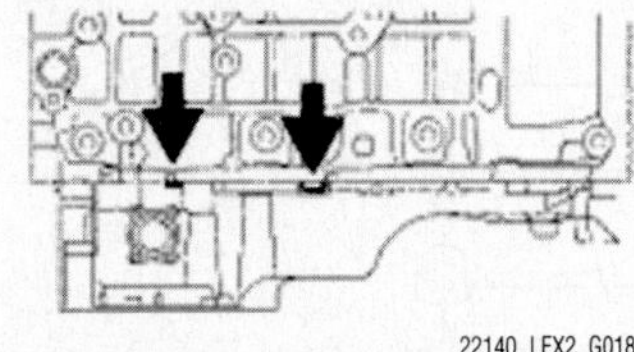

Fig. 103 Remove the oil pan by prying between the oil pan and cylinder block

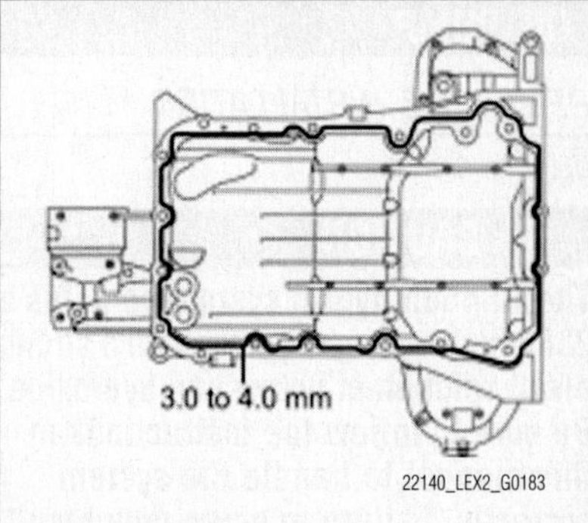

Fig. 104 Standard seal diameter: 0.118 to 0.156 inch (3.0 to 4.0 mm)

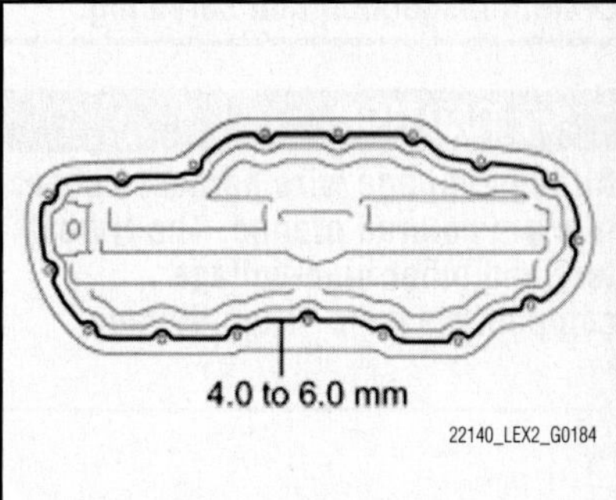

Fig. 105 Standard seal diameter: 0.156 to 0.236 inch (4.0 to 6.0 mm)

- Remove the 16 bolts and 2 nuts.
- Remove the oil pan by prying between the oil pan and cylinder block with a screwdriver.

To install:

➡Be sure to clean the bolts and stud bolts and check the threads for cracks or other damage.

7. Install oil pan sub-assembly as follows:
   - Apply seal packing in a continuous line as shown in the illustration. Toyota Genuine Seal Packing Block, Three Bond 1207B or equivalent.
   - Install the oil pan with the 16 bolts and 2 nuts. Tighten the bolts to 7 ft. lbs. (10 Nm). Tighten the nuts to 15 ft. lbs. (21 Nm).
8. Install the No. 2 oil pan sub-assembly as follows:
   - Apply seal packing in a continuous line as shown in the illustration.
   - Install the oil pan sub-assembly with the 15 bolts and 2 nuts. Tighten the bolts and nuts to 7 ft. lbs. (10 Nm).

OIL PUMP

REMOVAL & INSTALLATION

See Figure 106.

CAUTION

The GS450h hybrid system contains a 288V high-voltage system with a strong alkali solution of potassium hydroxide. Be sure to follow the instructions in this manual to handle the system correctly. Failure to do so may result in serious injury or electrocution. Engineer must undergo special training to be able to perform high-voltage system inspection and servicing.

CAUTION

All high-voltage wire harness connectors are colored orange. The HV battery and other high-voltage components have "High Voltage" caution labels. Do not carelessly touch these wires and components.

CAUTION

Before inspecting or servicing the high-voltage system, be sure to follow safety measures, such as wearing insulated gloves and removing the service plug to prevent electrocution. Carry the removed service plug in your pocket to prevent other technicians from reinstalling it while you are servicing the vehicle.

CAUTION

After removing the service plug, wait 10 minutes before touching any of the high-voltage connectors and terminals.

1. Remove the hybrid service plug grip.
2. Disconnect the negative battery cable.

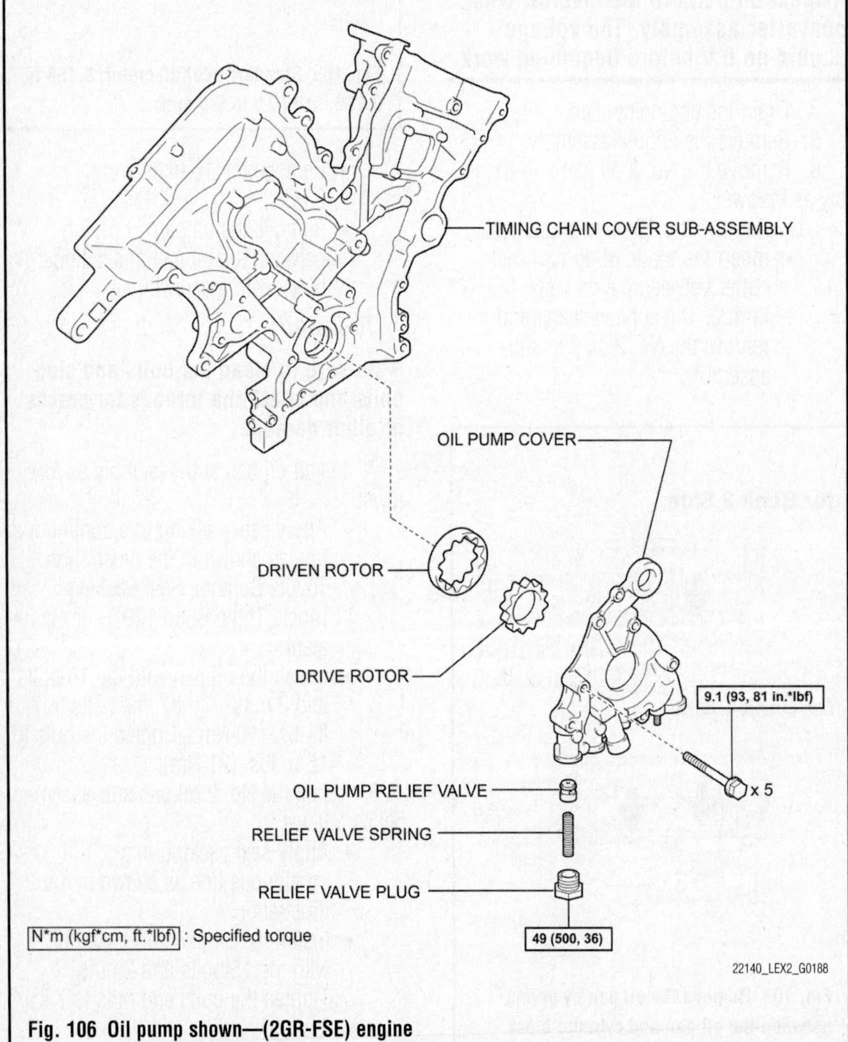

Fig. 106 Oil pump shown—(2GR-FSE) engine

CAUTION

Before inspecting the high-voltage system or disconnecting the low voltage connector of the inverter with converter assembly, take safety precautions, such as wearing insulated gloves and removing the service plug grip to prevent electrical shocks. After removing the service plug grip, put it in your pocket to prevent other technicians from accidentally reconnecting it while you are working on the high-voltage system.

CAUTION

After disconnecting the service plug grip, wait for at least 10 minutes before touching any of the high-voltage connectors or terminals. After waiting, check the voltage at the inspection point in the inverter with converter assembly. The voltage should be 0 V before beginning work.

3. Remove the service cover and plug.
4. Disconnect the negative battery cable.
5. Remove the drive belt.
6. Using SST, loosen the crankshaft pulley set bolt.
7. Remove the oil pump assembly from the timing cover.

To install:

8. Install the oil pump assembly to the timing cover.
9. Install the front crankshaft seal and drive it into the cover with a seal driver.
10. Align the pulley set key with the key groove of the pulley, and slide on the pulley.
11. Install the pulley bolt and tighten bolt to 184 ft. lbs. (240 Nm).
12. Install the drive belt.
13. Install the service cover and plug.
14. Connect the negative battery cable.

INSPECTION

See Figures 107 through 111.

1. Check the oil pump relief valve as follows:
   - Coat the relief valve with engine oil. Check that it falls smoothly into the valve hole by its own weight.
   - If it does not, replace the relief valve. If necessary, replace the oil pump assembly
2. Inspect oil pump rotor set as follows:
   - Install the rotors to the timing chain cover with the rotors' marks facing

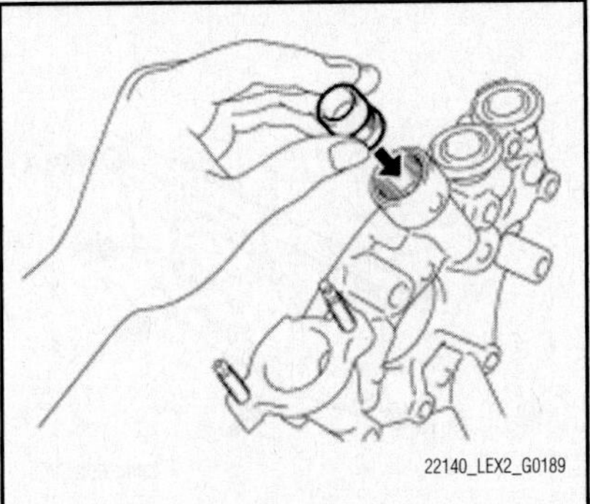

Fig. 107 Oil pump relief valve—(2GR- FSE) engine

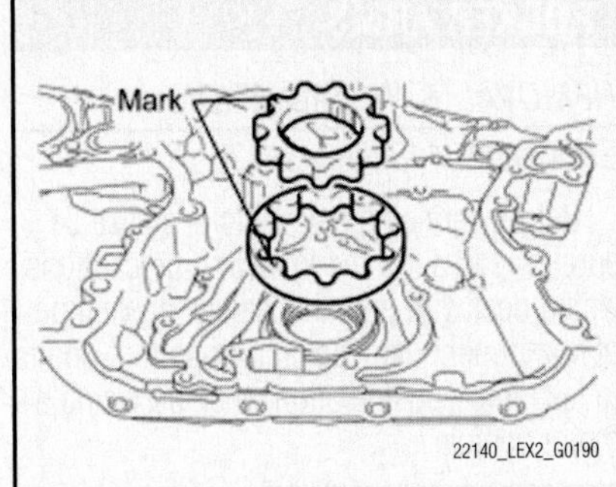

Fig. 108 Check that the rotor revolves smoothly.

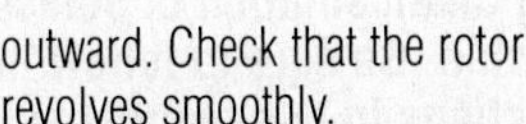

outward. Check that the rotor revolves smoothly.

3. Check the tip clearance as follows:
   - Using a feeler gauge, measure the clearance between the drive and driven rotor tips, as shown in the illustration. Standard tip clearance: 0.0024 to 0.0063 inch (0.060 to 0.160 mm) Maximum tip clearance: 0.0063 inch (0.160 mm)
   - If the tip clearance is greater than the maximum, replace the drive and driven rotors.

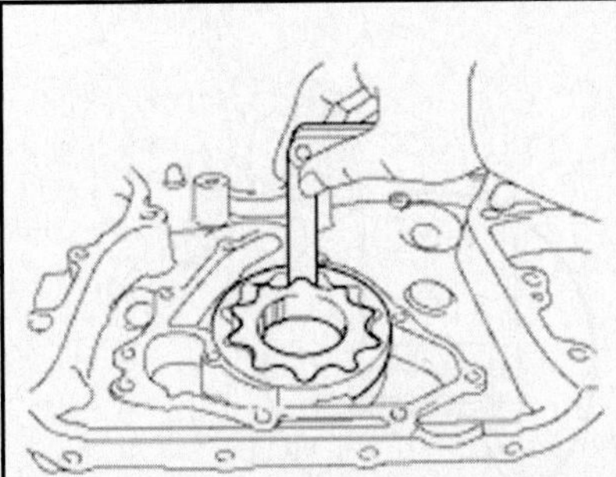

Fig. 109 Using a feeler gauge, measure the clearance between the drive and driven rotor tips

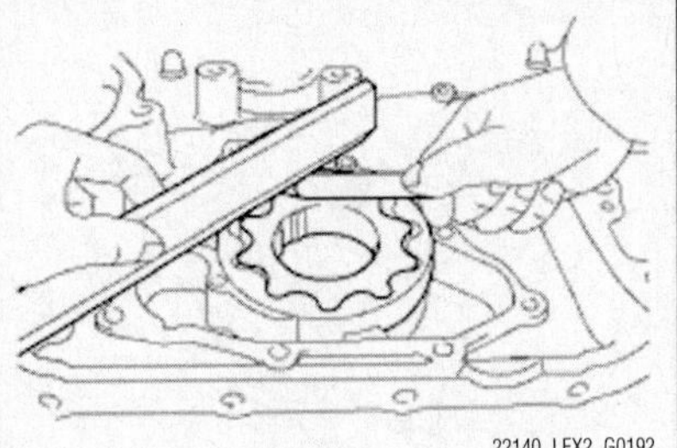

Fig. 110 Using a feeler gauge and precision straightedge, measure the clearance between the rotors and precision straightedge

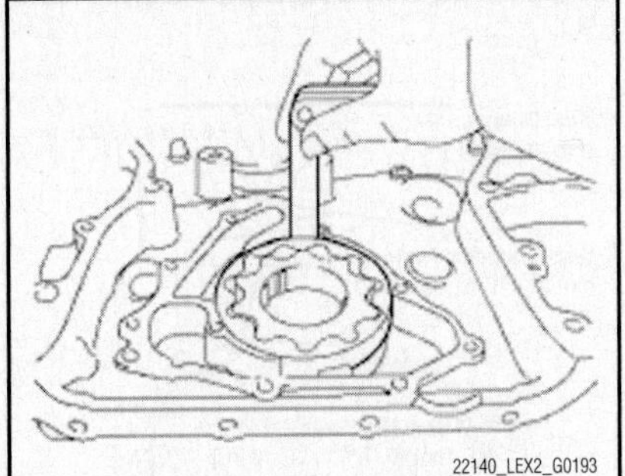

Fig. 111 Using a feeler gauge, measure the clearance between the timing chain cover and driven rotor

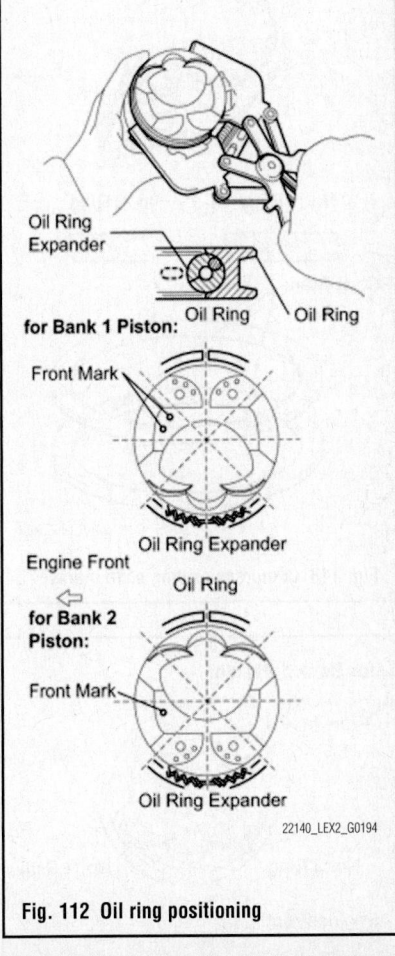

4. Check the side clearance as follows:
   - Using a feeler gauge and precision straightedge, measure the clearance between the rotors and precision straightedge, as shown in the illustration. Standard side clearance: 0.0012 to 0.0035 inch (0.030 to 0.090 mm) Maximum side clearance: 0.0035 inch (0.090 mm)
   - If the side clearance is greater than the maximum, replace the timing chain cover.
5. Check the body clearance as follows:
   - Using a feeler gauge, measure the clearance between the timing chain cover and driven rotor, as shown in the illustration. Standard body clearance: 0.0098 to 0.0128 inch (0.250 to 0.325 mm) Maximum side clearance: 0.0128 inch (0.325 mm)
   - If the body clearance is greater than the maximum, replace the timing chain cover.

PISTON AND RING

POSITIONING

See Figures 112 through 114.

Oil Ring Expander
for Bank 1 Piston:
Oil Ring
Oil Ring
Front Mark
Oil Ring Expander
Engine Front
Oil Ring
for Bank 2 Piston:
Front Mark
Oil Ring Expander
22140_LEX2_G0194

Fig. 112 Oil ring positioning

REAR MAIN SEAL

REMOVAL & INSTALLATION

See Figures 115 through 117.

✻✻ CAUTION

The GS450h hybrid system contains a 288V high-voltage system with a strong alkali solution of potassium hydroxide. Be sure to follow the instructions in this manual to handle the system correctly. Failure to do so may result in serious injury or electrocution. Engineer must undergo special training to be able to perform high-voltage system inspection and servicing.

✻✻ CAUTION

All high-voltage wire harness connectors are colored orange. The HV battery and other high-voltage components have "High Voltage" caution labels. Do not carelessly touch these wires and components.

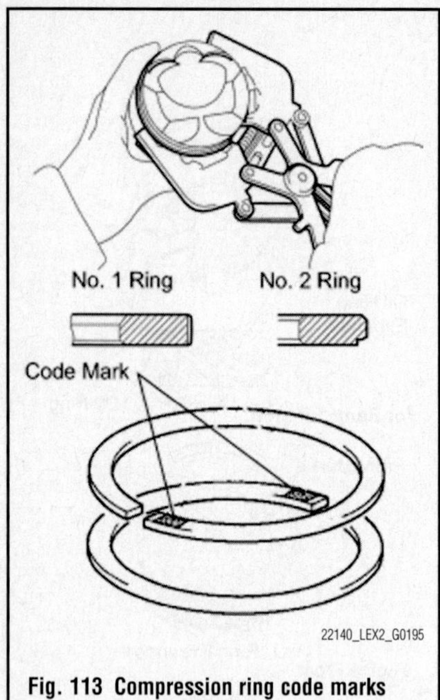

Fig. 113 Compression ring code marks

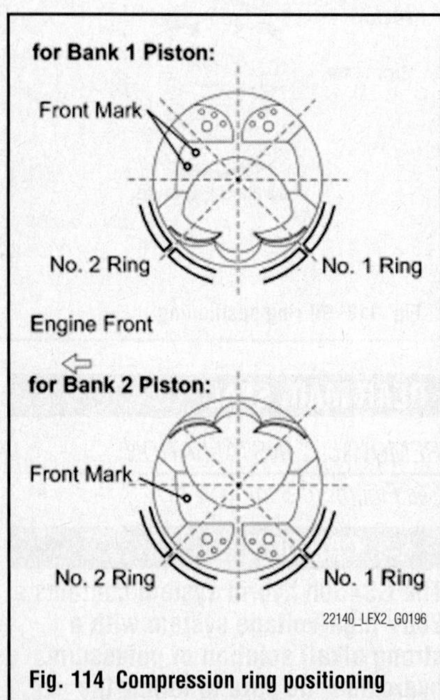

Fig. 114 Compression ring positioning

⁂ CAUTION

Before inspecting or servicing the high-voltage system, be sure to follow safety measures, such as wearing insulated gloves and removing the service plug to prevent electrocution. Carry the removed service plug in your pocket to prevent other technicians from reinstalling it while you are servicing the vehicle.

⁂ CAUTION

After removing the service plug, wait 10 minutes before touching any of the high-voltage connectors and terminals.

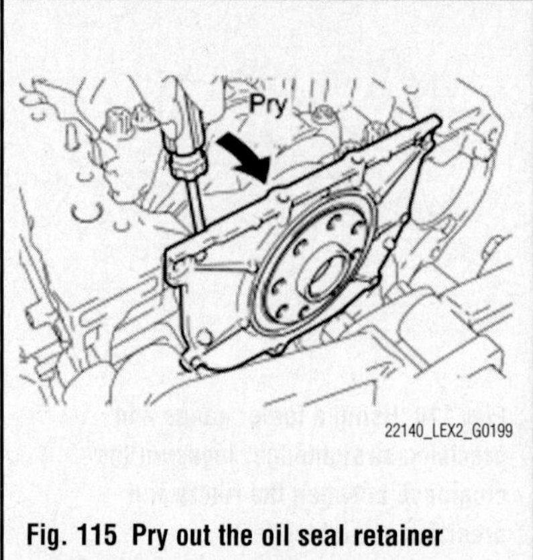

Fig. 115 Pry out the oil seal retainer

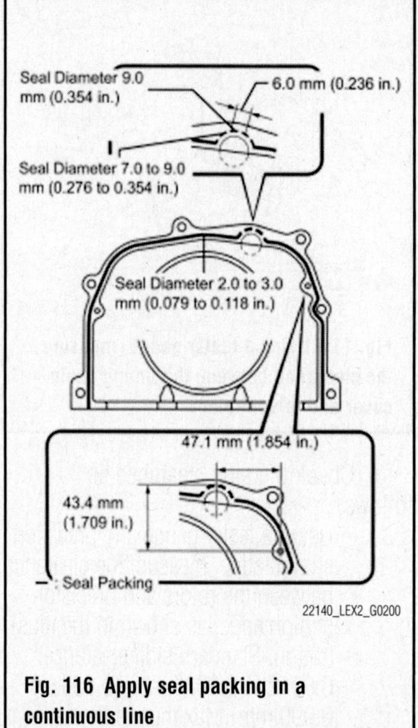

Fig. 116 Apply seal packing in a continuous line

1. Remove the hybrid service plug grip.
2. Disconnect the negative battery cable.
3. Remove hybrid transmission assembly.
4. Remove the flywheel.
5. Remove the rear oil seal retainer bolts.
6. Using a screwdriver, pry out the oil seal retainer.

To install:

7. Apply seal packing in a continuous line as shown.
8. Install the oil seal retainer with the 6 bolts and tighten to 7 ft. lbs. (10 Nm).

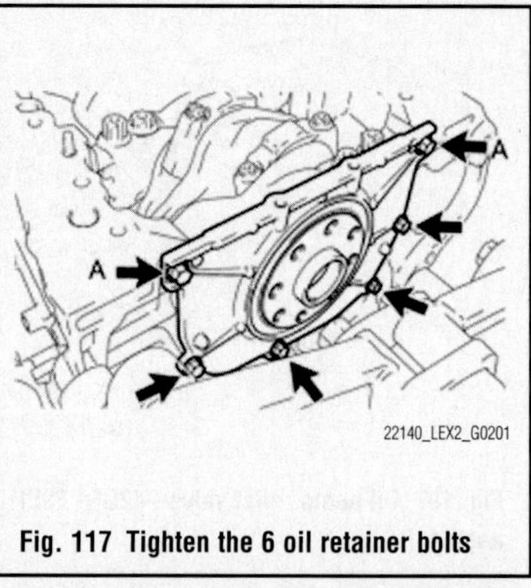

Fig. 117 Tighten the 6 oil retainer bolts

ROCKER ARMS

REMOVAL & INSTALLATION

See Figures 118 through 136.

Before servicing any vehicle, please be sure to read all of the following precautions, which deal with personal safety, prevention of component damage, and important points to take into consideration when servicing a motor vehicle:

⁂ CAUTION

The GS450h hybrid system contains a 288V high-voltage system with a strong alkali solution of potassium hydroxide. Be sure to follow the instructions in this manual to handle the system correctly. Failure to do so may result in serious injury or electrocution. Engineer must undergo special training to be able to perform high-voltage system inspection and servicing.

⁂ CAUTION

All high-voltage wire harness connectors are colored orange. The HV bat-

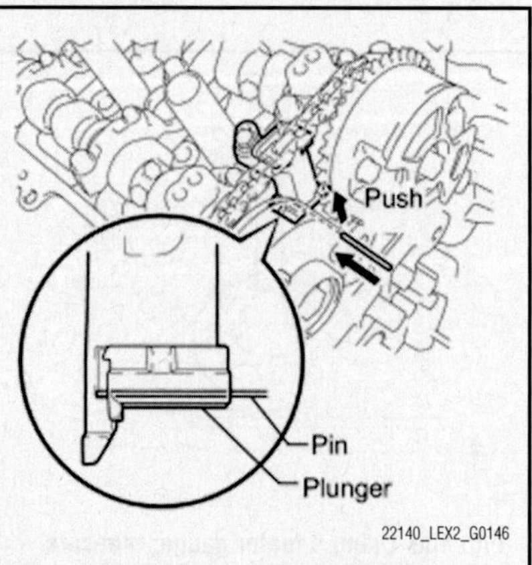

Fig. 118 Insert a pin of 0.039 inch (1.0 mm) into the hole to fix it in place

tery and other high-voltage components have "High Voltage" caution labels. Do not carelessly touch these wires and components.**

⁂ CAUTION

Before inspecting or servicing the high-voltage system, be sure to follow safety measures, such as wearing insulated gloves and removing the service plug to prevent electrocution. Carry the removed service plug in your pocket to prevent other technicians from reinstalling it while you are servicing the vehicle.

⁂ CAUTION

After removing the service plug, wait 10 minutes before touching any of the high-voltage connectors and terminals.

1. Discharge the fuel system.
2. Recover the refrigerant from the A/C system using a refrigerant recovery unit.
3. Place the front wheels in the straight ahead position.
4. Disconnect the negative battery cable.
5. Remove the hybrid service plug grip.

⁂ CAUTION

Before inspecting the high-voltage system or disconnecting the low voltage connector of the inverter with converter assembly, take safety precautions, such as wearing insulated gloves and removing the service plug grip to prevent electrical shocks. After removing the service plug grip, put it in your pocket to prevent other technicians from accidentally reconnecting it while you are working on the high-voltage system.

⁂ CAUTION

After disconnecting the service plug grip, wait for at least 10 minutes before touching any of the high-voltage connectors or terminals. After waiting, check the voltage at the inspection point in the inverter with converter assembly. The voltage should be 0 V before beginning work.

6. Remove the luggage floor mat.
7. Remove the service plug cover.
8. Remove the service plug.
9. Disconnect the negative battery cable.
10. Drain the engine oil and coolant.
11. Remove the cool air intake duct seal.
12. Remove the engine room side cover.
13. Remove the V-bank cover.
14. Remove hoses and electrical connectors.
15. Remove the air intake surge tank sub-assembly.
16. Remove the ignition coil connectors and coils.
17. Remove the ventilation valve.
18. Remove the 4 bolts and the 4 VVT sensors.
19. Remove the 4 bolts and 4 oil control valves.
20. Remove the oil check valve bolt, oil pipe union and oil pipe.
21. Remove valve covers.
22. Remove the drive belt.
23. Remove the crankshaft pulley.
24. Remove the timing cover.
25. Remove the timing components.
26. Remove the camshaft timing gear assembly (Bank 1) as follows:
    - While raising up the No. 2 chain tensioner, insert a pin of 0.039 inch (1.0 mm) into the hole to fix it in place.
    - Hold the hexagonal portion of the camshaft with a wrench, and remove the 2 bolts and 2 camshaft timing gears.
    - Remove the No. 2 chain.
    - Remove the No. 2 Chain tensioner
27. Remove the camshaft for (Bank 1) as follows:
    - Remove the 3 gaskets.
    - Make sure that the knock pin of the camshaft is positioned as shown in the illustration.
    - Uniformly loosen and remove the 9 bearing cap bolts in the sequence shown in the illustration.
    - Uniformly loosen and remove the 14 bearing cap bolts in the sequence shown in the illustration.
    - Remove the 6 bearing caps.
    - Remove the No. 1 and No. 2 camshafts.
28. Remove the camshaft timing gear assembly (Bank 2) as follows:
    - While raising up the No. 2 chain tensioner, insert a pin of 0.039 inch (1.0 mm) into the hole to fix it in place.
    - Hold the hexagonal portion of the camshaft with a wrench, and

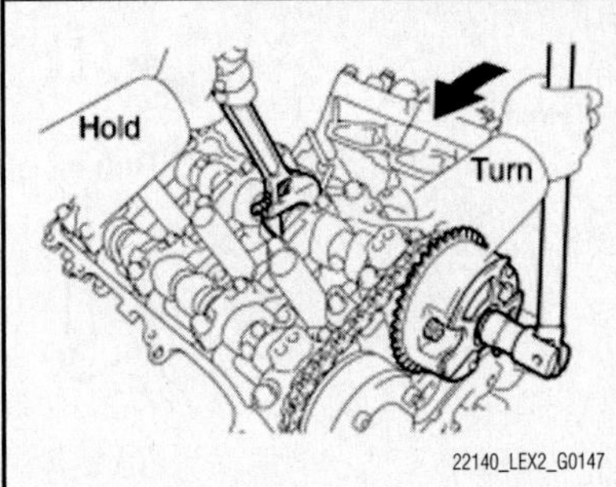

Fig. 119 Hold the hexagonal portion of the camshaft with a wrench, and remove the 2 bolts and 2 camshaft timing gears

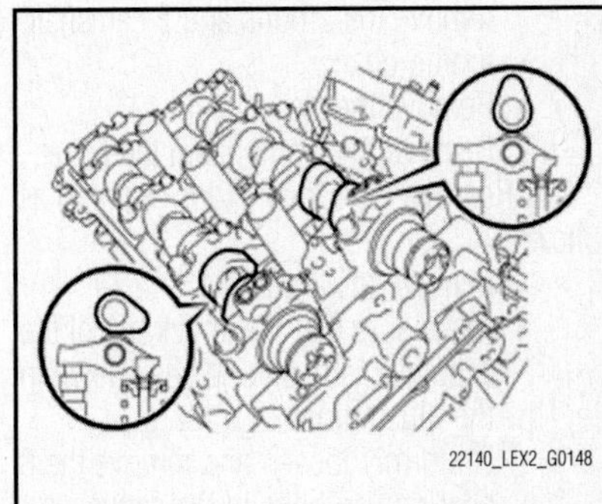

Fig. 120 Make sure that the knock pin of the camshaft is positioned as shown

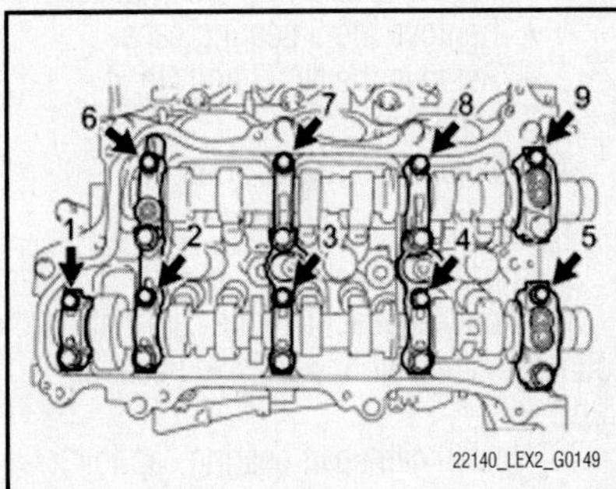

Fig. 121 Uniformly loosen and remove the 9 bearing cap bolts in the sequence shown

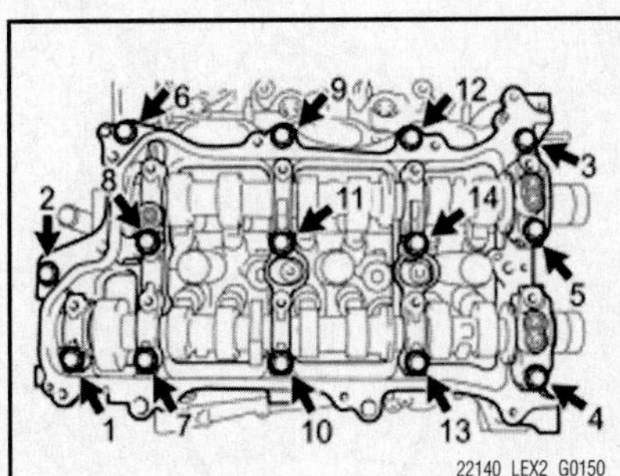

Fig. 122 Uniformly loosen and remove the 14 bearing cap bolts in the sequence shown

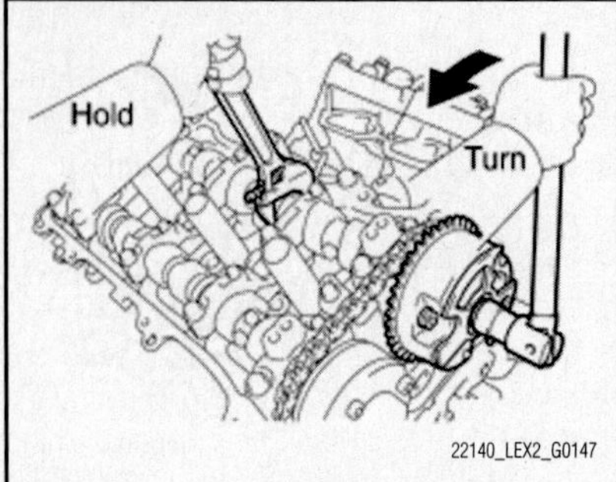

Fig. 123 Hold the hexagonal portion of the camshaft with a wrench, and remove the 2 bolts and 2 camshaft timing gears

remove the 2 bolts and 2 camshaft timing gears.
- Remove the No. 3 chain.
- Remove the No. 3 chain tensioner.

29. Remove the camshaft for (Bank 2) as follows:
- Remove the 3 gaskets.
- Make sure that the knock pin of the camshaft is positioned as shown in the illustration.
- Uniformly loosen and remove the 8 bearing cap bolts in the sequence shown in the illustration.
- Uniformly loosen and remove the 13 bearing cap bolts in the sequence shown in the illustration.
- Remove the 5 bearing caps.
- Remove the No. 3 and No. 4 camshafts.

30. Remove the rocker arms.

To install:

31. Install rocker arms.
32. Apply engine oil to the camshaft journals, camshaft housings and bearing caps.
33. Install camshaft bearing cap for (Bank 1).

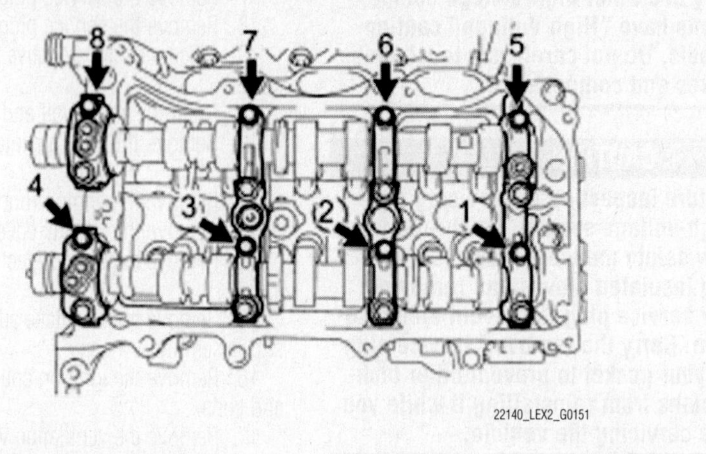

Fig. 125 Uniformly loosen and remove the 8 bearing cap bolts in the sequence shown

34. Install the camshaft and No. 2 camshaft to the camshaft housing.
35. Confirm the marks and numbers on the camshaft bearing caps and place them each in their proper position and direction.
36. Temporarily install the 8 bolts in the order shown in the illustration.
37. Tighten bolts to 7 ft. lbs. (10 Nm).
38. Install camshaft assembly sub-assembly.
39. Apply seal packing in a continuous line as shown in the illustration.
40. Install the camshaft housing, and install the 12 bolts in the order shown in the illustration. Tighten to 21 ft. lbs. (28 Nm).

➡Make sure that the knock pin of the camshaft is positioned as shown in the illustration before installing the camshaft housing.

41. Tighten the 9 bolts in the order shown in the illustration. Tighten bolts to 12 ft. lbs. (16 Nm).
42. Install 3 new gaskets as shown in the illustration.
43. Apply engine oil to the camshaft journals, camshaft housings and bearing caps.
44. Install camshaft bearing cap for (Bank 1).
45. Install the camshaft and No. 2 camshaft to the camshaft housing.
46. Confirm the marks and numbers on the camshaft bearing caps and place them each in their proper position and direction.
47. Install the camshaft No. 3 and camshaft No. 4 to the camshaft housing.
48. Confirm the marks and numbers on the camshaft bearing caps and place them each in their proper position and direction.
49. Temporarily install the 8 bolts in the order shown in the illustration. Tighten bolts to 7 ft. lbs. (10 Nm).
50. Install the camshaft sub-assembly.
51. Apply seal packing in a continuous line.
52. Install the camshaft housing and tighten the 13 bolts in the order shown in the illustration. Tighten to 21 ft. lbs. (28 Nm).

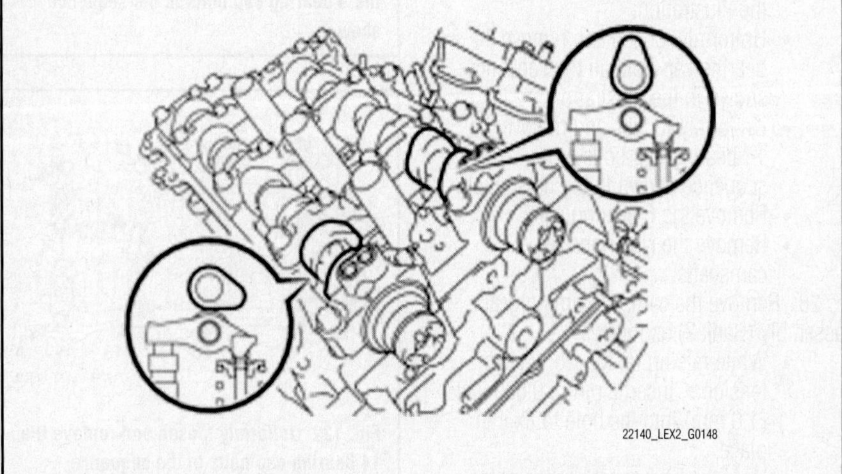

Fig. 124 Make sure that the knock pin of the camshaft is positioned as shown

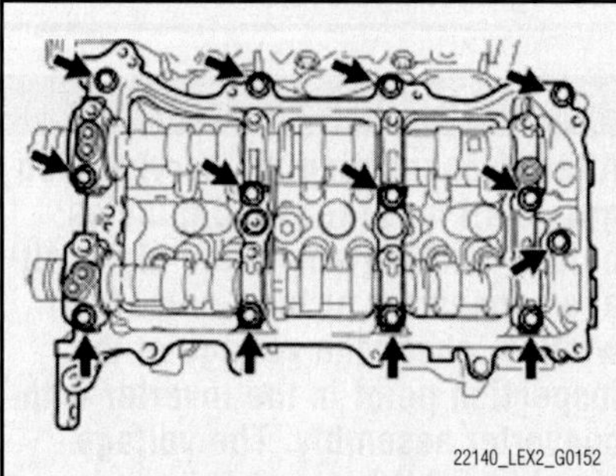

Fig. 126 Uniformly loosen and remove the 13 bearing cap bolts in the sequence shown

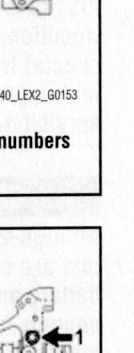

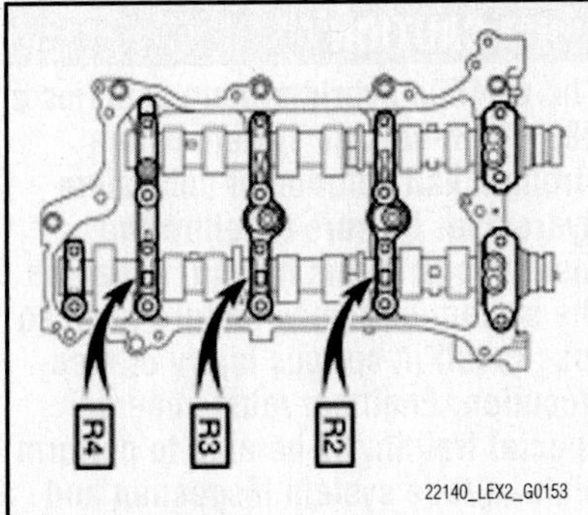

Fig. 127 Confirm the marks and numbers on the camshaft bearing caps

Fig. 128 Temporarily install the 8 bolts in the order shown

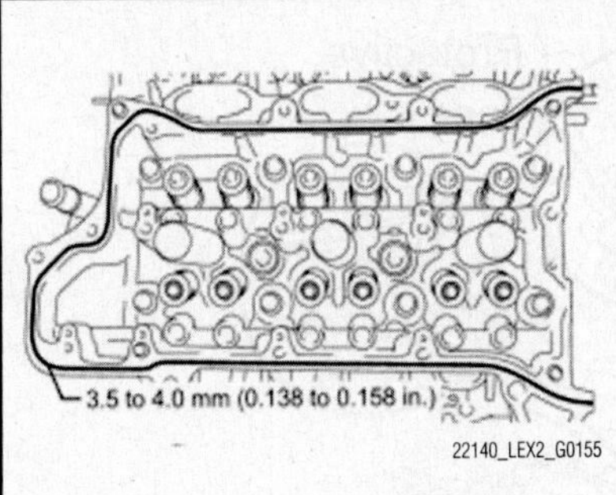

Fig. 129 Apply seal packing in a continuous line

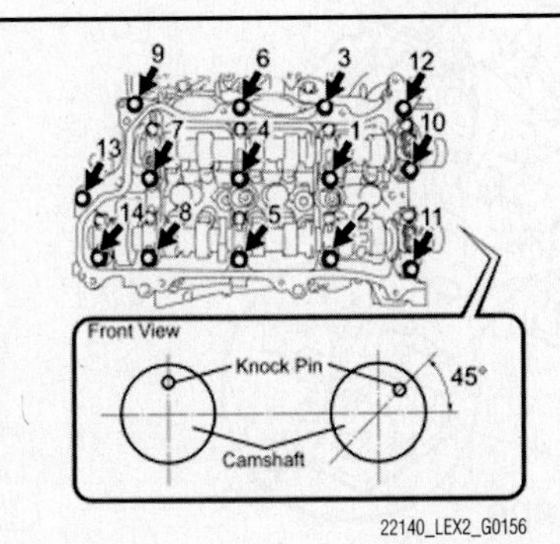

Fig. 130 Install the 12 bolts in the order shown

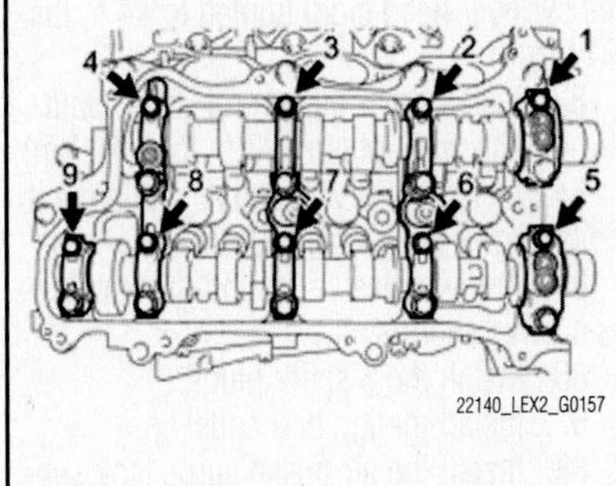

Fig. 131 Tighten the 9 bolts in the order shown

➡Make sure that the knock pin of the camshaft is positioned as shown in the illustration before installing the camshaft housing.

53. Tighten the 8 bolts in the order shown in the illustration. Tighten to 12 ft. lbs. (16 Nm).
54. Install the chain tensioner and tighten the mounting the bolt to 15 ft. lbs. (21 Nm).
55. While pushing in the tensioner, insert a pin of 1.0 mm (0.039 in.) into the hole to fix it in place.
56. Install the camshaft timing gears and No.2 chain for (Bank 1) as follows:
57. Align the mark plate (yellow) with the timing marks (1 dot mark) of the camshaft timing gears as shown in the illustration.
58. Apply a small amount of engine oil to the bolt threads and bolt-seating surface.
59. Align the knock pin of the camshaft with the pin hole of the camshaft timing gear. Install the camshaft timing gear and camshaft timing exhaust gear with the No. 2 chain installed
60. Hold the hexagonal portion of the camshaft with a wrench, and tighten the 2 bolts to 74 ft. lbs. (100 Nm).
61. Remove the pin from the No. 2 chain tensioner.
62. Install the chain tensioner and tighten the mounting the bolt to 15 ft. lbs. (21 Nm).
63. While pushing in the tensioner, insert a pin of 1.0 mm (0.039 in.) into the hole to fix it in place.
64. Install the camshaft timing gears No. 2 for (Bank 2) as follows:
65. Align the mark plate (yellow) with the timing marks (1 dot mark) of the camshaft timing gears as shown in the illustration.
66. Apply a small amount of engine oil to the bolt threads and bolt-seating surface.
67. Align the knock pin of the camshaft with the pin hole of the camshaft timing gear. Install the camshaft timing gear and camshaft timing exhaust gear with the No. 2 chain installed
68. Hold the hexagonal portion of the camshaft with a wrench, and tighten the 2 bolts to 74 ft. lbs. (100 Nm).
69. Remove the pin from the No. 2 chain tensioner.
70. Install the timing components.
71. Install the timing cover.
72. Install the crankshaft pulley.
73. Install the drive belt.

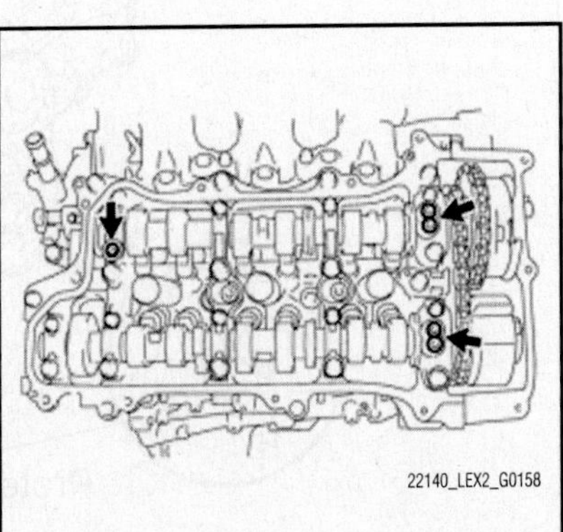

Fig. 132 Install 3 new gaskets as shown

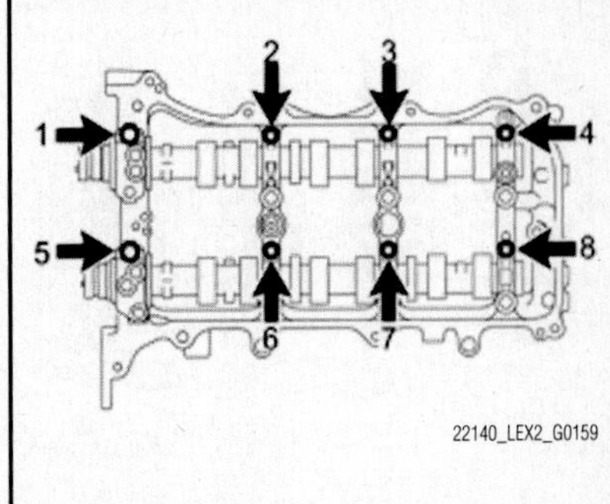

Fig. 133 Temporarily install the 8 bolts in the order shown

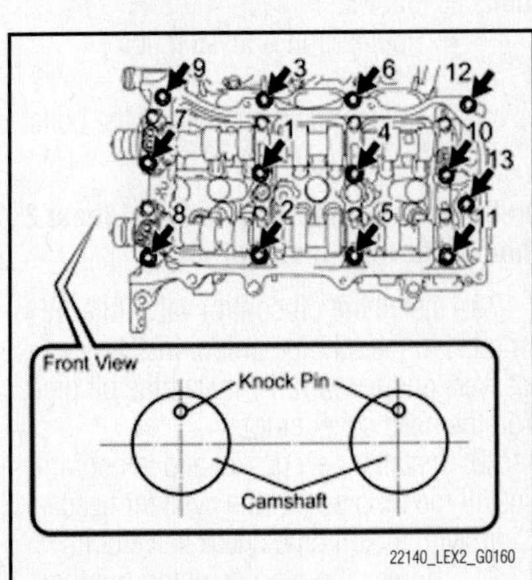

Fig. 134 Install the camshaft housing and tighten the 13 bolts in the order shown

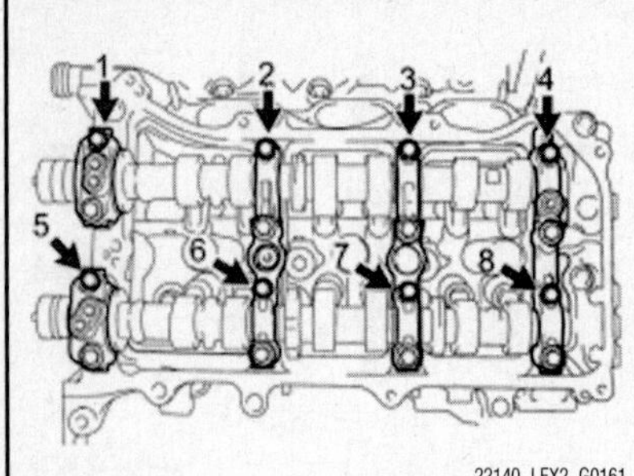

Fig. 135 Tighten the 8 bolts in the order shown

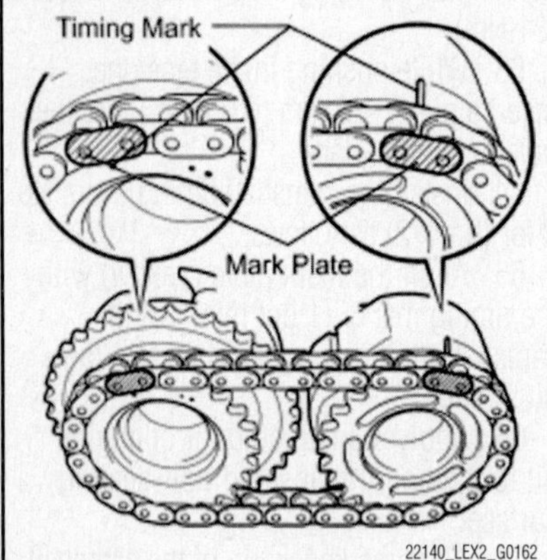

Fig. 136 Align the mark plate (yellow) with the timing marks (1 dot mark) of the camshaft timing gears

74. Install the 4 bolts and 4 oil control valves.
75. Install the 4 bolts and the 4 VVT sensors.
76. Install the oil check valve bolt, oil pipe union and oil pipe.
77. Install the ventilation valve.
78. Install the head cover with the 14 bolts as follows:

- Tighten bolt A to 15 ft. lbs. (21 Nm)
- Tighten the remainder of the bolts to 7ft. lbs. (10 Nm).

➡**Do not start the engine for at least 2 hours after the installation.**

79. Install the oil control valve filter RH and LH to the oil pipe union. Install new gaskets and temporarily install the oil pipe (on the head cover side).
80. Install a new gasket and temporarily install the oil pipe (on the cylinder head side) with the oil pipe check valve bolt.
81. Tighten the oil pipe union (on the head cover side) tighten to 44 ft. lbs. 60 Nm).
82. Tighten the oil check valve bolt (on the cylinder head side) tighten to 44 ft. lbs. 60 Nm).
83. Install the 4 oil control valves with the 4 bolts and tighten to 7 ft. lbs. (10 Nm).
84. Install the 4 VVT sensors with the 4 bolts. Tighten 7 ft. lbs. (10 Nm).
85. Apply adhesive around the ventilation valve and install.
86. Install the 6 spark plugs.
87. Install the ignition coils.
88. Install the air intake surge tank sub-assembly.
89. Install all hoses and electrical connectors.
90. Install the service plug and cover.
91. Connect the negative battery cable.
92. Install the luggage floor mat.

TIMING CHAIN & SPROCKETS

REMOVAL & INSTALLATION

See Figures 137 through 149.

Before servicing any vehicle, please be sure to read all of the following precautions, which deal with personal safety, prevention of component damage, and important points to take into consideration when servicing a motor vehicle:

⁂ CAUTION

The GS450h hybrid system contains a 288V high-voltage system with a strong alkali solution of potassium hydroxide. Be sure to follow the instructions in this manual to handle the system correctly. Failure to do so may result in serious injury or electrocution. Engineer must undergo special training to be able to perform high-voltage system inspection and servicing.

⁂ CAUTION

All high-voltage wire harness connectors are colored orange. The HV battery and other high-voltage components have "High Voltage" caution labels. Do not carelessly touch these wires and components.

⁂ CAUTION

Before inspecting or servicing the high-voltage system, be sure to follow safety measures, such as wearing insulated gloves and removing the service plug to prevent electrocution. Carry the removed service plug

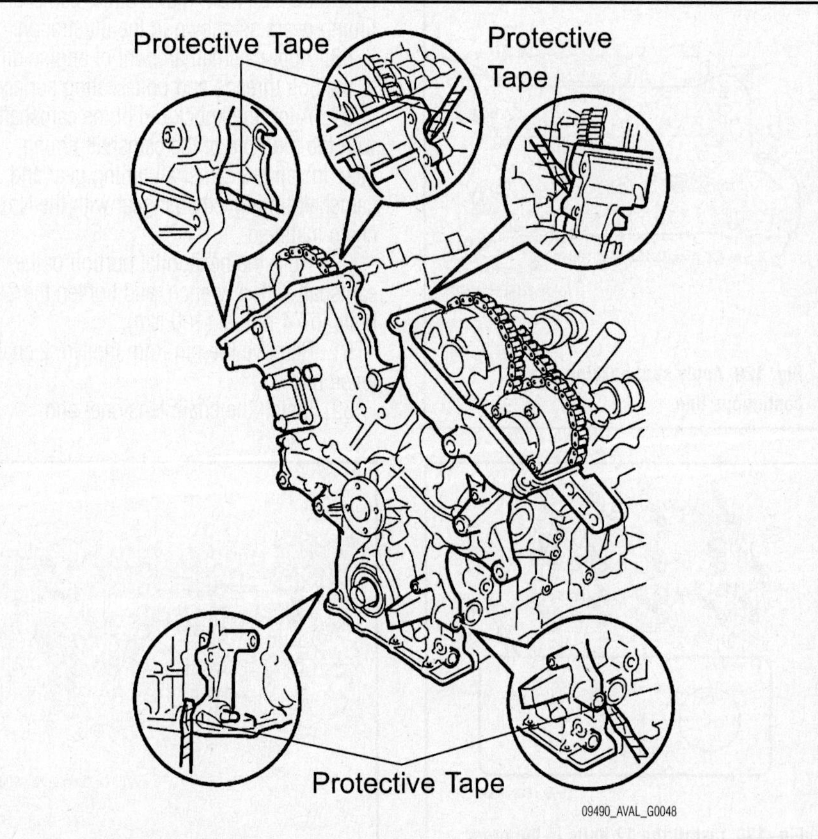

Fig. 137 Pry the front cover at the locations 3.5L (2GR-FSE) Engine

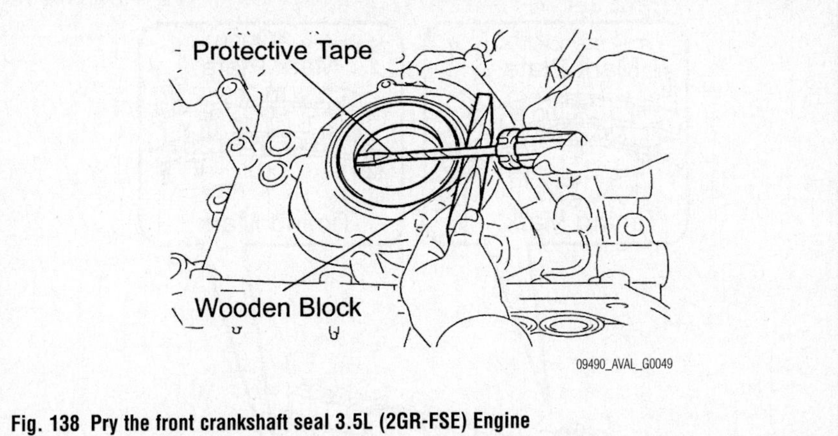

Fig. 138 Pry the front crankshaft seal 3.5L (2GR-FSE) Engine

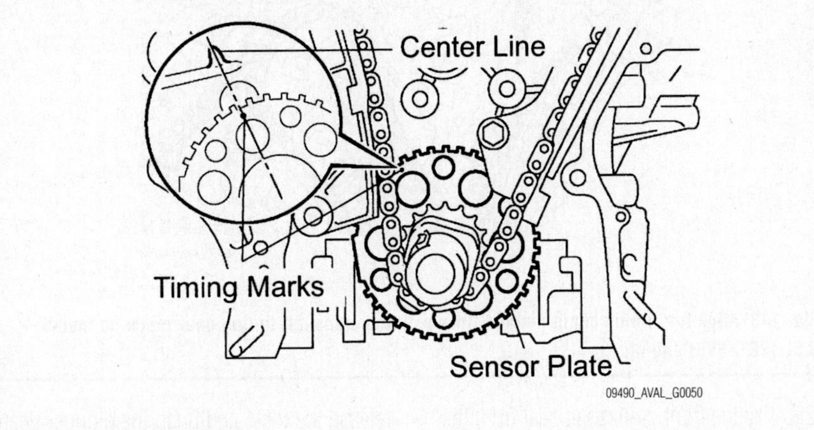

Fig. 139 Align the timing mark on the sensor plate and block to set the No. 1 cylinder at TDC—3.5L (2GR-FSE) engine

in your pocket to prevent other technicians from reinstalling it while you are servicing the vehicle.

CAUTION

After removing the service plug, wait 10 minutes before touching any of the high-voltage connectors and terminals.

1. Discharge the fuel system.
2. Recover the refrigerant from the A/C system using a refrigerant recovery unit.
3. Place the front wheels in the straight ahead position.
4. Disconnect the negative battery cable.
5. Remove the hybrid service plug grip.

CAUTION

Before inspecting the high-voltage system or disconnecting the low voltage connector of the inverter with converter assembly, take safety precautions, such as wearing insulated gloves and removing the service plug grip to prevent electrical shocks. After removing the service plug grip, put it in your pocket to prevent other technicians from accidentally reconnecting it while you are working on the high-voltage system.

CAUTION

After disconnecting the service plug grip, wait for at least 10 minutes before touching any of the high-voltage connectors or terminals. After waiting, check the voltage at the inspection point in the inverter with converter assembly. The voltage should be 0 V before beginning work.

6. Remove the luggage floor mat.
7. Remove the service plug cover.
8. Remove the service plug.
9. Disconnect the negative battery cable.
10. Drain the engine oil and coolant.
11. Remove the engine/transaxle assembly from the vehicle.
12. Remove the transaxle.
13. Remove the oil dipstick tube.
14. Remove the driveplate.
15. Install the engine to a suitable engine stand.

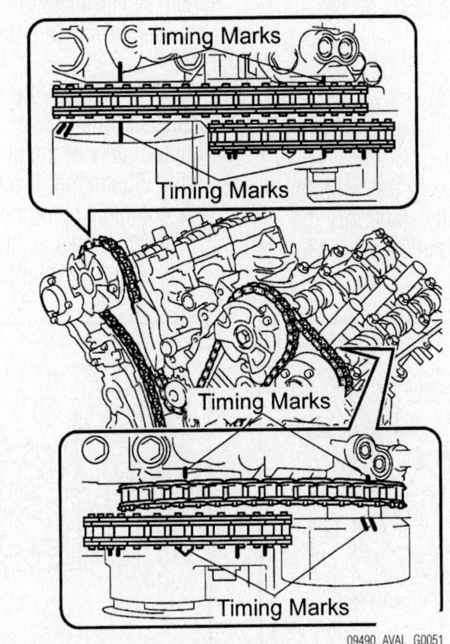

Fig. 140 Ensure the timing marks of the camshaft timing gears are aligned with the bearing cap timing marks—3.5L (2GR-FSE) engine

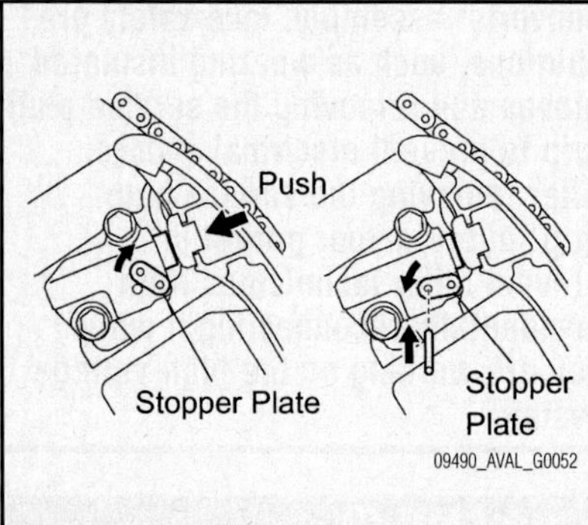

Fig. 141 Procedure to remove the timing chain tensioner—3.5L (2GR-FSE) engine

16. Remove or disconnect the following:
- Idler pulley
- Right-side engine mounting bracket
- Accessory drive belt tensioner
- Water pump pulley
- No. 2 timing gear cover
- Engine mounting stay and bracket
- Water inlet housing

17. Using Special Tool 09213-70011, hold the crankshaft pulley and loosen the pulley bolt.

18. Using the pulley bolt and Special Tool 09950-50013, remove the crankshaft pulley

19. Remove or disconnect the following:
- Upper and lower oil pans
- O-rings from the oil pump
- Air intake surge tank
- Ignition coils
- Oil pipes
- Cylinder head cover

20. Remove the mounting bolts from the front cover.

21. Using a suitable pry tool with the tip covered with protective tape, pry the front cover in the specified locations to remove

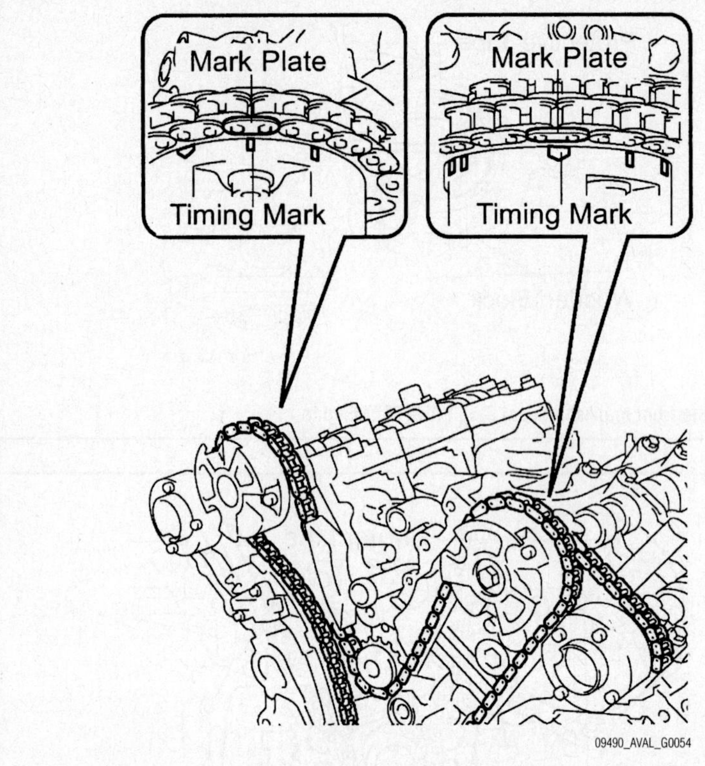

Fig. 143 Align the timing chain marked links with the camshaft timing gear marks to install—3.5L (2GR-FSE) engine

22. Pry the front crankshaft seal from the front cover.

23. Temporarily tighten the pulley set bolt.

24. Set the timing mark on the crank angle sensor plate to the right-hand block bore center line to put the No. 1 cylinder at TDC.

25. Check that the timing marks of the camshaft timing gears are aligned with the timing marks of the bearing caps as shown in the illustration. If not, turn the crankshaft one complete revolution (360°) and align the timing marks as shown.

26. Move the stopper plate upward to release the lock, and push the plunger deep into the tensioner.

27. Move the stopper plate downward to set the lock, and insert a hexagon wrench into the stopper plate's hole.

28. Remove the chain tensioner.

29. Remove the chain tensioner slipper

30. Turn the crankshaft 10° counterclockwise to loosen the chain off the crankshaft timing gear.

31. Remove the timing chain from the crank timing gear and place it on the crankshaft.

32. Turn the camshaft timing gear on the right-side bank clockwise (approximately

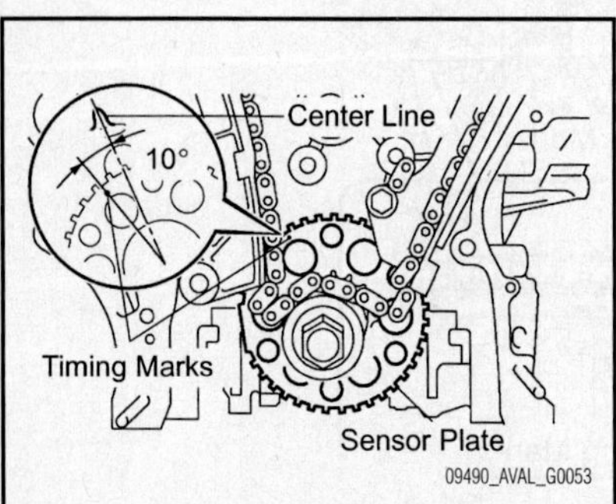

Fig. 142 Turn the crankshaft counterclockwise to loosen the timing chain—3.5L (2GR-FSE) engine

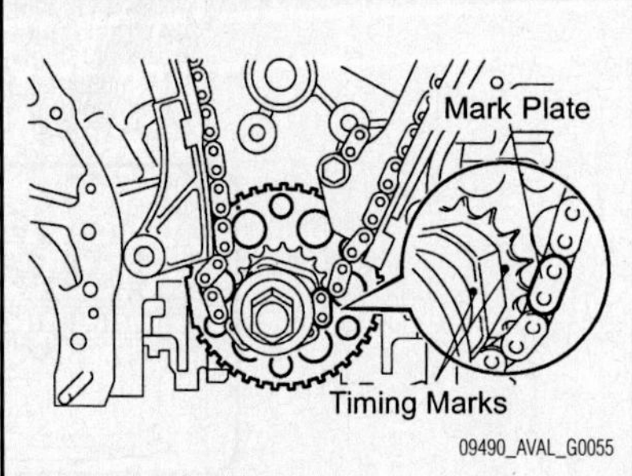

Fig. 144 Align the timing chain marked link with the crankshaft gear to install—3.5L (2GR-FSE) engine

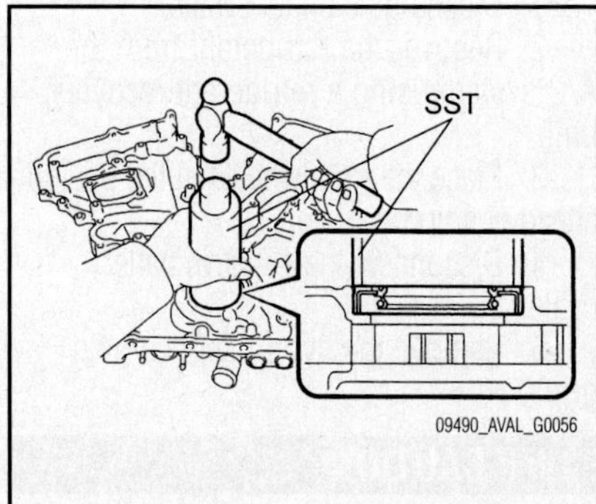

Fig. 145 Tap a new oil seal into place in the front cover

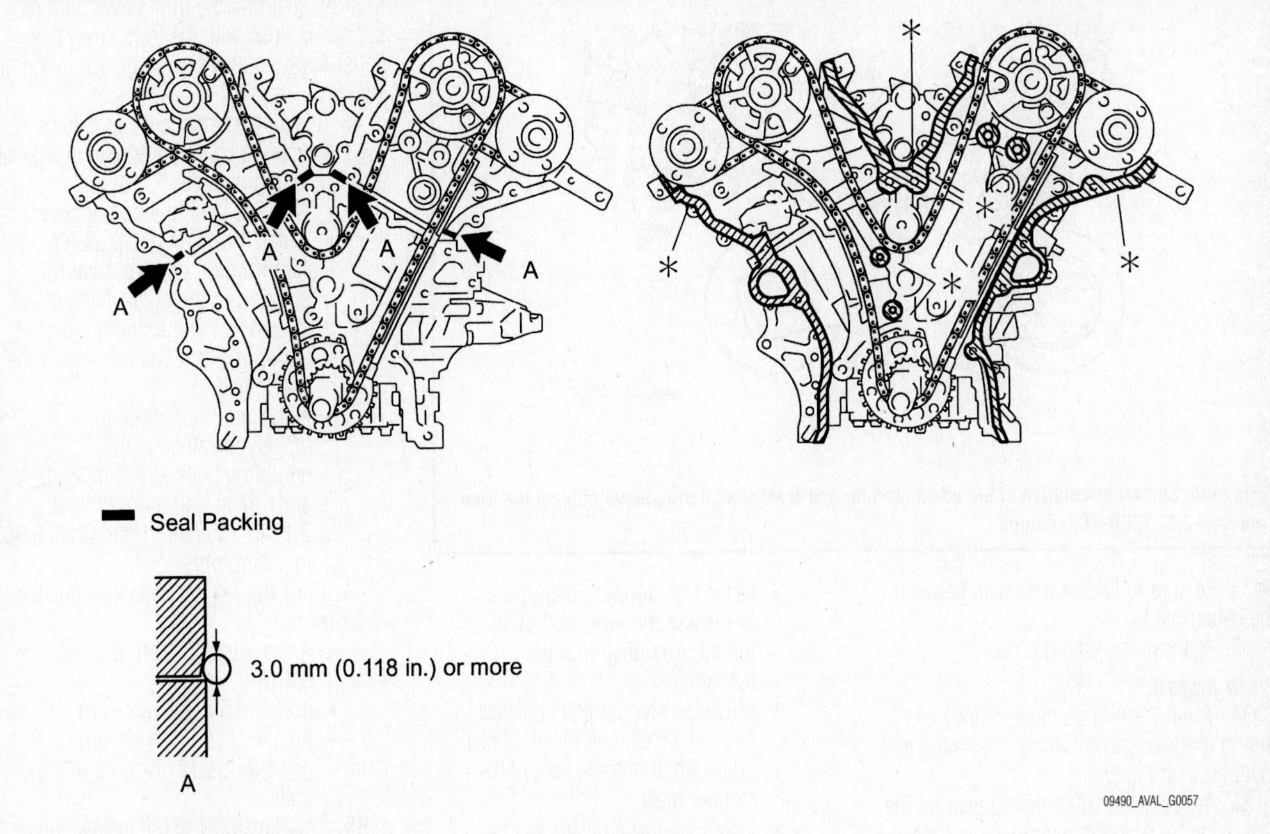

Fig. 146 Apply sealant to the engine block as shown

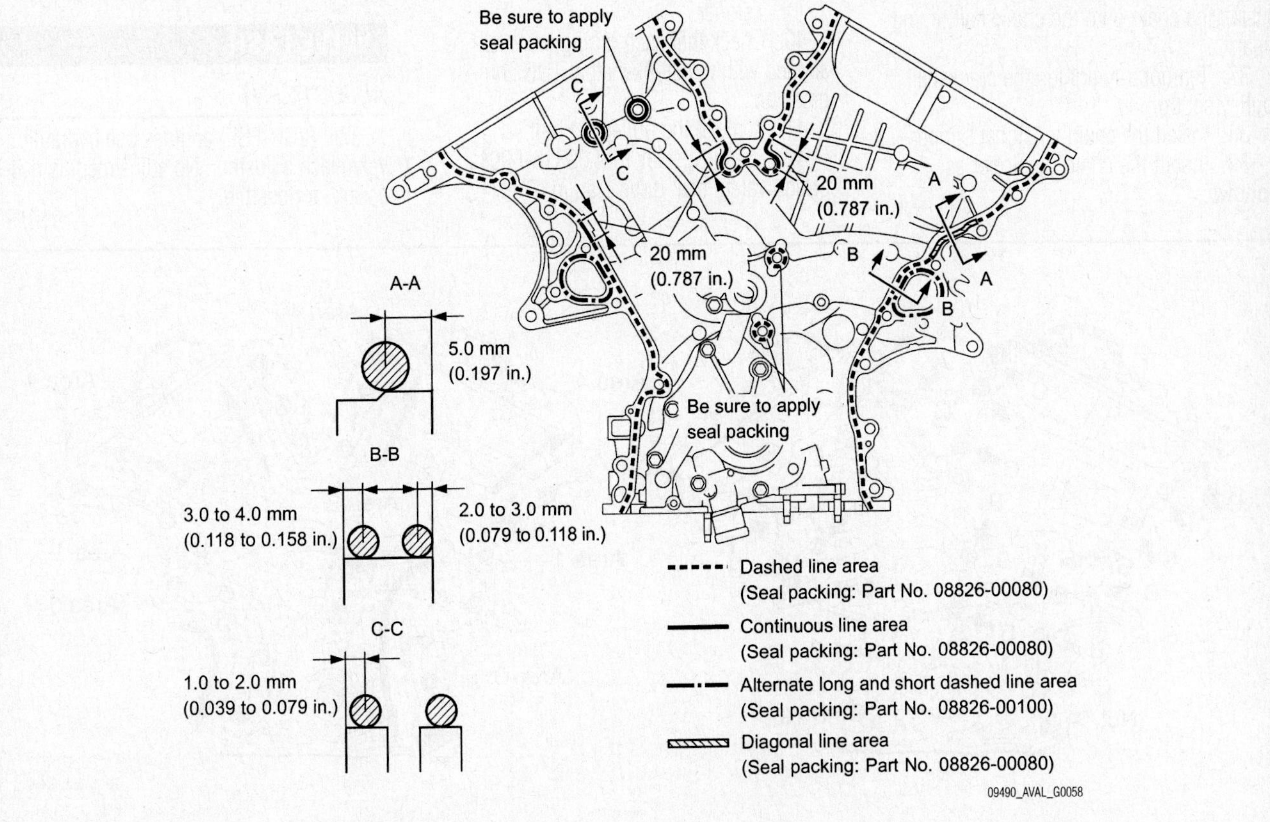

Fig. 147 Apply sealant to the engine front cover as shown

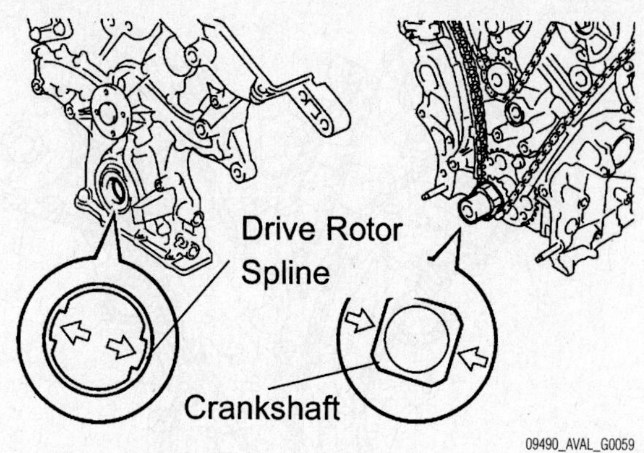

Fig. 148 Correct orientation of the oil pump rotor and crankshaft during installation of the front cover—3.5L (2GR-FSE) engine

60°). Be sure to loosen the chain between the center banks.

33. Remove the timing chain

To install:

34. Align the orange marked links and timing mark as shown in the illustration and install the timing chain.

35. Turn the camshaft timing gear on the right-side bank counterclockwise to tighten the chain between banks.

36. Align the yellow marked link and timing mark as shown in the illustration and install the chain onto the crankshaft timing gear.

37. Temporarily tighten the crankshaft pulley set bolt.

38. Install the chain tensioner slipper.

39. Install the chain tensioner as follows:

- Move the stopper plate upward to release the lock, and push the plunger deep into the tensioner.
- Move the stopper plate downward to set the lock, and insert a hexagon wrench into the hole of the stopper plate.
- Install the chain tensioner and tighten the bolts to 7 ft. lbs. (10 Nm).
- Remove the lock pin of chain tensioner.

40. Check that each timing mark is aligned with the crankshaft at TDC compression.

41. Remove the pulley set bolt.

42. Using Special Tool 09316-60011 or equivalent seal driver, tap in a new front oil seal into the front cover until its surface is flush with the front cover case edge. Apply multi-purpose grease to the oil seal lip.

43. Install the front cover as follows:

- Apply a continuous bead of sealant to engine block.
- Apply sealant to the front cover.
- Install a new oil pump gasket.
- Align the oil pump's drive rotor spline and the crankshaft as shown in the illustration. Install the spline and chain cover to the crankshaft.
- Install the front cover and loosely install all of the mounting bolts and nuts. Bolts A are 1.57 in. (40 mm); Bolts B are 2.17 in. (55 mm); Bolts C are 0.98 in. (25 mm).

44. Fully tighten the bolts in sequence as follows:

- Areas 1 and 2: 15 ft. lbs. (21 Nm)
- Area 3: 15 ft. lbs. (21 Nm)
- Area 4: 32 ft. lbs. (43 Nm) for Bolt A; 15 ft. lbs. (21 Nm) for all other bolts

45. The remainder of the installation is the reverse order of removal

46. Start the engine and check for leaks

VALVE LASH

ADJUSTMENT

3.5L (2GR-FSE) engines use hydraulic valve lash adjusters. No adjustment is necessary or possible.

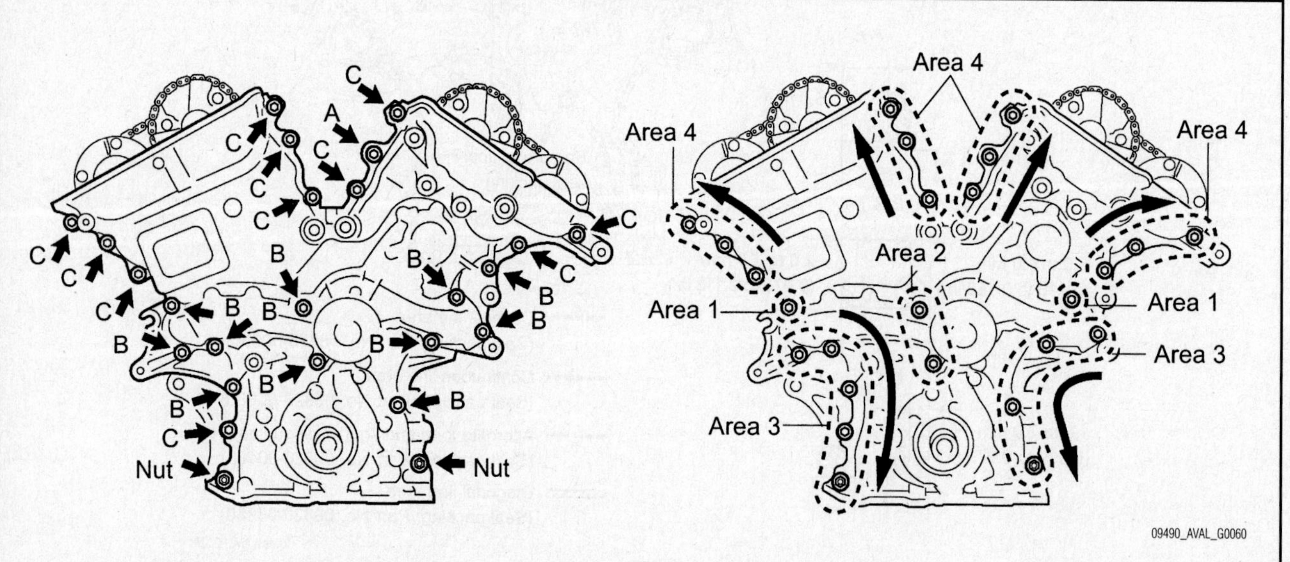

Fig. 149 Bolt identification and torque sequence for front timing cover—(2GR-FSE) engine

ENGINE PERFORMANCE & EMISSION CONTROLS

CAMSHAFT POSITION (CMP) SENSOR

REMOVAL & INSTALLATION

See Figures 150 through 153.

1. Disconnect the negative battery cable. Wait at least 90 seconds after disconnecting the cable from the negative (-) battery terminal to prevent airbag and seat belt pretensioner activation.
2. Remove the V-bank engine cover.
3. Disconnect the sensor connector.
4. Remove the bolt and sensor.

To install:

5. Install the sensor with the bolt and tighten to 7 ft. lbs. (10 Nm).
6. Connect the sensor connector.
7. Install the V-bank cover.
8. Connect the negative battery cable.
9. Certain systems need to be initialized after disconnecting and reconnecting the cable from the negative (-) battery terminal.

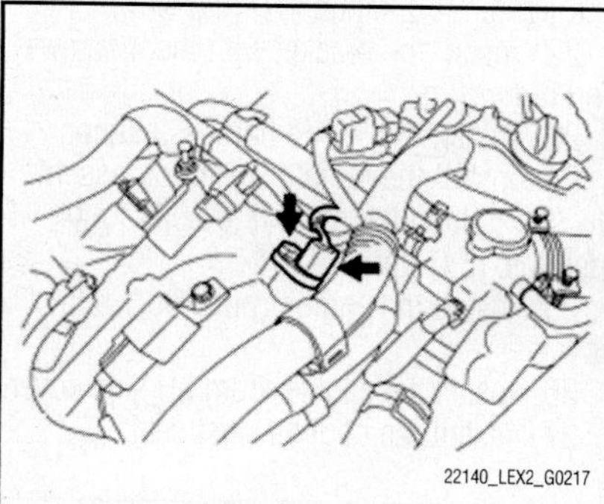

Fig. 150 Intake side bank 1 sensor location

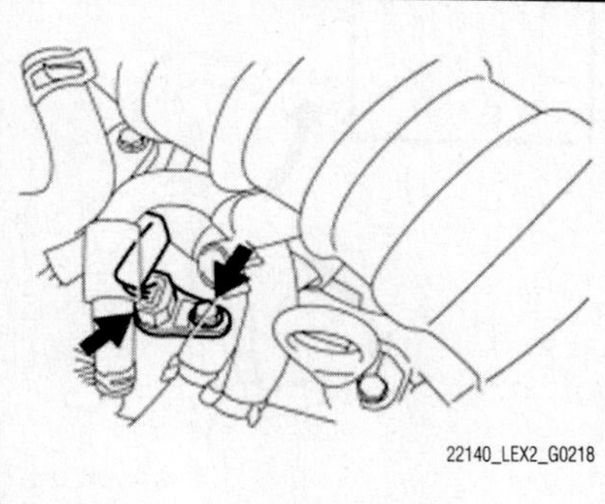

Fig. 151 Exhaust side bank 1 sensor location

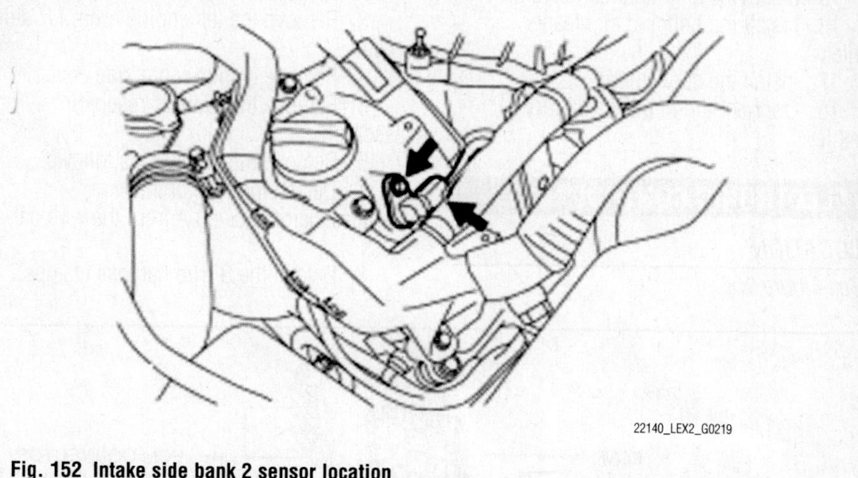

Fig. 152 Intake side bank 2 sensor location

Fig. 153 Exhaust side bank 2 sensor location

CRANKSHAFT POSITION (CKP) SENSOR

REMOVAL & INSTALLATION

See Figure 154.

1. Remove the service plug grip.
2. Disconnect the negative battery cable.
3. Remove the bolt and air cleaner inlet.
4. Evacuate the A/C system.
5. Remove the engine under covers.
6. Disconnect the A/C lines.
7. Remove the electric A/C inverter compressor.
8. Disconnect the sensor connector.
9. Remove the bolt and sensor.

To install:

10. Apply a coat of engine oil to the O-ring of the sensor.
11. Install the sensor with the bolt and tighten to 7 ft. lbs. (10 Nm).
12. Check compressor oil and add if low.
13. Install the electric A/C compressor and tighten the mounting bolts to 18 ft. (2.4 Nm).

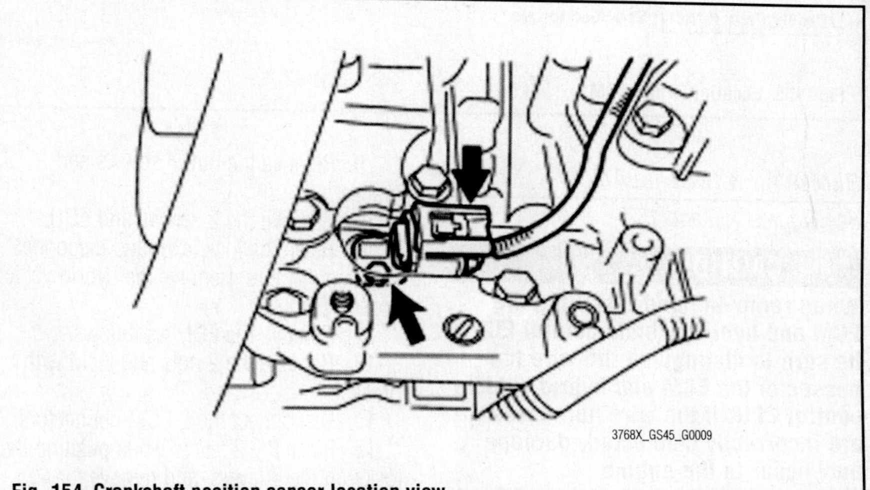

Fig. 154 Crankshaft position sensor location view

14. Install the A/C lines and tighten to 7 ft. lbs. (10 Nm).
15. Install the engine under covers.
16. Install the bolt and air cleaner inlet.
17. Install the service grip.
18. Connect the negative battery cable.

ELECTRONIC CONTROL UNIT

LOCATION

See Figure 155.

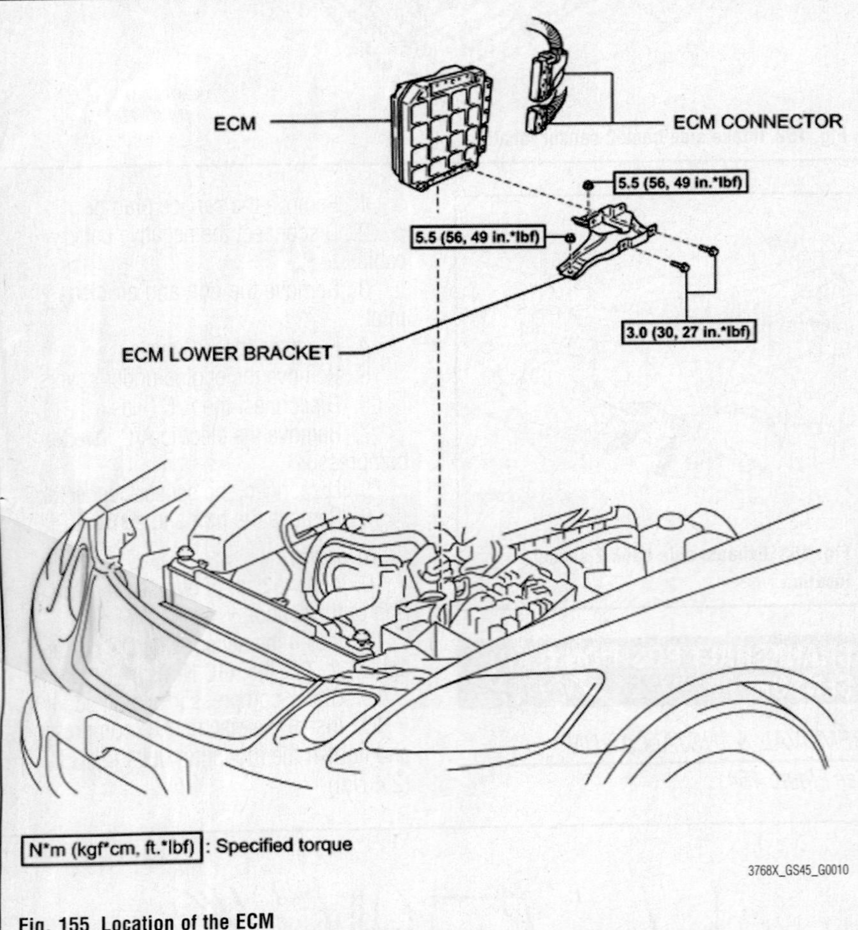

Fig. 155 Location of the ECM

REMOVAL & INSTALLATION

See Figures 156 and 157.

⁂ CAUTION

When removing and installing the ECM and hybrid vehicle control ECU, be sure to distinguish the wire harnesses of the ECM and hybrid vehicle control ECU. If the wire harnesses are incorrectly connected, damage may occur to the engine.

1. Disconnect the negative battery cable.
2. Remove the cool air intake duct seal.
3. Remove the left engine room LH side cover.
4. Remove engine room side cover.
5. Remove the radiator reservoir assembly.
6. Remove the HV ECU as follows:
7. Detach the side of the wire harness clamp labeled A from the wire harness.
8. Detach the 3 wire harness clamps.
9. Remove the nut, 4 screws and bracket.
10. Remove the 2 screws and ECU.
11. Raise the lock lever and disconnect the 2 connectors. Remove the hybrid vehicle control ECU.
12. Remove the ECM as follows:
13. Remove the 2 nuts and ECM with bracket.
14. Disconnect the 2 ECM connectors.
15. Raise the 2 levers while pushing the locks on the 2 levers, and remove the 2 ECM connectors.

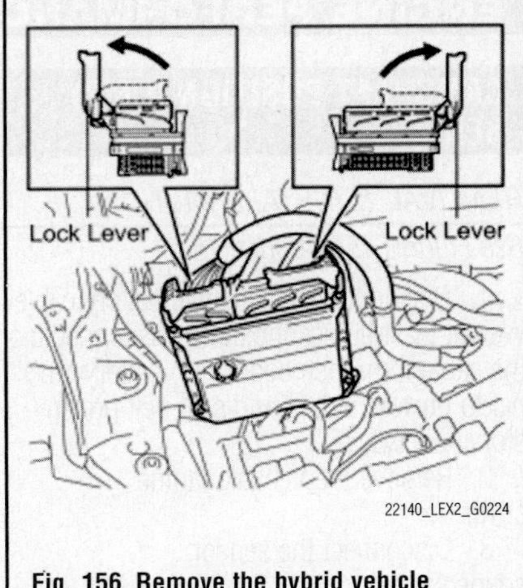

Fig. 156 Remove the hybrid vehicle control ECU

To install:

16. Install the ECM to the lower bracket and tighten to 27 inch lbs. (3.0 Nm).
17. Connect the 2 ECM connectors.
18. Install the ECM with bracket with the 2 nuts and tighten to 49 inch lbs. (5.5 Nm).
19. Install the HV control unit as follows:
20. Connect the 2 connectors to the hybrid vehicle control ECU and push each lock lever down to lock the connectors.
21. Install the ECU with the 2 screws and tighten to 27 inch lbs. (3.0 Nm).
22. Install the bracket with the 4 screws and nut.
23. Attach the 3 wire harness clamps.
24. Install the radiator reservoir assembly and tighten the mounting bolts to 44 inch lbs. (5.0 Nm).
25. Install the engine room No.1 relay cover.
26. Install the engine room LH side cover.
27. Install the cool air duct seal.

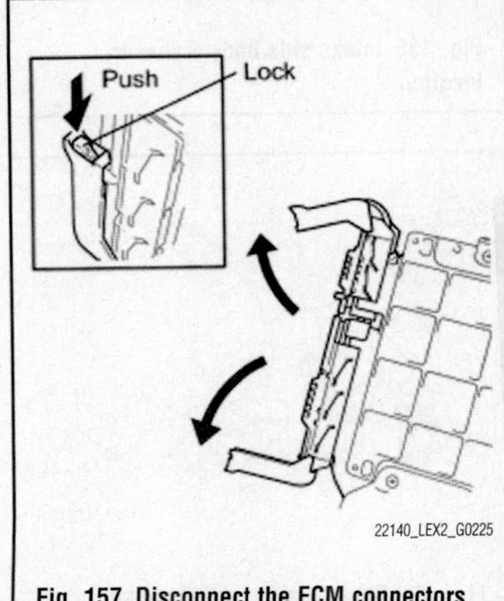

Fig. 157 Disconnect the ECM connectors as shown

28. Connect the negative battery cable.

29. Some vehicle systems require initialization after reconnecting the cable to the negative battery terminal.

ENGINE COOLANT TEMPERATURE (ECT) SENSOR

LOCATION

The Engine Coolant Temperature (ECT) Sensor is located at the rear of the engine in the cylinder head.

REMOVAL & INSTALLATION

See Figure 158.

1. Before servicing the vehicle, refer to the Precautions section.
2. Discharge the fuel system pressure.
3. Disconnect the negative battery cable.

CAUTION

Remove the service plug grip to interrupt a high voltage circuit at the time of the check. Keep the removed service plug grip in your pocket to prevent other technicians from accidentally reconnecting it while you are servicing the vehicle. All the high voltage wiring connectors are colored in orange.

4. Wear insulated gloves. Remove the service plug grip after sliding the lever of the service plug grip.
5. Remove or disconnect the following:
   - Cool air intake seal
   - LH and RH engine room cover
   - Engine V-bank cover
   - LH and RH front upper fender protector
   - LH and RH fender cowl seal
   - LH and RH wiper arm assembly
   - Cowl top ventilator louver
   - Drain engine coolant
   - No 2 ventilation hose
   - Air cleaner cap and hose
   - Throttle body
   - Intake air surge tank
   - Intake manifold
6. Disconnect the coolant temperature sensor connector.
7. Remove the coolant temperature sensor.

To install:

8. Install a new gasket to the sensor.
9. Install the sensor and tighten to 15 ft. lbs. (20 Nm).
10. Connect the sensor connector.
11. Install or Reconnect the following:
    - Intake manifold
    - Intake air surge tank
    - Throttle body
    - Air cleaner cap and hose
    - Drain engine coolant
    - Cowl top ventilator louver
    - LH and RH wiper arm assembly
    - LH and RH fender cowl seal
    - LH and RH front upper fender protector
    - Engine V-bank cover
    - LH and RH engine room cover
    - Cool air intake seal
12. Connect the negative battery cable.
13. Wear insulated gloves and install the service plug.
14. Install the engine coolant, bleed system and check for leaks.

HEATED OXYGEN (HO2S) SENSOR

LOCATION

See Figure 159.

Refer to the accompanying illustration.

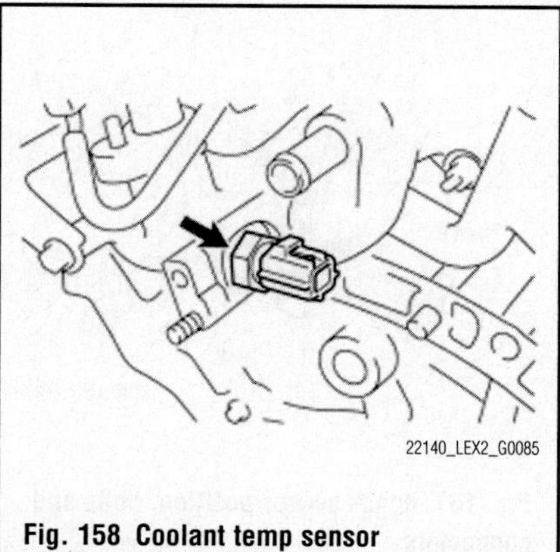

Fig. 158 Coolant temp sensor

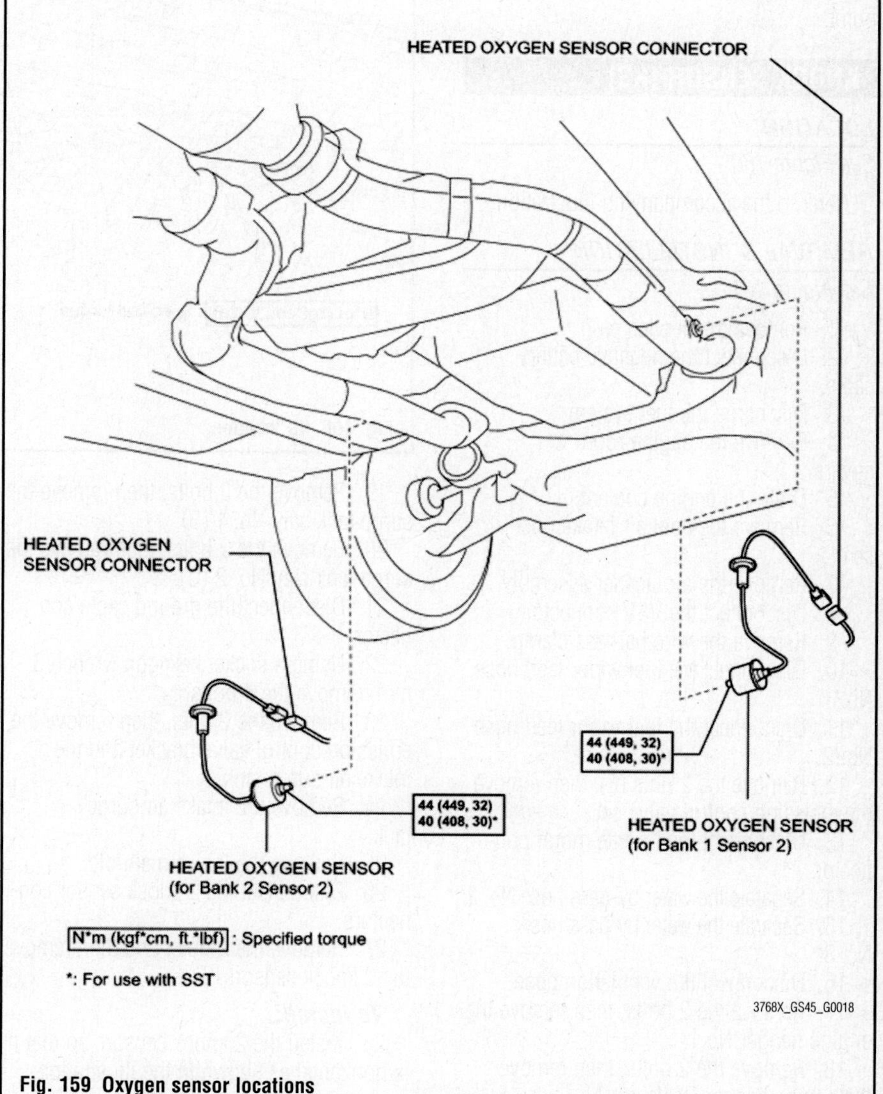

Fig. 159 Oxygen sensor locations

REMOVAL & INSTALLATION

1. Remove the service plug grip.
2. Disconnect the negative battery cable.
3. Remove front console upper garnish.
4. Remove the console box plate.
5. Remove the RH and LH instrument panel finish panel ends.
6. Remove the heated O2 sensor as follows:
   - Disconnect the sensor connector.
   - Remove the grommet and pass the sensor connector out of the cabin through the floor panel.
   - Using a O2 socket remove the sensor from the front exhaust pipe.

To install:

7. Install the O2 sensor and tighten to 30 ft. lbs. (40 Nm).
8. Reverse the removal procedure at this point.

KNOCK SENSOR (KS)

LOCATION

See Figure 160.

Refer to the accompanying illustration.

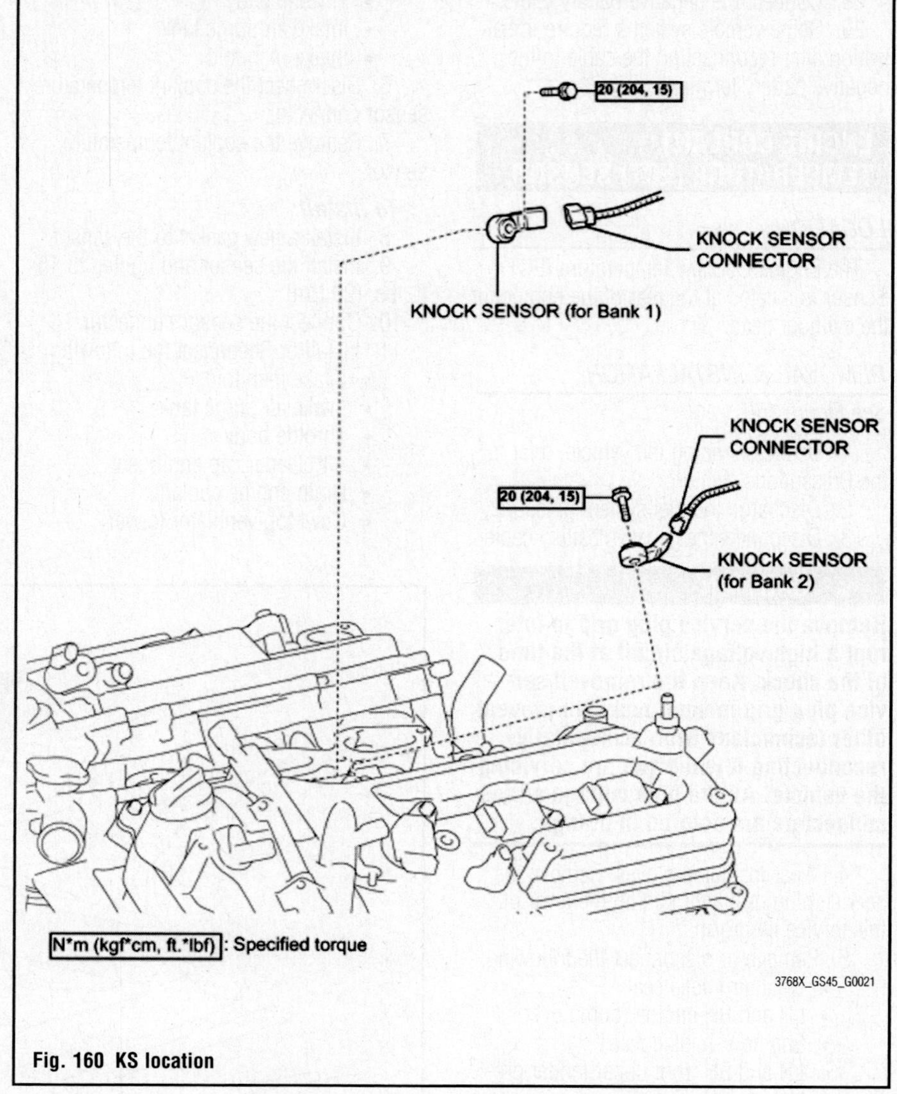

Fig. 160 KS location

REMOVAL & INSTALLATION

See Figure 161.

1. Remove the service plug.
2. Disconnect the negative battery cable.
3. Discharge the fuel system.
4. Remove the engine room covers.
5. Drain the engine coolant.
6. Remove the cool air intake duct seal.
7. Remove the air cleaner assembly.
8. Disconnect the VSV connector.
9. Remove the wire harness clamp.
10. Disconnect the fuel vapor feed hose No. 1.
11. Disconnect the fuel vapor feed hose No. 2.
12. Remove the 2 nuts (E), then remove the emission control valve set.
13. Disconnect the throttle motor connector.
14. Separate the water by-pass hose No. 2.
15. Separate the water by-pass hose No. 3.
16. Disconnect the ventilation hose.
17. Remove the 2 bolts, then remove the engine hanger No.1
18. Remove the 2 bolts, then remove the engine hanger No.1
19. Remove the 2 bolts, then remove the surge tank stay No. 1 (B).
20. Remove the 2 bolts, then remove the surge tank stay No. 2 (C).
21. Disconnect the ground cable connector.
22. Using a socket hexagon wrench 8 mm, remove the 4 bolts.
23. Remove the 2 nuts, then remove the emission control valve bracket and the intake air surge tank.
24. Remove the intake air surge tank.
25. Remove the intake manifold.
26. Disconnect the 2 knock sensor connectors.
27. Remove the 2 nuts, and then remove the 2 knock sensors.

To install:

28. Install the 2 knock sensors so that it is horizontal as shown in the illustration. Then install the 2 bolts and tighten to 15 ft. lbs. (20 Nm).
29. Connect the 2 knock sensor connectors.

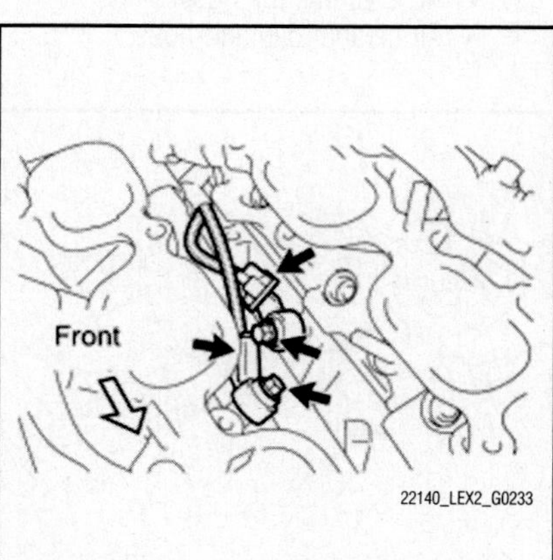

Fig. 161 Knock sensor position, bolts and connectors

30. Install the intake manifold.
31. Install the surge tank stay.
32. Install air filter assembly bracket and tighten to 14 ft. lbs. 20 (Nm).
33. Install the cool air intake duct.
34. Install engine coolant and bleed the system.
35. Install the service plug.
36. Connect the negative battery cable.
37. Some systems need initialization when reconnecting the battery cable.

MALFUNCTION INDICATOR LIGHT (MIL)

RESET PROCEDURE

1. Clear DTC (Using the Techstream) as follows:
    a. Connect the Techstream to the DLC3.
    b. Turn the ignition switch ON.
    c. Enter the following menus: Powertrain / Engine and ECT / Trouble Codes.
    d. Press the YES button.
2. CLEAR DTC (Without using Techstream)
    a. Perform either one of the following operations.
    b. Disconnect the negative battery cable for more than 1 minute.
    c. Remove the EFI and ETCS fuses from the engine room No. 1 junction block located inside the engine compartment for more than 1 minute.

MASS AIR FLOW (MAF) METER

REMOVAL & INSTALLATION

1. Disconnect the Mass Air Flow (MAF) meter connector.
2. Disconnect the 2 wire harness clamps from the air cleaner assembly.
3. Remove the 2 screws and Mass Air Flow (MAF) meter.

To install:

4. Install the Mass Air Flow (MAF) meter with the 2 screws.
5. Connect the Mass Air Flow (MAF) meter connector.
6. Connect the 2 wire harness clamps to the air cleaner assembly.

THROTTLE POSITION SENSOR (TPS)

REMOVAL & INSTALLATION

See Figure 162.

1. Disconnect the negative battery cable.
2. Remove the cool air intake duct seal.
3. Drain the engine coolant.
4. Remove the V-bank cover.
5. Remove the air cleaner inlet.
6. Disconnect the No. 2 ventilation hose.
7. Disconnect the air cleaner cap and hose.
8. Disconnect the MAF meter connector.
9. Disconnect the clamp from the air cleaner.
10. Disconnect the purge line hose.
11. Disconnect the 4 clamps.
12. Remove the hose clamp and air cleaner cap with air cleaner hose.
13. Remove the throttle body as follows:
    - Disconnect the throttle motor connector.
    - Remove the 4 bolts and disconnect the throttle body from the intake air surge tank.
    - Disconnect the 2 water by-pass hoses from the throttle body.
    - Remove the throttle body and gasket.

To install:

14. Install a new gasket to the intake air surge tank.
15. Connect the 2 water by-pass hoses to the throttle body.
16. Install the throttle body with the 4 bolts and tighten to 7 ft. lbs. (10 Nm).
17. Connect the throttle motor connector.
18. Install the air cleaner cap with air cleaner hose assembly with the 4 clamps and hose clamp.
19. Install the purge line hose to the air cleaner hose.
20. Connect the MAF meter connector and clamp to the air cleaner.
21. Connect the ventilation hose to the cylinder head cover with the clamp.
22. Add engine coolant.
23. Connect the negative battery cable.
24. Check the for coolant leaks.
25. Check the function of the throttle body.
26. Install the air cleaner inlet.
27. Install the engine room covers.
28. Install the cool air duct seal.
29. Certain systems need to be initialized after disconnecting and reconnecting the cable from the negative (-) battery terminal.

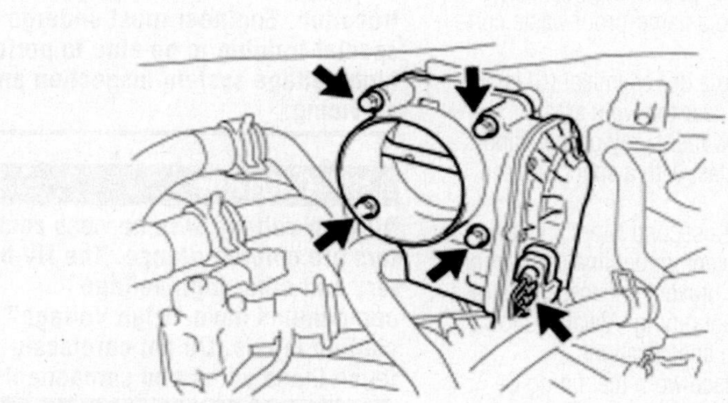

Fig. 162 Remove the 4 bolts and disconnect the throttle body

FUEL SYSTEM SERVICE PRECAUTIONS

Safety is the most important factor when performing not only fuel system maintenance, but any type of maintenance. Failure to conduct maintenance and repairs in a safe manner may result in serious personal injury or death. Work on a vehicle's fuel system components can be accomplished safely and effectively by adhering to the following rules and guidelines.

- To avoid the possibility of fire and personal injury, always disconnect the negative battery cable unless the repair or test procedure requires that battery voltage be applied.
- Always relieve the fuel system pressure prior to disconnecting any fuel system component (injector, fuel rail, pressure regulator, etc.) fitting or fuel line connection. Exercise extreme caution whenever relieving fuel system pressure to avoid exposing skin, face and eyes to fuel spray. Please be advised that fuel under pressure may penetrate the skin or any part of the body that it contacts.
- Always place a shop towel or cloth around the fitting or connection prior to loosening to absorb any excess fuel due to spillage. Ensure that all fuel spillage is quickly removed from engine surfaces. Ensure that all fuel-soaked cloths or towels are deposited into a flame-proof waste container with a lid.
- Always keep a dry chemical (Class B) fire extinguisher near the work area.
- Do not allow fuel spray or fuel vapors to come into contact with a spark or open flame.
- Always use a second wrench when loosening or tightening fuel line connection fittings. This will prevent unnecessary stress and torsion on fuel piping. Always follow the proper torque specifications.
- Always replace worn fuel fitting O-rings with new ones. Do not substitute fuel hose where rigid pipe is installed.

FUEL SYSTEM PRESSURE

RELIEVING

1. Connect the Techstream to the DLC3.
2. Set the vehicle to the "INSPECTION MODE
3. Start the engine.
4. Disconnect the fuel pump connector.
5. After the engine has stopped on its own, turn the power switch OFF.
6. Loosen the fuel tank cap, then discharge the pressure in the fuel tank completely.
7. Connect the fuel pump connector.

➡DTC P3190, P3191, P0171, P0172 and/or P0A0F may be set.

FUEL FILTER

REMOVAL & INSTALLATION

The fuel filter is integral to the fuel pump assembly inside the fuel tank.

FUEL LEVEL SENDING UNIT

REMOVAL & INSTALLATION

See Figures 163 through 165.

Before servicing any vehicle, please be sure to read all of the following precautions, which deal with personal safety, prevention of component damage, and important points to take into consideration when servicing a motor vehicle:

⁂ CAUTION

The GS450h hybrid system contains a 288V high-voltage system with a strong alkali solution of potassium hydroxide. Be sure to follow the instructions in this manual to handle the system correctly. Failure to do so may result in serious injury or electrocution. Engineer must undergo special training to be able to perform high-voltage system inspection and servicing.

⁂ CAUTION

All high-voltage wire harness connectors are colored orange. The HV battery and other high-voltage components have "High Voltage" caution labels. Do not carelessly touch these wires and components.

⁂ CAUTION

Before inspecting or servicing the high-voltage system, be sure to follow safety measures, such as wearing insulated gloves and removing the service plug to prevent electrocution. Carry the removed service plug in your pocket to prevent other technicians from reinstalling it while you are servicing the vehicle.

⁂ CAUTION

After removing the service plug, wait 10 minutes before touching any of the high-voltage connectors and terminals.

1. Discharge the fuel system.
2. Remove the service plug.
3. Disconnect the negative battery cable.
4. Remove the rear seat cushion assembly.
5. Remove the clip and room partition pad.
6. Remove the service hole cover and disconnect the 2 connectors.
7. Disconnect the fuel main tubes and fuel return vent tube.
8. Remove the 2 tube joint clips and fuel tubes as shown below.

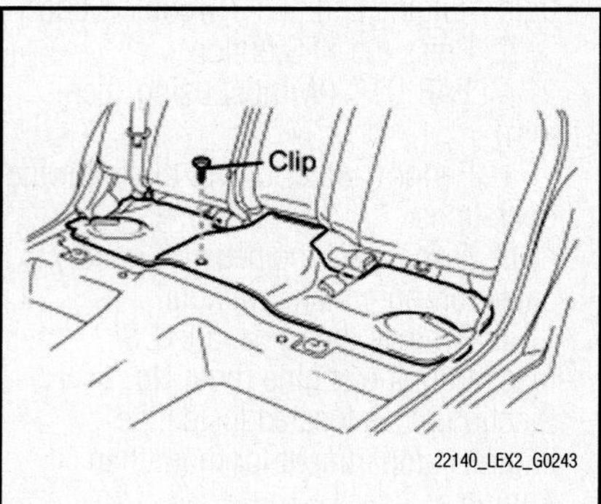

Fig. 163 Remove the clip and room partition pad

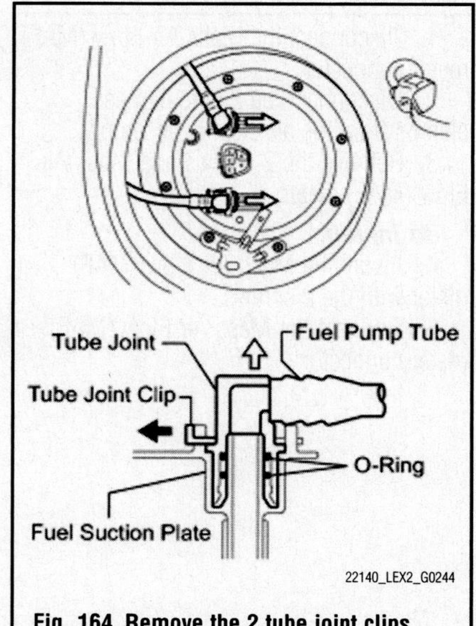

Fig. 164 Remove the 2 tube joint clips and fuel tubes

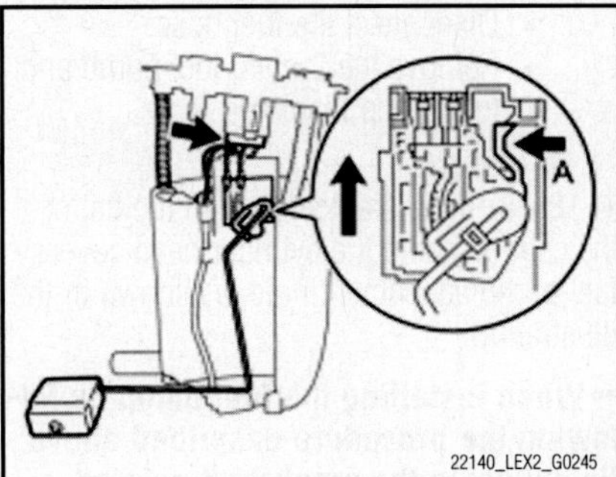

Fig. 165 Press down on the sender gauge claw labeled A. Then slide the sender gauge upward

9. Remove the 8 bolts and set plate.
10. Disconnect the fuel tube.
11. Remove the fuel suction tube with pump and gauge from the fuel tank.
12. Disconnect the fuel sender gauge connector.
13. Press down on the sender gauge claw labeled A. Then slide the sender gauge upward.

To install:

14. Set the fuel sender gauge to the sub-tank. Then slide the sender gauge downward to install it.
15. Connect the fuel sender gauge connector.
16. Install the fuel suction with pump and gauge tube assembly.
17. Install the service hole cover.
18. Install the partition pad.
19. Install the rear seat cushion assembly.
20. Install the service plug grip.
21. Connect the negative battery cable.
22. Check for fuel leaks.
23. Certain systems need to be initialized after disconnecting and reconnecting the cable from the negative (-) battery terminal.

FUEL PUMP MODULE

REMOVAL & INSTALLATION

See Figures 163 through 167.

Before servicing any vehicle, please be sure to read all of the following precautions, which deal with personal safety, prevention of component damage, and important points to take into consideration when servicing a motor vehicle:

CAUTION

The GS450h hybrid system contains a 288V high-voltage system with a strong alkali solution of potassium hydroxide. Be sure to follow the instructions in this manual to handle the system correctly. Failure to do so may result in serious injury or electrocution. Engineer must undergo special training to be able to perform high-voltage system inspection and servicing.

CAUTION

All high-voltage wire harness connectors are colored orange. The HV battery and other high-voltage components have "High Voltage" caution labels. Do not carelessly touch these wires and components.

CAUTION

Before inspecting or servicing the high-voltage system, be sure to follow safety measures, such as wearing insulated gloves and removing the service plug to prevent electrocution. Carry the removed service plug in your pocket to prevent other technicians from reinstalling it while you are servicing the vehicle.

CAUTION

After removing the service plug, wait 10 minutes before touching any of the high-voltage connectors and terminals.

1. Discharge the fuel system.
2. Remove the service plug.
3. Disconnect the negative battery cable.
4. Remove the rear seat cushion assembly.
5. Remove the clip and room partition pad.
6. Remove the service hole cover and disconnect the 2 connectors.
7. Disconnect the fuel main tubes and fuel return vent tube.
8. Remove the 2 tube joint clips and fuel tubes as shown below.
9. Remove the 8 bolts and set plate.
10. Disconnect the fuel tube.
11. Remove the fuel suction tube with pump and gauge from the fuel tank.
12. Disconnect the fuel sender gauge connector.
13. Press down on the sender gauge claw labeled A. Then slide the sender gauge upward.
14. Using needle nozzle pliers, remove the E-ring.

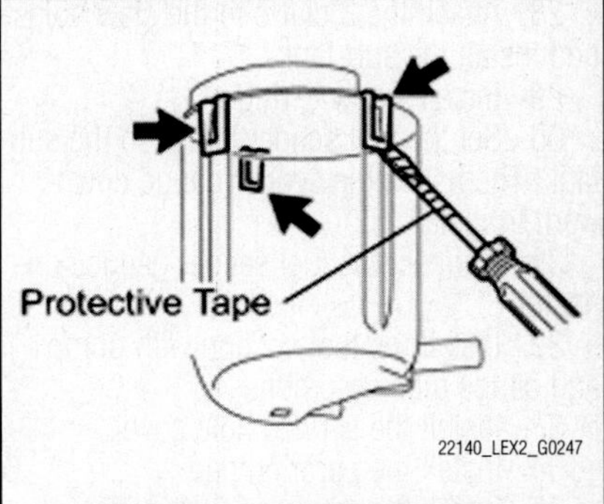

Fig. 166 Detach the 3 claws and remove the sub-tank

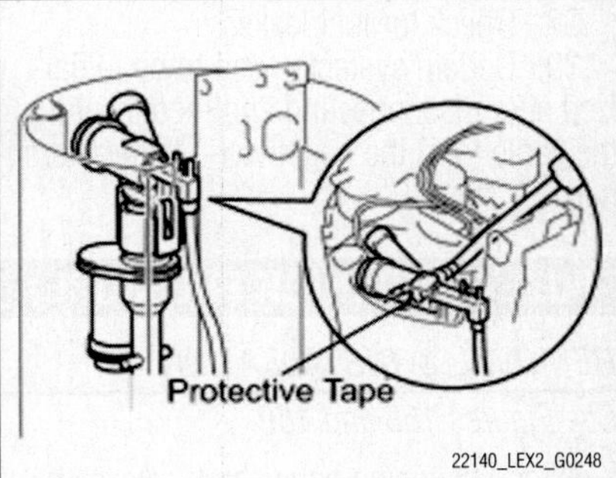

Fig. 167 Using a clip remover with the tip taped, remove the jet pump

15. Using a screwdriver with the tip taped, detach the claw of the jet pump nozzle.
16. Using a screwdriver with the tip taped, detach the 3 claws and remove the sub-tank.
17. Using a clip remover with the tip taped, remove the jet pump.
18. Disconnect the 2 connectors and terminal, and remove the fuel pump wire.
19. Using a screwdriver with the tip taped, detach the 2 claws from the claw holes and remove the fuel pump.
20. Remove the fuel pump seal from the fuel pump.

To install:

21. Install the fuel pump spacer to the fuel pump.
22. Apply a light coat of gasoline to a new seal, and install it to the fuel pump.
23. Install the fuel pump to the fuel filter.
24. Connect the 2 connectors and terminal.
25. Apply gasoline to a new O-ring and install it to the jet pump.
26. Install the jet pump while aligning it to the installation position of the sub-tank.
27. Connect the jet pump nozzle to the sub-tank.

28. Attach the 3 claws to the claw holes and install the sub-tank.

29. Install a new E-ring.

30. Set the fuel sender gauge to the sub-tank. Then slide the sender gauge downward to install it.

31. Connect the fuel sender gauge connector.

32. Install the fuel suction with pump and gauge tube assembly.

33. Install the service hole cover.

34. Install the partition pad.

35. Install the rear seat cushion assembly.

36. Install the service plug grip.

37. Connect the negative battery cable.

38. Check for fuel leaks.

39. Certain systems need to be initialized after disconnecting and reconnecting the cable from the negative (-) battery terminal.

FUEL PUMP (HIGH PRESSURE)

REMOVAL & INSTALLATION

See Figures 168 and 169.

Before servicing any vehicle, please be sure to read all of the following precautions, which deal with personal safety, prevention of component damage, and important points to take into consideration when servicing a motor vehicle:

CAUTION

The GS450h hybrid system contains a 288V high-voltage system with a strong alkali solution of potassium hydroxide. Be sure to follow the instructions in this manual to handle the system correctly. Failure to do so may result in serious injury or electrocution. Engineer must undergo special training to be able to perform high-voltage system inspection and servicing.

CAUTION

All high-voltage wire harness connectors are colored orange. The HV battery and other high-voltage components have "High Voltage" caution labels. Do not carelessly touch these wires and components.

CAUTION

Before inspecting or servicing the high-voltage system, be sure to follow safety measures, such as wearing insulated gloves and removing the service plug to prevent electrocution. Carry the removed service plug in your pocket to prevent other technicians from reinstalling it while you are servicing the vehicle.

CAUTION

After removing the service plug, wait 10 minutes before touching any of the high-voltage connectors and terminals.

1. Discharge fuel system.
2. Drain the engine coolant.
3. Remove the cool air intake duct seal.
4. Remove the engine room side cover.
5. Remove the V-bank cover.
6. Remove the air cleaner cap and hose.
7. Remove the wiper arm assembly LH and RH
8. Remove the cowl top ventilator louver sub-assembly.
9. Disconnect the No. 2 ventilator hose.
10. Remove the air cleaner cap with air cleaner hose.
11. Remove the throttle body with motor assembly.
12. Remove the air intake surge tank.
13. Remove the intake manifold.
14. Remove the water hose joint.
15. Remove the fuel pressure pulsation damper.
16. Remove the No. 1 fuel pipe.
17. Disconnect the No 2 fuel pipe as follows:
    - Disconnect the fuel high pressure side fuel pump connector.
    - Fix the union bolt on the fuel pump side in place with a 21 mm wrench. Using a 19 mm union nut wrench, loosen the union and remove the fuel pipe.
    - Remove the 2 bolts on the delivery pipe side.
    - Disconnect the fuel hose.
    - Remove the 2 nuts, fuel pump and fuel pump insulator.

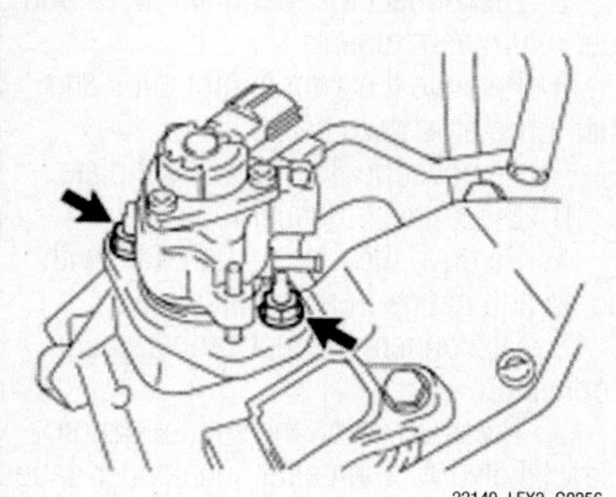

Fig. 168 Remove the mounting nuts and high pressure fuel pump.

To install:

18. Turn the crankshaft until the flat of the cam is facing the cylinder head cover's fuel pump attachment hole, as shown in the illustration.

➡When installing the fuel pump by following the procedure described above: By not using the crankshaft pointed side to push up the pump activation surface, it is easier to install the fuel pump and No. 2 fuel pipe later.

19. Pour 30 cc of engine oil through the cylinder head cover's fuel pump attachment hole into the cylinder head oil collector

20. Apply a coat of engine oil to the pump activation cam and pump lifter part.

21. Install a new fuel pump insulator to the cylinder head cover. Then pass the 2 stud bolts through the holes of the fuel pump and set them on the insulator.

22. Temporarily install the No. 2 fuel pipe sub-assembly to the fuel pump assembly.

23. Install the 2 nuts and tighten them in several passes to 18 ft. lbs. 25 (Nm).

24. Connect the fuel hose.

25. Install the No. 2 fuel pipe to the delivery pipe with the 2 bolts. Tighten the mounting bolts to 7ft. lbs. (10 Nm).

26. Using a 19 mm union nut wrench, connect the fuel pipe and tighten to 22 ft. lbs. (30 Nm).

27. Connect the connector to the fuel pump.

28. Install the No. 2 fuel pipe.

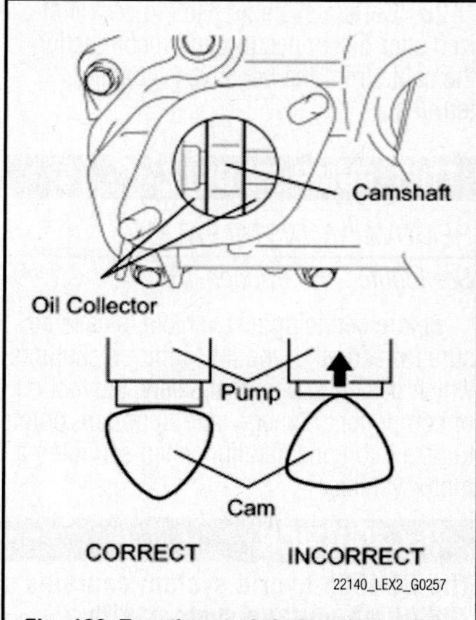

Fig. 169 Turn the crankshaft until the flat of the cam is facing the cylinder head

29. Apply gasoline to the O-ring and connect the fuel pipe to the delivery pipe.
30. Install the high pressure pump assembly.
31. Connect the No. 2 fuel pipe.
32. Install the No.1 fuel pipe and tighten the bolts to 7 ft. lbs. (10 Nm).
33. Install the fuel pressure pulsation damper.
34. Install the water hose joint.
35. Install the intake manifold.
36. Install the intake air surge tank.
37. Install the throttle body with motor assembly.
38. Connect the throttle motor connector.
39. Install the air cleaner cap with air cleaner hose.
40. Reconnect the No. 2 ventilator hose.
41. Install the cowl top ventilator louver sub-assembly.
42. Install the wiper arm assembly LH and RH
43. Install the air cleaner cap and hose.
44. Install the V-bank cover.
45. Install the engine room side cover.
46. Install the cool air intake duct seal.
47. Install and bleed engine coolant.

FUEL RAIL AND INJECTOR

REMOVAL & INSTALLATION

Direct Fuel Injection

CAUTION

The GS450h hybrid system contains a 288V high-voltage system with a strong alkali solution of potassium hydroxide. Be sure to follow the instructions in this manual to handle the system correctly. Failure to do so may result in serious injury or electrocution. Engineer must undergo special training to be able to perform high-voltage system inspection and servicing.

CAUTION

All high-voltage wire harness connectors are colored orange. The HV battery and other high-voltage components have "High Voltage" caution labels. Do not carelessly touch these wires and components.

CAUTION

Before inspecting or servicing the high-voltage system, be sure to follow safety measures, such as wearing insulated gloves and removing the service plug to prevent electrocution. Carry the removed service plug in your pocket to prevent other technicians from reinstalling it while you are servicing the vehicle.

CAUTION

After removing the service plug, wait 10 minutes before touching any of the high-voltage connectors and terminals.

CAUTION

Before inspecting the high-voltage system or disconnecting the low voltage connector of the inverter with converter assembly, take safety precautions, such as wearing insulated gloves and removing the service plug grip to prevent electrical shocks. After removing the service plug grip, put it in your pocket to prevent other technicians from accidentally reconnecting it while you are working on the high-voltage system.

CAUTION

After disconnecting the service plug grip, wait for at least 10 minutes before touching any of the high-voltage connectors or terminals. After waiting, check the voltage at the inspection point in the inverter with converter assembly. The voltage should be 0 V before beginning work.

1. Discharge the fuel system.
2. Remove the hybrid service plug grip.
3. Disconnect the negative battery cable.
4. Drain the engine coolant.
5. Remove the cool air intake duct seal.
6. Remove the engine room side cover.
7. Remove the V-bank cover.
8. Remove the air cleaner cap and hose.
9. Remove the wiper arm assembly LH and RH
10. Remove the cowl top ventilator louver sub-assembly.
11. Drain the engine coolant.
12. Disconnect the No. 2 ventilator hose.
13. Remove the air cleaner cap with air cleaner hose.
14. Remove the throttle body with motor assembly.
15. Remove the air intake surge tank.
16. Remove the intake manifold.
17. Remove the water hose joint.
18. Remove the fuel pressure pulsation damper.
19. Remove the No. 1 fuel pipe.
20. Disconnect the No 2 fuel pipe.
21. Remove the high pressure fuel pump assembly.
22. Remove the No.2 and 3 fuel pipe.
23. Remove the No.2 fuel delivery pipe.
24. With the connectors still connected, disconnect the No. 2 fuel delivery pipe.
25. Disconnect the 3 injector connectors.
26. Remove the injectors, and keep injectors in the original order.
27. Remove the 3 injector vibration insulators from the cylinder head.
28. Remove the No.1 fuel delivery pipe.
29. Disconnect the 2 wire harness clamps and fuel pressure sensor connector.
30. Remove the 2 bolts and 2 nuts.
31. With the connectors still connected, disconnect the No. 1 fuel delivery pipe.
32. Disconnect the 3 injector connectors.
33. Remove the 3 injector vibration insulators from the cylinder head.
34. Remove the injectors, and keep injectors in the original order.

To install:

35. Install the injectors with properly installed seals.
36. Install the injectors in their original order.
37. Install a new O-ring, new backup rings (No. 1, No. 2, No. 3) and new E-ring to the fuel injector.
38. Install the injector nozzle holder clamp.
39. Apply gasoline to the O-ring. Install the nozzle holder clamp by aligning the protruding part of the clamp to the notch of the delivery pipe.
40. Install a new injector vibration insulator to the cylinder head.
41. Apply lubricant to the installation hole of the injector.
42. Install the fuel delivery pipe by uniformly tightening the 2 bolts and 2 nuts in several passes in the order shown in the illustration. Tighten 15 ft. lbs. (21 Nm).
43. Connect the 3 connectors, 2 clamps and fuel pressure sensor clamp.
44. Install the No. 3 fuel pipe, apply gasoline to the O-rings.
45. Press the fuel pipe and delivery pipe together by hand until there is no gap between them. Then install the No. 3 fuel pipe with the 4 bolts and tighten to 7 ft. lbs. (10 Nm).
46. Install the No. 2 fuel pipe.
47. Apply gasoline to the O-ring and connect the fuel pipe to the delivery pipe.

48. Install the high pressure pump assembly.
49. Connect the No. 2 fuel pipe.
50. Install the No.1 fuel pipe and tighten the bolts to 7 ft. lbs. (10 Nm).
51. Install the fuel pressure pulsation damper.
52. Install the water hose joint.
53. Install the intake manifold.
54. Install the intake air surge tank.
55. Install the throttle body with motor assembly.
56. Connect the throttle motor connector.
57. Install the air cleaner cap with air cleaner hose.
58. Reconnect the No. 2 ventilator hose.
59. Install the cowl top ventilator louver sub-assembly.
60. Install the wiper arm assembly LH and RH
61. Install the air cleaner cap and hose.
62. Install the V-bank cover.
63. Install the engine room side cover.
64. Install the cool air intake duct seal.
65. Install and bleed engine coolant.

Port Injection

See Figures 170 through 172.

CAUTION

The GS450h hybrid system contains a 288V high-voltage system with a strong alkali solution of potassium hydroxide. Be sure to follow the instructions in this manual to handle the system correctly. Failure to do so may result in serious injury or electrocution. Engineer must undergo special training to be able to perform high-voltage system inspection and servicing.

CAUTION

All high-voltage wire harness connectors are colored orange. The HV battery and other high-voltage components have "High Voltage" caution labels. Do not carelessly touch these wires and components.

CAUTION

Before inspecting or servicing the high-voltage system, be sure to follow safety measures, such as wearing insulated gloves and removing the service plug to prevent electrocution. Carry the removed service plug in your pocket to prevent other technicians from reinstalling it while you are servicing the vehicle.

CAUTION

After removing the service plug, wait 10 minutes before touching any of the high-voltage connectors and terminals.

1. Discharge the fuel system.
2. Remove the hybrid service plug grip.
3. Disconnect the negative battery cable.

CAUTION

Before inspecting the high-voltage system or disconnecting the low voltage connector of the inverter with converter assembly, take safety precautions, such as wearing insulated gloves and removing the service plug grip to prevent electrical shocks. After removing the service plug grip, put it in your pocket to prevent other technicians from accidentally reconnecting it while you are working on the high-voltage system.

CAUTION

After disconnecting the service plug grip, wait for at least 10 minutes before touching any of the high-voltage connectors or terminals. After waiting, check the voltage at the inspection point in the inverter with converter assembly. The voltage should be 0 V before beginning work.

4. Drain the engine coolant.
5. Remove the cool air intake duct seal.
6. Remove the engine room side cover.
7. Remove the V-bank cover.
8. Remove the air cleaner cap and hose.
9. Remove the wiper arm assembly LH and RH
10. Remove the cowl top ventilator louver sub-assembly.
11. Drain the engine coolant.
12. Disconnect the No. 2 ventilator hose.
13. Remove the air cleaner cap with air cleaner hose.
14. Remove the throttle body with motor assembly.
15. Remove the air intake surge tank.
16. Remove the fuel delivery pipe sub-assembly as follows:
    - Pinch and pull the fuel tube connector to disconnect the connector from the delivery pipe.
    - Disconnect the 6 injector connectors and 2 clamps.

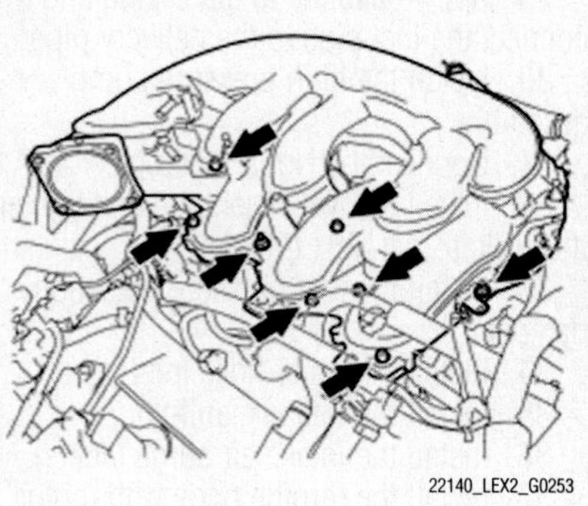

Fig. 170 Using a 5 mm hexagon socket wrench, remove the 6 bolts, 2 nuts and gasket from the air intake surge tank

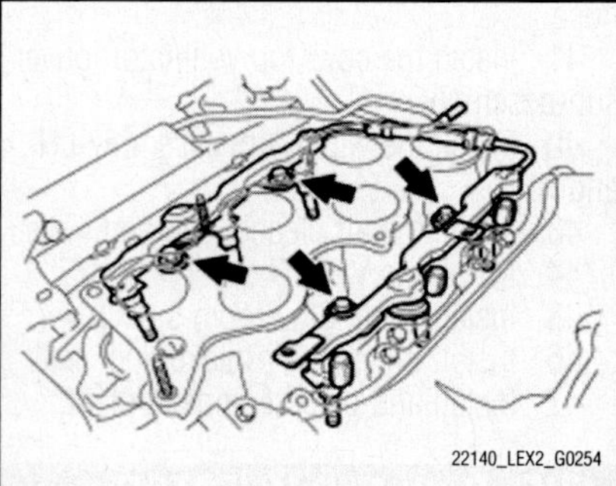

Fig. 171 Remove the 4 bolts and fuel delivery pipe sub-assembly

- Remove the 4 bolts and fuel delivery pipe sub-assembly.

➡When removing the delivery pipe, hold the pipe by both ends and pull it straight upward.

- Remove the 4 delivery pipe spacers from the intake manifold.
- Remove the 6 fuel injectors from the fuel delivery pipe. For reinstallation, attach a tag or label to the injector shaft.

To install:

17. Apply gasoline to 2 new O-rings and install them to the injector.
18. Install the fuel injector assemblies to the fuel delivery pipe.
19. Install the fuel delivery pipe as follows:
    - Install the 4 delivery pipe spacers to the intake manifold.
    - Install the delivery pipe (with injector) to the intake manifold.
    - Install the 4 bolts and tighten to 13 ft. lbs. (17 Nm).
    - Connect the 6 injector connectors and 2 clamps.
    - Push in the fuel tube connector to

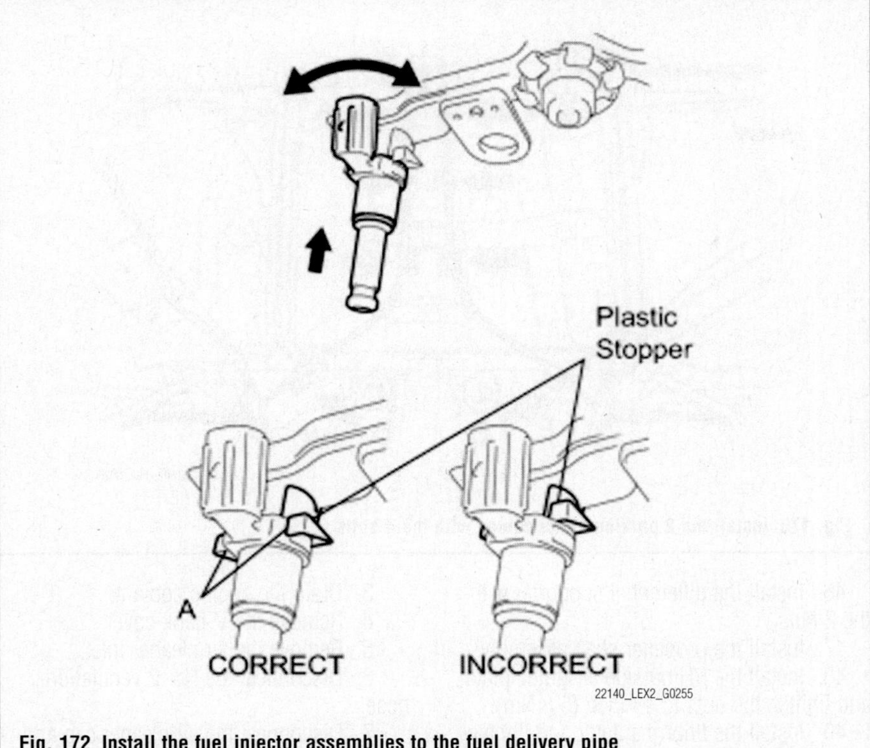

Fig. 172 Install the fuel injector assemblies to the fuel delivery pipe

the delivery pipe and push up the retainer to engage the claws.

- After connecting the fuel tube, align the paint marks on the delivery pipe.

20. Install the intake air surge tank.
21. Install the throttle body with motor assembly.
22. Connect the throttle motor connector.
23. Install the air cleaner cap with air cleaner hose.
24. Reconnect the No. 2 ventilator hose.
25. Install the cowl top ventilator louver sub-assembly.
26. Install the wiper arm assembly LH and RH
27. Install the air cleaner cap and hose.
28. Install the V-bank cover.
29. Install the engine room side cover.
30. Install the cool air intake duct seal.
31. Install and bleed engine coolant.

FUEL TANK

REMOVAL & INSTALLATION

See Figures 163, 173 through 175.

**** CAUTION**

The GS450h hybrid system contains a 288V high-voltage system with a strong alkali solution of potassium hydroxide. Be sure to follow the instructions in this manual to handle the system correctly. Failure to do so may result in serious injury or electrocution. Engineer must undergo special training to be able to perform high-voltage system inspection and servicing.

**** CAUTION**

All high-voltage wire harness connectors are colored orange. The HV battery and other high-voltage components have "High Voltage" caution labels. Do not carelessly touch these wires and components.

**** CAUTION**

Before inspecting or servicing the high-voltage system, be sure to follow safety measures, such as wearing insulated gloves and removing the service plug to prevent electrocution. Carry the removed service plug in your pocket to prevent other technicians from reinstalling it while you are servicing the vehicle.

**** CAUTION**

After removing the service plug, wait 10 minutes before touching any of the high-voltage connectors and terminals.

1. Discharge the fuel system.
2. Remove the service plug.
3. Disconnect the negative battery cable.
4. Remove the rear seat cushion assembly.
5. Remove the clip and room partition pad.
6. Remove the service hole cover and disconnect all the electrical connectors.
7. Disconnect the fuel main tubes and fuel return vent tube.
8. Remove the 2 tube joint clips and fuel tubes.
9. Remove the exhaust pipe.
10. Remove the 3 nuts, 2 clips and floor LH and RH covers.
11. Remove the 5 nuts and heat insulator.
12. Remove the RH outside air guide plate.
13. Remove the propeller shaft assembly.
14. Remove both differential support protectors.
15. Remove the RH rear floor side member cover.
16. Remove the RH and LH rear suspension member brace.
17. Detach the 2 parking brake cables from the 4 clamps.
18. Remove the 4 bolts and disconnect the 2 parking brake cables from the parking brake equalizer.
19. Disconnect the fuel tank return vent tube.
20. Disconnect the fuel tank main tube RH.
21. Disconnect the fuel tank main tube LH.
22. Remove the clamp and disconnect the vent hose.
23. Remove the fuel tube clamp and disconnect the canister tube.
24. Disconnect the fuel vapor containment valve connector and 2 clamps.
25. Loosen the 2 fuel filler pipe bolts.

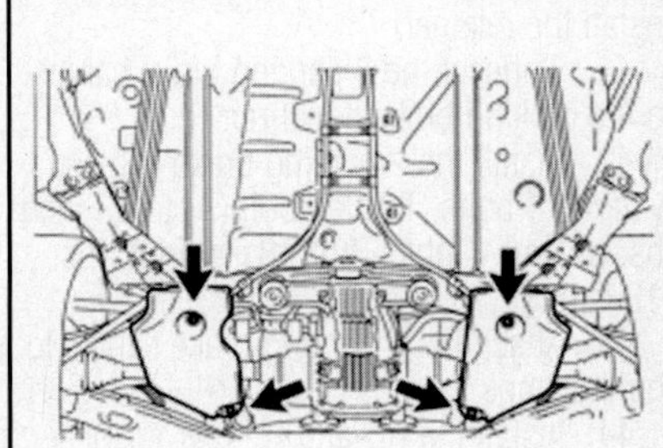

Fig. 173 Remove both differential support protectors

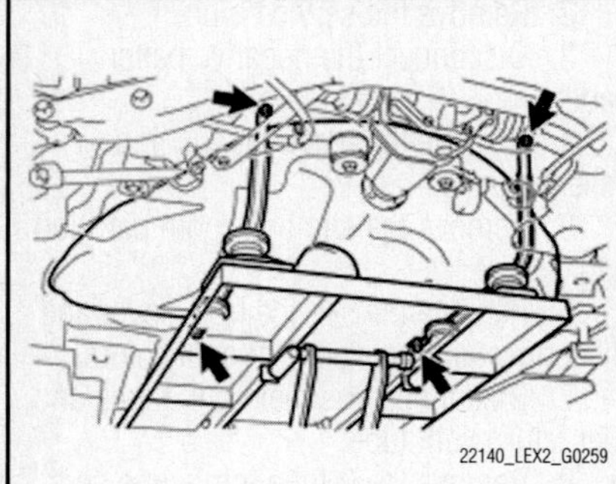

Fig. 174 Slowly lower the mission jack slightly

26. Place a mission jack under the fuel tank.
27. Remove the 4 bolts and 2 fuel tank bands.
28. Slowly lower the mission jack slightly.
29. Remove the clamp, disconnect the fuel filler pipe from the fuel tank, and lower the mission jack.
30. Remove the fuel tank from the mission jack.

To install:

31. Set the fuel tank on a mission jack and raise the fuel tank.
32. Install the clamp, and connect the fuel filler pipe to the fuel tank.
33. Install the 2 fuel tank bands with the 4 bolts and tighten to 29 ft. lbs. (39 Nm).
34. Tighten the 2 fuel filler pipe bolts to 13 ft. lbs. (18 Nm).
35. Connect the fuel vapor containment valve connector and 2 clamps.
36. Connect the canister tube and install the clamp.
37. Connect the fuel tank vent hose and install the clamp.
38. Connect the fuel tank main tube (LH).
39. Connect the fuel tank main tube (RH).
40. Connect the fuel return vent tube and install the retainer.
41. Connect the 2 parking brake cables to the parking brake equalizer.
42. Install the 2 parking brake cables with the 4 bolts. Tighten bolt (A) to 53 inch lbs. (6 Nm). Tighten bolt (B) to 14 ft. lbs. (19 Nm).
43. Attach the 2 parking brake cables to the 4 clamps.
44. Install the RH and LH rear suspension member brace with the bolts. Tighten to 14 ft. lbs. (19 Nm).
45. Install the LH and RH member covers with the 3 bolts.

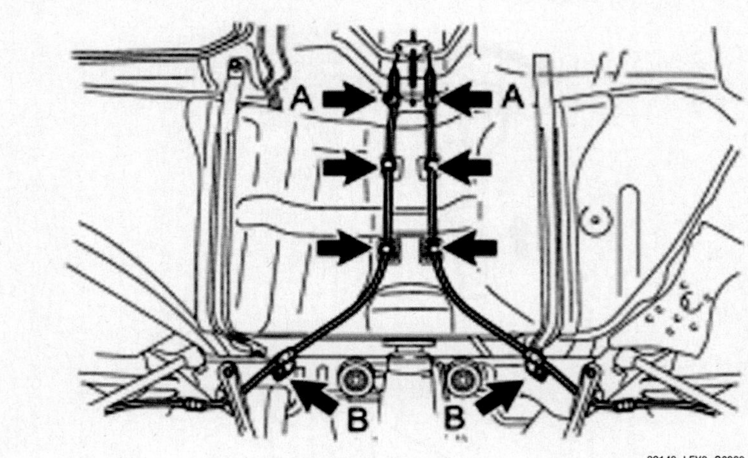

Fig. 175 Install the 2 parking brake cables with the 4 bolts

46. Install the differential supporter with the 2 nuts.
47. Install the propeller shaft assembly.
48. Install the RH outside air guide plate and tighten the nuts to 48 inch (5.8 Nm).
49. Install the floor insulator and tighten the nuts to 48 inch (5.8 Nm).
50. Install the LH and RH floor cover center with the 3 nuts.
51. Install 2 new gaskets and the exhaust pipe front ends to the exhaust manifolds with the 4 bolts and 2 nuts. Tighten to 29 ft. lbs. (39 Nm).
52. Install all fuel and electrical connections to top of the fuel tank.
53. Install the partition pad with the clip.
54. Install the seat cushion assembly.
55. Install the service plug grip.
56. Connect the negative battery cable.
57. Check for fuel leaks.
58. Certain systems need to be initialized after disconnecting and reconnecting the cable from the negative (-) battery terminal.

IDLE SPEED

ADJUSTMENT

Idle speed is maintained by the Engine Control Module (ECM). No adjustment is necessary or possible.

THROTTLE BODY

REMOVAL & INSTALLATION

1. Disconnect the negative battery cable.
2. Remove the cool air intake duct seal.
3. Drain the engine coolant.
4. Remove the V-bank cover.
5. Remove the air cleaner inlet.
6. Disconnect the No. 2 ventilation hose.
7. Disconnect the air cleaner cap and hose.
8. Disconnect the MAF meter connector.
9. Disconnect the clamp from the air cleaner.
10. Disconnect the purge line hose.
11. Disconnect the 4 clamps.
12. Remove the hose clamp and air cleaner cap with air cleaner hose.
13. Remove the throttle body as follows:
    - Disconnect the throttle motor connector.
    - Remove the 4 bolts and disconnect the throttle body from the intake air surge tank.
    - Disconnect the 2 water by-pass hoses from the throttle body.
    - Remove the throttle body and gasket.

To install:

14. Install a new gasket to the intake air surge tank.
15. Connect the 2 water by-pass hoses to the throttle body.
16. Install the throttle body with the 4 bolts and tighten to 7 ft. lbs. (10 Nm).
17. Connect the throttle motor connector.
18. Install the air cleaner cap with air cleaner hose assembly with the 4 clamps and hose clamp.
19. Install the purge line hose to the air cleaner hose.
20. Connect the MAF meter connector and clamp to the air cleaner.

21. Connect the ventilation hose to the cylinder head cover with the clamp.

22. Add engine coolant.

23. Connect the negative battery cable.

24. Check the for coolant leaks.

25. Check the function of the throttle body.

26. Install the air cleaner inlet.

27. Install the engine room covers.

28. Install the cool air duct seal.

29. Certain systems need to be initialized after disconnecting and reconnecting the cable from the negative (-) battery terminal.

HEATING & AIR CONDITIONING SYSTEM

BLOWER MOTOR

REMOVAL & INSTALLATION

See Figure 176.

1. Before servicing any vehicle, please be sure to read all of the precautions.
2. Remove the service plug
3. Disconnect the negative battery cable.
4. Remove the A/C unit.
5. Remove the blower unit assembly.
6. Disconnect the connector.
7. Remove the 3 screws and fan motor.

To install:

8. Install the fan motor with the 3 screws.
9. Reconnect the connector.
10. Install the blower unit assembly.
11. Install the A/C unit.
12. Install the service plug.
13. Connect the negative battery cable.

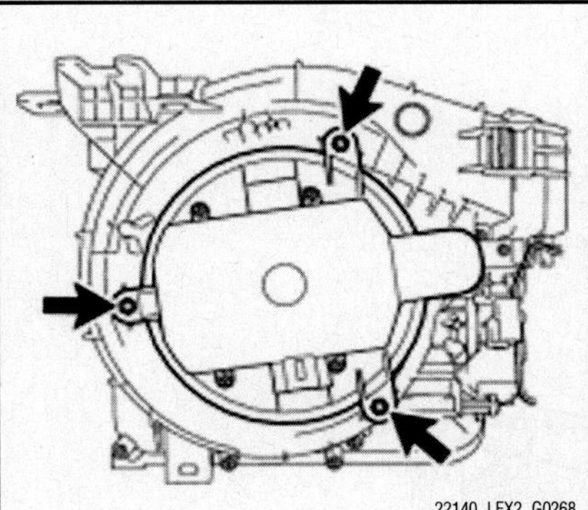

22140_LEX2_G0268

Fig. 176 Remove the 3 screws and fan motor

HEATER CORE

REMOVAL & INSTALLATION

See Figures 177 and 178.

Before servicing any vehicle, please be sure to read all of the following precautions, which deal with personal safety, prevention of component damage, and important points to take into consideration when servicing a motor vehicle.

CAUTION

The GS450h hybrid system contains a 288V high-voltage system with a strong alkali solution of potassium hydroxide. Be sure to follow the instructions in this manual to handle the system correctly. Failure to do so may result in serious injury or electrocution. Engineer must undergo special training to be able to perform high-voltage system inspection and servicing.

CAUTION

All high-voltage wire harness connectors are colored orange. The HV battery and other high-voltage components have "High Voltage" caution labels. Do not carelessly touch these wires and components.

CAUTION

Before inspecting or servicing the high-voltage system, be sure to follow safety measures, such as wearing insulated gloves and removing the service plug to prevent electrocution. Carry the removed service plug in your pocket to prevent other technicians from reinstalling it while you are servicing the vehicle.

CAUTION

After removing the service plug, wait 10 minutes before touching any of the high-voltage connectors and terminals.

1. Disable air bag system.
2. Remove the service plug.
3. Before servicing the vehicle, refer to the Precautions section.
   - Discharge the A/C system.
   - Set radio receiver assembly to shipment mode as follows:
   - Be sure that all discs and tapes have been removed from the unit.
   - Be sure that the engine switch off.
   - While simultaneously pressing the "SEEK UP" and "DISC" switches, turn the engine switch on (ACC).
   - Turn the engine switch off.
4. Disconnect the negative battery cable. Wait 90 seconds before doing any further work while the airbag system de-energizes.
5. Drain the cooling system into a clean container for reuse.
6. Align the front wheels facing straight ahead.
7. Remove or disconnect the following:
   - Cool air intake duct seal
   - Left and right engine room side covers
   - Left and right front pillar to front side seals, using a clip remover to detach the 3 claws
   - Left and right nut, windshield wiper arms and blades
   - Left and right front fender to cowl side seals by moving the component toward the center of the vehicle to detach the 2 claws
8. Remove cowl top ventilator louver assembly as follows:
   - Remove the 2 clips and detach the 5 claws.
   - Pull the ventilator louver in the direction indicated by the arrow in the illustration to detach the 10 claws and remove the ventilator louver.
9. Remove the windshield wiper motor and link assembly as follows:
   - Disconnect the connector. Then detach the 2 clamps and remove the wire harness from the cowl top panel.

➡There are 6 bolts total, however, 2 bolts cannot be removed from the wiper motor and link because they are integrated into the wiper motor and link.

   - Remove the 4 bolts and wiper motor and link.
10. Separate suction pipe sub-assembly as follows:
    - Remove the bolt, and slide the hook connector.
    - Disconnect the suction pipe sub-assembly.
    - Remove the O-ring from the suction pipe sub-assembly

⁂ WARNING

Seal the openings of the disconnected parts using vinyl tape to prevent moisture and foreign matter from entering.

11. Separate liquid tube sub-assembly as follows:
   - Disconnect the liquid tube sub-assembly.
   - Remove the O-ring from the liquid tube subassembly.
   - Inlet and outlet heater water outlet hoses
   - Instrument panel assembly.
   - A/C blower unit assembly
   - Mounting screw and No. 2 air duct
   - Upper and lower foot ducts
   - Connector, mounting screw and air conditioning amplifier assembly

12. Remove the heater core from the A/C-blower assembly unit as follows:
   - Disconnect the wiring connector, remove the 3 screws and right side air outlet control servo motor.
   - Disengage 3 clamps, connector and wire harness to right side air mix control servo motor.
   - Remove the 2 screws and heater piping cover.

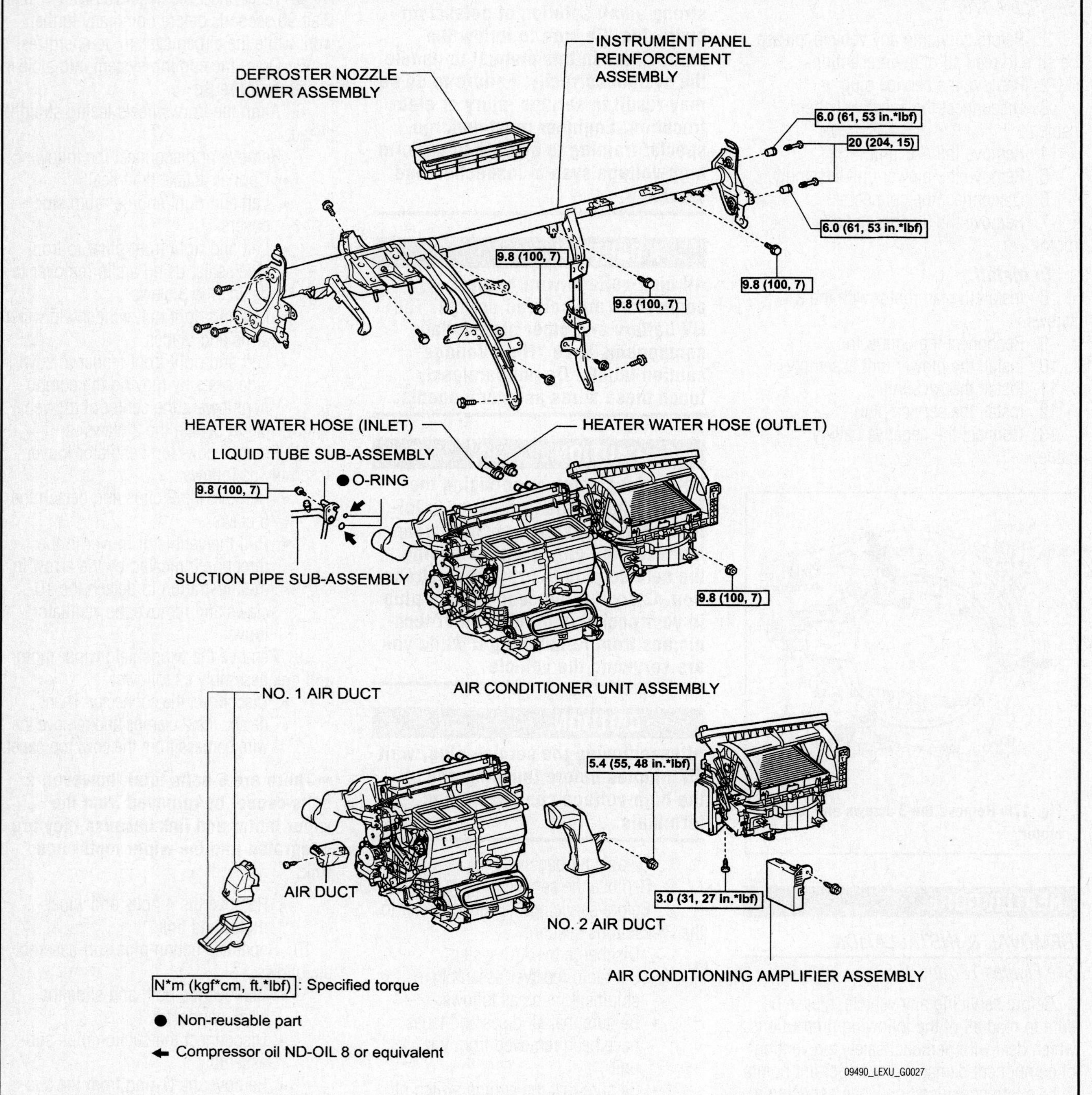

Fig. 177 Exploded view of the A/C unit assembly, instrument panel reinforcement and related components—GS450h

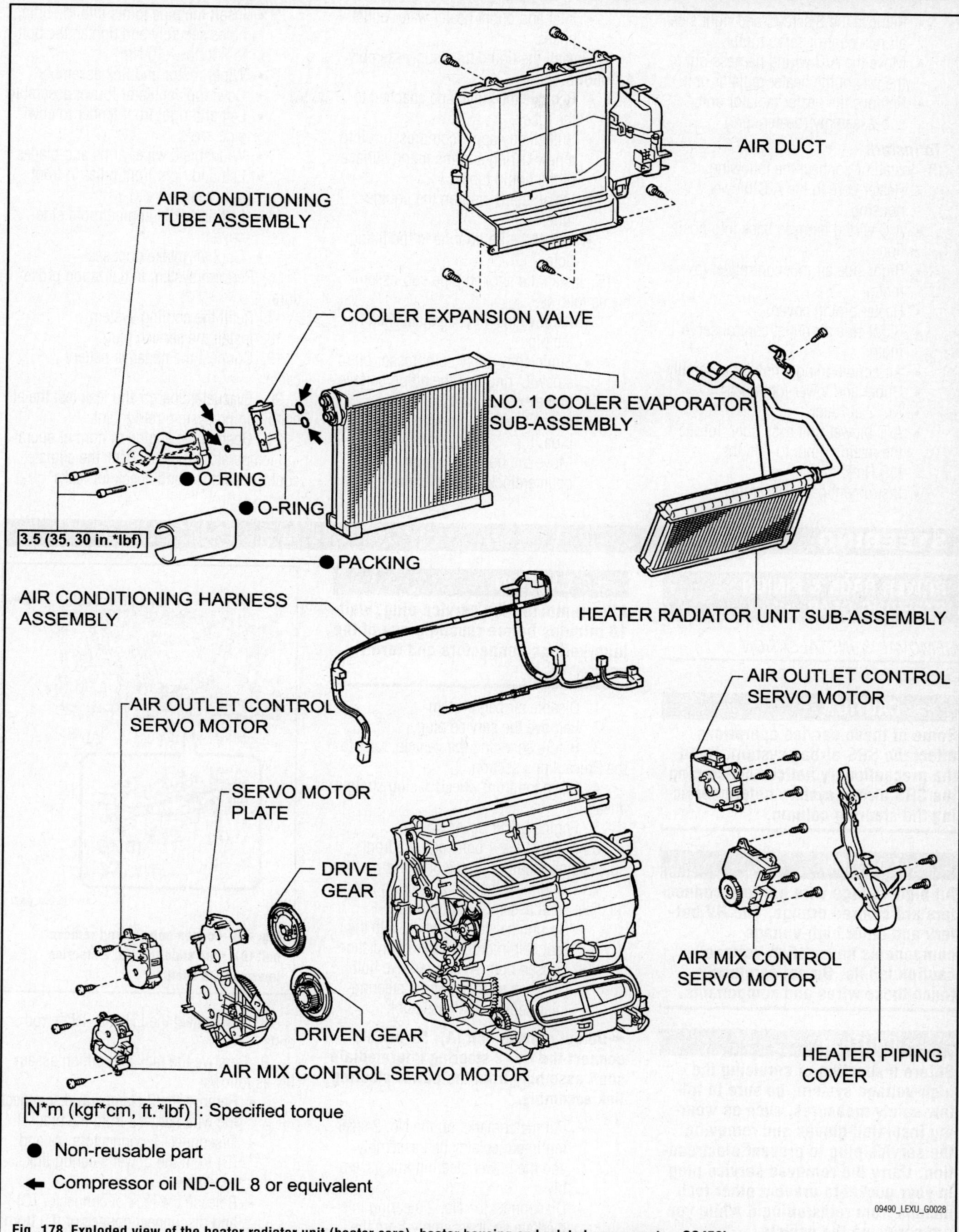

Fig. 178 Exploded view of the heater radiator unit (heater core), heater housing and related components—GS450h

- Remove the 3 screws and right side air mix control servo motor.
- Move the A/C wiring harness out of the way of the heater radiator unit.
- Remove the heater radiator unit sub-assembly (heater core)

To install:

13. Install or connect the following:
    - Heater core to the A/C blower housing
    - A/C wiring harness back into position
    - Right side air mix control servo motor
    - Heater piping cover
    - Right side air outlet control servo motor
    - Air conditioning amplifier assembly
    - Upper and lower foot ducts
    - No. 2 air duct
    - A/C blower unit assembly. Torque the retaining nut to 7 ft. lbs. (10 Nm).
    - Instrument panel assembly.
    - Inlet and outlet heater water outlet hoses
14. Install the liquid tube sub-assembly as follows:
    - Remove the vinyl tape attached to the tube.
    - Sufficiently apply compressor oil to a new O-ring and the fitting surface of the liquid tube.
    - Install the O-ring on the liquid tube.
    - Install the liquid tube to the fitting hole.
15. Install the suction pipe sub-assembly as follows:
    - Remove the vinyl tape attached to the pipe.
    - Sufficiently apply compressor oil to a new O-ring and the fitting surface of the suction pipe.
    - Install the O-ring on the suction pipe.
    - Move the hook connector in a counterclockwise direction.
    - Insert the pipe joints into the fitting holes securely and tighten the bolt to 7 ft. lbs. (10 Nm).
    - Wiper motor and link assembly
    - Cowl top ventilator louver assembly
    - Left and right front fender to cowl side seals
    - Windshield wiper arms and blades
    - Left and right front pillar to front side seals
    - Left and right engine room side covers
    - Cool air intake duct seal
16. Perform system initialization procedure.
17. Refill the cooling system.
18. Install the service plug.
19. Connect the negative battery cable.
20. Evacuate, charge and leak test the air conditioning system refrigerant.
21. Operate the engine to normal operating temperatures; then, check the climate control operation and check for leaks.

STEERING

POWER RACK & PINION STEERING GEAR

REMOVAL & INSTALLATION

See Figures 179 through 181.

CAUTION

Some of these service operations affect the SRS airbag system. Read the precautionary notices concerning the SRS airbag system before servicing the steering column.

CAUTION

All high-voltage wire harness connectors are colored orange. The HV battery and other high-voltage components have "High Voltage" caution labels. Do not carelessly touch these wires and components.

CAUTION

Before inspecting or servicing the high-voltage system, be sure to follow safety measures, such as wearing insulated gloves and removing the service plug to prevent electrocution. Carry the removed service plug in your pocket to prevent other technicians from reinstalling it while you are servicing the vehicle.

CAUTION

After removing the service plug, wait 10 minutes before touching any of the high-voltage connectors and terminals.

1. Disable air bag system.
2. Remove the service plug.
3. Before servicing the vehicle, refer to the Precautions section.
4. Place the front wheels facing straight ahead.
5. Engine under cover No. 2.
6. Remove the 4 bolts and the front suspension member protector lower.
7. Disconnect the No. 2 steering intermediate shaft assembly as follows:
    - Secure the steering wheel with the seat belt in order to prevent rotation.
    - Loosen bolt (A) and remove bolt (B), then slide the No. 2 steering intermediate shaft assembly.

➡Do not remove bolt (A). Do not disconnect the No. 2 steering intermediate shaft assembly from the power steering link assembly.

- Put matchmarks on the No. 2 steering intermediate shaft assembly and the power steering link assembly.
- Disconnect the No. 2 steering intermediate shaft assembly from the power steering link assembly.

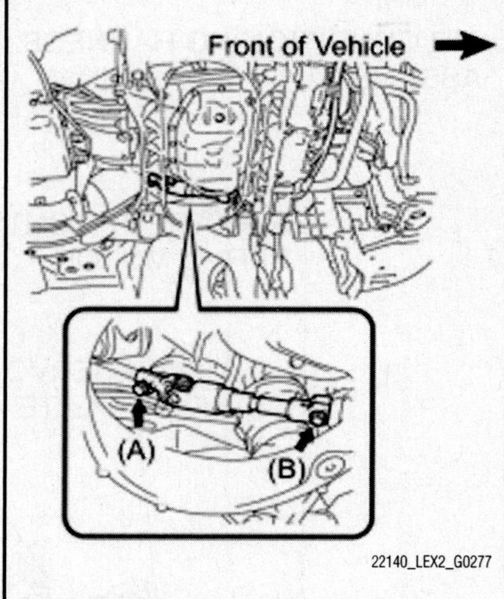

Fig. 179 Loosen bolt (A) and remove bolt (B), then slide the No. 2 steering intermediate shaft

8. Disconnect the LH and RH tie rod ends.
9. Remove the rack and pinion assembly as follows:
    - Remove the 2 clamps to disconnect the wire harness from the bracket.
    - Disconnect 2 connectors (A) and (B) from the power steering link assembly.
    - Release the lock of connector (C) and disconnect connector (C) from the power steering link assembly.

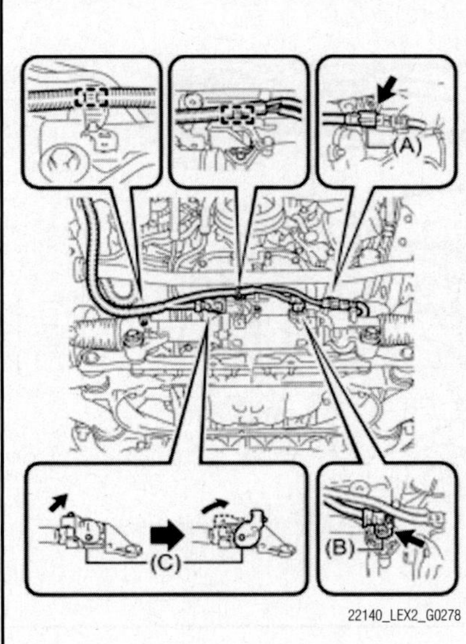

Fig. 180 Disconnect wire harness connector (A), (B) and (C) to the power steering

- Remove the 2 bolts, 2 washers, 2 nuts, and power steering link assembly from the front suspension crossmember.

To install:

10. Install the power steering link assembly with the 2 bolts, 2 washers and 2 nuts. Tighten to 87 ft. lbs. (118 Nm).

11. Connect wire harness connector (C) to the power steering link assembly and securely lock the connector.

12. Connect 2 wire harness connectors (A) and (B) to the power steering link assembly.

13. Install the 2 wire harness clamps to the power steering link assembly.

14. Connect the tie rod end LH and RH to the steering knuckle with the nut. Tighten to 50 ft. lbs. (65 Nm).

15. Install a new clip. If the holes for the clip are not aligned, tighten the nut up to 60° further.

16. Connect the No. 2 steering intermediate shaft as follow:
    - Align the matchmarks on the No. 2 intermediate shaft and the power steering link assembly.
    - Install bolt (A) and tighten the 2 bolts. Tighten to 26 ft. lbs. (35 Nm).
    - Install the front suspension member protector lower to the front suspension crossmember with the 4 bolts. Tighten to 71 inch lbs. (8 Nm).

17. Install the engine under cover.

18. Install the front wheels and tighten to 76 ft. lbs. (103 Nm).

19. Install service plug.

20. Connect the negative battery cable.

21. Inspect and adjust front wheel alignment.

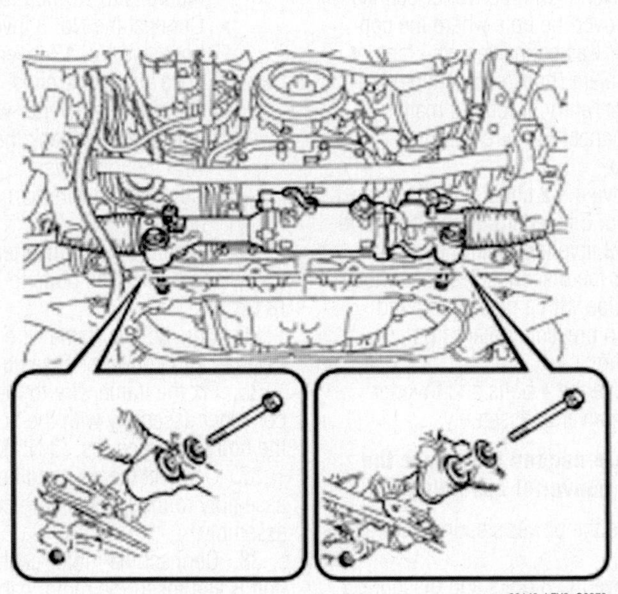

Fig. 181 Remove the 2 bolts, 2 washers, 2 nuts, and power steering link assembly from the front suspension crossmember.

22. Initialize the rotation angle sensor and calibrate torque sensor zero point as follow:
    - Inspection before calibration
    - Connect the Techstream to the DLC3.
    - Turn the power switch ON (IG).
    - Turn the Techstream on.
    - Check the IG power supply voltage on the Techstream. Enter the following menus: Chassis / EMPS / Data List.
    - Rotation angle sensor calibration value clear, rotation angle sensor initialization, and torque sensor zero point calibration

➡If DTC C1516 (Torque Sensor Zero Point Adjustment Incomplete) is stored, the torque sensor zero point cannot be calibrated. Clear the DTC before starting calibration. If DTC C1526 (Rotation Angle Sensor Initialization Incomplete) is stored, the rotation angle sensor cannot be initialized. Clear the DTC before starting initialization.

- Connect the Techstream to the DLC3.
- Turn the power switch ON (IG).
- Turn the Techstream on.
- Follow the procedures on the Techstream display to clear the rotation angle sensor calibration value, initialize the rotation angle sensor, and calibrate the torque sensor zero point. Enter the following menus: Chassis / EMPS / Utility / Torque Sensor Adjustment.

23. Perform variable gear ratio steering system calibration as follows:
    - Connect the Techstream to the DLC3.
    - Turn the power switch ON (IG).
    - Turn the Techstream on.
    - Enter the following menus: Chassis / VGRS / Utility / Steering Angle Adjust.
    - Perform the following procedures displayed on the Techstream.

POWER STEERING ECU

REMOVAL & INSTALLATION

See Figures 182 through 185.

1. Before servicing the vehicle, refer to the precautions.
2. Disconnect the negative battery cable.

✻✻ CAUTION

Before inspecting the high-voltage system or disconnecting the low voltage connector of the inverter with converter assembly, take safety precautions, such as wearing insulated gloves and removing the service plug grip to prevent electrical shocks. After removing the service plug grip, put it in your pocket to prevent other technicians from accidentally reconnecting it while you are working on the high-voltage system.

✻✻ CAUTION

After disconnecting the service plug grip, wait for at least 10 minutes before touching any of the high-voltage connectors or terminals. After waiting, check the voltage at the inspection point in the inverter with converter assembly. The voltage should be 0 V before beginning work.

3. Check for DTCs
4. Confirm that P0AA6 (High voltage insulation is unusual) is not output before doing removal or installation work inside the battery. If the DTC is output, perform troubleshooting procedures first.
5. Remove the luggage compartment floor mat.
6. Remove the battery service hole cover.
7. Remove the service plug grip.
8. Disconnect the negative battery cable.
9. Remove the cool air intake seal.
10. Remove the LH and RH engine room side cover.
11. Remove the LH and RH upper fender protector.
12. Remove the LH and RH wiper arm assembly.
13. Remove the front fender cowl seals.
14. Pull the ventilator louver in the direction indicated by the arrow in the illustration to detach the 10 claws and remove the ventilator louver.
15. Remove the engine under covers.
16. Drain the coolant for the inverter.
17. Remove the connector cover assembly.
18. Remove the 2 bolts and connector cover assembly.
19. Using the voltmeter, measure the voltage between the terminals of the 2 phase connectors (N-P).
20. Standard voltage: 0 volts.
21. Remove the 2 bolts (A).

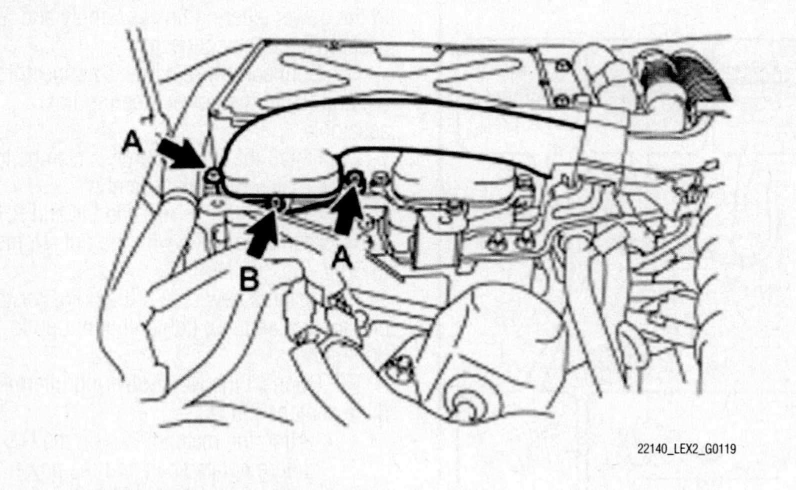

Fig. 182 Disconnect the generator cable

22. While loosening bolt B, disconnect the generator cable. Removing bolts A and B in a wrong order may cause the cable and the inverter with converter to break.
23. Separate the motor cable.
24. Disconnect the air conditioning harness assembly. Insulate the removed terminals with insulating tape.
25. Remove the bolt and disconnect the frame wire
26. Remove the 2 bolts and No. 6 inverter bracket
27. Remove the inverter with converter assembly as follows:
    - Raise the lock lever and disconnect the inverter with converter connector. Cover the hole where the connector was connected with tape or equivalent (non-residue type) to prevent entry of foreign matter.
    - Disconnect the wire harness clamp.
    - Remove the 2 clips, and disconnect the No. 8 inverter cooling hose and No. 13 inverter cooling hose.
    - Cover the disconnected water hose and pipe with a plastic bag and tape to prevent coolant from spilling.
    - Remove the 4 bolts and inverter with converter assembly.

➡2 people are needed to remove the inverter with converter assembly.

28. Remove the power steering ECU as follows:
    - Remove the 3 bolts and disconnect the ground cable terminal.
    - Release the locks of the 2 power steering ECU assembly connectors and disconnect the connectors.

To install:

29. Connect the 2 power steering ECU connectors and securely lock the connectors.
30. Install the ground cable terminal with bolt and tighten to 44 inch lbs. (5 Nm).
31. Install the power steering ECU to the battery tray with the 2 bolts. Tighten to 10 ft. lbs. (14 Nm).
32. Install the No. 1 battery tray support.
33. Install the inverter with converter assembly as follows:
    - Install the inverter with converter assembly with 4 new bolts. Tighten the bolts to 71 inch lbs. (8 Nm).
    - Connect the No. 8 inverter cooling hose and No. 13 inverter cooling hose with the 2 clips.
    - Connect the inverter with converter connector and lock the connector with the lock lever.
    - Connect the wire harness clamp.
34. Install the No. 6 inverter bracket with 2 new and tighten the bolts to 71 inch lbs. (8 Nm).
35. Connect the frame wire to the inverter with converter assembly.
36. Fix the frame wire to the inverter with converter assembly with the bolt. Tighten the bolt to 71 inch lbs. (8 Nm).
37. Connect the air conditioning harness assembly to the inverter with converter assembly.
38. Connect the motor cable. Tighten bolt B and insert the motor cable into the inverter with converter assembly.
39. Fix the motor cable to the inverter with converter assembly with the 2 bolts (A). Tighten the bolt to 71 inch lbs. (8 Nm).

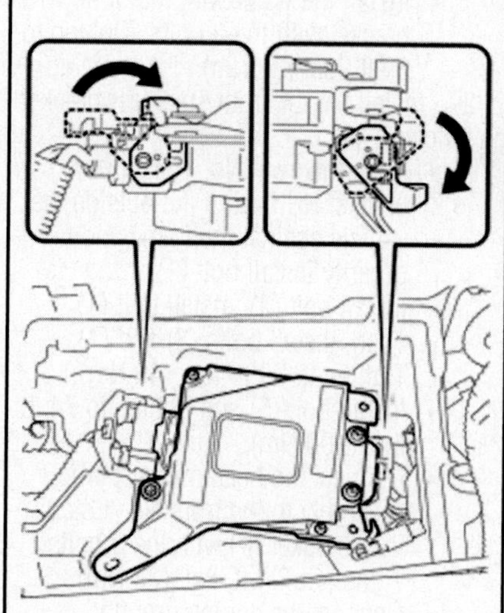
Fig. 183 Release the locks of the 2 power steering ECU assembly connectors

40. Install the inverter motor cable bracket with the 2 nuts (C). Tighten the bolt to 71 inch lbs. (8 Nm).

41. Connect the generator cable. Tighten bolt B and insert the generator cable into the inverter with converter.

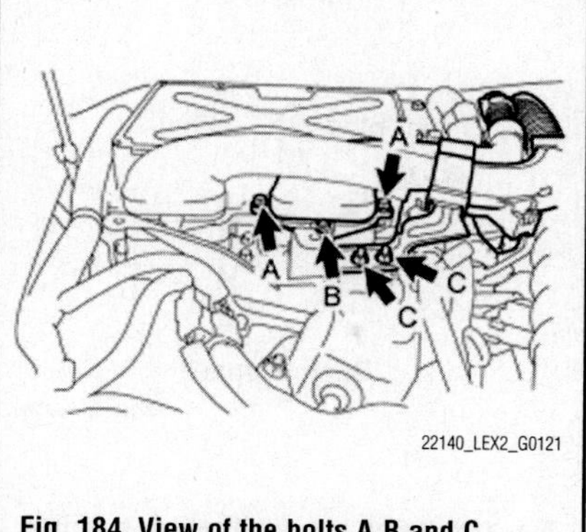

Fig. 184 View of the bolts A,B and C

42. Fix the generator cable to the inverter with converter assembly with the 2 bolts (A).

43. Install the connector cover assembly with the 2 bolts. Tighten the bolt to 71 inch lbs. (8 Nm).

44. Install the service plug grip.

45. Connect the negative battery cable.

46. Install the battery service hole cover.

47. Install the luggage compartment floor mat.

48. Install coolant and to the inverter and bleed the system.

49. Check for coolant leaks.

50. Install the engine under covers.

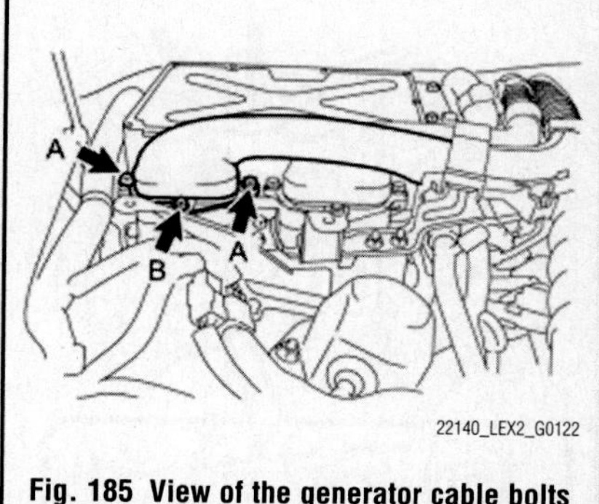

Fig. 185 View of the generator cable bolts A, and B

51. Install the cowl top ventilator sub-assembly

52. Install the LH and RH fender to cowl seal.

53. Install LH and RH wiper arm assembly.

54. Install both upper front fender protectors.

55. Engage the 3 clips and install the No. 2 engine room side RH cover.

56. Install the engine room covers.

57. Install the cool air intake duct seal.

58. Some vehicle systems require initialization after reconnecting the cable to the negative battery terminal.

SUSPENSION

FRONT SUSPENSION

ACTIVE STABILIZER CONTROL ACTUATOR

REMOVAL & INSTALLATION

See Figures 186 through 190.

1. Disconnect the negative battery cable.
2. Remove the front wheels.
3. Remove the engine under cover.
4. Remove the RH front fender liner.
5. Disconnect the front active stabilizer actuator control assembly as follows:
   - Remove the bolt and disconnect the ground wire.
   - Disconnect the 2 wire harness clamps from the bracket.
   - Using the procedures below, disconnect the ECU connector.
   - Release the lever's lock. (1)
   - Press the claw and move the lever in the direction of the arrow in the illustration (2).
   - Disconnect the ECU connector. (3)
6. Remove the LH and RH stabilizer links.
7. Remove the LH and RH rear under covers.

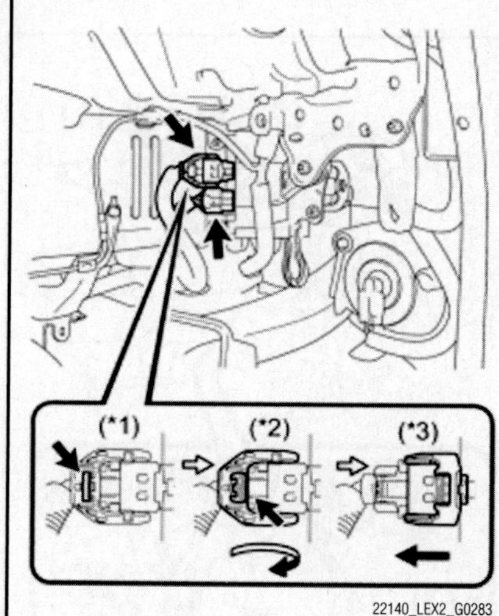

Fig. 186 Press the claw and move the lever in the direction of the arrow

8. Remove LH No. 2 stabilizer bar bracket.
9. Remove the EPS wire harness.
10. Remove the 2 bolts and bracket from the crossmember.
11. Remove RH No. 2 stabilizer bar bracket.
12. Remove the 2 bolts, bracket and control actuator.

➡The front active stabilizer control actuator is very heavy. Be careful not to drop it.

13. Remove the LH No. 1 stabilizer bracket as follows:
    - Remove the 2 bolts and water with motor and bracket from the front No. 2 stabilizer bracket LH

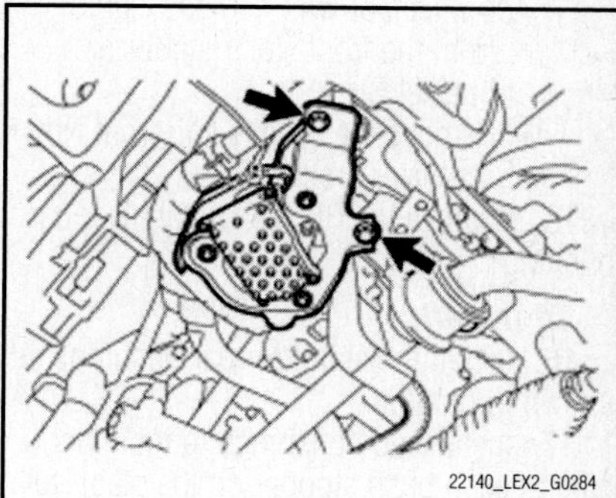
Fig. 187 Remove the 2 bolts and water with motor and bracket

Fig. 188 Disconnect the skid control sensor wire clamp

- Disconnect the skid control sensor wire clamp from the front No. 1 stabilizer bracket LH.
- Disconnect the 2 power steering link assembly wire harness clamps.
- Remove the 3 bolts and No. 1 ECU bracket.
- Remove the 5 bolts and front No. 1 stabilizer bracket LH from the frame.

14. Remove the RH No. 1 stabilizer bracket as follows:
    - Remove the 2 bolts and separate the heater water pump assembly from the front No. 1 stabilizer bracket RH.
    - Disconnect the engine ground wire clamp, heater water pump assembly wire harness clamp and skid control sensor wire harness clamp from the front No. 1 stabilizer bracket RH.
    - Remove the 5 bolts and front No. 1 stabilizer bracket RH.
15. Remove the RH No. 1 stabilizer bar bushing.

To install:

16. Install the stabilizer bar bushings as shown in graphic.
17. Install each bushings to the inner side of each bush stopper on the stabilizer bar.
18. Install each bushings with its slit facing the vehicle front side.

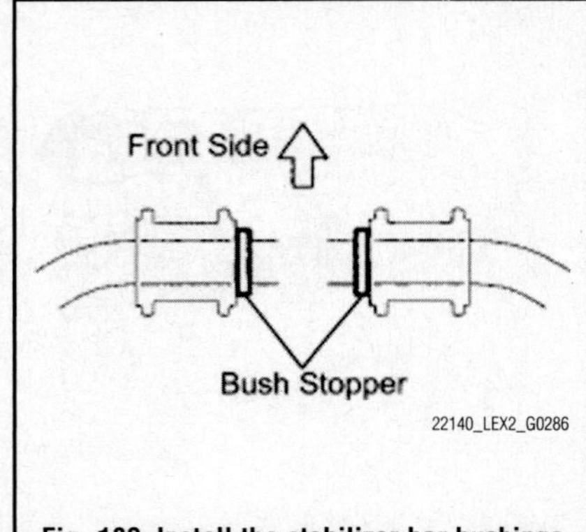

Fig. 189 Install the stabilizer bar bushings

19. Install the LH front stabilizer bracket as follows:
    - Push the front No. 1 stabilizer bracket LH toward the outside the vehicle against the frame and temporarily install bolt (1).
    - Install bolt (2). Install bolt (1). Then, install bolts (3) and (4). Tighten to 36 ft. lbs. (49 Nm).
    - Install bolt (5) and tighten to 74 ft. lbs. (100 Nm).
    - Connect the No. 1 ECU bracket to the front No. 1 stabilizer bracket LH. Tighten to 15 ft. lbs. (20 Nm).
    - Connect the 2 power steering link assembly wire harness clamps and skid control sensor wire clamp.

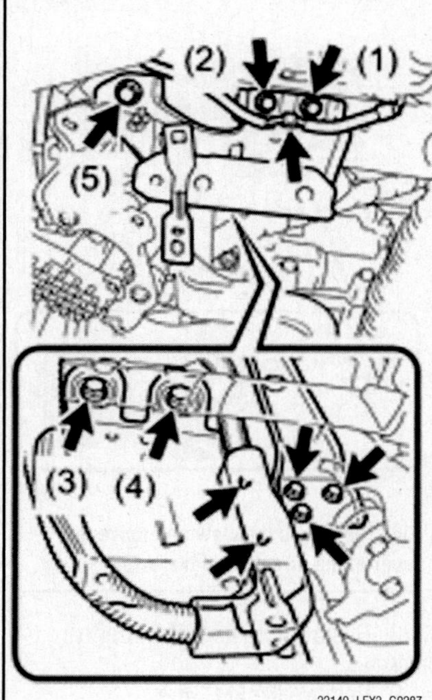

Fig. 190 LH front stabilizer bracket bolt view

- Install the water with motor and bracket with the 2 bolts. Tighten to 29 ft. lbs. (31 Nm).

20. Install the LH front stabilizer bracket as follows:
    - Push the front No. 1 stabilizer bracket RH toward the outside the vehicle against the frame and temporarily install bolt (1).
    - Install bolt (2). Install bolt (1). Then, install bolts (3) and (4). Tighten to 36 ft. lbs. (49 Nm).
    - Install bolt (5) and tighten to 74 ft. lbs. (100 Nm).
    - Connect the heater water pump assembly to the front No. 1 stabilizer bracket RH with the 2 bolts. Tighten to 24 ft. lbs. (33 Nm).
    - Connect the engine ground wire clamp, heater water pump assembly wire harness clamp and skid control sensor wire clamp to the front No. 1 stabilizer bracket RH.
21. Temporarily install the control actuator, 2 bushings and 2 brackets with the 4 bolts.
22. Install front LH and RH No. 2 stabilizer bracket and tighten to 36 ft. lbs. (49 Nm).
23. Install the rear RH and LH engine under covers.
24. Install the LH and RH stabilizer link assembly. Tighten to 62 ft. lbs. (84 Nm).
25. Connect the front active stabilizer control actuator assembly as follow:
    - Connect the connector to the front active stabilizer control ECU (1)
    - Rotate the lever in the direction of the arrow until a "click" sound is heard (2).
    - Lock the lever's lock. (3)
    - Install the wire clamp to the bracket.
    - Install the ground wire to the front active stabilizer control ECU with the bolt. Tighten to 7 ft. lbs. (10 Nm).
26. Install the RH front fender "liner.
27. Install the front RH fender wheel opening molding.
28. Install the engine under covers.
29. install and tighten the front wheels and to 76 ft. lbs. (103 Nm).
30. Connect the negative battery cable.
31. Certain systems need to be initialized after disconnecting and reconnecting the cable from the negative (-) battery terminal.

CONTROL LINKS

REMOVAL & INSTALLATION

See Figure 191.

1. Before servicing the vehicle, refer to the precautions section.
2. Remove the front wheels.
3. Remove engine under cover.
4. Remove the left and right rear engine under covers.
5. Remove the 2 nuts and the left front stabilizer link assembly. Repeat for the right side. If the ball joint turns together with the nut, use a hexagon (6mm) wrench to hold the stud.

To install:

6. Install the left front stabilizer link assembly with the 2 nuts and tighten to 62 ft. lbs. (84 Nm). Repeat for the right side. If the ball joint turns together with the nut, use a hexagon (6mm) wrench to hold the stud.
7. Install the left and right rear engine under covers.
8. Install engine under cover.
9. Install the front wheels.

LOWER BALL JOINT

REMOVAL & INSTALLATION

See Figure 192.

1. Before servicing the vehicle, refer to the precautions section.
2. Remove the front wheel.
3. Remove engine under cover.
4. Separate the ABS speed sensor.
5. For AWD, remove the clip and castle nut, then using SST 09628-00011, separate the upper ball joint from the steering knuckle.
6. Remove the clip and castle nut, then using SST 09628_62011, separate the tie rod end.
7. Remove the strut assembly.
8. Separate the front stabilizer link assembly.
9. Remove the engine under cover.
10. Remove the lower control arm.
11. Fix the front lower control arm in a vise using aluminum plates.
12. Remove the clip and castle nut.
13. Use SST 09950-40011 (09951-04010, 09952-04010, 09953-04020, 09954-04010, 09955-04051, 09957-04010, 09958-04011) to remove the front lower ball joint from the front lower control arm.
14. Inspect the lower control arm ball joint as follows:
    - Flip the ball joint stud back and forth 5 times.
    - Temporarily install the nut, and use a torque wrench to turn the nut continuously at a rate of 3 to 5 seconds per turn. Take the torque reading on the 5th turn.
    - Turning torque: 53 inch lbs (6 Nm)

    a. Check the dust boots for cracks or grease leakage. If the value is not within the specified range, replace the front lower ball joint with a new one.

To install:

15. Install the front lower ball joint to the front lower control arm with the nut. Ensure that the thread and taper are free of oil or other foreign matter. Tighten the nut to 120 ft. lbs. (162 Nm).
16. Install a new clip to the front lower ball joint. Further tighten the nut up to 60° if the holes for the cotter pin are not aligned.
17. Install the lower control arm.
18. Connect the front stabilizer link assembly.
19. Install the strut assembly, but temporarily tighten the bolts at the lower control arm.
20. Connect the tie rod end to the steering knuckle Tighten the nut to 50 ft. lbs. (65 Nm). Install a new cotter pin. If the holes for the clip are not aligned, tighten the nut up to 60° further.
21. Install the steering knuckle to the front suspension upper control arm, and tighten it with the castle nut to 64 ft. lbs. (87 Nm). Install a new clip to the steering knuckle. Further tighten the nut up to 60° if the holes for the cotter pin are not aligned.
22. Connect the ABS speed sensor.
23. Stabilize the suspension as follows:

    a. Install the front wheels. Tighten the lug nuts to 76 ft. lbs. (103 Nm).

    b. Lower the vehicle and bounce it up and down several times to stabilize the front suspension.

    c. Remove the front wheels.

    d. Jack up the front suspension lower arm placing a wooden block in between. Apply a load to the front suspension so that the front suspension lower arm is placed in a horizontal position.
24. Fully tighten the strut assembly bolts at the lower control arm.
25. Fully tighten the lower control arm.
26. Install engine under cover.
27. Install the front wheel.
28. Inspect and adjust wheel alignment.

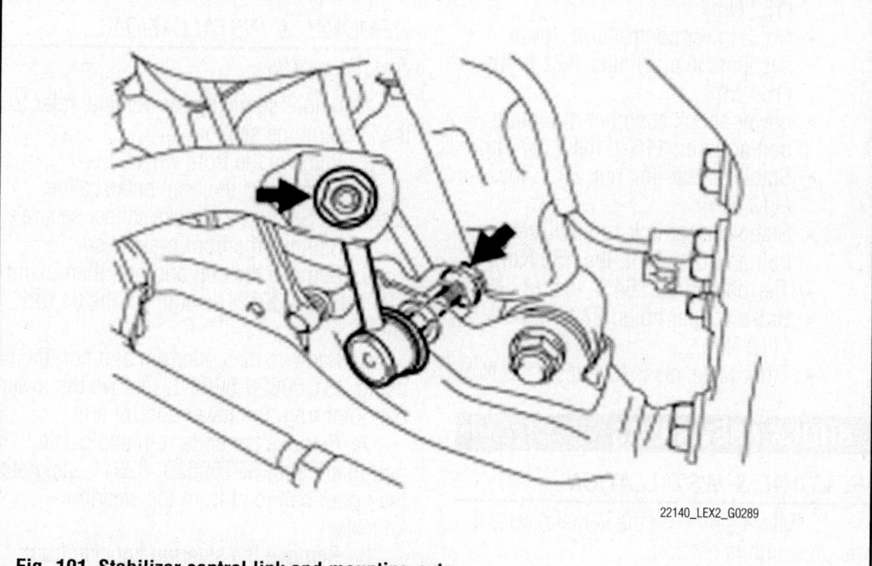

Fig. 191 Stabilizer control link and mounting nuts

LOWER CONTROL ARM

REMOVAL & INSTALLATION

See Figure 192.

1. Before servicing the vehicle, refer to the precautions section.
2. Raise the vehicle on a hoist, so that front suspension components are hanging and accessible.
3. Remove or disconnect the following:
    - Front wheels.
    - Engine under covers.
    - Brake caliper(s); Do NOT disconnect the brake hose; hang the

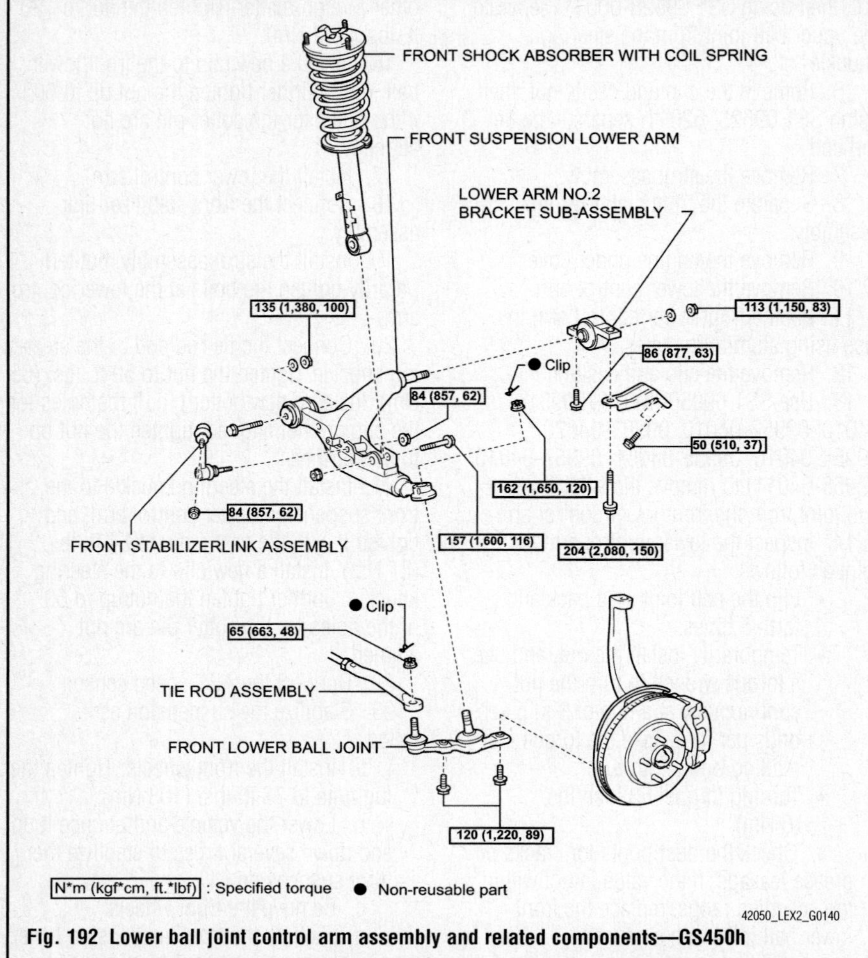

Fig. 192 Lower ball joint control arm assembly and related components—GS450h

caliper without stress on the hose
- Tie rod end from steering knuckle
- Stabilizer bar link from stabilizer bar
- Height control sensor link, if equipped, from shock absorber bracket
- Shock absorber lower mount
- Lower control arm set bolts (loosen only)
- Lower ball joint from No. 2 lower control arm (lower suspension arm)
- Steering gear assembly
- Strut bar bracket
- No. 1 lower control arm (lower suspension arm); matchmark adjusting cam to crossmember

To install:

4. To install, reverse the removal procedure, noting the following torque settings:
- No. 1 lower control arm (lower suspension arm) bolt to shock absorber bracket: 44 ft. lbs. (59 Nm)
- No. 1 lower control arm (lower suspension arm) adjusting cam bolt and nut to crossmember: 127 ft. lbs. (172 Nm)
- Strut bar bracket bolts: 43 ft. lbs. (58 Nm)
- Strut bar bracket nut: 112 ft. lbs. (152 Nm)
- No. 2 lower control arm (lower suspension arm) nuts: 122 ft. lbs. (164 Nm)
- Lower shock absorber mounting bolt and nut: 116 ft. lbs. (157 Nm)
- Stabilizer bar link nut: 83 ft. lbs. (113 Nm)
- Stabilizer bar link-to-stabilizer bar bolt and nut: 43 ft. lbs. (55 Nm)
- Tie rod end nut: 64 ft. lbs. (87 Nm)
- Brake caliper bolts: 87 ft. lbs. (118 Nm)
- Front wheel nuts: 76 ft. lbs. (103 Nm)

SHOCK ABSORBERS

REMOVAL & INSTALLATION

1. Before servicing the vehicle, refer to the precautions section.
2. Remove the front wheel.
3. Disconnect the front brake caliper assembly. Support the brake caliper securely.
4. Remove the front brake disc
5. Remove the clip and nut, then using SST 09610-20012, disconnect the tie rod end.
6. Remove the cotter pin and nut, then using SST 09628-6201 1, remove the lower ball joint from the lower control arm.
7. Remove the cotter pin and castle nut, then using SST09628_62011, separate the upper ball joint from the steering knuckle.
8. Remove the steering knuckle from the vehicle.
9. If necessary, remove the lower ball joint and hub and bearing assembly from the steering knuckle.

To install:

10. If necessary, install the hub and bearing assembly and lower ball joint to the steering knuckle.
11. Install the steering knuckle into the vehicle.
12. Connect the upper ball joint to the steering knuckle. Install castle nut and new cotter pin.
13. Connect the lower ball joint to lower control arm. Install castle nut and new cotter pin.
14. Connect the tie rod end to the steering knuckle. Install castle nut and new cotter pin.
15. Install the front brake disc
16. Install the front brake caliper assembly.
17. Install the front wheel.
18. Inspect and adjust wheel alignment.

STEERING KNUCKLE

REMOVAL & INSTALLATION

See Figure 193.

1. Before servicing the vehicle, refer to the Precautions section.
2. Remove the front wheel.
3. Disconnect the front brake caliper assembly. Support the brake caliper securely.
4. Remove the front brake disc
5. Remove the clip and nut, then using SST 09610-20012, disconnect the tie rod end.
6. Remove the cotter pin and nut, then using SST 09628-6201 1, remove the lower ball joint from the lower control arm.
7. Remove the cotter pin and castle nut, then using SST09628_62011, separate the upper ball joint from the steering knuckle.
8. Remove the steering knuckle from the vehicle.

Fig. 193 Remove the 2 bolts and steering knuckle from the lower ball joint—2WD models

9. If necessary, remove the lower ball joint and hub and bearing assembly from the steering knuckle.

To install:

10. If necessary, install the hub and bearing assembly and lower ball joint to the steering knuckle.
11. Install the steering knuckle into the vehicle.
12. Connect the upper ball joint to the steering knuckle. Install castle nut and new cotter pin. Tighten to 64 ft. lbs. (87 Nm).
13. Connect the lower ball joint to lower control arm. Install castle nut and new cotter pin. Tighten to 64 ft. lbs. (87 Nm).
14. Connect the tie rod end to the steering knuckle. Install castle nut and new cotter pin. Tighten to 48 ft. lbs. (65 Nm).
15. Install the front brake disc.
16. Install the front brake caliper assembly.
17. Install the front wheel.
18. Inspect and adjust wheel alignment.

STRUT

REMOVAL & INSTALLATION

See Figures 194.

1. Before servicing the vehicle, refer to the precautions section.
2. Remove or disconnect the following:
   - Negative battery cable.
   - Front wheel
   - Brake caliper, leaving the line attached
3. Loosen the 3 upper strut mounting nuts.
4. Loosen, but do not remove, the upper strut rod nut.

⁂ CAUTION

Do NOT remove the upper strut nut at this time.

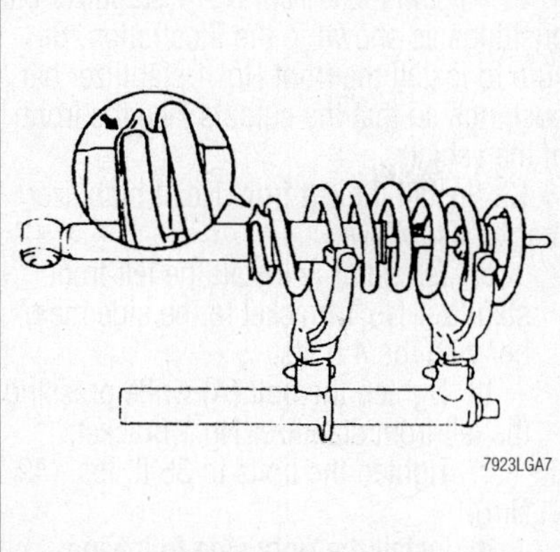

Fig. 194 Matching the spring to the seat

5. Remove or disconnect the following:
   - Anti-lock Brake System (ABS) speed sensor and harness
   - Upper suspension arm from the steering knuckle
   - Stabilizer bar from the link and remove the bracket
   - Strut from the lower suspension arm.
   - 3 upper strut mounting nuts and remove the strut
6. Compress the coil spring.
7. Remove or disconnect the following:
   - Piston rod locknut
   - Suspension support, coil spring and bumper
8. If disposing the strut, perform the following procedure:
   a. Fully extend the strut rod.
   b. Drill a hole near the bottom of the shock to remove the gas inside.

⁂ CAUTION

The gas is harmless, but be careful of chips that may fly up when the gas is released.

To install:

9. Install or connect the following:
   - Spring bumper
   - Coil spring
   - Suspension support to the rod and temporarily install a new nut
10. Turn the suspension support so one of the bolts on the support faces the same direction as shown in the illustration.

➡Align the bolt so a line drawn between the rod and bolt would be at 90° to the direction of the lower bushing.

11. Install or connect the following:
   - Spring compressor
   - Strut and tighten the upper retaining nuts to 41 ft. lbs. (56 Nm)
   - New upper strut rod nut to 20 ft. lbs. (27 Nm)
   - Strut to the lower arm and temporarily tighten the nut and bolt
   - Stabilizer bar bracket and tighten the bolts to 21 ft. lbs. (28 Nm)
   - The stabilizer bar to the link and tighten the bolts to 29 ft. lbs. (39 Nm)
   - Upper suspension arm to the steering knuckle. Tighten the nut to 64 ft. lbs. (87 Nm) and install a new cotter pin.
   - ABS speed sensor and tighten the bolt to 69 inch lbs. (8 Nm)
   - Caliper
   - Wheel
12. Bounce the vehicle several times to stabilize the suspension.
13. Tighten the lower strut bolt and nut to 116 ft. lbs. (157 Nm).
14. Check the front wheel alignment.

To install:

15. If necessary, install the hub and bearing assembly and lower ball joint to the steering knuckle.
16. Install the steering knuckle into the vehicle.
17. Connect the upper ball joint to the steering knuckle. Install castle nut and new cotter pin.
18. Connect the lower ball joint to lower control arm. Install castle nut and new cotter pin.
19. Connect the tie rod end to the steering knuckle. Install castle nut and new cotter pin.
20. Install the front brake disc
21. Install the front brake caliper assembly.
22. Install the front wheel.
23. Inspect and adjust wheel alignment.

STABILIZER BAR

REMOVAL & INSTALLATION

See Figures 195 through 197.

1. Before servicing the vehicle, refer to the precautions section.
2. Remove the front wheels.
3. Remove engine under cover.
4. Remove the left and right rear engine under covers.
5. Remove the 2 nuts and the left front stabilizer link assembly. Repeat for the right side. If the ball joint turns together with the nut, use a hexagon (6mm) wrench to hold the stud.
6. Remove the 2 bolts and the left front No. 2 stabilizer bracket from the front sus-

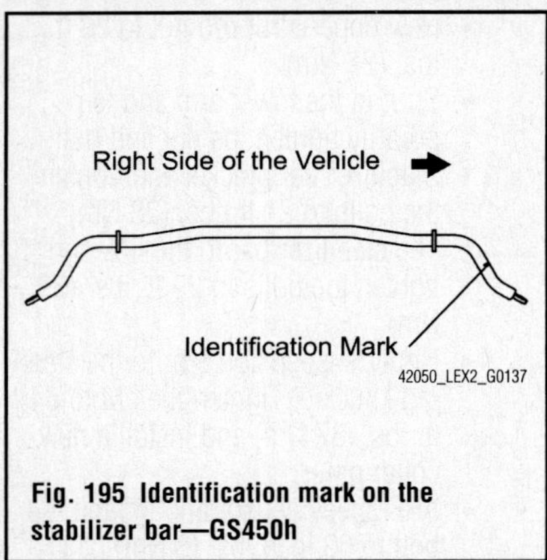

Fig. 195 Identification mark on the stabilizer bar—GS450h

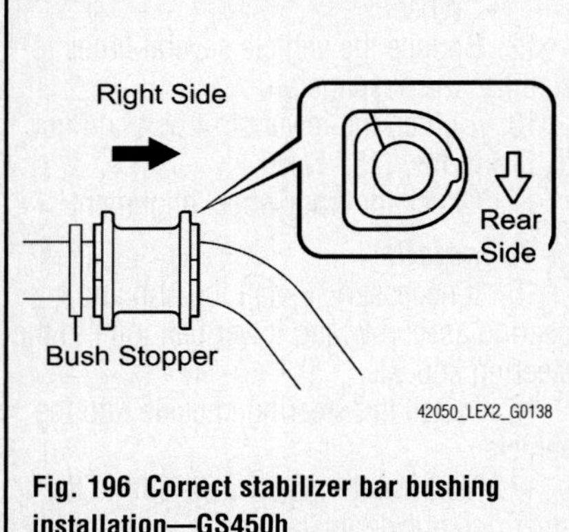

Fig. 196 Correct stabilizer bar bushing installation—GS450h

pension crossmember. Repeat for the right side.

7. Remove the 4 bolts and the left front No. 1 stabilizer bracket from the frame. Repeat for the right side.

8. Remove the 2 front No. 1 stabilizer bar bushings from the front stabilizer bar.

9. Remove the front stabilizer bar from the vehicle.

To install:

10. Install the front stabilizer bar to the vehicle. The identification mark must be on the right side of the vehicle when installing the front stabilizer bar.

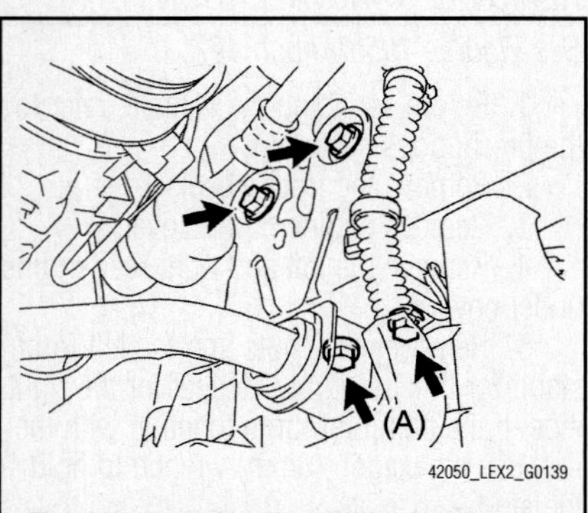

Fig. 197 Front No. 1 stabilizer bracket bolts—GS450h

11. Install the 2 front No. 1 stabilizer bar bushings as shown in the illustration. Be sure to install the front No. 1 stabilizer bar bushings so that the cutouts face the front of the vehicle.

12. Install the left front No. 1 stabilizer bracket as follows:

a. Temporarily install the left front stabilizer No.1 bracket to the side member with the 4 bolts.

b. Tighten the bolt (A) while pressing the left front stabilizer No.1 bracket.

c. Tighten the bolts to 36 ft. lbs. (49 Nm).

d. Install the right side following the same procedures as for the left side.

13. Install the left front No. 2 stabilizer bracket on the vehicle with the 2 bolts. Repeat for the right side. Tighten the bolts to 36 ft. lbs. (49 Nm).

14. Install the left front stabilizer link assembly with the 2 nuts and tighten to 62 ft. lbs. (84 Nm). Repeat for the right side. If the ball joint turns together with the nut, use a hexagon (6mm) wrench to hold the stud.

15. Install the left and right rear engine under covers.

16. Install engine under cover.

17. Install the front wheels.

18. Inspect and adjust the front wheel alignment.

UPPER BALL JOINT

REMOVAL & INSTALLATION

The upper ball joint is an integral part of the upper arm and is not replaced separately. The upper ball joint replacement is accomplished by replacing the upper arm.

UPPER CONTROL ARM

REMOVAL & INSTALLATION

See Figure 198.

1. Before servicing the vehicle, refer to the precautions section.

2. Remove or disconnect the following:

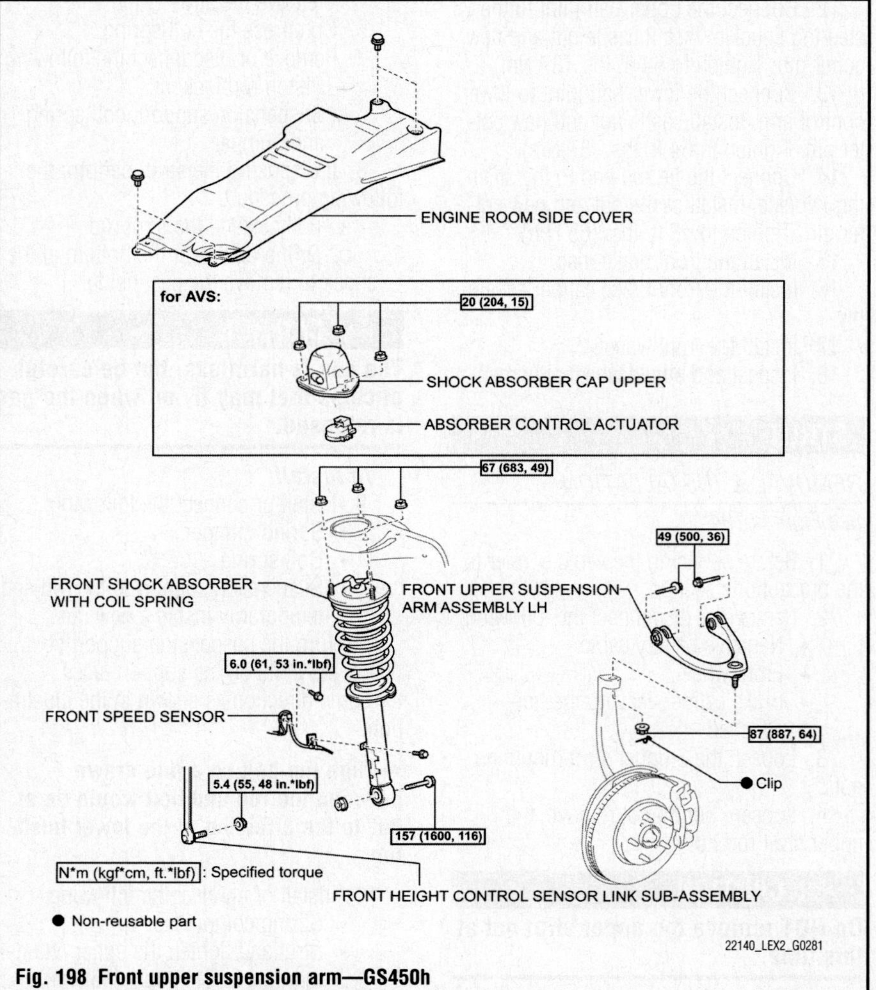

Fig. 198 Front upper suspension arm—GS450h

- Negative battery cable
- Wheel

3. Loosen the 3 upper strut mounting nuts.
4. Loosen, but do not remove, the upper strut rod nut.

CAUTION

Do NOT completely remove the upper strut nut at this time.

5. Remove or disconnect the following:
   - Brake caliper, leaving the line attached and secure it out of the way
   - Anti-lock Brake System (ABS) speed sensor and harness
   - Cotter pin and nut from the upper control arm
   - Upper control arm from the steering knuckle
   - Stabilizer bar from the link and remove the bracket
   - Cotter pin and nut from the lower control arm
   - Strut from the lower suspension arm
   - 3 upper strut mounting nuts and remove the strut
   - Mounting bolts holding the upper control arm to the frame
   - Upper control arm from the vehicle

To install:

6. Install or connect the following:
   - Upper suspension arm and tighten the mounting bolts to 39 ft. lbs. (53 Nm)
   - Strut and tighten the upper retaining nuts to 41 ft. lbs. (56 Nm). Tighten the new upper strut rod nut to 20 ft. lbs. (27 Nm).
   - Strut to the lower arm and temporarily tighten the nut and bolt
   - Stabilizer bar bracket and tighten the bolts to 21 ft. lbs. (28 Nm)
   - Stabilizer bar to the link and tighten the bolts to 29 ft. lbs. (39 Nm)
   - Upper suspension arm to the steering knuckle. Tighten the nut to 64 ft. lbs. (87 Nm) and install a new cotter pin.
   - ABS speed sensor and tighten the bolt to 69 inch lbs. (8 Nm)
   - Caliper
   - Front wheel
7. Lower the vehicle.
8. Bounce the vehicle several times to stabilize the suspension.
9. Tighten the lower strut bolt and nut to 116 ft. lbs. (157 Nm).
10. Check the front wheel alignment.

WHEEL HUB & BEARING

REMOVAL & INSTALLATION

See Figure 199.

1. Before servicing the vehicle, refer to the Precautions section.
2. Remove or disconnect the following:
   - Negative battery cable
   - Front wheel
   - Caliper, leaving the brake line connected and suspend it out of the way

WARNING

Never allow the brake caliper to hang freely from the brake hose.

   - Rotor
   - Anti-lock Brake System (ABS) speed sensor and harness
   - Tie rod from the arm on the lower ball joint
   - Upper suspension arm from the steering knuckle
   - Steering knuckle from the lower control arm
   - Ball joint from the steering knuckle
   - Front hub grease cap
3. Clamp the hub in a soft jaw vise.
4. Using a hammer and chisel, loosen the staked part of the locknut.
5. Remove or disconnect the following:
   - Locknut
   - ABS speed sensor rotor

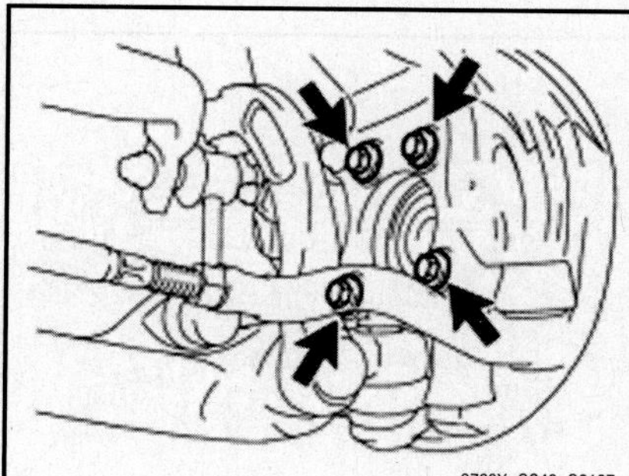

Fig. 199 Remove the 4 bolts, front axle hub sub-assembly and front disc brake dust cover

➡**Do NOT scratch the serrations of the sensor rotor.**

   - Brake dust cover bolts and shift the cover toward the outside.
   - Hub from the steering knuckle
   - Inner bearing race from the hub shaft
   - Oil seal from the knuckle
   - Bearing snapring from the steering knuckle
   - Bearing from the steering knuckle

To install:

6. Install or connect the following:
   - New bearing into the steering knuckle

➡**If the inner race and balls come loose from the bearing outer race, be sure to install them on the same side as before.**

   - Snapring
   - New outside inner race and tap in the new seal. Tap the seal until it is flush with the end surface of the steering knuckle.
   - Brake dust cover to the knuckle and tighten the bolts to 74 inch lbs. (8 Nm)
   - Hub into the steering knuckle
   - ABS speed sensor rotor
   - Axle hub locknut. Tighten the nut to 147 ft. lbs. (199 Nm) and stake it.
   - Grease cap to the steering knuckle by tapping lightly around the circumference of the cap with a hammer
   - Ball joint to the steering knuckle. Tighten the 2 bolts to 83 ft. lbs. (113 Nm).
   - Steering knuckle to the upper and lower suspension arms. Tighten the upper nut to 64 ft. lbs. (87 Nm) and the lower nut to 95 ft. lbs. (127 Nm). Install a new cotter pin on the lower nut. Install the clip on the upper suspension arm nut.
   - Tie rod end to the steering knuckle. Tighten the nut to 64 ft. lbs. (87 Nm) and install a new cotter pin.
   - Rotor, disc brake pads and the brake caliper
   - ABS speed sensor and harness. Tighten the sensor retaining bolt to 69 inch lbs. (8 Nm).
   - Wheel
7. Lower the vehicle and connect the negative battery cable.
8. Check the front wheel alignment.

ACTIVE STABILIZER CONTROL ACTUATOR

REMOVAL & INSTALLATION

See Figures 200 through 203.

1. Before servicing the vehicle, refer to the precautions section.
2. Disconnect the negative battery cable.
3. Remove the luggage compartment floor mat.
4. Remove the RH luggage compartment trim.
5. Remove the tool box.
6. Remove the active converter cover.
7. Disconnect the rear active stabilizer control actuator assembly as follows:
   - Using a screwdriver, release the lever's lock. (1)
   - Move the lever in the direction of the arrow in the illustration. (2)
   - Disconnect the ECU connector. (3)
   - Disconnect the connector (B) and connector (C).
   - Remove the 4 clamps and packing, and move the disconnected side of the wire harness out of the vehicle.
8. Remove the 2 bolts and 2 compression springs, and disconnect the front exhaust pipe from the tailpipe.
9. Remove the tailpipe from the 6 No. 4 exhaust pipe supports.
10. Remove the rear wheels.
11. Remove the 2 nuts and the support protector.
12. Remove the nut labeled A, and disconnect the link from the stabilizer.
13. Remove the bolt labeled B, and disconnect the link and bracket from the No. 2 suspension arm
14. Remove the 4 bolts, 2 brackets, 2 bushings and control actuator.

To install:

15. Install the 2 bushings.
16. Temporarily install the rear stabilizer control actuator with the bracket and 4 bolts so that the actuator is on the right side of the vehicle.
17. Tighten the 4 bolts to 24 ft. lbs. (32 Nm).
18. Install the LH and RH stabilizer link, bracket to the rear No. 2 suspension arm with the bolt and nut labeled (B). Tighten to 20 ft. lbs. (27 Nm).
19. Install the stabilizer link to the stabilizer with the nut labeled (A). Tighten to 66 ft. lbs. (89 Nm).

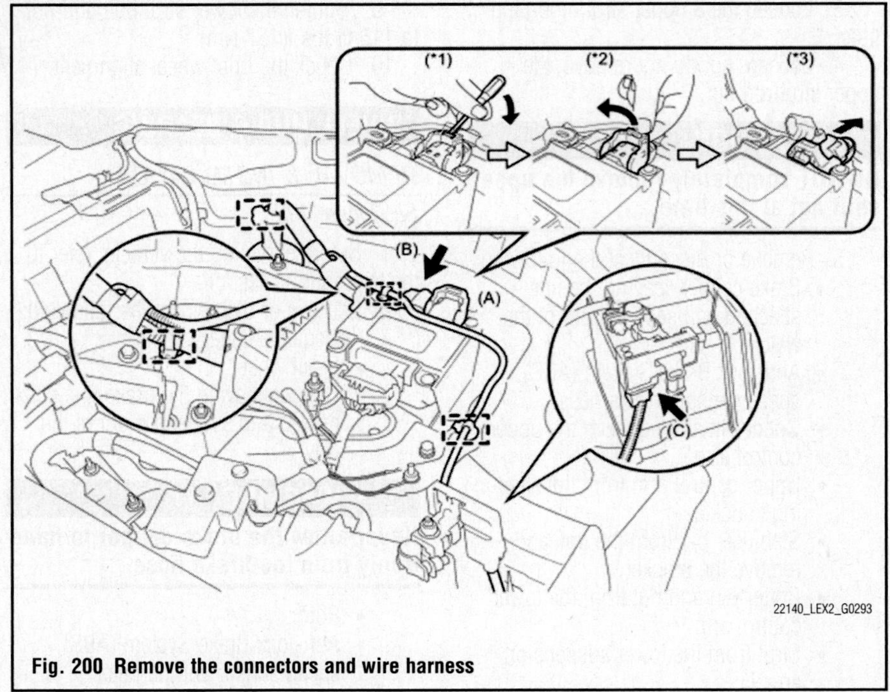

Fig. 200 Remove the connectors and wire harness

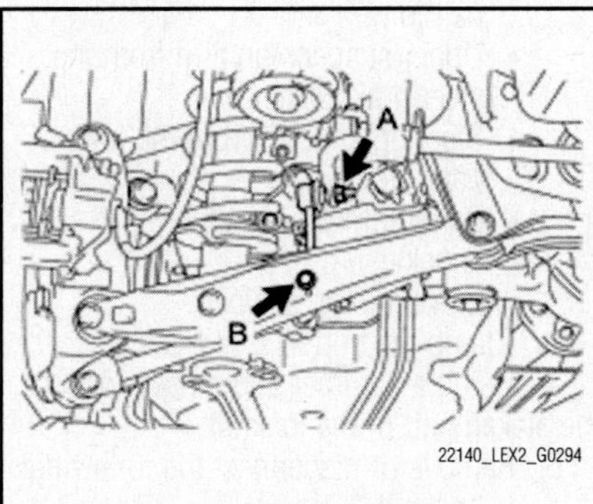

Fig. 201 Remove the nut labeled A and bolt labeled B

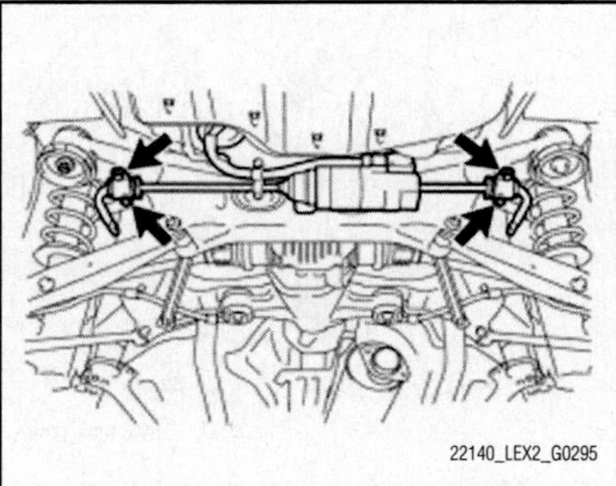

Fig. 202 Remove the 4 bolts, 2 brackets, 2 bushings and control actuator

20. Install the No. 2 support protector.
21. Install the rear wheels and tighten to 76 ft. lbs. (103 Nm).

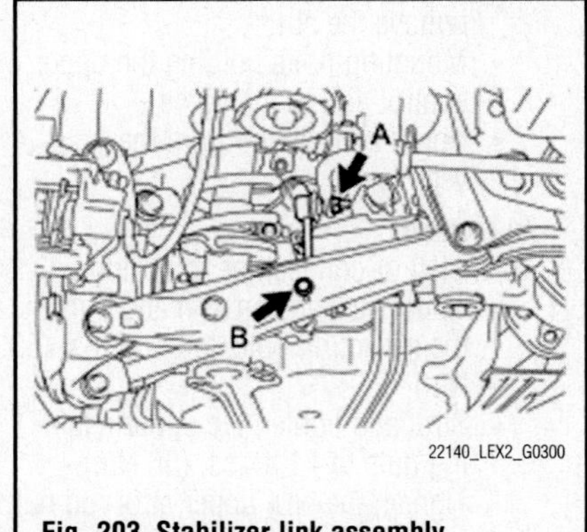

Fig. 203 Stabilizer link assembly

22. Install the tail pipe assembly.
23. Install the 4 clamps and wire harness packing.
24. Connect the connector (B) and connector (C).
25. Connect the connector (A).
26. Install the active stabilizer converter cover.
27. Install the tool box.
28. Install the RH luggage compartment trim side cover.
29. Install the luggage compartment trim cover.
30. Connect the negative battery cable.
31. Certain systems need to be initialized after disconnecting and reconnecting the cable from the negative (-) battery terminal

CONTROL ARMS/LINKS

REMOVAL & INSTALLATION

Stabilizer Link

See Figure 204.

1. Before servicing the vehicle, refer to the precautions section.
2. Before servicing the vehicle, refer to the precautions section.
3. Remove the stabilizer control arm link upper and lower nuts.

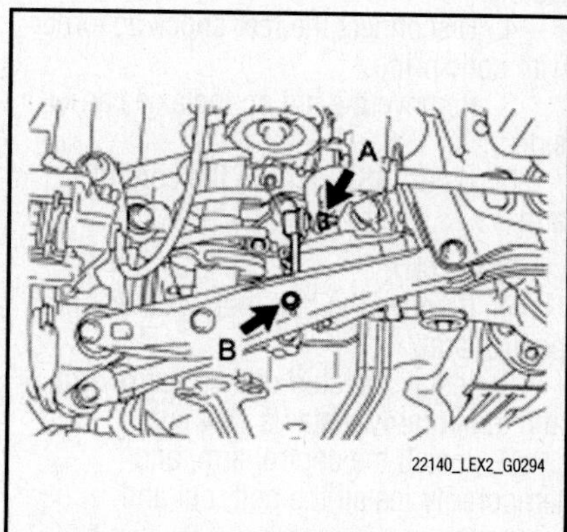

Fig. 204 Stabilizer link assembly

To install:

4. Install the LH and RH stabilizer link, bracket to the rear No. 2 suspension arm with the bolt and nut labeled (B). Tighten to 20 ft. lbs. (27 Nm).
5. Install the stabilizer link to the stabilizer with the nut labeled (A). Tighten to 66 ft. lbs. (89 Nm).

No. 1 Suspension Arm

See Figures 205 and 206.

1. Before servicing the vehicle, refer to the Precautions Section.
2. Remove the rear wheels.
3. Remove the differential support protector.

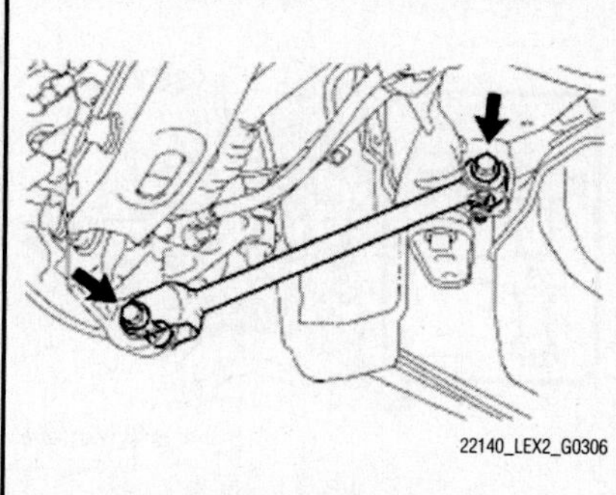

Fig. 205 Remove the 2 bolts, 2 nuts and suspension arm

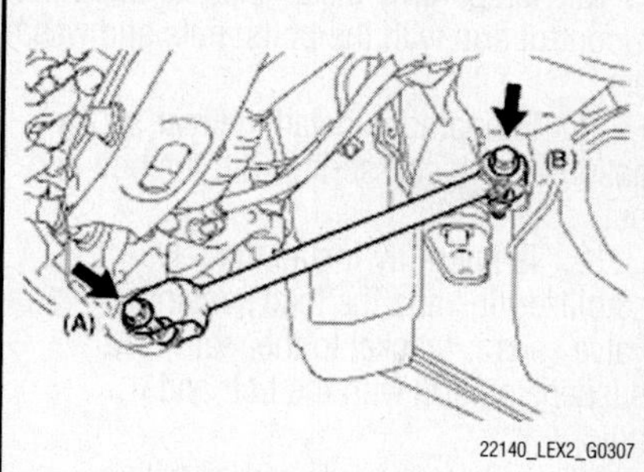

Fig. 206 Tighten the No. 1 rear suspension arm bolt (A) and (B)

4. Remove the LH rear side member cover.
5. Remove the 2 bolts, 2 nuts and suspension arm.

To install:

6. Temporarily install the suspension arm with the 2 bolts and 2 nuts.
7. Jack up the axle carrier, with a wooden block placed between the jack and axle carrier, to apply load to the rear suspension so that the rear drive shaft assembly becomes level.
8. Tighten the No. 1 rear suspension arm. Tighten bolt (A) to 70 ft. lbs. (95 Nm).
9. Install the member cover with the 2 screws.
10. Install the NO. 2 differential support protector.
11. Install the rear wheel and tighten to 76 ft. lbs. (103 Nm).
12. Inspect and adjust the rear wheel alignment.
13. Check the head light adjustment.

No. 2 Suspension Arm

See Figure 207.

1. Before servicing the vehicle, refer to the Precautions Section.
2. Remove the rear wheels.
3. Remove the differential support protector.
4. Remove the LH rear member brace.
5. Remove the rear No. 2 suspension arm as follows:
   - Remove the bolt and nut, and disconnect the rear stabilizer link and the load sensing valve sensor bracket.
   - Remove the bolt and nut, and disconnect the rear shock absorber with coil spring.
   - Remove the nut and bolt on the axle carrier side, and disconnect the rear No. 2 suspension arm.
   - Remove the bolt, nut, and washer on the rear suspension member side, and remove the rear No. 2 suspension arm.

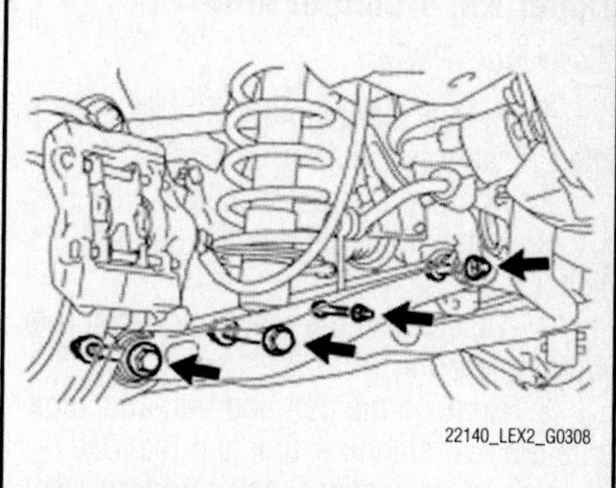

Fig. 207 Remove the rear No. 2 suspension arm

To install:

6. Temporarily install the rear suspension No. 2 arm with the bolt, nut and washer as follows:
   - Temporarily install the rear suspension No. 2 arm with the bolt and nut.
   - Temporarily install the rear shock absorber with coil spring with the bolt and nut.
   - Temporarily install the stabilizer link and the load sensing valve sensor bracket to the rear suspension No. 2 arm with the bolt and nut.
   - Jack up the axle carrier, with a wooden block placed between the jack and axle carrier, to apply load to the rear suspension so that the rear drive shaft becomes level.
7. Tighten the rear suspension arm as follows:
   - Tighten the bolt on the axle carrier side to 119 ft. lbs. (161 Nm).
   - Tighten the bolt holding the rear shock absorber with coil spring to 81 ft. lbs. (110 Nm).
   - Tighten the nut holding the rear stabilizer link to 20 ft. lbs. (27 Nm).
   - Tighten the nut on the rear suspension member side to 103 ft. lbs. (140 Nm).
8. Install the member cover with the 2 screws.
9. Install the No. 2 differential support protector.
10. Install the rear wheel and tighten to 76 ft. lbs. (103 Nm).
11. Inspect and adjust the rear wheel alignment.
12. Check the head light adjustment.

Upper No. 1 Control Arm

See Figure 208.

1. Before servicing the vehicle, refer to the Precautions Section.
2. Remove the rear wheels.
3. Remove the differential support protector.
4. Disconnect the rear shock absorber with coil spring.
5. Remove the bolt and nut, and separate the rear stabilizer link and the load sensing valve sensor bracket from the rear No. 2 suspension arm.
6. Remove the bolt and nut, and separate the rear shock absorber with coil spring from the rear No. 2 suspension arm.
7. Remove the bolts, nuts washers and the control arm from the rear axle carrier.
8. Remove the bolts, nuts, washers and control arm from the rear suspension member.

To install:

9. Temporarily install the control arm with the bolts, nuts and washers.
10. Temporarily install the rear upper No. 1 control arm with the bolts, nuts and washers.
11. Temporarily install the rear shock absorber with coil spring with the bolt and nut.
12. Temporarily install the rear stabilizer link and the load sensing valve sensor bracket to the rear No. 2 suspension arm with the bolt and nut.
13. Jack up the axle carrier, with a wooden block placed between the jack and axle carrier, to apply load to the suspension so that the rear drive shaft assembly becomes level
14. Tighten the nuts on the control arm to 119 ft. lbs. (161 Nm).
15. Tighten the bolt holding the rear shock absorber with coil spring to 81 ft. lbs. (110 Nm).
16. Tighten the nut holding the rear stabilizer link to 20 ft. lbs. (27 Nm).
17. Install the No. 2 differential support protector.
18. Install the rear wheel and tighten to 76 ft. lbs. (103 Nm).
19. Inspect and adjust the rear wheel alignment.
20. Check the head light adjustment.

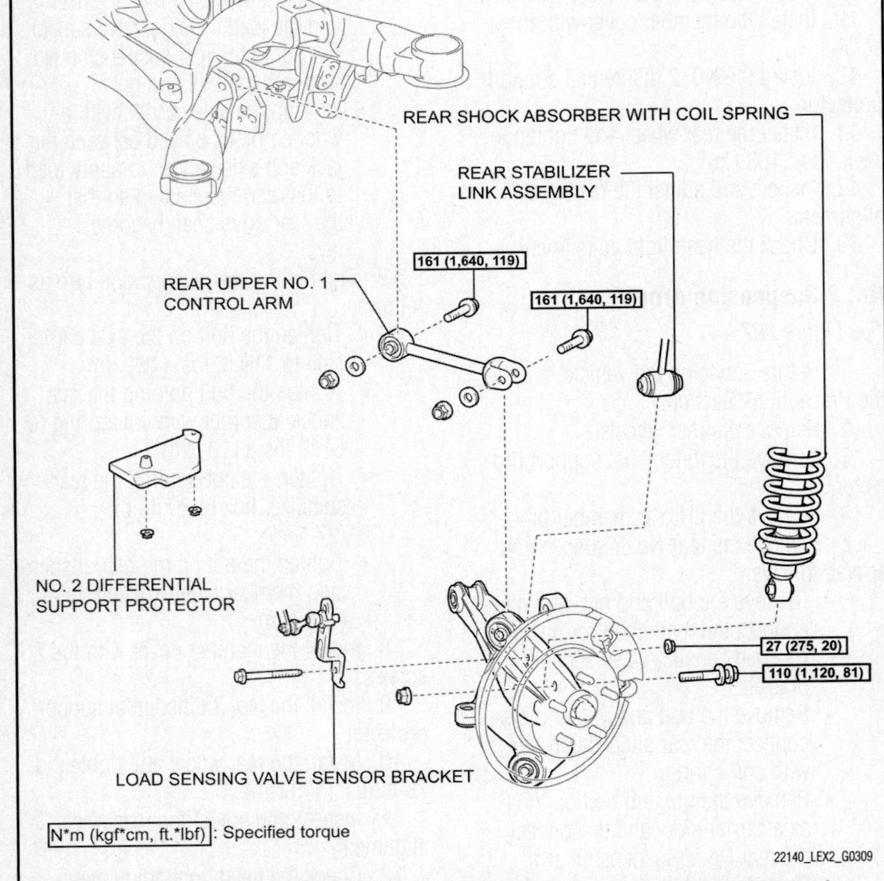

Fig. 208 Remove the bolts, nuts, washers and control arm from the rear suspension member

Upper No. 2 Control Arm

See Figure 209.

1. Before servicing the vehicle, refer to the Precautions Section.
2. Remove the rear wheels.
3. Remove the differential support protector.
4. Disconnect the rear shock absorber with coil spring.
5. Remove the nut on the axle carrier side.
6. Using SST, remove the control arm.

To install:

7. Remove the bolt, nut, washer, and control arm.
8. Install the stud of the control arm, and temporarily install a new nut.
9. Install the control arm, and temporarily install the bolt, nut and washer.
10. Tighten the nut of the control arm to 52 ft. lbs. (70 Nm).
11. Jack up the axle carrier, with a wooden block placed between the jack and axle carrier, to apply load to the suspension so that the rear drive shaft becomes level.
12. Tighten the nut on the rear suspension member side to 77 ft. lbs. (105 Nm).
13. Tighten the rear shock absorber with coil spring.
14. Install the No. 2 differential support protector.
15. Install the rear wheel and tighten to 76 ft. lbs. (103 Nm).
16. Inspect and adjust the rear wheel alignment.
17. Check the head light adjustment.

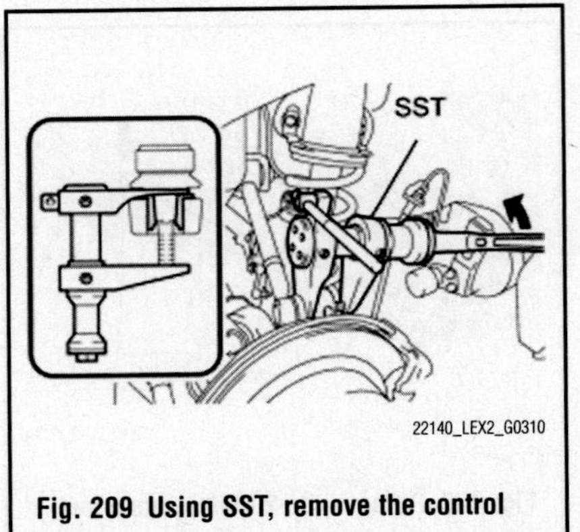

Fig. 209 Using SST, remove the control arm

SHOCK ABSORBER

REMOVAL & INSTALLATION

See Figures 210 through 212.

1. Before servicing the vehicle, refer to the precautions section.
2. Remove the luggage compartment floor mat.
3. Remove the LH and RH luggage compartment side trim covers.
4. Remove the tool box.
5. Remove the rear floor finish plate.
6. Remove the front luggage compartment trim cover.
7. Remove the luggage compartment trim cover inner LH.
8. Remove the rear wheel.
9. Remove the front luggage compartment trim cover.
10. Remove the 2 bolts and brace.
11. Remove the No. 2 suspension arm as follows:
    - Loosen the nut on the rear suspension member side.
    - Remove the bolt and nut and disconnect the rear stabilizer link and the load sensing valve sensor bracket.
    - Remove the bolt and nut and disconnect the rear shock absorber.
    - Remove the nut and bolt on the rear axle carrier side.
12. Remove rear shock absorber cap and disconnect the connector.
13. Turn the absorber control actuator counterclockwise approximately 40° to remove it.
14. Remove the 3 nuts on the upper side of the rear shock absorber with coil spring.
15. Remove the 2 bolts and the rear shock absorber with coil spring from the body.

To install:

16. Temporarily install the rear shock absorber with rear coil spring with the 2 bolts.
17. Install the 3 nuts on the upper side of the rear shock absorber with coil spring. Tighten to 55 ft. lbs. (74 Nm).
18. Tighten the lock nut to 15 ft. lbs. (18 Nm).
19. Tighten the 2 bolts 16 ft. lbs. (21 Nm).
20. Check that the control rod of the rear shock absorber is in the position shown in the illustration.
21. Install the absorber control actuator to the actuator support bracket. Turn the actuator clockwise approximately 40° until a click is felt.
22. Connect the connector and install the shock absorber cap.

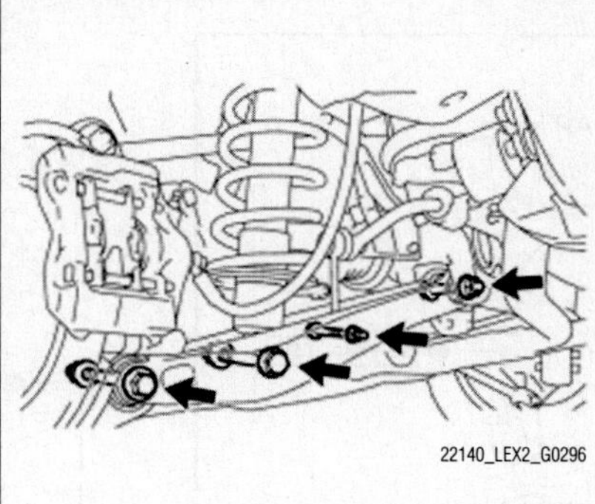

Fig. 210 Remove the No. 2 suspension arm

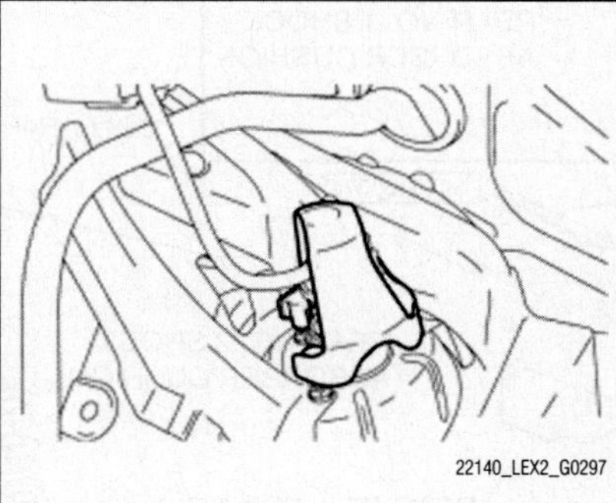

Fig. 211 Remove rear shock absorber cap and disconnect the connector

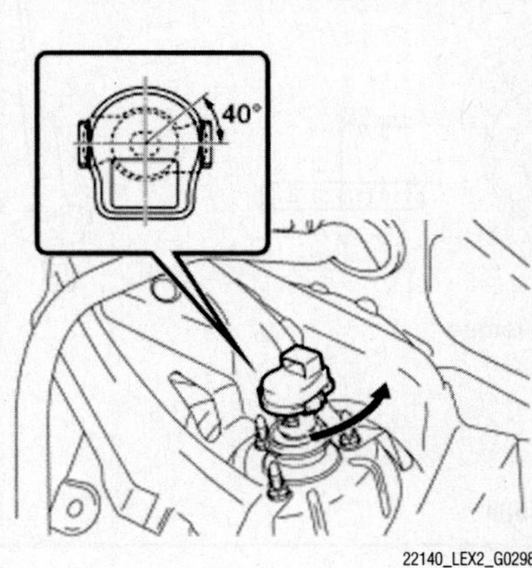

Fig. 212 Turn the absorber control actuator counterclockwise approximately 40° to remove it

23. Temporarily install the rear No. 2 suspension arm with the bolt and nut.
24. Temporarily install the rear shock absorber with rear coil spring with the bolt and nut.
25. Temporarily install the rear stabilizer link and load sensing valve sensor bracket to the rear No. 2 suspension arm with the bolt and nut.
26. Stabilize the suspension.
27. Tighten the No. 2 Suspension arm as follows:
    - Tighten the bolt on the axle carrier side. Tighten to 119 ft. lbs. (161 Nm).
    - Tighten the bolt holding the rear shock absorber with coil spring. Tighten to 81 ft. lbs. (110 Nm).
    - Tighten the nut holding the rear stabilizer link to 20 ft. lbs. (27 Nm).
    - Tighten the nut on the rear suspension member side. Tighten to 103 ft. lbs. (140 Nm).
28. Install the LH rear suspension brace.
29. Install the NO. 2 differential support protector.
30. Install the rear wheel and tighten to 76 ft. lbs. (103 Nm).
31. Inspect and adjust the rear wheel alignment.
32. Check the head light adjustment.

OVERHAUL

See Figures 213 through 215.

1. Before servicing the vehicle, refer to the Precautions Section.
2. Install SST to the rear coil spring so that the distance between the upper and lower hooks is as wide as possible within the installation area.
3. Compress the rear coil spring until it can be moved freely.
4. Remove the nut.
5. Remove the rear No. 1 shock absorber cushion washer, rear No. 1 shock absorber cushion, rear suspension support, rear coil spring insulator upper, collar, rear No. 2 shock absorber cushion, and rear No. 1 spring bumper.
6. Remove the rear coil spring and rear coil spring insulator lower from the rear shock absorber. Release the SST and remove the rear coil spring.

To install:

7. Using SST, compress the rear coil spring.
8. Compress the coil spring and fit the lower end of the spring into the spring seat gap.
9. Install the rear coil spring insulator lower and fit the rear coil spring end into the recessed part of the rear shock absorber lower seat.
10. Install the rear No. 1 spring bumper.
11. Install the rear coil spring insulator upper, rear No. 1 and No. 2 shock absorber cushions, and collar to the rear suspension support.
12. Install the rear suspension support and rear No. 1 shock absorber cushion washer.

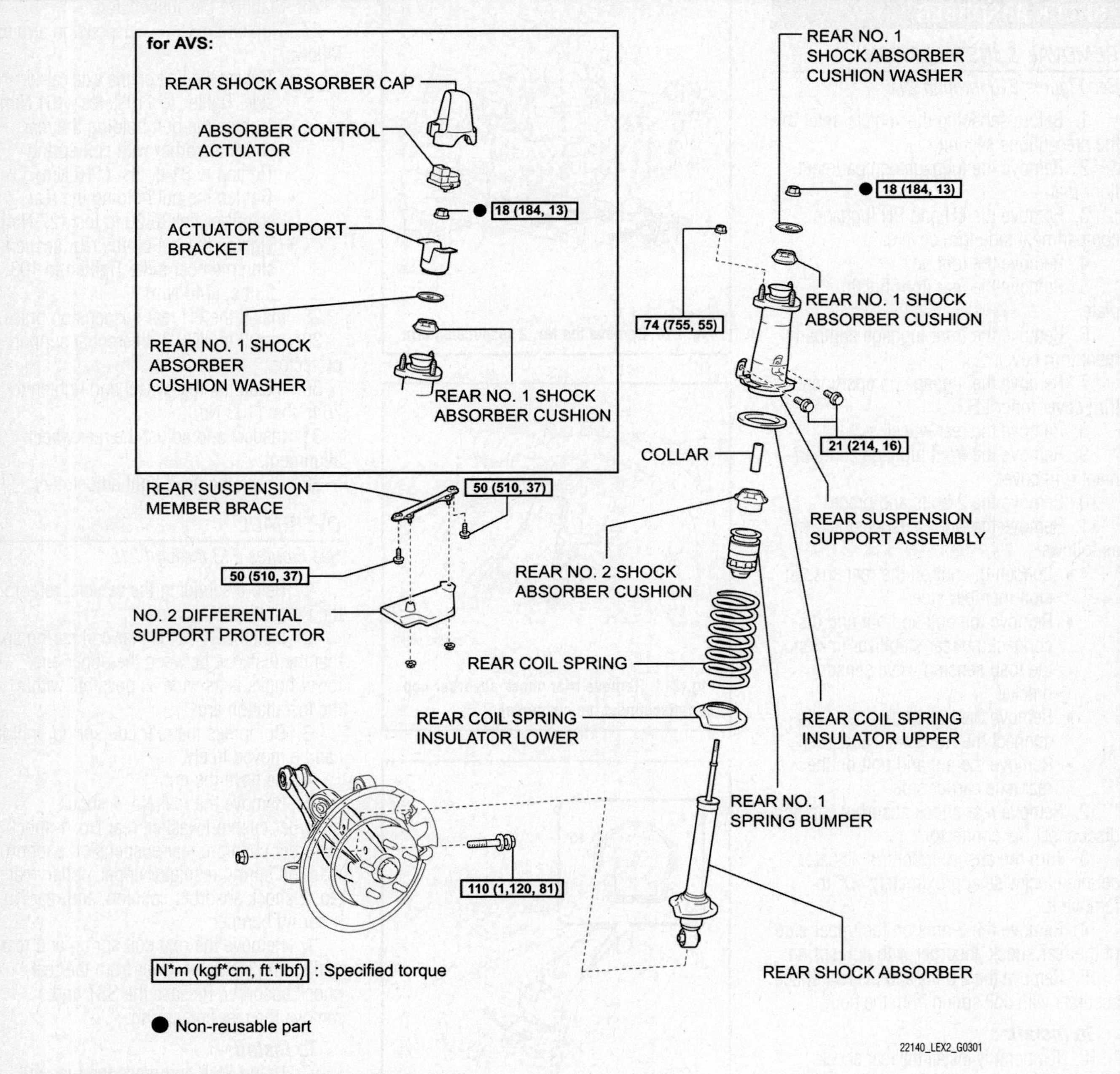

for AVS:
REAR SHOCK ABSORBER CAP
ABSORBER CONTROL ACTUATOR
18 (184, 13)
ACTUATOR SUPPORT BRACKET
REAR NO. 1 SHOCK ABSORBER CUSHION WASHER
REAR NO. 1 SHOCK ABSORBER CUSHION
REAR NO. 1 SHOCK ABSORBER CUSHION WASHER
18 (184, 13)
74 (755, 55)
REAR NO. 1 SHOCK ABSORBER CUSHION
21 (214, 16)
COLLAR
REAR SUSPENSION MEMBER BRACE
50 (510, 37)
50 (510, 37)
REAR SUSPENSION SUPPORT ASSEMBLY
REAR NO. 2 SHOCK ABSORBER CUSHION
NO. 2 DIFFERENTIAL SUPPORT PROTECTOR
REAR COIL SPRING
REAR COIL SPRING INSULATOR LOWER
REAR COIL SPRING INSULATOR UPPER
REAR NO. 1 SPRING BUMPER
110 (1,120, 81)
N*m (kgf*cm, ft.*lbf) : Specified torque
REAR SHOCK ABSORBER
● Non-reusable part
22140_LEX2_G0301

Fig. 213 Explode view of rear strut assembly —GS450h

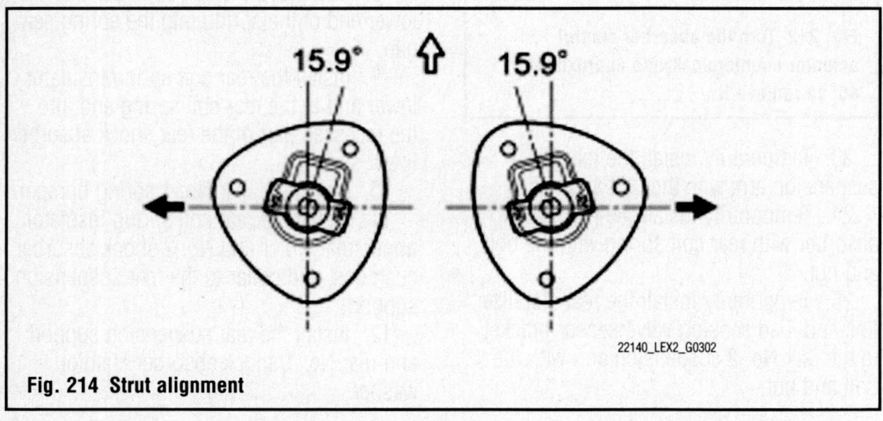

Fig. 214 Strut alignment

13. Align the width across flat on the piston rod end of the shock absorber and the width across flat on the suspension support bracket. Then align the center of the actuator support bracket and the stud bolt as shown in the illustration to install the rear suspension support to the rear shock absorber (for AVS).

14. Temporarily install the nut.

15. Adjust the rear suspension support to the installation position of the lower part of the rear shock absorber so that the studs are positioned as shown in the illustration.

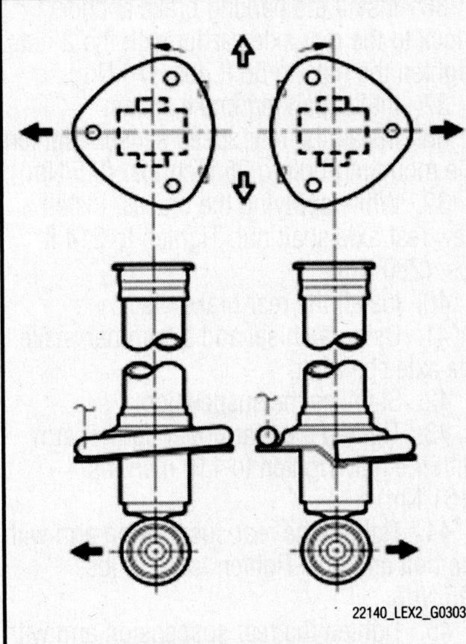

Fig. 215 Adjust the rear suspension support to the installation position of the lower part of the rear shock absorber

16. Release the SST and remove it from the rear coil spring.

STABILIZER BAR

REMOVAL & INSTALLATION

See Figures 216 and 217.

1. Before servicing the vehicle, refer to the Precautions Section.
2. Remove the rear wheels.
3. Remove the differential support protector.
4. Remove the LH and RH stabilizer links.
5. Remove the 4 bolts and the rear stabilizer bar.

To install:

6. Insert the rear stabilizer bar between the rear suspension member and exhaust tailpipe assembly so that the oval holes are on the right of the vehicle.
7. Install the rear stabilizer bar with the 4 bolts in order from 1 to 4. Tighten to 24 ft. lbs. (32 Nm).
8. Install the LH and RH stabilizer links. Tighten the top nut to 20 ft. lbs. (27 Nm). Tighten the bottom nut to 66 ft. lbs. (89 Nm).
9. Install the NO. 2 differential support protector.
10. Install the rear wheel and tighten to 76 ft. lbs. (103 Nm).
11. Inspect and adjust the rear wheel alignment.
12. Check the head light adjustment.

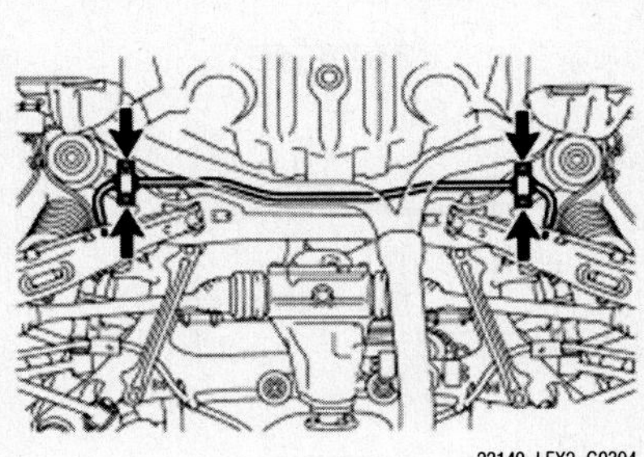

Fig. 216 Remove the 4 bolts and the rear stabilizer bar

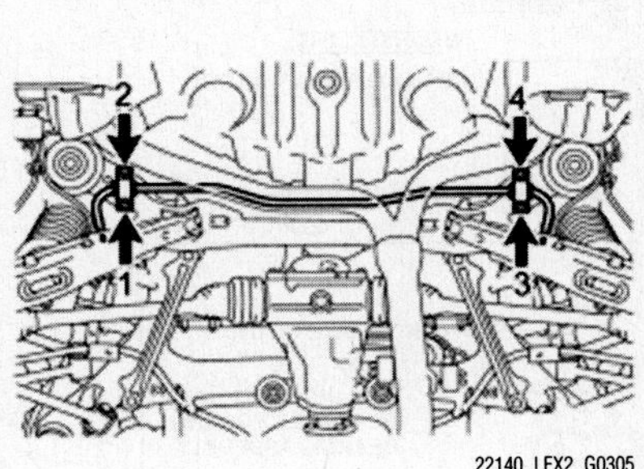

Fig. 217 Install the rear stabilizer bar with the 4 bolts in order from 1 to 4

WHEEL HUB & BEARING

REMOVAL & INSTALLATION

See Figures 218 through 220.

WARNING

When the brake caliper is removed from the brake disc, do not start the hybrid system. If the hybrid system is started, the brake fluid pressure may increase.

1. Before servicing the vehicle, refer to the precautions section.

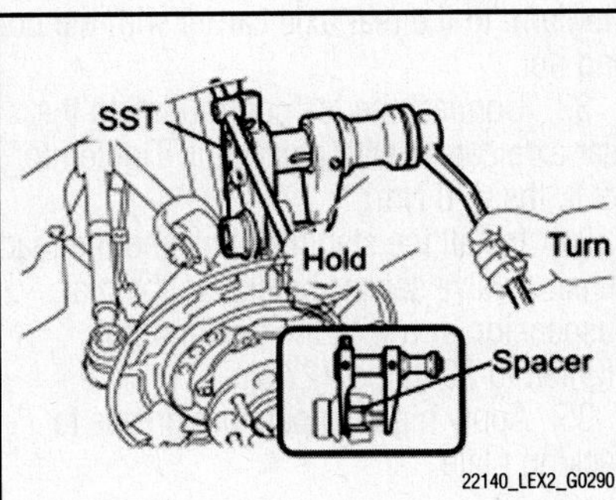

Fig. 218 Using SST, remove the rear upper control arm from the rear axle carrier.

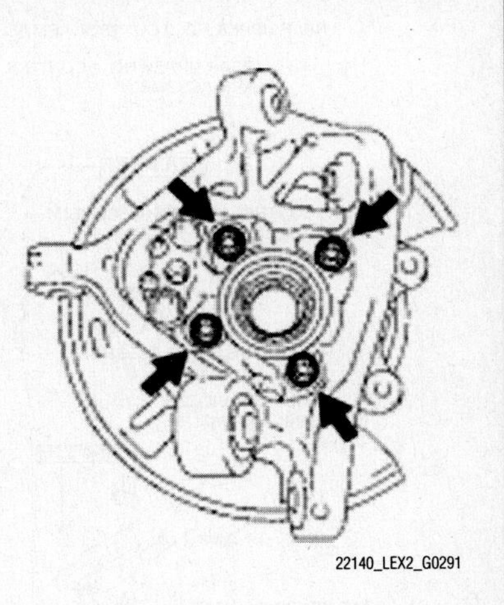

Fig. 219 Remove the 4 bolts and axle hub and bearing from the rear axle carrier

2. Turn the power switch OFF, and check that the "READY" indicator turns off.
3. Disconnect the negative battery cable.
4. Remove the rear wheel.
5. Remove the bolt, nut, load sensing valve sensor bracket and stabilizer link.
6. Using a punch and a hammer, release the staked part of the axle shaft nut.
7. While applying the brakes, remove the rear axle shaft nut.
8. Disconnect the rear brake caliper assembly.
9. Remove the rear brake disc.
10. Remove the rear speed sensor.
11. Remove the rear parking brake shoes.
12. Remove the 2 nuts and the parking brake anchor block from the rear axle carrier.
13. Remove the nut from the rear upper control arm.
14. Using SST, remove the rear upper control arm from the rear axle carrier.
15. Jack up the rear axle so that the bolt on the rear upper No. 1 control arm can be removed.
16. Remove the bolt, washer and nut, and the rear upper No. 1 control arm from the rear axle carrier.
17. Remove the bolt and nut, and separate the rear suspension arm from the rear axle carrier.
18. Remove the bolt and nut, and separate the rear No. 2 suspension arm from the rear axle carrier
19. Using SST, disconnect the toe control link from the rear axle carrier.

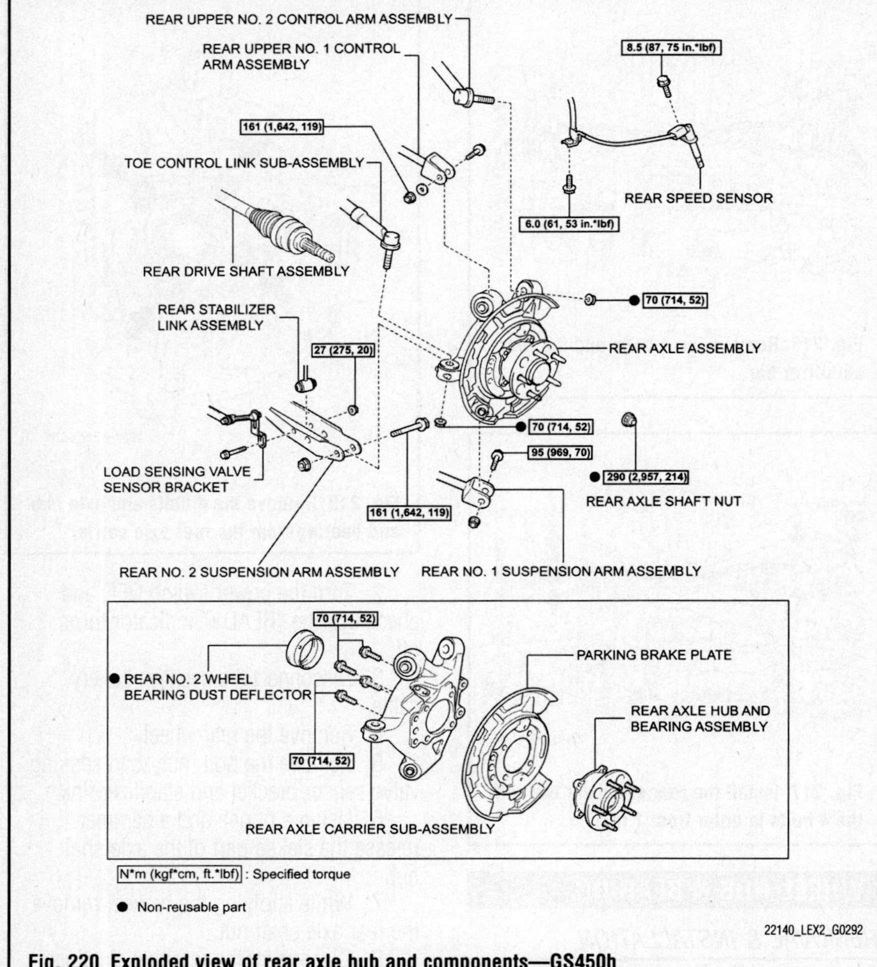

Fig. 220 Exploded view of rear axle hub and components—GS450h

20. Using a plastic-faced hammer, remove the drive shaft from the rear axle carrier.
21. Remove the rear axle.
22. Using a screwdriver, remove the bearing dust deflector from the rear axle carrier.
23. Hold the axle hub and bearing in a vise between aluminum plates.
24. Remove the 4 bolts and axle hub and bearing from the rear axle carrier.

To install:

25. Hold the axle hub and bearing in a vise between aluminum plates.
26. Install the axle hub and bearing to the rear axle carrier with the 4 bolts. Tighten to 52 ft. lbs. (70 Nm).
27. Using a seal driver and a hammer, install the No. 2 bearing dust deflector to the rear axle carrier.
28. Install the rear drive shaft to the rear axle.
29. Install the rear No. 2 upper control arm to the rear axle carrier with a new nut. Tighten to 52 ft. lbs. (70 Nm).
30. Temporarily tighten the rear upper control arm to the rear axle carrier with the bolt, washer and nut.
31. Temporarily tighten the rear suspension arm to the rear axle carrier with the bolt and nut.
32. Temporarily tighten the rear suspension arm to the rear axle carrier with the bolt and nut.
33. Connect the toe control link to the rear axle carrier with a new nut. Tighten to 52 ft. lbs. (70 Nm).
34. Install the stabilizer link and the load sensing valve sensor bracket to the rear suspension arm with the bolt and nut. Tighten to 20 ft. lbs. (27 Nm).
35. Apply high temperature grease to backing plate.
36. Install the parking brake anchor block to the rear axle carrier with the 2 nuts. Tighten the nuts to 56 ft. lbs. (76 Nm).
37. Install parking brake shoes.
38. Install the rear speed sensor. Tighten the mounting bolt to 75 inch lbs. (8.5 Nm).
39. While applying the brakes, install a new rear axle shaft nut. Tighten to 214 ft. lbs. (290 Nm).
40. Install the rear brake disc.
41. Using a chisel and a hammer, stake the axle shaft nut.
42. Stabilize the suspension.
43. Tighten the rear upper control arm with the nut. Tighten to 119 inch lbs. (161 Nm).
44. Tighten the rear suspension arm with the bolt and nut. Tighten to 70 ft. lbs. (95 Nm).
45. Tighten the rear suspension arm with the bolt and nut. Tighten 119 inch lbs. (161 Nm).
46. Install the rear wheel and tighten to 76 ft. lbs. (103 Nm).
47. Inspect and adjust parking brake travel.
48. Inspect and adjust the rear wheel alignment.
49. Verify rear speed sensor operation.

ADJUSTMENT

See Figure 221.

Check the backlash in bearing shaft direction and the axle hub deviation. Maximum for backlash should be 0.0020 in. (0.05mm) and for axle hub deviation 0.020 in. (0.05mm).

➡**The rear wheel bearings are non-adjustable. If the wheel bearing is out of specifications, replace the wheel bearing.**

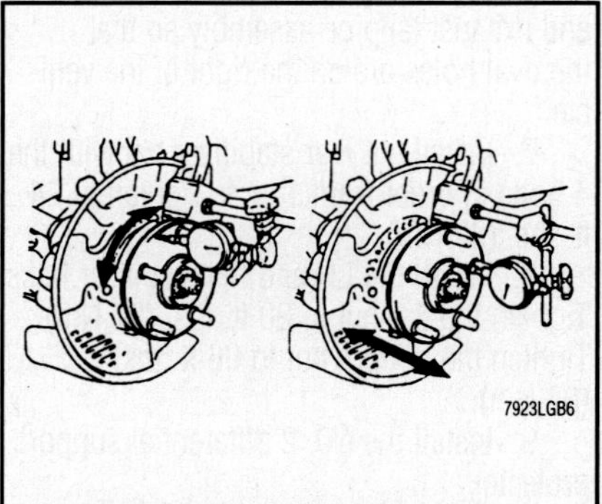

Fig. 221 Checking wheel bearings for excessive play

LEXUS

GX470

26

SPECIFICATIONS AND MAINTENANCE CHARTS

ENGINE AND VEHICLE IDENTIFICATION

Engine							Model Year	
Code ①	Liters (cc)	Cu. In.	Cyl.	Fuel Sys.	Engine Type	Eng. Mfg.	Code ②	Year
2UZ-FE	4.7 (4664)	285	8	SFI	DOHC	Toyota	9	2009

SFI: Sequential Fuel Injection

DOHC: Double Overhead Camshaft

① Stamped on the left side of the engine block

② 10th digit of the Vehicle Identification Number (VIN)

3768X_GX47_C0001

GENERAL ENGINE SPECIFICATIONS

Year	Model	Engine Displacement Liters	Engine Series ID	Net Horsepower @ rpm	Net Torque @ rpm (ft. lbs.)	Bore x Stroke (in.)	Com-pression Ratio	Oil Pressure @ rpm
2009	GX470	4.7	2UZ-FE	260@5400	306@3400	3.70x3.31	10:01	45-65@3000

3768X_GX47_C0002

ENGINE TUNE-UP SPECIFICATIONS

Year	Engine Displacement Liters	Engine ID	Spark Plug Gap (in.)	Ignition Timing (deg.)*	Fuel Pump (psi)	Idle Speed (rpm)		Valve Clearance	
						MT	AT	Intake	Exhaust
2009	4.7	2UZ-FE	0.043	NA	38-44	—	650-750	0.006-0.010	0.010-0.014

NOTE: The Vehicle Emission Control Information label often reflects specification changes made during production. The label figures must be used if they differ from those in this chart.

NA: Not available

B: Before top dead center

* With terminals TC and E1 connected to DLC1 or for 5.7L Terminal TC and CG of DLC3 connected

3768X_GX47_C0003

CAPACITIES

Year	Model	Engine Displacement Liters	Engine ID	Engine Oil with Filter (qts.)	Transmission (qts.)		Transfer Case (pts.)	Drive Axle		Fuel Tank (gal.)	Cooling System (qts.)
					5-Spd	Auto.*		Front (pts.)	Rear (pts.)		
2009	GX470	4.7	2UZ-FE	6.5	—	11.0	3.0	3.9	6.4	23.0	13.6

*After draining, add the following amounts, then fill to the cold full line

3768X_GX47_C0004

FLUID SPECIFICATIONS

Year	Model	Engine Displacement Liters	Engine ID/VIN	Engine Oil	Auto. Trans.	Power Steering Fluid	Brake Master Cylinder
2009	GX470	4.7	2UZ-FE	5W-30	Toyota Genuine ATF WS	ATF Dexron II Or III	DOT 3

DOT: Department Of Transpotation

3768X_GX47_C0013

VALVE SPECIFICATIONS

Year	Engine Displacement Liters	Engine ID	Seat Angle (deg.)	Face Angle (deg.)	Spring Test Pressure (lbs. @ in.)	Spring Installed Height (in.)	Stem-to-Guide Clearance (in.)		Stem Diameter (in.)	
							Intake	Exhaust	Intake	Exhaust
2009	4.7	2UZ-FE	45	44.5	47.2-50.7@ 1.378	1.380	0.0010-0.0024	0.0012-0.0026	0.2154-0.2159	0.2152-0.2157

3768X_GX47_C0005

CAMSHAFT AND BEARING SPECIFICATIONS CHART

All measurements are given in inches.

Year	Engine Displacement Liters	Engine ID/VIN	Journal Dia.	Brg. Oil Clearance	Shaft End-play	Runout	Journal Bore	Lobe Height Intake	Lobe Height Exhaust
2009	4.7	2UZ-FE	①	②	NA	0.0031	NA	1.8624-1.8664	1.8104-1.1843

NA: Not Available

① 1.0612-1.0618 in.

② 0.012-0.0028 in.

3768X_GX47_C0014

CRANKSHAFT AND CONNECTING ROD SPECIFICATIONS

All measurements are given in inches.

			Crankshaft				Connecting Rod		
Year	Engine Displacement Liters	Engine ID	Main Brg. Journal Dia.	Main Brg. Clearance	Shaft End-play	Thrust on No.	Journal Diameter	Oil Clearance	Side Clearance
2009	4.7	2UZ-FE	2.6373-2.6378	①	0.0008-0.0087	3	2.0465-2.0472	0.0011-0.0021	0.0063-0.0138

① Nos. 1 and 2: 0.0011-0.0018

All others: 0.0016-0.0023

3768X_GX47_C0006

PISTON AND RING SPECIFICATIONS

All measurements are given in inches.

				Ring Gap			Ring Side Clearance		
Year	Engine Displacement Liters	Engine ID	Piston Clearance	Top Comp.	Bottom Comp.	Oil Control	Top Comp.	Bottom Comp.	Oil Control
2009	4.7	2UZ-FE	0.0035-0.0044	0.0118-0.0157	0.0157-0.0217	0.0051-0.0150	0.0012-0.0031	0.0012-0.0028	SNUG

① No 1: 0.039, No 2: 0.043, No 3: 0.039

3768X_GX47_C0007

TORQUE SPECIFICATIONS

All readings in ft. lbs.

Year	Engine Displacement Liters	Engine ID	Cylinder Head Bolts	Main Bearing Bolts	Rod Bearing Bolts	Crankshaft Damper Bolts	Flywheel Bolts	Manifold Intake	Manifold Exhaust	Spark Plugs	Oil Pan Drain Plug
2009	4.7	2UZ-FE	①	②	③	181	④	13	33	13	29

① Step 1: 30 ft. lbs.
Step 2: Plus 90 degrees
Step 3: Plus 90 degrees

② Step 1: 20 ft. lbs.
Step 2: Plus 90 degrees

③ Step 1: 18 ft. lbs.
Step 2: Plus 90 degrees

④ Step 1: 22 ft. lbs.
Step 2: Plus 90 degrees

3768X_GX47_C0008

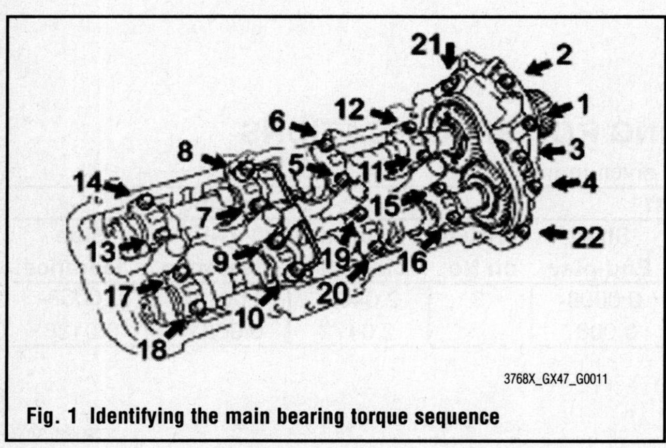

Fig. 1 Identifying the main bearing torque sequence

WHEEL ALIGNMENT

Year	Model	Caster Range (+/-Deg.)	Caster Preferred Setting (Deg.)	Camber Range (+/-Deg.)	Camber Preferred Setting (Deg.)	Toe-in (in.)	Steering Axis Inclination (Deg.)
2009	GX470	0.75	+3.28	0.75	-0.02	0.08+/-0.16	12.48+/-0.75

Note: All alignment specifications are based on nominal ride height and standard tires

① 2WD except air suspension +3.38
2WD with air suspension +3.55
4WD except air suspension +3.22
4WD with air suspension +3.37

② 2WD except air suspension -0.47
2WD with air suspension -0.50
4WD except air suspens
4WD with air suspension -0.17

③ 2WD except air suspension 12.97+/-.075
2WD with air suspension 13.00+/-.075
4WD except air suspension 12.65+/-.075
4WD with air suspension 12.67+/-.075

3768X_GX47_C0009

TIRE, WHEEL AND BALL JOINT SPECIFICATIONS

Year	Model	OEM Tires		Tire Pressures (psi)		Wheel	Ball Joint	Lug Nut Torque
		Standard	Optional	Front	Rear	Size	Inspection	(ft. lbs.)
2009	GX470	P265/65SR17	None	①	①	7.5JJ	②	83

OEM: Original Equipment Manufacturer

① See placard on vehicle

② Upper arm ball joint turning torque: 40 inch lbs.
Lower arm ball joint turning torque: 27 inch lbs.

3768X_GX47_C0010

BRAKE SPECIFICATIONS

All measurements in inches unless noted

Year	Model		Brake Disc			Minimum Lining Thickness	Brake Caliper	
			Original Thickness	Minimum Thickness	Maximum Runout		Bracket Bolts (ft. lbs.)	Mounting Bolts (ft. lbs.)
2009	GX 470	F	1.102	1.024	0.0020	0.039	—	91
		R	0.709	0.630	0.0079	0.039	77	65

F: Front

R: Rear

3768X_GX47_C0011

SCHEDULED MAINTENANCE INTERVALS

LEXUS GX470

TO BE SERVICED	TYPE OF SERVICE	VEHICLE MILEAGE INTERVAL (x1000)													
		5	10	15	20	25	30	35	40	45	50	55	60	90	120
Engine oil & filter	R	✓	✓	✓	✓	✓	✓	✓	✓	✓	✓	✓	✓	✓	✓
Automatic transmission fluid	S/I			✓			✓			✓			✓	✓	✓
Ball joints & dust covers	S/I			✓			✓			✓			✓	✓	✓
Bolts & nuts on chassis & body	S/I			✓			✓			✓			✓	✓	✓
Brake linings & drums	S/I	✓	✓	✓	✓	✓	✓	✓	✓	✓	✓	✓	✓	✓	✓
Brake line pipes & hoses	S/I			✓			✓			✓			✓	✓	✓
Brake pads & discs (front & rear)	S/I	✓	✓	✓	✓	✓	✓	✓	✓	✓	✓	✓	✓	✓	✓
Brake fluid	R						✓						✓	✓	✓
Rack and pinion assembly	S/I			✓			✓			✓			✓	✓	✓
Steering linkage & boots	S/I			✓			✓			✓			✓	✓	✓
Air cleaner filter	R						✓						✓	✓	✓
Spark plugs ①	R														✓
Drive belts	S/I												✓	✓	✓
Exhaust pipes & mountings	S/I			✓			✓			✓			✓	✓	✓
Fuel lines & connections	S/I						✓						✓	✓	✓
Engine coolant ②	S/I			✓			✓			✓			✓	✓	
Rear differential & transfer case oil	S/I			✓			✓			✓			✓	✓	✓
Fuel tank cap gasket	S/I						✓						✓	✓	✓
Rotate tires	S/I			✓			✓			✓			✓		✓
Clean air conditioning filter ③	S/I			✓			✓			✓			✓		✓
Axle shaft bolts	S/I			✓			✓			✓			✓	✓	✓
runout	S/I						✓						✓	✓	✓

R: Replace S/I: Service or Inspect

① Spark plugs are replaced at 120,000 miles

② Replace engine coolant at 100,000 miles and then inspect every 15,000 miles

③ Replace air conditioning filter every 30,000 miles

FREQUENT OPERATION MAINTENANCE (SEVERE SERVICE)

If a vehicle is operated under any of the following conditions it is considered severe service:

- Extremely dusty areas.

- 50% or more of the vehicle operation is in 32°C (90°F) or higher temperatures, or constant temperatures below 0°C (32°F).

- Prolonged idling (vehicle operation in stop and go traffic).

- Frequent short running periods (engine does not warm to normal operating temperatures).

- Police, taxi, delivery usage or trailer towing usage.

Air cleaner filter: service or inspect every 5000 miles

Rear differential & transfer case oil: replace every 15,000 miles.

Ball joints & dust covers: service or inspect every 5000 miles.

Bolts & nuts on chassis & body: service or inspect every 5000 miles.

Axle shaft bolts: service or inspect every 5000 miles.

Steering linkage: service or inspect every 5000 miles.

3768X_GX47_C0012

BRAKES — INFORMATION AND PRECAUTIONS

ANTI-LOCK SYSTEMS

- Certain components within the ABS system are not intended to be serviced or repaired individually.
- Do not use rubber hoses or other parts not specifically specified for and ABS system. When using repair kits, replace all parts included in the kit. Partial or incorrect repair may lead to functional problems and require the replacement of components.
- Lubricate rubber parts with clean, fresh brake fluid to ease assembly. Do not use shop air to clean parts; damage to rubber components may result.
- Use only DOT 3 brake fluid from an unopened container.
- If any hydraulic component or line is removed or replaced, it may be necessary to bleed the entire system.
- A clean repair area is essential. Always clean the reservoir and cap thoroughly before removing the cap. The slightest amount of dirt in the fluid may plug an orifice and impair the system function. Perform repairs after components have been thoroughly cleaned; use only denatured alcohol to clean components. Do not allow ABS components to come into contact with any substance containing mineral oil; this includes used shop rags.
- The Anti-Lock control unit is a microprocessor similar to other computer units in the vehicle. Ensure that the ignition switch is **OFF** before removing or installing controller harnesses. Avoid static electricity discharge at or near the controller.
- If any arc welding is to be done on the vehicle, the control unit should be unplugged before welding operations begin.

DISC AND DRUM SYSTEMS

CAUTION

Dust and dirt accumulating on brake parts during normal use may contain asbestos fibers from production or aftermarket brake linings. Breathing excessive concentrations of asbestos fibers can cause serious bodily harm. Exercise care when servicing brake parts. Do not sand or grind brake lining unless equipment used is designed to contain the dust residue. Do not clean brake parts with compressed air or by dry brushing. Cleaning should be done by dampening the brake components with a fine mist of water, then wiping the brake components clean with a dampened cloth. Dispose of cloth and all residue containing asbestos fibers in an impermeable container with the appropriate label. Follow practices prescribed by the Occupational Safety and Health Administration (OSHA) and the Environmental Protection Agency (EPA) for the handling, processing, and disposing of dust or debris that may contain asbestos fibers.

BRAKES — BLEEDING THE BRAKE SYSTEM

BLEEDING PROCEDURE

BLEEDING PROCEDURE

➡If any work is done on the brake system or if air is suspected in the brake lines, bleed the air from the system.

➡Do not let brake fluid remain on a painted surface. Wash it off immediately.

1. Before servicing the vehicle, refer to the Precautions section.
2. Check the fluid level in the reservoir after bleeding each wheel. Add DOT3 fluid, if necessary.
3. If the hydraulic brake booster was disassembled or if the reservoir becomes empty, bleed the air from the hydraulic brake booster as follows:

➡Perform this step only if the brake booster with accumulator pump assembly is removed and/or installed.

a. Turn the ignition switch OFF, depress the brake pedal 20 times or more to release the pressure from the accumulator.

b. Fully depress the brake pedal 10 times.

c. Turn the ignition switch to the ON position and start the brake booster pump.

d. Make sure the pump operates for 8 to 14 seconds.

➡If the pump does not operate as specified, repeat the above and recheck the operating time.

4. Bleeding Front Brake Lines:

a. Turn the ignition switch to the ON position and wait until the pump motor has stopped.

b. Connect the vinyl tube to the brake caliper.

c. Depress the brake pedal several times, then loosen the bleeder plug with the pedal held down.

d. At the point when the fluid stops coming out, tighten the bleeder plug, 8 ft. lbs. (11 Nm) then release the brake pedal.

e. Repeat procedure until all the air in the fluid has been bled out.

f. Repeat the above procedures to bleed the other brake line.

5. Bleeding Rear Brake Lines:

a. Turn the ignition switch to the ON position and depress the brake pedal.

b. Connect the vinyl tube to the brake caliper.

c. Loosen the bleeder plug and release air.

➡Brake fluid is sent through the pump, so keep the brake pedal depressed until the air is completely bled out.

d. When the air is completely bled out of the brake fluid through the bleeder plug, tighten the bleeder plug to 8 ft. lbs. (11 Nm) then release.

e. Repeat the above procedures to bleed the other brake line.

6. Bleeding Master Cylinder Solenoid is only possible with a Toyota proprietary scan system.

WHEEL SPEED SENSORS

REMOVAL & INSTALLATION

Front

See Figures 2 and 3.

➡Replacement of RH side is same as that of LH side.

1. Disconnect negative battery cable.
2. Remove front wheel.
3. Disconnect the speed sensor connector.
4. Using a hexagon wrench (5 mm), remove the bolt and front speed sensor from the steering knuckle.

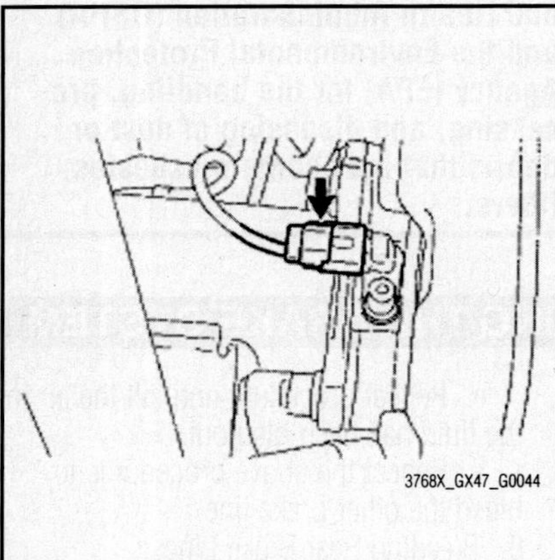

Fig. 2 Disconnecting the speed sensor connector

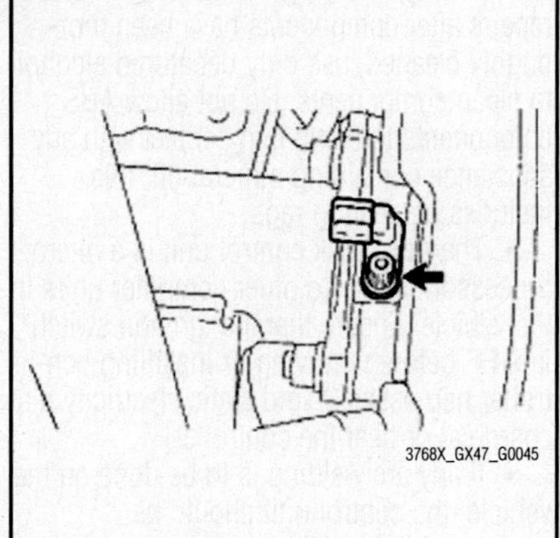

Fig. 3 Removing the bolt and front speed sensor from the steering knuckle/hub

➡Do not stick and foreign matter on the sensor tip. Do not let the foreign matter into the sensor installation hole.

To install:

➡Make sure the sensor tip is clean.

5. To install, Using a hexagon wrench (5 mm), install the front speed sensor with the bolt to the steering knuckle and torque to 73 inch lbs. (8.3 Nm).
6. Connect the speed sensor connector.
7. Install front wheel and tighten to 82 ft. lbs. (112 Nm).
8. Connect negative battery cable.
9. Check speed sensor signal.

Rear

See Figures 2 and 3.

➡Replacement of RH side is same as that of LH side.

1. Disconnect negative battery cable.
2. Remove rear wheel.
3. Disconnect the speed sensor connector.
4. Remove the bolt and front speed sensor from the axle hub.
5. Disconnect retaining clips.

➡Do not stick and foreign matter on the sensor tip. Do not let the foreign matter into the sensor installation hole.

To install:

➡Make sure the sensor tip is clean.

6. To install, install the front speed sensor with the bolt to the axle hub and torque to 73 inch lbs. (8.3 Nm).
7. Connect the speed sensor connector.
8. Connect retaining clips.
9. Install rear wheel and tighten to 82 ft. lbs. (112 Nm).
10. Connect negative battery cable.
11. Check speed sensor signal.

BRAKE CALIPER

REMOVAL & INSTALLATION

See Figure 4.

1. Remove the wheel.
2. Remove the anti-rattle spring from the caliper.
3. Remove the clips and anti-rattle pins.
4. Lift out the pads and shims.
5. If the caliper is being replaced, disconnect the brake line. Plug the line to prevent fluid loss.
6. Remove the caliper mounting bolts. Lift off the caliper.

To install:

7. Installation is the reverse of removal. Bleed the brakes. Observe the following torques:
   - Caliper mounting bolts: 91 ft. lbs. (123 Nm)
   - Brake line-to-caliper: 11 ft. lbs. (15 Nm)

DISC BRAKE PADS

REMOVAL & INSTALLATION

See Figure 4.

1. Raise the vehicle and support it safely.
2. Remove the wheels.
3. Remove the clip, pins and anti-rattle spring.
4. Withdraw the pads and remove the anti-squeal shims.

To install:

5. Before installing the new pads, check the disc thickness and disc runout.
6. Siphon out a small amount of brake fluid from the reservoir.
7. Press in the pistons with a hammer handle or equivalent.
8. Apply disc brake grease to both sides of the inner anti-squeal shim. Install the anti-squeal shims to the new pads.
9. Install the pads.
10. Install the anti-rattle springs and pins. Install the clip.
11. Install the wheels.
12. Check and adjust the fluid level. Apply the brake pedal several times.
13. Road-test the vehicle for proper operation.

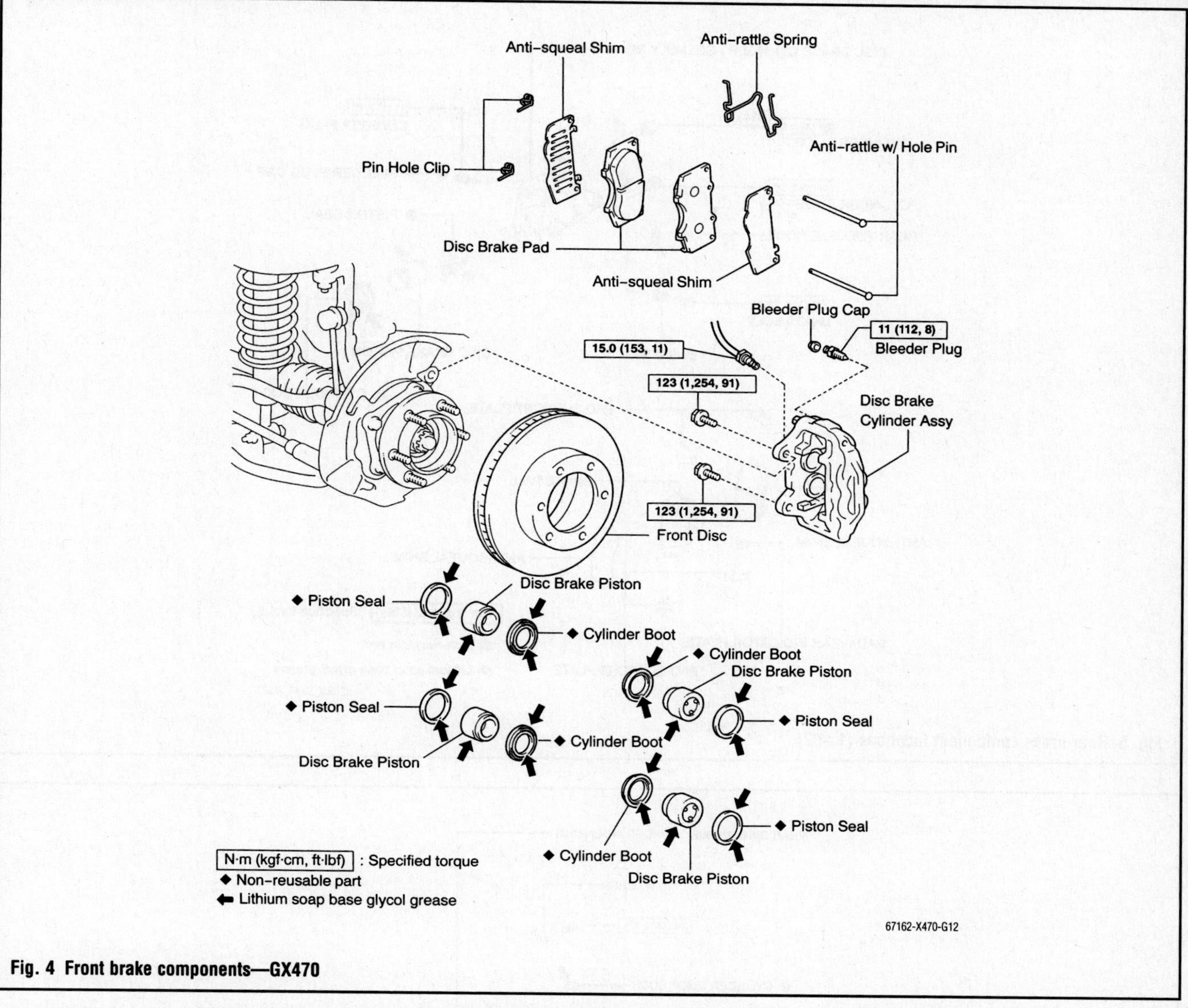

Fig. 4 Front brake components—GX470

BRAKES

REAR DISC BRAKES

BRAKE CALIPER

REMOVAL & INSTALLATION

See Figures 5 and 6.

1. Remove the wheel.
2. Remove the anti-rattle spring from the caliper.
3. Remove the clips and anti-rattle pins.
4. Lift out the pads and shims.
5. If the caliper is being replaced, disconnect the brake line. Plug the line to prevent fluid loss.
6. Remove the caliper mounting bolts. Lift off the caliper.

To install:

7. Installation is the reverse of removal. Bleed the brakes as necessary. Observe the following torques:
   - Caliper mounting bolts: 65 ft. lbs. (88 Nm)
   - Brake line-to-caliper: 23 ft. lbs. (31 Nm)

DISC BRAKE PADS

REMOVAL & INSTALLATION

See Figures 5 and 6.

1. Raise the vehicle and support it safely.
2. Remove the wheels.
3. Remove the brake caliper and suspend it so the hose is not stretched.
4. Remove the brake pads, anti-squeal shim, pad support plates and wear indicators.

To install:

5. Before installing the new pads, check the disc thickness and disc runout.
6. Install the pad support plates.
7. Install the pad wear indicator plates on each pad.
8. Install the anti-squeal shim to the outer pad. Install the pads.
9. Install the brake caliper and tighten to 65 ft. lbs. (88 Nm).
10. Install the wheels.
11. Apply the brake pedal several times.
12. Road-test the vehicle for proper operation.

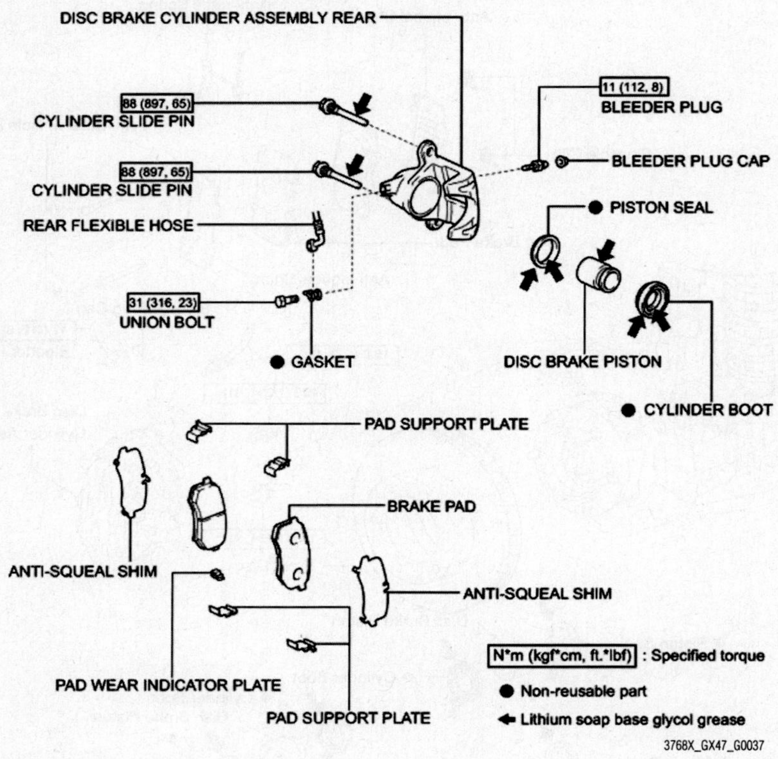

Fig. 5 Rear brake component locations (1 of 2)

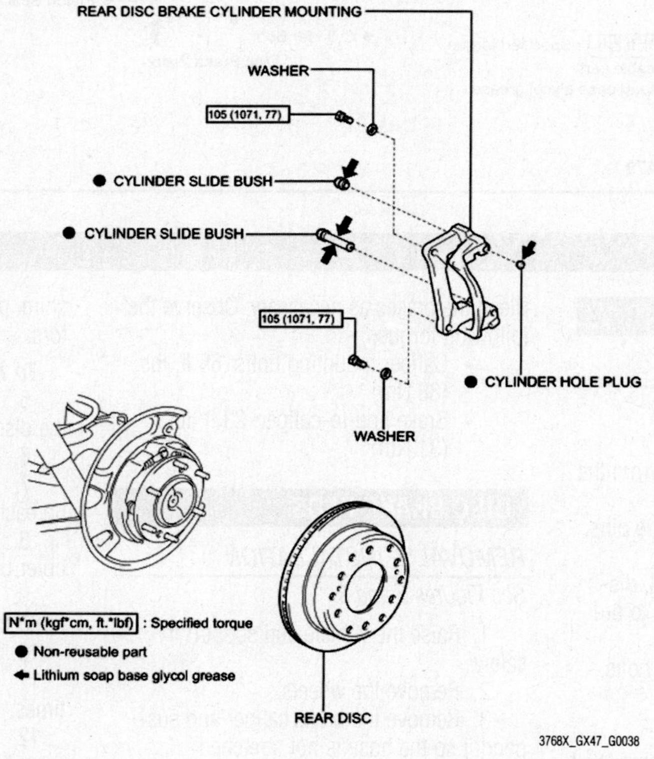

Fig. 6 Rear brake component locations (2 of 2)

PARKING BRAKE CABLES

ADJUSTMENT

See Figure 7.

1. Remove rear wheel.
2. Adjust parking brake shoe clearance.
3. Install rear wheel an tighten to 82 ft. lbs. (112 Nm)
4. Inspect parking brake lever travel.
5. Slowly depress the parking brake lever all the way, and count the number of clicks.
   a. Parking brake lever travel at 66 ft. lbs. (294 Nm) 5 to 7 clicks.

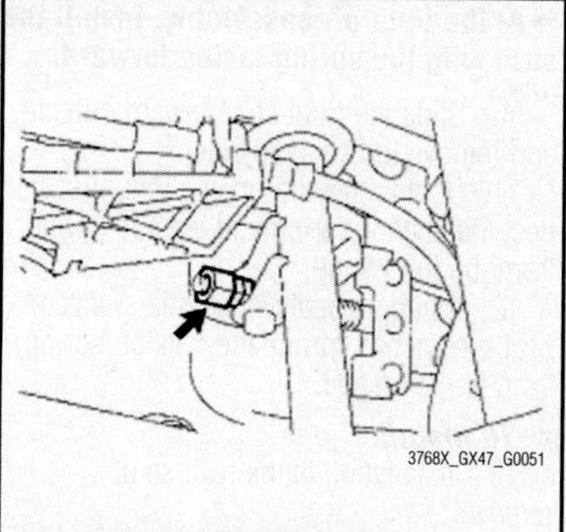

Fig. 7 Locating the adjusting nut

6. Adjust parking brake lever travel by removing the console panel upper.
7. Turn the adjusting nut until the parking brake lever travel becomes correct.
8. Check whether parking brake drags or not.
9. When operating the parking brake lever, check that the parking brake lever indicator light comes on.
10. Install the console panel upper.

PARKING BRAKE SHOES

REMOVAL & INSTALLATION

See Figure 8.

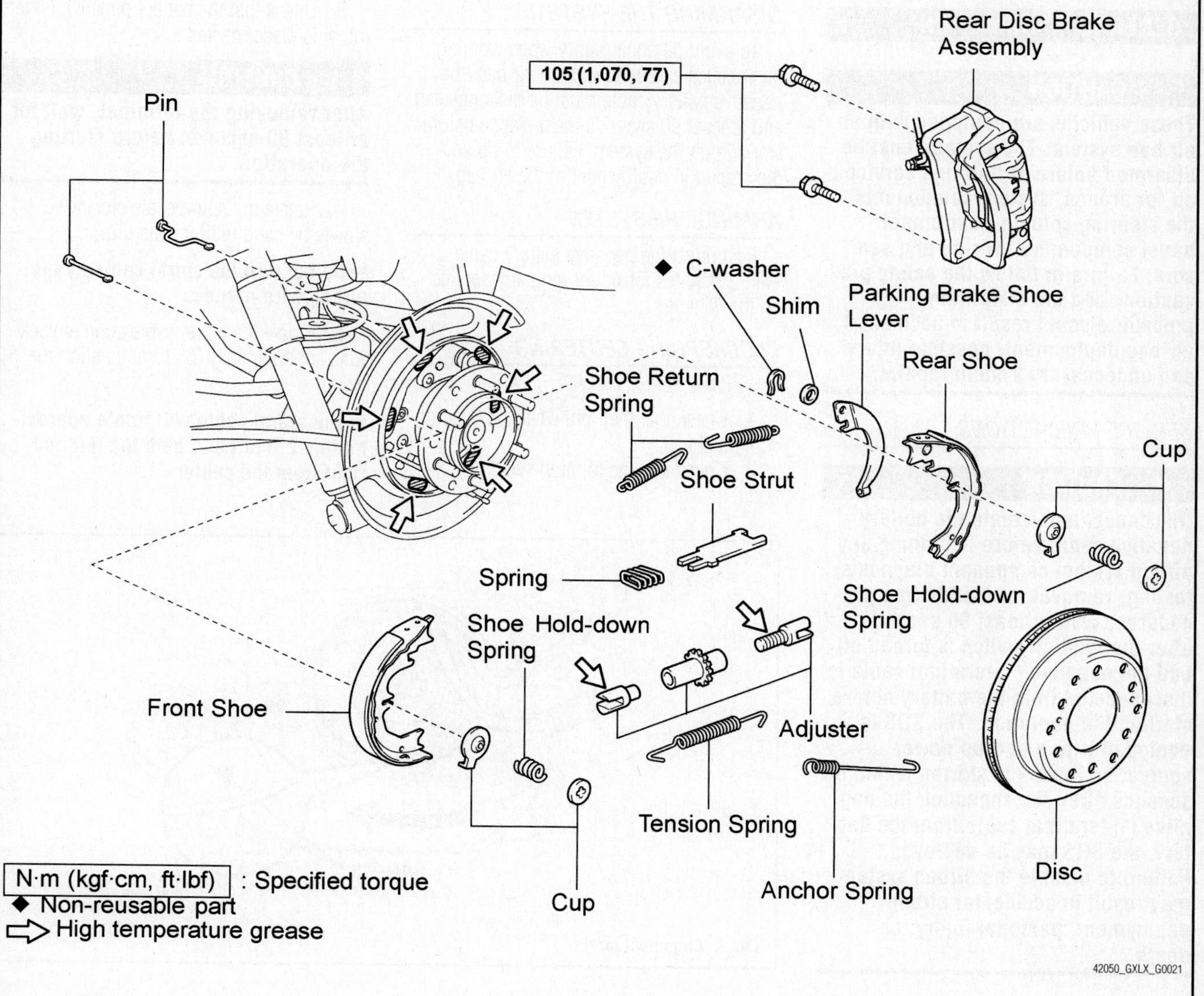

Fig. 8 Exploded view of the parking brake

1. Before servicing the vehicle, refer to the Precautions section.
2. Raise and safely support the vehicle.
3. Remove the rear wheel.
4. Remove the 2 mounting bolts and remove the disc brake assembly.
5. Suspend the disc brake securely and so the hose is not stretched.
6. Release the parking brake lever.
7. Place matchmarks on the disc and rear axle hub.
8. Remove the disc.

➡If the disc cannot be removed easily, turn the shoe adjuster until the wheel turns freely.

9. Using needle-nose pliers, remove the 2 shoe return springs.

➡At the time of reassembly, install the strut with the spring facing forward.

10. Slide the front shoe toward outside and remove the shoe adjuster.
11. Using a needle-nose pliers, disconnect the anchor spring and tension spring from the front shoe.
12. Using a needle-nose pliers, disconnect the anchor spring and tension spring from the rear shoe.

To install:

13. Installation is the reverse of removal.

ADJUSTMENT

1. Before servicing the vehicle, refer to the Precautions section.
2. Turn the adjuster and expand the shoes until the disc locks.
3. Return the adjuster 8 notches.
4. Depress the parking brake pedal with 33 ft. lbs (147 Nm).
5. Drive the vehicle at about 31 mph (50 km/h) on a safe, level and dry road for about 0.25 mile (400 meters) in this condition.
6. Repeat this procedure 2 or 3 times.

CHASSIS ELECTRICAL

AIR BAG (SUPPLEMENTAL RESTRAINT SYSTEM)

GENERAL INFORMATION

⁂ CAUTION

These vehicles are equipped with an air bag system. The system must be disarmed before performing service on, or around, system components, the steering column, instrument panel components, wiring and sensors. Failure to follow the safety precautions and the disarming procedure could result in accidental air bag deployment, possible injury and unnecessary system repairs.

SERVICE PRECAUTIONS

⁂ CAUTION

Disconnect and isolate the battery negative cable before beginning any airbag system component diagnosis, testing, removal, or installation procedures. Wait at least 90 seconds after the ignition switch is turned off and the negative (-) terminal cable is disconnected from the battery before starting the operation. The SRS is equipped with a backup power source, so if work is started within 90 seconds after disconnecting the negative (-) terminal cable from the battery, the SRS may be deployed. Failure to disable the airbag system may result in accidental airbag deployment, personal injury, or death.

DISARMING THE SYSTEM

To avoid personal injury when working on vehicles equipped with an air bag, the negative battery cable must be disconnected and at least 90 seconds must elapse before working on the system. Failure to do so may result in deployment of the air bag.

ARMING THE SYSTEM

Reconnect the negative battery cable. Wait 2 minutes for performing any service on the vehicle.

CLOCKSPRING CENTERING

See Figure 9.

1. Check that the front wheels are facing straight ahead.
2. Check that the ignition switch is off.
3. Check that the battery negative (-) terminal is disconnected

⁂ CAUTION

After removing the terminal, wait for at least 90 seconds before starting the operation.

4. Rotate the spiral cable clockwise slowly by hand until it feels firm.

➡Do not turn the spiral cable by the airbag wire harness

5. Rotate the spiral cable counterclockwise approximately 2.5 turns to align the marks

➡The spiral cable will rotate approximately 2.5 turns to both the left and right from the center

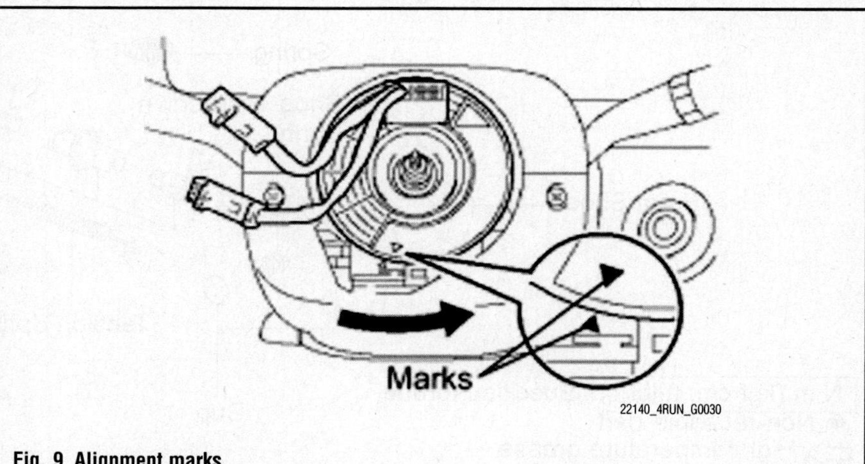

Fig. 9 Alignment marks

DRIVE TRAIN

TRANSFER CASE ASSEMBLY

REMOVAL & INSTALLATION

See Figure 10.

1. Before servicing the vehicle, refer to the Precautions section.
2. Drain the fluid.
3. Remove the skid plate.
4. Remove the transmission.
5. Remove the 8 bolts and 2 clamps from transfer assembly.
6. Separate the transfer case from the transmission.

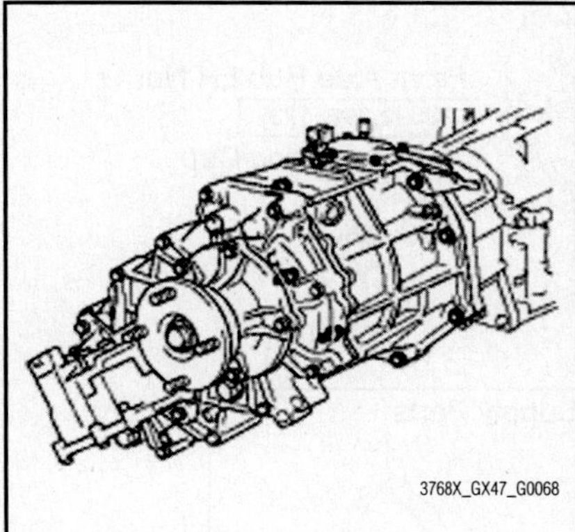

Fig. 10 Removing the transfer case

To install:

7. Installation is the reverse of removal. Torque the bolts to 17 ft. lbs. (24 Nm).

FRONT AXLE SHAFT, BEARING & SEAL

REMOVAL & INSTALLATION

Differential Side Seal

See Figure 11.

1. Remove front wheel.
2. Remove the 6 bolts and engine under cover assembly rear.
3. Remove the 4 bolts and no. 1 engine under cover sub-assembly.
4. Drain differential oil.
5. Remove front axle shaft LH nut.
6. Separate front speed sensor LH.
7. Separate tie rod end sub-assembly LH.
8. Separate no. 1 front suspension arm sub-assembly lower LH.
9. Remove front drive shaft assembly LH.
10. Using service tool, remove the oil seal.

To install:

11. Using SST and a hammer, install a new oil seal.
12. Coat the oil seal lip with MP grease.
13. Install front drive shaft assembly LH.
14. Connect no. 1 front suspension arm sub-assembly lower LH.
15. Connect tie rod end sub-assembly LH.
16. Connect front speed sensor LH.
17. Install front axle shaft LH nut and tighten to 173 ft. lbs. (235 Nm).
18. Fill up differential oil.
19. Install the no. 1 engine under cover sub-assembly with the 4 bolts.
20. Install the engine under cover assembly with the 6 bolts.
21. Install front wheel.
22. Inspect ABS speed sensor signal.

FRONT HALFSHAFT

REMOVAL & INSTALLATION

See Figures 12 and 13.

1. Before servicing the vehicle, refer to the Precautions section.
2. Remove the wheel.
3. Drain the differential oil.
4. Remove the cotter pin and cap, then remove the hub nut.
5. Remove the speed sensor wiring harness. Remove the sensor.

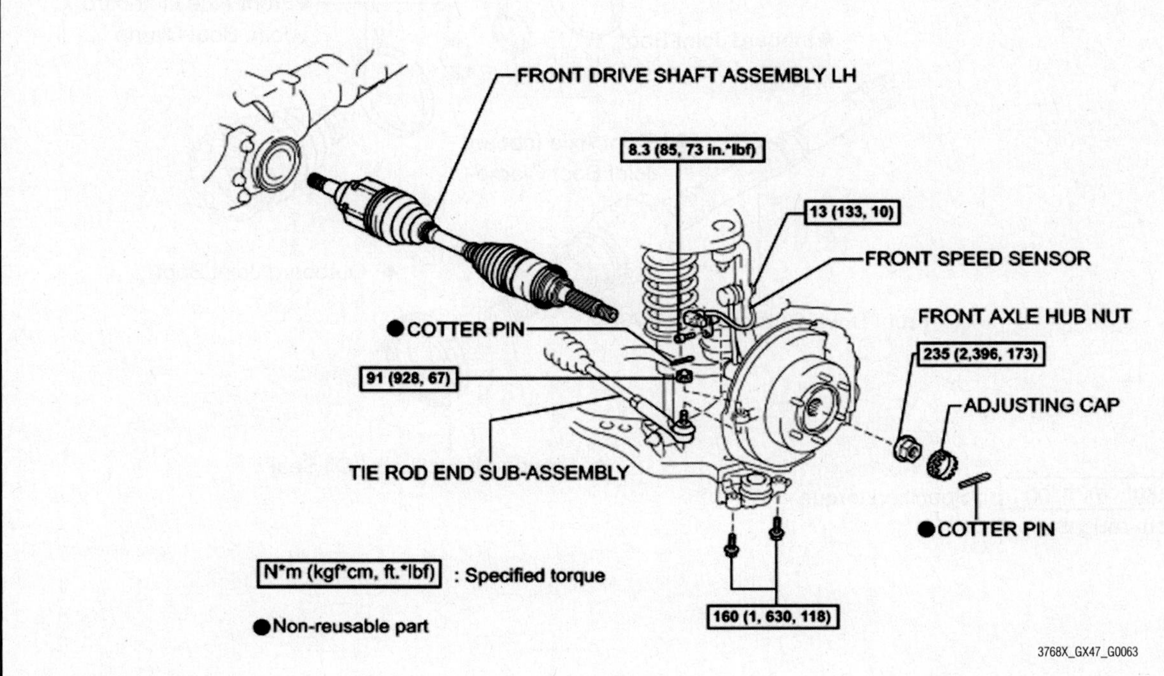

Fig. 11 Front drive shaft assembly components

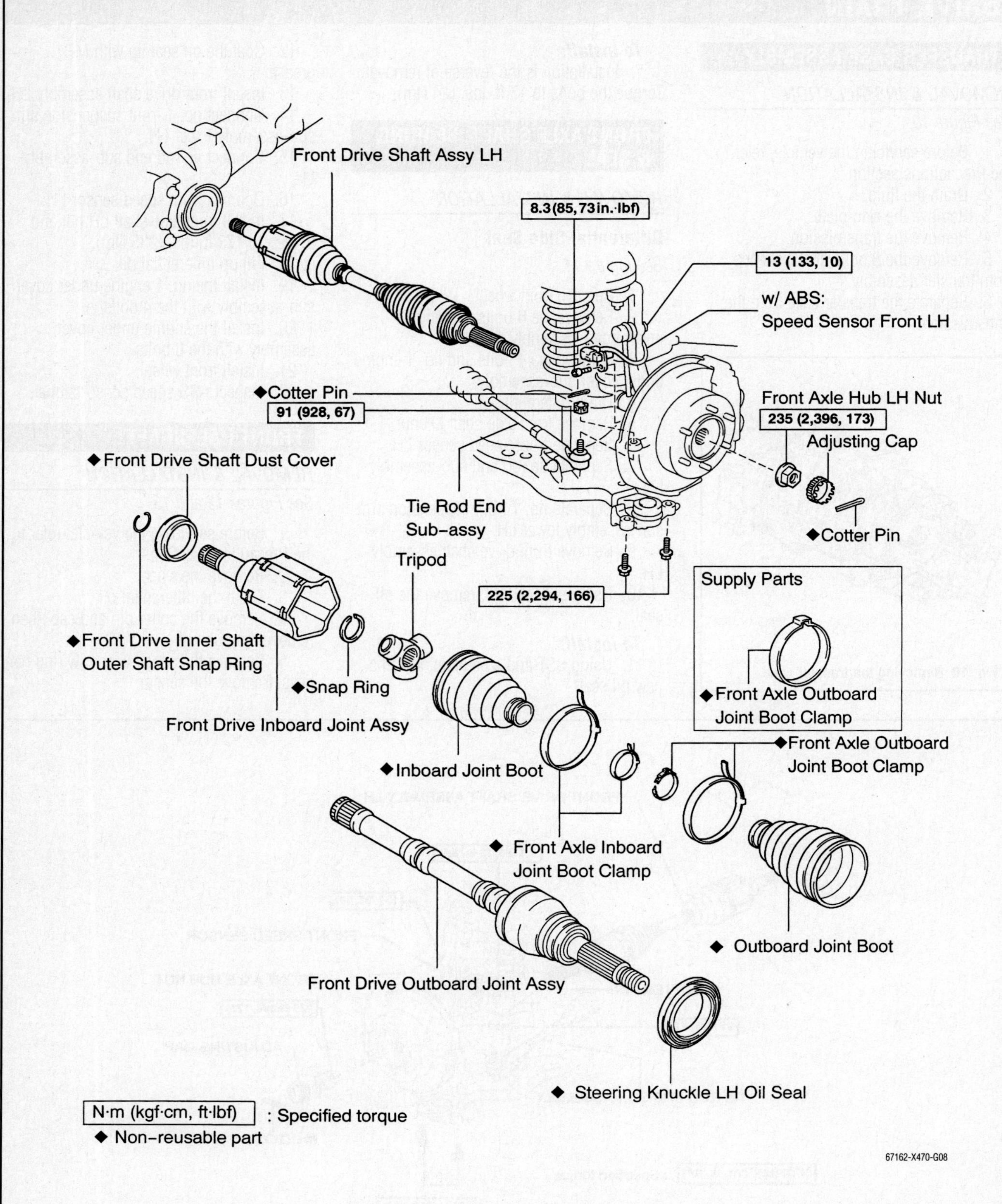

Front Drive Shaft Assy LH
8.3(85,73in.·lbf)
13 (133, 10)
w/ ABS:
Speed Sensor Front LH
◆Cotter Pin
91 (928, 67)
Front Axle Hub LH Nut
235 (2,396, 173)
Adjusting Cap
◆Front Drive Shaft Dust Cover
Tie Rod End
Sub-assy
◆Cotter Pin
Tripod
225 (2,294, 166)
Supply Parts
◆Front Drive Inner Shaft
Outer Shaft Snap Ring
◆Snap Ring
◆Front Axle Outboard
Joint Boot Clamp
Front Drive Inboard Joint Assy
◆Front Axle Outboard
Joint Boot Clamp
◆Inboard Joint Boot
◆ Front Axle Inboard
Joint Boot Clamp
◆ Outboard Joint Boot
Front Drive Outboard Joint Assy
◆ Steering Knuckle LH Oil Seal
N·m (kgf·cm, ft·lbf) : Specified torque
◆ Non-reusable part
67162-X470-G08

Fig. 12 Front halfshaft, left side shown—GX470

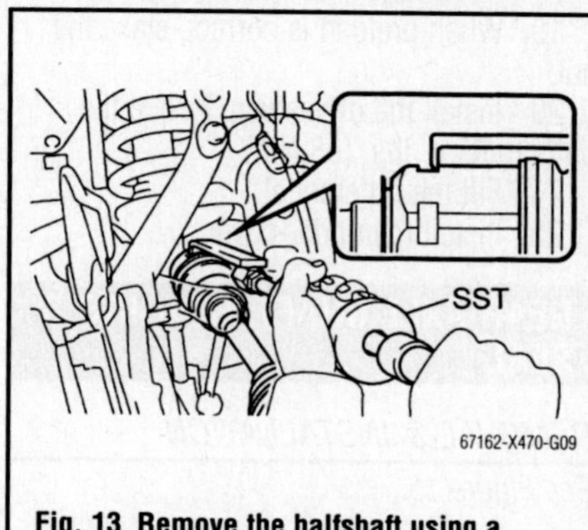

Fig. 13 Remove the halfshaft using a slide hammer and adapter

6. Remove the tie rod end from the knuckle.

7. Remove the 2 bolts and separate the lower arm from the ball joint.

8. Remove the halfshaft using a slide hammer and adapter. Keep the halfshaft level when carrying it.

To install:

9. Coat the inboard end splines of the halfshaft with clean ATF.

10. Align the splines and drive the halfshaft into place with a brass drift.

11. Install a new snapring with the opening facing down.

12. Install the sensor. Torque to 10 ft. lbs. (13 Nm). Connect the wire harness.

13. Connect the arm to the ball joint. Torque to 166 ft. lbs. (225 Nm).

14. Connect the tie rod end. Torque to 67 ft. lbs. (91 Nm). The nut can be advanced up to 60°to align the cotter pin hole.

15. Install the hub nut. Torque to 173 ft. lbs. (235 Nm). Install the cap and a new cotter pin.

16. Fill the differential.

17. Install the wheel. Torque to 83 ft. lbs. (112 Nm).

FRONT PINION SEAL

REMOVAL & INSTALLATION

See Figure 14.

1. Before servicing the vehicle, refer to the Precautions section.

2. Remove the wheels.

3. Remove the engine under-covers.

4. Remove the front driveshaft.

5. Remove the pinion nut.

6. Remove the companion flange with a puller.

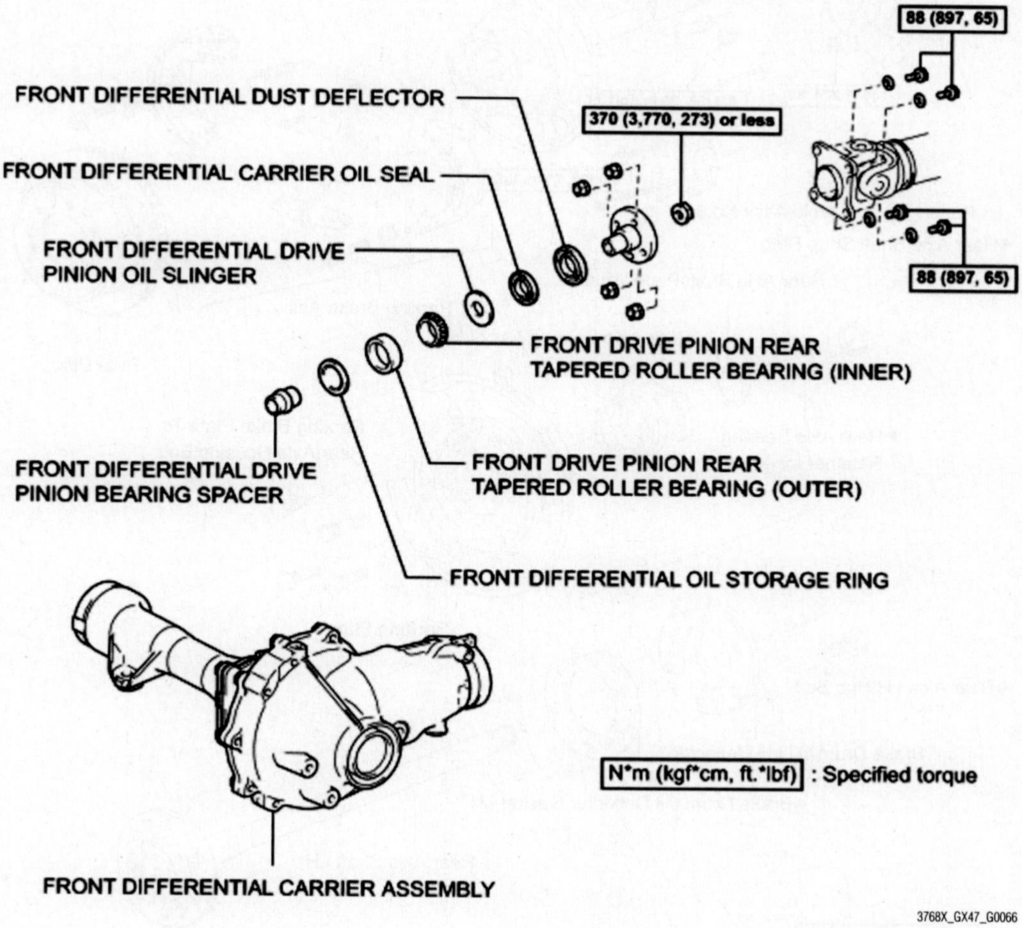

88 (897, 65)
FRONT DIFFERENTIAL DUST DEFLECTOR
370 (3,770, 273) or less
FRONT DIFFERENTIAL CARRIER OIL SEAL
FRONT DIFFERENTIAL DRIVE PINION OIL SLINGER
88 (897, 65)
FRONT DRIVE PINION REAR TAPERED ROLLER BEARING (INNER)
FRONT DIFFERENTIAL DRIVE PINION BEARING SPACER
FRONT DRIVE PINION REAR TAPERED ROLLER BEARING (OUTER)
FRONT DIFFERENTIAL OIL STORAGE RING
N*m (kgf*cm, ft.*lbf) : Specified torque
FRONT DIFFERENTIAL CARRIER ASSEMBLY
3768X_GX47_G0066

Fig. 14 Front differential carrier oil seal components

7. Remove the oil seal with a seal puller.
8. Remove the oil slinger.
9. Remove the bearing with a puller.
10. Remove the oil storage ring.
11. Remove the spacer and discard it.

To install:

12. Install a new spacer.
13. Install the oil storage ring using a brass drift.
14. Install the bearing.
15. Install the slinger.
16. Using a seal driver, install the new oil seal. Drive the seal into a depth of 4.35mm/0.45mm.
17. Install the companion flange. Coat the threads of a new flange nut with gear oil. Hold the flange and torque the nut to 273 ft. lbs. (370 Nm).
18. Using an inch-pound torque wrench, check the preload. Preload for a new bearing should be 9—14 inch lbs.; for a used bearing, 4.3–7 inch lbs. If not, a new spacer must be installed.
19. When preload is correct, stake the nut.
20. Install the driveshaft. Torque the bolts to 65 ft. lbs. (88 Nm).
21. Fill the differential.
22. Install the under-covers.

REAR AXLE SHAFT, BEARING & SEAL

REMOVAL & INSTALLATION

See Figure 15.

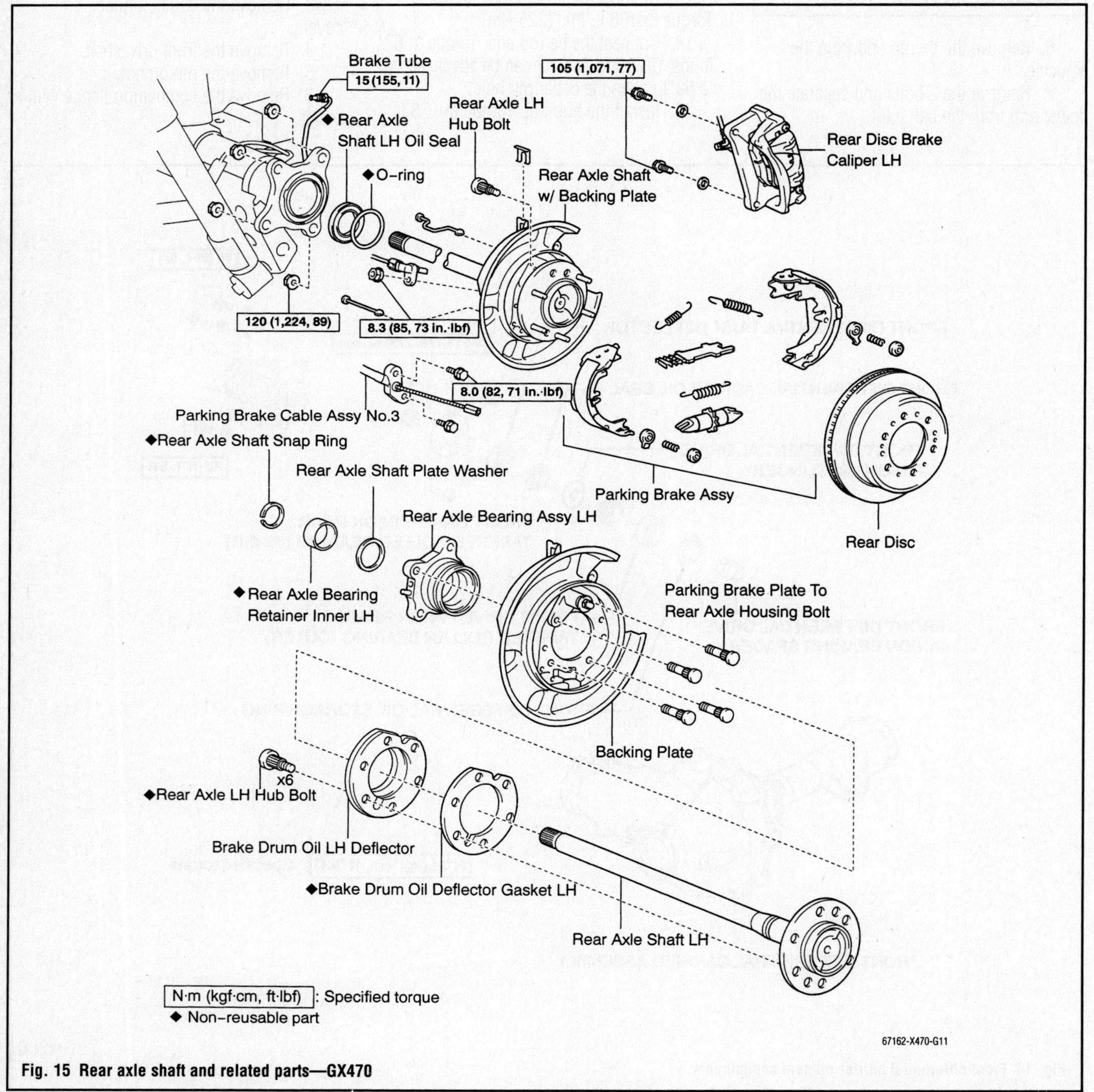

Fig. 15 Rear axle shaft and related parts—GX470

1. Remove the wheel.
2. Remove the speed sensor.
3. Remove the caliper.
4. Remove the rotor.
5. Remove the parking brake assembly.
6. Remove the 4 nuts and pull out the axle shaft with backing plate.
7. Remove the oil seal with a slide hammer.

To install:

8. Installation is the reverse of removal. Torque the nuts to 89 ft. lbs. (120 Nm).

REAR PINION SEAL

REMOVAL & INSTALLATION

See Figure 16.

1. Before servicing the vehicle, refer to the Precautions section.
2. Remove the wheels.
3. Remove the engine under-covers.
4. Remove the front driveshaft.
5. Remove the pinion nut.
6. Remove the companion flange with a puller.
7. Remove the oil seal with a seal puller.
8. Remove the oil slinger.
9. Remove the bearing with a puller.
10. Remove the oil storage ring.
11. Remove the spacer and discard it.

To install:

12. Install a new spacer.
13. Install the oil storage ring using a brass drift.
14. Install the bearing.
15. Install the slinger.
16. Using a seal driver, install the new oil seal. Drive the seal into a depth of 1.00mm / 0.45mm.
17. Install the companion flange. Coat the threads of a new flange nut with gear oil. Hold the flange and torque the nut to 273 ft. lbs. (370 Nm).
18. Using an inch-pound torque wrench, check the preload. Preload for a new bearing should be 9–15 inch lbs.; for a used bearing, 5–7.5 inch lbs. If not, a new spacer must be installed.
19. When preload is correct, stake the nut.
20. Install the driveshaft. Torque the bolts to 65 ft. lbs. (88 Nm).
21. Fill the differential.
22. Install the under-covers.

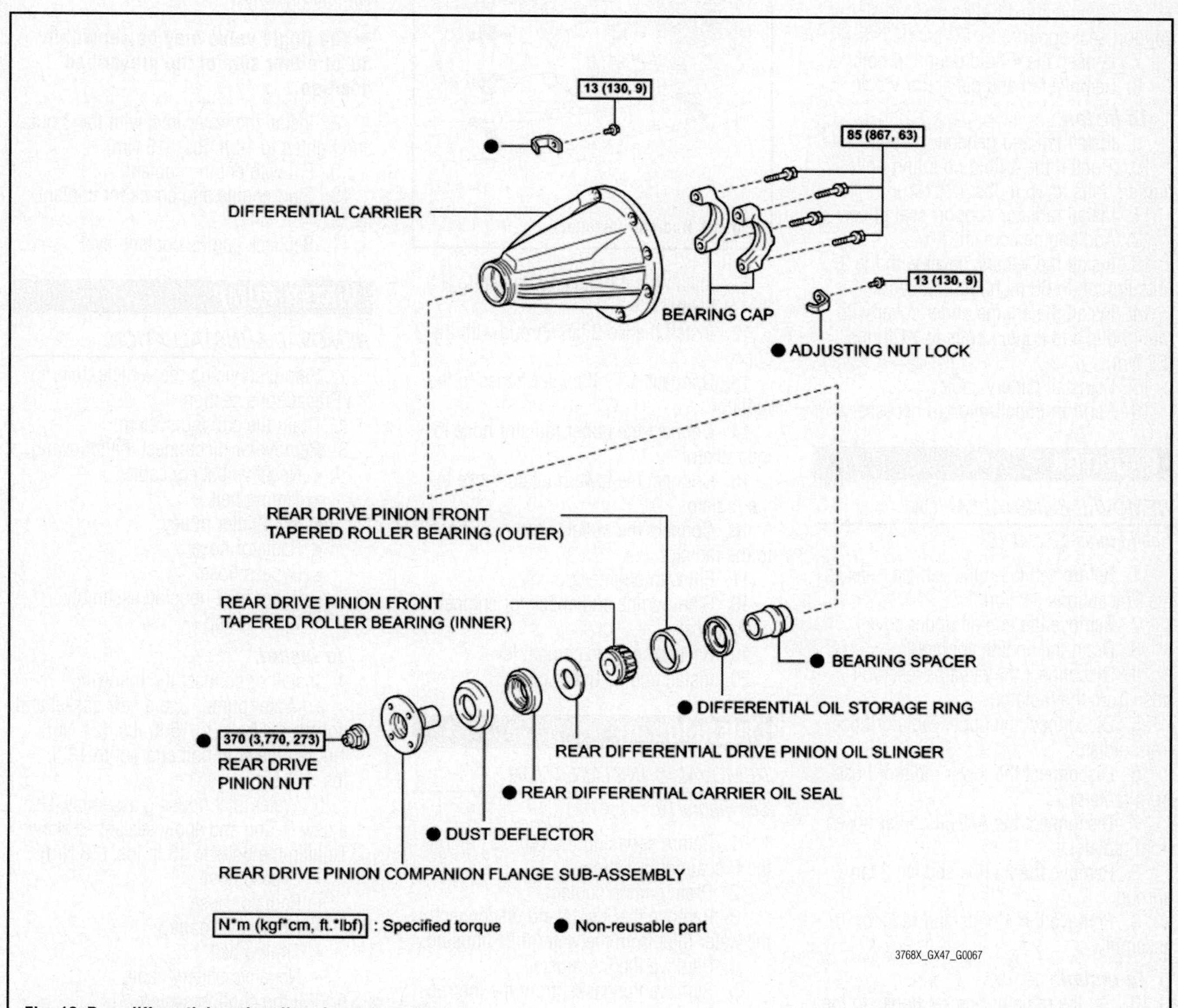

Fig. 16 Rear differential carrier oil seal components

ENGINE COOLING

ENGINE FAN

REMOVAL & INSTALLATION

1. Disconnect cable from negative battery terminal.

CAUTION

Wait at least 90 seconds after disconnecting the cable from negative (-) battery terminal to prevent airbag and seat belt pretensioner activation.

2. Remove no. 1 engine under cover.
3. Drain engine coolant.
4. Remove the 4 bolts and engine under cover.
5. Remove V-bank cover.
6. Remove the 11 clips and radiator support seal upper.
7. Loosen the 4 fluid coupling bolts.
8. Remove fan and generator v belt.

To install:

9. Install fan and generator v belt.
10. Tighten the 4 fluid coupling bolts tighten bolts to 15 ft. lbs. (21 Nm).
11. Install radiator support seal upper.
12. Add engine coolant.
13. Install the V-bank cover with the 2 nuts tighten to 66 inch lbs. (7.5 Nm).
14. Install the engine under cover with the 4 bolts and tighten bolts to 21 ft. lbs. (29 Nm).
15. Reinstall battery cable.
16. Perform initialization, if necessary.

RADIATOR

REMOVAL & INSTALLATION

See Figures 17 and 18.

1. Before servicing the vehicle, refer to the Precautions section.
2. Remove the engine under cover.
3. Drain the engine coolant.
4. Disconnect the radiator reservoir hose from the radiator.
5. Disconnect the upper radiator hose from radiator.
6. Disconnect the lower radiator hose from radiator.
7. Disconnect the A/T oil cooler hoses from radiator.
8. Remove the 2 clips and No.2 fan shroud.
9. Remove the 4 bolts and radiator assembly.

To install:

10. Set the radiator bracket hooks to the radiator support holes.

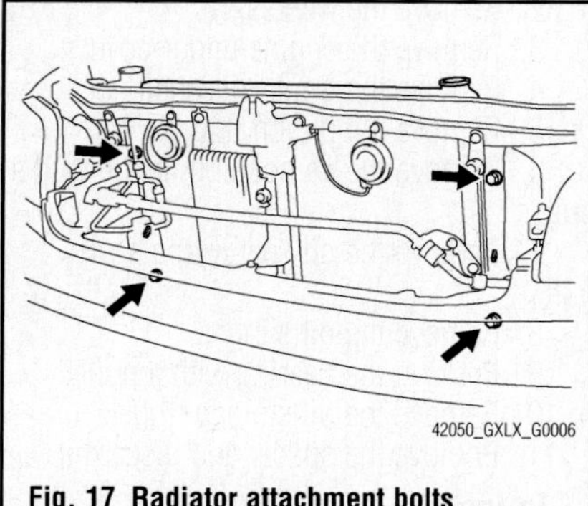

Fig. 17 Radiator attachment bolts

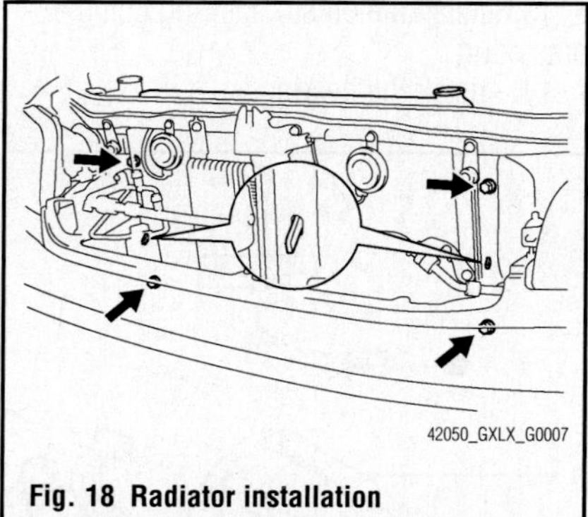

Fig. 18 Radiator installation

11. Install the 4 bolts and tighten to 9 ft. lbs. (12 Nm)
12. Install the No.2 fan shroud with the 2 clips.
13. Connect A/T oil cooler hoses to the radiator.
14. Connect the upper radiator hose to the radiator.
15. Connect the lower radiator hose to the radiator.
16. Connect the radiator reservoir hose to the radiator.
17. Fill with engine coolant.
18. Start engine and check for engine coolant leaks.
19. Recheck engine coolant level.
20. Install engine under cover.

THERMOSTAT

REMOVAL & INSTALLATION

See Figure 19.

1. Before servicing the vehicle, refer to the Precautions section.
2. Drain engine coolant.
3. Remove the 3 nuts and disconnect the water inlet from the water inlet housing.
4. Remove the thermostat.
5. Remove the gasket from the thermostat.

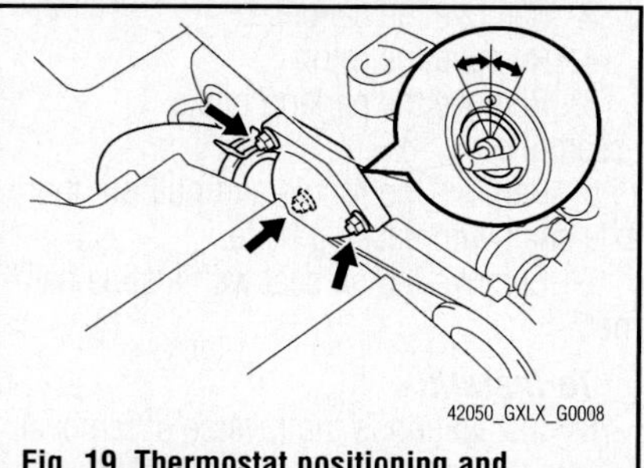

Fig. 19 Thermostat positioning and installation

To install:

6. Install a new gasket to the thermostat.
7. Insert the thermostat into the water inlet housing with the jiggle valve facing straight upward.

➡The jiggle valve may be set within 30°of either side of the prescribed position.

8. Install the water inlet with the 3 nuts and tighten to 14 ft. lbs. (19 Nm).
9. Fill with engine coolant.
10. Start engine and check for coolant leaks.
11. Recheck engine coolant level.

WATER PUMP

REMOVAL & INSTALLATION

1. Before servicing the vehicle, refer to the Precautions section.
2. Drain the cooling system.
3. Remove or disconnect the following:
   - Negative battery cable
   - Timing belt.
   - No. 2 idler pulley
   - Radiator hose
   - Bypass hose
   - Water inlet housing assembly
   - Water pump

To install:

4. Install or connect the following:
   a. Water pump. Use a new gasket and tighten the bolts to 15 ft. lbs. (21 Nm). Tighten the stud bolt and nut to 13 ft. lbs. (18 Nm).
   b. Water inlet housing assembly. Use a new O-ring and apply sealant as shown. Tighten the bolts to 13 ft. lbs. (18 Nm).
   - Bypass hose
   - Radiator hose
   - No. 2 idler pulley
   - Timing belt
   - Negative battery cable
5. Fill the cooling system.
6. Start the engine and check for leaks.

ENGINE ELECTRICAL — CHARGING SYSTEM

ALTERNATOR

REMOVAL & INSTALLATION

1. Before servicing the vehicle, refer to the Precautions section.
2. Disconnect the negative battery cable.
3. Remove the drive belt.
4. Disconnect the 2 oil cooler lines from the fan shroud, remove the fan shroud.
5. Remove the 4 nuts and remove the fan with the fluid coupling.
6. Disconnect the vane pump assembly.
7. Disconnect the alternator wiring.
8. Remove the nuts and bolts and remove the alternator.

To install:

9. Install the alternator. Tighten the bolt to 29 ft. lbs. (39 Nm), the upper nut to 29 ft. lbs. (39 Nm) and the side nut to 12 ft. lbs. (16 Nm).
10. Attach the alternator wiring.
11. Install the vane pump assembly.
12. Install the shroud and fluid coupling together and tighten shroud bolts. Tighten the fan coupling nuts to 21 ft. lbs. (29 Nm).

ENGINE ELECTRICAL — IGNITION SYSTEM

FIRING ORDER

4.0L Engine Firing order: 1–2–3–4–5–6.

4.7L Engine Firing order: 1–8–4–3–6–5–7–2.

IGNITION COIL

REMOVAL & INSTALLATION

1. Disconnect cable from negative battery terminal.

CAUTION

Wait at least 90 seconds after disconnecting the cable from the negative (-) battery terminal to prevent airbag and seat belt pretensioner activation.

2. Remove the V-bank cover.
3. Remove intake air connector pipe.
   a. Disconnect the 3 hoses.
   b. Remove the 2 bolts.
   c. Loosen the 2 clamp bolts and remove the intake air connector.
4. Disconnect the 2 engine wire clamps from the cylinder head cover LH side.
5. Disconnect the connector and remove ignition coil assembly.

To install:

6. Install the ignition coil with the bolt tighten bolt to 66 inch lbs. (7.5 Nm).
7. Connect the connector.
8. Connect the 2 engine wire clamps to the cylinder head cover LH side.
9. Install intake air connector pipe.
10. Install v-bank cover sub-assembly.
11. Connect cable to negative battery terminal.
12. Perform initialization if necessary.

IGNITION TIMING

INSPECTION

See Figures 20 and 21.

1. Warm up the engine.
2. Connect the terminals 13 (TC) and 4 (CG) of the DLC3.

➡Be sure not to connect the terminals wrongly. It causes breakage of the engine.

3. Remove the air cleaner cap sub-assembly.

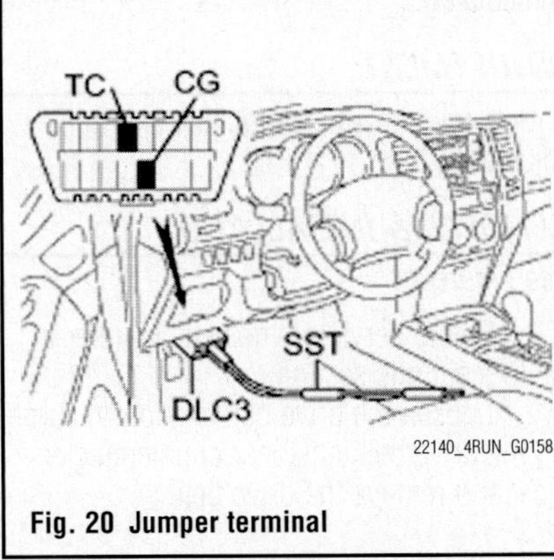

Fig. 20 Jumper terminal

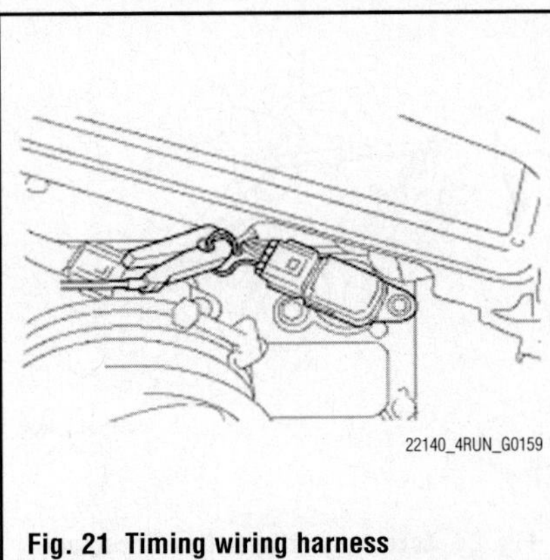

Fig. 21 Timing wiring harness

4. Pull out the wire harness as shown in the illustration.
5. Connect the tester probe of a timing light to the wire of the ignition coil connector for No. 1 cylinder.

➡Use a timing light that detects the first signal. After checking, be sure to wrap the wire harness with tape.

6. Inspect the ignition timing during idling. 8 to 12° BTDC during idling (Transmission in neutral position).
7. Remove the connector from the DLC3.
8. Inspect the ignition timing during idling. 7 to 24° BTDC during idling (Transmission in neutral position).
9. Disconnect the timing light from the engine.
10. Install the air cleaner cap sub-assembly.

ADJUSTMENT

The ignition timing is controlled by the Powertrain Control Module (PCM). No adjustment is necessary or possible.

SPARK PLUGS

REMOVAL & INSTALLATION

1. Remove the ignition coils.
2. Using a 16mm plug wrench, remove the spark plugs.
3. Clean the spark plugs.

To install:

4. Adjust the spark plug electrode gap. Electrode gap for new spark plug is 0.039–0.043 inch (1.0–1.1 mm).
5. Using a 16 mm plug wrench, install the spark plugs and tighten to 13 ft. lbs. (17.5 Nm).
6. Reinstall the ignition coils.

ENGINE ELECTRICAL

STARTER

REMOVAL & INSTALLATION

1. Before servicing the vehicle, refer to the Precautions section.
2. Drain the cooling system.
3. Relieve the fuel system pressure.
4. Remove or disconnect the following:
    - Negative battery cable
    - Engine appearance cover
    - Air intake tube
    - No1 & No2 fuel hoses
    - Ventilation hose
    - Purge VSV
    - Vacuum control valve
    - Engine wire harness
    - Throttle body water bypass hose
    - Intake manifold
    - Air pump and switching valve
    - Starter motor mounting bolts
    - Starter wiring connectors
    - Starter motor

To install:

5. Install or connect the following:
    - Starter motor
    - Starter wiring connectors: tighten the cable nut to 86 inch lbs. (10 Nm)
    - Starter motor mounting bolts: tighten the bolts to 29 ft. lbs. (39 Nm)
    - Air pump and switching valve
    - Intake manifold
    - Throttle body water bypass hose
    - Engine wire harness
    - Vacuum control valve
    - Purge VSV
    - Ventilation hose
    - No1 and No2 fuel hoses
    - Air intake tube
    - Engine appearance cover
    - Negative battery cable
6. Fill the cooling system.
7. Start the engine and check for leaks.
8. Perform initialization if necessary.

ENGINE MECHANICAL

ACCESSORY DRIVE BELTS

ACCESSORY BELT ROUTING

See Figures 22 and 23.

Refer to the accompanying illustrations for belt routing.

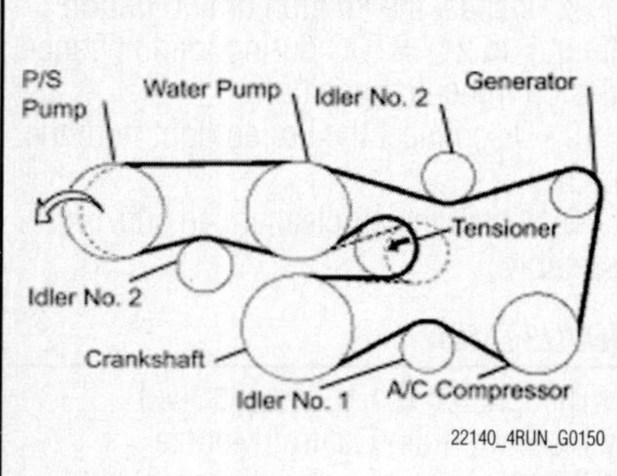

Fig. 22 Accessory drive belt routing —V6 engine

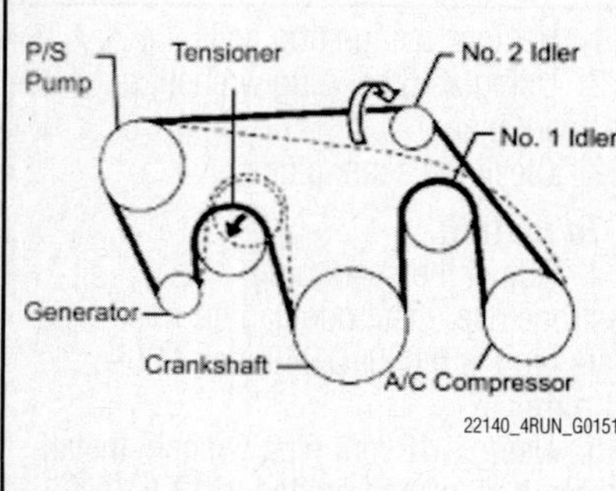

Fig. 23 Accessory drive belt routing —V8 engine

INSPECTION

Inspect the drive belt for signs of glazing or cracking. A glazed belt will be perfectly smooth from slippage, while a good belt will have a slight texture of fabric visible. Cracks will usually start at the inner edge of the belt and run outward. All worn or damaged drive belts should be replaced immediately.

ADJUSTMENT

Belt adjustment is automatic and non-adjustable.

REMOVAL & INSTALLATION

See Figure 24.

1. Before servicing the vehicle, refer to the Precautions section.
2. Loosen the drive belt tension by turning the drive belt tensioner counterclockwise, and remove the drive belt.

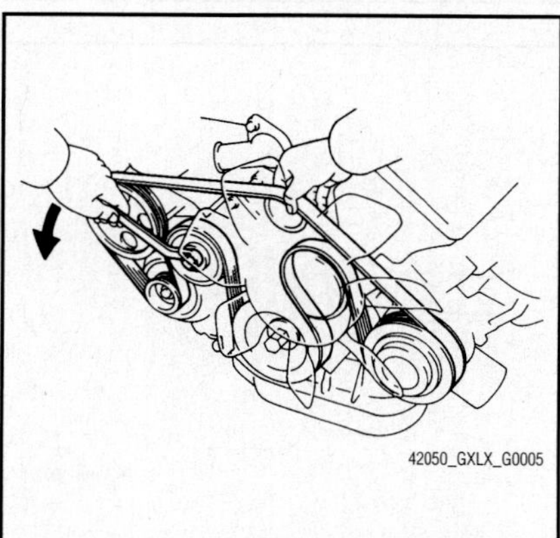

Fig. 24 Accessory drive belt replacement

To install:

3. Installation is the reverse of removal.

CAMSHAFT AND VALVE LIFTERS

REMOVAL & INSTALLATION

See Figures 25 through 39.

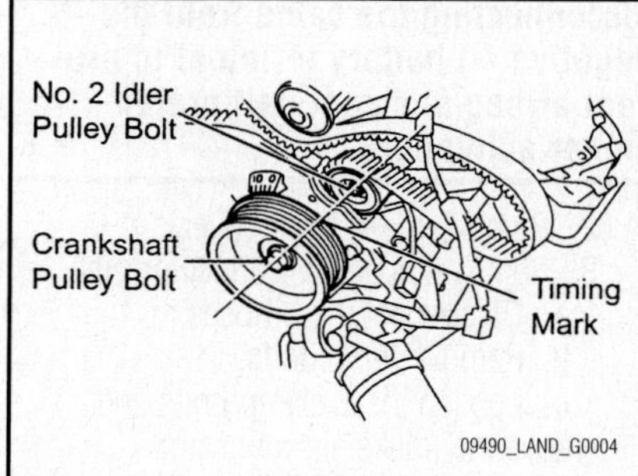

Fig. 25 Check the timing mark of the crankshaft pulley is aligned with the center(s) of the crankshaft pulley bolt and idler pulley bolt

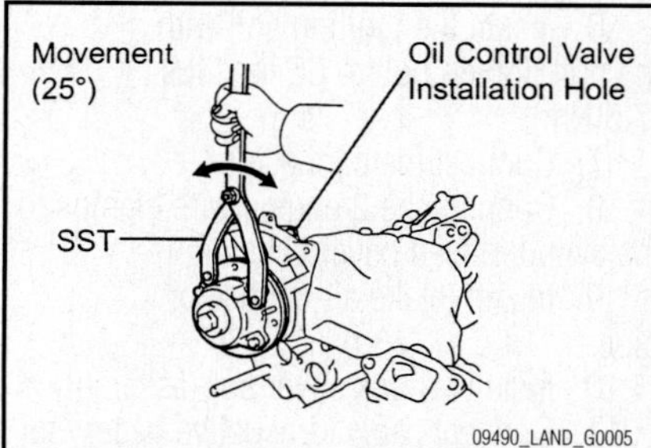

Fig. 26 Release the oil from the front bearing caps using the tool illustrated

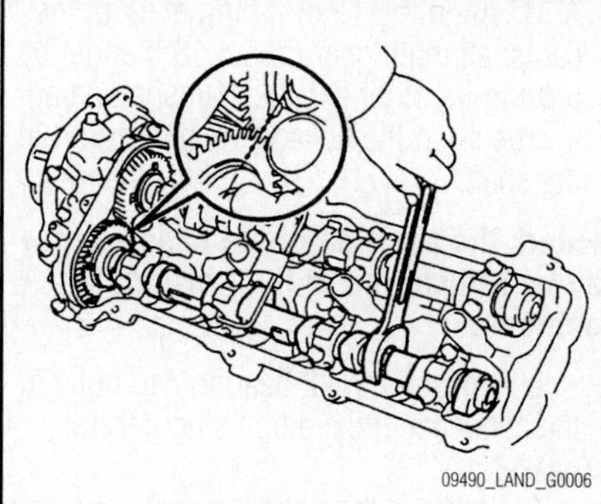

Fig. 27 Align the 2 dot timing mark of the left side camshaft by turning the left exhaust camshaft using a wrench on the hexagon head portion of the shaft

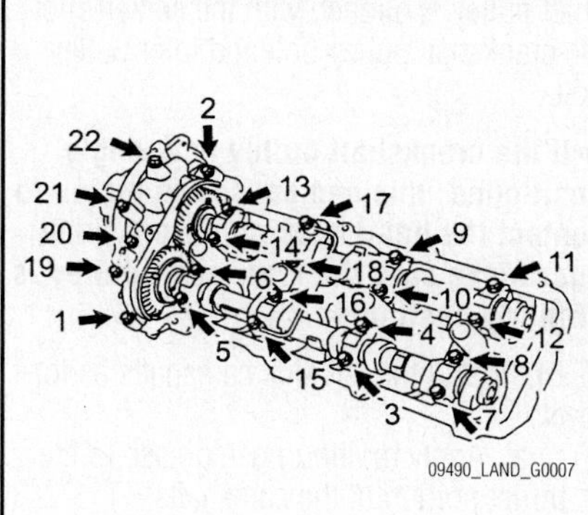

Fig. 28 Loosen the left side 22 bearing cap bolts in the sequence illustrated using several passes

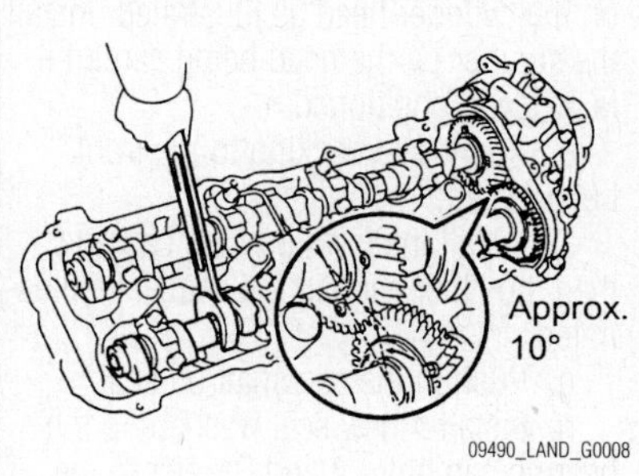

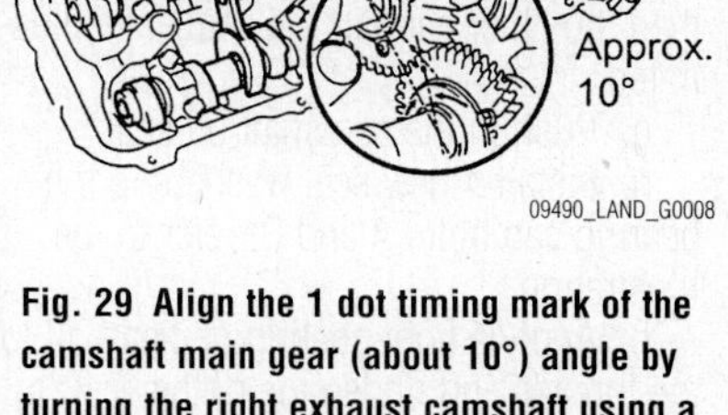

Fig. 29 Align the 1 dot timing mark of the camshaft main gear (about 10°) angle by turning the right exhaust camshaft using a wrench on the hexagon head portion of the shaft

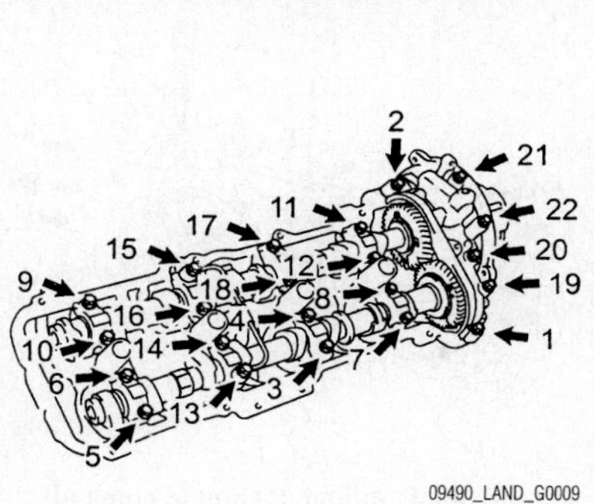

Fig. 30 Loosen the right side 22 bearing cap bolts in the sequence illustrated using several passes

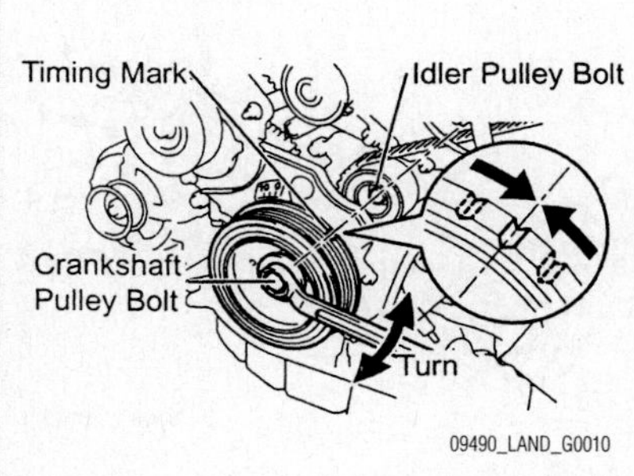

Fig. 31 Check the timing mark of the crankshaft pulley is aligned with the center(s) of the crankshaft pulley bolt and idler pulley bolt

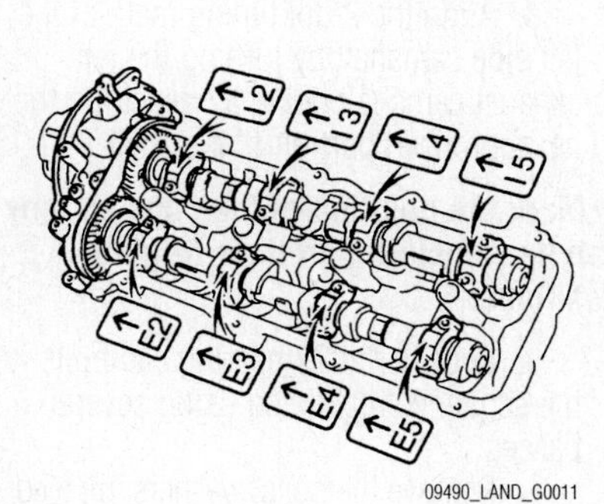

Fig. 32 Install the front bearing cap and then the other caps in the sequence illustrated on the left side camshafts

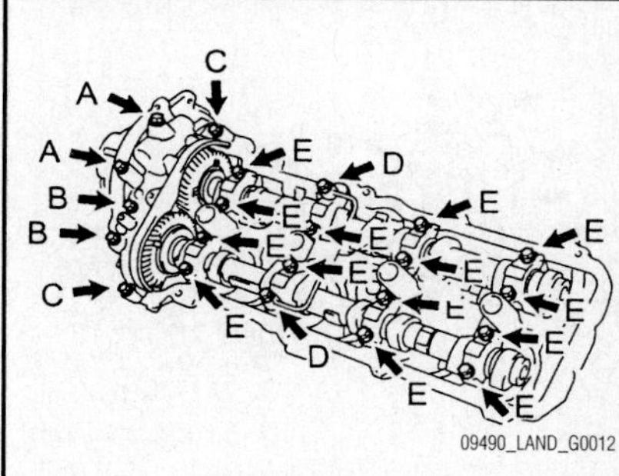

Fig. 33 Apply a light coating of clean oil to the threads and underside of the bolt heads D and E. make sure no oil gets under the heads of bolts A, B and C on the left side camshafts

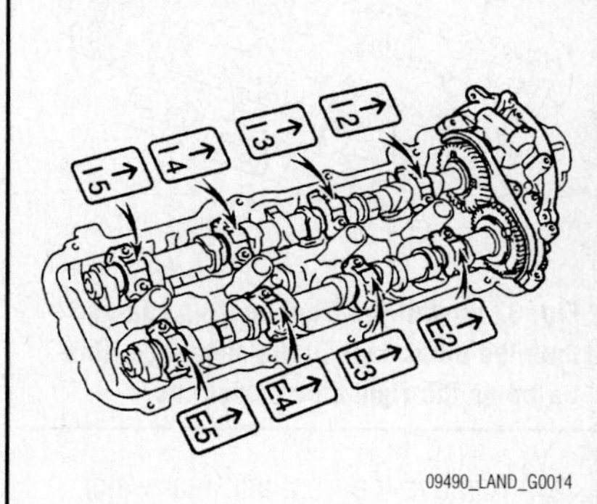

Fig. 34 Install the front bearing cap and then the other caps in the sequence illustrated on the left side camshafts

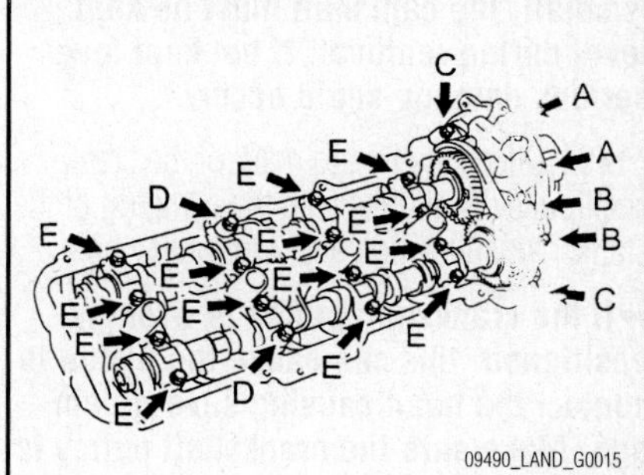

Fig. 35 Apply a light coating of clean oil to the threads and underside of the bolt heads D and E (make sure no oil gets under the heads of bolts A, B and C on the right side camshafts)

1. Before servicing the vehicle, refer to the Precautions section.
2. Drain the cooling system.
3. Relieve the fuel system pressure.
4. Remove the V bank cover.
5. Remove the timing belt.
6. Remove the camshaft pulleys.
7. Remove the Camshaft Position (CMP) Sensor.
8. Remove the power steering pump and set it aside with the lines still attached.
9. Remove the front exhaust pipe.
10. On models with an automatic transmission, remove the oil dipstick and tube.
11. Remove the ignition coils.
12. Remove the rear timing belt plates being careful not to drop anything.
13. Disconnect the fuel inlet hose.
14. Remove the intake manifold.
15. Remove the water inlet and inlet housing. Refer to water pump removal.

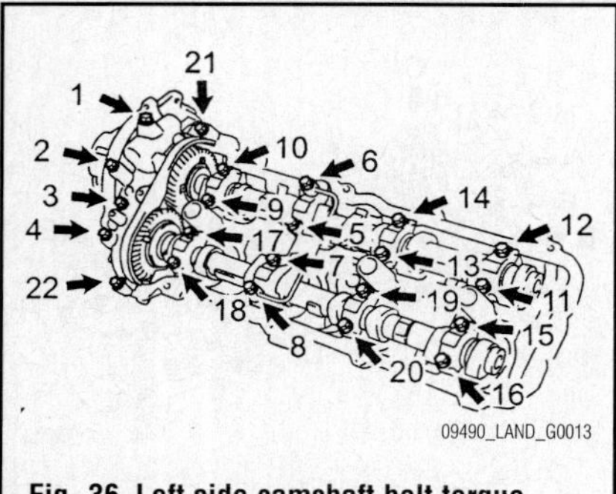

Fig. 36 Left side camshaft bolt torque sequence

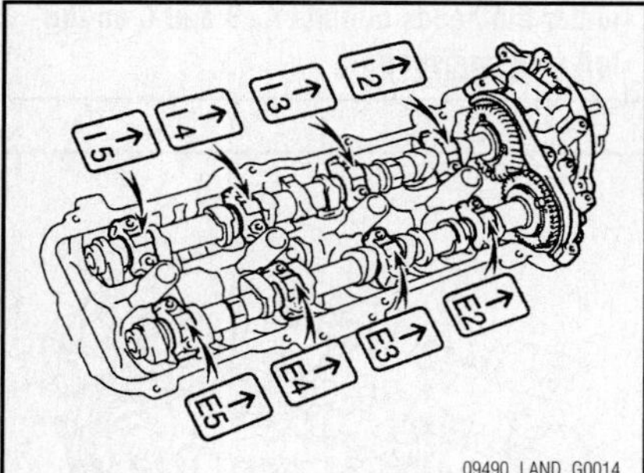

Fig. 37 Install the front bearing cap and then the other caps in the sequence illustrated on the right side camshafts

16. Remove the front and rear water bypass joint.

17. Remove the engine hangers and if needed the oil dipstick and tube.

18. Remove the valve covers.

➡Since the thrust level of the camshaft is small, the camshaft must be kept level during removal. If not kept level serious damage could occur.

19. Check the timing mark of the crankshaft pulley is aligned with the center(s) of the crankshaft pulley bolt and idler pulley bolt.

➡If the crankshaft pulley is wrongly positioned, this can cause the piston to contact the head causing severe damage. Make sure the crankshaft pulley is properly positioned.

20. Release the oil from the front bearing caps using the tool illustrated. Rotate the camshaft timing tube from left to right 2 to 3 times within its VVT-I range of 25° and collect the oil from the timing oil control valve installation hole using a rag.

21. Remove the left hand camshafts as follows:

a. Bring the service bolt of the sub gear up by turning the left exhaust camshaft using a wrench on the hexagon head portion of the shaft.

b. Secure the sub gear to the main gear

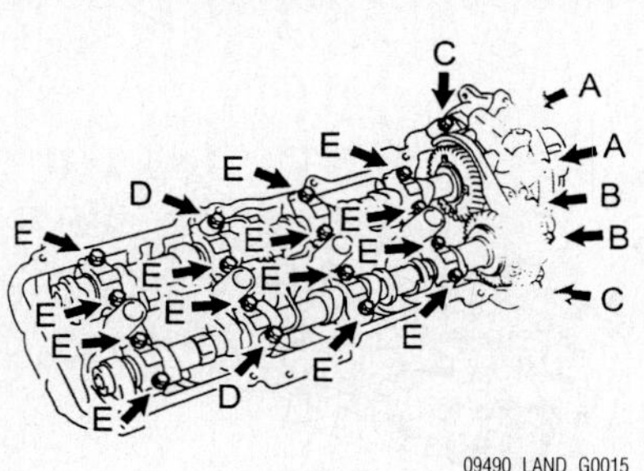

Fig. 38 Apply a light coating of clean oil to the threads and underside of the bolt heads D and E (make sure no oil gets under the heads of bolts A, B and C on the right side camshafts)

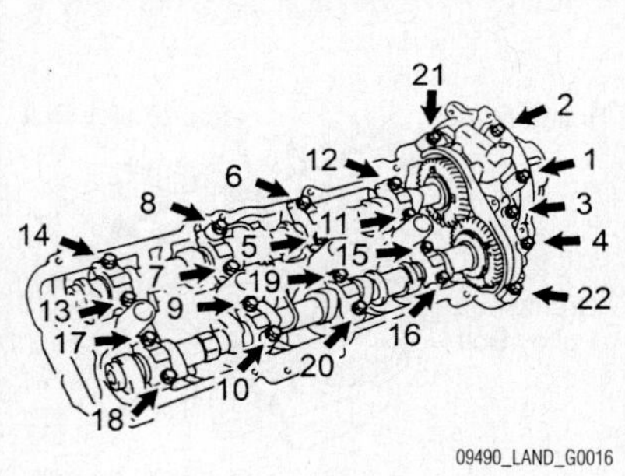

Fig. 39 Right side camshaft bolt torque sequence

using a 16—20 mm bolt with a diameter of 6mm and a thread pitch of 1mm.

c. Make sure the torsional force of the sub gear is retained by the bolt.

d. Align the 2 dot timing mark of the left side camshaft by turning the left exhaust camshaft using a wrench on the hexagon head portion of the shaft.

➡Mark the position of the caps so they can be reinstalled in their original positions.

e. Loosen the 22 bearing cap bolts in the sequence illustrated using several passes.

f. Remove the bolts, washers, oil feed pipe, bearing caps, camshaft housing plug, oil control valve filter and the camshafts.

22. Remove the right hand camshafts as follows:

a. Bring the service bolt of the sub gear up by turning the right exhaust camshaft using a wrench on the hexagon head portion of the shaft.

b. Secure the sub gear to the main gear using a 16—20 mm bolt with a diameter of 6mm and a thread pitch of 1mm.

c. Make sure the torsional force of the sub gear is retained by the bolt.

d. Align the 1 dot timing mark of the camshaft main gear (about 10°) angle by turning the right exhaust camshaft using a wrench on the hexagon head portion of the shaft.

➡Mark the position of the caps so they can be reinstalled in their original positions.

e. Loosen the 22 bearing cap bolts in the sequence illustrated using several passes.

f. Remove the bolts, washers, oil feed pipe, bearing caps, camshaft housing plug, oil control valve filter and the camshafts.

To install:

23. Check the timing mark of the crankshaft pulley is aligned with the center(s) of the crankshaft pulley bolt and idler pulley bolt.

➡If the crankshaft pulley is wrongly positioned, this can cause the piston to contact the head causing severe damage. Make sure the crankshaft pulley is properly positioned.

24. Install the left side camshafts as follows:

a. Apply multipurpose grease to the thrust portion of the camshafts.

b. Align the 2 dot timing mark of the camshaft drive and driven main gears and install the camshafts.

c. Apply seal packing to the camshaft housing plug.

d. Install the camshaft housing plug on the cylinder head as illustrated. Install the strainer on the head being careful it is properly positioned.

e. Apply seal packing to the front bearing cap.

f. Install the front bearing cap and then the other caps in the sequence illustrated.

g. Push in the camshaft oil seal.

h. Install 4 new seal washers to the bearing cap bolts A and B, refer to the illustration.

i. Apply a light coating of clean oil to the threads and underside of the bolt heads D and E (make sure no oil gets under the heads of bolts A, B and C).

j. The bolt lengths and positions are as follows. Refer to the illustration for bolt location:

- 94mm bolts A
- 72mm bolts B
- 25mm bolts C
- 52mm bolts D
- 38mm bolts E

k. Tighten the cap bolts using several passes. Tighten bolt C to 66 inch lbs. (7.5 Nm) an the remaining bolts to 12 ft. lbs. (16 Nm).

l. Remove the service bolt.

25. Install the right side camshafts as follows:

a. Apply multi-purpose grease to the thrust portion of the camshafts.

b. Align the 1 dot timing mark of the camshaft drive and driven main gears and install the camshafts.

c. Set the 1 dot timing mark of the camshaft drive and driven gears at a 10°angle.

d. Apply seal packing to the camshaft housing plug.

e. Install the camshaft housing plug on the cylinder head as illustrated. Install the strainer on the head being careful it is properly positioned.

f. Apply seal packing to the front bearing cap.

g. Install the front bearing cap and then the other caps in the sequence illustrated.

h. Push in the camshaft oil seal.

i. Install 4 new seal washers to the bearing cap bolts A and B, refer to the illustration.

j. Apply a light coating of clean oil to the threads and underside of the bolt heads D and E (make sure no oil gets under the heads of bolts A, B and C).

k. The bolt lengths and positions are as follows. refer to the illustration for bolt location:

- 94mm bolts A
- 72mm bolts B
- 25mm bolts C
- 52mm bolts D
- 38mm bolts E

l. Tighten the cap bolts using several passes. Tighten bolt C to 66 inch lbs. (7.5 Nm) an the remaining bolts to 12 ft. lbs. (16 Nm).

m. Remove the service bolt.

26. Check and adjust the valve clearance.

27. Install the camshaft timing control valve.

28. Install the 4 half moon plugs onto the cylinder heads.

29. Install the valve covers and tighten to 53 inch lbs. (6 Nm).

30. Install the engine hangers and tighten to 27 ft. lbs. (37 Nm).

31. Install the VVT sensors.

32. Install the oil dipstick tube and dipstick.

33. Install the ignition coils.

34. Install the water bypass joint and tighten the retainers to 13 ft. lbs. (18 Nm).

35. Install the water inlet and housing assembly.

36. Install the intake manifold.

37. Install the timing belt rear plates, right plates first, then left plates. Tighten the retainers to 66 inch lbs. (7 Nm).

38. Install the throttle body cover.

39. Install the front exhaust pipe, power steering pump.

40. Install the Camshaft Position (CMP) Sensor and camshaft timing pulleys, tighten to 25 ft. lbs. (34 Nm).

41. Install the timing belt.

42. Fill the cooling system and perform an oil change.

43. Start the vehicle and check for leaks.

CRANKSHAFT DAMPER

REMOVAL & INSTALLATION

See Figure 40.

1. Before servicing the vehicle, refer to the Precautions section.
2. Drain the cooling system.
3. Remove or disconnect the following:
   - Negative battery cable
   - Engine under cover
   - Engine appearance cover
   - Air intake assembly
   - Accessory drive belt
   - Cooling fan and pulley
   - Radiator
   - Drive belt idler pulley
   - Camshaft Position (CMP) sensor connector
   - Upper timing covers
   - Oil cooler pipe
   - Center timing cover
   - A/C compressor
   - Cooling fan bracket
   - Crankshaft pulley

To install:

4. Install of connect the following:
   - Crankshaft pulley: tighten the bolt to 181 ft. lbs. (245 Nm)
   - Cooling fan bracket: tighten the 12mm bolts to 12 ft. lbs. (16 Nm) and the 14mm bolts to 24 ft. lbs. (32 Nm)
   - A/C compressor
   - Center timing cover
   - Oil cooler pipe
   - Upper timing covers
   - CMP sensor connector
   - Drive belt idler pulley: tighten the bolt to 27 ft. lbs. (37 Nm)
   - Radiator
   - Cooling fan and pulley: tighten the nuts to 16 ft. lbs. (21 Nm)
   - Accessory drive belt
   - Air intake assembly
   - Engine appearance cover
   - Engine under cover
   - Negative battery cable
5. Fill the cooling system.
6. Start the engine and check for leaks.

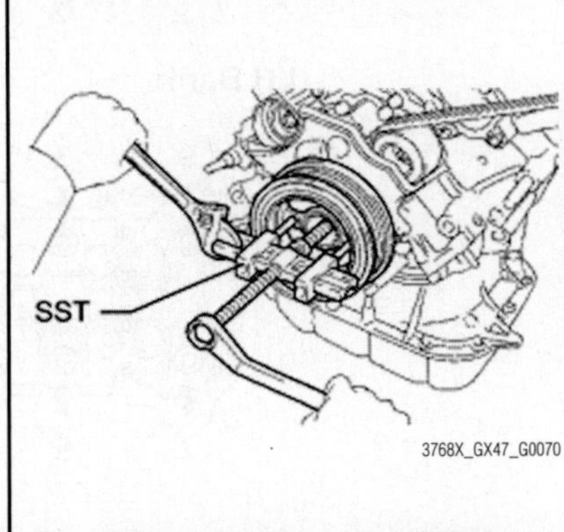

Fig. 40 Removing the crankshaft damper

CRANKSHAFT FRONT SEAL

REMOVAL & INSTALLATION

See Figure 41.

1. Before servicing the vehicle, refer to the Precautions section.
2. Drain the cooling system.
3. Remove or disconnect the following:
   - Negative battery cable
   - Engine under cover
   - Engine appearance cover
   - Air intake assembly
   - Accessory drive belt
   - Cooling fan and pulley
   - Radiator
   - Drive belt idler pulley
   - Camshaft Position (CMP) sensor connector
   - Upper timing covers
   - Oil cooler pipe
   - Center timing cover
   - A/C compressor
   - Cooling fan bracket

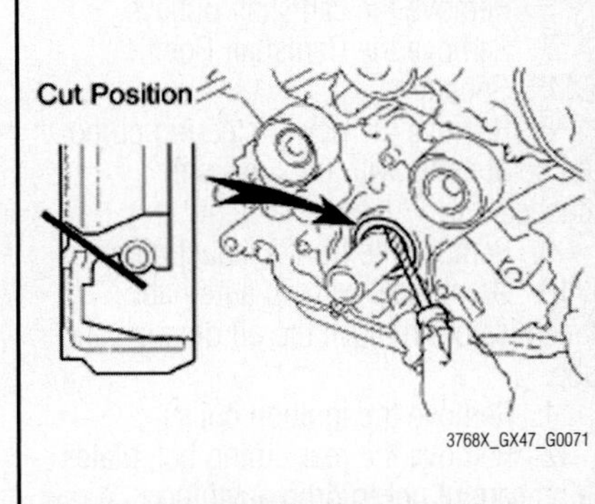

Fig. 41 Removing the front crankshaft seal

- Crankshaft pulley
- Lower timing cover
- Timing belt.
- Crankshaft timing sprocket
- Front crankshaft seal

To install:

4. Install the oil seal so that it is flush with the oil pump housing.
5. Install or connect the following:
   - Crankshaft timing sprocket
   - Timing belt
   - Lower timing cover
   - Crankshaft pulley: tighten the bolt to 181 ft. lbs. (245 Nm)
   - Cooling fan bracket: tighten the 12mm bolts to 12 ft. lbs. (16 Nm) and the 14mm bolts to 24 ft. lbs. (32 Nm)
   - A/C compressor
   - Center timing cover
   - Oil cooler pipe
   - Upper timing covers
   - CMP sensor connector
   - Drive belt idler pulley: tighten the bolt to 27 ft. lbs. (37 Nm)
   - Radiator
   - Cooling fan and pulley: tighten the nuts to 16 ft. lbs. (21 Nm)
   - Accessory drive belt
   - Air intake assembly
   - Engine appearance cover
   - Engine under cover
   - Negative battery cable
6. Fill the cooling system.
7. Start the engine and check for leaks.

CYLINDER HEAD

REMOVAL & INSTALLATION

See Figure 42.

1. Before servicing the vehicle, refer to the Precautions section.
2. Drain the cooling system.
3. Relieve the fuel system pressure.
4. Remove the V bank cover.
5. Remove the timing belt.
6. Remove the camshaft pulleys.
7. Remove the Camshaft Position (CMP) Sensor.
8. Remove the power steering pump and set it aside with the lines still attached.
9. Remove the front exhaust pipe.
10. On models with an automatic transmission, remove the oil dipstick and tube.
11. Remove the ignition coils.
12. Remove the rear timing belt plates being careful not to drop anything.
13. Disconnect the fuel inlet hose.
14. Remove the intake manifold.
15. Remove the water inlet and inlet housing. Refer to water pump removal.
16. Remove the front and rear water bypass joint.
17. Remove the engine hangers and if needed the oil dipstick and tube.
18. Remove the valve covers.

➡Since the thrust level of the camshaft is small, the camshaft must be kept level during removal. If not kept level serious damage could occur.

19. Check the timing mark of the crankshaft pulley is aligned with the center(s) of the crankshaft pulley bolt and idler pulley bolt.

➡If the crankshaft pulley is wrongly positioned, this can cause the piston to contact the head causing severe damage. Make sure the crankshaft pulley is properly positioned.

20. Release the oil from the front bearing caps using the tool illustrated. Rotate the camshaft timing tube from left to right 2–3 times within its VVT-I range of 25°and collect the oil from the timing oil control valve installation hole using a rag.
21. Remove the left hand camshafts as follows:
    a. Bring the service bolt of the sub gear up by turning the left exhaust camshaft using a wrench on the hexagon head portion of the shaft.
    b. Secure the sub gear to the main gear using a 16—20 mm bolt with a diameter of 6mm and a thread pitch of 1mm.
    c. Make sure the torsional force of the sub gear is retained by the bolt.
    d. Align the 2 dot timing mark of the left side camshaft by turning the left exhaust camshaft using a wrench on the hexagon head portion of the shaft.

➡Mark the position of the caps so they can be reinstalled in their original positions.

    e. Loosen the 22 bearing cap bolts in the sequence illustrated using several passes.
    f. Remove the bolts, washers, oil feed pipe, bearing caps, camshaft housing plug, oil control valve filter and the camshafts.
22. Remove the right hand camshafts as follows:
    a. Bring the service bolt of the sub gear up by turning the right exhaust camshaft using a wrench on the hexagon head portion of the shaft.
    b. Secure the sub gear to the main gear using a 16—20 mm bolt with a diameter of 6mm and a thread pitch of 1mm.
    c. Make sure the torsional force of the sub gear is retained by the bolt.
    d. Align the 1 dot timing mark of the camshaft main gear (about 10°) angle by turning the right exhaust camshaft using a wrench on the hexagon head portion of the shaft.

➡Mark the position of the caps so they can be reinstalled in their original positions.

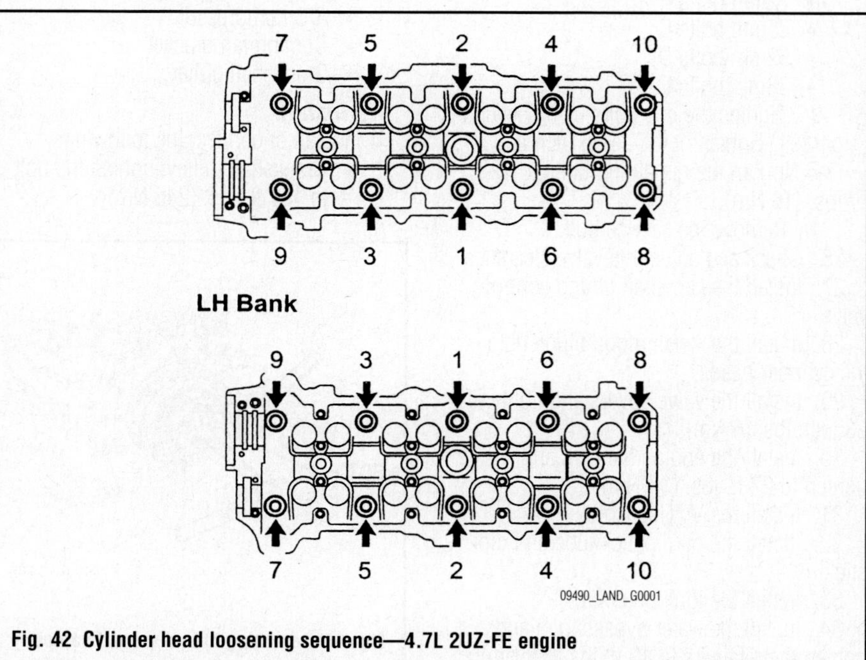

Fig. 42 Cylinder head loosening sequence—4.7L 2UZ-FE engine

e. Loosen the 22 bearing cap bolts in the sequence illustrated using several passes.

f. Remove the bolts, washers, oil feed pipe, bearing caps, camshaft housing plug, oil control valve filter and the camshafts.

23. Loosen the cylinder head bolts in the sequence shown, using several passes.

24. Remove the cylinder heads and exhaust manifolds together as an assembly.

To install:

25. Install new gaskets and the cylinder heads.

26. Tighten the bolts in sequence as follows:

a. Step 1: 30 ft. lbs. (40 Nm).

b. Step 2: Plus 90°.

c. Step 3: Plus 90°.

27. Check the timing mark of the crankshaft pulley is aligned with the center(s) of the crankshaft pulley bolt and idler pulley bolt.

➡If the crankshaft pulley is wrongly positioned, this can cause the piston to contact the head causing severe damage. Make sure the crankshaft pulley is properly positioned.

28. Install the left side camshafts as follows:

a. Apply multipurpose grease to the thrust portion of the camshafts.

b. Align the 2 dot timing mark of the camshaft drive and driven main gears and install the camshafts.

c. Apply seal packing to the camshaft housing plug.

d. Install the camshaft housing plug on the cylinder head as illustrated. Install the strainer on the head being careful it is properly positioned.

e. Apply seal packing to the front bearing cap.

f. Install the front bearing cap and then the other caps in the sequence illustrated.

g. Push in the camshaft oil seal.

h. Install 4 new seal washers to the bearing cap bolts A and B, refer to the illustration.

i. Apply a light coating of clean oil to the threads and underside of the bolt heads D and E (make sure no oil gets under the heads of bolts A, B and C).

j. The bolt lengths and positions are as follows. refer to the illustration for bolt location:

- 94mm bolts A
- 72mm bolts B
- 25mm bolts C
- 52mm bolts D
- 38mm bolts E

k. Tighten the cap bolts using several passes. Tighten bolt C to 66 inch lbs. (7.5 Nm) and the remaining bolts to 12 ft. lbs. (16 Nm).

l. Remove the service bolt.

29. Install the right side camshafts as follows:

a. Apply multipurpose grease to the thrust portion of the camshafts.

b. Align the 1 dot timing mark of the camshaft drive and driven main gears and install the camshafts.

c. Set the 1 dot timing mark of the camshaft drive and driven gears at a 10°angle.

d. Apply seal packing to the camshaft housing plug.

e. Install the camshaft housing plug on the cylinder head as illustrated. Install the strainer on the head being careful it is properly positioned.

f. Apply seal packing to the front bearing cap.

g. Install the front bearing cap and then the other caps in the sequence illustrated.

h. Push in the camshaft oil seal.

i. Install 4 new seal washers to the bearing cap bolts A and B, refer to the illustration.

j. Apply a light coating of clean oil to the threads and underside of the bolt heads D and E (make sure no oil gets under the heads of bolts A, B and C).

k. The bolt lengths and positions are as follows. refer to the illustration for bolt location:

- 94mm bolts A
- 72mm bolts B
- 25mm bolts C
- 52mm bolts D
- 38mm bolts E

l. Tighten the cap bolts using several passes. Tighten bolt C to 66 inch lbs. (7.5 Nm) and the remaining bolts to 12 ft. lbs. (16 Nm).

m. Remove the service bolt.

30. Check and adjust the valve clearance.

31. Install the camshaft timing control valve.

32. Install the 4 half moon plugs onto the cylinder heads.

33. Install the valve covers and tighten to 53 inch lbs. (6 Nm).

34. Install the engine hangers and tighten to 27 ft. lbs. (37 Nm).

35. Install the VVT sensors.

36. Install the oil dipstick tube and dipstick.

37. Install the ignition coils.

38. Install the water bypass joint and tighten the retainers to 13 ft. lbs. (18 Nm).

39. Install the water inlet and housing assembly.

40. Install the intake manifold.

41. Install the timing belt rear plates, right plates first, then left plates. Tighten the retainers to 66 inch lbs. (7 Nm).

42. Install the throttle body cover.

43. Install the front exhaust pipe, power steering pump.

44. Install the Camshaft Position (CMP) Sensor and camshaft timing pulleys, tighten to 25 ft. lbs. (34 Nm).

45. Install the timing belt.

46. Fill the cooling system and perform an oil change.

47. Start the vehicle and check for leaks.

EXHAUST MANIFOLD

REMOVAL & INSTALLATION

See Figures 43 and 44.

1. Before servicing the vehicle, refer to the Precautions section.

2. Attach a hoist to the engine lifting eyes.

3. Remove or disconnect the following:

- Negative battery cable

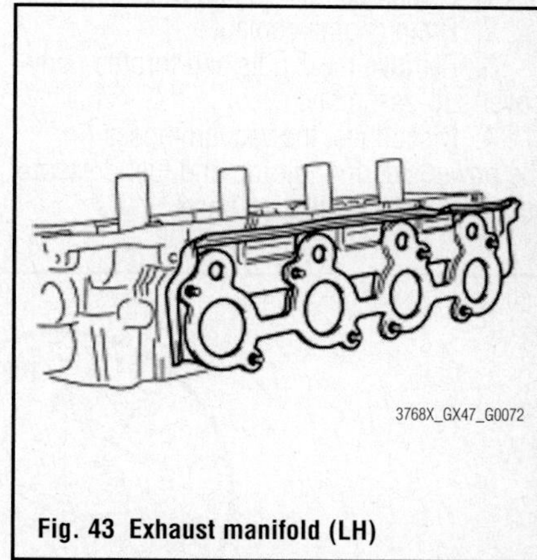

Fig. 43 Exhaust manifold (LH)

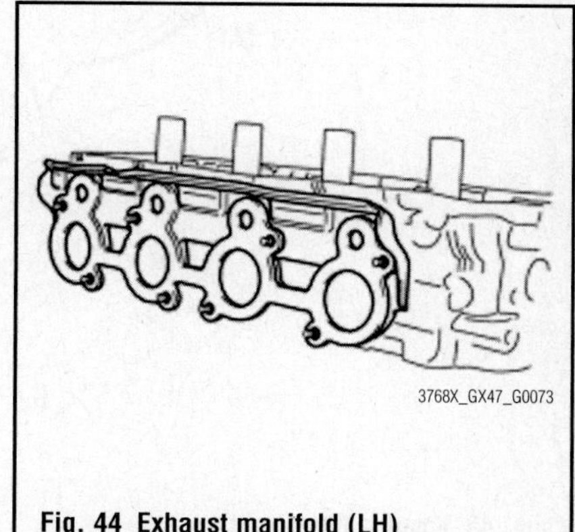

Fig. 44 Exhaust manifold (LH)

- Heated Oxygen (HO_2S) sensor connectors
- Exhaust manifold heat shield
- Exhaust front pipe
- Motor mount
- Motor mount bracket
- Exhaust manifold

To install:

➡Use new exhaust manifold nuts for assembly.

4. Install or connect the following:
   - Exhaust manifold: tighten the nuts to 32 ft. lbs. (44 Nm)
   - Motor mount bracket: tighten the bolts to 27 ft. lbs. (36 Nm)
   - Motor mount: tighten the fasteners to 22 ft. lbs. (30 Nm)
   - Exhaust front pipe: tighten the nuts to 46 ft. lbs. (62 Nm)
   - Exhaust manifold heat shield
   - HO_2S sensor connectors
   - Negative battery cable
5. Start the engine and check for leaks.

INTAKE MANIFOLD

REMOVAL & INSTALLATION

See Figure 45.

1. Discharge fuel system pressure.
2. Drain engine coolant.
3. Remove the 2 nuts and throttle body cover sub-assembly.
4. Disconnect the vacuum hoses (for the power steering idle-up and fuel pressure regulator) and ventilation hose.
5. Remove the air cleaner hose assembly.
6. Disconnect fuel hose.
7. Disconnect fuel hose No.2.
8. Disconnect the throttle control connector.
9. Disconnect the purge VSV.
10. Disconnect the 8 injector connectors.
11. Disconnect the ECT sensor connector.
12. Disconnect the 2 VSV connectors for the air injection system.
13. Disconnect the 8 ignition coil connectors.
14. Disconnect the 2 air fuel ratio sensor connectors.
15. Disconnect the vacuum hose from the fuel pressure regulator.
16. Disconnect the PCV hoses from the PCV valve on the LH cylinder head.
17. Disconnect the EVAP hose (from the charcoal canister) from the purge VSV.
18. Disconnect the 2 vacuum hoses from the VSV for the air injection system.
19. Disconnect the 2 water by-pass hoses from the throttle body.
20. Disconnect the 2 wire clamps from the wire clamp bracket on the RH delivery pipe.
21. Remove the bolt and nut holding the engine wire protector from the intake manifold and cylinder head.
22. Remove the 2 bolts and ground cables from the RH and LH cylinder heads.

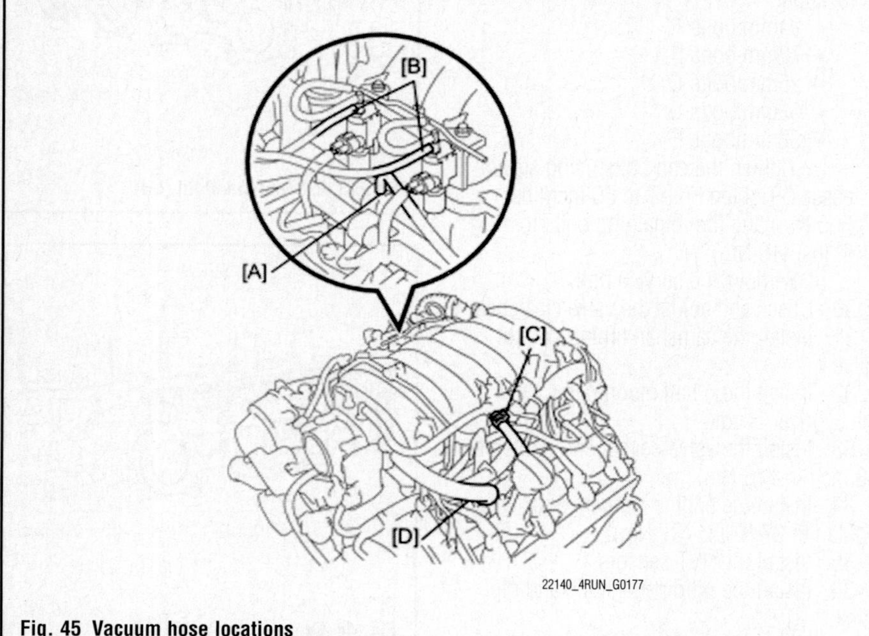

Fig. 45 Vacuum hose locations

23. Remove the bolt and V-bank cover bracket from the intake manifold.
24. Disconnect the engine wire from the engine hanger and wire bracket.
25. Remove the bolt and wire bracket from the intake manifold.
26. Remove the 6 bolts, 4 nuts, intake manifold assembly and 2 gaskets.
27. Remove air pump assembly w/ bracket.
28. Remove the 2 nuts and 2 knock sensors.

To install:

29. Install the 2 knock sensors with the 2 nuts and tighten to 15 ft. lbs. (20 Nm).
30. Connect the 2 knock sensor connectors.
31. Place 2 new gaskets on the intake manifold.
32. Place the intake manifold on the cylinder heads.
33. Install and uniformly tighten the 6 bolts and 4 nuts in several steps and tighten to 13 ft. lbs. (18 Nm).
34. Install the V-bank cover bracket to the intake manifold.
35. Install the wire bracket to the intake manifold with the bolt.
36. Connect the engine wire to the engine hanger and wire bracket.
37. Connect the wire protector to the intake manifold and cylinder heads with the bolt and nut.
38. Install the 2 ground cables with the 2 bolts to the RH and LH cylinder heads.
39. Connect the 2 water by-pass hoses to the throttle body.
40. Connect the 2 wire clamps to the wire clamp bracket on the RH delivery pipe.
41. Connect the vacuum hose to the fuel pressure regulator.
42. Connect the PCV hose to the PCV valve on the LH cylinder head.
43. Connect the EVAP hose (from the charcoal canister) to the purge VSV.
44. Connect the 2 vacuum hoses to the VSV for the air injection system.
45. Connect the throttle control connector.
46. Connect the 2 VSV connectors for the air injection system.
47. Connect the purge VSV connector.
48. Connect the 8 injector connectors.
49. Connect the ECT sensor connector.

50. Connect the 8 ignition coil connectors.
51. Connect the 2 air fuel ratio sensor connectors.
52. Install fuel hose No.2.
53. Install fuel hose.
54. Install throttle body cover sub-assembly.
55. Add engine coolant.
56. Check for leaks.

OIL PAN

REMOVAL & INSTALLATION

See Figures 46 through 48.

1. Before servicing the vehicle, refer to the Precautions section.
2. Remove the engine from the vehicle and mount it on a stand.
3. Remove or disconnect the following:
    - Oil dipstick tube
    - Lower oil pan
    - Oil pan baffle
    - Upper oil pan

To install:

4. The upper oil pan bolts are different lengths and are identified as follows:
    - A: 0.79 inch (20mm) w/10mm head
    - B: 0.98 inch (25mm) w/12mm head
    - C: 2.36 inch (60mm) w/12mm head
    - D: 1.38 inch (35mm) w/10mm head
5. Apply silicone sealant to the upper oil pan as shown.
6. Install the upper oil pan and tighten the fasteners in several passes to the following specifications:
    - 10mm: 66 inch lbs. (7.5 Nm)
    - 12mm: 21 ft. lbs. (28 Nm)
7. Install or connect the following:
    - Oil pan baffle: tighten the fasteners to 66 inch lbs. (7.5 Nm)

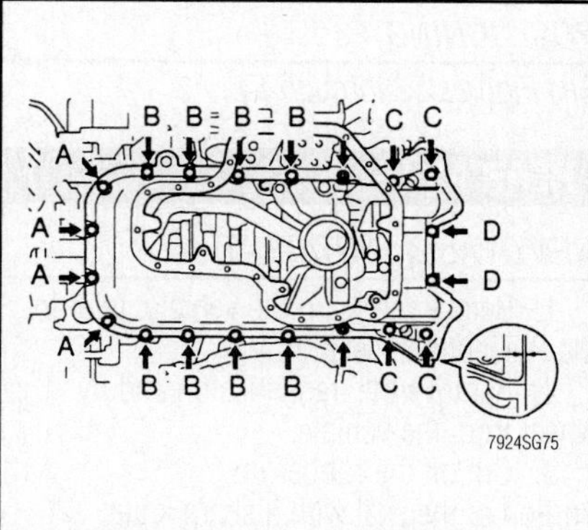

Fig. 46 Upper oil pan bolt location

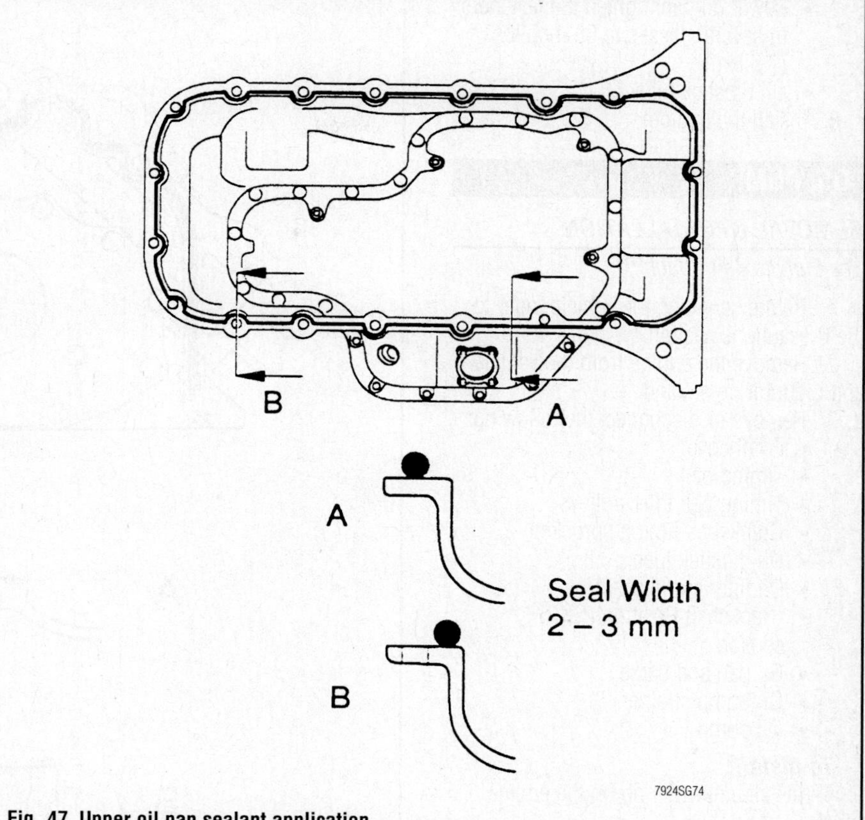

Fig. 47 Upper oil pan sealant application

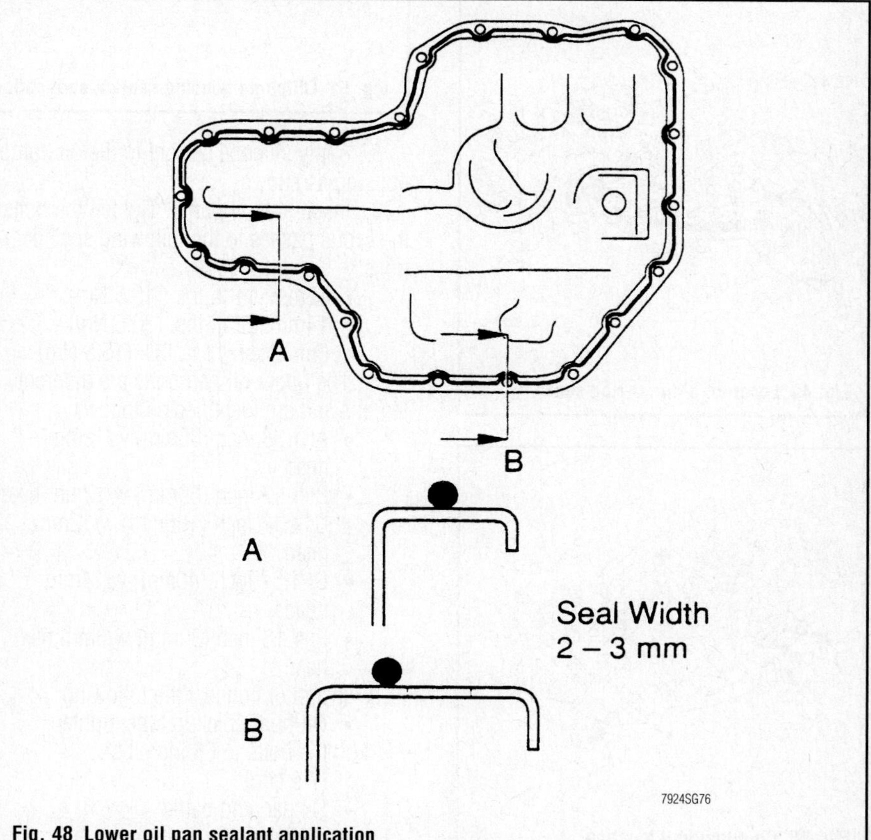

Fig. 48 Lower oil pan sealant application

- Lower oil pan: tighten the fasteners in several passes to 66 inch lbs. (7.5 Nm)
- Oil dipstick tube

8. Install the engine.

OIL PUMP

REMOVAL & INSTALLATION

See Figures 49 through 51.

1. Before servicing the vehicle, refer to the Precautions section.
2. Remove the engine from the vehicle and mount it on a stand.
3. Remove or disconnect the following:
   - Front cover
   - Timing belt
   - Timing belt idler pulleys
   - Crankshaft timing sprocket
   - Oil dipstick tube
   - Oil filter and bracket
   - Crankshaft Position (CKP) sensor
   - Oil pan and baffle
   - Oil pump strainer
   - Oil pump

To install:

4. Install a new O-ring on the engine block.

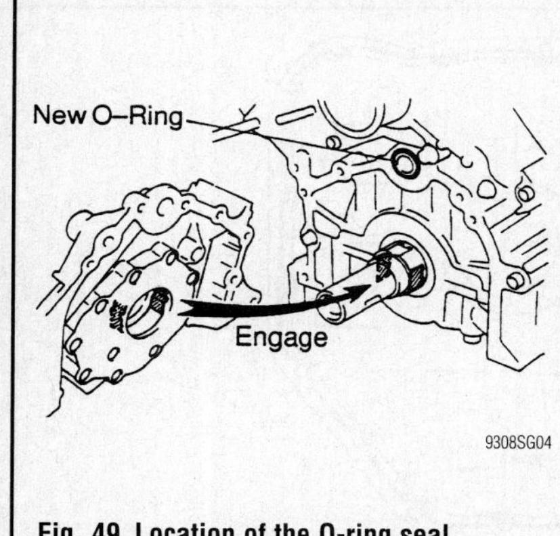

Fig. 49 Location of the O-ring seal

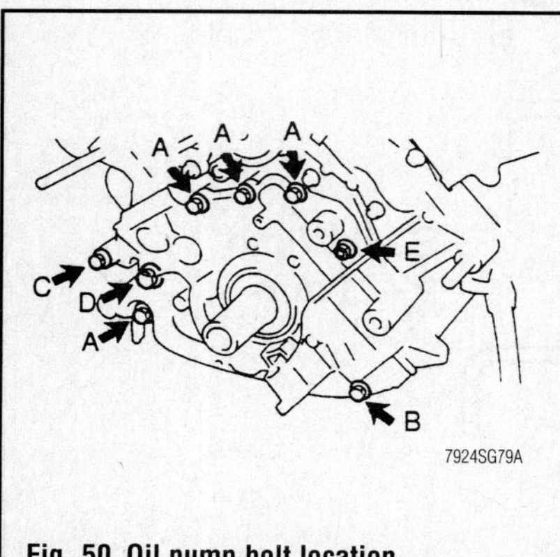

Fig. 50 Oil pump bolt location

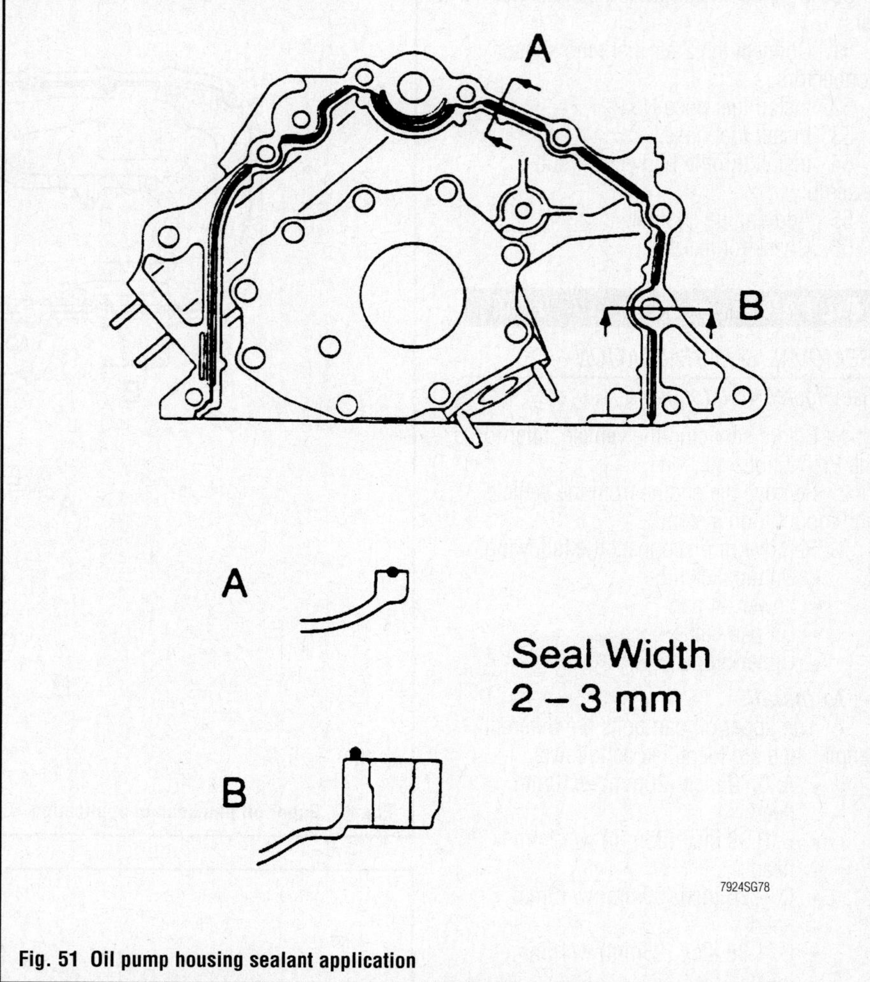

Fig. 51 Oil pump housing sealant application

5. Apply silicone sealant to the oil pump housing as shown.
6. Install the oil pump. Tighten the bolts in several passes to the following specifications:
   - 12mm: 11 ft. lbs. (15.5 Nm)
   - 14mm: 22 ft. lbs. (30.5 Nm)
   - 6mm Hex: 11 ft. lbs. (15.5 Nm)
7. The upper oil pan bolts are different lengths and are identified as follows:
   - A: 1.38 inch (35mm) w/12mm head
   - B: 1.97 inch (50mm) w/12mm head
   - C: 4.17 inch (106mm) w/12mm head
   - D: 1.57 inch (40mm) w/14mm head
   - E: 1.18 inch (30mm) w/6mm hex head
8. Install or connect the following:
   - Oil pump pickup tube: tighten the bolts to 66 inch lbs. (7.5 Nm)
   - Oil pan and baffle
   - CKP sensor
   - Oil filter and bracket: tighten the bolts to 13 ft. lbs. (18 Nm)
   - Oil dipstick tube
   - Crankshaft timing sprocket
   - Timing belt idler pulleys
   - Timing belt
   - Front cover
9. Install the engine.

PISTON AND RING

POSITIONING

See Figures 52 through 54.

REAR MAIN SEAL

REMOVAL & INSTALLATION

1. Before servicing the vehicle, refer to the Precautions section.
2. Remove the transmission and flywheel from the vehicle.
3. Cut off the rubber lip portion of the seal with a sharp knife.
4. Pry out the oil seal.

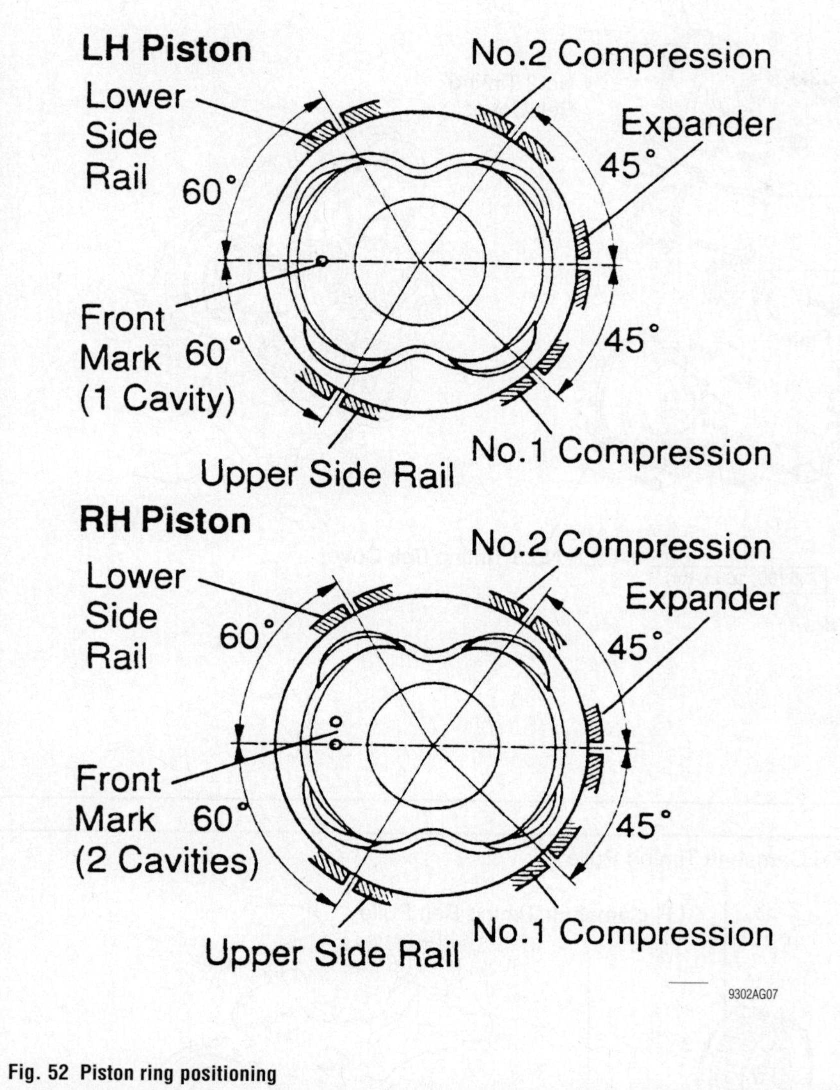

Fig. 52 Piston ring positioning

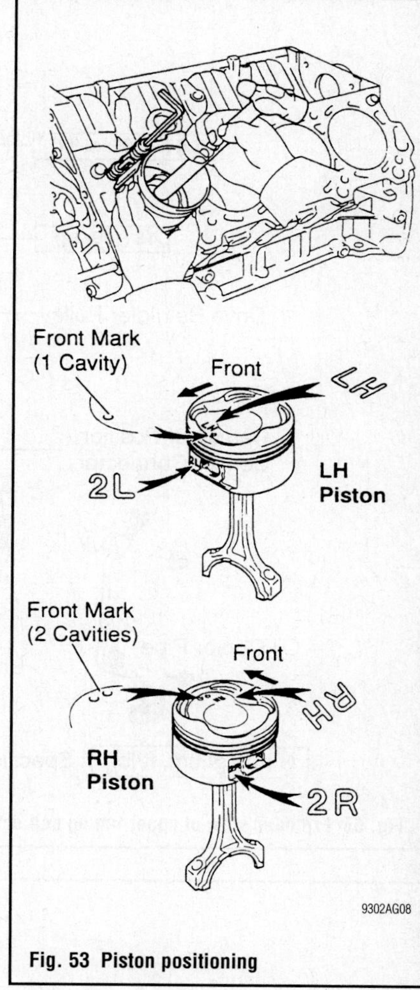

Fig. 53 Piston positioning

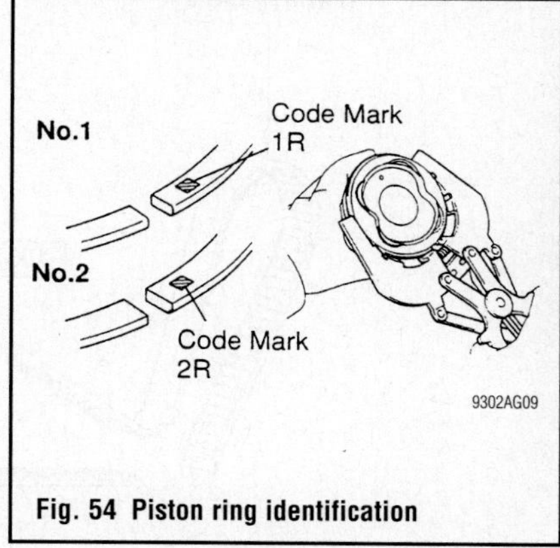

Fig. 54 Piston ring identification

To install:

5. Install the rear main seal so that it is flush with the seal retainer housing.
6. Install or connect the following:
   - Flywheel/driveplate: tighten the bolts to 35 ft. lbs. (48 Nm) plus a 90°turn
   - Transmission

TIMING BELT FRONT COVER

REMOVAL & INSTALLATION

See Timing Belt & Sprockets.

TIMING BELT & SPROCKETS

REMOVAL & INSTALLATION

See Figures 55 through 62.

1. Disconnect the negative battery cable.
2. Raise and safely support the vehicle.
3. Remove the oil pan protector and the engine under cover.
4. Drain the cooling system and store the coolant for refilling purposes.
5. Lower the vehicle and remove the battery clamp cover.
6. From the top of the engine, remove the fuel return hose, the engine cover nuts/bolts and the cover.
7. Remove the air cleaner and the intake air connector assembly.
8. Remove the cooling fan pulley by performing the following procedures:
   a. Loosen the 4 fan clutch-to-fan pulley nuts.
   b. Using a box-end wrench on the serpentine drive belt tensioner bolt, rotate the tensioner counterclockwise and remove the drive belt.

➡The serpentine drive belt tensioner bolt is a left-hand thread.

   c. Remove the fan clutch-to-fan pulley nuts, the fan, the clutch assembly and the fan pulley.
9. Remove the radiator by performing the following procedures:

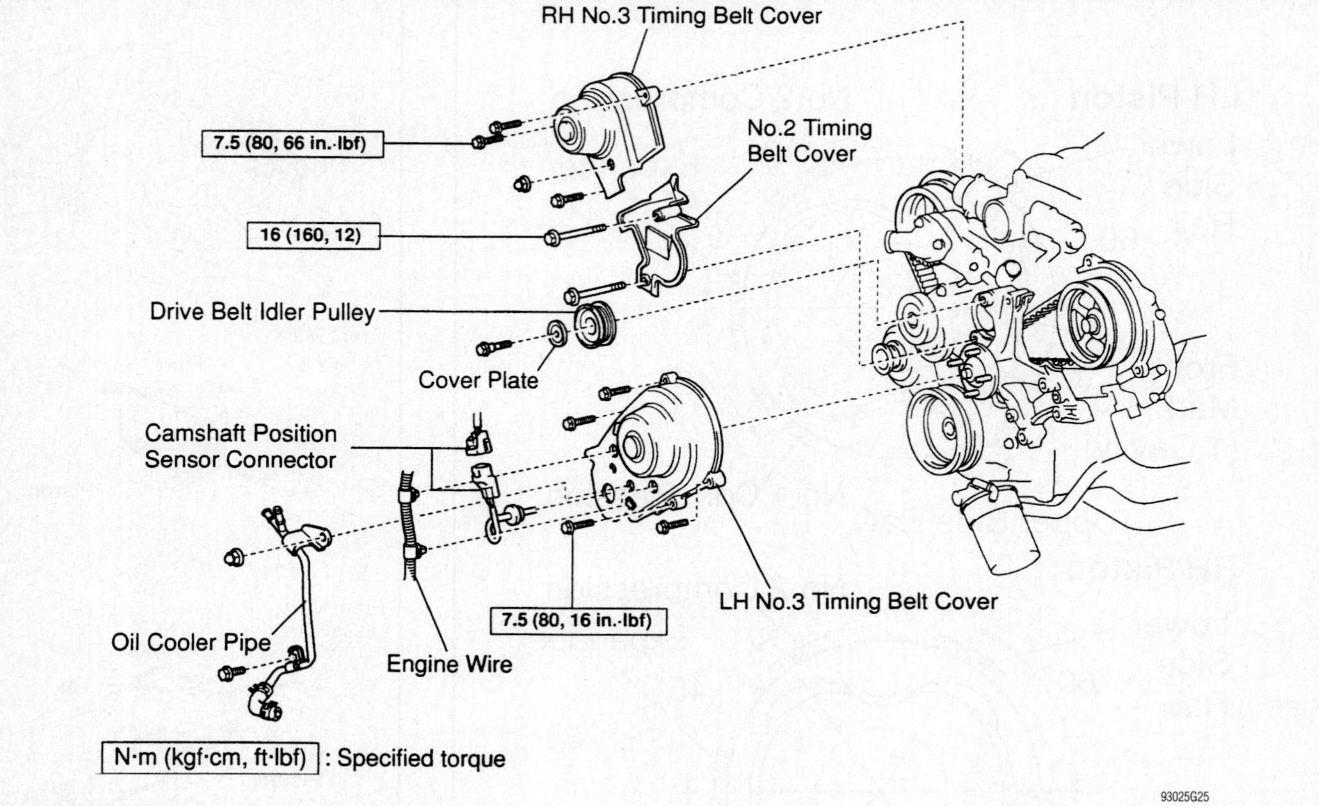

Fig. 55 Exploded view of upper timing belt covers

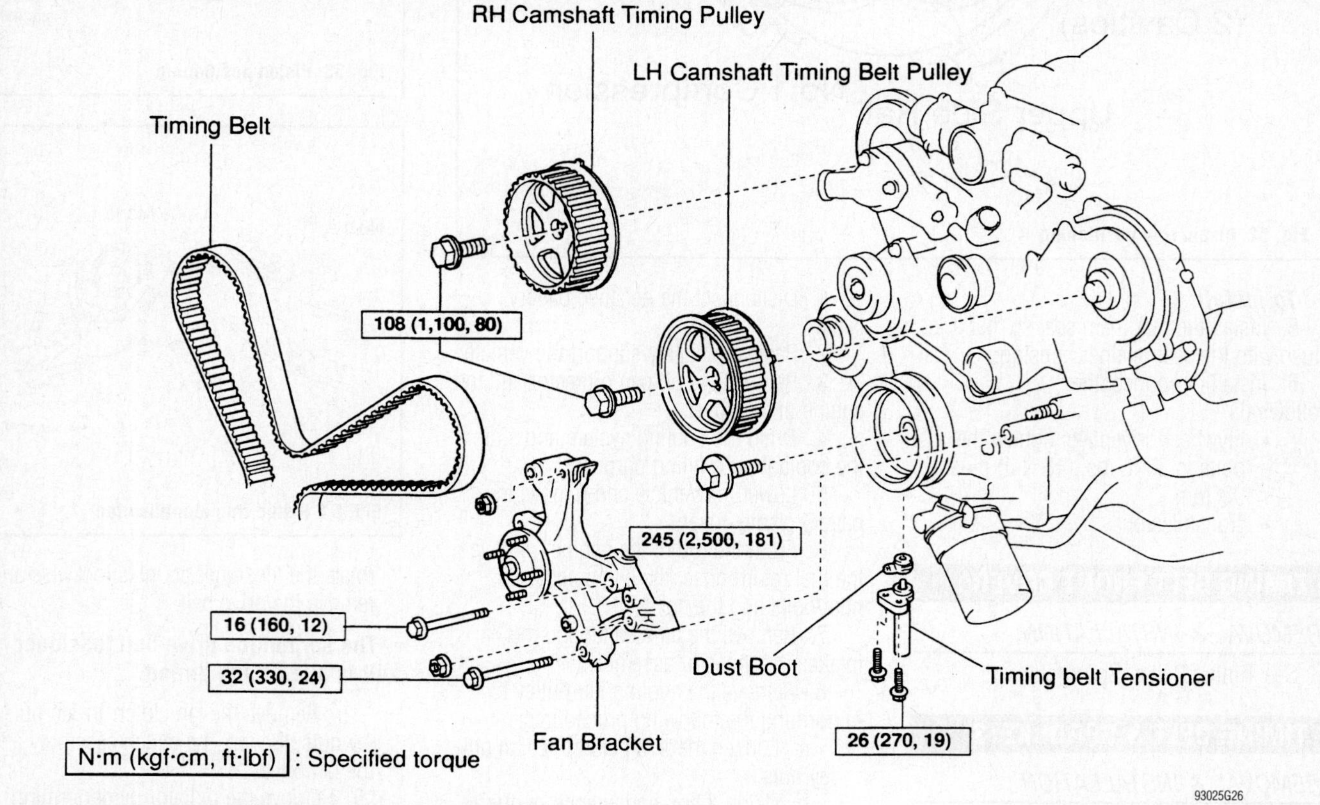

Fig. 56 Exploded view of upper timing sprockets and components

a. Disconnect the upper, lower and reservoir hoses from the radiator.

b. Disconnect and plug the automatic transmission oil cooler at the radiator. Disconnect the automatic transmission oil cooler hoses from the fan shroud clamp.

c. Remove the radiator reservoir tank.

d. Remove the fan shroud-to-radiator bolts and the shroud.

e. Remove the 2 upper radiator-to-chassis nuts.

f. Remove the middle radiator-to-chassis nut/bolts and brackets.

g. Carefully, lift the radiator from the vehicle.

10. Remove the serpentine drive belt idler pulley bolt, cover plate and pulley.

11. Remove the right side (No. 3) timing belt cover.

12. Remove the left side (No. 3) timing belt cover by performing the following procedures:

a. Disconnect the engine wire from both wire clamps.

b. Disconnect the Camshaft Position (CMP) Sensor wire from the wire clamp on the left-side (No.3) timing belt cover.

c. Disconnect the sensor connector from the connector bracket.

d. Disconnect the sensor connector.

e. Remove the wire grommet from the left-side (No. 3) timing belt cover.

f. Remove the oil cooler tube bolts and tube.

13. Remove the middle (No. 2) timing belt cover bolts and cover.

14. Remove the cooling fan bracket nuts/bolts and bracket.

➡If reusing the timing belt, make sure that there are 3 installation marks on the belt; if there are none, install them.

15. Using the Crankshaft Pulley Holding tool, Bolt tool and Companion Flange Holding tool, or equivalent, loosen the crankshaft pulley bolt.

16. Position the No. 1 cylinder to approximately 50°After Top Dead Center (ATDC) of the compression stroke by performing the following procedures:

a. Rotate the crankshaft pulley (CLOCKWISE) to align its groove with the timing mark "0" on the lower (No. 1) timing belt cover.

b. Check that the camshaft sprocket timing marks are aligned with the rear timing belt plate marks; if not, rotate the crankshaft 1 revolution (360°).

c. Rotate the crankshaft pulley approximately 50°(CLOCKWISE) and align the crankshaft pulley timing mark between the centers of the crankshaft pulley bolt and the idler pulley bolt.

✲✲ WARNING

If the timing belt is disengaged, having the crankshaft pulley in the wrong angle can cause the valve to come into contact with the piston when removing the camshaft pulley.

17. Remove the crankshaft pulley bolt.

➡If reusing the timing belt and the installation marks have disappeared, place new installation marks on the timing belt to match the camshaft timing sprocket marks.

➡To avoid meshing the timing sprocket and the timing belt, secure one with a string; then, place matchmarks on the timing belt and the right-side camshaft timing sprocket.

18. Remove the timing belt tensioner bolts and the tensioner.

19. Using the Camshaft Holding tool, or equivalent, slightly turn the left-side camshaft sprocket clockwise to loosen the tension spring. Then, disconnect the timing belt from the camshaft sprockets.

20. Remove the alternator by performing the following procedures:

a. Disconnect the electrical connector from the alternator.

b. Remove the rubber cap/nut and disconnect the battery wire from the alternator.

c. Disconnect the wire clamp from the alternator cord clip.

d. Remove the alternator-to-engine nuts/bolts and the alternator.

21. Remove the serpentine drive belt tensioner nuts/bolts and the tensioner.

22. Using the Crankshaft Puller Assembly tool, or equivalent, press the crankshaft pulley from the crankshaft.

✲✲ WARNING

DO NOT rotate the crankshaft pulley.

23. Remove the lower (No. 1) timing belt cover bolts and the cover.

24. Remove the timing belt guide, spacer and the timing belt.

To install:

➡With the timing belt removed, this is a perfect opportunity to inspect and/or replace the water pump.

25. Inspect the timing belt tensioner by performing the following procedures:

a. Inspect the seal for leakage; if leakage is suspected, replace the tensioner.

b. Using both hands to hold the tensioner facing upward, strongly press the pushrod against a solid surface. If the pushrod moves, replace the tensioner.

✲✲ WARNING

Never hold the tensioner with the pushrod facing downward.

c. Measure the pushrod protrusion from the housing end, it should be 0.413–0.453 inch (10.5–11.5mm). If the protrusion is not as specified, replace the tensioner.

26. Temporarily install the timing belt by performing the following procedures:

a. Align the timing belt's installation mark with the crankshaft timing sprocket.

b. Install the timing belt on the crankshaft timing sprocket, the No. 1 idler pulley and the No. 2 idler pulley.

27. Install the gasket to the timing belt cover spacer and install the cover spacer.

28. Install the timing belt guide with the cup side facing outward.

29. Install the lower (No. 1) timing belt cover.

30. Install the crankshaft pulley by performing the following procedures:

a. Align the crankshaft pulley with the crankshaft key.

b. Using the Crankshaft Installer tool, or equivalent, and a hammer, tap the crankshaft pulley into position.

31. Install the serpentine drive belt tensioner and torque the tensioner-to-engine bolts to 12 ft. lbs. (16 Nm).

➡To install the serpentine drive belt tensioner, use a bolt 4.18 inch (106mm) in length.

32. Check that the crankshaft pulley's timing mark is aligned with the centers of the idler pulley and crankshaft pulley bolts.

33. Install the alternator and torque the alternator-to-engine nuts/bolts to 29 ft. lbs. (39 Nm). Connect the alternator's electrical connectors and clip.

34. Install the timing belt to the left-side camshaft by performing the following procedures:

a. Rotate the left-side camshaft pulley to align the timing belt installation mark with the camshaft sprocket's timing mark and slide the belt onto the camshaft timing sprocket.

b. Using the Camshaft Holding tool, or equivalent, slightly turn the left-side camshaft sprocket counterclockwise to place tension on the timing belt between the crankshaft sprocket and the camshaft sprocket.

35. Rotate the right-side camshaft pulley to align the timing belt installation mark with the camshaft sprocket's timing mark and slide the belt onto the camshaft timing sprocket.

36. Using a vertical press, slowly press the pushrod into the housing using 200—2205 lbs. (981–9807 N) until the holes align, then, install a 1.27mm Allen® wrench to secure the pushrod and release the press. Install the dust boot on the tensioner housing.

37. Install the timing belt tensioner and torque the bolts to 19 ft. lbs. (26 Nm).

38. Using a pair of pliers, remove the Allen® wrench from the tensioner housing.

39. Check the valve timing by performing the following procedure:

a. Temporarily install the crankshaft pulley bolt.

b. Slowly, rotate the crankshaft pulley 2 revolutions (CLOCKWISE) and realign the TDC marks.

➡If the pulley/sprocket timing marks do not realign, remove the timing belt and reinstall it.

40. Using the Crankshaft Pulley Holding tool, Bolt tool and Companion Flange Holding tool, or equivalent, torque the crankshaft pulley bolt to 181 ft. lbs. (245 Nm).

41. Install the cooling fan bracket and torque the 12mm (head size) bolt to 12 ft. lbs. (16 Nm) and the 14mm (head size) bolt to 24 ft. lbs. (32 Nm).

42. Install the air conditioning compressor.

43. Install the middle (No. 2) timing belt cover and torque the bolts to 12 ft. lbs. (16 Nm).

44. Install the upper right-side (No. 3) timing belt cover and torque the bolts to 66 inch lbs. (7.5 Nm).

45. Install the upper left-side (No. 3) timing belt cover by performing the following procedures:

a. Install the oil cooler tube and bolt.

b. Feed the Camshaft Position (CMP) Sensor (CPS) through the left-side (No. 3) timing belt cover hole.

c. Install the left-side (No. 3) timing belt cover and torque the bolts to 66 inch lbs. (7.5 Nm).

d. Install the wire grommet to the left-side (No. 3) timing belt cover.

e. Install the sensor connector to the connector bracket and connect the sensor connector.

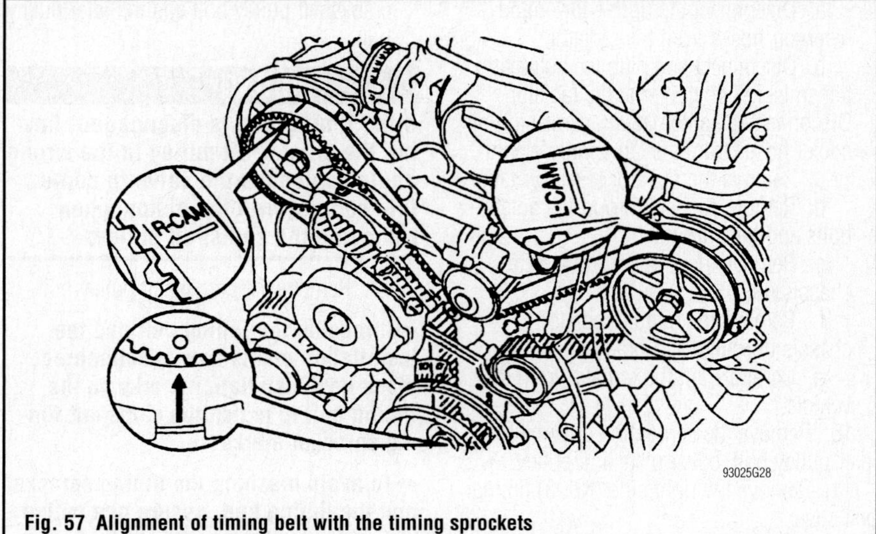

Fig. 57 Alignment of timing belt with the timing sprockets

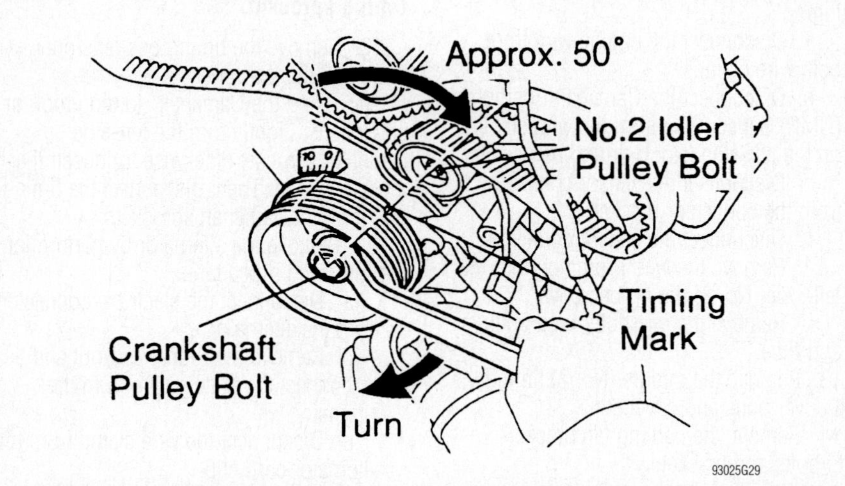

Fig. 58 Aligning of crankshaft pulley timing mark with the center line of the crankshaft pulley bolt and the idler pulley bolt

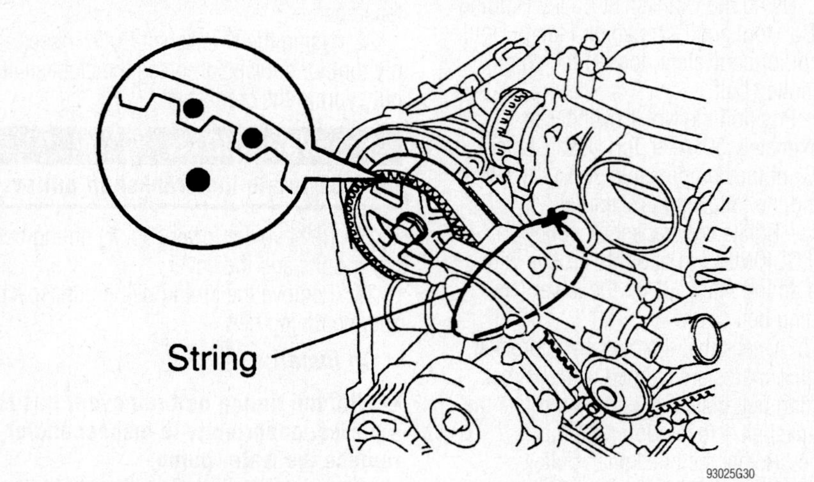

Fig. 59 Securing the timing belt with string and matchmarking the camshaft with the timing belt

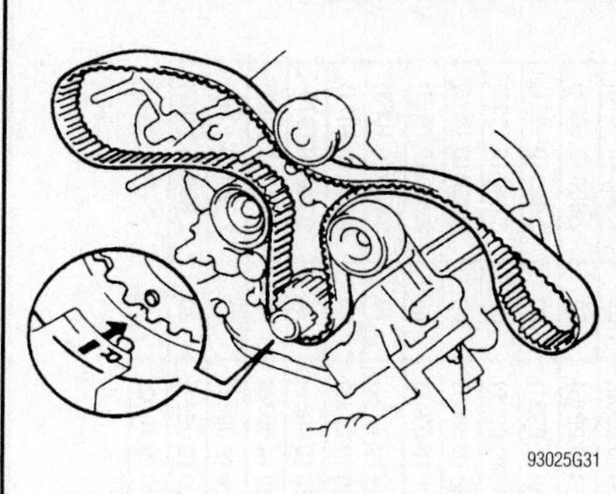

Fig. 60 Installing the timing belt on the crankshaft sprocket

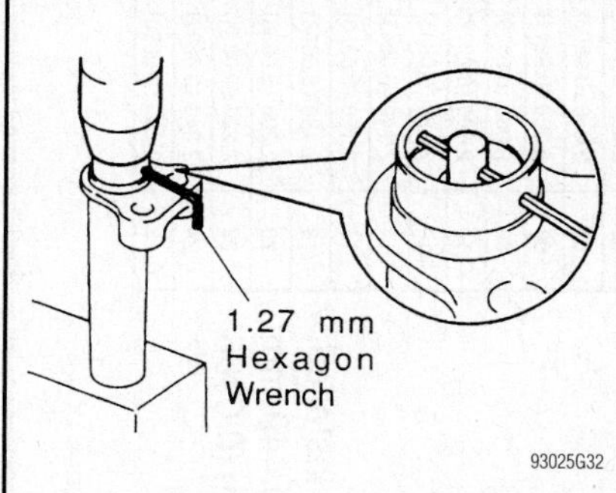

Fig. 61 Securing the timing belt tensioner pushrod

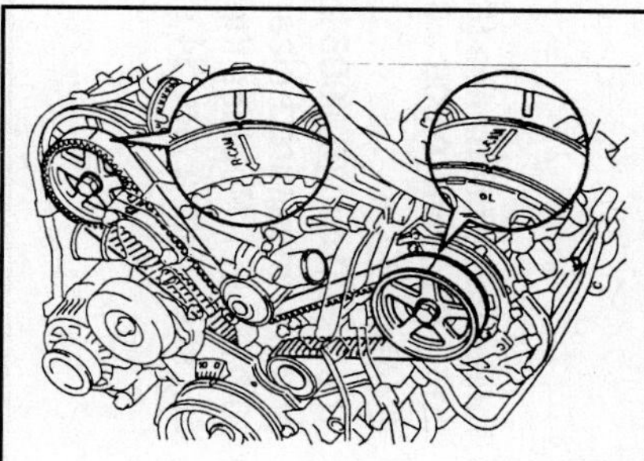

Fig. 62 Checking the TDC alignment marks after rotating the crankshaft 2 revolutions

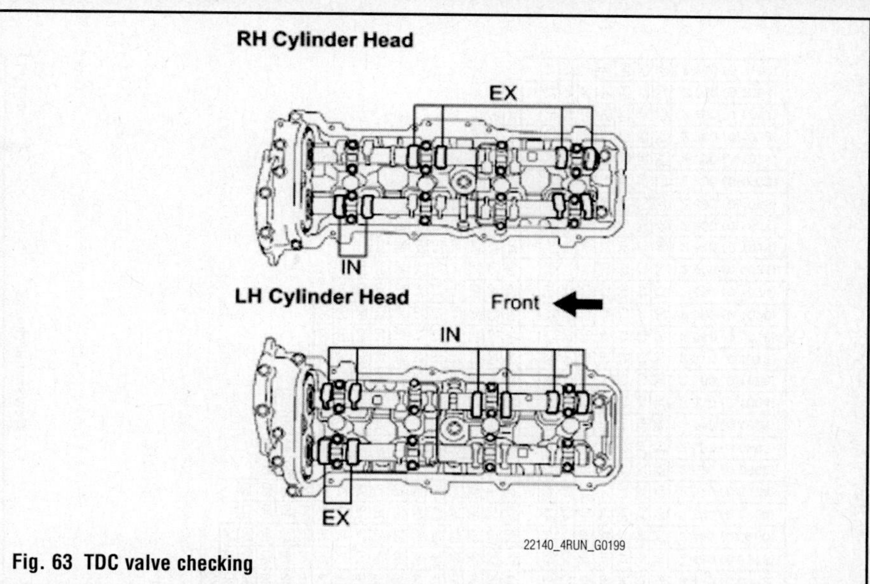

Fig. 63 TDC valve checking

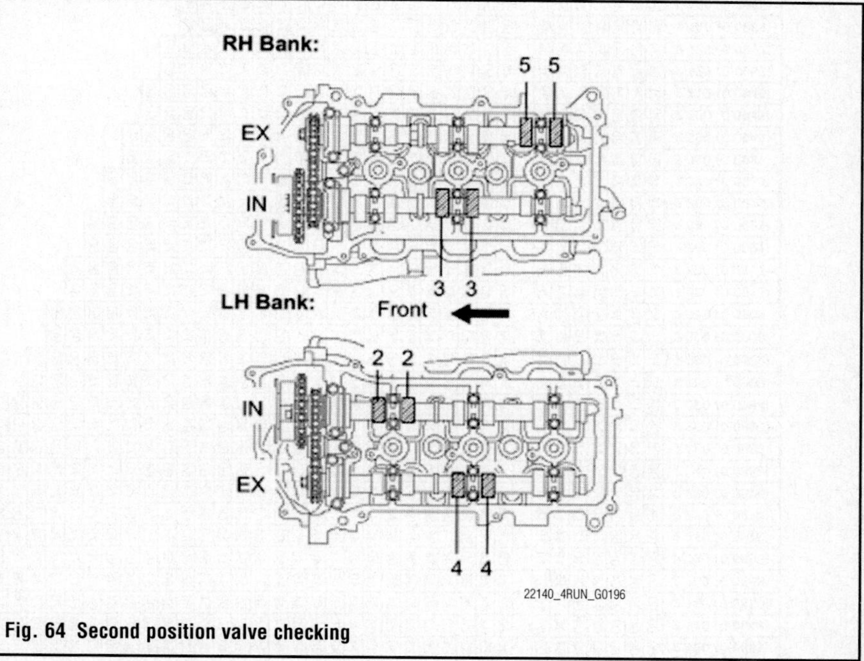

Fig. 64 Second position valve checking

f. Install the sensor wire and the engine wire to the clamps on the left-side (No. 3) timing belt cover.

46. Install the drive belt idler pulley and cover plate; then, torque the pulley bolt to 27 ft. lbs. (37 Nm).

47. To complete the installation, reverse the removal procedures.

48. Refill the cooling system and connect the negative battery cable.

VALVE LASH

ADJUSTMENT

See Figures 63 through 66.

➡Measure valve clearance with the engine cold.

1. Before servicing the vehicle, refer to the Precautions section.
2. Drain the cooling system.
3. Remove or disconnect the following:
    - Negative battery cable
    - Ignition coils
    - Valve covers
4. Turn the crankshaft pulley to align its notch with timing mark "0" of the No. 1 timing belt cover.
5. Check only the valves indicated.
6. Using a feeler gauge, measure the clearance between the valve lifter and camshaft.
7. Record the out-of-specification valve clearance measurements. They will be used later to determine the required replacement adjusting shim.
8. Turn the crankshaft 1 complete revolution (360°) and align the camshaft timing marks.
9. Check the valve clearance.

The valve clearance specifications are as follows:

- Intake: 0.006–0.010 inch (0.15–0.25mm)
- Exhaust: 0.010–0.014 inch (0.25–0.35mm)

10. Record the measurements for each valve.

Measured clearance mm (in.) \ Installed shim thickness mm (in.)	2.000 (0.0787)	2.020 (0.0795)	2.040 (0.0803)	2.060 (0.0811)	2.080 (0.0819)	2.100 (0.0827)	2.120 (0.0835)	2.140 (0.0843)	2.160 (0.0850)	2.180 (0.0858)	2.200 (0.0866)	2.210 (0.0870)	2.220 (0.0874)	2.230 (0.0878)	2.240 (0.0882)	2.250 (0.0886)	2.260 (0.0890)	2.270 (0.0894)	2.280 (0.0898)	2.290 (0.0902)	2.300 (0.0906)	2.310 (0.0909)	2.320 (0.0913)	2.330 (0.0917)	2.340 (0.0921)	2.350 (0.0925)	2.360 (0.0929)	2.370 (0.0933)	2.380 (0.0937)	2.390 (0.0941)	2.400 (0.0945)	2.410 (0.0949)	2.420 (0.0953)	2.430 (0.0957)	2.440 (0.0961)	2.450 (0.0965)	2.460 (0.0969)	2.470 (0.0972)	2.480 (0.0976)	2.490 (0.0980)	2.500 (0.0984)	2.510 (0.0988)	2.520 (0.0992)	2.530 (0.0996)	2.540 (0.1000)	2.550 (0.1004)	2.560 (0.1008)	2.570 (0.1012)	2.580 (0.1016)	2.590 (0.1020)	2.600 (0.1024)	2.620 (0.1031)	2.640 (0.1039)	2.660 (0.1047)	2.680 (0.1055)	2.700 (0.1063)	2.720 (0.1071)	2.740 (0.1079)	2.760 (0.1087)	2.780 (0.1094)	2.800 (0.1102)
0.000-0.030 (0.0000-0.0012)							00	00	00	00	02	02	04	04	06	06	08	08	10	10	12	12	14	14	16	16	18	18	20	20	22	22	24	24	26	26	28	28	30	30	32	32	34	34	36	36	38	38	40	40	42	44	46	48	50	52	54	56	58	60	62
0.031-0.050 (0.0012-0.0020)						00	00	00	00	02	04	06	06	08	08	10	10	12	12	14	14	16	16	18	18	20	20	22	22	24	24	26	26	28	28	30	30	32	32	34	34	36	36	38	38	40	40	42	42	44	44	46	48	50	52	54	56	58	60	62	64
0.051-0.070 (0.0020-0.0028)					00	00	00	00	02	04	06	08	08	10	10	12	12	14	14	16	16	18	18	20	20	22	22	24	24	26	26	28	28	30	30	32	32	34	34	36	36	38	38	40	40	42	42	44	44	46	46	48	50	52	54	56	58	60	62	64	66
0.071-0.090 (0.0028-0.0035)				00	00	00	00	02	04	06	08	10	10	12	12	14	14	16	16	18	18	20	20	22	22	24	24	26	26	28	28	30	30	32	32	34	34	36	36	38	38	40	40	42	42	44	44	46	46	48	48	50	52	54	56	58	60	62	64	66	68
0.091-0.110 (0.0036-0.0043)			00	00	00	00	02	04	06	08	10	12	12	14	14	16	16	18	18	20	20	22	22	24	24	26	26	28	28	30	30	32	32	34	34	36	36	38	38	40	40	42	42	44	44	46	46	48	48	50	50	52	54	56	58	60	62	64	66	68	70
0.111-0.130 (0.0044-0.0051)		00	00	00	00	02	04	06	08	10	12	14	14	16	16	18	18	20	20	22	22	24	24	26	26	28	28	30	30	32	32	34	34	36	36	38	38	40	40	42	42	44	44	46	46	48	48	50	50	52	52	54	56	58	60	62	64	66	68	70	72
0.131-0.149 (0.0052-0.0059)		00	00	00	02	04	06	08	10	12	14	16	16	18	18	20	20	22	22	24	24	26	26	28	28	30	30	32	32	34	34	36	36	38	38	40	40	42	42	44	44	46	46	48	48	50	50	52	52	54	54	56	58	60	62	64	66	68	70	72	74
0.150-0.250 (0.0059-0.0098)																																																													
0.251-0.270 (0.0099-0.0106)	06	08	10	12	14	16	18	20	22	24	26	28	28	30	30	32	32	34	34	36	36	38	38	40	40	42	42	44	44	46	46	48	48	50	50	52	52	54	54	56	56	58	58	60	60	62	62	64	64	66	66	68	70	72	74	76	78	80	80	80	
0.271-0.290 (0.0107-0.0114)	08	10	12	14	16	18	20	22	24	26	28	30	30	32	32	34	34	36	36	38	38	40	40	42	42	44	44	46	46	48	48	50	50	52	52	54	54	56	56	58	58	60	60	62	62	64	64	66	66	68	68	70	72	74	76	78	80	80	80		
0.291-0.310 (0.0115-0.0122)	10	12	14	16	18	20	22	24	26	28	30	32	32	34	34	36	36	38	38	40	40	42	42	44	44	46	46	48	48	50	50	52	52	54	54	56	56	58	58	60	60	62	62	64	64	66	66	68	68	70	70	72	74	76	78	80	80	80			
0.311-0.330 (0.0122-0.0130)	12	14	16	18	20	22	24	26	28	30	32	34	34	36	36	38	38	40	40	42	42	44	44	46	46	48	48	50	50	52	52	54	54	56	56	58	58	60	60	62	62	64	64	66	66	68	68	70	70	72	72	74	76	78	80	80	80				
0.331-0.350 (0.0130-0.0138)	14	16	18	20	22	24	26	28	30	32	34	36	36	38	38	40	40	42	42	44	44	46	46	48	48	50	50	52	52	54	54	56	56	58	58	60	60	62	62	64	64	66	66	68	68	70	70	72	72	74	74	76	78	80	80	80					
0.351-0.370 (0.0138-0.0146)	16	18	20	22	24	26	28	30	32	34	36	38	38	40	40	42	42	44	44	46	46	48	48	50	50	52	52	54	54	56	56	58	58	60	60	62	62	64	64	66	66	68	68	70	70	72	72	74	74	76	76	78	80	80	80						
0.371-0.390 (0.0146-0.0154)	18	20	22	24	26	28	30	32	34	36	38	40	40	42	42	44	44	46	46	48	48	50	50	52	52	54	54	56	56	58	58	60	60	62	62	64	64	66	66	68	68	70	70	72	72	74	74	76	76	78	78	80	80	80							
0.391-0.410 (0.0154-0.0161)	20	22	24	26	28	30	32	34	36	38	40	42	42	44	44	46	46	48	48	50	50	52	52	54	54	56	56	58	58	60	60	62	62	64	64	66	66	68	68	70	70	72	72	74	74	76	76	78	78	80	80	80	80								
0.411-0.430 (0.0162-0.0169)	22	24	26	28	30	32	34	36	38	40	42	44	44	46	46	48	48	50	50	52	52	54	54	56	56	58	58	60	60	62	62	64	64	66	66	68	68	70	70	72	72	74	74	76	76	78	78	80	80	80	80	80									
0.431-0.450 (0.0170-0.0177)	24	26	28	30	32	34	36	38	40	42	44	46	46	48	48	50	50	52	52	54	54	56	56	58	58	60	60	62	62	64	64	66	66	68	68	70	70	72	72	74	74	76	76	78	78	80	80	80	80	80	80										
0.451-0.470 (0.0178-0.0185)	26	28	30	32	34	36	38	40	42	44	46	48	48	50	50	52	52	54	54	56	56	58	58	60	60	62	62	64	64	66	66	68	68	70	70	72	72	74	74	76	76	78	78	80	80	80	80	80	80	80											
0.471-0.490 (0.0185-0.0193)	28	30	32	34	36	38	40	42	44	46	48	50	50	52	52	54	54	56	56	58	58	60	60	62	62	64	64	66	66	68	68	70	70	72	72	74	74	76	76	78	78	80	80	80	80	80	80	80													
0.491-0.510 (0.0193-0.0201)	30	32	34	36	38	40	42	44	46	48	50	52	52	54	54	56	56	58	58	60	60	62	62	64	64	66	66	68	68	70	70	72	72	74	74	76	76	78	78	80	80	80	80	80	80	80															
0.511-0.530 (0.0201-0.0209)	32	34	36	38	40	42	44	46	48	50	52	54	54	56	56	58	58	60	60	62	62	64	64	66	66	68	68	70	70	72	72	74	74	76	76	78	78	80	80	80	80	80	80	80																	
0.531-0.550 (0.0209-0.0217)	34	36	38	40	42	44	46	48	50	52	54	56	56	58	58	60	60	62	62	64	64	66	66	68	68	70	70	72	72	74	74	76	76	78	78	80	80	80	80	80	80	80																			
0.551-0.570 (0.0217-0.0224)	36	38	40	42	44	46	48	50	52	54	56	58	58	60	60	62	62	64	64	66	66	68	68	70	70	72	72	74	74	76	76	78	78	80	80	80	80	80	80	80																					
0.571-0.590 (0.0225-0.0232)	38	40	42	44	46	48	50	52	54	56	58	60	60	62	62	64	64	66	66	68	68	70	70	72	72	74	74	76	76	78	78	80	80	80	80	80	80	80																							
0.591-0.610 (0.0233-0.0240)	40	42	44	46	48	50	52	54	56	58	60	62	62	64	64	66	66	68	68	70	70	72	72	74	74	76	76	78	78	80	80	80	80	80	80	80																									
0.611-0.630 (0.0241-0.0248)	42	44	46	48	50	52	54	56	58	60	62	64	64	66	66	68	68	70	70	72	72	74	74	76	76	78	78	80	80	80	80	80	80	80																											
0.631-0.650 (0.0248-0.0256)	44	46	48	50	52	54	56	58	60	62	64	66	66	68	68	70	70	72	72	74	74	76	76	78	78	80	80	80	80	80	80	80																													
0.651-0.670 (0.0256-0.0264)	46	48	50	52	54	56	58	60	62	64	66	68	68	70	70	72	72	74	74	76	76	78	78	80	80	80	80	80	80	80																															
0.671-0.690 (0.0264-0.0272)	48	50	52	54	56	58	60	62	64	66	68	70	70	72	72	74	74	76	76	78	78	80	80	80	80	80	80	80																																	
0.691-0.710 (0.0272-0.0280)	50	52	54	56	58	60	62	64	66	68	70	72	72	74	74	76	76	78	78	80	80	80	80	80	80	80																																			
0.711-0.730 (0.0280-0.0287)	52	54	56	58	60	62	64	66	68	70	72	74	74	76	76	78	78	80	80	80	80	80	80	80																																					
0.731-0.750 (0.0288-0.0295)	54	56	58	60	62	64	66	68	70	72	74	76	76	78	78	80	80	80	80	80	80	80																																							
0.751-0.770 (0.0296-0.0303)	56	58	60	62	64	66	68	70	72	74	76	78	78	80	80	80	80	80	80	80																																									
0.771-0.790 (0.0304-0.0311)	58	60	62	64	66	68	70	72	74	76	78	80	80	80	80	80	80	80																																											
0.791-0.810 (0.0311-0.0319)	60	62	64	66	68	70	72	74	76	78	80	80	80	80	80	80																																													
0.811-0.830 (0.0319-0.0327)	62	64	66	68	70	72	74	76	78	80	80	80	80	80																																															
0.831-0.850 (0.0327-0.0335)	64	66	68	70	72	74	76	78	80	80	80	80																																																	
0.851-0.870 (0.0335-0.0343)	66	68	70	72	74	76	78	80	80	80																																																			
0.871-0.890 (0.0343-0.0350)	68	70	72	74	76	78	80	80	80																																																				
0.891-0.910 (0.0351-0.0358)	70	72	74	76	78	80	80	80																																																					
0.911-0.930 (0.0359-0.0366)	72	74	76	78	80	80	80																																																							
0.931-0.950 (0.0367-0.0374)	74	76	78	80	80	80																																																								
0.951-0.970 (0.0374-0.0382)	76	78	80	80	80																																																									
0.971-0.990 (0.0382-0.0390)	78	80	80	80																																																										
0.991-1.010 (0.0390-0.0398)	80	80	80																																																											
1.011-1.030 (0.0398-0.0406)	80	80																																																												
1.031-1.050 (0.0406-0.0413)	80																																																													

Intake valve clearance (Cold):
0.15 – 0.25 mm (0.006 – 0.010 in.)

EXAMPLE:
The 2.300 mm (0.0906 in.) shim is installed, and the measured clearance is 0.440 mm (0.0173 in.). Replace the 2.300 mm (0.0906 in.) shim with a No. 54 shim.

New shim thickness mm (in.)

Shim No.	Thickness	Shim No.	Thickness	Shim No.	Thickness
00	2.000 (0.0787)	28	2.280 (0.0898)	56	2.560 (0.1008)
02	2.020 (0.0795)	30	2.300 (0.0906)	58	2.580 (0.1016)
04	2.040 (0.0803)	32	2.320 (0.0913)	60	2.600 (0.1024)
06	2.060 (0.0811)	34	2.340 (0.0921)	62	2.620 (0.1031)
08	2.080 (0.0819)	36	2.360 (0.0929)	64	2.640 (0.1039)
10	2.100 (0.0827)	38	2.380 (0.0937)	66	2.660 (0.1047)
12	2.120 (0.0835)	40	2.400 (0.0945)	68	2.680 (0.1055)
14	2.140 (0.0843)	42	2.420 (0.0953)	70	2.700 (0.1063)
16	2.160 (0.0850)	44	2.440 (0.0961)	72	2.720 (0.1071)
18	2.180 (0.0858)	46	2.460 (0.0969)	74	2.740 (0.1079)
20	2.200 (0.0866)	48	2.480 (0.0976)	76	2.760 (0.1087)
22	2.220 (0.0874)	50	2.500 (0.0984)	78	2.780 (0.1094)
24	2.240 (0.0882)	52	2.520 (0.0992)	80	2.800 (0.1102)
26	2.260 (0.0890)	54	2.540 (0.1000)		

7924SG71

Fig. 65 Intake valve clearance shim selection chart

Measured clearance mm (in.) \ Installed shim thickness mm (in.)	2.000 (0.0787)	2.020 (0.0795)	2.040 (0.0803)	2.060 (0.0811)	2.080 (0.0819)	2.100 (0.0827)	2.120 (0.0835)	2.140 (0.0843)	2.160 (0.0850)	2.180 (0.0858)	2.200 (0.0866)	2.210 (0.0870)	2.220 (0.0874)	2.230 (0.0878)	2.240 (0.0882)	2.250 (0.0886)	2.260 (0.0890)	2.270 (0.0894)	2.280 (0.0898)	2.290 (0.0902)	2.300 (0.0906)	2.310 (0.0909)	2.320 (0.0913)	2.330 (0.0917)	2.340 (0.0921)	2.350 (0.0925)	2.360 (0.0929)	2.370 (0.0933)	2.380 (0.0937)	2.390 (0.0941)	2.400 (0.0945)
0.000–0.030 (0.0000–0.0012)												00	00	00	00	00	00	00	00	00	02	02	04	04	06	06	08	08	10	10	12
0.031–0.050 (0.0012–0.0020)											00	00	00	00	00	00	00	02	02	04	04	06	06	08	08	10	10	12	12	14	14
0.051–0.070 (0.0020–0.0028)										00	00	00	00	00	00	02	02	04	04	06	06	08	08	10	10	12	12	14	14	16	16
0.071–0.090 (0.0028–0.0035)									00	00	00	00	00	02	02	04	04	06	06	08	08	10	10	12	12	14	14	16	16	18	18
0.091–0.110 (0.0036–0.0043)								00	00	00	00	02	02	04	04	06	06	08	08	10	10	12	12	14	14	16	16	18	18	20	20
0.111–0.130 (0.0044–0.0051)							00	00	00	00	02	04	04	06	06	08	08	10	10	12	12	14	14	16	16	18	18	20	20	22	22
0.131–0.150 (0.0052–0.0059)						00	00	00	00	02	04	06	06	08	08	10	10	12	12	14	14	16	16	18	18	20	20	22	22	24	24
0.151–0.170 (0.0059–0.0067)					00	00	00	00	02	04	06	08	08	10	10	12	12	14	14	16	16	18	18	20	20	22	22	24	24	26	26
0.171–0.190 (0.0067–0.0075)				00	00	00	00	02	04	06	08	10	10	12	12	14	14	16	16	18	18	20	20	22	22	24	24	26	26	28	28
0.191–0.210 (0.0075–0.0083)			00	00	00	00	02	04	06	08	10	12	12	14	14	16	16	18	18	20	20	22	22	24	24	26	26	28	28	30	30
0.211–0.230 (0.0083–0.0091)		00	00	00	00	02	04	06	08	10	12	14	14	16	16	18	18	20	20	22	22	24	24	26	26	28	28	30	30	32	32
0.231–0.249 (0.0091–0.0098)		00	00	00	02	04	06	08	10	12	14	16	16	18	18	20	20	22	22	24	24	26	26	28	28	30	30	32	32	34	34
0.250–0.350 (0.0098–0.0138)																															
0.351–0.370 (0.0138–0.0146)	06	08	10	12	14	16	18	20	22	24	26	28	28	30	30	32	32	34	34	36	36	38	38	40	40	42	42	44	44	46	46
0.371–0.390 (0.0146–0.0154)	08	10	12	14	16	18	20	22	24	26	28	30	30	32	32	34	34	36	36	38	38	40	40	42	42	44	44	46	46	48	48
0.391–0.410 (0.0154–0.0161)	10	12	14	16	18	20	22	24	26	28	30	32	32	34	34	36	36	38	38	40	40	42	42	44	44	46	46	48	48	50	50
0.411–0.430 (0.0162–0.0169)	12	14	16	18	20	22	24	26	28	30	32	34	34	36	36	38	38	40	40	42	42	44	44	46	46	48	48	50	50	52	52
0.431–0.450 (0.0170–0.0177)	14	16	18	20	22	24	26	28	30	32	34	36	36	38	38	40	40	42	42	44	44	46	46	48	48	50	50	52	52	54	54
0.451–0.470 (0.0178–0.0185)	16	18	20	22	24	26	28	30	32	34	36	38	38	40	40	42	42	44	44	46	46	48	48	50	50	52	52	54	54	56	56
0.471–0.490 (0.0185–0.0193)	18	20	22	24	26	28	30	32	34	36	38	40	40	42	42	44	44	46	46	48	48	50	50	52	52	54	54	56	56	58	58
0.491–0.510 (0.0193–0.0201)	20	22	24	26	28	30	32	34	36	38	40	42	42	44	44	46	46	48	48	50	50	52	52	54	54	56	56	58	58	60	60
0.511–0.530 (0.0201–0.0209)	22	24	26	28	30	32	34	36	38	40	42	44	44	46	46	48	48	50	50	52	52	54	54	56	56	58	58	60	60	62	62
0.531–0.550 (0.0209–0.0217)	24	26	28	30	32	34	36	38	40	42	44	46	46	48	48	50	50	52	52	54	54	56	56	58	58	60	60	62	62	64	64
0.551–0.570 (0.0217–0.0224)	26	28	30	32	34	36	38	40	42	44	46	48	48	50	50	52	52	54	54	56	56	58	58	60	60	62	62	64	64	66	66
0.571–0.590 (0.0225–0.0232)	28	30	32	34	36	38	40	42	44	46	48	50	50	52	52	54	54	56	56	58	58	60	60	62	62	64	64	66	66	68	68
0.591–0.610 (0.0233–0.0240)	30	32	34	36	38	40	42	44	46	48	50	52	52	54	54	56	56	58	58	60	60	62	62	64	64	66	66	68	68	70	70
0.611–0.630 (0.0241–0.0248)	32	34	36	38	40	42	44	46	48	50	52	54	54	56	56	58	58	60	60	62	62	64	64	66	66	68	68	70	70	72	72
0.631–0.650 (0.0248–0.0256)	34	36	38	40	42	44	46	48	50	52	54	56	56	58	58	60	60	62	62	64	64	66	66	68	68	70	70	72	72	74	74
0.651–0.670 (0.0256–0.0264)	36	38	40	42	44	46	48	50	52	54	56	58	58	60	60	62	62	64	64	66	66	68	68	70	70	72	72	74	74	76	76
0.671–0.690 (0.0264–0.0272)	38	40	42	44	46	48	50	52	54	56	58	60	60	62	62	64	64	66	66	68	68	70	70	72	72	74	74	76	76	78	78
0.691–0.710 (0.0272–0.0280)	40	42	44	46	48	50	52	54	56	58	60	62	62	64	64	66	66	68	68	70	70	72	72	74	74	76	76	78	78	80	80
0.711–0.730 (0.0280–0.0287)	42	44	46	48	50	52	54	56	58	60	62	64	64	66	66	68	68	70	70	72	72	74	74	76	76	78	78	80	80	80	80
0.731–0.750 (0.0288–0.0295)	44	46	48	50	52	54	56	58	60	62	64	66	66	68	68	70	70	72	72	74	74	76	76	78	78	80	80	80	80	80	80
0.751–0.770 (0.0296–0.0303)	46	48	50	52	54	56	58	60	62	64	66	68	68	70	70	72	72	74	74	76	76	78	78	80	80	80	80	80	80	80	
0.771–0.790 (0.0304–0.0311)	48	50	52	54	56	58	60	62	64	66	68	70	70	72	72	74	74	76	76	78	78	80	80	80	80	80	80	80			
0.791–0.810 (0.0311–0.0319)	50	52	54	56	58	60	62	64	66	68	70	72	72	74	74	76	76	78	78	80	80	80	80	80	80	80					
0.811–0.830 (0.0319–0.0327)	52	54	56	58	60	62	64	66	68	70	72	74	74	76	76	78	78	80	80	80	80	80	80	80							
0.831–0.850 (0.0327–0.0335)	54	56	58	60	62	64	66	68	70	72	74	76	76	78	78	80	80	80	80	80	80	80									
0.851–0.870 (0.0335–0.0343)	56	58	60	62	64	66	68	70	72	74	76	78	78	80	80	80	80	80	80	80											
0.871–0.890 (0.0343–0.0350)	58	60	62	64	66	68	70	72	74	76	78	80	80	80	80	80	80	80													
0.891–0.910 (0.0351–0.0358)	60	62	64	66	68	70	72	74	76	78	80	80	80	80	80	80															
0.911–0.930 (0.0359–0.0366)	62	64	66	68	70	72	74	76	78	80	80	80	80	80																	
0.931–0.950 (0.0367–0.0374)	64	66	68	70	72	74	76	78	80	80	80	80																			
0.951–0.970 (0.0374–0.0382)	66	68	70	72	74	76	78	80	80	80																					
0.971–0.990 (0.0382–0.0390)	68	70	72	74	76	78	80	80	80																						
0.991–1.010 (0.0390–0.0398)	70	72	74	76	78	80	80	80																							
1.011–1.030 (0.0398–0.0406)	72	74	76	78	80	80	80																								
1.031–1.050 (0.0406–0.0413)	74	76	78	80	80	80																									
1.051–1.070 (0.0414–0.0421)	76	78	80	80	80																										
1.071–1.090 (0.0422–0.0429)	78	80	80	80																											
1.091–1.110 (0.0430–0.0437)	80	80	80																												
1.111–1.130 (0.0437–0.0445)	80	80																													
1.131–1.150 (0.0445–0.0453)	80																														

Measured clearance mm (in.) \ Installed shim thickness mm (in.)	2.410 (0.0949)	2.420 (0.0953)	2.430 (0.0957)	2.440 (0.0961)	2.450 (0.0965)	2.460 (0.0969)	2.470 (0.0972)	2.480 (0.0976)	2.490 (0.0980)	2.500 (0.0984)	2.510 (0.0988)	2.520 (0.0992)	2.530 (0.0996)	2.540 (0.1000)	2.550 (0.1004)	2.560 (0.1008)	2.570 (0.1012)	2.580 (0.1016)	2.590 (0.1020)	2.600 (0.1024)	2.620 (0.1031)	2.640 (0.1039)	2.660 (0.1047)	2.680 (0.1055)	2.700 (0.1063)	2.720 (0.1071)	2.740 (0.1079)	2.760 (0.1087)	2.780 (0.1094)	2.800 (0.1102)
0.000–0.030 (0.0000–0.0012)	12	14	14	16	16	18	18	20	20	22	22	24	24	26	26	28	28	30	30	32	34	36	38	40	42	44	46	48	50	52
0.031–0.050 (0.0012–0.0020)	16	16	18	18	20	20	22	22	24	24	26	26	28	28	30	30	32	32	34	34	36	38	40	42	44	46	48	50	52	54
0.051–0.070 (0.0020–0.0028)	18	18	20	20	22	22	24	24	26	26	28	28	30	30	32	32	34	34	36	36	38	40	42	44	46	48	50	52	54	56
0.071–0.090 (0.0028–0.0035)	20	20	22	22	24	24	26	26	28	28	30	30	32	32	34	34	36	36	38	38	40	42	44	46	48	50	52	54	56	58
0.091–0.110 (0.0036–0.0043)	22	22	24	24	26	26	28	28	30	30	32	32	34	34	36	36	38	38	40	40	42	44	46	48	50	52	54	56	58	60
0.111–0.130 (0.0044–0.0051)	24	24	26	26	28	28	30	30	32	32	34	34	36	36	38	38	40	40	42	42	44	46	48	50	52	54	56	58	60	62
0.131–0.150 (0.0052–0.0059)	26	26	28	28	30	30	32	32	34	34	36	36	38	38	40	40	42	42	44	44	46	48	50	52	54	56	58	60	62	64
0.151–0.170 (0.0059–0.0067)	28	28	30	30	32	32	34	34	36	36	38	38	40	40	42	42	44	44	46	46	48	50	52	54	56	58	60	62	64	66
0.171–0.190 (0.0067–0.0075)	30	30	32	32	34	34	36	36	38	38	40	40	42	42	44	44	46	46	48	48	50	52	54	56	58	60	62	64	66	68
0.191–0.210 (0.0075–0.0083)	32	32	34	34	36	36	38	38	40	40	42	42	44	44	46	46	48	48	50	50	52	54	56	58	60	62	64	66	68	70
0.211–0.230 (0.0083–0.0091)	34	34	36	36	38	38	40	40	42	42	44	44	46	46	48	48	50	50	52	52	54	56	58	60	62	64	66	68	70	72
0.231–0.249 (0.0091–0.0098)	36	36	38	38	40	40	42	42	44	44	46	46	48	48	50	50	52	52	54	54	56	58	60	62	64	66	68	70	72	74
0.250–0.350 (0.0098–0.0138)																														
0.351–0.370 (0.0138–0.0146)	48	48	50	50	52	52	54	54	56	56	58	58	60	60	62	62	64	64	66	66	68	70	72	74	76	78	80	80	80	
0.371–0.390 (0.0146–0.0154)	50	50	52	52	54	54	56	56	58	58	60	60	62	62	64	64	66	66	68	68	70	72	74	76	78	80	80	80		
0.391–0.410 (0.0154–0.0161)	52	52	54	54	56	56	58	58	60	60	62	62	64	64	66	66	68	68	70	70	72	74	76	78	80	80	80			
0.411–0.430 (0.0162–0.0169)	54	54	56	56	58	58	60	60	62	62	64	64	66	66	68	68	70	70	72	72	74	76	78	80	80	80				
0.431–0.450 (0.0170–0.0177)	56	56	58	58	60	60	62	62	64	64	66	66	68	68	70	70	72	72	74	74	76	78	80	80	80					
0.451–0.470 (0.0178–0.0185)	58	58	60	60	62	62	64	64	66	66	68	68	70	70	72	72	74	74	76	76	78	80	80	80						
0.471–0.490 (0.0185–0.0193)	60	60	62	62	64	64	66	66	68	68	70	70	72	72	74	74	76	76	78	78	80	80	80							
0.491–0.510 (0.0193–0.0201)	62	62	64	64	66	66	68	68	70	70	72	72	74	74	76	76	78	78	80	80	80	80								
0.511–0.530 (0.0201–0.0209)	64	64	66	66	68	68	70	70	72	72	74	74	76	76	78	78	80	80	80	80	80									
0.531–0.550 (0.0209–0.0217)	66	66	68	68	70	70	72	72	74	74	76	76	78	78	80	80	80	80	80	80										
0.551–0.570 (0.0217–0.0224)	68	68	70	70	72	72	74	74	76	76	78	78	80	80	80	80	80	80	80											
0.571–0.590 (0.0225–0.0232)	70	70	72	72	74	74	76	76	78	78	80	80	80	80	80	80	80													
0.591–0.610 (0.0233–0.0240)	72	72	74	74	76	76	78	78	80	80	80	80	80	80	80															
0.611–0.630 (0.0241–0.0248)	74	74	76	76	78	78	80	80	80	80	80	80	80																	
0.631–0.650 (0.0248–0.0256)	76	76	78	78	80	80	80	80	80	80	80																			
0.651–0.670 (0.0256–0.0264)	78	78	80	80	80	80	80	80	80																					
0.671–0.690 (0.0264–0.0272)	80	80	80	80	80	80	80																							
0.691–0.710 (0.0272–0.0280)	80	80	80	80	80																									
0.711–0.730 (0.0280–0.0287)	80	80	80																											
0.731–0.750 (0.0288–0.0295)	80																													

Exhaust valve clearance (Cold):
0.25 – 0.35 mm (0.010 – 0.014 in.)

EXAMPLE:
The 2.300 mm (0.0906 in.) shim is installed, and the measured clearance is 0.440 mm (0.0173 in.). Replace the 2.300 mm (0.0906 in.) shim with a No. 44 shim.

New shim thickness mm (in.)

Shim No.	Thickness	Shim No.	Thickness	Shim No.	Thickness
00	2.000 (0.0787)	28	2.280 (0.0898)	56	2.560 (0.1008)
02	2.020 (0.0795)	30	2.300 (0.0906)	58	2.580 (0.1016)
04	2.040 (0.0803)	32	2.320 (0.0913)	60	2.600 (0.1024)
06	2.060 (0.0811)	34	2.340 (0.0921)	62	2.620 (0.1031)
08	2.080 (0.0819)	36	2.360 (0.0929)	64	2.640 (0.1039)
10	2.100 (0.0827)	38	2.380 (0.0937)	66	2.660 (0.1047)
12	2.120 (0.0835)	40	2.400 (0.0945)	68	2.680 (0.1055)
14	2.140 (0.0843)	42	2.420 (0.0953)	70	2.700 (0.1063)
16	2.160 (0.0850)	44	2.440 (0.0961)	72	2.720 (0.1071)
18	2.180 (0.0858)	46	2.460 (0.0969)	74	2.740 (0.1079)
20	2.200 (0.0866)	48	2.480 (0.0976)	76	2.760 (0.1087)
22	2.220 (0.0874)	50	2.500 (0.0984)	78	2.780 (0.1094)
24	2.240 (0.0882)	52	2.520 (0.0992)	80	2.800 (0.1102)
26	2.260 (0.0890)	54	2.540 (0.1000)		

7924SG72

Fig. 66 Exhaust valve clearance shim selection chart

11. When all valve clearances have been measured, remove the camshafts.

12. Remove the valve shims and measure them. Note this measurement along with the clearance measurement recorded earlier.

13. Using the valve clearance and shim thickness measurements, find replacement shims in the Shim Selection charts.

To install:

14. Install or connect the following:

- Replacement valve shims
- Camshafts
- Valve covers
- Ignition coils
- Negative battery cable

15. Fill the cooling system.

16. Start the engine and check for leaks.

ENGINE PERFORMANCE & EMISSION CONTROLS

CAMSHAFT POSITION (CMP) SENSOR

LOCATION

See Figure 67.

Refer to the accompanying illustration.

REMOVAL & INSTALLATION

1. Drain engine coolant.
2. Remove V-bank covers sub-assembly.
3. Remove fan and generator v belt.
4. Remove oil cooler pipe.
5. Remove timing belt cover sub-assembly No. 3 LH.
6. Disconnect the Camshaft Position (CMP) Sensor connector.
7. Remove the bolt, stud bolt and Camshaft Position (CMP) Sensor.

To install:

8. Install the Camshaft Position (CMP) Sensor with the bolt and stud bolt. Tighten bolt to 66 inch lbs. (7.5 Nm).
9. Reconnect the Camshaft Position (CMP) Sensor connector.
10. Install timing belt cover sub-assembly no. 3 LH.
11. Install oil cooler pipe.
12. Install fan and generator v belt.
13. Add engine coolant.
14. Check for engine coolant leaks.
15. Install V-bank cover sub-assembly.

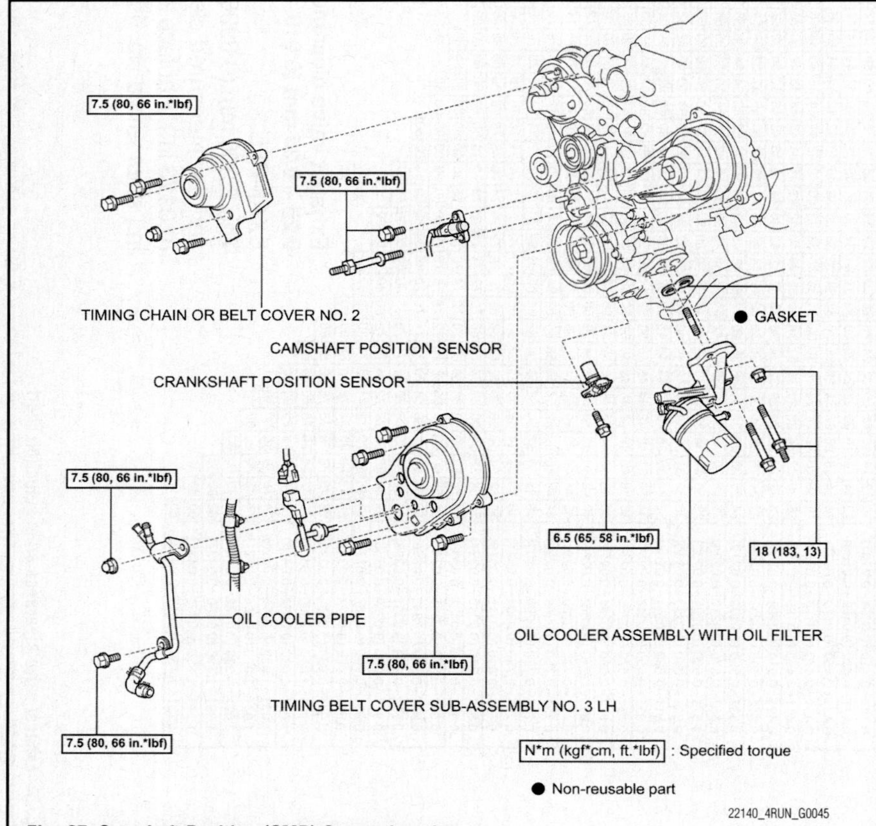

Fig. 67 Camshaft Position (CMP) Sensor location

CRANKSHAFT POSITION (CKP) SENSOR

LOCATION

See Figure 68.

Refer to the accompanying illustration.

REMOVAL & INSTALLATION

1. Disconnect cable from negative battery terminal.

➡Wait at least 90 seconds after disconnecting the cable from the negative (-) battery terminal to prevent airbag and seat belt pretensioner activation.

2. Remove No. 1 engine under cover.
3. For V6 engine, remove A/C compressor.
4. Disconnect the sensor connector.
5. Remove the bolt and sensor.

To install:

6. Install Crankshaft Position (CKP) Sensor.
7. Install the sensor with the bolt. Tighten bolt to 57 inch lbs. (6.5 Nm).
8. Connect the sensor connector.
9. For V6 engine, install A/C compressor and charge system.
10. Install no. 1 engine under cover.
11. Connect cable to negative battery terminal.
12. Perform initialization, if necessary.

ELECTRONIC CONTROL MODULE (ECM)

REMOVAL & INSTALLATION

1. Disconnect cable from negative battery terminal.

➡Wait at least 90 seconds after disconnecting the cable from the negative

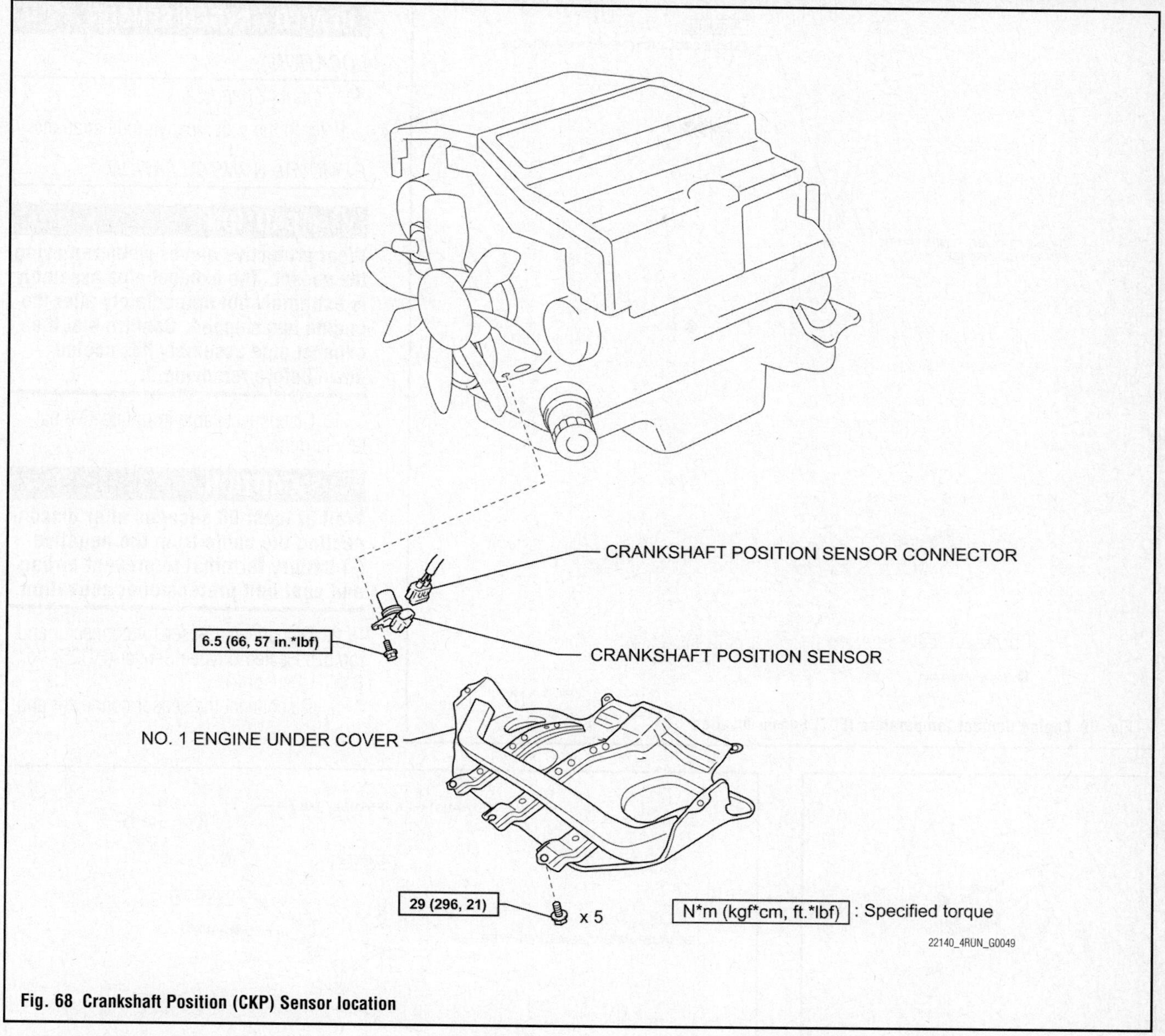

Fig. 68 Crankshaft Position (CKP) Sensor location

(-) battery terminal to prevent airbag and seat belt pretensioner activation.

2. Remove glove box compartment door.
3. Remove the 2 screws and glove compartment door.
4. Remove no. 2 finish panel lower.
5. Remove the 3 screws and no. 2 finish panel lower.
6. Disconnect the 5 ECM connectors.
7. Remove the 3 screws and ECM.

To install:

8. Install the ECM with the 3 screws, tighten to 49 inch lbs. (5.5 Nm).
9. Connect the 5 ECM connectors.

➡Be sure to securely connect the connectors.

10. Install the No. 2 finish panel lower with the 3 screws.
11. Install the glove compartment door with the 2 screws.
12. Connect cable to negative battery terminal.
13. Perform initialization, if necessary.

➡Certain systems need to be initialized after disconnecting and reconnecting the cable from the negative (-) battery terminal.

ENGINE COOLANT TEMPERATURE (ECT) SENSOR

LOCATION

See Figure 69.

Refer to the accompanying illustration.

REMOVAL & INSTALLATION

See Figure 70.

1. Disconnect cable from negative battery terminal.

➡Wait at least 90 seconds after disconnecting the cable from the negative (-) battery terminal to prevent airbag and seat belt pretensioner activation.

2. Drain engine coolant.
3. Remove V-bank cover sub-assembly.
4. Remove intake air connector pipe.
5. Remove throttle body.
6. Disconnect the sensor connector.
7. Remove the sensor.
8. Remove the gasket from the sensor.

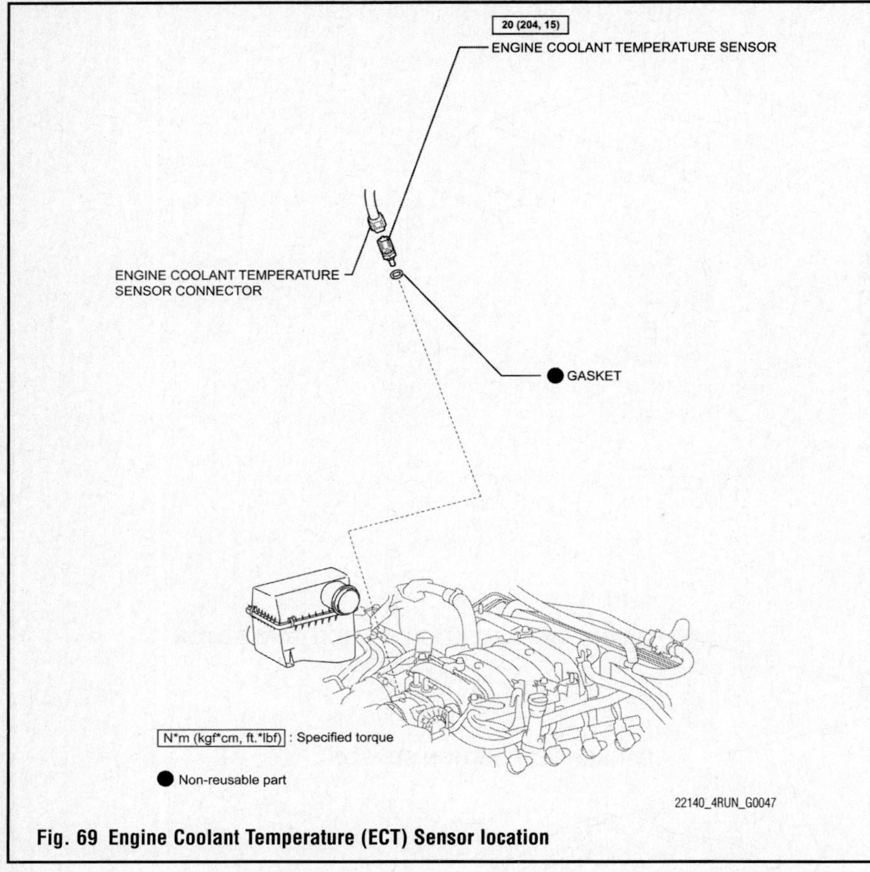

Fig. 69 Engine Coolant Temperature (ECT) Sensor location

HEATED OXYGEN SENSOR

LOCATION

See Figures 71 and 72.

Refer to the accompanying illustrations.

REMOVAL & INSTALLATION

CAUTION

Wear protective gloves when removing the sensor. The exhaust pipe assembly is extremely hot immediately after the engine has stopped. Confirm that the exhaust pipe assembly has cooled down before removing it.

1. Disconnect cable from negative battery terminal.

CAUTION

Wait at least 90 seconds after disconnecting the cable from the negative (-) battery terminal to prevent airbag and seat belt pretensioner activation.

2. Disconnect the sensor connector and remove Heated Oxygen Sensor (HO2S) (for Bank 1 Sensor 2).
3. Disconnect the sensor connector and

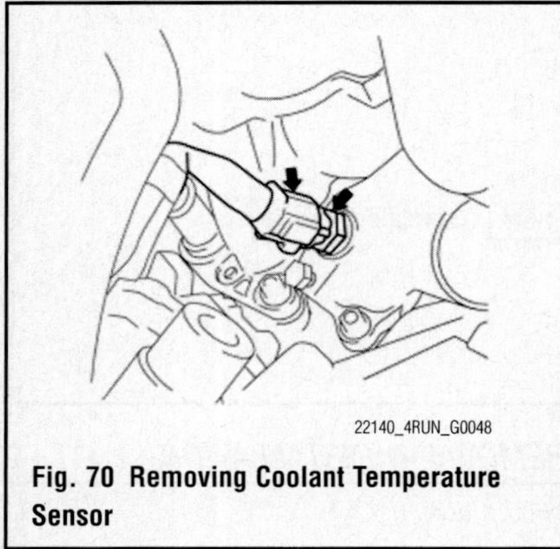

Fig. 70 Removing Coolant Temperature Sensor

To install:

9. Install a new gasket to the sensor.
10. Install the sensor. Tighten to 15 ft. lbs. (20 Nm).
11. Install throttle body.
12. Install intake air connector pipe.
13. Install V-bank cover sub-assembly.
14. Connect cable to negative battery terminal.
15. Add engine coolant.
16. Check for engine coolant leaks.
17. Perform initialization, if necessary.

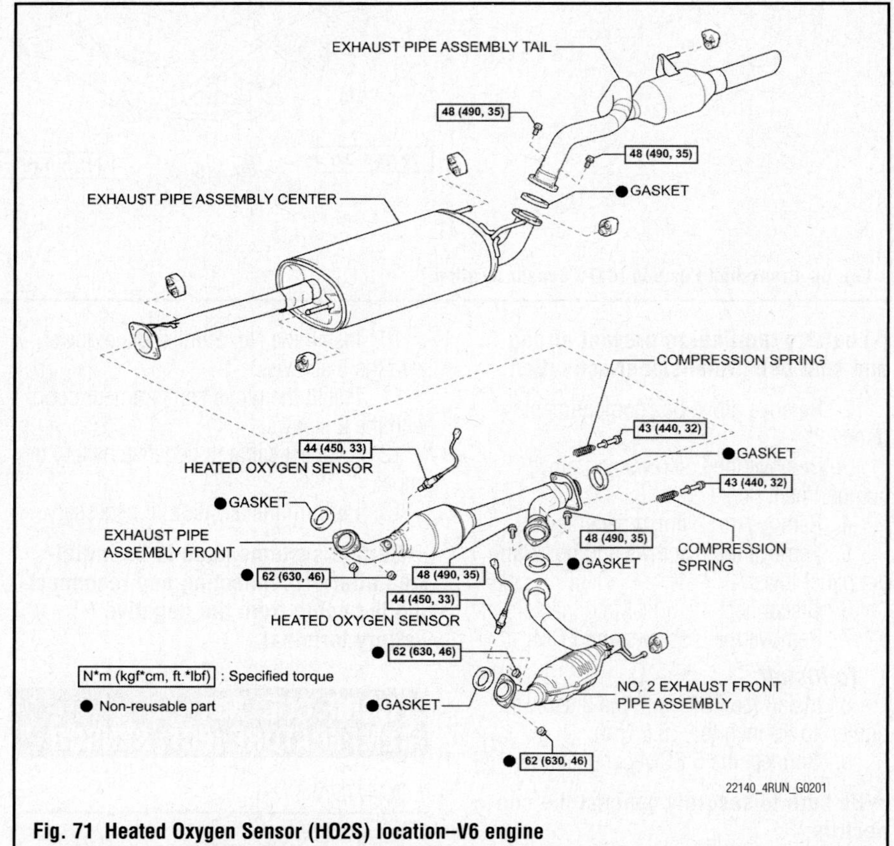

Fig. 71 Heated Oxygen Sensor (HO2S) location–V6 engine

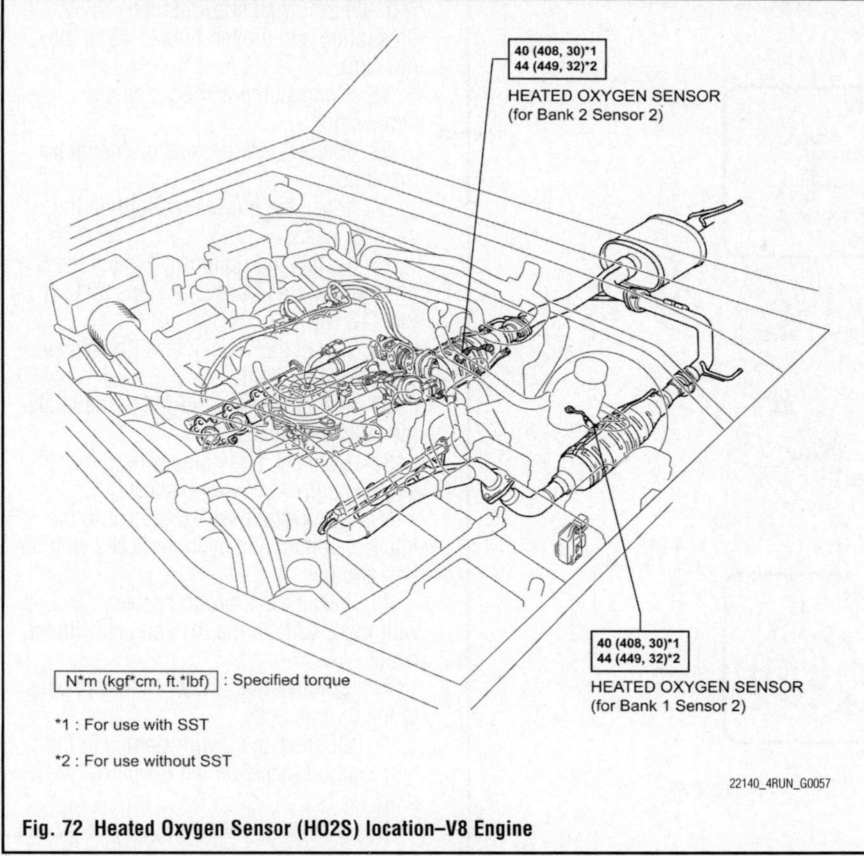

Fig. 72 Heated Oxygen Sensor (HO2S) location–V8 Engine

remove Heated Oxygen Sensor (HO2S) (for Bank 2 Sensor 2).

To install:

4. Install Heated Oxygen Sensor (HO2S) (for Bank 1 Sensor 2) and tighten to 32 ft. lbs. (44 Nm).

➡Use a torque wrench with a fulcrum length of 11.81 inch (30 cm).

5. Connect the sensor connector.
6. Install Heated Oxygen Sensor (HO2S) (for Bank 2 Sensor 2) and tighten to 32 ft. lbs. (44 Nm).

➡Use a torque wrench with a fulcrum length of 11.81 inch (30 cm).

7. Connect the sensor connector.
8. Connect cable to negative battery terminal.
9. Perform initialization, if necessary.

INTAKE AIR TEMPERATURE (IAT) SENSOR

LOCATION

The Intake Air Temperature (IAT) sensor, built into the Mass Air Flow (MAF) meter.

REMOVAL & INSTALLATION

See Mass Air Flow (MAF) meter.

KNOCK SENSOR (KS)

LOCATION

See Figure 73.

Refer to the accompanying illustration.

REMOVAL & INSTALLATION

See Figures 74 and 75.

1. Discharge fuel system pressure.
2. Drain engine coolant.
3. Remove V-bank cover sub-assembly.
4. Disconnect the vacuum hoses (for the power steering idle-up and fuel pressure regulator) and ventilation hose.
5. Remove the air cleaner hose assembly.
6. Disconnect fuel hose.
7. Disconnect fuel hose No. 2.
8. Disconnect the throttle control connector.
9. Disconnect the purge VSV connector.
10. Disconnect the 8 injector connectors.
11. Disconnect the ECT sensor connector.
12. Disconnect the 8 ignition coil connectors.
13. Disconnect the 2 VSV connectors for the air injection system.
14. Disconnect the 8 ignition coil connectors.

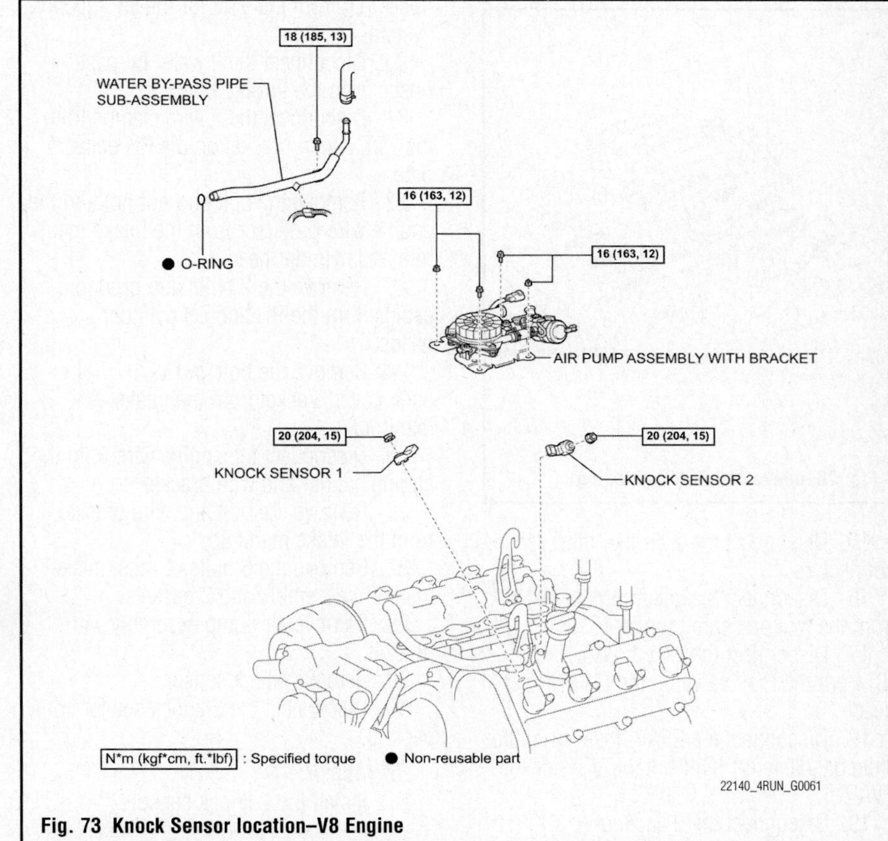

Fig. 73 Knock Sensor location–V8 Engine

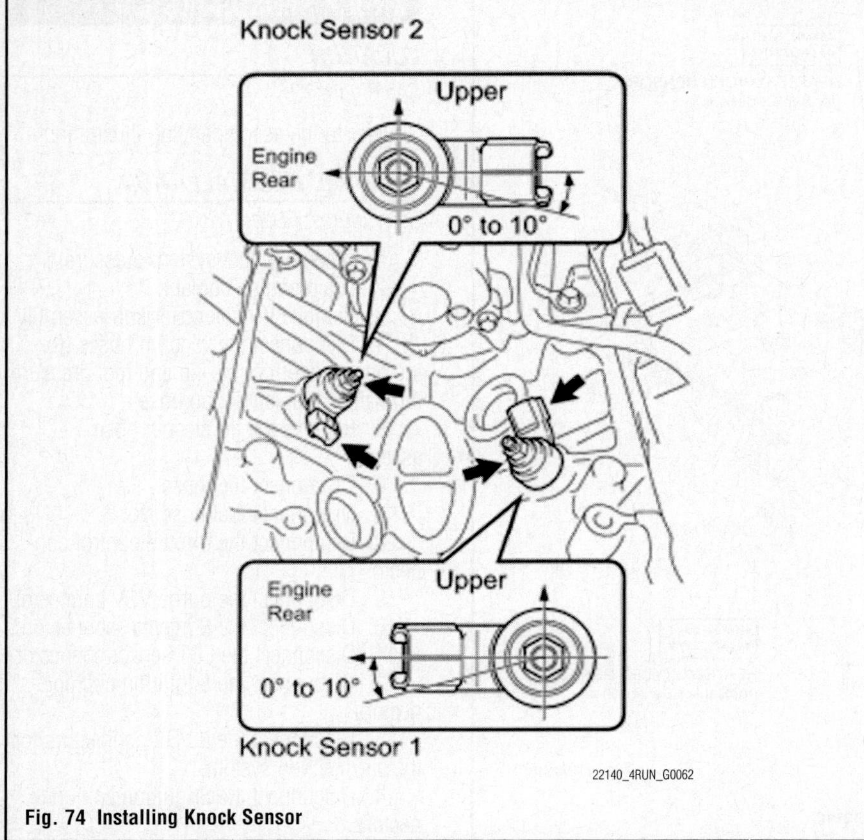

Fig. 74 Installing Knock Sensor

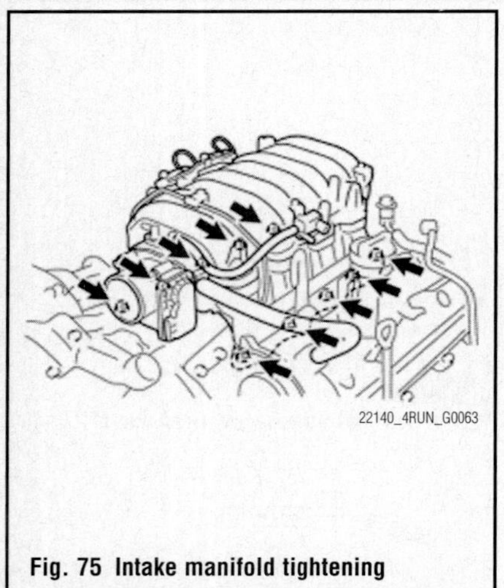

Fig. 75 Intake manifold tightening

15. Disconnect the 2 air fuel ratio sensor connectors.
16. Disconnect the vacuum hose [A] from the fuel pressure regulator.
17. Disconnect the PCV hoses [B] from the PCV valve on the LH cylinder head.
18. Disconnect the EVAP hose (from the charcoal canister) [C] from the VSV for the EVAP.
19. Disconnect the 2 vacuum hoses [D] from the VSV for the air injection system.
20. Disconnect the 2 water by-pass hoses from the throttle body.
21. Disconnect the 2 wire clamps from the wire clamp bracket on the RH delivery pipe.
22. Remove the bolt and nut holding the engine wire protector from the intake manifold and cylinder head.
23. Remove the 2 bolts and ground cables from the RH and LH cylinder heads.
24. Remove the bolt and V-bank cover bracket from the intake manifold.
25. Disconnect the engine wire from the engine hanger and wire bracket.
26. Remove the bolt and wire bracket from the intake manifold.
27. Remove the 6 bolts, 4 nuts, intake manifold assembly and 2 gaskets.
28. Remove air pump assembly with bracket.
29. Remove knock sensor.
30. Disconnect the 2 knock sensor connectors.

To install:

31. Install the 2 knock sensors with the 2 nuts as shown in the illustration and tighten nuts to 15 ft. lbs. (20 Nm).
32. Connect the 2 knock sensor connectors.
33. Place 2 new gaskets on the intake manifold.
34. Place the intake manifold on the cylinder heads.
35. Install and uniformly tighten the 6 bolts and 4 nuts in several steps to 13 ft. lbs. (18 Nm).
36. Install the V-bank cover bracket to the intake manifold.
37. Install the wire bracket to the intake manifold with the bolt.
38. Connect the engine wire to the engine hanger and wire bracket.
39. Connect the wire protector to the intake manifold and cylinder heads with the bolt and nut.
40. Install the 2 ground cables with the 2 bolts to the RH and LH cylinder heads.
41. Connect the 2 water by-pass hoses to the throttle body.
42. Connect the 2 wire clamps to the wire clamp bracket on the RH delivery pipe.
43. Connect the vacuum hose to the fuel pressure regulator.
44. Connect the PCV hose to the PCV valve on the LH cylinder head.
45. Connect the EVAP hose (from the charcoal canister) to the purge VSV.
46. Connect the 2 vacuum hoses to the VSV for the air injection system.
47. Connect the throttle control connector.
48. Connect the 2 VSV connectors for the air injection system.
49. Connect the purge VSV connector.
50. Connect the 8 injector connectors.
51. Connect the ECT sensor connector.
52. Connect the 8 ignition coil connectors.
53. Connect the 2 air fuel ratio sensor connectors.
54. Install fuel hose No. 2.
55. Install fuel hose.
56. Install V-bank cover sub-assembly.
57. Add engine coolant.
58. Check for engine coolant leaks.
59. Check for fuel leaks.

MALFUNCTION INDICATOR LIGHT (MIL)

RESET PROCEDURE

Clearing DTC codes resets MIL.

MASS AIR FLOW (MAF) SENSOR

LOCATION

The MAF sensor is located in the air intake snorkel.

REMOVAL & INSTALLATION

1. Disconnect connector.
2. Remove attaching screws and remove MAF sensor.

To install:

3. Installation is the reverse of the removal procedure.

THROTTLE POSITION SENSOR (TPS)

LOCATION

See Figure 76.

Refer to the accompanying illustration.

REMOVAL & INSTALLATION

1. Disconnect connector.
2. Remove attaching screws and remove TPS.

To install:

3. Installation is the reverse of the removal procedure.

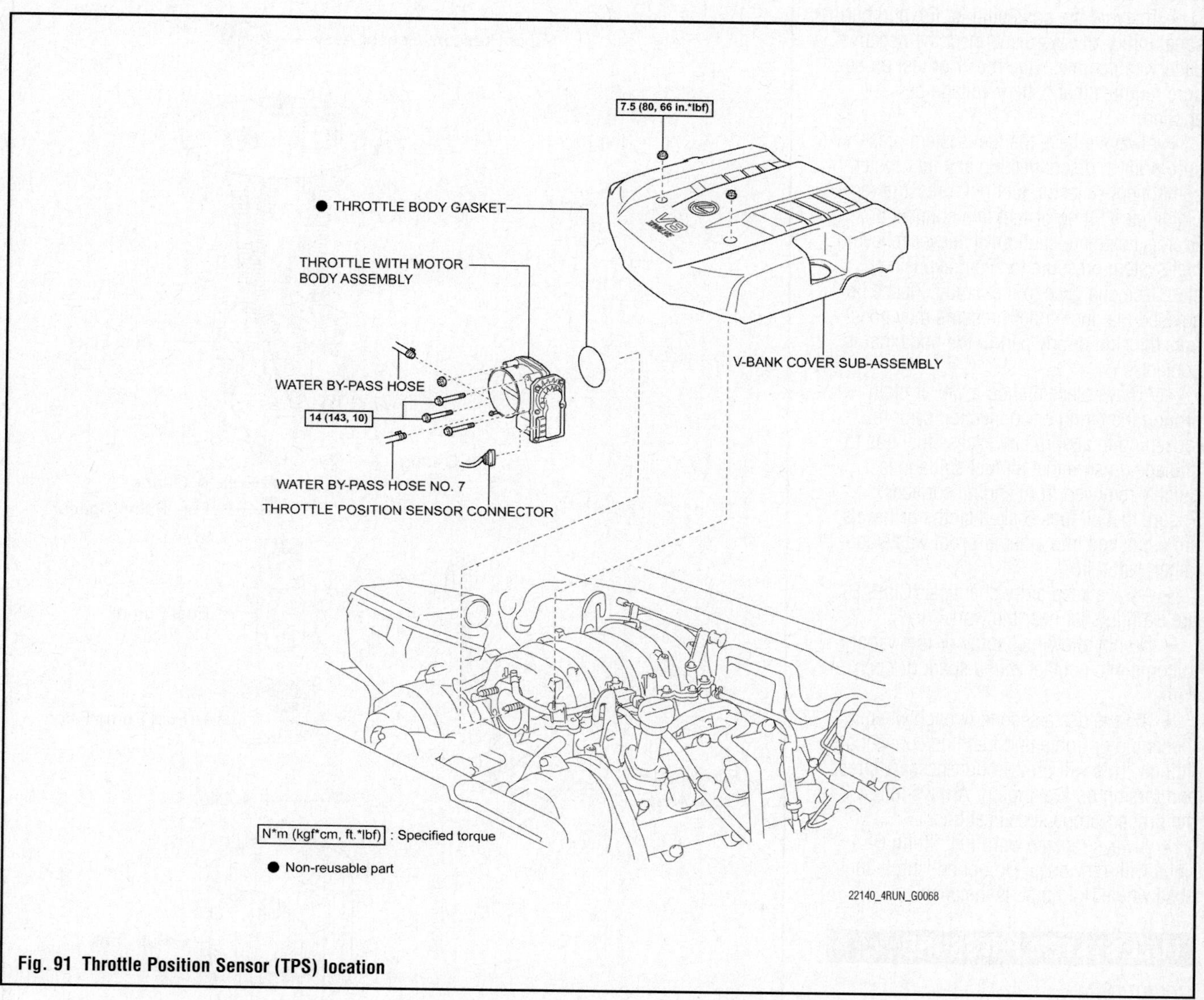

Fig. 91 Throttle Position Sensor (TPS) location

FUEL SYSTEM SERVICE PRECAUTIONS

Safety is the most important factor when performing not only fuel system maintenance, but any type of maintenance. Failure to conduct maintenance and repairs in a safe manner may result in serious personal injury or death. Work on a vehicle's fuel system components can be accomplished safely and effectively by adhering to the following rules and guidelines.

- To avoid the possibility of fire and personal injury, always disconnect the negative battery cable unless the repair or test procedure requires that battery voltage be applied.
- Always relieve the fuel system pressure prior to disconnecting any fuel system component (injector, fuel rail, pressure regulator, etc.) fitting or fuel line connection. Exercise extreme caution whenever relieving fuel system pressure to avoid exposing skin, face and eyes to fuel spray. Please be advised that fuel under pressure may penetrate the skin or any part of the body that it contacts.
- Always place a shop towel or cloth around the fitting or connection prior to loosening to absorb any excess fuel due to spillage. Ensure that all fuel spillage is quickly removed from engine surfaces. Ensure that all fuel-soaked cloths or towels are deposited into a flame-proof waste container with a lid.
- Always keep a dry chemical (Class B) fire extinguisher near the work area.
- Do not allow fuel spray or fuel vapors to come into contact with a spark or open flame.
- Always use a second wrench when loosening or tightening fuel line connection fittings. This will prevent unnecessary stress and torsion on fuel piping. Always follow the proper torque specifications.
- Always replace worn fuel fitting O-rings with new ones. Do not substitute fuel hose where rigid pipe is installed.

FUEL SYSTEM PRESSURE

RELIEVING

1. Remove the fuel pump relay from the engine compartment relay block.
2. Start the engine and let it run until it shuts off.
3. Turn the ignition to OFF.
4. Try to start the engine and make sure it won't start.
5. Disconnect the negative battery cable.
6. Install the relay.

FUEL FILTER

REMOVAL & INSTALLATION

See Figure 77.

The fuel filter is part of the fuel pump module unit and is not a normally replaced item.

FUEL LEVEL SENDING UNIT

REMOVAL & INSTALLATION

See Figures 78 through 80.

CAUTION

Do not smoke or work near an open flame when working on the fuel pump.

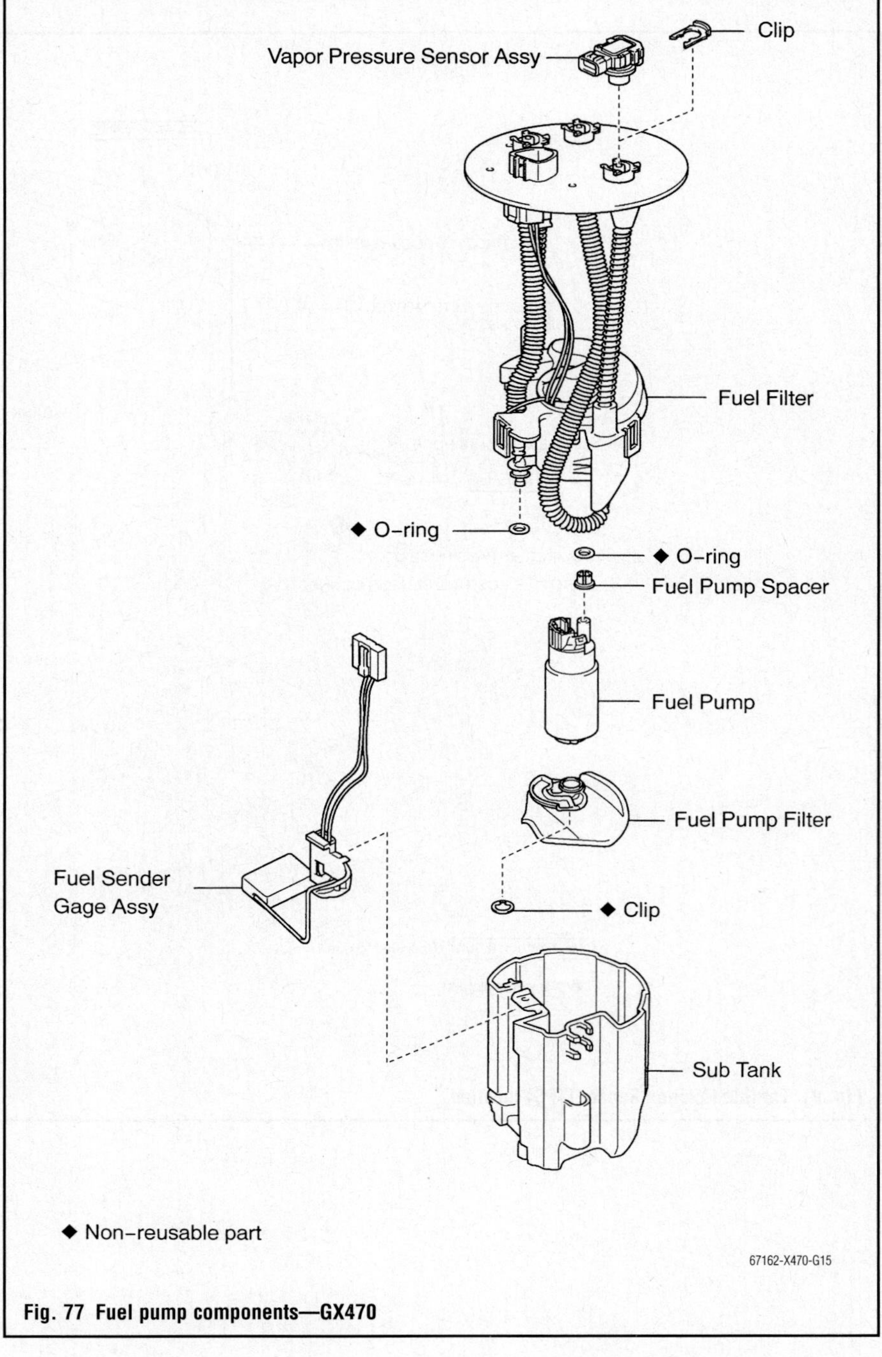

Fig. 77 Fuel pump components—GX470

1. Discharge fuel system pressure.
2. Disconnect cable from negative battery terminal

⁂ CAUTION

Wait at least 90 seconds after disconnecting the cable from the negative (-) battery terminal to prevent airbag and seat belt pretensioner activation.

3. Remove the 2 rear seats.
4. Remove rear door scuff plate.
5. Remove step plate.
6. Remove rear seat lock cover.
7. Take off the front and rear floor carpets.
8. Remove the 2 screws and floor service hole cover.

CAUTION

Prevent the retained pressure in the fuel line from splashing inside the vehicle compartment. When sealing the tube and suction plates with the O-ring of the quick connector, be careful not to damage any contact surfaces or allow foreign matter to contact any surface. Be sure to perform the disconnection by hand. Do not use tools. Do not bend or turn the nylon tube by force.

9. Disconnect fuel main tube and return tube.
10. Before the operation, remove foreign matter or dirt sticking to the tube joint clips.
11. Widen the tip of the clips with your fingers and pull them out for disconnection.
12. Pull out the fuel main tube and the return tube. If the nylon tube and the suction plate stick together, turn the nylon tube with your fingers and pull it out for disconnection.
13. After the disconnection, protect the connector with a plastic bag.
14. Remove the 8 bolts.
15. Pull out the fuel pump and sender gauge assembly.

➡Do not damage the fuel pump filter. Be careful that the arm of the sender gauge is not bent.

16. Remove sending unit.

To install:

17. Install sending unit.
18. Install a new gasket to the fuel suction plate.
19. Insert the fuel pump and sender gauge assembly into the fuel tank.
20. Install the fuel tank vent tube set plate with the 8 bolts tighten bolts to 31 inch lbs. (3.5 Nm).
21. Before installing the tube connectors, check for foreign matter on the connection between the nylon tube and the suction plate.
22. Attach the fuel tube connectors to the ports of the fuel suction plate and insert the clips until you hear a click.
23. After the connection, pull the clips to check that they are installed securely.
24. Connect cable to negative battery terminal.
25. Check for fuel leaks.
26. Install the service hole cover with the 2 screws.

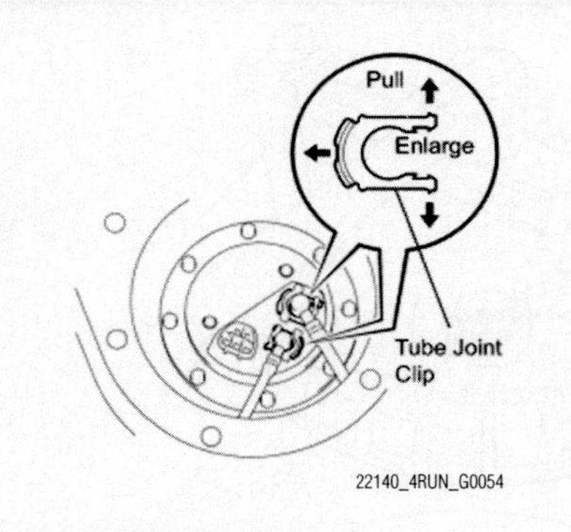

Fig. 78 Removing clip

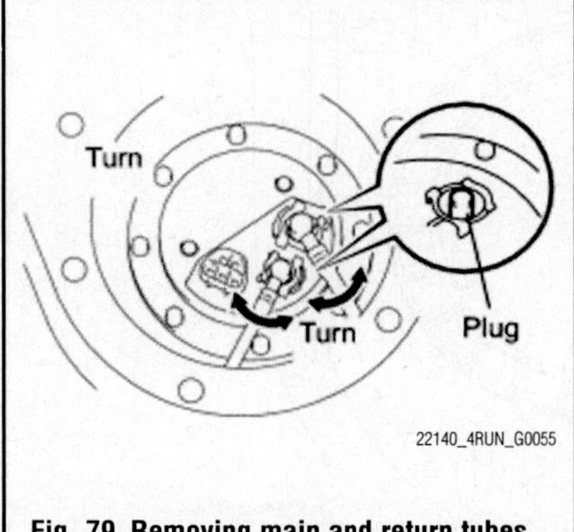

Fig. 79 Removing main and return tubes

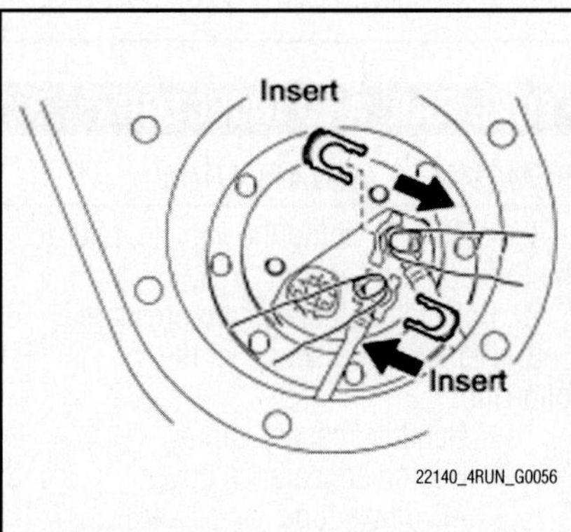

Fig. 80 Installing main and return tubes

27. Install the front and rear floor carpets.
28. Install rear seat lock cover.
29. Install step plate.
30. Install rear door scuff plate.
31. Install the 2 rear seats.
32. Perform initialization, if necessary.

FUEL PUMP

REMOVAL & INSTALLATION

See Figures 81 through 83.

1. Before servicing the vehicle, refer to the Precautions section.
2. Relieve the fuel system pressure.
3. Remove the spare tire.
4. Disconnect the fuel pump connector and remove the fuel tank protector.
5. Disconnect the main and fuel return tubes.
6. Disconnect the fuel tank vent hose.
7. Disconnect the inlet and breather hoses.
8. Support the fuel tank with a jack, loosen the tank strap bolts remove the straps and lower the tank.
9. Disconnect any necessary hoses and wiring from the pump.

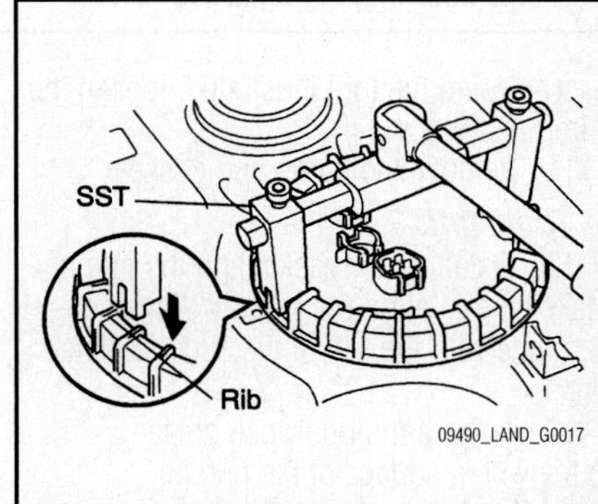

Fig. 81 Use the tool illustrated to remove the fuel pump retainer—GX470

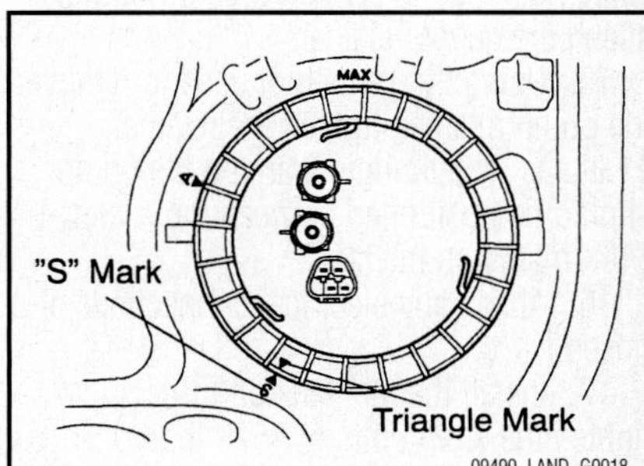

Fig. 82 Align the triangle mark on the new pump retainer with the S mark on the tank—GX470

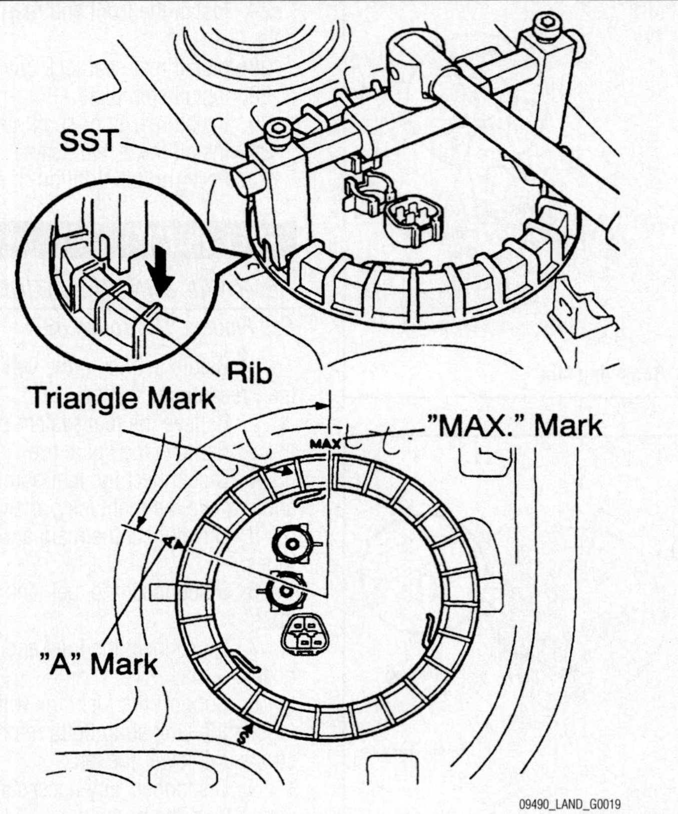

Fig. 83 The triangle mark on the pump should be positioned between the A and MAX marks on the tank when properly tightened

10. Using the tool illustrated, loosen the pump retainer.
11. Remove the pump and gasket.

To install:

12. Install a new gasket and the pump. Make sure to align the keyway of the suction tube with the key of the suction plate No. 1.
13. Apply a multipurpose grease to the whole surface of the pump retainer.
14. Align the triangle mark on the new pump retainer with the S mark on the tank while pushing the suction tube down and attach the gauge retainer.
15. Using the same tool used to remove the pump retainer, tighten the retainer 1½times. The triangle mark on the pump should be positioned between the A and MAX marks on the tank.
16. Attach any electrical connections and hoses.
17. Install the fuel tank and tighten the strap bolts to 45 ft. lbs. (62 Nm).
18. Install the remaining components.

FUEL RAIL AND INJECTOR

REMOVAL & INSTALLATION

1. Before servicing the vehicle, refer to the Precautions section.
2. Relieve the fuel system pressure.
3. Remove or disconnect the following:
   - Negative battery cable
   - Engine appearance cover
   - Air intake tube
   - Fuel lines
   - Fuel pulsation damper
   - Fuel pressure regulator vacuum line
   - Accelerator cable and bracket
   - Positive Crankcase Ventilation (PCV) valve and hose
   - Evaporative Emissions (EVAP) vacuum switching valve
   - Engine appearance cover brackets
   - Fuel injector harness connectors
   - Engine harness protector
   - Fuel supply manifold crossover pipe
   - Fuel supply manifolds with injectors attached
   - Fuel injectors

To install:

4. Install the fuel injectors to the supply manifold with new O-ring seals and new grommets.
5. Install new injector insulators to the intake manifold.
6. Install or connect the following:
   - Fuel supply manifolds with injectors attached: tighten the bolts to 66 inch lbs. (7.5 Nm)
   - Fuel supply manifold crossover pipe: tighten the bolts to 29 ft. lbs. (39 Nm)
   - Engine harness protector
   - Fuel injector harness connectors
   - Engine appearance cover brackets
   - EVAP vacuum switching valve
   - PCV valve and hose
   - Accelerator cable and bracket
   - Fuel pressure regulator vacuum line
   - Fuel pulsation damper
   - Fuel lines
   - Air intake tube
   - Engine appearance cover
   - Negative battery cable
7. Start the engine and check for leaks.

FUEL TANK

REMOVAL & INSTALLATION

See Figures 84 through 86.

1. Discharge fuel system pressure.
2. Disconnect cable from negative battery terminal.
3. Disconnect vent line tube.
4. Remove the bolt and bracket from the fuel tank band.
5. Remove the bolt and bracket from the body.
6. Disconnect the fuel main tube, return tube and fuel tube.
   a. With Fuel hose connector cover type disengage the lock claw by lifting up the cover.

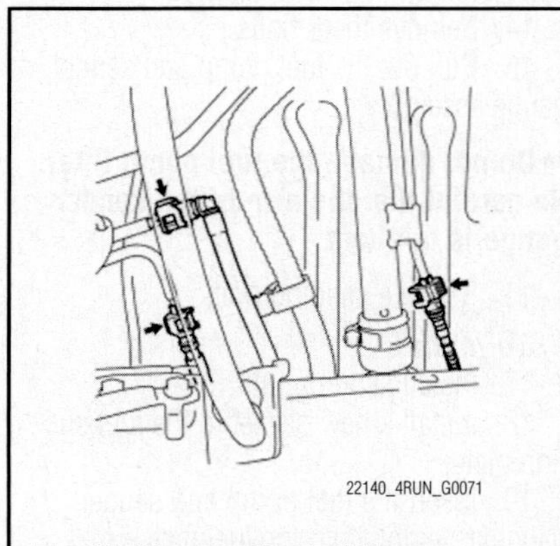

Fig. 84 Disconnecting fuel lines

b. Check for dirt or mud on the pipe and around the connector before disconnection. Clean if necessary.

c. Disconnect the connector and pipe by hand.

d. If the connector and the pipe stuck, pinch the connector, and push and pull the pipe to disconnect it.

➡**Do not use any tools.**

e. Check for dirt or mud on the seal surface of the disconnected pipe. Clean if necessary.

f. To protect the disconnected pipe and connector from damage and contamination, cover it with a plastic bag.

7. Loosen the bolt of the clamp and disconnect the fuel inlet hose from the fuel inlet pipe.

8. Set up a transmission jack under the fuel tank.

9. Remove the 2 bolts and disconnect the 2 fuel tank bands from the fuel tank.

10. Slightly lower the mission jack so that the fuel pump and sender gauge connector and 2 clamps can be removed.

➡**Do not lower the mission jack excessively as this may damage the connector.**

11. Operate the transmission jack and remove the fuel tank.

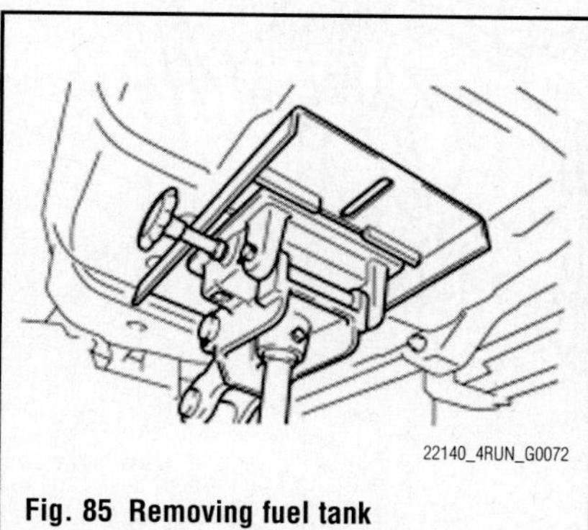

Fig. 85 Removing fuel tank

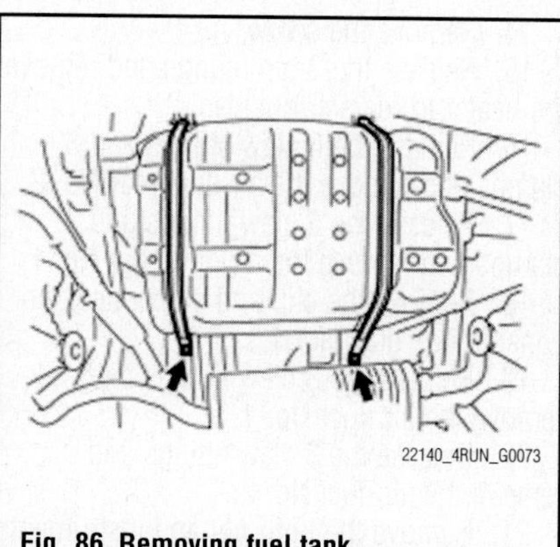

Fig. 86 Removing fuel tank

12. Remove fuel pump and sender gauge assembly.

13. Remove fuel inlet hose.

14. Remove fuel hose.

To install:

15. Install fuel hose.

16. Install fuel inlet hose.

17. Install fuel pump and sender gauge assembly.

18. Install fuel tank assembly.

19. Operate the transmission jack so that the fuel pump and sender gauge connector and 2 clamps can be installed. Then raise the transmission jack again to install the fuel tank.

20. Install the 2 fuel tank bands with the 2 bolts and tighten to 30 ft. lbs. (40 Nm).

21. Connect the fuel main tube, return tube and fuel hose.

a. Check that there is no damage or contamination in the connected part of the pipe.

b. Align the axis of the connector with the axis of the pipe. Push the pipe into the connector until the connector makes a "click" sound. If the connection is tight, apply a little amount of fresh engine oil on the tip of the pipe.

c. After having finished the connection, try to pull apart the pipe and the connector and confirm that they are securely connected.

d. With fuel hose connector cover type attach the lock claw by lifting up the cover, as shown in the illustration.

22. Connect the fuel inlet hose to the fuel inlet pipe and tighten the bolt of the clamp to 66 inch lbs. (7.5 Nm).

23. Install the bracket to the body with the bolt and tighten bolt to 11 ft. lbs. (15 Nm).

24. Install the bracket to the fuel tank band with the bolt and tighten bolt to 11 ft. lbs. (15 Nm).

25. Connect the vent line tube to the fuel tank.

26. Check for fuel leaks.

27. Connect cable to negative battery terminal.

28. Perform initialization, if necessary.

IDLE SPEED

ADJUSTMENT

Idle speed is maintained by the Powertrain Control Module (PCM). No adjustment is necessary or possible.

THROTTLE BODY

REMOVAL & INSTALLATION

See Figures 87 and 88.

1. Before servicing the vehicle, refer to the Precautions section.

2. Remove the 2 nuts, then remove the V-bank cover.

3. Remove throttle body cover.

4. Drain engine coolant.

5. Remove intake air connector.

6. Disconnect the throttle control connector.

7. Disconnect the 2 water bypass hoses from the throttle body.

8. Remove the nut and 3 bolts, and remove the throttle body from the intake manifold.

To install:

9. Install the throttle body with the nut and 3 bolts. Tighten them to 10 ft. lbs. (14 Nm).

10. Connect the 2 water bypass hoses to the throttle body.

11. Connect the throttle control connector.

12. Install intake air connector.

13. Fill with engine coolant.

14. Start engine and check for engine coolant leaks.

15. Install throttle body cover.

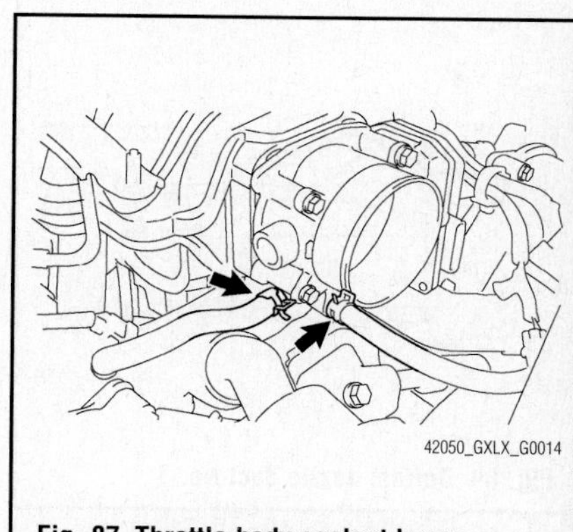

Fig. 87 Throttle body coolant hoses

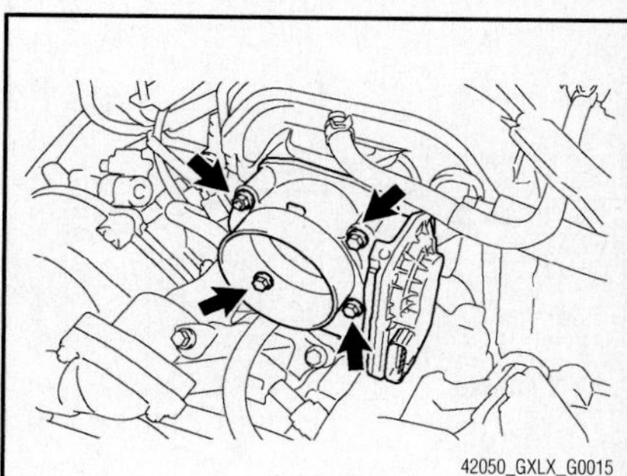

Fig. 88 Throttle body bolts and nut

HEATING & AIR CONDITIONING SYSTEM

BLOWER MOTOR

REMOVAL & INSTALLATION

1. Before servicing the vehicle, refer to the Precautions section.
2. Disconnect the connector.
3. Remove the three screws and the blower motor.

To install:

4. Install the blower motor with three screws.
5. Connect the connector.

HEATER CORE

REMOVAL & INSTALLATION

See Figures 89 through 98.

1. Discharge refrigerant from refrigeration system.
2. Disconnect cooler refrigerant suction pipe C.
3. Disconnect cooler refrigerant liquid pipe C.
4. Using pliers, grip the claws of the clip and slide the clip and disconnect the heater water outlet hose.

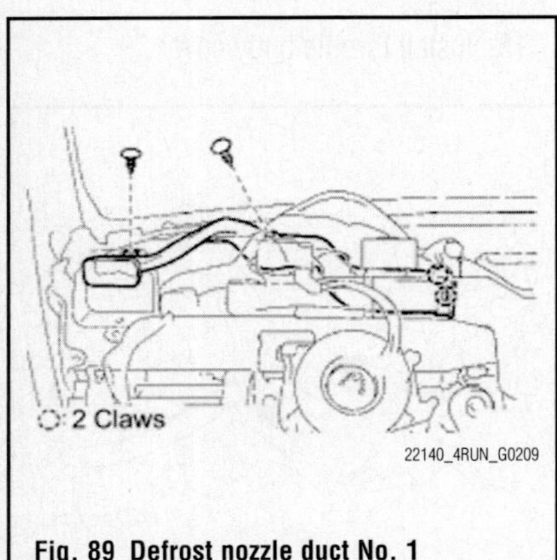

Fig. 89 Defrost nozzle duct No. 1

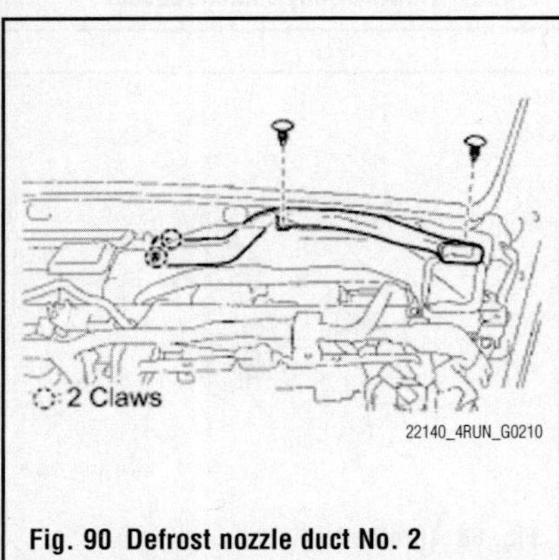

Fig. 90 Defrost nozzle duct No. 2

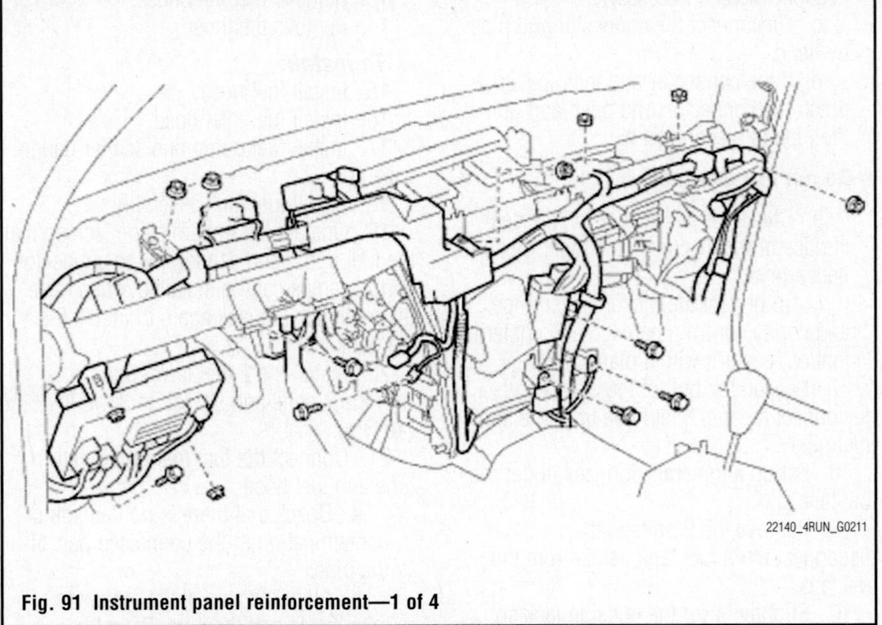

Fig. 91 Instrument panel reinforcement—1 of 4

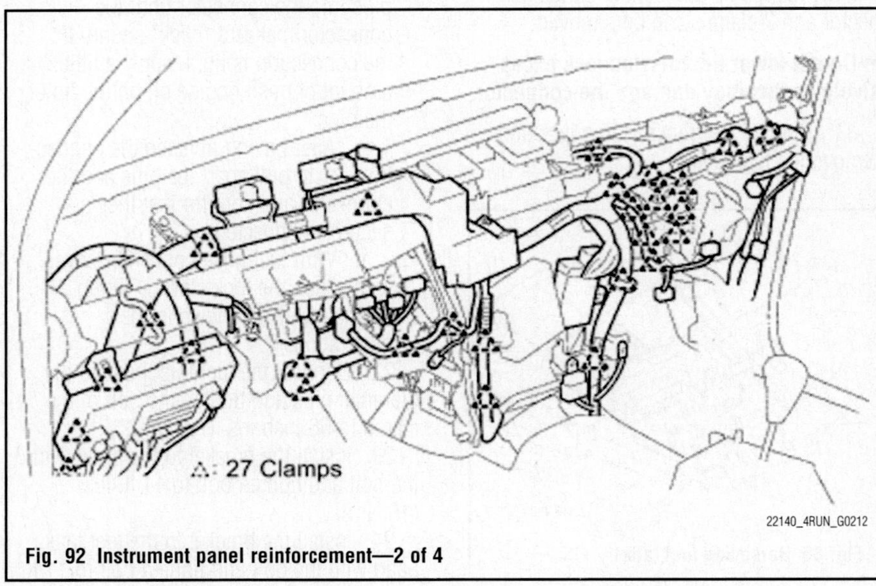

Fig. 92 Instrument panel reinforcement—2 of 4

5. Disconnect heater water outlet hose.
6. Remove instrument panel safety pad sub-assembly.
7. Remove air conditioning amplifier assembly.
8. Remove the 2 clips.
9. Release the 2 claw fittings and remove the side defroster nozzle duct No. 1.
10. Remove the 2 clips.
11. Release the 2 claw fittings and remove the side defroster nozzle duct No. 2.
12. Remove the screw.
13. Release the 2 pin fittings and remove the heater to register duct No. 1.
14. Remove the screw.
15. Release the 2 pin fittings and remove the heater to register duct No. 3.
16. Release the 6 claw fittings and 3 clamps and remove the air duct rear No. 2.
17. Release the 6 claw fittings and 3 clamps and remove the air duct rear No. 1.
18. Remove the clip and disconnect the console box duct No. 1.
19. Release the 3 claw fittings and remove the air duct No. 1.
20. Release the 3 claw fittings and remove the air duct No. 2.
21. Remove the bolt, nut and instrument panel brace mounting brackets.

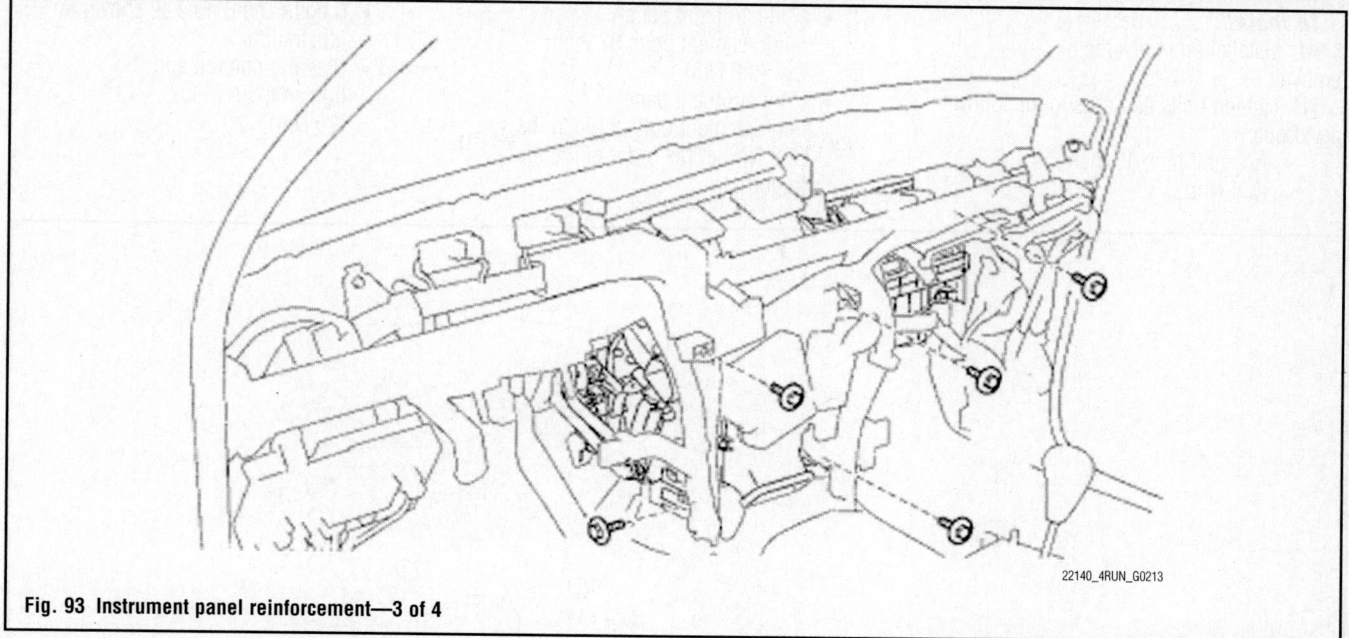

Fig. 93 Instrument panel reinforcement—3 of 4

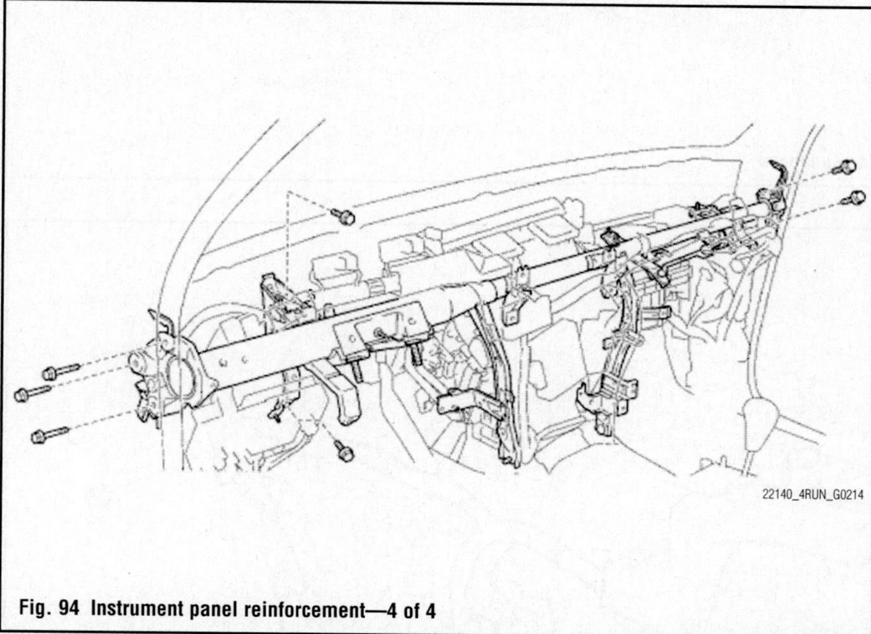

Fig. 94 Instrument panel reinforcement—4 of 4

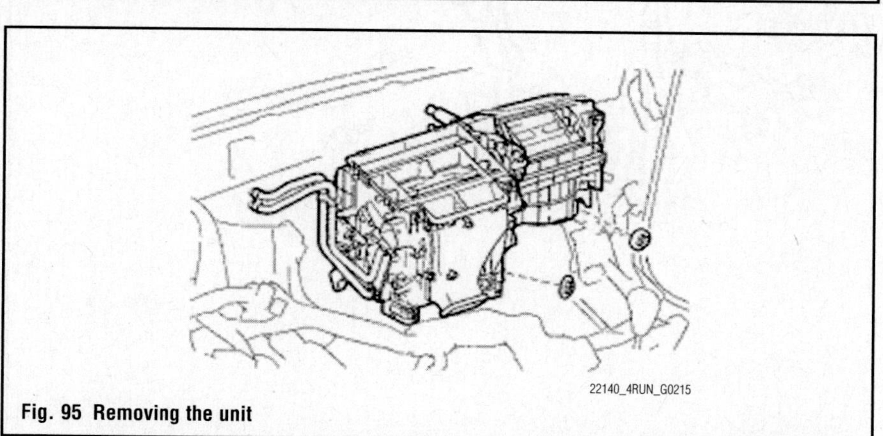

Fig. 95 Removing the unit

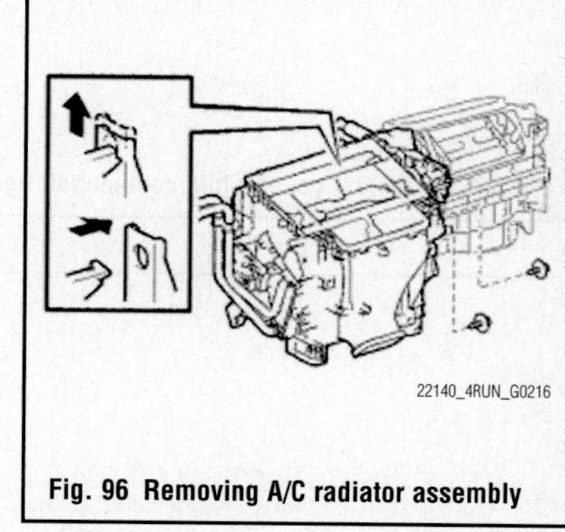

Fig. 96 Removing A/C radiator assembly

22. Release the 3 clamps and disconnect the connector.

23. Remove the 4 nuts and disconnect the steering column assembly.

24. Remove instrument panel reinforcement.

a. Remove the 6 bolts and 8 nuts.

b. Release the 27 clamps.

c. Disconnect the connectors.

d. Remove the 5 bolts.

e. Remove the 7 bolts and instrument panel reinforcement.

25. Release the 4 claw fittings and remove the heater to register duct center.

26. Release the 4 claw fittings and remove the defroster nozzle assembly lower.

27. Disconnect the connectors.

28. Remove the 2 nut and air conditioner unit assembly.

29. Remove the 2 screws and air conditioning radiator assembly.

To install:

30. Installation is reverse of removal.

31. Tighten bolts and nuts to the following torque:

- A/C unit nuts 48 inch lbs. (5.4 Nm)
- (7) Instrument panel reinforcement bolts 87 inch lbs. (9.8 Nm)
- (5) Instrument panel reinforcement bolts 87 inch lbs. (9.8 Nm) in the order of the illustration
- 6 bolts and 8 nuts as shown in illustration
- Steering column and tighten to 19 ft. lbs. (26 Nm)

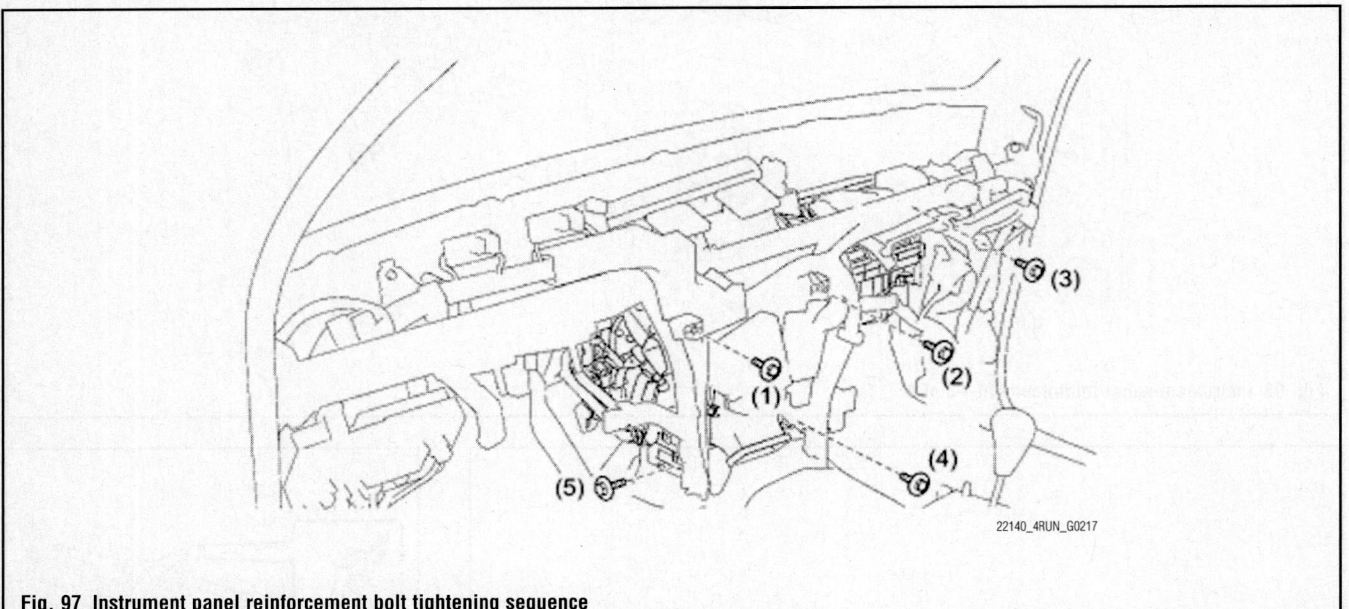

Fig. 97 Instrument panel reinforcement bolt tightening sequence

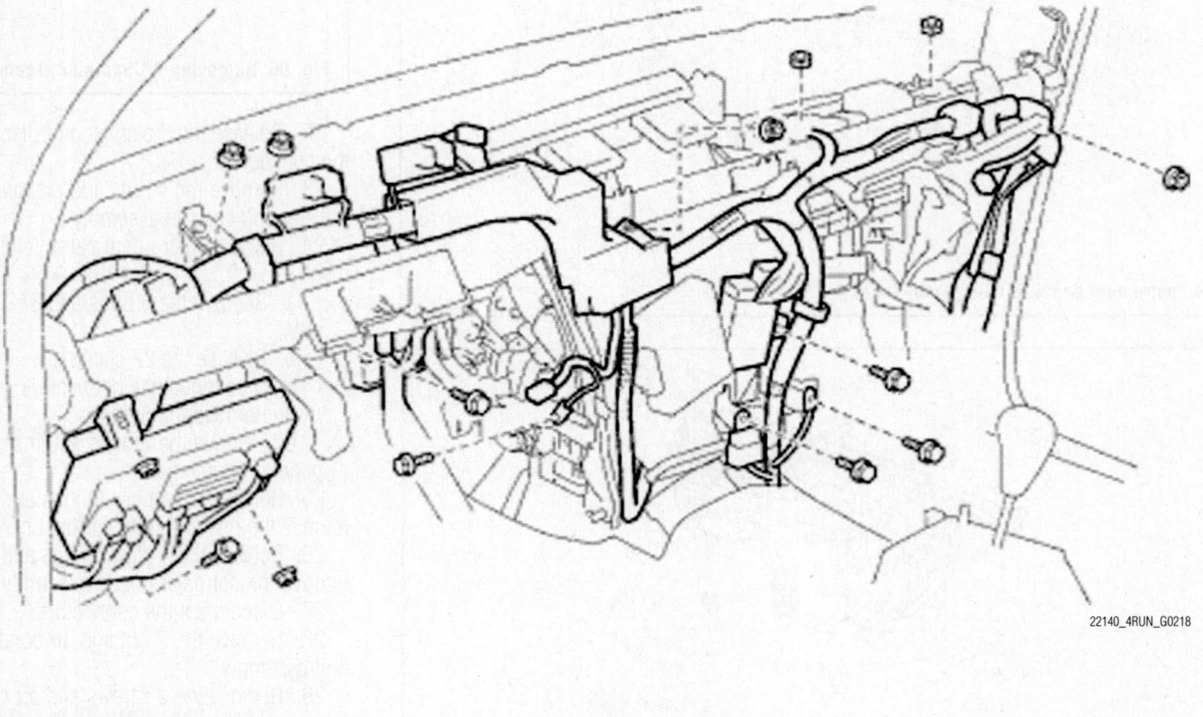

22140_4RUN_G0218

Fig. 98 Installing 6 bolts and 8 nuts

STEERING

POWER RACK & PINION STEERING GEAR

REMOVAL & INSTALLATION

See Figure 99.

1. Before servicing the vehicle, refer to the Precautions section.
2. Disconnect the battery ground cable.
3. Place the front wheels in the straight ahead position.
4. Remove the horn pad.
5. Remove the steering wheel.
6. Remove the lower steering column cover.
7. Remove the turn signal switch.

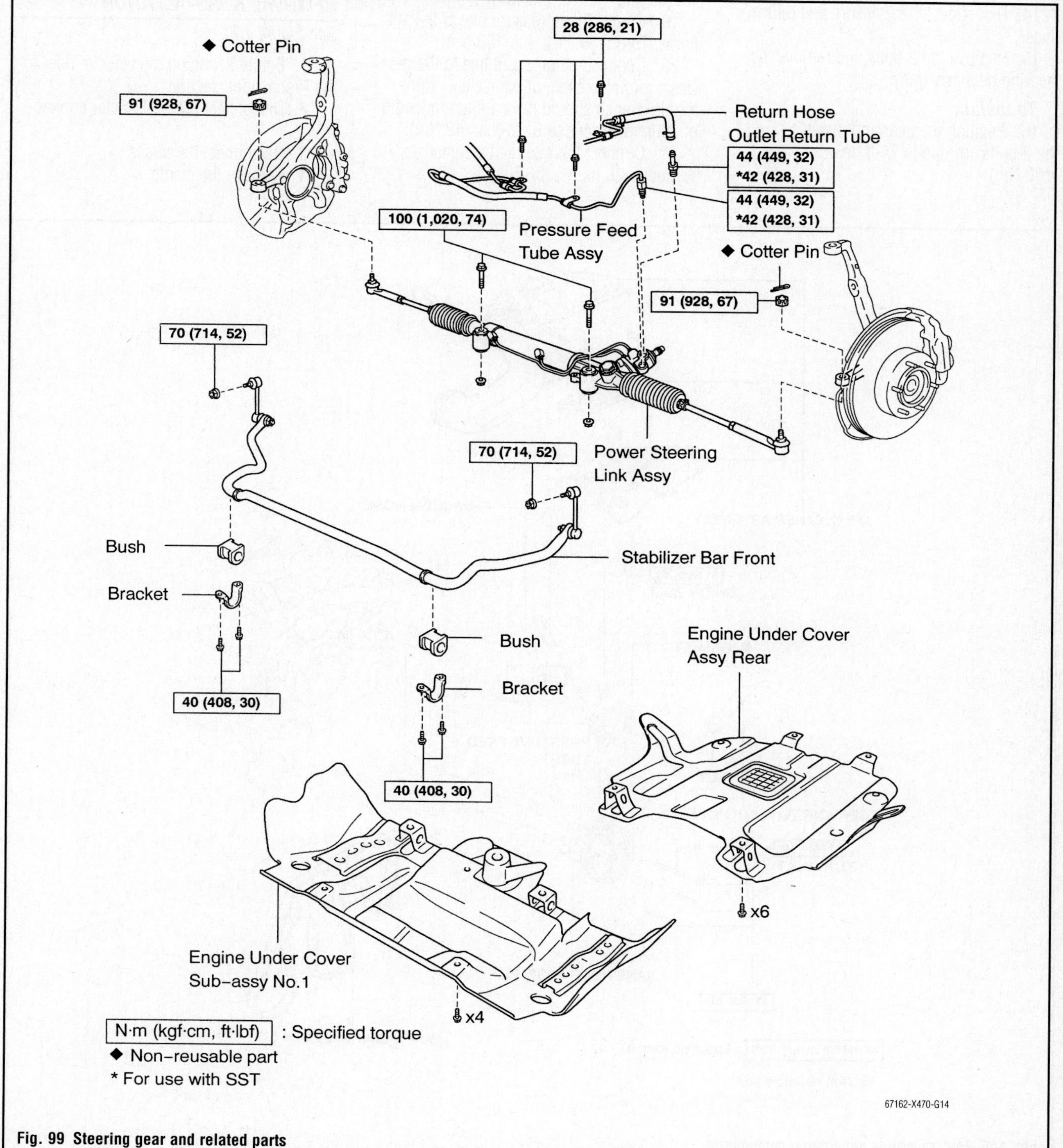

Fig. 99 Steering gear and related parts

8. Remove the spiral cable assembly.
9. Remove the front wheels.
10. Remove the engine under-covers.
11. Remove the stabilizer bar.
12. Remove the tie rod ends from the knuckle.
13. Remove the steering intermediate shaft.
14. Disconnect the pressure and return lines.
15. Remove the 2 bolts and remove the steering gear assembly.

To install:

16. Position the gear and install the 2 bolts. Torque to 74 ft. lbs. (100 Nm).

➡The nuts have detents. Never turn the nuts, just the bolts.

17. Install the stabilizer bar. Torque the end links to 52 ft. lbs. (70 Nm); the clamp bolts to 30 ft. lbs. (40 Nm).
18. Connect the return line. Use a torque wrench, or equivalent. The torque wrench should have a fulcrum length of 300mm. Torque to 31 ft. lbs. (42 Nm).
19. Connect the pressure line at the sub-frame. Torque to 21 ft. lbs. (28 Nm).
20. Connect the pressure line to the gear. Use a torque wrench, or equivalent. The torque wrench should have a fulcrum length of 300mm. Torque to 31 ft. lbs. (42 Nm).
21. Connect the intermediate shaft. Torque to 26 ft. lbs. (36 Nm).
22. Connect the tie rod ends. Torque to 67 ft. lbs. (91 Nm).
23. Install the under-covers.
24. The remainder of installation is the reverse of removal.

POWER STEERING PUMP

REMOVAL & INSTALLATION

See Figure 100.

1. Before servicing the vehicle, refer to the Precautions section.
2. Disconnect the MAF meter connector.
3. Disconnect the hoses.
4. Remove the clamp.

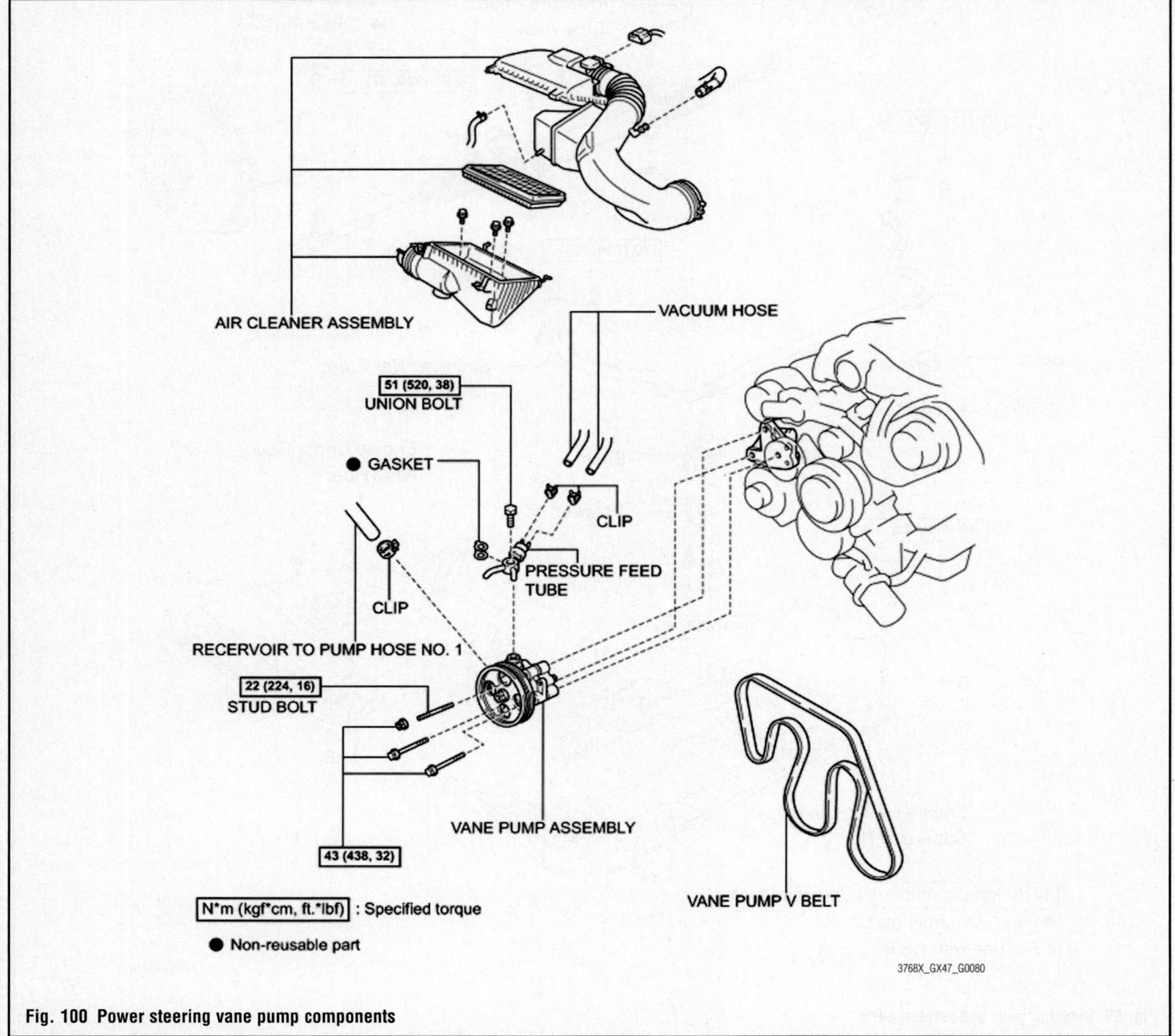

Fig. 100 Power steering vane pump components

5. Remove the 3 bolts and air cleaner assembly with air cleaner hose connected.

6. Loosen the drive belt tension by turning the drive belt tensioner counterclockwise, and remove the drive belt.

7. Remove the 2 clips and disconnect the 2 vacuum hoses.

8. Remove the clip and disconnect the return hose.

9. Remove the union bolt and gasket, disconnect the pressure feed tube.

10. Remove the 2 bolts, nut, stud bolt and power steering pump assembly.

To install:

11. Install the power steering pump assembly with the stud bolt.

12. Tighten the stud bolt to 16 ft. lbs. (22 Nm).

13. Install the 2 bolts and nut and tighten them to 33 ft. lbs. (44 Nm).

14. Install a new gasket and the union bolt on the pressure feed tube.

➡Make sure that the stopper of the pressure feed tube contacts the power steering pump body as shown in the illustration.

15. Tighten the union bolt to 34 ft. lbs. (46.5 Nm).

16. Connect the return hose with the clip.

17. Connect the 2 vacuum hoses and install the 2 clips.

18. Loosen the drive belt tension by turning the drive belt tensioner counterclockwise, and install the belt.

19. Install the air cleaner assembly with air cleaner hose and the 3 bolts.

20. Install the clamp.

21. Connect the MAF meter connector.

22. Fill with power steering fluid and bleed the system.

BLEEDING

1. Before servicing the vehicle, refer to the Precautions section.

2. Check fluid level.

3. Jack up front of vehicle and support it with stands.

4. With the engine stopped, turn the wheel slowly from lock to lock several times.

5. Lower the vehicle.

6. Start the engine and run at idle for a few minutes.

7. With the engine idling, turn the wheel left or right to the full lock position and keep it there for 2 to 3 seconds, then turn the wheel to the opposite full lock position and keep it there for 2 to 3 seconds. Repeat several times.

8. Stop the engine.

9. Check for foaming or emulsification of the power steering fluid.

10. If the system has to be bled twice specifically because of foaming or emulsification, check for fluid leaks in the system.

11. Check fluid level.

SUSPENSION

FRONT SUSPENSION

COIL SPRING

REMOVAL & INSTALLATION

See Figure 101.

1. Remove the strut.

2. Place the strut in a compressor, and compress the spring.

3. Hold the rod and remove the nut.

➡Don't use an impact wrench.

4. Remove the bushing retainer

5. Remove the upper bushing.

6. Remove the support.

7. Remove the lower bushing retainer.

8. Remove the spring.

9. Remove the lower bushing.

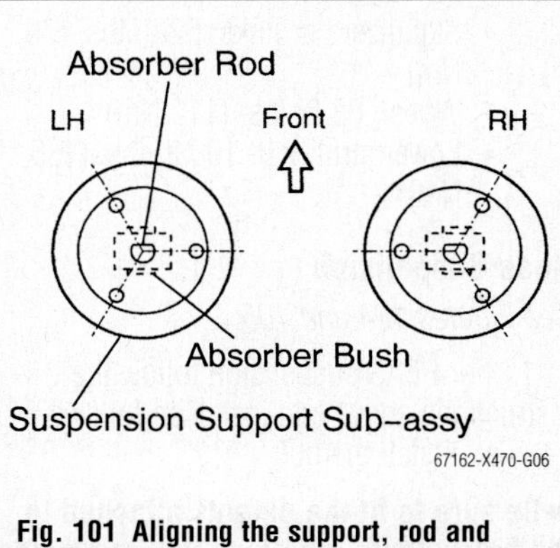

Fig. 101 Aligning the support, rod and bushing

To install:

10. Install the new lower bushing.

11. Compress the spring and install it.

12. Install the bushing retainer.

13. Install the suspension support.

14. Install the upper bushing.

15. Install the retainer.

16. Align the support, rod and bushing as shown. Install the locknut and torque to 18 ft. lbs. (25 Nm).

17. Release the spring from the compressor and check the alignment of the parts.

18. Install the strut.

CONTROL LINKS

REMOVAL & INSTALLATION

Non-KDSS

1. Remove front disc wheel.

2. Remove the 2 nuts and the front stabilizer link assembly.

➡If the ball joint turns together with the nut, use a hexagon (6 mm) wrench to hold the stud.

3. Remove front stabilizer link assembly.

To install:

4. Installation is reverse of removal. Tighten nuts to 52 ft. lbs. (70 Nm).

LOWER BALL JOINT

REMOVAL & INSTALLATION

The lower ball joint is serviced with the lower control arm as an assembly.

LOWER CONTROL ARM

REMOVAL & INSTALLATION

See Figure 102

1. Before servicing the vehicle, refer to the Precautions section.

2. Remove the wheel.

3. Support the lower arm with a jack.

4. Remove the lower strut bolt.

5. Remove the 2 bolts and separate the lower ball joint attachment from the knuckle.

6. Place matchmarks on the camber adjusting cam and toe adjusting cam.

7. Remove the 2 nuts and remove the arm along with the cams.

To install:

8. Installation is the reverse of removal. Align all matchmarks. Use new nuts and cotter pins. Don't fully tighten the control arm bolts until the vehicle is on the ground and the suspension jounced a few times. Observe the following torques:

- Lower ball joint stud: 103 ft. lbs. (140 Nm)
- Lower ball joint attachment bolts: 166 ft. lbs. (225 Nm)

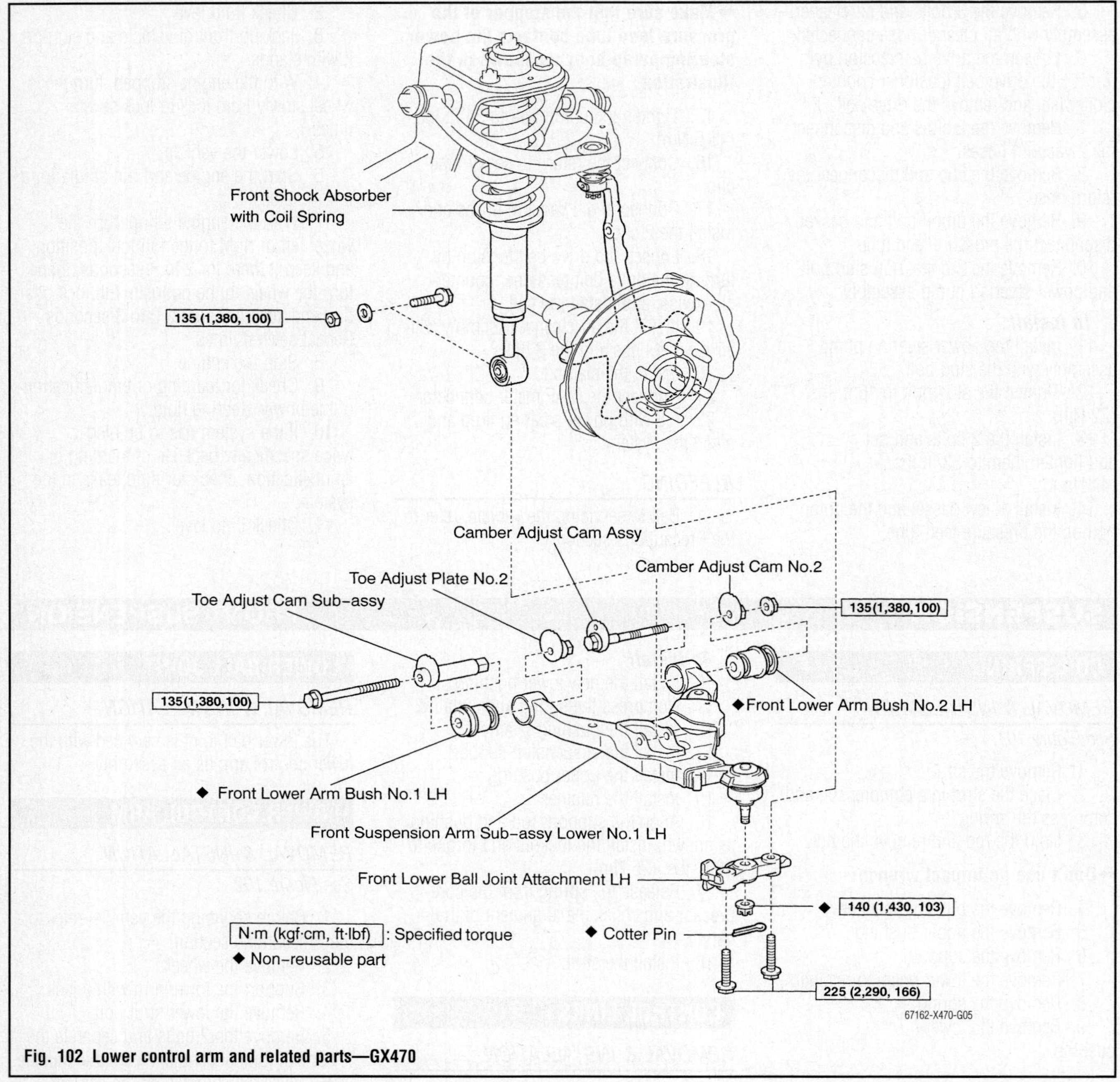

Fig. 102 Lower control arm and related parts—GX470

- Lower arm bolts: 100 ft. lbs. (135 Nm)

MACPHERSON STRUT

REMOVAL & INSTALLATION

Non-Reas Suspension

See Figure 103.

1. Before servicing the vehicle, refer to the Precautions section.
2. Remove the wheel.
3. Remove the stabilizer bar.
4. Remove the clamps and connector.
5. Remove the wire bracket.
6. Remove the lower strut bolt.
7. Remove the 3 upper strut nuts.
8. Remove the strut.

To install:

9. Installation is the reverse of removal. Do not fully tighten the lower strut bolt until the vehicle is resting on the ground and the suspension has been jounced a few times. Observe the following torques:
   - Upper nuts: 47 ft. lbs. (64 Nm)
   - Bracket nut: 11 ft. lbs. (15 Nm)
   - Stabilizer bar links: 52 ft. lbs. (70 Nm)
   - Wheel: 83 ft. lbs. (112 Nm)
   - Lower strut bolt: 100 ft. lbs. (135 Nm)

Reas Suspension

See Figures 104 and 105.

1. For REAS suspension follow the instructions above and install as follows:
   a. Install the bolt.

➡Be sure to fit the detents attached to the bracket into a hole on the frame side

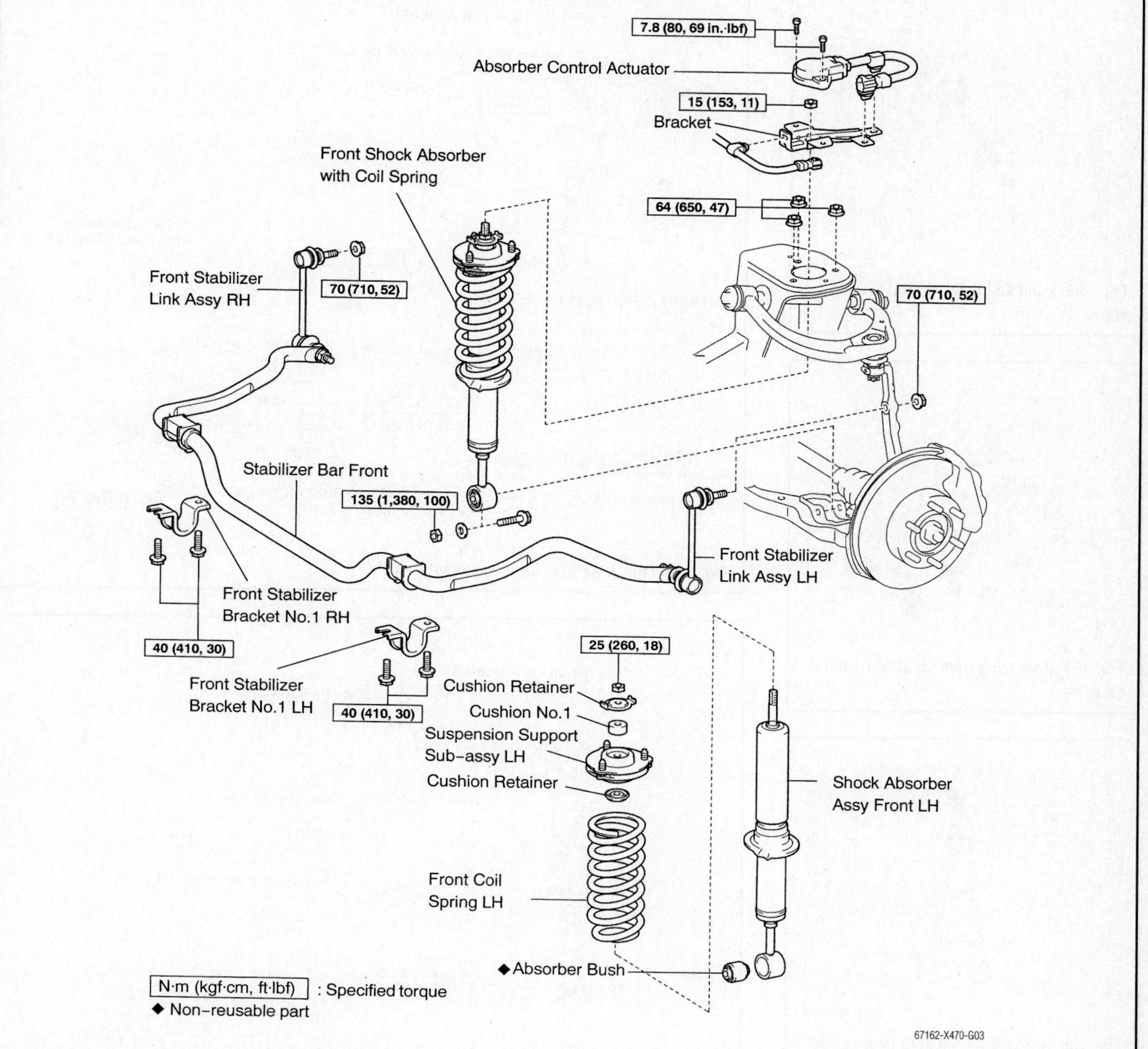

Fig. 103 Front strut and related components—GX470

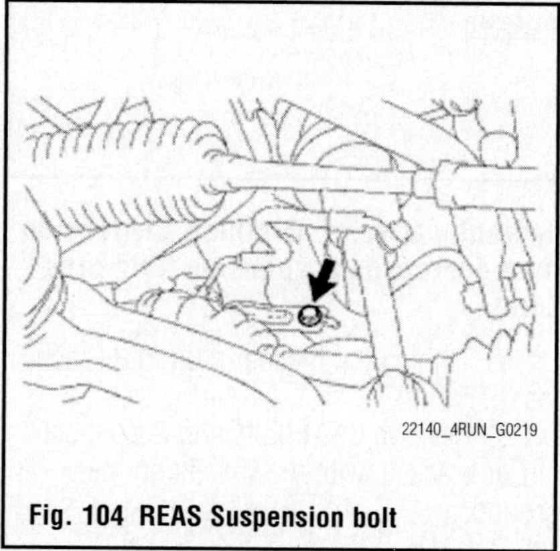

Fig. 104 REAS Suspension bolt

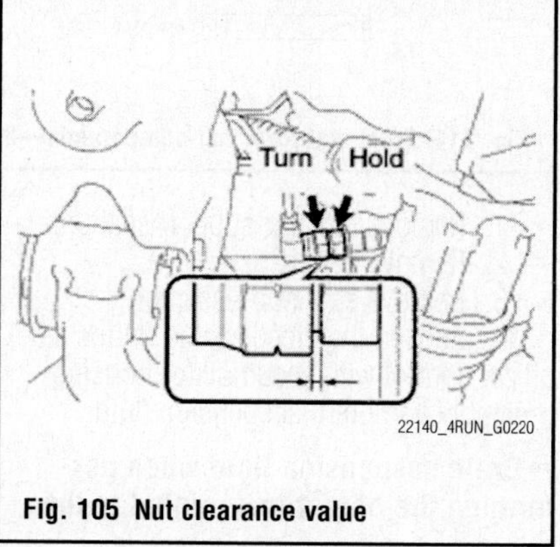

Fig. 105 Nut clearance value

2. As shown in the illustration, tighten the nut of clearance to standard value. Tighten to 18 ft. lbs. (25 Nm).

STABILIZER BAR

REMOVAL & INSTALLATION

With KDSS

See Figures 106 through 114.

➡Bleeding and testing system requires a proprietary service tool. Do not attempt to service system without proper equipment and training.

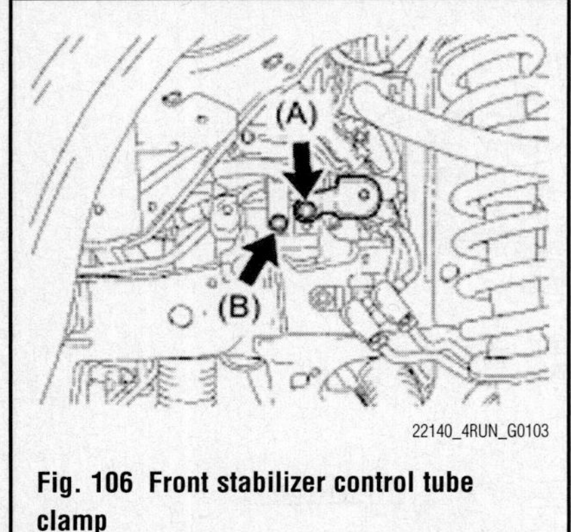

Fig. 106 Front stabilizer control tube clamp

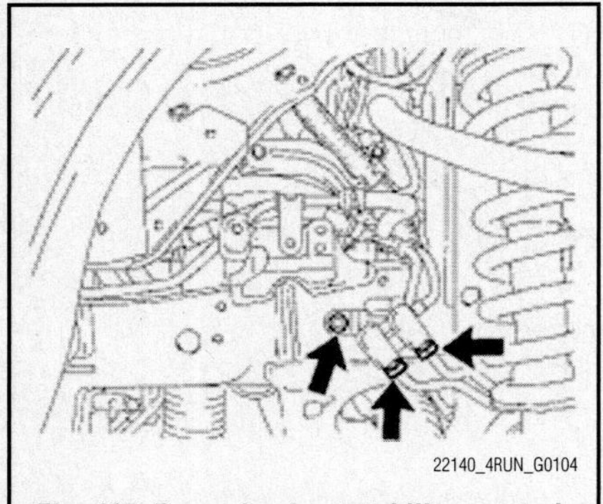

Fig. 107 Removing front stabilizer control tube

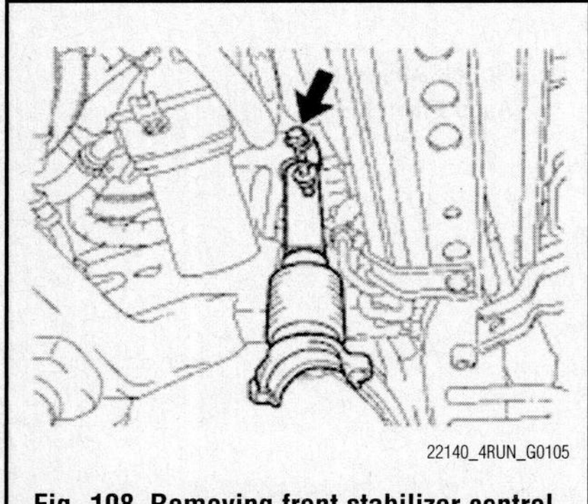

Fig. 108 Removing front stabilizer control & tube cylinder assembly

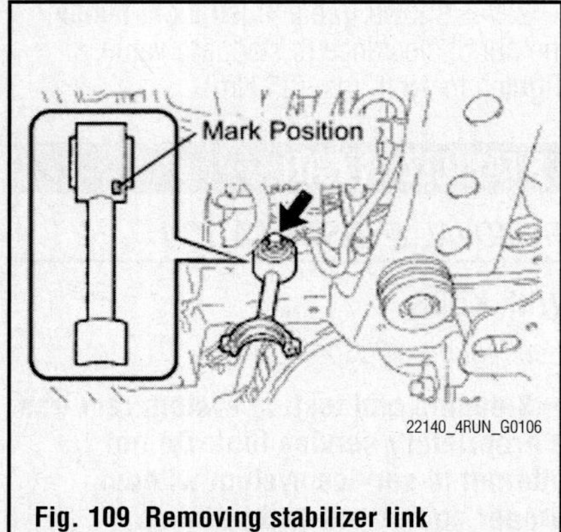

Fig. 109 Removing stabilizer link

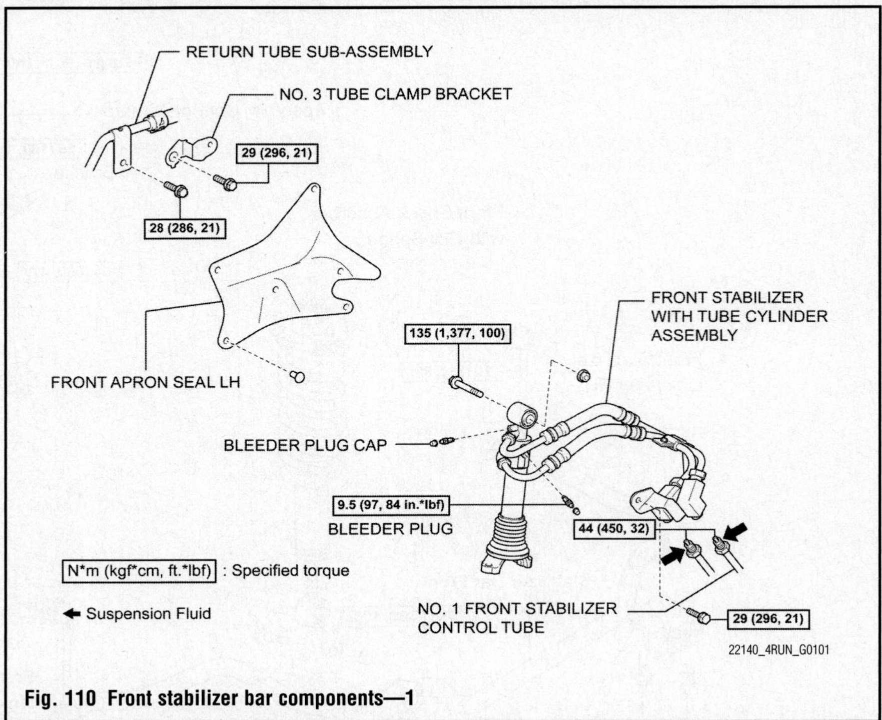

Fig. 110 Front stabilizer bar components—1

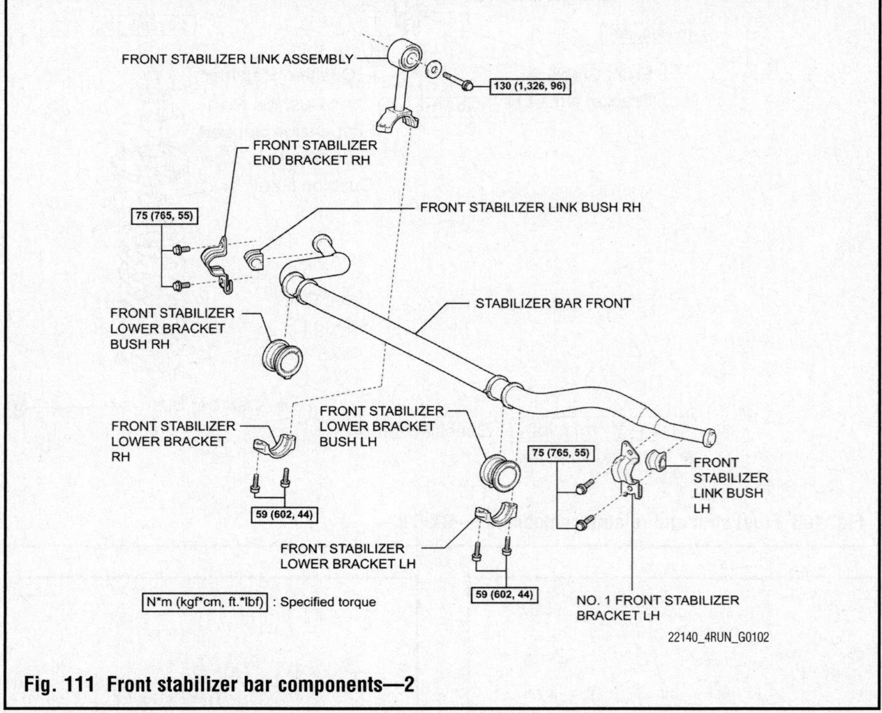

Fig. 111 Front stabilizer bar components—2

1. Remove radiator support seal upper.
2. Remove battery.
3. Remove step sub-assembly.
4. Loosen the bleeder plug on the stabilizer control with accumulator housing assembly and drain suspension fluid.

➡Drain suspension fluid when performing the operations related to the hydraulic circuits. Draining suspension fluid decreases suspension fluid pressure.

5. Tighten the bleeder plug to 84 inch lbs. (9.5 Nm)
6. Remove the 4 bolts and 2 front stabilizer brackets with the wheels on the ground.

7. Using a socket hexagon wrench (10 mm), remove the 4 bolts, 2 front stabilizer lower brackets and stabilizer bar front.

8. Remove the 2 front stabilizer link bushes and front stabilizer lower bracket bushes from the stabilizer bar front.

9. Remove front disc wheel.

10. Remove the 7 clips and front apron seal.

11. Remove bolt (A) and No. 3 tube clamp bracket.

12. Remove bolt (B) and separate the return tube sub-assembly.

13. Using a union nut wrench, separate the No. 1 front stabilizer control tube from the front stabilizer with tube cylinder assembly.

14. Remove the bolt.

15. Remove the bolt, nut and front stabilizer with tube cylinder assembly.

➡Turn the bolt while holding the nut. Do not loosen or remove flare nuts (A) and (B) shown in the illustration. Do not remove or hold the front stabilizer with tube cylinder assembly by the cylinder boot.

16. Remove the 2 bleeder plug caps and bleeder plugs from the front stabilizer with tube cylinder assembly.

17. Remove the bolt, spacer and front stabilizer link assembly.

To install:

18. Install the front stabilizer link assembly with the spacer by temporarily tightening the bolt. Ensure that the identification mark on the front stabilizer link assembly faces inward and to the front of the vehicle.

19. Install the 2 bleeder plugs and bleeder plug caps to the front stabilizer cylinder with tube cylinder assembly. Tighten to 84 inch lbs. (9.5 Nm).

20. Install the front stabilizer with tube cylinder assembly by temporarily tightening the nut and bolt.

➡Turn the bolt while holding the nut. Do not hold or install the front stabilizer with tube cylinder assembly by the cylinder boot. Pass the tube side of the front stabilizer with tube cylinder assembly under the return tube sub-assembly before installing.

21. Install the bracket by temporarily tightening the bolt.

22. Connect the No. 1 front stabilizer control tube.

23. Apply suspension fluid to the threads of the flare nuts.

Fig. 112 Installing clamp bracket bolts

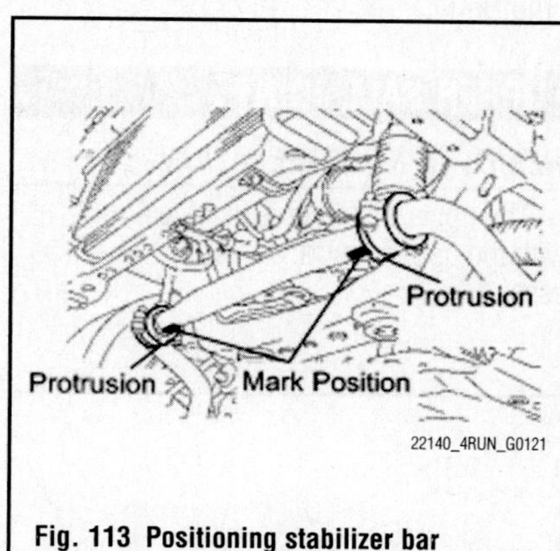

Fig. 113 Positioning stabilizer bar

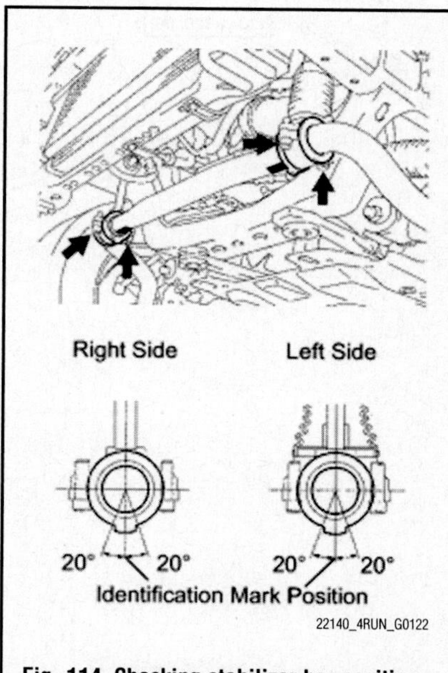

Fig. 114 Checking stabilizer bar position

24. Using a union nut wrench, connect the No. 1 front stabilizer control tube to the front stabilizer with tube cylinder assembly and tighten the flare nuts to 32 ft. lbs. (44 Nm).

25. Fully tighten the bracket bolt to 21 ft. lbs. (29 Nm)

26. Install the No. 3 tube clamp bracket with bolt (A) and tighten to 21 ft. lbs. (29 Nm).

27. Install the return tube sub-assembly with bolt (B) and tighten to 21 ft. lbs. (29 Nm).

28. Install the front apron seal LH with the 7 clips.

➡Bleeding system requires a proprietary service tool.

29. Install the 2 front stabilizer lower bracket bushes to the stabilizer bar front.

a. Align the protrusions on the bushes with the identification marks on the front stabilizer bar front with the protrusions facing inward.

➡Place the jack under the left side of the vehicle.

30. Support the stabilizer bar front with the identification marks facing down with a jack. Place a wooden block between the jack and the stabilizer bar front to prevent damage.

31. Using a socket hexagon wrench (10 mm), install the stabilizer bar front with the 4 bolts and 2 front stabilizer lower brackets and tighten to 44 ft. lbs. (59 Nm).

a. Check that the protrusions on the front stabilizer lower bracket bushes are positioned within 20°of the identification marks.

32. Install the 4 bolts, 2 front stabilizer brackets and 2 front stabilizer link bushes with the wheels on the ground and tighten to 5 ft. lbs. (75 Nm).

➡There are stamps on the front stabilizer brackets to distinguish between right and left.

33. Fully tighten the bolt on the front stabilizer with tube cylinder assembly to 100 ft. lbs. (135 Nm).

➡Tighten the bolt with the wheels on the ground.

34. Fully tighten the bolt on the front stabilizer link assembly and tighten to 96 ft. lbs. (130 Nm).

➡Tighten the bolt with the wheels on the ground.

➡System must be tested. Testing system requires a proprietary service tool.

35. Install no. 1 engine under cover sub-assembly.

36. Install spare disc wheel.

37. Install step sub-assembly LH.

38. Install battery

39. Install radiator support seal upper.

Without KDSS

1. Remove front disc wheel.
2. Remove the 2 nuts and the front stabilizer link assembly.
3. Remove front stabilizer link assembly.
4. Remove the 2 bolts and front stabilizer bracket.
5. Remove the 2 front stabilizer bar bush.
6. Remove stabilizer bar front.

To install:

7. Install the 2 front stabilizer bar bush.

➡Install the bushing to the inner side of the bushing stopper on the stabilizer bar. Install the stabilizer bush No. 1 as the protrusion to be on the inner side of the vehicle.

8. Install the front stabilizer bracket and tighten to 30 ft. lbs. (40 Nm).
9. Install the front stabilizer link assemblies.
10. Install front wheels.

TRACK BAR

REMOVAL & INSTALLATION

See Figure 115.

1. Remove the bolt.
2. Remove the bolt, nut and lateral control rod assembly.

To install:

3. Install the lateral control rod assembly with the bolt.
4. Install the bolt and the nut.
5. Stabilize suspension.
6. Fully tighten the 2 bolts to 96 ft. lbs. (130 Nm).

UPPER BALL JOINT

REMOVAL & INSTALLATION

The upper ball joint is serviced with the upper control arm as an assembly.

UPPER CONTROL ARM

REMOVAL & INSTALLATION

See Figure 116.

1. Before servicing the vehicle, refer to the Precautions section.
2. Remove the wheel.
3. Disconnect the skid control wire.
4. Support the lower arm with a jack.
5. Remove the cable bracket.
6. Disconnect the ball joint from the knuckle.
7. Remove the through-bolt, washers and nut.
8. Remove the arm.

To install:

9. Installation is the reverse of removal. Don't fully tighten the through-bolt until the vehicle is on the ground and the suspension is jounced a few times.
    - Ball joint nut: 81 ft. lbs. (110 Nm)
    - Through-bolt: 85 ft. lbs. (115 Nm)

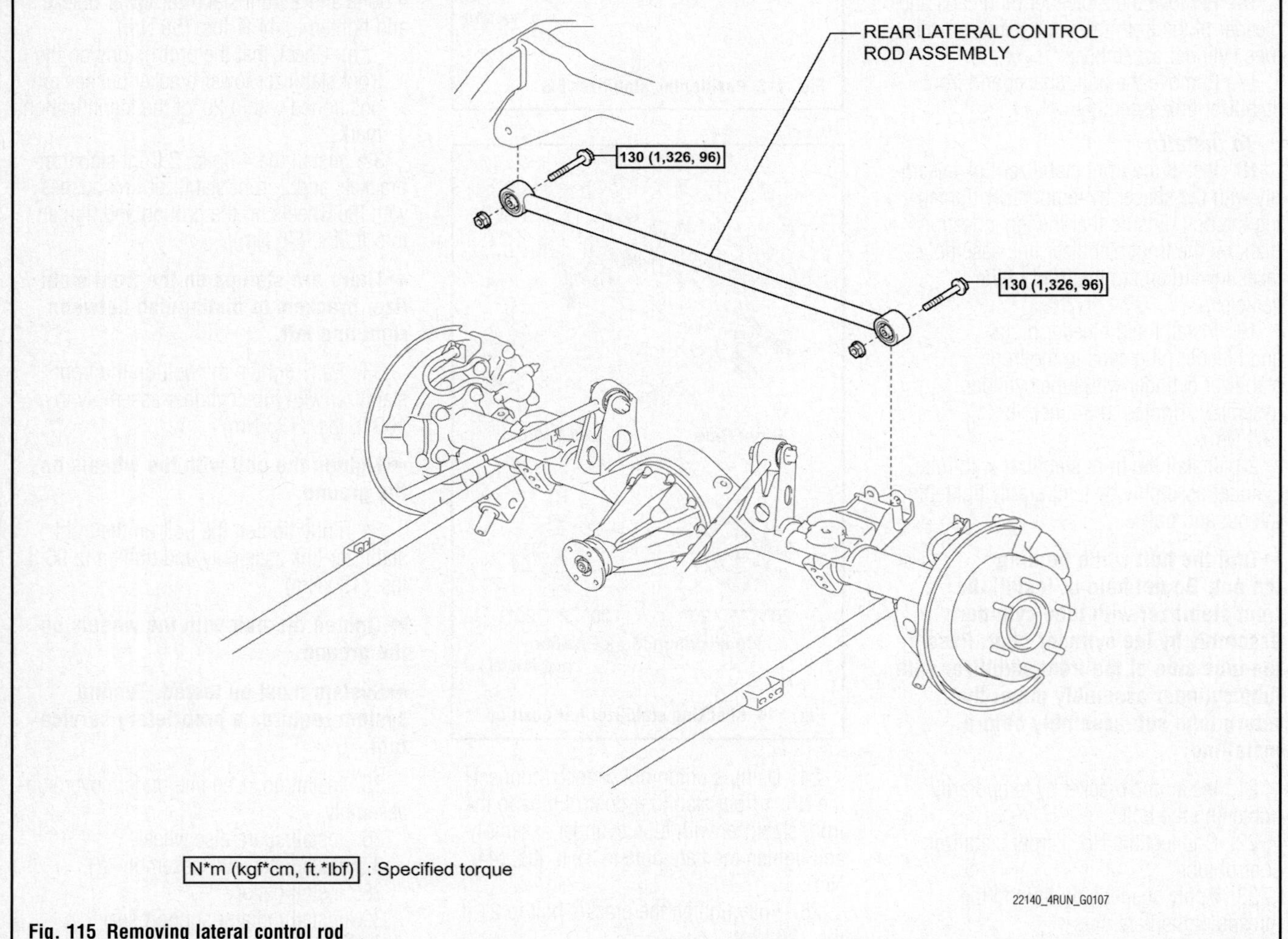

Fig. 115 Removing lateral control rod

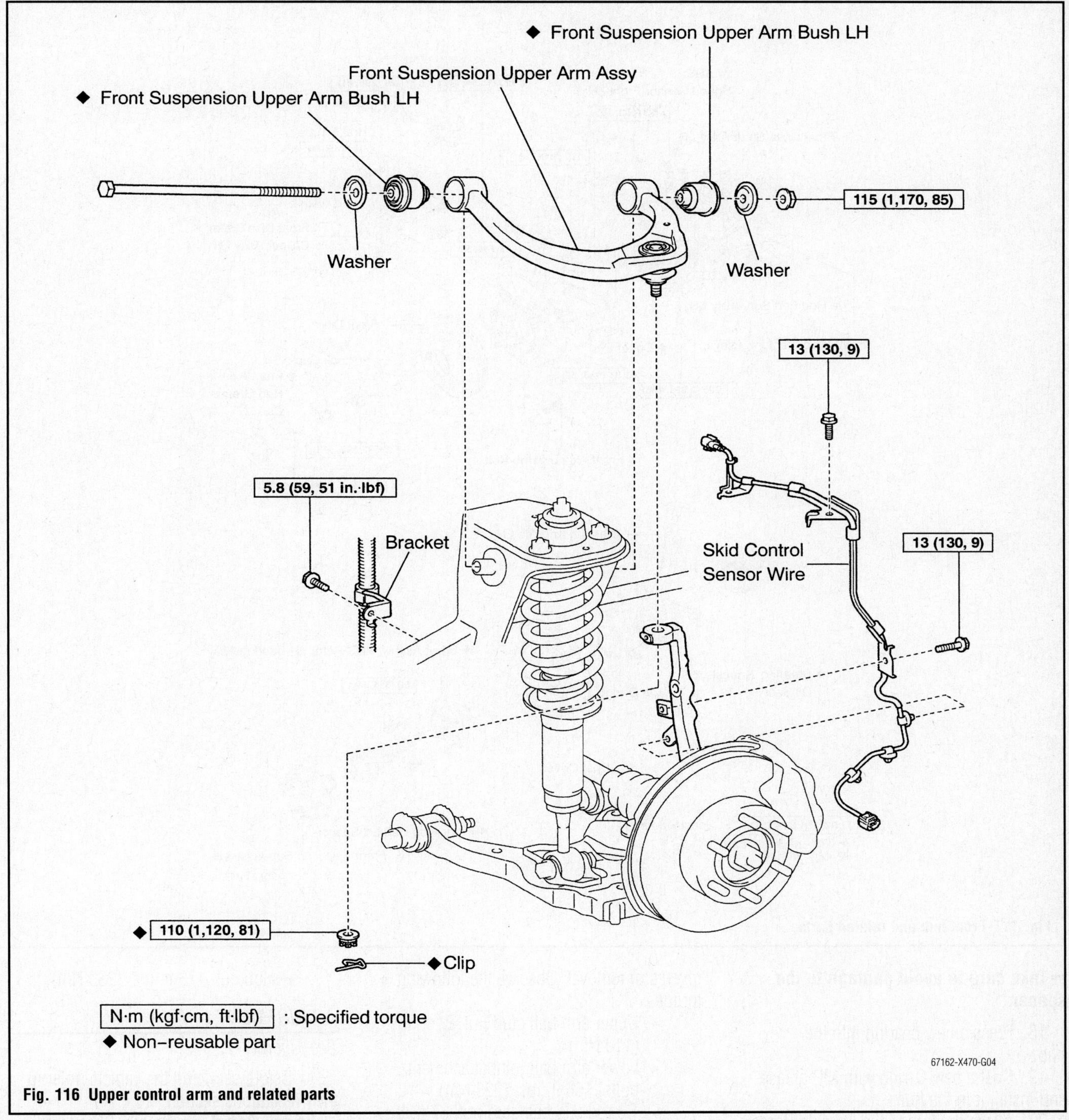

Fig. 116 Upper control arm and related parts

WHEEL BEARINGS

REMOVAL & INSTALLATION

See Figure 117.

1. Remove the wheel.
2. Remove the caliper.
3. Remove the hub grease cap.
4. Remove the cotter pin.
5. Remove the hub nut.
6. Remove the speed sensor.
7. Remove the stabilizer links from the knuckles.
8. Remove the tie rod end from the knuckle.
9. Remove the lower arm from the knuckle.
10. Remove the upper arm from the knuckle.
11. Remove the hub/knuckle assembly from the shaft.
12. Mount the assembly in a vise.
13. Remove the knuckle oil seal.
14. Remove the 4 bolts and remove the hub assembly from the knuckle.
15. Using the special tool and its components, remove the bearing from the hub.
16. Remove the oil seal.

To install:

17. Using a seal driver, install a new seal.

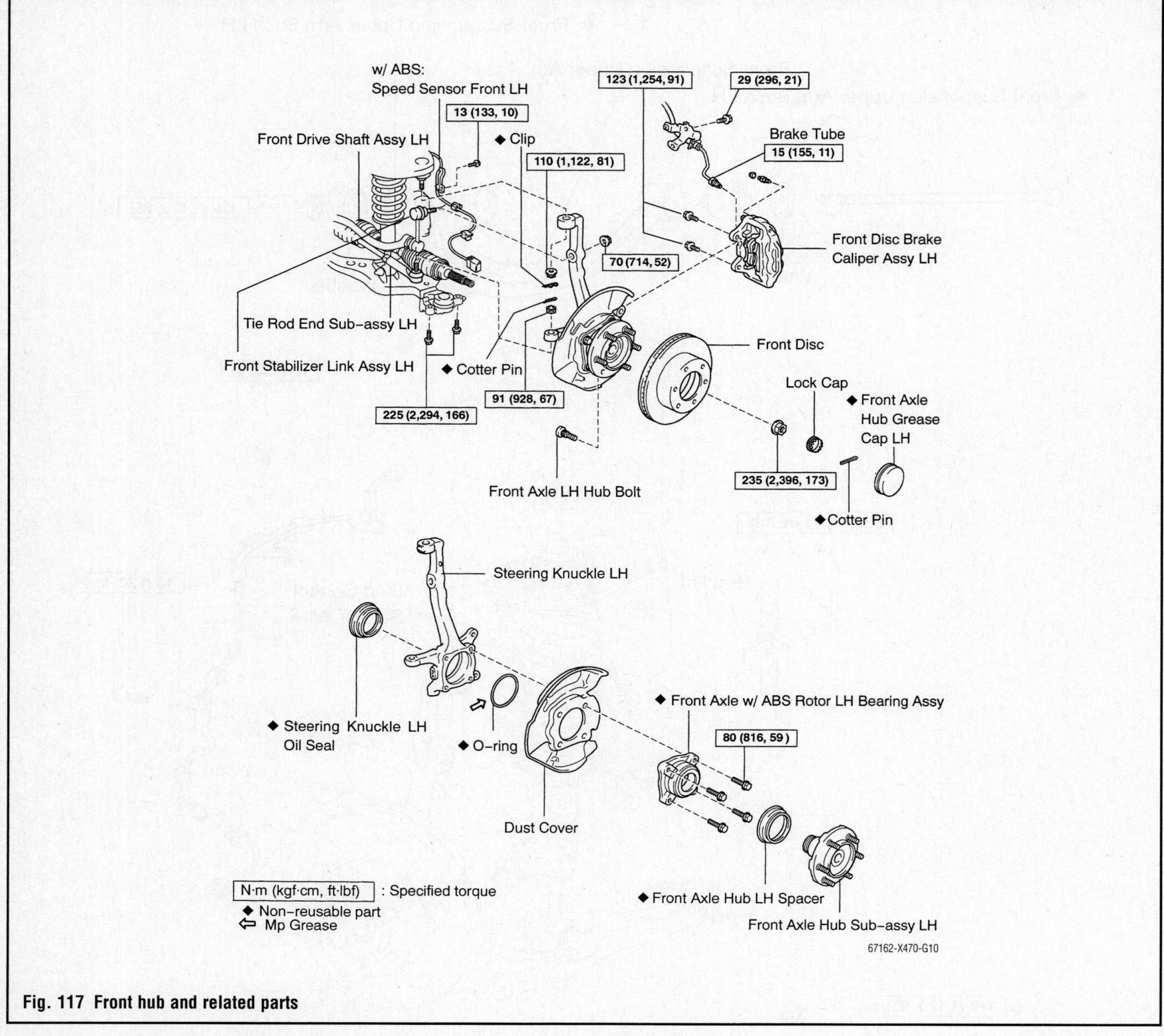

Fig. 117 Front hub and related parts

➡Take care to avoid damage to the spacer.

18. Press a new bearing into the hub.
19. Coat a new O-ring with MP grease and install it in the hub.
20. Attach the hub to the knuckle. Torque to 59 ft. lbs. (80 Nm).
21. Install a new knuckle oil seal.
22. The remainder of installation is the reverse of removal. Observe the following torques:
    - Upper arm ball stud nut: 81 ft. lbs. (110 Nm)
    - Lower arm ball joint attachment bolts: 166 ft. lbs. (225 Nm)
    - Tie rod end ball stud nut: 67 ft. lbs. (91 Nm)
    - Stabilizer end links: 52 ft. lbs. (70 Nm)
    - Hub nut: 173 ft. lbs. (235 Nm)

ADJUSTMENT

See Figure 117.

1. Before servicing the vehicle, refer to the Precautions section.

No adjustment is possible. Check for axle hub backlash and axle hub deviation. If either exceeds 0.0020 inch, replace the bearing.

CONTROL ARMS/LINKS

REMOVAL & INSTALLATION

Lower

See Figures 118 through 120.

1. Remove the rear wheel.
2. Separate the parking brake cable assembly No. 3.
3. Remove the lower control arm assembly.
   a. Remove the nut, washer and bolt from the rear axle housing.

➡While fixing the nut, turn and remove the bolt.

   b. Remove the nut, washer, bolt and the lower control arm assembly.

To install:

4. Install the lower control arm assembly.

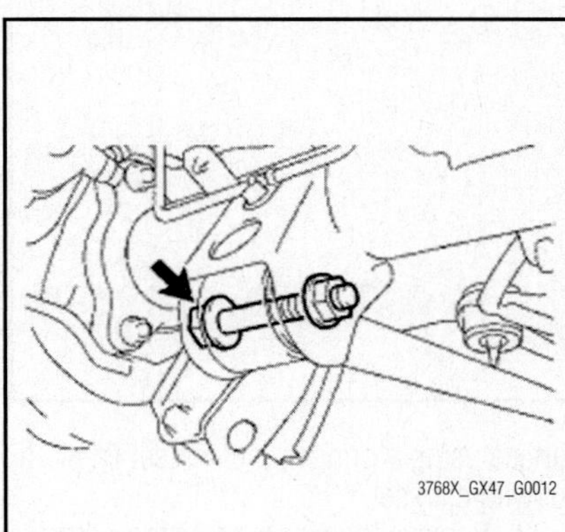

Fig. 118 Removing the nut, washer and bolt from the rear axle housing LH

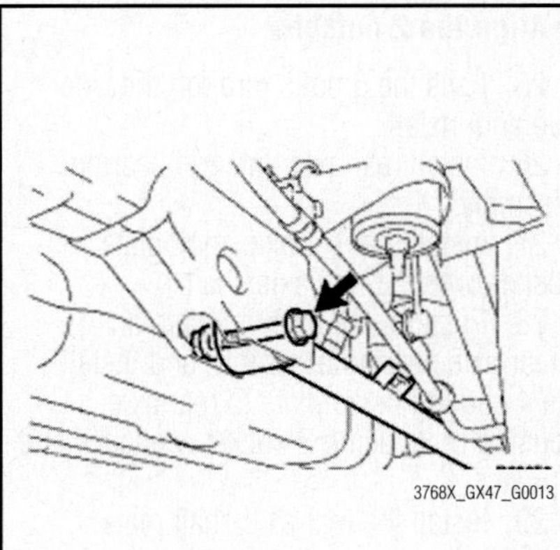

Fig. 119 Removing the nut, washer, bolt and the lower control arm assembly

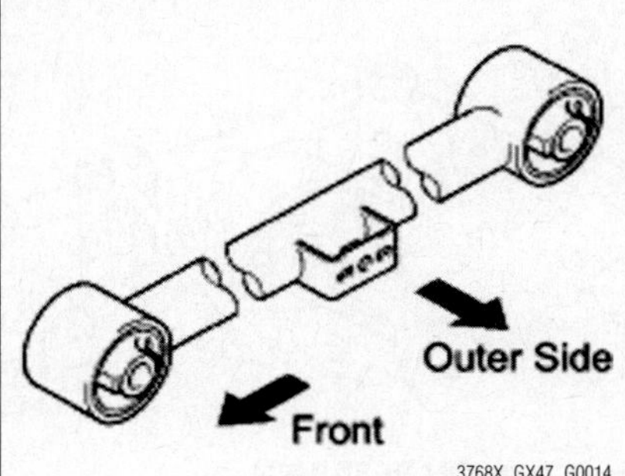

Fig. 120 Installing the lower control arm assembly

   a. Install the lower control arm assembly, temporarily tighten the bolt, washer and nut.
   b. Install the lower control arm assembly.
   c. Install the rear axle housing, temporarily tighten the bolt, washer and nut.
5. Connect the parking brake cable assembly No. 3. Tighten the bolt to 9 ft. lbs. (13 Nm).
6. Install the rear wheel. Tighten the wheel to 83 ft. lbs. (112 Nm).
7. Stabilize the suspension.
8. Fully tighten the lower control arm assembly. Tighten to 96 ft. lbs. (130 Nm).

Upper

See Figures 121 and 122.

1. Remove the rear wheel.
2. Disconnect the height control sensor link rear sub assembly.
3. Remove the upper control arm assembly.
   a. Remove the nut, washer and bolt from the rear axle housing.

Fig. 121 Removing the nut, washer and bolt from the rear axle housing

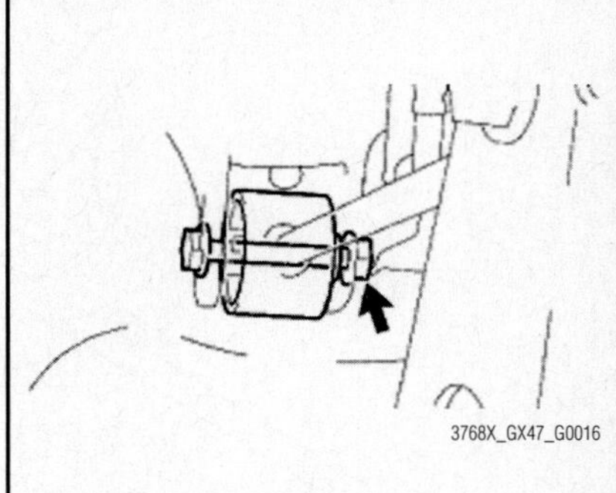

Fig. 122 Removing the nut, washer and bolt with the upper control arm assembly

➡While fixing the bolt, turn and remove the nut.

   b. Remove the nut, washer and bolt with the upper control arm assembly.

To install:

4. Install the upper control arm assembly.
   a. Install the upper control arm assembly, temporarily tighten the nut, washer and bolt.
   b. Install the rear axle housing, temporarily tighten the nut, washer and bolt.
5. Connect the height control sensor link sub assembly.
6. Install the rear wheel.
7. Stabilize the suspension.
8. Fully tighten the upper control arm assembly. Tighten the upper control arm assembly nuts to 59 ft. lbs. (80 Nm).

SHOCK ABSORBER

REMOVAL & INSTALLATION

See Figure 123.

1. Before servicing the vehicle, refer to the precautions section.
2. Support the axle with a jackstand.
3. Disconnect the actuator at the shock absorber.

➡Don't over extend the pneumatic shock.

4. Remove the lower shock bolt.
5. Remove the upper nut and remove the shock.

To install:

6. Installation is the reverse of removal. Don't fully tighten the lower bolt until the

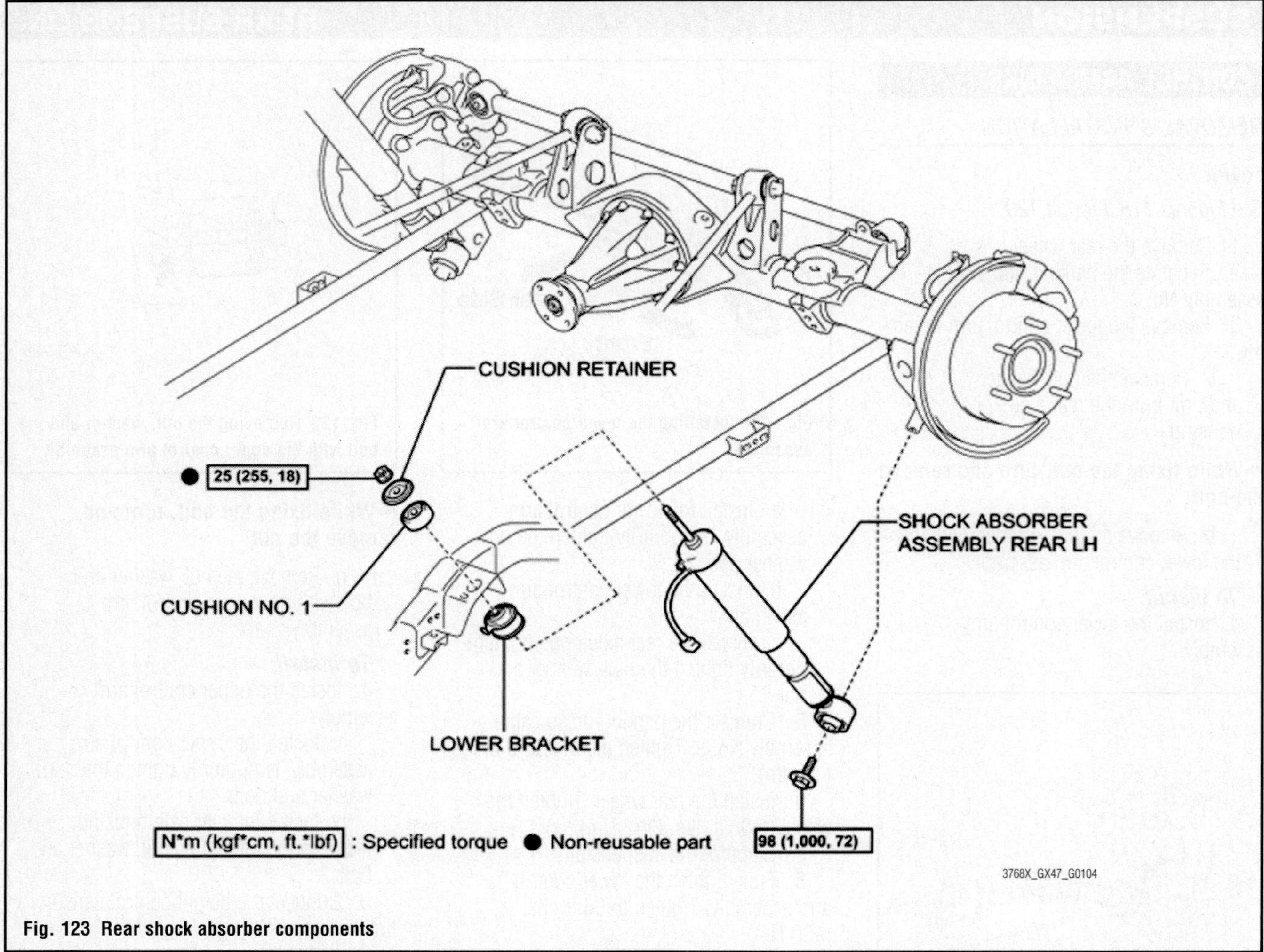

Fig. 123 Rear shock absorber components

vehicle is on the ground and the suspension jounced a few times. Torque the upper nut to 18 ft. lbs. (25 Nm); the lower bolt to 72 ft. lbs. (98 Nm).

WHEEL BEARINGS

REMOVAL & INSTALLATION

See Figures 124 through 128.

1. Remove rear wheel.
2. Remove rear speed sensor (w/ ABS).
3. Separate rear disc brake caliper assembly.
4. Remove rear disc.
5. Remove parking brake shoe return tension spring.
6. Remove parking brake shoe strut compression spring.
7. Remove parking brake shoe strut.
8. Remove parking brake shoe.
9. Remove rear axle shaft with backing plate.
10. Remove the 4 nuts and rear axle shaft with backing plate.
11. Remove the O-ring.
12. Using a snap ring expander, remove the snap ring.
13. Remove the rear axle shaft from the bearing.
14. Remove the rear axle bearing retainer inner from the rear axle bearing assembly.
15. Remove the rear axle shaft washer from the rear axle bearing assembly.
16. Attach the 4 nuts to the parking brake plate to rear axle housing bolts.
17. Using a hammer, remove the 4 parking brake plate to rear axle housing bolts and rear axle bearing assembly.

➡Do not reuse the nuts previously removed from the vehicle.

18. Remove the 6 hub bolts.
19. Remove brake drum oil deflector.
20. Remove brake drum oil deflector gasket.
21. Remove rear axle bearing oil seal.
22. Grind the rear axle bearing inner race surface using a grinder, then chisel them out with a chisel.
23. Remove the rear axle shaft oil seal from the rear axle shaft.

To install:

24. Install a new deflector gasket and deflector to the rear axle shaft.

➡Align the 2 notches.

25. Pass the 6 bolts through the axle hub and install.
26. Install rear axle hub and bearing assembly.
27. Install the rear axle shaft plate washer onto the rear axle shaft.
28. Position the backing plate on a rear axle bearing assembly, and install the 4 parking brake plate to rear axle housing bolts using 2 socket wrenches and a press.
29. Install the rear axle shaft plate washer onto the rear axle shaft.
30. Install a new rear axle bearing retainer inner to the rear axle shaft.

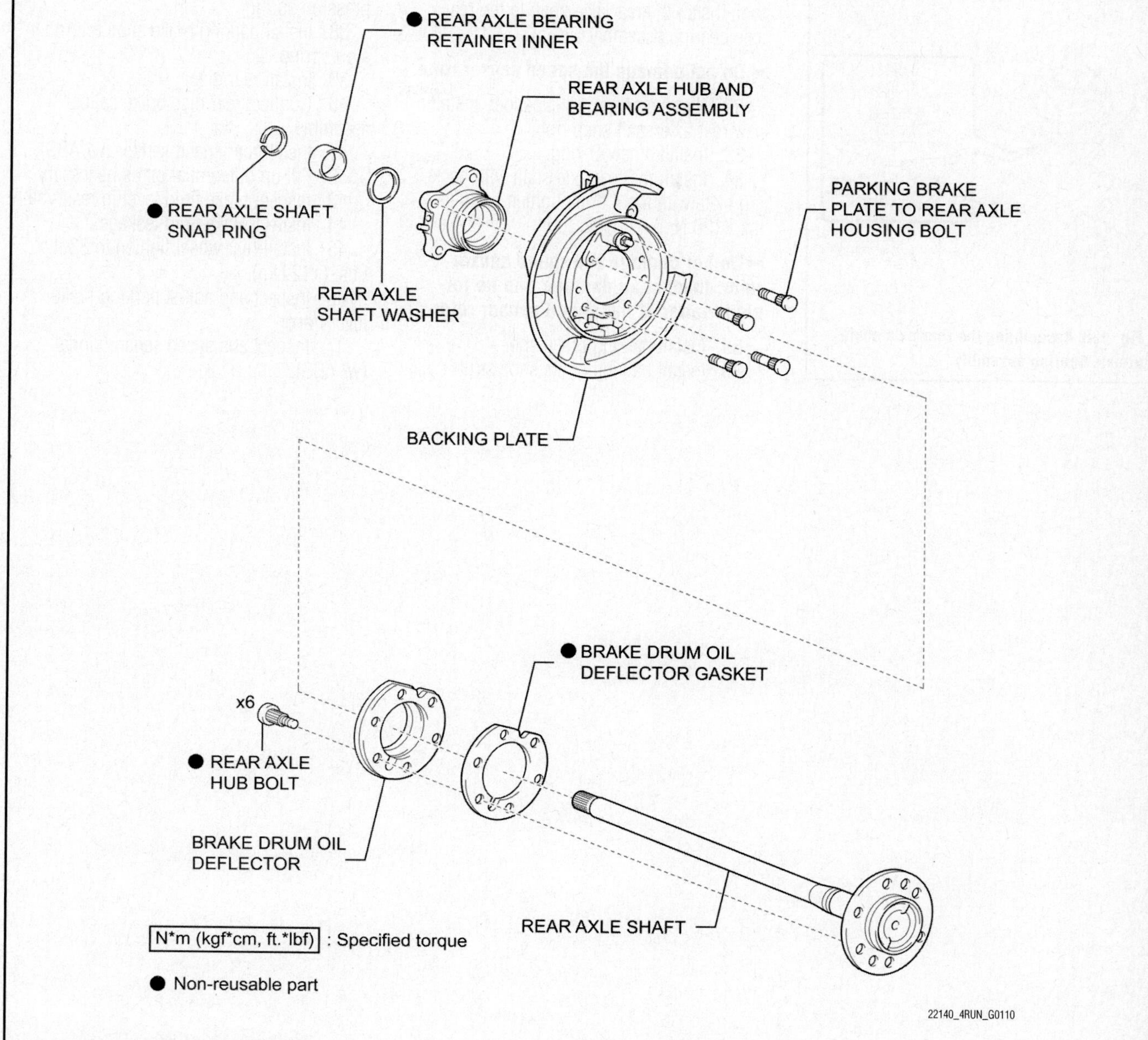

● REAR AXLE BEARING RETAINER INNER

REAR AXLE HUB AND BEARING ASSEMBLY

● REAR AXLE SHAFT SNAP RING

REAR AXLE SHAFT WASHER

PARKING BRAKE PLATE TO REAR AXLE HOUSING BOLT

BACKING PLATE

● BRAKE DRUM OIL DEFLECTOR GASKET

x6

● REAR AXLE HUB BOLT

BRAKE DRUM OIL DEFLECTOR

REAR AXLE SHAFT

N*m (kgf*cm, ft.*lbf) : Specified torque

● Non-reusable part

22140_4RUN_G0110

Fig. 124 Rear axle assembly components

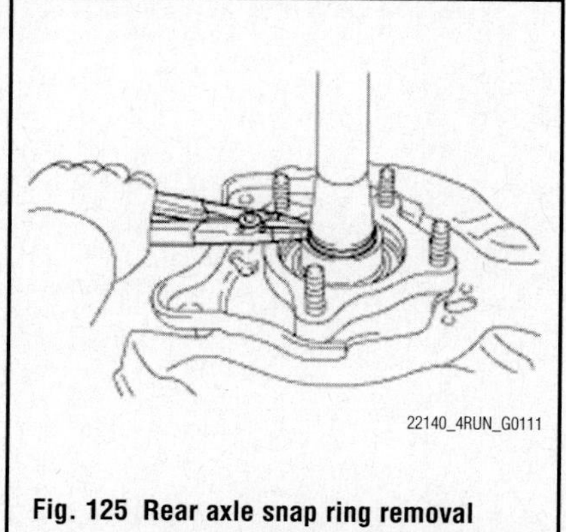

Fig. 125 Rear axle snap ring removal

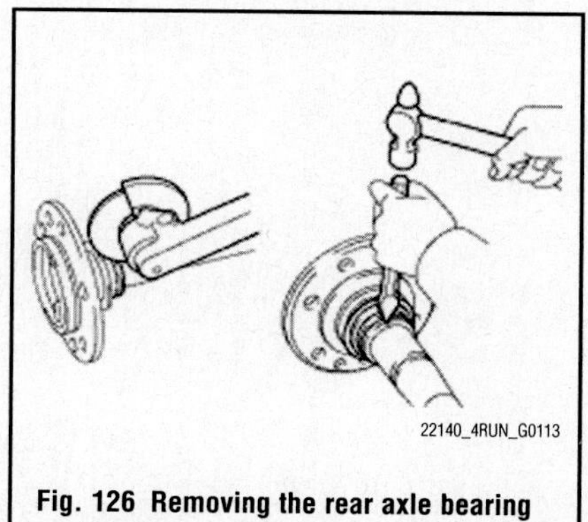

Fig. 126 Removing the rear axle bearing inner race

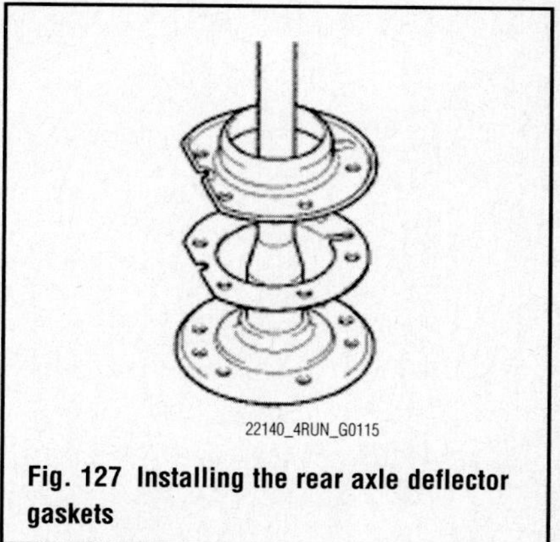

Fig. 127 Installing the rear axle deflector gaskets

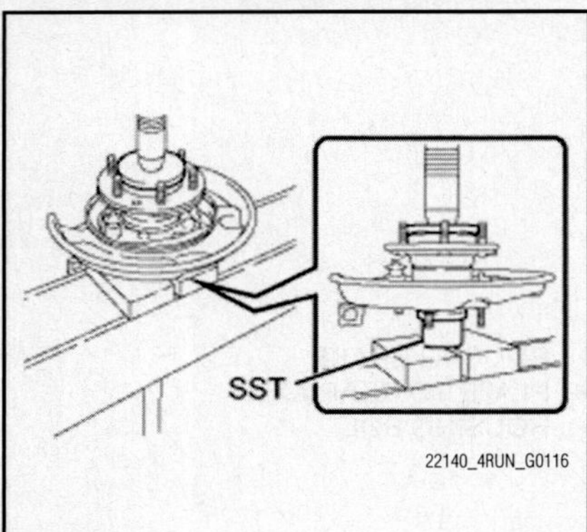

Fig. 128 Assembling the rear axle shaft-to-axle bearing assembly

31. With and press and the appropriate tool, install the rear axle shaft to the rear axle bearing assembly.

➡Do not damage the speed sensor rotor.

32. Using a snap ring expander, install a new rear axle shaft snap ring.
33. Install a new O-ring.
34. Install the rear axle shaft with backing plate with the 4 nuts, tighten to 89 ft. lbs. (120 Nm).

➡Do not damage the speed sensor rotor. Inspect no damage and no foreign matter at the speed sensor rotor.

35. Install parking brake shoe.
36. Install parking brake shoe strut.
37. Install parking brake shoe strut compression spring.
38. Install parking brake shoe return tension spring.
39. Install rear disc.
40. Connect rear disc brake caliper assembly.
41. Install rear speed sensor (w/ ABS).
42. Fill up differential oil as necessary.
43. Inspect brake fluid level in reservoir.
44. Inspect brake fluid leakage.
45. Install rear wheel tighten to 83 ft. Lbs. (112 Nm).
46. Inspect and adjust parking brake lever travel.
47. Inspect abs speed sensor signal (w/ ABS).

LEXUS

HS250h

27

SPECIFICATIONS AND MAINTENANCE CHARTS

ENGINE AND VEHICLE IDENTIFICATION

Engine							Model Year	
Code ①	Liters (cc)	Cu. In.	Cyl.	Fuel Sys.	Engine Type	Eng. Mfg.	Code ②	Year
2AZ-FXE	2.4 (2362)	144	4	SFI	DOHC	Toyota	A	2010

SFI: Sequential Fuel Injection

DOHC: Double Overhead Camshaft

① Stamped on the left side of the engine block

② 10th digit of the Vehicle Identification Number (VIN)

3768X_HS25_C0001

GENERAL ENGINE SPECIFICATIONS

All measurements are given in inches.

Year	Model	Engine Displacement Liters	Engine Series VIN	Net Horsepower @ rpm	Net Torque @ rpm (ft. lbs.)	Bore x Stroke (in.)	Com-pression Ratio	Oil Pressure @ rpm
2010	HS250h	2.4	2AZ-FXE	155@6000	158@4000	3.48x3.78	9.8:1	55@3000

3768X_HS25_C0002

GASOLINE ENGINE TUNE-UP SPECIFICATIONS

Year	Engine Displacement Liters	Engine VIN	Spark Plug Gap (in.)	Ignition Timing (deg.)	Fuel Pump (psi)	Idle Speed (rpm)	Valve Clearance (in.)	
							Intake	Exhaust
2010	2.4	2AZ-FXE	0.039-0.043	NA	44-50	①	0.0075-0.0114	0.0150-0.0189

NOTE: The Vehicle Emission Control Information label often reflects specification changes made during production.

The label figures must be used if they differ from those in this chart.

NA: Not available

① Manual transmission: 650 to 750 rpm, Automatic transmission: 610 to 710 rpm

3768X_HS25_C0003

CAPACITIES

Year	Model	Engine Displacement Liters	Engine VIN	Engine Oil with Filter (qts.)	Transmission (pts.) 5-Spd	Transmission (pts.) Auto.	Transfer Case (pts.)	Drive Axle Front (pts.)	Drive Axle Rear (pts.)	Fuel Tank (gal.)	Cooling System (qts.)
2010	HS250h	2.4	2AZ-FXE	4.5	—	3.7	—	—	—	14.5	6.4

NOTE: All capacities are approximate. Add fluid gradually and check to be sure a proper fluid level is obtained.

3768X_HS25_C0004

FLUID SPECIFICATIONS

Year	Model	Engine Displacement Liters	Engine ID/VIN	Engine Oil	Auto. Trans. ①	Drive Axle	Power Steering Fluid	Engine Coolant	Brake Master Cylinder
2010	HS250h	2.4	2AZ-FXE	0W-20	H V fluid	NA	N/S	②	DOT 3

DOT: Department Of Transpotation

NA: Not Available

N/S: Not Specified

① Hybrid Transaxle fluid.

② Toyota Super long life Coolant (SLLC)

3768X_HS25_C0005

VALVE SPECIFICATIONS

Year	Engine Displacement Liters	Engine VIN	Seat Angle (deg.)	Face Angle (deg.)	Spring Test Pressure (lbs. @ in.)	Spring Installed Height (in.)	Stem-to-Guide Clearance (in.) Intake	Stem-to-Guide Clearance (in.) Exhaust	Stem Diameter (in.) Intake	Stem Diameter (in.) Exhaust
2010	2.4	2AZ-FXE	45	44.5	NA	NA	0.0010-0.0031	0.0012-0.0039	0.2154-0.2159	0.2151-0.2157

NA: Not Available

3768X_HS25_C0006

CAMSHAFT AND BEARING SPECIFICATIONS CHART

All measurements are given in inches.

Year	Engine Displ. Liters	Engine ID/VIN	Journal Dia.	Brg. Oil Clearance	Shaft End-play	Runout	Journal Bore	Lobe Height Intake	Lobe Height Exhaust
2010	2.4	2AZ-FXE	①	②	NA	NA	NA	1.8624-1.8664	1.8104-1.1843

NA: Not Available

① No. 1 journal: 1.4162-1.4167

Other Journals: 0.9039-0.9045 in.

② No. 1 journal mark 1: 0.000276-0.00150 in.

No. 1 journal mark 2 and 3: 0.000315-0.00150in.

Other Maximum Journals: 0.00276 in.

3768X_HS25_C0007

CRANKSHAFT AND CONNECTING ROD SPECIFICATIONS

All measurements are given in inches.

Year	Engine Displacement Liters	Engine VIN	Crankshaft Main Brg. Journal Dia.	Crankshaft Main Brg. Oil Clearance	Crankshaft Shaft End-play	Crankshaft Thrust on No.	Connecting Rod Journal Diameter	Connecting Rod Oil Clearance	Connecting Rod Side Clearance
2010	2.4	2AZ-FXE	2.1649-2.1654	0.0007-0.0016	0.0016-0.0095	3	①	0.0013-0.0025	0.0063-0.0143

① Mark 1: 2.0079-2.0082 in.

Mark 2: 2.0082-2.0084 in.

Mark 3: 2.0084-2.0087 in.

3768X_HS25_C0008

PISTON AND RING SPECIFICATIONS

All measurements are given in inches.

Year	Engine Displ. Liters	Engine VIN	Piston Clearance	Ring Gap Top Compression	Ring Gap Bottom Compression	Ring Gap Oil Control	Ring Side Clearance Top Compression	Ring Side Clearance Bottom Compression	Ring Side Clearance Oil Control
2010	2.4	2AZ-FXE	0.0020-0.0029	0.0094-0.0122	0.0130-0.0169	0.0039-0.0118	0.0008-0.0028	0.0008-0.0024	0.00079-0.0028

3768X_HS25_C0009

TORQUE SPECIFICATIONS

All readings in ft. lbs.

Year	Engine Displacement Liters	Engine VIN	Cylinder Head Bolts	Main Bearing Bolts	Rod Bearing Bolts	Crankshaft Damper Bolts	Flywheel Bolts	Manifold Intake	Manifold Exhaust	Spark Plugs	Oil Pan Drain Plug
2010	2.4	2AZ-FXE	①	②	③	125	96	22	27	14	29

① Step 1: 52 ft. lbs.

Step 2: plus 90 degrees

② Step 1: 15 ft. lbs.

Step 2: 29 ft. lbs.

Step 3: Plus 90 degrees

3768X_HS25_C0010

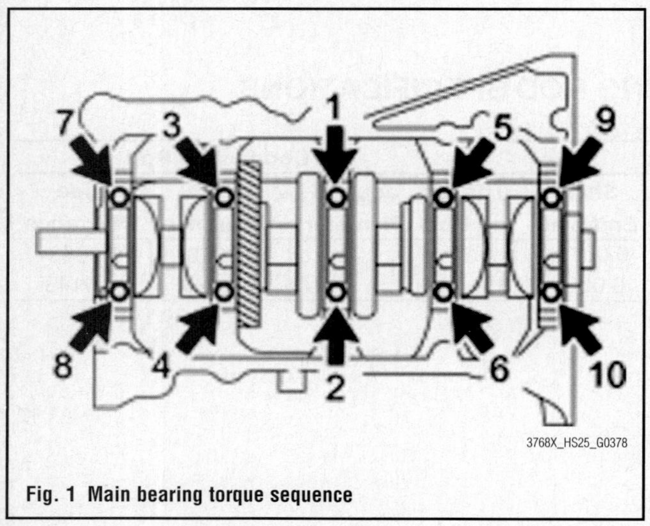

Fig. 1 Main bearing torque sequence

WHEEL ALIGNMENT

Year	Model		Caster Range (+/-Deg.)	Caster Preferred Setting (Deg.)	Camber Range (+/-Deg.)	Camber Preferred Setting (Deg.)	Toe-in (in.)	Steering Axis Inclination (Deg.)
2010	HS250h	Front	0.75	①	0.75	-0.22	0+/-0.09	12.02
		Rear	—	—	②	-1.15	0.16+/-0.08	—

① Tire size P215/55R17: 6.07 degrees

Tire size P225/45R18: 5.87 degrees

② Tire size P215/55R17: -1.02 +/- -0.75

Tire size P225/45R18: -0.85 +/- -0.75

3768X_HS25_C0011

TIRE, WHEEL AND BALL JOINT SPECIFICATIONS

Year	Model	OEM Tires		Tire Pressures (psi)		Wheel	Ball Joint	Lug Nut Torque
		Standard	Optional	Front	Rear	Size	Inspection	(ft. lbs.)
2010	HS250h	P215/55R17	P225/45R18	33	33/32	NA	①	76

NA: Not Available

OEM: Original Equipment Manufacturer

OPT: Optional

PSI: Pounds Per Square Inch

STD: Standard

① Replace if any measurable movement is found.

3768X_HS25_C0012

BRAKE SPECIFICATIONS

All measurements in inches unless noted

Year	Model		Brake Disc			Minimum	Brake Caliper	
			Original Thickness	Minimum Thickness	Maximum Runout	Lining Thickness	Bracket Bolts (ft. lbs.)	Mounting Bolts (ft. lbs.)
2010	HS250h	Front	0.984	0.866	0.0020	0.039	79	25
		Rear	0.472	0.413	0.0059	0.039	58	20

3768X_HS25_C0013

SCHEDULED MAINTENANCE INTERVALS

LEXUS HS250h

TO BE SERVICED	TYPE OF SERVICE	VEHICLE MILEAGE INTERVAL (x1000)													
		5	10	15	20	25	30	35	40	45	50	55	60	90	120
Engine oil & filter	R	✓	✓	✓	✓	✓	✓	✓	✓	✓	✓	✓	✓	✓	✓
Smart key battery	R		✓		✓		✓		✓		✓		✓	✓	✓
Automatic transmission fluid	S/I			✓			✓			✓			✓	✓	✓
Ball joints & dust covers	S/I		✓	✓	✓		✓		✓		✓		✓	✓	✓
Bolts & nuts on chassis & body	S/I	✓	✓	✓	✓	✓	✓	✓	✓	✓	✓	✓	✓	✓	✓
Brake linings & drums	S/I	✓	✓	✓	✓	✓	✓	✓	✓	✓	✓	✓	✓	✓	✓
Brake line pipes & hoses	S/I			✓			✓			✓			✓	✓	✓
Brake pads & discs (front & rear)	S/I	✓	✓	✓	✓	✓	✓	✓	✓	✓	✓	✓	✓	✓	✓
Brake fluid	R						✓						✓	✓	✓
Rack and pinion assembly	S/I			✓			✓			✓			✓	✓	✓
Steering linkage & boots	S/I	✓	✓	✓	✓	✓	✓	✓	✓	✓	✓	✓	✓	✓	✓
Steering gear box	S/I			✓			✓			✓			✓	✓	✓
Spark plugs ①	R														✓
Drive belts	S/I												✓	✓	✓
Exhaust pipes & mountings	S/I			✓			✓			✓			✓	✓	✓
Fuel lines & connections	S/I						✓						✓	✓	✓
Engine/inverter coolant ②	S/I			✓			✓			✓			✓	✓	✓
Radiator, condenser and/or	S/I			✓			✓			✓			✓	✓	✓
Fuel tank cap gasket	S/I						✓						✓	✓	✓
Rotate tires	S/I	✓	✓	✓	✓	✓	✓	✓	✓	✓	✓	✓	✓	✓	✓
Clean air conditioning filter ③	S/I						✓			✓			✓		✓
Axle shaft bolts	S/I			✓			✓			✓			✓	✓	✓
Brake pad thickness and rotor runout	S/I					✓					✓		✓	✓	✓

R: Replace S/I: Service or Inspect

① Spark plugs are replaced at 120,000 miles

② Replace engine coolant at 100,000 miles and then inspect every 15,000 miles

③ Replace air conditioning filter every 30,000 miles

FREQUENT OPERATION MAINTENANCE (SEVERE SERVICE)

If a vehicle is operated under any of the following conditions it is considered severe service:

- Extremely dusty areas.

- 50% or more of the vehicle operation is in 32°C (90°F) or higher temperatures, or constant temperatures below 0°C (32°F).

- Prolonged idling (vehicle operation in stop and go traffic).

- Frequent short running periods (engine does not warm to normal operating temperatures).

- Police, taxi, delivery usage or trailer towing usage.

Air cleaner filter: service or inspect every 5000 miles

Rear differential & transfer case oil: replace every 15,000 miles.

Ball joints & dust covers: service or inspect every 5000 miles.

Bolts & nuts on chassis & body: service or inspect every 5000 miles.

Axle shaft bolts: service or inspect every 5000 miles.

Steering linkage: service or inspect every 5000 miles.

3768X_HS25_C0014

BRAKES — INFORMATION AND PRECAUTIONS

ANTI-LOCK SYSTEMS

• Certain components within the ABS system are not intended to be serviced or repaired individually.

• Do not use rubber hoses or other parts not specifically specified for and ABS system. When using repair kits, replace all parts included in the kit. Partial or incorrect repair may lead to functional problems and require the replacement of components.

• Lubricate rubber parts with clean, fresh brake fluid to ease assembly. Do not use shop air to clean parts; damage to rubber components may result.

• Use only DOT 3 brake fluid from an unopened container.

• If any hydraulic component or line is removed or replaced, it may be necessary to bleed the entire system.

• A clean repair area is essential. Always clean the reservoir and cap thoroughly before removing the cap. The slightest amount of dirt in the fluid may plug an orifice and impair the system function. Perform repairs after components have been thoroughly cleaned; use only denatured alcohol to clean components. Do not allow ABS components to come into contact with any substance containing mineral oil; this includes used shop rags.

• The Anti-Lock control unit is a microprocessor similar to other computer units in the vehicle. Ensure that the ignition switch is **OFF** before removing or installing controller harnesses. Avoid static electricity discharge at or near the controller.

• If any arc welding is to be done on the vehicle, the control unit should be unplugged before welding operations begin.

DISC AND DRUM SYSTEMS

**** CAUTION**

Dust and dirt accumulating on brake parts during normal use may contain asbestos fibers from production or aftermarket brake linings. Breathing excessive concentrations of asbestos fibers can cause serious bodily harm. Exercise care when servicing brake parts. Do not sand or grind brake lining unless equipment used is designed to contain the dust residue. Do not clean brake parts with compressed air or by dry brushing. Cleaning should be done by dampening the brake components with a fine mist of water, then wiping the brake components clean with a dampened cloth. Dispose of cloth and all residue containing asbestos fibers in an impermeable container with the appropriate label. Follow practices prescribed by the Occupational Safety and Health Administration (OSHA) and the Environmental Protection Agency (EPA) for the handling, processing, and disposing of dust or debris that may contain asbestos fibers.

BRAKES — BLEEDING THE BRAKE SYSTEM

BLEEDING PROCEDURE

See Figures 2 through 4.

**** CAUTION**

The Techstream must be used for air bleeding. If not used, the air bleeding will be incomplete, which is hazardous and may lead to an accident.

➡Perform air bleeding with park (P) selected and the parking brake applied.

➡As brake fluid may overflow when bleeding, do not place the fluid can on the reservoir filler opening.

➡Perform air bleeding while maintaining the brake fluid level between the MIN and MAX lines on the brake fluid reservoir.

➡Air bleeding will be difficult if the following occurs:

a. The brake actuator hose (the hose between the brake booster pump assembly and brake fluid reservoir) is higher than the fluid level and air enters the hose.

b. During the air bleeding procedure, air enters the brake booster pump assembly while the pump motor is operating.

➡While performing air bleeding, the accumulator pressure drop may cause a buzzer to sound. As there is no problem, continue with the operation.

➡During air bleeding, DTCs for pressure sensor malfunctions, etc. may be stored. After air bleeding and if instructed in the procedures, clear the DTCs.

➡Release the parking brake before performing the linear valve offset calibration.

➡Do not allow brake fluid to adhere to any painted surface such as the vehicle body. If brake fluid leaks onto any painted surface, immediately clean it off.

Brake Line

1. Remove the center No. 1 cowl top ventilator louver.
   a. Disengage 2 claws and separate the hood to cowl top seal.
   b. Disengage 2 claws B and the 2 guides to remove the center No. 1 cowl top ventilator louver.
2. Bleed the brake line.
   a. Remove the brake master cylinder reservoir filler cap assembly.
   b. Add brake fluid into the reservoir between the MAX and MIN lines on the brake fluid reservoir. Brake fluid: SAE J1703 or FMVSS No. 116 DOT3

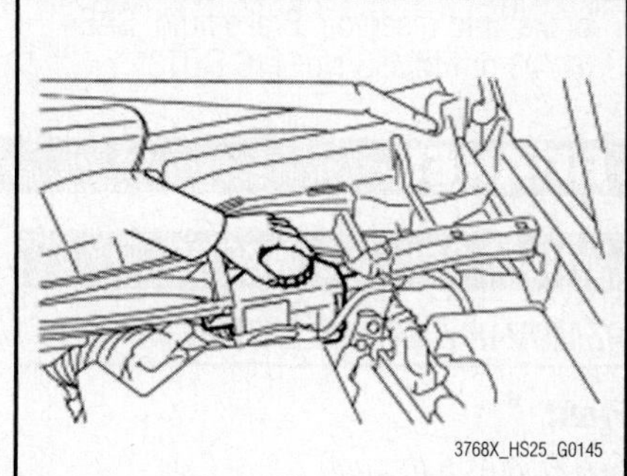

Fig. 2 Removing the master cylinder reservoir filler cap assembly

   c. Connect the Techstream to the DLC3 and turn the power switch on (IG).
   d. Turn the Techstream on and enter the following menus: Chassis / ABS/ VSC/TRC / Air Bleeding.
   e. Select "Usual air bleeding" on the Techstream display, and bleed air from the brake fluid following the instructions on the Techstream.
   f. After air bleeding, tighten each bleeder plug. Tighten to 73 inch lbs. (8.3 Nm).
   g. Clear the DTCs.

h. Turn the Techstream off and turn the power switch off.

3. Inspect for brake fluid leaks.

4. Install the brake master cylinder reservoir filler cap.

5. Install the center No. 1 cowl top ventilator louver.

a. Engage the 2 guides and 2 claws B to install the center No. 1 cowl top ventilator louver.

b. Engage 2 claws A to install the hood to cowl top seal.

Brake System

6. Remove the outer cowl top panel sub assembly.

7. Bleed the brake system.

a. Wait at least 2 minutes with the power switch off, and disconnect the reservoir level switch connector.

➡Do not depress the brake pedal or open/close the doors until the reservoir level switch connector is disconnected.

➡This procedure is not required if the reservoir level switch connector has been disconnected.

b. Remove the brake master cylinder reservoir filler cap assembly.

c. Add brake fluid into the reservoir between the MAX and MIN lines on the brake fluid reservoir. Brake fluid: SAE J1703 or FMVSS No. 116 DOT3.

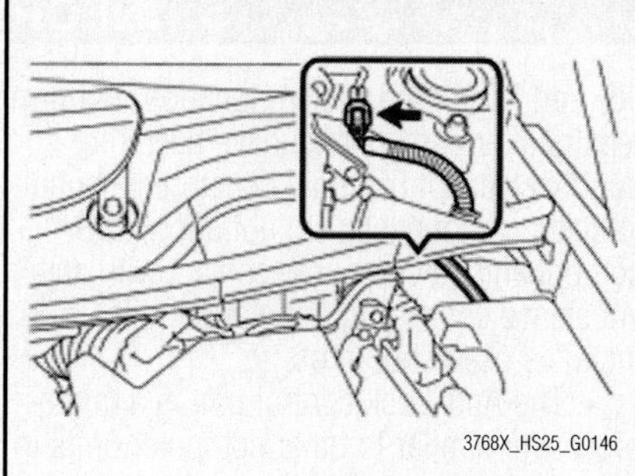

Fig. 3 Disconnecting the reservoir level switch connector

d. Connect the Techstream to the DLC3 and turn the power switch on (IG).

e. Turn the Techstream on and enter the following menus: Chassis / ABS/ VSC/TRC / Air Bleeding.

f. Select "ABS actuator has been replaced" on the Techstream display, and bleed air from the brake fluid following the instructions on the Techstream.

➡Before following the instructions on the Techstream to perform linear valve offset calibration, release the parking brake. When calibration is complete, immediately apply the parking brake.

g. After air bleeding, tighten each bleeder plug. Tighten the disc brake

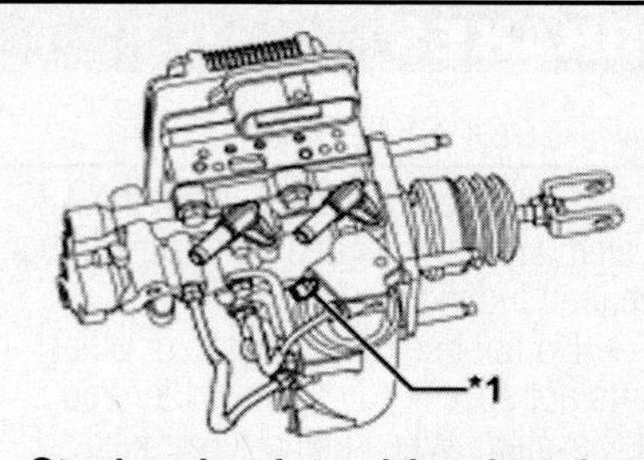

Fig. 4 Locating the stroke simulator bleeder plug

caliper bleeder plug to 73 inch lbs. (8.3 Nm). Tighten the stroke simulator bleeder plug to 75 inch lbs. (8.5 Nm).

➡The stroke simulator bleeder plug is positioned as shown in the illustration.

h. Clear the DTCs.

i. Turn the Techstream off and turn the power switch off.

8. Install the brake master cylinder reservoir filler cap.

9. Inspect for brake fluid leaks.

10. Install the outer cowl top panel sub-assembly.

BRAKES — ANTI-LOCK BRAKE SYSTEM (ABS)

WHEEL SPEED SENSORS

REMOVAL & INSTALLATION

Front

See Figures 5 through 7.

➡While the battery is connected, even if the power switch is off, the brake control system activates when the brake pedal is depressed or any door courtesy switch is turned on. Therefore, when servicing the brake system components, do not depress the brake pedal or open/close the doors while the battery is connected.

➡Use the same procedure for the LH side and RH side.

➡The following procedure is for the LH side.

➡If the sensor rotor needs to be replaced, replace it together with the front axle hub and bearing assembly.

1. Remove the luggage compartment floor mat.

⁂ CAUTION

After the power switch is turned off, the display and navigation module display (HDD navigation system) records various types of memory and settings. As a result, after turning the power switch off, make sure to wait at least 60 seconds before disconnecting the cable from the negative (-) battery terminal.

2. Disconnect the cable from the negative battery terminal.

3. Remove the front wheel.

4. Remove the front fender liner.

a. Remove the 9 clips, 10 screws, 3 grommets, bolt and front fender liner.

5. Remove the front speed sensor.

a. Remove the 2 front speed sensor wire harness clamp and disconnect the front speed sensor connector.

b. Remove the bolt and No. 2 sensor clamp from the body.

c. Remove the bolt, No. 1 sensor and the front brake flexible hose together from the front shock absorber assembly.

d. Remove the clamp from the front shock absorber assembly.

e. Remove the bolt and front speed sensor from the steering knuckle.

➡Prevent foreign matter from attaching to the front speed sensor tip.

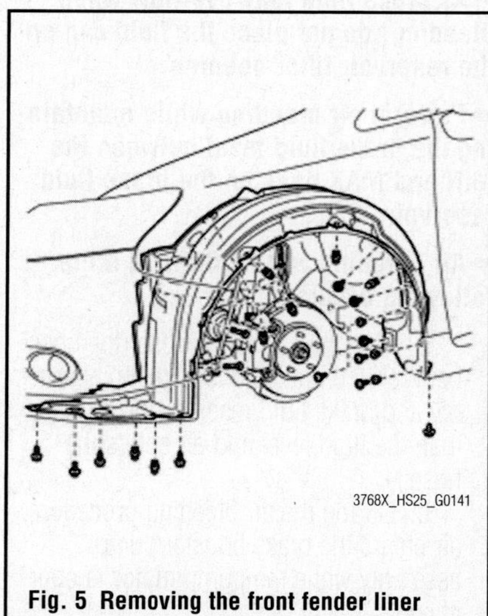

Fig. 5 Removing the front fender liner

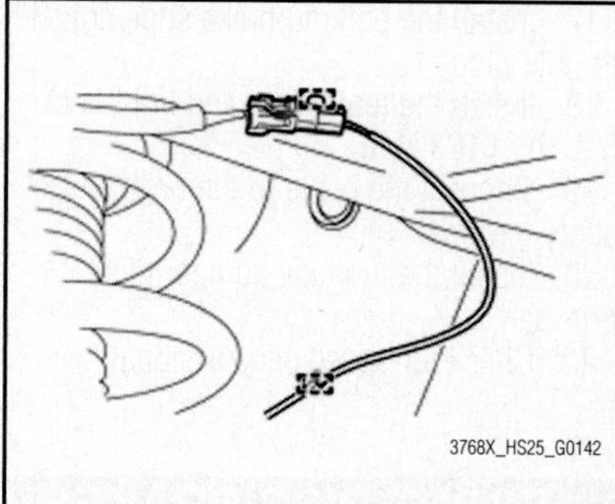

Fig. 6 Disconnecting the front speed sensor

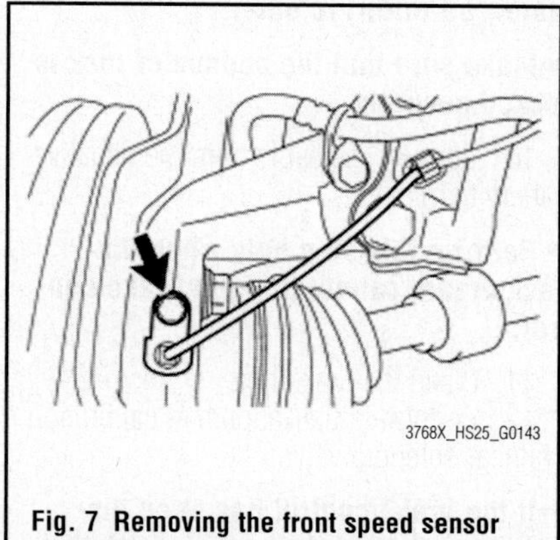

Fig. 7 Removing the front speed sensor

➡**Clean the front speed sensor installation hole and the contact surfaces every time the front speed sensor is removed.**

To install:

➡**Use the same procedure for the LH side and RH side.**

➡**The following procedure is for the LH side.**

➡**If the sensor rotor needs to be replaced, replace it together with the front axle hub and bearing assembly.**

6. Install the front speed sensor assembly. Tighten the bolt to 75 inch lbs. (8.5 Nm).

➡**Prevent foreign matter from attaching to the front speed sensor tip.**

➡**Firmly insert the front speed sensor body into the knuckle before tightening the bolt.**

➡**After installing the front speed sensor to the knuckle, make sure that there is no clearance between the front speed sensor stay and knuckle. Also make sure that no foreign matter is stuck between the parts.**

a. Install the clamp to the front shock absorber assembly.

b. Temporarily install the front brake flexible hose.

c. Install the front brake flexible hose and No. 1 sensor clamp together to the front shock absorber with the bolt. Tighten to 14 ft. lbs. (19 Nm).

➡**Do not twist the wire harness for the front speed sensor when installing the front speed sensor.**

➡**A bolt tightens the brake flexible hose and front speed sensor together. Make sure that the front speed sensor is positioned over the front brake flexible hose.**

d. Install the No. 2 sensor clamp to the body with the bolt. Tighten to 75 inch lbs. (8.5 Nm).

e. Connect the speed sensor connector.

f. Install the 2 front speed sensor wire harness clamps to the body.

7. Install the front fender liner.

a. Install the front fender liner with the 9 clips, 10 screws, bolt and 3 new grommets.

8. Install the front wheel. Tighten to 75 ft. lbs. (103 Nm).

9. Connect the cable to the negative battery terminal.

➡**When disconnecting the cable, some systems need to be initialized after the cable is reconnected.**

10. Install the luggage compartment floor mat.

11. Check for speed sensor signal.

Rear

See Figure 8.

➡**While the battery is connected, even if the power switch is off, the brake control system activates when the brake pedal is depressed or any door courtesy switch is turned on. Therefore, when servicing the brake system components, do not depress the brake pedal or open/close the doors while the battery is connected.**

➡**When the brake pedal is first depressed after replacing the brake pads or pushing back the disc brake piston, DTC C1214 may be output. As there is no malfunction, clear the DTC.**

➡**Use the same procedure for the LH side and RH side.**

➡**The following procedure is for the LH side.**

➡**If the sensor rotor needs to be replaced, replace it together with the rear axle hub and bearing assembly.**

➡**The rear speed sensor is a component of the rear axle hub and bearing assembly. If the rear speed sensor malfunctions, replace the rear axle hub and bearing assembly.**

1. Remove the luggage compartment floor mat.

➡**After the power switch is turned off, the display and navigation module display (HDD navigation system) records various types of memory and settings. As a result, after turning the power switch off, make sure to wait at least 60 seconds before disconnecting the cable from the negative (-) battery terminal.**

2. Disconnect the cable from the negative battery terminal.

3. Remove the rear wheel.

4. Remove the rear suspension arm cover.

5. Disconnect the rear speed sensor wire.

a. Using a screwdriver, disconnect the connector from the rear speed sensor.

➡**Be careful not to damage the rear speed sensor.**

6. Separate the rear disc brake caliper assembly.

7. Remove the parking brake shoe adjusting hole plug.

8. Remove the rear disc.

9. Remove the rear axle hub and bearing assembly.

➡**The rear speed sensor is a component of the rear axle hub and bearing**

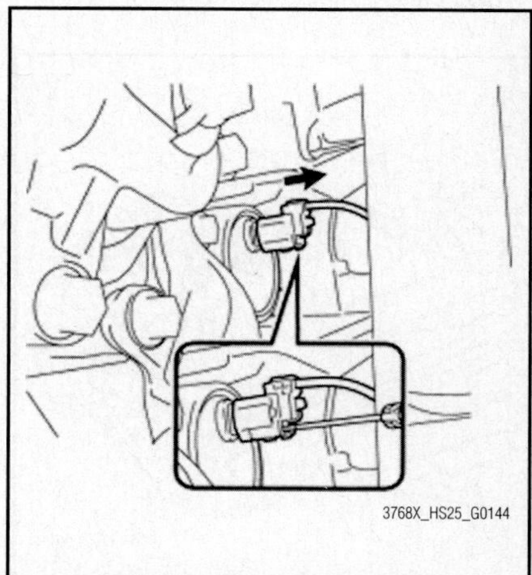

Fig. 8 Disconnecting the rear speed sensor

assembly. If the rear speed sensor malfunctions, replace the rear axle hub and bearing assembly.

➡**If the sensor rotor needs to be replaced, replace it together with the rear axle hub and bearing assembly.**

To install:

10. Install the rear axle hub and bearing assembly.
11. Inspect the rear axle hub bearing looseness.
12. Install the rear disc.
13. Install the rear disc brake caliper assembly.
14. Connect the rear speed sensor wire.
15. Install the rear suspension arm cover.
16. Adjust the parking brake shoe clearance and parking brake pedal travel.
17. Install the parking brake shoe adjusting hole plug.
18. Install the rear wheel and tighten to 76 ft. lbs. (103 Nm).
19. Connect the cable to the negative battery terminal.
20. Install the luggage compartment floor mat.
21. Check for speed sensor signal.

BRAKES

FRONT DISC BRAKES

BRAKE CALIPER

REMOVAL & INSTALLATION

See Figures 9 and 10.

➡**When the brake pedal is first depressed after replacing the brake pads or pushing back the disc brake piston, DTC C1214 may be output. As there is no malfunction, clear the DTC.**

➡**Use the same procedure for the LH side and RH side.**

➡**The following procedure is for the LH side.**

1. Disable the brake control.
2. Remove the front wheel.
3. Drain the brake fluid.

➡**If the brake fluid leaks onto any painted surface, immediately wash it off.**

4. Separate front flexible hose.
   a. Remove the union bolt and gasket, and separate the front flexible hose from the disc brake caliper assembly.
5. Remove the disc brake caliper assembly.
   a. Hold the front disc brake caliper slide pin, and remove the 2 bolts and disc brake caliper assembly.

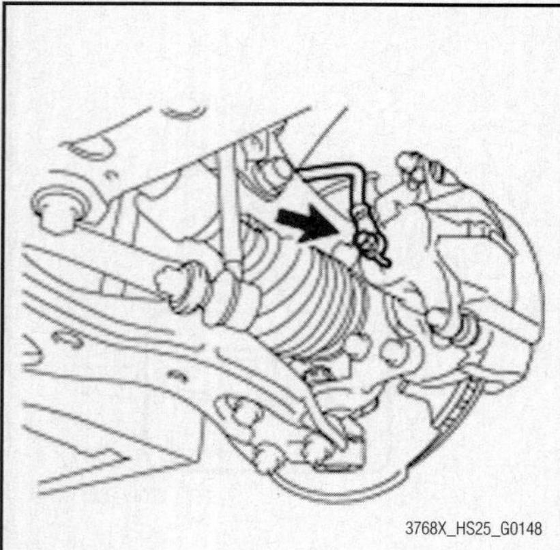

Fig. 9 Separating the front flexible hose

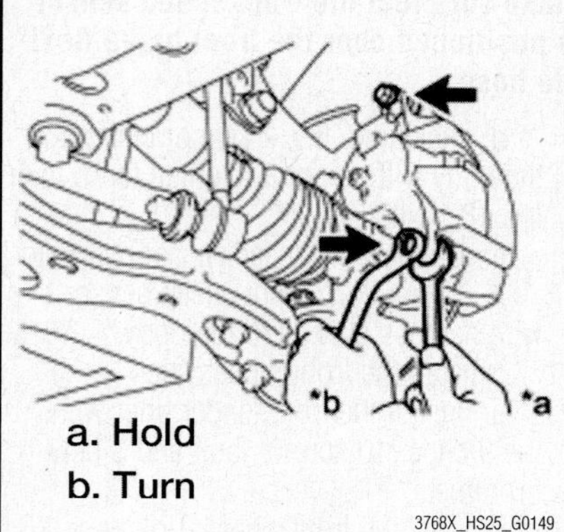

Fig. 10 Removing the disc brake caliper assembly

To install:

6. Install the disc brake caliper assembly.
   a. Hold the front disc brake caliper slide pin, and install the disc brake caliper assembly to the front disc brake caliper mounting with the 2 bolts. Tighten the bolts to 25 ft. lbs. (34 Nm).
7. Connect the front flexible hose.
   a. Connect the front flexible hose to the disc brake caliper assembly with a new union bolt and a new gasket. Tighten the bolt to 21 ft. lbs. (29 Nm).

➡**Install the flexible hose lock securely into the lock hole in the disc brake cylinder.**

8. Disconnect the cable from the negative battery terminal.

➡**Perform this step only when the Techstream cannot prohibit brake control.**

9. Connect the connector.

➡**Perform the following step only when the Techstream cannot prohibit brake control.**

   a. Connect the connector to the brake booster with the master cylinder assembly.

➡**Make sure that the connector can be connected smoothly. Do not allow water, oil or dirt to enter.**

➡**Make sure that the connector lock is locked securely.**

10. Connect the cable from the negative battery terminal.

➡**Perform this step only when the Techstream cannot prohibit brake control.**

11. Bleed the brake line.
12. Perform initialization and calibration of linear solenoid valve.

➡**If the brake control has been disabled, make sure to perform initialization and calibration of the linear solenoid valve.**

13. Install the front wheel. Tighten to 76 ft. lbs. (103 Nm).

DISC BRAKE PADS

REMOVAL & INSTALLATION

See Figures 9 and 10.

➡**When the brake pedal is first depressed after replacing the brake pads or pushing back the disc brake piston, DTC C1214 may be output. As there is no malfunction, clear the DTC.**

➡**Use the same procedure for the LH side and RH side.**

➡**The following procedure is for the LH side.**

1. Disable the brake control.
2. Remove the front wheel.
3. Drain the brake fluid.

➡**If the brake fluid leaks onto any painted surface, immediately wash it off.**

4. Separate front flexible hose.
   a. Remove the union bolt and gasket, and separate the front flexible hose from the disc brake caliper assembly.

5. Remove the disc brake caliper assembly.
a. Hold the front disc brake caliper slide pin, and remove the 2 bolts and disc brake caliper assembly.
6. Remove the front brake pad.
a. Remove the 2 front disc brake pads from the front disc brake caliper mounting.

To install:

7. Install the front disc brake pad.
a. Install the 2 front disc brake pads to the front disc brake caliper mounting.

➡There should be no oil or grease on the friction surfaces of the disc brake pads of the front disc.

8. Install the disc brake caliper assembly.
a. Hold the front disc brake caliper slide pin, and install the disc brake caliper assembly to the front disc brake caliper mounting with the 2 bolts. Tighten the bolts to 25 ft. lbs. (34 Nm).
9. Connect the front flexible hose.
a. Connect the front flexible hose to the disc brake caliper assembly with a new union bolt and a new gasket. Tighten the bolt to 21 ft. lbs. (29 Nm).

➡Install the flexible hose lock securely into the lock hole in the disc brake cylinder.

10. Disconnect the cable from the negative battery terminal.

➡Perform this step only when the Techstream cannot prohibit brake control.

11. Connect the connector.

➡Perform the following step only when the Techstream cannot prohibit brake control.

a. Connect the connector to the brake booster with the master cylinder assembly.

➡Make sure that the connector can be connected smoothly. Do not allow water, oil or dirt to enter.

➡Make sure that the connector lock is locked securely.

12. Connect the cable from the negative battery terminal.

➡Perform this step only when the Techstream cannot prohibit brake control.

13. Bleed the brake line.
14. Perform initialization and calibration of linear solenoid valve.

➡If the brake control has been disabled, make sure to perform initialization and calibration of the linear solenoid valve.

15. Install the front wheel. Tighten to 76 ft. lbs. (103 Nm).

BRAKES — REAR DISC BRAKES

BRAKE CALIPER

REMOVAL & INSTALLATION

See Figure 11.

➡When the brake pedal is first depressed after replacing the brake pads or pushing back the disc brake piston, DTC C1214 may be output. As there is no malfunction, clear the DTC.

➡Use the same procedure for the RH side and LH side.

➡The following procedure is for the LH side.

1. Disable brake control.
2. Remove rear wheel.
3. Drain brake fluid.

➡If brake fluid leaks onto any painted surface, immediately wash it off.

4. Disconnect rear flexible hose.
a. Remove the union bolt and gasket, and disconnect the rear flexible hose from the rear disc brake caliper assembly.
5. Remove the rear disc brake caliper assembly.
a. Hold the 2 rear disc brake caliper slide pins, and remove the 2 bolts and rear disc brake caliper assembly.

To install:

➡There should be no oil or grease on the friction surfaces of the disc brake pads or the rear disc.

6. Install the rear disc brake caliper assembly.
a. Hold the 2 rear disc brake caliper slide pins and install the rear disc brake caliper assembly to the rear disc brake caliper mounting with the 2 bolts. Tighten to 20 ft. lbs. (27 Nm).
7. Connect the rear flexible hose.
a. Connect the rear flexible hose to the rear disc brake caliper assembly with a new union bolt and a new gasket. Tighten to 29 ft. lbs. (39 Nm).
8. Disconnect the cable from the negative battery terminal.

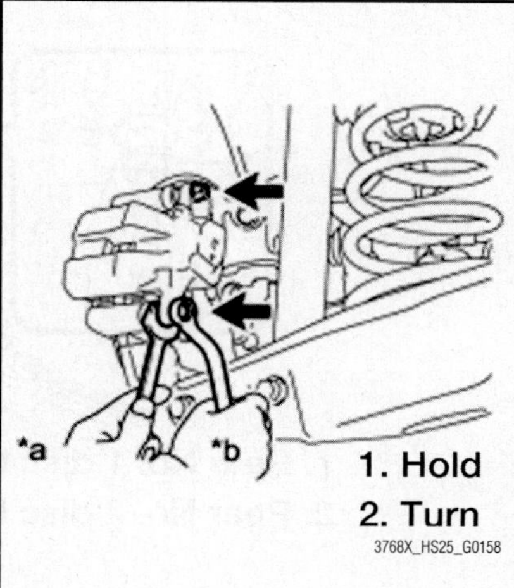

Fig. 11 Removing the rear disc brake caliper assembly

➡Perform this step only when the Techstream cannot prohibit brake control.

a. Disconnect the cable from the negative (-) battery terminal.
9. Connect the connector.

➡Perform this step only when the Techstream cannot prohibit brake control.

a. Connect the connector to the brake booster with master cylinder assembly.

➡Make sure that the connector can be connected smoothly. Do not allow water, oil or dirt to enter.

➡Make sure that the connector lock is locked securely.

10. Connect the cable from the negative battery terminal.

➡Perform this step only when the Techstream cannot prohibit brake control.

11. Bleed the brake line.
12. Perform initialization and calibration of the linear solenoid valve.

➡If the brake control has been disabled, make sure to perform initialization and calibration of the linear solenoid valve.

13. Adjust the parking brake show clearance and parking brake pedal travel.
14. Install the rear wheel and tighten to 76 ft. lbs. (103 Nm).

DISC BRAKE PADS

REMOVAL & INSTALLATION

See Figure 11.

➡When the brake pedal is first depressed after replacing the brake pads or pushing back the disc brake piston, DTC C1214 may be output. As there is no malfunction, clear the DTC.

➡Use the same procedure for the RH side and LH side.

➡The following procedure is for the LH side.

1. Disable brake control.
2. Remove rear wheel.
3. Drain brake fluid.

➡If brake fluid leaks onto any painted surface, immediately wash it off.

4. Disconnect rear flexible hose.
   a. Remove the union bolt and gasket, and disconnect the rear flexible hose from the rear disc brake caliper assembly.
5. Remove the rear disc brake caliper assembly.
   a. Hold the 2 rear disc brake caliper slide pins, and remove the 2 bolts and rear disc brake caliper assembly.
6. Remove the rear disc brake pad.

To install:

7. Install the rear disc brake pad.
   a. Install the 2 rear disc brake pads to the rear disc brake caliper mounting.

➡There should be no oil or grease on the friction surfaces of the disc brake pads or the rear disc.

8. Install the rear disc brake caliper assembly.
   a. Hold the 2 rear disc brake caliper slide pins and install the rear disc brake caliper assembly to the rear disc brake caliper mounting with the 2 bolts. Tighten to 20 ft. lbs. (27 Nm).
9. Connect the rear flexible hose.
   a. Connect the rear flexible hose to the rear disc brake caliper assembly with a new union bolt and a new gasket. Tighten to 29 ft. lbs. (39 Nm).
10. Disconnect the cable from the negative battery terminal.

➡Perform this step only when the Techstream cannot prohibit brake control.

   a. Disconnect the cable from the negative (-) battery terminal.

11. Connect the connector.

➡Perform this step only when the Techstream cannot prohibit brake control.

   a. Connect the connector to the brake booster with master cylinder assembly.

➡Make sure that the connector can be connected smoothly. Do not allow water, oil or dirt to enter.

➡Make sure that the connector lock is locked securely.

12. Connect the cable from the negative battery terminal.

➡Perform this step only when the Techstream cannot prohibit brake control.

13. Bleed the brake line.
14. Perform initialization and calibration of the linear solenoid valve.

➡If the brake control has been disabled, make sure to perform initialization and calibration of the linear solenoid valve.

15. Adjust the parking brake show clearance and parking brake pedal travel.
16. Install the rear wheel and tighten to 76 ft. lbs. (103 Nm).

PARKING BRAKE SHOES

ADJUSTMENT

See Figures 12 and 13.

1. Remove the No. 1 switch hole base.
2. Completely release the parking brake pedal.
3. Loosen the lock nut and the adjusting nut to completely release the parking brake cable.
4. Remove the rear wheels.
5. Temporarily install the hub nuts to the hub bolts.

➡Securely install the hub nuts to the rear disc.

6. Remove the parking brake shoe adjusting hole plug.
7. Turn the shoe adjuster and expand the shoe until the disc locks.
8. Turn and contract the shoe adjuster until the disc can rotate smoothly. Standard returns is 8 notches.
9. Check that there is no brake drag against the shoe.
10. Install the parking brake shoe adjusting hole plug.
11. Turn the adjusting nut until the parking brake pedal travel is corrected to be within the specified range. Parking brake pedal travel is 8 to 11 notches at 67.5 ft. lbs. (300 N).
12. Using a wrench or an equivalent

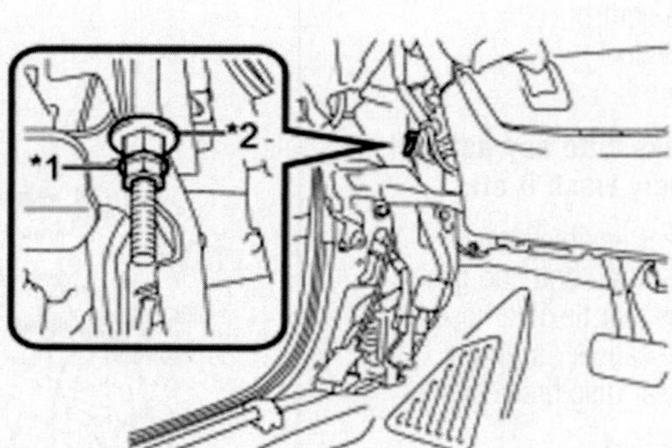

Fig. 12 Locating the lock nut and adjusting nut

tool, hold the adjusting nut and tighten the lock nut. Tighten the nuts to 62 inch lbs. (7 Nm).

13. Operate the parking brake pedal 3 to 4 times, and check the parking brake pedal travel.

14. Check that there is no brake drag against the shoe.

15. Remove the hub nuts from the hub bolts.

16. Install the rear wheels. Tighten to 76 ft. lbs. (103 Nm).

17. Install the No. 1 switch hole base.

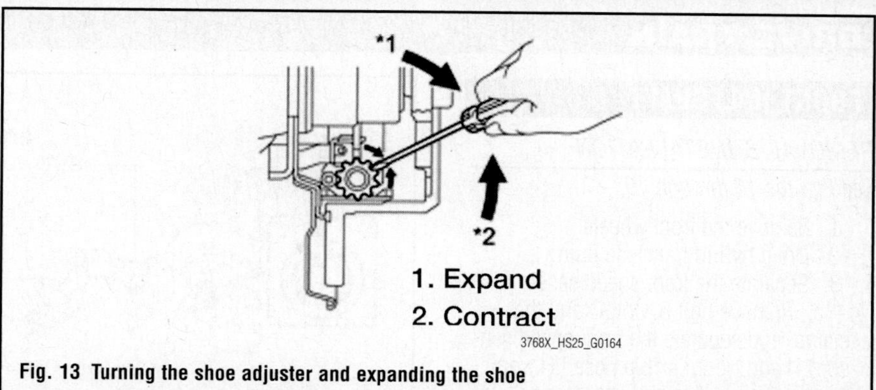

Fig. 13 Turning the shoe adjuster and expanding the shoe

CHASSIS ELECTRICAL

AIR BAG (SUPPLEMENTAL RESTRAINT SYSTEM)

GENERAL INFORMATION

The vehicle is equipped with a Supplemental Restraint System (SRS). It consists of a driver airbag, front passenger airbag, driver side knee airbag, front seat side airbag and curtain shield airbag. Failure to carry out service operations in the correct sequence could cause the SRS to unexpectedly deploy during servicing, possibly leading to a serious accident. Further, if a mistake is made in servicing the SRS, it is possible that the SRS may fail to operate when required. Before performing servicing (including removal or installation of parts, inspection or replacement), be sure to read the following carefully, then follow the correct procedures indicated in the repair manual.

SERVICE PRECAUTIONS

**** CAUTION**

Disconnect and isolate the battery negative cable before beginning any airbag system component diagnosis, testing, removal, or installation procedures. Wait at least 90 seconds after the ignition switch is turned off and the negative (-) terminal cable is disconnected from the battery before starting the operation. The SRS is equipped with a backup power source, so if work is started within 90 seconds after disconnecting the negative (-) terminal cable from the battery, the SRS may be deployed. Failure to disable the airbag system may result in accidental airbag deployment, personal injury, or death.

DISARMING THE SYSTEM

To avoid personal injury when working on vehicles equipped with an air bag, the negative battery cable must be disconnected and at least 90 seconds must elapse before working on the system. Failure to do so may result in deployment of the air bag.

ARMING THE SYSTEM

To arm the system after service is finished, connect the negative battery cable.

CLOCKSPRING CENTERING

See Figures 14 and 15.

1. Before servicing the vehicle, refer to the Precautions Section.
2. Check that the ignition switch is **OFF**.
3. Check that the battery negative (-) terminal is disconnected.

**** CAUTION**

After removing the terminal, wait for at least 90 seconds before starting the operation.

4. Rotate the spiral cable counterclockwise slowly by hand until it feels firm.
5. Rotate the spiral cable clockwise approximately 2.5 turns to align the marks.

➡Do not turn the spiral cable by the airbag wire harness.

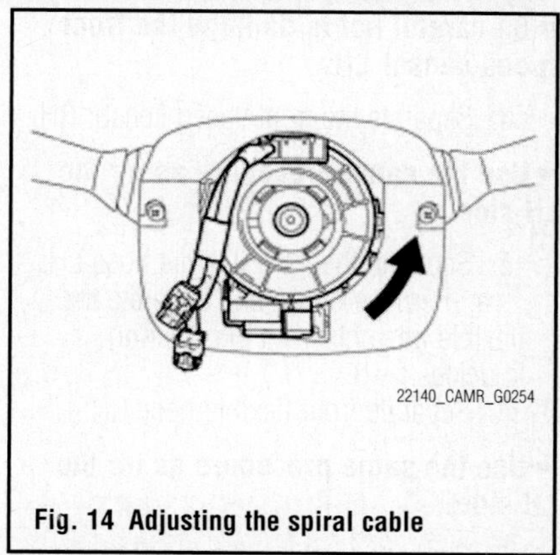

Fig. 14 Adjusting the spiral cable

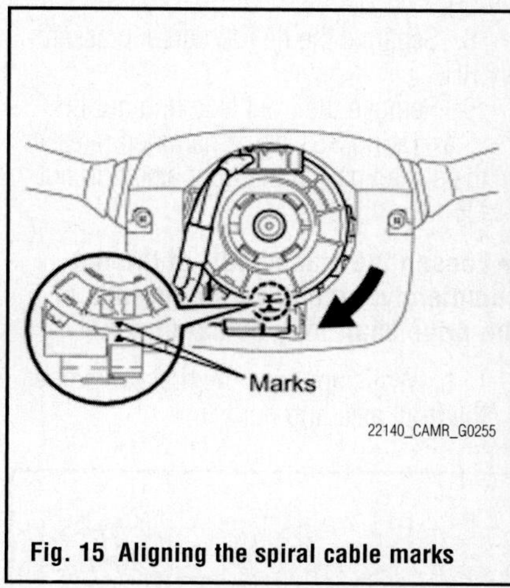

Fig. 15 Aligning the spiral cable marks

➡The spiral cable will rotate approximately 2.5 turns to both the left and right from the center.

FRONT HALFSHAFT

REMOVAL & INSTALLATION

See Figures 16 through 19.

1. Remove the front wheels.
2. Drain hybrid transaxle fluid.
3. Separate the front speed sensor LH.
   a. Remove bolt B, bolt C and the clamp, and separate the front speed sensor LH and front flexible hose LH from the shock absorber and the steering knuckle.

➡Prevent foreign matter from adhering to the front speed sensor LH.

➡Be careful not to damage the front speed sensor LH.

4. Separate the front speed sensor RH.

➡Use the same procedure as for the LH side.

5. Separate the front flexible hose LH.
   a. Remove bolt A and separate the flexible hose LH from the steering knuckles.
6. Separate front flexible hose RH.

➡Use the same procedure as for the LH side.

7. Separate the tie rod end sub assembly LH.
8. Separate the tie rod end sub assembly RH.
9. Remove the front axle hub nut LH.
   a. Using SST and a hammer, release the staked part of the front axle hub nut LH.

➡Loosen the staked part of the nut completely, otherwise the threads of the drive shaft may be damaged.

   b. While applying the brakes, remove the front axle hub nut LH.

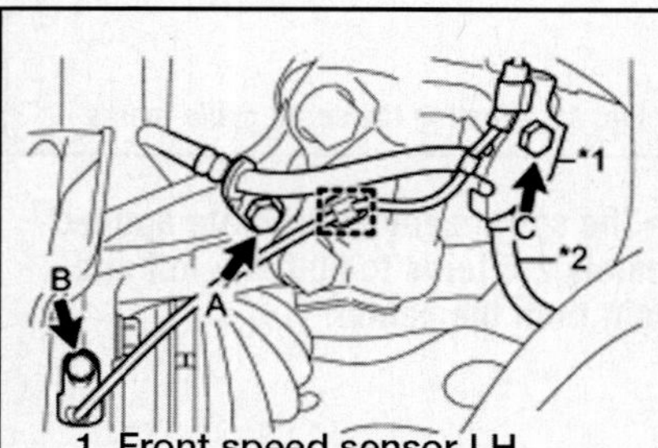

Fig. 16 Separating the front speed sensor LH

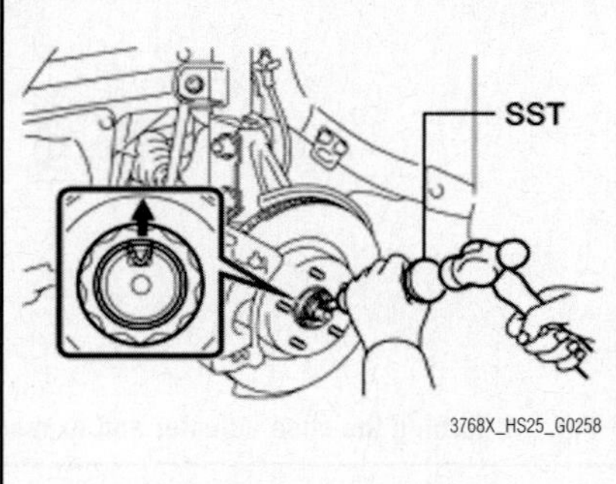

Fig. 17 Removing the front axle hub nut LH

10. Remove the front axle hub nut RH.

➡Use the same procedure as for the LH side.

11. Separate the front stabilizer link assembly LH.
   a. Remove the nut and separate the front stabilizer link assembly LH.

➡If the ball joint turns together with the nut, use a hexagon wrench (6 mm) to hold the stud.

12. Separate the front stabilizer link assembly RH.

➡Use the same procedure as for the LH side.

13. Separate the front No. 1 lower suspension arm sub assembly LH.
   a. Remove the bolt and 2 nuts, and separate the front No. 1 lower suspension arm sub-assembly LH from the lower ball joint LH.
14. Separate the front No. 1 lower suspension arm sub assembly RH.

➡Use the same procedure as for the LH side.

   a. Put matchmarks on the front drive shaft assembly LH and the front axle hub sub assembly LH.
   b. Using a plastic hammer, separate the front drive shaft assembly LH from the front axle assembly LH.

➡Be careful not to damage the front axle outboard joint boot and front speed sensor rotor.

15. Separate the front drive shaft assembly RH.

➡Use the same procedure as for the LH side.

16. Remove the front drive shaft assembly LH.
   a. Using SST, remove the front drive shaft assembly LH.

➡Do not damage the hybrid transaxle case oil seal.

➡Do not damage the front axle inboard joint boot.

➡Do not drop the front drive shaft assembly LH.

17. Remove the front drive shaft assembly RH.
   a. Remove the 2 bolts and pull out the front drive shaft assembly RH together with the drive shaft bearing case sub assembly.
   b. Remove the front drive shaft assembly RH from the hybrid transaxle.

➡Do not damage the hybrid transaxle case oil seal.

➡Do not damage the front axle inboard joint boot.

➡Do not drop the front drive shaft assembly RH.

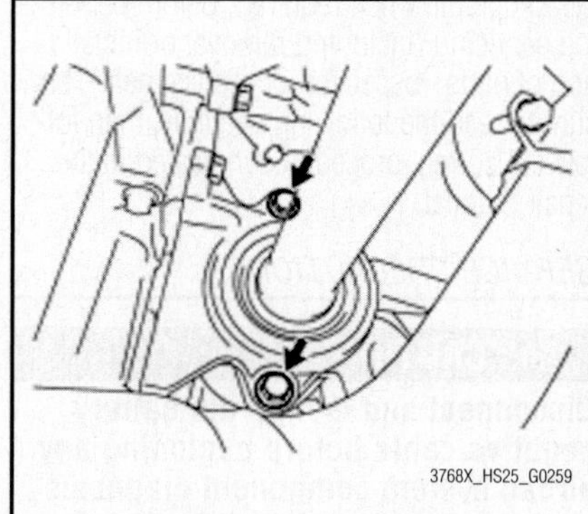

Fig. 18 Removing the front drive shaft assembly RH

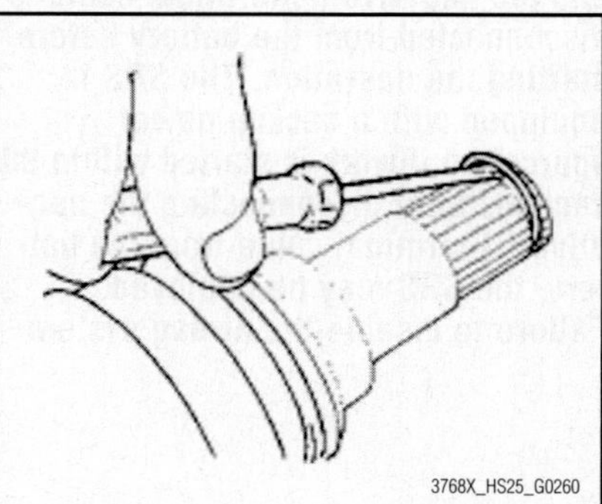

Fig. 19 Removing the front drive shaft LH hold snap ring (for LH side)

18. Remove the front drive shaft LH hole snap ring (for LH side).

a. Using a screwdriver, remove the front drive shaft LH hole snap ring.

To install:

19. Install the front drive shaft LH hole snap ring (for LH side).

a. Install a new front drive shaft LH hole snap ring to the front drive inboard joint assembly LH.

➡Face the end gap of the front drive inboard joint hole snap ring downward.

20. Install the front drive shaft assembly LH.

a. Coat the splines of the inboard joint shaft with ATF WS.

b. Align the inboard joint splines, and using a brass bar and a hammer, install the front drive shaft assembly LH.

➡Face the end gap of the front drive shaft LH hole snap ring downward.

➡Do not damage the hybrid transaxle case oil seal.

➡Do not damage the front axle inboard joint boot.

➡Make sure to center the front drive shaft assembly LH during installation to prevent damage to the front drive shaft LH hole snap ring.

➡Confirm whether the drive shaft is securely driven in by checking the reaction force and sound.

c. Align the matchmarks and install the front drive shaft assembly LH to the front axle hub sub-assembly LH.

➡Be careful not to damage the front axle outboard joint boot or speed sensor rotor.

21. Install the front drive shaft assembly RH.

a. Coat the splines of the inboard joint shaft assembly with ATF WS.

b. Align the shaft splines and securely insert the front drive shaft assembly RH.

➡Do not damage the hybrid transaxle case oil seal, inboard joint boot or front drive shaft dust cover.

c. Install the front drive shaft assembly RH with the 2 bolts. Tighten to 47 ft. lbs. (64 Nm).

d. Align the matchmarks and install the front drive shaft assembly RH to the front axle hub sub-assembly RH.

➡Be careful not to damage the outboard joint boot or speed sensor rotor.

22. Install the front stabilizer link assembly LH.

a. Install the front stabilizer link assembly LH to the front shock absorber LH with the nut. Tighten to 55 ft. lbs. (74 Nm).

➡If the ball joint turns together with the nut, use a hexagon wrench (6 mm) to hold the stud.

23. Install the front stabilizer link assembly RH.

➡Use the same procedure as for the LH side.

24. Install the front No. 1 lower suspension arm sub assembly LH.

a. Install the lower ball joint to the front No. 1 lower suspension arm sub-assembly LH with the bolt and 2 nuts. Tighten to 68 ft. lbs. (92 Nm).

25. Install the front NO. 1 lower suspension arm sub assembly RH.

➡Use the same procedure as for the LH side.

26. Install the tie rod end sub assemblies.

27. Install the front speed sensor LH to the steering knuckle. Tighten the bolt to 75 inch lbs. (8.5 Nm).

➡Prevent foreign matter from adhering to the front speed sensor.

➡Be careful not to damage the front speed sensor.

a. Install the flexible hose LH and the front speed sensor LH to the shock absorber with the bolt. Tighten to 14 ft. lbs. (19 Nm).

➡Be careful not to damage the speed sensor.

➡Prevent foreign matter from adhering to the front speed sensor.

➡Do not twist the sensor wire when installing the front speed sensor.

b. Install the clamp to the shock absorber.

28. Install the front speed sensor RH.

➡Use the same procedure as for the LH side.

29. Install the front flexible hose LH with the bolt to the steering knuckle. Tighten to 14 ft. lbs. (19 Nm).

30. Install the front flexible hose RH.

➡Use the same procedure as the LH.

31. Install the front axle hub nut LH.

a. Clean the threaded parts on the front drive shaft assembly LH and axle hub nut LH using a non residue solvent.

➡Be sure to perform this work for a new front drive shaft assembly.

➡Keep the threaded parts free of oil and foreign matter.

b. Using a socket wrench (30 mm), install a new front axle hub nut LH. Tighten to 159 ft. lbs. (216 Nm).

c. Using a chisel and hammer, stake the front axle hub nut LH.

32. Install the front axle hub nut RH.

➡Use the same procedure as for the LH side.

33. Add hybrid transaxle fluid.

34. Inspect the hybrid transaxle fluid.

35. Install the front wheels. Tighten to 76 ft. lbs. (103 Nm).

36. Adjust the front wheel alignment.

37. Check the speed sensor signal.

ENGINE COOLING

ENGINE FAN

REMOVAL & INSTALLATION

See Figures 20 and 21.

1. Remove the radiator assembly.
2. Remove the cooling fan motor insulator.
 a. Remove the 2 bolts and cooling fan motor insulator.
3. Remove the fan.
 a. Remove the nut and the fan.
4. Remove the No. 2 fan.
 a. Remove the nut and No. 2 fan.
5. Remove the cooling fan motor.
 a. Disconnect the clamp and cooling fan motor connector.
 b. Remove the 3 screws and cooling fan motor.
6. Remove the No. 2 cooling fan motor.
 a. Disconnect the clamp and No. 2 cooling fan motor connector.
 b. Remove the 3 screws and No. 2 cooling fan motor.

To install:

7. Install the No. 2 cooling fan motor.
 a. Install the No. 2 cooling fan motor with the 3 screws. Tighten to 35 inch lbs. (3.9 Nm).
 b. Connect the clamp and No. 2 cooling fan motor connector.
8. Install the cooling fan motor.
 a. Install the cooling fan motor with the 3 screws. Tighten to 35 inch lbs. (3.9 Nm).
 b. Connect the clamp and cooling fan motor connector.
9. Install the No. 2 fan.
 a. Install the No. 2 fan with the nut. Tighten to 56 inch lbs. (6.3 Nm).
10. Install the fan with the nut. Tighten to 56 inch lbs. (6.3 Nm).
11. Install the cooling fan motor insulator.
 a. Install the cooling fan motor insulator with the 2 bolts. Tighten to 23 inch lbs. (2.6 Nm).
12. Install the radiator assembly.

RADIATOR

REMOVAL & INSTALLATION

See Figures 22 through 29.

➡After the power switch is turned off, the display and navigation module display (HDD navigation system) records various types of memory and settings. As a result, after turning the power switch off, make sure to wait at least 60 seconds before disconnecting the cable from the negative (-) battery terminal.

1. Disconnect cable from negative battery terminal.
2. Remove cools air intake duct seal.
3. Remove No. 1 engine under cover.
4. Drain engine coolant (for Engine).
5. Remove front bumper assembly.
6. Remove low pitched horn assembly.
7. Remove high pitched horn assembly.
8. Remove millimeter wave radar sensor assembly (w/ Dynamic Radar Cruise Control System).
9. Remove hood lock assembly.
10. Remove radiator grille bracket.
11. Remove No. 5 inverter bracket.
12. Remove radiator support LH.
 a. Remove the 2 bolts and radiator support LH.
 b. Remove the radiator support cushion from the radiator support LH.
13. Remove the radiator support RH.
 a. Disconnect the No. 2 water by pass hose clamp from the radiator support RH.
 b. Remove the bolt and radiator support RH.
 c. Remove the radiator support cushion from the radiator support RH.
14. Remove the hood lock support sub assembly.
 a. Remove the 4 bolts and hood lock support sub assembly.
15. Disconnect the No. 3 radiator hose and the No. 3 water by pass hose.
16. Disconnect the No. 2 radiator hose.

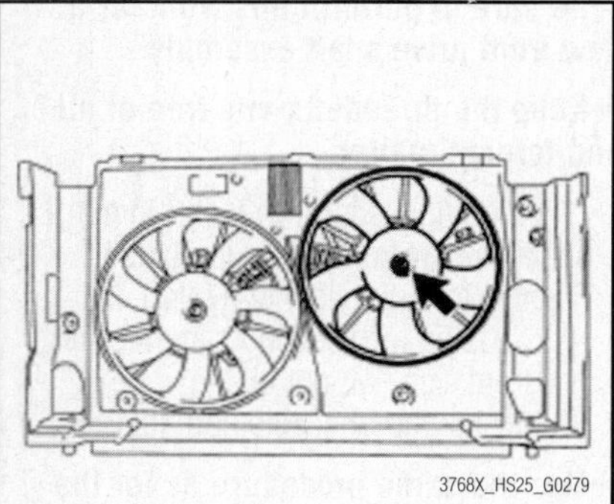

Fig. 20 Removing the fan

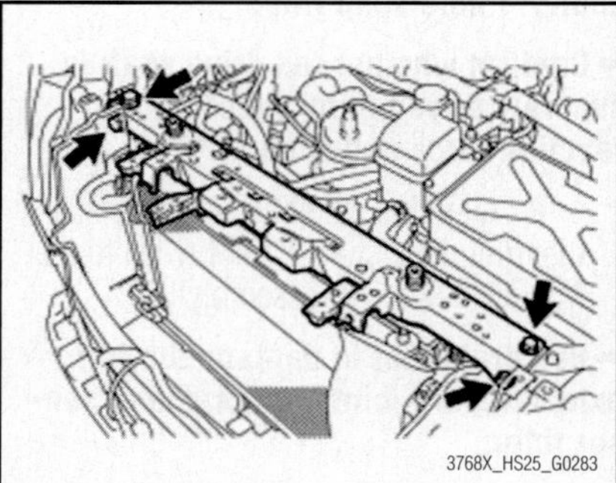

Fig. 23 Removing the hood lock support sub assembly

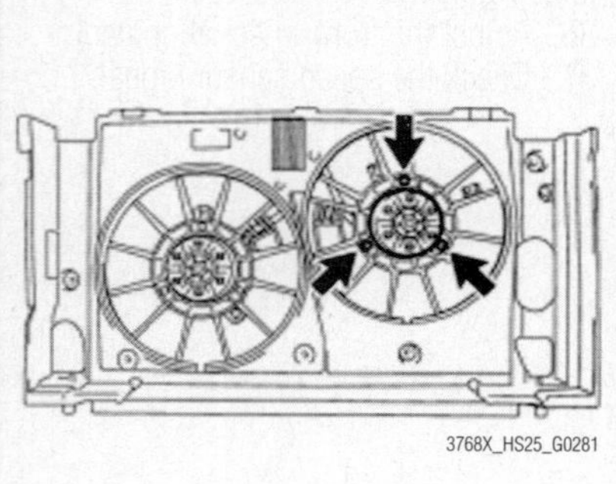

Fig. 21 Removing the fan motor

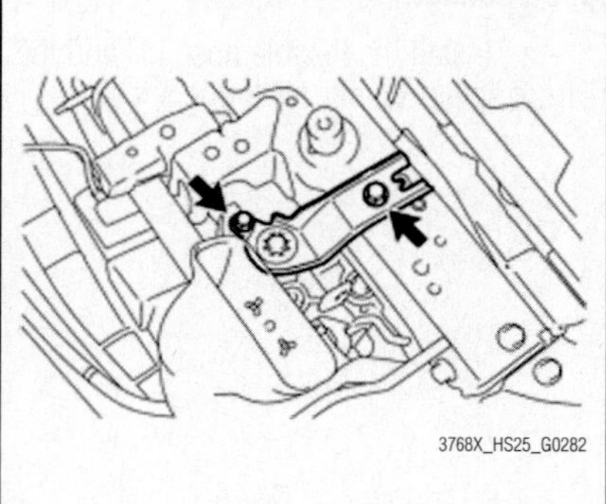

Fig. 22 Removing the radiator support LH

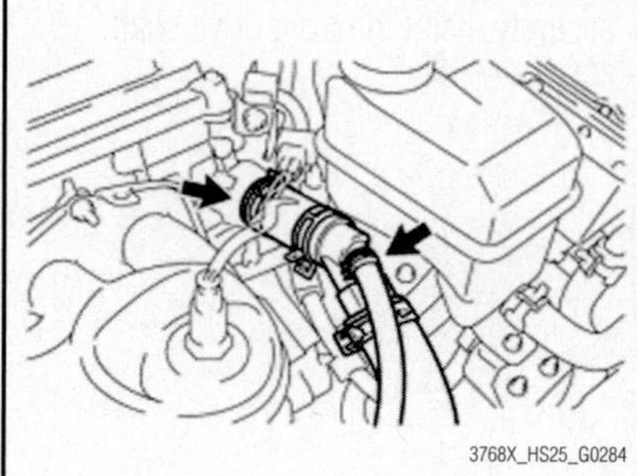

Fig. 24 Disconnecting the No. 3 radiator hose and No. 4 water by pass hose

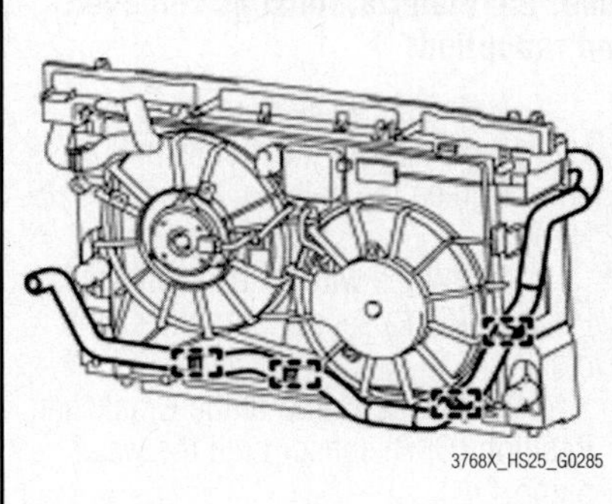

Fig. 25 **Disconnecting the No. 5 inverter cooling hose**

17. Disconnect the No. 5 inverter cooling hose.
 a. Disconnect the 4 clamps and No. 5 inverter cooling hose from the fan shroud.
18. Disconnect the No. 1 water by pass hose.
19. Disconnect the No. 3 water by pass hose.
 a. Disconnect the 4 clamps and No. 3 water by pass hose from the No. 2 fan shroud.
 b. Disconnect the No. 3 water by pass hose.
20. Remove the No. 2 fan shroud.
 a. Remove the 2 bolts, 2 claws and No. 2 fan shroud.
21. Remove the radiator side deflector RH.
 a. Disconnect the 3 clips and 2 claws, and remove the radiator side deflector RH.
22. Remove the radiator and fan assembly.

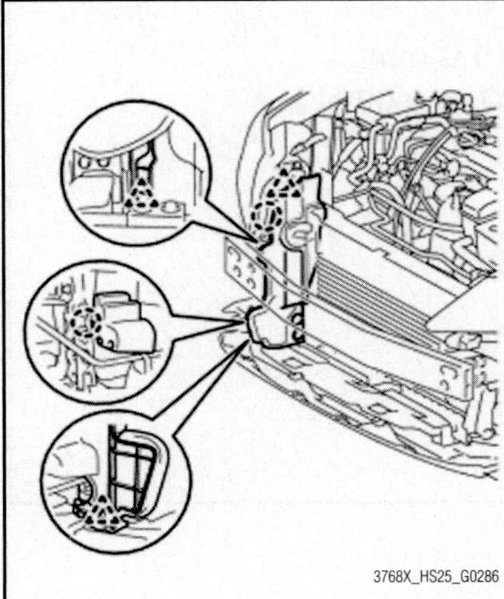

Fig. 26 **Removing the radiator side deflector**

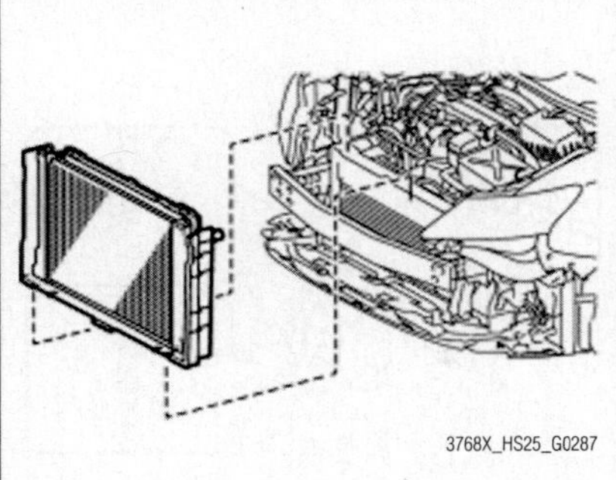

Fig. 27 **Removing the radiator and fan assembly**

 a. Disconnect the 3 wire harness clamps and connector.
 b. Remove the radiator assembly from the vehicle together with the cooling fan assembly.

➡For vehicles with the air conditioning system, do not apply any excessive force to the cooler condenser assembly or pipe when removing the radiator assembly.

 c. Remove the 2 lower radiator supports.

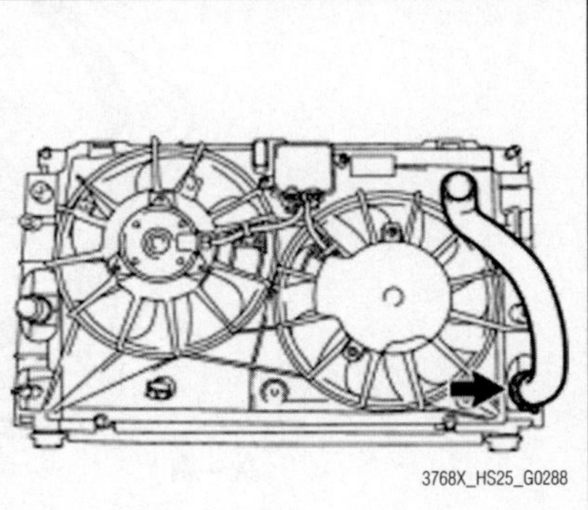

Fig. 28 **Removing the radiator hose**

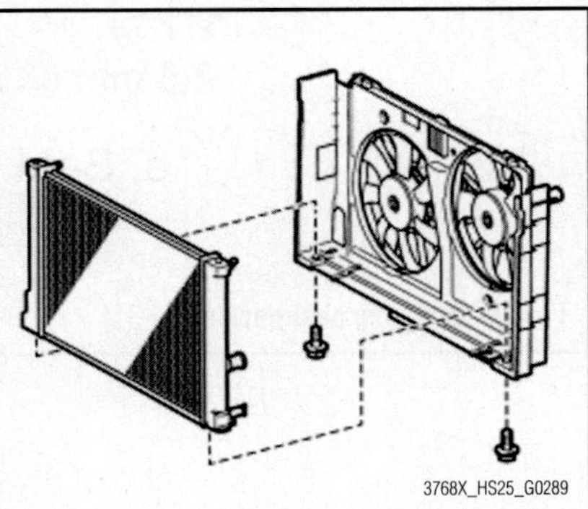

Fig. 29 **Removing the radiator assembly**

23. Remove the No. 1 radiator hose.
 a. Remove the No. 1 radiator hose from the radiator assembly.
24. Remove the No. 2 radiator hose.
 a. Remove the No. 2 radiator hose from the radiator assembly.
25. Remove the radiator assembly.
 a. Remove the 2 bolts and radiator assembly.

To install:

26. To install, reverse the removal procedure. Tighten the radiator assembly bolts to 62 inch lbs. (7 Nm). Tighten the No. 2 fan shroud bolts to 62 inch lbs. (7 Nm). Tighten the hood lock sub assembly bolts to 9 ft. lbs. (13 Nm). Tighten the radiator support RH to 14 ft. lbs. (19 Nm). Tighten the radiator support LH Bolt A to 14 ft. lbs. (19 Nm) and Bolt B to 62 inch lbs. (7 Nm).

THERMOSTAT

REMOVAL & INSTALLATION

See Figure 30.

1. Remove the No. 1 engine under cover.
2. Drain coolant.
3. Disconnect the No. 2 radiator hose.
4. Remove the water inlet.
 a. Remove the 2 nuts and the water inlet from the cylinder block.
5. Remove the thermostat and gasket.

To install:

6. To install, reverse the removal procedure. Install the thermostat with the jiggle valve facing upward. The jiggle valve may be set to within 10° on either side of the prescribed position. Tighten the water inlet nuts to 80 inch lbs. (9 Nm).

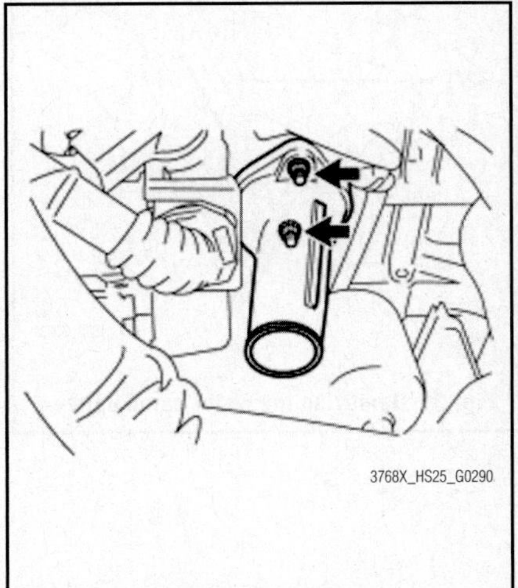

Fig. 30 **Removing the thermostat**

WATER PUMP

REMOVAL & INSTALLATION

See Figures 31 through 33.

1. Remove the No. 1 engine cover.
2. Drain the coolant.
3. Remove the v-ribbed belt.
4. Remove the water pump pulley.
   a. Using a special tool, remove the 4 bolts and water pump pulley.
5. Remove the water pump assembly.
   a. Remove the clamp of the Crankshaft Position (CKP) sensor from the water pump.
   b. Disconnect the wire of the Crankshaft Position (CKP) sensor from the clamp bracket.
   c. Remove the 4 bolts, 2 nuts and clamp bracket.
   d. Using a screwdriver, pry between the water pump and cylinder block, and then remove the water pump.

➡Tape the screwdriver tip before use.

➡Be careful not to damage the contact surfaces of the water pump and cylinder block.

To install:

6. Install the water pump assembly.
   a. Remove the old seal packing material from the contact surfaces.
   b. Apply a continuous line of seal packing.

Seal packing: Toyota Genuine Seal Packing 1232B, Three Bond 1282B or equivalent.

The standard diameter is 0.086-0.984 inch (2.2-2.5 mm).

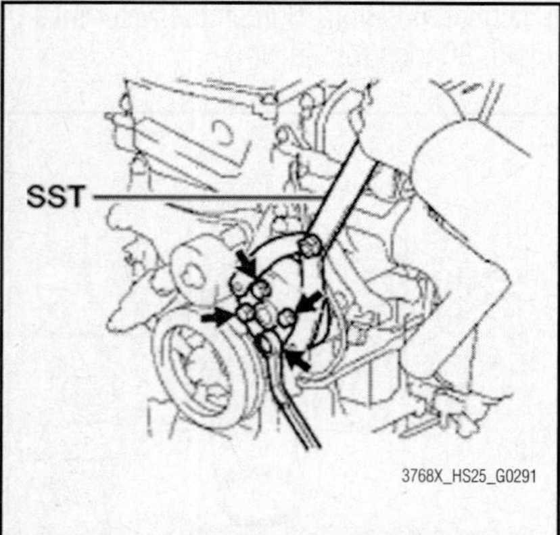

Fig. 31 Removing the water pump pulley

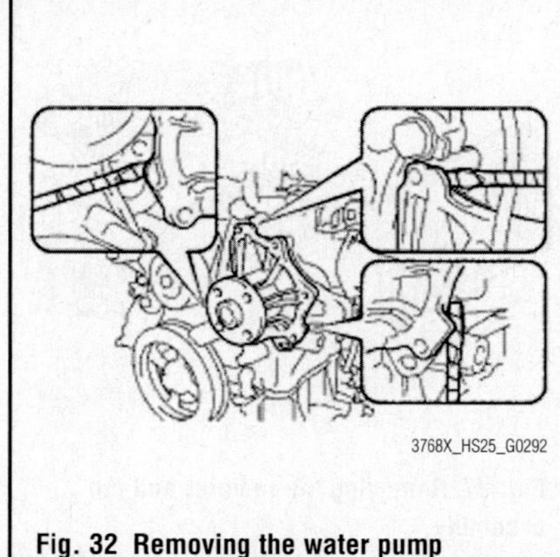

Fig. 32 Removing the water pump

➡Remove any oil from the contact surfaces.

➡The parts must be set within 3 minutes after applying seal packing. Otherwise, the material must be removed and reapplied.

   a. Install the water pump and clamp bracket with the 4 bolts and 2 nuts. Tighten the bolts to 80 inch lbs. (9 Nm).
   b. Install the wire of the Crankshaft Position (CKP) sensor onto the clamp bracket.
   c. Install the clamp of the Crankshaft Position (CKP) sensor onto the water pump.
7. Install the water pump pulley.
   a. Using the special too, install the water pump pulley with the 4 bolts. Tighten to 19 ft. lbs. (26 Nm).
8. Install the v-ribbed belt.
9. Add coolant.
10. Inspect for coolant leaks.
11. Install the No. 1 engine cover.

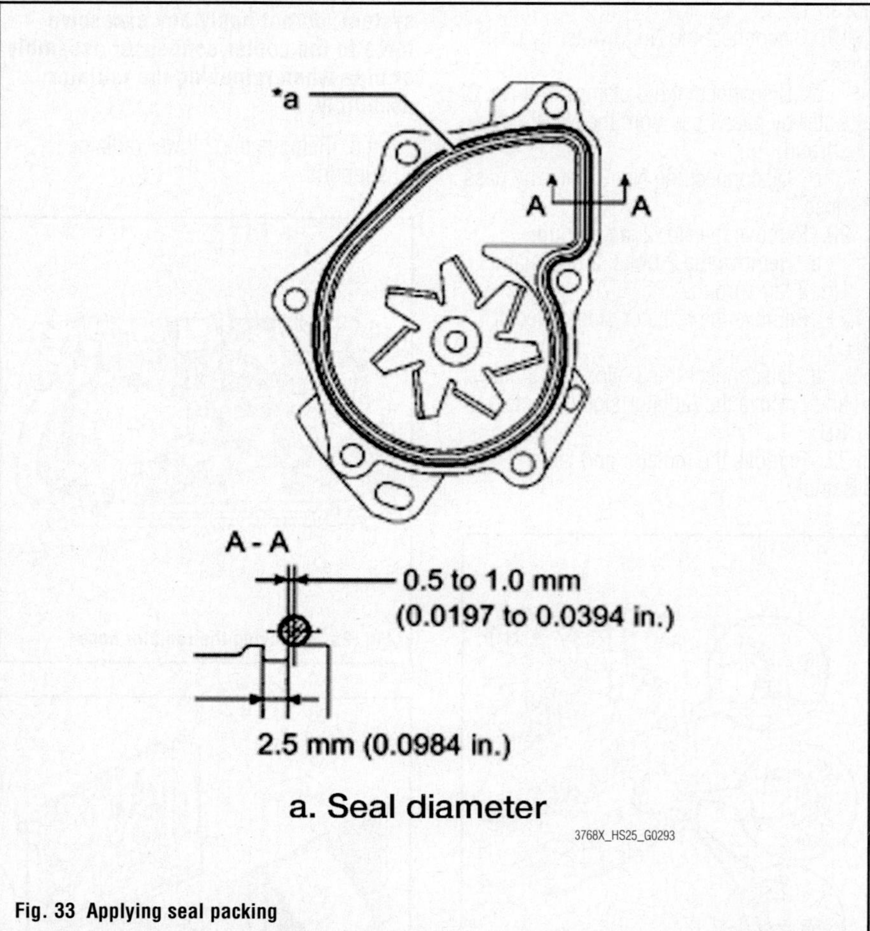

Fig. 33 Applying seal packing

ENGINE ELECTRICAL

IGNITION COIL

REMOVAL & INSTALLATION

See Figures 34 and 35.

1. Remove the No. 1 engine cover sub assembly.
2. Remove the ignition coil assembly.
   a. Disconnect the 4 ignition coil connectors.
   b. Remove the 4 bolts and 4 ignition coils.
3. Remove the spark plug.
   a. Remove the 4 spark plugs.

To install:

4. Install the 4 spark plugs.

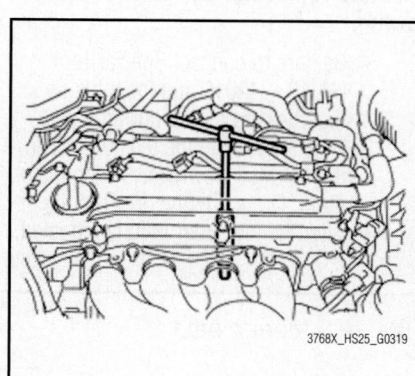

Tighten the spark plugs to 14 ft. lbs. (19 Nm).

5. Install the ignition coil assembly and bolts. Tighten to 80 inch lbs. (9 Nm).
   a. Connect the 4 ignition coil connectors.
6. Install the No. 1 engine cover sub assembly.

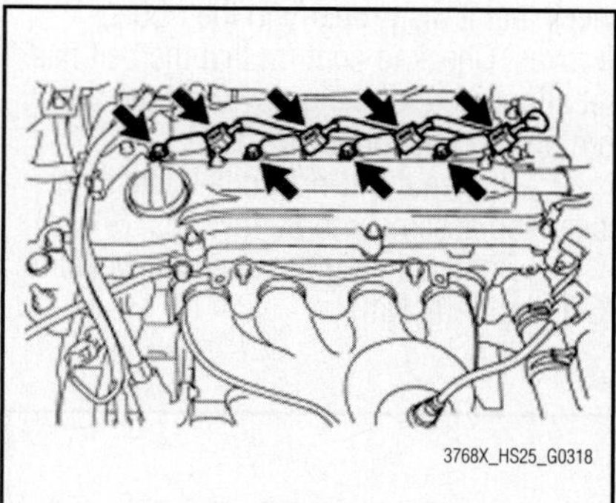

Fig. 34 Disconnecting the ignition coil connectors

Fig. 35 Removing the spark plugs

IGNITION TIMING

INSPECTION

All engines are equipped with a Distributorless Ignition System (DIS). No timing adjustment is possible.

SPARK PLUGS

REMOVAL & INSTALLATION

Refer to IGNITION COIL for removal and installation procedures.

ENGINE MECHANICAL

ACCESSORY DRIVE BELTS

ACCESSORY BELT ROUTING

See Figure 36.

Refer to the accompanying illustration.

INSPECTION

See Figure 37.

Visually check the V-ribbed belt for excessive wear, frayed cords, etc. If any defect has been found, replace the V-ribbed belt.

- Cracks on the rib side of a belt are considered acceptable. If the belt has chunks missing from the ribs, it should be replaced
- A "new belt" is a belt which has been used for less than 5 minutes with the engine running
- A "used belt" is a belt which has been used for 5 minutes or more with the engine running

ADJUSTMENT

This vehicle is equipped with an auto-tensioner and cannot be adjusted.

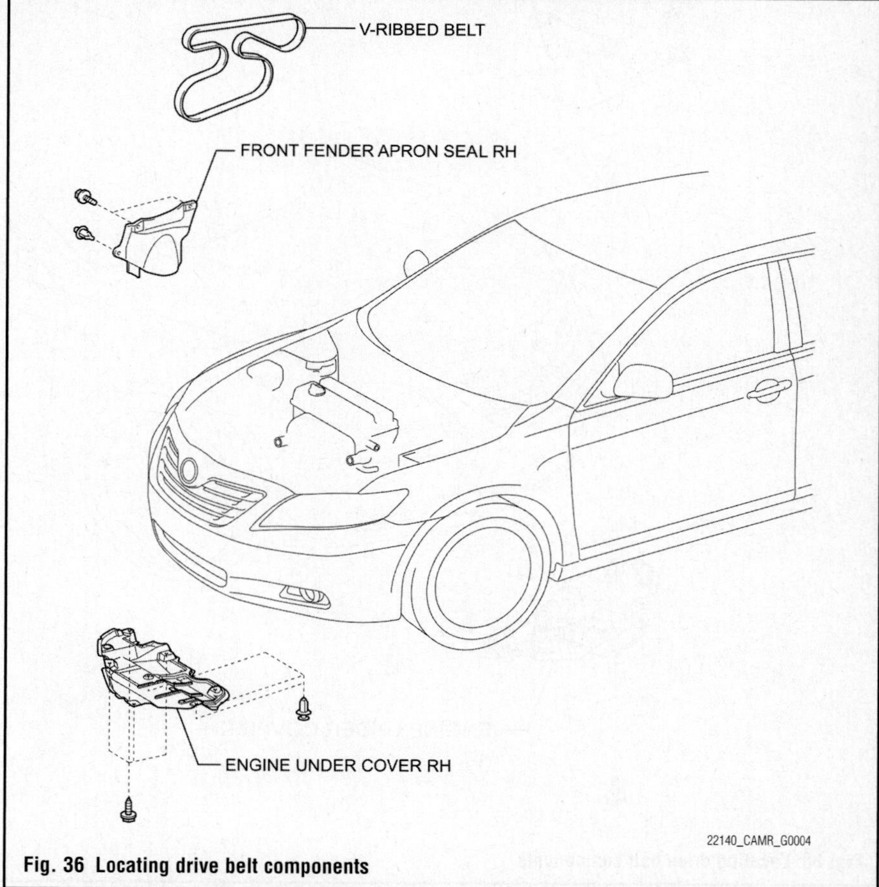

Fig. 36 Locating drive belt components

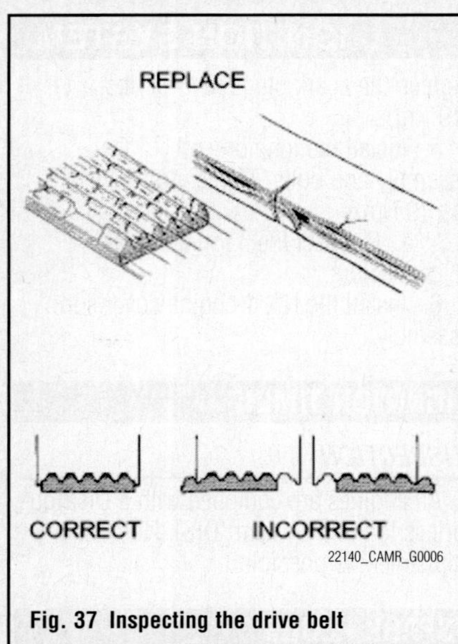

Fig. 37 Inspecting the drive belt

REMOVAL & INSTALLATION

See Figure 38.

1. Before servicing the vehicle, refer to the Precautions Section.
2. Remove the right hand front wheel.
3. Remove the right hand engine under cover.
4. Remove the right hand front fender apron seal.

➡Before removing, take note of the following:

- Be sure to connect Special Tool: 09216—42010 and the tools so that they are in line during use
- When retracting the tensioner, turn it clockwise slowly for 3 seconds or more. Do not apply force rapidly
- After the tensioner is fully retracted, do not apply force any more than necessary

5. Using the Special Tool and a 19 mm socket wrench, loosen the v-ribbed belt tensioner arm clockwise, then remove the v-ribbed belt.
6. Remove the v-ribbed belt.

To install:

7. To install, reverse removal procedure.
8. After installing the V-ribbed belt, check that it fits properly in the ribbed grooves. Check to confirm that the belt has not slipped out of the grooves on the bottom of the crank pulley by hand.
9. Tighten the V-ribbed belt tensioner cover sub assembly to 80 inch lbs. (9 Nm).
10. Tighten the right hand front wheel to: 76 ft. lbs. (103 Nm).

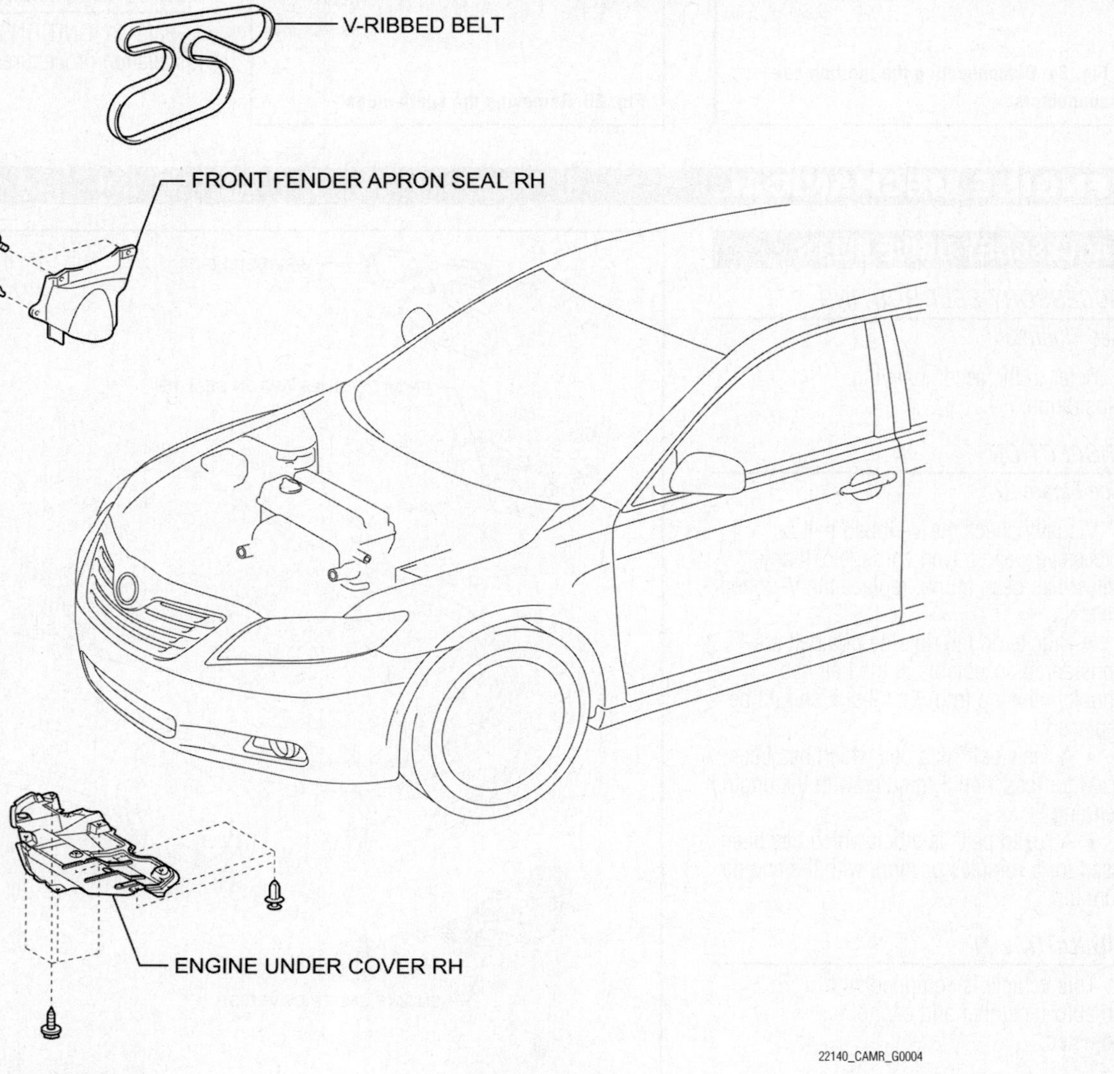

V-RIBBED BELT
FRONT FENDER APRON SEAL RH
ENGINE UNDER COVER RH
22140_CAMR_G0004

Fig. 38 Locating drive belt components

BALANCE SHAFT

REMOVAL & INSTALLATION

See Figures 39 through 46.

1. Before servicing the vehicle, refer to the Precautions Section.
2. Connect the negative battery cable.
3. Drain the engine oil.
4. Remove the oil pump. Refer to Oil Pump below for removal procedure.
5. Remove the No. 1 and No. 2 balance shaft sub-assembly. Remove the eight bolts in sequence.
6. Remove the No. 1 and No. 2 balance shafts.
7. Remove the balance shaft bearings if necessary.

To install:

➡Do not apply engine oil to the bearings and the contact surfaces.

8. Install the bearings in the crankcase and balance shaft housing.
9. Apply a light coat of engine oil to the bearings.

➡Confirm that the match marks on driven gears No. 1 and No. 2 are matched.

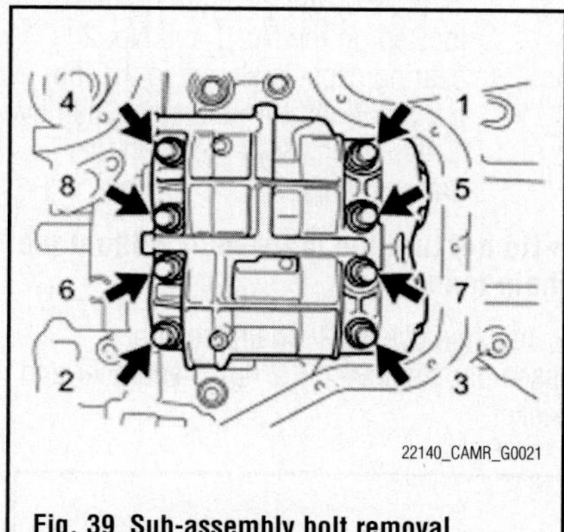

Fig. 39 Sub-assembly bolt removal sequence

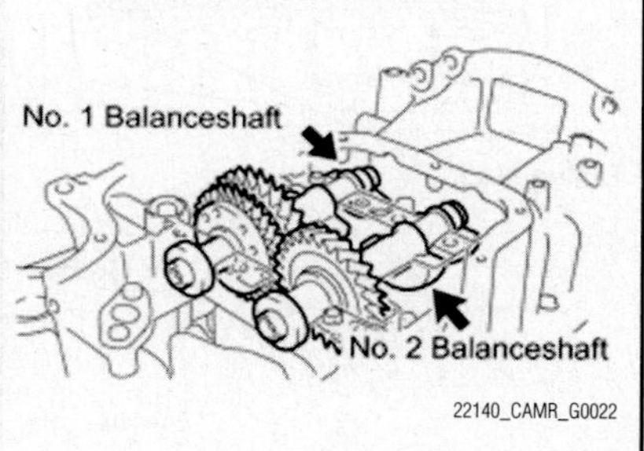

Fig. 40 Removing the No. 1 and No. 2 balance shafts

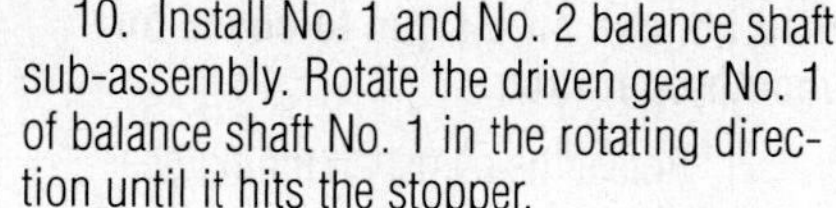

10. Install No. 1 and No. 2 balance shaft sub-assembly. Rotate the driven gear No. 1 of balance shaft No. 1 in the rotating direction until it hits the stopper.

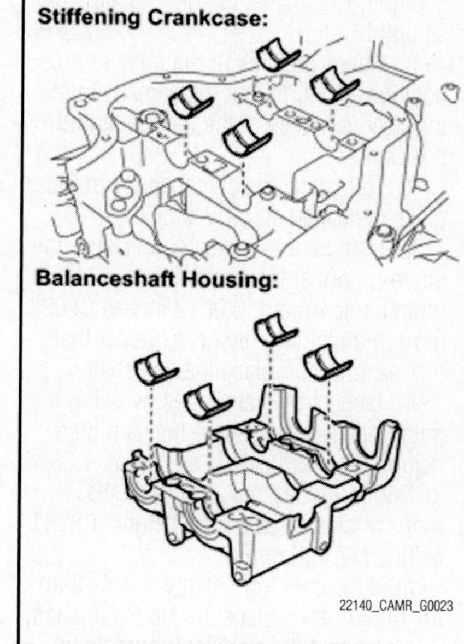

Fig. 41 Identifying and removing the balance shaft bearings

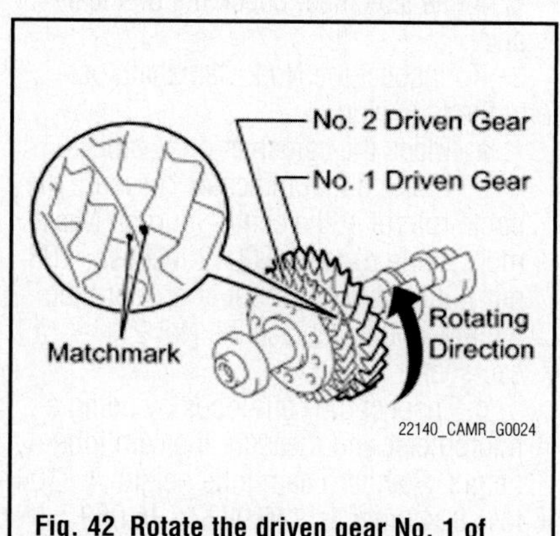

Fig. 42 Rotate the driven gear No. 1 of balance shaft No. 1

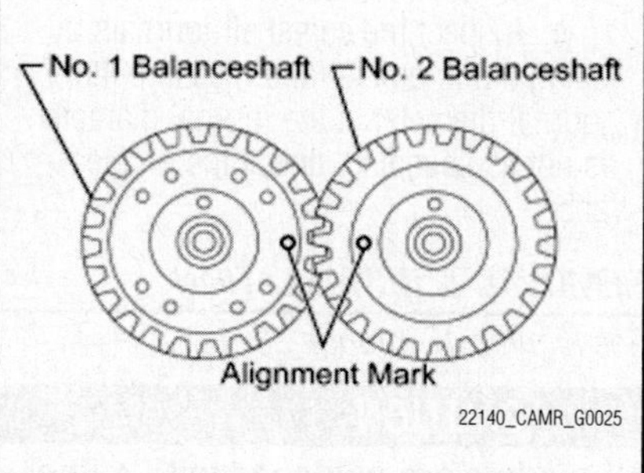

Fig. 43 Aligning marks of the No. 1 and No. 2 balance shafts

11. Align the alignment marks of the No. 1 and No. 2 balance shafts as shown.
12. Place the No. 1 and No. 2 balance shafts on the crankcase.
13. Apply a light coat of engine oil under the heads of the balance shaft housing bolts.
14. Install the balance shaft housing bolts. The balance shaft housing bolts should be tightened in 2 progressive steps as follows:

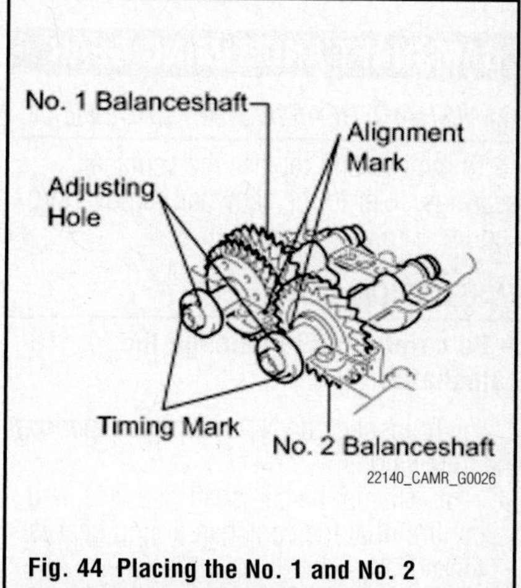

Fig. 44 Placing the No. 1 and No. 2 balance shafts on the crankcase

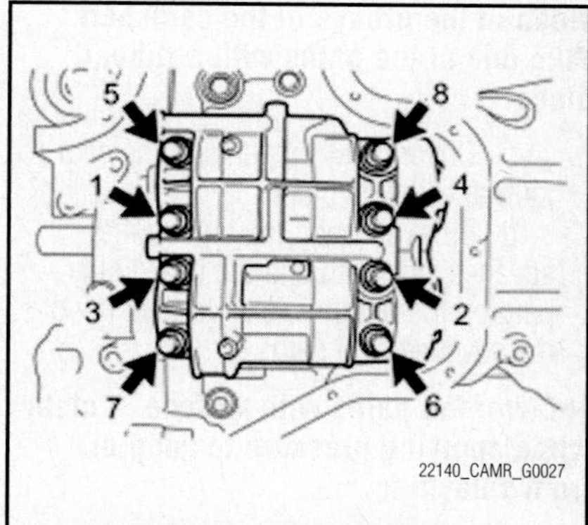

Fig. 45 Balance shaft housing bolt tightening sequence

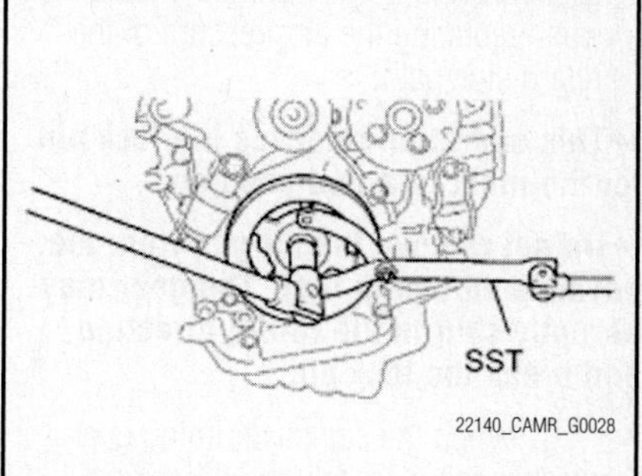

Fig. 46 Marking the bolt head and tightening procedure

a. Tighten the eight balance shaft housing bolts in sequence to: 16 ft. lbs. (22 Nm).

b. Mark the front side of each balance shaft housing bolt head with paint. Retighten the bolts by 90°. Check that the paint marks are now at a 90°angle to the front.

15. To complete installation, reverse remaining removal procedure.

16. Check the engine for leaks.

CAMSHAFT AND TIMING GEAR

CAMSHAFT BEARING REPLACEMENT

To remove and replace the camshaft bearings, refer to the camshaft removal procedure.

INSPECTION

➡**Be careful not to damage the camshafts.**

1. To inspect the No.1 camshaft, perform the following:

a. Clamp the camshaft in a vise, and confirm that the camshaft timing gear is locked.

b. Release the lock pin.

➡**The 2 advance side paths are provided in the groove of the camshaft. Plug one of the paths with a rubber piece.**

c. Cover the 4 oil paths of the cam journal with vinyl tape.

d. Break through the tape of the advance side path and the retard side path on the opposite side to the hole of the advance side path.

➡**Cover the paths with a piece of cloth when applying pressure to keep oil from splashing.**

e. Apply approximately 28 psi of air pressure to the two broken paths.

f. Check that the camshaft timing gear revolves in the advance direction when reducing the air pressure of the retard side path.

➡**This operation releases the lock pin for the most retarded position.**

➡**Do not remove the air gun from the advance side path first. The gear may abruptly shift in the retard direction and break the lock pin.**

g. When the camshaft timing gear reaches the most advanced position, remove the air gun from the retard side path and advance side path, in that order.

➡**Do not use an air gun to check for smooth operation.**

h. Rotate the camshaft timing gear within its movable range several times, but do not turn it to the most retarded position. Check that the gear rotates smoothly.

i. Check the lock in the most retarded position. Confirm that the camshaft timing gear is locked at the most retarded position.

j. To inspect the camshaft for runout, place the camshaft on V-blocks.

k. Using a dial indicator, measure the circle runout at the center journal. Maximum circle runout: 0.0012 inches (0.03 mm). If the circle runout is greater than the maximum, replace the camshaft.

l. Inspect the cam lobes by using a micrometer and measure the cam lobe height. Standard cam lobe height: 1.8624 to 1.8664 inches. (47.306 to 47.406 mm). Minimum cam lobe height: 1.8581 inches (47.196 mm).

m. If the cam lobe height is less than the minimum, replace the No.1 camshaft.

n. Inspect the camshaft journals by using a micrometer and measure the journal diameter. If the journal diameter is not as specified, check the oil clearance.

2. To inspect the No.2 Camshaft, perform the following:

a. Place the camshaft on V-blocks.

b. Using a dial indicator, measure the circle runout at the center journal. Maximum circle runout: 0.0012 inches. (0.03 mm). If the circle runout is greater than the maximum, replace the No. 2 camshaft.

c. Inspect the cam lobes by using a micrometer and measure the cam lobe height. Standard cam lobe height: 1.8104 to 1.8143 inches. (45.983 to 46.083 mm). Minimum cam lobe height: 1.8060 inches (45.873 mm).

d. If the cam lobe height is less than the minimum, replace the No.2 camshaft.

e. Inspect the camshaft journals by using a micrometer and measure the journal diameter. If the journal diameter is not as specified, check the oil clearance.

REMOVAL & INSTALLATION

See Figures 47 through 58.

CAUTION

All models are equipped with a Supplemental Restraint System (SRS), which uses an air bag. Whenever working near any of the SRS components, such as the impact sensors, the air bag module, steering column and instrument panel, disable the SRS.

1. Before servicing the vehicle, refer to the Precautions Section.
2. Disconnect the negative battery cable.
3. Loosen the lug nuts on the front right hand wheel.
4. Apply the parking brake, block the rear wheels, then raise and safely support the front of the vehicle securely on jackstands.
5. Remove the front right hand wheel.
6. Remove the left and right hand under cover.
7. Remove the No.1 engine cover subassembly and 2 nuts.
8. Remove the ignition coil assembly.
9. Remove the cylinder head cover subassembly
10. Set No.1 Cylinder to TDC/Compression by performing the following:
    - Turn the crankshaft pulley until its groove and the timing mark "0" of the timing chain cover are aligned
    - Check that each timing mark of the camshaft timing gear and sprocket is aligned with each timing mark located on the No. 1 and No. 2 bearing caps as shown in the illustration. If not, turn the crankshaft by 1 revolution (360°) to align the timing marks

➡**Do not turn the crankshaft without the chain tensioner.**

11. Remove No.1 Chain tensioner assembly. Remove the 2 nuts, tensioner and gasket.

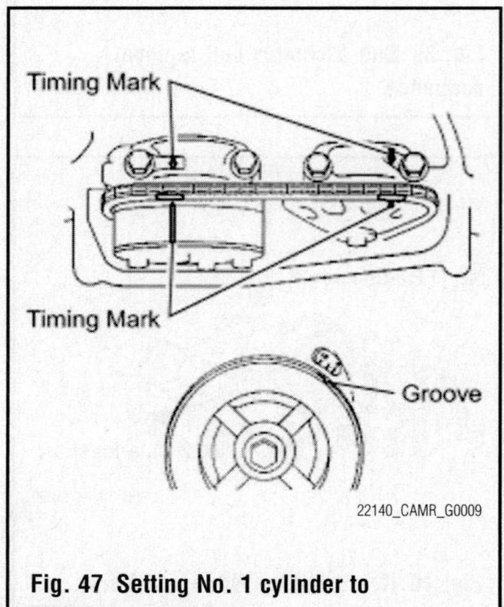

Fig. 47 Setting No. 1 cylinder to TDC/Compression

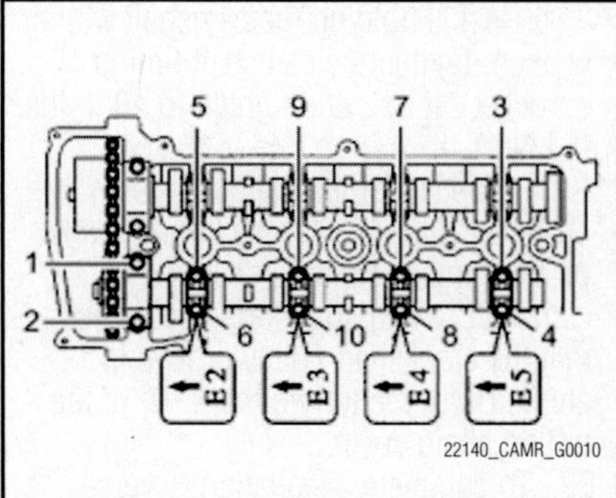

Fig. 48 No. 2 camshaft bearing cap bolt removal sequence

12. Remove the No. 2 camshaft by performing the following:
    a. While holding the camshaft with a wrench, loosen the camshaft timing set bolt.
    b. Using several steps, uniformly loosen and remove the 10 bearing cap bolts in the sequence shown.
    c. While holding the No. 2 camshaft by hand, remove the camshaft timing sprocket set bolt.
    d. Remove the camshaft timing sprocket from the No. 2 camshaft with the timing chain wrapped on the sprocket.
    e. Remove the camshaft timing sprocket from the timing chain.
13. Remove the No. 1 camshaft by performing the following:
    a. Using several steps, uniformly loosen and remove the 10 bearing cap bolts in the sequence shown.
    b. Remove the 5 bearing caps.
    c. Remove the camshaft and camshaft timing gear while holding the timing chain by hand.

➡Be careful not to drop anything inside the timing chain cover.

    d. Tie the timing chain with a string.

14. Remove the camshaft timing gear assembly by performing the following:
    a. Clamp the camshaft in a vise, and make sure that the camshaft timing gear does not rotate.
    b. Cover all the oil ports except the advance side port shown in the illustration with vinyl tape

➡Cover the paths with a shop rag or piece of cloth to avoid oil splashes.

➡Depending on the air pressure, the camshaft timing gear will turn to the advance angle side without applying force by hand. Also, if the pressure is difficult to apply because of air leakage from the port, the lock may be difficult to release.

    c. Apply air pressure of 14 psi to the oil path, then turn the camshaft timing gear in the advance direction (counterclockwise) by hand.
    d. Remove the flange bolt of the camshaft timing gear. Be sure not to remove the other four bolts. If planning to reuse the gear, be sure to release the straight pin lock before installing the gear.

To install:

15. Put the camshaft timing gear and camshaft together with the straight pin and key groove misaligned.

➡Be sure not to turn the camshaft timing gear to the retard angle side (the right angle).

16. Turn the camshaft timing gear as shown in the illustration while pushing it gently against the camshaft. Push further at the position where the pin fits into the groove.
17. Check that there is no clearance between the gear and camshaft.
18. Tighten the flange bolt with the camshaft timing gear fixed in place and tighten to 40 ft. lbs. (54 Nm).
19. Check that the camshaft timing gear can move to the retard angle side (the right direction) and is locked in the most retarded position.
20. To install the No.1 camshaft, perform the following:
    a. Apply a light coat of engine oil to the journal portion of the camshaft.
    b. Install the timing chain onto the camshaft timing gear with the paint mark aligned with the timing mark in the camshaft timing gear.

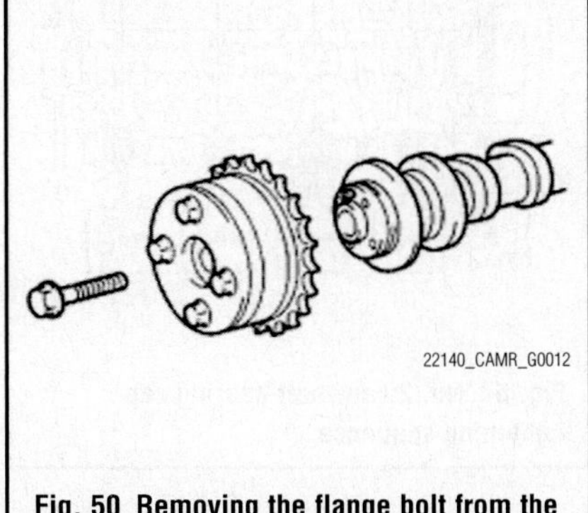

Fig. 50 Removing the flange bolt from the camshaft timing gear

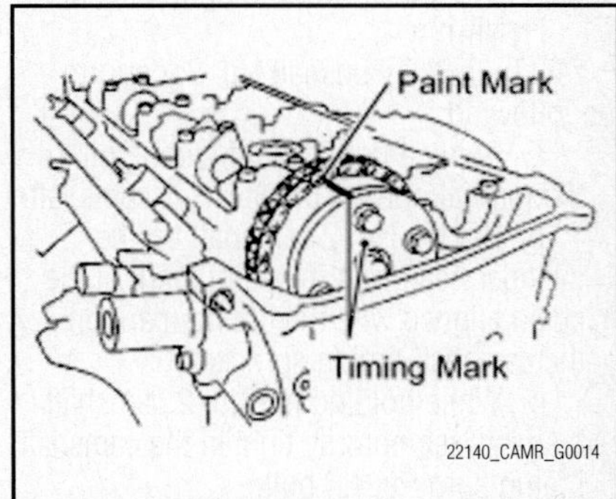

Fig. 52 Aligning No. 1 camshaft paint mark with the timing mark

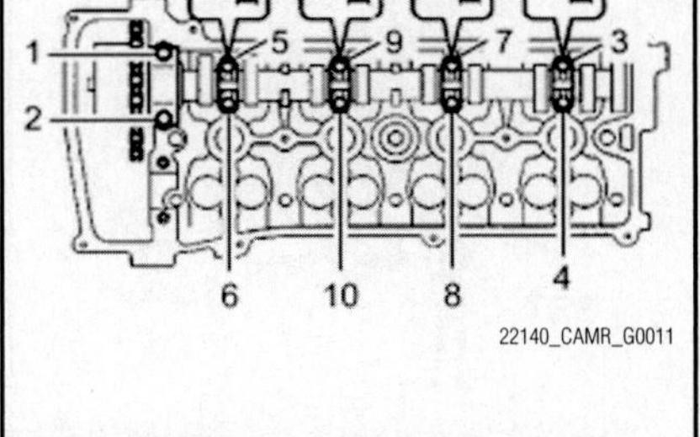

Fig. 49 No.1 camshaft bearing cap bolt removal sequence

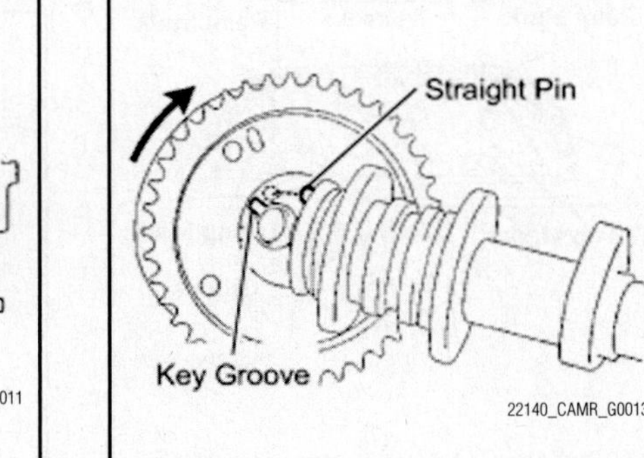

Fig. 51 No. 1 camshaft straight pin and key groove misaligned

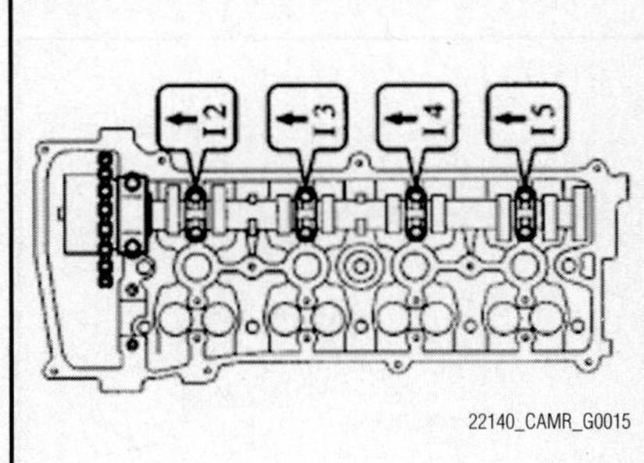

Fig. 53 Checking the No. 1 camshaft bearing cap front marks, numbers and order

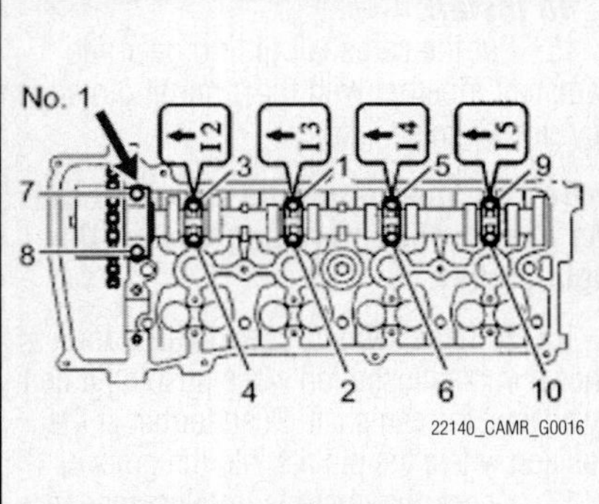

Fig. 54 No. 1 camshaft bearing cap tightening sequence

c. Examine the front marks and numbers, and check that the order is as shown in the illustration below. Then install the bearing caps into the cylinder head.

d. Apply a light coat of engine oil to the threads and under the heads of the bearing cap bolts.

e. Using several steps, uniformly tighten the 10 bearing cap bolts in the sequence shown. Tighten the following to specification:

- No. 1 Bearing cap: 22 ft. lbs. (30 Nm)
- No. 3 Bearing cap: 80 inch lbs. (9 Nm)

21. To install camshaft No. 2, perform the following:

a. Apply a light coat of engine oil to the journal portion of the No. 2 camshaft.

b. Put the No. 2 camshaft on the cylinder head with the paint mark of the chain aligned with the timing mark on the camshaft timing sprocket.

c. While holding the No. 2 camshaft by hand, temporarily tighten the camshaft timing sprocket set bolt.

d. Examine the front marks and numbers, and check that the order is as shown in the illustration. Then install the bearing caps onto the cylinder head.

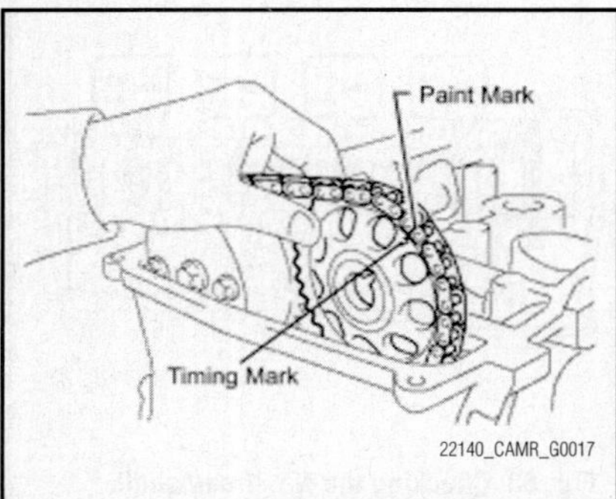

Fig. 55 Aligning No. 2 camshaft paint mark with the timing mark

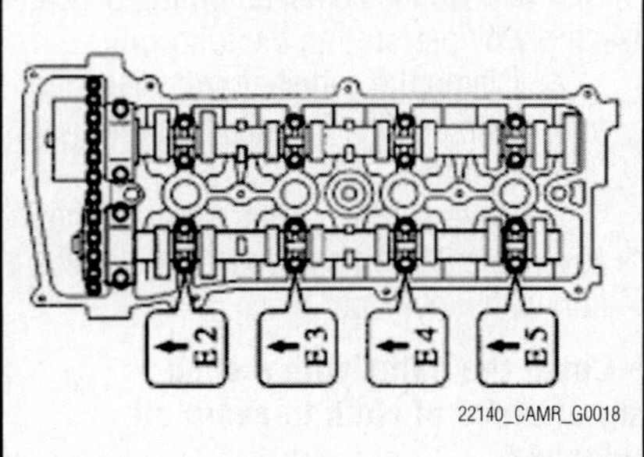

Fig. 56 Checking the No. 2 camshaft bearing cap front marks, numbers and order

e. Apply a light coat of engine oil to the threads and under the heads of the bearing cap bolts.

f. Using several steps, uniformly tighten the 10 bearing cap bolts in the sequence shown. Tighten the following to specification:

- No. 1 Bearing cap: 22 ft. lbs. (30 Nm)
- No. 3 Bearing cap: 80 inch lbs. (9 Nm)

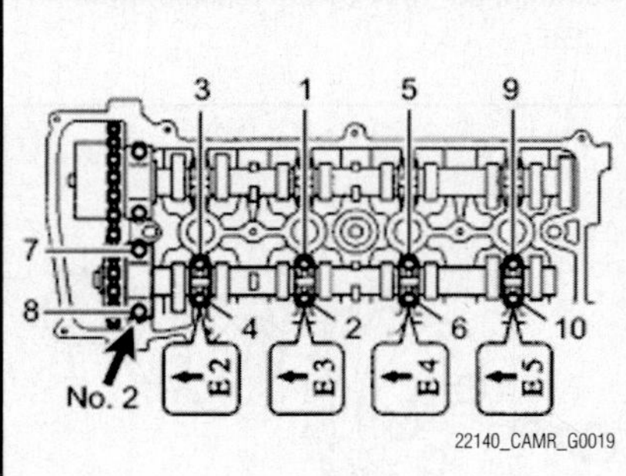

Fig. 57 No. 2 camshaft bearing cap tightening sequence

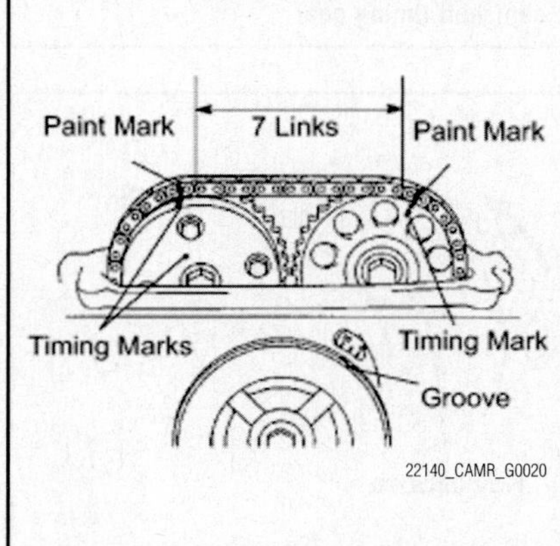

Fig. 58 Aligning paint marks on the chain with the timing marks on the camshaft timing gear and camshaft timing sprocket

g. While holding the camshaft with a wrench, tighten the camshaft timing sprocket set bolt and tighten to 40 ft. lbs. (54 Nm).

h. Check that the paint marks on the chain are aligned with the timing marks on the camshaft timing gear and camshaft timing sprocket. Also, check that the crankshaft pulley groove is aligned with the timing mark "0" of the timing chain cover.

22. To complete installation, reverse remaining removal procedure.

23. Check engine for oil leaks.

24. Connect the negative battery cable.

CRANKSHAFT DAMPER

REMOVAL & INSTALLATION

See Figure 59.

1. Before servicing the vehicle, refer to the Precautions Section.
2. Remove the engine assembly.
3. Install on engine stand.
4. Remove the oil filler cap and gasket.
5. Remove the spark plugs and ignition coil assembly.
6. Remove the drain plug and gasket.
7. Remove the ventilation valve.
8. Remove the 4 bolts and 4 Camshaft Position (CMP) sensors.
9. Remove the 4 bolts and 4 camshaft timing oil control valves.
10. Remove the bolt and Crankshaft Position (CKP) sensor.
11. Remove the No. 1 oil pipe.
12. Remove the oil pipe.
13. Remove the cylinder block water drain cock sub-assembly.
14. Remove the oil filter element.
15. Remove the crankshaft pulley, as follows:

a. Using SST (SST: 09213-70011, SST: 09330-00021) or equivalent, loosen the crankshaft pulley bolt.

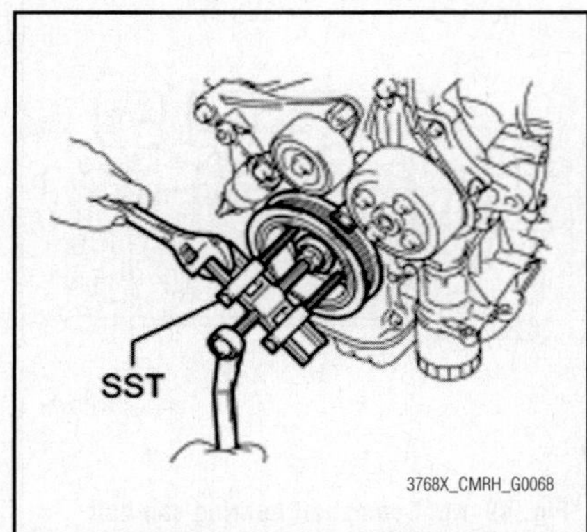

Fig. 59 Removing the crankshaft pulley

b. Using SST (SST: 09950-50013) or equivalent, remove the crankshaft pulley bolt and crankshaft pulley.

To install:

16. Install the crankshaft pulley, as follows:

a. Align the pulley set key with the key groove of the pulley, and slide on the pulley.

b. Using SST (SST: 09213-70011, SST: 09330-00021) or equivalent, install the pulley bolt and tighten to 184 ft. lbs. (250 Nm).

17. Install the oil filter element.
18. Install the cylinder block water drain cock sub-assembly.
19. Install the No. 1 oil pipe.
20. Install the oil pipe.
21. Install the Crankshaft Position (CKP) sensor.
22. Install the camshaft timing oil control valve assembly.
23. Install the Camshaft Position (CMP) sensor.
24. Install the ventilation valve sub-assembly.
25. Install the spark plugs and the ignition coil assembly.
26. Install the oil filler cap sub-assembly.
27. Remove the engine stand.
28. Install the engine assembly.

CRANKSHAFT FRONT SEAL

REMOVAL & INSTALLATION

See Figures 60 and 61.

1. Remove the front wheel RH.
2. Remove the engine under cover.
3. Remove the No. 1 engine under cover.
4. Remove the rear engine under cover RH.
5. Remove the rear side rail reinforcement sub assembly RH.

a. Remove the 4 bolts and rear side rail reinforcement sub assembly.

6. Remove the cool air intake duct seal.
7. Remove the V-ribbed belt tensioner cover sub assembly.
8. Remove the V-ribbed belt.
9. Remove the crankshaft pulley.

a. Using the special tool, secure the pulley in place and loosen the pulley bolt.

b. Using the special tool, remove the pulley bolt and pulley.

➡If necessary, remove the pulley and pulley bolt using the special tool.

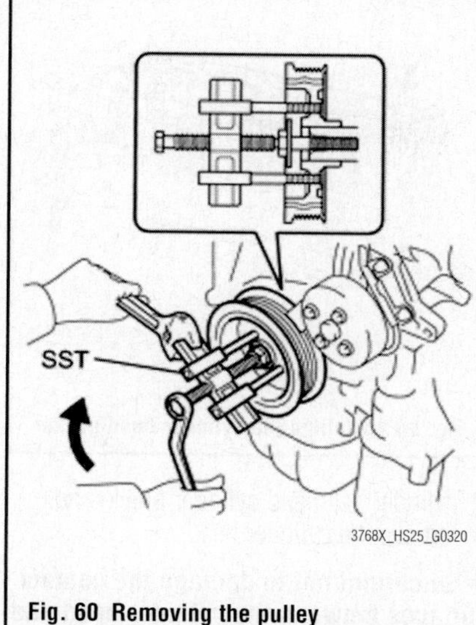

Fig. 60 Removing the pulley

10. Remove the timing chain case oil seal.

a. Using a knife, cut off the oil seal lip.

b. Using a screwdriver with the tip taped, pry out the oil seal.

➡After the removal, check the crankshaft for damage. If it is damaged, smooth the surface with 400 grit sandpaper.

To install:

11. Install the timing chain case oil seal.

a. Apply MP grease to a new oil seal lip.

➡Keep the lip free from the foreign matter.

b. Using the special tool and a hammer, tap in the oil seal until its surface is flush with the front oil seal retainer edger.

➡Wipe off extra grease from the crankshaft.

12. Install the crankshaft pulley.

a. Align the pulley set key with the key groove of the pulley.

b. Using the special tool, hold the pulley in place and tighten the bolt. Tighten to 133 ft. lbs. (180 Nm).

13. Install the V-ribbed belt.
14. Install the V-ribbed belt tensioner cover sub assembly.
15. Inspect for an oil leak.
16. Install the cool air intake duct seal.
17. Install the rear side rail reinforcement sub assembly RH.

a. Install the rear side rail reinforcement sub assembly RH with the 4 bolts. Tighten to 71 ft. lbs. (96 Nm).

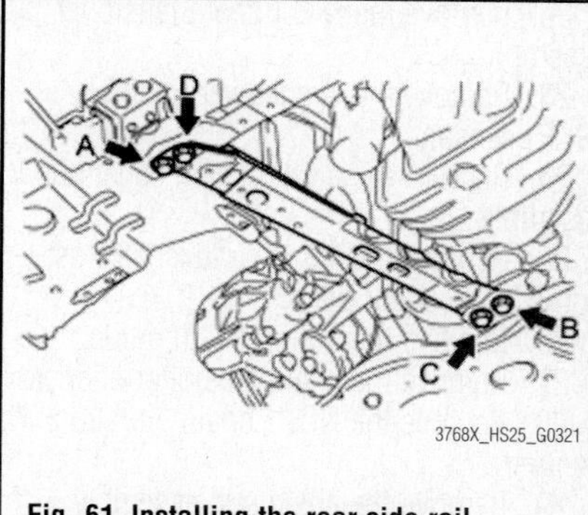

Fig. 61 Installing the rear side rail reinforcement sub assembly RH

➡Temporarily tighten Bolts A and B, and then fully tighten the 4 bolts in the order of C, B, D and A.

18. Install the rear engine under cover RH.
19. Install the No. 1 engine under cover.
20. Install the engine under cover.

CYLINDER HEAD

REMOVAL & INSTALLATION

See Figures 62 through 66.

1. Before servicing the vehicle, refer to the Precautions Section.
2. Disconnect the negative battery cable.
3. Remove the engine assembly with the transaxle.
4. Remove the V-ribbed belt tensioner cover sub assembly.
5. Remove the V-ribbed belt.
6. Remove the throttle body assembly.
7. Remove the ignition coil assembly.
8. Remove the spark plug.
9. Remove the engine oil level dipstick.
10. Remove the engine oil level dipstick guide.
11. Remove the engine delivery pipe sub assembly.
12. Remove the camshaft timing oil control valve assembly.
13. Remove the intake manifold assembly.
14. Remove the No. 1 intake manifold insulator.
15. Remove the manifold stay.
16. Remove the No. 2 manifold stay.
17. Remove the exhaust manifold converter sub assembly.
18. Remove the idler pulley sub assembly.
19. Remove the oil pan sub assembly.
20. Set No. 1 cylinder to TDC/compression.
21. Remove the crankshaft pulley.

22. Remove the No. 1 chain tensioner assembly.
23. Remove the Crankshaft Position (CKP) sensor.
24. Remove the timing chain cover sub assembly.
25. Remove the No. 1 Crankshaft Position (CKP) sensor plate.
26. Remove the timing chain guide.
27. Remove the chain tensioner slipper.
28. Remove the NO. 1 chain vibration damper.
29. Remove the chain sub assembly.
30. Remove the No. 2 camshaft.
31. Remove the camshaft.
32. Remove the No. 1 camshaft bearing.
33. Remove the No. 2 camshaft bearing.
34. Remove the cylinder head sub assembly.
   a. In several steps, uniformly loosen and remove the 10 cylinder head bolts and 10 washers with a10 mm bi-hexagon wrench in the sequence shown.

➡Head warpage or cracking could result from removing the bolts in the wrong order.

   b. Using a screwdriver with its tip wrapped with tape, pry between the cylinder head and cylinder block, and remove the cylinder head.

➡Be careful not to damage the contact surfaces between the cylinder head and the cylinder block.

35. Remove the cylinder head gasket.
36. Inspect the cylinder head set bolt.

To install:

37. Install the cylinder head gasket.
   a. Place a new cylinder head gasket on the cylinder block surface with the Lot No. stamp facing upward.

➡Remove any oil from the contact surfaces.

➡Be careful of the installation direction.

38. Install the cylinder head sub assembly.
   a. Place the cylinder head on the cylinder head gasket.

➡Place the cylinder head gently in order to avoid damaging the cylinder head gasket.

   b. Install the cylinder head bolts.

➡The cylinder head bolts are tightened in 2 successive steps.

   c. Apply a light coat of engine oil to the threads and under the heads of the cylinder head set bolts.
   d. Using several steps, uniformly install and tighten the 10 cylinder head set bolts and plate washers with a 10 mm bi-hexagon wrench in the order shown. Tighten to 52 ft. lbs. (70 Nm).
   e. Mark the front of the cylinder head bolts with paint.
   f. Further tighten the cylinder head bolts 90°.
   g. Check that the paint mark in now at 90°angle to the front.
39. Install the No. 1 camshaft bearing.
40. Install the No. 2 camshaft bearing.
41. Install the camshaft.
42. Install the No. 2 camshaft.
43. Install the No. 1 chain vibration damper.
44. Install the chain sub assembly.
45. Install the chain tensioner slipper.
46. Install the timing chain guide.
47. Install the No. 1 Crankshaft Position (CKP) sensor plate.
48. Install the timing chain cover sub assembly.
49. Install the crankshaft pulley.
50. Install the No. 1 chain tensioner assembly.
51. Install the oil pan sub assembly.
52. Install the cylinder head cover sub assembly.
53. Install the Crankshaft Position (CKP) sensor.
54. Install the idler pulley sub assembly.
55. Install the exhaust manifold converter sub assembly.
56. Install the No. 2 manifold stay.
57. Install the manifold stay.

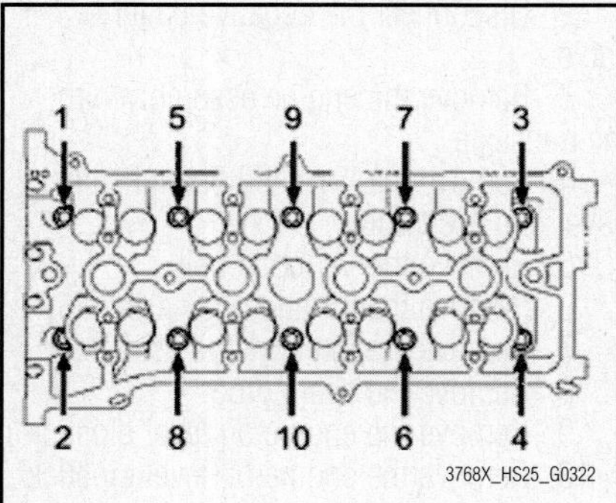

Fig. 62 Removing the cylinder head sub assembly

Fig. 63 Separating the cylinder head from the cylinder block

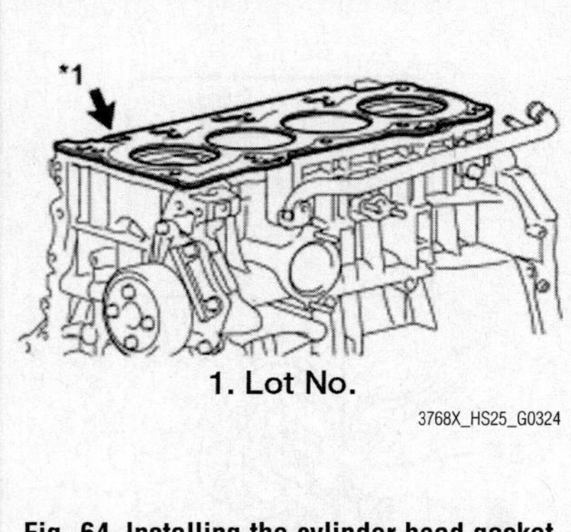

Fig. 64 Installing the cylinder head gasket

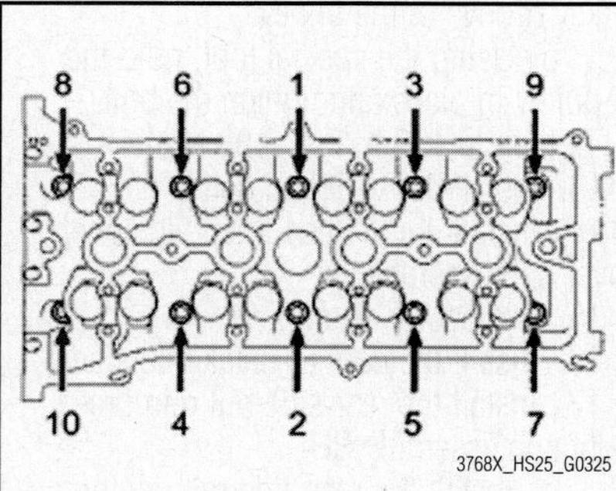

Fig. 65 Cylinder head bolt tightening sequence

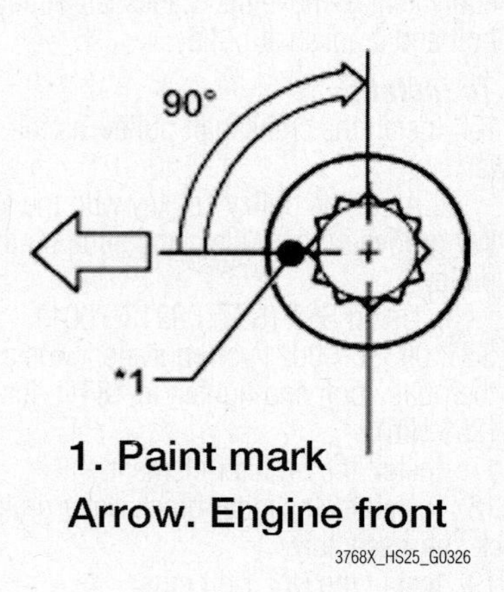

Fig. 66 Marking the front of the cylinder head bolts

58. Install the No. 1 intake manifold insulator.
59. Install the intake manifold assembly.
60. Install the camshaft timing oil control valve assembly.
61. Install the fuel delivery pipe sub assembly.
62. Install the engine oil level dipstick guide.
63. Install the engine oil level dipstick.
64. Install the spark plug.
65. Install the ignition coil assembly.
66. Install the throttle body assembly.
67. Install the V-ribbed belt.
68. Install the V-ribbed belt tensioner cover sub assembly.
69. Install the engine assembly with the transaxle.

EXHAUST MANIFOLD

REMOVAL & INSTALLATION

See Figures 67 through 72.

1. Remove the air fuel ratio sensor.
2. Remove the cool air intake duct seal.
3. Remove the engine under cover.
4. Remove the No. 1 engine under cover.
5. Remove the front No. 3 engine under cover.
6. Remove the front exhaust pipe assembly.
   a. Disconnect the oxygen sensor connector and wire harness clamp.
   b. Remove the 4 bolts, 2 compressions springs and front exhaust pipe assembly.
   c. Remove the 2 gaskets from the exhaust manifold converter sub-assembly and front exhaust pipe assembly.
7. Remove the No. 2 manifold converter insulator.

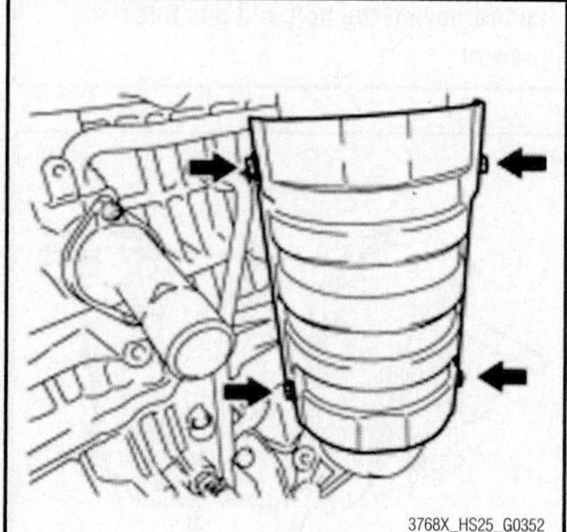

Fig. 67 Removing the No. 2 manifold converter insulator

Fig. 68 Removing the No. 2 manifold stay

Fig. 69 Removing the exhaust manifold

   a. Remove the 2 bolts and No. 2 manifold converter insulator.
8. Remove the No. 1 manifold converter insulator.
   a. Remove the 4 bolts and No. 1 manifold converter insulator.
9. Remove the manifold stay.
   a. Remove the bolt, nut and manifold stay.
10. Remove the No. 2 manifold stay.
   a. Remove the bolt, nut and No. 2 manifold stay.
11. Remove the exhaust manifold converter sub assembly.
   a. Using a 12 mm deep socket wrench, remove the 5 nuts and exhaust manifold converter sub assembly.
   b. Remove the exhaust manifold from the head gasket.

To install:

12. Install the exhaust manifold converter sub assembly.
   a. Install a new exhaust manifold to head gasket.
   b. Using a 12 mm deep socket wrench, install the exhaust manifold converter sub assembly and 5 nuts in the order shown. Tighten to 27 ft. lbs. (37 Nm).
13. Install the No. 2 manifold stay.

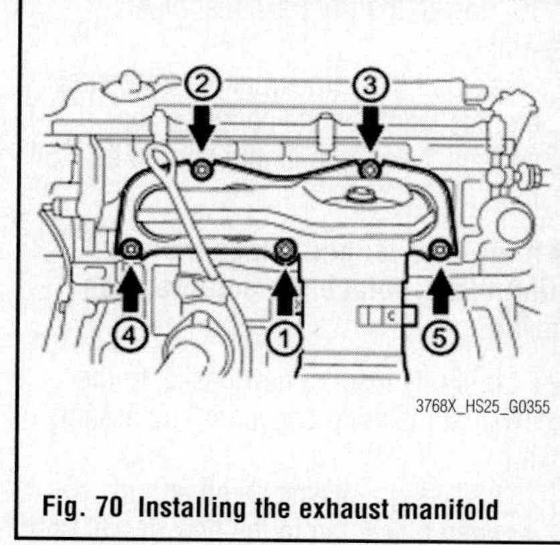

Fig. 70 Installing the exhaust manifold

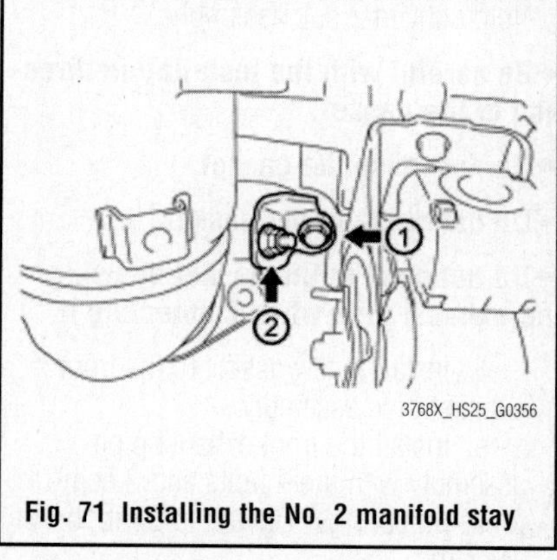

Fig. 71 Installing the No. 2 manifold stay

   a. Temporarily install the No. 2 manifold stay with the bolt and nut.
   b. Tighten the bolt and nut of the No. 2 manifold stay in the order shown. Tighten to 32 ft. lbs. (44 Nm).
14. Install the No. 1 manifold converter insulator with the 4 bolts. Tighten to 9 ft. lbs. (12 Nm).
15. Install the No. 2 manifold converter insulator with the 2 bolts. Tighten to 80 inch lbs. (9 Nm).

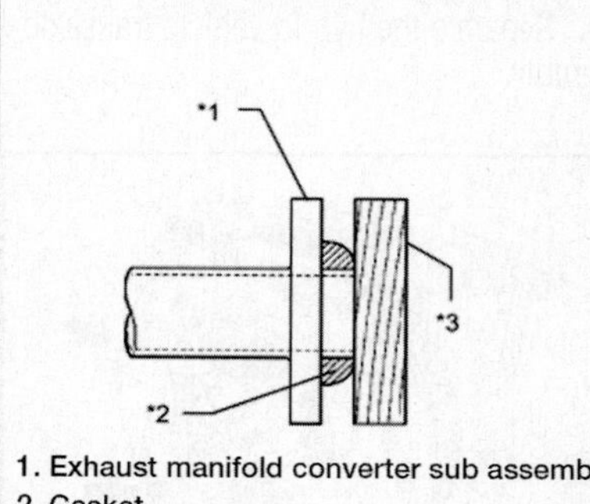

Fig. 72 Installing the new exhaust manifold converter sub assembly gasket

16. Install the front exhaust pipe assembly.
    a. Using a vernier caliper, measure the free length of the compression springs. Minimum length should be 1.64 inch (41.55 mm).

➡If the free length is less than the minimum, replace the compression spring.

    b. Fully insert a new gasket to the exhaust manifold converter sub assembly.
    c. Using a plastic hammer and wooden block, tap in the new gasket until its surface is flush with the exhaust manifold converter sub assembly.

➡Be careful with the installation direction of the gasket.

➡Do not reuse the gasket.

➡Do not damage the gasket.

➡Do not push in the gasket by using the exhaust pipe when connecting it.

    d. Install a new gasket to the front exhaust pipe assembly.
    e. Install the front exhaust pipe assembly with the 4 bolts and 2 compression springs. Tighten to 32 ft. lbs. (43 Nm).
17. Install the front No. 3 engine under cover.
18. Install the No. 1 engine under cover.
19. Install the engine under cover.
20. Install the cool air intake duct seal.
21. Install the air fuel ratio sensor.

FLYWHEEL

REMOVAL & INSTALLATION

See Figures 73 through 75.

1. Separate the hybrid vehicle transaxle assembly.

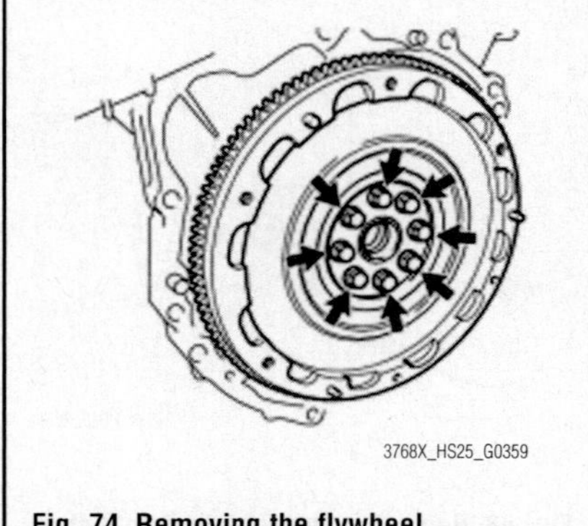

Fig. 74 Removing the flywheel

2. Remove the transmission input damper assembly.
    a. Using the special tool, hold the crankshaft.
    b. Remove the 6 bolts and transmission input damper from the flywheel.
3. Remove the flywheel sub assembly.
    a. Remove the 8 bolts and the flywheel.

To install:

4. Install the flywheel sub assembly.
    a. Using the special tool, hold the crankshaft.
    b. Clean the 8 bolts and the bolt holes.
    c. Install the flywheel with the 8 new bolts. Uniformly tighten the 8 new bolts in the sequence shown. Tighten to 96 ft. lbs. (130 Nm).
5. Install the transmission input damper assembly.
    a. Clean the 6 bolt holes.
    b. Install the transmission input damper assembly with the 6 new bolts. Tighten to 22 ft. lbs. (30 Nm).

➡Take care not to insert the transmission input damper in the wrong direction.

6. Install the hybrid vehicle transaxle assembly.

INTAKE MANIFOLD

REMOVAL & INSTALLATION

See Figures 76 through 79.

1. Remove the throttle body.
    a. Remove the front suspension crossmember sub assembly.
2. Remove the front exhaust pipe assembly.
3. Remove the rear No. 1 engine mounting insulator.
4. Remove the rear engine mounting bracket.
5. Remove the intake manifold.
    a. Disconnect the 2 air hoses and remove the bolt and gas filter bracket.
    b. Remove the No. 2 ventilation hose.
    c. Remove the 5 bolts, 2 nuts and intake manifold.
    d. Remove the No. 1 intake manifold to head gasket.

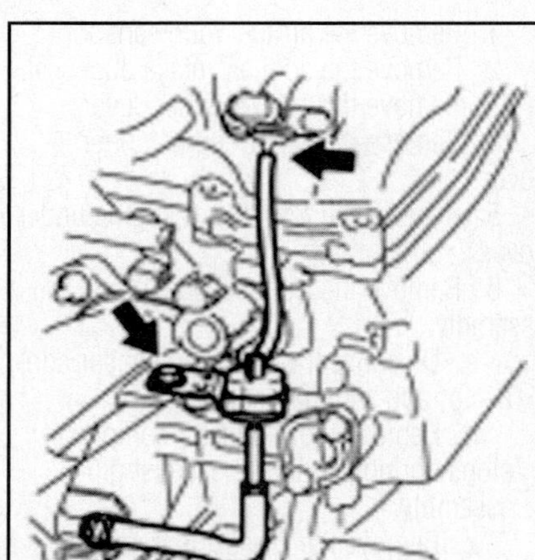

Fig. 76 Disconnecting the 2 air hoses and removing the bolt and gas filter bracket

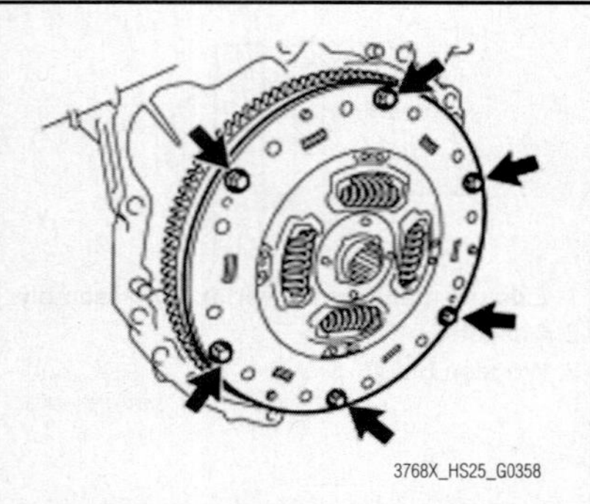

Fig. 73 Removing the transmission input damper from the flywheel

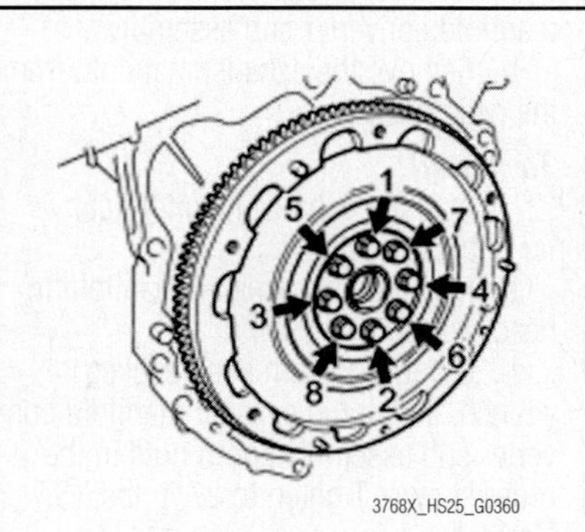

Fig. 75 Installing the flywheel

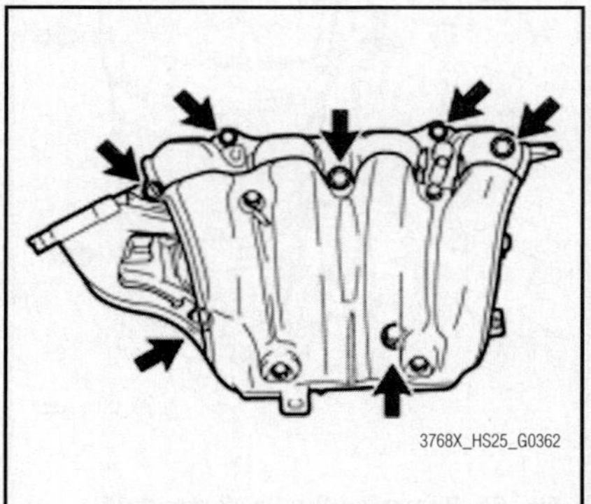

Fig. 77 Removing the intake manifold

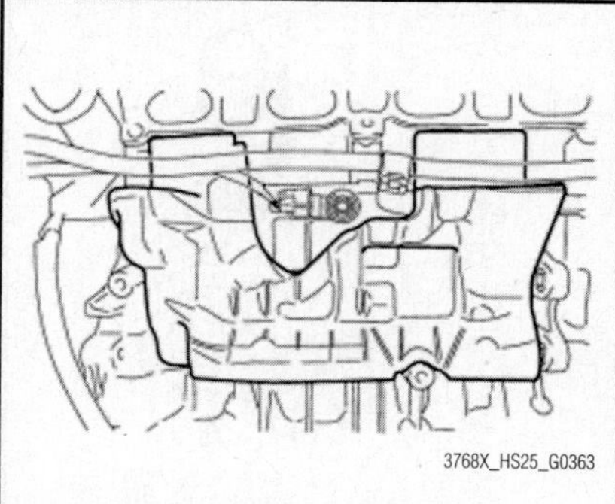

Fig. 78 Removing the No. 1 intake manifold insulator

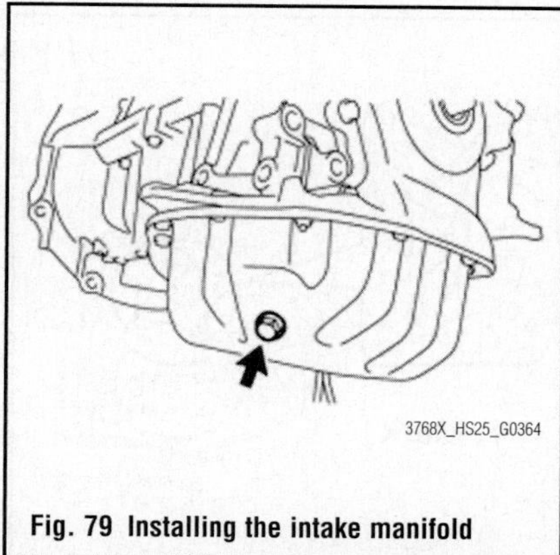

Fig. 79 Installing the intake manifold

e. Remove the No. 1 intake manifold insulator.

To install:

6. Install the intake manifold.
   a. Install the No. 1 intake manifold insulator.
   b. Install a new No. 1 intake manifold to the head gasket.
   c. Install the intake manifold with the 5 bolts and 2 nuts. Tighten to 22 ft. lbs. (30 Nm).
   d. Install the gas filter bracket with the bolt and connect the 2 air hoses. Tighten to 48 inch lbs. (5.4 Nm).
   e. Install the No. 2 ventilation hose.
7. Install the rear engine mounting bracket.
8. Install the rear No. 1 engine mounting insulator.
9. Install the front exhaust pipe assembly.
10. Install the front suspension crossmember sub assembly.
11. Install the throttle body.
12. Inspect for exhaust gas leaks.

OIL PAN

REMOVAL & INSTALLATION

See Figures 80 through 83.

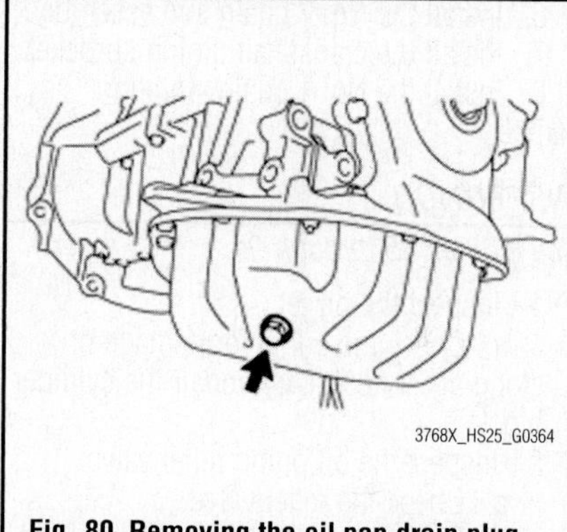

Fig. 80 Removing the oil pan drain plug

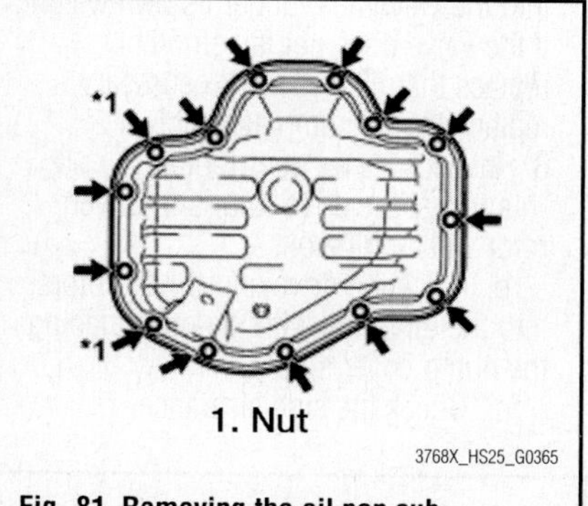

Fig. 81 Removing the oil pan sub assembly

1. Remove the oil pan drain plug.
   a. Remove the oil pan drain plug and gasket.
2. Remove the oil pan sub assembly.
   a. Remove the 12 bolts and 2 nuts.
   b. Insert the blade of an oil pan seal cutter between the crankcase, chain cover and oil pan, then cut through the applied sealer and remove the oil pan.

➡Be careful not to damage the contact surfaces of the crankcase, chain cover or oil pan.

To install:

3. Remove any old packing material and be careful not to drop any oil on the contact surfaces of the cylinder block or oil pan. Apply a continuous bead of seal packing 0.118-0.157 inch (3-4 mm). (Toyota Genuine Seal Packing Black, Three Bond 1207B or equivalent.

➡Remove any oil from the contact surfaces.

➡Install the oil pan within 3 minutes of applying seal packing.

➡Do not start the engine for at least 2 hours after installing the oil pan.

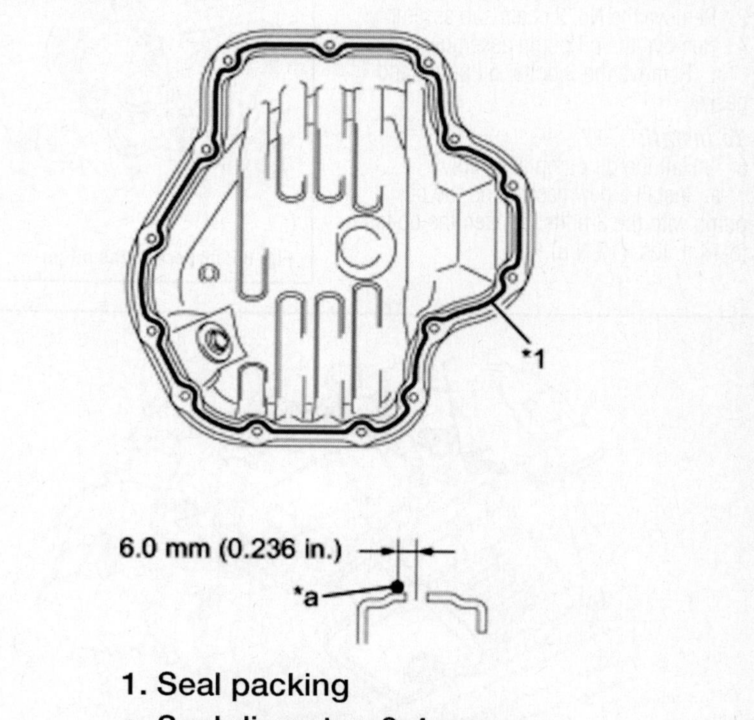

Fig. 82 Applying seal packing

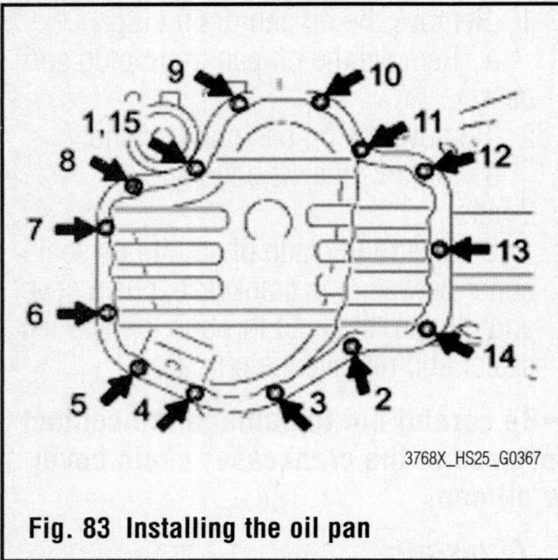

Fig. 83 Installing the oil pan

a. Install the oil pan sub-assembly to the cylinder block.

b. Uniformly tighten the 12 bolts and 2 nuts in sequence. Tighten to 80 inch lbs. (9 Nm).

4. Install the oil pan drain plug.

a. Install a new gasket and oil pan drain plug. Tighten to 29 ft. lbs. (40 Nm).

OIL PUMP

REMOVAL & INSTALLATION

See Figure 84.

1. Remove the chain sub assembly.
2. Remove the crankshaft timing sprocket.
3. Remove the No. 2 chain sub assembly.
4. Remove the oil pump assembly.

a. Remove the 3 bolts, oil pump and gasket.

To install:

5. Install the oil pump assembly.

a. Install a new gasket and the oil pump with the 3 bolts. Tighten the bolts to 14 ft. lbs. (19 Nm).

6. Install the No. 2 chain sub assembly.
7. Install the crankshaft timing sprocket.
8. Install the No. 1 chain vibration damper.

INSPECTION

See Figures 85 through 90.

1. Inspect the oil jet.

a. Check the oil jet for damage or clogging. If necessary, repair the cylinder block.

2. Inspect the oil pump relief valve.

a. Check the relief valve.

b. Coat the valve with engine oil, and then check that the valve falls smoothly into the valve hole under its own weight. If the valve does not fall smoothly, replace the relief valve. If necessary, replace the oil pump assembly.

3. Inspect the oil pump rotor.

a. Coat the drive rotor and driven rotor with engine oil.

b. Place the drive and driven rotors into the oil pump with the marks facing the pump cover side.

c. Check the side clearance.

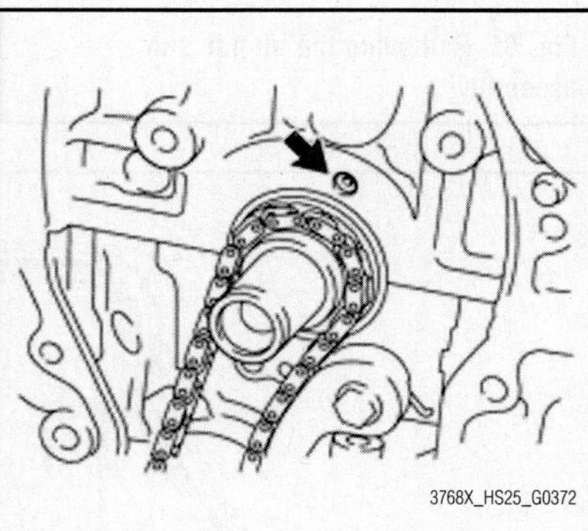

Fig. 85 Inspecting the oil jet

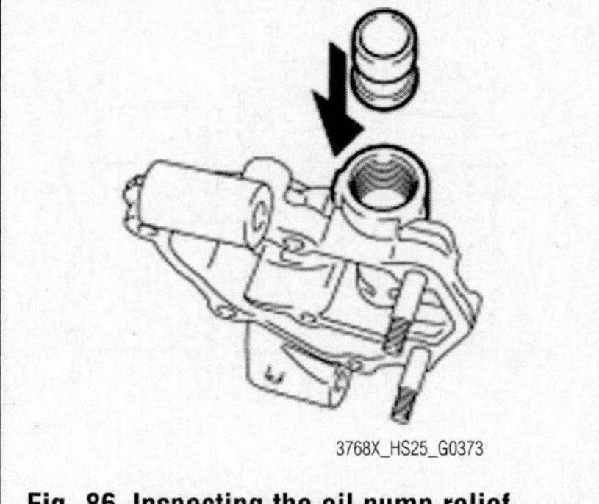

Fig. 86 Inspecting the oil pump relief valve

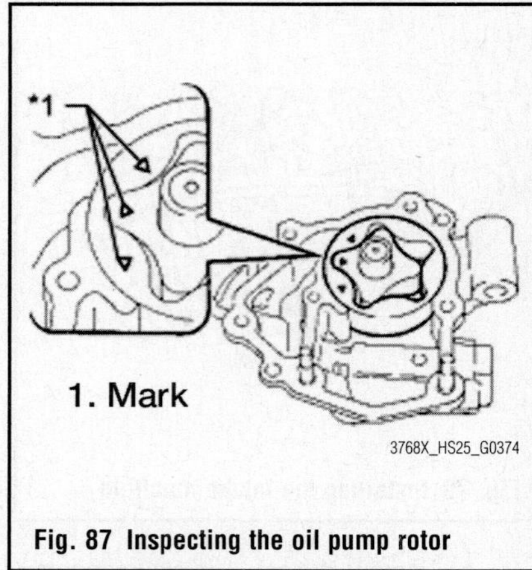

Fig. 87 Inspecting the oil pump rotor

d. Using a feeler gauge and precision straight edge, measure the clearance between the rotors and precision straight edge.

Standard clearance:

- 0.00118-0.00335 inch (0.030-0.085 mm)

Maximum clearance:

- 0.00630 inch (0.16 mm)

➡If the side clearance is greater than the maximum, replace the oil pump assembly.

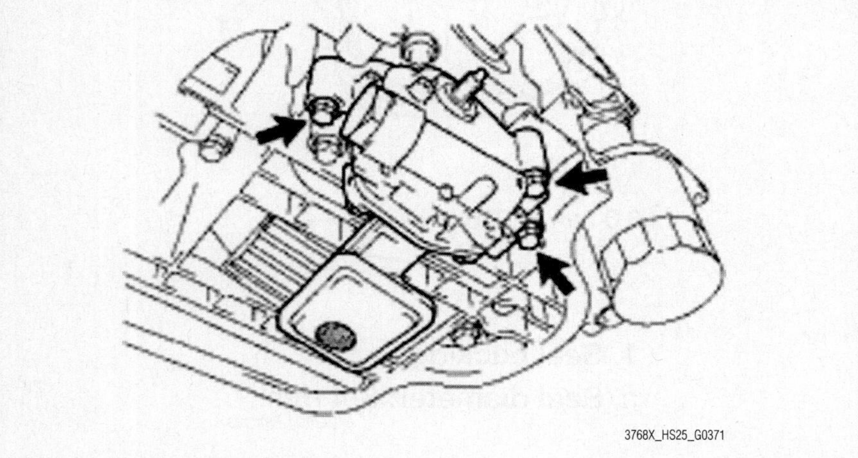

Fig. 84 Removing the oil pump assembly

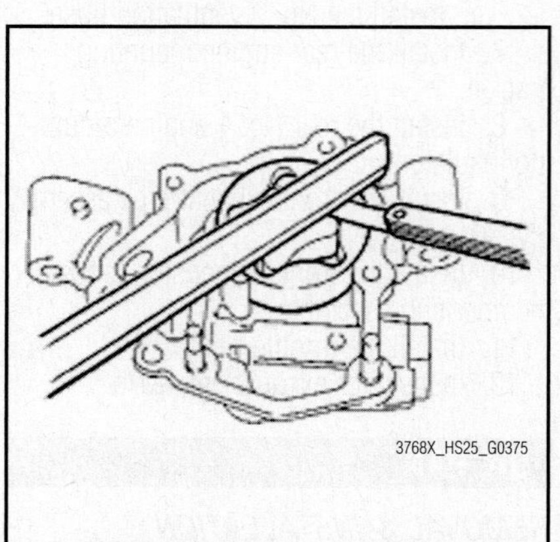

Fig. 88 Checking the side clearance

a. Check the tip clearance.

b. Using a feeler gauge, measure the clearance between the drive and driven rotor tips.

Standard clearance:

- 0.00315-0.00630 inch (0.080-0.160 mm)

Maximum clearance:

- 0.0138 inch (0.35 mm)

➡If the tip clearance is greater than the maximum, replace the oil pump assembly.

a. Check the body clearance.

b. Using a feeler gauge, measure the clearance between the driven rotor and pump body.

Standard clearance:

- 0.00394-0.00669 inch

Maximum clearance:

- 0.0128 inch (0.325 mm)

➡If the body clearance is greater than the maximum, replace the oil pump assembly.

4. Inspect the No. 2 chain sub assembly.
5. Inspect the oil pump drive gear.
6. Inspect the oil pump drive shaft gear.
7. Inspect the chain tensioner plate.

PISTON AND RING

POSITIONING

See Figures 91 through 93.

1. Install the oil ring expander and oil ring rail by hand.

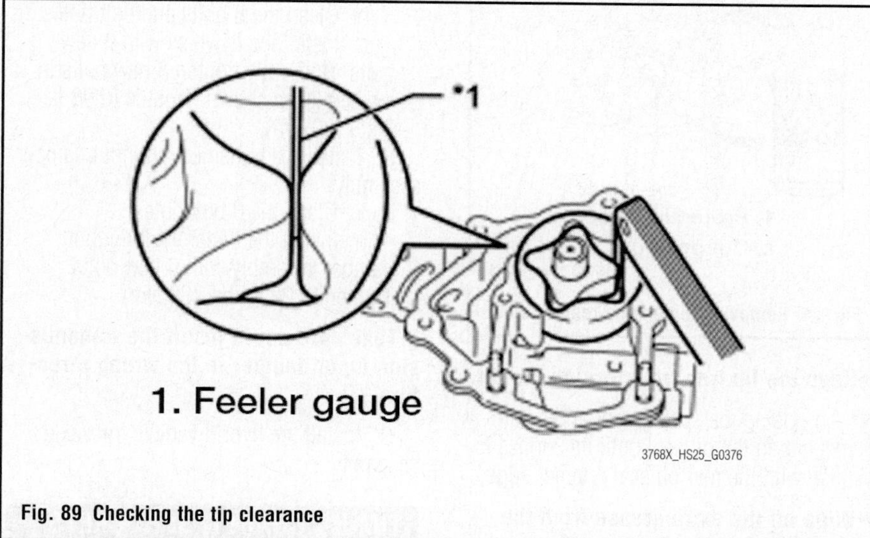

Fig. 89 Checking the tip clearance

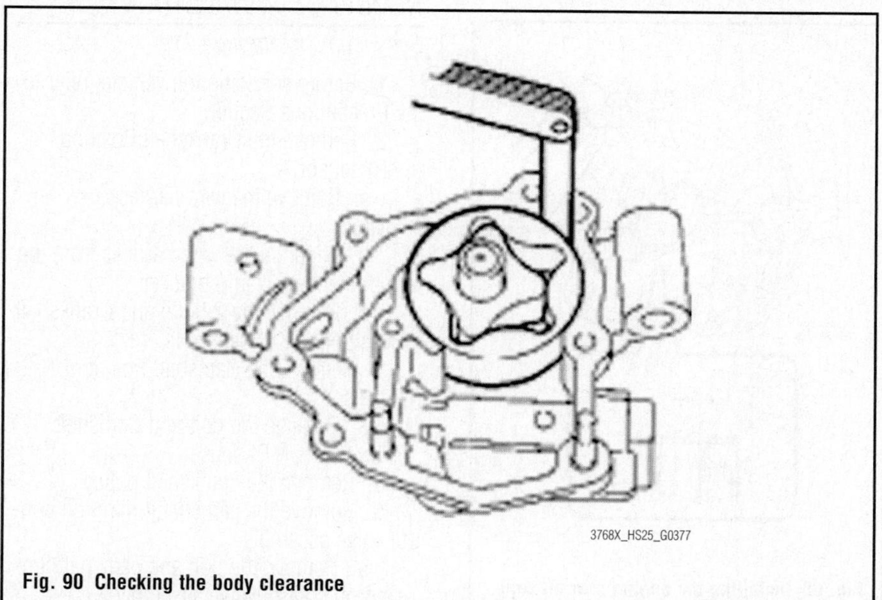

Fig. 90 Checking the body clearance

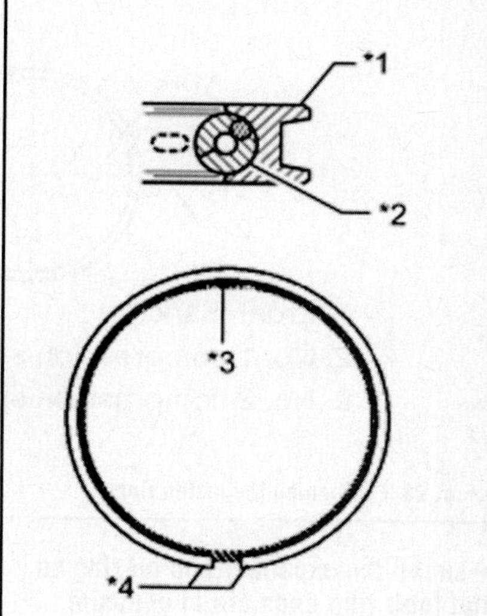

Fig. 91 Expanding the oil ring rail

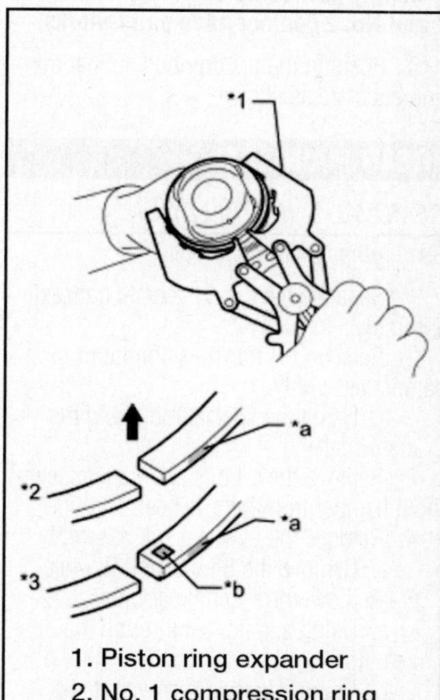

Fig. 92 Installing the 2 compression rings

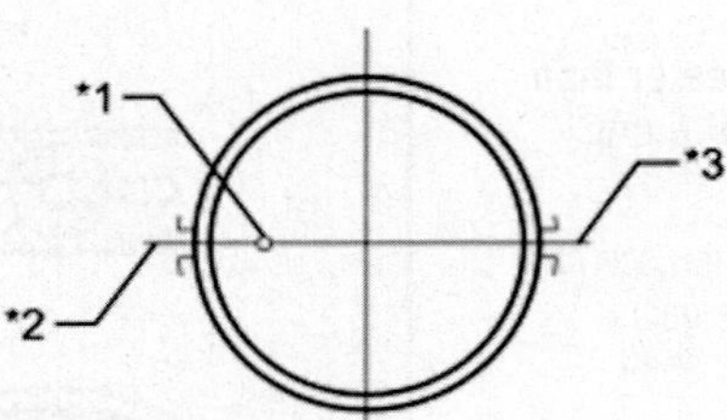

Fig. 93 Positioning the piston rings

➡Install the expander and oil ring so that their ring ends are at opposite sides.

➡Securely install the expander into the inner groove of the oil ring.

2. Using a piston ring expander, install the 2 compression rings so that the paint marks are positioned as shown.

➡Install the No. 2 compression ring with the code mark (2N) facing upward.

➡Supply part compression springs (No. 1 and No. 2) do not have paint marks.

3. Position the piston rings so that the ring ends are as shown.

REAR MAIN SEAL

REMOVAL & INSTALLATION

See Figures 94 through 96.

1. Separate the hybrid vehicle transaxle assembly.
2. Remove the transmission input damper assembly.
   a. Using the special tool, hold the crankshaft.
3. Remove the 6 bolts and transmission input damper from the flywheel.
4. Remove the flywheel sub assembly.
   a. Remove the 8 bolts and flywheel.
5. Remove the engine rear oil seal.
   a. Using a knife, cut through the oil seal lip.
   b. Using a screwdriver with its tip taped, pry out the oil seal.

➡After removal, check the crankshaft for damage. If damaged, smooth the surface with 400 grit sandpaper.

To install:

6. Install the engine rear oil seal.
   a. Apply MP grease to a new oil seal lip.

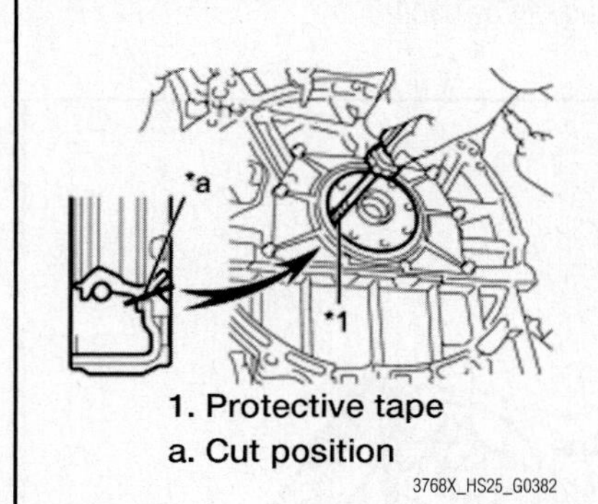

Fig. 94 Removing the engine rear oil seal

➡Keep the lip free from foreign matter.

b. Using the special tool and a hammer, tap in the oil seal until its surface is flush with the rear oil seal retainer edge.

➡Wipe off the extra grease from the crankshaft.

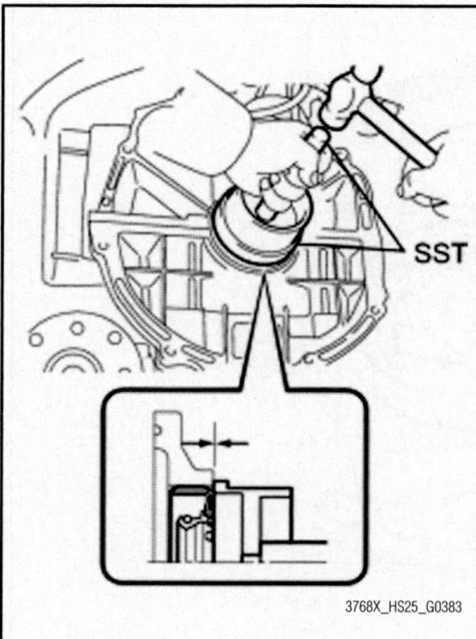

Fig. 95 Installing the engine rear oil seal

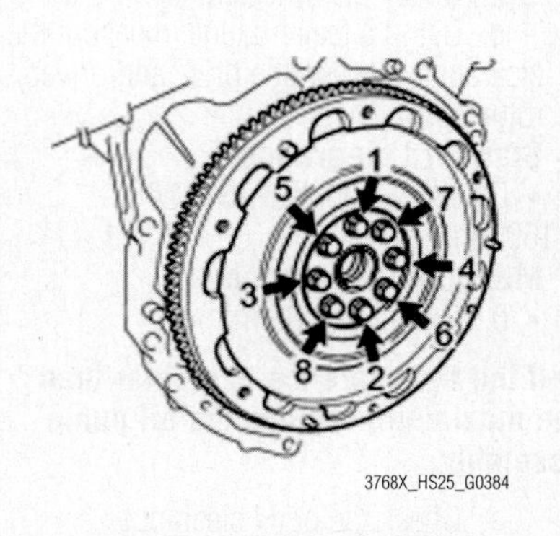

Fig. 96 Identifying the flywheel bolt tightening sequence

7. Install the flywheel sub assembly.
   a. Using the special tool, hold the crankshaft.
   b. Clean the 8 bolts and bolt holes.
   c. Install the flywheel with 8 new bolts. Uniformly tighten 8 new bolts in the sequence shown. Tighten to 96 ft. lbs. (130 Nm).
8. Install the transmission input damper assembly.
   a. Clean the 6 bolt holes.
   b. Install the transmission input damper assembly with 6 new bolts. Tighten to 22 ft. lbs. (30 Nm).

➡Take care not to insert the transmission input damper in the wrong direction.

9. Install the hybrid vehicle transaxle assembly.

TIMING CHAIN & SPROCKETS

REMOVAL & INSTALLATION

See Figures 97 through 119.

1. Before servicing the vehicle, refer to the Precautions Section.
2. Remove the Crankshaft Position (CKP) sensor.
   a. Remove the wire harness clamp.
   b. Separate the wire harness from the wire harness clamp bracket.
   c. Remove the 2 bolts and Crankshaft Position (CKP) sensor.
3. Remove the Camshaft Position (CMP) sensor.
   a. Remove the bolt and Camshaft Position (CMP) sensor.
4. Remove the crankshaft pulley.
5. Remove the camshaft timing oil control valve assembly.
   a. Remove the bolt and camshaft timing oil control valve assembly.

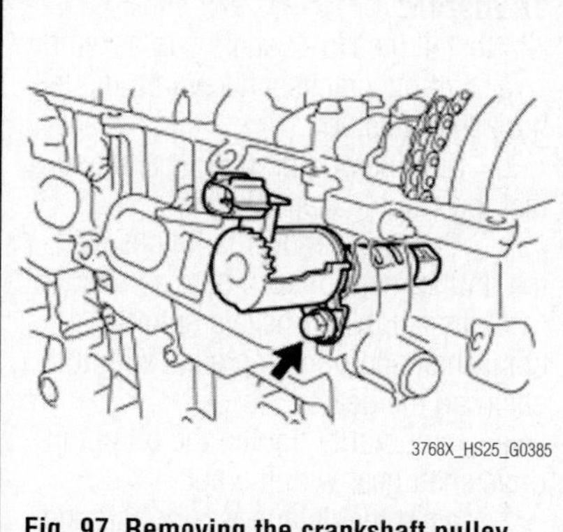

Fig. 97 Removing the crankshaft pulley

b. Remove the o-ring from the camshaft timing oil control valve assembly.

6. Remove the No. 1 chain tensioner assembly.

a. Remove the 2 nuts, No. 1 chain tensioner assembly and gasket.

➡Do not turn the crankshaft without the No. 1 chain tensioner assembly.

7. Remove the water pump pulley.
8. Remove the water pump assembly.
9. Remove the oil pan drain plug.
10. Remove the oil pan sub assembly.

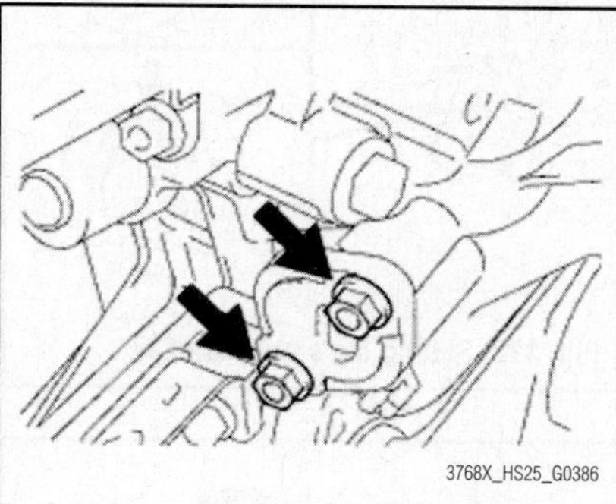

Fig. 98 Removing the No. 1 chain tensioner assembly

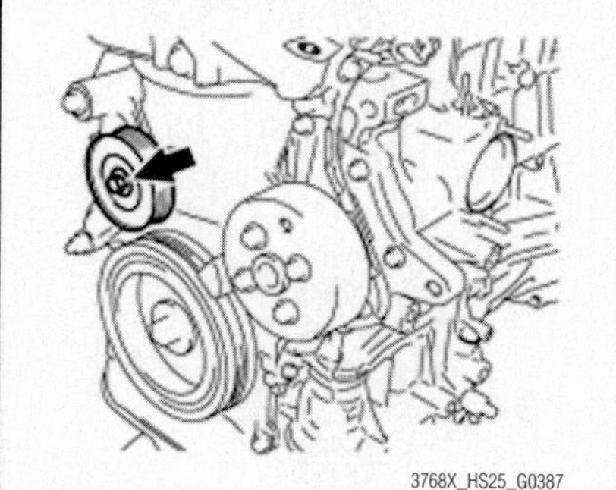

Fig. 99 Removing the idler pulley sub assembly

11. Remove the idler pulley sub assembly.
12. Remove the idler pulley bracket.
13. Remove the timing chain cover sub assembly.

a. Remove the 3 bolts and engine mounting bracket RH.

b. Using an E10 TORX®.$ socket, remove the stud bolt for the v-ribbed belt tensioner.

c. Remove the timing chain cover by prying the portions between the timing chain cover, cylinder head and cylinder block with a screwdriver.

➡Be careful not to damage the contact surfaces of the timing chain cover, cylinder head or cylinder block.

➡Tape the screwdriver tip before use.

14. Remove the timing chain case oil seal.

a. Using a screwdriver and a hammer, remove the oil seal.

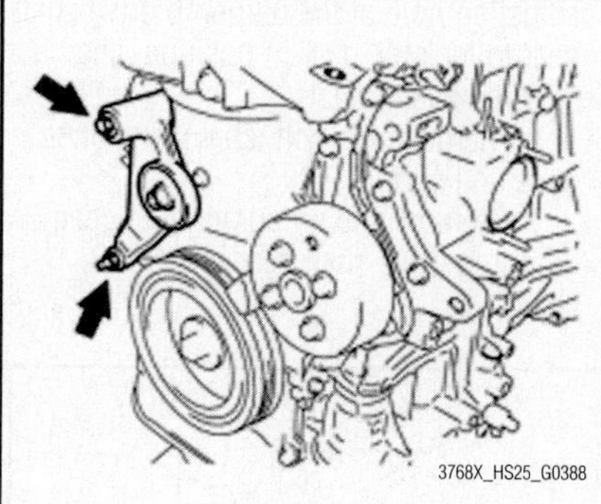

Fig. 100 Removing the idler pulley bracket

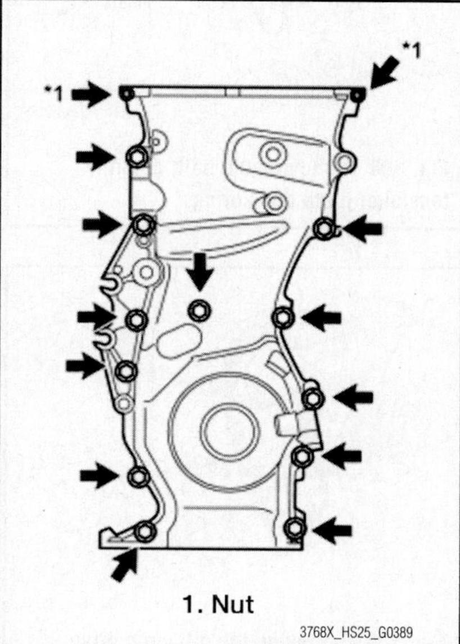

Fig. 101 Removing the time chain cover sub assembly bolts and nuts

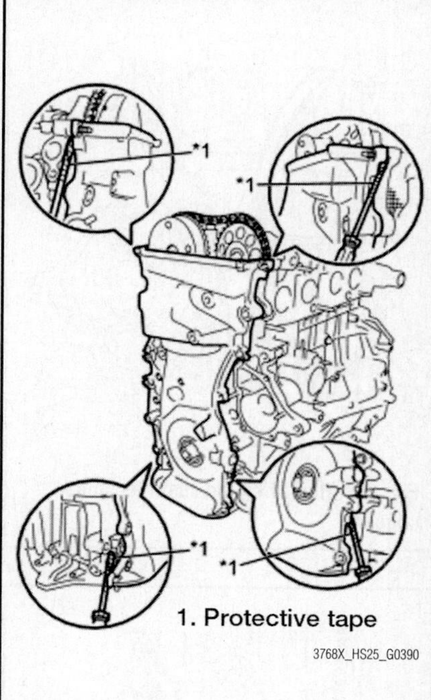

Fig. 102 Removing the timing chain cover

➡Tape the screwdriver tip before use.

15. Remove the No. 1 Crankshaft Position (CKP) sensor plate.
16. Remove the timing chain guide.
17. Remove the chain tensioner slipper.
18. Remove the No. 1 chain vibration damper.
19. Remove the chain sub assembly.
20. Remove the crankshaft timing sprocket.
21. Remove the No. 2 chain sub assembly.

a. Turn the crankshaft 90°counter-clockwise to align the adjusting hole of the oil pump drive shaft gear with the groove of the oil pump.

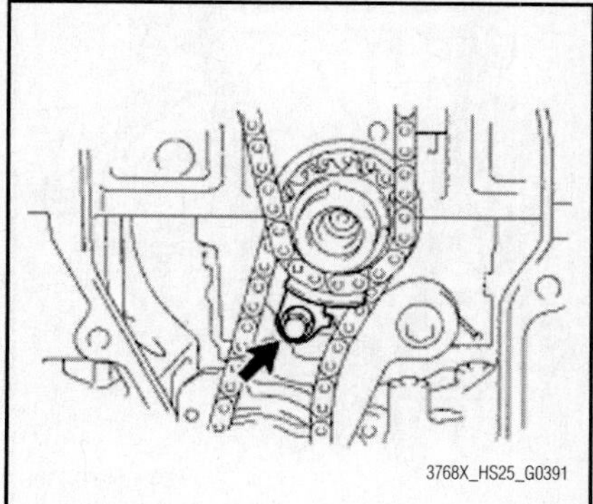

Fig. 103 Removing the timing chain guide

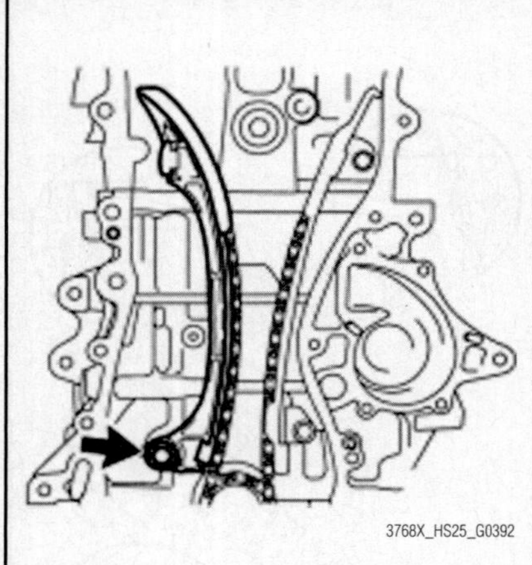

Fig. 104 Removing the chain tensioner slipper

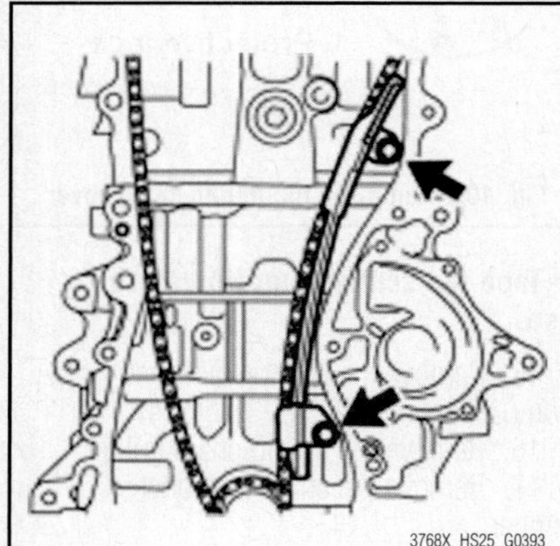

Fig. 105 Removing the No. 1 chain vibration damper

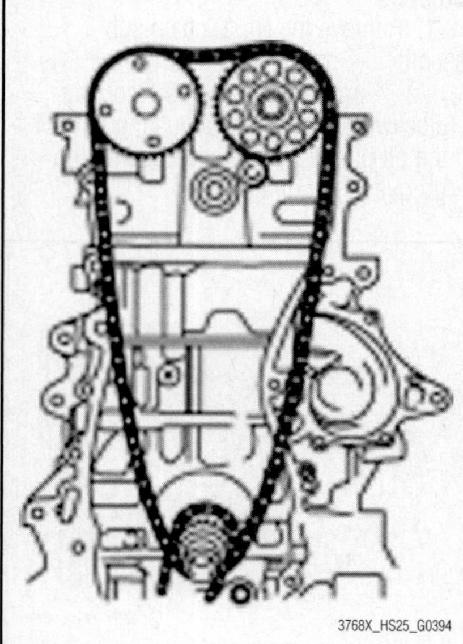

Fig. 106 Removing the chain sub assembly

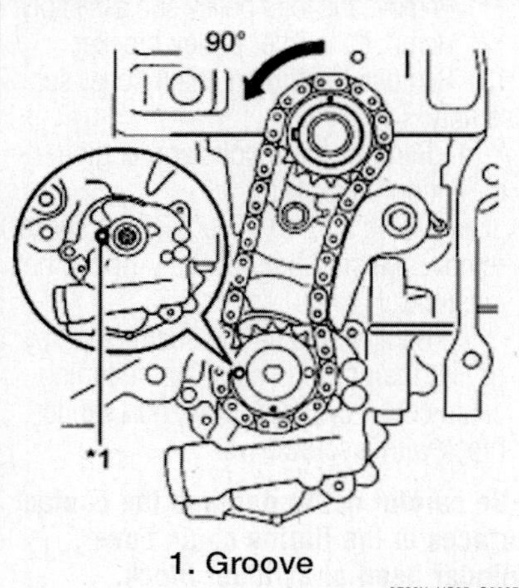

Fig. 107 Aligning the adjusting hole of the oil pump drive shaft gear with the groove of the oil pump

b. Insert a 4 mm diameter bar into the adjusting hole of the oil pump drive shaft gear to lock the gear in position, and then remove the nut.

c. Remove the bolt, chain tensioner plate and spring.

d. Remove the oil pump drive gear, oil pump drive shaft gear and No. 2 chain.

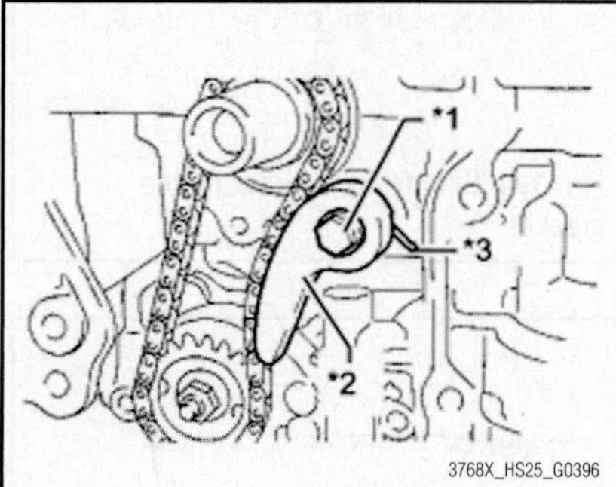

Fig. 108 Removing the bolt, chain tensioner plate and spring

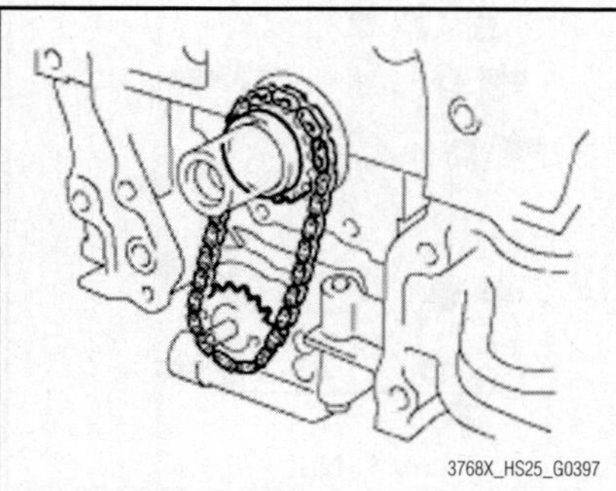

Fig. 109 Removing the oil pump drive gear, oil pump drive shaft gear and No. 2 chain

To install:

22. Install the No. 2 chain sub assembly.

a. Set the crankshaft key into the left horizontal position.

b. Turn the oil pump drive shaft so that the cutout faces upward.

c. Align the yellow mark links with the timing mark of each gear as shown.

d. Install the sprockets onto the crankshaft and oil pump shaft with the chain on the gears.

e. Temporarily tighten the oil pump drive shaft gear with the nut.

f. Insert the damper spring into the adjusting hole, and then install the chain tensioner plate with the bolt. Tighten to 9 ft. lbs. (12 Nm).

g. Align the adjusting hole of the oil pump drive shaft gear with the groove of the oil pump.

h. Insert a 4 mm diameter bar into the adjusting hole of the oil pump drive shaft gear to lock the gear in position, and then tighten the nut. Tighten to 22 ft. lbs. (30 Nm).

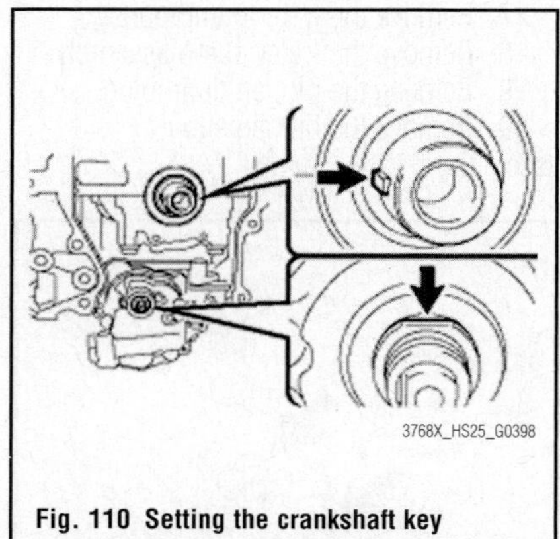

Fig. 110 Setting the crankshaft key

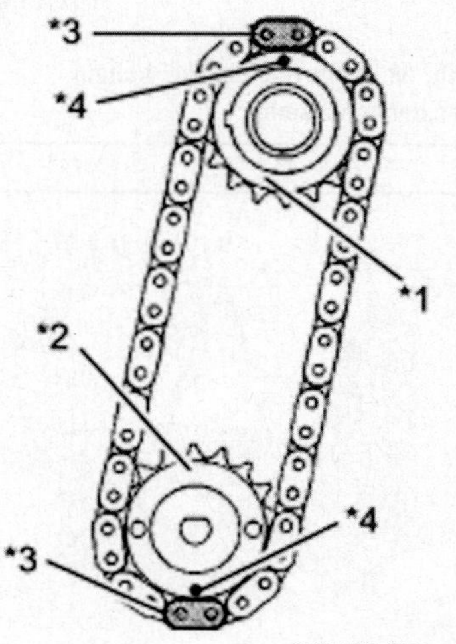

Fig. 111 Aligning the timing marks

i. Rotate the crankshaft 90° clockwise, and position the crankshaft key to face up as shown.

23. Install the crankshaft timing sprocket.

24. Install the No. 1 chain vibration damper with the 2 bolts. Tighten to 80 inch lbs. (9 Nm).

25. Install the chain sub assembly.

a. Set the No. 1 cylinder to TDC/compression.

b. Turn the camshafts with a wrench (using the hexagonal lobe) to align the timing marks of the camshaft timing gears with the timing marks located on the No. 1 and No. 2 bearing caps as shown.

c. Check that the crankshaft is positioned with the crankshaft key facing up.

d. Install the chain onto the crankshaft timing sprocket with the gold or pink mark link aligned with the timing mark on the crankshaft.

e. Using SST and a hammer, tap in the crankshaft timing sprocket.

f. Align the gold or yellow mark links with each timing mark located on the camshaft timing gear and sprocket, then install the chain.

26. Install the chain tensioner slipper. Tighten the bolt to 14 ft. lbs. (19 Nm).

27. Install the timing chain guide with the bolt. Tighten to 80 inch lbs. (9 Nm).

28. Install the No. 1 Crankshaft Position (CKP) sensor plate with the "F" mark facing forward.

29. Install the timing chain cover sub assembly.

a. Remove any old packing (FIPG) material and be careful not to drop any oil on the contact surfaces of the timing chain cover, cylinder head or cylinder block.

b. Apply seal packing as shown.

c. Apply seal packing in a continuous bead.

➡Remove any oil from the contact surfaces.

➡Install the chain cover within 3 minutes of applying seal packing.

➡Do not start the engine for at least 2 hours after installation.

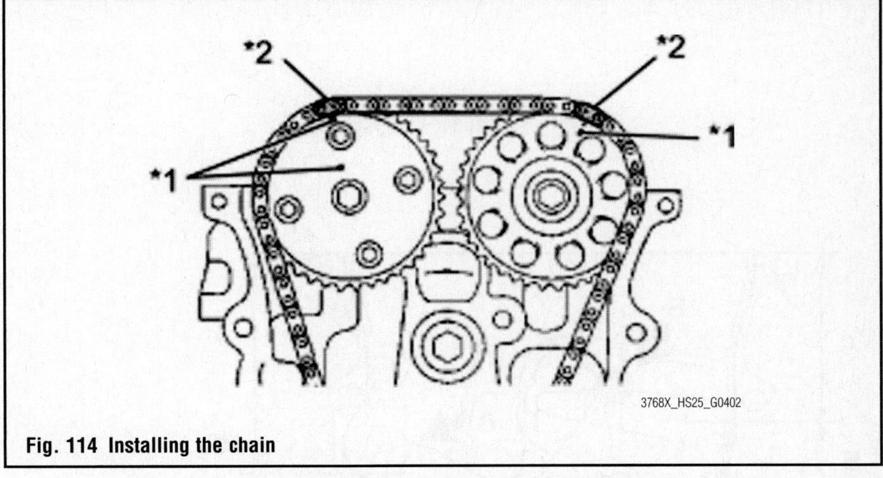

Fig. 114 Installing the chain

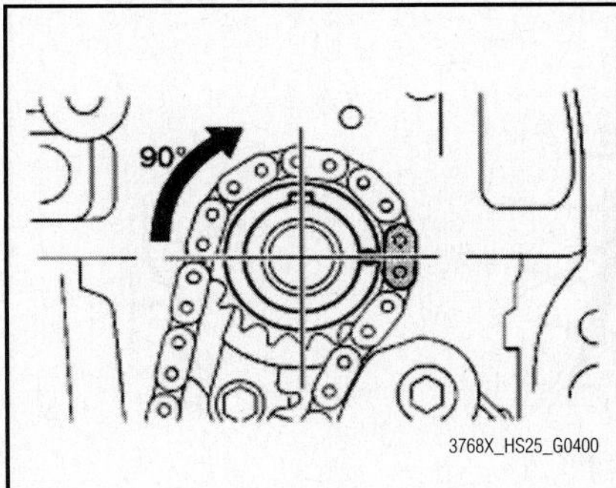

Fig. 112 Rotating the crankshaft key to face up

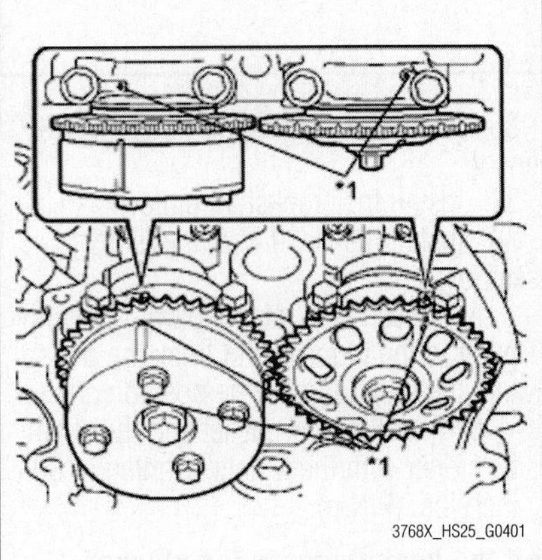

Fig. 113 Aligning the camshaft timing gear timing marks

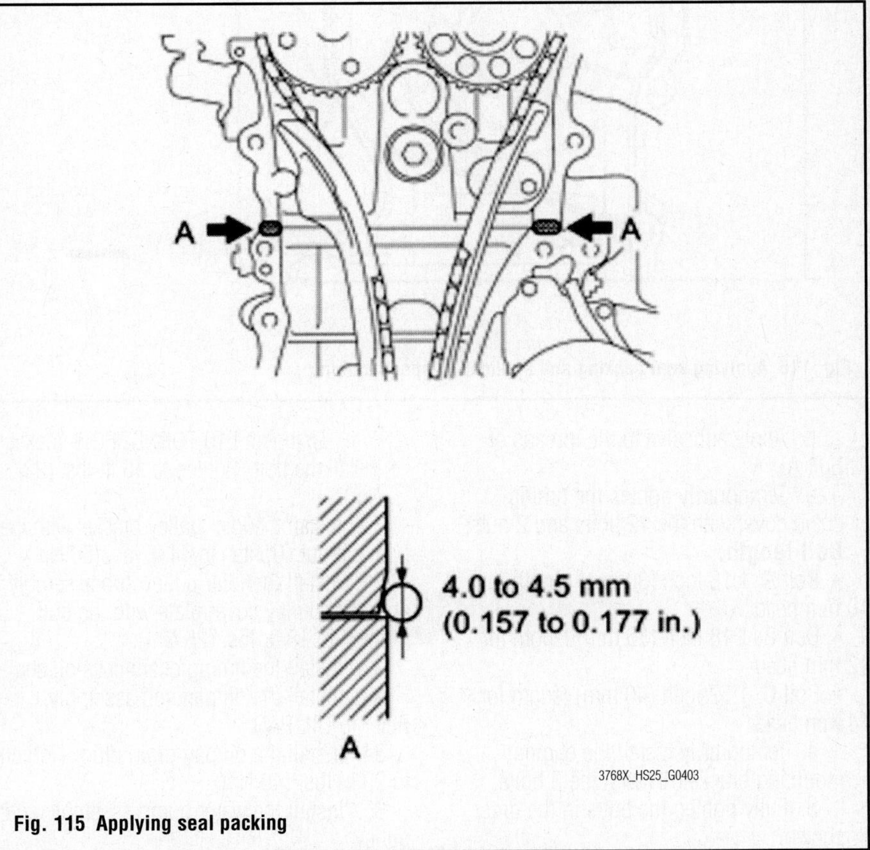

Fig. 115 Applying seal packing

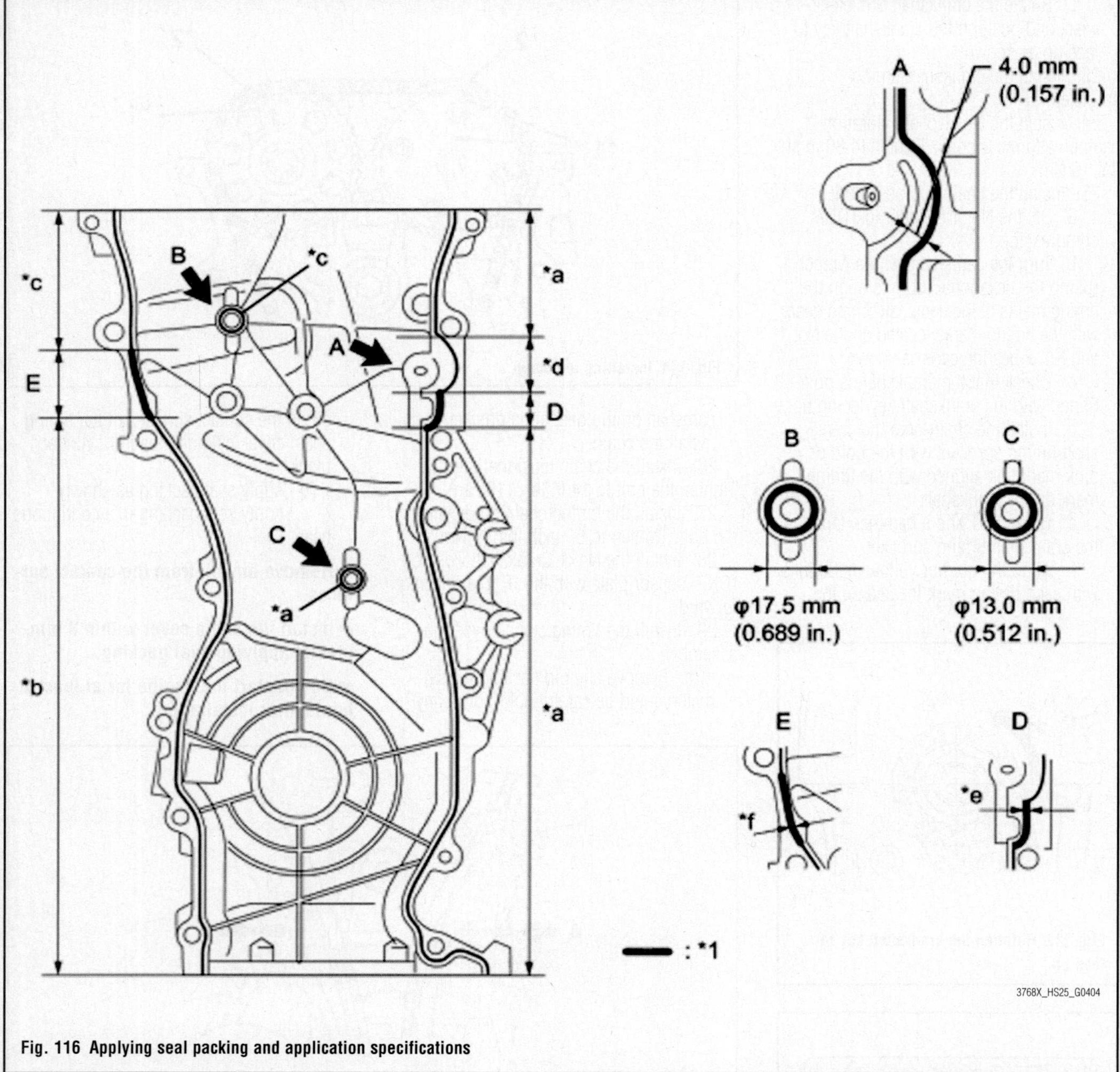

Fig. 116 Applying seal packing and application specifications

d. Apply adhesive to the threads of bolt A.

e. Temporarily tighten the timing chain cover with the 12 bolts and 2 nuts.

Bolt length:

- Bolt A: 1.18 inch (30 mm) length for 10 mm head
- Bolt B: 1.18 inch (30 mm) length for 12 mm head
- Bolt C: 1.57 inch (40 mm) length for 14 mm head

a. Temporarily install the engine mounting bracket RH with the 3 bolts.

b. Fully tighten the bolts in the order shown.

c. Using an E10 TORX®.$ socket, install the bolt. Tighten to 16 ft. lbs. (22 Nm).

30. Install the idler pulley bracket with the bolt and nut. Tighten to 44 ft. lbs. (60 Nm).

31. Install the idler pulley sub assembly and idler pulley cover plate with the bolt. Tighten to 18 ft. lbs. (25 Nm).

32. Install the timing chain case oil seal.

33. Install the oil pan sub assembly. Refer to OIL PAN.

34. Install the oil pan drain plug. Tighten to 29 ft. lbs. (40 Nm).

35. Install the water pump assembly and pulley.

36. Install the Crankshaft Position (CKP) sensor.

37. Install the crankshaft pulley.

38. Install the No. 1 chain tensioner assembly.

a. Release the ratchet pawl, then fully push in the plunger and hook the pin so that the plunger is in the position shown.

b. Install a new gasket and the chain tensioner with the 2 nuts. Tighten to 80 inch lbs. (9 Nm).

➡If the hook releases the plunger while the chain tensioner is being installed, set the hook again.

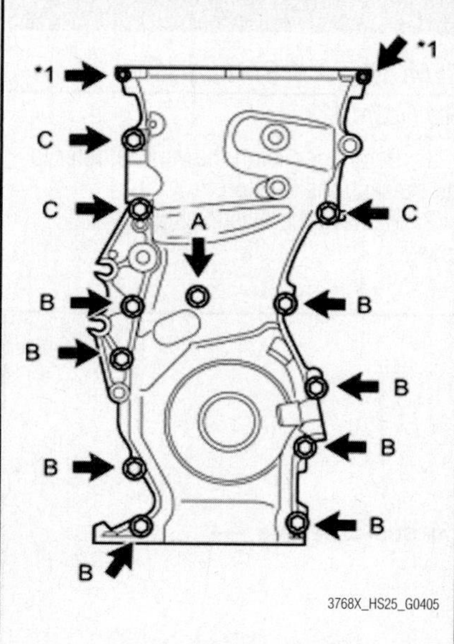

Fig. 117 Identifying different bolt and size locations

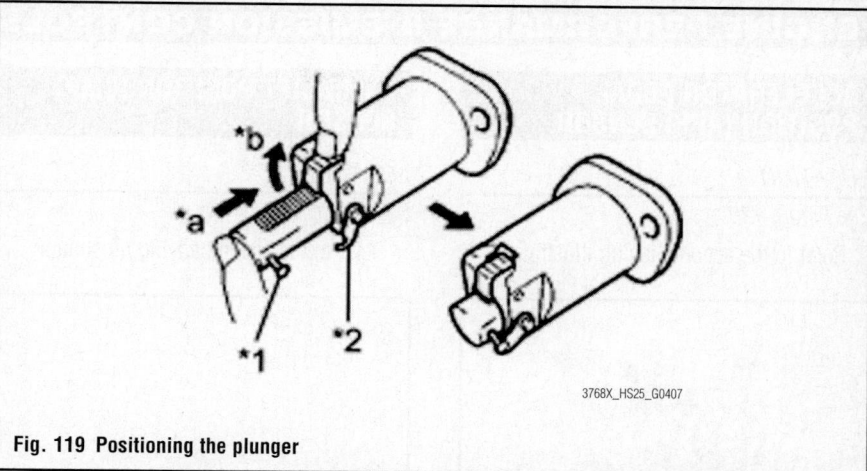

Fig. 119 Positioning the plunger

c. Turn the crankshaft counterclockwise, then disconnect the hook from the plunger knock pin.

d. Turn the crankshaft clockwise, then check that the plunger is extended.

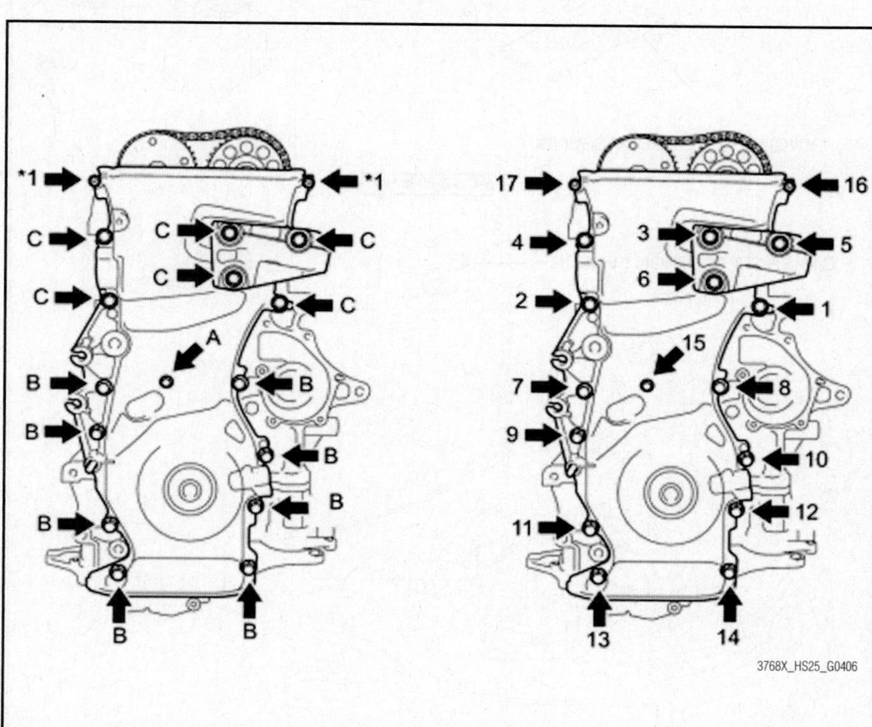

Fig. 118 Installing the timing chain sub assembly and tightening specifications

VALVE LASH

ADJUSTMENT

➡**Keep the lash adjuster free of dirt and foreign objects.**

➡**Only use clean engine oil.**

1. Place the lash adjuster into a container filled with engine oil.
2. Insert the SST's (SST: 09276-75010) tip into the lash adjuster's plunger and use the tip to press down on the check ball inside the plunger.
3. Squeeze the SST and lash adjuster together to move the plunger up and down 5 to 6 times.
4. Check the movement of the plunger and bleed the air. OK: Plunger moves up and down.

➡**When bleeding air from the high-pressure chamber, make sure that the tip of the SST is actually pressing the check ball as shown in the illustration. If the check ball is not pressed, air will not bleed.**

5. After bleeding the air, remove the SST. Then, try to press the plunger quickly and firmly with a finger. OK: Plunger is very difficult to move. If the result is not as specified, replace the lash adjuster.
6. Install the lash adjusters.

➡**Install the lash adjuster to the same place where it was removed from.**

ENGINE PERFORMANCE & EMISSION CONTROLS

ACCELERATOR PEDAL POSITION (APP) SENSOR

LOCATION

See Figure 120.

Refer to the accompanying illustration.

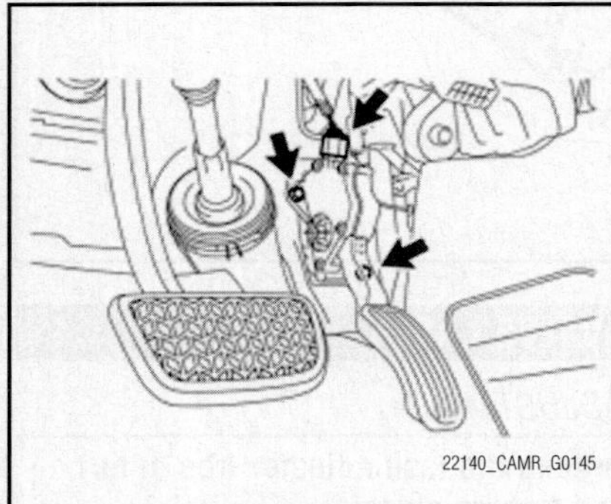

Fig. 120 Accelerator pedal assembly location

REMOVAL & INSTALLATION

See Figure 121.

1. Before servicing the vehicle, refer to the Precautions Section.
2. Remove left center floor carpet cover.
3. Disconnect the accelerator pedal connector.
4. Remove the 2 nuts and accelerator pedal assembly.

➡Avoid physical shock to the accelerator pedal assembly.

➡Do not disassemble the accelerator pedal assembly.

To install:

5. Installation is the reverse of the removal procedure. Tighten the accelerator pedal rod nuts to 43 inch lbs. (5.4 Nm).

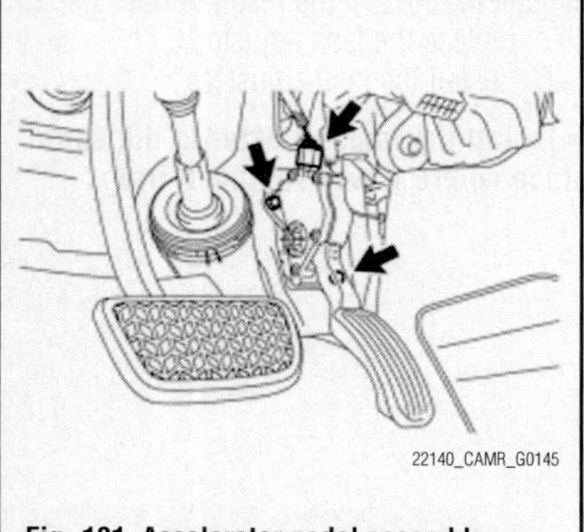

Fig. 121 Accelerator pedal assembly

CAMSHAFT POSITION (CMP) SENSOR

LOCATION

See Figure 122.

Refer to the accompanying illustration.

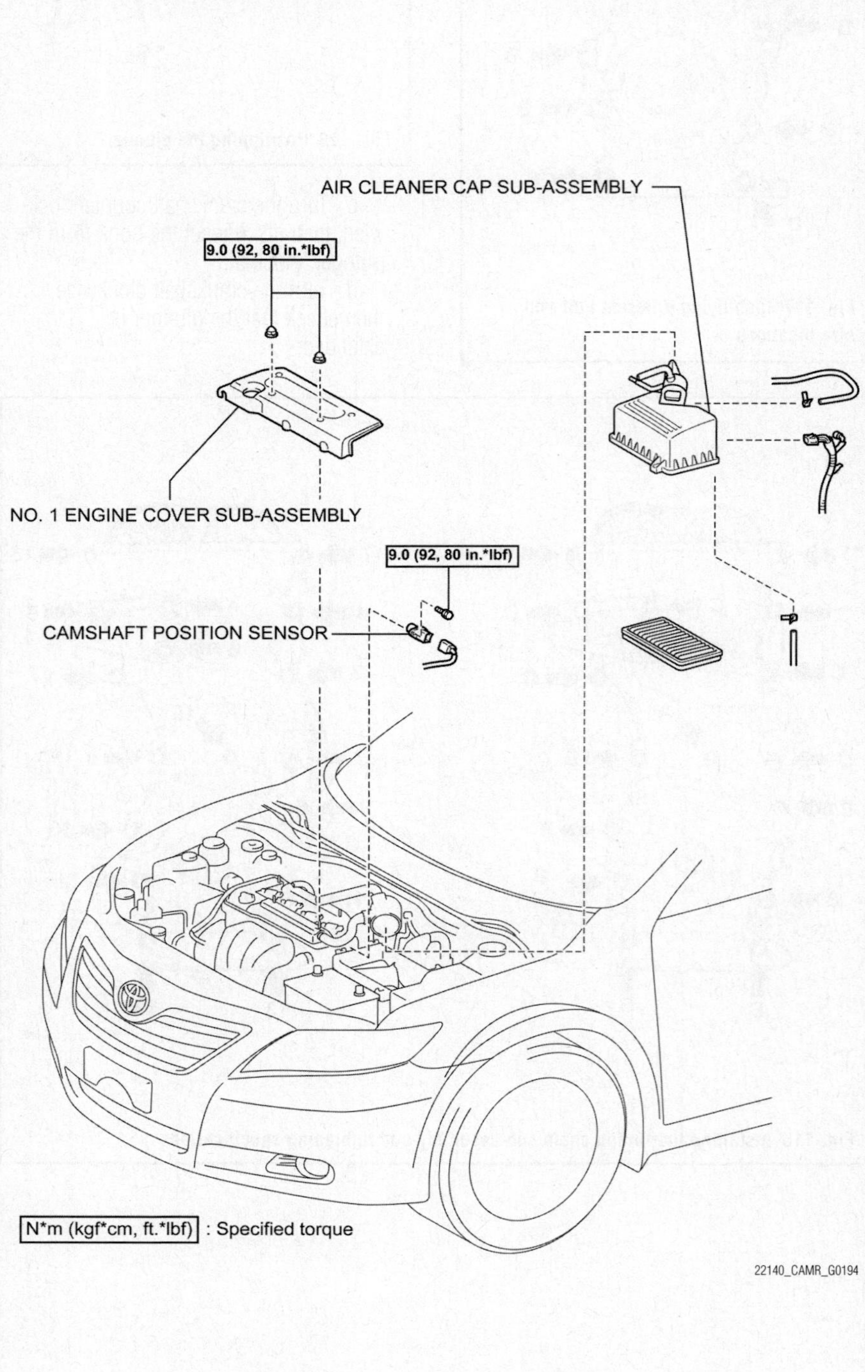

Fig. 122 Camshaft Position (CMP) sensor

REMOVAL & INSTALLATION

See Figure 123.

1. Before servicing the vehicle, refer to the Precautions Section.
2. Remove the plastic engine cover.

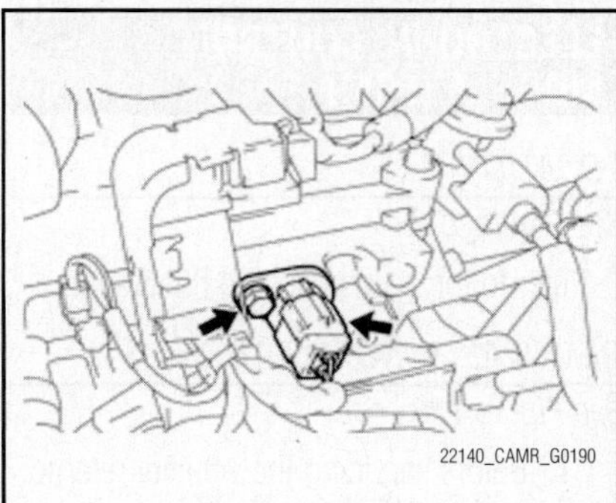

Fig. 123 Remove the bolt and Camshaft Position (CMP) sensor

3. Remove air cleaner cap sub-assembly.
4. Disconnect the Camshaft Position (CMP) sensor connector.
5. Remove the bolt and Camshaft Position (CMP) sensor.

To install:

➡Make sure that the O-ring is not cracked or jammed when installing it.

6. Apply a light coat of engine oil to the O-ring of the sensor.
7. Install the Camshaft Position (CMP) sensor with the bolt and tighten to: 80 inch lbs. (9 Nm).
8. To complete installation, reverse removal procedure.

CRANKSHAFT POSITION (CKP) SENSOR

LOCATION

See Figure 124.

Refer to the accompanying illustration.

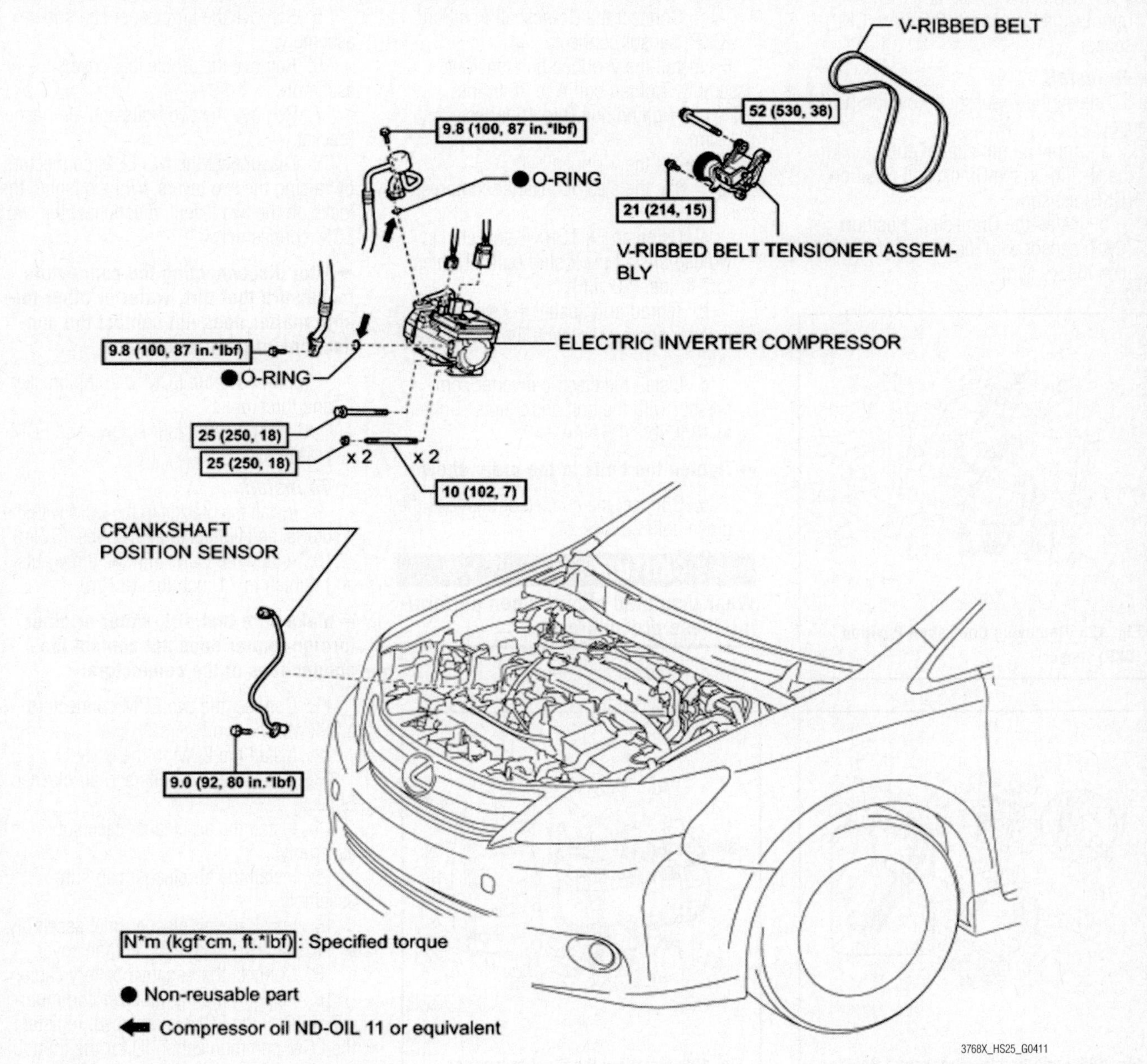

Fig. 124 Crankshaft Position (CKP) sensor location

REMOVAL & INSTALLATION

See Figures 125 through 128.

1. Remove the electric inverter compressor.
2. Remove the v-ribbed belt.
3. Remove the v-ribbed belt tensioner assembly.
4. Remove the Crankshaft Position (CKP) sensor.
   a. Disconnect the Crankshaft Position (CKP) sensor connector.
   b. Remove the connector clamp and wire harness clamp.
   c. Remove the wire harness clamp bracket from the wire harness.
   d. Remove the bolt, and then remove the Crankshaft Position (CKP) sensor.

To install:

5. Install the Crankshaft Position (CKP) sensor.
   a. Apply a light coat of engine oil to the O-ring on the Crankshaft Position (CKP) sensor.
   b. Install the Crankshaft Position (CKP) sensor with the bolt. Tighten to 80 inch lbs. (9 Nm).

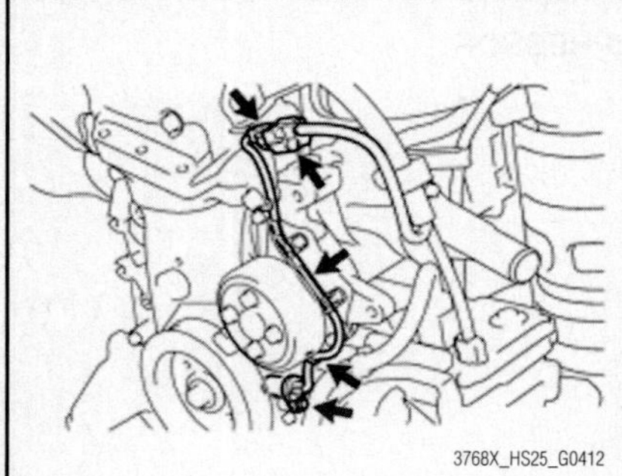

Fig. 125 Removing Crankshaft Position (CKP) sensor

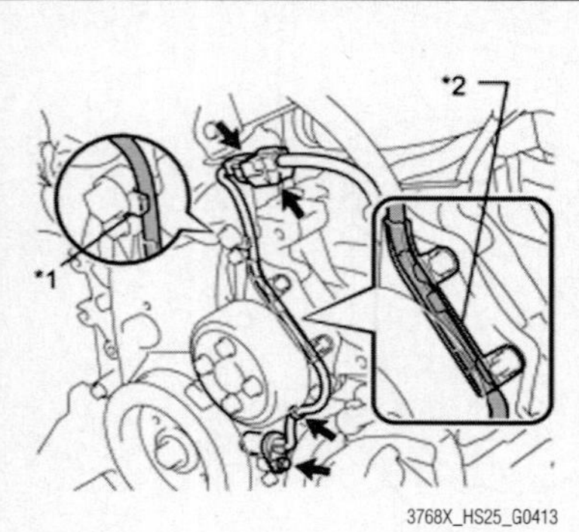

Fig. 126 Installing the Crankshaft Position (CKP) sensor

Fig. 127 Installing the v-ribbed belt tensioner

   c. Connect the Crankshaft Position (CKP) sensor connector.
6. Install the v-ribbed belt tensioner assembly. Tighten bolt A to 38 ft. lbs. (52 Nm). Tighten bolt B to 15 ft. lbs. (21 Nm).
7. Install the v-ribbed belt.
8. Install the electric inverter compressor.
   a. Using an E8 TORX® socket wrench, install the 2 stud bolts. Tighten to 7 ft. lbs. (10 Nm).
   b. Temporarily install the electric inverter compressor with the bolt and 2 nuts.
   c. Install the electric inverter compressor with the bolt and 2 nuts. Tighten to 18 ft. lbs. (25 Nm).

➡Tighten the bolts in the order shown.

   d. Connect the connector and lock the green colored lock.

✲✲ CAUTION

Wear insulated gloves when performing these procedures.

   e. Connect the connector.

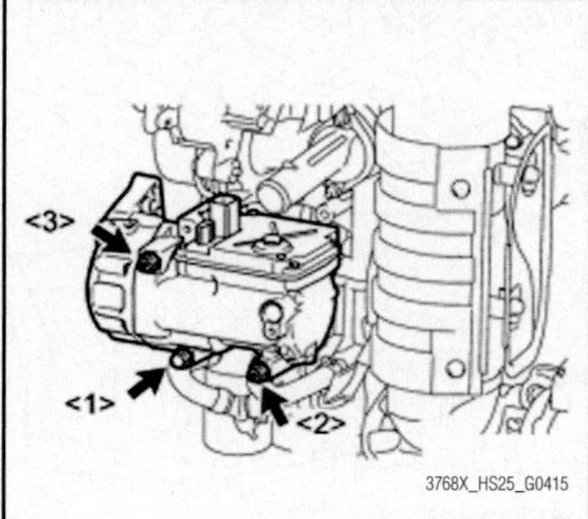

Fig. 128 Installing the electric inverter compressor

ELECTRONIC CONTROL MODULE (ECM)

LOCATION

See Figure 129.

Refer to the accompanying illustration.

REMOVAL & INSTALLATION

See Figure 130.

1. Before servicing the vehicle, refer to the Precautions Section.
2. Disconnect the negative battery cable.
3. Remove the plastic engine cover.
4. Remove the air cleaner inlet assembly.
5. Remove the air cleaner cap sub-assembly.
6. Remove the air cleaner case sub-assembly.
7. Remove the two bolts and air cleaner bracket.
8. Disconnect the two ECM connectors by raising the two levers. While pushing the locks on the two levers, disconnect the two ECM connectors.

➡After disconnecting the connectors, make sure that dirt, water or other foreign matter does not contact the connections of the connectors.

   a. Remove the ECM with the bracket and three bolts.
   b. Remove the four screws and ECM brackets.

To install:

9. Install the bracket to the ECM with the 4 screws, and tighten to 27 inch lbs. (3 Nm).
10. Attach the ECM with the three nuts and tighten to 71 inch lbs. (8 Nm).

➡Make sure that dirt, water or other foreign matter does not contact the connections of the connectors.

11. Connect the two ECM connectors and lower the two levers.
12. Install the ECM to the body.
13. Install the two bolts and air cleaner bracket.
14. Install the air cleaner case sub-assembly.
15. Install the air cleaner cap sub-assembly.
16. Install the air cleaner inlet assembly.
17. Remove the plastic engine cover.
18. Connect the negative battery cable.
19. Register the immobilizer communication ID. If the ECM is replaced, register the ECM communication ID for the immobilizer system (refer to the Service Bulletin for registration).

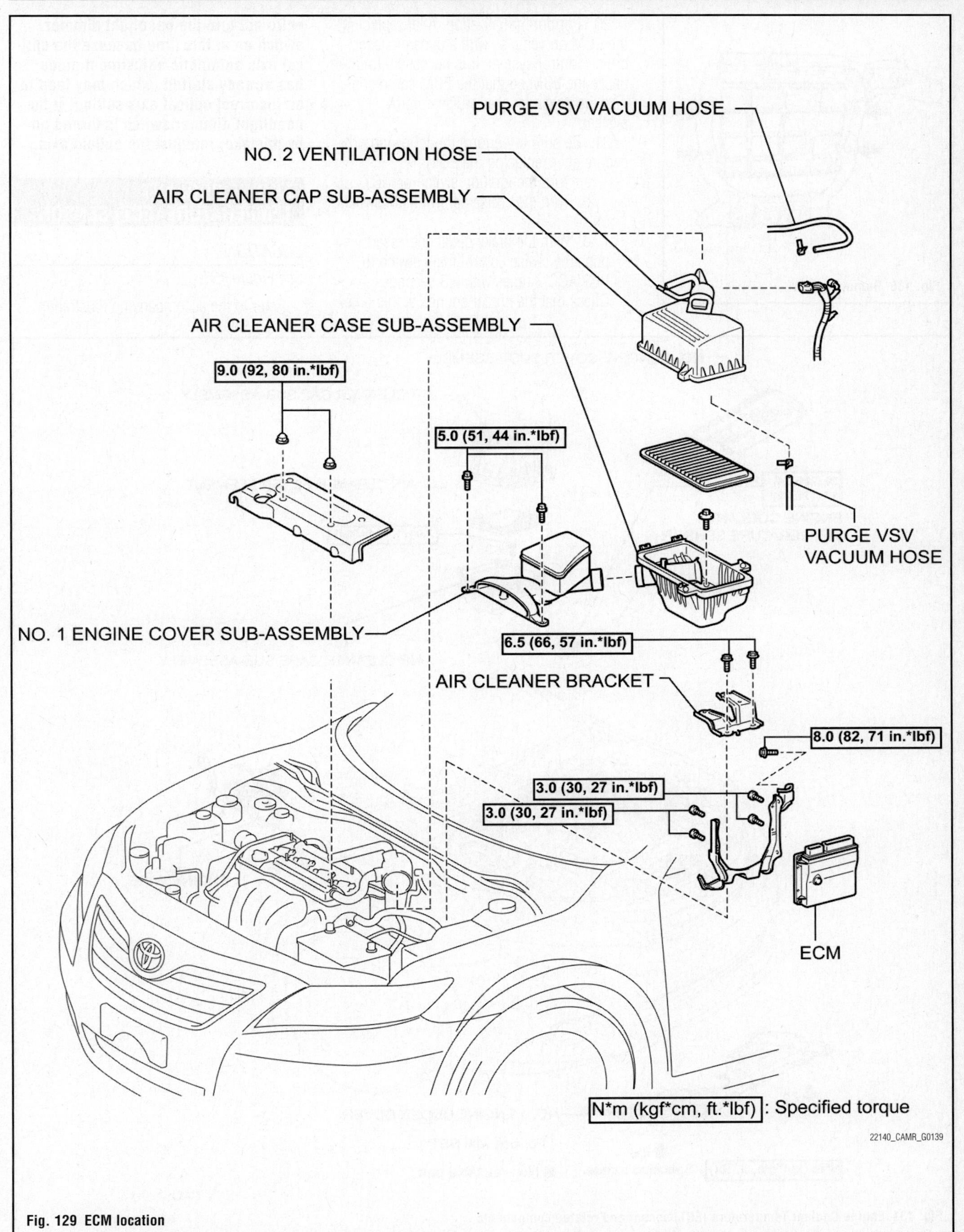

Fig. 129 ECM location

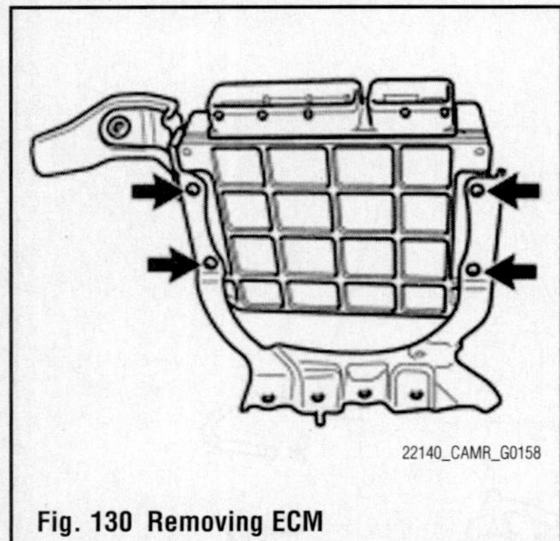

Fig. 130 Removing ECM

20. Perform initialization. After replacing the ECM on vehicles with a dynamic laser cruise control system, it is necessary to initialize the ECM so that the ECM can recognize the dynamic laser cruise control system.

21. Be sure to perform the following procedure after replacing the ECM:

a. Turn the ignition switch on (IG).

b. Turn the cruise control main switch on.

c. With the brake pedal depressed, push the cruise control main switch to RES/ACC 3 times within 3 seconds. Check that the buzzer sounds at this time.

➡Do not turn the headlight dimmer switch on at this time because the optical axis automatic adjustment mode has already started, which may lead to an incorrect optical axis setting. If the headlight dimmer switch is turned on by mistake, readjust the optical axis.

ENGINE COOLANT TEMPERATURE (ECT) SENSOR

LOCATION

See Figure 131.

Refer to the accompanying illustration.

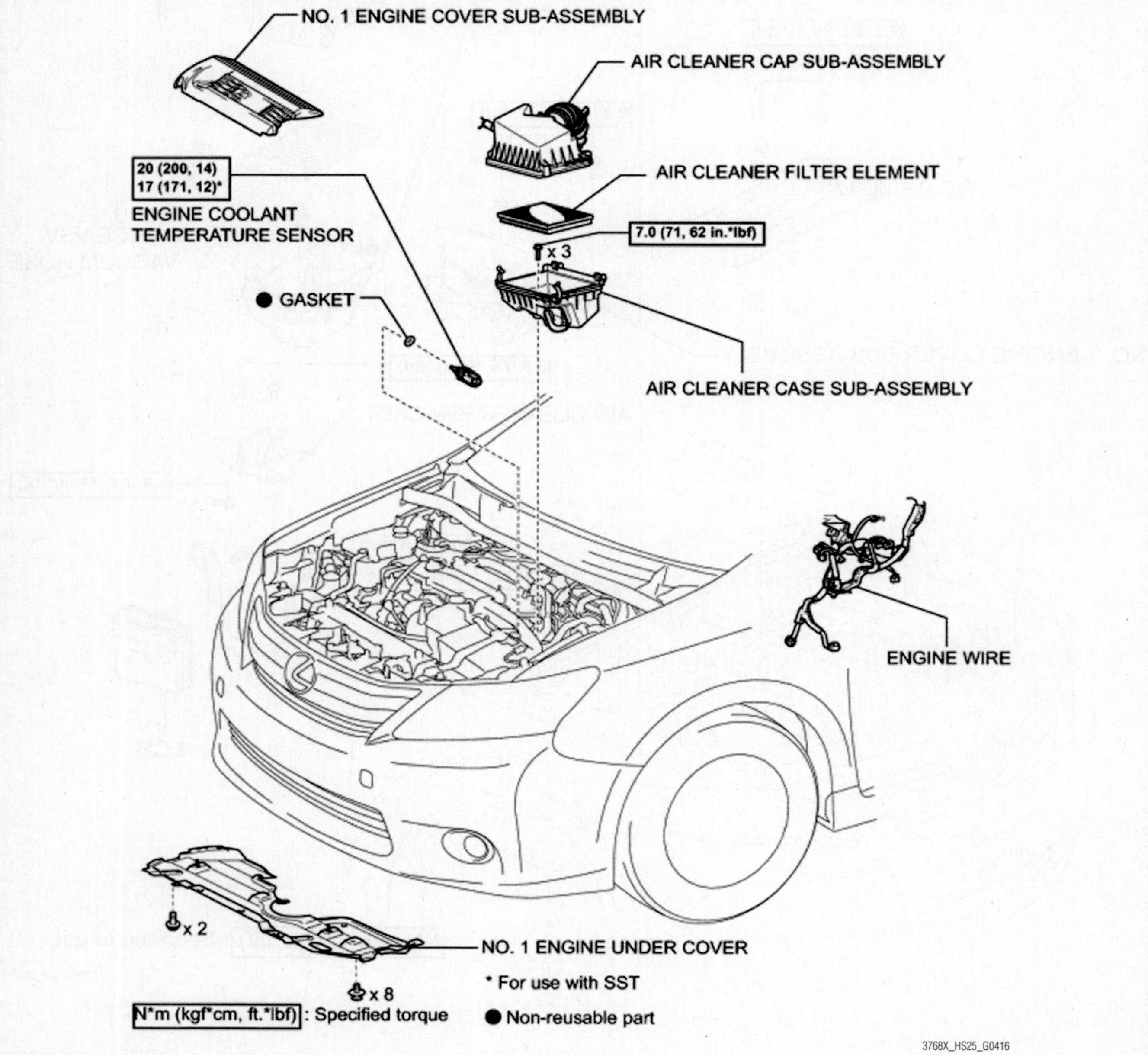

Fig. 131 Engine Coolant Temperature (ECT) sensor and related components

REMOVAL & INSTALLATION

See Figure 132.

1. Remove the No. 1 engine under cover.
2. Drain the engine coolant.
3. Remove the front wiper motor and link.
4. Remove the water guard plate RH.
5. Remove the No. 2 heater air duct splash shield seal.
6. Remove the cowl body mounting reinforcement RH.

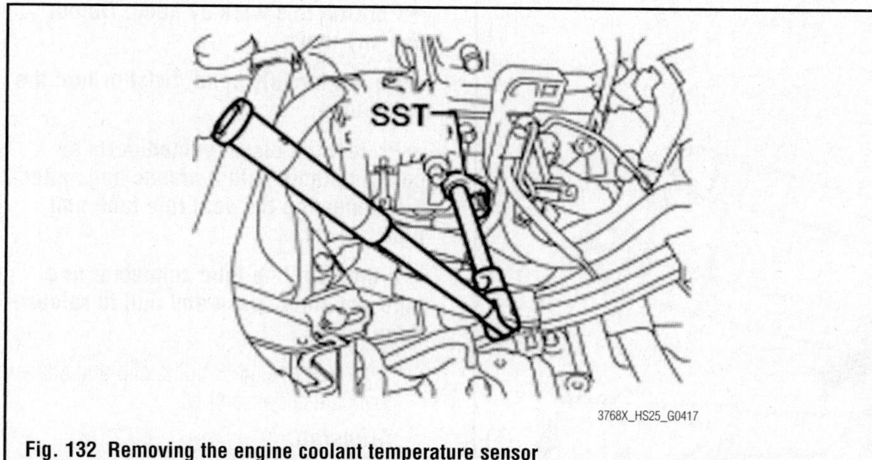

Fig. 132 Removing the engine coolant temperature sensor

7. Remove the outer cowl top panel sub assembly.
8. Remove the No. 1 engine cover sub assembly.
9. Remove the air cleaner cap sub assembly.
10. Remove the air cleaner case sub assembly.
11. Separate the engine wire.
    a. Remove the 2 bolts and disconnect the 2 ground cables.
    b. Disconnect the 4 connectors.
    c. Disconnect the connector clamp from the bracket.
12. Remove the engine coolant temperature sensor.
    a. Using the special tool, remove the engine coolant temperature sensor.

To install:

13. Install a new gasket onto the engine coolant temperature sensor.
    a. Using the special tool, install the engine coolant temperature sensor. Using the special tool tighten to 14 ft. lbs. (20 Nm). Without the special tool, tighten the sensor to 12 ft. lbs. (17 Nm).

➡The "with special tool" torque value is effective when using the special tool with a fulcrum length of 1.18 inch (30 mm).

➡The "with special tool" torque value is effective when using a torque wrench with a fulcrum length of 7.09 inch (180 mm).

➡The torque value is effective when the special tool is parallel to the torque wrench.

    b. Connect the engine coolant temperature sensor connector.

14. Install the engine wire.
    a. Install the connector clamp from the bracket.
    b. Connect the 4 connectors.
    c. Install the 2 bolts and disconnect the 2 ground cables.
15. Install the air cleaner case sub assembly.
16. Install the air cleaner cap sub assembly.
17. Add engine coolant.
18. Inspect for engine coolant leak.
19. Install the No. 1 engine under cover.
20. Install the outer cowl top panel sub assembly.
21. Install the cowl body mounting reinforcement RH.
22. Install the No. 2 heater air duct splash shield seal.
23. Install the water guard plate RH.
24. Install the front wiper motor and link.
25. Install the No. 1 engine cover sub assembly.

EVAPORATIVE EMISSIONS (EVAP) CANISTER

LOCATION

See Figure 133.

Refer to the accompanying illustration.

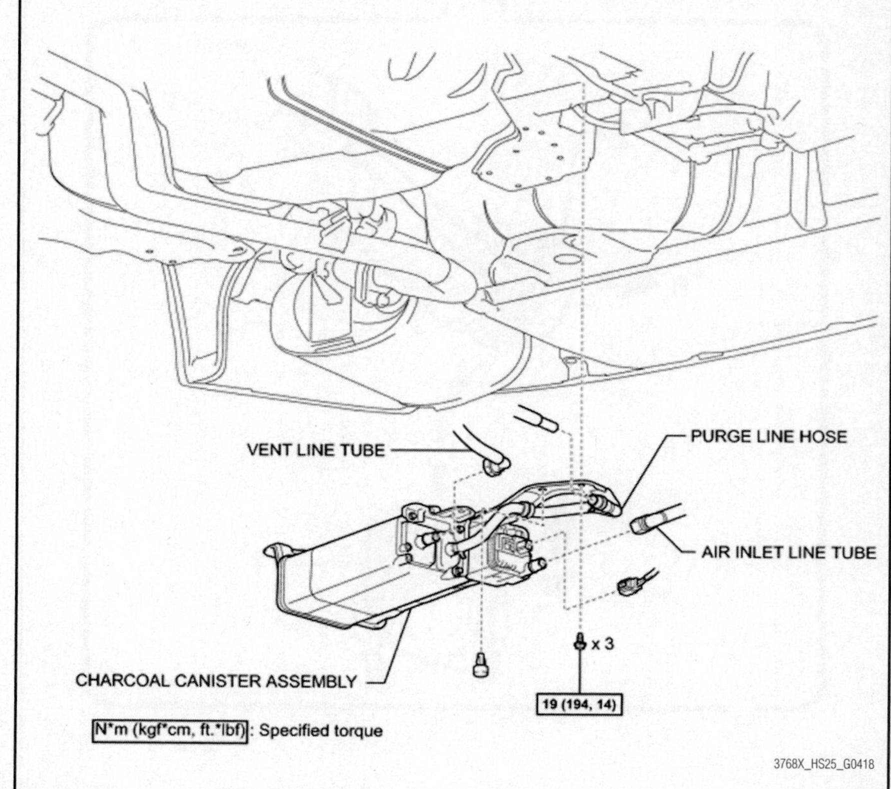

Fig. 133 Evaporative emissions canister

REMOVAL & INSTALLATION

See Figure 134.

➡**After the power switch is turned off, the display and navigation module display (HDD navigation system) records various types of memory and settings. As a result, after turning the power switch off, make sure to wait at least 60 seconds before disconnecting the cable from the negative (-) battery terminal.**

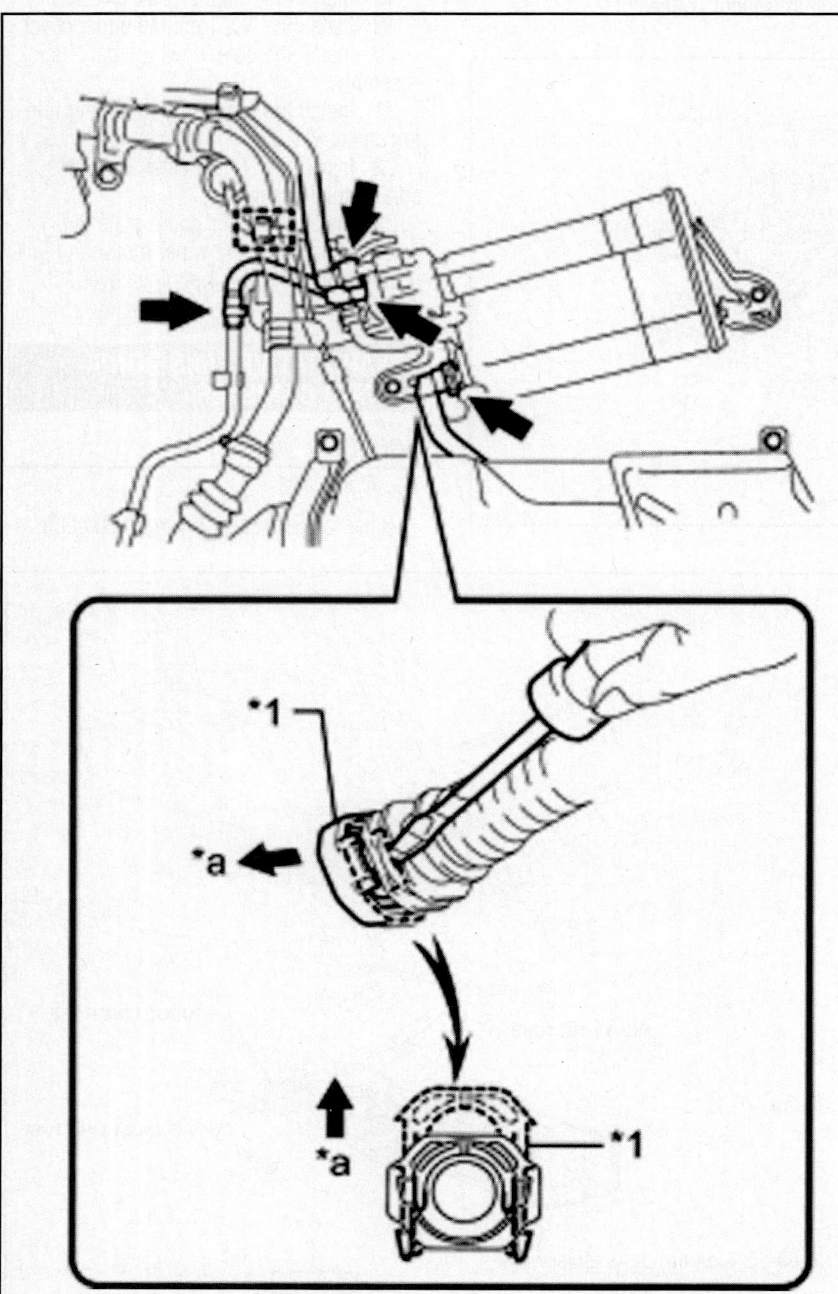

Fig. 134 Removing the charcoal canister assembly

1. Remove the rear suspension member sub assembly.
2. Disconnect the cable from the negative battery terminal.
3. Remove the charcoal canister assembly.
    a. Using a screwdriver, pry up the retainer.

➡**Do not remove the retainer.**

➡**Tape the screwdriver tip before use.**

b. Disconnect the purge line hose, air inlet tube and wire harness connector.
c. Disconnect the vent line tube.

➡**Remove any dirt or foreign matter on the vent line tube connector before performing this work.**

➡**Do not allow any scratches or foreign matter on the parts when disconnecting them as the vent line tube connector has an O-ring that seals the pipe.**

➡**Perform this work by hand. Do not use any tools.**

➡**Do not forcibly bend, twist or turn the nylon tube.**

➡**Protect the disconnected parts by covering them with a plastic bags after disconnecting the vent line tube and pipe.**

➡**If the vent line tube connector and pipe are stuck, push and pull to release them.**

d. Remove the 3 bolts, clip and charcoal canister assembly.

To install:

4. Install the charcoal canister assembly with the 3 bolts and a clip. Tighten to 14 ft. lbs. (19 Nm).
    a. Connect the vent line tube.
    b. Connect the pipe to the vent line tube connector. Then push up the retainer to lock the claws.

➡**Check that there are no scratches or foreign matter around the connected parts of the vent line tube connector and pipe before performing this work.**

➡**After connecting the vent line tube, check that the vent line tube is securely connected by pulling the vent line tube connector and pipe.**

c. Connect the purge line hose, air inlet line tube and wire harness connector.

5. Connect the cable to the negative battery terminal.
6. Install the rear suspension member sub assembly.

HEATED OXYGEN SENSOR (HO2S)

LOCATION

See Figure 135.

Refer to the accompanying illustration.

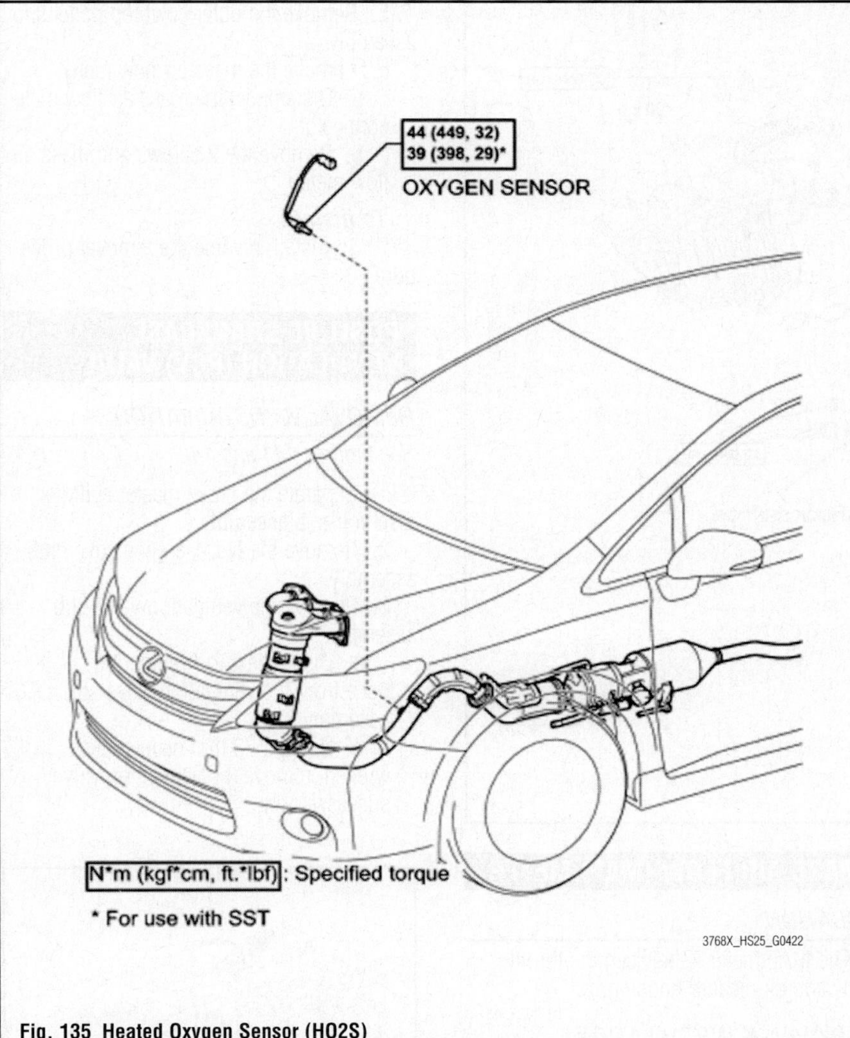

Fig. 135 Heated Oxygen Sensor (HO2S)

REMOVAL & INSTALLATION

See Figure 136.

1. Remove the oxygen sensor.
   a. Disconnect the oxygen sensor connector.
   b. Using the special tool, remove the oxygen sensor from the front exhaust pipe assembly.

➡**Do not damage the oxygen sensor.**

To install:

   c. Temporarily tighten the oxygen sensor.

➡**Do not damage the oxygen sensor.**

   d. Using the special tool, fully tighten the oxygen sensor. Using the special tool, tighten to 29 ft. lbs. (39 Nm). To tighten without the special tool, tighten to 32 ft. lbs. (44 Nm).
   e. Connect the oxygen sensor connector.
2. Inspect for exhaust gas leaks.

INTAKE AIR TEMPERATURE (IAT) SENSOR

REMOVAL & INSTALLATION

See Mass Air Flow Meter.

KNOCK SENSOR (KS)

LOCATION

See Figure 137.

Refer to the accompanying illustration.

REMOVAL & INSTALLATION

See Figures 138 and 139.

1. Remove the intake manifold.
2. Remove the knock sensor.
   a. Disconnect the knock sensor connector.
   b. Remove the nut and knock sensor.

To install:

3. Install the knock sensor with the nut. Tighten to 15 ft. lbs. (20 Nm).

➡**Make sure that the knock sensor is in the correct position.**

   a. Connect the knock sensor connector.
4. Install the intake manifold.

MALFUNCTION INDICATOR LIGHT (MIL)

RESET PROCEDURE

Clear the DTC codes.

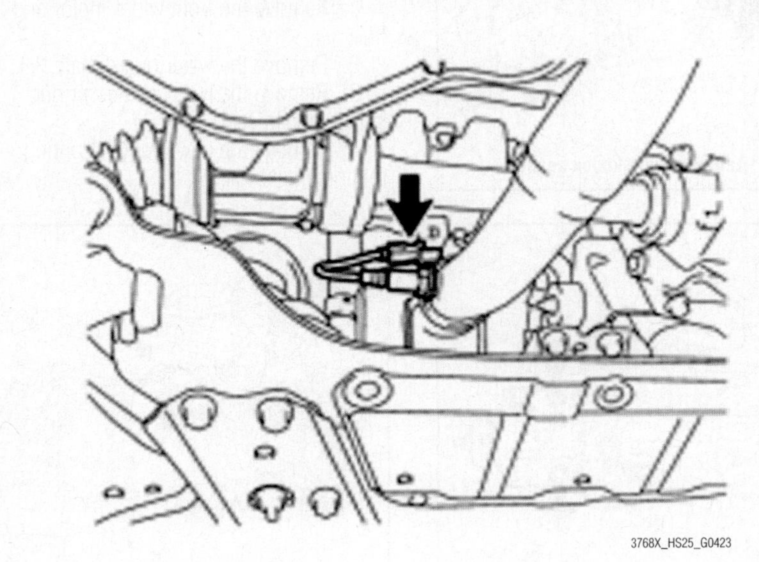

Fig. 136 Disconnecting the oxygen sensor connector

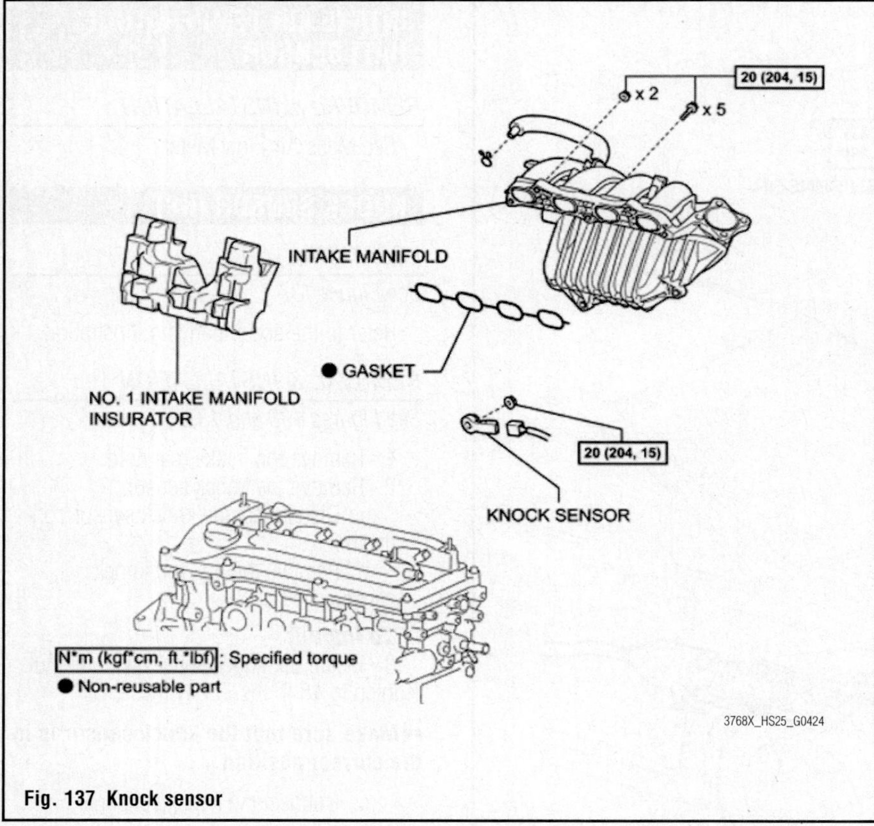

Fig. 137 Knock sensor

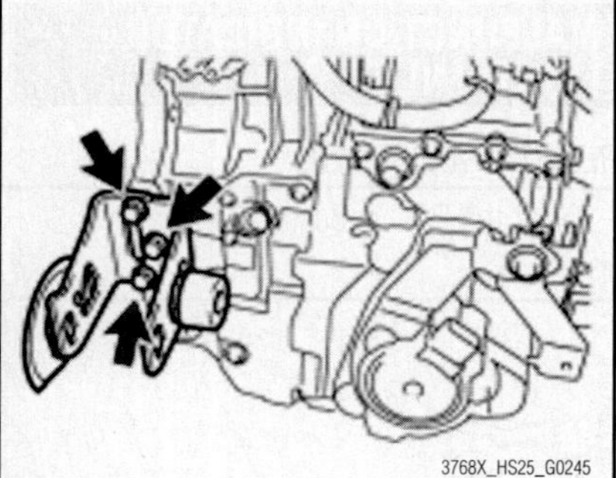

Fig. 138 Removing the knock sensor

MASS AIR FLOW (MAF) METER

LOCATION

The MAF meter is between the throttle body and air cleaner housing.

REMOVAL & INSTALLATION

See Figure 140.

1. Remove the front wiper motor and link.
2. Remove the water guard plate RH.
3. Remove the No. 2 heater air duct splash shield seal.
4. Remove the cowl body mounting reinforcement RH.

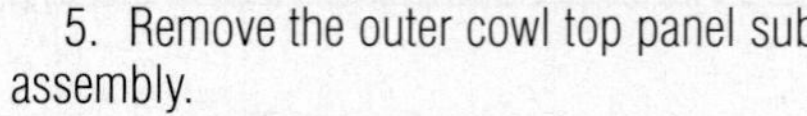

5. Remove the outer cowl top panel sub assembly.
6. Remove the mass air flow meter.
   a. Disconnect the mass air flow meter connector.
   b. Remove the 2 screws and mass air flow meter.

To install:

7. To install, reverse the removal procedure.

POSITIVE CRANKCASE VENTILATION (PCV) VALVE

REMOVAL & INSTALLATION

See Figures 141 and 142.

1. Separate the brake master cylinder reservoir sub assembly.
2. Remove the No. 1 engine cover sub assembly.
3. Remove the ventilation valve sub assembly.
   a. Disconnect the ventilation hose from the ventilation valve sub assembly.
   b. Using a 22 mm deep socket wrench, remove the ventilation valve sub-assembly.

Fig. 141 Disconnecting the ventilation hose from the ventilation valve sub assembly

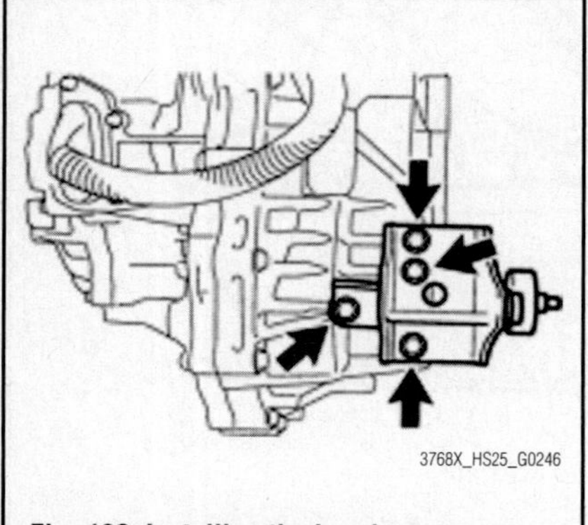

Fig. 139 Installing the knock sensor

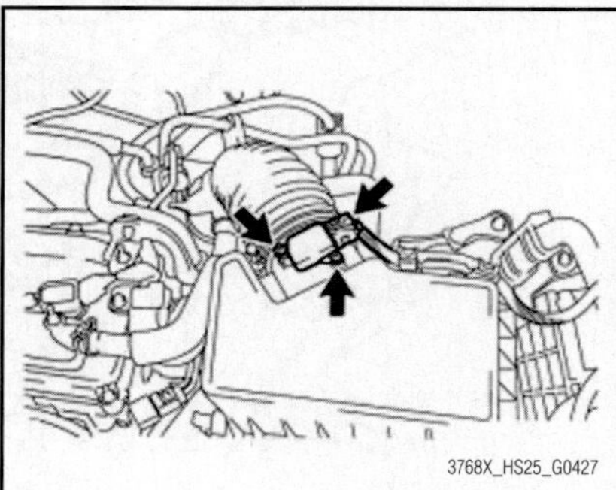

Fig. 140 Removing the Mass Air Flow (MAF) meter

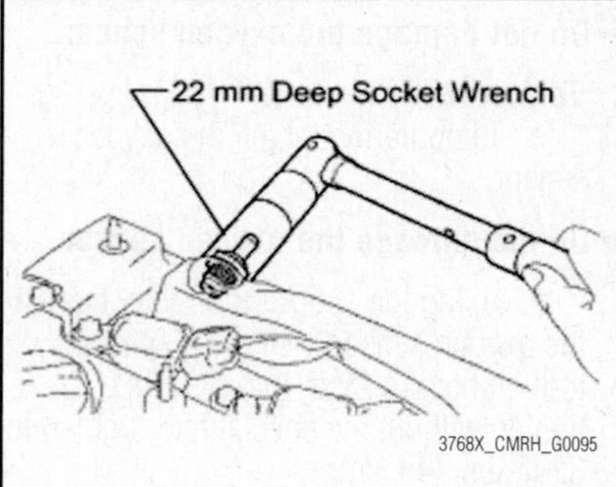

Fig. 142 Removing the ventilation valve sub assembly

To install:

⁂ CAUTION

Do not start the engine within 1 hour after installation.

4. To install, reverse the removal procedure. Tighten the ventilation valve to 14 ft. lbs. (19 Nm).

THROTTLE POSITION SENSOR (TPS)

LOCATION

Refer to the Throttle Body removal and installation procedures.

REMOVAL & INSTALLATION

Refer to the Throttle Body removal and installation procedure in Fuel System..

FUEL

GASOLINE FUEL INJECTION SYSTEM

FUEL SYSTEM SERVICE PRECAUTIONS

Safety is the most important factor when performing not only fuel system maintenance, but any type of maintenance. Failure to conduct maintenance and repairs in a safe manner may result in serious personal injury or death. Work on a vehicle's fuel system components can be accomplished safely and effectively by adhering to the following rules and guidelines.

- To avoid the possibility of fire and personal injury, always disconnect the negative battery cable unless the repair or test procedure requires that battery voltage be applied.
- Always relieve the fuel system pressure prior to disconnecting any fuel system component (injector, fuel rail, pressure regulator, etc.) fitting or fuel line connection. Exercise extreme caution whenever relieving fuel system pressure to avoid exposing skin, face and eyes to fuel spray. Please be advised that fuel under pressure may penetrate the skin or any part of the body that it contacts.
- Always place a shop towel or cloth around the fitting or connection prior to loosening to absorb any excess fuel due to spillage. Ensure that all fuel spillage is quickly removed from engine surfaces. Ensure that all fuel-soaked cloths or towels are deposited into a flame-proof waste container with a lid.
- Always keep a dry chemical (Class B) fire extinguisher near the work area.
- Do not allow fuel spray or fuel vapors to come into contact with a spark or open flame.
- Always use a second wrench when loosening or tightening fuel line connection fittings. This will prevent unnecessary stress and torsion on fuel piping. Always follow the proper torque specifications.
- Always replace worn fuel fitting O-rings with new ones. Do not substitute fuel hose where rigid pipe is installed.

FUEL SYSTEM PRESSURE

RELIEVING

⁂ CAUTION

Perform the following procedures to prevent fuel from spilling out before removing any fuel system parts.

⁂ CAUTION

Pressure will still remain in the fuel line even after performing the following procedures. When disconnecting the fuel line, cover it with a shop rag or a piece of cloth to prevent fuel from spraying or coming out.

1. Disconnect the fuel pump connector:
   a. Remove the rear seat cushion assembly.
   b. Remove the rear floor service hole cover.
   c. Disconnect the fuel pump connector.
   d. Start the engine.
   e. After the engine stops, turn the ignition switch off.

➡DTC P0171/25 (fuel problem) may be detected.

   f. Crank the engine again. Check that the engine does not start.
   g. Remove the fuel tank cap to discharge pressure from the fuel tank.
   h. Disconnect the cable from the negative (-) battery terminal.
   i. Reconnect the fuel pump connector.
   j. Install the rear floor service hole cover.
   k. Install the rear seat.
2. Check that there are no fuel leaks from the fuel system after doing any maintenance or repairs.

FUEL FILTER

REMOVAL & INSTALLATION

See Figure 143.

1. Before servicing the vehicle, refer to the precautions section.
2. Remove the fuel pump from the vehicle.
3. Remove the fuel pump filter, as follows:

➡Do not damage the fuel pump filter. Do not remove the suction filter.

   a. Using a screwdriver, pry out the clips.
   b. Pull out the fuel pump filter from the fuel pump.

To install:

4. Install the fuel pump filter with a new clip.
5. Install the fuel pump.

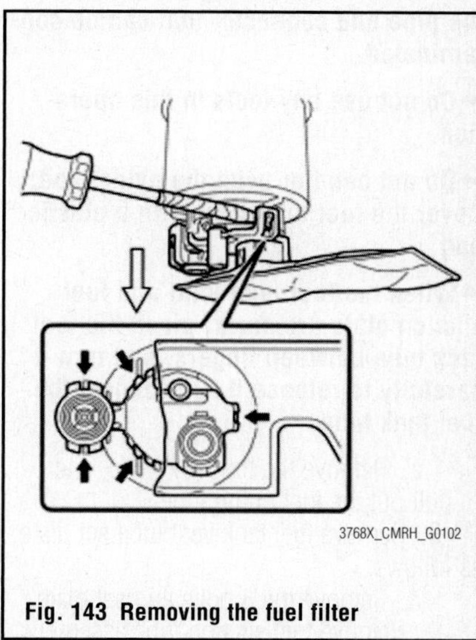

Fig. 143 Removing the fuel filter

FUEL LEVEL SENDING UNIT

REMOVAL & INSTALLATION

1. Before servicing the vehicle, refer to the precautions section.
2. Remove the fuel pump from the vehicle.

3. Remove the fuel sender gauge assembly, as follows:
   a. Disconnect the fuel sender gauge connector.
   b. Unlock the fuel sender gauge, and slide it to remove.

To install:

4. Slide the fuel sender gauge to engage with the claw.
   a. Connect the fuel sender gauge connector.
5. Install the fuel pump.

FUEL PUMP

REMOVAL & INSTALLATION

See Figures 144 and 145.

1. Before servicing the vehicle, refer to the precautions section.
2. Discharge fuel system pressure.
3. Disconnect battery negative cable.
4. Remove the rear seat cushion assembly.
5. Remove the rear floor service hole cover.
6. Disconnect the fuel pump connector.
7. Separate the fuel pump tube sub-assembly, as follows:

➡Check if there is any dirt or mud around the connector before this operation and remove the dirt as necessary.

➡Be careful of mud because the quick connector has an O-ring which seals the pipe and connector that can be contaminated.

➡Do not use any tools in this operation.

➡Do not bend or twist the nylon tube. Cover the fuel tube joint with a plastic bag.

➡When the fuel tube joint and fuel suction plate are stuck, pinch the fuel tank tube between fingers, and turn it carefully to release it. Disconnect the fuel tank tube.

   a. Remove the tube joint clip, and pull out the fuel pump tube.
8. Remove fuel tank vent tube set plate, as follows:
   a. Remove the 8 bolts and set plate.
9. Remove fuel suction tube assembly with pump and gauge, as follows:
   a. Pull out the fuel suction tube from the fuel tank.

➡Do not damage the fuel pump filter.

➡Be careful not to bend the arm of the fuel sender gauge.

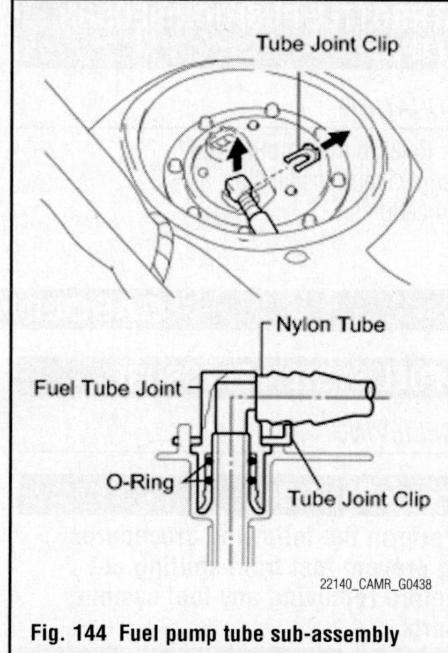

Fig. 144 Fuel pump tube sub-assembly

   b. Remove the gasket from the fuel suction tube.

To install:

10. Install a new gasket to the fuel suction tube.
11. Install the fuel suction tube.

➡Do not damage the fuel pump filter.

➡Be careful not to bend the arm of the fuel sender gauge.

12. Install the fuel tank vent tube set plate, as follows:
   a. Align the mark of the set plate with the fuel suction tube.
   b. Install the set plate with the 8 bolts and tighten to 52 inch lbs. (5.9 Nm).
13. Install the fuel pump tube with the tube joint clip.

➡Check that there is no scratches or foreign objects on the connecting part.

➡Check that the fuel tube joint is inserted securely.

➡Check that the tube joint clip is on the collar of the fuel tube joint.

➡After installing the tube joint clip, check that the fuel tube joint is pulled off.

14. Connect battery negative cable.
15. Inspect for fuel leak.
16. Install the rear floor service hole cover.
17. Install the rear seat cushion assembly.

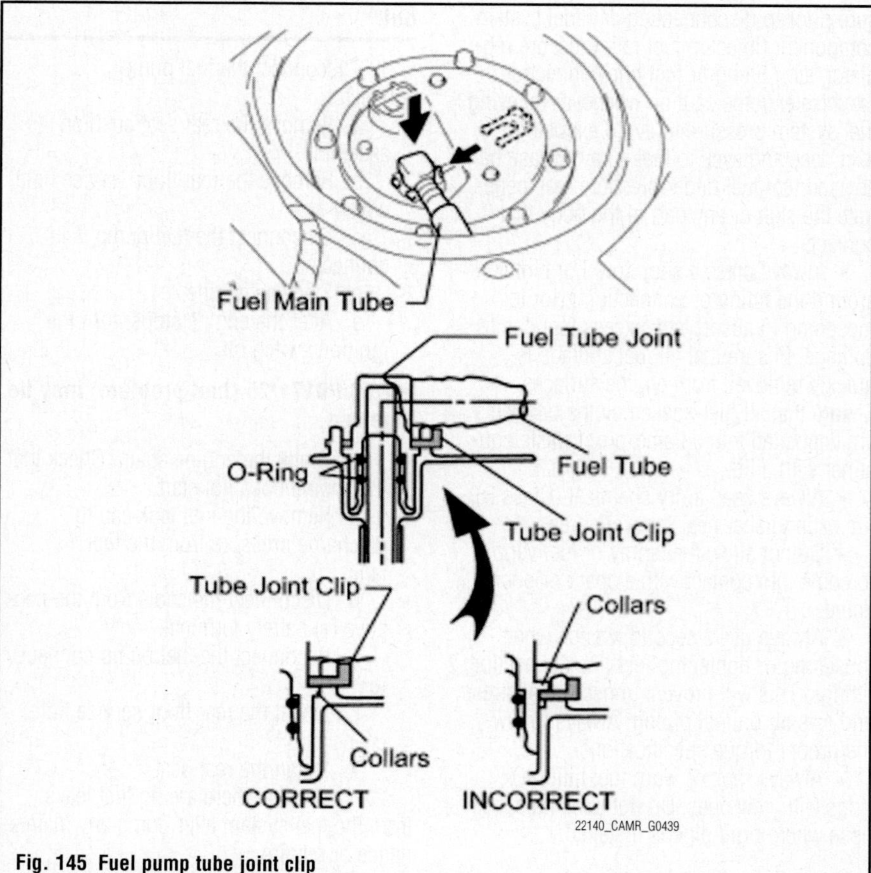

Fig. 145 Fuel pump tube joint clip

FUEL RAIL AND INJECTOR

REMOVAL & INSTALLATION

See Figures 146 and 147.

➡After the power switch is turned off, the display and navigation module display (HDD navigation system) records various types of memory and settings. As a result, after turning the power switch off, make sure to wait at least 60 seconds before disconnecting the cable from the negative (-) battery terminal.

1. Discharge the fuel system pressure.
2. Remove the luggage compartment floor mat.
3. Disconnect the cable from the negative battery terminal.
4. Remove the windshield wiper motor and link assembly.
5. Remove the water guard plate RH.
6. Remove the No. 2 heater air duct splash shield seal.
7. Remove the cowl body mounting reinforcement RH.
8. Remove the outer cowl top panel sub assembly.
9. Remove the No. 1 engine cover sub assembly.
10. Remove the air cleaner cap sub assembly.
11. Disconnect the fuel tube sub assembly.
    a. Release the claw and remove the No. 1 fuel pipe clamp.

➡Check for foreign matter on the pipe and around the connector before disconnecting the quick connector. Clean the connector if necessary.

b. Pinch the tube connector, and then pull the tube connector off of the pipe.

➡Check for foreign matter in the fuel tube around the fuel tube connector.

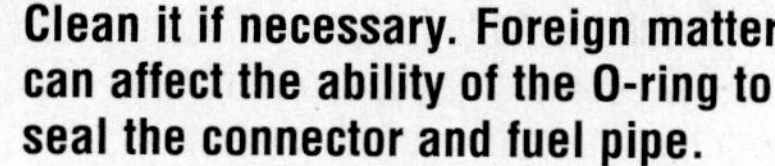

Clean it if necessary. Foreign matter can affect the ability of the O-ring to seal the connector and fuel pipe.

➡Do not use any tools to separate the connector and pipe.

➡Do not forcefully bend, kink or twist the hose.

➡Keep the connector and pipe free from foreign matter.

➡If the connector and pipe are stuck together, pinch the connector and turn it carefully to disconnect it.

➡Put the connector in a plastic bag to prevent damage and contamination.

c. Separate the fuel tube from the fuel hose clamp.

12. Disconnect the No. 2 ventilation hose from the ventilation valve.
13. Remove the fuel delivery pipe sub assembly.
    a. Remove the 2 wire harness clamps.
    b. Disconnect the 4 fuel injectors.
    c. Remove the 2 bolts, and then remove the fuel delivery pipe together with the 4 fuel injectors.

➡Be careful not to drop the fuel injectors when removing the fuel delivery pipe.

d. Remove the 2 spacers from the cylinder head.
e. Remove the 4 insulators from the cylinder head.

14. Remove the fuel injector assembly.
    a. Pull out the 4 injectors from the delivery pipe.
    b. Remove the o-ring from each fuel injector.
    c. For reinstallation, attach tag or label to each injector shaft.

➡Prevent entry of foreign objects by covering the fuel injectors with plastic bags.

To install:

15. Install the fuel injector assembly.
    a. Install a new insulator and o-ring to each fuel injector assembly.
    b. Apply a light coat of gasoline or spindle oil to the contact surfaces of the new o-ring on each fuel injector assembly.
    c. Apply a light coat of gasoline or spindle oil to the part of the fuel delivery pipe which comes into contact with the o-ring of the fuel injector.
    d. Push and twist each fuel injector to install them into the fuel delivery pipe.

➡Make sure that the O-ring is not cracked or jammed before installing the injector.

e. Check that the fuel injector rotates smoothly. If the fuel injector does not rotate, replace the O-ring.

16. Install a fuel delivery pipe sub assembly.
    a. Install the 2 spacers onto the cylinder head.
    b. Install the fuel delivery pipe together with the 4 fuel injectors, then temporarily tighten the 2 bolts.

➡Be careful not to drop the fuel injectors when installing the fuel delivery pipe.

c. Check that the fuel injector rotates smoothly. If the fuel injector does not rotate smoothly, replace the O-ring.
d. Tighten the 2 bolts to 15 ft. lbs. (20 Nm).
e. Connect the 4 fuel injector connectors.
f. Install the 2 wire harness clamps.

17. Install the No. 2 ventilation hose.
    a. Connect the ventilation hose to the ventilation valve.

➡Make sure that the paint mark and hose clamp are positioned as shown after connecting the hose.

18. Connect the fuel tube sub assembly.
    a. Install the fuel tube to the fuel hose clamp.
    b. Push in the fuel tube connector to the fuel main tube until the fuel tube connector makes a "click" sound.

➡Check that there is no damage or foreign objects on the fuel pipe connectors.

➡After disconnecting, check that the fuel tube connector and the pipe are securely connected by pulling on them.

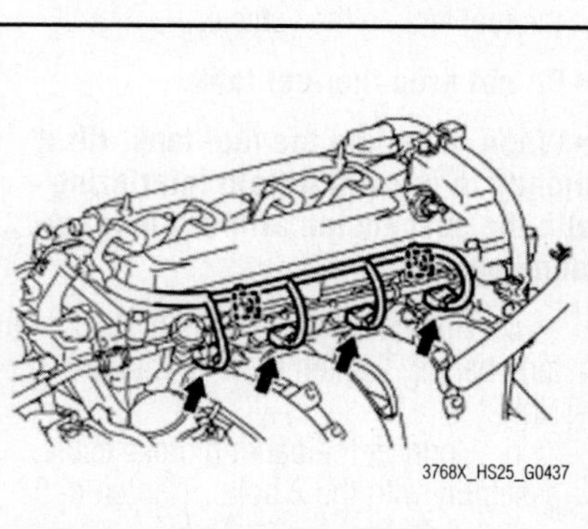

Fig. 146 Removing the fuel injector connectors

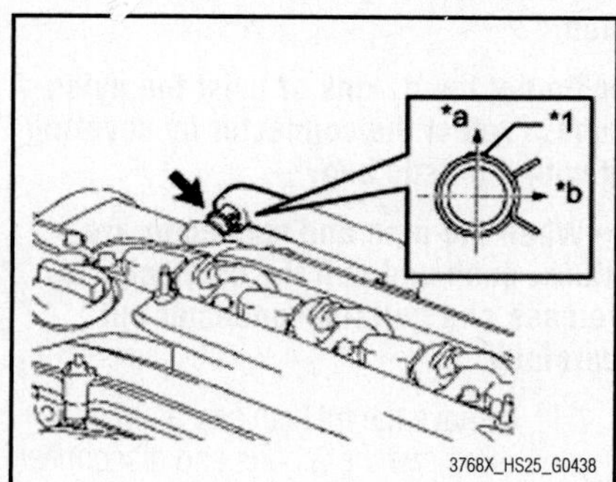

Fig. 147 Installing the No. 2 ventilation hose

c. Install the No. 1 fuel pipe clamp.

19. Install the air cleaner cap sub assembly.

20. Connect the cable to the negative battery terminal.

21. Install the luggage compartment floor mat.

22. Inspect for a fuel leak.

23. Install the No. 1 engine cover sub assembly.

24. Install the outer cowl top pane sub assembly.

25. Install the cowl body mounting reinforcement RH.

26. Install the No. 2 heater air duct splash shield seal.

27. Install the water guard plate RH.

28. Install the windshield wiper motor and link assembly.

FUEL TANK

REMOVAL & INSTALLATION

See Figures 148 through 151.

1. Remove the fuel suction tube assembly with the pump and gauge.
2. Drain the fuel.
3. Remove the center exhaust pipe assembly.
4. Remove the rear floor side member cover LH.
5. Remove the rear floor side member cover RH.
6. Remove the No. 1 fuel tank protector.
   a. Remove the bolt, 2 nuts and No. 1 fuel tank protector.
7. Disconnect the fuel tank breather tube.
   a. Pinch the tabs of the retainer to remove the lock claws and pull it down.
   b. Pull out the fuel tank breather tube.

➡Check that there is no dirt or other foreign objects around the connector before this operation and clean the connector as necessary.

➡It is necessary to prevent mud or dirt from entering the connector. If mud or dirt gets in the connector, the O-rings may not seal properly.

➡Do not use any tools in this operation.

➡Do not bend, kink or twist the nylon tube. Protect the connector by covering it with a plastic bag.

➡When the pipe and connector are stuck, push and pull the connector to release and pull the connector out carefully.

8. Disconnect the fuel cut off tube.
   a. Disconnect the fuel cut off tube from the charcoal canister assembly.

➡Do not remove the retainer.

➡Remove any dirt or foreign matter on the fuel cut-off tube connector before performing this work.

➡Do not allow any scratches or foreign matter on the parts when disconnecting them as the fuel cut-off tube connector has an O-ring that seals the pipe.

➡Perform this work by hand. Do not use any tools.

➡Do not forcibly bend, twist or turn the fuel cut-off tube.

➡Protect the disconnected part by covering it with a plastic bag after disconnecting the fuel cut-off tube.

➡If the vent hose connector and pipe are stuck, push and pull to release them.

9. Disconnect the fuel tank to filler pipe hose.
   a. Loosen the clamp, then disconnect the fuel tank to filler pipe hose from the fuel tank.
10. Remove the fuel tank main tube sub assembly.
   a. Pinch the tabs of the retainer to remove the lock claws and pull it down as shown in the illustration.
   b. Pull out and remove the fuel tank main tube sub-assembly.

➡Check that there is no dirt or other foreign objects around the connector before this operation and clean the connector as necessary.

➡It is necessary to prevent mud or dirt from entering the connector. If mud or dirt gets in the connector, the O-rings may not seal properly.

➡Do not use any tools in this operation.

➡Do not bend, kink or twist the nylon tube. Protect the connector by covering it with a plastic bag.

➡When the pipe and connector are stuck, push and pull the connector to release and pull the connector out carefully.

11. Remove the fuel tank assembly.
   a. Remove the 2 bolts and disconnect the parking brake cable assembly.
   b. Support the fuel tank using an engine lifter.

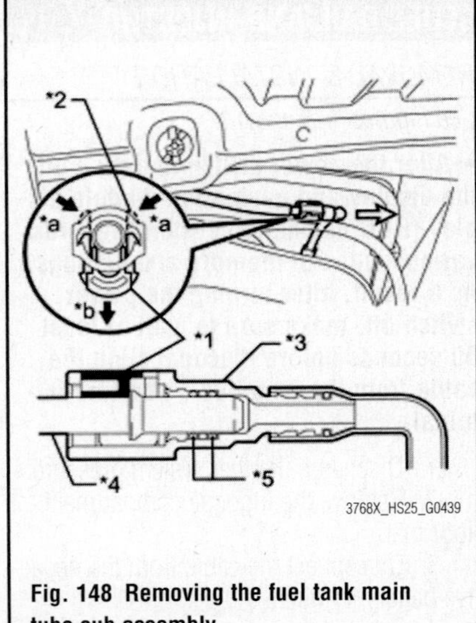

Fig. 148 Removing the fuel tank main tube sub assembly

   c. Remove the 4 set bolts of the 2 fuel tank bands.
   d. Lower the engine lifter to remove the fuel tank.

➡Slowly operate the engine lifter to lower the fuel tank.

➡Do not drop the fuel tank.

➡When removing the fuel tank, tilt it slightly to prevent it from interfering with the suspension arm or other surrounding parts.

12. Remove the fuel tank cushion.
   a. Remove the 3 No. 1 fuel tank cushions and 2 No. 2 fuel tank cushions.

To install:

13. Install the fuel tank cushion.
   a. Install the 3 new No. 1 fuel tank cushions and 2 new No. 2 fuel tank cushions.
14. Install the fuel tank assembly.
   a. Support the fuel tank using an engine lifter.
   b. Raise the engine lifter, then install the fuel tank to the vehicle.

➡Do not drop the fuel tank.

➡When installing the fuel tank, tilt it slightly to prevent it from interfering with the suspension arm or other surrounding parts.

   c. Tighten the 4 set bolts of the 2 fuel tank bands. Tighten to 29 ft. lbs. (39 Nm).
   d. Connect the parking brake cable assembly with the 2 bolts. Tighten to 53 inch lbs. (6 Nm).
15. Connect the fuel tank main tube sub assembly.

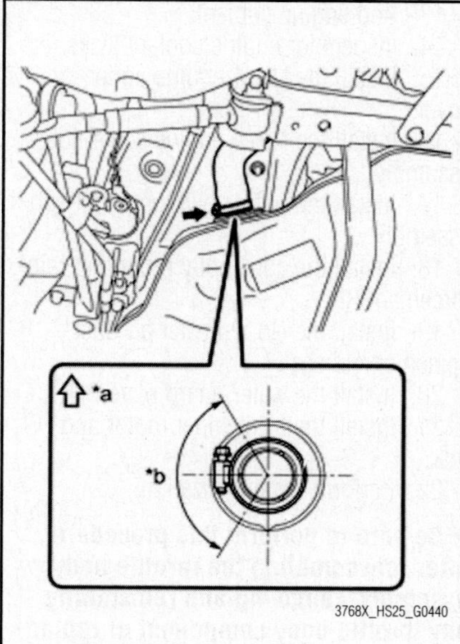

Fig. 149 Connecting the fuel tank to the filler pipe hose

a. Push in the fuel tank main tube connector to the pipe and push up the retainer so that the claws engage.

➡Check that there are no scratches or foreign objects around the connected parts of the fuel tube connector and pipe before starting this step.

➡After connecting the fuel tank main tube sub-assembly, check that the fuel tank main tube sub-assembly is securely connected by pulling on the fuel tube connector.

16. Connect the fuel tank to the filler pipe hose.
  a. Connect the fuel tank to the filler pipe hose to the fuel tank, then fit it with the clamp.
17. Connect the fuel tank breather tube.
  a. Push in the fuel tank breather tube connector to the pipe and push up the retainer so that the claws engage.

➡Check that there are no scratches or foreign objects around the connected parts of the fuel tube connector and pipe before starting this step.

➡After connecting the fuel tank main tube sub-assembly, check that the fuel tank main tube sub-assembly is securely connected by pulling on the fuel tube connector.

18. Connect the fuel cut off tube.
  a. Push in the fuel cut-off tube connector to the charcoal canister and push up the retainer so that the claws engage.

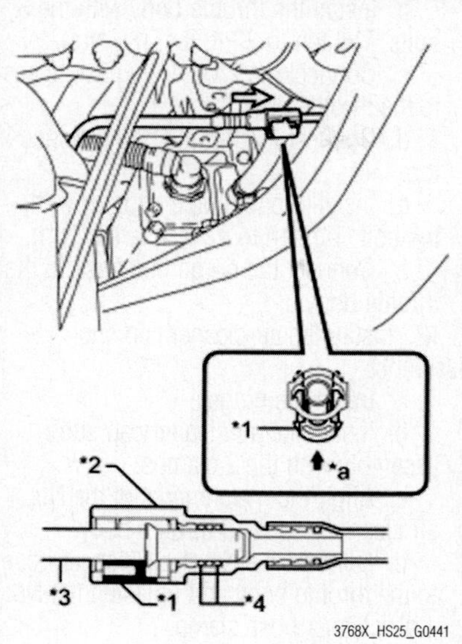

Fig. 150 Connecting the fuel tank breather tube

➡Check that there are no scratches or foreign matter around the connected parts of the fuel cut-off tube connector and pipe before performing this work.

➡After connecting the fuel cut-off tube, check that the fuel cut-off tube is securely connected by pulling the fuel cut-off tube connector and the charcoal canister.

19. Install the No. 1 fuel tank protector.
  a. Install the No. 1 fuel tank protector with the bolt and 2 nuts. Tighten to 49 inch lbs. (5.5 Nm).
20. install the rear floor side member cover LH.

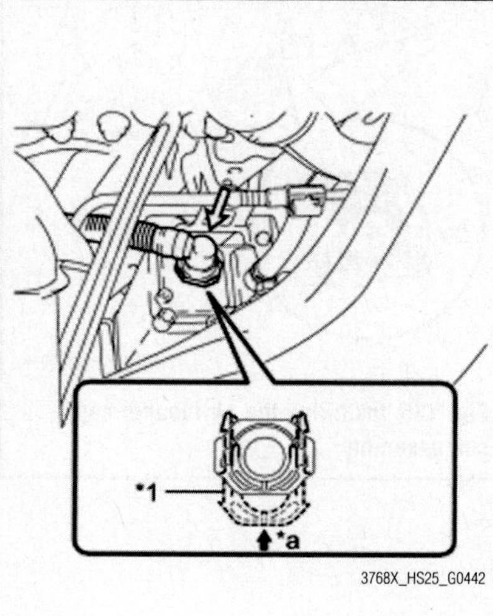

Fig. 151 Connecting the fuel cut off tube

21. Install the rear floor side member cover RH.
22. Install the center exhaust pipe assembly.
23. Add fuel.
24. Install the fuel suction tube assembly with the pump and gauge.

IDLE SPEED

ADJUSTMENT

Idle speed is maintained by the ECM. No adjustment is necessary or possible.

THROTTLE BODY

REMOVAL & INSTALLATION

See Figures 152 through 155.

1. Remove the No. 1 engine under cover.
2. Drain the engine coolant.
3. Remove the front wiper motor and link.
4. Remove the water guard plate RH.
5. Remove the No. 2 heater air duct splash shield seal.
6. Remove the cowl body mounting reinforcement RH.
7. Remove the outer cowl top pane sub assembly.
8. Remove the No. 1 engine cover sub assembly.
9. Remove the air cleaner cap sub assembly.
  a. Disconnect the mass air flow meter connector and clamp.

➡Push down on the tab to disengage the claw, and then slide the harness toward the rear of the vehicle to disengage it.

  b. Disconnect the No. 2 ventilation hose.
  c. Disconnect the purge VSV connector and wire harness clamp.
  d. Disconnect the purge VSV vacuum hose.
  e. Disconnect the purge line hose.
  f. Lock the No. 1 air cleaner hose clamp, and then disconnect the No. 1 air cleaner hose from the throttle body.
  g. Remove the 2 clamps and air cleaner cap sub assembly.
10. Remove the throttle body assembly.
  a. Remove the purge VSV vacuum hose.
  b. Remove the bolt and separate the fuel pipe support from the throttle body.
  c. Disconnect the throttle body connector.
  d. Disconnect the 2 water by pass hoses from the throttle body.

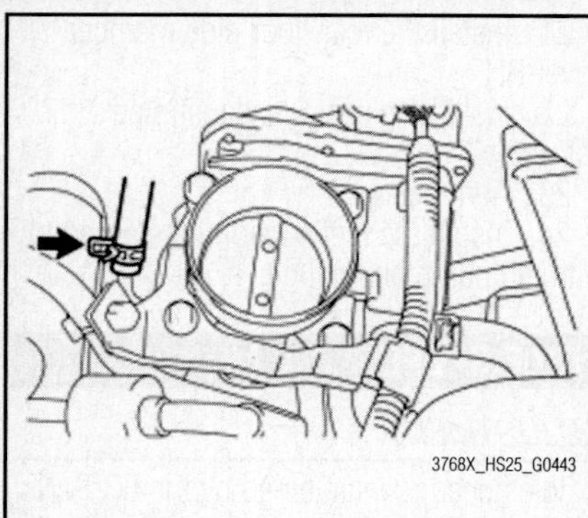

Fig. 152 Removing the purge VSV vacuum hose

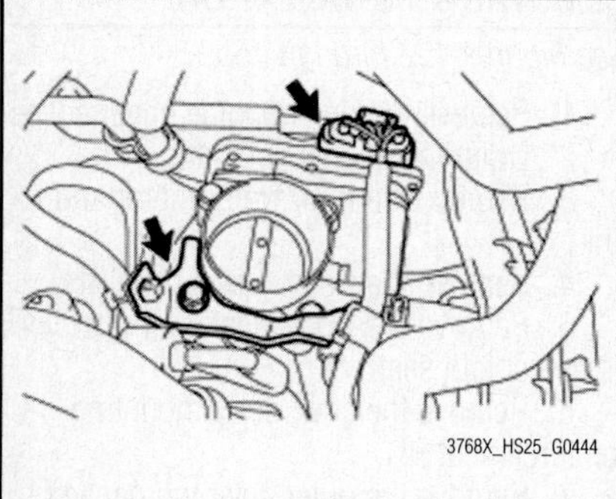

Fig. 153 Separating the fuel pipe support from the throttle body

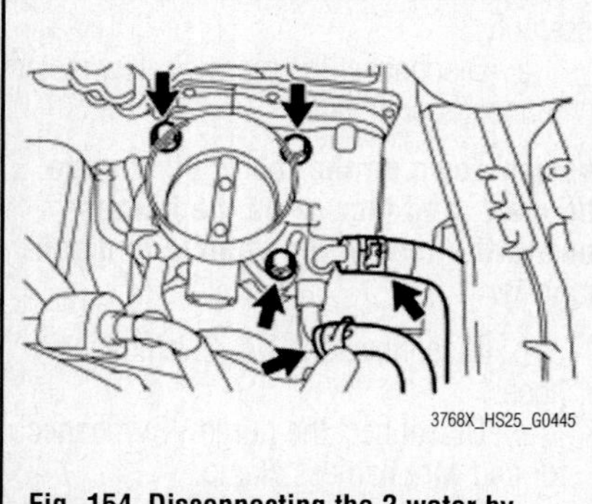

Fig. 154 Disconnecting the 2 water by pass hoses from the throttle body

e. Remove the 3 bolts and throttle body.

f. Remove the gasket from the intake manifold.

To install:

11. Install the throttle body assembly.

a. Install a new gasket onto the intake manifold.

b. Install the throttle body with the 3 bolts. Tighten to 22 ft. lbs. (30 Nm).

c. Connect the 2 water by pass hoses to the throttle body.

d. Connect the throttle body connector.

e. Install the fuel pipe support with the bolt. Tighten to 22 ft. lbs. (30 Nm).

f. Connect the purge line hose to the throttle body.

12. Install the air cleaner cap sub assembly.

a. Insert the hinges.

b. Install the air cleaner cap sub assembly with the 2 clamps.

c. Align the matchmarks of the No. 1 air cleaner hose and throttle body.

d. Connect the No. 1 air cleaner hose to the throttle body and unfasten the No. 1 air cleaner hose clamp.

➡Make sure that the hose clamp is at the correct angle.

e. Connect the purge line hose.

f. Connect the purge VSV vacuum hose.

g. Connect the purge VSV connector.

h. Connect the No. 2 ventilation hose to the hose.

i. Connect the mass air flow meter connector and clamp.

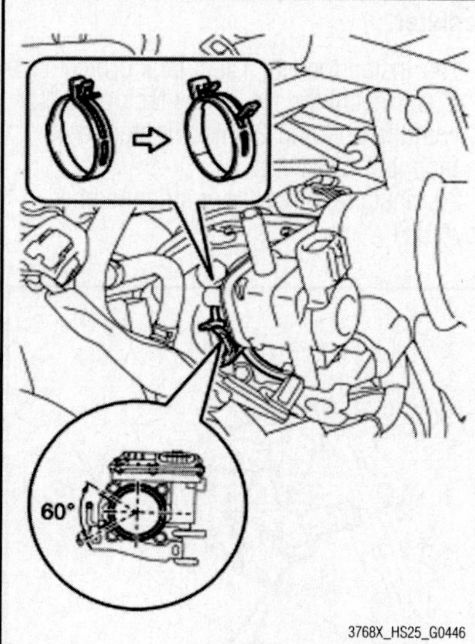

Fig. 155 Installing the air cleaner cap sub assembly

13. Add engine coolant.

14. Inspect for engine coolant leaks.

15. Install the No. 1 engine under cover.

16. Install the No. 1 engine cover sub assembly.

17. Install the outer cowl top panel sub assembly.

18. Install the cowl body mounting reinforcement RH.

19. Install the No. 2 heater air duct splash shield seal.

20. Install the water guard plate RH.

21. Install the front wiper motor and link.

22. Perform the initialization.

➡Be sure to perform this procedure after reassembling the throttle body assembly, removing and reinstalling any throttle body component or replacing the hybrid vehicle control ECU.

a. Disconnect the cable from the negative (-) battery terminal.

➡After the power switch is turned off, the display and navigation module display (HDD navigation system) records various types of memory and settings. As a result, after turning the power switch off, make sure to wait at least 60 seconds before disconnecting the cable from the negative (-) battery terminal.

b. Connect the Techstream to the DLC3 and clear the DTCs.

c. Set the vehicle to inspection mode.

d. Start the engine without operating the accelerator pedal and check that the MIL is not illuminated and that the idle speed is within the specified range when the air conditioning is switched off after the engine is warmed up.

Standard:

- A/C switched off: 850-950 rpm

➡If the accelerator pedal is operated, perform the above steps again.

➡Be sure to perform this step with all accessories off.

➡Make sure that park (P) is selected.

a. Perform a road test and confirm that there are no abnormalities.

HEATING & AIR CONDITIONING SYSTEM

BLOWER MOTOR

REMOVAL & INSTALLATION

See Figure 156.

1. Remove the No. 2 instrument panel under cover sub assembly.
2. Remove the front blower motor sub assembly.
    a. Disconnect the connector.
    b. Remove the 3 screws and front blower motor sub assembly.

➡Do not remove the front blower motor sub assembly if it has been damaged or impacted.

Fig. 156 Removing the front blower motor

To install:

3. Install the front blower motor sub assembly with the 3 screws.

➡Do not install the front blower motor sub assembly if it has been damaged or impacted.

    a. Connect the connector.
4. Install the No. 2 instrument panel under cover sub assembly.

STEERING

POWER RACK & PINION STEERING GEAR

REMOVAL & INSTALLATION

See Figures 157 through 160.

1. Place the front wheels facing straight ahead.
2. Secure the steering wheel.
    a. Secure the steering wheel with the seat belt in order to prevent rotation.

➡This operation is useful to prevent damage to the spiral cable.

3. Remove the column hole cover silencer sheet.
4. Separate the No. 2 steering intermediate shaft assembly.
5. Separate the No. 1 steering column hole cover sub assembly.
    a. Remove clip A, detach clip B from the body and disconnect the No. 1 steering column hole cover sub assembly.

➡Do not damage clips A and B.

6. Remove the front wheels.
7. Remove the No. 1 engine under cover.
8. Remove the No. 2 engine under cover.
9. Remove the front No. 3 engine under cover.

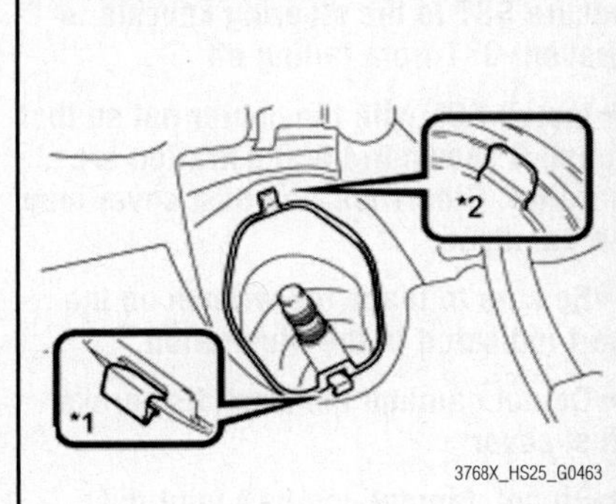

Fig. 157 Separating the No. 1 steering column hole cover sub assembly

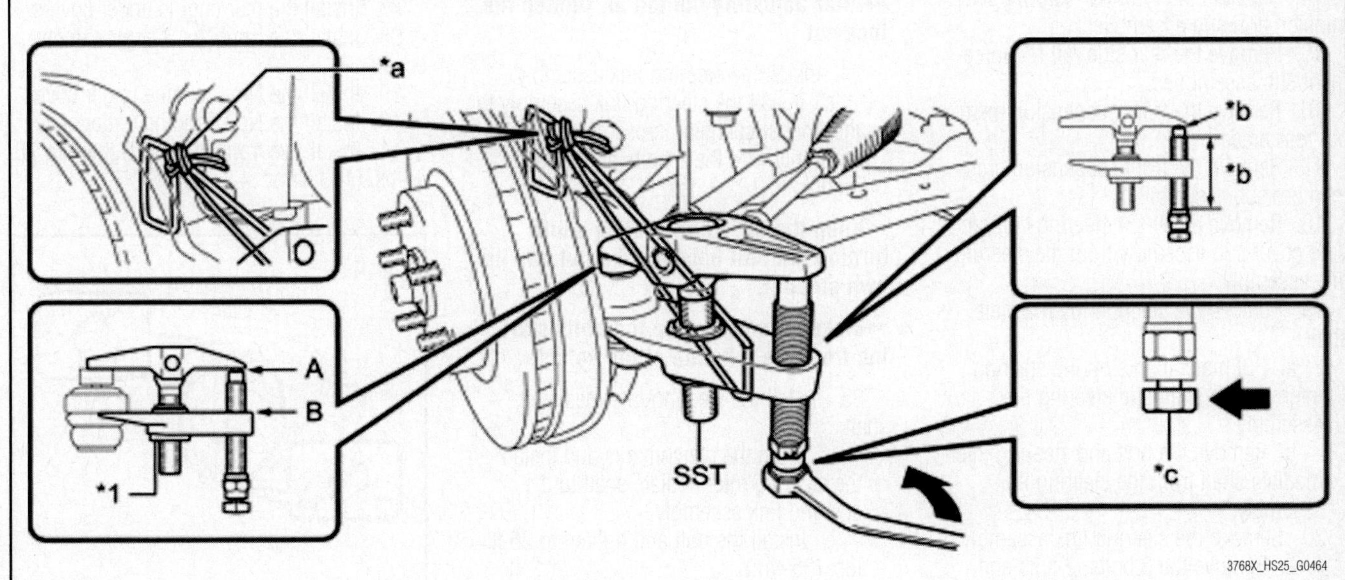

Fig. 158 Separating the tie rod end from the steering knuckle

10. Remove the rear engine under covers.

11. Separate the front stabilizer link assemblies.

12. Separate the tie rod end sub assembly LH.

a. Remove the cotter pin and nut.

b. Install the special tool to the tie rod end.

➡Make sure that the upper ends of the tie rod end and special tool are aligned.

c. Using the special tool, separate the tie rod end from the steering knuckle.

⁂ CAUTION

Apply grease to the bolt threads and the tip of the special tool.

➡Be sure to tighten the string firmly to secure SST to the steering knuckle to prevent SST from falling off.

➡Install SST with the center nut so that A and B shown in the illustration are parallel. Otherwise, the dust cover may be damaged.

➡Be sure to place the wrench on the part indicated in the illustration.

➡Do not damage the front disc brake dust cover.

➡Do not damage the ball joint dust cover.

➡Do not damage the steering knuckle.

13. Separate the tie rod end sub assemblies.

14. Separate the front No. 1 lower suspension arm sub assemblies.

15. Remove the rear side rail reinforcement sub assemblies.

16. Remove the front suspension member rear braces.

17. Remove the front suspension crossmember sub assembly.

18. Remove the No. 1 steering column hole cover sub assembly from the steering link assembly.

19. Remove the steering intermediate shaft.

a. Put matchmarks on the steering intermediate shaft and steering link assembly.

b. Remove the bolt and steering intermediate shaft from the steering link assembly.

20. Remove the steering link assembly.

a. Remove the 2 bolts, 2 nuts and steering link assembly from the front suspension crossmember sub-assembly.

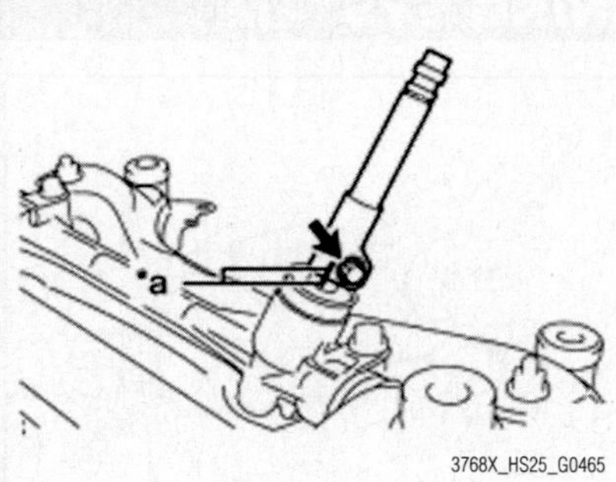

Fig. 159 Marking the steering intermediate shaft

➡Keep the nut from rotating while turning the bolt because the nut has its own stopper.

21. Secure the steering link assembly.

a. Using the special tool secure the steering link assembly in a vise.

➡Tape the special tool before use.

22. Remove the tie rod end sub assemblies.

a. Put matchmarks on the tie rod end sub assemblies and the steering gear assembly.

b. Remove the tie rod end sub assemblies and the lock nuts.

To install:

23. Install the tie rod end sub assemblies.

a. Install the lock nuts and tie rod end sub assemblies to the steering gear assembly until the matchmarks are aligned.

➡After adjusting the toe in, tighten the lock nut.

24. Install the steering link assembly.

a. Install the steering link assembly to the front suspension crossmember sub-assembly with the 2 bolts and 2 nuts. Tighten to 102 ft. lbs. (138 Nm).

➡Keep the nut from rotating while turning the bolt because the nut has its own stopper.

➡Make sure to tighten the bolts starting from the left side of the vehicle.

25. Install the steering intermediate shaft.

a. Align the matchmarks and install the steering intermediate shaft to the steering link assembly.

b. Install the bolt and tighten to 26 ft. lbs. (35 Nm).

26. Install the No. 1 steering column hole cover sub assembly.

a. Align the round hole in the No. 1 steering column hole cover sub-assembly with the protrusion of the steering link assembly to install the cover.

27. Install the front suspension crossmember sub assembly.

28. Install the front suspension member rear braces.

29. Install the rear side rail reinforcement sub assemblies.

30. Connect the front No. 1 lower suspension arm sub assemblies.

31. Connect the tie rod end sub assemblies.

a. Connect the tie rod end sub assembly to the steering knuckle with the nut. Tighten to 36 ft. lbs. (49 Nm).

➡Further tighten the nut up to 60°if the holes for the cotter pin are not aligned.

b. Install a new cotter pin.

32. Connect the front stabilizer link assemblies.

33. Connect the No. 1 steering column hole cover sub assembly.

a. Place clip A as shown to engage clip B to the body to connect the No. 1 steering column hole cover sub assembly.

➡Make sure that the lips of the No. 1 steering column hole cover sub assembly are not damaged.

34. Connect the No. 2 steering intermediate shaft assembly.

35. Place the front wheels facing straight ahead.

36. Install the column hole cover silencer sheet.

37. Install the rear engine under covers.

38. Install the front No. 3 engine under cover.

39. Install the No. 2 engine under cover.

40. Install the No. 1 engine under cover.

41. Install the front wheels. Tighten to 76 ft. lbs. (103 Nm).

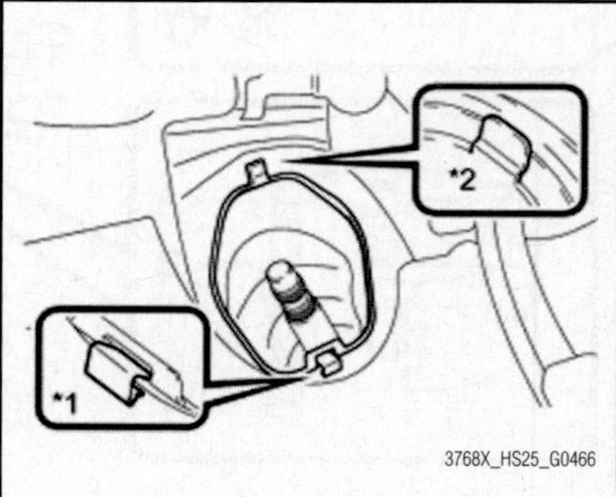

Fig. 160 Connecting the No. 1 steering column hole cover sub assembly

42. Stabilize the suspension.
43. Inspect and adjust the front wheel alignment.

POWER STEERING PUMP

REMOVAL & INSTALLATION

See Figures 161 through 164.

1. Before servicing the vehicle, refer to the precautions section.
2. Drain power steering fluid.
3. Remove RH engine under cover.
4. Remove RH front fender apron seal.
5. Remove fan and generator v belt.
6. Slide the clip and disconnect the No. 1 fluid reservoir to pump hose from the vane pump assembly.
7. Disconnect pressure feed tube assembly, as follows:
   a. Remove the union bolt and disconnect the pressure feed tube assembly from the vane pump assembly.
   b. Remove the gasket from the pressure feed tube assembly.
8. Disconnect the power steering fluid pressure switch connector.
9. Remove vane pump assembly, as follows:

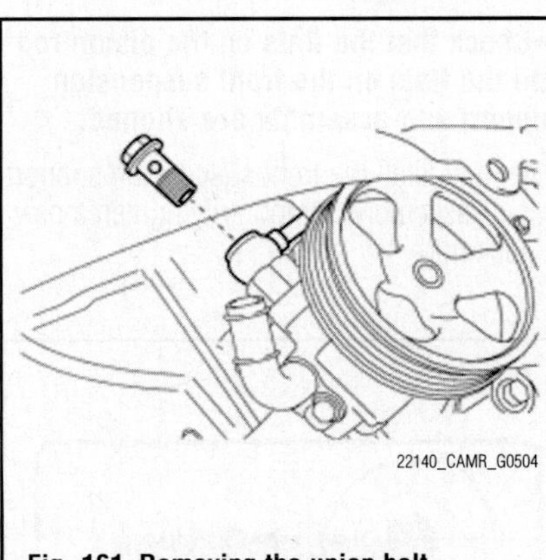

Fig. 161 Removing the union bolt

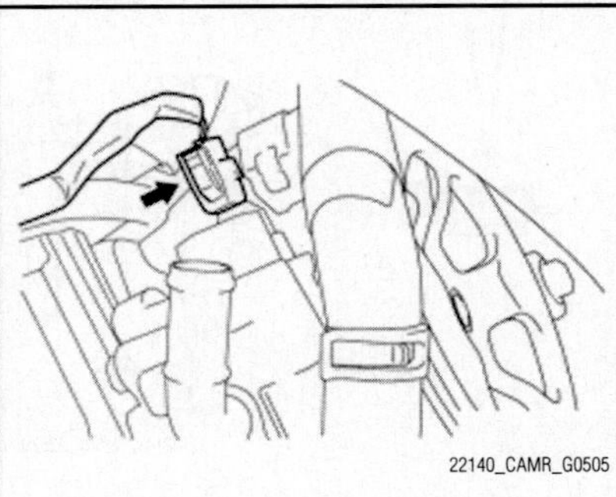

Fig. 162 Disconnecting the power steering fluid pressure switch connector

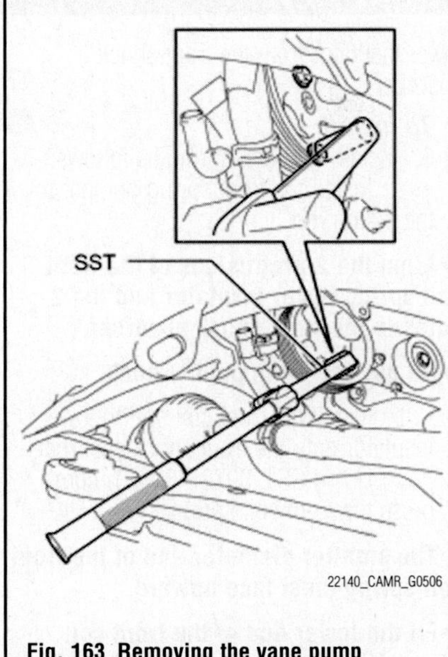

Fig. 163 Removing the vane pump assembly

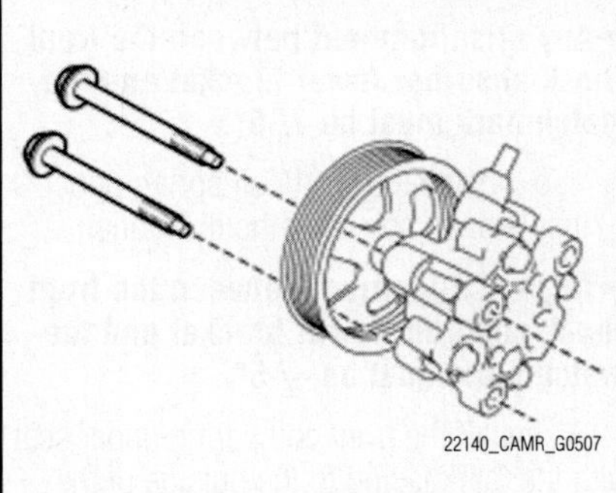

Fig. 164 Removing 2 bolts from the vane pump assembly

   a. Using SST: 09249-63010, loosen the 2 bolts and remove the vane pump assembly.
   b. Remove the 2 bolts from the vane pump assembly.

To install:

10. Install vane pump assembly, as follows:
   a. Temporarily install the 2 bolts to the vane pump assembly.
   b. Install the vane pump assembly.

➡Use a torque wrench with a fulcrum length of 11.81 inches (300 mm).

➡This torque value is effective when SST is parallel to the torque wrench.

   c. Using SST: 09249-63010, tighten the 2 bolts to 32 ft. lbs. (43 Nm).

11. Connect the connector to the power steering fluid pressure switch.
12. Connect pressure feed tube assembly, as follows:
   a. Install a new gasket to the pressure feed tube assembly.

➡Make sure that the stopper of the pressure feed tube assembly contacts the vane pump assembly securely as shown in the illustration.

   b. Connect the pressure feed tube assembly to the vane pump assembly with the union bolt. Tighten to 37 ft. lbs. (50 Nm).

13. Connect No. 1 fluid reservoir to pump hose, as follows:

➡Connect the No. 1 oil reservoir to pump hose with the paint mark facing toward the rear of the vehicle.

➡Push the No. 1 oil reservoir to pump hose as far as it will go.

➡Install the clip at the position specified in the illustration.

   a. Connect the No. 1 fluid reservoir to pump hose to the vane pump assembly with the clip.

14. To complete installation, reverse removal procedure.

BLEEDING

1. Before servicing the vehicle, refer to the precautions section.
2. Check the fluid level.
3. Jack up the front of the vehicle and support it with stands.
4. With the engine stopped, turn the wheel slowly from lock to lock several times.
5. Lower the vehicle.
6. Start the engine.
7. Run the engine at idle for a few minutes.
8. With the engine idling, turn the wheel left or right to the full lock position and keep it there for 2 to 3 seconds, then turn the wheel to the opposite full lock position and keep it there for 2 to 3 seconds.
9. Repeat the above steps several times.
10. Stop the engine.
11. Check for foaming or emulsification. If the system has to be bled twice because of foaming or emulsification, check for fluid leaks in the system.
12. Check the fluid level.

COIL SPRING

REMOVAL & INSTALLATION

See Figures 165 through 170.

1. Before servicing the vehicle, refer to the precautions section.
2. Remove the front shock absorber.
3. As shown in the illustration, secure the front shock absorber with coil spring in a vise using aluminum plates by clamping onto a double nutted bolt affixed to the bracket at the bottom of the absorber.

➡Do not use an impact wrench.

➡If the front coil spring is compressed at an angle, using 2 SST will make the work easier.

4. Using SST: 09727-30021, compress the front coil spring.
5. Remove the front suspension support sub-assembly, front suspension support bearing, front coil spring upper seat, front coil spring upper insulator, front coil spring, front spring bumper, and front coil spring lower insulator from the front shock absorber.

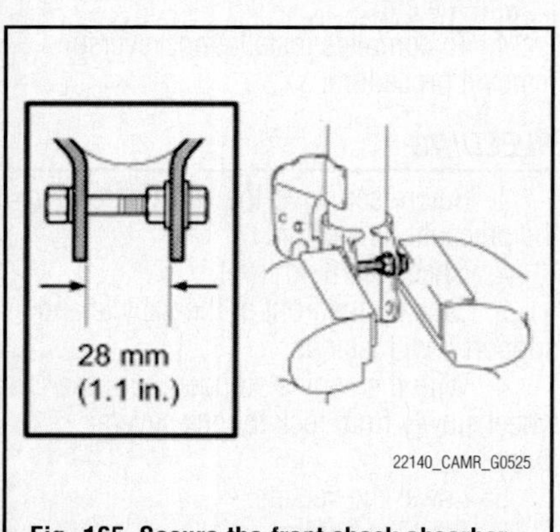

Fig. 165 Secure the front shock absorber

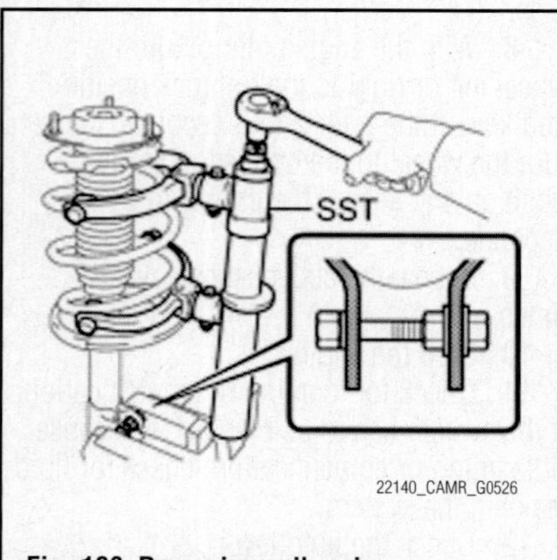

Fig. 166 Removing coil spring components

To install:

6. Install front coil spring as follows:
   a. Install the front spring bumper to the piston rod.

➡Align the 2 protrusions of the front coil spring lower insulator and the 2 holes in the front shock absorber.

➡Do not use an impact wrench.

   b. Install the front coil spring lower insulator onto the front shock absorber.
   c. Using SST: 09727-30021, compress the front coil spring.

➡The smaller diameter end of the front coil spring must face upward.

➡Fit the lower end of the front coil spring into the gap of the insulator.

   d. Install the front coil spring to the front shock absorber.

➡Any misalignment between the front shock absorber lower bracket and the matchmark must be +/-5°.

   e. Install the front coil spring upper insulator as shown in the illustration.

➡Any misalignment between the front shock absorber lower bracket and the matchmark must be +/-5°.

7. Install the front coil spring upper seat with the mark facing to the outside of the vehicle.

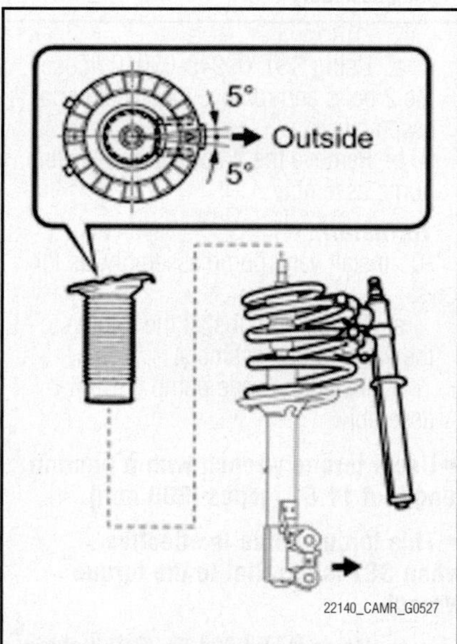

Fig. 167 Installing the front coil spring upper insulator

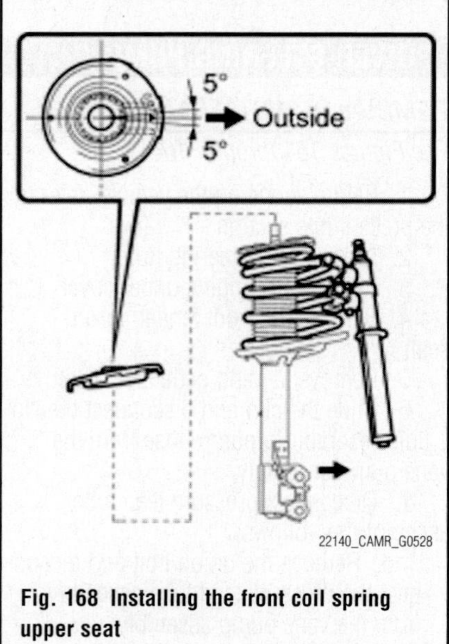

Fig. 168 Installing the front coil spring upper seat

➡If there is foreign matter inside the front suspension support bearing, replace it with a new one.

   a. Install a new front suspension support bearing.

➡Check that the flats on the piston rod and the flats on the front suspension support sub-assembly are aligned.

   b. Install the front suspension support sub-assembly. Temporarily tighten a new lock nut.

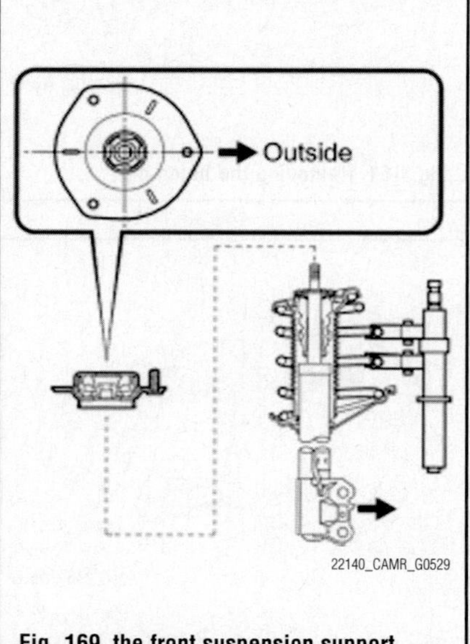

Fig. 169 the front suspension support sub-assembly

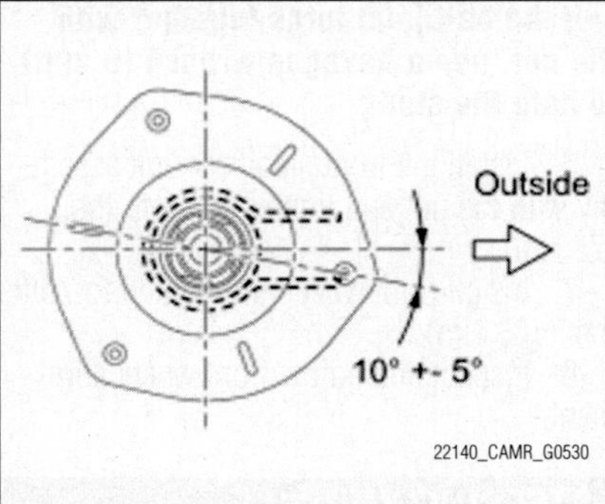

Fig. 170 Aligning the front shock absorber lower bracket and arrows

➡**Do not use an impact wrench.**

➡**Any misalignment between the front shock absorber lower bracket and the matchmark must be +/-5°.**

c. Remove the SST slowly in order to release the coil spring.

CONTROL LINKS

REMOVAL & INSTALLATION

See Front Stabilizer Bar.

LOWER BALL JOINT

REMOVAL & INSTALLATION

See Figure 171.

1. Before servicing the vehicle, refer to the precautions section.
2. Remove the front wheel.
3. Remove the front axle hub nut.
4. Separate the front speed sensor.
5. Separate the front disc the brake caliper assembly.
6. Remove front disc.
7. Separate the tie rod assembly.
8. Separate the No. 1 front lower suspension arm.
9. Remove the front axle assembly.
10. Remove front wheel No. 1 bearing dust deflector.
11. Remove front axle hub hole snap ring.
12. Remove front axle hub.
13. Remove front disc brake dust cover.
14. Remove the front lower ball joint assembly, as follows:
    a. Secure the steering knuckle in a vise using aluminum plates.
    b. Remove the cotter pin and castle nut.

➡**Do not damage the dust cover of the ball joint.**

➡**Do not damage the steering knuckle.**

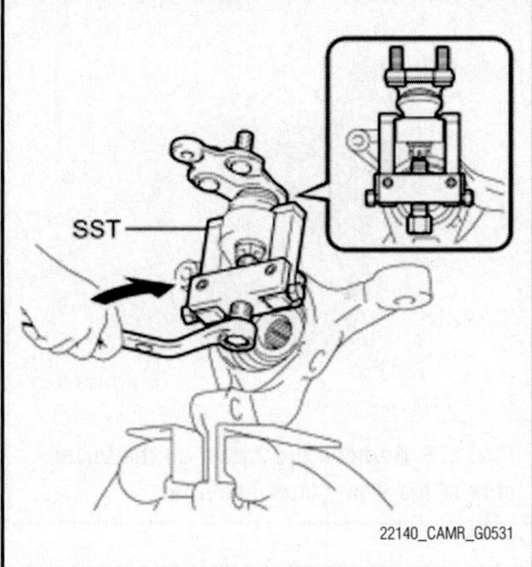

Fig. 171 Remove the front lower ball joint assembly

c. Using SST (SST: 09628-62011) or equivalent, remove the front lower ball joint assembly.

To install:

15. Installation is the reverse of the removal procedure, noting the following:
    a. Install the front lower ball joint assembly to the steering knuckle with the castle nut and tighten to 91 ft. lbs. (123 Nm). Further tighten the nut up to 60°if the holes for the cotter pin are not aligned.
    b. Inspect and adjust the front wheel alignment.
    c. Inspect the ABS speed sensor signal.

LOWER CONTROL ARM

REMOVAL & INSTALLATION

See Figures 172 through 174.

1. Before servicing the vehicle, refer to the precautions section.
2. Remove the engine assembly with transaxle.

➡**Use the same procedures for the RH side and the LH side. The procedures listed below are for the LH side.**

3. Remove the 3 nuts and the engine mounting insulator.
4. Remove the 3 bolts and the nut on the front suspension lower No. 1 arm and remove it from the front frame assembly.
5. Remove the front lower arm bushing stopper.

To install:

6. Install the front lower arm bushing stopper.
7. Install the front suspension lower No. 1 arm to the front frame assembly with the 3 bolts and the nut, but do not tighten them yet.

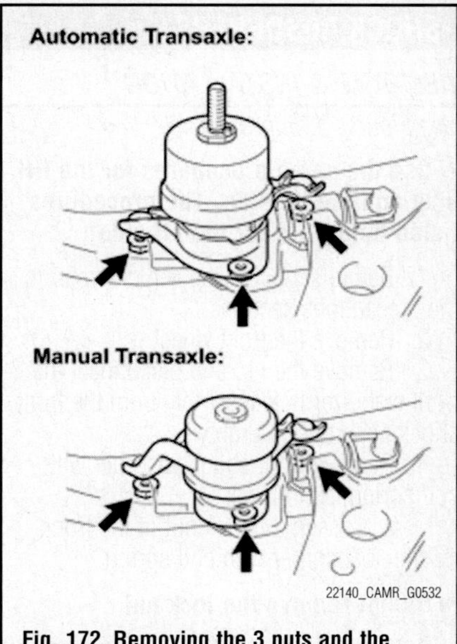

Fig. 172 Removing the 3 nuts and the engine mounting insulator

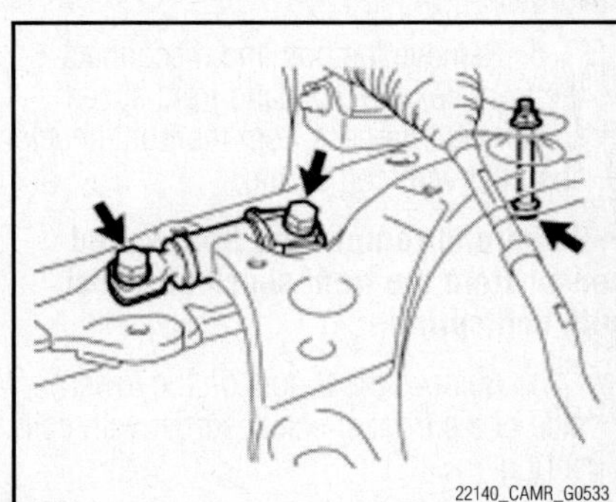

Fig. 173 Removing the 3 bolts and the nut on the front suspension lower No. 1 arm

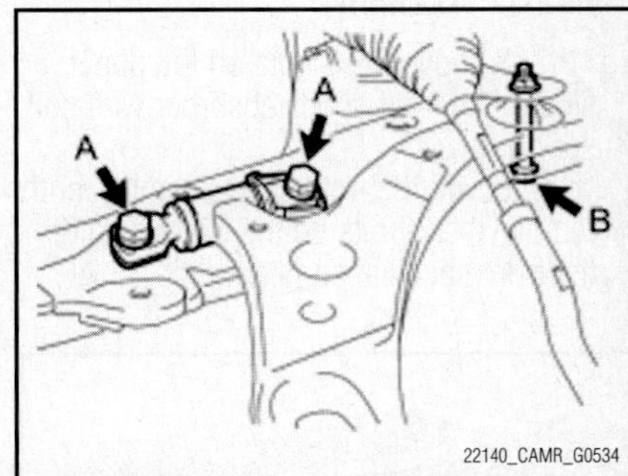

Fig. 174 Tightening the 3 bolts and the nut on the front suspension lower No. 1 arm

8. Tighten bolts "A" to 148 ft. lbs. (200 Nm). Tighten bolts "B" to 152 ft. lbs. (206 Nm).
9. Install the engine mounting insulator with the 3 nuts and tighten to 64 ft. lbs. (87 Nm).
10. Install the engine assembly with transaxle.

SHOCK ABSORBERS

REMOVAL & INSTALLATION

See Figures 175 through 177.

➡Use the same procedures for the RH side and the LH side. The procedures listed below are for the LH side.

1. Before servicing the vehicle, refer to the precautions section.
2. Remove the front wheel.
3. Remove the nut and disconnect the front stabilizer link assembly from the front shock absorber assembly.
4. Remove front shock absorber with coil spring, as follows:
   a. Loosen the lock nut of the front shock absorber with coil spring.

➡Do not remove the lock nut.

➡Only loosen the nut when disassembling the front shock absorber with coil spring.

   b. Remove the bolt and disconnect the front flexible hose and front speed sensor wire harness from the front shock absorber with coil spring.

➡Be sure to remove the front speed sensor from the front shock absorber with coil spring.

   c. Remove the 2 nuts on the lower side of the front shock absorber with coil spring.

➡When removing the nuts, keep the bolts from rotating.

➡Keep the bolts inserted to secure the front axle assembly.

   d. Remove the 3 nuts on the upper side of the front shock absorber with coil spring.
   e. Lower the front axle assembly, and remove the 2 bolts on the lower side of the front shock absorber.

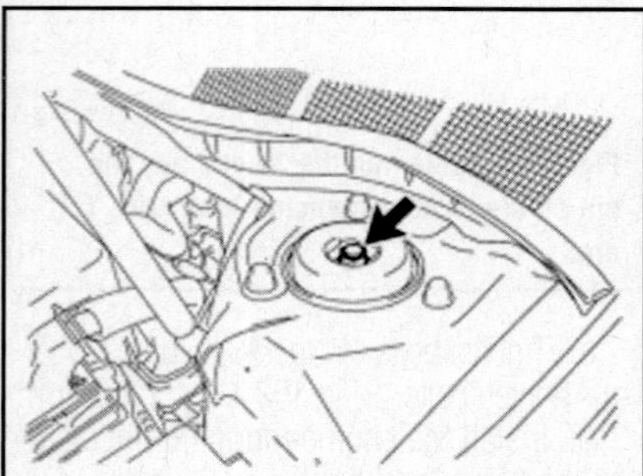

Fig. 175 Loosening the lock nut of the front shock absorber arm

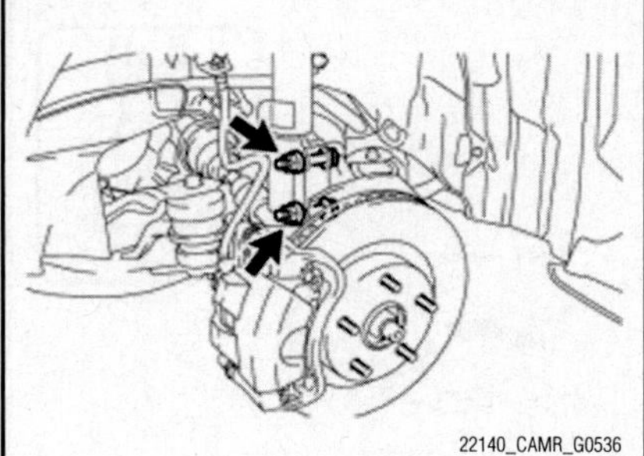

Fig. 176 Remove the 2 nuts on the lower side of the front shock absorber

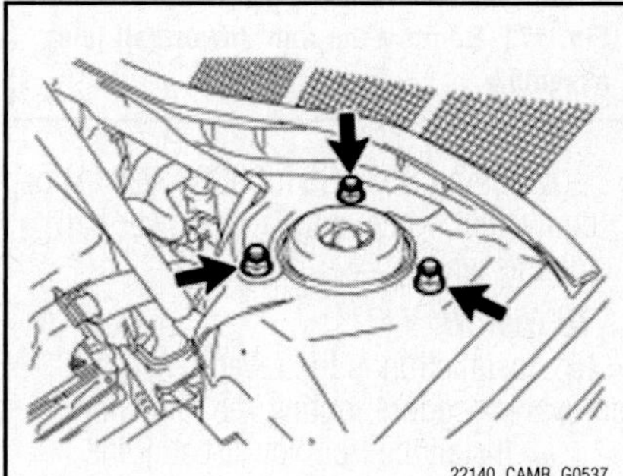

Fig. 177 Removing the 3 nuts on the upper side of the front shock absorber

➡Make sure that the front speed sensor is disconnected from the front shock absorber with coil spring.

   f. Remove the front shock absorber with coil spring.

To install:

5. Install front shock absorber with coil spring, as follows:
   a. Install the front shock absorber with coil spring to the front axle assembly and insert the 2 bolts from the front side of the vehicle.
   b. Slowly jack up the vehicle using a wooden block and install the front shock absorber with coil spring (upper side) to the vehicle.
   c. Install the 3 nuts to the upper side of the front shock absorber with coil spring and tighten to 63 ft. lbs. (85 Nm).

➡When installing the nuts, keep the bolts from rotating.

   d. Install the 2 nuts to the lower side of the front shock absorber with coil spring and tighten to 155 ft. lbs. (210 Nm).
   e. Install the front flexible hose and front speed sensor wire harness with the bolt and tighten to 14 ft. lbs. (19 Nm).
   f. Fully tighten the lock nut and tighten to 52 ft. lbs. (70 Nm).

➡If the ball joint turns together with the nut, use a hexagon wrench (6 mm) to hold the stud.

6. Install the front stabilizer link assembly with the nut and tighten to 55 ft. lbs. (74 Nm).
7. Install front wheel and tighten to 76 ft. lbs. (103 Nm).
8. Inspect and adjust front wheel alignment.

STEERING KNUCKLE

REMOVAL & INSTALLATION

See Wheel Hub and Bearing.

STABILIZER BAR & LINKS

REMOVAL & INSTALLATION

See Figures 178 through 180.

1. Before servicing the vehicle, refer to the precautions section.
2. Remove the front wheels.
3. Separate steering intermediate shaft assembly.
4. Separate tie rod end sub-assembly.

➡If the ball joint turns together with the nut, use a hexagon wrench (6 mm) to hold the stud.

5. Remove the 2 nuts and the front stabilizer link assembly
6. Remove the engine assembly with transaxle.
7. Remove the bolts and the left and right No. 1 stabilizer brackets.
8. Remove the engine assembly with transaxle.
9. Remove the bolts and the left and right No. 1 stabilizer brackets.
10. Remove the 2 front No. 1 stabilizer bar bushings from the front stabilizer bar.
11. Remove the front stabilizer bar from the vehicle.

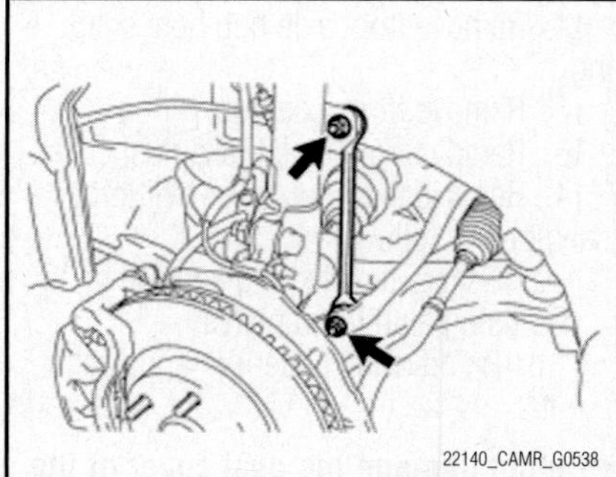

Fig. 178 Remove the 2 nuts and front stabilizer link assembly (left hand shown)

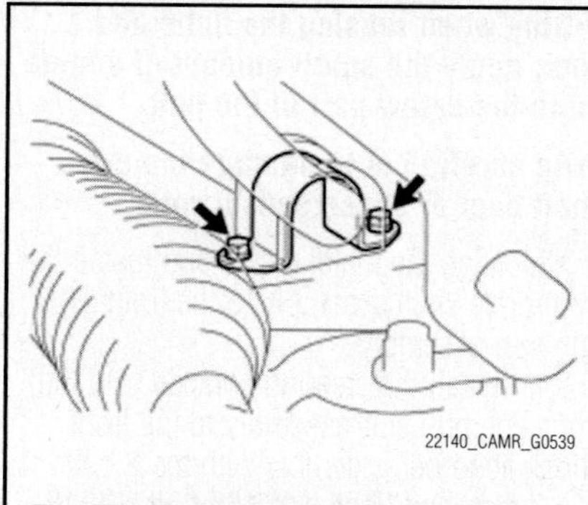

Fig. 179 Remove the 2 bolts and No. 1 stabilizer bracket (left hand shown)

To install:

➡Make sure that the cutout of the front stabilizer bar bushing No. 1 faces the rear side as shown.

12. Install the 2 front stabilizer bar bushings No. 1 to the outside of the bushing stopper on the front stabilizer bar.
13. Install the No. 1 left front stabilizer bracket with the 2 bolts and tighten to 20 ft. lbs. (27 Nm).
14. Install the No. 1 right front stabilizer bracket with the 2 bolts and tighten to 20 ft. lbs. (27 Nm).
15. Install the engine assembly with transaxle.
16. Install the left front stabilizer link assembly with the 2 nuts and tighten to 55 ft. lbs. (74 Nm).
17. Install the right front stabilizer link assembly with the 2 nuts and tighten to 55 ft. lbs. (74 Nm).

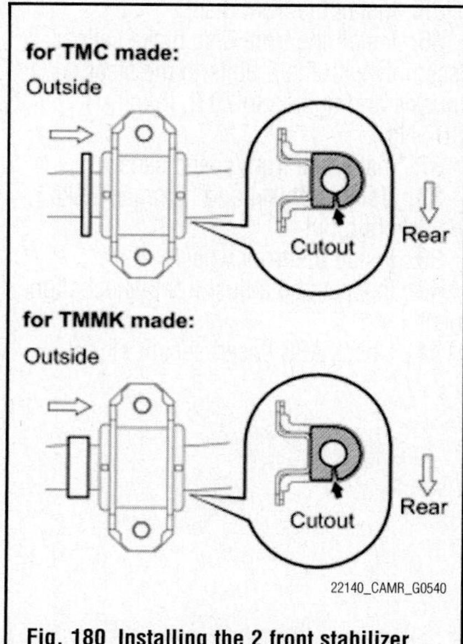

Fig. 180 Installing the 2 front stabilizer bar bushings

18. To complete installation, reverse removal procedure.
19. Inspect and adjust the front wheel alignment.

WHEEL HUB & BEARING

REMOVAL & INSTALLATION

See Figures 181 through 187.

1. Before servicing the vehicle, refer to the Precautions Section.
2. Remove front wheel.
3. Remove front axle hub nut.
4. Separate front speed sensor.
5. Remove the 2 bolts and separate the front disc brake caliper assembly from the steering knuckle. Use wire or an equivalent tool to keep the brake caliper from hanging down by the flexible hose.
6. Remove front disc.
7. Separate tie rod end sub-assembly.
8. Separate front suspension lower no. 1 arm.
9. Remove front axle assembly.
10. Using a screwdriver with its tip wrapped with vinyl tape, remove the No. 1 front wheel bearing dust deflector. Be careful not to damage the steering knuckle.
11. Using snap ring pliers, remove the front axle hub hole snap ring.
12. Remove front axle hub sub-assembly by performing the following:
    a. Hold the front axle assembly between aluminum plates in a vise.

➡Do not overtighten the vise.

    b. Using SST 09520-00031, remove the front axle hub sub-assembly.

➡Be careful not to drop the front axle hub sub-assembly.

    c. Using SST 09555-55010, SST: 09950-60010 and SST: 09950-70010 and a press, remove the bearing inner race (outside) from the front axle hub sub-assembly.

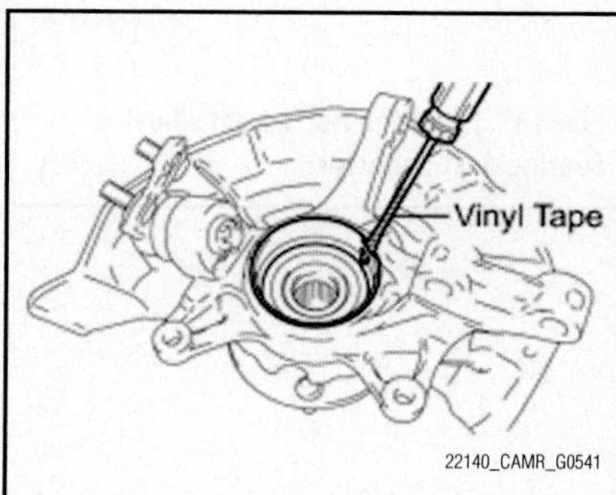

Fig. 181 Remove the No. 1 front wheel bearing dust deflector

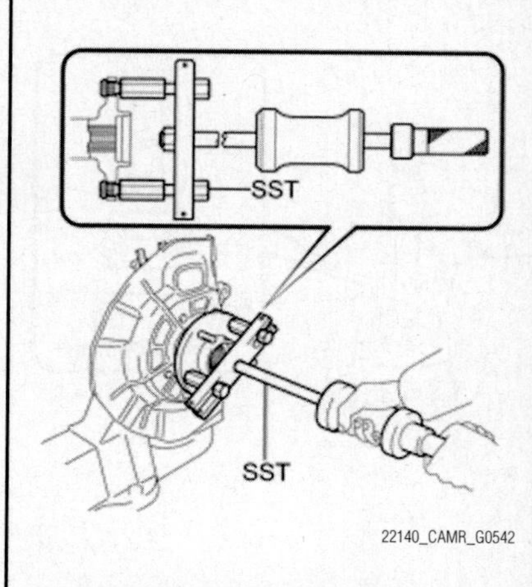

Fig. 182 Remove the front axle hub sub-assembly

13. Remove the 4 bolts and disc brake dust cover from the steering knuckle.
14. Remove front lower ball joint assembly.
15. Remove front axle hub bearing by performing the following:
    a. Place the bearing inner race (outside) on the front axle hub bearing.
    b. Using SST 09527-17011, SST: 09950-60010 and a press, press the front axle hub bearing until it contacts the SST: 09950-70010.
    c. Using SST: 09527-20011, SST: 09950-60010 to make the steering knuckle horizontal, fix it to the V-block.
16. Using SST: 09950-70010 and a press, remove the front axle hub bearing from the steering knuckle.

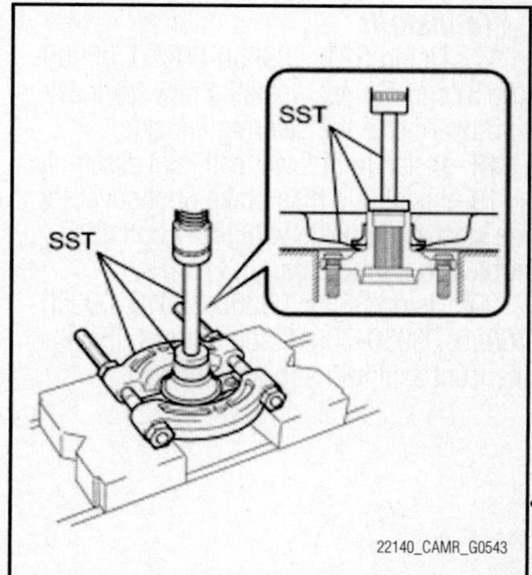

Fig. 183 Remove the bearing inner race (outside) from the front axle hub sub-assembly

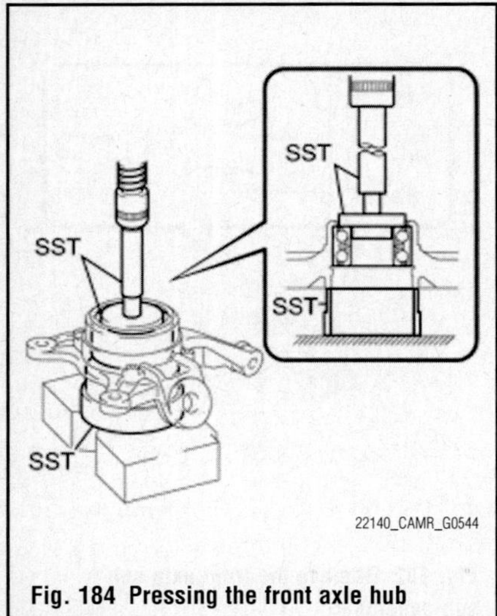

Fig. 184 Pressing the front axle hub bearing

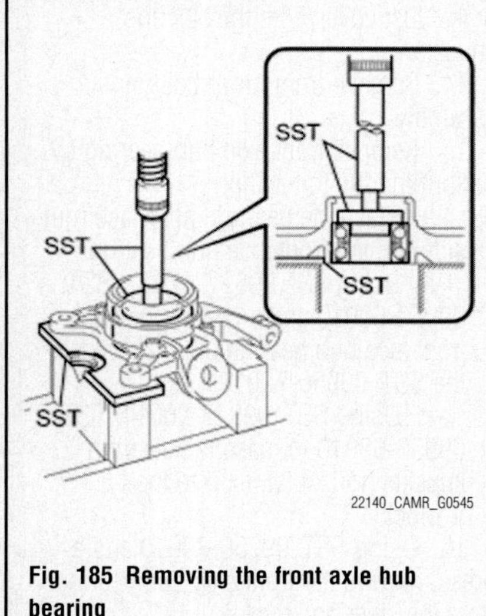

Fig. 185 Removing the front axle hub bearing

To install:

17. Using SST's: 09950-60020, 09950-70010 and a press, install a new front axle hub bearing to the steering knuckle.
18. Install front lower ball joint assembly.
19. Install the disc brake dust cover to the steering knuckle with the 4 bolts and tighten to 73 inch lbs. (8.3 Nm).
20. Using SST's: 09608-32010, 09950-60020, 09950-70010 and a press, install the front axle hub sub-assembly.
21. Using snap ring pliers, install a new front axle hub hole snap ring.

➡Align the hole for the speed sensor in the No. 1 front wheel bearing dust deflector with the steering knuckle.

22. Using SST's: 09316-60011, 09608-32010 and a hammer, install a new No. 1 front wheel bearing dust deflector.

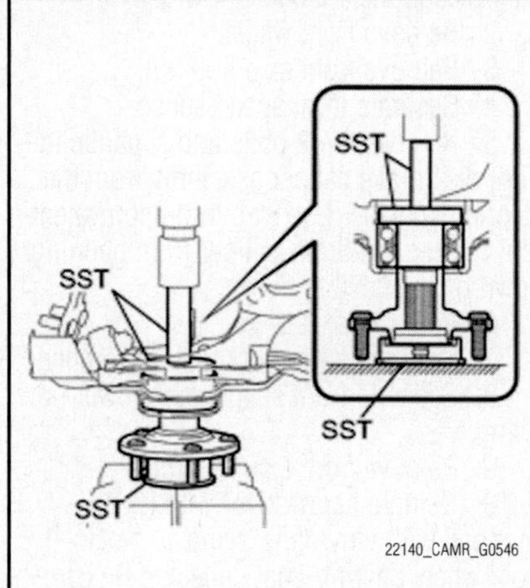

Fig. 186 Installing the front axle hub sub-assembly

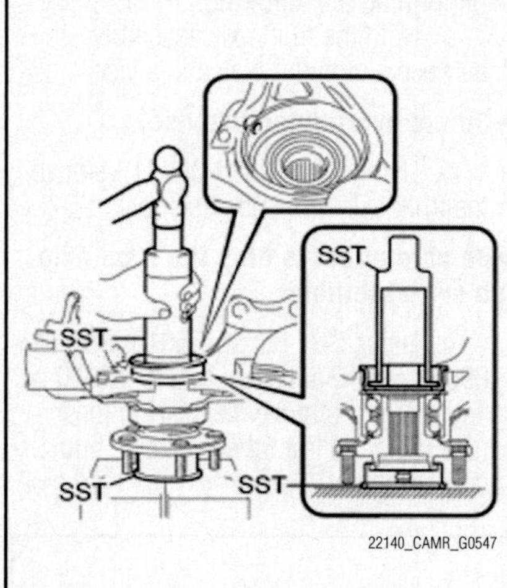

Fig. 187 Installing No. 1 front wheel bearing dust deflector

➡Only when reusing the bolts and nuts, apply the small amount of engine oil to the screw part of the nuts.

➡Be careful not to damage the drive shaft boot or speed sensor rotor.

23. Align the matchmarks and install the front drive shaft assembly to the front axle hub sub-assembly.
24. Install the steering knuckle with the front axle hub sub-assembly to the front shock absorber assembly with the 2 bolts and 2 nuts and tighten to 155 ft. lbs. (210 Nm).
25. Install the lower No. 1 front suspension arm sub-assembly.
26. Install the tie rod end sub-assembly.
27. Install the front disc.
28. Install the front disc brake caliper assembly with the 2 bolts to the steering knuckle and tighten to 79 ft. lbs. (107 Nm).
29. Clean the threaded parts on the drive shaft and axle hub nut using a non-residue solvent.

➡Be sure to perform this work for a new drive shaft.

➡Keep the threaded parts free of oil and foreign objects.

30. Using a 30 mm socket wrench, install the front axle hub nut and tighten to 217 ft. lbs. (294 Nm).
31. Remove the 2 bolts and separate the front disc brake caliper assembly from the steering knuckle.
32. Remove the front disc.
33. Inspect front axle hub bearing looseness.
34. Inspect front axle hub runout.
35. Install the front disc.
36. Install the front disc brake caliper assembly with the 2 bolts to the steering knuckle and tighten to 79 ft. lbs. (107 Nm).
37. Install the front speed sensor.
38. Using a chisel and hammer, stake the axle hub nut.
39. Install the front wheel.
40. Inspect and adjust front wheel alignment.
41. Check ABS speed sensor signal.

COIL SPRING

REMOVAL & INSTALLATION

See Figures 188 through 190.

1. Before servicing the vehicle, refer to the precautions section.
2. Secure the rear shock absorber with coil spring in a vise using aluminum plates by closing the vise onto the double netted bolt affixed to the bracket at the bottom of the absorber.

➡Do not use an impact wrench.

➡If the rear coil spring is compressed at an angle, using 2 SST will make the work easier.

3. Using SST: 09727-30021, compress the rear coil spring
4. Remove the nut, rear shock absorber collar and rear suspension support assembly.
5. Remove the rear coil spring, rear No. 1 spring bumper, and rear coil spring lower insulator.

To install:

6. Install the rear No. 1 spring bumper to the piston rod.
7. Install the rear coil spring lower insulator onto the rear shock absorber.

➡Do not use an impact wrench.

8. Using SST: 09727-30021, compress the rear coil spring.

➡The smaller diameter end must face upward.

➡Fit the lower end of the rear coil spring into the gap of the lower seat.

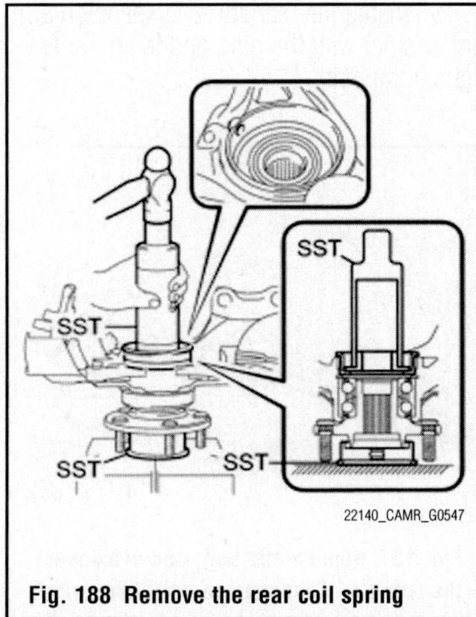

Fig. 188 Remove the rear coil spring

➡If the front coil spring is compressed at an angle, using 2 SST will make the work easier.

9. Install the rear coil spring to the rear shock absorber.

➡Align the notches of the piston rod and the rear suspension support assembly as shown in the illustration before installing the rear suspension support assembly.

10. Install the rear suspension support assembly.
11. Align the notches of the shock absorber with the notch of the rear suspension support assembly so that the notches face the outside of the vehicle.
12. Install the rear shock absorber collar.
13. Loosely tighten a new lock nut to the rear suspension piston rod.

➡Do not use an impact wrench.

➡When lining up the rear suspension support assembly's stud bolts at the middle point between the two sides of the bracket, the maximum permissible degree of error is plus or minus 5°.

14. Release the spring while adjusting the rear suspension support assembly to the position shown in the illustration, and remove the SST from the rear coil spring.

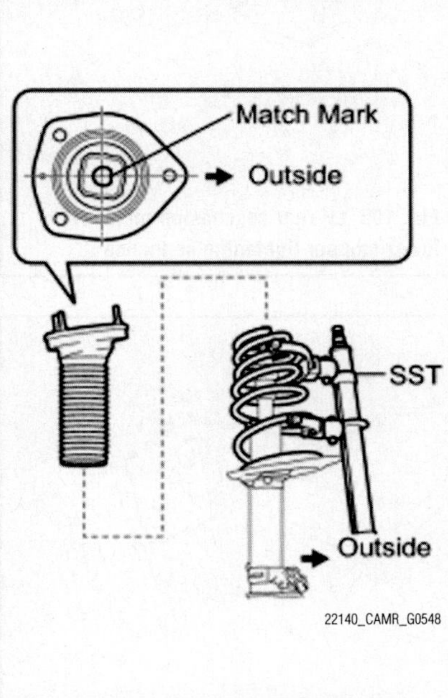

Fig. 189 Align the notches of the shock absorber

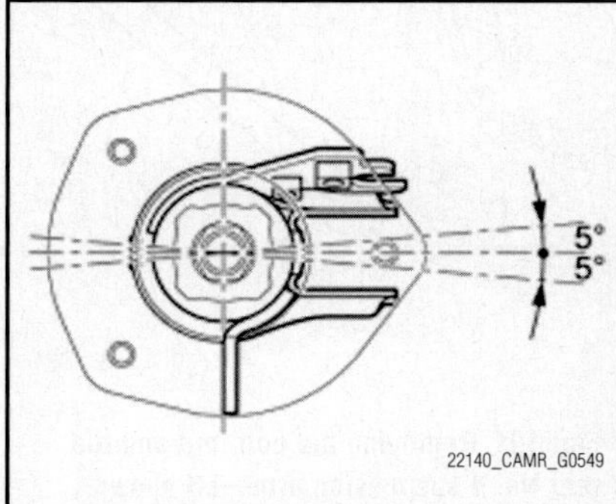

Fig. 190 Lining up the rear suspension support assembly's stud bolts

CONTROL ARMS/LINKS

REMOVAL & INSTALLATION

No. 1 Suspension Arm

See Figures 191 through 196.

1. Before servicing the vehicle, refer to the precautions section.

➡Check if an old gasket still remains on the pipe. If so, remove it. Also, check if any bolts or nuts are rusted. If so, replace them.

2. Remove rear wheel.
3. Remove center exhaust pipe assembly.
4. Remove tail exhaust pipe assembly.
5. Separate both rear stabilizer link assemblies.
6. Remove rear stabilizer bar no. 2 and no. 1 bracket.
7. Remove rear stabilizer bar.
8. Remove rear stabilizer bushing.
9. Separate rear strut rod.

➡When removing the bolt, keep the nut from rotating.

10. Remove the bolt, nut and separate the rear suspension No. 2 arm (outer side) from the rear axle carrier.

➡When removing the bolt, keep the nut from rotating.

11. Remove the bolt, nut and the rear No. 1 suspension arm (outer side) from the rear axle carrier.
12. Remove the 2 nuts and the LH rear suspension member lower stopper.
13. Remove the 2 nuts and the RH rear suspension member lower stopper.
14. Support the rear suspension member with a jack.
15. Remove the 2 bolts, and the rear suspension member sub-assembly.

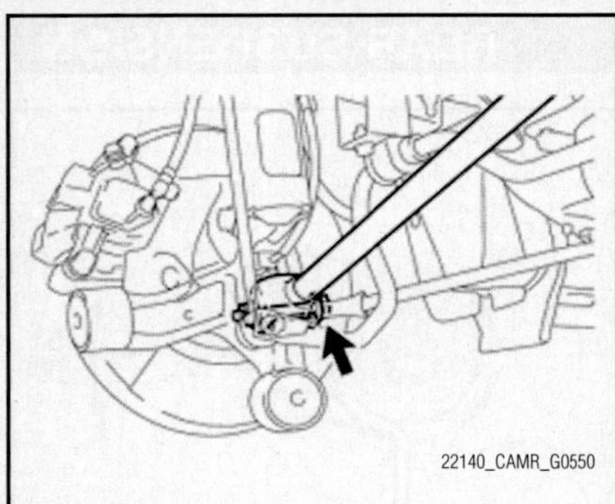

Fig. 191 Removing the bolt, nut and the rear No. 1 suspension arm—LH shown

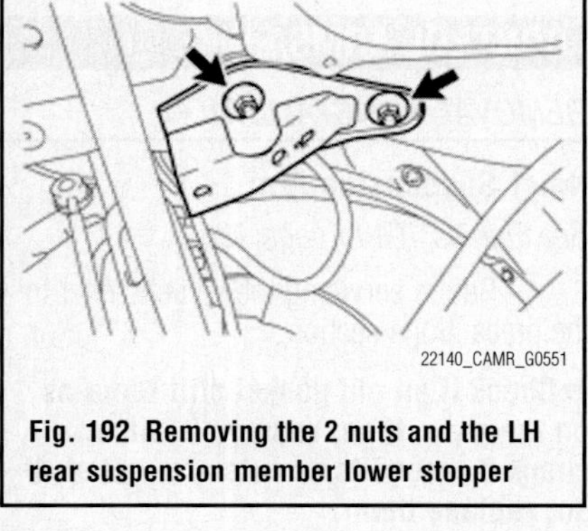

Fig. 192 Removing the 2 nuts and the LH rear suspension member lower stopper

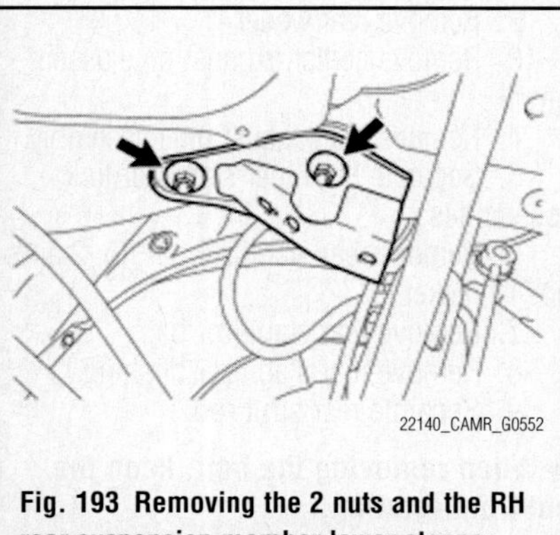

Fig. 193 Removing the 2 nuts and the RH rear suspension member lower stopper

16. Remove the bolt and rear No. 1 suspension arm assembly.

To install:

17. Install the No. 1 rear suspension arm (inner side) with the bolt, and temporarily tighten the bolt.
18. Install the rear No. 1 suspension arm so that the bracket leans toward the front side of the vehicle.
19. Ensure that the paint mark faces the rear side of the vehicle.
20. Set the rear No.1 suspension arm in the position shown in the illustration, and fully tighten the bolt to 74 ft. lbs. (100 Nm).

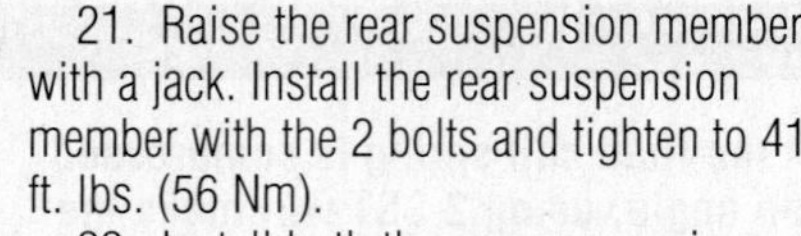

21. Raise the rear suspension member with a jack. Install the rear suspension member with the 2 bolts and tighten to 41 ft. lbs. (56 Nm).
22. Install both the rear suspension member lower stoppers with the 2 nuts and tighten to:
    a. Nut A: 41 ft. lbs. (55 Nm).
    b. Nut B: 28 ft. lbs. (38 Nm).

➡Insert the bolt from the front of the vehicle and temporarily install the bolt.

23. Connect the rear No.1 suspension arm (outer side) to the rear axle carrier with

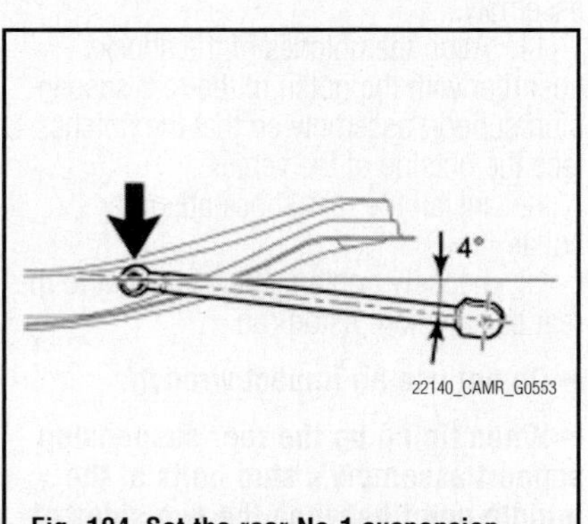

Fig. 194 Set the rear No.1 suspension arm

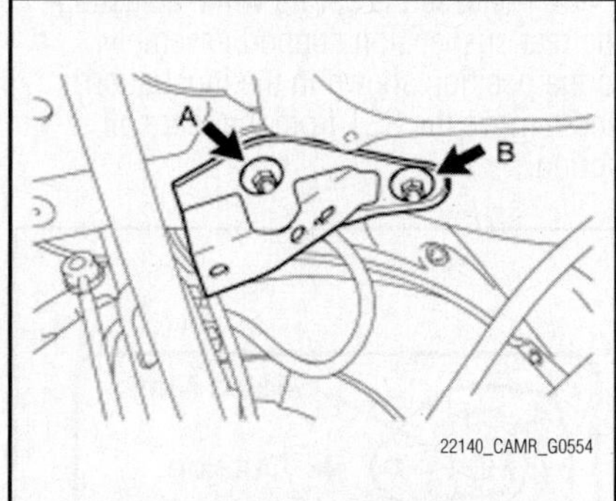

Fig. 195 LH rear suspension member lower stopper tightening sequence

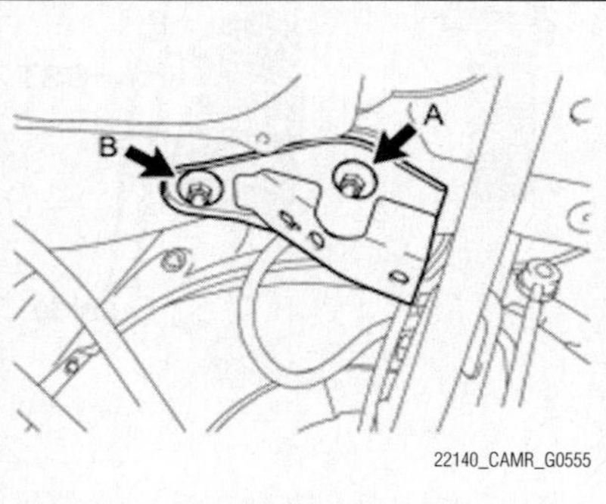

Fig. 196 LH rear suspension member lower stopper tightening sequence

the bolt and nut and temporarily tighten the bolt and nut. When temporarily tightening the bolt, keep the nut from rotating.

➡Insert the bolt from the inside of the vehicle and temporarily install the bolt.

24. Connect the strut rod assembly rear to the axle carrier with the bolt and nut and temporarily tighten the bolt. When temporarily tightening the bolt, keep the nut from rotating.
25. Jack up the rear axle carrier, placing a wooden block to avoid damage. Apply load to the suspension so that the installed bolt of the rear No. 1 suspension arm (inner side) is horizontally aligned with the center of the rear axle hub.
26. Fully tighten rear No. 1 suspension arm and tighten the bolt to 74 ft. lbs. (100 Nm).
27. Fully tighten rear No. 2 suspension arm and tighten the bolt to 74 ft. lbs. (100 Nm).
28. To complete installation, reverse removal procedure.

No. 2 Suspension Arm

See Figures 197 and 198.

1. Before servicing the vehicle, refer to the precautions section.
2. Remove the rear wheel.
3. Remove the bolt, and disconnect the rear No. 2 suspension arm (inner side).

➡When removing the bolt, keep the nut from rotating.

4. Remove the bolt, nut and the rear No. 2 suspension arm (outer side) from the rear axle carrier.

To install:

➡Ensure that the paint mark faces to the rear of the vehicle.

5. Install the rear No. 2 suspension arm (inner side) with the bolt, and temporarily tighten the bolt.

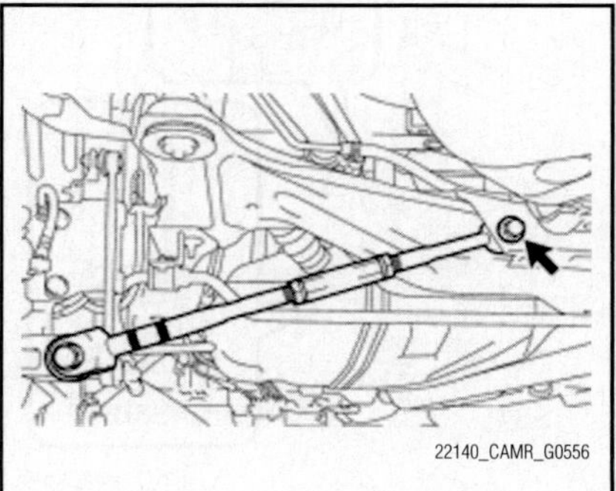

Fig. 197 Remove the bolt, and disconnect the rear No. 2 suspension arm (inner side)

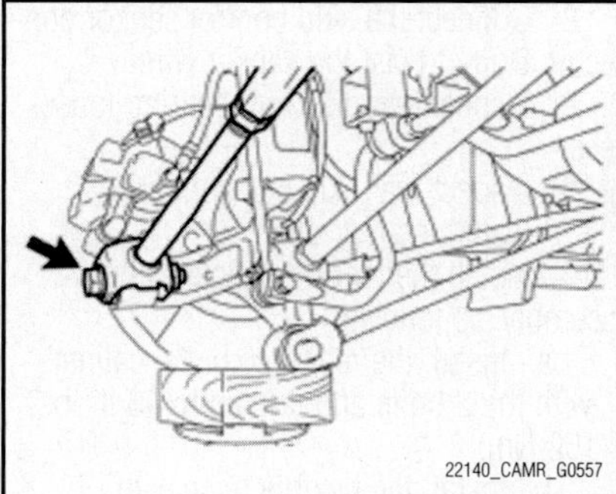

Fig. 198 Removing the bolt, nut and the rear No. 2 suspension arm (outer side)

➡When temporarily tightening the bolt, keep the nut from rotating.

6. Connect the rear No. 2 suspension arm (outer side) to the rear axle carrier with the bolt and nut, and temporarily tighten the bolt.
7. Stabilize suspension.
8. Fully tighten the rear No. 2 suspension arm bolt (inner side) to 74 ft. lbs. (100 Nm).
9. Fully tighten the rear No. 2 suspension arm bolt (outer side) to 74 ft. lbs. (100 Nm).
10. Install the rear wheel.
11. Inspect and adjust the rear wheel alignment.

SHOCK ABSORBER

REMOVAL & INSTALLATION

See Figures 199 and 200.

1. Before servicing the vehicle, refer to the precautions section.
2. Remove the rear seat cushion assembly.
3. Remove rear seat headrest plate cover.
4. Remove rear seat headrest assembly.
5. Remove the rear seatback assembly.
6. Remove the rear wheel.
7. Separate LH rear stabilizer link assembly.
8. Remove the 2 bolts, and disconnect the rear brake flexible hose and rear speed sensor from the rear shock absorber with coil spring and rear axle carrier.
9. Remove the 4 claws and the rear suspension support No. 1 cover.
10. Remove the rear shock absorber with coil spring, as follows:

➡Do not remove the lock nut.

➡Only loosen the nut when disassembling the rear shock absorber with coil spring.

a. Loosen the lock nut of the rear shock absorber with coil spring.

➡When removing the nuts, keep the bolts from rotating.

➡Keep one bolt inserted to secure the hub and disc rotor.

b. Remove the 2 nuts and 2 bolts on the lower side of the rear shock absorber with coil spring.
c. Remove the 3 nuts on the upper side of the rear shock absorber with coil spring.

➡Make sure that the rear speed sensor is disconnected from the rear shock absorber with coil spring.

d. Lower the rear axle carrier, and remove the 2 bolts on the lower side of the rear shock absorber with coil spring.

To install:

11. Install the rear shock absorber with coil spring to the rear axle carrier assembly and insert the 2 bolts from the rear of the vehicle.
12. Slowly jack up the vehicle using a wooden block and install the rear shock absorber with coil spring (upper side) to the vehicle.

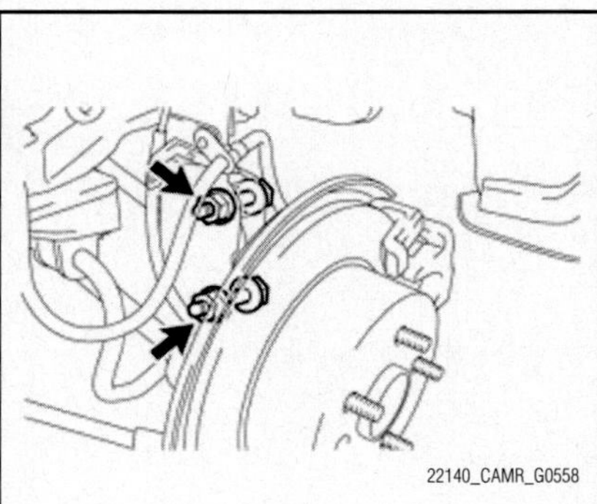

Fig. 199 Loosen the 2 nuts on the lower side of the shock absorber

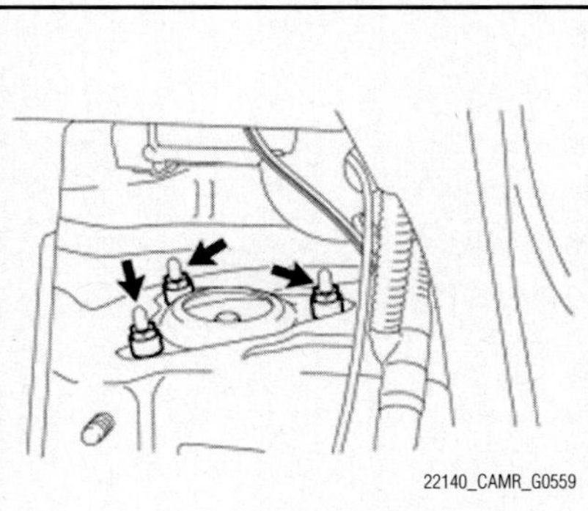

Fig. 200 Remove the 3 nuts

13. Install the 3 nuts to the upper side of the rear shock absorber with coil spring and tighten to 29 ft. lbs. (39 Nm).

➡When installing the nuts, keep the bolts from rotating.

14. Install the 2 nuts and 2 bolts to the lower side of the rear shock absorber with coil spring and tighten 133 ft. lbs. (180 Nm).
15. Fully tighten the lock nut to 41 ft. lbs. (55 Nm).
16. Connect rear speed sensor.
17. Install the LH rear stabilizer link assembly.
18. Engage the 4 claws and install the rear suspension support No. 1 cover.
19. To complete installation, reverse remaining removal.
20. Check abs speed sensor signal.
21. Inspect and adjust rear wheel alignment.

TESTING

1. Inspect the shock absorber.
   a. Compress and extend the shock absorber rod 4 or more times.

➡A normal shock has no abnormal resistance or sound and operation resistance is normal.

➡If there is any abnormality, replace the rear shock absorber with a new one.

WHEEL BEARINGS

REMOVAL & INSTALLATION

See Figure 201.

➡Use the same procedures for the RH side and LH side.

➡The procedures listed below are for the LH side.

1. Before servicing the vehicle, refer to the precautions section.
2. Remove the rear wheel.
3. Separate the rear disc brake caliper assembly, as follows:
   a. Remove the bolt and separate the flexible hose from the shock absorber.
   b. Remove the 2 bolts and separate the rear disc brake caliper assembly.
4. Remove the rear disc.
5. Disconnect the skid control sensor connector.
6. Remove the 4 bolts and the rear axle hub and bearing assembly.

To install:

7. Install the hub and bearing assembly with the 4 bolts and tighten to 59 ft. lbs. (80 Nm).

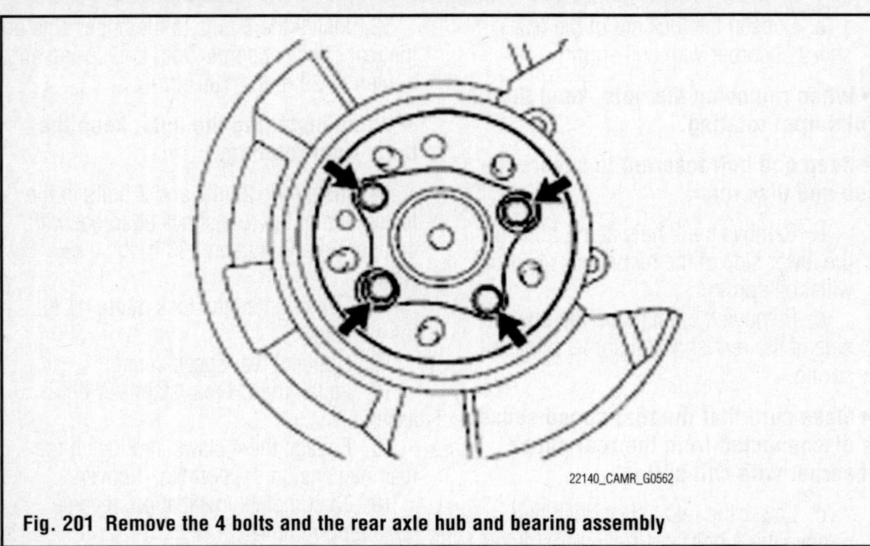

Fig. 201 Remove the 4 bolts and the rear axle hub and bearing assembly

8. Connect the skid control sensor connector. Do not twist the sensor wire.
9. Inspect rear axle hub bearing looseness.
10. Inspect rear axle hub runout.
11. Install the rear disc.
12. Install the rear disc brake caliper assembly, as follows:
    a. Install the rear disc brake caliper with the 2 bolts and tighten to 46 ft. lbs. (62 Nm).
    b. Install the flexible hose with the bolt and tighten to 14 ft. lbs. (19 Nm).
13. Install the rear wheel.
14. Inspect and adjust the rear wheel alignment.
15. Check ABS speed sensor signal.

LEXUS

IS250 • IS350

28

SPECIFICATIONS AND MAINTENANCE CHARTS

ENGINE AND VEHICLE IDENTIFICATION

Engine						
Code ①	Liters (cc)	Cu. In.	Cyl.	Fuel Sys.	Engine Type	Eng. Mfg.
4GR-FSE	2.5 (2500)	153	V6	DI	DOHC	Toyota
2GR-FSE	3.5 (3456)	211	V6	DI	DOHC	Toyota

Model Year	
Code ②	Year
7	2007
8	2008
9	2009
A	2010

SFI: Sequential Multi-port Fuel Injection

DI: Direct Injection

DOHC: Double Overhead Camshaft

① Located on the timing belt cover

② 10th digit of the VIN

3768X_IS25_C0001

GENERAL ENGINE SPECIFICATIONS

All measurements are given in inches.

Year	Model	Engine Displacement Liters (cc)	Engine ID/VIN	Fuel System Type	Net Horsepower @ rpm	Net Torque @ rpm (ft. lbs.)	Bore x Stroke (in.)	Com-pression Ratio	Oil Pressure @ rpm
2007	IS 250	2.5 (2500)	4GR-FSE	DI	204@6400	185@4800	3.27x3.03	12.0:1	55.5@6000
	IS 350	3.5 (3456)	2GR-FSE	DI	306@6400	277@4800	3.70x3.27	11.8:1	55.5@6000
2008	IS 250	2.5 (2500)	4GR-FSE	DI	204@6400	185@4800	3.27x3.03	12.0:1	55.5@6000
	IS 350	3.5 (3456)	2GR-FSE	DI	306@6400	277@4800	3.70x3.27	11.8:1	55.5@6000
2009	IS 250	2.5 (2500)	4GR-FSE	DI	204@6400	185@4800	3.27x3.03	12.0:1	55.5@6000
	IS 350	3.5 (3456)	2GR-FSE	DI	306@6400	277@4800	3.70x3.27	11.8:1	55.5@6000
2010	IS 250	2.5 (2500)	4GR-FSE	DI	204@6400	185@4800	3.27x3.03	12.0:1	55.5@6000
	IS 350	3.5 (3456)	2GR-FSE	DI	306@6400	277@4800	3.70x3.27	11.8:1	55.5@6000

DI : Direct Injection

3768X_IS25_C0002

ENGINE TUNE-UP SPECIFICATIONS

Year	Engine Displacement Liters	Engine ID/VIN	Spark Plug Gap (in.)	Ignition Timing (deg.)	Fuel Pump (psi)	Idle Speed (rpm)	Valve Clearance Intake	Valve Clearance Exhaust
2007	2.5	4GR-FSE	0.039-0.043	8-12B ③	28-85	650-750	HYD	HYD
	3.5	2GR-FSE	0.039-0.043	8-12B ③	28-85	600-700	HYD	HYD
2008	2.5	4GR-FSE	0.039-0.043	8-12B ③	28-85	650-750	HYD	HYD
	3.5	2GR-FSE	0.039-0.043	8-12B ③	28-85	600-700	HYD	HYD
2009	2.5	4GR-FSE	0.039-0.043	8-12B ③	28-85	650-750	HYD	HYD
	3.5	2GR-FSE	0.039-0.043	8-12B ③	28-85	600-700	HYD	HYD
2010	2.5	4GR-FSE	0.039-0.043	8-12B ③	28-85	650-750	HYD	HYD
	3.5	2GR-FSE	0.039-0.043	8-12B ③	28-85	600-700	HYD	HYD

NOTE: The Vehicle Emission Control Information label often reflects specification changes made during production. The label figures must be used if they differ from those in this chart.

B: Before top dead center

HYD: Hydraulic Valve Lifters

① Terminals TE1 and E1 of check connector must be connected

② Terminals TC and E1 of check connector must be connected

③ Terminals TC and CG of check connector must be connected

④ Terminals TE and E1 of check connector must be connected

3768X_IS25_C0003

CAPACITIES

Year	Model	Engine Displacement Liters	Engine ID/VIN	Engine Oil with Filter	Transmission (pts.) ① Auto.	Transmission (pts.) ① Manual	Drive Axle (pts.)	Fuel Tank (gal.)	Cooling System (qts.)
2007	IS 250	2.5	4GR-FSE	②	③	3.8	④	17.2	9.6
	IS 350	3.5	2GR-FSE	6.6	3.6	3.8	2.8	17.2	9.6
2008	IS 250	2.5	4GR-FSE	②	③	3.8	④	17.2	9.6
	IS 350	3.5	2GR-FSE	6.6	3.6	3.8	2.8	17.2	9.6
2009	IS 250	2.5	4GR-FSE	②	③	3.8	④	17.2	9.6
	IS 350	3.5	2GR-FSE	6.6	3.6	3.8	2.8	17.2	9.6
2010	IS 250	2.5	4GR-FSE	②	③	3.8	④	17.2	9.6
	IS 350	3.5	2GR-FSE	6.6	3.6	3.8	2.8	17.2	9.6

NOTE: All capacities are approximate. Add fluid gradually and check to be sure a proper fluid level is obtained.

① Specification is for transmission drain and refill, not overhaul.

② 2WD models: 6.2 qts.
AWD models: 6.3 qts.

③ 2WD models: 3.2 pts.
AWD models: 5.6 pts.

④ 2WD models: 2.4 pts.
AWD models: 2.2 pts. for the rear; 1.48 pts. for the front

3768X_IS25_C0004

FLUID SPECIFICATIONS

Year	Model	Engine Displacement Liters	Engine Oil	Auto. Trans.	Drive Axle	Power Steering Fluid	Brake Master Cylinder
2009	IS250	2.5	①	ATF World Standard	②	NA	DOT 3
	IS350	3.5	5W-30	ATF World Standard	③	NA	DOT 3
2010	IS250	2.5	①	ATF World Standard	②	NA	DOT 3
	IS350	3.5	5W-30	ATF World Standard	③	NA	DOT 3

DOT: Department Of Transportation

NA: Not Available

Note: If specification disagrees with specification in owners manual, use specification in owners manaual

① 0W-20 above 40 degrees F. 5W-20 below 40 degrees F

② API GL-5 SAE 90 above 0 degrees F. API GL-5 SAE 80W-90 below 0 degrees

③ API GL-5 SAE 85W-90 above 0 degrees F. API GL-5 SAE 80W-90 below 0 degrees

3768X_IS25_C0013

VALVE SPECIFICATIONS

Year	Engine Displacement Liters	Engine ID/VIN	Seat Angle (deg.)	Face Angle (deg.)	Spring Test Pressure (lbs. @ in.)	Spring Free-Length (in.)	Stem-to-Guide Clearance (in.)		Stem Diameter (in.)	
							Intake	Exhaust	Intake	Exhaust
2007	2.5	4GR-FSE	45	NA	NA	1.831	0.0010-0.0024	0.0012-0.0026	0.2154-0.2159	0.2151-0.2157
	3.5	2GR-FSE	45	NA	NA	2.035	0.0010-0.0024	0.0012-0.0026	0.2154-0.2159	0.2151-0.2157
2008	2.5	4GR-FSE	45	NA	NA	1.831	0.0010-0.0024	0.0012-0.0026	0.2154-0.2159	0.2151-0.2157
	3.5	2GR-FSE	45	NA	NA	2.035	0.0010-0.0024	0.0012-0.0026	0.2154-0.2159	0.2151-0.2157
2009	2.5	4GR-FSE	45	NA	NA	1.831	0.0010-0.0024	0.0012-0.0026	0.2154-0.2159	0.2151-0.2157
	3.5	2GR-FSE	45	NA	NA	2.035	0.0010-0.0024	0.0012-0.0026	0.2154-0.2159	0.2151-0.2157
2010	2.5	4GR-FSE	45	NA	NA	1.831	0.0010-0.0024	0.0012-0.0026	0.2154-0.2159	0.2151-0.2157
	3.5	2GR-FSE	45	NA	NA	2.035	0.0010-0.0024	0.0012-0.0026	0.2154-0.2159	0.2151-0.2157

NA: Not Available

3768X_IS25_C0005

CAMSHAFT SPECIFICATIONS

All measurements in inches unless noted

Year	Engine Displacement Liters	Engine Code/VIN	Journal Dia.	Brg. Oil Clearance	Shaft End-play	Circle Runout	Lobe Height Intake	Lobe Height Exhaust
2009	2.5	4GR-FSE	①	③	NA	NA	1.8624-1.8664	1.8104-1.8143
	3.5	2GR-FSE	②	③	0.0031 0.0051	0.0016	1.7447-1.7487	1.7426-1.7465
2010	2.7	4GR-FSE	①	③	NA	NA	1.8624-1.8664	1.8104-1.8143
	3.5	2GR-FSE	②	③	0.0031 0.0051	0.0016	1.7447-1.7487	1.7426-1.7465

① No1: 1.4152-1.4157

All others: 1.0220-1.0226

③ No. 1: 0.0012-0.0031

All others: 0.00098-0.0024

3768X_IS25_C0014

CRANKSHAFT AND CONNECTING ROD SPECIFICATIONS

All measurements are given in inches.

Year	Engine Displacement Liters	Engine ID/VIN	Crankshaft Main Brg. Journal Dia.	Crankshaft Main Brg. Oil Clearance	Crankshaft Shaft End-play	Crankshaft Thrust on No.	Connecting Rod Journal Diameter	Connecting Rod Oil Clearance	Connecting Rod Side Clearance
2007	2.5	4GR-FSE	2.4011-2.4016	0.0010-0.0019	0.0008-0.0087	2	1.8894-1.8898	0.0017-0.0027	0.0098-0.0157
	3.5	2GR-FSE	2.4011-2.4016	①	0.0016-0.0094	NA	2.0863-2.0866	0.0018-0.0026	0.0059-0.0157
2008	2.5	4GR-FSE	2.4011-2.4016	0.0010-0.0019	0.0008-0.0087	2	1.8894-1.8898	0.0017-0.0027	0.0098-0.0157
	3.5	2GR-FSE	2.4011-2.4016	①	0.0016-0.0094	NA	2.0863-2.0866	0.0018-0.0026	0.0059-0.0157
2009	2.5	4GR-FSE	2.4011-2.4016	0.0010-0.0019	0.0008-0.0087	2	1.8894-1.8898	0.0017-0.0027	0.0098-0.0157
	3.5	2GR-FSE	2.4011-2.4016	①	0.0016-0.0094	NA	2.0863-2.0866	0.0018-0.0026	0.0059-0.0157
2010	2.5	4GR-FSE	2.4011-2.4016	0.0010-0.0019	0.0008-0.0087	2	1.8894-1.8898	0.0017-0.0027	0.0098-0.0157
	3.5	2GR-FSE	2.4011-2.4016	①	0.0016-0.0094	NA	2.0863-2.0866	0.0018-0.0026	0.0059-0.0157

NA: Not Available

① Journal No. 1 and 4: 0.0010 - 0.0019 inch

Journal No. 2 and 3: 0.0013 - 0.0021 inch

3768X_IS25_C0008

PISTON AND RING SPECIFICATIONS

All measurements are given in inches.

Year	Engine Displacement Liters	Engine ID/VIN	Piston Clearance	Ring Gap Top Compression	Ring Gap Bottom Compression	Ring Gap Oil Control	Ring Side Clearance Top Compression	Ring Side Clearance Bottom Compression	Ring Side Clearance Oil Control
2007	2.5	4GR-FSE	0.0006-0.0014	0.0087-0.0126	0.0236-0.0276	0.0039-0.0138	0.0008-0.0028	0.0012-0.0028	0.0008-0.0026
	3.5	2GR-FSE	0.0008-0.0020	0.0091-0.0130	0.0138-0.0177	0.0039-0.0157	0.0008-0.0028	0.0008-0.0024	0.0008-0.0028
2008	2.5	4GR-FSE	0.0006-0.0014	0.0087-0.0126	0.0236-0.0276	0.0039-0.0138	0.0008-0.0028	0.0012-0.0028	0.0008-0.0026
	3.5	2GR-FSE	0.0008-0.0020	0.0091-0.0130	0.0138-0.0177	0.0039-0.0157	0.0008-0.0028	0.0008-0.0024	0.0008-0.0028
2009	2.5	4GR-FSE	0.0006-0.0014	0.0087-0.0126	0.0236-0.0276	0.0039-0.0138	0.0008-0.0028	0.0012-0.0028	0.0008-0.0026
	3.5	2GR-FSE	0.0008-0.0020	0.0091-0.0130	0.0138-0.0177	0.0039-0.0157	0.0008-0.0028	0.0008-0.0024	0.0008-0.0028
2010	2.5	4GR-FSE	0.0006-0.0014	0.0087-0.0126	0.0236-0.0276	0.0039-0.0138	0.0008-0.0028	0.0012-0.0028	0.0008-0.0026
	3.5	2GR-FSE	0.0008-0.0020	0.0091-0.0130	0.0138-0.0177	0.0039-0.0157	0.0008-0.0028	0.0008-0.0024	0.0008-0.0028

3768X_IS25_C0007

TORQUE SPECIFICATIONS

All readings in ft. lbs.

Year	Engine Displacement Liters	Engine ID/VIN	Cylinder Head Bolts	Main Bearing Bolts	Rod Bearing Bolts	Crankshaft Damper Bolts	Flywheel Bolts	Manifold Intake	Manifold Exhaust	Spark Plugs	Oil Pan Drain Plug
2007	2.5	4GR-FSE	①	②	③	184	④	15	15	18	30
	3.5	2GR-FSE	①	②	⑤	184	61	15	15	18	30
2008	2.5	4GR-FSE	①	②	③	184	④	15	15	18	30
	3.5	2GR-FSE	①	②	⑤	184	61	15	15	18	30
2009	2.5	4GR-FSE	①	②	③	184	④	15	15	18	30
	3.5	2GR-FSE	①	②	⑤	184	61	15	15	18	30
2010	2.5	4GR-FSE	①	②	③	184	④	15	15	18	30
	3.5	2GR-FSE	①	②	⑤	184	61	15	15	18	30

① Step 1: 27 ft. lbs.
Step 2: Tighten an additional 90 degrees
Step 3: Tighten an additional 90 degrees
14mm head bolt: 22 ft. lbs.

② 12mm head bolts: 19 ft. lbs.
16-point bolts:
A74 Step 1: 45 ft. lbs.
Step 2: Plus an additional 90 degrees

③ Step 1: 28 ft. lbs.
Step 2: Plus 90 degrees

④ Automatic: 61 ft. lbs.
Manual: 54 ft. lbs.

⑤ Step 1: 30 ft. lbs.
Step 2: Plus 90 degrees

3768X_IS25_C0006

WHEEL ALIGNMENT

Year	Model		Caster Range (+/-Deg.)	Caster Preferred Setting (Deg.)	Camber Range (+/-Deg.)	Camber Preferred Setting (Deg.)	Toe-in (in.)	Steering Axis Inclination (Deg.)
2007	IS 250 2WD	F	0.75	①	0.75	-0.38	0.04 +/- 0.08	10.68
		R	—	—	0.75	-1.23	0.12 +/- 0.08	—
	IS 250 AWD	F	0.75	+4.63	0.75	-0.42	0.04 +/- 0.08	11.27
		R	—	—	0.75	-0.83	0.12 +/- 0.08	—
	IS 350	F	0.75	①	0.75	-0.38	0.04 +/- 0.08	10.68
		R	—	—	0.75	-1.23	0.12 +/- 0.08	—
2008	IS 250 2WD	F	0.75	①	0.75	-0.38	0.04 +/- 0.08	10.68
		R	—	—	0.75	-1.23	0.12 +/- 0.08	—
	IS 250 AWD	F	0.75	+4.63	0.75	-0.42	0.04 +/- 0.08	11.27
		R	—	—	0.75	-0.83	0.12 +/- 0.08	—
	IS 350	F	0.75	①	0.75	-0.38	0.04 +/- 0.08	10.68
		R	—	—	0.75	-1.23	0.12 +/- 0.08	—
2009	IS 250 2WD	F	0.75	①	0.75	-0.38	0.04 +/- 0.08	10.68
		R	—	—	0.75	-1.23	0.12 +/- 0.08	—
	IS 250 AWD	F	0.75	+4.63	0.75	-0.42	0.04 +/- 0.08	11.27
		R	—	—	0.75	-0.83	0.12 +/- 0.08	—
	IS 350	F	0.75	①	0.75	-0.38	0.04 +/- 0.08	10.68
		R	—	—	0.75	-1.23	0.12 +/- 0.08	—
2010	IS 250 2WD	F	0.75	①	0.75	-0.38	0.04 +/- 0.08	10.68
		R	—	—	0.75	-1.23	0.12 +/- 0.08	—
	IS 250 AWD	F	0.75	+4.63	0.75	-0.42	0.04 +/- 0.08	11.27
		R	—	—	0.75	-0.83	0.12 +/- 0.08	—
	IS 350	F	0.75	①	0.75	-0.38	0.04 +/- 0.08	10.68
		R	—	—	0.75	-1.23	0.12 +/- 0.08	—

① 16 inch wheels: +8.12
17 inch wheels: +8.18
18 inch wheels: +8.07

3768X_IS25_C0009

TIRE, WHEEL AND BALL JOINT SPECIFICATIONS

Year	Model	OEM Tires		Tire Pressures (psi)		Wheel	Ball Joint	Lug Nut
		Standard	Optional	Front	Rear	Size	Inspection	(ft. lbs.
2007	IS 250/ IS 350	205/55R16 89W	225/45R17 90W 245/45R17 95W 225/45R17 91V 225/45R17 95V 245/45R17 95V 225/40R18 88Y {F} 255/40R18 95Y {R}	35	38	Std: 7-JJ Opt: 8-JJ Opt: 8-J {F} Opt: 8.5-J {R}	L: 4.5-53 in. ①	76
2008	IS 250/ IS 350	205/55R16 89W	225/45R17 90W 245/45R17 95W 225/45R17 91V 225/45R17 95V 245/45R17 95V 225/40R18 88Y {F} 255/40R18 95Y {R}	35	38	Std: 7-JJ Opt: 8-JJ Opt: 8-J {F} Opt: 8.5-J {R}	L: 4.5-53 in. ①	76
2009	IS 250/ IS 350	205/55R16 89W	225/45R17 90W 245/45R17 95W 225/45R17 91V 225/45R17 95V 245/45R17 95V 225/40R18 88Y {F} 255/40R18 95Y {R}	35	38	Std: 7-JJ Opt: 8-JJ Opt: 8-J {F} Opt: 8.5-J {R}	L: 4.5-53 in. ①	76
2010	IS 250/ IS 350	205/55R16 89W	225/45R17 90W 245/45R17 95W 225/45R17 91V 225/45R17 95V 245/45R17 95V 225/40R18 88Y {F} 255/40R18 95Y {R}	35	38	Std: 7-JJ Opt: 8-JJ Opt: 8-J {F} Opt: 8.5-J {R}	L: 4.5-53 in. ①	76

OEM: Original Equipment Manufacturer

PSI: Pounds Per Square Inch

STD: Standard

OPT: Optional

L: Lower

U: Upper

① Torque required in inch lbs. to rotate ball joint when removed from the knuckle

3768X_IS25_C0010

BRAKE SPECIFICATIONS

All measurements in inches unless noted

		Front Brake Disc			Rear Brake Disc			Minimum	Brake Caliper	
		Original	Minimum	Maximum	Original	Minimum	Maximum	Lining	Bracket Bolts	Mounting Bolts
Year	Model	Thickness	Thickness	Run-out	Thickness	Thickness	Run-out	Thickness	(ft. lbs.)	(ft. lbs.)
2007	IS 250	1.102	0.983	0.0020	0.413	0.039	0.0020	0.0390	58	①
	IS 350	1.181	1.063	0.0020	0.709	0.650	0.0020	0.0390	—	①
2008	IS 250	1.102	0.983	0.0020	0.413	0.039	0.0020	0.0390	58	①
	IS 350	1.181	1.063	0.0020	0.709	0.650	0.0020	0.0390	—	①
2009	IS 250	1.102	0.983	0.0020	0.413	0.039	0.0020	0.0390	58	①
	IS 350	1.181	1.063	0.0020	0.709	0.650	0.0020	0.0390	—	①
2010	IS 250	1.102	0.983	0.0020	0.413	0.039	0.0020	0.0390	58	①
	IS 350	1.181	1.063	0.0020	0.709	0.650	0.0020	0.0390	—	①

① Front: 58 ft. lbs.
Rear: 40 ft. lbs.

3768X_IS25_C0011

SCHEDULED MAINTENANCE INTERVALS

Lexus—IS250 & IS350

TO BE SERVICED	TYPE OF SERVICE	VEHICLE MILEAGE INTERVAL (x1000)												
		7.5	15	22.5	30	37.5	45	52.5	60	67.5	75	82.5	90	97.5
Engine oil & filter	R	✓	✓	✓	✓	✓	✓	✓	✓	✓	✓	✓	✓	✓
A/C filter (if equipped) ①	S/I	✓	✓	✓	✓	✓	✓	✓	✓	✓	✓	✓	✓	✓
Automatic transaxle fluid & filter	S/I		✓		✓		✓		✓		✓		✓	
Ball joints & dust covers	S/I		✓		✓		✓		✓		✓		✓	
Bolts & nuts on chassis & body	S/I		✓		✓		✓		✓		✓		✓	
Brake fluid ②	S/I		✓		✓		✓		✓		✓		✓	
Brake line pipes & hoses	S/I		✓		✓		✓		✓		✓		✓	
Brake pads & discs (front & rear)	S/I		✓		✓		✓		✓		✓		✓	
Differential oil	S/I		✓		✓		✓		✓		✓		✓	
Driveshaft boots (if equipped)	S/I		✓		✓		✓		✓		✓		✓	
Steering gear housing oil	S/I		✓		✓		✓		✓		✓		✓	
Steering linkage	S/I		✓		✓		✓		✓		✓		✓	
Air filter	R				✓				✓				✓	
Exhaust pipes & mountings	S/I				✓				✓				✓	
Fuel lines & connections	S/I				✓				✓				✓	
Engine coolant	R						✓					✓		
Fuel tank cap gasket	R								✓					
Spark plugs	R								✓					
Charcoal canister	S/I								✓					
Drive belts	S/I								✓					
Valve clearance	S/I								✓					

R: Replace S/I: Service or Inspect

① Replace at 15,000 miles.

② Replace at 30,000 miles (unless previously replaced).

FREQUENT OPERATION MAINTENANCE (SEVERE SERVICE)

If a vehicle is operated under any of the following conditions it is considered severe service

- Extremely dusty areas.

- 50% or more of the vehicle operation is in 32°C (90°F) or higher temperatures, or constant operation in temperatures below 0°C (32°F).

- Prolonged idling (vehicle operation in stop and go traffic).

- Frequent short running periods (engine does not warm to normal operating temperatures).

- Police, taxi, delivery usage or trailer towing usage.

Oil & oil filter: change every 3750 miles.

Ball joints & dust covers: service or inspect every 7500 miles.

Bolts & nuts on chassis & body: service or inspect every 7500 miles.

Brake pads & discs (front & rear): service or inspect every 7500 miles.

Driveshaft boots (if equipped): service or inspect every 7500 miles.

Steering linkage: service or inspect every 7500 miles.

Air filter: service or inspect every 15,000 miles.

Automatic transmission fluid & filter: replace every 15,000 miles.

Differential oil: replace every 15,000 miles.

Exhaust pipes & mountings: service or inspect every 15,000 miles.

Drive belts: service or inspect at 60,000 miles & every 7500 miles thereafter.

Timing belts: replace every 60,000 miles.

3768X_IS25_C0012

BRAKES — INFORMATION AND PRECAUTIONS

ANTI-LOCK SYSTEMS

• Certain components within the ABS system are not intended to be serviced or repaired individually.

• Do not use rubber hoses or other parts not specifically specified for and ABS system. When using repair kits, replace all parts included in the kit. Partial or incorrect repair may lead to functional problems and require the replacement of components.

• Lubricate rubber parts with clean, fresh brake fluid to ease assembly. Do not use shop air to clean parts; damage to rubber components may result.

• Use only DOT 3 brake fluid from an unopened container.

• If any hydraulic component or line is removed or replaced, it may be necessary to bleed the entire system.

• A clean repair area is essential. Always clean the reservoir and cap thoroughly before removing the cap. The slightest amount of dirt in the fluid may plug an orifice and impair the system function. Perform repairs after components have been thoroughly cleaned; use only denatured alcohol to clean components. Do not allow ABS components to come into contact with any substance containing mineral oil; this includes used shop rags.

• The Anti-Lock control unit is a microprocessor similar to other computer units in the vehicle. Ensure that the ignition switch is **OFF** before removing or installing controller harnesses. Avoid static electricity discharge at or near the controller.

• If any arc welding is to be done on the vehicle, the control unit should be unplugged before welding operations begin.

DISC AND DRUM SYSTEMS

✱✱ CAUTION

Dust and dirt accumulating on brake parts during normal use may contain asbestos fibers from production or aftermarket brake linings. Breathing excessive concentrations of asbestos fibers can cause serious bodily harm. Exercise care when servicing brake parts. Do not sand or grind brake lining unless equipment used is designed to contain the dust residue. Do not clean brake parts with compressed air or by dry brushing. Cleaning should be done by dampening the brake components with a fine mist of water, then wiping the brake components clean with a dampened cloth. Dispose of cloth and all residue containing asbestos fibers in an impermeable container with the appropriate label. Follow practices prescribed by the Occupational Safety and Health Administration (OSHA) and the Environmental Protection Agency (EPA) for the handling, processing, and disposing of dust or debris that may contain asbestos fibers.

BRAKES — BLEEDING THE BRAKE SYSTEM

BLEEDING PROCEDURE

BLEEDING PROCEDURE

See Figure 1.

➡If any work is done on the brake system or if air in the brake lines is suspected, bleed the system of air.

1. Fill the reservoir with brake fluid.

➡If the master cylinder has been disassembled or if the reservoir becomes empty, bleed the air from the master cylinder.

2. Bleed the brake master cylinder as follows:
   a. Disconnect the brake lines from the master cylinder.
   b. Slowly depress the brake pedal and hold it.
   c. Block off the outer holes with your fingers, and release the brake pedal.
   d. Repeat the previous 2 steps 3 or 4 times.
3. Bleed the brake line as follows:
   a. Connect the vinyl tube to the brake caliper.
   b. Depress the brake pedal several times, then loosen the bleeder plug with the pedal held down.
   c. At the point when fluid stops coming out, tighten the bleeder plug, then release the brake pedal.
   d. Repeat the previous 2 steps until all the air in the fluid has been bled out.
   e. Repeat the above procedure to bleed the air out of the brake line for each wheel.
   f. Tighten the bleeder plug to 8 ft. lbs. (11 Nm).
4. Bleed the brake actuator as follows:
   a. Remove the reservoir cap.
   b. Install the SST 09992-00242, 09992-00350 to the reservoir.
   c. Connect the vinyl tube to the bleeder plug of the brake actuator.

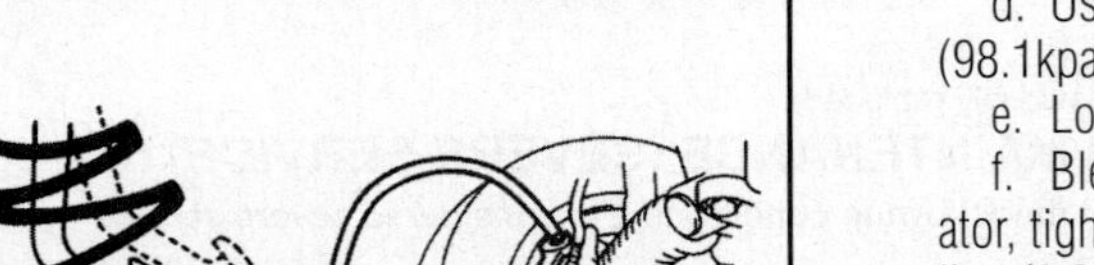

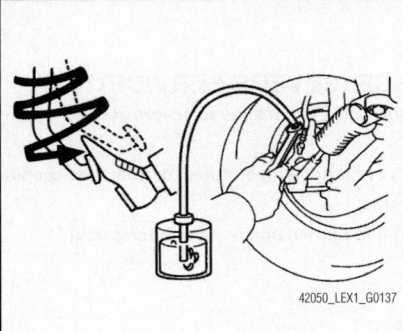

42050_LEX1_G0137

Fig. 1 Bleeding the brake line

   d. Using SST, apply the 14.2 psi (98.1kpa) of pressure to the reservoir.
   e. Loosen the bleeder plug.
   f. Bleed the air out of the brake actuator, tighten the bleeder plug to 74 inch lbs. (8.3 Nm).
5. Check the fluid level and add fluid if necessary.

BLEEDING THE ABS SYSTEM

With VSC

➡After performing the usual air bleeding in the brake system, if the height or feel of the brake pedal cannot be obtained, perform air bleeding in the brake actuator assembly with a hand–held tester by following procedures below. Make sure that the brake fluid in the master cylinder reservoir tank does not become empty.

1. Depress the brake pedal more than 20 times with the engine off.
2. Connect the hand–held tester to the DLC3, and then turn the ignition switch to the ON position.
3. Do not start the engine.
4. Select "AIR BLEEDING" on the hand–held tester. Please refer to the Hand–Held Tester Operator's Manual for further details.

5. Bleed the air out of the regular brake line in"Step1: Increase" on the hand–held tester display. Perform the air bleeding by following the steps displayed on the hand–held tester. Make sure that the brake fluid in the master cylinder reservoir tank does not become empty.

6. Connect the vinyl tube to either one of the bleeder plugs.

7. Depress the brake pedal several times, and then loosen the bleeder plug of one of the above wheels with the pedal depressed.

8. When fluid stops coming out, tighten the bleeder plug, then release the brake pedal.

9. Repeat (2) and (3) until all air in the fluid is completely bled out.

10. Tighten the bleeder plug to 73 inch lbs. (8.3 Nm).

11. Repeat the above procedure to bleed the air out of the brake line for each wheel.

12. Bleed the air out of the suction line in"Step2: Inhalation" on the hand–held tester display.

13. Connect the vinyl tube to the bleeder plug at the right front wheel or the right rear wheel and loosen the bleeder plug.

14. Operate the brake actuator assembly using the hand–held tester to bleed the air.

15. Check that the operation has stopped, by referring to the hand–held tester display.

16. Repeat (2) and (3) until all the air in the fluid is completely bled out.

17. Tighten the bleeder plug.

18. For the rest of the wheels, bleed the air in the same way as stated in the above procedure.

19. Bleed the air out of the pressure reduction line in "Step3: Decrease" on the hand–held tester display.

20. Connect a vinyl tube to either one of the bleeder plugs.

21. Loosen the bleeder plug.

22. Using the hand–held tester, operate the brake actuator assembly using hand–held tester, completely depress the brake pedal and keep it.

➡The operation stops automatically in 4 seconds. When performing this procedure continuously, an interval of at least 20 seconds is required. When the operation is completed, the brake pedal slightly goes down. This is a normal phenomenon caused when the solenoid opens. During this procedure, the pedal seems heavy, but completely depress it so that the brake fluid comes out from the bleeder plug. Be sure to keep depressing the brake pedal. Never depress and release the pedal repeatedly.

23. Tighten the bleeder plug, then release the brake pedal.

24. Repeat 3 previous steps until all the air in the fluid is completely bled out.

25. Tighten the bleeder plug.

26. Repeat the above procedure to bleed the air out of the brake line for each wheel.

27. Bleed the air out of the regular brake line again in "Step4: Increase" on the hand–held tester display.

28. Connect the vinyl tube to either one of the bleeder plug.

29. Depress the brake pedal several times, then loosen the bleeder plug of one of the above wheels with the pedal depressed.

30. When fluid stops coming out, tighten the bleeder plug, then release the brake pedal.

31. Repeat the previous 2 steps until all the air in the fluid is completely bled out.

32. Tighten the bleeder plug.

33. Repeat the above procedure to bleed the air out of the brake line for each wheel.

WHEEL SPEED SENSORS

REMOVAL & INSTALLATION

Mounted In The Hub & Bearing Assembly

See Figure 2.

1. Before servicing the vehicle, refer to the Precautions section.
2. Disconnect the negative battery cable.
3. Remove the wheel.
4. Disconnect the connector from the speed sensor.
5. Remove the hub and bearing assembly.
6. Using a pin punch and hammer, drive out the 2 pins and remove the 2 attachments from SST 09520-00031.
7. Mount the front axle hub in a soft jaw vise.
8. Using SST 09520-00031 (09520-00040, 09521-00020), 09950-00020and 2 bolts (Diameter: 12mm, Pitch: 1.5mm), remove the speed sensor.

To install:

9. Clean the contacting surface of the axle hub and a new speed sensor.

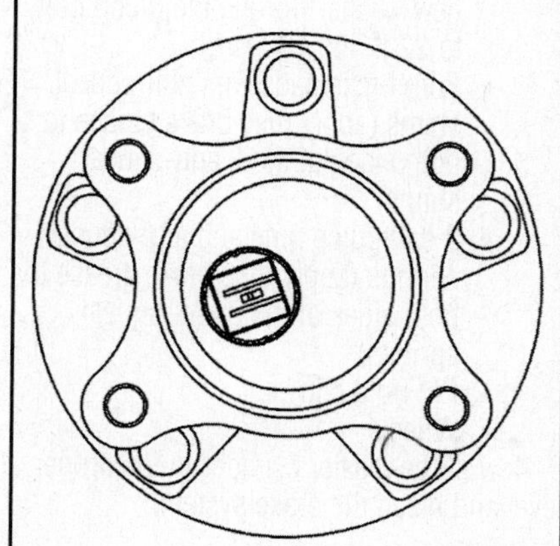

Fig. 2 Correct installation of the speed sensor in the hub assembly

10. Place the speed sensor on the axle hub so that the connector will be positioned as shown in the illustration under the on-vehicle condition.
11. Using SST 09214-7601 1 and a press, install a new speed sensor to the axle hub.

⁂ CAUTION

Do not tap the speed sensor with a hammer directly. Check that there should be no foreign objects on the speed sensor detection portion.

12. Install the speed sensor to the front axle.
13. Connect the speed sensor connector.
14. Install fender liner.
15. Install front wheel.
16. Connect the negative battery cable.
17. Check for speed sensor signal.

Mounted Separately

1. Before servicing the vehicle, refer to the Precautions section.
2. Disconnect the negative battery cable.
3. Remove the wheel.
4. Remove fender liner, if necessary.
5. Disconnect the connector from the speed sensor.

6. Remove the clamp bolts holding the sensor harness and clamp from the body and shock absorber.

7. Remove the bolt and speed sensor. Do not stick and foreign matter on the sensor tip.

To install:

8. Install the speed sensor with the bolt and tighten to 71 inch lbs. (8 Nm). Make sure the sensor tip is clean.

9. Install the sensor harness clamps

10. Connect the speed sensor connector and clamp.

11. Install fender liner.

12. Install front wheel.

13. Connect the negative battery cable.

14. Check for ABS signal.

BRAKES — FRONT DISC BRAKES

BRAKE CALIPER

REMOVAL & INSTALLATION

1. Before servicing the vehicle, refer to the Precautions section.
2. Remove or disconnect the following:
   - Wheels
   - Brake hose from the caliper
   - Bolts that attach the caliper to the mounting bracket
   - Caliper assembly by lifting the bottom

To install:

3. Grease the caliper slides and bolts with lithium grease.
4. Install or connect the following:
   - Caliper. Hold the sliding pin and tighten the mounting bolts to 25 ft. lbs. (34 Nm).
   - Brake hose to the caliper using 2 new washers. Torque the union bolt to 22 ft. lbs. (30 Nm)
   - Wheels
5. Fill the master cylinder to the proper level and bleed the brake system.

DISC BRAKE PADS

REMOVAL & INSTALLATION

1. Before servicing the vehicle, refer to the Precautions section.
2. Remove or disconnect the following:
   - Wheels
   - 2 bolts and remove the disc brake caliper assembly

➡Support the caliper. Do NOT allow to hang by the brake hose.

   - Pads with anti-squeal shims

To install:

3. Install or connect the following:
   - Anti-squeal shims to each pad
4. Using a suitable tool, compress the piston carefully in the cylinder bores
   a. Inner and outer pads with the wear indicator plates facing correct position.
   b. Tighten the calipers 2 bolts to 25 ft. lbs. (34 Nm).
   c. Install the front wheel.

BRAKES — REAR DISC BRAKES

BRAKE CALIPER

REMOVAL & INSTALLATION

See Figure 3.

1. Before servicing the vehicle, refer to the Precautions section.
2. Remove or disconnect the following:
   - Wheels
   - Pin hold clip
   - 2 pad guide pins and anti-squeal spring
   - Both brake pads with anti-squeal shims (remove the 4 anti-squeal shims from each pad)
   - Brake hose from the caliper
   - Caliper support pins
   - Caliper assembly by lifting the bottom

To install:

3. Install or connect the following:
   - Caliper assembly to the caliper support bracket
   - Caliper support pins and torque to 18 ft. lbs. (25 Nm)
   - Brake hose to the caliper using a new washer. Torque the union bolt to 22 ft. lbs. (30 Nm)
   - Both brake pads with anti-squeal shims (apply disc brake grease to both sides of the 2 anti-squeal shims)
   - 2 pad guide pins and anti-squeal springs (apply disc brake grease to both sides of the 2 anti-squeal springs)
   - Pin hold clip
   - Wheels
4. Fill the master cylinder to the proper level and bleed the brake system.

DISC BRAKE PADS

REMOVAL & INSTALLATION

1. Before servicing the vehicle, refer to the Precautions section.
2. Remove or disconnect the following:
   - Wheels
   - Pin hold clip
   - 2 pad guide pins and anti-squeal spring
   - Both brake pads with anti-squeal shims (remove the 4 anti-squeal shims from each pad)

To install:

3. Install or connect the following:
   - Compress the caliper pistons
   - Both brake pads with anti-squeal shims (apply disc brake grease to both sides of the 2 anti-squeal shims)
   - 2 pad guide pins and anti-squeal springs (apply disc brake grease to both sides of the 2 anti-squeal springs)
   - Pin hold clip
   - Wheels

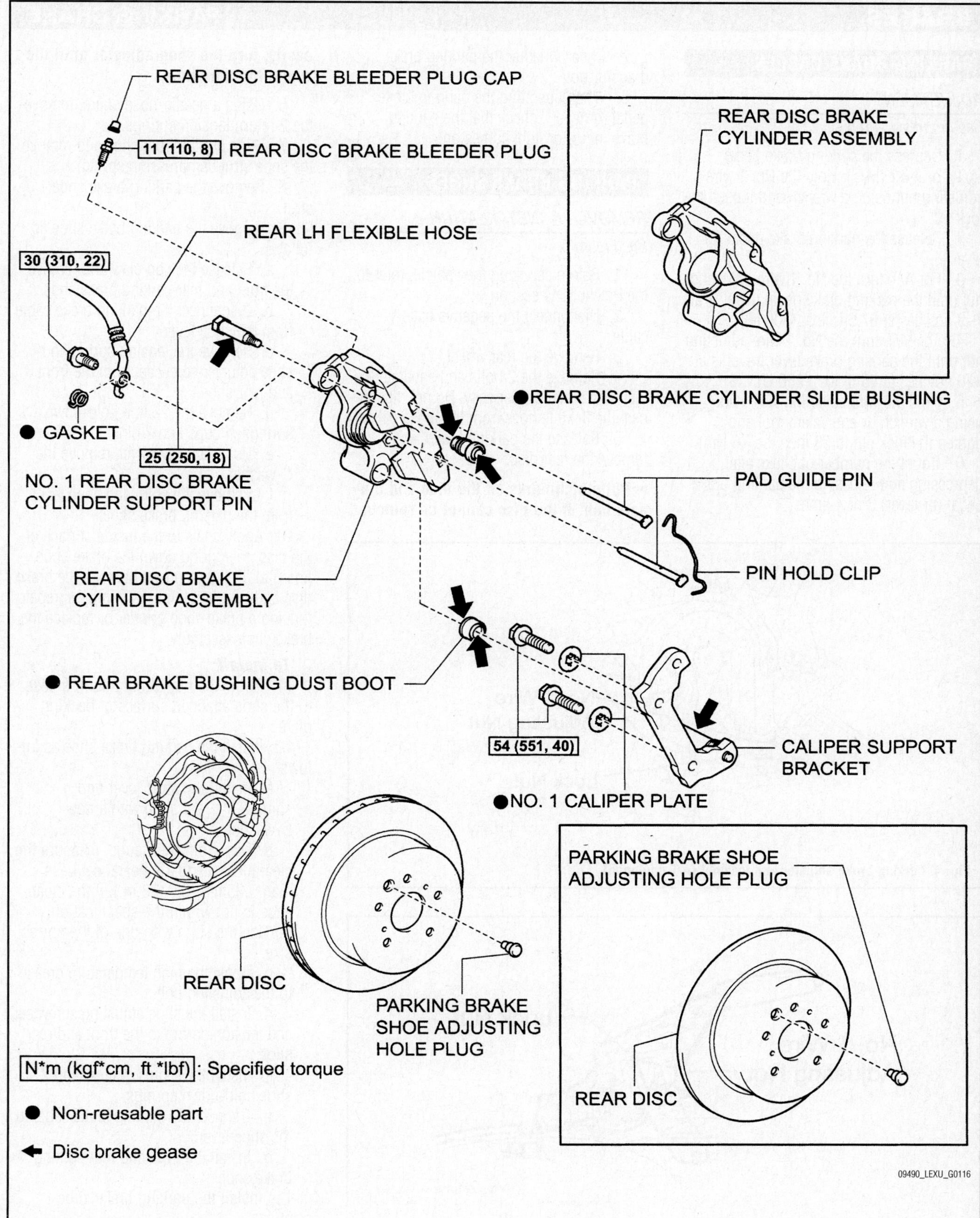

Fig. 3 Exploded view of the rear disc brakes

PARKING BRAKE CABLES

ADJUSTMENT

See Figures 4 and 5.

1. Depress the parking brake pedal (A/T), or lever (M/T). Hold the No. 1 wire adjusting nut using a wrench and loosen the lock nut.
2. Release the parking brake pedal (or lever).
3. For A/T, turn the No. 1 wire adjusting nut until the parking brake pedal travel is at 7–9 notches at 67.5 ft. lbs. (300 N).
4. For M/T, turn the No. 1 wire adjusting nut until the parking brake lever travel is at 4–6 notches at 45 ft. lbs. (200 N).
5. Hold the wire adjusting No. 1 nut using a wrench or equivalent tool and tighten the lock nut to 53 inch lbs. (6 Nm).
6. Count the number of clicks after depressing and releasing the parking brake pedal (or lever) 3 or 4 times.
7. Check whether the parking brake drags or not.
8. When operating the parking brake pedal (or lever), check that the parking brake indicator light comes on.

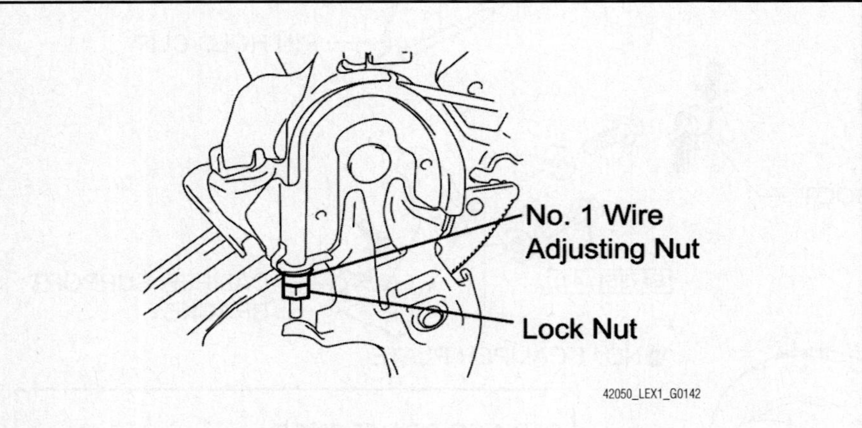

Fig. 4 Parking brake adjuster and lock nut (A/T equipped)

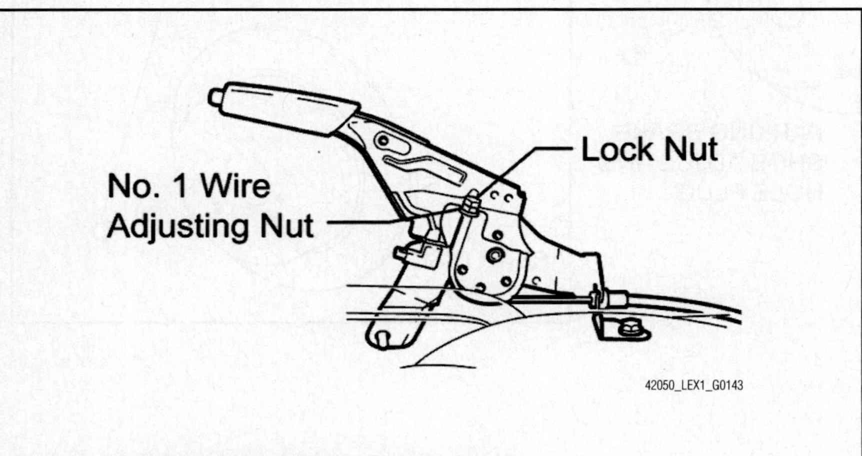

Fig. 5 Parking brake adjuster and lock nut (M/T equipped)

PARKING BRAKE SHOES

REMOVAL & INSTALLATION

See Figure 6.

1. Before servicing the vehicle, refer to the Precautions section.
2. Disconnect the negative battery cable.
3. Remove the rear wheel.
4. Remove the 2 bolts and separate the rear disc brake caliper assy. Do not disconnect the flexible hose from the brake caliper.
5. Release the parking brake, and remove the rear disc.

➡Put matchmarks on the disc and the axle hub. If the disc cannot be removed easily, turn the shoe adjuster until the wheel turns freely.

6. Using a needle-nose pliers, remove the 2 return tension springs.
7. Slide out the front shoe and remove the shoe strut compression spring.
8. Remove the parking brake shoe strut.
9. Remove the parking brake shoe as follows:
   a. Release the cup claw and remove the front and rear parking brake shoe.
   b. Disconnect the parking brake cable from the shoe lever.
   c. Remove the tension spring and shoe adjuster screw set from the front and rear shoe.
   d. Remove the 2 shoe hold-down springs, 4 cups and 2 pins.
   e. Using a screwdriver, remove the C-washer.
   f. Remove the shim and shoe lever from the parking brake shoe.
10. Apply chalk to the inside surface of the disc, then grind down the brake shoe lining to fit. If the contact between the brake disc and the shoe lining is improper, repair it using a brake shoe grinder or replace the brake shoe assembly.

To install:

11. Apply the high temperature grease to the shoe attached surface of backing plate.
12. Install the parking brake shoe as follows:
    a. Install the shoe lever and shim to the rear shoe with a new C-washer.
    b. Using a feeler gauge, measure the clearance. Standard clearance: Less than 0.35mm (0.0138 in.). If the clearance is not within the specification, replace the shim with one of the correct size.
    c. Apply the high temperature grease to the adjusting bolt.
    d. Install the shoe adjusting screw set and tension spring to the front and rear shoe.
    e. Install the 2 pins, 4 cups and 2 shoe hold-down springs.
    f. Connect the parking brake cable to the shoe lever.
    g. Install the front and rear parking brake shoe.
13. Install the parking brake shoe strut.
14. Install the parking brake shoe strut compression spring.

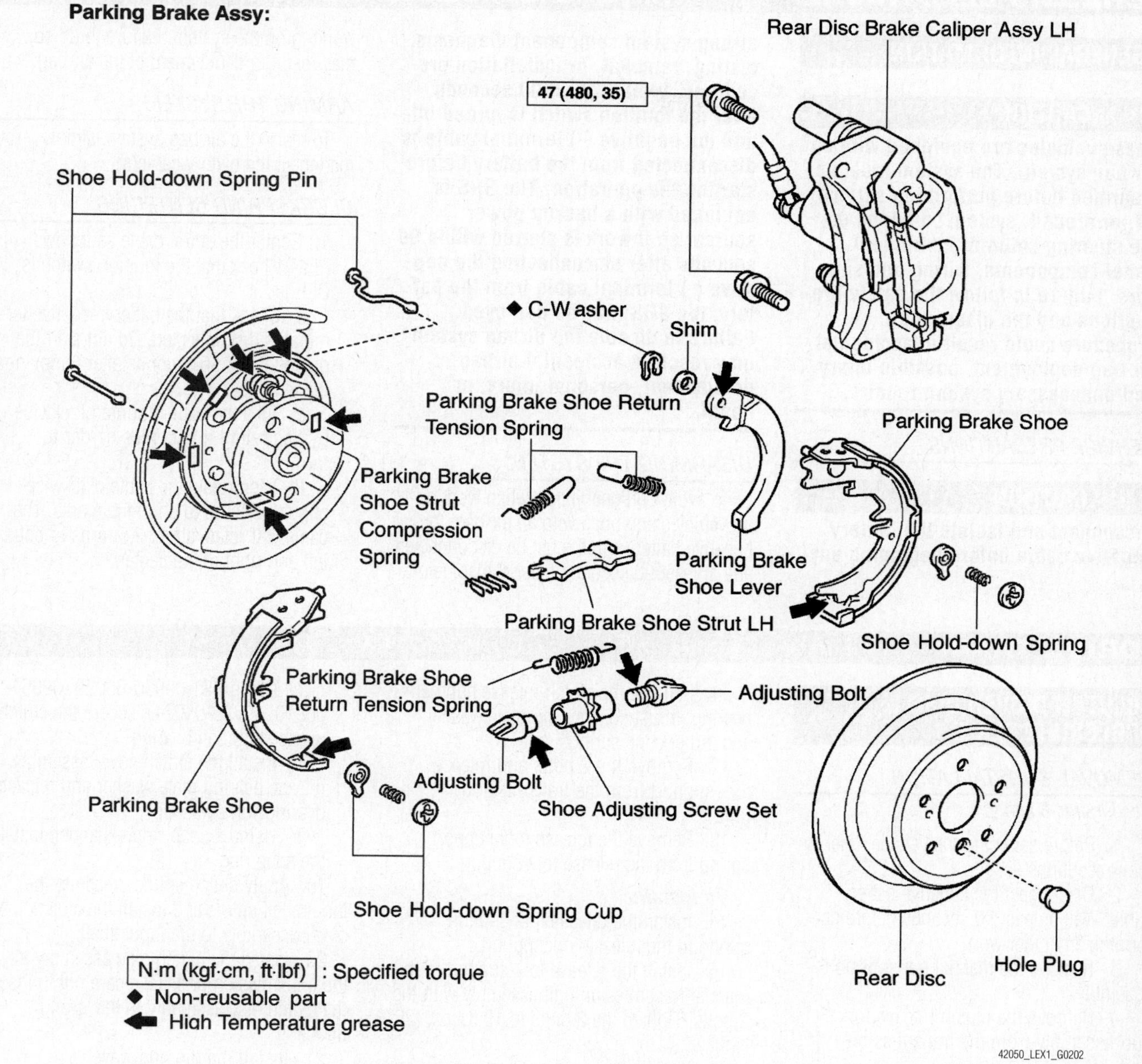

Fig. 6 Exploded view of typical parking brake shoes assembly

15. Using a needle nose pliers, install the 2 return tension springs.

16. Check that each part is installed properly.

➡There should be no oil or grease adhering to the friction surface of the shoe lining and disc.

17. Install the rear disc.

18. Adjust the parking brake shoe clearance.

19. Install the rear disc brake caliper assy.

20. Install the rear wheel.

21. Inspect and adjust parking brake pedal or lever travel.

ADJUSTMENT

See Figure 7.

1. Temporarily install the hub nuts.
2. Remove the hole plug, and turn the adjuster and expand the shoes until the disc locks.
3. Contract the shoe adjuster until the disc can rotate smoothly. Standard: Return 7–8 notches
4. Check shoe is no brake drag.
5. Install the hole plug.

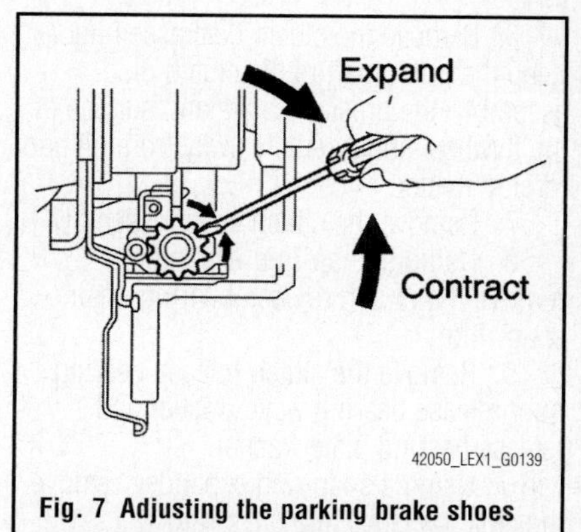

Fig. 7 Adjusting the parking brake shoes

GENERAL INFORMATION

⁂ CAUTION

These vehicles are equipped with an air bag system. The system must be disarmed before performing service on, or around, system components, the steering column, instrument panel components, wiring and sensors. Failure to follow the safety precautions and the disarming procedure could result in accidental air bag deployment, possible injury and unnecessary system repairs.

SERVICE PRECAUTIONS

⁂ CAUTION

Disconnect and isolate the battery negative cable before beginning any airbag system component diagnosis, testing, removal, or installation procedures. Wait at least 90 seconds after the ignition switch is turned off and the negative (-) terminal cable is disconnected from the battery before starting the operation. The SRS is equipped with a backup power source, so if work is started within 90 seconds after disconnecting the negative (-) terminal cable from the battery, the SRS may be deployed. Failure to disable the airbag system may result in accidental airbag deployment, personal injury, or death.

DISARMING THE SYSTEM

To avoid personal injury when working on vehicles equipped with an air bag, the negative battery cable must be disconnected and at least 90 seconds must elapse before working on the system. Failure to do so may result in deployment of the air bag.

ARMING THE SYSTEM

To rearm the air bag system, simply reconnect the battery cable(s).

CLOCKSPRING CENTERING

1. Center the spiral cable as follows:
   a. Check that the ignition switch is OFF.
   b. Check that the battery negative terminal is disconnected. Do not start the operation for 90 seconds after removing the terminal.
   c. Turn the cable counterclockwise by hand until it becomes harder to turn.
   d. Then rotate the cable clockwise about 2.5 turns to align the marks. The cable will rotate about 2.5 turns to both left and right of the center.

DRIVE TRAIN

CLUTCH DRIVEN DISC & PRESSURE PLATE

REMOVAL & INSTALLATION

See Figures 8 and 9.

1. Before servicing the vehicle, refer to the Precautions section.
2. Disconnect the negative battery cable. Wait at least 90 seconds before performing any other work.
3. Remove the manual transmission assembly.
4. Remove the clutch release fork sub-assembly from the transmission unit.
5. Remove the E-ring, pin and wave washer from the clutch release fork.
6. Remove the clutch disc assembly. Keep the lining part of the clutch disc assembly, the pressure plate and surface of the flywheel subassembly away from oil and foreign matter.
7. Remove the clutch cover assembly.
8. Using a snap ring expander, remove the clutch release bearing shaft snap ring.
9. Remove the clutch release bearing hub, release bearing wave washer and release bearing plate washer.
10. Using a snap ring expander, remove the clutch release hub snap ring.
11. Remove the clutch release hub ball bearing, thrust cone spring plate washer and thrust cone spring.
12. Remove the 2 bolts and release fork support from the transmission unit.
13. Remove the release fork support spring from the release fork support.

To install:

14. Install the release fork support spring to the release fork support.
15. Install the release fork support to the manual transmission unit assembly with the 2 bolts. Tighten the 2 bolts to 19 ft. lbs. (26 Nm).
16. Install the thrust cone spring, thrust cone spring plate washer and clutch release hub ball bearing to the clutch release hub as shown in the illustration.
17. Using a snap ring expander, install the snap ring.
18. Install the clutch release bearing hub as follows:
    a. Fill the groove inside the release bearing hub with grease.
    b. Apply grease to the contact surface between the release hub bearing and the clutch cover. Be sure to apply an even and thin layer of grease. Excessive grease or a lump of grease on the contact surface between the release hub bearing and the clutch cover can splatter and result in trouble.
    c. Using SST 09950-60020 (09951-00810, 09726-05021), secure the clutch release hub ball bearing.
    d. Install the clutch cover assembly, release bearing plate washer and release bearing wave washer.
    e. Using a snap ring expander, install the snap ring.
19. Apply clutch spline grease to the input shaft spline and install the clutch cover assembly to the input shaft.
20. Install the clutch disc assembly to the transmission unit. Take care not to insert the clutch disc assembly in the wrong direction.
21. Install the pin and wave washer to the clutch release fork with a new E-ring.
22. Apply release hub grease to the contact surface between, the release fork and release bearing hub, the contact surface between the release fork and push rod, and release fork pivot points as shown in the illustration.
23. Install the clutch release fork sub-assembly to the release fork support. Do not use excess force to install the clutch release fork sub-assembly. After installation, move the fork back and forth to check that the clutch release bearing hub slides smoothly.
24. Install the manual transmission assembly.
25. Connect the negative battery cable.

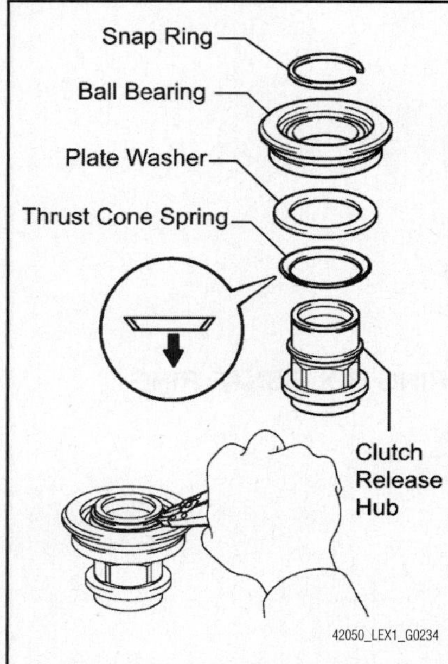

Fig. 8 Install the thrust cone spring, thrust cone spring plate washer and clutch release hub ball bearing to the clutch release hub

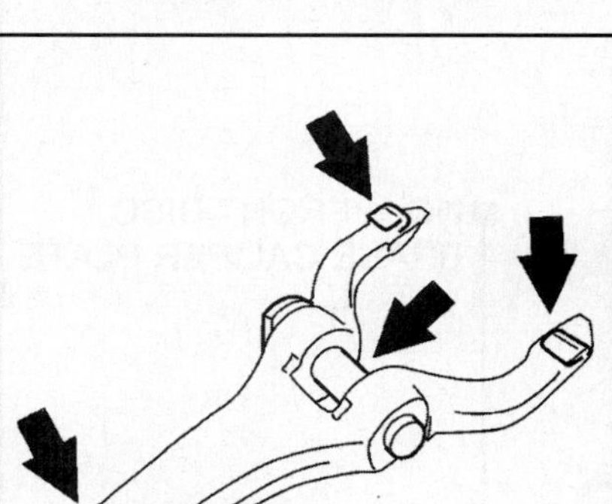

Fig. 9 Apply release hub grease to the release fork pivot points as shown

HYDRAULIC SYSTEM BLEEDING

BLEEDING PROCEDURE

See Figure 11.

If any work is performed on the clutch system or if air in the clutch lines is suspected, bleed air from the brake system.

➡Wash off clutch fluid immediately if it comes in contact with any painted surface.

1. Remove the bleeder plug cap.
2. Connect a vinyl tube to the bleeder plug.

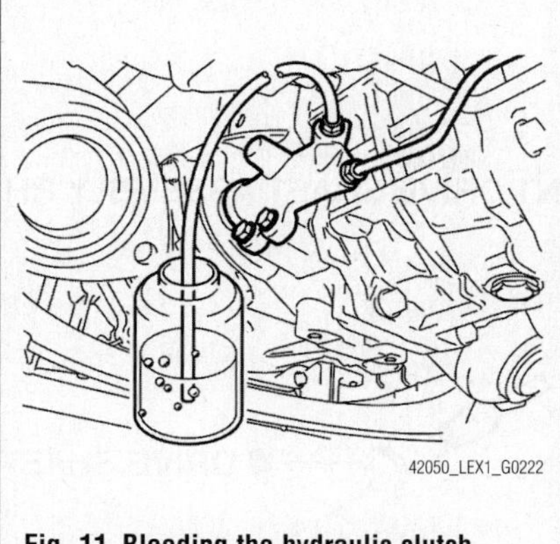

Fig. 11 Bleeding the hydraulic clutch system

3. Depress the clutch pedal several times, and then loosen the bleeder plug with the pedal depressed.
4. When fluid no longer comes out, tighten the bleeder plug, and then release the clutch pedal.
5. Repeat the previous 2 steps until all the air in the fluid is completely bled.
6. Tighten the bleeder plug to 8 ft. lbs. (11 Nm).
7. Install the bleeder plug cap.
8. Check that all the air has been bled from the clutch line.
9. Check fluid level in reservoir.

FRONT DIFFERENTIAL CASE OIL SEAL

REMOVAL & INSTALLATION

With AWD

See Figures 12 and 13.

1. Before servicing the vehicle, refer to the Precautions section.
2. Remove the halfshaft assemblies.
3. Using SST 09308-00010, remove the

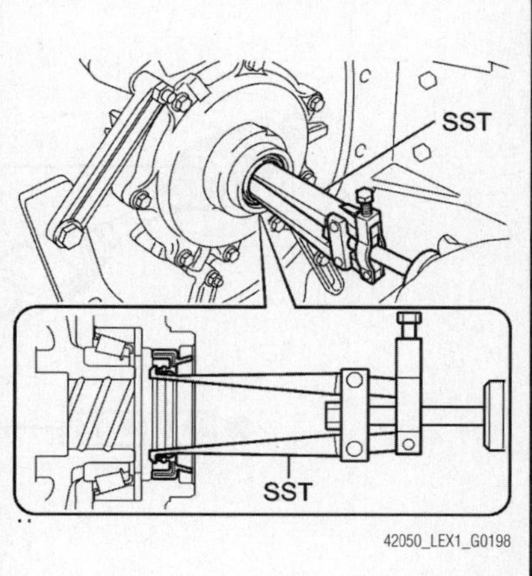

Fig. 12 Using SST 09308-00010 to remove the oil seals from the differential carrier

2 (left and right side) oil seals from the differential carrier.

To install:

4. Using SST 09613-22011, 09950-60010 (09951-00560) and a hammer, install 2 new oil seals to the following installation depth:
   a. Left Side: 0.039–0.079 inch (1–2mm)
   b. Right Side: 0.295–0.335 inch (7.5–8.5mm)
5. Apply MP grease to the oil seal lips.
6. Install the halfshaft assemblies.

FRONT HALFSHAFT

REMOVAL & INSTALLATION

With AWD

See Figure 14.

1. Before servicing the vehicle, refer to the Precautions section.

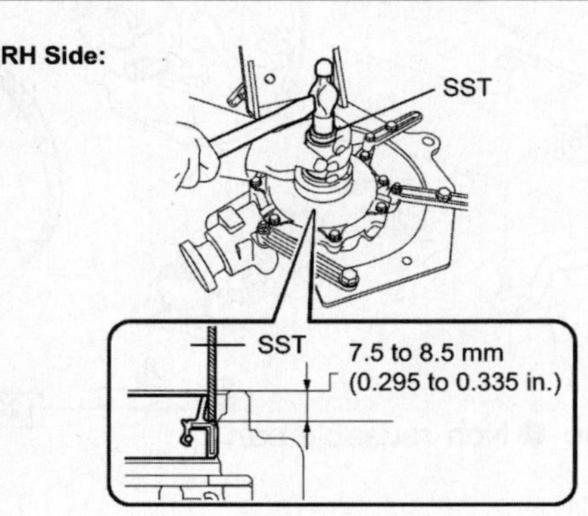

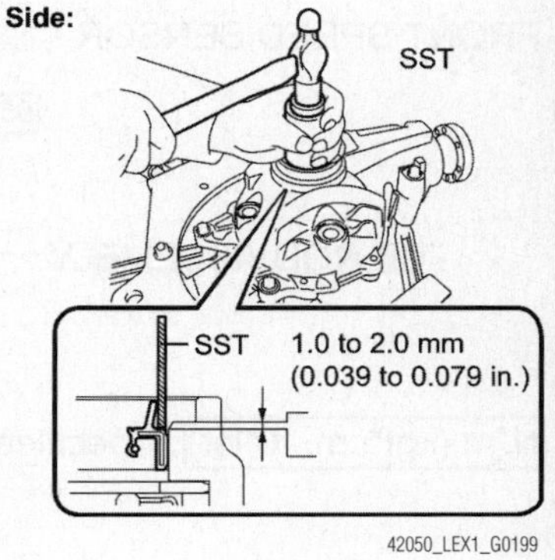

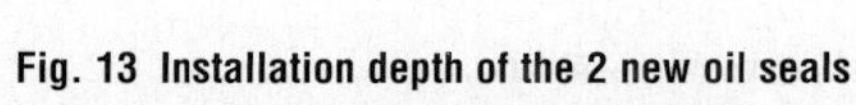

Fig. 13 Installation depth of the 2 new oil seals

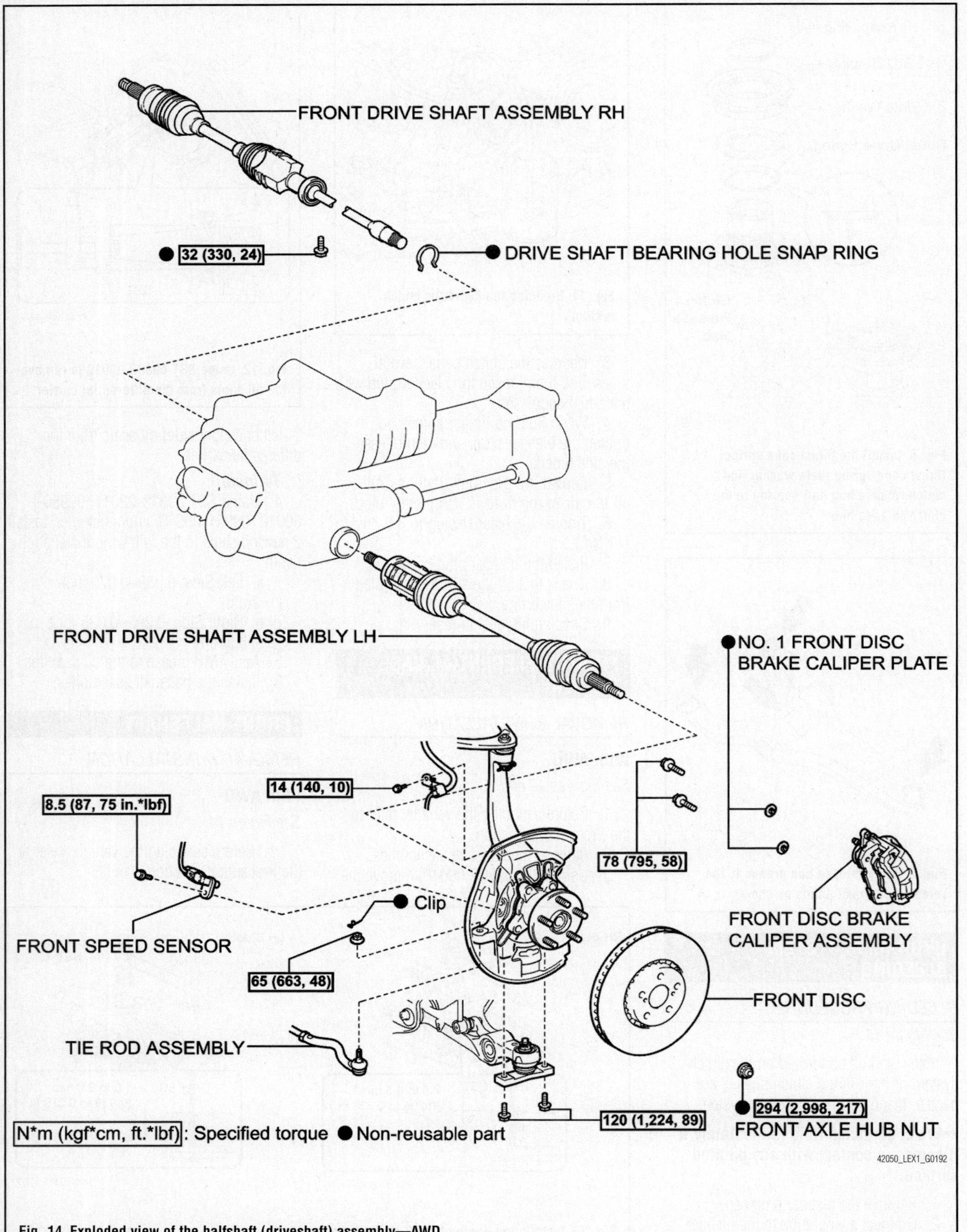

Fig. 14 Exploded view of the halfshaft (driveshaft) assembly—AWD

2. Disconnect the negative battery cable. Wait at least 90 seconds before performing any other work.
3. Remove the front wheel.
4. Separate the front speed sensor.
5. Separate the front disc brake caliper assembly.
6. Remove the front brake rotor.
7. Using SST 09930-00010 and a hammer, release the staked part of the front axle hub nut and remove. Release the staked part of the nut completely, otherwise the threads of the halfshaft may be damaged.
8. Separate the front lower ball joint assembly.
9. Separate the tie rod assembly.
10. Using a plastic hammer, separate the front axle assembly from the halfshaft assembly. Be careful not to damage the halfshaft boot, dust cover and oil seal

➡Be careful not to drop the halfshaft assembly.

11. Using SST 09520-01010, 09520-24010 (09520-32040), remove the left front halfshaft assembly.
12. Using water pump pliers, remove the bearing bracket hole snap ring.
13. Remove the bolt and right front halfshaft assembly from the halfshaft bearing bracket.
14. Inspect the halfshaft as follows:
  a. Check that there is no excessive play in the outboard joint.
  b. Check that the inboard joint slides smoothly in the thrust direction.
  c. Check that there is no excessive play in the radial directions of the inboard joint.
  d. Check the boot for damage.

To install:

15. Install the left front halfshaft assembly as follows:
16. Coat the spline of the inboard joint shaft assembly with gear oil.
17. Set the shaft snap ring with the opening side facing down.
18. Align the shaft splines and install the halfshaft assembly with a brass bar and hammer. Move the driveshaft assembly while keeping it level.

➡It is possible to determine if the inboard joint shaft is properly engaged (the shaft is in contact with the pinion shaft, and the snap ring is engaged in the pinion gear) based on the sound or feeling when the shaft is driven in.

19. Install the right front halfshaft assembly as follows:
20. Coat the spline of the inboard joint shaft assembly with gear oil.
21. Install the halfshaft assembly.
22. Using water pump pliers, install a new bearing bracket hole snap ring.
23. Install a new bolt and tighten it to 24 ft. lbs. (32 Nm).
24. Install the front halfshaft assembly to the front axle assembly. Be careful not to damage the halfshaft boot.
25. Install the tie rod assembly.
26. Install the front lower ball joint assembly.
27. Install the front brake rotor.
28. Install the front disc brake caliper assembly.
29. Using a socket wrench (30mm), install a new axle hub nut and tighten it to 217 ft. lbs. (294 Nm).
30. Using a chisel and hammer, stake the front axle hub nut.
31. Install the front speed sensor.
32. Install the front wheel.
33. Inspect and adjust the front wheel alignment.
34. Connect the negative battery cable.
35. Check ABS speed sensor signal.

FRONT PINION SEAL

REMOVAL & INSTALLATION

With AWD

1. Before servicing the vehicle, refer to the Precautions section.
2. Remove the differential and place the differential carrier in an overhaul stand.
3. Using SST 09930-00010 and a hammer, unstake the staked part of the drive pinion nut. Be sure to use the SST with the tapered surface facing the shaft. Completely loosen the staked part of the nut when removing it.
4. Using SST 09330-00021, 09950-30012 (09955-03030) to hold the flange, remove the drive pinion nut.
5. Using SST 09950-30012 (09951-03010, 09953-03010, 09954-03010, 09955-03030, 09956-03020), remove the companion flange. Apply grease to the threads and the tip of the SST center bolt before use.
6. Using SST 09950-00020, 09950-60010 (09951-00430), 09950-70010 (09951-07100) and a press, remove the dust deflector. Perform this procedure only when the dust deflector is damaged.
7. Using SST 09308-00010, remove the oil seal from the differential carrier.

To install:

8. Install the new pinion seal as follows:
  a. Join the 2 SST 09309-36010, 09502-12010 and secure them with vinyl tape.
  b. Using SST 09309-36010, 09502-12010 and a hammer, install a new oil seal. Oil seal installation depth: 0.028–0.051 inch (0.7–1.3mm)
  c. Apply MP grease to the oil seal lip.
9. Install the differential side bearing retainer dust deflector with the bolt and tighten to 62 inch lbs. (7 Nm).
10. Install the front drive pinion companion flange as follows:
  a. Using SST 09950-30012 (09951-03010, 09953-03010, 09954-03010, 09955-03030, 09956-03020), install the companion flange to the drive pinion.

➡Apply grease to the threads and tip of the SST center bolt before use.

  b. Coat the threads of a new drive pinion nut with hypoid gear oil LSD.
  c. Using SST 09330-00021, 09950-30012 (09955-03030), to hold the flange, tighten the drive pinion nut to 80–173 ft. lbs. (108–235 Nm).
11. Adjust the differential drive pinion preload as follows:
  a. Using a torque wrench, measure the preload of the drive pinion. Preload (at starting): New bearing: 8.7–13.9 inch lbs. (0.98–1.57 Nm). Reused bearing: 4.3–6.9 inch lbs. (0.49–0.78 Nm).
  b. If the preload is less than the specified minimum value, check the preload while retightening the drive pinion nut by 5–10° to adjust it into the specified range.
  c. If the preload is less than the specified minimum value even when the tightening torque of the drive pinion nut is greater than the specified maximum value, loosen the nut and check that the threads of the drive pinion nut and drive pinion are not stripped.
  d. If the threads are not stripped, replace the bearing spacer. Apply hypoid gear oil LSD to the threads of the drive pinion and repeat the procedure.

12. Install the differential case to the differential carrier. Do not damage the case bearing and ring gear.

13. Install the front drive pinion companion flange front nut. Using a chisel and hammer, stake the drive pinion nut.

REAR HALFSHAFTS

REMOVAL & INSTALLATION

See Figure 15.

1. Before servicing the vehicle, refer to the Precautions section.

2. Remove or disconnect the following:
   a. Negative battery cable
   b. Rear tire and wheel assembly
   c. Cotter pin, locknut cap, and locknut
   d. Height control sensor, if equipped
   e. 2 exhaust pipe support brackets, if necessary

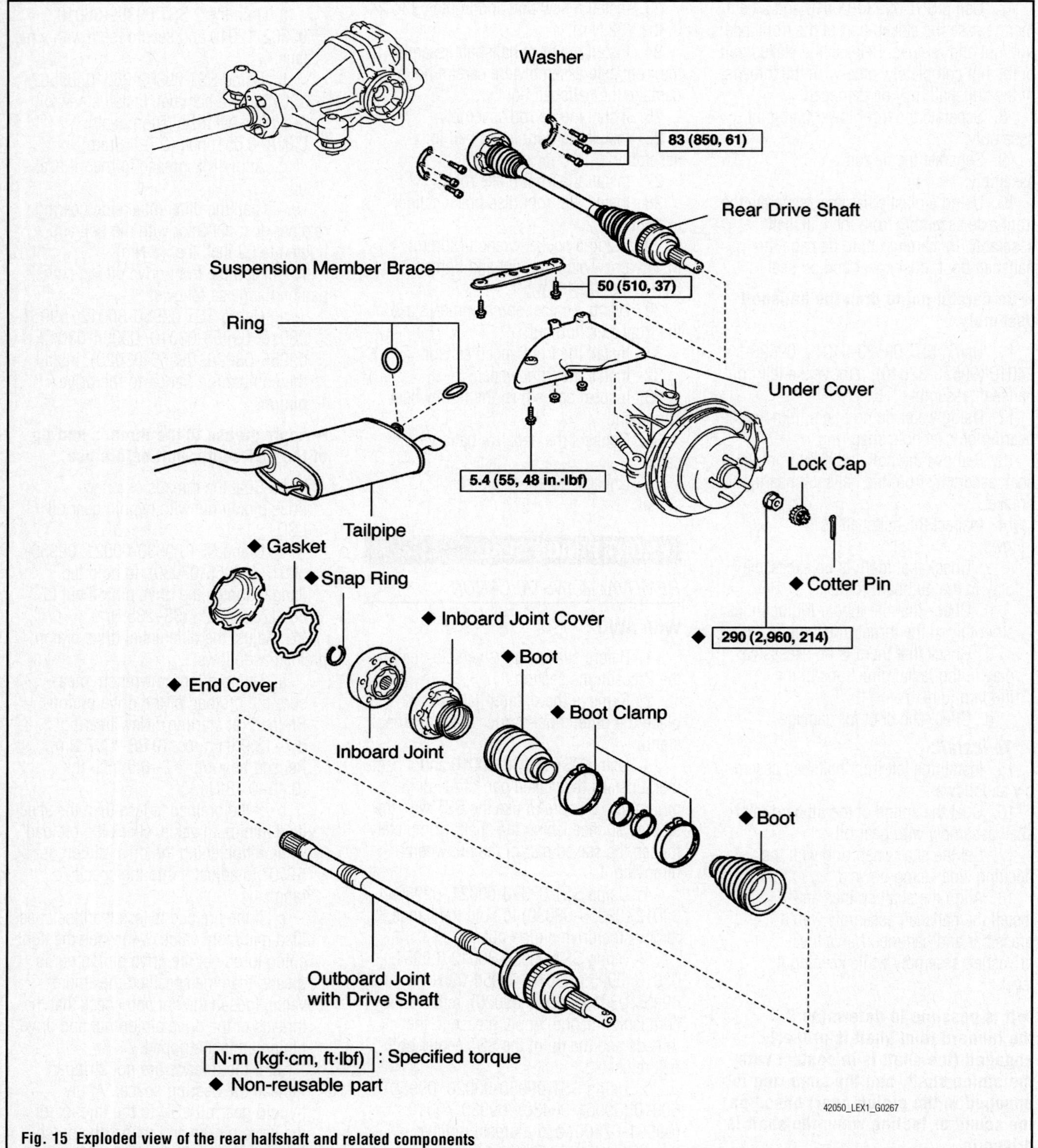

Fig. 15 Exploded view of the rear halfshaft and related components

3. Place matchmarks on the halfshaft and the side gear shaft. Remove the 6 hex bolts and 2 washers.

4. Hold the inboard joint side of the halfshaft so the outboard joint side does not bend too much. Tap the end of the halfshaft with a rubber mallet to loosen it from the axle hub and remove the halfshaft.

To install:

5. Insert the outboard joint side of the halfshaft through the axle hub. Align the matchmarks on the side gear shaft and the halfshaft.

6. Coat the threads with clean oil and install the hex bolts. Tighten the bolts to 61 ft. lbs. (83 Nm).

7. Install or connect the following:
- Exhaust pipe support brackets, if removed, and tighten to 14 ft. lbs. (19 Nm)
- Bearing locknut, if removed, and have a helper apply the brakes. Tighten the locknut to 213 ft. lbs. (289 Nm).
- Lock cap and a new cotter pin
- Height control sensor, if removed
- Rear tire and wheel assembly
- Negative battery cable

REAR PINION SEAL

REMOVAL & INSTALLATION

1. Before servicing the vehicle, refer to the Precautions section.

2. Remove the differential and place the differential carrier in an overhaul stand.

3. Using SST 09930-00010 and a hammer, unstake the staked part of the drive pinion nut. Be sure to use the SST with the tapered surface facing the shaft. Completely loosen the staked part of the nut when removing it.

4. Using SST 09330-00021, 09950-30012 (09955-03030) to hold the flange, remove the drive pinion nut.

5. Using SST 09950-30012 (09951-03010, 09953-03010, 09954-03010, 09955-03030, 09956-03020), remove the companion flange. Apply grease to the threads and the tip of the SST center bolt before use.

6. Using SST 09950-00020, 09950-60010 (09951-00430), 09950-70010 (09951-07100) and a press, remove the dust deflector. Perform this procedure only when the dust deflector is damaged.

7. Using SST 09308-00010, remove the oil seal from the differential carrier.

To install:

8. Install the new pinion seal as follows:

a. Join the 2 SST 09309-36010, 09502-12010 and secure them with vinyl tape.

b. Using SST 09309-36010, 09502-12010 and a hammer, install a new oil seal. Oil seal installation depth: 0.028–0.051 inch (0.7–1.3mm)

c. Apply MP grease to the oil seal lip.

9. Install the differential side bearing retainer dust deflector with the bolt and tighten to 62 inch lbs. (7 Nm).

10. Install the front drive pinion companion flange as follows:

a. Using SST 09950-30012 (09951-03010, 09953-03010, 09954-03010, 09955-03030, 09956-03020), install the companion flange to the drive pinion.

➡Apply grease to the threads and tip of the SST center bolt before use.

b. Coat the threads of a new drive pinion nut with hypoid gear oil LSD.

c. Using SST 09330-00021, 09950-30012 (09955-03030), to hold the flange, tighten the drive pinion nut to 74 ft. lbs. (100 Nm).

11. Adjust the differential drive pinion preload as follows:

a. Using a torque wrench, measure the preload of the drive pinion. Preload (at starting):

b. New bearing: 9.9–15 inch lbs. (1.12–1.70 Nm)

c. Reused bearing: 3.5–6.2 inch lbs. (0.40–0.70 Nm)

d. If the preload is less than the specified minimum value, check the preload while retightening the drive pinion nut by 5–10° to adjust it into the specified range.

e. If the preload is less than the specified minimum value even when the tightening torque of the drive pinion nut is greater than the specified maximum value, loosen the nut and check that the threads of the drive pinion nut and drive pinion are not stripped.

If the threads are not stripped, replace the bearing spacer. Apply hypoid gear oil LSD to the threads of the drive pinion and repeat the procedure.

12. Install the differential case to the differential carrier. Do not damage the case bearing and ring gear.

13. Install the drive pinion companion flange front nut. Using a chisel and hammer, stake the drive pinion nut.

TRANSFER CASE ASSEMBLY

REMOVAL & INSTALLATION

1. Before servicing the vehicle, refer to the Precautions section.

2. Disconnect the negative battery cable. Wait at least 90 seconds before performing any other work.

3. Remove or disconnect the following:

a. Automatic transmission assembly.

b. Transfer case oil pan.

4. Remove the transfer valve body assembly as follows:

a. Remove the 4 bolts and the transfer valve body assembly

b. Disconnect the transfer wire connector from the transfer control solenoid. Remove the bolt and the transfer wire. Transfer extension housing using SST 09950-40011. Remove the 8 bolts (in order), clamp and transfer case. Use a plastic hammer to tap the transfer case to remove it. Transfer rear output shaft assembly.

5. Remove the flange yoke as follows:

a. Using a hammer and SST 09930-00010, release the staked part of the nut.

b. Using SST 09330-00021 and 09213-54015, hold the yoke. Remove the nut and the flange yoke. Remove the 10 bolts and rear transfer chain case. Use a plastic hammer to tap the rear transfer chain case to remove it. Transfer front drive chain, transfer drive sprocket and transfer front driven clutch sleeve. Remove the 6 bolts (in order) and front transfer chain case. Use a plastic hammer to tap the front transfer chain case to remove it.

To install:

6. Install front transfer chain case as follows:

7. Clean the front mating surface of any residual sealant.

a. Apply new sealant in a width of 0.04–0.06 inches (1.0–1.5mm) all around to the front transfer chain case front mating surface.

b. Temporarily tighten the 6 bolts in several steps, and then tighten them to 25 ft. lbs. (34 Nm).

8. Install or connect the following:

a. Transfer front drive chain, transfer drive sprocket and transfer front driven clutch sleeve

9. Install rear transfer chain case as follows:

a. Clean the front mating surface of any residual sealant.

b. Apply new sealant in a width of 0.04–0.06 inches (1.0–1.5mm) all around to the rear transfer chain case front mating surface.

c. Temporarily tighten the 10 bolts in several steps, and then tighten them to 25 ft. lbs. (34 Nm). Tighten the bolt in the middle of the case cover to 42 ft. lbs. (57 Nm).

10. Install the flange yoke as follows:

a. Install the flange yoke.

b. Using SST 09330-00021 and 09213-54015, hold the yoke. Torque the new nut to 91 ft. lbs. (123 Nm). Transfer rear output shaft assembly. Transfer extension housing using SST 09950-40011.

11. Install transfer case as follows:

a. Clean the front mating surface of any residual sealant.

b. Apply new sealant in a width of 0.04–0.06 inches (1.0–1.5mm) all around to the transfer case front mating surface.

c. Temporarily tighten the 8 bolts and clamp in several steps, and then tighten them to 25 ft. lbs. (34 Nm).

12. Install the transfer valve body assembly as follows:

a. Coat the transfer wire O-ring with ATF.

b. Install the transfer wire with the bolt.

c. Connect the transfer wire connector to the transfer control solenoid.

d. Install the transfer valve body assembly. Torque the 4 bolts to 87 inch lbs. (10 Nm).

13. Install transfer case oil pan as follows:

a. Clean and install the transfer case oil cleaner magnet to the pan.

b. Clean the pan contact surface of any residual sealant.

c. Apply new sealant in a width of 0.12–0.14 inches (3.0–3.5mm) all around the contact surface of the transfer case oil pan.

d. Temporarily tighten the 9 bolts in several steps, and then tighten them to 65 inch lbs. (7.5 Nm). Automatic transmission assembly.

ENGINE COOLING

ELECTRIC FAN SWITCH

REMOVAL & INSTALLATION

1. Before servicing the vehicle, refer to the Precautions section.
2. Disconnect the negative battery cable. Wait at least 90 seconds before performing any other work.
3. Disconnect the electric fan switch (temperature switch) connector for electric cooling fan.
4. Using a switch socket or open end wrench, remove the switch from the vehicle.

To install:

5. Installation is the reverse of the removal procedure.

TESTING

1. Using an ohmmeter, check for continuity between the terminals.
2. If the ohmmeter reading is above 208°F (98°C), there is continuity.
3. If the ohmmeter reading is below 190°F (88°C), there is no continuity.
4. If the result is not as specified, replace the temperature switch.

ENGINE FAN

REMOVAL & INSTALLATION

1. Before servicing the vehicle, refer to the Precautions section.
2. Disconnect the negative battery cable. Wait at least 90 seconds before performing any work.
3. Drain the engine coolant.
4. Remove the radiator assembly (the fan assembly is attached).
5. Separate the fan assembly from the radiator.

To install:

6. Install the fan assembly to the radiator and tighten the fasteners to 44 inch lbs. (5 Nm).
7. Install the radiator and fan assembly.
8. Refill the engine cooling system.
9. Connect the negative battery cable.
10. Start the engine and check system for coolant leaks.
11. Install the engine under cover No. 1.

RADIATOR

REMOVAL & INSTALLATION

See Figures 18 through 20.

1. Before servicing the vehicle, refer to the Precautions section.
2. Disconnect the negative battery cable. Wait at least 90 seconds before performing any work.
3. Remove the cool air intake duct seal.
4. Remove the left and right engine room side covers.
5. Remove V-bank cover sub-assembly.
6. Remove the No. 1 inlet air cleaner.
7. Disconnect the No. 2 ventilation hose.
8. Remove the air cleaner cap with air cleaner hose.
9. Remove engine under cover.
10. Drain engine coolant.
11. Remove the front bumper assembly as follows:

a. Using a screwdriver, turn the pin 90° and remove the 2 pin hold clips. Tape the screwdriver tip before use. Use the same procedures for the RH side and LH side.

12. Put protective tape around the front bumper assembly.

a. Using a clip remover, remove the 2 clips

b. Remove the 2 radiator grille protectors.

c. Remove the bolt and 10 screws.

d. Disengage the 5 claws and disconnect the front bumper assembly as shown in the illustration. Use the same procedures for the RH side and LH side.

e. Disconnect the headlight washer hose. (w/ headlight cleaner system)

f. Disconnect the connector and remove the front bumper assembly.

13. Remove the front bumper energy absorber.
14. Disengage the 2 clips and remove the radiator support opening cover.
15. Disconnect the connector, remove the 3 bolts and the millimeter wave radar sensor assembly (w/ Dynamic Radar Cruise Control System).
16. Remove the 3 screws, clamp and the hood lock control cable cover.
17. Disconnect the hood lock assembly as follows:

a. Disconnect the connector.

b. Remove the hood lock nut cap and the hood lock nut.

c. Remove the 2 bolts, and separate the hood lock assembly from the upper radiator support.

➡Do not forcibly bend the hood lock control cable.

18. Remove the upper radiator support as follows:

a. Separate the 4 wire harness clamps and the 3 connectors.

b. Separate the clamp and the cooling fan motor connector.

c. Separate the clamp and the No.2 cooling fan motor connector.

d. Separate the 6 clamps and the No.2 engine room wire.

e. Remove the 5 bolts and the upper radiator support.

19. Remove the 2 clips and the radiator inlet hose from the radiator assembly.

20. Remove the 2 clamps and the radiator outlet hose from the radiator assembly.

21. Remove the 4 bolts, separate the cooler condenser assembly from the radiator assembly.

22. Remove the engine room ECU outlet duct from the engine room ECU box.

23. Separate the reserve tank cap sub-assembly from the reserve tank.

24. Remove the radiator assembly from the vehicle together with the cooling fan assembly.

➡Make sure that the cooler condenser assembly and radiator assembly do not come into contact with each other.

25. Remove the 3 claws and the cooling fan assembly from the radiator assembly.

26. Remove the 2 radiator support cushions and the 2 lower radiator supports.

To install:

27. Install the 2 radiator support cushions and the 2 lower radiator supports.

28. Install the cooling fan assembly to the radiator assembly by engaging them at the bottom. Engage the 3 claws to secure the cooling fan assembly to the radiator assembly.

29. Install the radiator assembly to the vehicle together with the cooling fan assembly.

➡Make sure that the cooler condenser assembly and radiator assembly do not come into contact with each other.

30. Install the reserve tank cap sub-assembly to the reserve tank.

31. Install the engine room ECU outlet duct to the engine room ECU box.

32. Install the 4 bolts and the cooler condenser assembly to the radiator assembly. Tighten the 4 bolts to 44 inch lbs. (5 Nm).

33. Install the radiator outlet hose to the radiator assembly and secure it with the clips. Make sure that the claws on the clips are facing downward.

34. Install the radiator inlet hose to the radiator assembly and secure it with the clips.

35. Install the upper radiator support as follows:

a. Install the 5 bolts and the upper radiator support. Tighten the bolts to 71 inch lbs. (8 Nm).

b. Install the 6 wire harness clamps and the No.2 engine room wire.

c. Connect the clamp and the No.2 cooling fan motor connector.

d. Connect the clamp and the cooling fan motor connector.

e. Connect the 4 wire harness clamps and the 3 connectors.

36. Install the hood lock assembly as follows:

a. Install the 2 bolts and the hood lock assembly to the upper radiator support. Tighten the bolts to 71 inch lbs. (8 Nm).

b. Install the hood lock nut and the hood lock nut cap. Tighten the nut to 71 inch lbs. (8 Nm).

c. Connect the connector.

37. Connect the clamp, and install the 3 screws and the hood lock control cable cover.

38. Install the millimeter wave radar sensor assembly (w/ Dynamic Radar Cruise Control System).

39. Install the air cleaner cap with air cleaner hose.

40. Connect the No. 2 ventilation hose.

41. Install the No. 1 inlet air cleaner.

42. Connect the 2 clips, and install the radiator support opening cover.

43. Install the front bumper energy absorber.

44. Install the front bumper assembly.

45. Refill the engine cooling system.

46. Connect the negative battery cable.

47. Start the engine and check system for coolant leaks.

⁂ WARNING

Exposure to radio frequency emissions is hazardous to your health. It is hazardous to your health to be within 7.9 inches (20cm) of the device's radio frequency aperture.

➡Perform measurements on a level surface. Make sure that no large pieces of metal are within a 10m (32.81 ft) x 14m (45.93 ft) area in front of the vehicle. If possible, the surrounding area should also be free of large metal objects.

48. Adjust the millimeter wave radar sensor assembly as follows:

a. Check the tire pressure and adjust it if necessary and remove all excess weight from the vehicle (luggage, heavy objects, etc.).

b. Check and adjust the vertical direction of the radar sensor.

c. Remove dust, oil, and foreign matter from the radar sensor's level rack.

d. Set a level on the radar sensor's level rack.

e. Check that the air bubble is within the red frame on the level. OK: The air bubble is within the red frame on the level. If the bubble is not within the red frame, use a hexagon wrench to adjust screw A until the air bubble is within the red frame. The adjustable range within the red frame on the level is + - 0.2°. The target angle is +0.2°(upward angle of 0.2°).

f. Adjust the reflector height.

g. Adjust the reflector so that the center of the SST 09870-60000 (09870-60010, 09870- 60040) reflector is the same height as the millimeter wave radar sensor.

h. Prepare a 10m (32.81 ft) string, a string with a sharp-pointed weight (plumb bob), and a 5m (16.41 ft) tape measure.

i. Place the reflector.

j. Hang the string (with weight) from the center of the vehicle's rear emblem. Mark the vehicle's rear center point on the ground. Repeat the same procedure for the front of the vehicle.

k. Set one end of the 10m (32.82 ft) string on the vehicle's rear center point. Run the string over the vehicle's front center point to a position 5m (16.41 ft) beyond the vehicle front center point, as shown in the illustration. Mark the 5m (16.41 ft) position.

l. Using a tape measure, measure

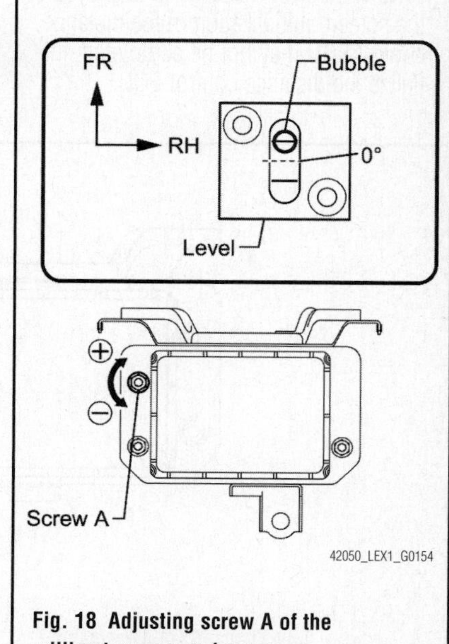

Fig. 18 Adjusting screw A of the millimeter wave radar sensor

DISTANCE

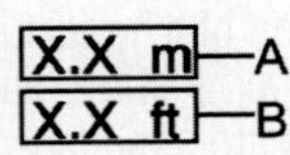

PRESS [ENTER]

42050_LEX1_G0156

Fig. 19 Display screen for adjusting the millimeter wave radar sensor

12mm (0.47 in.) to the left of the 5m (16.41 ft) position. Place the reflector at that position.

➡Perform the operation as precisely as possible.

m. Adjust the radar beam axis by first connecting the intelligent tester to the Controller Area Network Vehicle Interface Module (CAN VIM). Then connect the CAN VIM to the Data Link Connectors 3 (DLC3).

n. Turn the engine switch on (IG).

o. Turn the intelligent tester main switch on, and turn the cruise control main switch on.

p. Select "1: OBD/MOBD".

q. Press the "ENTER" key.

r. Select "8: RADAR CRUISE".

s. Press the "ENTER" key.

t. Select "5: BEAM AXIS ADJUST".

u. Press the "ENTER" key. Pressing the "ENTER" key will make the ECM transfer to BEAM AXIS ADJUST MODE.

v. If an error message is displayed on the screen, initialization of the distance control ECU may not be completed. Initialize the distance control ECU.

w. When the ECM transfers to BEAM AXIS ADJUST MODE, the buzzer sounds for 1 second.

x. Press the "ENTER" key.

y. Press the "ENTER" key.

z. After the CRUISE main indicator light illuminates, press the "ENTER" key.

aa. Press the "ENTER" key.

bb. Check that a distance of 5 m (16.41 ft) is displayed on the tester.

cc. If the distance is 0 m (0 ft), the sensor cannot detect the target. Reconfirm that there is no metal in the surrounding area.

dd. Check that a value between 0.0 and 6.3 is displayed in the meter display area A.

ee. Check that a value between 0.0 and 20.67 is displayed in the feet display area B.

ff. Check that LEFT or RIGHT is displayed in area C.

gg. Check that an angle value between 0.0 and 6.3 is displayed in area D. While using the intelligent tester BEAM AXIS ADJUST MODE, the actual direction and angle of the radar sensor may be different from the intelligent tester's data. In such a case, the deviation is displayed on the combination meter's multi-information display.

hh. Check and adjust the horizontal direction of the radar sensor.

ii. Check that the divergence of the radar beam axis is 0°. Standard: 0° (Both right and left) If the axis is not as specified, use a hexagon wrench to adjust screw B until the divergence of the radar beam axis is 0°.

jj. Based on the measured divergence of the beam axis, turn and adjust screw B for horizontal adjustment of the millimeter wave radar sensor using a hexagon wrench.

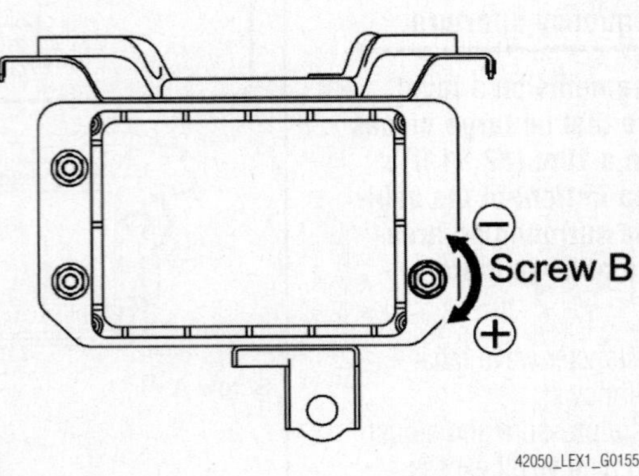

Fig. 20 Adjusting screw B of the millimeter wave radar sensor

kk. If "LEFT SIDE: 1.0°" is displayed, the divergence is 1.0°in the left direction. Turn screw B approximately 3 turns to the negative (-) side.

ll. If the value does not change to 0°, it is possible that the sensor is aiming at something different. Reconfirm that there are no reflective materials in the surrounding area.

mm. Reset the radar sensor's driving learning values. Prepare a type of metal that can block radio waves, such as aluminum foil. Cover the radar sensor's left half with the metal for 10 seconds.

nn. Be sure to keep the reflector in place and make sure that there is nothing between the sensor's left half and the reflector.

oo. When the reset is completed, the buzzer sounds for 10 seconds.

pp. Disconnect the intelligent tester from the DLC3.

qq. Recheck and readjust the vertical direction of the radar sensor.

rr. Set a level on the radar sensor's level rack.

ss. Check that the level's air bubble is within the red frame. OK: Level's air bubble is within red frame. If the bubble is not within the red frame, use a hexagon wrench to adjust screw A until the level's air bubble is within the red frame.

tt. The adjustable range within the red frame is + -0.2°.

uu. The target angle is +0.2°(upward angle of 0.2°).

49. Start the engine and charge the battery.
50. Aim and adjust the fog lights.
51. Install engine under cover.
52. Install V-bank cover sub-assembly.
53. Install the left and right engine room side covers.
54. Install the cool air intake duct seal.

THERMOSTAT

REMOVAL & INSTALLATION

See Figure 23.

1. Before servicing the vehicle, refer to the Precautions section.
2. Disconnect the negative battery cable. Wait at least 90 seconds before performing any other work.
3. Remove the cool air intake duct seal.
4. Remove the No. 1 inlet air cleaner.
5. Remove the V-bank cover sub-assembly.
6. Drain the engine coolant.

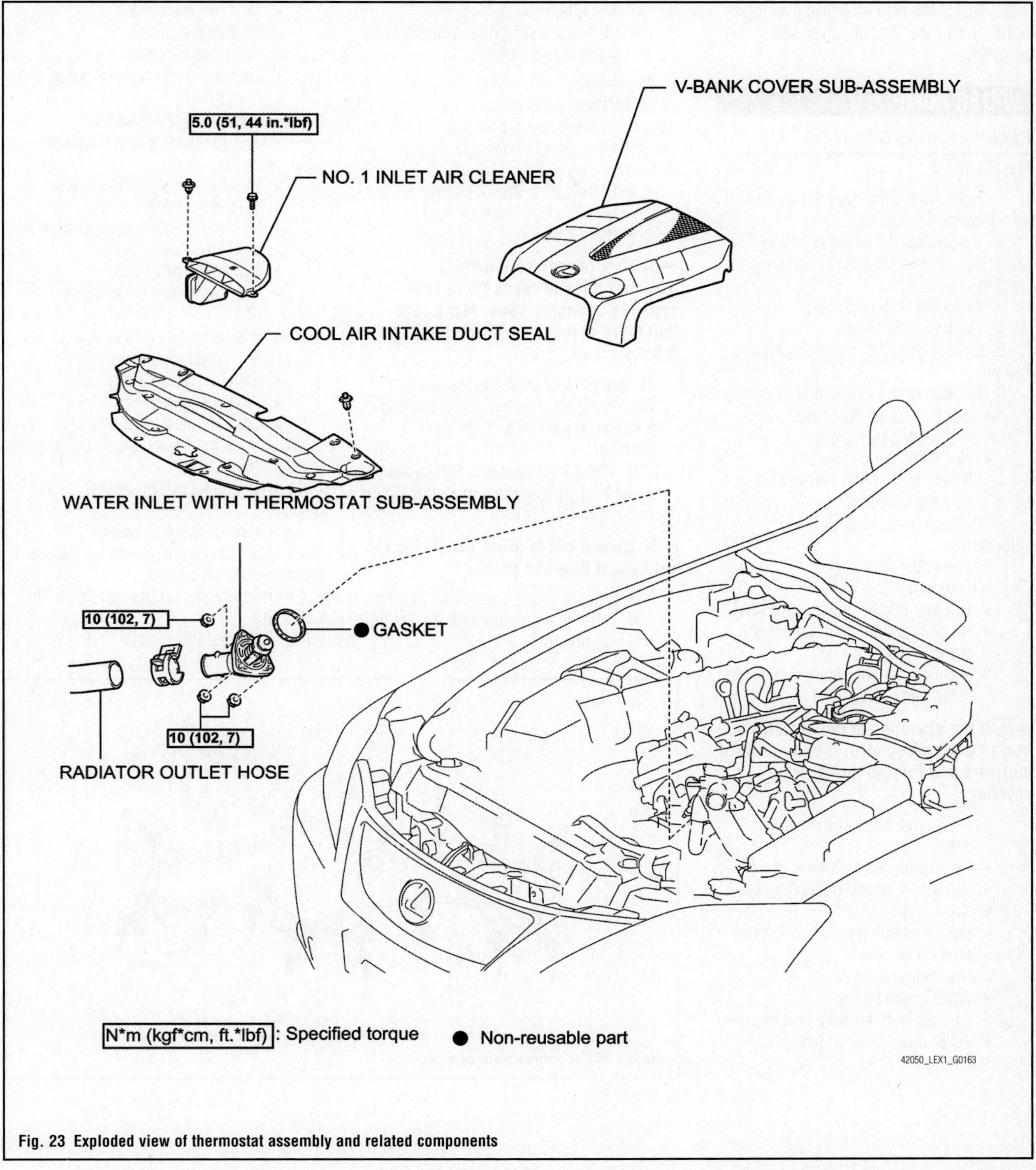

Fig. 23 Exploded view of thermostat assembly and related components

7. Disconnect the hose from the water inlet with thermostat.

8. Remove the 3 nuts, water inlet with thermostat and gasket.

To install:

9. Install a new gasket and the water inlet with thermostat with the 3 nuts. Tighten the 3 nuts to 7 ft. lbs. (10 Nm).

10. Connect the radiator outlet hose to the radiator assembly and secure it with the clip. Make sure that the claws on the clip are positioned outward with space between the hose end and the clip end should be less than 1–5mm. Hose contact with the stopper should be less than 2.5mm.

11. Refill the engine coolant.

12. Connect the negative battery cable.

13. Start the engine and check cooling system for leaks.

14. Install the No. 1 inlet air cleaner.

15. Install the cool air intake duct seal.
16. Install the V-bank cover sub-assembly.

WATER PUMP

REMOVAL & INSTALLATION

See Figure 24.

1. Before servicing the vehicle, refer to the Precautions section.
2. Disconnect the negative battery cable. Wait at least 90 seconds before performing any work.
3. Drain the engine coolant.
4. Drain the engine oil.
5. Discharge refrigerant from A/C system.
6. Remove or disconnect the following:
   - Cool air intake duct seal
   - V-bank engine cover
   - No. 1 inlet air cleaner
7. Loosen, but do not remove water pump pulley mounting bolts.
8. Remove or disconnect the following:
   - Serpentine drive belt
   - Engine under cover
   - Left rear engine under cover (2WD)
   - Oil filter and mounting bracket assembly (AWD)
   - No. 1 cooler refrigerant suction and discharge hoses

➡Seal the openings of the disconnected parts using vinyl tape to prevent moisture and foreign matter from entering.

   - A/C compressor and magnetic clutch
   - Serpentine belt tensioner
   - Radiator inlet and outlet hoses
   - No. 1 engine cover
   - Bolt, 2 retaining nuts and injector driver unit
   - No. 2 engine cover
   - Water pump pulley
   - 5 hoses from the water inlet housing
   - Water inlet housing, gasket and O-ring
   - No. 2 idler pulley
   - 16 bolts, water pump assembly and water pump gasket

To install:

Install a new water pump gasket and water pump assembly with the 16 mounting bolts and tighten as follows:

- Bolt A: 15 ft. lbs. (21 Nm)
- Bolt B: 81 inch lbs. (9 Nm)
- Bolt C: 81 inch lbs. (9 Nm)

➡Be sure to replace 2 bolts C with new ones or reuse them after applying adhesive 1344. Make sure that there is no oil on the threads of the A bolts.

9. Install or connect the following:
   a. No. 2 idler pulley, cover plate and bolt. Torque the bolt to 32 ft. lbs. (43 Nm).
   b. Water inlet housing, new gasket and O-ring. Torque the 4 mounting bolts to 7 ft. lbs. (10 Nm).

➡Be careful not to allow the O-ring to get caught between parts.

   - 5 hoses to the water inlet housing
   - Water pump pulley with 4 bolts finger tight only
   - No. 2 engine cover
   - Injector driver unit
   - No. 1 engine cover
   - Inlet and outlet radiator hoses and new clamps
   - Serpentine belt tensioner
   - A/C compressor and magnetic clutch
   - Oil filter and mounting bracket assembly (AWD)
   - No. 1 cooler refrigerant suction and discharge hoses
   - Serpentine drive belt
   - Left rear engine under cover (2WD)
   - Engine under cover
   - No. 1 inlet air cleaner
   - Negative battery cable
   - V-bank engine cover
   - Cool air intake duct seal
10. Torque the 4 water pump pulley bolts to 15 ft. lbs. (21 Nm).
11. Add engine oil and engine coolant.
12. Recharge the A/C system.
13. Start the engine, check for leaks and bleed the cooling system.
14. Recheck all fluid levels and add if necessary.
15. Road test the vehicle.

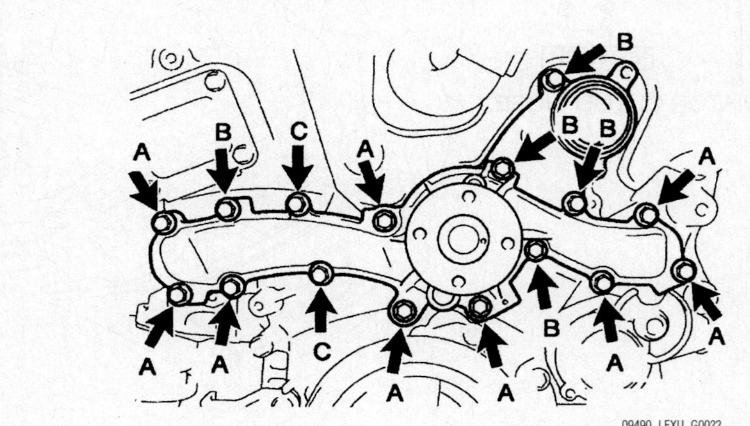

Fig. 24 Water pump mounting bolts

ENGINE ELECTRICAL — CHARGING SYSTEM

ALTERNATOR

REMOVAL & INSTALLATION

1. Before servicing the vehicle, refer to the Precautions section.
2. Remove or disconnect the following:
   - Negative battery cable. Wait at least 90 seconds before performing any other work.
   - Cool air intake duct seal
   - Left engine room side cover
   - V-bank cover sub-assembly
   - No. 1 inlet air cleaner
   - Serpentine drive belt
   - Engine coolant
   - Radiator inlet hose
   - No. 2 engine cover
   - No. 2 idler pulley sub-assembly
   - Wiring harness
   - Terminal cap
   - Alternator wiring harnesses
   - Bolt and clamp bracket
   - Alternator connector, and 2 clamps
   - Nut and alternator bracket at the engine
   - 2 mounting bolts and the alternator
   - 2 bolts and 2 alternator mounting brackets

To install:

3. Install or connect the following:
   - 2 alternator mounting brackets and 2 bolts. Torque to 15 ft. lbs. (20 Nm).
   - Alternator and 2 mounting bolts. Torque to 32 ft. lbs. (43 Nm).
   - Alternator bracket and nut at the engine. Torque to 15 ft. lbs. (20 Nm).
   - Alternator wire to terminal. Torque nut to 87 inch lbs. (10 Nm).
   - Terminal cap
   - Clamp bracket and bolt. Torque to 7 ft. lbs. (10 Nm).
   - 2 clamps, and connect alternator connector
   - Wire harness with the 3 nuts. Torque to 7 ft. lbs. (10 Nm).
   - 2 alternator connectors.
   - No. 2 idler pulley sub-assembly
   - No. 2 engine cover
   - Radiator inlet hose
   - Serpentine drive belt
   - No. 1 inlet air cleaner
   - V-bank cover sub-assembly
   - Left engine room side cover
   - Cool air intake duct seal
   - Negative battery cable
4. Add engine coolant.

ENGINE ELECTRICAL — IGNITION SYSTEM

FIRING ORDER

See Figure 25.

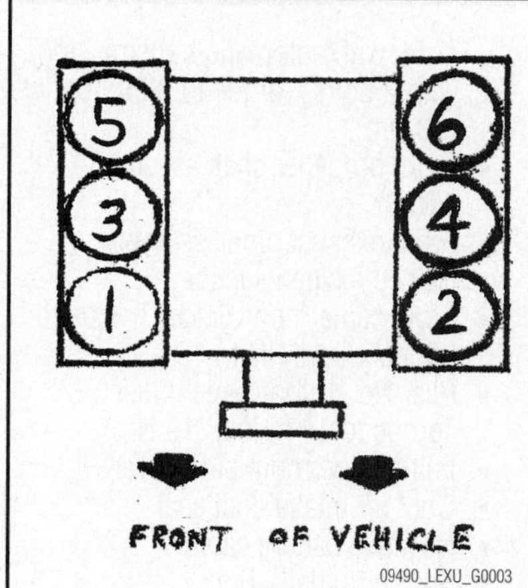

Fig. 25 Firing order: 1-2-3-4-5-6 Distributorless ignition system (one coil per cylinder)

IGNITION COIL

REMOVAL & INSTALLATION

1. Before servicing the vehicle, refer to the Precautions section.
2. Disconnect the negative battery cable. Wait at least 90 seconds before performing any other work.
3. Remove V-bank cover sub-assembly.
4. Remove cool air intake duct seal.
5. Remove the right and left engine room side covers.
6. Disconnect the ventilation hose from the cylinder head.
7. Remove the air cleaner cap with air cleaner hose.
8. Remove intake air surge tank assembly.
9. Disconnect the 6 ignition coil connectors.
10. Remove the 6 bolts and the 6 ignition coil assemblies.

To install:

11. Install the 6 ignition coils with the 6 bolts and torque to 7 ft. lbs. (10 Nm).
12. Connect the 6 ignition coil connectors.
13. Install the intake air surge tank assembly.
14. Install the air cleaner cap with air cleaner hose.
15. Connect the ventilation hose to the cylinder head cover with the clamp.
16. Install the right and left engine room side covers.
17. Install cool air intake duct seal.
18. Install V-bank cover sub-assembly.
19. Connect the negative battery cable.
20. Perform system initialization procedure.

IGNITION TIMING

INSPECTION

See Figures 27 and 28.

1. Warm up the engine and stop the engine.

➡A warmed up engine should have an engine coolant temperature of over 176°F (80°C), have an engine oil temperature of 140°F (60°C), and the engine rpm should be stabilized.

2. When using the intelligent tester, perform the following:
   a. Connect the intelligent tester to the DLC3.
   b. Start the engine and idle it.
   c. Push the intelligent tester main switch ON.
   d. Enter the following items: Powertrain / Engine and ECT / Data list / IGN Advance.
   e. Ignition timing should measure 5–15°BTDC at idle.

➡Refer to the intelligent tester operator's manual for further details

3. When not using the intelligent tester, perform the following:
   a. Remove the V-bank cover sub-assembly.
   b. Connect the tester probe of a timing light to the wire of the ignition con-

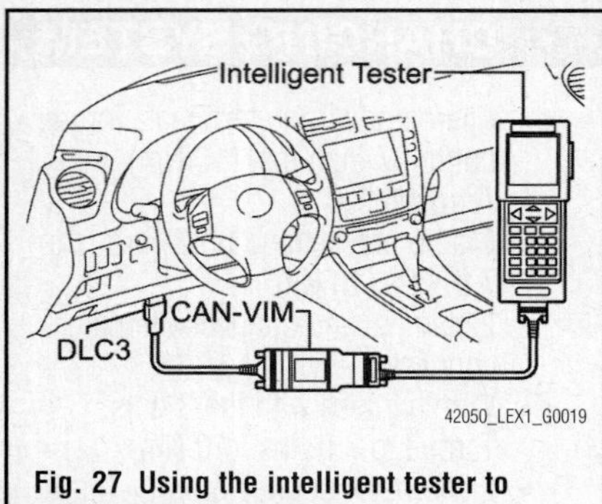

Fig. 27 Using the intelligent tester to check the ignition timing

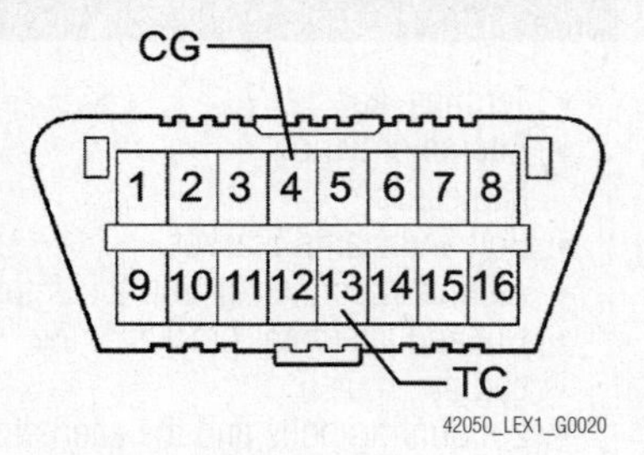

Fig. 28 Terminals 13 (TC) and 4 (CG) of DLC3

nector for No. 1 cylinder. Use a timing light which can detect the primary current.

c. Using SST 09843-18040, connect terminals TC and CG of the DLC3.

⁂ CAUTION

Confirm the terminal numbers before connecting them. Connecting the wrong terminals can damage the engine. When checking the ignition timing, the transmission should be in neutral.

d. Using a timing light, check the ignition timing. Ignition timing should measure 8–12°BTDC at idle.

e. Remove the SST from the DLC3.

f. Check the ignition timing. Ignition timing should measure 5–15°BTDC at idle.

g. Check that the ignition timing advances immediately when the engine speed is increased.

h. Disconnect the timing light from the engine.

4. Install the V-bank cover sub-assembly.

ADJUSTMENT

The engines covered in this section are equipped with a Distributorless Ignition System (DIS). No timing adjustments are possible.

SPARK PLUGS

REMOVAL & INSTALLATION

1. Before servicing the vehicle, refer to the Precautions section.
2. Disconnect the negative battery cable. Wait at least 90 seconds before performing any other work.
3. Remove the ignition coil(s).
4. Using a 0.63 inch (16mm) plug wrench, remove the spark plug(s).

To install:

5. Install the spark plug(s). Torque the spark plugs to 18 ft. lbs. (25 Nm).
6. Install the ignition coil(s).
7. Connect the negative battery cable.
8. Perform system initialization procedure.

ENGINE ELECTRICAL — STARTING SYSTEM

STARTER

REMOVAL & INSTALLATION

1. Before servicing the vehicle, refer to the Precautions section.
2. Remove or disconnect the following:
   - Negative battery cable. Wait at least 90 seconds before performing any other work.
   - Cool air intake duct seal
   - Left engine room side cover
   - Rear No. 1 floor panel brace
   - Front center floor brace
   - Heated oxygen sensor
   - Front exhaust pipe assembly
   - Front propeller shaft assembly (AWD)
   - No. 1 exhaust pipe support bracket (2.5L with automatic transmission)
   - Left exhaust manifold
   - Terminal 50 connector from starter
   - Terminal cap
   - Nut and wire harness from terminal 30
   - 2 bolts and starter motor

To install:

3. Install or connect the following:
   - Starter assembly with 2 bolts. Torque to 43 ft. lbs. (58 Nm).
   - Starter wires
   - Left exhaust manifold
   - No. 1 exhaust pipe support bracket (2.5L with automatic transmission) with 2 bolts. Torque to 32 ft. lbs. (43 Nm).
   - Front propeller shaft assembly (AWD)
   - Front exhaust pipe assembly
   - Heated oxygen sensor
   - Front center floor brace. Torque to 65 inch lbs. (7 Nm).
   - Rear No. 1 floor panel brace. Torque to 14 ft. lbs. (19 Nm).
   - Left engine room side cover
   - Cool air intake duct seal
   - Negative battery cable
   - System initialization
   - Start the vehicle and check for exhaust leaks.

ENGINE MECHANICAL

ACCESSORY DRIVE BELTS

ACCESSORY BELT ROUTING

See Figure 29.

Refer to the accompanying illustration for belt routing.

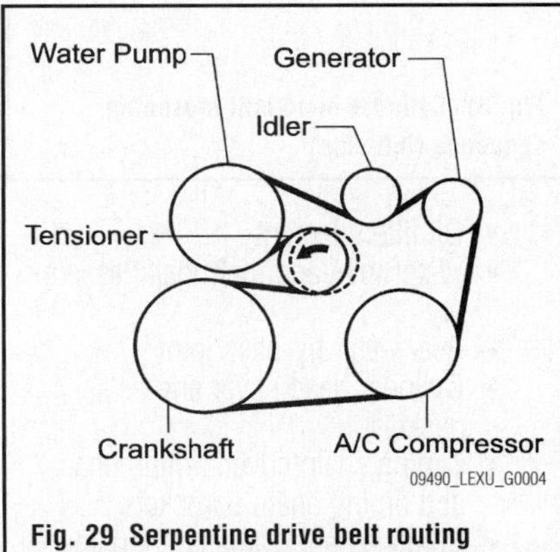

Fig. 29 Serpentine drive belt routing

INSPECTION

Inspect the drive belt for signs of glazing or cracking. A glazed belt will be perfectly smooth from slippage, while a good belt will have a slight texture of fabric visible. Cracks will usually start at the inner edge of the belt and run outward. All worn or damaged drive belts should be replaced immediately.

REMOVAL & INSTALLATION

See Figures 30 and 31.

1. Before servicing the vehicle, refer to the Precautions section.
2. Disconnect the negative battery cable. Wait at least 90 seconds before performing any other work.
3. Remove the cool air intake duct seal.
4. Remove the inlet air cleaner.
5. Remove the V-bank cover.

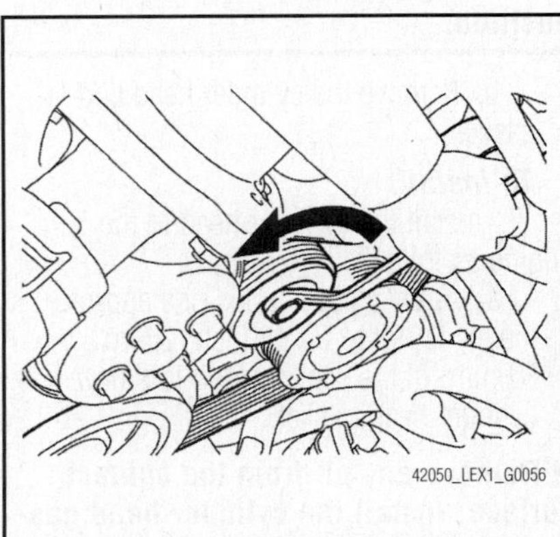

Fig. 30 Loosening the drive belt tension

Fig. 31 Fix the belt tensioner in place by inserting a 5mm bihexagon wrench into the holes

6. While releasing the belt tension by turning the belt tensioner counterclockwise, and remove the drive belt from the belt tensioner.
7. While turning the belt tensioner counterclockwise, align with its holes, and then insert the 5mm bihexagon wrench into the holes to fix the belt tensioner.
8. Visually check the drive belt for excessive wear, frayed cords, chunks missing from its ribs, etc. If any defect has been found, replace the belt.
9. Check that nothing gets caught in the tensioner by turning it clockwise and counterclockwise. If a malfunction exists, replace the tensioner.

To install:

10. Install the drive belt.
11. While turning the belt tensioner counterclockwise, remove the bar.
12. Put the backside of the drive belt on the tensioner pulley and idler pulley. Check that the belt is properly set to each pulley.
13. If it is difficult to install the drive belt, perform the following procedure:
    a. Put the belt on every part except the tensioner pulley as shown in the routing illustration.
    b. While releasing the belt tension by turning the belt tensioner counterclockwise, put the belt on the tensioner pulley.
14. Install the V-bank cover.
15. Install the inlet air cleaner.
16. Install the cool air intake duct seal.
17. Connect the negative battery cable.

CRANKSHAFT DAMPER

REMOVAL & INSTALLATION

See Figure 32.

1. Before servicing the vehicle, refer to the Precautions section.

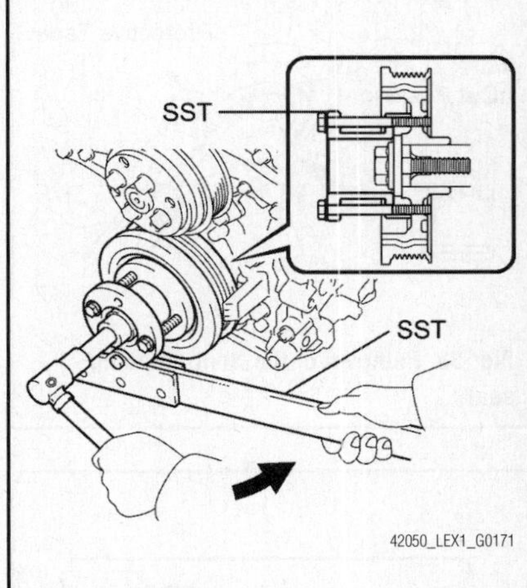

Fig. 32 Using special tools to remove the crankshaft pulley

2. Disconnect the negative battery cable. Wait at least 90 seconds before performing any other work.
3. It may be necessary to remove the radiator assembly for access.
4. Remove the accessory drive belt(s).
5. Using SST 09213-54015 (91651-60855), 09330-00021, loosen the crankshaft pulley bolt.
6. Using SST 09950-50013 (09951-05010, 09952-05010, 09953-05010, 09954-05030) and the pulley bolt, remove the crankshaft pulley.

➡Before using SST, apply lubricating oil on the threads and tip of the center bolt.

To install:

7. Align the pulley set key with the key groove of the pulley, and slide on the pulley.
8. Using SST 09213-54015 (91651-60855), 09330-00021, install the pulley bolt.
9. Tighten the crankshaft pulley bolt to 184 ft. lbs. (250 Nm).
10. Install the accessory drive belt(s).
11. Install the radiator, if necessary.
12. Connect the negative battery cable.

CRANKSHAFT FRONT SEAL

REMOVAL & INSTALLATION

See Figures 33 and 34.

1. Before servicing the vehicle, refer to the Precautions section.
2. Disconnect the negative battery

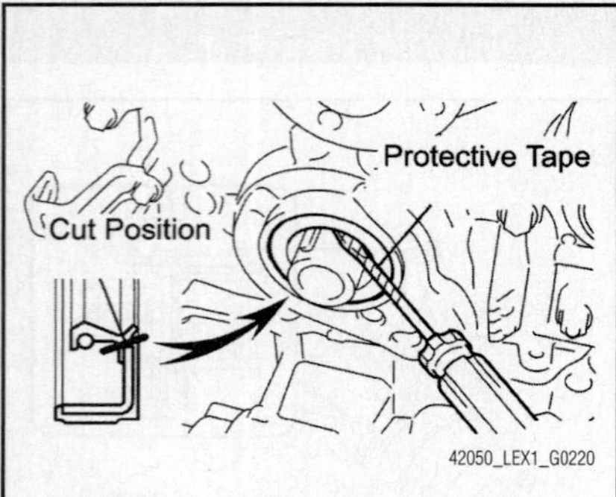

Fig. 33 Removal of the front oil pump seal

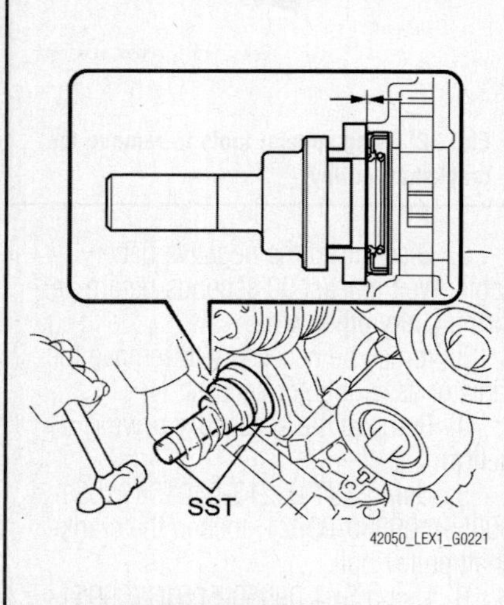

Fig. 34 Installation of the front oil pump seal

cable. Wait at least 90 seconds before performing any other work.

3. Remove the radiator assembly.
4. Remove the accessory drive belt.
5. Remove the crankshaft pulley.
6. Using a screwdriver with its tip taped, pry out the timing chain case oil seal.

➡After the removal, check the crankshaft for damage. If it is damaged, smooth the surface with 400-grit sandpaper.

To install:

7. Using SST 09223-22010, 09506-35010 and a hammer, tap in the oil seal until its surface is flush with the rear oil seal retainer edge. Keep the lip free of foreign matter. Do not tap the oil seal at an angle.
8. Install the crankshaft pulley.
9. Install the accessory drive belt.
10. Install the radiator assembly.
11. Inspect for oil leaks.
12. Connect the negative battery cable.

CYLINDER HEAD

REMOVAL & INSTALLATION

See Figures 35 through 42.

1. Before servicing the vehicle, refer to the Precautions section.
2. Drain the cooling system.
3. Drain the engine oil.
4. Relieve the fuel system pressure.
5. Remove or disconnect the following:
   - Negative battery cable. Wait at least 90 seconds before performing any other work
   - Oil filler cap and gasket
   - Radiator cap
   - Spark plugs
   - Ventilation valve
   - 4 camshaft position sensors
   - 4 camshaft timing oil control valves
   - Crankshaft position sensor
   - Left and right side oil check valve bolt, oil pipe union and oil pipe
   - Left and right side oil control valve filter and gaskets
   - Cylinder block water drain cocks

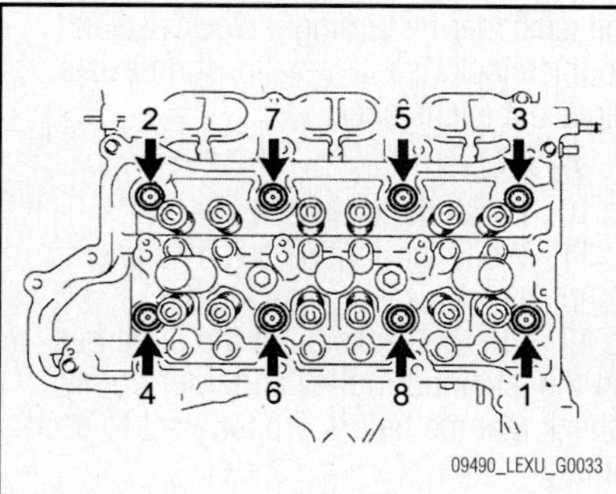

Fig. 35 Cylinder head bolt loosening sequence (right side)

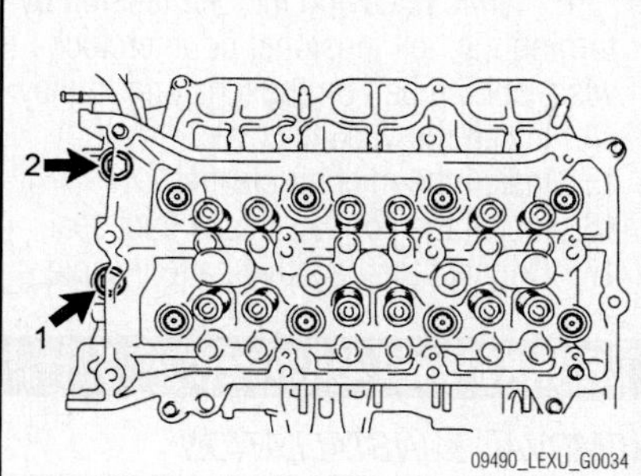

Fig. 36 Cylinder head 14mm bolt loosening sequence (left side)

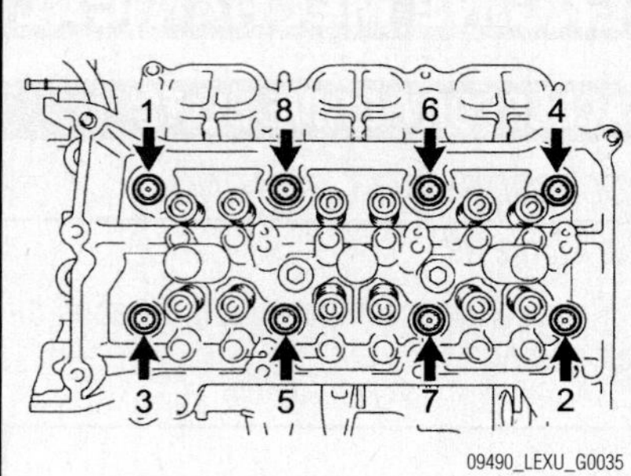

Fig. 37 Cylinder head bolt loosening sequence (left side)

   - Oil filter element
   - Water inlet and thermostat assembly
   - Rear water by-pass joint
   - Cylinder head cover and gaskets
   - Timing chain cover, timing chain and timing chain sprockets
   - Camshaft and camshaft housing assembly
   - Valve rocker arms. Arrange the removed rocker arms in the correct order.
   - Valve lash adjusters. Arrange the removed valve lash adjusters in the correct order.
6. Remove the cylinder head (left or right) as follows:
   a. Using a 10mm bi-hexagon wrench, uniformly loosen the 8 bolts in the sequence shown in the illustration. Remove the 8 cylinder head bolts and plate washers.

⁂ WARNING

Be careful not to drop washers into the cylinder head. Cylinder head warpage or cracking could result from removing bolts in an incorrect order. Be sure to keep separate the removed parts for each installation position.

   b. Remove the cylinder head and gasket.

To install:

7. Install the cylinder head to the engine as follows:
   a. Apply a continuous line approximately 0.098 to 0.118 inch (2.5 to 3.0mm) of the seal packing to a new cylinder head gasket.

➡Remove any oil from the contact surface. Install the cylinder head gasket within 3 minutes after applying

the seal packing. Install the cylinder head bolt within 15 minutes after applying the seal packing. Do not apply engine oil within 2 hours of installation.

b. Place the cylinder head gasket on the cylinder block surface with the Lot No. stamp upward.

⁂ WARNING

Be careful of the installation direction. Gently place the cylinder head in order not to damage the gasket with the bottom part of the head.

c. Place the cylinder head on the cylinder block.

⁂ WARNING

Be careful not to allow oil to adhere to the bottom part of the cylinder head.

d. Apply a light coat of engine oil to the threads and under the heads of the cylinder head bolts.

e. Using a 10mm bi-hexagon wrench, install and uniformly tighten the 8 cylinder head bolts with the plate washers to 27 ft. lbs. (36 Nm) in the sequence shown in the illustration. If any of the bolts does not meet the torque, replace it.

f. Mark the forward edge of each bolt with paint, then retighten each bolt, in proper sequence, an additional 90 degrees. Check that each painted mark is now at a 90 degrees angle to the front. The paint mark should have been applied to the bolt in the 9 o'clock position and should now be in the 12 o'clock position.

g. Tighten each bolt again, in proper sequence, an additional 90 degrees. Check that each painted mark is now facing rearward.

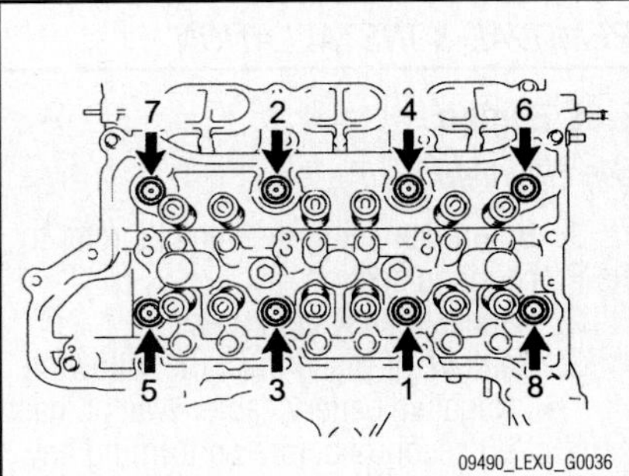

Fig. 38 Cylinder head bolt tightening sequence (right side)

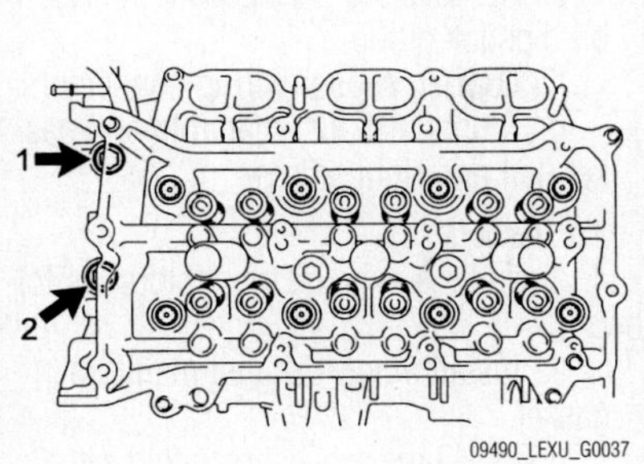

Fig. 39 Cylinder head 14mm bolt tightening sequence (left side)

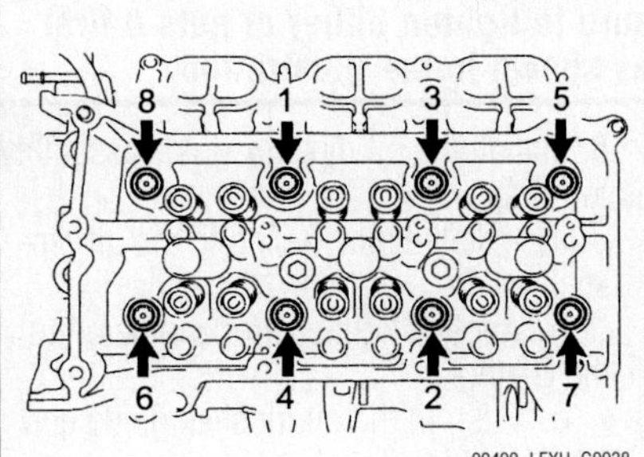

Fig. 40 Cylinder head bolt tightening sequence (left side)

h. Tighten the 2 bolts on the left cylinder head in the order shown in the illustration. Torque to 22 ft. lbs. (30 Nm).

➡Do not use the tightening procedure for a plastic region bolt (if equipped) when tightening bolts 1 and 2 shown in the illustration.

i. Seal packing will seep out on the engine's front side. Thoroughly wipe off seeped out seal packing.

8. Install or connect the following:
   - Valve lash adjusters
   - Valve rocker arms

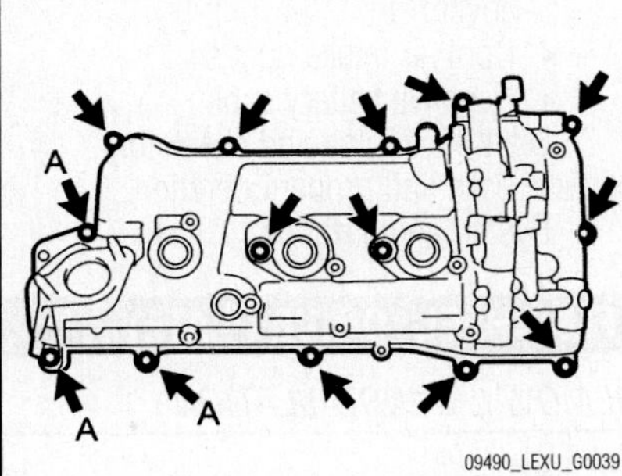

Fig. 41 Cylinder head cover bolt tightening sequence (right side)

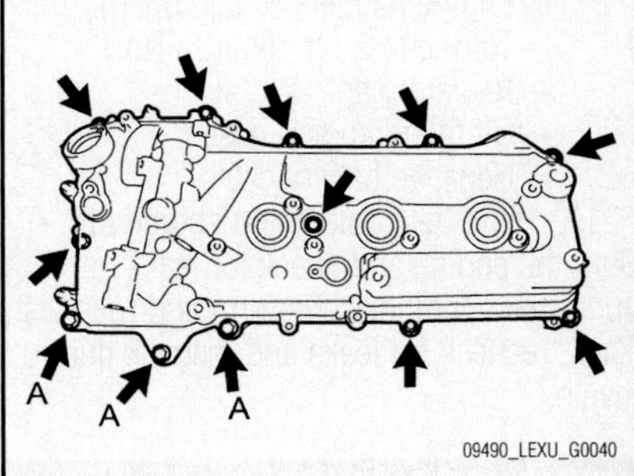

Fig. 42 Cylinder head cover bolt tightening sequence (left side)

- Camshaft and camshaft housing assembly
- Timing chain cover, timing chain and timing chain sprockets

9. Install the cylinder head cover and gaskets as follows:

a. Apply seal packing where the cylinder head meets the timing cover.

➡Remove any oil from the contact surface. Install the head cover within 3 minutes after applying seal packing. Do not start the engine for at least 2 hours after installing.

b. Install all new gaskets.

c. Install the head cover along with the retaining bolts. Torque the A bolts to 15 ft. lbs. (21 Nm); and remaining bolts to 7 ft. lbs. (10 Nm).

- New gaskets, O-ring and rear water by-pass joint. Torque to 7 ft. lbs. (10 Nm).
- New gasket, water inlet and thermostat assembly. Torque to 7 ft. lbs. (10 Nm).
- Oil filter element
- Adhesive sealer, left and right side cylinder block water drain cocks. Torque to 22 ft. lbs. (30 Nm).
- Left and right side oil control valve filter and gaskets
- Left and right side oil check valve bolt, oil pipe union and oil pipe. Torque to 44 ft. lbs. (60 Nm).
- Crankshaft position sensor. Torque to 7 ft. lbs. (10 Nm).
- 4 camshaft timing oil control valves
- 4 camshaft position sensors
- Adhesive and ventilation valve
- Spark plugs. Torque to 18 ft. lbs. (24 Nm) for 2.5L and 3.5L engines; 13 ft. lbs. (18 Nm) for 3.0L engine.

- Engine hangers (if equipped). Torque to 24 ft. lbs. (33 Nm).
- Radiator cap
- Oil filler cap and gasket
- Negative battery cable

10. Refill the coolant and engine oil. Start the engine and check for leaks or abnormal conditions. Perform and road test. Then, recheck for leaks and recheck fluid levels.

EXHAUST MANIFOLD

REMOVAL & INSTALLATION

See Figure 43.

1. Before servicing the vehicle, refer to the Precautions section.
2. Remove or disconnect the following:
   - Negative battery cable. Wait at least 90 seconds before performing any other work
   - Cool air intake duct seal
   - Left and right engine room side covers
   - V-bank cover
   - Engine under cover
   - No. 2 engine under cover (2WD)
   - Left and right rear engine under cover (2WD)
   - Front exhaust pipe assembly
   - Front lower suspension member protector (2WD)
   - Exhaust pipe No. 1 support bracket (4GR-FSE with automatic transmission)
3. Remove the oil dipstick guide assembly as follows:
   a. Remove the oil level gauge.
   b. Remove the bolt, then remove the No. 2 oil dipstick guide.
   c. Remove the O-ring from the No. 2 oil dipstick guide.
   d. Remove the bolt and clamp (2WD), then remove the No. 1 oil dipstick guide.

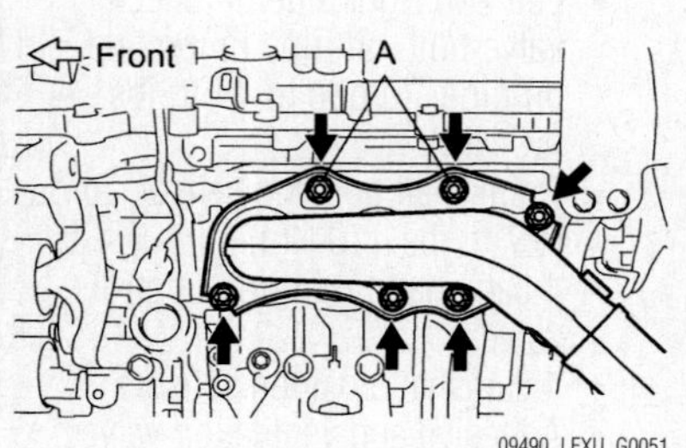

Fig. 43 When tightening the exhaust manifold nuts, start with nuts A as shown (left side illustrated, right side the same

   e. Remove the O-ring from the No. 1 oil dipstick guide.
   f. Oxygen sensor wiring connectors
   g. 6 nuts, exhaust manifold and gasket (left and right sides)

To install:

4. Install the exhaust manifold(s) as follows:
   a. Install a new exhaust manifold gasket.
   b. Install the exhaust manifold with 6 new nuts and torque to 15 ft. lbs. (21 Nm).

⁂ WARNING

Do not damage the stud bolt when installing the exhaust manifold. Be sure to tighten either of nuts A first as shown in the illustration.

5. Install the oil dipstick guide assembly as follows:
   a. Install a new O-ring to the oil dipstick guide.
   b. Apply a light coat of engine oil to the O-ring.
   c. Push in the oil dipstick guide end into the guide hole.
   d. Install the No. 1 oil dipstick guide with the bolt and torque to 7 ft. lbs. (10 Nm).
   e. Connect the clamp (2WD).
   f. Install the No. 2 oil dipstick guide with the bolt and torque to 15 ft. lbs. (21 Nm).
   g. Install the oil dipstick.
6. Install or connect the following:
   - Oxygen sensor wiring connectors
   - Front lower suspension member protector (2WD) and 4 bolts. Torque the 4 bolts to 71 inch lbs. (8 Nm).
   - Front exhaust pipe assembly
   - Left and right rear engine under cover (2WD)
   - No. 2 engine under cover (2WD)
   - Engine under cover
   - V-bank cover
   - Left and right engine room side covers
   - Cool air intake duct seal
   - Negative battery cable
7. Start the engine and check for exhaust leaks and proper operation.
8. System initialization.

FLYWHEEL (DRIVE PLATE)

REMOVAL & INSTALLATION

See Figure 44.

1. Before servicing the vehicle, refer to the Precautions section.
2. Disconnect the negative battery

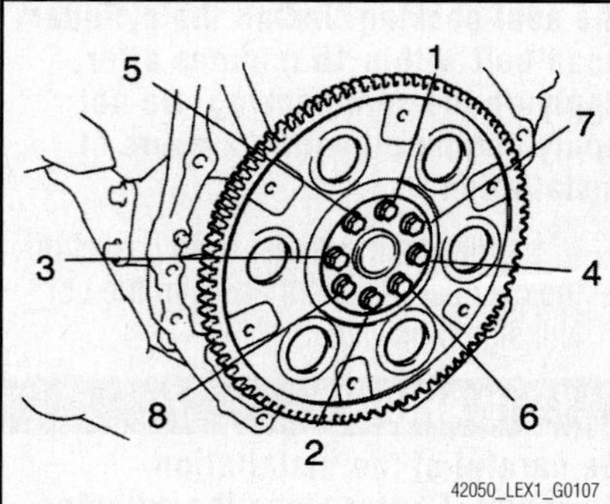

Fig. 44 Bolt tightening sequence for flywheel (drive plate)

cable. Wait at least 90 seconds before performing any other work
3. Remove the engine or transmission assembly.
4. Using SST 09213-54015 (91651-60855), 09330-00021, hold the crankshaft.
5. Remove the 8 bolts and remove the rear spacer, drive plate and front spacer (A/T) or flywheel (M/T).

To install:

6. Using SST 09213-54015 (91651-60855), 09330-00021, hold the crankshaft.
7. Clean the bolts and bolt holes.
8. Apply adhesive (Part No. 08833-00070, THREE BOND 1324 or equivalent) to 2 or 3 threads of the bolts.
9. Install the rear spacer, drive plate and front spacer (A/T) or flywheel (M/T).
10. Install and uniformly tighten the 8 bolts in several passes, in the sequence shown, to the following torque values:
    - A/T: 61 ft. lbs. (83 Nm)
    - M/T: 64 ft. lbs. (73 Nm)

➡Do not start the engine within an hour after installing.

11. Install the engine or transmission assembly.
12. Connect the negative battery cable.

INTAKE MANIFOLD

REMOVAL & INSTALLATION

2.5L Engine

See Figure 45.

1. Before servicing the vehicle, refer to the Precautions section.
2. Relieve fuel system pressure.
3. Remove or disconnect the following:
   - Negative battery cable. Wait at least 90 seconds before performing any other work.
   - Coolant
   - Cool air intake duct seal

- Left and right engine room side covers
- V-bank cover
- Left and right front upper fender protectors
- Left and right roof drip side finish moldings using a moulding removal tool
- Front wiper arm head cap
- Left and right windshield wiper arm and blade assemblies
- Cowl top ventilator louver assembly
- No. 2 ventilation hose
- Air cleaner cap with air cleaner hose
- Throttle body assembly
- Cold start injector

4. Remove intake air surge tank as follows:

5. Disconnect the vacuum hose from the intake air surge tank.

a. Remove the bolt and disconnect the No. 1 vacuum switching valve assembly from the intake air surge tank.

b. Disconnect the wire harness and hose from the surge tank.

c. Disconnect the ventilation hose from the intake air surge tank.

d. Remove the bolt and water hose joint from the intake air surge tank.

e. Remove the bolt and disconnect the surge tank stay.

f. Using a 5mm hexagon socket wrench, remove the 7 bolts, 2 nuts and gasket.

6. Remove intake manifold as follows:

a. Disconnect the connector for the SCV.

b. Disconnect the SCV position sensor connector.

c. Remove the 4 bolts, 4 nuts, intake manifold and gasket.

To install:

7. Place new gaskets onto the intake manifold and position the intake manifold between the cylinder heads. Tighten the nuts and bolts to 15 ft. lbs. (21 Nm).

8. Install or connect the following:

- SCV position sensor connector
- DC motor connector for the SCV

9. Install the intake air surge tank as follows:

a. Install a new gasket to the intake air surge tank.

b. Using a 5mm hexagon socket wrench, install the 6 bolts. Torque all bolts, except A, to 13 ft. lbs. (18 Nm).

c. Install the bolt and 2 nuts to the intake air surge tank. Torque bolt A to 15 ft. lbs. (21 Nm) and the nuts to 12 ft. lbs. (16 Nm).

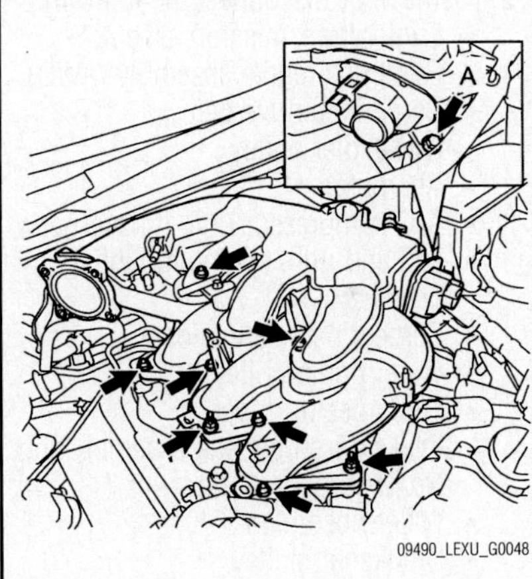

Fig. 45 Intake air surge tank mounting bolt locations

d. Install the surge tank stay to the intake air surge tank. Torque the bolt to 15 ft. lbs. (21 Nm).

e. Connect the water hose joint with the bolt and torque to 7 ft. lbs. (10 Nm).

f. Connect the ventilation hose to the intake air surge tank.

g. Connect the wire harness and hose to the intake air surge tank.

10. Connect the No. 1 vacuum switching valve assembly to the intake air surge tank. Torque the fasteners to 13 ft. lbs. (18 Nm).

a. Connect the vacuum hose to the intake air surge tank.

- Cold start injector. Torque the bolts to 7 ft. lbs. (10 Nm).
- New gasket and throttle body assembly. Torque the 4 bolts to 7 ft. lbs. (10 Nm)
- Air cleaner cap with air cleaner hose
- No. 2 ventilation hose
- Cowl top ventilator louver assembly
- Left and right windshield wiper arm and blade assemblies
- Front wiper arm head cap
- Left and right roof drip side finish moldings
- Left and right front upper fender protectors
- V-bank cover
- Left and right engine room side covers
- Cool air intake duct seal
- Negative battery cable

11. Refill the cooling system. Start the engine and check for leaks and proper operation.

12. Check the function of the throttle body unit.

13. Perform the system initialization.

3.5L Engine

1. Before servicing the vehicle, refer to the Precautions section.

2. Relieve fuel system pressure.

3. Remove or disconnect the following:

- Negative battery cable. Wait at least 90 seconds before performing any other work.
- Coolant
- Cool air intake duct seal
- Left and right engine room side covers
- V-bank cover
- Left and right front upper fender protectors
- Left and right roof drip side finish moldings using a moulding removal tool
- Front wiper arm head cap
- Left and right windshield wiper arm and blade assemblies
- Cowl top ventilator louver assembly
- No. 2 ventilation hose
- Air cleaner cap with air cleaner hose
- Throttle body assembly

4. Remove intake air surge tank as follows:

a. Disconnect the vacuum hose from the intake air surge tank.

b. Remove the bolt and disconnect the No. 1 vacuum switching valve assembly from the intake air surge tank.

c. Disconnect the ventilation hose, union to check valve hose and water by-pass hose from the surge tank.

d. Disconnect the 4 wire harness clamps from the intake air surge tank.

e. Remove the bolt and water hose joint from the intake air surge tank.

f. Remove the bolt and disconnect the surge tank stay.

g. Using a 5mm hexagon socket wrench, remove the 6 bolts, 2 nuts and gasket.

5. Remove intake manifold as follows:

a. Disconnect the fuel tube from the delivery pipe sub-assembly.

b. Disconnect the 4 connectors.

c. Remove the 4 bolts, 4 nuts, intake manifold and gasket.

To install:

6. Place new gaskets onto the intake manifold and position the intake manifold between the cylinder heads. Tighten the nuts and bolts to 15 ft. lbs. (21 Nm).

7. Install or connect the following:

- 4 connectors
- Fuel tube to the delivery pipe sub-assembly

8. Install the intake air surge tank as follows:

a. Install a new gasket to the intake air surge tank.

b. Install the intake air surge tank with the 2 nuts. Torque the 2 nuts to 12 ft. lbs. (16 Nm).

c. Using a 5mm hexagon socket wrench, install the 6 bolts. Torque the bolts to 13 ft. lbs. (18 Nm).

d. Install the surge tank stay to the intake air surge tank. Torque the bolt to 15 ft. lbs. (21 Nm).

e. Connect the water hose joint with the bolt and torque to 7 ft. lbs. (10 Nm).

f. Connect the 4 wire harness clamps to the intake air surge tank.

g. Connect the ventilation hose, union to check valve hose and water by-pass hose to the intake air surge tank.

h. Connect the No. 1 vacuum switching valve assembly to the intake air surge tank. Torque the bolt to 13 ft. lbs. (18 Nm).

i. Connect the vacuum hose to the intake air surge tank.

- New gasket and throttle body assembly. Torque the 4 bolts to 7 ft. lbs. (10 Nm)
- Air cleaner cap with air cleaner hose
- No. 2 ventilation hose
- Cowl top ventilator louver assembly
- Left and right windshield wiper arm and blade assemblies
- Front wiper arm head cap
- Left and right roof drip side finish moldings
- Left and right front upper fender protectors
- V-bank cover
- Left and right engine room side covers
- Cool air intake duct seal
- Negative battery cable

9. Refill the cooling system. Start the engine and check for leaks and proper operation.

10. Check the function of the throttle body unit.

11. Perform the system initialization.

OIL PUMP

REMOVAL & INSTALLATION

See Figures 47 through 52.

➡The oil pump cannot be removed with the engine in the vehicle. The engine and transmission must be removed as a unit, then separated.

1. Before servicing the vehicle, refer to the Precautions section.

2. Remove or disconnect the following:

- Engine/transmission assembly
- Front differential assembly (AWD)
- Serpentine drive belt
- No. 2 idler pulley
- Alternator
- A/C compressor unit, if necessary
- Left and right engine mounting brackets
- Serpentine belt tensioner
- Water pump pulley
- Fuel injector driver
- Intake air surge tank assembly and No. 2 surge tank stay
- Water hose joint
- Crankshaft pulley
- Water inlet
- Oil pan assembly
- Oil strainer
- No. 1 and 2 fuel pipes
- High pressure side fuel pump
- Ignition coil assembly
- No. 1 and 2 oil pipes
- Left and right cylinder head covers

3. Remove the timing chain cover assembly as follows:

a. Remove bolt and wiring harness clamp bracket.

Fig. 47 Location of mounting bolts for oil pump cover behind the timing chain cover

Drive Rotor Spline
Crankshaft
09490_LEXU_G0090

Fig. 48 Align the oil pump drive rotor spline and the crankshaft

b. Remove 25 mounting bolts and 2 mounting nuts.

c. Remove the timing chain cover by prying between the timing chain cover and cylinder head or cylinder block with a screwdriver.

d. Remove the gasket.

➡The oil pump assembly is incorporated into the back of the timing chain cover. The oil pump assembly can be disassembled from the back of the timing chain cover for inspection purposes.

4. Install the timing chain cover assembly as follows:

a. Apply seal packing in a continuous line of 0.197–0.217 inches (5.0–5.5mm) in diameter to the engine at the seam where the cylinder head meets the camshaft bearing cap assembly and the cylinder head meets the cylinder block.

➡Be sure to clean, degrease and dry the contact surfaces before applying the seal packing. Install the component within 3 minutes after applying seal packing. Do not start the engine for at least 2 hours after installing.

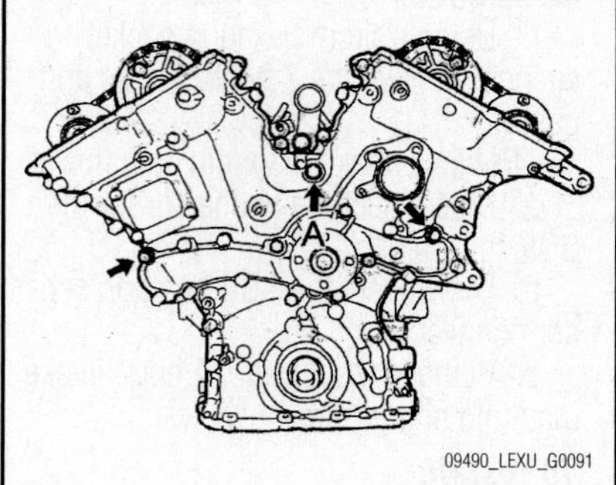

Fig. 49 Location of 3 bolts to be tightened first (location of bolt A shown)

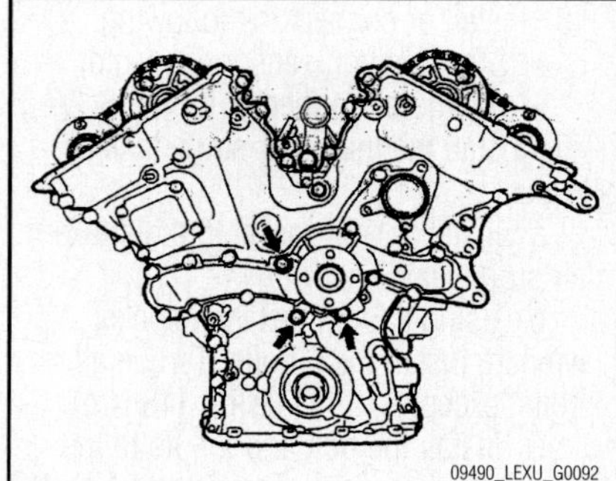

Fig. 50 Location of 3 bolts to be tightened second

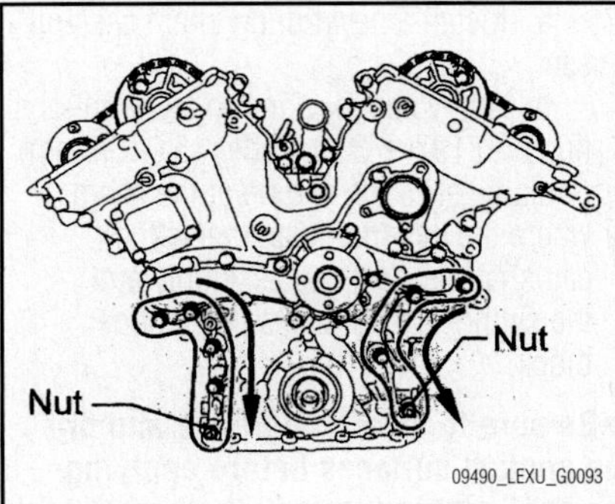

Fig. 51 Location of 7 bolts and 2 nuts to be tightened third

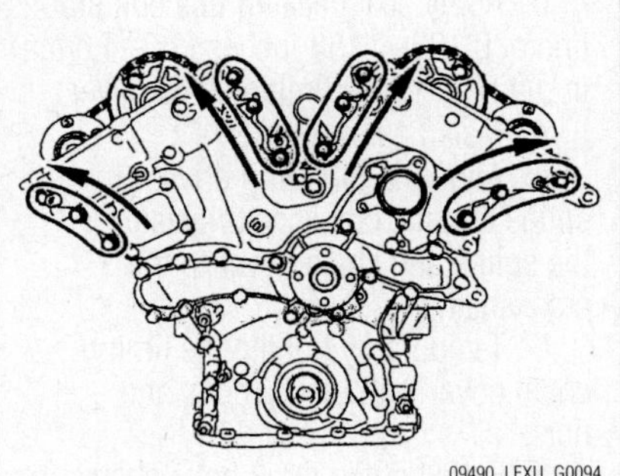

Fig. 52 Location of 12 bolts to be tightened fourth

b. Apply seal packing in a continuous line of 0.138–0.158 inches (3.5–4.0mm) in diameter to the timing chain cover.

c. Install a new gasket.

d. Align the oil pump drive rotor spline and the crankshaft. Install the spline and chain cover to the crankshaft.

e. Temporarily tighten the timing chain cover with the 25 bolts and nuts.

f. Fully tighten the 3 bolts shown in the illustration. Torque bolt A to 32 ft. lbs. (43 Nm). Torque the 2 remaining bolts to 15 ft. lbs (21 Nm).

g. Fully tighten the 3 bolts shown in the illustration. Torque the bolts to 15 ft. lbs (21 Nm).

h. Fully tighten the 7 bolts and 2 nuts shown in the illustration. Torque the bolts to 15 ft. lbs (21 Nm).

➡Be sure to tighten the bolts and nuts in order of upper to lower.

i. Fully tighten the 12 bolts shown in the illustration. Torque the bolts to 15 ft. lbs (21 Nm).

➡Be sure to tighten the bolts in order of lower to upper.

j. Install the bolt and wiring harness bracket. Torque the bolts to 7 ft. lbs (10 Nm).

5. Install or connect the following:
   - Left and right cylinder head covers
   - No. 1 and 2 oil pipes
   - Ignition coil assembly
   - High pressure side fuel pump
   - No. 1 and 2 fuel pipes
   - Oil strainer
   - Oil pan assembly
   - Water inlet
   - Crankshaft pulley. Torque the bolt to 192 ft. lbs (260 Nm).
   - Water hose joint
   - Intake air surge tank assembly and No. 2 surge tank stay
   - Fuel injector driver
   - Water pump pulley. Torque the bolt to 15 ft. lbs (21 Nm).
   - Serpentine belt tensioner
   - Left and right engine mounting brackets
   - A/C compressor unit, if necessary
   - Alternator
   - No. 2 idler pulley
   - Serpentine drive belt
   - Front differential assembly (AWD)
   - Engine/transmission assembly

INSPECTION

Oil Pump Relief Valve

See Figure 53.

1. Inspect the oil pump relief valve as follows:

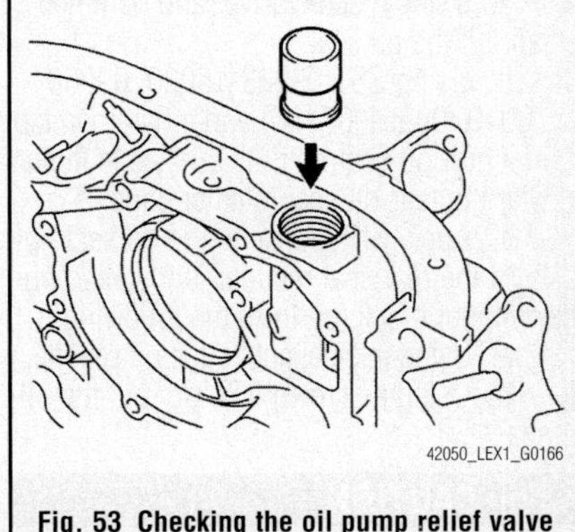

Fig. 53 Checking the oil pump relief valve

a. Remove the plug, spring and relief valve.

b. Apply a light coat of engine oil.

c. Check that it falls smoothly into the valve hole by its own weight. If it doesn't, replace the relief valve. If necessary, replace the oil pump assembly.

PISTON AND RING

POSITIONING

See Figure 54.

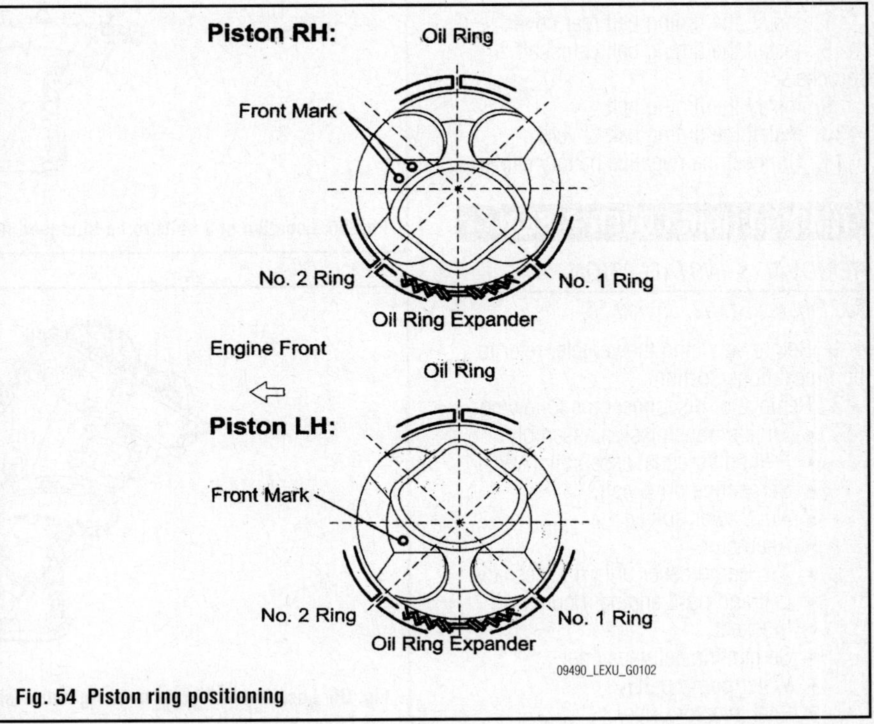

Fig. 54 Piston ring positioning

REAR MAIN SEAL

REMOVAL & INSTALLATION

1. If the rear oil seal retainer is removed from the cylinder block, perform the following:

a. Using a screwdriver and hammer, tap out the oil seal.

b. Using SST 09223-15030, 09950-70010 (09951-07100) and a hammer, tap in a new oil seal until its surface is flush with the rear oil seal retainer edge.

c. Apply MP grease to the oil seal lip.

2. If the rear seal retainer is installed on the cylinder block, perform the following:

a. Using a knife, cut off the oil seal lip.

b. Using a screwdriver, pry out the oil seal.

⁂ CAUTION

Be careful not to damage the crankshaft. Tape the screwdriver tip.

To install:

c. Apply MP grease to a new oil seal lip.

d. Using SST 09223-15030, 09950-70010 (09951-07100) and a hammer, tap in the oil seal until its surface is flush with the rear oil seal retainer edge.

TIMING BELT REAR COVER

REMOVAL & INSTALLATION

1. Before servicing the vehicle, refer to the Precautions section.
2. Disconnect the negative battery cable.
3. Remove the timing belt covers.
4. Remove the timing belt.
5. Remove the timing belt camshaft sprockets.
6. Remove the timing belt rear cover.

To install:

7. Install the timing belt rear cover.
8. Install the timing belt camshaft sprockets.
9. Install the timing belt.
10. Install the timing belt covers.
11. Connect the negative battery cable.

TIMING CHAIN COVER & SEAL

REMOVAL & INSTALLATION

See Figures 51, 52, 55 and 56.

1. Before servicing the vehicle, refer to the Precautions section.
2. Remove or disconnect the following:
   - Engine/transmission assembly
   - Front differential assembly (AWD)
   - Serpentine drive belt
   - No. 2 idler pulley
   - Alternator
   - A/C compressor unit, if necessary
   - Left and right engine mounting brackets
   - Serpentine belt tensioner
   - Water pump pulley
   - Fuel injector driver
   - Intake air surge tank assembly and No. 2 surge tank stay
   - Water hose joint
   - Crankshaft pulley
   - Water inlet
   - Oil pan assembly
   - Oil strainer
   - No. 1 and 2 fuel pipes
   - High pressure side fuel pump
   - Ignition coil assembly
   - No. 1 and 2 oil pipes
   - Left and right cylinder head covers
3. Remove the timing chain cover assembly as follows:

a. Remove bolt and wiring harness clamp bracket.

b. Remove 25 mounting bolts and 2 mounting nuts.

c. Remove the timing chain cover by prying between the timing chain cover and cylinder head or cylinder block with a screwdriver.

d. Remove the gasket.

e. Remove the timing chain case oil seal.

To install:

4. Install the timing chain cover assembly as follows:

a. Install a new timing chain case oil seal.

b. Apply seal packing in a continuous line of 0.197–0.217 inches (5.0–5.5mm) in diameter to the engine at the seam where the cylinder head meets the camshaft bearing cap assembly and the cylinder head meets the cylinder block.

➡Be sure to clean, degrease and dry the contact surfaces before applying the seal packing. Install the component within 3 minutes after applying seal packing. Do not start the engine for at least 2 hours after installing.

c. Apply seal packing in a continuous line of 0.138–0.158 inches (3.5–4.0mm) in diameter to the timing chain cover

d. Install a new gasket

e. Align the oil pump drive rotor spline and the crankshaft. Install the spline and chain cover to the crankshaft.

f. Temporarily tighten the timing chain cover with the 25 bolts and nuts.

g. Fully tighten the 3 bolts shown in

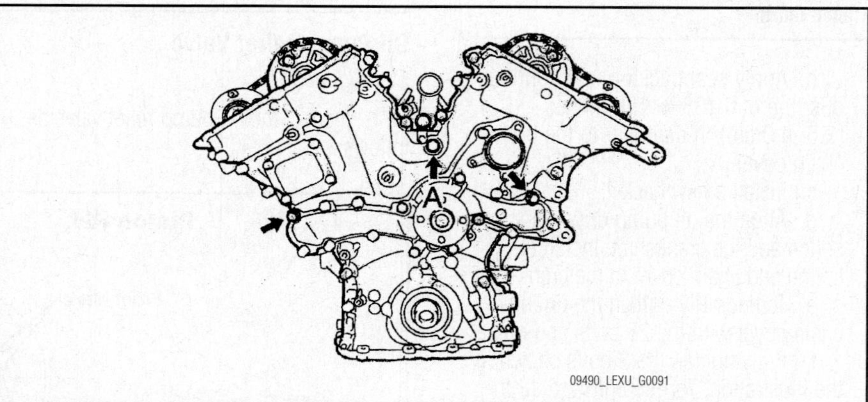

Fig. 55 Location of 3 bolts to be tightened first (location of bolt A shown)

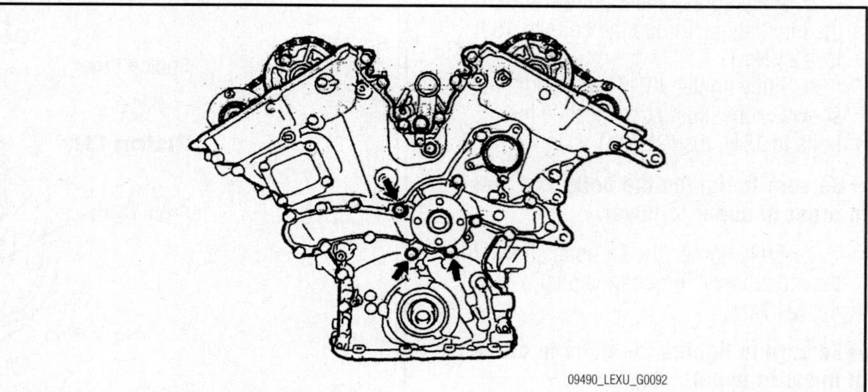

Fig. 56 Location of 3 bolts to be tightened second

the illustration. Torque bolt A to 32 ft. lbs. (43 Nm). Torque the 2 remaining bolts to 15 ft. lbs (21 Nm).

h. Fully tighten the 3 bolts shown in the illustration. Torque the bolts to 15 ft. lbs (21 Nm).

i. Fully tighten the 7 bolts and 2 nuts shown in the illustration. Torque the bolts to 15 ft. lbs (21 Nm).

➡Be sure to tighten the bolts and nuts in order of upper to lower.

j. Fully tighten the 12 bolts shown in the illustration. Torque the bolts to 15 ft. lbs (21 Nm).

➡Be sure to tighten the bolts in order of lower to upper.

k. Install the bolt and wiring harness bracket. Torque the bolts to 7 ft. lbs (10 Nm).

5. Install or connect the following:
- Left and right cylinder head covers
- No. 1 and 2 oil pipes
- Ignition coil assembly
- High pressure side fuel pump
- No. 1 and 2 fuel pipes
- Oil strainer
- Oil pan assembly
- Water inlet
- Crankshaft pulley. Torque the bolt to 192 ft. lbs (260 Nm).
- Water hose joint
- Intake air surge tank assembly and No. 2 surge tank stay
- Fuel injector driver
- Water pump pulley. Torque the bolt to 15 ft. lbs (21 Nm).
- Serpentine belt tensioner
- Left and right engine mounting brackets
- A/C compressor unit, if necessary
- Alternator
- No. 2 idler pulley
- Serpentine drive belt
- Front differential assembly (AWD)
- Engine/transmission assembly

TIMING CHAIN & SPROCKETS

REMOVAL & INSTALLATION

See Figures 57 through 59.

➡The timing chain cannot be removed with the engine in the vehicle. The engine and transmission must be removed as a unit, then separated.

1. Before servicing the vehicle, refer to the Precautions section.
2. Remove or disconnect the following:
- Engine/transmission assembly
- Front differential assembly (AWD)

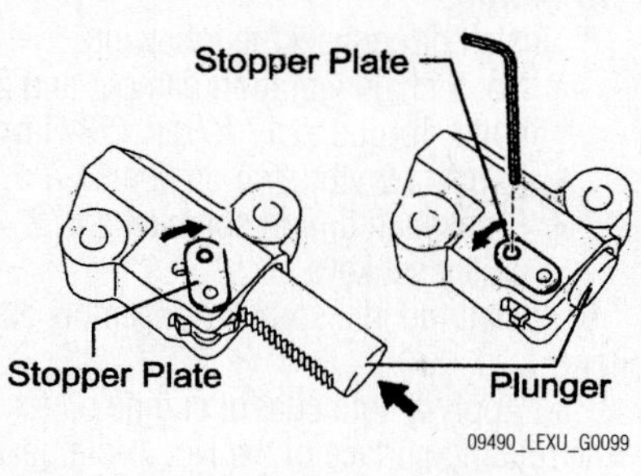

Fig. 57 Aligning the timing marks at the block bore centerline and camshaft bearing caps

- Serpentine drive belt
- No. 2 idler pulley
- Alternator
- A/C compressor unit, if necessary
- Left and right engine mounting brackets
- Serpentine belt tensioner
- Water pump pulley
- Fuel injector driver
- Intake air surge tank assembly and No. 2 surge tank stay
- Water hose joint
- Crankshaft pulley
- Water inlet
- Oil pan assembly
- Oil strainer
- No. 1 and 2 fuel pipes
- High pressure side fuel pump
- Ignition coil assembly
- No. 1 and 2 oil pipes
- Left and right cylinder head covers

3. Remove the timing chain cover and seal.
4. Set the No. 1 cylinder to TDC/compression as follows:

a. Temporarily tighten the pulley set bolt.

b. Set the timing mark on the crank angle sensor plate to the right block bore center line (TDC/compression).

c. Check that the timing marks of the camshaft timing gears are aligned with the timing marks of the bearing cap. If

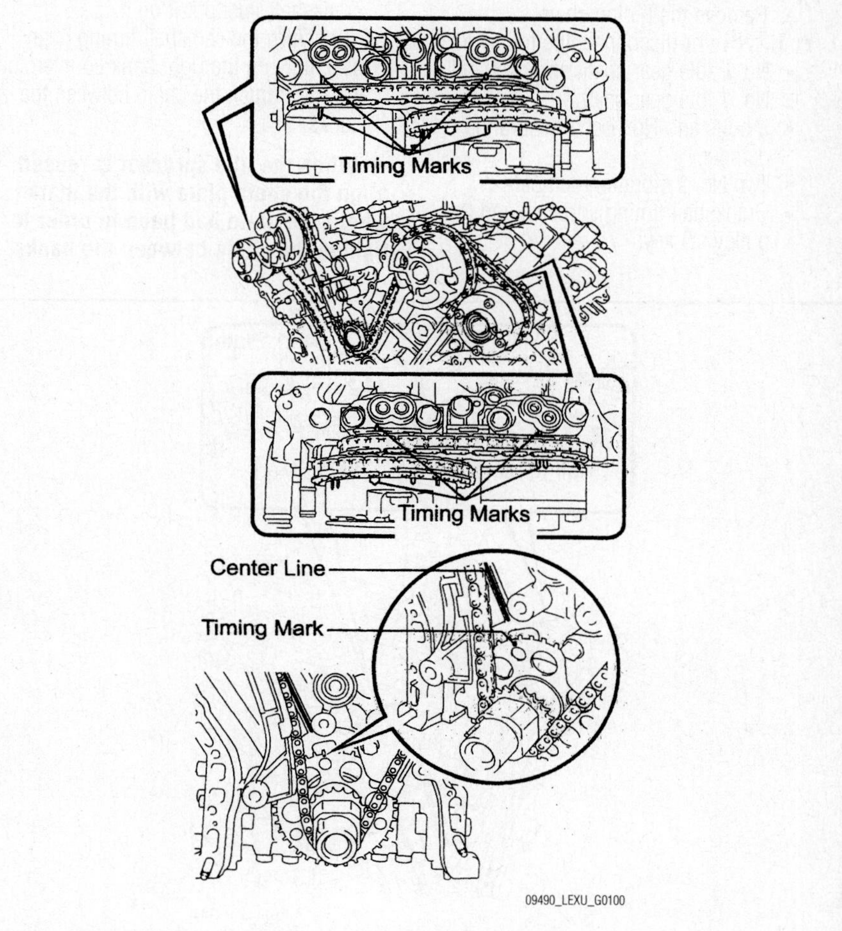

Fig. 58 Chain tensioner component showing stopper plate and plunger

not, turn the crankshaft 1 revolution (360 degrees) and align the timing marks.

5. Remove the No. 1 chain tensioner assembly as follows:
   a. Move the stopper plate upward to release the lock, and push the plunger deep into the tensioner.
   b. Move the stopper plate downward to set the lock, and insert a hexagon wrench into the stopper plate's hole.
   c. Remove the 2 bolts and chain tensioner.
   d. Remove the chain tensioner slipper.
6. Remove the timing chain as follows:
   a. Turn the crankshaft counterclockwise 10 degrees to loosen the chain of the crankshaft timing sprocket.
   b. Remove the pulley set bolt.
   c. Remove the chain from the crankshaft timing sprocket and place it on the crankshaft.
   d. Turn the camshaft timing gear assembly on the right bank clockwise (approx. 60 degrees). Be sure to loosen the chain between the banks.
   e. Remove the timing chain.
7. Remove or disconnect the following:
   - No. 2 idle gear shaft, sprocket and No. 1 idle gear shaft
   - 2 bolts and No. 1 chain vibration damper
   - Two No. 2 vibration dampers
   - Crankshaft timing sprocket and 2 pulley set keys

To install:

8. Install or connect the following:
   - No. 1 chain vibration damper and 2 bolts. Torque to 17 ft. lbs. (23 Nm).
   - Two No. 2 vibration dampers
   - Crankshaft timing sprocket and 2 pulley set keys
9. Install the idle sprocket assembly as follows:
   a. Apply a light coat of engine oil to the rotating surface of the No. 1 idle gear shaft.
   b. Temporarily install the No. 1 idle gear shaft and idle sprocket with the No. 2 idle gear shaft while aligning the knock pin of the No. 1 idle gear with the knock pin groove of the cylinder block.

➡Be careful of the idle gear direction.

   c. Using a 10mm hexagon wrench, tighten the No. 2 idle gear shaft to 44 ft. lbs. (60 Nm). Check that the idle sprocket turns smoothly.
10. Install the timing chain as follows:
   a. Align the mark plate and timing mark and install the chain.
   b. Do not pass the chain over the crankshaft, just put it on it.
   c. Turn the camshaft timing gear assembly on the right bank counterclockwise to tighten the chain between the banks.

➡ When the idle sprocket is reused, align the chain plate with the mark where the plate had been in order to tighten the chain between the banks.

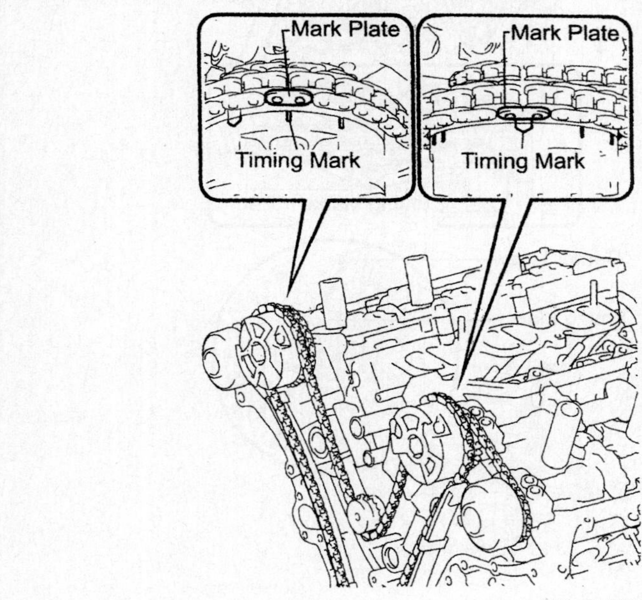

Fig. 59 Aligning the mark plates to the timing marks

   d. Align the mark plate and timing mark and install the chain onto the crankshaft timing sprocket.
   e. Temporarily tighten the pulley set bolt.
   f. Turn the crankshaft clockwise to set it to the right block bore center line (TDC/compression).
   - Chain tensioner slipper
11. Install the No. 1 chain tensioner assembly as follows:
   a. Move the stopper plate upward to release the lock, and push the plunger deep into the tensioner.
   b. Move the stopper plate downward to set the lock, and insert a hexagon wrench into the hole of the stopper plate.
   c. Install the chain tensioner with the 2 bolts and torque to 7 ft. lbs. (10 Nm).
   d. Remove the lock pin of the chain tensioner. Check that each timing mark is aligned with the crankshaft at the TDC/compression.
   e. Remove the pulley set bolt.
12. Install or connect the following:
   - Timing chain cover assembly
   - Left and right cylinder head covers
   - No. 1 and 2 oil pipes
   - Ignition coil assembly
   - High pressure side fuel pump
   - No. 1 and 2 fuel pipes
   - Oil strainer
   - Oil pan assembly
   - Water inlet
   - Crankshaft pulley. Torque the bolt to 192 ft. lbs (260 Nm).
   - Water hose joint
   - Intake air surge tank assembly and No. 2 surge tank stay
   - Fuel injector driver
   - Water pump pulley. Torque the bolt to 15 ft. lbs (21 Nm).
   - Serpentine belt tensioner
   - Left and right engine mounting brackets
   - A/C compressor unit, if necessary
   - Alternator
   - No. 2 idler pulley
   - Serpentine drive belt
   - Front differential assembly (AWD)
   - Engine/transmission assembly

VALVE LASH

ADJUSTMENT

The 2.5L (4GR-FSE) and 3.5L (2GR-FSE) engines are equipped with hydraulic valves which are not adjustable.

ENGINE PERFORMANCE & EMISSION CONTROLS

CRANKSHAFT POSITION (CKP) SENSOR

LOCATION

See Figures 62 and 63.

Refer to the accompanying illustrations.

REMOVAL & INSTALLATION

2.5L Engine

See Figure 64.

1. Remove the cool air intake duct seal.
2. Remove the engine room side cover LH.
3. Remove the engine room side cover RH.
4. Remove the v-bank cover sub assembly.
5. Discharge the refrigerant from the refrigeration.
6. Remove the No. 1 inlet air cleaner.
7. Remove the v-ribbed belt.
8. Remove the engine under cover.
9. Remove the rear engine under cover LH.
10. Remove the oil filter bracket (4WD).
11. Remove the oil filter bracket sub assembly (4WD).
12. Disconnect the No. 1 cooler refrigerant suction hose.
13. Disconnect the discharge hose sub assembly.

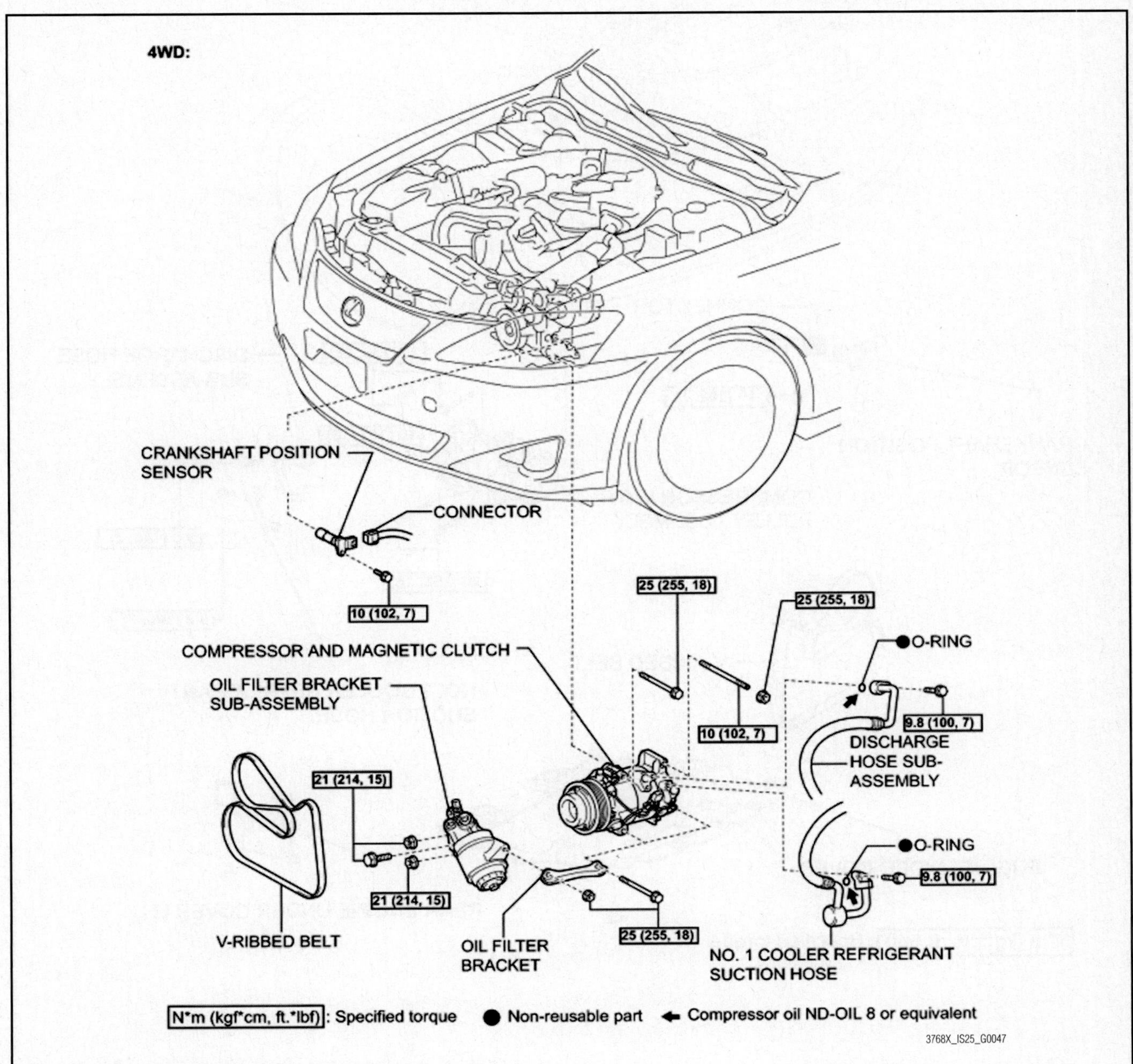

Fig. 62 Crankshaft Position (CKP) sensor location—2.5L engine

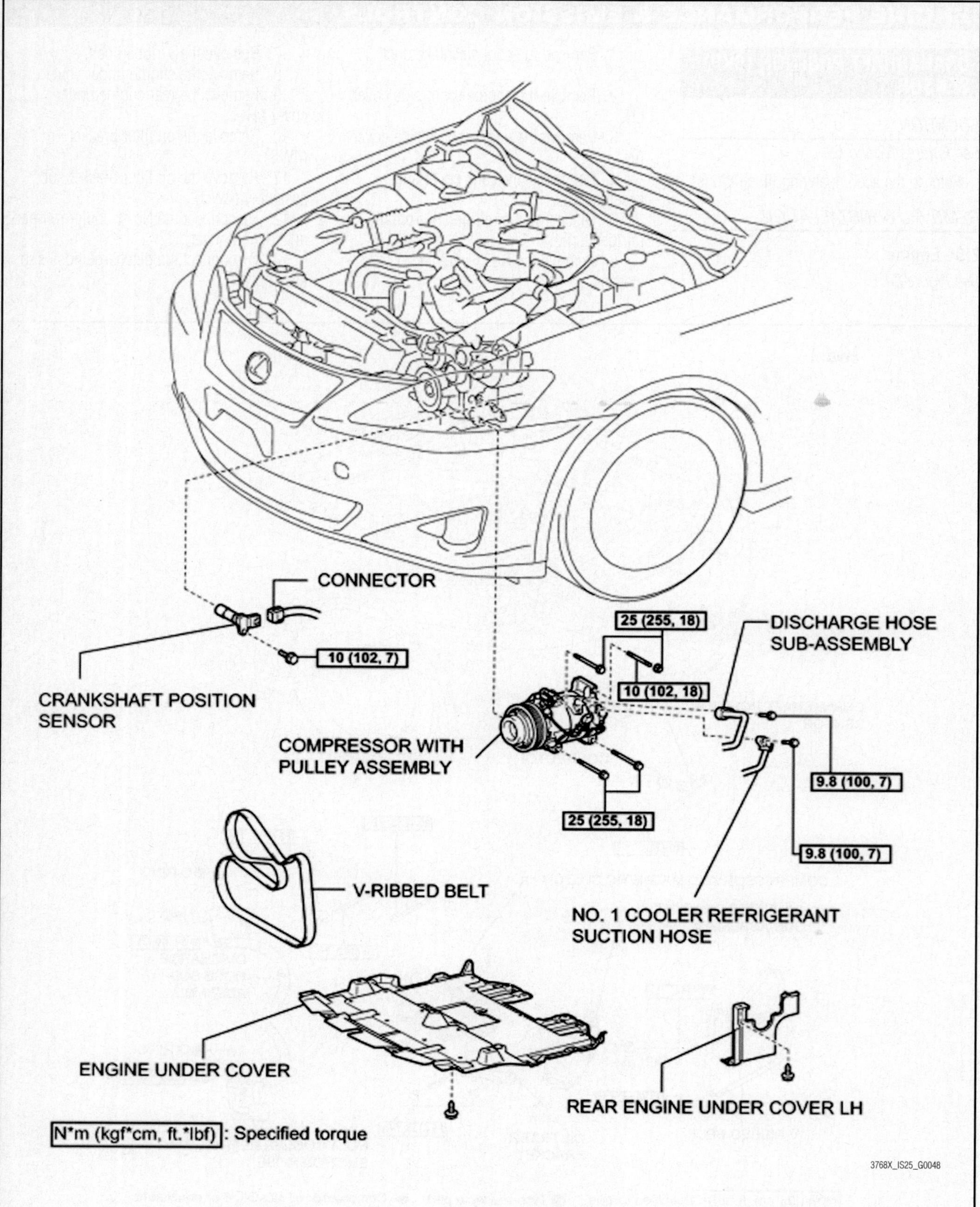

Fig. 63 Crankshaft Position (CKP) sensor location—3.5L engine

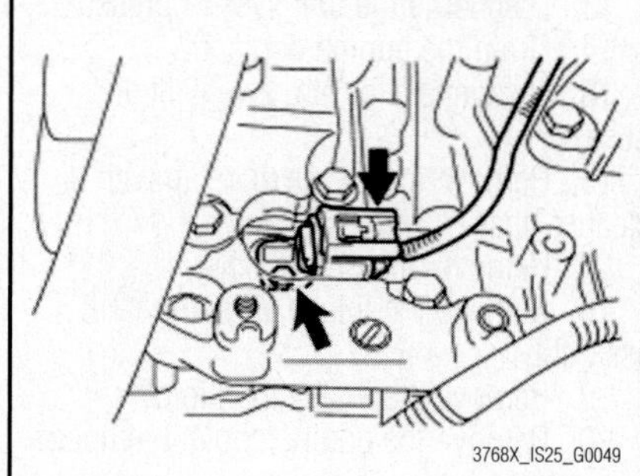

Fig. 64 Removing the crankshaft position sensor—2.5L engine

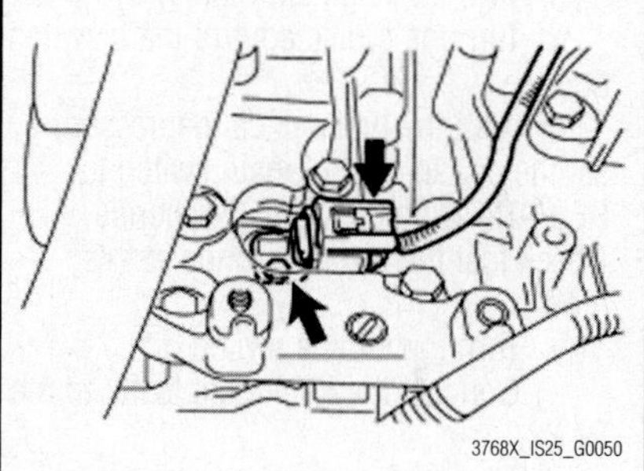

Fig. 65 Removing the crankshaft position sensor—3.5L engine

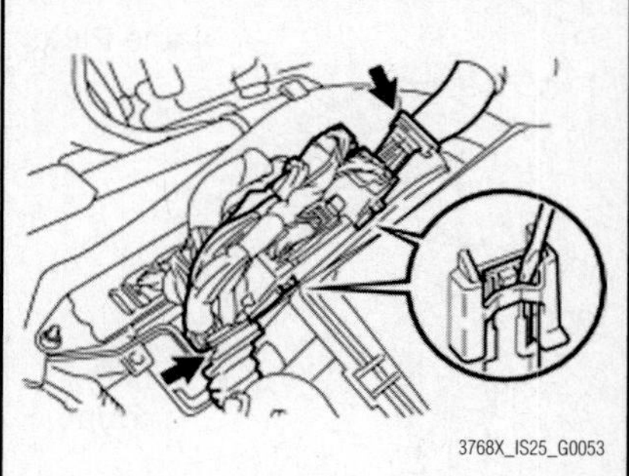

Fig. 67 Disconnecting the connector holder

14. Remove the compressor and magnetic clutch.
15. Remove the crankshaft position sensor.
    a. Disconnect the crankshaft position sensor connector.
    b. Remove the bolt and crankshaft position sensor.

To install:

16. To install, reverse the removal procedure. Tighten the crankshaft position sensor to 7 ft. lbs. (10 Nm).

3.5L Engine

See Figure 67.

1. Remove the cool air intake duct seal.
2. Remove the v-bank cover sub assembly.
3. Remove the No. 1 inlet air cleaner.
4. Discharge the refrigerant from the refrigeration system.
5. Remove the v-ribbed belt.
6. Remove the engine under cover.
7. Remove the rear engine under cover LH.
8. Disconnect the No. 1 cooler refrigerant suction hose.
9. Disconnect the discharge hose sub assembly.
10. Remove the compressor with the pulley assembly.
11. Remove the crankshaft position sensor.
    a. Disconnect the crankshaft position sensor connector.
    b. Remove the bolt and crankshaft position sensor.

To install:

12. To install, reverse the removal procedure. Tighten the crankshaft position sensor to 7 ft. lbs. (10 Nm).

ELECTRONIC CONTROL MODULE (ECM)

REMOVAL & INSTALLATION

See Figures 66 through 69.

1. Disconnect the cable from negative battery terminal.
2. Remove the cool air intake duct seal.
3. Remove the engine room side cover LH.
4. Remove the ECM.
    a. Remove the 3 bolts and ECM cover.

➡Remove all water on and around the ECM cover.

➡Perform the inspection indoors to avoid rain.

➡Be sure to prevent water intrusion to the ECM (connectors and screw parts).

5. Remove the ECM.
    a. Disconnect the connectors. Using a screwdriver disconnect the connector holder. Disconnect the 6 ECM connectors.
    b. Remove the 2 nuts and ECM.

To install:

6. Install the ECM to the ECM box.

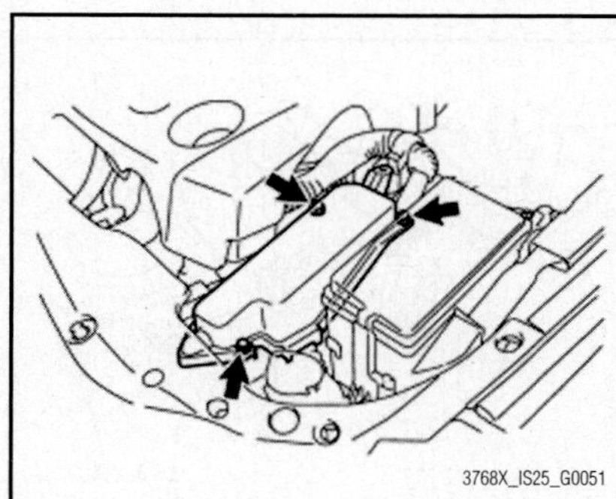

Fig. 66 Removing the ECM cover—3.5L engine

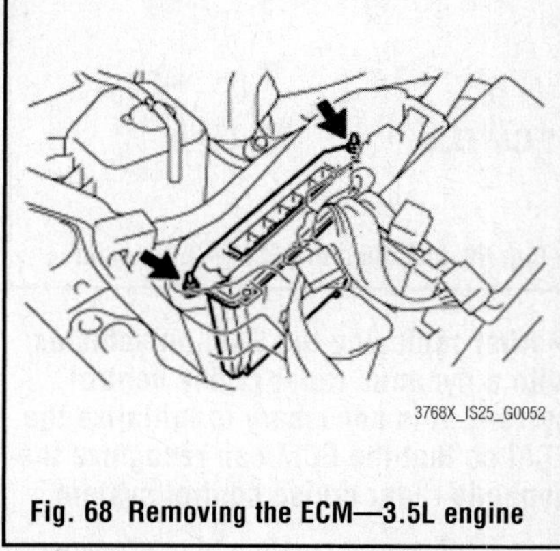

Fig. 68 Removing the ECM—3.5L engine

➡Install the ECM on the ECM box with the name plate facing the inside of the vehicle.

➡Make sure that there is no foreign matter on the gasket surface of the ECM box.

➡Insert the stud bolt on the vehicle rear side to the ECM first.

➡Make sure that stud bolts are securely installed into the front and rear bolt holes.

    a. Install the 2 nuts. Tighten to 49 inch lbs. (5.5 Nm).

➡Be sure to first tighten the nut on the vehicle rear.

    b. Connect the ECM connectors.

➡Be sure to securely connect the connectors.

7. Install the ECM cover. Tighten the nuts to 49 inch lbs. (5.5 Nm).

➡Make sure that the wire harness does not get caught between the parts.

8. Install the engine room side cover LH.
9. Install the cool air intake duct seal.
10. Initialized ECM.

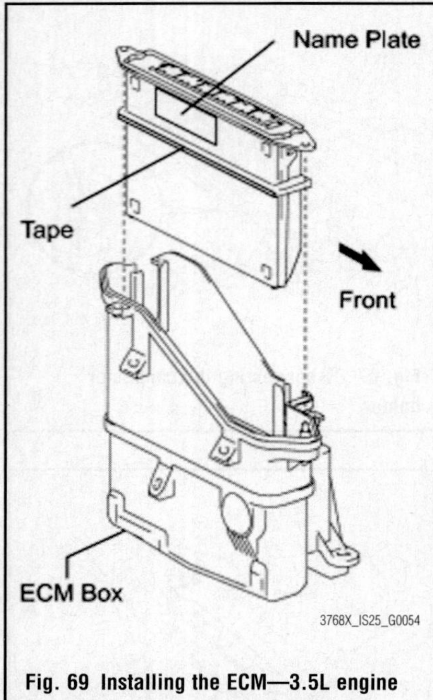

Fig. 69 Installing the ECM—3.5L engine

➡After replacing the ECM on vehicles with a dynamic radar cruise control system, it is necessary to initialize the ECM so that the ECM can recognize the dynamic radar cruise control system.

a. Be sure to perform the following procedures after replacing the ECM.

b. Turn the engine switch on (IG).

c. Turn the cruise main switch on.

d. With the brake pedal depressed, push the cruise control main switch to RES/ACC 3 times within 3 seconds. Check that the buzzer sounds at this time.

➡Do not turn the headlight dimmer switch on at this time because the optical axis automatic adjustment mode has already started, which may lead to an incorrect optical axis setting. If the headlight dimmer switch is turned on by mistake, readjust the optical axis.

11. Perform registration.

➡The VIN must be input into the replacement ECM.

12. Set the ECM for w/optional dynamic radar cruise control system.

➡When replacing the ECM of vehicles with the optional dynamic radar cruise control system, it is necessary to set the ECM for this option.

a. Be sure perform the following procedures after replacing the ECM.

b. Turn the engine switch on (IG).

c. Turn the cruise control main switch on.

d. With the brake pedal depressed, set the cruise control main switch to RES/ACC 3 times within 3 seconds. Check that the buzzer sounds at this time.

e. Turn the engine switch off.

f. Connect the intelligent tester to the DLC3.

g. Turn the engine switch and intelligent tester main switch on.

h. Select the item L/C Option Flag in the Date List.

i. Check that YES is displayed.

ENGINE COOLANT TEMPERATURE (ECT) SENSOR

REMOVAL & INSTALLATION

See Figure 70.

1. Remove the cool air intake duct seal.
2. Remove the engine room side cover LH.
3. Remove the engine room side cover RH.
4. Remove the v-bank cover sub assembly.
5. Remove the front fender protector upper LH.
6. Remove the front fender protector upper RH.
7. Remove the roof drip side finish moulding LH.
8. Remove the roof drip side finish moulding RH.
9. Remove the front wiper arm head cap.
10. Remove the windshield wiper arm and blade assembly LH.
11. Remove the windshield wiper arm and blade assembly RH.
12. Remove the cowl top ventilator louver sub assembly.
13. Discharge the fuel system pressure.
14. Drain the engine coolant.
15. Disconnect the No. 2 ventilation hose.
16. Remove the air cleaner cap with air cleaner hose.
17. Remove the throttle body.
18. Remove the intake air surge tank assembly.
19. Remove the intake manifold.
20. Remove the engine coolant temperature sensor.
    a. Disconnect the engine coolant temperature sensor connector.
    b. Remove the engine coolant temperature sensor.
    c. Remove the gasket from the engine coolant temperature sensor.

To install:

21. To install, reverse the removal procedure. Tighten the engine coolant temperature sensor to 15 ft. lbs. (20 Nm).

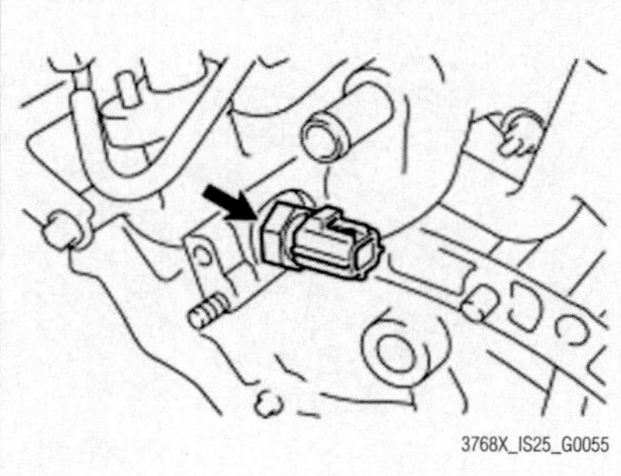

Fig. 70 Removing the engine coolant temperature sensor

EVAPORATIVE EMISSIONS (EVAP) CANISTER

REMOVAL & INSTALLATION

See Figures 71 through 74.

1. Disconnect the cable from the negative battery terminal.
2. Remove the floor under cover.
    a. Remove the 5 nuts, 4 clips and floor under cover.
3. Remove the charcoal canister assembly.
    a. Disconnect the wire harness from the 2 clamps.
    b. Disconnect the 2 quick connectors, hose and connector.
    c. Replace the 4 nuts, 2 spacers and charcoal canister assembly.

To install:

4. To install, reverse the removal procedure. Tighten the charcoal canister to 49 inch lbs. (5.5 Nm).

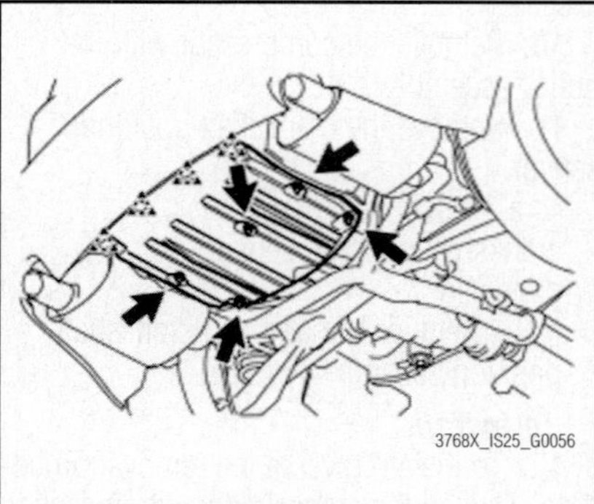

Fig. 71 Removing the floor under cover

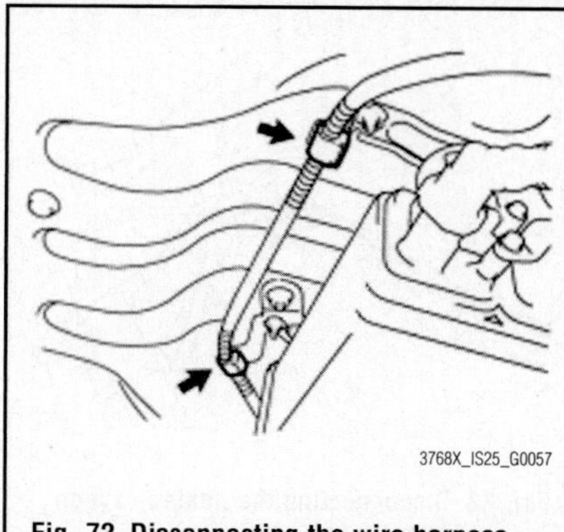

Fig. 72 Disconnecting the wire harness clamps

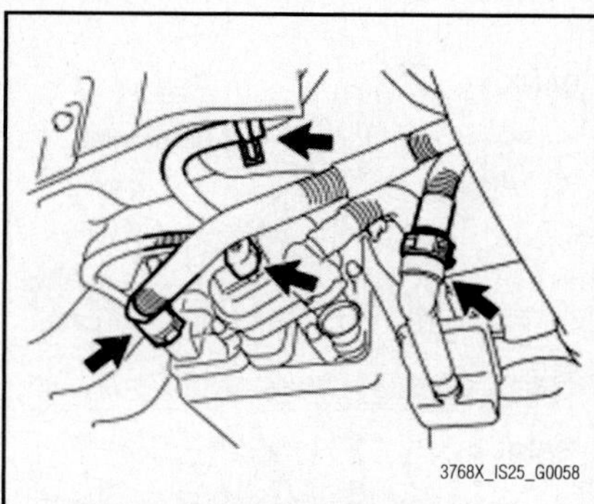

Fig. 73 Disconnecting the 2 quick connectors, hose and connector

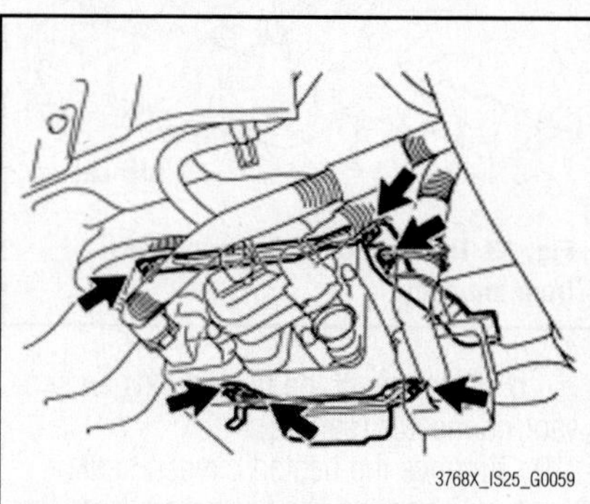

Fig. 74 Removing the 4 nuts, 2 spacers and charcoal canister assembly

HEATED OXYGEN (HO2S) SENSOR

REMOVAL & INSTALLATION

2.5L Engine

See Figures 75 through 80.

1. Disconnect the cable from the negative battery terminal.

⁂ CAUTION

Wait at least 90 seconds after disconnecting the cable from the negative (-) battery terminal to prevent airbag and seat belt pretensioner deployment.

2. Remove the shift lever knob sub assembly.
3. Remove the upper console panel No. 1 garnish (A/T).
4. Remove the upper console panel No. 2 garnish (A/T).
5. Remove the console panel sub assembly (A/T).
6. Remove the rear console panel sub assembly (M/T).
7. Remove the front console panel sub assembly (M/T).
8. Remove the front ash receptacle sub assembly.
9. Remove the console box register assembly.
10. Remove the console box.
11. Remove the center lower instrument cluster finish panel.
12. Remove the front seat track bracket cover outer LH (power seat, 4WD).
13. Remove the front seat track bracket cover inner LH (power seat, 4WD).
14. Remove the front seat track cover rear outer LH (power seat, 4WD).
15. Remove the front seat track cover rear inner LH (power seat, 4WD).
16. Remove the front seat assembly (power seat, 4WD).
17. Remove the front seat inner track bracket cover (manual seat, 4WD).
18. Remove the front seat outer track bracket cover (manual seat, 4WD).
19. Remove the rear seat outer track bracket cover (manual seat, 4WD).
20. Remove the rear seat inner track bracket cover (manual seat, 4WD).
21. Remove the front seat assembly (manual seat, 4WD).
22. Disconnect the connector.

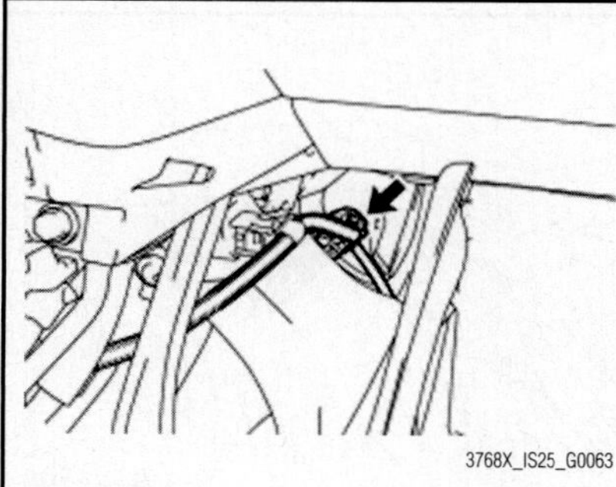

Fig. 75 Disconnecting the oxygen sensor connector (1 of 2)

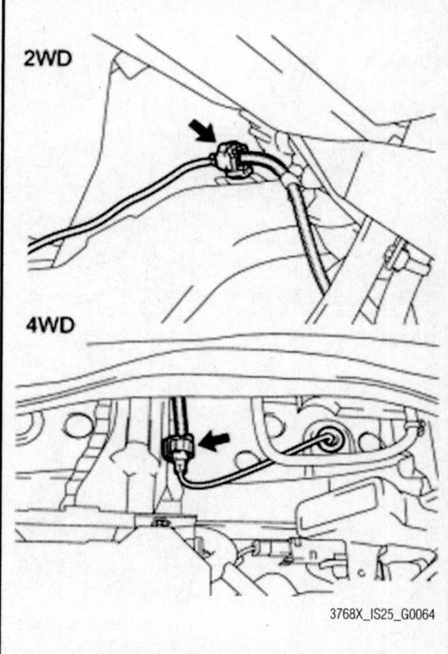

Fig. 76 Disconnecting the oxygen sensor connector (2 of 2)

a. Disconnect the oxygen sensor connector.

b. Disconnect the oxygen sensor connector.

23. Remove the oxygen sensor (2WD).

a. Disconnect the grommet to the body.

b. Using the special tool (09224-00010) remove the oxygen sensor to the exhaust pipe assembly.

24. Remove the oxygen sensor (4WD).

a. Remove the grommet and 2 clips.

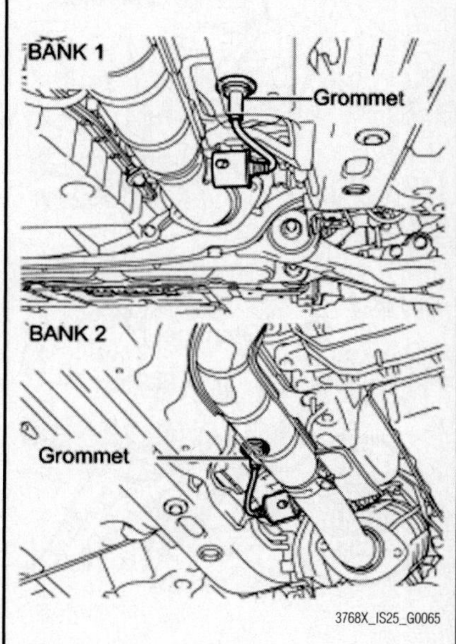

Fig. 77 Disconnecting the grommet to the body

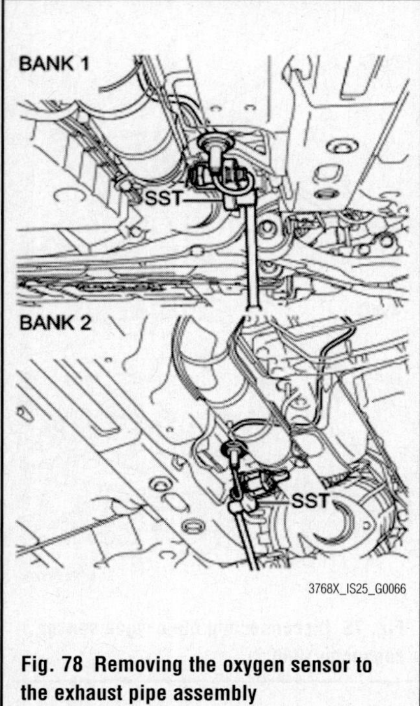

Fig. 78 Removing the oxygen sensor to the exhaust pipe assembly

b. Using the special tool (09224-00010), remove the heated oxygen sensor.

To install:

25. To install, reverse the removal procedure. Tighten the heated oxygen sensor to 30 ft. lbs. (40 Nm).

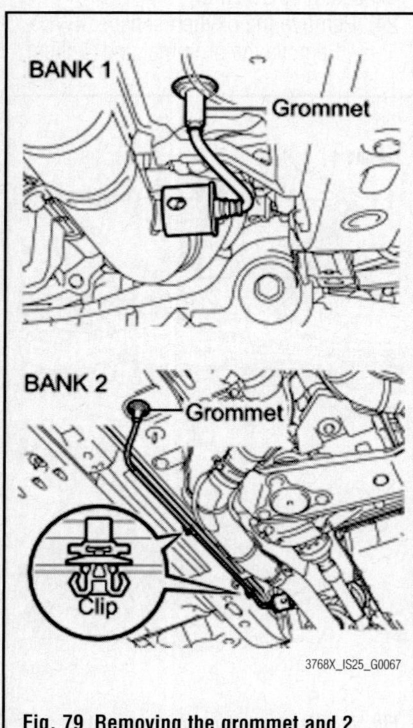

Fig. 79 Removing the grommet and 2 clips

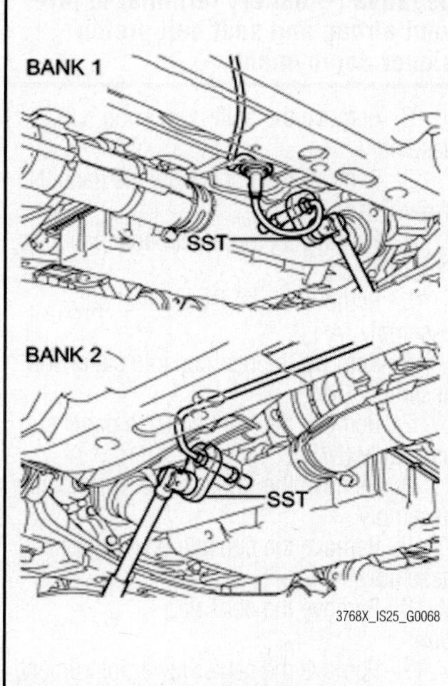

Fig. 80 Removing the heated oxygen sensor (4WD)

3.5L Engine

See Figures 81 through 84.

1. Remove the shift lever knob sub assembly.
2. Remove the No. 1 console upper panel garnish.
3. Remove the No. 2 console upper panel garnish.
4. Remove the console panel sub assembly.
5. Remove the front ash receptacle sub assembly.
6. Remove the console box register assembly.
7. Remove the console box.
8. Disconnect the heated oxygen sensor connector.

a. Disconnect the heated oxygen sensor connector (bank 1).

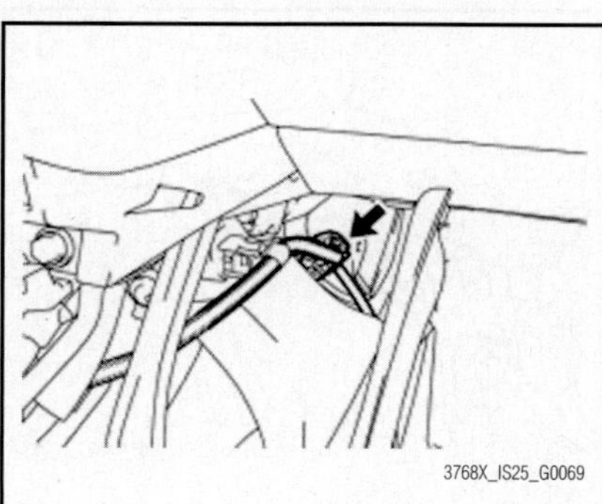

Fig. 81 Disconnecting the heated oxygen sensor connector (bank 1)

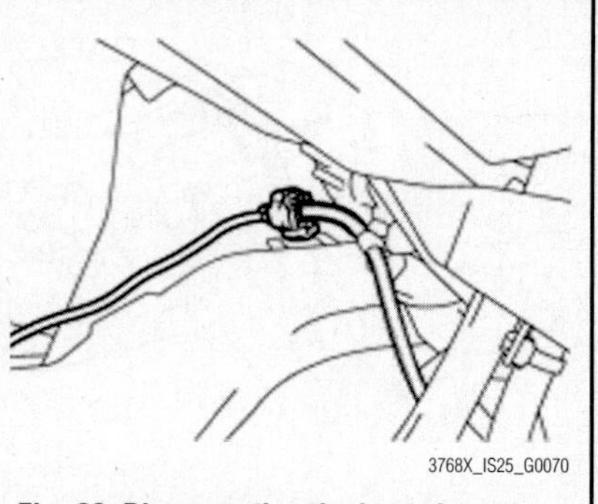

Fig. 82 Disconnecting the heated oxygen sensor connector (bank 2)

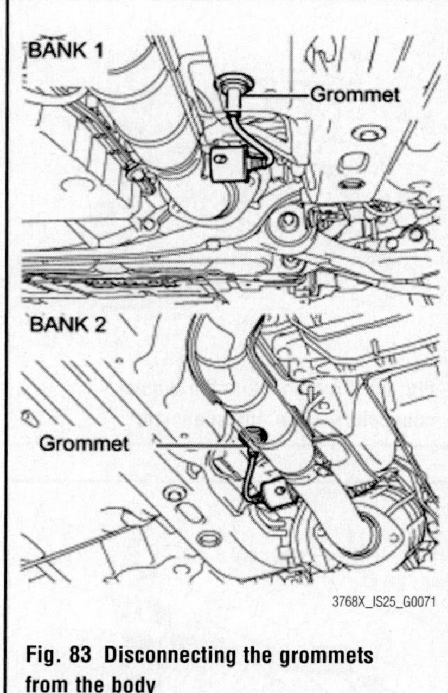

Fig. 83 Disconnecting the grommets from the body

b. Disconnect the heated oxygen sensor connector (bank 2).

9. Remove the heated oxygen sensor.

a. Disconnect the grommets from the body.

b. Using the special tool (09224-00010), remove the heated oxygen sensor from the exhaust pipe assembly.

To install:

10. Install the heated oxygen sensor.

a. Using the special tool (09224-00010), install the heated oxygen sensor.

b. Pass the heated oxygen sensor connectors through the floor panel and install the grommets.

11. Connect the heated oxygen sensor connector (bank 2).

a. Connect the heated oxygen sensor connector (bank 1).

12. Install the console box.

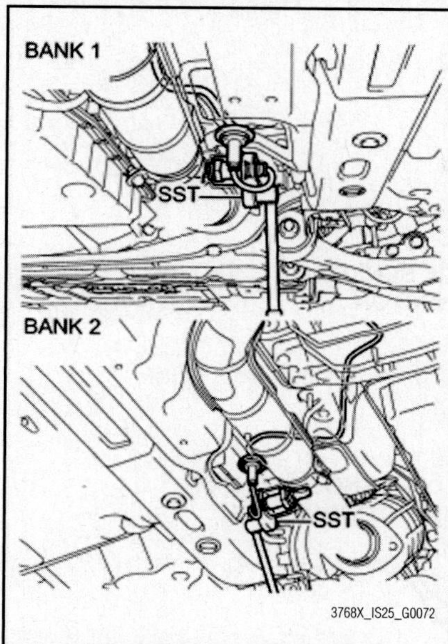

Fig. 84 Removing the heated oxygen sensor from the exhaust pipe assembly

13. Install the console box register assembly.
14. Install the front ash receptacle sub assembly.
15. Install the console panel sub assembly.
16. Install the No. 2 console upper panel garnish.
17. Install the No. 1 console upper panel garnish.
18. Install the shift lever knob sub assembly.

KNOCK SENSOR (KS)

LOCATION

See Figures 85 and 86.

Refer to the accompanying illustrations.

REMOVAL & INSTALLATION

2.5L Engine

See Figures 87 and 88.

CAUTION

Do not allow fuel to spray when removing the pipe between the high pressure side fuel pump and the fuel injectors. The fuel in the pipe is highly pressurized.

1. Remove the cool air intake duct seal.
2. Remove the engine room side cover LH.
3. Remove the engine room side cover RH.
4. Remove the v-bank cover sub assembly.
5. Remove the front fender protector upper LH.
6. Remove the front fender protector upper RH.
7. Remove the roof drip side finish moulding LH.
8. Remove the roof drip side finish moulding RH.

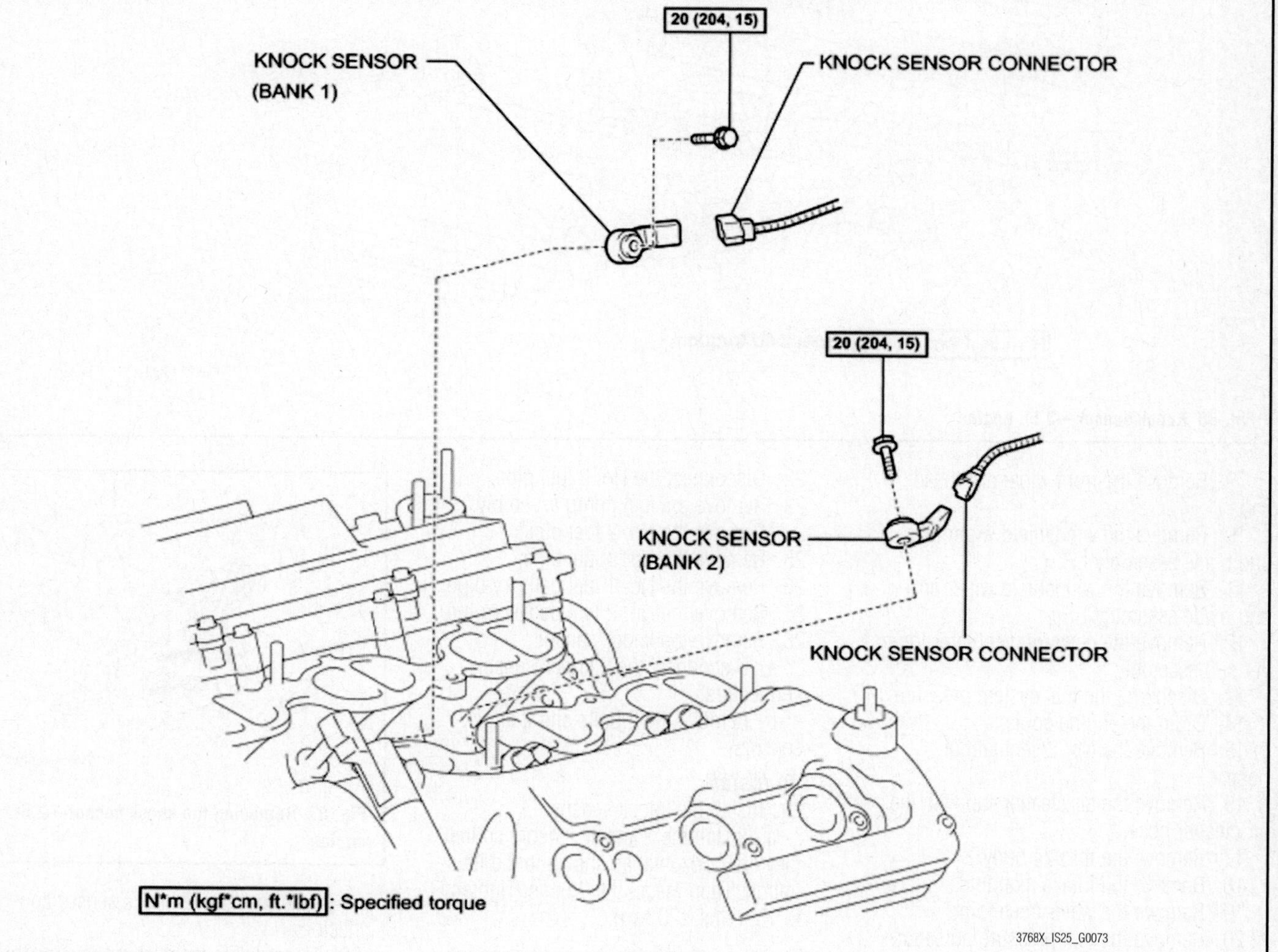

Fig. 85 Knock sensor—2.5L engine

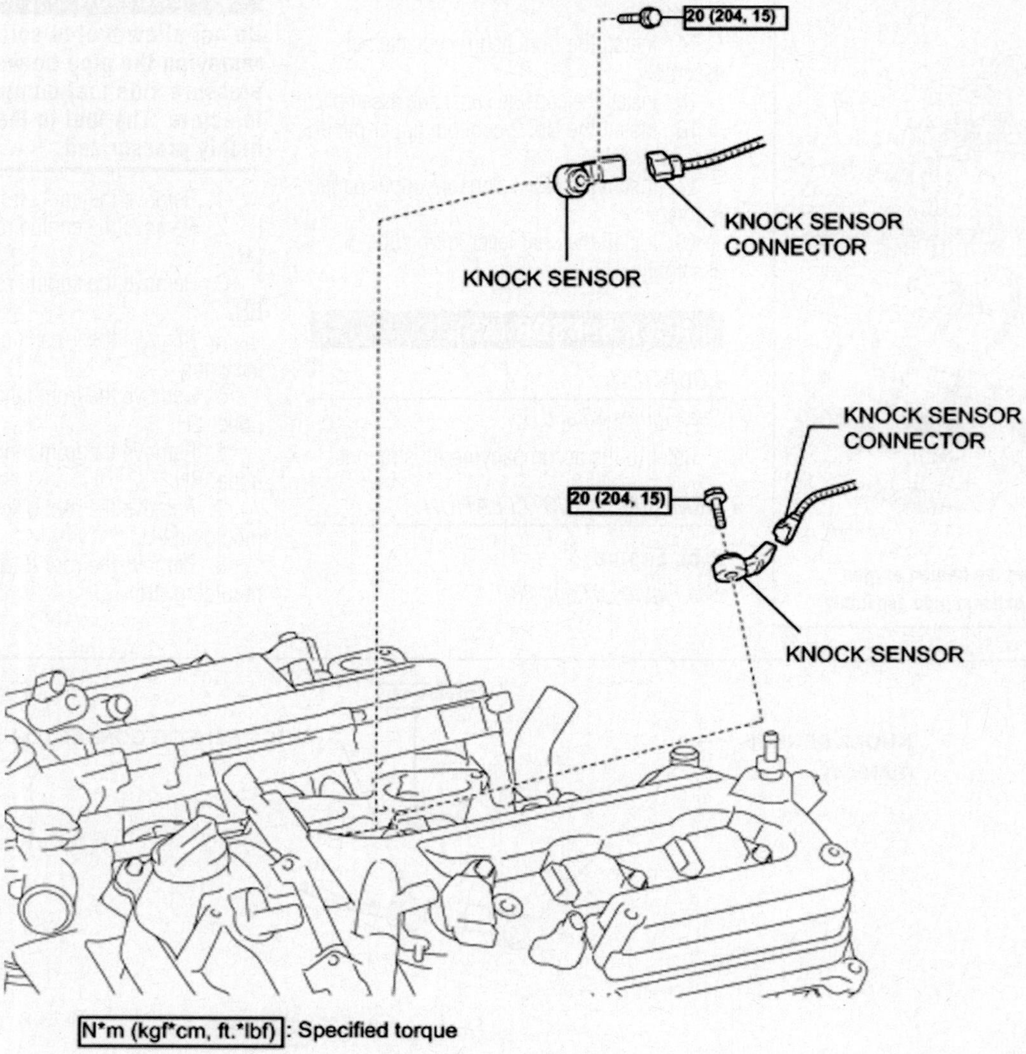

Fig. 86 Knock sensor—3.5L engine

9. Remove the front wiper arm head cap.
10. Remove the windshield wiper arm and blade assembly LH.
11. Remove the windshield wiper arm and blade assembly RH.
12. Remove the cowl top ventilator louver sub assembly.
13. Discharge the fuel system pressure.
14. Drain the engine coolant.
15. Remove the No. 2 ventilation hose.
16. Remove the air cleaner cap with the air cleaner hose.
17. Remove the throttle body.
18. Remove the intake manifold.
19. Remove the water hose joint.
20. Remove the fuel pressure pulsation damper.
21. Remove the No. 1 fuel pipe.
22. Disconnect the No. 2 fuel pipe.
23. Remove the fuel pump assembly.
24. Remove the No. 2 fuel pipe.
25. Remove the No. 3 fuel pipe.
26. Remove the No. 1 fuel delivery pipe.
27. Remove the No. 2 fuel delivery pipe.
28. Remove the knock sensor.
   a. Disconnect the 2 knock sensor connectors.
   b. Remove the 2 bolts and 2 knock sensors.

To install:

29. Install the knock sensor.
   a. Install the 2 knock sensors so that they are horizontal as shown in the illustration. Then install the 2 bolts. Tighten to 15 ft. lbs. (20 Nm).

➡It is acceptable for the knock sensor to be tilted +10°to -5°.

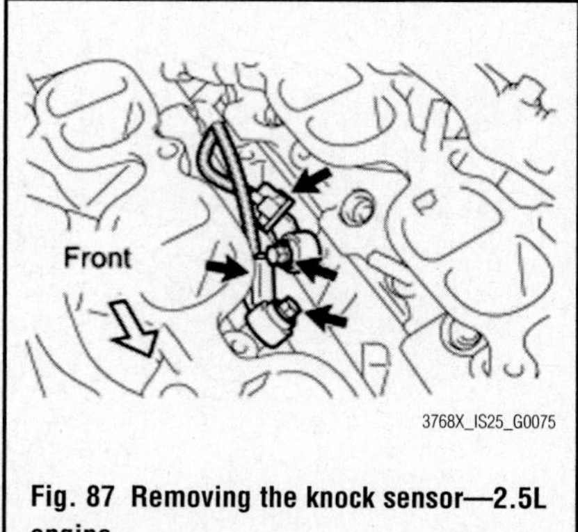

Fig. 87 Removing the knock sensor—2.5L engine

   b. Connect the 2 knock sensor connectors.
30. To complete installation, reverse the remaining removal procedure.

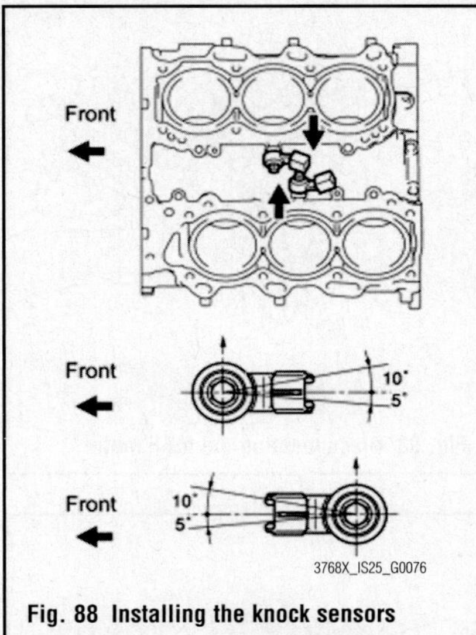

Fig. 88 Installing the knock sensors

3.5L Engine

See Figures 89 and 90.

CAUTION

Do not allow fuel to spray when removing the pipe between the high pressure side fuel pump and the fuel injectors. The fuel in the pipe is highly pressurized.

1. Remove the cool air intake duct seal.
2. Remove the engine room side cover LH.
3. Remove the engine room side cover RH.
4. Remove the v-bank cover sub assembly.
5. Remove the front fender protector upper LH.
6. Remove the front fender protector upper RH.
7. Remove the roof drip side finish moulding LH.
8. Remove the roof drip side finish moulding RH.
9. Remove the front wiper arm head cap.
10. Remove the windshield wiper arm and blade assembly LH.
11. Remove the windshield wiper arm and blade assembly RH.
12. Discharge the fuel system pressure.
13. Drain the engine coolant.
14. Remove the No. 2 ventilation hose.
15. Remove the air cleaner cap with the air cleaner hose.
16. Remove the throttle body.
17. Remove the intake air surge tank assembly.

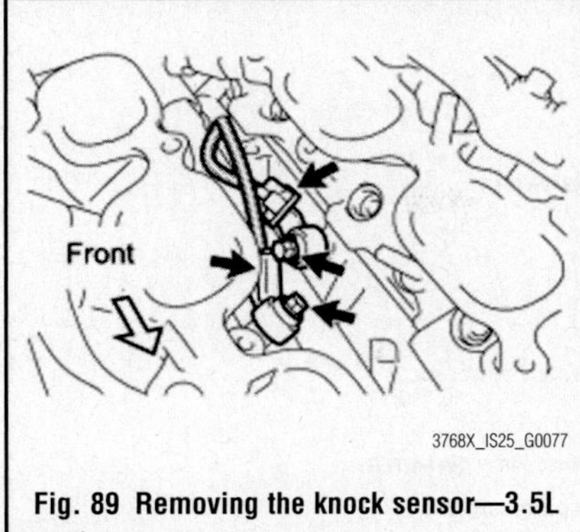

Fig. 89 Removing the knock sensor—3.5L engine

18. Remove the intake manifold.
19. Remove the water hose joint.
20. Remove the fuel pressure pulsation damper.
21. Remove the No. 1 fuel pipe.
22. Disconnect the No. 2 fuel pipe.
23. Remove the fuel pump assembly.
24. Remove the No. 2 fuel pipe.
25. Remove the No. 3 fuel pipe.
26. Remove the No. 1 fuel delivery pipe.
27. Remove the No. 2 fuel delivery pipe.
28. Remove the knock sensor.
    a. Disconnect the 2 knock sensor connectors.
    b. Remove the 2 bolts and 2 knock sensors.

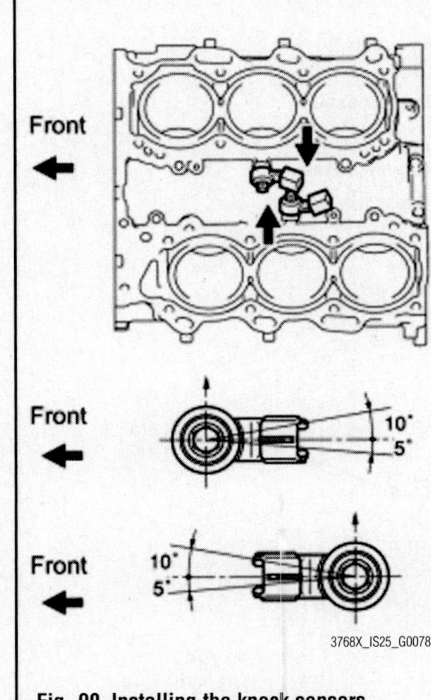

Fig. 90 Installing the knock sensors

To install:

29. Install the knock sensor.
    a. Install the 2 knock sensors so that they are horizontal as shown in the illustration. Then install the 2 bolts. Tighten to 15 ft. lbs. (20 Nm).

➡It is acceptable for the knock sensor to be tilted +10°to -5°.

    b. Connect the 2 knock sensor connectors.
30. To complete installation, reverse the remaining removal procedure.

MASS AIR FLOW (MAF) METER

LOCATION

See Figures 91 and 92.

Refer to the accompanying illustrations.

REMOVAL & INSTALLATION

See Figures 93 and 94.

1. Remove the cool air intake duct seal.
2. Remove the engine room side cover RH.
3. Remove the Mass Air Flow (MAF) meter.
    a. Disconnect the MAF meter connector.
    b. Remove the 2 screws and MAF meter.
    c. Remove the o-ring from the MAF meter.

To install:

4. To install, reverse the removal procedure.

VEHICLE SPEED SENSOR (VSS)

REMOVAL & INSTALLATION

A760E

See Figures 95 and 96.

1. Remove the transmission revolution sensor (NT).
    a. Disconnect the transmission revolution sensor connector.
    b. Remove the bolt and transmission revolution sensor.
    c. Remove the o-ring from the transmission revolution sensor.
2. Remove the transmission revolution sensor (SP2).
    a. Disconnect the transmission revolution sensor connector.
    b. Remove the bolt and transmission revolution sensor.
    c. Remove the o-ring from the transmission revolution sensor.

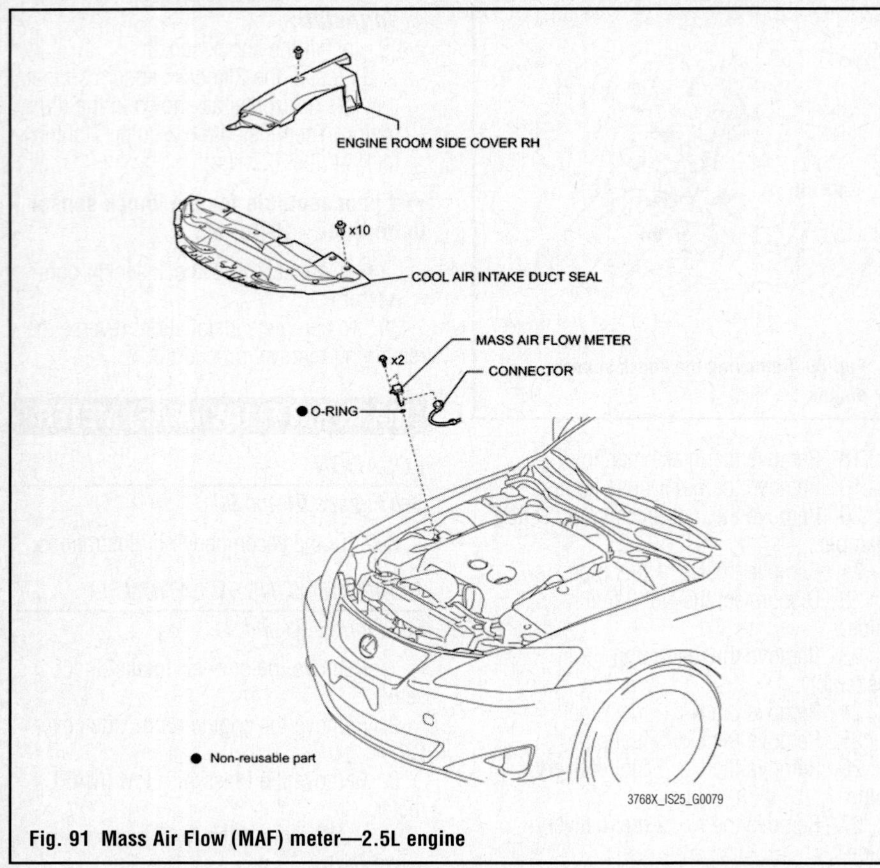

Fig. 91 Mass Air Flow (MAF) meter—2.5L engine

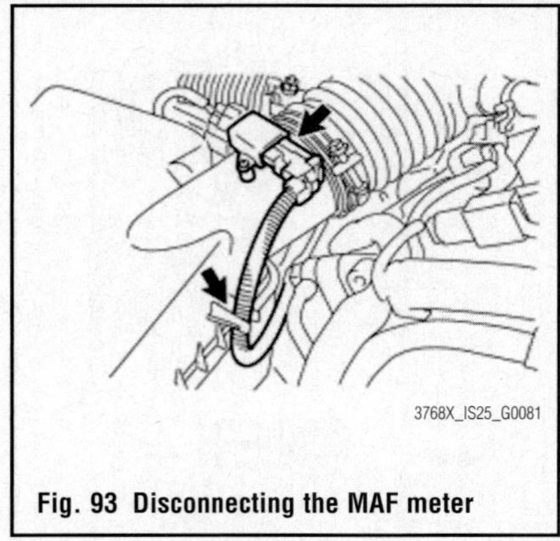

Fig. 93 Disconnecting the MAF meter

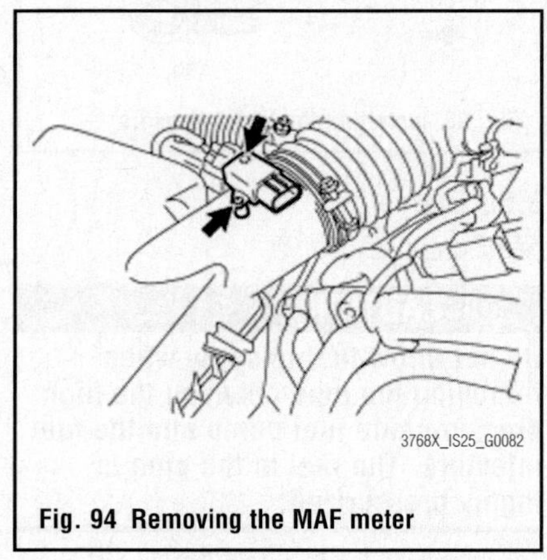

Fig. 94 Removing the MAF meter

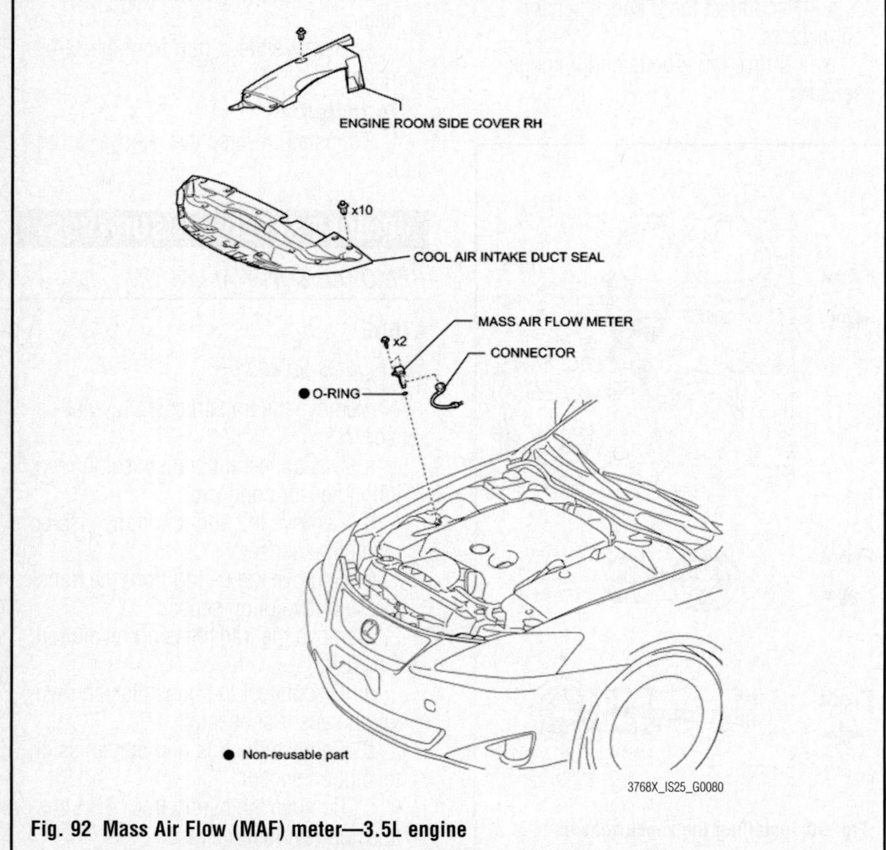

Fig. 92 Mass Air Flow (MAF) meter—3.5L engine

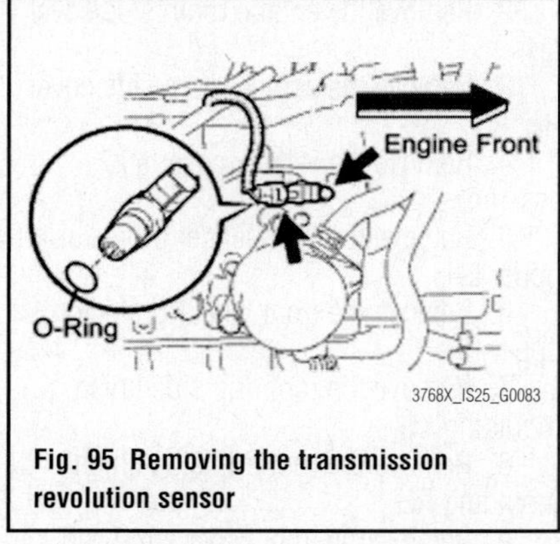

Fig. 95 Removing the transmission revolution sensor

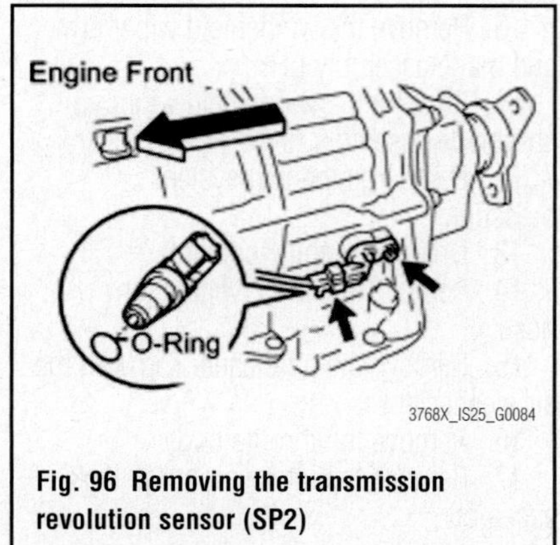

Fig. 96 Removing the transmission revolution sensor (SP2)

To install:

3. To install, reverse the removal procedure. Coat new o-rings with ATF WS. Tighten the sensors to 48 inch lbs. (5.4 Nm).

A760H

See Figures 97 and 98.

1. Remove the transmission revolution sensor (NT).
   a. Disconnect the transmission revolution sensor connector.
   b. Remove the bolt and transmission revolution sensor.
   c. Remove the o-ring from the transmission revolution sensor.
2. Remove the transmission revolution sensor (SP2).
   a. Disconnect the transmission revolution sensor connector.

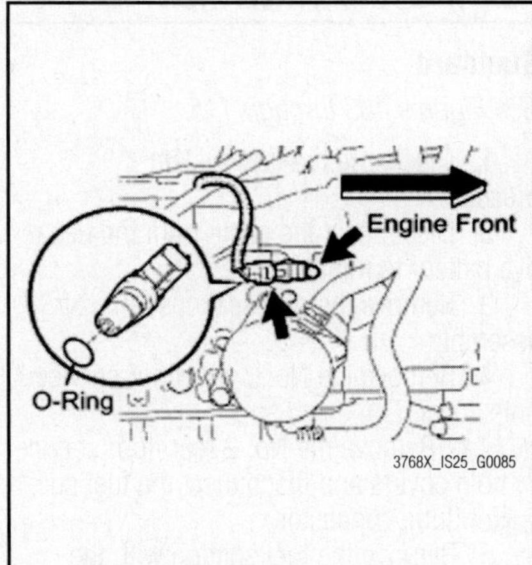

Fig. 97 Removing the transmission revolution sensor (NT)

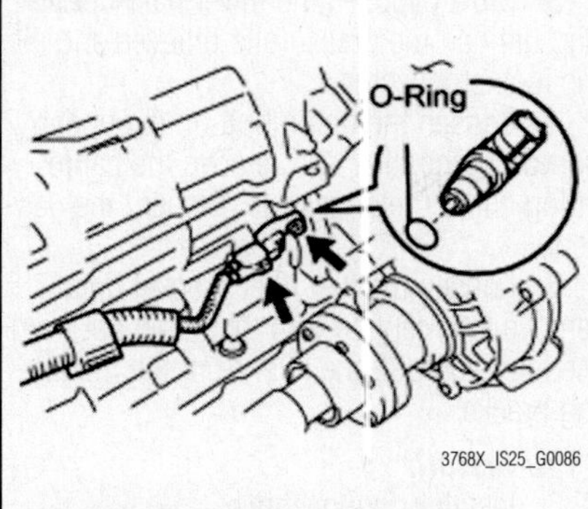

Fig. 98 Removing the transmission revolution sensor (SP2)

   b. Remove the bolt and transmission revolution sensor.
   c. Remove the o-ring from the transmission revolution sensor.

To install:

3. To install, reverse the removal procedure. Coat new o-rings with ATF WS. Tighten the sensors to 48 inch lbs. (5.4 Nm).

A960E

See Figures 99 and 100.

1. Remove the transmission revolution sensor (NT).
   a. Disconnect the transmission revolution sensor connector.
   b. Remove the bolt and transmission revolution sensor.
   c. Remove the o-ring from the transmission revolution sensor.
2. Remove the transmission revolution sensor (SP2).
   a. Disconnect the transmission revolution sensor connector.

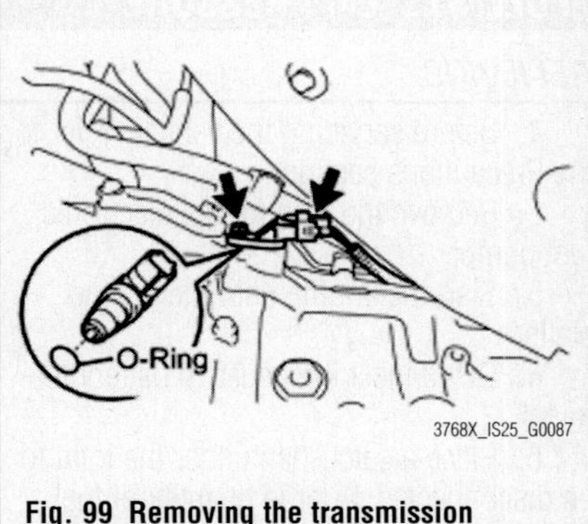

Fig. 99 Removing the transmission revolution sensor (NT)

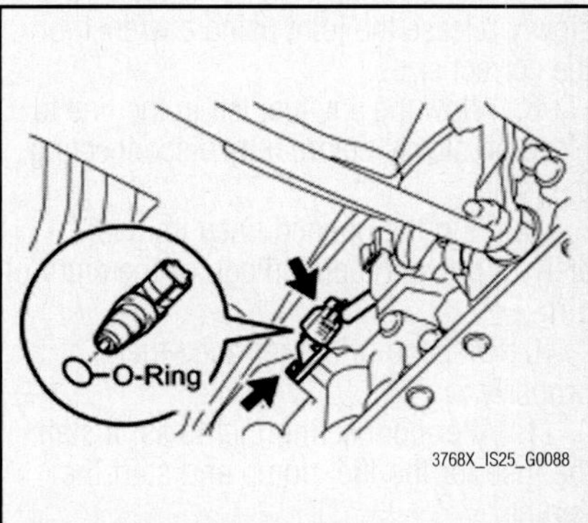

Fig. 100 Removing the transmission revolution sensor (SP2)

   b. Remove the bolt and transmission revolution sensor.
   c. Remove the o-ring from the transmission revolution sensor.

To install:

3. To install, reverse the removal procedure. Coat new o-rings with ATF WS. Tighten the sensors to 48 inch lbs. (5.4 Nm).

FUEL SYSTEM SERVICE PRECAUTIONS

Safety is the most important factor when performing not only fuel system maintenance, but any type of maintenance. Failure to conduct maintenance and repairs in a safe manner may result in serious personal injury or death. Work on a vehicle's fuel system components can be accomplished safely and effectively by adhering to the following rules and guidelines.

- To avoid the possibility of fire and personal injury, always disconnect the negative battery cable unless the repair or test procedure requires that battery voltage be applied.
- Always relieve the fuel system pressure prior to disconnecting any fuel system component (injector, fuel rail, pressure regulator, etc.) fitting or fuel line connection. Exercise extreme caution whenever relieving fuel system pressure to avoid exposing skin, face and eyes to fuel spray. Please be advised that fuel under pressure may penetrate the skin or any part of the body that it contacts.
- Always place a shop towel or cloth around the fitting or connection prior to loosening to absorb any excess fuel due to spillage. Ensure that all fuel spillage is quickly removed from engine surfaces. Ensure that all fuel-soaked cloths or towels are deposited into a flame-proof waste container with a lid.
- Always keep a dry chemical (Class B) fire extinguisher near the work area.
- Do not allow fuel spray or fuel vapors to come into contact with a spark or open flame.
- Always use a second wrench when loosening or tightening fuel line connection fittings. This will prevent unnecessary stress and torsion on fuel piping. Always follow the proper torque specifications.
- Always replace worn fuel fitting O-rings with new ones. Do not substitute fuel hose where rigid pipe is installed.

FUEL SYSTEM PRESSURE

RELIEVING

1. Before servicing the vehicle, refer to the Precautions section.
2. Remove the fuse for the electronic fuel pump.
3. Start the engine until the engine stalls.
4. Disconnect the negative battery terminal.
5. Place a catch-pan under the joint to be disconnected. A large quantity of fuel may be released when the joint is opened.
6. Wear eye or full-face protection.
7. Place a shop towel over the area and slowly release the joint using a wrench of the correct size.
8. Allow the any fuel left in the line to bleed off slowly before fully disconnecting the joint.
9. Plug the opened lines immediately to prevent fuel spillage or the entry of dirt.
10. Dispose of the released fuel properly.
11. After connecting fuel lines, install the fuse for the fuel pump and start the engine.
12. Check for leaks and repair as needed.

FUEL FILTER

REMOVAL & INSTALLATION

See Figure 101.

The fuel filter is an integral component of the in-tank fuel pump assembly. Refer to the Fuel Pump Removal procedure.

1. Before servicing the vehicle, refer to the Precautions section.
2. Disconnect the negative battery cable. Wait at least 90 seconds before performing any other work.
3. Slowly loosen the lower flare nut fitting until all the pressure is relieved and all the fuel is collected.
4. Loosen the union bolt on the upper portion of the filter and remove the banjo fitting and 2 metal gaskets. Discard the gaskets.
5. Loosen the fuel filter bracket bolt, remove the fuel line with the flared nut from the filter and pull the filter from the mounting bracket.

To install:

6. Install or connect the following:
   - A new fuel filter to the vehicle and tighten the bracket bolt
   - Banjo fitting with a new metal gasket on each side
   - Union bolt. Tighten the union bolt to 22 ft. lbs. (30 Nm).
   - Flare nut to the lower connection. Tighten the flare nut to 22 ft. lbs. (30 Nm).
   - Lower the vehicle if raised.
   - Remove the drain pan and/or rags and connect the negative battery cable.
   - Start the engine and visually inspect the upper and lower connections for leaks.

FUEL LEVEL SENDING UNIT

REMOVAL & INSTALLATION

See Figure 102.

1. Before servicing the vehicle, refer to the Precautions section.
2. Relieve the fuel system pressure.
3. Disconnect the negative battery cable. Wait at least 90 seconds before performing any other work
4. Remove the fuel pump and sending gauge assembly.
5. Remove the fuel sending gauge from the fuel pump assembly as follows:
   a. Disconnect the fuel sender gauge connector.
   b. Press down on the sender gauge claw. Then slide the sender gauge upward.

To install:

6. Install the fuel sending gauge to the fuel pump assembly as follows:
   a. Set the fuel sender gauge to the No. 1 fuel sub-tank. Then slide the sender gauge downward to install it.
   b. Connect the fuel sender gauge connector.
7. Install the fuel pump and sending gauge assembly.
8. Connect the negative battery cable.

FUEL PUMP

REMOVAL & INSTALLATION

Standard

See Figures 103 through 115.

1. Discharge the fuel system pressure.
2. Disconnect the cable from the negative battery terminal.
3. Remove the rear seat cushion assembly.
4. Remove the No. 2 rear floor service hole cover.
   a. Remove the No. 2 rear floor service hole covers and disconnect the fuel suction tube connector.
5. Remove the fuel suction with the pump and gauge tube assembly.
   a. Disconnect the connector from the fuel suction with pump and gauge assembly.
   b. Disconnect fuel pump tube.

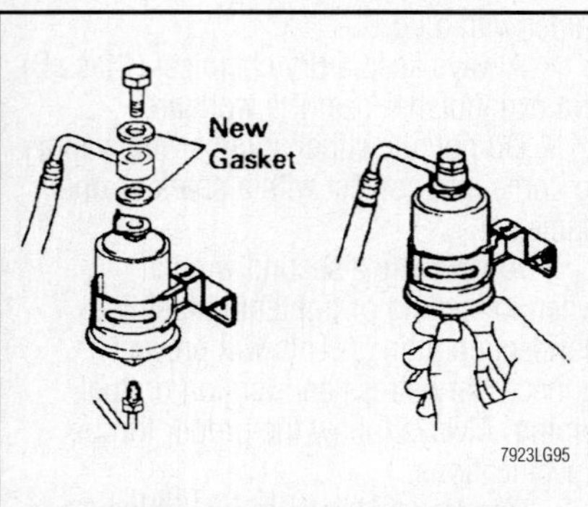

Fig. 101 Exploded view of a typical fuel line connection at the filter

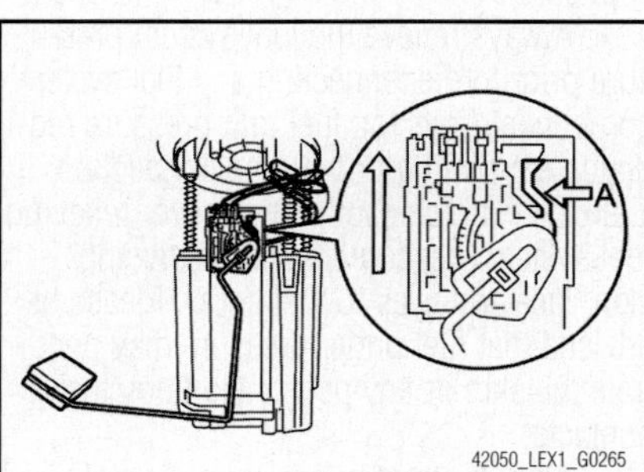

Fig. 102 Press down on the sender gauge claw (A), then slide the sender gauge upward

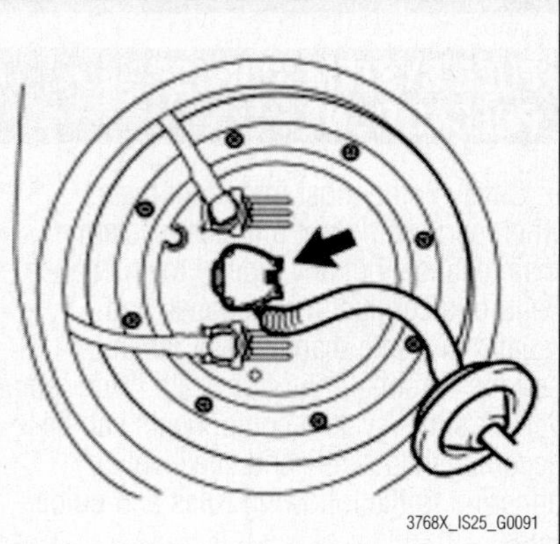

Fig. 103 Disconnecting the connector from the fuel suction with pump and gauge assembly

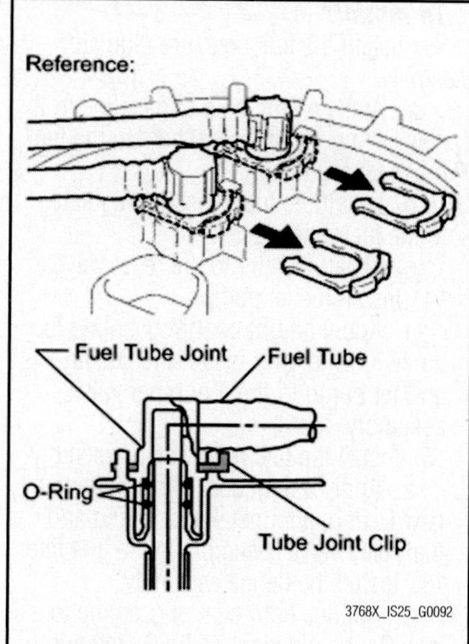

Fig. 104 Removing the 2 tube joint clips, fuel tank main tube and fuel tank return vent tube

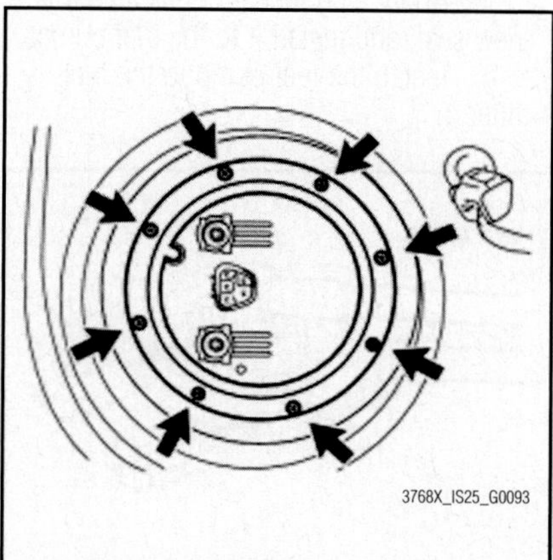

Fig. 105 Removing the 8 bolts and the fuel tank vent tube set plate

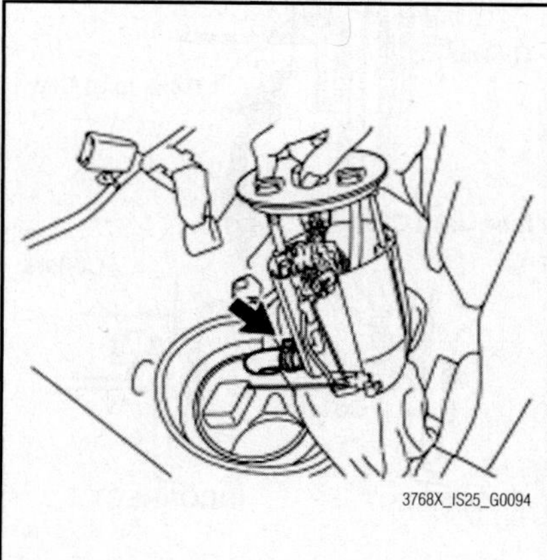

Fig. 106 Disconnecting the fuel tube

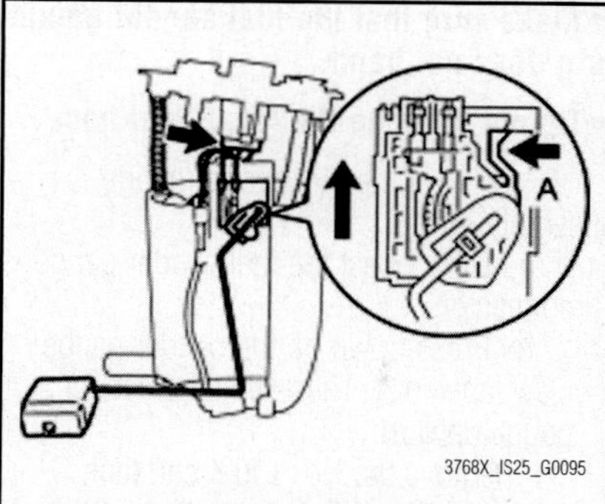

Fig. 107 Removing the fuel sender gauge assembly

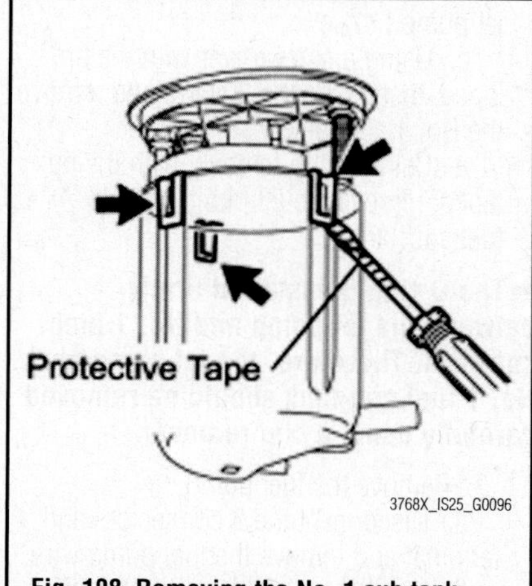

Fig. 108 Removing the No. 1 sub tank

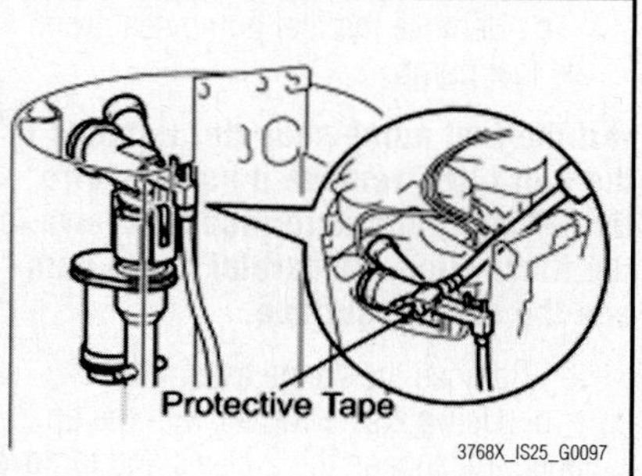

Protective Tape

3768X_IS25_G0097

Fig. 109 Removing the jet pump and the No. 1 fuel sub tank

➡Before beginning this procedure, check for foreign matter on the joint clips. Clean if necessary.

c. Remove the 2 tube joint clips, fuel tank main tube and fuel tank return vent tube.

➡Keep the O-ring free of foreign matter, as it becomes contaminated easily.

➡Do not use any tools in this procedure.

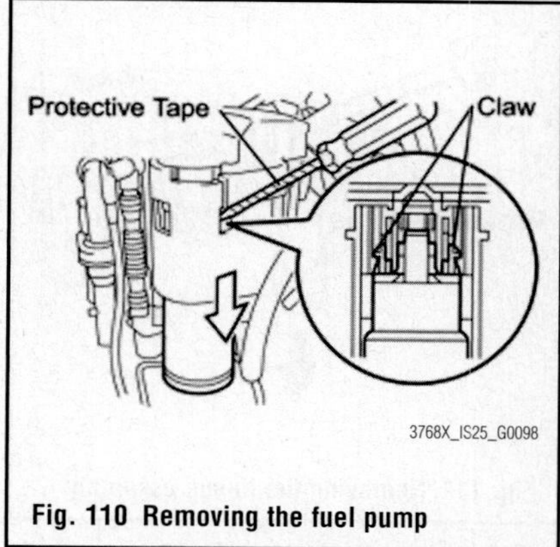

Fig. 110 Removing the fuel pump

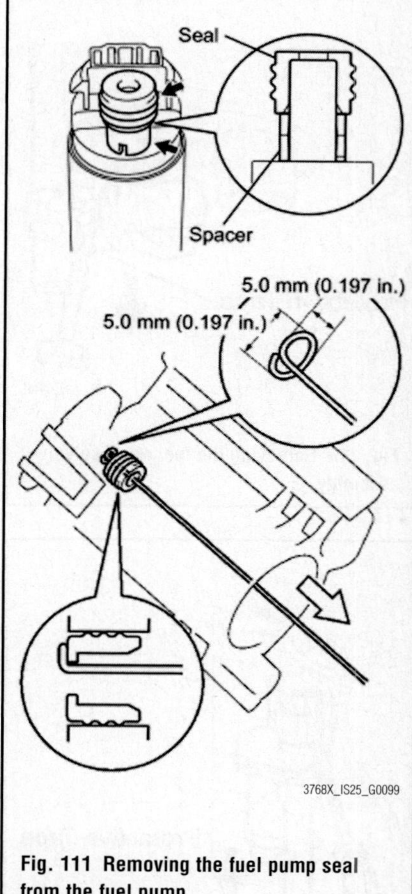

Fig. 111 Removing the fuel pump seal from the fuel pump

➡Do not forcefully bend or twist the tube.

➡Put the tube in a plastic bag to prevent damage and contamination.

➡If the fuel suction plate and tube are stuck together, pinch the tube and turn it carefully to disconnect it.

➡Be careful not to damage any clips. If a clip is damaged, replace it.

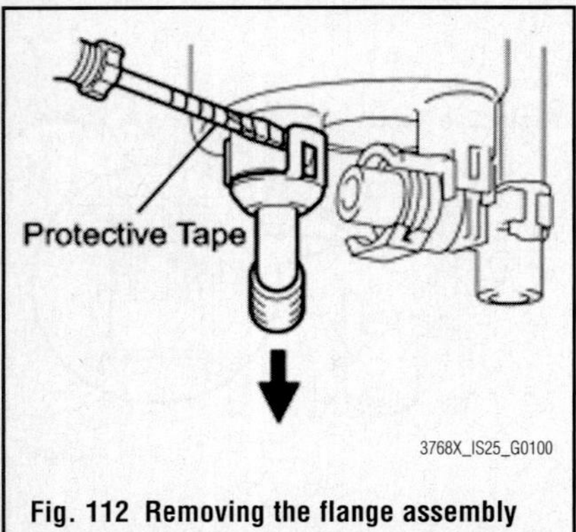

Fig. 112 Removing the flange assembly

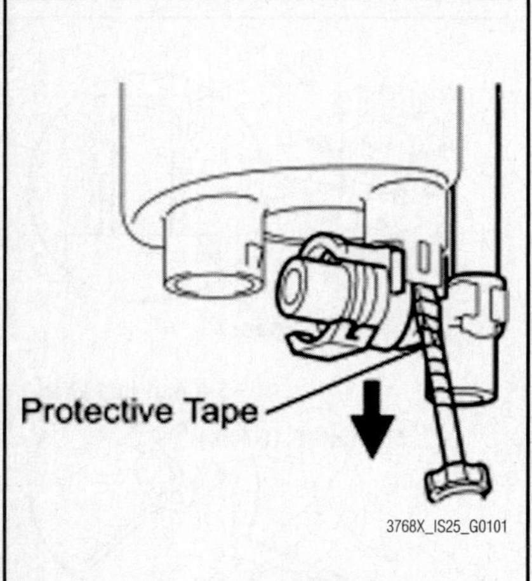

Fig. 113 Removing the fuel relief valve assembly

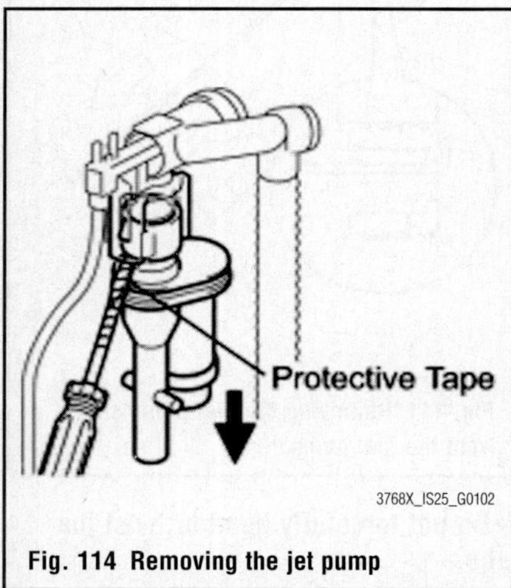

Fig. 114 Removing the jet pump

d. Remove the 8 bolts and the fuel tank vent tube set plate.

➡While holding the fuel suction tube by hand, remove the fuel tank vent tube set plate.

e. Disconnect the fuel tube.

f. Remove the fuel suction tube from the fuel tank.

➡Make sure that the fuel sender gauge arm does not bend.

➡Do not damage the No. 2 fuel hose.

6. Remove the fuel sender gauge assembly.
   a. Disconnect the fuel sender gauge connector.
   b. Press down on the sender gauge claw labeled A. Then slide the sender gauge upward.
7. Remove the No. 1 fuel sub tank.
   a. Using needle nozzle pliers, remove the E-ring.
   b. Using a screwdriver with the tip taped disengage the claw and remove the jet pump nozzle.
   c. Using a screwdriver with the tip taped, disengage the 3 claws and remove the No. 1 sub tank.
   d. Using a clip remover with the tip taped, remove the jet pump and No. 1 fuel sub-tank.

➡The O-ring is installed firmly between the jet pump and No. 1 fuel sub-tank. Therefore, the jet pump and No. 1 fuel sub-tank should be removed carefully using a clip remover.

8. Remove the fuel pump.
   a. Disconnect the 2 connectors and terminal and remove the fuel pump wire.
   b. Using a screwdriver with the tip taped, detach the 2 claws from the claw holes and remove the fuel pump.
   c. Remove the fuel pump seal from the fuel pump.

➡If the fuel pump seal still remains in the fuel filter, remove it using a wire tip (φ1 mm) that is formed as shown in the illustration. Be careful not to damage the sealing surface.

9. Remove the flange assembly.
   a. Using a screwdriver with the tip taped, disengage the 2 claws and remove the flange assembly.
10. Remove the fuel relief valve assembly.
    a. Using a screwdriver with the tip taped, disengage the claw and remove the fuel relief valve assembly.
11. Remove the jet pump.
    a. Using a screwdriver with the tip taped, disengage the 2 claws and remove the jet pump.
12. Remove the fuel pressure regulator assembly.
    a. Remove the clip and fuel pressure regulator from the relief valve.
    b. Remove the 2 O-rings from the fuel pressure regulator.

To install:

13. Install the fuel pressure regulator assembly.
    a. Apply a light coat of gasoline to 2 new O-rings, and install them to the fuel pressure regulator.
    b. Install the fuel pressure regulator to the fuel relief valve.
    c. Install the clip to the relief valve.
14. Install the jet pump.
    a. Apply a light coat of gasoline to a new O-ring, and install the O-ring and jet pump to the fuel relief valve assembly.
15. Install the fuel relief valve assembly.
    a. Apply a light coat of gasoline to a new O-ring, and install the O-ring and fuel relief valve assembly to the fuel filter.
16. Install the flange assembly.
    a. Apply a light coat of gasoline to a new O-ring, and install the O-ring and flange assembly to the fuel filter.
17. Install the fuel pump.
    a. Install the fuel pump spacer to the fuel pump.
    b. Apply a light coat of gasoline to a new seal, and install it to the fuel pump.
    c. Install the fuel pump to the fuel filter.

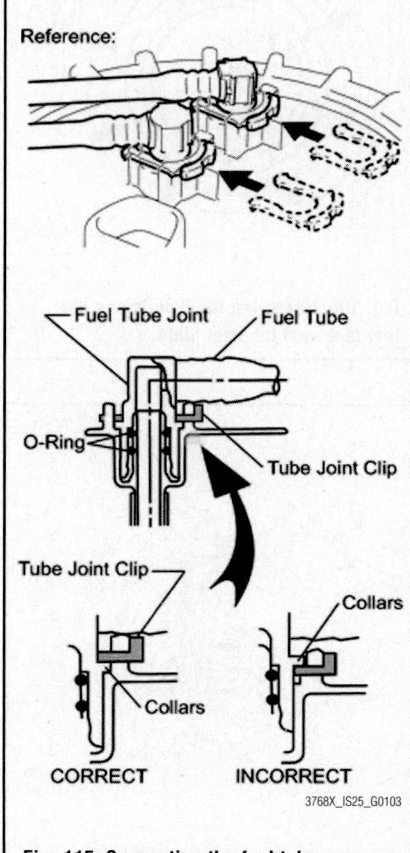

Fig. 115 Connecting the fuel tube

d. Connect the 2 connectors and terminal.

18. Install the No. 1 fuel sub tank.

a. Apply a light coat of gasoline to a new O-ring, and install the O-ring and No. 1 fuel sub-tank to the fuel filter with flange assembly.

b. Connect the jet pump nozzle to the No. 1 fuel sub-tank.

c. Install a new E-ring.

19. Install a fuel sender gage assembly.

a. Set the fuel sender gauge to the No. 1 fuel sub-tank. Then slide the sender gauge downward to install it.

b. Connect the fuel sender gauge connector.

20. Install the fuel suction with the pump and gauge tube assembly.

a. Install a new gasket onto the fuel tank.

b. Connect the fuel hose with the clip.

c. Set the fuel suction tube to the fuel tank.

➡Make sure that the fuel sender gauge arm does not bend.

➡Do not damage the No. 2 fuel hose.

d. Align the protrusion of the fuel suction tube and the cutout of the fuel tank vent tube set plate.

e. While holding the fuel suction tube by hand, install the fuel tank vent tube to the fuel tank with the 8 bolts. Tighten to 53 inch lbs. (6 Nm).

f. Connect the fuel tube. Push the fuel tube joint in the plug of the fuel suction plate, then install the 2 tube joint clips.

➡Check that there are no scratches or foreign objects on the connecting parts.

➡Check that the fuel tube joint is inserted securely.

➡Check that the tube joint clips are on the collars of the fuel tube joints.

➡After installing the tube joint clips, check that the fuel tube joints have not been pulled off.

➡Be careful not to damage any clips. If a clip is damaged, replace it.

g. Connect the fuel suction tube connector.

21. Connect the cable to the negative battery terminal.

22. Check for fuel leaks.

23. Install the No. 2 rear floor service hole cover.

a. Install the No. 2 rear floor service hole cover with new butyl tape.

24. Install the rear seat cushion assembly.

25. Perform the initialization.

➡Certain systems need to be initialized after reconnecting the cable to the negative (-) battery terminal.

High Pressure

See Figures 116 through 118.

1. Remove the cool air intake duct seal.
2. Remove the engine room side cover LH.
3. Remove the engine room side cover RH.
4. Remove the v-bank cover sub assembly.
5. Remove the front upper fender protector LH.
6. Remove the front upper fender protector RH.
7. Remove the roof drip side finish moulding LH.
8. Remove the roof drip side finish moulding RH.
9. Remove the front wiper arm head cap.
10. Remove the front wiper arm and blade assembly LH.
11. Remove the front wiper arm and blade assembly RH.
12. Remove the cowl top ventilator louver sub assembly.
13. Discharge the fuel system pressure.
14. Disconnect the cable from the negative battery terminal.
15. Drain the engine coolant.
16. Disconnect the No. 2 ventilation hose.
17. Remove the air cleaner cap with the air cleaner hose.
18. Remove the throttle with the motor body assembly.
19. Remove the intake air surge tank.
20. Remove the intake manifold.

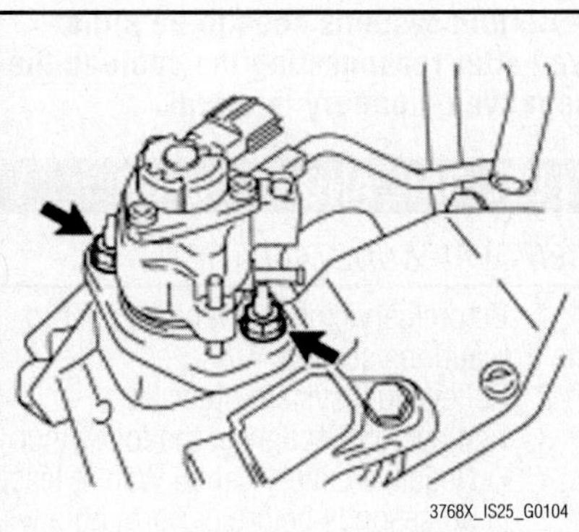

Fig. 116 Removing the fuel pump assembly

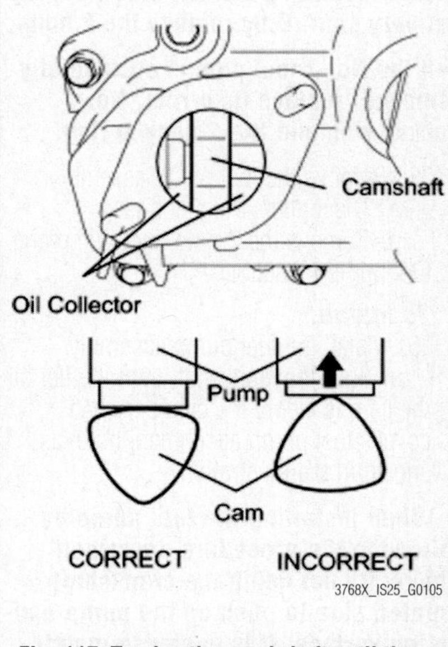

Fig. 117 Turning the crankshaft until the flat of the cam is facing the cylinder head cover's fuel pump attachment hole

21. Disconnect the water hose joint.
22. Remove the fuel pressure pulsation damper.
23. Remove the No. 1 fuel pipe.
24. Disconnect the No. 2 fuel pipe.

a. Disconnect the fuel high pressure side fuel pump connector.

b. Fix the union bolt on the fuel pump side in place with a 21 mm wrench. Using a 19 mm union nut wrench, loosen the union and remove the fuel pipe.

➡There must be absolutely no free play in the union on the fuel pump side.

➡If the union on the fuel pumps side has free play, replace the fuel pump.

c. Remove the 2 bolts on the delivery pipe side.

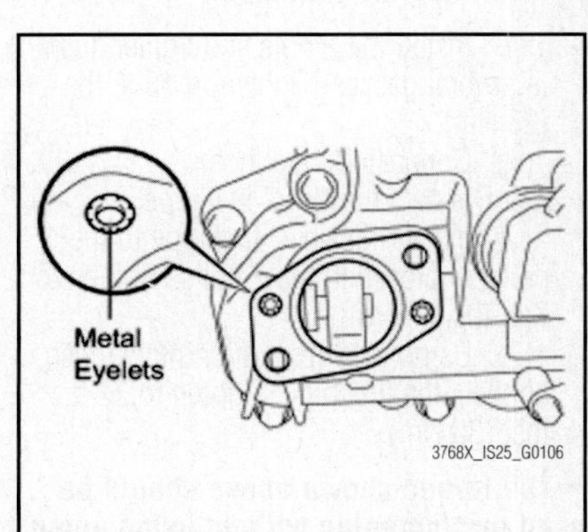

Fig. 118 Installing a new fuel pump insulator to the cylinder head cover

➡Do not remove the fuel pipe from the delivery pipe. Only remove the 2 bolts.

➡If the No. 2 fuel pipe is accidentally removed, replace its o-ring, No. 1 backup ring and No. 2 backup ring.

25. Remove the fuel pump assembly.
   a. Disconnect the fuel hose.
   b. Remove the 2 nuts, fuel pump and fuel pump insulator.

To install:

26. Install the fuel pump assembly.
   a. Turn the crankshaft until the flat of the cam is facing the cylinder head cover's fuel pump attachment hole, as shown in the illustration.

➡When installing the fuel pump by following the procedure described above: By not using the crankshaft pointed side to push up the pump activation surface, it is easier to install the fuel pump and No. 2 fuel pipe later.

   b. Pour 30 cc of engine oil through the cylinder head cover's fuel pump attachment hole into the cylinder head oil collector.
   c. Apply a coat of engine oil to the pump activation cam and pump lifter part.
   d. Install a new fuel pump insulator to the cylinder head cover. Then pass the 2 stud bolts through the holes of the fuel pump and set them on the insulator.

➡Install the insulator so that the open sides of the metal eyelets are facing outward, as shown in the illustration.

   e. Temporarily install the No.2 fuel pipe sub-assembly to the fuel pump assembly.

➡Be careful not to damage the sealing surface of the fuel pipe when temporarily installing the fuel pipe.

   f. Install the 2 nuts and tighten them in several passes. Tighten to 18 ft. lbs. (25 Nm).
   g. Connect the fuel hose.
27. Connect the No. 2 fuel pipe.
   a. Install the No. 2 fuel pipe to the delivery pipe with the 2 bolts. Tighten to 7 ft. lbs. (10 Nm).
   b. Using a 19 mm union nut wrench, connect the fuel pipe. Tighten to 22 ft. lbs. (30 Nm).

➡The torque shown above should be used for tightening without using union nut wrench. When the union nut wrench is used for tightening, the torque should be calculated based on the length of the union nut wrench.

   c. Connect the connector to the fuel pump.
28. Install the No. 1 fuel pipe.
29. Install a fuel pressure pulsation damper.
30. Connect the water hose joint.
31. Install the intake manifold.
32. Install the intake air surge tank.
33. Install the throttle with the motor body assembly.
34. Install the air cleaner cap with the air cleaner hose.
35. Connect the No. 2 ventilation hose.
36. Connect the cable to the negative battery terminal.
37. Add engine coolant.
38. Check for engine coolant leaks.
39. Check for fuel leaks.
40. Install the cowl top ventilator louver sub assembly.
41. Install the front wiper arm and blade assembly LH.
42. Install the front wiper arm and blade assembly RH.
43. Install the front wiper arm head cap.
44. Install the roof drip side finish moulding LH.
45. Install the roof drip side finish moulding RH.
46. Install the front upper fender protector LH.
47. Install the front upper fender protector RH.
48. Install the v-bank cover sub assembly.
49. Install the engine room side cover LH.
50. Install the engine room side cover RH.
51. Install the cool air intake duct seal.
52. Check the function of the throttle with the motor body assembly.
53. Perform initialization.

➡Certain systems need to be initialized after reconnecting the cable to the negative (-) battery terminal.

FUEL RAIL AND INJECTOR

REMOVAL & INSTALLATION

1. Before servicing the vehicle, refer to the Precautions section.
2. Relieve fuel system pressure.
3. Remove or disconnect the following:
   - Negative battery cable. Wait at least 90 seconds before performing any other work.
   - Coolant
   - Cool air intake duct seal
   - Left and right engine room side covers
   - V-bank cover
   - Left and right front upper fender protectors
   - Left and right roof drip side finish moldings using a moulding removal tool
   - Front wiper arm head cap
   - Left and right windshield wiper arm and blade assemblies
   - Cowl top ventilator louver assembly
   - No. 2 ventilation hose
   - Air cleaner cap with air cleaner hose
   - Throttle body assembly
   - Cold start injector
   - Intake air surge tank
   - Intake manifold
   - Water hose joint
   - Fuel pressure pulsation damper
   - No. 3 fuel hose from the No. 1 fuel pipe and remove fuel pipe
   - No. 2 and 3 fuel pipes
   - No. 1 and 2 fuel delivery pipes
   - Fuel injectors, O-rings and seals

To install:

4. Install 2 new seals to each injector.
5. Install the fuel injectors as follows:
   - Install a new O-ring, new backup rings (No. 1, No. 2, No. 3) and new E-ring to the fuel injector.

⁂ WARNING

Check that there is no foreign matter or damaged areas in the injector's O-ring groove. Check that the installation direction of the No. 1 and No. 2 backup ring are correct. Make sure the backup rings and O-ring are installed in the correct order. Check that the alignment openings of the backup rings are not overlapped or stretched. After installing the O-ring, check that it is not contaminated with foreign matter and is not damaged.

   - Install the injector nozzle holder clamp.
   - Apply gasoline to the O-ring. Install the nozzle holder clamp by aligning the protruding part of the clamp to the notch of the delivery pipe.

⁂ WARNING

Make sure there is no gap between the delivery pipe and clamp. Check that there is no foreign matter or damage in the injector insertion hole

of the delivery pipe. Insert the injector straight into the delivery pipe without tilting it.

- No. 1 and 2 fuel delivery pipes. Torque the fasteners to 15 ft. lbs. (21 Nm).
- No. 2 and 3 fuel pipes and new O-rings. Torque No. 3 fuel pipe fastener to 7 ft. lbs. (10 Nm).
- No. 1 fuel pipe. Torque fasteners to 7 ft. lbs. (10 Nm).
- Fuel pressure pulsation damper and new gasket. Torque to 28 ft. lbs. (40 Nm).
- Water hose joint. Torque fasteners to 7 ft. lbs. (10 Nm).
- Intake manifold
- Intake air surge tank
- Cold start injector. Torque the bolts to 7 ft. lbs. (10 Nm).
- New gasket and throttle body assembly. Torque the 4 bolts to 7 ft. lbs. (10 Nm)
- Air cleaner cap with air cleaner hose
- No. 2 ventilation hose
- Cowl top ventilator louver assembly
- Left and right windshield wiper arm and blade assemblies
- Front wiper arm head cap
- Left and right roof drip side finish moldings
- Left and right front upper fender protectors
- V-bank cover
- Left and right engine room side covers
- Cool air intake duct seal
- Negative battery cable

6. Refill the cooling system. Start the engine and check for coolant and fuel leaks and proper operation.
7. Check the function of the throttle body unit
8. System initialization

FUEL TANK

REMOVAL & INSTALLATION

See Figures 119 through 125.

1. Discharge fuel system pressure.
2. Disconnect the cable from the negative battery terminal.
3. Remove the rear seat cushion assembly.
4. Remove the No. 2 rear floor service hole cover.
5. Remove the fuel suction with the pump and gauge tube assembly.
6. Drain the fuel.
7. Remove the floor service hole cover.
   a. Remove the rear floor service hole cover.
8. Remove the fuel sender gauge assembly.
   a. Disconnect the fuel sender gauge connector.
   b. Remove the 5 bolts and fuel sender gauge.
   c. Remove the fuel sender gauge from the fuel tank.

➡Be careful not to bend the arm of the fuel sender gauge.

9. Remove the front exhaust pipe assembly.
10. Remove the propeller shaft with the center bearing assembly.
11. Remove the No. 2 differential support protector.
12. Remove the No. 1 differential support protector.

➡Removal procedure of the No. 1 differential support protector is the same as that of the No. 2 differential support protector.

13. Remove the rear floor side member cover LH.
14. Remove the rear floor side member cover RH.
15. Remove the rear suspension member brace LH.
    a. Remove the 4 bolts and suspension member brace lowers LH.
16. Remove the rear suspension member brace RH.
    a. Remove the 4 bolts and suspension member brace lower RH.
17. Disconnect the parking brake cable assembly.
    a. Remove the 2 parking brake cables from the 4 clamps.
    b. Remove the 4 bolts and disconnect the 2 parking brake cables.
18. Disconnect the fuel tube sub assembly.
    a. Disconnect the fuel tank main tube (LH side).

➡Remove any dirt and foreign objects on the fuel tube connector before performing this work.

➡Do not allow any scratches or foreign objects on the parts when disconnecting, as the fuel tube connector has O-rings that seal the pipe.

➡Perform this work by hand. Do not use any tools.

➡Do not forcibly bend, twist or turn the nylon tube.

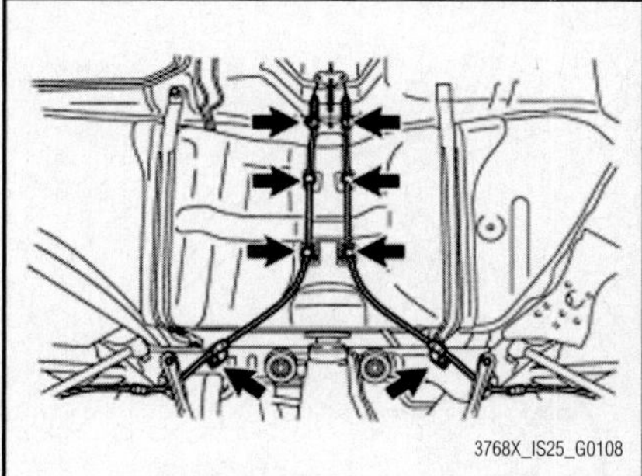

Fig. 119 Disconnecting the parking brake cable assembly

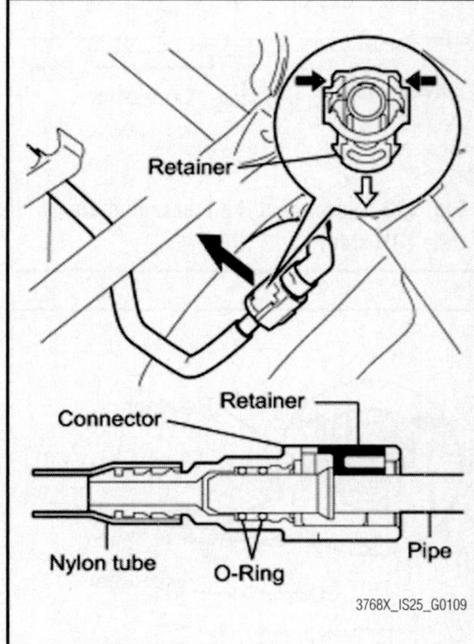

Fig. 120 Disconnecting the fuel tank main tube (LH side)

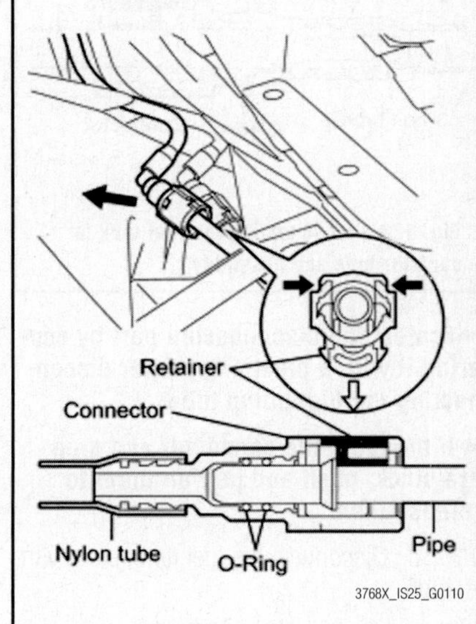

Fig. 121 Disconnect the fuel tank return vent tube

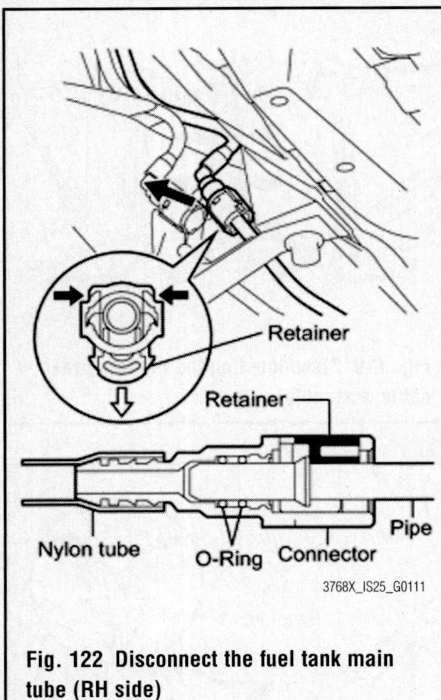

Fig. 122 Disconnect the fuel tank main tube (RH side)

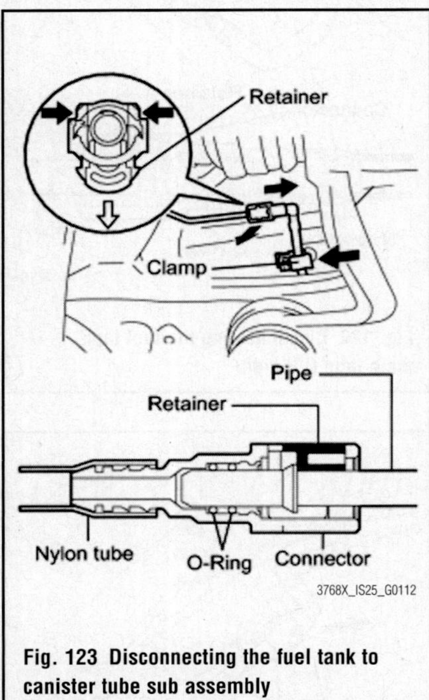

Fig. 123 Disconnecting the fuel tank to canister tube sub assembly

➡Protect the disconnected part by covering it with a plastic bag after disconnecting the fuel pump tube.

➡If the fuel tube connector and pipe are stuck, push and pull on them to release them.

b. Disconnect the fuel tank return vent tube.

➡Remove any dirt and foreign objects on the fuel tube connector before performing this work.

➡Do not allow any scratches or foreign objects on the parts when disconnecting, as the fuel tube connector has O-rings that seal the pipe.

➡Perform this work by hand. Do not use any tools.

➡Do not forcibly bend, twist or turn the nylon tube.

➡Protect the disconnected part by covering it with a plastic bag after disconnecting the fuel pump tube.

➡If the fuel tube connector and pipe are stuck, push and pull on them to release them.

c. Disconnect the fuel tank main tube (RH side).

➡Remove any dirt and foreign objects on the fuel tube connector before performing this work.

➡Do not allow any scratches or foreign objects on the parts when disconnecting, as the fuel tube connector has O-rings that seal the pipe.

➡Perform this work by hand. Do not use any tools.

➡Do not forcibly bend, twist or turn the nylon tube.

➡Protect the disconnected part by covering it with a plastic bag after disconnecting the fuel pump tube.

➡If the fuel tube connector and pipe are stuck, push and pull on them to release them.

19. Disconnect the fuel tank to filler pipe hose.

a. Loosen the hose clamp bolt and disconnect the fuel tank to filler pipe hose.

20. Disconnect the No. 3 fuel hose sub assembly.

a. Disconnect the fuel hose clamp.

b. Remove the clip and disconnect the No. 3 fuel sub assembly.

21. Disconnect the fuel tank to canister tube sub assembly.

a. Disconnect the fuel tube clamp.

b. Disconnect the fuel tank to canister tube sub assembly.

➡Remove any dirt and foreign objects on the fuel tube connector before performing this work.

➡Do not allow any scratches or foreign objects on the parts when disconnecting, as the fuel tube connector has O-rings that seal the pipe.

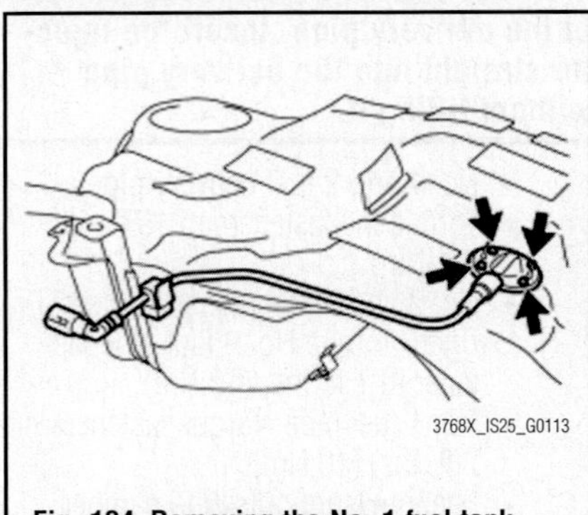

Fig. 124 Removing the No. 1 fuel tank breather tube sub assembly

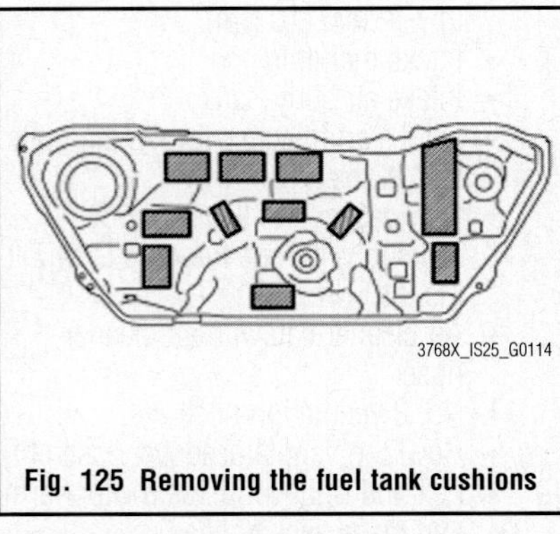

Fig. 125 Removing the fuel tank cushions

➡Perform this work by hand. Do not use any tools.

➡Do not forcibly bend, twist or turn the nylon tube.

➡Protect the disconnected part by covering it with a plastic bag after disconnecting the fuel pump tube.

➡If the fuel tube connector and pipe are stuck, push and pull on them to release them.

22. Remove the fuel tank assembly.

a. Place an engine lifter under the fuel tank assembly.

b. Remove the 4 bolts, fuel tank bands and fuel tank assembly.

23. Remove the No. 1 fuel tank breather tube sub assembly.

a. Remove the 4 bolts, No. 1 fuel tank breather tube sub assembly and gasket.

24. Remove the fuel tube sub assembly.

a. Disconnect the clamp and remove the fuel tank main tube and fuel tank return vent tube.

b. Disconnect the 2 clamps and remove the edge protector.

25. Remove the fuel tank to filler pipe hose.

a. Loosen the hose clamp bolt and remove the fuel tank pipe hose from the fuel tank.

26. Remove the fuel tank vent hose.
    a. Push the fuel tank vent hose deep into the fuel tank to release the lock pin.
    b. Pinch portion A.
    c. Pull out the fuel tank vent hose.
    d. Remove the fuel vapor containment valve from the fuel tank.
27. Remove the fuel tank cushion.
    a. Remove the 11 fuel tank cushions.
28. Remove the No. 2 fuel tank protector.
    a. Remove the No. 2 fuel tank protector from the fuel tank assembly.

To install:

29. To install, reverse the removal procedure. Tighten the fuel tank breather tube sub assembly bolts to 13 inch lbs. (1.5 Nm). Tighten the fuel tank bands to 29 ft. lbs. (39 Nm).

IDLE SPEED

ADJUSTMENT

Idle speed is maintained by the Powertrain Control Module (PCM). No adjustment is necessary or possible.

THROTTLE BODY

REMOVAL & INSTALLATION

See Figure 126.

1. Before servicing the vehicle, refer to the Precautions section.
2. Disconnect the negative battery cable. Wait at least 90 seconds before performing any other work.
3. Remove the cool air intake duct seal.

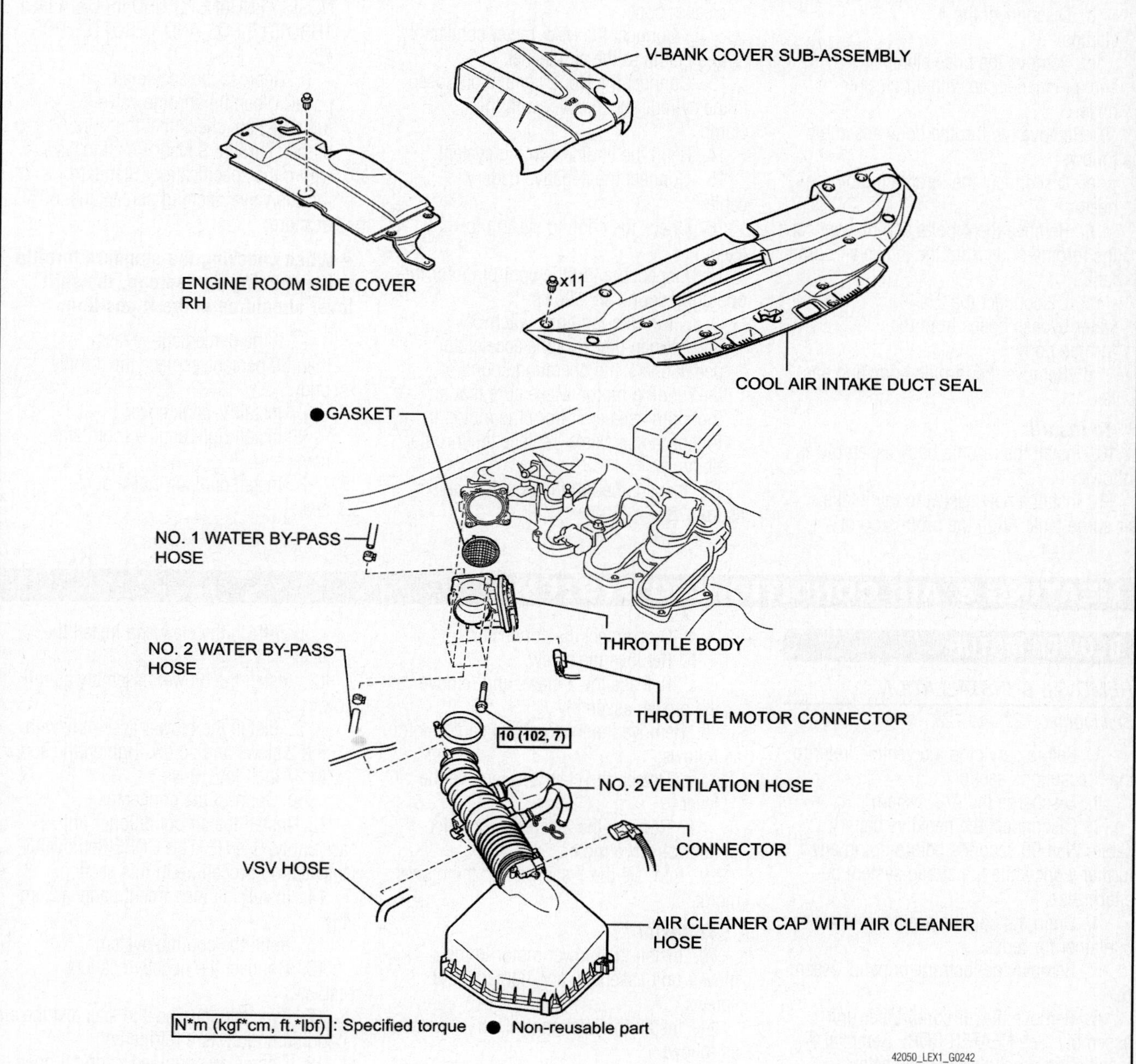

Fig. 126 Exploded view of the throttle body assembly and related components

4. Drain the engine coolant.
5. Remove the right engine room side cover.
6. Remove the V-bank cover.
7. Disconnect the No. 2 ventilation hose from the cylinder head.
8. Remove the air cleaner cap with air cleaner hose as follows:
   a. Disconnect the MAF meter connector.
   b. Disconnect the clamp from the air cleaner.
   c. Disconnect the VSV hose.
   d. Disconnect the 4 clamps.
   e. Remove the hose clamp and air cleaner cap with air cleaner hose.
9. Remove the throttle body assembly as follows:
   a. Disconnect the throttle motor connector.
   b. Remove the 4 bolts and disconnect the throttle body from the intake air surge tank.
   c. Disconnect the 2 water bypass hoses from the throttle body.
   d. Remove the throttle body and gasket.

To install:

10. Install the throttle body assembly as follows:
11. Install a new gasket to the intake air surge tank. Align the protrusion of the gasket on the intake air surge tank.
   a. Connect the 2 water by-pass hoses to the throttle body.
   b. Install the throttle body with the 4 bolts. Tighten the 4 bolts to 7 ft. lbs. (10 Nm).
   c. Connect the throttle motor connector.
12. Install air cleaner cap with air cleaner hose as follows:
   a. Install the air cleaner cap with air cleaner hose assembly with the 4 clamps and hose clamp.
   b. Install the VSV hose to the air cleaner hose.
   c. Connect the MAF meter connector and clamp to the air cleaner.
13. Connect the No. 2 ventilation hose to the cylinder head cover with the clamp.
14. Refill the engine cooling system.
15. Connect the negative battery cable.
16. Check the cooling system for leaks.
17. Inspect the throttle control motor for operating sound as follows:
   a. Turn the ignition switch ON.
   b. When pressing the accelerator pedal, check the operating sound of the running motor. Make sure that no friction noises emit from the motor. If friction noise exists, replace the throttle body.
18. Inspect the throttle position sensor as follows:
   a. Connect the hand-held tester or OBD II scan tool to the DLC3.
   b. Turn the ignition switch ON.
   c. Check that the check engine warning light does not light up.
   d. Check that the throttle valve opening percentage (THROTTLE POS) of the CURRENT DATA shown the standard value.
   e. Standard throttle valve opening percentage: 60 percent or more
   f. Turn the intelligent tester main switch ON.
   g. Enter the following menus: DIAGNOSIS / ENHANCED OBD II / DATA LIST / THROTTLE POS AND THROTTLE POS #2.
   h. Depress the accelerator pedal. When the throttle valve is fully opened, check that the value of the "Throttle Sensor Position" is within the specification. Standard throttle valve opening percentage: 60% or more.

➡When checking the standard throttle valve opening percentage, the shift lever should be in the N position.

   i. If the percentage is less than 60 percent, replace the throttle body.
   j. Install V-bank cover.
   k. Install right engine room side cover.
   l. Install cool air intake duct seal.

HEATING & AIR CONDITIONING SYSTEM

BLOWER MOTOR

REMOVAL & INSTALLATION

See Figures 127 and 128.

1. Before servicing the vehicle, refer to the Precautions section.
2. Discharge the A/C system.
3. Disconnect the negative battery cable. Wait 90 seconds before doing any further work while the airbag system de-energizes.
4. Drain the cooling system into a clean container for reuse.
5. Remove the instrument panel assembly.
6. Remove the air conditioner unit assembly. See HEATER CORE Removal & Installation procedure in this section.
7. Remove the blower assembly as follows:
   a. Disconnect the connector.
   b. Remove the screw.
   c. Release the 2 claws and remove the blower assembly.
8. Remove the inlet control servo motor as follows:
   a. Detach the claw and remove the lever.
   b. Remove the 2 screws and inlet control servo motor.
9. Remove the 3 screws and the blower motor.

To install:

10. Install the blower motor to the blower unit assembly and tighten the 3 screws.
11. Install the inlet control servo motor as follows:
   a. Install the inlet control servo motor with the 2 screws.
   b. Attach the claw and install the lever.
12. Install the blower assembly as follows:
   a. Install the blower assembly with the 2 claws and screw. Tighten the screw to 27 inch lbs. (3 Nm).
   b. Connect the connector.
13. Install the air conditioner unit assembly. See HEATER CORE Removal & Installation procedure in this section.
14. Install the instrument panel assembly.
15. Refill the cooling system.
16. Connect the negative battery cable.
17. Evacuate, charge and leak test the air conditioning system refrigerant.
18. Operate the engine to normal operating temperatures; then, check the climate control operation and check for leaks.

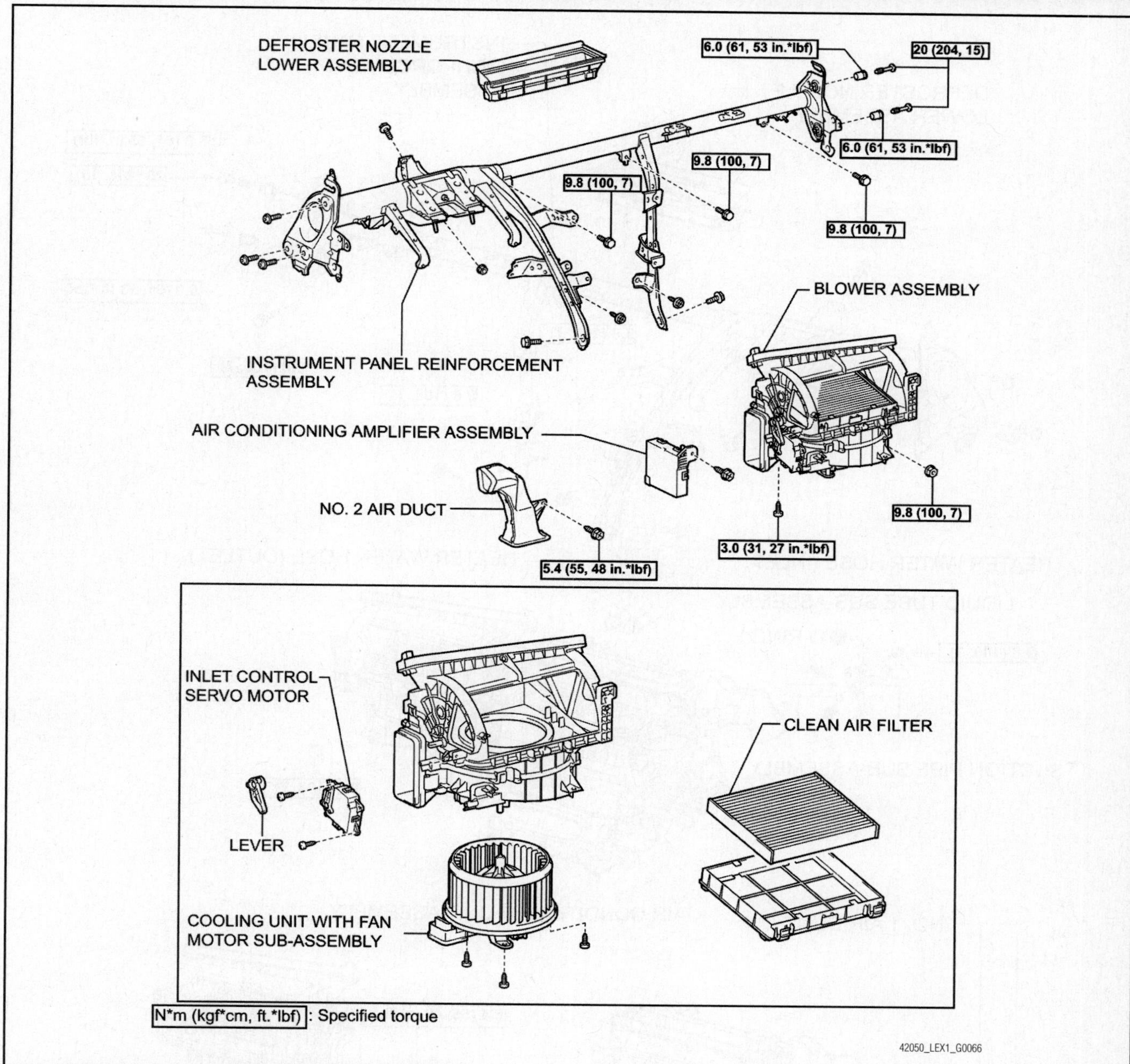

DEFROSTER NOZZLE LOWER ASSEMBLY
6.0 (61, 53 in.*lbf)
20 (204, 15)
6.0 (61, 53 in.*lbf)
9.8 (100, 7)
9.8 (100, 7)
9.8 (100, 7)
BLOWER ASSEMBLY
INSTRUMENT PANEL REINFORCEMENT ASSEMBLY
AIR CONDITIONING AMPLIFIER ASSEMBLY
NO. 2 AIR DUCT
9.8 (100, 7)
3.0 (31, 27 in.*lbf)
5.4 (55, 48 in.*lbf)
INLET CONTROL SERVO MOTOR
CLEAN AIR FILTER
LEVER
COOLING UNIT WITH FAN MOTOR SUB-ASSEMBLY
N*m (kgf*cm, ft.*lbf) : Specified torque

42050_LEX1_G0066

Fig. 127 Exploded view of the blower assembly components

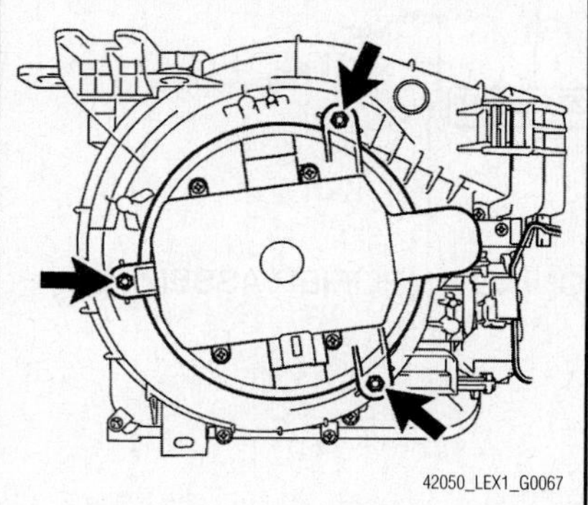

42050_LEX1_G0067

Fig. 128 Blower motor screw locations

HEATER CORE

REMOVAL & INSTALLATION

See Figures 129 and 130.

1. Before servicing the vehicle, refer to the Precautions section.
2. Discharge the A/C system.
3. Disconnect the negative battery cable. Wait 90 seconds before doing any further work while the airbag system de-energizes.
4. Drain the cooling system into a clean container for reuse.
5. Align the front wheels facing straight ahead.
6. Remove or disconnect the following:
    a. Cool air intake duct seal
    b. Left and right engine room side covers
7. Remove the left and right front upper fender protector as follows:
    a. Using a clip remover, separate the clip on the rubber portion of the cowl top ventilator louver subassembly from the front upper fender protector.
    b. Disengage the 3 clips and the claw to remove the front upper fender protector.
8. Remove the left and right roof drip side finish moulding as follows:

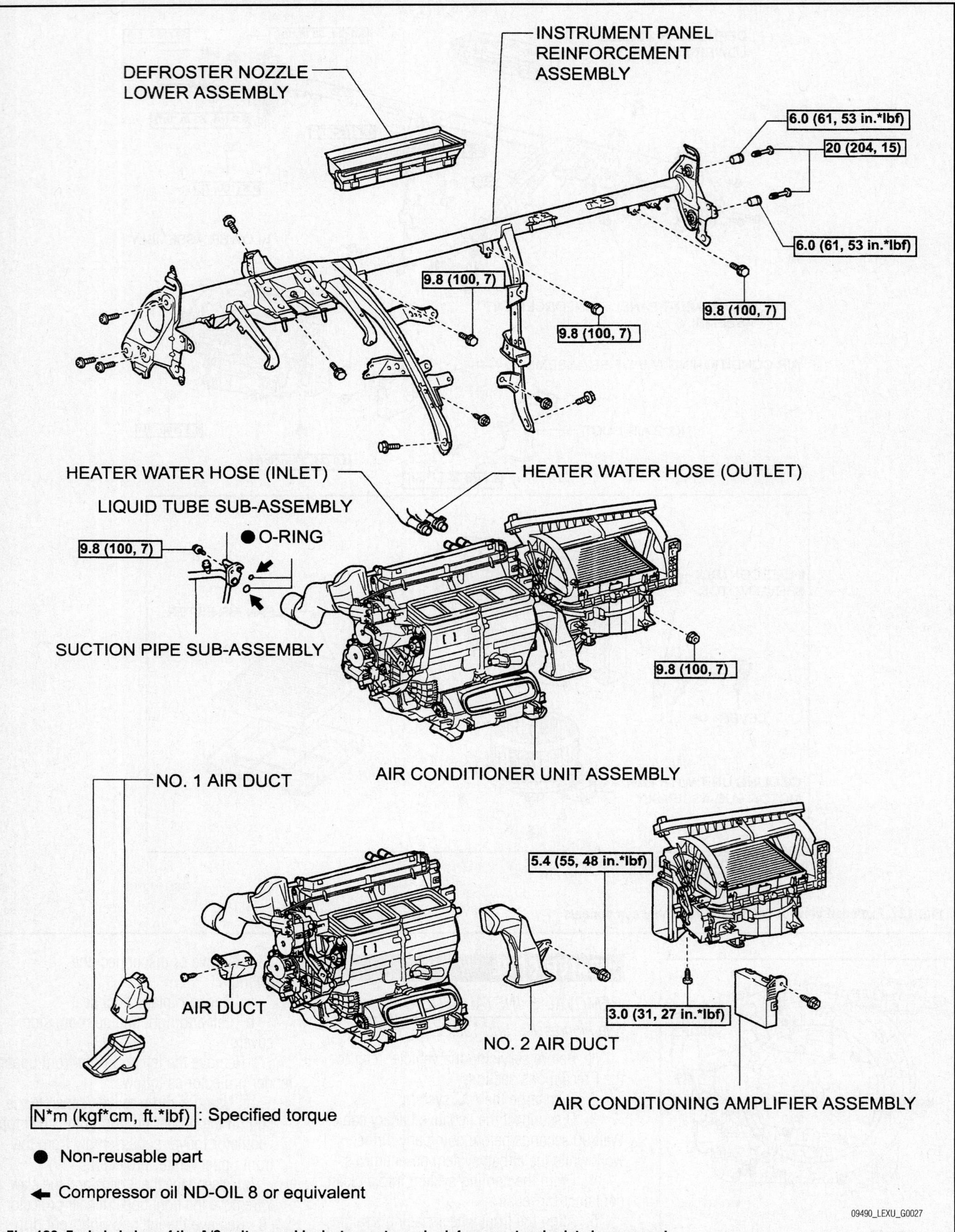

Fig. 129 Exploded view of the A/C unit assembly, instrument panel reinforcement and related components

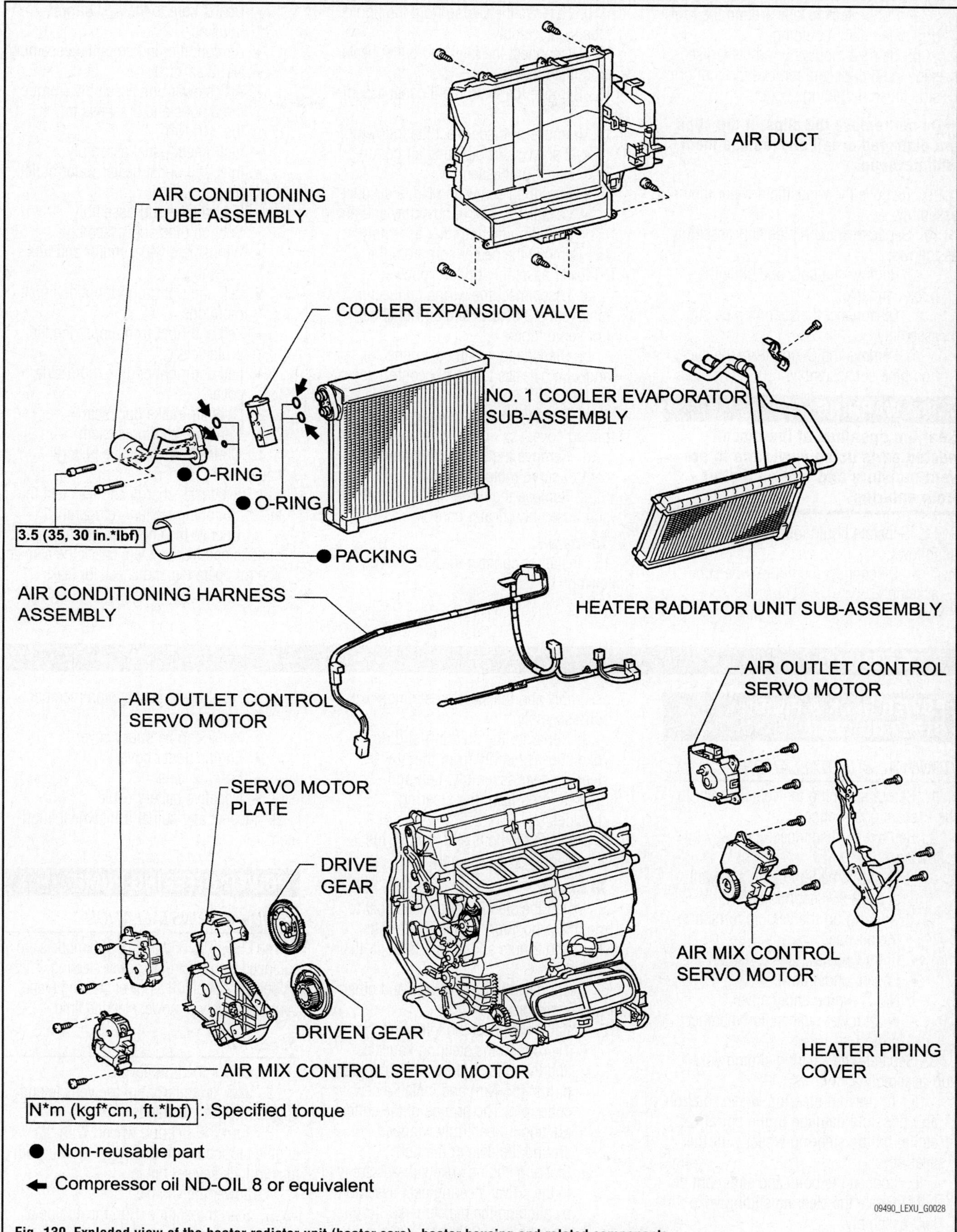

Fig. 130 Exploded view of the heater radiator unit (heater core), heater housing and related components

a. Put protective tape around the roof drip side finish moulding.

b. Using a moulding remover, disengage the 6 clips and remove the roof drip side finish moulding.

➡Do not remove the clips. If the clips are damaged or fall off, replace them with new clips.

9. Remove the windshield wiper motor assembly.

10. Separate suction pipe sub-assembly as follows:

a. Remove the bolt, and slide the hook connector.

b. Disconnect the suction pipe sub-assembly.

c. Remove the O-ring from the suction pipe subassembly.

⁂ WARNING

Seal the openings of the disconnected parts using vinyl tape to prevent moisture and foreign matter from entering.

11. Separate liquid tube sub-assembly as follows:

a. Disconnect the liquid tube sub-assembly.

b. Remove the O-ring from the liquid tube subassembly.

12. Disconnect the inlet and outlet heater water outlet hoses

13. Remove the instrument panel assembly.

14. Remove or disconnect the following:

- 2 screws, 3 bolts, and air conditioner unit assembly
- Mounting screw and No. 2 air duct
- Connector, mounting screw and air conditioning amplifier assembly

15. Remove the heater core from the A/C-blower assembly unit as follows:

a. Disconnect the wiring connector, remove the 3 screws and air outlet control servo motor.

b. Disengage 4 clamps, connector and wire harness to air mix control servo motor.

c. Remove the 3 screws and heater piping cover.

d. Remove the 3 screws and air mix control servo motor.

e. Remove the heater radiator unit sub-assembly (heater core).

To install:

16. Install or connect the following:

- Heater core to the A/C blower housing
- Air conditioning amplifier assembly
- No. 2 air duct
- A/C blower unit assembly. Torque the 3 bolts and 2 screws to 7 ft. lbs. (10 Nm).
- Instrument panel assembly
- Inlet and outlet heater water outlet hoses
- Liquid tube sub-assembly
- Suction pipe sub-assembly
- Windshield wiper motor and link assembly
- Left and right roof drip side finish moldings
- Left and right front upper fender protectors
- Left and right engine room side covers
- Cool air intake duct seal

a. Refill the cooling system.

b. Connect the negative battery cable.

c. Evacuate, charge and leak test the air conditioning system refrigerant.

d. Operate the engine to normal operating temperatures; then, check the climate control operation and check for leaks.

STEERING

POWER RACK & PINION STEERING GEAR

REMOVAL & INSTALLATION

1. Before servicing the vehicle, refer to the Precautions section.

2. Remove or disconnect the following:

- Negative battery cable and wait at least 90 seconds before working on the vehicle to disarm the air bag.
- Front wheels
- Engine under cover
- No. 2 engine under cover
- Front lower suspension member protector

3. Separate the steering sliding yoke sub-assembly as follows:

a. To prevent steering wheel rotation and possible damage to the spiral cable, fix the steering wheel with the seat belt.

b. Loosen the bolts and slide, but do not separate the steering sliding yoke sub-assembly.

c. Matchmark the power steering gear assembly and the steering sliding yoke sub-assembly.

d. Separate the steering sliding yoke sub-assembly from the power steering gear assembly. Left and right tie rods from the steering knuckles

4. Remove the bolts and remove the power steering gear assembly.

To install:

5. Install the steering gear assembly to its mounting position. Install the 2 bolts, washers and 2 nuts and torque to 87 ft. lbs. (118 Nm).

6. Reconnect the ground wire and other wiring connectors.

7. Install or connect the following:

- Tie rods to the steering knuckles and the nuts. Tighten the nut to 50 ft. lbs. (65 Nm) and install a new cotter pin. The prongs of the cotter pin should be firmly wrapped around the flats of the nut.
- Steering sliding yoke sub-assembly to the power steering gear assembly by aligning the matchmarks. Torque the 2 bolts to 26 ft. lbs. (35 Nm).
- Front lower suspension member protector
- No. 2 engine under cover
- Engine under cover
- Front wheels
- Negative battery cable

8. Inspect and adjust front wheel alignment.

POWER STEERING PUMP

REMOVAL & INSTALLATION

The LEXUS IS 250 and IS 350 models are equipped with electronic power steering. These models do not utilize a power steering (vane) pump assembly or power steering fluid.

BLEEDING

1. Check the fluid level.

2. Jack up the front of the vehicle and support it with the stands.

3. Turn the steering wheel. With the engine stopped, turn the wheel slowly from lock to lock several times.

4. Lower the vehicle.

5. Start the engine. Run the engine at idle for a few minutes.

6. Turn the steering wheel. With the engine idling, turn the wheel to left or right full lock position and keep it there for 2–3 seconds, then turn the wheel to the opposite full lock position and keep it there for 2–3 seconds.

7. Repeat last step several times.

8. Stop the engine.

9. Check for foaming or emulsification. If the system has to be bled twice specifically because of foaming or emulsification, check for fluid leaks in the system.

10. Check the fluid level.

SUSPENSION

FRONT SUSPENSION

LOWER BALL JOINT

REMOVAL & INSTALLATION

1. Before servicing the vehicle, refer to the Precautions section.
2. Remove the lower control arm.
3. Fix the front lower control arm in a vise using aluminum plates.
4. Remove the clip and castle nut.
5. Use SST 09950-40011 (09951-04010, 09952-04010, 09953-04020, 09954-04010, 09955-04051, 09957-04010, 09958-04011) to remove the front lower ball joint from the front lower suspension arm.
6. Inspect the lower control arm ball joint as follows:
    a. Flip the ball joint stud back and forth 5 times.
    b. Temporarily install the nut, and use a torque wrench to turn the nut continuously at a rate of 3 to 5 seconds per turn. Take the torque reading on the 5th turn. Tighten to 4.4–53 inch lbs (0.5–6 Nm)
    c. Check the dust boots for cracks or grease leakage. If the value is not within the specified range, replace the front lower ball joint with a new one.

To install:

7. Install the front lower ball joint to the front lower control arm with the nut. Ensure that the thread and taper are free of oil or other foreign matter. Tighten the nut to 120 ft. lbs. (162 Nm) for RWD, and 92 ft. lbs. (125 Nm) for AWD.
8. Install a new clip to the front lower ball joint. Further tighten the nut up to 60° if the holes for the cotter pin are not aligned.
9. Install the lower control arm.

LOWER CONTROL ARM

REMOVAL & INSTALLATION

See Figure 131.

1. Before servicing the vehicle, refer to the Precautions section.
2. Remove or disconnect the following:
    - Negative battery cable
    - Front wheel(s)
    - Front speed sensor
    - Tie rod assembly
    - Height control sensor link assembly
    - Lower part of strut from lower control arm
    - Stabilizer link assembly
    - Engine under cover
3. Remove the lower control arm as follows:
    a. Support front suspension crossmember with a transmission jack and a block of wood
    b. Remove the 2 bolts from the front lower ball joint.
    c. Loosen, but do not remove the nut of the lower No. 2 arm bracket assembly.
    d. Remove the bolt, washer and nut on the front of the front lower control arm.
    e. Remove the 4 bolts, side rail plate and front lower suspension arm with the lower No. 2 arm bracket assembly.

To install:

4. Install and temporarily tighten the front lower control arm as follows:
    a. Torque bolt 1 to 150 ft. lbs. (204 Nm); bolt 2 to 63 ft. lbs. (86 Nm); bolt 3 to 37 ft. lbs. (50 Nm).
    b. Install bolt from the front and temporarily tighten the bolt, washer and nut.
    c. Install the front lower ball joint, with 2 bolts and tighten to 89 ft. lbs. (120 Nm).
5. Install or connect the following:
    a. Lower part of strut to lower control arm and temporarily tighten the bolt

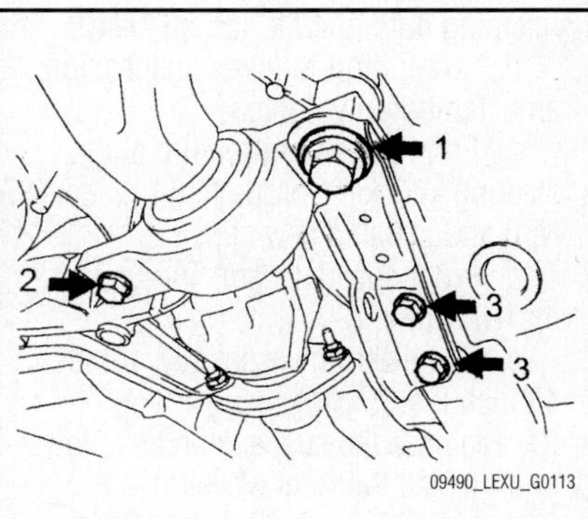

Fig. 131 Lower control arm and side rail plate bolt locations

    b. Stabilizer link assembly. Torque to 48 inch lbs. (5 Nm).
    c. Tie rods to the steering knuckles and the nuts. Tighten the nut to 50 ft. lbs. (65 Nm) and install a new cotter pin. The prongs of the cotter pin should be firmly wrapped around the flats of the nut.
    d. Front speed sensor. Torque to 4 ft. lbs. (6 Nm).
    e. Height control sensor link assembly. Torque to 48 inch lbs. (5 Nm).
6. Stabilize the suspension as follows:
    a. Install the front wheel(s).
    b. Lower the vehicle and bounce it up and down several times to stabilize the front suspension.
    c. Raise the vehicle.
    d. Remove the front wheel.
    e. Jack up the front lower suspension arm placing a wood block in between. Apply a load to the front suspension so that the front lower suspension arm is placed in a horizontal position.
7. Fully tighten the lower strut to lower suspension arm nut and bolt to 116 ft. lbs. (157 Nm).
8. Fully tighten the bolt on the front of the front lower control arm to 100 ft. lbs. (135 Nm).
9. Fully tighten the nut on the lower No. 2 arm bracket assembly to 83 ft. lbs. (113 Nm).
    - Engine under cover
    - Front wheel
10. Inspect and adjust front wheel alignment.

STEERING KNUCKLE

REMOVAL & INSTALLATION

1. Before servicing the vehicle, refer to the Precautions section.
2. Remove the front wheel.
3. Disconnect the front brake caliper assembly. Support the brake caliper securely.
4. Remove the front brake disc
5. Remove the clip and nut, then using SST 09610-20012, disconnect the tie rod end.

6. Remove the cotter pin and nut, then using SST 09628-6201 1, remove the lower ball joint from the lower control arm.

7. Remove the cotter pin and castle nut, then using SST09628_62011, separate the upper ball joint from the steering knuckle.

8. Remove the steering knuckle from the vehicle.

9. If necessary, remove the lower ball joint and hub and bearing assembly from the steering knuckle.

To install:

10. If necessary, install the hub and bearing assembly and lower ball joint to the steering knuckle.

11. Install the steering knuckle into the vehicle.

12. Connect the upper ball joint to the steering knuckle. Install castle nut and new cotter pin.

13. Connect the lower ball joint to lower control arm. Install castle nut and new cotter pin.

14. Connect the tie rod end to the steering knuckle. Install castle nut and new cotter pin.

15. Install the front brake disc

16. Install the front brake caliper assembly.

17. Install the front wheel.

18. Inspect and adjust wheel alignment.

STRUT

REMOVAL & INSTALLATION

See Figure 132.

1. Before servicing the vehicle, refer to the Precautions section.

2. Remove or disconnect the following:

- Negative battery cable
- Tire and wheel assembly
- Front speed sensor
- Stabilizer link assembly
- Front upper control arm at the steering knuckle

3. Loosen, but do not remove the bolt securing the strut to the lower suspension arm. Support lower arm with a floor jack and block of wood.

- Engine room side cover
- Loosen the top strut lock nut
- 3 upper mounting nuts from the strut tower
- Spring support reinforcement

4. Slowly lower the jack.

- Lower strut bolt
- Strut assembly

⁂ CAUTION

Do NOT remove the center nut to the strut at this time. The spring on the strut is under high pressure and can cause serious injury.

Front Suspension Upper Side View

Outer Side of Vehicle — Outer Side of Vehicle

Front

Paint Mark — Paint Mark

30°+-2° LH — 30°+-2° RH

09490_LEXU_G0112

Fig. 132 Align the out mark of the upper spring seat with the mark on the upper insulator

5. Secure the strut in a vise.

6. Compress the coil spring

7. Remove or disconnect the following:

- Upper strut retaining nut
- Suspension support
- Upper insulator
- Bumper
- Coil spring
- Insulator

To install:

8. Install or connect the following:

- Lower insulator
- Coil spring end into the step of the lower seat
- Bumper to the piston rod
- Upper insulator
- Upper support to the piston rod, aligning it with the groove in the strut rod

9. Adjust the front suspension support assembly so that the bolts come to the positions shown in the illustration.

a. Front spring support reinforcement. Tighten the 3 new upper strut retaining nuts to 49 ft. lbs. (67 Nm).

b. Lower strut to lower suspension arm. Temporarily tighten.

c. Front upper control arm at the steering knuckle. Torque to 64 ft. lbs. (87 Nm) and install a new clip.

d. Front speed sensor. Torque to 4 ft. lbs. (6 Nm).

e. Stabilizer link assembly. Torque to 48 inch lbs. (5 Nm).

10. Stabilize the suspension as follows:

a. Install the front wheel(s).

b. Lower the vehicle and bounce it up and down several times to stabilize the front suspension.

c. Raise the vehicle.

d. Remove the front wheel.

e. Jack up the front lower suspension arm placing a wood block in between. Apply a load to the front suspension so that the front lower suspension arm is placed in a horizontal position.

11. Fully tighten the lower strut to lower suspension arm nut and bolt to 116 ft. lbs. (157 Nm).

- Engine room side cover
- Tire and wheel assembly
- Negative battery cable

12. Check the front alignment.

The strut removal procedure also includes the separation of the coil spring.

OVERHAUL

The strut removal procedure also includes the separation of the coil spring.

STABILIZER BAR

REMOVAL & INSTALLATION

AWD

See Figures 133 and 134.

1. Before servicing the vehicle, refer to the Precautions section.

2. Disconnect the negative battery cable. Wait at least 90 seconds before performing any other work

3. Remove the engine assembly with transmission

4. Remove the 2 nuts and the left front stabilizer link assembly. Repeat for the right side. If the ball joint turns together with the nut, use a hexagon (6mm) wrench to hold the stud.

5. Remove the 2 bolts and the left front No. 2 stabilizer bracket from the front suspension crossmember. Repeat for the right side.

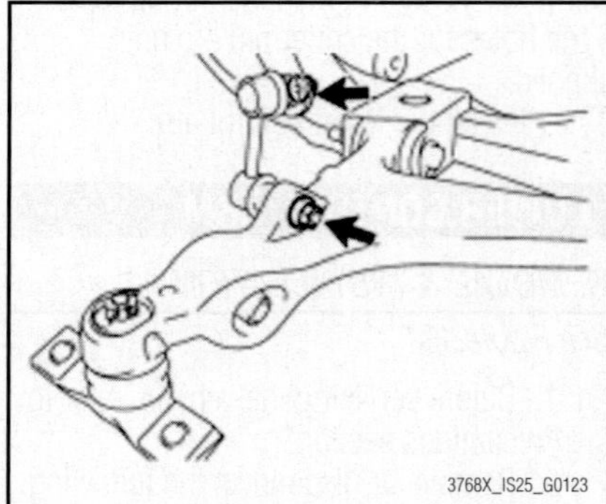

Fig. 133 Removing the front stabilizer link assembly

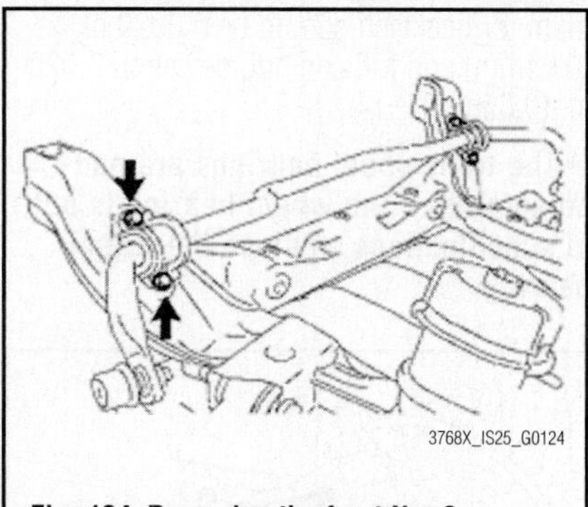

Fig. 134 Removing the front No. 2 stabilizer bracket

6. Remove the 2 front No. 1 stabilizer bar bushings from the front stabilizer bar.

7. Remove the front stabilizer bar from the vehicle.

To install:

8. Install the front stabilizer bar to the vehicle. The identification mark must be on the right side of the vehicle when installing the front stabilizer bar.

9. Install the 2 front No. 1 stabilizer bar bushings outside the bush stoppers on the front stabilizer bar as shown in the illustration. Be sure to install the front No. 1 stabilizer bar bushings so that the cutouts face the front of the vehicle.

10. Install the left front No. 2 stabilizer bracket on the vehicle with the 2 bolts. Repeat for the right side.

11. Tighten the bolts to 57 ft. lbs. (78 Nm).

12. Install the left front stabilizer link assembly with the 2 nuts and tighten to 62 ft. lbs. (84 Nm). Repeat for the right side.

13. Install the engine assembly with transmission

14. Connect the negative battery cable.

RWD

See Figures 135 and 136.

1. Before servicing the vehicle, refer to the Precautions section.

2. Remove the front wheels.

3. Remove engine under cover.

4. Remove the left and right rear engine under covers.

5. Disconnect the front height control sensor link assembly.

6. Remove the 2 nuts and the left front stabilizer link assembly. Repeat for the right side. If the ball joint turns together with the nut, use a hexagon (6mm) wrench to hold the stud.

7. Remove the 2 bolts and the left front No. 2 stabilizer bracket from the front suspension crossmember. Repeat for the right side.

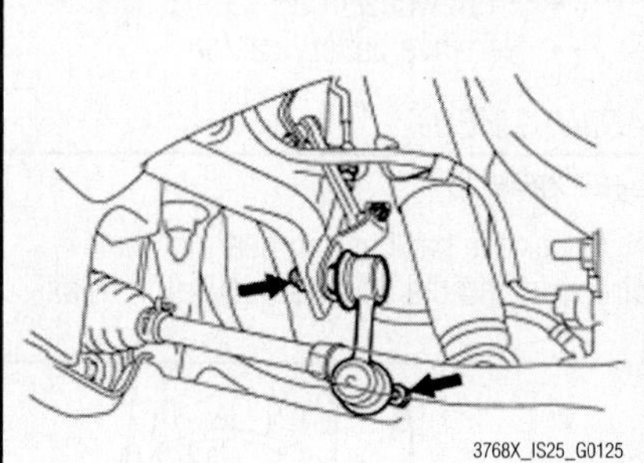

Fig. 135 Removing the front stabilizer link assembly

Fig. 136 Removing the front No. 1 stabilizer bracket

8. Remove the 4 bolts and the left front No. 1 stabilizer bracket from the frame. Repeat for the right side.

9. Remove the 2 front No. 1 stabilizer bar bushings from the front stabilizer bar.

10. Remove the front stabilizer bar from the vehicle.

To install:

11. Install the front stabilizer bar to the vehicle. The identification mark must be on the right side of the vehicle when installing the front stabilizer bar.

12. Install the 2 front No. 1 stabilizer bar bushings as shown in the illustration. Be sure to install the front No. 1 stabilizer bar bushings so that the cutouts face the front of the vehicle.

13. Install the left front No. 1 stabilizer bracket as follows:

a. Press the left front No. 1 stabilizer bracket against the frame toward the outside of the vehicle and temporarily tighten bolt A.

b. Install bolt B; fully tighten bolt A, and then install bolts C and D.

c. Install the skid control sensor wire clamp to the left front No. 1 stabilizer bracket.

d. Tighten the bolts to 36 ft. lbs. (49 Nm).

e. Install the right side following the same procedures as for the left side.

14. Install the left front No. 2 stabilizer bracket on the vehicle with the 2 bolts. Repeat for the right side. Tighten the bolts to 36 ft. lbs. (49 Nm).

15. Install the left front stabilizer link assembly with the 2 nuts and tighten to 62 ft. lbs. (84 Nm). Repeat for the right side. If the ball joint turns together with the nut, use a hexagon (6mm) wrench to hold the stud.

16. Connect the front height control sensor link assembly.

17. Install the left and right rear engine under covers.

18. Install engine under cover.

19. Install the front wheels.

20. Inspect and adjust the front wheel alignment.

UPPER BALL JOINT

REMOVAL & INSTALLATION

The upper ball joint is an integral part of the upper arm and is not replaced separately. The upper ball joint replacement is accomplished by replacing the upper arm.

UPPER CONTROL ARM

REMOVAL & INSTALLATION

1. Before servicing the vehicle, refer to the Precautions section.

2. Remove or disconnect the following:
   - Negative battery cable
   - Wheel
   - Front strut assembly
   - Front upper control arm

To install:

3. Install or connect the following:
   - Upper suspension arm and tighten the mounting bolts to 36 ft. lbs. (49 Nm)
   - Front strut assembly
   - Wheel
   - Negative battery cable

4. Lower the vehicle.

5. Check the front wheel alignment.

WHEEL HUB & BEARING

REMOVAL & INSTALLATION

1. Before servicing the vehicle, refer to the Precautions section.

2. Remove or disconnect the following:

- Negative battery cable
- Front wheels
- Speed sensor
- Disc brake caliper
- Brake disc
- Front axle hub nut (AWD)
- 4 bolts and front axle hub

➡On AWD models, use a plastic hammer to tap the hub unit away from the driveshaft.

To install:

3. Install or connect the following:
   - Front axle hub. Torque the mounting bolts to 51 ft. lbs. (69 Nm).
   - New front axle hub nut (AWD). Torque to 217 ft. lbs. (294 Nm).
   - Brake disc
   - Disc brake caliper
4. Stake the front axle hub nut using a hammer and chisel.
   - Speed sensor
   - Front wheels
   - Negative battery cable

ADJUSTMENT

See Figure 137.

Check the backlash in bearing shaft direction and the axle hub deviation. Maximum for backlash should be 0.0020 in. (0.05mm) and for axle hub deviation 0.020 in. (0.05mm).

➡The front wheel bearings are non-adjustable. If the wheel bearing is out of specifications, replace the wheel bearing.

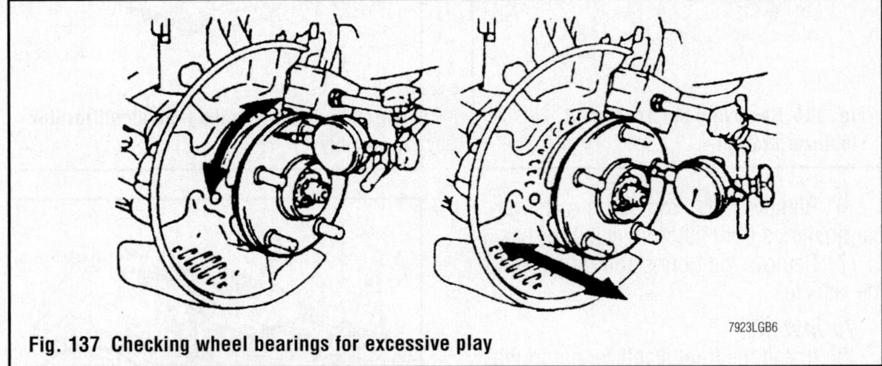

Fig. 137 Checking wheel bearings for excessive play

SUSPENSION

REAR SUSPENSION

STABILIZER BAR

REMOVAL & INSTALLATION

See Figure 138.

1. Remove the rear wheels.
2. Remove the 2 nuts and left stabilizer bar link. If the ball joint turns together with the nut, use a hexagon wrench (5 mm) to hold the stud. Repeat for the right side.
3. Remove the 8 bolts, 2 No. 1 brackets and 2 bushings. Two types of bolts are used, so make sure the correct bolts are installed.
4. Remove the stabilizer bar.

To install:

5. Install the stabilizer bar.
6. Install the bushing and bracket with the 2 bolts and tighten to 14 ft. lbs. (19 Nm).
7. Install the bushing to the inner side of the bushing stopper on the stabilizer bar.
8. Connect the stabilizer bar links and tighten the nuts to 29 ft. lbs. (39 Nm). Repeat for the right side.
9. Install the rear wheels.
10. Inspect and adjust rear wheel alignment.

WHEEL HUB & BEARING

REMOVAL & INSTALLATION

1. Before servicing the vehicle, refer to the Precautions section.
2. Raise and safely support the vehicle.
3. Remove or disconnect the following:
   - Rear tire and wheel assembly
   - Rear stabilizer link
   - Rear axle shaft nut
   - Rear disc brake caliper and disc
   - Speed sensor
   - Parking brake assembly at the wheel hub
   - No. 1 and 2 upper control arm assembly
   - No. 1 and 2 rear suspension arm assembly
   - Toe control link
   - Rear axle from hub assembly using a plastic hammer
   - No. 2 rear wheel bearing dust deflector using a screwdriver
   - 4 bolts and rear axle hub and bearing assembly from the axle carrier assembly

To install:

4. Install or connect the following:
   - Hub on the carrier and tighten the bolts to 52 ft. lbs. (70 Nm)
   - No. 2 rear wheel bearing dust deflector using a screwdriver
   - Rear axle into hub assembly
   - No. 1 and 2 upper control arm assembly. Torque the new nut to the No. 2 rear upper control arm to 52 ft. lbs. (70 Nm). Temporarily tighten the No. 1 arm bolt.
   - No. 1 and 2 rear suspension arm assembly. Temporarily tighten.
   - Toe control link. Torque new nut to 52 ft. lbs. (70 Nm).
   - Rear stabilizer link. Torque to 20 ft. lbs. (27 Nm).
   - Parking brake assembly at the wheel hub
   - Rear speed sensor
   - Rear axle shaft nut. Torque to 214 ft. Lbs. (290 Nm) and stake the nut with a chisel and hammer.
   - Rear brake disc and disc brake caliper
5. Adjust the parking brake.
6. Stabilize the suspension as follows:
   a. Install the wheel(s).
   b. Lower the vehicle and bounce it up and down several times to stabilize the rear suspension.
   c. Raise the vehicle.
   d. Remove the wheel.
   e. Jack up the rear lower suspension arm placing a wood block in between. Apply a load to the suspension so that the lower suspension arm is placed in a horizontal position.
7. Fully tighten the No. 1 rear upper control arm assembly to 119 ft. lbs. (161 Nm).
8. Fully tighten the No. 1 rear suspension arm assembly to 70 ft. lbs. (95 Nm).

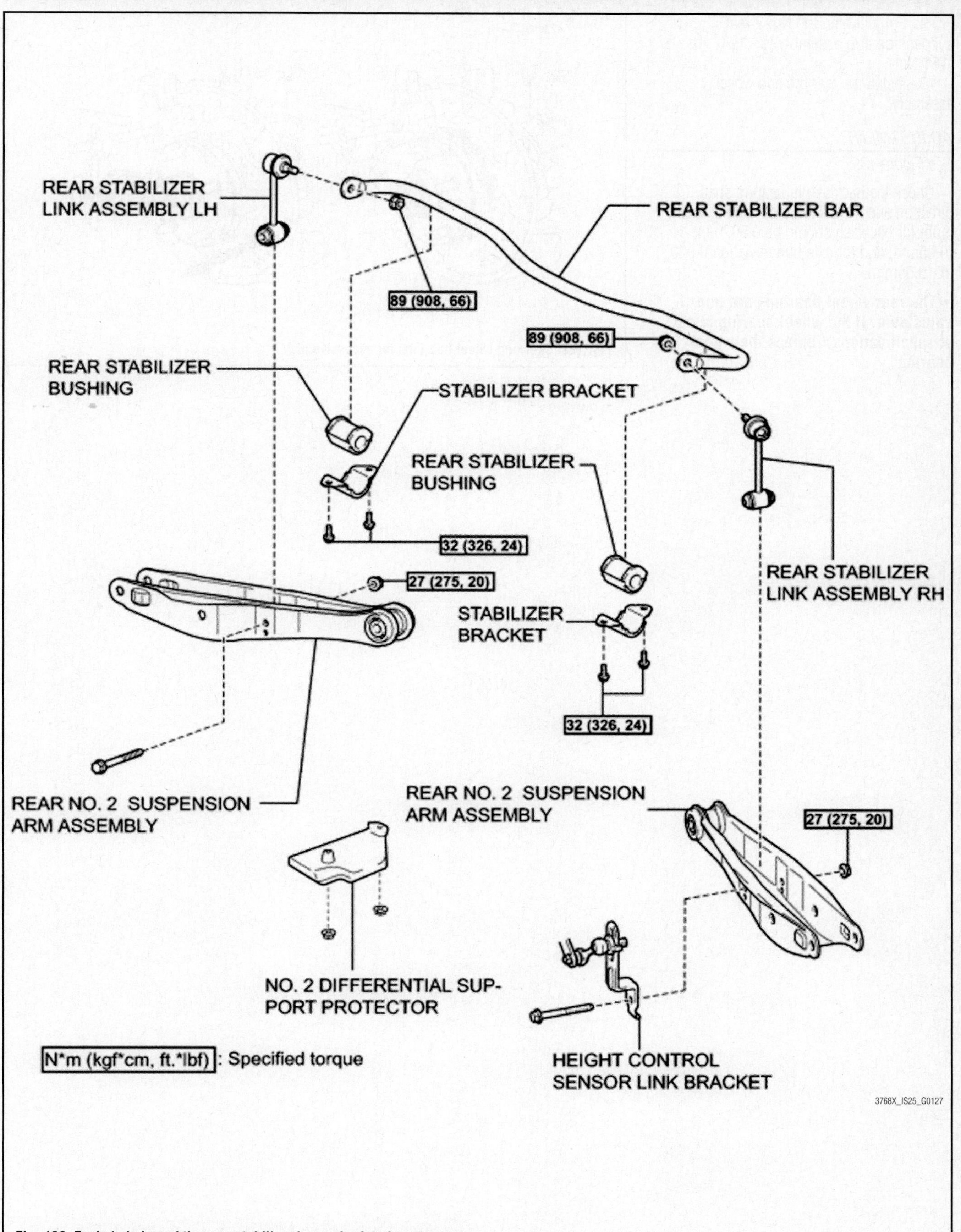

Fig. 138 Exploded view of the rear stabilizer bar and related components

9. Fully tighten the No. 2 rear suspension arm assembly to 119 ft. lbs. (161 Nm).

10. Install the rear tire and wheel assembly.

ADJUSTMENT

See Figure 139.

Check the backlash in bearing shaft direction and the axle hub deviation. Maximum for backlash should be 0.0020 in. (0.05mm) and for axle hub deviation 0.020 in. (0.05mm).

➡The rear wheel bearings are non-adjustable. If the wheel bearing is out of specifications, replace the wheel bearing.

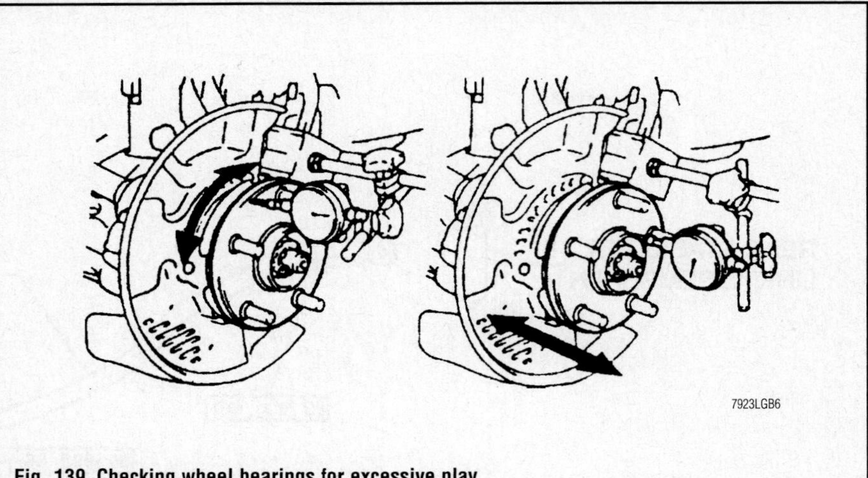

Fig. 139 Checking wheel bearings for excessive play

LEXUS

LS460

29

SPECIFICATIONS AND MAINTENANCE CHARTS

ENGINE AND VEHICLE IDENTIFICATION

Engine							Model Year	
Code ①	Liters (cc)	Cu. In.	Cyl.	Fuel Sys.	Engine Type	Eng. Mfg.	Code ②	Year
1UR-FSE	4.6 (4608)	282	V8	SFI D4S	DOHC	Toyota	9	2009
							A	2010

SFI: Sequential Fuel Injection

DOHC: Double Overhead Camshaft

NA: Information not available

① Stamped on the left side of the engine block

② 10th digit of the Vehicle Identification Number (VIN)

3768X_LS46_C0001

GENERAL ENGINE SPECIFICATIONS

Year	Model	Engine Displacement Liters	Engine Series ID	Net Horsepower @ rpm	Net Torque @ rpm (ft. lbs.)	Bore x Stroke (in.)	Com-pression Ratio	Oil Pressure @ rpm
2009	LS460	4.6	1UR-FSE	380@6400	367@4100	3.70x3.27	11.8:1	31 plus@2500
2010	LS460	4.6	1UR-FSE	380@6400	367@4100	3.70x3.27	11.8:1	31 plus@2500

3768X_LS46_C0002

ENGINE TUNE-UP SPECIFICATIONS

Year	Engine Displacement Liters	Engine ID	Spark Plug Gap (in.)	Ignition Timing (deg.)	Fuel Pump (psi)	Idle Speed (rpm)	Valve Clearance	
							Intake	Exhaust
2009	4.6	1UR-FSE	0.039-0.043	8-12B ①	②	700-800	HYD	HYD
2010	4.6	1UR-FSE	0.039-0.043	8-12B ①	②	700-800	HYD	HYD

NOTE: The Vehicle Emission Control Information label often reflects specification changes made during production. The label figures must be used if they differ from those in this chart.

HYD: Hydraulic Valve Lifters

① Terminals TC and CG of check connector must be connected

② Low pressure system 28-85 psi. High pressure system 508-653 psi.

3768X_LS46_C0003

CAPACITIES

Year	Model	Engine Disp. Liters	Engine ID	Engine Oil with Filter (qts.)	Transmission (pts.) Auto. ①	Transfer Case (pts.)	Drive Axle Front (pts.)	Drive Axle Rear (pts.)	Fuel Tank (gal.)	Cooling System (qts.)
2009	LS460 2WD	4.6	1UR-FSE	9.1	21.7	NA	NA	3.0	22.1	11.6
	LS460 AWD	4.6	1UR-FSE	9.5	22.6	1.48	1.58	3.0	22.1	11.7
2010	LS460 2WD	4.6	1UR-FSE	9.1	21.7	NA	NA	3.0	22.1	11.6
	LS460 AWD	4.6	1UR-FSE	9.5	22.6	1.48	1.58	3.0	22.1	11.7

NA: Not Applicable

① Specification is for transmission drain and refill, not overhaul.

3768X_LS46_C0004

FLUID SPECIFICATIONS

Year	Model	Engine Displ. Liters	Engine Oil	Auto. Trans.	Drive Axle ② Front	Drive Axle ② Rear	Transfer Case	Power Steering Fluid ③	Brake Master Cylinder	Cooling System
2009	LS460	4.6	①	Toyota ATF WS	75W-85	75W-85	80W-90	NE	DOT 3	SLLC ④
2010	LS460	4.6	①	Toyota ATF WS	75W-85	75W-85	80W-90	NE	DOT 3	SLLC ④

NE: Not Equipped

DOT: Department Of Transpotation

① 0W-20 or 5W-20 ILSAC multigrade engine oil

② Oil grade: Hypoid gear oil API GL-5

③ The LS460 is not equipped with a power steering pump. It utilizes a power steering motor.

④ Toyota Super Long Life Coolant

3768X_LS46_C0013

VALVE SPECIFICATIONS

Year	Engine Displacement Liters	Engine ID	Seat Angle (deg.)	Face Angle (deg.)	Spring Test Pressure (lbs. @ in.)	Spring Free-Length (in.)	Stem-to-Guide Clearance (in.) Intake	Stem-to-Guide Clearance (in.) Exhaust	Stem Diameter (in.) Intake	Stem Diameter (in.) Exhaust
2009	4.6	1UR-FSE	45	44.5	NA	2.035	0.0010-0.0024	0.0012-0.0026	0.2154-0.2159	0.2152-0.2157
2010	4.6	1UR-FSE	45	44.5	NA	2.035	0.0010-0.0024	0.0012-0.0026	0.2154-0.2159	0.2152-0.2157

NA: Information not available

3768X_LS46_C0005

CAMSHAFT AND BEARING SPECIFICATIONS

All measurements are given in inches.

Year	Engine Displacement Liters	Engine VIN	Journal Diameter	Brg. Oil Clearance	Shaft End-play	Runout	Journal Bore	Lobe Lift Intake	Lobe Lift Exhaust
2009	4.6	1UR-FSE	①	②	NA	0.0016	NA	1.7438-1.7497	1.7447-1.7506
2010	4.6	1UR-FSE	①	②	NA	0.0016	NA	1.7438-1.7497	1.7447-1.7506

NA: Not Available

① No. 1 journal: 1.1794 to 1.1799 inches
Other journals: 1.0220 to 1.0226 inches

② No. 1 journal: 0.00118 to 0.00256 inches
Other journals: 0.0010 to 0.00024 inches

3768X_LS46_C0014

CRANKSHAFT AND CONNECTING ROD SPECIFICATIONS

All measurements are given in inches.

Year	Engine Displacement Liters	Engine ID	Crankshaft Main Brg. Journal Dia.	Crankshaft Main Brg. Oil Clearance	Crankshaft Shaft End-play	Crankshaft Thrust on No.	Connecting Rod Journal Diameter	Connecting Rod Oil Clearance	Connecting Rod Side Clearance
2009	4.6	1UR-FSE	2.6373-2.6378	①	0.0008-0.0087	3	2.0863-2.0866	0.0010-0.0020	0.0059-0.0217
2010	4.6	1UR-FSE	2.6373-2.6378	①	0.0008-0.0087	3	2.0863-2.0866	0.0010-0.0020	0.0059-0.0217

① Journal No. 1 and 5: 0.0007 - 0.0012 inch
Remaining journals: 0.0009 - 0.0015 inch

3768X_LS46_C0006

PISTON AND RING SPECIFICATIONS

All measurements are given in inches.

Year	Engine Displ. Liters	Engine ID	Piston Clearance	Ring Gap Top Comp.	Ring Gap Bottom Comp.	Ring Gap Oil Control	Ring Side Clearance Top Comp.	Ring Side Clearance Bottom Comp.	Ring Side Clearance Oil Control
2009	4.6	1UR-FSE	0.0014-0.0022	0.0091-0.0130	0.0138-0.0177	0.0039-0.0157	0.0008-0.0028	0.0008-0.0024	0.0008-0.0028
2010	4.6	1UR-FSE	0.0014-0.0022	0.0091-0.0130	0.0138-0.0177	0.0039-0.0157	0.0008-0.0028	0.0008-0.0024	0.0008-0.0028

3768X_LS46_C0007

TORQUE SPECIFICATIONS

All readings in ft. lbs.

Year	Engine Disp. Liters	Engine ID	Cylinder Head Bolts	Main Bearing Bolts	Rod Bearing Bolts	Crankshaft Damper Bolts	Flywheel Bolts	Manifold Intake	Manifold Exhaust	Spark Plugs	Oil Pan Drain Plug
2009	4.6	1UR-FSE	①	②	③	221	22	15	15	13	30
2010	4.6	1UR-FSE	①	②	③	221	22	15	15	13	30

① Step 1: 27 ft. lbs.
Step 2: Tighten an additional 90 degrees
Step 3: Tighten an additional 90 degrees
12 mm bolt: 15 ft. lbs.

② Inside position: 45 ft. lbs.
Outside position Step 1: 20 ft. lbs.
Outside position Step 2: Tighten an additional 90 degrees
Cylinder block Side postion: 33 ft. lbs.

③ Step 1: 30 ft. lbs.
Step 2: Plus 90 degrees

3768X_LS46_C0008

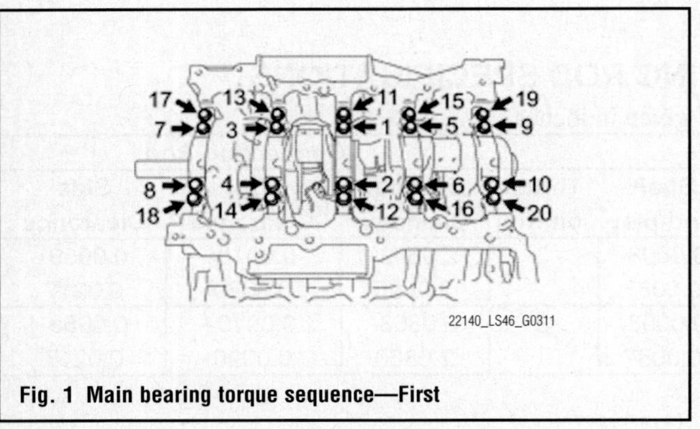

Fig. 1 Main bearing torque sequence—First

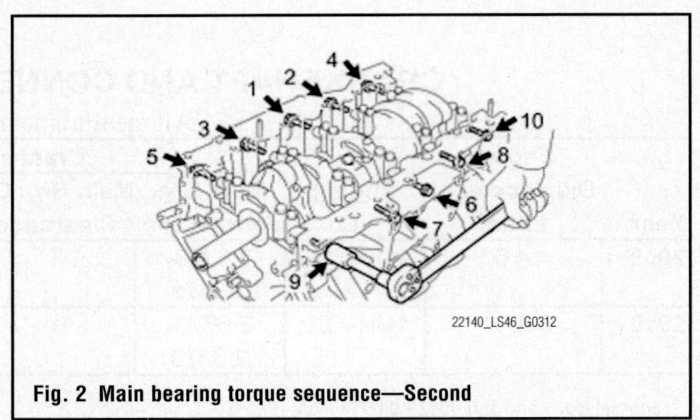

Fig. 2 Main bearing torque sequence—Second

WHEEL ALIGNMENT

Year	Model		Caster Range (+/-Deg.)	Caster Preferred Setting (Deg.)	Camber Range (+/-Deg.)	Camber Preferred Setting (Deg.)	Toe-in (in.)	Steering Axis Inclination (Deg.)
2009	LS460	Front	0.75	+6.62	0.75	-0.28	0.04+/-0.08	9.12+/-0.75
	w/o air susp	Rear	—	—	0.75	-1.63	0.12+/-0.08	—
	with air	Front	0.75	+7.02	0.75	-0.43	0+/-0.08	9.3+/-0.75
	suspension	Rear	—	—	0.75	-1.83	0.12+/-0.08	—
2010	LS460	Front	0.75	+6.62	0.75	-0.28	0.04+/-0.08	9.12+/-0.75
	w/o air susp	Rear	—	—	0.75	-1.63	0.12+/-0.08	—
	with air	Front	0.75	+7.02	0.75	-0.43	0+/-0.08	9.3+/-0.75
	suspension	Rear	—	—	0.75	-1.83	0.12+/-0.08	—

3768X_LS46_C0009

TIRE, WHEEL AND BALL JOINT SPECIFICATIONS

Year	Model	OEM Tires Standard	OEM Tires Optional	Tire Pressures (psi) Front	Tire Pressures (psi) Rear	Wheel Size	Ball Joint Inspection	Lug Nut Torque (ft. lbs.)
2009	LS460	P235/50R18	P235/50R18	33	33	7.5-J	①	103
2010	LS460	P235/50R18	P235/50R18	33	33	7.5-J	①	103

OEM: Original Equipment Manufacturer

PSI: Pounds Per Square Inch

STD: Standard

OPT: Optional

① Replace if any measurable movement is found.

3768X_LS46_C0010

BRAKE SPECIFICATIONS

All measurements in inches unless noted

Year	Model		Brake Disc Original Thickness	Brake Disc Minimum Thickness	Brake Disc Maximum Runout	Minimum Lining Thickness	Brake Caliper Bracket Bolts (ft. lbs.)	Brake Caliper Mounting Bolts (ft. lbs.)
2009	LS460	F	①	②	0.0016	0.039	100	NA
		R	③	④	0.0016	0.039	63	NA
2010	LS460	F	①	②	0.0016	0.039	100	NA
		R	③	④	0.0016	0.039	63	NA

F: Front

R: Rear

NA: Not Available

① 13.14 inch disc: 1.181 inches
14.06 inch disc: 1.339 inches

② 13.14 inch disc: 1.063 inches
14.06 inch disc: 1.221 inches

③ 12.40 inch disc: 0.787 inches
13.19 inch disc: 0.866 inches

④ 12.40 inch disc: 0.709 inches
13.19 inch disc: 0.887 inches

3768X_LS46_C0011

SCHEDULED MAINTENANCE INTERVALS

LEXUS—LS460

TO BE SERVICED	TYPE OF SERVICE	VEHICLE MILEAGE INTERVAL (x1000)													
		5	10	15	20	25	30	35	40	45	50	55	60	90	120
Engine oil & filter	R	✓	✓	✓	✓	✓	✓	✓	✓	✓	✓	✓	✓	✓	✓
Automatic transmission fluid	S/I			✓			✓			✓			✓	✓	✓
Ball joints & dust covers	S/I			✓			✓			✓			✓	✓	✓
Bolts & nuts on chassis & body	S/I			✓			✓			✓			✓	✓	✓
Brake linings & drums	S/I	✓	✓	✓	✓	✓	✓	✓	✓	✓	✓	✓	✓	✓	✓
Brake line pipes & hoses	S/I			✓			✓			✓			✓	✓	✓
Brake pads & discs (front & rear)	S/I	✓	✓	✓	✓	✓	✓	✓	✓	✓	✓	✓	✓	✓	✓
Brake fluid	R						✓						✓	✓	✓
Rack and pinion assembly	S/I			✓			✓			✓			✓	✓	✓
Steering linkage & boots	S/I			✓			✓			✓			✓	✓	✓
Air cleaner filter	R						✓						✓	✓	✓
Spark plugs ①	R														✓
Drive belts	S/I						✓						✓	✓	✓
Exhaust pipes & mountings	S/I			✓			✓			✓			✓	✓	✓
Fuel lines & connections	S/I						✓						✓	✓	✓
Engine coolant ②	S/I			✓			✓			✓			✓	✓	
Rear differential	S/I			✓			✓			✓			✓	✓	✓
Fuel tank cap gasket	S/I						✓						✓	✓	✓
Rotate tires	S/I			✓			✓			✓			✓		✓
Clean air conditioning filter ③	S/I			✓			✓			✓			✓		✓
Axle shaft bolts	S/I			✓			✓			✓			✓	✓	✓
Brake pad thickness and rotor runout	S/I						✓						✓	✓	✓

R: Replace S/I: Service or Inspect

① Spark plugs are replaced at 120,000 miles

② Replace engine coolant at 100,000 miles and then inspect every 15,000 miles

③ Replace air conditioning filter every 30,000 miles

FREQUENT OPERATION MAINTENANCE (SEVERE SERVICE)

If a vehicle is operated under any of the following conditions it is considered severe service:

- Extremely dusty areas.

- 50% or more of the vehicle operation is in 32°C (90°F) or higher temperatures, or constant temperatures below 0°C (32°F).

- Prolonged idling (vehicle operation in stop and go traffic).

- Frequent short running periods (engine does not warm to normal operating temperatures).

- Police, taxi, delivery usage or trailer towing usage.

Air cleaner filter: service or inspect every 5000 miles

Rear differential & transfer case oil: replace every 15,000 miles.

Ball joints & dust covers: service or inspect every 5000 miles.

Bolts & nuts on chassis & body: service or inspect every 5000 miles.

Axle shaft bolts: service or inspect every 5000 miles.

Steering linkage: service or inspect every 5000 miles.

3768X_LS46_C0012

BRAKES — INFORMATION AND PRECAUTIONS

ANTI-LOCK SYSTEMS

• Certain components within the ABS system are not intended to be serviced or repaired individually.

• Do not use rubber hoses or other parts not specifically specified for and ABS system. When using repair kits, replace all parts included in the kit. Partial or incorrect repair may lead to functional problems and require the replacement of components.

• Lubricate rubber parts with clean, fresh brake fluid to ease assembly. Do not use shop air to clean parts; damage to rubber components may result.

• Use only DOT 3 brake fluid from an unopened container.

• If any hydraulic component or line is removed or replaced, it may be necessary to bleed the entire system.

• A clean repair area is essential. Always clean the reservoir and cap thoroughly before removing the cap. The slightest amount of dirt in the fluid may plug an orifice and impair the system function. Perform repairs after components have been thoroughly cleaned; use only denatured alcohol to clean components. Do not allow ABS components to come into contact with any substance containing mineral oil; this includes used shop rags.

• The Anti-Lock control unit is a microprocessor similar to other computer units in the vehicle. Ensure that the ignition switch is **OFF** before removing or installing controller harnesses. Avoid static electricity discharge at or near the controller.

• If any arc welding is to be done on the vehicle, the control unit should be unplugged before welding operations begin.

DISC AND DRUM SYSTEMS

※※ CAUTION

Dust and dirt accumulating on brake parts during normal use may contain asbestos fibers from production or aftermarket brake linings. Breathing excessive concentrations of asbestos fibers can cause serious bodily harm. Exercise care when servicing brake parts. Do not sand or grind brake lining unless equipment used is designed to contain the dust residue. Do not clean brake parts with compressed air or by dry brushing. Cleaning should be done by dampening the brake components with a fine mist of water, then wiping the brake components clean with a dampened cloth. Dispose of cloth and all residue containing asbestos fibers in an impermeable container with the appropriate label. Follow practices prescribed by the Occupational Safety and Health Administration (OSHA) and the Environmental Protection Agency (EPA) for the handling, processing, and disposing of dust or debris that may contain asbestos fibers.

BRAKES — BLEEDING THE BRAKE SYSTEM

BLEEDING PROCEDURE

Proprietary testing equipment is required. The Techstream must be used during fluid replacement. If not used, the fluid replacement will be incomplete, which is hazardous and may lead to an accident.

BRAKES — ANTI-LOCK BRAKE SYSTEM (ABS)

WHEEL SPEED SENSORS

REMOVAL & INSTALLATION

Front

See Figures 3 through 7.

➡The front speed sensor is a component of the front axle hub sub-assembly. If the sensor malfunctions, replace the front axle hub sub-assembly.

※※ WARNING

While the battery is connected, even if the engine switch is off, the brake control system activates when the brake pedal is depressed or the door courtesy switch turns on. Therefore during servicing of the brake system components, do not operate the brake pedal and open/close the doors while the battery is connected.

※※ WARNING

After the engine switch is turned off, the HDD navigation system requires approximately 6 minutes to record various types of memory and settings. As a result, after turning the engine switch off, wait 6 minutes or more before disconnecting the cable from the negative battery terminal.

1. Remove the right cowl top ventilator louver.
2. Disconnect cable from negative battery terminal.

※※ CAUTION

Wait at least 90 seconds after disconnecting the cable from the negative battery terminal to prevent airbag and seat belt pretensioner activation.

3. Remove the front wheel.
4. Remove the skid control sensor wire, as follows:
    a. Disconnect the front fender liner so that the skid control sensor wire connector (labeled A) can be seen.
    b. Disconnect the sensor connector (labeled A).
    c. Detach the sensor clip (labeled B).
    d. Remove the bolt (labeled C) and sensor clamp (labeled D).
    e. Remove the 2 bolts (labeled E) and 2 sensor clamps (labeled F).

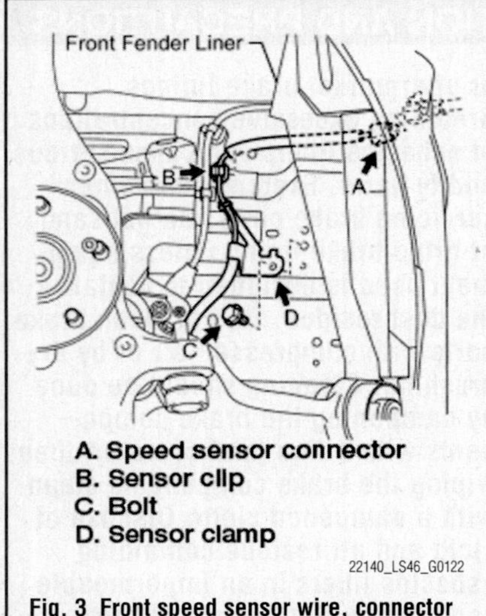

Fig. 3 Front speed sensor wire, connector clip, bolt, and clamp

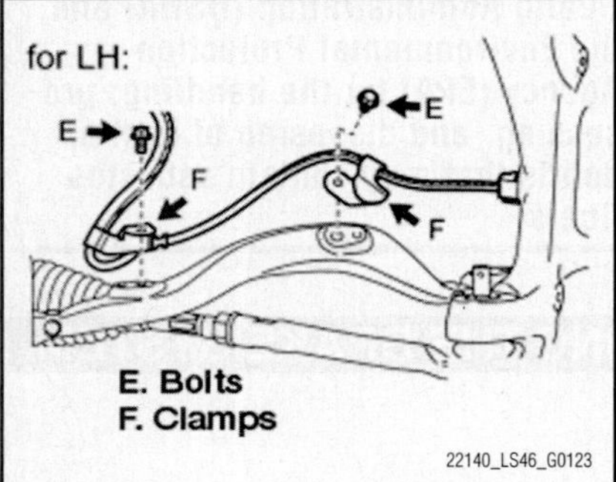

Fig. 4 Front speed sensor wire bolts and clamps—left hand

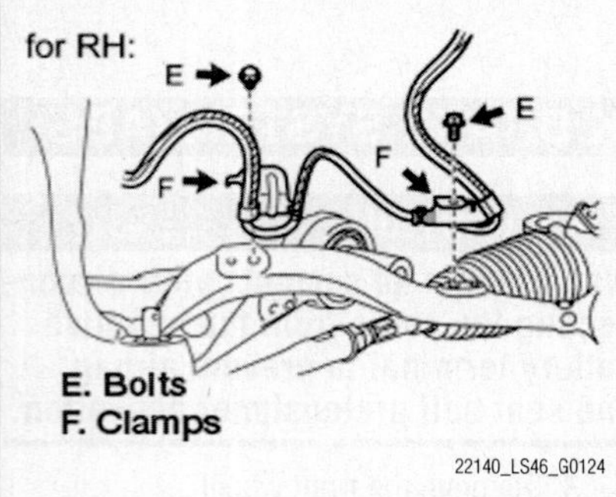

Fig. 5 Front speed sensor wire bolts and clamps—right hand

f. Disconnect the pad wear indicator connector (labeled G).

g. Remove the bolt (labeled H) and sensor clamp (labeled I).

h. Disconnect the connector (labeled J) from the front speed sensor.

➡Be careful not to damage the speed sensor. Prevent foreign matter from adhering to the speed sensor.

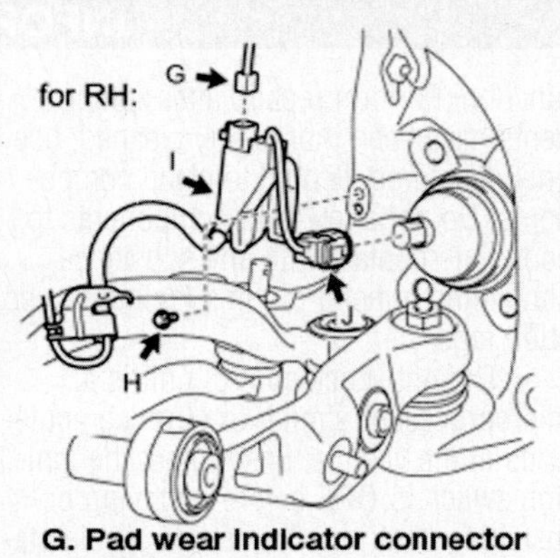

Fig. 6 Front pad wear indicator connectors, bolt, and clamp—right hand

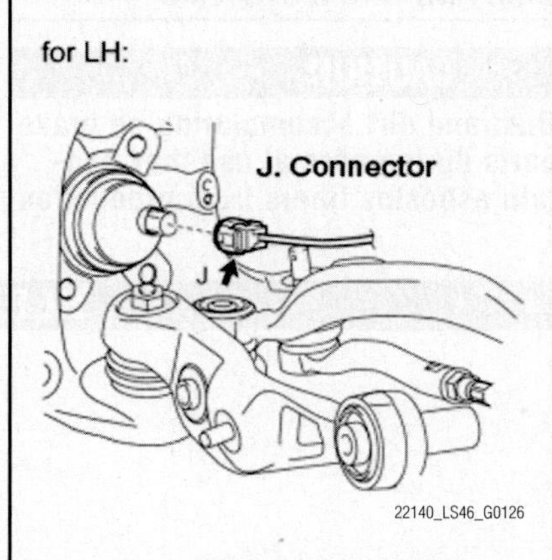

Fig. 7 Front pad wear indicator connector—left hand

5. Disconnect front disc brake caliper assembly.
6. Remove front disc.
7. Remove the front axle hub.

To install:

8. Installation is the reverse of removal. Tighten the skid control sensor wire bolts to 75 inch lbs. (8.5 Nm).
9. Perform initialization.
10. Check the speed sensor signal.

Rear

See Figures 8 through 12.

1. Remove the right cowl top ventilator louver.
2. Disconnect cable from negative battery terminal.

⁂ CAUTION

Wait at least 90 seconds after disconnecting the cable from the negative battery terminal to prevent airbag and seat belt pretensioner activation.

➡After the engine switch is turned off, the HDD navigation system requires approximately 6 minutes recording various types of memory and settings. As a result, after turning the engine switch off, wait 6 minutes or more before disconnecting the cable from the negative battery terminal.

3. Remove the luggage compartment mat sub-assembly.
4. Remove the rope hook assembly and rope hook.
5. Remove the left and right deck trim side board.
6. Remove the luggage compartment No. 1 light assembly.
7. Remove the luggage compartment No. 2 trim hook (without rear cooler).
8. Remove the No. 1 cooler cover (with rear cooler).
9. Remove the front luggage compartment trim cover.
10. Remove rear floor finish plate.

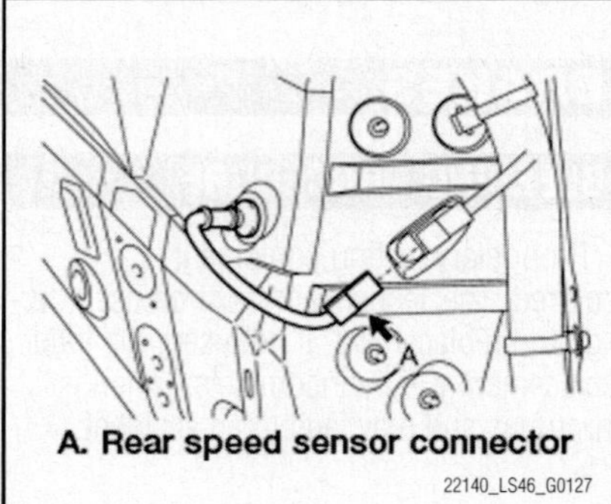

Fig. 8 Rear speed sensor connector

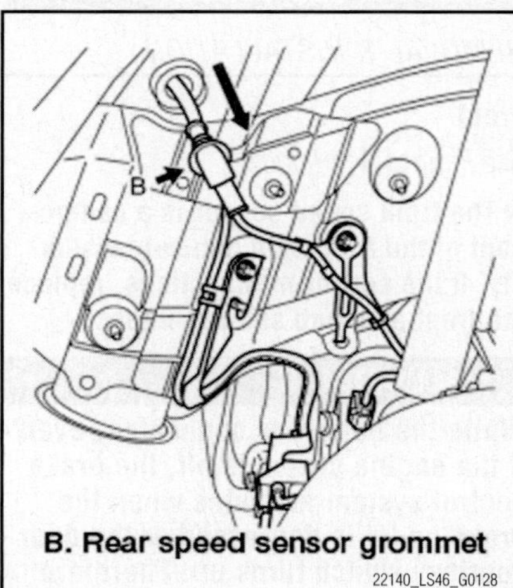

Fig. 9 Rear speed sensor grommet

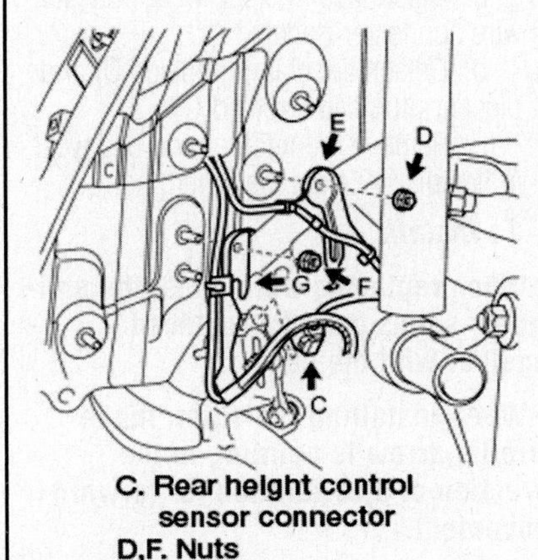

Fig. 10 Rear height control sensor connector, nuts, and clamp

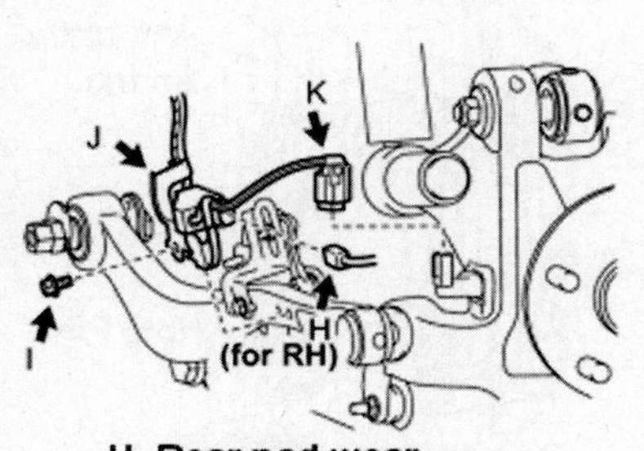

Fig. 11 Rear pad wear indicator connector, bolt, and sensor clamp and connector

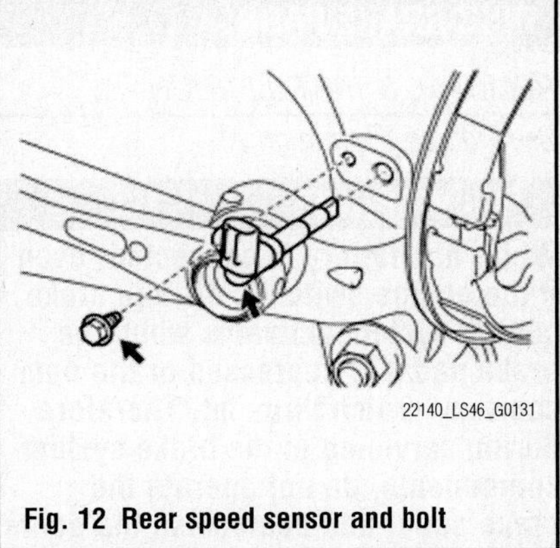

Fig. 12 Rear speed sensor and bolt

11. Remove the left and right luggage compartment trim cover assembly.
12. Remove rear wheel.
13. Remove skid control sensor wire, as follows:
    a. Disconnect the sensor connector (labeled A).
    b. Disconnect the grommet (labeled B) of the speed sensor wire from the hole of the wheel house.
    c. Except for the right hand side of models with coil spring type suspension: Disconnect the rear height control sensor connector (labeled C) and remove the nut (labeled D) and sensor clamp (labeled E).
    d. Remove the nut (labeled F) and sensor clamp (labeled G).
    e. For right hand: Disconnect the pad wear indicator connector (labeled H).
    f. Remove the bolt (labeled I) and sensor clamp (labeled J).
    g. Disconnect the connector (labeled K) from the rear speed sensor.

➡Be careful not to damage the speed sensor. Prevent foreign matter from adhering to the speed sensor.

14. Remove the bolt and speed sensor. The rear speed sensor is easily damaged. When pulling out the rear speed sensor from the rear axle hub, do not use excessive force to rotate and remove it. Prevent foreign matter from attaching to the sensor tip.

To install:

15. Installation is the reverse of removal. Tighten the skid control sensor wire bolts to 75 inch lbs. (8.5 Nm), and sensor clamp nuts to 49 inch lbs. (5.5 Nm).
16. Perform initialization.
17. Check the speed sensor signal.

BRAKE CALIPER

REMOVAL & INSTALLATION

See Figure 13.

⁂ WARNING

While the battery is connected, even if the engine switch is off, the brake control system activates when the brake pedal is depressed or the door courtesy switch turns on. Therefore during servicing of the brake system components, do not operate the brake pedal and open/close the doors while the battery is connected.

⁂ WARNING

After the engine switch is turned off, the HDD navigation system requires approximately 6 minutes to record various types of memory and settings. As a result, after turning the engine switch off, wait 6 minutes or more before disconnecting the cable from the negative battery terminal.

1. Remove the right cowl top ventilator louver.
2. Disconnect cable from negative battery terminal.

⁂ CAUTION

Wait at least 90 seconds after disconnecting the cable from the negative battery terminal to prevent airbag and seat belt pretensioner activation.

3. Remove the front wheel.
4. Remove front disc brake pad kit.
5. Drain brake fluid.
6. Remove the union bolt and gasket from the disc brake caliper, then disconnect the flexible hose from the disc brake caliper.
7. Remove the 2 bolts and disc brake caliper from the knuckle.

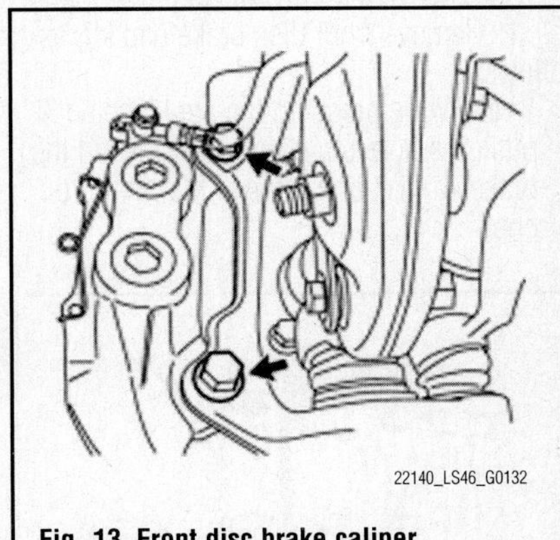

Fig. 13 Front disc brake caliper

To install:

8. Installation is the reverse of removal, noting the following:
    a. Tighten the brake caliper bolts to 100 ft. lbs. (135 Nm).
    b. Tighten the flexible hose union bolt to 29 ft. lbs. (39 Nm).
9. Fill reservoir and bleed brake system.

DISC BRAKE PADS

REMOVAL & INSTALLATION

See Figures 14 through 21.

⁂ WARNING

While the battery is connected, even if the engine switch is off, the brake control system activates when the brake pedal is depressed or the door courtesy switch turns on. Therefore during servicing of the brake system components, do not operate the brake pedal and open/close the doors while the battery is connected.

⁂ WARNING

After the engine switch is turned off, the HDD navigation system requires approximately 6 minutes to record various types of memory and settings. As a result, after turning the engine switch off, wait 6 minutes or more before disconnecting the cable from the negative battery terminal.

1. Disconnect cable from negative battery terminal.

⁂ CAUTION

Wait at least 90 seconds after disconnecting the cable from the negative battery terminal to prevent airbag and seat belt pretensioner activation.

2. Remove the front wheel.
3. Remove front disc brake pad kit, as follows:
   a. While pressing the area labeled A, push the hole pin (labeled B) toward the brake caliper, and remove the pin hold clip.

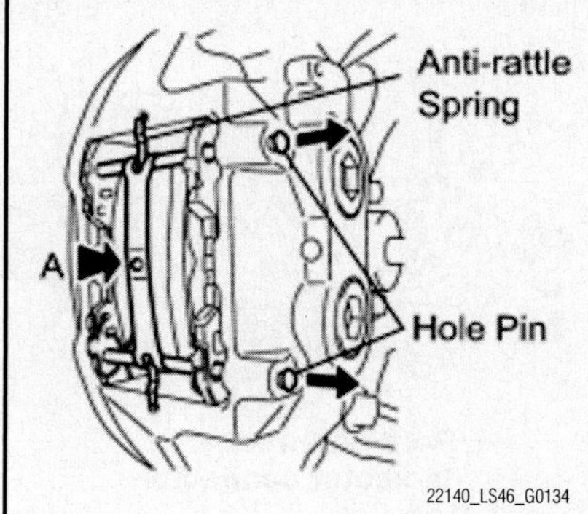

Fig. 15 Press on area "A" and remove the 2 front disc brake anti-rattles with hole pins and anti-rattle spring

➡The pin hold clip can be used again if it has sufficient rebound; no deformation or wear; and has had all rust, dirt and foreign particles cleaned off.

   b. While pressing the area labeled A, remove the 2 front disc brake anti-rattles with hole pins and front disc brake anti-rattle spring.

➡The anti-rattle spring can be used again if it has sufficient rebound; no deformation, cracks or wear; and has had all rust, dirt and foreign particles cleaned off.

   c. Remove the front disc brake anti-rattle spring.
   d. Remove the 2 pads from the disc brake caliper.
   e. Remove the No. 1 and No. 2 anti-squeal shims and anti-rattle spring from each pad.

4. Remove the pad wear indicator wire as follows:
   a. Disconnect the pad wear indicator wire connector (labeled A).
   b. Detach the clamp (labeled B) and bleeder plug cap (labeled C).
   c. Remove the retainer and pad wear indicator wire from the inner pad.

To install:

➡When replacing worn pads, the anti-squeal shims must be replaced together with the pad.

➡When installing the shim, make sure its arrow is pointing in the direction of disc rotation for forward movement.

➡Install each shim in the correct position and direction.

➡Install each pad as shown in the illustration.

➡There should be no oil or grease on the friction surface of the pads and the disc.

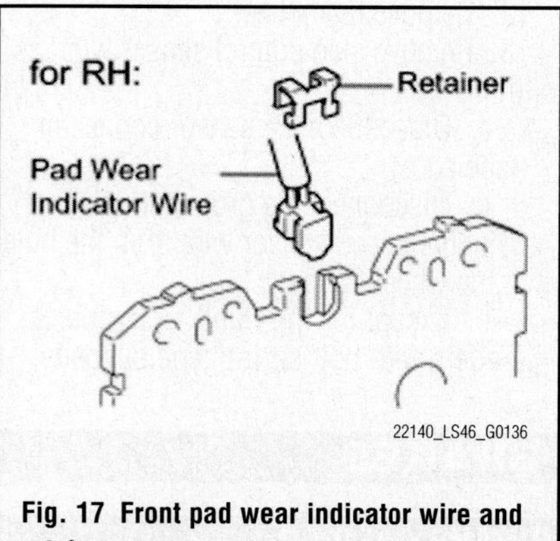

Fig. 17 Front pad wear indicator wire and retainer

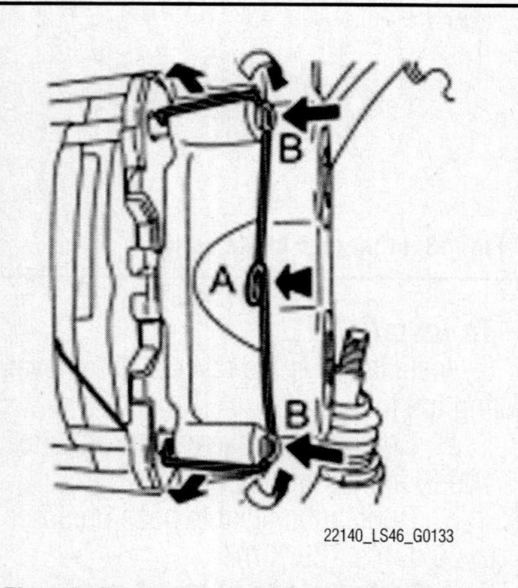

Fig. 14 Press on area "A", push the hole pin "B", and remove pin hole clip

for RH:
B
C
A
22140_LS46_G0135

Fig. 16 Front pad wear indicator, clamp, and bleeder plug cap

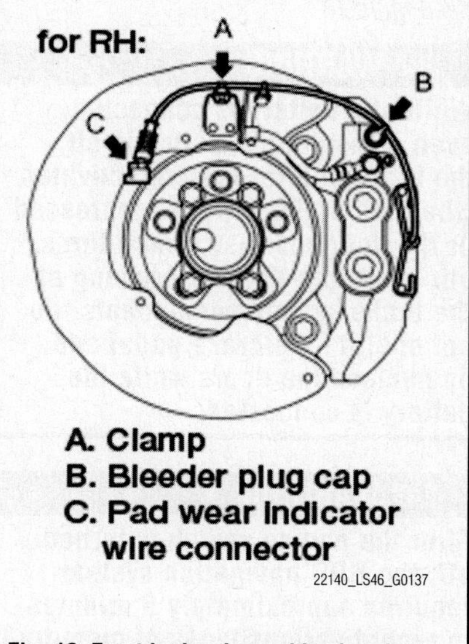

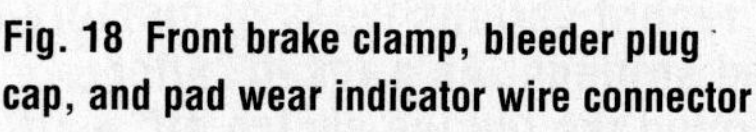

Fig. 18 Front brake clamp, bleeder plug cap, and pad wear indicator wire connector

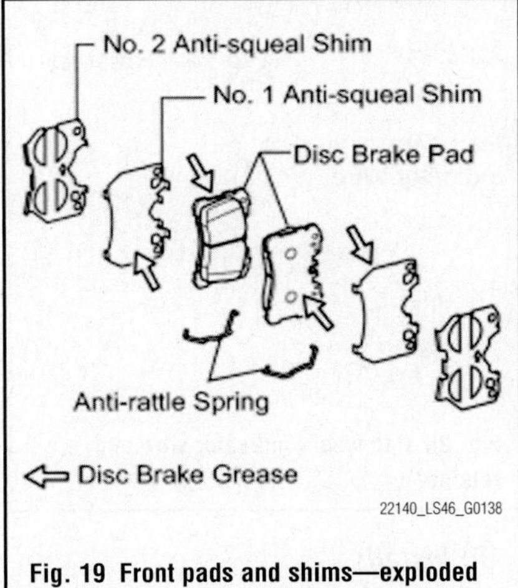

Fig. 19 Front pads and shims—exploded view

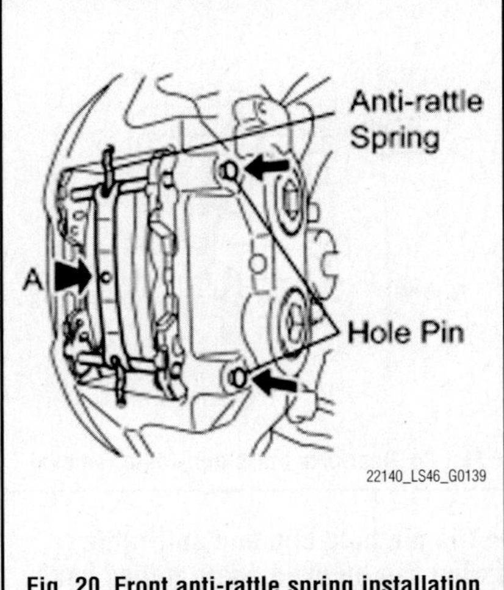

Fig. 20 Front anti-rattle spring installation

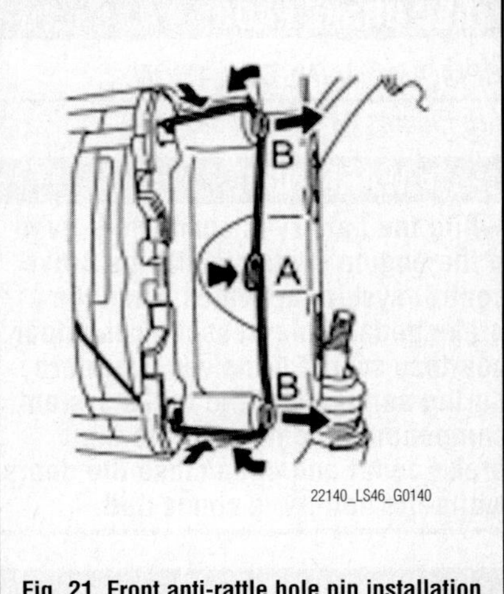

Fig. 21 Front anti-rattle hole pin installation

➡**When the brake pedal is first depressed after replacing the brake pad, DTC C1341, C1342, C1343 and/or C1344 may be output. As there is no malfunction, delete the DTC(s).**

5. Install the pad wear indicator wire as follows:
   a. Install the pad wear indicator wire and a new retainer to the inner pad.
6. Attach the clamp (labeled A) and bleeder plug cap (labeled B).
7. Connect the pad wear indicator wire connector (labeled C).
8. Install the anti-rattle spring to each pad. Install the spring lock securely in the groove of the pad.
9. Apply disc brake grease to the sides of the 2 No. 1 anti-squeal shims that contact the disc brake pad.

➡**Do not apply grease to the sides of the 2 No. 1 anti-squeal shims that contact the No. 2 anti-squeal shims.**

10. Install the No. 1 and No. 2 anti-squeal shims to each pad.
11. Install the 2 pads to the disc brake caliper.
12. Install the front disc brake anti-rattle spring between the 2 pads.

➡**The anti-rattle spring and pin hold clip can be used again if they have sufficient rebound; no deformation, cracks or wear; and have had all rust, dirt and foreign particles cleaned off.**

13. Install the front disc brake anti-rattle spring to the disc brake caliper.
14. While pressing the area labeled A, install the 2 front disc brake anti-rattles with hole pins.
15. While pressing the area labeled A, slightly pull out the hole pin (labeled B) from the brake caliper, and install the pin hold clip.
16. Install the front wheel.

BRAKES — REAR DISC BRAKES

BRAKE CALIPER

REMOVAL & INSTALLATION

See Figure 22.

⁂ WARNING

While the battery is connected, even if the engine switch is off, the brake control system activates when the brake pedal is depressed or the door courtesy switch turns on. Therefore during servicing of the brake system components, do not operate the brake pedal and open/close the doors while the battery is connected.

⁂ WARNING

After the engine switch is turned off, the HDD navigation system requires approximately 6 minutes to record various types of memory and settings. As a result, after turning the engine switch off, wait 6 minutes or more before disconnecting the cable from the negative battery terminal.

1. Remove the right cowl top ventilator louver.
2. Disconnect cable from negative battery terminal.

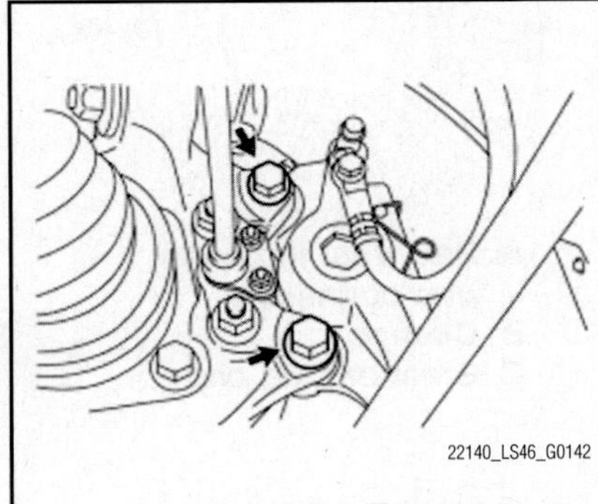

Fig. 22 Rear disc brake caliper and bolts

⁂ CAUTION

Wait at least 90 seconds after disconnecting the cable from the negative battery terminal to prevent airbag and seat belt pretensioner activation.

3. Remove the rear wheel.
4. Remove disc brake pad kit.
5. Drain brake fluid.
6. Remove the union bolt and gasket from the disc brake caliper, then disconnect the flexible hose.
7. Remove the 2 bolts and disc brake caliper from the knuckle.

To install:

8. The remainder of installation is the reverse of removal, noting the following:
   a. Tighten the brake caliper bolts to 63 ft. lbs. (86 Nm).
   b. Tighten the flexible hose union bolt to 29 ft. lbs. (39 Nm).
9. Fill reservoir and bleed brake system.

DISC BRAKE PADS

REMOVAL & INSTALLATION

See Figures 23 through 30.

⁂ WARNING

While the battery is connected, even if the engine switch is off, the brake control system activates when the brake pedal is depressed or the door courtesy switch turns on. Therefore during servicing of the brake system components, do not operate the brake pedal and open/close the doors while the battery is connected.

⁂ WARNING

After the engine switch is turned off, the HDD navigation system requires approximately 6 minutes to record various types of memory and settings. As a result, after turning the engine switch off, wait 6 minutes or more before disconnecting the cable from the negative battery terminal.

1. Remove the right cowl top ventilator louver.
2. Disconnect cable from negative battery terminal.

⁂ CAUTION

Wait at least 90 seconds after disconnecting the cable from the negative battery terminal to prevent airbag and seat belt pretensioner activation.

3. Remove the rear wheel.
4. Remove disc brake pad kit, as follows:
5. While pressing the area labeled A, push the hole pin (labeled B) toward the brake caliper, and remove the pin hold clip.

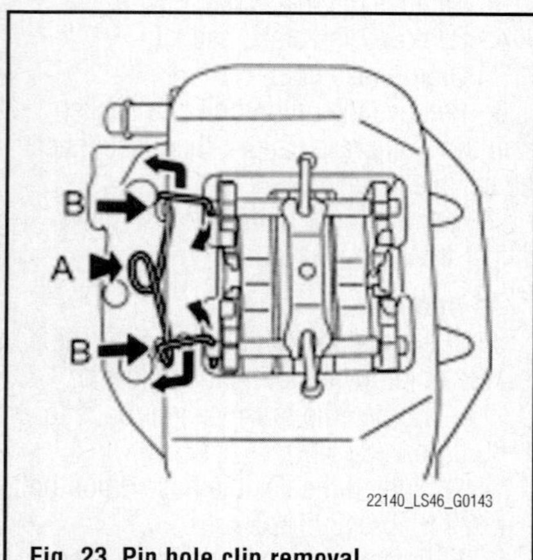

Fig. 23 Pin hole clip removal

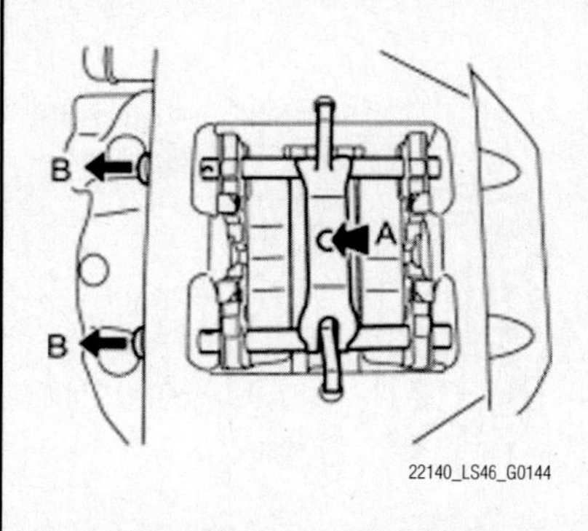

Fig. 24 Rear disc brake anti-rattle removal

➡The pin hold clip and anti-rattle spring can be used again if they have sufficient rebound; no deformation or wear; and have had all rust, dirt and foreign particles cleaned off.

6. While pressing the area labeled A, remove the 2 rear disc brake anti-rattles with hole pins (labeled B).
7. Remove the rear disc brake anti-rattle spring.
8. Remove the 2 pads from the disc brake caliper.
9. Remove the 2 anti-squeal shims and anti-rattle spring from each pad.
10. Remove the pad wear indicator wire as follows:
    a. Disconnect the pad wear indicator wire connector (labeled A).
    b. Detach the 3 clamps (labeled B) and bleeder plug cap (labeled C).
    c. Remove the retainer and pad wear indicator wire from the inner pad.

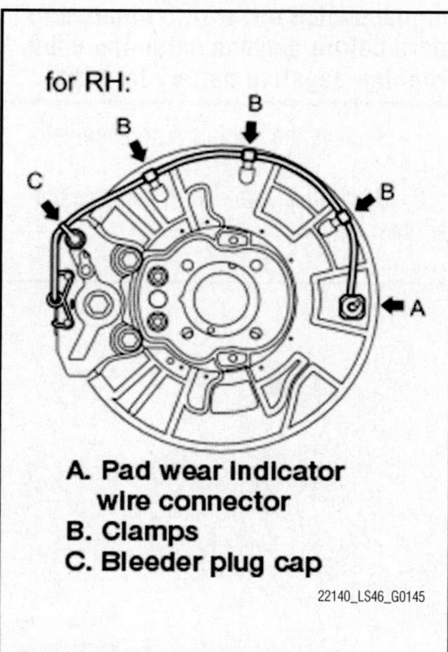

Fig. 25 Pad wear indicator wire connector, clamps, and bleeder plug cap

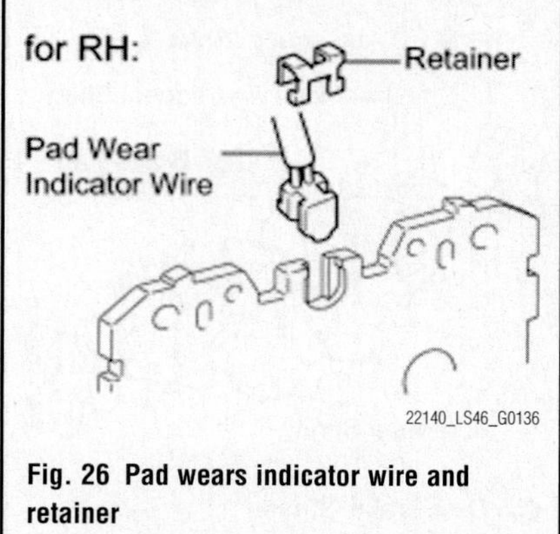

Fig. 26 Pad wears indicator wire and retainer

To install:

➡When replacing worn pads, the anti-squeal shims must be replaced together with the pad.

➡When installing the shim, make sure its arrow is pointing in the direction of disc rotation for forward movement.

➡Install each shim in the correct position and direction.

➡Install each pad as shown in the illustration.

➡There should be no oil or grease on the friction surface of the pads and the disc.

➡When the brake pedal is first depressed after replacing the brake pad, DTC C1341, C1342, C1343 and/or C1344 may be output. As there is no malfunction, delete the DTC(s).

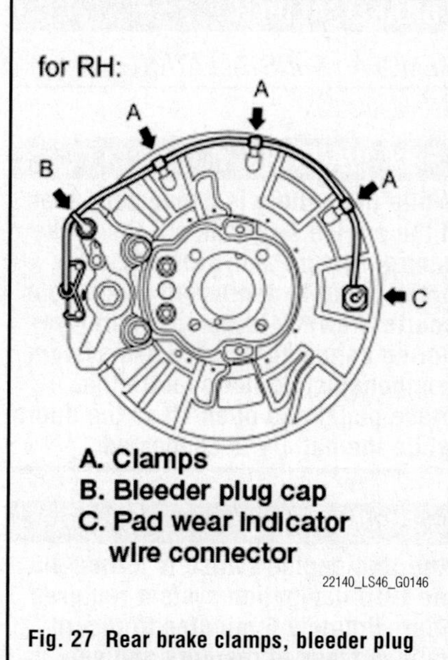

Fig. 27 Rear brake clamps, bleeder plug cap, and pad wear indicator connector

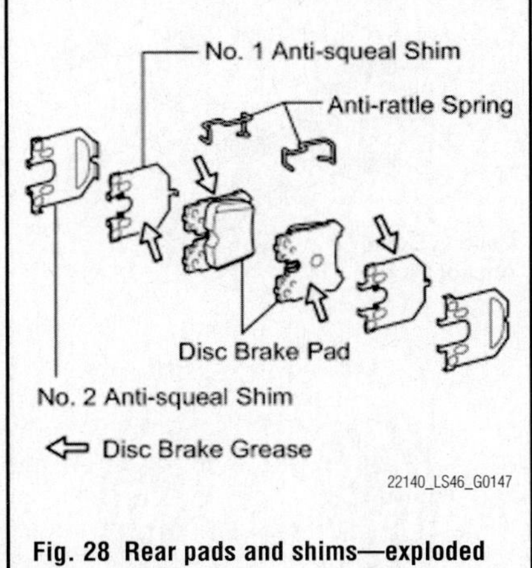

Fig. 28 Rear pads and shims—exploded view

11. Install the pad wear indicator wire as follows:
   a. Install the pad wear indicator wire and a new retainer to the inner pad.
12. Attach the 3 clamps (labeled A) and bleeder plug cap (labeled B).
13. Connect the pad wear indicator wire connector (labeled C).
14. Install the anti-rattle spring to each pad. Install the spring lock securely in the groove of the pad.

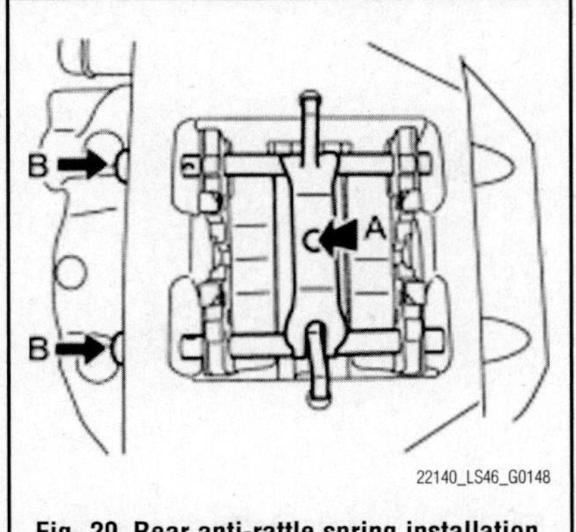

Fig. 29 Rear anti-rattle spring installation

15. Apply disc brake grease to the sides of the 2 No. 1 anti-squeal shims that contact the disc brake pad.

➡Do not apply grease to the sides of the 2 No. 1 anti-squeal shims that contact the No. 2 anti-squeal shims.

16. Install the No. 1 and No. 2 anti-squeal shims to each pad.
17. Install the 2 pads to the disc brake caliper.

➡The anti-rattle spring and pin hold clip can be used again if they have sufficient rebound; no deformation, cracks or wear; and have had all rust, dirt and foreign particles cleaned off.

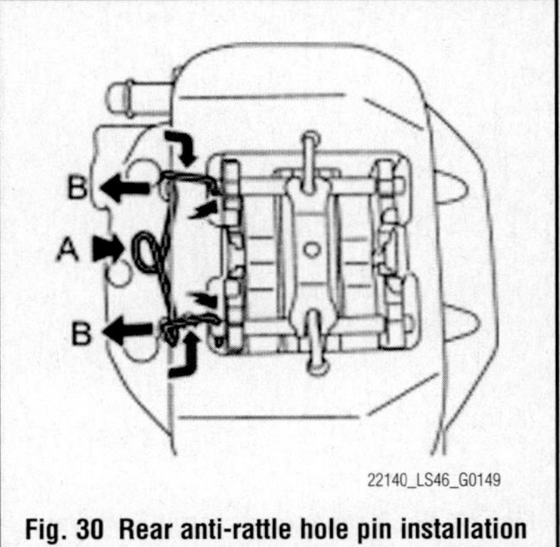

Fig. 30 Rear anti-rattle hole pin installation

18. Install the rear disc brake anti-rattle spring to the disc brake caliper.
19. While pressing the area labeled A, install the 2 rear disc brake anti-rattles with hole pins (labeled B).
20. While pressing the area labeled A, slightly pull out the hole pin (labeled B) from the brake caliper, and install the pin hold clip.
21. Install the front wheel.

BRAKES — **PARKING BRAKE**

PARKING BRAKE SHOES

REMOVAL & INSTALLATION

See Figures 31 through 40.

⁂ WARNING

While the battery is connected, even if the engine switch is off, the brake control system activates when the brake pedal is depressed or the door courtesy switch turns on. Therefore during servicing of the brake system components, do not operate the brake pedal and open/close the doors while the battery is connected.

⁂ WARNING

After the engine switch is turned off, the HDD navigation system requires approximately 6 minutes recording various types of memory and settings. As a result, after turning the engine switch off, wait 6 minutes or more before disconnecting the cable from the negative battery terminal.

1. Remove the right cowl top ventilator louver.
2. Disconnect cable from negative battery terminal.

⁂ CAUTION

Wait at least 90 seconds after disconnecting the cable from the negative battery terminal to prevent airbag and seat belt pretensioner activation.

3. Remove the 2 bolts and disconnect the rear disc brake caliper. Hang the caliper with wire or equivalent.

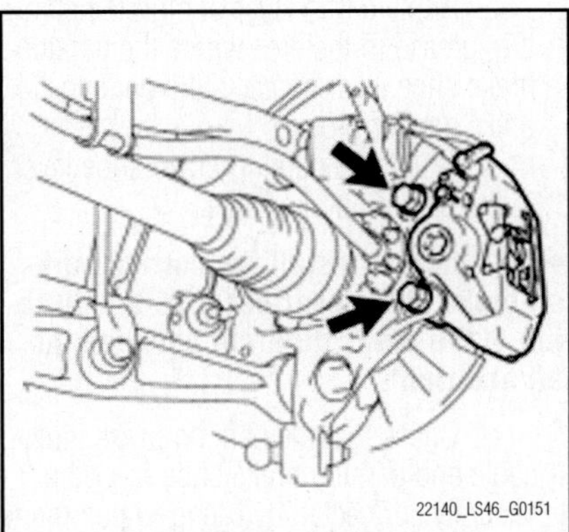

Fig. 31 Disconnect rear caliper

4. Put matchmarks on the rear disc and rear axle hub.
5. Remove the rear disc.

➡If the rear disc is difficult to remove, turn the adjustment screw in the contraction direction as shown in the illustration to make the disc easier to remove.

6. Using SST (SST: 09703-30011), remove the No. 2 and No. 1 parking brake shoe return tension springs.

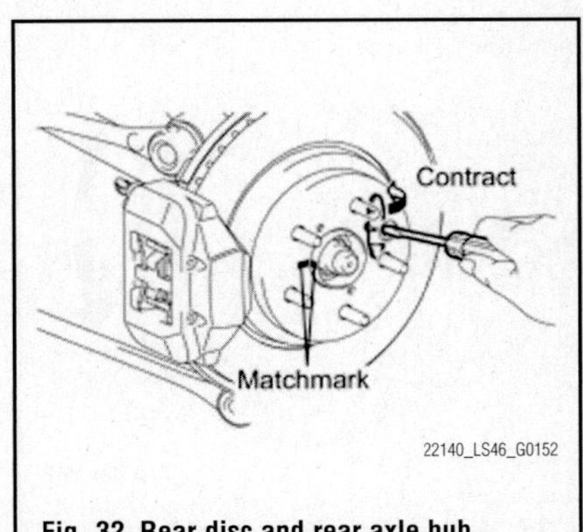

Fig. 32 Rear disc and rear axle hub matchmarks

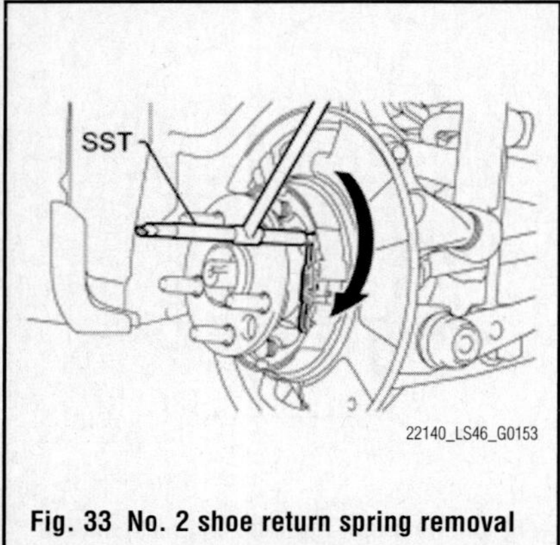

Fig. 33 No. 2 shoe return spring removal

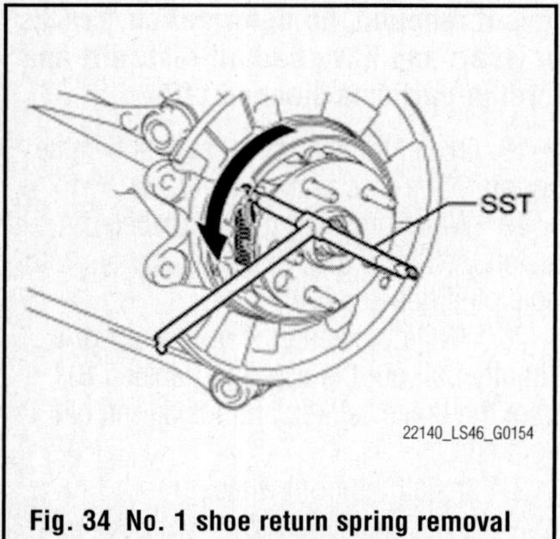

Fig. 34 No. 1 shoe return spring removal

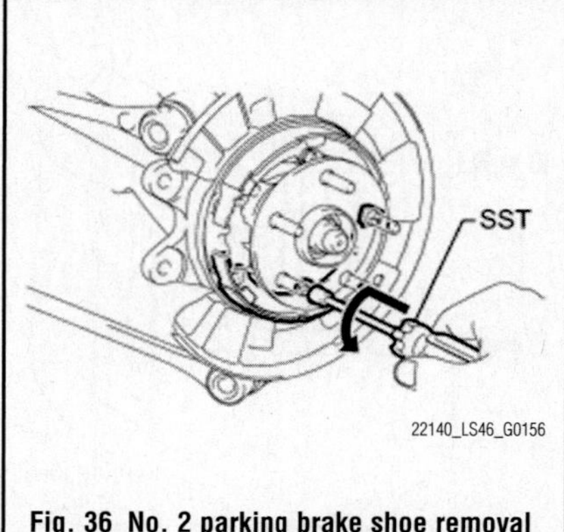

Fig. 36 No. 2 parking brake shoe removal

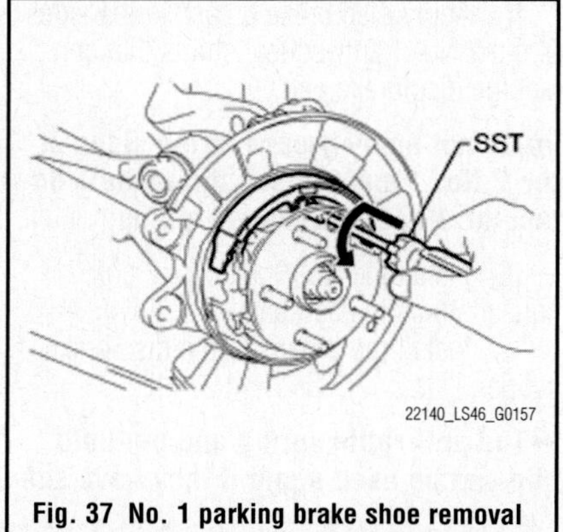

Fig. 37 No. 1 parking brake shoe removal

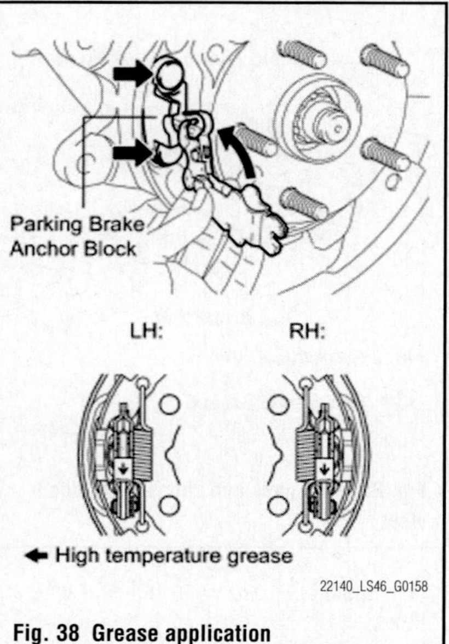

Fig. 38 Grease application

7. Slide the parking brake shoe, and remove the parking brake shoe adjusting screw set.

8. Using SST (SST: 09718-00010), remove the No. 2 and No. 1 shoe hold down spring cups, No. 2 and No. 1 compression springs and No. 1 shoe hold down spring pin. Use the service hole to retain the No. 1 shoe hold down spring pin with your finger.

9. Remove the No. 2 parking brake shoe

To install:

10. Using SST (SST: 09718-00010), install the No. 2 and No. 1 parking brake shoes with the No. 2 and No. 1 shoe hold down spring cups, No. 2 and No. 1 compression springs and No. 1 shoe hold down spring pin. Use the service hole to retain the No. 1 shoe hold down spring pin with your finger.

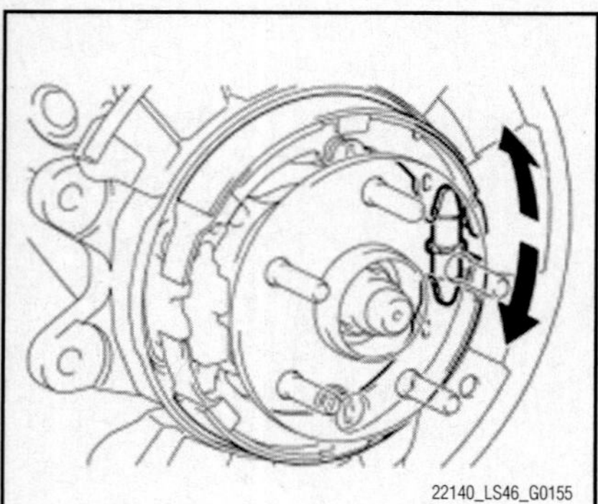

Fig. 35 Parking brake shoe adjusting screw set removal

11. Apply high temperature grease to the thread and all joining areas of the parking brake shoe adjusting screw set.

12. Install the parking brake shoe adjusting screw set.

13. Install the No. 2 parking brake shoe return spring.

14. Install the parking brake shoe lever, as follows:

a. Apply a thin layer of high temperature grease to the area where the parking brake shoe lever contacts the parking brake anchor block.

b. Install the parking brake shoe lever to the parking brake cable.

➡Take care to install the correct parking brake shoe lever because the direction of the pin is different between the left and right.

c. Check that the parking brake inner cable and parking brake shoe lever are securely connected by trying to pull them apart.

d. Install the parking brake cable to the support bracket with the 2 nuts and tighten to 71 inch lbs. (8 Nm).

e. With the vehicle on the ground, check that the gap between the parking brake cable and drive shaft, and the gap between the parking brake cable and shaft boot are 10 mm (0.394 in.) or more. Make sure that the cable is not twisted.

15. Install the No. 1 parking brake shoe return spring.

16. Check parking brake installation. If necessary, reinstall the parts properly.

17. Align the matchmarks on the rear disc and rear axle hub, and install the rear disc.

18. Connect the rear disc brake caliper with 2 new bolts to63 ft. lbs. (86 Nm).

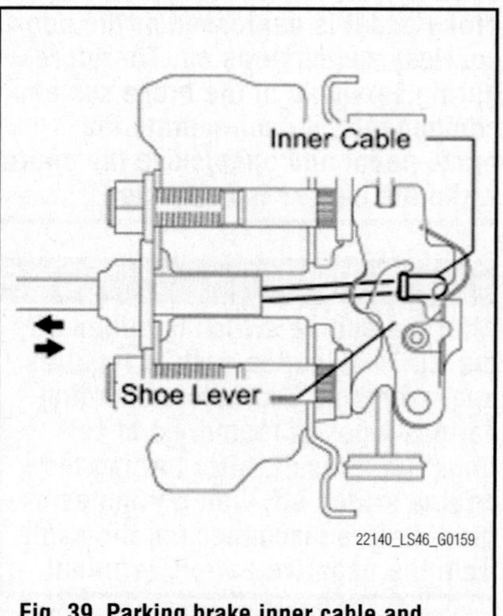

Fig. 39 Parking brake inner cable and shoe lever connection

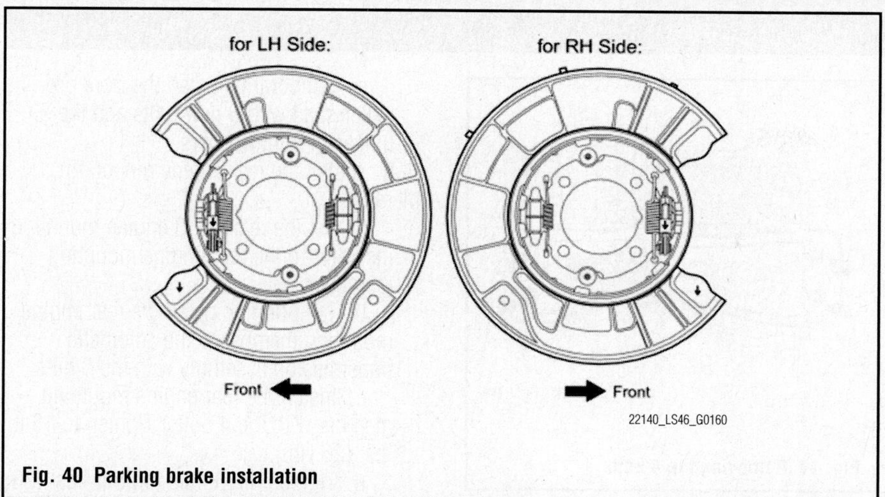

Fig. 40 Parking brake installation

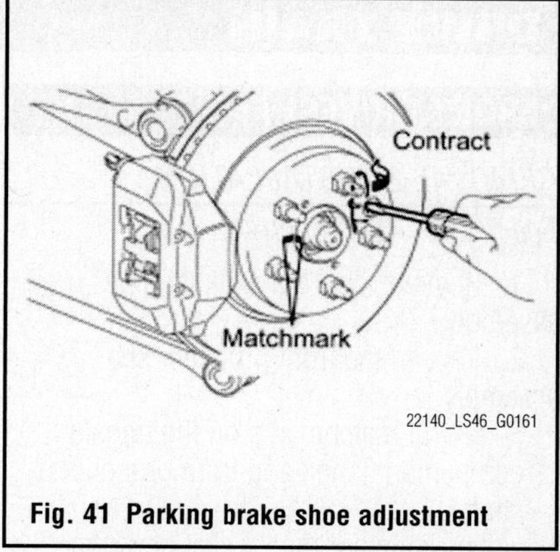

Fig. 41 Parking brake shoe adjustment

➡Do not twist the brake hose, Make sure that the bolts are free from damage and foreign matter. Do not overtighten the bolts.

19. Adjust parking brake shoe clearance.
20. Install the rear wheel
21. Perform parking brake shoe bedding.

ADJUSTMENT

See Figure 41.

1. Remove the rear wheel.
2. Temporarily install the hub nuts.
3. With the engine switch on, operate the electric parking brake switch to release the parking brake, and then turn the engine switch off.
4. Remove the shoe adjusting hole plug, and rotate the rear disc so that the service hole is aligned with the adjusting screw.
5. Using a screwdriver, turn the adjusting screw of the parking brake shoe in the expansion direction until the disc locks.
6. With the engine switch on (IG), operate the electric parking brake switch to lock and release the parking brake. Repeat again. Then turn the engine switch off.

➡Make sure that the parking brake is released.

7. Turn the adjusting screw again in the expansion direction to lock the disc.
8. Loosen the adjusting screw so that the rear disc can rotate slightly.

➡The standard number of return notches is 7.

9. Check that there is no brake drag.
10. Install the shoe adjusting hole plug and remove the hub nuts.
11. Install the rear wheel.

CHASSIS ELECTRICAL — AIR BAG (SUPPLEMENTAL RESTRAINT SYSTEM)

GENERAL INFORMATION

CAUTION

These vehicles are equipped with an air bag system. The system must be disarmed before performing service on, or around, system components, the steering column, instrument panel components, wiring and sensors. Failure to follow the safety precautions and the disarming procedure could result in accidental air bag deployment, possible injury and unnecessary system repairs.

SERVICE PRECAUTIONS

CAUTION

Disconnect and isolate the battery negative cable before beginning any airbag system component diagnosis, testing, removal, or installation procedures. Wait at least 90 seconds after the ignition switch is turned off and the negative (-) terminal cable is disconnected from the battery before starting the operation. The SRS is equipped with a backup power source, so if work is started within 90 seconds after disconnecting the negative (-) terminal cable from the battery, the SRS may be deployed. Failure to disable the airbag system may result in accidental airbag deployment, personal injury, or death.

DISARMING THE SYSTEM

To avoid personal injury when working on vehicles equipped with an air bag, the negative battery cable must be disconnected and at least 90 seconds must elapse before working on the system. Failure to do so may result in deployment of the air bag.

After the engine switch is turned off, the HDD navigation system requires approximately 6 minutes to record various types of memory and settings. As a result, after turning the engine switch off, wait 6 minutes or more before disconnecting the cable from the negative (-) battery terminal.

When the cable is disconnected from the negative battery terminal, the memory settings of each system will be cleared. Because of this, be sure to write down the settings of each system before starting work. When work is finished, reset the settings of each system as before. Never use a backup power supply from outside the vehicle to avoid erasing the memory in a system.

ARMING THE SYSTEM

To arm the system after service is finished, connect the negative battery cable.

DRIVE TRAIN

FRONT DRIVESHAFT

REMOVAL & INSTALLATION

See Figures 42 through 45.

1. Remove the front exhaust pipe assembly.
2. Loosen the front propeller shaft assembly.
    a. Put matchmarks on the transfer companion flange and front propeller shaft.
    b. Using a 6 mm socket hexagon wrench, loosen the 6 bolts.

➡Be careful not to damage the front propeller shaft.

    c. Put matchmarks on the front differential companion flange and front propeller shaft.
    d. Using a 6 mm socket hexagon wrench, loosen the 6 bolts.

➡Be careful not to damage the front propeller shaft.

3. Remove the rear engine mounting member.
    a. Support the automatic transmission assembly with a transmission jack.

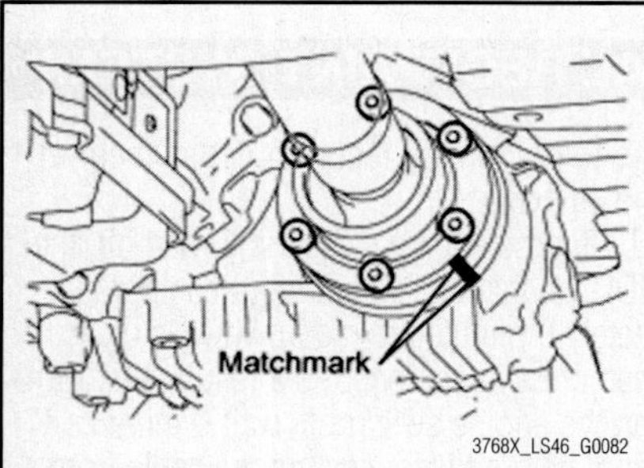

Fig. 42 Putting matchmarks on the transfer companion flange and the front propeller shaft

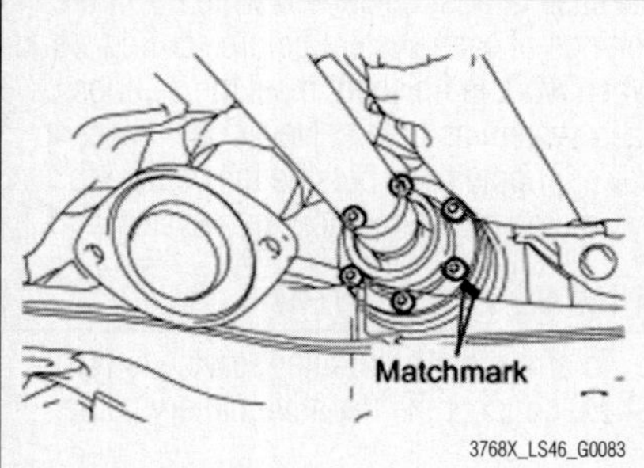

Fig. 43 Putting matchmarks on the front differential companion flange and the front propeller shaft

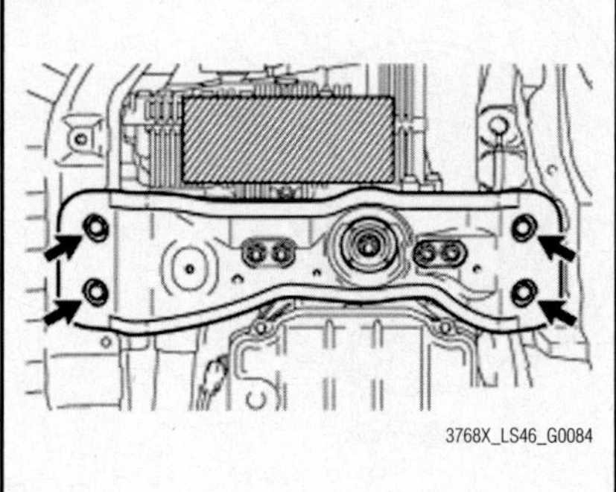

Fig. 44 Removing the 4 bolts

➡Keep the transmission jack level with the weight properly supported.

    b. Remove the 4 bolts.
    c. Remove the 5 nuts and rear No. 3 engine mounting insulator, and remove the rear engine mounting member from the automatic transmission assembly.

4. Remove the front propeller shaft assembly.
    a. Remove the 12 bolts, 4 universal joint washers and the front propeller shaft assembly.

➡Do not bend the front propeller shaft at an excessive angle.

To install:

5. Temporarily install the front propeller shaft assembly.
    a. Align the matchmarks on the differential companion flange and the front propeller shaft.
    b. Temporarily install the front propeller shaft with 6 new bolts and the 2 universal joint washers.
    c. Align the matchmarks on the transfer companion flange and the front propeller shaft.

Fig. 45 Removing the rear engine mounting member from the automatic transmission assembly

    d. Temporarily install the front propeller shaft with 6 new bolts and the 2 universal joint washers.

6. Install the rear engine mounting member.
    a. Set the rear No. 3 engine mounting insulator to the rear engine mounting member.
    b. Temporarily install the rear engine mounting member to the automatic transmission assembly with the 5 nuts.
    c. Install the rear engine mounting member with the 4 bolts. Tighten to 19 ft. lbs. (26 Nm).
    d. Tighten the 5 nuts to 28 ft. lbs. (38 Nm).
7. Tighten the front propeller shaft assembly.
    a. Using a 6 mm socket hexagon wrench, tighten the front propeller shaft with the 6 bolts. Tighten to 18 ft. lbs. (25 Nm).

➡Be careful not to damage the front propeller shaft.

8. Install the front exhaust pipe assembly.
9. Inspect for exhaust gas leaks.

FRONT DIFFERENTIAL CARRIER

REMOVAL & INSTALLATION

AWD Models

See Figures 46 and 47.

1. Place the front wheels facing straight ahead.
2. Drain the differential oil.
3. Remove the engine assembly.
4. Install the engine on an engine stand.
5. Remove the front differential carrier assembly.
    a. Support the rear differential carrier with a jack.

CAUTION

As the differential carrier assembly is very heavy, securely support it with the jack.

    b. Remove the 4 bolts and differential carrier.

➡Do not damage the installation surface when removing the differential carrier.

➡The remaining oil may leak out when removing the differential carrier.

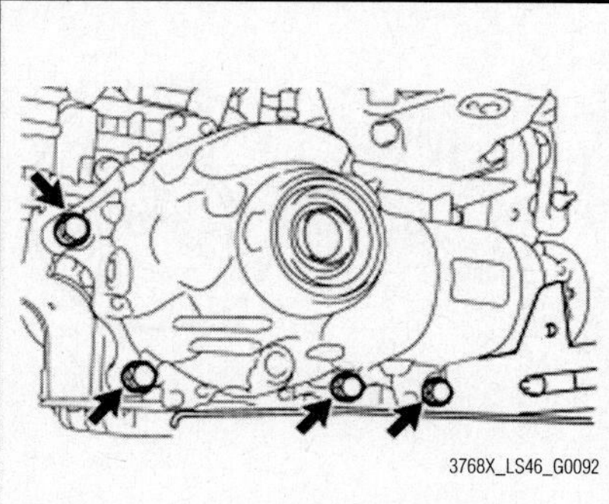

Fig. 46 Removing the front differential carrier

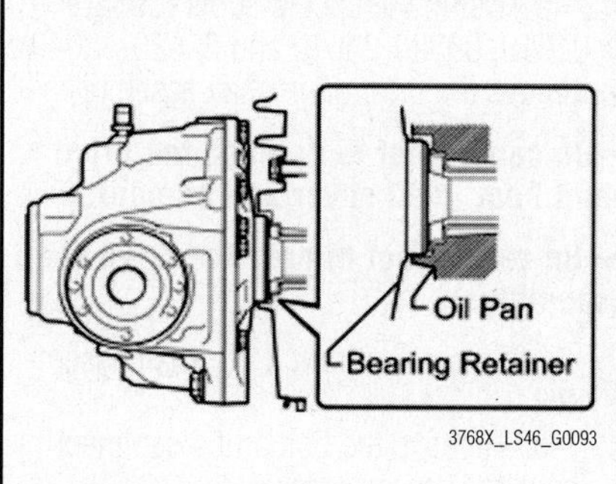

Fig. 47 Aligning the side bearing retainer and the oil pan

To install:

6. Install the front differential carrier assembly.

a. Support the differential carrier assembly with a jack and temporarily install it to the oil pan with the 4 bolts.

CAUTION

As the differential carrier assembly is very heavy, securely support it with the jack.

b. Align the side bearing retainer and oil pan so that their ends are aligned and tighten the 4 bolts to install the differential carrier assembly. Tighten to 60 ft. lbs. (81 Nm).

7. Install the engine assembly to the vehicle.

a. Remove the engine from the engine stand and then install it to the vehicle.

8. Add differential oil.

FRONT DIFFERENTIAL CARRIER OIL SEAL

REPLACEMENT

See Figures 48 through 53.

1. Place the front wheels facing straight ahead.
2. Drain the differential oil.
3. Remove the engine assembly.
4. Install the engine on an engine stand.
5. Remove the front differential carrier assembly.
6. Fix the front differential carrier assembly to an overhaul stand.
7. Remove the front differential carrier retainer sub assembly.
8. Remove the No. 1 front differential case sub assembly.
9. Fix the differential carrier.
10. Remove the front drive pinion companion flange front nut.
11. Remove the front drive pinion companion flange front sub assembly with the deflector.
12. Remove the front differential carrier oil seal.

a. Using the special tool (09308-00010), remove the oil seal from the differential carrier.

13. Remove the front differential drive pinion oil slinger.
14. Remove the differential drive pinion with the bearing inner race.

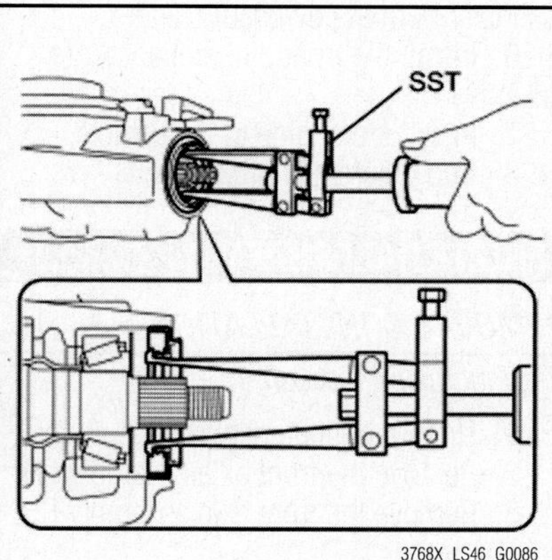

Fig. 48 Removing the oil seal from the differential carrier

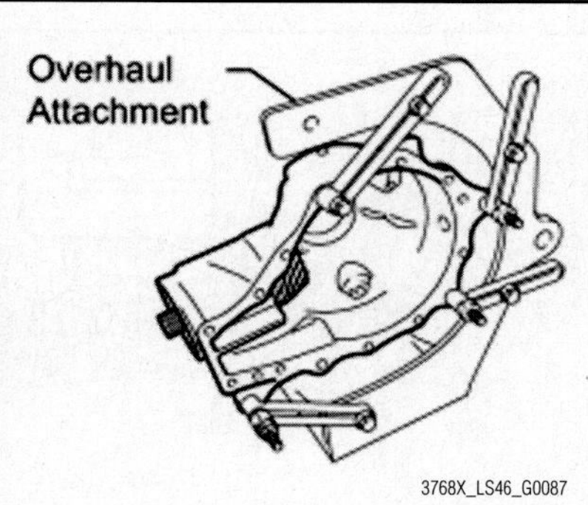

Fig. 49 Removing the differential carrier together with the overhaul attachment

a. Remove the differential carrier together with the overhaul attachment from the engine stand.

b. Install the differential carrier to the attachment.

c. Using a press, remove the drive pinion from the differential carrier.

➡**Do not drop the drive pinion or bearing inner race.**

To install:

15. Replace the front differential drive pinion bearing spacer.

a. Replace the drive pinion bearing spacer with a new one and install it to the drive pinion.

16. Install the differential drive pinion with the bearing inner race.

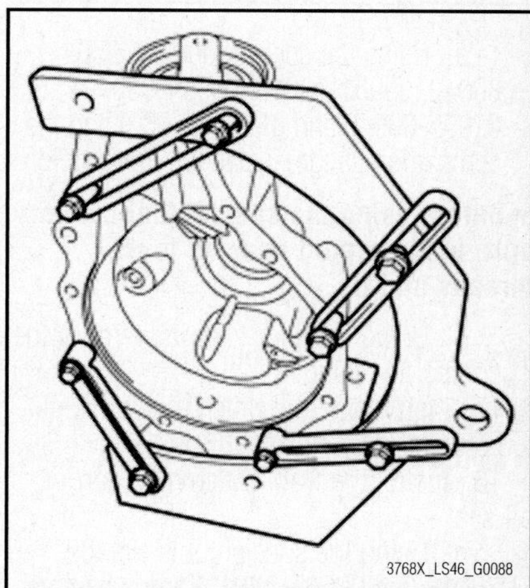

Fig. 50 Installing the differential carrier to the attachment

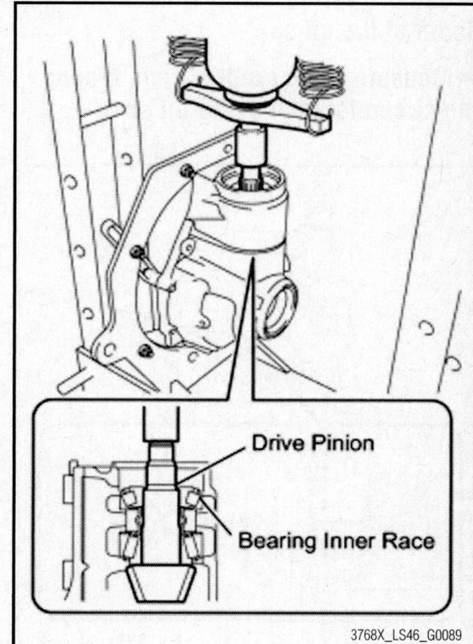

Fig. 51 Removing the drive pinion from the differential carrier

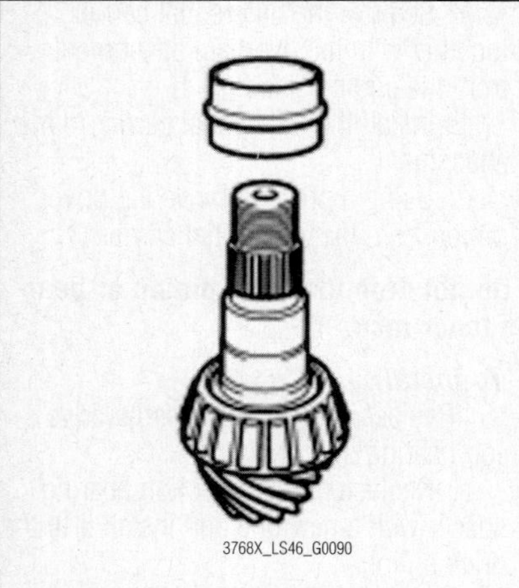

Fig. 52 Replacing the drive pinion bearing spacer

a. Using the special tools (09950-30012, 09951-03010, 09954-03010, 09955-03040 and 09956-03060) and the companion flange, install the drive pinion.

➡Before using the special tool center bolt, apply hypoid gear oil to its threads and tip.

b. Using the special tools, remove the companion flange.

17. Install the front differential drive pinion oil slinger.

18. Install the front differential carrier oil seal.

a. Using the special tools (09309-36010 and 09502-24010) and a hammer, tap in a new oil seal. The standard depth is 0.0244–0.0283 inch (3.2–7.2 mm).

➡Using a vernier caliper, measure the depth of the oil seal.

➡Measure at 3 or more areas around the circumference of the oil seal

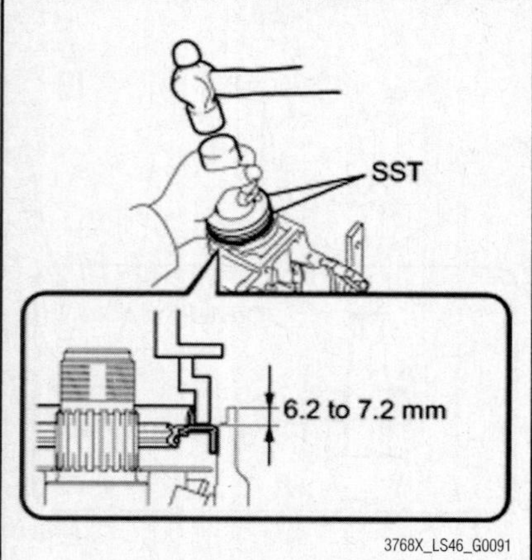

Fig. 53 Installing the front differential carrier oil seal

➡Make sure the difference between the maximum and minimum measured values is less than 0.0295 inch (0.75 mm), as a greater difference may lead to oil leaks.

➡Tap the oil seal uniformly so that the oil seal is straight.

➡Do not excessively tap the oil seal.

➡First, uniformly tap in the oil seal until it is flush with the edge of the carrier, then tap it little by little until the depth is within the standard range.

b. Apply MP grease to the lip of the oil seal.

19. Install the front drive pinion companion flange front sub assembly with the deflector.

20. Inspect the drive pinion preload.

21. Install the No. 1 front differential case sub assembly.

22. Install the front differential carrier retainer sub assembly.

23. Inspect the total preload.

24. Stake the front drive pinion companion flange front nut.

25. Remove the front differential carrier assembly from the overhaul stand.

26. Install the front differential carrier assembly.

27. Install the engine to the vehicle.

28. Add differential oil.

FRONT HALFSHAFT

REMOVAL & INSTALLATION

See Figures 54 through 57.

1. Drain the differential oil.
2. Remove the front axle assembly LH.
3. Remove the front axle assembly RH.

➡Use the same procedures described for the LH side.

4. Remove the front drive shaft assembly LH.

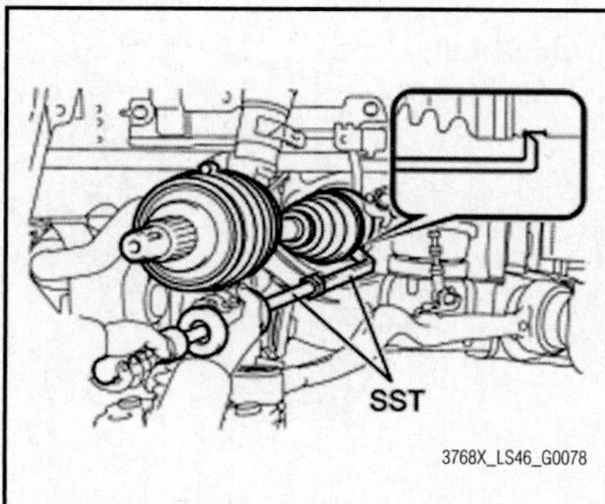

Fig. 54 Removing the front drive shaft assembly LH

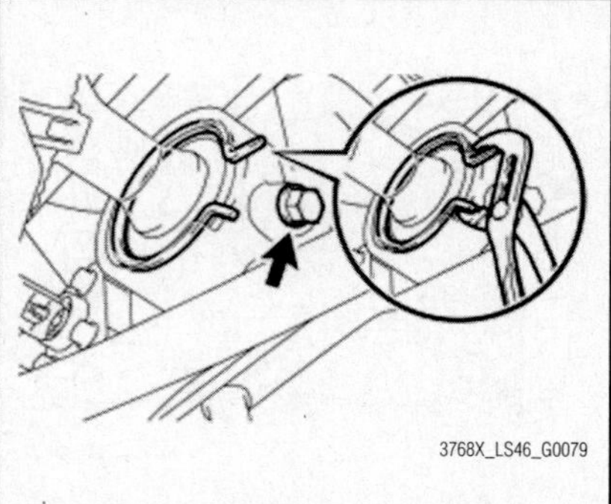

Fig. 55 Removing the bearing bracket hole snap ring

a. Using the special tools (09520-01010, 09520-24010 and 09520-32040), remove the front drive shaft assembly.

➡Be careful not to damage the drive shaft boot, dust cover and oil seal.

➡ Be careful not to drop the drive shaft assembly.

5. Remove the front drive shaft assembly RH.

a. Remove the bolt and disconnect the wire harness bracket.

b. Using water pump pliers, remove the bearing bracket hole snap ring.

c. Remove the bolt and front drive shaft assembly from the drive shaft bearing bracket.

➡Be careful not to damage the drive shaft boot, dust cover and oil seal.

➡Be careful not to drop the drive shaft assembly.

6. Inspect the front drive shaft.

a. Check that there are no excessive play in the outboard joint.

b. Check that the inboard joint slides smoothly in the thrust direction.

c. Check that there is no excessive

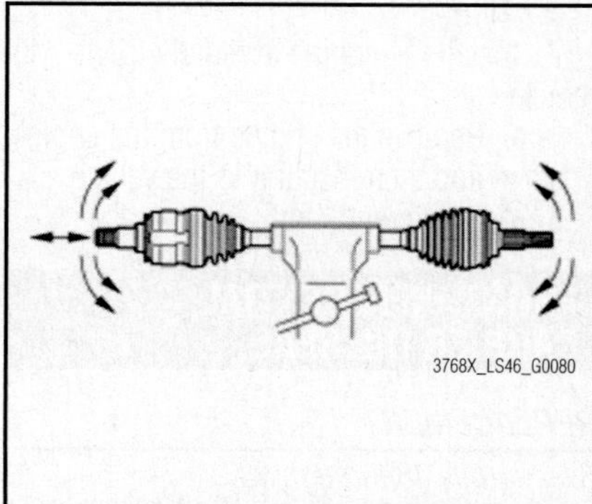

Fig. 56 Checking for excessive play in the outboard joint

play in the radial directions of the inboard joint.

d. Check the boot for damage.

To install:

7. Install the front drive shaft assembly LH.

a. Coat the spline of the inboard joint shaft assembly with gear oil.

b. Set the shaft snap ring with the opening side facing down.

c. Align the shaft splines and tap in the drive shaft assembly with a brass bar and hammer.

➡Be careful not to damage the drive shaft dust cover, boot and oil seal.

➡Move the drive shaft assembly while keeping it level.

➡Bending or sliding the drive shaft excessively may cause the tripod joint to come out of its groove. Therefore, be careful when handing or installing the drive shaft.

8. Install the front drive shaft assembly RH.

a. Coat the spline of the inboard joint shaft assembly with gear oil.

b. Install the drive shaft assembly.

c. Using water pump pliers, install a new bearing bracket hole snap ring.

➡Do not damage the boot and oil seal.

➡Move the drive shaft assembly while keeping it level.

d. Install a new bolt and tighten to 24 ft. lbs. (32 Nm).

➡Before installing the bolt, check that there is rubber on the tip of the bolt. If there is no rubber on the bolt, use a new one.

e. Install the new wire harness bracket with the bolt. Tighten to 9 ft. lbs. (12 Nm).

9. Install the front axle assembly LH.

10. Install the front axle assembly RH.

11. Add differential oil.

FRONT PINION SEAL

REMOVAL & INSTALLATION

See Figures 58 through 67.

1. Fix the front differential carrier assembly to the overhaul stand.

2. Inspect the differential ring gear backlash.

a. Set a lever probe to a dial indicator. Insert the probe into the drain plug hole and set it onto the edge of one of the teeth of the ring gear at a right angle.

b. Using the special tool (09564-50010), turn the differential case clockwise and counterclockwise and measure the ring gear backlash while holding the companion flange with your hand. The standard backlash is 0.00512–0.00905 inch (0.13–0.23 mm).

➡Measure at 3 or more areas around the circumference of the ring gear.

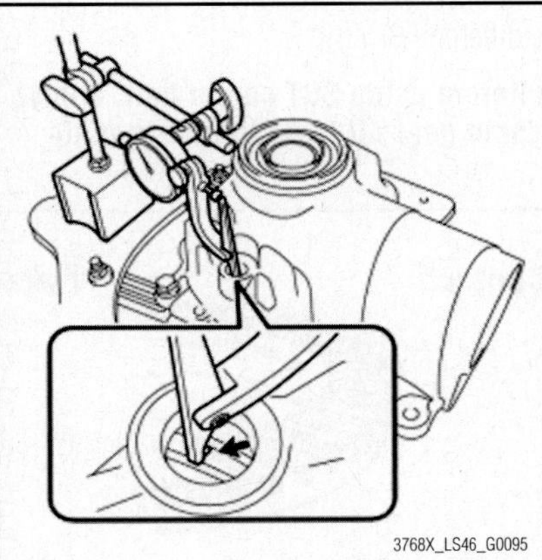

Fig. 58 Setting a lever probe to a dial indicator.

➡For reassembly purposes, record the result before disassembly.

c. If the backlash is not within the specified range, adjust the ring gear backlash or repair as necessary.

3. Remove the front differential breather plug from the differential carrier.

4. Remove the front differential case oil seal LH using the special tool (09308-00010) from the differential carrier.

a. Using the special tool (098308-00010), remove the oil seal RH from the differential carrier.

5. Remove the front differential carrier retainer sub assembly.

a. Loosen the 11 bolts.

b. Remove the differential carrier from the overhaul attachment.

c. Remove the 11 bolts.

d. Insert the blade of an oil pan seal cutter between the differential carrier and differential carrier retainer. Cut through the applied seal packing.

➡Do not damage the installation surface of the differential carrier.

e. Using a brass bar and hammer, lightly tap the retainer to remove it from the differential carrier.

6. Remove the No. 1 front differential case sub assembly.

a. Remove the differential case from the differential carrier.

➡Do not damage the differential case bearing and ring gear.

7. Inspect the tooth contact between the ring gear and the drive pinion.

8. Fix the differential carrier to the overhaul stands.

9. Remove the front differential breather plug oil deflector.

a. Remove the 2 bolts and the differential breather plug oil deflector.

10. Remove the front drive pinion companion flange front nut.

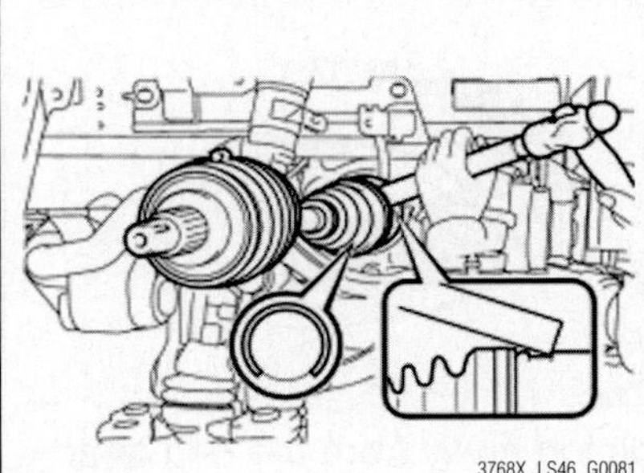

Fig. 57 Installing the front drive shaft assembly LH

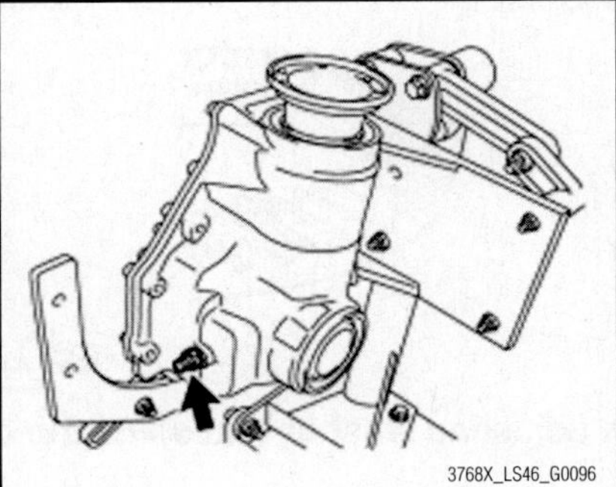

Fig. 59 Removing the front differential breather plug from the differential carrier

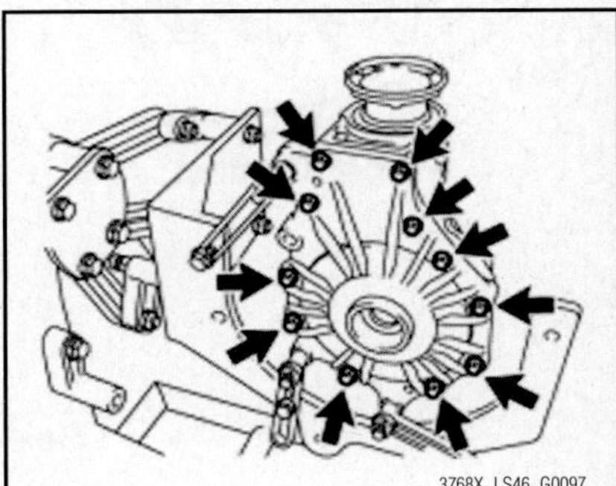

Fig. 60 Locating the front differential carrier retainer sub assembly bolts

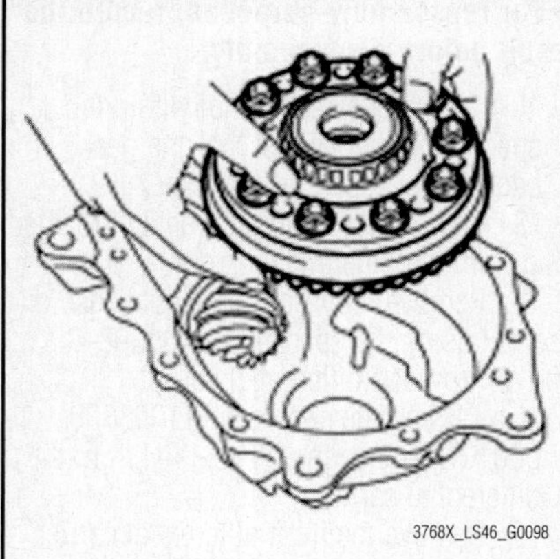

Fig. 61 Removing the No. 1 front differential case sub assembly

a. Using the special tool (09930-00010) and a hammer, unstake the staked part of the drive pinion companion flange nut.

➡**Be sure to use SST with the tapered surface facing the shaft.**

➡**Do not grind the tip of SST with a grinder etc.**

➡**Completely loosen the staked part of the nut when removing it.**

➡**Do not damage the threads of the drive pinion.**

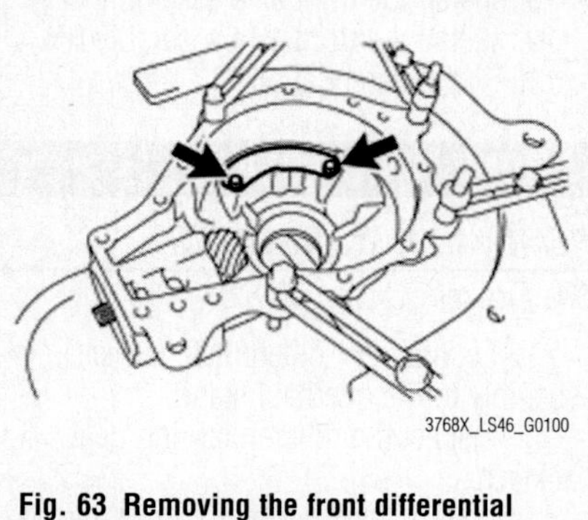

Fig. 63 Removing the front differential breather plug oil deflector

b. Using the special tools (09213-58013, 09229-55010 and 09330-00021) to hold the flange, remove the drive pinion companion flange nut.

➡**Perform the removal while supporting the overhaul attachment.**

11. remove the front drive pinion companion flange front sub assembly.

a. Using the special tools, (09951-03010, 09953-03010, 09954-03010, 09955-03030 and 09956-03020), remove the companion flange from the differential carrier.

➡**Before using SST center bolt, apply hypoid gear oil to its threads and tip.**

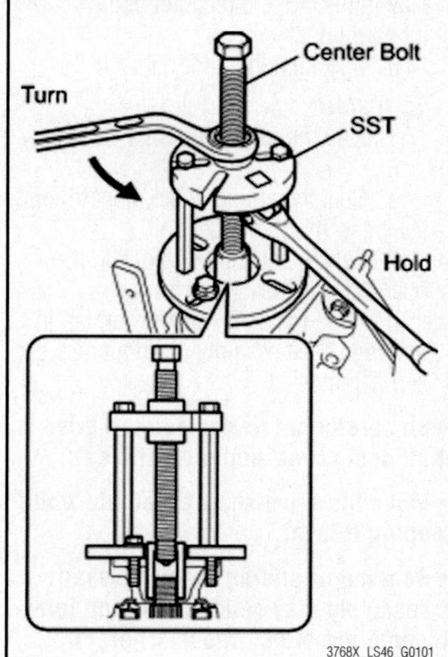

Fig. 64 Removing the front drive pinion companion flange front sub assembly

12. Remove the front differential dust deflector.

a. Using the special tools (09950-00020, 09950-60010, 09951-00430, 09950-70010 or 09951-07100) and a press, press out

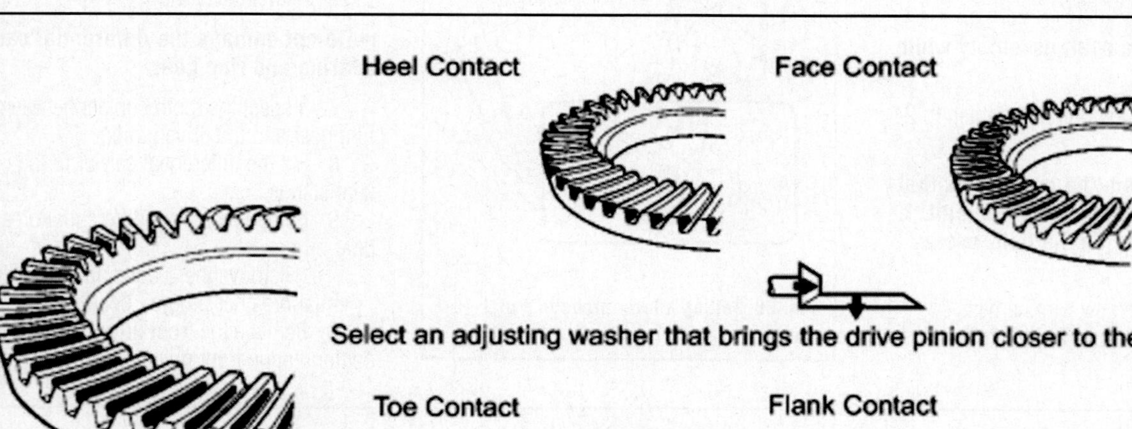

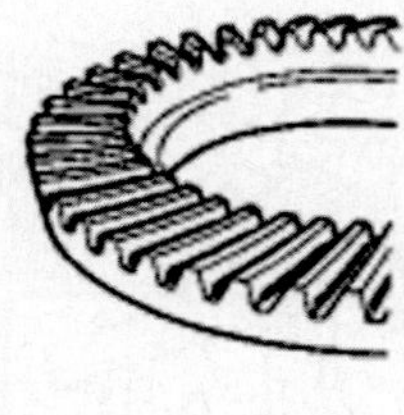

Fig. 62 Identifying tooth contact between the ring gear and the drive pinion

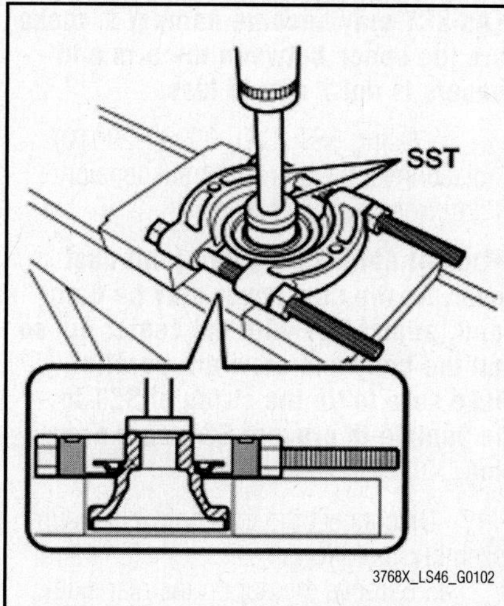

Fig. 65 Removing the front differential dust deflector

the dust deflector from the companion flange.

➡**Perform this procedure while tightening the special tool nut to secure the contact surfaces of the special tool and the dust deflector.**

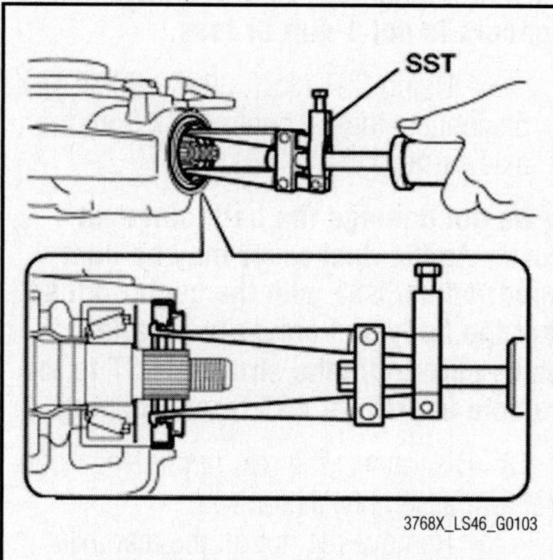

Fig. 66 Removing the differential carrier oil seal

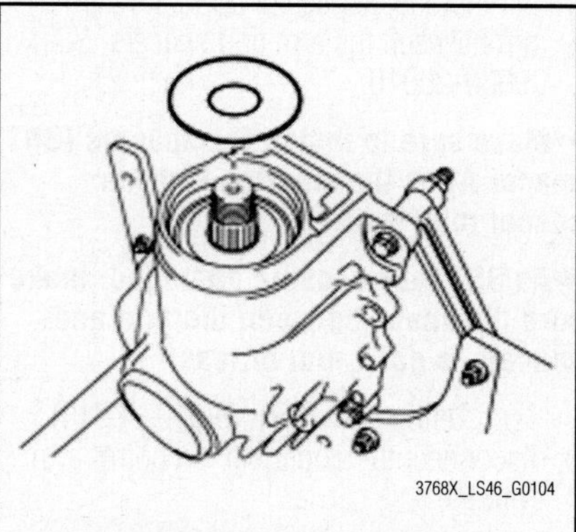

Fig. 67 Removing the front differential drive pinion oil slinger

➡**Do not drop the companion flange.**

13. Remove the differential carrier oil seal.
    a. Using the special tool (09308-00010) remove the oil seal from the differential carrier.
14. Remove the front differential drive pinion oil slinger.
    a. Remove the drive pinion oil slinger from the differential carrier.

To install:

15. To install, reverse the removal procedure.

REAR AXLE SHAFT, BEARING & SEAL

REMOVAL & INSTALLATION

See Figures 68 through 74.

1. Remove the rear wheel.
2. Remove the bolt and load sensing valve sensor bracket from the toe control link sub-assembly.
3. Remove the bolt and nut, and then disconnect the rear stabilizer link assembly from the rear No. 2 suspension arm assembly.
4. Using SST (SST: 09930-00010) and a hammer, release the staked part of the rear axle shaft nut. Release the staked part of the nut completely; otherwise the screw of the drive shaft may be damaged.
5. While applying the brakes, remove the rear axle shaft nut.
6. Remove the 2 bolts and separate the rear disc brake caliper assembly.

➡**Hang the caliper with wire or equivalent. Do not damage the brake hose.**

7. Disconnect the rear speed sensor.
8. Remove the rear disc.
9. Remove the No. 2 and No. 1 parking brake shoe return tension springs.

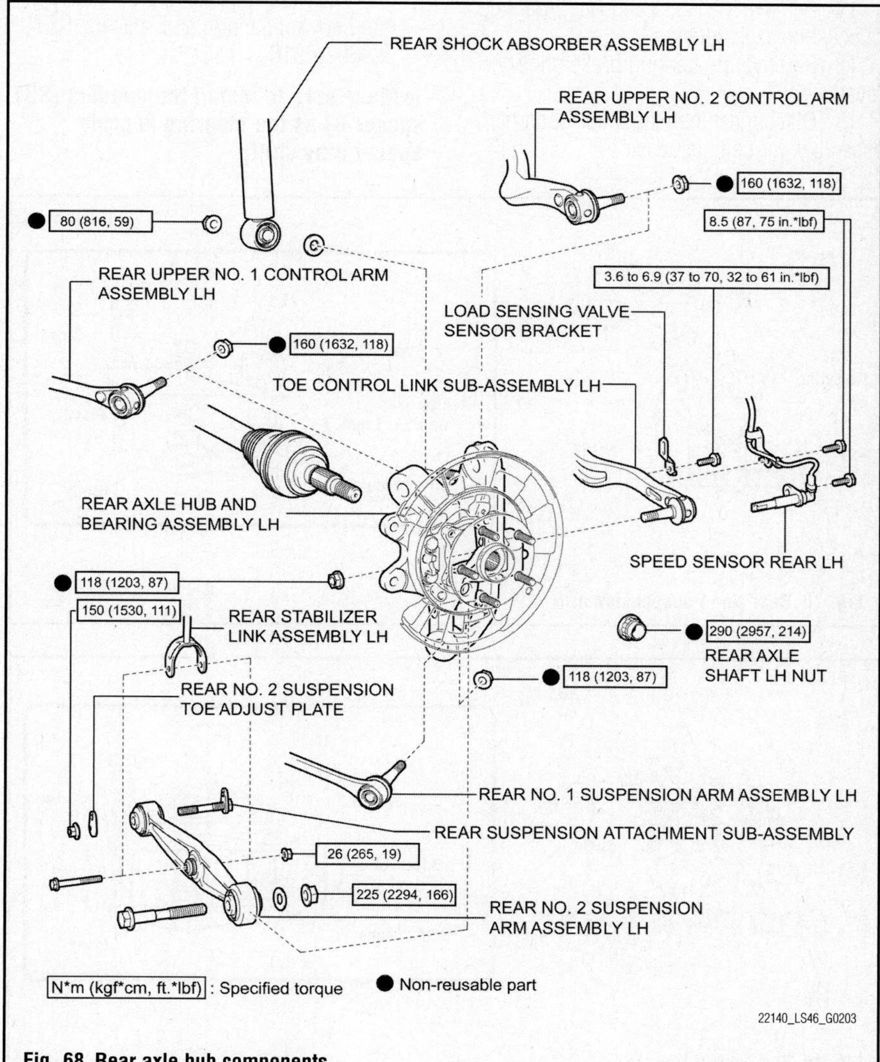

Fig. 68 Rear axle hub components

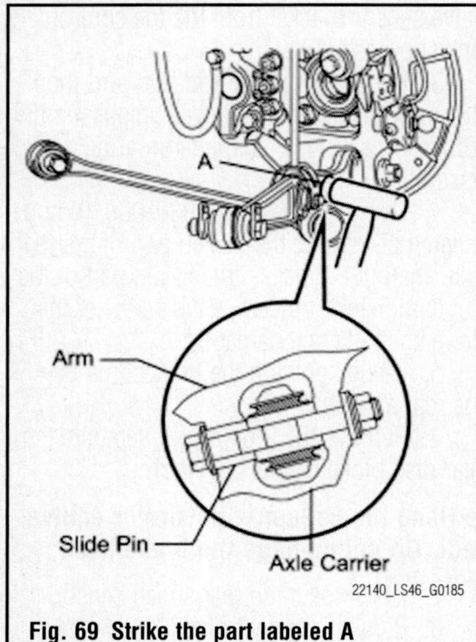

Fig. 69 Strike the part labeled A

10. Remove the parking brake shoe adjusting screw set.
11. Remove the No. 1 and No. 2 parking brake shoe assemblies.
12. Remove the parking brake shoe lever.
13. Disconnect the rear shock absorber assembly, for coil suspension.
14. Disconnect the pneumatic cylinder with rear shock absorber assembly, for air suspension.
15. Remove the rear No. 2 suspension arm assembly, as follows:
    a. Using a plastic-faced hammer or equivalent, strike the part labeled A from the rear of the vehicle to maintain the clearance at the slide pin area.
    b. Remove the nut, washer, bolt and No. 2 suspension arm from the axle carrier.
    c. Remove the nut, rear suspension attachment sub-assembly, and rear No. 2 suspension toe adjust plate, and then remove the rear No. 2 suspension arm assembly.
16. Remove the rear No. 1 suspension arm assembly, as follows:
    a. Remove the nut from the rear No. 1 suspension arm.
    b. Install 2 spacers (SST spacer B) onto the rear No. 1 suspension arm so that there is a space of approximately 1 mm between the arm and spacers, SST: 09960-20010.

➡Make sure to install the spacers (SST spacer B) as the steering knuckle spacer may shift.

➡As SST may become damaged, make sure the space between the arm and spacers is not 1 mm or less.

    c. Using SST (SST: 09960-20010), disconnect the rear No. 1 suspension arm from the axle carrier.

➡Do not damage the ball joint dust cover. As the dust cover may be damaged, adjust SST with the center nut so that the body and crow are parallel. Make sure to tie the string of SST to the vehicle to prevent SST from dropping.

17. Disconnect the toe control link sub-assembly, as follows:
    a. Remove the nut on the rear axle carrier side.
    b. Install 2 spacers (SST spacer B) onto the toe control link so that there is a space of approximately 1 mm between the arm and spacers. SST: 09960-20010.

➡Make sure to install the spacers (SST spacer B) as the steering knuckle spacer may shift.

➡As SST may become damaged, make sure the space between the arm and spacers is not 1 mm or less.

    c. Using SST (SST: 09960-20010), disconnect the toe control link from the axle carrier.

➡Do not damage the ball joint dust cover. As the dust cover may be damaged, adjust SST with the center nut so that the body and crow are parallel. Make sure to tie the string of SST to the vehicle to prevent SST from dropping.

18. Disconnect the rear upper No. 1 control arm assembly, as follows:
    a. Remove the nut on the rear axle carrier side.
    b. Install 2 spacers (SST spacer A) onto the rear upper No. 1 control arm so that there is a space of approximately 1 mm between the arm and spacers. SST: 09960-20010.

➡Make sure to install the spacers (SST spacer A) as the steering knuckle spacer may shift.

➡As SST may become damaged, make sure the space between the arm and spacers is not 1 mm or less.

    c. Using SST (SST: 09960-20010), disconnect the upper No. 1 control arm from the axle carrier.

➡Do not damage the ball joint dust cover. As the dust cover may be

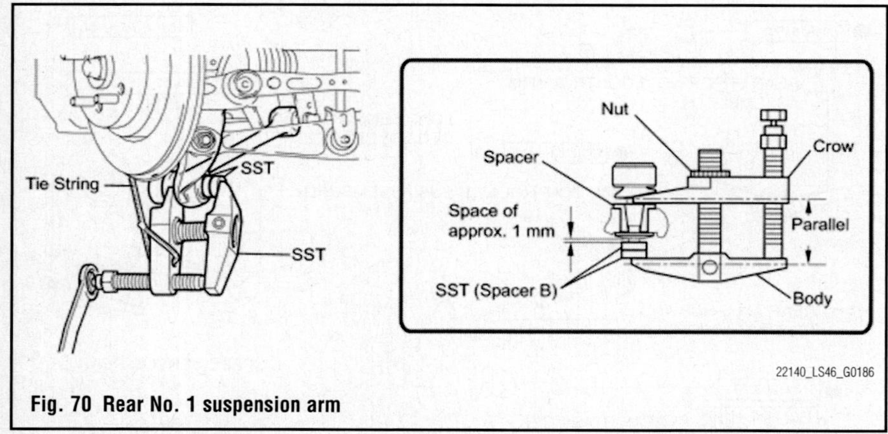

Fig. 70 Rear No. 1 suspension arm

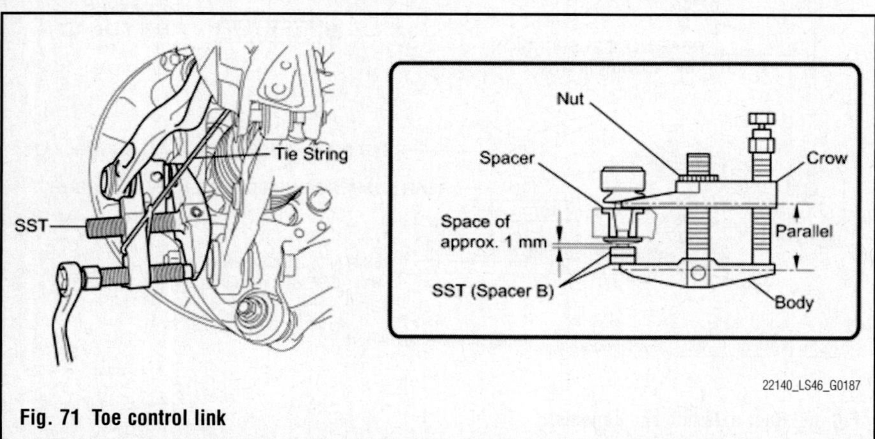

Fig. 71 Toe control link

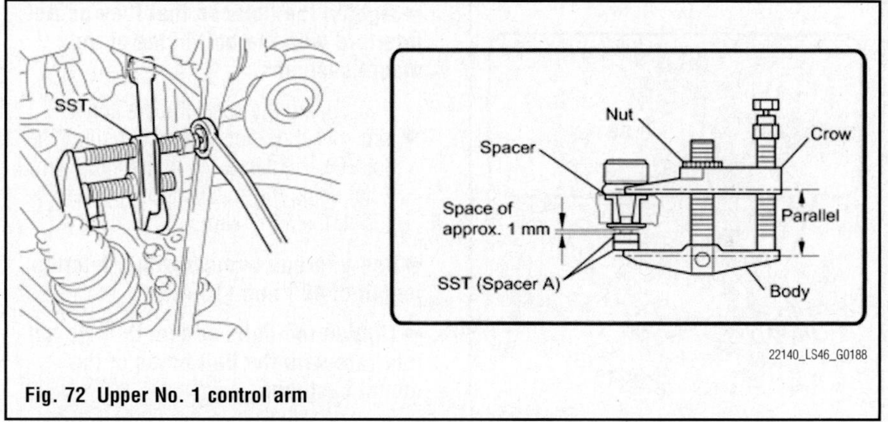

Fig. 72 Upper No. 1 control arm

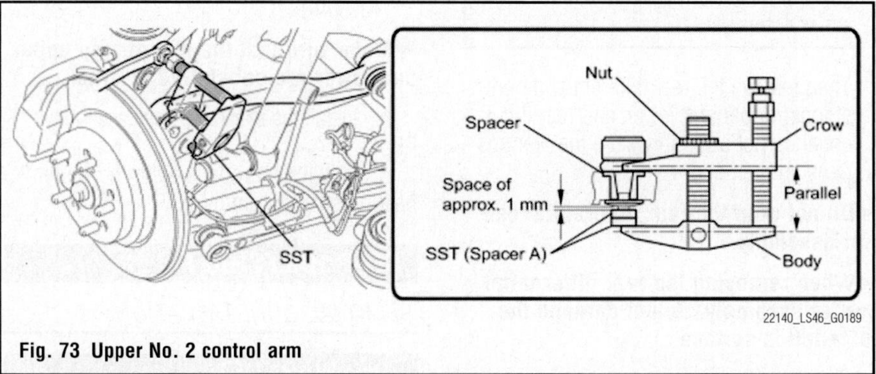

Fig. 73 Upper No. 2 control arm

damaged, adjust SST with the center nut so that the body and crow are parallel. Make sure to tie the string of SST to the vehicle to prevent SST from dropping.

19. Using the same procedure, disconnect the rear upper No. 2 control arm assembly.

20. Using a plastic-faced hammer, separate the drive shaft from the rear axle carrier sub-assembly and remove the rear axle assembly.

➡Be careful not to damage the boots. Use a wire or an equivalent to keep the rear drive shaft assembly from hanging down.

21. Using a screwdriver, remove the No. 1 wheel bearing dust deflector from the rear axle carrier.

22. Hold the axle hub and bearing in a vise between aluminum plates. Do not overtighten the vise.

23. Remove the 4 bolts and axle hub and bearing from the rear axle carrier.

24. Remove the 2 nuts, parking brake anchor block, parking brake cable support bracket, and parking brake plate from the rear axle carrier.

To install:

25. Install the parking brake plate, cable support bracket and parking brake anchor block to the rear axle carrier with the 2 nuts and tighten to 56 ft. lbs. (76 Nm).

26. Hold the axle hub and bearing in a vise between aluminum plates. Do not overtighten the vise.

27. Install the axle hub and bearing to the rear axle carrier with the 4 bolts and tighten to 72 ft. lbs. (97 Nm).

28. Using SST (SST: 09950-70010. SST: 09951-01000) and a hammer, install the No. 1 bearing dust deflector to the rear axle carrier.

29. Align the hole for the speed sensor in the No. 1 bearing dust deflector with the rear axle carrier.

30. Engage the spline part of the rear axle carrier to the spline part of the driver shaft assembly.

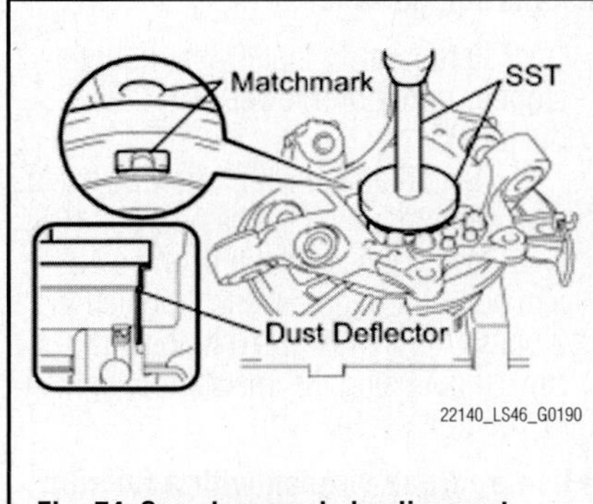

Fig. 74 Speed sensor hole alignment

31. Using a jack, lift up the rear axle carrier, and align the installation positions of each arm.

➡Place a wooden block between the jack and rear axle carrier to prevent damage.

32. Temporarily connect rear shock absorber assembly, for coil suspension.

33. Temporarily connect pneumatic cylinder with rear shock absorber assembly, for air suspension.

34. Connect the upper No. 1 control arm with a new nut to the axle carrier and tighten to 118 ft. lbs. (160 Nm).

35. Connect the upper No. 2 control arm with a new nut to the axle carrier and tighten to 118 ft. lbs. (160 Nm).

36. Connect the control link with a new nut to the axle carrier and tighten to 87 ft. lbs. (118 Nm).

37. Temporarily tighten the No. 2 suspension arm, as follows:

a. Install the stud of the suspension arm, and temporarily install a washer, nut and the bolt to the axle carrier.

b. Temporarily install the No. 2 suspension arm with the No. 2 suspension toe adjust plate, rear suspension attachment and nut to the suspension member.

38. Connect the rear No. 1 suspension arm with a new nut to the axle carrier and tighten to 87 ft. lbs. (118 Nm).

39. Apply high temperature grease.

40. Install the parking brake shoe lever.

41. Install the No. 1 and No. 2 parking brake shoe assembly.

42. Check parking brake installation.

43. Install a new rear axle shaft nut and tighten to 214 ft. lbs. (290 Nm).

44. Inspect rear axle hub bearing looseness.

45. Inspect rear axle hub runout.

46. Install rear disc.

47. Adjust parking brake shoe clearance.

48. Connect rear speed sensor.

49. Connect the rear disc brake caliper assembly with new 2 bolts and tighten to 63 ft. lbs. (86 Nm).

➡Do not twist the flexible hose. Make sure the screw parts are free from foreign matter and are not damaged. Be careful not to overtighten the bolts, as the rear axle carrier is made of aluminum.

50. Using a chisel and a hammer, stake the axle shaft nut.

51. Install the load sensing valve sensor bracket to the toe control link sub-assembly

with the bolt. Tighten to 32 to 61 inch lbs. (3.6 to 6.9 Nm).

52. Stabilize the suspension.

53. Connect the rear shock absorber assembly, for coil suspension.

54. Connect the pneumatic cylinder with rear shock absorber assembly, for air suspension.

55. Insert the bolt from the front of the vehicle, and then install the rear stabilizer link assembly with the nut and tighten to 19 ft. lbs. (26 Nm).

56. Tighten the rear No. 2 suspension arm, as follows:

a. Tighten the nut on the rear suspension member side to 111 ft. lbs. (150 Nm).

b. Tighten the nut on the rear axle carrier side to 166 ft. lbs. (225 Nm).

57. Install the rear wheel.

58. Perform parking brake shoe bedding.

59. Inspect and adjust rear wheel alignment.

60. Check the ABS speed sensor signal.

61. Adjust headlight aiming.

62. Adjust the object recognition camera.

REAR DIFFERENTIAL CARRIER

REMOVAL & INSTALLATION

2WD Models Vehicles

See Figure 75.

1. Remove the propeller with the center bearing shaft assembly.

2. Remove the rear drive shaft assembly.

3. Remove the rear differential carrier assembly.

a. Support the rear differential carrier assembly with a transmission jack.

CAUTION

As the rear differential carrier assembly is very heavy, securely support it with the transmission jack.

CAUTION

Perform this procedure with several people supporting the rear differential carrier assembly so that it does not tilt or fall.

b. Using the special tool or a 12 mm socket hexagon wrench, remove the 3 bolts (A) from the rear No. 1 and No. 2 differential mount cushions.

c. Remove the 2 bolts (B).

d. Slowly lower the transmission jack.

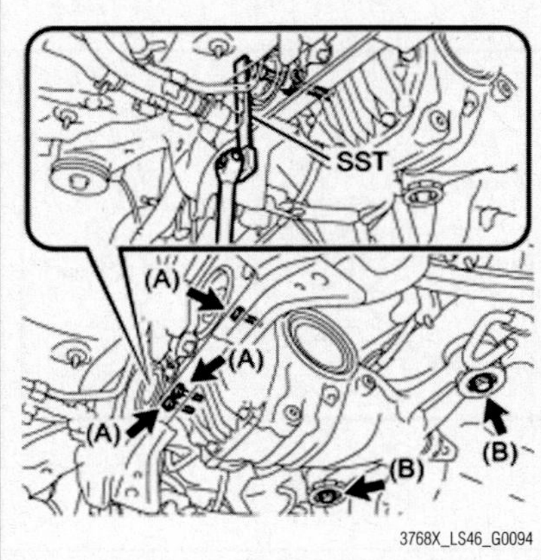

Fig. 75 Removing the rear differential carrier assembly

Then remove the rear differential mount stopper upper and lower, and rear differential carrier assembly from the rear suspension member.

➡Do not drop the rear differential carrier assembly.

➡When removing the rear differential carrier assembly, do not damage the installation surface.

To install:

4. Install the rear differential carrier assembly.

a. Place the rear differential carrier assembly on the transmission jack.

➡As the rear differential carrier assembly is very heavy, securely support it with the transmission jack.

➡Perform this procedure with several people supporting the rear differential carrier assembly so that it does not tilt or fall.

b. Slowly raise the transmission jack so that the rear differential carrier is at its installation position.

➡When installing the rear differential carrier assembly, do not damage the installation surface.

c. Temporarily install the mount stopper upper and lower with 2 new bolts (B).

d. Temporarily install 2 new bolts (A).

e. Temporarily install a new bolt (C).

f. Using the special tool and a 12 mm socket hexagon wrench, tighten the 2 bolts (A). Tighten to 76 ft. lbs. (103 Nm). If not using the special tool tighten to 105 ft. lbs. (142 Nm).

➡Use a torque wrench with a fulcrum length of 400 mm (16 in.)

➡Tighten the bolts so that they do not interfere with the bolt holes of the mount cushions.

g. Using the special tools and a 12 mm socket hexagon wrench, tighten the bolt (C). Tighten to 76 ft. lbs. (103 Nm). If not using the special tools tighten to 105 ft. lbs. (142 Nm).

➡Use a torque wrench with a fulcrum length of 400 mm (16 in.)

➡Tighten the bolts so that they do not interfere with the bolt holes of the mount cushions.

h. Tighten the 2 bolts (B) to 94 ft. lbs. (127 Nm).

➡Make sure that the mount stop upper and lower are not shifted or tilted.

5. Install the rear drive shaft assembly RH.

6. Install the propeller with the center bearing shaft assembly.

REAR HALFSHAFT

REMOVAL & INSTALLATION

WARNING

After the engine switch is turned off, the HDD navigation system requires approximately 6 minutes to record various types of memory and settings. As a result, after turning the engine switch off, wait 6 minutes or more before disconnecting the cable from the negative battery terminal.

1. Remove the right cowl top ventilator louver.

2. Disconnect cable from negative battery terminal.

CAUTION

Wait at least 90 seconds after disconnecting the cable from the negative battery terminal to prevent airbag and seat belt pretensioner activation.

3. Remove the rear wheel.

4. Inspect the rear upper No. 1 and No. 2 control arm ball joint rattle.

5. Inspect the rear No. 1 suspension arm ball joint rattle.

6. Inspect the toe control link ball joint rattle.

7. Remove the bolt and disconnect the left and right load sensing valve sensor brackets, with air suspension.

8. Remove the nuts and No. 2 and No. 1 differential support protectors.

9. Remove the bolts and the left and right rear suspension member brace.
10. Remove the right and left rear axle shaft nuts.
11. Remove the bolts and rear speed sensors.
12. Remove the right pad wear indicator wire.
13. Remove parking brake shoe adjusting hole plug.
14. Disconnect the right and left rear disc brake caliper assemblies.
15. Remove the rear disc.
16. Remove the No. 2 and No. 1 parking brake shoe return springs.
17. Remove the parking brake shoe adjusting screw set.
18. Remove the No. 2 and No. 1 parking brake shoe assemblies.
19. Remove the parking brake shoe lever.
20. Remove the 2 nuts and parking cable support bracket.
21. Disconnect the left and right rear stabilizer link assemblies.
22. Disconnect the left and right rear shock absorber assemblies, with coil suspension.
23. Disconnect the left and right pneumatic cylinder with rear shock absorber assemblies, with air suspension.
24. Disconnect the right and left rear No. 2 suspension arm assemblies.
25. Disconnect the right and left rear No. 1 suspension arm assemblies.
26. Disconnect the left and right toe control link sub-assemblies.
27. Disconnect the right and left upper rear No. 1 control arm assemblies.
28. Disconnect the right and left upper rear No. 2 control arm assemblies.
29. Using a plastic-faced hammer, separate the drive shaft from the left and right rear axle assembly, and then remove the rear axle carrier.

➡Be careful not to damage the boot. Use a wire or an equivalent to keep the rear drive shaft assembly from hanging down.

30. Using SST (SST: 09520-01010, SST: 09520-24010), remove the left and right rear drive shafts.

➡Be careful not to damage the oil seal, inboard joint boot and drive shaft dust cover, or drop the drive shaft.

To install:

31. Install the left and right rear drive shaft assemblies, as follows:
a. Coat the spline of the inboard joint shaft with gear oil.
b. Set the shaft snap ring with the opening side facing downward.
c. Align the shaft splines and install the drive shaft with a brass bar and hammer.

➡Be careful not to damage the drive shaft dust cover, boot and oil seal. Move the drive shaft while keeping it level.

32. Install the rear drive shaft to the left and right rear axle assemblies.

➡Be careful not to damage the drive shaft boot.

33. Connect the right and left upper rear No. 1 control arm assemblies.
34. Connect the right and left upper rear No. 2 control arm assemblies.
35. Temporarily connect the right and left rear No. 2 suspension arm assemblies.
36. Connect the left and right toe control link sub-assemblies.
37. Connect the left and right rear No. 1 suspension arm assemblies.
38. Temporarily connect the right and left rear shock absorber assemblies, with coil suspension.
39. Temporarily connect the right and left rear pneumatic cylinder with shock absorber assemblies, with air suspension.
40. Connect the left and right rear stabilizer link assemblies.
41. Install the parking brake anchor block and parking cable support bracket with the 2 nuts and tighten to 56 ft. lbs. (76 Nm).
42. Apply high temperature grease.
43. Install the parking brake shoe lever.
44. Install the No. 1 and No. 2 parking brake shoe assemblies.
45. Install the parking brake shoe adjusting screw set.
46. Install the No. 1 and No. 2 parking brake shoe return springs.
47. Check parking brake installation.
48. Install the right and left rear axle shaft nuts.
49. Inspect rear axle hub bearing looseness.
50. Inspect rear axle hub runout.
51. Install rear disc.
52. Adjust parking brake shoe clearance.
53. Install parking brake shoe adjusting hole plug.
54. Connect the right and left rear disc brake caliper assemblies.
55. Install the rear speed sensors with the bolts and tighten to 75 inch lbs. (8.5 Nm).
56. Install the right pad wear indicator wire.
57. Stake the right and left rear axle shaft nuts.
58. Stabilize the suspension.
59. Tighten the left and right rear No. 2 suspension arm assemblies.
60. Connect the right and left rear shock absorber assemblies.
61. Connect the right and left rear pneumatic cylinder with shock absorber assemblies, with air suspension.
62. Install the right and left rear suspension member brace with the bolts and tighten to 37 ft. lbs. (50 Nm).
63. Install the No. 2 and No. 1 differential support protector with the nuts and tighten to 48 inch lbs. (5.4 Nm).
64. Connect the left and right load sensing valve sensor brackets with the bolts and tighten to 32 to 61 inch lbs. (3.6 to 6.9 Nm).
65. Install the rear wheel.
66. Connect cable to negative battery terminal.
67. Install the cowl top ventilator louver protector.
68. Perform parking brake shoe bedding.
69. Perform initialization.
70. Check the suspension control system for air suspension.
71. Inspect and adjust the rear wheel alignment.
72. Check ABS speed sensor signal.
73. Adjust the headlight.
74. Adjust the object recognition camera.

REAR PINION SEAL

REMOVAL & INSTALLATION

See Figures 76 through 86.

✲✲ WARNING

After the engine switch is turned off, the HDD navigation system requires approximately 6 minutes to record various types of memory and settings. As a result, after turning the engine switch off, wait 6 minutes or more before disconnecting the cable from the negative battery terminal.

1. Remove the right cowl top ventilator louver.
2. Disconnect cable from negative battery terminal.

✲✲ CAUTION

Wait at least 90 seconds after disconnecting the cable from the negative battery terminal to prevent airbag and seat belt pretensioner activation.

3. Drain differential oil.
4. Remove the rear wheel.
5. Remove the propelier with center bearing shaft assembly.

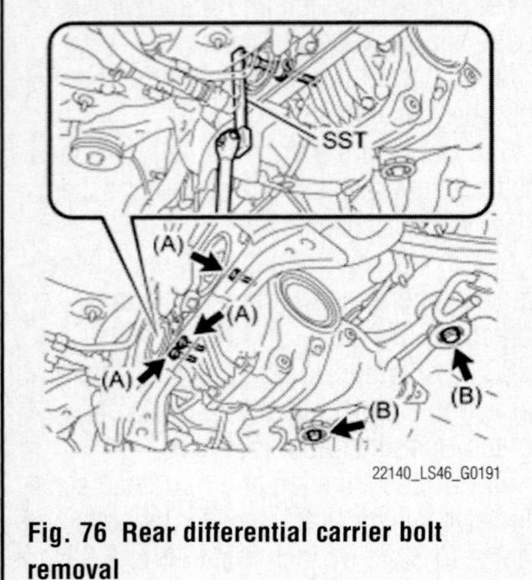

Fig. 76 Rear differential carrier bolt removal

6. Remove the rear drive shaft assembly.

7. Remove the rear differential carrier assembly, as follows:

a. Support the rear differential carrier assembly with a transmission jack.

CAUTION

As the rear differential carrier assembly is very heavy, securely support it with the transmission jack.

CAUTION

Perform this procedure with several people supporting the rear differential carrier assembly so that it does not tilt or fall.

b. Using SST or a 12 mm socket hexagon wrench, remove the 3 bolts (A) from the rear No. 1 and No. 2 differential mount cushions.

c. Remove the 2 bolts (B).

d. Slowly lower the transmission jack. Then remove the rear differential mount stopper upper and lower, and rear differential carrier assembly from the rear suspension member.

➡Do not drop the rear differential carrier assembly.

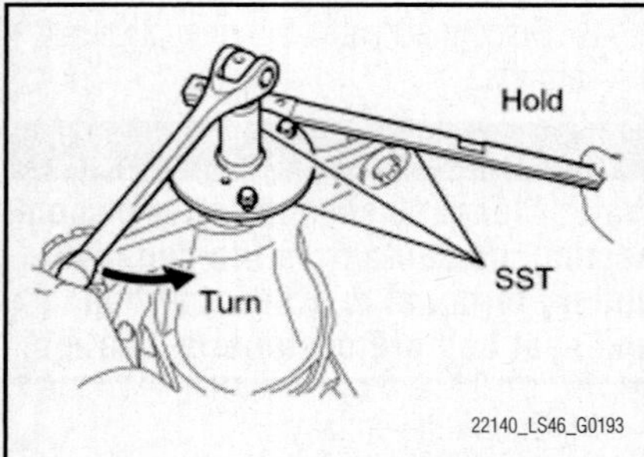

Fig. 77 Secure companion flange and remove drive pinion nut

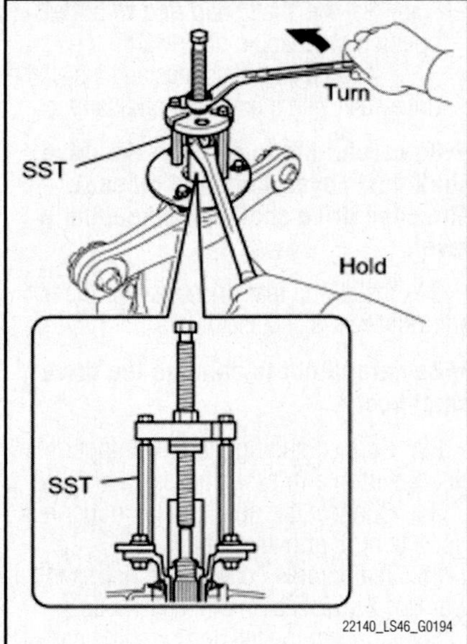

Fig. 78 Companion flange removal

➡When removing the rear differential carrier assembly, do not damage the installation surface.

8. Remove the rear drive pinion nut, as follows:

a. Using SST (SST: 09930-00010) and a hammer, unstake the staked part of the drive pinion nut.

➡Be sure to use SST with the tapered surface facing the shaft. Do not grind the tip of SST with a grinder, etc. Completely unstake the drive pinion nut to prevent damaging the threads of the drive pinion.

b. Using SST (SST: 09330-00021, SST: 09950-30012) secure the companion flange in place.

c. Perform the removal while supporting the overhaul attachment. SST: 09229-55010.

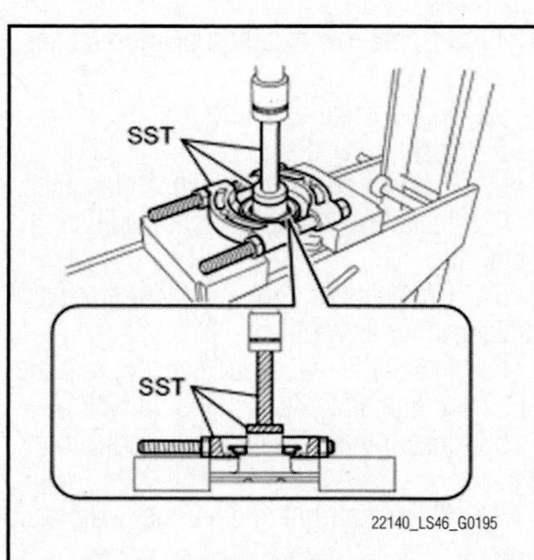

Fig. 79 Dust deflector removal

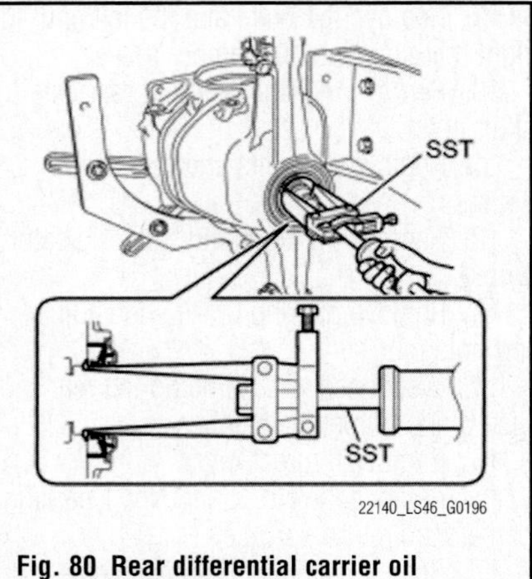

Fig. 80 Rear differential carrier oil seal removal

9. Using SST (SST: 09950-30012), remove the rear drive pinion companion flange. Apply grease to the threads and tip of SST center bolt before use.

10. Using SST (SST: 09950-00020, SST: 09950-60010, SST: 09950-70010) and a press, press out the rear differential dust deflector.

➡Perform this procedure while tightening SST nut to secure the contact surfaces of SST and the dust deflector. Do not drop the companion flange.

11. Using SST (SST: 09308-00010), remove the oil seal from the differential carrier.

To install:

12. Using SST (SST: 09316-60011, SST: 09710-04101) and a hammer, tap in a new oil seal. Tap the oil seal uniformly so that the oil seal is straight. Do not excessively tap the oil seal.

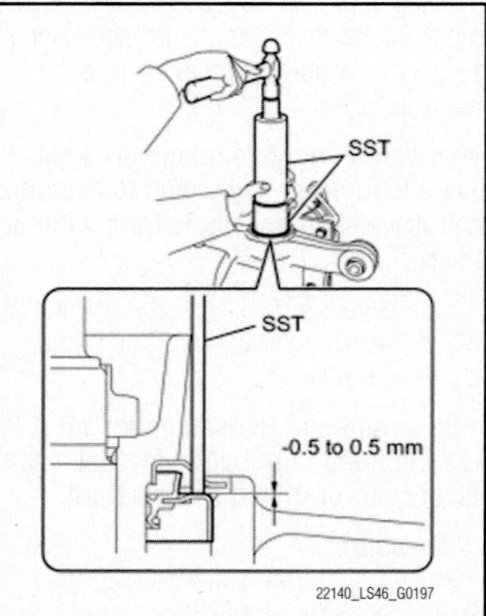

Fig. 81 Rear differential carrier oil seal installation

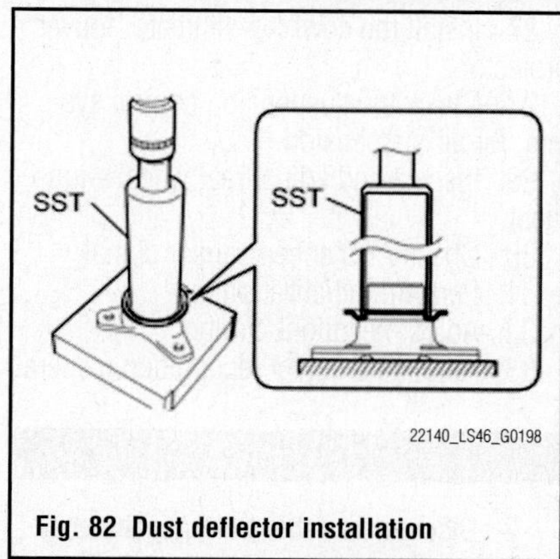

Fig. 82 Dust deflector installation

13. Using a vernier caliper, measure the depth of the oil seal. Standard depth: 0 to 0.020 in. (0 to 0.5 mm).

➡Measure at 3 or more areas around the circumference of the oil seal, Make sure difference between the maximum and minimum measured values is less than 0.65 mm, as a greater difference may lead to oil leaks.

14. Apply MP grease to the lip of the oil seal.

➡Perform this procedure only when replacing the dust deflector.

15. Using SST (SST: 09316-6001) and a press, insert a new dust deflector into the companion flange.

➡Slowly press in the dust deflector but not excessively. If any burrs remain after pressing in the deflector, remove them.

16. Using SST (SST: 09950-30012),

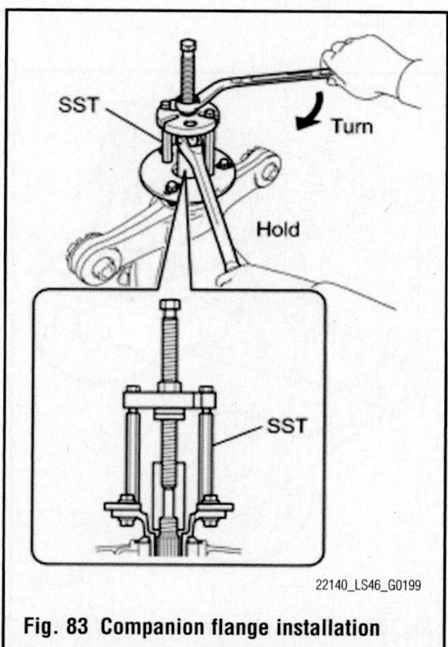

Fig. 83 Companion flange installation

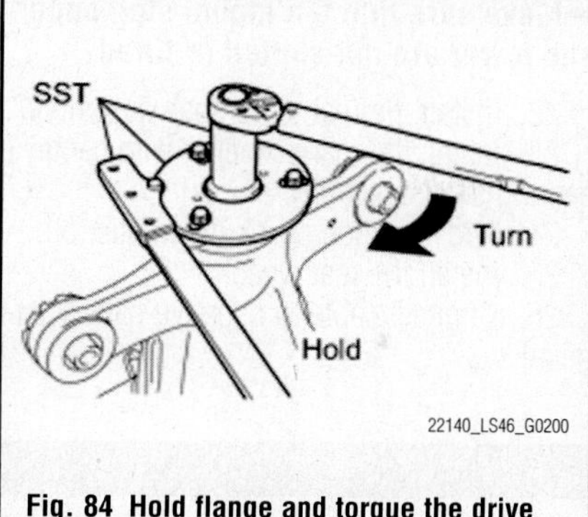

Fig. 84 Hold flange and torque the drive pinion nut

install the companion flange to the drive pinion.

➡Perform this procedure after aligning the companion flange with the spline of the drive pinion. Apply grease to the threads and tip of SST center bolt before use.

17. Coat the threads of a new drive pinion nut with hypoid gear oil LSD.

18. Use SST (SST: 09950-30012), to hold the flange and torque the drive pinion nut.

19. Using SST (SST: 09229-55010, SST: 09330-00021) and a torque wrench, tighten the drive pinion nut while checking the starting torque of the drive pinion. Maximum torque: T=490 Nm (5000 kgfcm).

➡Tighten the nut with a force of 100 N*m (1020 kgf*cm) while checking the starting torque of the drive pinion.

➡Do not overtighten the nut, as the threads will become damaged.

➡Apply hypoid gear oil LSD to the threads of the nut and drive pinion.

➡Perform the removal while supporting the overhaul attachment.

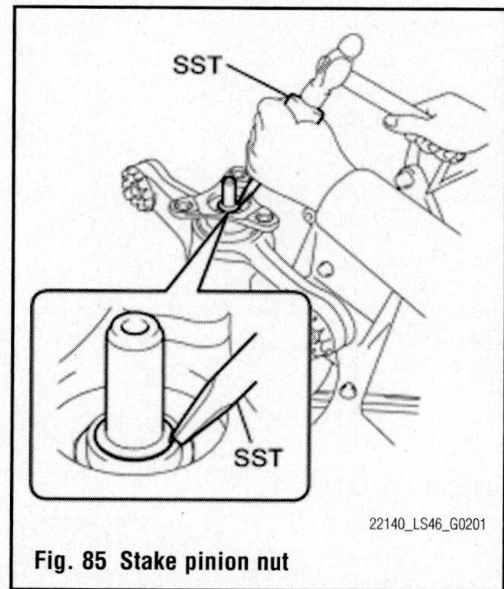

Fig. 85 Stake pinion nut

20. Using a chisel and a hammer, stake the drive pinion nut. SST: 09930-00010.

21. Install the rear differential carrier assembly, as follows:

a. Place the rear differential carrier assembly on a transmission jack.

CAUTION

As the rear differential carrier assembly is very heavy, securely support it with the transmission jack.

CAUTION

Perform this procedure with several people supporting the rear differential carrier assembly so that it does not tilt or fall.

b. Slowly raise the transmission jack so that the rear differential carrier is at its installation position.

➡When installing the rear differential carrier assembly, do not damage the installation surface.

c. Temporarily install the mount stopper upper and lower with 2 new bolts (B).

d. Temporarily install 2 new bolts (A).

e. Temporarily install a new bolt (C).

f. Using SST (SST: 09249-63010) and a 12 mm socket hexagon wrench, tighten the 2 bolts (A) to 76 ft. lbs. (103 Nm), or 105 ft. lbs. (142 Nm) without SST.

➡Use a torque wrench with a fulcrum length of 16 inches (400 mm). Tighten the bolts so that they do not interfere with the bolt holes of the mount cushions.

g. Using SST (SST: 09249-63010) and a 12 mm socket hexagon wrench, tighten

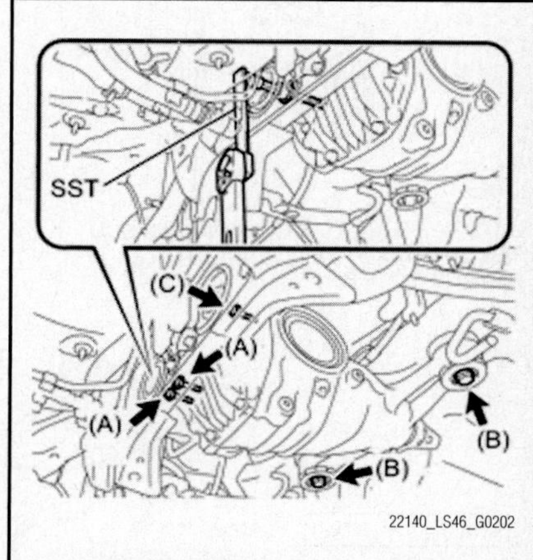

Fig. 86 Rear differential carrier assembly bolt installation

the bolt (C) to 76 ft. lbs. (103 Nm), or 105 ft. lbs. (142 Nm) without SST.

➡**Use a torque wrench with a fulcrum length of 16 inches (400 mm). Tighten the bolts so that they do not interfere with the bolt holes of the mount cushions.**

h. Tighten the 2 bolts (B) to 94 ft. lbs. (127 Nm).

➡**Make sure that the mount stop upper and lower are not shifted or tilted.**

22. Install the rear drive shaft assembly.
23. Install the rear propeller with center bearing shaft assembly.
24. Add and then check differential oil.
25. Install the rear wheel.
26. Connect cable to negative battery terminal.
27. Install the cowl top ventilator louver protector.
28. Check the suspension control system, for air suspension.
29. Inspect and adjust rear wheel alignment.
30. Check ABS speed sensor signal.
31. Perform initialization.
32. Adjust headlight aiming.
33. Adjust the object recognition camera.

ENGINE COOLING

ENGINE FAN

REMOVAL & INSTALLATION

2WD Models

See Figures 87 through 91.

WARNING

After the engine switch is turned off, the HDD navigation system requires approximately 6 minutes to record various types of memory and settings. As a result, after turning the engine switch off, wait 6 minutes or more before disconnecting the cable from the negative battery terminal.

1. Remove the right cowl top ventilator louver.
2. Disconnect cable from negative battery terminal.

CAUTION

Wait at least 90 seconds after disconnecting the cable from the negative battery terminal to prevent airbag and seat belt pretensioner activation.

3. Remove the V-bank cover sub-assembly.
4. Remove the air cleaner inlet cover.
5. Remove the No. 1 air cleaner inlet.
6. Remove the right and left engine room side cover.
7. Remove the No. 1 engine under cover.
8. Drain engine coolant.
9. Disconnect the No.1 and No. 2 radiator hoses.
10. Remove the ECM outlet duct.
11. Remove the radiator reservoir assembly.
12. Remove the fan shroud with fan and motor, as follows:
    a. Disconnect the cooling fan ECU's connector.
    b. Using a clip remover, detach the 3 clamps.

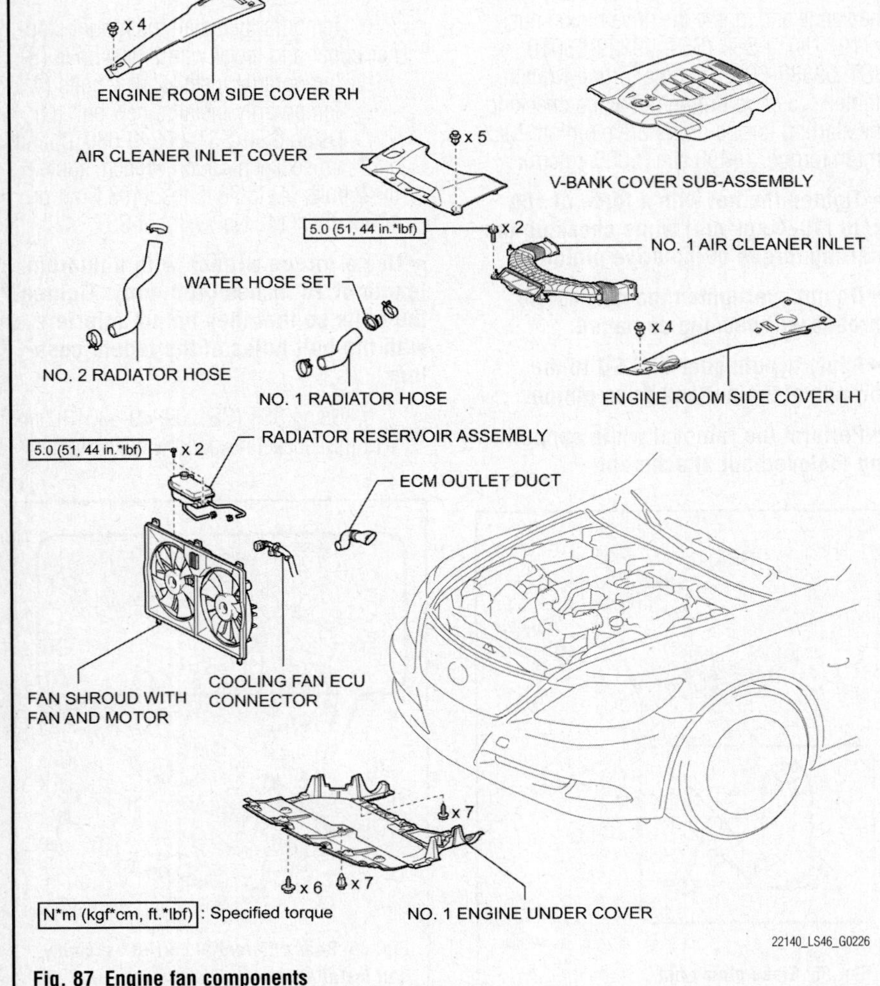

Fig. 87 Engine fan components

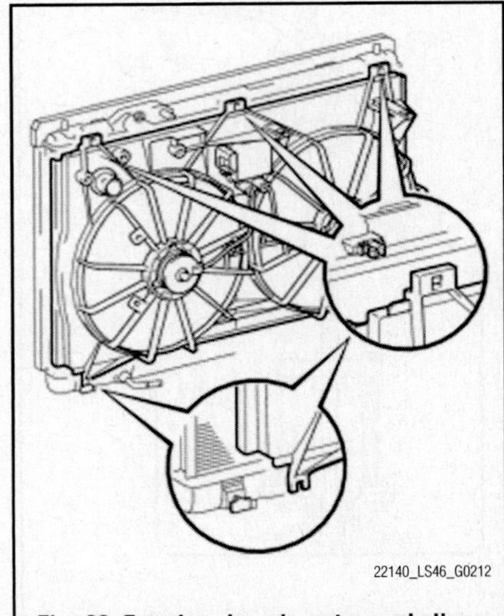

Fig. 88 Fan, fan shroud, motor, and clips

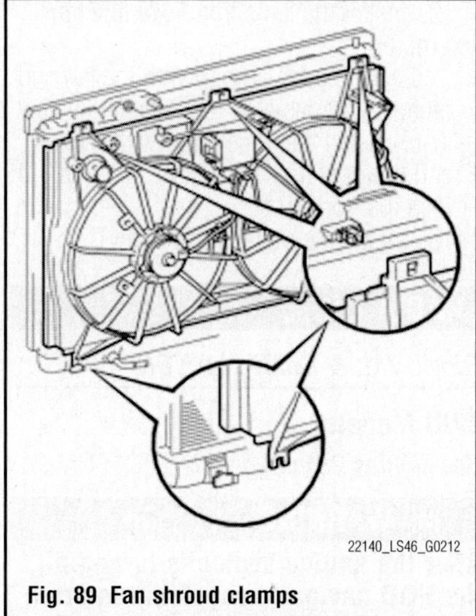

Fig. 89 Fan shroud clamps

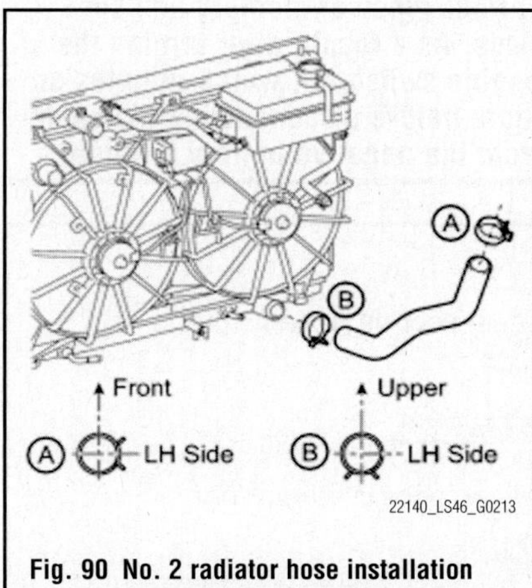

Fig. 90 No. 2 radiator hose installation

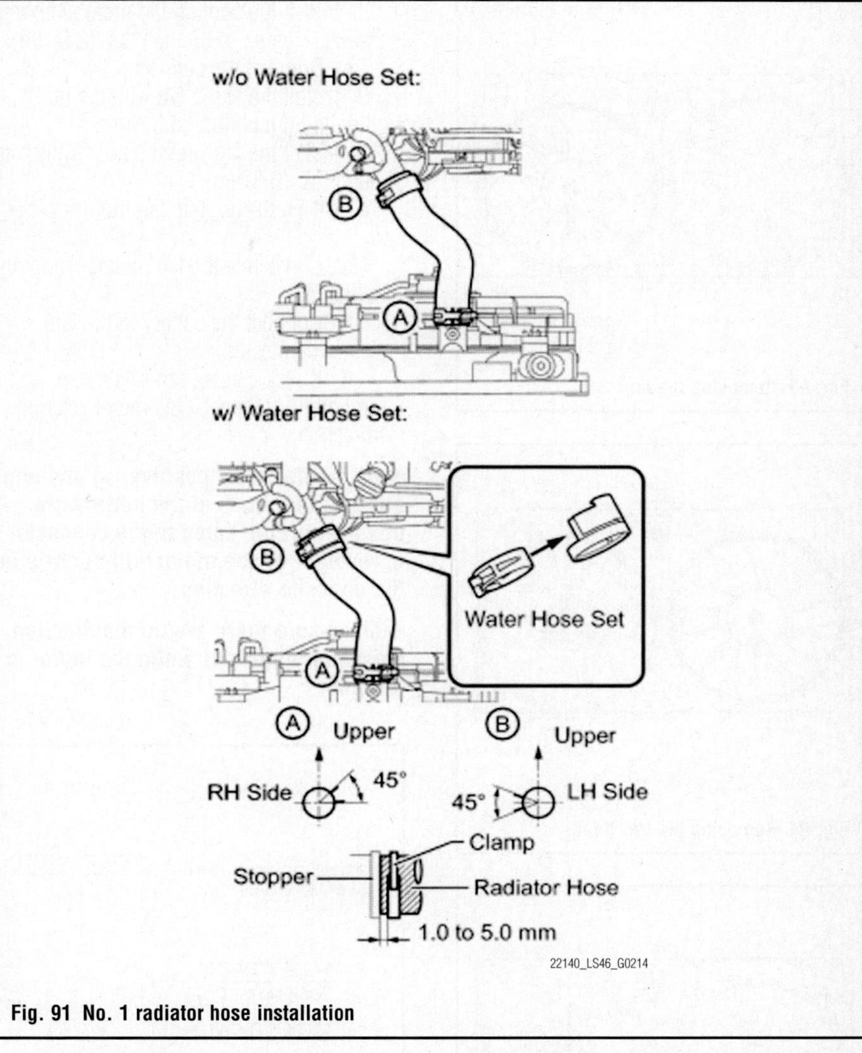

Fig. 91 No. 1 radiator hose installation

c. Detach the 3 claws, and then remove the fan shroud with fan and motor.

To install:

13. Install the fan shroud with fan and motor to the radiator, and attach the 3 claws.
14. Attach the 3 clamps to the fan shroud, and connect the cooling fan ECU's connector.
15. Install the radiator reservoir assembly.
16. Connect the No. 2 radiator hose.
17. Connect the No. 1 radiator hose to the radiator and water outlet, and then secure it with the hose clamps. The direction of the hose clamp is indicated in the illustration. Insert the radiator hose into the stopper. Set the hose clamp so that the clearance between the hose clamp and the stopper is within 1.0 to 5.0 mm (0.039 to 0.20 in.).

➡If the vehicle is equipped with a water hose set on the water outlet side, assemble the water hose set before use.

18. Add coolant and inspect for coolant leak.
19. The remainder of installation is the reverse of installation.
20. Perform initialization.

AWD Models

See Figures 92 through 96.

1. Remove the radiator assembly from the vehicle.
2. Remove the fan shroud with the fan and motor.
   a. Remove the 2 bolts.
   b. Detach the claw, and then remove the fan shroud with the fan and motor.
3. Remove the nut and fan.
4. Remove the No. 2 nut and fan.
5. Remove the cooling fan motor.
   a. Disconnect the connector.
   b. Remove the 3 screws and motor.
6. Remove the No. 2 cooling fan motor.
   a. Disconnect the connector.
   b. Remove the 3 screws and the motor.

To install:

7. Install the No. 2 cooling fan motor with the 3 screws. Tighten to 35 inch lbs. (3.9 Nm).
   a. Connect the connector.

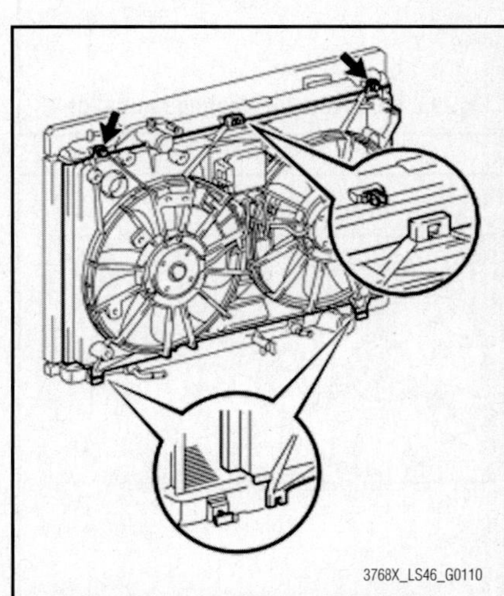

Fig. 92 Removing the fan shroud with the fan and motor

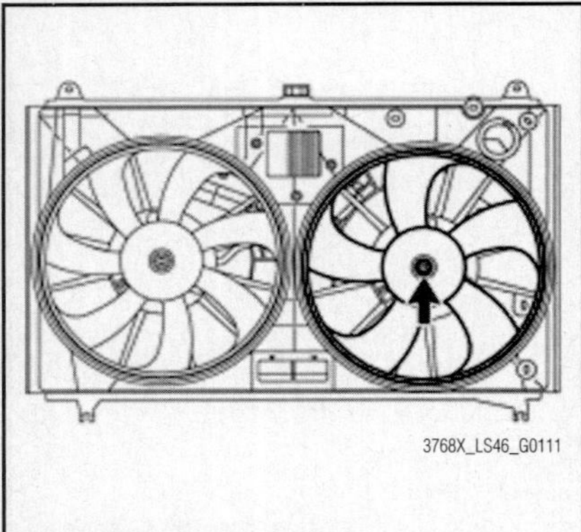

Fig. 93 Removing the fan

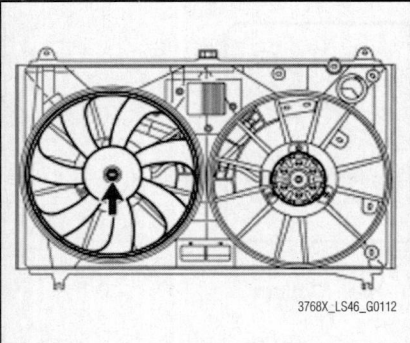

3768X_LS46_G0112

Fig. 94 Removing the No. 2 fan

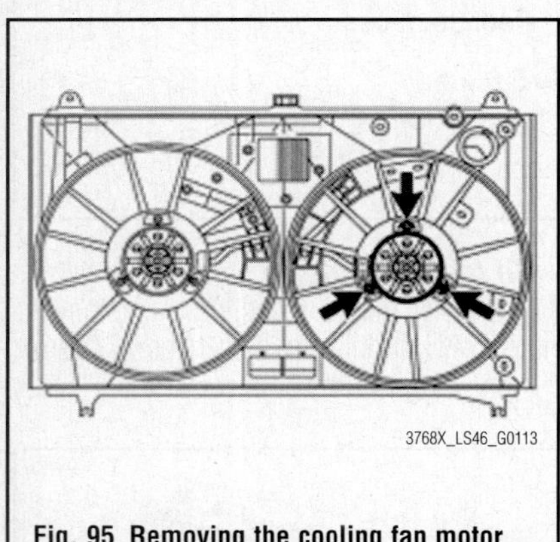

Fig. 95 Removing the cooling fan motor

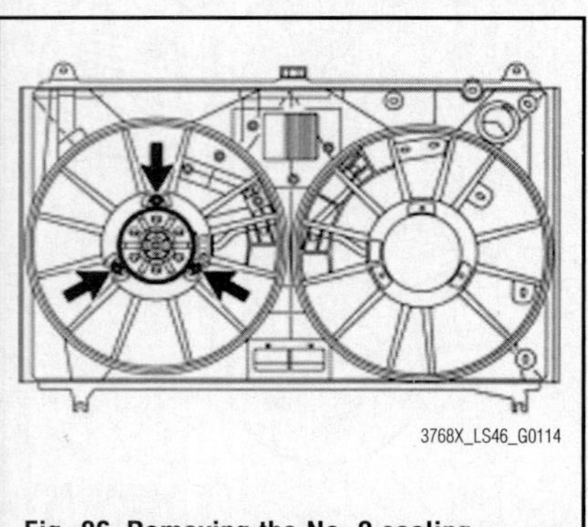

Fig. 96 Removing the No. 2 cooling fan motor

8. Install the cooling fan motor with the 3 screws. Tighten to 35 inch lbs. (3.9 Nm).

a. Connect the connector.

9. Install the No. 2 fan with the nut. Tighten to 56 inch lbs. (6.3 Nm).

10. Install the fan with the nut. Tighten to 56 inch lbs. (6.3 Nm).

11. Check the cooling fan motor operation.

a. Check that the fan rotates smoothly by hand.

b. Connect the battery to the fan motor connector.

c. Check that the fan rotates smoothly, and check the motor rotation direction.

➡**Be careful if the positive (+) and negative (-) terminals of the battery are mistakenly connected to the connector in reverse, as the motor will operate in the opposite direction.**

➡**Make sure there are no malfunction noises or vibrations while the motor is rotating.**

12. Install the fan shroud with the fan and motor.

a. Set the fan shroud with the fan and motor to the radiator, and attach the claw.

b. Install the fan shroud with the fan and motor with the 2 bolts. Tighten to 44 inch lbs. (5 Nm).

13. Install the radiator assembly.

RADIATOR

REMOVAL & INSTALLATION

2WD Models

See Figures 97 through 102.

⁂ WARNING

After the engine switch is turned off, the HDD navigation system requires approximately 6 minutes to record various types of memory and settings. As a result, after turning the engine switch off, wait 6 minutes or more before disconnecting the cable from the negative battery terminal.

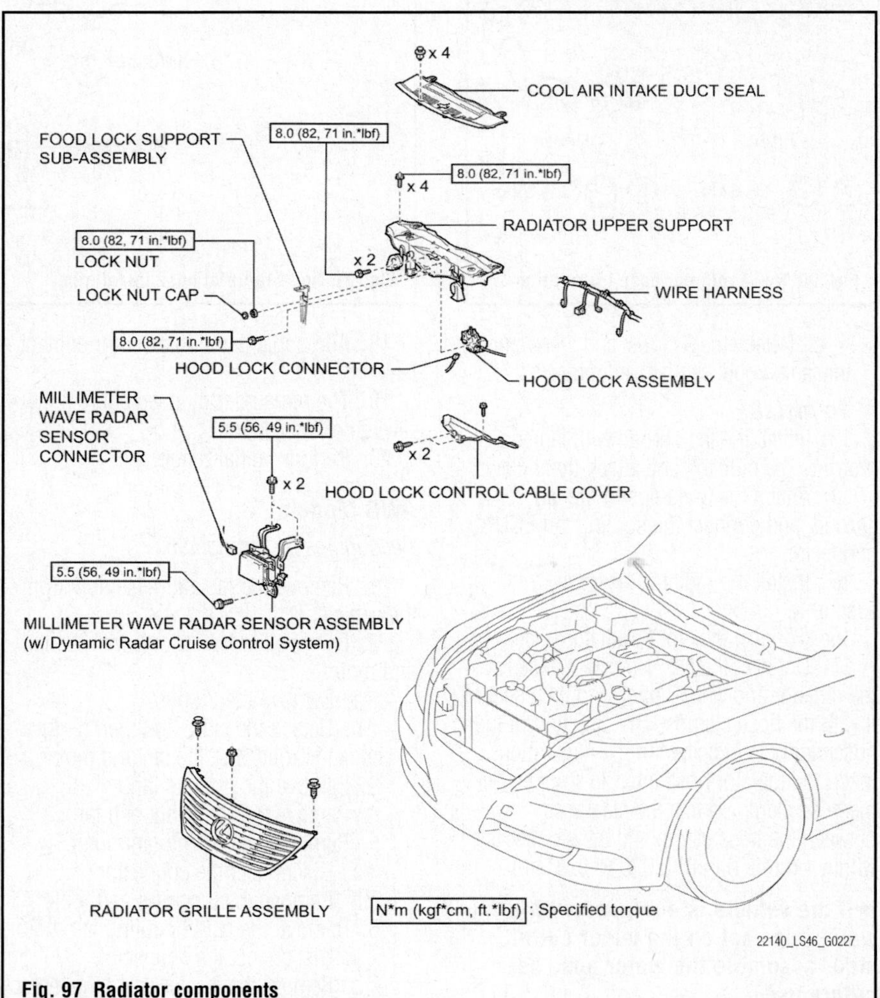

Fig. 97 Radiator components

1. Remove the right cowl top ventilator louver.
2. Disconnect cable from negative battery terminal.

CAUTION

Wait at least 90 seconds after disconnecting the cable from the negative battery terminal to prevent airbag and seat belt pretensioner activation.

3. Remove the No. 1 engine under cover.
4. Drain engine coolant.
5. Remove the V-bank cover sub-assembly.
6. Remove the air cleaner inlet cover.
7. Remove the No. 1 air cleaner inlet.
8. Remove the right and left engine room side cover.
9. Remove the 4 clips and cool air intake duct seal.
10. Remove the radiator grill.
11. Disconnect the No.1 and No. 2 radiator hoses.
12. Disconnect the oil cooler inlet and outlet hoses.
13. Remove the ECM outlet duct.
14. Remove the radiator reservoir assembly.
15. Remove millimeter wave radar sensor assembly (with Dynamic Radar Cruise Control System), as follows:
   a. Disconnect the connector.
   b. Remove the 3 bolts and sensor.
16. Remove the hood lock control cable cover, as follows:
   a. Remove the 3 screws.
   b. Detach the claw and remove the cover.
17. Remove the hood lock assembly.
18. Remove the radiator upper support, as follows:

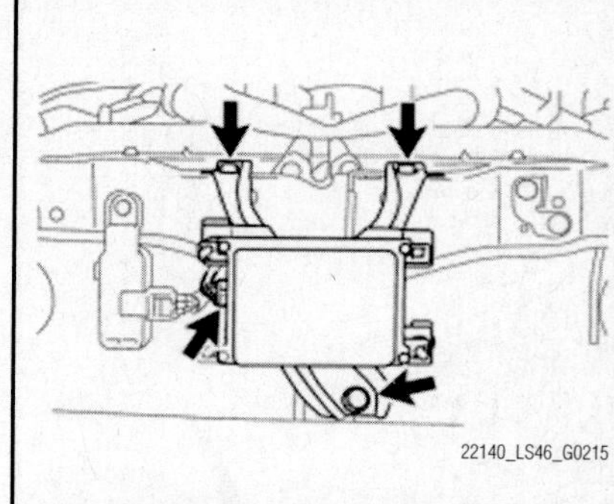

Fig. 99 Millimeter wave radar sensor assembly and bolts

   a. Disconnect the 3 connectors and detach the 2 clamps.
   b. Detach the 5 clamps and disconnect the wire harness.
   c. Remove the 5 bolts and upper support.
19. Remove the radiator assembly, as follows:
   a. Detach the 3 clamps and disconnect the ECU's connector.
   b. Remove the 4 bolts and separate the radiator from the A/C condenser.
20. Detach the 3 claws and remove the fan shroud with fan and motor.
21. Remove the 2 radiator support cushions.
22. Remove the 2 lower radiator supports.

To install:

23. Install the 2 lower radiator supports.
24. Install the 2 radiator support cushions.
25. Set the fan shroud with fan and motor to the radiator, and attach the 3 claws to install the fan shroud.
26. Install the radiator to the front cross-member. Install the lower packing.
27. Align the radiator with the A/C condenser, and then install the 4 bolts and tighten to 44 inch lbs. (5 Nm).

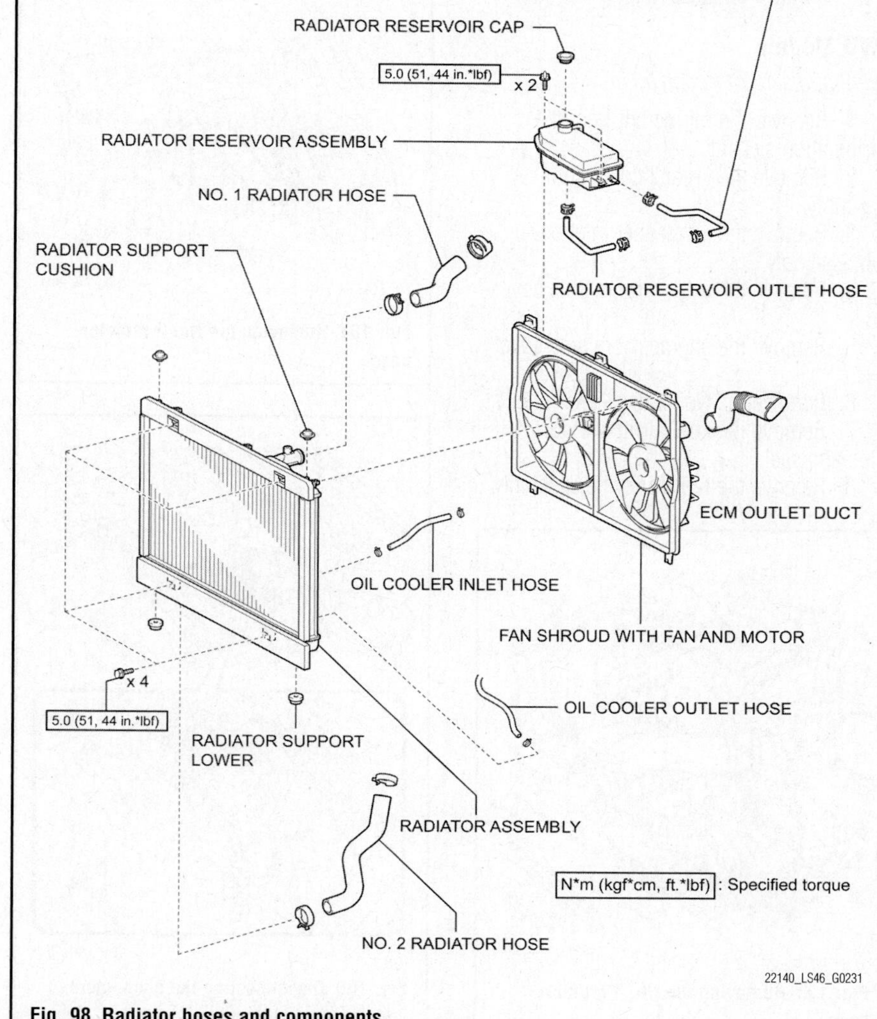

Fig. 98 Radiator hoses and components

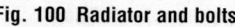

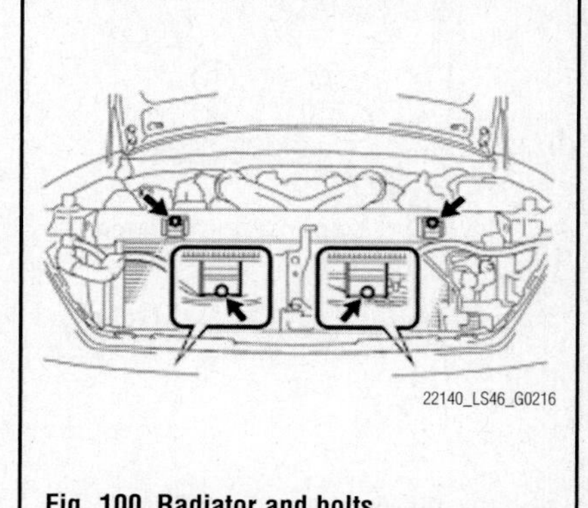

Fig. 100 Radiator and bolts

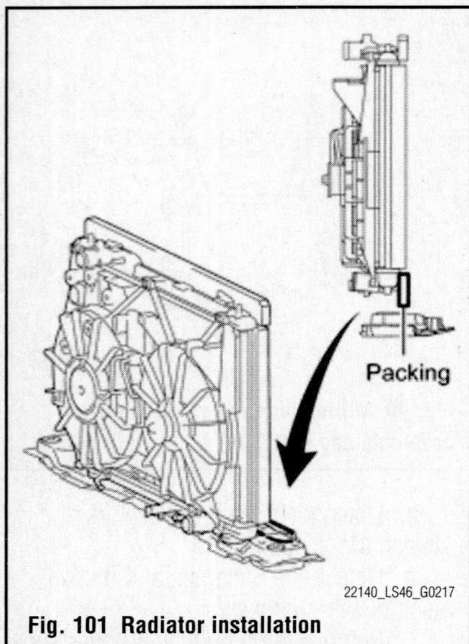

Fig. 101 Radiator installation

28. Attach the 3 clamps and connect the ECU's connector.
29. Attach the wire harness to the upper radiator support with the 5 clamps.
30. Install the upper support with the 5 bolts and tighten to 71 inch lbs. (8 Nm).
31. Attach the 2 clamps to the right and left side radiator support.
32. Install the hood lock assembly.
33. Attach the claw to install the hood lock control cable cover, and install the 3 screws.
34. Install the millimeter wave radar sensor assembly sensor with the 3 bolts (with Dynamic Radar Cruise Control System), tighten to 49 inch lbs. (5.5 Nm), and connect the connector.
35. Install the radiator reservoir assembly.
36. Install the ECM outlet duct.
37. Connect the oil cooler inlet and outlet hoses. Insert the outlet hose into the stopper.

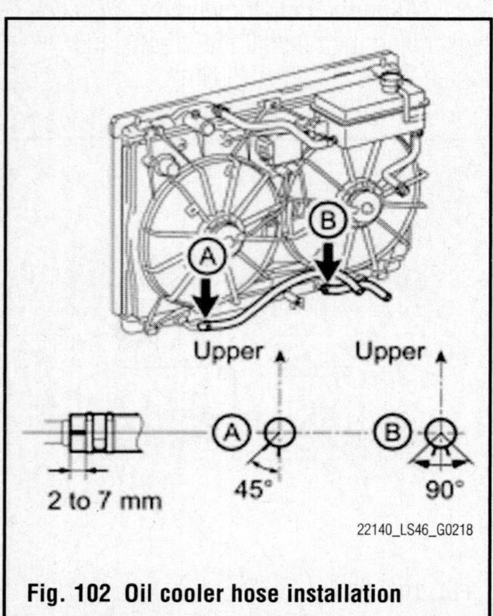

Fig. 102 Oil cooler hose installation

Set the hose clamp so that the clearance between the hose clamp and the stopper is within 2 to 7 mm (0.08 to 0.28 in.).

38. Connect the No.1 and No. 2 radiator hoses.
39. Install the cool air intake duct seal with the 4 clips.
40. Install the No. 1 air cleaner inlet.
41. Connect the cable to the negative battery terminal.
42. Install the right cowl top ventilator louver.
43. Add coolant and inspect for coolant leak.
44. Adjust the millimeter wave radar sensor assembly sensor (with Dynamic Radar Cruise Control System).
45. Install the radiator grill.
46. Install the right and left engine room side cover.
47. Install the air cleaner inlet cover.
48. Install the V-bank cover sub-assembly.
49. Install the No. 1 engine under cover.
50. Perform initialization.

AWD Models

See Figures 103 through 117.

1. Recover the refrigerant from the refrigeration system.
2. Remove the v-bank cover sub assembly.
3. Remove the air cleaner inlet cover sub assembly.
4. Remove the engine room side cover RH.
5. Remove the engine room side cover LH.
6. Remove the No. 1 air cleaner inlet.
7. Remove the cool air intake duct 4 clips and seal.
8. Remove the radiator grille assembly.

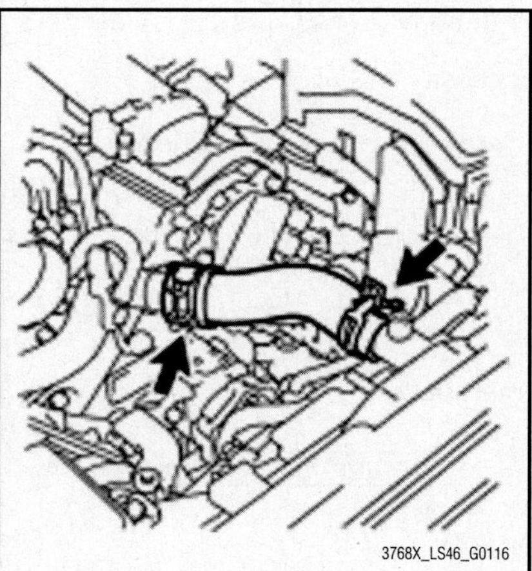

Fig. 103 Removing the No. 1 radiator hose

9. Remove the No. 1 engine under cover.
10. Drain the engine coolant.
11. Remove the front fender wheel opening moulding LH.
12. Remove the front fender wheel opening moulding RH.
13. Remove the front wheel opening extension pad LH.
14. Remove the front wheel opening extension pad RH.
15. Remove the front bumper cover.
16. Remove the front bumper energy absorber.
17. Remove the front bumper reinforcement.
18. Remove the radiator reservoir assembly.
19. Remove the No. 1 radiator hose.
20. Remove the No. 2 radiator hose.
21. Disconnect the inlet oil cooler hose.
22. Disconnect the outlet oil cooler hose.
23. Remove the outlet engine room ECU duct.

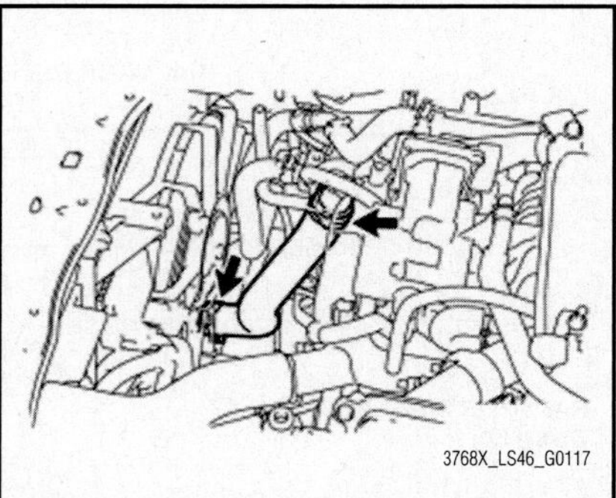

Fig. 104 Removing the No. 2 radiator hose

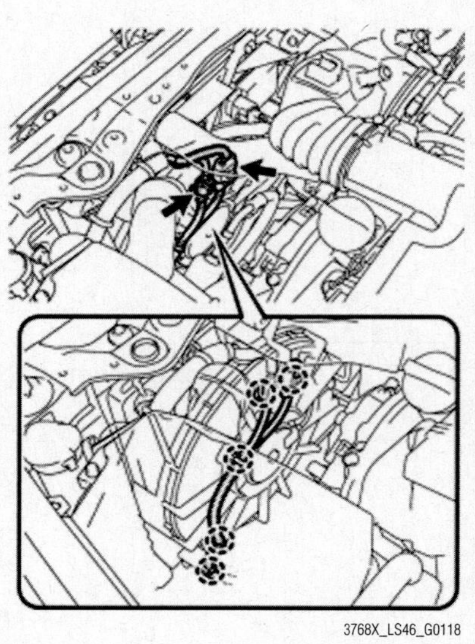

Fig. 105 Disconnecting the connectors and detaching the clamps

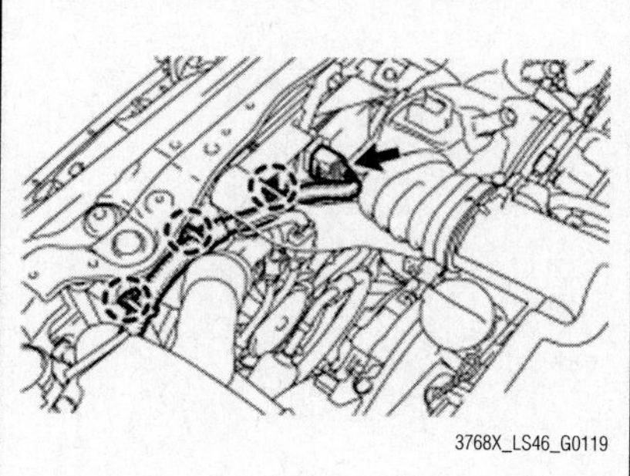

Fig. 106 Detaching the 3 clamps and disconnecting the ECU connector

Fig. 107 Removing the sensor

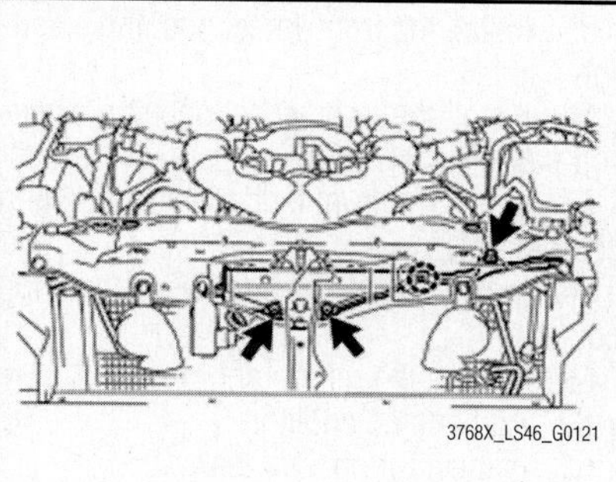

Fig. 108 Removing the hood lock control cable cover

24. Remove the wire harness.
   a. Disconnect the 2 connectors.
   b. Detach the 5 clamps and disconnect the wire harness.
   c. Detach the 3 clamps and disconnect the ECU connector.
25. Remove the millimeter wave radar sensor assembly (with dynamic radar cruise control system).
   a. Disconnect the connector.
   b. Remove the 3 bolts and sensor.
26. Remove the hood lock control cable cover.
   a. Remove the 3 screws.

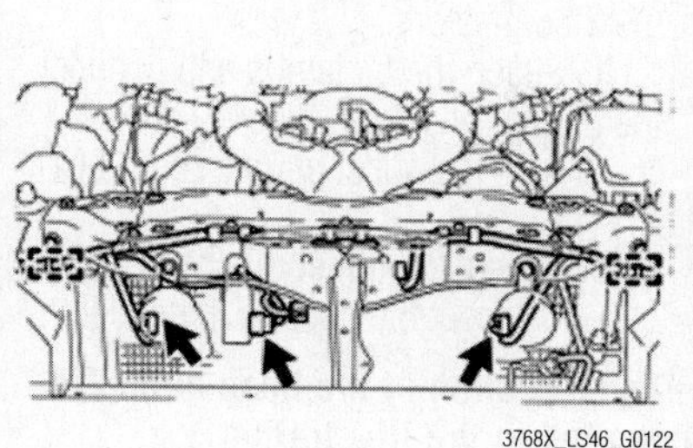

Fig. 109 Disconnecting the 3 connectors and detaching the 2 clamps

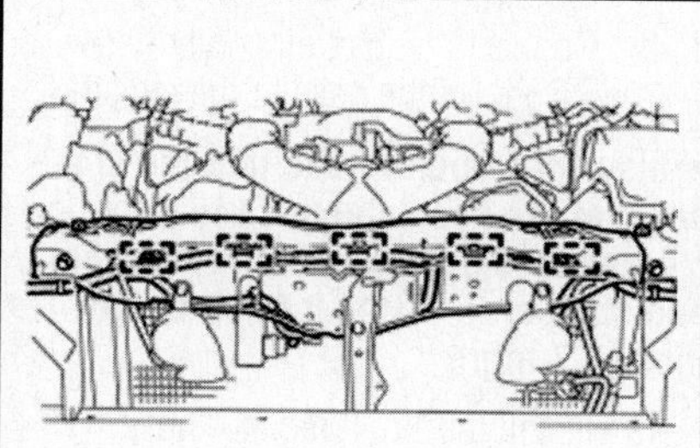

Fig. 110 Detaching the 5 clamps and disconnecting the wire harness

Fig. 111 Removing the upper support

   b. Detach the claw and remove the cover.
27. Remove the hood lock assembly.
28. Remove the upper radiator support sub assembly.
   a. Disconnect the 3 connectors and detach the 2 clamps.
   b. Detach the 5 clamps and disconnect the wire harness.
   c. Remove the 5 bolts and upper support.
29. Disconnect the discharge hose sub assembly.
30. Disconnect the liquid tube sub assembly.

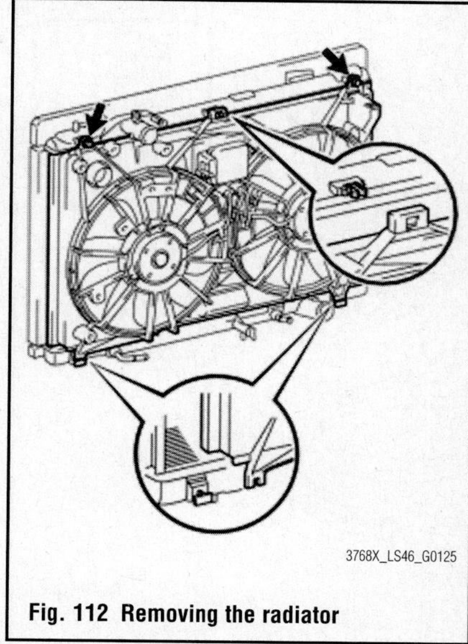

Fig. 112 Removing the radiator

31. Remove the radiator assembly.
   a. Remove the radiator from the crossmember.
   b. Remove the 2 bolts.
   c. Detach the claw and remove the fan shroud with the fan and motor from the radiator.
32. Remove the cooler condenser assembly.
   a. Remove the 4 bolts and separate the condenser from the radiator.

➡Do not damage the core of the A/C condenser.

33. remove the 2 radiator support cushions.
34. Remove the 2 lower radiator supports.

To install:

35. Install the lower radiator supports.
36. Install the radiator support cushions.
37. Install the cooler condenser assembly to the radiator with the 4 bolts. Tighten to 44 inch lbs. (5 Nm).

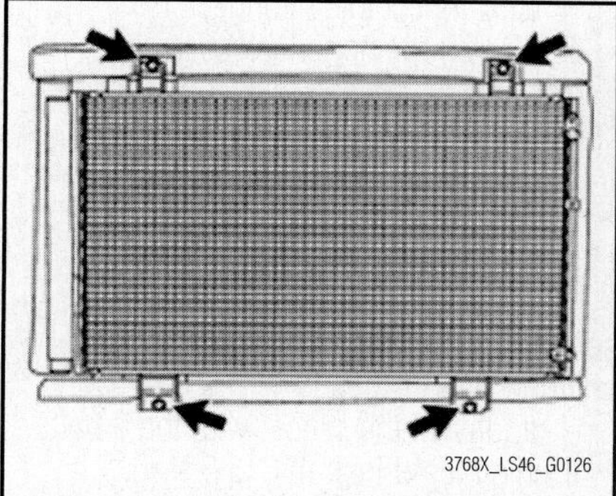

Fig. 113 Removing the cooler condenser assembly

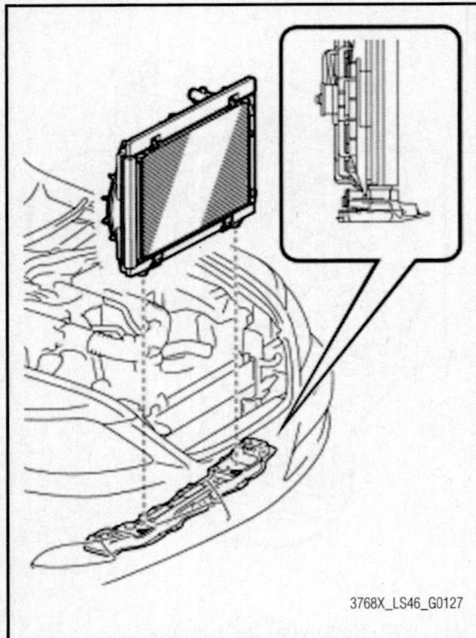

Fig. 114 Installing the radiator to the front crossmember

38. Install the radiator assembly.
    a. Set the fan shroud with the fan and motor to the radiator, and attach the claw.
    b. Install the fan shroud to the radiator with the 2 bolts. Tighten to 44 inch lbs. (5 Nm).
    c. Install the radiator to the front crossmember.

➡Install the radiator to the front side holes of the radiator's lower support.

39. Connect the liquid tube sub assembly A.
40. Connect the discharge hose sub assembly.
41. Install the upper radiator support sub assembly.
    a. Attach the wire harness to the upper support with the 5 clamps.
    b. Install the upper support with the 5 bolts. Tighten to 71 inch lbs. (8 Nm).
    c. Attach the 2 clamps to the radiator support LH and RH side.
    d. Connect the 3 connectors.
42. Install the hood lock assembly.
43. Install the hood lock control cable cover.
    a. Attach the claw to install the cover.
    b. Install the 3 screws.
44. Install the millimeter wave radar sensor assembly (with dynamic radar cruise control system).
    a. Install the sensor with the 3 bolts. Tighten to 49 inch lbs. (5.5 Nm).
    b. Connect the connector.
45. Connect the wire harness.
    a. Attach the 5 clamps and connect the 2 connectors.
    b. Attach the 3 clamps and connect the ECU connector.
46. Install the outlet engine room ECU duct.
47. Connect the outlet oil cooler hose.
    a. Connect the hose to the radiator.

➡The direction of the hose clamp is indicated in the illustration.

➡Insert the cooler hose into the stopper. Set the hose clamp so that the clearance is between the hose clamp and stopper is within 0.0787-0.276 inch (2-7 mm).

48. Connect the inlet oil cooler hose.
    a. Connect the hose to the radiator.

➡Insert the cooler hose into the stopper. Set the hose clamp so that the clearance is between the hose clamp and stopper is within 0.0787-0.276 inch (2-7 mm).

49. Install the No. 2 radiator hose.

➡The direction of the hose clamp is indicated in the illustration.

50. Install the No. 1 radiator hose.
    a. Connect the No. 1 radiator hose to the radiator and water outlet, and then secure it with the hose clamps.

➡If the vehicle is equipped with a water hose set on the water outlet side, assembly the water hose set before use.

➡The direction of the hose clamp is indicated in the illustration.

➡Insert the radiator hose into the stopper. Set the hose clamp so that the clearance between the hose clamp and the stopper is within 0.039-0.20 inch (1-5 mm).

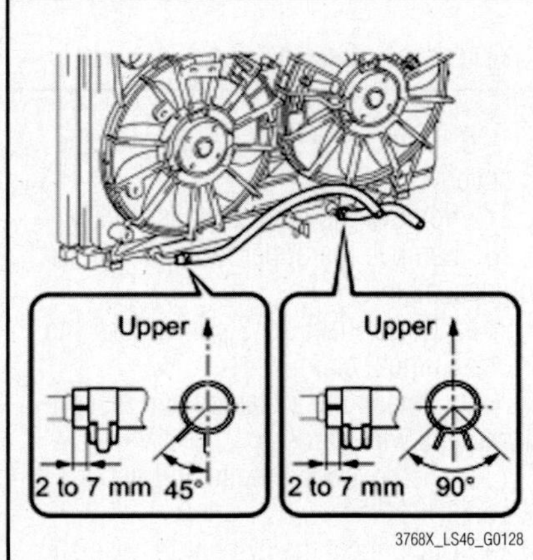

Fig. 115 Connecting the outlet oil cooler hose

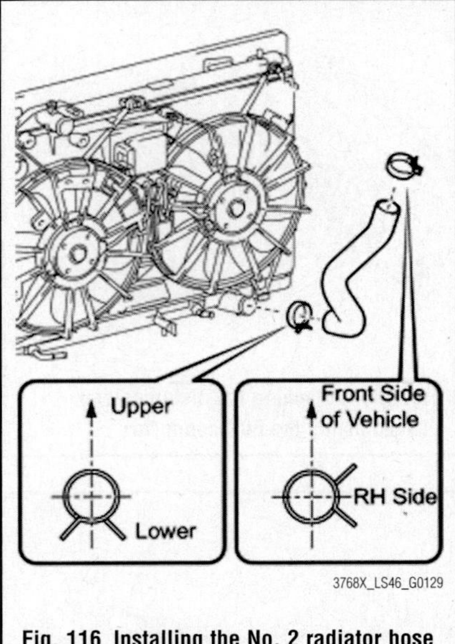

Fig. 116 Installing the No. 2 radiator hose

51. Install the radiator reservoir assembly.
52. Install the front bumper reinforcement.
53. Install the front bumper energy absorber.
54. Install the front bumper cover.
55. Install the front wheel opening extension pad LH.
56. Install the front wheel opening extension pad RH.
57. Install the front fender wheel opening moulding LH.
58. Install the front fender wheel opening moulding RH.
59. Charge the refrigerant.
60. Add engine coolant.
61. Inspect for coolant leaks.
62. Add automatic transmission fluid.
63. Inspect for refrigerant leaks.
64. Install the cool air intake duct seal with the 4 clips.
65. Adjust the millimeter wave radar sensor assembly (with dynamic radar cruise control system).
66. Install the radiator grille assembly.
67. Install the No. 1 engine under cover.
68. Install the No. 1 air cleaner inlet.
69. Install the engine room side cover RH.
70. Install the engine room side cover LH.
71. Install the air cleaner inlet cover sub assembly.
72. Install the v-bank cover sub assembly.

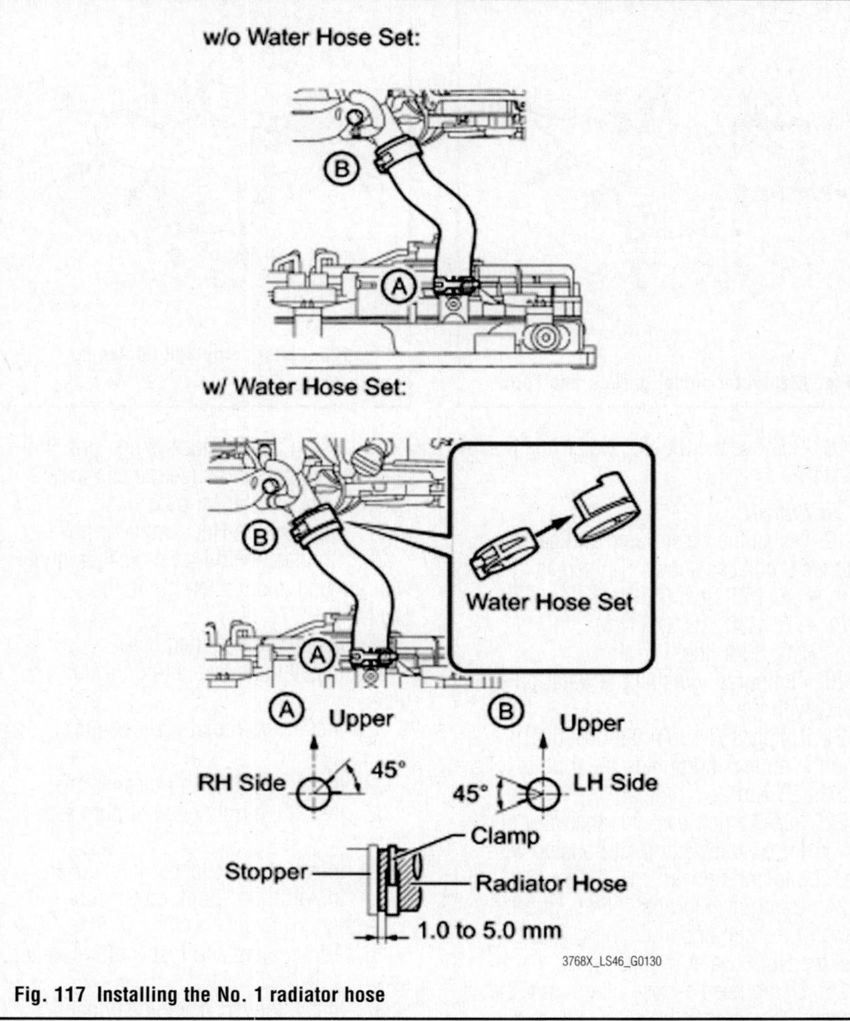

Fig. 117 Installing the No. 1 radiator hose

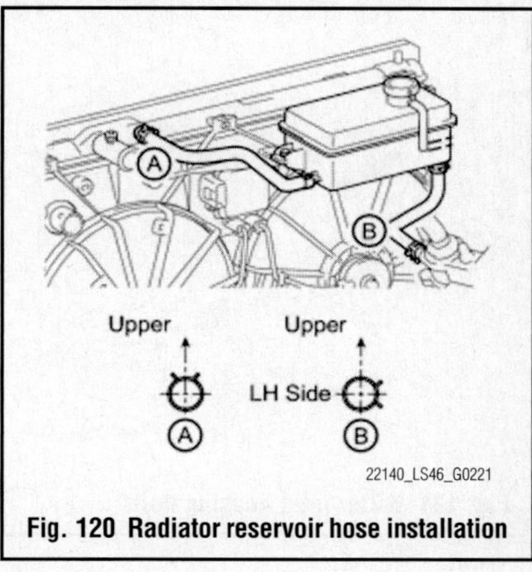

Fig. 120 Radiator reservoir hose installation

THERMOSTAT

REMOVAL & INSTALLATION

See Figures 118 through 120.

1. Remove the V-bank cover sub-assembly.
2. Remove the air cleaner inlet cover.
3. Remove the No. 1 air cleaner inlet.

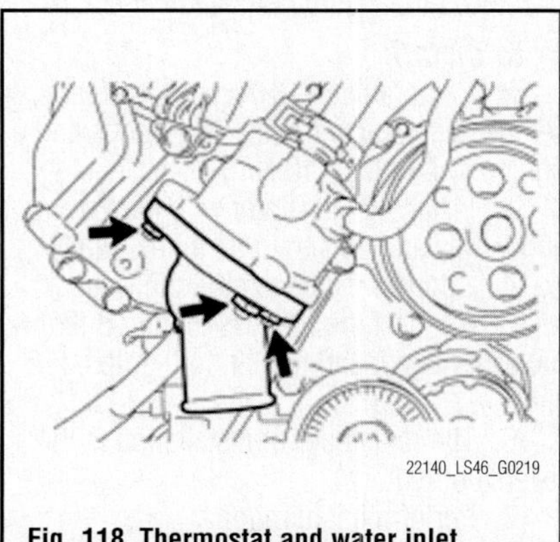

Fig. 118 Thermostat and water inlet

4. Remove the right engine room side cover.
5. Remove the No. 1 engine under cover.
6. Drain engine coolant.

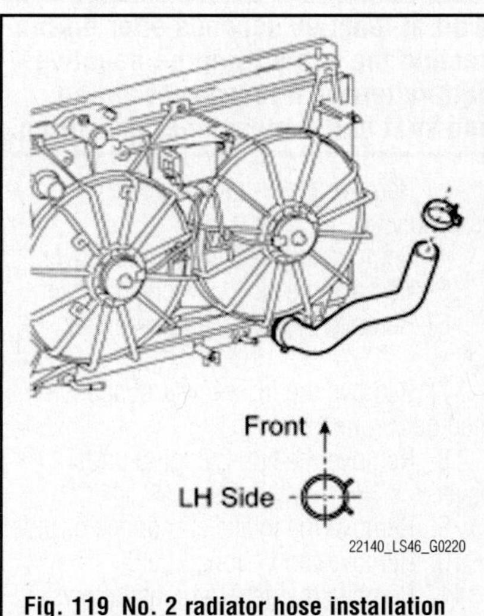

Fig. 119 No. 2 radiator hose installation

7. Remove the radiator reservoir assembly, as follows:
   a. Disconnect the 2 reservoir hoses.
   b. Remove the 2 bolts and reservoir.
8. Disconnect the No. 2 radiator hose.
9. Remove the 3 nuts, water inlet with thermostat and gasket.

To install:

10. Install a new gasket and the water inlet with thermostat with the 3 nuts and tighten to 7 ft. lbs. (10 Nm).
11. Connect the No. 2 radiator hose. The direction of the hose clamp is indicated in the illustration.
12. Install the radiator reservoir with the 2 bolts and tighten to 44 inch lbs. (5 Nm).
13. Connect the 2 reservoir hoses. The direction of the hose clamp is indicated in the illustration.
14. Add coolant and inspect for coolant leak.
15. Install the No. 1 air cleaner inlet.
16. Install the right engine room side cover.
17. Install the air cleaner inlet cover.
18. Install the V-bank cover sub-assembly.
19. Install the No. 1 engine under cover.

WATER PUMP

REMOVAL & INSTALLATION

See Figures 121 through 123.

1. Remove the No. 1 engine under cover.
2. Drain engine coolant.
3. Remove the V-bank cover sub-assembly.
4. Remove the air cleaner inlet cover.
5. Remove the right engine room side cover.
6. Remove the No. 1 air cleaner inlet.
7. Remove the radiator reservoir assembly.

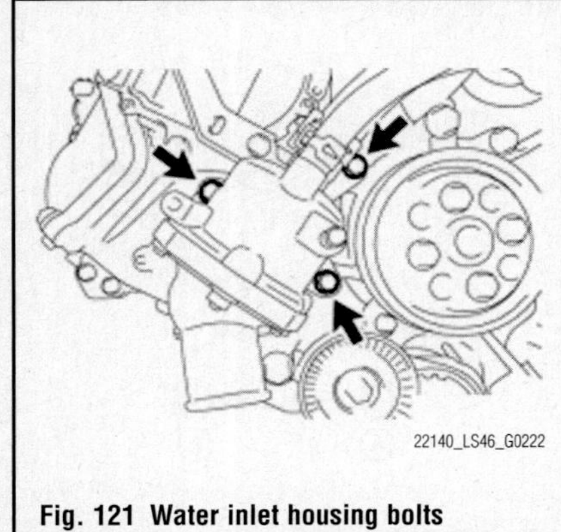

Fig. 121 Water inlet housing bolts

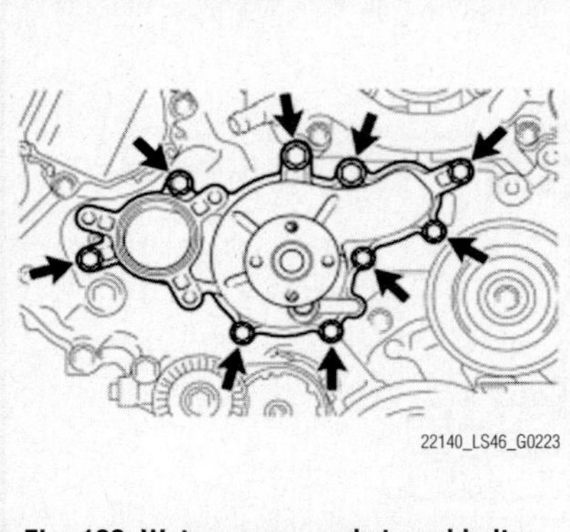

Fig. 122 Water pump, gasket, and bolts

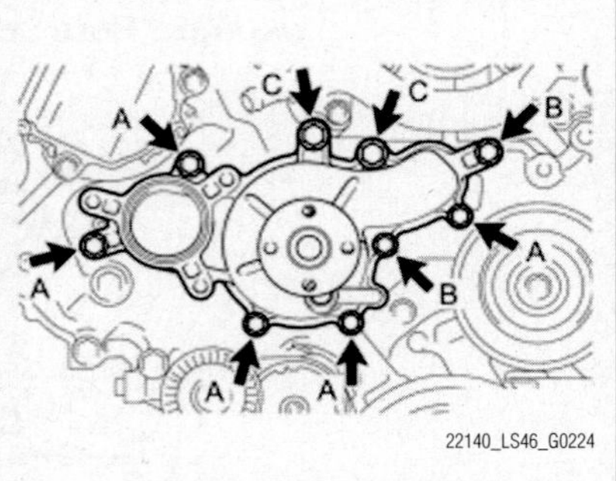

Fig. 123 Water pump bolt tightening sequence

8. Remove the intake air connector pipe.
9. Remove the V-ribbed belt.
10. Disconnect the bolt and the oil level dipstick guide.
11. Disconnect the No. 2 radiator hose.
12. Using needle-nose pliers, grip the claws of the clip and slide the clip to disconnect the No. 5 water by-pass hose.
13. Using needle-nose pliers, grip the claws of the clip and slide the clip to disconnect the water inlet hose.
14. Using needle-nose pliers, grip the claws of the clip and slide the clip to disconnect the No. 3 water by-pass hose.
15. Remove the 3 bolts, water inlet housing and gasket.
16. Using SST (SST: 09960-10010), hold the water pump pulley.
17. Remove the 4 bolts and water pump pulley.
18. Remove the 9 bolts, water pump and gasket.

To install:

19. Install the water pump and gasket with the 9 bolts as shown. Tighten as follows:
- A: 15 ft. lbs. (20 Nm)
- B: 17 ft. lbs. (23 Nm)
- C: 35 ft. lbs. (47 Nm)

20. Temporarily install the water pump pulley with the 4 bolts.
21. Using SST (SST: 09960-10010), hold the pulley and tighten the 4 bolts to15 ft. lbs. (21 Nm).
22. Install a new gasket to the water pump.
23. Install the water inlet housing with the 3 bolts and tighten to15 ft. lbs. (21 Nm).
24. Using needle-nose pliers, grip the claws of the clip and slide the clip to connect the No. 3 water by-pass hose.
25. Using needle-nose pliers, grip the claws of the clip and slide the clip to connect the water inlet hose.
26. Using needle-nose pliers, grip the claws of the clip and slide the clip to connect the No. 5 water by-pass hose.
27. Connect the No. 2 radiator hose.
28. Connect the oil level dipstick guide with the bolt and tighten to7 ft. lbs. (10 Nm).
29. Install the V-ribbed belt.
30. Install the intake air connector pipe.
31. Install the radiator reservoir assembly.
32. Install the No. 1 air cleaner inlet.
33. Install the right engine room side cover.
34. Install the air cleaner inlet cover.
35. Install the V-bank cover sub-assembly.
36. Add coolant and inspect for coolant leak.
37. Install the No. 1 engine under cover.

ENGINE ELECTRICAL — CHARGING SYSTEM

ALTERNATOR

REMOVAL & INSTALLATION

⁂ WARNING

After the engine switch is turned off, the HDD navigation system requires approximately 6 minutes to record various types of memory and settings. As a result, after turning the engine switch off, wait 6 minutes or more before disconnecting the cable from the negative battery terminal.

1. Remove the right cowl top ventilator louver.
2. Disconnect cable from negative battery terminal.

⁂ CAUTION

Wait at least 90 seconds after disconnecting the cable from the negative battery terminal to prevent airbag and seat belt pretensioner activation.

3. Remove the V-bank cover sub-assembly.
4. Remove the air cleaner inlet cover.
5. Remove the No. 1 air cleaner inlet.
6. Remove the No. 1 engine under cover.
7. Remove the front lower suspension member protector.
8. Remove the No. 2 engine under cover.
9. Remove the intake air connector pipe.
10. Remove the V-ribbed belt.
11. Remove the generator, as follows:
   a. Remove the nut, and disconnect the harness from the +B terminal.
   b. Disconnect the generator connector.
   c. Remove the 2 bolts and 2 nuts.
   d. Using an E8 TORX® socket wrench, remove the 2 stud bolts and generator.

To install:

12. Using an E8 TORX® socket wrench, set the generator with the 2 stud bolts and tighten to 7 ft. lbs. (10 Nm).
13. Install the generator with the 2 bolts and 2 nuts and tighten to 32 ft. lbs. (43 Nm).
14. Connect the generator connector.
15. Connect the harness to the +B terminal with the nut and tighten to 9 ft. lbs. (12 Nm).
16. The remainder of installation is the reverse of removal.
17. Perform initialization.

ENGINE ELECTRICAL — IGNITION SYSTEM

FIRING ORDER

Firing order for 4.6L (1UR-FSE) engine: 1–8–4–3–6–5–7–2

IGNITION COIL

REMOVAL & INSTALLATION

See Figures 124 and 125.

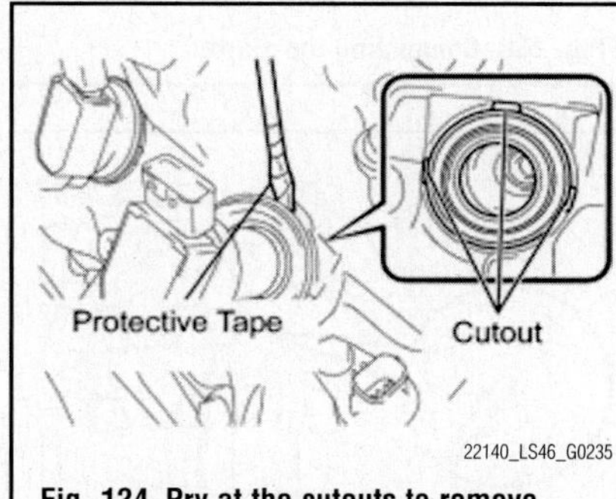

Fig. 124 Pry at the cutouts to remove ignition coils

⁂ WARNING

After the engine switch is turned off, the HDD navigation system requires approximately 6 minutes recording various types of memory and settings. As a result, after turning the engine switch off, wait 6 minutes or more before disconnecting the cable from the negative battery terminal.

1. Remove the right cowl top ventilator louver.
2. Disconnect cable from negative battery terminal.

⁂ CAUTION

Wait at least 90 seconds after disconnecting the cable from the negative battery terminal to prevent airbag and seat belt pretensioner activation.

3. Remove the V-bank cover sub-assembly.
4. Remove the air cleaner inlet cover.
5. Remove the right and left engine room side cover.
6. Remove the No. 1 air cleaner inlet.
7. Remove the skid control ECU.
8. Remove the skid control ECU bracket.
9. Remove the battery and tray.
10. Remove the right and left air cleaner assembly.
11. Remove the ignition coil assembly, as follows:
    a. Disconnect the 8 ignition coil connectors.
    b. Remove the 8 bolts.
    c. Using a screwdriver with the tip taped, pry at the cutouts to remove the 8 ignition coils together with the 8 spark plug tube gaskets. Do not damage the cylinder head cover when removing the spark plug tube gasket.
    d. Using a 16 mm plug wrench, remove the 8 spark plugs.

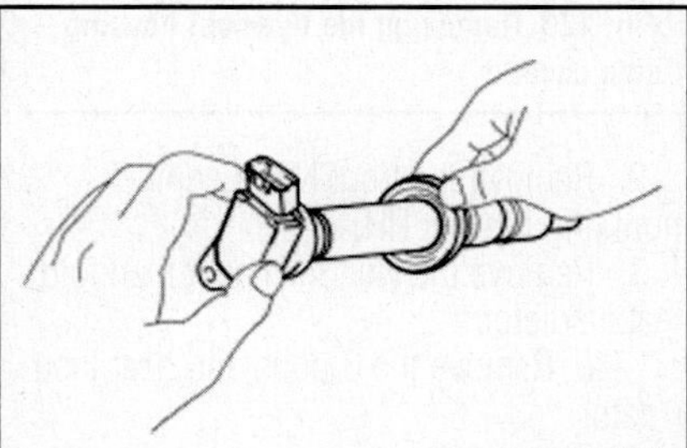

Fig. 125 Spark plug tube gasket installation

To install:

12. Using a 16 mm plug wrench, install the 8 spark plugs and tighten to 13 ft. lbs. (18 Nm).
13. Perform a visual inspection on the spark plug tube gasket. There should be not scratches or deformation on the upper surface or inner or outer lip.
14. Slide a new spark plug tube gasket onto the ignition coil.
15. After installing the spark plug tube gasket, firmly insert the ignition coil.
16. Install the 8 bolts and tighten to 7 ft. lbs. (10 Nm).
17. Connect the 8 ignition coil connectors.
18. The remainder of installation is the reverse of removal.
19. Perform initialization.

IGNITION TIMING

ADJUSTMENT

The engines covered in this section are equipped with a Distributorless Ignition System (DIS). No timing adjustments are possible.

SPARK PLUGS

REMOVAL & INSTALLATION

See Ignition Coil Pack. The recommended spark plug is DENSO made, FK20HBR11.

ENGINE ELECTRICAL — STARTING SYSTEM

STARTER

REMOVAL & INSTALLATION

2WD Models

⁂ WARNING

After the engine switch is turned off, the HDD navigation system requires approximately 6 minutes to record various types of memory and settings. As a result, after turning the engine switch off, wait 6 minutes or more before disconnecting the cable from the negative battery terminal.

1. Remove the right cowl top ventilator louver.
2. Disconnect cable from negative battery terminal.

⁂ CAUTION

Wait at least 90 seconds after disconnecting the cable from the negative battery terminal to prevent airbag and seat belt pretensioner activation.

3. Remove the generator.
4. Remove the front exhaust pipe.
5. Remove the right and left rear engine under cover.
6. Disconnect the front stabilizer bar.
7. Remove the screw, clip and right engine under cover.
8. Remove the 4 bolts and the right front suspension member reinforcement.
9. Remove the engine oil level dipstick guide.
10. Remove the No. 1 exhaust manifold heat insulator.
11. Remove the right exhaust manifold sub-assembly.
12. Remove the 3 bolts and the No. 3 exhaust manifold heat insulator.
13. Remove the starter assembly, as follows:

14. Detach the 11 claws and remove the terminal upper cover.

15. Disconnect the starter connector.

16. Remove the nut and disconnect the wire harness.

17. Remove the 2 bolts and starter.

18. Remove the nut and starter terminal lower cover.

19. Remove the flywheel housing side cover.

To install:

20. Installation is the reverse of removal, noting the following:

a. Tighten the starter terminal lower cover nut to 7 ft. lbs. (10 Nm).

b. Tighten the starter bolts to 27 ft. lbs. (37 Nm).

c. Tighten the wire harness nut to 88 inch lbs. (9.8 Nm).

d. Perform initialization.

AWD Models

See Figures 126 through 131.

1. Remove the exhaust manifold sub assembly RH.

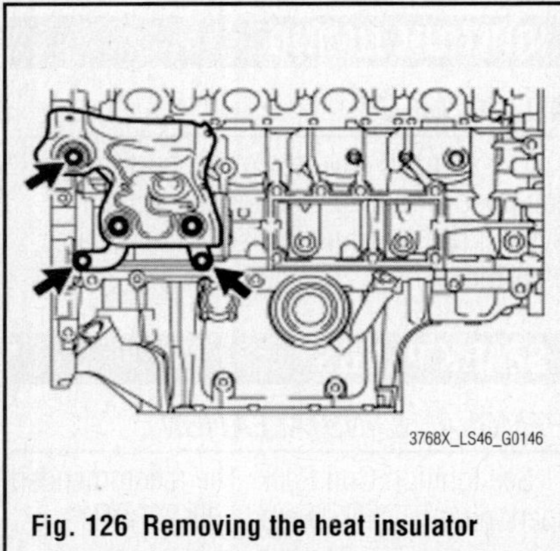

Fig. 126 Removing the heat insulator

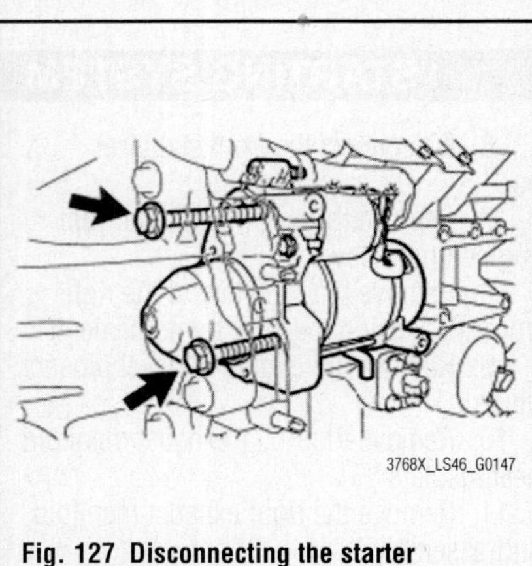

Fig. 127 Disconnecting the starter

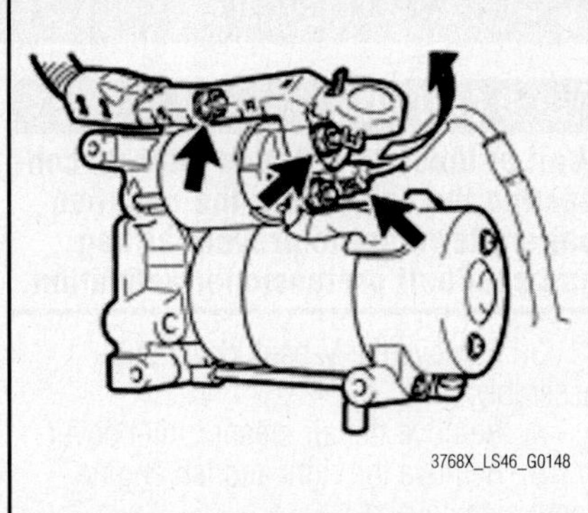

Fig. 128 Opening the terminal cap

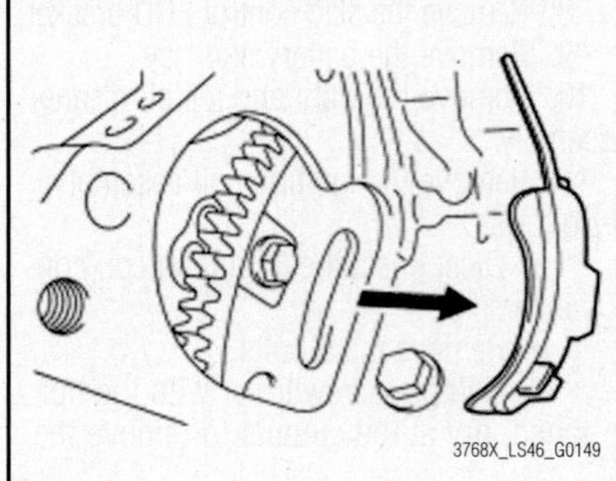

Fig. 129 Removing the flywheel housing side cover

2. Remove the front No. 2 engine mounting bracket RH.

3. Remove the No. 3 exhaust manifold heat insulator.

a. Remove the 3 bolts and heat insulator.

4. Remove the starter assembly.

a. Remove the 2 bolts and disconnect the starter.

b. Open the terminal cap.

c. Remove the 2 nuts and disconnect the starter wire.

d. Disconnect the starter connector and remove the starter.

e. Remove the flywheel housing side cover.

To install:

5. Install the starter.

a. Install the flywheel housing side cover.

b. Connect the starter connector.

c. Connect the starter wire with the 2 nuts. Tighten nut A to 7 ft. lbs. (10 Nm). Tighten nut B to 88 inch lbs. (9.8 Nm).

d. Close the terminal cap.

e. Install the starter with the 2 bolts. Tighten to 27 ft. lbs. (37 Nm).

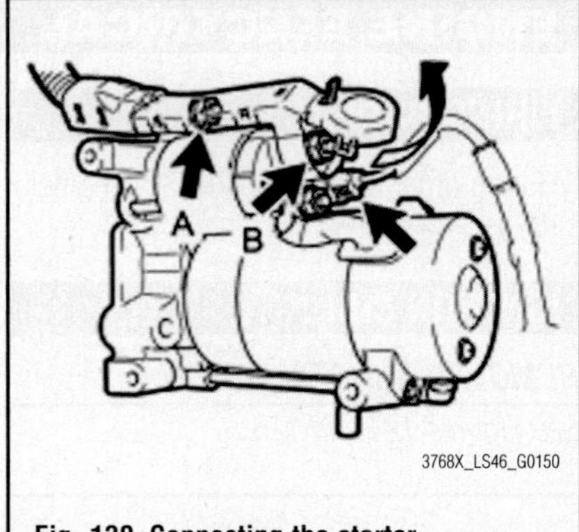

Fig. 130 Connecting the starter

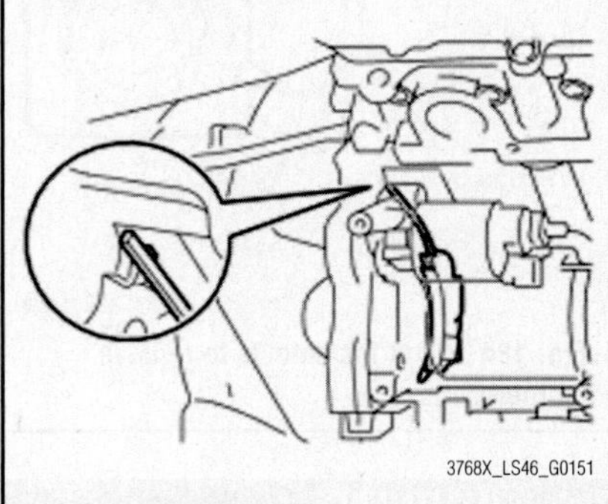

Fig. 131 Checking the flywheel cover installation

➡Make sure the flywheel housing cover is as shown in the illustration.

6. Install the No. 3 exhaust manifold heat insulator with the 3 bolts. Tighten to 7 ft. lbs. (10 Nm).

7. Install the front No. 2 engine mounting bracket RH.

8. Install the exhaust manifold sub assembly RH.

SOLENOID OR RELAY REPLACEMENT

1. Remove the starter.

2. Remove the nut, and disconnect the lead wire from the starter magnetic switch.

3. Remove the 2 nuts and starter magnetic switch assembly.

To install:

4. Temporarily install the magnetic switch by hooking its tip to the upper side of the pinion drive lever.

5. Install the 2 nuts and tighten to 66 inch lbs. (7.5 Nm).

6. Connect the lead wire to the terminal with the nut and tighten to 57 inch lbs. (5.5 Nm).

7. Install the starter.

ENGINE MECHANICAL

ACCESSORY DRIVE BELTS

ACCESSORY BELT ROUTING

See Figure 132.

Refer to the accompanying illustration.

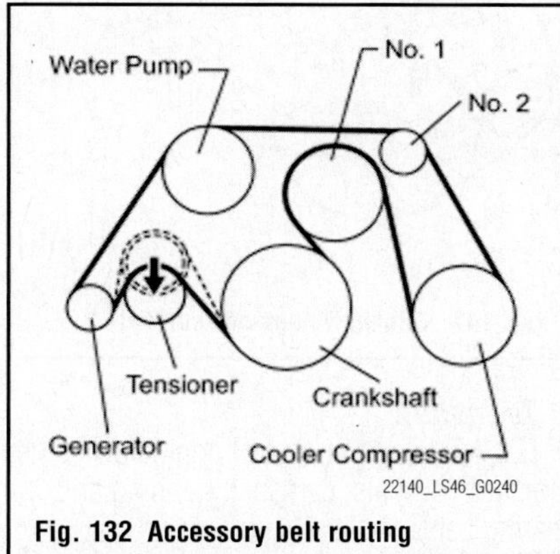

Fig. 132 Accessory belt routing

INSPECTION

Inspect the drive belt for signs of glazing or cracking. A glazed belt will be perfectly smooth from slippage, while a good belt will have a slight texture of fabric visible. Cracks will usually start at the inner edge of the belt and run outward. All worn or damaged drive belts should be replaced immediately.

ADJUSTMENT

These engines are equipped with automatic belt tensioners. Adjusting the belt tension is not possible or necessary.

REMOVAL & INSTALLATION

See Figure 132.

1. Remove the V-bank cover sub-assembly.
2. Remove the air cleaner inlet cover.
3. Remove the No. 1 air cleaner inlet.
4. Remove the intake air connector pipe.
5. Remove the V-ribbed belt, as follows:
6. Rotate the tensioner pulley counterclockwise to loosen the belt tension. The pulley bolt for the belt tensioner has a left-handed thread.
7. While turning the belt tensioner counterclockwise, align the holes. Insert a bar of φ5 mm (0.20 in.) into the holes to fix the belt tensioner in place.
8. Remove the V belt.

To install:

9. Install the V belt as shown. Check that the drive belt is properly set to each pulley.
10. Rotate the tensioner pulley counterclockwise, and then remove the fix bar.
11. Install the intake air connector pipe.
12. Install the No. 1 air cleaner inlet.
13. Install the air cleaner inlet cover.
14. Install the V-bank cover sub-assembly.

CAMSHAFT AND VALVE LIFTERS

INSPECTION

See Figures 133 and 134.

1. To inspect the camshaft, place the camshaft on V-blocks.
   a. Using a dial indicator, measure the circle runout at the center journal.
   b. Maximum runout: 0.0016 inches (0.04 mm).
   c. If the runout is greater than the maximum, replace the camshaft.

➡Check the oil clearance after replacing the camshaft.

2. Using a micrometer, measure the cam lobe height.

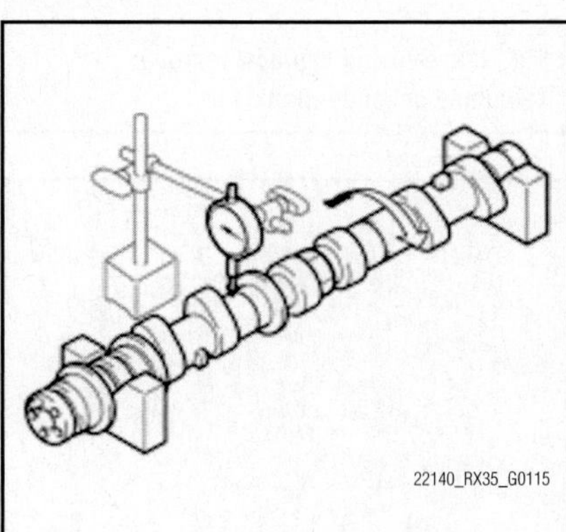

Fig. 133 Measuring camshaft runout

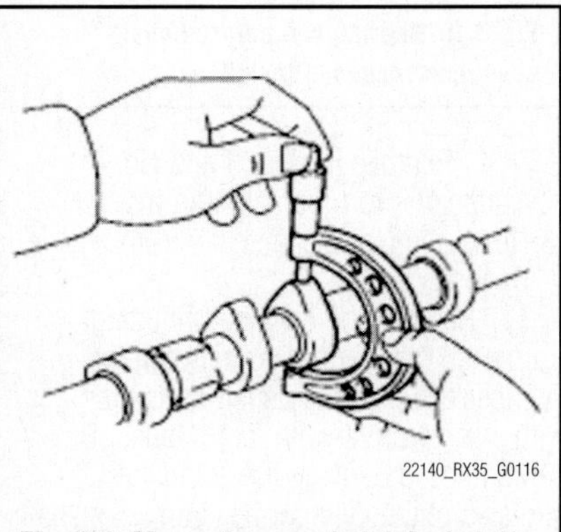

Fig. 133 Measuring camshaft lobe height

3. Standard lobe height intake camshafts: 1.7438 to 1.7497 inches (44.293 mm to 44.443 mm)
4. Standard lobe height exhaust camshafts: 1.7447 to 1.7506 inches (44.316 mm to 44.466 mm)
5. If the cam lobe height is less than the minimum, replace the camshaft
6. Using a micrometer, measure the journal diameter.
7. Standard No.1 journal diameter: 1.1794 to 1.1799 inches (29.956 mm to 29.970 mm)
8. Standard other journal diameter: 1.0220 to 1.0226 inches (25.959 mm to 25.975 mm)
   a. If the journal diameter is not as specified, check the oil clearance.

REMOVAL & INSTALLATION

See Figures 135 through 148.

⁂ WARNING

After the engine switch is turned off, the HDD navigation system requires approximately 6 minutes to record various types of memory and settings. As a result, after turning the engine switch off, wait 6 minutes or more before disconnecting the cable from the negative battery terminal.

1. Remove the right cowl top ventilator louver.
2. Disconnect cable from negative battery terminal.

⁂ CAUTION

Wait at least 90 seconds after disconnecting the cable from the negative battery terminal to prevent airbag and seat belt pretensioner activation.

3. Discharge fuel system pressure.
4. Remove timing chain cover sub-assembly.
5. Set No. 1 cylinder to TDC/compression.
6. Remove No. 1 chain tensioner assemblies.
7. Remove No. 1 chain vibration damper.
8. Remove chain sub-assembly.
9. Remove No. 3 chain tensioner assembly.
10. Remove camshaft bearing caps:
    a. Make sure that the knock pin of the camshaft is positioned as shown in the illustration.

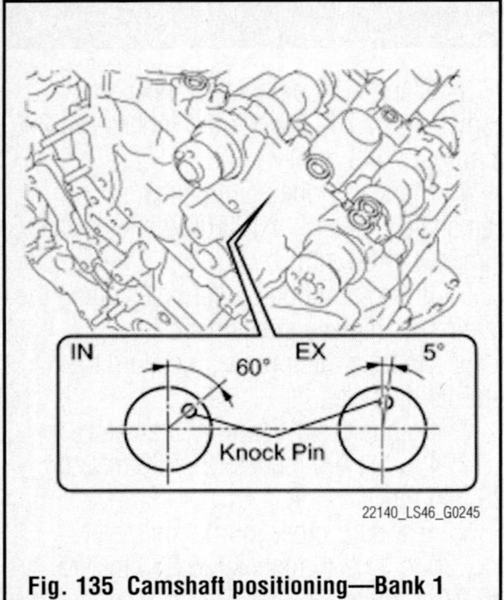

Fig. 135 Camshaft positioning—Bank 1

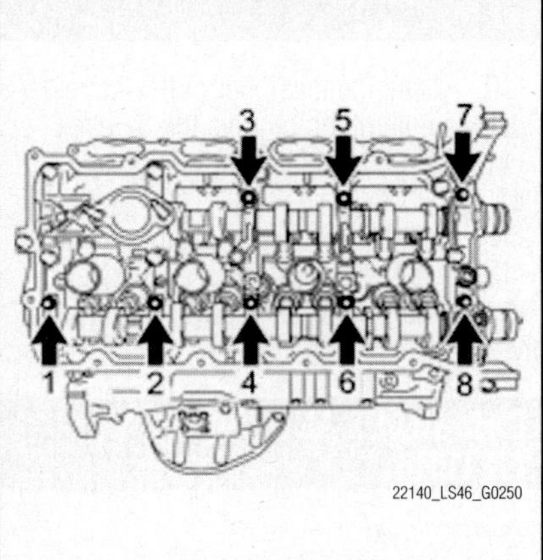

Fig. 138 Bearing cap bolt removal sequence first—Bank 2

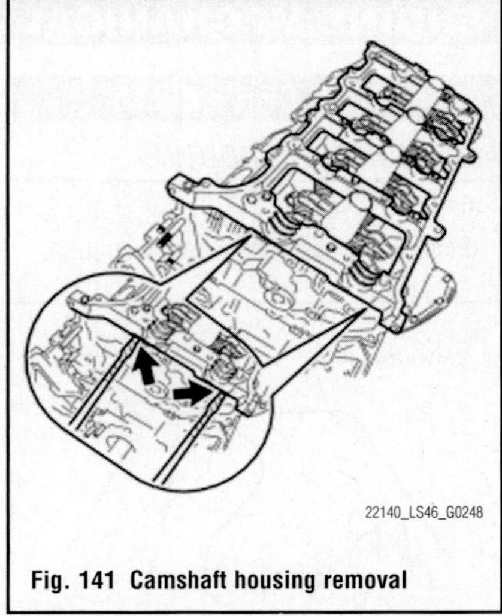

Fig. 141 Camshaft housing removal

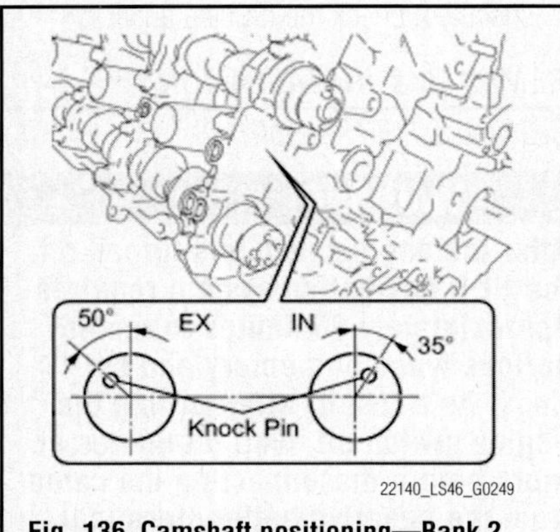

Fig. 136 Camshaft positioning—Bank 2

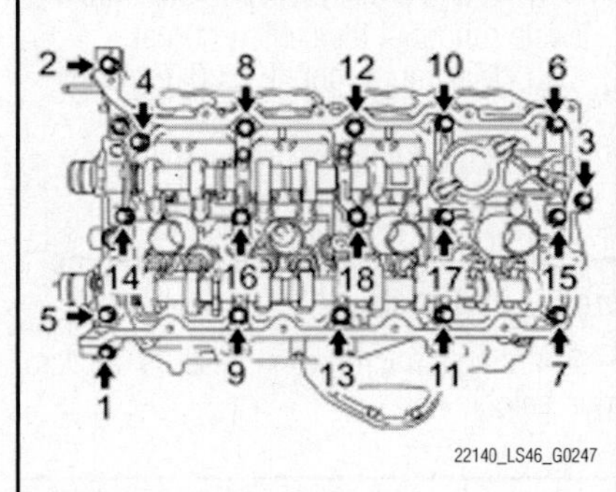

Fig. 139 Bearing cap bolt removal sequence second—Bank 1

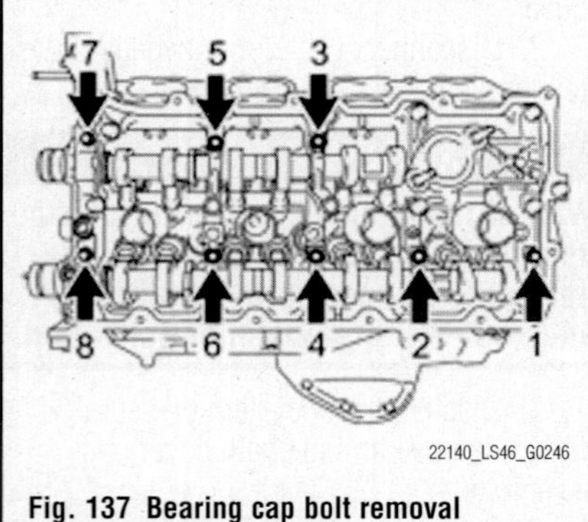

Fig. 137 Bearing cap bolt removal sequence first—Bank 1

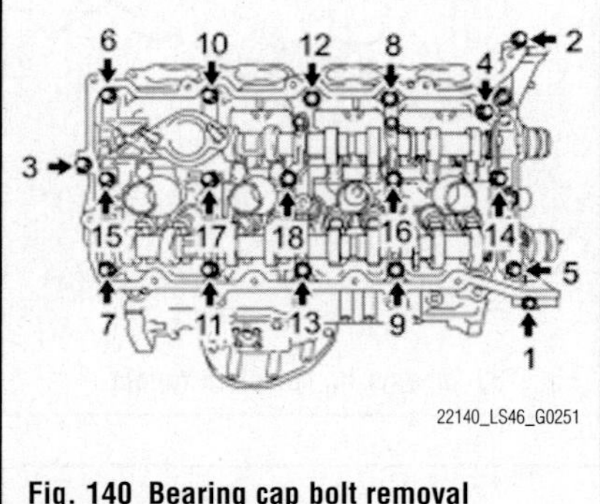

Fig. 140 Bearing cap bolt removal sequence second—Bank 2

b. Uniformly loosen and remove the 8 bearing cap bolts in the sequence shown in the illustration

c. Uniformly loosen and remove the 18 bearing cap bolts in the sequence shown in the illustration

➡Uniformly loosen the bolts while keeping the camshaft level

d. Remove the 7 bearing caps.

e. Remove the No. 3 and No. 4 camshafts from Bank 1, and the No. 1 and No. 2 camshafts from Bank 2.

11. Remove the camshaft housing from each side by prying between the cylinder head and camshaft housing with a screwdriver with its tip taped. Be careful not to damage the contact surfaces of the cylinder head and camshaft housing.

To install:

12. Apply a light coat of engine oil to the camshaft journals, camshaft housing and bearing caps.

13. Install the No. 3 and No. 4 camshafts to the Bank 1 camshaft housing, and the No. 1 and No. 2 camshafts to the Bank 2 camshaft housing.

14. Confirm the marks and numbers on the camshaft bearing caps and place them in their proper positions and directions.

15. Temporarily install the 8 bolts in the order shown in the illustration.

16. Make sure that the valve rocker arms are installed correctly.

17. Apply seal packing in a continuous line as shown. Seal packing: Toyota Genuine Seal Packing Black, Three Bond 1207B or equivalent. Standard seal diameter: 3.5 to 4.0 mm (0.138 to 0.158 in.)

➡Remove any oil from the contact surface. Install the camshaft housing within 3 minutes and tighten the bolts within 15 minutes after applying seal packing. Do not start the engine for at least 2 hours after the installation.

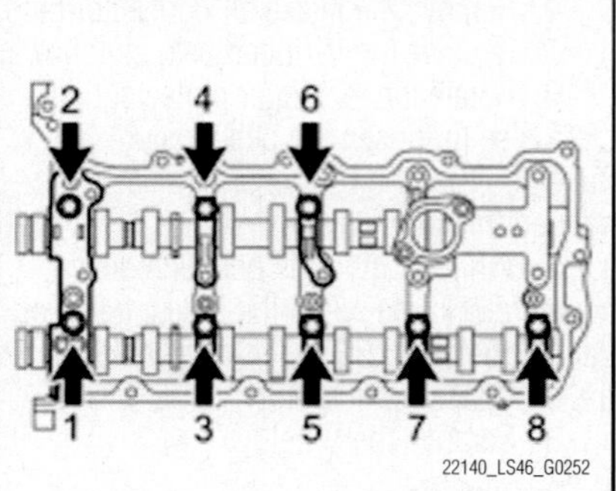

Fig. 142 Camshaft bearing cap bolt installation sequence—Bank 1

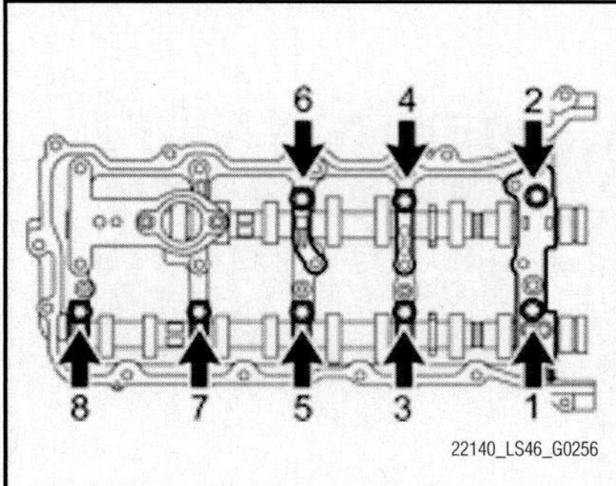

Fig. 143 Camshaft bearing cap bolt installation sequence—Bank 2

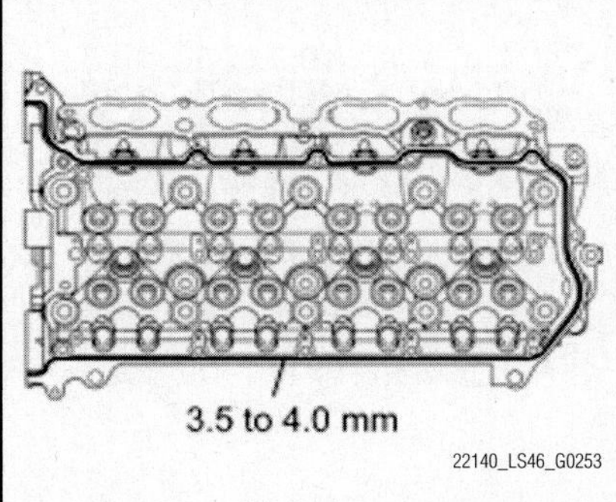

Fig. 144 Sealant packing location, Bank 1 shown, Bank 2 similar

18. Install the camshaft housing, and install the 12 bolts and tighten to 7 ft. lbs. (10 nm) and 22 ft. lbs. (30 Nm), in the order shown.

➡Make sure that each knock pin of the camshafts is positioned as shown in the illustration before installing the camshaft housing.

19. Tighten the 8 bolts to 12 ft. lbs. (16 Nm) in the order shown in the illustration

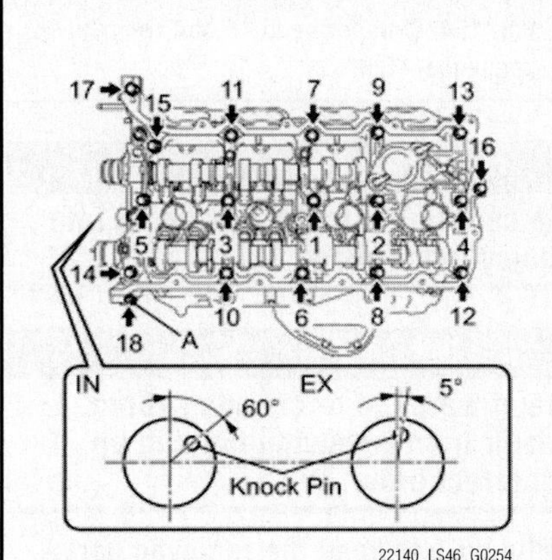

Fig. 145 Camshaft housing bolt tightening sequence—Bank 1

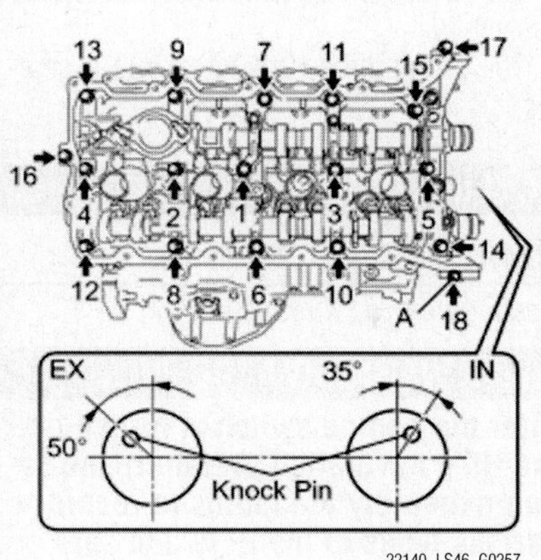

Fig. 146 Camshaft housing bolt tightening sequence—Bank 2

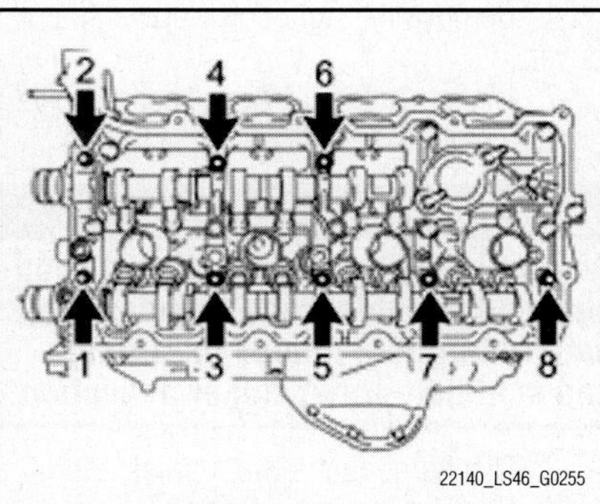

Fig. 147 Camshaft bearing cap bolt tightening sequence—Bank 1

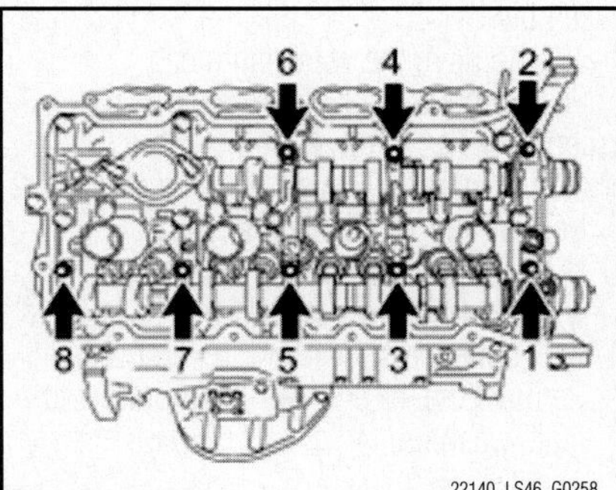

Fig. 148 Camshaft bearing cap bolt tightening sequence—Bank 2

20. Install No. 3 chain tensioner assembly.
21. Install timing chain.
22. Install chain tensioner slipper.
23. Install No. 1 chain tensioner assembly.
24. Install No. 1 chain vibration damper.
25. Tighten camshaft timing gear assembly.
26. Check No. 1 cylinder to TDC/compression.
27. Install timing chain cover.
28. Connect cable from negative battery terminal.

CRANKSHAFT DAMPER

REMOVAL & INSTALLATION

See Figures 149 and 150.

1. Before servicing the vehicle, refer to the Precautions Section.
2. Remove the V-bank cover sub-assembly.
3. Remove the air cleaner inlet cover sub-assembly.
4. Remove the No. 1 air cleaner inlet.
5. Remove the intake air connector pipe.
6. Remove the V-ribbed belt.
7. Remove the resonator bracket sub-assembly.
8. Remove the crankshaft pulley, as follows:
   a. Using SST (SST: 09213-54015, SST: 09330-00021), loosen the crankshaft pulley set bolt.
   b. Using the pulley set bolt and SST (09950-50013), remove the crankshaft pulley.

To install:

9. Align the pulley set key with the key groove of the pulley, and slide on the pulley.
10. Using SST (SST: 09213-54015, SST: 09330-00021), install the pulley bolt and tighten to 221 ft. lbs. (300 Nm).
11. The remainder of installation is the reverse of removal.

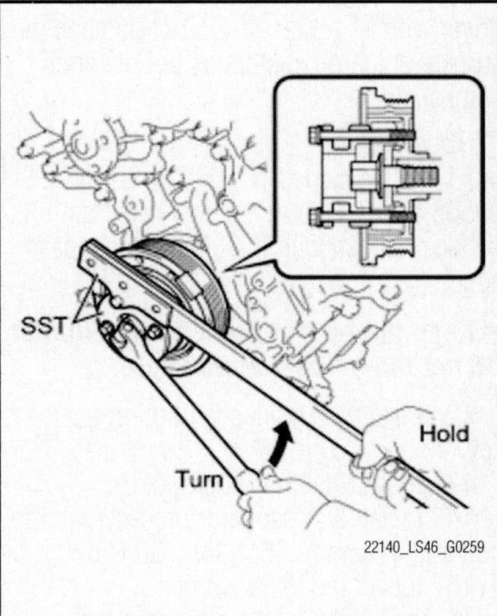

Fig. 149 Loosen the crankshaft pulley set bolt

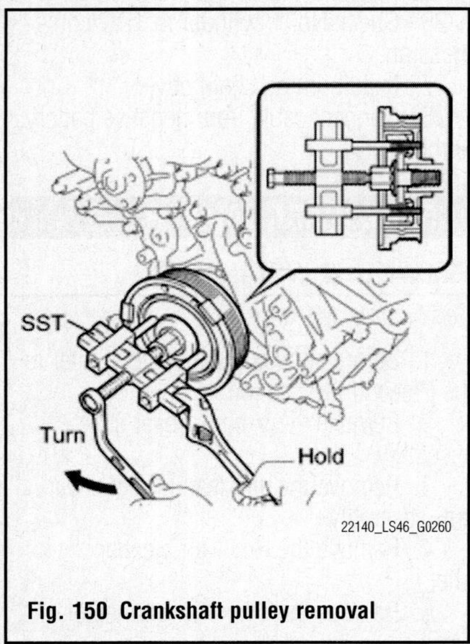

Fig. 150 Crankshaft pulley removal

CRANKSHAFT FRONT SEAL

REMOVAL & INSTALLATION

1. Before servicing the vehicle, refer to the Precautions Section.
2. Remove the V-bank cover sub-assembly.
3. Remove the air cleaner inlet cover sub-assembly.
4. Remove the No. 1 air cleaner inlet.
5. Remove the intake air connector pipe.
6. Remove the V-ribbed belt.
7. Remove the resonator bracket sub-assembly.
8. Remove the crankshaft pulley.
9. Remove the crankshaft timing gear key from the crankshaft.
10. Remove the oil seal with a screwdriver with its tip taped. Do not damage the surface of the oil seal press fit hole and crankshaft.

To install:

11. Using SST (SST: 09223-22010, SST: 09506-35010) tap in a new oil seal until its surface is flush with the timing chain case edge.

➡Keep the lip free from foreign matter. Do not tap oil seal at an angle.

12. Install the crankshaft timing gear key.
13. Install the crankshaft pulley.
14. Install the resonator bracket with the bolt and tighten to 15 ft. lbs. (20 Nm).
15. Install the V-ribbed belt.
16. Install the intake air connector pipe.
17. Install the No. 1 air cleaner inlet.
18. Install the air cleaner inlet cover sub-assembly.
19. Install the V-bank cover sub-assembly.

CYLINDER HEAD

REMOVAL & INSTALLATION

See Figures 151 through 162.

✲✲ WARNING

After the engine switch is turned off, the HDD navigation system requires approximately 6 minutes to record various types of memory and settings. As a result, after turning the engine switch off, wait 6 minutes or more before disconnecting the cable from the negative battery terminal.

1. Remove the right cowl top ventilator louver.
2. Disconnect cable from negative battery terminal.

✲✲ CAUTION

Wait at least 90 seconds after disconnecting the cable from the negative battery terminal to prevent airbag and seat belt pretensioner activation.

3. Discharge fuel system pressure.
4. Remove the camshafts.
5. Remove the No. 1 valve rocker arm sub-assembly.
6. Remove the valve lash adjuster assembly.
7. Remove the valve stem cap.
8. Remove the cylinder head sub-assemblies, as follows:
   a. Uniformly loosen and remove the 2 bolts in the sequence shown.
   b. Using a 10 mm bi-hexagon wrench, uniformly loosen the 10 bolts in the sequence shown in the illustration. Remove the 10 cylinder head bolts and plate washers.

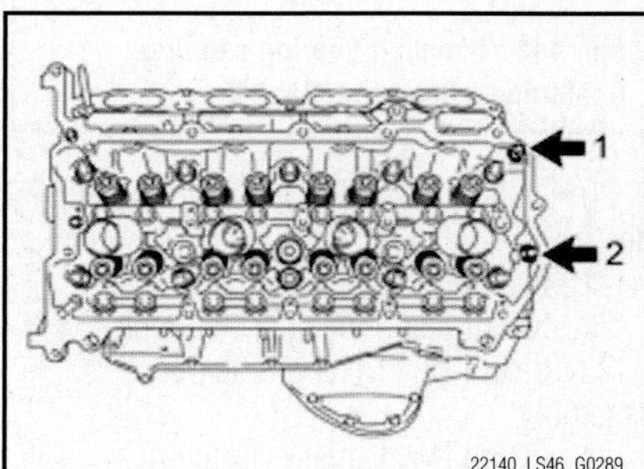

Fig. 151 Cylinder head 2 bolt removal sequence—Bank 1

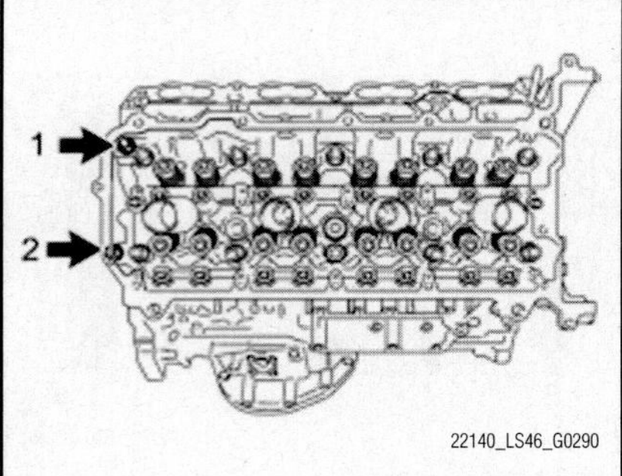

Fig. 152 Cylinder head 2 bolt removal sequence—Bank 2

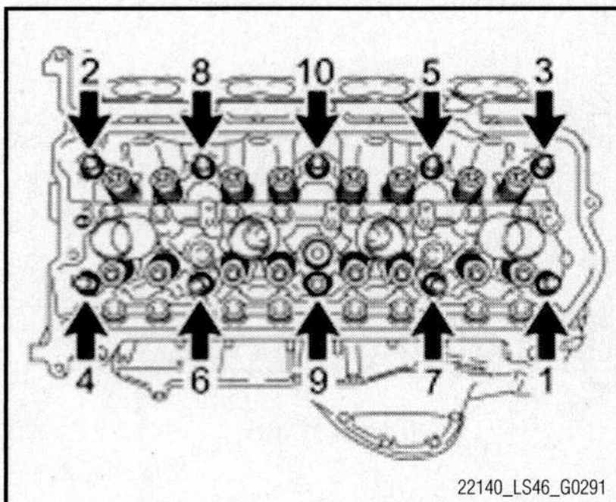

Fig. 153 Cylinder head 10 bolt removal sequence—Bank 1

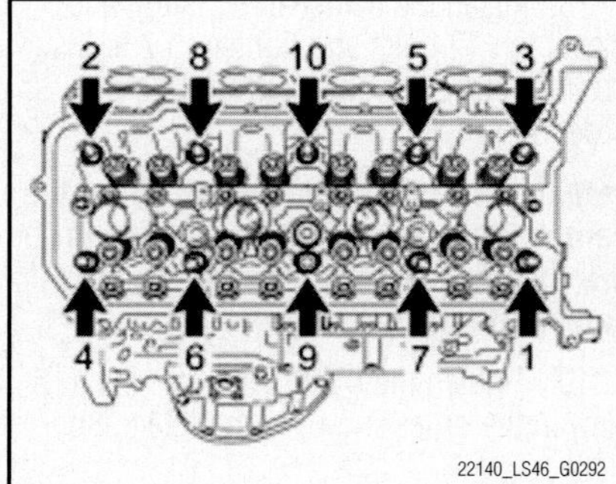

Fig. 154 Cylinder head 10 bolt removal sequence—Bank 2

✲✲ WARNING

Be careful not to drop washers into the cylinder head.

✲✲ WARNING

Head warpage or cracking could result from removing bolts in an incorrect order.

➡Be sure to keep the removed parts separate for each installation position.

   c. Remove the cylinder head and gasket.

To install:

9. Inspect cylinder head set bolt.
10. Inspect cylinder head sub-assembly.
11. Check the piston protrusions for each cylinder:
    a. Clean the cylinder block with solvent.
    b. Set the piston of the cylinder to be measured to slightly ATDC.
12. Place the cylinder head gasket on the cylinder block surface with the front face of the Lot No. stamp upward.

✲✲ WARNING

Be careful of the installation direction.

✲✲ WARNING

Gently place the cylinder head in order not to damage the gasket with the bottom part of the head. Place the cylinder head on the cylinder block.

➡Ensure that no oil is on the mounting surface of the cylinder head.

13. Apply a light coat of engine oil to the threads and under the heads of the cylinder head bolts

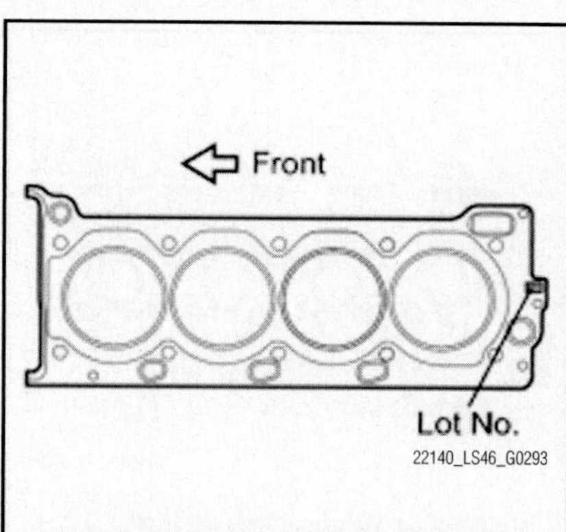

Fig. 155 Position the cylinder head gasket with Lot No. stamp upward—Bank 1

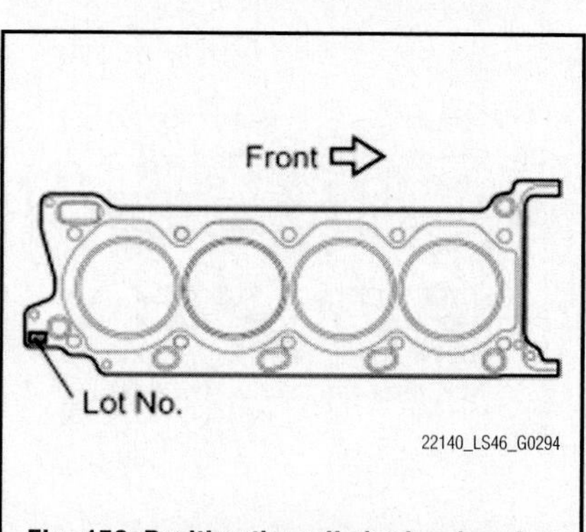

Fig. 156 Position the cylinder head gasket with Lot No. stamp upward—Bank 2

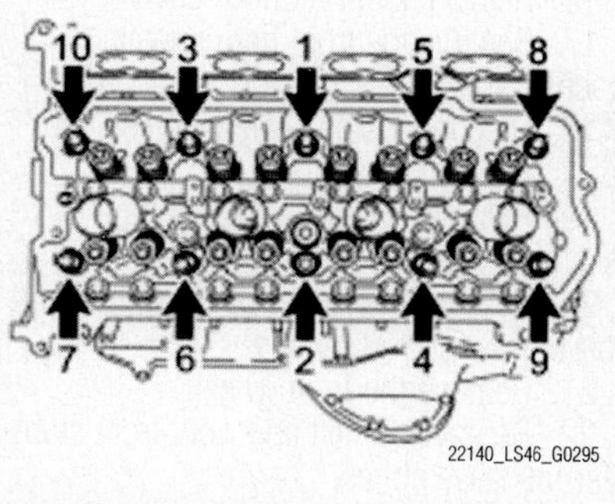

Fig. 157 Cylinder head 10 bolt tightening sequence—Bank 1

14. The cylinder head bolts are tightened in 3 progressive steps:
    a. Using a 10 mm bi-hexagon wrench, install and uniformly tighten the 10 cylinder head bolts with the plate washers in several steps, in the sequence shown in the illustration. Tighten to 27 ft. lbs. (38 Nm).
    b. Mark the cylinder head bolt head with paint as shown in the illustration.
    c. Tighten the cylinder head bolts another 90°in the sequence shown above.

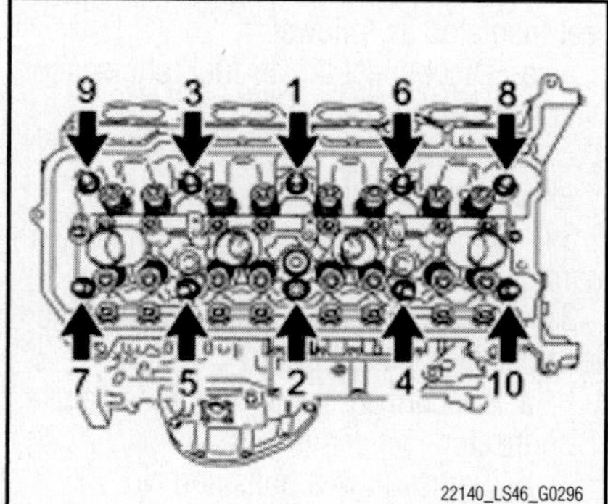

Fig. 158 Cylinder head 10 bolt tightening sequence—Bank 2

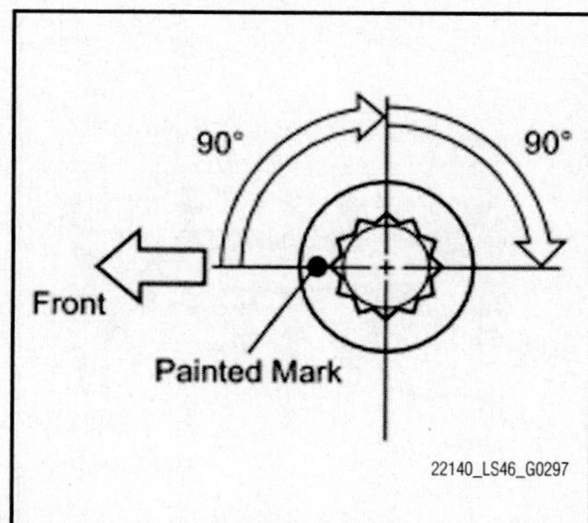

Fig. 159 Mark the cylinder head bolt head

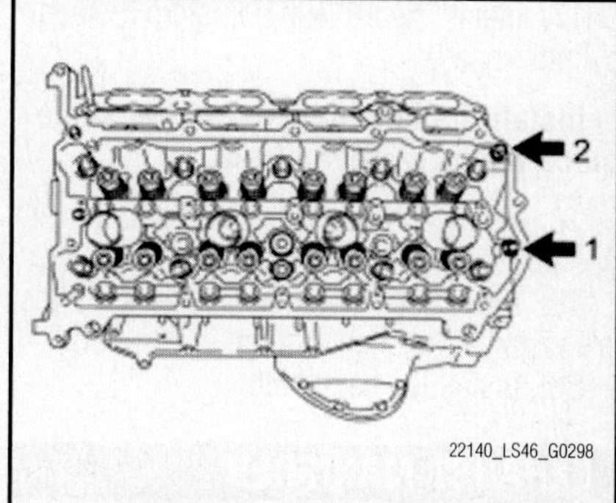

Fig. 160 Cylinder head 2 bolt tightening sequence—Bank 1

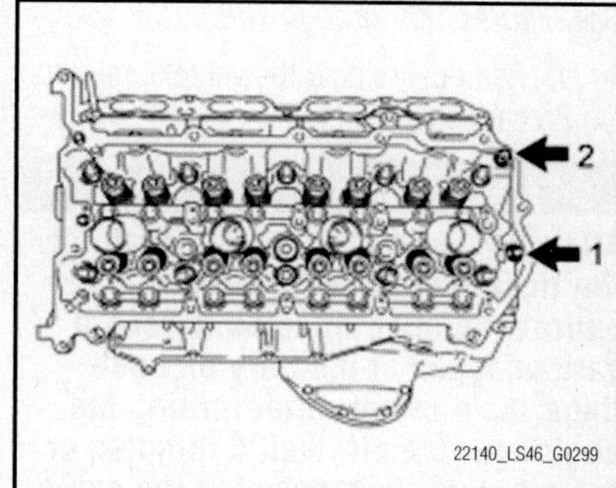

Fig. 161 Cylinder head 2 bolt tightening sequence—Bank 2

    d. Again, tighten the cylinder head bolts by an additional 90°in the sequence shown above
    e. Check that the painted marks are now facing rearward.
15. Uniformly install the 2 bolts in the sequence shown in the illustration. Tighten to 15 ft. lbs. (21 Nm).
16. Apply a light coat of engine oil to the valve stem caps and install the 16 valve stem caps to each cylinder head.

➡Be sure to inspect the valve lash adjuster before installing it.

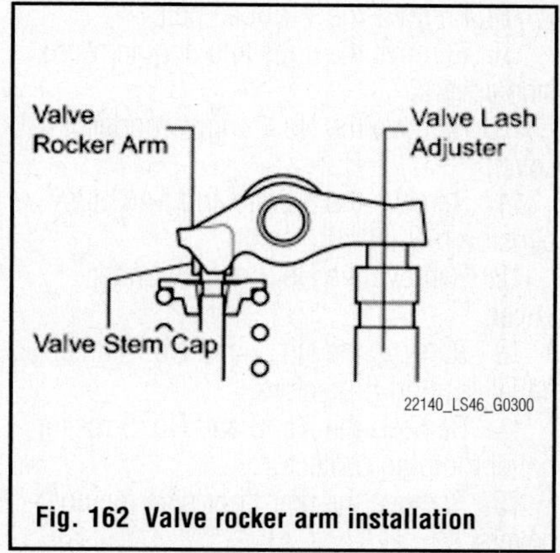

Fig. 162 Valve rocker arm installation

17. Install the 16 lash adjusters to each cylinder head.

➡**Install the lash adjuster at the same place it was removed from.**

18. Apply engine oil to the lash adjuster tips and valve stem cap ends.
19. Make sure that the valve rocker arms are installed as shown in the illustration.
20. Install the camshaft.

EXHAUST MANIFOLD

REMOVAL & INSTALLATION

2WD Models

See Figures 163 through 168.

1. Before servicing the vehicle, refer to the Precautions Section.

WARNING

After the engine switch is turned off, the HDD navigation system requires approximately 6 minutes to record various types of memory and settings. As a result, after turning the engine switch off, wait 6 minutes or more before disconnecting the cable from the negative battery terminal.

2. Remove the right cowl top ventilator louver.
3. Disconnect cable from negative battery terminal.

CAUTION

Wait at least 90 seconds after disconnecting the cable from the negative battery terminal to prevent airbag and seat belt pretensioner activation.

4. Remove the V-bank cover sub-assembly.
5. Remove the air cleaner inlet cover.
6. Remove the No. 1 air cleaner inlet.
7. Remove the intake air connector pipe.
8. Remove the V-ribbed belt.
9. Remove the clips and engine room side covers.
10. Remove the No 1 engine under cover.
11. Remove the 8 bolts and front suspension member protector.
12. Remove the No 2 engine under cover.
13. Remove the No. 1 and No. 2 differential support protectors.
14. Remove the No. 6 and No. 5 rocker panel molding protector.
15. Remove the rear floor side member covers.
16. Remove the front floor covers.
17. Remove the front floor center brace.
18. Disconnect the heated oxygen sensor.
19. Remove the front exhaust pipe assembly.
20. Remove the No. 1 exhaust pipe support bracket sub-assembly.
21. Remove the front wheel.
22. Remove the left and right front stabilizer link assemblies.
23. Remove the left and right engine under covers.
24. Remove the left and right rear engine under covers.
25. Remove the left and right front stabilizer No. 1 brackets.
26. Remove the front stabilizer bar.
27. Remove the front stabilizer bar No. 1 bush.
28. Remove the left and right front suspension member reinforcement.
29. Remove the generator.
30. Remove the steering sliding yoke with shaft sub-assembly.
31. Remove the No. 2 steering intermediate shaft assembly.
32. Remove the engine oil level dipstick guide.
33. Remove the No. 1 exhaust manifold heat insulator, as follows:
   a. Disconnect the air fuel ratio sensor connector.
   b. Remove the 3 bolts and No. 1 exhaust manifold heat insulator.
34. Remove the 8 nuts, right exhaust manifold, and gasket.
35. Remove the No. 2 exhaust manifold heat insulator, as follows:
   a. Disconnect the air fuel ratio sensor connector.
   b. Remove the 3 bolts and No. 2 exhaust manifold heat insulator.
36. Remove the 8 nuts, left exhaust manifold, and gasket.

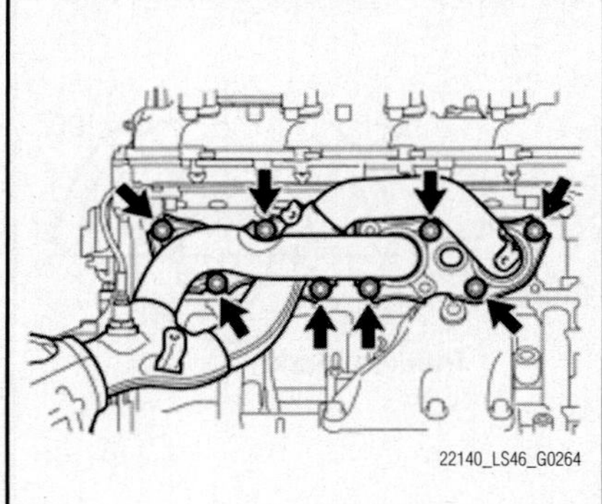

Fig. 163 Right exhaust manifold

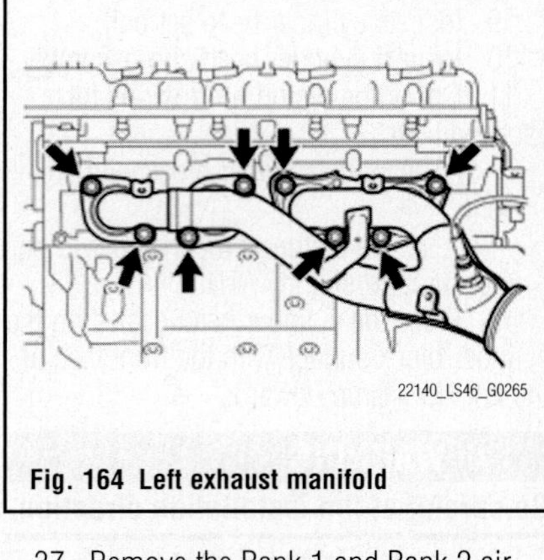

Fig. 164 Left exhaust manifold

37. Remove the Bank 1 and Bank 2 air fuel ratio sensors.

To install:

38. Install the Bank 1 and Bank 2 air fuel ratio sensors.
39. Install the left exhaust manifold sub-assembly, as follows:
   a. Install a new gasket.
   b. Install the left exhaust manifold, install 8 new nuts in the order shown in the illustration, and tighten to 15 ft. lbs. (21 Nm).
40. Install the No. 2 exhaust manifold heat insulator, as follows:

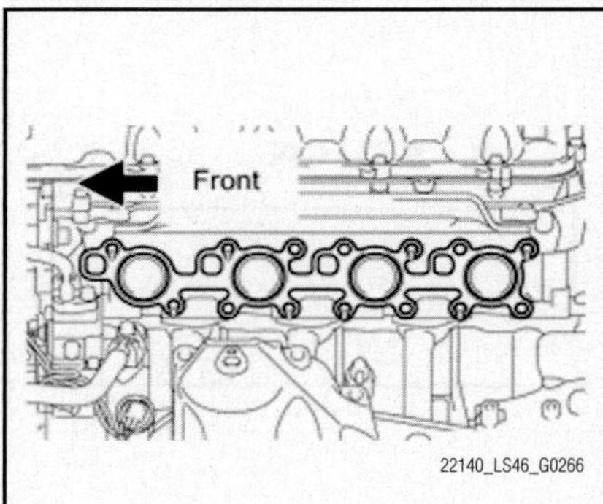

Fig. 165 Left exhaust manifold gasket installation

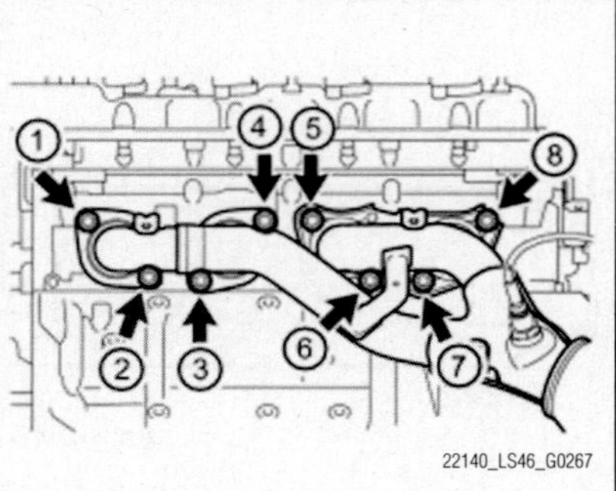

Fig. 166 Left exhaust manifold nut tightening sequence

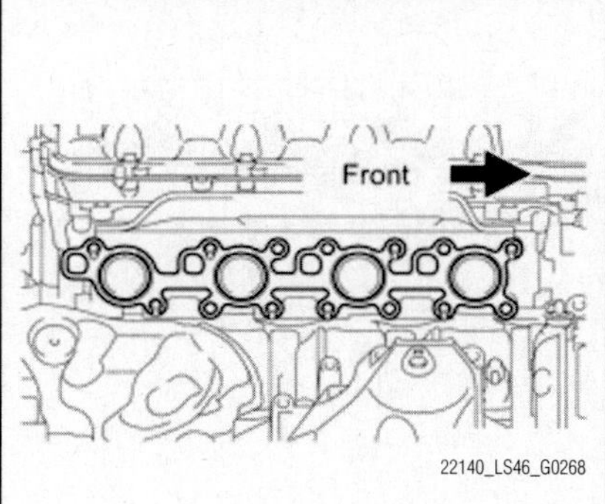

Fig. 167 Right exhaust manifold gasket installation

a. Install the exhaust manifold heat insulator with the 3 bolts and tighten to 7 ft. lbs. (10 Nm).

b. Connect the air fuel ratio sensor connector.

41. Install the right exhaust manifold sub-assembly, as follows:

a. Install a new gasket.

b. Install the right exhaust manifold, install 8 new nuts in the order shown in the illustration, and tighten to 15 ft. lbs. (21 Nm).

42. Install the No. 1 exhaust manifold heat insulator, as follows:

a. Install the No 1 exhaust manifold heat insulator with the 3 bolts and tighten to 7 ft. lbs. (10 Nm).

b. Connect the air fuel ratio sensor connector.

43. Install the engine oil level dipstick guide.

44. Install the No. 2 steering intermediate shaft assembly.

45. Install the steering sliding yoke with shaft sub-assembly.

46. Install the generator.

47. Install the V-ribbed belt.

48. Install the left and right front suspension member reinforcement.

49. Install the front stabilizer bar No. 1 bush.

50. Temporarily tighten the front stabilizer bar.

51. Tighten the left and right front stabilizer No. 1 brackets.

52. Install the left and right rear engine under covers.

53. Install the left and right engine under covers.

54. Install the front exhaust pipe assembly.

55. Install the front floor center brace.

56. Connect the heated oxygen sensor.

57. Connect the cable to the negative battery terminal.

58. Install the right cowl top ventilator louver.

59. Inspect for exhaust gas leak.

60. Install the front floor covers.

61. Install the rear floor side member covers.

62. Install the No. 6 and No. 5 rocker panel molding protector.

63. Install the No. 1 and No. 2 differential support protectors.

64. Install the No 2 engine under cover.

65. Install the 8 bolts and front suspension member protector.

66. Install the No 1 engine under cover.

67. Install the front wheel.

68. Install the clips and engine room side covers.

69. Install the intake air connector pipe.

70. Install the No 1 air cleaner inlet.

71. Install the air cleaner inlet cover.

72. Install the V-bank cover sub-assembly.

73. Perform initialization.

AWD Models

See Figures 169 through 179.

1. Remove the engine and transmission.

2. Remove the front frame assembly.

3. Remove the engine oil level dipstick guide.

4. Remove the No. 1 exhaust manifold heat insulator.

a. Remove the 3 bolts and the heat insulator.

5. Remove the No. 2 exhaust manifold heat insulator.

a. Remove the 3 bolts and the heat insulator.

6. Remove the front engine mounting insulator.

7. Remove the front No. 1 engine mounting bracket RH.

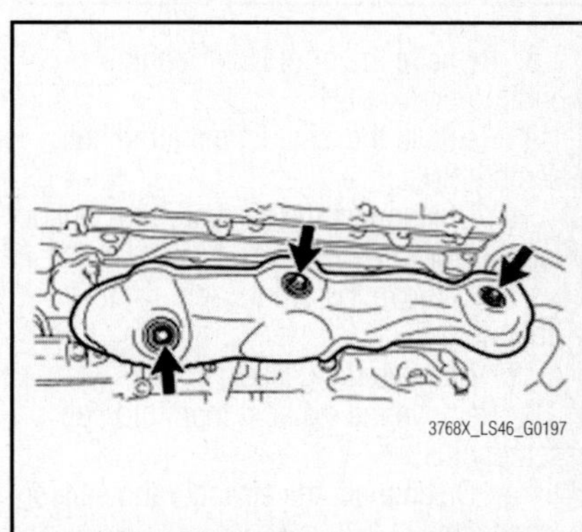

Fig. 170 Removing the No. 2 exhaust manifold heat insulator

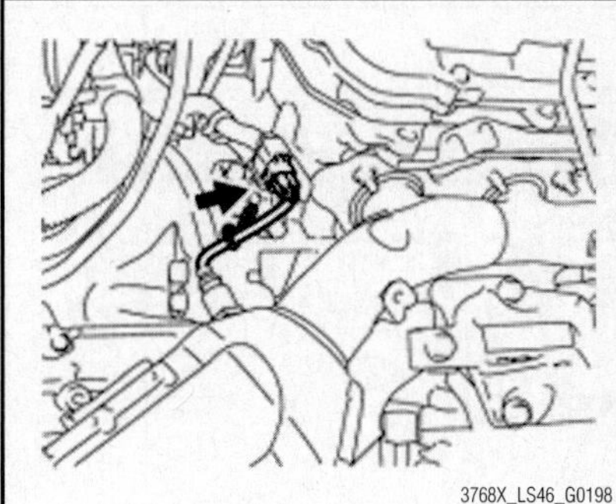

Fig. 171 Disconnecting the air fuel ratio sensor RH

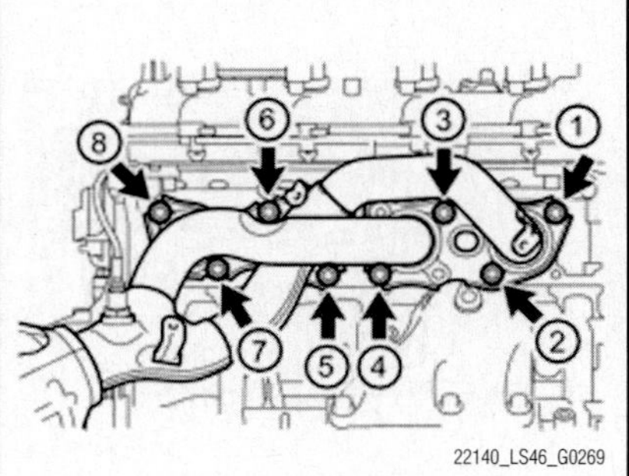

Fig. 168 Right exhaust manifold nut tightening sequence

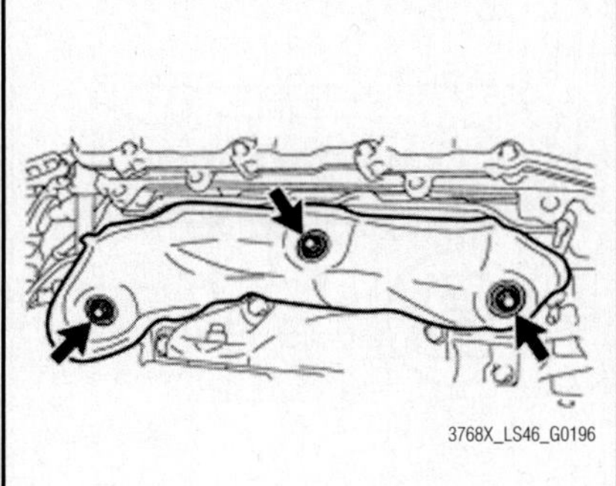

Fig. 169 Removing the heat insulator

Fig. 172 Removing the RH exhaust manifold

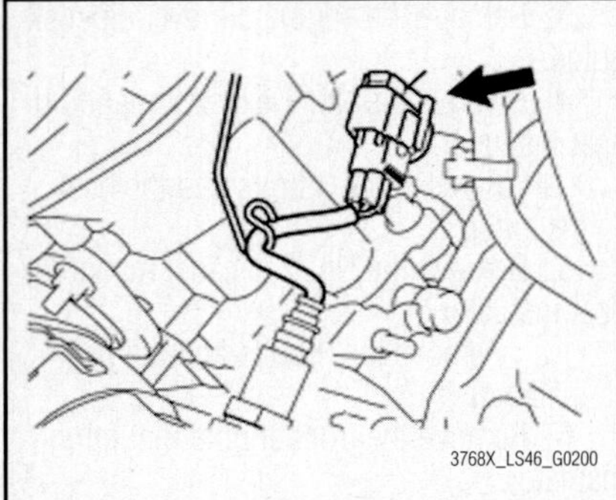

Fig. 173 Disconnecting the air fuel ratio sensor connector LH

8. Remove the front No. 1 engine mounting bracket LH.
9. Remove the exhaust manifold sub assembly RH.
   a. Disconnect the air fuel ratio sensor connector.
   b. Remove the 8 nuts and exhaust manifold RH.
   c. Remove the gasket.
10. Remove the exhaust manifold sub assembly LH.
   a. Disconnect the air fuel ratio sensor connector.
   b. Remove the 8 nuts and exhaust manifold LH.

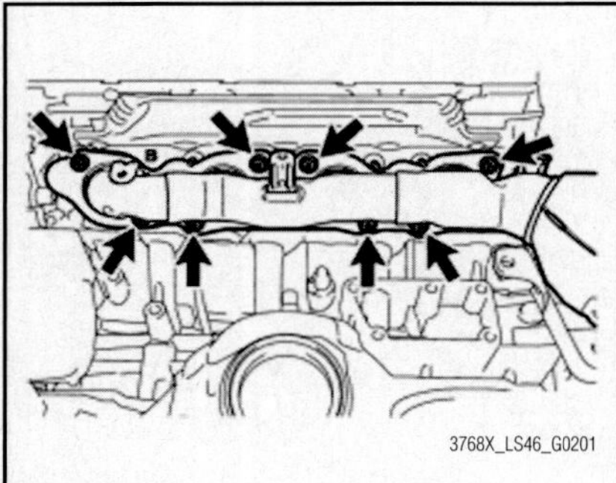

Fig. 174 Removing the LH exhaust manifold

   c. Remove the gasket.
11. Remove the air fuel ratio sensor (bank 1 sensor 1).
   a. Using the special tool (09224-00010), remove the air fuel ratio sensor from the exhaust manifold LH.
12. Remove the air fuel ratio sensor (bank 2 sensor 1).
   a. Using the special tool (09224-00010), remove the air fuel ratio sensor from the exhaust manifold RH.

To install:

13. Install the air fuel ratio sensor (bank 1 sensor 1).
   a. Using the special tool (09224-00010), install the fuel ratio sensor to the exhaust manifold LH. If using the special tool, tighten to 30 ft. lbs. (40 Nm). If not using the special tool, tighten to 32 ft. lbs. (44 Nm).

➡Use a torque wrench with a fulcrum length of 11.8 inches (300 mm).

➡Make sure the special tool and the torque wrench are connected in a straight line.

14. Install the air fuel ratio sensor (bank 2 sensor 1).
   a. Using the special tool (09224-00010), install the air fuel ratio sensor to the exhaust manifold RH. If using the special tool, tighten to 30 ft. lbs. (40 Nm). If not using the special tool, tighten to 32 ft. lbs. (44 Nm).

➡Use a torque wrench with a fulcrum length of 11.8 inches (300 mm).

➡Make sure the special tool and the torque wrench are connected in a straight line.

15. Install the exhaust manifold sub assembly LH.
   a. Place a new gasket on the cylinder head with the "B" mark facing the manifold side.

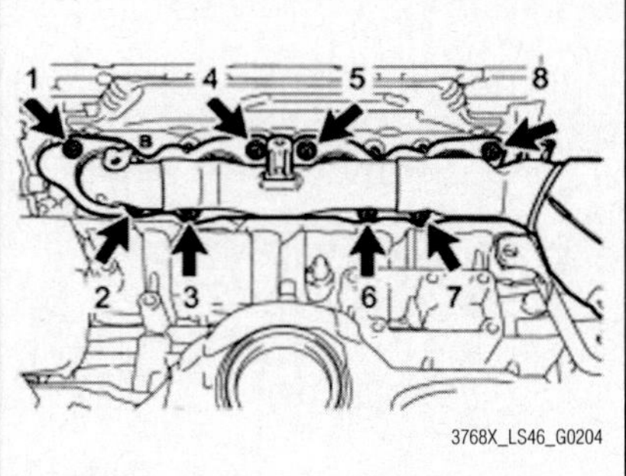

Fig. 177 LH exhaust manifold bolt tightening sequence

➡Make sure that the gasket is installed facing the proper direction.

   b. Install the exhaust manifold with the new nuts by uniformly tightening the nuts in several steps, in the sequence shown. Tighten to 15 ft. lbs. (21 Nm).
   c. Connect the air fuel ratio sensor connector.
   d. Hook the wire harness to the bracket.
16. Install the exhaust manifold sub assembly RH.
   a. Place a new gasket on the cylinder head with the "B" mark facing the manifold side.

➡Make sure that the gasket is installed facing the proper direction.

   b. Install the exhaust manifold with the new nuts by uniformly tightening the nuts in several steps, in the sequence shown.
   c. Connect the air fuel ratio sensor connector.
   d. Hook the wire harness to the bracket.
17. Install the front No. 1 engine mounting bracket LH.
18. Install the front No. 1 engine mounting bracket RH.

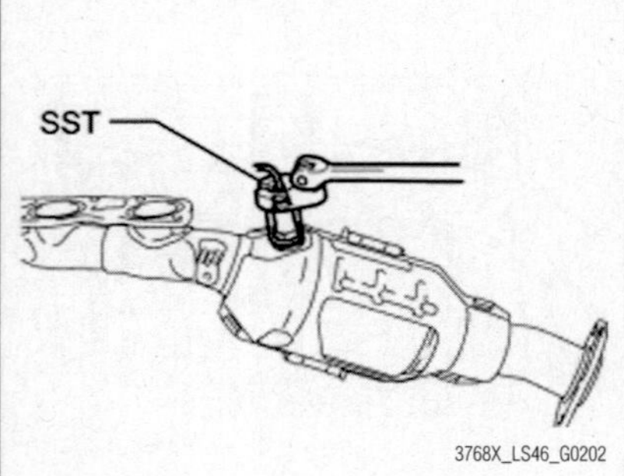

Fig. 175 Removing the air fuel ratio sensor (bank 1 sensor 1)

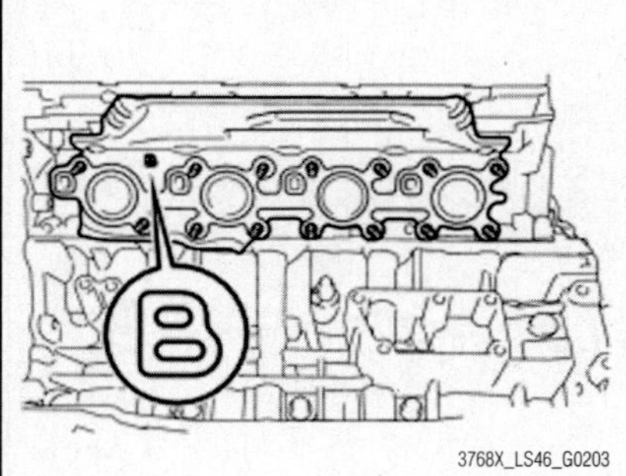

Fig. 176 Installing the new gasket to the LH cylinder head

Fig. 178 Installing the new gasket on the RH cylinder head

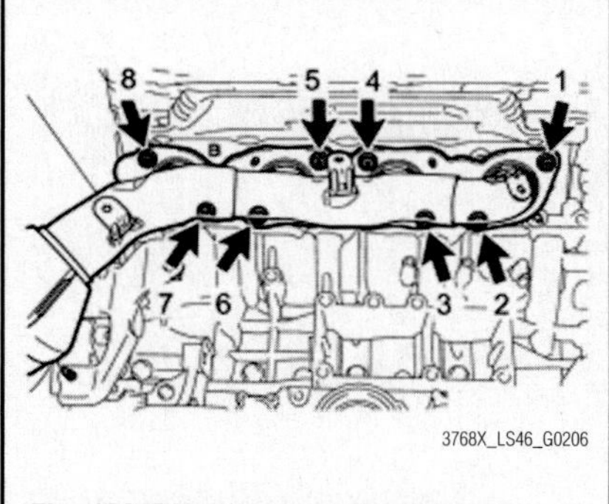

Fig. 179 RH exhaust manifold bolt tightening sequence

19. Install the front engine mounting insulator.
20. Install the No. 2 exhaust manifold heat insulator.
 a. Install the heat insulator with the 3 bolts. Tighten to 7 ft. lbs. (10 Nm).
21. Install the No. 1 exhaust manifold heat insulator.
 a. Install the heat insulator with the 3 bolts. Tighten to 7 ft. lbs. (10 Nm).
22. Install the engine oil level dipstick guide.
23. Install the front frame assembly.
24. Install the engine and transmission.

FLYWHEEL

REMOVAL & INSTALLATION

See Figures 180 through 182.

1. Before servicing the vehicle, refer to the Precautions Section.

WARNING

After the engine switch is turned off, the HDD navigation system requires approximately 6 minutes to record various types of memory and settings. As a result, after turning the engine switch off, wait 6 minutes or more before disconnecting the cable from the negative battery terminal.

2. Remove the right cowl top ventilator louver.
3. Disconnect cable from negative battery terminal.

CAUTION

Wait at least 90 seconds after disconnecting the cable from the negative battery terminal to prevent airbag and seat belt pretensioner activation.

4. Remove the transmission assembly.
5. Remove the drive plate and ring gear sub-assembly, as follows:

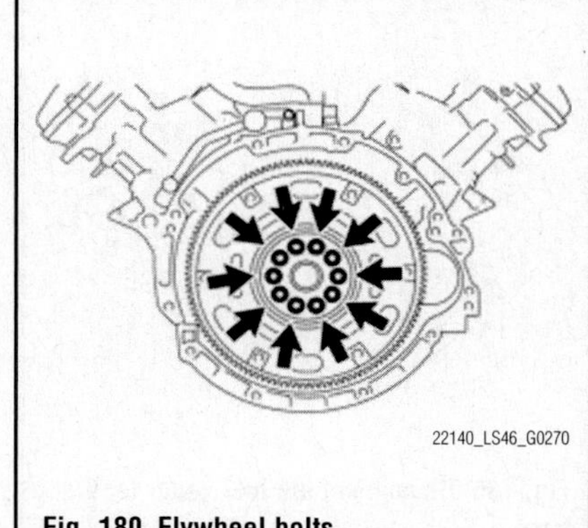

Fig. 180 Flywheel bolts

 a. Using SST (SST: 09213-54015, SST: 09330-00021) hold the crankshaft.
 b. Remove the 10 bolts, spacer plate, ring gear and sensor rotor.

To install:

6. Using SST (SST: 09213-54015, SST: 09330-00021) hold the crankshaft.
7. Install the sensor rotor, ring gear and spacer plate on the crankshaft.
8. Uniformly install and tighten 10 new bolts in the sequence shown in the illustration to 22 ft. lbs. (30 Nm).

➡Do not reuse the flywheel installation bolts.

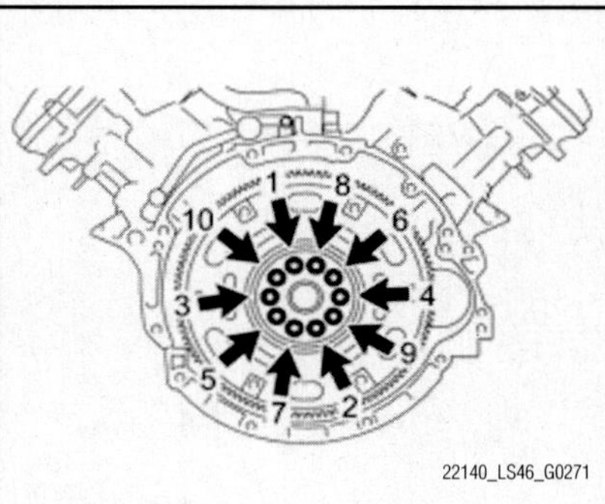

Fig. 181 Flywheel bolt tightening sequence

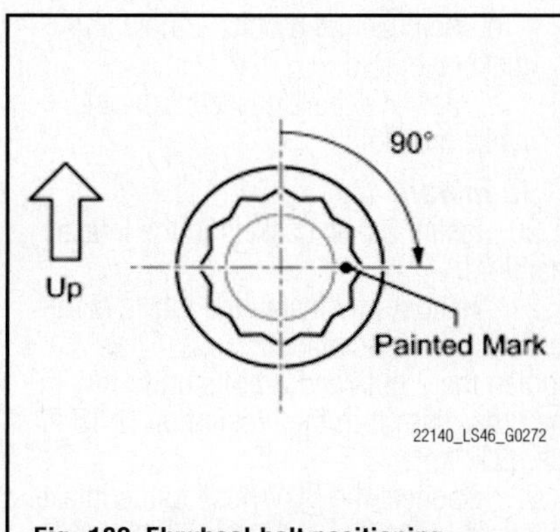

Fig. 182 Flywheel bolt positioning

➡Do not impact or damage the flywheel installation bolts. Be sure to handle them carefully.

9. Mark the upside of each flywheel installation bolt with paint.
10. Retighten the flywheel installation bolts by 90°as shown.
11. Check that the painted marks are now at a 90°angle to the upside.
12. Install the transmission.
13. Connect the cable to the negative battery terminal.

INTAKE MANIFOLD

REMOVAL & INSTALLATION

See Figures 183 through 192.

1. Before servicing the vehicle, refer to the Precautions Section.

WARNING

After the engine switch is turned off, the HDD navigation system requires approximately 6 minutes to record various types of memory and settings. As a result, after turning the engine switch off, wait 6 minutes or more before disconnecting the cable from the negative battery terminal.

2. Remove the right cowl top ventilator louver.
3. Disconnect cable from negative battery terminal.

CAUTION

Wait at least 90 seconds after disconnecting the cable from the negative battery terminal to prevent airbag and seat belt pretensioner activation.

4. Remove the V-bank cover sub-assembly.
5. Remove the air cleaner inlet cover.
6. Remove the right engine room side cover.
7. Remove the No. 1 air cleaner inlet.
8. Remove the No. 1 engine under cover.
9. Remove the front lower suspension member protector.
10. Remove the No. 2 engine under cover.
11. Drain engine coolant.
12. Remove the intake air connector pipe.
13. Remove the throttle body.
14. Remove the air cleaner assemblies.
15. Disconnect the engine wire.
16. Remove the injector driver, as follows:

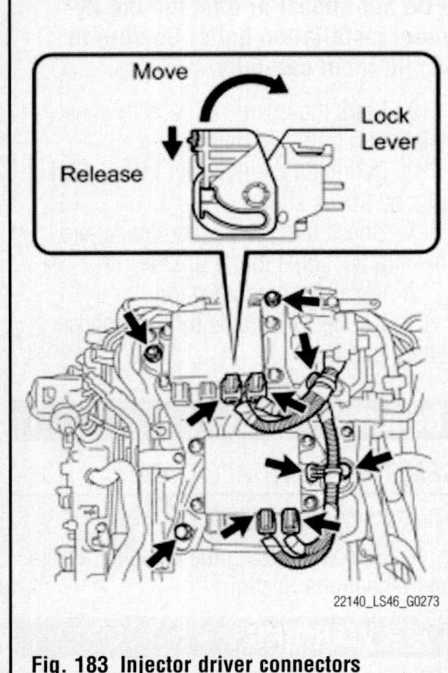

Fig. 183 Injector driver connectors

a. Disconnect the 4 injector driver connectors. To disconnect the injector driver connectors, push the claw downward and move the lock lever to release the lock.

b. Disconnect the 2 clamps from the injector driver.

c. Remove the 2 bolts, 2 nuts and injector driver from the intake manifold.

17. Slide the 4 clamps, and disconnect the heater water inlet hose, heater water outlet hose, water inlet hose, and No. 3 water by-pass hose from the water by-pass pipe and remove the 2 bolts and water by-pass pipe.

18. Disconnect the fuel vapor feed hose and No. 2 fuel vapor feed hose and remove the bolt and purge VSV.

19. Remove the intake manifold, as follows:

a. Disconnect the PCV hose from intake manifold.

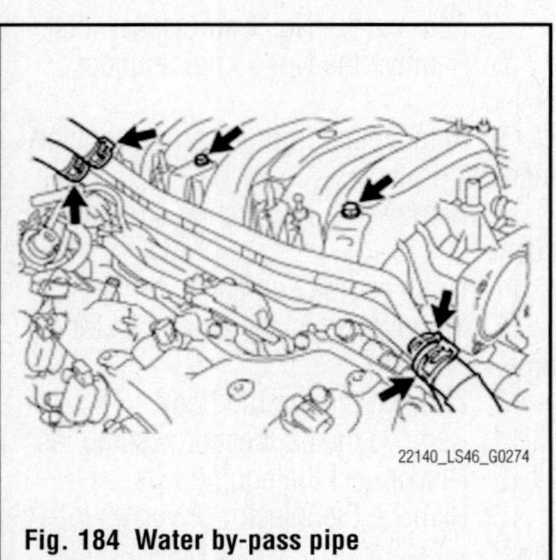

Fig. 184 Water by-pass pipe

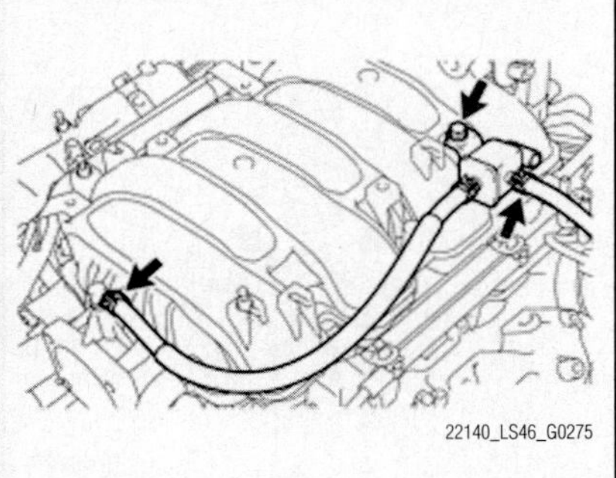

Fig. 185 Disconnect the fuel vapor feed hose

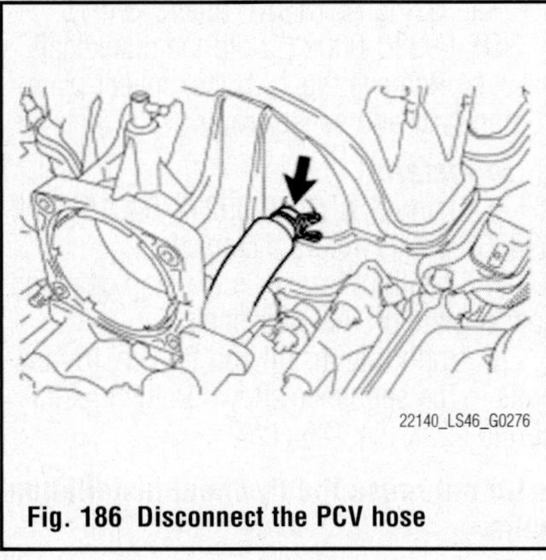

Fig. 186 Disconnect the PCV hose

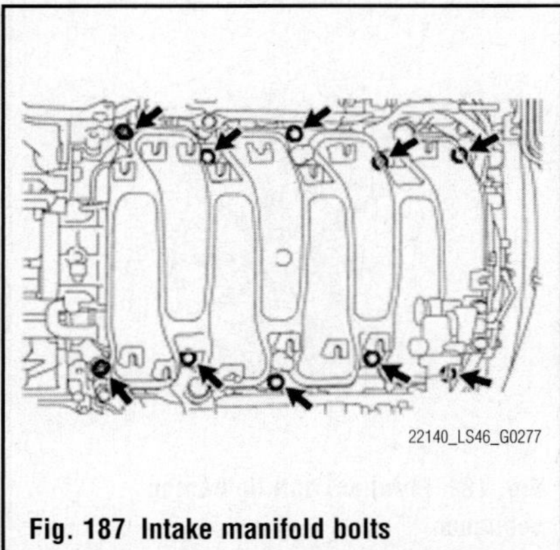

Fig. 187 Intake manifold bolts

b. Remove the 8 bolts, 2 nuts and intake manifold.

c. Remove the 2 gaskets from the intake manifold.

To install:

20. Install 2 new gaskets to the intake manifold.

21. Temporarily install the intake manifold with the 2 nuts and 8 bolts. Then tighten the 2 nuts and 8 bolts uniformly in the order shown in the illustration to 15 ft. lbs. (21 Nm).

22. Connect the PCV hose to the intake manifold.

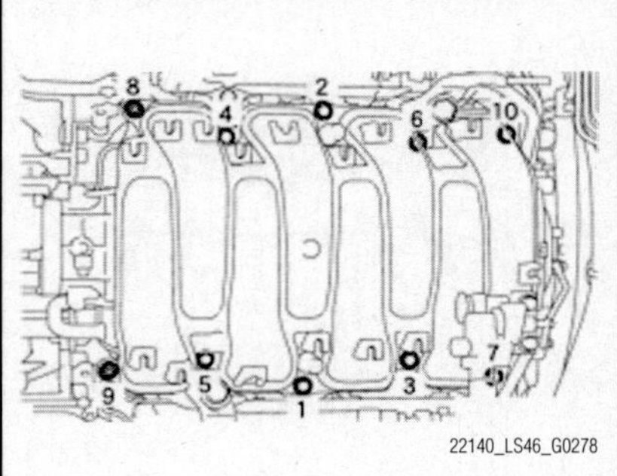

Fig. 188 Intake manifold bolt tightening sequence

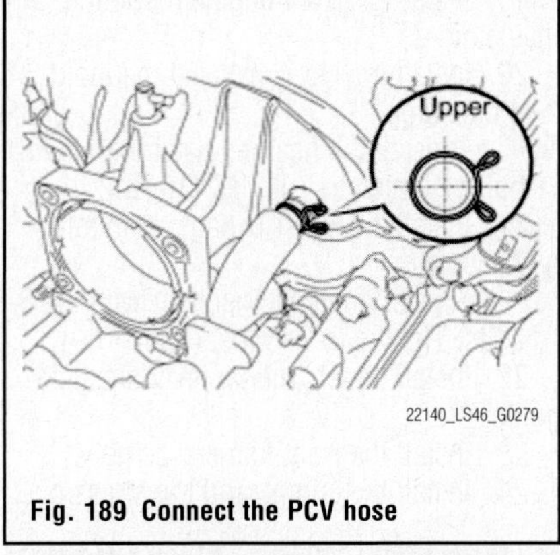

Fig. 189 Connect the PCV hose

23. Install the purge VSV with the bolt and tighten to 15 ft. lbs. (21 Nm).

24. Connect the fuel vapor feed hose to the intake manifold.

25. Connect the No. 2 fuel vapor feed hose to the purge VSV.

26. Install the water by-pass pipe to the intake manifold with the 2 bolts.

27. Connect the heater water inlet hose, heater water outlet hose, water inlet hose, and No. 3 water by-pass hose to the water by-pass pipe with the 4 clamps.

28. Install the injector driver to the intake manifold by installing the 2 bolts and 2 nuts

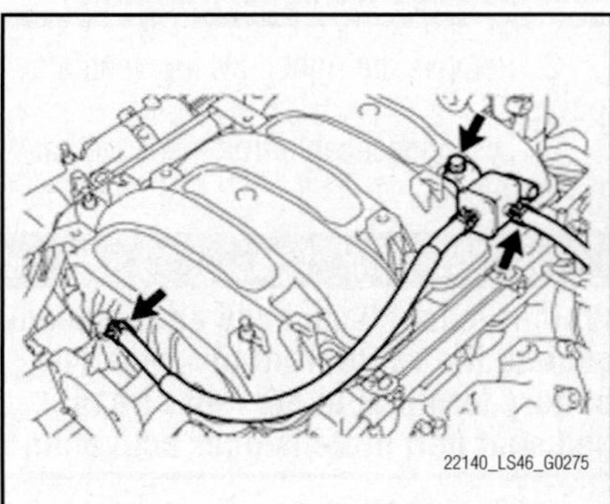

Fig. 190 Connect the No. 2 fuel vapor feed hose

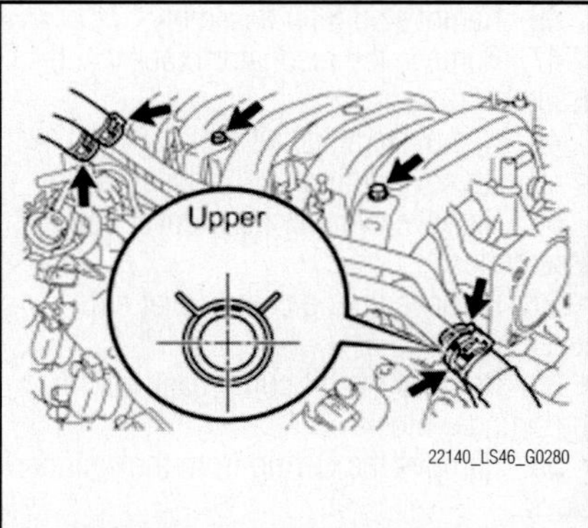

Fig. 191 Water by-pass pipe installation

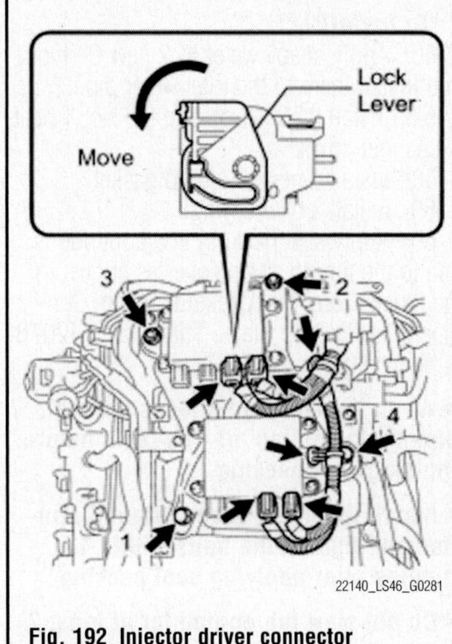

Fig. 192 Injector driver connector installation

in the order shown in the illustration and tighten to 7 ft. lbs. (10 Nm).

29. Connect the 4 wire harness connectors to the injector driver. Then move the lock lever as shown in the illustration to lock the connectors.
30. Connect the engine wire.
31. Install the air cleaner assemblies.
32. Install the throttle body.
33. Install the intake air connector pipe.
34. Install the No. 1 air cleaner inlet.
35. Connect the cable to the negative battery terminal.
36. Add engine coolant and inspect for coolant leak.
37. Install the right engine room side cover.
38. Install the air cleaner inlet cover.
39. Install the V-bank cover sub-assembly.
40. Install the No. 2 engine under cover.
41. Install the front lower suspension member protector.
42. Install the No. 1 engine under cover.
43. Install the right cowl top ventilator louver.
44. Perform initialization.

OIL PAN

REMOVAL & INSTALLATION

See Figures 193 through 197.

➡It may be possible to remove the oil pan from the vehicle while the engine is still in the vehicle. If not. the engine and transmission must be removed as a unit, and then separated.

1. Before servicing the vehicle, refer to the precautions in the beginning of this section.
2. Remove engine/transmission assembly.
3. Remove the oil pan protector with the 2 bolts.
4. Remove the 15 bolts and 2 nuts and remove the No. 2 oil pan.
5. Insert the blade of an oil pan seal cutter between the oil pans. Cut through the applied sealer and remove the No. 2 oil pan.
6. Remove the 14 bolts and 2 nuts and remove the No. 1 oil pan.
7. Remove the oil pan by prying between the oil pan and cylinder block with a screwdriver with its tip taped.

➡Be sure to clean the bolts and stud bolts, and check the threads for cracks or other damage.

➡Be careful not to damage the contact surfaces of the oil pans.

To install:

8. Apply the seal packing to oil pan No. 1, as shown. Seal packing: Toyota Genuine Seal Packing Black, Three Bond 1207B or equivalent. Standard seal diameter: 3.0 to 4.0 mm (0.118 to 0.156 in.). Application position from inside edge of oil pan. 6.0 mm (0.236 in.).

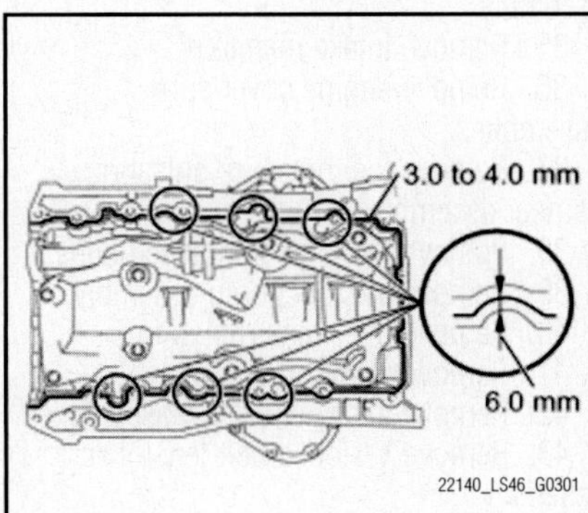

Fig. 193 Oil pan No. 1 sealant application

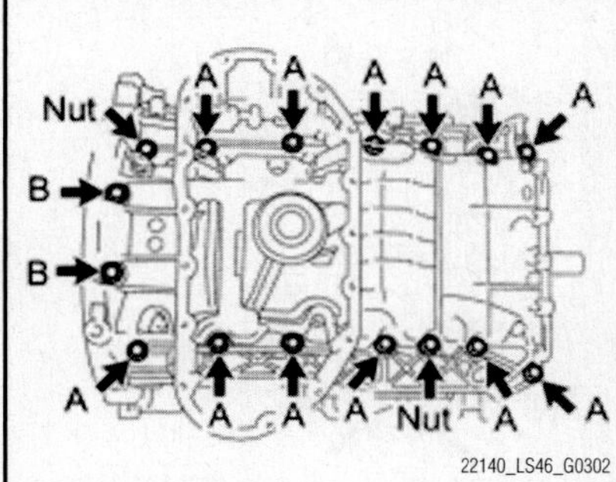

Fig. 194 No. 1 oil pan bolt tightening sequence

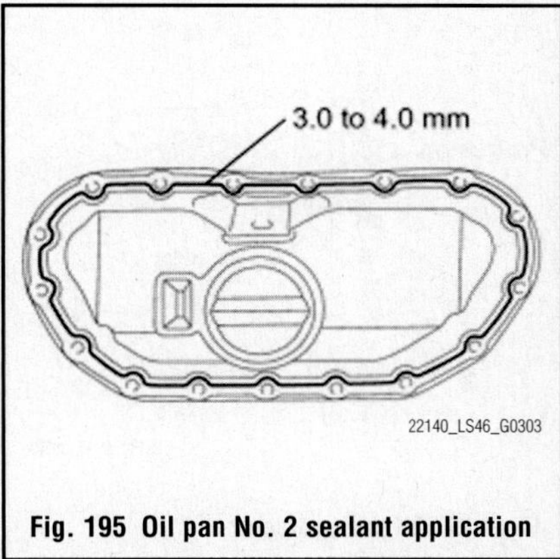

Fig. 195 Oil pan No. 2 sealant application

➡Remove any oil from the contact surface.

➡Install the oil pan within 3 minutes and tighten the bolts and nuts within 15 minutes after applying seal packing.

➡Do not start the engine for at least 2 hours after installing.

9. Install the oil pan sub-assembly with the 14 bolts and 2 nuts.
10. Uniformly tighten the bolts and nuts in several passes to the following:
    - Bolts A to 7 ft. lbs. (10 Nm).
    - Bolts B 26 ft. lbs. (35 Nm).
    - Nut to 26 ft. lbs. (35 Nm).
11. Apply seal packing to oil pan No. 2, as shown.
12. Install the oil pan with the 15 bolts and 2 nuts. Uniformly tighten the bolts and nuts in several passes to 7 ft. lbs. (10 Nm).
13. Install the oil pan protector with the 2 nuts and tighten to 44 inch lbs. (5 Nm). Install the oil pan protector with its protrusion facing the bottom of the engine.
14. Install engine/transaxle assembly, including adding fluids.

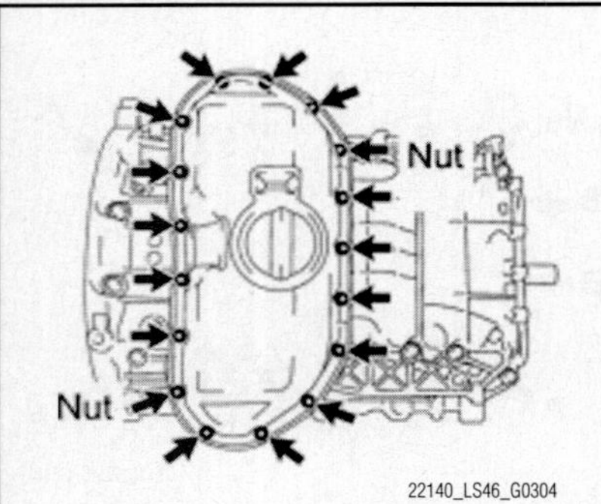

Fig. 196 No. 2 oil pan bolt tightening sequence

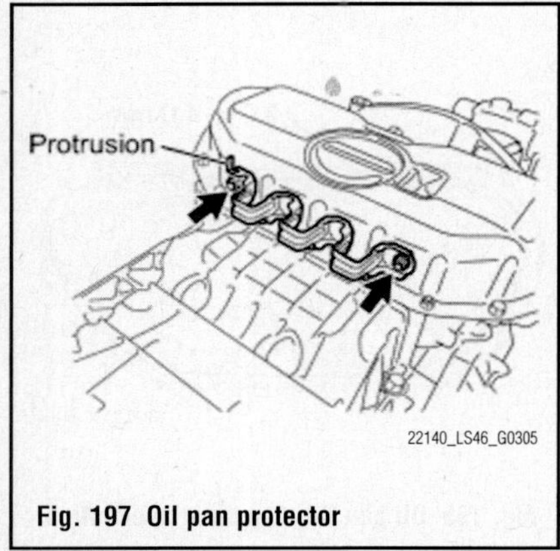

Fig. 197 Oil pan protector

OIL PUMP

REMOVAL & INSTALLATION

2WD Models

See Figures 198 through 204.

1. Before servicing the vehicle, refer to the Precautions Section.

⁂ WARNING

After the engine switch is turned off, the HDD navigation system requires approximately 6 minutes to record various types of memory and settings. As a result, after turning the engine switch off, wait 6 minutes or more before disconnecting the cable from the negative battery terminal.

2. Remove the right cowl top ventilator louver.
3. Disconnect cable from negative battery terminal.

⁂ CAUTION

Wait at least 90 seconds after disconnecting the cable from the negative battery terminal to prevent airbag and seat belt pretensioner activation.

4. Remove the V-bank cover sub-assembly.
5. Remove the air cleaner inlet cover.
6. Remove the No. 1 air cleaner inlet.
7. Remove the engine room side covers.
8. Remove the battery, nut, clamp, bolts, insulator, and tray.
9. Remove the No. 1 and No. 2 engine under covers.
10. Remove the front lower suspension member protector.
11. Remove the rear engine under covers.
12. Drain engine oil and coolant.
13. Remove the intake air connector pipe.
14. Remove the air cleaner assemblies.
15. Remove the radiator reservoir.
16. Remove the V-ribbed belt.
17. Disconnect the radiator hoses.
18. Remove engine room ECU outlet duct.
19. Remove skid control ECU.
20. Remove skid control ECU bracket.
21. Disconnect engine wire.
22. Remove front exhaust pipe assembly.
23. Disconnect front stabilizer bar.
24. Remove the left front suspension member reinforcement.
25. Remove steering sliding yoke with shaft sub-assembly.
26. Remove No. 2 steering intermediate shaft assembly.
27. Remove No. 2 exhaust manifold heat insulator.
28. Remove exhaust manifold sub-assembly.
29. Disconnect cooler compressor assembly.
30. Remove oil level dipstick guide.
31. Remove generator.
32. Remove injector driver.
33. Remove No. 1 engine cover sub-assembly.
34. Remove water by-pass pipe sub-assembly.
35. Remove intake manifold.
36. Remove engine cover sub-assemblies.
37. Remove fuel pressure pulsation damper assembly.
38. Remove fuel pipe sub-assemblies.
39. Remove the fuel pump assembly.
40. Remove inlet water housing.
41. Remove water pump pulley.
42. Remove idler pulley sub-assemblies.
43. Remove V-ribbed belt tensioner assembly.
44. Remove front water by-pass joint.
45. Remove the Bank 1 and Bank 2 camshaft timing control motor assembly.
46. Remove oil filter assembly.
47. Remove the resonator bracket sub-assembly.
48. Remove crankshaft pulley.
49. Remove ignition coil assembly.
50. Remove cylinder head cover sub-assemblies.
51. Remove timing chain cover sub-assembly.
52. Remove the oil pump gasket from the cylinder block.
53. Remove the O-ring from the cylinder block.
54. Remove the inlet water pipe.
55. Remove the 2 O-rings from the inlet water pipe.

To install:

56. Apply soapy water to 2 new O-rings and install them to the inlet water pipe.
57. Install the inlet pipe to the No. 1 heat exchanger cover.
58. Install a new oil pump gasket.
59. Install a new O-ring.
60. Apply seal packing in a continuous line to the timing chain cover as shown in the illustration. Seal packing: Toyota Genuine Seal Packing Black, Three Bond 1207B or equivalent.

➡When the contact surfaces are wet, wipe them with an oil-free cloth before applying seal packing.

➡Install the chain cover within 3 minutes and tighten the bolts within 10 minutes after applying seal packing.

➡Do not start the engine for at least 2 hours after installing.

61. Align the oil pump's drive rotor spline and the crankshaft as shown in the illustration. Install the spline and chain cover to the crankshaft.
62. Temporarily tighten the timing chain cover with the 30 bolts and nut.
63. Tighten the 11 bolts in several steps, in the sequence shown in the illustration to 35 ft. lbs. (47 Nm).
64. Temporarily tighten the belt tensioner with the standard bolt and 6 mm hexagon wrench bolt.
65. Tighten the 21 bolts and nut in several steps, in the sequence shown in the illustration to 17 ft. lbs. (23 Nm).
66. Install 2 new gaskets and 2 plugs and tighten to 34 ft. lbs. (46 Nm).
67. Install cylinder head cover sub-assemblies.
68. Install ignition coil assembly.
69. Install crankshaft pulley.
70. Install the resonator bracket sub-assembly.

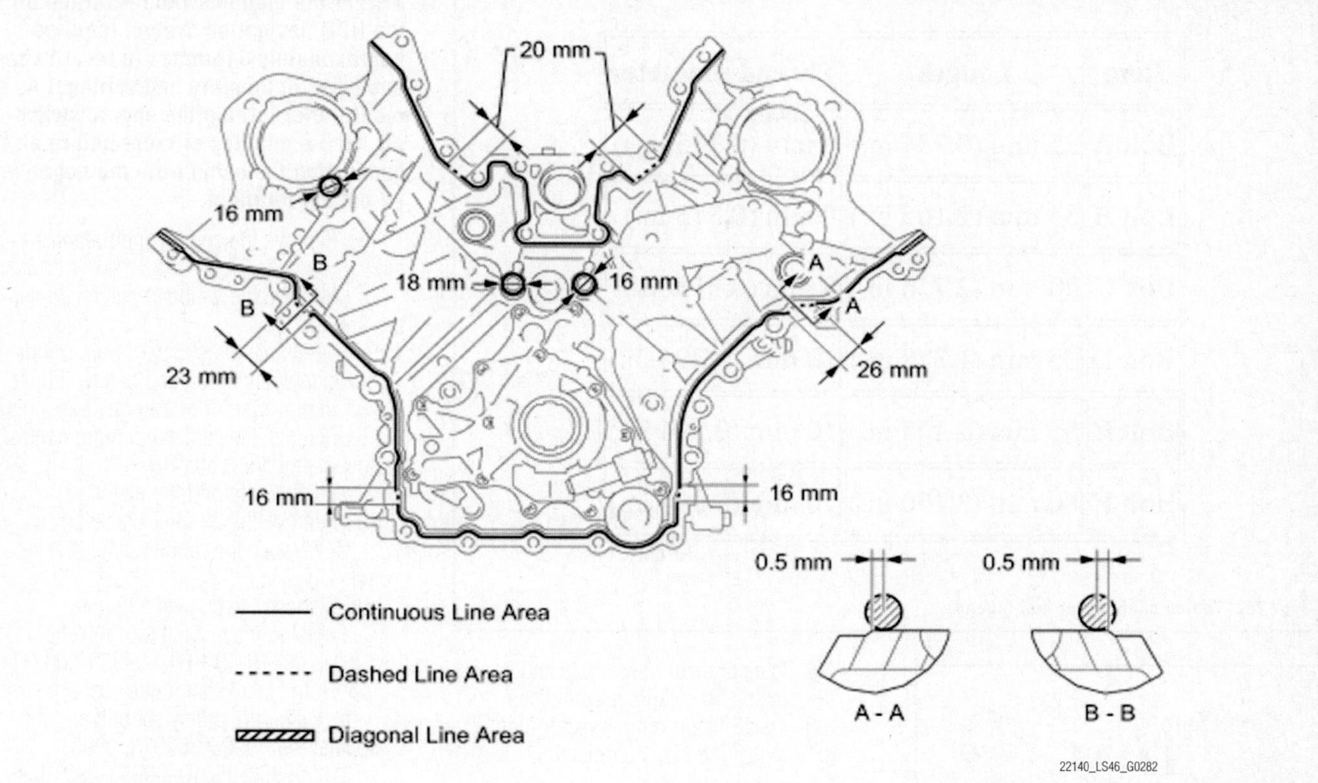

20 mm
16 mm
18 mm
16 mm
B
B
A
A
23 mm
26 mm
16 mm
16 mm
0.5 mm
0.5 mm
A - A
B - B
Continuous Line Area
Dashed Line Area
Diagonal Line Area
22140_LS46_G0282

Fig. 198 Timing chain cover sealant application

• Area	Seal packing diameter	Application position from inside edge of cover
Continuous Line Area	3.0 to 4.0 mm (0.1181 to 0.1575 in.)	2.5 mm (0.098 in.)
Dashed Line Area	6.4 mm (0.2520 in.) or more, or within OK area shown in illustration	0.5 mm (0.020 in.)
Diagonal Line Area	3.0 to 4.0 mm (0.1181 to 0.1575 in.)	5.5 mm (0.217 in.)

22140_LS46_G0283

Fig. 199 Sealant application specifications

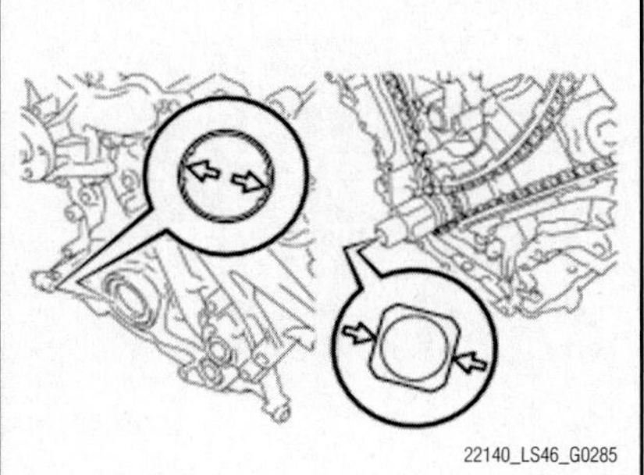

Fig. 200 Timing chain cover oil pump alignment

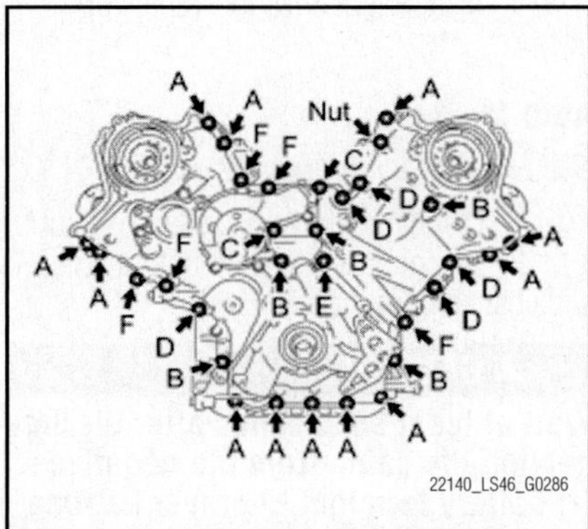

Fig. 201 Timing chain cover bolt position

71. Install oil filter assembly.
72. Install camshaft timing control motor assembly.
73. Install front water by-pass joint.
74. Install idler pulleys.
75. Install water pump pulley.
76. Install inlet water housing.
77. Install fuel pump assemblies.
78. Install fuel pipes.
79. Install fuel pressure pulsation damper assembly.
80. Install engine covers.
81. Install intake manifold.
82. Install water by-pass pipe sub-assembly.
83. Install injector driver.
84. Install generator.
85. Install oil level dipstick guide.
86. Connect cooler compressor assembly.
87. Install exhaust manifold sub-assembly.
88. Install No. 2 exhaust manifold heat insulator.
89. Install No. 2 steering intermediate shaft assembly.
90. Install steering sliding yoke with shaft sub-assembly.
91. Install left front suspension member reinforcement.

Item	Length	Thread diameter
Bolt A	25 mm (0.984 in.)	8 mm (0.315 in.)
Bolt B	55 mm (2.165 in.)	8 mm (0.315 in.)
Bolt C	70 mm (2.756 in.)	8 mm (0.315 in.)
Bolt D	35 mm (1.378 in.)	10 mm (0.394 in.)
Bolt E	55 mm (2.165 in.)	10 mm (0.394 in.)
Bolt F	80 mm (3.150 in.)	10 mm (0.394 in.)

22140_LS46_G0284

Fig. 202 Timing chain cover bolt legend

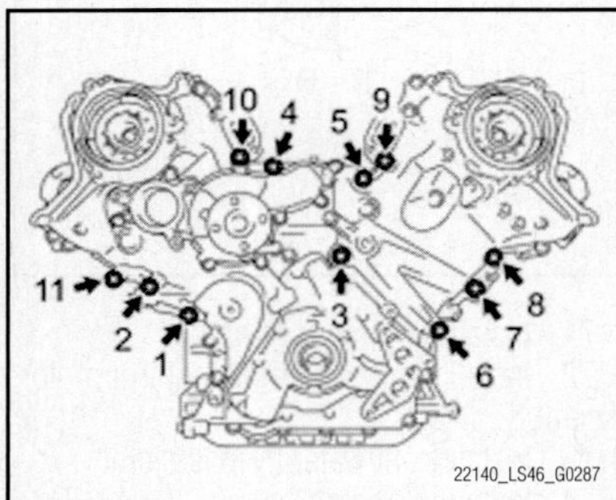

Fig. 203 Timing chain cover bolt tightening sequence 1

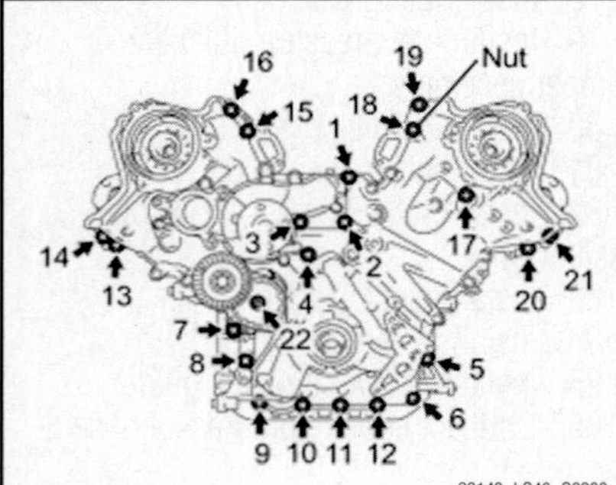

Fig. 204 Timing chain cover bolt tightening sequence 2

92. Connect front stabilizer bar.
93. Install front exhaust pipe assembly.
94. Connect engine wire.
95. Install engine room ECU outlet duct.
96. Install skid control ECU.
97. Install V-ribbed belt.
98. Install radiator assembly and hoses.
99. Install air cleaner assemblies.
100. Install intake air connector pipe.
101. Install the No. 1 air cleaner inlet.
102. Install rear engine under covers.
103. Install battery, nut, clamp, bolts, insulator, and tray.
104. Add engine oil and coolant.
105. Connect cable to negative battery terminal.
106. Perform initialization.
107. Inspect for oil, coolant, fuel, and exhaust gas leaks.
108. Check engine oil level.
109. Install front lower suspension member protector.
110. Install No. 1 and No. 2 engine under covers.
111. Install the engine room side covers.
112. Install air cleaner inlet cover.
113. Install V-bank cover sub-assembly.
114. Install right cowl top ventilator louver.

AWD Models

See Figures 205 through 227.

1. Discharge the fuel system pressure.
2. Disconnect the cable from the negative battery terminal.

✲✲ CAUTION

Wait at least 90 seconds after disconnecting the cable from the negative (-) battery terminal to prevent airbag and seat belt pretensioner activation.

➡After the engine switch is turned off, the HDD navigation system requires approximately 6 minutes to record various types of memory and settings. As a result, after turning the engine switch off, wait 6 minutes or more before disconnecting the cable from the negative (-) battery terminal.

3. Remove the engine and transmission.
4. Remove the v-ribbed belt tensioner assembly.
   a. Remove the standard bolt, 6 mm hexagon wrench bolt and belt tensioner.
5. Remove the front water by pass joint.
6. Remove the camshaft timing control motor assembly LH and RH.
7. Remove the oil filter element.
8. Remove the oil filter bracket.
   a. Remove the 3 bolts, filter bracket and 2 gaskets.
9. Remove the crankshaft pulley.
   a. Using the special tool (09213-54015, 90119-08216 or 09330-00021), loosen the crankshaft pulley set bolt.
   b. Using the pulley set bolt and the special tool (09950-50013, 09951-05010, 09952-05010, 09953-05010 or 09954-05011), remove the crankshaft pulley.
10. Remove the ignition coil assembly.
11. Remove the cylinder head cover sub assembly LH.
   a. Remove the 15 bolts, 2 seal washers, cylinder head cover and gasket.

➡Make sure the removed parts are returned to the same places they were removed from.

   b. Remove the 4 gaskets and 2 O-rings from the camshaft bearing caps (No. 2, No. 3, No. 7).
12. Remove the cylinder head cover sub assembly RH.
   a. Remove the 15 bolts, 2 seal washers, cylinder head cover and gasket.

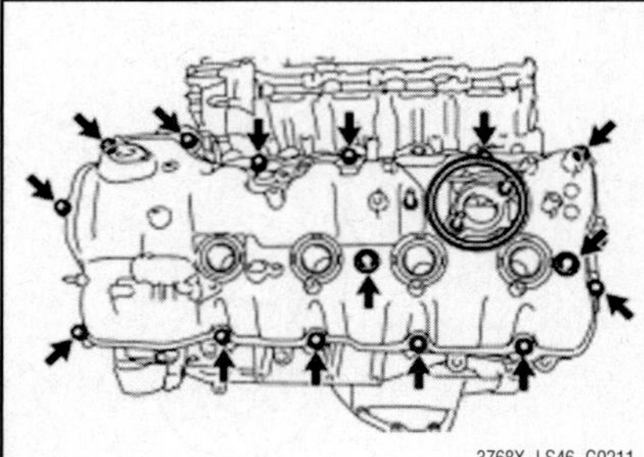

Fig. 205 Removing the cylinder head cover sub assembly LH bolts, washers and gasket

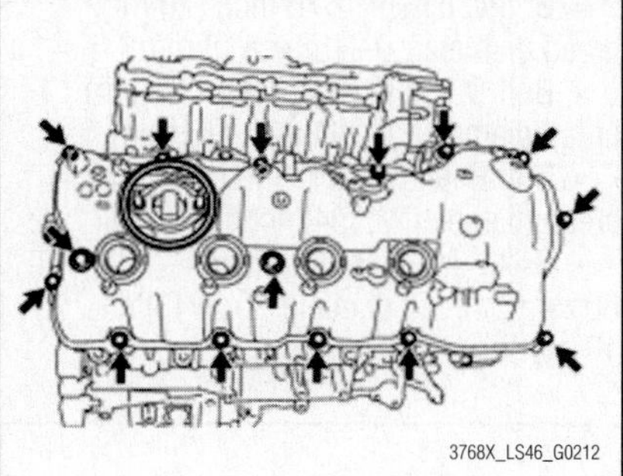

Fig. 206 Removing the 4 gaskets and 2 o-rings (LH)

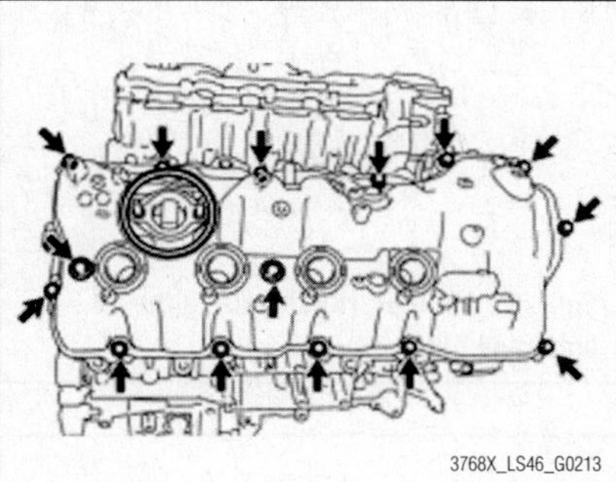

Fig. 207 Removing the cylinder head cover sub assembly RH

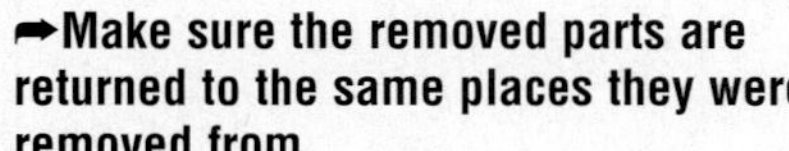

➡Make sure the removed parts are returned to the same places they were removed from.

b. Remove the 4 gaskets and 2 O-rings from the camshaft bearing caps (No. 1, No. 3, No. 6).

13. Remove the timing chain cover sub assembly.

a. Remove the 2 plugs and 2 gaskets.

b. Remove the 30 bolts and nut shown in the illustration.

c. Remove the timing chain cover by prying between the timing chain cover and cylinder head and cylinder block with a screwdriver as shown in the illustration.

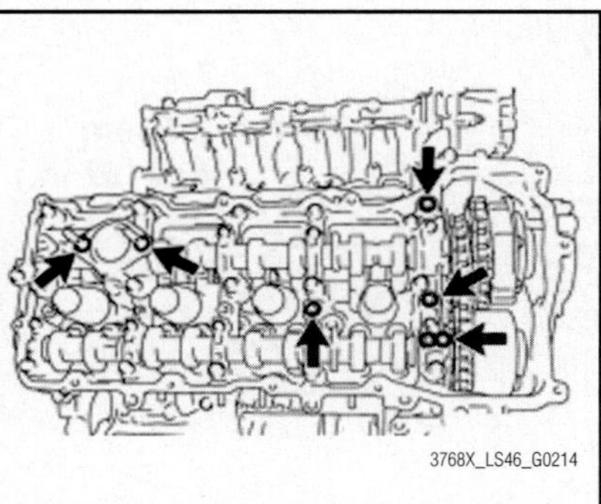

Fig. 208 Removing the 4 gaskets and 3 o-rings (RH)

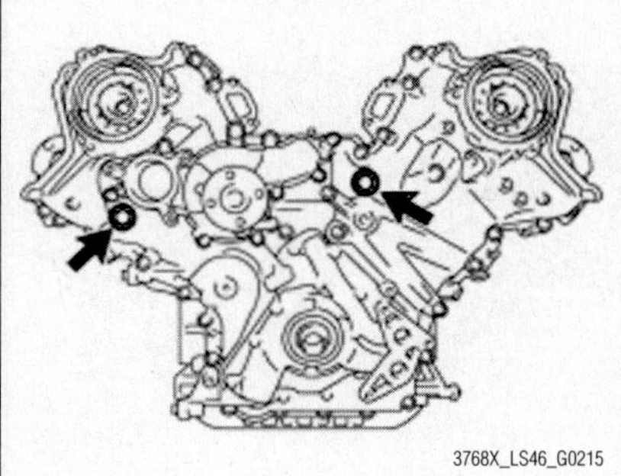

Fig. 209 Removing the 2 plugs and gaskets

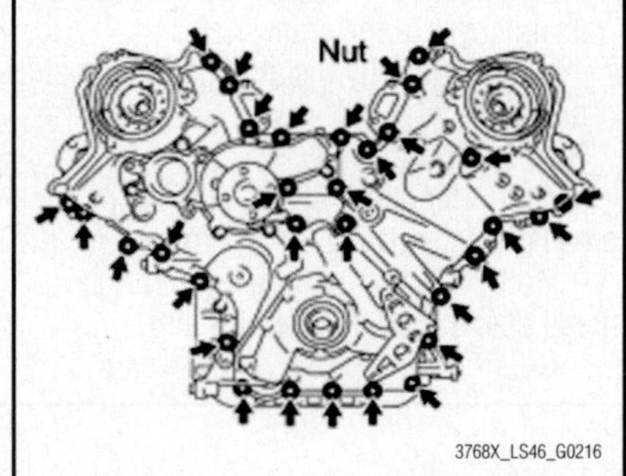

Fig. 210 Removing the timing chain cover bolts and nut

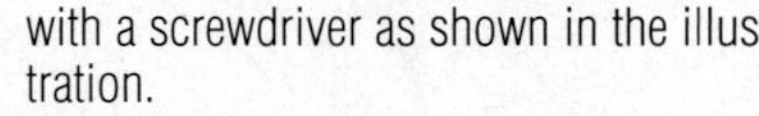

➡Tape the screwdriver tip before use.

➡Be careful not to damage the contact surfaces of the cylinder head, cylinder block and chain cover.

d. Remove the oil pump gasket from the cylinder block.

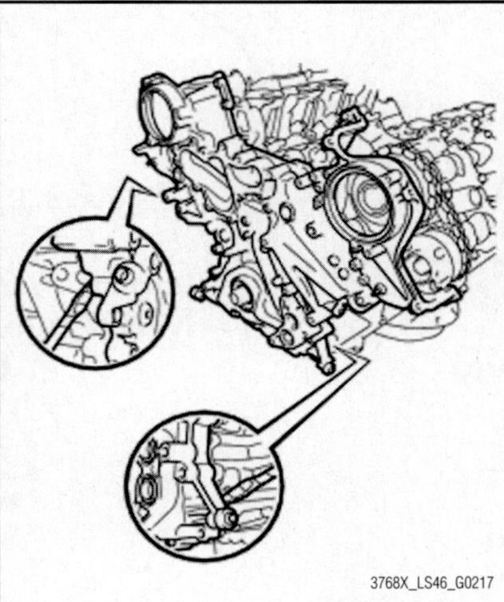
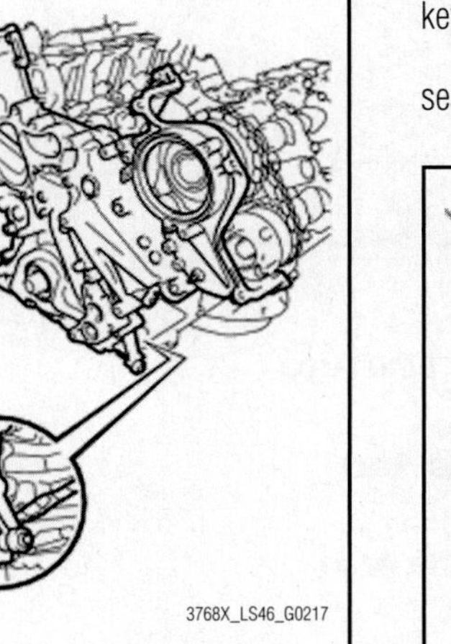

Fig. 211 Removing the timing chain cover

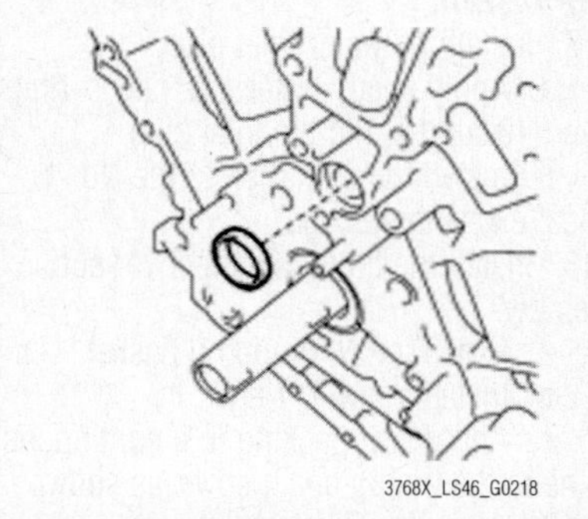

Fig. 212 Removing the oil pump gasket from the cylinder block

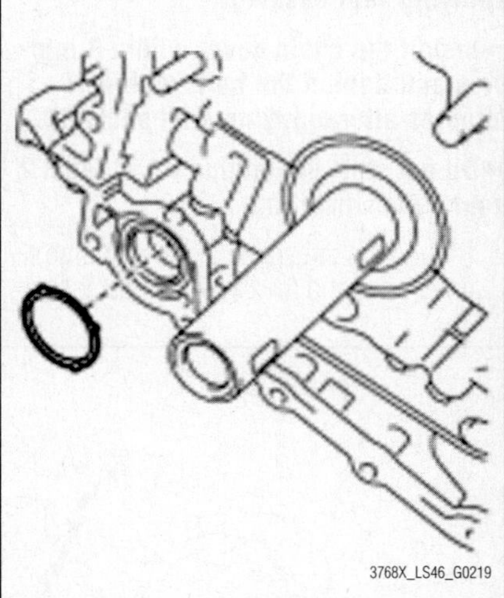

Fig. 213 Removing the o-ring from the cylinder block

e. Remove the O-ring from the cylinder block.

14. Remove the water inlet pipe.

a. Remove the 2 o-rings from the water inlet pipe.

15. Remove the crankshaft timing gear key.

16. Remove the timing chain case oil seal.

Fig. 214 Removing the water inlet pipe and o-rings

To install:

17. Install the water inlet pipe.
    a. Apply soapy water to 2 new o-rings and install them to the inlet pipe.
    b. Install the inlet pipe to the No. 1 heat exchanger cover.
18. Install the timing chain cover sub assembly.
    a. Install a new oil pump gasket.
    b. Install a new o-ring.
    c. Apply seal packing in a continuous line to the timing chain cover as shown in the following illustration.

➡When the contact surfaces are wet, wipe them with an oil-free cloth before applying seal packing.

➡Install the chain cover within 3 minutes and tighten the bolts within 10 minutes after applying seal packing.

➡Do not start the engine for at least 2 hours after installing.

    d. Align the spline of the oil pump's drive rotor and the crankshaft as shown in the illustration. Install the spline and chain cover to the crankshaft.

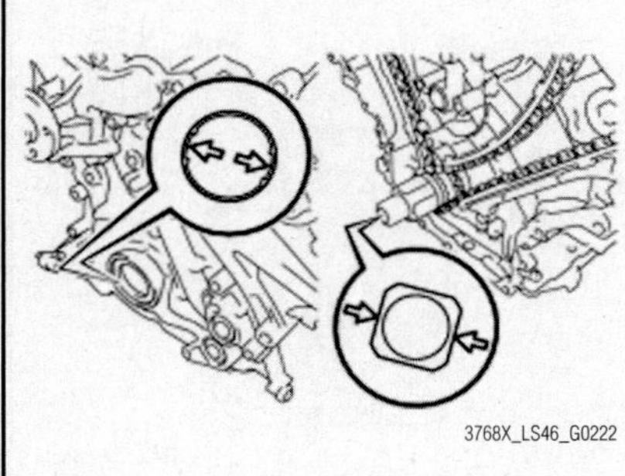

Fig. 216 Aligning the spline of the oil pump's drive rotor and the crankshaft

    e. Temporarily install the 30 bolts and nut.

Bolt Length:

- Bolt A length; 0.984 inch (25 mm), thread diameter; 0.315 inch (8 mm)
- Bolt B length; 2.18 inch (55 mm), thread diameter; 0.315 inch (8 mm)
- Bolt C length; 2.76 inch (70 mm), thread diameter; 0.315 inch (8 mm)
- Bolt D length; 1.38 inch (35 mm), thread diameter; 0.394 inch (10 mm)
- Bolt E length; 2.17 inch (55 mm), thread diameter; 0.394 inch (10 mm)
- Bolt F length; 3.15 inch (80 mm), thread diameter; 0.394 inch (10 mm)

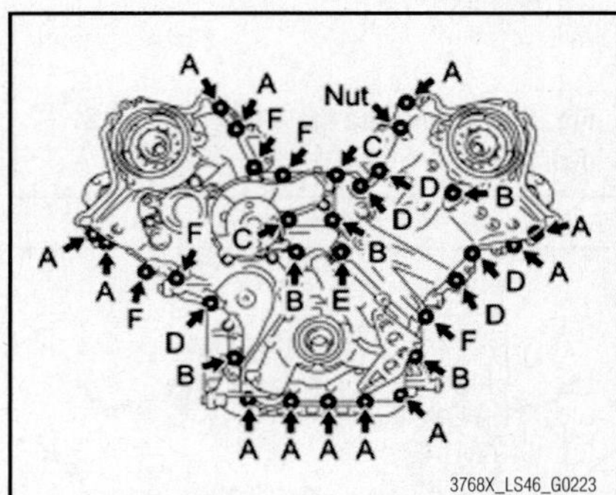

Fig. 217 Temporarily installing the 30 bolts and nut

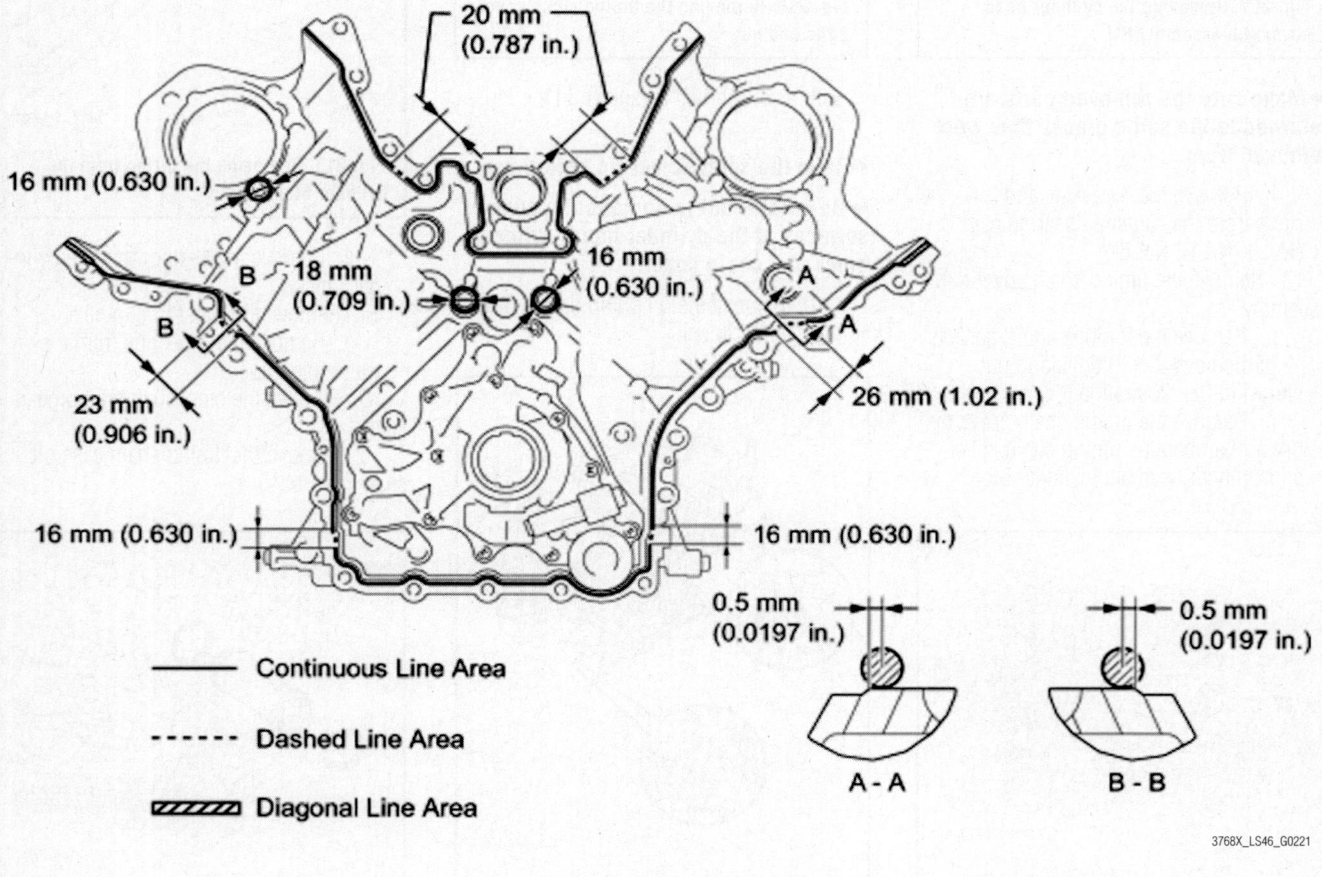

Fig. 215 Applying seal packing

➨Make sure there is no oil on the bolt threads.

a. Tighten the 11 bolts in several steps, in the sequence shown. Tighten to 35 ft. lbs. (47 Nm).

b. Temporarily install the belt tensioner with the standard bolt and 6 mm hexagon wrench bolt.

c. Tighten the 21 bolts and nut in several steps, in the sequence shown in the illustration. Tighten to 17 ft. lbs. (23 Nm).

➨After the installation, if the seal packing has seeped out at the areas labeled A shown in the illustration, wipe it off.

d. Install 2 new gaskets and 2 plugs. Tighten to 34 ft. lbs. (46 Nm).

19. Install the cylinder head cover sub assembly LH.

a. Install 4 new gaskets and 2 new o-rings to the camshaft bearing caps (No. 2, No. 3, No. 7).

b. Install a new gasket to the cylinder head cover.

➨Remove any oil from the contact surface.

c. Apply seal packing as shown.

➨Remove any oil from the contact surface.

➨Install the cylinder head cover within 3 minutes and tighten the bolts within 15 minutes after applying seal packing.

➨Do not start the engine for at least 2 hours after the installation.

d. Install the cylinder head cover with 2 new seal washers and 15 bolts. Tighten all bolts except bolt A to 9 ft. lbs. (12 Nm). Tighten bolt A to 15 ft. lbs. (21 Nm).

20. Install the cylinder head cover sub assembly RH.

a. Install 4 new gaskets and 2 new o-rings to the camshaft bearing caps (No. 1, No. 3, No. 6).

b. Install a new gasket to the cylinder head cover.

➨Remove any oil from the contact surface.

c. Apply seal packing as shown.

➨Remove any oil from the contact surface.

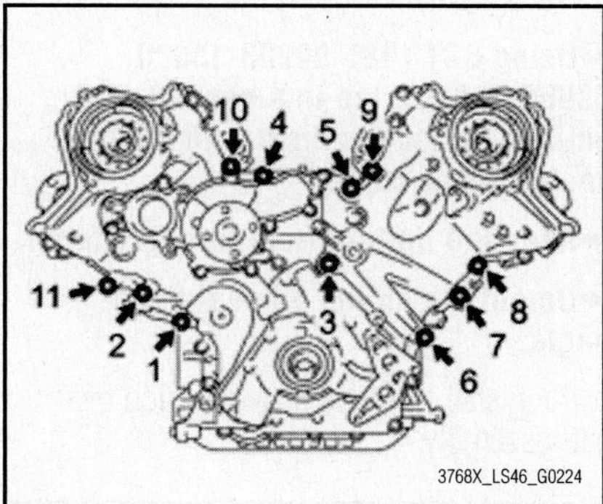

Fig. 218 Identifying the bolt tightening sequence

Fig. 221 Installing 2 new gaskets and plugs

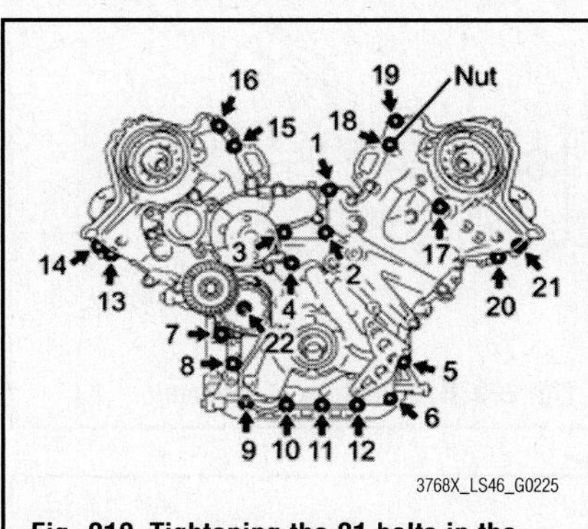

Fig. 219 Tightening the 21 bolts in the sequence provided

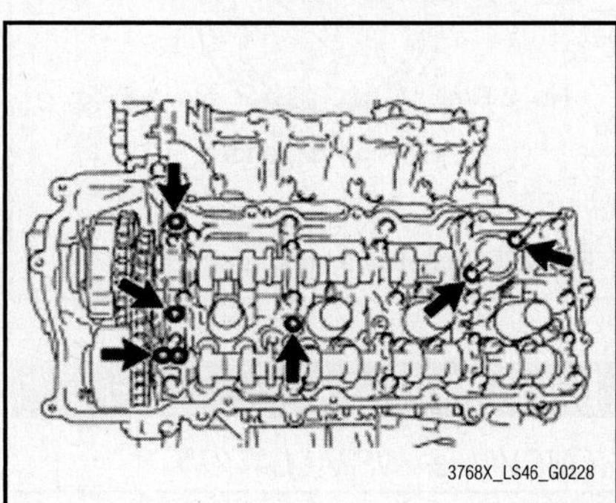

Fig. 222 Installing 4 new gaskets and 2 new o-rings to the camshaft bearing caps

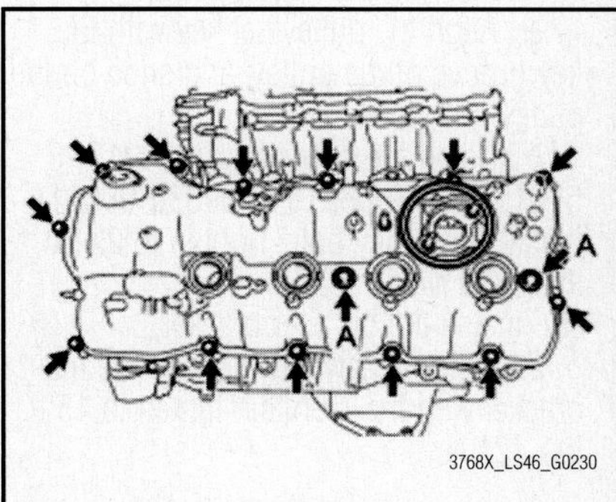

Fig. 224 Installing the cylinder head cover (LH)

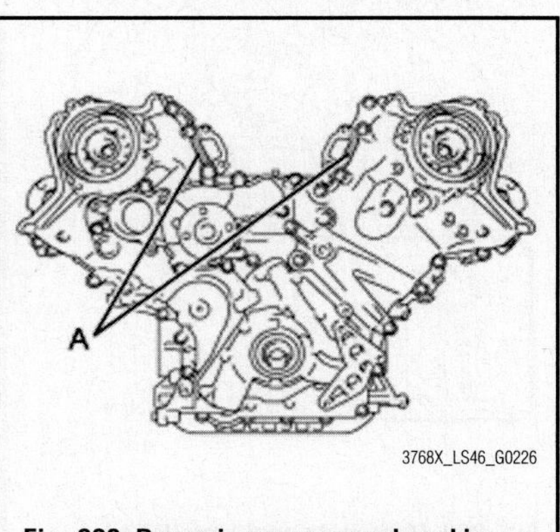

Fig. 220 Removing excess seal packing

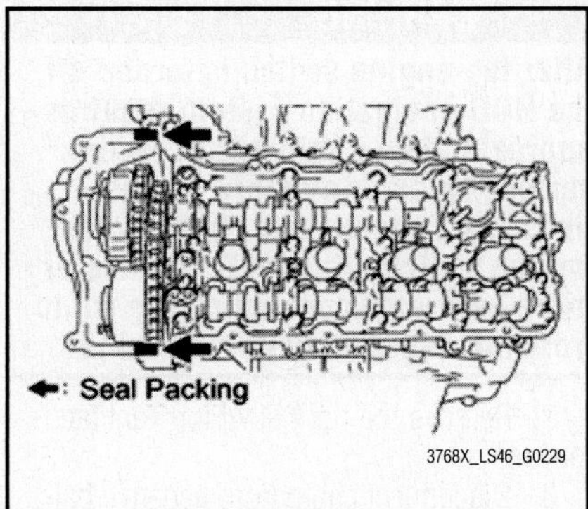

Fig. 223 Applying seal packing

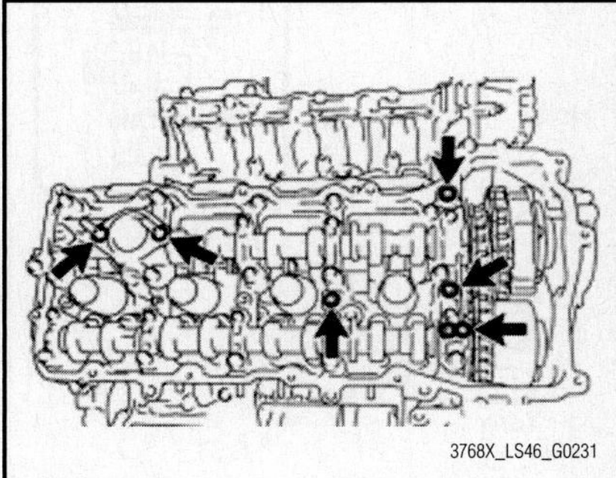

Fig. 225 Installing 4 new gaskets and 2 new o-rings to the camshaft bearing caps

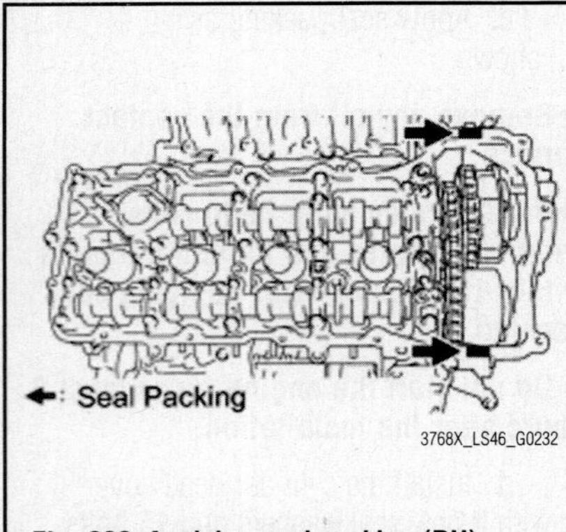

Fig. 226 Applying seal packing (RH)

➡Install the cylinder head cover within 3 minutes and tighten the bolts within 15 minutes after applying seal packing.

➡Do not start the engine for at least 2 hours after the installation.

d. Install the cylinder head cover with 2 new seal washers and the 15 bolts. Tighten all bolts except for bolt A to 9 ft. lbs. (12 Nm). Tighten bolt A to 15 ft. lbs. (21 Nm).

21. Install the ignition coil assembly.
22. Install the timing chain case oil seal.
23. Install the crankshaft timing gear key.
24. Install the crankshaft pulley.

a. Align the pulley set key with the key groove of the pulley, and slide on the pulley.

b. Using the special tool (09213-54015, 90119-08216 or 09330-00021), install the pulley bolt. Tighten to 221 ft. lbs. (300 Nm).

25. Install the oil filter bracket.

a. Install 2 new gaskets and the filter bracket with the 3 bolts. Tighten to 15 ft. lbs. (21 Nm).

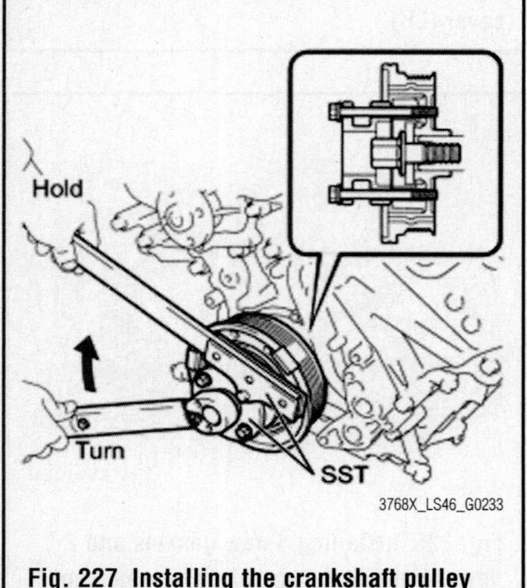

Fig. 227 Installing the crankshaft pulley

26. Install the oil filter element.
27. Install the camshaft timing control motor assembly LH.
28. Install the camshaft timing control motor assembly RH.
29. Install the front water by pass joint.
30. Install the engine and transmission.
31. Perform initialization.

PISTON AND RING

POSITIONING

See Figure 228.

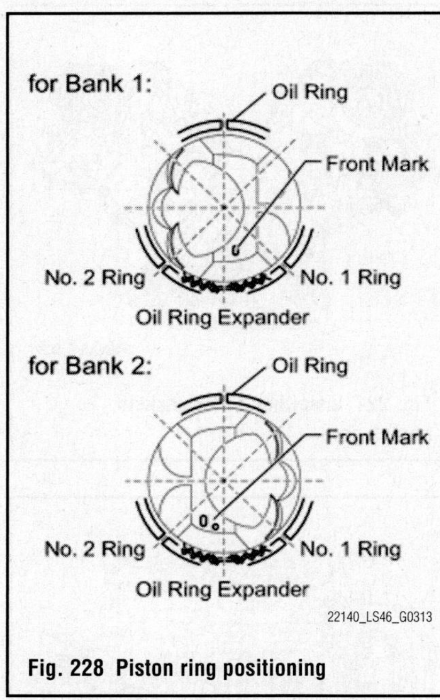

Fig. 228 Piston ring positioning

REAR MAIN SEAL

REMOVAL & INSTALLATION

See Figures 229 and 230.

1. Before servicing the vehicle, refer to the Precautions Section.

WARNING

After the engine switch is turned off, the HDD navigation system requires approximately 6 minutes to record various types of memory and settings. As a result, after turning the engine switch off, wait 6 minutes or more before disconnecting the cable from the negative battery terminal.

2. Remove the right cowl top ventilator louver.
3. Disconnect cable from negative battery terminal.

CAUTION

Wait at least 90 seconds after disconnecting the cable from the negative battery terminal to prevent airbag and seat belt pretensioner activation.

4. Remove the automatic transmission.
5. Remove the drive plate and ring gear sub-assembly.
6. Remove the oil seal with a screwdriver with its tip taped.

➡Do not damage the surface of the oil seal press fit hole and crankshaft.

To install:

➡Using SST (SST 09223-15030, 09950-70010), tap in a new oil seal until its surface is flush with the oil seal retainer edge.

➡Keep the lip free from foreign matter.

➡Do not tap on the oil seal at an angle.

7. Install the drive plate and ring gear sub-assembly.

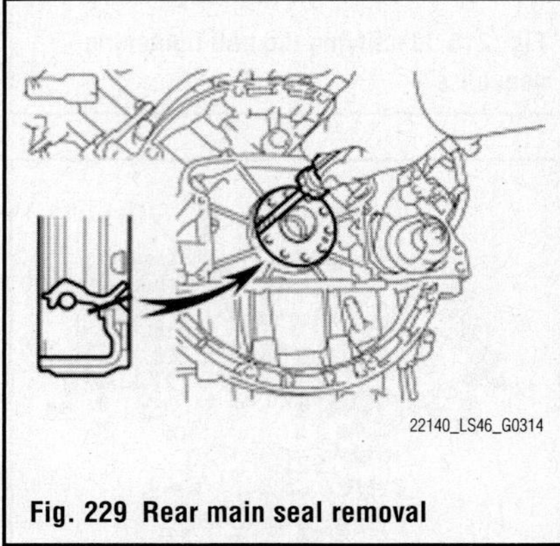

Fig. 229 Rear main seal removal

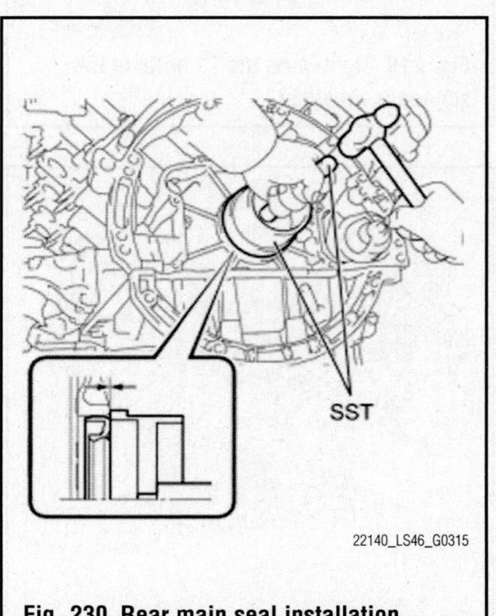

Fig. 230 Rear main seal installation

8. Install the automatic transmission.
9. Connect cable to negative battery terminal.

TIMING CHAIN COVER AND SEAL

REMOVAL & INSTALLATION

See Figures 231 through 239.

1. Before servicing the vehicle, refer to the Precautions Section.

WARNING

After the engine switch is turned off, the HDD navigation system requires approximately 6 minutes to record various types of memory and settings. As a result, after turning the engine switch off, wait 6 minutes or more before disconnecting the cable from the negative battery terminal.

2. Remove the right cowl top ventilator louver.
3. Disconnect cable from negative battery terminal.

CAUTION

Wait at least 90 seconds after disconnecting the cable from the negative battery terminal to prevent airbag and seat belt pretensioner activation.

4. Remove the V-bank cover sub-assembly.
5. Remove the air cleaner inlet cover.
6. Remove the No. 1 air cleaner inlet.
7. Remove the engine room side covers.
8. Remove the battery, nut, clamp, bolts, insulator, and tray.
9. Remove the No. 1 and No. 2 engine under covers.
10. Remove the front lower suspension member protector.

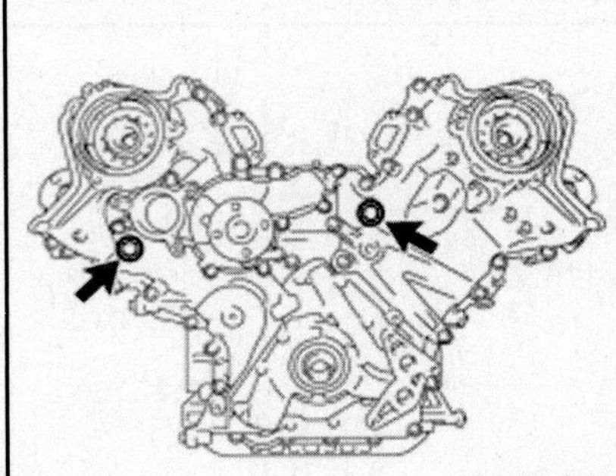

Fig. 231 Timing chain cover plugs and gaskets

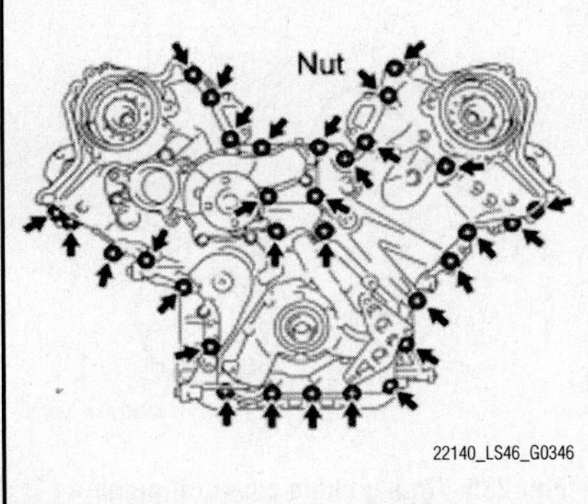

Fig. 232 Timing chain cover bolt removal

11. Remove the rear engine under covers.
12. Drain engine oil and coolant.
13. Remove the intake air connector pipe.
14. Remove the air cleaner assemblies.
15. Remove the radiator reservoir.
16. Remove the V-ribbed belt.
17. Disconnect the radiator hoses.
18. Remove engine room ECU outlet duct.
19. Remove skid control ECU.
20. Remove skid control ECU bracket.
21. Disconnect engine wire.
22. Remove front exhaust pipe assembly.
23. Disconnect front stabilizer bar.
24. Remove the left front suspension member reinforcement.
25. Remove steering sliding yoke with shaft sub-assembly.
26. Remove No. 2 steering intermediate shaft assembly.
27. Remove No. 2 exhaust manifold heat insulator.
28. Remove exhaust manifold sub-assembly.

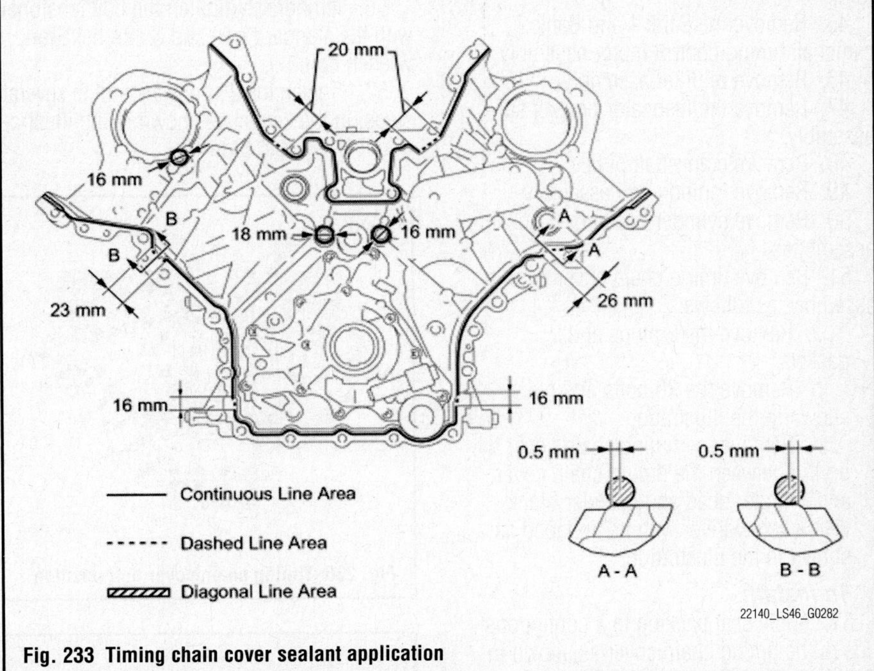

Fig. 233 Timing chain cover sealant application

• Area	Seal packing diameter	Application position from inside edge of cover
Continuous Line Area	3.0 to 4.0 mm (0.1181 to 0.1575 in.)	2.5 mm (0.098 in.)
Dashed Line Area	6.4 mm (0.2520 in.) or more, or within OK area shown in illustration	0.5 mm (0.020 in.)
Diagonal Line Area	3.0 to 4.0 mm (0.1181 to 0.1575 in.)	5.5 mm (0.217 in.)

22140_LS46_G0283

Fig. 234 Sealant application specifications

29. Disconnect cooler compressor assembly.
30. Remove oil level dipstick guide.
31. Remove generator.
32. Remove injector driver.
33. Remove No. 1 engine cover sub-assembly.
34. Remove water by-pass pipe sub-assembly.
35. Remove intake manifold.
36. Remove engine cover sub-assemblies.
37. Remove fuel pressure pulsation damper assembly.
38. Remove fuel pipe sub-assemblies.
39. Remove the fuel pump assembly.
40. Remove inlet water housing.
41. Remove water pump pulley.
42. Remove idler pulley sub-assemblies.
43. Remove V-ribbed belt tensioner assembly.
44. Remove front water by-pass joint.
45. Remove the Bank 1 and Bank 2 camshaft timing control motor assembly.
46. Remove oil filter assembly.
47. Remove the resonator bracket sub-assembly.
48. Remove crankshaft pulley.
49. Remove ignition coil assembly.
50. Remove cylinder head cover sub-assemblies.
51. Remove timing chain cover sub-assembly, as follows:

a. Remove the 2 plugs and 2 gaskets.

b. Remove the 30 bolts and nut shown in the illustration.

c. Remove the timing chain cover by prying between the timing chain cover and cylinder head and cylinder block with a screwdriver with its tip taped as shown in the illustration.

To install:

52. Apply seal packing in a continuous line to the timing chain cover as shown in the illustration. Seal packing: Toyota Genuine Seal Packing Black, Three Bond 1207B or equivalent.

➡When the contact surfaces are wet, wipe them with an oil-free cloth before applying seal packing.

➡Install the chain cover within 3 minutes and tighten the bolts within 10 minutes after applying seal packing.

➡Do not start the engine for at least 2 hours after installing.

53. Align the oil pump's drive rotor spline and the crankshaft as shown in the illustration. Install the spline and chain cover to the crankshaft.

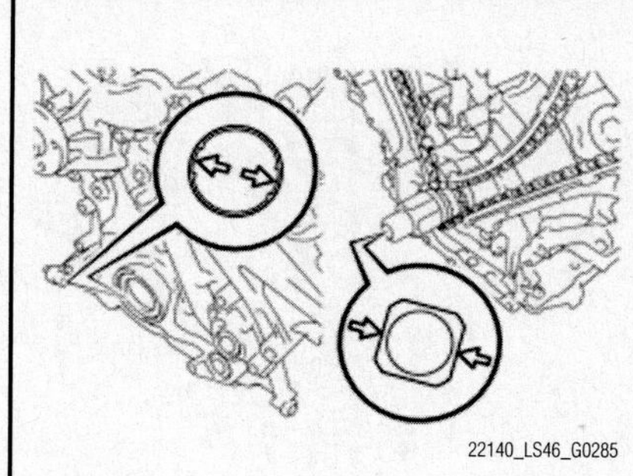

Fig. 235 Timing chain cover oil pump alignment

54. Temporarily tighten the timing chain cover with the 30 bolts and nut.
55. Tighten the 11 bolts in several steps, in the sequence shown in the illustration to 35 ft. lbs. (47 Nm).
56. Temporarily tighten the belt tensioner with the standard bolt and 6 mm hexagon wrench bolt.
57. Tighten the 21 bolts and nut in several steps, in the sequence shown in the illustration to 17 ft. lbs. (23 Nm).

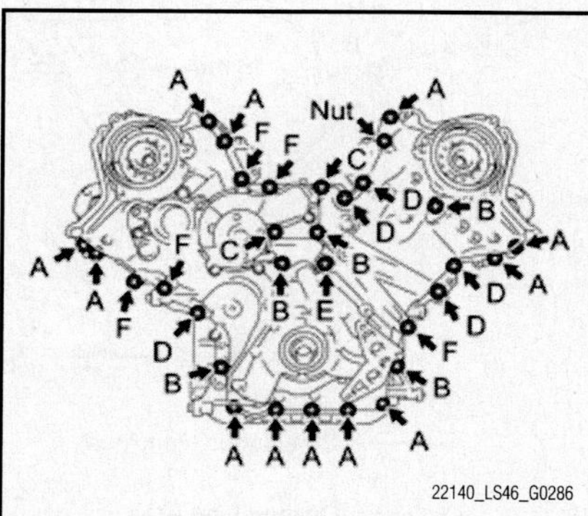

Fig. 236 Timing chain cover bolt position

Item	Length	Thread diameter
Bolt A	25 mm (0.984 in.)	8 mm (0.315 in.)
Bolt B	55 mm (2.165 in.)	8 mm (0.315 in.)
Bolt C	70 mm (2.756 in.)	8 mm (0.315 in.)
Bolt D	35 mm (1.378 in.)	10 mm (0.394 in.)
Bolt E	55 mm (2.165 in.)	10 mm (0.394 in.)
Bolt F	80 mm (3.150 in.)	10 mm (0.394 in.)

22140_LS46_G0284

Fig. 237 Timing chain cover bolt legend

58. Install 2 new gaskets and 2 plugs and tighten to 34 ft. lbs. (46 Nm).
59. Install cylinder head cover sub-assemblies.
60. Install ignition coil assembly.
61. Install crankshaft pulley.
62. Install the resonator bracket sub-assembly.
63. Install oil filter assembly.
64. Install camshaft timing control motor assembly.
65. Install front water by-pass joint.
66. Install idler pulleys.
67. Install water pump pulley.
68. Install inlet water housing.
69. Install fuel pump assemblies.
70. Install fuel pipes.
71. Install fuel pressure pulsation damper assembly.
72. Install engine covers.
73. Install intake manifold.
74. Install water by-pass pipe sub-assembly.
75. Install injector driver.
76. Install generator.
77. Install oil level dipstick guide.
78. Connect cooler compressor assembly.
79. Install exhaust manifold sub-assembly.

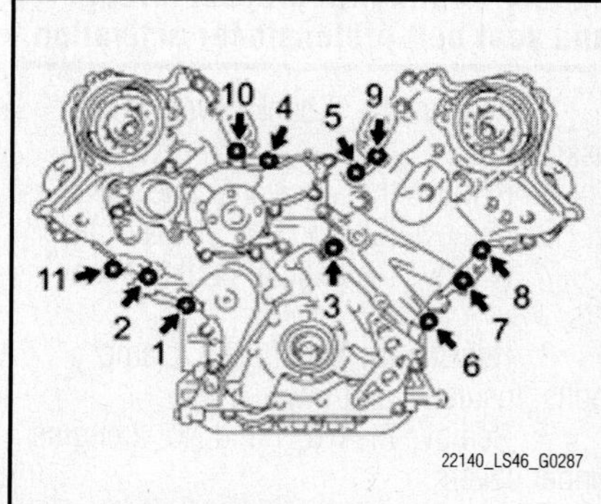

Fig. 238 Timing chain cover bolt tightening sequence 1

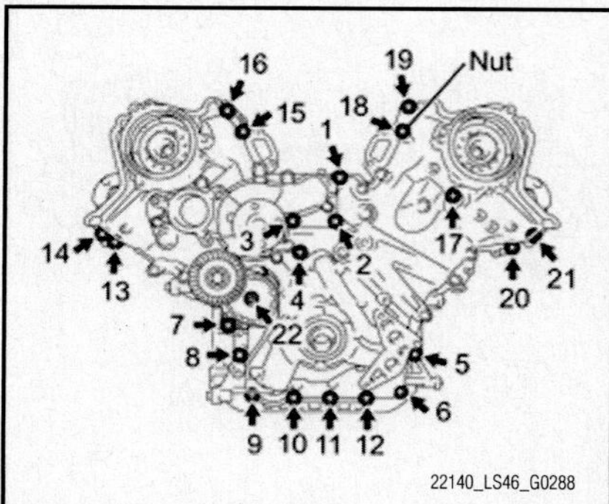

Fig. 239 Timing chain cover bolt tightening sequence 2

80. Install No. 2 exhaust manifold heat insulator.
81. Install No. 2 steering intermediate shaft assembly.
82. Install steering sliding yoke with shaft sub-assembly.
83. Install left front suspension member reinforcement.
84. Connect front stabilizer bar.
85. Install front exhaust pipe assembly.
86. Connect engine wire.
87. Install engine room ECU outlet duct.
88. Install skid control ECU.
89. Install V-ribbed belt.
90. Install radiator assembly and hoses.
91. Install air cleaner assemblies.
92. Install intake air connector pipe.
93. Install the No. 1 air cleaner inlet.
94. Install rear engine under covers.
95. Install battery, nut, clamp, bolts, insulator, and tray.
96. Add engine oil and coolant.
97. Connect cable to negative battery terminal.
98. Perform initialization.
99. Inspect for oil, coolant, fuel, and exhaust gas leaks.
100. Check engine oil level.
101. Install front lower suspension member protector.
102. Install No. 1 and No. 2 engine under covers.
103. Install the engine room side covers.
104. Install air cleaner inlet cover.
105. Install V-bank cover sub-assembly.
106. Install right cowl top ventilator louver.

TIMING CHAIN & SPROCKETS

REMOVAL & INSTALLATION

See Figures 240 through 272.

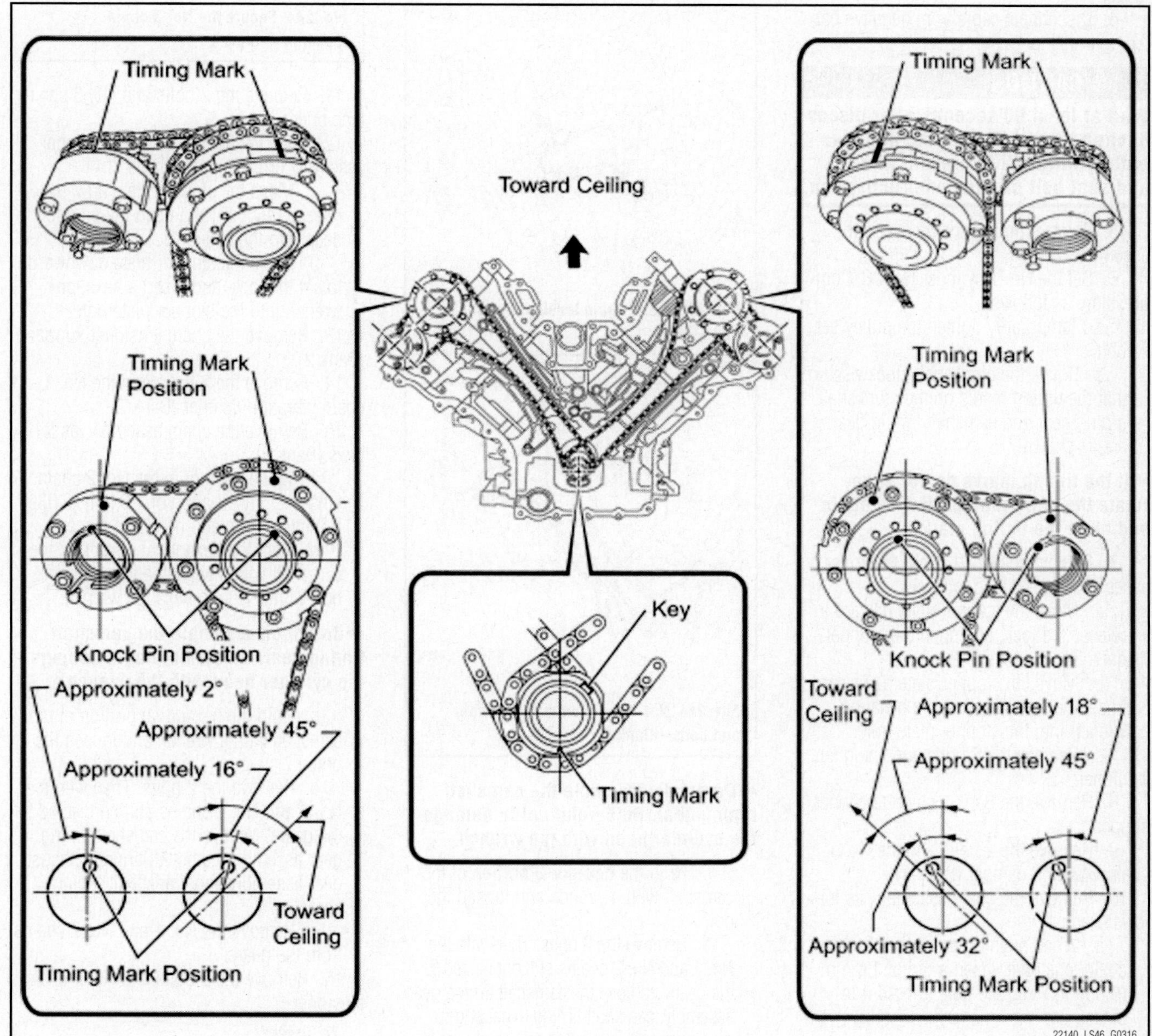

Fig. 240 Timing chain alignment

1. Before servicing the vehicle, refer to the Precautions Section.
2. Discharge fuel system pressure.

WARNING

After the engine switch is turned off, the HDD navigation system requires approximately 6 minutes to record various types of memory and settings. As a result, after turning the engine switch off, wait 6 minutes or more before disconnecting the cable from the negative battery terminal.

3. Remove the right cowl top ventilator louver.
4. Disconnect cable from negative battery terminal.

CAUTION

Wait at least 90 seconds after disconnecting the cable from the negative battery terminal to prevent airbag and seat belt pretensioner activation.

5. Remove timing chain cover sub-assembly.
6. Set the No. 1 cylinder to TDC/Compression, as follows:
   a. Temporarily tighten the pulley set bolt.
   b. Rotate the crankshaft clockwise so that the timing marks on the crankshaft timing gear and camshaft timing gears are as shown.

➡If the timing marks do not align, rotate the crankshaft clockwise again and align the timing marks.

7. Remove the No. 1 chain tensioner assembly (Bank 1), as follows:
   a. Move the stopper plate upward to release the lock, and push the plunger deep into the tensioner.
   b. Move the stopper plate downward to set the lock, and insert a hexagon wrench into the stopper plate hole.
   c. Remove the 2 bolts and chain tensioner.
8. Remove the Bank 1 chain tensioner slipper.
9. Remove the 2 bolts and the No. 1 chain vibration damper (Bank 1).
10. Remove the chain assembly, as follows (Bank 1):
   a. While pushing down the No. 3 chain tensioner, insert a pin of 1.0 mm (0.039 in.) into the hole to hold it in place.
   b. Hold the hexagonal portion of the camshaft with a wrench and loosen the bolt with a 12mm hexagon wrench.

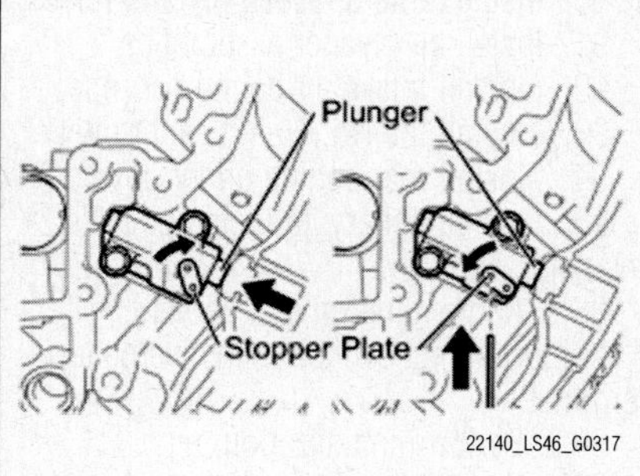

Fig. 241 No. 1 chain tensioner stopper plate and plunger—Bank 1

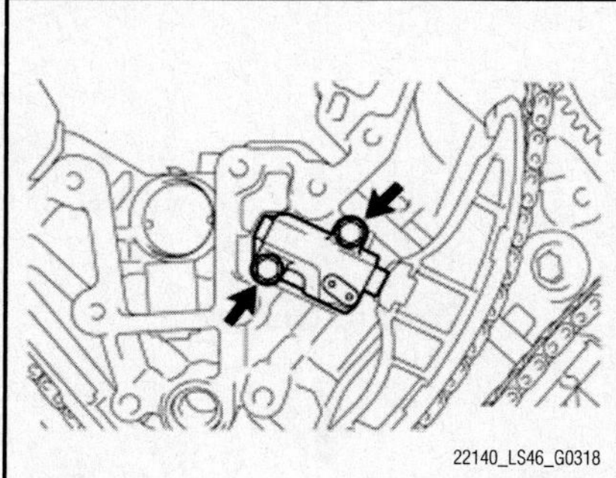

Fig. 242 No. 1 chain tensioner and bolts—Bank 1

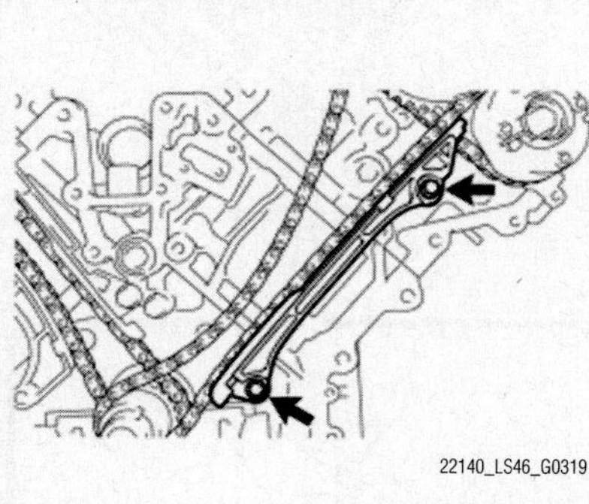

Fig. 243 No. 1 chain vibration damper and bolts—Bank 1

➡Do not disassemble the camshaft timing gear. Be careful not to damage the cylinder head with the wrench.

   c. Hold the hexagonal portion of the camshaft with a wrench and loosen the bolt.
   d. Remove the 2 bolts. Then with the No. 1 and No. 2 chains still connected to the gears, remove the camshaft timing gear assembly, camshaft timing exhaust gear assembly and crankshaft timing sprocket.
   e. Remove the No. 1 and No. 2 chains from the gears.

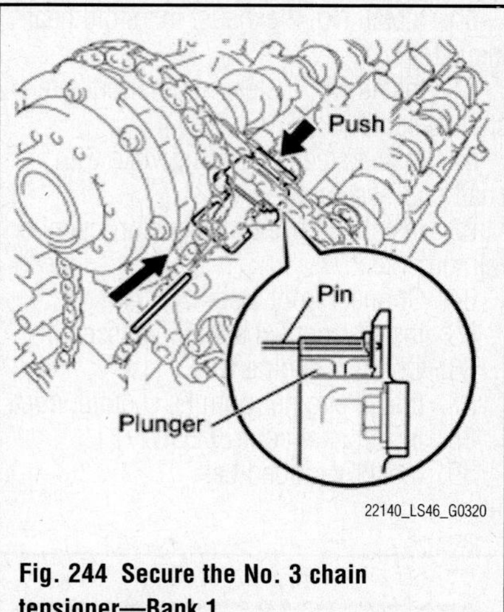

Fig. 244 Secure the No. 3 chain tensioner—Bank 1

11. Remove the 2 bolts and No. 3 chain tensioner.
12. Remove the No. 1 chain tensioner assembly (Bank 2), as follows:
   a. Move the stopper plate upward to release the lock, and push the plunger deep into the tensioner.
   b. Move the stopper plate downward to set the lock, and insert a hexagon wrench into the stopper plate hole.
13. Remove the chain tensioner slipper (Bank 2).
14. Remove the 2 bolts and the No. 1 chain vibration damper (Bank 2).
15. Remove the chain assembly, as follows (Bank 2):
   a. While raising up the No. 2 chain tensioner, insert a pin of 1.0 mm (0.039 in.) into the hole to hold it in place.
   b. Hold the hexagonal portion of the camshaft with a wrench and loosen the bolt with a 12 mm hexagon wrench.

➡Do not disassemble the camshaft timing gear. Be careful not to damage the cylinder head with the wrench.

   c. Hold the hexagonal portion of the camshaft with a wrench and loosen the bolt.
   d. Remove the 2 bolts. Then with the No. 1 and No. 2 chains still connected to the gears, remove the camshaft timing gear assembly, camshaft timing exhaust gear assembly and crankshaft timing sprocket.
   e. Remove the No. 1 and No. 2 chains from the gears.
16. Remove the 2 bolts and No. 2 chain tensioner.

To install:

17. Install the No. 2 chain tensioner, as follows:

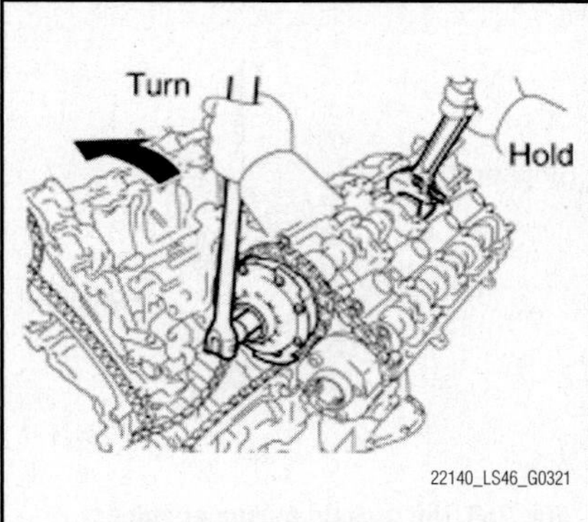

Fig. 245 Camshaft bolt loosening—Bank 1

a. Install the No. 2 chain tensioner with the 2 bolts and tighten to 7 ft. lbs. (10 Nm).

b. While raising up the No. 2 chain tensioner, insert a pin of 1.0 mm (0.039 in.) into the hole to hold it in place.

18. Install the No. 1 chain sub-assembly (Bank 2), as follows:

a. Align the No. 1 chain's orange mark plates with the camshaft timing gear's timing mark, and attach the chain to the gear as shown in the illustration.

b. Align the No. 1 chain's orange mark plate with the crankshaft timing

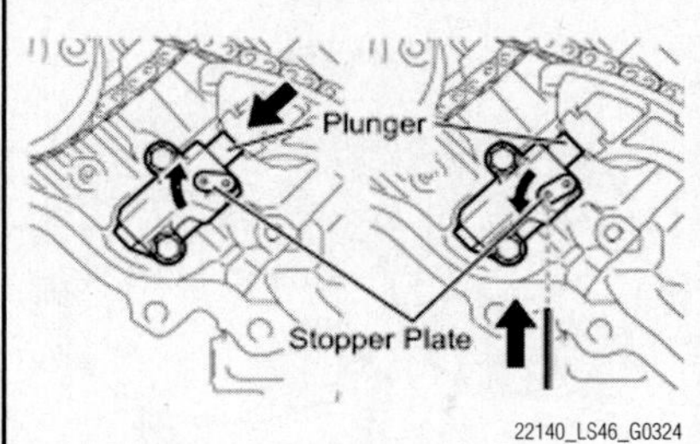

Fig. 248 No. 1 chain tensioner stopper plate and plunger—Bank 2

gear's timing mark, and attach the chain to the gear as shown in the illustration.

c. Align the No. 2 chain's mark plates (yellow) with the timing marks of the camshaft timing gear assembly and camshaft timing exhaust gear assembly, and attach the No. 2 chain to the gears as shown in the illustration.

➡The crankshaft timing gear and camshaft exhaust gear assembly will be installed with the No. 1 and No. 2 chains connected to the gears.

d. Install the crankshaft timing gear to the crankshaft.

e. Align and attach the knock pin of the No. 1 camshaft with the pin hole of the camshaft timing gear assembly.

f. Using the hexagonal portion of the No. 2 camshaft, align and attach the knock pin of the No. 2 camshaft with the pin hole of the camshaft timing exhaust gear assembly.

g. Remove the pin from the No. 2 chain tensioner.

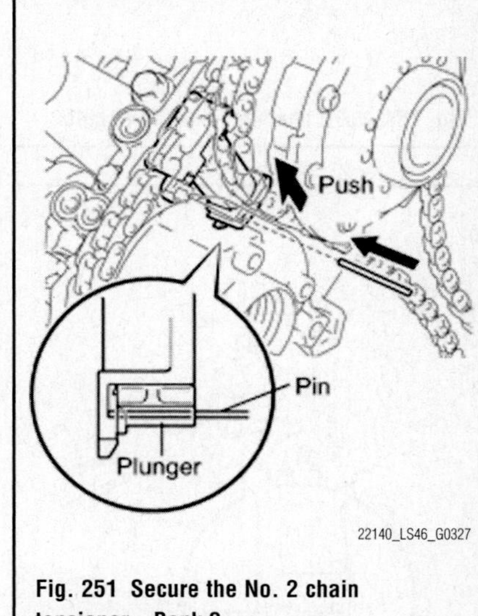

Fig. 251 Secure the No. 2 chain tensioner—Bank 2

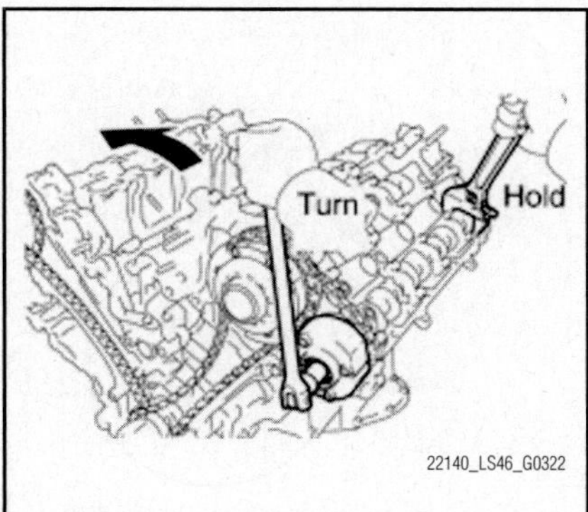

Fig. 246 Loosen camshaft bolt—Bank 1

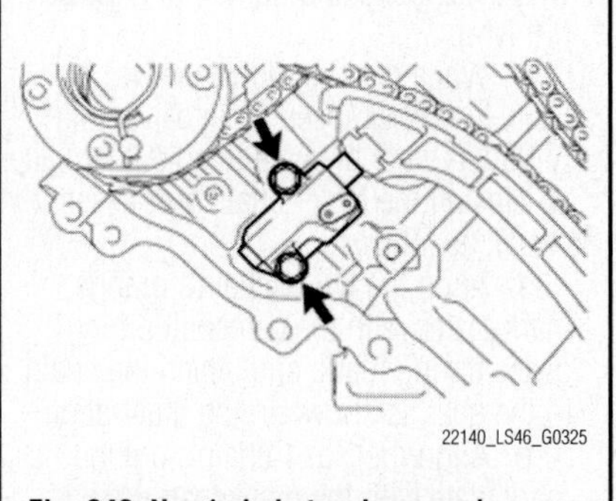

Fig. 249 No. 1 chain tensioner and bolts—Bank 2

Fig. 252 Camshaft bolt loosening—Bank 2

Fig. 247 No. 3 chain tensioner and bolts

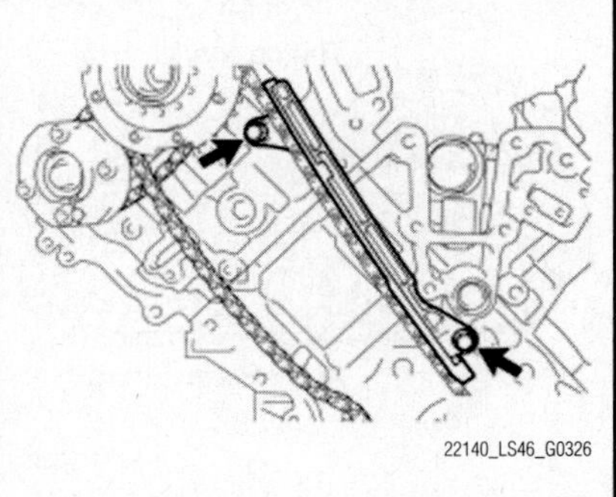

Fig. 250 No. 1 chain vibration damper and bolts—Bank 2

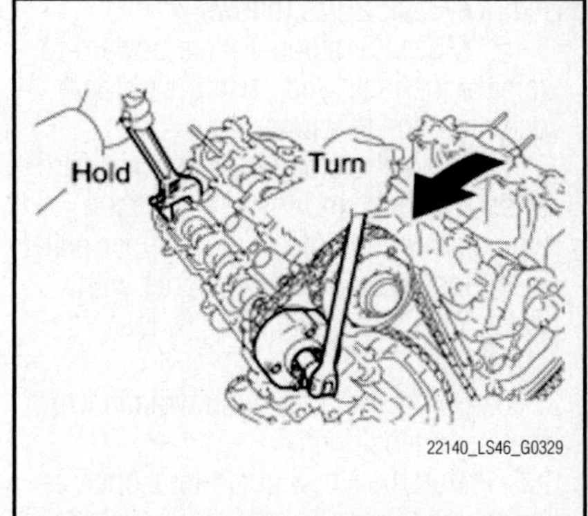

Fig. 253 Loosen camshaft bolt—Bank 2

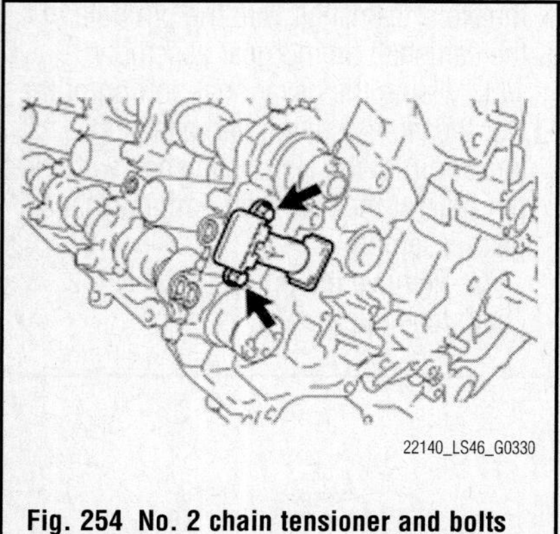
Fig. 254 No. 2 chain tensioner and bolts

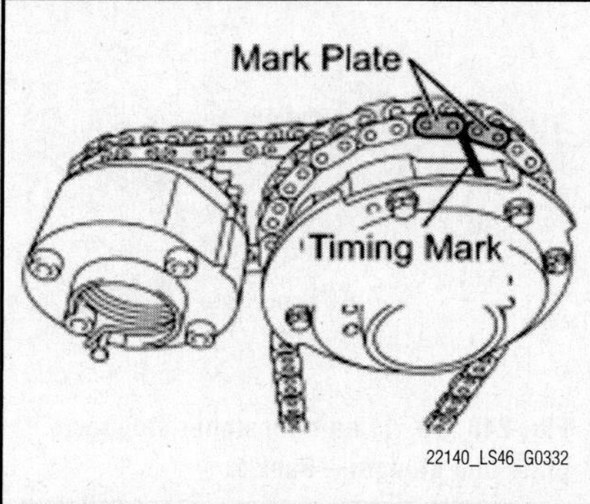

Fig. 256 No. 1 chain and camshaft timing gear alignment—Bank 2

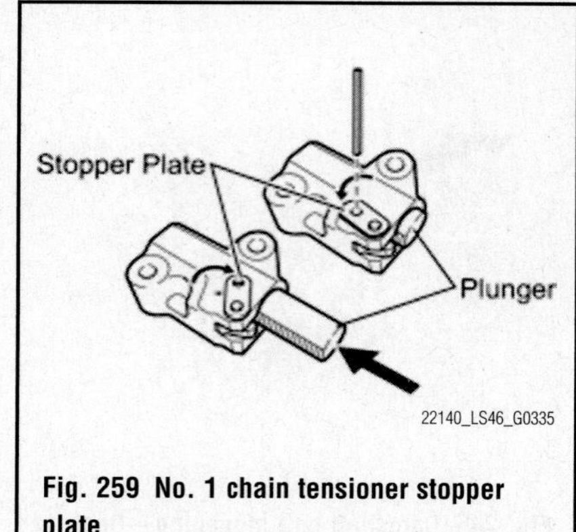

Fig. 259 No. 1 chain tensioner stopper plate

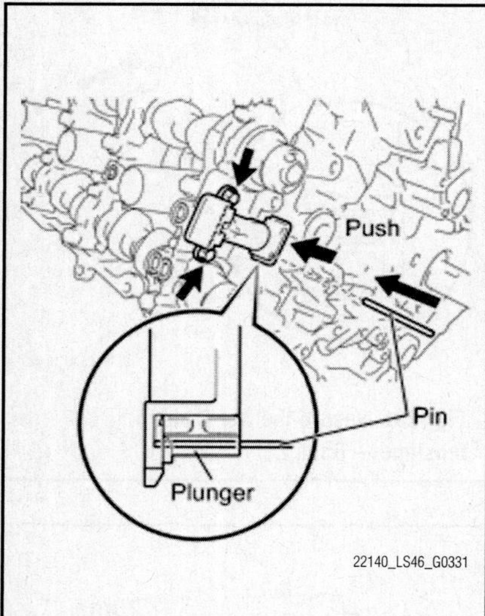

Fig. 255 No. 2 chain tensioner installation

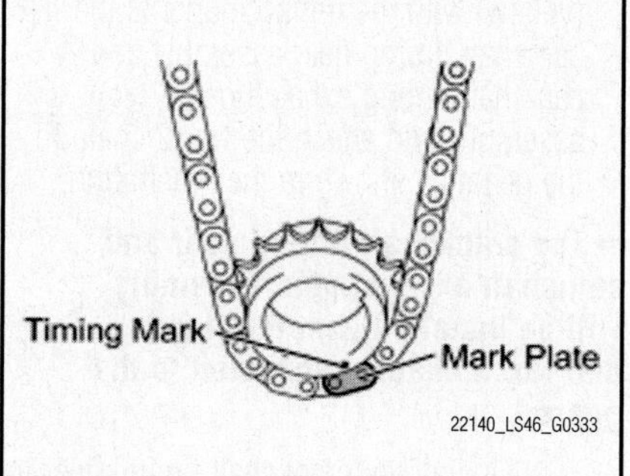

Fig. 257 Chain and camshaft timing gear alignment—Bank 2

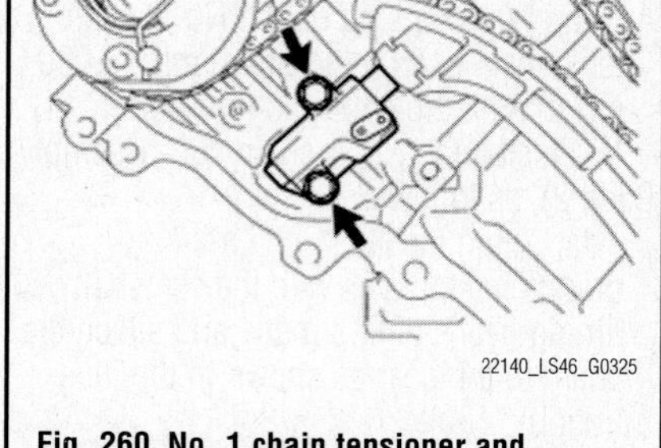
Fig. 260 No. 1 chain tensioner and bolts—Bank 2

19. Install the No. 1 chain vibration damper (Bank 2) with the 2 bolts and tighten to 15 ft. lbs. (21 Nm).

20. Install the chain tensioner slipper (Bank 2). If you cannot install the chain tensioner slipper due to the tension of the chain, use the hexagonal portion of the camshaft to loosen the chain, and then install the chain tensioner slipper.

21. Install the No. 1 chain tensioner assembly (Bank 2), as follows:

a. Move the stopper plate upward to release the lock, and push the plunger deep into the tensioner.

b. Move the stopper plate downward to set the lock, and insert a hexagon wrench into the hole of the stopper plate.

c. Install the chain tensioner with the 2 bolts and tighten to 7 ft. lbs. (10 Nm).

d. Remove the hexagon wrench from the chain tensioner.

22. Install the No. 3 chain tensioner, as follows:

a. Install the No. 3 chain tensioner with the 2 bolts and tighten to 7 ft. lbs. (10 Nm).

b. While pushing down the No. 3 chain tensioner, insert a pin of 1.0 mm (0.039 in.) into the hole to hold it in place.

23. Install the No. 1 chain sub-assembly (Bank 1), as follows:

a. Align the No. 1 chain's orange mark plates with the camshaft timing gear's timing mark, and attach the chain to the gear as shown in the illustration.

b. Align the No. 1 chain's orange mark plate with the crankshaft timing gear's timing mark, and attach the chain to the gear as shown in the illustration.

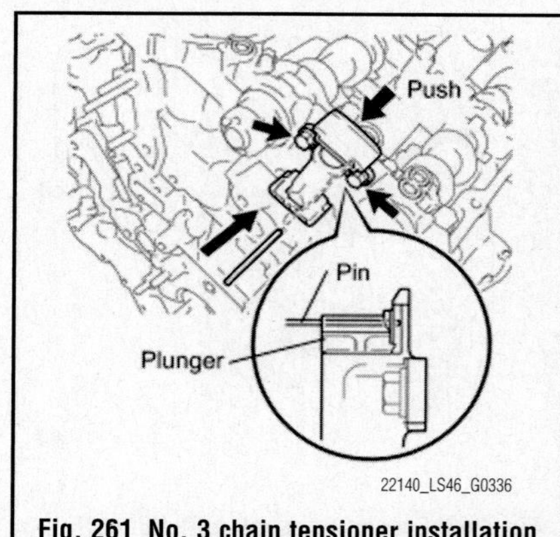

Fig. 261 No. 3 chain tensioner installation

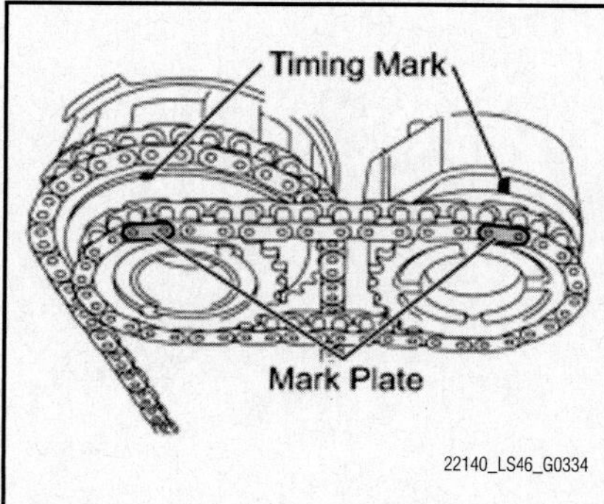

Fig. 258 No. 2 chain and camshaft timing gear alignment—Bank 2

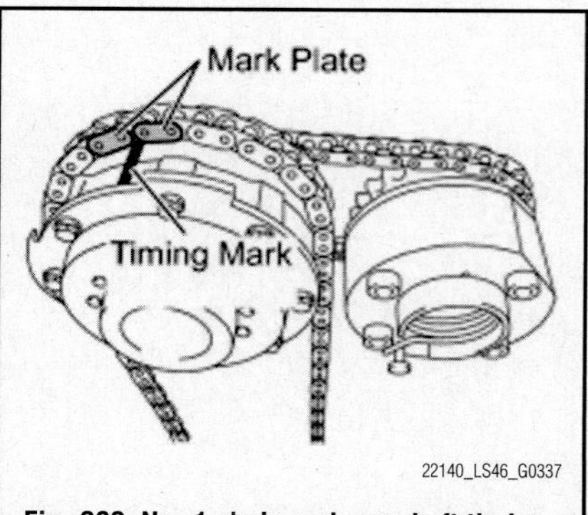

Fig. 262 No. 1 chain and camshaft timing gear alignment—Bank 1

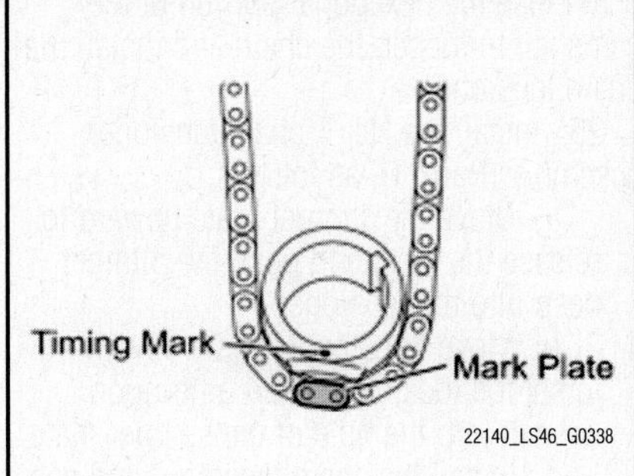

Fig. 263 Chain and camshaft timing gear alignment—Bank 1

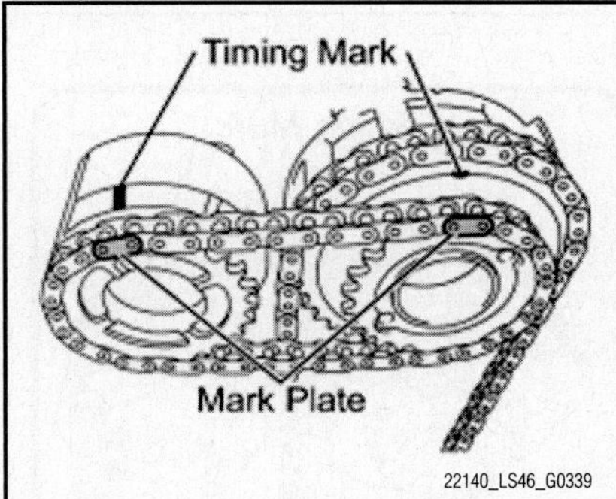

Fig. 264 No. 2 chain and camshaft timing gear alignment—Bank 1

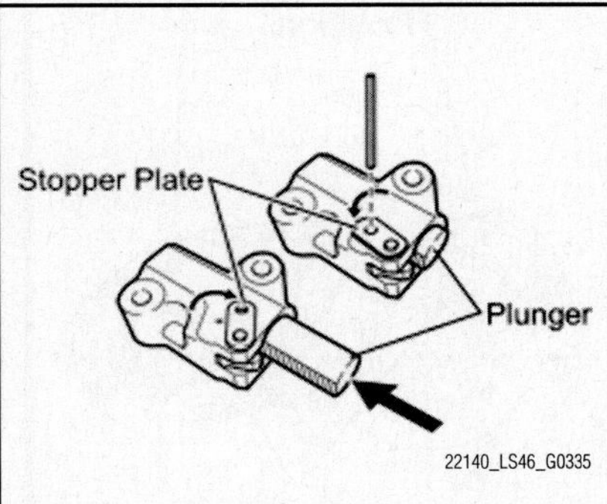

Fig. 265 No. 1 chain tensioner stopper plate

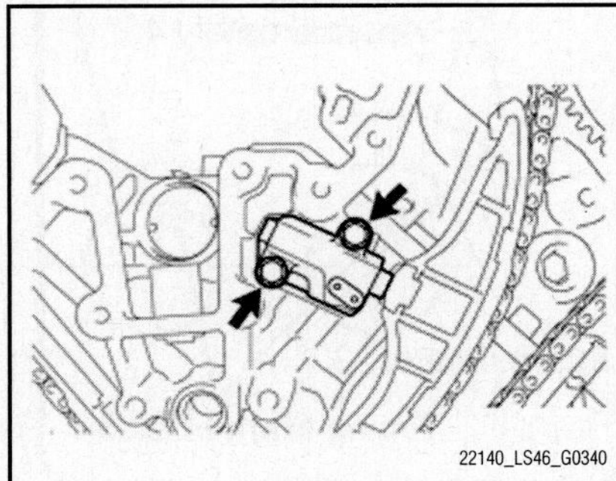

Fig. 266 No. 1 chain tensioner and bolts—Bank 1

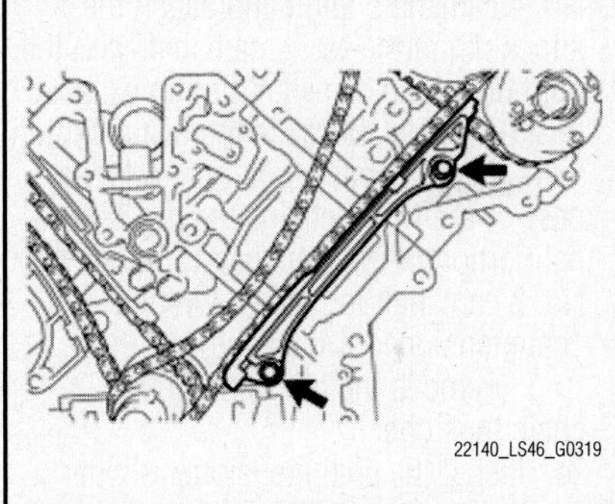

Fig. 267 No. 1 chain vibration damper and bolts—Bank 1

c. Align the No. 2 chain's mark plates (yellow) with the timing marks of the camshaft timing gear assembly and camshaft timing exhaust gear assembly, and attach the No. 2 chain to the gears as shown in the illustration.

➡The crankshaft timing gear and camshaft exhaust gear assembly will be installed with the No. 1 and No. 2 chains connected to the gears.

d. Install the crankshaft timing gear to the crankshaft.

e. Align and attach the knock pin of the No. 3 camshaft with the pin hole of the camshaft timing gear assembly.

f. Using the hexagonal portion of the

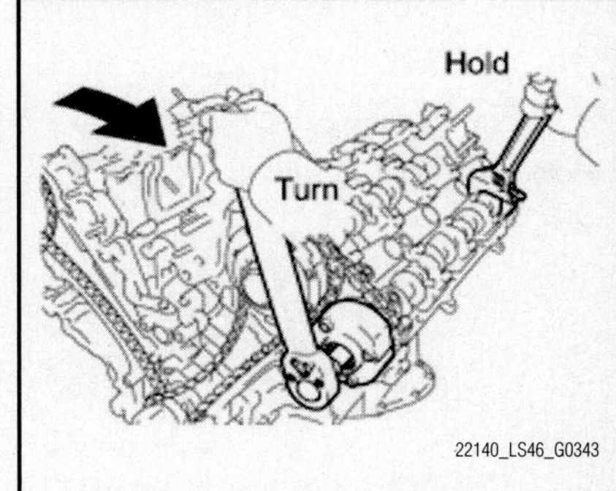

Fig. 270 Tighten the camshaft timing exhaust gear assembly—Bank 1

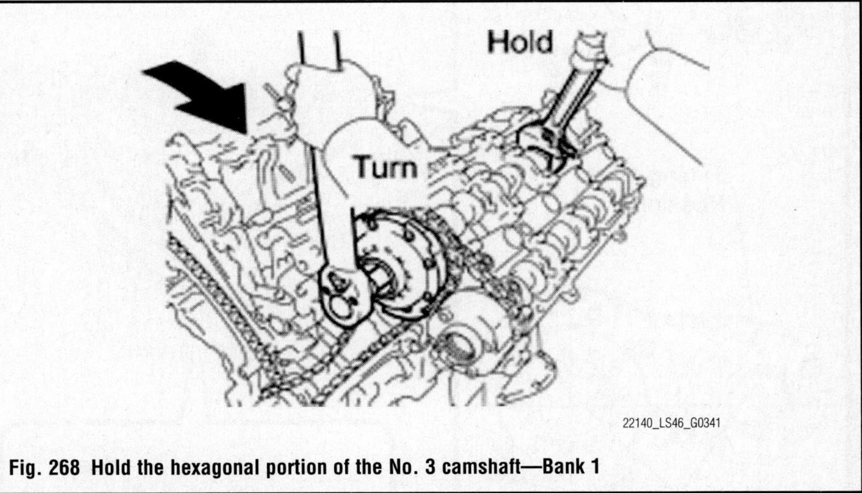

Fig. 268 Hold the hexagonal portion of the No. 3 camshaft—Bank 1

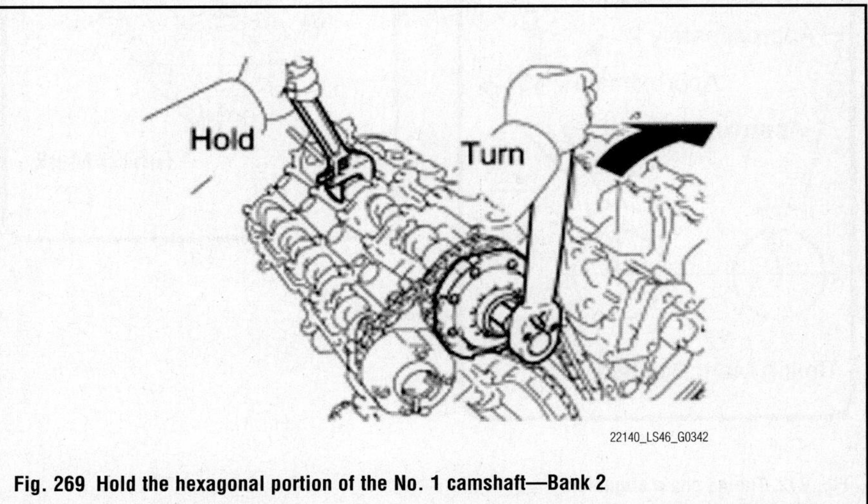

Fig. 269 Hold the hexagonal portion of the No. 1 camshaft—Bank 2

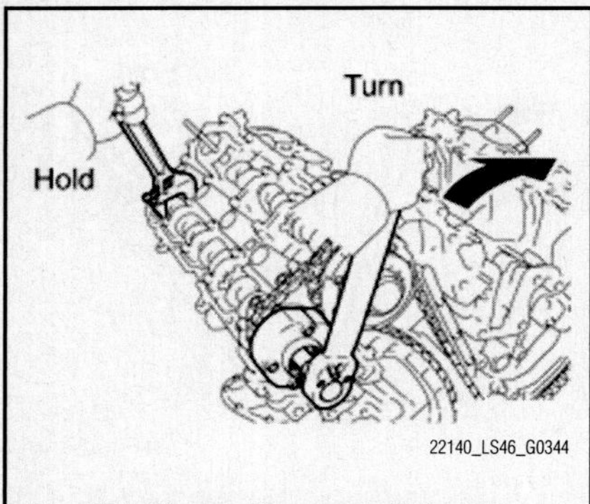

Fig. 271 Tighten the camshaft timing exhaust gear assembly—Bank 2

No. 4 camshaft, align and attach the knock pin of the No. 4 camshaft with the pin hole of the camshaft timing exhaust gear assembly. Because the gears' timing mark positions may shift due to looseness of the No. 1 chain, use the hexagonal portion of the camshaft to hold the No. 3 camshaft in place until the No. 1 chain tensioner is installed.

g. Remove the pin from the No. 2 chain tensioner.

24. Install the chain tensioner slipper (Bank 1). If you cannot install the chain tensioner slipper due to the tension of the chain, use the hexagonal portion of the camshaft to loosen the chain and install the chain tensioner.

25. Install the No. 1 chain tensioner assembly (Bank 1), as follows:

a. Move the stopper plate upward to release the lock, and push the plunger deep into the tensioner.

b. Move the stopper plate downward to set the lock, and insert a hexagon wrench into the hole of the stopper plate.

c. Install the chain tensioner and gasket with the 2 bolts and tighten 7 ft. lbs. (10 Nm).

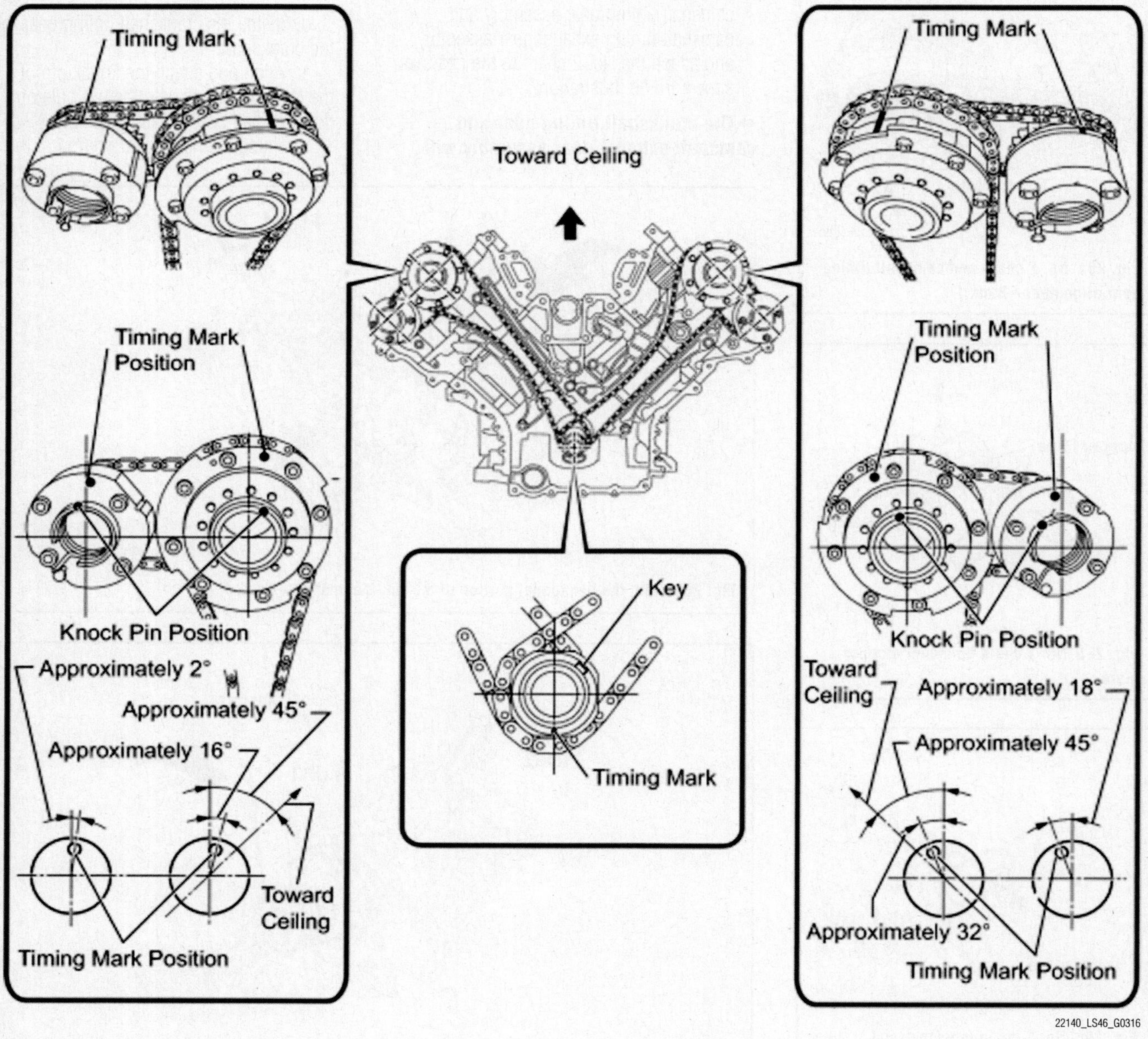

Fig. 272 Timing chain alignment

26. Install the No. 1 chain vibration damper (Bank 1), as follows:

a. Install the vibration damper with the 2 bolts and tighten to 15 ft. lbs. (21 Nm).

b. Remove the hexagon wrench from the No. 1 chain tensioner.

27. Tighten the camshaft timing gear assembly, as follows:

a. Using a wrench, hold the hexagonal portion of the No. 3 camshaft (for Bank 1).

b. Using a wrench, hold the hexagonal portion of the No. 1 camshaft (for Bank 2).

c. Using a 12 mm socket hexagon wrench, tighten the camshaft timing gear assembly with a new bolt to 58 ft. lbs. (79 Nm).

d. Using a wrench to hold the hexagonal portion of the No. 4 camshaft (for Bank 1), tighten the camshaft timing exhaust gear assembly with the bolt to 74 ft. lbs. (100 Nm).

e. Using a wrench to hold the hexagonal portion of the No. 2 camshaft (for Bank 2), tighten the camshaft timing exhaust gear assembly with the bolt to 74 ft. lbs. (100 Nm).

28. Set the No. 1 cylinder to TDC/Compression, as follows:

a. Temporarily tighten the pulley bolt.

b. Rotate the crankshaft clockwise and check that the timing marks on the crankshaft timing gear and camshaft timing gears are as shown in the illustration.

c. Remove the crankshaft pulley bolt.

29. Install the timing chain cover.

30. Connect cable to negative battery terminal.

VALVE LASH

ADJUSTMENT

No adjustment is necessary.

ENGINE PERFORMANCE & EMISSION CONTROLS

CAMSHAFT POSITION (CMP) SENSOR

LOCATION

See Figure 273.

Refer to the accompanying illustration for sensor location.

REMOVAL & INSTALLATION

1. Remove V-bank cover sub-assembly.
2. Disconnect the Camshaft Position (CMP) sensor connector.
3. Remove the bolt and Camshaft Position (CMP) sensor.

To install:

4. Installation is the reverse of removal. Tighten the Camshaft Position (CMP) sensor bolt to 7 ft. lbs. (10 Nm).

CRANKSHAFT POSITION (CKP) SENSOR

LOCATION

See Figure 274.

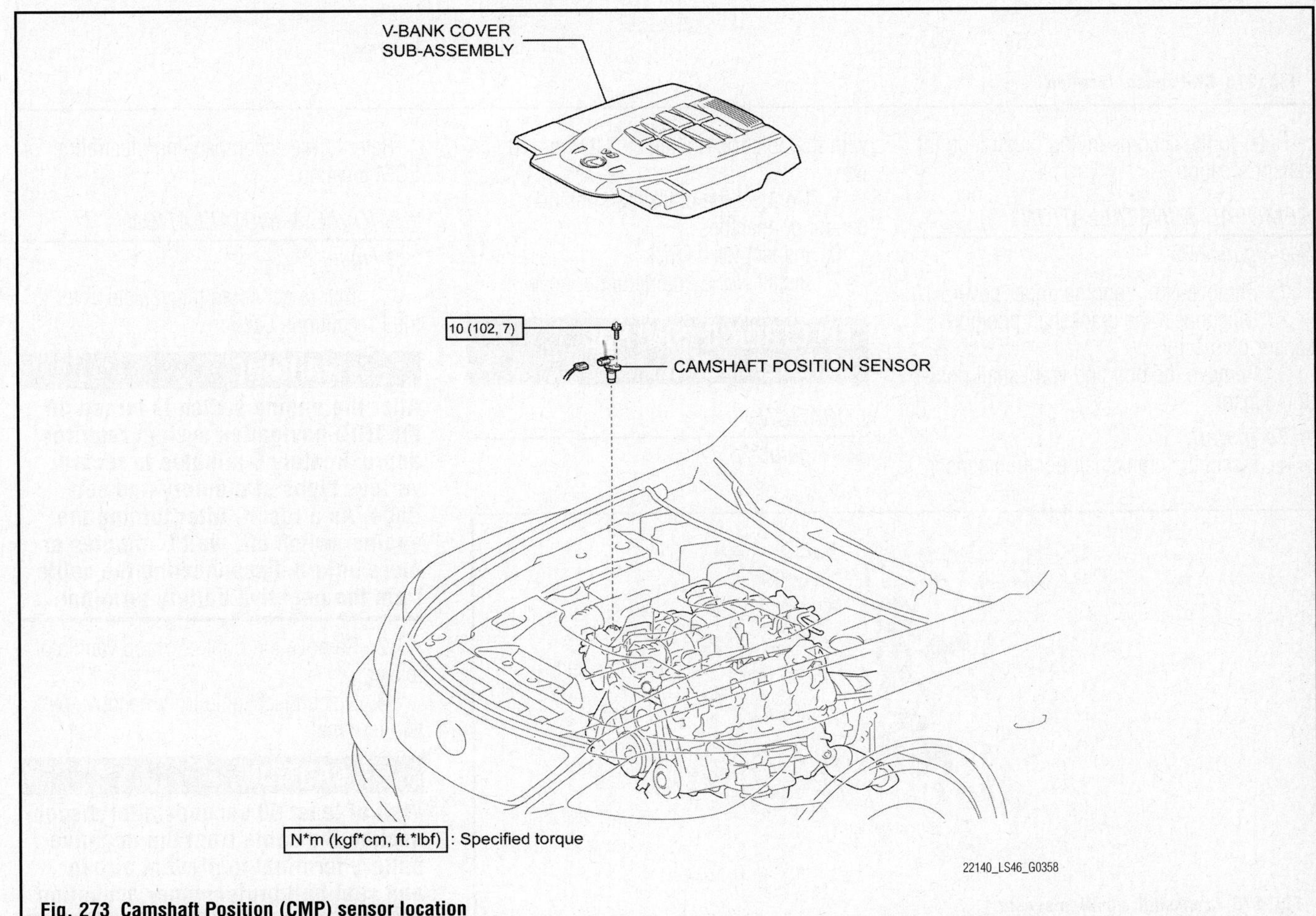

Fig. 273 Camshaft Position (CMP) sensor location

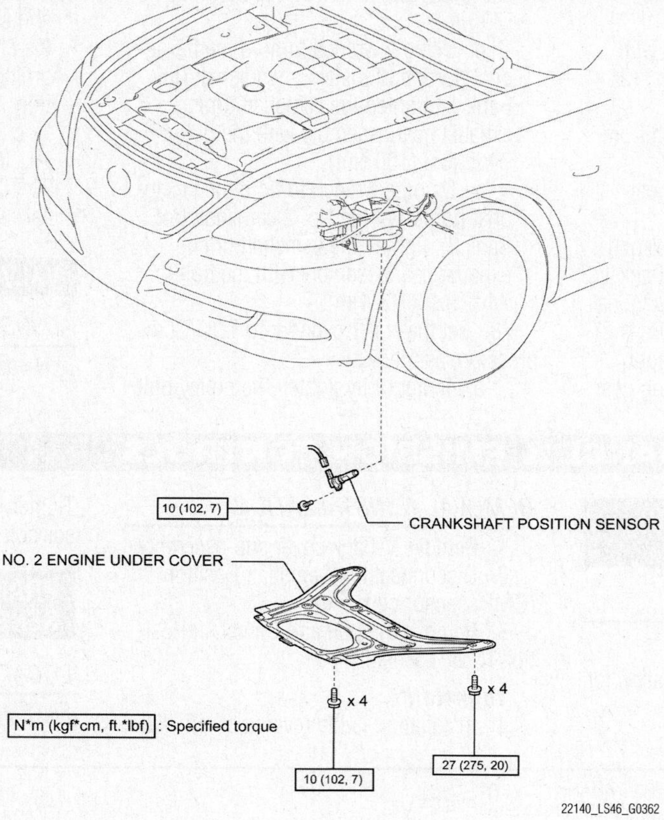

Fig. 274 CKP sensor location

Refer to the accompanying illustration for sensor location.

REMOVAL & INSTALLATION

See Figure 275.

1. Remove No. 2 engine under cover.
2. Disconnect the crankshaft position sensor connector.
3. Remove the bolt and crankshaft position sensor.

To install:

4. Install the crankshaft position sensor with the bolt and tighten to 7 ft. lbs. (10 Nm).
5. Connect the crankshaft position sensor connector.
6. Inspect for oil leak.
7. Install No. 2 engine under cover.

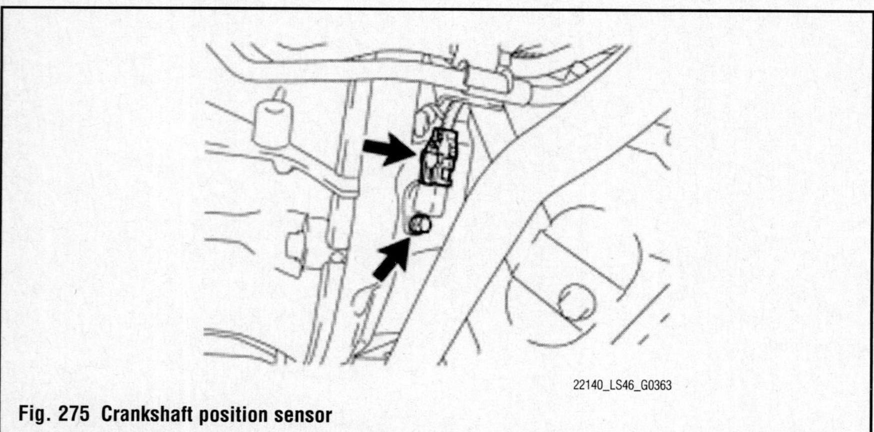

Fig. 275 Crankshaft position sensor

ELECTRONIC CONTROL MODULE (ECM)

LOCATION

See Figure 276.

Refer to the accompanying illustration for ECM location.

REMOVAL & INSTALLATION

See Figure 277.

1. Before servicing the vehicle, refer to the Precautions Section.

WARNING

After the engine switch is turned off, the HDD navigation system requires approximately 6 minutes to record various types of memory and settings. As a result, after turning the engine switch off, wait 6 minutes or more before disconnecting the cable from the negative battery terminal.

2. Remove the right cowl top ventilator louver.
3. Disconnect cable from negative battery terminal.

CAUTION

Wait at least 90 seconds after disconnecting the cable from the negative battery terminal to prevent airbag and seat belt pretensioner activation.

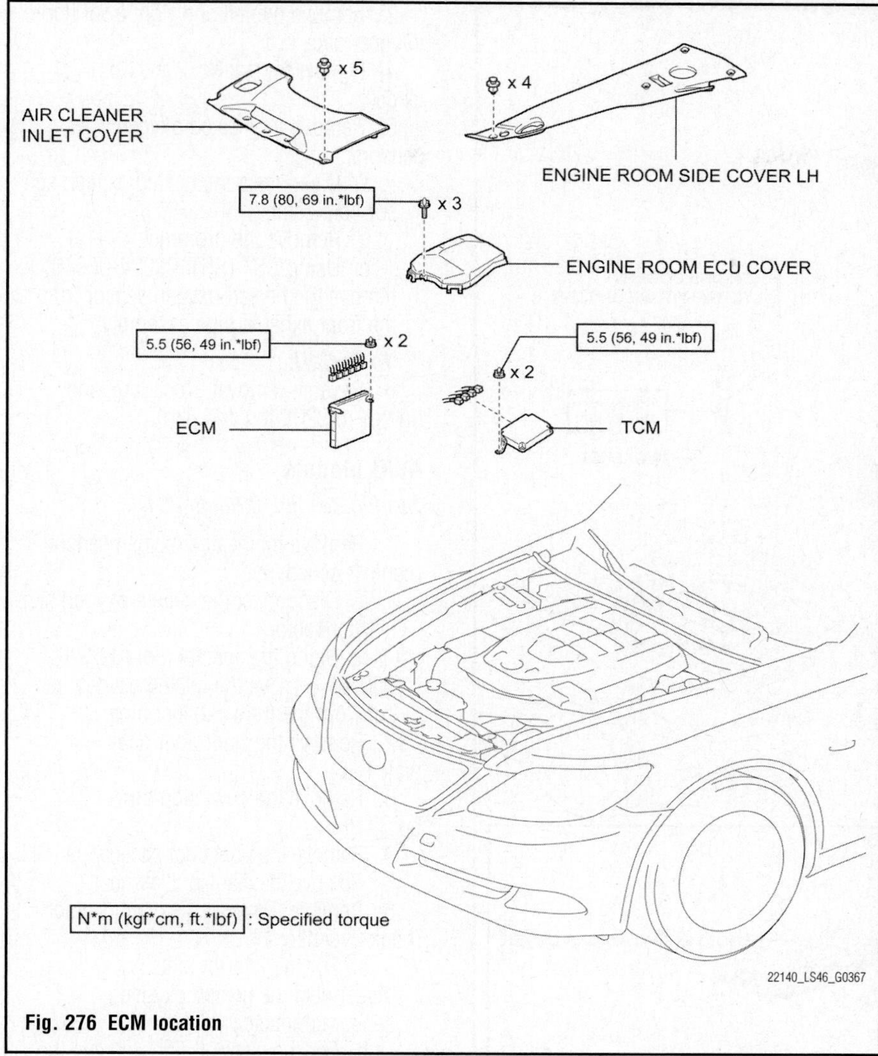

Fig. 276 ECM location

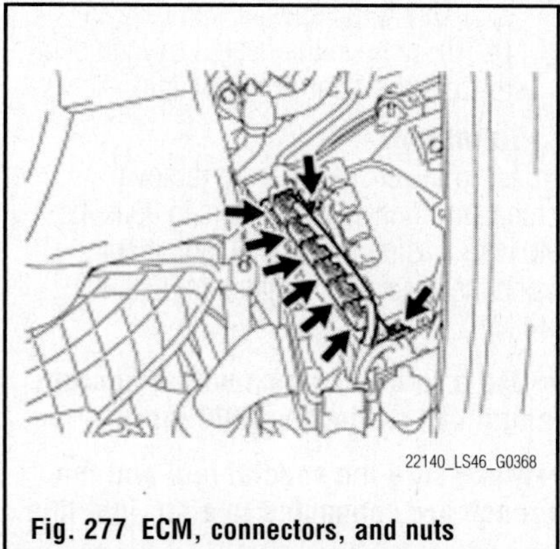

Fig. 277 ECM, connectors, and nuts

4. Remove the air cleaner inlet cover.
5. Remove the left engine room side cover.

➡Wipe off any water on or around the engine room ECU cover.

➡Perform these procedures in a dry place away from rain, etc.

➡Do not allow water to enter the ECM through its connectors areas, screw areas, etc.

6. Remove the 3 bolts and engine room ECU cover.
7. Disconnect the 4 TCM connectors.
8. Remove the 2 nuts and TCM.
9. Disconnect the 6 ECM connectors.
10. Remove the 2 nuts and ECM.

To install:

11. Install the ECM with the 2 nuts and tighten to 49 inch lbs. (5.5 Nm).
12. Connect the 6 ECM connectors.
13. Install the TCM with the 2 nuts and tighten to 49 inch lbs. (5.5 Nm).
14. Connect the 4 TCM connectors.
15. Install the engine room ECU cover with the 3 bolts and tighten to 69 inch lbs. (7.8 Nm).
16. Install the left engine room side cover.
17. Install the air cleaner inlet cover.
18. Connect cable to negative battery terminal.
19. Install the right cowl top ventilator louver.
20. Perform reset memory. Perform the RESET MEMORY (AT initialization) when replacing the automatic transmission assembly, engine assembly or ECM. Initialization cannot be completed by only removing the battery.
21. Perform initialization.

ENGINE COOLANT TEMPERATURE (ECT) SENSOR

LOCATION

See Figure 278.

Refer to the accompanying illustration for sensor location.

REMOVAL & INSTALLATION

See Figure 279.

1. Remove the v-bank cover sub assembly.
2. Remove the air cleaner inlet cover sub assembly.
3. Remove the No. 1 air cleaner inlet.
4. Remove the intake air connector pipe.
5. Remove the No. 1 engine under cover.
6. Drain the engine coolant.
7. Remove the Engine Coolant Temperature sensor.
   a. Disconnect the engine coolant temperature sensor connector.
   b. Using a 19 mm ball joint lock nut wrench, remove the sensor and gasket.
   c. Remove the gasket from the ECT sensor.

To install:

8. To install, reverse the removal procedure. Tighten the sensor to 10 ft. lbs. (13 Nm) using the ball joint lock nut wrench. If not using the ball joint lock nut wrench, tighten to 14 ft. lbs. (20 Nm).

➡Do not use a tool such as an impact wrench or equivalent.

➡Use a torque wrench with a fulcrum length of 11.8 inches (300 mm).

➡Make sure the ball joint lock nut wrench and torque wrench are connected in a straight line.

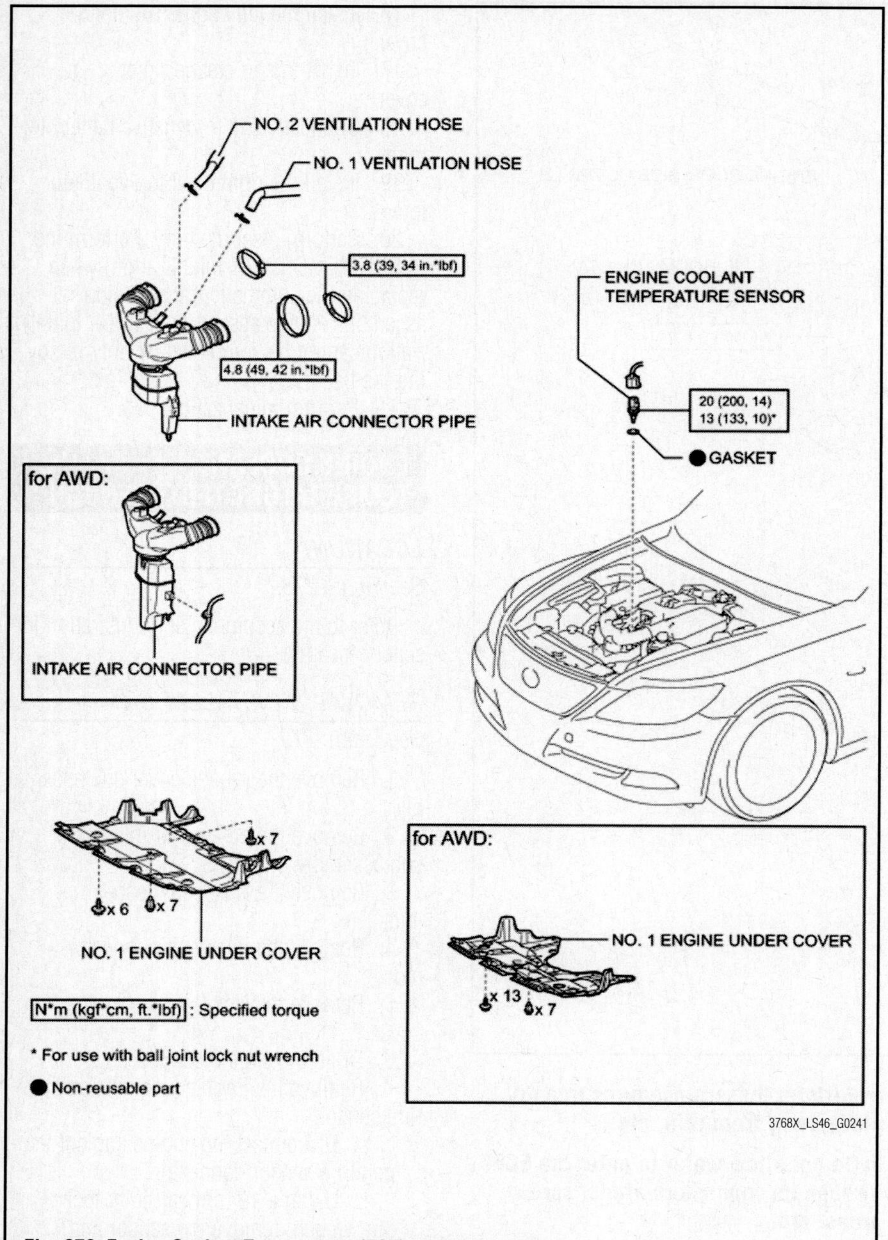

Fig. 278 Engine Coolant Temperature (ECT) sensor location

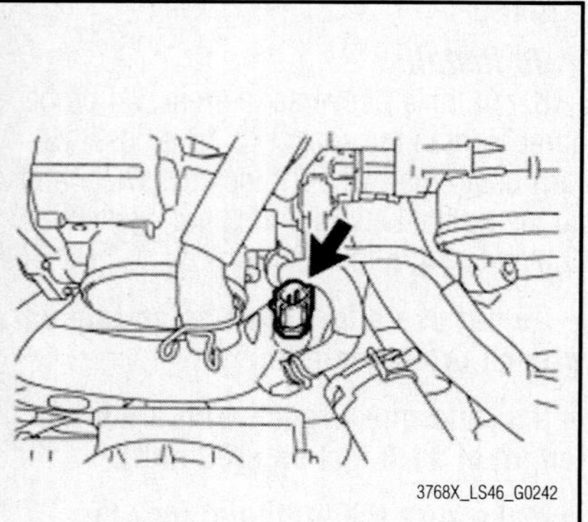

Fig. 279 Removing the ECT sensor

HEATED OXYGEN (HO2S) SENSOR

LOCATION

See Figures 280 and 281.

Refer to the accompanying illustrations for sensor location.

REMOVAL & INSTALLATION

2WD Models

1. Remove the left and right front door scuff plates.
2. Remove the left and right cowl side trim boards.
3. Remove the left and right front floor silencer pads.
4. Remove the rear No. 2 and No. 1 air duct.
5. Remove the heated oxygen sensors:
    a. Disconnect the heated oxygen sensor connector.
    b. Remove the grommet.
    c. Using SST (SST: 09224-00010), remove the heated oxygen sensor from the front exhaust pipe assembly.

To install:

6. Reverse removal procedure, and tighten to 32 ft. lbs. (44 Nm).

AWD Models

See Figures 282 through 284.

1. Remove the heated oxygen sensor (bank 1, sensor 2).
    a. Disconnect the heated oxygen sensor connector.
    b. Using the special tool (09224-00010), remove the heated oxygen sensor from the front exhaust pipe.
2. Remove the front door scuff plate RH.
3. Remove the cowl side trim board RH.
4. Remove the front floor silencer pad RH.
5. Remove the rear No. 2 air duct.
6. Remove the heated oxygen sensor (bank 2, sensor 2).
    a. For the interior side, disconnect the heated oxygen sensor connector.
    b. For the exterior side, remove the grommet.
    c. Using the special tool (09224-00010), remove the heated oxygen sensor from the front exhaust pipe.

To install:

7. To install, reverse the removal procedure. Tighten the sensors to 30 ft. lbs. (40 Nm) if using the special tool. If not using the special tool tighten to 32 ft. lbs. (44 Nm).

➡Use a torque wrench with a fulcrum length of 11.8 inches (300 mm).

➡Make sure the special tool and the wrench are connected in a straight line.

INTAKE AIR TEMPERATURE (IAT) SENSOR

LOCATION

The Intake Air Temperature (IAT) sensor is mounted on the Mass Air Flow (MAF) meter.

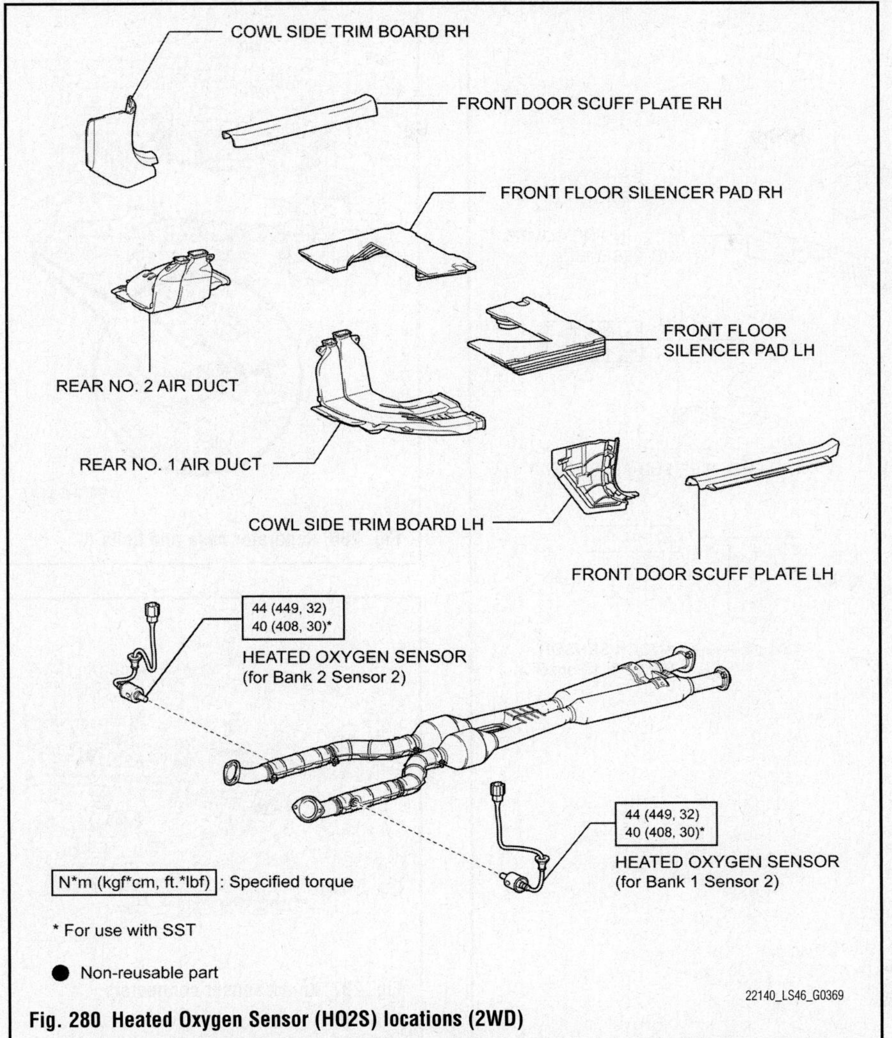

Fig. 280 Heated Oxygen Sensor (HO2S) locations (2WD)

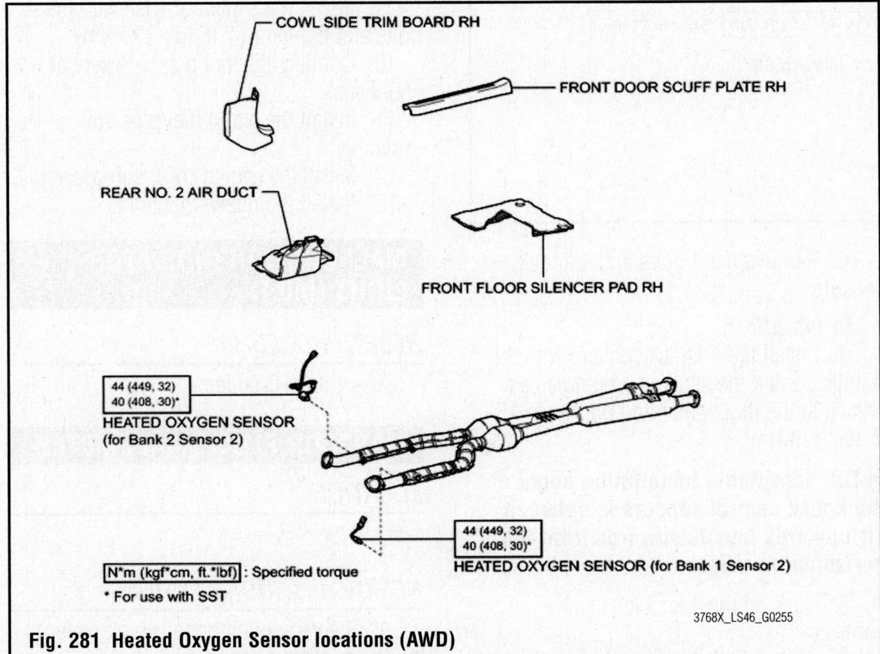

Fig. 281 Heated Oxygen Sensor locations (AWD)

Fig. 282 Removing the heated oxygen sensor (bank 1, sensor 2)

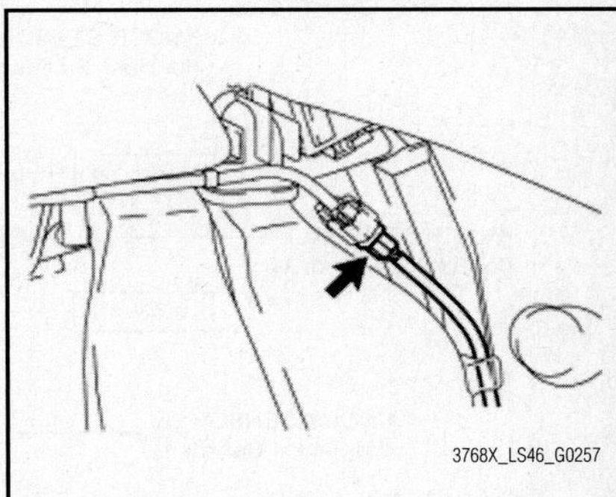

Fig. 283 Disconnecting the heated oxygen sensor connector (bank 2, sensor 2)

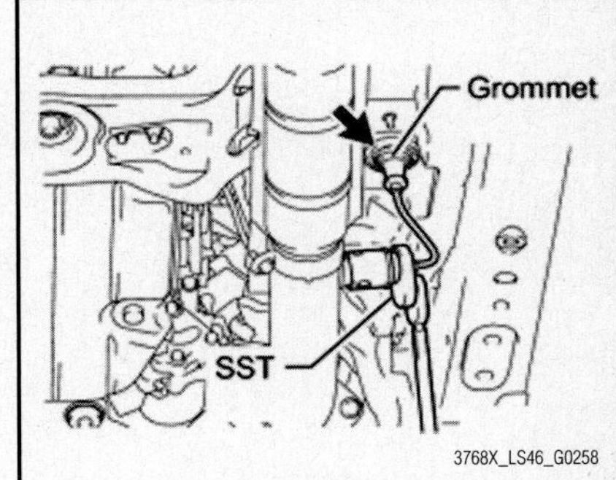

Fig. 284 Removing the heated oxygen sensor (bank 2, sensor 2)

REMOVAL & INSTALLATION

1. Remove air flow meter.
2. Remove Intake Air Temperature (IAT) sensor.

To install:

3. Reverse removal procedure.

KNOCK SENSOR (KS)

LOCATION

See Figure 285.

Refer to the accompanying illustration for sensor location.

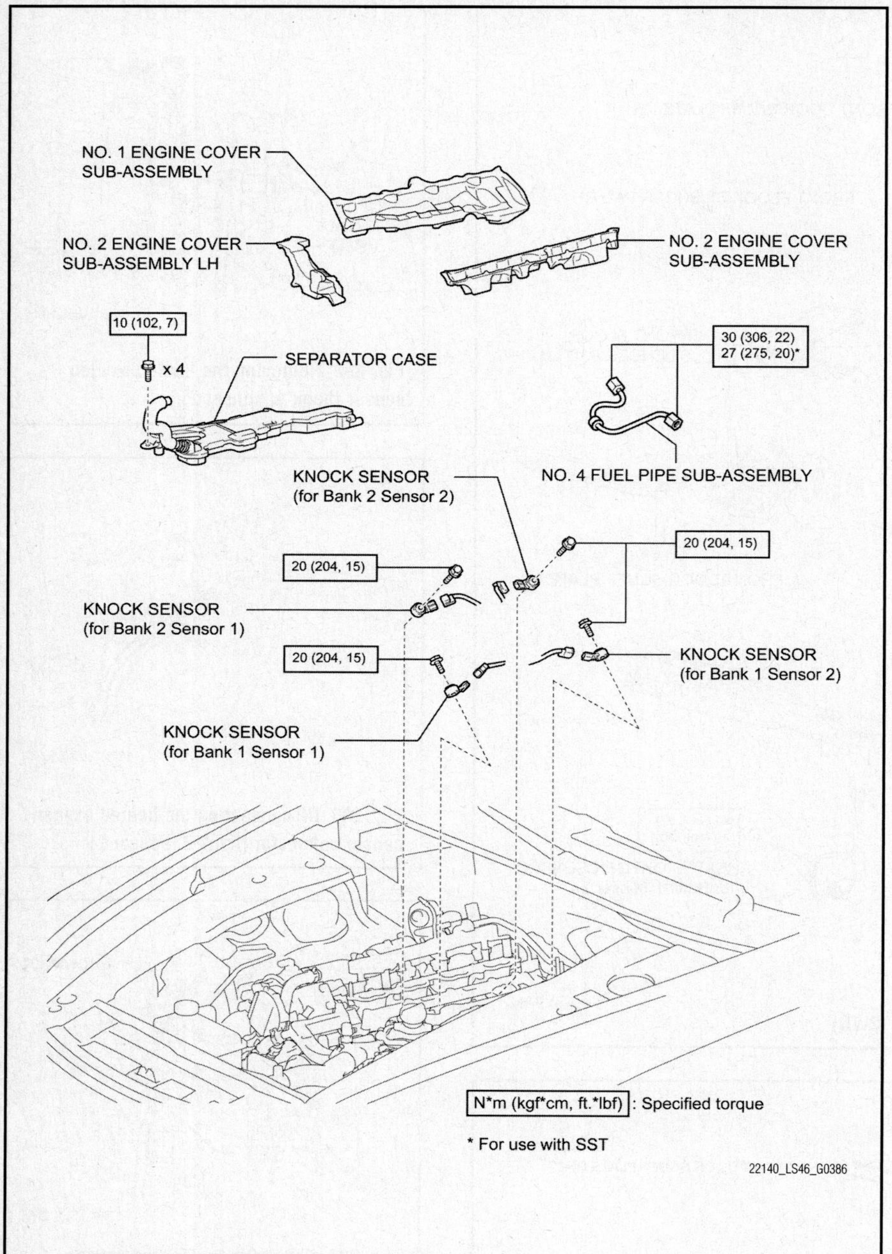

Fig. 285 Knock Sensor (KS) location

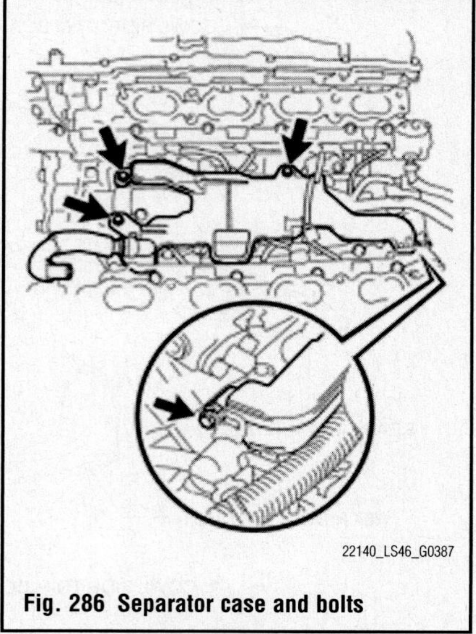

Fig. 286 Separator case and bolts

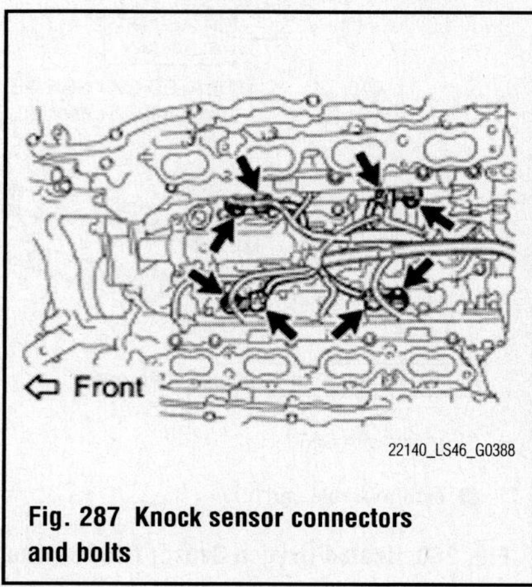

Fig. 287 Knock sensor connectors and bolts

REMOVAL & INSTALLATION

See Figures 286 through 288.

1. Remove the intake manifold.
2. Remove the engine cover sub-assembly.
3. Remove the No. 4 fuel pipe sub-assembly.
4. Remove the separator case, as follows:
   a. Disconnect the fuel pressure sensor connector.
   b. Remove the 4 bolts and separator case.
5. Disconnect the 4 knock sensor connectors.
6. Remove the 4 bolts and 4 knock sensors.

To install:

7. Install the 4 knock sensors with the 4 bolts so that the sensors are angled as shown in the illustration and tighten to 15 ft. lbs. (20 Nm).

➡The acceptable installation angle of the knock control sensors is between 10°upwards and downwards from the horizontal position.

8. Connect the 4 knock sensor connectors.
9. Install the separator case with the 4 bolts and tighten to 7 ft. lbs. (10 Nm).
10. Connect the fuel pressure sensor connector.
11. Install the No. 4 fuel pipe sub-assembly.
12. Install the engine cover sub-assembly.
13. Install the intake manifold.

MALFUNCTION INDICATOR LIGHT (MIL)

RESET PROCEDURE

Clear the DTC codes.

MASS AIR FLOW (MAF) METER

LOCATION

See Figure 289.

REMOVAL & INSTALLATION

1. Before servicing the vehicle, refer to the Precautions Section.

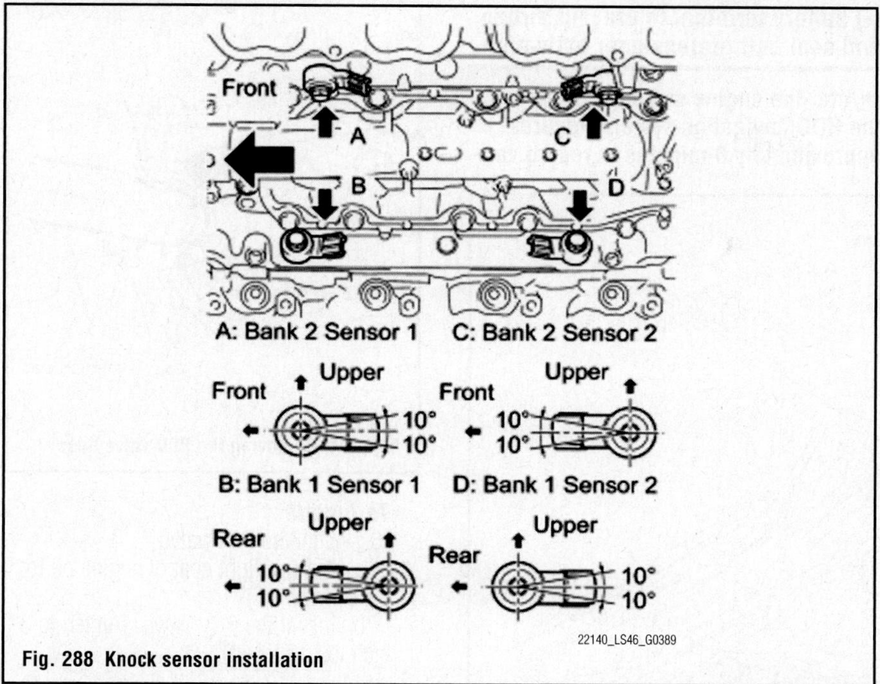

Fig. 288 Knock sensor installation

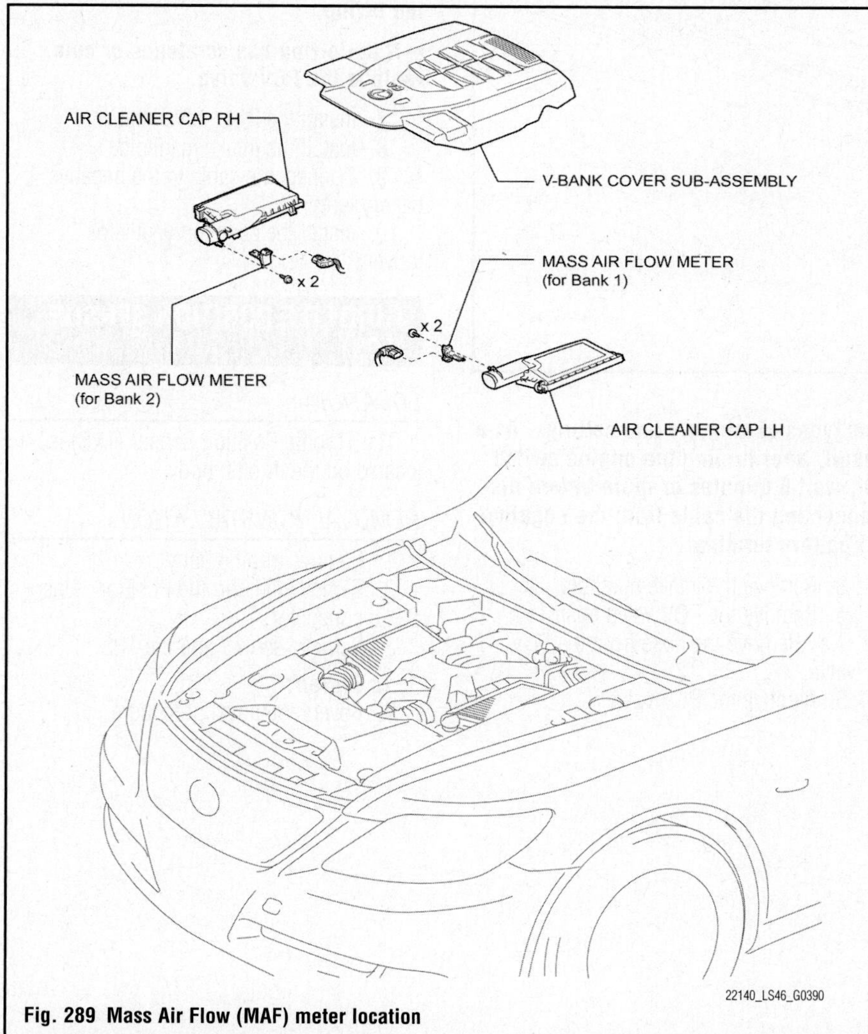

Fig. 289 Mass Air Flow (MAF) meter location

WARNING

After the engine switch is turned off, the HDD navigation system requires approximately 6 minutes recording various types of memory and settings. As a result, after turning the engine switch off, wait 6 minutes or more before disconnecting the cable from the negative battery terminal.

2. Remove the right cowl top ventilator louver.
3. Disconnect cable from negative battery terminal.

CAUTION

Wait at least 90 seconds after disconnecting the cable from the negative battery terminal to prevent airbag and seat belt pretensioner activation.

4. Remove the V-bank cover sub-assembly.
5. Remove the left and right air cleaner caps:
   a. Disconnect the MAF meter connector.
   b. Remove the clamps, loosen the hose clamps and remove the air cleaner caps.
6. Remove the 2 screws and MAF meter from Bank 1.
7. Remove the 2 screws and MAF meter from Bank 2.

To install:

8. Install each MAF meter with the 2 screws and tighten to 9 inch lbs. (1 Nm).

➡If the screw is tightened excessively, the screw hole may be damaged. Make sure the O-ring is not pinched.

9. Install the right and left air cleaner caps to the air cleaner hose assembly with the hose clamps and tighten to 34 inch lbs. (3.8 Nm). Insert the protrusion of the air cleaner hose into the hole of the hose clamp.

➡The hose clamp can be tightened within the range of 18 inch lbs. (2 Nm) to 49 inch lbs. (5.5 Nm).

10. Install the air cleaner caps to the air cleaner case with 2 clips each.
11. Connect the MAF meter connector.
12. Install the V-bank cover sub-assembly.
13. Connect cable to negative battery terminal.
14. Install the right cowl top ventilator louver.
15. Perform initialization.

POSITIVE CRANKCASE VENTILATION (PCV) VALVE

LOCATION

See Figure 290.

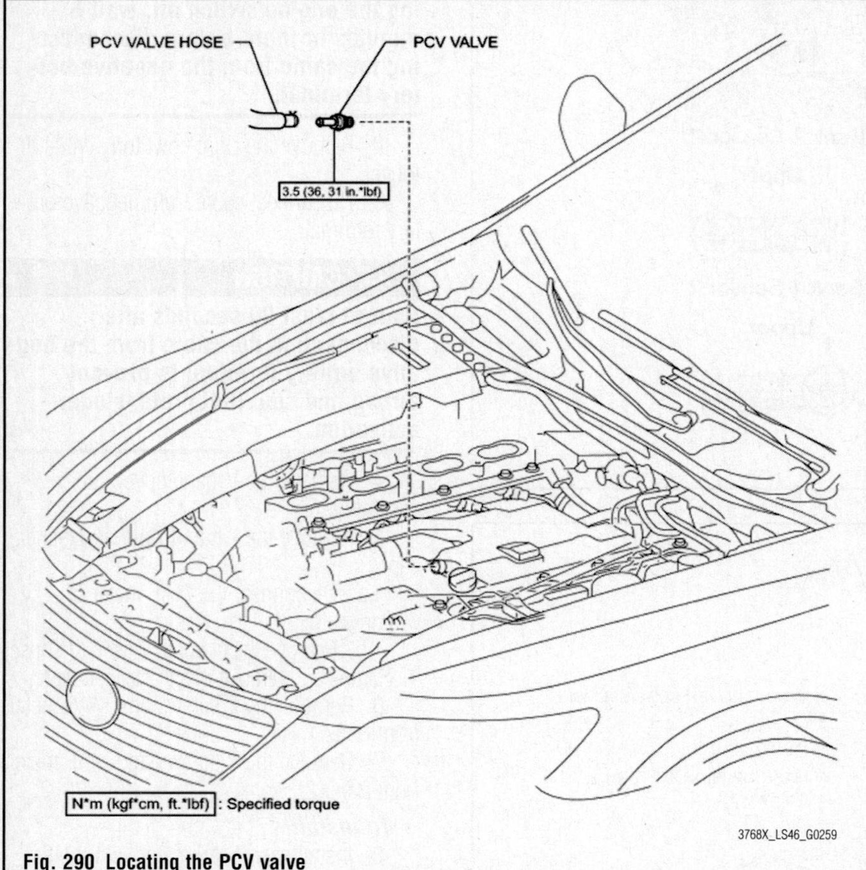

Fig. 290 Locating the PCV valve

REMOVAL & INSTALLATION

See Figure 291.

1. Remove the cowl top ventilator louver RH.
2. Disconnect the cable from the negative battery terminal.

CAUTION

Wait at least 90 seconds after disconnecting the cable from the negative (-) battery terminal to prevent airbag and seat belt pretensioner activation.

➡After the engine switch is turned off, the HDD navigation system requires approximately 6 minutes to record various types of memory and settings. As a result, after turning the engine switch off, wait 6 minutes or more before disconnecting the cable from the negative (-) battery terminal.

3. Remove the intake manifold.
4. Remove the PCV valve hose.
   a. Remove the hose from the PCV valve.
5. Remove the PCV valve.

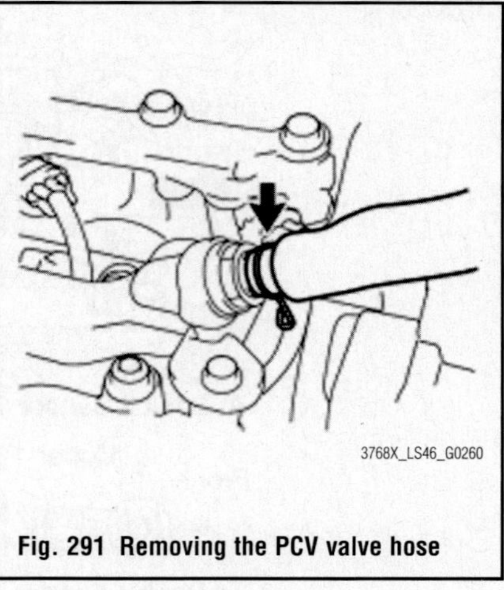

Fig. 291 Removing the PCV valve hose

To install:

6. Install the PCV valve.
   a. Apply a light coat of engine oil to the o-ring.
   b. Install the PCV valve. Tighten to 31 inch lbs. (3.5 Nm).

➡When reusing the PCM valve, inspect the o-ring.

➡If the o-ring has scratches or cuts, replace the PCV valve.

7. Install the PCV valve hose.
8. Install the intake manifold.
9. Connect the cable to the negative battery terminal.
10. Install the cowl top ventilator louver RH.

THROTTLE POSITION SENSOR (TPS)

LOCATION

The Throttle Position Sensor (TPS) is located on the throttle body.

REMOVAL & INSTALLATION

1. Remove throttle body.
2. Remove the Throttle Position Sensor (TPS) connector.
3. Remove screws and the TPS.

To install:

4. Reverse removal procedure.

FUEL SYSTEM SERVICE PRECAUTIONS

Safety is the most important factor when performing not only fuel system maintenance but any type of maintenance. Failure to conduct maintenance and repairs in a safe manner may result in serious personal injury or death. Maintenance and testing of the vehicle's fuel system components can be accomplished safely and effectively by adhering to the following rules and guidelines.

- To avoid the possibility of fire and personal injury, always disconnect the negative battery cable unless the repair or test procedure requires that battery voltage be applied.
- Always relieve the fuel system pressure prior to disconnecting any fuel system component (injector, fuel rail, pressure regulator, etc.), fitting or fuel line connection. Exercise extreme caution whenever relieving fuel system pressure to avoid exposing skin, face and eyes to fuel spray. Please be advised that fuel under pressure may penetrate the skin or any part of the body that it contacts.
- Always place a shop towel or cloth around the fitting or connection prior to loosening to absorb any excess fuel due to spillage. Ensure that all fuel spillage (should it occur) is quickly removed from engine surfaces. Ensure that all fuel soaked cloths or towels are deposited into a suitable waste container.
- Always keep a dry chemical (Class B) fire extinguisher near the work area.
- Do not allow fuel spray or fuel vapors to come into contact with a spark or open flame.
- Always use a back-up wrench when loosening and tightening fuel line connection fittings. This will prevent unnecessary stress and torsion to fuel line piping.
- Always replace worn fuel fitting O-rings with new Do not substitute fuel hose or equivalent where fuel pipe is installed.

Before servicing the vehicle, make sure to also refer to the precautions in the beginning of this section as well.

FUEL SYSTEM PRESSURE

RELIEVING

✻✻ CAUTION

Perform the following procedures to prevent fuel from spilling out before removing any fuel system parts.

✻✻ CAUTION

Do not disconnect any part of the fuel system until you have discharged the fuel system pressure.

✻✻ CAUTION

Pressure will still remain in the fuel line even after performing the following procedures. Even after discharging the fuel pressure, place a piece of cloth or equivalent over fittings as you separate them to reduce the risk of fuel spray on yourself or in the engine compartment.

✻✻ CAUTION

Allow the engine to cool down before performing the following procedures, as the high pressure side's fuel pressure does not decrease until the engine cools down.

➡Perform these procedures with the engine coolant temperature at 140°F (60°C) or less.

1. Disconnect the fuel pump connector.
2. Remove the left rear seat assembly.
3. Remove the rear floor service hole cover.
4. Disconnect the fuel pump connector.
5. Start the engine.
6. After the engine stops, turn the ignition switch off.

➡DTC P3190, P3191, P0171, P0172 and/or POAOF may be set.

7. Loosen the fuel tank cap, then discharge the pressure in the fuel tank completely.
8. Reconnect the fuel pump connector.
9. Install the rear seat.
10. Install the rear floor service hole cover.
11. Check that there are no fuel leaks from the fuel system after doing any maintenance or repairs.

FUEL FILTER

REMOVAL & INSTALLATION

Fuel filters are in the fuel tank along with the fuel pump.

FUEL LEVEL SENDING UNIT

REMOVAL & INSTALLATION

See Figures 292 and 293.

1. Before servicing the vehicle, refer to the Precautions Section.
2. Remove the fuel pump from the vehicle.
3. Remove the fuel sender gauge assembly, as follows:
   a. Disconnect the fuel sender gauge connector.
   b. Press down on the sender gauge claw labeled A. Then slide the sender gauge upward.

➡Do not touch the sender plate's resistance plate area or the contact area.

To install:

4. Set the fuel sender gauge to the fuel sub-tank. Then slide the sender gauge downward to install it.

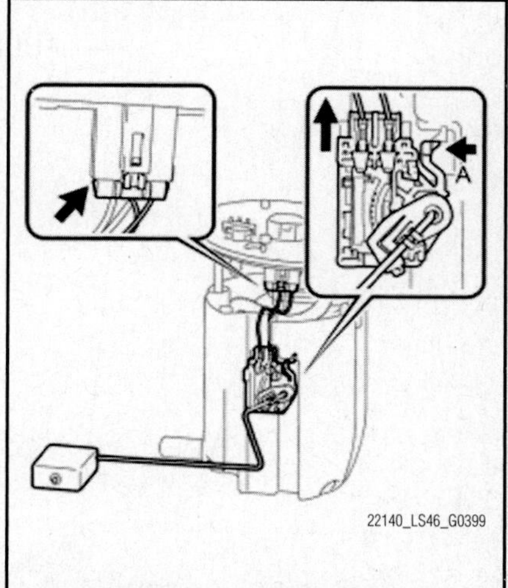

Fig. 292 Fuel sender gauge removal

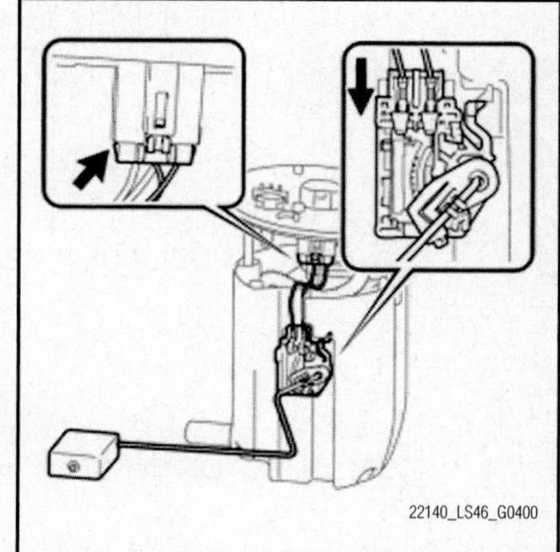

Fig. 293 Fuel sender gauge installation

a. Connect the fuel sender gauge connector to the fuel suction plate.

➡Do not touch the sender plate's resistance plate area or the contact area.

5. Install the fuel pump.

FUEL PUMP

REMOVAL & INSTALLATION

See Figures 294 through 298.

1. Before servicing the vehicle, refer to the Precautions Section.
2. Discharge fuel system pressure.

⁂ WARNING

After the engine switch is turned off, the HDD navigation system requires approximately 6 minutes to record various types of memory and settings. As a result, after turning the engine switch off, wait 6 minutes or more before disconnecting the cable from the negative battery terminal.

3. Remove the right cowl top ventilator louver.
4. Disconnect cable from negative battery terminal.

⁂ CAUTION

Wait at least 90 seconds after disconnecting the cable from the negative battery terminal to prevent airbag and seat belt pretensioner activation.

5. Remove the left rear seat assembly, for power seat or ottoman.
6. Remove the rear seat cushion assembly, for fixed seat.
7. Remove the rear floor No. 2 service hole cover and disconnect the connector from the fuel suction with pump and gauge tube assembly,

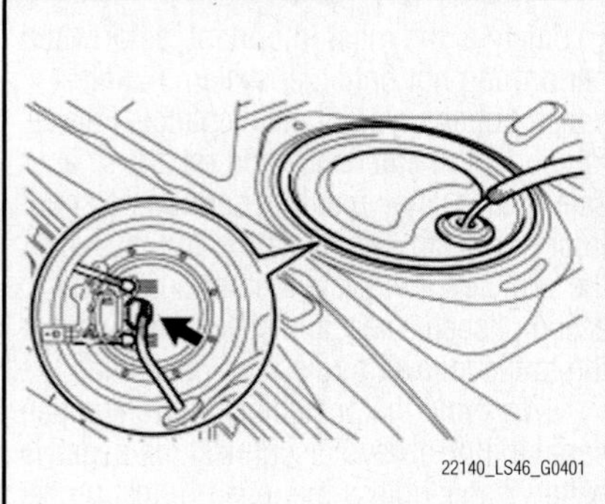

Fig. 295 Rear floor No. 2 service hole cover

FUEL INJECTOR (for Port Injection)
FUEL PUMP (for High Pressure)
INJECTOR DRIVER
FUEL PUMP
FUEL PRESSURE REGULATOR
FUEL PUMP RESISTOR
FUEL PUMP (for High Pressure)
FUEL INJECTOR (for Direct Injection)
22140_LS46_G0409

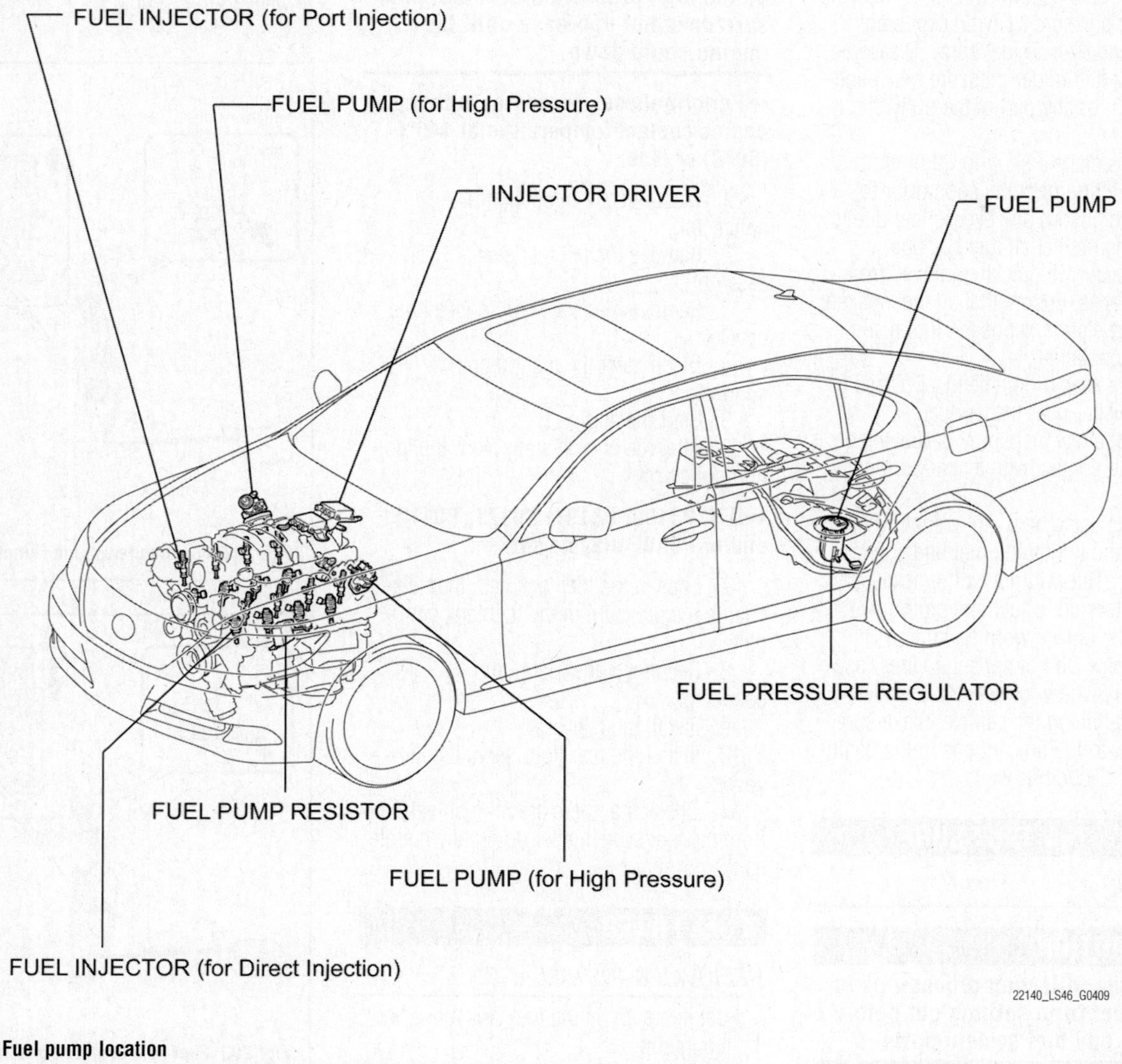

Fig. 294 Fuel pump location

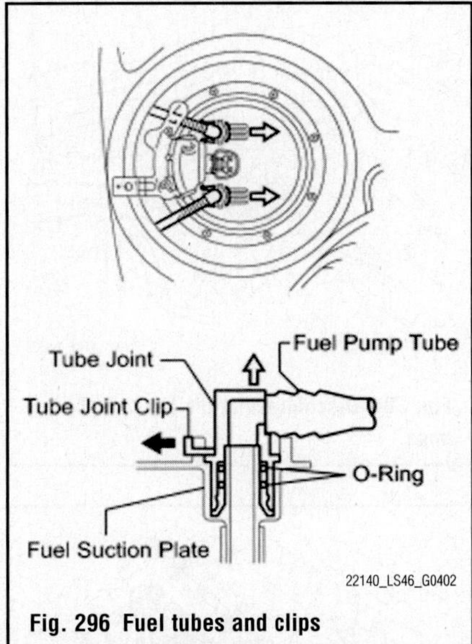

Fig. 296 Fuel tubes and clips

8. Remove the fuel suction with pump and gauge tube assembly, as follows:
   a. Disconnect the fuel main tube and fuel return tube. Remove the 2 tube joint clips and 2 fuel tubes.

➡**Remove any dirt and foreign matter on the fuel tube joint.**

➡**Do not allow any scratches or foreign matter on the parts when disconnecting them, as the fuel tube joint contains the O-rings that seal the plug.**

➡**Perform this work by hand. Do not use any tools.**

➡**Do not forcibly bend, twist or turn the nylon tube.**

➡**Protect the disconnected part by covering it with a plastic bag and tape after disconnecting the fuel tubes.**

   b. Remove the 8 bolts and set plate.
   c. Disconnect the fuel hose.
   d. Remove the fuel suction with pump and gauge tube from the fuel tank.

➡**Make sure that the sender gauge arm does not bend. Do not damage the fuel suction with pump and gauge tube.**

   e. Remove the gasket.

To install:

9. Apply a light coat of gasoline to a new gasket, and install it to the fuel tank.
10. Connect the fuel hose, and set the fuel suction with pump and gauge.
11. Install the set plate with the 8 bolts and tighten to 53 inch lbs. (6 Nm).

➡**Align the protrusion of the set plate with the cutout of the fuel suction pump and gauge.**

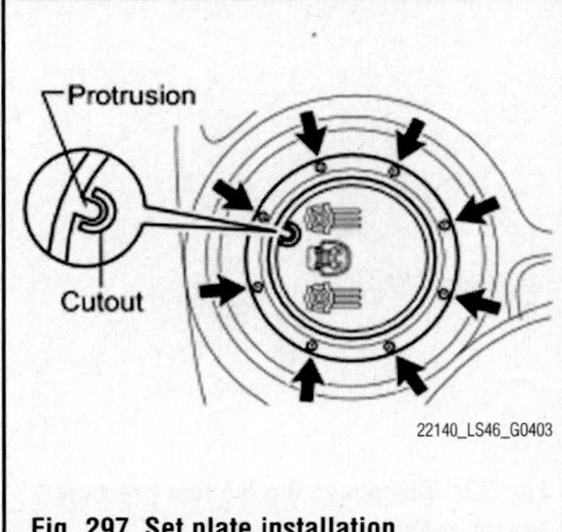

Fig. 297 Set plate installation

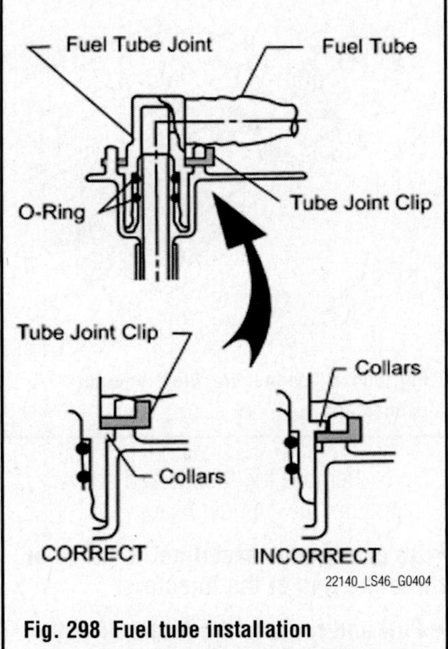

Fig. 298 Fuel tube installation

12. Install the fuel main tube and fuel return tube with the 2 tube joint clips.

➡**Check that there are no scratches or foreign objects on the connecting parts, the fuel tube joint is inserted securely, and the tube joint clips are on the collars of the fuel tube joints.**

13. Connect the connector to the fuel suction with pump and gauge tube assembly.
14. The remainder of installation is the reverse of removal.
15. Perform initialization.
16. Check for fuel leaks.

FUEL RAIL AND INJECTOR

REMOVAL & INSTALLATION

Direct Injection

See Figures 299 through 324.

✱✱ CAUTION

Do not smoke or be near an open flame when working on the fuel system.

✱✱ CAUTION

Keep gasoline away from rubber or leather parts.

✱✱ CAUTION

Do not allow fuel to spray when removing the pipe between the high pressure side fuel pump and the fuel injector. The fuel in the pipe is highly pressurized.

1. Discharge the fuel system pressure.
2. Disconnect the cable from the negative battery terminal.

➡**After the engine switch is turned off, the HDD navigation system requires approximately 6 minutes to record various types of memory and settings. As a result, after turning the engine switch off, wait 6 minutes or more before disconnecting the cable from the negative (-) battery terminal.**

3. Remove the fuel pressure pulsation damper assembly.
4. Remove the No. 3 fuel pipe sub assembly.
5. Remove the No. 2 fuel pipe sub assembly.
6. Remove the No. 1 engine cover sub assembly.
7. Remove the No. 2 engine cover sub assembly.
8. Remove the No. 2 engine cover sub assembly LH
9. Remove the No. 4 fuel pipe sub assembly.
   a. Using a 19 mm union nut wrench, remove the fuel pipe.

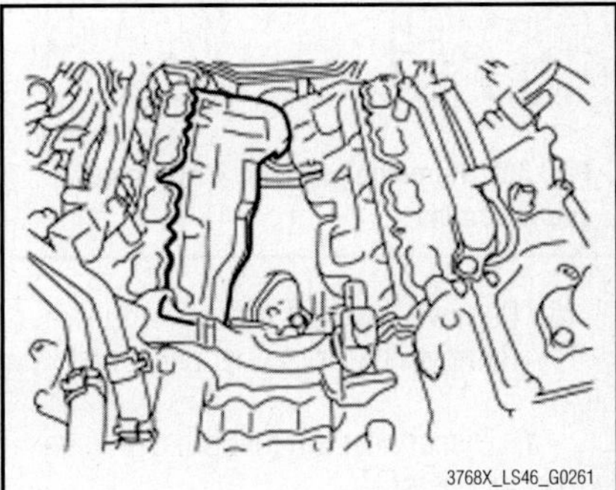

Fig. 299 Removing the No. 1 engine cover sub assembly

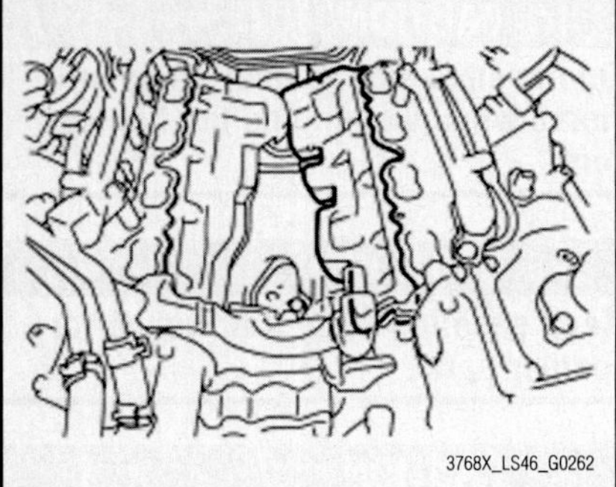

Fig. 300 Removing the No. 2 engine cover sub assembly

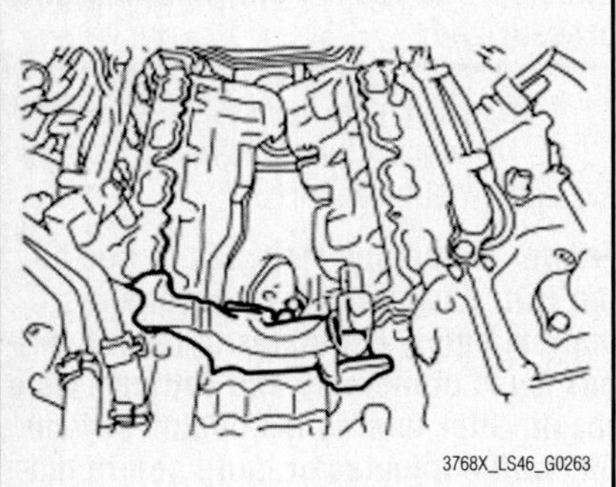

Fig. 301 Removing the No. 2 engine cover sub assembly LH

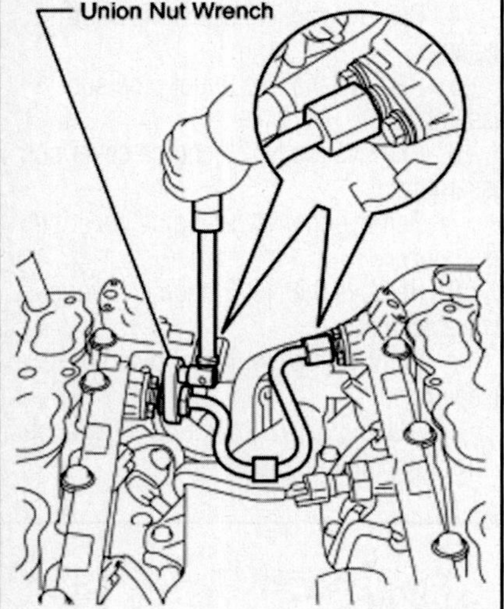

Fig. 302 Removing the No. 4 fuel pipe sub assembly

10. Remove the PCV hose.
11. Remove the No. 2 fuel delivery pipe.
    a. Disconnect the fuel pressure sensor connector.
    b. Disconnect the 4 injector connectors.

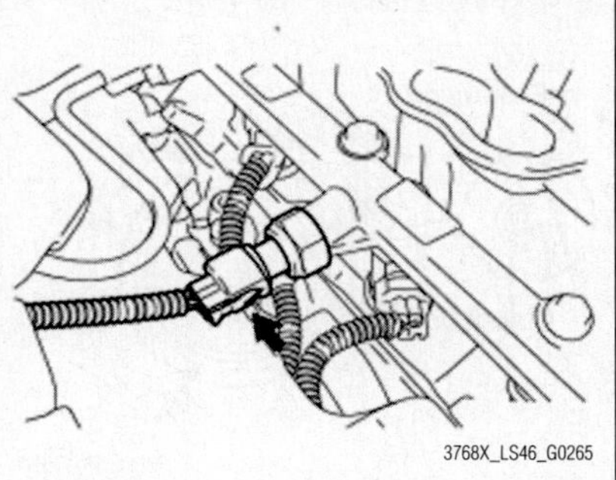

Fig. 303 Disconnecting the fuel pressure sensor connector

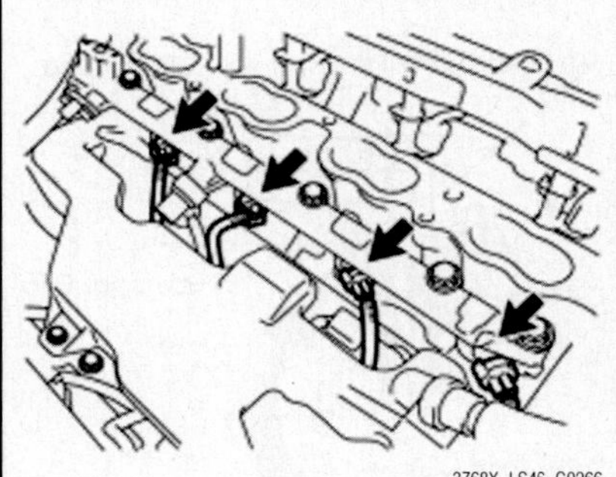

Fig. 304 Disconnecting the 4 injector connectors

c. Remove the 5 bolts and delivery pipe from the cylinder head.

➡Be extremely careful not to touch or strike the tips of the injectors.

➡Pull and remove the fuel delivery pipe in a straight line without tilting it.

d. Remove the 4 injector vibration insulators from the cylinder head.

12. Remove the fuel delivery pipe.
    a. Disconnect the No. 3 fuel hose.
    b. Disconnect the 4 injector connectors.

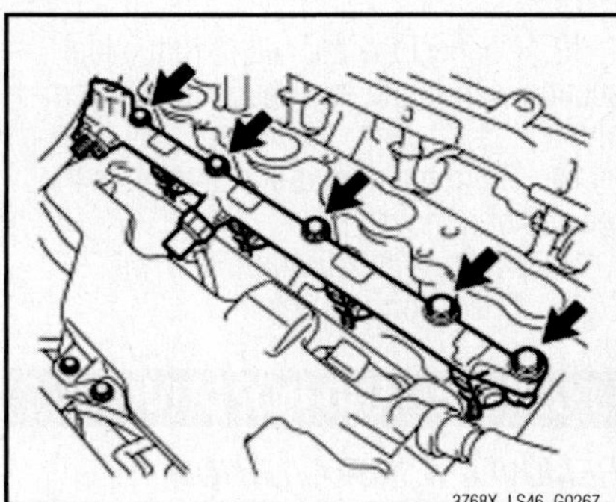

Fig. 305 Removing the delivery pipe from the cylinder head

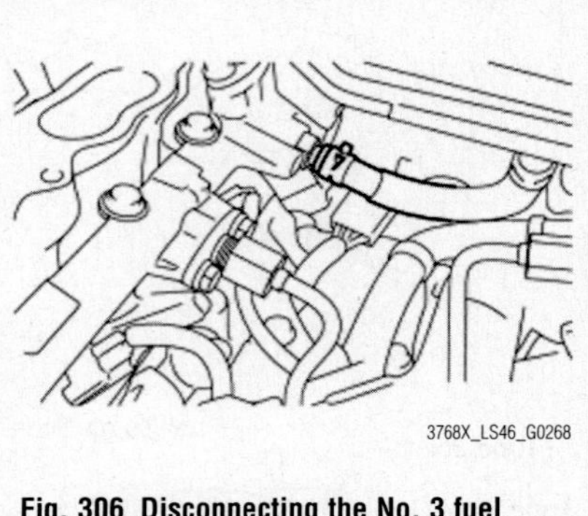

Fig. 306 Disconnecting the No. 3 fuel hose

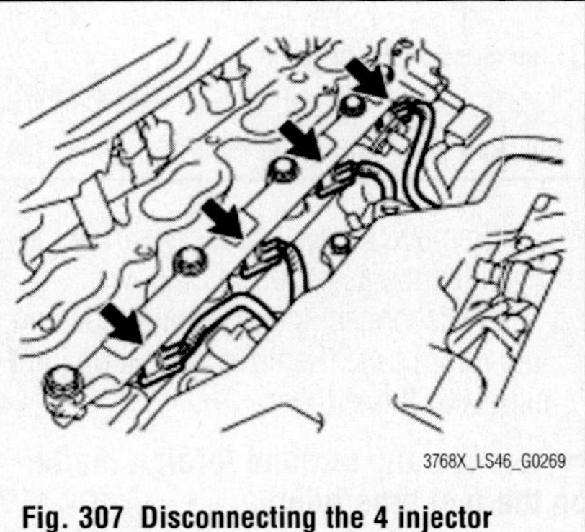

Fig. 307 Disconnecting the 4 injector connectors

c. Remove the 5 bolts and delivery pipe from the cylinder head.

➡Be extremely careful not to touch or strike the tips of the injectors.

➡Pull and remove the fuel delivery pipe in a straight line without tilting it.

13. Remove the fuel injector assembly.
    a. Secure the delivery pipe between aluminum plates in a vise and pull out the injector in a straight line.

➡Pull and remove the injector in a straight line to avoid damage to the seal surface of the delivery pipe o-ring.

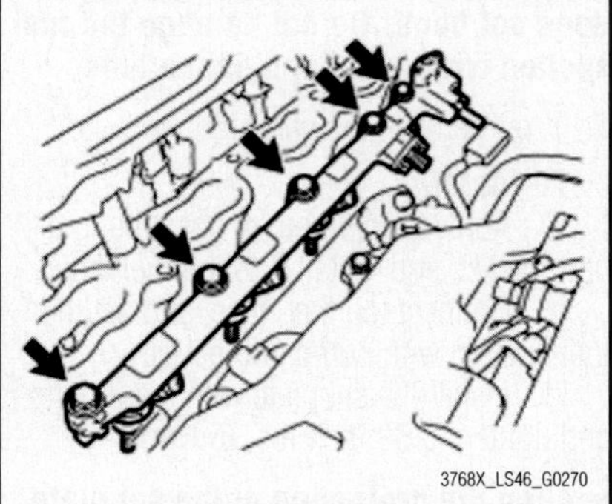

Fig. 308 Removing the delivery pipe front the cylinder head

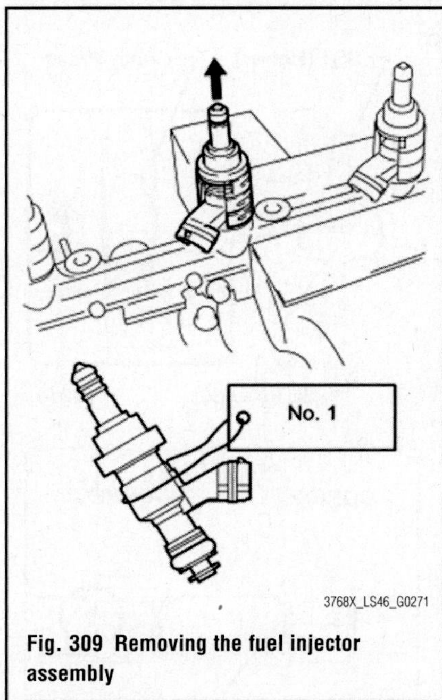

Fig. 309 Removing the fuel injector assembly

➡**For reinstallation, attach a tag or label to the injector shaft.**

b. Remove the nozzle holder clamp from the injector.

c. Remove the o-ring, backup rings and E-ring from the fuel injector.

14. Remove the fuel injector seal.

a. Using the tips of needle-nose pliers, pinch and pull one of the 2 injector seals at several points to stretch it. Repeat this for the other injector seal.

➡**Excessively pinching the injector seal may damage the groove of the injector.**

➡**If an injector is dropped or the tips of the injectors are struck, replace it with a new one.**

b. Remove the 2 injector seals from the injector.

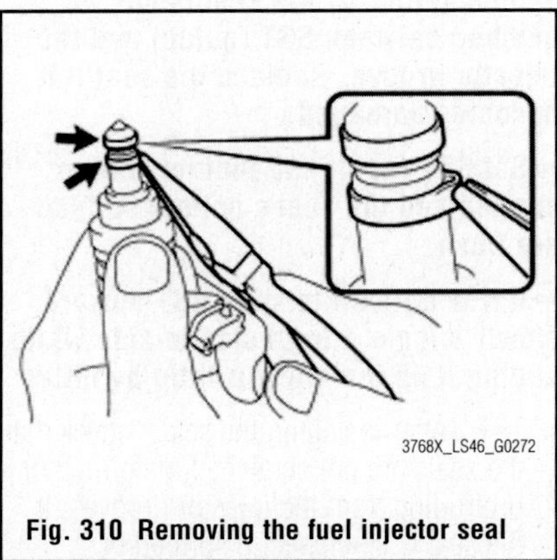

Fig. 310 Removing the fuel injector seal

To install:

15. Install the fuel injector seal.

a. Apply engine conditioner to the clean area shown in the illustration. Using a piece of cloth, clean carbon deposits from the injector and its grooves.

➡**Do not clean the tip of the injector.**

➡**Do not use a wire brush to clean the injector.**

➡**If an injector is dropped or the tips of the injectors is struck, replace it with a new one.**

b. Apply engine oil to the injector contact surface of SST (09260-39015) (guide). Then attach SST (guide) to the injector with the tapered inner portion facing the tip of the injector, as shown in the illustration.

c. Install a new injector seal to the special tool (09260-39015) (holder).

➡**Be careful not to install the injector seal to SST (holder) at an angle. Doing so will stretch the seal and correcting this problem is very complicated.**

d. Install SST (holder with injector seal) to the tip of the injector. Slide the seal downward into the injector groove (injector connector side) with your fingers, as shown in the illustration.

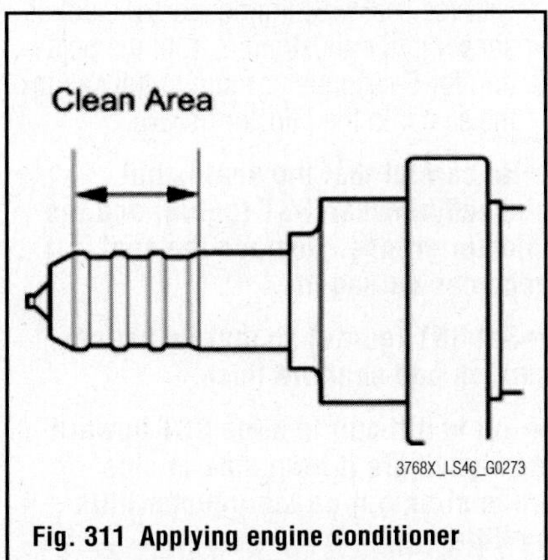

Fig. 311 Applying engine conditioner

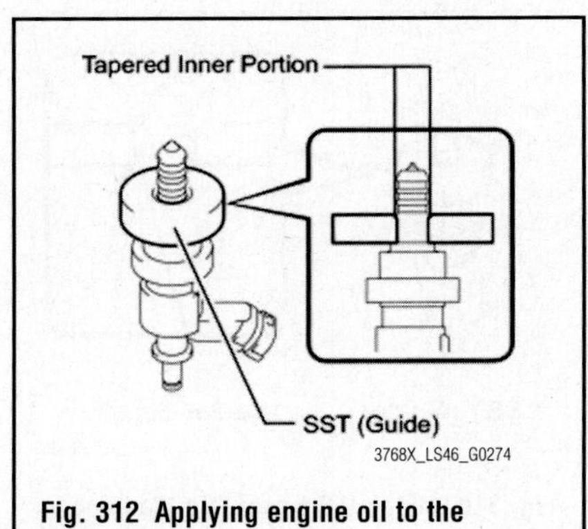

Fig. 312 Applying engine oil to the injector

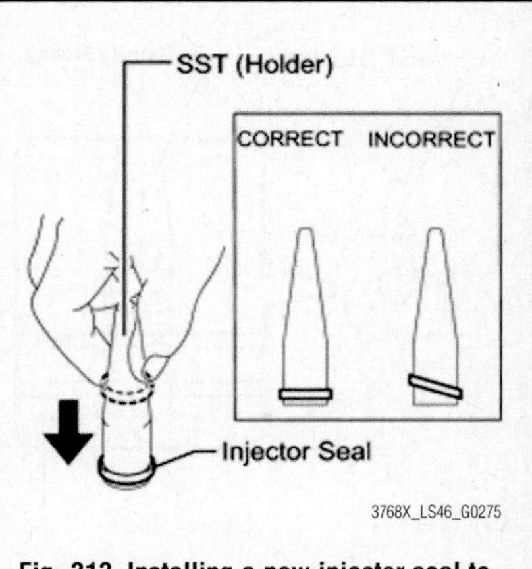

Fig. 313 Installing a new injector seal to the special tool

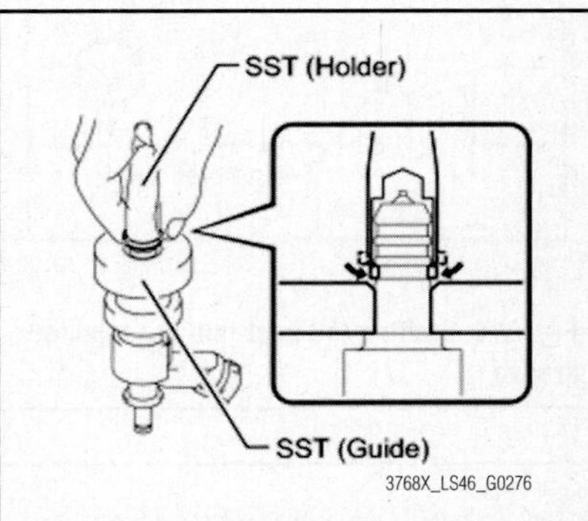

Fig. 314 Installing the special tool to the tip of the injector

➡**Check that the seal covers the circumference of the injector groove as shown in the illustration.**

e. Using SST (holder), gently press downward on the injector seal (injector connector side). Then slowly slide SST (guide) towards the injector tip to settle the seal into the injector groove.

➡**Be careful that the seal is not pinched between SST (guide) and the injector groove. Replace the seal if it becomes damaged.**

➡**When using SST (guide) to settle the seal into the groove, SST (guide) only needs to be slid upward to the position labeled A in the illustration.**

➡**After using SST (guide) to settle the seal into the groove, return SST (guide) to its position labeled B in the illustration.**

f. Install a new injector seal to the special tool (holder).

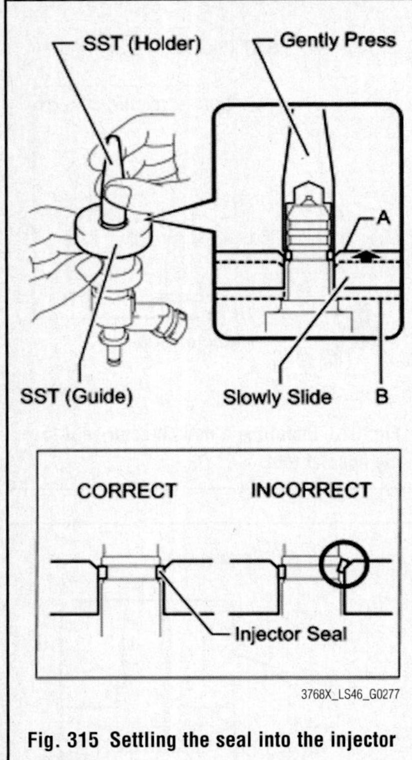

Fig. 315 Settling the seal into the injector groove

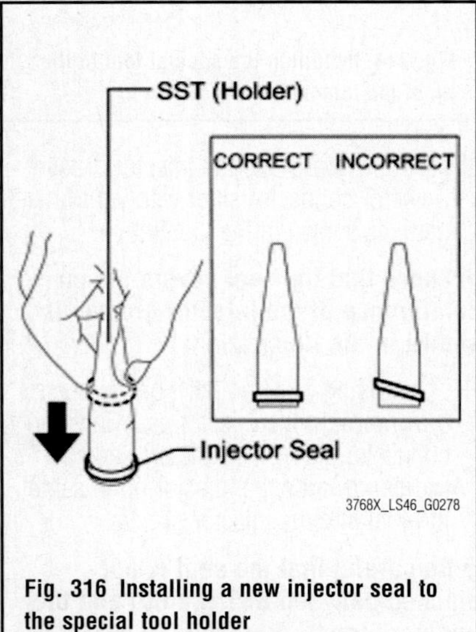

Fig. 316 Installing a new injector seal to the special tool holder

➡Be careful not to install the injector seal to SST (holder) at an angle. Doing so will stretch the seal and correcting this problem is very complicated.

g. Install SST (holder with injector seal) to the tip of the injector. Slide the seal downward into the injector groove (injector tip side) with your fingers, as shown in the illustration.

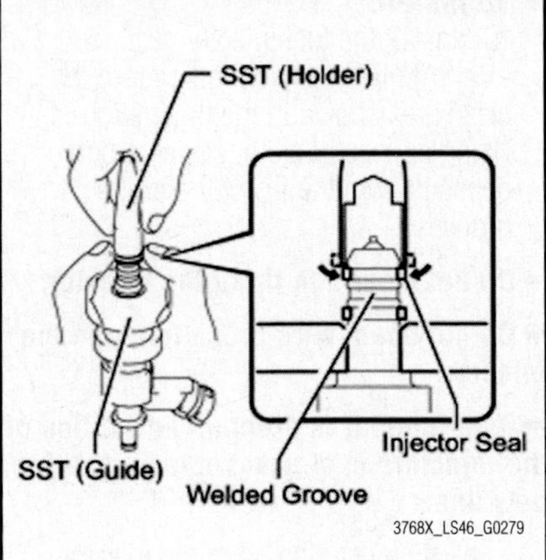

Fig. 317 Installing the special tool (holder with injector seal) to the tip of the injector

➡Make sure that the seal does not slip into the welded groove of the injector shown in the illustration. If it does, replace it with a new one.

➡Check that the seal covers the circumference of the injector groove as shown in the illustration.

h. Slowly slide SST (guide) towards the tip of the injector. When the injector contact surface of SST (guide) aligns with the seal (injector connector side) as shown in the illustration, hold the position for 5 seconds or more to fully align the seal into the injector groove.

➡Be careful that the seal is not pinched between SST (guide) and the injector groove. Replace the seal if it becomes damaged.

➡Set SST (guide) so that its bottom surface and seal are flush.

➡If it is difficult to slide SST upward, slowly wiggle it from side to side while sliding it up the injector little by little.

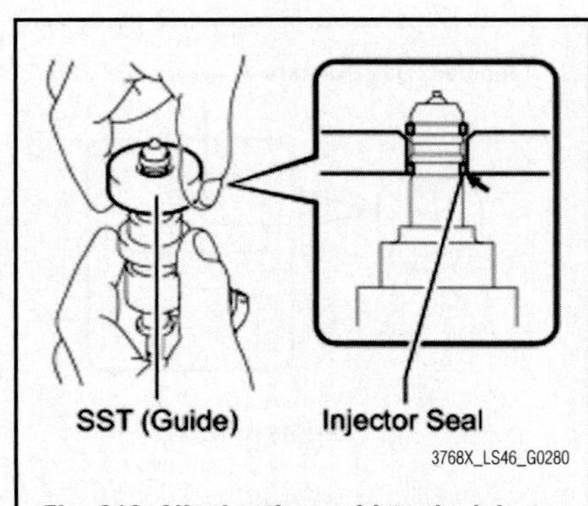

Fig. 318 Aligning the seal into the injector groove

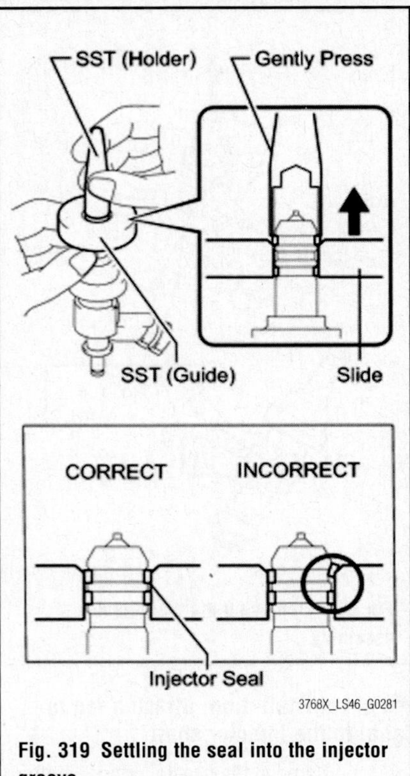

Fig. 319 Settling the seal into the injector groove

i. Using SST (holder), gently press downward on the injector seal (injector tip side). Then slowly slide SST (guide) towards the injector tip to settle the seal into the injector groove.

➡Be careful that the seal is not pinched between SST (guide) and the injector groove. Replace the seal if it becomes damaged.

j. Slowly slide SST (guide) towards the tip of the injector. When the injector contact surface of SST (guide) aligns with the seal (injector tip side) as shown in the illustration, hold the position for 5 seconds or more to fully align the seal into the injector groove.

➡Be careful that the seal is not pinched between SST (guide) and the injector groove. Replace the seal if it becomes damaged.

➡Set SST (guide) so that its bottom surface and the seal's bottom surface are flush.

➡If it is difficult to slide SST upward, slowly wiggle it from side to side while sliding it up the injector little by little.

k. After installing the seals, check that the seals are not scratched, deformed or protruding from the injector grooves. If the seal is scratched, deformed or

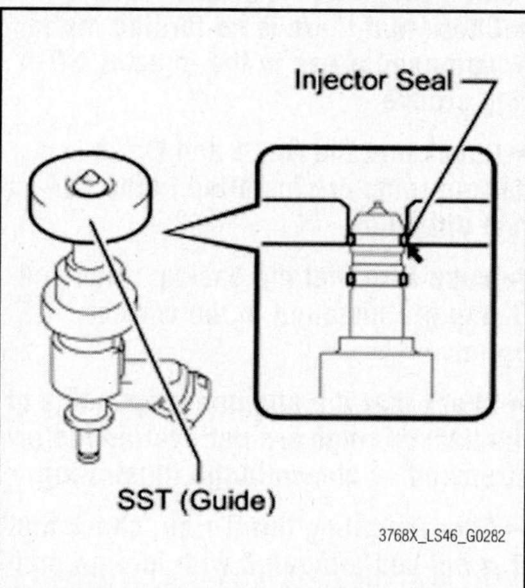

Fig. 320 Aligning the seal into the injector groove

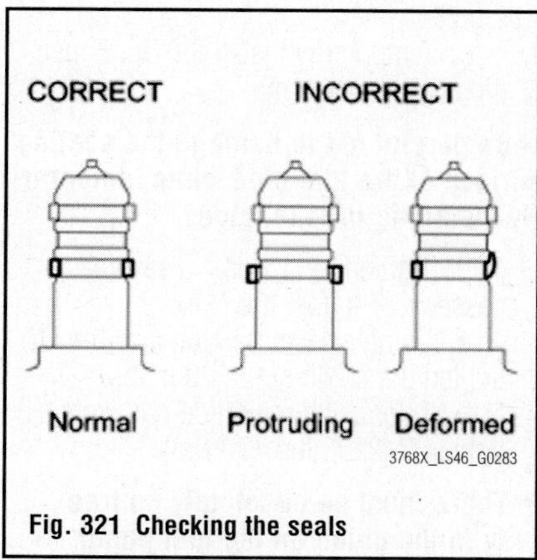

Fig. 321 Checking the seals

protruding from the groove, replace it with a new one.

16. Install the fuel injector assembly.
   a. Install a new o-ring, backup rings (No. 1, No. 2 and No. 3) and a new E-ring to the fuel injector.

➡Check that there is no foreign matter or damaged areas in the injector O-ring groove.

➡Check that the No. 1 and No. 2 backup rings are installed in the correct direction.

➡Make sure that the backup rings and O-ring are installed in the correct order.

➡Check that the alignment openings of the backup rings are not overlapped or stretched as shown in the illustration.

➡After installing the O-ring, check that it is not contaminated with foreign matter and is not damaged.

   b. install the injector nozzle holder clamp. Apply gasoline to the o-ring. Install the nozzle holder clamp by aligning the protruding part of the clamp to the notch of the delivery pipe.

➡The injector connectors can be differentiated by the color of the insulators, which are black, yellow, green or brown. Install the injectors so that insulators are the same color for a given bank.

➡Make sure that there is no gap between the delivery pipe and clamp.

➡Check that there is no foreign matter or damage to the injector insertion hole of the delivery pipe.

➡Insert the injector straight into the delivery pipe without tilting it.

17. Install the No. 2 fuel delivery pipe.
   a. Install 4 new injector vibration insulators to the cylinder head.
   b. Apply lubricant to the injector seal and installation holes.
   c. Install the delivery pipe (with injectors) to the cylinder head with the 5 bolts. Tighten to 15 ft. lb s. (21 Nm).

➡If an injector is dropped or the tips of the injectors is struck, replace it with a new one.

➡Check that there is no foreign matter or damage to the injector insertion hole of the cylinder head.

➡When inserting the fuel delivery pipe, push it in evenly without tilting it.

   d. Connect the 4 injector connectors.
   e. Connect the fuel pressure sensor connector.

18. Install the fuel delivery pipe.
   a. Install 4 new injector vibration insulators to the cylinder head.
   b. Apply lubricant to the injector seal and installation holes.

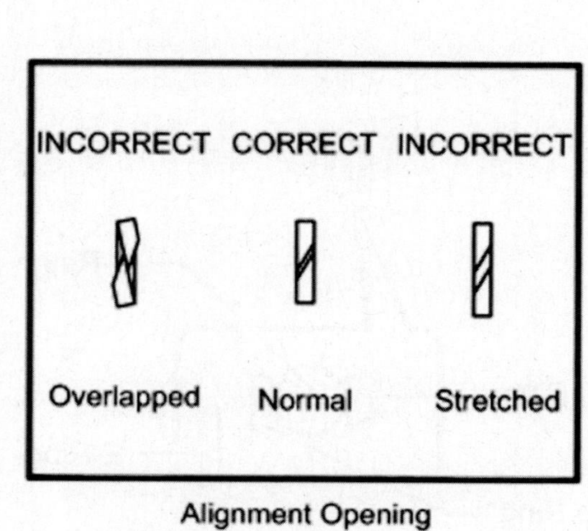

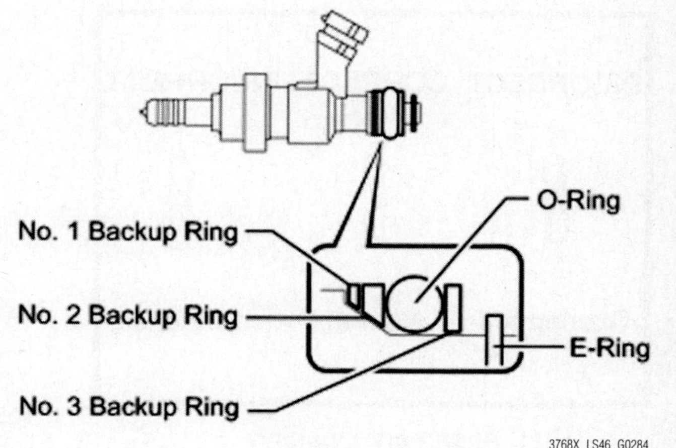

Fig. 322 Installing new o-rings, backup rings and a new E-ring

c. Install the delivery pipe (with injectors) to the cylinder head with the 5 bolts. Tighten to 15 ft. lbs. (21 Nm).

➡If an injector is dropped or the tips of the injectors is struck, replace it with a new one.

➡Check that there is no foreign matter or damage to the injector insertion hole of the cylinder head.

➡When inserting the fuel delivery pipe, push it in evenly without tilting it.

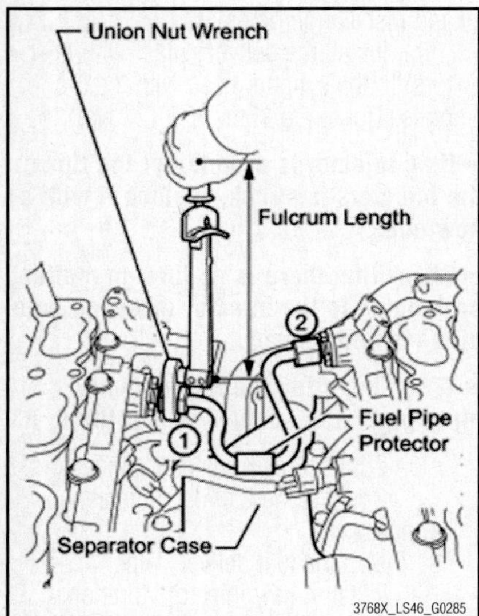

Fig. 323 Installing the No. 4 fuel pipe sub assembly

d. Connect the 4 injector connectors.

e. Connect the No. 3 fuel hose.

19. Install the PCV hose.

20. Install the No. 4 fuel pipe sub assembly.

a. Temporarily install the fuel pipe.

b. Using a 19 mm union nut wrench, tighten the fuel pipe in the order shown in the illustration. If tightening without a union nut wrench, tighten to 22 ft. lbs. (30 Nm). If tightening with a union nut wrench, tighten to 20 ft. lbs. (27 Nm).

➡After installing the fuel pipe, check that the fuel pipe protector contacts with the separator case.

➡Use a torque wrench with a fulcrum length of 11.8 inch (30 cm).

➡Make sure to use a union nut wrench and torque wrench are connected in a straight line.

21. Install the No. 2 engine cover sub assembly LH.

22. Install the No. 2 engine cover sub assembly.

23. Install the No. 1 engine cover sub assembly.

24. Install the No. 3 fuel pipe sub assembly.

a. Install a new O-ring, new backup rings (No. 1 and No. 2) and new E-ring to the fuel injector as shown in the illustration.

➡Check that there is no foreign matter or damaged areas in the injector's O-ring groove.

➡Check that the No. 1 and No. 2 backup rings are installed in the correct direction.

➡Make sure that the backup rings and O-ring are installed in the correct order.

➡Check that the alignment openings of the backup rings are not overlapped or stretched as shown in the illustration.

➡After installing the O-ring, check that it is not contaminated with foreign matter and is not damaged.

b. Temporarily install the No. 3 fuel pipe to the delivery pipe with the 2 bolts.

c. Temporarily install the No. 3 fuel pipe to the fuel pump.

➡Be careful not to damage the sealing surface of the fuel pipe when temporarily installing the fuel pipe.

d. Tighten the 2 bolts in several passes to 7 ft. lbs. (10 Nm).

e. Using a 19 mm union nut wrench, tighten the union nut to 20 ft. lbs. (27 Nm). If not using the union nut wrench tighten to 22 ft. lbs. (30 Nm).

➡There must be absolutely no free play in the union on the fuel pump

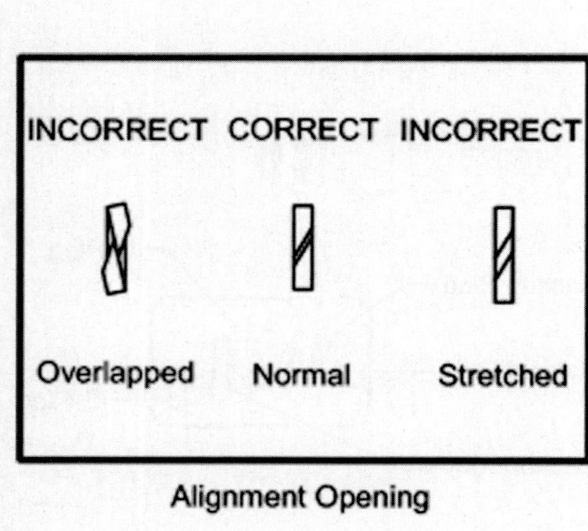

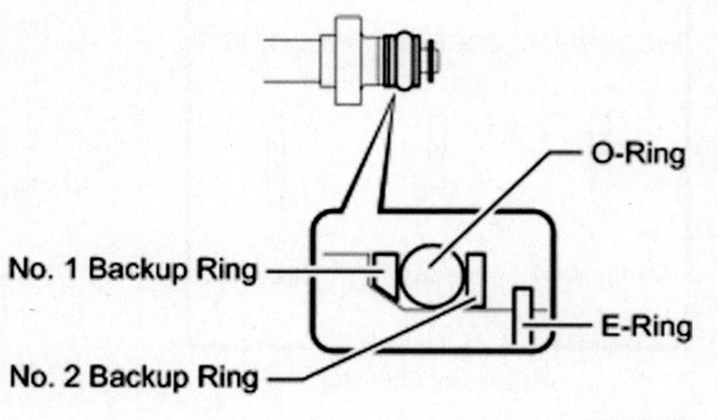

Fig. 324 Installing the No. 3 fuel pipe sub assembly

side. If the union on the fuel pump side has free play, replace the fuel pump.

➡Use a torque wrench with a fulcrum length of 11.8 inches (30 cm).

➡Make sure the union nut wrench and torque wrench are connected in a straight line.

f. Connect the fuel pump connector.

25. Install the No. 2 fuel pipe sub assembly.

➡The installation procedures are the same as the No. 3 fuel pipe.

26. Install the fuel pressure pulsation damper assembly.
27. Connect the cable to the negative battery terminal.
28. Perform the initialization.
29. Add the engine coolant.
30. Inspect for coolant leaks.
31. Inspect for fuel leaks.

Port Injection

See Figures 325 through 330.

➡Do not smoke or be near an open flame when working on the fuel system.

➡Keep gasoline away from rubber or leather parts.

➡Do not allow fuel to spray when removing the pipe between the high pressure side fuel pump and the fuel injector. The fuel in the pipe is highly pressurized.

1. Discharge the fuel system pressure.
2. Disconnect the cable from the negative battery terminal.

➡After the engine switch is turned off, the HDD navigation system requires approximately 6 minutes to record various types of memory and settings. As a result, after turning the engine switch off, wait 6 minutes or more before disconnecting the cable from the negative (-) battery terminal.

3. Remove the intake manifold.
4. Disconnect the No. 3 fuel hose.
   a. Remove the pipe clamp.
   b. Pinch the tube connector and then pull out the fuel hose.

➡Check for any dirt and foreign matter contamination in the pipe and around the connector. Clean if necessary. Foreign matter may damage the O-rings or cause leaks in the seal between the pipe and connector.

➡Do not use any tools to separate the pipe and connector.

➡Do not forcefully bend or twist the nylon tube.

➡Check for any dirt and foreign matter on the pipe seal surface. Clean if necessary.

➡Put the pipe and connector ends in plastic bags to prevent damage and dirt contamination.

➡If the pipe and connector are stuck together, pinch the tube between your fingers and turn it carefully to free it. Then disconnect the hose.

5. Remove the fuel tube sub assembly.
   a. Lift up the retainer to release its lock.
   b. Pinch the tube connector and then pull out the fuel tube.

➡Check for any dirt and foreign matter contamination in the pipe and around the connector. Clean if necessary. Foreign matter may damage the O-rings or cause leaks in the seal between the pipe and connector.

➡Do not use any tools to separate the pipe and connector.

➡Do not forcefully bend or twist the nylon tube.

➡Check for any dirt and foreign matter on the pipe seal surface. Clean if necessary.

➡Put the pipe and connector ends in plastic bags to prevent damage and dirt contamination.

3768X_LS46_G0287

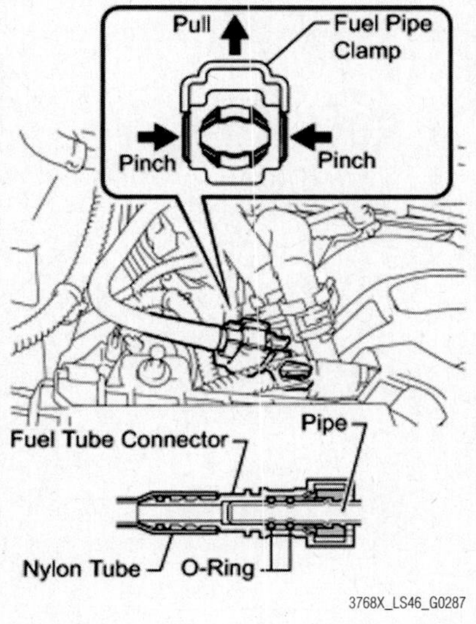

Fig. 325 Disconnecting the No. 3 fuel hose

➡If the pipe and connector are stuck together, pinch the tube between your fingers and turn it carefully to free it. Then disconnect the tube.

6. Remove the fuel delivery pipe.
   a. Disconnect the 2 delivery pipe connectors.
   b. Disconnect the 4 wire harness clamps.
   c. Remove the 4 bolts and 2 fuel delivery pipes.

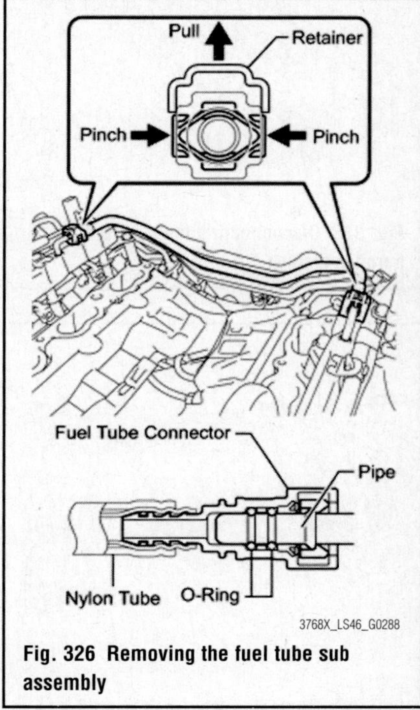

Fig. 326 Removing the fuel tube sub assembly

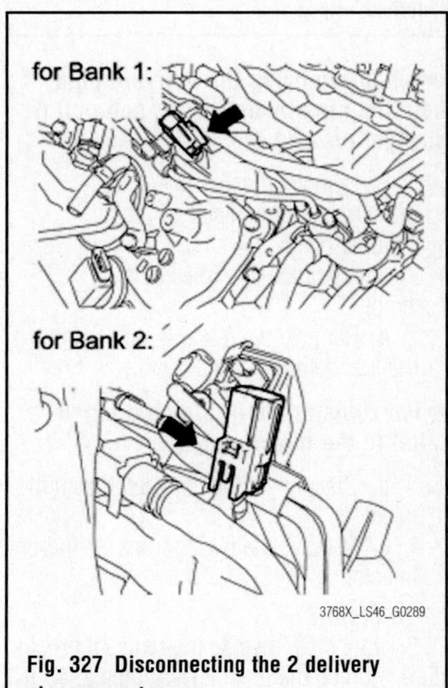

Fig. 327 Disconnecting the 2 delivery pipe connectors

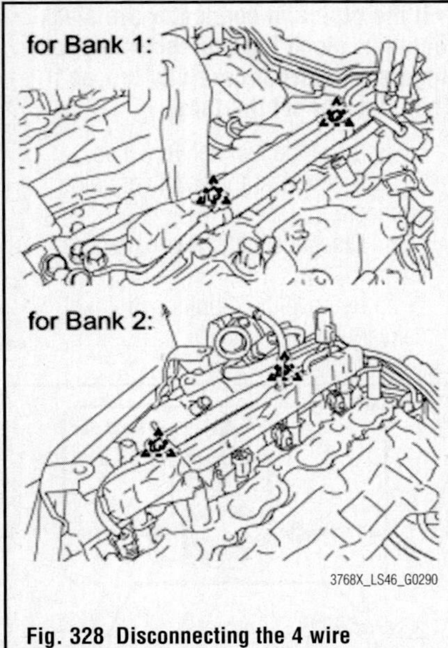

Fig. 328 Disconnecting the 4 wire harness clamps

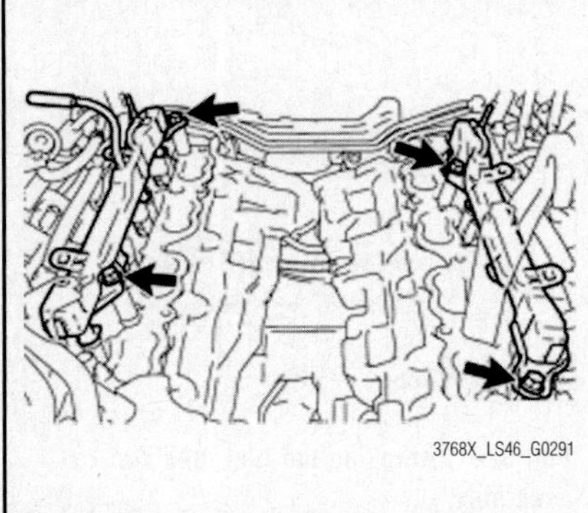

Fig. 329 Removing the 4 bolts and 2 fuel delivery pipes

➡**When removing the delivery pipe, hold the pipe by both ends and pull it straight upward.**

d. Remove the 4 delivery pipe spacers and 8 insulators from the cylinder head.

7. Remove the fuel injector assembly.

a. Remove the 8 fuel injectors from the fuel delivery pipes.

➡**For reinstallation, attach a tag or label to the injector shaft.**

b. Disconnect the connector from the injector.

c. Remove the o-ring from the fuel injector.

To install:

8. To install, reverse the removal procedure. Tighten the fuel delivery pipe to 15 ft. lbs. (21 Nm).

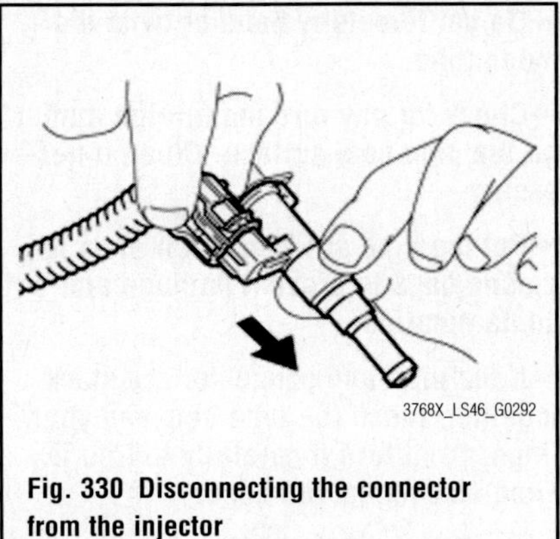

Fig. 330 Disconnecting the connector from the injector

FUEL TANK

REMOVAL & INSTALLATION

See Figures 331 through 336.

1. Before servicing the vehicle, refer to the Precautions Section.

2. Discharge fuel system pressure.

⁂ WARNING

After the engine switch is turned off, the HDD navigation system requires approximately 6 minutes to record various types of memory and settings. As a result, after turning the engine switch off, wait 6 minutes or more before disconnecting the cable from the negative battery terminal.

3. Remove the right cowl top ventilator louver.

4. Disconnect cable from negative battery terminal.

⁂ CAUTION

Wait at least 90 seconds after disconnecting the cable from the negative battery terminal to prevent airbag and seat belt pretensioner activation.

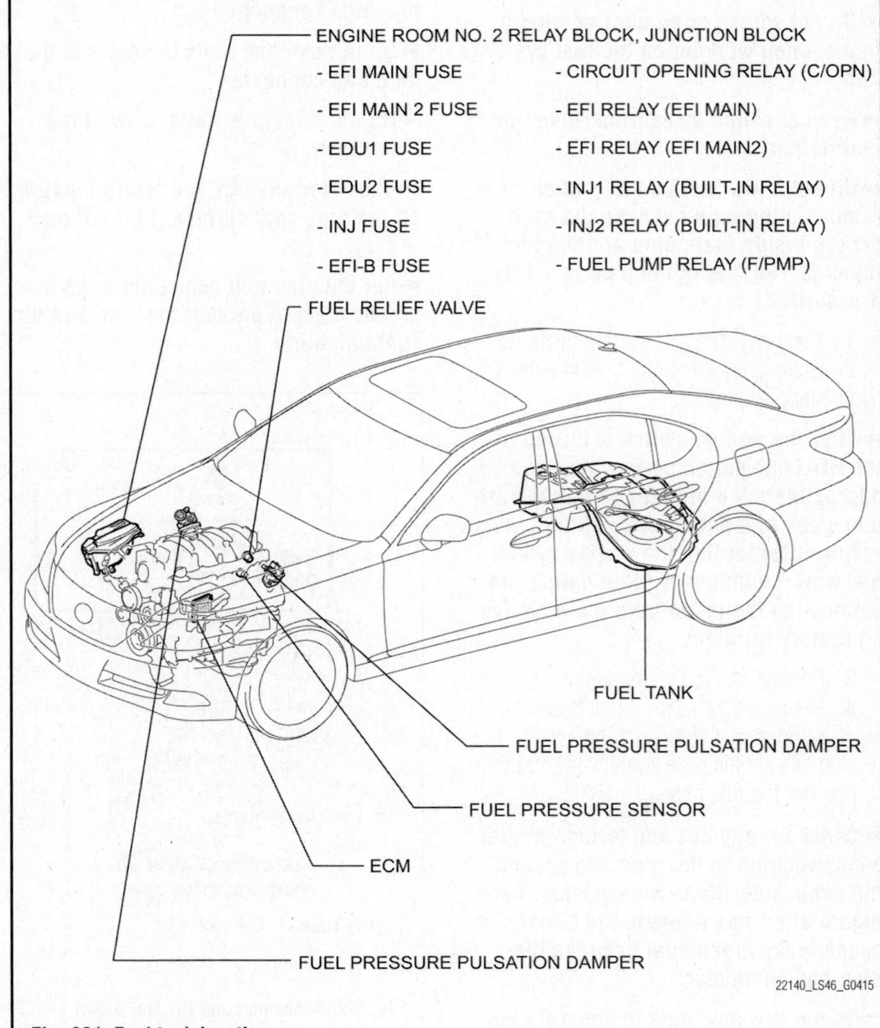

Fig. 331 Fuel tank location

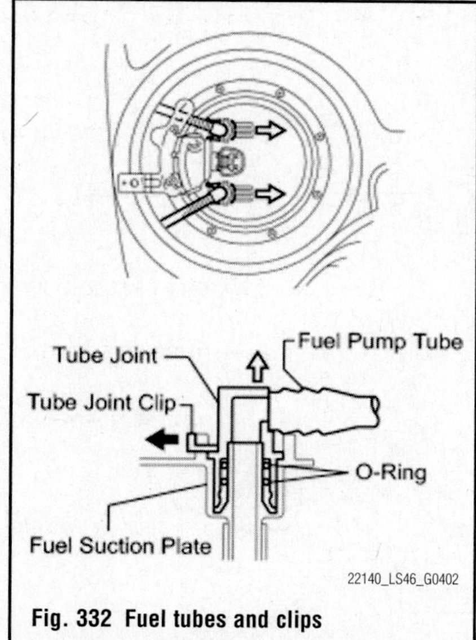

Fig. 332 Fuel tubes and clips

5. Remove the left rear seat assembly, for power seat or ottoman.

6. Remove the rear seat cushion assembly, for fixed seat.

7. Remove the rear floor No. 2 service hole cover and disconnect the connector from the fuel suction with pump and gauge tube assembly,

8. Disconnect the fuel main tube and fuel return tube. Remove the 2 tube joint clips and 2 fuel tubes.

➡Remove any dirt and foreign matter on the fuel tube joint.

➡Do not allow any scratches or foreign matter on the parts when disconnecting them, as the fuel tube joint contains the O-rings that seal the plug.

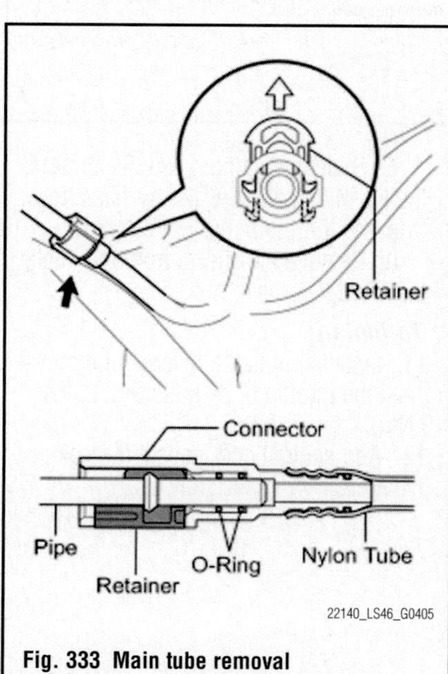

Fig. 333 Main tube removal

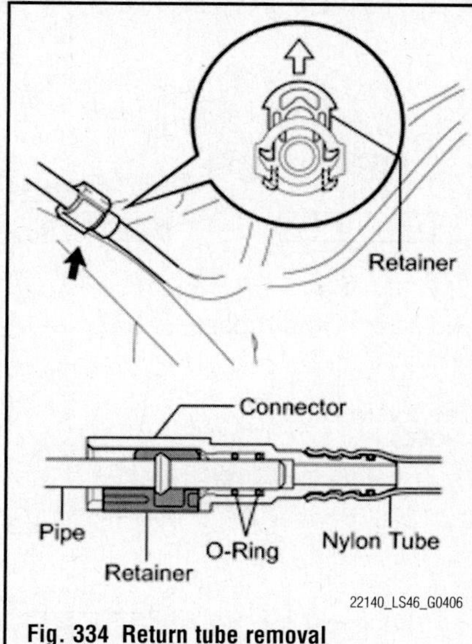

Fig. 334 Return tube removal

➡Perform this work by hand. Do not use any tools.

➡Do not forcibly bend, twist or turn the nylon tube.

➡Protect the disconnected part by covering it with a plastic bag and tape after disconnecting the fuel tubes.

9. Remove the rear service hole cover and disconnect the fuel sender gauge connector.

10. Remove the exhaust pipe.

11. Remove the propeller with center bearing shaft.

12. Disconnect the fuel main tube.

13. Disconnect the fuel return tube.

➡Check for any dirt and foreign matter contamination in the pipe, around the connector, and on the pipe seal surface. Foreign matter may damage the O-ring or cause leaks in the seal between the pipe and connector.

➡Do not use any tools to separate the pipe and connector.

➡Do not allow any scratches or foreign matter on the parts when disconnecting them, as the fuel tube joint contains the O-rings that seal the plug.

➡Do not forcibly bend, twist or turn the nylon tube.

➡Protect the disconnected part by covering it with a plastic bag and tape after disconnecting the fuel tubes.

➡If the pipe and connector are stuck together, pinch the tube between your fingers and turn it carefully to free it. Then disconnect the main tube.

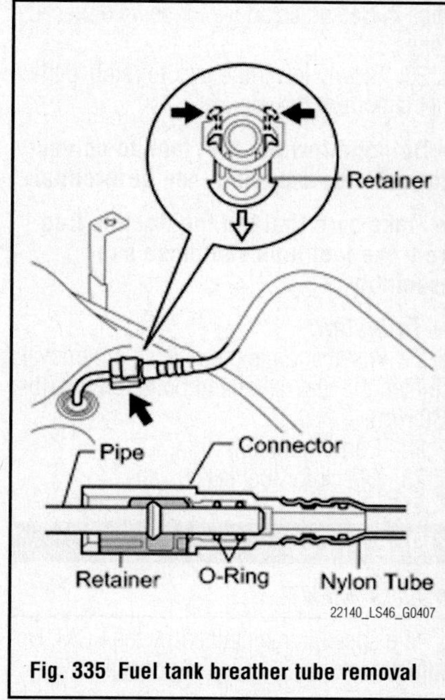

Fig. 335 Fuel tank breather tube removal

14. Loosen the clamp and disconnect the filler pipe hose.

15. Place a jack under the fuel tank.

16. Remove the 4 bolts and 2 fuel tank bands.'

17. Slowly lower the jack so that the fuel tank breather tube can be disconnected. Make sure that no load is applied to the fuel filler pipe.

18. Disconnect the fuel tank breather tube sub-assembly. See notes above for service precautions.

19. Disconnect the fuel tank vent hose sub-assembly. Pinch the retainer and then

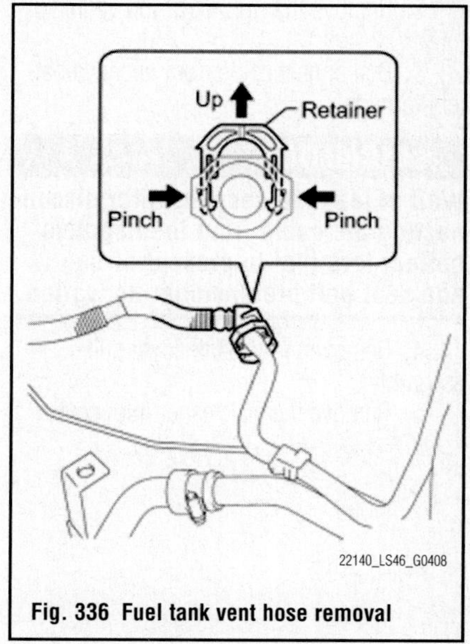

Fig. 336 Fuel tank vent hose removal

raise it. See notes above for service precautions.

20. Slowly lower the jack to remove the fuel tank from the vehicle.

➡Do not allow the fuel tank to contact the vehicle, especially the differential.

➡Make sure that the fuel tank is free from the fuel tank vent hose sub-assembly.

To install:

21. Installation is the reverse of removal. Tighten the fuel tank band bolts to 29 ft. lbs. (39 Nm).

22. Perform initialization.

23. Check for fuel leaks.

IDLE SPEED

ADJUSTMENT

Idle speed is maintained by the ECM. No adjustment is necessary or possible.

THROTTLE BODY

REMOVAL & INSTALLATION

See Figure 337.

1. Before servicing the vehicle, refer to the Precautions Section.

WARNING

After the engine switch is turned off, the HDD navigation system requires approximately 6 minutes to record various types of memory and settings. As a result, after turning the engine switch off, wait 6 minutes or more before disconnecting the cable from the negative battery terminal.

2. Remove the right cowl top ventilator louver.

3. Disconnect cable from negative battery terminal.

CAUTION

Wait at least 90 seconds after disconnecting the cable from the negative battery terminal to prevent airbag and seat belt pretensioner activation.

4. Remove the V-bank cover sub-assembly.

5. Remove the air cleaner inlet cover.

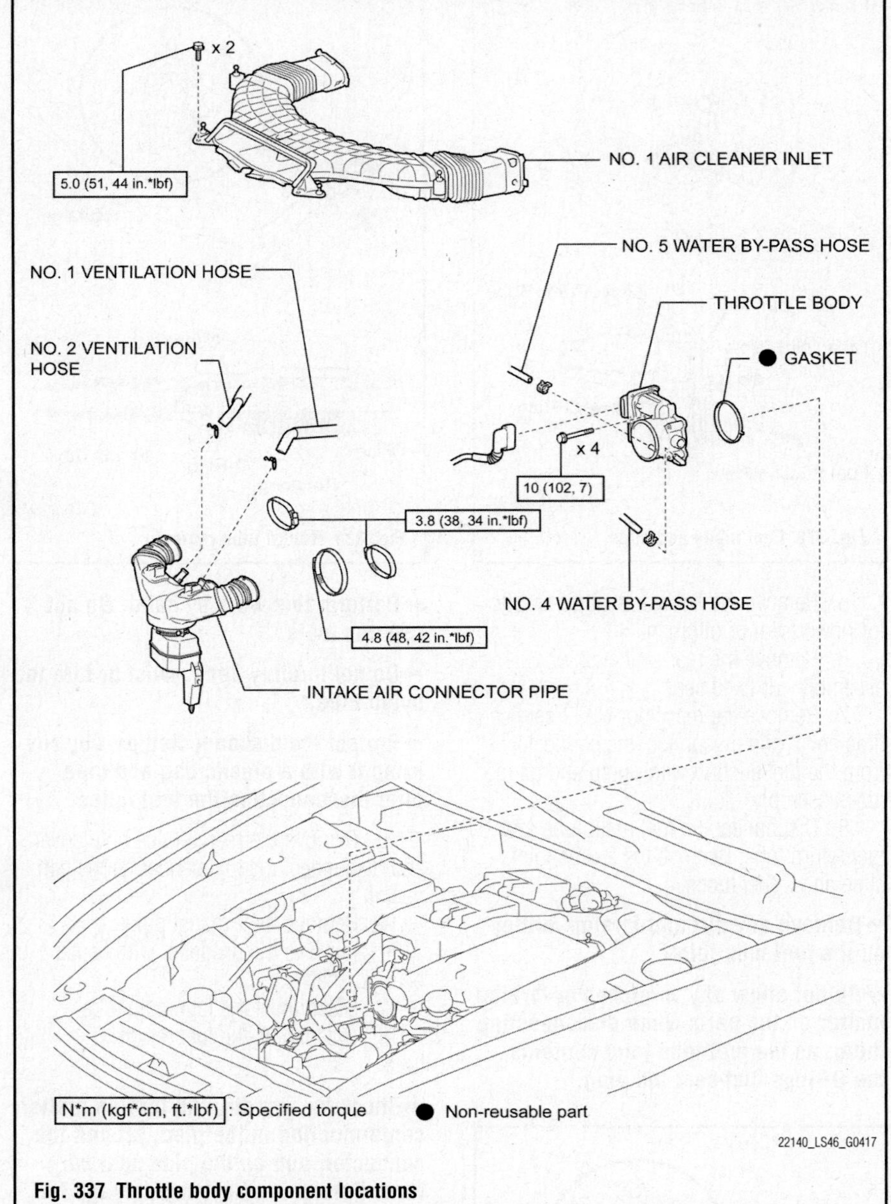

Fig. 337 Throttle body component locations

6. Remove the No. 1 air cleaner inlet.

7. Remove the No. 1 engine under cover.

8. Drain engine coolant.

9. Remove the intake air connector pipe.

10. Remove the throttle body, as follows:

a. Disconnect the throttle motor connector.

b. Remove the 4 bolts and throttle body.

c. Slide the clamps, and disconnect the No. 4 and No. 5 water by-pass hoses from the throttle body.

d. Remove the gasket from the intake manifold.

To install:

11. Installation is the reverse of removal. Tighten the throttle body bolts to 7 ft. lbs. (10 Nm).

12. Add coolant and inspect for leak.

13. Perform initialization.

HEATING & AIR CONDITIONING SYSTEM

BLOWER MOTOR

REMOVAL & INSTALLATION

See Figure 338.

1. Before servicing the vehicle, refer to the Precautions Section.
2. Remove the air conditioning unit.
3. Remove the blower assembly, as follows:
   a. Remove the screw.
   b. Disconnect the connector.
   c. Detach the claw and remove the blower unit.
4. Remove the 3 screws and damper servo.
5. Remove the 3 screws and blower with fan motor.

To install:

6. Installation is the reverse of removal.

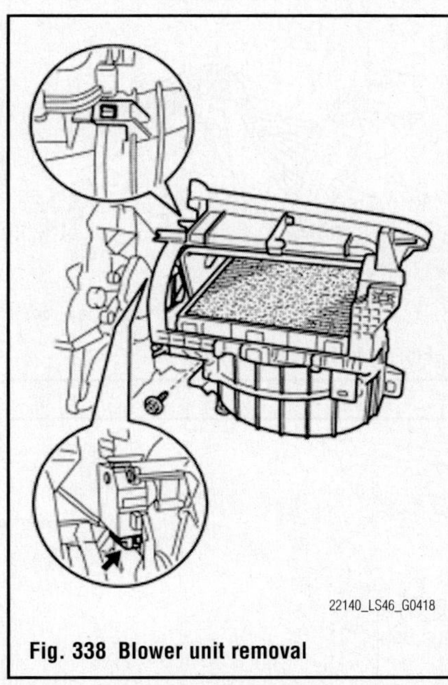

Fig. 338 Blower unit removal

HEATER CORE

REMOVAL & INSTALLATION

See Figures 339 through 348.

1. Before servicing the vehicle, refer to the Precautions Section.
2. Recover refrigerant from system.

WARNING

After the engine switch is turned off, the HDD navigation system requires approximately 6 minutes to record various types of memory and settings. As a result, after turning the engine switch off, wait 6 minutes or more before disconnecting the cable from the negative battery terminal.

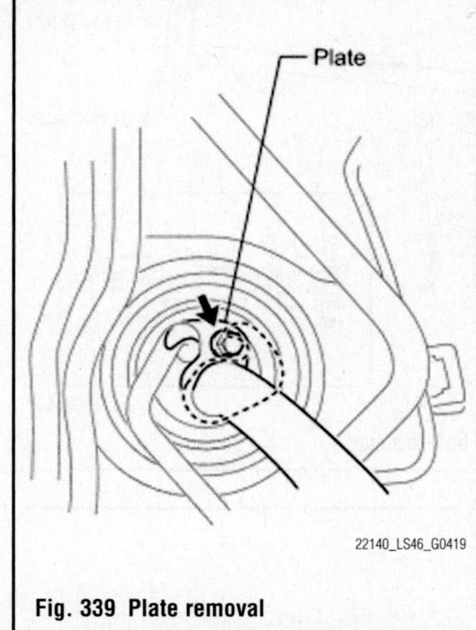

Fig. 339 Plate removal

3. Remove the right cowl top ventilator louver.
4. Disconnect cable from negative battery terminal.

CAUTION

Wait at least 90 seconds after disconnecting the cable from the negative battery terminal to prevent airbag and seat belt pretensioner activation.

5. Remove tube sub-assembly, as follows:
   a. Remove the bolt.
   b. Remove plate.
   c. Disconnect the suction tube and liquid tube.

➡Do not use a screwdriver or similar tool to disconnect the tube. Seal the opening of the disconnected parts to prevent moisture and foreign matter from entering them.

6. Using pliers, grip the claws of the clip and slide the clip, and disconnect the heater water inlet and outlet hoses.
7. Remove the instrument panel safety pad.
8. Remove the bolt, screw detach the claws and remove the air ducts.
9. Remove the steering column.
10. Remove the windshield wiper motor.
11. Remove the instrument panel reinforcement assembly, as follows:
    a. Disconnect the clamps, connectors and wire harness.
    b. Remove the bolts and nuts, and disconnect the ground wire and junction block.
    c. Remove the 9 bolts and 2 screws.
    d. Using a T40 TORX® socket, remove the 2 TORX® bolts (passenger side) and 3 bolts (driver side).
    e. For passenger side, using a 12 mm hexagon wrench, remove the 2 collars and instrument panel reinforcement with spacer.

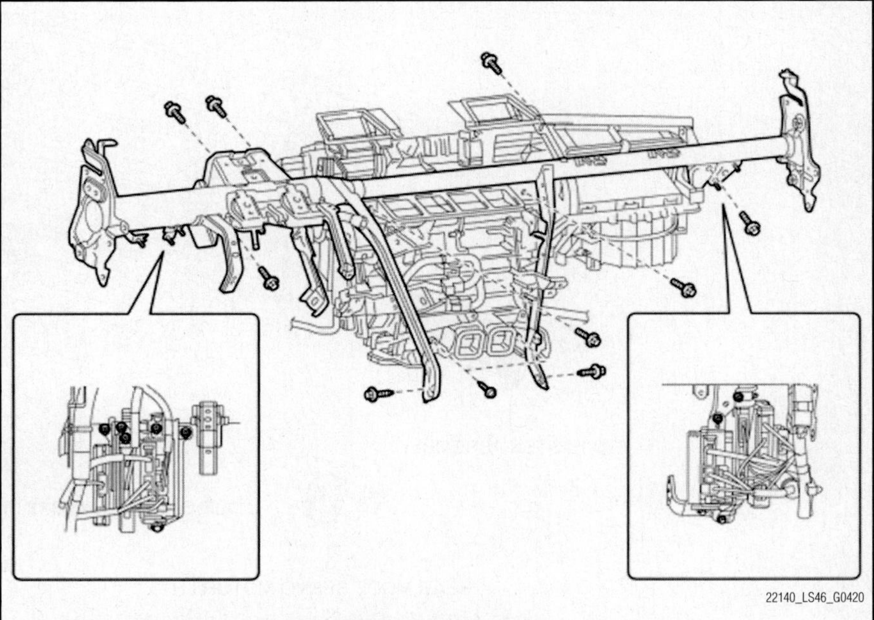

Fig. 340 Instrument panel reinforcement assembly removal

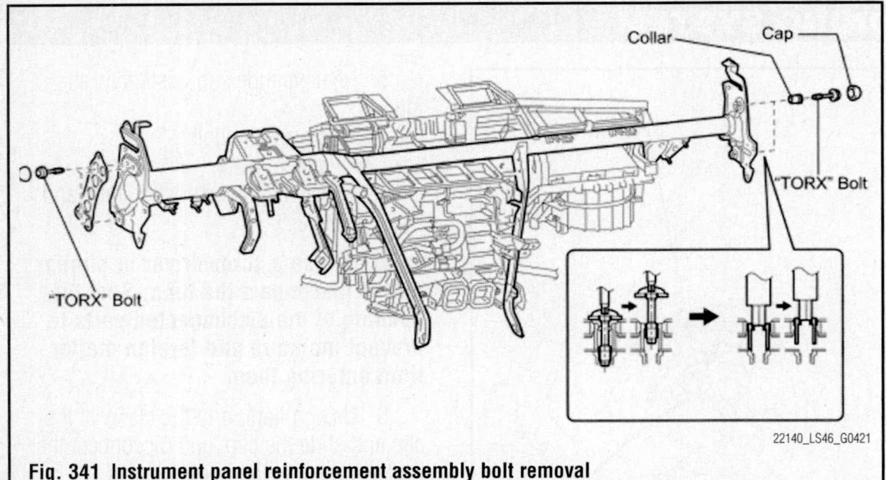

Fig. 341 Instrument panel reinforcement assembly bolt removal

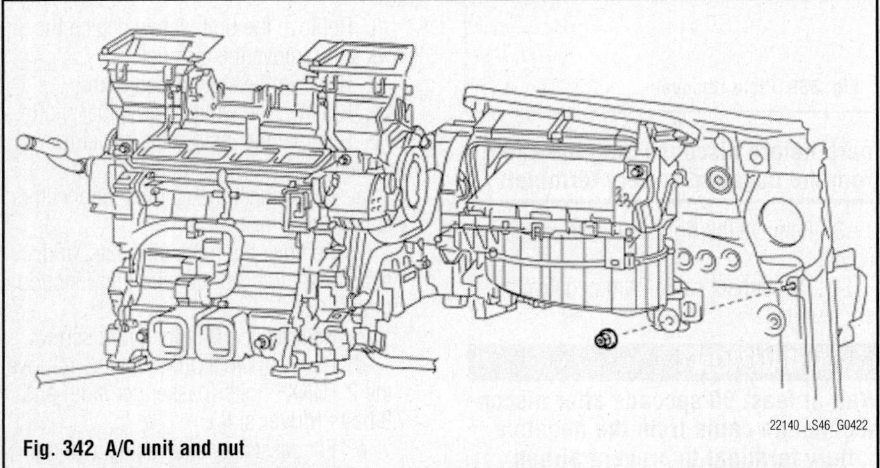

Fig. 342 A/C unit and nut

➡The TORX® bolts on the passenger side can be removed with the collar for adjustment. Remove the nut and air conditioning unit.

12. Remove the rear left-hand mode servo motor.
13. Remove the 2 screws, 3 clamps and PTC heater.

To install:

14. Install the PTC heater with the 2 screws and 3 clamps.
15. Install the rear left-hand mode servo motor.
16. Install the air conditioning unit with the nut and tighten to 7 ft. lbs. (9.8 Nm).
17. Install the instrument panel reinforcement assembly, as follows:
    a. Install the instrument panel reinforcement with spacer.
    b. Using a T40 TORX® socket, install the 2 TORX® bolts (passenger side) and

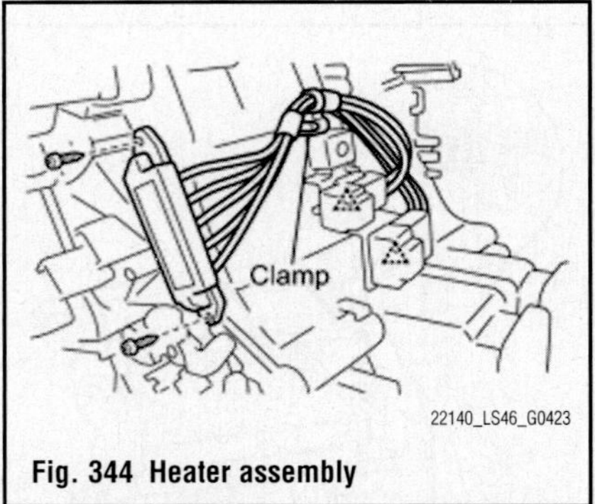

Fig. 344 Heater assembly

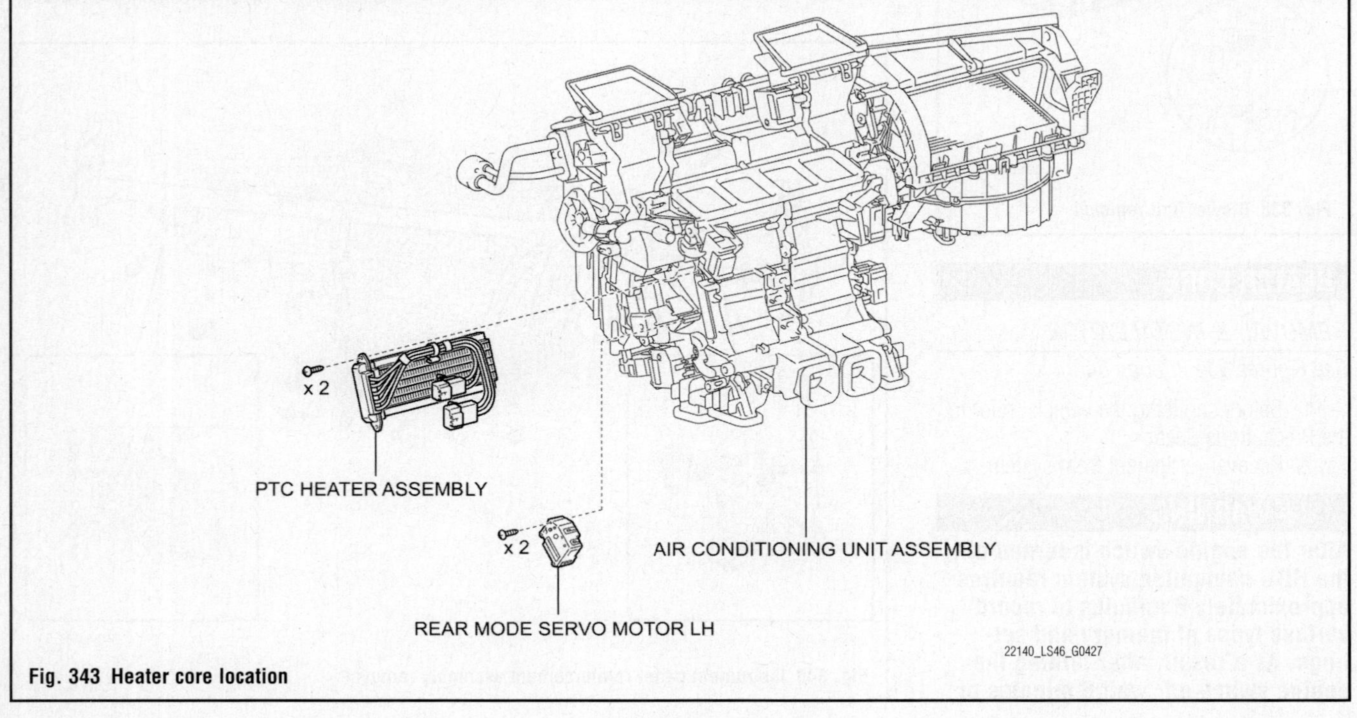

Fig. 343 Heater core location

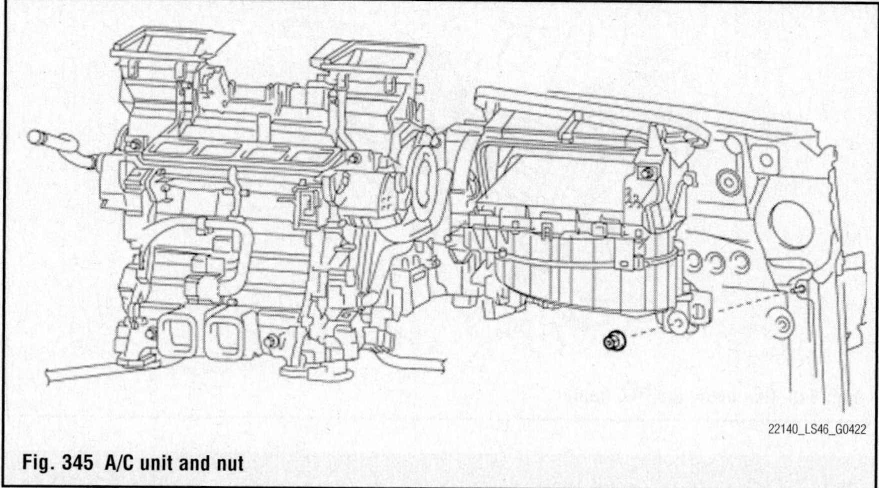
Fig. 345 A/C unit and nut

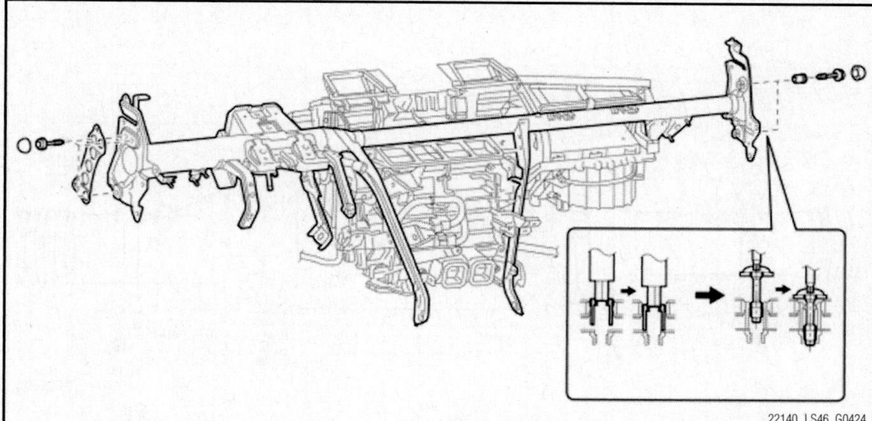
Fig. 346 Instrument panel reinforcement assembly bolt installation

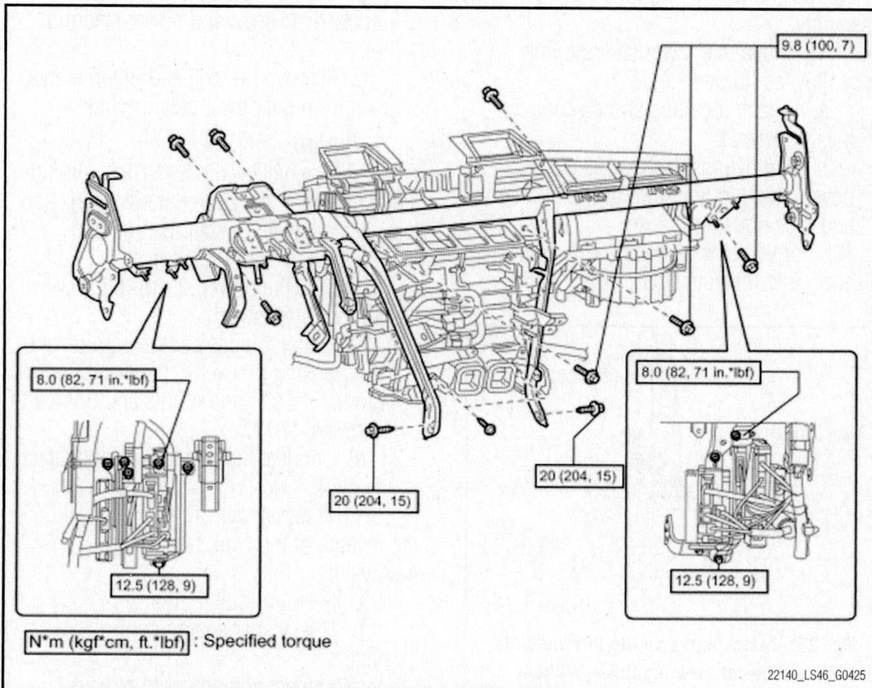

Fig. 347 Instrument panel reinforcement assembly installation

3 bolts (driver side) and tighten to 15 ft. lbs. (20 Nm).

c. For passenger side, using a 12 mm hexagon wrench, install the 2 collars and tighten to 15 ft. lbs. (20 Nm).

d. Install the 9 bolts and 2 screws.

e. Install the ground wires and junction block with the bolts and nuts.

f. Attach the clamps and connectors to the wire harness.

18. Install the windshield wiper motor.
19. Attach the claws, install the screw and bolt, tighten to 7 ft. lbs. (9.8 Nm), and install the ducts.
20. Install the instrument panel safety pad.
21. Connect the water inlet and outlet hoses and attach the clips.
22. Connect the tube sub-assembly, as follows:

a. Remove the attached vinyl tape from the pipe.

b. Sufficiently apply compressor oil (ND-OIL 8 or equivalent) to 2 new O-rings and the fitting surface of the suction tube and liquid tube.

c. Install the 2 O-rings on the suction tube and liquid tube.

d. Connect the suction tube and liquid tube. After the connection, check that the claw of the piping clamp is engaged.

e. Install the plate with the bolt and tighten to 7 ft. lbs. (9.8 Nm).

23. Connect the cable to negative battery terminal.
24. Install the right cowl top ventilator louver.
25. Add engine coolant and compressor oil.
26. Charge refrigerant.
27. Warm up the engine.
28. Check for coolant and refrigerant leaks.
29. Check the SRS warning light.
30. Perform initialization.

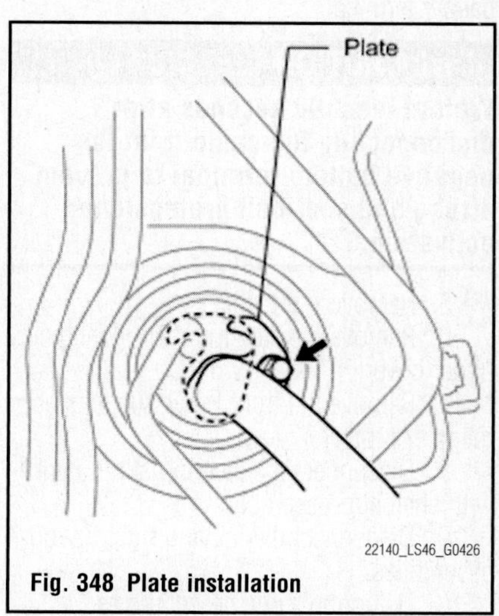

Fig. 348 Plate installation

HEATER UNIT

REMOVAL & INSTALLATION

See Figure 349.

1. Remove the air conditioning unit assembly.
2. Remove the rear mode servo motor LH.
3. Remove the PTC heater assembly.
   a. Remove the 2 screws, 3 clamps and PTC heater.

To install:

4. To install, reverse the removal procedure.

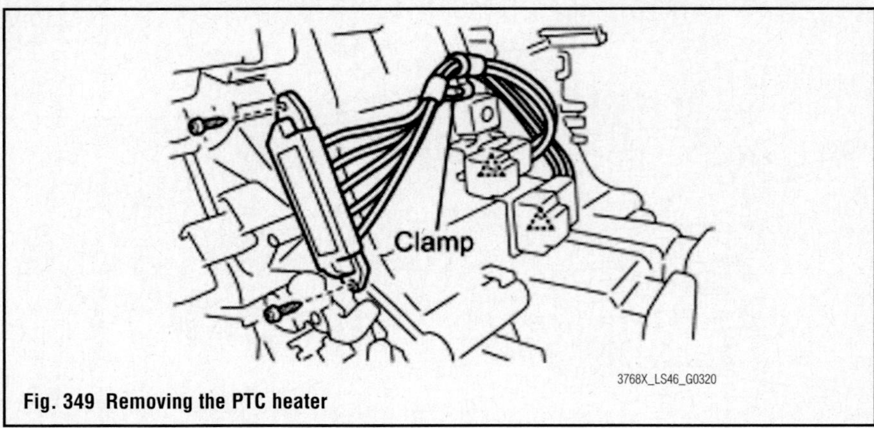

Fig. 349 Removing the PTC heater

STEERING

POWER RACK & PINION STEERING GEAR

REMOVAL & INSTALLATION

See Figures 350 through 352.

1. Before servicing the vehicle, refer to the Precautions Section.
2. Position the front wheels straight ahead.

⁂ WARNING

After the engine switch is turned off, the HDD navigation system requires approximately 6 minutes to record various types of memory and settings. As a result, after turning the engine switch off, wait 6 minutes or more before disconnecting the cable from the negative battery terminal.

3. Remove the right cowl top ventilator louver.
4. Disconnect cable from negative battery terminal.

⁂ CAUTION

Wait at least 90 seconds after disconnecting the cable from the negative battery terminal to prevent airbag and seat belt pretensioner activation.

5. Remove front wheels.
6. Remove the No. 1 and No. 2 engine under cover.
7. Remove the front lower suspension member protector.
8. Disconnect the steering sliding yoke with shaft sub-assembly.
9. Disconnect the left and right tie rod assemblies.
10. Using the same procedure as above, disconnect the right tie rod assembly.
11. Remove the power steering link assembly, as follows:
    a. Detach the clips and disconnect the connectors.
    b. Remove the 4 bolts, 4 nuts and power steering link from the front frame and rack housing bracket.
12. For vehicles without VGRS, remove steering intermediate shaft, as follows:
    a. Put the matchmarks on the intermediate shaft and power steering link.
    b. Remove the bolt and intermediate shaft from the power steering link.

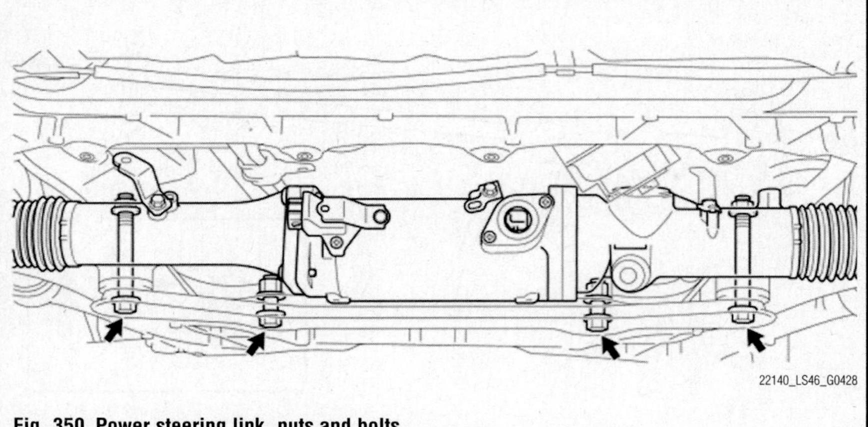

Fig. 350 Power steering link, nuts and bolts

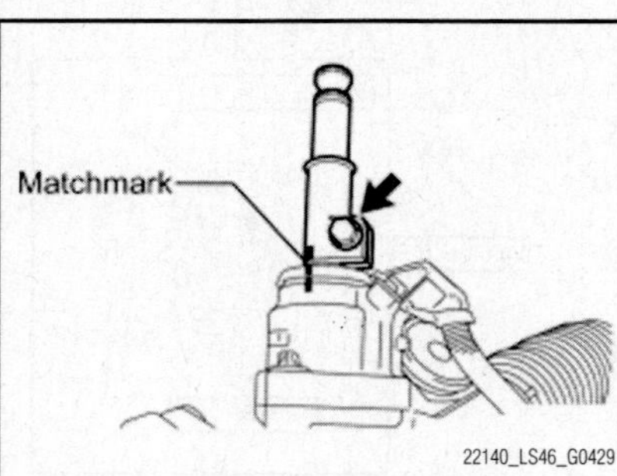

Fig. 351 Matchmarks on the intermediate shaft and power steering link—Without VGRS

To install:

13. For vehicles without VGRS, align the matchmarks on the power steering link and intermediate shaft, install the bolt and tighten to 26 ft. lbs. (35 Nm).
14. Install the power steering link assembly, as follows:
    a. Install the power steering link and rack housing bracket to the front frame with the 4 bolts and 4 nuts and tighten to 52 ft. lbs. (70 Nm).
    b. Connect the connectors and attach the clips.
15. The remainder of installation is the reverse of removal, noting the following:
    a. Perform initialization.
    b. Check suspension control system.
    c. Inspect and adjust front wheel alignment.

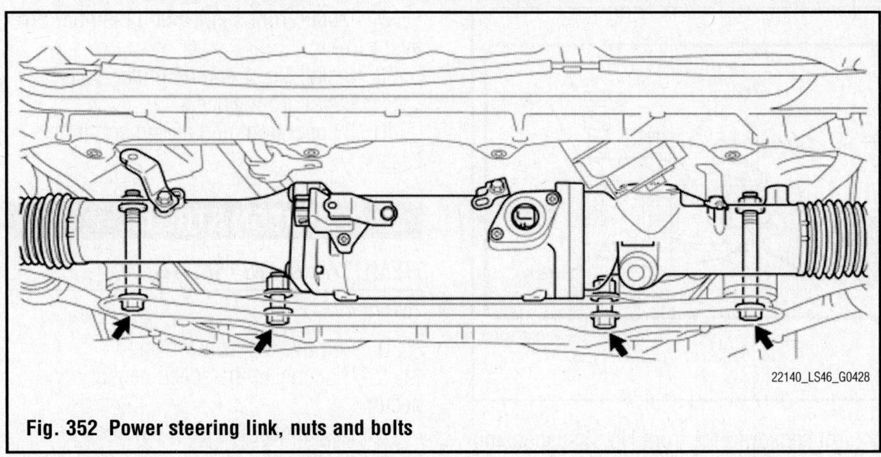

Fig. 352 Power steering link, nuts and bolts

d. Adjust headlight aiming.
e. Inspect the SRS warning light.
f. Adjust object recognition camera.
g. Initialize rotation angle sensor and calibrate torque sensor zero point.
h. Perform variable gear ratio steering system calibration, if necessary.

POWER STEERING PUMP

REMOVAL & INSTALLATION

This vehicle is equipped with electronic power steering. These models do not utilize a power steering (vane) pump assembly or power steering fluid.

SUSPENSION

FRONT SUSPENSION

LOWER BALL JOINT

REMOVAL & INSTALLATION

See Figure 353.

1. Remove the front wheels.
2. Remove the No. 1 engine under cover.
3. Remove the lower front suspension member protector.
4. Remove the clip and nut.
5. Install 2 spacers (Service Tool spacer A) onto the front No. 1 suspension lower arm assembly so that there is a space of approximately 1 mm (0.0394 in.) between the arm and spacers.

➡Make sure to install the spacers (Service Tool spacer A) as the steering knuckle spacer may shift, as Service Tool may become damaged, make sure the space between the arm and spacers is not 1 mm (0.0394 in.) or less.

6. Using Service Tool, disconnect the front No. 1 suspension lower arm assembly from the steering knuckle.

To install:

7. Install the front No. 1 suspension lower arm to the steering knuckle with the nut and tighten to 107 ft. lbs. (145 nm).
8. Install a new clip.

➡If it is necessary to align the holes for the clips after installing the nuts, the nuts can be tightened up to 60°more.

9. Install the lower front suspension member protector.
10. Install the No. 1 engine under cover.
11. Install the front wheels.
12. Inspect front end alignment, if necessary.

LOWER CONTROL ARM

REMOVAL AND & INSTALLATION

See Figure 354.

1. Disconnect cable from negative battery terminal.
2. Remove the front wheels.
3. Remove the No. 1 engine under cover.
4. Remove the lower front suspension member protector.
5. Remove the No. 2 engine under cover.
6. Disconnect the steering sliding yoke with shaft sub-assembly.
7. Disconnect the tie rod assembly.
8. Remove the power steering link assembly.
9. Disconnect the speed sensor connector.
10. Disconnect the front disc brake caliper assembly.
11. Remove the front disc.
12. Remove the front disc brake dust cover.
13. Remove the front No. 1 suspension lower arm assembly.
14. Remove the clip and nut.
    a. Install 2 spacers (Service Tool spacer A) onto the front No. 1 suspension lower arm assembly so that there is a space of approximately 1 mm (0.0394 in.) between the arm and spacers.

➡Make sure to install the spacers (Service Tool spacer A) as the steering knuckle spacer may shift, as Service Tool may become damaged, make sure the space between the arm and spacers is not 1 mm (0.0394 in.) or less.

    b. Using Service Tool, disconnect the front No. 1 suspension lower arm assembly from the steering knuckle.
    c. Remove the bolt and nut, and then remove the No. 1 suspension lower arm.
15. Remove the bolt, and then remove the bracket of the front height control sensor.
16. Remove the 2 nuts and front stabilizer link assembly.

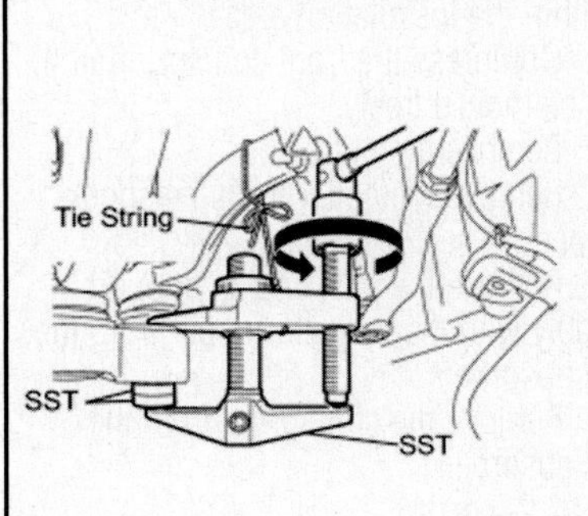

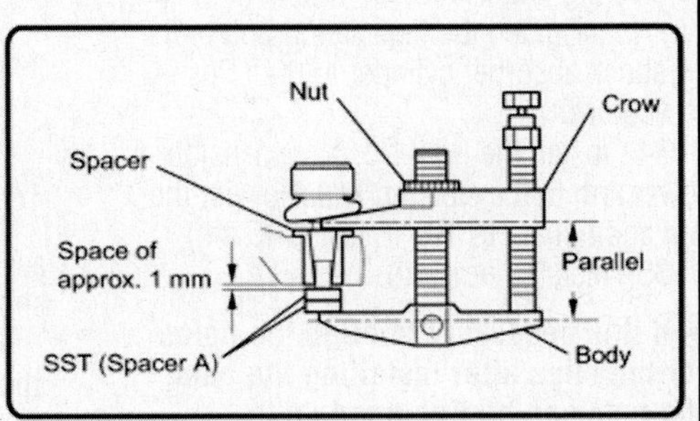

22140_LS46_G0501

Fig. 353 Remove lower ball joint

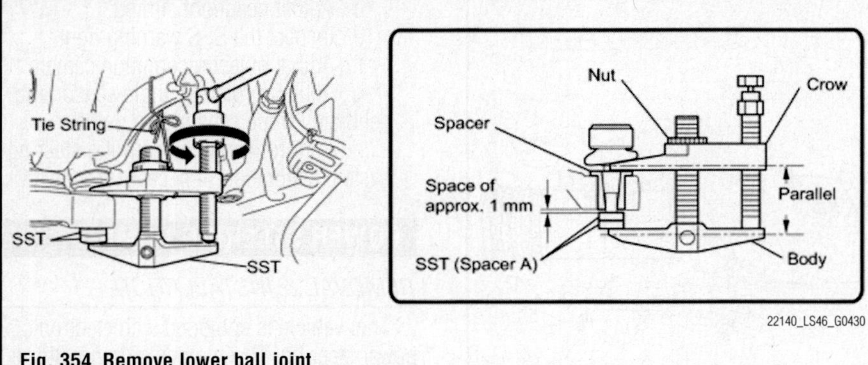

Fig. 354 Remove lower ball joint

a. Remove the front No. 2 suspension lower arm assembly.

b. Support the steering knuckle with a jack and wooden block.

c. Remove the bolt and nut, and then disconnect the bottom side of the pneumatic front with shock absorber cylinder assembly from the front No. 2 suspension lower arm.

d. Remove the nut and clip.

e. Install 2 spacers (Service Tool spacer A) onto the front No. 2 suspension lower arm assembly so that there is a space of approximately 1 mm (0.0394 in.) between the arm and spacers.

➡Make sure to install the spacers (Service Tool spacer A) as the steering knuckle spacer may shift, as Service Tool may become damaged, make sure the space between the arm and spacers is not 1 mm (0.0394 in.) or less.

f. Using Service Tool, disconnect the front No. 2 suspension lower arm assembly from the steering knuckle.

17. Remove the bolt and nut, and then remove the No. 2 suspension lower arm.

To install:

18. Temporarily tighten the front No. 2 suspension lower arm assembly.

a. Insert the bolt from the front of the vehicle. Then temporarily install the front No. 2 suspension lower arm with the nut.

b. Temporarily install the front No. 2 suspension lower arm to the steering knuckle with the nut.

19. Install the bottom side of the pneumatic front with shock absorber cylinder assembly to the No. 2 suspension lower arm. Then insert the bolt from the rear of the vehicle and temporarily install it with the nut.

20. Install the front stabilizer link with the 2 nuts and tighten to 62 ft. lbs. (84 Nm).

21. Install the bracket of the front height control sensor to the front No. 2 suspension lower arm with the bolt and tighten to 48 inch lbs. (5.4 Nm).

22. Temporarily tighten front No. 1 suspension lower arm assembly.

a. Insert the bolt from the back of the vehicle. Temporarily install the front No. 1 suspension lower arm with the nut.

b. Temporarily install the front No. 1 suspension lower arm to the steering knuckle with the nut.

➡Tighten the nuts after the vehicle is stabilized.

23. Install front disc brake dust cover.

24. Install front disc.

25. Connect front disc brake caliper assembly.

26. Connect speed sensor connector.

27. Install power steering link assembly.

28. Connect tie rod assembly.

29. Connect steering sliding yoke with shaft sub-assembly.

30. Connect cable to negative battery terminal.

31. Install cowl top ventilator louver.

32. Stabilize suspension.

33. Tighten front No. 2 suspension lower arm assembly.

a. Tighten the installation nut of the front No. 2 suspension lower arm assembly to 107 ft. lbs. (145 Nm).

b. Tighten the installation nut of the steering knuckle to 107 ft. lbs. (145 Nm).

c. Tighten the installation bolt of the shock absorber cylinder to 80 ft. lbs. (108 Nm).

34. Install the front No. 1 suspension lower arm to the steering knuckle with the nut and tighten to 107 ft. lbs. (145 nm).

35. Install a new clip.

➡If it is necessary to align the holes for the clips after installing the nuts, the nuts can be tightened up to 60°more.

36. Install No. 2 engine under cover.

37. Install front suspension member protector lower.

38. Install No. 1 engine under cover.

39. Install front wheels.

40. Inspect front end alignment, if necessary.

MACPHERSON STRUT

REMOVAL & INSTALLATION

See Figure 355.

1. Remove the front wheel.

2. Disconnect the speed sensor connector.

3. Disconnect the front disc brake caliper assembly.

4. Remove the front disc.

5. Remove the front disc brake dust cover.

6. Disconnect the tie rod assembly.

7. Disconnect the front No. 1 suspension lower arm assembly.

8. Disconnect the front height control sensor sub-assembly.

9. Remove the front stabilizer link assembly.

10. Disconnect the front No. 2 suspension lower arm assembly.

11. Remove the air cleaner inlet cover sub-assembly

12. Remove the engine room side cover.

13. Remove the front shock absorber assembly.

a. Loosen the lock nut of the front shock absorber.

➡Do not remove the lock nut. Loosen the lock nut only when disassembling the front shock absorber with coil spring.

b. Remove the 3 nuts from the upper side of the front suspension support.

c. Slowly lower the jack. Remove the bolt from the lower side to remove the front shock absorber with coil spring.

14. Install Service Tool to the front coil spring so that the distance between the upper and lower hooks is as wide as possible within the installation area.

15. Compress the front coil spring until it can be moved freely.

16. Remove the nut.

17. Remove the collar from the shock absorber.

18. Remove the suspension support assembly with the insulator upper from the shock absorber.

19. Remove the coil spring from the shock absorber.

To install:

20. Using Service Tool, compress the coil spring.

21. Install the coil spring so that the end comes to the stepped portion of the lower coil spring insulator.

22. Align the bolt of the front suspension support and the cutout of the front coil spring insulator, and install the front coil spring insulator on the front suspension support.

23. Align the width across flat of the piston rod end and of the front suspension support to install the front shock absorber.

24. Install the suspension support assembly with the upper insulator to the shock absorber.

25. Install the collar to the shock absorber.

26. Temporarily install a new lock nut to the front shock absorber.

27. Adjust the front suspension support so that the bolts come to the positions as shown in the illustration, and remove Service Tool from the front coil spring.

28. Install the front shock absorber on the vehicle by installing the 3 nuts on the suspension support side and tighten nuts to 49 ft. lbs. (67 Nm).

29. Tighten the lock nut to 20 ft. lbs. (28 Nm).

30. Install the engine room side cover.

31. Install the air cleaner inlet cover sub-assembly.

32. Temporarily tighten the front No. 2 suspension lower arm assembly.

33. Install the front stabilizer link assembly.

34. Connect the front height control sensor sub-assembly.

35. Install the front No. 1 suspension lower arm assembly.

36. Connect the tie rod assembly.

37. Install the front disc brake dust cover.

38. Install the front disc.

39. Connect the front disc brake caliper assembly.

40. Connect the speed sensor connector.

41. Stabilize the suspension:

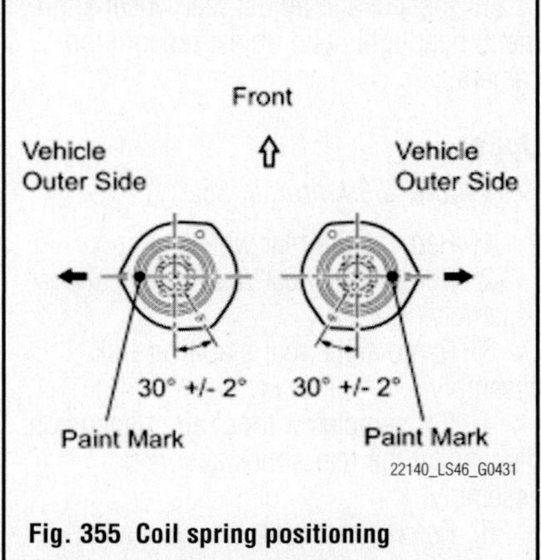

Fig. 355 Coil spring positioning

a. Install the front tires.

b. Lower the vehicle and bounce it up and down several times to stabilize the front suspension.

c. Remove the front tires.

d. Jack up the front suspension lower arm with a wooden block between the jack and front suspension lower arm. Apply a load to the front suspension so that the front suspension lower arm is placed in a horizontal position.

42. Tighten the front No. 2 suspension lower arm assembly.

43. Install the front wheel.

44. Inspect and adjust the front wheel and headlight alignment.

STEERING KNUCKLE

REMOVAL & INSTALLATION

1. Remove the front wheel.
2. Disconnect the speed sensor connector.
3. Disconnect the front disc brake caliper assembly.
4. Remove the front disc.
5. Remove the front disc brake dust cover.
6. Disconnect the tie rod assembly.
7. Disconnect the front No. 1 suspension lower arm assembly.
8. Disconnect the front height control sensor sub-assembly.
9. Remove the front stabilizer link assembly.
10. Disconnect the front shock absorber assembly.
11. Disconnect the front No. 2 suspension lower arm assembly.
12. Remove the steering knuckle sub-assembly.
13. Remove the front axle hub sub-assembly.

To install:

14. Install the front axle hub sub-assembly.
15. Install the steering knuckle sub-assembly.
16. Connect the front No. 2 suspension lower arm assembly.
17. Connect the front shock absorber assembly.
18. Install the front stabilizer link assembly.
19. Connect the front height control sensor sub-assembly.
20. Connect the front No. 1 suspension lower arm assembly.
21. Connect the tie rod assembly.
22. Install the front disc brake dust cover.
23. Install the front disc.
24. Connect the front disc brake caliper assembly.
25. Connect the speed sensor connector.
26. Install the front wheel.

STABILIZER BAR

REMOVAL & INSTALLATION

See Figures 356 and 357.

1. Remove the front wheel.
2. Remove the 8 bolts and the No. 2 engine under cover.
3. Remove the 2 nuts and the front stabilizer link.
4. Remove the front stabilizer link assembly.
5. Remove the 2 screws and the engine under cover sub-assembly from the front suspension member reinforcement.
6. Remove the engine under cover sub-assembly.
7. Remove the 2 screws and the engine under cover rear.
8. Remove the 2 bolts and the front No. 1 stabilizer bracket from the front suspension crossmember.
9. Remove the front No. 1 stabilizer bracket.

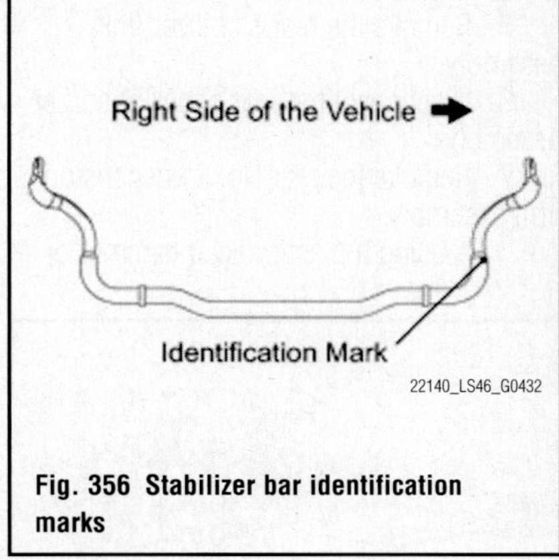

Fig. 356 Stabilizer bar identification marks

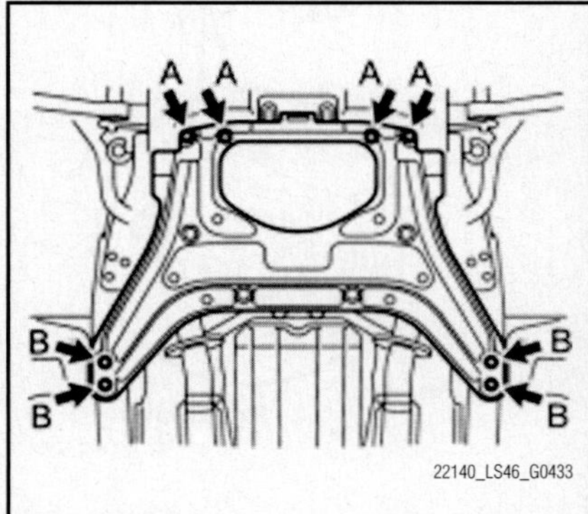

Fig. 357 Engine under cover bolt locations

10. Remove the front stabilizer bar from the vehicle.

To install:

11. Temporarily install the front stabilizer bar with the 4 bolts and the front No. 1 stabilizer bracket.

➡Install the front stabilizer bar so that the identification mark face vehicle right side.

12. Tighten the front No. 1 stabilizer bracket with the 2 bolts to 36 ft. lbs. (49 Nm).
13. Tighten the front No. 1 stabilizer bracket.
14. Install the engine under cover rear with the 2 screws.
15. Install the engine under cover rear.
16. Install the engine under cover sub-assembly with the 2 screws.
17. Install the engine under cover sub-assembly.
18. Install the front stabilizer link with the 2 nuts and tighten to 62 ft. lbs. (84 nm).
19. Install the No. 2 engine under cover with the 8 bolts.
20. Install front wheel.
21. Inspect and adjust front wheel and headlight alignment.

WHEEL HUB & BEARING

REMOVAL & INSTALLATION

1. Remove the front wheel.
2. Remove the steering knuckle sub-assembly.
3. Remove the 4 bolts and front axle hub from the steering knuckle.

To install:

4. Install the front axle hub to the steering knuckle with the 4 bolts and tighten to 48 ft. lbs. (65 Nm).
5. Install the steering knuckle sub-assembly.
6. Install the front wheel.

SUSPENSION

REAR SUSPENSION

CONTROL ARMS/LINKS

REMOVAL & INSTALLATION

Lower

See Figure 358.

1. Remove the rear wheel.
2. Remove the No. 2 differential support protector.
3. Remove the rear speed sensor.
4. Remove the load sensing valve sensor bracket.
5. Remove the rear stabilizer link assembly.
6. Disconnect the rear shock absorber assembly.
7. Remove the rear No. 2 suspension arm assembly.
    a. Using a plastic-faced hammer or equivalent, strike the part labeled from the rear of the vehicle to maintain the clearance at the slide pin area.
    b. Remove the nut, washer, bolt and no. 2 suspension arm from the axle carrier.
    c. Remove the nut, no. 2 suspension toe adjust plate, rear suspension attachment and no. 2 suspension arm from the suspension member.

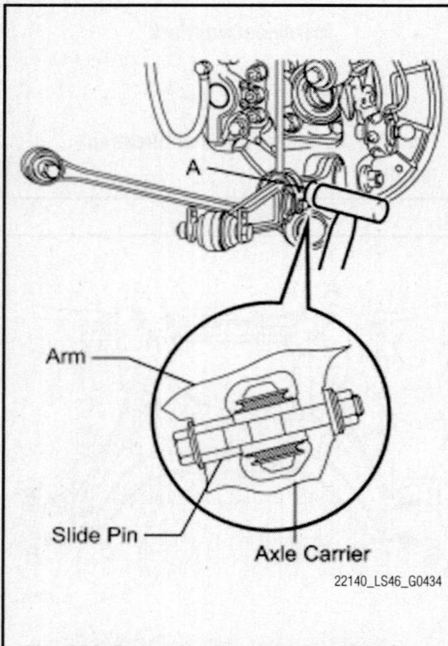

Fig. 358 Removing No. 2 suspension arm

8. Remove the rear No. 1 suspension arm assembly.
    a. Remove the nut on the rear axle carrier side.
    b. Install 2 spacers (Service Tool spacer B) onto the rear No. 1 suspension arm so that there is a space of approximately 1 mm (0.039 in.) between the arm and spacers.

➡Make sure to install the spacers (Service Tool spacer B) as the axle carrier spacer may shift, as Service Tool may become damaged, make sure the space between the arm and spacers is not 1 mm (0.039 in.) or less.

    c. Using Service Tool, disconnect the rear No. 1 control arm from the axle carrier.
    d. Remove the bolt, nut and No. 1 suspension arm from the suspension member.

To install:

9. Temporarily install the rear No. 2 suspension arm assembly.
    a. Temporarily install the No. 2 suspension arm with the no. 2 suspension toe adjust plate, rear suspension attachment and nut to the suspension member.
    b. Install the stud of the suspension arm, and temporarily install a washer, nut and the bolt to the axle carrier.
10. Temporarily install the rear No. 1 suspension arm assembly.
    a. Temporarily install the No. 1 suspension arm with the bolt and nut to the suspension member.
    b. Install the stud of the suspension arm, and temporarily install a new nut to the axle carrier.
11. Tighten rear No. 1 suspension arm assembly bolt to 53 ft. lbs. (72 Nm) and the nut to 87 ft. lbs. (118 Nm).
12. Temporarily connect the rear shock absorber assembly.
13. Install the rear stabilizer link assembly.
14. Stabilize the suspension.
15. Tighten the rear No. 2 suspension arm assembly to 111 ft. lbs. (150 Nm) on member side and 166 ft. lbs. (225 Nm) for carrier side.
16. Tighten connect rear shock absorber assembly to 59 ft. lbs. (80 Nm).
17. Install the rear speed sensor.
18. Install the load sensing valve sensor bracket.
19. Install the No. 2 differential support protector.
20. Install the rear wheel.
21. Inspect and adjust rear wheel alignment, headlights and object recognition camera.

Upper

See Figures 359 through 362.

1. Remove the rear wheel.
2. Remove the rear disc brake caliper assembly.
3. Remove the rear stabilizer link assembly.
4. For vehicles without air suspension, disconnect the rear shock absorber assembly.
5. For vehicles with air suspension,

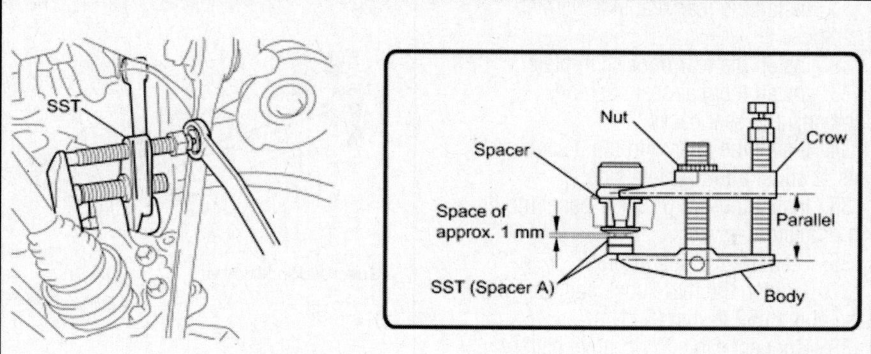

Fig. 359 Install 2 spacers onto the rear upper control arm—No. 1 control arm

Fig. 360 Install 2 spacers onto the rear upper control arm—No. 2 control arm

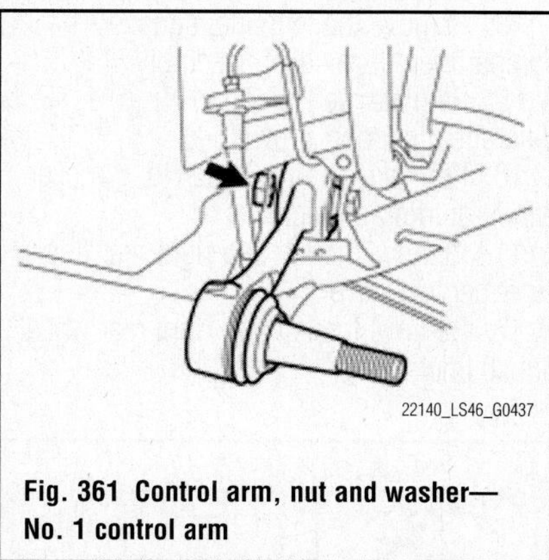

Fig. 361 Control arm, nut and washer—No. 1 control arm

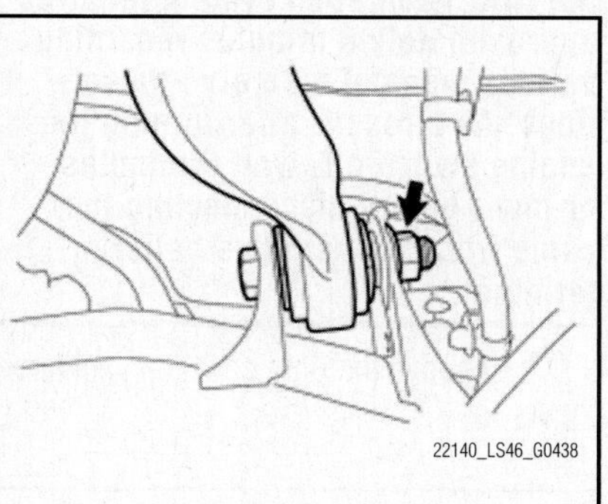

Fig. 362 Control arm, nut and washer—No. 2 control arm

disconnect the pneumatic cylinder with rear shock absorber assembly.

6. Remove the rear No. 2 suspension arm assembly.

7. Disconnect the rear No. 1 suspension arm assembly.

8. Disconnect the toe control link sub-assembly.

9. Remove the rear upper control arm assembly (use the same procedure for the No. 1 and No. 2 control arms), as follows:

a. Remove the nut on the rear axle carrier side.

b. Install 2 spacers (SST spacer A) onto the rear upper control arm so that there is a space of approximately 1 mm (0.039 in.) between the arm and spacers. SST: 09960-20010.

➡Make sure to install the spacers (SST spacer A) as the axle carrier spacer may shift.

➡As SST may become damaged, make sure the space between the arm and spacers is not 1 mm (0.039 in.) or less.

c. Using SST (SST: 09960-20010), disconnect the rear upper control arm from the axle carrier.

➡Do not damage the dust cover. As the dust cover may be damaged, adjust SST with the center nut so that the body and crow are parallel. Make sure to tie the string of to the vehicle to prevent SST from dropping.

d. Remove the nut and washer, bolt and control arm from the rear suspension member.

To install:

10. For vehicles without air suspension, temporarily connect the rear shock absorber assembly.

11. For vehicles with air suspension, temporarily connect the pneumatic cylinder with rear shock absorber assembly.

12. Temporarily install the rear upper control arm assembly (use the same procedure for the No. 1 and No. 2 control arms), as follows:

a. Temporarily install the control arm with the bolt, nut and washer to the suspension member.

b. Install the stud of the control arm, and temporarily install a new nut to the axle carrier.

➡Push the axle carrier downward.

13. Tighten the nuts on the rear upper No. 1 control arm to 111 ft. lbs. (150 Nm) for suspension member side, and 118 ft. lbs. (160 Nm) for axle carrier side.

14. Tighten the nuts on the rear upper No. 2 control arm to 166 ft. lbs. (225 Nm) for suspension member side, and 118 ft. lbs. (160 Nm) for axle carrier side.

15. Temporarily install the rear No. 2 suspension arm.

16. Connect the toe control link sub-assembly.

17. Connect the rear No. 1 suspension arm.

18. Install rear disc brake caliper assembly.

19. Stabilize the suspension.

20. Tighten the rear No. 2 suspension arm assembly.

21. For vehicles without air suspension, connect the rear shock absorber assembly.

22. For vehicles with air suspension, connect the pneumatic cylinder with rear shock absorber assembly.

23. Install the rear stabilizer link assembly.

24. Install the rear wheel.

25. Check the suspension control system.

26. Inspect and adjust the rear wheel alignment.

27. Adjust the headlight.

28. Adjust the object recognition camera.

MACPHERSON STRUTS

REMOVAL & INSTALLATION

1. Remove the cowl top ventilator louver.
2. Disconnect cable from negative battery terminal.
3. Remove the rear wheel.
4. Remove the rear seat cushions, ottomans and seat backs.
5. Remove No. 6 rocker panel moulding protector.
6. Remove the 3 screws, 11 nuts and 2 clips from the liner.
7. Remove the rear door scuff plates.
8. Remove the rear seat side garnish moldings.
9. Remove the package tray trim panel assembly.
10. Remove the rear speed sensor.
11. Remove the load sensing valve sensor bracket.
12. Disconnect the rear upper No. 2 control arm assembly.
13. Disconnect the rear upper No. 1 control arm assembly.
14. Disconnect the toe control link sub-assembly.
15. Remove the rear axle shaft nut.
16. Remove the 3 nuts and cap.
17. Remove the 3 nuts on the upper side of the shock absorber.
18. Remove the nut from the shock absorber lower side.
19. Angle the axle carrier's upper tip toward the vehicle's outer side, and remove the shock absorber.

To install:

20. Install the rear shock absorber assembly and tighten to 47 ft. lbs. (64 Nm).
21. Temporarily install the rear shock absorber lower side with a new nut and washer to the axle carrier.
22. Tighten the new lock nut to 20 ft. lbs. (27 Nm).
23. Install the cap with the 3 nuts to the shock absorber and tighten to 10 ft. lbs. (14 Nm).
24. Connect the toe control link sub-assembly.
25. Connect the rear upper no. 2 control arm assembly.
26. Connect the rear upper no. 1 control arm assembly.
27. Tighten the rear axle shaft nut to 214 ft. lbs. (290 Nm) and stake.
28. Install the load sensing valve sensor bracket.
29. Install the rear speed sensor.
30. Install the package tray trim panel assembly.
31. Install the rear seat side garnish moldings.
32. Install the rear door scuff plate.
33. Install the rear seat cushions, ottomans and seat backs.
34. Install the liner with the 3 screws, 11 nuts, 2 clips to the vehicle side.
35. Install the No. 6 rocker panel moulding protector.
36. Stabilize the suspension.
37. Tighten the rear shock absorber assembly to 59 ft. lbs. (80 Nm).
38. Connect cable to negative battery terminal.
39. Install the cowl top ventilator louver.
40. Perform initialization, if necessary.
41. Install the rear wheel.
42. Inspect and adjust rear wheel alignment, and headlights.

STABILIZER BAR

REMOVAL & INSTALLATION

See Figures 363 through 367.

1. Before servicing the vehicle, refer to the Precautions Section.

**** WARNING**

After the engine switch is turned off, the HDD navigation system requires approximately 6 minutes recording various types of memory and settings. As a result, after turning the engine switch off, wait 6 minutes or more before disconnecting the cable from the negative battery terminal.

2. Remove the right cowl top ventilator louver.

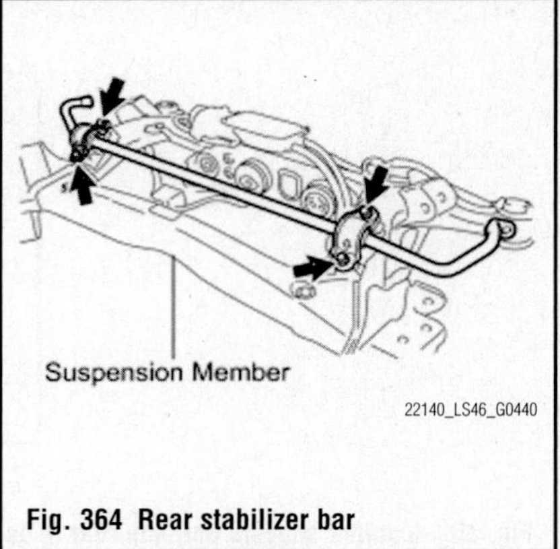

Fig. 364 Rear stabilizer bar

3. Disconnect cable from negative battery terminal.

**** CAUTION**

Wait at least 90 seconds after disconnecting the cable from the negative battery terminal to prevent airbag and seat belt pretensioner activation.

4. Remove the rear wheel.
5. Remove the luggage compartment mat sub-assembly.
6. Remove the left deck trim side board.
7. Remove rear floor finish plate.
8. Remove the left side luggage compartment trim cover assembly.
9. Remove the No. 2 and No. 1 differential support protector.
10. Remove the No. 6 and No. 5 rocker panel molding protector.
11. Remove the left and right rear floor side member cover.
12. Remove the left and right rear wheel house liners.

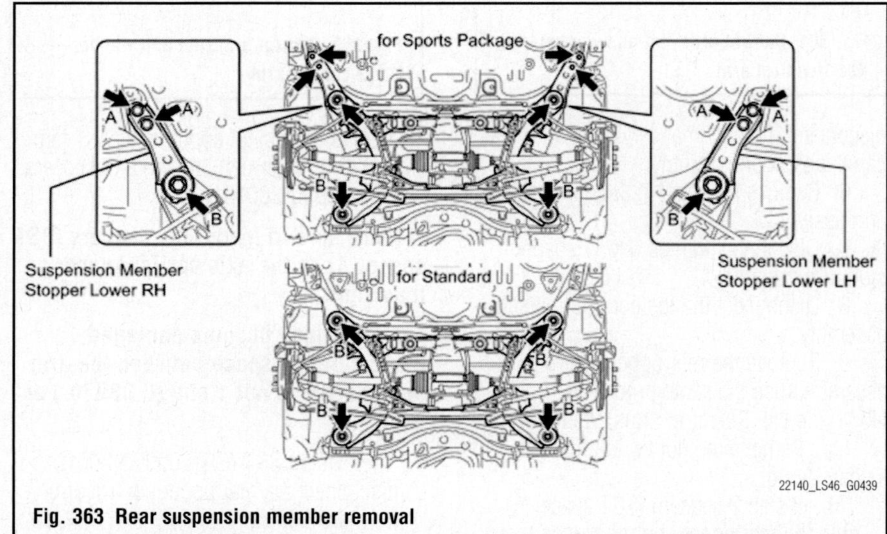

Fig. 363 Rear suspension member removal

13. Remove the No. 1 floor under cover.

14. Remove the left and right rear speed sensors.

15. Disconnect the No. 1 actuator harness clamp.

16. Remove the parking brake cable.

17. Remove the front floor center brace.

18. Remove the heated oxygen sensor (for Sensor 2).

19. Remove the left and right tailpipes.

20. Remove the No. 1 front floor heat insulator.

21. Remove the propeller shaft heat insulator.

22. Remove the propeller with center bearing shaft assembly.

23. Remove the left and right rear disc brake caliper assemblies.

24. Remove the left rear stabilizer link assembly, as follows:

a. Remove the nut and disconnect the stabilizer link from the stabilizer bar.

b. Remove the bolt, nut and stabilizer link from the rear No. 2 suspension arm.

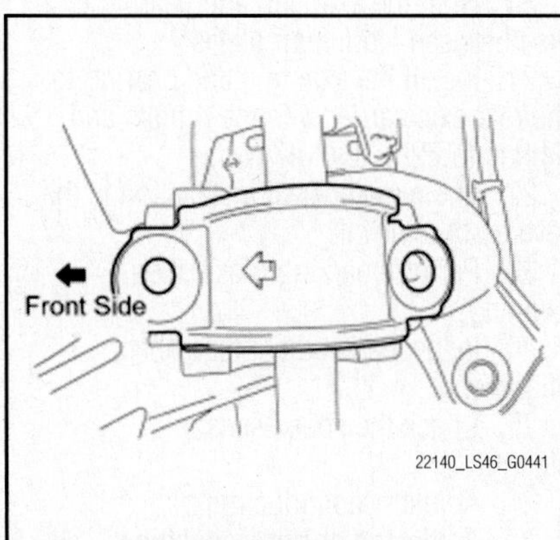

Fig. 365 Rear stabilizer bar bracket installation

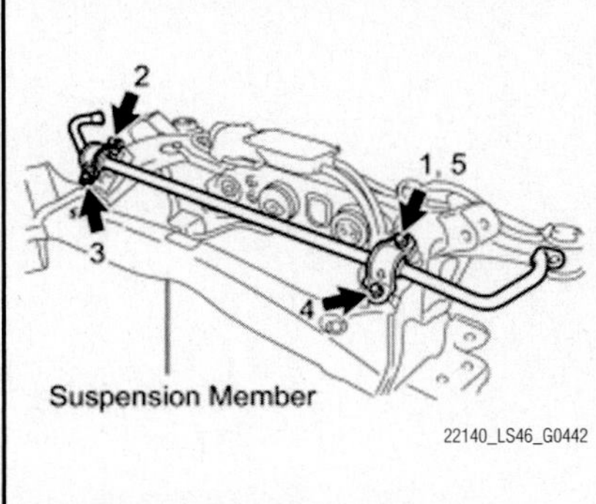

Fig. 366 Rear stabilizer bar bolt tightening sequence

25. Using the same procedure as above, remove the right rear stabilizer link assembly.

26. For vehicles without air suspension, disconnect the left and right rear shock absorber assemblies.

27. For vehicles with air suspension, disconnect the left and right pneumatic cylinder with rear shock absorber assemblies.

28. Support the rear suspension cross-member with the jack.

29. Remove the rear suspension member sub-assembly for sports package:

a. Remove the 4 bolts labeled A.

b. Remove the 4 bolts labeled B and 2 suspension member stoppers.

c. Slowly lower the jack, and disconnect the suspension member from the vehicle.

30. Remove the rear suspension member sub-assembly for standard package:

a. Remove the 4 bolts.

31. Slowly lower the jack, and disconnect the suspension member from the vehicle.

32. Remove the 4 bolts and the rear stabilizer bar from the suspension member.

➡The stabilizer bracket and bush are built onto the stabilizer bar. If the bracket and/or bush detach from the bar, replace the bar.

To install:

33. Install the rear stabilizer bar, as follows:

a. Make sure the bracket's arrows face the front of the vehicle.

b. First temporarily install bolt 1. Then install bolts 2, 3 and 4. Then tighten bolt 5. Tighten the bolts to 35 ft. lbs. (48 Nm).

34. Install the rear stabilizer link assemblies with the nuts and bolts and tighten as follows:

- Stabilizer bar side: 66 ft. lbs. (89 Nm).
- No. 2 suspension arm side: 19 ft. lbs. (26 Nm).

35. Slowly raise the suspension member jack, and install the suspension member to the body.

36. For sports package:

a. Install the 2 member stoppers with the 4 bolts labeled B, and install the 4 bolts labeled A, and tighten the A bolts to 14 ft. lbs. (19 Nm), and the B bolts to 94 ft. lbs. (127 Nm).

37. For standard package:

a. Install the 4 bolts labeled B and tighten to 94 ft. lbs. (127 Nm).

38. For vehicles with air suspension, connect the left and right pneumatic cylinder with rear shock absorber assemblies.

39. For vehicles without air suspension, connect the left and right rear shock absorber assemblies.

40. Install the left and right rear disc brake caliper assemblies.

41. Install the propeller with center bearing shaft assembly.

42. Inspect and adjust No. 2 and No. 3 joint angle.

43. Install the propeller shaft heat insulator.

44. Install the No. 1 front floor heat insulator.

45. Install the left and right tailpipes.

46. Install the front exhaust pipe assembly.

47. Install the front floor center brace.

48. Install the heated oxygen sensor (for Sensor 2).

49. Inspect for gas leak.

50. Install the parking brake cable.

51. Connect the No. 1 actuator harness clamp.

52. Install the left and right rear speed sensors.

53. Install the left and right rear wheel house liners.

54. Install the left and right rear floor side member cover.

55. Install the No. 6 and No. 5 rocker panel molding protector.

56. Install the No. 2 and No. 1 differential support protector.

57. Install the No. 1 floor under cover.

58. Install the left side luggage compartment trim cover assembly.

59. Install the rear floor finish plate.

60. Install the left deck trim side board.

61. Install the luggage compartment mat sub-assembly.

62. Install the rear wheel.

63. Connect cable to negative battery terminal.

64. Install the right cowl top ventilator louver.

65. Perform initialization.

66. Check the suspension control system.

67. Inspect and adjust the rear wheel alignment.

68. Check the speed sensor signal.

69. Adjust the headlight.

70. Adjust the object recognition camera.

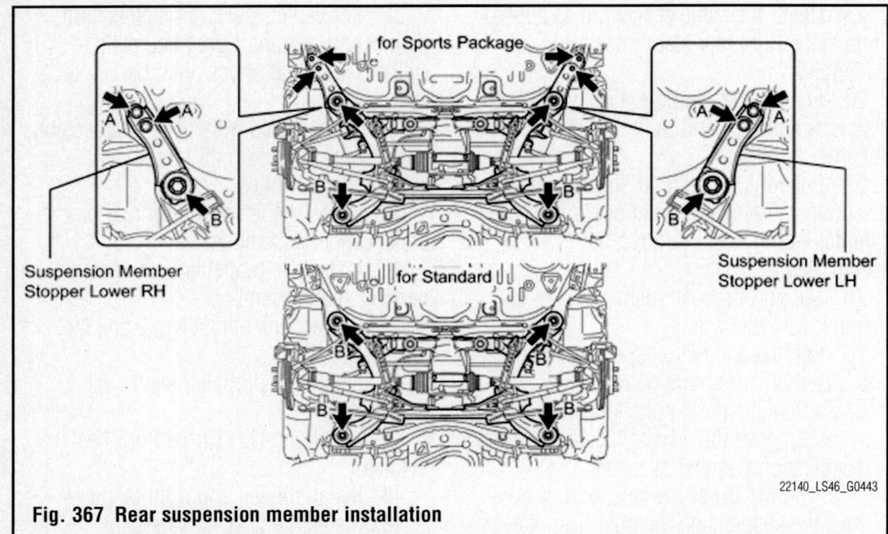

Fig. 367 Rear suspension member installation

WHEEL HUB & BEARING

REMOVAL & INSTALLATION

See Figure 368.

1. Remove the rear wheel.

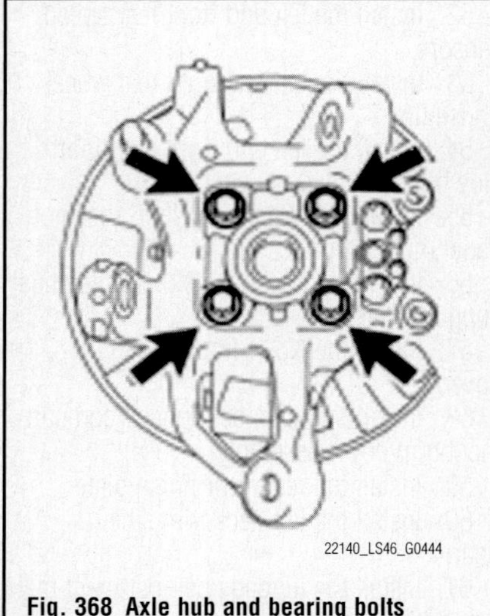

Fig. 368 Axle hub and bearing bolts

2. Remove the bolt and load sensing valve sensor bracket from the toe control link sub-assembly.
3. Remove the bolt and nut, and then disconnect the rear stabilizer link assembly from the rear No. 2 suspension arm assembly.
4. Using SST (SST: 09930-00010) and a hammer, release the staked part of the rear axle shaft nut, and while applying the brakes, remove the rear axle shaft nut.

⁂ WARNING

Release the staked part of the nut completely, otherwise the screw of the drive shaft may be damaged.

5. Remove the 2 bolts and separate the rear disc brake caliper assembly. Hang the caliper with wire or equivalent. Do not damage the brake hose.
6. Disconnect the rear speed sensor.
7. Remove the rear disc.
8. Remove the No. 2 and No. 1 Parking brake shoe return tension spring.
9. Remove the parking brake shoe adjusting screw set.
10. Remove the No. 1 and No. 2 parking brake shoe assemblies.
11. Remove the parking brake shoe lever.
12. For coil suspension, disconnect the rear shock absorber assembly.
13. For air suspension, disconnect the pneumatic cylinder with rear shock absorber assembly.
14. Remove the rear No. 2 suspension arm assembly.
15. Disconnect the rear No. 1 suspension arm assembly.
16. Disconnect the toe control link sub-assembly.
17. Disconnect the rear upper No. 1 and No. 2 control arm assembly.
18. Remove the rear axle assembly.
19. Remove the rear No. 1 wheel bearing dust deflector
    a. Remove the rear axle hub and bearing assembly, as follows:
    b. Hold the axle hub and bearing in a vise between aluminum plates. Do not overtighten the vise.
    c. Remove the 4 bolts and axle hub and bearing from the rear axle carrier.

To install:

20. Hold the axle hub and bearing in a vise between aluminum plates.
21. Install the axle hub and bearing to the rear axle carrier with the 4 bolts and tighten to 72 ft. lbs. (97 Nm).
22. The remainder of installation is the reverse of removal.
23. Perform parking brake shoe bedding.
24. Inspect and adjust rear wheel alignment.
25. Check the speed sensor signal.
26. Adjust headlight aiming.
27. Adjust the object recognition camera.

30

LEXUS

RX350

SPECIFICATIONS AND MAINTENANCE CHARTS

ENGINE AND VEHICLE IDENTIFICATION

Engine						
Code ①	Liters (cc)	Cu. In.	Cyl.	Fuel Sys.	Engine Type	Eng. Mfg.
2GR-FE	3.5 (3456)	210	6	SFI	DOHC	Toyota

Model Year	
Code ②	Year
9	2009
A	2010

SFI: Sequential Fuel Injection

DOHC: Double Overhead Camshaft

NA: Information not available

① Stamped on the left side of the engine block

② 10th digit of the Vehicle Identification Number (VIN)

3768X_RX35_C0001

GENERAL ENGINE SPECIFICATIONS

Year	Model	Engine Displacement Liters	Engine Series ID	Net Horsepower @ rpm	Net Torque @ rpm (ft. lbs.)	Bore x Stroke (in.)	Com-pression Ratio	Oil Pressure @ rpm
2009	RX350	3.5	2GR-FE	270@6200	251@4700	3.70x3.27	10.8:1	36-78@3000
2010	RX350	3.5	2GR-FE	270@6200	251@4700	3.70x3.27	10.8:1	55@3000

3768X_RX35_C0002

ENGINE TUNE-UP SPECIFICATIONS

Year	Engine Displacement Liters	Engine ID	Spark Plug Gap (in.)	Ignition Timing (deg.)*	Fuel Pump (psi)	Idle Speed (rpm)	Valve Clearance	
							Intake	Exhaust
2009	3.5	2GR-FE	0.039-0.043	N/A	44-50	650-750	N/A	N/A
2010	3.5	2GR-FE	0.039-0.043	N/A	44-50	650-750	N/A	N/A

NOTE: The Vehicle Emission Control Information label often reflects specification changes made during production.

The label figures must be used if they differ from those in this chart.

N/A: Not available

3768X_RX35_C0003

CAPACITIES

Year	Model	Engine Displacement Liters	Engine ID	Engine Oil with Filter (qts.)	Transmission (pts.) Auto.*	Transfer Case (pts.)	Drive Axle Front (pts.)	Drive Axle Rear (pts.)	Fuel Tank (gal.)	Cooling System (qts.)
2009	RX350	3.5	2GR-FE	6.4	3.7	2.0	NE	2.0	19.2	①
2010	RX350	3.5	2GR-FE	6.4	3.7	2.0	NE	2.0	19.2	①

NE: Not Equipped

*After draining, add the following amounts, then, fill to the cold full line.

① Non-towing pkg. 7.3 qt. w/o rear air, 9.6 qts. w/ rear air

Towing pkg. 8 qt. w/o rear air, 10.4 qts. w/ rear air

3768X_RX35_C0004

FLUID SPECIFICATIONS

Year	Model	Engine Displ. Liters	Engine Oil	Auto. Trans.	Drive Axle Rear ①	Transfer Case	Power Steering Fluid ②	Brake Master Cylinder	Cooling System
2009	RX350	3.5	5W-30	ATF WS	80W-90	80W-90	NE	DOT 3	SLLC ③
2010	RX350	3.5	5W-30	ATF WS	80W-90	80W-90	NE	DOT 3	SLLC ③

NE: Not Equipped

DOT: Department Of Transpotation

① Oil grade: Hypoid gear oil API GL-5

② The RX350 is not equipped with a power steering pump. It utilizes a power steering motor.

③ Toyota Super Long Life Coolant

3768X_RX35_C0005

VALVE SPECIFICATIONS

Year	Engine Displacement Liters	Engine ID	Seat Angle (deg.)	Face Angle (deg.)	Spring Test Pressure (lbs. @ in.)	Spring Installed Height (in.)	Stem-to-Guide Clearance (in.) Intake	Stem-to-Guide Clearance (in.) Exhaust	Stem Diameter (in.) Intake	Stem Diameter (in.) Exhaust
2009	3.5	2GR-FE	45	44.5	NA	NA	0.0010-0.0024	0.0012-0.0026	0.2154-0.2159	0.2151-0.2157
2010	3.5	2GR-FE	45	44.5	NA	NA	0.0010-0.0024	0.0012-0.0026	0.2154-0.2159	0.2151-0.2157

NA: Information not available

3768X_RX35_C0006

CAMSHAFT AND BEARING SPECIFICATIONS

All measurements are given in inches.

Year	Engine Displacement Liters	Engine VIN	Journal Diameter	Brg. Oil Clearance	Shaft End-play	Runout	Journal Bore	Lobe Lift Intake	Lobe Lift Exhaust
2009	3.5	2GR-FE	①	②	NA	0.0016	NA	1.7447-1.7487	1.7426-1.7465
2010	3.5	2GR-FE	①	②	NA	0.0016	NA	1.7447-1.7487	1.7426-1.7465

NA: Not Available

① No. 1 journal: 1.4152 to 1.4157 inches
Other journals: 1.0220 to 1.0226 inches

② No. 1 journal: 0.0016 to 0.0031 inches
Other journals: 0.00098 to 0.0024 inches

3768X_RX35_C0007

CRANKSHAFT AND CONNECTING ROD SPECIFICATIONS

All measurements are given in inches.

Year	Engine Displacement Liters	Engine ID	Crankshaft Main Brg. Journal Dia.	Crankshaft Main Brg. Oil Clearance	Crankshaft Shaft End-play	Crankshaft Thrust on No.	Connecting Rod Journal Diameter	Connecting Rod Oil Clearance	Connecting Rod Side Clearance
2009	3.5	2GR-FE	2.4011-2.4016	0.0010-0.0019	0.0016-0.0095	2	2.0863-2.0866	0.0018-0.0026	0.0059-0.0157
2010	3.5	2GR-FE	2.4011-2.4016	0.0010-0.0019	0.0016-0.0095	2	2.0863-2.0866	0.0018-0.0026	0.0059-0.0157

3768X_RX35_C0008

PISTON AND RING SPECIFICATIONS

All measurements are given in inches.

Year	Engine Displacement Liters	Engine ID	Piston Clearance	Ring Gap Top Comp.	Ring Gap Bottom Comp.	Ring Gap Oil Control	Ring Side Clearance Top Comp.	Ring Side Clearance Bottom Comp.	Ring Side Clearance Oil Control
2009	3.5	2GR-FE	0.0018-0.0020	0.0098-0.0138	0.0197-0.0236	0.0039-0.0157	0.0008-0.0028	0.0008-0.0024	0.0028-0.0059
2010	3.5	2GR-FE	0.0018-0.0020	0.0098-0.0138	0.0197-0.0236	0.0039-0.0157	0.0008-0.0028	0.0008-0.0024	0.0028-0.0059

3768X_RX35_C0009

TORQUE SPECIFICATIONS

All readings in ft. lbs.

Year	Engine Displacement Liters	Engine ID	Cylinder Head Bolts	Main Bearing Bolts	Rod Bearing Bolts	Crankshaft Damper Bolts	Flywheel Bolts	Manifold Intake	Manifold Exhaust	Spark Plugs	Oil Pan Drain Plug
2009	3.5	2GR-FE	①	②	③	184	61	15	15	18	33
2010	3.5	2GR-FE	①	②	③	184	61	15	15	18	33

① Step 1: 10 mm bolts to 27 ft. lbs.
Step 2: 10mm point cap bolts plus 90 degrees
Step 3: 10mm point cap bolts plus 90 degrees
Step 4: Front bolts to 22 ft. lbs.

② Step 1: 16 cap cap bolts 45 ft. lbs.
Step 2: 16 cap bolts plus 90 degrees
Step 3: 8 side bolts to 38 ft. lbs.

② Step 1: 18 ft. lbs.
Step 2: Plus 90 degrees

3768X_RX35_C0010

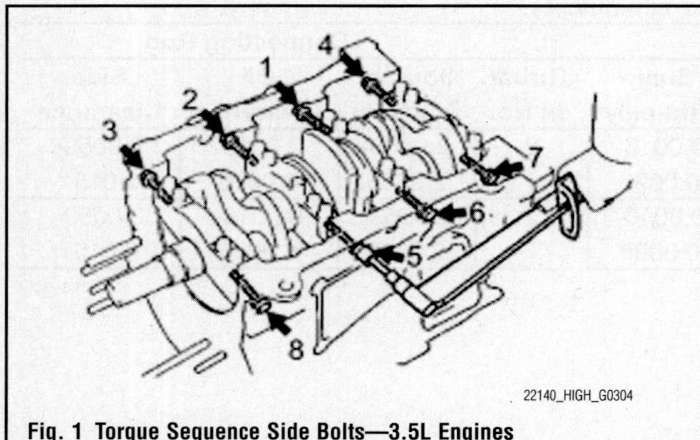

Fig. 1 Torque Sequence Side Bolts—3.5L Engines

WHEEL ALIGNMENT

Year	Model		Caster Range (+/-Deg.)	Caster Preferred Setting (Deg.)	Camber Range (+/-Deg.)	Camber Preferred Setting (Deg.)	Toe-in (in.)	Steering Axis Inclination (Deg.)
2009	RX350 without air suspension	2WD F	0.75	+2.75	0.75	-0.67	0+/-0.2	10.75+/-0.75
		4WD F	0.75	+2.75	0.75	-0.58	0+/-0.2	10.75+/-0.75
		2WD R	NA	NA	0.75	-1.33	0.24+/-0.16	—
		4WD R	NA	NA	0.75	-0.83	0.24+/-0.16	—
	with air suspension	2WD F	0.75	+2.85	0.75	-0.67	0+/-0.2	10.7+/-0.75
		4WD F	0.75	+2.83	0.75	-0.62	0+/-0.2	10.58+/-0.75
		2WD R	NA	NA	0.75	-1.35	0.3+/-0.2	—
		4WD R	NA	NA	0.75	-0.92	0.3+/-0.2	—
2010	RX350 without air suspension	2WD F	0.75	+2.75	0.75	-0.67	0+/-0.2	10.75+/-0.75
		4WD F	0.75	+2.75	0.75	-0.58	0+/-0.2	10.75+/-0.75
		2WD R	NA	NA	0.75	-1.33	0.24+/-0.16	—
		4WD R	NA	NA	0.75	-0.83	0.24+/-0.16	—
	with air suspension	2WD F	0.75	+2.85	0.75	-0.67	0+/-0.2	10.7+/-0.75
		4WD F	0.75	+2.83	0.75	-0.62	0+/-0.2	10.58+/-0.75
		2WD R	NA	NA	0.75	-1.35	0.3+/-0.2	—
		4WD R	NA	NA	0.75	-0.92	0.3+/-0.2	—

3768X_RX35_C0011

TIRE, WHEEL AND BALL JOINT SPECIFICATIONS

Year	Model	OEM Tires Standard	OEM Tires Optional	Tire Pressures (psi) Front	Tire Pressures (psi) Rear	Wheel Size	Ball Joint Inspection	Lug Nut Torque (ft. lbs.)
2009	RX350	P225/65R17	P235/55R18	30	30	6.5-J	①	76
2010	RX350	P225/65R17	R235/55R18	30	30	6.5-J	①	76

OEM: Original Equipment Manufacturer

PSI: Pounds Per Square Inch

STD: Standard

OPT: Optional

① Replace if any measurable movement is found.

3768X_RX35_C0012

BRAKE SPECIFICATIONS

All measurements in inches unless noted

Year	Model		Brake Disc Original Thickness	Brake Disc Minimum Thickness	Brake Disc Maximum Runout	Minimum Lining Thickness	Brake Caliper Bracket Bolts (ft. lbs.)	Brake Caliper Mounting Bolts (ft. lbs.)
2009	RX350	F	1.102	1.024	0.0020	0.039	77	25
		R	0.394	0.335	0.0059	0.039	58	32
2010	RX350	F	1.102	1.024	0.0020	0.039	77	25
		R	0.394	0.335	0.0059	0.039	58	32

F: Front

R: Rear

3768X_RX35_C0013

SCHEDULED MAINTENANCE INTERVALS

LEXUS—RX350

TO BE SERVICED	TYPE OF SERVICE	VEHICLE MILEAGE INTERVAL (x1000)																	
		5	10	15	20	25	30	35	40	45	50	55	60	65	70	75	80	85	90
Engine oil & filter	R	✓	✓	✓	✓	✓	✓	✓	✓	✓	✓	✓	✓	✓	✓	✓	✓	✓	✓
Automatic transmission fluid	S/I			✓			✓			✓			✓			✓			✓
Ball joints & dust covers	S/I			✓			✓			✓			✓			✓			✓
Bolts & nuts on chassis & body	S/I			✓			✓			✓			✓			✓			✓
Brake line pipes & hoses	S/I			✓			✓			✓			✓			✓			✓
Brake pads & discs (front & rear)	S/I	✓	✓	✓	✓	✓	✓	✓	✓	✓	✓	✓	✓	✓	✓	✓	✓	✓	✓
Brake fluid	R						✓						✓			✓			✓
Rack and pinion assembly	S/I			✓			✓			✓			✓			✓			✓
Steering linkage & boots	S/I			✓			✓			✓			✓			✓			✓
Air cleaner filter	R						✓						✓			✓			✓
Spark plugs ①	R	Replace at 120,000 miles																	
Drive belts	S/I												✓						✓
Exhaust pipes & mountings	S/I			✓			✓			✓			✓			✓			✓
Fuel lines & connections	S/I						✓						✓			✓			✓
Engine coolant ②	S/I			✓			✓			✓			✓			✓			✓
Rear differential & transfer case oil	S/I			✓			✓			✓			✓			✓			✓
Fuel tank cap gasket	S/I						✓						✓			✓			✓
Rotate tires	S/I	✓	✓	✓	✓	✓	✓	✓	✓	✓	✓	✓	✓	✓	✓	✓	✓	✓	✓
Clean air conditioning filter ③	S/I			✓			✓			✓			✓						✓
Axle shaft bolts	S/I			✓			✓			✓			✓						✓
Brake pad thickness and rotor runout	S/I						✓						✓						✓

R: Replace S/I: Service or Inspect

① Spark plugs are replaced at 120,000 miles

② Replace engine coolant at 100,000 miles and then inspect every 15,000 miles

③ Replace air conditioning filter every 30,000 miles

FREQUENT OPERATION MAINTENANCE (SEVERE SERVICE)

If a vehicle is operated under any of the following conditions it is considered severe service:

- Extremely dusty areas.

- 50% or more of the vehicle operation is in 32°C (90°F) or higher temperatures, or constant temperatures below 0°C (32°F).

- Prolonged idling (vehicle operation in stop and go traffic).

- Frequent short running periods (engine does not warm to normal operating temperatures).

- Police, taxi, delivery usage or trailer towing usage.

Air cleaner filter: service or inspect every 5000 miles

Rear differential & transfer case oil: replace every 15,000 miles.

Ball joints & dust covers: service or inspect every 5000 miles.

Bolts & nuts on chassis & body: service or inspect every 5000 miles.

Axle shaft bolts: service or inspect every 5000 miles.

Steering linkage: service or inspect every 5000 miles.

3768X_RX35_C0014

BRAKES — INFORMATION AND PRECAUTIONS

ANTI-LOCK SYSTEMS

- Certain components within the ABS system are not intended to be serviced or repaired individually.
- Do not use rubber hoses or other parts not specifically specified for and ABS system. When using repair kits, replace all parts included in the kit. Partial or incorrect repair may lead to functional problems and require the replacement of components.
- Lubricate rubber parts with clean, fresh brake fluid to ease assembly. Do not use shop air to clean parts; damage to rubber components may result.
- Use only DOT 3 brake fluid from an unopened container.
- If any hydraulic component or line is removed or replaced, it may be necessary to bleed the entire system.
- A clean repair area is essential. Always clean the reservoir and cap thoroughly before removing the cap. The slightest amount of dirt in the fluid may plug an orifice and impair the system function. Perform repairs after components have been thoroughly cleaned; use only denatured alcohol to clean components. Do not allow ABS components to come into contact with any substance containing mineral oil; this includes used shop rags.
- The Anti-Lock control unit is a microprocessor similar to other computer units in the vehicle. Ensure that the ignition switch is **OFF** before removing or installing controller harnesses. Avoid static electricity discharge at or near the controller.
- If any arc welding is to be done on the vehicle, the control unit should be unplugged before welding operations begin.

DISC AND DRUM SYSTEMS

⁂ CAUTION

Dust and dirt accumulating on brake parts during normal use may contain asbestos fibers from production or aftermarket brake linings. Breathing excessive concentrations of asbestos fibers can cause serious bodily harm. Exercise care when servicing brake parts. Do not sand or grind brake lining unless equipment used is designed to contain the dust residue. Do not clean brake parts with compressed air or by dry brushing. Cleaning should be done by dampening the brake components with a fine mist of water, then wiping the brake components clean with a dampened cloth. Dispose of cloth and all residue containing asbestos fibers in an impermeable container with the appropriate label. Follow practices prescribed by the Occupational Safety and Health Administration (OSHA) and the Environmental Protection Agency (EPA) for the handling, processing, and disposing of dust or debris that may contain asbestos fibers.

BRAKES — BLEEDING THE BRAKE SYSTEM

BLEEDING PROCEDURE

BLEEDING & FLUID FILL PROCEDURE

When any work is done on the brake system that includes disconnecting fluid lines, or if air in the brake lines is suspected, bleed the air from the system.

⁂ WARNING

Do not let brake fluid remain on painted surfaces - it will eat away the paint if left on too long. Wash it off immediately.

Before proceeding, fill the brake fluid reservoir with brake fluid: SAE J1703 or FMVSS no. 116 DOT3

Bleeding the Master Cylinder

See Figure 2.

If the master cylinder has been disassembled or if the reservoir becomes empty, bleed the air from the master cylinder.

1. Remove the air cleaner assembly with hose.
2. Disconnect the brake lines from the master cylinder, using 12mm union nut wrench, or a suitable brake line wrench.
3. Have an assistant slowly depress the brake pedal and hold it.
4. Cover the outer holes with your fingers, and have your assistant release the brake pedal.
5. Repeat steps 3 and 4 several times.
6. Connect the brake lines. Tighten as follows:
   a. For 2006–07 models, tighten to 11 ft. lbs. (15 Nm).
   b. For 2008 models, use a torque wrench with a fulcrum length of 250 mm (9.84 inches) and tighten to 14 ft. lbs. (19 Nm). This torque value is effective when the union nut wrench is parallel to the torque wrench.
7. Install the air cleaner assembly with hose.

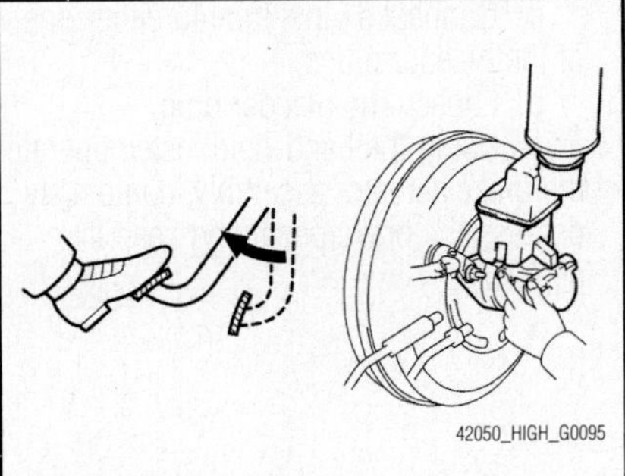

Fig. 2 Use your fingers to cover the outer holes, then release the brake pedal

Bleeding the Brake Lines

See Figure 3.

1. Raise and safely support the vehicle.
2. Connect a piece of vinyl tubing to the brake caliper.
3. Have an assistant depress the brake pedal several times, then loosen the bleeder plug while the pedal is depressed.
4. When fluid stops coming out, tighten the bleeder plug, then release the brake pedal.
5. Repeat steps 2 and 3 until all the air in the fluid has been bled out.
6. Tighten the brake bleeder plug to 73 inch lbs. (8.3 Nm).

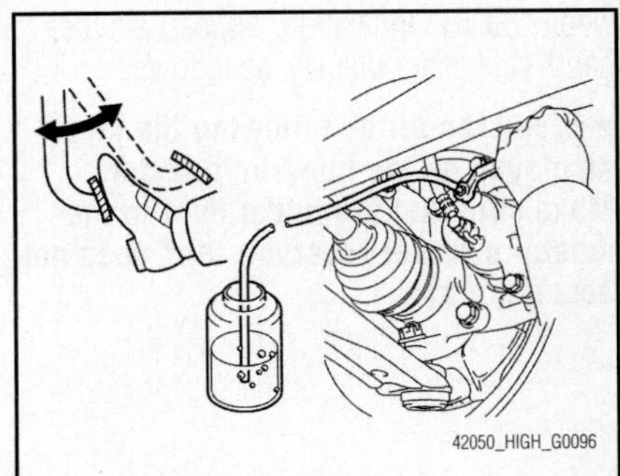

Fig. 3 Bleeding the brake lines at each wheel

7. Repeat the above steps to bleed the air out of the brake line for each wheel.

Bleeding the Actuator Assembly

See Figure 4.

➡After bleeding the air from the brake system, if the height or feel of the brake pedal cannot be obtained, perform air bleeding in the brake actuator assembly with a hand-held tester by following the procedures below.

1. Depress the brake pedal more than 20 times with the engine off.
2. Connect the hand-held tester to the DLC3, then turn the ignition switch to the **ON** position, but do NOT start the engine.
3. Select "AIR BLEEDING" on the hand-held tester.

➡Refer to the hand-held tester operator's manual for more details.

4. Bleed the air out of the regular brake line when "Step 1: Increase" appears on the hand-held tester display, as follows:

➡Bleed the air by following the steps displayed on the hand-held tester. Make sure that the brake fluid in the master cylinder reservoir tank does not become empty.

a. Connect the vinyl tube to either one of the bleeder plugs.
b. Have an assistant depress the brake pedal several times, then loosen the bleeder plug connected to the vinyl tube with the pedal depressed.
c. When fluid stops coming out, tighten the bleeder plug and release the brake pedal.
d. Repeat the previous 2 steps until all the air in the fluid is completely bled out.
e. Tighten the bleeder plug completely to 73 inch lbs. (8.3 Nm).
f. Repeat the above procedures for each wheel to bleed the air out of the brake line.

5. Bleed the air out of the suction line when "Step 2: Inhalation" appears on the hand-held tester display, as follows:

➡Bleed the air by following the steps displayed on the hand-held tester. Make sure that the brake fluid in the master cylinder reservoir tank does not become empty.

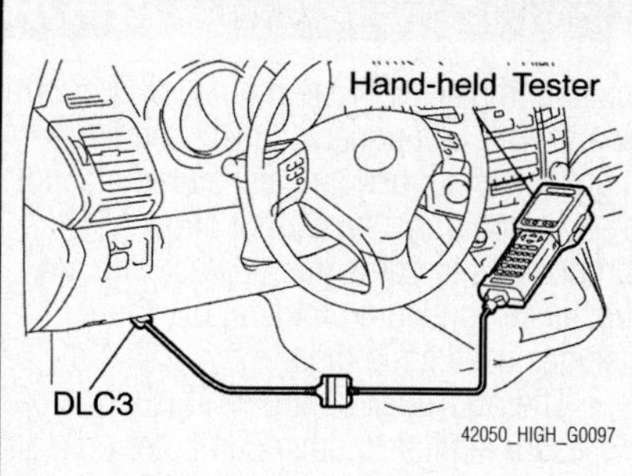

Fig. 4 Connect a suitable hand-held tester to the DLC3 to bleed the actuator assembly

a. Connect the vinyl tube to the bleeder plug at the right front wheel or the right rear wheel and loosen the bleeder plug.
b. Operate the brake actuator assembly to bleed the air using the hand-held tester.

➡This operation stops automatically after 4 seconds. At this time, be sure to release the brake pedal.

c. Check if the operation has stopped by referring to the hand-held tester display.
d. Repeat the previous 2 steps until all air in the fluid is completely bled out.
e. Tighten the bleeder plug completely to 73 inch lbs. (8.3 Nm).
f. Repeat the above procedures to bleed the air out of the brake line for each wheel.

6. Bleed the air out of the pressure reduction line when "Step 3: Decrease" appears on the hand-held tester display, as follows:

➡Bleed the air by following the steps displayed on the hand-held tester. Make sure that the brake fluid in the master cylinder reservoir tank does not become empty.

a. Connect a vinyl tube to either one of the bleeder plugs.
b. Loosen the bleeder plug.
c. Using the hand-held tester, operate the brake actuator assembly, completely depress the brake pedal and keep it depressed.

➡The operation stops automatically after 4 seconds. When performing this procedure continuously, set an interval of at least 20 seconds. When the operation is complete, the brake pedal goes down slightly. This is a normal phenomenon caused when the solenoid opens. During this procedure, the pedal will feel heavy, but completely depress it so that the brake fluid comes out from the bleeder plug. Be sure to keep depressing the brake pedal. Do not depress and release the pedal repeatedly.

d. Tighten the bleeder plug, then release the brake pedal.
e. Repeat the previous 3 steps until all the air in the fluid is completely bled out.
f. Tighten the bleeder plug completely to 73 inch lbs. (8.3 Nm).
g. Repeat the above procedures for each wheel to bleed the air out of the brake line.

7. Bleed the air out of the regular brake line again when "Step 4: Increase" appears on the hand-held tester display, as follows:

➡Bleed the air by following the steps displayed on the hand-held tester. Make sure that the brake fluid in the master cylinder reservoir tank does not become empty.

a. Connect the vinyl tube to either one of the bleeder plugs.
b. Depress the brake pedal several times, then loosen the bleeder plug connected to the vinyl tube with the pedal depressed.

8. When fluid stops coming out, tighten the bleeder plug, then release the brake pedal.

a. Repeat the previous 2 steps until all the air in the fluid is completely bled out.
b. Tighten the bleeder plug completely to 73 inch lbs. (8.3 Nm).
c. Repeat the above procedures for each wheel to bleed the air out of the brake line.

9. Check the fluid level and add fluid if necessary. Use SAE J1703 or FMVSS No. 116 DOT3 Brake fluid.

BRAKE CALIPER

REMOVAL & INSTALLATION

2009 Models

See Figure 5.

1. Before servicing the vehicle, refer to the precautions section.
2. Remove front wheel.
3. Drain brake fluid.

➡Wash the brake fluid off immediately if it adheres to any painted surfaces.

4. Remove the union bolt and the gasket from the disc brake caliper sub-assembly, then disconnect the flexible hose.
5. Hold the front disc brake caliper slide pin, slide pin No. 2 and remove the 2 bolts and brake cylinder assembly.

To install:

6. Install the disc brake caliper assembly with the 2 bolts and tighten to 25 ft. lbs. (34 Nm).

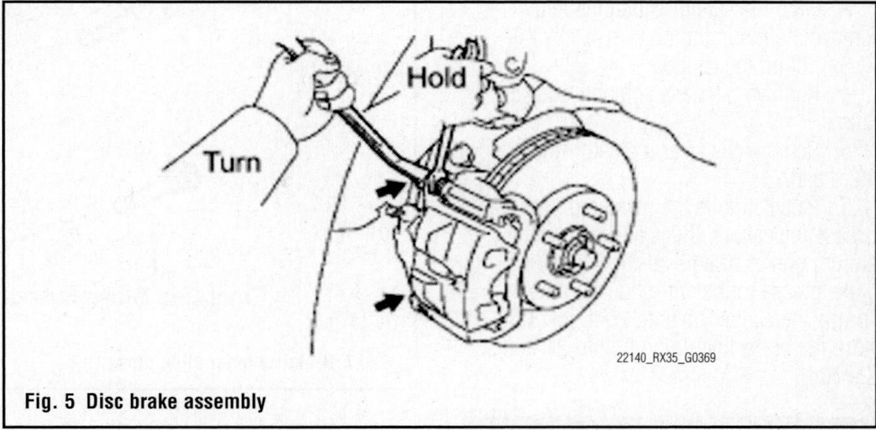

Fig. 5 Disc brake assembly

7. Connect the flexible hose with the union bolt and a new gasket and tighten to 21 ft. lbs. (29 Nm).

➡Install the flexible hose lock securely in the lock hole in the disc brake caliper sub-assembly.

8. Fill reservoir with brake fluid.
9. Bleed master cylinder.
10. Bleed brake line.
11. Check fluid level in reservoir.
12. Check brake fluid leakage.
13. Install front wheel.

2010 Models

See Figure 6.

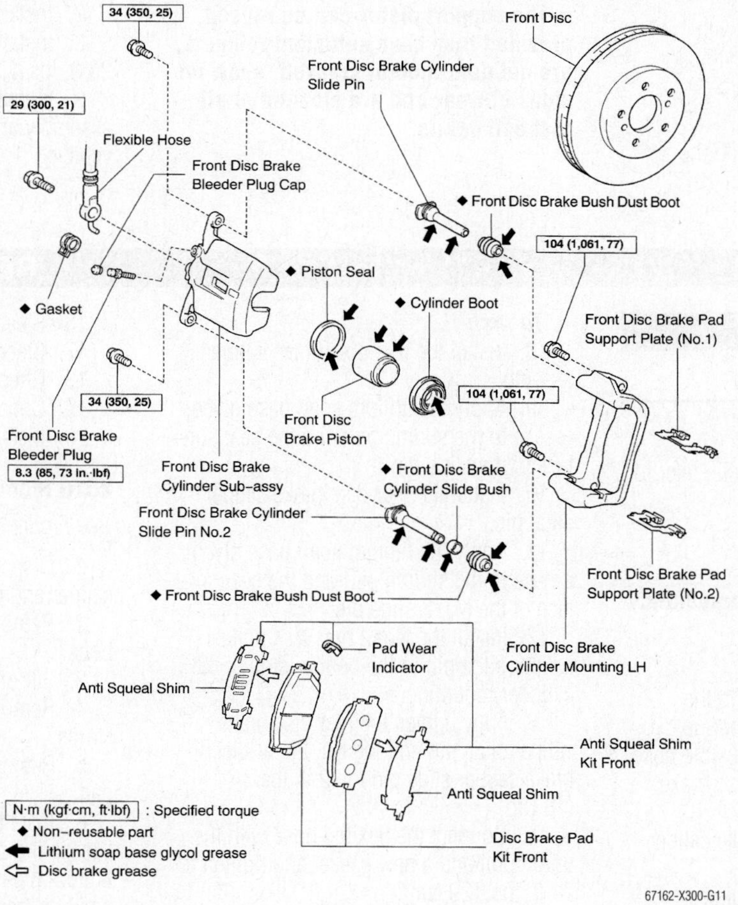

Fig. 6 Front disc brake components

1. Disconnect the brake line from the caliper and plug it.
2. Hold the caliper slide pins and remove the mounting bolts.
3. Lift off the caliper.
4. Remove the pads and anti-squeal shims.
5. Remove the wear indicator from the inner pad.
6. Installation is the reverse of removal. Grease the caliper slides and bolts with lithium grease or equivalent. Apply disc brake grease to the anti-squeal shims. Torque the caliper bolts to 25 ft. lbs. (34 Nm); the brake line union bolt to 21 ft. lbs. (29 Nm).

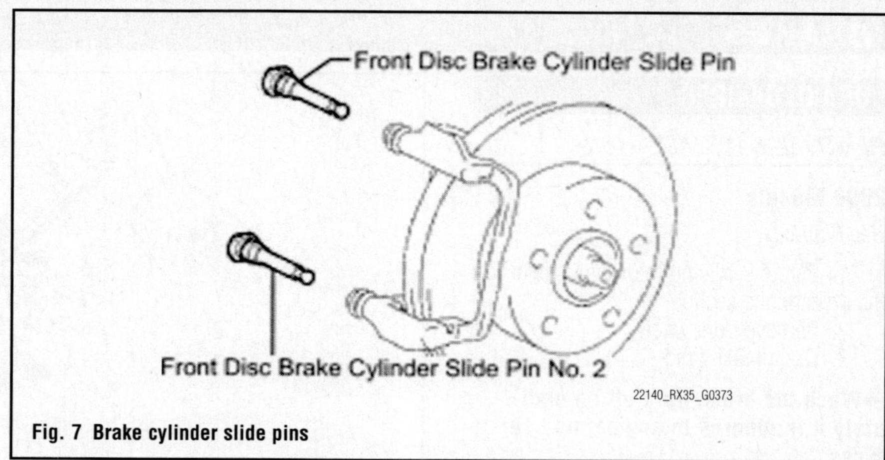

Fig. 7 Brake cylinder slide pins

DISC BRAKE PADS

REMOVAL & INSTALLATION

2009 Models

See Figure 7.

1. Before servicing the vehicle, refer to the Precautions Section.
2. Remove the disc brake caliper slide pins (upper and lower) from the disc brake caliper mounting.
3. Remove brake cylinder.
4. Remove the 2 brake pads with the anti squeal shims from the cylinder mounting.

To install:

5. Install the 2 brake pads with the anti squeal shims to the disc brake caliper mounting.

2010 Models

See Figure 6.

1. Hold the sliding pin and remove the lower bolt.
2. Lift the caliper up and secure it.
3. Remove the pads, 4 shims and wear indicator plate. Remove the 2 pad support plates.

➡The support plates can be reused, provided they have sufficient rebound, are not deformed or cracked, show no signs of wear and are cleaned of all rust and debris.

To install:

➡Always use new shims and wear indicators, even when re-installing the original pads.

4. Install a wear indicator plate on the inner pad.
5. Apply disc brake grease to both sides of the inner anti-squeal shims and install the shims.
6. Install the inner pad with the wear indicator plate facing upwards.
7. Install the outer pad.
8. Install the caliper. Torque the bolt to 25 ft. lbs. (34 Nm).
9. Install the wheel and tire assembly and carefully lower the vehicle.

BRAKES

REAR DISC BRAKES

BRAKE CALIPER

REMOVAL & INSTALLATION

2009 Models

1. Before servicing the vehicle, refer to the precautions section.
2. Remove rear wheel.
3. Drain brake fluid.

➡Wash the brake fluid off immediately if it adheres to any painted surfaces.

4. Remove the union bolt and the gasket from the disc brake caliper sub-assembly, then disconnect the flexible hose.
5. Remove the No. 2 rear disc brake caliper slide pin.
6. Remove the rear disc brake caliper slide pin.
7. Remove the disc brake caliper assembly.

To install:

8. Install the rear disc brake caliper assembly.
9. Apply the lithium soap base glycol grease to the sliding part and the seal surface of the slide pin.
10. Install the rear disc brake caliper slide pin.
11. Apply the lithium soap base glycol grease to the sliding part and the seal surface of the No. 2 slide pin.
12. Install the No. 2 rear disc brake caliper slide pin to the bottom side of the cylinder mounting.
13. Fully tighten the rear disc brake caliper slide pin and the No. 2 rear disc brake caliper slide pin to 32 ft. lbs. (43 Nm).
14. Connect the flexible hose with the union bolt and a new gasket and tighten to 21 ft. lbs. (29 Nm).
15. Fill reservoir with brake fluid.
16. Bleed master cylinder.
17. Bleed brake line.
18. Check fluid level in reservoir.
19. Check brake fluid leakage.
20. Install rear wheel.

2010 Models

See Figure 8.

1. Disconnect the brake line from the caliper and plug it.
2. Remove the caliper mounting bolts.
3. Lift off the caliper.
4. Remove the pads and anti-squeal shims.
5. Remove the wear indicators from each pad.
6. Installation is the reverse of removal. Grease the caliper slides and bolts with lithium grease or equivalent. Apply disc brake grease to the anti-squeal shims. For 2007 models, torque the caliper bolts to 32

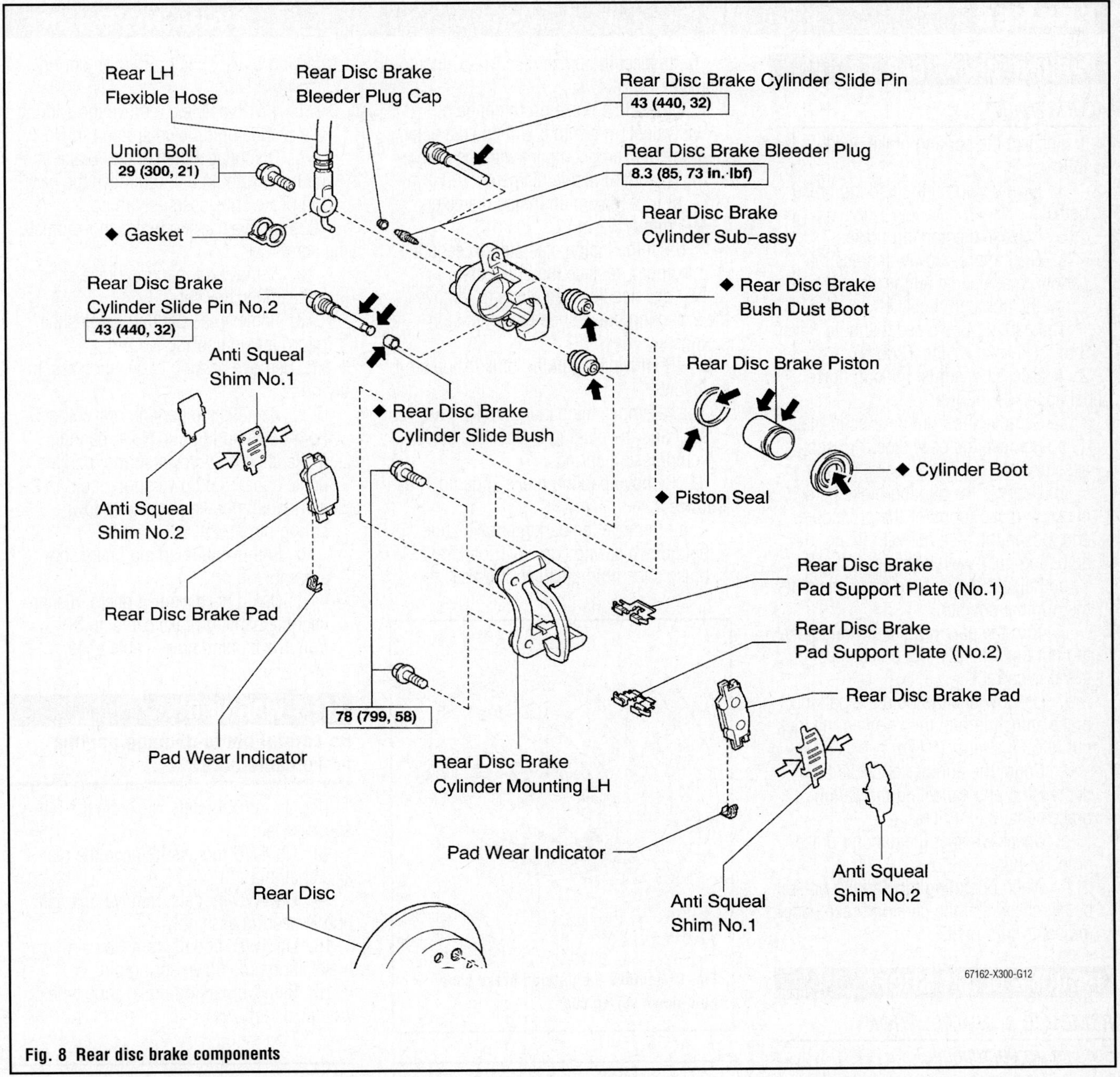

Fig. 8 Rear disc brake components

ft. lbs. (43 Nm); the brake line union bolt to 21 ft. lbs. (29 Nm). For 2008 models, torque the caliper bolts to 25 ft. lbs. (34 Nm); the brake line union bolt to 24 ft. lbs. (33 Nm).

DISC BRAKE PADS

REMOVAL & INSTALLATION

See Figure 8.

1. Disconnect the brake line from the caliper and plug it.
2. Remove the caliper mounting bolts.
3. Lift off the caliper.
4. Remove the pads and anti-squeal shims.
5. Remove the wear indicators from each pad.
6. Installation is the reverse of removal.

Grease the caliper slides and bolts with lithium grease or equivalent. Apply disc brake grease to the anti-squeal shims. For 2007 models, torque the caliper bolts to 32 ft. lbs. (43 Nm); the brake line union bolt to 21 ft. lbs. (29 Nm). For 2008 models, torque the caliper bolts to 25 ft. lbs. (34 Nm); the brake line union bolt to 24 ft. lbs. (33 Nm).

PARKING BRAKE CABLES

ADJUSTMENT

1. Inspect the parking brake pedal travel, as follows:
   a. Firmly step on the parking brake pedal.
   b. Release the parking brake.
   c. Once more, slowly depress the parking brake pedal all the way, and count the number of clicks.
   The parking brake pedal should travel 5 to 7 clicks at 67 lbs. (300 N).
2. If necessary, adjust parking brake pedal travel, as follows:
   a. Remove the front door scuff plate.
   b. Remove the cowl side trim sub-assembly.
   c. Depress the parking brake pedal 5 clicks to make room for the procedure, and loosen the lock nut with fixing adjusting nut by wrench.
   d. Release the parking brake pedal to the original position.
   e. Turn the parking brake wire adjusting nut until the parking brake pedal travel is correct.
   f. Use a wrench to hold the parking brake adjusting nut, then tighten the lock nut to 53 inch lbs. (6 Nm).
   g. Count the number of clicks after depressing and canceling the parking brake pedal 3 to 4 times.
   h. Check whether the parking brake drags or not.
   i. When operating the parking brake pedal, check that the parking brake pedal indicator light is lit.

PARKING BRAKE SHOES

REMOVAL & INSTALLATION

See Figures 9 through 16.

1. Raise and safely support the vehicle.
2. Remove the rear wheel and tire assemblies.
3. Unbolt and remove the rear caliper, but do not disconnect the fluid line. Suspend the caliper out of the way with a piece of wire.
4. Matchmark the brake disc (rotor) to the axle hub.
5. Make sure the parking brake is fully released, then remove the rear brake disc (rotor).

➡If the rotor cannot be easily removed, turn the shoe adjuster until the wheel turns freely.

6. Inspect the brake disc (rotor) inside diameter, as follows:
   a. Using a brake drum gauge or equivalent, measure the inside diameter of the disc and compare with the following: Standard inside diameter: 190 mm (7.48 in.). Maximum inside diameter: 191 mm (7.52 in.)
   b. If the inside diameter exceeds the maximum, replace the brake disc.
7. Use needle-nose pliers to remove the 2 parking brake shoe return tension springs.
8. Remove the parking brake shoe strut, as follows:
   a. Remove the parking brake shoe strut and the parking brake shoe strut compression spring.
9. Remove parking brake shoe no.1, as follows:
   a. Remove the parking brake shoe hold down spring cup No.1, parking brake shoe hold down spring and parking brake shoe hold down spring cup no.2.
   b. FWD vehicles, remove the parking brake shoe hold down spring pin No.1.
   c. Disconnect the parking brake shoe return spring no.2 and remove the parking brake shoe assembly lh no.1.

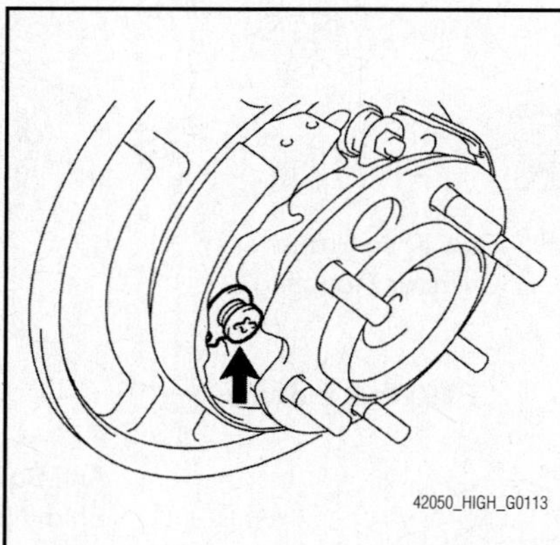

Fig. 9 Remove the parking brake shoe hold-down spring cups

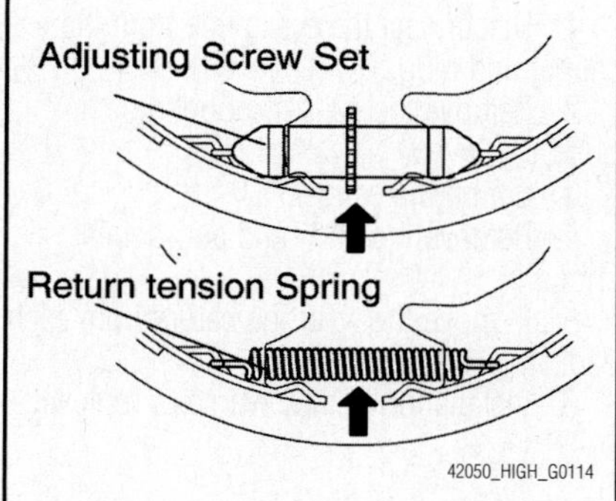

Fig. 10 Remove the parking brake shoe adjusting screw set and shoe return spring

10. Remove parking brake shoe adjusting screw set:
    a. Remove the parking brake shoe adjusting screw set.
    b. Remove the parking brake shoe return tension spring No.2.
11. Remove parking brake shoe assembly no.2:
    a. Remove the parking brake shoe hold down spring cup No.1, parking brake shoe hold down spring, parking brake shoe hold down spring cup no.2 and parking brake shoe hold down spring pin no.2.
    b. Remove the parking brake shoe assembly lh no.2.
    c. Using needle-nose pliers, disconnect the parking brake cable no.3 from the parking brake cable shoe lever.

⁂ WARNING

Be careful not to damage parking brake cable no.3.

12. On 4WD models, separate the rear speed sensor.
13. On 4WD models, remove the rear axle shaft nut.
14. On 4WD models, remove rear axle hub & bearing assembly
15. On 4WD models, remove parking brake shoe hold down spring pin.
16. Remove parking brake shoe type C-washer, as follows:

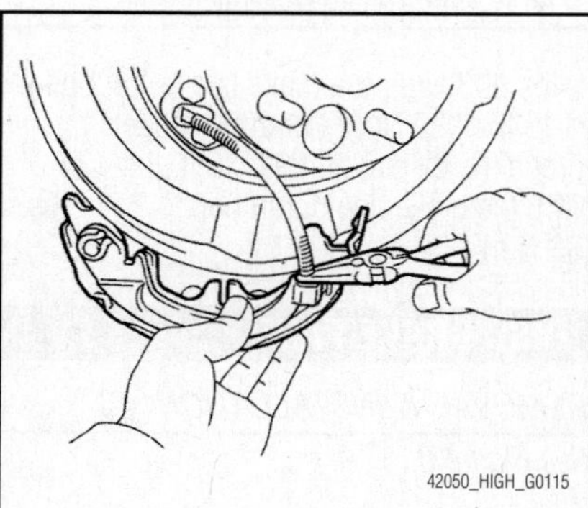

Fig. 11 Using needle-nose pliers, disconnect the parking brake cable no.3 from the parking brake cable shoe lever

a. Using a screwdriver, remove the c-washer.

b. Remove the shim and parking brake shoe lever from the parking brake shoe no.2.

17. Inspect parking brake shoe lining thickness:

a. Using a ruler, measure the thickness of the shoe lining. Standard thickness is 2.5 mm (0.098 in.) and minimum thickness is 1.0 mm (0.039 in.). If the lining thickness is less than or equal to the minimum, or

If there is severe or uneven wear, replace the brake shoe.

18. Inspect brake disc and parking brake shoe lining for proper contact

a. Apply chalk to the inside surface of the disc, then grind down the brake shoe lining to fit disc.

b. If the contact between the brake disc and the shoe lining is improper, repair it using a brake shoe grinder or replace the brake shoe assembly.

To install:

19. Install the parking brake shoe type C-washer, as follows

a. Using a feeler gauge, measure the clearance. Standard clearance: less than 0.35 mm (0.014 in.).

If the clearance is not within the specifications, replace the shim with one of the correct size. The shim sizes: 0.3 mm (0.012 in.), 0.9 mm (0.035 in.) or 0.6 mm (0.024 in.).

b. Using pliers, install the parking brake shoe lever and the shim with a new C-washer.

20. Apply high temperature grease to the shaded parts shown in the illustration of the backing plate which make contact with the shoe.

21. On 4WD models, perform the following:

a. Install the parking brake shoe hold down spring pin.

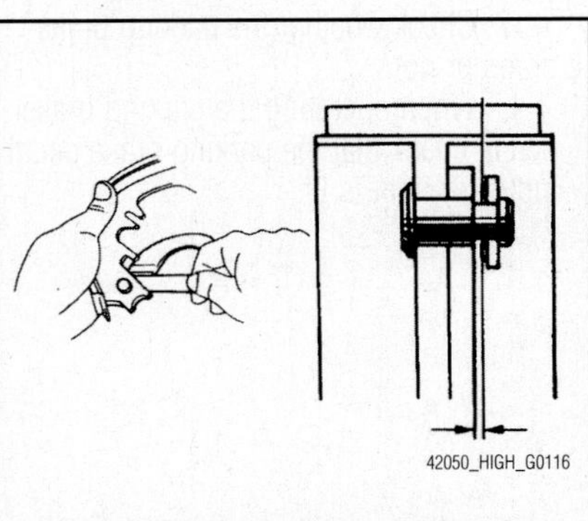

Fig. 12 Installing the parking brake shoe type C-washer

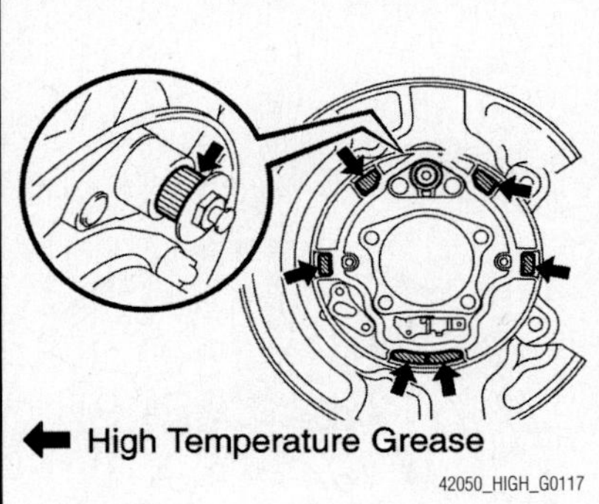

Fig. 13 Apply high temperature grease to the shaded parts of the backing plate which make contact with the shoe

b. Install the rear axle hub & bearing.

c. Install the rear axle shaft nut.

d. Install the rear speed sensor.

22. Install parking brake shoe no.2, as follows:

a. Using needle-nose pliers, connect the parking brake cable no.3 to the parking brake cable shoe lever.

➡Be careful not to damage the parking brake cable no.3.

b. Install the parking brake shoe no.2 with the parking brake shoe hold down spring, parking brake shoe hold down spring cup no.1, parking brake shoe hold down spring cup no.2 and parking brake shoe hold down spring pin no.2.

23. Install the parking brake shoe adjusting screw set, as follows:

a. Apply high temperature grease to the parking brake shoe adjusting bolt and piece.

b. Attach the parking brake shoe return tension spring no.2 to the parking brake shoe no.1 and parking brake shoe assembly no.2.

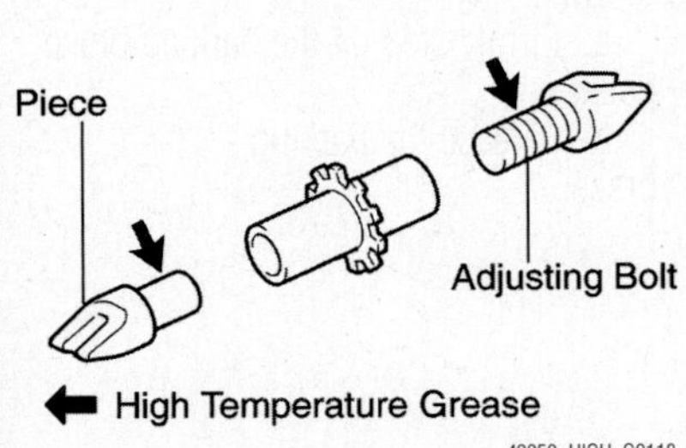

Fig. 14 Apply high temperature grease to the parking brake shoe adjusting bolt and piece

c. Attach the parking brake shoe adjusting screw set to the parking brake shoe no.1 and parking brake shoe no.2.

24. Install parking brake shoe no.1:

a. For FWD models, install the parking brake shoe hold down spring pin no.1.

b. Install the parking brake shoe no.1 with the parking brake shoe hold down spring, parking brake shoe hold down spring cup no.2, parking brake shoe hold down spring cup no.2.

25. attach the parking brake shoe strut and the parking brake shoe strut compression spring to parking brake shoe no.2 and parking brake shoe no.1.

26. Install parking brake shoe return tension spring using needle-nose pliers as shown in the illustration.

➡First install the front side spring then the rear side spring.

27. Check that the parking brake components are properly installed.

✲✲ WARNING

There should be no oil or grease on the friction surface of the shoe lining and disc.

28. For 4WD models, inspect the bearing backlash and axle hub deviation.

29. Install the rear disc (rotor), aligning the matchmarks made during removal.

30. Adjust parking brake shoe clearance, as follows:

a. Temporarily install the hub nuts.

b. Remove the hole plug, turn the adjuster and expand the shoes until the disc locks.

c. Contract the shoe adjuster until the disc rotates smoothly. Standard : return 8 notches

d. Check that the shoe has no brake drag.

e. Install the hole plug.

31. Install the caliper, as outlined earlier in this section.

32. Install the rear wheel and tire assembly and tighten the lug nuts to 76 ft. lbs. (103 Nm).

33. Inspect and adjust the parking brake pedal travel, as outlined in this section.

34. For 4WD models, check the ABS speed sensor signal.

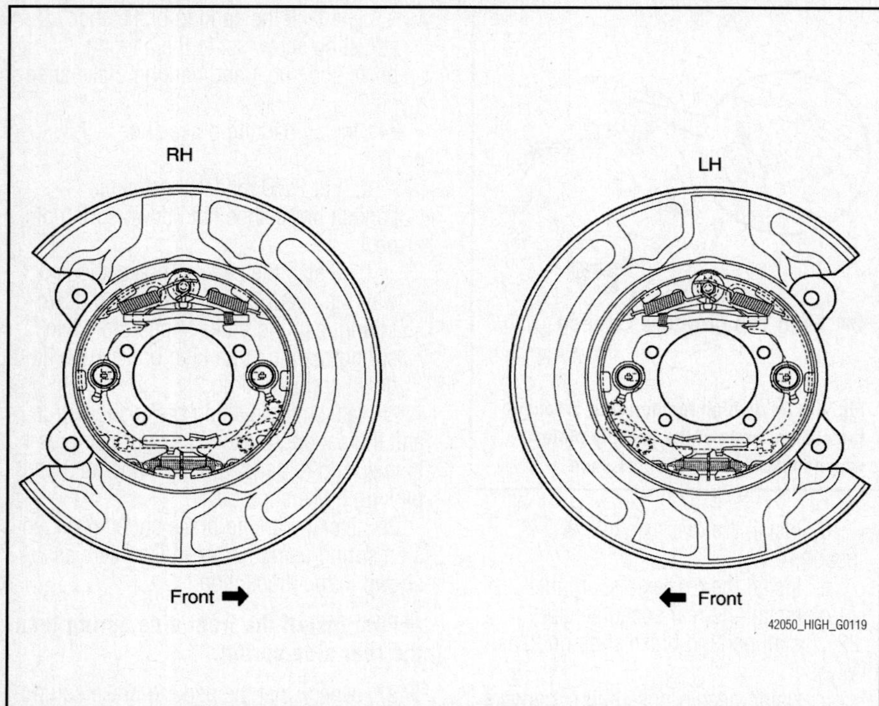

Fig. 15 Check that the parking brake components are properly installed. There should be no oil or grease on the friction surface of the shoe lining and disc

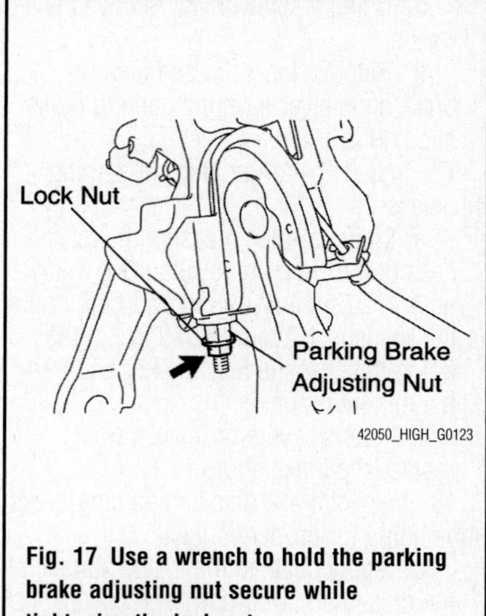

Fig. 17 Use a wrench to hold the parking brake adjusting nut secure while tightening the lock nut

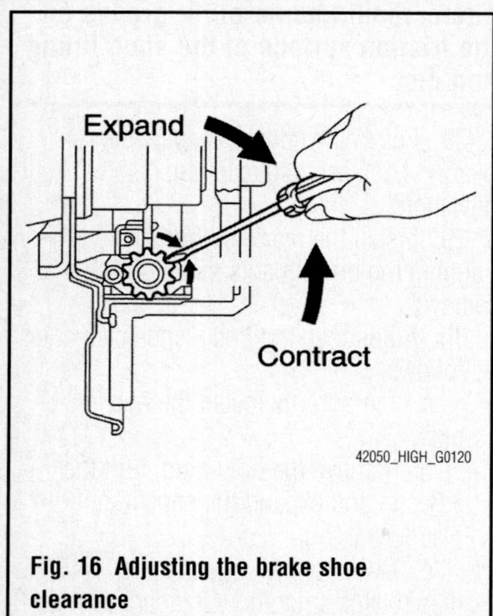

Fig. 16 Adjusting the brake shoe clearance

ADJUSTMENT

See Figures 16 and 17.

1. Raise and safely support the vehicle.
2. Remove the rear wheel and tire assemblies.
3. Adjust parking brake shoe clearance, as follows:
   a. Temporarily install the hub nuts.
   b. Remove the hole plug, turn the adjuster and expand the shoes until the disc locks.
   c. Contract the shoe adjuster until the disc rotates smoothly. Standard : return 8 notches
   d. Check that the shoe has no brake drag.
   e. Install the hole plug.
4. Install the rear wheel and tire assembly and tighten the lug nuts to 76 ft. lbs. (103 Nm).
5. Inspect the parking brake pedal travel, as follows:
   a. Firmly step on the parking brake pedal.
   b. Release the parking brake.
   c. Once more, slowly depress the parking brake pedal all the way, and count the number of clicks. The parking brake pedal should travel 8 to 10 clicks at 67 lbs. (300 N).
6. If necessary, adjust parking brake pedal travel, as follows:
   a. Remove the lower instrument panel finish panel sub-assembly.
   b. Remove the lower instrument panel insert sub- assembly.
   c. Depress the parking brake pedal 5 clicks to make room for the procedure, and loosen the lock nut with fixing adjusting nut by wrench.
   d. Release the parking brake pedal to the original position.
   e. Turn the parking brake wire adjusting nut until the parking brake pedal travel is correct.
   f. Use a wrench to hold the parking brake adjusting nut, then tighten the lock nut to 62 inch lbs. (7 Nm).
   g. Count the number of clicks after depressing and canceling the parking brake pedal 3 to 4 times.
   h. Check whether the parking brake drags or not.
   i. When operating the parking brake pedal, check that the parking brake pedal indicator light is lit.

CHASSIS ELECTRICAL — AIR BAG (SUPPLEMENTAL RESTRAINT SYSTEM)

GENERAL INFORMATION

CAUTION

These vehicles are equipped with an air bag system. The system must be disarmed before performing service on, or around, system components, the steering column, instrument panel components, wiring and sensors. Failure to follow the safety precautions and the disarming procedure could result in accidental air bag deployment, possible injury and unnecessary system repairs.

SERVICE PRECAUTIONS

CAUTION

Disconnect and isolate the battery negative cable before beginning any airbag system component diagnosis, testing, removal, or installation procedures. Wait at least 90 seconds after the ignition switch is turned off and the negative (-) terminal cable is disconnected from the battery before starting the operation. The SRS is equipped with a backup power source, so if work is started within 90 seconds after disconnecting the negative (-) terminal cable from the battery, the SRS may be deployed. Failure to disable the airbag system may result in accidental airbag deployment, personal injury, or death.

DISARMING THE SYSTEM

To avoid personal injury when working on vehicles equipped with an air bag, the negative battery cable must be disconnected and at least 90 seconds must elapse before working on the system. Failure to do so may result in deployment of the air bag.

ARMING THE SYSTEM

To arm the system after service is finished, connect the negative battery cable.

CLOCKSPRING CENTERING

See Figure 18.

1. Check that the ignition switch is **OFF**.
2. Check that the battery negative (-) terminal is disconnected.

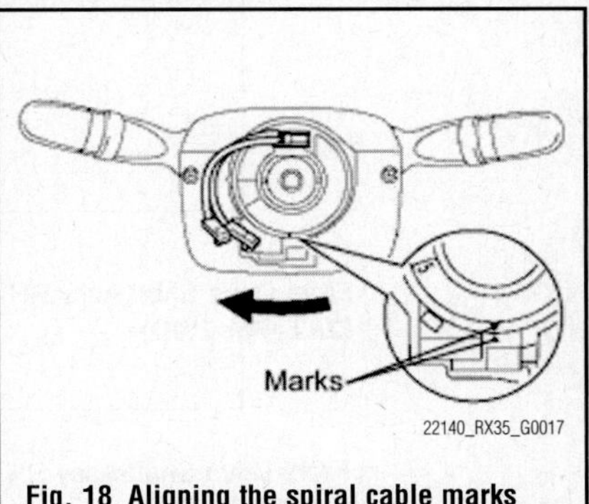

Fig. 18 Aligning the spiral cable marks

CAUTION

After removing the terminal, wait for at least 90 seconds before starting the operation.

3. Rotate the spiral cable counterclockwise slowly by hand until it feels firm.
4. Rotate the spiral cable clockwise approximately 2.5 turns to align the marks.

➡Do not turn the spiral cable by the airbag wire harness.

➡The spiral cable will rotate approximately 2.5 turns to both the left and right from the center.

DRIVE TRAIN

FRONT DRIVESHAFT

REMOVAL & INSTALLATION

Refer to Driveshaft in Rear Drive Axle in this section.

FRONT HALFSHAFT

REMOVAL & INSTALLATION

See Figures 19 and 20.

1. Remove the engine under cover assembly.
2. Remove the No. 2 engine under cover.
3. Drain automatic transaxle fluid.
   a. Remove the drain plug, gasket and drain ATF
   b. Install a new gasket and the drain plug and tighten to 36 ft. lbs. (49 Nm).
4. Drain transfer oil
5. Remove the front wheel.
6. Remove the left front axle hub nut.
   a. Using SST(SST: 09930-00010) or equivalent and a hammer, unstake the staked part of the left axle hub nut.

➡Loosen the staked part of the nut completely, otherwise the screw of the driveshaft may be damaged.

   b. While applying the brakes, remove the left lock axle hub nut.
7. Remove the nut and separate the left stabilizer link assembly.

➡If the ball joint turns together with the nut, use a hexagon wrench (6 mm) to hold the stud.

8. Separate the left front speed sensor, as follows:
   a. Remove the bolt and clip, and separate the sensor wire and hose from the shock absorber.

➡Be careful not to damage the speed sensor.

   b. Remove the bolt, and separate the left speed sensor from the steering knuckle.

➡Do not allow foreign matter to adhere to the speed sensor.

9. Separate the left tie rod end sub-assembly, as follows:
   a. Remove the cotter pin and nut.
10. Separate the No. 1 lower front suspension arm sub-assembly, as follows:
   a. Remove the bolt and 2 nuts, and separate the No. 1 front suspension arm sub-assembly lower from the lower ball joint.
11. Using a plastic hammer, separate the driveshaft from the axle hub.

➡Be careful not to damage the boot and speed sensor rotor.

12. Using SST(SST: 09520-01010, SST: 09520-24010) or equivalent, remove the front driveshaft assembly.

➡Be careful not to damage the transaxle case oil seal, inboard joint

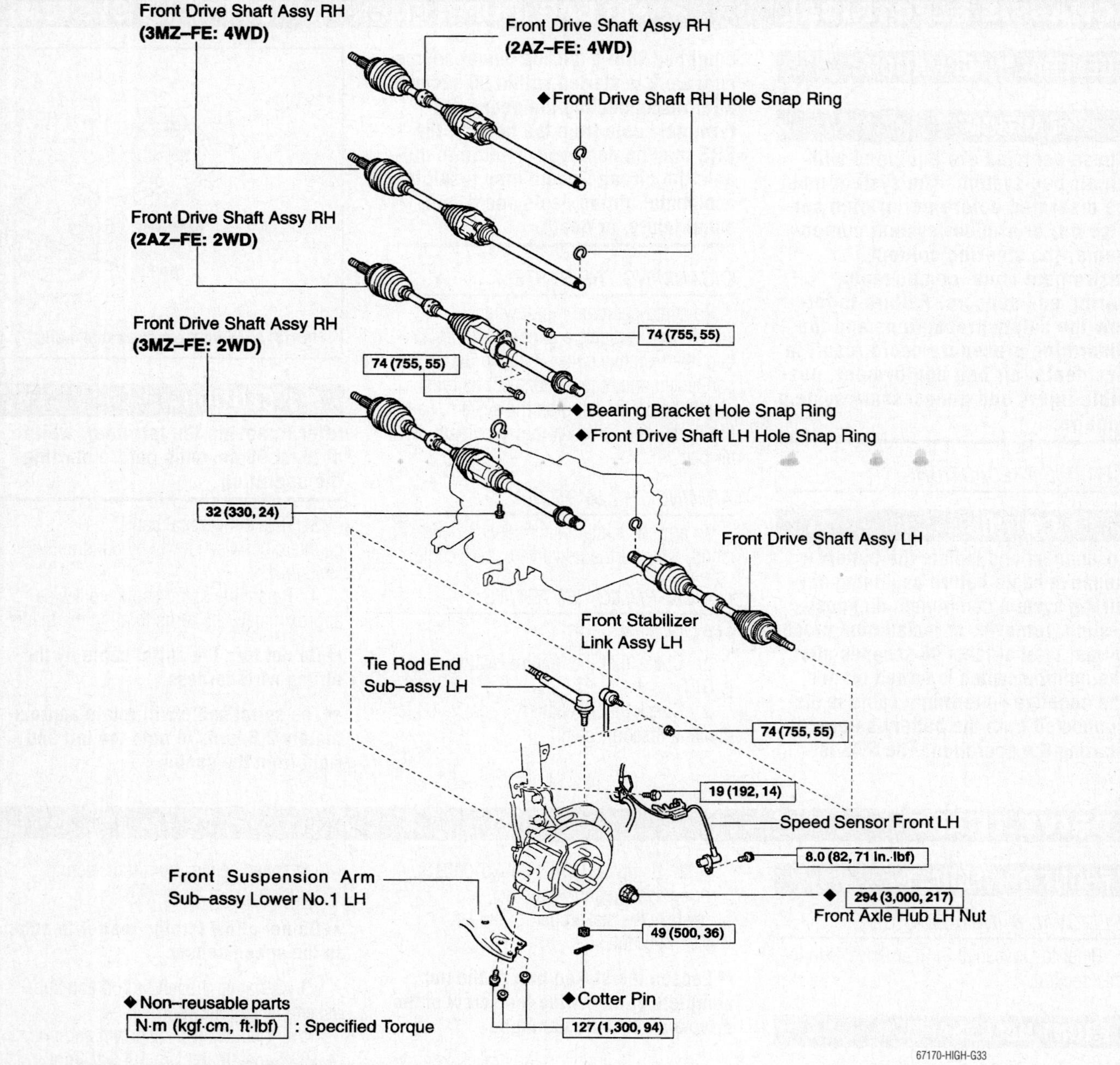

Front Drive Shaft Assy RH (3MZ–FE: 4WD)
Front Drive Shaft Assy RH (2AZ–FE: 4WD)
◆ Front Drive Shaft RH Hole Snap Ring
Front Drive Shaft Assy RH (2AZ–FE: 2WD)
Front Drive Shaft Assy RH (3MZ–FE: 2WD)
74 (755, 55)
74 (755, 55)
◆ Bearing Bracket Hole Snap Ring
◆ Front Drive Shaft LH Hole Snap Ring
32 (330, 24)
Front Drive Shaft Assy LH
Front Stabilizer Link Assy LH
Tie Rod End Sub–assy LH
74 (755, 55)
19 (192, 14)
Speed Sensor Front LH
8.0 (82, 71 in.·lbf)
Front Suspension Arm Sub–assy Lower No.1 LH
◆ 294 (3,000, 217)
Front Axle Hub LH Nut
49 (500, 36)
◆ Non–reusable parts
N·m (kgf·cm, ft·lbf) : Specified Torque
◆ Cotter Pin
127 (1,300, 94)
67170-HIGH-G33

Fig. 19 Front halfshaft and related parts—3.5L engine

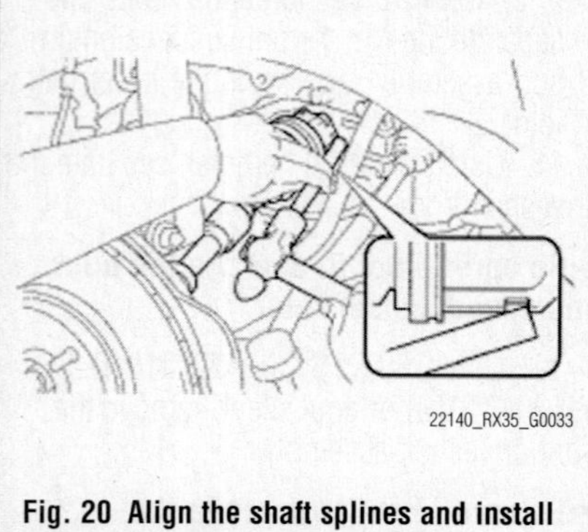

Fig. 20 Align the shaft splines and install the driveshaft assembly

boot and driveshaft dust cover or drop the driveshaft assembly.

13. For 2WD vehicles, remove the right front driveshaft assembly, as follows:

a. Using a screwdriver, remove the bearing brake hole snap ring.

b. Remove the bolt and front driveshaft assembly from the driveshaft bearing bracket.

14. For 4WD vehicles, remove the right front driveshaft assembly, as follows:

a. Using SST(SST: 09520-01010, SST: 09520-24010) or equivalent, remove the front driveshaft assembly.

➡When removing and installing the right front driveshaft assembly in 4WD vehicle, be sure to first drain all the transaxle oil and transfer oil. If removal and installation is carried out without draining these oils, the transfer oil will flow into the transaxle side. Extensive cleaning will be required if the two oils mix.

➡Do not damage the oil seal and dust cover.

➡Move the driveshaft assembly while keeping it level.

To install:

15. Install the left front driveshaft assembly, as follows:
  a. Coat the spline of the inboard joint shaft assembly with ATF.
  b. Align the shaft splines and install the driveshaft assembly with a brass bar and hammer.

➡Set the snap ring with the opening side facing down.

➡Be careful not to damage the driveshaft dust cover, boot and oil seal.

➡Move the driveshaft assembly while keeping it level.

16. For 2WD vehicles, install the right front driveshaft assembly, as follows:
  a. Using a screwdriver, install a new bearing bracket hole snap ring.

➡Do not damage the oil seal and boot.

  b. Install the bolt and tighten to 24 ft. lbs. (32 Nm).

17. For 4WD vehicles, install the right front driveshaft assembly by using the same procedures as for the left side, as described above.

➡Set the snap ring with the opening side facing downward.

➡Be careful not to damage the transaxle case oil seal, inboard joint boot and driveshaft dust cover.

18. Install the left driveshaft assembly to the left front axle assembly.

➡Be careful not to damage the outboard joint boot or speed sensor rotor.

19. Install the lower ball joint to the lower front suspension arm sub-assembly with the bolt and nuts and tighten to 94 ft. lbs. (127 Nm).

20. Install the left tie rod end sub-assembly, as follows:
  a. Install the tie rod end to the steering knuckle with the nut and tighten to 36 ft. lbs. (49 Nm).
  b. Install a new cotter pin.

➡If the holes for the cotter pin are not are not aligned, tighten the nut up to 60° further.

21. Install the left front speed sensor to the steering knuckle with the bolt and tighten to 71 inch lbs. (8 Nm).

➡Prevent foreign matter from adhering to the speed sensor.

22. Install the flexible hose and the speed sensor to the shock absorber with the bolt and set the clip of sensor on knuckle and tighten to 14 ft. lbs. (19 Nm).

➡Be careful not to damage the speed sensor, allow foreign matter to adhere to the speed sensor, or twist the sensor wire when installing the speed sensor.

23. Install the left stabilizer link assembly with the nut and tighten to 55 ft. lbs. (74 Nm).

➡If the ball joint turns together with nut. use a hexagon (6 mm) wrench to hold the stud.

24. Using s socket wrench (30 mm), install a new left axle hub nut and tighten to 217 ft. lbs. (294 Nm).
25. Using a chisel and hammer, stake the left axle hub nut.
26. Install the front wheel and tighten the lug nuts to 76 ft. lbs. (103 Nm).
27. Add automatic transaxle fluid.
28. Inspect automatic transaxle fluid.
29. For 4wd vehicles, add transfer oil.
30. Inspect transfer oil.
31. Adjust front wheel alignment.
32. Install the No. 2 engine under cover.
33. Install the engine under cover assembly.
34. Check the ABS speed sensor signal.

REAR DRIVESHAFT

REMOVAL & INSTALLATION

See Figures 21 through 27.

1. Depress the brake pedal and hold it.
2. Using a hexagon wrench (6 mm), loosen the cross groove joint set bolts 1/2 turn.

➡Put a piece of cloth or equivalent into the inside of the universal joint cover so that the boot does not touch the inside of the universal joint cover.

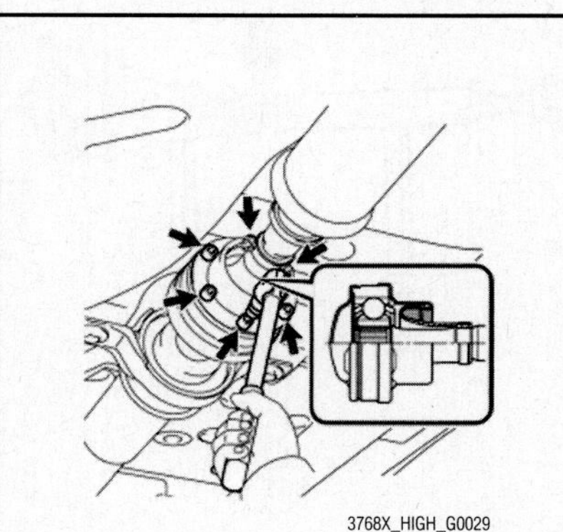

Fig. 21 Locating the cross groove joint set bolts

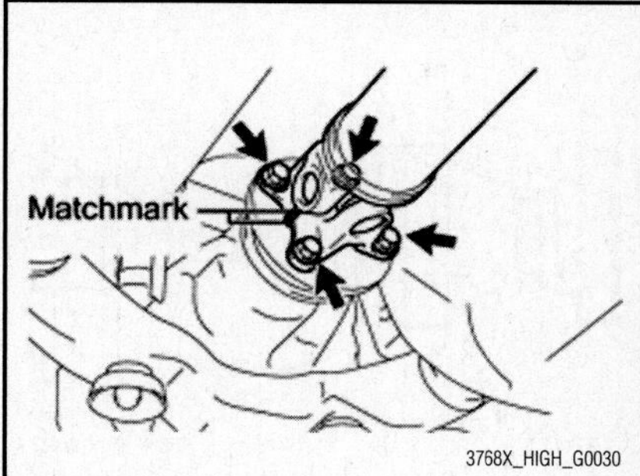

Fig. 22 Placing the matchmark and remove the 4 nuts, 4 bolts and 4 washers

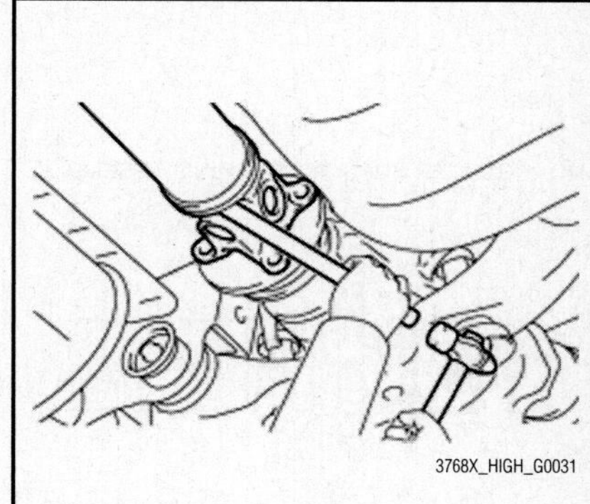

Fig. 23 Using a brass bar and a hammer, remove the propeller shaft with center bearing shaft assembly

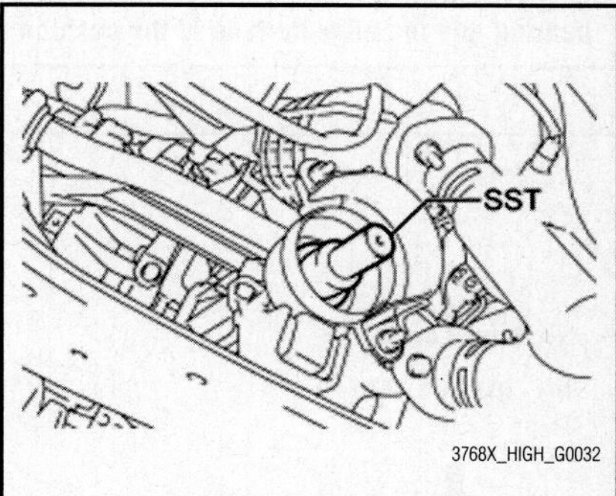

Fig. 24 Insert SST 09325-20010 into the transfer to prevent oil leakage

3. Place matchmarks on the rear propeller shaft and rear drive pinion flange sub-assembly.
4. Remove the 4 nuts, 4 bolts and 4 washers.
5. Remove the 4 bolts and 4 adjusting shims.
6. Using a brass bar and a hammer, remove the propeller shaft with center bearing shaft assembly.
7. Insert SST 09325-20010 into the transfer to prevent oil leakage.

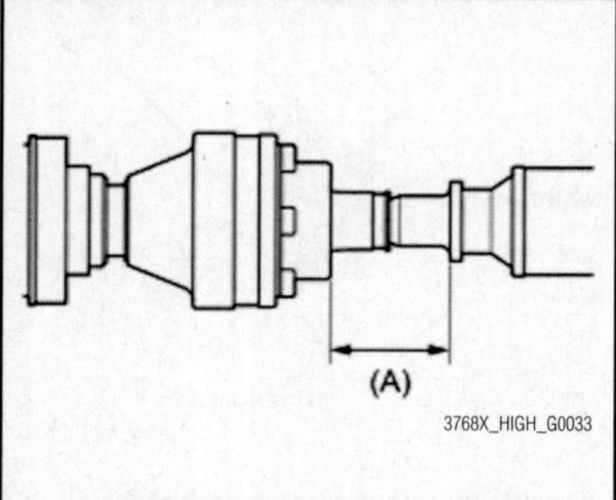

Fig. 25 Adjust the dimension between the rear side of the cover and shaft

To install:

8. Align the matchmarks on the propeller shaft flange and differential companion flange, and connect the shaft with the 4 bolts, 4 washers and 4 nuts.
9. Remove SST from the transaxle.
10. Insert the yoke into the transaxle.
11. Install the 4 adjusting shims and propeller shaft with center bearing, and temporarily tighten the 4 bolts.
12. Tighten the 4 bolts to 54 ft. lbs. (74 Nm).
13. Fully tighten propeller with center bearing shaft assembly:

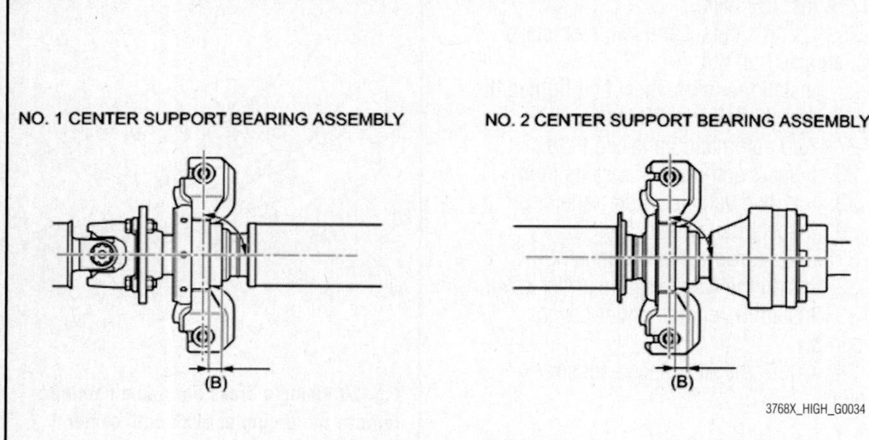

Fig. 26 Adjust the front and rear dimensions between the edge surface of the center support bearing and the edge surface of the cushion

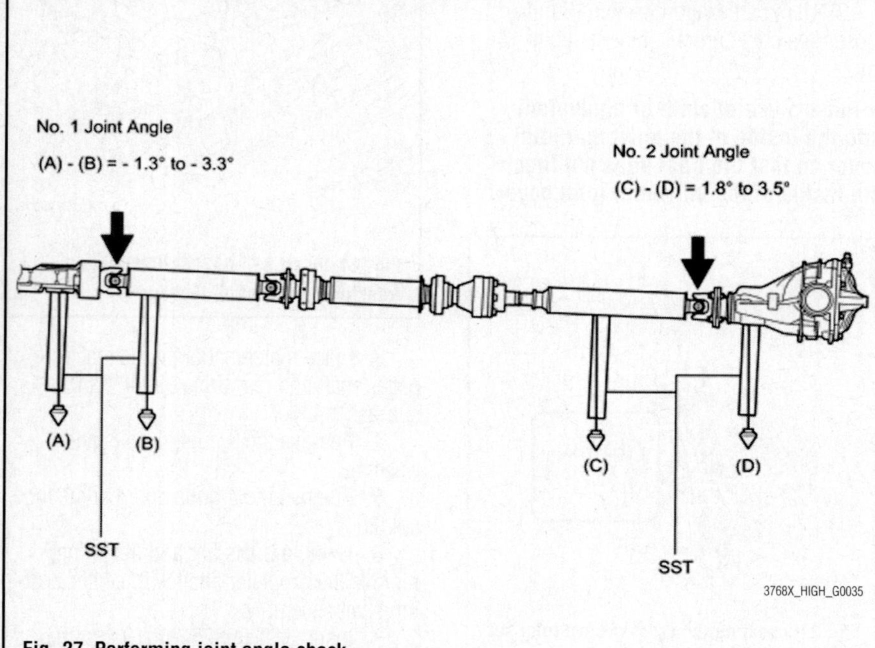

Fig. 27 Performing joint angle check

a. Remove the piece of cloth from the joint.

b. Using a hexagon wrench (6 mm), tighten the 6 bolts. Tighten to 19 ft. lbs. (26 Nm).

c. With the vehicle unloaded, adjust the dimension between the rear side of the cover and shaft as shown in the illustration. (A): 58.0 +/- 0.5 mm (2.283 +/- 0.02 in.)

d. Under the same condition as above, adjust the front and rear dimensions between the edge surface of the center support bearing and the edge surface of the cushion respectively as shown, and then tighten the bolts. (B): 12.5 +/- 1.0 mm (0.492 +/- 0.039 in.). Tighten to 27 ft. lbs. (37 Nm).

e. Check that the center line of the bracket is at the right angle in the shaft axial direction.

f. If any vibration or noise occurs, perform joint angle check as follows and replace the adjusting shim with a proper one.

- Turn the propeller shaft several times by hand to stabilize the center support bearings.
- Using a jack, raise and lower the differential to stabilize the differential mounting cushion.
- Remove the transfer dynamic damper.
- Using SST 09370-50010, measure the transfer installation angle (A) and front propeller shaft installation angle (B). No. 1 joint angle: (A) - (B) = -1.3° to -3.3°
- Using SST 09370-50010, measure the rear propeller shaft installation angle (C) and rear differential shaft installation angle (D). No. 2 joint angle: (C) - (D) = 1.8° to 3.5°

➡If the measured angle is not within the specification, adjust it with the center support bearing adjusting shim.

14. Install the transfer dynamic damper. Tighten to 19 ft. lbs. (26 Nm).
15. Inspect and adjust the transfer oil.

REAR HALFSHAFT

REMOVAL & INSTALLATION

See Figures 28 through 32.

1. Raise and support the vehicle.
2. Remove the rear wheel.
3. Using a hexagon wrench (10 mm), remove the rear differential filler plug

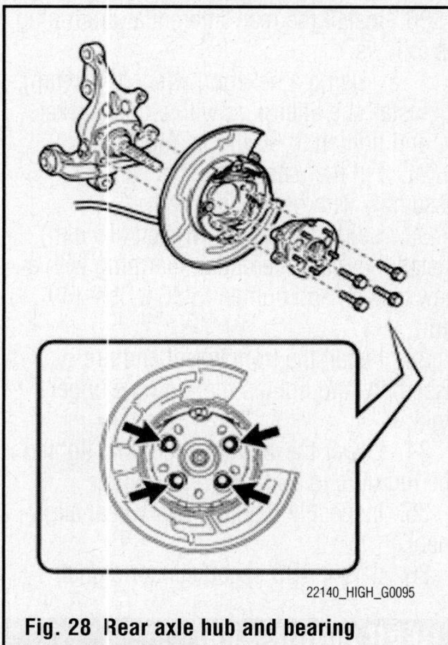

Fig. 28 Rear axle hub and bearing assembly and bolts

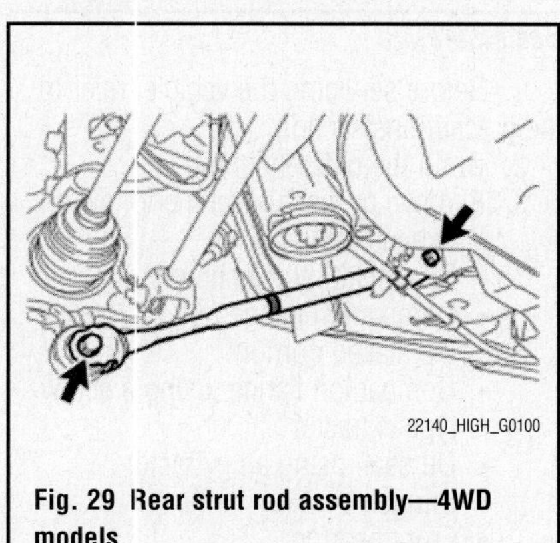

Fig. 29 Rear strut rod assembly—4WD models

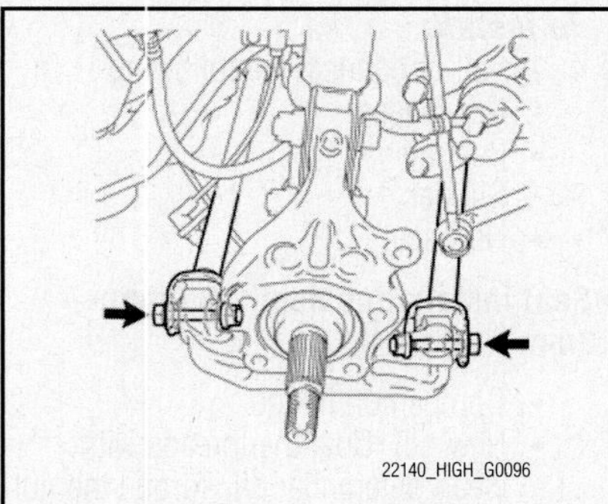

Fig. 30 Rear axle carrier sub-assembly bolts

and rear differential filler plug gasket.

4. Using a hexagon wrench (10 mm), remove the rear differential drain plug and rear differential drain plug gasket, and drain the oil.

5. Remove the bolt and separate the rear speed sensor from the rear axle carrier sub-assembly.

➡Keep the sensor tip and rear speed sensor installation hole free from foreign matter.

6. Using Special Service Tool (SST) 09930-00010 or equivalent, along with a hammer, release the staked part of the rear axle shaft nut.

➡Loosen the staked part of the nut completely, otherwise the threads of the driveshaft may be damaged.

7. While applying the brakes, remove the rear axle shaft nut.

8. Remove the 2 caliper bracket bolts and separate the rear disc brake caliper assembly.

➡Use wire or an equivalent tool to keep the brake caliper from hanging down by the flexible hose.

9. Put matchmarks on the rear disc and the axle hub.

10. Release the parking brake and remove the rear disc.

➡If the disc cannot be removed easily, turn and press firmly the shoe adjuster until the wheel comes free.

11. Remove rear axle hub and bearing assembly, as follows:

a. Put matchmarks on the driveshaft and axle hub.

➡Do not punch the marks.

b. Remove the 4 bolts and the rear axle hub and bearing assembly.

➡Do not rotate the driveshaft with the rear axle hub and bearing assembly removed.

➡Use wire or an equivalent tool to keep the parking brake assembly from hanging down by the parking brake cable assembly.

12. Remove the bolt and the nut, and separate the No. 3 parking brake cable assembly.

13. Remove the 2 bolts, the 2 nuts, and the rear strut rod assembly.

➡Since stopper nuts are used, loosen the bolts.

14. Remove the rear axle carrier sub-assembly, as follows:

a. Loosen the 2 bolts.

➡Since stopper nuts are used, loosen the bolts.

b. Remove the 2 bolts and 2 nuts, and separate the rear shock absorber with coil spring (lower side) from the rear axle carrier sub-assembly.

➡Be careful not to damage the outboard joint boot or the speed sensor rotor.

c. Remove the 2 bolts, the 2 nuts, and the rear axle carrier sub-assembly.

➡Be careful not to damage the outboard joint boot or the speed sensor rotor.

15. Using a slide hammer (SST: 09520-01010, SST: 09520-24010 or equivalent), remove the rear driveshaft assembly (halfshaft) as shown in the illustration.

➡Remove the rear driveshaft assembly while keeping it level.

To install:

16. Align the shaft splines and install the rear driveshaft assembly (halfshaft) with a brass bar and hammer.

➡Set the snap ring with the opening facing downward.

➡Be careful not to damage the oil seal, boot, or dust cover.

➡Install the driveshaft assembly while keeping it level.

17. Temporarily install the rear axle carrier sub-assembly with the 2 bolts and the 2 nuts.

➡Be careful not to damage the outboard joint boot.

➡Be careful not to damage the speed sensor rotor.

➡Prevent foreign matter from adhering to the speed sensor rotor.

18. Install the rear axle carrier sub-assembly with the 2 bolts and the 2 nuts, and tighten to 213 ft. lbs. (290 Nm).

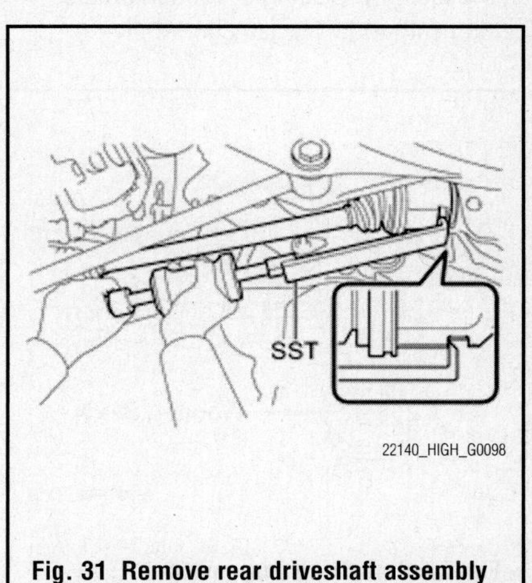

Fig. 31 Remove rear driveshaft assembly

➡**Do not rotate the driveshaft with the rear axle hub and bearing assembly removed.**

➡**Insert the bolts from the rear side.**

19. Check that the identification mark of the rear strut rod assembly is positioned on the inner side of the vehicle.
20. Temporarily install the rear strut rod assembly with the 2 bolts and the 2 nuts.

➡**Since stopper nuts are used, temporarily tighten the bolts.**

21. Install the rear axle hub and bearing assembly, as follows:
    a. Align the matchmarks on the driveshaft and rear axle hub.

➡**Do not rotate the driveshaft.**

    b. Install the parking brake assembly and the rear axle hub and bearing assembly with the 4 bolts, and tighten to 55 ft. lbs. (75 Nm).
22. Align the matchmarks and install the rear disc.

➡**When replacing the rear disc with a new one, select the installation position where the rear disc has minimal runout.**

23. Install the rear disc brake caliper assembly with the 2 bolts, and tighten to 57 ft. lbs. (78 Nm).
24. Install the rear speed sensor to the rear axle carrier sub-assembly with the bolt, and tighten to 71 inch lbs. (8 Nm).

➡**Keep the rear speed sensor tip and sensor installation hole free from foreign matter.**

➡**Do not twist the rear speed sensor wire when installing.**

25. Jack up the rear axle carrier sub-assembly, placing a wooden block underneath to avoid damage. Apply load to the suspension so that the rear driveshaft assembly is positioned horizontally.

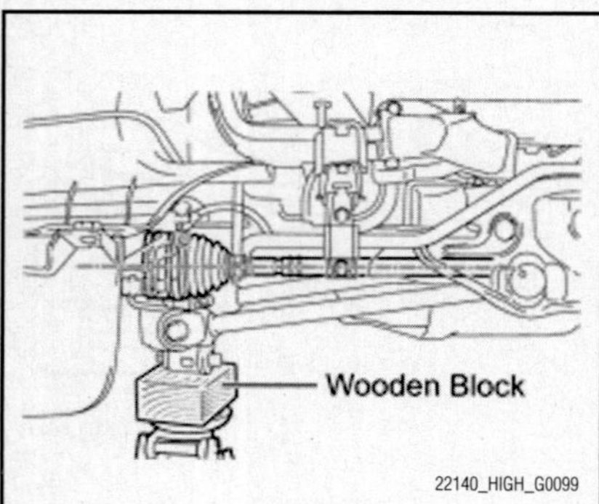

Fig. 32 Stabilizing suspension

CAUTION

Do not jack up the rear axle carrier sub-assembly too high as the vehicle may fall.

➡**Do not bend the brake dust cover.**

➡**If the rear driveshaft assembly cannot be positioned horizontally as shown in the illustration even when the rear axle carrier sub-assembly is jacked up, apply additional load to the vehicle such as by having a person sit in the rear seat.**

➡**Use the same procedures for the RH side and LH side.**

26. Fully tighten the rear No. 1 and 2 suspension arm assemblies with the bolts and nuts, and tighten to 82 ft. lbs. (112 Nm).

➡**Since a stopper nut is used, tighten the bolt.**

➡**The final torque must be applied under standard vehicle height conditions.**

27. Complete the installation of the rear strut rod assembly, tightening the bolts to 59 ft. lbs. (80 Nm).

➡**Since a stopper nut is used, fully tighten the bolt.**

➡**The final torque must be applied under standard vehicle height conditions.**

28. Install the No. 3 parking brake cable assembly with the bolt and the nut, and tighten to 29 ft. lbs. (39 Nm) and 53 inch lbs. (6 Nm).

➡**Do not twist the No. 3 parking brake cable assembly when installing it.**

29. Install the rear axle shaft nut, as follows:
    a. Clean the threaded parts on the driveshaft and axle hub nut using a non-residue solvent.

➡**Be sure to perform this work for a new driveshaft.**

➡**Keep the threaded parts free of oil and foreign objects.**

    b. Install a new rear axle shaft nut, and tighten to 216 ft. lbs. (294 Nm).
    c. Using a chisel and hammer, stake the rear axle shaft nut.
30. Install the rear differential drain plug, as follows:
    a. Using a hexagon wrench (10 mm), install the filler plug with a new gasket, and tighten to 36 ft. lbs. (49 Nm).
31. Fill the rear differential carrier assembly with hypoid gear oil.
32. Using a hexagon wrench (10 mm), install the rear differential filler plug with a new gasket, and tighten to 36 ft. lbs. (49 Nm).
33. Install the front wheel and tire assembly and tighten the lug nuts finger-tight.
34. Lower the vehicle, then final tighten the lug nuts to 76 ft. lbs. (103 Nm).
35. Inspect and adjust rear wheel alignment.
36. Check ABS speed sensor signal.

REAR PINION SEAL

REMOVAL & INSTALLATION

See Figure 33.

1. Before servicing the vehicle, refer to the precautions section.
2. Drain the differential oil.
3. Remove or disconnect the following:
    - Exhaust pipe
    - Driveshaft by matchmarking it
    - Companion flange nut, by loosen the staked portion
    - Companion flange, using a screw-type extractor
    - Oil seal, using an extractor
    - Slinger
    - Front bearing
    - Spacer

To install:

4. Install or connect the following
    - New spacer
    - Bearing
    - Slinger
    - New seal

➡**Seal installation depth: 2.0mm +/- 0.3mm**

    - Companion flange
    - New nut. Coat the threads with clean differential oil. Torque the nut to 80 ft. lbs. (108 Nm).
5. The remainder of installation is the reverse of removal.

TRANSFER CASE ASSEMBLY

REMOVAL & INSTALLATION

See Figures 34 and 35.

1. Remove the engine/transaxle assembly.

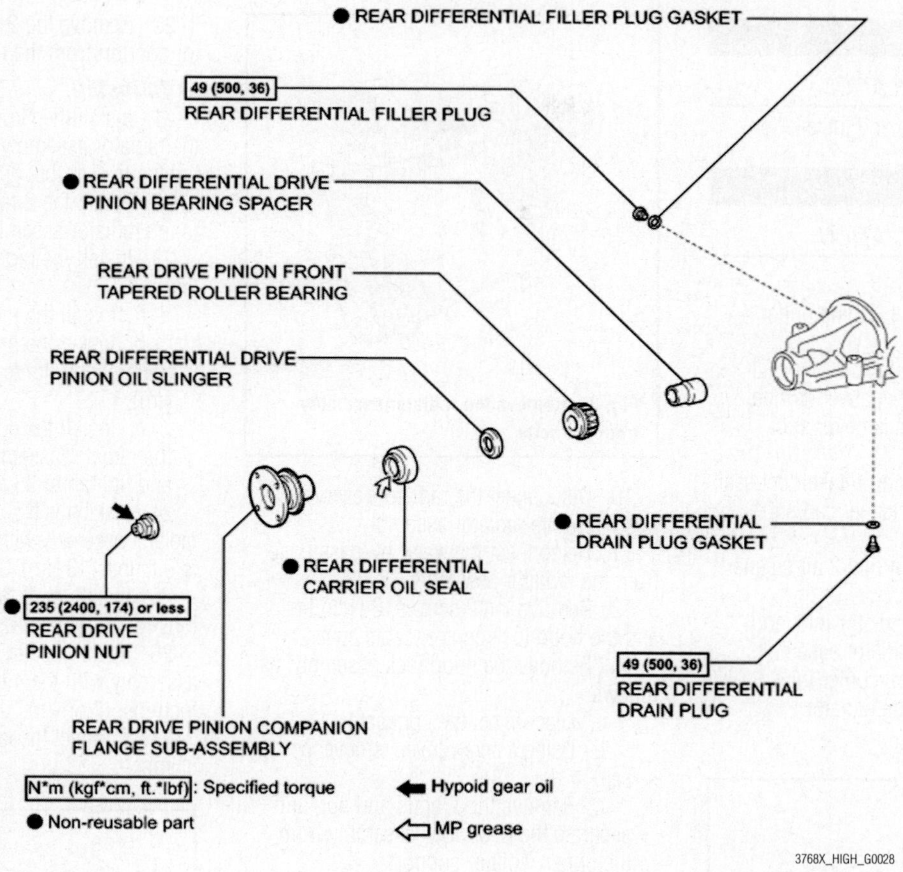

● REAR DIFFERENTIAL FILLER PLUG GASKET
49 (500, 36)
REAR DIFFERENTIAL FILLER PLUG
● REAR DIFFERENTIAL DRIVE PINION BEARING SPACER
REAR DRIVE PINION FRONT TAPERED ROLLER BEARING
REAR DIFFERENTIAL DRIVE PINION OIL SLINGER
● REAR DIFFERENTIAL DRAIN PLUG GASKET
● REAR DIFFERENTIAL CARRIER OIL SEAL
● 235 (2400, 174) or less
REAR DRIVE PINION NUT
49 (500, 36)
REAR DIFFERENTIAL DRAIN PLUG
REAR DRIVE PINION COMPANION FLANGE SUB-ASSEMBLY
N*m (kgf*cm, ft.*lbf): Specified torque
● Non-reusable part
Hypoid gear oil
MP grease
3768X_HIGH_G0028

Fig. 33 Exploded view of the pinion seal and components

67170-HIGH-G31

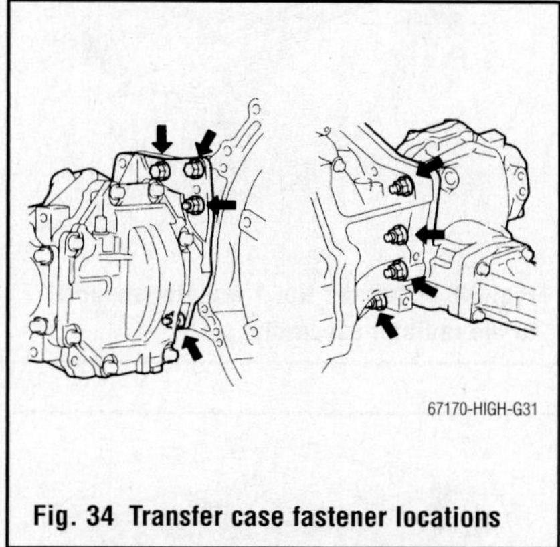

Fig. 34 Transfer case fastener locations

2. Drain the transaxle.
3. Separate the engine and transaxle.
4. Remove the 2 bolts and 6 nuts and separate the transfer case from the transaxle. It will be necessary to break it loose with a plastic mallet.

Driving Plug
Sealant
Gasket
67170-HIGH-G32

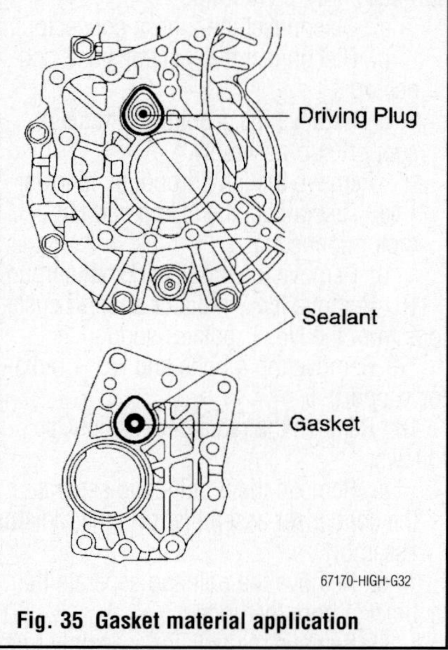

Fig. 35 Gasket material application

➡Keep the transfer case level during removal. Don't grasp the oil seals.

To install:

5. Clean all grease from the mating surfaces.
6. Apply a continuous 1.2mm diameter bead of silicone gasket material to the transaxle and transfer case as shown.
7. Join the transfer case to the transaxle within 10 minutes of gasket material application. If not, remove the material and start again.
8. Torque the nuts and bolts to 51 ft. lbs. (69 Nm).
9. The remainder of installation is the reverse of removal. Observe the following torques:
   - Engine mount bracket: 47 ft. lbs. (64 Nm)
   - Stiffener plate: 25 ft. lbs. (34 Nm)
   - Drain plug: 36 ft. lbs. (49 Nm)

ENGINE COOLING

ENGINE FAN

REMOVAL & INSTALLATION

See Radiator removal and installation.

RADIATOR

REMOVAL & INSTALLATION

See Figures 36 through 40.

1. Remove the No. 1 engine under cover.
2. Drain engine coolant.
3. Remove battery negative terminal.
4. Remove the V-bank cover sub-assembly.
5. Remove the cool air intake duct seal.
6. Remove the air cleaner cap sub-assembly.
7. Remove the No. 1 and 2 air cleaner inlets.
8. Disconnect the radiator inlet and outlet hoses from the radiator assembly.
9. Disconnect the oil cooler inlet and outlet hoses from the radiator assembly.

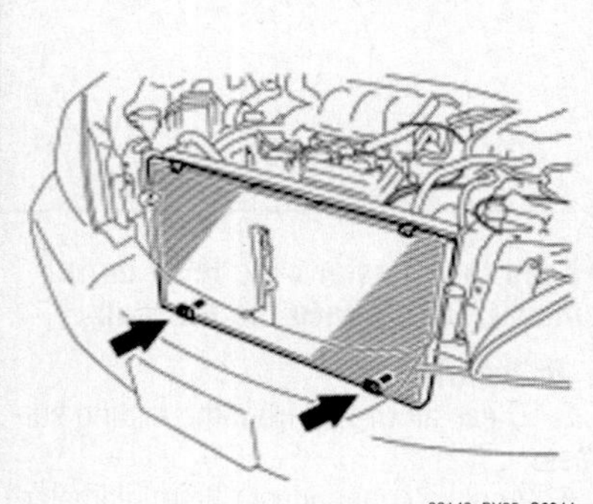

Fig. 36 Remove the 2 bolts and separate the condenser assembly from the radiator assembly

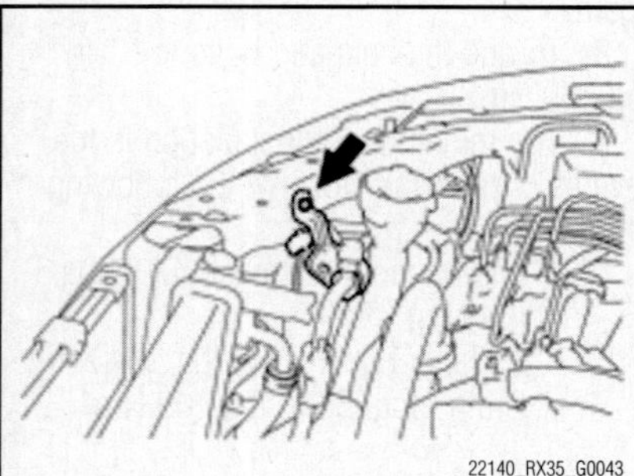

Fig. 37 Remove the bolt and separate the bracket from the body

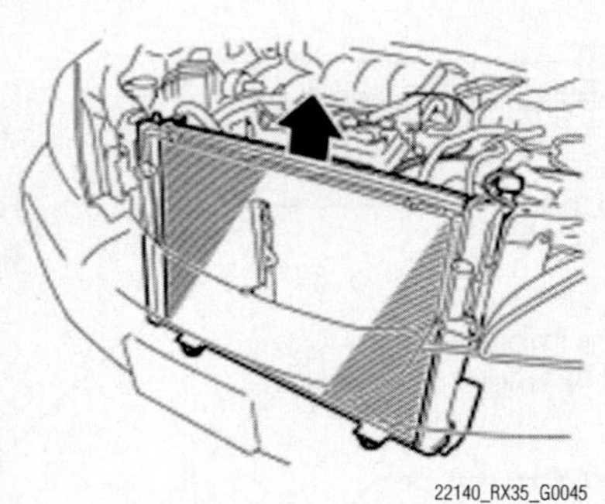

Fig. 38 Remove the radiator assembly from the body

10. Disconnect the radiator reserve tank hose from the radiator assembly.
11. Disconnect the water by-pass hose from the radiator assembly.
12. Remove the 2 screws, 2 snap fits and the hood lock control cable cover.
13. Remove the hood lock assembly, as follows:
    a. Disconnect the connector.
    b. Using a screwdriver, remove the cap.
    c. Remove the 2 bolts and nut, and separate the hood lock assembly from the upper radiator support.
    d. Remove the hood lock control cable from the hood lock assembly.
14. Remove the upper radiator support sub-assembly, as follows:
    a. Disconnect the 2 horn connectors.
    b. Disconnect the 3 clamps and connector.
    c. Remove the 5 bolts and radiator support sub-assembly.
15. Remove the fan shroud, as follows:
    a. Disconnect the clamp and fan motor connector.
    b. Remove the 4 bolts and fan shroud.
16. Remove the 2 radiator support cushions from the No. 1 radiator support.
17. Remove the 4 bolts and No. 1 radiator support.
18. Remove the radiator assembly, as follows:
    a. Remove the 2 bolts and separate the condenser assembly from the radiator assembly.
    b. Remove the bolt and separate the bracket from the body.
    c. Remove the radiator assembly from the body.
19. Remove the 2 radiator support lowers from the radiator assembly.
20. Remove the 2 bolts and No. 2 radiator support from the radiator assembly.

To install:

21. Install the No. 2 radiator support to the radiator assembly with the 2 bolts and tighten to 9 ft. lbs. (13 Nm).
22. Install the 2 radiator support lowers to the radiator assembly.
23. Install the radiator assembly as follows:
    a. Install the radiator to the body.
    b. Install the bracket to the body with the bolt and tighten to 87 inch lbs. (9.8 Nm).
    c. Install the condenser assembly to the radiator assembly with the 2 bolts and tighten to 35 inch lbs. (3.9 Nm).
24. Install the No. 1 radiator support to the radiator assembly with the 4 bolts and tighten to 9 ft. lbs. (13 Nm), 35 inch lbs. (3.9 Nm).
25. Install the 2 radiator support cushions to the radiator assembly.
26. Install the fan shroud to the radiator assembly with the 4 bolts and tighten to 66 inch lbs. (7.5 Nm).
27. Connect the fan motor connector and clamp.

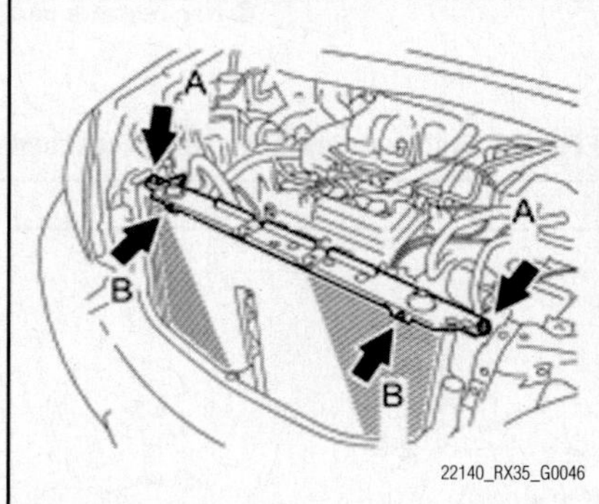

Fig. 39 Install the No. 1 radiator support to the radiator assembly

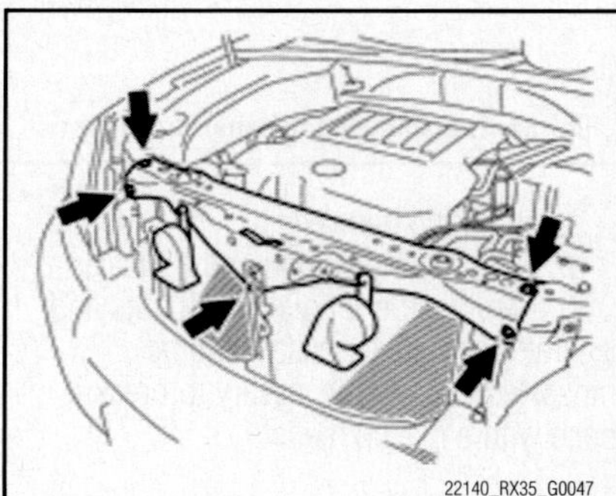

Fig. 40 Install the radiator support sub-assembly upper with the 5 bolts

28. Install the radiator support sub-assembly upper as follows:
    a. Install the radiator support sub-assembly upper with the 5 bolts and tighten to 69 inch lbs. (7.8 Nm).
    b. Connect the 3 clamps and connector.
    c. Connect the 2 horn connectors.
29. Install the hood lock assembly as follows:
    a. Install the hood lock control cable to the hood lock assembly.
    b. Install the 2 bolts and the hood lock assembly to the radiator support sub-assembly upper and tighten to 71 inch lbs. (8 Nm).
    c. Install the hood lock nut and the hood lock nut cap and tighten to 71 inch lbs. (8 Nm).
    d. Connect the connector.
30. Connect the 2 clips, and install the 2 screws and the hood lock control cable cover.
31. Connect the water by-pass hose No. 2 to the radiator assembly with the clip.
32. Connect the radiator reserve tank hose to the radiator assembly with the clip.
33. Connect the oil cooler inlet and outlet hoses to the radiator assembly.
34. Connect the radiator inlet and outlet hoses to the radiator assembly.
35. Install the No. 1 and 2 air cleaner inlets.
36. Install the air cleaner cap sub-assembly.
37. Install the cool air intake duct seal.
38. Install the V-bank cover sub-assembly.
39. Connect battery negative terminal.
40. Add engine coolant.
41. Inspect for coolant leak.
42. Install the No. 1 engine under cover.
43. Perform initialization procedure.

THERMOSTAT

REMOVAL & INSTALLATION

See Figures 41 and 42.

1. Drain engine coolant.
2. Remove the V-bank cover sub-assembly.
3. Remove the side engine room cover.
4. Remove the right front fender seal.
5. Remove the right No. 2 engine mounting stay.
6. Remove the engine moving control rod sub-assembly.
7. Remove the V-ribbed belt.
8. Remove the bolt, No. 2 idler pulley cover plate and No. 2 idler pulley.
9. Slide the clamp and disconnect the radiator hose outlet.
10. Remove the 2 nuts and water inlet.
11. Remove the gasket from the thermostat and remove thermostat.

To install:

12. Install a new gasket to the thermostat.
13. Install the thermostat with the jiggle valve facing up.

➡The jiggle valve may be set within 10° on either side of the prescribed position.

14. Install the water inlet with the 2 nuts and tighten to 7 ft. lbs. (10 Nm).
15. Connect the radiator hose outlet with the clamp.
16. Install the No. 2 idler pulley cover plate and No. 2 idler pulley with the bolt and tighten to 32 ft. lbs. (43 Nm).
17. Install the V-ribbed belt.
18. Install the engine moving control rod sub-assembly.
19. Install the right No. 2 engine mounting stay.
20. Install the right front fender seal.
21. Add engine coolant.
22. Inspect for coolant leak.
23. Install the side engine room cover.
24. Install the V-bank cover sub-assembly.

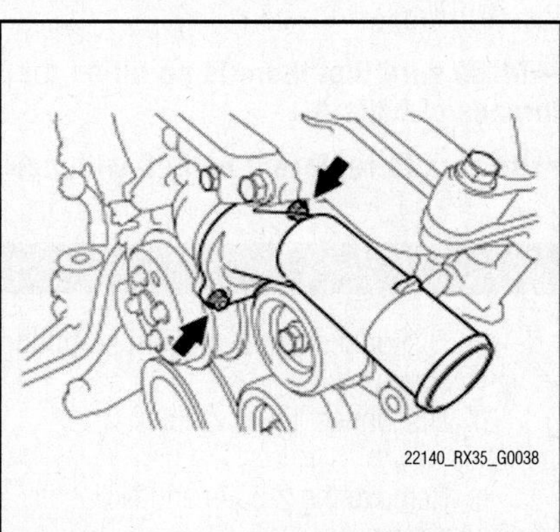

Fig. 41 Remove the 2 nuts and water inlet

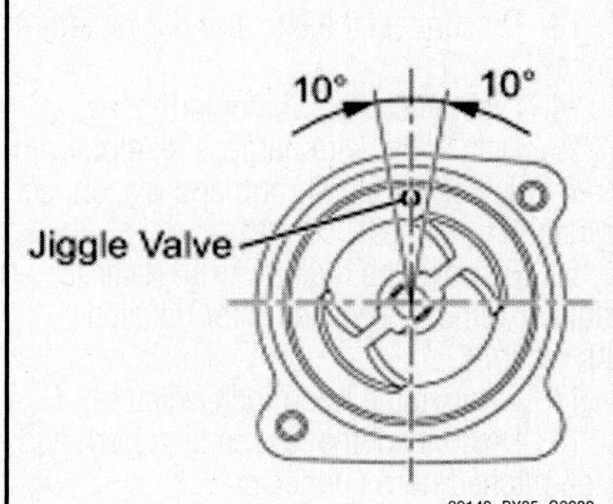

Fig. 42 Install the thermostat

WATER PUMP

REMOVAL & INSTALLATION

See Figures 43 through 46.

1. Remove the engine assembly and transaxle.
2. Secure the engine.
3. Remove the left front No. 1 engine mounting bracket.

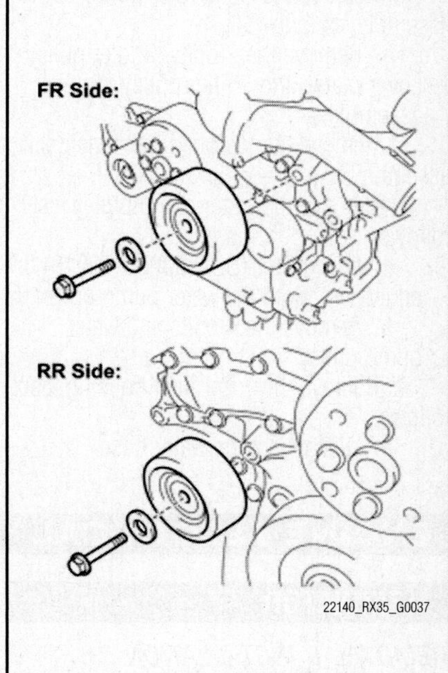

Fig. 43 Remove the 2 bolts, 2 idler pulley cover plates and idler pulley sub-assemblies

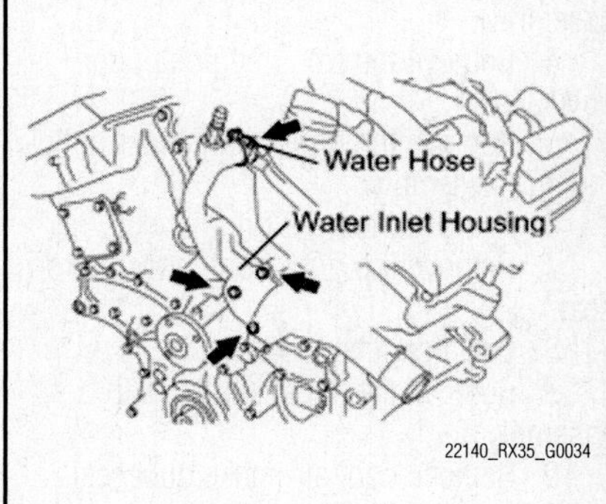

Fig. 44 Remove the 2 bolts, nut and water inlet housing

Fig. 45 Remove the 16 bolts, water pump assembly and water pump gasket

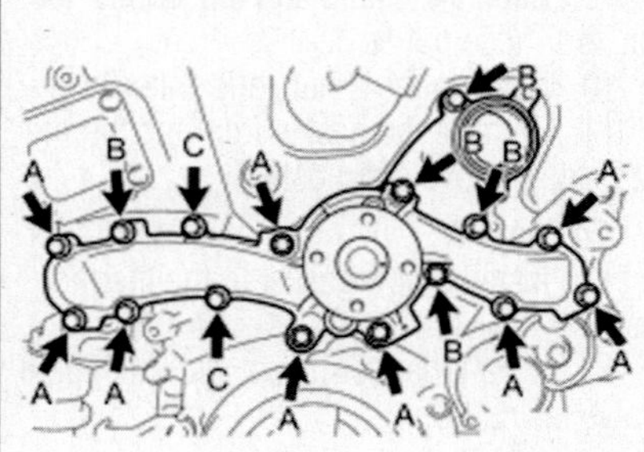

Fig. 46 Install a new water pump gasket and the water pump assembly with the 16 bolts

4. Remove the No. 2 idler pulley sub-assembly, as follows:
   a. Remove the 2 bolts, 2 idler pulley cover plates and 2 idler pulley sub-assemblies.
5. Remove the V-ribbed belt tensioner assembly.
6. Remove the water pump pulley, as follows:
   a. Using SST (SST: 09960-10010) or equivalent, hold the water pump pulley.
   b. Remove the 4 bolts and water pump pulley.
7. Remove the water inlet housing, as follows:
   a. Disconnect the water hose.
   b. Remove the 2 bolts, nut and water inlet housing.
   c. Remove the water inlet housing gasket and water outlet pipe O-ring.
8. Remove the 16 bolts, water pump assembly and water pump gasket.

To install:

9. Install a new water pump gasket and the water pump assembly with the 16 bolts and tighten to 15 ft. lbs. (21 Nm), 81 inch lbs. (9.1 Nm).

➡Make sure that there is no oil on the threads of bolts A.

➡Be sure to replace 2 bolts C with new ones or reuse them after applying adhesive 1344. Adhesive: Toyota Genuine Seal Packing Black, Three Bond 1207B or equivalent.

10. Install a new water inlet housing gasket and water outlet pipe O-ring.
11. Install the water inlet housing with the 2 bolts and nut and tighten to 7 ft. lbs. (10 Nm).

➡Be careful not to allow the O-ring to get caught between the parts.

12. Connect the water hose.
13. Install the water pump pulley, as follows:
   a. Temporarily install the water pump pulley with the 4 bolts.
   b. Using SST (SST: 09960-10010) or equivalent, hold the water pump pulley.
   c. Tighten the 4 bolts to 15 ft. lbs. (21 Nm).
14. Install the V-ribbed belt tensioner assembly.
15. Install the 2 idler pulley cover plates and idler pulley sub-assemblies with the 2 bolts and tighten to 32 ft. lbs. (43 Nm).
16. Install the left front No. 1 engine mounting bracket.
17. Remove engine stand.
18. Install the engine assembly and transaxle.

ENGINE ELECTRICAL

CHARGING SYSTEM

ALTERNATOR

REMOVAL & INSTALLATION

See Figure 47.

1. Raise and safely support the vehicle.
2. Remove the front wheel.
3. Remove the engine under cover assembly.
4. Remove the No. 1 engine under cover.
5. Remove the right front fender molding sub-assembly.
6. Remove the right front fender liner.
7. Remove the right front fender apron seal.
8. Drain engine coolant.
9. Remove the V-bank cover sub-assembly
10. Remove cool air intake duct seal.
11. Remove the battery.
12. Remove the No. 1 and 2 air cleaner inlets, as follows:
   a. Disconnect the 2 vacuum switching valve clamps.
   b. Disconnect the 2 vacuum hoses.
   c. Remove the 2 bolts and No. 2 air cleaner inlet.
   d. Disconnect the 2 vacuum hoses, and remove the 2 bolts and No. 1 air cleaner inlet.
13. Disconnect the No. 1 and 2 radiator hoses.
14. Disconnect the oil cooler hoses.
15. Detach the wire harness clamps from both sides of the fan shroud and disconnect the cooling fan ECU connector.
16. Remove the radiator grill. Refer to radiator removal procedure for detailed instructions.
17. Remove the hood lock assembly.
18. Disconnect the low pitched horn and high pitched horn connectors.
19. Detach the hood lock control cable clamp and remove the 6 bolts and upper radiator support sub-assembly.
20. Remove the 4 bolts and move the cooler condenser assembly.
21. Remove the radiator assembly and fan assembly with motor.
22. Remove the bolt and the No. 2 oil level dipstick guide.
23. Remove the V-ribbed belt.
24. Remove the alternator assembly, as follows:

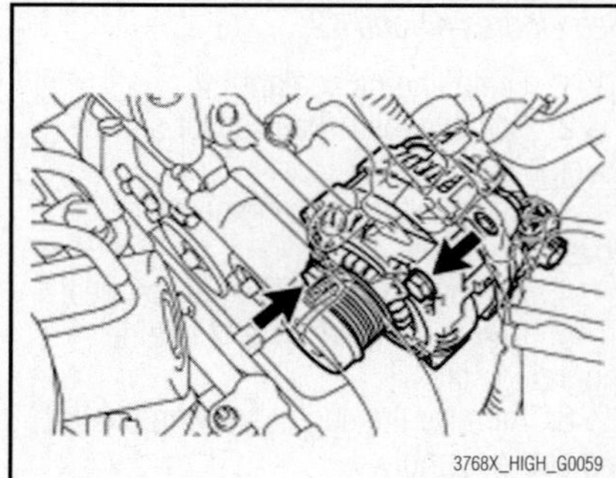

Fig. 47 Remove the 2 bolts and alternator—3.5L Engine

a. Remove the terminal cap.

b. Remove the nut and disconnect the wire harness from terminal B.

c. Disconnect the alternator connector from the alternator assembly.

d. Disconnect the connector from the compressor and magnetic clutch.

e. Disconnect the 3 wire harness clamps.

f. Remove the 2 bolts, and then disconnect the bracket.

g. Remove the 2 bolts and the alternator assembly.

h. Disconnect the wire harness clamp, and then remove the alternator bracket.

i. Remove the bolt and the wire harness clamp stay.

To install:

25. Install the alternator assembly, as follows:

a. Install the wire harness clamp stay with the bolt, and tighten to 15 ft. lbs. (20 Nm).

b. Connect the alternator bracket with the wire harness clamp.

c. Install alternator assembly with the 2 bolts, and tighten to 32 ft. lbs. (43 Nm).

d. Temporally install the 2 bolts., then fully tighten the 2 bolts to 15 ft. lbs. (20 Nm).

e. Connect the alternator connector to the alternator assembly.

f. Install the alternator wire with the nut, and tighten to 87 ft. lbs. (9.8 Nm).

g. Install the terminal cap.

h. Connect the 3 wire harness clamps.

i. Connect the magnetic clutch connector to the compressor and magnetic clutch.

26. Install the V-ribbed belt.

27. Install the No. 2 oil level dipstick guide, as follows:

a. Install a new O-ring to the No. 2 oil level dipstick guide.

b. Apply a light coat of engine oil to the O-ring.

c. Push in the No. 2 oil level dipstick guide end into the No. 1 oil level dipstick guide.

d. Install the No. 2 oil level dipstick guide with the bolt, and tighten to 15 ft. lbs. (20 Nm).

28. Install the radiator assembly and fan assembly with motor.

29. Install the cooler condenser assembly.

30. Install the upper radiator support sub-assembly.

31. Install the hood lock assembly.

32. Install the radiator grill.

33. Connect the cooling fan ECU connector.

34. Connect the oil cooler hose.

35. Connect the No. 1 and 2 radiator hose.

36. Install the No. 1 and 2 air cleaner inlet.

37. Install the battery.

38. Install the cool air intake duct seal.

39. Install the right front fender apron seal.

40. Install the right front fender liner.

41. Install the right front fender molding sub-assembly.

42. Add engine coolant. Refer to radiator installation procedure for detailed instructions.

43. Inspect for coolant leak. Refer to radiator installation procedure for detailed instructions.

44. Inspect automatic transaxle fluid.

45. Inspect for oil leaks.

46. Install the No. 1 engine under cover.

47. Install the engine under cover assembly.

48. Install the right front wheel. Tighten the lug nuts to 76 ft. lbs. (103 Nm).

49. Lower the vehicle.

50. Install the V-bank cover sub-assembly.

ENGINE ELECTRICAL

IGNITION SYSTEM

FIRING ORDER

Firing order for 3.5L engine: 1–2–3–4–5–6

IGNITION COIL

REMOVAL & INSTALLATION

See Figures 48 and 49.

1. Disconnect the negative battery cable.
2. Remove the No. 1 engine under cover.
3. Remove the engine room side cover.
4. Remove the V-bank cover sub-assembly.
5. Drain engine coolant.
6. Remove the windshield wiper motor and link assembly.
7. Remove the cowl top panel outer sub-assembly.
8. Remove the air cleaner cap sub-assembly.
9. Remove the intake air surge tank assembly.

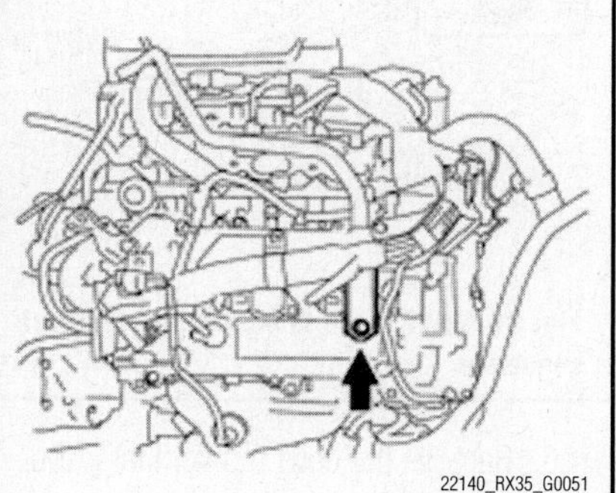

Fig. 48 Remove the bolt and No. 1 surge tank stay

10. Remove the No. 1 surge tank stay, as follows:

a. Remove the bolt and disconnect the harness clamp.

b. Remove the bolt and No. 1 surge tank stay.

11. Remove the ignition coil assembly, as follows:

a. Disconnect the 6 ignition coil connectors.

b. Remove the 6 bolts and 6 ignition coils.

To install:

12. Installation is the reverse of the removal procedure, noting the following:

a. Tighten the ignition coil bolts to 7 ft. lbs. (10 Nm).

b. Tighten the No. 1 surge tank stay bolt to 15 ft. lbs. (21 Nm).

c. Tighten the surge tank bolt and clamp to 62 inch lbs. (7 Nm).

d. After adding engine coolant, check for leaks.

e. After completing installation, perform initialization procedure.

IGNITION TIMING

ADJUSTMENT

All engines are equipped with a Distributorless Ignition System (DIS). No timing adjustment is possible.

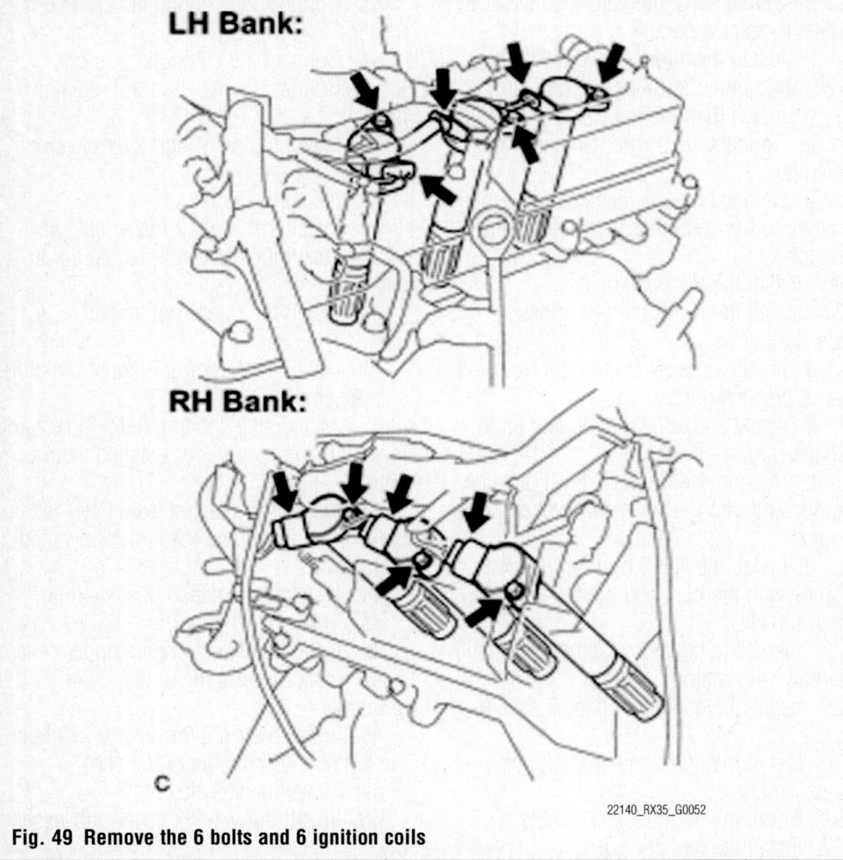

Fig. 49 Remove the 6 bolts and 6 ignition coils

SPARK PLUGS

REMOVAL & INSTALLATION

See Figures 50 through 56.

1. Remove the engine under cover assembly.
2. Remove the No. 1 engine under cover.
3. Drain the engine coolant.
4. Remove the V-bank cover sub-assembly.
5. Remove both the front wiper arms and blade assemblies.

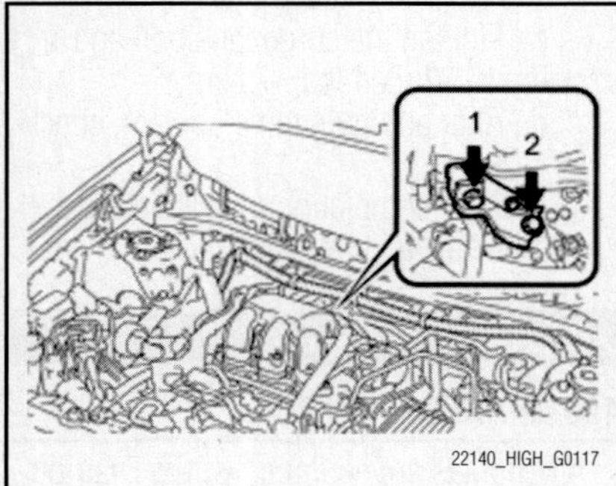

Fig. 50 Throttle body bracket bolts removal sequence—3.5L engine

Fig. 51 No. 1 surge tank stay bolt removal sequence

6. Remove the cowl top ventilator louver sub-assembly.
7. Remove the windshield wiper motor and link assembly.
8. Remove the outer cowl top panel sub-assembly.
9. Disconnect the engine room main wire.
10. Remove the 2 bolts in the order shown in the illustration and remove the throttle body bracket.
11. Remove the 2 bolts in the order shown in the illustration and remove the No. 1 surge tank stay.
12. Remove the air cleaner cap sub-assembly.
13. Remove the intake air surge tank assembly, as follows:
    a. Disconnect the 4 hoses.
    b. Disconnect the throttle body connector and clamp.
    c. Disconnect the connector.
    d. Disconnect the 2 water by-pass hoses from the throttle body.
    e. Remove the 4 bolts and 2 nuts in the order shown in the illustration.

➡Use a 5 mm hexagon socket wrench to remove the 4 bolts.

    f. Remove the gasket from the intake air surge tank.
14. Remove the ignition coil assembly, as follows:
    a. Remove the 4 bolts.
    b. Remove the nut.
    c. Disconnect the 2 harness clamps.
    d. Disconnect the 6 ignition coil connectors.
    e. Remove the 6 bolts and 6 ignition coils
15. Remove the 6 spark plugs.

To install:

16. Install the 6 spark plugs and tighten to 13 ft. lbs. (18 Nm).
17. Install the ignition coil assembly as follows:
    a. Install the 6 ignition coils with the 6 bolts and tighten to 7 ft. lbs. (10 Nm).
    b. Connect the 6 ignition coil connectors.
    c. Install the 4 bolts and tighten to 73 inch lbs. (8.3 Nm).
    d. Install the nut and tighten to 73 inch lbs. (8.3 Nm).
    e. Install the 2 clamps.
18. Install the intake air surge tank assembly as follows:
    a. Install the surge tank with the 4 bolts and 2 nuts in the order shown in the illustration, and tighten to 12 ft. lbs. (16 Nm) and 13 ft. lbs. (18 Nm), using a 5 mm hexagon socket wrench. **DO NOT** apply oil to the bolts.
    b. Connect the 2 water by-pass hoses to the throttle with motor body assembly.
    c. Connect the connector.
    d. Install the clamp and connect the throttle with motor body assembly connector.
    e. Connect the 4 hoses.
19. Temporarily install the No. 1 surge tank stay as follows:
    a. Temporarily install the intake air surge tank assembly with 3 new gaskets on the intake manifold.

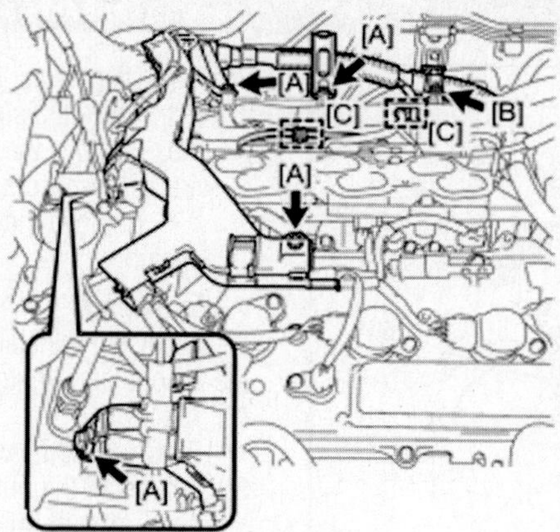

Fig. 52 Ignition coil assembly harness clamps

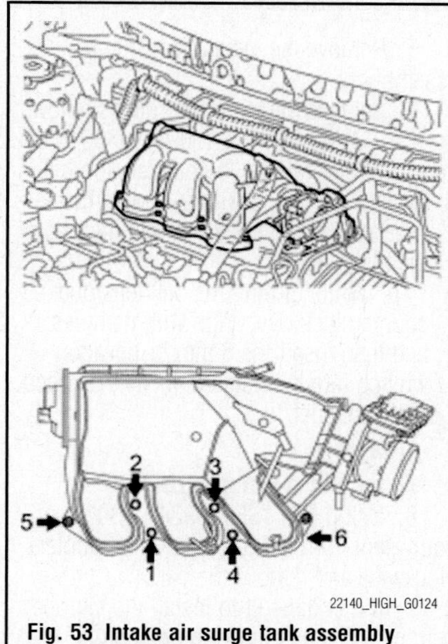

Fig. 53 Intake air surge tank assembly installation sequence

➡**Do not allow the gaskets to slip out of place during installation.**

b. Temporarily install the No. 1 surge tank stay with the 2 bolts. **DO NOT** apply oil to the bolts.

20. Temporarily install the throttle body bracket with the 2 bolts. **DO NOT** apply oil to the bolts.

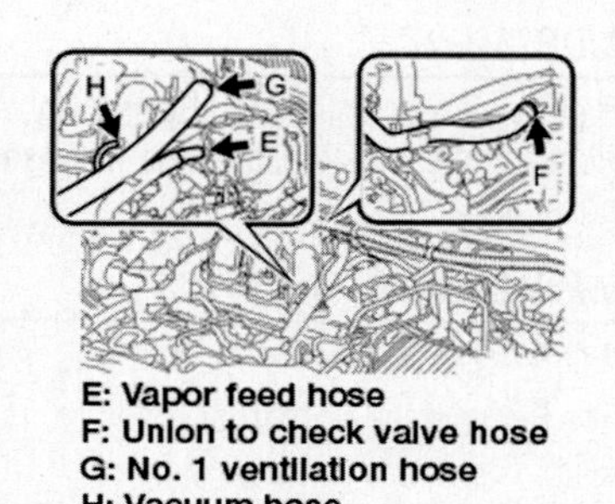

Fig. 54 Intake air surge tank assembly hoses

Fig. 55 No. 1 surge tank stay bolt tightening sequence

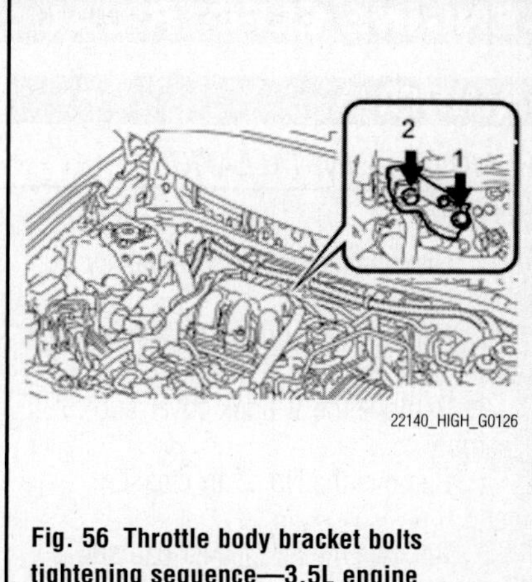

Fig. 56 Throttle body bracket bolts tightening sequence—3.5L engine

21. Fully tighten the No. 1 surge tank stay, as follows:
   a. Fully tighten the 2 bolts in the order shown in the illustration. Tighten to 15 ft. lbs. (21 Nm). **DO NOT** apply oil to the bolts.
22. Fully tighten the throttle body bracket, as follows:
   a. Fully tighten the 2 bolts in the order shown in the illustration. Tighten to 15 ft. lbs. (21 Nm). **DO NOT** apply oil to the bolts.
23. Connect the engine room main wire.
24. Install the air cleaner cap sub-assembly.
25. Install the outer cowl top panel sub-assembly, as follows:
   a. Install the outer cowl top panel sub-assembly with the 8 bolts and 6 nuts.
   b. Engage the 4 clamps.
26. Install the windshield wiper motor and link assembly.
27. Install the cowl top ventilator louver sub-assembly.
28. Install both front wiper arm and blade assemblies.
29. Install the V-bank cover sub-assembly.
30. Add engine coolant.
31. Inspect for engine coolant leak.
32. Install the No. 1 engine under cover.
33. Install the engine under cover assembly.

STARTER

REMOVAL & INSTALLATION

See Figure 57.

1. Disconnect the negative battery cable.
2. Remove the cool air intake duct seal.
3. Remove the V-bank cover sub-assembly.
4. Remove the No. 2 air cleaner inlet.
5. Remove the air cleaner cap sub-assembly.
6. Remove the air cleaner case sub-assembly.
7. Remove the No. 1 air cleaner inlet.

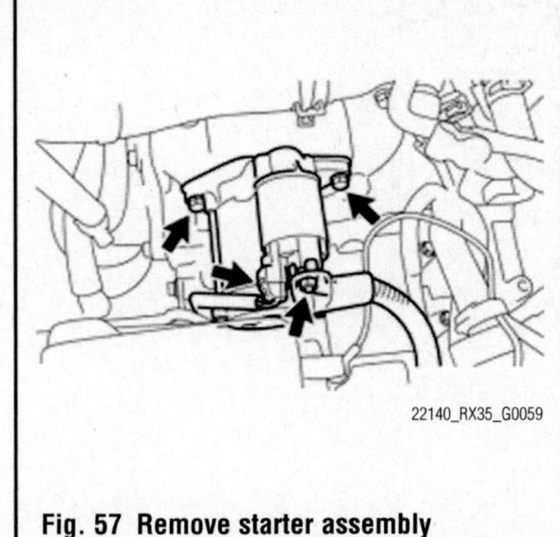

Fig. 57 Remove starter assembly

8. Remove the starter assembly, as follows:
   a. Disconnect the terminal 50 connector from the starter assembly.
   b. Remove the nut and disconnect the wire harness from terminal 30.
   c. Remove the 2 bolts and starter assembly.

To install:

9. Install the starter assembly with the 2 bolts and tighten to 28 ft. lbs. (37 Nm).
10. Connect the wire harness to terminal 30 and install the nut, and then attach the terminal cap and tighten to 87 inch lbs. (9.8 Nm).
11. Connect the terminal 50 connector to the starter assembly.
12. The remainder of installation is the reverse of the removal procedure.

ENGINE MECHANICAL

ACCESSORY DRIVE BELTS

ACCESSORY BELT ROUTING

See Figure 58.

Refer to the accompanying illustration for belt routing.

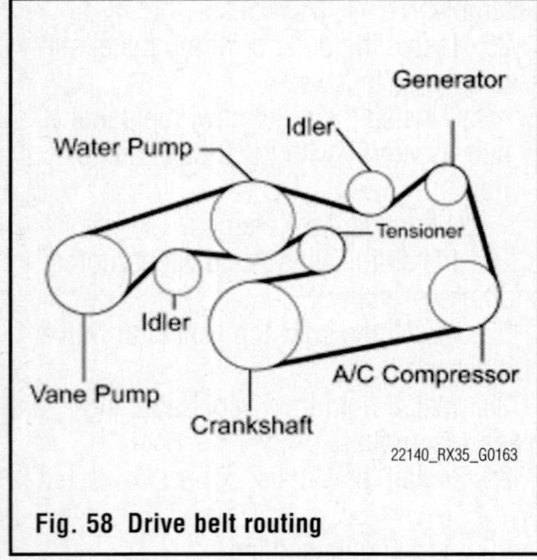

Fig. 58 Drive belt routing

INSPECTION

Visually check the V-ribbed belt for excessive wear, frayed cords, etc. All worn or damaged drive belts should be replaced immediately. Cracks on the rib side of a V-ribbed belt are considered acceptable, If the drive belt has chunks missing from its ribs, it should be replaced. After installing the V-ribbed belt, check that it fits properly in the ribbed grooves. Check to confirm that the belt has not slipped out of the grooves on the bottom of the crank pulley by hand.

ADJUSTMENT

Belt tension is maintained by an automatic tensioner. No adjustment is necessary or possible.

REMOVAL & INSTALLATION

See Figures 59 through 61.

1. Remove the right front wheel.
2. Separate the right front fender splash shield sub-assembly.
3. Remove the right front fender apron seal.
4. Remove the engine room side cover.

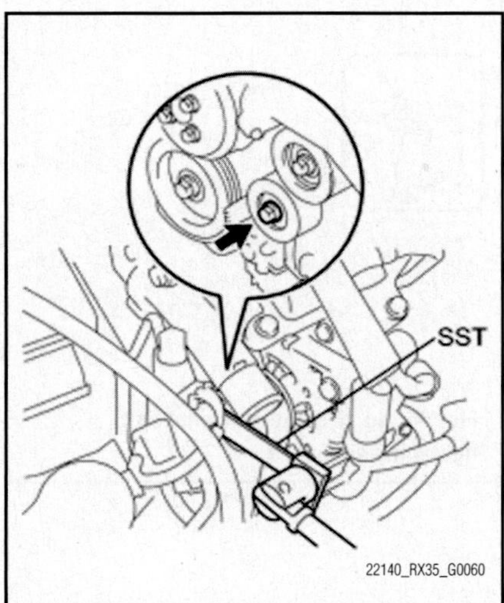

Fig. 59 Release the belt tension by turning the belt tensioner counterclockwise

5. Remove the V-bank cover sub-assembly.
6. Remove the V-ribbed belt, as follows:
   a. Using SST (SST: 09249-63010) or equivalent, release the belt tension by turning the belt tensioner counterclockwise, and remove the V-ribbed belt from the belt tensioner.
   b. While turning the belt tensioner counterclockwise, align with its holes, and then insert the 5 mm bi-hexagon wrench into the holes to fix the V-ribbed belt tensioner.

To install:

7. Install the V-ribbed belt.
8. Using SST (SST: 09249-63010) or equivalent, turn the belt tensioner counterclockwise and remove the bar.
9. If it is difficult to install the V-ribbed belt, perform the following procedure:
   a. Put the V-ribbed belt on every pulley except the tensioner pulley as shown in the illustration.
   b. While releasing the belt tension by turning the belt tensioner counterclockwise, put the V-ribbed belt on the tensioner pulley.

➡Put the backside of the V-ribbed belt on the tensioner pulley and idler pulley.

➡Check that the V-ribbed belt is properly set to each pulley.

   c. After installing the V-ribbed belt, check that it fits properly in the ribbed grooves. Confirm that the belt has not

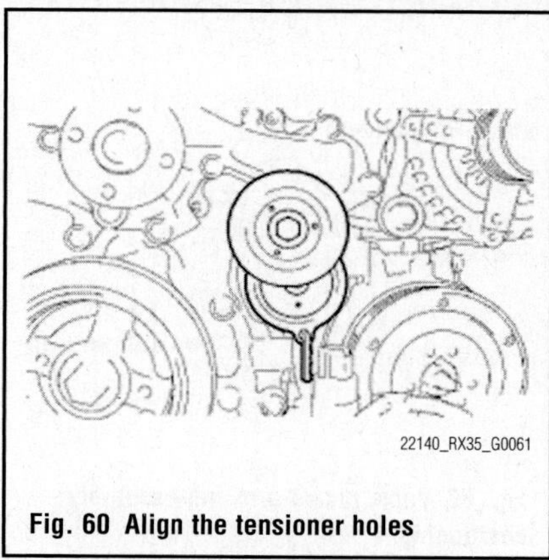
Fig. 60 Align the tensioner holes

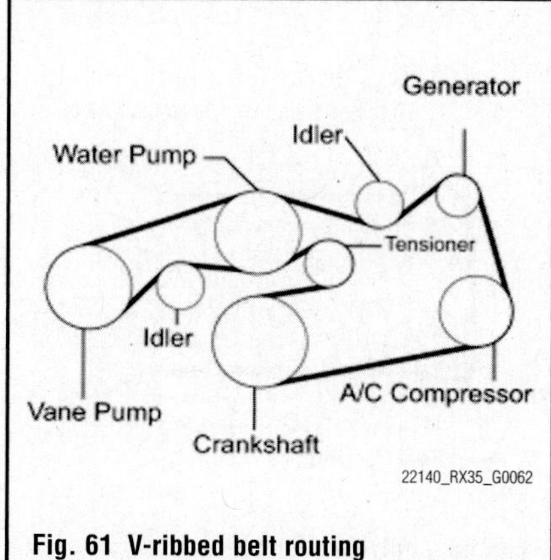

Fig. 61 V-ribbed belt routing

slipped out of the grooves on the bottom of the crank pulley by hand.

10. Install the V-bank cover sub-assembly.
11. Install the engine room side cover.
12. Install the right front fender apron seal.
13. Install the right front fender splash shield sub-assembly.
14. Install the right front wheel and tighten the lug nuts to 76 ft. lbs. (103 Nm).

CAMSHAFT AND VALVE LIFTERS

REMOVAL & INSTALLATION

See Figures 62 through 79.

1. Remove the engine assembly.
2. Install on engine stand.
3. The following must be removed:
   a. Remove ignition coil assembly.
   b. Remove the right hand No. 2 engine mounting stay.
   c. Remove the intake manifold.
   d. Remove the right hand exhaust manifold sub-assembly.
   e. Remove the No. 2 engine oil level dipstick guide.
   f. Remove the No. 2 manifold stay.
   g. Remove the No. 2 exhaust manifold heat insulator.
   h. Remove the left hand exhaust manifold sub-assembly.
   i. Remove the transverse engine mounting bracket.
   j. Remove the generator assembly.
   k. Remove the V-ribbed belt tensioner assembly.
   l. Remove the No. 2 timing gear cover.
   m. Remove the No. 2 idler pulley sub-assembly.
   n. Remove the left hand No. 1 engine front mounting bracket.
   o. Remove the left hand 6 bolts and No. 1 front engine mounting bracket.
   p. Remove the radio setting condenser.
   q. Remove the No. 1 vacuum switching valve.
   r. Remove the knock control sensor wire.
   s. Remove the knock control sensor.
   t. Remove the Crankshaft Position (CKP) sensor.
   u. Remove the No. 1 oil pipe.
   v. Remove the oil pipe.
   w. Remove the crankshaft pulley.
   x. Remove the oil cooler assembly, if necessary.
   y. Remove the No. 1 oil cooler bracket, if necessary.
   z. Remove the water inlet housing.
   aa. Remove the water outlet.
   bb. Remove the cylinder head cover sub-assembly (for Bank 1).
   cc. Remove the cylinder head cover sub-assembly (for Bank 2).
   dd. Remove the No. 2 oil pan sub-assembly.
   ee. Remove the oil strainer sub-assembly.
   ff. Remove the oil pan sub-assembly.
   gg. Remove the timing chain cover sub-assembly
   hh. Remove the timing chain case oil seal.
   ii. Set the No. 1 cylinder to TDC/compression.
   jj. Remove the No. 1 chain tensioner assembly.
   kk. Remove the chain tensioner slipper.
   ll. Remove the chain sub-assembly.
   mm. Remove the idle sprocket assembly.
   nn. Remove the camshaft timing gears and No. 2 chain (for Bank 1).
   oo. While raising the No. 2 chain tensioner assembly, insert a pin of 1.0 mm (0.039 in.) diameter into the hole to fix the No. 2 chain tensioner assembly.
   pp. Hold the hexagonal portion of the camshaft with a wrench, and remove the 2 bolts and 2 camshaft timing gear assemblies.

➡Be careful not to damage the cylinder head with the wrench. Do not disassemble the camshaft timing gear assemblies.

   qq. Remove the No. 2 chain assembly.
   rr. Remove the bolt and No. 2 chain tensioner assembly.
   ss. Check that the camshafts are positioned as shown in the illustration.
   tt. Uniformly loosen and remove the 8 bearing cap bolts in several steps and in the sequence shown in the illustration.
   uu. Uniformly loosen and remove the 12 bearing cap bolts in several steps and in the sequence shown in the illustration.

➡Uniformly loosen the bolts while keeping the camshaft level.

   vv. Remove the 5 camshaft bearing caps.
   ww. Remove the camshaft.
   xx. Remove the No. 2 camshaft.
   yy. Remove the right hand camshaft housing sub-assembly by prying between the cylinder head and the right hand camshaft housing sub-assembly with a screwdriver.

➡Be careful not to damage the contact surfaces of the cylinder head and the right hand camshaft housing sub-assembly.

To install:

4. Install the following:
   a. Apply engine oil to the camshaft journals, camshaft housing sub-assembly RH and camshaft bearing caps.
   b. Install the camshaft and No. 2 camshaft to the right hand camshaft housing sub-assembly.
   c. Make sure of the marks and numbers on the camshaft bearing caps and place them in each proper position and direction.
   d. Temporarily tighten the 8 bearing cap bolts to 7 ft. lbs. (10 Nm) in the order shown in the illustration.
   e. Make sure that the No. 1 valve rocker arm sub-assembly is installed as shown in the illustration.
   f. Apply seal packing in a continuous line as shown in the illustration.

➡Remove any oil from the contact surface. Install the right hand camshaft housing sub-assembly within

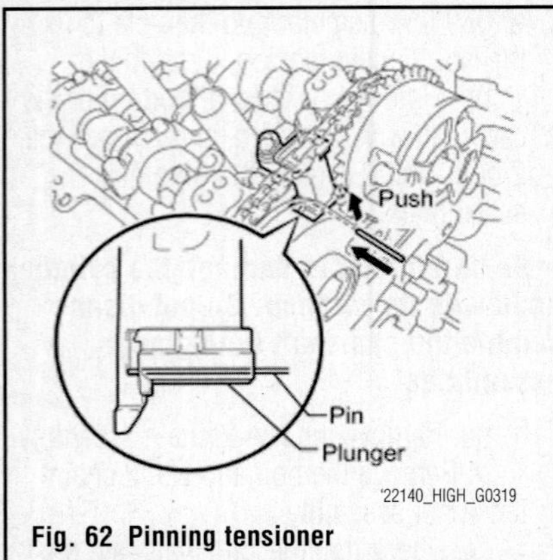

Fig. 62 Pinning tensioner

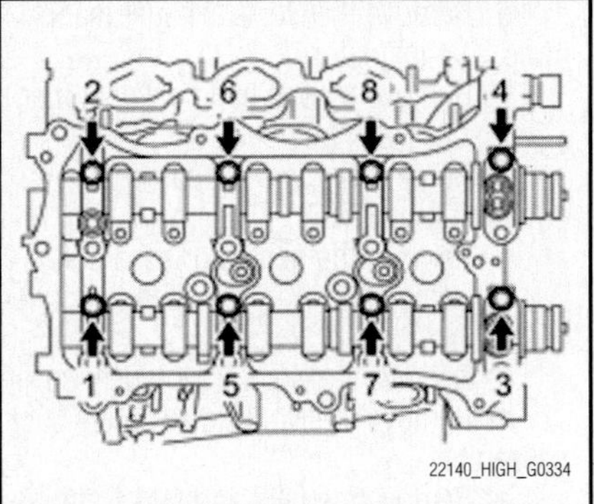

Fig. 65 Camshafts bearing cap removal sequence

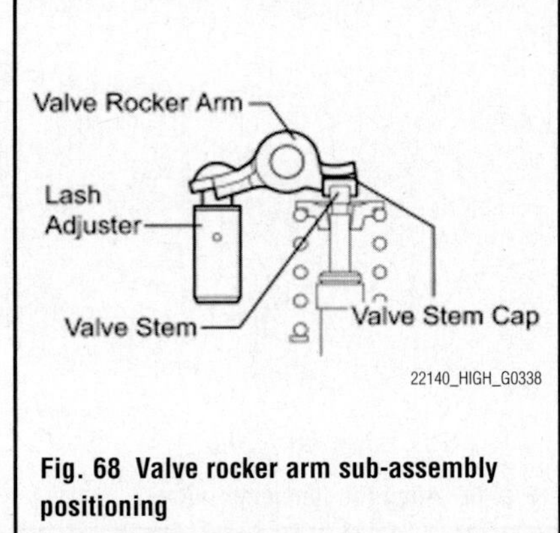

Fig. 68 Valve rocker arm sub-assembly positioning

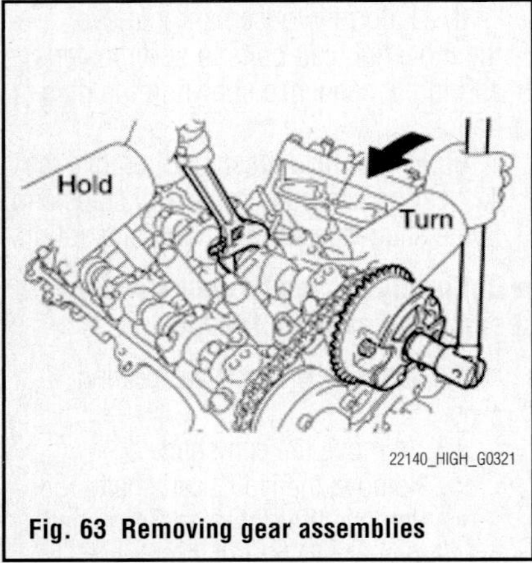

Fig. 63 Removing gear assemblies

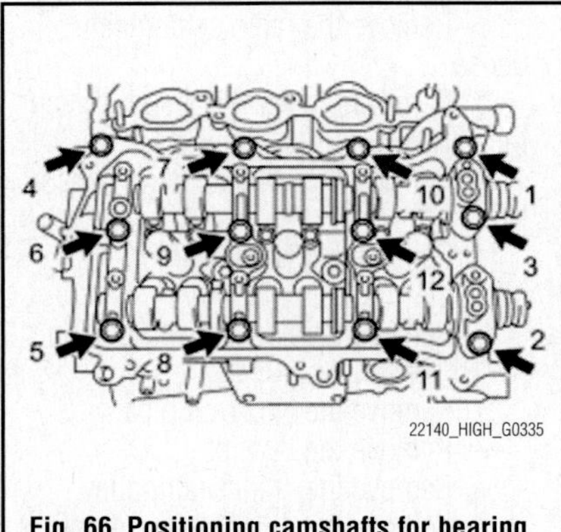

Fig. 66 Positioning camshafts for bearing cap removal

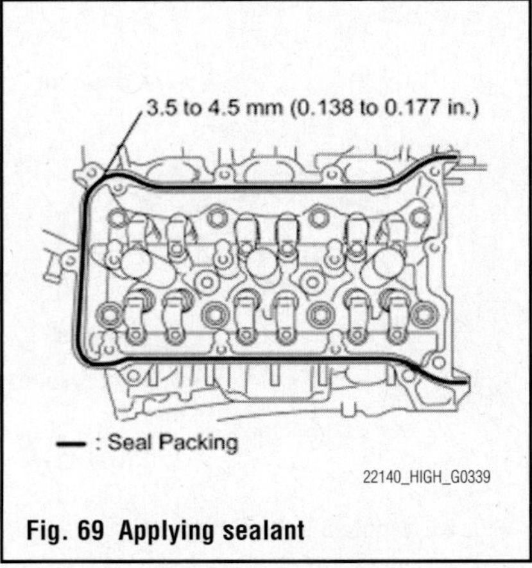

Fig. 69 Applying sealant

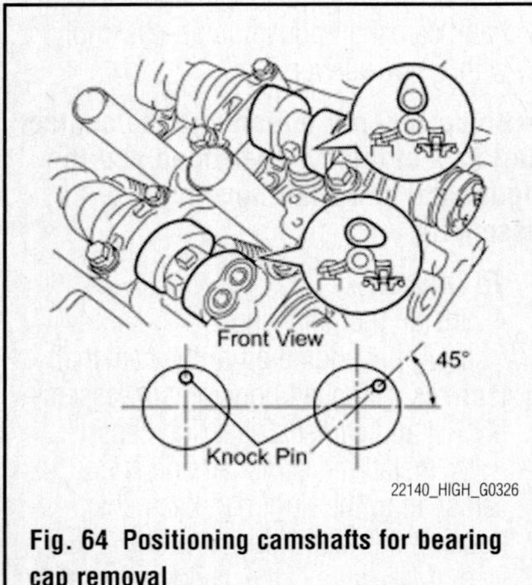

Fig. 64 Positioning camshafts for bearing cap removal

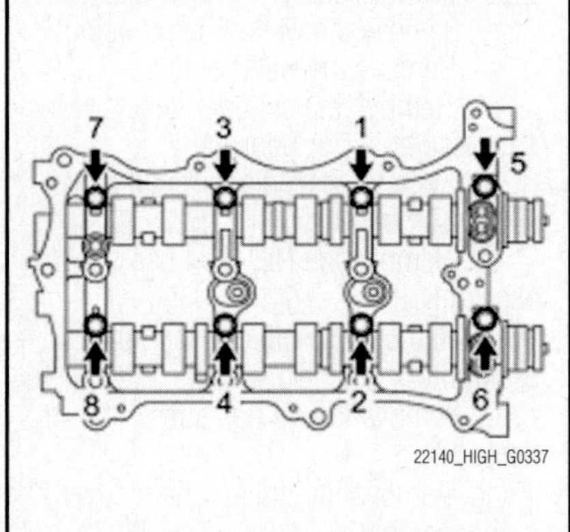

Fig. 67 Camshafts bearing cap tightening sequence

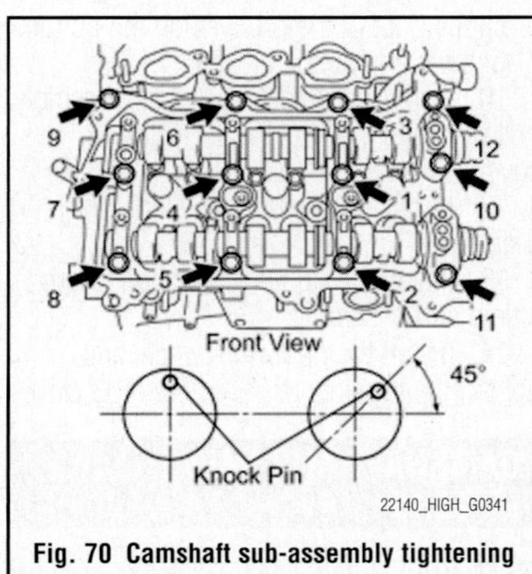

Fig. 70 Camshaft sub-assembly tightening sequence

3 minutes. Do not start the engine for at least 2 hours after installing.

g. Install the right hand camshaft housing sub-assembly and tighten the 12 bolts to 21 ft. lbs. (28 Nm) in the order shown in the illustration.

➡When installing the right hand camshaft housing, it is necessary to correctly position the camshafts as shown in the removal illustration. If the camshaft housing sub-assembly is removed because any of the bolts are loosened during installation, make sure that the previously applied seal packing does not enter any oil passages.

h. Complete the tightening of the 8 bolts to 12 ft. lbs. (16 Nm) in the order shown above.

i. Install the No. 2 chain tensioner assembly with the bolt and tighten to 15 ft. lbs. (21 Nm).

j. While pushing in the tensioner, insert a pin of 1.0 mm (0.039 in.) diameter into the hole to fix it.

k. Align the mark plate with the timing marks of the camshaft timing gear.

l. Apply a light coat of engine oil to the bolt threads and bolts seating surface.

m. Align the knockpin of the camshaft with pinhole of the camshaft timing gear assembly. Install the camshaft timing

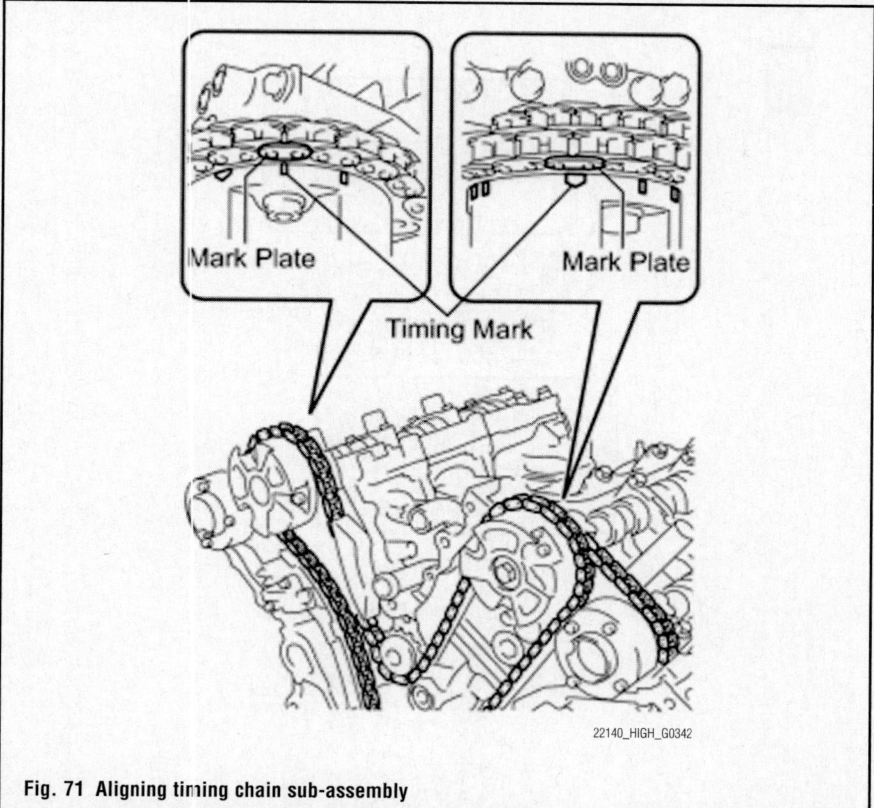

Fig. 71 Aligning timing chain sub-assembly

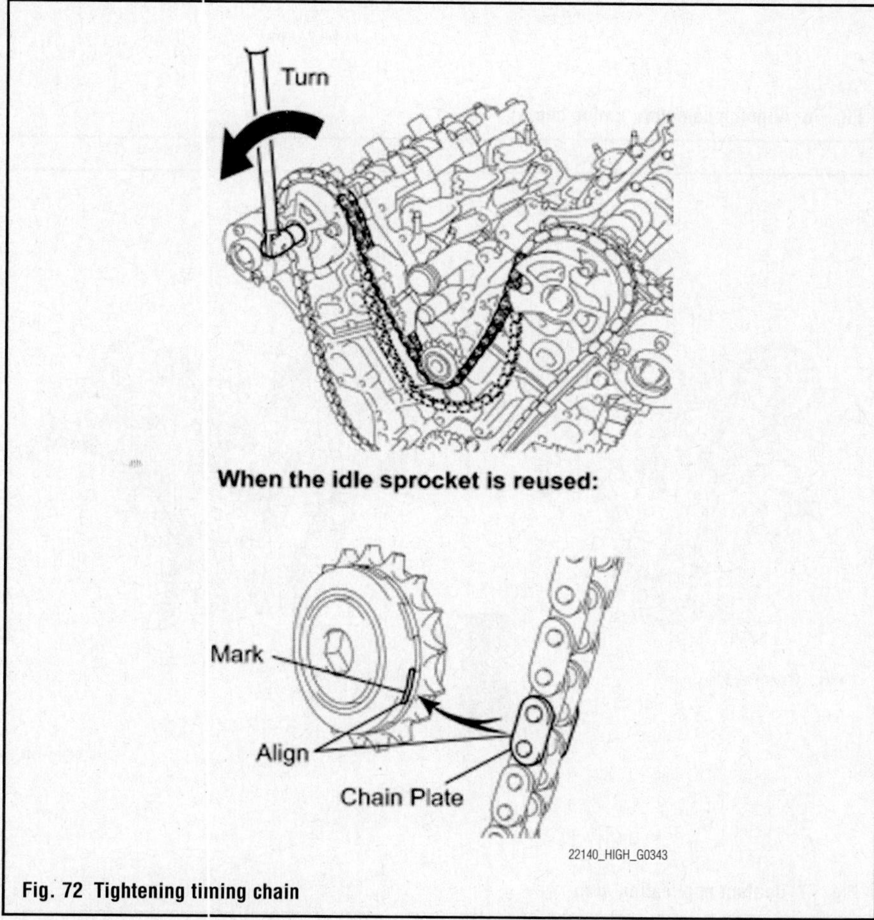

Fig. 72 Tightening timing chain

gear assembly and camshaft timing exhaust timing gear assembly and camshaft timing exhaust gear with the No. 2 chain sub-assembly installed.

n. Hold the hexagonal portion of the camshaft with the wrench and tighten the two bolts and camshaft timing gear assemblies to 74 ft. lbs. (100 Nm).

o. Remove the pan from the No 2 chain tensioner assembly.

p. Install idle sprocket assembly and tighten to 44 ft. lbs. (60 Nm).

q. Install chain sub-assembly.

r. Align the mark plate and timing marks as shown in the illustration and install the chain.

➡The camshaft mark plates are orange.

s. Do not pass the chain over the crankshaft, just temporarily place it on the crankshaft.

t. Turn the camshaft timing gear assembly on bank 1 counterclockwise to tighten the chain between the banks.

➡When the idle sprocket assembly is reused, align the timing chain plate with the mark where the plate has been in order to tighten the chain between the banks.

u. Align the mark plate and timing marks as shown in the illustration and install the chain onto the crankshaft timing sprocket. The crankshaft to mark plate is yellow.

v. Turn the crankshaft clockwise to set it to the right-hand block bore more centerline. (TDC Compression).

w. Install chain tensioner slipper.

x. Move the stopper plate upward to release the lock, and push the plunger deep into the tensioner.

y. Move the stopper plate downward to set the loss, and insert a hexagon wrench into the hole of the stopper plate.

z. Install No. 1 chain tensioner assembly and tighten bolts to 7 ft. lbs. (10 Nm).

aa. Remove the hexagon wrench from the No. 1 chain tensioner assembly. Check that the each timing mark is aligned with the crankshaft at TDC compression.

bb. Remove the pulley set bolt.

cc. Install timing chain case oil seal.

dd. Install sealant to timing chain cover sub-assembly.

ee. Install new O ring gasket on cylinder block.

ff. Align the oil pump's drive rotor spline and the crankshaft as shown in the

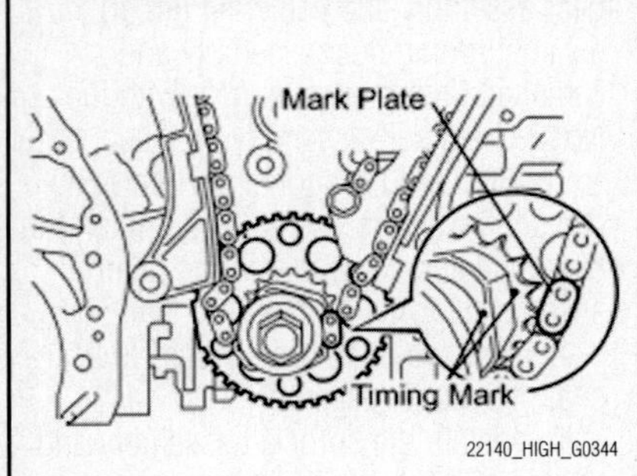

Fig. 73 Installing timing chain on crankshaft

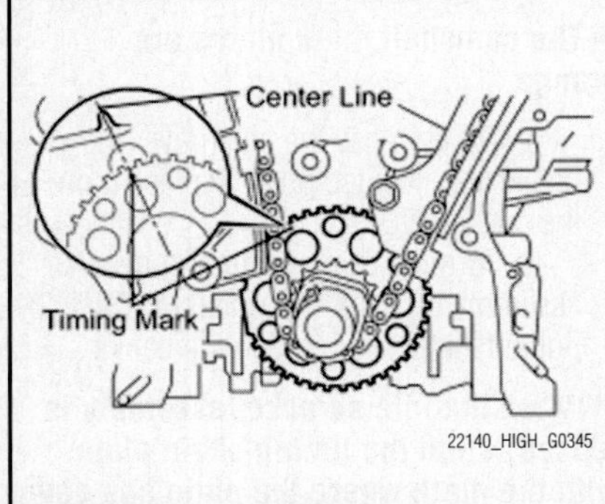

Fig. 74 Aligning timing chain on crankshaft

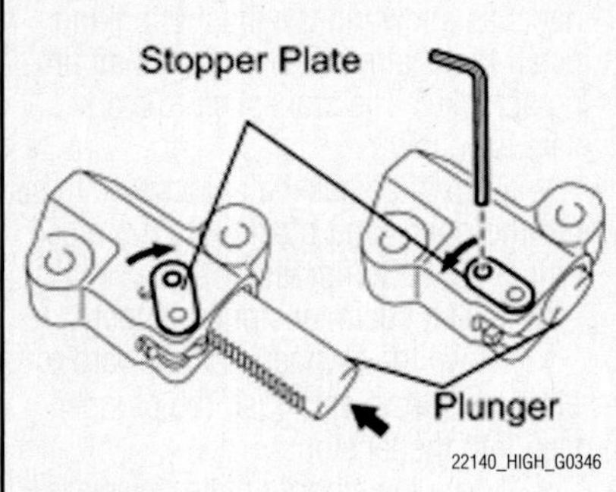

Fig. 75 Aligning timing chain on crankshaft

illustration. Install the spline and chain cover to the crankshaft.

gg. Temporarily tighten the timing chain cover with the 23 bolts and 2 nuts.

- Tighten bolts in area 1 and 2: 15 ft. lbs. (21 Nm).
- Tighten bolt in area 3: 15 ft. lbs. (21 Nm).

➡First tighten the upper bolts and nuts, followed by the lower bolts and nuts as shown.

- Tighten bolt in area 4: 32 ft. lbs. (43 Nm)

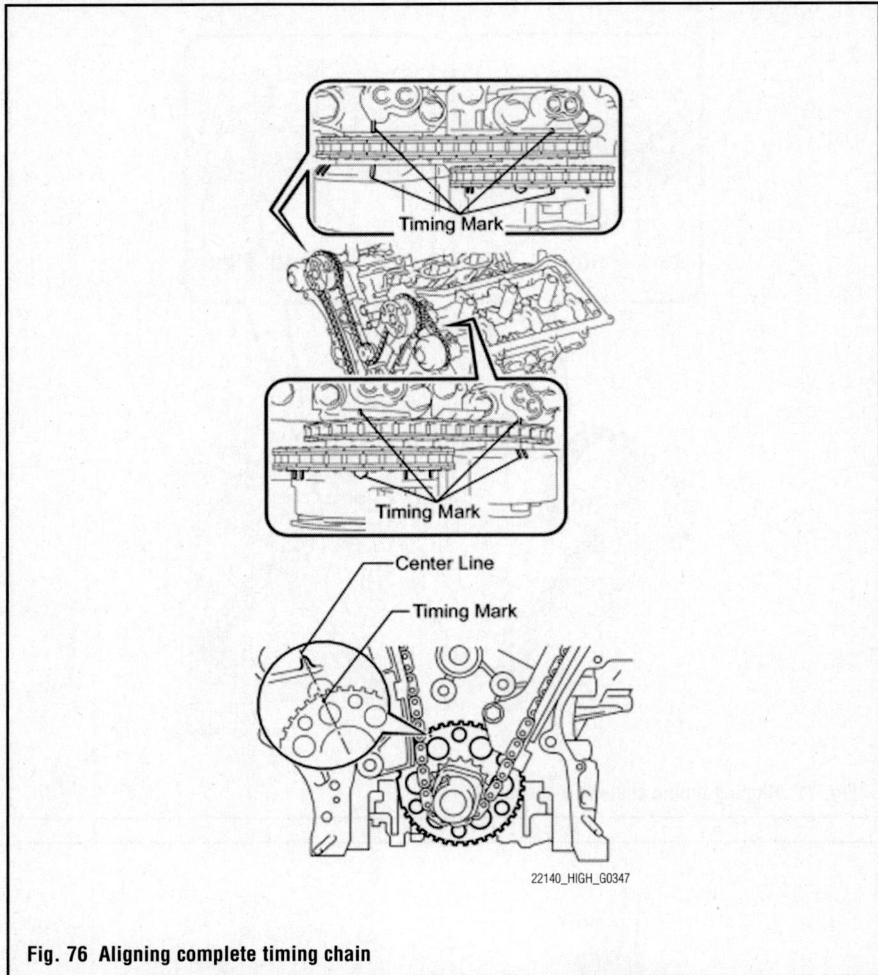

Fig. 76 Aligning complete timing chain

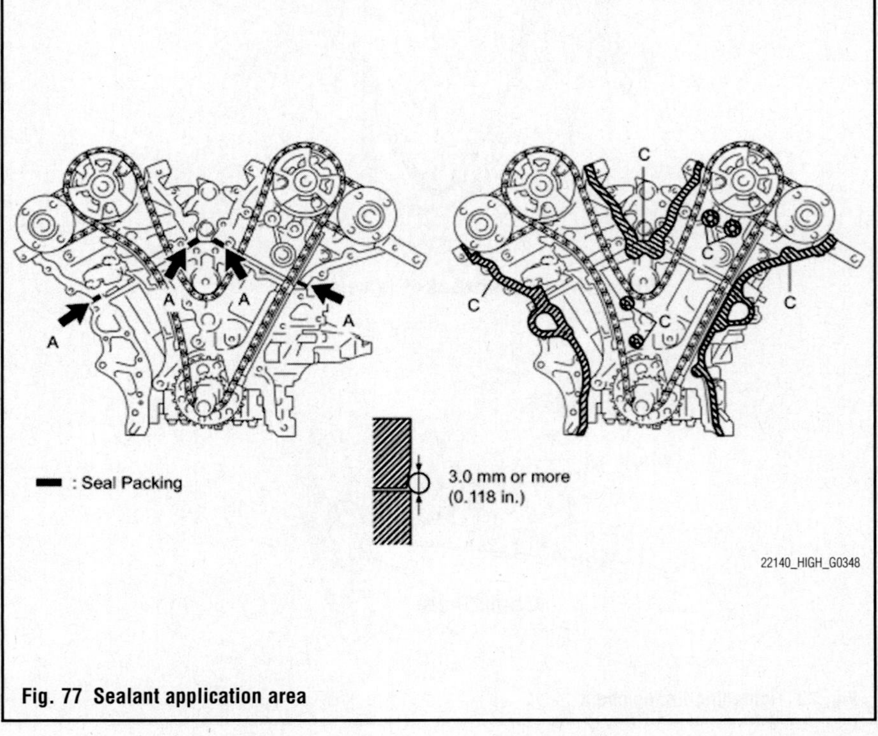

Fig. 77 Sealant application area

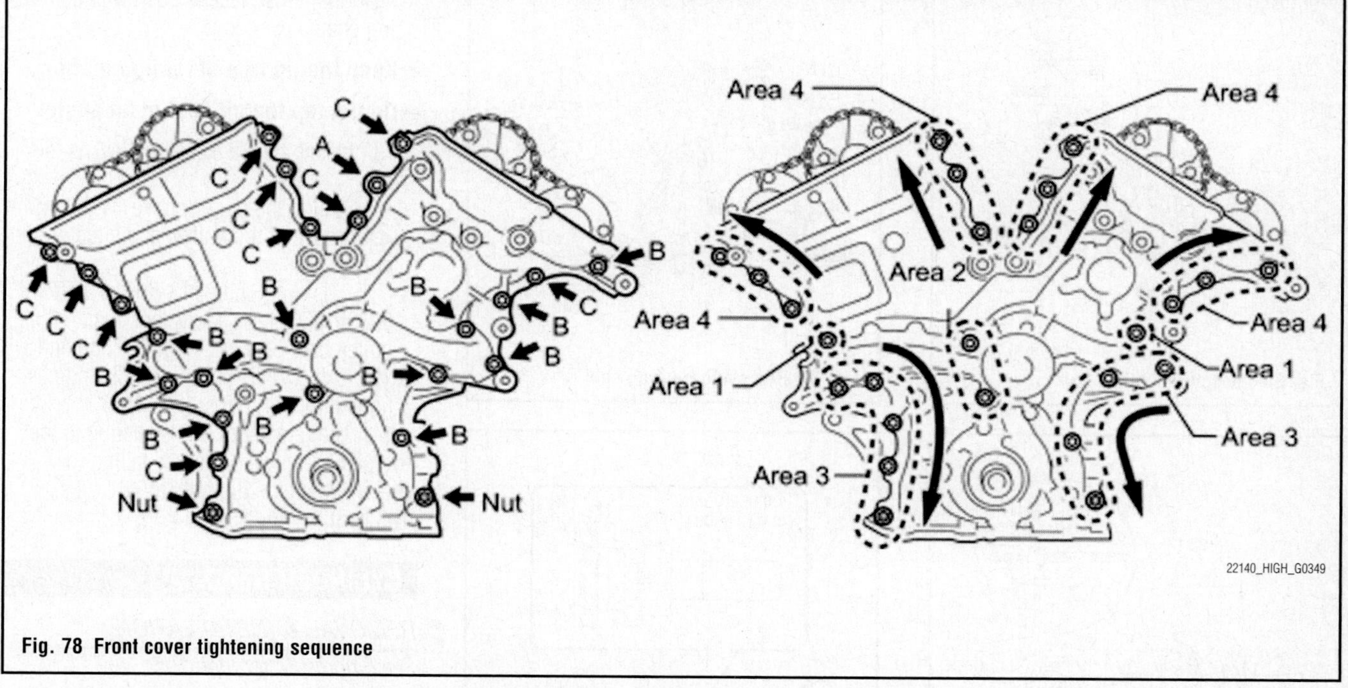

Fig. 78 Front cover tightening sequence

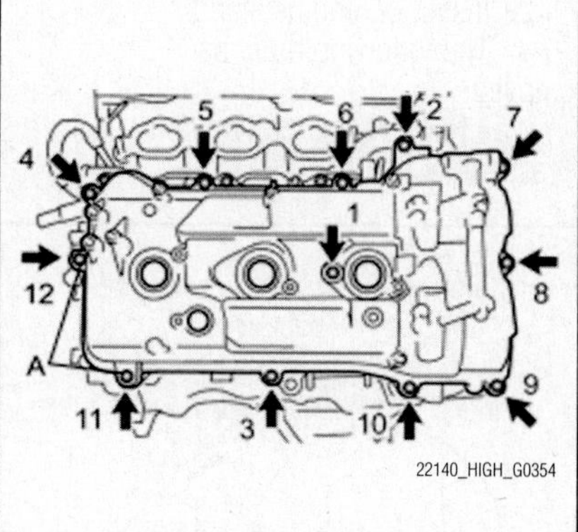

Fig. 79 Cylinder head cover tightening sequence

- Tighten bolt in area 4: 15 ft. lbs. (21 Nm)

hh. Install oil pan sub-assembly and tighten 16 bolts and 2 nuts to 7 ft. lbs. (10 Nm) and 15 ft lbs. (21 Nm).

ii. Install oil strainers sub-assembly and tighten bolts and nuts to 7 ft. lbs. (10 Nm).

jj. Install No.2 oil pan sub-assembly and tighten 16 bolts and 2 nuts to 7 ft. lbs. (10 Nm).

kk. Install cylinder head cover sub-assemblies and tighten to 7 ft. lbs. (10 Nm) and 15 ft. lbs. (21 Nm).

ll. Install the water inlet housing.

mm. Install No. 1 oil cooler bracket (w/ oil cooler).

nn. Install the oil cooler assembly (w/ oil cooler).

oo. Install the crankshaft pulley.

pp. Install the oil pipe.

qq. Install the No. 1 oil pipe.

rr. Install the Crankshaft Position (CKP) sensor.

ss. Install the knock control sensor.

tt. Install the knock control sensor wire.

uu. Install the No. 1 vacuum switching valve.

vv. Install the radio setting condenser.

ww. Install the No. 1 left front engine mounting bracket with the 6 bolts and tighten to 40 ft. lbs. (54 Nm).

➡Install the water inlet and mounting bracket within 15 minutes after installing the chain cover. Do not start the engine for at least 2 hours after installation.

xx. Install the No. 2 idler pulley sub-assembly.

yy. Install the No. 2 timing gear cover

zz. Install the V-ribbed belt tensioner assembly.

53. Install the generator assembly.
54. Install the transverse engine mounting bracket.
55. Install the left hand exhaust manifold sub-assembly and tighten to 15 ft. lbs. (21 Nm).
56. Install the No. 2 exhaust manifold heat insulator.
57. Install the No. 2 manifold stay.
58. Install the No. 2 engine oil level dipstick guide.
59. Install the right hand exhaust manifold sub-assembly and tighten to 15 ft. lbs. (21 Nm).
60. Install the intake manifold and tighten the 6 bolts and 4 nuts uniformly in several steps to 15 ft. lbs. (21 Nm).
61. Install the right hand No. 2 engine mounting stay.
62. Install the ignition coil assembly.
63. Remove the engine stand.
64. Install the engine assembly.

CATALYTIC CONVERTER

REMOVAL & INSTALLATION

The catalytic converter is integrated with the exhaust manifold. Refer to Exhaust Manifold in this section.

CRANKSHAFT DAMPER

REMOVAL & INSTALLATION

See Figures 80 and 81.

1. Raise and support the vehicle.
2. Remove the right front wheel.
3. Remove the engine under cover assembly.
4. Remove the No. 1 engine under cover.
5. Remove the right front fender molding sub-assembly.
6. Remove the right front fender liner.
7. Remove the right front fender apron seal.
8. Remove the V-ribbed belt.
9. Using a special service tool (SST: 09213-70011, SST: 09330-00021 or equivalent), loosen the crankshaft pulley bolt.

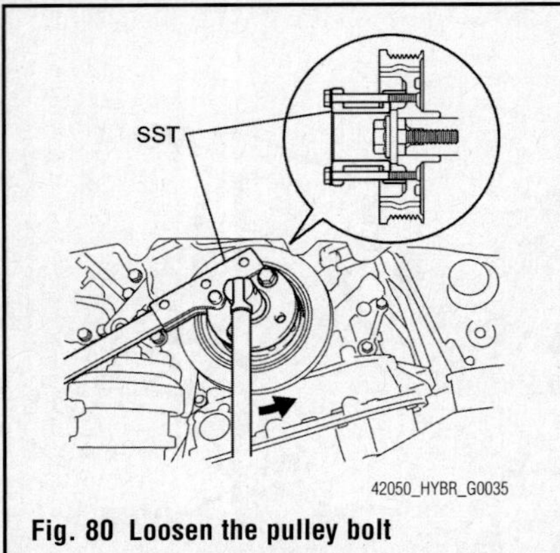

Fig. 80 Loosen the pulley bolt

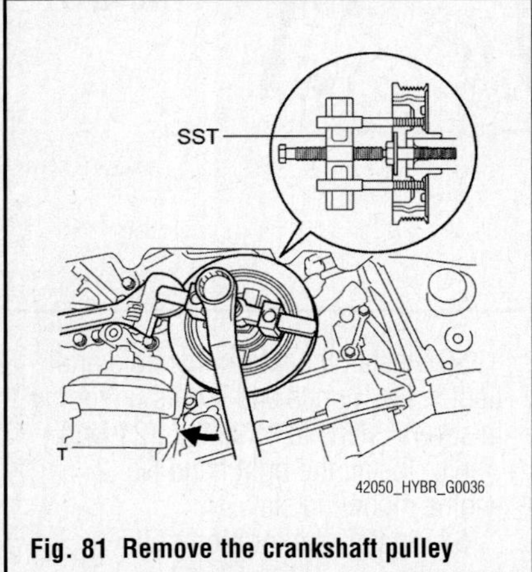

Fig. 81 Remove the crankshaft pulley

10. Using SST: 09950-50013 or equivalent, remove the crankshaft pulley bolt and crankshaft pulley.

To install:

11. Install the crankshaft pulley, as follows:
   a. Align the pulley set key with the key groove of the pulley, and slide on the pulley.
   b. Using SST: 09213-70011, SST: 09330-00021, or equivalent, install the pulley bolt. Tighten to 184 ft. lbs. (250 Nm).
12. The remainder of installation is the reverse of removal. When installing the wheel, tighten the lug nuts to 76 ft. lbs. (103 Nm).

CRANKSHAFT FRONT SEAL

REMOVAL & INSTALLATION

See Figures 82 and 83.

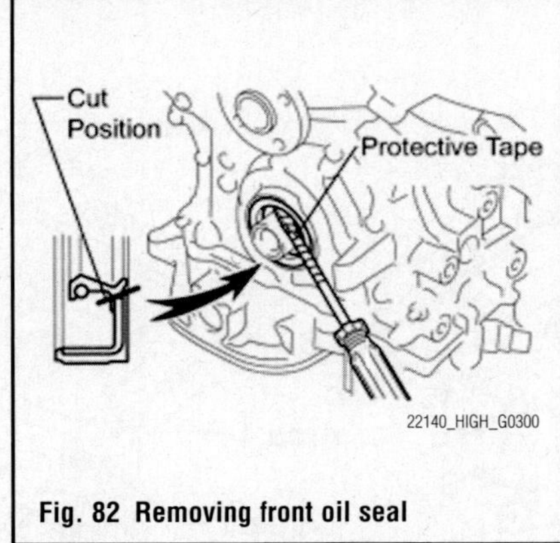

Fig. 82 Removing front oil seal

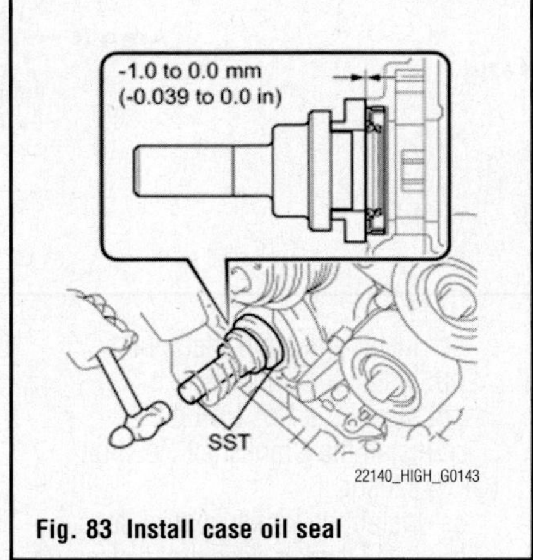

Fig. 83 Install case oil seal

1. Raise and support the vehicle.
2. Remove the right front wheel.
3. Remove the engine under cover assembly.
4. Remove the No. 1 engine under cover.
5. Remove the right front fender molding sub-assembly.
6. Remove the right front fender liner.
7. Remove the right front fender apron seal.
8. Remove the V-ribbed belt.
9. Using a special service tool (SST: 09213-70011, SST: 09330-00021 or equivalent), loosen the crankshaft pulley bolt.
10. Using SST: 09950-50013 or equivalent, remove the crankshaft pulley bolt and crankshaft pulley.
11. Using a screwdriver, pry out the timing chain case oil seal.

➡Tape the screwdriver tip before use.

➡After the removal, check the crankshaft for damage. If it is damaged, smooth the surface with 400-grit sandpaper.

To install:

12. Install timing chain case oil seal, as follows:
   a. Apply MP grease to a new oil seal lip.
   b. Using a Special Service Tool (SST: 09223-22010, SST: 09506-35010 or equivalent) and a hammer, tap in the oil seal until its surface is flush with the timing chain cover edge.

➡Keep the lip free of foreign matter.

➡Do not tap the oil seal at an angle.

13. Install the crankshaft pulley, as follows:
   a. Align the pulley set key with the key groove of the pulley, and slide on the pulley.
   b. Using SST: 09213-70011, SST: 09330-00021, or equivalent, install the pulley bolt. For 2.7L engines, tighten to 192 ft. lbs. (260 Nm). For 3.5L engines, tighten to 184 ft. lbs. (250 Nm).
14. The remainder of installation is the reverse of removal. When installing the wheel, tighten the lug nuts to 76 ft. lbs. (103 Nm).

CYLINDER HEAD

REMOVAL & INSTALLATION

See Figures 84 through 105.

1. Remove the engine assembly.
2. Install on engine stand.
3. The following must be removed:
   a. Remove the ignition coil assembly.

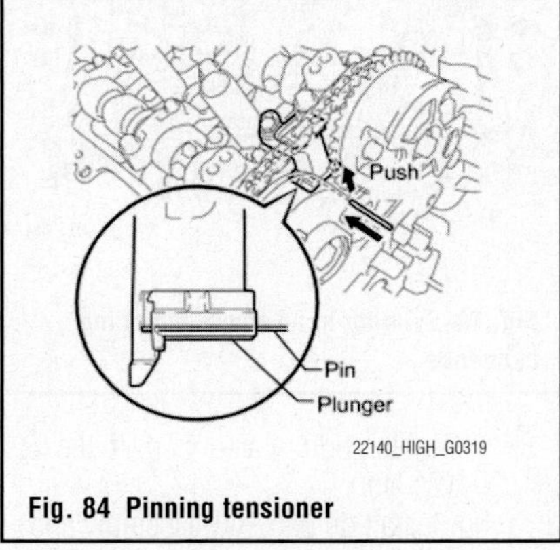

Fig. 84 Pinning tensioner

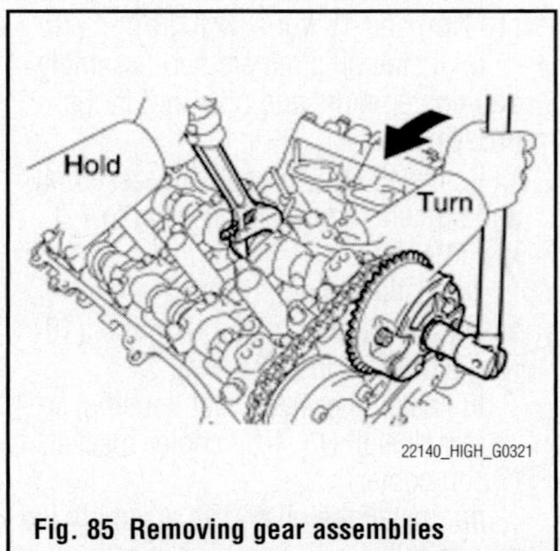

Fig. 85 Removing gear assemblies

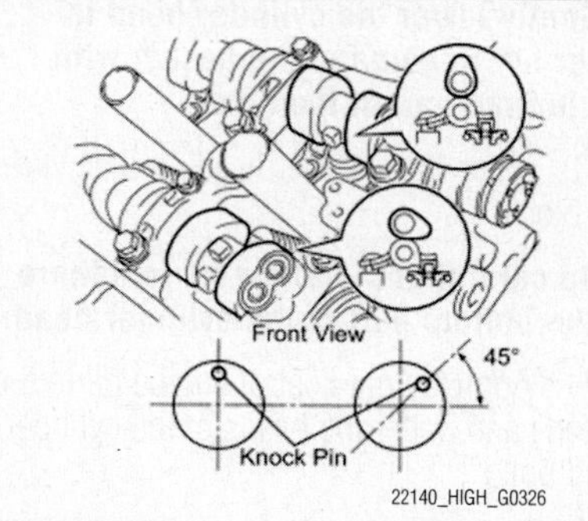

Fig. 86 Positioning camshafts for bearing cap removal

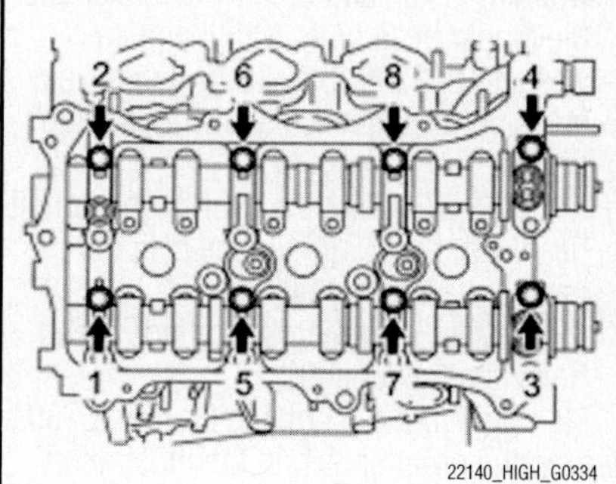

Fig. 87 Camshafts bearing cap removal sequence

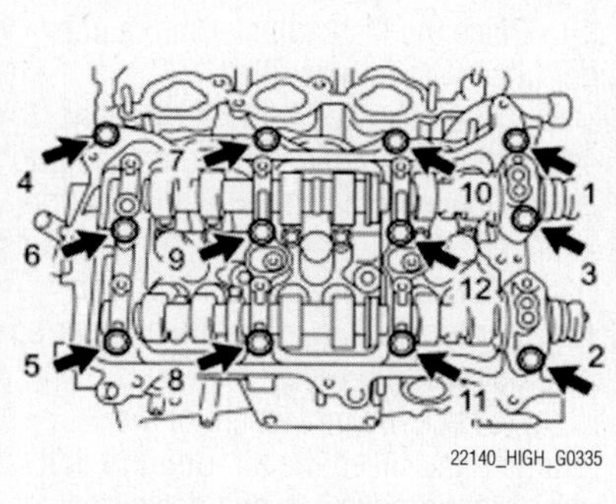

Fig. 88 Camshaft bearing cap loosening sequence

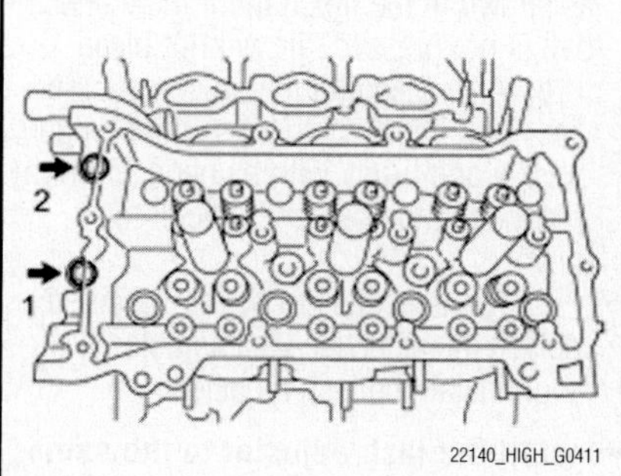

Fig. 89 Loosening cylinder head bolts LH shown

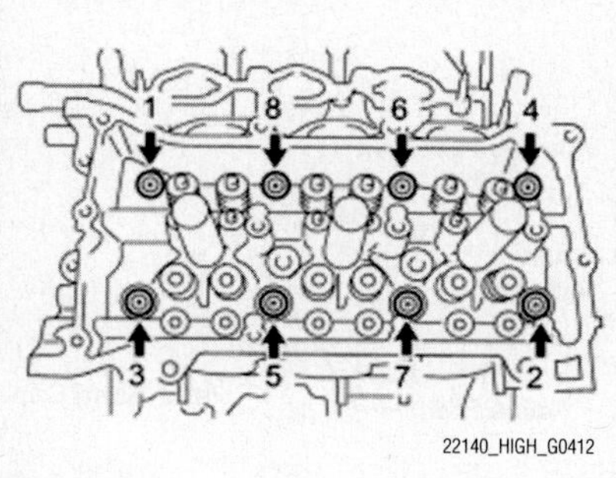

Fig. 90 Cylinder head bolt loosening sequence LH shown

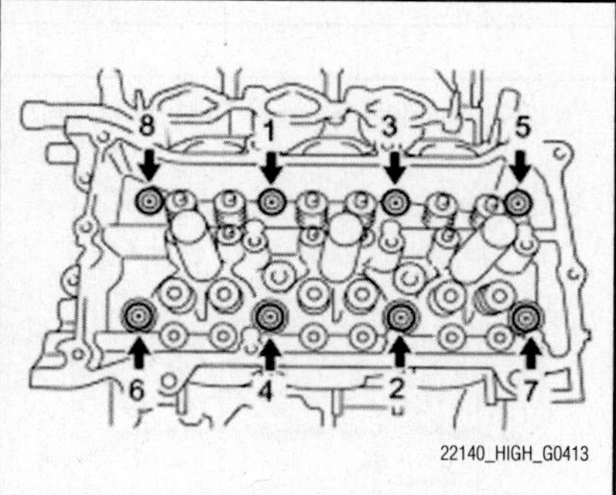

Fig. 91 Cylinder head bolt tightening sequence LH shown

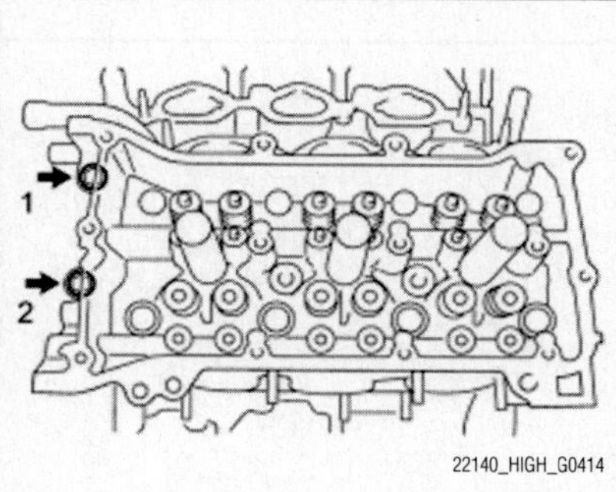

Fig. 92 Tightening cylinder head bolts LH shown

b. Remove right hand No. 2 engine mounting stay.

c. Remove the intake manifold.

d. Remove the right hand exhaust manifold sub-assembly.

e. Remove the No. 2 engine oil level dipstick guide.

f. Remove the No. 2 manifold stay.

g. Remove the No. 2 exhaust manifold heat insulator.

h. Remove the left exhaust manifold sub-assembly.

i. Remove the transverse engine mounting bracket.

j. Remove the generator assembly.

k. Remove the V-ribbed belt tensioner assembly.

l. Remove the No. 2 timing gear cover.

m. Remove the No. 2 idler pulley sub-assembly.

n. Remove the left No. 1 engine front mounting bracket.

o. Remove the 6 bolts and left hand No. 1 front engine mounting bracket.

p. Remove the radio setting condenser.

q. Remove the No. 1 vacuum switching valve.

r. Remove the knock control sensor wire.

s. Remove the knock control sensor.

t. Remove the Crankshaft Position (CKP) sensor.

u. Remove the No. 1 oil pipe.

v. Remove the oil pipe.

w. Remove the crankshaft pulley.

x. Remove the oil cooler assembly, if necessary.

y. Remove the No. 1 oil cooler bracket, if necessary.

z. Remove the water inlet housing.

aa. Remove the water outlet.

bb. Remove the cylinder head cover sub-assembly (for bank 1).

cc. Remove the cylinder head cover sub-assembly (for bank 2).

dd. Remove the No. 2 oil pan sub-assembly.

ee. Remove the oil strainer sub-assembly.

ff. Remove the oil pan sub-assembly.

gg. Remove the timing chain cover sub-assembly.

hh. Remove the timing chain case oil seal.

ii. Set the No. 1 cylinder to TDC/compression.

jj. Remove the No. 1 chain tensioner assembly.

kk. Remove the chain tensioner slipper.

ll. Remove the chain sub-assembly.

mm. Remove the idle sprocket assembly.

nn. Remove the camshaft timing gears and No. 2 chain (for bank 1).

oo. While raising the No. 2 chain tensioner assembly, insert a pin of 1.0 mm (0.039 in.) diameter into the hole to fix the No. 2 chain tensioner assembly.

pp. Hold the hexagonal portion of the camshaft with a wrench, and remove the 2 bolts and 2 camshaft timing gear assemblies.

➡Be careful not to damage the cylinder head with the wrench. Do not disassemble the camshaft timing gear assemblies.

qq. Remove the No. 2 chain assembly.

rr. Remove the bolt and No. 2 chain tensioner assembly.

ss. Check that the camshafts are positioned as shown in the illustration.

tt. Uniformly loosen and remove the 8 bearing cap bolts in several steps and in the sequence shown in the illustration.

uu. Uniformly loosen and remove the 12 bearing cap bolts in several steps and in the sequence shown in the illustration.

➡Uniformly loosen the bolts while keeping the camshaft level.

vv. Remove the 5 camshaft bearing caps.

ww. Remove the camshaft.

xx. Remove the No. 2 camshaft.

yy. Remove the right camshaft housing sub-assembly by prying between the cylinder head and right camshaft housing sub-assembly with a screwdriver.

➡Be careful not to damage the contact surfaces of the cylinder head and the right camshaft housing sub-assembly.

4. Remove the 24 valve lash adjuster assemblies from the cylinder head.

➡Arrange the removed parts in the correct order.

5. Uniformly loosen and remove the 2 cylinder head set bolts in several steps and in the sequence shown in the illustration.

➡Be careful not to drop washers into the cylinder head. Cylinder head warpage or cracking could result from removing bolts in an incorrect order. Be sure to keep separate the removed parts for each installation position.

6. Using a 10 mm bi-hexagon wrench, uniformly loosen the 8 bolts in the sequence shown in the illustration. Remove the 8 cylinder head bolts and plate washers.

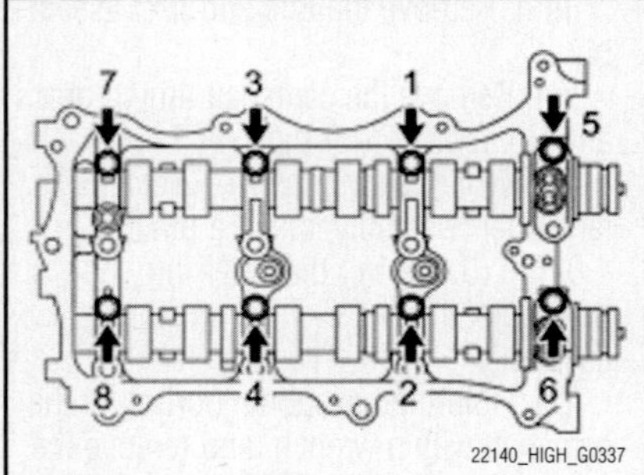

Fig. 93 Camshafts bearing cap tightening sequence

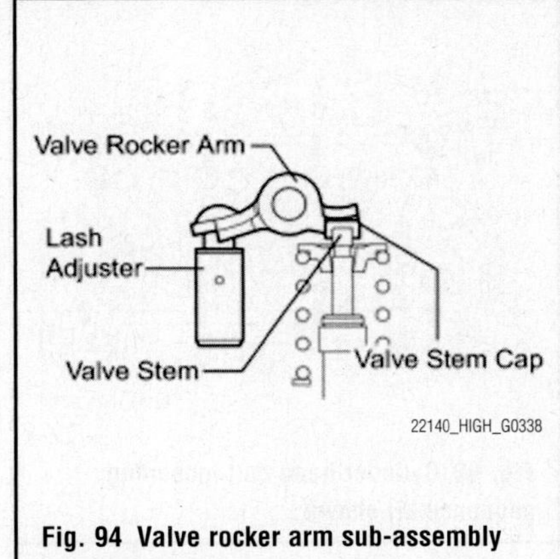

Fig. 94 Valve rocker arm sub-assembly positioning

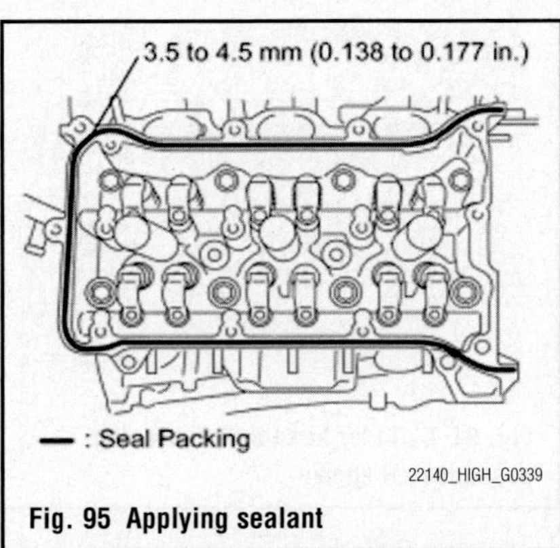

Fig. 95 Applying sealant

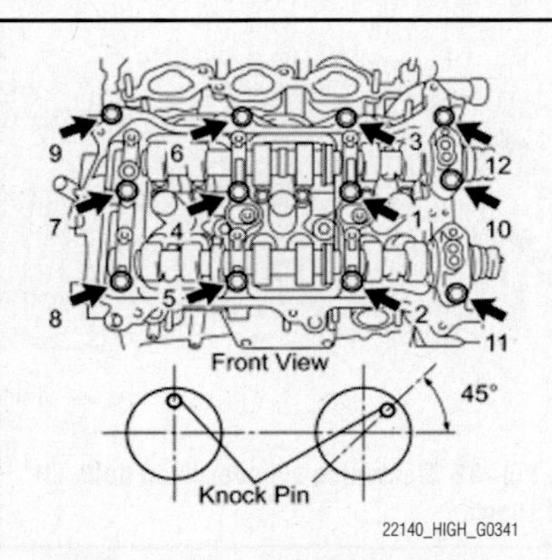

Fig. 96 Camshaft sub-assembly tightening sequence

7. Remove the cylinder head sub-assembly.

8. Remove the No. 2 cylinder head gasket.

To install:

9. Place the No. 2 cylinder head gasket on the cylinder block surface with the Lot No. stamp upward.

➡Gently lower the cylinder head in order not to damage the gasket with the bottom part of the head.

10. Place the cylinder head on the cylinder block.

➡Be careful not to allow oil to adhere to the bottom part of the cylinder head.

11. Apply a light coat of engine oil to the threads and under the heads of the cylinder head bolts.

➡The cylinder head bolts are tightened in 3 progressive steps.

a. Step 1: Using a 10 mm bi-hexagon wrench, install and uniformly tighten the 8 cylinder head bolts with the plate washers to 27 ft. lbs. (36 Nm) in several steps in the sequence shown in the illustration.

b. Step 2: Mark the cylinder head bolt head with paint. Tighten the cylinder head bolts another 90°.

c. Step 3: Tighten the cylinder head bolts an additional 90°.

12. Tighten the 2 bolts to 22 ft. lbs. (30 Nm) in the order shown in the illustration.

13. Install the 12 valve stem caps.

➡Keep the lash adjuster free of dirt and foreign objects. Only use clean engine oil.

14. Place the lash adjuster into a container filled with engine oil.

a. Insert Service Tool's tip into the lash adjuster's plunger and use the tip to press down on the check ball inside the plunger.

b. Squeeze Service Tool and lash adjuster together to move the plunger up and down 5 to 6 times.

c. Check the movement of the plunger and bleed the air. Confirm that the plunger moves up and down freely.

d. When bleeding air from the high-pressure chamber, make sure that the tip of SST is actually pressing the check ball as shown in the illustration. If the check ball is not pressed, air will not bleed.

e. After bleeding the air, remove Service Tool. Then, try to press the plunger quickly and firmly with by hand. Confirm that the plunger is very difficult to move.

f. If the results are not as specified, replace the defective lash adjuster.

g. Install the lash adjusters.

➡Install the lash adjuster to the same place where it was removed from.

15. Install the following:

a. Apply engine oil to the camshaft journals, right camshaft housing sub-assembly, and camshaft bearing caps.

b. Install the camshaft and No. 2 camshaft to the right camshaft housing sub-assembly.

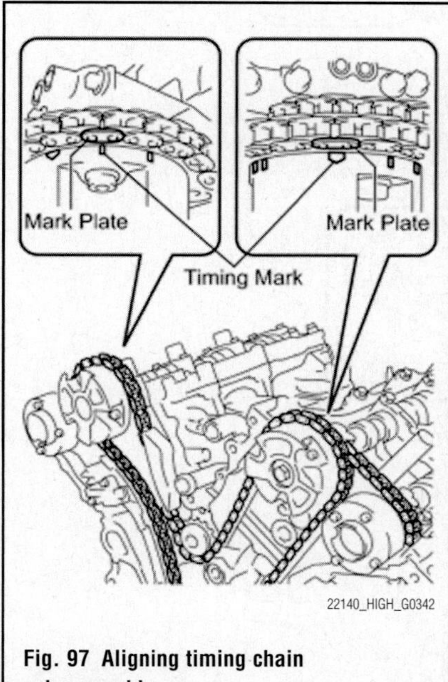

Fig. 97 Aligning timing chain sub-assembly

c. Make sure of the marks and numbers on the camshaft bearing caps and place them in each proper position and direction.

d. Temporarily tighten the 8 bearing cap bolts to 7 ft. lbs. (10 Nm) in the order shown in the illustration.

e. Make sure that the No. 1 valve rocker arm sub-assembly is installed as shown in the illustration.

f. Apply seal packing in a continuous line as shown in the illustration.

➡Remove any oil from the contact surface. Install the camshaft housing sub-assembly RH within 3 minutes. Do not start the engine for at least 2 hours after installing.

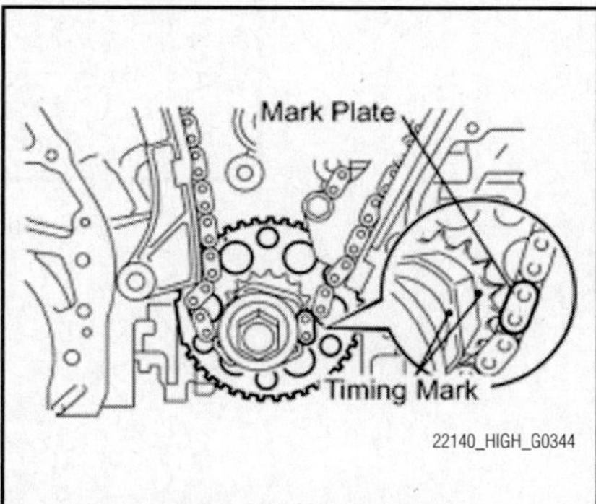

Fig. 99 Installing timing chain on crankshaft

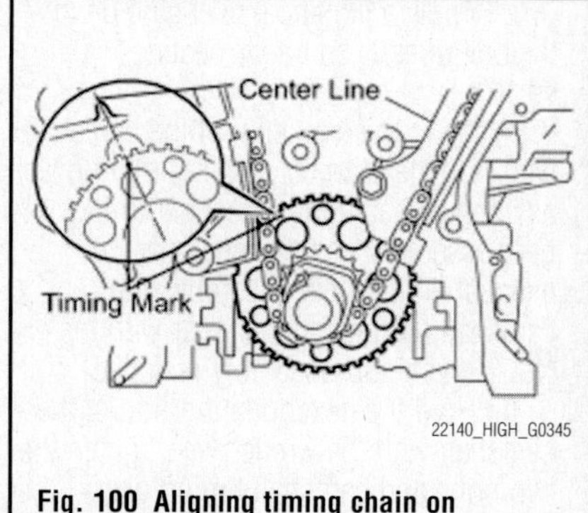

Fig. 100 Aligning timing chain on crankshaft

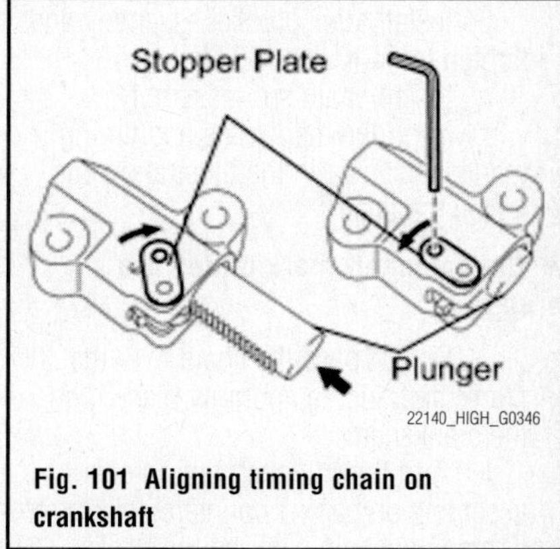

Fig. 101 Aligning timing chain on crankshaft

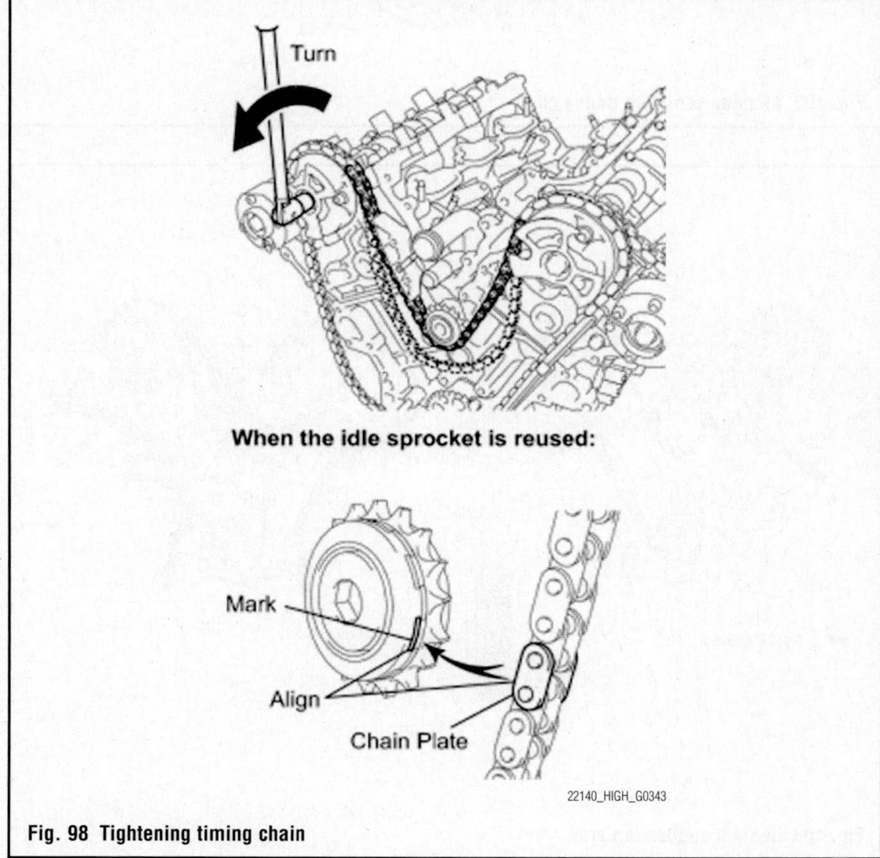

Fig. 98 Tightening timing chain

g. Install the right camshaft housing sub-assembly and tighten the 12 bolts to 21 ft. lbs. (28 Nm) in the order shown in the illustration.

➡When installing the camshaft housing RH, it is necessary to correctly position the camshafts as shown in the removal illustration. If the camshaft housing sub-assembly is removed because any of the bolts are loosened during installation, make sure that the previously applied seal packing does not enter any oil passages.

h. Complete the tightening of the 8 bolts to 12 ft. lbs. (16 Nm) in the order shown above.

i. Install the No. 2 chain tensioner assembly with the bolt and tighten to 15 ft. lbs. (21 Nm).

j. While pushing in the tensioner, insert a pin of 1.0 mm (0.039 in.) diameter into the hole to fix it.

k. Align the mark plate with the timing marks of the camshaft timing gear.

l. Apply a light coat of engine oil to the bolt threads and bolts seating surface.

m. Align the knockpin of the camshaft with pinhole of the camshaft timing gear assembly. Install the camshaft timing gear assembly and camshaft timing exhaust timing gear assembly and camshaft timing exhaust gear with the No. 2 chain sub-assembly installed.

n. Hold the hexagonal portion of the camshaft with the wrench and tighten the two bolts and camshaft timing gear assemblies to 74 ft. lbs. (100 Nm).

o. Remove the pan from the No 2 chain tensioner assembly.

p. Install idle sprocket assembly and tighten to 44 ft. lbs. (60 Nm).

q. Install chain sub-assembly.

r. Align the mark plate and timing marks as shown in the illustration and install the chain.

➡The camshaft mark plates are orange.

s. Do not pass the chain over the crankshaft, just temporarily place it on the crankshaft.

t. Turn the camshaft timing gear assembly on bank 1 counterclockwise to tighten the chain between the banks.

➡When the idle sprocket assembly is reused, align the timing chain plate with the mark where the plate has been in order to tighten the chain between the banks.

u. Align the mark plate and timing marks as shown in the illustration and install the chain onto the crankshaft timing sprocket. The crankshaft to mark plate is yellow.

v. Turn the crankshaft clockwise to set it to the right-hand block bore more centerline. (TDC Compression)

w. Install chain tensioner slipper.

x. Move the stopper plate upward to release the lock, and push the plunger deep into the tensioner.

y. Move the stopper plate downward to set the loss, and insert a hexagon wrench into the hole of the stopper plate.

z. Install No. 1 chain tensioner assembly and tighten bolts to 7 ft. lbs. (10 Nm).

aa. Remove the hexagon wrench from the No. 1 chain tensioner assembly. Check that the each timing mark is aligned with the crankshaft at TDC compression.

bb. Remove the pulley set bolt.

cc. Install timing chain case oil seal.

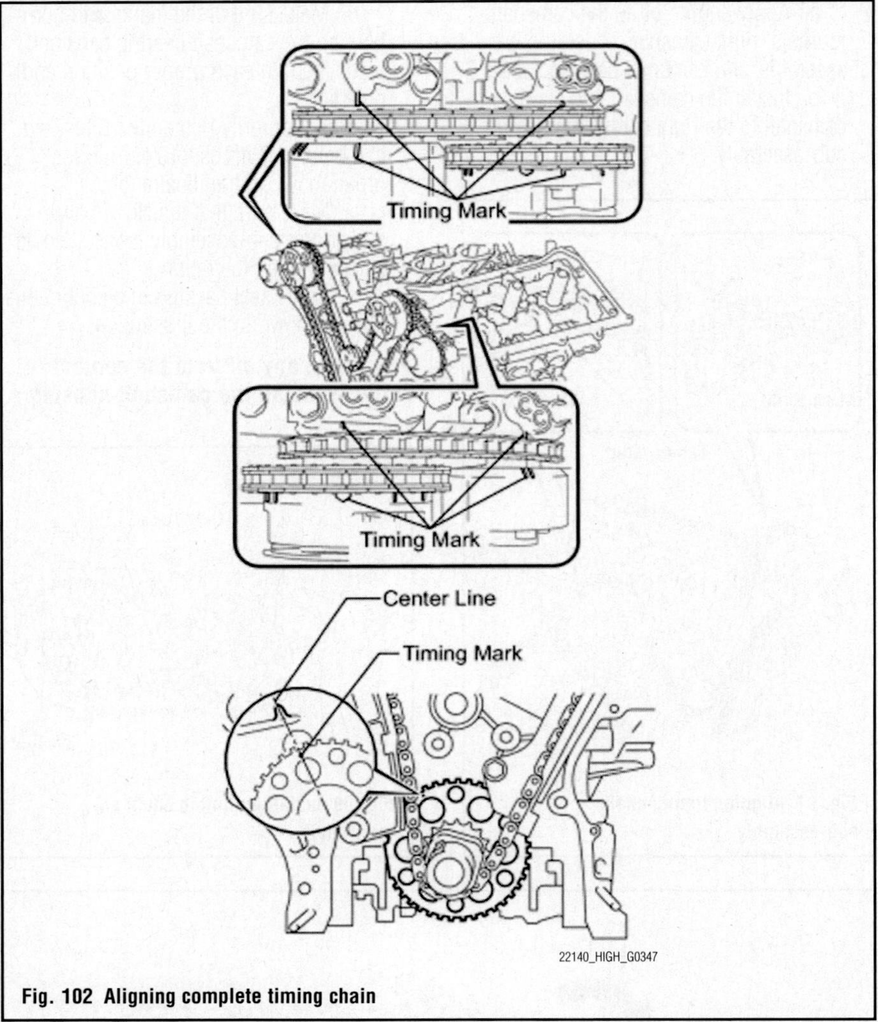

Fig. 102 Aligning complete timing chain

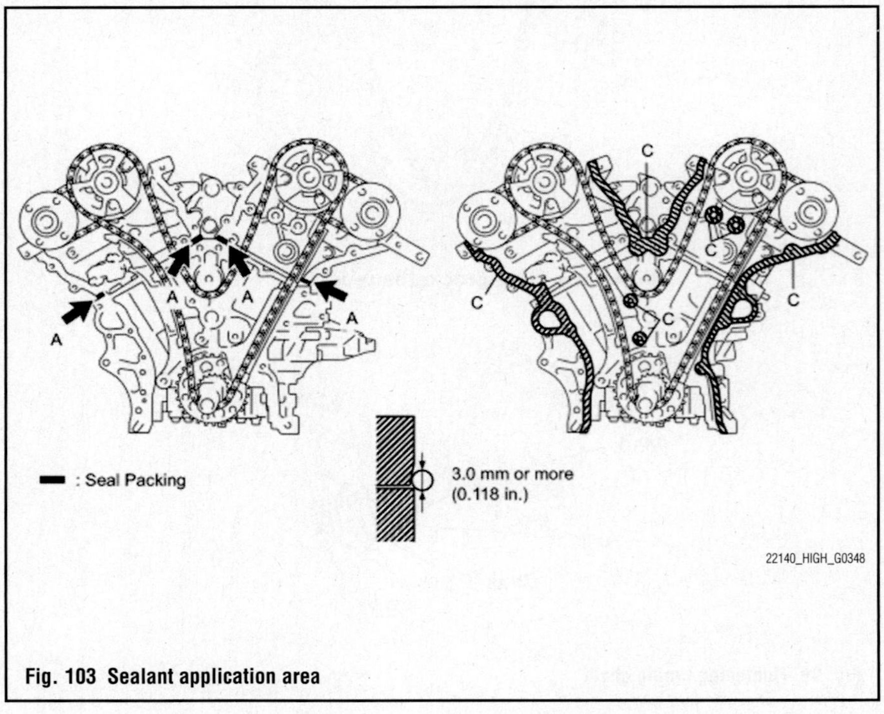

Fig. 103 Sealant application area

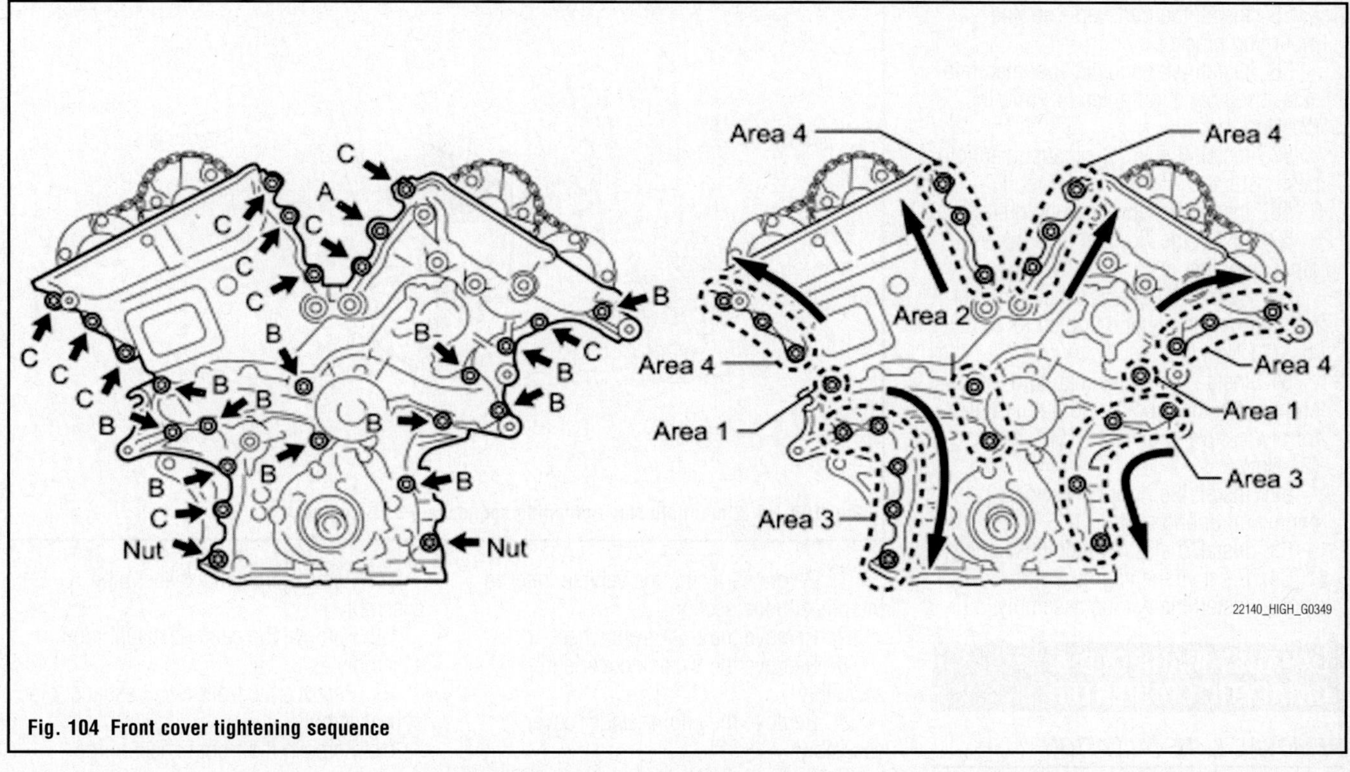

Fig. 104 Front cover tightening sequence

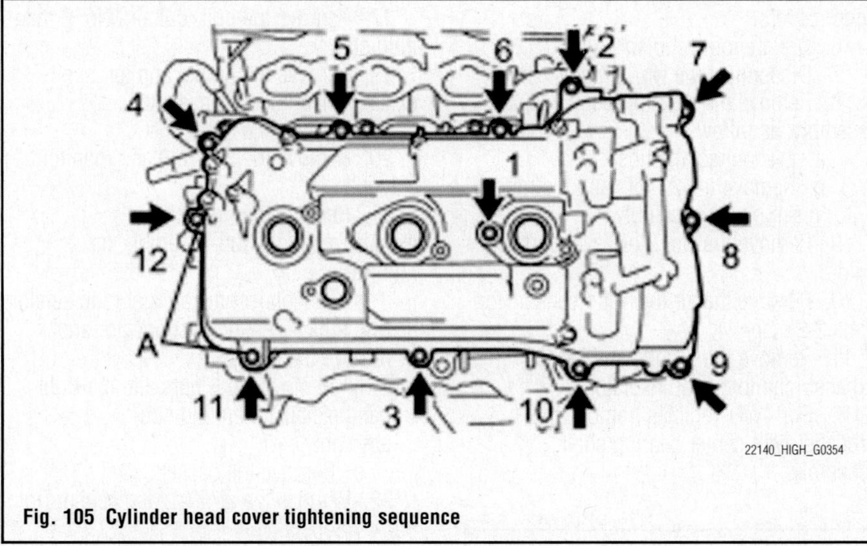

Fig. 105 Cylinder head cover tightening sequence

dd. Install sealant to timing chain cover sub-assembly.

ee. Install new O ring gasket on cylinder block.

ff. Align the oil pump's drive rotor spline and the crankshaft as shown in the illustration. Install the spline and chain cover to the crankshaft.

gg. Temporarily tighten the timing chain cover with the 23 bolts and 2 nuts.

- Tighten bolts in area 1 and 2 15 ft. lbs. (21 Nm)
- Tighten bolt in area 3 15 ft. lbs. (21 Nm)

➡First tighten the upper bolts and nuts followed by the lower bolts and nuts as shown.

- Tighten bolt in area 4 32 ft. lbs. (43 Nm)
- Tighten bolt in area 4 15 ft. lbs. (21 Nm)

hh. Install oil pan subassembly and tighten 16 bolts and 2 nuts to 7 ft. lbs. (10 Nm) and 15 ft lbs. (21 Nm).

ii. Install oil strainers sub-assembly and tighten bolts and nuts to 7 ft. lbs. (10 Nm).

jj. Install No.2 oil pan sub-assembly and tighten 16 bolts and 2 nuts to 7 ft. lbs. (10 Nm).

kk. Install cylinder head cover sub-assemblies and tighten to 7 ft. lbs. (10 Nm) and 15 ft. lbs. (21 Nm).

ll. Install water inlet housing

mm. Install the No. 1 oil cooler bracket (w/ oil cooler).

nn. Install the oil cooler assembly (w/ oil cooler).

oo. Install the crankshaft pulley.

pp. Install the oil pipe.

qq. Install the No. 1 oil pipe.

rr. Install the Crankshaft Position (CKP) sensor.

ss. Install the knock control sensor.

tt. Install the knock control sensor wire.

uu. Install the No. 1 vacuum switching valve.

vv. Install radio setting condenser.

ww. Install the left hand No. 1 front engine mounting bracket with the 6 bolts and tighten to 40 ft. lbs. (54 Nm).

➡Install the water inlet and mounting bracket within 15 minutes after installing the chain cover. Do not xx. start the engine for at least 2 hours after installation.

yy. Install the No. 2 idler pulley sub-assembly.

zz. Install the No. 2 timing gear cover.

53. Install the V-ribbed belt tensioner assembly.

54. Install the generator assembly.

55. Install the transverse engine mounting bracket.

56. Install left hand exhaust manifold sub-assembly and tighten to 15 ft. lbs. (21 Nm).

57. Install the No. 2 exhaust manifold heat insulator.

58. Install the No. 2 manifold stay.

59. Install the No. 2 engine oil level dipstick guide.

60. Install right hand exhaust manifold sub-assembly and tighten to 15 ft. lbs. (21 Nm).

61. Install the intake manifold and tighten the 6 bolts and 4 nuts uniformly in several steps to 15 ft. lbs. (21 Nm).

62. Install the right hand No. 2 engine mounting stay.

63. Install the ignition coil assembly.

64. Remove the engine stand.

65. Install the engine assembly.

EXHAUST MANIFOLD & CATALYTIC CONVERTER

REMOVAL & INSTALLATION

See Figures 106 through 109.

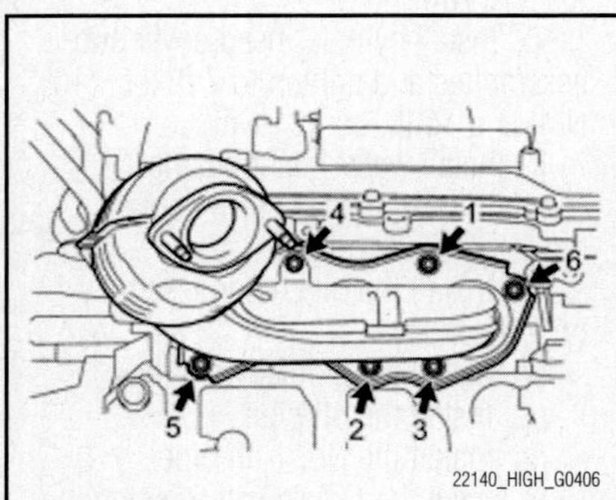

Fig. 106 Right exhaust manifold sub-assembly tightening sequence—3.5L engine

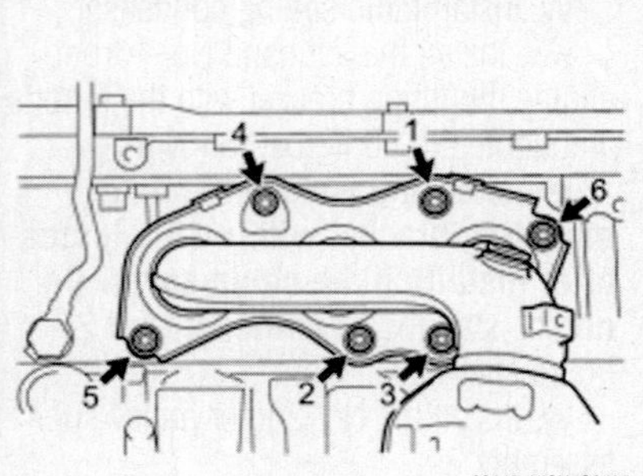

Fig. 107 Left exhaust manifold sub-assembly tightening sequence—3.5L engine

Fig. 108 No. 2 manifold stay tightening sequence—3.5L engine

1. Before servicing the vehicle, refer to the precautions section.

2. Remove the right front wheel.

3. Remove the V-bank cover sub-assembly.

4. Remove the engine under cover assembly.

5. Remove the No. 1 and 2 engine under covers.

6. Drain engine coolant.

7. Disconnect the No. 1 radiator hose.

8. Remove the radiator reserve tank assembly, as follows:

a. Disconnect the hose.

b. Remove the 2 bolts and the radiator reserve tank assembly.

9. Remove the No. 2 oil level dipstick guide.

10. Remove the air fuel ratio sensor (for Bank 2 Sensor 1).

11. Remove the 3 bolts and No. 2 exhaust manifold heat insulator.

12. For 4WD vehicles, remove the propeller with center bearing shaft assembly.

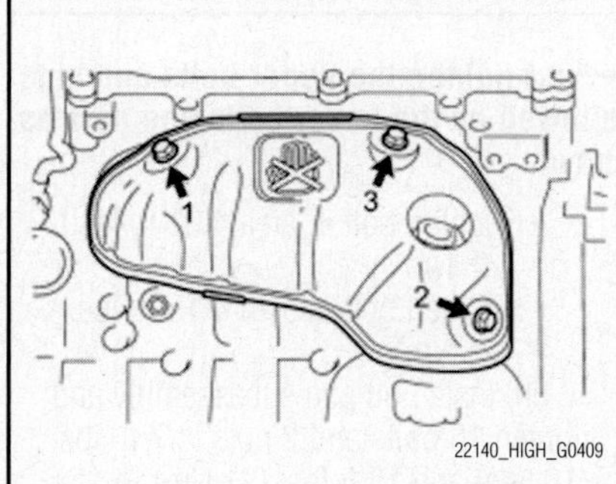

Fig. 109 No. 2 exhaust manifold heat insulator tightening sequence—3.5L engine

13. Remove the tail exhaust pipe assembly.

14. Remove the center exhaust pipe assembly.

15. Remove the front No. 3 exhaust pipe sub-assembly.

16. Remove the front exhaust pipe assembly.

17. Remove the bolt, nut and No. 2 manifold stay.

18. Remove the 6 nuts and left hand exhaust manifold sub-assembly.

19. Remove the gasket.

20. Remove the bolt, nut and manifold stay.

21. Remove the right hand exhaust manifold sub-assembly, as follows:

a. Disconnect the air fuel ratio sensor (for bank 1 sensor 1) connector and remove the clamp.

b. Remove the 6 nuts and the right hand exhaust manifold sub-assembly.

c. Remove the gasket.

22. Remove the air fuel ratio sensor (for Bank 1 Sensor 1).

To install:

23. Install the air fuel ratio sensor (for Bank 1 Sensor 1).

24. Install the right hand exhaust manifold sub-assembly, as follows:

a. Install a new gasket.

b. Install the right hand exhaust manifold sub-assembly by tightening the 6 nuts in the order shown to 15 ft. lbs (21 Nm).

c. Connect the air fuel ratio sensor (for Bank 1 Sensor 1) connector and install the clamp.

25. Install the manifold stay with the bolt and nut and tighten to 25 ft. lbs (34 Nm), 26 ft. lbs (35 Nm).

26. Install the left hand exhaust manifold sub-assembly, as follows:

a. Install a new gasket.

b. Install the left hand exhaust manifold sub-assembly by tightening the 6 nuts in the order shown to 15 ft. lbs (21 Nm).

27. Install the No. 2 manifold stay by tightening the bolt and nut in the order shown to 25 ft. lbs (34 Nm).

28. Install the front exhaust pipe assembly.

29. Install the front No. 3 exhaust pipe sub-assembly.

30. Install the center exhaust pipe assembly.

31. Install the tail exhaust pipe assembly.

32. For 4WD vehicles, temporarily tighten the propeller with center bearing shaft assembly.

33. For 4WD vehicles, fully tighten the propeller with center bearing shaft assembly.

34. Install the No. 2 exhaust manifold heat insulator by tightening the 3 bolts in the order shown to 75 inch lbs (8.5 Nm).

35. Install the air fuel ratio sensor (for Bank 2 Sensor 1).

36. Install the No. 2 oil level dipstick guide.

37. Install the radiator reserve tank assembly with the 2 bolts and tighten to 48 inch lbs (5.4 Nm).

38. Connect the hose.

39. Connect the No. 1 radiator hose.

40. Add engine coolant.

41. Inspect for coolant leak.

42. Inspect for gas leak, and repair as necessary.

43. For 4WD vehicles, inspect and adjust transfer oil.

44. Install the No. 1 and 2 engine under covers.

45. Install the engine under cover assembly.

46. Install the V-bank cover sub-assembly.

47. Install the right front wheel.

FLEXPLATE

REMOVAL & INSTALLATION

See Figure 110.

1. Remove automatic transaxle assembly

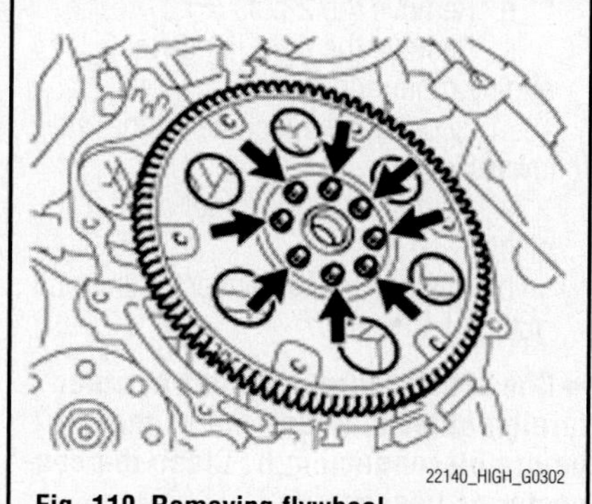

Fig. 110 Removing flywheel

2. Hold the crankshaft and remove the 8 bolts, front spacer, drive plate and rear spacer.

To install:

3. Installation is reverse of removal.

INTAKE MANIFOLD

REMOVAL & INSTALLATION

See Figures 111 through 123.

1. Discharge fuel system pressure.

2. Remove the engine under cover assembly.

3. Remove the No. 1 engine under cover.

4. Drain engine coolant.

5. Remove the V-bank cover sub-assembly.

6. Remove both front wiper arm and blade assemblies.

7. Remove the cowl top ventilator louver sub-assembly.

8. Remove the windshield wiper motor and link assembly.

9. Remove the outer cowl top panel sub-assembly.

10. Remove the air cleaner cap sub-assembly.

Fig. 111 Throttle body bracket bolt removal sequence—3.5L engine

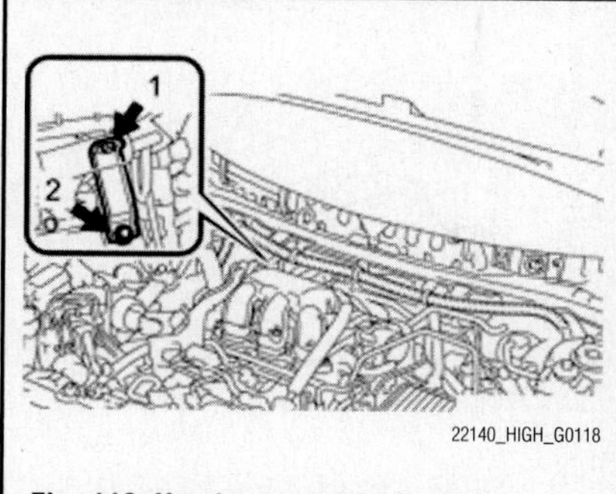

Fig. 112 No. 1 surge tank stay bolt removal sequence—3.5L engine

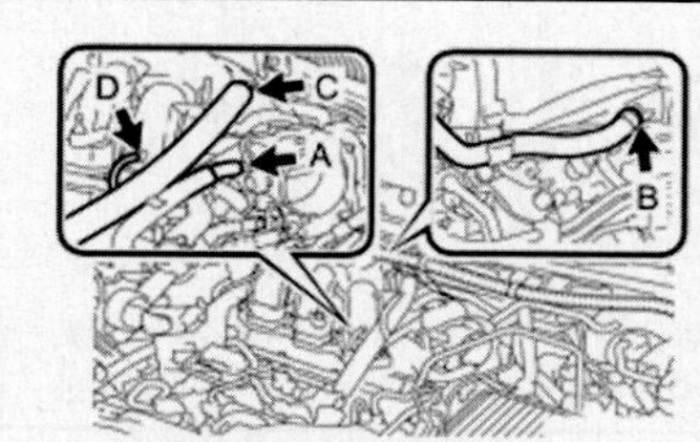

Fig. 113 Intake air surge tank assembly hoses—3.5L engine

Fig. 114 Throttle body connector and clamp—3.5L engine

11. Disconnect the engine room main wire, as follows:

a. Disconnect the 5 harness clamps.

12. Remove throttle body bracket:

a. Remove the 2 bolts in the order shown in the illustration and remove the throttle body bracket.

13. Remove the 2 bolts in the order shown in the illustration and remove the No. 1 surge tank stay.

14. Remove the intake air surge tank assembly, as follows:

a. Disconnect the 4 hoses.

b. Disconnect the throttle body connector and clamp.

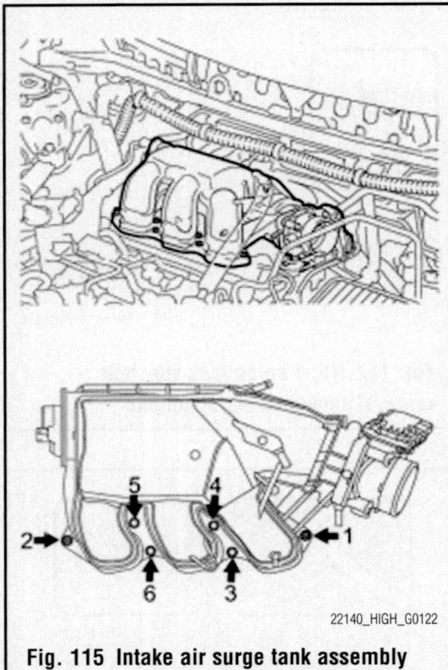

Fig. 115 Intake air surge tank assembly bolt removal sequence—3.5L engine

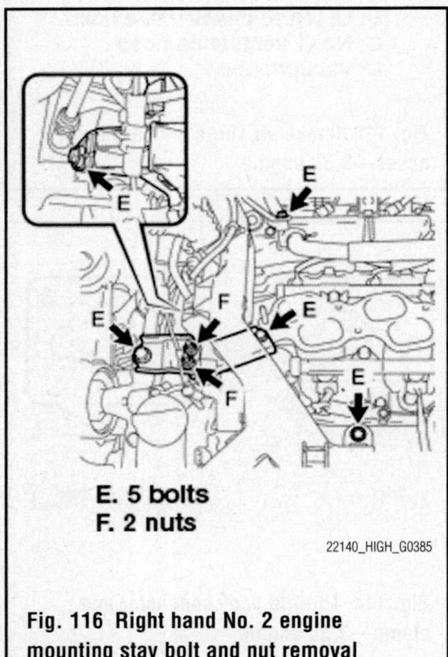

Fig. 116 Right hand No. 2 engine mounting stay bolt and nut removal sequence—3.5L engine

c. Disconnect the connector.

d. Disconnect the 2 water by-pass hoses from the throttle body.

e. Remove the 4 bolts and 2 nuts in the order shown in the illustration.

➡Use a 5 mm socket hexagon wrench to remove the 4 bolts

f. Remove the gasket from the intake air surge tank.

15. Remove the right hand No. 2 engine mounting stay, as follows:

a. Remove the 5 bolts (E).

b. Remove the 2 nuts (F).

c. Remove the right hand No. 2 engine mounting stay.

16. Disconnect the fuel main tube, as follows:

a. Remove the No. 2 fuel pipe clamp.

b. Pinch the tube connector and pull out the fuel pipe.

➡Check that there is no dirt or other foreign objects around the connector before disconnecting it. Clean the connector as necessary.

➡It is necessary to prevent dirt or foreign objects from entering the quick connector. If dirt or foreign objects enter the connector, the O-rings may not seal properly.

➡Only disconnect the quick connector by hand.

➡Do not bend, kink or twist the nylon tubes. Protect the connector by covering it with a plastic bag.

➡If the pipe and the connector are stuck, carefully try wiggling or pushing and pulling on the connector to release it. Pull the connector off the pipe carefully.

17. Remove the fuel delivery pipe sub-assembly, as follows:

a. Disconnect the 6 fuel injector connectors.

18. Remove the 5 bolts and fuel delivery pipe sub-assembly together with the 6 fuel injectors.

➡Be careful not to drop the fuel injectors when removing the fuel delivery pipe sub-assembly.

a. Remove the 6 injector vibration insulators from the intake manifold.

19. Remove the intake manifold, as follows:

a. Remove the 6 bolts and 4 nuts in the order shown in the illustration and remove the intake manifold.

20. Remove the 2 No. 1 intake manifold to head gaskets.

To install:

21. Install the intake manifold, as follows:

a. Set 2 new gaskets on each cylinder head.

➡Align the port holes of the gaskets and cylinder head.

➡Make sure that the gaskets are installed in the correct direction.

b. Set the intake manifold on the cylinder heads.

c. Install the intake manifold with the 6 bolts and 4 nuts in the order shown in the illustration and tighten to 15 ft. lbs. (21 Nm). **DO NOT** apply oil to the bolts.

22. Install the right hand No. 2 engine mounting stay.

23. Install the bolt (A), and tighten to 15 ft. lbs. (21 Nm). **DO NOT** apply oil to the bolt.

24. Install the 2 nuts (B), and tighten to 17 ft. lbs. (23 Nm).

25. Install the bolt (C), and tighten to 28 ft. lbs. (38 Nm). **DO NOT** apply oil to the bolt.

26. Install the 3 bolts (D), and tighten to 73 inch lbs. (8.3 Nm). **DO NOT** apply oil to the bolts.

27. Install the fuel delivery pipe sub-assembly, as follows:

a. Install 6 new insulators to the intake manifold.

b. Place the fuel delivery pipe which has the 6 fuel injectors installed to it in position on the intake manifold.

➡Be careful not to drop the fuel injectors when installing the fuel delivery pipe.

c. Temporarily install the 5 bolts which are used to hold the fuel delivery pipe to the intake manifold. **DO NOT** apply oil to the bolts.

➡After installing the fuel injectors, check that they turn smoothly. If not, reinstall the injectors with new O-rings.

d. Tighten the 5 bolts which are used to hold the fuel delivery pipe to the intake manifold to 15 ft. lbs. (21 Nm). **DO NOT** apply oil to the bolts.

e. Connect the 6 fuel injector connectors.

28. Connect the fuel main tube, as follows:

a. Push in the tube connector onto the pipe until the tube connector clicks.

➡Before connecting the tube, make sure that it is not damaged. Make sure that there is no dirt present on the connecting surfaces.

➡After connecting, check that the fuel tube connector and the pipe are securely connected by pulling on them.

b. Install the No. 2 fuel pipe clamp.

29. Temporarily install the No. 1 surge tank stay, as follows:

a. Temporarily install the intake air

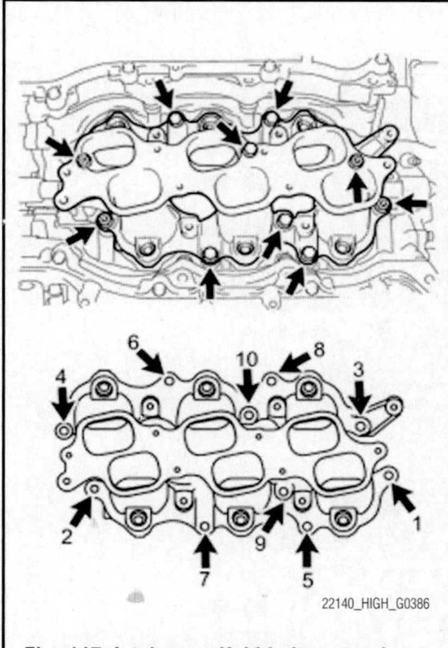

Fig. 117 Intake manifold bolt removal sequence—3.5L engine

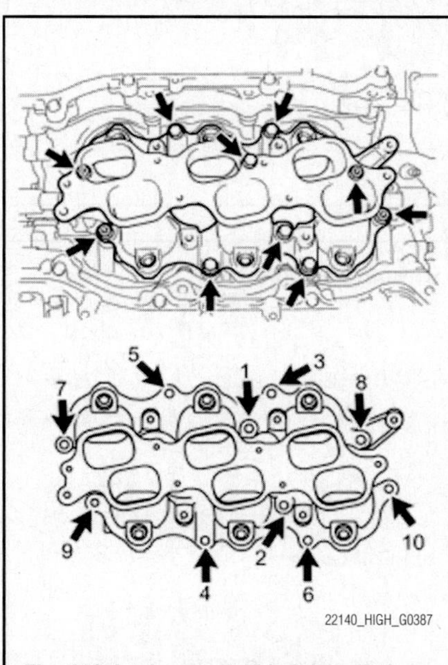

Fig. 118 Intake manifold bolt installation sequence—3.5L engine

surge tank assembly with 3 new gaskets on the intake manifold.

➡Do not allow the gaskets to slip out of place during installation.

b. Temporarily install the No. 1 surge tank stay with the 2 bolts. **DO NOT** apply oil to the bolts.

30. Temporarily install the throttle body bracket with the 2 bolts. **DO NOT** apply oil to the bolts.

31. Install the intake air surge tank assembly, as follows:

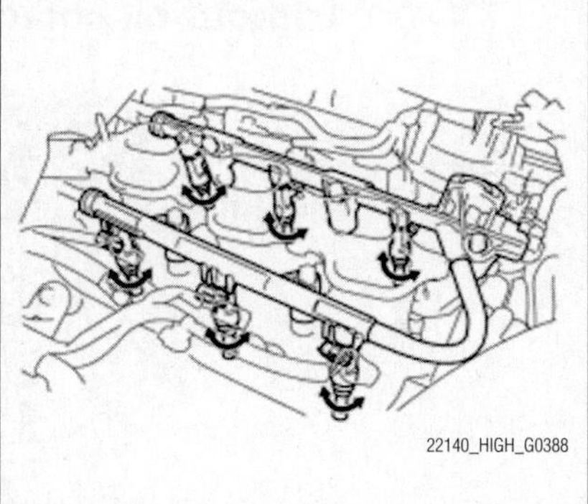

Fig. 119 Place fuel delivery pipe on intake manifold—3.5L engine

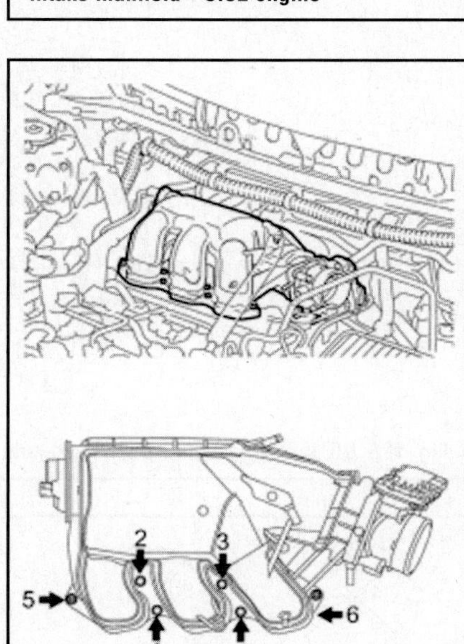

Fig. 120 Surge tank tightening sequence—3.5L engine

a. Install the surge tank with the 4 bolts and 2 nuts in the order shown in the illustration and tighten to 12 ft. lbs. (16 Nm) and 13 ft. lbs. (18 Nm). **DO NOT** apply oil to the bolts.

➡Use a 5 mm hexagon socket wrench to tighten the 4 bolts.

b. Connect the 2 water by-pass hoses to the throttle with motor body assembly.

c. Connect the connector.

d. Install the clamp and connect the throttle with motor body assembly connector.

e. Connect the 4 hoses.

32. Fully tighten the No. 1 surge tank stay, as follows:

a. Fully tighten the 2 bolts in the order shown in the illustration to 15 ft. lbs. (21 Nm). **DO NOT** apply oil to the bolts.

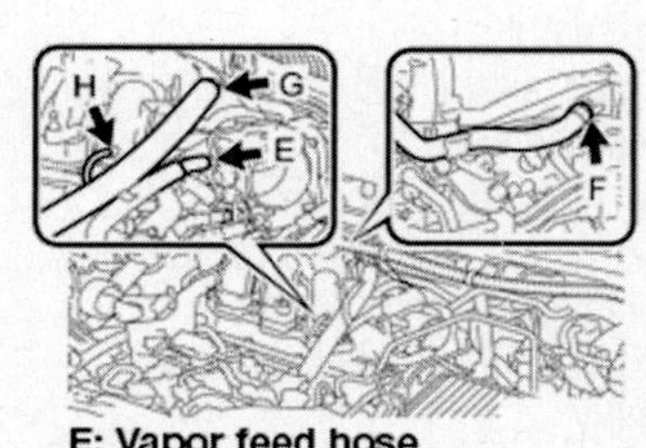

Fig. 121 Intake air surge tank assembly hoses—3.5L engine

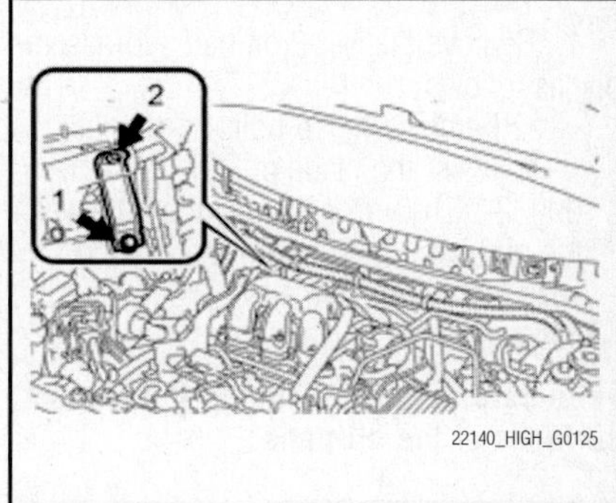

Fig. 122 No. 1 surge tank stay tightening sequence—3.5L engine

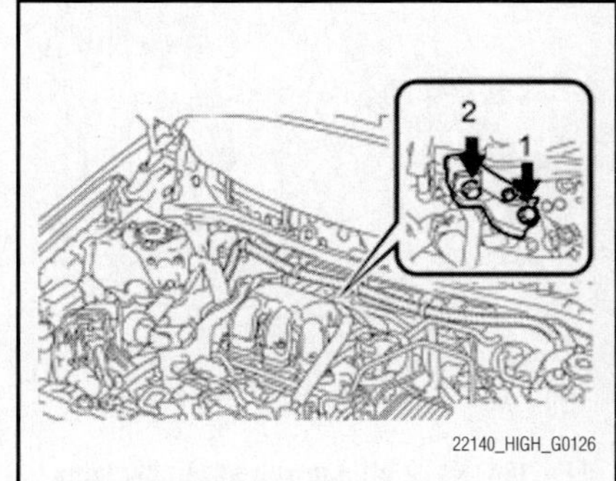

Fig. 123 Throttle body bracket tightening sequence—3.5L engine

33. Fully tighten the throttle body bracket, as follows:

a. Fully tighten the 2 bolts in the order shown in the illustration to 15 ft. lbs. (21 Nm). **DO NOT** apply oil to the bolts.

34. Connect the 5 engine room main wire harness clamps.

35. Install the air cleaner cap sub-assembly.

36. Install the outer cowl top panel sub-assembly.

37. Install the windshield wiper motor and link assembly.

38. Install the cowl top ventilator louver sub-assembly.
39. Install the both front wiper arm and blade assemblies.
40. Add engine coolant.
41. Inspect for engine coolant leak.
42. Inspect for fuel leak.
43. Install the V-bank cover sub-assembly.
44. Install the No. 1 engine under cover.
45. Install the engine under cover assembly.

OIL PAN

REMOVAL & INSTALLATION

See Figures 124 through 133.

1. Remove the No. 2 oil pan sub-assembly, as follows:
   a. Remove the 16 bolts and 2 nuts.
   b. Insert the blade of SST (SST: 09032-00100) or equivalent tool between the oil pans. Cut through the applied sealer and remove the No. 2 oil pan sub-assembly.

➡Be careful not to damage the contact surfaces of the oil pans.

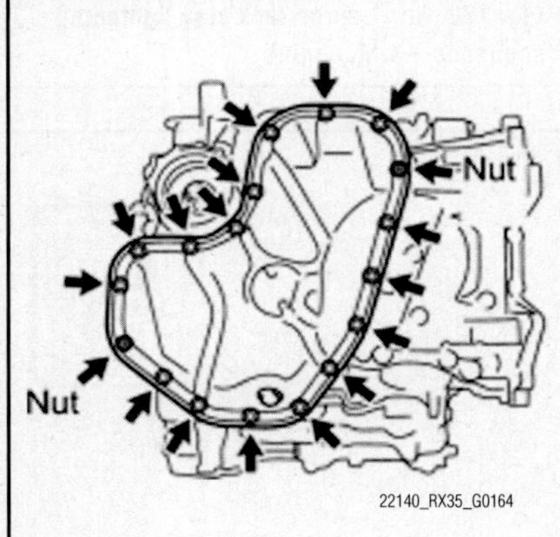

Fig. 124 No. 2 oil pan sub-assembly bolts and nuts

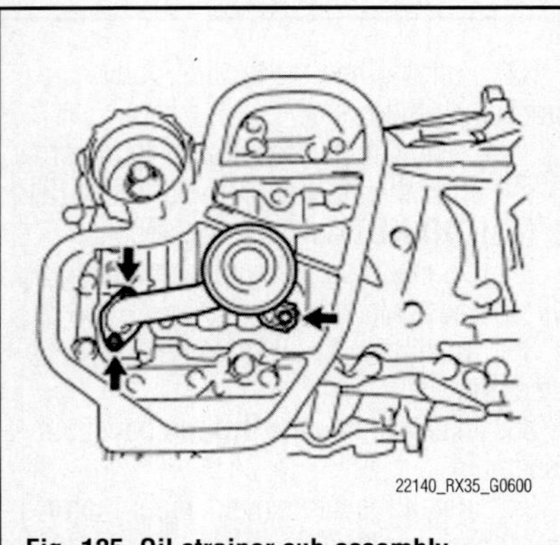

Fig. 125 Oil strainer sub-assembly

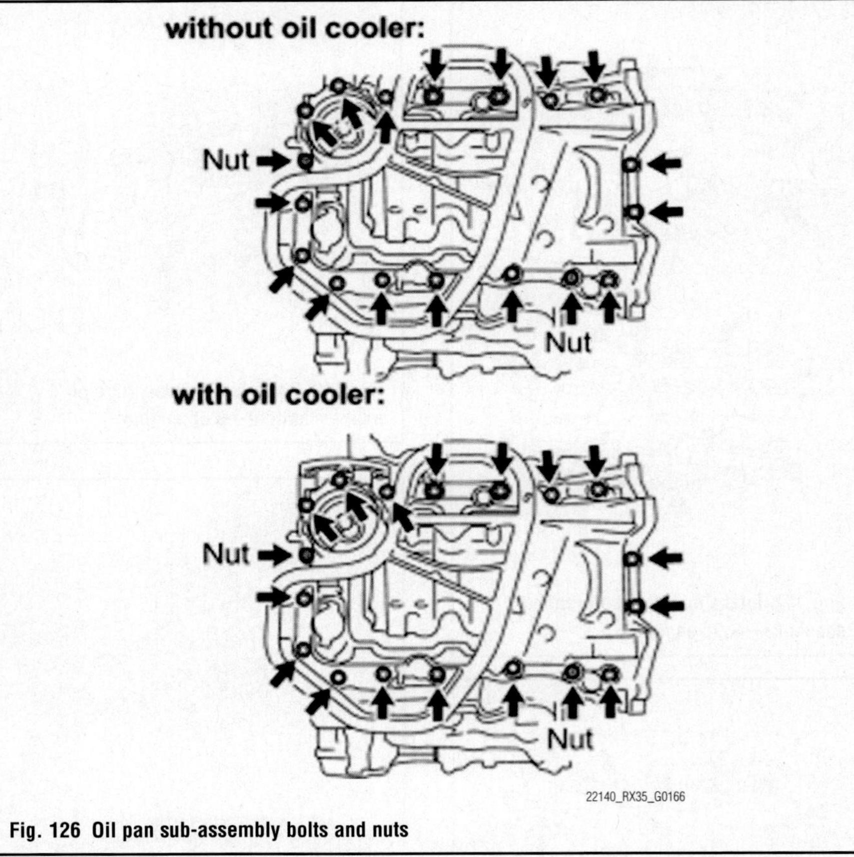

Fig. 126 Oil pan sub-assembly bolts and nuts

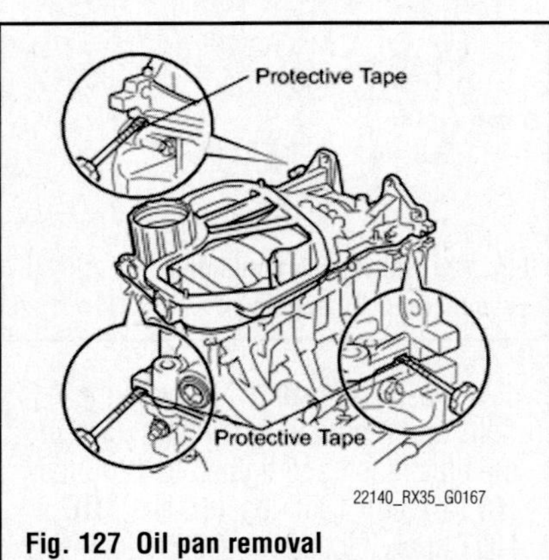

Fig. 127 Oil pan removal

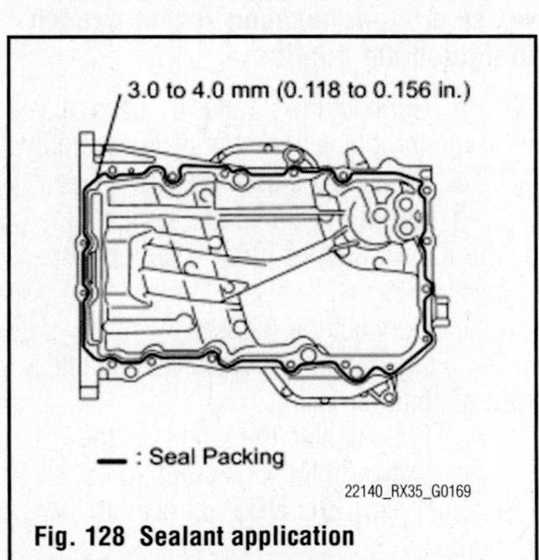

Fig. 128 Sealant application

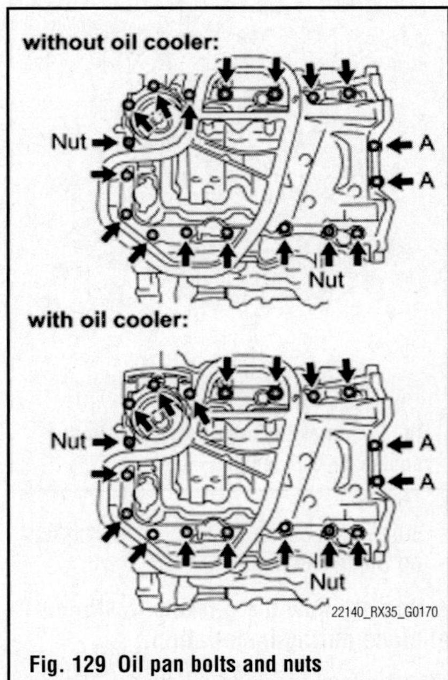

Fig. 129 Oil pan bolts and nuts

   c. Using a TORX® socket wrench E6, remove the 2 stud bolts.

2. Remove the oil strainer sub-assembly, as follows:
   a. Remove the bolt, 2 nuts, oil strainer and gasket.

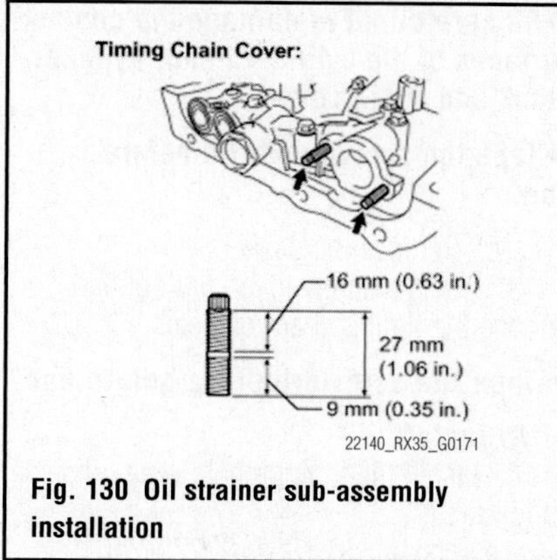

Fig. 130 Oil strainer sub-assembly installation

b. Using a TORX® socket wrench E6, remove the 2 stud bolts.

3. Remove the oil pan sub-assembly, as follows:

a. Remove the 16 bolts and 2 nuts.

➡Be sure to clean the bolts and stud bolts and check the threads for cracks or other damage.

b. Remove the oil pan by prying between the oil pan and cylinder block with a taped screwdriver.

➡Be careful not to damage the contact surfaces of the cylinder block and oil pan.

c. Remove the 2 O-rings.

d. Using a TORX® socket wrench E8, remove the 2 stud bolts. (without oil cooler).

e. Using a TORX® socket wrench E8, remove the 4 stud bolts. (with oil cooler).

To install:

4. Install the oil pan sub-assembly, as follows:

a. When replacing a stud bolt, install it by using an E8 TORX® socket wrench. Tighten to 7 ft. lbs (10 Nm).

b. Install 2 new O-rings.

c. Apply seal packing in a continuous line as shown in the illustration. Seal packing: Toyota Genuine Seal Packing Black, Three Bond 1207B or equivalent. Seal diameter: 3.0 to 4.0 mm (0.118 to 0.156 in.).

➡Remove any oil from the contact surface.

➡Install the oil pan within 3 minutes after applying seal packing.

➡Do not start the engine for at least 2 hours after installing.

d. Install the oil pan with the 16 bolts and 2 nuts and tighten to 7 ft. lbs (10 Nm), and 15 ft. lbs (21 Nm).

5. Install the oil strainer sub-assembly, as follows:

a. Using an E6 TORX® socket, install the stud bolts as shown in the illustration and tighten to 35 inch lbs (4 Nm).

b. Install a new gasket and the oil strainer with the bolt and 2 nuts and tighten to 7 ft. lbs (10 Nm).

6. Install the No. 2 oil pan sub-assembly, as follows:

a. Using an E6 TORX® socket, install the stud bolts as shown in the illustration and tighten to 35 inch lbs (4 Nm).

b. Apply seal packing in a continuous line as shown in the illustration. Seal packing: Toyota Genuine Seal Packing Black, Three Bond 1207B or equivalent. Seal diameter: 3.0 to 4.0 mm (0.118 to 0.156 in.).

➡Remove any oil from the contact surface.

➡Install the oil pan No. 2 within 3 minutes after applying seal packing.

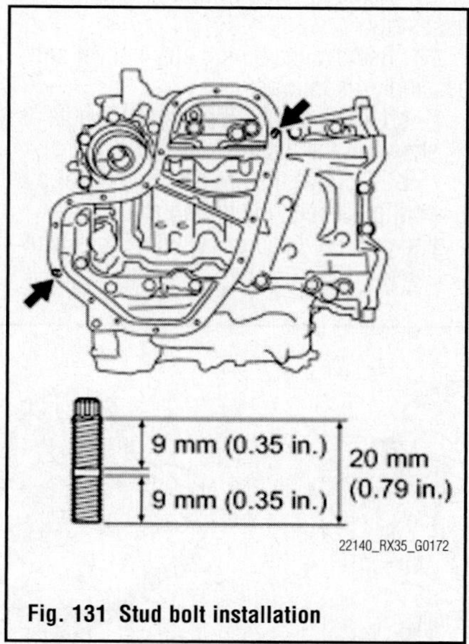

Fig. 131 Stud bolt installation

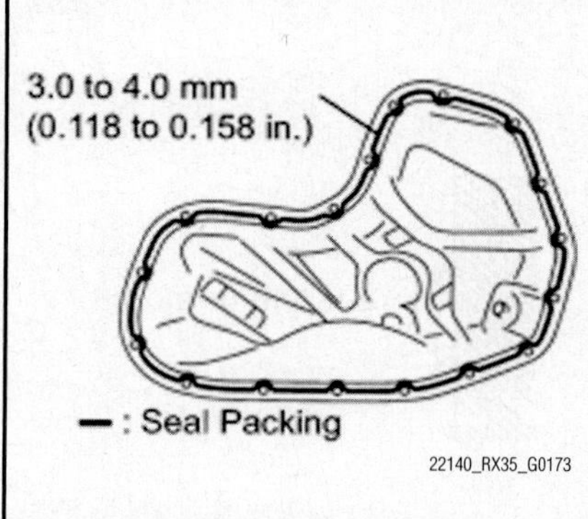

Fig. 132 Sealant application

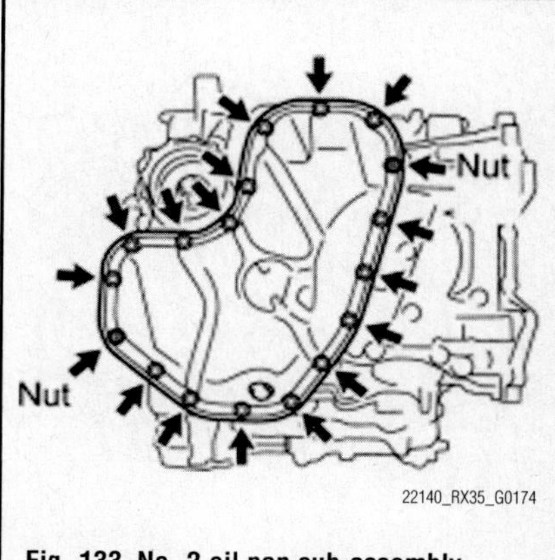

Fig. 133 No. 2 oil pan sub-assembly installation

➡Do not start the engine for at least 2 hours after installing.

c. Install the oil pan with the 16 bolts and 2 nuts and tighten to 7 ft. lbs (10 Nm).

OIL PUMP

REMOVAL & INSTALLATION

See Figures 134 through 141.

1. Remove the engine assembly with transaxle.
2. Secure the engine.
3. Remove the No. 1 oil pipe, as follows:

a. Remove the 2 oil pipe unions, gaskets and No. 1 oil pipe.

b. Remove the left oil control valve filter and gaskets.

4. Remove the oil pipe, as follows:

a. Remove the bolt.

b. Remove the 2 oil pipe unions and oil pipe.

c. Remove the right oil control valve filter and gaskets.

5. Using a special service tool (SST: 09213-70011, SST: 09213-70011 or

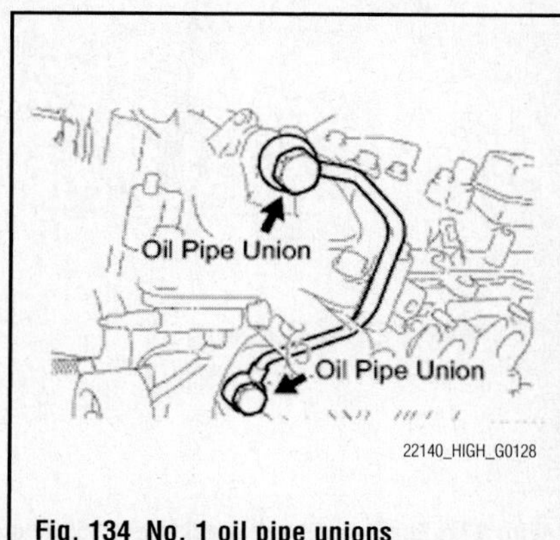

Fig. 134 No. 1 oil pipe unions

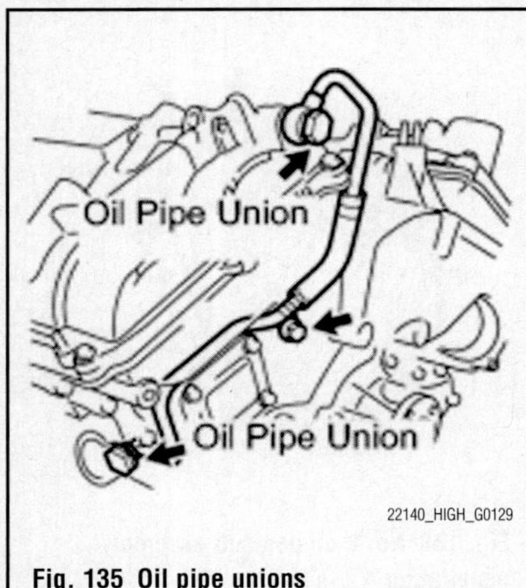

Fig. 135 Oil pipe unions

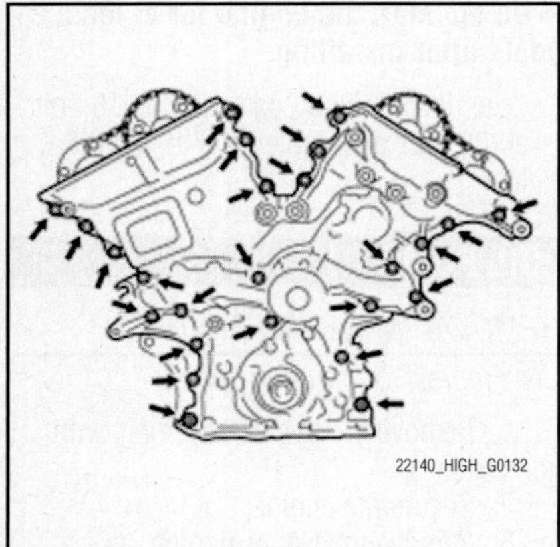

Fig. 136 Timing chain cover sub-assembly bolts and nuts

equivalent), loosen the crankshaft pulley bolt.

6. Using SST: 09950-50013 or equivalent, remove the crankshaft pulley bolt and crankshaft pulley.

7. Separate the oil cooler pipe (w/ oil cooler), as follows:

a. Remove the bolt and 2 nuts, and disconnect the oil cooler pipe from the oil pan sub-assembly.

b. Remove the gasket from the oil pan sub-assembly.

8. Remove the water inlet housing, as follows:

a. Disconnect the water hose.

b. Remove the 2 bolts, nut and water inlet housing.

c. Remove the water inlet housing gasket and water outlet pipe O-ring.

9. Remove the Bank 1 and Bank 2 cylinder head cover sub-assemblies.

10. Remove the No. 2 oil pan sub-assembly.

11. Remove the oil strainer sub-assembly.

12. Remove the timing chain cover sub-assembly, as follows:

a. Remove the 23 bolts and 2 nuts as shown in the illustration.

b. Remove the timing chain cover by prying between the timing chain cover and cylinder head or cylinder block with a screwdriver.

➡Be careful not to damage the contact surfaces of the cylinder head, cylinder block and chain cover.

➡Tape the screwdriver tip before use.

c. Remove the gasket.

13. Using a screwdriver, pry out and remove the timing chain oil seal.

➡Tape the screwdriver tip before use.

To install:

14. Install the timing chain case oil seal, as follows:

a. Using SST (SST: 09223-22010, SST: 09506-35010) tap in a new oil seal until its surface is flush with the timing chain case edge.

➡Keep the lip free from foreign matter.

➡Make sure that the oil seal edge does not stick out of the timing chain case.

➡Do not tap on the oil seal at an angle.

15. Install the timing chain cover sub-assembly, as follows:

a. Apply seal packing in a continuous line to the engine unit as shown in the illustration. Seal packing: Toyota Seal Packing Black, Three Bond 1207B or equivalent. Seal diameter: 3.0 mm (0.118 in.).

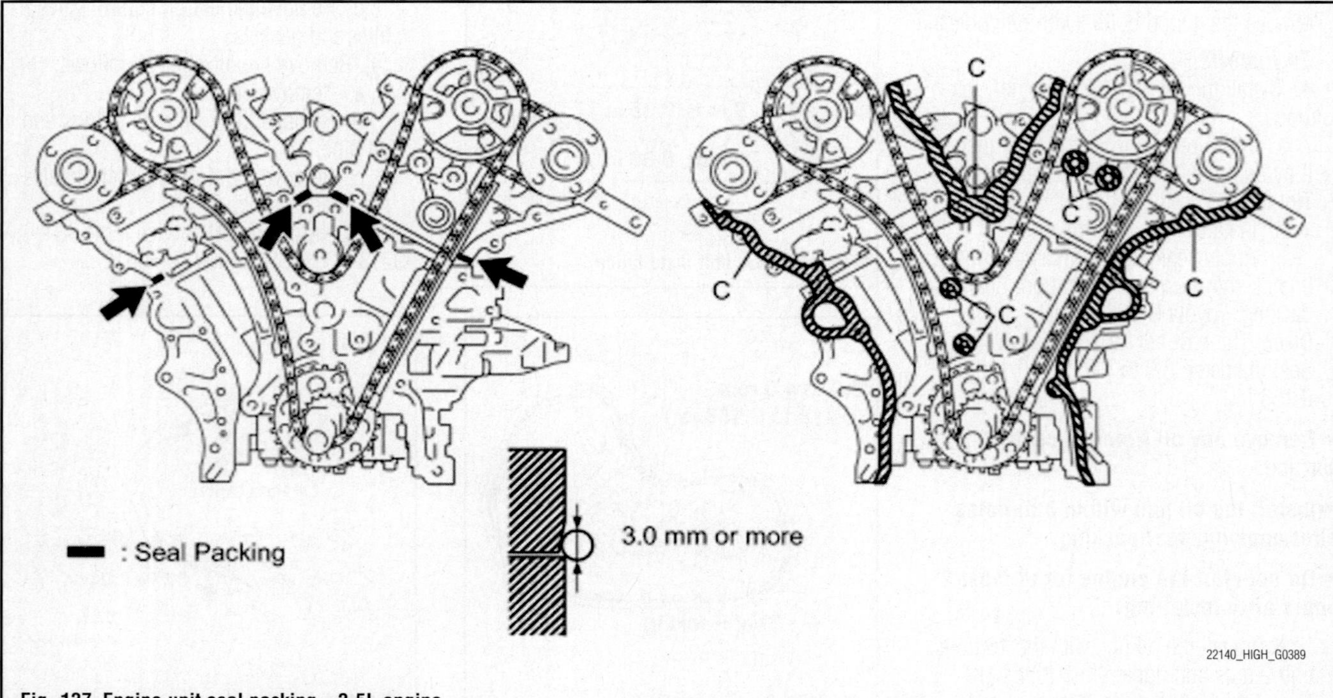

Fig. 137 Engine unit seal packing—3.5L engine

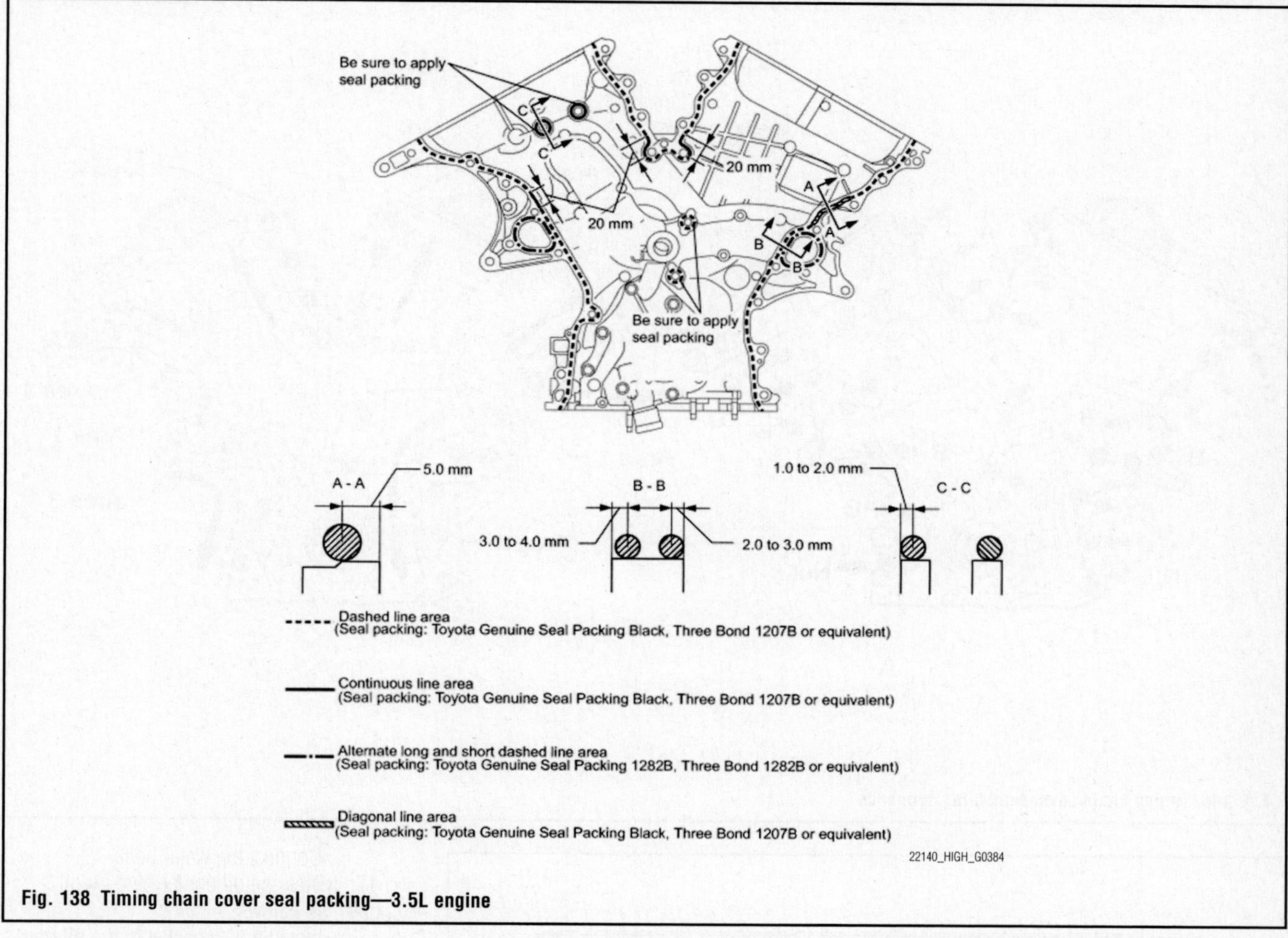

Fig. 138 Timing chain cover seal packing—3.5L engine

Area	Seal Packing Diameter	Application Position from Inside Seal Line
Continuous Line Area	4.5 mm (0.177 in.) or more	3.0 to 4.0 mm (0.118 to 0.158 in.)
Alternate Long and Short Dashed Line Area	3.5 mm (0.138 in.) or more	2.0 to 3.0 mm (0.079 to 0.118 in.)
Dashed Line Area	3.5 mm (0.138 in.) or more	3.0 to 4.0 mm (0.118 to 0.158 in.)
Diagonal Line Area	6.0 mm (0.236 in.) or more	5.0 mm (0.197 in.)

22140_HIGH_G0380

Fig. 139 Seal packing specifications—3.5L engine

➡Be sure to clean and degrease the contact surfaces, especially the surfaces indicated by C in the illustration.

➡If the contact surfaces are wet, wipe them with an oil-free cloth before applying seal packing.

➡Install the chain cover sub-assembly within 3 minutes after applying seal packing.

➡Do not start the engine for at least 2 hours after installing the chain cover sub-assembly.

b. Apply seal packing in a continuous line to the timing chain cover as shown in the illustration. Seal packing: Toyota Seal Packing Black, Three Bond 1207B, Three Bond 1282B or equivalent.

➡If the contact surfaces are wet, wipe them with an oil-free cloth before applying seal packing.

➡Install the chain cover sub-assembly within 3 minutes and tighten the bolts within 15 minutes after applying seal packing.

➡Do not start the engine for at least 2 hours after installing the chain cover sub-assembly.

c. Apply seal packing as follows:

d. Install a new gasket.

e. Align the oil pump's drive rotor spline and the crankshaft as shown in the illustration. Install the spline and chain cover to the crankshaft.

f. Temporarily tighten the timing chain cover with the 23 bolts and 2 nuts.

➡Make sure that there is no oil on the bolt threads.

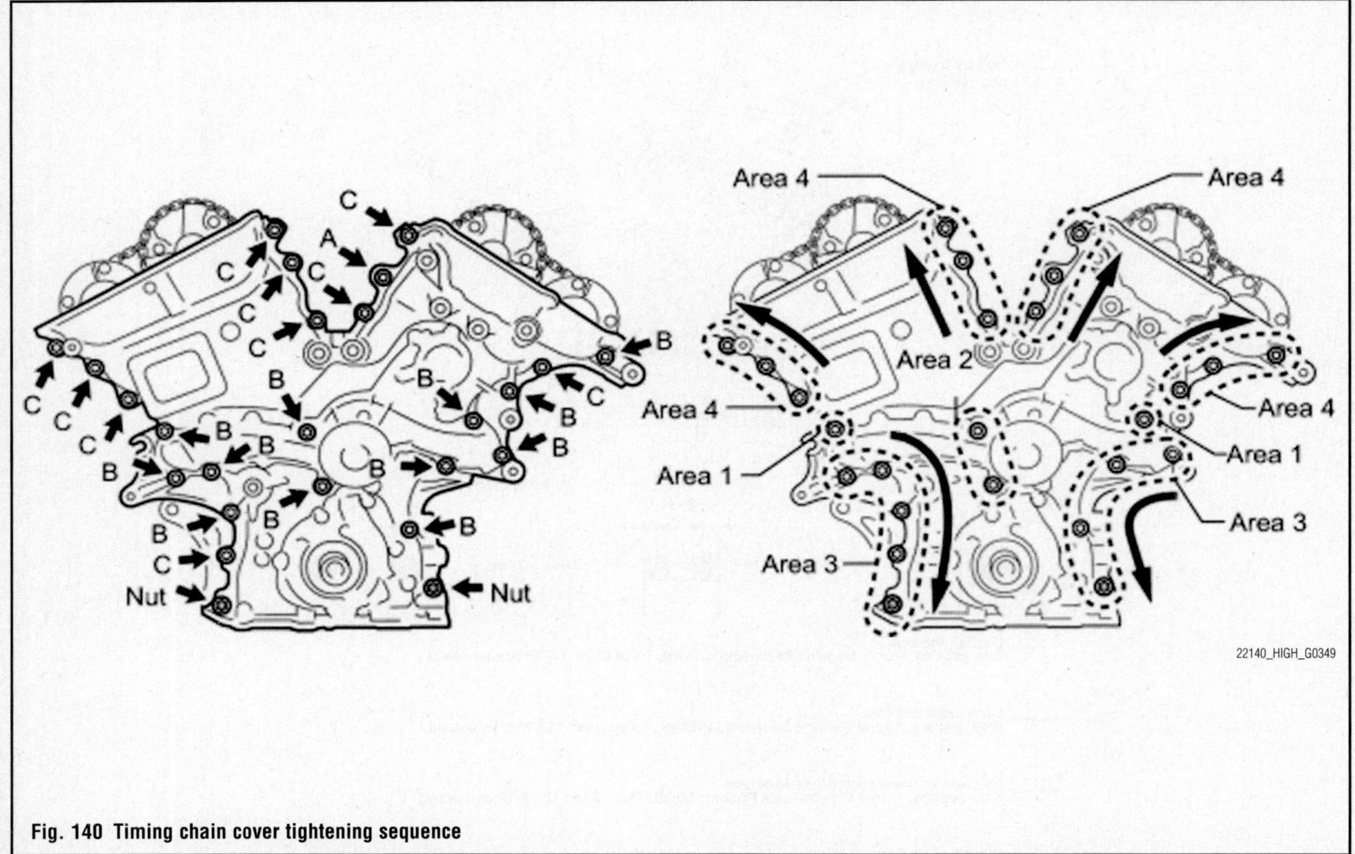

Fig. 140 Timing chain cover tightening sequence

Item	Length
Bolt A	40 mm (1.57 in.)
Bolt B	55 mm (2.17 in.)
Bolt C	25 mm (0.98 in.)

22140_HIGH_G0133

Fig. 141 Timing chain cover bolt length

g. Fully tighten the bolts in area 1 and area 2 (from top to bottom as shown) to 15 ft. lbs. (21 Nm).

h. Fully tighten the bolts and nuts in area 3 (from top to bottom as shown) to 15 ft. lbs. (21 Nm).

➡**Tighten the bolts and nuts from top to bottom as shown in the illustration.**

i. Fully tighten the bolts in area 4 (from bottom to top as shown) to 32 ft. lbs. (43 Nm), 15 ft. lbs. (21 Nm).

j. Install the oil pan sub-assembly.

k. Install the oil strainer sub-assembly.

l. Install the No. 2 oil pan sub-assembly.

m. Install the cylinder head cover sub-assembly.

16. Install the water inlet housing, as follows:

a. Install a new water inlet housing No. 1 gasket and water outlet pipe O-ring.

b. Install the water inlet housing with the 2 bolts and nut and tighten to 7 ft. lbs. (10 Nm).

➡**Be careful not to allow the O-ring to get caught between the parts.**

c. Connect the water hose.

17. Install the oil cooler pipe (w/oil cooler), as follows:

a. Install a new gasket to the oil pan sub-assembly.

b. Install the oil cooler pipe with the bolt and 2 nuts and tighten to 15 ft. lbs. (21 Nm).

18. Install the crankshaft pulley.
19. Install oil pipe.
20. Install the No. 1 oil pipe.
21. Install the engine hangers.
22. Remove engine stand.
23. Install engine assembly with transaxle.

PISTON AND RING

POSITIONING

See Figure 142.

REAR MAIN SEAL

REMOVAL & INSTALLATION

See Figures 143 through 147.

➡**This procedure requires a variety of special tools.**

1. Remove the automatic transaxle assembly (2WD) or

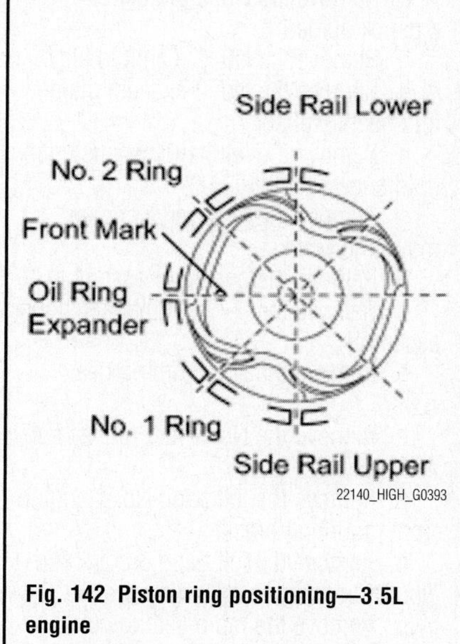

Fig. 142 Piston ring positioning—3.5L engine

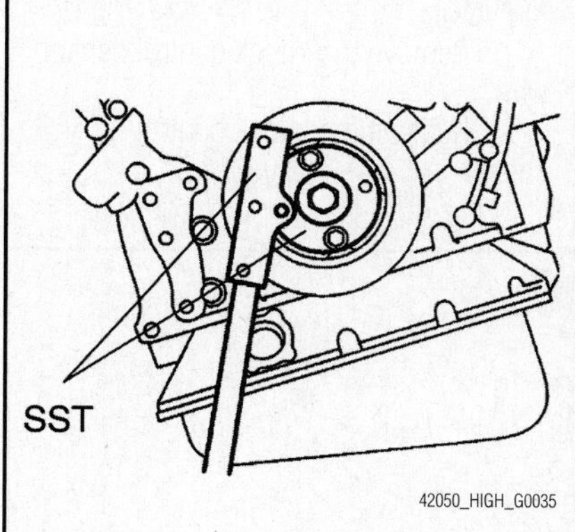

Fig. 143 Hold the crankshaft in place with the special tools as shown

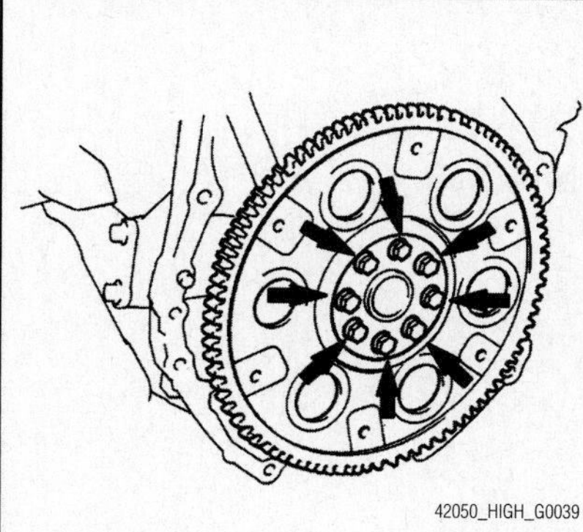

Fig. 144 Remove the 8 bolts, rear spacer, drive plate and front spacer

automatic transaxle and transfer assembly (4WD), as outlined in the Drive Train Section.

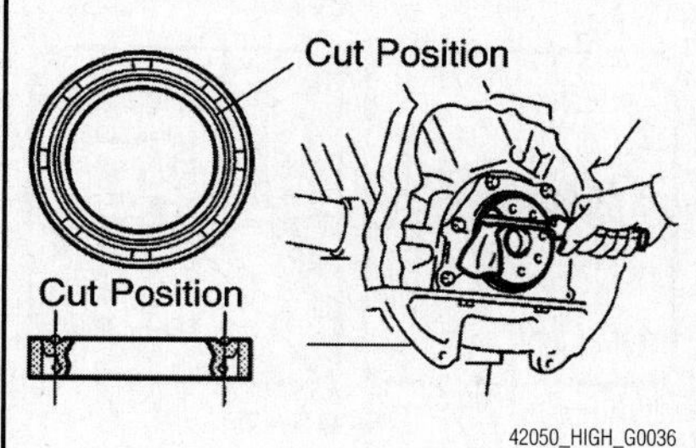

Fig. 145 If using a screwdriver wrapped in tape to remove the rear main seal, be very carefully not to damage the crankshaft

2. Remove the drive plate and ring gear, as follows:

a. Secure the crankshaft with Special Service Tool (SST) 09213-54015 (91651-60855), 09330-00021 or their equivalents.

b. Remove the 8 bolts, rear spacer, drive plate and front spacer.

3. Remove the rear main seal, as follows:

a. Using a knife, carefully cut off the oil seal lip.

b. Use a suitable prytool, or taped screwdriver, pry out the oil seal.

⁂ WARNING

After removing the seal, make sure the crankshaft is not damaged or scratched. If it is, you can mend it with a fine grit (No. 400) sandpaper.

To install:

4. Install engine rear oil seal, as follows:

a. Apply suitable grease to a new oil seal lip. Make sure to keep the lip away from foreign materials to avoid picking up contamination or debris.

b. Using SST 09223-15030, 09950-70010 (09951-07100), or their equivalent, and a hammer, tap in the oil seal until its surface is flush with the rear oil seal retainer edge. Tap the seal in squarely to make sure it seats properly

➡Wipe the extra grease off of the crankshaft.

5. Install drive plate and ring gear, as follows:

a. Fix the crankshaft with SST 09213-54015 (91651-60855), 09330-00021, or their equivalents.

b. Clean the bolts and the bolt holes.

c. Apply a suitable adhesive (part no. 08833-00070, three bond or equivalent) to 2 or 3 threads of the bolt end.

d. Install and uniformly tighten the

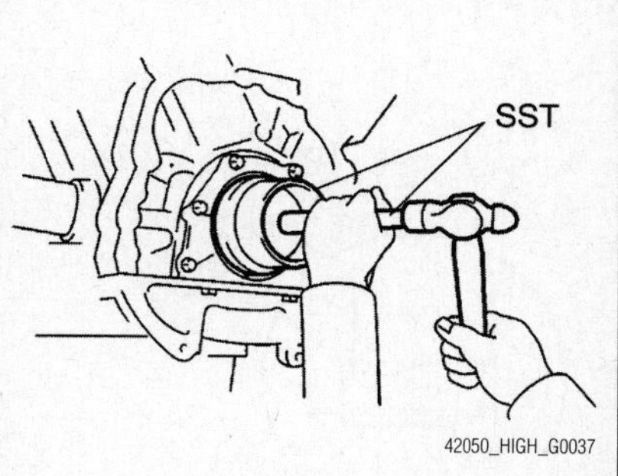

Fig. 146 Use a seal installation tool and hammer to install the rear main seal until its surface is flush with the seal retainer edge

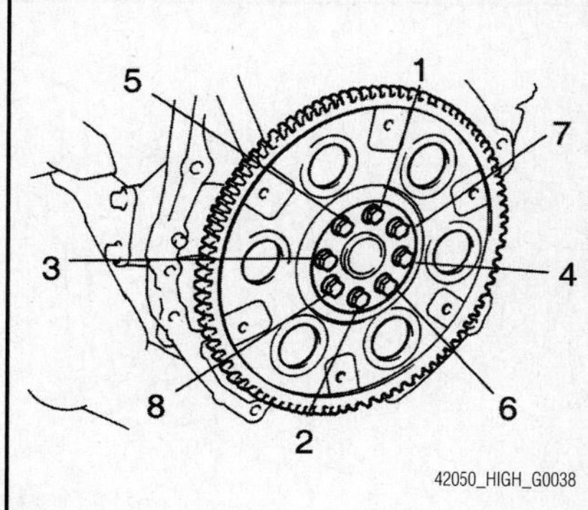

Fig. 147 Drive plate bolt tightening sequence—3.3L engines

8 bolts, in several passes in the sequence shown in the accompanying illustration to a final torque of 61 ft. lbs. (83 Nm).

⁂ WARNING

Do not start the engine for AT LEAST one hour after installing the seal!

6. Install the automatic transaxle assembly (2WD) or automatic transaxle and transfer assembly (4WD), as outlined in the Drive Train Section.

TIMING CHAIN FRONT COVER

REMOVAL & INSTALLATION

Refer to the timing chain and sprocket procedure.

TIMING CHAIN & SPROCKETS

REMOVAL & INSTALLATION

See Figures 148 through 159.

1. Remove the engine assembly.

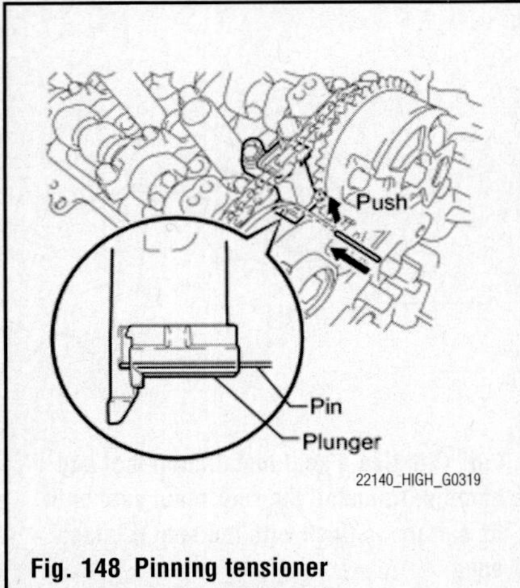

Fig. 148 Pinning tensioner

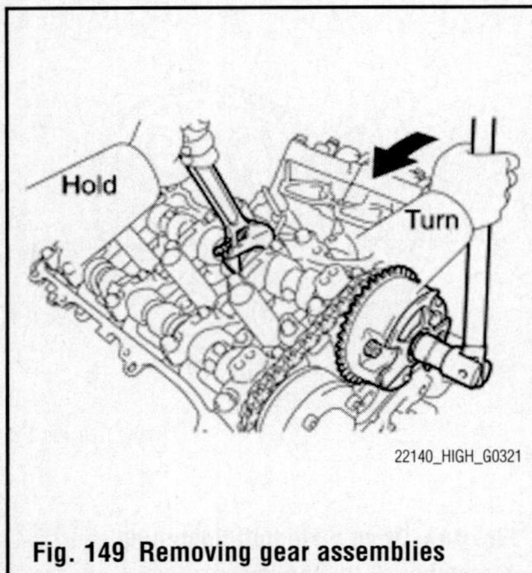

Fig. 149 Removing gear assemblies

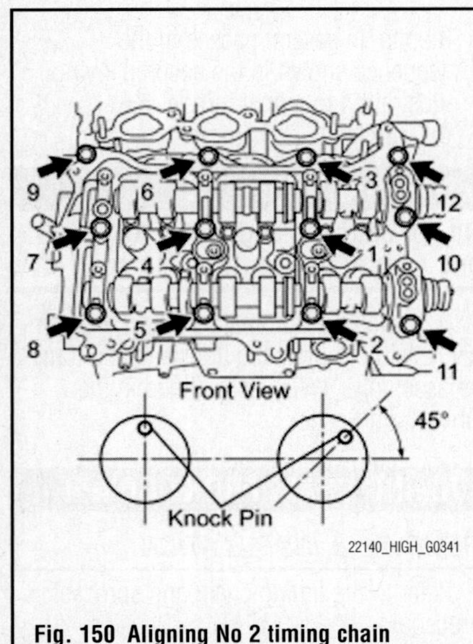

Fig. 150 Aligning No 2 timing chain

2. Install on engine stand.
3. The following must be removed:
   a. Remove ignition coil assembly.
   b. Remove the right hand No. 2 engine mounting stay.

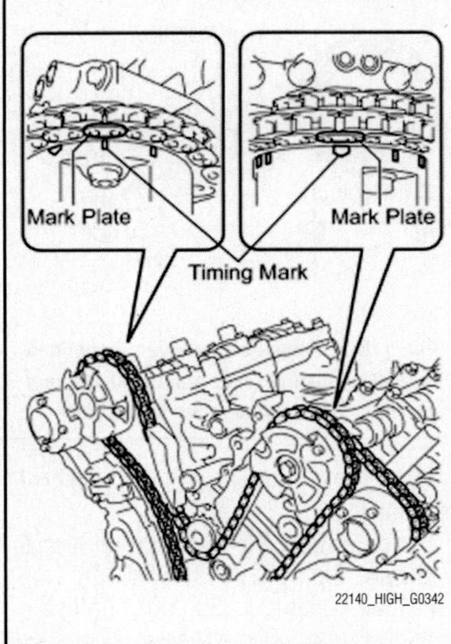

Fig. 151 Aligning timing chain sub-assembly

   c. Remove the intake manifold.
   d. Remove the right hand exhaust manifold sub-assembly.
   e. Remove the No. 2 engine oil level dipstick guide.
   f. Remove the No. 2 manifold stay.
   g. Remove the No. 2 exhaust manifold heat insulator.
   h. Remove the left hand exhaust manifold sub-assembly.
   i. Remove the transverse engine mounting bracket.
   j. Remove the generator assembly.
   k. Remove the V-ribbed belt tensioner assembly.
   l. Remove the No. 2 timing gear cover.
   m. Remove the No. 2 idler pulley sub-assembly.
   n. Remove the left hand No. 1 engine front mounting bracket.
   o. Remove the left hand 6 bolts and No. 1 front engine mounting bracket.
   p. Remove the radio setting condenser.
   q. Remove the No. 1 vacuum switching valve.
   r. Remove the knock control sensor wire.
   s. Remove the knock control sensor.

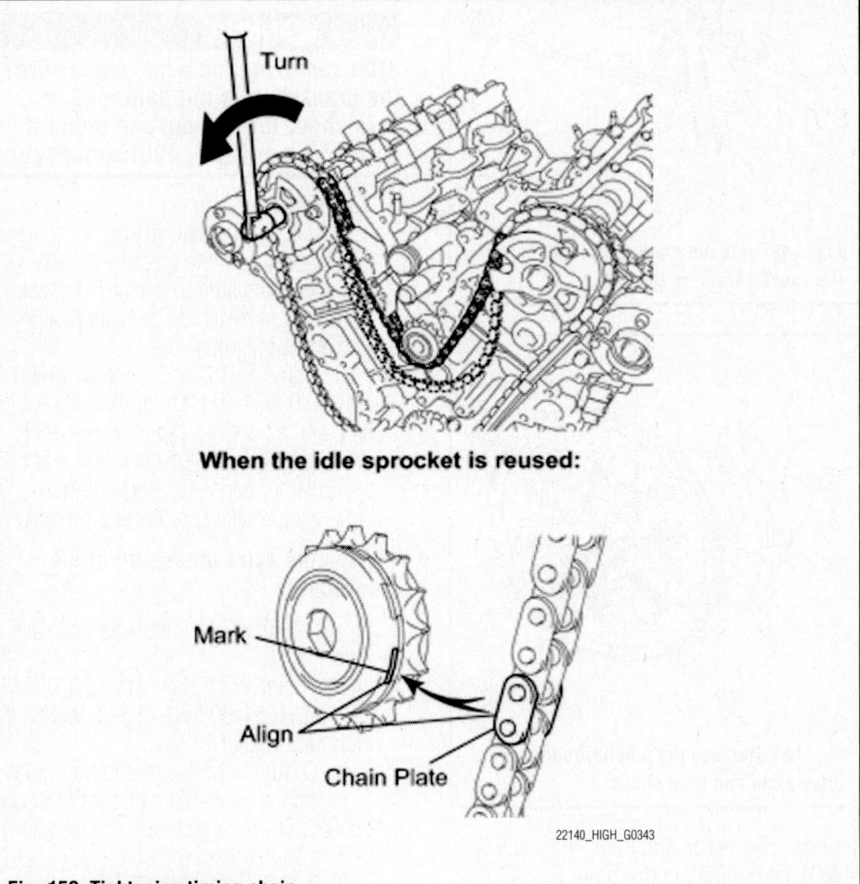

Fig. 152 Tightening timing chain

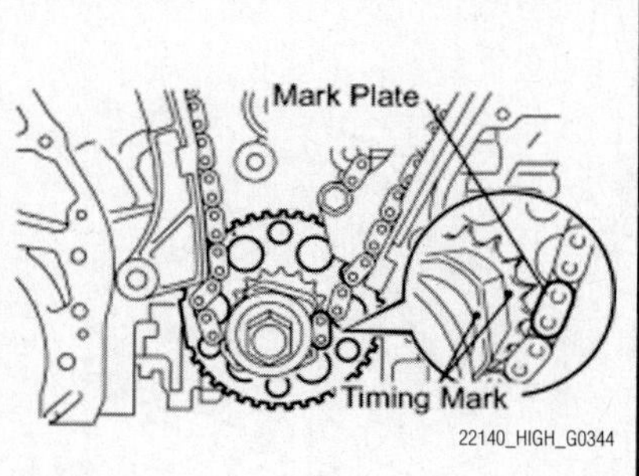

Fig. 153 Installing timing chain on crankshaft

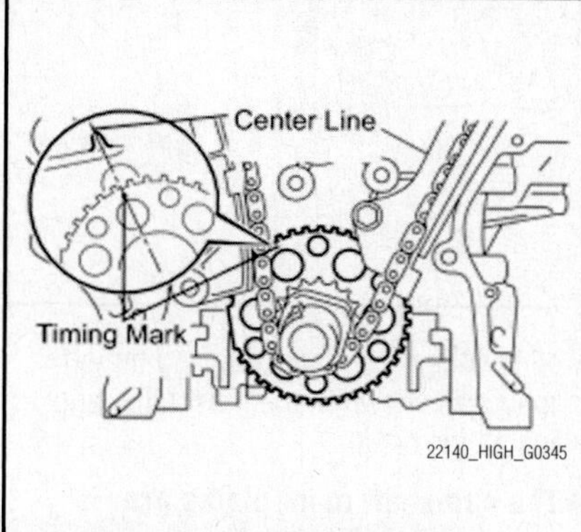

Fig. 154 Aligning timing chain on crankshaft

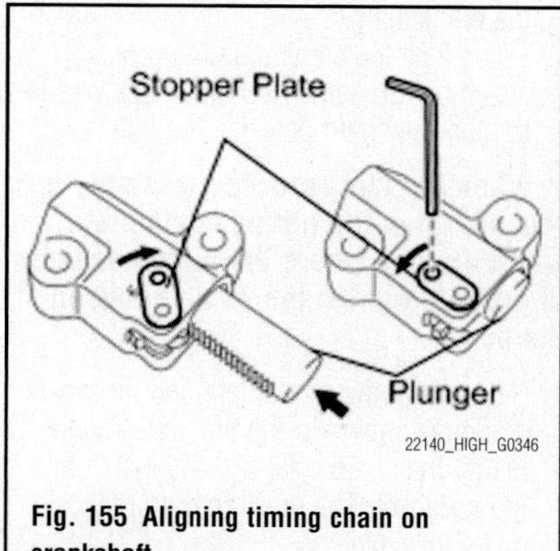

Fig. 155 Aligning timing chain on crankshaft

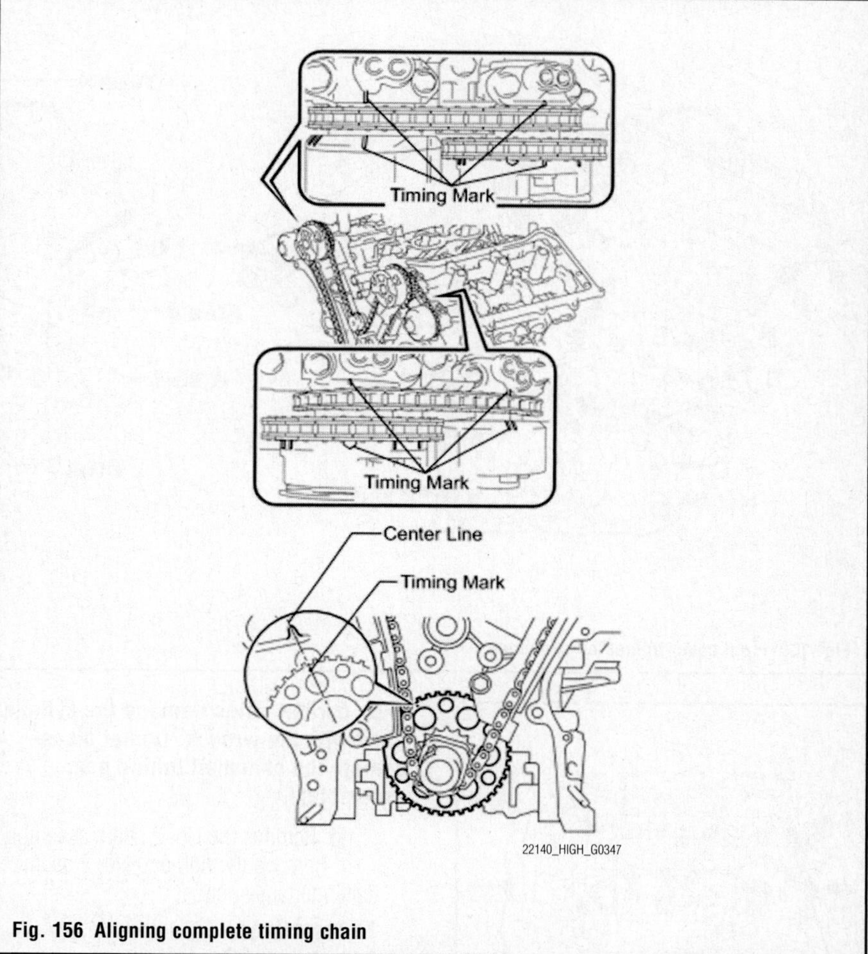

Fig. 156 Aligning complete timing chain

Fig. 157 Sealant application area

t. Remove the Crankshaft Position (CKP) sensor.

u. Remove the No. 1 oil pipe.

v. Remove the oil pipe.

w. Remove the crankshaft pulley.

x. Remove the oil cooler assembly, if necessary.

y. Remove the No. 1 oil cooler bracket, if necessary.

z. Remove the water inlet housing.

aa. Remove the water outlet.

bb. Remove the cylinder head cover sub-assembly (for Bank 1).

cc. Remove the cylinder head cover sub-assembly (for Bank 2).

dd. Remove the No. 2 oil pan sub-assembly.

ee. Remove the oil strainer sub-assembly.

ff. Remove the oil pan sub-assembly.

gg. Remove the timing chain cover sub-assembly

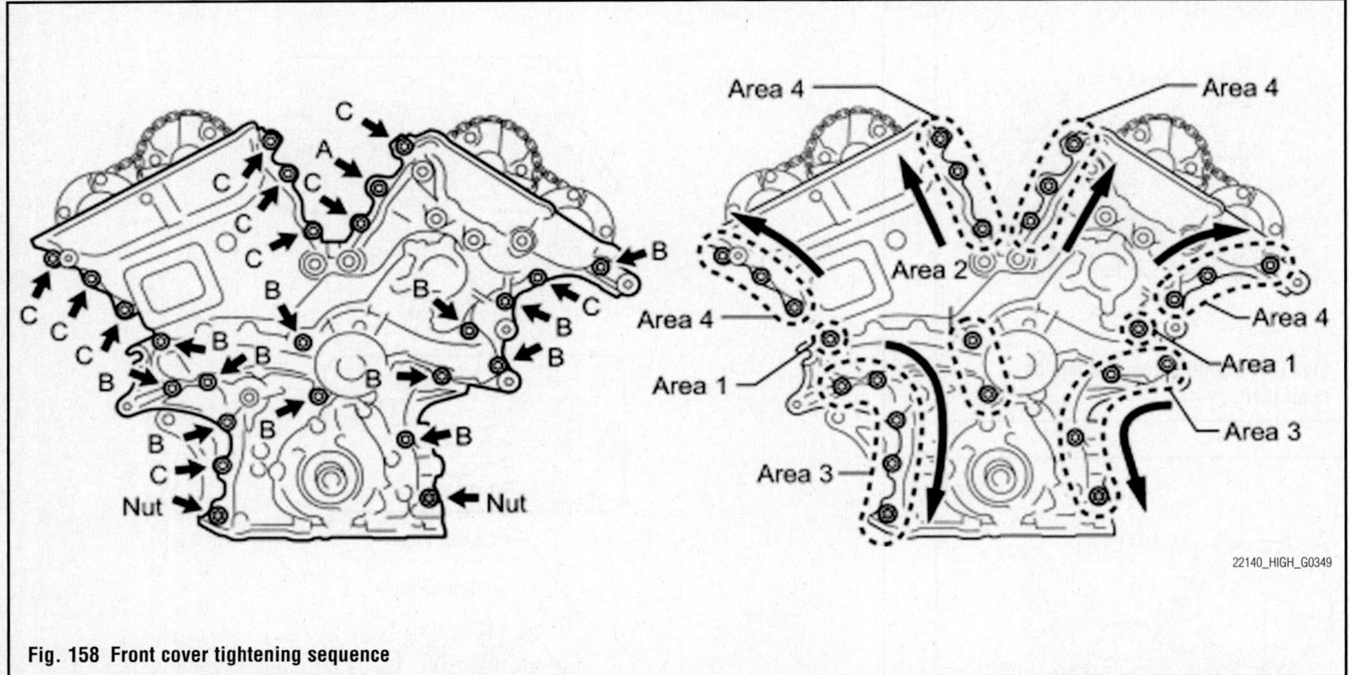

Fig. 158 Front cover tightening sequence

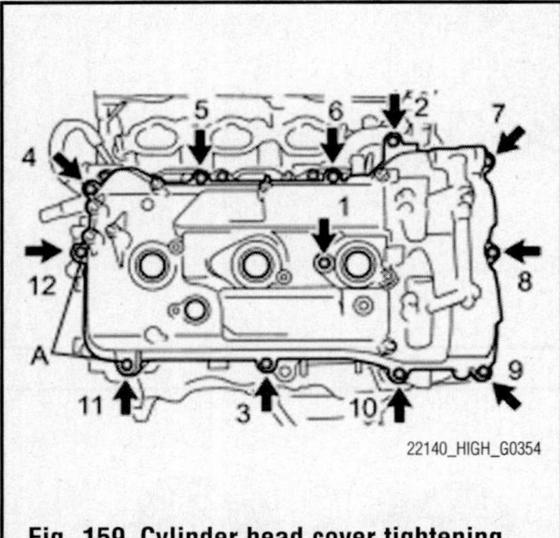

Fig. 159 Cylinder head cover tightening sequence

hh. Remove the timing chain case oil seal.

ii. Set the No. 1 cylinder to TDC/compression.

jj. Remove the No. 1 chain tensioner assembly.

kk. Remove the chain tensioner slipper.

ll. Remove the chain sub-assembly.

mm. Remove the idle sprocket assembly.

nn. Remove the camshaft timing gears and No. 2 chain (for Bank 1).

oo. While raising the No. 2 chain tensioner assembly, insert a pin of 1.0 mm (0.039 in.) diameter into the hole to fix the No. 2 chain tensioner assembly.

pp. Hold the hexagonal portion of the camshaft with a wrench, and remove the 2 bolts and 2 camshaft timing gear assemblies.

➡Be careful not to damage the cylinder head with the wrench. Do not disassemble the camshaft timing gear assemblies.

qq. Remove the No. 2 chain assembly.

rr. Remove the bolt and No. 2 chain tensioner assembly.

To install:

4. Install the following:

a. Install the No. 2 chain tensioner assembly with the bolt and tighten to 15 ft. lbs. (21 Nm).

b. While pushing in the tensioner, insert a pin of 1.0 mm (0.039 in.) diameter into the hole to fix it.

c. Align the mark plate with the timing marks of the camshaft timing gear as shown.

d. Apply a light coat of engine oil to the bolt threads and bolts seating surface.

e. Align the knockpin of the camshaft with pinhole of the camshaft timing gear assembly. Install the camshaft timing gear assembly and camshaft timing exhaust timing gear assembly and camshaft timing exhaust gear with the No. 2 chain sub-assembly installed.

f. Hold the hexagonal portion of the camshaft with the wrench and tighten the two bolts and camshaft timing gear assemblies to 74 ft. lbs. (100 Nm).

g. Remove the pan from the No 2 chain tensioner assembly.

h. Install idle sprocket assembly and tighten to 44 ft. lbs. (60 Nm).

i. Install chain sub-assembly.

j. Align the mark plate and timing marks as shown in the illustration and install the chain.

➡The camshaft mark plates are orange.

k. Do not pass the chain over the crankshaft, just temporarily place it on the crankshaft.

l. Turn the camshaft timing gear assembly on bank 1 counterclockwise to tighten the chain between the banks.

➡When the idle sprocket assembly is reused, align the timing chain plate with the mark where the plate has been in order to tighten the chain between the banks.

m. Align the mark plate and timing marks as shown in the illustration and install the chain onto the crankshaft timing sprocket. The crankshaft to mark plate is yellow.

n. Turn the crankshaft clockwise to set it to the right-hand block bore more centerline. (TDC Compression).

o. Install chain tensioner slipper.

p. Move the stopper plate upward to release the lock, and push the plunger deep into the tensioner.

q. Move the stopper plate downward to set the loss, and insert a hexagon wrench into the hole of the stopper plate.

r. Install No. 1 chain tensioner assembly and tighten bolts to 7 ft. lbs. (10 Nm).

s. Remove the hexagon wrench from

the No. 1 chain tensioner assembly. Check that the each timing mark is aligned with the crankshaft at TDC compression.

t. Remove the pulley set bolt.

u. Install timing chain case oil seal.

v. Install sealant to timing chain cover sub-assembly.

w. Install new O ring gasket on cylinder block.

x. Align the oil pump's drive rotor spline and the crankshaft as shown in the illustration. Install the spline and chain cover to the crankshaft.

y. Temporarily tighten the timing chain cover with the 23 bolts and 2 nuts.

- Tighten bolts in area 1 and 2: 15 ft. lbs. (21 Nm).
- Tighten bolt in area 3: 15 ft. lbs. (21 Nm).

➡First tighten the upper bolts and nuts, followed by the lower bolts and nuts as shown.

- Tighten bolt in area 4: 32 ft. lbs. (43 Nm)
- Tighten bolt in area 4: 15 ft. lbs. (21 Nm)

z. Install oil pan sub-assembly and tighten 16 bolts and 2 nuts to 7 ft. lbs. (10 Nm) and 15 ft lbs. (21 Nm).

aa. Install oil strainers sub-assembly and tighten bolts and nuts to 7 ft. lbs. (10 Nm).

bb. Install No.2 oil pan sub-assembly and tighten 16 bolts and 2 nuts to 7 ft. lbs. (10 Nm).

cc. Install cylinder head cover sub-assemblies and tighten to 7 ft. lbs. (10 Nm) and 15 ft. lbs. (21 Nm).

dd. Install the water inlet housing.

ee. Install No. 1 oil cooler bracket (w/ oil cooler).

ff. Install the oil cooler assembly (w/ oil cooler).

gg. Install the crankshaft pulley.

hh. Install the oil pipe.

ii. Install the No. 1 oil pipe.

jj. Install the Crankshaft Position (CKP) sensor.

kk. Install the knock control sensor.

ll. Install the knock control sensor wire.

mm. Install the No. 1 vacuum switching valve.

nn. Install the radio setting condenser.

oo. Install the No. 1 left front engine mounting bracket with the 6 bolts and tighten to 40 ft. lbs. (54 Nm).

➡Install the water inlet and mounting bracket within 15 minutes after installing the chain cover. Do not start the engine for at least 2 hours after installation.

pp. Install the No. 2 idler pulley sub-assembly.

qq. Install the No. 2 timing gear cover

rr. Install the V-ribbed belt tensioner assembly.

ss. Install the generator assembly.

tt. Install the transverse engine mounting bracket.

uu. Install the left hand exhaust manifold sub-assembly and tighten to 15 ft. lbs. (21 Nm).

vv. Install the No. 2 exhaust manifold heat insulator.

ww. Install the No. 2 manifold stay.

xx. Install the No. 2 engine oil level dipstick guide.

yy. Install the right hand exhaust manifold sub-assembly and tighten to 15 ft. lbs. (21 Nm).

zz. Install the intake manifold and tighten the 6 bolts and 4 nuts uniformly in several steps to 15 ft. lbs. (21 Nm).

53. Install the right hand No. 2 engine mounting stay.
54. Install the ignition coil assembly.
55. Remove the engine stand.
56. Install the engine assembly.

VALVE LASH

ADJUSTMENT

See Figure 160.

➡Keep the lash adjuster free of dirt and foreign objects.

➡Only use clean engine oil.

1. Place the lash adjuster into a container filled with engine oil.
2. Insert the SST's (SST: 09276-75010) tip into the lash adjuster's plunger and use the tip to press down on the check ball inside the plunger.
3. Squeeze the SST and lash adjuster together to move the plunger up and down 5 to 6 times.
4. Check the movement of the plunger and bleed the air. OK: Plunger moves up and down.

➡When bleeding air from the high-pressure chamber, make sure that the tip of the SST is actually pressing the check ball as shown in the illustration. If the check ball is not pressed, air will not bleed.

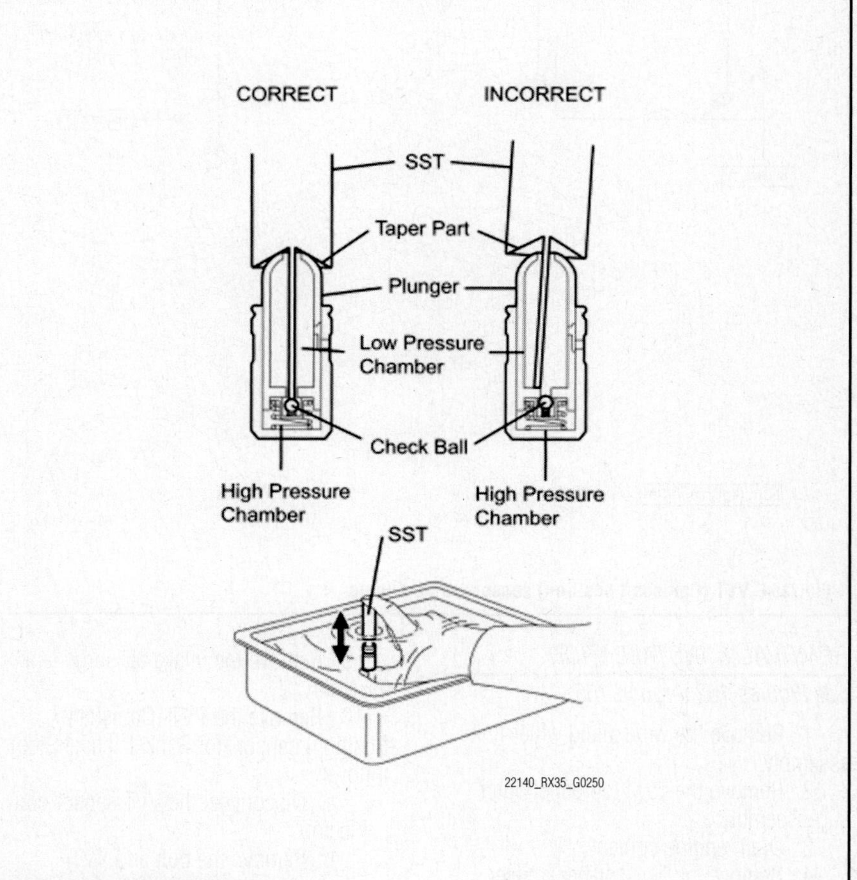

Fig. 160 Valve lash adjuster

5. After bleeding the air, remove the SST. Then, try to press the plunger quickly and firmly with a finger. OK: Plunger is very difficult to move. If the result is not as specified, replace the lash adjuster.

6. Install the lash adjusters.

➡Install the lash adjuster to the same place where it was removed from.

ENGINE PERFORMANCE & EMISSION CONTROLS

CAMSHAFT POSITION (CMP) SENSOR

LOCATION

See Figure 161.

Refer to the accompanying illustration.

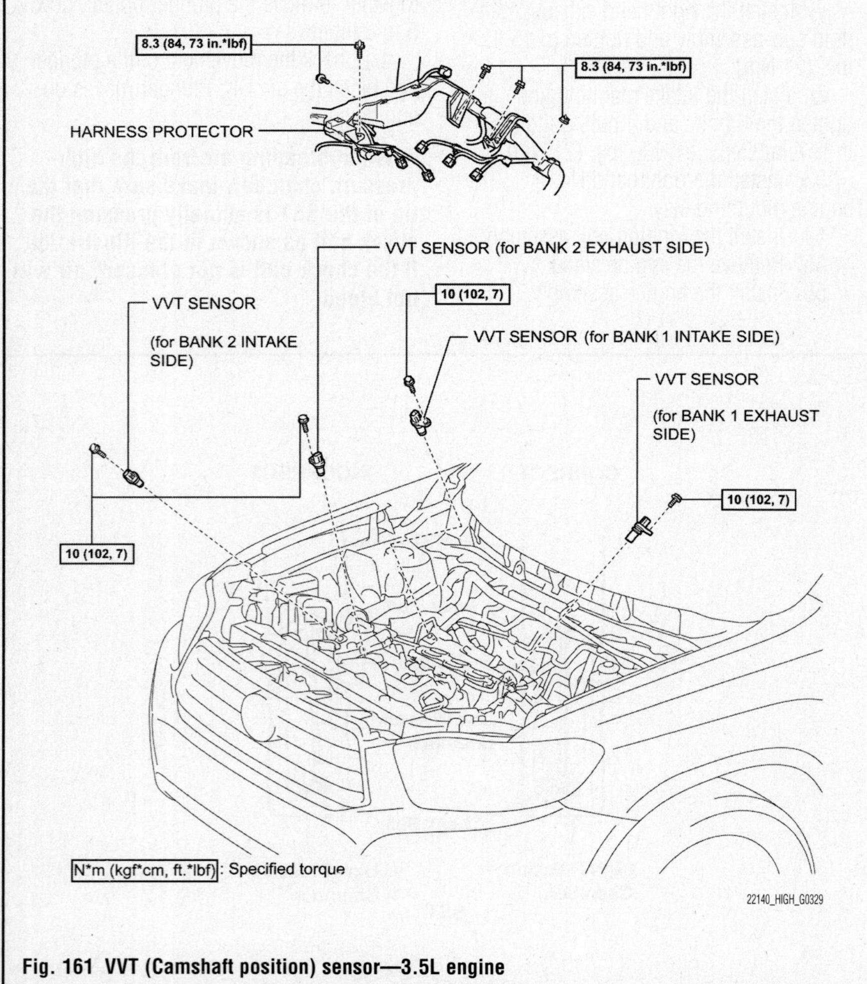

Fig. 161 VVT (Camshaft position) sensor—3.5L engine

REMOVAL & INSTALLATION

See Figures 162 through 165.

1. Remove the windshield wiper link assembly.
2. Remove the cowl top panel outer sub-assembly.
3. Drain engine coolant.
4. Remove the No. 1 V-bank cover bracket.
5. Remove No. 2 air cleaner inlet.
6. Remove No. 1 air cleaner inlet.
7. Remove the air cleaner cap sub-assembly.
8. Remove the air cleaner case sub-assembly.
9. Remove the intake air surge tank assembly.
10. Remove the VVT (Camshaft Position) sensor (for Bank 1 Intake Side), as follows:
    a. Disconnect the VVT sensor connector.
    b. Remove the bolt and VVT sensor.

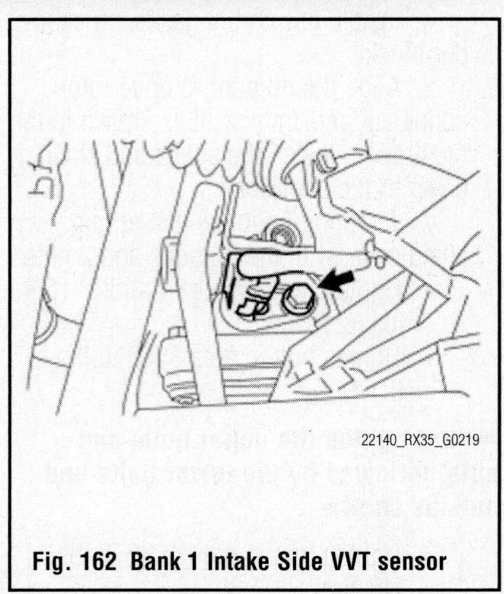

Fig. 162 Bank 1 Intake Side VVT sensor

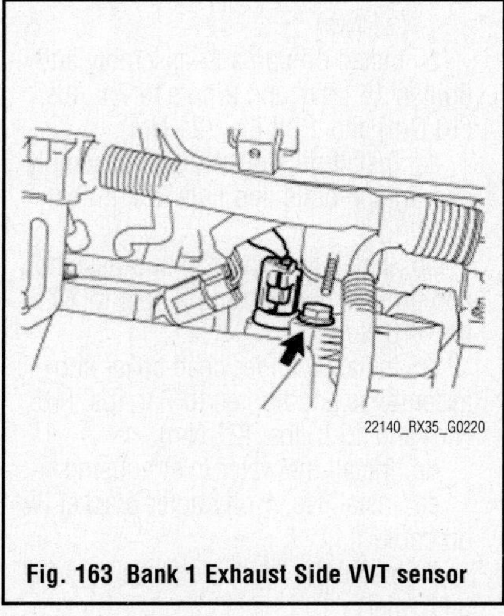

Fig. 163 Bank 1 Exhaust Side VVT sensor

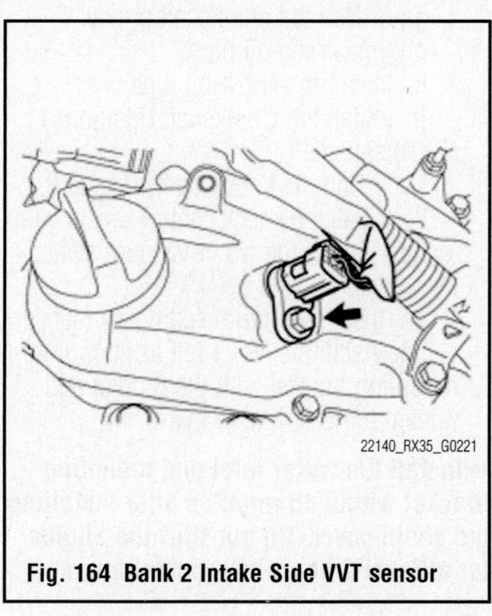

Fig. 164 Bank 2 Intake Side VVT sensor

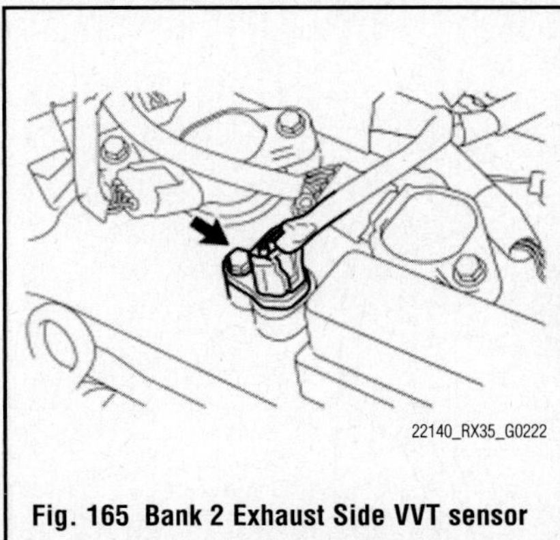

Fig. 165 Bank 2 Exhaust Side VVT sensor

11. Remove the VVT sensor (for Bank 1 Exhaust Side), as follows:
    a. Disconnect the VVT sensor connector.
    b. Remove the bolt and VVT sensor.
12. Remove the VVT sensor (for Bank 2 Intake Side), as follows:
    a. Disconnect the VVT sensor connector.
    b. Remove the bolt and VVT sensor.
13. Remove the VVT sensor (for Bank 2 Exhaust Side), as follows:
    a. Disconnect the VVT sensor connector.
    b. Remove the bolt and VVT sensor.

CRANKSHAFT POSITION (CKP) SENSOR

LOCATION

See Figure 166.

Refer to the accompanying illustration.

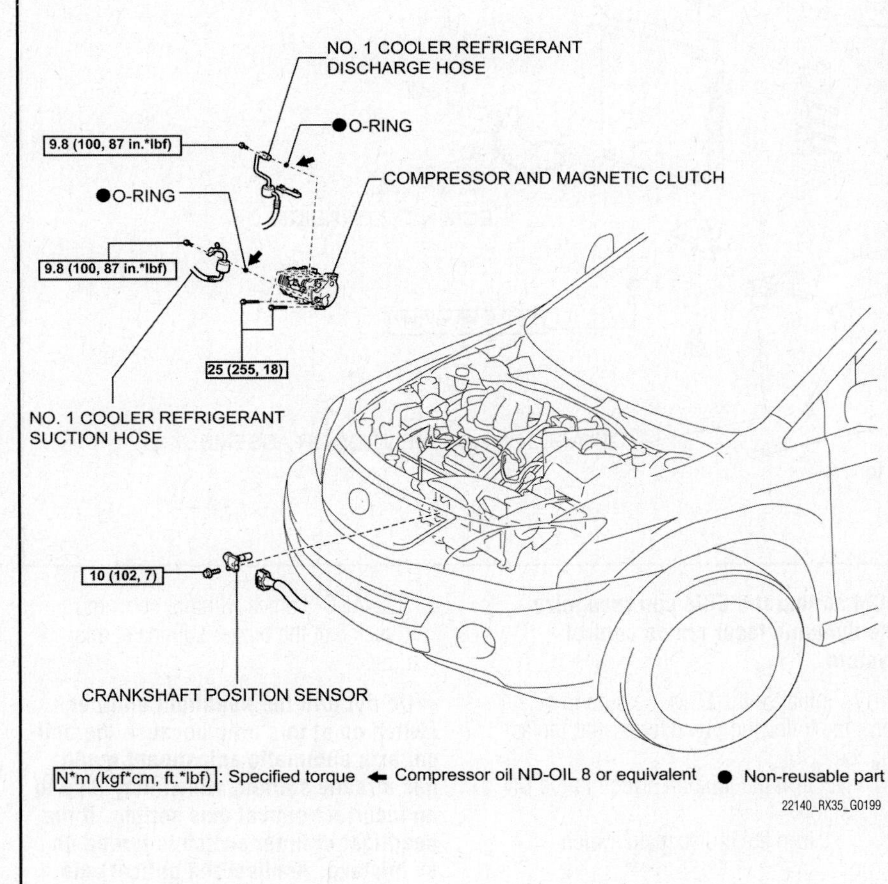

Fig. 166 Crankshaft Position (CKP) sensor location

REMOVAL & INSTALLATION

See Figure 167.

1. Remove alternator assembly.
2. Remove compressor and magnetic clutch.
3. Disconnect the Crankshaft Position (CKP) sensor connector.
4. Remove the bolt, and then remove the Crankshaft Position (CKP) sensor.

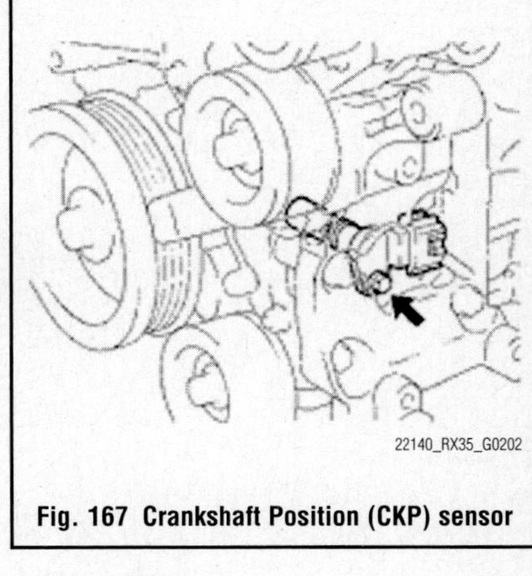

Fig. 167 Crankshaft Position (CKP) sensor

To install:

5. Apply a light coat of engine oil to the O-ring on the Crankshaft Position (CKP) sensor.
6. Install the Crankshaft Position (CKP) sensor with the bolt and tighten to 7 ft. lbs. (10 Nm).
7. Connect the Crankshaft Position (CKP) sensor connector.
8. Install compressor and magnetic clutch.
9. Install alternator assembly.

ELECTRONIC CONTROL MODULE (ECM)

LOCATION

See Figures 168 and 169.

Refer to the accompanying illustrations.

REMOVAL & INSTALLATION

2009 Models

See Figure 170.

1. Disconnect the negative battery cable.
2. Remove glove compartment door assembly.
3. Disconnect the 5 ECM connectors.
4. Disconnect the wire harness clamp.
5. Remove the 2 nuts and ECM.
6. Remove the 2 screws and the ECM No. 1 bracket from the ECM.
7. Remove the 2 screws and the ECM No. 2 bracket from the ECM.
8. Remove the 2 screws and the ECM No. 3 bracket from the ECM.

To install:

9. Install the ECM No. 1 bracket to the ECM with the 2 screws.
10. Install the ECM No. 2 bracket to the ECM with the 2 screws.
11. Install the ECM No. 3 bracket to the ECM with the 2 screws.

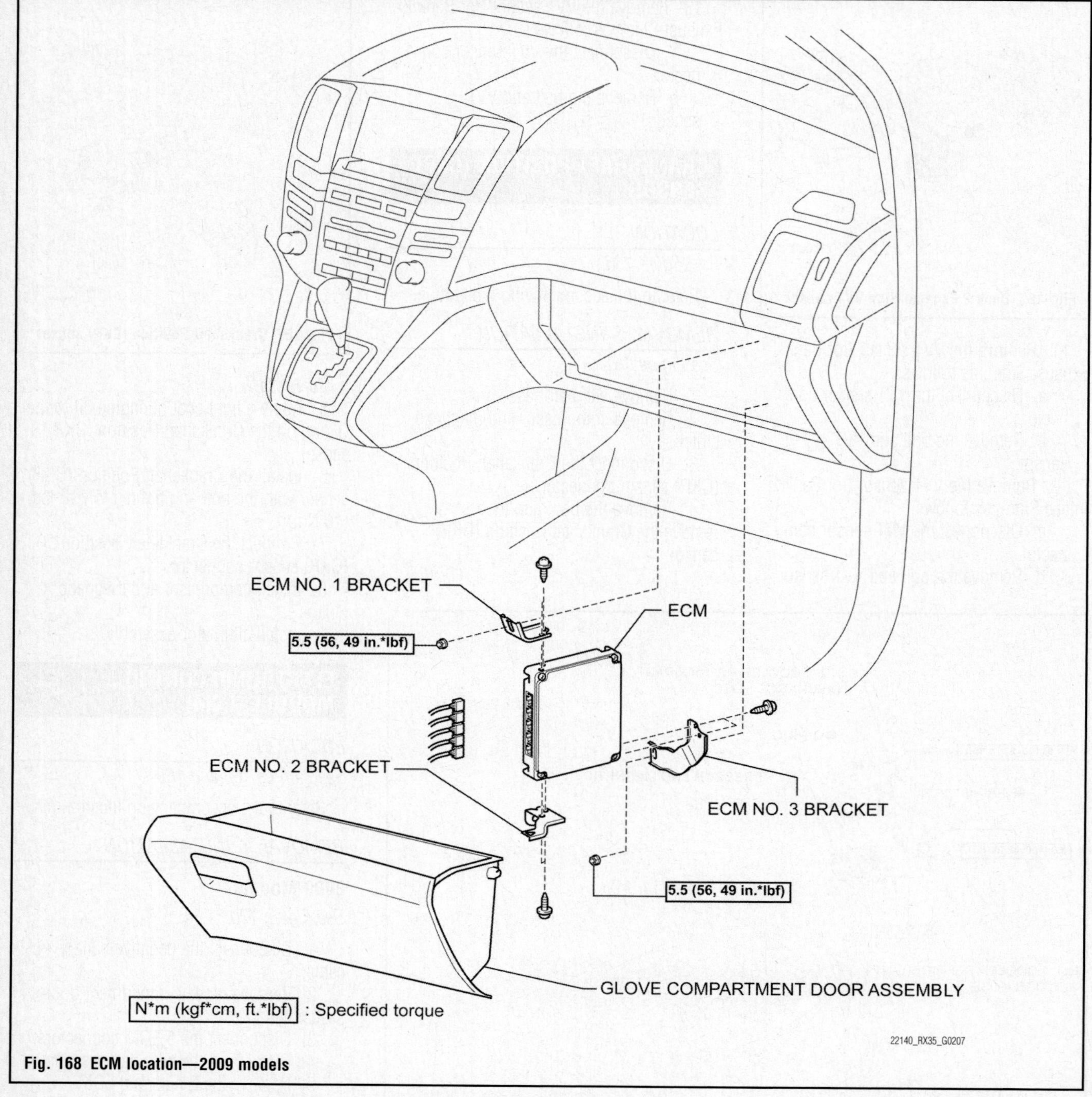

Fig. 168 ECM location—2009 models

12. Install the ECM with the 2 nuts and tighten to 49 inch lbs. (5.5 Nm).
13. Install the wire harness clamp.
14. Connect the 5 ECM connectors.
15. Install glove compartment door assembly.
16. Connect the negative battery cable.

➡After replacing the ECM on vehicles with a dynamic laser cruise control system, it is necessary to initialize the ECM so that the ECM can recognize the dynamic laser cruise control system.

17. Initialize the ECM. Be sure to perform the following procedures after replacing the ECM:
    a. Turn the ignition switch to the ON position.
    b. Turn the cruise main switch on.
    c. With the brake pedal depressed, push the cruise control main switch to RES/ACC 3 times within 3 seconds. Check that the buzzer sounds at this time.

➡Do not turn the headlight dimmer switch on at this time because the optical axis automatic adjustment mode has already started, which may lead to an incorrect optical axis setting. If the headlight dimmer switch is turned on by mistake, readjust the optical axis.

    d. Turn the ignition switch off.

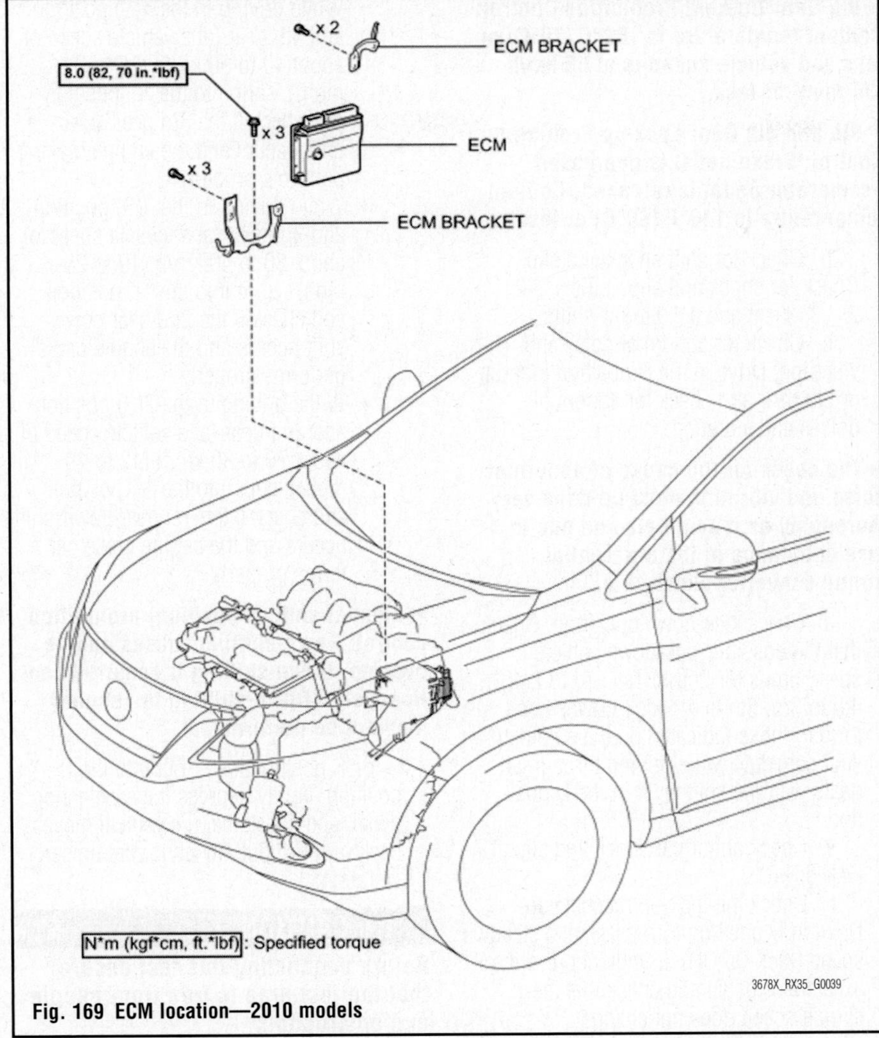

Fig. 169 ECM location—2010 models

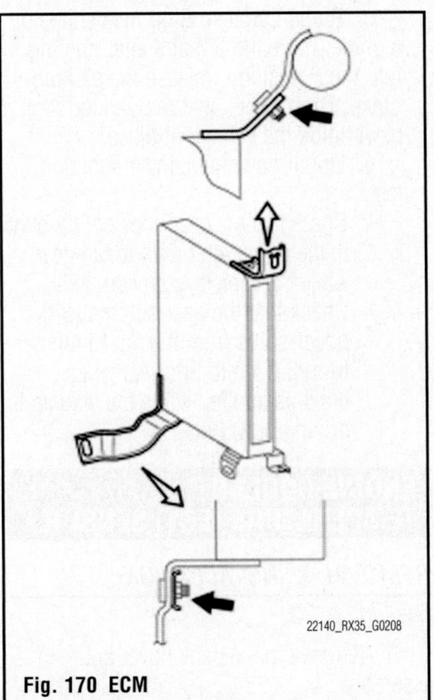

Fig. 170 ECM

2010 Models

See Figure 171.

➡The Vehicle Identification Number (VIN) must be input into a replacement ECM.

➡After the engine switch is turned off, the display and navigation module display (HDD navigation system) records various types of memory and settings. As a result, after turning the engine switch off, make sure to wait at least 60 seconds before disconnecting the cable from the negative (-) battery terminal.

1. Disconnect the negative battery cable.
2. Remove the windshield wiper motor and link assembly.
3. Remove front shock absorber cap (with air suspension).
4. Remove outer cowl top panel sub-assembly.
5. Remove v-bank cover sub-assembly.

⁂ CAUTION

Wait at least 90 seconds after disconnecting the cable from the negative (-) battery terminal to disable the SRS system.

➡When disconnecting the cable, some systems need to be initialized after the cable is reconnected.

6. Remove no. 2 air cleaner inlet.
7. Remove no. 1 air cleaner inlet .
8. Remove battery.
9. Remove air cleaner assembly
10. Remove ECM:
    a. Separate the wire harness clamp.
    b. Separate the wire harness clamp.
    c. Disconnect the 2 ECM connectors and remove the ECM with brackets. Push in the locks on the 2 levers, raise the levers, and disconnect the 2 ECM connectors.

➡After disconnecting the connectors, make sure that dirt, water or other foreign matter does not contact the connecting part of the connectors.

    d. Remove the 3 bolts and the ECM with bracket.
    e. Remove the 5 screws and 2 ECM brackets.

To install:

11. Installation is the reverse of the removal procedure. After connecting the negative battery cable, perform the following procedures:
    a. Register the transmitter ID.
    b. Inspect the tire pressure warning system.
    c. Initialize tire pressure warning system.

➡Be sure to register the transmitter IDs of all tires in the ECU before initialization.

➡Be sure to inflate all tires to the proper inflation pressure before initialization.

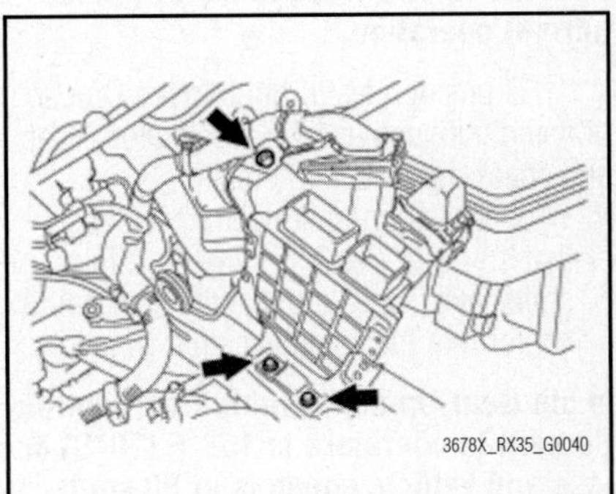

Fig. 171 Removing the 3 bolts and the ECM with bracket

d. Register immobilizer communication ID.

➡If the ECM is replaced, register the ECU communication ID for the immobilizer system.

e. Perform initialization.

➡If the ECM is replaced, perform RESET MEMORY (at initialization).

RESET

➡Perform the RESET MEMORY (AT initialization) when replacing the automatic transaxle assembly, engine assembly or ECM. The RESET MEMORY can be performed only with Techstream.

➡The ECM memorizes the condition that the ECT controls the automatic transaxle assembly and engine assembly according to those characteristics. Therefore, when the automatic transaxle assembly, engine assembly, or ECM has been replaced, it is necessary to reset the memory so that the ECM can memorize the new information.

1. Reset procedure is as follows:
   a. Turn the ignition switch off.
   b. Connect Techstream to the DLC3.
   c. Turn the ignition switch on (IG) and push the Techstream main switch on.
   d. Enter the following menu: Powertrain / Engine and ECT / Utility / Reset Memory. Then, press "Next".
   e. Perform the reset memory procedure from the main menu.

⁂ CAUTION

After performing the RESET MEMORY, be sure to perform the ROAD TEST.

ROAD TESTING

➡Perform the test at the ATF temperature 122 to 176°F (50 to 80°C) in the normal operation.

1. D position test: Shift into the D position and fully depress the accelerator pedal and check the following points:
   a. Check up-shift operation: Check that 1 _ 2, 2 _ 3, 3 _ 4 and 4 _ 5th up-shifts take place, and that the shift points conform to the automatic shift schedule.

➡5th Gear Up-shift Prohibition Control: Coolant temperature is 158°F (70°C) or less and vehicle speed is at 80 km/h (50 mph) or less. ATF temperature is 28°F (-2°C) or less.

➡4th Gear Up-shift Prohibition Control: Coolant temperature is 158°F (70°C) or less and vehicle speed is at 55 km/h (34 mph) or less.

➡5th and 4th Gear Lock-up Prohibition Control: Brake pedal is depressed. Accelerator pedal is released. Coolant temperature is 140°F (60°C) or less.

b. Check for shift shock and slip. Check for shock and slip at the 1 _ 2, 2 _ 3, 3 _ 4 and 4 _ 5th up-shifts.

c. Check for abnormal noise and vibration. Drive in the D position lock-up or 5th gear and check for abnormal noises and vibration.

➡The check for the cause of abnormal noise and vibration must be done very thoroughly as it could also be due to loss of balance in the differential, torque converter clutch, etc.

d. Check kick-down operation. Check that the possible kick-down vehicle speed limits for 2nd to 1st, 3rd to 2nd, 4th to 3rd, 5th to 4th kick-downs conform to those indicated on the automatic shift schedule while driving through all gears with the shift lever in the D position.

e. Check abnormal shock and slip at kick-down.

f. Check the lock-up mechanism. Drive in D position (5th gear), at a steady speed (lock-up ON). Lightly depress the accelerator pedal and check that the engine speed does not change abruptly.

➡

- There is no lock-up in the 1st, 2nd, and 3rd gear.
- 3rd lock-up operates while uphill control is active.
- If there is a big jump in engine speed, there is no lock-up.

2. S position test:
   a. Shift to the S position, depress the accelerator pedal and check the following points. Check shift operation:
   - While driving in the D position and 5th gear, shift into the D position and S position and back to the D position. Check that the gear change 5 _ 4 down-shift and 4 _ 5 up-shift can be performed
   - With the shift lever in the S position (with the vehicle stopped), shift into the "+" position to check that the shift position on the combination meter changes as follows : 1 _ 2, 2 _ 3, 3 _ 4 and 4 _ 5
   - While driving in the 4(S) position and 4th gear (at a vehicle speed of about 40 to 50 km/h (25 to 31 mph)), shift into the "-" position and check if the 3rd gear down-shift occurs and the engine brake performs properly
   - While driving in the 3(S) position and 3rd gear (at a vehicle speed of about 30 to 40 km/h (19 to 25 mph)), shift into the "-" position and check if the 2nd gear down-shift occurs and the engine brake performs properly
   - While driving in the 2(S) position and 2nd gear (at a vehicle speed of about 20 to 30 km/h (12 to 19 mph)), shift into the "-" position and check if the 1st gear down-shift occurs and the engine brake performs properly

➡Manual shift (S position) prohibition control: Down-shifting causes engine overrun. Down-shifting is required continuously. (Down-shifting to 1st gear may not be performed.)

b. R position test: Shift into the R position, lightly depress the accelerator pedal, and check that the vehicle moves backward without any abnormal noise or vibration.

⁂ CAUTION

Before conducting this test ensure that the test area is free from people and obstruction.

c. P position test: Stop the vehicle on a grade (more the 5°) and after shifting into the P position, release the parking brake. Then, check that the parking lock pawl holds the vehicle in place.

d. Uphill/downhill control function test:

- Check that the gear does not up-shift to the 3rd to 4th or 4th to 5th gear while the vehicle is driving uphill.
- Check that the gear automatically down-shifts from the 5th to 4th or from the 4th to 3rd gear when brake is applied while the vehicle is driving downhill.

ENGINE COOLANT TEMPERATURE (ECT) SENSOR

REMOVAL & INSTALLATION

See Figure 172.

1. Remove the V-bank cover sub-assembly.

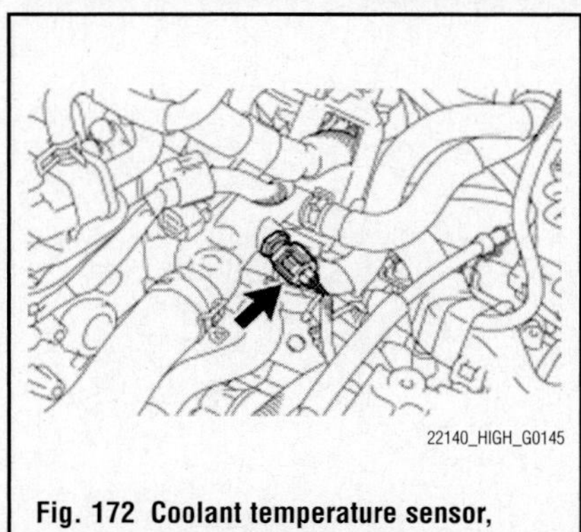

Fig. 172 Coolant temperature sensor, connector, and gasket—3.5L engine

2. Remove the engine under cover assembly.
3. Remove the No. 1 engine under cover.
4. Drain engine coolant.
5. Remove the cool air intake duct seal.
6. Remove the battery.
7. Remove the No. 1 and 2 air cleaner inlets.
8. Remove the air cleaner cap and case sub-assemblies.
9. Disconnect the engine coolant temperature sensor connector.
10. Remove the engine coolant temperature sensor and gasket.

To install:

11. Install a new gasket onto the engine coolant temperature sensor.
12. Install the engine coolant temperature sensor and tighten to 15 ft. lbs. (20 Nm).
13. The remainder of installation is the reverse of the removal procedure.

HEATED OXYGEN SENSOR (HO2S)

LOCATION

See Figure 173.

Refer to the accompanying illustration.

REMOVAL & INSTALLATION

See Figures 174 through 176.

1. Remove the oxygen sensor (for Bank 1):
   a. Disconnect the oxygen sensor connector and 2 clamps.

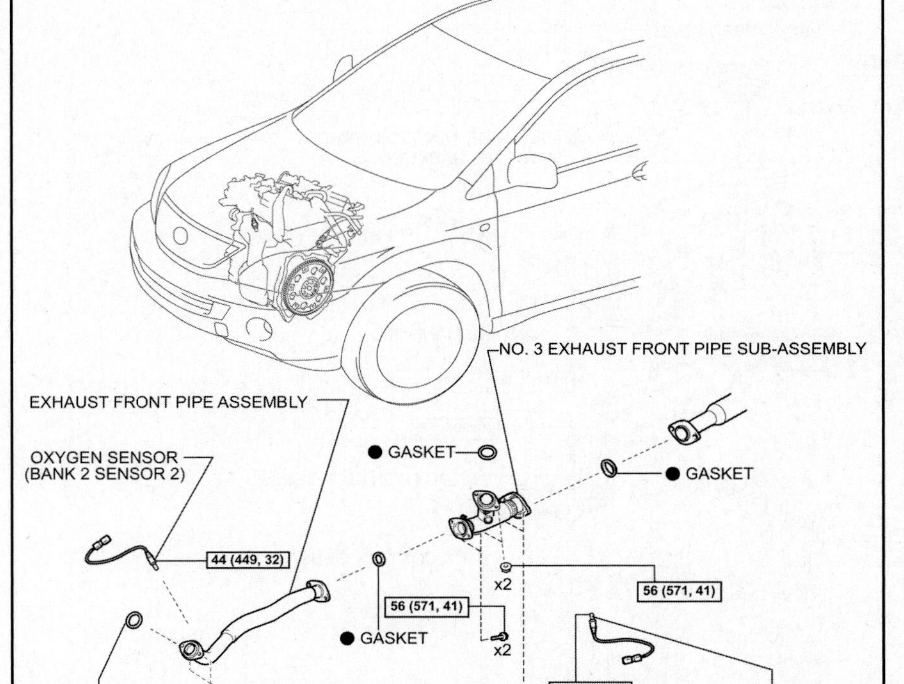

Fig. 173 Heated Oxygen Sensor (HO2S) location

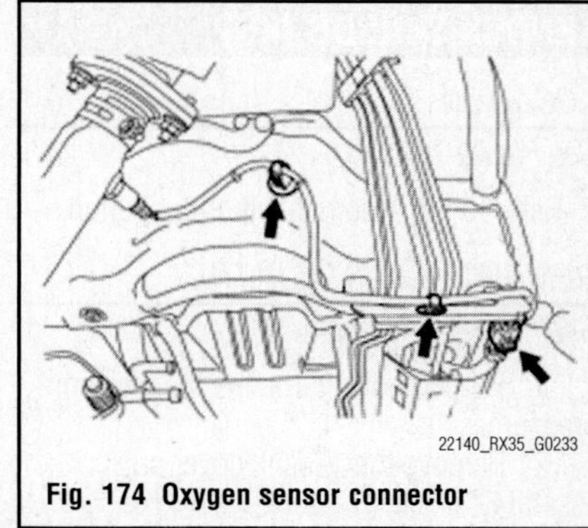

Fig. 174 Oxygen sensor connector

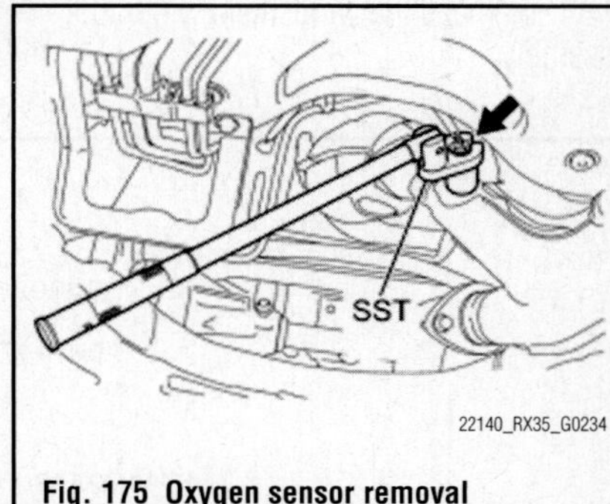

Fig. 175 Oxygen sensor removal

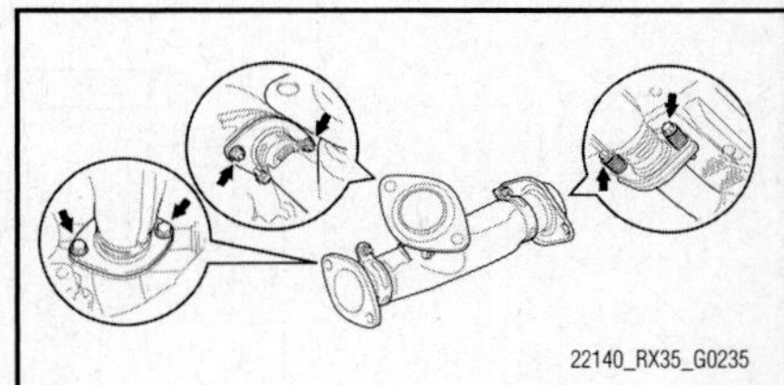

Fig. 176 No. 3 exhaust front pipe sub-assembly

   b. Using SST (SST: 09224-00010) or equivalent, remove the oxygen sensor from the No. 3 exhaust front pipe sub-assembly.

➡Do not damage the oxygen sensor.

2. Remove the No. 3 exhaust front pipe sub-assembly (for Bank 2):
   a. Remove the 4 bolts, 2 nuts, 2 compression springs and No. 3 exhaust front pipe sub-assembly.
3. Remove the exhaust front pipe assembly (for Bank 2).
4. Remove the oxygen sensor (for Bank 2):
   a. Using SST (SST: 09224-00010)) or equivalent, remove the oxygen sensor from the exhaust front pipe assembly.

➡Do not damage the oxygen sensor.

To install:

5. Installation is the reverse of the removal procedure. Tighten the oxygen sensors to 32 ft. lbs. (44 Nm).

KNOCK SENSOR (KS)

LOCATION

See Figure 177.

Refer to the accompanying illustration.

REMOVAL & INSTALLATION

See Figures 178 and 179.

1. Properly discharge the fuel system pressure.
2. Remove the V-bank cover sub-assembly.
3. Drain engine coolant.
4. Remove the windshield wiper link assembly.
5. Remove the cowl top panel outer sub-assembly.
6. Remove the air cleaner cap sub-assembly.
7. Remove the air cleaner case sub-assembly.
8. Remove the intake air surge tank assembly.
9. Remove the fuel tube sub-assembly.
10. Remove the intake manifold.
11. Disconnect the 2 Knock Sensor (KS) connectors.
12. Remove the 2 bolts and then remove the 2 knock sensors.

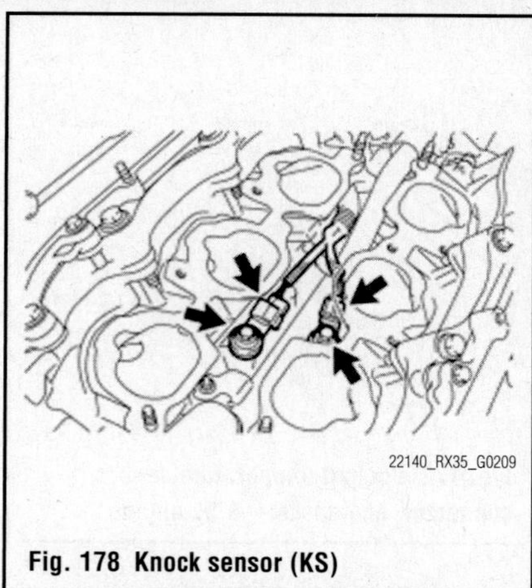

Fig. 178 Knock sensor (KS)

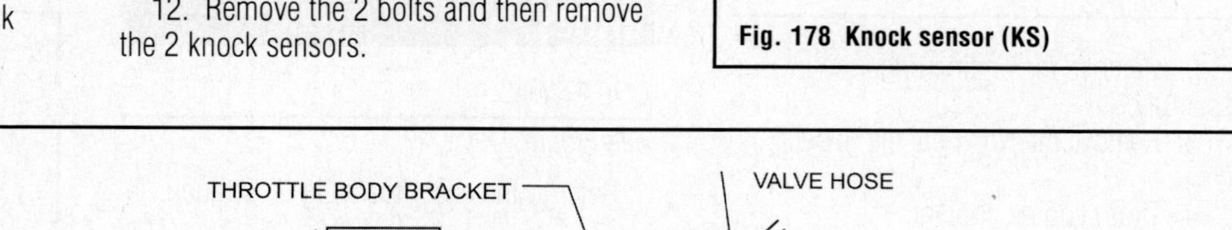

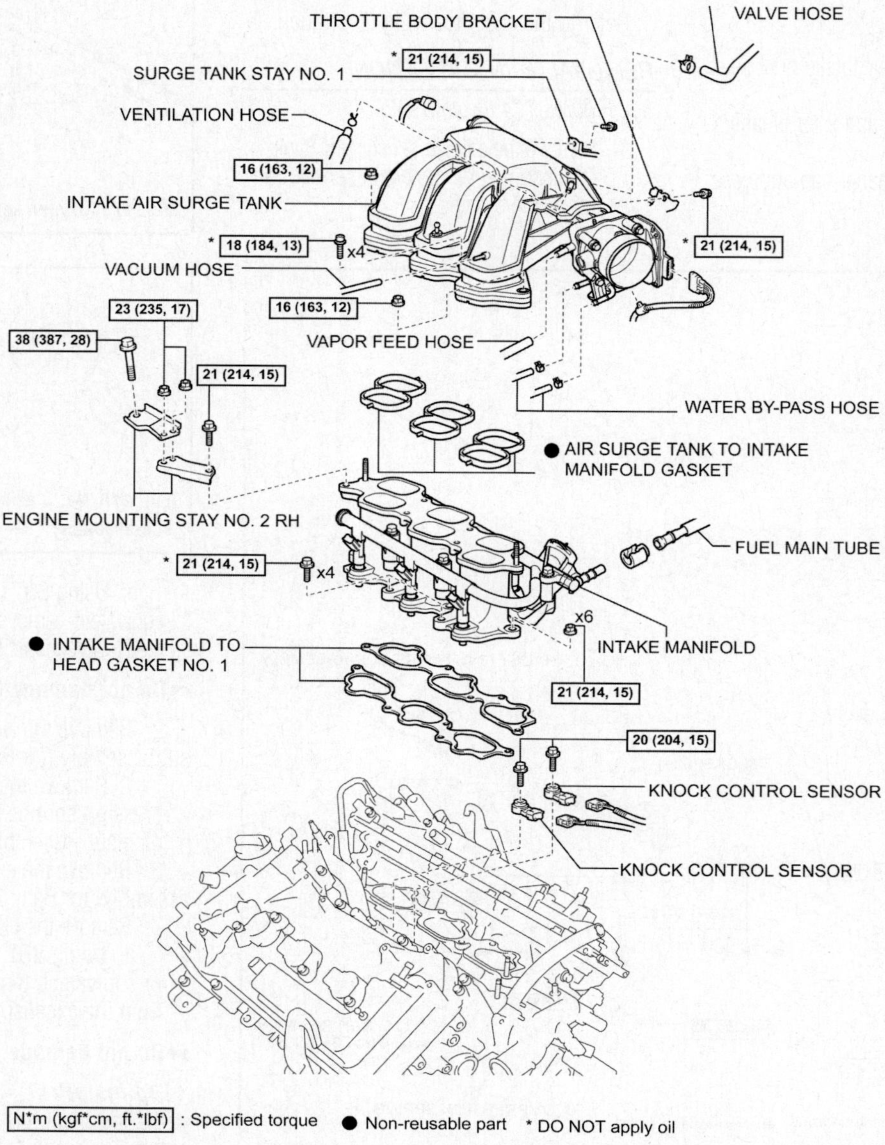

Fig. 177 Knock Sensor (KS) location

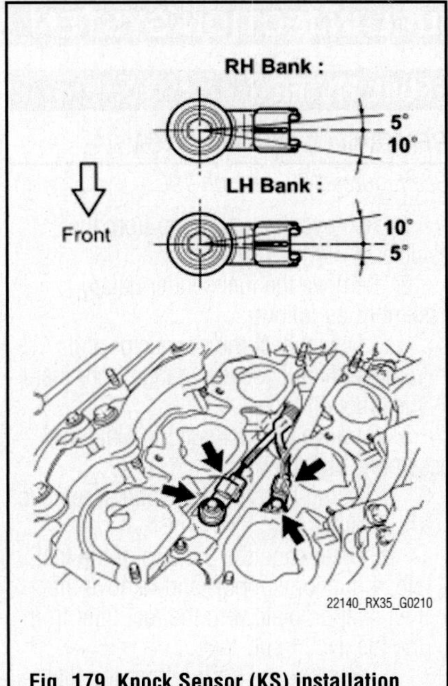

Fig. 179 Knock Sensor (KS) installation

To install:

13. Install the 2 knock control sensors with the 2 bolts as shown in the illustration and tighten to 15 ft. lbs. (20 Nm).
14. Connect the 2 knock control sensor connectors.
15. The remainder of installation is the reverse of the removal procedure.
16. After adding coolant, check for coolant and fuel leaks.

MALFUNCTION INDICATOR LIGHT (MIL)

RESET PROCEDURE

Using a ODB II scan tool, clear the DTC codes.

MASS AIR FLOW (MAF) METER

LOCATION

See Figures 180 and 181.

Refer to the accompanying illustrations.

REMOVAL & INSTALLATION

See Figure 182.

1. Disconnect the Mass Air Flow (MAF) meter connector.
2. Remove the 2 screws and the MAF meter..

To install:

3. Installation is the reverse of the removal procedure.

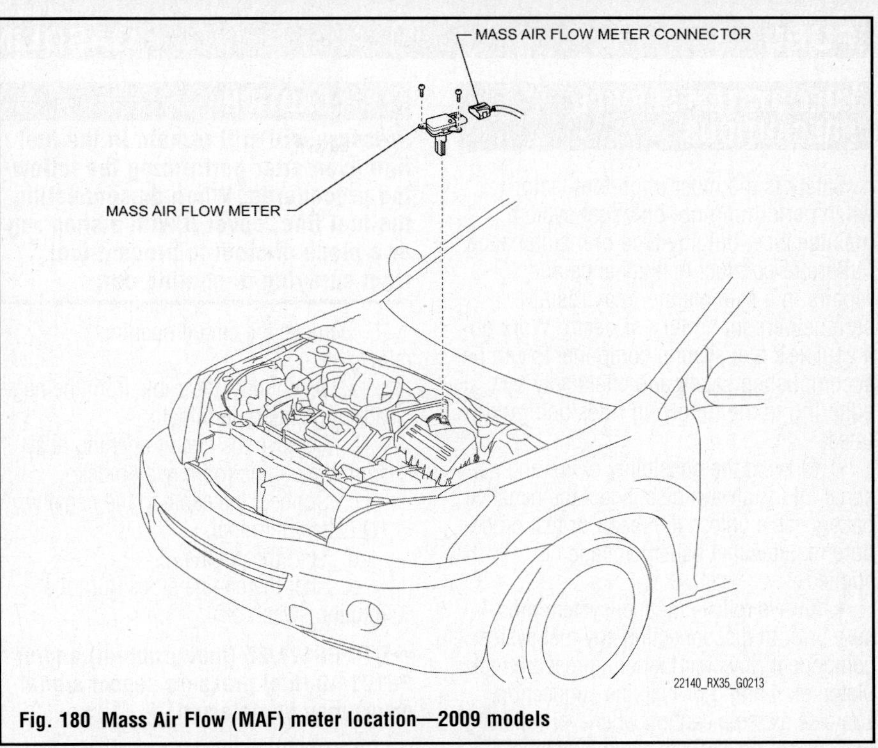

Fig. 180 Mass Air Flow (MAF) meter location—2009 models

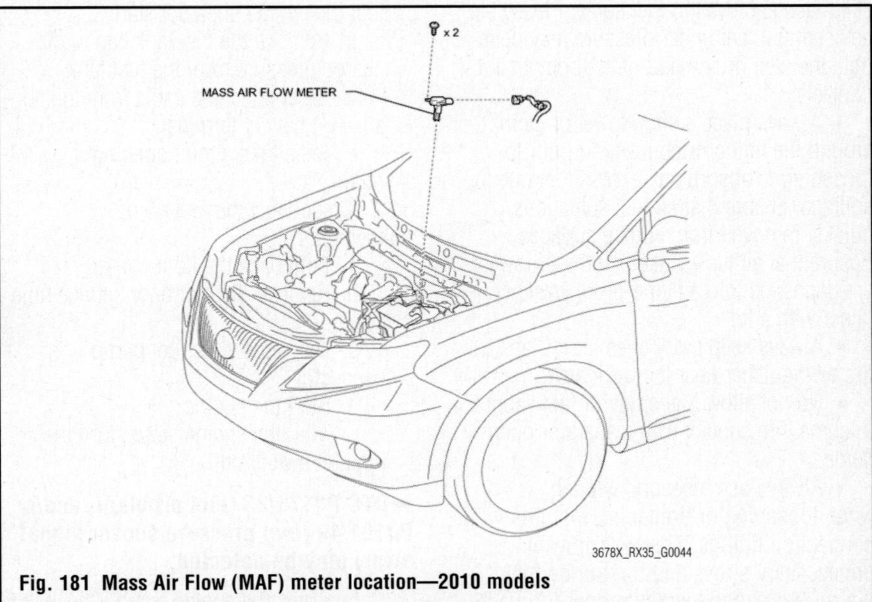

Fig. 181 Mass Air Flow (MAF) meter location—2010 models

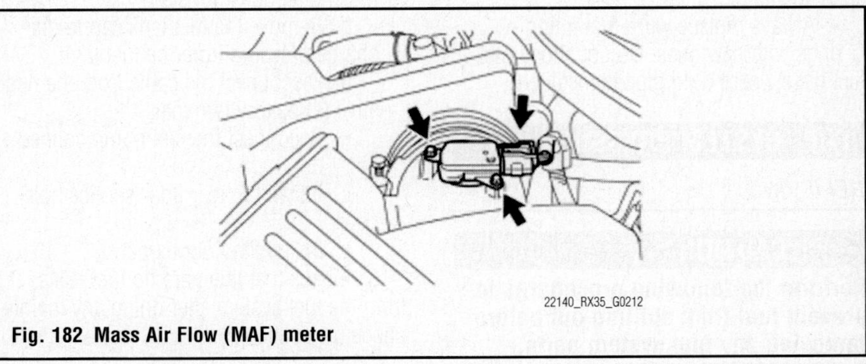

Fig. 182 Mass Air Flow (MAF) meter

FUEL SYSTEM SERVICE PRECAUTIONS

Safety is the most important factor when performing not only fuel system maintenance, but any type of maintenance. Failure to conduct maintenance and repairs in a safe manner may result in serious personal injury or death. Work on a vehicle's fuel system components can be accomplished safely and effectively by adhering to the following rules and guidelines.

- To avoid the possibility of fire and personal injury, always disconnect the negative battery cable unless the repair or test procedure requires that battery voltage be applied.
- Always relieve the fuel system pressure prior to disconnecting any fuel system component (injector, fuel rail, pressure regulator, etc.) fitting or fuel line connection. Exercise extreme caution whenever relieving fuel system pressure to avoid exposing skin, face and eyes to fuel spray. Please be advised that fuel under pressure may penetrate the skin or any part of the body that it contacts.
- Always place a shop towel or cloth around the fitting or connection prior to loosening to absorb any excess fuel due to spillage. Ensure that all fuel spillage is quickly removed from engine surfaces. Ensure that all fuel-soaked cloths or towels are deposited into a flame-proof waste container with a lid.
- Always keep a dry chemical (Class B) fire extinguisher near the work area.
- Do not allow fuel spray or fuel vapors to come into contact with a spark or open flame.
- Always use a second wrench when loosening or tightening fuel line connection fittings. This will prevent unnecessary stress and torsion on fuel piping. Always follow the proper torque specifications.
- Always replace worn fuel fitting O-rings with new ones. Do not substitute fuel hose where rigid pipe is installed.

FUEL SYSTEM PRESSURE

RELIEVING

CAUTION

Perform the following procedures to prevent fuel from spilling out before removing any fuel system parts.

CAUTION

Pressure will still remain in the fuel line even after performing the following procedures. When disconnecting the fuel line, cover it with a shop rag or a piece of cloth to prevent fuel from spraying or coming out.

1. Remove the circuit opening relay/;
   a. Disconnect the cable from the negative (-) battery terminal.
   b. Remove the circuit opening relay from the engine room relay block.
   c. Connect the cable to the negative (-) battery terminal.
   d. Start the engine.
   e. After the engine stops, turn the ignition switch off.

➡DTC P0171/25 (fuel problem) and/or P0191/49 (fuel pressure sensor signal error) may be detected.

   f. Crank the engine again. Check that the engine does not start.
   g. Remove the fuel tank cap to discharge pressure from the fuel tank.
   h. Disconnect the cable from the negative (-) battery terminal.
   i. Install the circuit opening relay.
2. Disconnect the fuel pump connector:
   a. Fold back the floor carpet.
   b. Remove the rear floor service hole cover.
   c. Disconnect the fuel pump connector.
   d. Start the engine.
   e. After the engine stops, turn the ignition switch off.

➡DTC P0171/25 (fuel problem) and/or P0191/49 (fuel pressure sensor signal error) may be detected.

   f. Crank the engine again. Check that the engine does not start.
   g. Remove the fuel tank cap to discharge pressure from the fuel tank.
   h. Disconnect the cable from the negative (-) battery terminal.
   i. Reconnect the fuel pump connector.
   j. Install the rear floor service hole cover.
   k. Install the floor carpet.
3. Check that there are no fuel leaks from the fuel system after doing any maintenance or repairs.

FUEL FILTER

REMOVAL & INSTALLATION

See Figures 183 through 190.

1. Remove the fuel pump from the vehicle.
2. Remove the fuel sender gauge assembly, as follows:
   a. Disconnect the connector and remove the fuel sender gauge from the fuel suction tube.
3. Remove the fuel suction plate sub-assembly, as follows:
   a. Using needle nozzle pliers, remove the E-ring.
   b. Disengage the 2 claws of the fuel No. 1 suction support and remove the fuel suction plate with the fuel filter from the fuel No. 1 sub-tank.
   c. Remove the spring from the fuel suction plate.
   d. Disengage the claw of the jet pump nozzle.
   e. Separate the fuel pump filter hose.
4. Disconnect the fuel pump harness.

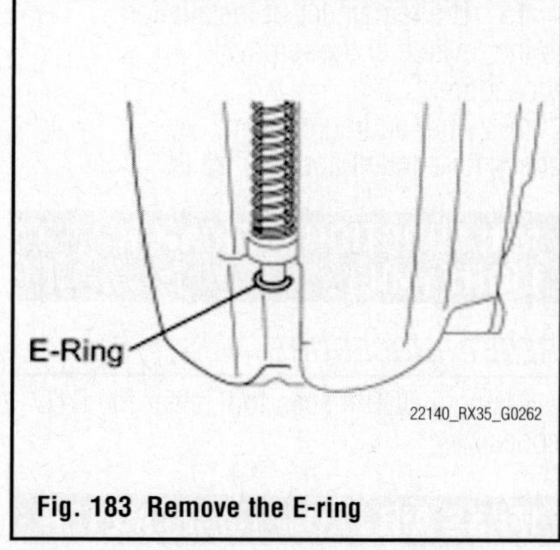

Fig. 183 Remove the E-ring

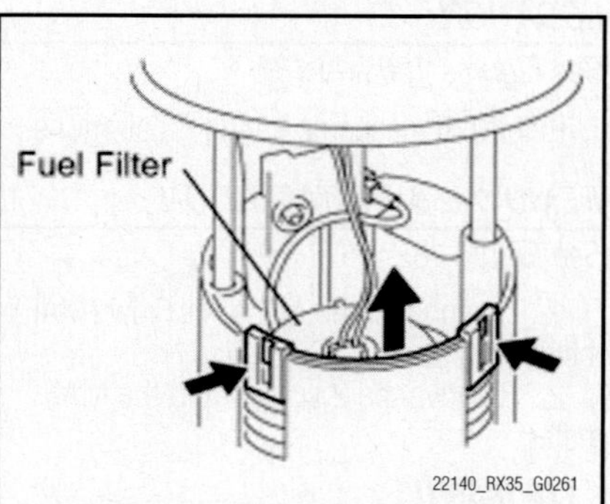

Fig. 184 Remove the fuel suction plate with the fuel filter from the fuel No. 1 sub-tank

5. Using a screwdriver with its tip wrapped in protective tape, disengage the 5 claws on the filter and remove the fuel pump from the fuel filter.

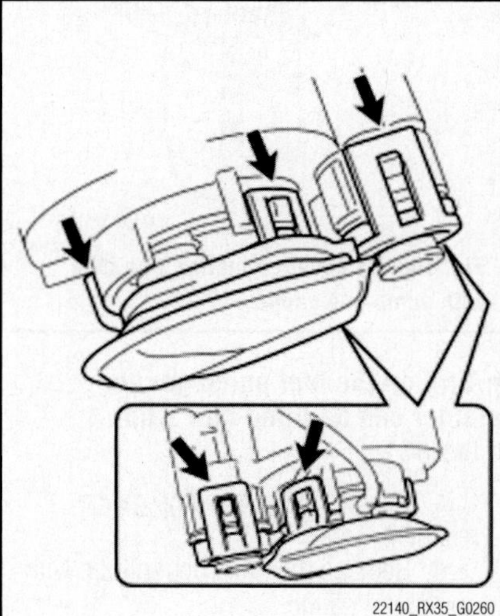

Fig. 185 Disengage the 5 claws on the filter and remove the fuel pump from the fuel filter

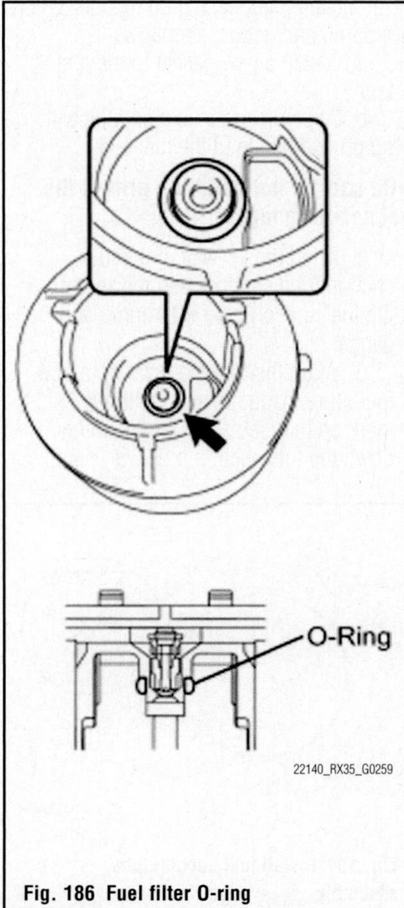

Fig. 186 Fuel filter O-ring

➡Do not damage the fuel filter.

➡Do not remove the suction filter.

➡Do not use either the fuel pump or the suction filter if the suction filter is removed from the fuel pump.

6. Remove the O-ring from the fuel filter.

To install:

7. Apply gasoline to a new O-ring and install it to the fuel filter.
8. Engage the 5 claws on the fuel filter and install the fuel pump with the pump filter.
9. Connect the fuel pump harness connector.
10. Install the fuel suction plate sub-assembly, as follows:
    a. Install the fuel pump filter tube while aligning it to the installation position of the fuel No. 1 sub-tank.
    b. Connect the jet pump nozzle.
    c. Make sure that the fuel tube passes under the protrusion of the fuel filter, and engage the claw of the fuel No. 1 suction support.
    d. Install the spring to the fuel suction plate shaft and install it to the fuel No. 1 sub-tank.
    e. Install a new E-ring.

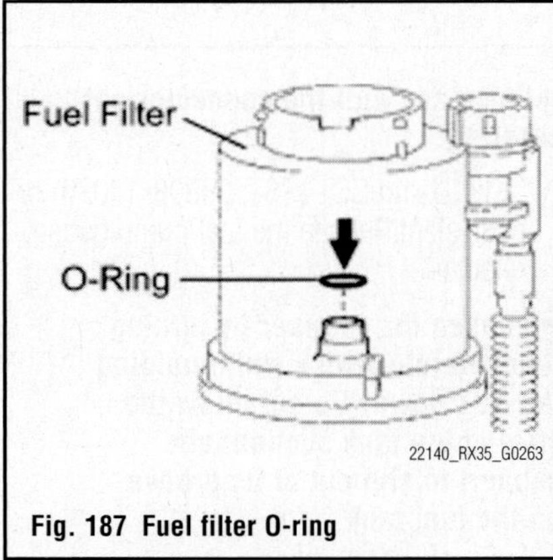

Fig. 187 Fuel filter O-ring

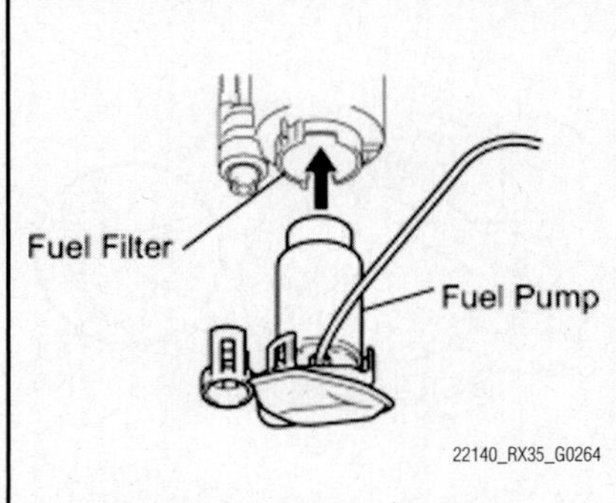

Fig. 188 Install the fuel pump with the pump filter

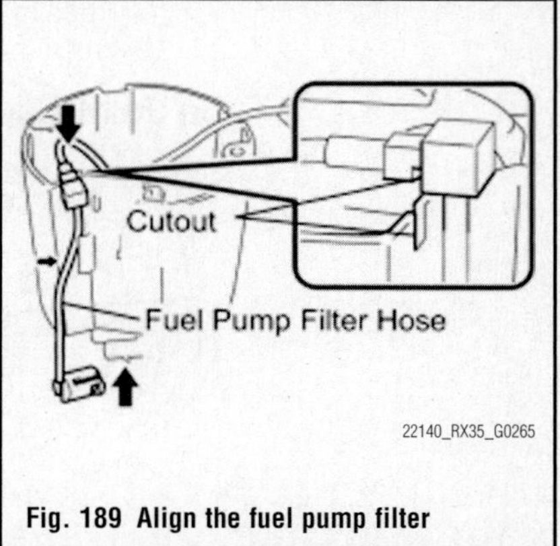

Fig. 189 Align the fuel pump filter

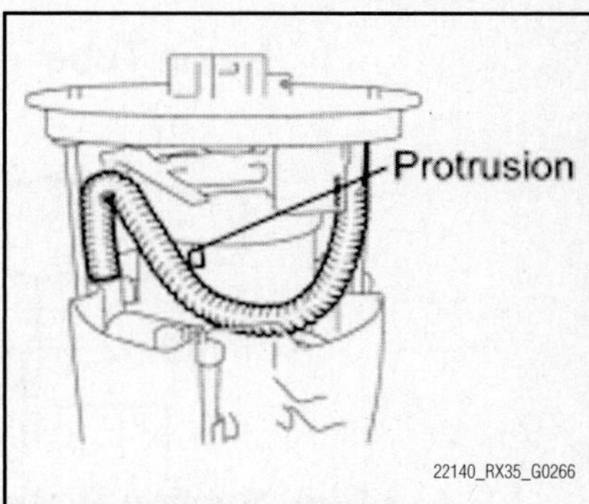

Fig. 190 Make sure that the fuel tube passes under the protrusion of the fuel filter

11. Install the fuel sender gauge assembly, as follows;
    a. Slide the fuel sender gauge to fit the claw.
    b. Connect the fuel sender gauge connector.
12. Install fuel pump to vehicle.

FUEL PUMP/ FUEL PUMP MODULE/FUEL TANK MODULE

REMOVAL & INSTALLATION

See Figures 191 through 197.

1. Discharge fuel system pressure.
2. Remove the deck board sub-assembly, as follows:
    a. Disengage the 5 clips and turn up the front side of the deck board.
3. Remove the right and left rear seat assemblies.
4. Remove the left rear door scuff plate.
5. Remove the left deck side trim cover.
6. Remove the left rear seat side cover.
7. Remove the rear floor service hole cover, as follows:
    a. Using a clip remover, remove the 2 clips and tear off the front floor carpet.

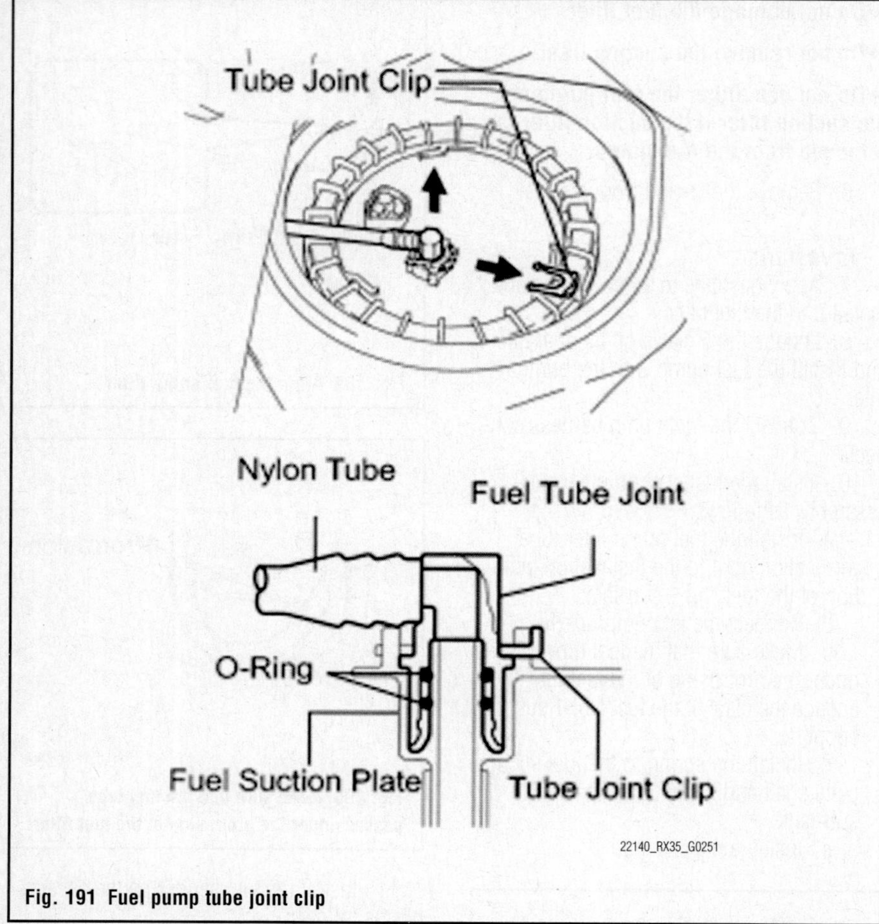

Fig. 191 Fuel pump tube joint clip

b. Remove the rear floor service hole cover.

c. Disconnect the fuel pump connector.

8. Remove the fuel suction tube assembly with pump and gauge, as follows:

a. Disconnect the fuel pump tube by removing the tube joint clip, and pull out the fuel pump tube.

➡**Check if there is any dirt or mud around the connector before this operation and remove the dirt as necessary.**

➡**Be careful of mud because the quick connector has an O-ring which seals the pipe and connector that can be contaminated.**

➡**Do not use any tool in this operation.**

➡**Do not bend or twist the nylon tube. Protect the connector by covering it with a vinyl or plastic bag.**

➡**When the pipe and connector are stuck, push and pull the connector to release and pull the connector out carefully.**

b. Using SST (SST: 09808-14020) or equivalent, loosen the fuel pump gauge retainer.

➡**Loosen the retainer by turning it counterclockwise while holding the SST down. Do not allow the claw of the tank suction tube support to slip out of its groove on the fuel tank.**

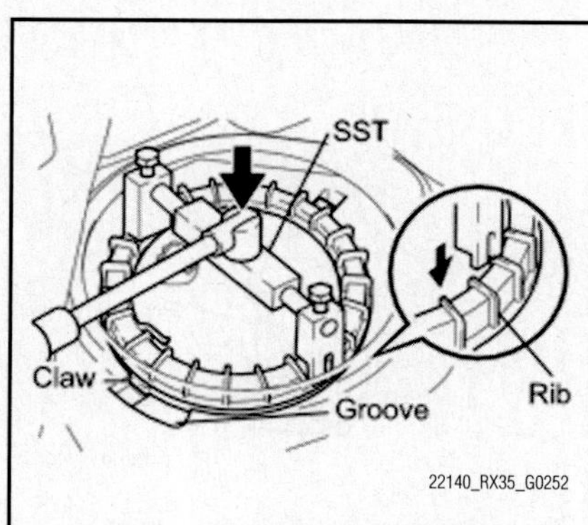

Fig. 192 Loosen fuel pump gauge retainer

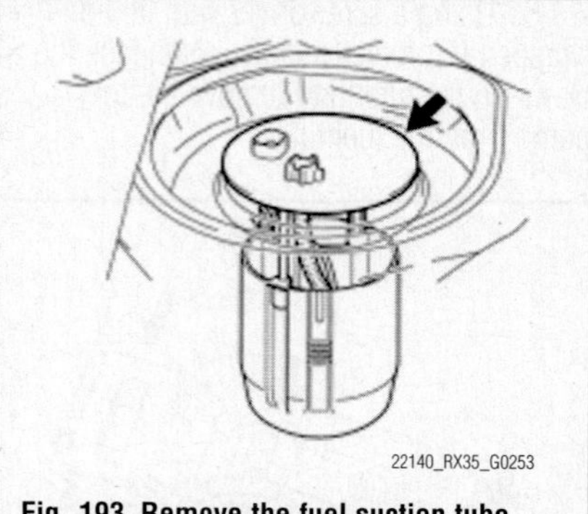

Fig. 193 Remove the fuel suction tube with pump and gauge

➡**A rib on the fuel pump gauge retainer can be fitted into a tip of the SST.**

c. Remove the fuel pump gauge retainer.

d. Remove the fuel suction tube with pump and gauge.

➡**Be careful not to bend the arm of the fuel sender gauge.**

e. Remove the gasket from the fuel tank.

To install:

9. Install the fuel suction tube assembly with pump and gauge, as follows:

a. Install a new gasket to the fuel tank.

b. Attach the fuel suction tube with pump and gauge to the fuel tank.

➡**Be careful not bend the arm of the fuel sender gauge.**

c. Align the keyway of the fuel suction tube support with the key of the fuel suction tube with pump and gauge.

d. Align the triangle mark on a new fuel pump gauge retainer with the "S" mark on the fuel tank while pushing down the fuel suction tube with pump

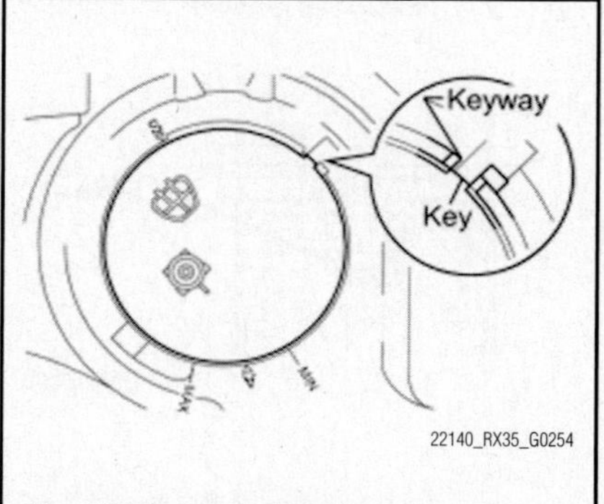

Fig. 194 Install fuel suction tube assembly

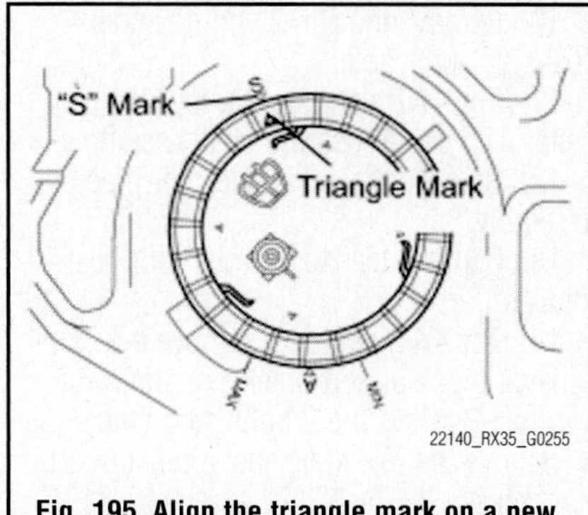

Fig. 195 Align the triangle mark on a new fuel pump

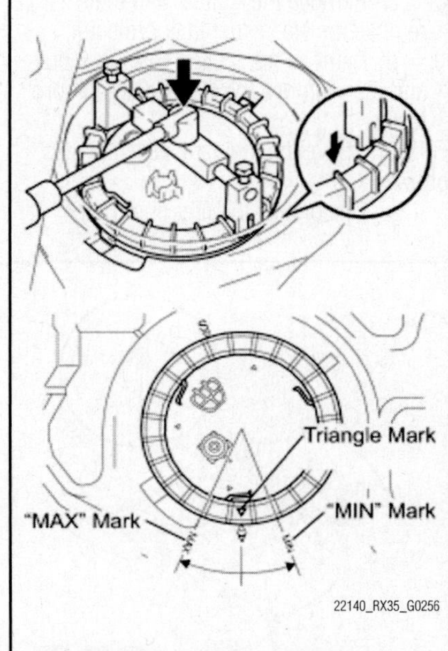

Fig. 196 Properly align the triangle mark

and gauge, and attach the fuel pump gauge retainer.

e. Rotate the fuel pump gauge retainer by hand, then tighten it one complete turn and another half turn using SST (SST: 09808-14020). The triangle mark on the fuel pump gauge retainer must be positioned between the "MIN" and "MAX" marks on the fuel tank.

➡Do not use other tools in this operation. Damage to the fuel pump gauge retainer and the fuel tank may result.

➡A rib on the fuel pump gauge retainer can be fitted into a tip of the SST.

f. Connect the fuel pump tube by installing the fuel pump tube and the tube joint clip.

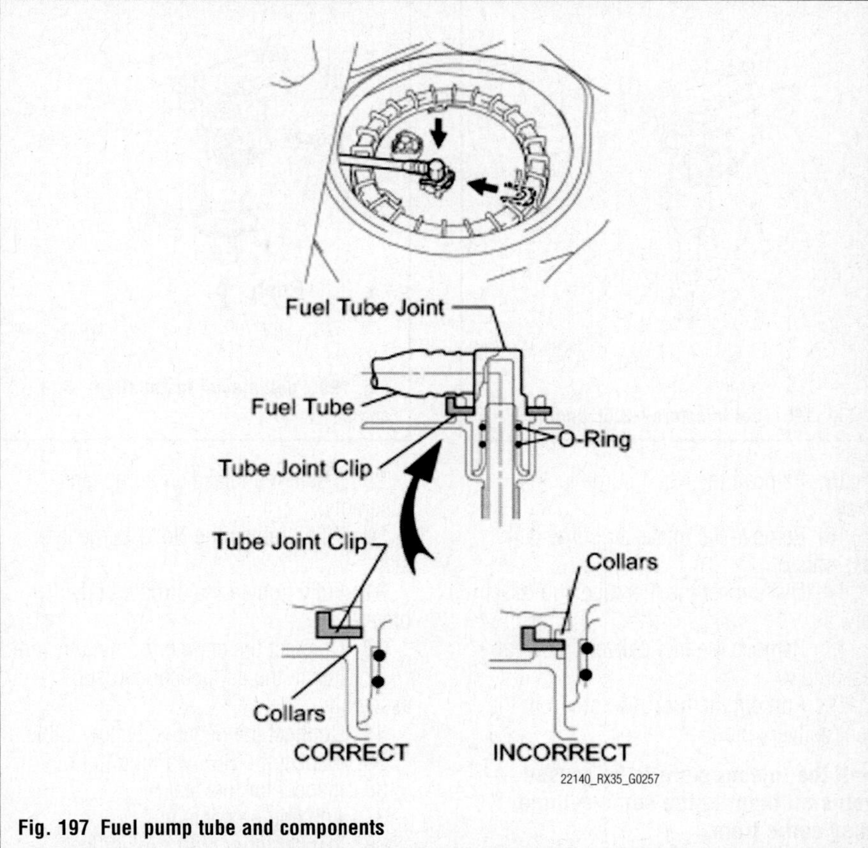

Fig. 197 Fuel pump tube and components

➡Check that there are no scratch or foreign objects on the connecting part.

➡Check that the fuel tube joint is inserted securely.

➡Check that the tube joint clip is on the collar of the fuel tube joint.

➡After installing the tube joint clip, check that the fuel tube joint has not been pulled off.

10. Inspect for fuel leak.
11. Install the rear floor service hole cover, as follows:
    a. Install a new butyl tape to the rear floor service hole cover.
    b. Connect the fuel pump connector.
    c. Install the rear floor service hole cover.
    d. Install the front floor carpet with the 3 clips.
12. Install the left rear seat side cover.
13. Install the left deck side trim cover.
14. Install the left rear door scuff plate.
15. Install the right and left rear seat assemblies.
16. Install the deck board sub-assembly.

FUEL RAIL AND INJECTOR

REMOVAL & INSTALLATION

See Figures 198 and 199.

1. Before servicing the vehicle, refer to the precautions section.
2. Relieve the fuel system pressure.
3. Disconnect the negative battery cable.
4. Remove the engine under cover assembly.
5. Remove the No. 1 engine under cover.
6. Drain the coolant.
7. Remove the front wiper arm and blade assemblies.
8. Remove the cowl top ventilator louver sub-assembly.
9. Remove the wiper motor and link assembly.
10. Remove the outer cowl top panel sub-assembly.
11. Remove the V-bank cover sub-assembly.
12. Remove the air cleaner cap sub-assembly.
13. Disconnect the engine room main wire.
14. Remove the throttle body bracket.

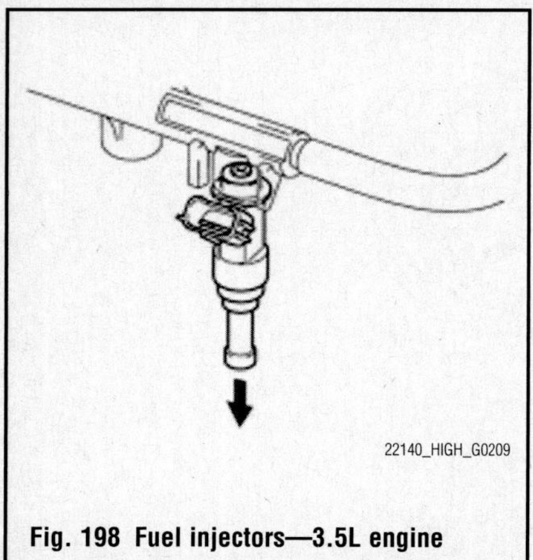

Fig. 198 Fuel injectors—3.5L engine

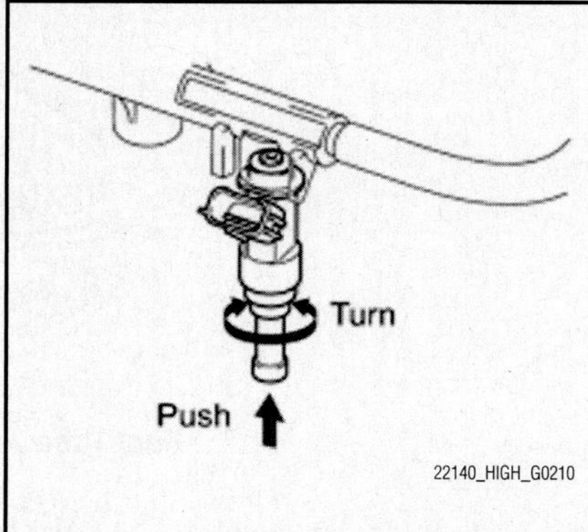

Fig. 199 Fuel injector installation—3.5L engine

15. Remove the No. 1 surge tank stay.
16. Remove the intake air surge tank assembly.
17. Disconnect the fuel tube sub-assembly.
18. Remove the fuel delivery pipe sub-assembly.
19. Pull out the fuel injectors from the fuel delivery pipe.

➡If the injectors are to be reused, reinstall them to the same cylinder they came from.

20. Remove the 6 O-rings from the injectors.

To install:

21. Apply a light coat of spindle oil or gasoline to new O-rings, and install them to each injector.

➡The wound or the foreign body must not adhere in the ditch of O-ring.

22. Apply a light coat of spindle oil or gasoline where the fuel delivery pipe contacts the O-ring.
23. Push the fuel injector while turning it to install the injector in the fuel delivery pipe.
24. Position the fuel injector connector outward.

➡Be careful not to twist the O-ring.

➡After installing the fuel injector, check that it turns smoothly. If not, reinstall it with a new O-ring.

25. Install the fuel delivery pipe sub-assembly.
26. Connect the fuel tube sub-assembly.
27. Temporarily install the No. 1 surge tank stay.
28. Temporarily install the throttle body bracket.
29. Install the intake air surge tank assembly.
30. Fully tighten the No. 1 surge tank stay.
31. Fully tighten the throttle body bracket.
32. Connect the engine room main wire.
33. Install the air cleaner cap sub-assembly.
34. Connect the negative battery cable.
35. Inspect the SRS warning light.
36. Inspect for fuel leak.
37. Add engine coolant.
38. Inspect for engine coolant leak.
39. Install the V-bank cover sub-assembly.
40. Install the outer cowl top panel sub-assembly.
41. Install the windshield wiper motor and link assembly.
42. Install the cowl top ventilator louver sub-assembly.
43. Install both front wiper arm and blade assemblies.
44. Install the No. 1 engine under cover.
45. Install the engine under cover assembly.

FUEL TANK

REMOVAL & INSTALLATION

See Figures 200 through 214.

1. Discharge fuel system pressure.
2. Remove the deck board sub-assembly.
3. Remove the right and left rear seat assemblies.
4. Remove the left rear door scuff plate.
5. Remove the left deck side trim cover.
6. Remove the left rear seat side cover.
7. Remove the rear floor service hole cover.
8. Remove the fuel suction tube assembly with pump and gauge, as follows:
9. Drain fuel.
10. Remove the No. 2 engine under cover, as applicable.
11. For 4WD vehicles, remove the propeller with center bearing shaft assembly.
12. Remove the exhaust center pipe assembly.
13. Remove the No. 3 front floor heat insulator.
14. For 4WD vehicles, remove the No. 4 exhaust pipe support bracket, as follows:
    a. Remove the 2 bolts, and then remove the No. 4 exhaust pipe support bracket.
    b. Remove the 2 clips and unfasten the claw, and then remove the No. 1 fuel tube protector.
    c. Remove the 4 nuts, and then remove the No. 1 fuel tank protector.
    d. Remove the 3 clips (A) and 6 nuts, and then remove the No. 1 fuel tank protector.
15. Remove the fuel tank assembly, as follows:
    a. Disconnect the fuel pump tube:

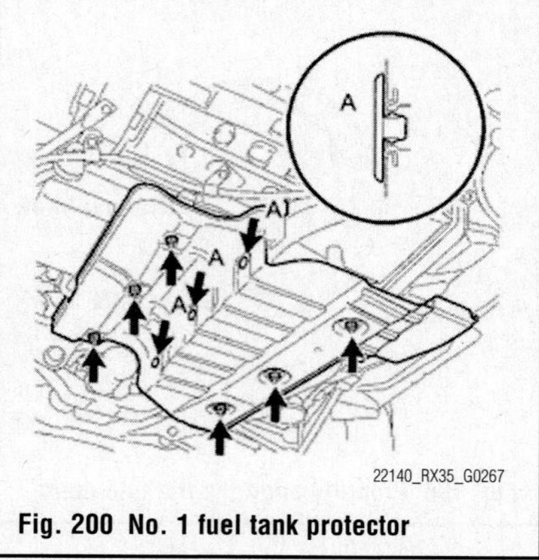

Fig. 200 No. 1 fuel tank protector

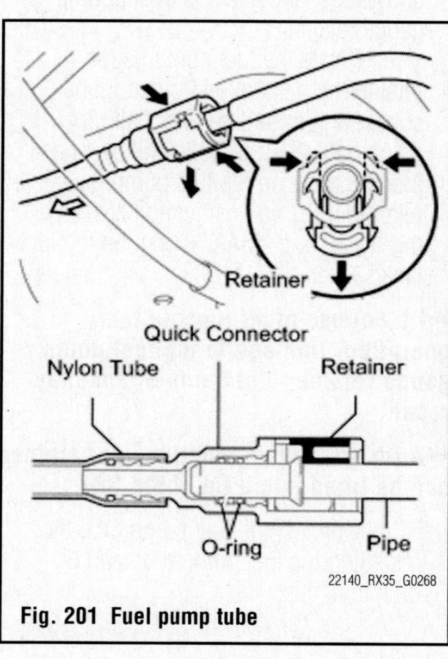

Fig. 201 Fuel pump tube

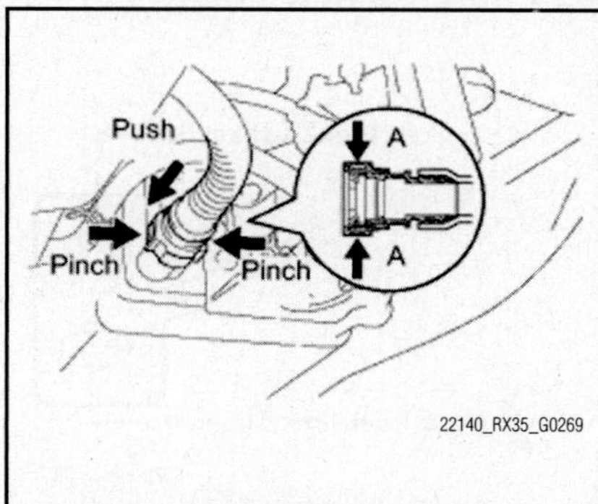

Fig. 202 Disconnect the fuel tank vent hose

- Pinch the projection of the retainer to remove the lock claws and pull it down as shown in the illustration
- Pull out the fuel pump tube

➡**Check if there is any dirt or mud around the connector before this operation and remove the dirt if necessary.**

➡**Be careful of mud because the quick connector can be contaminated.**

➡**Do not use any tool in this operation.**

➡**Do not bend or twist the nylon tube. Protect the connector by covering it with a vinyl or plastic bag.**

➡**When the pipe and connector are stuck, push and pull the connector to release and pull the connector out carefully.**

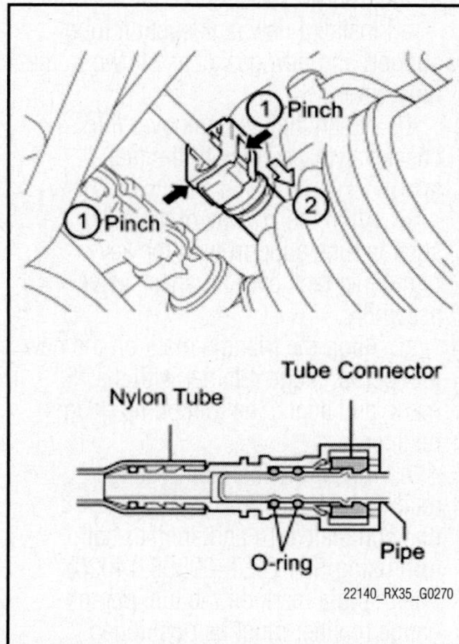

Fig. 203 Disconnect the No. 3 fuel tank breather

b. Disconnect the fuel tank vent hose:

- Deeply push the connector to release the locking tab
- Pinch portion A
- Pull out the connector

c. Disconnect the No. 3 fuel tank breather:

- Pinch the tube connector and then pull out the No. 3 fuel tank breather tube.

➡**Check if there is any dirt or mud around the connector before this operation and remove the dirt as necessary.**

➡**Be careful of mud because the quick connector has an O-ring which seals the pipe and connector that can be contaminated.**

➡**Do not use any tool in this operation.**

➡**Do not bend or twist the nylon tube. Protect the connector by covering it with a vinyl or plastic bag.**

➡**When the pipe and connector are stuck, push and pull the connector to release and pull the connector out carefully.**

d. Loosen the hose clamp bolt and disconnect the fuel tank to filler pipe hose.

e. Set up a transmission jack under the fuel tank.

f. Remove the 4 bolts, and then remove the 2 fuel tank bands.

g. Remove the 2 nuts.

h. Operate the transmission jack to remove the fuel tank.

16. Unfasten the 2 claws, and then remove the fuel pump tube.

17. Loosen the hose clamp bolt, and then remove the fuel tank to filler pipe hose.

18. Remove the fuel tank vent hose, as follows:

a. Completely push the connector to release the locking tab (*1).

b. Pinch portion A (*2).

c. Pull out the connector.

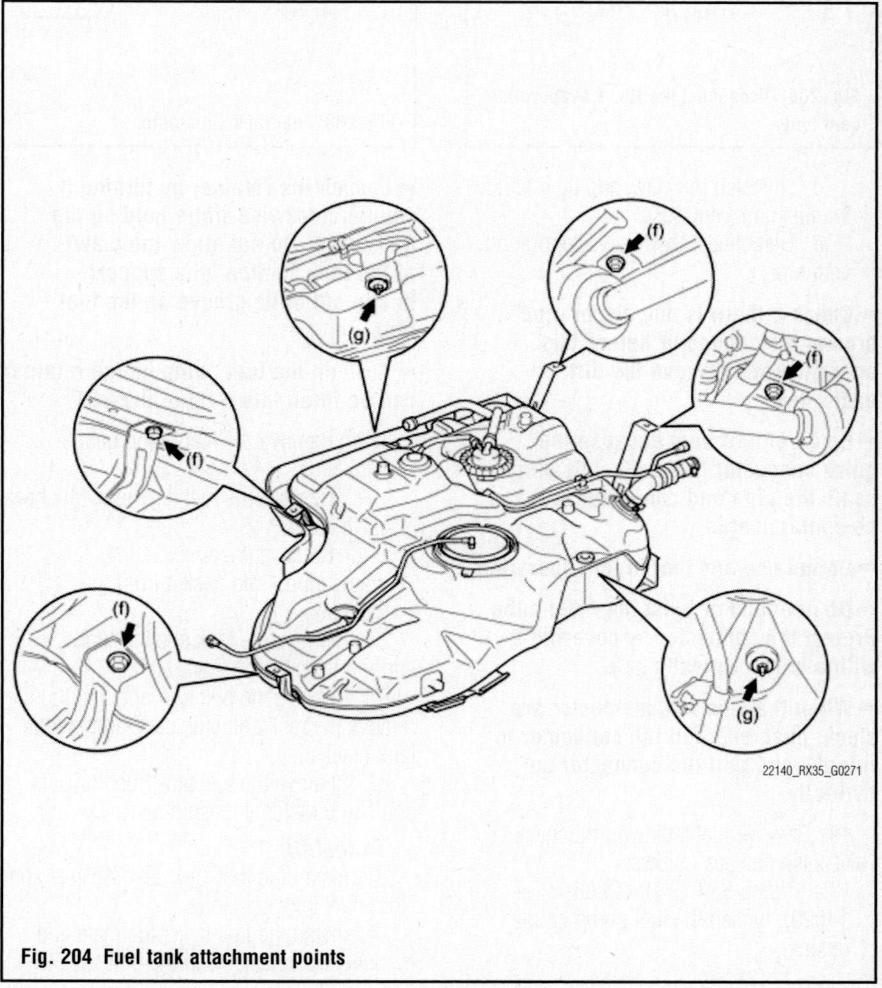

Fig. 204 Fuel tank attachment points

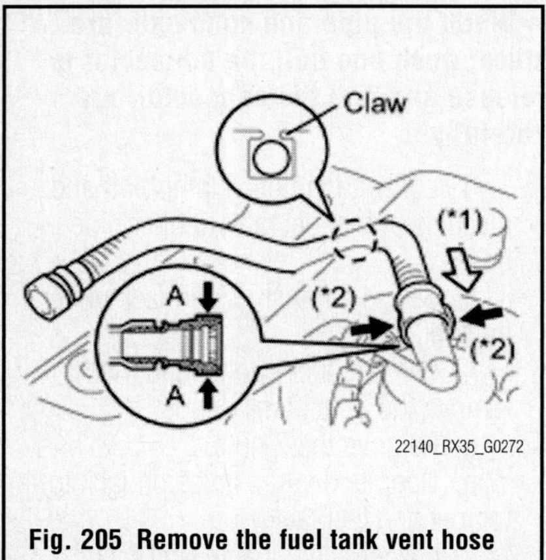

Fig. 205 Remove the fuel tank vent hose

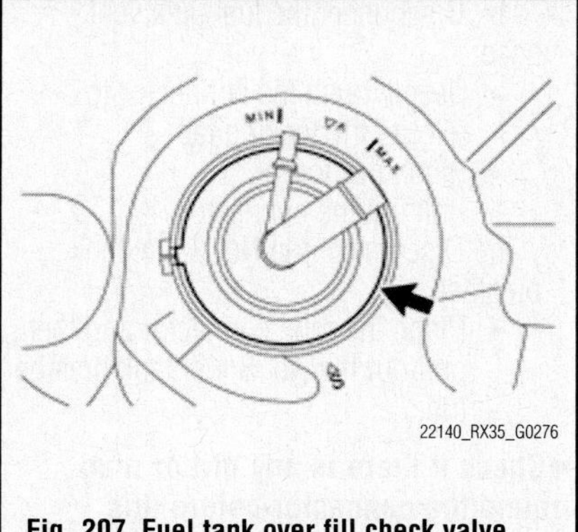

Fig. 207 Fuel tank over fill check valve assembly

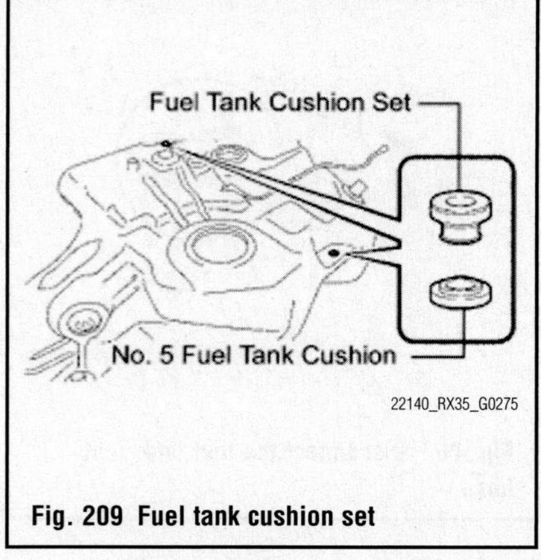

Fig. 209 Fuel tank cushion set

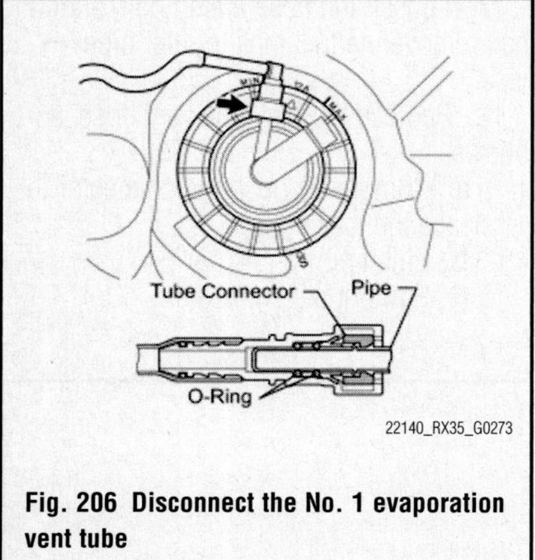

Fig. 206 Disconnect the No. 1 evaporation vent tube

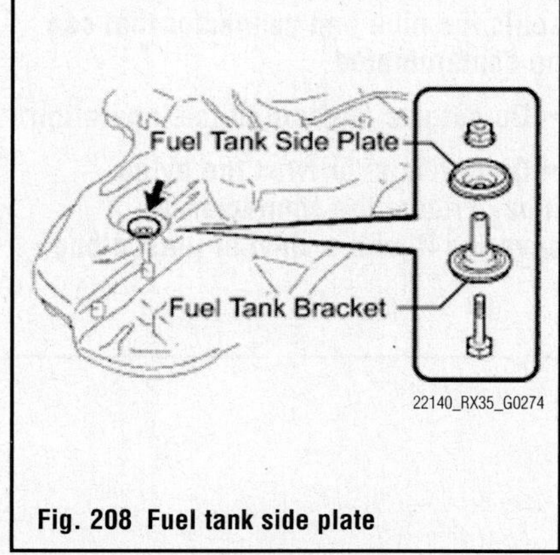

Fig. 208 Fuel tank side plate

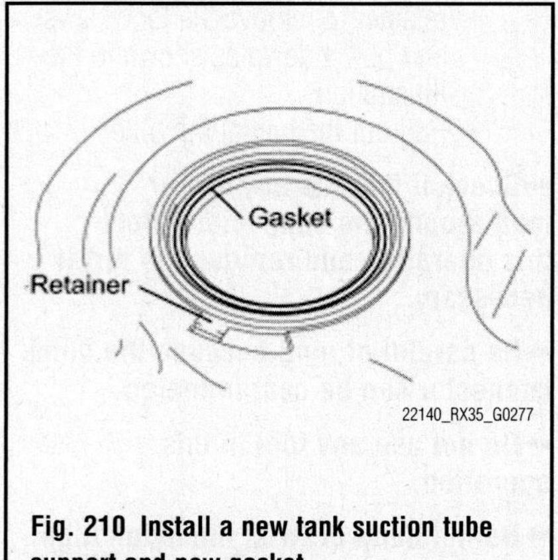

Fig. 210 Install a new tank suction tube support and new gasket

d. Unfasten the claw, and then remove the fuel tank vent hose.

e. Disconnect the No. 1 evaporation vent tube.

➡**Check if there is any dirt or mud around the connector before this operation and remove the dirt as necessary.**

➡**Be careful of mud because the quick connector has an O-ring which seals the pipe and connector that can be contaminated.**

➡**Do not use any tool in this operation.**

➡**Do not bend or twist the nylon tube. Protect the connector by covering it with a vinyl or plastic bag.**

➡**When the pipe and connector are stuck, push and pull the connector to release and pull the connector out carefully.**

19. Remove fuel tank over fill check valve assembly, as follows:

a. Using SST (SST: 09808-14020), loosen the fuel pump gauge retainer.

➡**Loosen the retainer by turning it counterclockwise while holding the SST down. Do not allow the claw of the tank suction tube support to slip out of its groove on the fuel tank.**

➡**A rib on the fuel pump gauge retainer can be fitted into a tip of the SST.**

b. Remove the fuel pump gauge retainer.

c. Remove the fuel tank over fill check valve assembly.

d. Remove the tank suction tube support and gasket from the fuel tank.

20. Remove the tank suction tube support from the fuel tank.

21. Remove the bolt and nut, and then remove the fuel tank side plate and the fuel tank bracket.

22. Remove the fuel tank cushion sets and No. 5 fuel tank cushions.

To install:

23. Install the fuel tank cushion sets and No. 5 fuel tank cushions.

24. Install the fuel tank side plate and the fuel tank bracket with the bolt and nut and tighten to 22 ft. lbs. (30 Nm).

25. Install a new tank suction tube support as shown in the illustration.

26. Install the fuel tank over fill check valve assembly, as follows:

a. Install a new tank suction tube support and new gasket as shown in the illustration.

b. Attach the fuel tank over fill check valve assembly to the fuel tank.

c. Align the keyway of the fuel suction tube support with the key of the fuel tank over fill check valve assembly.

d. Align the triangle mark on the new fuel pump gauge retainer with the "S" mark, and attach the fuel pump gauge retainer.

e. Rotate the fuel pump gauge retainer by hand, then tighten it one complete turn and another half turn using SST (SST: 09808-14020). The triangle mark on the fuel pump gauge retainer must be positioned between the "MIN" and "MAX" marks on the fuel tank.

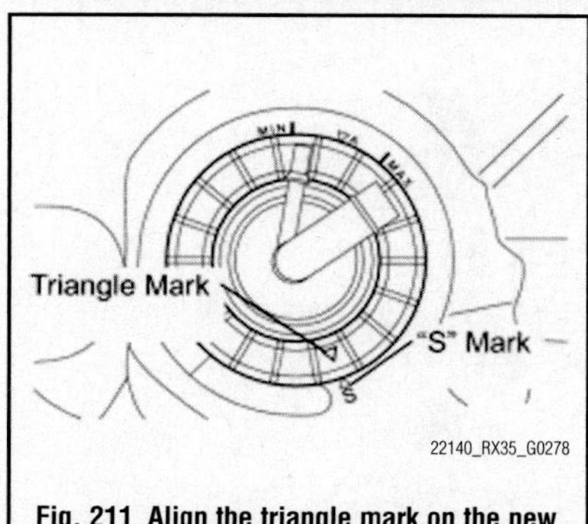

Fig. 211 Align the triangle mark on the new fuel pump gauge retainer with the "S" mark

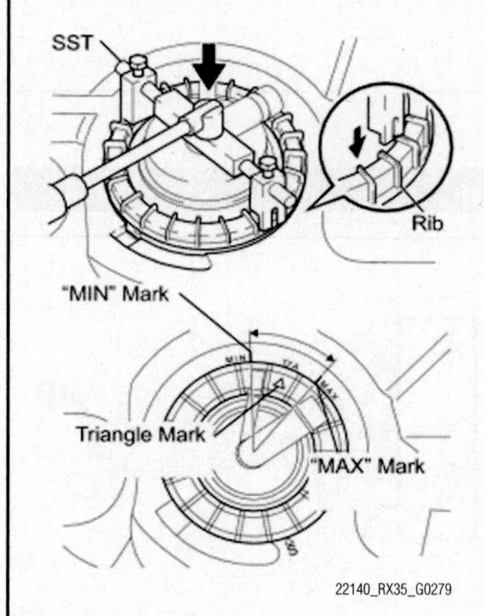

Fig. 212 Positioning fuel pump gauge retainer triangle mark

➡Do not use other tools in this operation. Damage to the fuel pump gauge retainer and the fuel tank may result.

➡A rib on the fuel pump gauge retainer can be fitted into a tip of the SST.

27. Install the fuel tank vent hose, as follows:

a. Connect the No. 1 evaporation vent tube:

- Align the axis of the connector with the axis of the pipe. Push the pipe into the connector until the connector makes a click sound. If the connection is too tight, apply a small amount of fresh engine oil to the tip of the pipe.

➡Check for damage or foreign objects on the parts that are to be connected.

➡After connecting, check that the fuel tube connector and the pipe are securely connected by pulling on them.

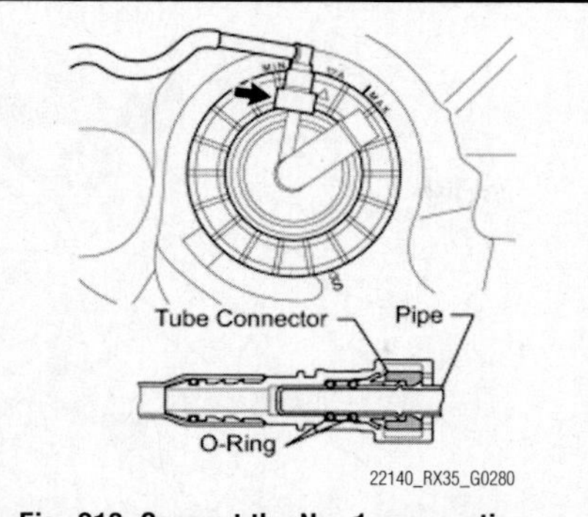

Fig. 213 Connect the No. 1 evaporation vent tube

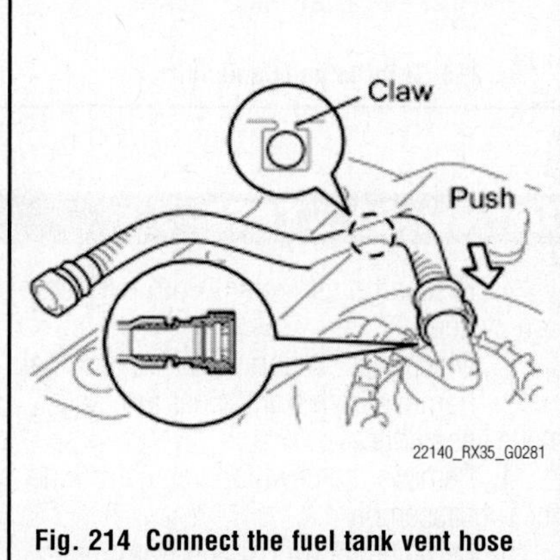

Fig. 214 Connect the fuel tank vent hose

b. Connect the fuel tank vent hose:

- Align the axis of the connector with the axis of the pipe. Push the pipe into the connector until the connector makes a click sound. If the connection is too tight, apply a small amount of fresh engine oil to the tip of the pipe.

➡Check for damage or foreign objects on the parts that are to be connected.

➡After connecting, check that the fuel tube connector and the pipe are securely connected by pulling on them.

- Install the fuel tank vent hose, and fasten the claw.

28. Install the fuel tank to filler pipe hose with the hose clamp.

29. Install the fuel pump tube, and fasten the 2 claws.

30. Install the fuel tank assembly, as follows:

a. Set up the fuel tank to the transmission jack.

b. Operating the transmission jack, install the fuel tank.

c. Tighten the 2 nuts to 14 ft. lbs. (20 Nm).

d. Install the 2 fuel tank bands with the 4 bolts and tighten to 29 ft. lbs. (39 Nm).

e. Connect the fuel tank to filler pipe hose.

f. Connect the No. 3 fuel tank breather tube:

- Push the quick connector to the pipe until it makes a "click" sound.

➡Check if there is any damage or foreign objects on the connected part.

➡After connecting, check if the quick connector and the pipe are securely connected by pulling on them.

g. Connect the fuel tank vent hose.

h. Connect the fuel pump tube:

- Push the quick connector and push up the retainer to lock the claws.

➡Check if there is any damage or foreign objects on the connected part.

➡After connecting, check if the quick connector and the pipe are securely connected by pulling on them.

31. Install the No. 1 fuel tank protector sub-assembly, as follows:

a. Install the No. 1 fuel tank protector with the 6 nuts and tighten to 49 inch lbs. (5.5 Nm).

b. Install 3 new clips. (A).

32. Install the No. 1 fuel tank protector with the 4 nuts and tighten to 49 inch lbs. (5.5 Nm).

33. For 4WD vehicles, install the No. 4 exhaust pipe support bracket with the 2 bolts and tighten to 16 ft. lbs. (22 Nm).

34. The remainder of installation is the reverse of the removal procedure.

35. Add fuel, inspect for fuel leak and exhaust gas leak.

IDLE SPEED

ADJUSTMENT

Idle speed is maintained by the Powertrain Control Module (PCM). No adjustment is necessary or possible.

THROTTLE BODY

REMOVAL & INSTALLATION

See Figure 215.

1. Remove the windshield wiper link assembly.
2. Remove the cowl top panel outer sub-assembly.
3. Drain engine coolant.
4. Remove the V-bank cover sub-assembly.
5. Remove the No. 2 air cleaner inlet.

6. Remove the No. 1 air cleaner inlet.
7. Remove the air cleaner cap sub-assembly.
8. Remove the air cleaner case sub-assembly.
9. Remove the throttle body, as follows:
   a. Disconnect the throttle body connector and clamp.
   b. Disconnect the 2 water by-pass hoses from the throttle Body.
   c. Remove the 4 bolts and throttle body.
   d. Remove the throttle body gasket from the intake air surge tank.

To install:

10. Installation is the reverse of the removal procedure. Torque the throttle body bolts to 7 ft. lbs. (10 Nm).
11. Inspect for coolant leak.

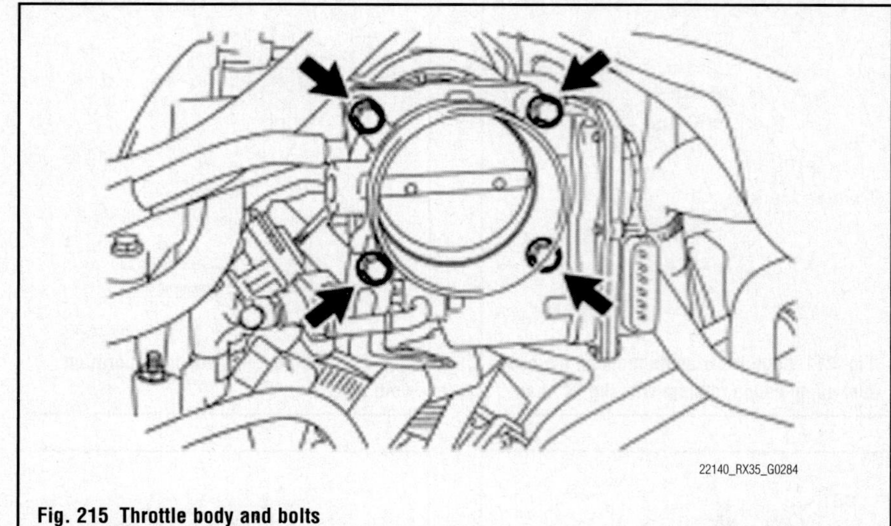

Fig. 215 Throttle body and bolts

HEATING & AIR CONDITIONING SYSTEM

BLOWER MOTOR

REMOVAL & INSTALLATION

See Figure 216.

1. Remove the No. 2 instrument panel under cover sub-assembly.
2. Remove front blower motor sub-assembly, as follows:
   a. Disconnect the connector.
   b. Remove the 3 screws and the front blower motor sub-assembly.

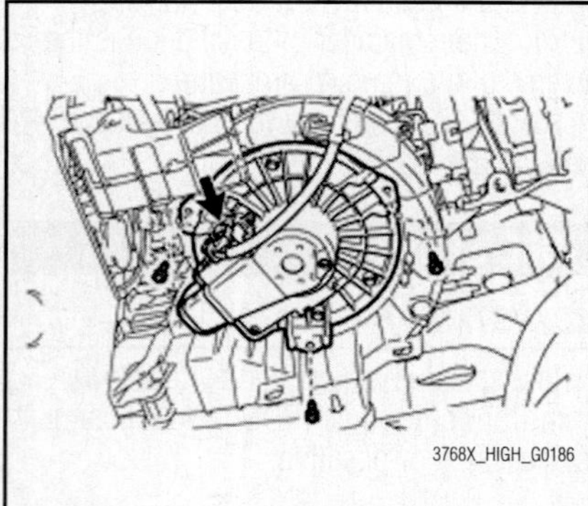

Fig. 216 Remove the 3 screws and the front blower motor sub-assembly

To install:

3. Installation is the reverse of the removal procedure.

HEATER CORE

REMOVAL & INSTALLATION

See Figures 217 through 229.

1. Discharge refrigerant from refrigeration system.
2. Disconnect battery negative terminal.
3. Remove both front wiper arm and blade assemblies.
4. Remove the cowl top ventilator louver sub-assembly.
5. Remove the windshield wiper link assembly.
6. Remove the cowl top panel outer sub-assembly.
7. Disconnect the air conditioning tube and accessory assembly, as follows:
   a. Install SST (SST: 09870-00025) on the piping clamp.

➡Make sure the direction of the piping clamp and SST by seeing the illustration shown on the caution label.

   b. Push down SST and release the clamp lock. Do not deform the tube when pushing the SST.
   c. Pull the SST slightly and push the release lever, and then remove the piping clamp with SST.
   d. Disconnect the air conditioning tube and accessory assembly by hand.

➡Do not use tools like a screwdriver to remove the tube.

➡Cap the open fittings immediately to keep moisture or dirt out of the system.

   e. Remove the 2 O-rings from the air conditioning tube and accessory assembly.

8. Disconnect the No. 1cooler refrigerant suction hose, using SST: 09870-00015, and the same procedure as for removing the air conditioning tube and accessory assembly.

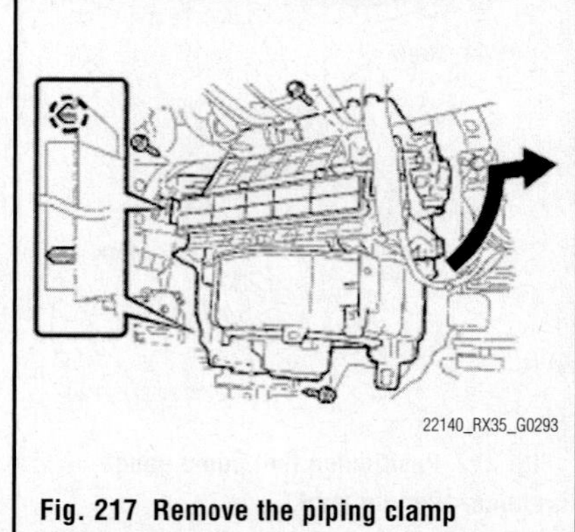

Fig. 217 Remove the piping clamp

9. Using pliers, grip the claws of the clip and slide the clip, and then disconnect the heater water outlet hose. Do not apply any excessive force to the heater water outlet hose.

➡Prepare a drain pan or cloth for cooling water leaks.

10. Using the same procedure as for the heater water outlet hose, disconnect the heater water inlet hose.
11. Remove the instrument panel assembly.
12. Remove the air conditioner amplifier assembly.
13. Turn back the floor carpet, release the claw and remove the rear No. 1 and 2air ducts.
14. Remove the bolt, release the 3 claws and remove the No. 1 air duct.

15. Disconnect the transmission control cable assembly.
16. Remove the shift lever.
17. Remove the 4 nuts and instrument panel bracket No. 4.
18. Remove instrument panel brace No. 1 sub-assembly, as follows:
    a. Remove the clamp and disconnect the connector.
    b. Remove the 3 bolts and 2 nuts and instrument panel brace No. 1 sub-assembly.
19. Remove instrument panel brace No. 2 sub-assembly, as follows:
    a. Remove the clamp and disconnect the connector.
    b. Remove the 3 bolts and 2 nuts and instrument panel brace No. 2 sub-assembly.
20. Separate steering intermediate shaft sub-assembly.
21. Remove steering column assembly.
22. Remove the 3 nuts and air duct No. 1 sub-assembly.
23. Remove instrument panel reinforcement assembly, as follows:
    a. Remove the 4 bolts and 4 nuts.
    b. Remove the clamp.
    c. Disconnect the connectors.
    d. Remove the 9 bolts and the 2 nuts and then remove the instrument panel reinforcement while holding the air conditioner unit assembly.

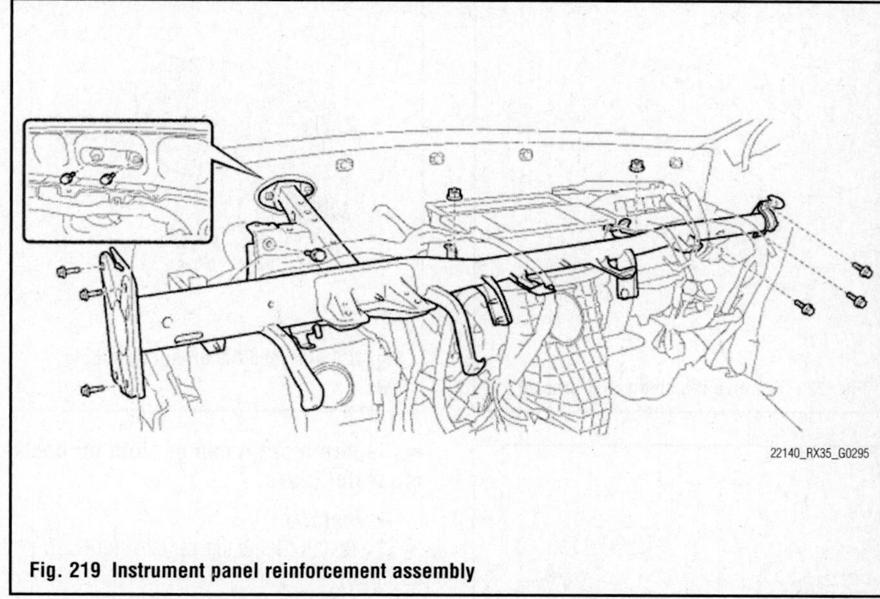

Fig. 219 Instrument panel reinforcement assembly

➡Make sure to hold the air conditioner unit assembly securely as its bracket installation parts may be damaged.

24. Remove the 2 nuts and air conditioner unit assembly.

WARNING

Make sure to hold the air conditioner unit assembly securely as its bracket installation parts may be damaged.

25. Remove the air duct No. 2, as follows:
    a. Remove the screw.
    b. Release the 2 claws and remove the air duct No. 2.
26. Remove the wiring air indicator harness No .2 sub-assembly.
27. Remove the blower assembly.
28. Release the 4 claws and remove the center heater to register duct.
29. Remove the 3 screws and air outlet control servomotor.
30. Remove the air mix control servomotor.
31. Remove the evaporator temperature sensor.
32. Remove the heater radiator unit sub-assembly, as follows:
    a. Remove the 2 screws and piping cover.
    b. Remove the screw and bracket.
    c. Remove the heater radiator unit sub-assembly.

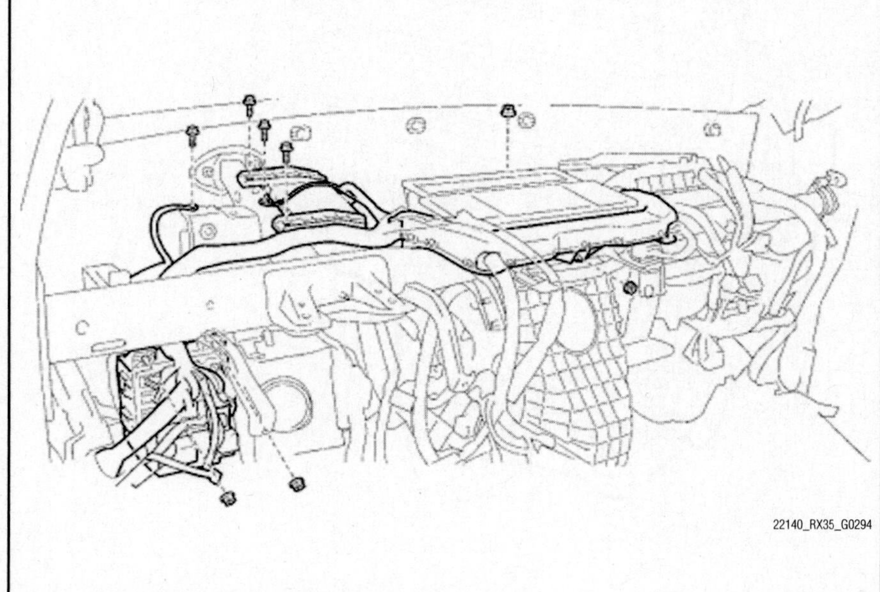

Fig. 218 Instrument panel reinforcement assembly

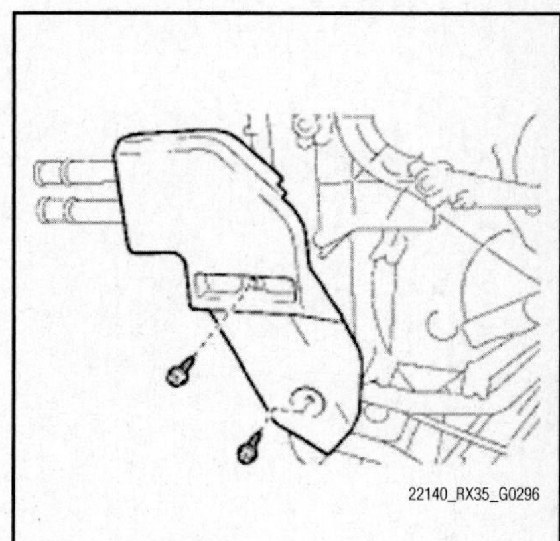

Fig. 220 Remove the 2 screws and piping cover

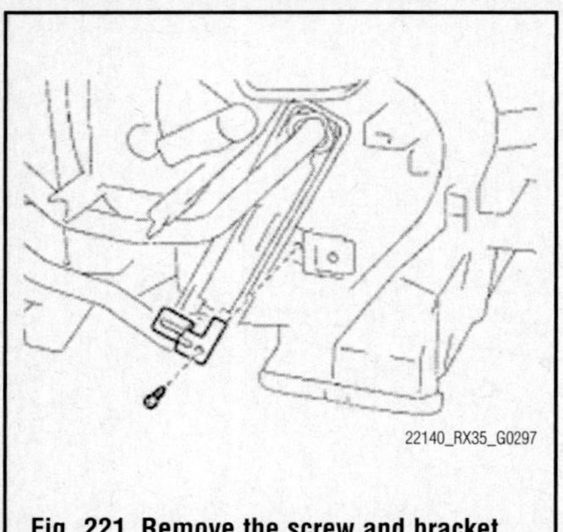

Fig. 221 Remove the screw and bracket

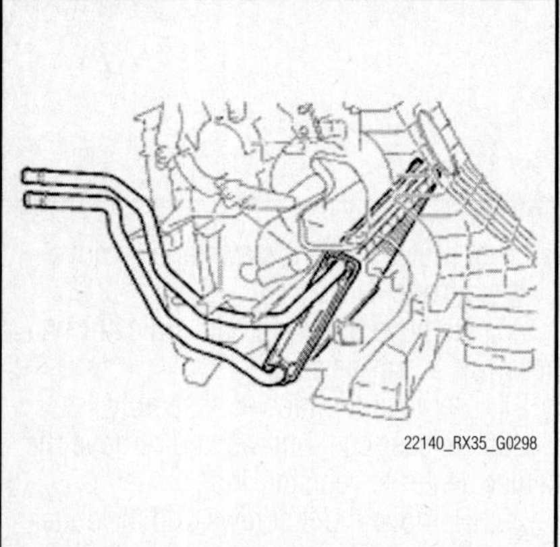

Fig. 222 Remove the heater radiator unit sub-assembly

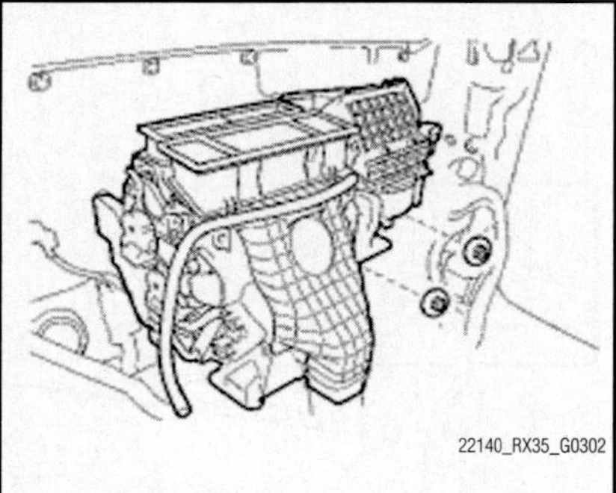

Fig. 223 Air conditioner unit assembly nuts

➡Prepare a drain pan or cloth for cooling water leaks.

To install:

33. Install the heater radiator unit sub-assembly.
34. Install the bracket with the screw.
35. Install the piping cover with the 2 screws.
36. Engage the 2 claws to install the evaporator temperature sensor.
37. Engage the clamp to install the connector.
38. Install the air mix control servomotor with the 3 screws.
39. Install the air outlet control servomotor with the 3 screws.
40. Engage the 4 clamps to install the heater to register duct center.
41. Engage the claw to install the blower assembly.
42. Install the wiring air indicator harness No. 2 sub-assembly with the 3 screws.
43. Connect the connectors and clamp.
44. Install the air duct No. 2 with the screw and tighten to 22 inch lbs. (2.5 Nm).
45. Temporarily tighten the air conditioner unit assembly with the 2 nuts. Hold the A/C unit securely to prevent damaging its bracket installation parts.
46. Install the instrument panel reinforcement assembly, as follows:
    a. Install the instrument panel reinforcement with the 8 bolts and tighten to 15 ft. lbs. (20 Nm).
    b. Temporarily tighten the air conditioner unit assembly with the 2 nuts and bolt.
    c. Install the 4 bolts and 4 nuts.
    d. Install the clamps.
47. Connect the connectors.
48. Install the instrument panel brace No. 2 sub-assembly with the 3 bolts and 2 nuts.
49. Install the clamps and connect the connectors.
50. Install the instrument panel brace No. 1 sub-assembly with the 3 bolts and 2 nuts.
51. Install the clamps and connect the connectors.
52. Install the No. 4 instrument panel bracket.

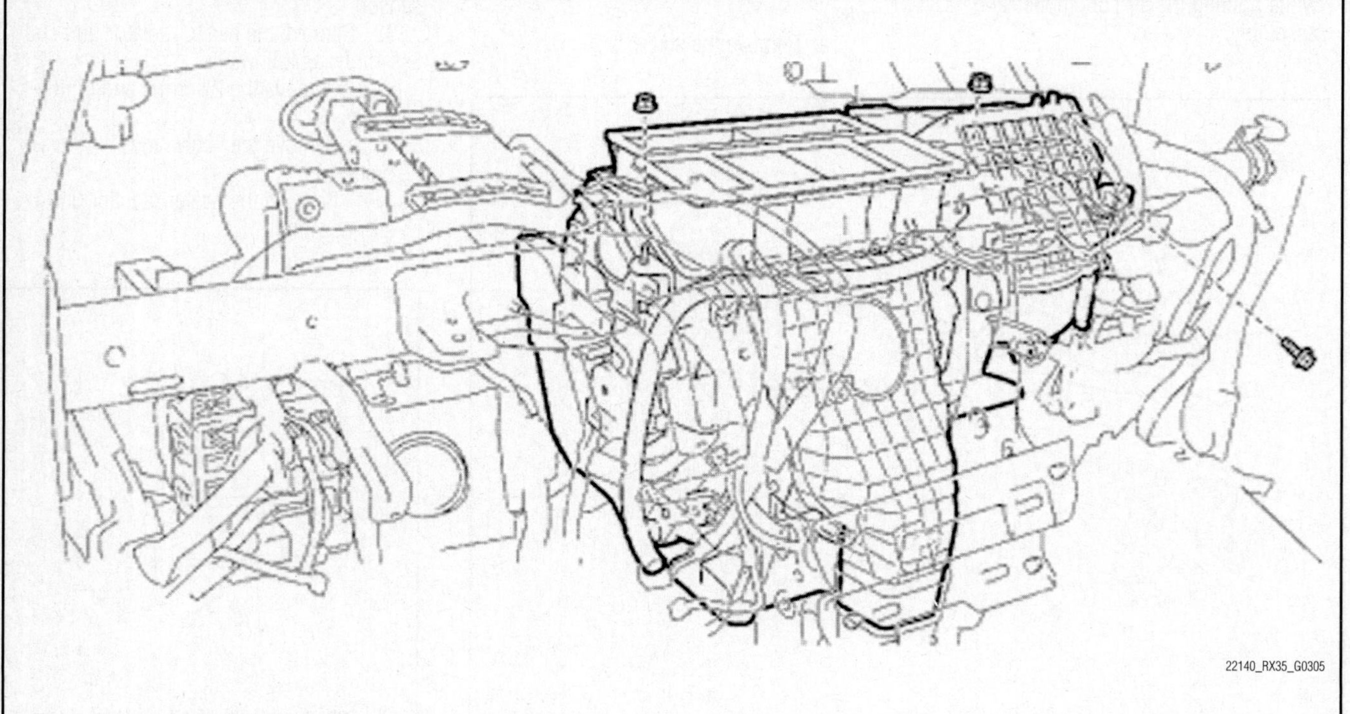

Fig. 224 Temporarily tighten the air conditioner unit assembly

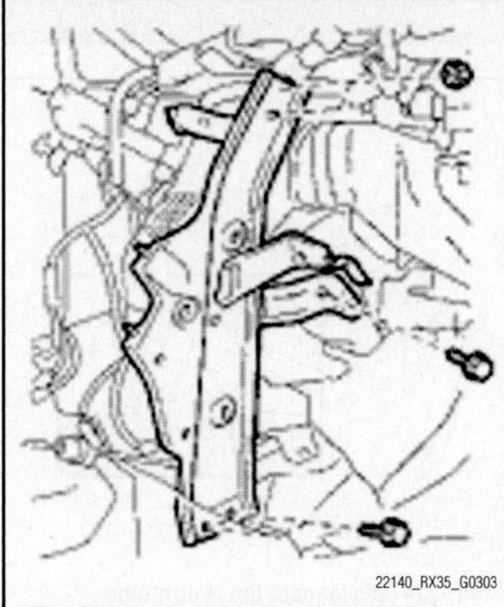
Fig. 225 Instrument panel brace No. 2 sub-assembly

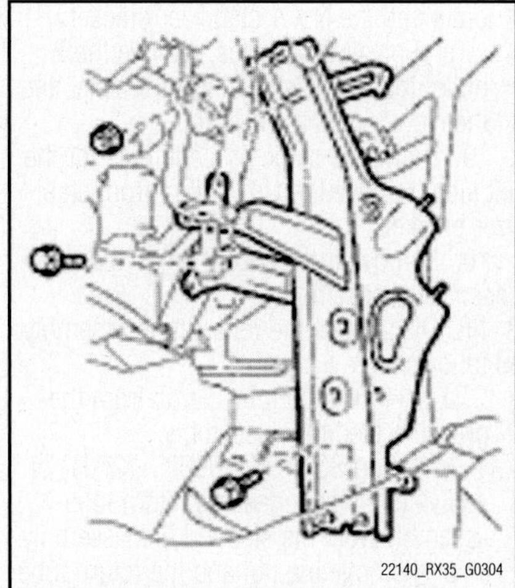
Fig. 226 Instrument panel brace No. 1 sub-assembly

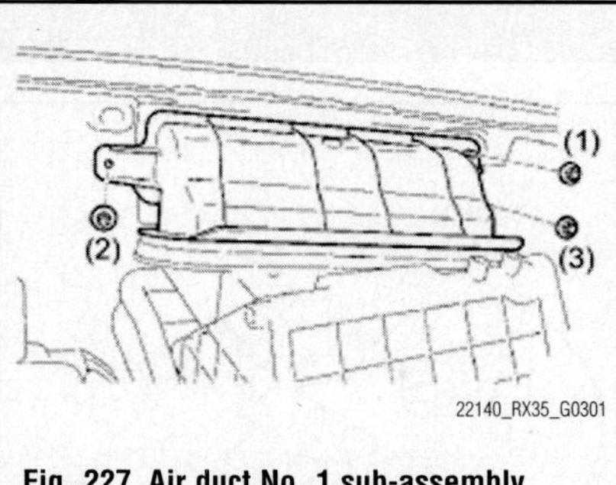

Fig. 227 Air duct No. 1 sub-assembly tightening sequence

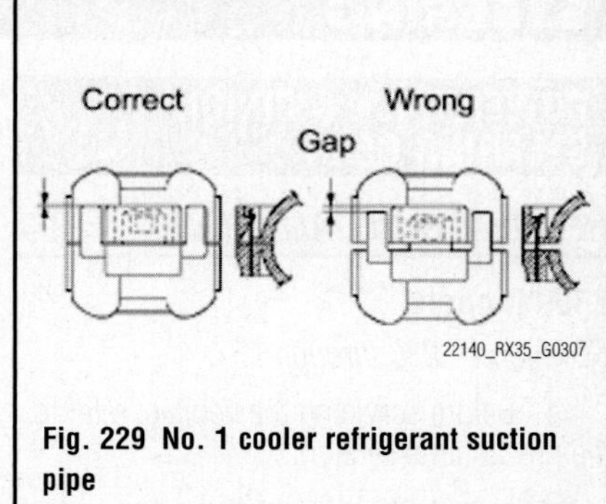

Fig. 229 No. 1 cooler refrigerant suction pipe

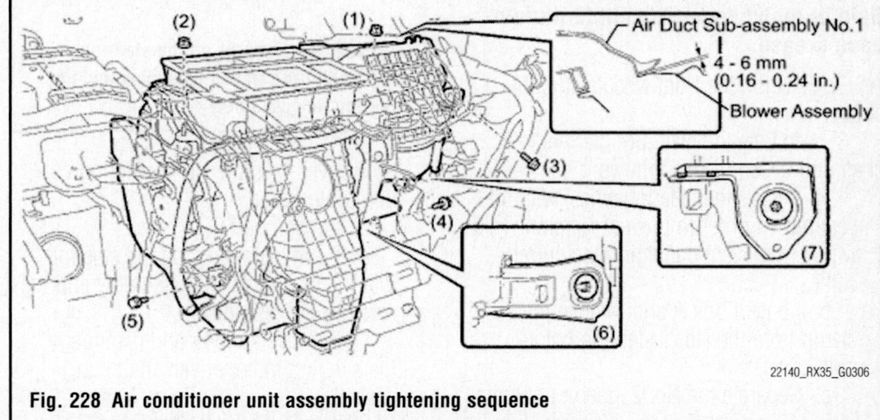

Fig. 228 Air conditioner unit assembly tightening sequence

53. Install the air duct No. 1 sub-assembly with the 3 nuts and tighten to 87 inch lbs. (9.8 Nm).

54. Fully tighten the air conditioner unit assembly with the 3 bolts and 4 nuts to 87 inch lbs. (9.8 Nm) in the order shown. Install the air conditioner unit assembly so that there is no clearance between the duct and blower assemblies.

55. Install the steering column assembly.

56. Install the steering intermediate shaft sub-assembly

57. Install the air duct No. 1 with the bolt and tighten to 87 inch lbs. (9.8 Nm).

58. Install the air conditioner amplifier assembly.

59. Install the shift lever.

60. Connect the transmission control cable assembly.

61. Inspect shift lever position.

62. Install the rear No. 1 air duct.

63. Install the rear No. 2 air duct.

64. Install the instrument panel assembly.

65. Install the heater water inlet and outlet hoses.

66. Install the No. 1 cooler refrigerant suction pipe, as follows:

a. Remove the attached vinyl tape from the hose.

b. Coat a new O-ring with compressor oil and install it to the hose. Compressor oil: ND-OIL 11 or equivalent.

WARNING

Do not use any compressor oil other than ND-OIL 11 or equivalent. If any compressor oil other than ND-OIL 11 or equivalent is used, compressor motor insulation performance may decrease, resulting in leakage of electric power.

c. Install the No. 1 cooler refrigerant suction pipe and piping clamp. Be sure to connect the hose securely, and check the fitting for the claw of the piping clamp.

67. Install the air conditioning tube and accessory assembly, as follows:

a. Remove the attached vinyl tape from the hose.

b. Coat a new O-ring with compressor oil and install it to the pipe. Compressor oil: ND-OIL 11 or equivalent.

WARNING

Do not use any compressor oil other than ND-OIL 11 or equivalent. If any compressor oil other than ND-OIL 11 or equivalent is used, compressor motor insulation performance may decrease, resulting in leakage of electric power.

c. Install the air conditioning tube and accessory assembly and piping clamp. Be sure to connect the hose securely, and check the fitting for the claw of the piping clamp.

68. Install cowl top panel outer sub-assembly.

69. Install windshield wiper link assembly.

70. Install cowl top ventilator louver sub-assembly.

71. Install both front wiper arm and blade assemblies.

72. Add engine coolant.

73. Connect cable to negative battery terminal.

74. Charge refrigerant.

75. Warm up compressor.

76. Check for engine coolant leaks.

77. Check for refrigerant leaks.

78. Perform initialization.

POWER RACK & PINION STEERING GEAR

REMOVAL & INSTALLATION

2009 Models

See Figures 230 through 237.

1. Before servicing the vehicle, refer to the precautions section.

➡When installing, coat the parts indicated by arrows with power steering fluid or molybdenum disulfide lithium base grease.

2. Position the front wheels straight ahead.
3. Separate the steering intermediate shaft sub-assembly, as follows:
   a. Secure the steering wheel with the seat belt in order to prevent rotation. This will help prevent damaging the spiral cable.
   b. Loosen bolt A and remove the clamp from the No. 1 steering column hole cover.
   c. Separate the No. 2 steering column hole cover from the No. 1 steering column hole cover.
   d. Loosen bolt B.
   e. Put matchmarks on the steering intermediate shaft sub-assembly and the steering link assembly.
   f. Remove the bolt and disengage the steering intermediate shaft sub-assembly.
4. Remove the front wheel.
5. Separate the right and left tie rod assemblies.
6. Separate the right and left front stabilizer link assemblies.
7. Remove the front exhaust pipe assembly.
8. Remove the No. 1 left front stabilizer bracket:
   a. For 2 WD vehicles, remove the 2 bolts and the No. 1 stabilizer bracket.
   b. For 4WD vehicles, remove the 2 bolts, the No. 1 left stabilizer bracket and the No. 2 stabilizer bracket.
9. Using the same procedure as for the left side, remove the No. 1 right front stabilizer bracket.
10. Remove the left front height control sensor sub-assembly.
11. Disconnect the return tube assembly, as follows:
    a. Remove the tube clamp from the pressure feed tube assembly.
    b. Using SST (SST: 09023-12701) or equivalent, disconnect the return tube assembly from the steering link assembly.
    c. Remove the nut and the return tube clamp.
12. Disconnect the pressure feed tube assembly, as follows:
    a. Using SST (SST: 09023-12701) or equivalent, disconnect the pressure feed tube assembly from the steering link assembly.

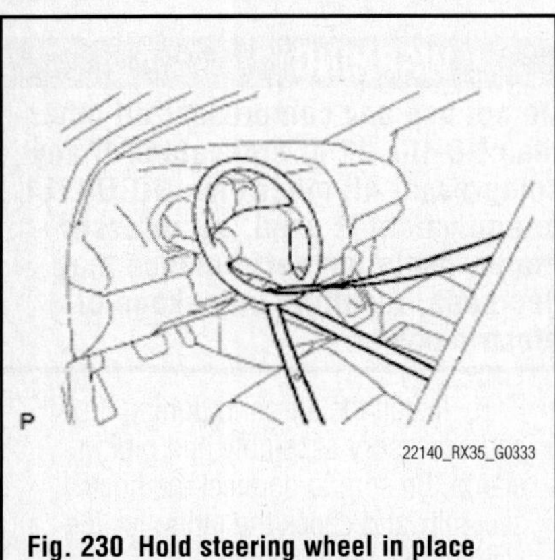

Fig. 230 Hold steering wheel in place

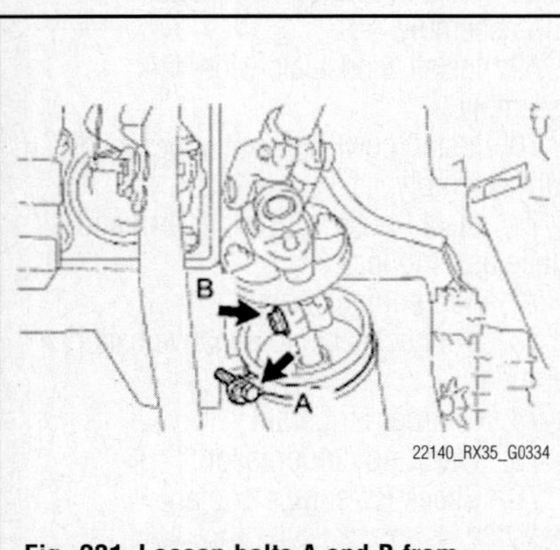

Fig. 231 Loosen bolts A and B from steering column hole covers

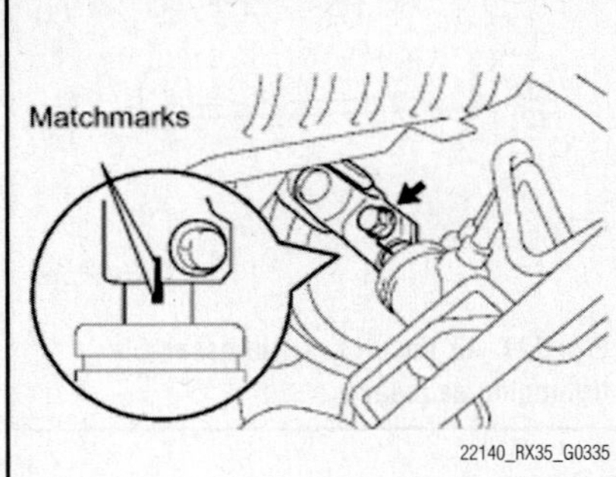

Fig. 232 Matchmarks on the steering intermediate shaft sub-assembly and the steering link assembly

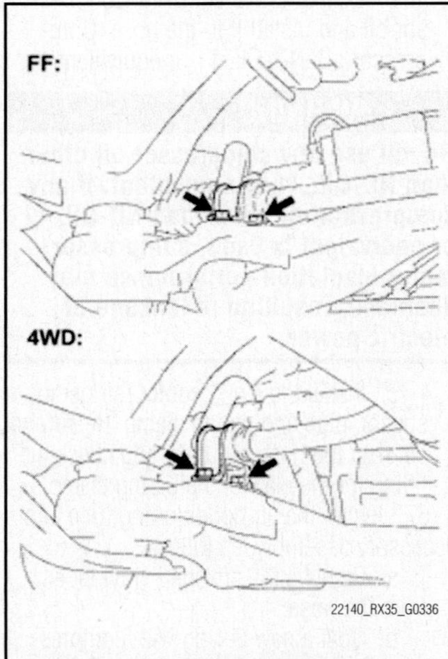

Fig. 233 Front No. 1 stabilizer bracket

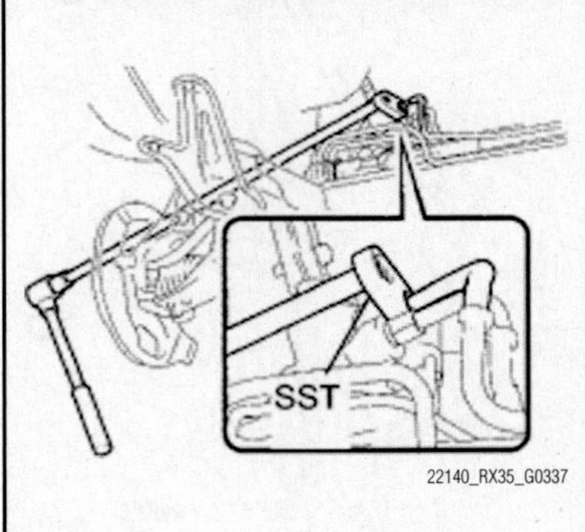

Fig. 234 Disconnect the return tube assembly

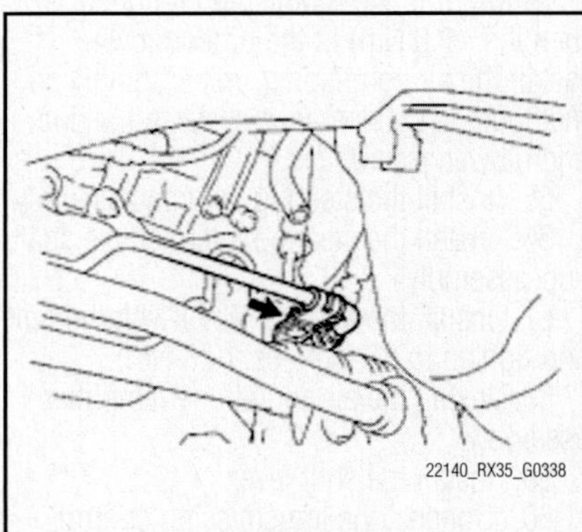

Fig. 235 Remove the nut and the return tube clamp

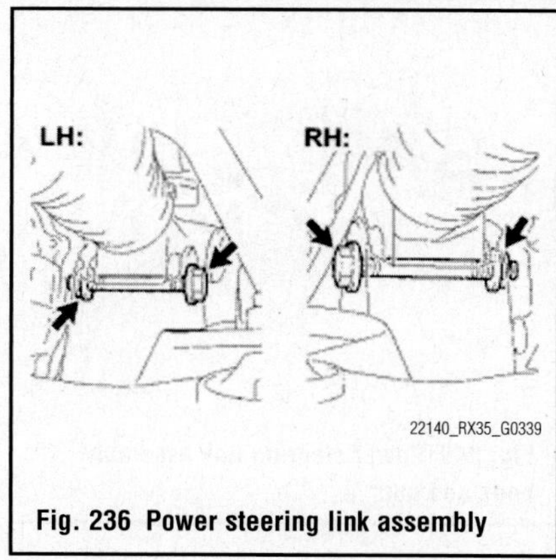

Fig. 236 Power steering link assembly

b. Remove the bolt and the pressure feed tube clamp.

13. Remove the 2 bolts, the nuts and the steering link assembly.

To install:

14. Install the power steering link assembly with the 2 bolts and the nuts and tighten to 52 ft. lbs. (70 Nm).

15. Using SST (SST: 09023-12701) or equivalent, connect the pressure feed tube assembly to the steering link assembly and tighten to 16 ft. lbs. (22 Nm). Use a torque wrench with a fulcrum length of 300 mm (11.81 in.). This torque value is effective when SST is parallel to a torque wrench.

16. Install the pressure feed tube assembly clamp with the bolt and tighten to 87 inch lbs. (9.8 Nm).

17. Using SST, connect the return tube assembly to the steering link assembly and tighten to 16 ft. lbs. (22 Nm). Use a torque wrench with a fulcrum length of 300 mm (11.81 in.). This torque value is effective when SST is parallel to a torque wrench.

18. Install the tube clamp to the pressure feed tube assembly.

19. Install the pressure feed tube clamp with the nut and tighten to 87 inch lbs. (9.8 Nm).

20. Install the left front height control sensor sub-assembly.

21. Install the No. 1 left front stabilizer bracket:

a. For 2WD vehicles, install the No. 1 left stabilizer bracket with the 2 bolts and tighten to 12 ft. lbs. (16 Nm).

b. For 4WD vehicles, install the No. 1 left stabilizer bracket and the No. 2 stabilizer bracket with the 2 bolts and tighten to 12 ft. lbs. (16 Nm).

22. Using the same procedure as for the left side, install the No. 1 right front stabilizer bracket.

23. Install the front exhaust pipe assembly.

24. Install the right and left front stabilizer link assemblies.

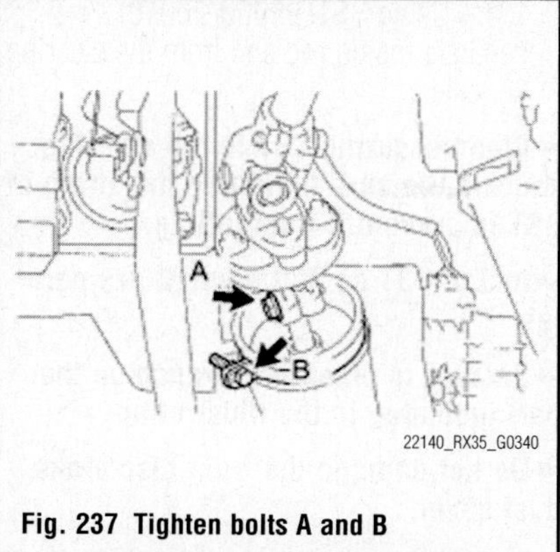

Fig. 237 Tighten bolts A and B

25. Install the right and left tie rod assemblies.

26. Install the front wheel.

27. Connect the steering intermediate shaft sub-assembly, as follows:

a. Align the matchmarks on the intermediate shaft sub-assembly and the steering link assembly.

b. Install the bolt and tighten to 26 ft. lbs. (35 Nm).

c. Tighten the bolt A to 26 ft. lbs. (35 Nm).

d. Install the steering column hole cover No. 2 to the No. 1 steering hole cover.

e. Install the clamp to the No. 1 steering column hole cover and tighten the bolt B.

28. Bleed power steering fluid.

29. Inspect for power steering fluid leak.

30. Inspect for gas leak.

31. Inspect the steering wheel center point.

2010 Models

See Figures 238 through 240.

1. Before servicing the vehicle, refer to the precautions section.

2. Discharge fuel system pressure.

3. Recover refrigerant from refrigeration system.

4. Remove the cool air intake duct seal.

5. Remove the battery.

6. Place the front wheels facing straight ahead.

7. Secure the steering wheel with the seat belt in order to prevent it from rotating.

WARNING

This operation is necessary to prevent damage to the spiral cable.

8. Remove the front wheels.

9. Remove the engine under cover assembly.

10. Remove the No. 1 and 2 engine under covers.

11. Remove the left hand floor under cover.

12. Remove both front fender molding sub-assemblies.

13. Remove both front fender liners.

14. Remove both front fender apron seals.

15. Drain engine oil.

16. Drain engine coolant.

17. Drain automatic transaxle fluid.

18. Remove both front wiper arm and blade assemblies.

19. Remove the cowl top ventilator louver sub-assembly.

20. Remove the windshield wiper motor and link assembly.

21. Remove the outer cowl top panel sub-assembly.

22. Remove the V-bank cover sub-assembly.

23. Remove the No. 1 and 2 air cleaner inlets.

24. Remove the air cleaner cap sub-assembly.

25. Remove the air cleaner filter element sub-assembly.

26. Remove the air cleaner case sub-assembly.

27. Remove the air cleaner bracket.

28. Separate the brake master cylinder reservoir assembly.

29. Remove the reservoir bracket.

30. Remove the right No. 2 engine mounting stay.

31. Remove the engine moving control rod.

32. Disconnect the No. 1 fuel vapor feed hose.

33. Disconnect the No. 1 and 2 radiator hose.

34. Disconnect the heater water outlet hose B.

35. Disconnect the heater water inlet hose B.

36. Disconnect the fuel tube sub-assembly.

37. Disconnect the oil cooler inlet and outlet hoses.

38. Disconnect the transmission control cable assembly.

39. Disconnect engine wire.

40. Disconnect the union to check valve hose.

41. For 4WD vehicles, remove the propeller with center bearing shaft assembly.

42. Remove the tail exhaust pipe assembly.

43. Remove the center exhaust pipe assembly.

44. Remove the front No. 3 exhaust pipe sub-assembly.

45. Remove the front exhaust pipe assembly.

46. Separate both front stabilizer link assemblies.
47. Remove both front axle hub nuts.
48. Disconnect both front speed sensors.
49. Separate the steering intermediate shaft assembly, as follows:
a. Remove the bolt and slide the steering intermediate shaft assembly.

➡Do not separate the steering intermediate shaft assembly from the power steering link assembly.

b. Put matchmarks on the steering intermediate shaft assembly and the power steering link assembly.
c. Separate the steering intermediate shaft assembly from the power steering link assembly.
50. Separate both tie rod assemblies:
a. Remove the cotter pin and the nut.
b. Install SST: 09960-20010 or equivalent to the tie rod end.

➡Make sure that the upper ends of the tie rod end and SST are aligned.

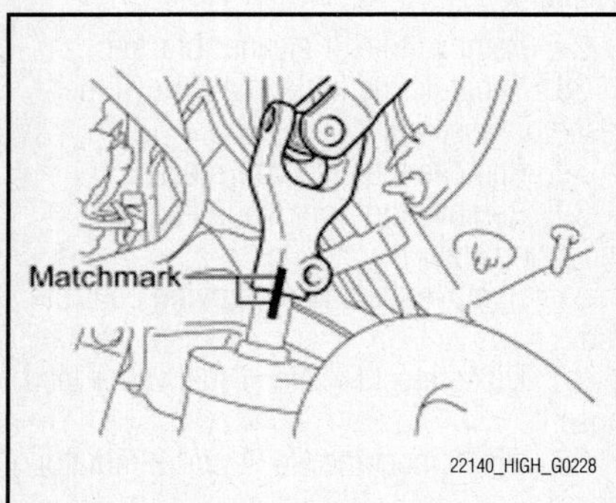

Fig. 238 Matchmarks on the steering intermediate shaft assembly and power steering link assembly

c. Using SST: 09960-20010, separate the tie rod end from the steering knuckle.

➡When securing SST to the steering knuckle, be sure to tighten the string of SST to prevent it from falling.

➡Install SST so that A and B are parallel.

➡Be sure to place the wrench on the part indicated in the illustration.

➡Do not damage the front disc brake dust cover.

➡Do not damage the ball joint dust cover.

➡Do not damage the steering knuckle.

51. Separate the No. 1 and 2 front suspension lower arms.
52. Separate both front axle assemblies.
53. Disconnect the discharge hose subassembly.
54. Disconnect the suction hose subassembly.
55. Remove the engine assembly with transaxle.
56. Remove both front No. 1 stabilizer brackets.
57. Remove the front stabilizer bar.
58. Remove the 2 bolts, 2 nuts, and power steering link assembly.

➡Because the nut has its own stopper, do not turn the nut. Loosen the bolt with the nut fixed.

59. Put matchmarks on the tie rod assemblies and the steering rack end subassemblies.
60. Loosen the lock nuts, and remove the tie rod assemblies and the lock nut.

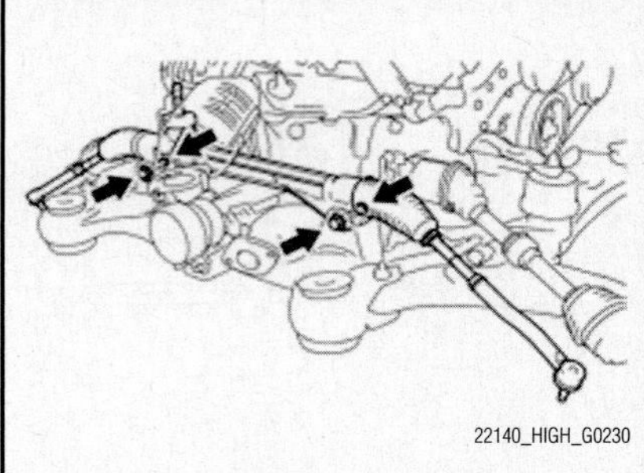

Fig. 240 Power steering link assembly bolts and nuts

To install:

61. Install the lock nuts and the tie rod assemblies to the steering rack end subassembly until the matchmarks are aligned.

➡After adjusting toe-in, torque the lock nut.

62. Install the power steering link assembly with the 2 bolts and 2 nuts and tighten to 51 ft. lbs. (70 Nm).

➡Make sure to tighten the bolts starting from the left side of the vehicle.

➡Because the nut has its own stopper, do not turn the nut. Tighten the bolt with the nut fixed.

63. Install the front stabilizer bar.
64. Install both front No. 1 stabilizer brackets.
65. Install the engine assembly with transaxle.
66. Reconnect the suction hose subassembly.
67. Reconnect the discharge hose subassembly.
68. Install both front axle assemblies.

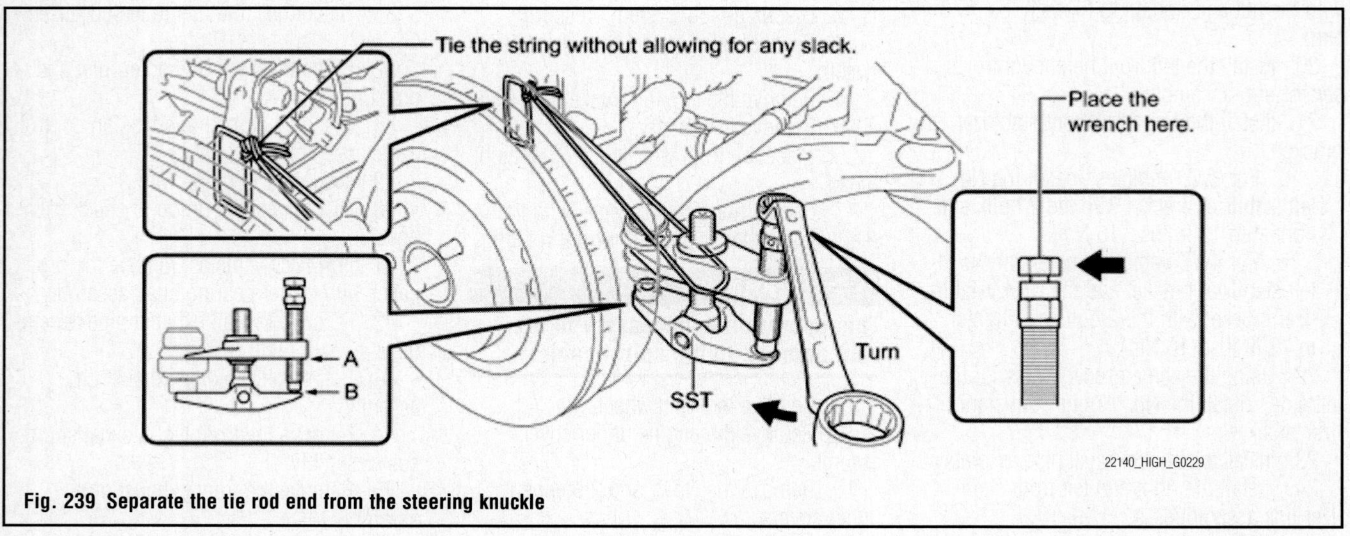

Fig. 239 Separate the tie rod end from the steering knuckle

69. Install both No. 1 front suspension lower arms.

70. Connect both tie rod assemblies and tighten to 36 ft. lbs. (49 Nm).

71. Install a new cotter pin.

➡Further tighten the nut up to 60° if the holes for the cotter pin are not aligned.

72. Align the matchmarks on the steering intermediate shaft assembly and the power steering link assembly.

73. Install the bolt and tighten to 26 ft. lbs. (35 Nm).

74. Install both front speed sensors.

75. Install both front axle hub nuts.

76. Install both front stabilizer link assemblies.

77. Install the front exhaust pipe assembly.

78. Install the front No. 3 exhaust pipe sub-assembly.

79. Install the center exhaust pipe assembly.

80. Install the tail exhaust pipe assembly.

81. For 4WD vehicles, temporarily tighten the propeller with center bearing shaft assembly.

82. For 4WD vehicles, fully tighten the propeller with center bearing shaft assembly.

83. Connect engine wire.

84. Connect the transmission control cable assembly.

85. Connect the fuel tube sub-assembly.

86. Connect the oil cooler inlet and outlet hoses.

87. Connect the heater water inlet hose B.

88. Connect the heater water outlet hose B.

89. Install the No. 1 and 2 radiator hoses.

90. Connect the union to check valve hose.

91. Connect the No. 1 fuel vapor feed hose.

92. Install the engine moving control rod.

93. Install the right No. 2 engine mounting stay.

94. Install the reservoir bracket.

95. Install the brake master cylinder reservoir assembly.

96. Install the air cleaner bracket.

97. Install battery.

98. Install the air cleaner case sub-assembly.

99. Install the air cleaner filter element sub-assembly.

100. Install the air cleaner cap sub-assembly.

101. Install the No. 1 and 2 air cleaner inlets.

102. Connect vacuum hoses.

103. Install the outer cowl top panel sub-assembly.

104. Install the windshield wiper motor and link assembly.

105. Install the cowl top ventilator louver sub-assembly.

106. Install both front wiper arm and blade assemblies.

107. Install front wheels. Tighten lug nuts to 76 ft. lbs. (103 Nm).

108. Add engine oil.

109. Add engine coolant.

110. Add automatic transaxle fluid.

111. Check automatic transaxle fluid.

112. Inspect for fuel leak.

113. Inspect for engine oil leak.

114. Inspect for coolant leak.

115. Inspect for exhaust gas leak.

116. Check shift lever position.

117. Place front wheels facing straight ahead.

118. Inspect and adjust front wheel alignment.

119. Check ignition timing.

120. Check engine idle speed.

121. Check CO/HC.

122. Check function of throttle body assembly.

123. Install both front fender apron seals.

124. Install both front fender liners.

125. Install both front fender molding sub-assemblies.

126. Install the left hand floor under cover.

127. Install the No. 1 and 2 engine under cover.

128. Install the engine under cover assembly.

129. Install the V-bank cover sub-assembly.

130. Check the ABS speed sensor signal.

131. Reset memory.

POWER STEERING PUMP

REMOVAL & INSTALLATION

See Figures 241 through 248.

1. Remove the right front wheel.
2. Remove the engine under cover assembly.
3. Remove the No. 2 engine under cover.
4. Drain power steering fluid.
5. Remove fan and alternator V-belt.
6. Slide the clip and disconnect the No. 1 fluid reservoir to pump hose from the vane pump (power steering pump) assembly.
7. Disconnect the pressure feed tube assembly, as follows:

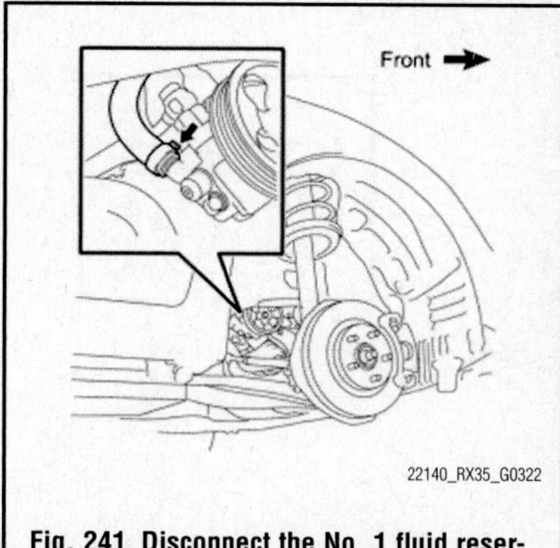

Fig. 241 Disconnect the No. 1 fluid reservoir to pump hose

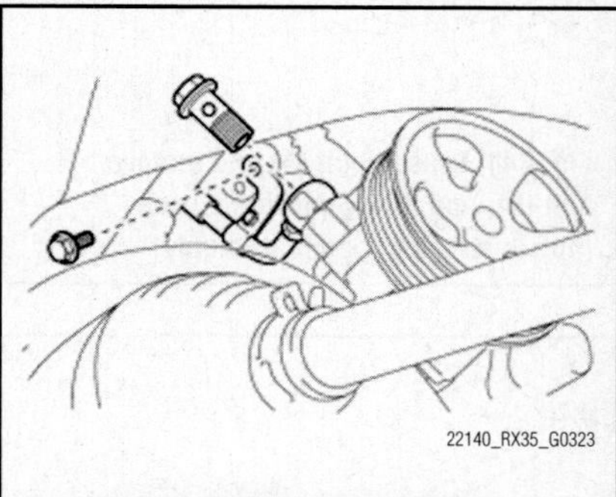

Fig. 242 Remove the pressure feed tube assembly bolts

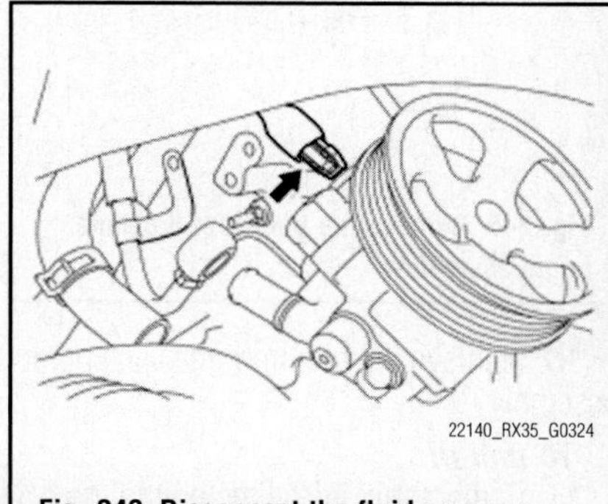

Fig. 243 Disconnect the fluid pressure switch connector

a. Remove the union bolt and disconnect the pressure feed tube assembly from the vane pump assembly.

b. Remove the bolt and separate the pressure feed tube clamp.

c. Remove the gasket from the pressure feed tube assembly.

8. Disconnect the fluid pressure switch connector.
9. Using SST (SST: 09249-63010), loosen bolt (A) and remove bolt (B), and then remove the vane pump (power steering pump) assembly.

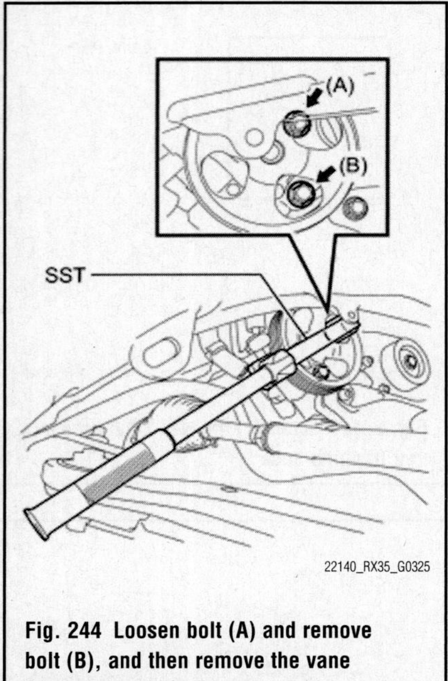

Fig. 244 Loosen bolt (A) and remove bolt (B), and then remove the vane (power steering) pump assembly

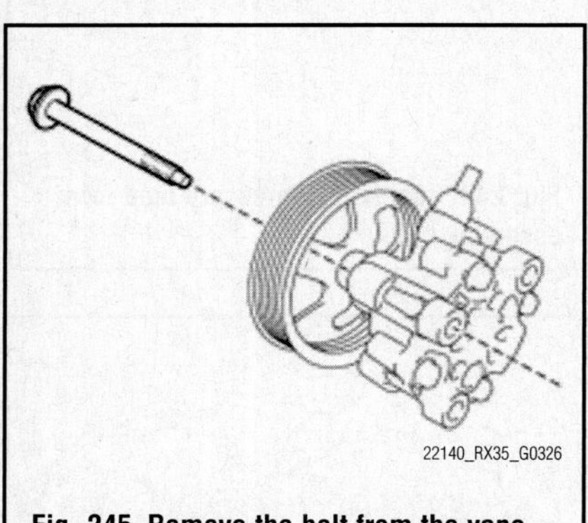

Fig. 245 Remove the bolt from the vane pump assembly

10. Remove the bolt from the vane pump assembly.

To install:

11. Temporarily install the bolt to the vane pump assembly.
12. Install the vane pump assembly.
13. Using SST (SST: 09249-63010), install bolt (B) and tighten the 2 bolts to 32 ft. lbs. (43 Nm), and 21 ft. lbs. (29 Nm). This torque value is accurate when SST is parallel to the torque wrench.

➡Use a torque wrench with a fulcrum length of 300 mm (11.81 in.).

14. Connect the connector to the power steering fluid pressure switch.
15. Install a new gasket to the pressure feed tube assembly.
16. Temporarily connect the pressure feed tube assembly to the vane pump assembly with the union bolt.

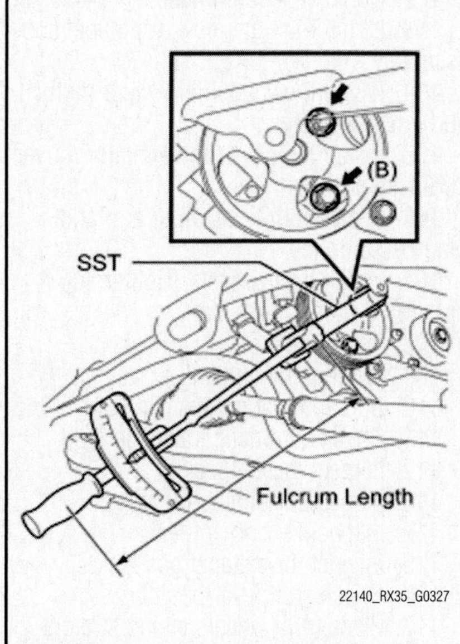

Fig. 246 Install bolt (B) and tighten the 2 bolts

17. Install the pressure feed tube assembly clamp with the bolt and tighten to 69 inch lbs. (7.8 Nm). Install the pressure feed tube assembly clamp in the correct position.
18. Fully tighten the union bolt to 37 ft. lbs. (50 Nm). Make sure that the stopper of the pressure feed tube assembly contacts the vane pump assembly securely as shown in the illustration.
19. Connect the No. 1 fluid reservoir to pump hose to the vane pump assembly with the clip.
20. After installing the No. 1 fluid reservoir to pump hose with its paint mark and the claw of the clip facing the right of the vehicle as shown in the illustration, turn the hose counterclockwise 90 degrees so that the paint mark and claw face the front of the vehicle.

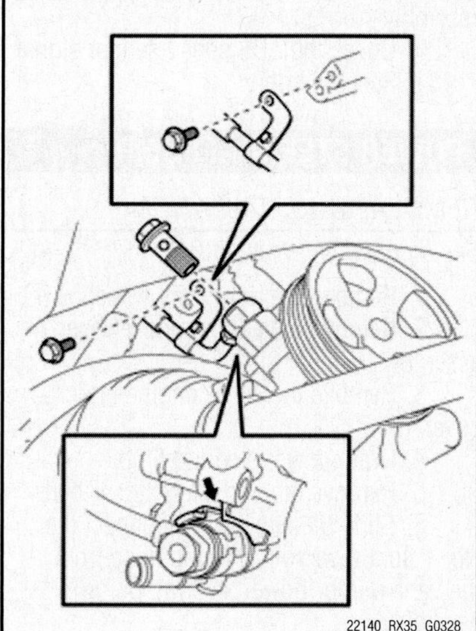

Fig. 247 Install pressure feed tube bolts

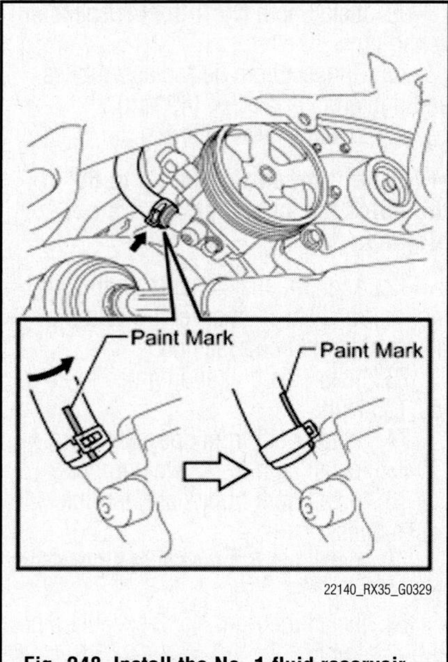

Fig. 248 Install the No. 1 fluid reservoir to pump hose

21. Install the No. 1 fluid reservoir to pump hose pushed all the way on as shown in the illustration.
22. Install fan and alternator V-belt.
23. Add power steering fluid.
24. Bleed power steering fluid.
25. Check power steering fluid level and inspect for leak.
26. Install the No. 2 engine under cover.
27. Install the engine under cover assembly.
28. Install the right front wheel.

BLEEDING

See Figure 249.

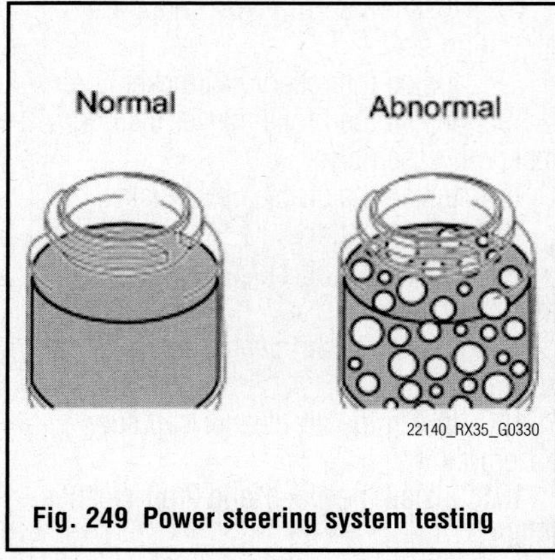

Fig. 249 Power steering system testing

1. Check the fluid level.
2. Jack up the front of the vehicle and support it with stands.
3. Turn the steering wheel:
   a. With the engine stopped, turn the steering wheel slowly from lock to lock several times.
4. Lower the vehicle.
5. Start the engine:
   a. Run the engine at idle for a few minutes.
6. Turn the steering wheel:
   a. With the engine idling, turn the steering wheel left or right to the full lock position and keep it in that position for 2 to 3 seconds, then turn the steering wheel to the opposite full lock position and keep it there for 2 to 3 seconds.
   b. Repeat this procedure several times.
7. Stop the engine.
8. Check for foaming or emulsification.
9. Be sure to check for fluid leaks in the system especially if the system has to be bled twice because of forming or emulsification.
10. Check the fluid level.

SUSPENSION

FRONT SUSPENSION

CONTROL LINKS

REMOVAL & INSTALLATION

See Figure 250.

➡Perform the same procedure on each side.

1. Remove wheel and tire assembly.
2. For 3.5L engines, remove the nut and separate the front stabilizer link assembly.

➡If the ball joint turns together with the nut, use a hexagon wrench (6 mm) to hold the stud bolt.

To install:

3. Install the front stabilizer link assembly with the nut and tighten to 55 ft. lbs. (74 Nm).
4. Install the wheel and tire assembly.

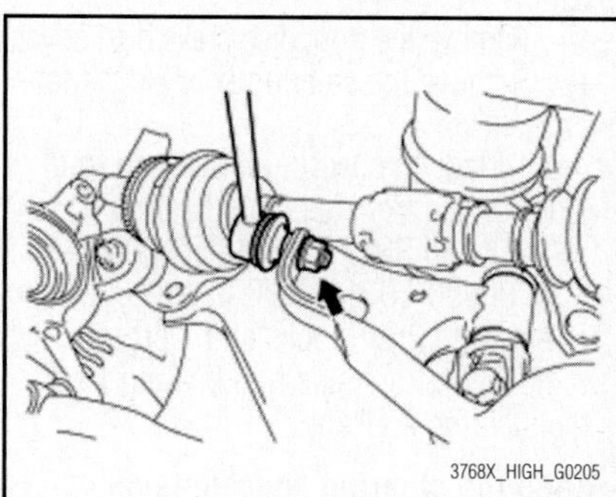

Fig. 250 Remove the nut and separate the front stabilizer link assembly

LOWER BALL JOINT

REMOVAL & INSTALLATION

See Figures 251 through 254.

1. Remove the front wheel.
2. Remove the front axle hub nut.
3. Remove the bolt and resin clamp, and separate the front speed sensor.

➡Be sure to completely separate the front speed sensor from the front shock absorber with coil spring.

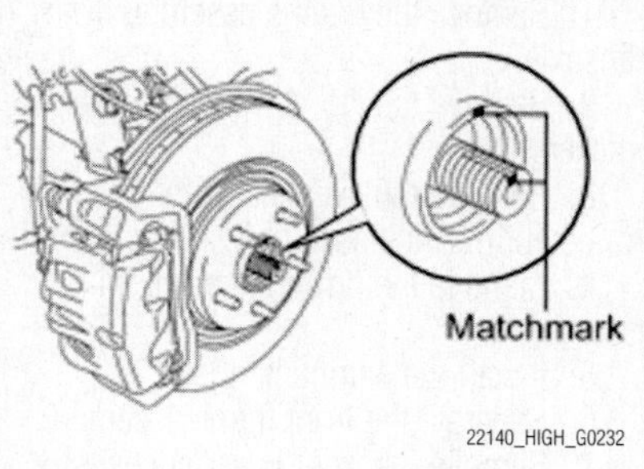

Matchmark

Fig. 251 Matchmark the front driveshaft assembly and the front axle hub sub-assembly

➡Clean the installation hole and the surface for the speed sensor every time the speed sensor is removed.

➡Be careful not to damage the front speed sensor.

4. Put matchmarks on the front driveshaft assembly and the front axle hub sub-assembly.
5. Using a plastic hammer, separate the front driveshaft assembly from the front axle assembly.

➡Loosen the staked part of the front axle hub nut completely, otherwise the threads of the driveshaft may be damaged.

6. Remove the 2 bolts and separate the front disc brake caliper assembly.

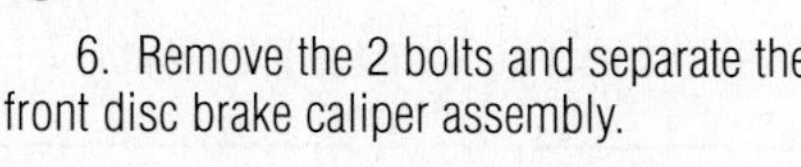

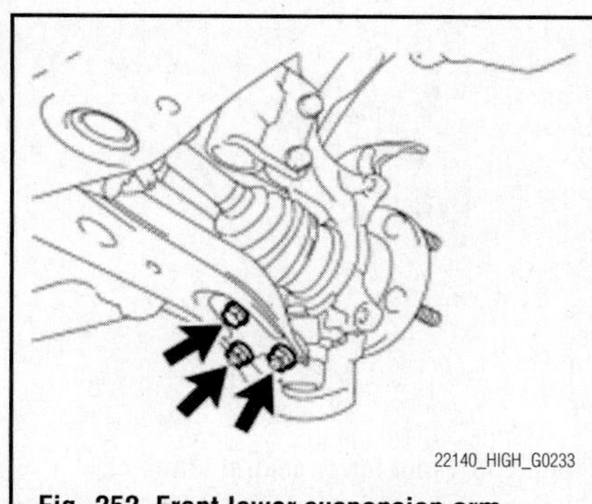

Fig. 252 Front lower suspension arm

➡Use wire or an equivalent tool to keep the brake caliper from hanging down by the flexible hose.

7. Remove the front disc.
8. Separate the tie rod assembly.
9. Remove the bolt, 2 nuts, and separate the front lower suspension arm from the lower ball joint.
10. Remove the 2 bolts, 2 nuts and front axle assembly.
11. Remove the front lower ball joint, as follows:
12. Secure the front axle assembly in a vise using aluminum plates.

➡When using a vise, do not overtighten it.

13. Remove the cotter pin and nut.
14. Install SST: 09960-20010 or equivalent to the front lower ball joint.
15. Using SST: 09960-20010 or equivalent, remove the front lower ball joint from the front axle assembly.

➡Install SST so that A and B are parallel.

➡Be sure to place a wrench on the part indicated in the illustration.

➡Do not damage the front lower ball joint dust cover.

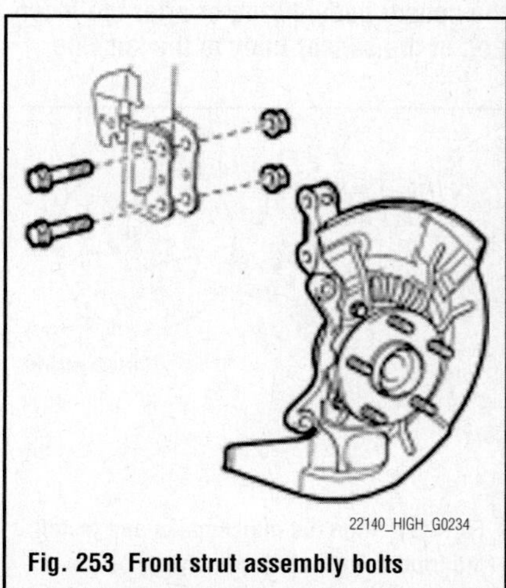

Fig. 253 Front strut assembly bolts

To install:

16. Install the front lower ball joint to the steering knuckle with the nut and tighten to 91 ft. lbs. (123 Nm).

➡Prevent oil from adhering to the screw and tapered parts.

17. Install a new cotter pin.

➡If the holes for the cotter pin are not aligned, tighten the nut further up to 60°.

18. Install the front axle assembly to the front shock absorber with the 2 bolts and 2 nuts and tighten to 213 ft. lbs. (290 Nm).

➡Only when reusing the bolts and nuts, apply a small amount of engine oil to the threads of the nuts.

19. Align the matchmarks and install the front driveshaft assembly to the front axle hub sub-assembly.
20. Install the front lower suspension arm to the front lower ball joint with the bolt and 2 nuts 68 ft. lbs. (92 Nm).
21. Connect the tie rod assembly.
22. Install the front disc.
23. Install the front disc brake caliper assembly to the steering knuckle with the 2 bolts and tighten to 77 ft. lbs. (104 Nm).
24. Install the clamp and front speed sensor with the bolt and tighten to 71 inch lbs. (8 Nm).

➡Prevent foreign matter from attaching to the sensor tip.

➡Firmly insert the sensor body into the knuckle before tightening the bolt.

➡After installing the sensor to the knuckle, make sure that there is no clearance between the sensor stay and knuckle. Also make sure that no foreign matter is stuck between the parts.

➡To prevent interference between the sensor and magnetic rotor, do not rotate the sensor body during or after the insertion of the sensor body to the knuckle.

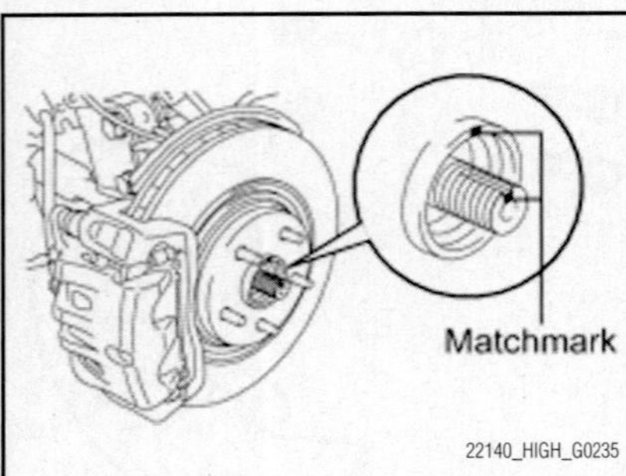

Fig. 254 Align the matchmarks and install the front driveshaft assembly

25. Install the front axle hub nut.
26. Check the ABS speed sensor signal

➡Check the ABS speed sensor signal.

27. Install front wheel and tighten lug nuts to 76 ft. lbs. (103 Nm).
28. Inspect and adjust front wheel alignment.

LOWER CONTROL ARM

REMOVAL & INSTALLATION

See Figure 255.

1. Remove the engine assembly with transaxle.
2. Remove the No. 1 front stabilizer brackets.
3. Remove the front stabilizer bar with front stabilizer link assembly.
4. Remove the power steering link assembly.
5. Install the engine hangers.
6. Separate the front frame assembly.
7. Remove the front lower suspension arm, as follows:
    a. Remove the 3 bolts, nut, and the front lower suspension arm from the front frame assembly.
    b. Remove the front lower arm bushing stopper from the front lower suspension arm.

To install:

8. Install the front lower suspension arm, as follows:
    a. Install the front lower arm bushing stopper to the front lower suspension arm.
    b. Install the front lower suspension arm to the front frame assembly with the 3 bolts and nut, but do not tighten them yet.
    c. Tighten the 3 bolts in numerical order shown. Tighten to 147 ft. lbs. (200 Nm), and 152 ft. lbs. (206 Nm).

➡Start installing the bolts from the front side of the vehicle.

9. Connect the front frame assembly.

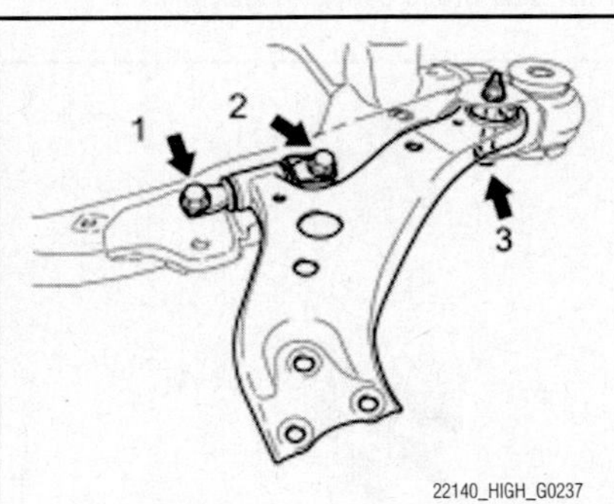

Fig. 255 Front lower control arm bolts tightening sequence

10. Remove the engine hangers.
11. Install the power steering link assembly.
12. Install the front stabilizer bar with front stabilizer link assembly.
13. Install the No. 1 front stabilizer brackets.
14. Install the engine assembly with transaxle.

STEERING KNUCKLE

REMOVAL & INSTALLATION

See Figure 256.

1. Remove the front wheel.
2. Remove the front axle hub nut.
3. Separate the front speed sensor.
4. Separate the front disc brake caliper assembly.
5. Remove the front disc.
6. Separate tie rod assembly.
7. Separate the front lower suspension arm.
8. Separate the front driveshaft assembly.
9. Remove the front axle assembly.
10. Remove the front lower ball joint.
11. Remove the No. 1 front wheel bearing dust deflector.
12. Remove the front axle hub hole snap ring.
13. Remove the front axle hub sub-assembly.
14. Remove the front disc brake dust cover.
15. Remove the steering knuckle, as follows:
    a. Place the bearing inner race (outside) on the front axle hub bearing.
    b. Using SST (SST: 09950-60010, SST: 09950-70010, SST: 09950-60020, or equivalent),V-blocks and a press, remove the front axle hub bearing from the steering knuckle.

➡Keep the steering knuckle level.

To install:

16. Using SST (SST: 09950-70010, SST: 09950-60020, or equivalent), install a new front axle hub bearing to the steering knuckle.
17. Install the front disc brake dust cover.
18. Install the front axle hub sub-assembly.
19. Install the front axle hub hole snap ring.
20. Install the No. 1 front wheel bearing dust deflector.
21. Install the front lower ball joint.
22. Install the front axle assembly.
23. Install the front driveshaft assembly.
24. Install the front lower suspension arm.
25. Connect tie rod assembly.
26. Install the front disc.

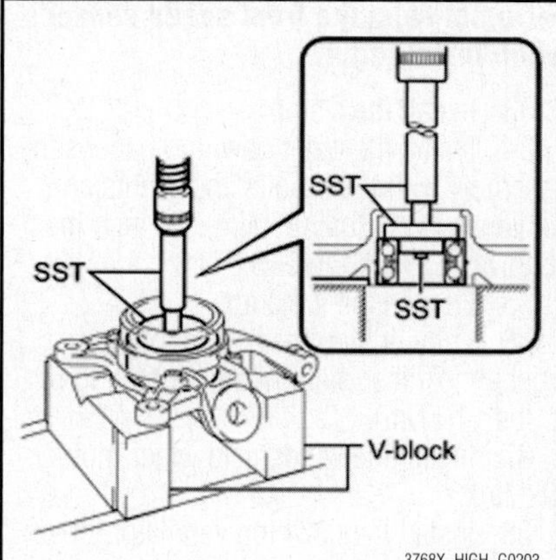

Fig. 256 Using SST (SST: 09950-60010, SST: 09950-70010, SST: 09950-60020, or equivalent),V-blocks and a press, remove the front axle hub bearing from the steering knuckle

27. Install the front disc brake caliper assembly.
28. Install the front axle hub nut.
29. Separate the front disc brake caliper assembly.
30. Remove the front disc.
31. Inspect the front axle bearing looseness.
32. Inspect front axle hub runout.
33. Install the front disc.
34. Install the front disc brake caliper assembly.
35. Install the front speed sensor.
36. Stake front axle hub nut.
37. Install the front wheel.
38. Inspect and adjust front wheel alignment.
39. Check for speed sensor signal.

STRUT

REMOVAL & INSTALLATION

See Figures 257 through 261.

1. Remove the front wheel.
2. Remove the front wiper arm and blade assemblies.
3. Loosen the front support to front shock absorber nut of the front shock absorber.

➡**Do not remove the front support to front shock absorber nut.**

➡**Loosen the nut only when the front shock absorber with coil spring needs to be disassembled.**

4. Remove the cowl top ventilator louver sub-assembly.
5. Remove the windshield wiper motor and link.

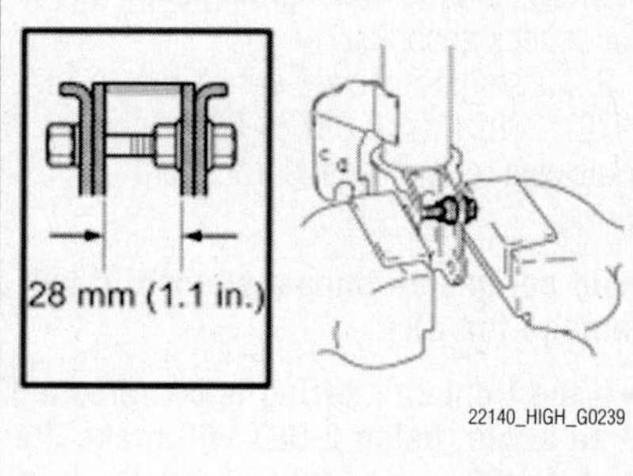

Fig. 257 Secure the front shock absorber with coil spring in a vise

6. Remove the outer cowl top panel sub-assembly, as follows:
   a. Disengage the 4 clamps and separate the wiper wire harness from the outer cowl top panel sub-assembly.
   b. Remove the 8 bolts, 6 nuts, and the outer cowl top panel sub-assembly.
7. Remove the bolt and clamp, and separate the front speed sensor and front flexible hose.
8. Remove the nut and separate the front stabilizer link assembly from the front shock absorber.

➡**If the ball joint turns together with the nut, use a hexagon wrench (6 mm) to hold the stud.**

9. Remove the front shock absorber with coil spring, as follows:
   a. Support the front axle using a jack and wooden block.
   b. Remove the 2 bolts and 2 nuts, and separate the front shock absorber with coil spring (lower side) from the steering knuckle.'

➡**When removing the nuts, keep the bolts from rotating.**

   c. Remove the nut and 2 spacers on the upper side of the front shock absorber with coil spring.

➡**Make sure that the front speed sensor is completely separated from the front shock absorber with coil spring.**

10. Secure the front shock absorber with coil spring, as follows:
    a. As shown in the illustration, secure the front shock absorber with coil spring in a vise using aluminum plates by clamping onto a double nutted bolt affixed to the bracket at the bottom of the absorber.
11. Remove the front support to front shock absorber nut, as follows:
    a. Using A Special Service Tool (SST: 09727-30021, SST: 09727-30021), compress the front coil spring.

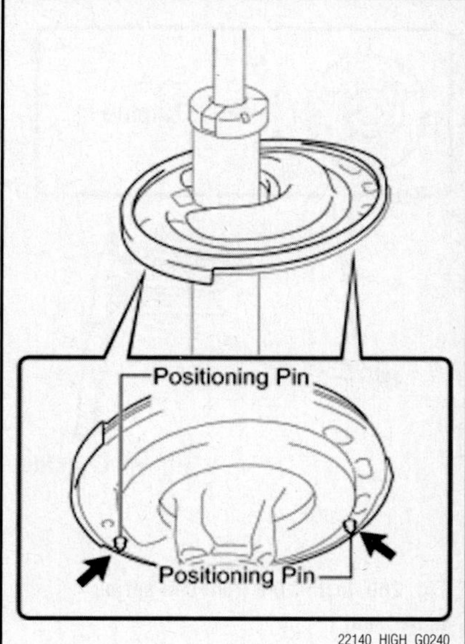

Fig. 258 Front coil spring lower insulator positioning pin

➡**Do not use an impact wrench. It will damage the SST.**

➡**If the front coil spring is compressed at an angle, using 2 SST will make the work easier.**

    b. Check that the front coil spring is fully compressed.
    c. Remove the front support to front shock absorber nut.
12. Remove the front suspension support sub-assembly.
13. Remove the front suspension support bearing.

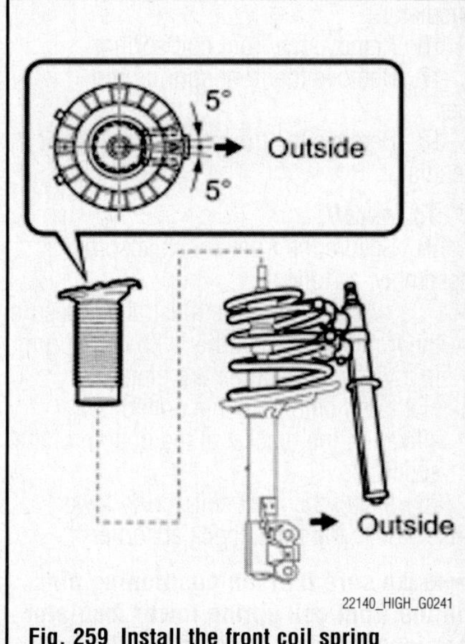

Fig. 259 Install the front coil spring upper insulator

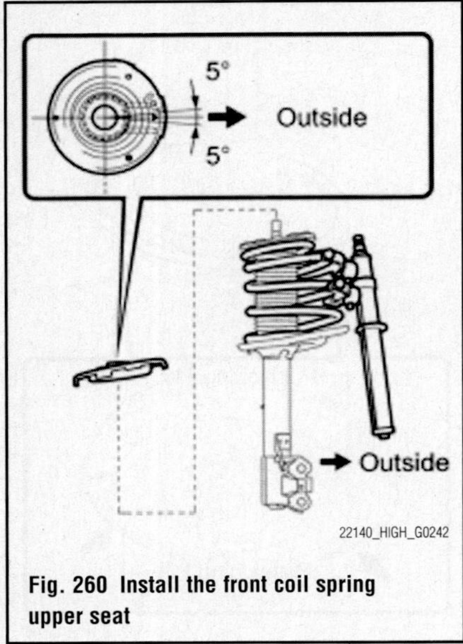

Fig. 260 Install the front coil spring upper seat

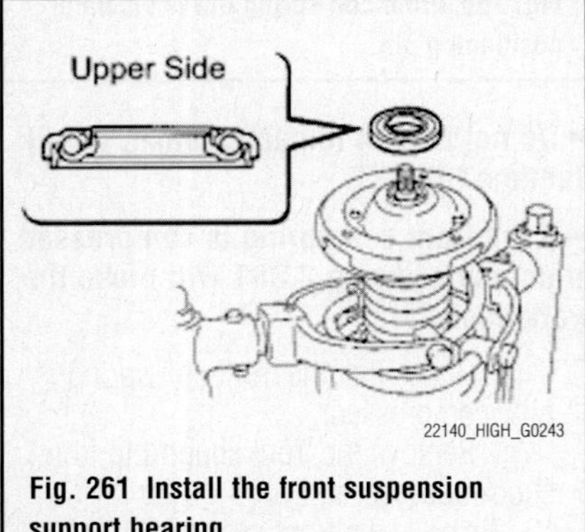

Fig. 261 Install the front suspension support bearing

14. Remove the front coil spring upper seat.
15. Remove the front coil spring upper insulator.
16. Remove the front coil spring.
17. Remove the front spring bum per.
18. Remove the front coil spring lower insulator.

To install:

19. Secure the front shock absorber assembly, as follows:
    a. As shown in the illustration, secure the front shock absorber with coil spring in a vise using aluminum plates by clamping onto a double nutted bolt affixed to the bracket at the bottom of the absorber.
20. Install the front coil spring lower insulator to the front shock absorber.

➡Make sure that the positioning pins on the front coil spring lower insulator are inserted into the holes in the front shock absorber.

21. Install the front spring bumper to the front shock absorber.
22. Using a Special Service Tool (SST: 09727-30021, SST: 09727-00050) or equivalent, compress the front coil spring.

➡Do not use an impact wrench. It will damage the SST.

➡If the front coil spring is compressed at an angle, using 2 SST will make the work easier.

23. Install the front coil spring to the front shock absorber.

➡Make sure that the end of the front coil spring is positioned in the depression of the lower spring seat.

24. Install the front coil spring upper insulator as shown in the illustration.

➡Any misalignment between the front shock absorber lower bracket and the alignment mark must be +/- 5°.

25. Install the front coil spring upper seat with the mark facing to the outside of the vehicle.

➡Any misalignment between the front shock absorber lower bracket and the alignment mark must be +/- 5°.

26. Install the front suspension support bearing as shown in the illustration.
27. Install the front suspension support sub-assembly as shown in the illustration.

➡Check that the slot on the piston rod and the slot on the front suspension support sub-assembly are aligned.

28. Temporarily tighten a new front support to front shock absorber nut.
29. Install the front shock absorber with coil spring (upper side) with the nut and 2 spacers and tighten to 63 ft. lbs. (85 Nm).
30. Install the front shock absorber with coil spring (lower side) to the steering knuckle and insert the 2 bolts and 2 nuts and tighten to 214 ft. lbs. (290 Nm).

➡When installing the nuts, keep the bolts from rotating.

31. Install the front stabilizer link assembly to the front shock absorber with the nut and tighten to 55 ft. lbs. (74 Nm).

➡If the ball joint turns together with the nut, use a hexagon wrench (6 mm) to hold the stud bolt.

32. Install the front speed sensor and front flexible hose with the bolt and tighten to 14 ft. lbs. (19 Nm).

➡Do not twist the front speed sensor when installing it.

33. Install the clamp.
34. Install the outer cowl top panel sub-assembly with the 8 bolts and 6 nuts and tighten to 63 ft. lbs. (85 Nm), 78 inch lbs. (8.8 Nm), 78 inch lbs. (8.8 Nm).
35. Engage the 4 clamps.
36. Fully tighten the front support to front shock absorber nut and tighten to 52 ft. lbs. (70 Nm).
37. Install the windshield wiper motor and link.
38. Install the cowl top ventilator louver sub-assembly.
39. Install the front wiper arm and blade assemblies.
40. Install the front wheel and tighten lug nuts to 76 ft. lbs. (103 Nm).
41. Inspect and adjust front wheel alignment.

STABILIZER BAR

REMOVAL & INSTALLATION

See Figure 262.

1. Remove the engine assembly with transaxle.
2. Remove the nuts and separate both front stabilizer link assemblies.

➡If the ball joint turns together with the nut, use a hexagon wrench (6 mm) to hold the stud.

3. Remove the 2 bolts and both No. 1 front stabilizer brackets from the front frame assembly.
4. Remove the front stabilizer bar.
5. Remove both No. 2 front stabilizer brackets from the front stabilizer bar bushing.
6. Remove the 2 No. 1 front stabilizer bar bushings from the front stabilizer bar.

To install:

7. Install the 2 No. 1 front stabilizer bar bushings to the front stabilizer bar as shown in the illustration.

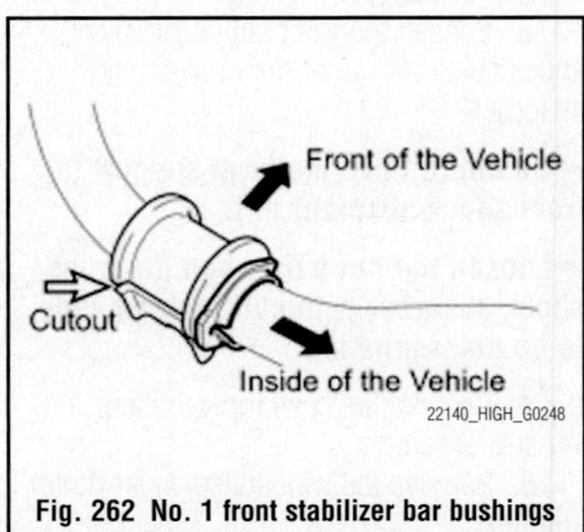

Fig. 262 No. 1 front stabilizer bar bushings

➡**Install the No. 1 front stabilizer bar bushings so that the cutout faces the rear of the vehicle.**

8. Install both No. 2 front stabilizer brackets to the No. 1 front stabilizer bar bushing.
9. Install the front stabilizer bar by inserting it from the right side of the vehicle.
10. Install both No. 1 front stabilizer brackets to the front frame assembly with the 2 bolts and tighten to 21 ft. lbs. (29 Nm).
11. Install both front stabilizer link assemblies with the nuts and tighten to 55 ft. lbs. (74 Nm).

➡**If the ball joint turns together with the nut, use a hexagon wrench (6 mm) to hold the stud bolt.**

12. Install the engine assembly with transaxle.
13. Inspect and adjust the front wheel alignment.

WHEEL HUB & BEARING

REMOVAL & INSTALLATION

See Steering Knuckle procedure.

SUSPENSION

REAR SUSPENSION

CONTROL ARMS/LINKS

REMOVAL & INSTALLATION

2009 Models

2WD Vehicles—No. 1

See Figures 263 and 264.

➡**Press the height control switch to stop the vehicle height control operation before jacking up or lifting up the vehicle with air suspension.**

➡**Press the height control switch to operate the vehicle height control after jacking down or lifting down the vehicle with air suspension.**

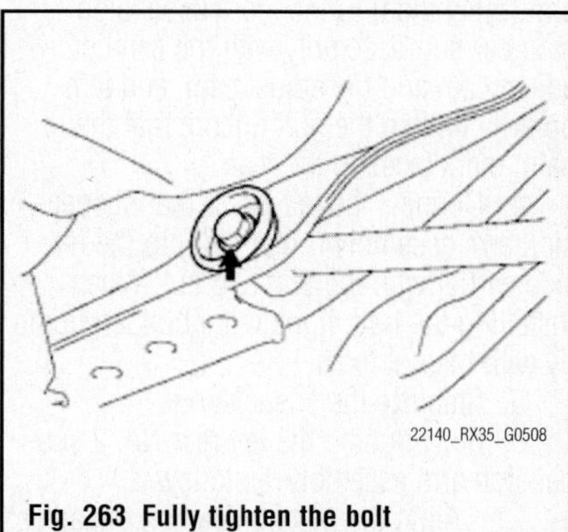

Fig. 263 Fully tighten the bolt

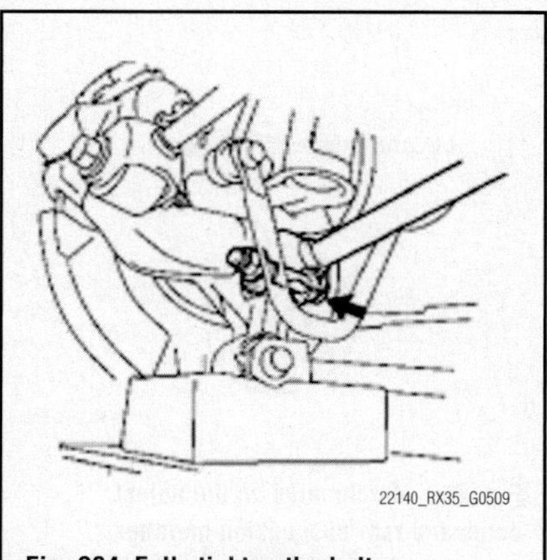

Fig. 264 Fully tighten the bolt

➡**Support the rear axle carrier with a jack.**

1. Remove the rear wheel.
2. Remove the rear stabilizer bar.
3. Remove the left rear No. 1 suspension arm assembly, as follows:
4. Remove the bolt and disconnect the rear suspension arm assembly No. 1 (inner side).
5. Remove the bolt, nut and the rear suspension arm assembly No. 1 (outer side) from the rear axle carrier. When removing the bolt, keep the nut from rotating.

To install:

6. Install the left rear No. 1 suspension arm assembly (inner side) with the bolt, and temporarily tighten the bolt. Ensure that the paint mark faces to the rear.
7. Connect the left rear No. 1 suspension arm assembly (outer side) to the rear axle carrier with the bolt and nut, and temporarily tighten the bolt and nut. When installing the bolt, tighten the bolt temporarily with the nut fixed.
8. Stabilize the suspension.
9. Fully tighten the left rear No. 1 suspension arm assembly, as follows:
   a. Fully tighten the bolt to 89 ft. lbs. (120 Nm).
   b. Fully tighten the bolt to 83 ft. lbs. (112 Nm).
10. Install the rear stabilizer bar.
11. Install the rear wheel.
12. Inspect the rear wheel alignment.

2WD Vehicles—No. 2

See Figures 265 and 266.

➡**Press the height control switch to stop the vehicle height control operation before jacking up or lifting up the vehicle with air suspension.**

➡**Press the height control switch to operate the vehicle height control after jacking down or lifting down the vehicle with air suspension.**

➡**Support the rear axle carrier with a jack.**

1. Remove the rear wheel.
2. Remove the left rear No. 2 suspension arm assembly, as follows:
   a. Remove the nut, and separate the height control sensor sub-assembly from the No. 2 suspension arm assembly (w/ height control sensor sub-assembly).
   b. Remove the bolt, nut and disconnect the No. 2 rear suspension arm assembly (inner side).
   c. Remove the bolt, nut and the No. 2 rear suspension arm assembly (outer side) from the rear axle carrier. When removing the bolt, keep the nut from rotating.

To install:

3. Install the rear No. 2 suspension arm assembly (inner side) with the bolt, and

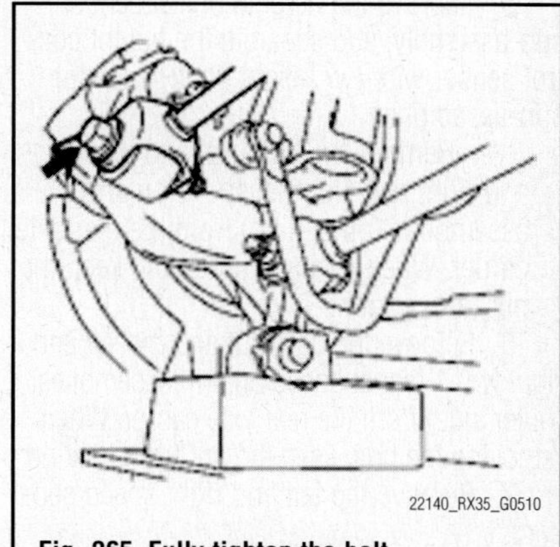

Fig. 265 Fully tighten the bolt

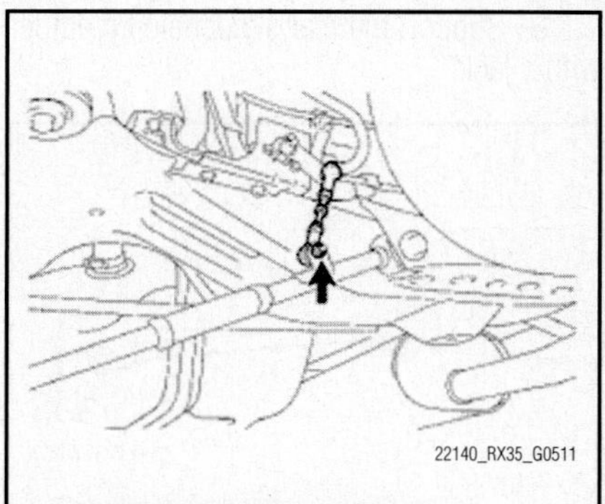

Fig. 266 Fully tighten the bolt

temporarily tighten the bolt. Ensure that the paint mark faces to the rear.

4. Connect the rear No. 2 suspension arm assembly (outer side) to the rear axle carrier with the bolt and nut, and temporarily tighten the bolt. When installing the bolt, tighten the bolt temporarily with the nut fixed.

5. Stabilize the suspension.

6. Fully tighten the left rear No. 2 suspension arm assembly, as follows:

a. Fully tighten the bolt to 89 ft. lbs. (120 Nm).

b. Fully tighten the bolt to 83 ft. lbs. (112 Nm).

7. Install the height control sensor sub-assembly and nut to the suspension arm assembly No. 2 (w/ height control sensor sub-assembly) and tighten to 48 inch lbs. (5.4 Nm).

8. Install the rear wheel.

9. Inspect the rear wheel alignment.

10. Adjust vehicle height.

11. Adjust headlight aim only.

4WD Vehicles—No. 1

See Figure 267.

1. Remove rear wheel.

2. Remove exhaust pipe assembly.

3. Remove propeller w/center bearing shaft assembly.

4. Remove rear strut rod assembly.

5. Remove the height control sensor sub-assembly, and separate the height control sensor wire (w/ height control sensor sub-assembly).

a. Remove the bolt, nut and the left and right rear No. 2 suspension arm assemblies (outer side) from the rear axle carrier. When removing the bolt, keep the nut from rotating.

6. Remove the bolt, nut and the left and right No. 1 rear suspension arm assemblies (outer side) from the rear axle carrier. When removing the bolt, keep the nut from rotating.

7. Remove the left and right speed sensors.

8. Remove the right and left rear driveshaft assemblies.

9. Support the rear suspension member with a jack.

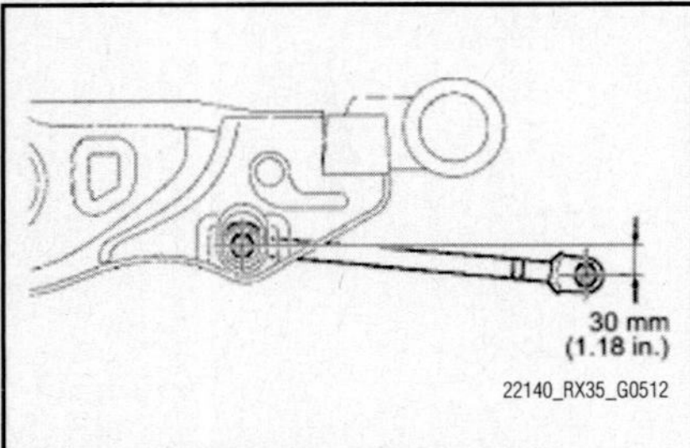

Fig. 267 Suspension arm placement

10. Remove the 4 nuts, 2 bolts and 2 retainers from the rear suspension member.

11. Lower the rear suspension member.

12. Remove the bolt, nuts and left rear No. 1 suspension arm assembly. When removing the bolt, keep the nut from rotating.

To install:

13. Install the left rear No. 1 suspension arm assembly with the bolt and nut, and temporarily tighten the bolt. Ensure that the paint mark faces to the rear.

14. Set the No. 1 rear suspension arm assembly as shown in the illustration, and fully tighten the bolt and tighten to 59 ft. lbs. (80 Nm).

15. Raise the rear suspension member with a jack.

16. Install the rear suspension member with the 4 nuts, 2 bolts and 2 retainers. Front side: Torque: 85 ft. lbs. (115 Nm). Rear side: 134 ft. lbs. (181 Nm).

17. Install the right and left rear driveshaft assemblies.

18. Install the left and right speed sensors.

19. Install the left and right No. 1 rear suspension arm assemblies (outside) to the rear axle carrier, and temporarily tighten the bolt. When installing the bolt, tighten the bolt temporarily with the nut fixed.

20. Install the left and right No. 2 rear suspension arms (outside) to the rear axle carrier, and temporarily tighten the bolt. When installing the bolt, tighten the bolt temporarily with the nut fixed.

21. Temporarily tighten the rear strut rod assembly.

22. Stabilize the suspension.

23. Fully tighten the left and right rear No. 1 suspension arm assemblies, as follows:

a. Fully tighten the bolt to 83 ft. lbs. (112 Nm).

24. Fully tighten the left and right rear No. 2 suspension arm assemblies, as follows:

a. Fully tighten the bolt to 48 inch lbs. (5.4 Nm).

25. Fully tighten the rear strut rod assembly.

26. Install the rear wheel.

27. Temporarily tighten the propeller w/center bearing shaft assembly.

28. Install the exhaust pipe assembly.

29. Check for exhaust gas leak.

30. Inspect the rear wheel alignment.

31. Adjust vehicle height.

32. Adjust headlight aim only.

33. Check the ABS speed sensor signal.

4WD Vehicles—No. 2

See Figures 268 and 269.

➡Press the height control switch to stop the vehicle height control operation before jacking up or lifting up the vehicle with air suspension.

➡Press the height control switch to operate the vehicle height control after jacking down or lifting down the vehicle with air suspension.

➡Support the rear axle carrier with a jack.

1. Remove the rear wheel.

2. Remove the left rear No. 2 suspension arm assembly, as follows:

a. Remove the nut, and separate the height control sensor sub-assembly from the No. 2 suspension arm assembly (w/ height control sensor sub-assembly).

b. Place matchmarks on the adjust cams and rear suspension member sub-assembly.

c. Remove the nut, camber adjust cam and toe adjust cam, and disconnect the No. 2 rear suspension arm assembly (inner side). When removing the nut, keep the bolt from rotating.

d. Remove the bolt, nut and the No. 2 rear suspension arm assembly (outer side) from the rear axle carrier. When removing the bolt, keep the nut from rotating.

To install:

3. Install the left No. 2 rear suspension arm (inner side) to the rear suspension member sub-assembly with the camber adjust cam and toe adjust cam, and temporarily tighten the nut. Ensure that the paint mark faces to the rear.

4. Connect the left No. 2 rear suspension arm assembly (outer side) to the rear axle carrier with the bolt and nut. When installing the bolt, tighten the bolt temporarily with the nut fixed.

5. Stabilize the suspension.

6. Fully tighten the left rear No. 2 suspension arm assembly, as follows:

a. Align the matchmarks on the adjust

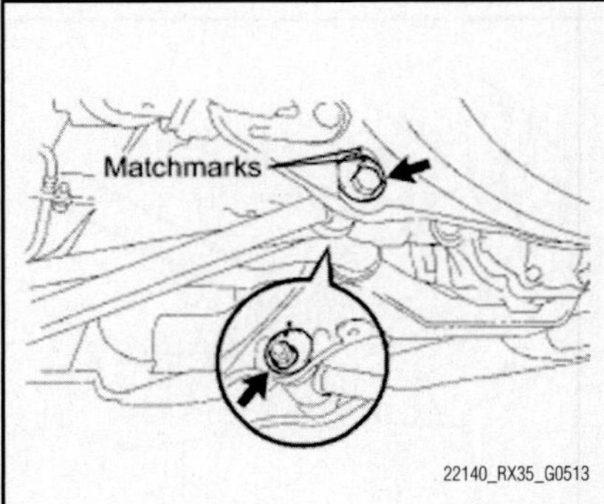

Fig. 268 Matchmarks on the adjust cams and rear suspension member sub-assembly

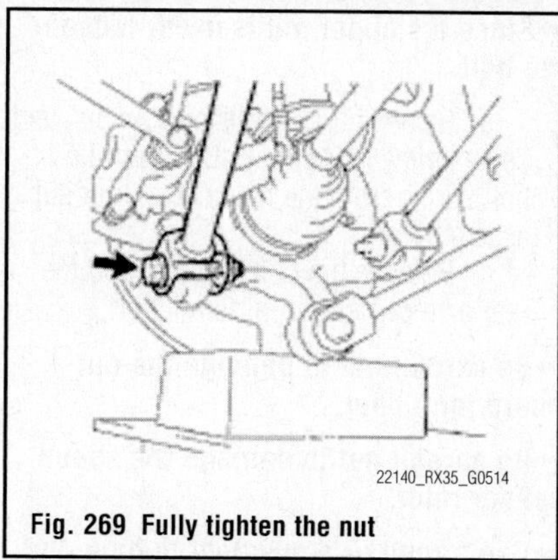

Fig. 269 Fully tighten the nut

cams and rear suspension member sub-assembly.

b. Fully tighten the nut to 74 ft. lbs. (100 Nm).

c. Fully tighten the nut to 83 ft. lbs. (112 Nm).

d. Install the height control sensor sub-assembly and nut to the No. 2 suspension arm assembly (w/ height control sensor sub-assembly) to 48 inch lbs. (5.4 Nm).

7. Install the rear wheel.
8. Inspect the rear wheel alignment.
9. Adjust vehicle height.
10. Adjust headlight aim only.

2010 Models

2WD Vehicles

See Figures 270 and 271.

➡Use the same procedures for the RH side and LH side.

1. Remove the deck board assembly.
2. Remove the No. 2 and 3 deck board sub-assemblies, as applicable.
3. Remove the tonneau cover assembly, as applicable.
4. Remove the rear mat.
5. Remove the deck trim service hole cover.
6. Remove the lower spare wheel carrier hinge cover, as applicable.
7. Remove the spare tire.
8. Remove the spare wheel carrier lock cover, as applicable.
9. Remove rear wheel.
10. Remove the nuts and separate the rear stabilizer link assemblies from the rear stabilizer bar.

➡If the ball joint turns together with the nut, use a hexagon wrench (5 mm) to hold the stud bolt.

11. Remove the 4 bolts and rear stabilizer bar.
12. Remove the rear No. 2 suspension arm assembly, as follows:

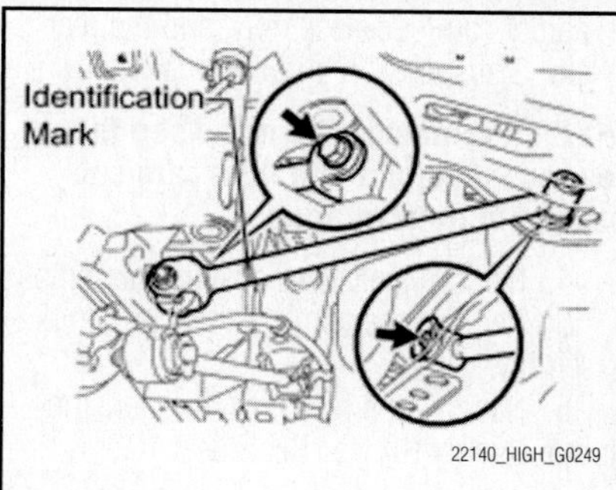

Fig. 270 Temporarily install the rear No. 1 and 2 suspension arm assemblies

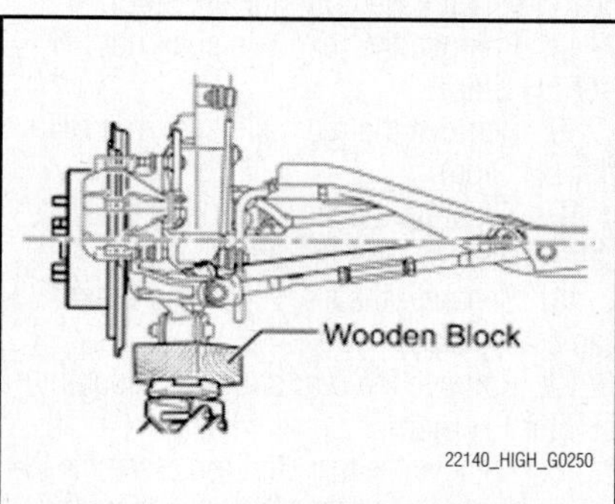

Fig. 271 Jack up the rear axle carrier

a. Remove the bolt and the nut, and separate the rear No. 2 suspension arm assembly from the rear axle carrier sub-assembly.

➡Since a stopper nut is used, loosen the bolt.

b. Remove the bolt and the rear No. 2 suspension arm assembly.

13. Remove the rear No. 1 suspension arm assembly, as follows:

a. Remove the bolt and the nut, and separate the rear No. 1 suspension arm assembly from the rear axle carrier sub-assembly.

➡Since a stopper nut is used, loosen the bolt.

b. Remove the bolt and the rear No. 1 suspension arm assembly.

To install:

14. Temporarily install the rear No. 1 and 2 suspension arm assemblies to the rear suspension member with the bolts.

➡Ensure that the identification marks face the rear side of the vehicle.

15. Temporarily install the rear No. 1 and 2 suspension arm assemblies to the rear axle carrier sub-assembly with the bolts and the nuts.

➡Since a stopper nut is used, temporarily tighten the bolts.

16. Jack up the rear axle carrier, placing a wooden block underneath to avoid damage. Apply load to the suspension so that the installed bolt of the rear No. 1 suspension arm (inner side) is horizontally aligned with the center of the rear axle hub.

CAUTION

Do not jack up the rear axle carrier sub-assembly too high as the vehicle may fall.

➡Do not bend the brake dust cover.

➡If the rear driveshaft assembly cannot be positioned horizontally as shown in the illustration even when the rear axle carrier sub-assembly is jacked up, apply additional load to the vehicle such as by having a person sit in the rear seat.

➡Use the same procedures for the RH side and LH side.

17. Fully tighten the rear No. 1 suspension arm assembly, as follows:

a. Using SST: 09961-00950 or equivalent and a socket wrench (19 mm), fully tighten the bolt to 88 ft. lbs. (120 Nm), 65 ft. lbs. (89 Nm).

➡Use a torque wrench with a fulcrum length of 425 mm (16.73 in.).

➡This torque value is effective when SST is parallel to the torque wrench.

➡The final torque must be applied under standard vehicle height conditions.

b. Fully tighten the bolt to 82 ft. lbs. (112 Nm).

➡Since a stopper nut is used, fully tighten the bolt.

➡The final torque must be applied under standard vehicle height conditions.

18. Fully tighten the rear No. 2 suspension arm assembly, as follows:

a. Fully tighten the bolts to 88 ft. lbs. (120 Nm), 82 ft. lbs. (112 Nm).

➡Since a stopper nut is used, fully tighten the bolt.

➡The final torque must be applied under standard vehicle height conditions.

19. Install the rear stabilizer bar with the 4 bolts and tighten to 14 ft. lbs. (19 Nm).
20. Install both rear stabilizer link assemblies to the rear stabilizer bar with the nuts and tighten to 29 ft. lbs. (39 Nm).

➡If the ball joint turns together with the nut, use a hexagon wrench (6 mm) to hold the stud bolt.

21. Install the wheels and tighten lug nuts to 76 ft. lbs. (103 Nm).
22. Inspect and adjust rear wheel alignment.
23. The remainder of installation is the reverse of the removal procedure.

4WD Vehicles

See Figures 272 through 276.

➡The removal procedures for the LH and RH sides are different.

➡When removing RH side components, it is not necessary to follow the steps with (for LH Side).

➡When removing LH side components, it is not necessary to follow the steps with (for RH side).

1. Remove the rear wheel (for RH Side).
2. Remove the rear wheels (for LH Side).
3. Remove the rear no. 2 suspension arm assembly LH (for LH Side), as follows:
   a. Put matchmarks on the adjust cams and the rear suspension member sub-assembly.
   b. Remove the bolt and the nut, and separate the rear No. 2 suspension arm assembly LH from the rear axle carrier sub-assembly LH.

➡Since a stopper nut is used, loosen the bolt.

   c. Remove the nut, the No. 2 camber adjust cam, the rear suspension toe adjust cam sub-assembly, and the rear No. 2 suspension arm assembly LH.

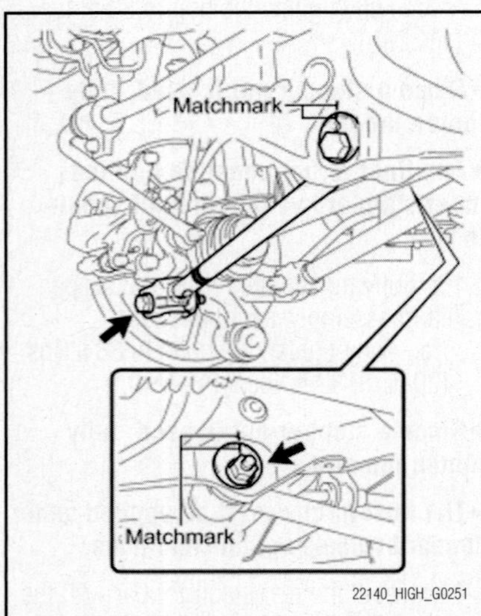

Fig. 272 Put matchmarks on the adjust cams and the rear suspension member sub-assembly

➡When removing the nut, keep the rear suspension toe adjust cam sub-assembly from rotating.

4. Remove the tail exhaust pipe assembly.
5. Remove the center exhaust pipe assembly.
6. Remove the rear no. 2 suspension arm assembly RH (for RH Side).

➡Perform the same procedure as the rear No. 2 suspension arm assembly LH.

7. Remove the propeller with center bearing shaft assembly (for LH Side).
8. Remove the rear axle shaft nut LH (for LH Side).
9. Remove the rear axle shaft nut RH (for LH Side).
10. Separate the No. 3 parking brake cable assembly (for LH Side).
11. Separate the No. 2 parking brake cable assembly.
12. Remove the rear strut rod assembly LH (for LH Side).
13. Remove the rear strut rod assembly RH.
14. Remove the rear No. 1 suspension arm assembly RH (for RH Side). As follows:
   a. Remove the bolt and the nut, and separate the rear No. 1 suspension arm assembly RH from the rear axle carrier sub-assembly RH.
   b. Remove the bolt, the nut, and the rear No. 1 suspension arm assembly RH from the rear suspension member sub-assembly

➡Since stopper nuts are used, loosen the bolts.

15. Remove the rear speed sensor LH (for LH Side).
16. Remove the rear speed sensor RH (for LH Side).
17. Separate the rear disc brake caliper assembly LH (for LH Side).
18. Separate the rear disc brake caliper assembly RH (for LH Side).
19. Remove the rear disc (for LH Side), as follows:
   a. Remove the rear disc from the rear axle hub and bearing assembly LH.
   b. Remove the rear disc from the rear axle hub and bearing assembly RH.

➡Perform the same procedure as the LH side.

20. Remove the rear axle hub and bearing assembly LH (for LH Side).
21. Remove the rear axle hub and bearing assembly RH (for LH Side).
22. Remove the rear axle carrier sub-assembly LH (for LH Side), as follows:
   a. Loosen the bolt.

➡Since a stopper nut is used, loosen the bolt.

   b. Remove the 2 bolts and 2 nuts, and separate the rear shock absorber with coil spring from the rear axle carrier sub-assembly LH.
   c. Remove the bolt, the nut, and the rear axle carrier sub-assembly LH.

➡Be careful not to damage the outboard joint boot.

➡Be careful not to damage the speed sensor rotor.

➡Use a rope or equivalent to hang the rear driveshaft assembly.

Remove rear axle carrier sub-assembly RH (for LH Side), as follows:

   d. Loosen the 2 bolts.
   e. Remove the 2 bolts and 2 nuts, and separate the rear shock absorber with coil spring from the rear axle carrier sub-assembly RH.
   f. Remove the 2 bolts, the 2 nuts, and the rear axle carrier sub-assembly RH.

➡Since stopper nuts are used, loosen the bolts.

➡Be careful not to damage the outboard joint boot.

➡Be careful not to damage the speed sensor rotor.

➡Use a rope or equivalent to hang the rear driveshaft assembly.

23. Remove the rear differential filler plug (for LH Side).
24. Remove the rear differential drain plug (for LH Side).
25. Remove the rear driveshaft assembly LH (for LH Side).
26. Remove the rear driveshaft snap ring LH (for LH Side).
27. Remove the rear driveshaft assembly RH (for LH Side).
28. Remove the rear driveshaft snap ring RH (for LH Side).
29. Remove the rear suspension member (for LH Side).
30. Remove the rear No. 1 suspension arm assembly LH (for LH Side), as follows:
   a. Remove the bolt, the nut, and the rear No. 1 suspension arm assembly LH from the rear suspension member sub-assembly.

➡Since a stopper nut is used, loosen the bolt.

To install:

31. Temporarily install rear No. 1 suspension arm assembly LH (for LH Side), as follows:

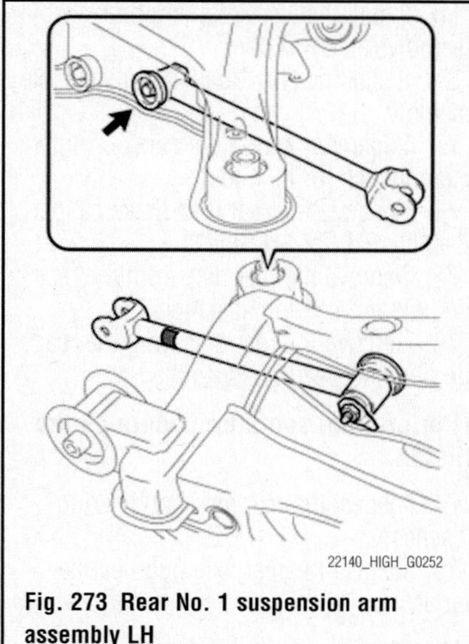
Fig. 273 Rear No. 1 suspension arm assembly LH

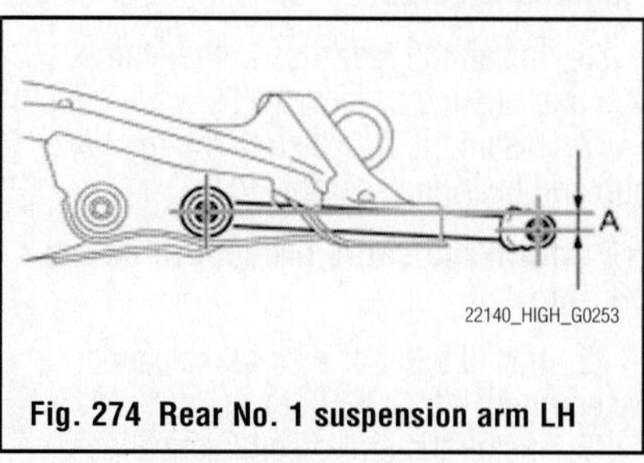

Fig. 274 Rear No. 1 suspension arm LH

a. Temporarily install the rear No. 1 suspension arm assembly LH to the rear suspension member sub-assembly with the bolt and the nut.

➡Ensure that the identification mark faces the rear side of the vehicle.

b. Set the rear No. 1 suspension arm LH in the position shown. Length A: 20 mm (0.787 in.)

c. Fully tighten the bolt to 59 ft. lbs. (80 Nm).

➡Since stopper nuts are used, temporarily tighten the bolts.

32. Install the rear suspension member (for LH Side), as follows:

a. Support the rear suspension member with a jack using a wooden block.

➡Use a properly sized wooden block to keep the jack and suspension member level.

➡Support the suspension member until retightening of the suspension member is complete.

b. Raise the rear suspension member with a jack.

c. Temporarily install the rear suspension member, 2 rear upper suspension member stoppers, and 2 rear lower suspension member stopper retainers with the 4 nuts and 2 bolts.

d. Fully tighten the 2 nuts 85 ft. lbs. (115 Nm).

e. Using SST: 09961-00950 or equivalent and a socket wrench (19 mm), fully tighten the nut (LH side) to 133 ft. lbs. (181 Nm), 98 ft. lbs. (134 Nm).

f. Using the same tools, fully tighten the nut (RH side) to the same specifications.

➡Use a torque wrench with a fulcrum length of 425 mm (16.73 in.).

➡These torque values are effective when SST is parallel to the torque wrench .

33. Install the rear driveshaft snap ring LH (for LH Side).

34. Install the rear driveshaft assembly LH (for LH Side).

35. Install the rear driveshaft snap ring RH (for LH Side).

36. Install the rear driveshaft assembly RH (for LH Side).

37. Install the rear differential drain plug (for LH Side).

38. Add differential oil (for LH Side).

39. Inspect differential oil (for LH Side).

40. Install the rear differential filler plug (for LH Side).

41. Install the rear axle carrier sub-assembly LH (for LH Side), as follows:

a. Temporarily install the rear axle carrier sub-assembly LH with the bolt and the nut.

b. Install the rear axle carrier sub-assembly LH with the 2 bolts and the 2 nuts and tighten to 213 ft. lbs. (290 Nm).

➡Be careful not to damage the outboard joint boot.

➡Be careful not to damage the speed sensor rotor.

➡Prevent foreign matter from adhering to the speed sensor rotor.

➡Do not rotate the rear driveshaft assembly without the rear axle hub and bearing assembly installed.

➡Insert the bolts from the rear side.

42. Install the rear axle carrier sub-assembly RH (for LH Side), as follows:

a. Temporarily install the rear axle carrier sub-assembly LH with the 2 bolts and the 2 nuts.

b. Install the rear axle carrier sub-assembly LH with the 2 bolts and the 2 nuts and tighten to 213 ft. lbs. (290 Nm).

➡Be careful not to damage the outboard joint boot.

➡Be careful not to damage the speed sensor rotor.

➡Prevent foreign matter from adhering to the speed sensor rotor.

➡Do not rotate the rear driveshaft assembly without the rear axle hub and bearing assembly installed.

➡Insert the bolts from the rear side.

43. Temporarily install the rear No. 1 suspension arm assembly RH to the rear suspension member sub-assembly with the bolt and the nut.

➡Ensure that the identification mark faces the rear side of the vehicle.

44. Temporarily install the rear No. 1 suspension arm assembly RH to the rear axle carrier sub-assembly RH with the bolt and the nut.

➡Since stopper nuts are used, temporarily tighten the bolts.

45. Temporarily install the rear strut rod assembly LH (for LH Side).

46. Temporarily install the rear strut rod assembly RH.

47. Temporarily install the rear No. 2 suspension arm assembly LH (for LH Side).

a. Temporarily install the rear No. 2 suspension arm assembly LH to the rear suspension member sub-assembly with the rear suspension toe adjust cam sub-assembly, the No. 2 camber adjust cam and the nut.

➡Ensure that the identification mark faces the rear side of the vehicle.

➡When temporarily tightening the nut, keep the rear suspension toe adjust cam sub-assembly from rotating.

b. Temporarily install the rear No. 2 suspension arm assembly LH to the rear axle carrier sub-assembly LH with the bolt and the nut.

➡Since a stopper nut is used, temporarily tighten the bolt.

48. Temporarily install the rear No. 2 suspension arm assembly RH (for LH Side).

➡Perform the same procedure as the rear No. 2 suspension arm assembly LH.

49. Install the rear axle hub and bearing assembly LH (for LH Side).

50. Install the rear axle hub and bearing assembly RH (for LH Side).

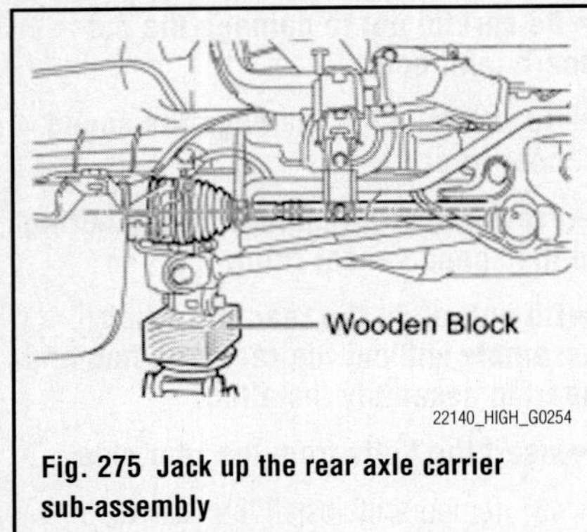

Fig. 275 Jack up the rear axle carrier sub-assembly

51. Install the rear disc to the rear axle hub and bearing assembly LH.
52. Install the rear disc to the rear axle hub and bearing assembly RH.

➡Perform the same procedure as the LH side.

53. Install the rear disc brake caliper assembly LH (for LH Side).
54. Install the rear disc brake caliper assembly RH (for LH Side).
55. Temporarily the install rear axle shaft nut LH (for LH Side).
56. Temporarily the install rear axle shaft nut RH (for LH Side).
57. Install the rear speed sensor LH (for LH Side).
58. Install the rear speed sensor RH (for LH Side).
59. Jack up the rear axle carrier sub-assembly, placing a wooden block underneath to avoid damage. Apply load to the suspension so that the rear driveshaft assembly is positioned horizontally.

⁂ CAUTION

Do not jack up the rear axle carrier sub-assembly too high as the vehicle may fall.

➡Do not bend the brake dust cover.

➡If the rear driveshaft assembly cannot be positioned horizontally as shown in the illustration even when the rear axle carrier sub-assembly is jacked up, apply additional load to the vehicle such as by having a person sit in the rear seat.

➡Use the same procedures for the RH side and LH side.

60. Fully tighten the rear No. 1 suspension arm assembly LH (for LH Side).
61. Fully tighten the rear No. 1 suspension arm assembly RH (for LH Side).

➡Perform the same procedure as the rear No. 1 suspension arm assembly LH.

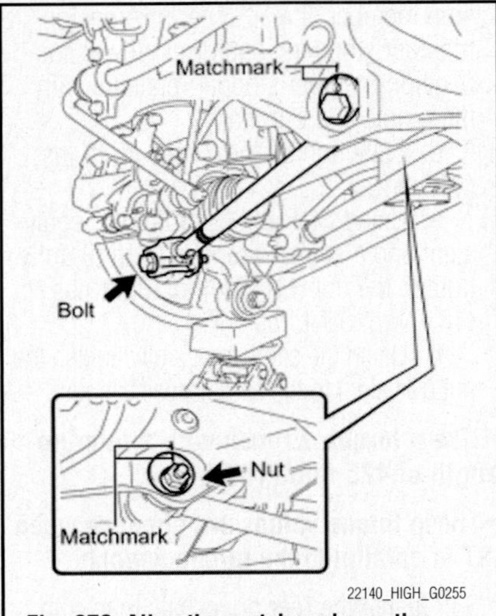

Fig. 276 Align the matchmarks on the adjust cams and rear suspension member sub-assembly

62. Fully tighten the rear No. 1 suspension arm assembly RH (for RH Side).
63. Fully tighten the 2 bolts to 59 ft. lbs. (80 Nm), 82 ft. lbs. (112 Nm).

➡Since a stopper nut is used, temporarily tighten the bolt.

➡The final torque must be applied under standard vehicle height conditions.

64. Fully tighten the rear No. 2 suspension arm assembly LH (for LH Side).
   a. Align the matchmarks on the adjust cams and rear suspension member sub-assembly.
   b. Fully tighten the nut to 74 ft. lbs. (100 Nm).

➡The final torque must be applied under standard vehicle height conditions.

➡When fully tightening the nut, keep the rear suspension toe adjust cam sub-assembly from rotating.

   c. Fully tighten the bolt to 82 ft. lbs. (112 Nm).

➡Since a stopper nut is used, temporarily tighten the bolt.

➡The final torque must be applied under standard vehicle height conditions.

65. Fully tighten the rear No. 2 suspension arm assembly RH (for RH Side).

➡Perform the same procedure as the rear No. 2 suspension arm assembly LH.

66. Fully tighten rear strut rod assembly LH (for LH Side).
67. Fully tighten rear strut rod assembly RH.
68. Install the No. 3 parking brake cable assembly (for LH Side).
69. Install the No. 2 parking brake cable assembly.
70. Separate the rear disc brake caliper assembly LH (for LH Side).
71. Separate the rear disc brake caliper assembly RH (for LH Side).
72. Remove the rear disc from the rear axle hub and bearing assembly LH.
73. Remove the rear disc from the rear axle hub and bearing assembly RH.

➡Perform the same procedure as the LH side.

74. Inspect the rear axle hub bearing looseness.
75. Inspect the rear axle hub bearing runout.

➡Use the same procedures for the RH side and LH side.

76. Install the rear disc to the rear axle hub and bearing assembly LH.
77. Install the rear disc to the rear axle hub and bearing assembly RH.

➡Perform the same procedure as the LH side.

78. Install the rear disc brake caliper assembly LH (for LH Side).
79. Install the rear disc brake caliper assembly RH (for LH Side).
80. Install the rear axle shaft nut LH (for LH Side).
81. Install the rear axle shaft nut RH (for LH Side).
82. Temporarily tighten the propeller with center bearing shaft assembly (for LH Side).
83. Fully tighten the propeller with center bearing shaft assembly (for LH Side).
84. Inspect and adjust transfer oil (for LH Side).
85. Install the center exhaust pipe assembly.
86. Install the tail exhaust pipe assembly.
87. Inspect for exhaust gas leak
88. Install the wheels and tighten lug nuts to 76 ft. lbs. (103 Nm).
89. Inspect and adjust rear wheel alignment.
90. Check for rear speed sensor signal (for LH Side).

STABILIZER BAR

REMOVAL & INSTALLATION

2WD Vehicles

See Figure 277.

1. Remove the nut and separate the rear stabilizer link assembly from the rear stabilizer bar.

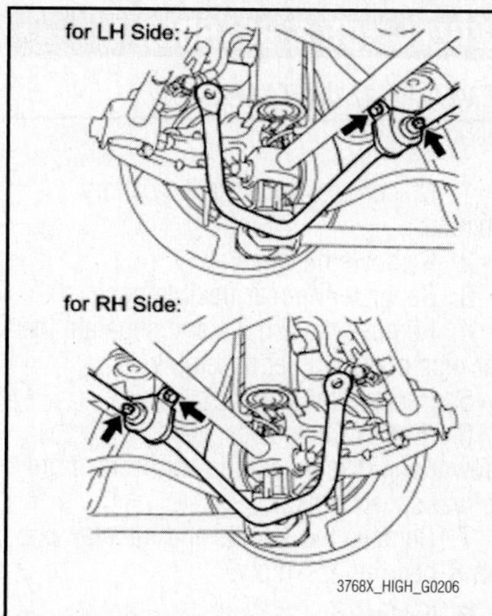

Fig. 277 Remove the 4 bolts and rear stabilizer bar—2WD vehicles

➡If the ball joint turns together with the nut, use a hexagon wrench (5 mm) to hold the stud bolt.

2. Remove the nut and the rear stabilizer link assembly from the rear shock absorber with coil spring.
3. Remove the 4 bolts and rear stabilizer bar.
4. Remove the rear No. 1 stabilizer bar bracket.
5. Remove the rear stabilizer bushing.

To install:

6. To install, reverse the removal procedure.
7. Tighten the rear stabilizer bar bolts to 14 ft. lbs. (19 Nm).
8. Tighten the rear stabilizer link assembly nut to 29 ft. lbs. (39 Nm).

4WD Vehicles

See Figure 278.

1. Remove deck board assembly.
2. Remove no. 3 deck board sub-assembly (w/ Tonneau Cover).
3. Remove no. 2 deck board sub-assembly (w/ Tonneau Cover).
4. Remove tonneau cover assembly (w/ Tonneau Cover).
5. Remove rear mat.
6. Remove deck trim service hole cover.
7. Remove lower spare wheel carrier hinge cover (w/ cover).
8. Remove spare tire.
9. Remove spare wheel carrier lock cover (w/ cover).
10. Remove rear wheels.
11. Remove tail exhaust pipe assembly.
12. Remove rear stabilizer link assembly.
13. Remove the 2 bolts and the rear stabilizer bar bracket.

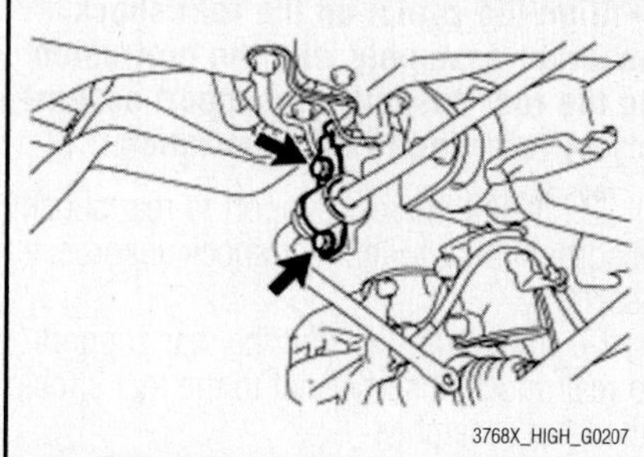

Fig. 278 Remove the 2 bolts and the rear stabilizer bar bracket

14. Remove the rear stabilizer bushing.
15. Remove the bolt and the rear stabilizer bar bracket.

To install:

16. Temporarily install the rear stabilizer bar with the identification mark positioned on the left side of the vehicle.
17. Temporarily install the LH and RH rear stabilizer bar bracket (front side) with the bolt.
18. Loosely tighten the bolt so that the bracket can be moved by hand.
19. Install the rear stabilizer bushings to the rear stabilizer bar.

➡Make sure that the cutout of the rear stabilizer bushing is positioned towards the rear of the vehicle.

20. Install the rear stabilizer bar bracket LH and RH (rear side) with the 2 bolts. Tighten to 14 ft. lbs. (19 Nm).
21. Fully tighten the LH rear stabilizer bar brackets. Tighten to 40 ft. lbs. (54 Nm).
22. Fully tighten the RH rear stabilizer bar brackets. Tighten to 14 ft. lbs. (19 Nm).
23. Install the rear stabilizer link assembly RH and LH to the rear shock absorber with coil spring RH and LH with the nut. Tighten to 29 ft. lbs. (39 Nm).
24. To complete installation, reverse remaining removal procedure.

STRUT

REMOVAL & INSTALLATION

See Figures 279 and 280.

1. Remove the wheel.
2. Remove the deck side trim cover.
3. Remove the deck side trim.
4. For 4WD vehicles, remove the bolt and separate the rear flexible hose from the rear shock absorber with coil spring.
5. Remove the bolt and separate the rear speed sensor from the rear shock absorber with coil spring.
6. Remove the nut and separate the rear stabilizer link assembly from the rear shock absorber with coil spring.

➡If the ball joint turns together with the nut, use a hexagon wrench (5 mm) to hold the stud bolt.

7. Disengage the 4 claws and remove the rear No. 1 suspension support cover.
8. Using a jack and wooden block, support the rear axle carrier sub-assembly.

➡Do not deform the dust cover.

➡Support the rear axle carrier sub-assembly until reinstallation of the rear shock absorber with coil spring is complete.

9. Loosen the rear support to rear shock absorber nut.

➡Do not remove the rear support to rear shock absorber nut.

➡Loosen the nut only when the rear shock absorber with coil spring needs to be disassembled.

10. Remove the 2 bolts and 2 nuts, and separate the rear shock absorber with coil spring from the rear axle carrier sub-assembly,

➡When removing the nuts, keep the bolts from rotating.

11. Remove the 3 nuts and rear shock absorber with coil spring.

➡Make sure that the rear speed sensor and rear flexible hose are disconnected from the rear shock absorber with coil spring.

12. Remove the rear support to rear shock absorber nut, as follows:
    a. Using SST: 09727-30021, compress the rear coil spring.

➡Do not use an impact wrench. It will damage SST.

    b. Check that the front coil spring is fully compressed.
    c. Hold the rear suspension support assembly and remove the rear support to rear shock absorber nut from the rear shock absorber assembly.
13. Remove the rear support to rear shock absorber collar from the rear shock absorber assembly.
14. Remove the rear suspension support assembly from the rear shock absorber assembly.
15. Remove the rear coil spring together with SST from the rear shock absorber assembly.

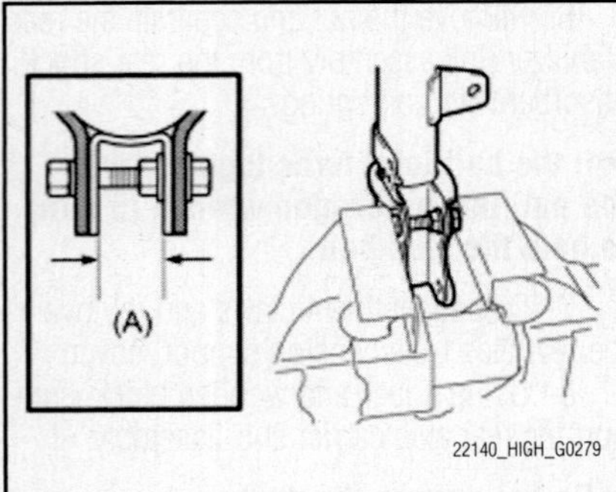

Fig. 279 Install the bolt and nut to the rear shock absorber with coil spring 2WD

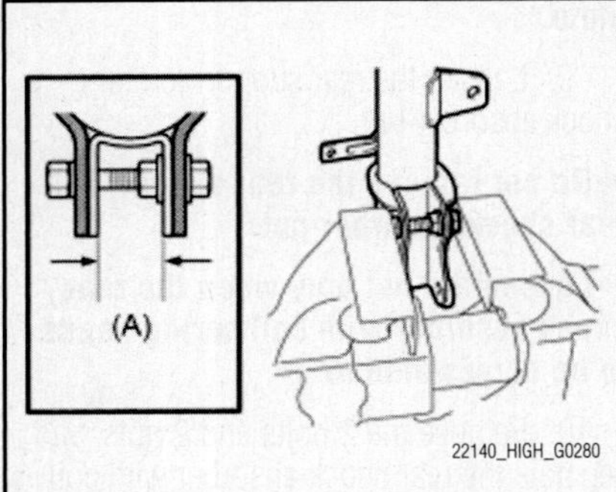

Fig. 280 Install the bolt and nut to the rear shock absorber with coil spring 4WD

16. Remove the rear No. 1 spring bumper from the rear shock absorber assembly.
17. Remove the rear lower coil spring insulator from the rear shock absorber assembly.

To install:

18. Install the rear lower coil spring insulator onto the rear shock absorber assembly.

➡Fit the recessed part of the rear lower coil spring insulator into the recession on the shock absorber assembly.

19. Install the rear No. 1 spring bumper to the rear shock absorber assembly.
20. Temporarily install rear coil spring, as follows:
    a. Using SST: 09727-30021 or equivalent, compress the rear coil spring.

➡Do not use an impact wrench. It will damage the SST.

    b. Temporarily install the rear coil spring together with SST to the rear shock absorber assembly.
21. Install the rear suspension support assembly to the rear shock absorber assembly.

➡Align the cutout on the rear shock absorber assembly with the protrusion on the rear suspension support assembly by referring to the illustration.

22. Install the rear support to rear shock absorber collar to the rear shock absorber assembly.
23. Temporarily install the rear support to rear shock absorber nut to the rear shock absorber assembly.
24. Install the rear coil spring.

➡Do not use an impact wrench. It will damage the SST.

➡Make sure that the end of the rear coil spring is positioned in the depression of the rear lower coil spring insulator.

➡Ensure that the stud bolt is positioned 3.5° to the outside of the vehicle as shown in the illustration. The deviation should be within +-5°.

25. Install the rear shock absorber with coil spring with the 3 nuts and tighten to 43 ft. lbs. (58 Nm).
26. Install the rear shock absorber with coil spring with the 2 bolts and 2 nuts and tighten to 213 ft. lbs. (290 Nm).

➡When installing the nuts, keep the bolts from rotating.

27. Fully tighten the rear support to rear shock absorber nut 40 ft. lbs. (55 Nm).
28. Install the rear No. 1 suspension support cover.
29. Install the rear stabilizer link assembly to the rear shock absorber with coil spring with the nut and tighten to 29 ft. lbs. (39 Nm).

➡If the ball joint turns together with the nut, use a hexagon wrench (5 mm) to hold the stud bolt.

30. Install the rear speed sensor wire to the rear shock absorber with coil spring with the bolt and tighten to 44 inch lbs. (5 Nm).

➡Do not twist the rear speed sensor wire when installing it.

31. Install the rear flexible hose to the rear shock absorber with coil spring with the bolt and tighten to 14 ft. lbs. (19 Nm).

➡Do not twist the rear flexible hose when installing it.

32. Install deck side trim.
33. Install deck side trim cover.
34. Install the wheels and tighten lug nuts to 76 ft. lbs. (103 Nm).
35. Inspect and adjust rear wheel alignment.

WHEEL HUB & BEARING

REMOVAL & INSTALLATION

See Figure 281.

1. Disconnect the negative battery cable.
2. Remove the wheel.
3. Separate the rear flexible hose.
4. Remove the 2 bolts and separate the rear disc brake caliper assembly.
5. Remove the rear disc.
6. For 4WD vehicles: Using a screwdriver, disconnect the connector from the rear speed sensor.
7. Remove the 4 bolts and the rear axle hub & bearing assembly.

To install:

8. Install the rear axle hub and bearing assembly with the 4 bolts and tighten to 55 ft. lbs. (75 Nm).
9. Inspect rear axle hub bearing looseness.
10. Inspect rear axle hub runout.
11. Connect the connector to the rear speed sensor.
12. Install the rear disc.
13. Install the rear disc brake caliper assembly with the 2 bolts and tighten to 57 ft. lbs. (78 Nm).
14. Install the rear flexible hose to the shock absorber with coil spring with the bolt and tighten to 14 ft. lbs. (19 Nm).
15. Install the wheel and tighten lug nuts to 76 ft. lbs. (103 Nm).
16. Connect the negative battery cable.
17. Inspect and adjust rear wheel alignment.
18. Check the ABS speed sensor signal.

Fig. 281 Remove the 4 bolts and the rear axle hub & bearing assembly

LEXUS

RX450h

31

BRAKES 31-9

CHASSIS ELECTRICAL 31-17

DRIVE TRAIN 31-18

ENGINE COOLING 31-20

ENGINE ELECTRICAL 31-22

ENGINE MECHANICAL......31-37

ENGINE PERFORMANCE & EMISSION CONTROLS 31-62

SPECIFICATIONS AND MAINTENANCE CHARTS

ENGINE AND VEHICLE IDENTIFICATION

Engine							Model Year	
Code	Liters (cc)	Cu. In.	Cyl.	Fuel Sys.	Engine Type	Eng. Mfg.	Code ①	Year
2GR-FXE	3.5 (3456)	210	6	SFI	DOHC	Toyota	A	2010

SFI: Sequential Fuel Injection

DOHC: Double Overhead Camshaft

① 10th digit of the VIN

3768X_RX45_C0001

GENERAL ENGINE SPECIFICATIONS

Year	Model	Engine Displacement Liters	Engine Series ID	Net Horsepower @ rpm	Net Torque @ rpm (ft. lbs.)	Bore x Stroke (in.)	Com-pression Ratio	Oil Pressure @ rpm
2010	RX450h	3.5	2GR-FXE	245@6000	234@4800	3.70x3.27	10.8:1	55@3000

3768X_RX45_C0002

ENGINE TUNE-UP SPECIFICATIONS

Year	Engine Displacement Liters	Engine ID	Spark Plug Gap (in.)	Ignition Timing (deg.)*	Fuel Pump (psi)	Idle Speed (rpm)	Valve Clearance	
							Intake	Exhaust
2010	3.5	2GR-FXE	0.043	8-12B	44-50	600-700	NA	NA

NA: Not Available

NOTE: The Vehicle Emission Control Information label often reflects specification changes made during production.

The label figures must be used if they differ from those in this chart.

B: Before top dead center

* With terminals TC and CG connected to DLC3 (ODB II connector)

3768X_RX45_C0003

CAPACITIES

Year	Model	Engine Displacement Liters	Engine ID	Engine Oil with Filter (qts.)	Transmission (pts.) Auto.*	Transfer Case (pts.)	Drive Axle Front (pts.)	Drive Axle Rear (pts.)	Fuel Tank (gal.)	Cooling System (qts.)
2010	RX450h	3.5	2GR-FXE	6.4	3.7	2.0	NE	2.0	19.1	①

NE: Not Equipped

*After draining, add the following amounts, then, fill to the cold full line.

① Non-towing pkg. 7.3 qt. w/o rear air, 9.6 qts. w/ rear air

Towing pkg. 8 qt. w/o rear air, 10.4 qts. w/ rear air

3768X_RX45_C0004

FLUID SPECIFICATIONS

Year	Model	Engine Displ. Liters	Engine Oil	Auto. Trans.	Drive Axle Rear ①	Transfer Case	Power Steering Fluid ②	Brake Master Cylinder	Cooling System
2010	RX450h	3.5	5W-30	ATF WS	80W-90	80W-90	NE	DOT 3	SLLC ③

NE: Not Equipped

DOT: Department Of Transpotation

① Oil grade: Hypoid gear oil API GL-5

② The RX450h is not equipped with a power steering pump. It utilizes a power steering motor.

③ Toyota Super Long Life Coolant

3768X_RX45_C0005

VALVE SPECIFICATIONS

Year	Engine Displacement Liters	Engine ID	Seat Angle (deg.)	Face Angle (deg.)	Spring Test Pressure (lbs. @ in.)	Spring Installed Height (in.)	Stem-to-Guide Clearance (in.) Intake	Stem-to-Guide Clearance (in.) Exhaust	Stem Diameter (in.) Intake	Stem Diameter (in.) Exhaust
2010	3.5	2GR-FXE	45	44.5	NA	NA	0.0010-0.0024	0.0012-0.0026	0.2154-0.2159	0.2152-0.2157

NA: Information not available

3768X_RX45_C0006

CAMSHAFT AND BEARING SPECIFICATIONS

All measurements are given in inches.

Year	Engine Displacement Liters	Engine VIN	Journal Diameter	Brg. Oil Clearance	Shaft End-play	Runout	Journal Bore	Lobe Lift	
								Intake	Exhaust
2010	3.5	2GR-FXE	①	②	NA	0.0016	NA	1.7447-1.7487	1.7426-1.7465

NA: Not Available

① No. 1 journal: 1.4152 to 1.4157 inches

Other journals: 1.0220 to 1.0226 inches

② No. 1 journal: 0.0016 to 0.0031 inches

Other journals: 0.00098 to 0.0024 inches

3768X_RX45_C0007

CRANKSHAFT AND CONNECTING ROD SPECIFICATIONS

All measurements are given in inches.

Year	Engine Displacement Liters	Engine ID	Crankshaft				Connecting Rod		
			Main Brg. Journal Dia.	Main Brg. Oil Clearance	Shaft End-play	Thrust on No.	Journal Diameter	Oil Clearance	Side Clearance
2010	3.5	2GR-FXE	2.4011-2.4016	0.0010-0.0019	0.0016-0.0095	2	2.0863-2.0866	0.0018-0.0026	0.0059-0.0157

NA: Not Available

3768X_RX45_C0008

PISTON AND RING SPECIFICATIONS

All measurements are given in inches.

Year	Engine Displacement Liters	Engine ID	Piston Clearance	Ring Gap			Ring Side Clearance		
				Top Comp.	Bottom Comp.	Oil Control	Top Comp.	Bottom Comp.	Oil Control
2010	3.5	2GR-FXE	0.0018-0.0020	0.0098-0.0138	0.0197-0.0236	0.0039-0.0157	0.0008-0.0028	0.0008-0.0024	0.0028-0.0059

3768X_RX45_C0009

TORQUE SPECIFICATIONS

All readings in ft. lbs.

Year	Engine Displacement Liters	Engine ID	Cylinder Head Bolts	Main Bearing Bolts	Rod Bearing Bolts	Crankshaft Damper Bolts	Flywheel Bolts	Manifold Intake	Manifold Exhaust	Spark Plugs	Oil Pan Drain Plug
2010	3.5	2GR-FXE	①	②	③	184	61	15	15	18	33

① Step 1: 10mm bolts to 27 ft. lbs.
Step 2: 10mm point cap bolts plus 90 degrees
Step 3: 10mm point cap bolts plus 90 degrees
Step 4: Front bolts to 22 ft. lbs.

② Step 1: 16 cap bolts to 45 ft. lbs.
Step 2: 16 cap bolts plus 90 degrees
Step 3: 8 side bolts to 38 ft. lbs.

③ Step 1: 18 ft. lbs.
Step 2: Plus 90 degrees

3768X_RX45_C0010

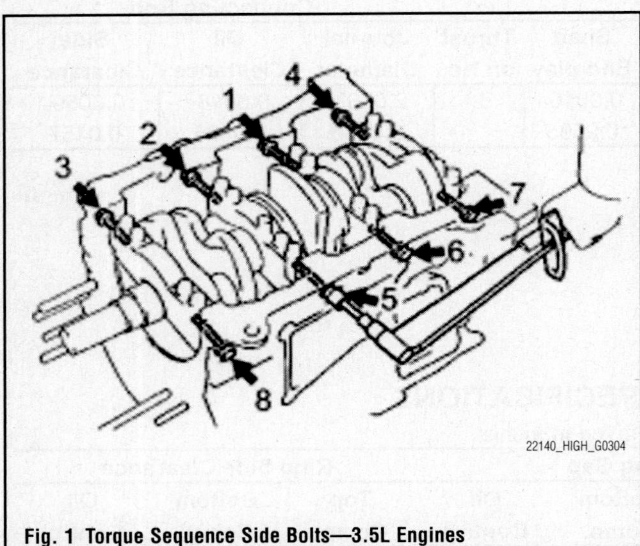

Fig. 1 Torque Sequence Side Bolts—3.5L Engines

WHEEL ALIGNMENT

Year	Model		Caster Range (+/-Deg.)	Caster Preferred Setting (Deg.)	Camber Range (+/-Deg.)	Camber Preferred Setting (Deg.)	Toe-in (in.)	Steering Axis Inclination (Deg.)
2010	RX450h	Front	0.75	+2.62	0.75	-0.63	0+/-0.08	11.02+/-0.75
		Rear	NA	NA	0.75	-1.00	0.12+/-0.08	NA

NA: Not Available

3768X_RX45_C0011

TIRE, WHEEL AND BALL JOINT SPECIFICATIONS

		OEM Tires		Tire Pressures (psi)		Wheel	Ball Joint	Lug Nut Torque
Year	Model	Standard	Optional	Front	Rear	Size	Inspection	(ft. lbs.)
2010	RX450h	P245/65R17	P245/55R19	30	30	7.5-J	①	76

OEM: Original Equipment Manufacturer

PSI: Pounds Per Square Inch

① Replace if any measurable movement is found.

3768X_RX45_C0012

BRAKE SPECIFICATIONS

All measurements in inches unless noted

			Brake Disc			Minimum	Brake Caliper	
			Original	Minimum	Maximum	Lining	Bracket Bolts	Mounting Bolts
Year	Model		Thickness	Thickness	Runout	Thickness	(ft. lbs.)	(ft. lbs.)
2010	RX450h	F	1.102	1.024	0.0020	0.039	76	25
		R	0.394	0.335	0.0059	0.039	56	25

F: Front

R: Rear

3768X_RX45_C0013

SCHEDULED MAINTENANCE INTERVALS

LEXUS—RX450h

TO BE SERVICED	TYPE OF SERVICE	VEHICLE MILEAGE INTERVAL (x1000)													
		5	10	15	20	25	30	35	40	45	50	55	60	90	120
Engine oil & filter	R	✓	✓	✓	✓	✓	✓	✓	✓	✓	✓	✓	✓	✓	✓
Automatic transmission fluid	S/I			✓			✓			✓			✓	✓	✓
Ball joints & dust covers	S/I			✓			✓			✓			✓	✓	✓
Bolts & nuts on chassis & body	S/I			✓			✓			✓			✓	✓	✓
Brake linings & drums	S/I	✓	✓	✓	✓	✓	✓	✓	✓	✓	✓	✓	✓	✓	✓
Brake line pipes & hoses	S/I			✓			✓			✓			✓	✓	✓
Brake pads & discs (front & rear)	S/I	✓	✓	✓	✓	✓	✓	✓	✓	✓	✓	✓	✓	✓	✓
Brake fluid	R						✓						✓	✓	✓
Rack and pinion assembly	S/I			✓			✓			✓			✓	✓	✓
Steering linkage & boots	S/I			✓			✓			✓			✓	✓	✓
Air cleaner filter	R						✓						✓	✓	✓
Spark plugs ①	R														✓
Drive belts	S/I												✓	✓	✓
Exhaust pipes & mountings	S/I			✓			✓			✓			✓	✓	✓
Fuel lines & connections	S/I						✓						✓	✓	✓
Engine coolant ②	S/I			✓			✓			✓			✓	✓	
Rear differential & transfer case oil	S/I			✓			✓			✓			✓	✓	✓
Fuel tank cap gasket	S/I						✓						✓	✓	✓
Rotate tires	S/I			✓			✓			✓			✓		✓
Clean air conditioning filter ③	S/I			✓			✓			✓			✓		✓
Axle shaft bolts	S/I			✓			✓			✓			✓	✓	✓
Brake pad thickness and rotor runout	S/I						✓						✓	✓	✓

R: Replace S/I: Service or Inspect

① Spark plugs are replaced at 120,000 miles

② Replace engine coolant at 100,000 miles and then inspect every 15,000 miles

③ Replace air conditioning filter every 30,000 miles

FREQUENT OPERATION MAINTENANCE (SEVERE SERVICE)

If a vehicle is operated under any of the following conditions it is considered severe service:

- Extremely dusty areas.

- 50% or more of the vehicle operation is in 32°C (90°F) or higher temperatures, or constant temperatures below 0°C (32°F).

- Prolonged idling (vehicle operation in stop and go traffic).

- Frequent short running periods (engine does not warm to normal operating temperatures).

- Police, taxi, delivery usage or trailer towing usage.

Air cleaner filter: service or inspect every 5000 miles

Rear differential & transfer case oil: replace every 15,000 miles.

Ball joints & dust covers: service or inspect every 5000 miles.

Bolts & nuts on chassis & body: service or inspect every 5000 miles.

Axle shaft bolts: service or inspect every 5000 miles.

Steering linkage: service or inspect every 5000 miles.

3768X_RX45_C0014

BRAKES — INFORMATION AND PRECAUTIONS

ANTI-LOCK SYSTEMS

- Certain components within the ABS system are not intended to be serviced or repaired individually.
- Do not use rubber hoses or other parts not specifically specified for and ABS system. When using repair kits, replace all parts included in the kit. Partial or incorrect repair may lead to functional problems and require the replacement of components.
- Lubricate rubber parts with clean, fresh brake fluid to ease assembly. Do not use shop air to clean parts; damage to rubber components may result.
- Use only DOT 3 brake fluid from an unopened container.
- If any hydraulic component or line is removed or replaced, it may be necessary to bleed the entire system.
- A clean repair area is essential. Always clean the reservoir and cap thoroughly before removing the cap. The slightest amount of dirt in the fluid may plug an orifice and impair the system function. Perform repairs after components have been thoroughly cleaned; use only denatured alcohol to clean components. Do not allow ABS components to come into contact with any substance containing mineral oil; this includes used shop rags.
- The Anti-Lock control unit is a microprocessor similar to other computer units in the vehicle. Ensure that the ignition switch is **OFF** before removing or installing controller harnesses. Avoid static electricity discharge at or near the controller.
- If any arc welding is to be done on the vehicle, the control unit should be unplugged before welding operations begin.

DISC AND DRUM SYSTEMS

CAUTION

Dust and dirt accumulating on brake parts during normal use may contain asbestos fibers from production or aftermarket brake linings. Breathing excessive concentrations of asbestos fibers can cause serious bodily harm. Exercise care when servicing brake parts. Do not sand or grind brake lining unless equipment used is designed to contain the dust residue. Do not clean brake parts with compressed air or by dry brushing. Cleaning should be done by dampening the brake components with a fine mist of water, then wiping the brake components clean with a dampened cloth. Dispose of cloth and all residue containing asbestos fibers in an impermeable container with the appropriate label. Follow practices prescribed by the Occupational Safety and Health Administration (OSHA) and the Environmental Protection Agency (EPA) for the handling, processing, and disposing of dust or debris that may contain asbestos fibers.

BRAKES — BLEEDING THE BRAKE SYSTEM

BLEEDING PROCEDURE

BLEEDING PROCEDURE

See Figures 2 and 3.

➡This procedure requires specialized tools. Please read through the procedure and make sure you have access to the proper equipment before beginning the bleeding procedure.

CAUTION

Never bleed air from the brake hydraulic system without using the intelligent tester. failure to use the intelligent tester could cause serious injury or an accident.

Note the following before bleeding the brake system:

- Move the shift lever to the P position and apply the parking brake before bleeding.
- Add brake fluid carefully and check that the reservoir level remains between the min and max lines while bleeding the brakes.
- Do not stand the fluid can on the reservoir inlet when bleeding the brake actuator. doing so will cause brake fluid to overflow.
- The actuator pump motor and solenoid can be operated by the driver even if the ignition switch is off.
- If the pump motor operates while air still remains inside the brake actuator hose, air will enter the actuator, making it more difficult to bleed the brakes. If there is concern about air remaining in the actuator hose, remove the two motor relays (skid control relay no.2) until instructed to reinstall them.
    - Although a buzzer may sound due to a decline in the accumulator pressure while bleeding, it is not necessary to stop bleeding.
    - Daces indicating a malfunction in the motor relays (skid control relay no.2) or the pressure sensor are stored after bleeding. clear the Daces when instructed during or after bleeding.

1. Add SAE J1703 or FMVSS no. 116 DOT 3 brake fluid to the max line in the reservoir.

CAUTION

Add brake fluid carefully and check that the reservoir level remains between the min and max lines while bleeding the brakes. Do not stand the fluid can on the reservoir inlet when bleeding the brake actuator. Doing so will cause brake fluid to overflow.

2. Disable the brake control (ECB). When using the intelligent tester:

➡When using the intelligent tester, refer to the intelligent tester operator's manual for further details. Bleed the air by following the steps displayed on the intelligent tester.

a. Move the shift lever to the P position and apply the parking brake.

b. Connect the intelligent tester to the DLC3 with the ignition switch **OFF** as shown in the illustration.

c. Turn the ignition switch to the ON position and turn on the intelligent tester.

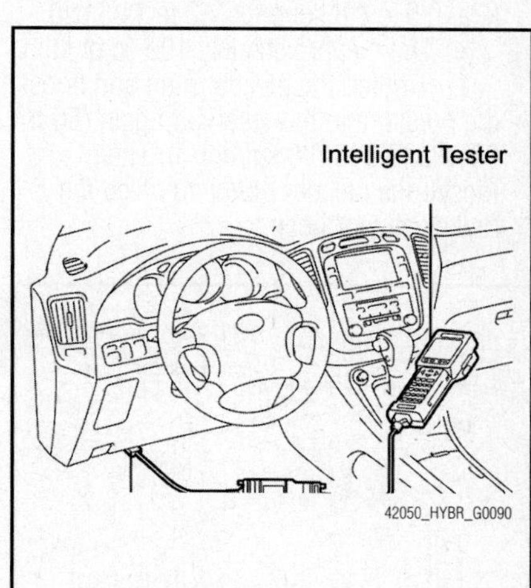

Fig. 2 Connect the intelligent tester to the DLC3 with the ignition switch OFF as shown in the illustration

➡Do not start the engine.

d. Enter the following menus: DIAGNOSIS / OBD/ MOBD / ABS/TRAC/VSC / ECB UTILITY / ECB INVALID.

⁂ WARNING

If the pump motor operates while air remains inside the brake actuator hose, air will enter the actuator, and this will make bleeding the brakes more difficult.

e. When removing the ABS motor relay: Remove the 2 ABS motor relays with the ignition switch off in order to disable brake control.

⁂ WARNING

If the pump motor operates while air remains inside the brake actuator hose, air will enter the actuator, and this will make bleeding the brakes more difficult.

➡After the brake actuator assembly has been replaced, remove the ABS motor relay

before bleeding the brakes.

3. **Bleed the brake actuator hose, as follows:**

a. Connect Special Tool 09992-00242, 0992-00350 or equivalent, to the reservoir with the brake reservoir pressure adapter.

b. Using Special Tool 09023-00101, loosen the bleeder plug of the actuator.

c. Connect a vinyl tube to the bleeder plug of the actuator.

d. Use the SST to boost pressure in the reservoir. Standard pressure is 50 to 80 kPa (0.5 to 0.8 kgf/cm2, 7.3 to 11.6 psi)

e. Drain approximately 100 cc of fluid.

f. Tighten the bleeder plug and boost the pressure in the reservoir again (50 to 80 kPa (0.5 to 0.8 kgf/cm2)). Then, loosen the bleeder plug and bleed the brake actuator hose.

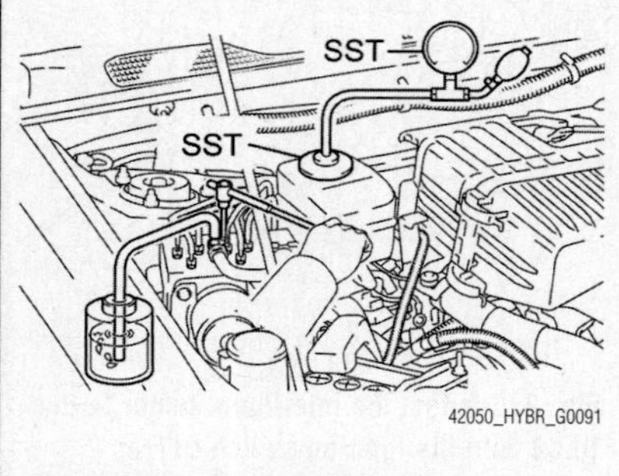

Fig. 3 View of the special tools

➡Repeat this procedure at least 5 times.

g. When air is completely bled out from the hose between the reservoir and the actuator, tighten the bleeder plug to 74 inch lbs. (8.3 Nm).

4. **Bleed the master cylinder, as follows:**

➡If the master cylinder has been disassembled or if the reservoir becomes empty, bleed the air from the master cylinder.

a. Enter the following menus: DIAGNOSIS / OBD/MOBD / ABS/TRAC/VSC / AIR BLEEDING.

b. Select "USUAL" if the front/rear brakes are removed, installed or disassembled.

c. Select "ACTUATOR" if the actuator is removed, installed or replaced.

d. Select "MASTER CYLINDER" if the brake master cylinder or the brake stroke simulator is removed, installed or replaced.

e. Disconnect the brake lines from the master cylinder.

f. Slowly depress and hold the brake pedal (Procedure A).

g. Cover the outer holes with fingers, and release the brake pedal (Procedure B).

h. Repeat procedure A and B 3 or 4 times.

i. Connect the brake lines to the master cylinder and tighten to 11 ft. lbs. (15 Nm).

5. **Bleed the front brake system, as follows:**

➡Air can be easily bled from the front brake system if air has been bled from the master cylinder when replacing the brake master cylinder assembly.

⁂ WARNING

If brake fluid leaks onto any painted surface of the vehicle, wash or otherwise remove it completely.

➡Bleed the air by following the steps displayed on the intelligent tester.

(a) Depress the brake pedal several times and bleed the front brake system from the bleeder plugs on the front brake cylinder RH and LH.

➡Repeat the procedure until air is completely bled from the front brake system.

6. Tighten the bleeder plugs to 74 inch lbs. (8.3 Nm) after bleeding.

7. Cancel brake control (ECB) disable

a. Install the 2 motor relays (skid control relay No.2) if they have been removed.

b. Complete brake control prevention following the prompts on the tester screen. (If brake control has been prevented using the intelligent tester.)

8. Clear the DTC(s).

9. Bleed the rear brake system, as follows:

⁂ WARNING

Never bleed air from the brake hydraulic system without using the intelligent tester. Failure to use the intelligent tester could cause serious injury or an accident.

➡Bleed the air by following the steps displayed on the intelligent tester.

a. Connect the intelligent tester to the DLC3 with the ignition switch off.

b. Check that the parking brake is applied and turn the ignition switch to the **ON** position.

c. Enter the following menus: DIAGNOSIS / OBD/ MOBD / ABS/TRAC/VSC / ECB UTILITY / ECB INVALID.

d. With the brake pedal depressed, bleed the rear brake system from the bleeder plug on the rear disc brake cylinder LH while the pump motor and solenoid are operating.

⁂ WARNING

Keep the fluid inside the reservoir above the LOW level by replenishing.

➡Depress and hold the brake pedal. After the solenoid operates for approximately 30 seconds, release the brake pedal to stop the solenoid. Repeat the procedures until air is completely bled from the rear brake system. The ECB warning light comes on and the buzzer sounds while bleeding, but they do not indicate a malfunction.

e. Tighten the bleeder plug to 74 inch lbs. (8.3 Nm) after bleeding.

f. Enter the following menus: DIAGNOSIS / OBD/ MOBD / ABS/TRAC/VSC / ECB UTILITY / ECB INVALID.

g. With the brake pedal depressed, bleed the rear brake system from the bleeder plug on the rear disc brake cylinder RH while the pump motor and solenoid are operating.

⁂ WARNING

Keep the fluid inside the reservoir above the LOW level by replenishing.

➡Depress and hold the brake pedal. After the solenoid operates for approximately 30 seconds, release the brake

pedal to stop the solenoid. Repeat the procedures until air is completely bled from the rear brake system. The ECB warning light comes on and the buzzer sounds while bleeding, but they do not indicate a malfunction.

h. Tighten the bleeder plug to 74 inch lbs. (8.3 Nm) after bleeding.

10. Perform the accumulator zero down:

WARNING

Never bleed air from the brake hydraulic system without using the intelligent tester. Failure to use the intelligent tester could cause serious injury or an accident. Be sure to perform this procedure before replacement, removal, or installation of the actuator.

➡**Perform accumulator zero down by following the steps displayed on the intelligent tester.**

a. Connect the intelligent tester to the DLC3 with the ignition switch **OFF**.

b. Depressurize the accumulator:

- Check that the parking brake is applied and turn the ignition switch to the **ON** position.
- Enter the following menus: DIAGNOSIS / OBD/ MOBD / ABS/TRAC/ VSC / ECB UTILITY / ZERO DOWN.
- When the buzzer sounds, turn the ignition switch **OFF**.

c. Circulate the fluid in the accumulator.

d. Depressurize the accumulator 5 times.

➡**Accumulator pressure is released and accumulated repeatedly, which circulates the fluid inside the accumulator, when repeating accumulator zero down. The pump motor rotates and the accumulator is pressurized every time the ignition switch is turned from off to on.**

11. Check the brake fluid level:

a. After performing accumulator zero down (accumulator depressurizing), return the fluid in the accumulator back to the reservoir and then adjust the fluid level in the master cylinder reservoir to the MAX level.

➡**After performing accumulator zero down (accumulator depressurizing), fluid is built up in the accumulator by turning the ignition switch to the ON position and the fluid level of the reservoir lowers. If the fluid level is adjusted without performing accumulator zero down (accumulator depressurizing), fluid is sent from the accumulator to the reservoir. The fluid level may exceed the MAX level, but it is normal.**

12. Clear the DTC(s).

13. When the brake actuator assembly is replaced, perform linear valve offset learning after bleeding is completed.

BRAKES

FRONT DISC BRAKES

BRAKE CALIPER

REMOVAL & INSTALLATION

See Figure 4.

➡**While the battery is connected, even if the power switch is off, the brake control system activates when the brake pedal is depressed or the door courtesy switch turns on. Therefore, during servicing of the brake system components, do not operate the brake pedal or open/close the doors while the battery is connected.**

1. Carefully raise the vehicle.
2. Remove the front wheel.
3. Drain the brake fluid.
4. Remove the bolt and gasket, and disconnect the front flexible hose from the disc brake caliper assembly.
5. Hold the front disc brake caliper slide pins and remove the 2 bolts and disc brake caliper assembly.

To install:

6. Hold the front disc brake caliper slide pins and install the disc brake caliper assembly with the 2 bolts. Tighten the bolts to 25 ft. lbs. 34 (Nm).
7. Connect the front flexible hose with the union bolt and a new gasket to the disc brake caliper assembly. Tighten the banjo bolt to 20 ft. lbs. (30 Nm).

➡**Install the front flexible hose lock securely in the lock hole in the disc brake caliper assembly.**

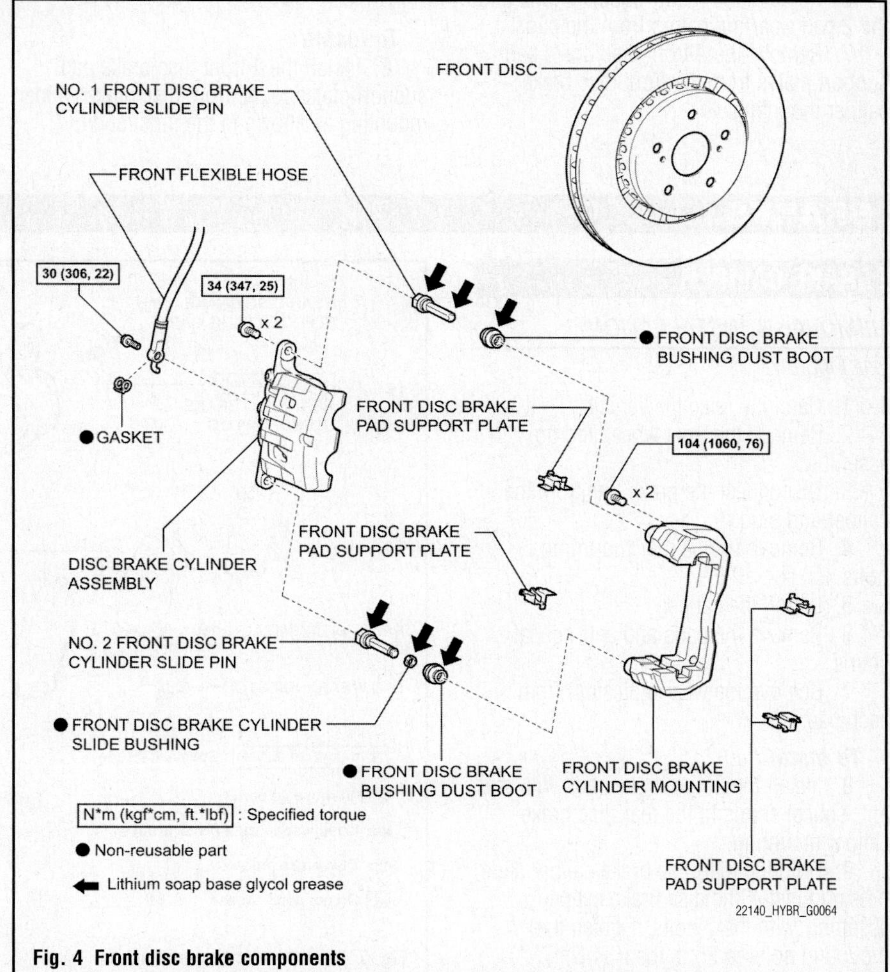

Fig. 4 Front disc brake components

8. Fill the master cylinder reservoir with brake fluid.
9. Bleed the brake system.
10. Inspect for brake fluid leaks.
11. Perform the accumulator zero down procedure.
12. Inspect the brake fluid level.
13. Install the front wheel and tighten to 76 ft. lbs. (103 Nm).
14. Clear the Daces
15. Check for Daces. If any DTC is output, perform the troubleshooting for that DTC.

DISC BRAKE PADS

REMOVAL & INSTALLATION

See Figures 5 and 6.

1. Carefully raise the vehicle.
2. Remove the front wheel.
3. Hold the front disc brake caliper slide pins and remove the 2 bolts and disc brake caliper assembly.
4. Remove the 2 anti-squeal springs.
5. Remove the 2 brake pads from the front disc brake caliper mounting.
6. Remove the 2 anti-squeal shims and the 2 pad wear indicators from the pads.
7. Remove the 4 front disc brake pad support plates from the front disc brake caliper mounting.

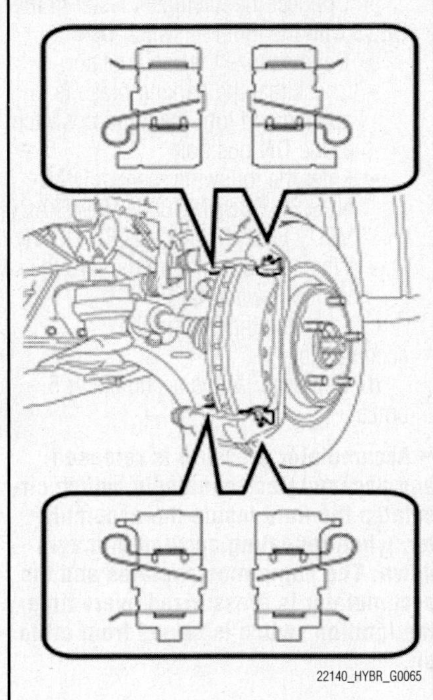

Fig. 5 Install the 4 front disc brake pad support plates as shown

To install:

8. Install the 4 front disc brake pad support plates to the front disc brake caliper mounting as shown in the illustration.

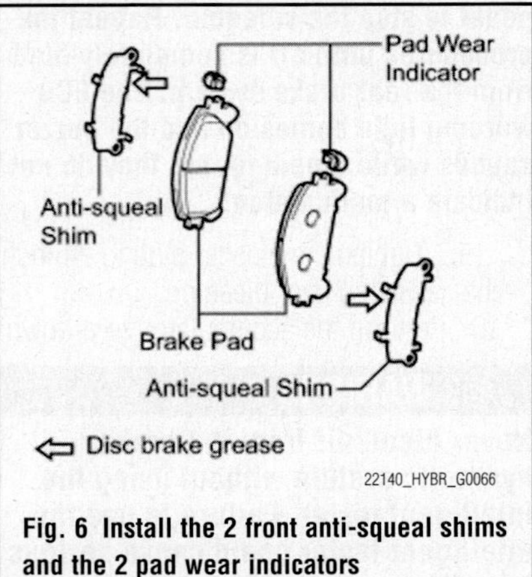

Fig. 6 Install the 2 front anti-squeal shims and the 2 pad wear indicators

9. Apply disc brake grease to the inside of the 2 front anti-squeal shims.
10. Install the 2 front anti-squeal shims and the 2 pad wear indicators to the pads.
11. Install the 2 brake pads with the front anti-squeal shims to the front disc brake caliper mounting.
12. Install the 2 anti-squeal springs to the front disc brake pads.
13. Hold the front disc brake caliper slide pins and install the disc brake caliper assembly with the 2 bolts. Tighten the bolts to 25 ft. lbs. (34 Nm).
14. Tighten the wheel to 76 ft. lbs. (103 Nm).

BRAKE CALIPER

REMOVAL & INSTALLATION

See Figure 7.

1. Carefully raise the vehicle.
2. Remove the front wheel and tire assembly.
3. Disconnect the brake line from the caliper and plug it.
4. Remove the caliper mounting bolts.
5. Lift off the caliper.
6. Remove the pads and anti-squeal shims.
7. Remove the wear indicators from each pad.

To install:

8. Install the 2 rear brake pads with rear anti-squeal shims to the rear disc brake caliper mounting.
9. Hold the rear disc brake caliper slide pins and install the disc brake caliper assembly with the 2 bolts. Tighten the mounting bolts to 25 ft. lbs. (34 Nm).

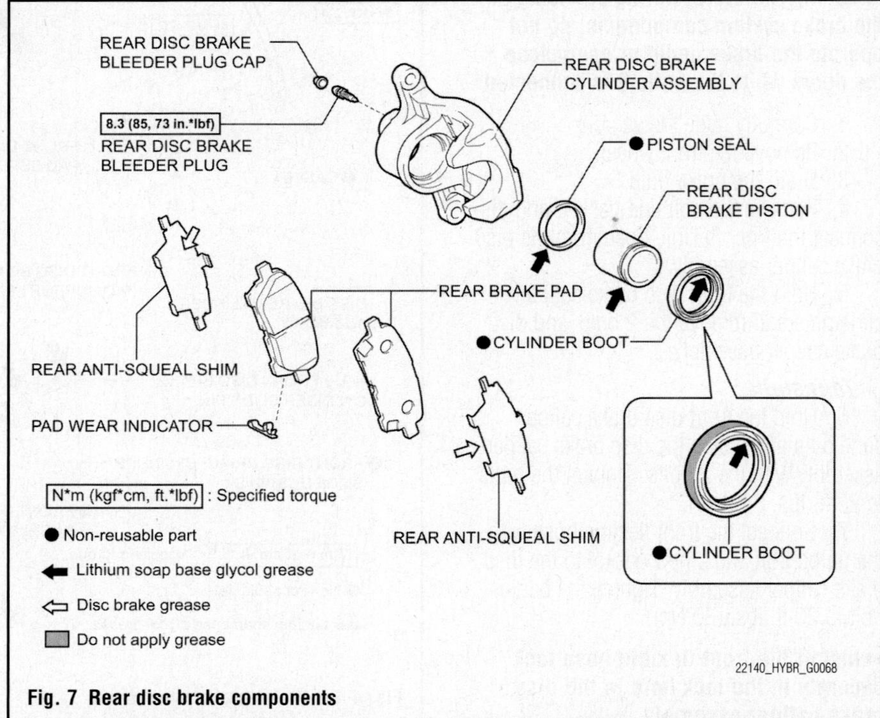

Fig. 7 Rear disc brake components

10. Install the brake hose to the caliper, tighten the banjo bolt to 24 ft. lbs. (33 Nm).

11. Tighten the wheel to 76 ft. lbs. (103 Nm).

12. Fill the master cylinder reservoir with brake fluid.

13. Bleed the brake system.

14. Inspect for brake fluid leaks.

15. Perform the accumulator zero down procedure.

16. Inspect the brake fluid level.

17. Install the front wheel and tighten to 76 ft. lbs. (103 Nm).

18. Clear the faces.

DISC BRAKE PADS

REMOVAL & INSTALLATION

See Figure 7.

1. Disconnect the brake line from the caliper and plug it.
2. Remove the caliper mounting bolts.
3. Lift off the caliper.
4. Remove the pads and anti-squeal shims.
5. Remove the wear indicators from each pad.
6. Installation is the reverse of removal. Grease the caliper slides and bolts with lithium grease or equivalent. Apply disc brake grease to the anti-squeal shims. Tighten the caliper bolts to 32 ft. lbs. (43 Nm); the brake line union bolt to 24 ft. lbs. (33 Nm).

PARKING BRAKE CABLES

ADJUSTMENT

See Figures 8 and 9.

1. Inspect parking brake pedal travel, as follows:
   a. Fully depress the parking brake pedal to engage the parking brake.
   b. Depress the pedal again to disengage the parking brake.
   c. Slowly depress the parking brake pedal using the specified force, and count the number of clicks. Parking brake pedal travel: 8 to 10 notches at 67 lbs (300 N). If the parking brake pedal travel is not as specified, adjust the parking brake shoe clearance and parking brake pedal travel.
2. Adjust parking brake shoe clearance and parking brake pedal travel, as follows:
   a. Remove the driver side knee airbag.
   b. Completely release the parking brake pedal.
   c. Loosen the lock nut and the adjusting nut to completely release the parking brake cable.

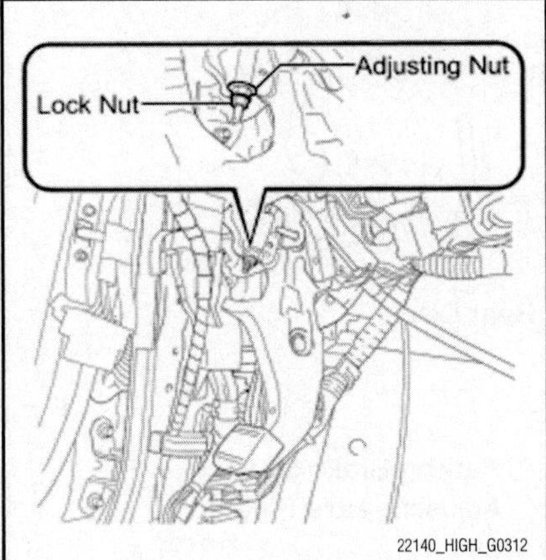

Fig. 8 Parking brake lock nut and adjusting nut

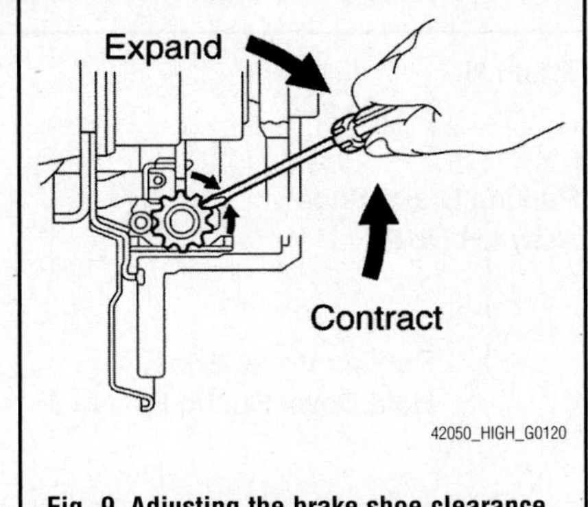

Fig. 9 Adjusting the brake shoe clearance

   d. Remove the rear wheel.
   e. Temporarily install the hub nuts.
   f. Remove the shoe adjusting hole plug.
   g. Turn the shoe adjuster and expand the shoe until the disc locks.
   h. Turn and contract the shoe adjuster until the disc can rotate smoothly. Standard: Return 8 notches.
   i. Check that there is no brake drag against the shoe.
   j. Install the shoe adjusting hole plug.
   k. Turn the adjusting nut until the parking brake pedal travel is corrected to be within the specified range. Parking brake pedal travel: 8 to 10 notches at 67 lbs (300 N).
   l. Using a wrench or an equivalent tool, hold the adjusting nut and tighten the lock nut and tighten to 62 inch lbs. (7 Nm).
   m. Operate the parking brake pedal 3 to 4 times, and check the parking brake pedal travel.
   n. Check that there is no brake drag against the shoe.
   o. Remove the hub nuts.
   p. Install the rear wheel and tighten the lug nuts to 76 ft. lbs. (103 Nm).
   q. Install the driver side knee airbag.
3. When operating the parking brake pedal, check that the brake warning light illuminates. Standard: the brake warning light always illuminates at the first click.

PARKING BRAKE SHOES

REMOVAL & INSTALLATION

See Figures 10 through 19.

1. Raise and safely support the vehicle.
2. Remove the rear wheel and tire assemblies.
3. Unbolt and remove the rear caliper, but do not disconnect the fluid line. Suspend the caliper out of the way with a piece of wire.

Fig. 10 Remove the parking brake shoe hold-down spring cups

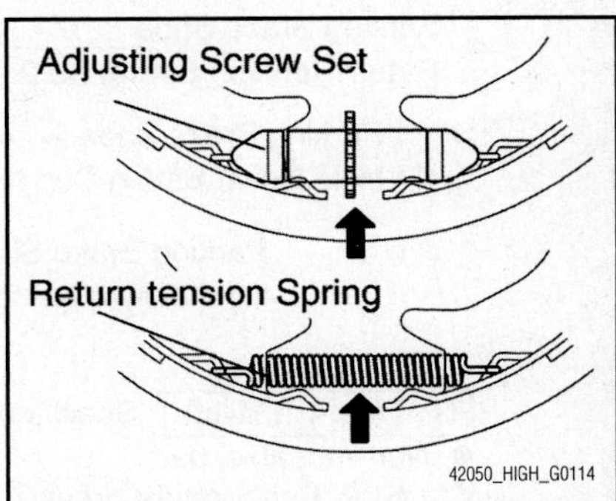

Fig. 11 Remove the parking brake shoe adjusting screw set and shoe return spring

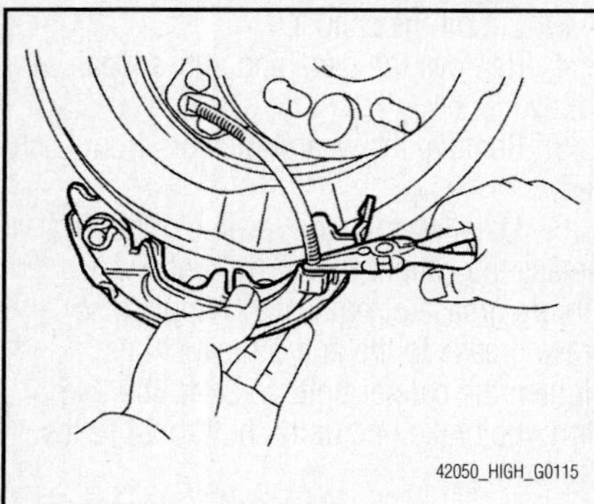

Fig. 12 Using needle-nose pliers, disconnect the parking brake cable no. 3 from the parking brake cable shoe lever

4. Matchmark the brake disc (rotor) to the axle hub.

5. Make sure the parking brake is fully released, then remove the rear brake disc (rotor).

➡**If the rotor cannot be easily removed, turn the shoe adjuster until the wheel turns freely.**

6. Inspect the brake disc (rotor) inside diameter, as follows:

a. Using a brake drum gauge or equivalent, measure the inside diameter of the disc and compare with the following: Standard inside diameter: 190 mm (7.48 in.). Maximum inside diameter: 191 mm (7.52 in.)

b. If the inside diameter exceeds the maximum, replace the brake disc.

7. Use needle-nose pliers to remove the 2 parking brake shoe return tension springs.

8. Remove the parking brake shoe strut, as follows:

a. Remove the parking brake shoe strut and the parking brake shoe strut compression spring.

9. Remove parking brake shoe no.1, as follows:

a. Remove the parking brake shoe hold down spring cup No.1, parking brake shoe hold down spring and parking brake shoe hold down spring cup no.2.

Parking Brake Shoe Strut LH
Rear Disc Brake Caliper Assy LH
Parking Brake Shoe Assy LH No.2
78 (800, 58)
Parking Brake Shoe Strut Compression Spring
Parking Brake Shoe Hold Down Spring Pin No.2
Parking Brake Shoe Hold Down Spring Cup No.2
Parking Brake Shoe Return Tension Spring No.1
Shim
Parking Brake Shoe Hold Down Spring
◆C-Washer
Parking Brake Shoe Lever
Parking Brake Shoe Hold Down Spring Cup No.1
Parking Brake Shoe Hold Down Spring Pin No.1
Parking Brake Shoe Assy LH No.1
Parking Brake Shoe Adjusting Screw Set
Parking Brake Shoe Return Tension Spring No.2
Parking Brake Shoe Hold Down Spring Cup No.2
Parking Brake Shoe Hold Down Spring
Parking Brake Shoe Hold Down Spring Cup No.1
Rear Disc
Parking Brake Shoe Adjusting Hole Plug
N·m (kgf·cm, ft·lbf) : Specified torque
◆ Non-reusable part
⇦ High Temperature grease

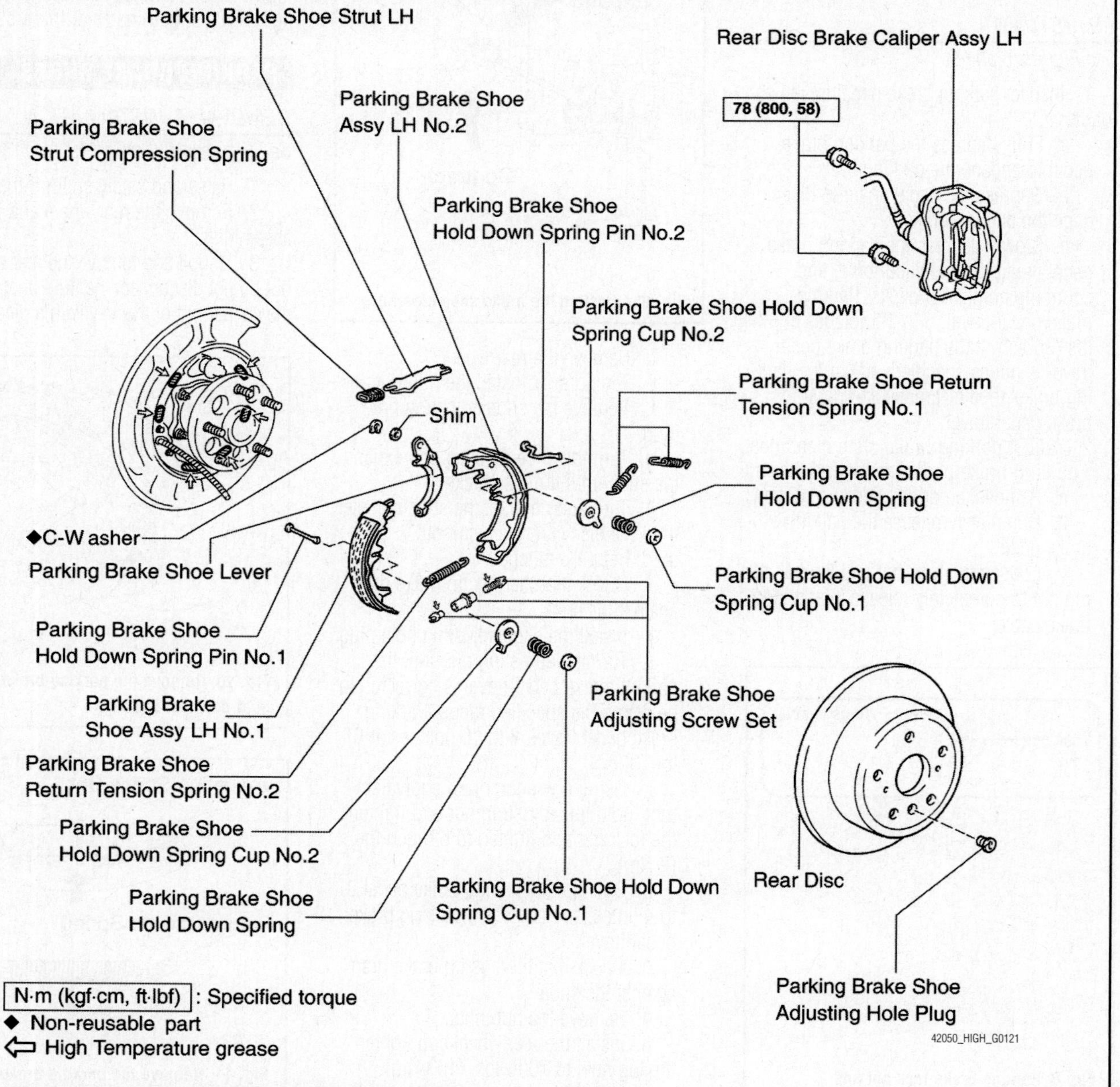

Fig. 13 Exploded view of the parking brake components—2WD vehicles

b. 2WD vehicles, remove the parking brake shoe hold down spring pin No.1.

c. Disconnect the parking brake shoe return spring no.2 and remove the parking brake shoe assembly lh no.1.

10. Remove parking brake shoe adjusting screw set:

a. Remove the parking brake shoe adjusting screw set.

b. Remove the parking brake shoe return tension spring No.2.

11. Remove parking brake shoe assembly no.2:

a. Remove the parking brake shoe hold down spring cup No.1, parking brake shoe hold down spring, parking brake shoe hold down spring cup no.2 and parking brake shoe hold down spring pin no.2.

b. Remove the parking brake shoe assembly lh no.2.

c. Using needle-nose pliers, disconnect the parking brake cable no.3 from the parking brake cable shoe lever.

WARNING

Be careful not to damage parking brake cable no.3.

12. On 4WD models, separate the rear speed sensor.

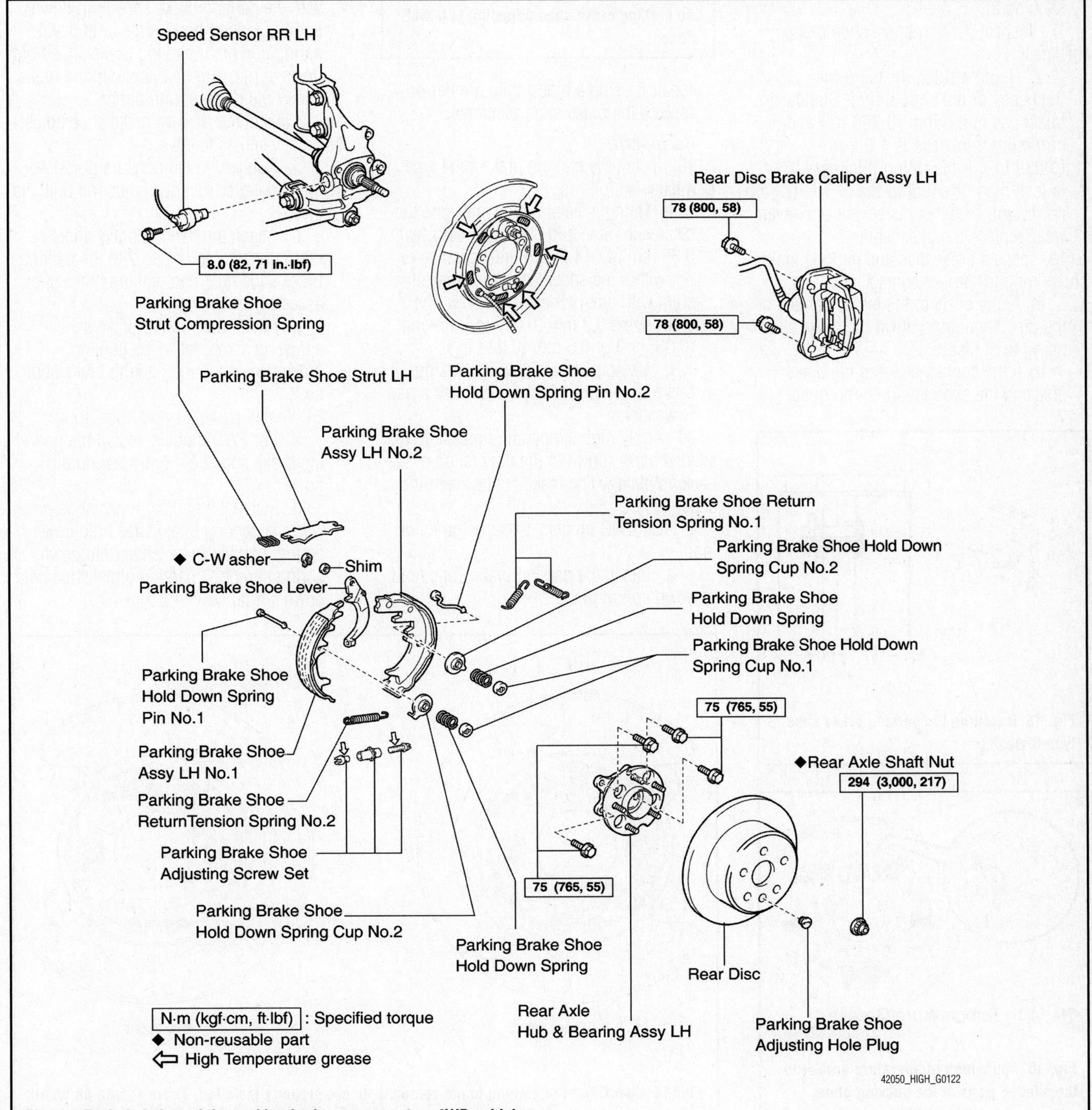

Fig. 14 Exploded view of the parking brake components—4WD vehicles

13. On 4WD models, remove the rear axle shaft nut.

14. On 4WD models, remove rear axle hub & bearing assembly

15. On 4WD models, remove parking brake shoe hold down spring pin.

16. Remove parking brake shoe type C-washer, as follows:

a. Using a screwdriver, remove the c-washer.

b. Remove the shim and parking brake shoe lever from the parking brake shoe no.2.

17. Inspect parking brake shoe lining thickness:

a. Using a ruler, measure the thickness of the shoe lining. Standard thickness is 2.5 mm (0.098 in.) and minimum thickness is 1.0 mm (0.039 in.). If the lining thickness is less than or equal to the minimum, or If there is severe or uneven wear, replace the brake shoe.

18. Inspect brake disc and parking brake shoe lining for proper contact

a. Apply chalk to the inside surface of the disc, then grind down the brake shoe lining to fit disc.

b. If the contact between the brake disc and the shoe lining is improper, repair it using a brake shoe grinder or replace the brake shoe assembly.

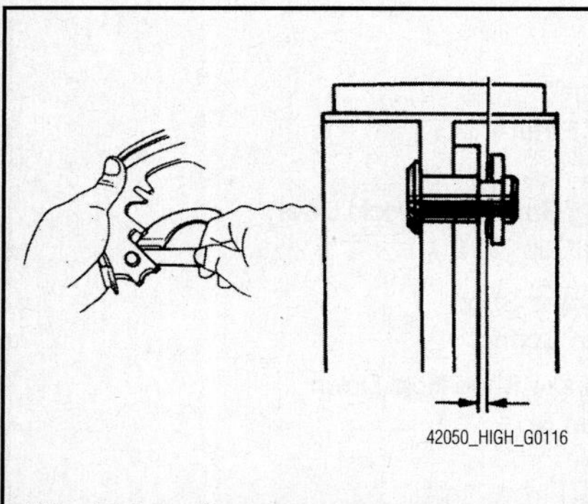

Fig. 15 Installing the parking brake shoe type C-washer

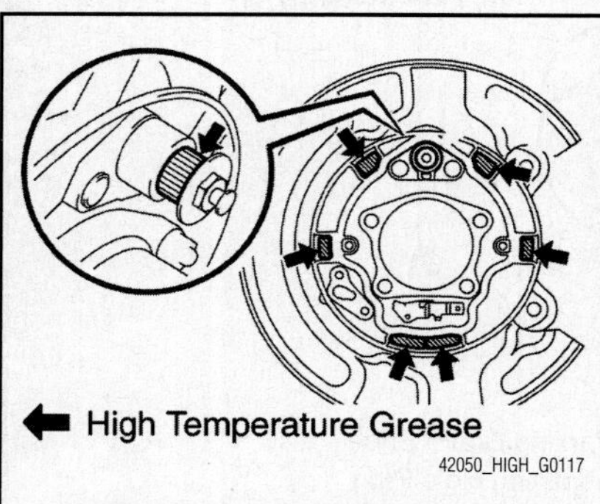

Fig. 16 Apply high temperature grease to the shaded parts of the backing plate which make contact with the shoe

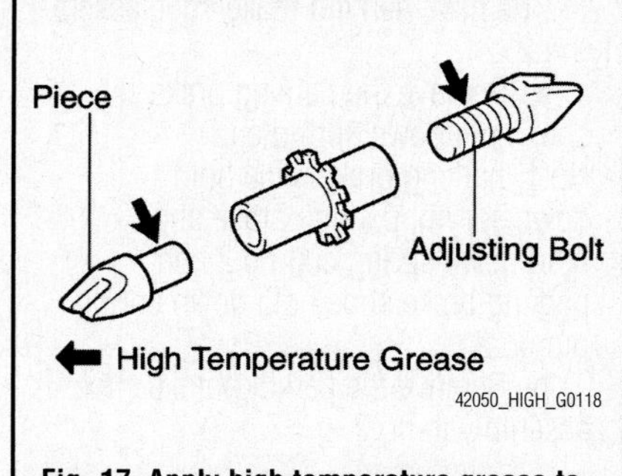

Fig. 17 Apply high temperature grease to the parking brake shoe adjusting bolt and piece

To install:

19. Install the parking brake shoe type C-washer, as follows

a. Using a feeler gauge, measure the clearance. Standard clearance: less than 0.35 mm (0.014 in.). If the clearance is not within the specifications, replace the shim with one of the correct size. The shim sizes: 0.3 mm (0.012 in.), 0.9 mm (0.035 in.) or 0.6 mm (0.024 in.).

b. Using pliers, install the parking brake shoe lever and the shim with a new C-washer.

20. Apply high temperature grease to the shaded parts shown in the illustration of the backing plate which make contact with the shoe.

21. On 4WD models, perform the following:

a. Install the parking brake shoe hold down spring pin.

b. Install the rear axle hub & bearing.

c. Install the rear axle shaft nut.

d. Install the rear speed sensor.

22. Install parking brake shoe no.2, as follows:

a. Using needle-nose pliers, connect the parking brake cable no.3 to the parking brake cable shoe lever.

➡Be careful not to damage the parking brake cable no.3.

b. Install the parking brake shoe no.2 with the parking brake shoe hold down spring, parking brake shoe hold down spring cup no.1, parking brake shoe hold down spring cup no.2 and parking brake shoe hold down spring pin no.2.

23. Install the parking brake shoe adjusting screw set, as follows:

a. Apply high temperature grease to the parking brake shoe adjusting bolt and piece.

b. Attach the parking brake shoe return tension spring no.2 to the parking brake shoe no.1 and parking brake shoe assembly no.2.

c. Attach the parking brake shoe adjusting screw set to the parking brake shoe no.1 and parking brake shoe no.2.

24. Install parking brake shoe no.1:

a. For 2WD models, install the parking brake shoe hold down spring pin no.1.

b. Install the parking brake shoe no.1 with the parking brake shoe hold down spring, parking brake shoe hold down spring cup no.2, parking brake shoe hold down spring cup no.2.

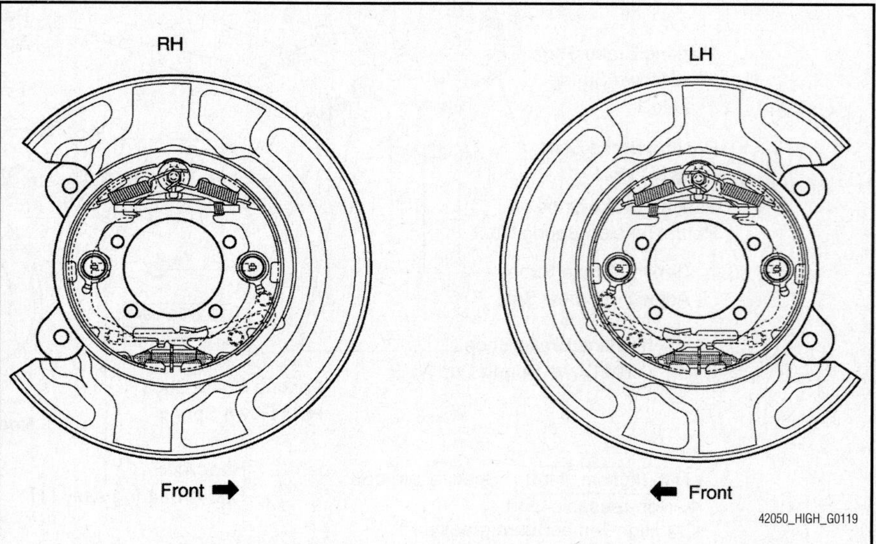

Fig. 18 Check that the parking brake components are properly installed. There should be no oil or grease on the friction surface of the shoe lining and disc

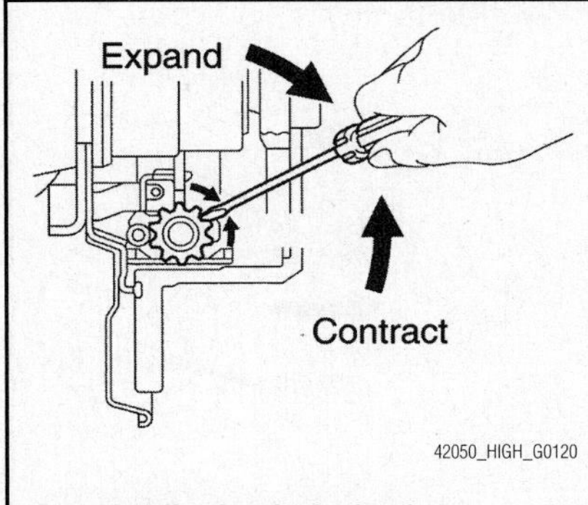

Fig. 19 Adjusting the brake shoe clearance

25. attach the parking brake shoe strut and the parking brake shoe strut compression spring to parking

brake shoe no.2 and parking brake shoe no.1.

26. Install parking brake shoe return tension spring using needle-nose pliers as shown in the illustration.

➡**First install the front side spring then the rear side spring.**

27. Check that the parking brake components are properly installed.

WARNING

There should be no oil or grease on the friction surface of the shoe lining and disc.

28. For 4WD models, inspect the bearing backlash and axle hub deviation.

29. Install the rear disc (rotor), aligning the matchmarks made during removal.

30. Adjust parking brake shoe clearance, as follows:

a. Temporarily install the hub nuts.

b. Remove the hole plug, turn the adjuster and expand the shoes until the disc locks.

c. Contract the shoe adjuster until the disc rotates smoothly. Standard : return 8 notches

d. Check that the shoe has no brake drag.

e. Install the hole plug.

31. Install the caliper, as outlined earlier in this section.

32. Install the rear wheel and tire assembly and tighten the lug nuts to 76 ft. lbs. (103 Nm).

33. Inspect and adjust the parking brake pedal travel, as outlined in this section.

34. For 4WD models, check the ABS speed sensor signal.

ADJUSTMENT

See Figures 19 and 20.

1. Raise and safely support the vehicle.

2. Remove the rear wheel and tire assemblies.

3. Adjust parking brake shoe clearance, as follows:

a. Temporarily install the hub nuts.

b. Remove the hole plug, turn the adjuster and expand the shoes until the disc locks.

c. Contract the shoe adjuster until the disc rotates smoothly. Standard : return 8 notches

d. Check that the shoe has no brake drag.

e. Install the hole plug.

4. Install the rotor and caliper.

5. Install the rear wheel and tire assembly and tighten the lug nuts to 76 ft. lbs. (103 Nm).

6. Inspect the parking brake pedal travel, as follows:

a. Firmly step on the parking brake pedal.

b. release the parking brake.

c. Once more, slowly depress the parking brake pedal all the way, and count the number of clicks. The parking brake pedal should travel 5 to 7 clicks at 67 lbs. (300 N).

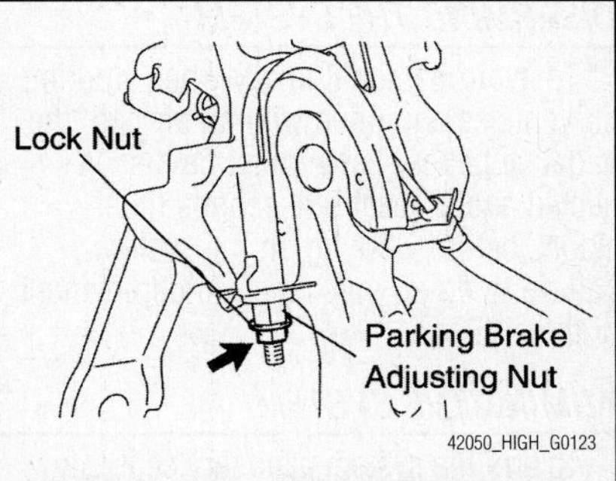

Fig. 20 Use a wrench to hold the parking brake adjusting nut secure while tightening the lock nut

7. If necessary, adjust parking brake pedal travel, as follows:

a. Remove the lower instrument panel finish panel sub-assembly.

b. Remove the lower instrument panel insert sub- assembly.

c. Depress the parking brake pedal 5 clicks to make room for the procedure, and loosen the lock nut with fixing adjusting nut by wrench.

d. Release the parking brake pedal to the original position.

e. Turn the parking brake wire adjusting nut until the parking brake pedal travel is correct.

f. Use a wrench to hold the parking brake adjusting nut, then tighten the lock nut to 53 inch lbs. (6 Nm).

g. Count the number of clicks after depressing and canceling the parking brake pedal 3 to 4 times.

h. Check whether the parking brake drags or not.

i. When operating the parking brake pedal, check that the parking brake pedal indicator light is lit.

CHASSIS ELECTRICAL

AIR BAG (SUPPLEMENTAL RESTRAINT SYSTEM)

GENERAL INFORMATION

CAUTION

These vehicles are equipped with an air bag system. The system must be disarmed before performing service on, or around, system components, the steering column, instrument panel components, wiring and sensors. Failure to follow the safety precautions and the disarming procedure could result in accidental air bag deployment, possible injury and unnecessary system repairs.

SERVICE PRECAUTIONS

CAUTION

Disconnect and isolate the battery negative cable before beginning any airbag system component diagnosis, testing, removal, or installation procedures. Wait at least 90 seconds after the ignition switch is turned off and the negative (-) terminal cable is disconnected from the battery before starting the operation. The SRS is equipped with a backup power source, so if work is started within 90 seconds after disconnecting the negative (-) terminal cable from the battery, the SRS may be deployed. Failure to disable the airbag system may result in accidental airbag deployment, personal injury, or death.

DISARMING THE SYSTEM

To avoid personal injury when working on vehicles equipped with an air bag, the negative battery cable must be disconnected and at least 90 seconds must elapse before working on the system. Failure to do so may result in deployment of the air bag.

ARMING THE SYSTEM

To arm the system after service is completed, connect the negative battery cable. If necessary, perform the initialization procedure, as outlined in the Chassis Electrical System.

CLOCKSPRING CENTERING

See Figure 21.

1. Check that the ignition switch is **OFF**.
2. Check that the battery negative (-) terminal is disconnected.

CAUTION

After removing the terminal, wait for at least 90 seconds before starting the operation.

3. Rotate the spiral cable counterclockwise slowly by hand until it feels firm.
4. Rotate the spiral cable clockwise approximately 2.5 turns to align the marks.

➡Do not turn the spiral cable by the airbag wire harness.

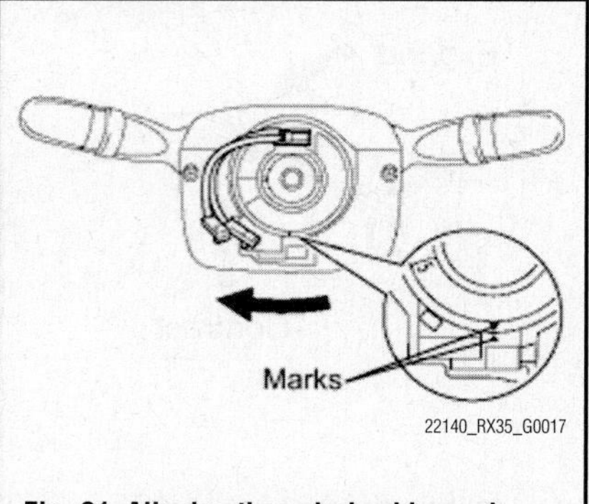

Fig. 21 Aligning the spiral cable marks

➡The spiral cable will rotate approximately 2.5 turns to both the left and right from the center.

DRIVE TRAIN

FRONT HALFSHAFT

REMOVAL & INSTALLATION

See Figures 22 through 24.

1. Before servicing the vehicle, refer to the Precautions Section.
2. Remove or disconnect the following:
   - Engine undercovers
   - Front wheels
   - Drain HV transaxle fluid
   - Cotter pin and hub nut
   - Front speed sensor
   - Brake caliper
   - Brake disc
   - Tie rod end, from the steering knuckle
   - Steering knuckle, from the lower control arm

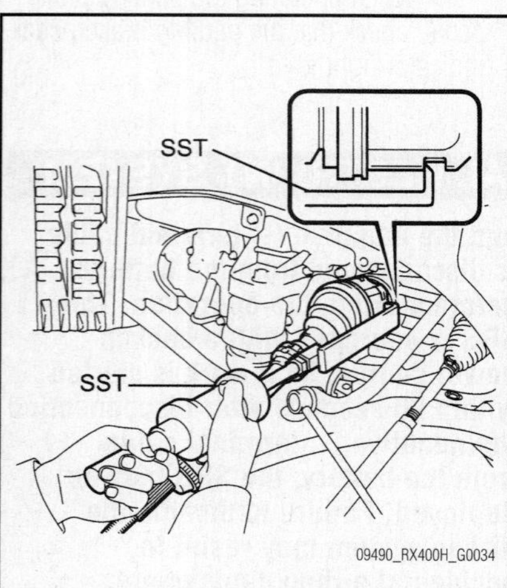

Fig. 22 Use the Special Tool to remove the halfshaft from the transaxle.

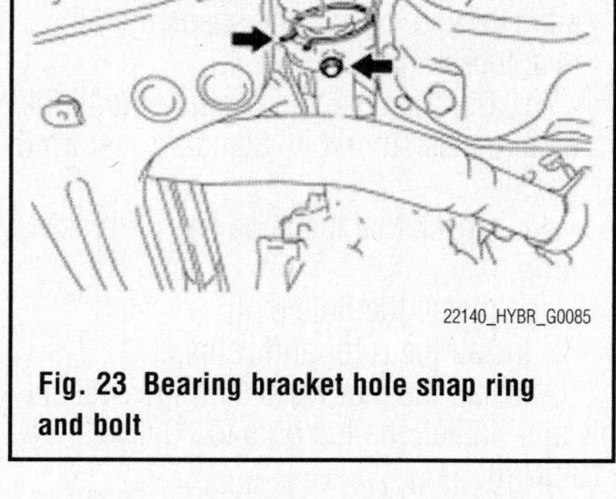

Fig. 23 Bearing bracket hole snap ring and bolt

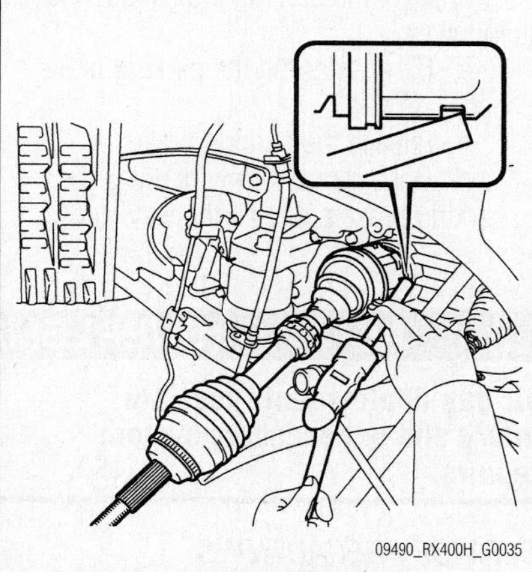

Fig. 24 Insert a brass drift into the groove to install the halfshaft.

   - Halfshaft from the axle hub, using a plastic hammer
   - Stabilizer link
3. Using Special Tool 095020-01010, remove the halfshaft from the transaxle.
4. For RH axle remove the bearing bracket hole snap ring from the drive shaft bearing bracket.
5. Remove the bolt and front drive shaft assembly RH from the drive shaft bearing bracket.

To install:

6. Install a new halfshaft hole snapring.
7. Coat the splines of the inboard joint shaft assembly with ATF.
8. Align the shaft splines and install the halfshaft assembly with a brass drift and hammer.
9. For RH side axle install the bracket hole snap ring and bolt. Tighten the bolt to 24 ft. lbs. (32 Nm).
10. Tighten the axle shaft nut and tighten to 216 ft. lbs. (294 Nm).
11. The remainder of installation is the reverse order of removal.
12. Fill the HV transaxle with gear oil, install the engine undercovers, check front end alignment and test drive.

➡If the cotter pin holes do not align, always correct by tightening the nut until the next hole aligns.

13. Install a new cotter pin.
14. Tighten the front wheels to 76 ft. lbs. (103 Nm).

REAR HALFSHAFT

REMOVAL & INSTALLATION

4WD Vehicles

See Figure 25.

1. Before servicing the vehicle, refer to the Precautions Section.
2. Remove the rear wheel.
3. Disconnect and remove the speed sensor.
4. Remove the brake disc and brake caliper assembly.
5. Remove the axle shaft nut.
6. Separate the rear axle hub and bearing assembly.
7. Separate the parking brake cable.
8. Disconnect the strut rod.
9. Separate the rear suspension arms.
10. Remove the rear carrier sub assembly.
11. Put matchmarks on the rear drive shaft assembly and differential side gear shaft.
12. Remove the 4 nuts, washers and rear drive shaft assembly.

To install:

13. Align the matchmarks.
14. Install the rear drive shaft assembly with the 4 nuts and washers. Tighten to 41 ft. lbs. (56 Nm).
15. Install the rear carrier sub assembly. Tighten the 2 mounting strut bolts to 133 ft. lbs. (180 Nm).
16. Temporarily tighten the rear suspension arm assembly No.2 with the bolt and nut.
17. Temporarily tighten the rear suspension arm assembly No.1 with the bolt and nut.
18. Temporarily tighten the strut rod assembly rear with the bolt and nut.
19. Install the hub and bearing assembly with the 4 bolts and tighten to 55 ft. lbs. (75 Nm).
20. Install the rear brake disc.
21. Install the rear disc brake caliper assembly with the 2 bolts to the rear axle carrier. Tighten the bolts to 58 ft. lbs. (78 Nm).
22. Install the speed sensor and tighten the mounting bolt to 71 inch lbs. (8 Nm).
23. Using a socket wrench (30mm), install a new rear axle shaft nut. Tighten the axle nut to 217 ft. lbs. (294 Nm).
24. Using a chisel and hammer, stake the rear axle shaft nut.
25. Stabilize the suspension. Then tighten the control arms to 83 ft. lbs. (112 Nm). Tighten the strut rod to 133 ft. lbs. (180 Nm).
26. Install the parking brake cable assembly No.3 with the nut. Tighten the nut to 53 inch lbs. (6 Nm)
27. Install the rear wheel and tighten to 76 ft. lbs. (103 Nm)
28. Check rear wheel alignment.
29. Verify speeds sensor operation.

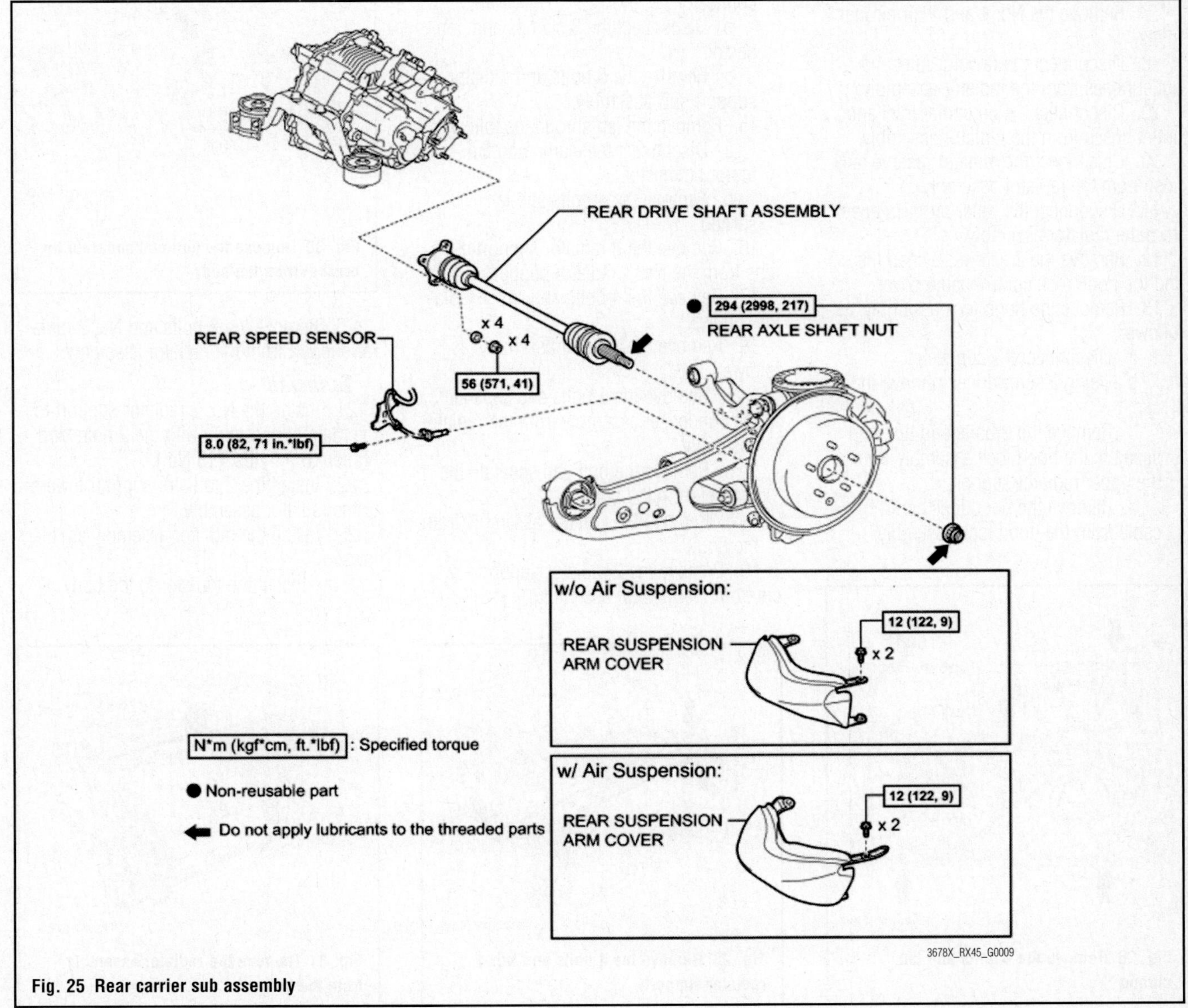

Fig. 25 Rear carrier sub assembly

ENGINE COOLING

ENGINE FAN

REMOVAL & INSTALLATION

See Radiator removal and installation.

RADIATOR

REMOVAL & INSTALLATION

See Figures 26 through 33.

1. Remove the No. 1 engine under-cover.
2. Drain engine coolant.
3. Remove battery negative terminal.
4. Remove the V-bank cover sub-assembly.
5. Remove the cool air intake duct seal.
6. Remove the air cleaner cap sub-assembly.
7. Remove the No. 1 and 2 air cleaner inlets.
8. Disconnect the radiator inlet and outlet hoses from the radiator assembly.
9. Disconnect the oil cooler inlet and outlet hoses from the radiator assembly.
10. Disconnect the radiator reserve tank hose from the radiator assembly.
11. Disconnect the water by-pass hose from the radiator assembly.
12. Remove the 2 screws, 2 snap fits and the hood lock control cable cover.
13. Remove the hood lock assembly, as follows:
    a. Disconnect the connector.
    b. Using a screwdriver, remove the cap.
    c. Remove the 2 bolts and nut, and separate the hood lock assembly from the upper radiator support.
    d. Remove the hood lock control cable from the hood lock assembly.

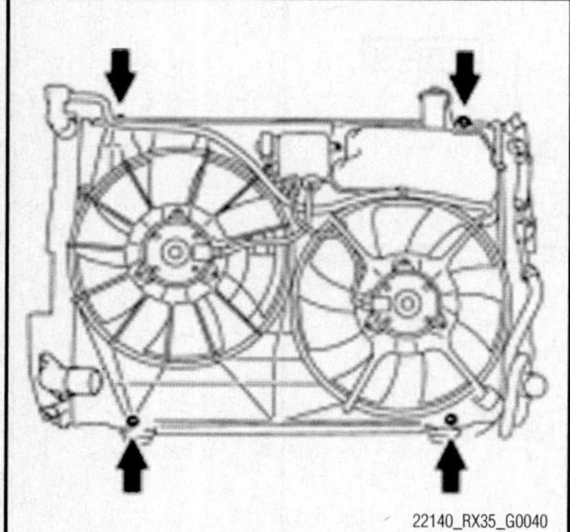

Fig. 26 Remove the 4 bolts and fan shroud

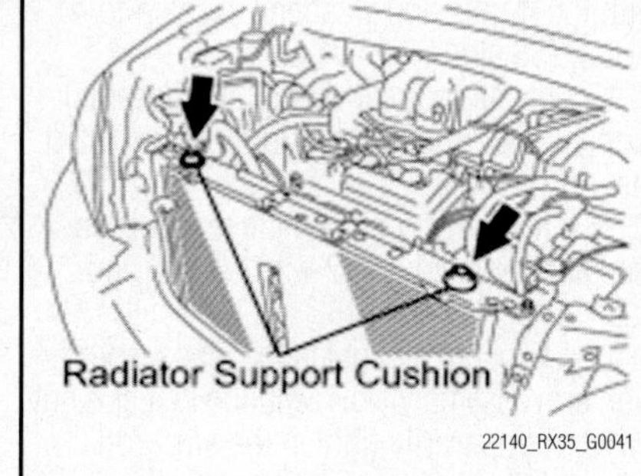

Fig. 27 Remove the 2 radiator support cushions

14. Remove the upper radiator support sub-assembly, as follows:
    a. Disconnect the 2 horn connectors.
    b. Disconnect the 3 clamps and connector.
    c. Remove the 5 bolts and radiator support sub-assembly.
15. Remove the fan shroud, as follows:
    a. Disconnect the clamp and fan motor connector.
    b. Remove the 4 bolts and fan shroud.
16. Remove the 2 radiator support cushions from the No. 1 radiator support.
17. Remove the 4 bolts and No. 1 radiator support.
18. Remove the radiator assembly, as follows:
    a. Remove the 2 bolts and separate the condenser assembly from the radiator assembly.
    b. Remove the bolt and separate the bracket from the body.
    c. Remove the radiator assembly from the body.
19. Remove the 2 radiator support lowers from the radiator assembly.

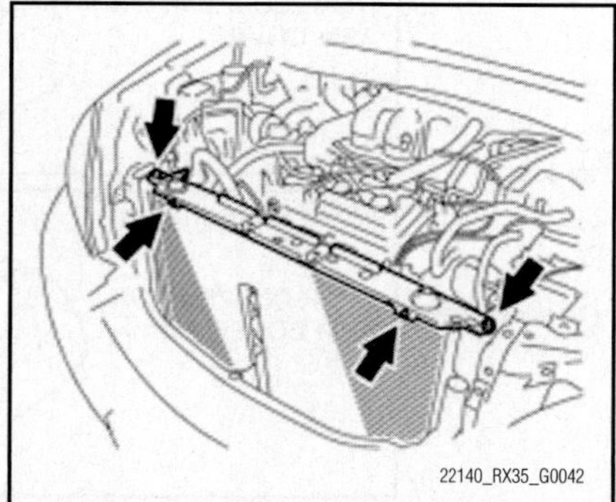

Fig. 28 Remove the 4 bolts and No. 1 radiator support

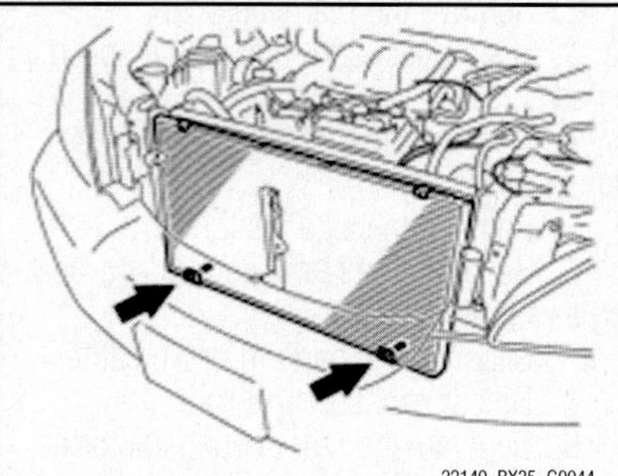

Fig. 29 Remove the 2 bolts and separate the condenser assembly from the radiator assembly

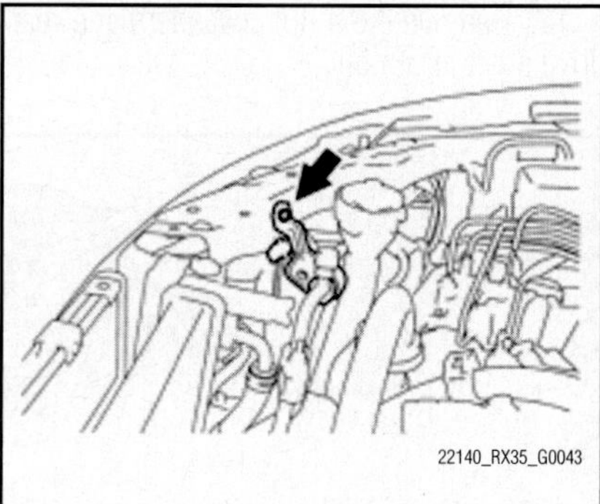

Fig. 30 Remove the bolt and separate the bracket from the body

20. Remove the 2 bolts and No. 2 radiator support from the radiator assembly.

To install:

21. Install the No. 2 radiator support to the radiator assembly with the 2 bolts and tighten to 9 ft. lbs. (13 Nm).
22. Install the 2 radiator support lowers to the radiator assembly.
23. Install the radiator assembly as follows:
    a. Install the radiator to the body.

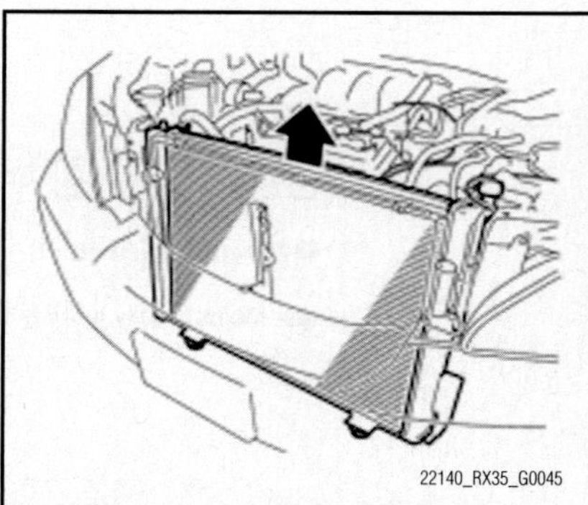

Fig. 31 Remove the radiator assembly from the body

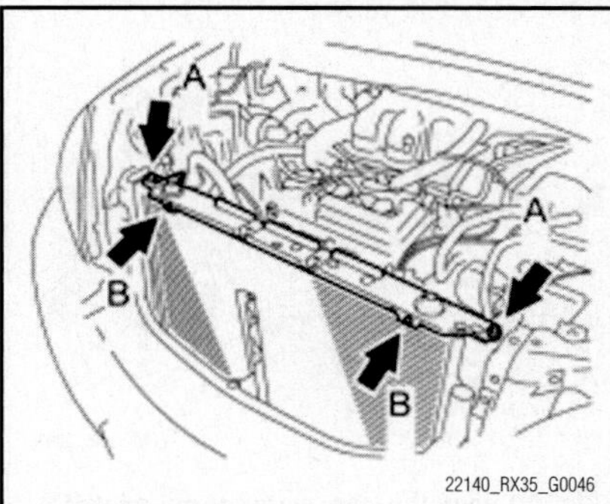

Fig. 32 Install the No. 1 radiator support to the radiator assembly

b. Install the bracket to the body with the bolt and tighten to 87 inch lbs. (9.8 Nm).

c. Install the condenser assembly to the radiator assembly with the 2 bolts and tighten to 35 inch lbs. (3.9 Nm).

24. Install the No. 1 radiator support to the radiator assembly with the 4 bolts and tighten to 9 ft. lbs. (13 Nm), 35 inch lbs. (3.9 Nm).

25. Install the 2 radiator support cushions to the radiator assembly.

26. Install the fan shroud to the radiator assembly with the 4 bolts and tighten to 66 inch lbs. (7.5 Nm).

27. Connect the fan motor connector and clamp.

28. Install the radiator support sub-assembly upper as follows:

a. Install the radiator support sub-assembly upper with the 5 bolts and tighten to 69 inch lbs. (7.8 Nm).

b. Connect the 3 clamps and connector.

c. Connect the 2 horn connectors.

29. Install the hood lock assembly as follows:

a. Install the hood lock control cable to the hood lock assembly.

b. Install the 2 bolts and the hood lock assembly to the radiator support sub-assembly upper and tighten to 71 inch lbs. (8 Nm).

c. Install the hood lock nut and the hood lock nut cap and tighten to 71 inch lbs. (8 Nm).

d. Connect the connector.

30. Connect the 2 clips, and install the 2 screws and the hood lock control cable cover.

31. Connect the water by-pass hose No. 2 to the radiator assembly with the clip.

32. Connect the radiator reserve tank hose to the radiator assembly with the clip.

33. Connect the oil cooler inlet and outlet hoses to the radiator assembly.

34. Connect the radiator inlet and outlet hoses to the radiator assembly.

35. Install the No. 1 and 2 air cleaner inlets.

36. Install the air cleaner cap sub-assembly.

37. Install the cool air intake duct seal.

38. Install the V-bank cover sub-assembly.

39. Connect battery negative terminal.

40. Add engine coolant.

41. Inspect for coolant leak.

42. Install the No. 1 engine undercover.

43. Perform initialization procedure.

THERMOSTAT

REMOVAL & INSTALLATION

See Figures 34 and 35.

1. Drain engine coolant.
2. Remove the V-bank cover sub-assembly.
3. Remove the side engine room cover.
4. Remove the right front fender seal.
5. Remove the right No. 2 engine mounting stay.
6. Remove the engine moving control rod sub-assembly.
7. Remove the V-ribbed belt.

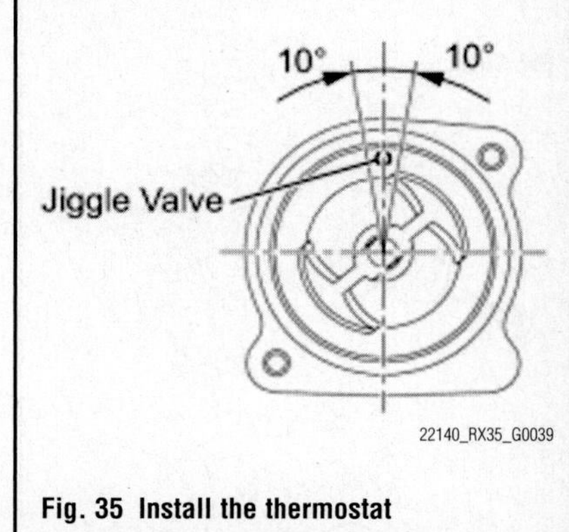

Fig. 35 Install the thermostat

8. Remove the bolt, No. 2 idler pulley cover plate and No. 2 idler pulley.
9. Slide the clamp and disconnect the radiator hose outlet.
10. Remove the 2 nuts and water inlet.
11. Remove the gasket from the thermostat and remove thermostat.

To install:

12. Install a new gasket to the thermostat.
13. Install the thermostat with the jiggle valve facing up.

➡The jiggle valve may be set within 10° on either side of the prescribed position.

14. Install the water inlet with the 2 nuts and tighten to 7 ft. lbs. (10 Nm).
15. Connect the radiator hose outlet with the clamp.
16. Install the No. 2 idler pulley cover plate and No. 2 idler pulley with the bolt and tighten to 32 ft. lbs. (43 Nm).
17. Install the V-ribbed belt.
18. Install the engine moving control rod sub-assembly.
19. Install the right No. 2 engine mounting stay.
20. Install the right front fender seal.
21. Add engine coolant.
22. Inspect for coolant leak.
23. Install the side engine room cover.
24. Install the V-bank cover sub-assembly.

WATER PUMP

REMOVAL & INSTALLATION

See Figures 36 through 39.

1. Remove the engine assembly and transaxle.
2. Secure the engine.
3. Remove the left front No. 1 engine mounting bracket.
4. Remove the No. 2 idler pulley sub-assembly, as follows:

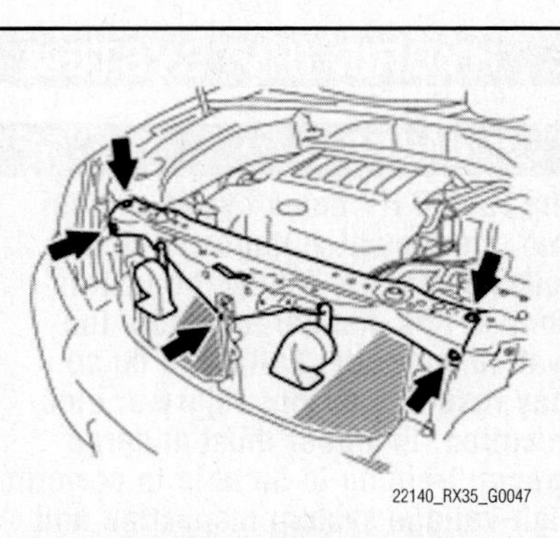

Fig. 33 Install the radiator support sub-assembly upper with the 5 bolts

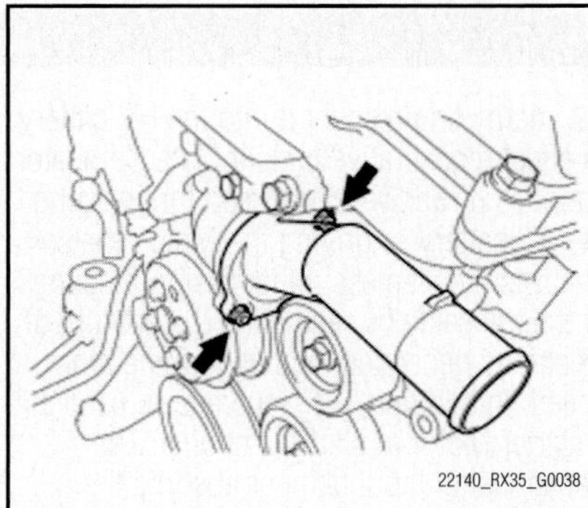

Fig. 34 Remove the 2 nuts and water inlet

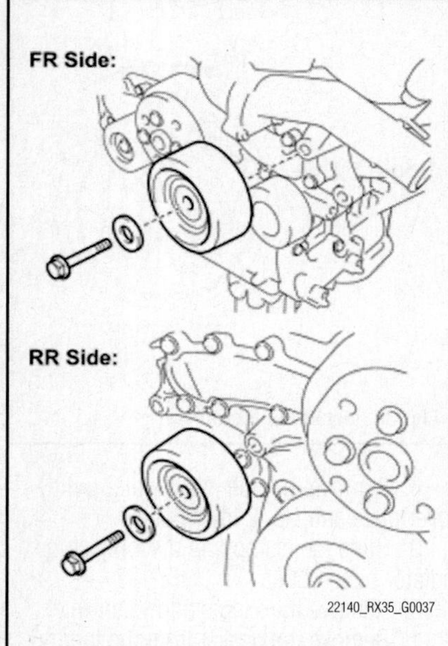

Fig. 36 Remove the 2 bolts, 2 idler pulley cover plates and idler pulley sub-assemblies

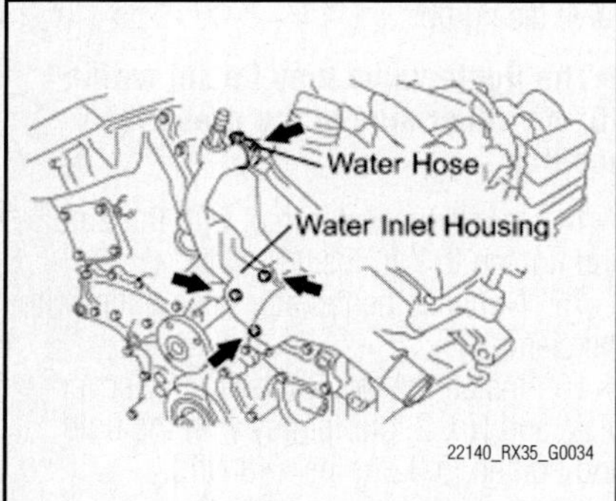

Fig. 37 Remove the 2 bolts, nut and water inlet housing

a. Remove the 2 bolts, 2 idler pulley cover plates and 2 idler pulley sub-assemblies.

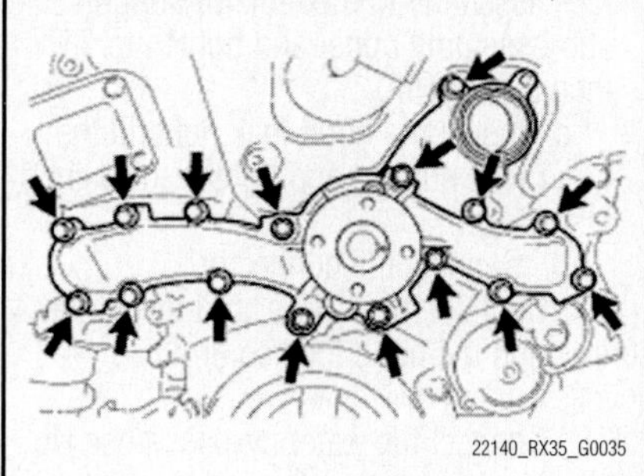

Fig. 38 Remove the 16 bolts, water pump assembly and water pump gasket

5. Remove the V-ribbed belt tensioner assembly.

6. Remove the water pump pulley, as follows:

a. Using SST (SST: 09960-10010) or equivalent, hold the water pump pulley.

b. Remove the 4 bolts and water pump pulley.

7. Remove the water inlet housing, as follows:

a. Disconnect the water hose.

b. Remove the 2 bolts, nut and water inlet housing.

c. Remove the water inlet housing gasket and water outlet pipe O-ring.

8. Remove the 16 bolts, water pump assembly and water pump gasket.

To install:

9. Install a new water pump gasket and the water pump assembly with the 16 bolts and tighten to 15 ft. lbs. (21 Nm), 81 inch lbs. (9.1 Nm).

➡Make sure that there is no oil on the threads of bolts A.

➡Be sure to replace 2 bolts C with new ones or reuse them after applying adhesive 1344. Adhesive: Toyota Genuine Seal Packing Black, Three Bond 1207B or equivalent.

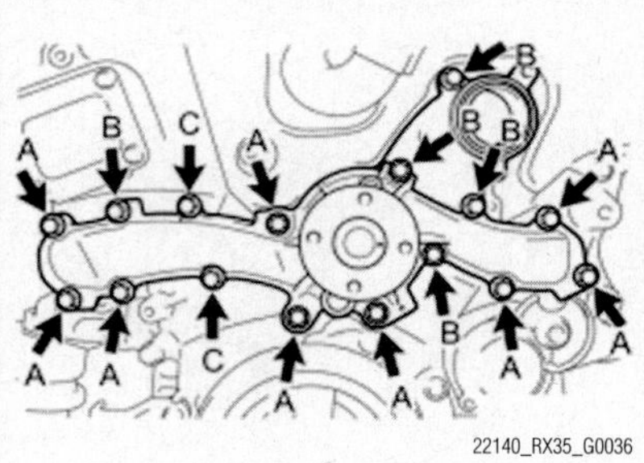

Fig. 39 Install a new water pump gasket and the water pump assembly with the 16 bolts

10. Install a new water inlet housing gasket and water outlet pipe O-ring.

11. Install the water inlet housing with the 2 bolts and nut and tighten to 7 ft. lbs. (10 Nm).

➡Be careful not to allow the O-ring to get caught between the parts.

12. Connect the water hose.

13. Install the water pump pulley, as follows:

a. Temporarily install the water pump pulley with the 4 bolts.

b. Using SST (SST: 09960-10010) or equivalent, hold the water pump pulley.

c. Tighten the 4 bolts to 15 ft. lbs. (21 Nm).

14. Install the V-ribbed belt tensioner assembly.

15. Install the 2 idler pulley cover plates and idler pulley sub-assemblies with the 2 bolts and tighten to 32 ft. lbs. (43 Nm).

16. Install the left front No. 1 engine mounting bracket.

17. Remove engine stand.

18. Install the engine assembly and transaxle.

ENGINE ELECTRICAL

HYBRID (HV) SYSTEM

HV PRECAUTIONS

The HV transaxle assembly consists of the planetary gear unit, Motor, and Generator. The gear unit uses the planetary gear to split engine output in accordance with a driving request while the vehicle is driven or the HV battery is charged. The Motor assists engine output while increasing vehicle driving force. The Motor also converts the energy, which is consumed in the form of heat during normal braking, into electrical energy and recover it into the HV battery to effect regenerative braking. The Generator supplies power, which is used for charging the HV battery or driving the Motor. It also controls the stepless transmission function of the transaxle by regulating the amount of electricity generated to change Generator speed. In addition, the Generator is used as a starter Motor to start the engine. The transmission input damper absorbs the shock generated when the driving force from the engine is transmitted.

✱✱ CAUTION

The RX450 HV has a hybrid system that operates at voltages up to 650 volts. Be sure to follow the instructions in this manual to handle the system correctly. Failure to do so may result in serious injury or electrocution. Engineer must undergo special training to be able to perform high-voltage system inspection and servicing.

⁂ CAUTION

All high-voltage wire harness connectors are colored orange. The HV battery and other high-voltage components have "High Voltage" caution labels. Do not carelessly touch these wires and components.

⁂ CAUTION

Before inspecting or servicing the high-voltage system, be sure to follow safety measures, such as wearing insulated gloves and removing the service plug to prevent electrocution. Carry the removed service plug in your pocket to prevent other technicians from reinstalling it while you are servicing the vehicle.

⁂ CAUTION

After removing the service plug, wait 5 minutes before touching any of the high-voltage connectors and terminals.

⁂ CAUTION

The RX450 HV has a hybrid system that operates at voltages up to 650 volts. Be sure to follow the instructions in this manual to handle the system correctly. Failure to do so may result in serious injury or electrocution. Engineer must undergo special training to be able to perform high-voltage system inspection and servicing.

⁂ CAUTION

All high-voltage wire harness connectors are colored orange. The HV battery and other high-voltage components have "High Voltage" caution labels. Do not carelessly touch these wires and components.

⁂ CAUTION

Before inspecting or servicing the high-voltage system, be sure to follow safety measures, such as wearing insulated gloves and removing the service plug to prevent electrocution. Carry the removed service plug in your pocket to prevent other technicians from reinstalling it while you are servicing the vehicle.

⁂ CAUTION

After removing the service plug, wait 5 minutes before touching any of the high-voltage connectors and terminals.

ALTERNATOR

The 3.5L engine has a DC electric converter and therefore does not use a standard alternator.

BATTERY BLOWER

REMOVAL & INSTALLATION

See Figure 40.

⁂ CAUTION

The RX450h HV has a hybrid system that operates at voltages up to 650 volts. Be sure to follow the instructions in this manual to handle the system correctly. Failure to do so may result in serious injury or electrocution. Engineer must undergo special training to be able to perform high-voltage system inspection and servicing.

⁂ CAUTION

All high-voltage wire harness connectors are colored orange. The HV battery and other high-voltage components have "High Voltage" caution labels. Do not carelessly touch these wires and components.

➡Wear insulating gloves and protective glasses.

1. Before inspecting or servicing the high-voltage system, be sure to follow safety measures, such as wearing insulated gloves and removing the service plug to prevent electrocution. Carry the removed service plug in your pocket to prevent other technicians from reinstalling it while you are servicing the vehicle.
2. After removing the service plug, wait 5 minutes before touching any of the high-voltage connectors and terminals.
3. To access the inverter cover for voltage verification. Remove the following:
    - Service plug grip
    - LH engine room side cover
    - Cool air intake duct seal
    - Air cleaner cap and case assembly
    - Inverter reserve tank
    - Inverter cover
4. Using the voltmeter, measure the voltage between the terminals of the 2 phase connectors (N-P). Standard voltage: 0 volts. Use measuring range of DC 750 V or more on the voltmeter.
5. Install the inverter cover and tighten the mounting bolts to 7 ft. lbs. (10 Nm).
6. Install the inverter reserve tank and tighten the mounting bolts to 7 ft. lbs. (10 Nm).
7. Install air cleaner case and cap assembly.
8. Install the cool air intake duct seal.
9. Install the LH engine room side cover.
10. To access the HV battery blower remove the following:
    - Rear center seat assembly
    - LH and RH rear seat head rest
    - LH and RH seat track bracket cover
    - LH and RH outer seat track bracket cover
    - Rear seat leg side cover
    - Disconnect the seat lock cables
    - LH and RH seat assembly
    - Deck board assembly No. 2 and No. 3
    - Tonneau cover if equipped
    - Rear No. 1 floor board LH (w/o rear No. 2 seat)
    - LH and RH rear seat side cover (w/ rear No. 2 seat)
    - Jack carrier support
    - Jack carrier cushion
    - Jack assembly and carrier assembly
    - RH deck side trim box
    - Deck floor board assembly (w/o rear No. 2 seat)
    - Rear No. 2 seat belt assembly (w/ rear No. 2 seat)
    - Rear seat belt lap assembly LH and RH (w/ rear No. 2 seat)
    - Rear No. 2 seat assembly (w/ rear No. 2 seat)
    - Rear floor finish plate
    - Rear deck side trim and cover
    - LH side trim cover (w/o rear seat entertainment system)
    - Power outlet socket (w/ rear seat entertainment system)
    - LH rear combination light service cover
    - Power socket assembly and cover
    - Rear deck trim cover (w/o rear seat entertainment system)
    - Reclining remote control lever bezel LH (w/ remote folding function)
    - LH rope hook assembly
    - No.2 deck side trim hook
    - LH front deck side trim cover

Fig. 40 Battery cooling blower assemblies

- Rear LH No. 1 outer seat belt assembly
- Deck trim side panel LH
- Quarter pillar garnish LH
- Roof side inner garnish assembly
- Rear seat side garnish cap
- Right side deck trim and cover
- Rear room temperature sensor (if equipped)
- Rear combination light service cover RH
- Rope hook assembly RH
- No. 1 luggage compartment trim hook
- RH front deck side trim cover
- RH rear No. 1 outer seat belt assembly
- RH deck trim side panel
- Air intake covers
- Turn back the front floor carpet assembly
- No. 1 HV battery tray
- Rear No.1. 2 and 3 floor boards
- Rear center seat inner belt assembly
- Battery cover lock striker
- Battery service hole cover.
- Battery cover sub-assembly

⁂ CAUTION

Be sure to wear insulated gloves and protective goggles.

11. Remove the battery cooling blower assembly as follows:
    a. Disconnect each battery cooling blower assembly connector and clamp.
    b. Remove the 9 nuts and 3 battery cooling blower assemblies.
12. Reverse the removal procedure and note the following:
    a. Install the 3 battery cooling blower assemblies with the 9 nuts. Tighten to 40 inch lbs. (4.5 Nm)
    b. Refer to HV battery removal for additional information.

HV BATTERY FUSE

REMOVAL & INSTALLATION

See Figure 41.

⁂ CAUTION

The RX450h HV has a hybrid system that operates at voltages up to 650 volts. Be sure to follow the instructions in this manual to handle the system correctly. Failure to do so may result in serious injury or electrocution. Engineer must undergo special training to be able to perform high-voltage system inspection and servicing.

⁂ CAUTION

All high-voltage wire harness connectors are colored orange. The HV battery and other high-voltage components have "High Voltage" caution labels. Do not carelessly touch these wires and components.

1. Before inspecting or servicing the high-voltage system, be sure to follow safety measures, such as wearing insulated gloves and removing the service plug to prevent electrocution. Carry the removed service plug in your pocket to prevent other technicians from reinstalling it while you are servicing the vehicle.
2. After removing the service plug, wait 5 minutes before touching any of the high-voltage connectors and terminals.

➡Wear insulating gloves and protective glasses.

3. To access the inverter cover for voltage verification. Remove the following:
    - Service plug grip
    - LH engine room side cover
    - Cool air intake duct seal
    - Air cleaner cap and case assembly
    - Inverter reserve tank
    - Inverter cover
4. Using the voltmeter, measure the voltage between the terminals of the 2 phase connectors (N-P). Standard voltage: 0 volts. Use measuring range of DC 750 V or more on the voltmeter.
5. Install the inverter cover and tighten the mounting bolts to 7 ft. lbs. (10 Nm).
6. Install the inverter reserve tank and tighten the mounting bolts to 7 ft. lbs. (10 Nm).
7. Install air cleaner case and cap assembly.
8. Install the cool air intake duct seal.
9. Install the LH engine room side cover.
10. To access the HV battery fuse remove the following:
    - Rear center seat assembly
    - LH and RH rear seat head rest
    - LH and RH seat track bracket cover
    - LH and RH outer seat track bracket cover
    - Rear seat leg side cover
    - Disconnect the seat lock cables
    - LH and RH seat assembly
    - Deck board assembly No. 2 and No. 3
    - Tonneau cover if equipped
    - Rear No. 1 floor board LH (w/o rear No. 2 seat)
    - LH and RH rear seat side cover (w/ rear No. 2 seat)
    - Jack carrier support
    - Jack carrier cushion
    - Jack assembly and carrier assembly
    - RH deck side trim box
    - Deck floor board assembly (w/o rear No. 2 seat)
    - Rear No. 2 seat belt assembly (w/ rear No. 2 seat)
    - Rear seat belt lap assembly LH and RH (w/ rear No. 2 seat)
    - Rear No. 2 seat assembly (w/ rear No. 2 seat)
    - Rear floor finish plate
    - Rear deck side trim and cover
    - LH side trim cover (w/o rear seat entertainment system)
    - Power outlet socket (w/ rear seat entertainment system)
    - LH rear combination light service cover
    - Power socket assembly and cover
    - Rear deck trim cover (w/o rear seat entertainment system)
    - Reclining remote control lever bezel LH (w/ remote folding function)
    - LH rope hook assembly
    - No.2 deck side trim hook

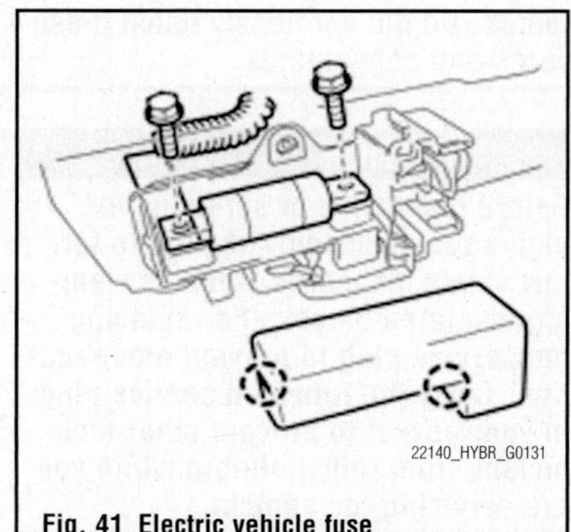

Fig. 41 Electric vehicle fuse

- LH front deck side trim cover
- Rear LH No. 1 outer seat belt assembly
- Deck trim side panel LH
- Quarter pillar garnish LH
- Roof side inner garnish assembly
- Rear seat side garnish cap
- Right side deck trim and cover
- Rear room temperature sensor (if equipped)
- Rear combination light service cover RH
- Rope hook assembly RH
- No. 1 luggage compartment trim hook
- RH front deck side trim cover
- RH rear No. 1 outer seat belt assembly
- RH deck trim side panel
- Air intake covers
- Turn back the front floor carpet assembly
- Release the 2 claws and remove the fuse block cover.
- Remove the 2 bolts and electric vehicle fuse.

11. Reverse the removal procedure and note the following:

a. Tighten the fuse mounting bolts to 48 inch lbs. (5.4 Nm).

b. Refer to HV battery removal for additional information.

BATTERY SMART UNIT

REMOVAL & INSTALLATION

See Figure 42.

⁂ CAUTION

The RX450 HV has a hybrid system that operates at voltages up to 650 volts. Be sure to follow the instructions in this manual to handle the system correctly. Failure to do so may result in serious injury or electrocution. Engineer must undergo special training to be able to perform high-voltage system inspection and servicing.

⁂ CAUTION

All high-voltage wire harness connectors are colored orange. The HV battery and other high-voltage components have "High Voltage" caution labels. Do not carelessly touch these wires and components.

➡Wear insulating gloves and protective glasses.

1. Before inspecting or servicing the high-voltage system, be sure to follow safety measures, such as wearing insulated gloves and removing the service plug to prevent electrocution. Carry the removed service plug in your pocket to prevent other technicians from reinstalling it while you are servicing the vehicle.
2. After removing the service plug, wait 5 minutes before touching any of the high-voltage connectors and terminals.
3. To access the inverter cover for voltage verification. Remove the following:
   - Service plug grip
   - LH engine room side cover
   - Cool air intake duct seal
   - Air cleaner cap and case assembly
   - Inverter reserve tank
   - Inverter cover
4. Using the voltmeter, measure the voltage between the terminals of the 2 phase connectors (N-P). Standard voltage: 0 volts. Use measuring range of DC 750 V or more on the voltmeter.
5. Install the inverter cover and tighten the mounting bolts to 7 ft. lbs. (10 Nm).
6. Install the inverter reserve tank and tighten the mounting bolts to 7 ft. lbs. (10 Nm).
7. Install air cleaner case and cap assembly.
8. Install the cool air intake duct seal.
9. Install the LH engine room side cover.
10. To access the HV battery smart unit remove the following:
    - Rear center seat assembly
    - LH and RH rear seat head rest
    - LH and RH seat track bracket cover
    - LH and RH outer seat track bracket cover
    - Rear seat leg side cover
    - Disconnect the seat lock cables
    - LH and RH seat assembly
    - Deck board assembly No. 2 and No. 3
    - Tonneau cover if equipped
    - Rear No. 1 floor board LH (w/o rear No. 2 seat)
    - LH and RH rear seat side cover (w/ rear No. 2 seat)
    - Jack carrier support
    - Jack carrier cushion
    - Jack assembly and carrier assembly
    - RH deck side trim box
    - Deck floor board assembly (w/o rear No. 2 seat)
    - Rear No. 2 seat belt assembly (w/ rear No. 2 seat)
    - Rear seat belt lap assembly LH and RH (w/ rear No. 2 seat)

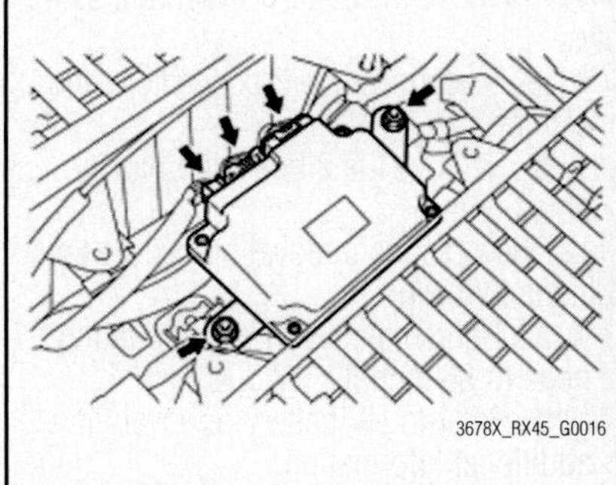

Fig. 42 Remove the 2 bolts and battery smart unit

- Rear No. 2 seat assembly (w/ rear No. 2 seat)
- Rear floor finish plate
- Rear deck side trim and cover
- LH side trim cover (w/o rear seat entertainment system)
- Power outlet socket (w/ rear seat entertainment system)
- LH rear combination light service cover
- Power socket assembly and cover
- Rear deck trim cover (w/o rear seat entertainment system)
- Reclining remote control lever bezel LH (w/ remote folding function)
- LH rope hook assembly
- No.2 deck side trim hook
- LH front deck side trim cover
- Rear LH No. 1 outer seat belt assembly
- Deck trim side panel LH
- Quarter pillar garnish LH
- Roof side inner garnish assembly
- Rear seat side garnish cap
- Right side deck trim and cover
- Rear room temperature sensor (if equipped)
- Rear combination light service cover RH
- Rope hook assembly RH
- No. 1 luggage compartment trim hook
- RH front deck side trim cover
- RH rear No. 1 outer seat belt assembly
- RH deck trim side panel
- Air intake covers
- Turn back the front floor carpet assembly
- No. 1 HV battery tray
- Rear No.1. 2 and 3 floor boards
- Rear center seat inner belt assembly
- Battery cover lock striker
- Battery service hole cover.
- Battery cover sub- assembly

11. Remove the battery smart unit as follows:

a. Disconnect the 3 connectors and 2 clamps.

b. Remove the 2 bolts and battery smart unit.

12. Reverse the removal procedure and note the following:

a. Tighten the smart unit mounting bolts to 66 inch lbs. (7.5 Nm).

b. Refer to HV battery removal for additional information.

HV BATTERY

REMOVAL & INSTALLATION

See Figures 43 through 46.

⁂ CAUTION

The RX450 HV has a hybrid system that operates at voltages up to 650 volts. Be sure to follow the instructions in this manual to handle the system correctly. Failure to do so may result in serious injury or electrocution. Engineer must undergo special training to be able to perform high-voltage system inspection and servicing.

⁂ CAUTION

All high-voltage wire harness connectors are colored orange. The HV battery and other high-voltage components have "High Voltage" caution labels. Do not carelessly touch these wires and components.

1. Before inspecting or servicing the high-voltage system, be sure to follow safety measures, such as wearing insulated gloves and removing the service plug to prevent electrocution. Carry the removed service plug in your pocket to prevent other technicians from reinstalling it while you are servicing the vehicle.

2. After removing the service plug, wait 5 minutes before touching any of the high-voltage connectors and terminals.

➡Wear insulating gloves and protective glasses.

3. Before inspecting or servicing the high-voltage system, be sure to follow safety measures, such as wearing insulated gloves and removing the service plug to prevent electrocution. Carry the removed service plug in your pocket to prevent other technicians from reinstalling it while you are servicing the vehicle.

4. After removing the service plug, wait 5 minutes before touching any of the high-voltage connectors and terminals.

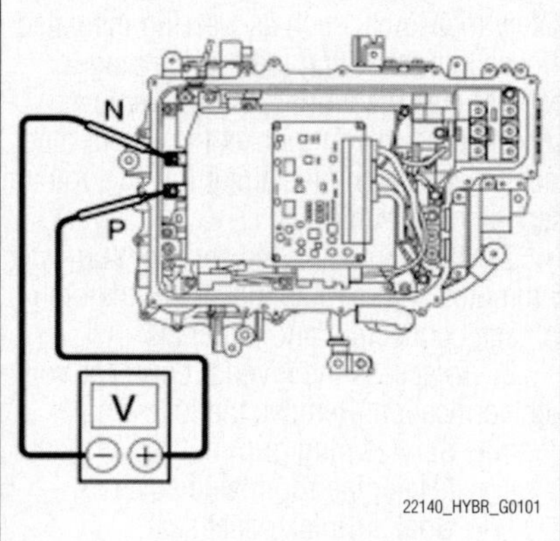

Fig. 43 Measure the voltage between the terminals of the 2 phase connectors

5. Check for Daces and confirm that P0AA6 (High voltage insulation is unusual) is not output before doing removal or installation inside the battery. If the DTC is output, perform troubleshooting first.

6. To access the inverter cover for voltage verification. Remove the following:

- Service plug grip
- LH engine room side cover
- Cool air intake duct seal
- Air cleaner cap and case assembly
- Inverter reserve tank
- Inverter cover

7. Using the voltmeter, measure the voltage between the terminals of the 2 phase connectors (N-P). Standard voltage: 0 volts. Use measuring range of DC 750 V or more on the voltmeter.

8. Install the inverter cover and t ighten the mounting bolts to 7 ft. lbs. (10 Nm).

9. Install the inverter reserve tank and tighten the mounting bolts to 7 ft. lbs. (10 Nm).

10. Install air cleaner case and cap assembly.

11. Install the cool air intake duct seal.

12. Install the LH engine room side cover.

13. To access the HV battery remove the following:

- Rear center seat assembly
- LH and RH rear seat head rest
- LH and RH seat track bracket cover
- LH and RH outer seat track bracket cover
- Rear seat leg side cover
- Disconnect the seat lock cables
- LH and RH seat assembly
- Deck board assembly No. 2 and No. 3
- Tonneau cover if equipped
- Rear No. 1 floor board LH (w/o rear No. 2 seat)
- LH and RH rear seat side cover (w/ rear No. 2 seat)
- Jack carrier support
- Jack carrier cushion
- Jack assembly and carrier assembly
- RH deck side trim box
- Deck floor board assembly (w/o rear No. 2 seat)
- Rear No. 2 seat belt assembly (w/ rear No. 2 seat)
- Rear seat belt lap assembly LH and RH (w/ rear No. 2 seat)
- Rear No. 2 seat assembly (w/ rear No. 2 seat)
- Rear floor finish plate
- Rear deck side trim and cover
- LH side trim cover (w/o rear seat entertainment system)
- Power outlet socket (w/ rear seat entertainment system)
- LH rear combination light service cover
- Power socket assembly and cover
- Rear deck trim cover (w/o rear seat entertainment system)
- Reclining remote control lever bezel LH (w/ remote folding function)
- LH rope hook assembly
- No.2 deck side trim hook
- LH front deck side trim cover
- Rear LH No. 1 outer seat belt assembly
- Deck trim side panel LH
- Quarter pillar garnish LH
- Roof side inner garnish assembly
- Rear seat side garnish cap
- Right side deck trim and cover
- Rear room temperature sensor (if equipped)
- Rear combination light service cover RH
- Rope hook assembly RH
- No. 1 luggage compartment trim hook
- RH front deck side trim cover
- RH rear No. 1 outer seat belt assembly
- RH deck trim side panel
- Air intake covers
- Turn back the front floor carpet assembly
- No. 1 HV battery tray
- Rear No.1. 2 and 3 floor boards
- Rear center seat inner belt assembly
- Battery cover lock striker
- Battery service hole cover.
- Battery cover sub- assembly
- Battery smart unit

CAUTION

Be sure to wear insulated gloves and protective goggles.

14. Remove the HV battery as follows:

a. Insulate the removed terminals with insulating tape

b. Remove the battery room ventilation hose between the HV battery module LH and center HV battery module.

c. Disconnect the wire harness clamp of the battery thermo sensor.

d. Remove the nut and main battery cable from the HV battery module LH.

e. Remove the nut, then disconnect the EV battery plug from the HV battery module LH.

f. Remove the carry belts from the center HV battery module and install them to the HV battery module LH.

g. Insert the ends of the carry belts into the installation holes. Pull the carry belts upward to securely install them.

h. Remove the HV battery module LH.

i. Remove the carry belts from the HV battery module LH.

j. Remove the center HV battery module.

k. Disconnect the 2 connectors from the HV relay assembly.

l. Disconnect the 2 connectors and clamp.

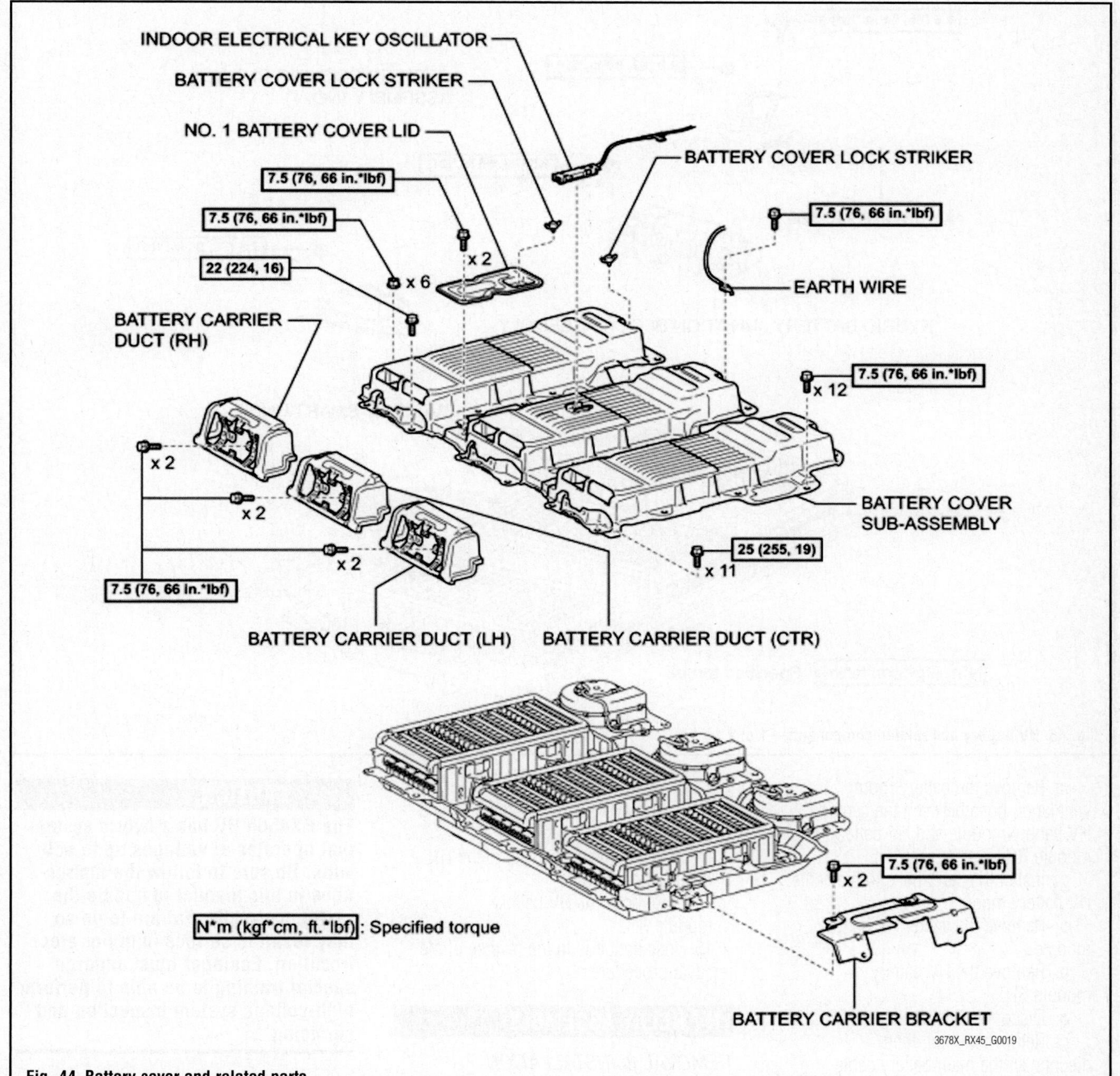

Fig. 44 Battery cover and related parts

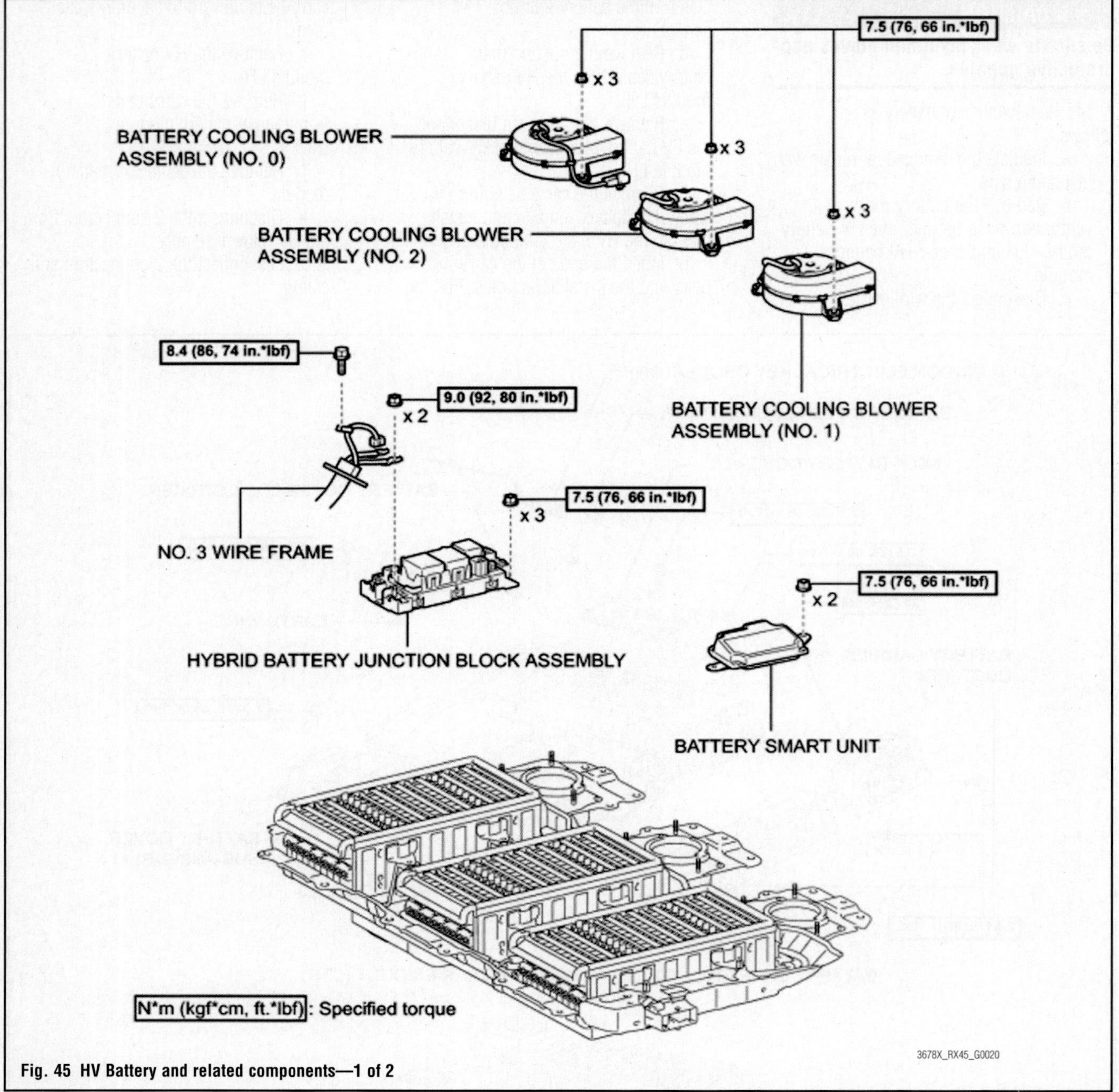

Fig. 45 HV Battery and related components—1 of 2

m. Remove the battery room ventilation hose between the center HV battery module and HV battery module RH.

n. Install the carry belts to the center HV battery module.

o. Remove the center HV battery module.

p. Remove the HV battery module RH.

q. Disconnect the 2 clamps.

r. Remove the nut, then disconnect the main battery cable connector.

s. Disconnect the clamp.

t. Remove the battery room ventilation hose.

u. Install the carry belts to the HV battery module RH.

v. Remove the HV battery module RH.

15. Installation is in the reverse of the removal procedure.

HV RELAY ASSEMBLY

REMOVAL & INSTALLATION

See Figure 47.

CAUTION

The RX450h HV has a hybrid system that operates at voltages up to 650 volts. Be sure to follow the instructions in this manual to handle the system correctly. Failure to do so may result in serious injury or electrocution. Engineer must undergo special training to be able to perform high-voltage system inspection and servicing.

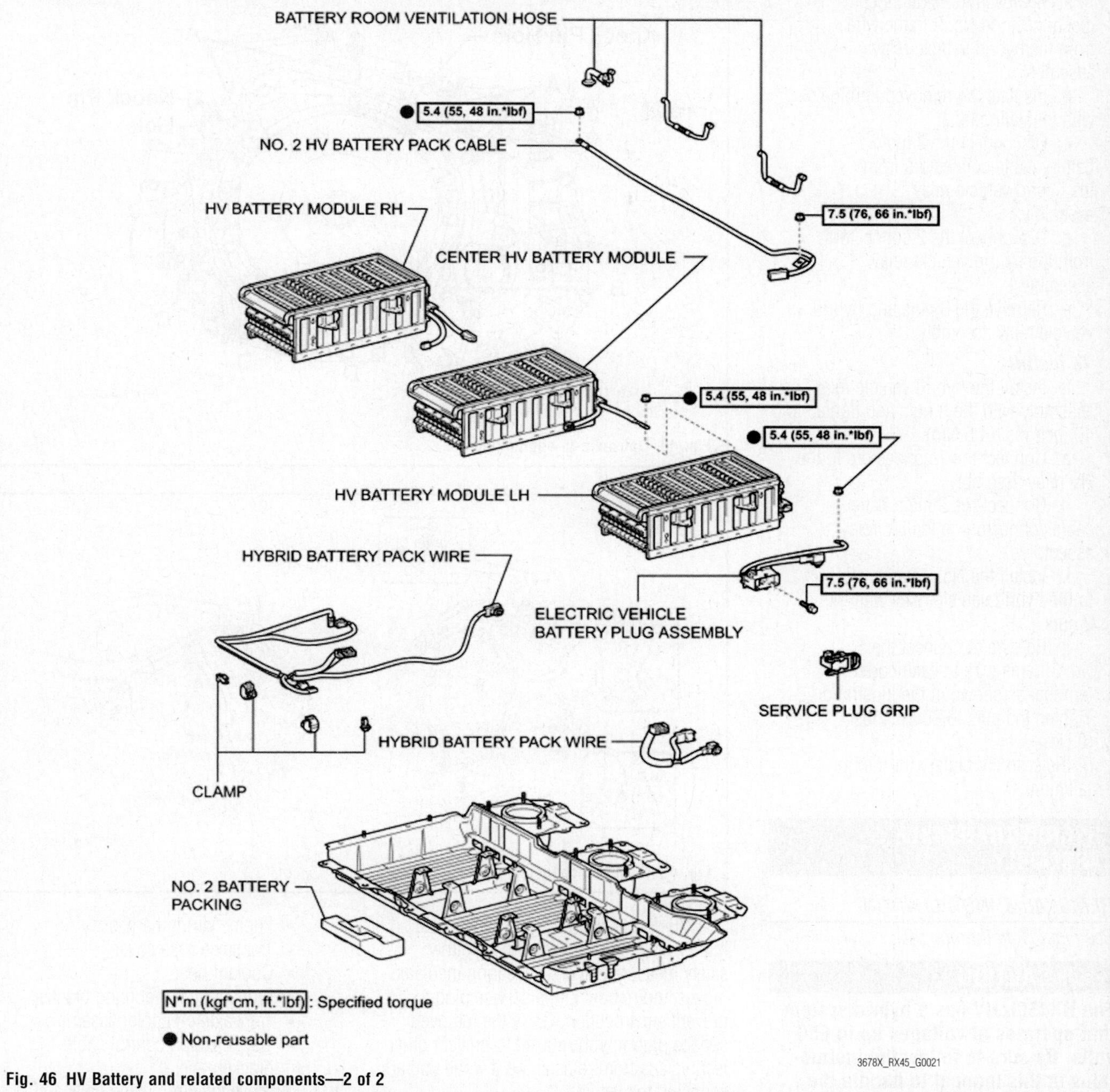

Fig. 46 HV Battery and related components—2 of 2

⁂ CAUTION

All high-voltage wire harness connectors are colored orange. The HV battery and other high-voltage components have "High Voltage" caution labels. Do not carelessly touch these wires and components.

1. Before inspecting or servicing the high-voltage system, be sure to follow safety measures, such as wearing insulated gloves and removing the service plug to prevent electrocution. Carry the removed service plug in your pocket to prevent other technicians from reinstalling it while you are servicing the vehicle.

2. After removing the service plug, wait 5 minutes before touching any of the high-voltage connectors and terminals.

3. Refer to HV battery removal for interior removal.

⁂ CAUTION

Be sure to wear insulated gloves and protective goggles.

4. Remove the HV relay assembly as follows:

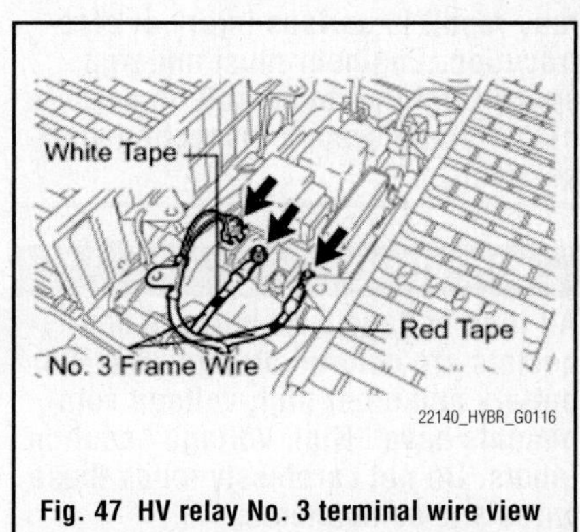

Fig. 47 HV relay No. 3 terminal wire view

a. Remove the 2 nuts, then disconnect the No. 3 frame wire from the hybrid vehicle relay assembly.

b. Insulate the removed terminals with insulating tape.

c. Disconnect the 2 main battery cable connectors from the hybrid vehicle relay assembly.

d. Disconnect the 2 connectors from the hybrid vehicle relay assembly.

e. Remove the 3 nuts and hybrid vehicle relay assembly.

To install:

f. Install the hybrid vehicle relay assembly with the 3 nuts and tighten to 40 inch lbs. (4.5 Nm)

g. Connect the 2 connectors to the HV relay assembly.

h. Connect the 2 main battery cable connectors to the HV relay assembly.

i. Install the No. 3 frame wire to the hybrid vehicle relay with the 2 nuts.

j. Be sure to connect the No. 3 frame wire to each correct terminal as shown in the illustration. Tighten the nuts to 80 inch lbs. (9 Nm).

5. Refer to HV battery for interior installation.

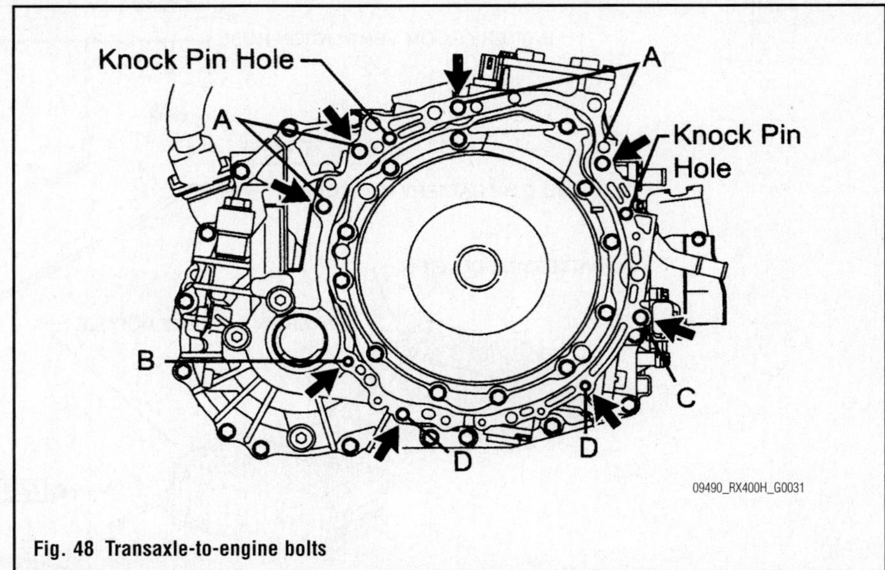

Fig. 48 Transaxle-to-engine bolts

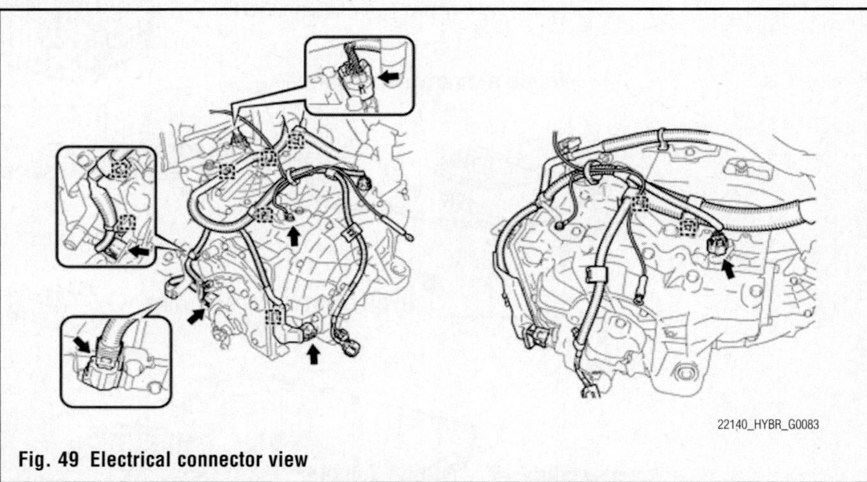

Fig. 49 Electrical connector view

HYBRID TRANSAXLE ASSEMBLY

REMOVAL & INSTALLATION

See Figures 48 through 50.

⁂ CAUTION

The RX450h HV has a hybrid system that operates at voltages up to 650 volts. Be sure to follow the instructions in this manual to handle the system correctly. Failure to do so may result in serious injury or electrocution. Engineer must undergo special training to be able to perform high-voltage system inspection and servicing.

⁂ CAUTION

All high-voltage wire harness connectors are colored orange. The HV battery and other high-voltage components have "High Voltage" caution labels. Do not carelessly touch these wires and components.

1. Before inspecting or servicing the high-voltage system, be sure to follow safety measures, such as wearing insulated gloves and removing the service plug to prevent electrocution. Carry the removed service plug in your pocket to prevent other technicians from reinstalling it while you are servicing the vehicle.

2. After removing the service plug, wait 5 minutes before touching any of the high-voltage connectors and terminals.

3. Before servicing the vehicle, refer to the Precautions Section.

4. Remove or disconnect the following:

- Engine/transaxle assembly. See Engine Removal and Installation.
- Manifold stay
- Transaxle damper
- Front frame assembly
- Halfshafts
- Flywheel housing undercover
- Engine wiring harnesses
- Transaxle case cover
- Coolant hose
- Front engine mounting bracket
- Transaxle oil cooler assembly
- Transmission control cable bracket

5. Remove the 8 mounting bolts and separate the transaxle assembly from the vehicle.

6. Installation is the reverse of removal. Observe the following torques:

- Transaxle-to-engine: Bolts A to 47 ft. lbs (64 Nm); Bolt B to 34 ft. lbs. (46 Nm); Bolts C to 47 ft. lbs. (64 Nm); Bolts D to 27 ft. lbs. (37 Nm)

➡Do not reuse Bolt B.

- Front engine mounting bracket: 47 ft. lbs. (64 Nm)
- Transaxle case cover: 74 inch lbs. (8.4 Nm)
- Undercover: 69 inch lbs. (8 Nm)

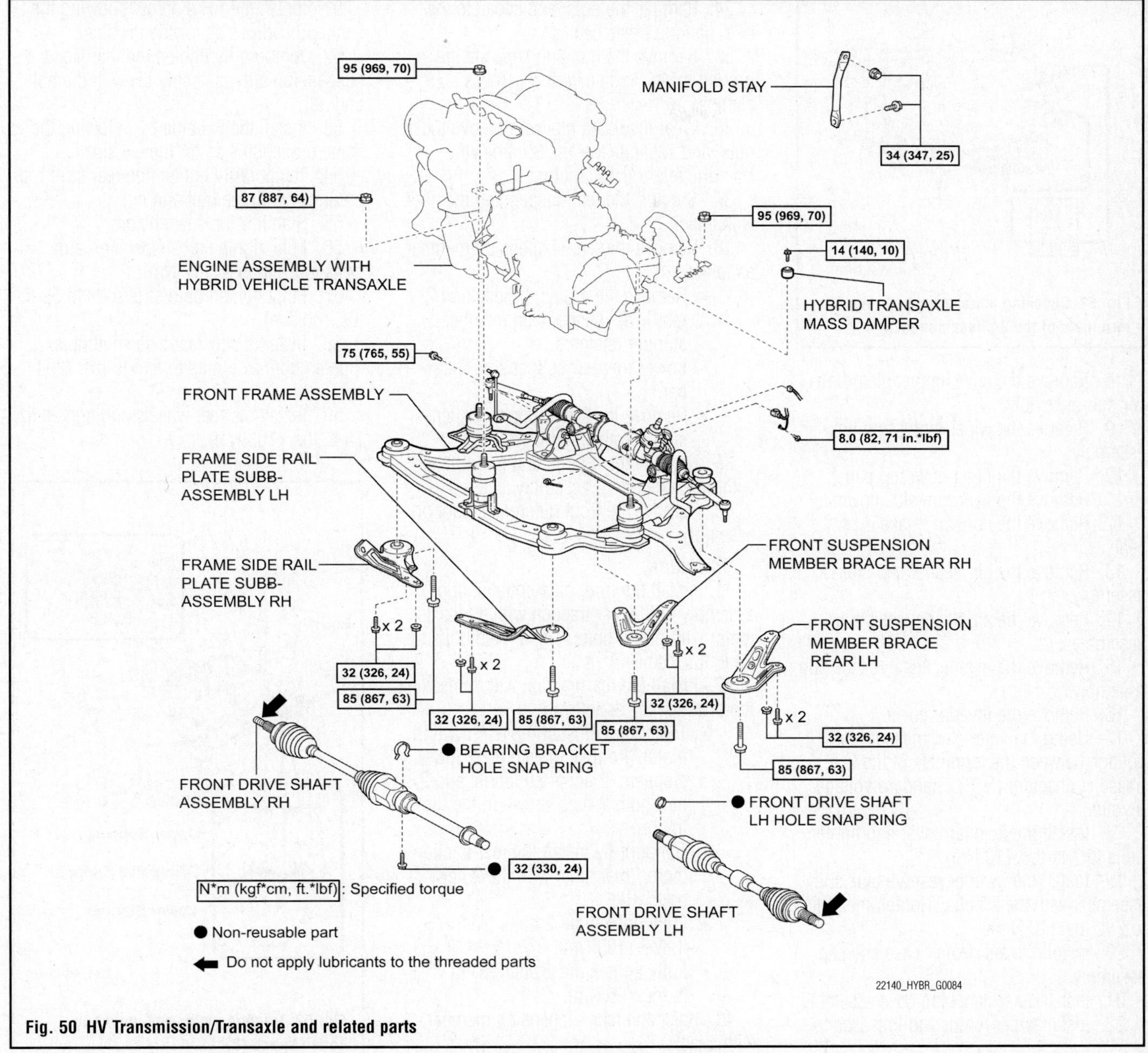

Fig. 50 HV Transmission/Transaxle and related parts

REAR TRACTION MOTOR

REMOVAL & INSTALLATION

See Figures 51 through 53.

CAUTION

The RX450 HV has a hybrid system that operates at voltages up to 650 volts. Be sure to follow the instructions in this manual to handle the system correctly. Failure to do so may result in serious injury or electrocution. Engineer must undergo special training to be able to perform high-voltage system inspection and servicing.

CAUTION

All high-voltage wire harness connectors are colored orange. The HV battery and other high-voltage components have "High Voltage" caution labels. Do not carelessly touch these wires and components.

1. Before inspecting or servicing the high-voltage system, be sure to follow safety measures, such as wearing insulated gloves and removing the service plug to prevent electrocution. Carry the removed service plug in your pocket to prevent other technicians from reinstalling it while you are servicing the vehicle.
2. After removing the service plug, wait 5 minutes before touching any of the high-voltage connectors and terminals.
3. Before servicing the vehicle, refer to the Precautions Section.
4. Disconnect the negative battery cable.
5. When disconnecting the cable, some systems need to be initialized after the cable is reconnected.
6. Remove the service plug grip, found underneath the Battery Service cover on the rear seat. Wait 5 minutes to discharge the high voltage capacitor.
7. Remove the LH and RH wiper arm assembly.

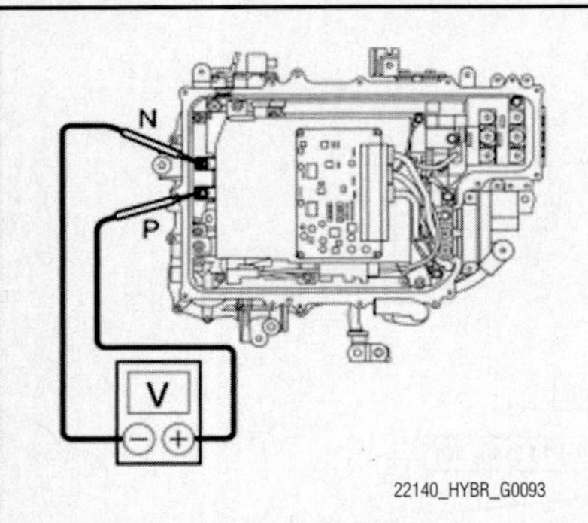

Fig. 51 Checking voltage between the terminals of the 2 phase connectors

8. Remove the cowl top ventilator louver sub-assembly.
9. Remove the wiper motor and link assembly.
10. Remove the outer cowl top panel.
11. Remove the LH room side cover.
12. Remove the cool air intake duct seal.
13. Remove the air cleaner cap sub assembly.
14. Remove the air cleaner case sub-assembly.
15. Remove the inverter reserve tank sub assembly.
16. Remove the inverter cover.
17. Using the voltmeter, measure the voltage between the terminals of the 2 phase connectors (N-P). Standard voltage: 0 volts.
18. Install the inverter cover and tighten bolts to 7 ft. lbs. (10 Nm).
19. Install the inverter reserve tank sub-assembly with the 2 bolts. Tighten the bolts to 7 ft. lbs. (10 Nm).
20. Install the air cleaner case and cap assembly.
21. Install outer panel top cowl assembly.
22. Install wiper motor and link assembly.
23. Install the cowl top ventilator assembly.
24. Install the LH and RH wiper arm assembly.
25. Install the cool air duct seal.
26. Install the engine room side cover.
27. Drain the rear traction motor fluid.
28. Remove the rear wheels.
29. Remove exhaust pipe assembly.
30. Remove the RH and LH axle shaft nuts.
31. Remove the LH and RH strut rod assembly.
32. Remove the LH and RH suspension arm assembly. No.1 and No. 2.
33. Remove axle assembly for RH and LH side.
34. Remove the nuts, and separate the both parking brake cables.
35. Remove the nut, and separate the ground cable. And all wiring harness clamps.
36. Wear insulated gloves. Remove the 2 nuts, and separate the No. 3 frame wire from the rear traction motor.
37. Support the rear suspension member with a jack.
38. Remove the rear suspension member as follows:
- Remove the 4 nuts, 2 bolts and 2 rear lower suspension member stopper retainers.
- Lower the rear suspension member.
- Remove the 2 rear upper suspension member stoppers.

39. Remove the 4 bolts and rear traction with transaxle motor assembly.
40. Remove the front differential support assembly.

To install:

41. Install the front differential support assembly to the rear traction with transaxle motor with new 2 bolts. Tighten the bolts to 59 ft. lbs. (80 Nm).
42. Install the rear traction with transaxle motor assembly as follows :
- Temporarily install the rear traction motor (front side) with the 2 lower stoppers, 2 upper supports, and 2 new bolts (A) as shown in the illustration.
- Temporarily install the rear traction motor (rear side) with the 2 new bolts (B).
- Fully tighten the 2 bolts (A) to 76 ft. lbs. (103 Nm).
- Fully tighten the 2 bolts (B) to 77 ft. lbs. (95 Nm).

43. Raise the rear suspension member with a jack.
44. Temporarily install the rear suspension member, the 2 rear upper suspension member stoppers and rear lower suspension member stopper retainers with the 4 nuts and the 2 bolts.
45. Fully tighten the rear suspension member to 133 ft lbs. (181 Nm).
46. Install the wiring harness and connectors.
47. Install both parking brake cables.
48. Install RH and LH axles.
49. Jack up the rear axle carrier, placing a wooden block to avoid damage.
50. Temporarily tighten the rear No. 2 suspension arm assembly LH with the bolt and nut.
51. Install the RH side by following the same procedures as for the LH side
52. Temporarily tighten the rear No. 1 suspension arm assembly LH with the bolt and nut.
53. Install the RH side by following the same procedures as for the LH side
54. Temporarily tighten the rear strut rod assembly with the bolt and nut.
55. Stabilize the suspension.
56. Fully tighten all suspension arm bolts to 82 ft. lbs. (112 Nm).
57. Fully tighten both strut rods to 59 ft. lbs. (80 Nm).
58. Install a new rear axle shaft nuts. Tighten both axle nuts to 216 ft. lbs. (294 Nm).
59. Install the rear wheels and tighten to 76 ft. lbs. (103 Nm).

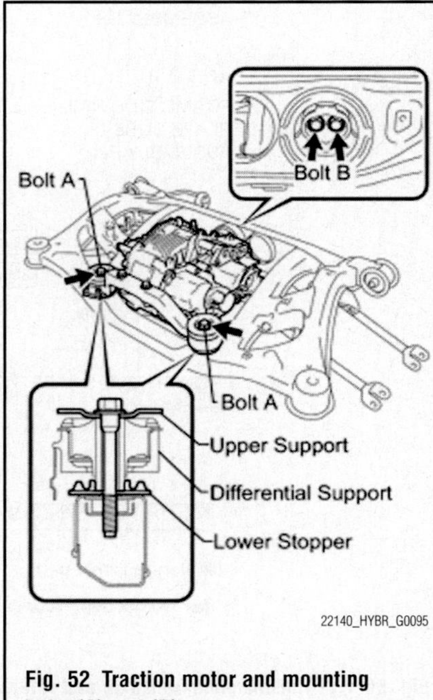

Fig. 52 Traction motor and mounting bolts (A) and (B)

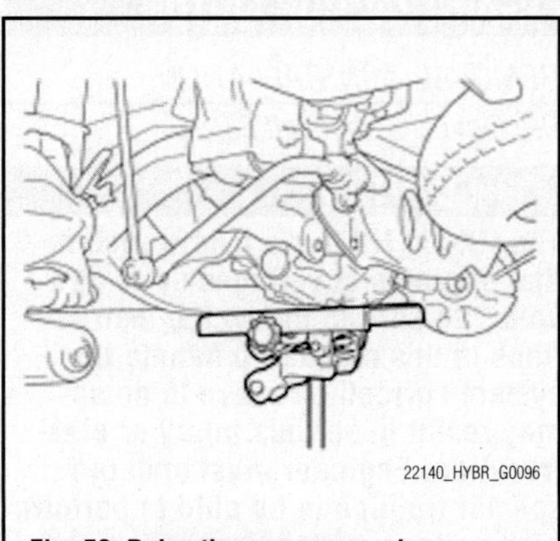

Fig. 53 Raise the rear suspension member with a jack

60. Check and adjust rear wheel alignment.

61. Check speed sensor operation.

ELECTRIC WATER PUMP WITH MOTOR

REMOVAL & INSTALLATION

See Figure 54.

⁂ CAUTION

The RX450 HV has a hybrid system that operates at voltages up to 650 volts. Be sure to follow the instructions in this manual to handle the system correctly. Failure to do so may result in serious injury or electrocution. Engineer must undergo special training to be able to perform high-voltage system inspection and servicing.

⁂ CAUTION

All high-voltage wire harness connectors are colored orange. The HV battery and other high-voltage components have "High Voltage" caution labels. Do not carelessly touch these wires and components.

1. Before inspecting or servicing the high-voltage system, be sure to follow safety measures, such as wearing insulated gloves and removing the service plug to prevent electrocution. Carry the removed service plug in your pocket to prevent other technicians from reinstalling it while you are servicing the vehicle.
2. After removing the service plug, wait 5 minutes before touching any of the high-voltage connectors and terminals.
3. Before servicing the vehicle, refer to the Precautions Section.
4. Disconnect the negative battery cable.
5. Remove the left engine room side cover.
6. Remove the transaxle side reserve tank.
7. Loosen the bleeder plug and drain the coolant from inverter cooler.
8. Loosen the bleeder plug and drain the coolant from inverter.
9. Remove the engine undercover.
10. Drain transaxle fluid if equipped with oil cooler.
11. Remove the front bumper if equipped with oil cooler.
12. Remove the frame side rail Plate sub-assembly as follows:
    - Using a transmission jack, hold the front frame.
    - Remove the 3 bolts, nut and frame side rail plate sub-assembly.

➡Be sure to position the transmission jack to properly support the front frame.

13. Disconnect the connector and 2 water hoses from the water with motor and bracket pump assembly.
14. Remove the bolt, nut and water with motor and bracket pump assembly.
15. If equipped with a oil cooler remove the 2 hoses.

To install:

16. Install the water with motor and bracket pump assembly with the bolt and nut. Tighten to 53 inch lbs. (6 Nm).
17. Connect the connector and 2 water hoses to the water with motor and bracket pump assembly.
18. Install bumper if removed for oil cooler. Inspect fluid level for hybrid transaxle.
19. Add engine coolant to inverter.
20. Connect the negative battery cable.
21. Check for coolant leaks.
22. Check oil cooler lines if removed.
23. Install the engine undercover.
24. Install the engine room left side cover.

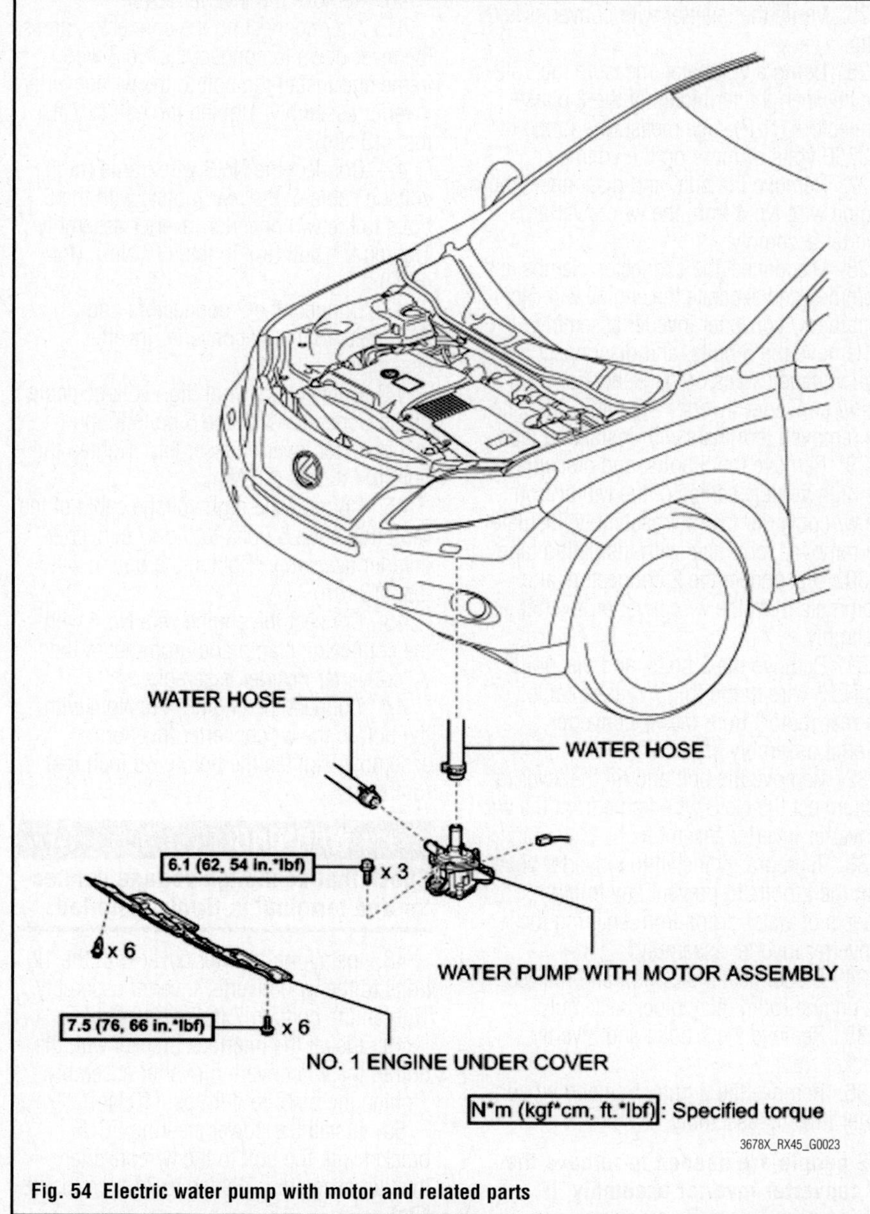

Fig. 54 Electric water pump with motor and related parts

INVERTER WITH CONVERTER

REMOVAL & INSTALLATION

See Figure 55.

✲✲ CAUTION

The RX450 HV has a hybrid system that operates at voltages up to 650 volts. Be sure to follow the instructions in this manual to handle the system correctly. Failure to do so may result in serious injury or electrocution. Engineer must undergo special training to be able to perform high-voltage system inspection and servicing.

✲✲ CAUTION

All high-voltage wire harness connectors are colored orange. The HV battery and other high-voltage components have "High Voltage" caution labels. Do not carelessly touch these wires and components.

1. Before inspecting or servicing the high-voltage system, be sure to follow safety measures, such as wearing insulated gloves and removing the service plug to prevent electrocution. Carry the removed service plug in your pocket to prevent other technicians from reinstalling it while you are servicing the vehicle.
2. After removing the service plug, wait 5 minutes before touching any of the high-voltage connectors and terminals.
3. Disconnect the negative battery cable.
4. Remove the engine room left side cover.
5. Remove the engine room cover.
6. Drain the coolant for the inverter.
7. Remove the wiper arm assembly LH and RH.
8. Remove the cowl top ventilator top louver sub-assembly.
9. Remove the wiper motor and link assembly.
10. Remove the cowl top outer panel sub-assembly.
11. Remove the cool air intake duct seal.
12. Remove the air cleaner cap with inlet.
13. Remove the air cleaner with resonator.
14. Remove the bolt and inverter bracket No.5.
15. Remove bolt and ground cable terminal from power steering ECU assembly.
16. Move the outer section to the wire harness side as illustrated, then disconnect the circuit breaker sensor No.1.
17. Remove the nut from the engine room wire No.2.
18. Remove the 2 bolts and inverter reserve tank sub-assembly.
19. Slide the 2 clamps, and disconnect the 2 water hoses from the inverter reserve tank sub-assembly.
20. Slide the clamp, and disconnect the water hose from the w/ converter inverter assembly.
21. Remove the bolt, and disconnect the power steering ECU bracket.
22. Remove the bolt and interlock bracket.
23. Insulate the removed terminal with insulating tape.

➡Make sure that the terminal does not stick out from the insulating tape.

24. Remove the 12 bolts and inverter cover.
25. Verify the inverter with converter is 0 volts.
26. Using a voltmeter, measure the voltage between the terminals of the 2 phase connectors (N-P). Use measuring range of DC 750 volts or more on the voltmeter.
27. Remove the bolt, and disconnect the engine wire No.4 from the w/ converter inverter assembly.
28. Disconnect the connector, clamps and grommet, and separate the engine wire No.4 from the w/ converter inverter assembly.

Remove the 5 bolts, and disconnect the high voltage cables of the Generator from the w/ converter inverter assembly. Insulate the removed terminals with insulating tape.

29. Remove the 5 bolts, and disconnect the high voltage cables of the Motor from the w/ converter inverter assembly. Insulate the removed terminals with insulating tape.
30. Disconnect the 2 connectors and grommets from the w/ converter inverter assembly.
31. Remove the 5 bolts, and disconnect the No.3 wire frame (high voltage cables of the rear motor) from the w/ converter inverter assembly. (for 4WD)
32. Remove the bolt and lift the lever to disconnect the No.3 wire frame from the w/ converter inverter assembly.
33. Temporarily install the inverter cover with the 2 bolts to prevent any foreign objects or water drops from entering the w/ converter inverter assembly.
34. Disconnect the clamp, and release the engine room relay block assembly.
35. Remove the 2 bolts and inverter bracket No.4.
36. Remove the 2 nuts, bolt and w/ converter inverter assembly.

➡2 people are needed to remove the w/ converter inverter assembly. If removing and storing the w/ converter inverter assembly, make sure to install the inverter cover to prevent any foreign objects or water drops from entering the w/ converter inverter assembly. Attach tape or equivalent (any adhesive should not be remained) to the holes for the connectors and grommets to prevent foreign matter or water from entering.

To install:

37. Install the w/ converter inverter assembly with the 2 nuts and bolt. Tighten to 15 ft lbs. (21 Nm).
38. Install the inverter bracket No.4 with the 2 bolts and tighten to 15 ft lbs. (21 Nm).
39. Install the engine room relay block assembly.
40. Remove the inverter cover.
41. After connecting the connector, press the lever down to connect the No.3 wire frame and install the bolt to the w/ converter inverter assembly. Tighten the bolt to 7 ft. lbs. (10 Nm).
42. Connect the No.3 wire frame (high voltage cable of the rear motor) with the 5 bolts to the w/ converter inverter assembly. Tighten the bolt to 7 ft. lbs. (10 Nm). (for 4WD)
43. Connect the 2 connectors and 2 grommets to the w/ converter inverter assembly.
44. Connect the front high voltage cable of the Generator with the 5 bolts to the w/ converter inverter assembly. Tighten the bolt to 7 ft. lbs. (10 Nm).
45. Connect the high voltage cable of the Motor with the 5 bolts to the w/ converter inverter assembly. Tighten the bolt to 7 ft. lbs. (10 Nm).
46. Connect the engine wire No.4 with the connector, clamp and grommet to the w/ converter inverter assembly.
47. Connect the engine wire No.4 with the bolt to the w/ converter inverter assembly. Tighten the bolt to 48 inch lbs. (5.4 Nm).

✲✲ WARNING

Check that each high voltage connector and terminal is firmly installed.

48. Install the inverter cover with the 12 bolts to the w/ converter inverter assembly. Tighten the bolts to 7 ft. lbs. (10 Nm).
49. Install the interlock bracket with the bolt to the w/ converter inverter assembly. Tighten the bolts to 7 ft. lbs. (10 Nm).
50. Install the power steering ECU bracket with the bolt to the w/ converter inverter assembly. Tighten to 71 inch lbs. (8 Nm).

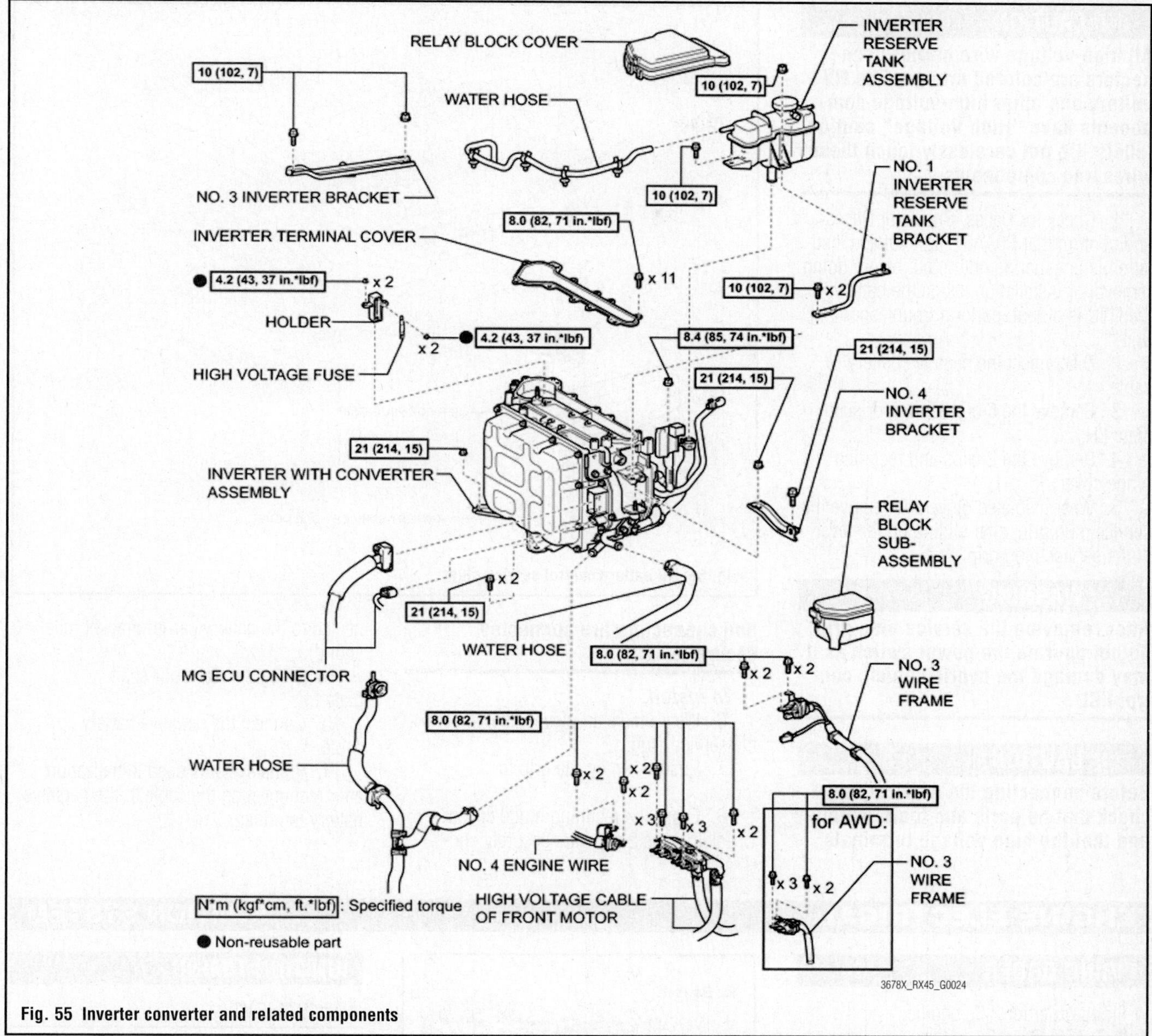

Fig. 55 Inverter converter and related components

51. Connect the water hose with the clamp to the w/ converter inverter assembly.
52. Connect the 2 water hoses with the 2 clamps to the inverter reserve tank sub-assembly.
53. Install the inverter reserve tank with the 2 bolts to the w/ converter inverter assembly. Tighten the mounting bolts to 7 ft. lbs. (10 Nm).
54. Connect the engine room wire No.2 with the nut and tighten to 75 inch lbs. (8.5 Nm).
55. Connect the circuit breaker sensor No.1 connector.
56. Install the power steering assembly.
57. Install the inverter bracket No.5 with the bolt and tighten to 7 ft. lbs. (10 Nm).
58. Install the air cleaner with resonator.
59. Install the air cleaner cap with inlet.
60. Install the cool air intake duct seal.
61. Install the cowl top panel sub-assembly outer.
62. Install wiper motor and link assembly.
63. Install the cowl top ventilator louver sub-assembly.
64. Install the RH and LH wiper arm assembly.
65. Install the engine room covers.
66. Install the service plug grip.
67. Connect the negative battery cable.
68. Add engine coolant to inverter.
69. Some systems need initialization when disconnecting the cable from the negative battery terminal.

SERVICE PLUG GRIP

REMOVAL & INSTALLATION

See Figure 56.

⁂ CAUTION

The hybrid system contains a 288V high-voltage system with a strong alkali solution of potassium hydroxide. Be sure to follow the instructions in this manual to handle the system correctly. Failure to do so may result in serious injury or electrocution. Engineer must undergo special training to be able to perform high-voltage system inspection and servicing.

⁂ CAUTION

All high-voltage wire harness connectors are colored orange. The HV battery and other high-voltage components have "High Voltage" caution labels. Do not carelessly touch these wires and components.

1. Check for Daces. Check for Daces and confirm that P0AA6 (High voltage insulation is unusual) is not output before doing removal or installation inside the battery. If the DTC is output, perform troubleshooting first.
2. Disconnect the negative battery cable.
3. Remove the 5 clips and door scuff plate LH.
4. Remove the 2 clips and reclining hinge cover.
5. Wear insulated glove and remove the service plug grip, after sliding up the lever of the service plug grip.

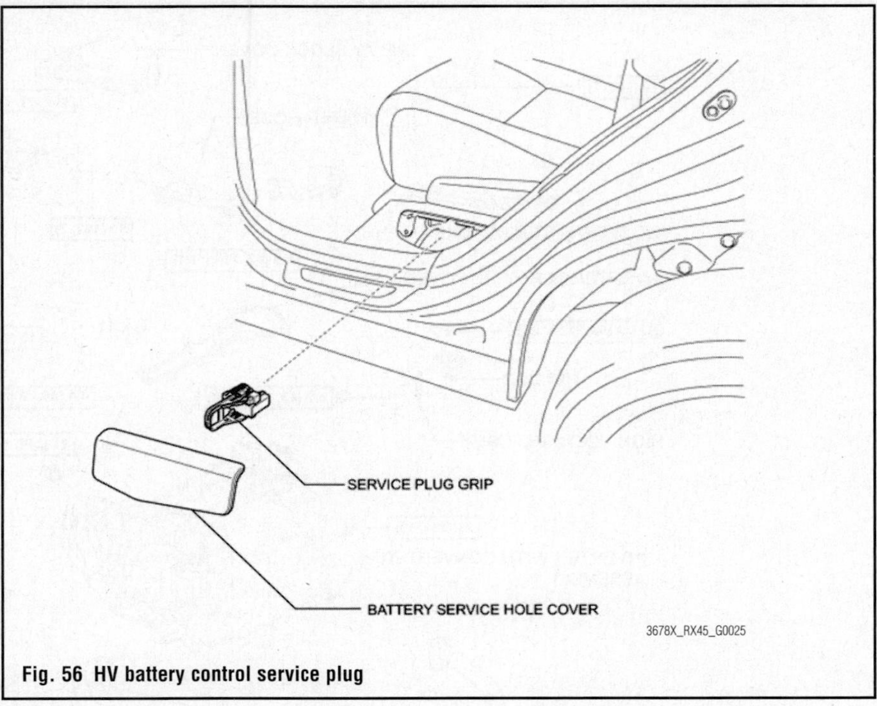

Fig. 56 HV battery control service plug

⁂ WARNING

After removing the service plug grip, do not operate the power switch as it may damage the hybrid vehicle control ECU.

⁂ CAUTION

Before connecting the service plug, check that no parts and tools remain and that the high voltage terminals and connectors are connected securely.

To install:

6. Wear insulated gloves, then insert the service plug.
7. Push down on the grip to lock.
8. Close the reclining hinge cover. Check that the 2 clips are securely connected to the battery carrier bracket (click sound).
9. Close the rear door scuff plate LH.
10. Connect the negative battery cable.
11. Some systems need initialization after reconnecting the cable to the negative battery terminal.

ENGINE ELECTRICAL

IGNITION SYSTEM

FIRING ORDER

Firing order for 3.5L engine:
1–2–3–4–5–6

IGNITION COIL

REMOVAL & INSTALLATION

See Figure 57.

1. Disconnect the negative battery cable.
2. Remove the EGR cooler assembly.
3. Disconnect the 6 ignition coil connectors.
4. Remove the 6 bolts and 6 ignition coils.
5. To install, reverse removal procedure.

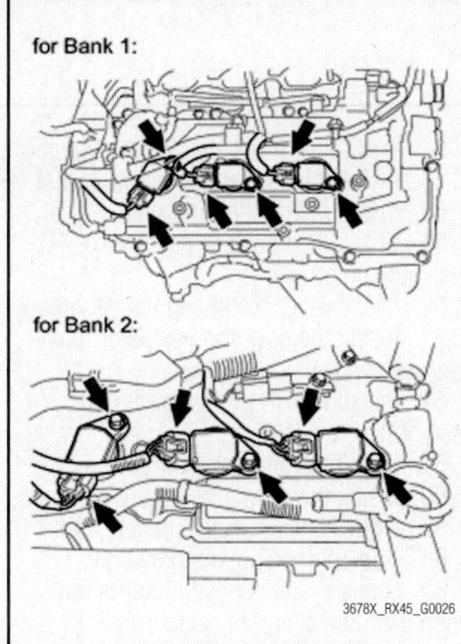

Fig. 57 Removing the ignition coils

IGNITION TIMING

ADJUSTMENT

All engines are equipped with a Distributorless Ignition System (DIS). No timing adjustment is possible.

INSPECTION

See Figure 58.

1. Warm up the engine.
2. Using SST: 09843-18040, connect terminals 13 (TC) and 4 (CG) of the DLC3.

➡Confirm the terminal numbers before connecting them. Connecting the wrong terminals can damage the engine.

➡Turn off all electrical systems before connecting the terminals.

➡Perform this inspection after the cooling fan motor is turned off.

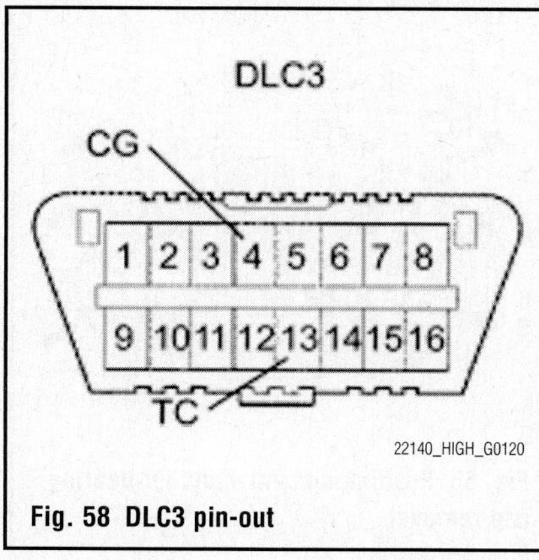

Fig. 58 DLC3 pin-out

3. Remove the V-bank cover sub-assembly.
4. Pull out the red lead wire harness.
5. Connect the tester terminal of the timing light to the red lead wire as shown in the illustration.

➡**Use a timing light which detects the No. 1 cylinder ignition signal.**

6. Check the ignition timing at idle. Standard ignition timing: 8 to 12° BTDC at idle.

➡**When checking the ignition timing, the transmission should be in the neutral position.**

➡**Run the engine at 1000 to 1300 RPM for 5 seconds, and then check that the engine RPM returns to idle speed.**

7. Disconnect terminals 13 (TC) and 4 (CG) of the DLC3.
8. Check the ignition timing at idle.

Standard ignition timing: 7 to 24° BTDC at idle.

9. Confirm that the ignition timing advances immediately when the engine RPM is increased.
10. Remove the timing light from the engine.

SPARK PLUGS

REMOVAL & INSTALLATION

1. Disconnect the negative battery cable.
2. Remove the ignition coils.
3. Remove the 6 spark plugs.
4. To install, tighten the spark plugs to 13 ft. lbs. (18 Nm).
5. To complete installation, reverse remaining removal procedure.

ENGINE MECHANICAL

ACCESSORY DRIVE BELTS

ACCESSORY BELT ROUTING

See Figure 59.

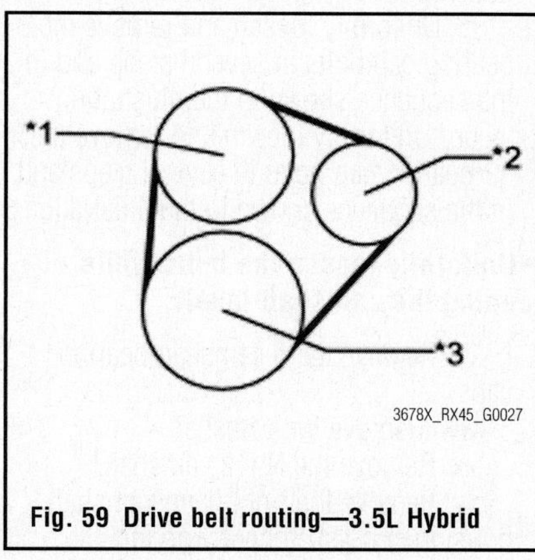

Fig. 59 Drive belt routing—3.5L Hybrid

INSPECTION

Visually check the V-ribbed belt for excessive wear, frayed cords, etc. All worn or damaged drive belts should be replaced immediately. Cracks on the rib side of a V-ribbed belt are considered acceptable, If the drive belt has chunks missing from its ribs, it should be replaced. After installing the V-ribbed belt, check that it fits properly in the ribbed grooves. Check to confirm that the belt has not slipped out of the grooves on the bottom of the crank pulley by hand.

ADJUSTMENT

Belt tension is maintained by an automatic tensioner. No adjustment is necessary or possible.

REMOVAL & INSTALLATION

See Figure 60.

1. Remove the right front wheel.
2. Separate the right front fender splash shield sub-assembly.
3. Remove the right front fender apron seal.
4. Remove the engine room side cover.
5. Remove the V-bank cover sub-assembly.
6. Remove the V-ribbed belt, as follows:
   a. Release the V-ribbed belt tension by turning the V-ribbed belt tensioner counterclockwise, and remove the V-ribbed belt from the V-ribbed belt tensioner.
   b. While turning the V-ribbed belt tensioner counterclockwise, align with its holes, and then insert a 5 mm bi-hexagon wrench into the holes to secure the V-ribbed belt tensioner.

Fig. 60 Removing the drive belt

To install:

7. Install the V-ribbed belt.
   a. Turn the V-ribbed belt tensioner counterclockwise and remove a 5 mm bi-hexagon wrench.
   b. After installing the V-ribbed belt, check that it fits properly in the ribbed grooves. Confirm that the belt has not slipped out of the grooves on the bottom of the crank pulley by hand.
8. Install the V-bank cover sub-assembly.
9. Install the engine room side cover.
10. Install the right front fender apron seal.
11. Install the right front fender splash shield sub-assembly.
12. Install the right front wheel and tighten the lug nuts to 76 ft. lbs. (103 Nm).

CAMSHAFT AND VALVE LIFTERS

REMOVAL & INSTALLATION

See Figures 61 through 78.

1. Remove the engine assembly.
2. Install on engine stand.
3. The following must be removed:
   a. Remove ignition coil assembly.
   b. Remove the right hand No. 2 engine mounting stay.
   c. Remove the intake manifold.
   d. Remove the right hand exhaust manifold sub-assembly.
   e. Remove the No. 2 engine oil level dipstick guide.
   f. Remove the No. 2 manifold stay.

g. Remove the No. 2 exhaust manifold heat insulator.

h. Remove the left hand exhaust manifold sub-assembly.

i. Remove the transverse engine mounting bracket.

j. Remove the generator assembly.

k. Remove the V-ribbed belt tensioner assembly.

l. Remove the No. 2 timing gear cover.

m. Remove the No. 2 idler pulley sub-assembly.

n. Remove the left hand No. 1 engine front mounting bracket.

o. Remove the left hand 6 bolts and No. 1 front engine mounting bracket.

p. Remove the radio setting condenser.

q. Remove the No. 1 vacuum switching valve.

r. Remove the knock control sensor wire.

s. Remove the knock control sensor.

t. Remove the crankshaft position sensor.

u. Remove the No. 1 oil pipe.

v. Remove the oil pipe.

w. Remove the crankshaft pulley.

x. Remove the oil cooler assembly, if necessary.

y. Remove the No. 1 oil cooler bracket, if necessary.

z. Remove the water inlet housing.

aa. Remove the water outlet.

bb. Remove the cylinder head cover sub-assembly (for Bank 1).

cc. Remove the cylinder head cover sub-assembly (for Bank 2).

dd. Remove the No. 2 oil pan sub-assembly.

ee. Remove the oil strainer sub-assembly.

ff. Remove the oil pan sub-assembly.

gg. Remove the timing chain cover sub-assembly

hh. Remove the timing chain case oil seal.

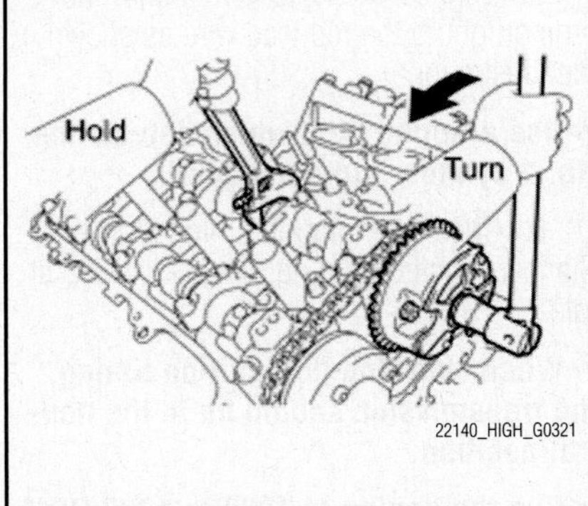

Fig. 62 Removing gear assemblies

ii. Set the No. 1 cylinder to TDC/compression.

jj. Remove the No. 1 chain tensioner assembly.

kk. Remove the chain tensioner slipper.

ll. Remove the chain sub-assembly.

mm. Remove the idle sprocket assembly.

nn. Remove the camshaft timing gears and No. 2 chain (for Bank 1).

oo. While raising the No. 2 chain tensioner assembly, insert a pin of 1.0 mm (0.039 in.) diameter into the hole to fix the No. 2 chain tensioner assembly.

pp. Hold the hexagonal portion of the camshaft with a wrench, and remove the 2 bolts and 2 camshaft timing gear assemblies.

➡Be careful not to damage the cylinder head with the wrench. Do not disassemble the camshaft timing gear assemblies.

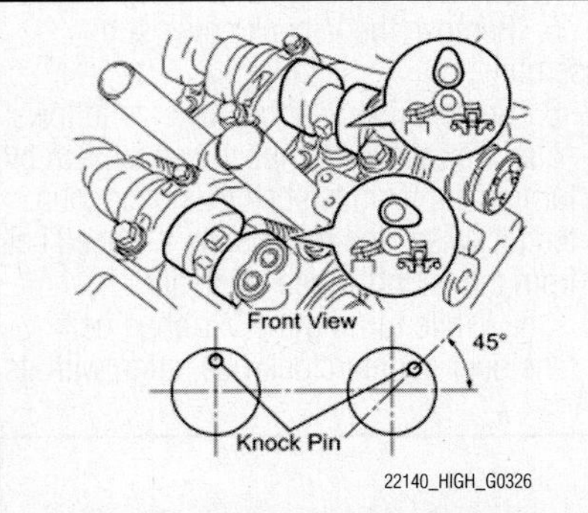
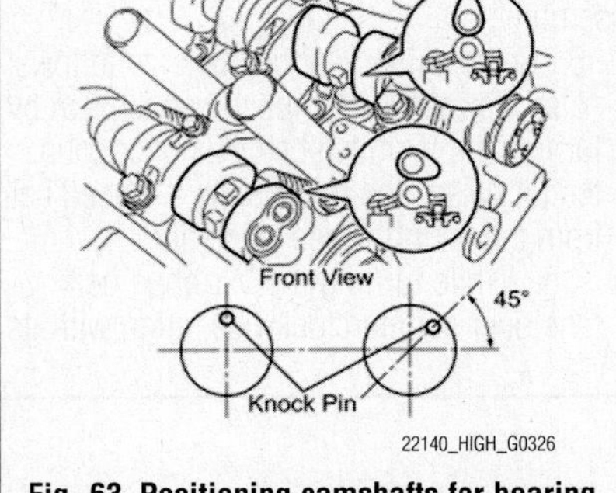

Fig. 63 Positioning camshafts for bearing cap removal

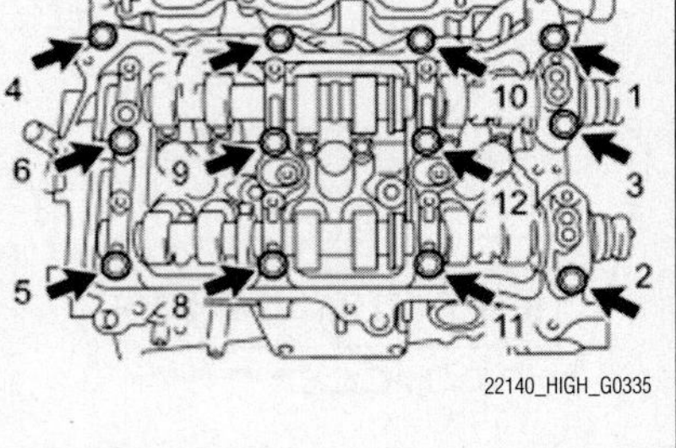

Fig. 65 Positioning camshafts for bearing cap removal

qq. Remove the No. 2 chain assembly.

rr. Remove the bolt and No. 2 chain tensioner assembly.

ss. Check that the camshafts are positioned as shown in the illustration.

tt. Uniformly loosen and remove the 8 bearing cap bolts in several steps and in the sequence shown in the illustration.

uu. Uniformly loosen and remove the 12 bearing cap bolts in several steps and in the sequence shown in the illustration.

➡Uniformly loosen the bolts while keeping the camshaft level.

vv. Remove the 5 camshaft bearing caps.

ww. Remove the camshaft.

xx. Remove the No. 2 camshaft.

yy. Remove the right hand camshaft housing sub-assembly by prying between the cylinder head and the right hand camshaft housing sub-assembly with a screwdriver.

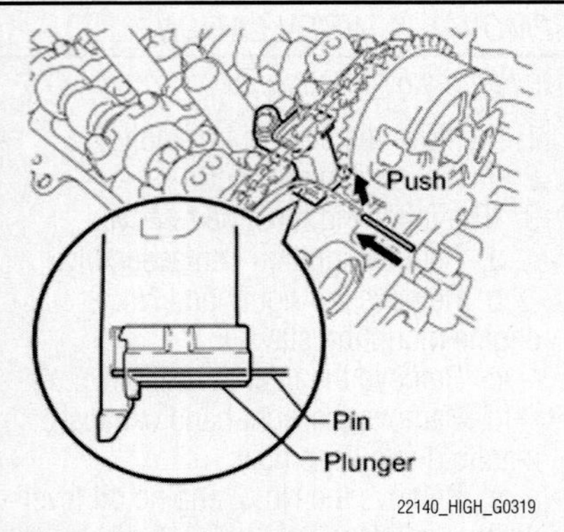

Fig. 61 Pinning tensioner

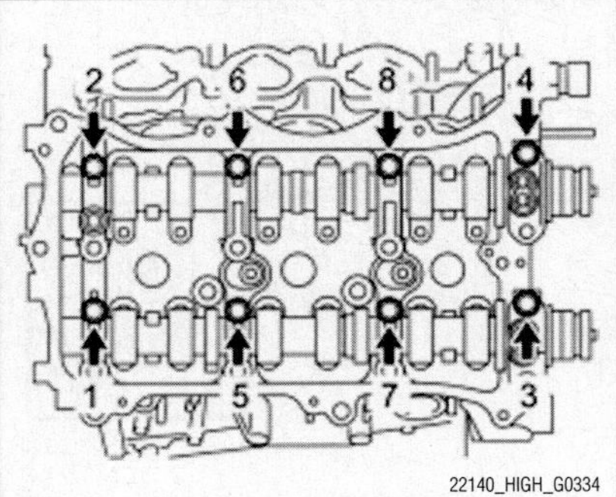

Fig. 64 Camshafts bearing cap removal sequence

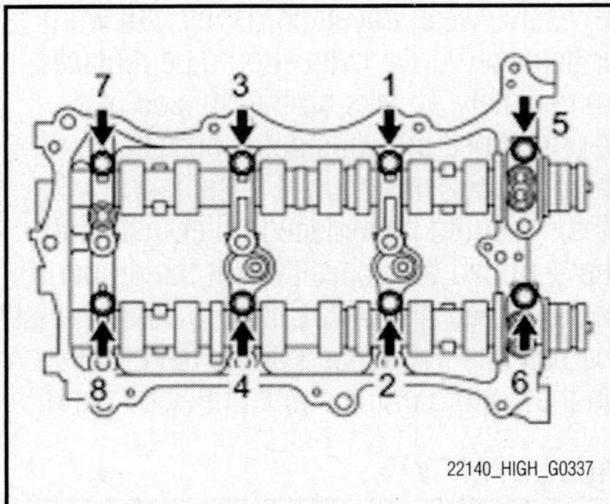

Fig. 66 Camshafts bearing cap tightening sequence

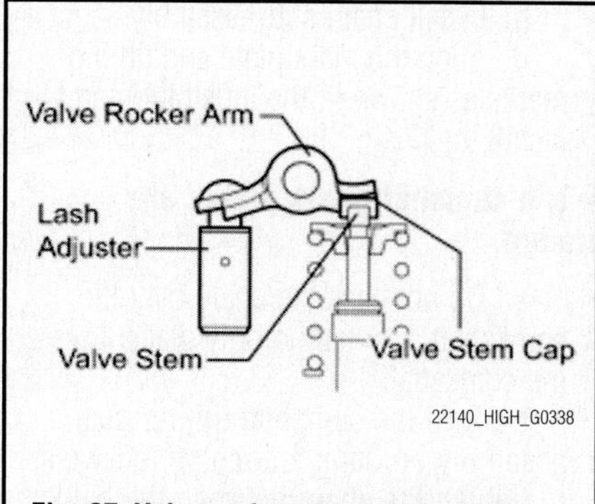

Fig. 67 Valve rocker arm sub-assembly positioning

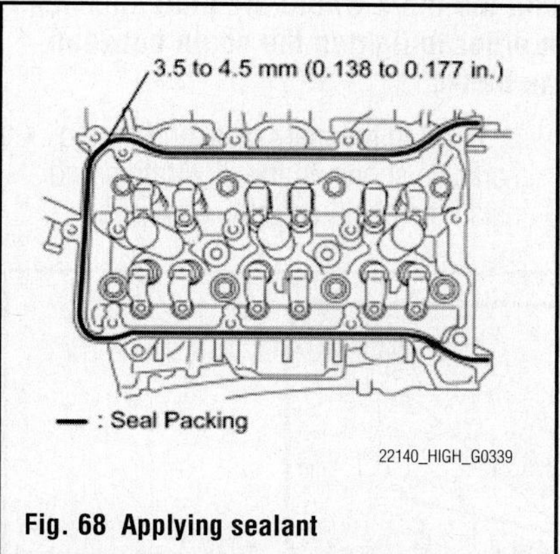

Fig. 68 Applying sealant

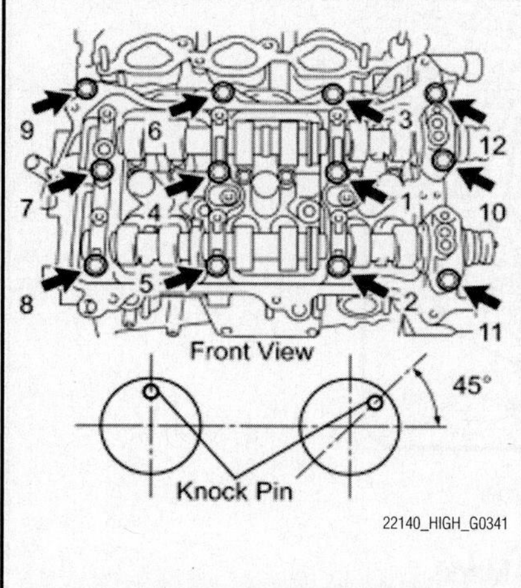

Fig. 69 Camshaft sub-assembly tightening sequence

➡**Be careful not to damage the contact surfaces of the cylinder head and the right hand camshaft housing sub-assembly.**

To install:

4. Install the following:

a. Apply engine oil to the camshaft journals, camshaft housing sub-assembly RH and camshaft bearing caps.

b. Install the camshaft and No. 2 camshaft to the right hand camshaft housing sub-assembly.

c. Make sure of the marks and numbers on the camshaft bearing caps and place them in each proper position and direction.

d. Temporarily tighten the 8 bearing cap bolts to 7 ft. lbs. (10 Nm) in the order shown in the illustration.

e. Make sure that the No. 1 valve rocker arm sub-assembly is installed as shown in the illustration.

f. Apply seal packing in a continuous line as shown in the illustration.

➡**Remove any oil from the contact surface. Install the right hand camshaft housing sub-assembly within 3 minutes. Do not start the engine for at least 2 hours after installing.**

g. Install the right hand camshaft housing sub-assembly and tighten the 12 bolts to 21 ft. lbs. (28 Nm) in the order shown in the illustration.

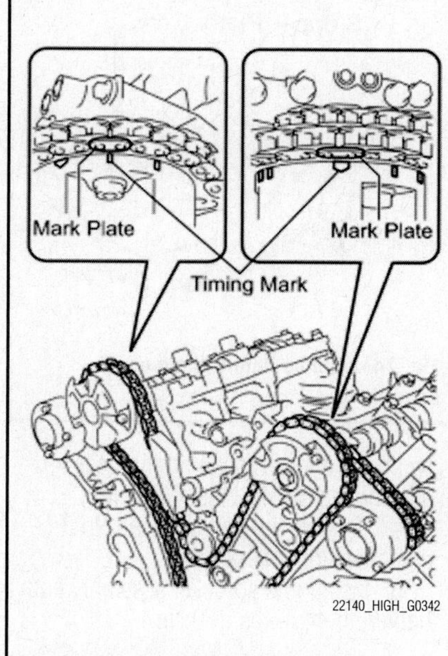

Fig. 70 Aligning timing chain sub-assembly

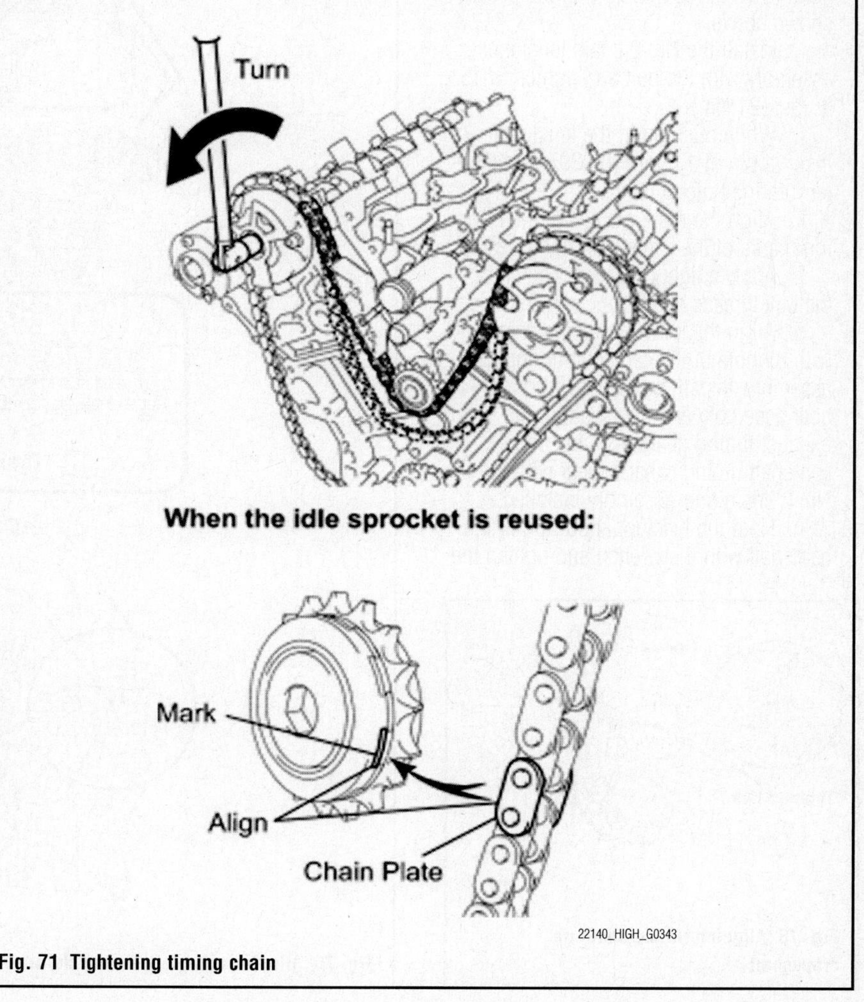

Fig. 71 Tightening timing chain

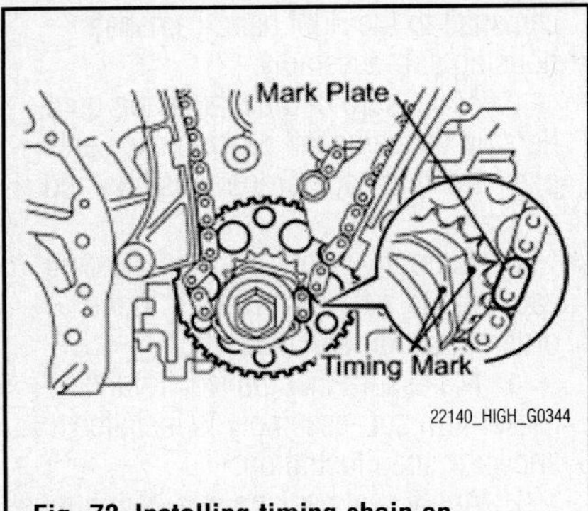

Fig. 72 Installing timing chain on crankshaft

➡**When installing the right hand camshaft housing, it is necessary to correctly position the camshafts as shown in the removal illustration. If the camshaft housing sub-assembly is removed because any of the bolts are loosened during installation, make sure that the previously applied seal packing does not enter any oil passages.**

h. Complete the tightening of the 8 bolts to 12 ft. lbs. (16 Nm) in the order shown above.

i. Install the No. 2 chain tensioner assembly with the bolt and tighten to 15 ft. lbs. (21 Nm).

j. While pushing in the tensioner, insert a pin of 1.0 mm (0.039 in.) diameter into the hole to fix it.

k. Align the mark plate with the timing marks of the camshaft timing gear.

l. Apply a light coat of engine oil to the bolt threads and bolts seating surface.

m. Align the knockpin of the camshaft with pinhole of the camshaft timing gear assembly. Install the camshaft timing gear assembly and camshaft timing exhaust timing gear assembly and camshaft timing exhaust gear with the No. 2 chain sub-assembly installed.

n. Hold the hexagonal portion of the camshaft with the wrench and tighten the two bolts and camshaft timing gear assemblies to 74 ft. lbs. (100 Nm).

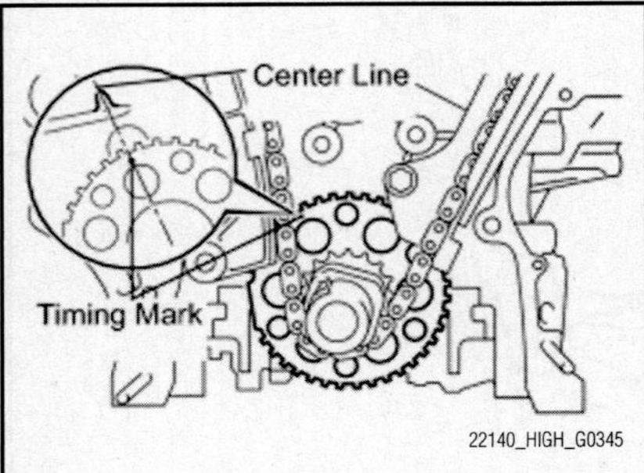

Fig. 73 Aligning timing chain on crankshaft

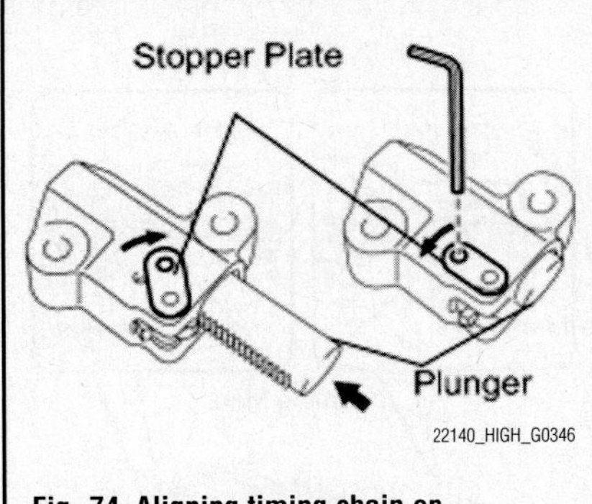

Fig. 74 Aligning timing chain on crankshaft

o. Remove the pan from the No 2 chain tensioner assembly.

p. Install idle sprocket assembly and tighten to 44 ft. lbs. (60 Nm).

q. Install chain sub-assembly.

r. Align the mark plate and timing marks as shown in the illustration and install the chain.

➡**The camshaft mark plates are orange.**

s. Do not pass the chain over the crankshaft, just temporarily place it on the crankshaft.

t. Turn the camshaft timing gear assembly on bank 1 counterclockwise to tighten the chain between the banks.

➡**When the idle sprocket assembly is reused, align the timing chain plate with the mark where the plate has been in order to tighten the chain between the banks.**

u. Align the mark plate and timing marks as shown in the illustration and install the chain onto the crankshaft

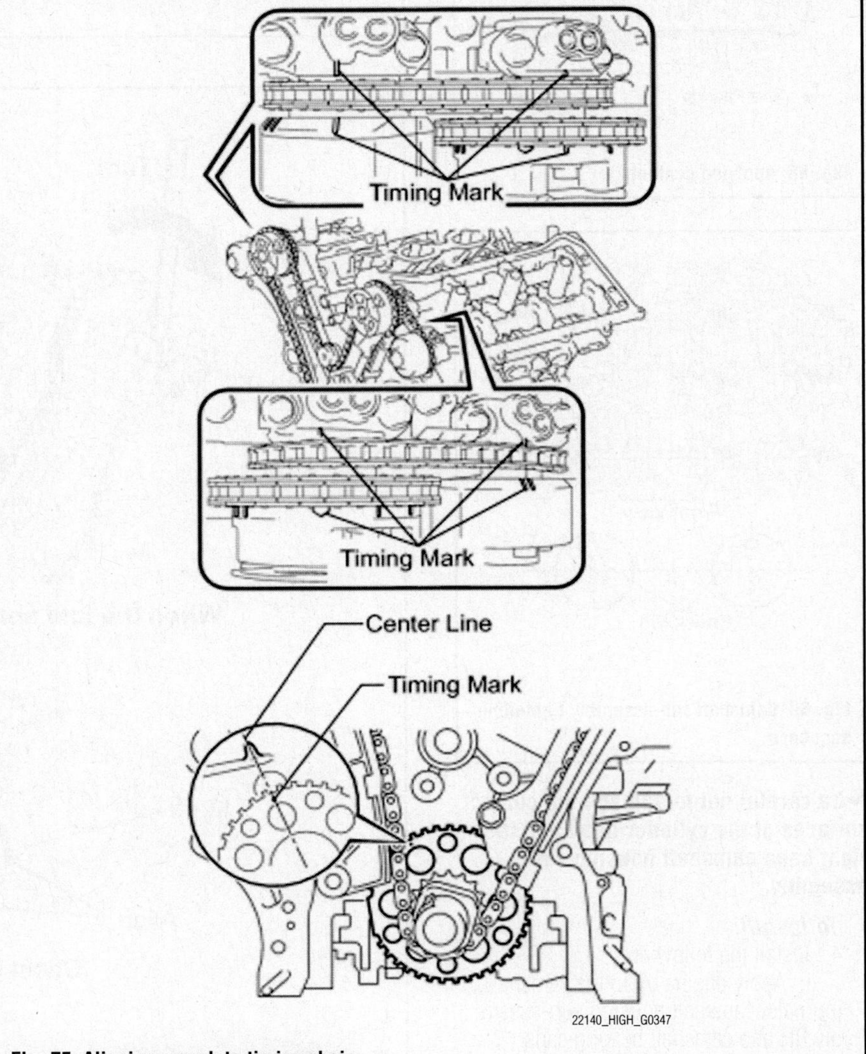

Fig. 75 Aligning complete timing chain

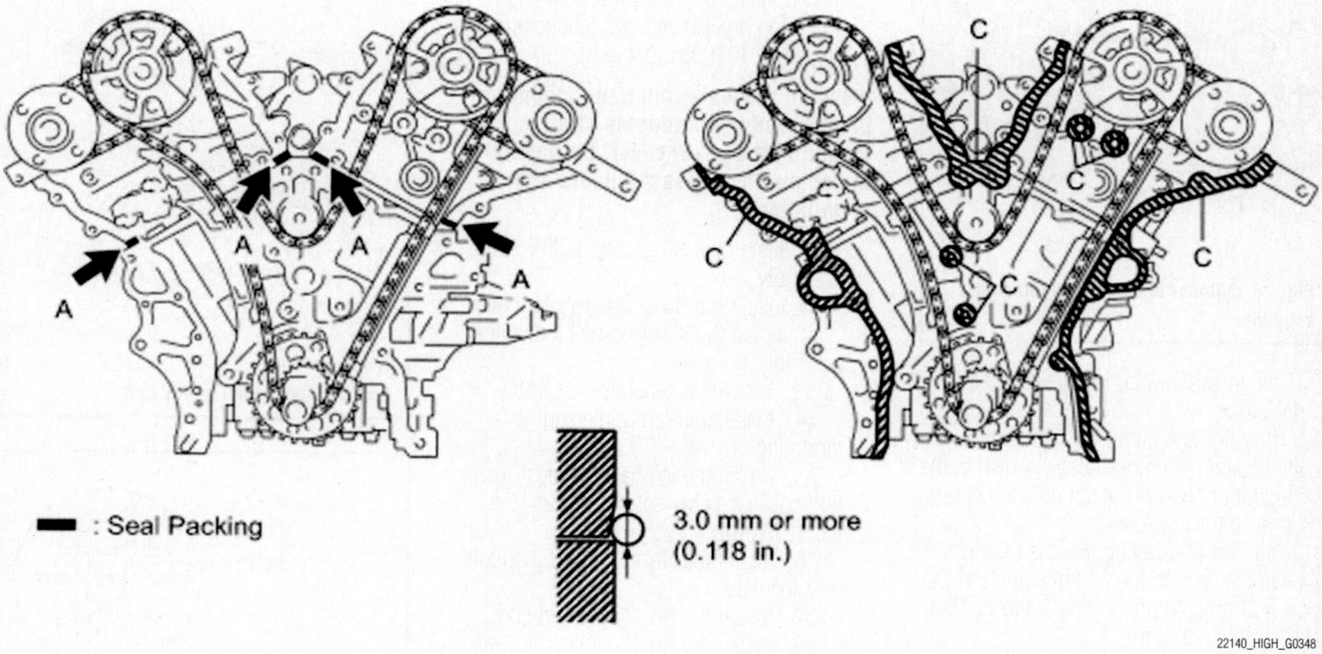

: Seal Packing

3.0 mm or more (0.118 in.)

22140_HIGH_G0348

Fig. 76 Sealant application area

timing sprocket. The crankshaft to mark plate is yellow.

v. Turn the crankshaft clockwise to set it to the right-hand block bore more centerline. (TDC Compression).

w. Install chain tensioner slipper.

x. Move the stopper plate upward to release the lock, and push the plunger deep into the tensioner.

y. Move the stopper plate downward to set the loss, and insert a hexagon wrench into the hole of the stopper plate.

z. Install No. 1 chain tensioner assembly and tighten bolts to 7 ft. lbs. (10 Nm).

aa. Remove the hexagon wrench from the No. 1 chain tensioner assembly. Check that the each timing mark is aligned with the crankshaft at TDC compression.

bb. Remove the pulley set bolt.

cc. Install timing chain case oil seal.

dd. Install sealant to timing chain cover sub-assembly.

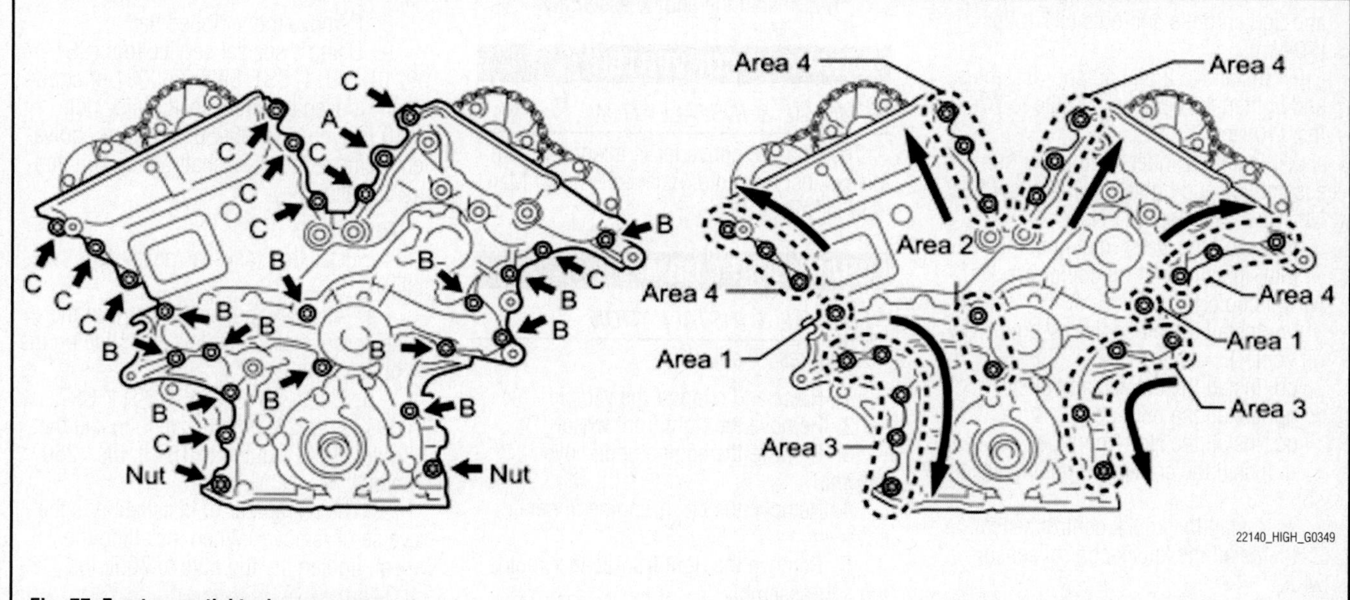

Fig. 77 Front cover tightening sequence

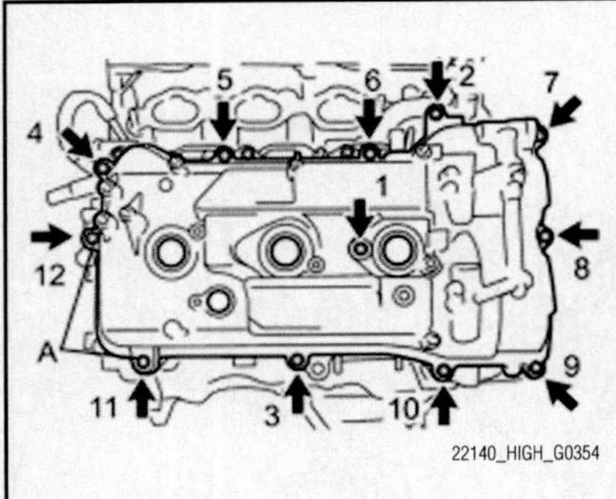

Fig. 78 Cylinder head cover tightening sequence

ee. Install new O ring gasket on cylinder block.

ff. Align the oil pump's drive rotor spline and the crankshaft as shown in the illustration. Install the spline and chain cover to the crankshaft.

gg. Temporarily tighten the timing chain cover with the 23 bolts and 2 nuts.

- Tighten bolts in area 1 and 2: 15 ft. lbs. (21 Nm).
- Tighten bolt in area 3: 15 ft. lbs. (21 Nm).

➡First tighten the upper bolts and nuts, followed by the lower bolts and nuts as shown.

- Tighten bolt in area 4: 32 ft. lbs. (43 Nm)
- Tighten bolt in area 4: 15 ft. lbs. (21 Nm)

hh. Install oil pan sub-assembly and tighten 16 bolts and 2 nuts to 7 ft. lbs. (10 Nm) and 15 ft lbs. (21 Nm).

ii. Install oil strainers sub-assembly and tighten bolts and nuts to 7 ft. lbs. (10 Nm).

jj. Install No.2 oil pan sub-assembly and tighten 16 bolts and 2 nuts to 7 ft. lbs. (10 Nm).

kk. Install cylinder head cover sub-assemblies and tighten to 7 ft. lbs. (10 Nm) and 15 ft. lbs. (21 Nm).

ll. Install the water inlet housing.

mm. Install No. 1 oil cooler bracket (w/ oil cooler).

nn. Install the oil cooler assembly (w/ oil cooler).

oo. Install the crankshaft pulley.

pp. Install the oil pipe.

qq. Install the No. 1 oil pipe.

rr. Install the crankshaft position sensor.

ss. Install the knock control sensor.

tt. Install the knock control sensor wire.

uu. Install the No. 1 vacuum switching valve.

vv. Install the radio setting condenser.

ww. Install the No. 1 left front engine mounting bracket with the 6 bolts and tighten to 40 ft. lbs. (54 Nm).

➡Install the water inlet and mounting bracket within 15 minutes after installing the chain cover. Do not start the engine for at least 2 hours after installation.

xx. Install the No. 2 idler pulley sub-assembly.

yy. Install the No. 2 timing gear cover

zz. Install the V-ribbed belt tensioner assembly.

53. Install the generator assembly.
54. Install the transverse engine mounting bracket.
55. Install the left hand exhaust manifold sub-assembly and tighten to 15 ft. lbs. (21 Nm).
56. Install the No. 2 exhaust manifold heat insulator.
57. Install the No. 2 manifold stay.
58. Install the No. 2 engine oil level dipstick guide.
59. Install the right hand exhaust manifold sub-assembly and tighten to 15 ft. lbs. (21 Nm).
60. Install the intake manifold and tighten the 6 bolts and 4 nuts uniformly in several steps to 15 ft. lbs. (21 Nm).
61. Install the right hand No. 2 engine mounting stay.
62. Install the ignition coil assembly.
63. Remove the engine stand.
64. Install the engine assembly.

CATALYTIC CONVERTER

REMOVAL & INSTALLATION

The catalytic converter is integrated with the exhaust manifold. Refer to Exhaust Manifold in this section.

CRANKSHAFT DAMPER

REMOVAL & INSTALLATION

See Figures 79 and 80.

1. Raise and support the vehicle.
2. Remove the right front wheel.
3. Remove the engine undercover assembly.
4. Remove the No. 1 engine undercover.
5. Remove the right front fender molding sub-assembly.
6. Remove the right front fender liner.

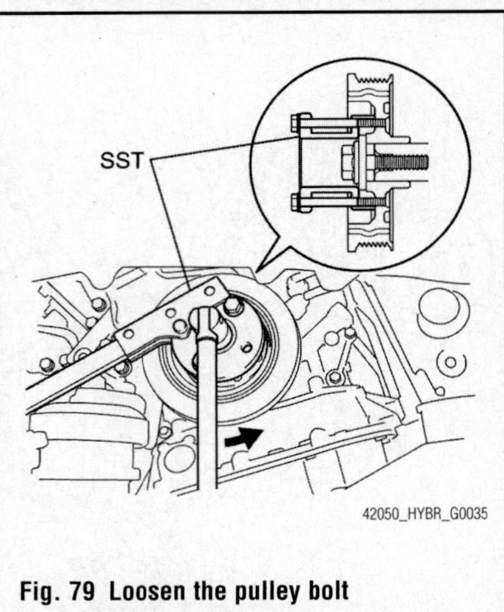

Fig. 79 Loosen the pulley bolt

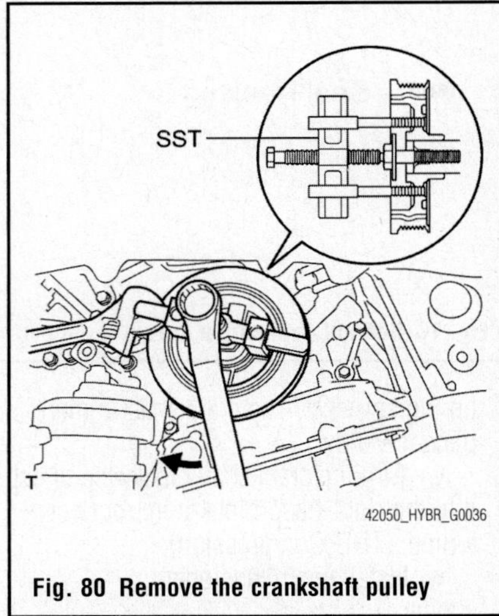

Fig. 80 Remove the crankshaft pulley

7. Remove the right front fender apron seal.
8. Remove the V-ribbed belt.
9. Using a special service tool (SST: 09213-70011, SST: 09330-00021 or equivalent), loosen the crankshaft pulley bolt.
10. Using SST: 09950-50013 or equivalent, remove the crankshaft pulley bolt and crankshaft pulley.

To install:

11. Install the crankshaft pulley, as follows:
    a. Align the pulley set key with the key groove of the pulley, and slide on the pulley.
    b. Using SST: 09213-70011, SST: 09330-00021, or equivalent, install the pulley bolt. Tighten to 184 ft. lbs. (250 Nm).
12. The remainder of installation is the reverse of removal. When installing the wheel, tighten the lug nuts to 76 ft. lbs. (103 Nm).

CRANKSHAFT FRONT SEAL

REMOVAL & INSTALLATION

See Figures 81 and 82.

1. Raise and support the vehicle.
2. Remove the right front wheel.
3. Remove the engine undercover assembly.
4. Remove the No. 1 engine undercover.
5. Remove the right front fender molding sub-assembly.
6. Remove the right front fender liner.
7. Remove the right front fender apron seal.
8. Remove the V-ribbed belt.
9. Using a special service tool (SST: 09213-70011, SST: 09330-00021 or equivalent), loosen the crankshaft pulley bolt.
10. Using SST: 09950-50013 or equivalent, remove the crankshaft pulley bolt and crankshaft pulley.
11. Using a screwdriver, pry out the timing chain case oil seal.

➡Tape the screwdriver tip before use.

➡After the removal, check the crankshaft for damage. If it is damaged, smooth the surface with 400-grit sandpaper.

To install:

12. Install timing chain case oil seal, as follows:
    a. Apply MP grease to a new oil seal lip.
    b. Using a Special Service Tool (SST: 09223-22010, SST: 09506-35010 or equivalent) and a hammer, tap in the oil seal until its surface is flush with the timing chain cover edge.

➡Keep the lip free of foreign matter.

➡Do not tap the oil seal at an angle.

13. Install the crankshaft pulley, as follows:

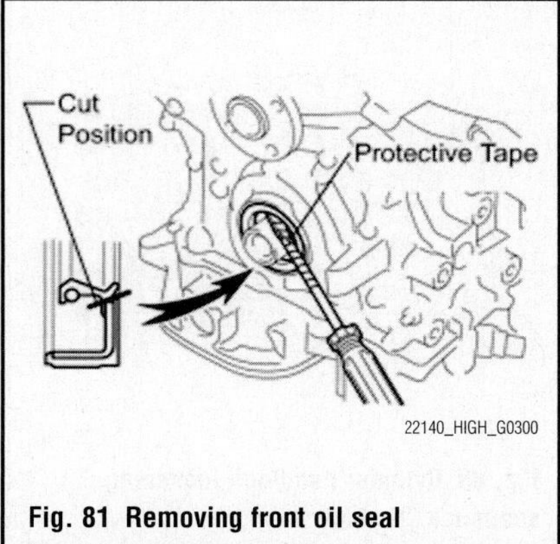

Fig. 81 Removing front oil seal

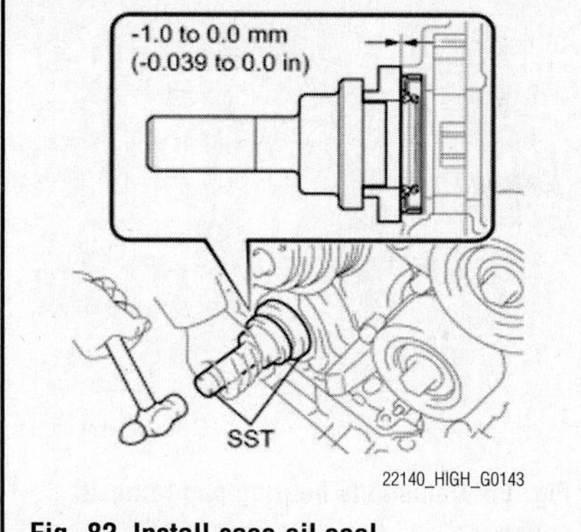

Fig. 82 Install case oil seal

a. Align the pulley set key with the key groove of the pulley, and slide on the pulley.

b. Using SST: 09213-70011, SST: 09330-00021, or equivalent, install the pulley bolt. For 2.7L engines, tighten to 192 ft. lbs. (260 Nm). For 3.5L engines, tighten to 184 ft. lbs. (250 Nm).

14. The remainder of installation is the reverse of removal. When installing the wheel, tighten the lug nuts to 76 ft. lbs. (103 Nm).

CYLINDER HEAD

REMOVAL & INSTALLATION

See Figures 83 through 104.

1. Remove the engine assembly.
2. Install on engine stand.
3. The following must be removed:
    a. Remove the ignition coil assembly.
    b. Remove right hand No. 2 engine mounting stay.
    c. Remove the intake manifold.
    d. Remove the right hand exhaust manifold sub-assembly.
    e. Remove the No. 2 engine oil level dipstick guide.
    f. Remove the No. 2 manifold stay.
    g. Remove the No. 2 exhaust manifold heat insulator.
    h. Remove the left exhaust manifold sub-assembly.
    i. Remove the transverse engine mounting bracket.
    j. Remove the generator assembly.
    k. Remove the V-ribbed belt tensioner assembly.
    l. Remove the No. 2 timing gear cover.
    m. Remove the No. 2 idler pulley sub-assembly.
    n. Remove the left No. 1 engine front mounting bracket.
    o. Remove the 6 bolts and left hand No. 1 front engine mounting bracket.
    p. Remove the radio setting condenser.
    q. Remove the No. 1 vacuum switching valve.
    r. Remove the knock control sensor wire.
    s. Remove the knock control sensor.
    t. Remove the crankshaft position sensor.
    u. Remove the No. 1 oil pipe.
    v. Remove the oil pipe.
    w. Remove the crankshaft pulley.
    x. Remove the oil cooler assembly, if necessary.
    y. Remove the No. 1 oil cooler bracket, if necessary.
    z. Remove the water inlet housing.
    aa. Remove the water outlet.
    bb. Remove the cylinder head cover sub-assembly (for bank 1).
    cc. Remove the cylinder head cover sub-assembly (for bank 2).
    dd. Remove the No. 2 oil pan sub-assembly.
    ee. Remove the oil strainer sub-assembly.
    ff. Remove the oil pan sub-assembly.
    gg. Remove the timing chain cover sub-assembly.
    hh. Remove the timing chain case oil seal.
    ii. Set the No. 1 cylinder to TDC/compression.
    jj. Remove the No. 1 chain tensioner assembly.
    kk. Remove the chain tensioner slipper.
    ll. Remove the chain sub-assembly.
    mm. Remove the idle sprocket assembly.

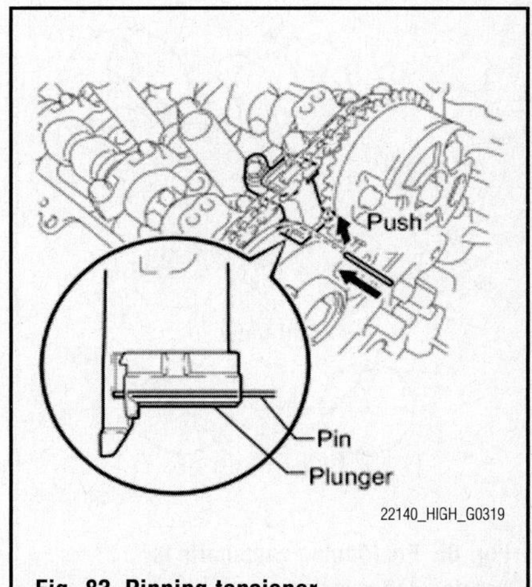

Fig. 83 Pinning tensioner

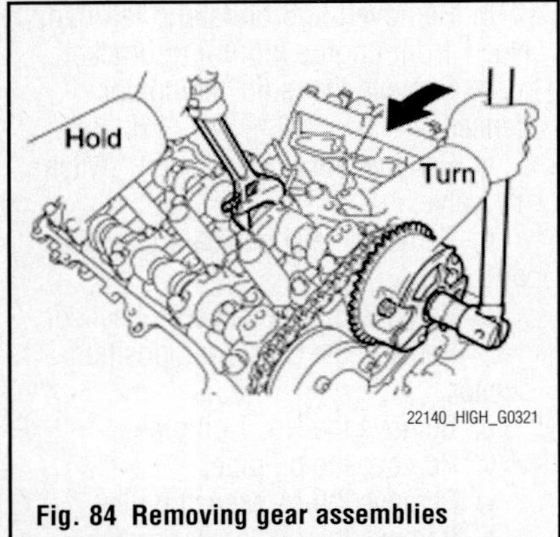

Fig. 84 Removing gear assemblies

nn. Remove the camshaft timing gears and No. 2 chain (for bank 1).

oo. While raising the No. 2 chain tensioner assembly, insert a pin of 1.0 mm (0.039 in.) diameter into the hole to fix the No. 2 chain tensioner assembly.

pp. Hold the hexagonal portion of the camshaft with a wrench, and remove the 2 bolts and 2 camshaft timing gear assemblies.

➡**Be careful not to damage the cylinder head with the wrench. Do not disassemble the camshaft timing gear assemblies.**

qq. Remove the No. 2 chain assembly.

rr. Remove the bolt and No. 2 chain tensioner assembly.

ss. Check that the camshafts are positioned as shown in the illustration.

tt. Uniformly loosen and remove the 8 bearing cap bolts in several steps and in the sequence shown in the illustration.

uu. Uniformly loosen and remove the 12 bearing cap bolts in several steps and in the sequence shown in the illustration.

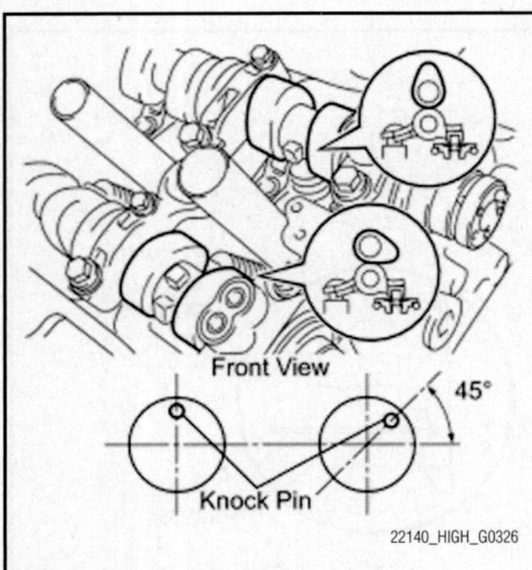

Fig. 85 Positioning camshafts for bearing cap removal

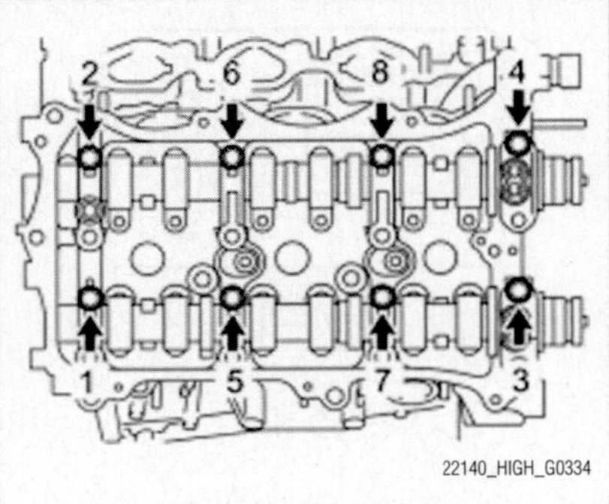

Fig. 86 Camshafts bearing cap removal sequence

➡**Uniformly loosen the bolts while keeping the camshaft level.**

vv. Remove the 5 camshaft bearing caps.

ww. Remove the camshaft.

xx. Remove the No. 2 camshaft.

yy. Remove the right camshaft housing sub-assembly by prying between the cylinder head and right camshaft housing sub-assembly with a screwdriver.

➡**Be careful not to damage the contact surfaces of the cylinder head and the right camshaft housing sub-assembly.**

4. Remove the 24 valve lash adjuster assemblies from the cylinder head.

➡**Arrange the removed parts in the correct order.**

5. Uniformly loosen and remove the 2 cylinder head set bolts in several steps and in the sequence shown in the illustration.

➡**Be careful not to drop washers into the cylinder head. Cylinder head warpage or cracking could result from removing bolts in an incorrect order. Be sure to keep separate the removed parts for each installation position.**

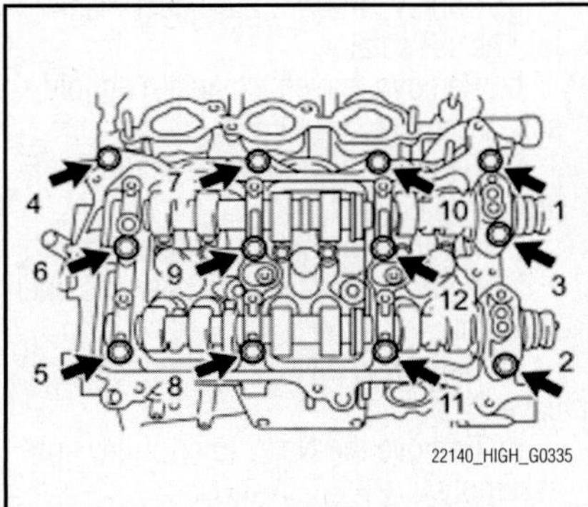

Fig. 87 Camshaft bearing cap loosening sequence

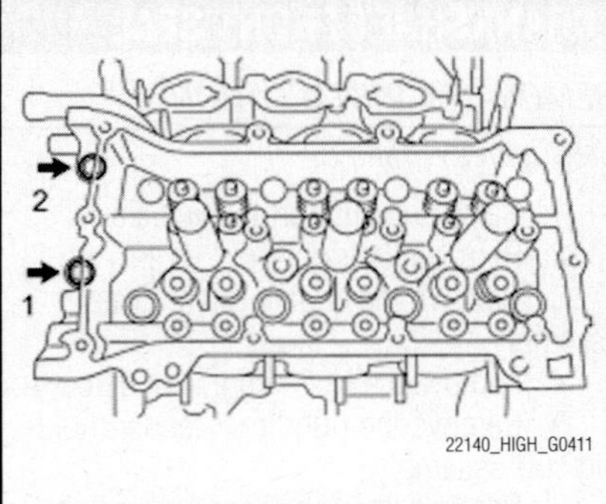

Fig. 88 Loosening cylinder head bolts LH shown

6. Using a 10 mm bi-hexagon wrench, uniformly loosen the 8 bolts in the sequence shown in the illustration. Remove the 8 cylinder head bolts and plate washers.

7. Remove the cylinder head sub-assembly.

8. Remove the No. 2 cylinder head gasket.

To install:

9. Place the No. 2 cylinder head gasket on the cylinder block surface with the Lot No. stamp upward.

➡**Gently lower the cylinder head in order not to damage the gasket with the bottom part of the head.**

10. Place the cylinder head on the cylinder block.

➡**Be careful not to allow oil to adhere to the bottom part of the cylinder head.**

11. Apply a light coat of engine oil to the threads and under the heads of the cylinder head bolts.

➡**The cylinder head bolts are tightened in 3 progressive steps.**

a. Step 1: Using a 10 mm bi-hexagon wrench, install and uniformly tighten the 8 cylinder head bolts with the plate

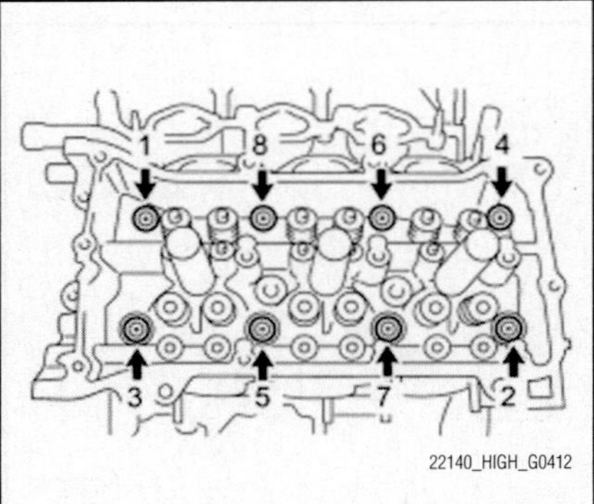

Fig. 89 Cylinder head bolt loosening sequence LH shown

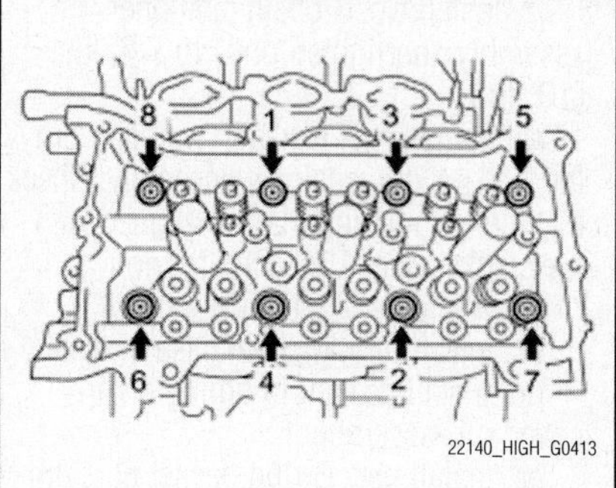

Fig. 90 Cylinder head bolt tightening sequence LH shown

washers to 27 ft. lbs. (36 Nm) in several steps in the sequence shown in the illustration.

b. Step 2: Mark the cylinder head bolt head with paint. Tighten the cylinder head bolts another 90°.

c. Step 3: Tighten the cylinder head bolts an additional 90°.

12. Tighten the 2 bolts to 22 ft. lbs. (30 Nm) in the order shown in the illustration.

13. Install the 12 valve stem caps.

➡Keep the lash adjuster free of dirt and foreign objects. Only use clean engine oil.

14. Place the lash adjuster into a container filled with engine oil.

a. Insert Service Tool's tip into the lash adjuster's plunger and use the tip to press down on the check ball inside the plunger.

b. Squeeze Service Tool and lash adjuster together to move the plunger up and down 5 to 6 times.

c. Check the movement of the plunger and bleed the air. Confirm that the plunger moves up and down freely.

d. When bleeding air from the high-pressure chamber, make sure that the tip of SST is actually pressing the check ball as shown in the illustration. If the check ball is not pressed, air will not bleed.

e. After bleeding the air, remove Service Tool. Then, try to press the plunger quickly and firmly with by hand. Confirm that the plunger is very difficult to move.

f. If the results are not as specified, replace the defective lash adjuster.

g. Install the lash adjusters.

➡Install the lash adjuster to the same place where it was removed from.

15. Install the following:

a. Apply engine oil to the camshaft journals, right camshaft housing sub-assembly, and camshaft bearing caps.

b. Install the camshaft and No. 2 camshaft to the right camshaft housing sub-assembly.

c. Make sure of the marks and numbers on the camshaft bearing caps and place them in each proper position and direction.

d. Temporarily tighten the 8 bearing cap bolts to 7 ft. lbs. (10 Nm) in the order shown in the illustration.

e. Make sure that the No. 1 valve rocker arm sub-assembly is installed as shown in the illustration.

f. Apply seal packing in a continuous line as shown in the illustration.

➡Remove any oil from the contact surface. Install the camshaft housing sub-assembly RH within 3 minutes. Do not start the engine for at least 2 hours after installing.

g. Install the right camshaft housing sub-assembly and tighten the 12 bolts to 21 ft. lbs. (28 Nm) in the order shown in the illustration.

➡When installing the camshaft housing RH, it is necessary to correctly position the camshafts as shown in the removal illustration. If the camshaft housing sub-assembly is removed because any of the bolts are loosened during installation, make sure that the previously applied seal packing does not enter any oil passages.

h. Complete the tightening of the 8 bolts to 12 ft. lbs. (16 Nm) in the order shown above.

i. Install the No. 2 chain tensioner assembly with the bolt and tighten to 15 ft. lbs. (21 Nm).

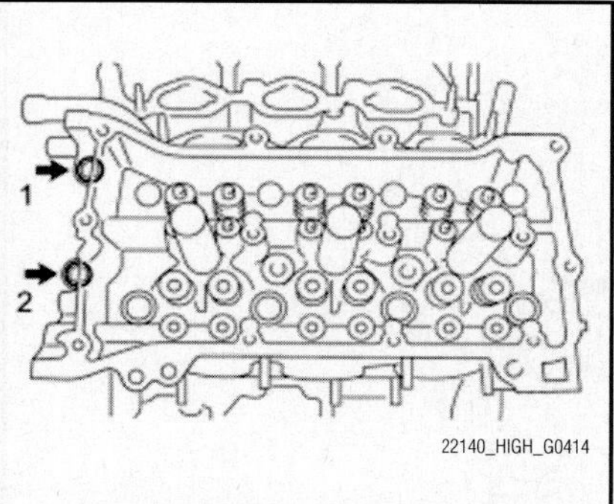

Fig. 91 Tightening cylinder head bolts LH shown

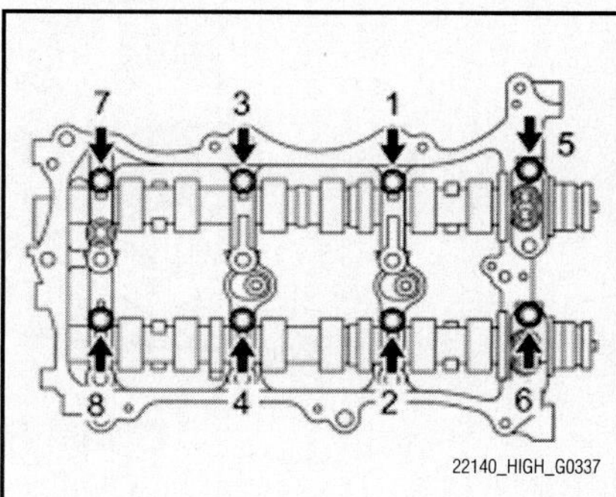

Fig. 92 Camshafts bearing cap tightening sequence

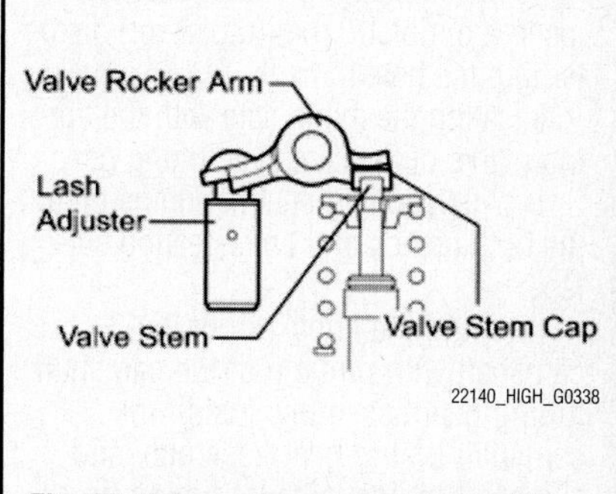

Fig. 93 Valve rocker arm sub-assembly positioning

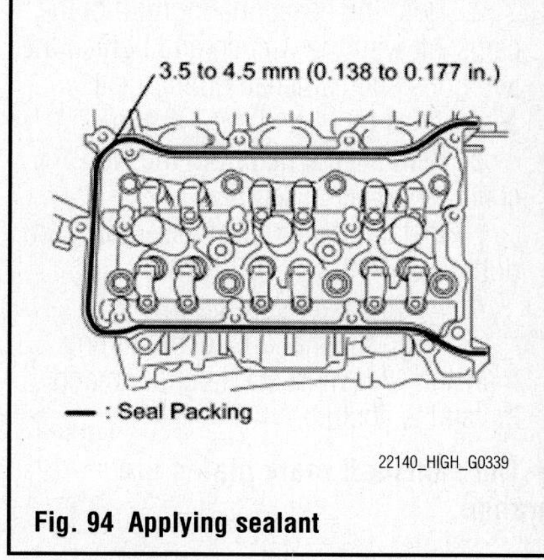

Fig. 94 Applying sealant

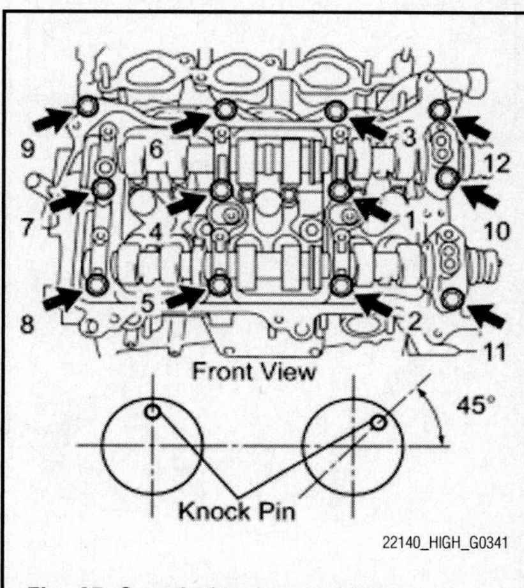

Fig. 95 Camshaft sub-assembly tightening sequence

j. While pushing in the tensioner, insert a pin of 1.0 mm (0.039 in.) diameter into the hole to fix it.

k. Align the mark plate with the timing marks of the camshaft timing gear.

l. Apply a light coat of engine oil to the bolt threads and bolts seating surface.

m. Align the knockpin of the camshaft with pinhole of the camshaft timing gear assembly. Install the camshaft timing gear assembly and camshaft timing exhaust timing gear assembly and camshaft timing exhaust gear with the No. 2 chain sub-assembly installed.

n. Hold the hexagonal portion of the camshaft with the wrench and tighten the two bolts and camshaft timing gear assemblies to 74 ft. lbs. (100 Nm).

o. Remove the pan from the No 2 chain tensioner assembly.

p. Install idle sprocket assembly and tighten to 44 ft. lbs. (60 Nm).

q. Install chain sub-assembly.

r. Align the mark plate and timing marks as shown in the illustration and install the chain.

➡The camshaft mark plates are orange.

s. Do not pass the chain over the crankshaft, just temporarily place it on the crankshaft.

t. Turn the camshaft timing gear assembly on bank 1 counterclockwise to tighten the chain between the banks.

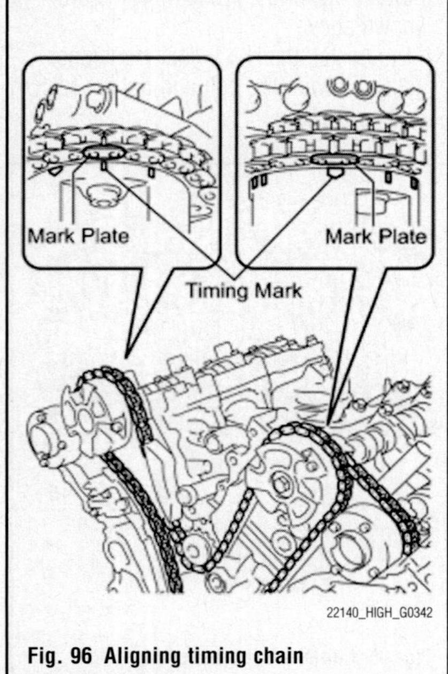

Fig. 96 Aligning timing chain sub-assembly

➡When the idle sprocket assembly is reused, align the timing chain plate with the mark where the plate has been in order to tighten the chain between the banks.

u. Align the mark plate and timing marks as shown in the illustration and install the chain onto the crankshaft timing sprocket. The crankshaft to mark plate is yellow.

v. Turn the crankshaft clockwise to set it to the right-hand block bore more centerline. (TDC Compression)

w. Install chain tensioner slipper.

x. Move the stopper plate upward to release the lock, and push the plunger deep into the tensioner.

y. Move the stopper plate downward to set the loss, and insert a hexagon wrench into the hole of the stopper plate.

z. Install No. 1 chain tensioner assembly and tighten bolts to 7 ft. lbs. (10 Nm).

aa. Remove the hexagon wrench from the No. 1 chain tensioner assembly. Check that the each timing mark is aligned with the crankshaft at TDC compression.

bb. Remove the pulley set bolt.

cc. Install timing chain case oil seal.

dd. Install sealant to timing chain cover sub-assembly.

ee. Install new O ring gasket on cylinder block.

ff. Align the oil pump's drive rotor spline and the crankshaft as shown in the illustration. Install the spline and chain cover to the crankshaft.

gg. Temporarily tighten the timing chain cover with the 23 bolts and 2 nuts.

- Tighten bolts in area 1 and 2 15 ft. lbs. (21 Nm)

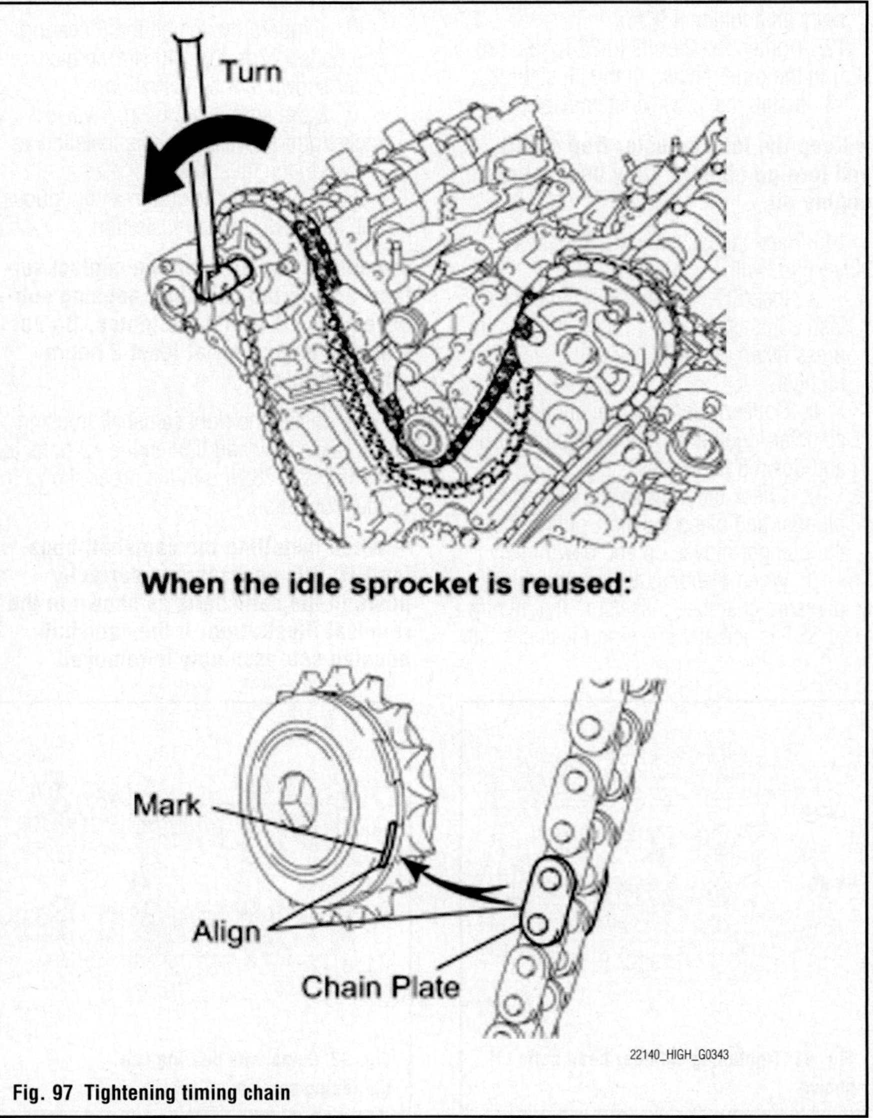

Fig. 97 Tightening timing chain

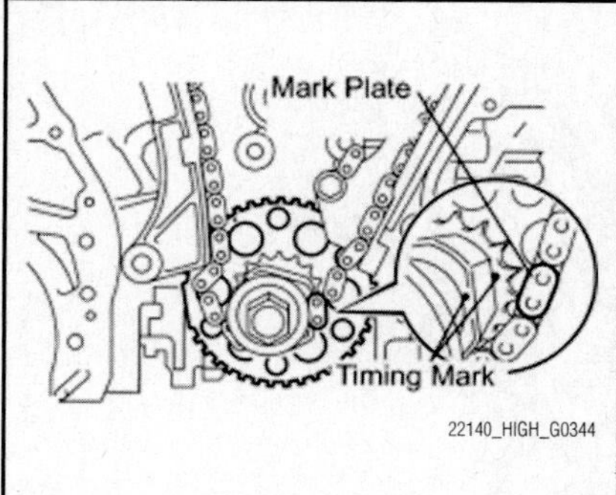

Fig. 98 Installing timing chain on crankshaft

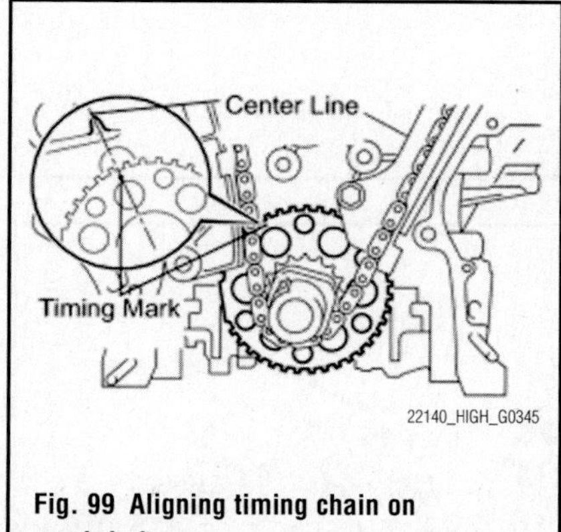

Fig. 99 Aligning timing chain on crankshaft

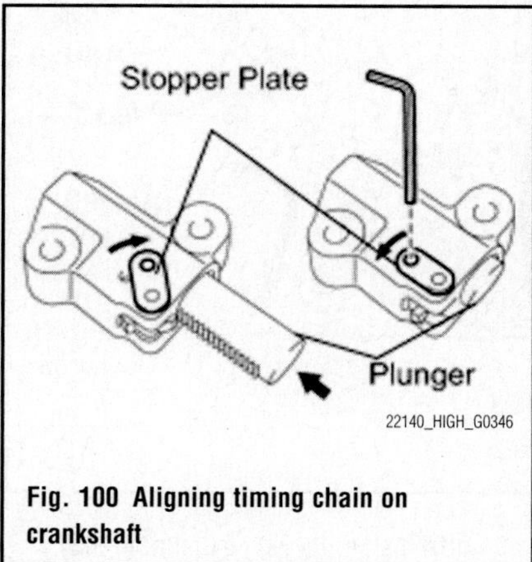

Fig. 100 Aligning timing chain on crankshaft

- Tighten bolt in area 3 15 ft. lbs. (21 Nm)

➡First tighten the upper bolts and nuts followed by the lower bolts and nuts as shown.

- Tighten bolt in area 4 32 ft. lbs. (43 Nm)
- Tighten bolt in area 4 15 ft. lbs. (21 Nm)

hh. Install oil pan subassembly and tighten 16 bolts and 2 nuts to 7 ft. lbs. (10 Nm) and 15 ft lbs. (21 Nm).

ii. Install oil strainers sub-assembly and tighten bolts and nuts to 7 ft. lbs. (10 Nm).

jj. Install No.2 oil pan sub-assembly and tighten 16 bolts and 2 nuts to 7 ft. lbs. (10 Nm).

kk. Install cylinder head cover sub-assemblies and tighten to 7 ft. lbs. (10 Nm) and 15 ft. lbs. (21 Nm).

ll. Install water inlet housing

mm. Install the No. 1 oil cooler bracket (w/ oil cooler).

nn. Install the oil cooler assembly (w/ oil cooler).

oo. Install the crankshaft pulley.

pp. Install the oil pipe.

qq. Install the No. 1 oil pipe.

rr. Install the crankshaft position sensor.

ss. Install the knock control sensor.

tt. Install the knock control sensor wire.

uu. Install the No. 1 vacuum switching valve.

vv. Install radio setting condenser.

ww. Install the left hand No. 1 front engine mounting bracket with the 6 bolts and tighten to 40 ft. lbs. (54 Nm).

➡Install the water inlet and mounting bracket within 15 minutes after installing the chain cover. Do not

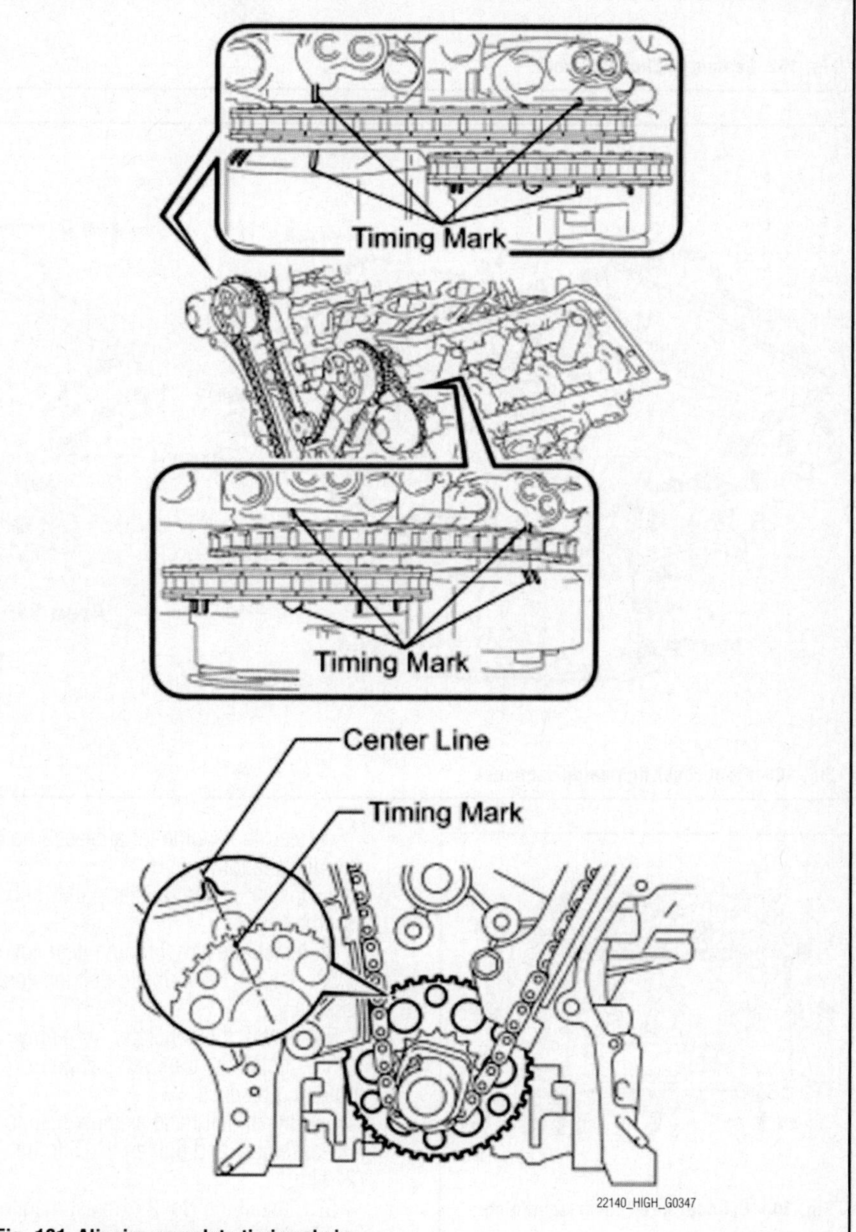

Fig. 101 Aligning complete timing chain

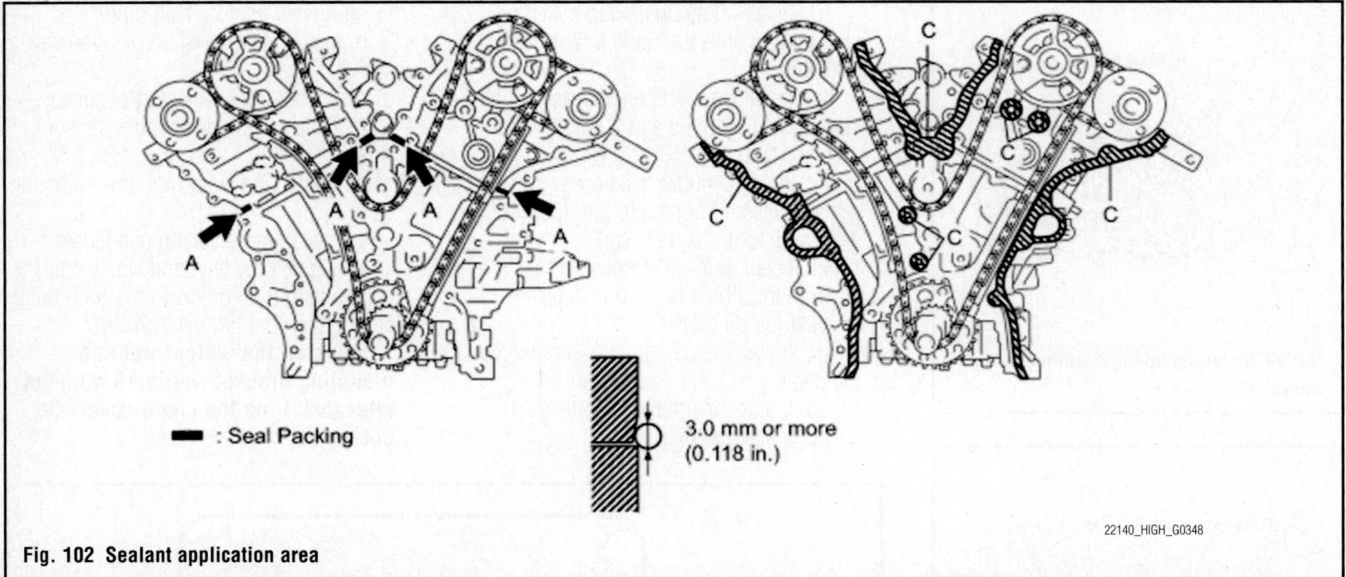

Fig. 102 Sealant application area

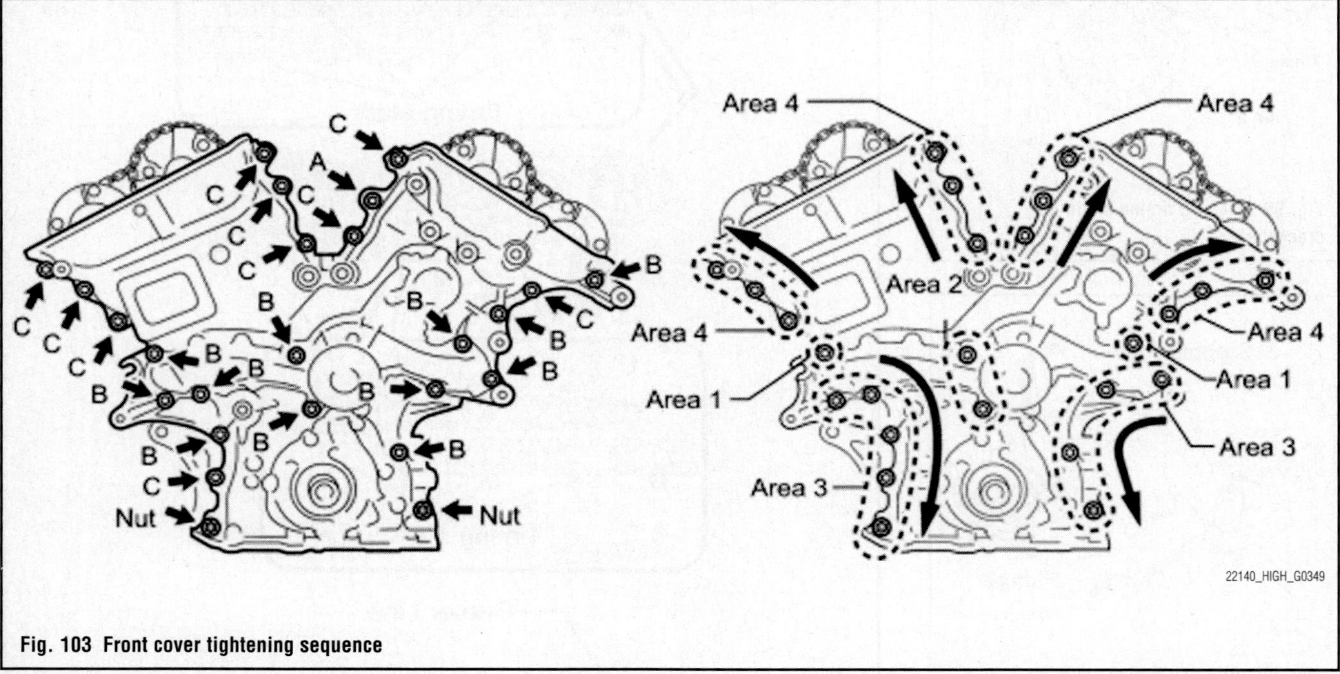

Fig. 103 Front cover tightening sequence

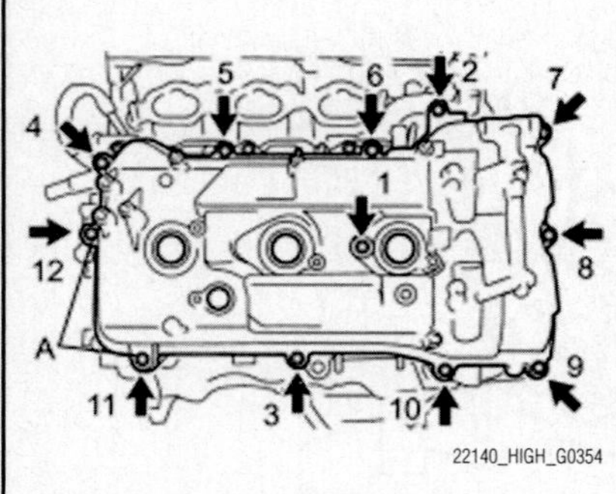

Fig. 104 Cylinder head cover tightening sequence

xx. start the engine for at least 2 hours after installation.

yy. Install the No. 2 idler pulley sub-assembly.

zz. Install the No. 2 timing gear cover.

53. Install the V-ribbed belt tensioner assembly.
54. Install the generator assembly.
55. Install the transverse engine mounting bracket.
56. Install left hand exhaust manifold sub-assembly and tighten to 15 ft. lbs. (21 Nm).
57. Install the No. 2 exhaust manifold heat insulator.
58. Install the No. 2 manifold stay.
59. Install the No. 2 engine oil level dipstick guide.
60. Install right hand exhaust manifold sub-assembly and tighten to 15 ft. lbs. (21 Nm).
61. Install the intake manifold and tighten the 6 bolts and 4 nuts uniformly in several steps to 15 ft. lbs. (21 Nm).
62. Install the right hand No. 2 engine mounting stay.
63. Install the ignition coil assembly.
64. Remove the engine stand.
65. Install the engine assembly.

EXHAUST MANIFOLD

REMOVAL & INSTALLATION

See Figures 105 and 106.

1. Disconnect the negative battery cable.
2. Remove the windshield wiper motor and link assembly.
3. Remove front shock absorber caps.
4. Remove outer cowl top panel sub-assembly.
5. Remove the No. 2 radiator assembly.
6. Remove front floor cover LH.
7. Remove front no. 3 exhaust pipe sub-assembly:
    a. Disconnect the 2 clamps and oxygen sensor connector (for Bank 1 Sensor 2).
    b. Remove the 4 bolts, 2 nuts, 2 compression springs and No. 3 exhaust pipe sub-assembly.
    c. Remove the 3 gaskets from the No. 3 exhaust pipe sub-assembly.
8. Remove the bolt, nut and manifold stay.
9. Remove the bolt, 4 nuts, No. 1 EGR pipe and 2 gaskets.
10. Remove exhaust manifold sub-assembly RH:
    a. Disconnect the 2 clamps and air fuel ratio sensor connector (for Bank 1 Sensor 1).
    b. Using a 12 mm deep socket wrench, remove the 6 nuts and exhaust manifold sub-assembly RH.
11. Remove the exhaust manifold to head gasket from the cylinder head sub-assembly.
12. Remove air fuel ratio sensor (for bank 1 sensor 1).
13. Remove front exhaust pipe assembly.
14. Remove the bolt, nut and No. 2 manifold stay.
15. Remove no. 2 engine oil level dipstick guide.
16. Disconnect the air fuel ratio sensor connector (for Bank 2 Sensor 1).
17. Disconnect the sensor wire from the radiator pipe clamp.
18. Remove the 3 bolts and No. 2 exhaust manifold heat insulator.
19. Using a 12 mm deep socket wrench, remove the 6 nuts and exhaust manifold sub-assembly LH.
20. Remove exhaust manifold to head gasket LH: Remove the exhaust manifold to head gasket LH from the cylinder head sub-assembly.
21. Remove air fuel ratio sensor (for bank 2 sensor 1).

To install:

22. To install, reverse the removal procedure while observing the following:

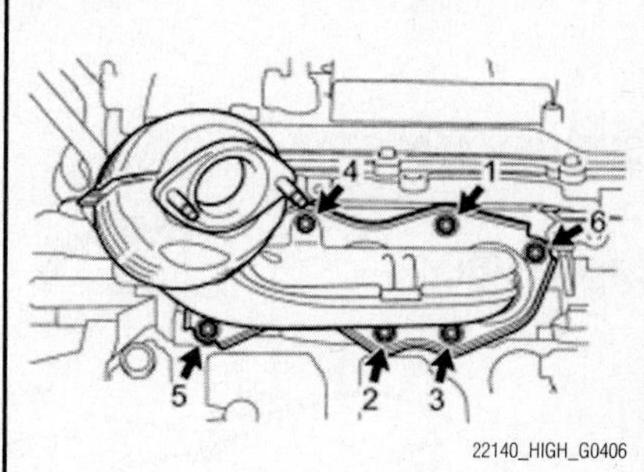

Fig. 105 Right exhaust manifold sub-assembly tightening sequence—3.5L engine

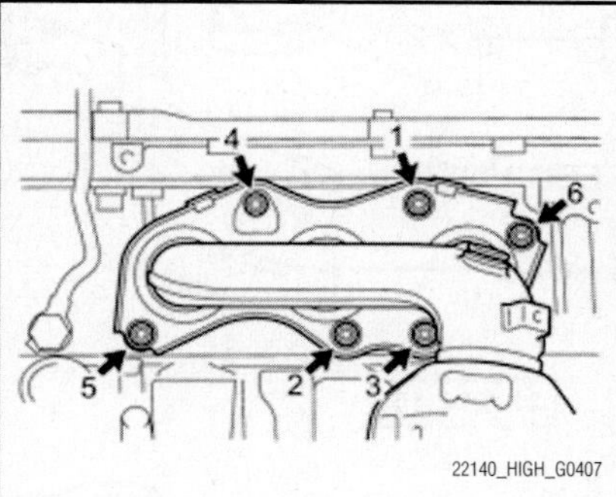

Fig. 106 Left exhaust manifold sub-assembly tightening sequence—3.5L engine

    a. Use new exhaust manifold gaskets.
    b. Tighten the exhaust manifolds in sequence as shown in the illustrations.

23. Inspect for gas leak, and repair as necessary.

FLYWHEEL

REMOVAL & INSTALLATION

See Figure 107.

Fig. 107 Removing flywheel

1. Remove automatic transaxle assembly
2. Hold the crankshaft and remove the 8 bolts, front spacer, drive plate and rear spacer.

To install:

3. Installation is reverse of removal.

INTAKE MANIFOLD

REMOVAL & INSTALLATION

See Figures 108 and 109.

1. Remove the air cleaner assembly with hose.
2. Discharge fuel system pressure.
3. Remove intake air surge tank assembly:
    a. Disconnect the throttle body assembly connector and wire harness clamp.
    b. Disconnect the fuel vapor feed hose.
    c. Disconnect the 2 water by-pass hoses.
    d. Disconnect the 5 clamps and separate the main wire.
    e. Disconnect the ventilation hose.
    f. Disconnect the connector from the manifold absolute pressure sensor.
    g. Remove the bolt and separate the No. 1 surge tank stay from the intake air surge tank assembly.
    h. Remove the bolt and separate the throttle body bracket from the intake air surge tank assembly.
    i. Remove the 2 nuts from the intake air surge tank assembly.
    j. Using a 5 mm socket hexagon wrench, remove the 4 bolts.
    k. Remove the intake air surge tank assembly and 3 air surge tank to intake manifold gaskets.
4. Remove no. 2 EGR pipe.
5. Remove EGR delivery chamber:
    a. Disconnect the 2 water hoses and connector.
    b. Remove the EGR delivery chamber.
    c. Remove the No. 2 intake manifold gasket.
6. Remove no. 2 engine mounting stay RH and LH.
7. Disconnect fuel tube sub-assembly.
8. Remove fuel injector assembly.
9. Remove the 6 bolts, 4 nuts and intake manifold.

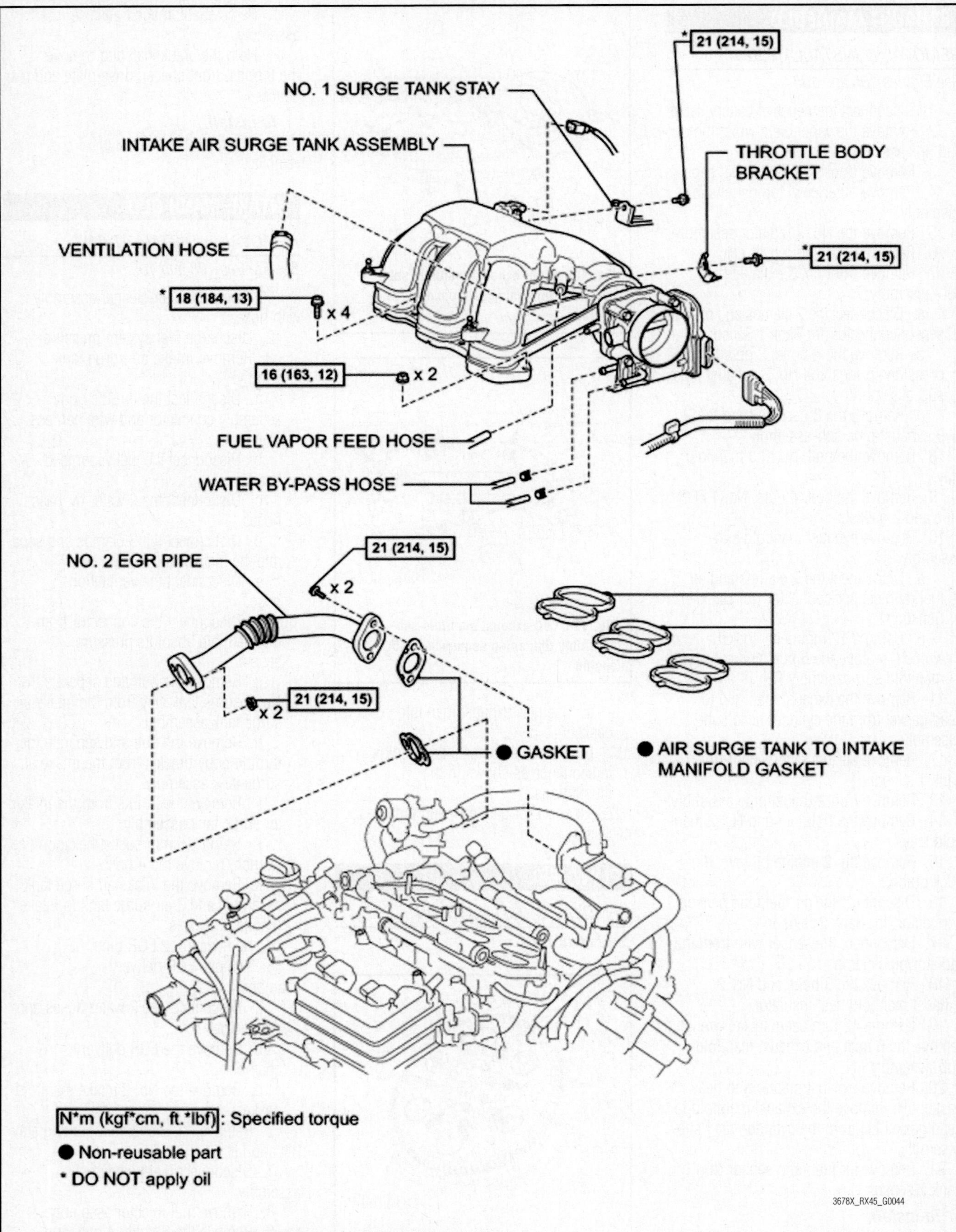

Fig. 108 Exploded view of the intake manifold and related components—1 of 2

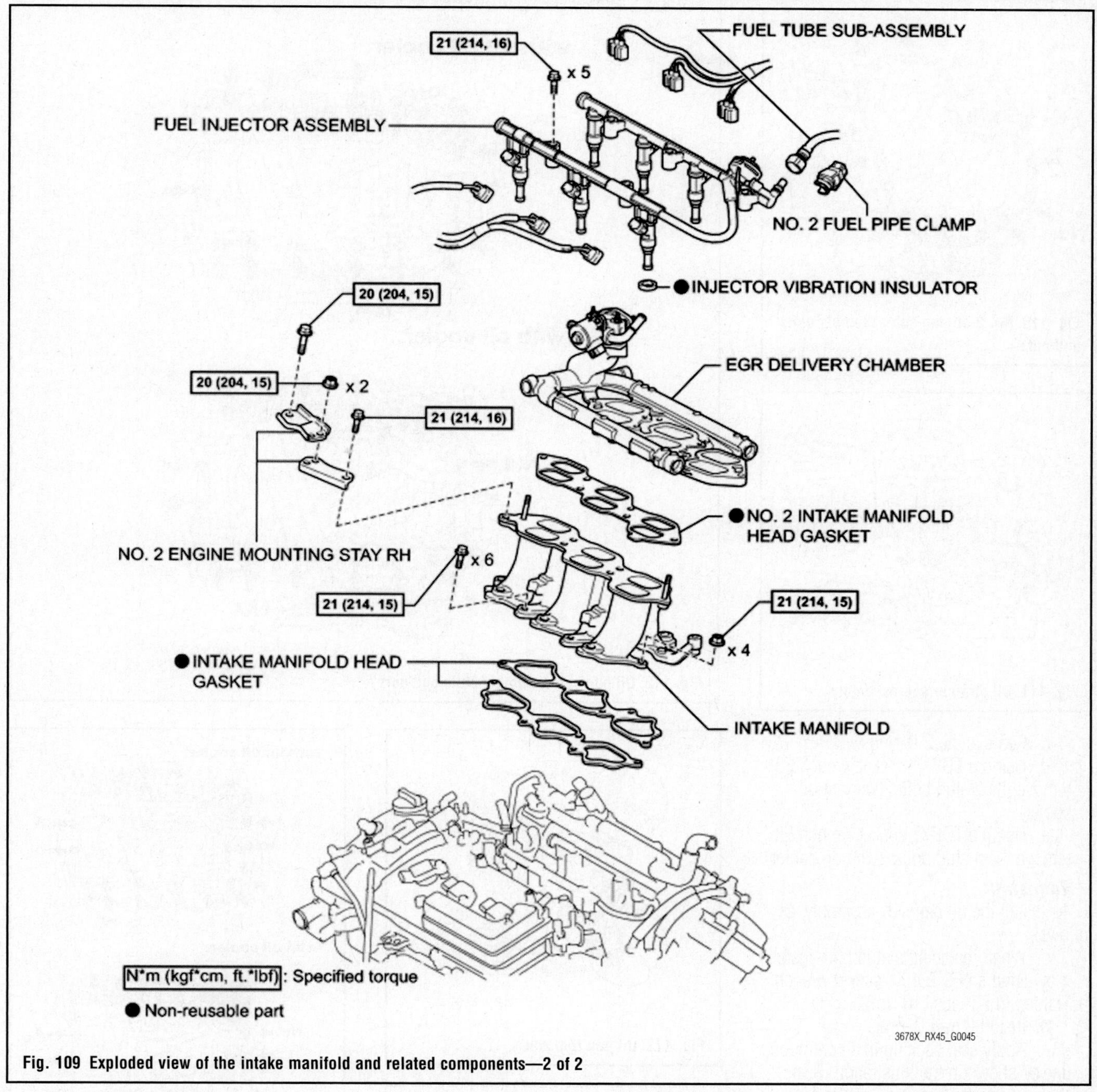

Fig. 109 Exploded view of the intake manifold and related components—2 of 2

10. Remove the 2 intake manifold gaskets.
11. To install, reverse the removal procedure. Refer to the illustrations for torque values.
12. Use new intake manifold gaskets upon installation.

OIL PAN

REMOVAL & INSTALLATION

See Figures 110 through 119.

1. Remove the No. 2 oil pan sub-assembly, as follows:
   a. Remove the 16 bolts and 2 nuts.
   b. Insert the blade of SST (SST: 09032-00100) or equivalent tool between the oil pans. Cut through the applied sealer and remove the No. 2 oil pan sub-assembly.

➡Be careful not to damage the contact surfaces of the oil pans.

   c. Using a TORX® socket wrench E6, remove the 2 stud bolts.
2. Remove the oil strainer sub-assembly, as follows:
   a. Remove the bolt, 2 nuts, oil strainer and gasket.
   b. Using a TORX® socket wrench E6, remove the 2 stud bolts.
3. Remove the oil pan sub-assembly, as follows:
   a. Remove the 16 bolts and 2 nuts.

➡Be sure to clean the bolts and stud bolts and check the threads for cracks or other damage.

   b. Remove the oil pan by prying between the oil pan and cylinder block with a taped screwdriver.

➡Be careful not to damage the contact surfaces of the cylinder block and oil pan.

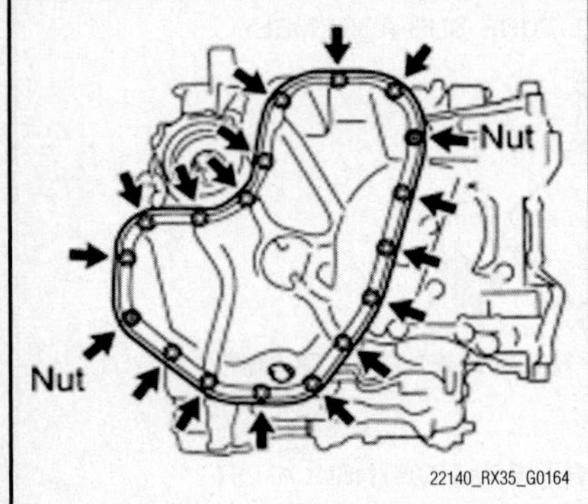

Fig. 110 No. 2 oil pan sub-assembly bolts and nuts

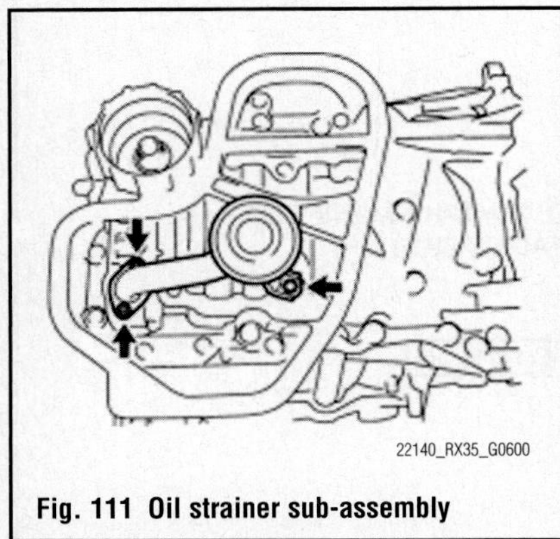

Fig. 111 Oil strainer sub-assembly

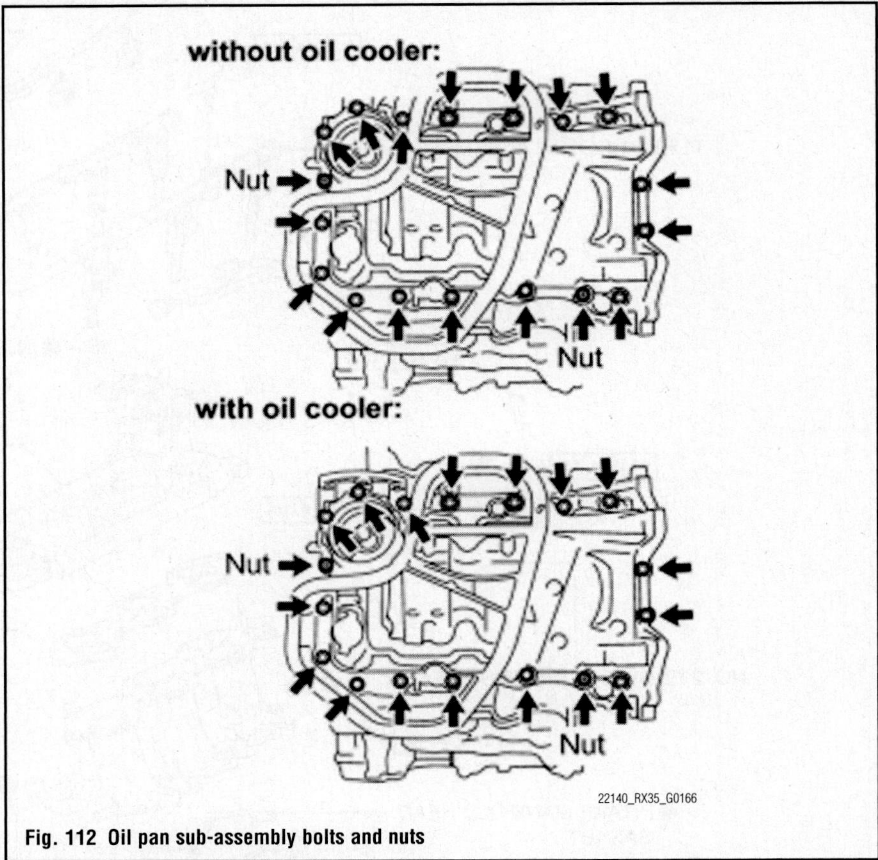

Fig. 112 Oil pan sub-assembly bolts and nuts

c. Remove the 2 O-rings.

d. Using a TORX® socket wrench E8, remove the 2 stud bolts. (without oil cooler).

e. Using a TORX® socket wrench E8, remove the 4 stud bolts. (with oil cooler).

To install:

4. Install the oil pan sub-assembly, as follows:

a. When replacing a stud bolt, install it by using an E8 TORX® socket wrench. Tighten to 7 ft. lbs (10 Nm).

b. Install 2 new O-rings.

c. Apply seal packing in a continuous line as shown in the illustration. Seal packing: Toyota Genuine Seal Packing Black, Three Bond 1207B or equivalent. Seal diameter: 3.0 to 4.0 mm (0.118 to 0.156 in.).

➡Remove any oil from the contact surface.

➡Install the oil pan within 3 minutes after applying seal packing.

➡Do not start the engine for at least 2 hours after installing.

d. Install the oil pan with the 16 bolts and 2 nuts and tighten to 7 ft. lbs (10 Nm), and 15 ft. lbs (21 Nm).

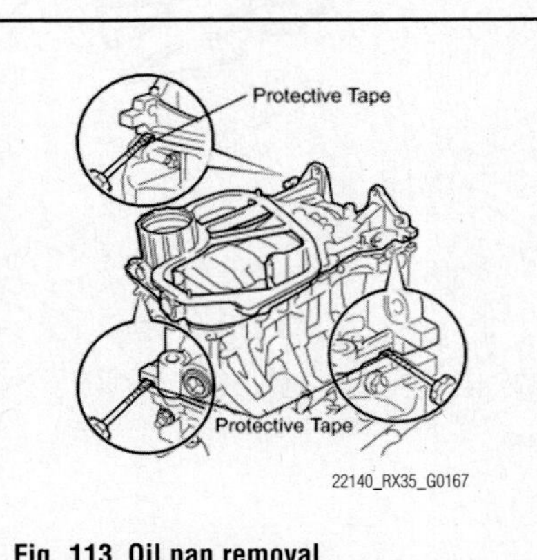

Fig. 113 Oil pan removal

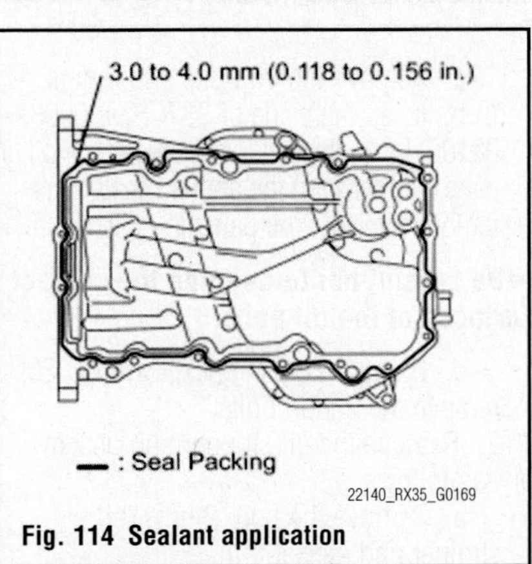

Fig. 114 Sealant application

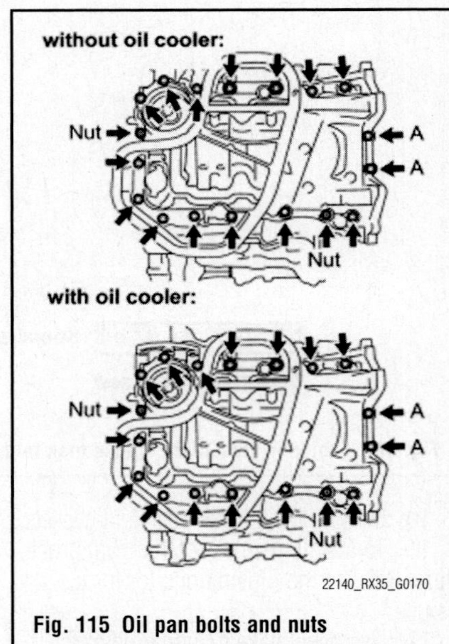

Fig. 115 Oil pan bolts and nuts

5. Install the oil strainer sub-assembly, as follows:

a. Using an E6 TORX® socket, install the stud bolts as shown in the illustration and tighten to 35 inch lbs (4 Nm).

b. Install a new gasket and the oil strainer with the bolt and 2 nuts and tighten to 7 ft. lbs (10 Nm).

6. Install the No. 2 oil pan sub-assembly, as follows:

a. Using an E6 TORX® socket, install the stud bolts as shown in the illustration and tighten to 35 inch lbs (4 Nm).

b. Apply seal packing in a continuous line as shown in the illustration. Seal packing: Toyota Genuine Seal Packing Black, Three Bond 1207B or equivalent. Seal diameter: 3.0 to 4.0 mm (0.118 to 0.156 in.).

➡Remove any oil from the contact surface.

➡Install the oil pan No. 2 within 3 minutes after applying seal packing.

➡Do not start the engine for at least 2 hours after installing.

c. Install the oil pan with the 16 bolts and 2 nuts and tighten to 7 ft. lbs (10 Nm).

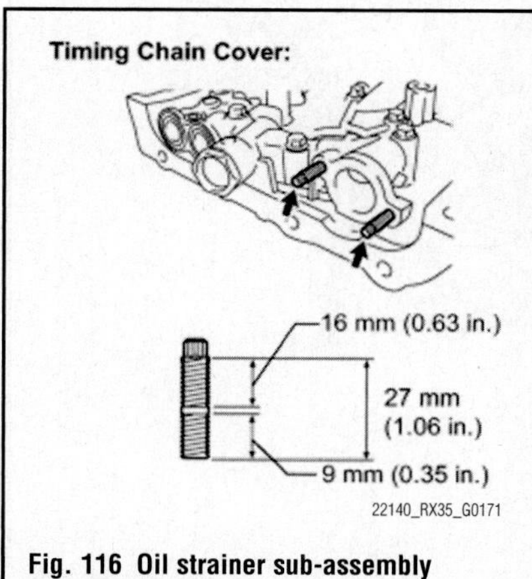

Fig. 116 Oil strainer sub-assembly installation

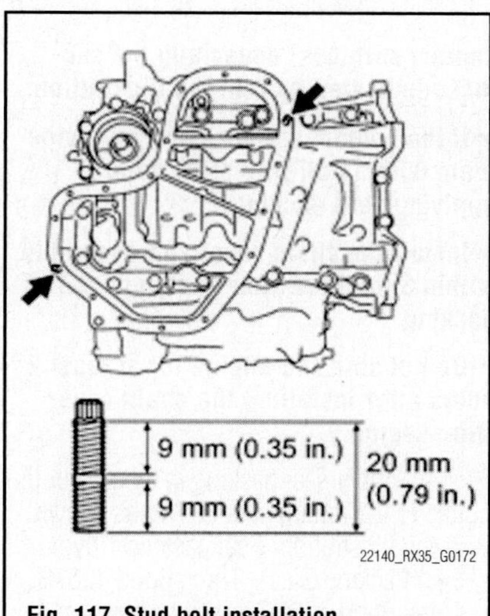

Fig. 117 Stud bolt installation

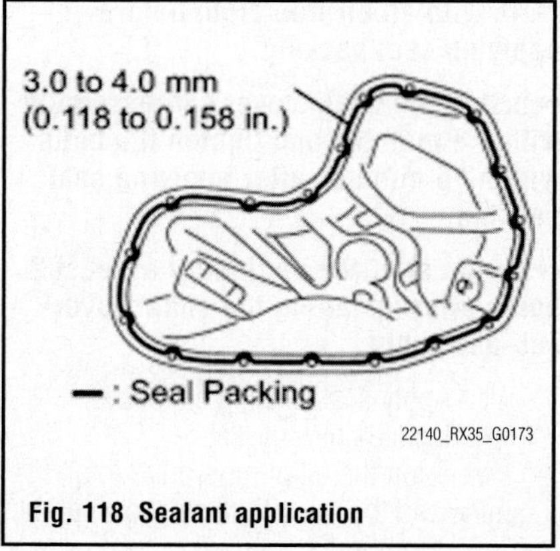

Fig. 118 Sealant application

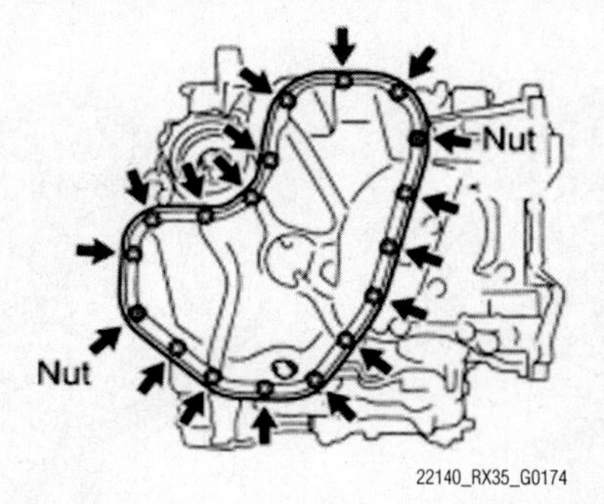

Fig. 119 No. 2 oil pan sub-assembly installation

OIL PUMP

REMOVAL & INSTALLATION

See Figures 120 through 129.

1. Remove the engine assembly with transaxle.
2. Secure the engine.
3. Remove the No. 1 oil pipe, as follows:

a. Remove the 2 oil pipe unions, gaskets and No. 1 oil pipe.

b. Remove the left oil control valve filter and gaskets.

4. Remove the oil pipe, as follows:

a. Remove the bolt.

b. Remove the 2 oil pipe unions and oil pipe.

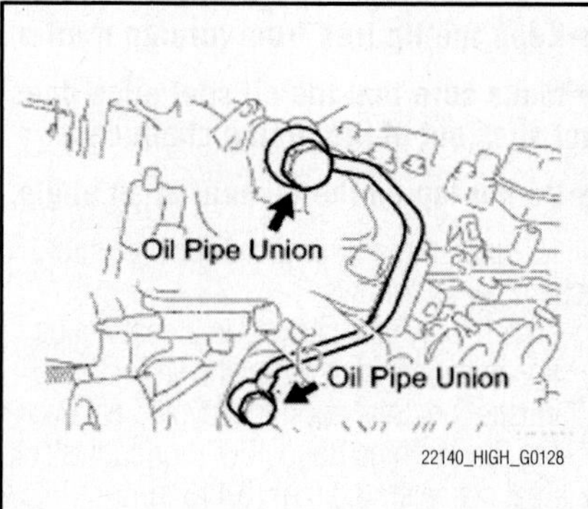

Fig. 120 No. 1 oil pipe unions

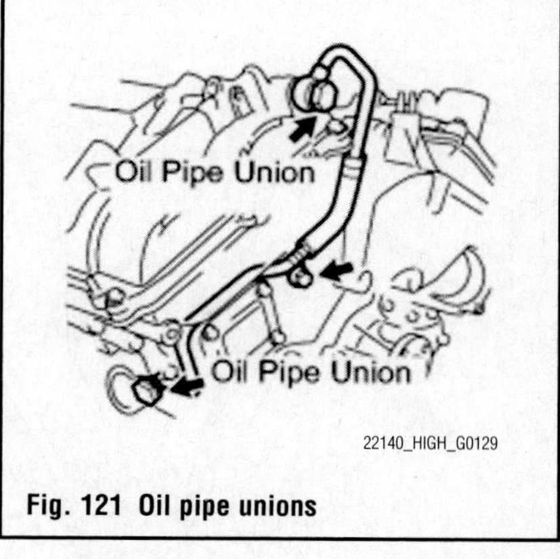

Fig. 121 Oil pipe unions

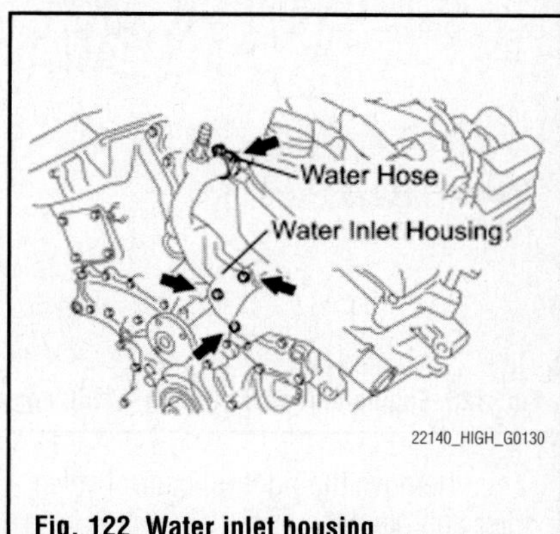

Fig. 122 Water inlet housing

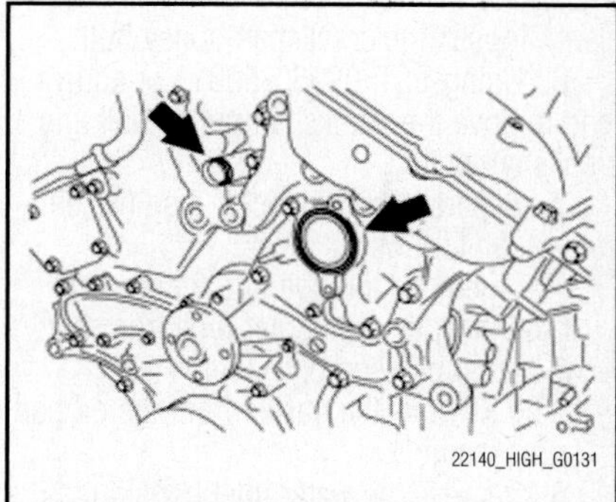

Fig. 123 Water inlet housing gasket and water outlet pipe O-ring

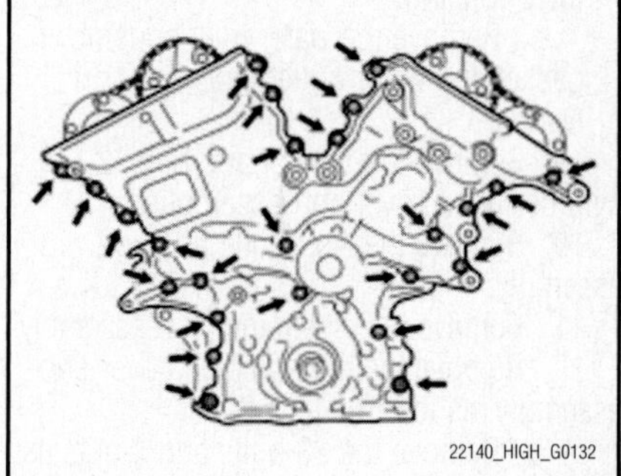

Fig. 124 Timing chain cover sub-assembly bolts and nuts

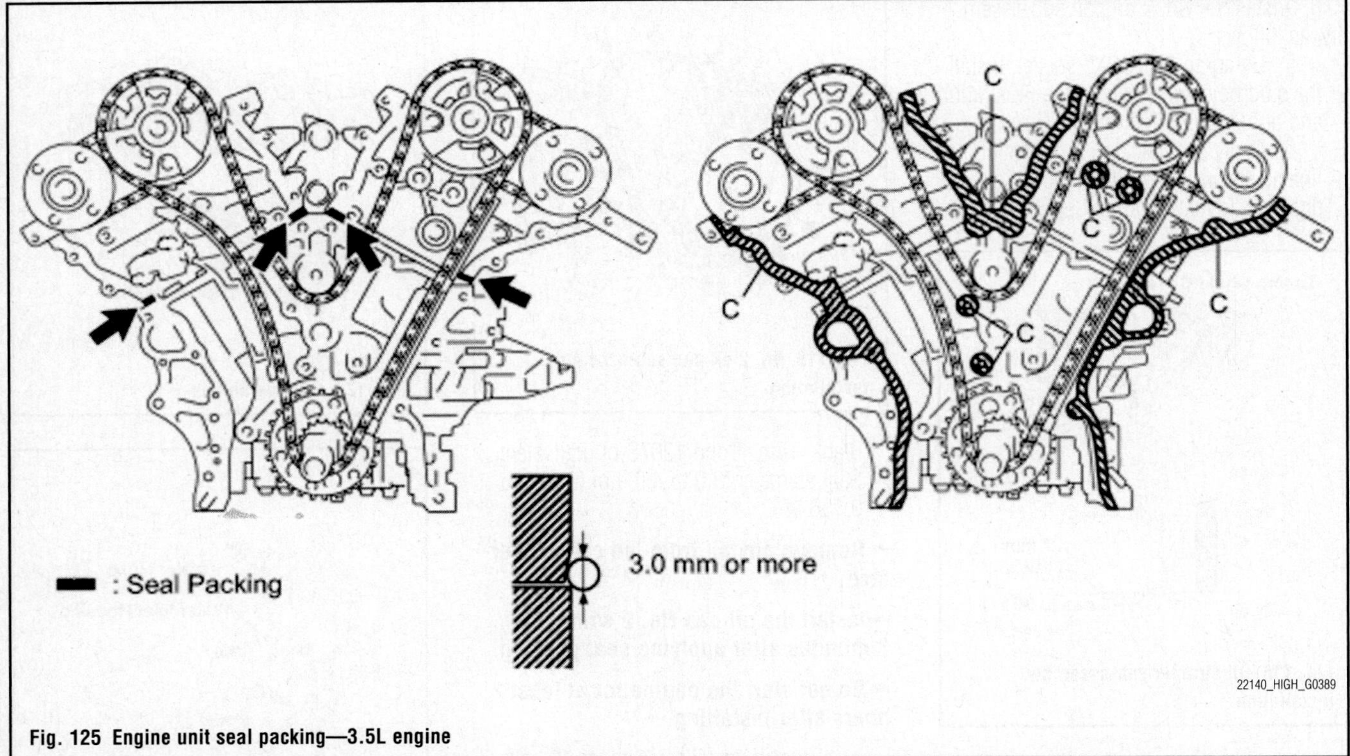

Fig. 125 Engine unit seal packing—3.5L engine

c. Remove the right oil control valve filter and gaskets.

5. Using a special service tool (SST: 09213-70011, SST: 09213-70011 or equivalent), loosen the crankshaft pulley bolt.

6. Using SST: 09950-50013 or equivalent, remove the crankshaft pulley bolt and crankshaft pulley.

7. Separate the oil cooler pipe (w/ oil cooler), as follows:

a. Remove the bolt and 2 nuts, and disconnect the oil cooler pipe from the oil pan sub-assembly.

b. Remove the gasket from the oil pan sub-assembly.

8. Remove the water inlet housing, as follows:

a. Disconnect the water hose.

b. Remove the 2 bolts, nut and water inlet housing.

c. Remove the water inlet housing gasket and water outlet pipe O-ring.

9. Remove the Bank 1 and Bank 2 cylinder head cover sub-assemblies.

10. Remove the No. 2 oil pan sub-assembly.

11. Remove the oil strainer sub-assembly.

12. Remove the timing chain cover sub-assembly, as follows:

a. Remove the 23 bolts and 2 nuts as shown in the illustration.

b. Remove the timing chain cover by prying between the timing chain cover and cylinder head or cylinder block with a screwdriver.

➡Be careful not to damage the contact surfaces of the cylinder head, cylinder block and chain cover.

➡Tape the screwdriver tip before use.

c. Remove the gasket.

13. Using a screwdriver, pry out and remove the timing chain oil seal.

➡Tape the screwdriver tip before use.

To install:

14. Install the timing chain case oil seal, as follows:

a. Using SST (SST: 09223-22010, SST: 09506-35010) tap in a new oil seal until its surface is flush with the timing chain case edge.

➡Keep the lip free from foreign matter.

➡Make sure that the oil seal edge does not stick out of the timing chain case.

➡Do not tap on the oil seal at an angle.

15. Install the timing chain cover sub-assembly, as follows:

a. Apply seal packing in a continuous line to the engine unit as shown in the illustration. Seal packing: Toyota Seal Packing Black, Three Bond 1207B or equivalent. Seal diameter: 3.0 mm (0.118 in.).

➡Be sure to clean and degrease the contact surfaces, especially the surfaces indicated by C in the illustration.

➡If the contact surfaces are wet, wipe them with an oil-free cloth before applying seal packing.

➡Install the chain cover sub-assembly within 3 minutes after applying seal packing.

➡Do not start the engine for at least 2 hours after installing the chain cover sub-assembly.

b. Apply seal packing in a continuous line to the timing chain cover as shown in the illustration. Seal packing: Toyota Seal Packing Black, Three Bond 1207B, Three Bond 1282B or equivalent.

➡If the contact surfaces are wet, wipe them with an oil-free cloth before applying seal packing.

➡Install the chain cover sub-assembly within 3 minutes and tighten the bolts within 15 minutes after applying seal packing.

➡Do not start the engine for at least 2 hours after installing the chain cover sub-assembly.

c. Apply seal packing as follows:

d. Install a new gasket.

e. Align the oil pump's drive rotor spline and the crankshaft as shown in the

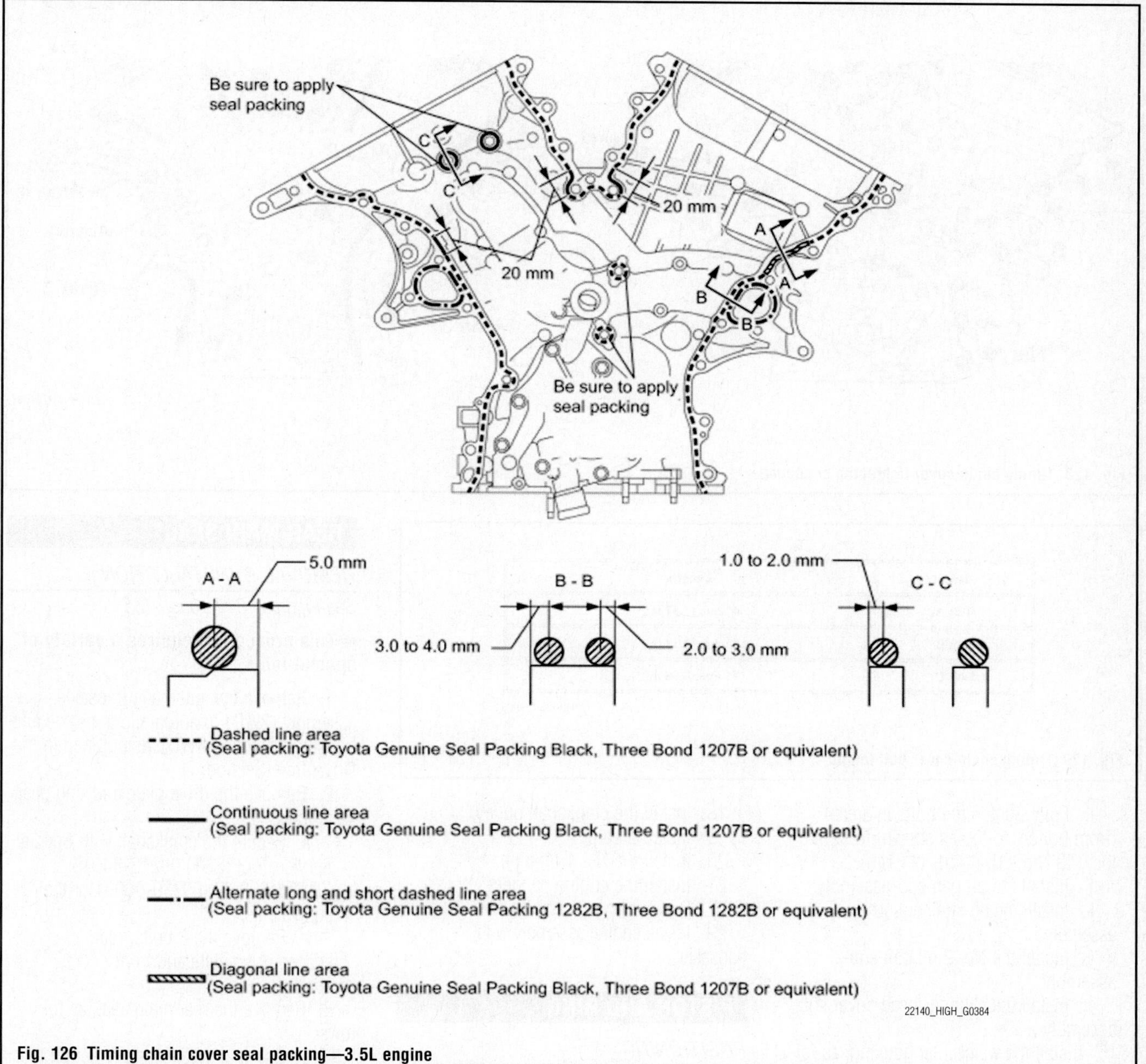

Fig. 126 Timing chain cover seal packing—3.5L engine

Area	Seal Packing Diameter	Application Position from Inside Seal Line
Continuous Line Area	4.5 mm (0.177 in.) or more	3.0 to 4.0 mm (0.118 to 0.158 in.)
Alternate Long and Short Dashed Line Area	3.5 mm (0.138 in.) or more	2.0 to 3.0 mm (0.079 to 0.118 in.)
Dashed Line Area	3.5 mm (0.138 in.) or more	3.0 to 4.0 mm (0.118 to 0.158 in.)
Diagonal Line Area	6.0 mm (0.236 in.) or more	5.0 mm (0.197 in.)

22140_HIGH_G0380

Fig. 127 Seal packing specifications—3.5L engine

illustration. Install the spline and chain cover to the crankshaft.

f. Temporarily tighten the timing chain cover with the 23 bolts and 2 nuts.

➡Make sure that there is no oil on the bolt threads.

g. Fully tighten the bolts in area 1 and area 2 (from top to bottom as shown) to 15 ft. lbs. (21 Nm).

h. Fully tighten the bolts and nuts in area 3 (from top to bottom as shown) to 15 ft. lbs. (21 Nm).

➡Tighten the bolts and nuts from top to bottom as shown in the illustration.

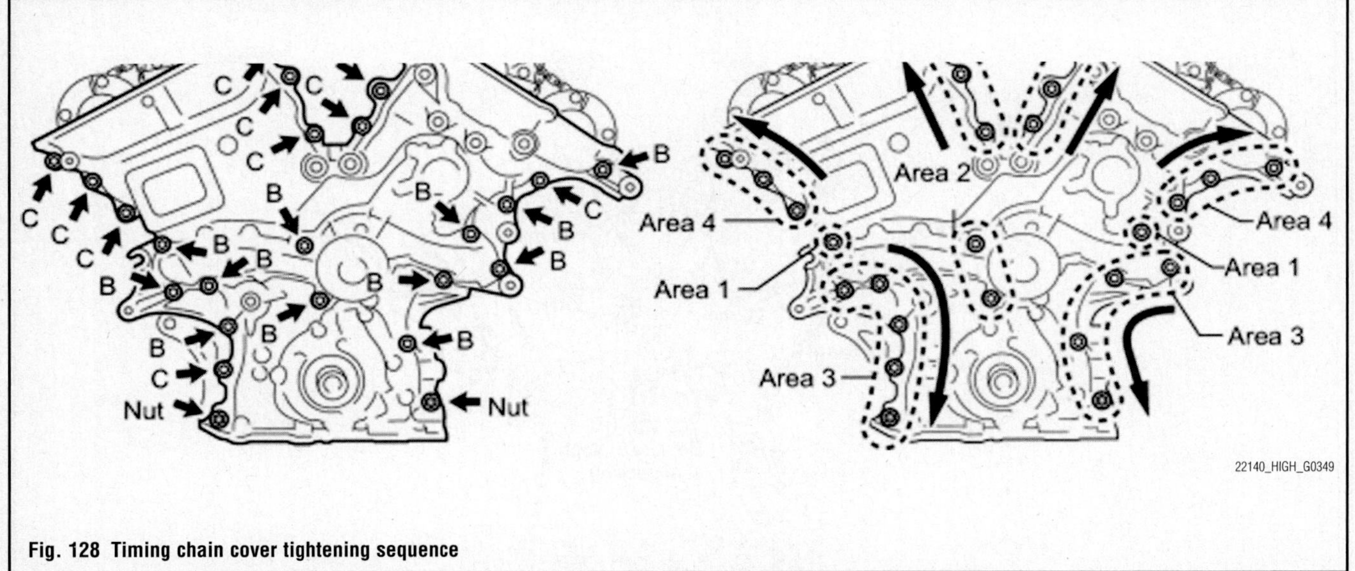

Fig. 128 Timing chain cover tightening sequence

Item	Length
Bolt A	40 mm (1.57 in.)
Bolt B	55 mm (2.17 in.)
Bolt C	25 mm (0.98 in.)

22140_HIGH_G0133

Fig. 129 Timing chain cover bolt length

i. Fully tighten the bolts in area 4 (from bottom to top as shown) to 32 ft. lbs. (43 Nm), 15 ft. lbs. (21 Nm).

j. Install the oil pan sub-assembly.

k. Install the oil strainer sub-assembly.

l. Install the No. 2 oil pan sub-assembly.

m. Install the cylinder head cover sub-assembly.

16. Install the water inlet housing, as follows:

a. Install a new water inlet housing No. 1 gasket and water outlet pipe O-ring.

b. Install the water inlet housing with the 2 bolts and nut and tighten to 7 ft. lbs. (10 Nm).

➡Be careful not to allow the O-ring to get caught between the parts.

c. Connect the water hose.

17. Install the oil cooler pipe (w/oil cooler), as follows:

a. Install a new gasket to the oil pan sub-assembly.

b. Install the oil cooler pipe with the bolt and 2 nuts and tighten to 15 ft. lbs. (21 Nm).

18. Install the crankshaft pulley.

19. Install oil pipe.

20. Install the No. 1 oil pipe.

21. Install the engine hangers.

22. Remove engine stand.

23. Install engine assembly with transaxle.

PISTON AND RING

POSITIONING

See Figure 130.

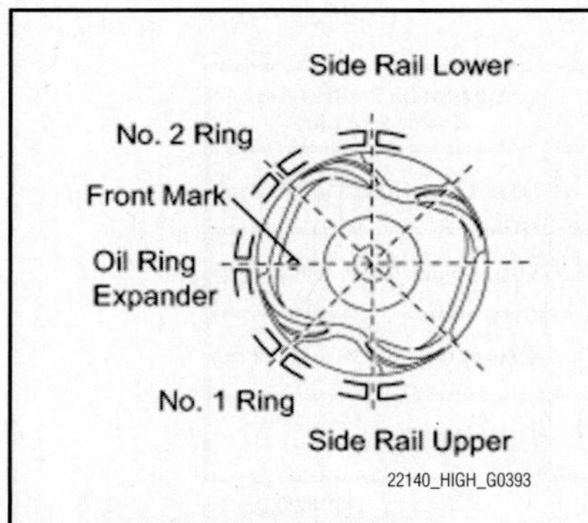

Fig. 130 Piston ring positioning—3.5L engine

REAR MAIN SEAL

REMOVAL & INSTALLATION

See Figures 131 through 135.

➡This procedure requires a variety of special tools.

1. Remove the automatic transaxle assembly (2WD) or automatic transaxle and transfer assembly (4WD), as outlined in the Drive Train Section.

2. Remove the drive plate and ring gear, as follows:

a. Secure the crankshaft with Special Service Tool (SST) 09213-54015 (91651-60855), 09330-00021 or their equivalents.

b. Remove the 8 bolts, rear spacer, drive plate and front spacer.

3. Remove the rear main seal, as follows:

a. Using a knife, carefully cut off the oil seal lip.

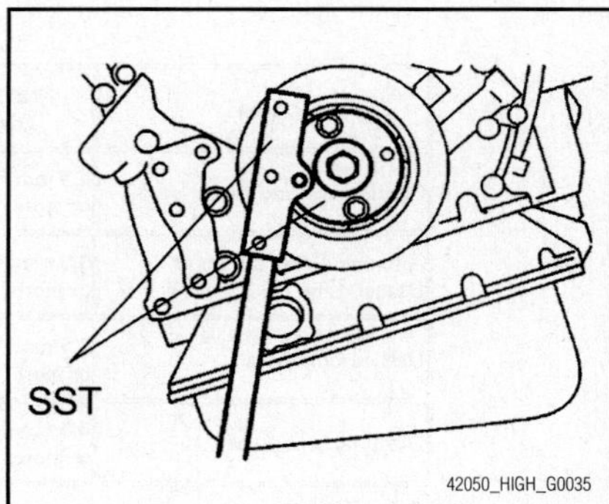

Fig. 131 Hold the crankshaft in place with the special tools as shown

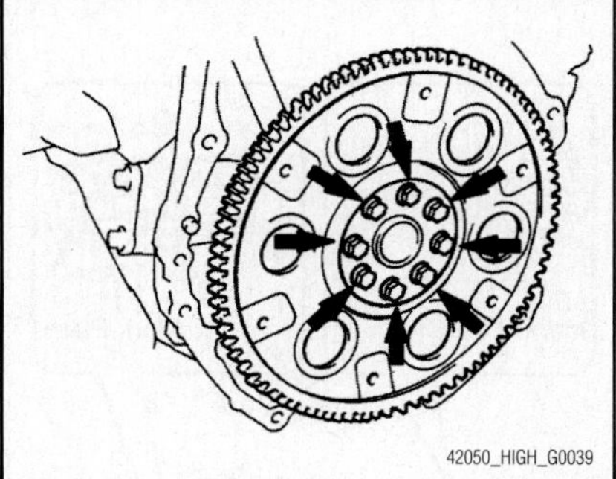

Fig. 132 Remove the 8 bolts, rear spacer, drive plate and front spacer

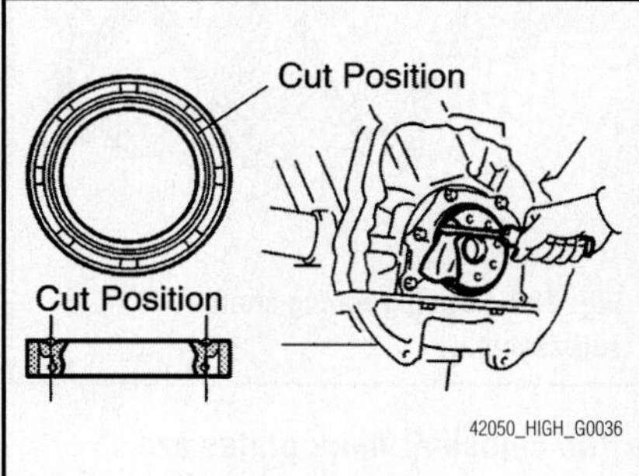

Fig. 133 If using a screwdriver wrapped in tape to remove the rear main seal, be very carefully not to damage the crankshaft

b. Use a suitable prytool, or taped screwdriver, pry out the oil seal.

WARNING

After removing the seal, make sure the crankshaft is not damaged or scratched. If it is, you can mend it with a fine grit (No. 400) sandpaper.

To install:

4. Install engine rear oil seal, as follows:

a. Apply suitable grease to a new oil seal lip. Make sure to keep the lip away from foreign materials to avoid picking up contamination or debris.

b. Using SST 09223-15030, 09950-70010 (09951-07100), or their equivalent, and a hammer, tap in the oil seal until its surface is flush with the rear oil seal retainer edge. Tap the seal in squarely to make sure it seats properly

➡Wipe the extra grease off of the crankshaft.

5. Install drive plate and ring gear, as follows:

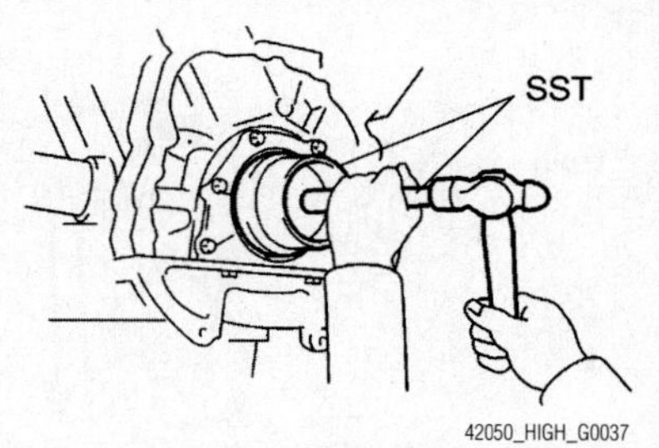

Fig. 134 Use a seal installation tool and hammer to install the rear main seal until its surface is flush with the seal retainer edge

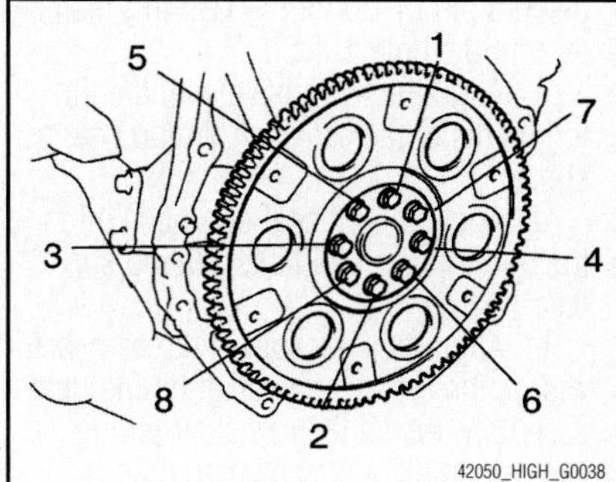

Fig. 135 Drive plate bolt tightening sequence—3.3L engines

a. Fix the crankshaft with SST 09213-54015 (91651-60855), 09330-00021, or their equivalents.

b. Clean the bolts and the bolt holes.

c. Apply a suitable adhesive (part no. 08833-00070, three bond or equivalent) to 2 or 3 threads of the bolt end.

d. Install and uniformly tighten the 8 bolts, in several passes in the sequence shown in the accompanying illustration to a final torque of 61 ft. lbs. (83 Nm).

WARNING

Do not start the engine for AT LEAST one hour after installing the seal!

6. Install the automatic transaxle assembly (2WD) or automatic transaxle and transfer assembly (4WD), as outlined in the Drive Train Section.

TIMING CHAIN FRONT COVER

REMOVAL & INSTALLATION

Refer to the timing chain and sprocket procedure.

TIMING CHAIN & SPROCKETS

REMOVAL & INSTALLATION

See Figures 136 through 147.

1. Remove the engine assembly.
2. Install on engine stand.
3. The following must be removed:

a. Remove ignition coil assembly.

b. Remove the right hand No. 2 engine mounting stay.

c. Remove the intake manifold.

d. Remove the right hand exhaust manifold sub-assembly.

e. Remove the No. 2 engine oil level dipstick guide.

f. Remove the No. 2 manifold stay.

g. Remove the No. 2 exhaust manifold heat insulator.

h. Remove the left hand exhaust manifold sub-assembly.

i. Remove the transverse engine mounting bracket.

j. Remove the generator assembly.

k. Remove the V-ribbed belt tensioner assembly.

l. Remove the No. 2 timing gear cover.

m. Remove the No. 2 idler pulley sub-assembly.

n. Remove the left hand No. 1 engine front mounting bracket.

o. Remove the left hand 6 bolts and No. 1 front engine mounting bracket.

p. Remove the radio setting condenser.

q. Remove the No. 1 vacuum switching valve.

r. Remove the knock control sensor wire.

s. Remove the knock control sensor.

t. Remove the crankshaft position sensor.

u. Remove the No. 1 oil pipe.

v. Remove the oil pipe.

w. Remove the crankshaft pulley.

x. Remove the oil cooler assembly, if necessary.

y. Remove the No. 1 oil cooler bracket, if necessary.

z. Remove the water inlet housing.

aa. Remove the water outlet.

bb. Remove the cylinder head cover sub-assembly (for Bank 1).

cc. Remove the cylinder head cover sub-assembly (for Bank 2).

dd. Remove the No. 2 oil pan sub-assembly.

ee. Remove the oil strainer sub-assembly.

ff. Remove the oil pan sub-assembly.

gg. Remove the timing chain cover sub-assembly

hh. Remove the timing chain case oil seal.

ii. Set the No. 1 cylinder to TDC/compression.

jj. Remove the No. 1 chain tensioner assembly.

kk. Remove the chain tensioner slipper.

ll. Remove the chain sub-assembly.

mm. Remove the idle sprocket assembly.

nn. Remove the camshaft timing gears and No. 2 chain (for Bank 1).

oo. While raising the No. 2 chain tensioner assembly, insert a pin of 1.0 mm (0.039 in.) diameter into the hole to fix the No. 2 chain tensioner assembly.

pp. Hold the hexagonal portion of the camshaft with a wrench, and remove the 2 bolts and 2 camshaft timing gear assemblies.

➡Be careful not to damage the cylinder head with the wrench. Do not disassemble the camshaft timing gear assemblies.

qq. Remove the No. 2 chain assembly.

rr. Remove the bolt and No. 2 chain tensioner assembly.

To install:

4. Install the following:

a. Install the No. 2 chain tensioner assembly with the bolt and tighten to 15 ft. lbs. (21 Nm).

b. While pushing in the tensioner,

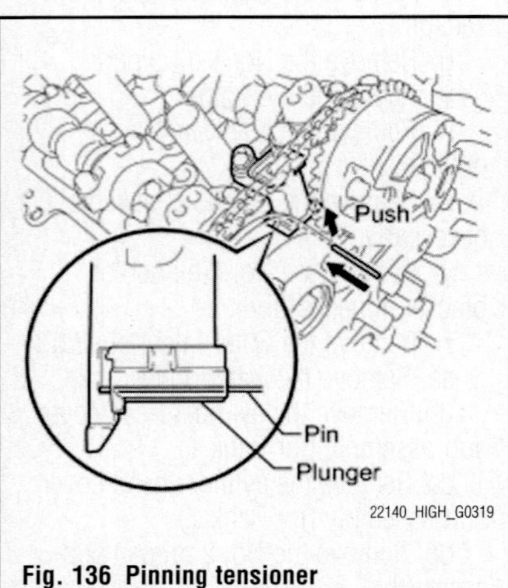

Fig. 136 Pinning tensioner

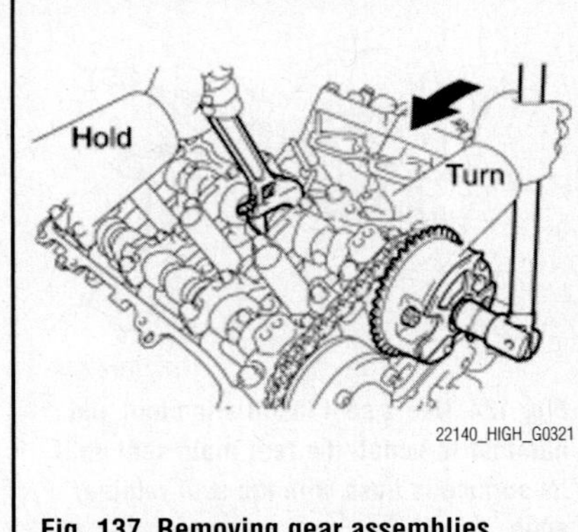

Fig. 137 Removing gear assemblies

insert a pin of 1.0 mm (0.039 in.) diameter into the hole to fix it.

c. Align the mark plate with the timing marks of the camshaft timing gear as shown.

d. Apply a light coat of engine oil to the bolt threads and bolts seating surface.

e. Align the knockpin of the camshaft with pinhole of the camshaft timing gear assembly. Install the camshaft timing gear assembly and camshaft timing exhaust timing gear assembly and camshaft timing exhaust gear with the No. 2 chain sub-assembly installed.

f. Hold the hexagonal portion of the camshaft with the wrench and tighten the two bolts and camshaft timing gear assemblies to 74 ft. lbs. (100 Nm).

g. Remove the pan from the No 2 chain tensioner assembly.

h. Install idle sprocket assembly and tighten to 44 ft. lbs. (60 Nm).

i. Install chain sub-assembly.

j. Align the mark plate and timing marks as shown in the illustration and install the chain.

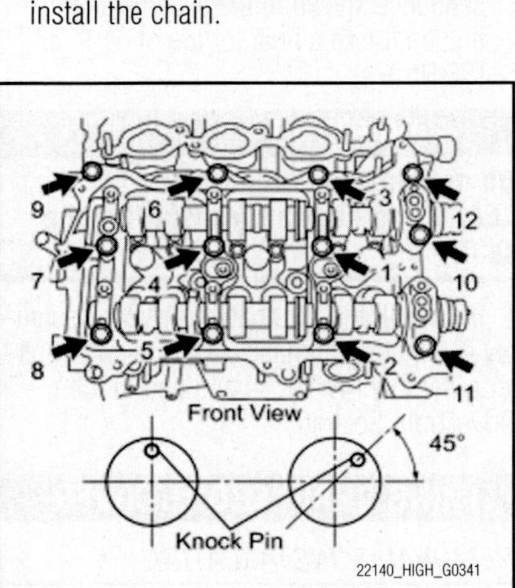

Fig. 138 Aligning No 2 timing chain

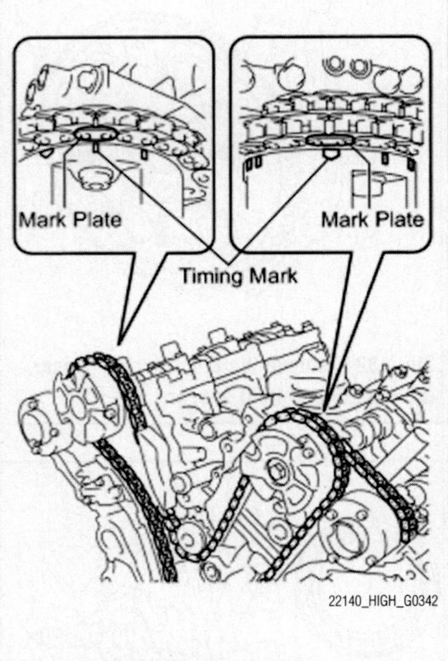

Fig. 139 Aligning timing chain sub-assembly

➡The camshaft mark plates are orange.

k. Do not pass the chain over the crankshaft, just temporarily place it on the crankshaft.

l. Turn the camshaft timing gear assembly on bank 1 counterclockwise to tighten the chain between the banks.

➡When the idle sprocket assembly is reused, align the timing chain plate with the mark where the plate has been in order to tighten the chain between the banks.

m. Align the mark plate and timing marks as shown in the illustration and install the chain onto the crankshaft timing sprocket. The crankshaft to mark plate is yellow.

n. Turn the crankshaft clockwise to set it to the right-hand block bore more centerline. (TDC Compression).

o. Install chain tensioner slipper.

p. Move the stopper plate upward to release the lock, and push the plunger deep into the tensioner.

q. Move the stopper plate downward to set the loss, and insert a hexagon wrench into the hole of the stopper plate.

r. Install No. 1 chain tensioner assembly and tighten bolts to 7 ft. lbs. (10 Nm).

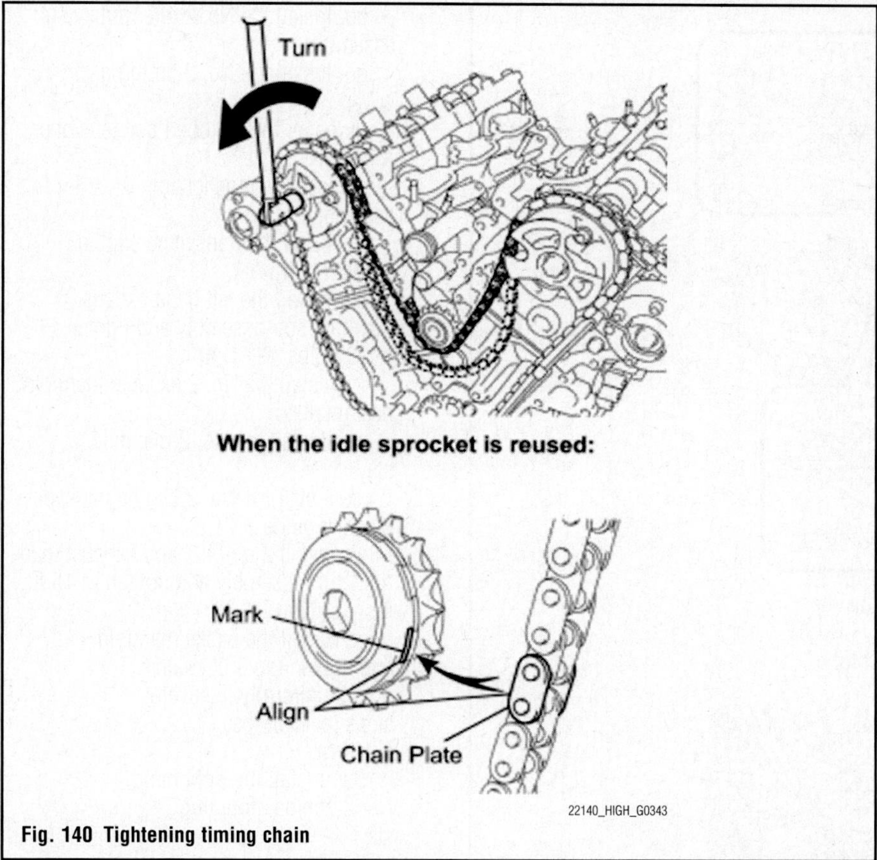

Fig. 140 Tightening timing chain

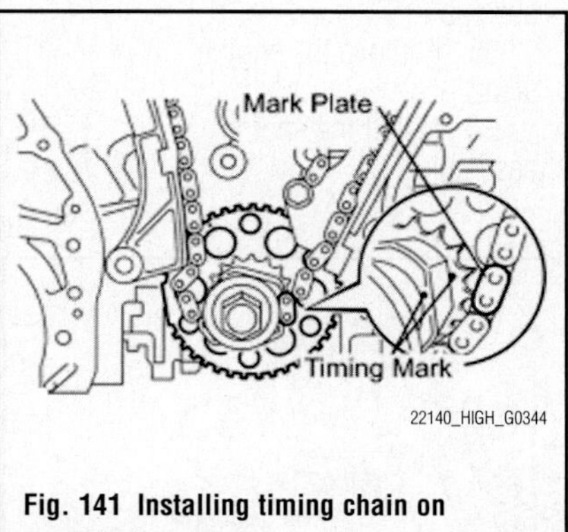

Fig. 141 Installing timing chain on crankshaft

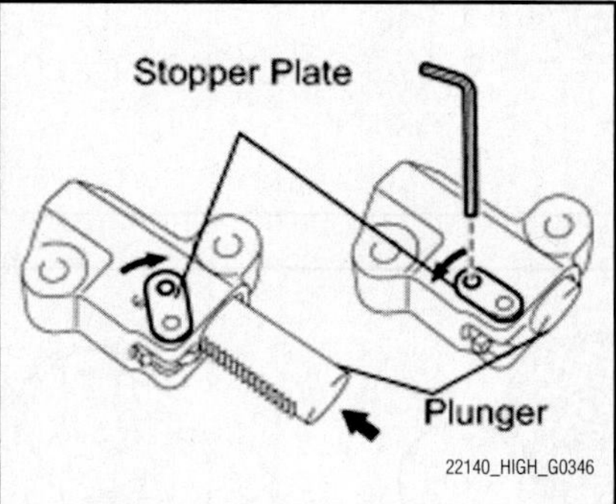

Fig. 143 Aligning timing chain on crankshaft

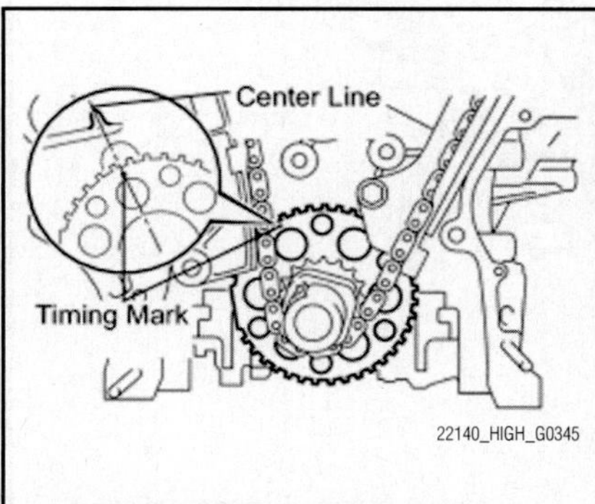

Fig. 142 Aligning timing chain on crankshaft

s. Remove the hexagon wrench from the No. 1 chain tensioner assembly. Check that the each timing mark is aligned with the crankshaft at TDC compression.

t. Remove the pulley set bolt.

u. Install timing chain case oil seal.

v. Install sealant to timing chain cover sub-assembly.

w. Install new O ring gasket on cylinder block.

x. Align the oil pump's drive rotor spline and the crankshaft as shown in the illustration. Install the spline and chain cover to the crankshaft.

y. Temporarily tighten the timing chain cover with the 23 bolts and 2 nuts.

- Tighten bolts in area 1 and 2: 15 ft. lbs. (21 Nm).
- Tighten bolt in area 3: 15 ft. lbs. (21 Nm).

➡First tighten the upper bolts and nuts, followed by the lower bolts and nuts as shown.

- Tighten bolt in area 4: 32 ft. lbs. (43 Nm)
- Tighten bolt in area 4: 15 ft. lbs. (21 Nm)

z. Install oil pan sub-assembly and tighten 16 bolts and 2 nuts to 7 ft. lbs. (10 Nm) and 15 ft lbs. (21 Nm).

aa. Install oil strainers sub-assembly and tighten bolts and nuts to 7 ft. lbs. (10 Nm).

bb. Install No.2 oil pan sub-assembly and tighten 16 bolts and 2 nuts to 7 ft. lbs. (10 Nm).

cc. Install cylinder head cover sub-assemblies and tighten to 7 ft. lbs. (10 Nm) and 15 ft. lbs. (21 Nm).

dd. Install the water inlet housing.

ee. Install No. 1 oil cooler bracket (w/ oil cooler).

ff. Install the oil cooler assembly (w/ oil cooler).

gg. Install the crankshaft pulley.

hh. Install the oil pipe.

ii. Install the No. 1 oil pipe.

jj. Install the crankshaft position sensor.

kk. Install the knock control sensor.

ll. Install the knock control sensor wire.

mm. Install the No. 1 vacuum switching valve.

nn. Install the radio setting condenser.

oo. Install the No. 1 left front engine mounting bracket with the 6 bolts and tighten to 40 ft. lbs. (54 Nm).

➡Install the water inlet and mounting bracket within 15 minutes after installing the chain cover. Do not start the engine for at least 2 hours after installation.

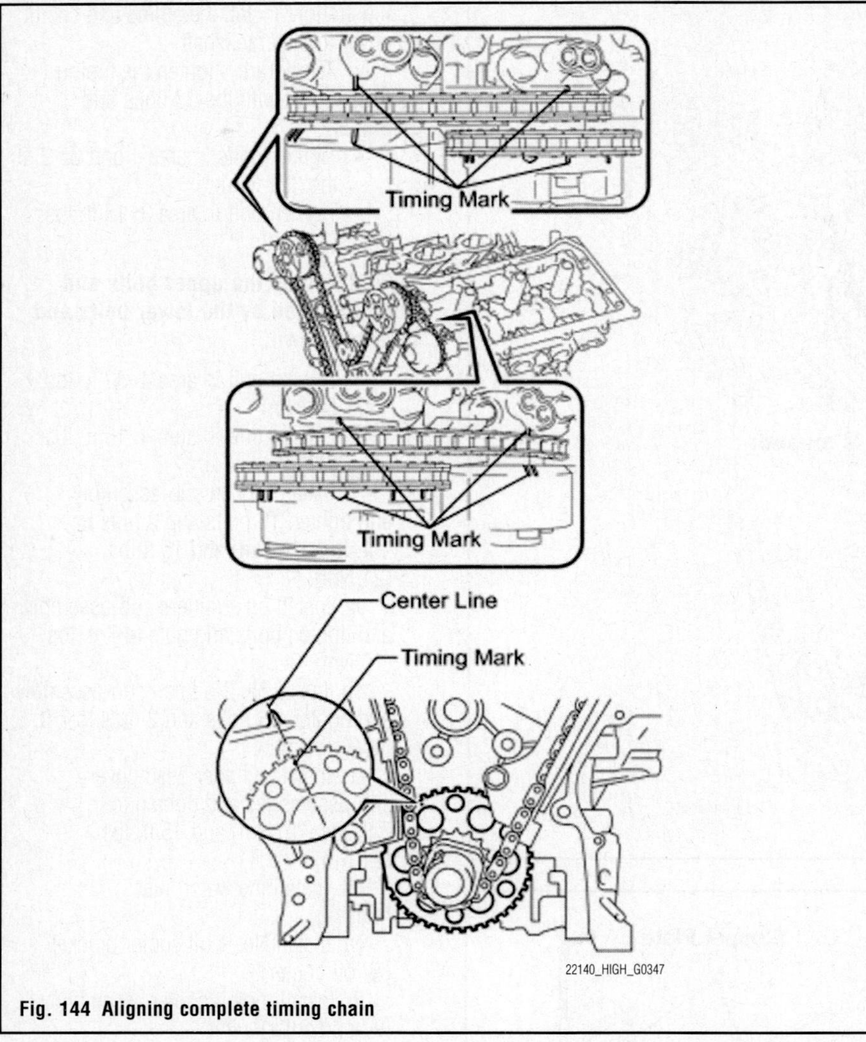

Fig. 144 Aligning complete timing chain

pp. Install the No. 2 idler pulley sub-assembly.

qq. Install the No. 2 timing gear cover

rr. Install the V-ribbed belt tensioner assembly.

ss. Install the generator assembly.

tt. Install the transverse engine mounting bracket.

uu. Install the left hand exhaust manifold sub-assembly and tighten to 15 ft. lbs. (21 Nm).

vv. Install the No. 2 exhaust manifold heat insulator.

ww. Install the No. 2 manifold stay.

xx. Install the No. 2 engine oil level dipstick guide.

yy. Install the right hand exhaust manifold sub-assembly and tighten to 15 ft. lbs. (21 Nm).

zz. Install the intake manifold and tighten the 6 bolts and 4 nuts uniformly in several steps to 15 ft. lbs. (21 Nm).

53. Install the right hand No. 2 engine mounting stay.

54. Install the ignition coil assembly.

55. Remove the engine stand.

56. Install the engine assembly.

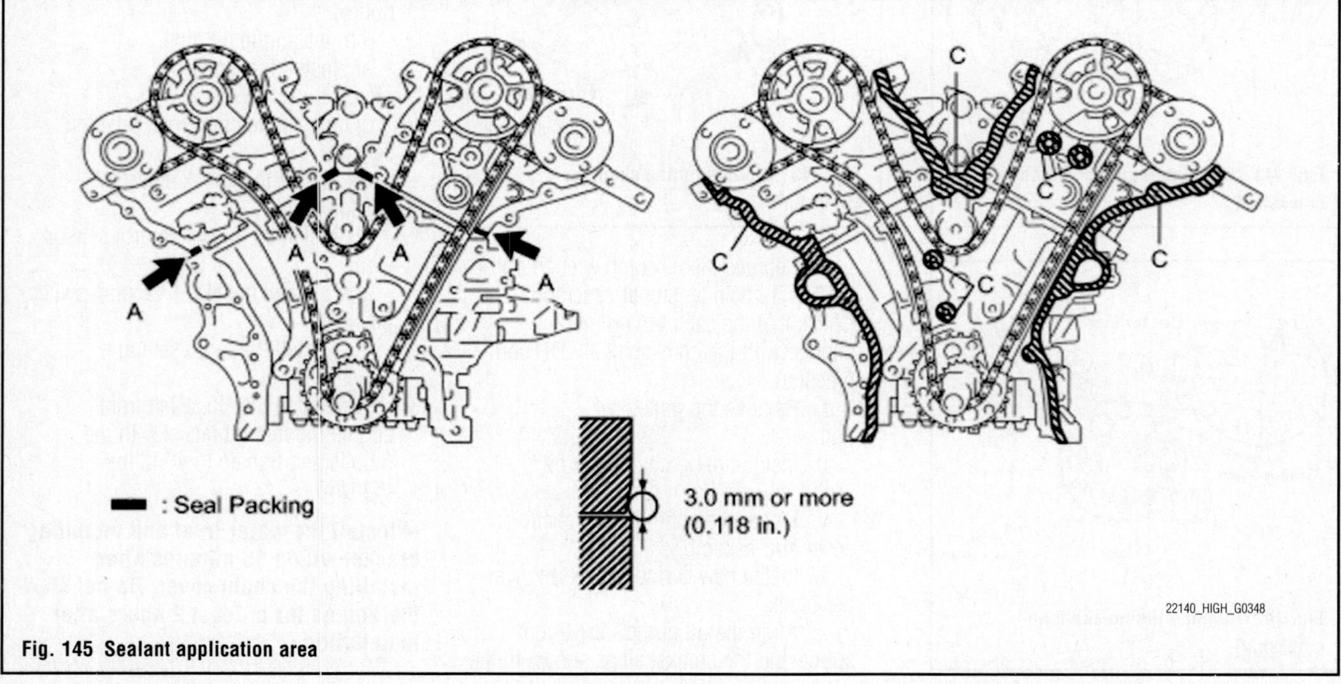

Fig. 145 Sealant application area

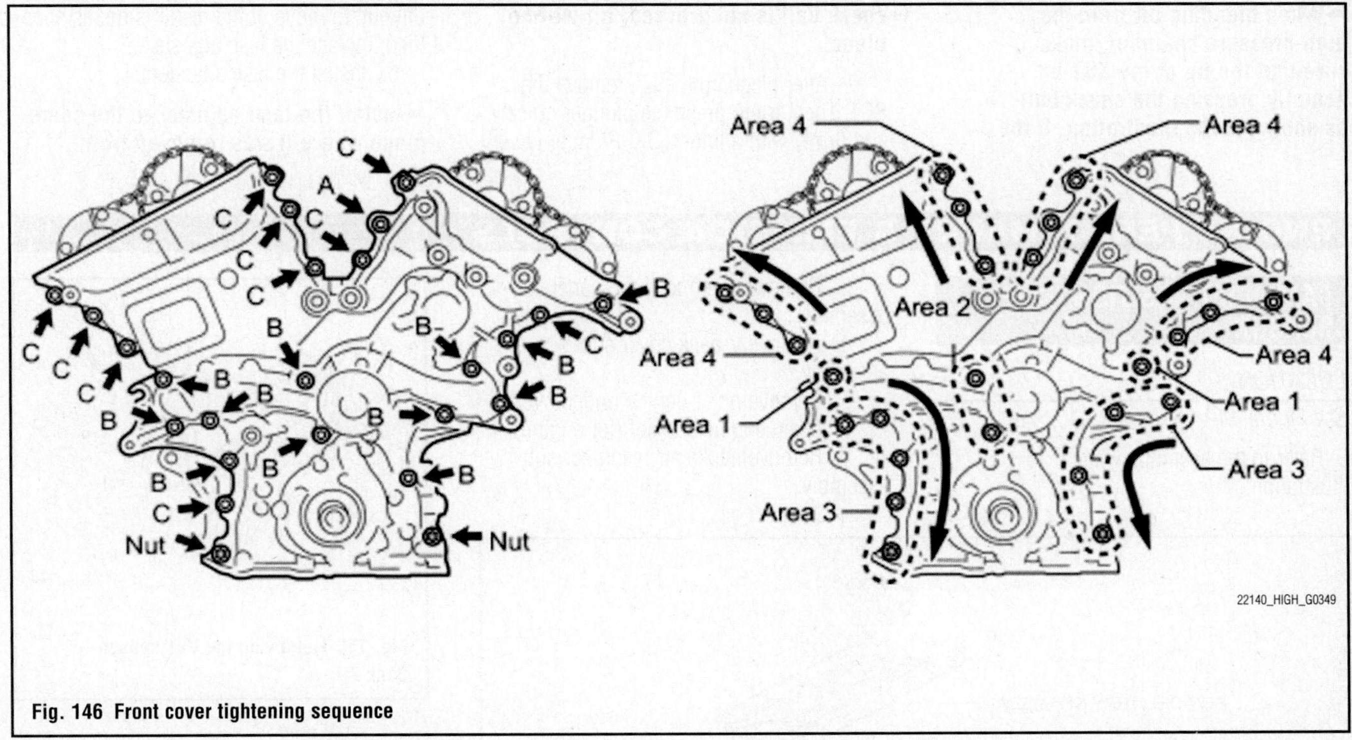

Fig. 146 Front cover tightening sequence

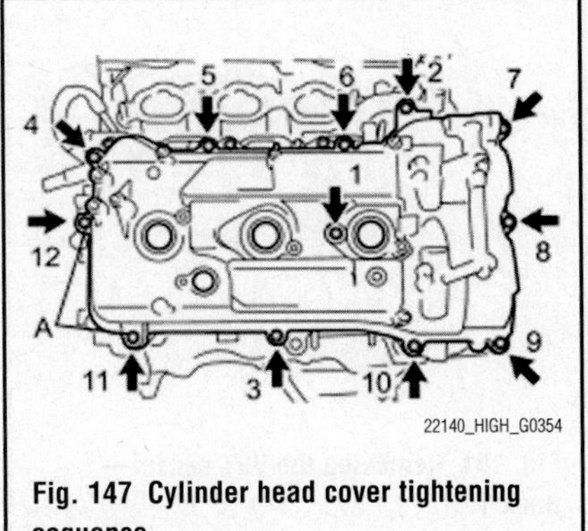

Fig. 147 Cylinder head cover tightening sequence

VALVE LASH

ADJUSTMENT

See Figure 148.

➡**Keep the lash adjuster free of dirt and foreign objects.**

➡**Only use clean engine oil.**

1. Place the lash adjuster into a container filled with engine oil.
2. Insert the SST's (SST: 09276-75010) tip into the lash adjuster's plunger and use the tip to press down on the check ball inside the plunger.
3. Squeeze the SST and lash adjuster together to move the plunger up and down 5 to 6 times.
4. Check the movement of the plunger and bleed the air. OK: Plunger moves up and down.

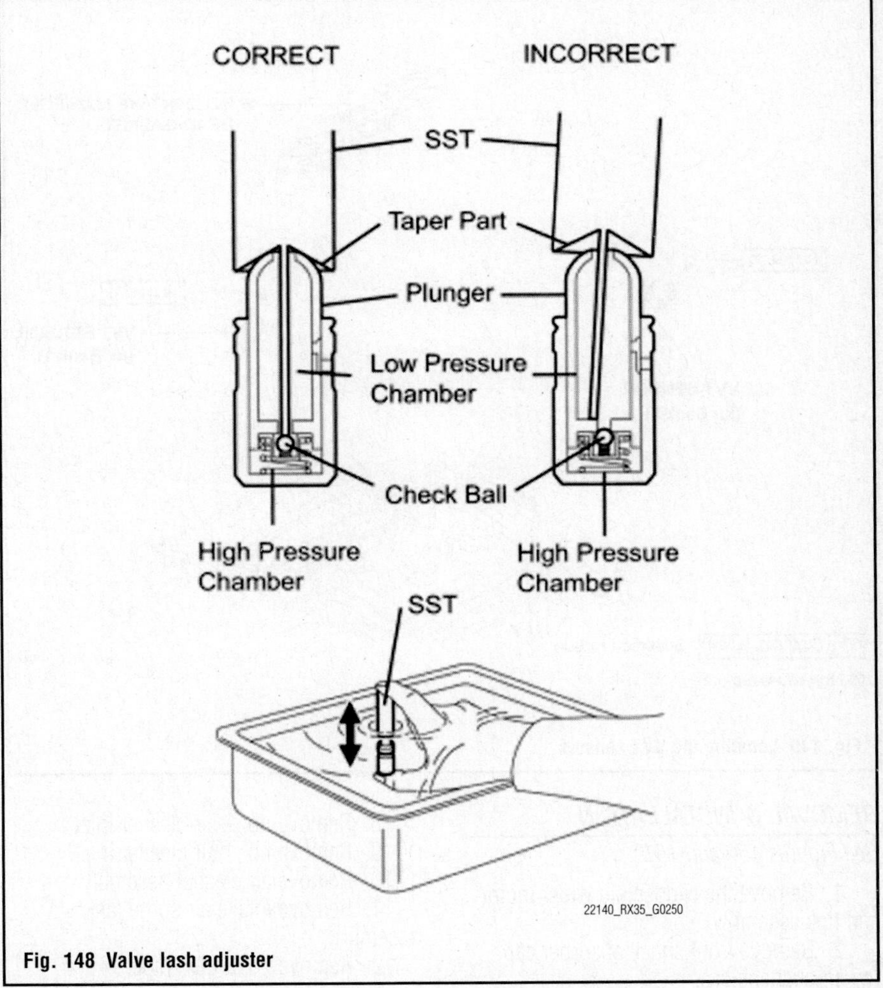

Fig. 148 Valve lash adjuster

➡When bleeding air from the high-pressure chamber, make sure that the tip of the SST is actually pressing the check ball as shown in the illustration. If the check ball is not pressed, air will not bleed.

5. After bleeding the air, remove the SST. Then, try to press the plunger quickly and firmly with a finger. OK: Plunger is very difficult to move. If the result is not as specified, replace the lash adjuster.

6. Install the lash adjusters.

➡Install the lash adjuster to the same place where it was removed from.

ENGINE PERFORMANCE & EMISSION CONTROLS

CAMSHAFT POSITION (CMP) SENSOR

LOCATION

See Figure 149.

Refer to the accompanying illustration.

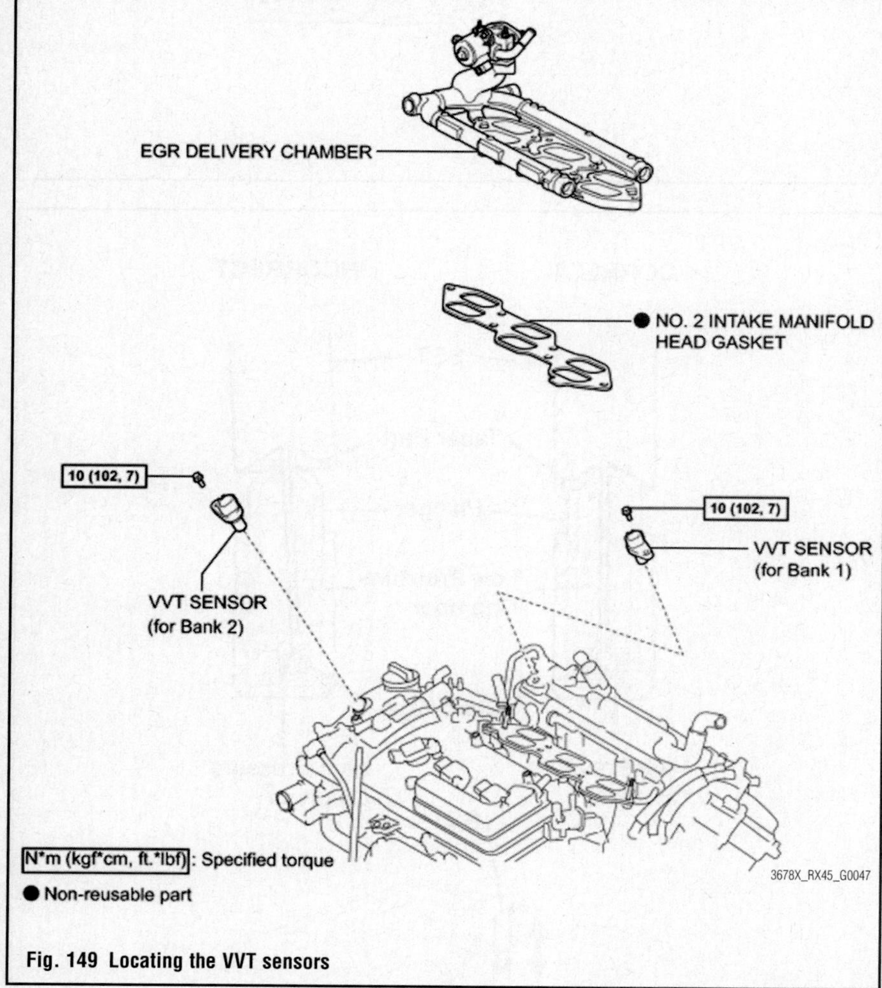

Fig. 149 Locating the VVT sensors

REMOVAL & INSTALLATION

See Figures 150 and 151.

1. Remove the windshield wiper motor and link assembly.
2. Remove front shock absorber cap (w/ air suspension).
3. Remove outer cowl top panel sub-assembly.
4. Remove v-bank cover sub-assembly.
5. Remove no. 1 engine undercover.
6. Drain engine coolant (for engine).
7. Remove intake air resonator sub-assembly.
8. Remove no. 2 air cleaner inlet.
9. Remove no. 1 air cleaner inlet.
10. Remove air cleaner assembly.
11. Remove intake air surge tank assembly.
12. Remove no. 2 EGR pipe.

Fig. 150 Removing the VVT sensor—Bank 2

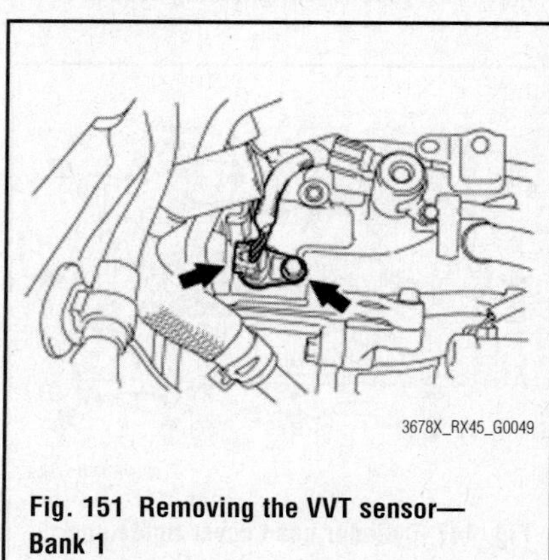

Fig. 151 Removing the VVT sensor—Bank 1

13. Remove EGR delivery chamber.
14. Remove VVT sensor:
    a. Disconnect the sensor connector.
    b. Remove the bolt and sensor.
15. To install, reverse removal procedure.
16. Apply a light coat of engine oil to the O-ring of the sensor.
17. Tighten the sensor bolt to 7 ft. lbs. (10 Nm).

CRANKSHAFT POSITION (CKP) SENSOR

LOCATION

See Figure 152.

Refer to the accompanying illustration.

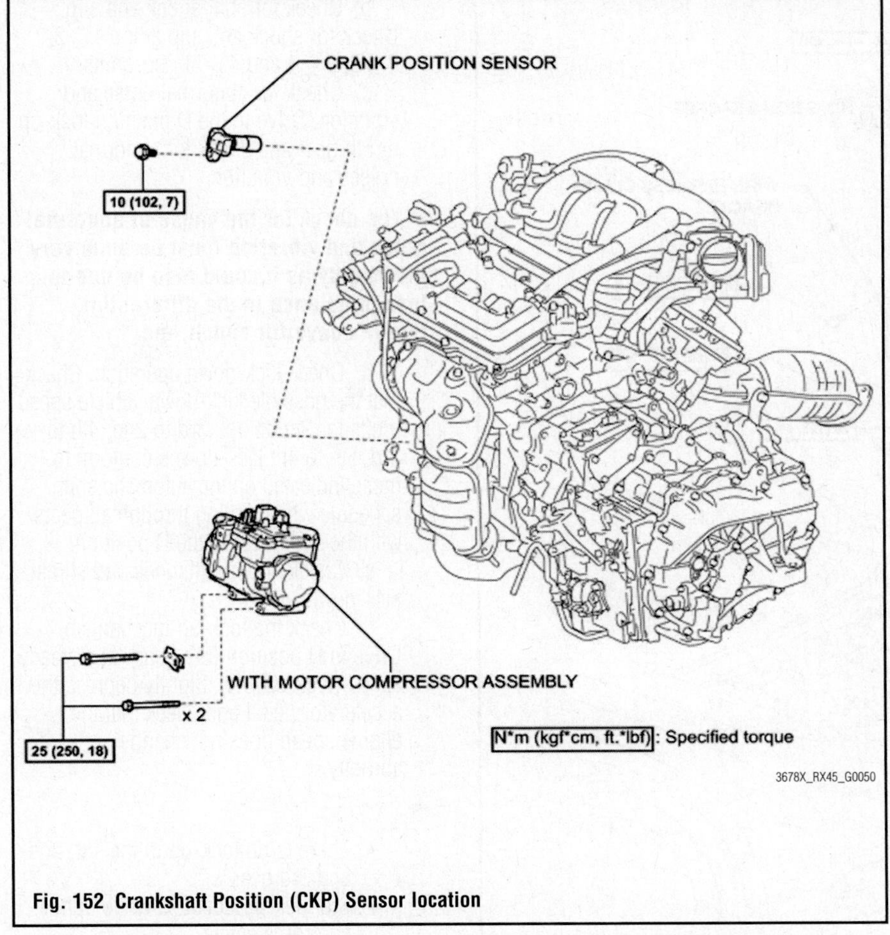

Fig. 152 Crankshaft Position (CKP) Sensor location

REMOVAL & INSTALLATION

See Figure 153.

1. Remove compressor and magnetic clutch.
2. Disconnect the crankshaft position sensor connector.
3. Remove the bolt, and then remove the crankshaft position sensor.

To install:

4. Apply a light coat of engine oil to the O-ring on the crankshaft position sensor.

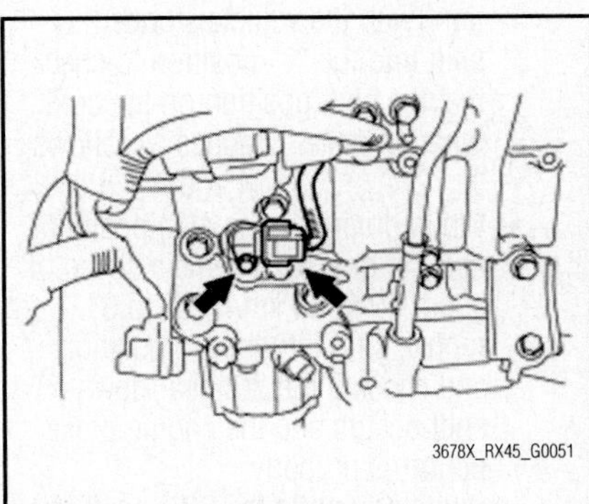

Fig. 153 Remove the bolt, and then remove the crankshaft position sensor

5. Install the crankshaft position sensor with the bolt and tighten to 7 ft. lbs. (10 Nm).
6. Connect the crankshaft position sensor connector.
7. Install compressor and magnetic clutch.
8. Install alternator assembly.

ELECTRONIC CONTROL MODULE (ECM)

LOCATION

See Figure 154.

Refer to the accompanying illustration.

REMOVAL & INSTALLATION

See Figure 154.

➡After the power switch is turned off, the display and navigation module display (HDD navigation system) records various types of memory and settings. As a result, after turning the power switch off, make sure to wait at least 60 seconds before disconnecting the cable from the negative (-) battery terminal.

1. Remove rear deck floor box.
2. Disconnect the negative battery cable.

CAUTION

Wait at least 90 seconds after disconnecting the cable from the negative (-) battery terminal to disable the SRS system.

3. Remove engine room side cover LH.
4. Remove engine room side cover.
5. Remove cool air intake duct seal.
6. Remove ECM:
   a. Separate the 2 wire harness clamps.
   b. Disconnect the 2 ECM connectors and remove the ECM with brackets.
   c. Push in the locks on the 2 levers, raise the levers, and disconnect the 2 ECM connectors.

➡After disconnecting the connectors, make sure that dirt, water or other foreign matter does not contact the connecting part of the connectors.

   d. Remove the bolt and separate the cooler pipe.
   e. Remove the 3 bolts and the ECM with bracket.

➡Do not deform the cooler pipe when moving it to access the bolts below.

   f. Remove the 2 bolts and the wire harness bracket.
   g. Remove the 4 bolts, and remove the No. 1 ECM bracket and No. 2 ECM bracket.
   h. Unlock the 3 claws, and remove the engine room ECM cover and ECM.
7. To install, reverse the removal procedure.

RESET

➡Perform the RESET MEMORY (AT initialization) when replacing the automatic transaxle assembly, engine assembly or ECM. The RESET MEMORY can be performed only with Techstream.

➡The ECM memorizes the condition that the ECT controls the automatic transaxle assembly and engine assembly according to those characteristics. Therefore, when the automatic transaxle assembly, engine assembly, or ECM has been replaced, it is necessary to reset the memory so that the ECM can memorize the new information.

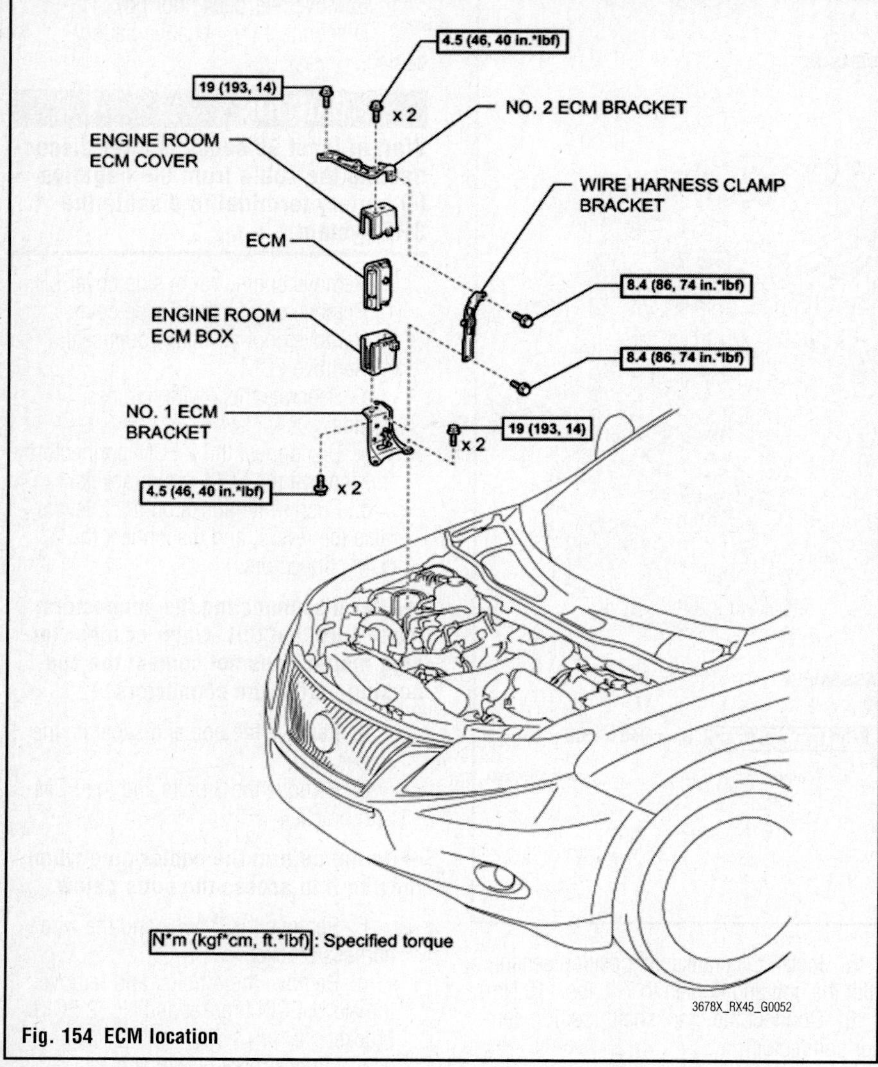

Fig. 154 ECM location

1. Reset procedure is as follows:
   a. Turn the ignition switch off.
   b. Connect Techstream to the DLC3.
   c. Turn the ignition switch on (IG) and push the Techstream main switch on.
   d. Enter the following menu: Powertrain / Engine and ECT / Utility / Reset Memory. Then, press "Next".
   e. Perform the reset memory procedure from the main menu.

CAUTION

After performing the RESET MEMORY, be sure to perform the ROAD TEST.

ROAD TESTING

➡Perform the test at the ATF temperature 122 to 176°F (50 to 80°C) in the normal operation.

1. D position test: Shift into the D position and fully depress the accelerator pedal and check the following points:
   a. Check up-shift operation: Check that 1 _ 2, 2 _ 3, 3 _ 4 and 4 _ 5th up-shifts take place, and that the shift points conform to the automatic shift schedule.

➡5th Gear Up-shift Prohibition Control: Coolant temperature is 158°F (70°C) or less and vehicle speed is at 80 km/h (50 mph) or less. ATF temperature is 28°F (-2°C) or less.

➡4th Gear Up-shift Prohibition Control: Coolant temperature is 158°F (70°C) or less and vehicle speed is at 55 km/h (34 mph) or less.

➡5th and 4th Gear Lock-up Prohibition Control: Brake pedal is depressed. Accelerator pedal is released. Coolant temperature is 140°F (60°C) or less.

   b. Check for shift shock and slip. Check for shock and slip at the 1 _ 2, 2 _ 3, 3 _ 4 and 4 _ 5th up-shifts.
   c. Check for abnormal noise and vibration. Drive in the D position lock-up or 5th gear and check for abnormal noises and vibration.

➡The check for the cause of abnormal noise and vibration must be done very thoroughly as it could also be due to loss of balance in the differential, torque converter clutch, etc.

   d. Check kick-down operation. Check that the possible kick-down vehicle speed limits for 2nd to 1st, 3rd to 2nd, 4th to 3rd, 5th to 4th kick-downs conform to those indicated on the automatic shift schedule while driving through all gears with the shift lever in the D position.
   e. Check abnormal shock and slip at kick-down.
   f. Check the lock-up mechanism. Drive in D position (5th gear), at a steady speed (lock-up ON). Lightly depress the accelerator pedal and check that the engine speed does not change abruptly.

➡

- There is no lock-up in the 1st, 2nd, and 3rd gear.
- 3rd lock-up operates while uphill control is active.
- If there is a big jump in engine speed, there is no lock-up.

2. S position test:
   a. Shift to the S position, depress the accelerator pedal and check the following points. Check shift operation:
   - While driving in the D position and 5th gear, shift into the D position and S position and back to the D position. Check that the gear change 5 _ 4 down-shift and 4 _ 5 up-shift can be performed
   - With the shift lever in the S position (with the vehicle stopped), shift into the "+" position to check that the shift position on the combination meter changes as follows : 1 _ 2, 2 _ 3, 3 _ 4 and 4 _ 5
   - While driving in the 4(S) position and 4th gear (at a vehicle speed of about 40 to 50 km/h (25 to 31 mph)), shift into the "-" position and check if the 3rd gear down-shift occurs and the engine brake performs properly
   - While driving in the 3(S) position and 3rd gear (at a vehicle speed of about 30 to 40 km/h (19 to 25

mph)), shift into the "-" position and check if the 2nd gear down-shift occurs and the engine brake performs properly

- While driving in the 2(S) position and 2nd gear (at a vehicle speed of about 20 to 30 km/h (12 to 19 mph)), shift into the "-" position and check if the 1st gear down-shift occurs and the engine brake performs properly

➡Manual shift (S position) prohibition control: Down-shifting causes engine overrun. Down-shifting is required continuously. (Down-shifting to 1st gear may not be performed.)

b. R position test: Shift into the R position, lightly depress the accelerator pedal, and check that the vehicle moves backward without any abnormal noise or vibration.

⁂ CAUTION

Before conducting this test ensure that the test area is free from people and obstruction.

c. P position test: Stop the vehicle on a grade (more the 5°) and after shifting into the P position, release the parking brake. Then, check that the parking lock pawl holds the vehicle in place.

d. Uphill/downhill control function test:

- Check that the gear does not up-shift to the 3rd to 4th or 4th to 5th gear while the vehicle is driving uphill.
- Check that the gear automatically down-shifts from the 5th to 4th or from the 4th to 3rd gear when brake is applied while the vehicle is driving downhill.

ENGINE COOLANT TEMPERATURE (ECT) SENSOR

REMOVAL & INSTALLATION

See Figure 155.

1. Remove the V-bank cover sub-assembly.
2. Remove the engine undercover assembly.
3. Remove the No. 1 engine undercover.
4. Drain engine coolant.
5. Remove the cool air intake duct seal.
6. Remove the battery.

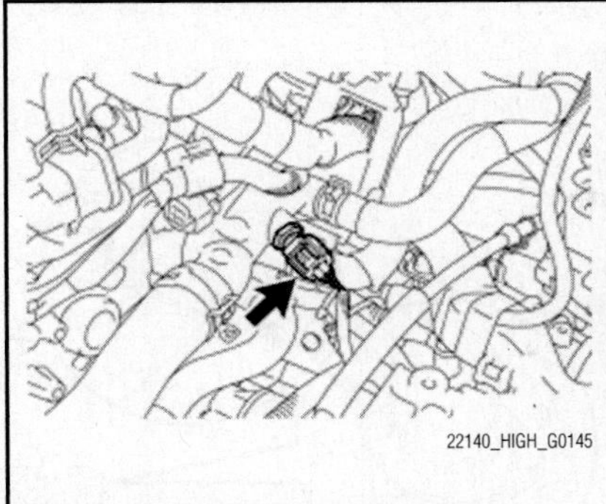

Fig. 155 Coolant temperature sensor, connector, and gasket—3.5L engine

7. Remove the No. 1 and 2 air cleaner inlets.
8. Remove the air cleaner cap and case sub-assemblies.
9. Disconnect the engine coolant temperature sensor connector.
10. Remove the engine coolant temperature sensor and gasket.

To install:

11. Install a new gasket onto the engine coolant temperature sensor.
12. Install the engine coolant temperature sensor and tighten to 15 ft. lbs. (20 Nm).
13. The remainder of installation is the reverse of the removal procedure.

HEATED OXYGEN SENSOR (HO2S)

LOCATION

See Figure 156.

Refer to the accompanying illustration.

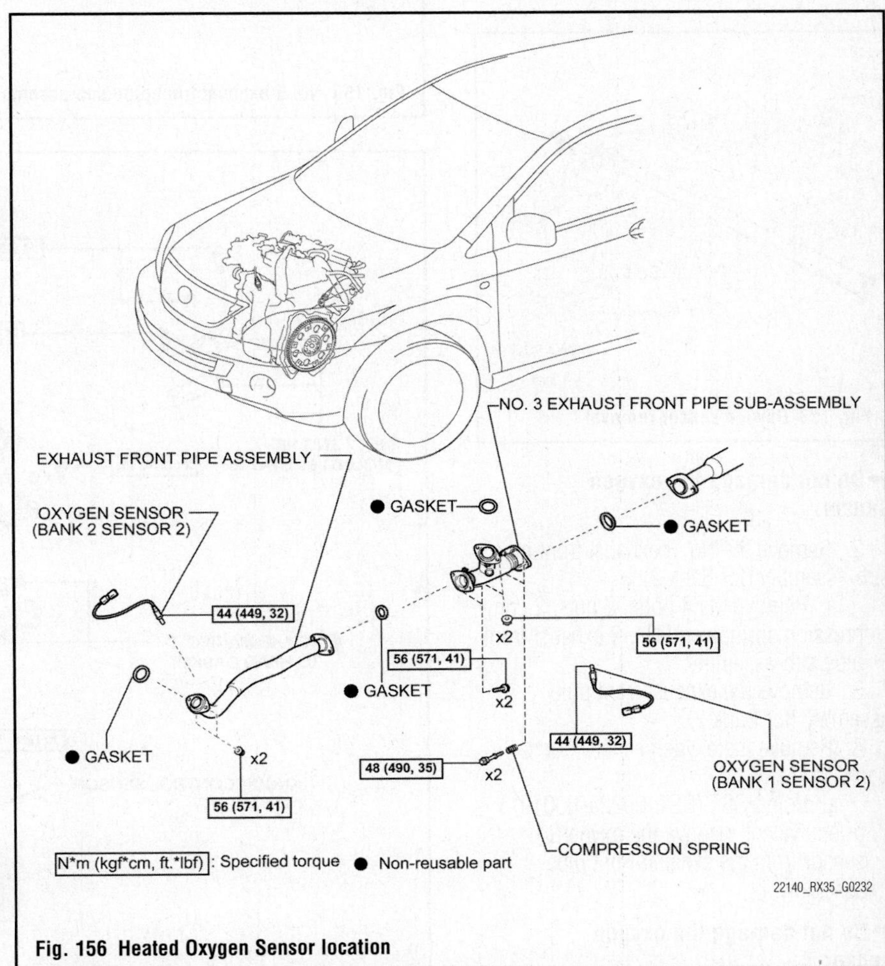

Fig. 156 Heated Oxygen Sensor location

REMOVAL & INSTALLATION

See Figures 157 through 159.

1. Remove the oxygen sensor (for Bank 1):

a. Disconnect the oxygen sensor connector and 2 clamps.

b. Using SST (SST: 09224-00010) or equivalent, remove the oxygen sensor from the No. 3 exhaust front pipe sub-assembly.

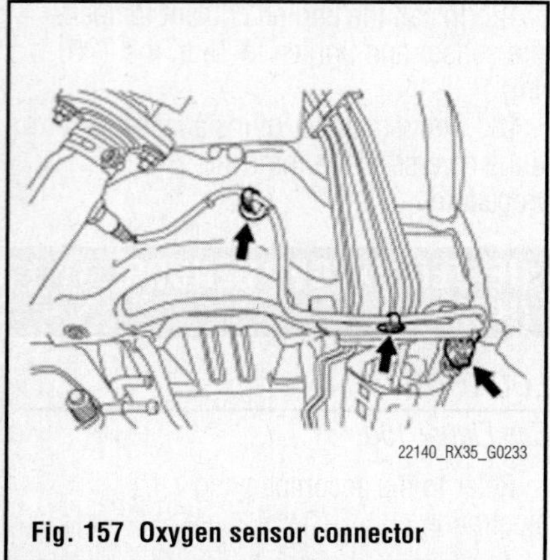

Fig. 157 Oxygen sensor connector

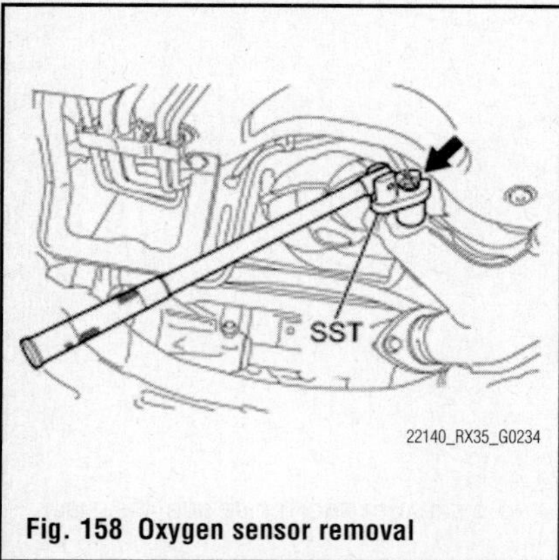

Fig. 158 Oxygen sensor removal

➡Do not damage the oxygen sensor.

2. Remove the No. 3 exhaust front pipe sub-assembly (for Bank 2):
   a. Remove the 4 bolts, 2 nuts, 2 compression springs and No. 3 exhaust front pipe sub-assembly.
3. Remove the exhaust front pipe assembly (for Bank 2).
4. Remove the oxygen sensor (for Bank 2):
   a. Using SST (SST: 09224-00010)) or equivalent, remove the oxygen sensor from the exhaust front pipe assembly.

➡Do not damage the oxygen sensor.

To install:

5. Installation is the reverse of the removal procedure. Tighten the oxygen sensors to 32 ft. lbs. (44 Nm).

KNOCK SENSOR (KS)

LOCATION

See Figure 160.

Refer to the accompanying illustration.

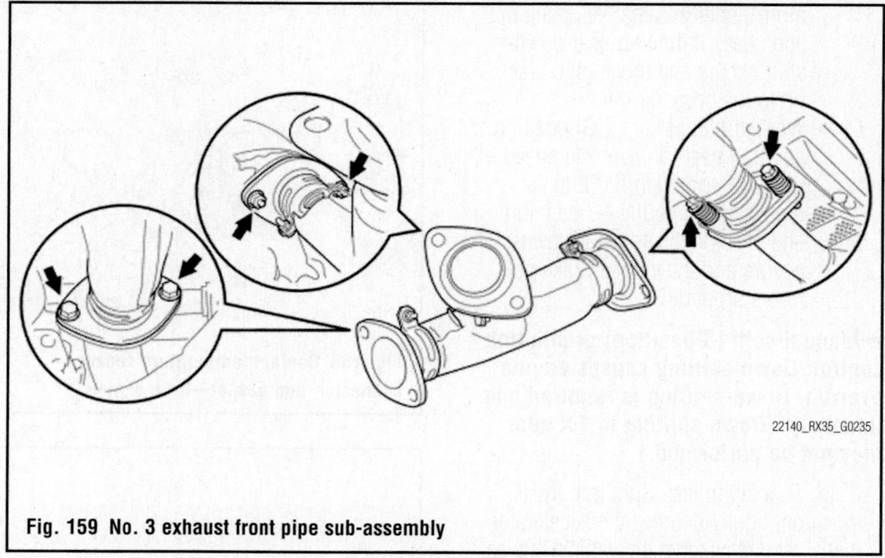

Fig. 159 No. 3 exhaust front pipe sub-assembly

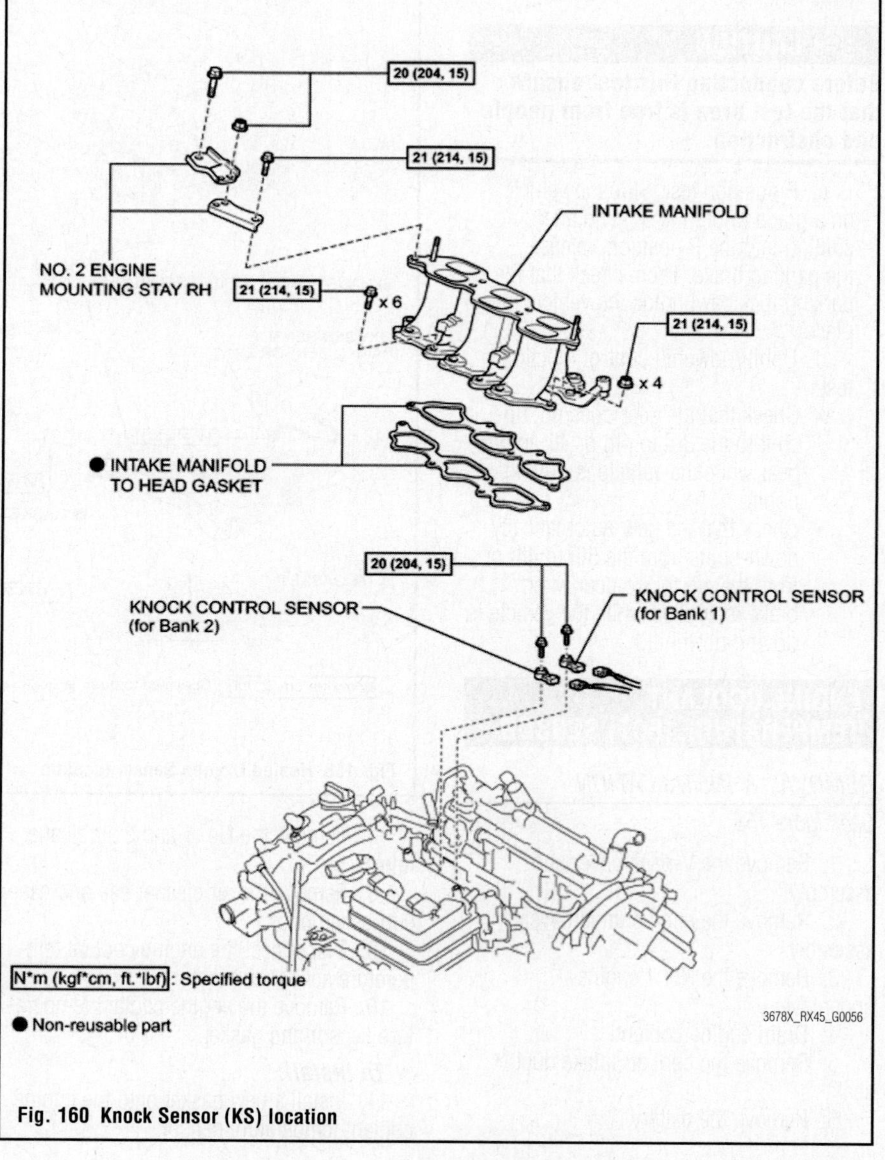

Fig. 160 Knock Sensor (KS) location

REMOVAL & INSTALLATION

See Figures 161 and 162.

1. Remove the intake manifold.
2. Disconnect the 2 knock control sensor connectors.
3. Remove the 2 bolts and 2 knock control sensors.

To install:

4. Install the 2 knock control sensors with the 2 bolts as shown in the illustration and tighten to 15 ft. lbs. (20 Nm).
5. Connect the 2 knock control sensor connectors.
6. The remainder of installation is the reverse of the removal procedure.
7. After adding coolant, check for coolant and fuel leaks.

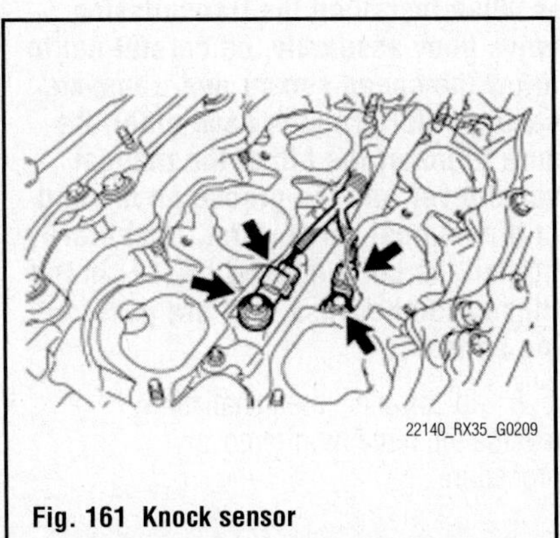

Fig. 161 Knock sensor

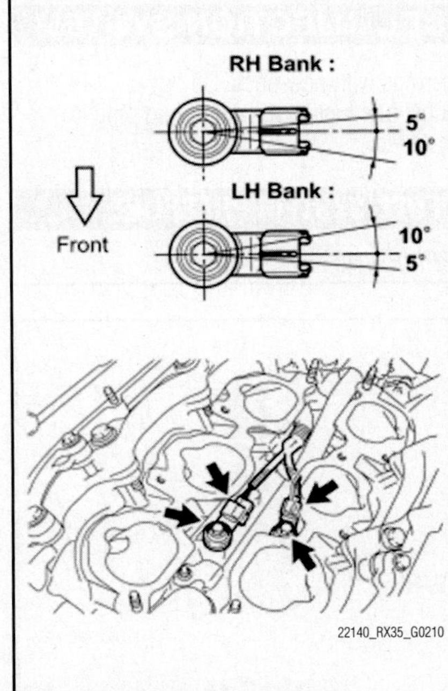

Fig. 162 Knock sensor installation

MALFUNCTION INDICATOR LIGHT (MIL)

RESET PROCEDURE

Using a ODB II scan tool, clear the DTC codes.

MASS AIR FLOW (MAF) METER

LOCATION

See Figure 163.

Refer to the accompanying illustration.

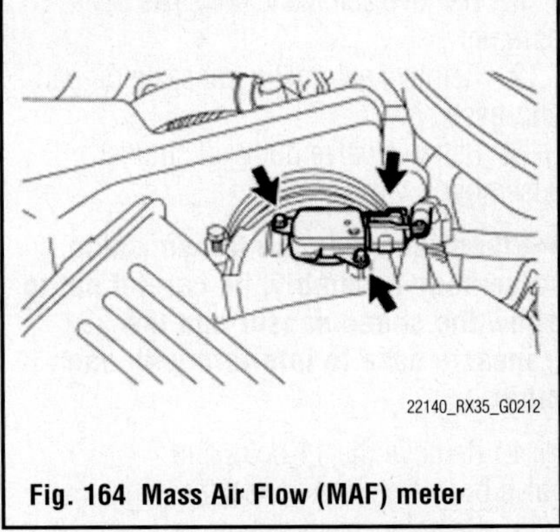

Fig. 164 Mass Air Flow (MAF) meter

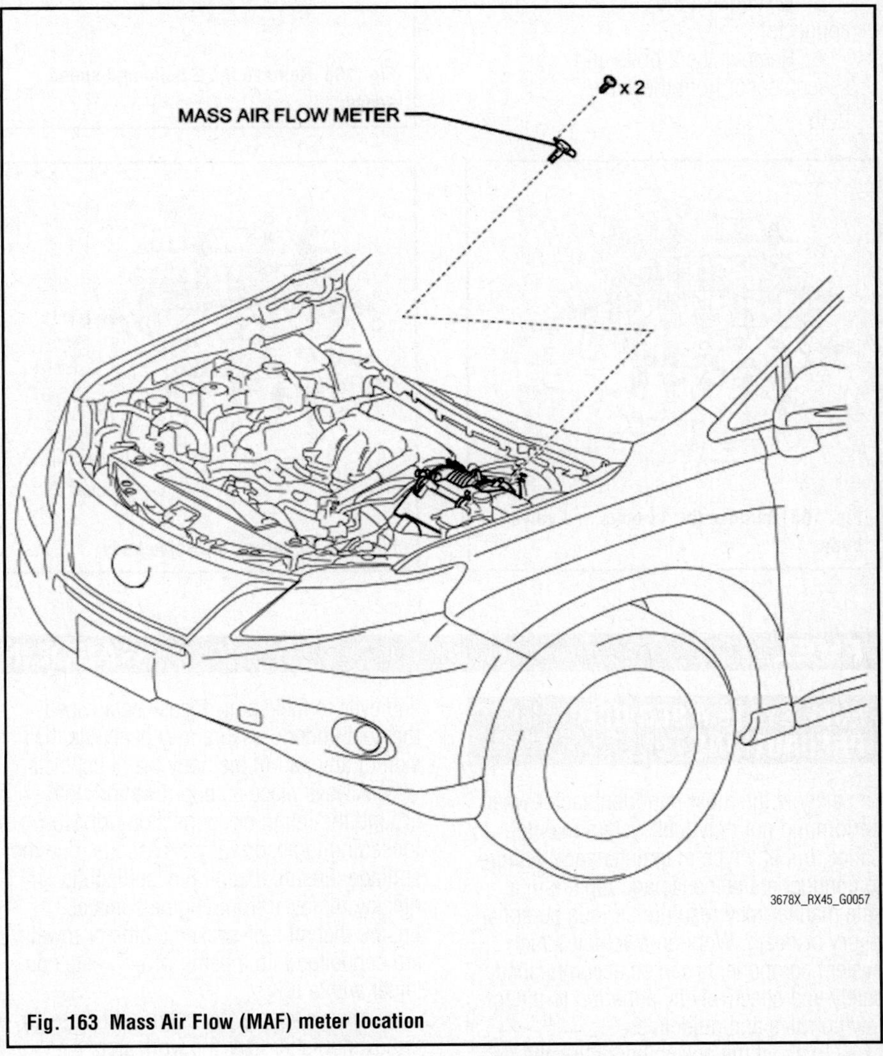

Fig. 163 Mass Air Flow (MAF) meter location

REMOVAL & INSTALLATION

See Figure 164.

1. Disconnect the Mass Air Flow (MAF) meter connector.
2. Remove the 2 screws and Mass Air Flow (MAF) meter.

To install:

3. Installation is the reverse of the removal procedure.

VEHICLE SPEED SENSOR (VSS)

REMOVAL & INSTALLATION

See Figures 165 through 167.

1. Remove automatic transaxle assembly.
2. Remove automatic transaxle oil pan sub-assembly
3. Remove valve body oil strainer assembly.

➡When removing the transmission valve body assembly, be careful not to allow the speed sensor and the transaxle case to interfere with each other.

4. Remove the 11 bolts and valve body from the transaxle.
5. Remove speed sensor:
   a. Disconnect the connector.
   b. Remove the 2 bolts and speed sensor from the valve body.

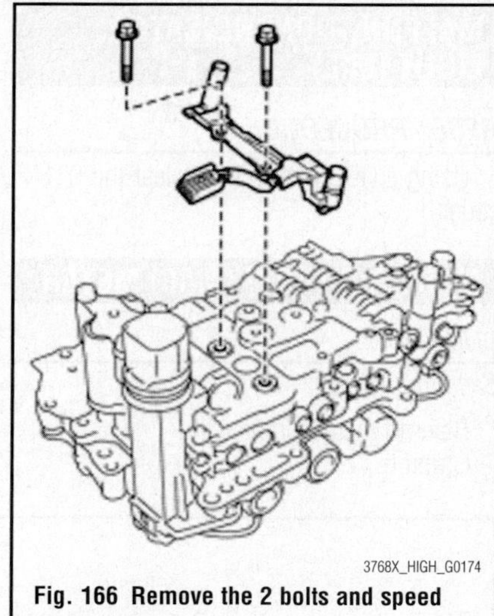

Fig. 166 Remove the 2 bolts and speed sensor

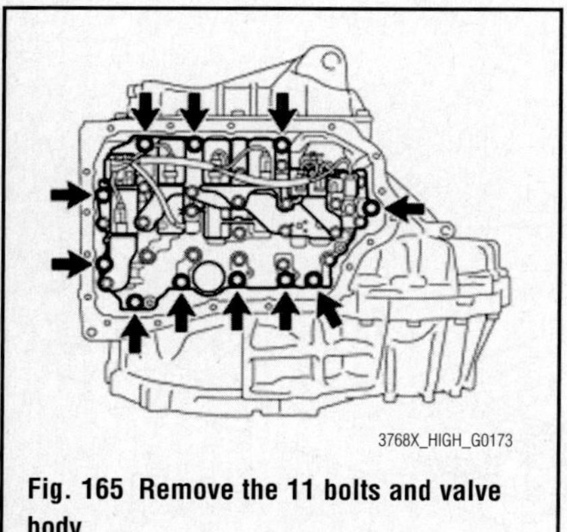

Fig. 165 Remove the 11 bolts and valve body

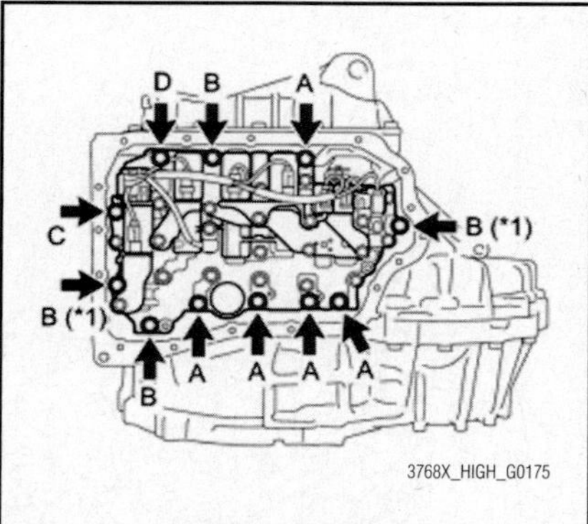

Fig. 167 Installing the valve body

To install:

6. Install the speed sensor to the valve body with the 2 bolts. Tighten to 8 ft. lbs. (11 Nm). Connect the connector.
7. Install transmission valve body assembly:
   a. Coat the O-ring of the transmission wire with ATF.
   b. Confirm that the manual valve lever is positioned as shown in the illustration and install the valve body assembly to the transaxle case with the 11 bolts.
      - Bolt A, B, C: Tighten to 8 ft. lbs. (11 Nm).
      - Bolt D: 85 inch lbs. (7 Nm)
   c. Bolt A: 25 mm (0.984 in.)
   d. Bolt B: 30 mm (1.18 in.)
   e. Bolt C, D: 35 mm (1.38 in.)

➡When installing the transmission valve body assembly, be careful not to allow the speed sensor and transaxle case to interfere with each other. Be sure to insert the pin of the manual valve lever into the groove on the end of the manual valve. First, temporarily tighten the bolts marked by (*1) in the illustration because they are positioning bolts.

8. To complete the installation, reverse the remaining removal procedure.

FUEL — GASOLINE FUEL INJECTION SYSTEM

FUEL SYSTEM SERVICE PRECAUTIONS

Safety is the most important factor when performing not only fuel system maintenance, but any type of maintenance. Failure to conduct maintenance and repairs in a safe manner may result in serious personal injury or death. Work on a vehicle's fuel system components can be accomplished safely and effectively by adhering to the following rules and guidelines.

- To avoid the possibility of fire and personal injury, always disconnect the negative battery cable unless the repair or test procedure requires that battery voltage be applied.
- Always relieve the fuel system pressure prior to disconnecting any fuel system component (injector, fuel rail, pressure regulator, etc.) fitting or fuel line connection. Exercise extreme caution whenever relieving fuel system pressure to avoid exposing skin, face and eyes to fuel spray. Please be advised that fuel under pressure may penetrate the skin or any part of the body that it contacts.
- Always place a shop towel or cloth around the fitting or connection prior to loosening to absorb any excess fuel due to spillage. Ensure that all fuel spillage is quickly removed from engine surfaces. Ensure that all fuel-soaked cloths or towels are deposited into a flame-proof waste container with a lid.
- Always keep a dry chemical (Class B) fire extinguisher near the work area.
- Do not allow fuel spray or fuel vapors to come into contact with a spark or open flame.
- Always use a second wrench when loosening or tightening fuel line connection fittings. This will prevent unnecessary stress and torsion on fuel piping. Always follow the proper torque specifications.
- Always replace worn fuel fitting O-rings with new ones. Do not substitute fuel hose where rigid pipe is installed.

FUEL SYSTEM PRESSURE

RELIEVING

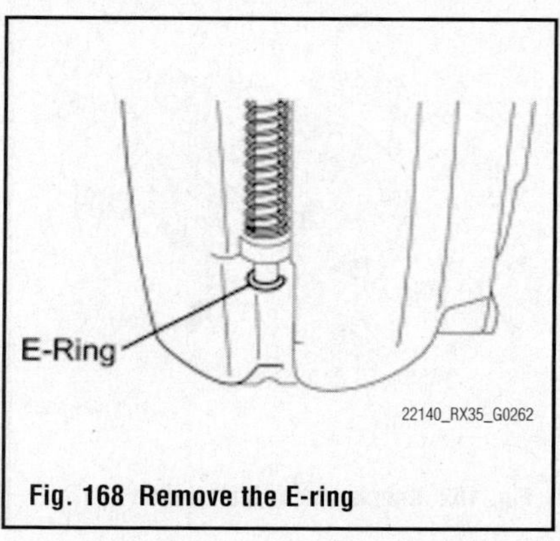

Fig. 168 Remove the E-ring

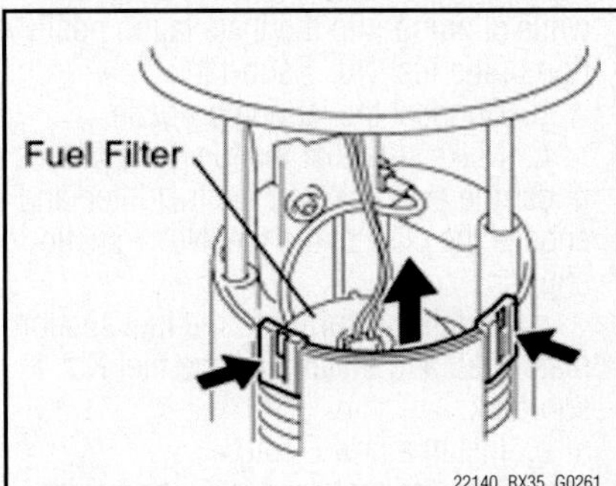

Fig. 169 Remove the fuel suction plate with the fuel filter from the fuel No. 1 sub-tank

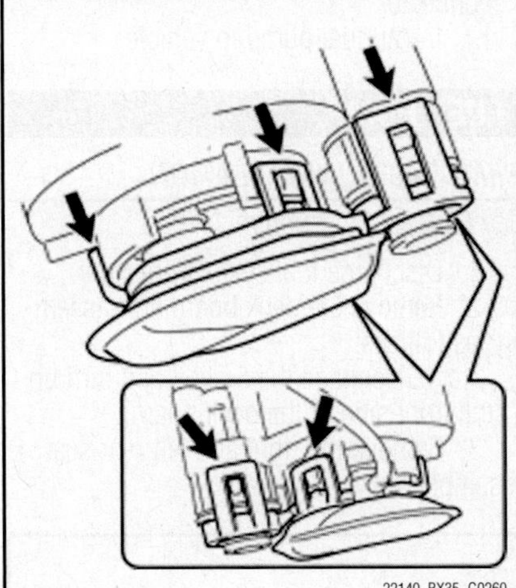

Fig. 170 Disengage the 5 claws on the filter and remove the fuel pump from the fuel filter

⁂ CAUTION

Perform the following procedures to prevent fuel from spilling out before removing any fuel system parts.

⁂ CAUTION

Pressure will still remain in the fuel line even after performing the following procedures. When disconnecting the fuel line, cover it with a shop rag or a piece of cloth to prevent fuel from spraying or coming out.

1. Remove the circuit opening relay/;
   a. Disconnect the cable from the negative (-) battery terminal.
   b. Remove the circuit opening relay from the engine room relay block.
   c. Connect the cable to the negative (-) battery terminal.
   d. Start the engine.
   e. After the engine stops, turn the ignition switch off.

➡DTC P0171/25 (fuel problem) and/or P0191/49 (fuel pressure sensor signal error) may be detected.

   f. Crank the engine again. Check that the engine does not start.
   g. Remove the fuel tank cap to discharge pressure from the fuel tank.
   h. Disconnect the cable from the negative (-) battery terminal.
   i. Install the circuit opening relay.
2. Disconnect the fuel pump connector:
   a. Fold back the floor carpet.
   b. Remove the rear floor service hole cover.
   c. Disconnect the fuel pump connector.
   d. Start the engine.
   e. After the engine stops, turn the ignition switch off.

➡DTC P0171/25 (fuel problem) and/or P0191/49 (fuel pressure sensor signal error) may be detected.

   f. Crank the engine again. Check that the engine does not start.
   g. Remove the fuel tank cap to discharge pressure from the fuel tank.
   h. Disconnect the cable from the negative (-) battery terminal.
   i. Reconnect the fuel pump connector.
   j. Install the rear floor service hole cover.
   k. Install the floor carpet.
3. Check that there are no fuel leaks from the fuel system after doing any maintenance or repairs.

FUEL FILTER

REMOVAL & INSTALLATION

See Figures 168 through 175.

1. Remove the fuel pump from the vehicle.
2. Remove the fuel sender gauge assembly, as follows:
   a. Disconnect the connector and remove the fuel sender gauge from the fuel suction tube.
3. Remove the fuel suction plate sub-assembly, as follows:
   a. Using needle nozzle pliers, remove the E-ring.
   b. Disengage the 2 claws of the fuel No. 1 suction support and remove the fuel suction plate with the fuel filter from the fuel No. 1 sub-tank.
   c. Remove the spring from the fuel suction plate.

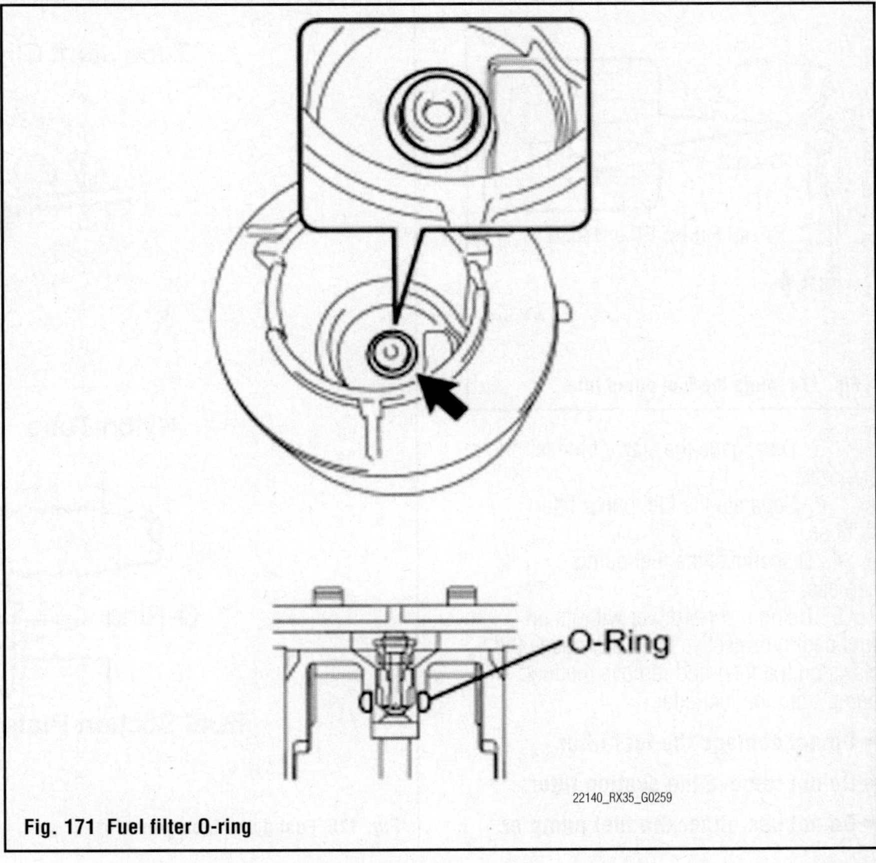

Fig. 171 Fuel filter O-ring

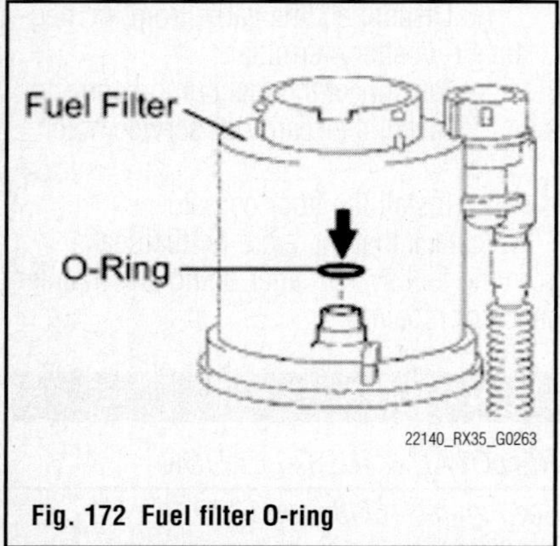

Fig. 172 Fuel filter O-ring

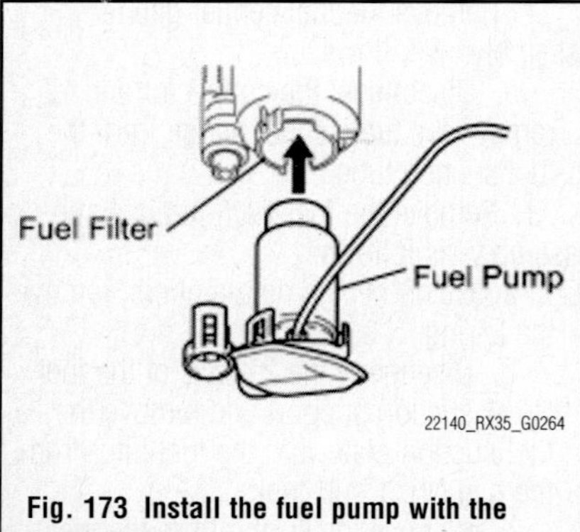

Fig. 173 Install the fuel pump with the pump filter

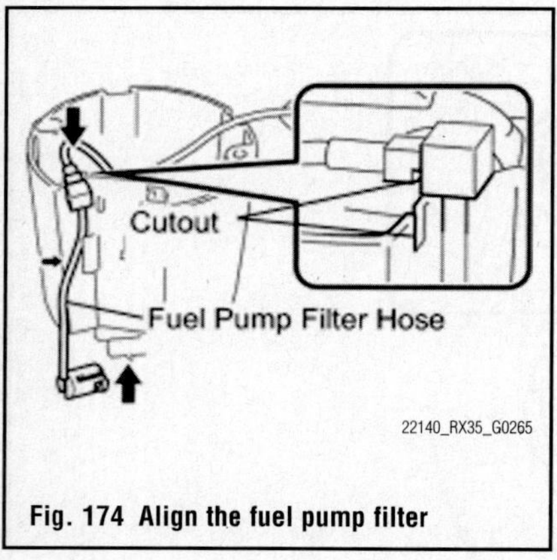

Fig. 174 Align the fuel pump filter

d. Disengage the claw of the jet pump nozzle.

e. Separate the fuel pump filter hose.

4. Disconnect the fuel pump harness.

5. Using a screwdriver with its tip wrapped in protective tape, disengage the 5 claws on the filter and remove the fuel pump from the fuel filter.

➡**Do not damage the fuel filter.**

➡**Do not remove the suction filter.**

➡**Do not use either the fuel pump or the suction filter if the suction filter is removed from the fuel pump.**

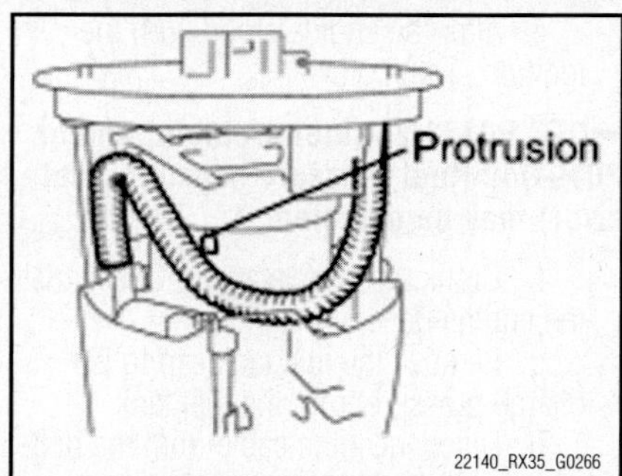

Fig. 175 Make sure that the fuel tube passes under the protrusion of the fuel filter

6. Remove the O-ring from the fuel filter.

To install:

7. Apply gasoline to a new O-ring and install it to the fuel filter.

8. Engage the 5 claws on the fuel filter and install the fuel pump with the pump filter.

9. Connect the fuel pump harness connector.

10. Install the fuel suction plate sub-assembly, as follows:

a. Install the fuel pump filter tube while aligning it to the installation position of the fuel No. 1 sub-tank.

b. Connect the jet pump nozzle.

c. Make sure that the fuel tube passes under the protrusion of the fuel filter, and engage the claw of the fuel No. 1 suction support.

d. Install the spring to the fuel suction plate shaft and install it to the fuel No. 1 sub-tank.

e. Install a new E-ring.

11. Install the fuel sender gauge assembly, as follows;

a. Slide the fuel sender gauge to fit the claw.

b. Connect the fuel sender gauge connector.

12. Install fuel pump to vehicle.

FUEL PUMP

REMOVAL & INSTALLATION

See Figures 176 through 182.

1. Discharge fuel system pressure.

2. Remove the deck board sub-assembly, as follows:

a. Disengage the 5 clips and turn up the front side of the deck board.

3. Remove the right and left rear seat assemblies.

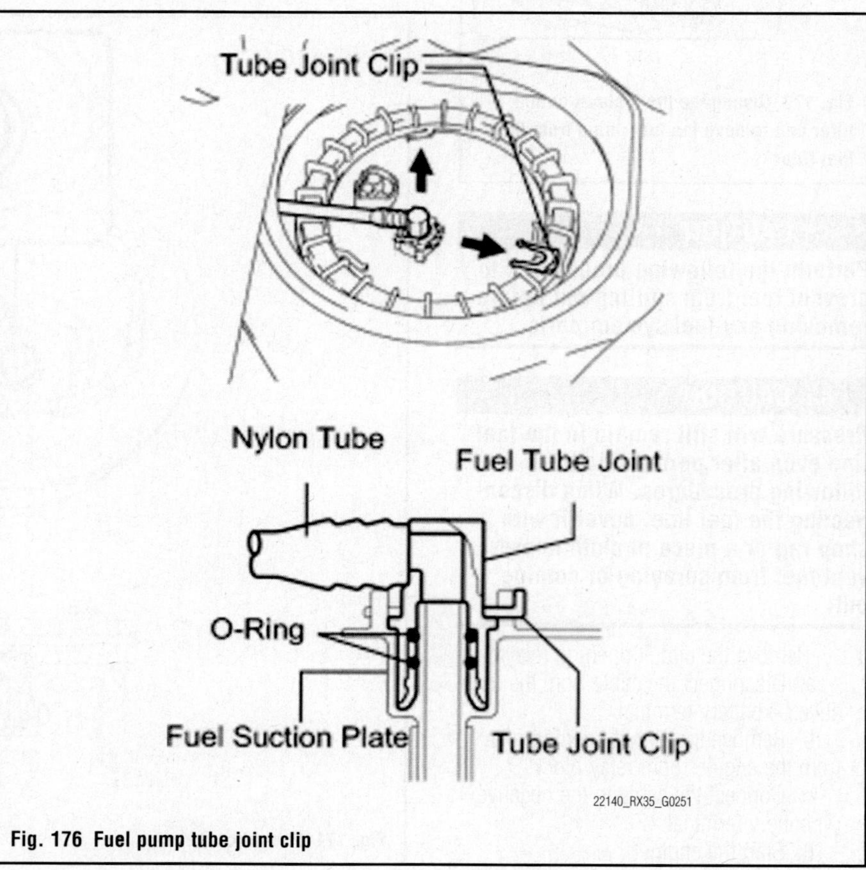

Fig. 176 Fuel pump tube joint clip

4. Remove the left rear door scuff plate.
5. Remove the left deck side trim cover.
6. Remove the left rear seat side cover.
7. Remove the rear floor service hole cover, as follows:
a. Using a clip remover, remove the 2 clips and tear off the front floor carpet.
b. Remove the rear floor service hole cover.
c. Disconnect the fuel pump connector.
8. Remove the fuel suction tube assembly with pump and gauge, as follows:
a. Disconnect the fuel pump tube by removing the tube joint clip, and pull out the fuel pump tube.

➡Check if there is any dirt or mud around the connector before this operation and remove the dirt as necessary.

➡Be careful of mud because the quick connector has an O-ring which seals the pipe and connector that can be contaminated.

➡Do not use any tool in this operation.

➡Do not bend or twist the nylon tube. Protect the connector by covering it with a vinyl or plastic bag.

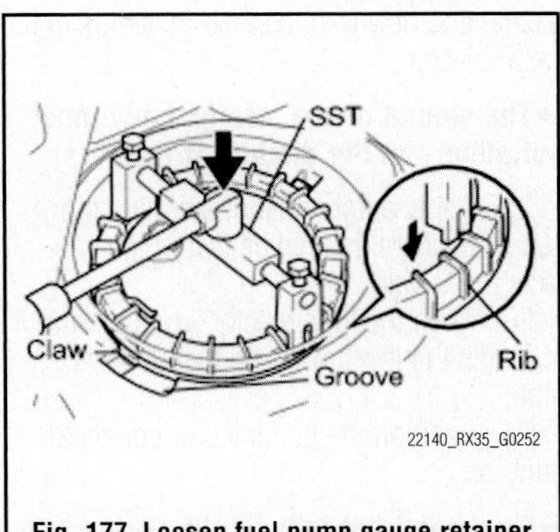

Fig. 177 Loosen fuel pump gauge retainer

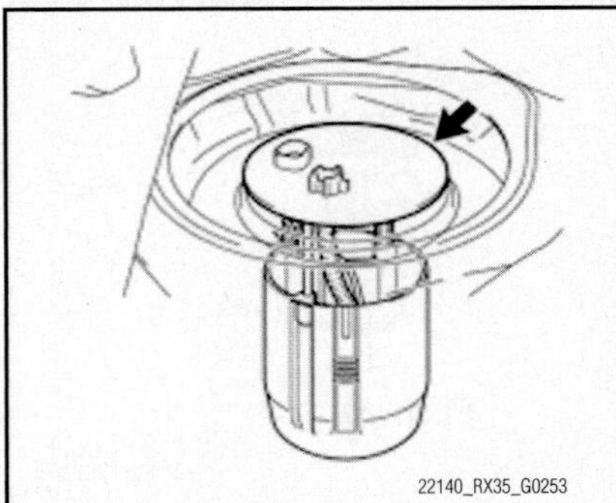

Fig. 178 Remove the fuel suction tube with pump and gauge

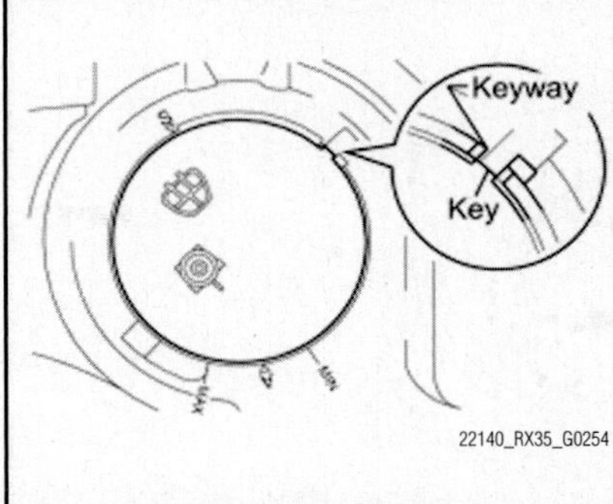

Fig. 179 Install fuel suction tube assembly

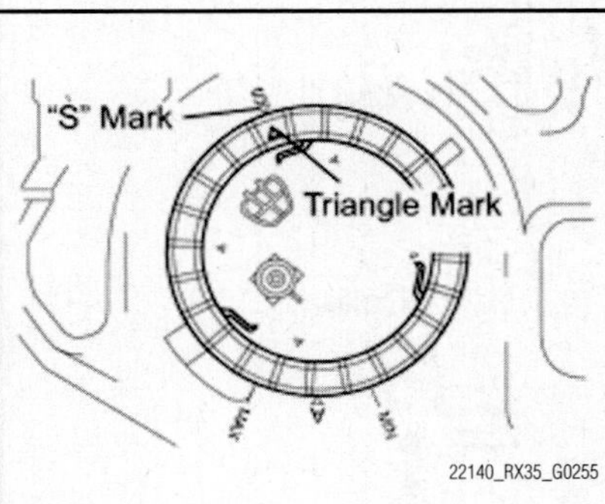

Fig. 180 Align the triangle mark on a new fuel pump

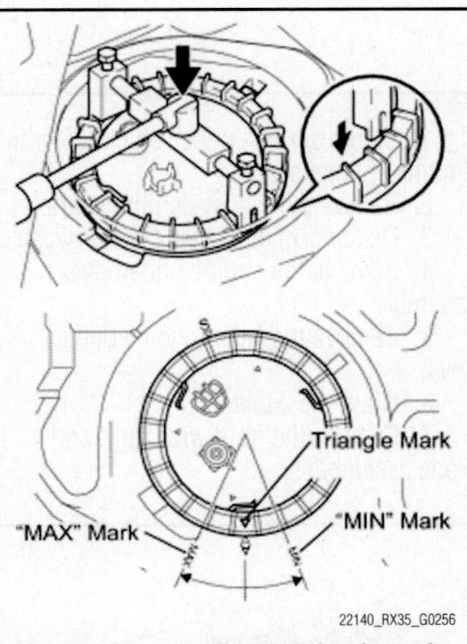

Fig. 181 Properly align the triangle mark

➡When the pipe and connector are stuck, push and pull the connector to release and pull the connector out carefully.

b. Using SST (SST: 09808-14020) or equivalent, loosen the fuel pump gauge retainer.

➡Loosen the retainer by turning it counterclockwise while holding the SST down. Do not allow the claw of the tank suction tube support to slip out of its groove on the fuel tank.

➡A rib on the fuel pump gauge retainer can be fitted into a tip of the SST.

c. Remove the fuel pump gauge retainer.
d. Remove the fuel suction tube with pump and gauge.

➡Be careful not to bend the arm of the fuel sender gauge.

e. Remove the gasket from the fuel tank.

To install:

9. Install the fuel suction tube assembly with pump and gauge, as follows:
a. Install a new gasket to the fuel tank.
b. Attach the fuel suction tube with pump and gauge to the fuel tank.

➡Be careful not bend the arm of the fuel sender gauge.

c. Align the keyway of the fuel suction tube support with the key of the fuel suction tube with pump and gauge.
d. Align the triangle mark on a new fuel pump gauge retainer with the "S" mark on the fuel tank while pushing down the fuel suction tube with pump and gauge, and attach the fuel pump gauge retainer.
e. Rotate the fuel pump gauge retainer by hand, then tighten it one complete turn and another half turn using SST (SST: 09808-14020). The triangle mark on the fuel pump gauge retainer must be positioned between the "MIN" and "MAX" marks on the fuel tank.

➡Do not use other tools in this operation. Damage to the fuel pump gauge retainer and the fuel tank may result.

➡A rib on the fuel pump gauge retainer can be fitted into a tip of the SST.

f. Connect the fuel pump tube by installing the fuel pump tube and the tube joint clip.

➡Check that there are no scratch or foreign objects on the connecting part.

➡Check that the fuel tube joint is inserted securely.

➡Check that the tube joint clip is on the collar of the fuel tube joint.

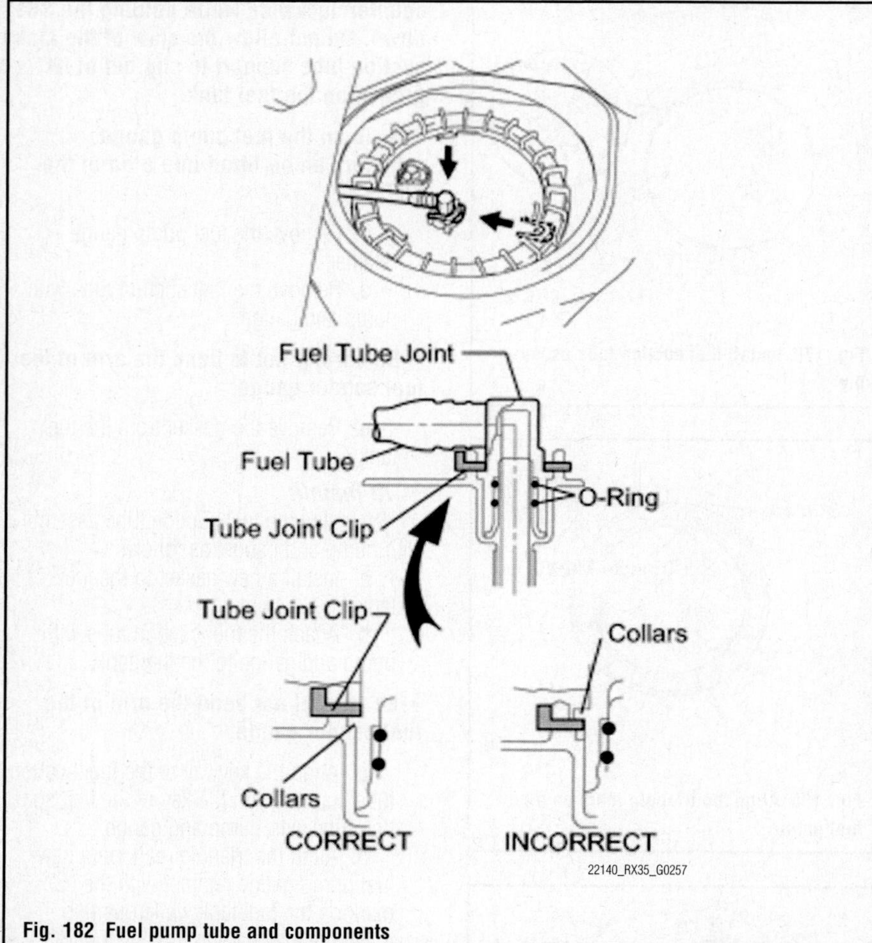

Fig. 182 Fuel pump tube and components

➡**After installing the tube joint clip, check that the fuel tube joint has not been pulled off.**

10. Inspect for fuel leak.
11. Install the rear floor service hole cover, as follows:
    a. Install a new butyl tape to the rear floor service hole cover.
    b. Connect the fuel pump connector.
    c. Install the rear floor service hole cover.
    d. Install the front floor carpet with the 3 clips.
12. Install the left rear seat side cover.
13. Install the left deck side trim cover.
14. Install the left rear door scuff plate.
15. Install the right and left rear seat assemblies.
16. Install the deck board sub-assembly.

FUEL RAIL AND INJECTOR

REMOVAL & INSTALLATION

See Figures 183 and 184.

1. Before servicing the vehicle, refer to the precautions section.
2. Relieve the fuel system pressure.
3. Disconnect the negative battery cable.
4. Remove the engine undercover assembly.
5. Remove the No. 1 engine under-cover.
6. Drain the coolant.
7. Remove the front wiper arm and blade assemblies.
8. Remove the cowl top ventilator louver sub-assembly.
9. Remove the wiper motor and link assembly.
10. Remove the outer cowl top panel sub-assembly.
11. Remove the V-bank cover sub-assembly.
12. Remove the air cleaner cap sub-assembly.
13. Disconnect the engine room main wire.
14. Remove the throttle body bracket.
15. Remove the No. 1 surge tank stay.
16. Remove the intake air surge tank assembly.
17. Disconnect the fuel tube sub-assembly.
18. Remove the fuel delivery pipe sub-assembly.
19. Pull out the fuel injectors from the fuel delivery pipe.

➡**If the injectors are to be reused, reinstall them to the same cylinder they came from.**

20. Remove the 6 O-rings from the injectors.

To install:

21. Apply a light coat of spindle oil or gasoline to new O-rings, and install them to each injector.

➡**The wound or the foreign body must not adhere in the ditch of O-ring.**

22. Apply a light coat of spindle oil or gasoline where the fuel delivery pipe contacts the O-ring.
23. Push the fuel injector while turning it to install the injector in the fuel delivery pipe.
24. Position the fuel injector connector outward.

➡**Be careful not to twist the O-ring.**

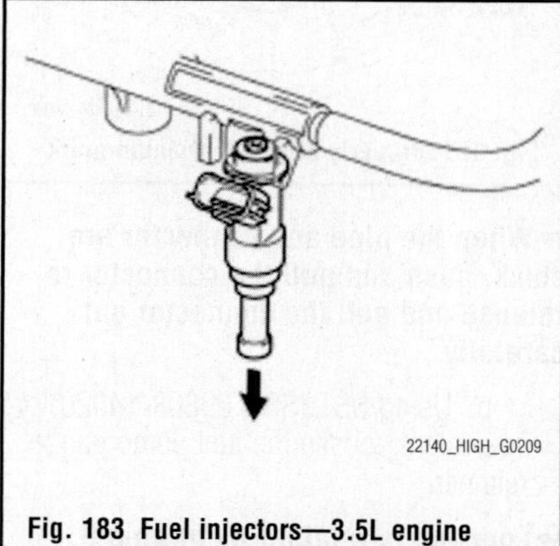

Fig. 183 Fuel injectors—3.5L engine

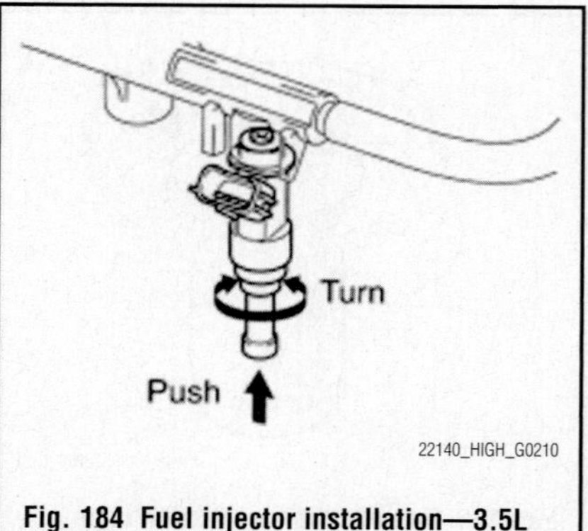

Fig. 184 Fuel injector installation—3.5L engine

➡After installing the fuel injector, check that it turns smoothly. If not, reinstall it with a new O-ring.

25. Install the fuel delivery pipe sub-assembly.
26. Connect the fuel tube sub-assembly.
27. Temporarily install the No. 1 surge tank stay.
28. Temporarily install the throttle body bracket.
29. Install the intake air surge tank assembly.
30. Fully tighten the No. 1 surge tank stay.
31. Fully tighten the throttle body bracket.
32. Connect the engine room main wire.
33. Install the air cleaner cap sub-assembly.
34. Connect the negative battery cable.
35. Inspect the SRS warning light.
36. Inspect for fuel leak.
37. Add engine coolant.
38. Inspect for engine coolant leak.
39. Install the V-bank cover sub-assembly.
40. Install the outer cowl top panel sub-assembly.
41. Install the windshield wiper motor and link assembly.
42. Install the cowl top ventilator louver sub-assembly.
43. Install both front wiper arm and blade assemblies.
44. Install the No. 1 engine undercover.
45. Install the engine undercover assembly.

FUEL TANK

REMOVAL & INSTALLATION

1. Before servicing the vehicle, refer to the Precautions Section.
2. Discharge the fuel pressure.
3. Disconnect the negative battery cable.
4. Remove fuel from fuel tank
5. Remove the center exhaust pipe.
6. Remove the front floor heat insulator.
7. Remove the fuel tank protector.
8. Remove the nut and disconnect the parking brake cable assembly
9. Separate the fuel tank wire connector from the bracket.
10. Disconnect the 2 connectors.
11. Disconnect the clamp from the charcoal canister protector.
12. Remove the 3 bolts and charcoal canister protector.
13. Disconnect the charcoal canister fuel hose from the charcoal canister assembly.
14. Pinch the retainer and pull out the quick connector with the quick connector pushed to the pipe side to disconnect the charcoal canister fuel hose from the charcoal canister assembly.
15. Disconnect the main fuel supply line. Pinch the tabs of the retainer to disengage the lock claws and pull the retainer down as shown in the illustration.
16. Pinch the retainer of the breather lower tube connector, and pull out the breather lower tube connector to disconnect the breather lower tube from the fuel tank.
17. Set a transmission jack to the fuel tank.
18. Remove the fuel filler pipe clamp and fuel tube connector from the fuel tank inlet pipe.
19. Remove the 4 bolts and the fuel tank bands.
20. Operate the transmission jack, and disconnect the fuel inlet pipe.
21. Operate the transmission jack, and remove the fuel tank.

To install:

22. Install new fuel tank cushions.
23. Install the 2 fuel tube clamps of the fuel tank main tube to the fuel tank.
24. Install the fuel tank main fuel assembly.
25. Install the fuel tank wire.
26. Connect the fuel tank vent hose to the fuel tank
27. Install the 2 nuts to the fuel vapor containment valve, and tighten the nuts to 71 inch lbs. (8 Nm).
28. Connect the connector and clamp.
29. Install a new gasket on the fuel pump assembly.
30. Install the fuel pump assembly.
31. Install the fuel tank vent tube set plate and fuel pump connector plate with the 8 bolts. Tighten the bolts to 53 inch lbs. (6 Nm).
32. Connect the fuel tank main tube with the tube joint clip and clamp.
33. Connect the fuel tank wire connector.
34. Connect the vapor pressure sensor connector and clamp.
35. Set the fuel tank on a transmission jack.
36. Operate the transmission jack, and install the fuel tank to the vehicle.
37. Operate the transmission jack, and connect the fuel tank inlet pipe with the tube connector and clamp.
38. Install the 2 fuel tank bands with the 4 bolts and tighten to 29 ft. lbs. (39 Nm).
39. Connect the breather lower tube. Push in the tube connector to the pipe until the tube connector makes a "click" sound.
40. Install the checker to the pipe.
41. Connect the fuel tank main supply line. Push in the fuel tube connector to the pipe and push up the retainer to engage the claws.
42. Align the quick connector with the pipe, then push in the quick connector until the retainer makes a "click" sound to connect the charcoal canister fuel hose to the charcoal canister assembly.
43. Connect the clamp to the charcoal canister protector.
44. Install the charcoal canister protector with the 3 bolts. Tighten the bolts to 48 inch lbs. (5.4 Nm).
45. Connect the fuel tank wire connectors, and install them to the bracket.
46. Install the parking brake cable assembly No. 3 with the nut.
47. Install the fuel tank protector.
48. Install the front floor heat insulator and tighten the retaining nuts to 43 inch lbs. (4.9 Nm).
49. Install the center pipe exhaust assembly.
50. Connect the negative battery cable.
51. Add fuel and inspect for fuel leaks.
52. Inspect for exhaust leaks.
53. Some systems need initialization after reconnecting the cable to the negative battery terminal.

IDLE SPEED

ADJUSTMENT

Idle speed is maintained by the Powertrain Control Module (PCM). No adjustment is necessary or possible.

THROTTLE BODY

REMOVAL & INSTALLATION

See Figure 185.

1. Remove the windshield wiper link assembly.
2. Remove the cowl top panel outer sub-assembly.
3. Drain engine coolant.
4. Remove the V-bank cover sub-assembly.
5. Remove the No. 2 air cleaner inlet.
6. Remove the No. 1 air cleaner inlet.
7. Remove the air cleaner cap sub-assembly.

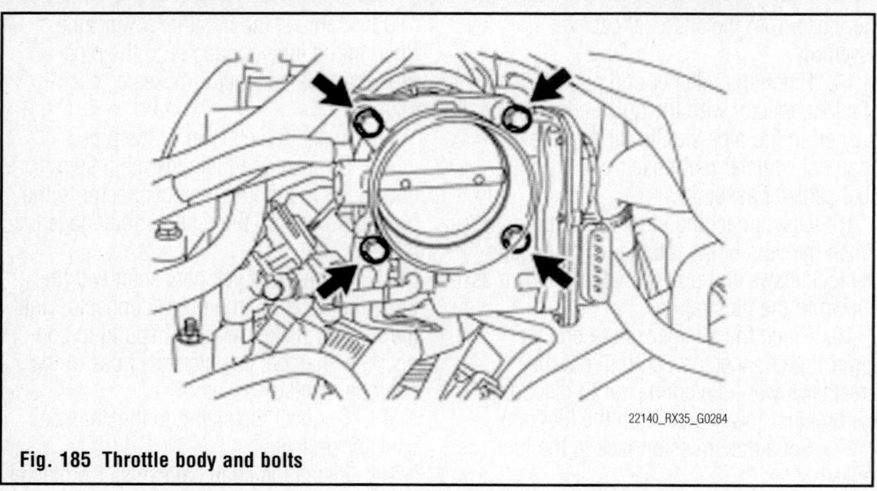

Fig. 185 Throttle body and bolts

8. Remove the air cleaner case sub-assembly.

9. Remove the throttle body, as follows:

a. Disconnect the throttle body connector and clamp.

b. Disconnect the 2 water by-pass hoses from the throttle Body.

c. Remove the 4 bolts and throttle body.

d. Remove the throttle body gasket from the intake air surge tank.

To install:

10. Installation is the reverse of the removal procedure. Torque the throttle body bolts to 7 ft. lbs. (10 Nm).

11. Inspect for coolant leak.

HEATING & AIR CONDITIONING SYSTEM

BLOWER MOTOR

REMOVAL & INSTALLATION

See Figure 186.

1. Remove the No. 2 instrument panel undercover sub-assembly.

2. Remove front blower motor sub-assembly, as follows:

a. Disconnect the connector.

b. Remove the 3 screws and the front blower motor sub-assembly.

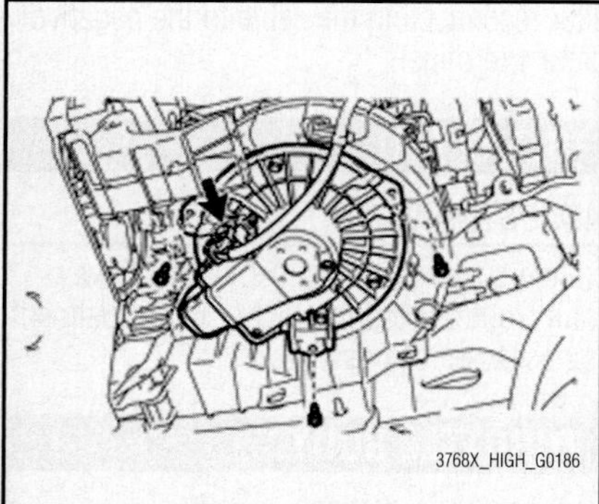

Fig. 186 Remove the 3 screws and the front blower motor sub-assembly

To install:

3. Installation is the reverse of the removal procedure.

HEATER CORE

REMOVAL & INSTALLATION

See Figures 187 through 201.

1. Discharge refrigerant from refrigeration system.

2. Disconnect battery negative terminal.

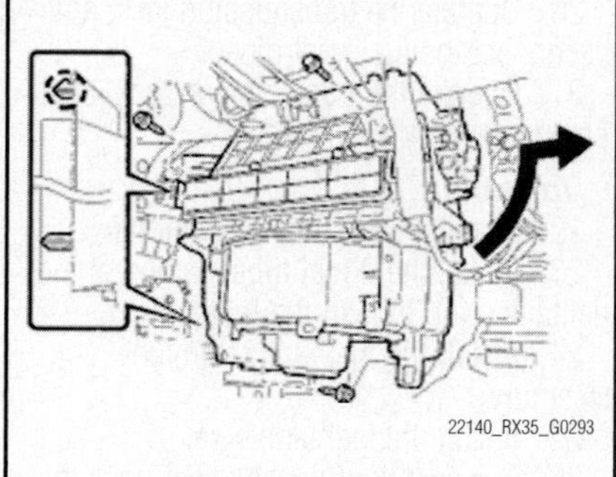

Fig. 187 Remove the piping clamp

3. Remove both front wiper arm and blade assemblies.

4. Remove the cowl top ventilator louver sub-assembly.

5. Remove the windshield wiper link assembly.

6. Remove the cowl top panel outer sub-assembly.

7. Disconnect the air conditioning tube and accessory assembly, as follows:

a. Install SST (SST: 09870-00025) on the piping clamp.

➡Make sure the direction of the piping clamp and SST by seeing the illustration shown on the caution label.

b. Push down SST and release the clamp lock. Do not deform the tube when pushing the SST.

c. Pull the SST slightly and push the release lever, and then remove the piping clamp with SST.

d. Disconnect the air conditioning tube and accessory assembly by hand.

➡Do not use tools like a screwdriver to remove the tube.

➡Cap the open fittings immediately to keep moisture or dirt out of the system.

e. Remove the 2 O-rings from the air conditioning tube and accessory assembly.

8. Disconnect the No. 1cooler refrigerant suction hose, using SST: 09870-00015, and the same procedure as for removing the air conditioning tube and accessory assembly.

9. Using pliers, grip the claws of the clip and slide the clip, and then disconnect the heater water outlet hose. Do not apply any excessive force to the heater water outlet hose.

➡Prepare a drain pan or cloth for cooling water leaks.

10. Using the same procedure as for the heater water outlet hose, disconnect the heater water inlet hose.

11. Remove the instrument panel assembly.

12. Remove the air conditioner amplifier assembly.

13. Turn back the floor carpet, release the claw and remove the rear No. 1 and 2air ducts.

14. Remove the bolt, release the 3 claws and remove the No. 1 air duct.

15. Disconnect the transmission control cable assembly.

16. Remove the shift lever.

17. Remove the 4 nuts and instrument panel bracket No. 4.

18. Remove instrument panel brace No. 1 sub-assembly, as follows:

a. Remove the clamp and disconnect the connector.

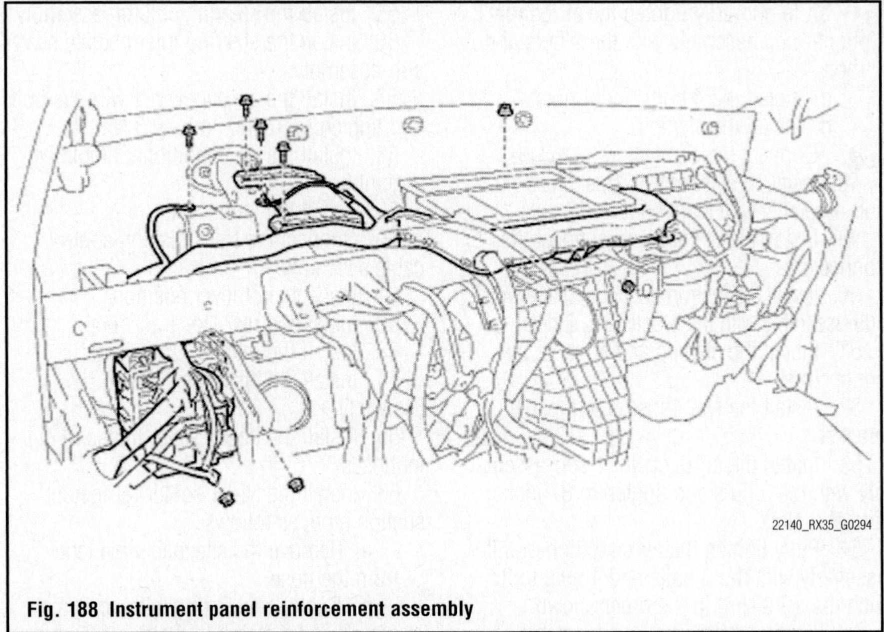

Fig. 188 Instrument panel reinforcement assembly

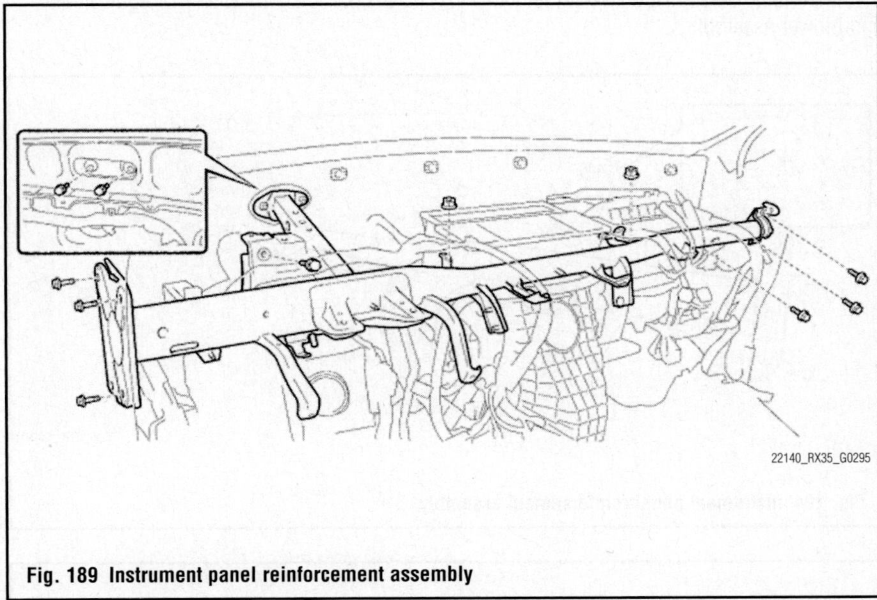

Fig. 189 Instrument panel reinforcement assembly

b. Remove the 3 bolts and 2 nuts and instrument panel brace No. 1 sub-assembly.

19. Remove instrument panel brace No. 2 sub-assembly, as follows:

a. Remove the clamp and disconnect the connector.

b. Remove the 3 bolts and 2 nuts and instrument panel brace No. 2 sub-assembly.

20. Separate steering intermediate shaft sub-assembly.

21. Remove steering column assembly.

22. Remove the 3 nuts and air duct No. 1 sub-assembly.

23. Remove instrument panel reinforcement assembly, as follows:

a. Remove the 4 bolts and 4 nuts.

b. Remove the clamp.

c. Disconnect the connectors.

d. Remove the 9 bolts and the 2 nuts and then remove the instrument panel reinforcement while holding the air conditioner unit assembly.

➡**Make sure to hold the air conditioner unit assembly securely as its bracket installation parts may be damaged.**

24. Remove the 2 nuts and air conditioner unit assembly.

⁂ WARNING

Make sure to hold the air conditioner unit assembly securely as its bracket installation parts may be damaged.

25. Remove the air duct No. 2, as follows:

a. Remove the screw.

b. Release the 2 claws and remove the air duct No. 2.

26. Remove the wiring air indicator harness No .2 sub-assembly.

27. Remove the blower assembly.

28. Release the 4 claws and remove the center heater to register duct.

29. Remove the 3 screws and air outlet control servomotor.

30. Remove the air mix control servomotor.

31. Remove the evaporator temperature sensor.

32. Remove the heater radiator unit sub-assembly, as follows:

a. Remove the 2 screws and piping cover.

b. Remove the screw and bracket.

c. Remove the heater radiator unit sub-assembly.

➡**Prepare a drain pan or cloth for cooling water leaks.**

To install:

33. Install the heater radiator unit sub-assembly.

34. Install the bracket with the screw.

35. Install the piping cover with the 2 screws.

36. Engage the 2 claws to install the evaporator temperature sensor.

37. Engage the clamp to install the connector.

38. Install the air mix control servomotor with the 3 screws.

39. Install the air outlet control servomotor with the 3 screws.

40. Engage the 4 clamps to install the heater to register duct center.

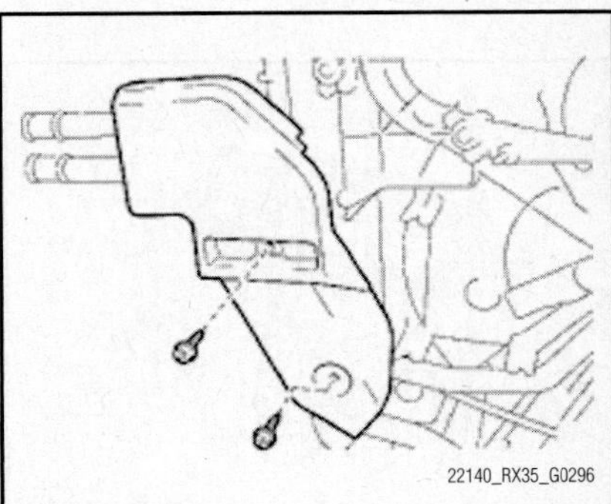

Fig. 190 Remove the 2 screws and piping cover

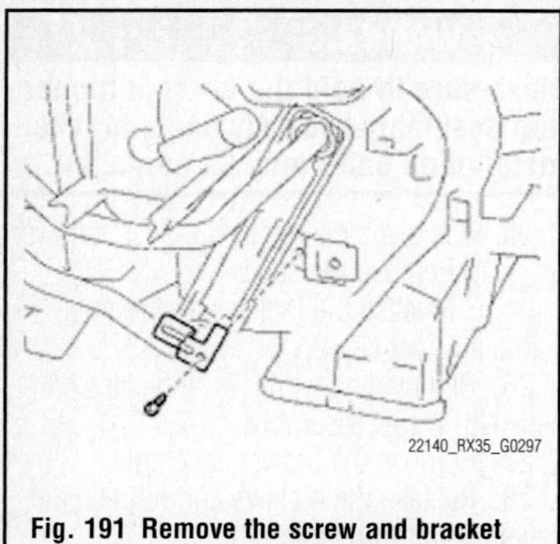

Fig. 191 Remove the screw and bracket

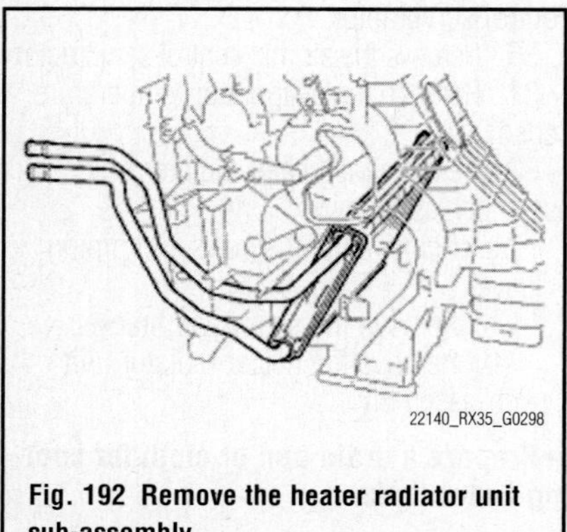

Fig. 192 Remove the heater radiator unit sub-assembly

41. Engage the claw to install the blower assembly.

42. Install the wiring air indicator harness No. 2 sub-assembly with the 3 screws.

43. Connect the connectors and clamp.

44. Install the air duct No. 2 with the screw and tighten to 22 inch lbs. (2.5 Nm).

45. Temporarily tighten the air conditioner unit assembly with the 2 nuts. Hold the A/C unit securely to prevent damaging its bracket installation parts.

46. Install the instrument panel reinforcement assembly, as follows:

a. Install the instrument panel reinforcement with the 8 bolts and tighten to 15 ft. lbs. (20 Nm).

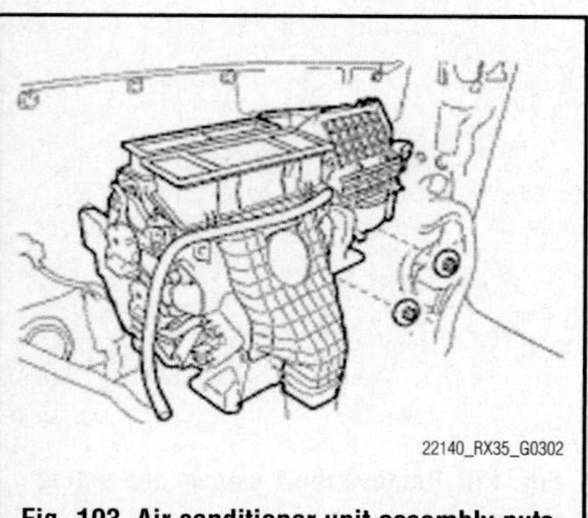

Fig. 193 Air conditioner unit assembly nuts

b. Temporarily tighten the air conditioner unit assembly with the 2 nuts and bolt.

c. Install the 4 bolts and 4 nuts.

d. Install the clamps.

47. Connect the connectors.

48. Install the instrument panel brace No. 2 sub-assembly with the 3 bolts and 2 nuts.

49. Install the clamps and connect the connectors.

50. Install the instrument panel brace No. 1 sub-assembly with the 3 bolts and 2 nuts.

51. Install the clamps and connect the connectors.

52. Install the No. 4 instrument panel bracket.

53. Install the air duct No. 1 sub-assembly with the 3 nuts and tighten to 87 inch lbs. (9.8 Nm).

54. Fully tighten the air conditioner unit assembly with the 3 bolts and 4 nuts to 87 inch lbs. (9.8 Nm) in the order shown. Install the air conditioner unit assembly so that there is no clearance between the duct and blower assemblies.

55. Install the steering column assembly.

56. Install the steering intermediate shaft sub-assembly

57. Install the air duct No. 1 with the bolt and tighten to 87 inch lbs. (9.8 Nm).

58. Install the air conditioner amplifier assembly.

59. Install the shift lever.

60. Connect the transmission control cable assembly.

61. Inspect shift lever position.

62. Install the rear No. 1 air duct.

63. Install the rear No. 2 air duct.

64. Install the instrument panel assembly.

65. Install the heater water inlet and outlet hoses.

66. Install the No. 1 cooler refrigerant suction pipe, as follows:

a. Remove the attached vinyl tape from the hose.

b. Coat a new O-ring with compressor oil and install it to the hose. Compressor oil: ND-OIL 11 or equivalent.

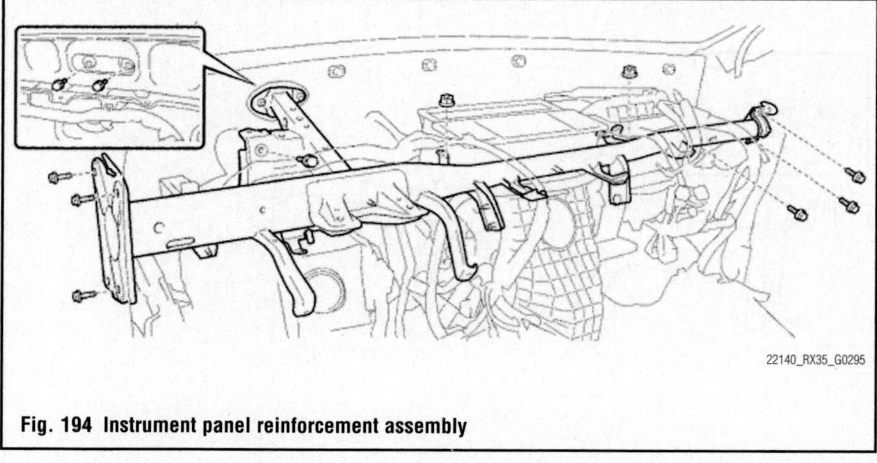

Fig. 194 Instrument panel reinforcement assembly

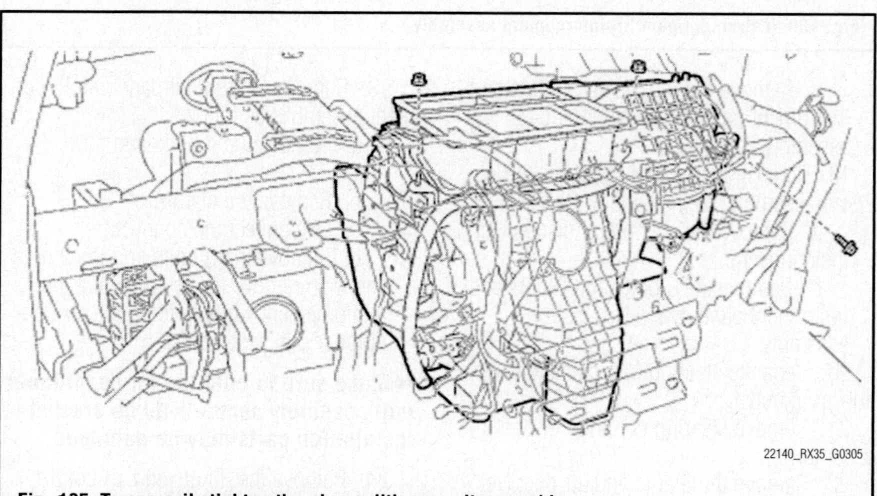

Fig. 195 Temporarily tighten the air conditioner unit assembly

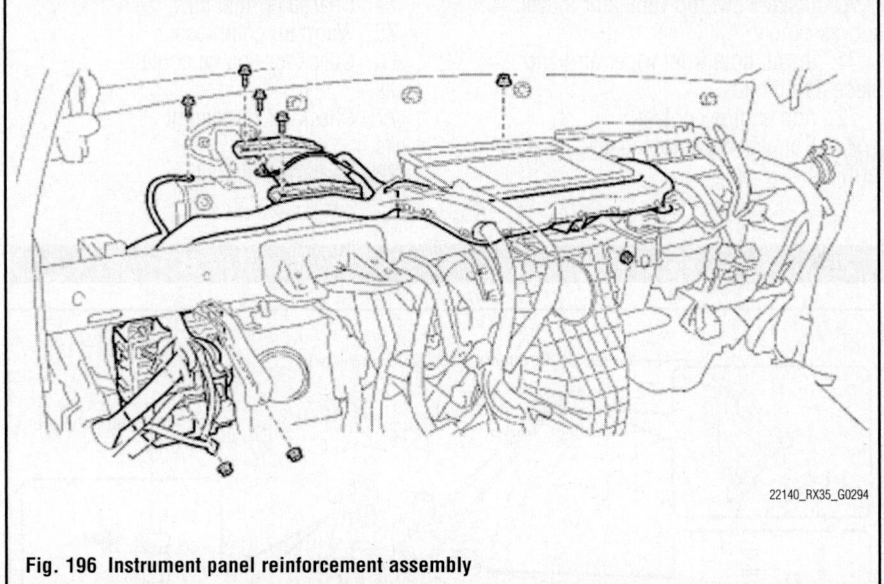

Fig. 196 Instrument panel reinforcement assembly

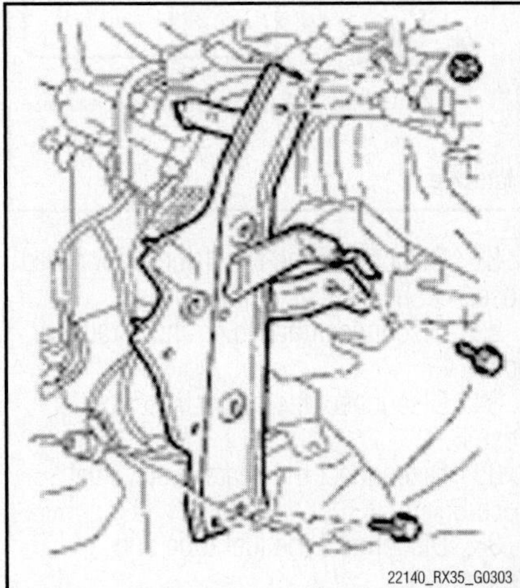

Fig. 197 Instrument panel brace No. 2 sub-assembly

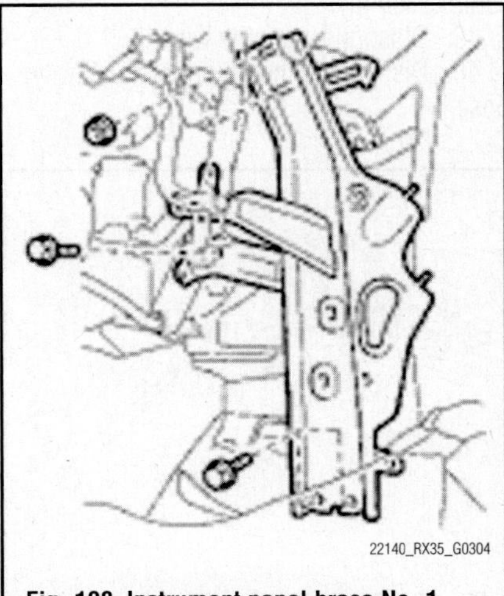

Fig. 198 Instrument panel brace No. 1 sub-assembly

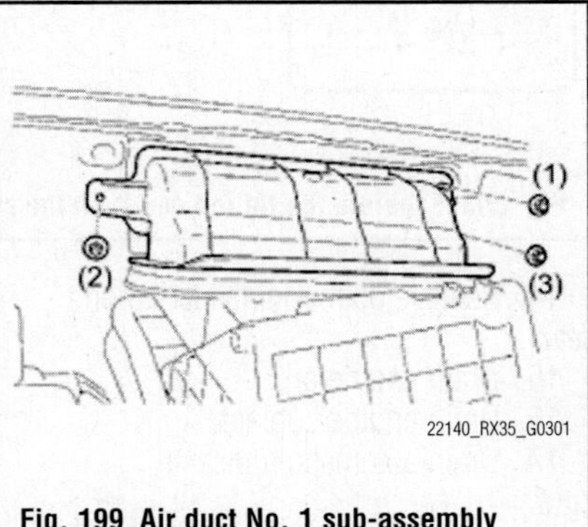

Fig. 199 Air duct No. 1 sub-assembly tightening sequence

WARNING

Do not use any compressor oil other than ND-OIL 11 or equivalent. If any compressor oil other than ND-OIL 11 or equivalent is used, compressor motor insulation performance may decrease, resulting in leakage of electric power.

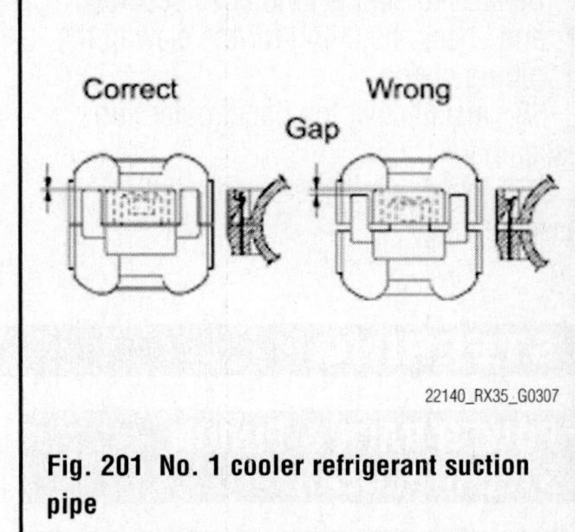

Fig. 201 No. 1 cooler refrigerant suction pipe

c. Install the No. 1 cooler refrigerant suction pipe and piping clamp. Be sure to connect the hose securely, and check the fitting for the claw of the piping clamp.

67. Install the air conditioning tube and accessory assembly, as follows:

a. Remove the attached vinyl tape from the hose.

b. Coat a new O-ring with compressor oil and install it to the pipe. Compressor oil: ND-OIL 11 or equivalent.

WARNING

Do not use any compressor oil other than ND-OIL 11 or equivalent. If any compressor oil other than ND-OIL 11 or equivalent is used, compressor motor insulation performance may decrease, resulting in leakage of electric power.

c. Install the air conditioning tube and accessory assembly and piping clamp.

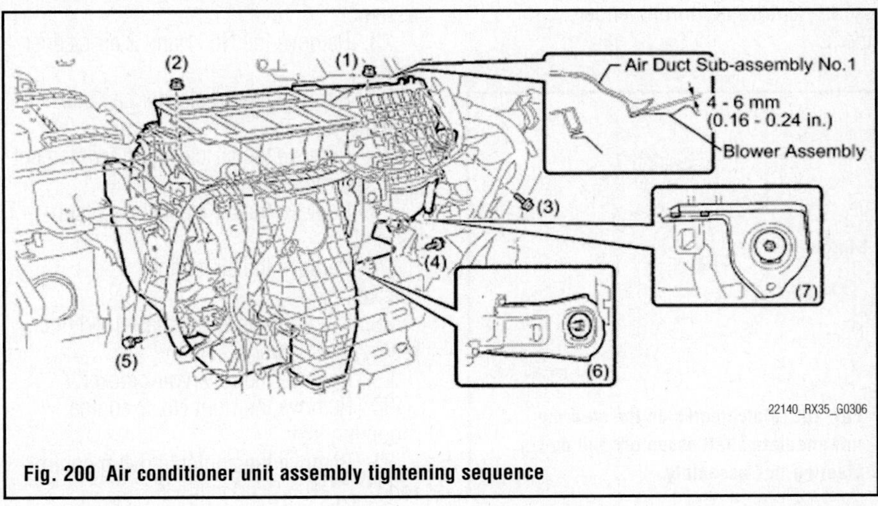

Fig. 200 Air conditioner unit assembly tightening sequence

Be sure to connect the hose securely, and check the fitting for the claw of the piping clamp.

68. Install cowl top panel outer sub-assembly.
69. Install windshield wiper link assembly.
70. Install cowl top ventilator louver sub-assembly.
71. Install both front wiper arm and blade assemblies.
72. Add engine coolant.
73. Connect cable to negative battery terminal.
74. Charge refrigerant.
75. Warm up compressor.
76. Check for engine coolant leaks.
77. Check for refrigerant leaks.
78. Perform initialization.

STEERING

POWER RACK & PINION STEERING GEAR

REMOVAL & INSTALLATION

See Figures 202 through 204.

1. Before servicing the vehicle, refer to the precautions section.
2. Discharge fuel system pressure.
3. Recover refrigerant from refrigeration system.
4. Remove the cool air intake duct seal.
5. Remove the battery.
6. Place the front wheels facing straight ahead.
7. Secure the steering wheel with the seat belt in order to prevent it from rotating.

**** WARNING**

This operation is necessary to prevent damage to the spiral cable.

8. Remove the front wheels.
9. Remove the engine undercover assembly.
10. Remove the No. 1 and 2 engine undercovers.
11. Remove the left hand floor undercover.
12. Remove both front fender molding sub-assemblies.
13. Remove both front fender liners.

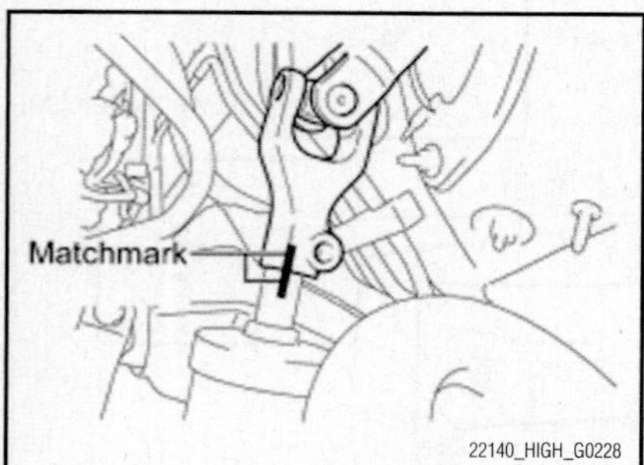

Fig. 202 Matchmarks on the steering intermediate shaft assembly and power steering link assembly

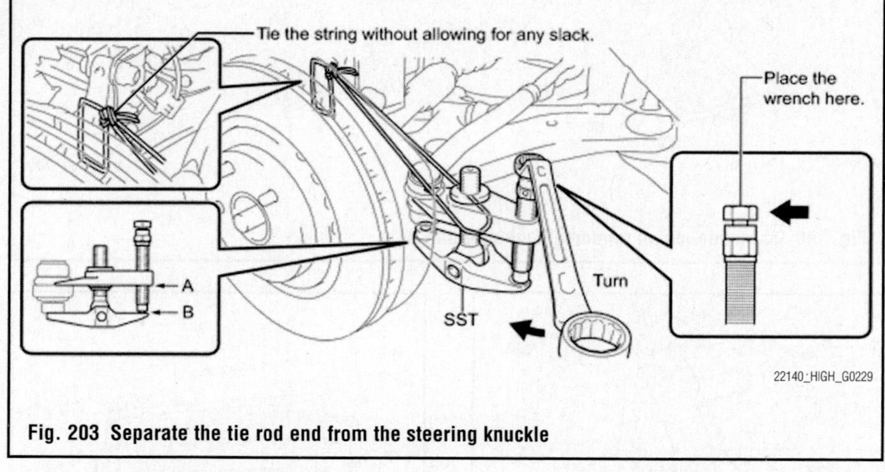

Fig. 203 Separate the tie rod end from the steering knuckle

14. Remove both front fender apron seals.
15. Drain engine oil.
16. Drain engine coolant.
17. Drain automatic transaxle fluid.
18. Remove both front wiper arm and blade assemblies.
19. Remove the cowl top ventilator louver sub-assembly.
20. Remove the windshield wiper motor and link assembly.
21. Remove the outer cowl top panel sub-assembly.
22. Remove the V-bank cover sub-assembly.
23. Remove the No. 1 and 2 air cleaner inlets.
24. Remove the air cleaner cap sub-assembly.
25. Remove the air cleaner filter element sub-assembly.
26. Remove the air cleaner case sub-assembly.
27. Remove the air cleaner bracket.
28. Separate the brake master cylinder reservoir assembly.
29. Remove the reservoir bracket.
30. Remove the right No. 2 engine mounting stay.
31. Remove the engine moving control rod.
32. Disconnect the No. 1 fuel vapor feed hose.
33. Disconnect the No. 1 and 2 radiator hose.
34. Disconnect the heater water outlet hose B.
35. Disconnect the heater water inlet hose B.
36. Disconnect the fuel tube sub-assembly.
37. Disconnect the oil cooler inlet and outlet hoses.
38. Disconnect the transmission control cable assembly.
39. Disconnect engine wire.
40. Disconnect the union to check valve hose.

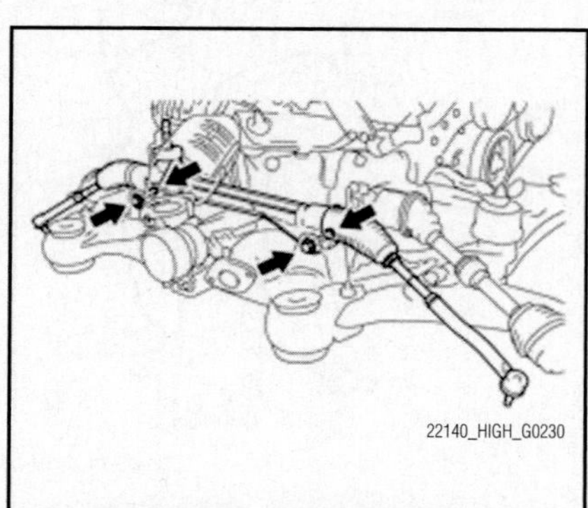

Fig. 204 Power steering link assembly bolts and nuts

41. For 4WD vehicles, remove the propeller with center bearing shaft assembly.

42. Remove the tail exhaust pipe assembly.

43. Remove the center exhaust pipe assembly.

44. Remove the front No. 3 exhaust pipe sub-assembly.

45. Remove the front exhaust pipe assembly.

46. Separate both front stabilizer link assemblies.

47. Remove both front axle hub nuts.

48. Disconnect both front speed sensors.

49. Separate the steering intermediate shaft assembly, as follows:

a. Remove the bolt and slide the steering intermediate shaft assembly.

➡Do not separate the steering intermediate shaft assembly from the power steering link assembly.

b. Put matchmarks on the steering intermediate shaft assembly and the power steering link assembly.

c. Separate the steering intermediate shaft assembly from the power steering link assembly.

50. Separate both tie rod assemblies:

a. Remove the cotter pin and the nut.

b. Install SST: 09960-20010 or equivalent to the tie rod end.

➡Make sure that the upper ends of the tie rod end and SST are aligned.

c. Using SST: 09960-20010, separate the tie rod end from the steering knuckle.

➡When securing SST to the steering knuckle, be sure to tighten the string of SST to prevent it from falling.

➡Install SST so that A and B are parallel.

➡Be sure to place the wrench on the part indicated in the illustration.

➡Do not damage the front disc brake dust cover.

➡Do not damage the ball joint dust cover.

➡Do not damage the steering knuckle.

51. Separate the No. 1 and 2 front suspension lower arms.

52. Separate both front axle assemblies.

53. Disconnect the discharge hose sub-assembly.

54. Disconnect the suction hose sub-assembly.

55. Remove the engine assembly with transaxle.

56. Remove both front No. 1 stabilizer brackets.

57. Remove the front stabilizer bar.

58. Remove the 2 bolts, 2 nuts, and power steering link assembly.

➡Because the nut has its own stopper, do not turn the nut. Loosen the bolt with the nut fixed.

59. Put matchmarks on the tie rod assemblies and the steering rack end sub-assemblies.

60. Loosen the lock nuts, and remove the tie rod assemblies and the lock nut.

To install:

61. Install the lock nuts and the tie rod assemblies to the steering rack end sub-assembly until the matchmarks are aligned.

➡After adjusting toe-in, torque the lock nut.

62. Install the power steering link assembly with the 2 bolts and 2 nuts and tighten to 51 ft. lbs. (70 Nm).

➡Make sure to tighten the bolts starting from the left side of the vehicle.

➡Because the nut has its own stopper, do not turn the nut. Tighten the bolt with the nut fixed.

63. Install the front stabilizer bar.

64. Install both front No. 1 stabilizer brackets.

65. Install the engine assembly with transaxle.

66. Reconnect the suction hose sub-assembly.

67. Reconnect the discharge hose sub-assembly.

68. Install both front axle assemblies.

69. Install both No. 1 front suspension lower arms.

70. Connect both tie rod assemblies and tighten to 36 ft. lbs. (49 Nm).

71. Install a new cotter pin.

➡Further tighten the nut up to 60° if the holes for the cotter pin are not aligned.

72. Align the matchmarks on the steering intermediate shaft assembly and the power steering link assembly.

73. Install the bolt and tighten to 26 ft. lbs. (35 Nm).

74. Install both front speed sensors.

75. Install both front axle hub nuts.

76. Install both front stabilizer link assemblies.

77. Install the front exhaust pipe assembly.

78. Install the front No. 3 exhaust pipe sub-assembly.

79. Install the center exhaust pipe assembly.

80. Install the tail exhaust pipe assembly.

81. For 4WD vehicles, temporarily tighten the propeller with center bearing shaft assembly.

82. For 4WD vehicles, fully tighten the propeller with center bearing shaft assembly.

83. Connect engine wire.

84. Connect the transmission control cable assembly.

85. Connect the fuel tube sub-assembly.

86. Connect the oil cooler inlet and outlet hoses.

87. Connect the heater water inlet hose B.

88. Connect the heater water outlet hose B.

89. Install the No. 1 and 2 radiator hoses.

90. Connect the union to check valve hose.

91. Connect the No. 1 fuel vapor feed hose.

92. Install the engine moving control rod.

93. Install the right No. 2 engine mounting stay.

94. Install the reservoir bracket.

95. Install the brake master cylinder reservoir assembly.

96. Install the air cleaner bracket.

97. Install battery.

98. Install the air cleaner case sub-assembly.

99. Install the air cleaner filter element sub-assembly.

100. Install the air cleaner cap sub-assembly.

101. Install the No. 1 and 2 air cleaner inlets.

102. Connect vacuum hoses.

103. Install the outer cowl top panel sub-assembly.

104. Install the windshield wiper motor and link assembly.

105. Install the cowl top ventilator louver sub-assembly.

106. Install both front wiper arm and blade assemblies.

107. Install front wheels. Tighten lug nuts to 76 ft. lbs. (103 Nm).

108. Add engine oil.

109. Add engine coolant.

110. Add automatic transaxle fluid.

111. Check automatic transaxle fluid.

112. Inspect for fuel leak.

113. Inspect for engine oil leak.

114. Inspect for coolant leak.

115. Inspect for exhaust gas leak.

116. Check shift lever position.

117. Place front wheels facing straight ahead.

118. Inspect and adjust front wheel alignment.

119. Check ignition timing.

120. Check engine idle speed.

121. Check CO/HC.

122. Check function of throttle body assembly.

123. Install both front fender apron seals.

124. Install both front fender liners.

125. Install both front fender molding sub-assemblies.

126. Install the left hand floor undercover.

127. Install the No. 1 and 2 engine undercover.

128. Install the engine undercover assembly.

129. Install the V-bank cover sub-assembly.

130. Check the ABS speed sensor signal.

131. Reset memory.

POWER STEERING ECU

REMOVAL & INSTALLATION

See Figure 205.

1. Before servicing the vehicle, refer to the Precautions Section.

2. Place wheels in straight ahead position.

3. Disconnect the negative battery cable.

4. Remove the left front door scuff plate.

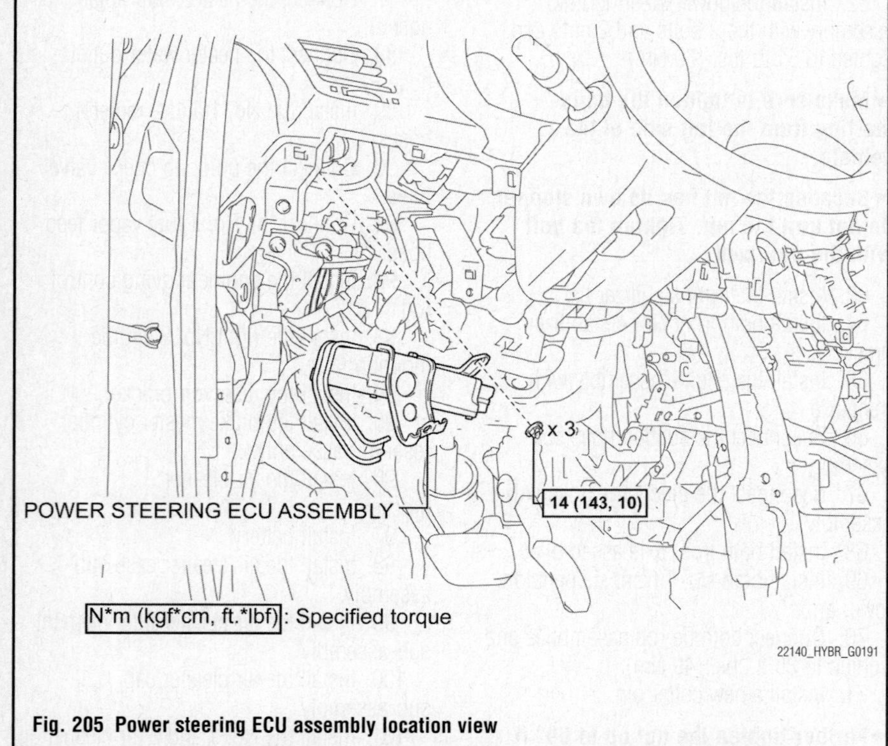

Fig. 205 Power steering ECU assembly location view

5. Remove the left side cowl trim.

6. Disconnect the hood lock control cable assembly and remove the lower instrument panel finish panel sub-assembly.

7. Remove the driver's side air bag assembly.

8. Remove the instrument panel junction block assembly.

9. Disconnect the 4 connectors from the power steering ECU assembly.

10. Remove the 3 nuts and the power steering ECU assembly.

To install:

11. Install the power steering ECU assembly with the 3 nuts, tighten the nuts to 10 ft. lbs. (14 Nm).

12. Check that the connector lever is at the fully unlocked position before installation.

13. Connect the 4 connectors to the power steering ECU assembly.

14. Connect the connectors to the back of the instrument panel junction block assembly.

15. Engage the wire harness clamp onto the instrument panel junction block assembly.

16. Install the instrument panel junction block assembly with the 3 nuts, tighten the nuts to 74 inch lbs. (8.4 Nm).

17. Connect the connectors to the instrument panel junction block assembly.

18. Engage the wire harness clamp onto the instrument panel junction block assembly.

19. Install the driver side knee airbag assembly with the 4 bolts. Tighten the bolts to 7 ft. lbs. (10 Nm).

20. Install the left cowl side trim.

21. Install the left front door scuff plate.

22. Connect the negative battery cable.

➡When disconnecting the cable, some systems need to be initialized after the cable is reconnected

23. Inspect the SRS warning light.

CONTROL LINKS

REMOVAL & INSTALLATION

See Figure 206.

➡Perform the same procedure on each side.

1. Remove wheel and tire assembly.
2. For 3.5L engines, remove the nut and separate the front stabilizer link assembly.

➡If the ball joint turns together with the nut, use a hexagon wrench (6 mm) to hold the stud bolt.

To install:

3. Install the front stabilizer link assembly with the nut and tighten to 55 ft. lbs. (74 Nm).
4. Install the wheel and tire assembly.

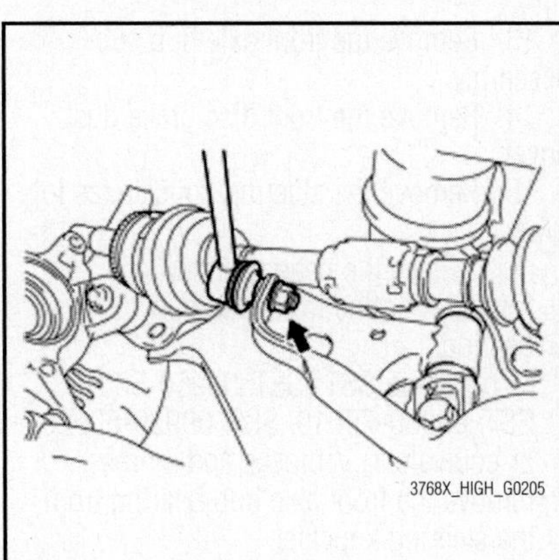

Fig. 206 Remove the nut and separate the front stabilizer link assembly—3.5L Engines

LOWER BALL JOINT

REMOVAL & INSTALLATION

See Figures 207 through 210.

1. Remove the front wheel.
2. Remove the front axle hub nut.
3. Remove the bolt and resin clamp, and separate the front speed sensor.

➡Be sure to completely separate the front speed sensor from the front shock absorber with coil spring.

➡Clean the installation hole and the surface for the speed sensor every time the speed sensor is removed.

➡Be careful not to damage the front speed sensor.

4. Put matchmarks on the front drive shaft assembly and the front axle hub sub-assembly.
5. Using a plastic hammer, separate the front drive shaft assembly from the front axle assembly.

➡Loosen the staked part of the front axle hub nut completely, otherwise the threads of the drive shaft may be damaged.

6. Remove the 2 bolts and separate the front disc brake caliper assembly.

➡Use wire or an equivalent tool to keep the brake caliper from hanging down by the flexible hose.

7. Remove the front disc.
8. Separate the tie rod assembly.
9. Remove the bolt, 2 nuts, and separate the front lower suspension arm from the lower ball joint.
10. Remove the 2 bolts, 2 nuts and front axle assembly.
11. Remove the front lower ball joint, as follows:
12. Secure the front axle assembly in a vise using aluminum plates.

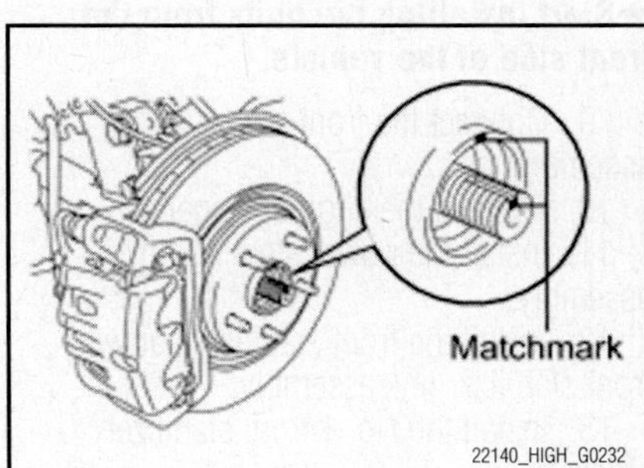

Fig. 207 Matchmark the front drive shaft assembly and the front axle hub sub-assembly

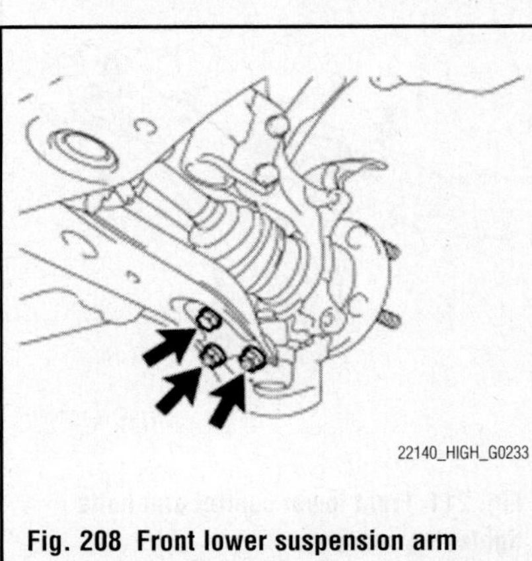

Fig. 208 Front lower suspension arm

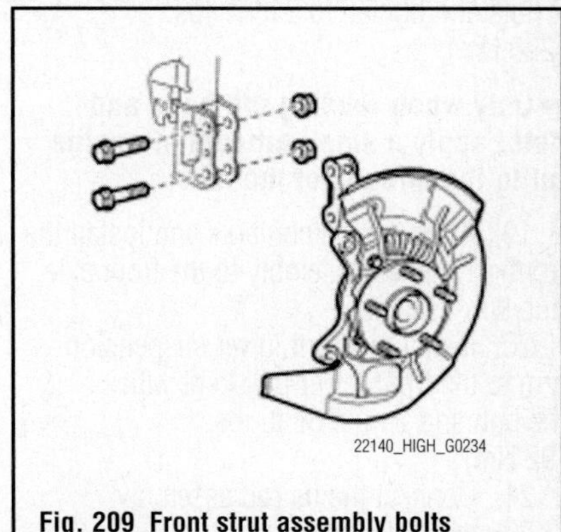

Fig. 209 Front strut assembly bolts

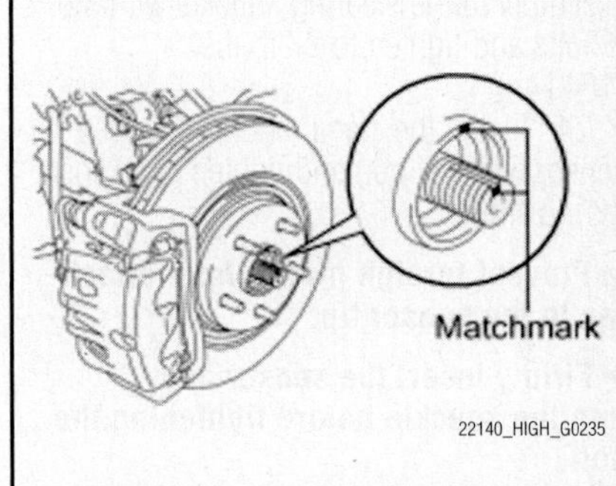

Fig. 210 Align the matchmarks and install the front drive shaft assembly

➡When using a vise, do not overtighten it.

13. Remove the cotter pin and nut.
14. Install SST: 09960-20010 or equivalent to the front lower ball joint.
15. Using SST: 09960-20010 or equivalent, remove the front lower ball joint from the front axle assembly.

➡Install SST so that A and B are parallel.

➡Be sure to place a wrench on the part indicated in the illustration.

➡Do not damage the front lower ball joint dust cover.

To install:

16. Install the front lower ball joint to the steering knuckle with the nut and tighten to 91 ft. lbs. (123 Nm).

➡Prevent oil from adhering to the screw and tapered parts.

17. Install a new cotter pin.

➡If the holes for the cotter pin are not aligned, tighten the nut further up to 60°.

18. Install the front axle assembly to the front shock absorber with the 2 bolts and 2 nuts and tighten to 213 ft. lbs. (290 Nm).

➡Only when reusing the bolts and nuts, apply a small amount of engine oil to the threads of the nuts.

19. Align the matchmarks and install the front drive shaft assembly to the front axle hub sub-assembly.

20. Install the front lower suspension arm to the front lower ball joint with the bolt and 2 nuts 68 ft. lbs. (92 Nm).

21. Connect the tie rod assembly.

22. Install the front disc.

23. Install the front disc brake caliper assembly to the steering knuckle with the 2 bolts and tighten to 77 ft. lbs. (104 Nm).

24. Install the clamp and front speed sensor with the bolt and tighten to 71 inch lbs. (8 Nm).

➡Prevent foreign matter from attaching to the sensor tip.

➡Firmly insert the sensor body into the knuckle before tightening the bolt.

➡After installing the sensor to the knuckle, make sure that there is no clearance between the sensor stay and knuckle. Also make sure that no foreign matter is stuck between the parts.

➡To prevent interference between the sensor and magnetic rotor, do not rotate the sensor body during or after the insertion of the sensor body to the knuckle.

25. Install the front axle hub nut.

26. Check the ABS speed sensor signal

➡Check the ABS speed sensor signal.

27. Install front wheel and tighten lug nuts to 76 ft. lbs. (103 Nm).

28. Inspect and adjust front wheel alignment.

LOWER CONTROL ARM

REMOVAL & INSTALLATION

See Figure 211.

1. Remove the engine assembly with transaxle.

2. Remove the No. 1 front stabilizer brackets.

3. Remove the front stabilizer bar with front stabilizer link assembly.

4. Remove the power steering link assembly.

5. Install the engine hangers.

6. Separate the front frame assembly.

7. Remove the front lower suspension arm, as follows:

a. Remove the 3 bolts, nut, and the front lower suspension arm from the front frame assembly.

b. Remove the front lower arm bushing stopper from the front lower suspension arm.

To install:

8. Install the front lower suspension arm, as follows:

a. Install the front lower arm bushing stopper to the front lower suspension arm.

b. Install the front lower suspension arm to the front frame assembly with the 3 bolts and nut, but do not tighten them yet.

c. Tighten the 3 bolts in numerical order shown. Tighten to 147 ft. lbs. (200 Nm), and 152 ft. lbs. (206 Nm).

➡Start installing the bolts from the front side of the vehicle.

9. Connect the front frame assembly.

10. Remove the engine hangers.

11. Install the power steering link assembly.

12. Install the front stabilizer bar with front stabilizer link assembly.

13. Install the No. 1 front stabilizer brackets.

14. Install the engine assembly with transaxle.

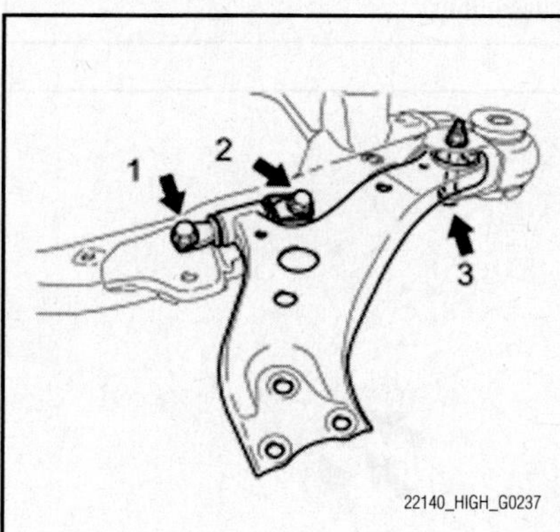

Fig. 211 Front lower control arm bolts tightening sequence

STEERING KNUCKLE

REMOVAL & INSTALLATION

See Figure 212.

1. Remove the front wheel.

2. Remove the front axle hub nut.

3. Separate the front speed sensor.

4. Separate the front disc brake caliper assembly.

5. Remove the front disc.

6. Separate tie rod assembly.

7. Separate the front lower suspension arm.

8. Separate the front drive shaft assembly.

9. Remove the front axle assembly.

10. Remove the front lower ball joint.

11. Remove the No. 1 front wheel bearing dust deflector.

12. Remove the front axle hub hole snap ring.

13. Remove the front axle hub sub-assembly.

14. Remove the front disc brake dust cover.

15. Remove the steering knuckle, as follows:

a. Place the bearing inner race (outside) on the front axle hub bearing.

b. Using SST (SST: 09950-60010, SST: 09950-70010, SST: 09950-60020, or equivalent),V-blocks and a press, remove the front axle hub bearing from the steering knuckle.

➡Keep the steering knuckle level.

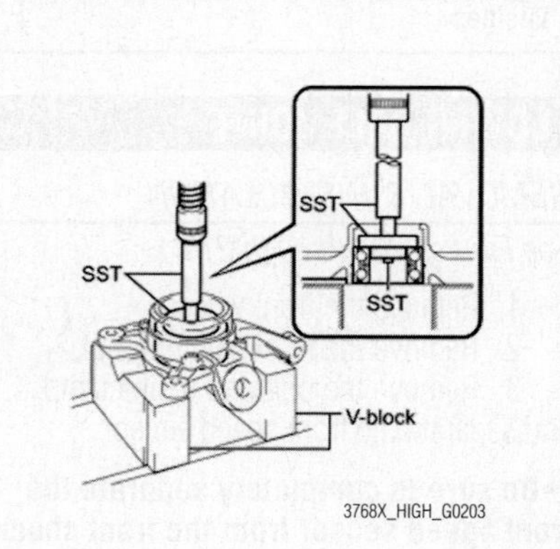

Fig. 212 Using SST (SST: 09950-60010, SST: 09950-70010, SST: 09950-60020, or equivalent),V-blocks and a press, remove the front axle hub bearing from the steering knuckle

To install:

16. Using SST (SST: 09950-70010, SST: 09950-60020, or equivalent), install a new front axle hub bearing to the steering knuckle.
17. Install the front disc brake dust cover.
18. Install the front axle hub sub-assembly.
19. Install the front axle hub hole snap ring.
20. Install the No. 1 front wheel bearing dust deflector.
21. Install the front lower ball joint.
22. Install the front axle assembly.
23. Install the front drive shaft assembly.
24. Install the front lower suspension arm.
25. Connect tie rod assembly.
26. Install the front disc.
27. Install the front disc brake caliper assembly.
28. Install the front axle hub nut.
29. Separate the front disc brake caliper assembly.
30. Remove the front disc.
31. Inspect the front axle bearing looseness.
32. Inspect front axle hub runout.
33. Install the front disc.
34. Install the front disc brake caliper assembly.
35. Install the front speed sensor.
36. Stake front axle hub nut.
37. Install the front wheel.
38. Inspect and adjust front wheel alignment.
39. Check for speed sensor signal.

STRUT

REMOVAL & INSTALLATION

See Figures 213 through 217.

1. Remove the front wheel.
2. Remove the front wiper arm and blade assemblies.
3. Loosen the front support to front shock absorber nut of the front shock absorber.

➡Do not remove the front support to front shock absorber nut.

➡Loosen the nut only when the front shock absorber with coil spring needs to be disassembled.

4. Remove the cowl top ventilator louver sub-assembly.

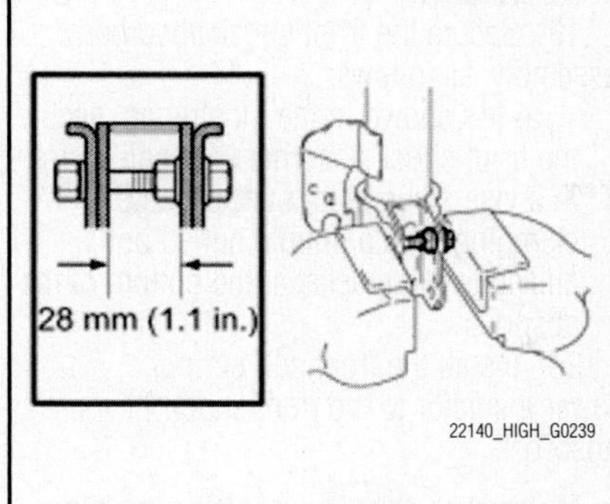

Fig. 213 Secure the front shock absorber with coil spring in a vise

5. Remove the windshield wiper motor and link.
6. Remove the outer cowl top panel sub-assembly, as follows:
   a. Disengage the 4 clamps and separate the wiper wire harness from the outer cowl top panel sub-assembly.
   b. Remove the 8 bolts, 6 nuts, and the outer cowl top panel sub-assembly.
7. Remove the bolt and clamp, and separate the front speed sensor and front flexible hose.
8. Remove the nut and separate the front stabilizer link assembly from the front shock absorber.

➡If the ball joint turns together with the nut, use a hexagon wrench (6 mm) to hold the stud.

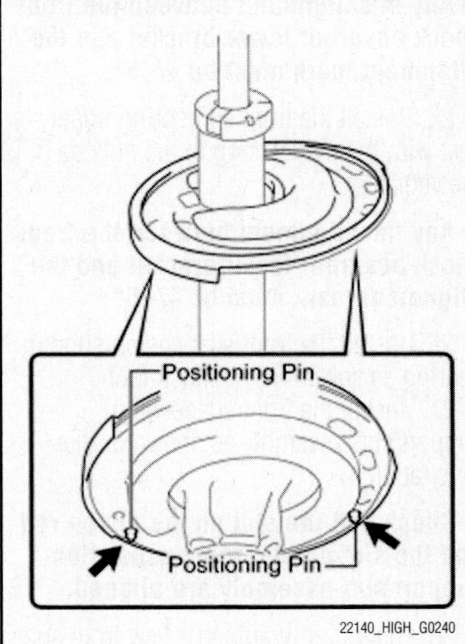

Fig. 214 Front coil spring lower insulator positioning pin

9. Remove the front shock absorber with coil spring, as follows:
   a. Support the front axle using a jack and wooden block.
   b. Remove the 2 bolts and 2 nuts, and separate the front shock absorber with coil spring (lower side) from the steering knuckle.'

➡When removing the nuts, keep the bolts from rotating.

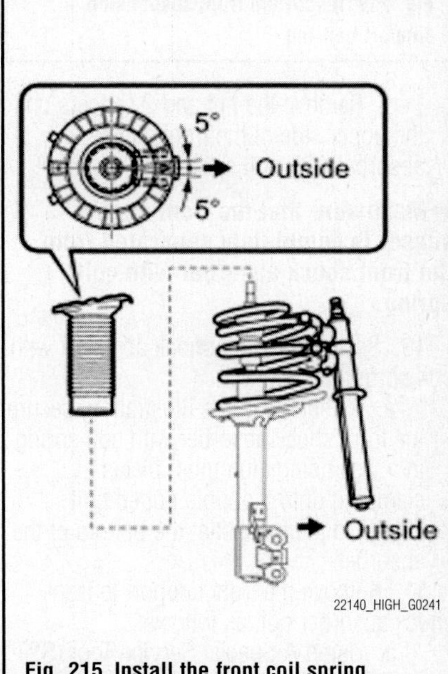

Fig. 215 Install the front coil spring upper insulator

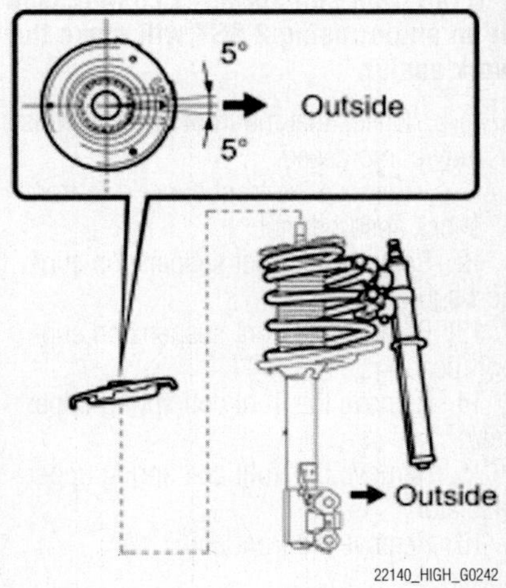

Fig. 216 Install the front coil spring upper seat

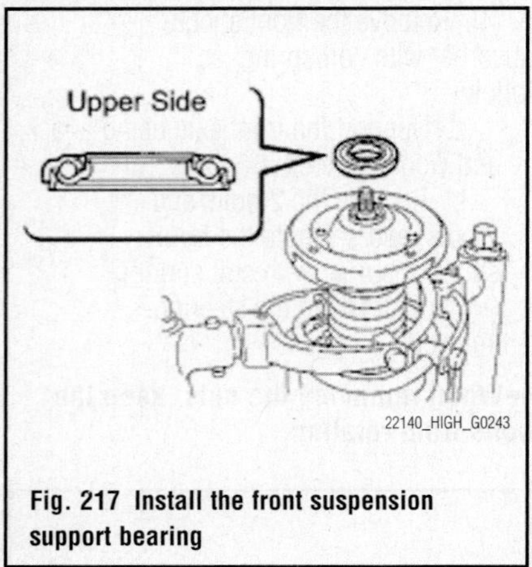

Fig. 217 Install the front suspension support bearing

c. Remove the nut and 2 spacers on the upper side of the front shock absorber with coil spring.

➡Make sure that the front speed sensor is completely separated from the front shock absorber with coil spring.

10. Secure the front shock absorber with coil spring, as follows:

a. As shown in the illustration, secure the front shock absorber with coil spring in a vise using aluminum plates by clamping onto a double nutted bolt affixed to the bracket at the bottom of the absorber.

11. Remove the front support to front shock absorber nut, as follows:

a. Using A Special Service Tool (SST: 09727-30021, SST: 09727-30021), compress the front coil spring.

➡Do not use an impact wrench. It will damage the SST.

➡If the front coil spring is compressed at an angle, using 2 SST will make the work easier.

b. Check that the front coil spring is fully compressed.

c. Remove the front support to front shock absorber nut.

12. Remove the front suspension support sub-assembly.

13. Remove the front suspension support bearing.

14. Remove the front coil spring upper seat.

15. Remove the front coil spring upper insulator.

16. Remove the front coil spring.

17. Remove the front spring bumper.

18. Remove the front coil spring lower insulator.

To install:

19. Secure the front shock absorber assembly, as follows:

a. As shown in the illustration, secure the front shock absorber with coil spring in a vise using aluminum plates by clamping onto a double nutted bolt affixed to the bracket at the bottom of the absorber.

20. Install the front coil spring lower insulator to the front shock absorber.

➡Make sure that the positioning pins on the front coil spring lower insulator are inserted into the holes in the front shock absorber.

21. Install the front spring bumper to the front shock absorber.

22. Using a Special Service Tool (SST: 09727-30021, SST: 09727-00050) or equivalent, compress the front coil spring.

➡Do not use an impact wrench. It will damage the SST.

➡If the front coil spring is compressed at an angle, using 2 SST will make the work easier.

23. Install the front coil spring to the front shock absorber.

➡Make sure that the end of the front coil spring is positioned in the depression of the lower spring seat.

24. Install the front coil spring upper insulator as shown in the illustration.

➡Any misalignment between the front shock absorber lower bracket and the alignment mark must be +/- 5°.

25. Install the front coil spring upper seat with the mark facing to the outside of the vehicle.

➡Any misalignment between the front shock absorber lower bracket and the alignment mark must be +/- 5°.

26. Install the front suspension support bearing as shown in the illustration.

27. Install the front suspension support sub-assembly as shown in the illustration.

➡Check that the slot on the piston rod and the slot on the front suspension support sub-assembly are aligned.

28. Temporarily tighten a new front support to front shock absorber nut.

29. Install the front shock absorber with coil spring (upper side) with the nut and 2 spacers and tighten to 63 ft. lbs. (85 Nm).

30. Install the front shock absorber with coil spring (lower side) to the steering knuckle and insert the 2 bolts and 2 nuts and tighten to 214 ft. lbs. (290 Nm).

➡When installing the nuts, keep the bolts from rotating.

31. Install the front stabilizer link assembly to the front shock absorber with the nut and tighten to 55 ft. lbs. (74 Nm).

➡If the ball joint turns together with the nut, use a hexagon wrench (6 mm) to hold the stud bolt.

32. Install the front speed sensor and front flexible hose with the bolt and tighten to 14 ft. lbs. (19 Nm).

➡Do not twist the front speed sensor when installing it.

33. Install the clamp.

34. Install the outer cowl top panel sub-assembly with the 8 bolts and 6 nuts and tighten to 63 ft. lbs. (85 Nm), 78 inch lbs. (8.8 Nm), 78 inch lbs. (8.8 Nm).

35. Engage the 4 clamps.

36. Fully tighten the front support to front shock absorber nut and tighten to 52 ft. lbs. (70 Nm).

37. Install the windshield wiper motor and link.

38. Install the cowl top ventilator louver sub-assembly.

39. Install the front wiper arm and blade assemblies.

40. Install the front wheel and tighten lug nuts to 76 ft. lbs. (103 Nm).

41. Inspect and adjust front wheel alignment.

STABILIZER BAR

REMOVAL & INSTALLATION

See Figure 218.

1. Remove the engine assembly with transaxle.

2. Remove the nuts and separate both front stabilizer link assemblies.

➡If the ball joint turns together with the nut, use a hexagon wrench (6 mm) to hold the stud.

3. Remove the 2 bolts and both No. 1 front stabilizer brackets from the front frame assembly.

4. Remove the front stabilizer bar.

5. Remove both No. 2 front stabilizer brackets from the front stabilizer bar bushing.

6. Remove the 2 No. 1 front stabilizer bar bushings from the front stabilizer bar.

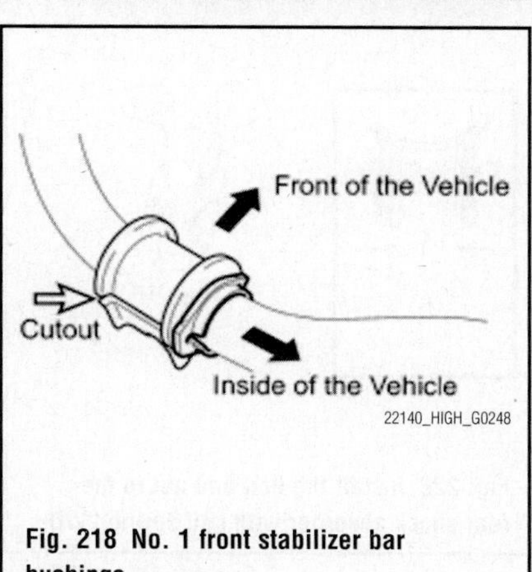

Fig. 218 No. 1 front stabilizer bar bushings

To install:

7. Install the 2 No. 1 front stabilizer bar bushings to the front stabilizer bar as shown in the illustration.

➡Install the No. 1 front stabilizer bar bushings so that the cutout faces the rear of the vehicle.

8. Install both No. 2 front stabilizer brackets to the No. 1 front stabilizer bar bushing.
9. Install the front stabilizer bar by inserting it from the right side of the vehicle.
10. Install both No. 1 front stabilizer brackets to the front frame assembly with the 2 bolts and tighten to 21 ft. lbs. (29 Nm).
11. Install both front stabilizer link assemblies with the nuts and tighten to 55 ft. lbs. (74 Nm).

➡If the ball joint turns together with the nut, use a hexagon wrench (6 mm) to hold the stud bolt.

12. Install the engine assembly with transaxle.
13. Inspect and adjust the front wheel alignment.

WHEEL HUB & BEARING (SEALED UNIT)

REMOVAL & INSTALLATION

See Steering Knuckle procedure.

SUSPENSION — REAR SUSPENSION

CONTROL ARMS/LINKS

REMOVAL & INSTALLATION

See Figures 219 and 220.

➡Use the same procedures for the RH side and LH side.

1. Remove the deck board assembly.
2. Remove the No. 2 and 3 deck board sub-assemblies, as applicable.
3. Remove the tonneau cover assembly, as applicable.
4. Remove the rear mat.
5. Remove the deck trim service hole cover.
6. Remove the lower spare wheel carrier hinge cover, as applicable.
7. Remove the spare tire.
8. Remove the spare wheel carrier lock cover, as applicable.
9. Remove rear wheel.
10. Remove the nuts and separate the rear stabilizer link assemblies from the rear stabilizer bar.

➡If the ball joint turns together with the nut, use a hexagon wrench (5 mm) to hold the stud bolt.

11. Remove the 4 bolts and rear stabilizer bar.
12. Remove the rear No. 2 suspension arm assembly, as follows:
    a. Remove the bolt and the nut, and separate the rear No. 2 suspension arm assembly from the rear axle carrier sub-assembly.

➡Since a stopper nut is used, loosen the bolt.

    b. Remove the bolt and the rear No. 2 suspension arm assembly.

13. Remove the rear No. 1 suspension arm assembly, as follows:
    a. Remove the bolt and the nut, and separate the rear No. 1 suspension arm assembly from the rear axle carrier sub-assembly.

➡Since a stopper nut is used, loosen the bolt.

    b. Remove the bolt and the rear No. 1 suspension arm assembly.

To install:

14. Temporarily install the rear No. 1 and 2 suspension arm assemblies to the rear suspension member with the bolts.

➡Ensure that the identification marks face the rear side of the vehicle.

15. Temporarily install the rear No. 1 and 2 suspension arm assemblies to the rear axle carrier sub-assembly with the bolts and the nuts.

➡Since a stopper nut is used, temporarily tighten the bolts.

16. Jack up the rear axle carrier, placing a wooden block underneath to avoid damage. Apply load to the suspension so that the installed bolt of the rear No. 1 suspension arm (inner side) is horizontally aligned with the center of the rear axle hub.

⁂ CAUTION

Do not jack up the rear axle carrier sub-assembly too high as the vehicle may fall.

➡Do not bend the brake dust cover.

➡If the rear drive shaft assembly cannot be positioned horizontally as shown in the illustration even when the rear axle carrier sub-assembly is jacked up, apply additional load to the vehicle such as by having a person sit in the rear seat.

➡Use the same procedures for the RH side and LH side.

17. Fully tighten the rear No. 1 suspension arm assembly, as follows:

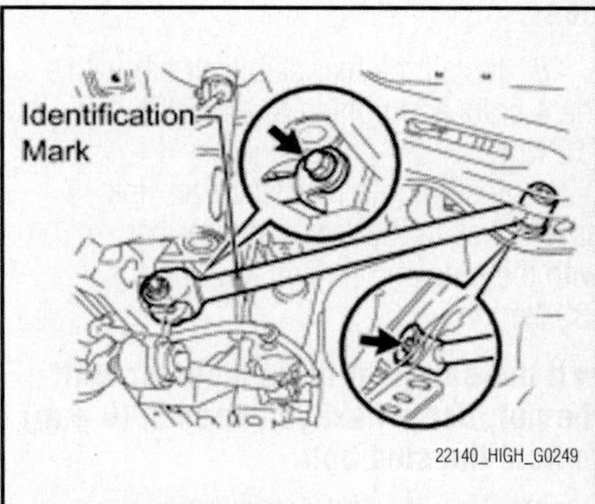

Fig. 219 Temporarily install the rear No. 1 and 2 suspension arm assemblies

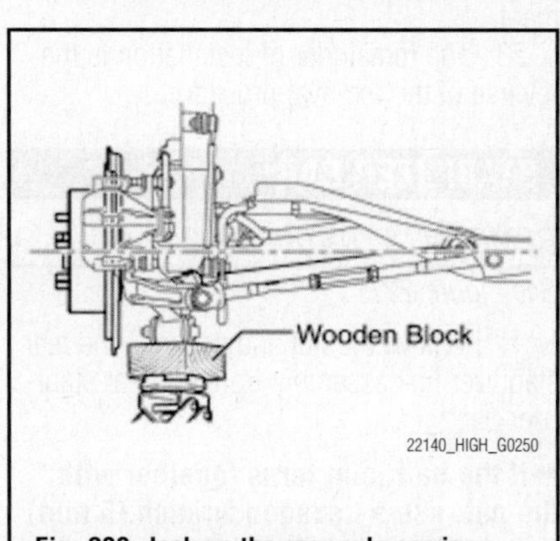

Fig. 220 Jack up the rear axle carrier

a. Using SST: 09961-00950 or equivalent and a socket wrench (19 mm), fully tighten the bolt to 88 ft. lbs. (120 Nm), 65 ft. lbs. (89 Nm).

➡Use a torque wrench with a fulcrum length of 425 mm (16.73 in.).

➡This torque value is effective when SST is parallel to the torque wrench.

➡The final torque must be applied under standard vehicle height conditions.

b. Fully tighten the bolt to 82 ft. lbs. (112 Nm).

➡Since a stopper nut is used, fully tighten the bolt.

➡The final torque must be applied under standard vehicle height conditions.

18. Fully tighten the rear No. 2 suspension arm assembly, as follows:

a. Fully tighten the bolts to 88 ft. lbs. (120 Nm), 82 ft. lbs. (112 Nm).

➡Since a stopper nut is used, fully tighten the bolt.

➡The final torque must be applied under standard vehicle height conditions.

19. Install the rear stabilizer bar with the 4 bolts and tighten to 14 ft. lbs. (19 Nm).

20. Install both rear stabilizer link assemblies to the rear stabilizer bar with the nuts and tighten to 29 ft. lbs. (39 Nm).

➡If the ball joint turns together with the nut, use a hexagon wrench (6 mm) to hold the stud bolt.

21. Install the wheels and tighten lug nuts to 76 ft. lbs. (103 Nm).

22. Inspect and adjust rear wheel alignment.

23. The remainder of installation is the reverse of the removal procedure.

STABILIZER BAR

REMOVAL & INSTALLATION

See Figure 221.

1. Remove the nut and separate the rear stabilizer link assembly from the rear stabilizer bar.

➡If the ball joint turns together with the nut, use a hexagon wrench (5 mm) to hold the stud bolt.

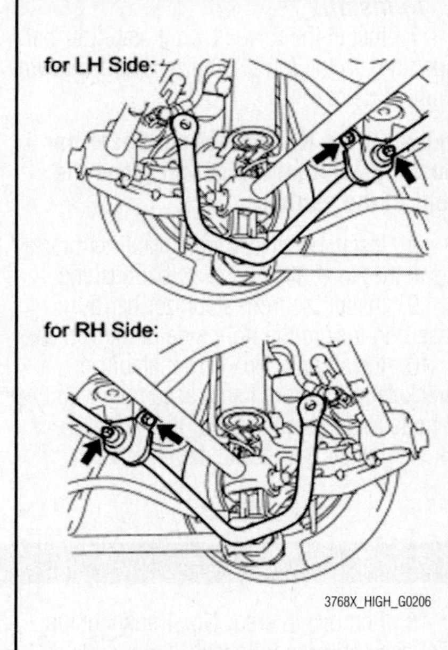

Fig. 221 Remove the 4 bolts and rear stabilizer bar—2WD vehicles

2. Remove the nut and the rear stabilizer link assembly from the rear shock absorber with coil spring.

3. Remove the 4 bolts and rear stabilizer bar.

4. Remove the rear No. 1 stabilizer bar bracket.

5. Remove the rear stabilizer bushing.

To install:

6. To install, reverse the removal procedure.

7. Tighten the rear stabilizer bar bolts to 14 ft. lbs. (19 Nm).

8. Tighten the rear stabilizer link assembly nut to 29 ft. lbs. (39 Nm).

STRUT

REMOVAL & INSTALLATION

See Figures 222 and 223.

1. Remove the wheel.

2. Remove the deck side trim cover.

3. Remove the deck side trim.

4. For 4WD vehicles, remove the bolt and separate the rear flexible hose from the rear shock absorber with coil spring.

5. Remove the bolt and separate the rear speed sensor from the rear shock absorber with coil spring.

6. Remove the nut and separate the rear stabilizer link assembly from the rear shock absorber with coil spring.

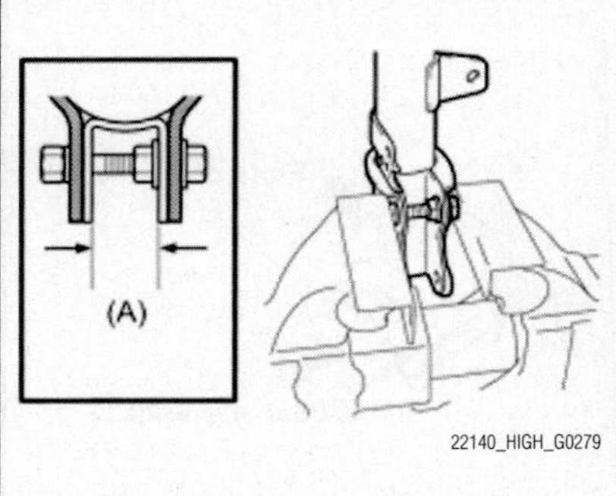

Fig. 222 Install the bolt and nut to the rear shock absorber with coil spring 2WD

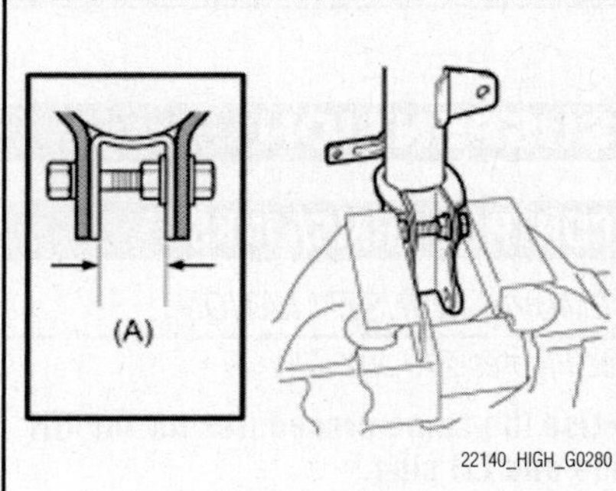

Fig. 223 Install the bolt and nut to the rear shock absorber with coil spring 4WD

➡If the ball joint turns together with the nut, use a hexagon wrench (5 mm) to hold the stud bolt.

7. Disengage the 4 claws and remove the rear No. 1 suspension support cover.

8. Using a jack and wooden block, support the rear axle carrier sub-assembly.

➡Do not deform the dust cover.

➡Support the rear axle carrier sub-assembly until reinstallation of the rear shock absorber with coil spring is complete.

9. Loosen the rear support to rear shock absorber nut.

➡Do not remove the rear support to rear shock absorber nut.

➡Loosen the nut only when the rear shock absorber with coil spring needs to be disassembled.

10. Remove the 2 bolts and 2 nuts, and separate the rear shock absorber with coil spring from the rear axle carrier sub-assembly,

➡When removing the nuts, keep the bolts from rotating.

11. Remove the 3 nuts and rear shock absorber with coil spring.

➡**Make sure that the rear speed sensor and rear flexible hose are disconnected from the rear shock absorber with coil spring.**

12. Remove the rear support to rear shock absorber nut, as follows:
   a. Using SST: 09727-30021, compress the rear coil spring.

➡**Do not use an impact wrench. It will damage SST.**

   b. Check that the front coil spring is fully compressed.
   c. Hold the rear suspension support assembly and remove the rear support to rear shock absorber nut from the rear shock absorber assembly.

13. Remove the rear support to rear shock absorber collar from the rear shock absorber assembly.
14. Remove the rear suspension support assembly from the rear shock absorber assembly.
15. Remove the rear coil spring together with SST from the rear shock absorber assembly.
16. Remove the rear No. 1 spring bumper from the rear shock absorber assembly.
17. Remove the rear lower coil spring insulator from the rear shock absorber assembly.

To install:

18. Install the rear lower coil spring insulator onto the rear shock absorber assembly.

➡**Fit the recessed part of the rear lower coil spring insulator into the recession on the shock absorber assembly.**

19. Install the rear No. 1 spring bumper to the rear shock absorber assembly.
20. Temporarily install rear coil spring, as follows:
   a. Using SST: 09727-30021 or equivalent, compress the rear coil spring.

➡**Do not use an impact wrench. It will damage the SST.**

   b. Temporarily install the rear coil spring together with SST to the rear shock absorber assembly.

21. Install the rear suspension support assembly to the rear shock absorber assembly.

➡**Align the cutout on the rear shock absorber assembly with the protrusion on the rear suspension support assembly by referring to the illustration.**

22. Install the rear support to rear shock absorber collar to the rear shock absorber assembly.
23. Temporarily install the rear support to rear shock absorber nut to the rear shock absorber assembly.
24. Install the rear coil spring.

➡**Do not use an impact wrench. It will damage the SST.**

➡**Make sure that the end of the rear coil spring is positioned in the depression of the rear lower coil spring insulator.**

➡**Ensure that the stud bolt is positioned 3.5° to the outside of the vehicle as shown in the illustration. The deviation should be within +-5°.**

25. Install the rear shock absorber with coil spring with the 3 nuts and tighten to 43 ft. lbs. (58 Nm).
26. Install the rear shock absorber with coil spring with the 2 bolts and 2 nuts and tighten to 213 ft. lbs. (290 Nm).

➡**When installing the nuts, keep the bolts from rotating.**

27. Fully tighten the rear support to rear shock absorber nut 40 ft. lbs. (55 Nm).
28. Install the rear No. 1 suspension support cover.
29. Install the rear stabilizer link assembly to the rear shock absorber with coil spring with the nut and tighten to 29 ft. lbs. (39 Nm).

➡**If the ball joint turns together with the nut, use a hexagon wrench (5 mm) to hold the stud bolt.**

30. Install the rear speed sensor wire to the rear shock absorber with coil spring with the bolt and tighten to 44 inch lbs. (5 Nm).

➡**Do not twist the rear speed sensor wire when installing it.**

31. Install the rear flexible hose to the rear shock absorber with coil spring with the bolt and tighten to 14 ft. lbs. (19 Nm).

➡**Do not twist the rear flexible hose when installing it.**

32. Install deck side trim.
33. Install deck side trim cover.
34. Install the wheels and tighten lug nuts to 76 ft. lbs. (103 Nm).
35. Inspect and adjust rear wheel alignment.

WHEEL HUB & BEARING

REMOVAL & INSTALLATION

See Figure 224.

1. Disconnect the negative battery cable.
2. Remove the wheel.
3. Separate the rear flexible hose.
4. Remove the 2 bolts and separate the rear disc brake caliper assembly.
5. Remove the rear disc.
6. For 4WD vehicles: Using a screwdriver, disconnect the connector from the rear speed sensor.
7. Remove the 4 bolts and the rear axle hub & bearing assembly.

To install:

8. Install the rear axle hub and bearing assembly with the 4 bolts and tighten to 55 ft. lbs. (75 Nm).
9. Inspect rear axle hub bearing looseness.
10. Inspect rear axle hub runout.
11. Connect the connector to the rear speed sensor.
12. Install the rear disc.

Fig. 224 Remove the 4 bolts and the rear axle hub & bearing assembly

13. Install the rear disc brake caliper assembly with the 2 bolts and tighten to 57 ft. lbs. (78 Nm).

14. Install the rear flexible hose to the shock absorber with coil spring with the bolt and tighten to 14 ft. lbs. (19 Nm).

15. Install the wheel and tighten lug nuts to 76 ft. lbs. (103 Nm).

16. Connect the negative battery cable.

17. Inspect and adjust rear wheel alignment.

18. Check the ABS speed sensor signal.

LEXUS

SC430

32

SPECIFICATIONS AND MAINTENANCE CHARTS

ENGINE AND VEHICLE IDENTIFICATION

Engine						
Code ①	Liters	Cu. In.	Cyl.	Fuel Sys.	Engine Type	Eng. Mfg.
3UZ-FE	4.3	262	V8	SFI	DOHC	Toyota

Model Year	
Code ②	Year
9	2009
A	2010

SFI: Sequential Multi-port Fuel Injection

DOHC: Double Overhead Camshaft

① Located on the timing belt cover

② 10th digit of the VIN

3768X_SC43_C0001

GENERAL ENGINE SPECIFICATIONS

All measurements are given in inches.

Year	Model	Engine Displacement Liters (cc)	Engine ID/VIN	Fuel System Type	Net Horsepower @ rpm	Net Torque @ rpm (ft. lbs.)	Bore x Stroke (in.)	Com-pression Ratio	Oil Pressure @ rpm
2009	SC430	4.3 (4293)	3UZ-FE	SFI	288@5600	317@3400	3.58x3.25	10.5:1	43-85@3000
2010	SC430	4.3 (4293)	3UZ-FE	SFI	288@5600	317@3400	3.58x3.25	10.5:1	43-85@3000

SFI : Sequential Multi-port Fuel Injection

3768X_SC43_C0002

ENGINE TUNE-UP SPECIFICATIONS

Year	Engine Displacement Liters	Engine ID/VIN	Spark Plug Gap (in.)	Ignition Timing (deg.)	Fuel Pump (psi)	Idle Speed (rpm)	Valve Clearance Intake	Valve Clearance Exhaust
2009	4.3	3UZ-FE	0.039-0.043	8-12B	44-50	700-800	0.006-0.010	0.010-0.014
2010	4.3	3UZ-FE	0.039-0.043	8-12B	44-50	700-800	0.006-0.010	0.010-0.014

NOTE: The Vehicle Emission Control Information label often reflects specification changes made during production. The label figures must be used if they differ from those in this chart.

B: Before top dead center

HYD: Hydraulic Valve Lifters

① Terminals TC and CG of check connector must be connected

3768X_SC43_C0003

CAPACITIES

Year	Model	Engine Displacement Liters	Engine ID/VIN	Engine Oil with Filter	Transmission (pts.) ① Auto.	Drive Axle (pts.)	Fuel Tank (gal.)	Cooling System (qts.)
2009	SC430	4.3	3UZ-FE	5.5	3.6	2.8	19.8	10.5
2010	SC430	4.3	3UZ-FE	5.5	3.6	2.8	19.8	10.5

NOTE: All capacities are approximate. Add fluid gradually and check to be sure a proper fluid level is obtained.

① Specification is for transmission drain and refill, not overhaul.

3768X_SC43_C0005

FLUID SPECIFICATIONS

Year	Model	Engine Displacement Liters	Engine ID/VIN	Engine Oil	Auto. Trans. ①	Drive Axle ②	Power Steering Fluid	Brake Master Cylinder	Engine Coolant ③
2009	SC430	4.3	3UZ-FE	5W-30	ATF-WS	75W-90	Dexron II or III	DOT 3	Toyota coolant
2010	SC430	4.3	3UZ-FE	5W-30	ATF-WS	75W-90	Dexron II or III	DOT 3	Toyota coolant

DOT: Department Of Transpotation

① The use of genuine Toyota ATF-WS is recommended

② Synthetic GL-5 (75W-90) or equivalent

③The use of genuine Toyota engine coolant is recommended or similar ethylene glycol based non-silicate, non-amine, non- nitrite, and non- borat coolant

3768X_SC43_C0004

VALVE SPECIFICATIONS

Year	Engine Displacement Liters	Engine ID/VIN	Seat Angle (deg.)	Face Angle (deg.)	Spring Test Pressure (lbs. @ in.)	Spring Free-Length (in.)	Stem-to-Guide Clearance (in.) Intake	Stem-to-Guide Clearance (in.) Exhaust	Stem Diameter (in.) Intake	Stem Diameter (in.) Exhaust
2009	4.3	3UZ-FE	45	44.5	45.9-50.7@ 1.3795	2.130	0.0010-0.0024	0.0012-0.0026	0.2154-0.2159	0.2152-0.2157
2010	4.3	3UZ-FE	45	44.5	45.9-50.7@ 1.3795	2.130	0.0010-0.0024	0.0012-0.0026	0.2154-0.2159	0.2152-0.2157

3768X_SC43_C0007

CAMSHAFT AND BEARING SPECIFICATIONS CHART

All measurements are given in inches.

Year	Engine Displ. Liters	Engine ID/VIN	Journal Dia.	Brg. Oil Clearance	Shaft End-play	Runout	Journal Bore	Lobe Height Intake	Lobe Height Exhaust
2009	4.3	3UZ-FE	1.0612-1.0618	0.0039	0.0024-0.0039	0.0031	NA	1.7303-1.7342	1.6783-1.6823
2010	4.3	3UZ-FE	1.0612-1.0618	0.0039	0.0024-0.0039	0.0031	NA	1.7303-1.7342	1.6783-1.6823

NA: Not Available

① Intake Journal 1: 0.0016 - 0.0031 inch.

All Others: 0.00098 - 0.0024 inch.

3768X_SC43_C0006

CRANKSHAFT AND CONNECTING ROD SPECIFICATIONS

All measurements are given in inches.

Year	Engine Displacement Liters	Engine ID/VIN	Crankshaft Main Brg. Journal Dia.	Crankshaft Main Brg. Oil Clearance	Crankshaft Shaft End-play	Crankshaft Thrust on No.	Connecting Rod Journal Diameter	Connecting Rod Oil Clearance	Connecting Rod Side Clearance
2009	4.3	3UZ-FE	2.6373-2.6378	①	0.0008-0.0087	3	2.0465-2.0472	0.0008-0.0019	0.0063-0.0138
2010	4.3	3UZ-FE	2.6373-2.6378	①	0.0008-0.0087	3	2.0465-2.0472	0.0008-0.0019	0.0063-0.0138

NA: Not Available

① Journal No. 1 and 5: 0.0007 - 0.0013 inch

Remaining journals: 0.0011 - 0.0018 inch

3768X_SC43_C0010

PISTON AND RING SPECIFICATIONS

All measurements are given in inches.

Year	Engine Displacement Liters	Engine ID/VIN	Piston Clearance	Ring Gap Top Compression	Ring Gap Bottom Compression	Ring Gap Oil Control	Ring Side Clearance Top Compression	Ring Side Clearance Bottom Compression	Ring Side Clearance Oil Control
2009	4.3	3UZ-FE	0.0023-0.0040	0.0118-0.0197	0.0157-0.0236	0.0059-0.0197	0.0012-0.0031	0.0008-0.0024	SNUG
2010	4.3	3UZ-FE	0.0023-0.0040	0.0118-0.0197	0.0157-0.0236	0.0059-0.0197	0.0012-0.0031	0.0008-0.0024	SNUG

3768X_SC43_C0009

TORQUE SPECIFICATIONS

All readings in ft. lbs.

Year	Engine Displacement Liters	Engine ID/VIN	Cylinder Head Bolts	Main Bearing Bolts	Rod Bearing Bolts	Crankshaft Damper Bolts	Flywheel Bolts	Manifold Intake	Manifold Exhaust	Spark Plugs
2009	4.3	3UZ-FE	①	②	③	181	④	13	32	13
2010	4.3	3UZ-FE	①	②	③	181	④	13	32	13

① Step 1: 44 ft. lbs.
Step 2: Plus 90 degrees

② Nuts:
Step 1: 20 ft. lbs.
Step 2: Plus 90 degrees
Bolts: 36 ft. lbs.

③ Step 1: 18 ft. lbs.
Step 2: Plus 90 degrees

④ Step 1: 36 ft. lbs.
Step 2: Plus 90 degrees

3768X_SC43_C0008

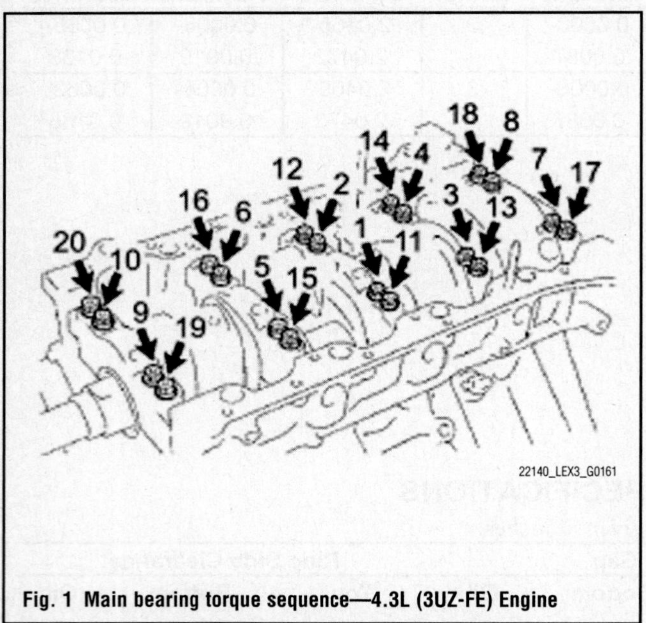

Fig. 1 Main bearing torque sequence—4.3L (3UZ-FE) Engine

WHEEL ALIGNMENT

Year	Model		Caster Range (+/-Deg.)	Caster Preferred Setting (Deg.)	Camber Range (+/-Deg.)	Camber Preferred Setting (Deg.)	Toe-in (in.)	Steering Axis Inclination (Deg.)
2009	SC 430	F	0.75	+7.92	0.75	-0.58	0.06 +/- 0.08	9.16
		R	—	—	0.50	-1.17	0.06 +/- 0.08	—
2010	SC 430	F	0.75	+7.92	0.75	-0.58	0.06 +/- 0.08	9.16
		R	—	—	0.50	-1.17	0.06 +/- 0.08	—

3768X_SC43_C0011

TIRE, WHEEL AND BALL JOINT SPECIFICATIONS

Year	Model	OEM Tires Standard	Tire Pressures (psi) Front	Tire Pressures (psi) Rear	Wheel Size	Ball Joint Inspection	Lug Nuts
2009	SC430	245/40ZR18	33	33	8-JJ	①	76
2010	SC430	245/40ZR18	33	33	8-JJ	①	76

OEM: Original Equipment Manufacturer

PSI: Pounds Per Square Inch

① Lower ball joint: 0.016 inch. Upper ball joint: standard turnig torque 9-30 inch.

3768X_SC43_C0012

BRAKE SPECIFICATIONS

All measurements in inches unless noted

Year	Model	Front Brake Disc Original Thickness	Front Brake Disc Minimum Thickness	Front Brake Disc Maximum Run-out	Rear Brake Disc Original Thickness	Rear Brake Disc Minimum Thickness	Rear Brake Disc Maximum Run-out	Minimum Lining Thickness	Brake Caliper Bracket Bolts (ft. lbs.)
2009	SC430	1.260	1.181	0.0020	0.472	0.413	0.0020	0.0390	①
2010	SC430	1.260	1.181	0.0020	0.472	0.413	0.0020	0.0390	①

① Front: 87 ft. lbs.
Rear: 77 ft. lbs.

3768X_SC43_C0013

SCHEDULED MAINTENANCE INTERVALS
Lexus SC430

TO BE SERVICED	TYPE OF SERVICE	VEHICLE MILEAGE INTERVAL (x1000)													
		5	10	15	20	25	30	35	40	45	50	55	60	90	120
Engine oil & filter	R	✓	✓	✓	✓	✓	✓	✓	✓	✓	✓	✓	✓	✓	✓
Automatic transmission fluid	S/I			✓			✓			✓			✓	✓	✓
Ball joints & dust covers	S/I			✓			✓			✓			✓	✓	✓
Bolts & nuts on chassis & body	S/I			✓			✓			✓			✓	✓	✓
Brake linings & drums	S/I	✓	✓	✓	✓	✓	✓	✓	✓	✓	✓	✓	✓	✓	✓
Brake line pipes & hoses	S/I			✓			✓			✓			✓	✓	✓
Brake pads & discs (front & rear)	S/I	✓	✓	✓	✓	✓	✓	✓	✓	✓	✓	✓	✓	✓	✓
Brake fluid	R						✓						✓	✓	✓
Rack and pinion assembly	S/I			✓			✓			✓			✓	✓	✓
Steering linkage & boots	S/I			✓			✓			✓			✓	✓	✓
Air cleaner filter	R												✓	✓	✓
Spark plugs ①	R														✓
Drive belts	S/I												✓	✓	✓
Exhaust pipes & mountings	S/I			✓			✓			✓			✓	✓	✓
Fuel lines & connections	S/I						✓						✓	✓	✓
Engine coolant ②	S/I			✓			✓			✓			✓	✓	
Rear differential & transfer case oil	S/I			✓			✓			✓			✓	✓	✓
Fuel tank cap gasket	S/I						✓						✓	✓	✓
Rotate tires	S/I			✓			✓			✓			✓		✓
Clean air conditioning filter ③	S/I			✓			✓			✓			✓		✓
Axle shaft bolts	S/I			✓			✓			✓			✓	✓	✓
Brake pad thickness and rotor	S/I						✓						✓	✓	✓

R: Replace S/I: Service or Inspect

① Spark plugs are replaced at 120,000 miles

② Replace engine coolant at 100,000 miles and then inspect every 15,000 miles

③ Replace air conditioning filter every 30,000 miles

FREQUENT OPERATION MAINTENANCE (SEVERE SERVICE)

If a vehicle is operated under any of the following conditions it is considered severe service:

- Extremely dusty areas.

- 50% or more of the vehicle operation is in 32°C (90°F) or higher temperatures, or constant temperatures below 0°C (32°F).

- Prolonged idling (vehicle operation in stop and go traffic).

- Frequent short running periods (engine does not warm to normal operating temperatures).

- Police, taxi, delivery usage or trailer towing usage.

Air cleaner filter: service or inspect every 5000 miles

Rear differential & transfer case oil: replace every 15,000 miles.

Ball joints & dust covers: service or inspect every 5000 miles.

Bolts & nuts on chassis & body: service or inspect every 5000 miles.

Axle shaft bolts: service or inspect every 5000 miles.

Steering linkage: service or inspect every 5000 miles.

Air filter: service or inspect every 15,000 miles.

Drive belts: service or inspect at 60,000 miles & every 7500 miles thereafter.

Timing belts: replace every 60,000 miles.

3768X_SC43_C0014

BRAKES — INFORMATION AND PRECAUTIONS

ANTI-LOCK SYSTEMS

• Certain components within the ABS system are not intended to be serviced or repaired individually.

• Do not use rubber hoses or other parts not specifically specified for and ABS system. When using repair kits, replace all parts included in the kit. Partial or incorrect repair may lead to functional problems and require the replacement of components.

• Lubricate rubber parts with clean, fresh brake fluid to ease assembly. Do not use shop air to clean parts; damage to rubber components may result.

• Use only DOT 3 brake fluid from an unopened container.

• If any hydraulic component or line is removed or replaced, it may be necessary to bleed the entire system.

• A clean repair area is essential. Always clean the reservoir and cap thoroughly before removing the cap. The slightest amount of dirt in the fluid may plug an orifice and impair the system function. Perform repairs after components have been thoroughly cleaned; use only denatured alcohol to clean components. Do not allow ABS components to come into contact with any substance containing mineral oil; this includes used shop rags.

• The Anti-Lock control unit is a microprocessor similar to other computer units in the vehicle. Ensure that the ignition switch is **OFF** before removing or installing controller harnesses. Avoid static electricity discharge at or near the controller.

• If any arc welding is to be done on the vehicle, the control unit should be unplugged before welding operations begin.

DISC SYSTEMS

CAUTION

Dust and dirt accumulating on brake parts during normal use may contain asbestos fibers from production or aftermarket brake linings. Breathing excessive concentrations of asbestos fibers can cause serious bodily harm. Exercise care when servicing brake parts. Do not sand or grind brake lining unless equipment used is designed to contain the dust residue. Do not clean brake parts with compressed air or by dry brushing. Cleaning should be done by dampening the brake components with a fine mist of water, then wiping the brake components clean with a dampened cloth. Dispose of cloth and all residue containing asbestos fibers in an impermeable container with the appropriate label. Follow practices prescribed by the Occupational Safety and Health Administration (OSHA) and the Environmental Protection Agency (EPA) for the handling, processing, and disposing of dust or debris that may contain asbestos fibers.

BRAKES — BLEEDING THE BRAKE SYSTEM

BLEEDING PROCEDURE

See Figure 2.

Except ABS System

If any work is done on the brake system or if air in the brake lines is suspected, bleed the system of air.

➡Do not let brake fluid remain on painted surfaces. Wash it off immediately.

1. Fill the reservoir with brake fluid.

➡If the master cylinder has been disassembled or if the reservoir becomes empty, bleed the air from the master cylinder.

2. Bleed the brake master cylinder as follows:
   - Disconnect the brake lines from the master cylinder.
   - Slowly depress the brake pedal and hold it.
   - Block off the outer holes with your fingers, and release the brake pedal.
   - Repeat the previous 2 steps 3 or 4 times.
3. Bleed the brake line as follows:
   - Connect the vinyl tube to the brake caliper.
   - Depress the brake pedal several times, then loosen the bleeder plug with the pedal held down.
   - At the point when fluid stops coming out, tighten the bleeder plug, then release the brake pedal.
   - Repeat the previous 2 steps until all the air in the fluid has been bled out.
   - Repeat the above procedure to bleed the air out of the brake line for each wheel.
   - Tighten the bleeder plug to 8 ft. lbs. (11 Nm).
4. Bleed the brake actuator as follows:
   - Remove the reservoir cap.
   - Install the SST 09992-00242, 09992-00350 to the reservoir.
   - Connect the vinyl tube to the bleeder plug of the brake actuator.
   - Using SST, apply the 14.2 psi (98.1kpa) of pressure to the reservoir.
   - Loosen the bleeder plug.
   - Bleed the air out of the brake actuator, tighten the bleeder plug to 74 inch lbs. (8.3 Nm).
5. Check the fluid level and add fluid if necessary.

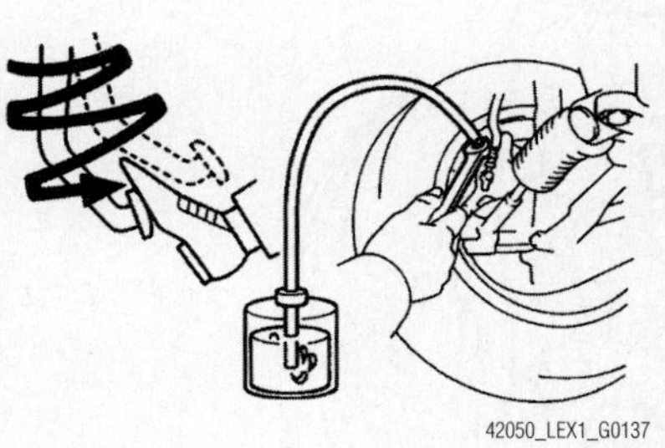

42050_LEX1_G0137

Fig. 2 Bleeding the brake line

ABS System

➡After performing the usual air bleeding in the brake system, if the height or feel of the brake pedal cannot be obtained, perform air bleeding in the brake actuator assembly with a hand held tester by following procedures below. Make sure that the brake fluid in the master cylinder reservoir tank does not become empty.

1. Depress the brake pedal more than 20 times with the engine off.

2. Connect the hand held tester to the DLC3, then turn the ignition switch to the ON position.

3. Do not start the engine.

4. Select "AIR BLEEDING" on the hand held tester. Please refer to the Hand Held Tester Operator's Manual for further details.

5. Bleed the air out of the regular brake line in "Step1: Increase" on the hand held tester display. Perform the air bleeding by following the steps displayed on the hand held tester. Make sure that the brake fluid in the master cylinder reservoir tank does not become empty.

6. Connect the vinyl tube to either one of the bleeder plugs.

7. Depress the brake pedal several times, then loosen the bleeder plug of one of the above wheels with the pedal depressed.

8. When fluid stops coming out, tighten the bleeder plug, then release the brake pedal.

9. Repeat (2) and (3) until all air in the fluid is completely bled out.

10. Tighten the bleeder plug to 73 inch lbs. (8.3 Nm).

11. Repeat the above procedure to bleed the air out of the brake line for each wheel.

12. Bleed the air out of the suction line in "Step2: Inhalation" on the hand held tester display.

13. Connect the vinyl tube to the bleeder plug at the right front wheel or the right rear wheel and loosen the bleeder plug.

14. Operate the brake actuator assembly using the hand held tester to bleed the air.

➡The operation stops automatically in 4 seconds. At this time, be sure to release the brake pedal.

15. Check that the operation has stopped, by referring to the hand held tester display.

16. Repeat (2) and (3) until all the air in the fluid is completely bled out.

17. Tighten the bleeder plug.

18. For the rest of the wheels, bleed the air in the same way as stated in the above procedure.

19. Bleed the air out of the pressure reduction line in "Step3: Decrease" on the hand held tester display.

20. Connect a vinyl tube to either one of the bleeder plugs.

21. Loosen the bleeder plug.

22. Using the hand held tester, operate the brake actuator assembly using hand held tester, completely depress the brake pedal and keep it.

➡The operation stops automatically in 4 seconds. When performing this procedure continuously, an interval of at least 20 seconds is required. When the operation is completed, the brake pedal slightly goes down. This is a normal phenomenon caused when the solenoid opens. During this procedure, the pedal seems heavy, but completely depress it so that the brake fluid comes out from the bleeder plug. Be sure to keep depressing the brake pedal. Never depress and release the pedal repeatedly.

23. Tighten the bleeder plug, then release the brake pedal.

24. Repeat 3 previous steps until all the air in the fluid is completely bled out.

25. Tighten the bleeder plug.

26. Repeat the above procedure to bleed the air out of the brake line for each wheel.

27. Bleed the air out of the regular brake line again in "Step4: Increase" on the hand-held tester display.

28. Connect the vinyl tube to either one of the bleeder plug.

29. Depress the brake pedal several times, then loosen the bleeder plug of one of the above wheels with the pedal depressed.

30. When fluid stops coming out, tighten the bleeder plug, then release the brake pedal.

31. Repeat the previous 2 steps until all the air in the fluid is completely bled out.

32. Tighten the bleeder plug.

33. Repeat the above procedure to bleed the air out of the brake line for each wheel.

WHEEL SPEED SENSORS

REMOVAL & INSTALLATION

Front

See Figure 3.

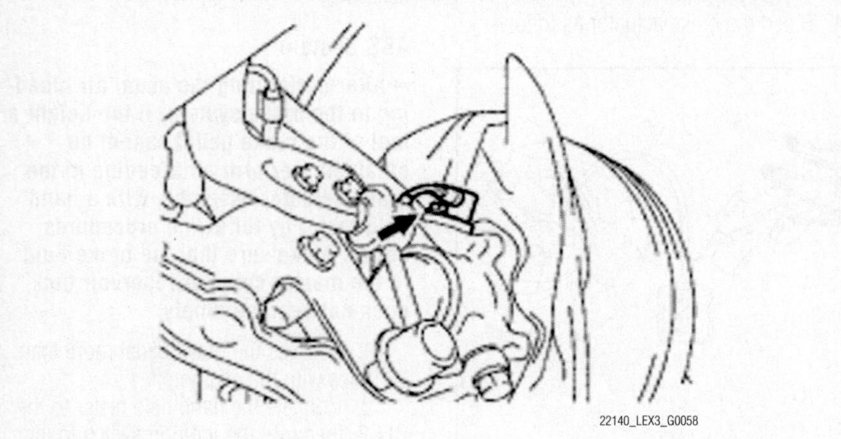

Fig. 3 Left rear speed sensor location—SC430 model

1. Remove the front wheel.

2. Remove the front fender liner.

3. Disconnect the sensor connector.

4. Remove the front speed sensor.

5. Remove the resin clip and 2 clamp bolts holding the sensor harness.

6. Remove the bolt and sensor.

➡Prevent foreign matter from attaching to the sensor tip.

To install:

7. Install the sensor with the bolt. Tighten bolt to 71 inch lbs. (8.0 Nm).

8. Install the sensor harness with the 2 clamp bolts and resin clip. Tighten the bolts to 44 inch lbs. (5.0 Nm).

9. Connect the speed sensor connector.

10. Install front fender liner.

11. Install front wheel. Tighten to 76 ft. lbs. (103 Nm).

Rear

See Figures 4 and 5.

1. Remove the rear seat cushion as follows:

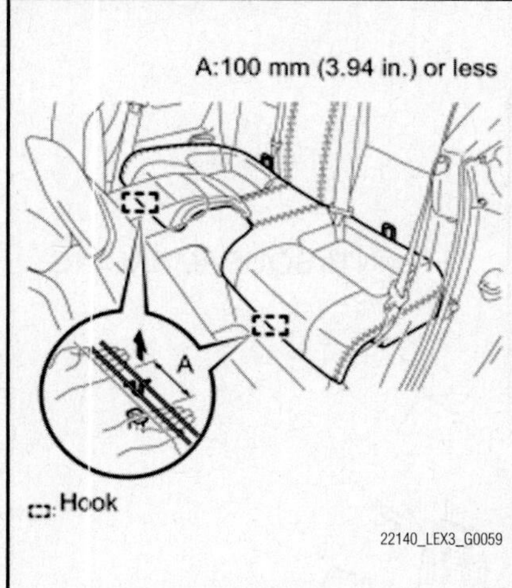

Fig. 4 Rear seat cushion assembly removal—SC430 model

- Detach the seat cushions 2 front hooks from the vehicle body.
- Choose a hook to detach first. Place your hands near the hook as shown in the illustration. Then lift the seat cushion to detach the hook.
- Repeat for the other hook.
- Detach the seat cushions 2 rear hooks from the seatback.
- Remove the seat cushion.

2. Using a screwdriver, detach the 8 claws and remove the 2 rear seat head rest plate covers.
3. Remove the 4 bolts and deflector board.

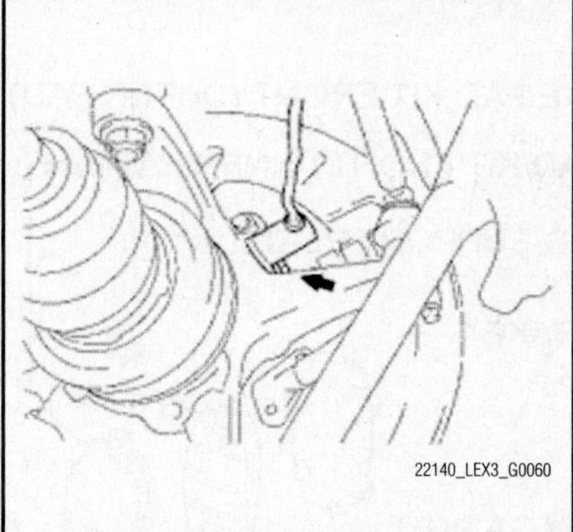

Fig. 5 Left rear speed sensor—SC430 model

4. Using a screwdriver, detach the 4 clips and 8 claws. Then remove the 2 headrest plates.
5. Remove the 2 nuts, 2 bolts and seatback assembly.
6. Remove rear speed sensor as follows:
   - Disconnect the sensor and HID connectors.
   - Remove the bolt and 2 nuts, and then disconnect the No. 2 and No. 3 clamps.
   - Remove the resin clamp and clamp bolt, and then disconnect the HID connector.
   - Remove the bolt and then disconnect the No. 1 clamp.
   - Remove the bolt and sensor.

➡Prevent foreign matter from attaching to the sensor tip.

To install:

7. Install the speed sensor with the bolt. Tighten to 71 inch lbs. (8.0 Nm).
8. Install the No. 1 clamp with the bolt. Tighten to 44 inch lbs. (5.0 Nm).
9. Install the No. 2 and No. 3 clamps with the bolt and 2 nuts. Tighten to 44 inch lbs. (5.0 Nm).
10. Connect the HID connector, and then install the resin clamp and clamp bolt. Tighten to 44 inch lbs. (5.0 Nm).
11. Connect the sensor and HID connector.
12. Install rear seat back assembly, tighten the bolts to 13 ft. lbs. (18 Nm).
13. Install the 2 headrests with the 4 nuts. Tighten to 44 inch lbs. (5.0 Nm).
14. Install the 2 headrest plates.
15. Install the deflector board with the 4 bolts. Tighten to 44 inch lbs. (5.0 Nm).
16. Attach the seat cushions 2 rear hooks to the seatback
17. Attach the seat cushions 2 front hooks to the vehicle body.
18. Confirm that the seat cushion is firmly installed.
19. Install the rear wheel. Tighten to 76 ft. lbs. (103 Nm).
20. Install the ornament with the 10 bolts. Tighten to 15 ft. lbs. (20 Nm).
21. Check speed sensor signal.

BRAKES

FRONT DISC BRAKES

BRAKE CALIPER

REMOVAL & INSTALLATION

See Figure 6.

1. Before servicing the vehicle, refer to the Precautions section.
2. Remove or disconnect the following:
   - Front wheels
   - Brake line at the caliper
   - 2 bolts to the holding the caliper to the steering knuckle
   - Caliper assembly

To install:

3. Install or connect the following:
   - Caliper. Tighten the 2 bolts to 87 ft. lbs. (118 Nm)
   - Brake line with 2 new gaskets and tighten the union to 22 ft. lbs. (30 Nm)
4. Refill the reservoir as necessary and bleed the brake system.
5. Install the rear wheel. Tighten to 76 ft. lbs. (103 Nm).
6. Install the ornament with the 10 bolts. Tighten to 15 ft. lbs. (20 Nm).
7. Road test vehicle.

DISC BRAKE PADS

REMOVAL & INSTALLATION

See Figure 7.

1. Before servicing the vehicle, refer to the Precautions section.
2. Remove or disconnect the following:
   - Front wheel
   - 2 bolts and remove the disc brake caliper assembly

➡Support the caliper. Do NOT allow to hang by the brake hose

   - 2 anti-squeal springs
   - Pads with anti-squeal shims
   - Disc brake support plates
   - Slide pins

To install:

3. Install or connect the following:
   - Slide pins
   - Disc brake pad support plate
   - Anti-squeal shims to each pad

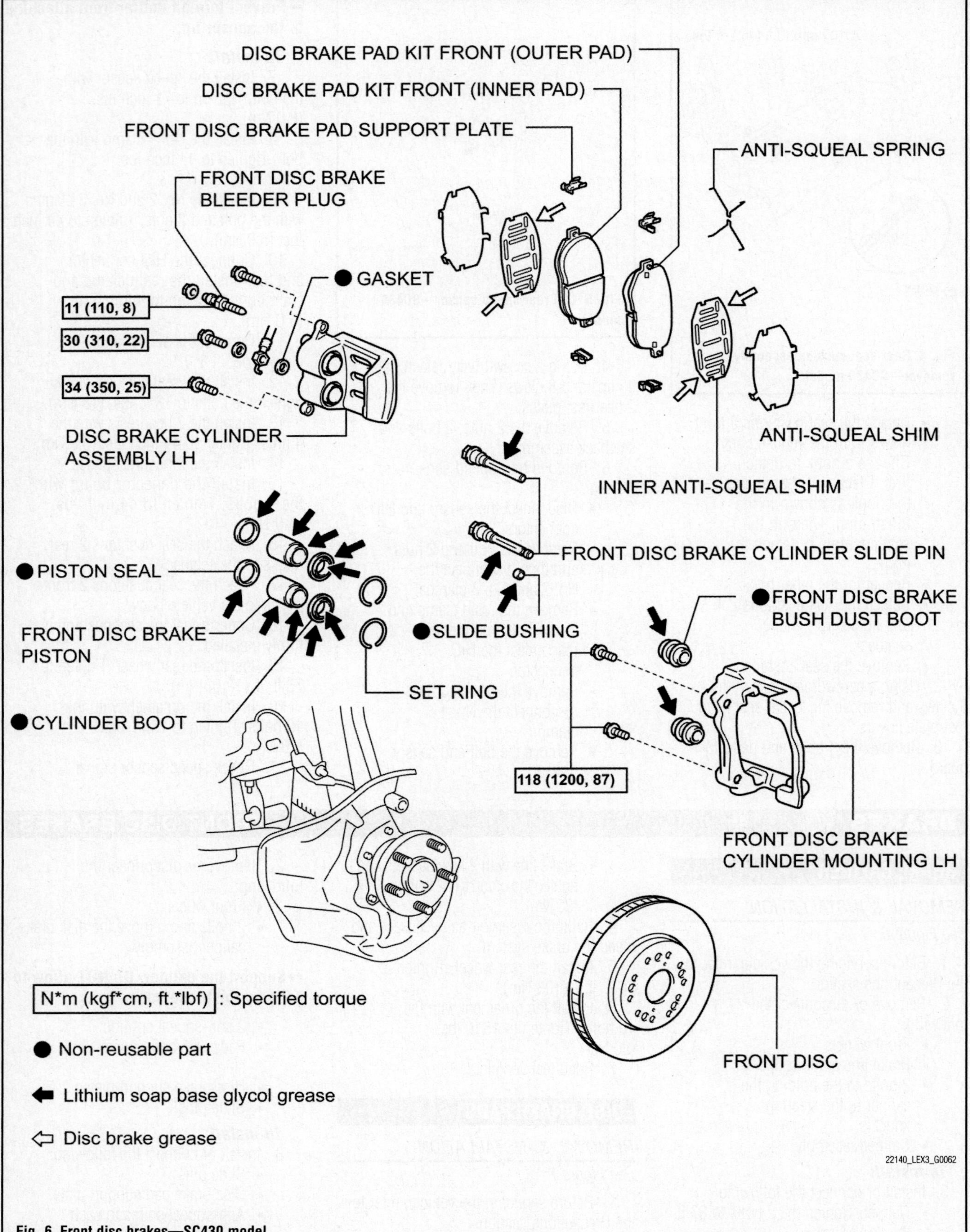

Fig. 6 Front disc brakes—SC430 model

4. Using a suitable tool, compress the piston carefully in the cylinder bores
- Inner pad with the wear indicator plate facing upward
- Install outer pad
- Caliper. Tighten the 2 bolts to 25 ft. lbs. (34 Nm)
- Front wheel

5. Install the rear wheel. Tighten to 76 ft. lbs. (103 Nm).
6. Install the ornament with the 10 bolts. Tighten to 15 ft. lbs. (20 Nm).
7. Road test vehicle.

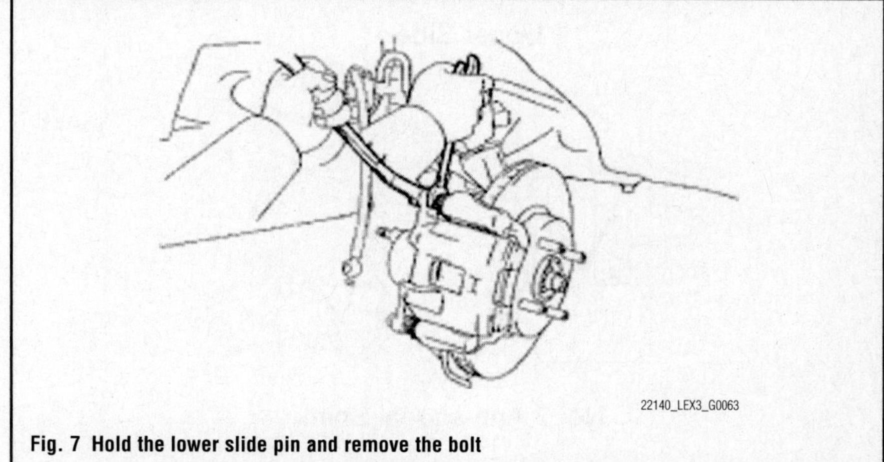

Fig. 7 Hold the lower slide pin and remove the bolt

BRAKES — REAR DISC BRAKES

BRAKE CALIPER

REMOVAL & INSTALLATION

See Figure 8.

1. Before servicing the vehicle, refer to the Precautions section.
2. Remove or disconnect the following:
   - Rear wheels
   - Brake line at the caliper, then plug it
   - Mounting bolts and the caliper assembly

To install:

3. Install the caliper on the torque plate with the 2 installation bolts. Tighten the mounting bolts to 77 ft. lbs. (104 Nm).
4. Connect the brake line with 2 new gaskets and tighten the union to 22 ft. lbs. (30 Nm).
5. Install the rear wheel. Tighten to 76 ft. lbs. (103 Nm).
6. Install the ornament with the 10 bolts. Tighten to 15 ft. lbs. (20 Nm).
7. Road test vehicle.
8. Refill the reservoir as necessary and bleed the brake system.
9. Road text vehicle and burnish brake linings.

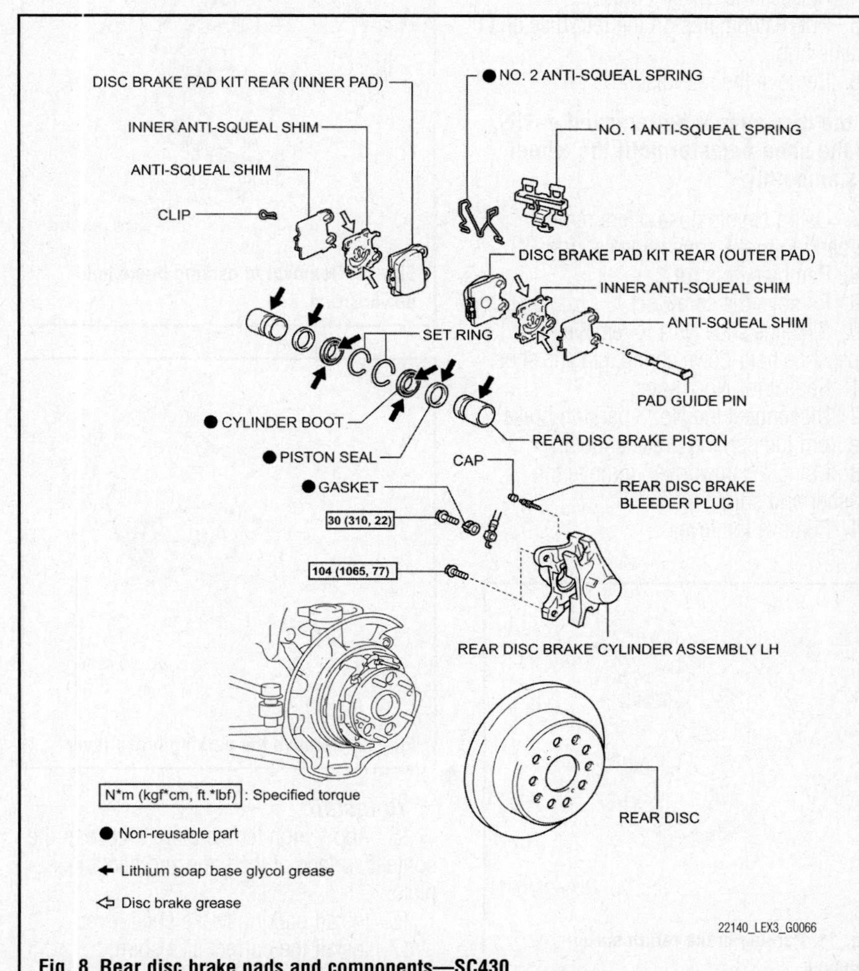

Fig. 8 Rear disc brake pads and components—SC430

DISC BRAKE PADS

REMOVAL & INSTALLATION

See Figure 9.

1. Before servicing the vehicle, refer to the Precautions section.
2. Remove or disconnect the following:
   - Rear wheel
   - Anti-squeal springs
   - Clip and guide pin
   - Disc pads
   - 4 anti-squeal shims from each pad

To install:

3. Install or connect the following:

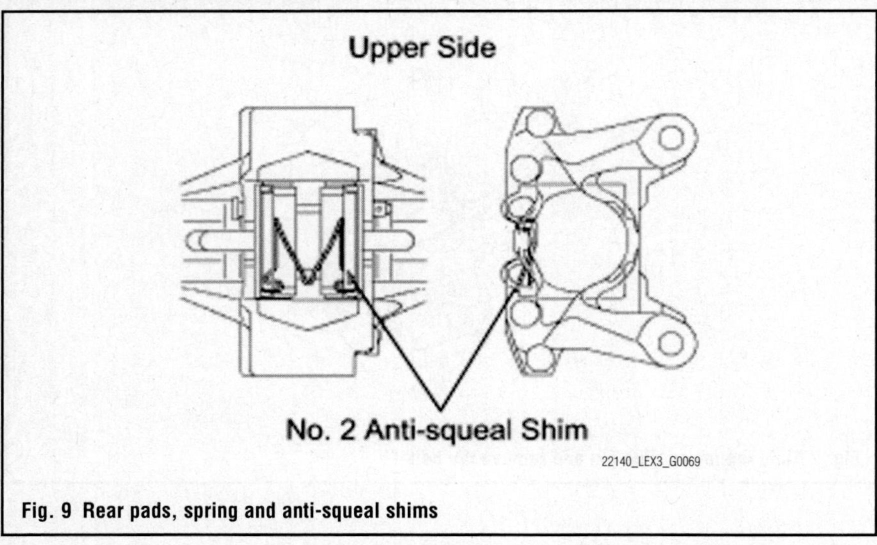

Fig. 9 Rear pads, spring and anti-squeal shims

- Apply disc brake grease to both sides of the inner anti-squeal shims
- Install 2 shims on each pad

➡Make sure that the arrows on the anti-squeal shims face the direction of wheel rotation.

4. Using a suitable tool, compress the piston carefully in the cylinder bores
   - Install pads
   - Anti-squeal springs
5. Install the rear wheel. Tighten to 76 ft. lbs. (103 Nm).
6. Install the ornament with the 10 bolts. Tighten to 15 ft. lbs. (20 Nm).
7. Road test vehicle.

BRAKES — PARKING BRAKE

PARKING BRAKE CABLES

ADJUSTMENT

See Figure 10.

1. Remove the instrument panel safety pad No. 1.
2. Loosen the lock nut and turn the adjusting nut until the pedal travel is correct at 44 lbs. (300 N) at 7–9 clicks.
3. Tighten the lock nut to 48 inch lbs. (5.4 Nm).
4. Install the instrument panel safety pad No. 1.

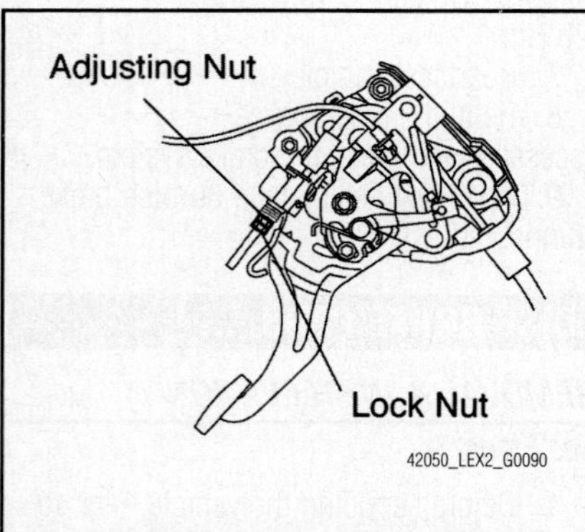

Fig. 10 Parking brake adjuster and lock nut—SC430

PARKING BRAKE SHOES

REMOVAL & INSTALLATION

See Figures 11 through 14.

1. Remove the rear wheels.
2. Remove the 2 bolts and separate the rear disc brake caliper assembly.

➡Hang the caliper with wire or equivalent.

3. Remove the caliper mounting bolts.
4. Release the parking brake.
5. Put matchmarks on the rear disc and rear axle hub.
6. Remove the rear disc.

➡If the disc cannot be removed easily, turn the shoe adjuster until the wheel turns smoothly.

7. Using needle nose pliers remove both parking brake shoe return springs.
8. Remove the strut.
9. Remove the screw set.
10. Slide the shoe No.1 to remove the 2 cups, shoe hold-down spring, pin and shoe.
11. Repeat for No 2 shoe.
12. Disconnect the No. 3 parking brake cable from the parking brake shoe lever.
13. Using a screwdriver, remove the C-washer and shim.
14. Remove the lever.

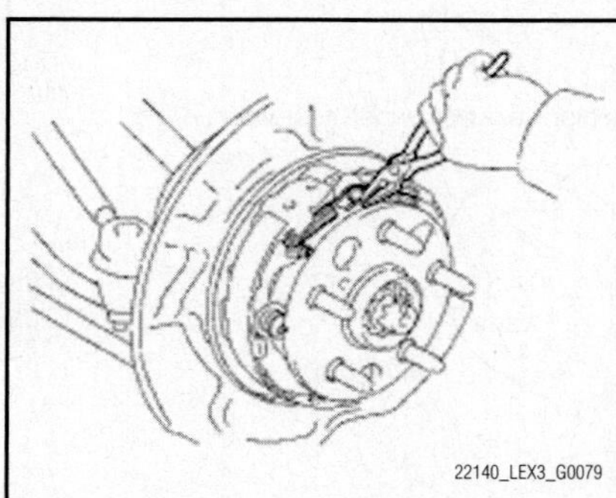

Fig. 11 Parking brake return spring removal

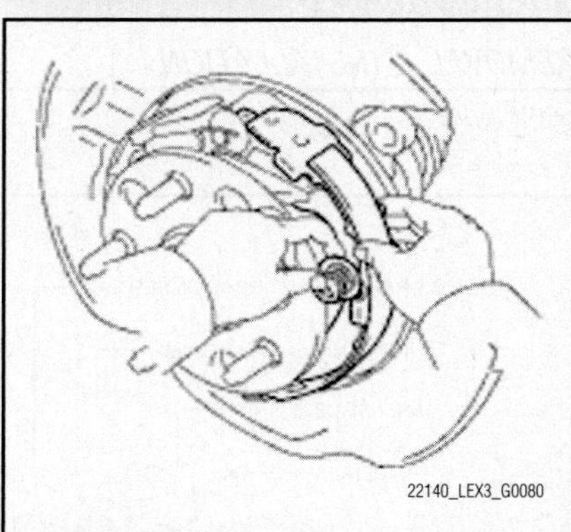

Fig. 12 Removal of parking brake hold down spring

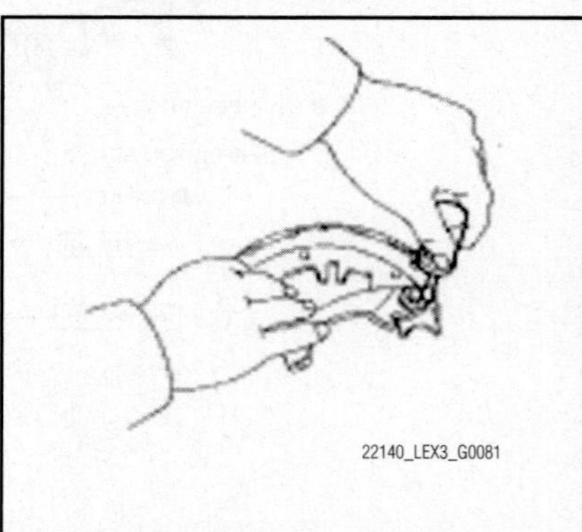

Fig. 13 Remove the parking brake lever

To install:

15. Apply high temperature grease to the contact surface of the shoe and backing plate.
16. Install parking brake shoe lever.
17. Install the correct size shim.
18. Install a new C-washer.

19. Connect the No. 3 parking brake cable to the parking brake shoe lever.

20. Install the No. 2 shoe with the 2 cups, shoe hold-down spring and pin.

21. Repeat for No. 1 shoe.

22. Apply high temperature grease to the adjusting bolt.

23. Install the parking brake adjusting screw.

24. Using needle-nose pliers, install the tension spring.

25. Apply high temperature grease to the contact surface of the strut and shoe return spring.

26. Install the strut.

27. Using needle-nose pliers, install the 2 shoe return springs.

28. Check that each part is installed properly

➡There should be no oil or grease adhering to the friction surface of the shoe lining and disc.

29. Align the matchmarks and install the disc.

30. Temporarily install the hub nuts.

31. Remove the hole plug, and then turn the adjuster to expand the shoe adjuster until the disc locks.

32. Contract the shoe adjuster until the disc can rotate smoothly. Standard : Return 8 notches.

33. Check that the shoe has no brake drag.

34. Install the hole plug.

35. Depress the parking brake pedal.

36. Drive the vehicle at approximately 31 mph (50 km/h) for approximately 0.25 miles.

37. Repeat the previous 2 steps 2 or 3 times.

38. Install rear brake caliper and tighten mounting bolts to 77 ft. lbs. (104 Nm).

39. Install the rear wheel. Tighten to 76 ft. lbs. (103 Nm).

40. Install the ornament with the 10 bolts. Tighten to 15 ft. lbs. (20 Nm).

41. Inspect parking brake travel and adjust as needed.

42. Depress the parking brake pedal.

43. Drive the vehicle at approximately 31 mph (50 km/h) for approximately 0.25 miles.

44. Repeat the previous 2 steps 2 or 3 times.

45. Check parking brake travel and adjust parking brake clearance if needed.

ADJUSTMENT

See Figures 15 and 16.

1. Temporarily install the hub nuts.
2. Remove the hole plug, and turn the adjuster and expand the shoes until the disc locks.
3. Contract the shoe adjuster until the disc can rotate smoothly. Standard: Return 7–8 notches
4. Check shoe is no brake drag.
5. Install the hole plug.

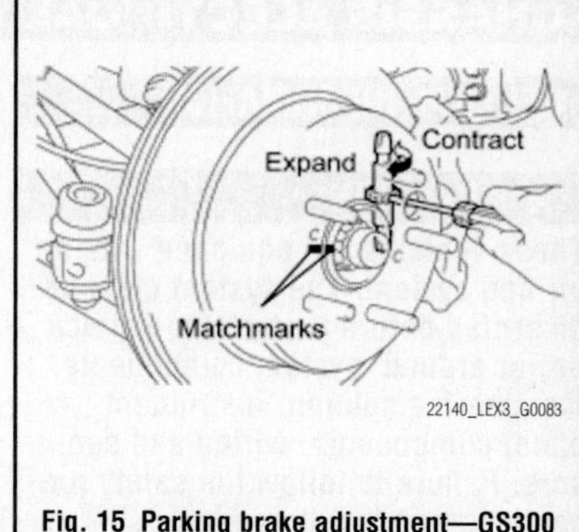

Fig. 15 Parking brake adjustment—GS300 model

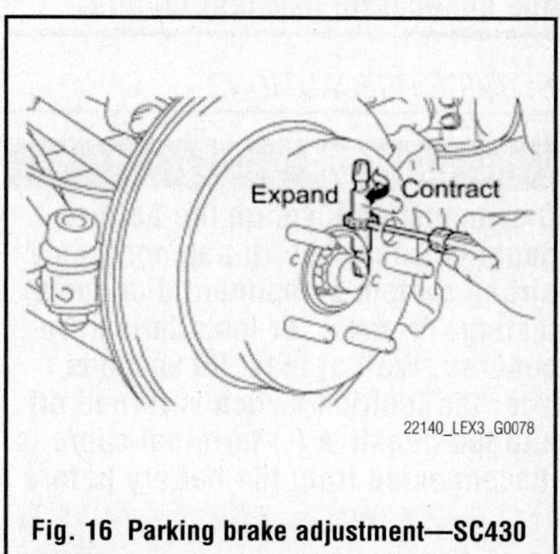

Fig. 16 Parking brake adjustment—SC430 model

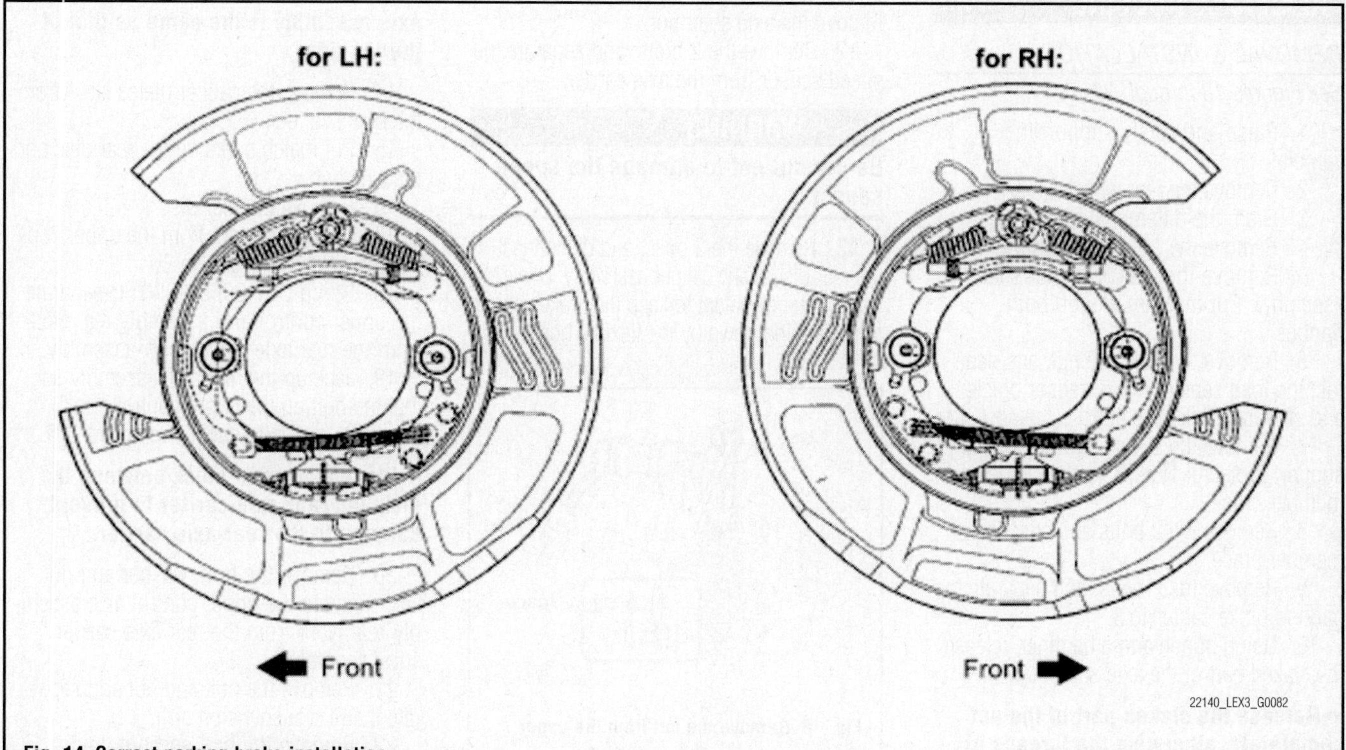

Fig. 14 Correct parking brake installation

GENERAL INFORMATION

CAUTION

These vehicles are equipped with an air bag system. The system must be disarmed before performing service on, or around, system components, the steering column, instrument panel components, wiring and sensors. Failure to follow the safety precautions and the disarming procedure could result in accidental air bag deployment, possible injury and unnecessary system repairs.

SERVICE PRECAUTIONS

CAUTION

Disconnect and isolate the battery negative cable before beginning any airbag system component diagnosis, testing, removal, or installation procedures. Wait at least 90 seconds after the ignition switch is turned off and the negative (-) terminal cable is disconnected from the battery before starting the operation. The SRS is equipped with a backup power source, so if work is started within 90 seconds after disconnecting the negative (-) terminal cable from the battery, the SRS may be deployed. Failure to disable the airbag system may result in accidental airbag deployment, personal injury, or death.

DISARMING THE SYSTEM

To avoid personal injury when working on vehicles equipped with an air bag, the negative battery cable must be disconnected and at least 90 seconds must elapse before working on the system. Failure to do so may result in deployment of the air bag.

ARMING THE SYSTEM

To rearm the air bag system, simply reconnect the battery cable(s).

CLOCKSPRING CENTERING

See Figure 17.

1. Center the clockspring spiral cable as follows:

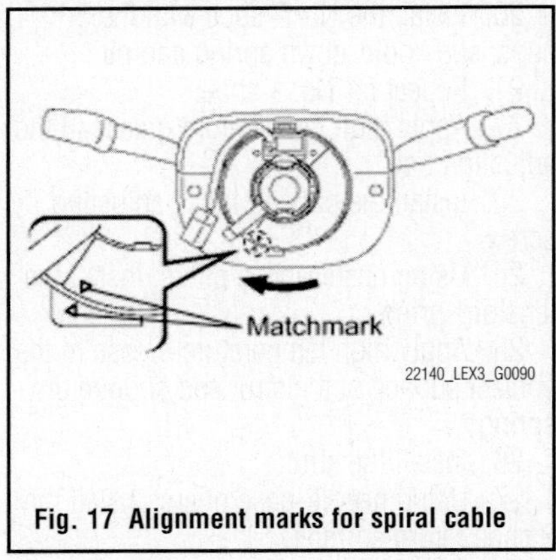

Fig. 17 Alignment marks for spiral cable

- Check that the ignition switch is OFF.
- Check that the battery negative terminal is disconnected. Do not start the operation for 90 seconds after removing the terminal.
- Turn the cable counterclockwise by hand until it becomes harder to turn.
- Then rotate the cable clockwise about 2.5 turns to align the marks. The cable will rotate about 2.5 turns to both left and right of the center.

DRIVE TRAIN

REAR AXLE HOUSING

REMOVAL & INSTALLATION

See Figures 18 through 22.

1. Raise and safely support the vehicle.
2. Remove the rear wheels.
3. Drain the differential oil.
4. Remove the exhaust assembly.
5. Remove the rear propeller shaft assembly. Put matchmarks on both flanges
6. Remove the bolt and nut, and separate the load sensing valve sensor bracket and stabilizer link assembly.
7. Remove the 2 nuts and differential support protector No.2 from the suspension member brace.
8. Remove the 2 bolts and suspension member brace.
9. Remove the 2 bolts, and separate the parking brake cable No.3.
10. Using punch and a hammer, release the staked part of the axle shaft nut.

➡Release the staked part of the nut completely, otherwise the threads of the drive shaft may be damaged.

11. While depressing the brake pedal, remove the axle shaft nut.
12. Remove the 2 bolts, and separate the speed sensor from the axle carrier.

WARNING

Be careful not to damage the speed sensor.

13. Remove the 2 bolts, and disconnect the rear disc brake caliper assembly. Use a wire or an equivalent to keep the brake caliper from hanging down by the flexible hose.

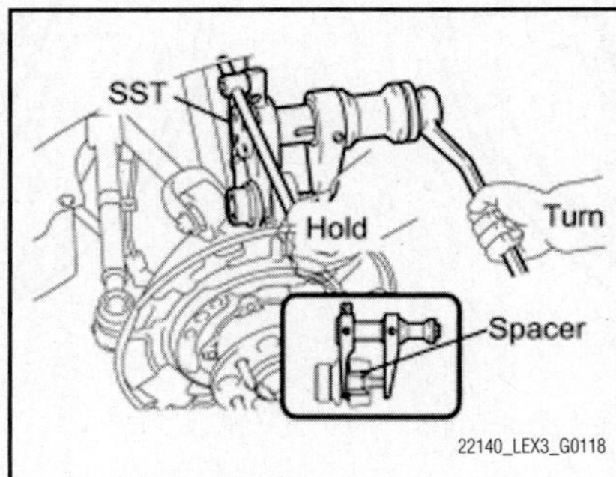

Fig. 18 Remove the nut from the upper control arm assembly rear No.2.

➡Removal procedure of the RH side axle assembly is the same as that of the LH side.

14. Remove the caliper plates No.1 from the brake caliper.
15. Put matchmarks on the rear disc and the axle hub.
16. Remove the rear disc.
17. Remove the nut from the upper control arm assembly rear No.2.
18. Using SST 09628-00011, separate the upper control arm assembly rear No.2 from the rear axle carrier sub-assembly.
19. Jack up the rear axle assembly so that the bolt on the upper control arm assembly rear No.1 can be removed.

➡Place a wooden block between the jack and rear axle carrier to prevent damage to the rear axle carrier.

20. Remove the bolt, washer and nut, and separate the upper control arm assembly rear No.1 from the rear axle carrier sub-assembly.
21. Remove the bolt and nut and separate the rear suspension arm.
22. Remove the bolt and nut, and separate the rear suspension arm assembly

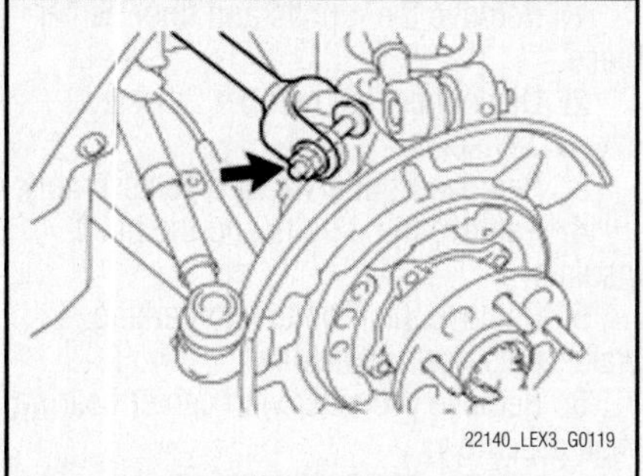

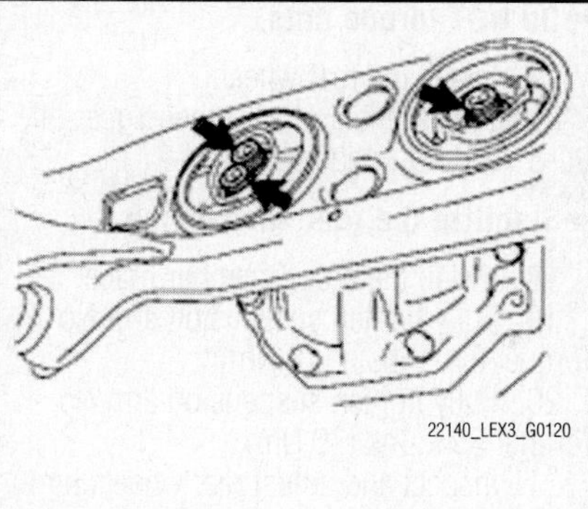

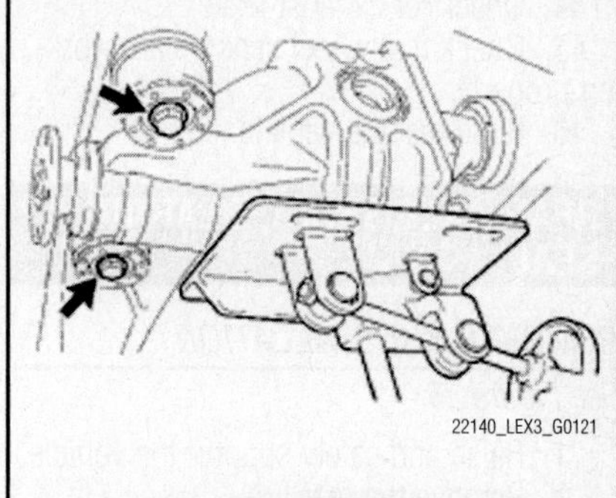

Fig. 19 Remove the bolt, washer and nut, and separate the upper control arm assembly

Fig. 20 Remove the axle carrier 3 bolts

Fig. 21 Remove the differential mount stopper bolts

No.2 from the rear axle carrier sub-assembly.

23. Using SST 09628-00011, separate the toe control link sub-assembly from the rear axle carrier sub-assembly.

24. Using a plastic hammer, separate the drive shaft from the rear axle carrier sub-assembly.

25. Using a slide hammer and adapter carefully, remove the rear drive shaft assembly.

26. Support the differential carrier with a jack.

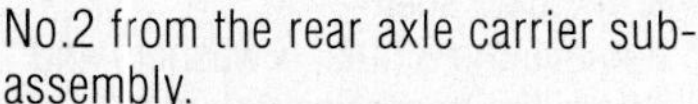

WARNING

Do not drop the differential carrier assembly.

27. Using a hexagon wrench (12 mm), remove the 3 bolts.

28. Remove the 2 bolts, 2 rear differential mount stopper uppers, 2 differential mount stopper lowers and rear differential carrier assembly.

To install:

29. Support the rear differential carrier with a jack and temporary install the lower mount stoppers and 2 new front side set bolts.

➡Do not let the rear differential carrier interfere with the drive shaft.

30. Using a 12 mm hexagon wrench, install the 3 rear side set hexagon bolts. Tighten to 105 ft. lbs. (142 Nm).

31. Using a 10 mm hexagon wrench, install the drain plug with a new gasket. Tighten to 39 ft. lbs. (49 Nm).

32. Add G-L5 75w90 gear oil.

33. Install rear differential fill plug and tighten to tighten to 39 ft. lbs. (49 Nm).

34. Install right and left drive shaft assembly, install rear wheels.

➡Whether or not the inboard joint shaft is in contact with the pinion shaft can be known from the sound or feeling when driving it in.

35. Insert the yoke of the intermediate shaft into the transmission.

36. Install the 2 center support bearing washers and center support bearing, and temporarily tighten the 2 bolts.

37. Align the matchmarks on the propeller shaft flange and differential companion flange, and connect the shaft with the 4 bolts, washers and nuts. Tighten to 54 ft. lbs. (74 Nm).

38. Check that the center line of the bracket is at right angles to the shaft axial direction.

39. Tighten the 2 intermediate shaft mounting bolts to 36 ft. lbs. (49 Nm).

40. Install the heat insulator No.1 with the 4 nuts and grommet. Tighten to 48 inch lbs. (5.4 Nm).

41. Install the tail pipe to connect the 6 exhaust pipe supports.

42. Install the support bracket with the 2 bolts. Tighten to 32 ft. lbs. (43 Nm).

43. Inspect and adjust rear wheel alignment.

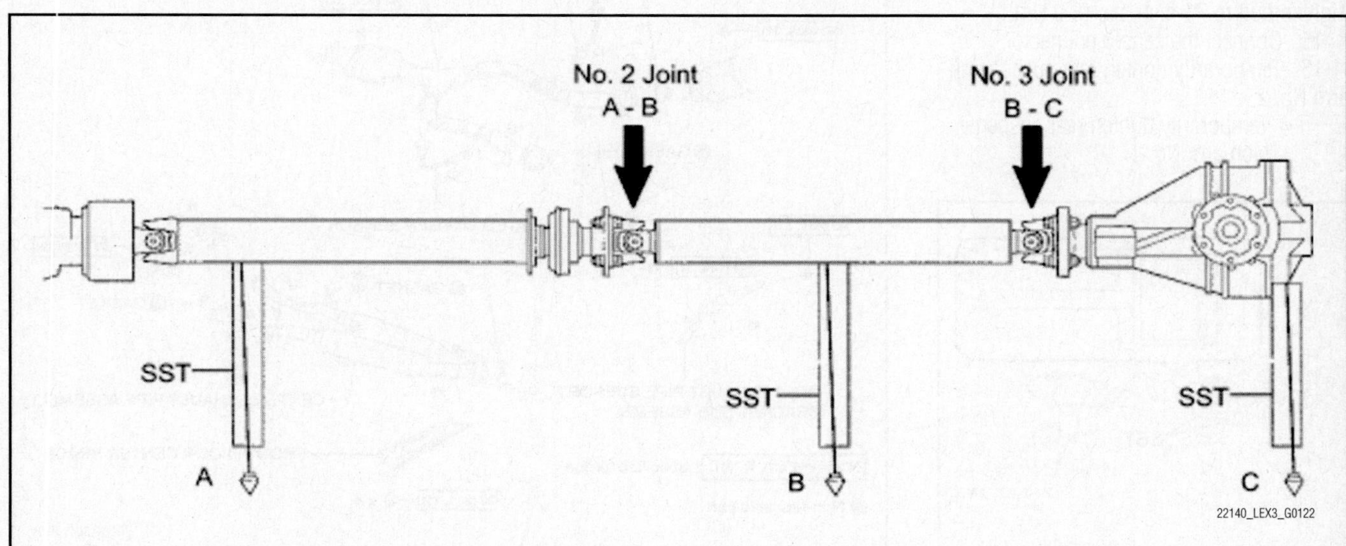

Fig. 22 Measure the installation angle by placing the SST 09370-50010 in the position shown in the illustration.

44. Check for exhaust leaks.
45. Check that wheel sensors are working properly.
46. Road test the vehicle.

REAR AXLE SHAFT, BEARING & SEAL

REMOVAL & INSTALLATION

See Figure 23.

1. Raise and safely support the vehicle.
2. Remove the rear tire.
3. Drain the differential oil, install plug and tighten to 36 ft. lbs. (49 Nm).
4. Remove right rear suspension member brace.
5. Remove the bolt and nut, and disconnect the stabilizer link and the height control link from the rear No. 2 suspension arm.
6. Remove the 2 bolts and nuts, and disconnect the No. 1 suspension arm.
7. Disconnect the speed sensor connector and mounting bolt.
8. Push the rear axle carrier toward the outside of the vehicle. Using a plastic-faced hammer, disconnect the rear drive shaft from the rear axle carrier.
9. Using seal removal adapter and slide hammer. Remove rear side gear shaft oil seal.

To install:

10. Using seal driver and a hammer, install new seal.
11. Apply MP grease to the oil seal lip.
12. Align the shaft splines and install the drive shaft assembly with a brass bar and hammer.
13. Install new rear axle shaft nut and tighten to 214 ft. lbs. (290 Nm).
14. Install the sensor with the bolt. Tighten bolt to 71 inch lbs. (8.0 Nm).
15. Connect the sensor connector.
16. Temporarily tighten rear suspension arm No. 2.
    - Temporarily tighten rear suspension arm No. 1.

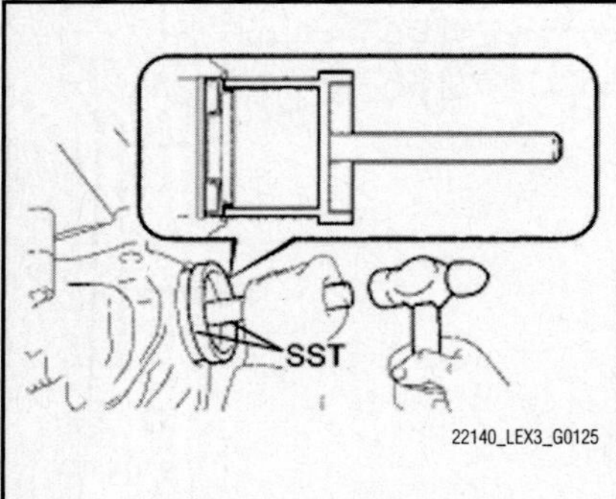

Fig. 23 Seal Installation on—SC430 rear differential shown

➡**Do NOT torque nuts.**

17. Install the rear wheel.

Install differential oil. Synthetic gear oil GL-5 75W-90 or equivalent

➡**Stabilize the rear suspension**

18. Install the cross member brace
19. Fully tighten suspension arm No. 2. Torque: 81 ft. lbs. (110 Nm).
20. Fully tighten suspension arm No. 1. Torque: 55 ft. lbs (75 Nm).
21. Inspect and adjust rear wheel alignment as needed.
22. Check that speed sensors are working properly.

REAR DRIVESHAFT

REMOVAL & INSTALLATION

See Figures 24 and 25.

1. Remove the 4 bolts and floor center brace.
2. Disconnect the number 2 heated oxygen sensor.
3. Remove front exhaust pipe assembly.
4. Remove the 4 bolts and the no. 1 insulator.
5. Remove the 4 bolts and parking brake cable heat insulator.
6. Remove propeller with center bearing shaft assembly:
    a. Using SST 09922-10010, loosen the adjusting nut until it can be turned by hand.
    b. Place matchmarks on the transmission companion flange, flexible coupling and intermediate shaft.
    c. Remove the 3 nuts, 3 washers and 3 bolts from the transmission side.

➡**The propeller shaft should not be disconnected from the flexible coupling.**

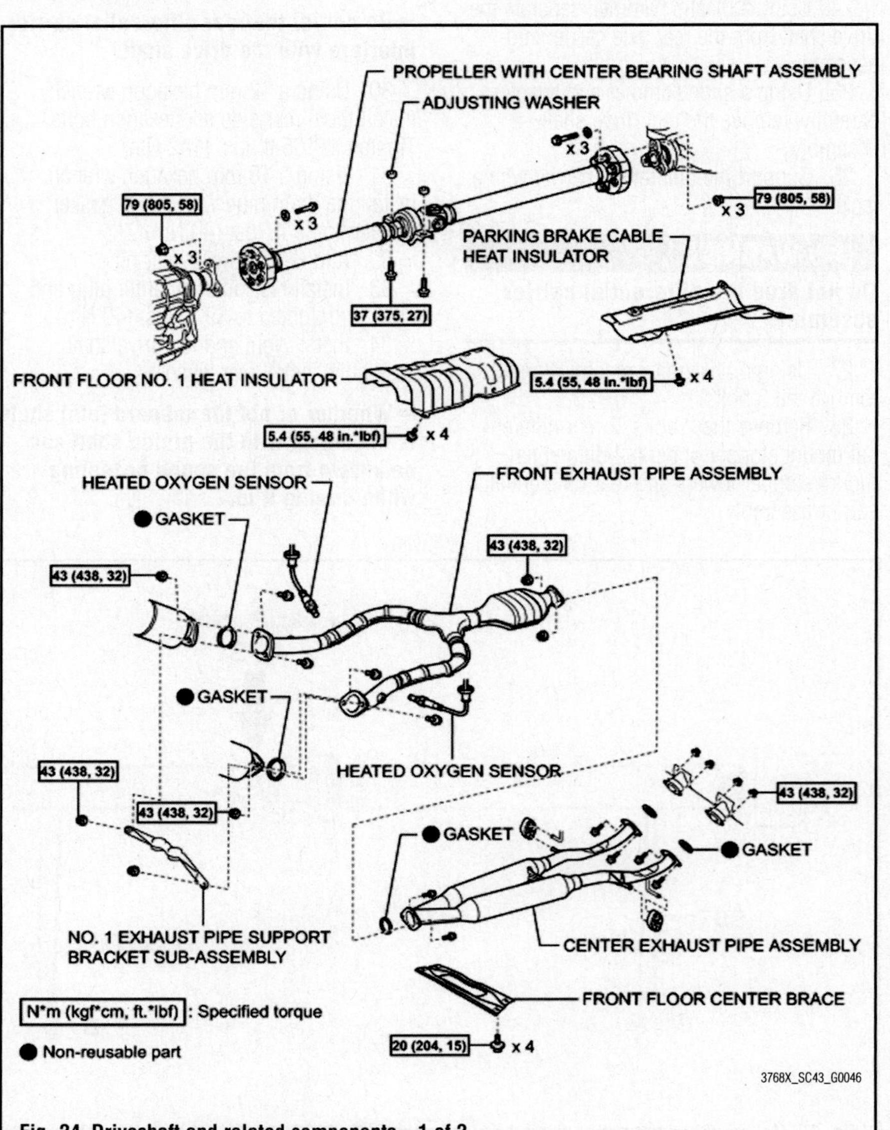

Fig. 24 Driveshaft and related components—1 of 2

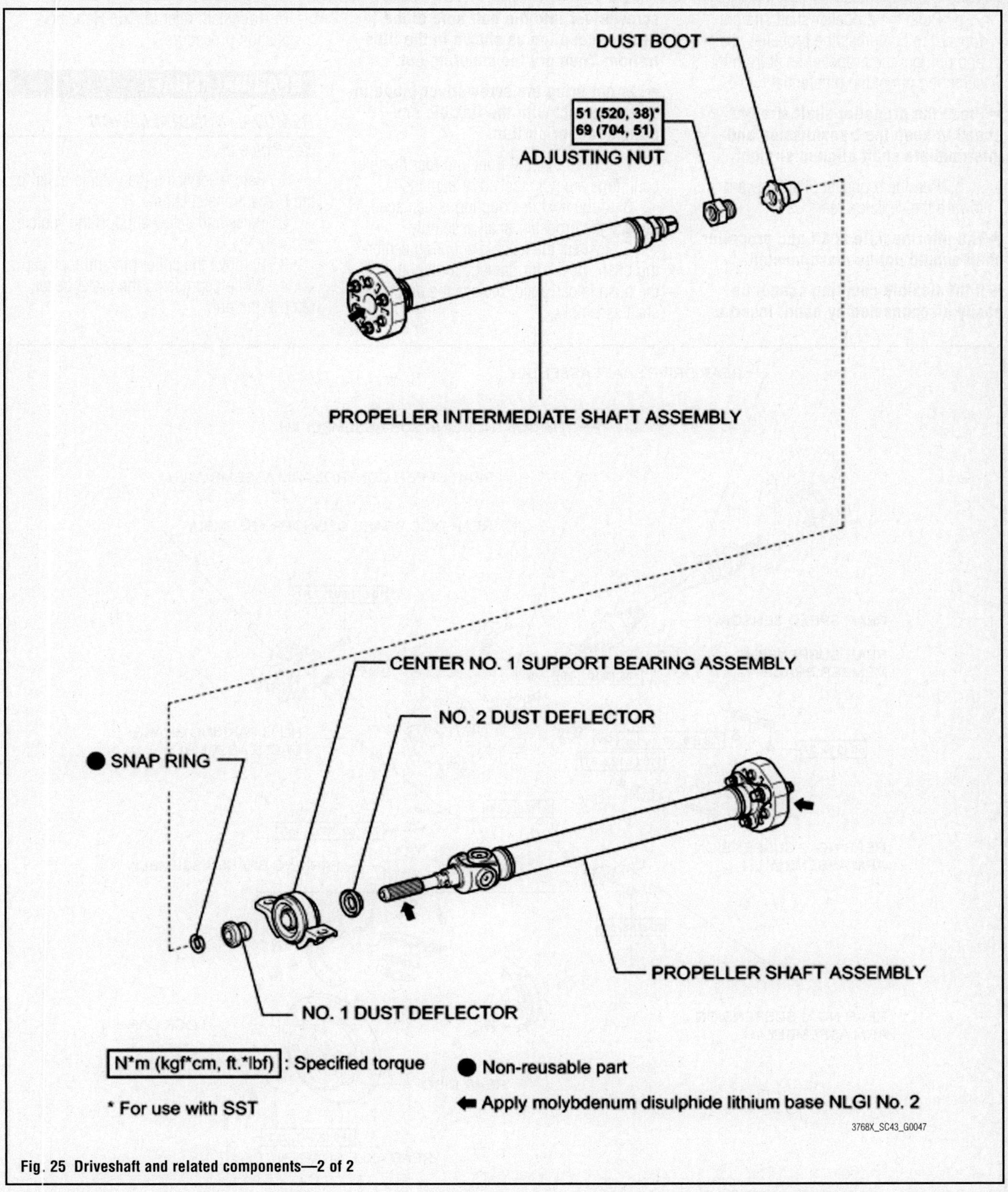

Fig. 25 Driveshaft and related components—2 of 2

d. Place matchmarks on the differential companion flange, flexible coupling and propeller shaft.

e. Remove the 3 nuts, 3 washers and 3 bolts from the differential side.

f. Remove the 2 center support bearing set bolts and adjusting washers. Maximum joint angle: 5°

➡**When removing the set bolts, support the center support bearing by hand so that the transmission and intermediate shaft, and propeller shaft and differential remain in a straight line.**

➡**Some vehicles are not equipped with an adjusting washer.**

g. Push the propeller shaft straight forward to compress the propeller shaft and pull out the propeller shaft from the centering pin of the differential.

➡Press the propeller shaft straight ahead to keep the transmission and intermediate shaft aligned straight.

h. Pull the propeller shaft outward toward the vehicle's rear.

➡The intermediate shaft and propeller shaft should not be disconnected.

➡If the flexible coupling cannot be easily disconnected by hand, insert a screwdriver into the bolt hole of the flexible coupling as shown in the illustration. Then pry the coupling out.

➡Do not bring the screwdriver blade in direct contact with the flexible coupling's rubber portion.

7. Check that the front and rear flexible couplings are not cracked or damaged.

8. If the flexible coupling is damaged, replace the propeller shaft assembly.

9. Inspect the flexible coupling centering bush. Check for damage to the bush. If the bush is damaged, replace the propeller shaft assembly.

10. To install, refer to Axle Housing installation procedure.

REAR HALFSHAFT

REMOVAL & INSTALLATION

See Figure 26.

1. Before servicing the vehicle, refer to the Precautions section.
2. Raise and safely support the vehicle securely.
3. Remove the cotter pin and lock cap.
4. While depressing the brake pedal, remove the nut.

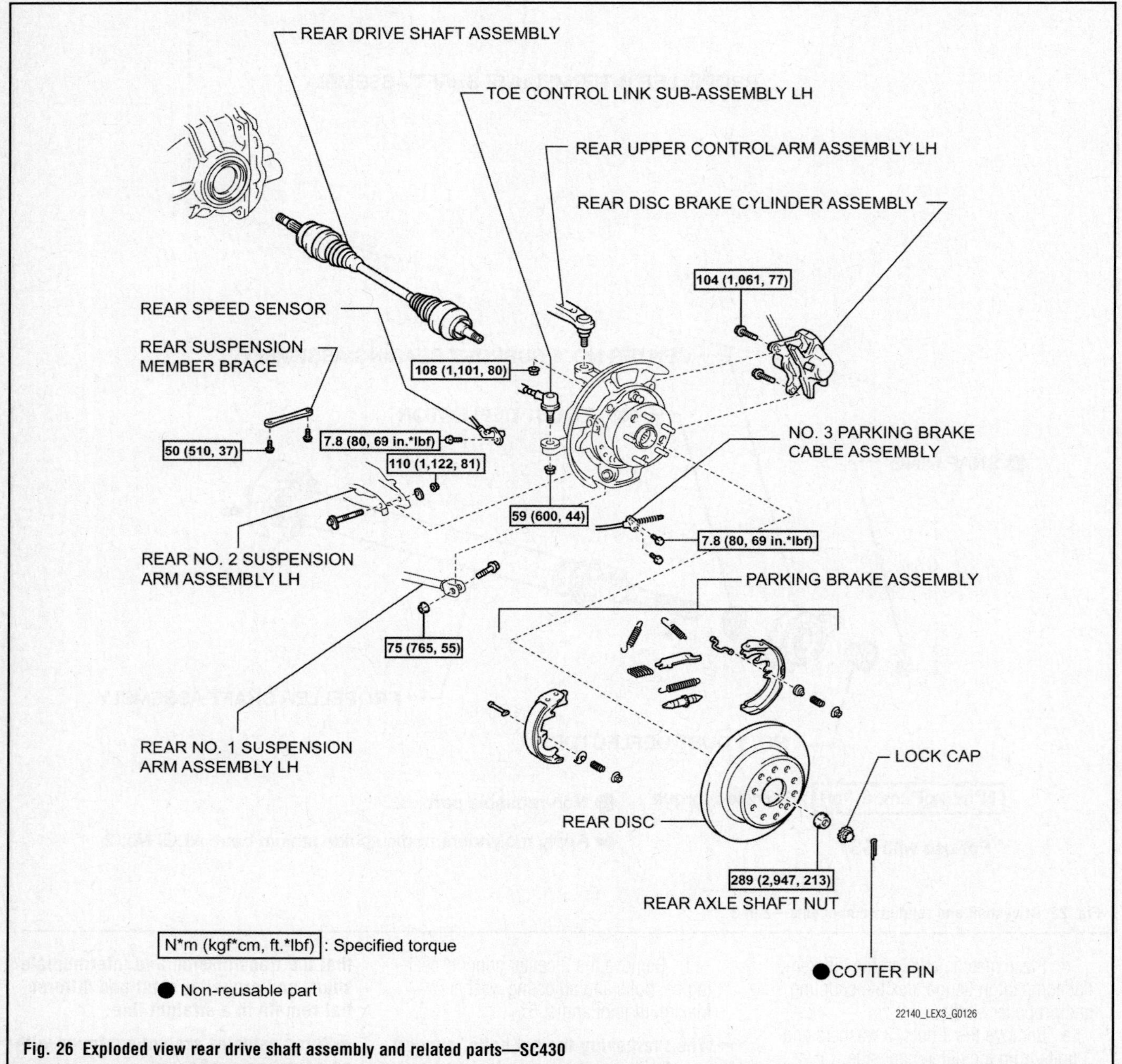

Fig. 26 Exploded view rear drive shaft assembly and related parts—SC430

5. Remove the 2 bolts, and disconnect the rear disc brake cylinder.

➡Use a wire or equivalent to keep the brake cylinder from hanging down by the flexible hose.

6. Remove the rear brake disc.
7. Remove the bolts and disconnect the speed sensor from the axle carrier.
8. Remove the 2 bolts, and disconnect the parking brake cable assembly.
9. Using SST 09610-20012, disconnect the toe control link from the axle carrier.
10. Remove the bolt and nut, and disconnect the No.1 rear suspension arm.
11. Remove the nut to the rear upper control arm assembly.
12. Using SST 09628-00011, disconnect the upper control arm from the axle carrier.
13. Connect the rear axle carrier to the upper control arm, and loosely install the nut.
14. Using a plastic-faced hammer, tap the rear drive shaft assembly's tip to detach the rear drive shaft assembly.
15. Remove the nut and disconnect the upper control arm.
16. Place matchmarks on the adjusting cam and rear suspension arm.
17. Remove the nut, No. 2 camber adjusting cam and No. 1 camber adjusting cam. Then disconnect the rear suspension arm and remove the axle carrier assembly.
18. Remove the 2 bolts and the rear suspension member brace.
19. Push the rear axle carrier toward the outside of the vehicle. Using a plastic-faced hammer, disconnect the rear drive shaft from the rear axle carrier.
20. Using a slide hammer and adapter remove the axle assembly.

To install:

21. Insert the outboard joint side of the halfshaft through the axle hub. Align the matchmarks on the side gear shaft and the halfshaft.
22. Coat the threads with clean oil and install the hex bolts.
23. Install axle shaft assembly.
24. Install the rear No. 2 suspension arm to the rear axle carrier.
25. Align the matchmarks of the camber adjust cam and rear axle carrier, and loosely install the nut.
26. Install the rear suspension member brace with the 2 bolts. Tighten the bolts to 37 ft. lbs. (50 Nm).
27. Connect the upper control arm to the rear axle carrier, and loosely install a new nut.
28. Connect the toe control link LH to the rear axle carrier, and install a new nut. Tighten to 44 ft. lbs. (59 Nm).
29. Tighten the rear upper control arm assembly to 80 ft. lbs. (108 Nm).
30. Connect the rear suspension arm and loosely install the bolt and a new nut.
31. Install the parking brake cable assembly and tighten backing plate bolts to 69 inch lbs. (7.8 Nm).
32. Inspect and adjust parking brake pedal travel.
33. Connect the brake cylinder with the 2 bolts and tighten to 77 ft. lbs. (104 Nm).
34. Install the sensor and connector tighten the mounting bolt to 71 inch lbs. (8.0 Nm).
35. While depressing the brake pedal, install the axle shaft nut. Tighten to 214 ft. lbs. (290 Nm).
36. Install the lock cap and a new cotter pin.
37. Lower the vehicle.
38. Press down on the vehicle several times to stabilize the suspension.
39. Tighten the No. 2 suspension arm to 81 ft. lbs. (110 Nm).
40. Tighten the No. 1 suspension arm to 55 ft. lbs. (74 Nm).
41. Install rear wheel and tighten to 76 ft. lbs. (103 Nm).
42. Inspect and adjust the rear wheel alignment.
43. Check ABS speed sensor signal.

REAR PINION SEAL

REMOVAL & INSTALLATION

See Figure 27.

1. Drain the gear oil.
2. Remove the rear propeller shaft.
3. Using a torque wrench, measure the total preload and record.

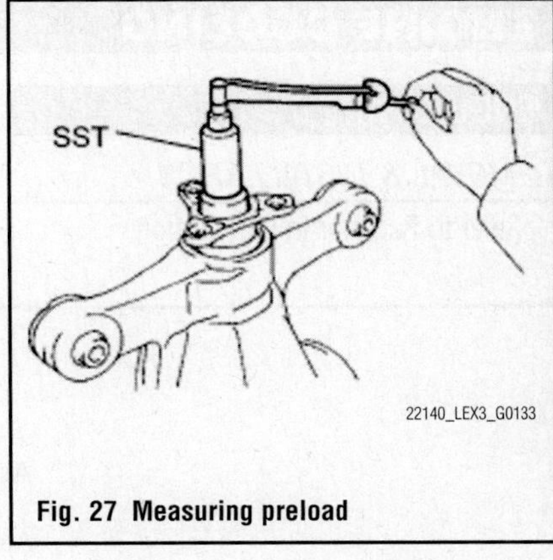

Fig. 27 Measuring preload

4. Using a punch and a hammer, unstake the staked part of the drive pinion nut.
5. Using SST 09330-00021 to hold the flange, remove the drive pinion nut.
6. With a puller remove the rear pinion flange.
7. Remove and replace rear dust deflector if it is damaged.
8. Carefully remove the oil seal from the differential carrier.

To install:

9. Install the companion flange with a flange installer.
10. Coat the threads of the drive pinion nut with hypoid gear oil LSD.
11. Using SST 09330-00021 to hold the flange, tighten the drive pinion nut.
12. Tighten the nut to approximately 72 ft. lbs. (98 Nm), and tighten it further while checking the preload.
13. Turn the bearing clockwise and counterclockwise several times to stabilize it
14. Using a torque wrench, measure the preload.
15. Drive pinion preload (at starting):
    a. New bearing 13.0 to 17.8 inch lbs. (1.5 to 2.0 Nm)
    b. Reused bearing 4.3 to 6.9 inch lbs. (0.5to 0.8 Nm)

➡Record the preload for total preload measurement.

16. If the preload is not within the specified range, adjust the differential drive pinion preload or repair as necessary.
17. Fill the differential with synthetic gear oil GL-5 75W-90 or equivalent.

ENGINE COOLING

ENGINE FAN

REMOVAL & INSTALLATION

Refer to Radiator in this section.

RADIATOR

REMOVAL & INSTALLATION

See Figure 28.

1. Before servicing the vehicle, refer to the Precautions section.
2. Disconnect the negative battery cable. Wait at least 90 seconds before performing any work.

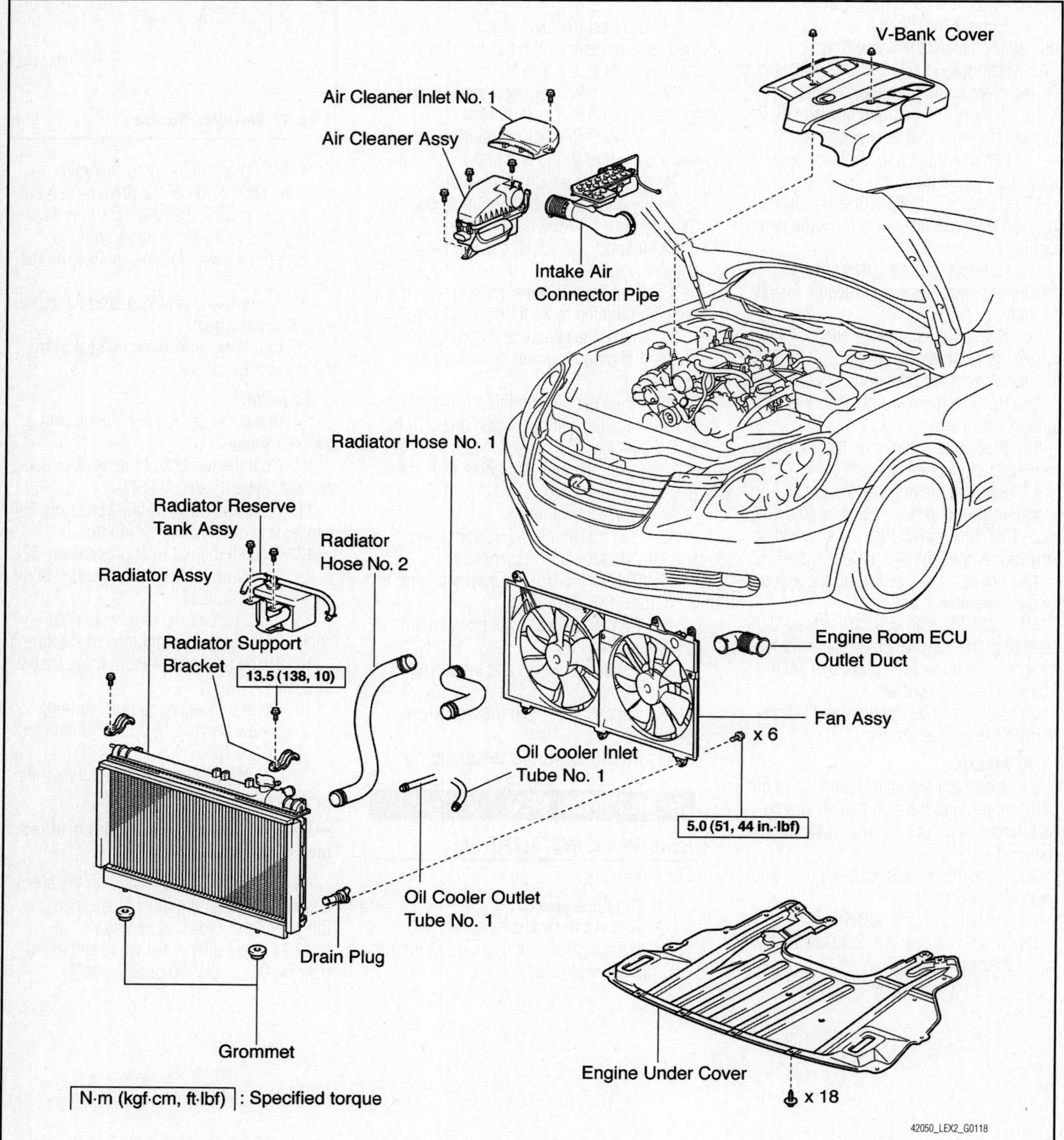

Fig. 28 Exploded view of the radiator assembly and related components—SC430

3. Remove the engine undercover.
4. Remove air cleaner inlet No. 1.
5. Drain engine coolant.
6. Remove the radiator reserve tank assembly.
7. Remove the V-bank cover.
8. Remove the intake air connector pipe.
9. Remove the air cleaner assembly.
10. Disconnect the radiator hoses No. 1 and No. 2.
11. Disconnect oil cooler inlet tube No. 1.
12. Disconnect oil cooler outlet tube No. 1.
13. Disconnect the cooling fan ECU connector.
14. Remove the 2 nuts and 2 radiator support brackets.
15. Remove the radiator and fan.
16. Remove the 2 grommets.
17. Disconnect the radiator reservoir hose from the radiator.
18. Remove the 6 bolts and fan assembly from the radiator.

To install:

19. Install the fan assembly to the radiator with the 6 bolts and tighten to 44 inch lbs. (5 Nm).
20. Connect the radiator reservoir hose to the radiator.
21. Install the 2 grommets to the radiator.
22. Attach the 2 grommets on the radiator to the body bracket.
23. Install the radiator and fan with the 2 radiator support brackets and 2 nuts. Tighten the nuts to 10 ft. lbs. (13.5 Nm).
24. Connect the cooling fan ECU connector.
25. Connect oil cooler outlet tube No. 1.
26. Connect oil cooler inlet tube No. 1.
27. Connect the radiator hoses No. 1 and No. 2.
28. Install the air cleaner assembly.
29. Install the intake air connector pipe.
30. Install the radiator reserve tank assembly.
31. Refill the cooling system.
32. Connect the negative battery cable.
33. Start the engine and check for leaks.
34. Install air cleaner inlet No. 1.
35. Remove the V-bank cover.
36. Install the engine undercover.
37. Perform the system initialization procedure.

THERMOSTAT

REMOVAL & INSTALLATION

See Figure 29.

1. Before servicing the vehicle, refer to the Precautions section.
2. Disconnect the negative battery cable. Wait at least 90 seconds before performing any other work.
3. Remove the air cleaner inlet No. 1.
4. Drain the engine coolant.
5. Remove the 3 nuts and disconnect the water inlet from the water inlet housing.
6. Remove the thermostat.
7. Remove the gasket from the thermostat.

To install:

8. Install a new gasket to the thermostat.
9. Insert the thermostat into the water inlet housing with the jiggle valve facing straight upward. The jiggle valve may be set within 30° of either side of the prescribed position.
10. Install the water inlet with the 3 nuts and tighten to 14 ft. lbs. (19 Nm).
11. Refill the engine coolant.
12. Connect the negative battery cable.
13. Start the engine and check cooling system for leaks.
14. Install the air cleaner inlet No. 1.
15. Check the coolant level and add if necessary.

WATER PUMP

REMOVAL & INSTALLATION

See Figure 30.

1. Before servicing the vehicle, refer to the Precautions section.
2. Disconnect the negative battery cable.
3. Drain the cooling system.
4. Remove or disconnect the following:
   - Engine undercover
   - Timing belt
   - Water inlet housing
   - Timing belt idler sub assembly
   - Water pump

To install:

5. Install or connect the following:
   - New gasket, water pump with 5 bolts and 2 stud bolts and nuts. Tighten bolt to 15 ft lbs (21 Nm) and stud nuts to 13 ft. lbs (18 Nm).
   - Water inlet housing with new O-ring and seal packing

➡If O-ring contacts engine oil it must be replaced

   - Timing belt idler sub assembly
   - Timing belt assembly
   - Radiator assembly
   - Engine coolant
   - Negative battery cable
6. Check for coolant leaks
7. Initialize the power windows and seat control systems
   - Engine undercover

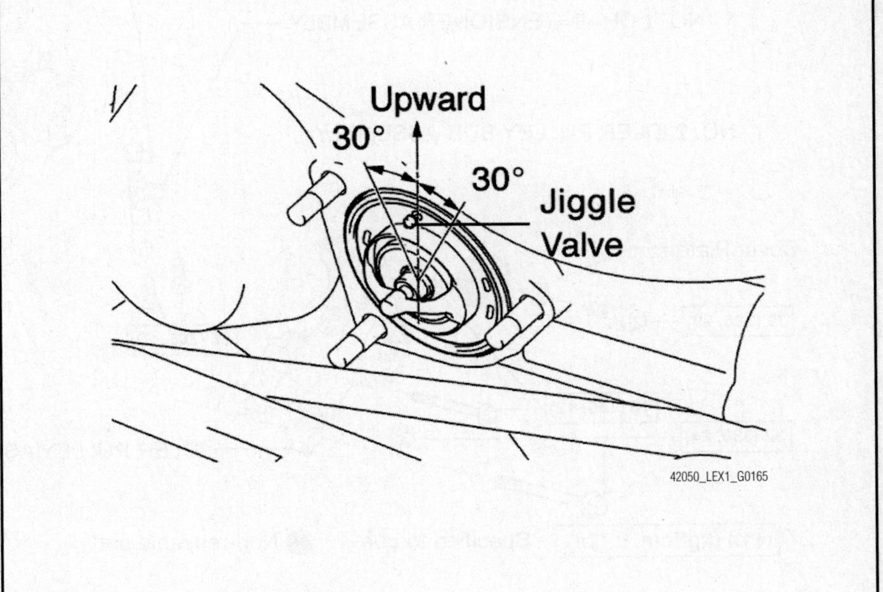

Fig. 29 Correct alignment of the thermostat jiggle valve—3UZ-FE engine

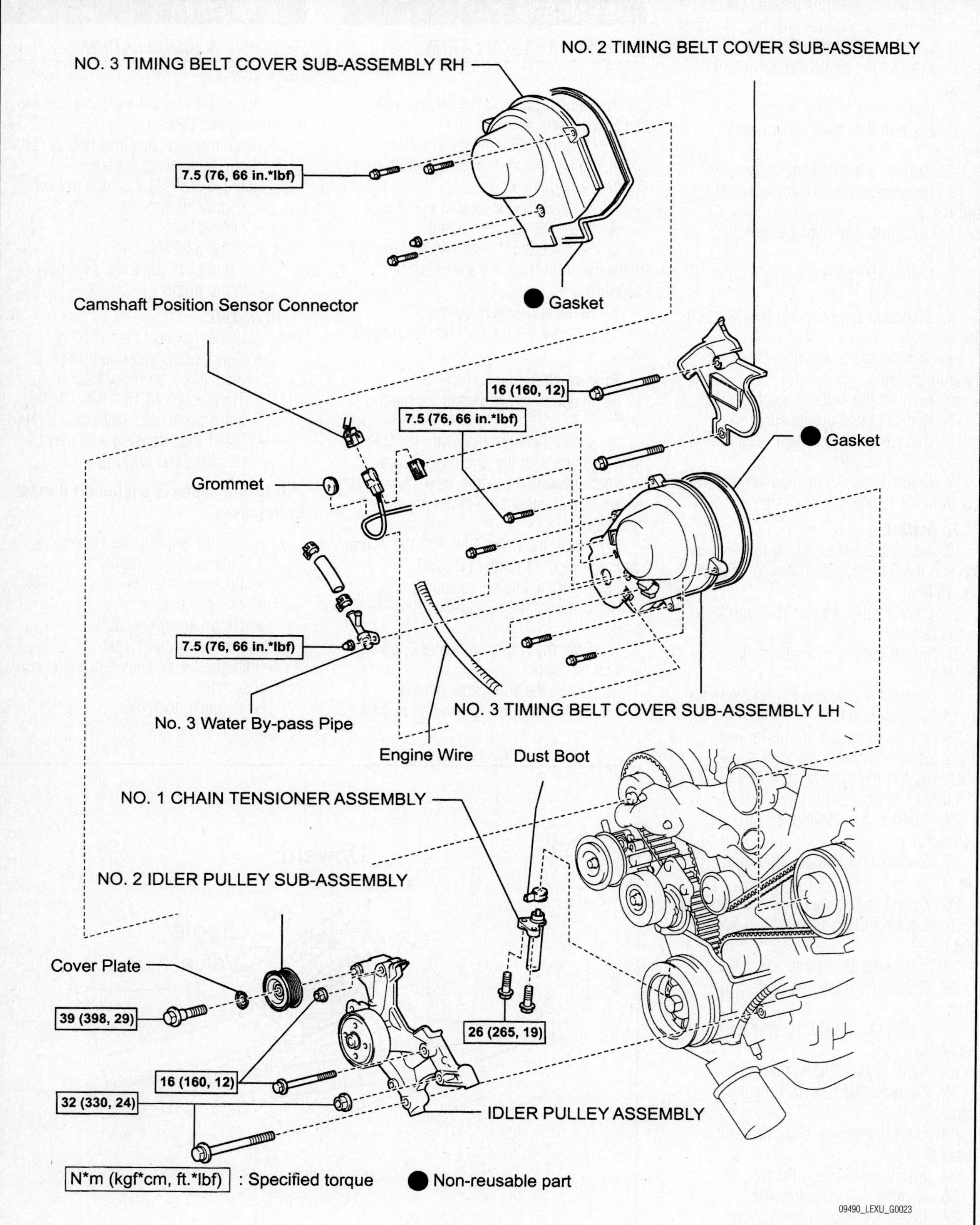

Fig. 30 Exploded view of water pump assembly—3UZ-FE Engine

ALTERNATOR

REMOVAL & INSTALLATION

See Figures 31 through 34.

1. Before servicing the vehicle, refer to the Precautions section.
2. Disconnect the negative battery cable.
3. Wait at least 90 seconds before performing any other work.
4. Remove the bolt and air cleaner inlet.

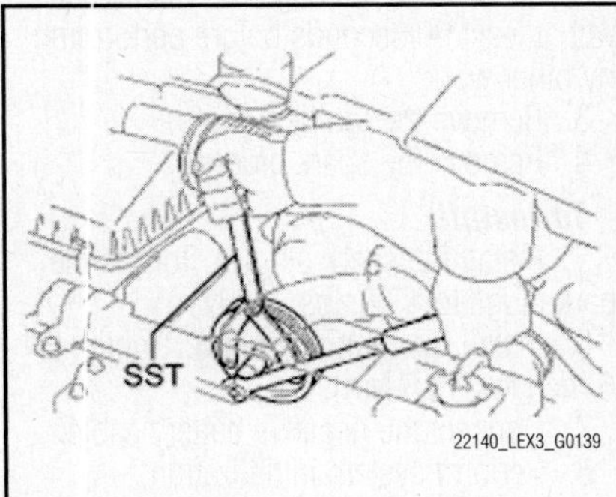

Fig. 31 Remove the vane pump pulley using the SST 09960-10010

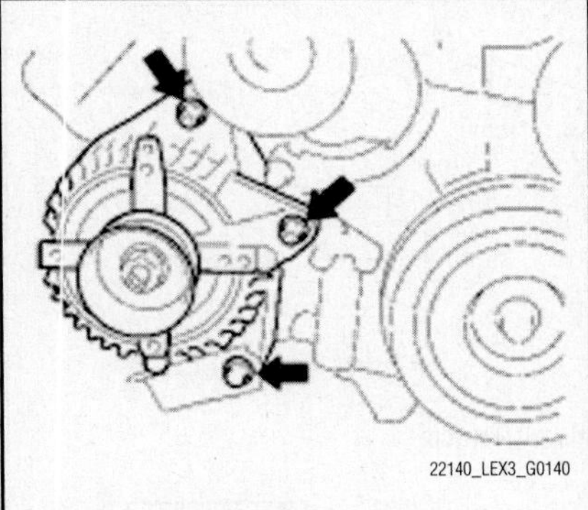

Fig. 32 Remove the 2 nuts, bolt and the alternator

5. Remove the 2 nuts and V-bank cover.
6. Loosen the belt tension by turning the belt tensioner counterclockwise, and remove the belt.

➡The tension pulley has a left hand thread.

7. Using SST 09960-10010, remove the nut and vane pump pulley.
8. Disconnect the generator connector.
9. Remove the rubber cap and nut, and disconnect the generator wire.
10. Remove the bolt and disconnect the engine wire bracket.
11. Remove the 2 nuts, bolt and the alternator.

To install:

12. Install the generator with the 2 nuts and bolt.
    a. Tighten the 2 nuts (B) to 11 ft. lbs. (15 Nm).
    b. Tighten the bolt (A) to 29 ft. lbs. (39 Nm).
13. Connect the engine wire bracket with the bolt.
14. Using SST, install the vane pump pulley with the nut. Tighten the nut to 32 ft. lbs. (43 Nm).
15. Set the V-belt to everything except the No. 2 idler pulley, as shown in the illustration.
16. Loosen the V-belt by turning the belt tensioner counterclockwise.
17. Then set the V-belt to the idler pulley.
18. Install engine undercover.
19. Install the V-bank cover with the 2 nuts. Tighten the nuts to 44 inch lbs. (5.0 Nm).
20. Install the air cleaner inlet with the bolt and tighten to 44 inch lbs. (5.0 Nm).
21. Connect the negative battery cable.
22. Certain systems need to be initialized after disconnecting and reconnecting the cable from the negative (-) battery terminal.

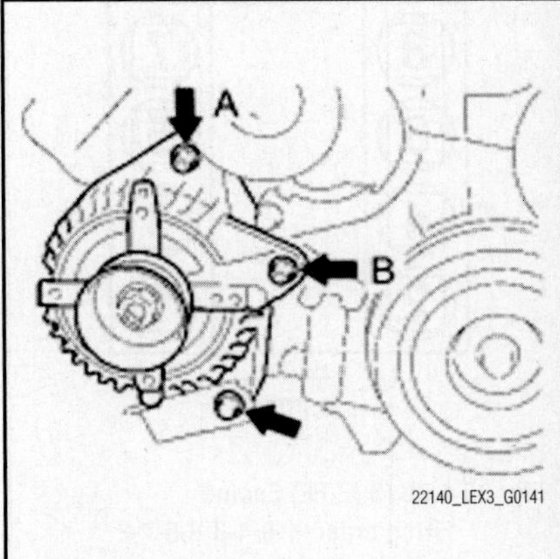

Fig. 33 Alternator mounting bolt (A) and nuts (B)

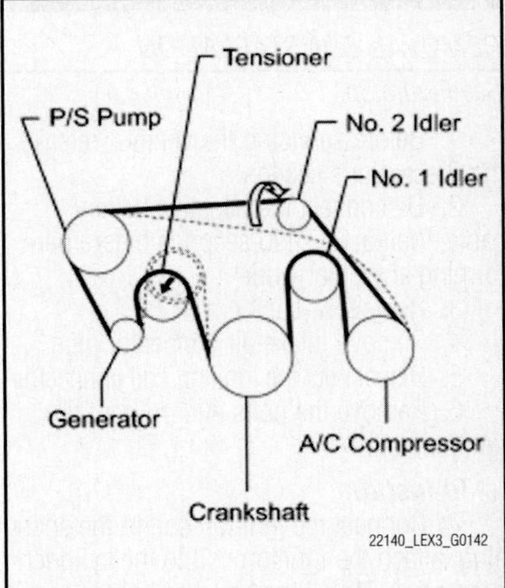

Fig. 34 V-belt diagram for 4.3L (3UZ-FE) Engine

FIRING ORDER

See Figure 35.

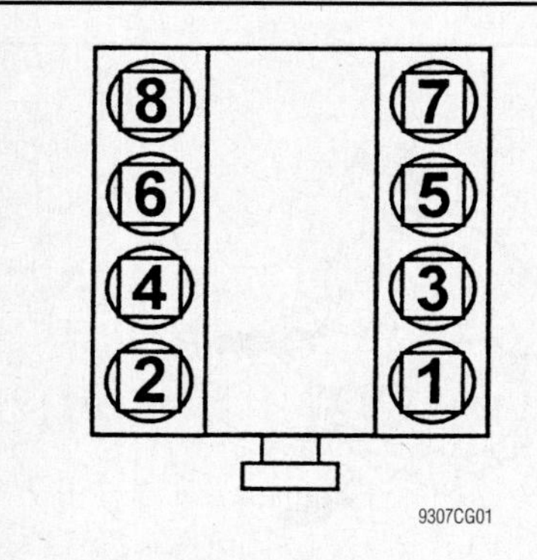

Fig. 35 4.3L (3UZ-FE) Engine Firing order: 1-8-4-3-6-5-7-2 Distributorless ignition system (one coil on each cylinder)

IGNITION COIL

REMOVAL & INSTALLATION

See Figure 36.

1. Before servicing the vehicle, refer to the Precautions section.
2. Disconnect the negative battery cable. Wait at least 90 seconds before performing any other work.
3. Remove V-bank cover.
4. Remove intake air connector pipe.
5. Disconnect the ignition coil connector.
6. Remove the bolts and pull out the ignition coil.

To install:

7. Connect the ignition coil to the spark plug, attach the ignition coil to the cylinder head cover. Install the bolt and tighten to 66 inch lbs. (7.5 Nm).
8. Connect the ignition coil connector.

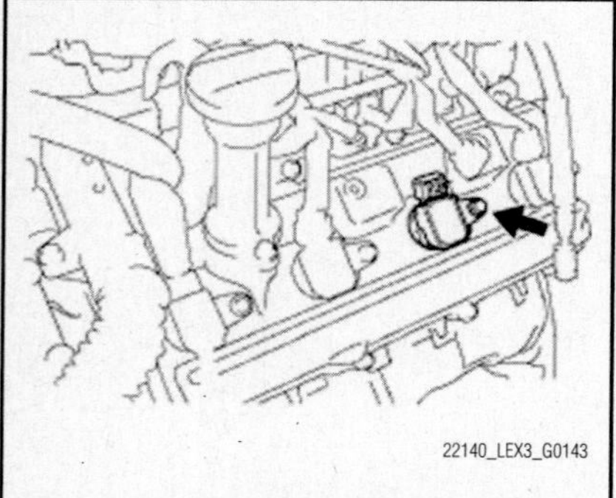

Fig. 36 Ignition coil and mounting bolt view

9. Install intake air connector pipe.
10. Install V-bank cover.
11. Connect the negative battery cable.
12. Perform system initialization procedure.

IGNITION TIMING

INSPECTION

1. Warm up the engine and stop the engine.

➡A warmed up engine should have an engine coolant temperature of over 80°C (176°F), have an engine oil temperature of 60°C (140°F), and the engine rpm should be stabilized.

2. Connect a hand-held tester to the DLC3.
3. Enter DATA LIST MODE on the hand-held tester. Please refer to the hand-held tester operator's manual for further details.
4. Start and idle the engine.
5. Check the ignition timing. The ignition timing should measure 8–12°BTDC
6. Disconnect the hand-held tester from the DLC3.

ADJUSTMENT

The ignition timing is controlled by the Powertrain Control Module (PCM). No adjustment is necessary or possible.

SPARK PLUGS

REMOVAL & INSTALLATION

See Figure 37.

1. Before servicing the vehicle, refer to the Precautions section.
2. Disconnect the negative battery cable. Wait at least 90 seconds before performing any other work.
3. Remove the ignition coil(s).
4. Remove the spark plug(s).

To install:

5. Install the spark plug(s). Torque the spark plugs to 13 ft. lbs. (19 Nm).
6. Install the ignition coil(s). Tighten to 66 inch lbs. (7.5 Nm).
7. Connect the negative battery cable.
8. Perform system initialization procedure.

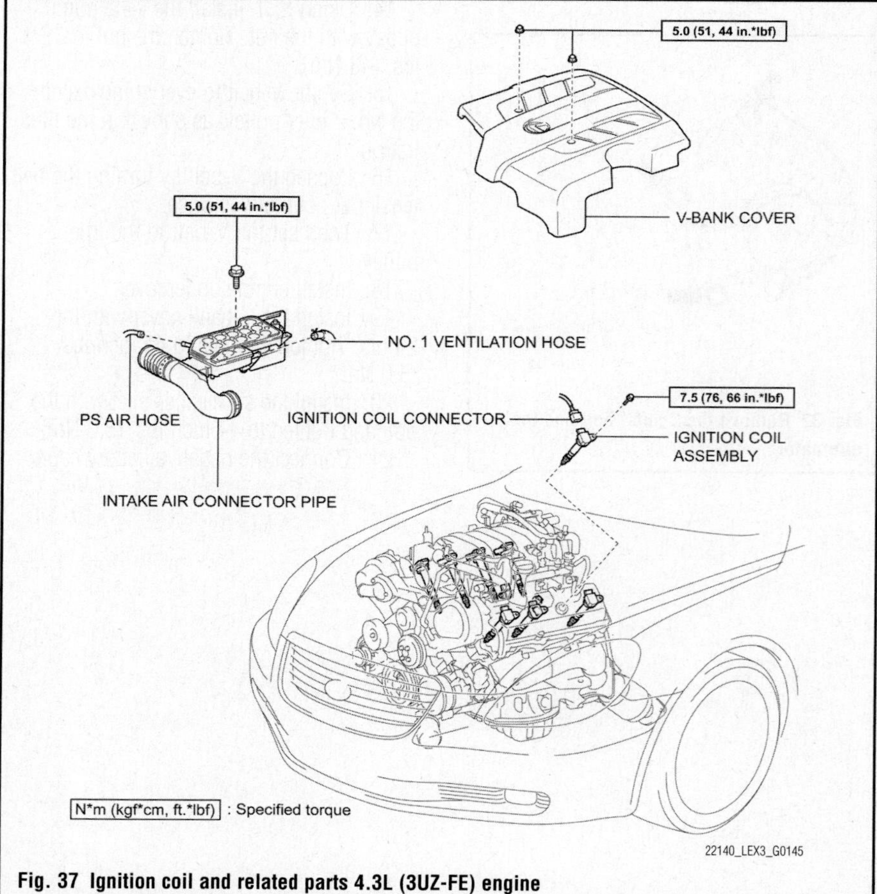

Fig. 37 Ignition coil and related parts 4.3L (3UZ-FE) engine

STARTER

REMOVAL & INSTALLATION

See Figure 38.

1. Before servicing the vehicle, refer to the Precautions section.
2. Remove or disconnect the following:
   - Negative battery cable. Wait at least 90 seconds before performing any other work
   - Drain engine coolant
   - V-bank cover
   - Accelerator cable
   - Intake air connector
   - Throttle Body
   - Upper and lower intake manifold assembly
   - Rear water bypass joint
   - Water bypass pipe
   - Water bypass pipe from the water pump
   - Wire clamp from the bracket on the water bypass pipe
   - O-ring from the water bypass pipe
   - Water bypass pipe bracket from the water bypass pipe
   - 2 bolts holding the starter to the cylinder block
   - Starter connector
   - Starter from the cylinder block
   - Nut, and disconnect the starter wire
   - Starter

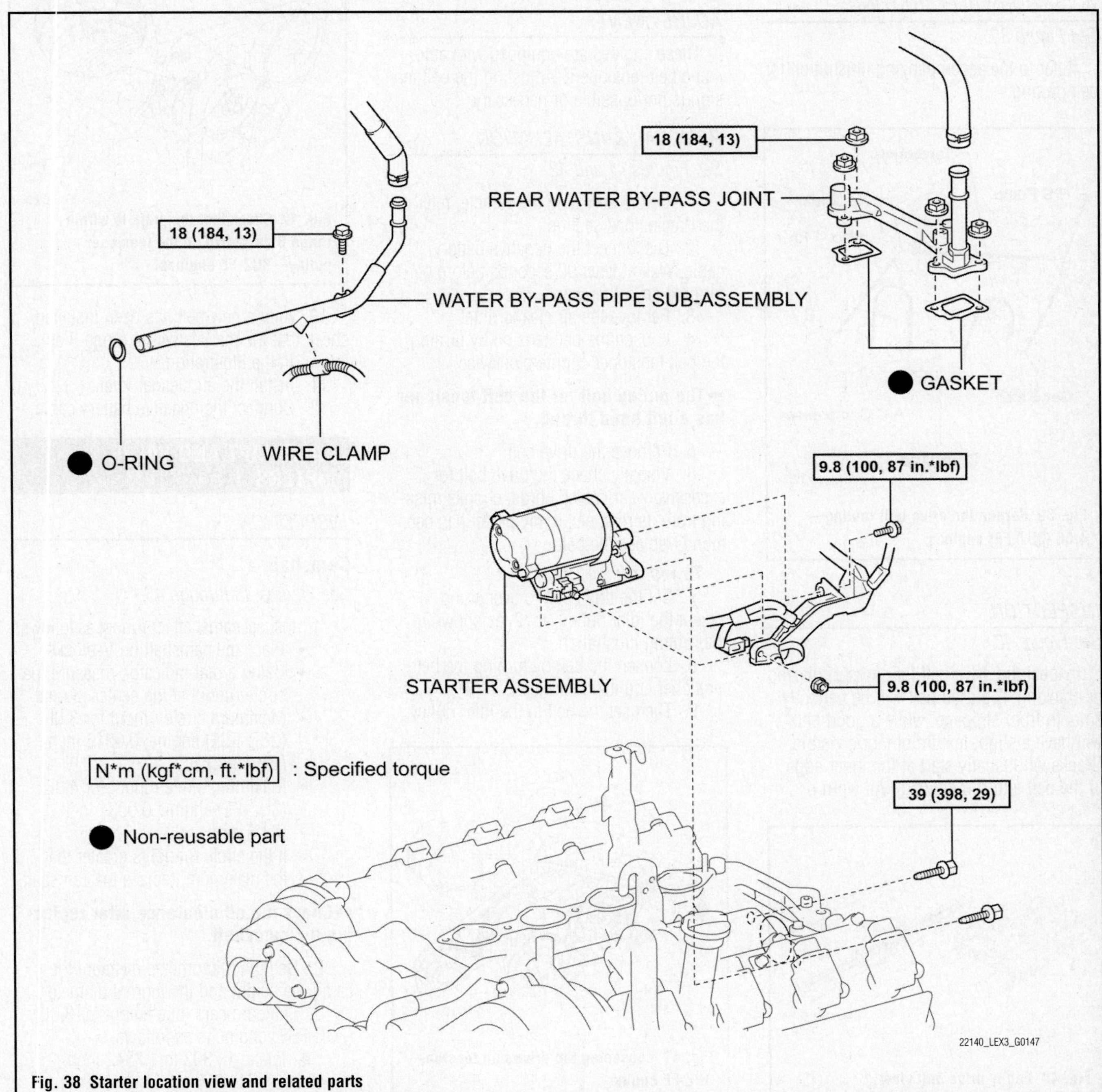

Fig. 38 Starter location view and related parts

To install:

3. Install or connect the following:
    - Wire clamp to the wire bracket with the bolt. Tighten to 87 inch lbs. (9.8 Nm).
    - Starter wire with the nut. Tighten to 87 inch lbs.
    - Starter connector
    - Starter with the 2 bolts. Torque the bolts to 29 ft. lbs. (39 Nm).
    - Water bypass pipe bracket to the water bypass pipe
    - O-ring to the water bypass pipe
    - Water bypass pipe
    - Wire clamp to the bracket on the water bypass pipe
    - Water bypass pipe bolts. Torque the bolts to 13 ft. lbs. (18 Nm).
    - Rear water bypass joint
    - Intake manifold assembly
    - Throttle body
    - Intake air connector
    - Accelerator cable
    - V-bank cover

ENGINE MECHANICAL

ACCESSORY DRIVE BELTS

ACCESSORY BELT ROUTING

See Figure 39.

Refer to the accompanying illustration for belt routing.

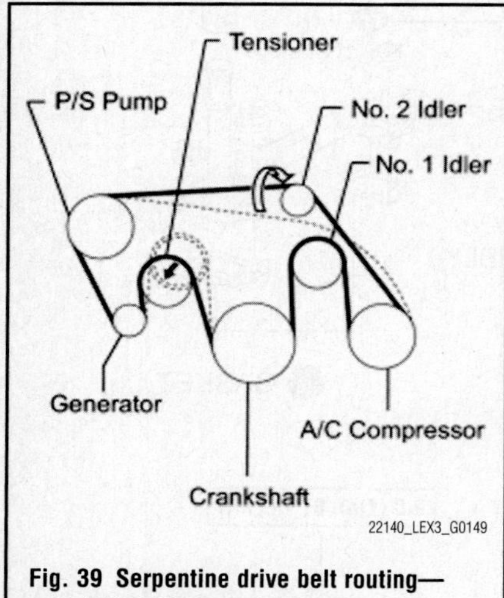

Fig. 39 Serpentine drive belt routing—4.3L (3UZ-FE) engines

INSPECTION

See Figure 40.

Inspect the drive belt for signs of glazing or cracking. A glazed belt will be perfectly smooth from slippage, while a good belt will have a slight texture of fabric visible. Cracks will usually start at the inner edge of the belt and run outward. All worn or damaged drive belts should be replaced immediately.

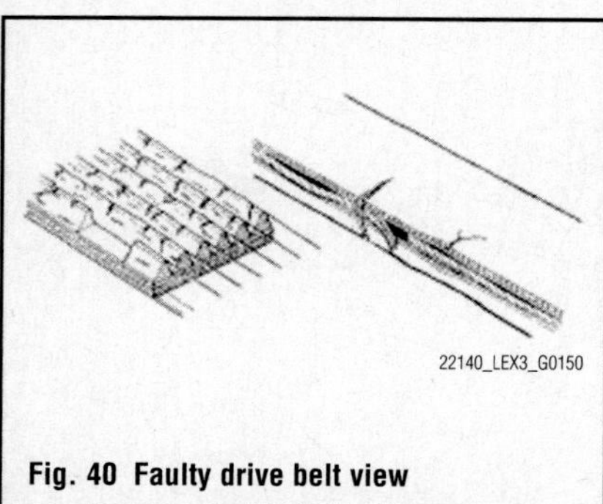

Fig. 40 Faulty drive belt view

ADJUSTMENT

These engines are equipped with automatic belt tensioners. Adjusting the belt tension is not possible or necessary.

REMOVAL & INSTALLATION

See Figures 41 and 42.

1. Before servicing the vehicle, refer to the Precautions section.
2. Disconnect the negative battery cable. Wait at least 90 seconds before performing any other work.
3. Remove the air cleaner inlet.
4. Loosen the belt tension by turning the belt tensioner counterclockwise.

➡The pulley bolt for the belt tensioner has a left hand thread.

5. Remove the drive belt.
6. Visually check the drive belt for excessive wear, frayed cords, chunks missing from its ribs, etc. If any defect has been found, replace the belt.

To install:

7. Set the drive belt to everything except the idler pulley No. 2, as shown in the routing illustration.
8. Loosen the belt by turning the belt tensioner counterclockwise.
9. Then set the belt to the idler pulley.

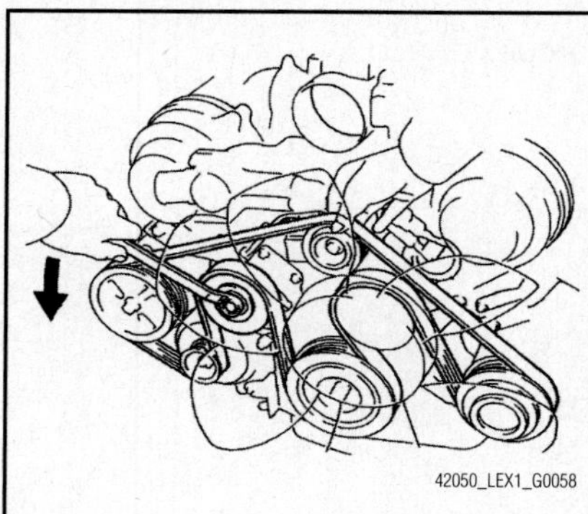

Fig. 41 Loosening the drive belt tension—3UZ-FE engine

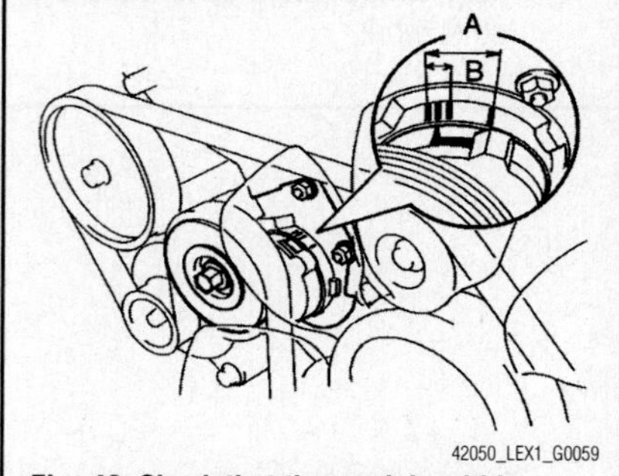

Fig. 42 Check that the mark is within range B as shown on the tensioner pulley—3UZ-FE engines

10. After a new belt has been installed, check that the mark is within range B as shown in the illustration.
11. Install the air cleaner inlet.
12. Connect the negative battery cable.

CAMSHAFT AND VALVE LIFTERS

INSPECTION

Camshaft

See Figures 43 through 45.

1. Inspect camshaft for runout as follows:
    - Place the camshaft on V-blocks.
    - Using a dial indicator, measure the circle runout at the center journal.
    - Maximum circle runout for 3.0L (3GR-FSE) engine: 0.0016 inch. (0.04 mm)
    - Maximum circle runout for 4.3L (3UZ-FE) engine: 0.0031 inch. (0.08 mm)
    - If the circle runout is greater than the maximum, replace the camshaft.

➡Check the oil clearance after replacing the camshaft.

2. Using a micrometer, measure the cam lobe height and the journal diameter.
3. Standard cam lobe height for 3.0L (3GR-FSE) engine is as follows:
    a. Intake 1.7303 to 1.7342 inch. (43.950 to 44.050 mm).

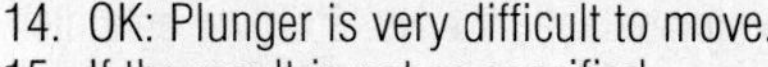

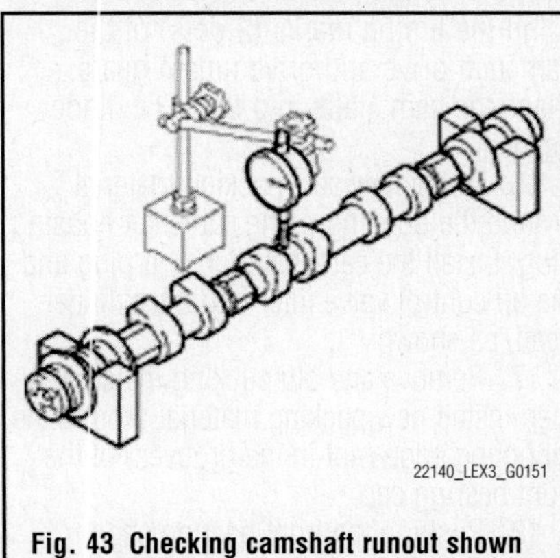

Fig. 43 Checking camshaft runout shown

b. Exhaust 1.7366 to 1.74.5 inch. (44.110 to 44.210 mm).

4. Standard cam lobe height for 4.3L (3UZ-FE) engine is as follows:

a. 1.6776 to 1.6815 inch. (42.610 to 42.710 mm).

5. Standard journal diameter for 3.0L (3GR-FSE) engine is as follows:

a. No.1 Journal 1.4152 to 1.4157 inch. (35.946 to 35.960 mm).

b. All other journals 1.0220 to 1.0226 inch. (25.959 to 25.975 mm).

6. Standard journal diameter for 4.3L (3UZ-FE) engine is as follows:

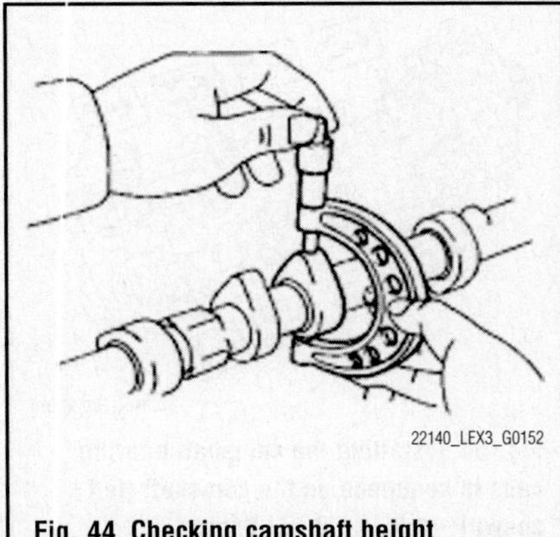

Fig. 44 Checking camshaft height

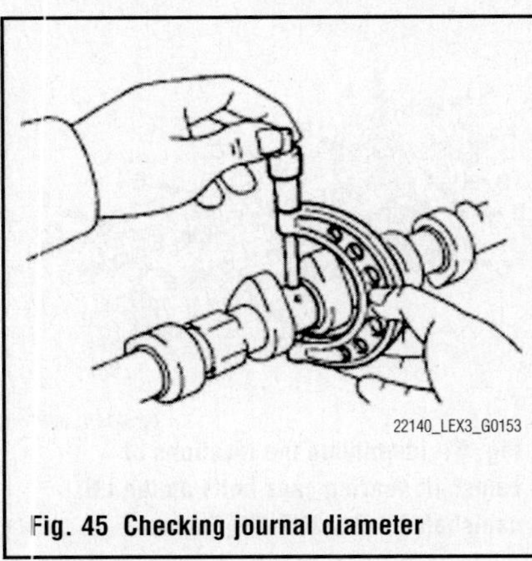

Fig. 45 Checking journal diameter

a. 1.0612 to 1.0618 inch. (26.954 to 26.970).

7. If the journal diameter is not as specified, check the oil clearance.

Valve Lash Adjusters

See Figure 46.

8. Place the lash adjuster into a container full of engine oil.

9. Insert SST 09276-75010 tip into the lash adjuster's plunger and use the tip to press down on the check ball inside the plunger.

10. Squeeze the SST and lash adjuster together to move the plunger up and down 5 to 6 times.

11. Check the movement of the plunger and bleed the air.

12. OK: Plunger moves up and down.

➡When bleeding high-pressure air from the compression chamber, make sure that the tip of the SST is actually pressing the checkball as shown in the illustration. If the checkball is not pressed, air will not bleed.

13. After bleeding the air, remove SST. Then try to quickly and firmly press the plunger with a finger.

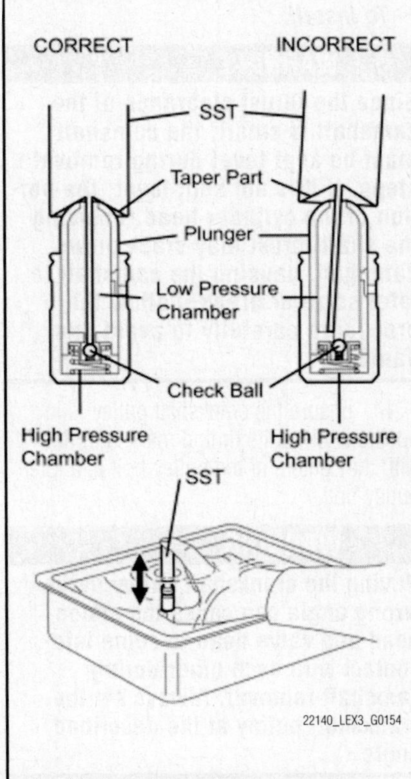

Fig. 46 Valve lash adjuster inspection shown

14. OK: Plunger is very difficult to move.

15. If the result is not as specified, replace the lash adjuster.

REMOVAL & INSTALLATION

See Figures 47 through 53.

1. Before servicing the vehicle, refer to the Precautions section.

2. Relieve the fuel pressure from the fuel lines.

3. Remove or disconnect the following:

- Engine undercover
- Drain engine coolant and engine oil
- V-bank cover
- Air cleaner assembly
- Intake air pipe
- Radiator
- Throttle body
- Upper and lower intake manifold assembly
- Timing belt from camshaft pulleys
- Camshaft Position (CMP) sensor and LH timing belt rear plates
- RH timing belt rear plates

⁂ WARNING

Do NOT drop anything inside timing belt cover during this procedure. Keep oil, water and dust from timing belt.

- Power steering pump from engine mount (Do NOT disconnect hoses)
- Catalytic converters
- Water inlet housing assembly
- Water bypass pipe, front bypass joint, and rear bypass joint
- Ignition coils
- Variable valve timing (VVT) sensors
- Engine hangers
- Oil dipsticks and guides for engine oil and transmission fluid
- Cylinder head covers
- Spark plugs
- Semi-circular plugs and camshaft housing plugs
- Camshaft timing oil control valve

⁂ WARNING

Since the thrust clearance of the camshaft is small, the camshaft must be kept level during removal steps. If it is not kept level, the portion of the cylinder head receiving the shaft thrust may crack or be damaged, causing the camshaft to later seize or break. Follow the procedure carefully to avoid this damage.

4. Check the crankshaft pulley position and ensure the timing mark of the pulley is aligned with the centers of the crankshaft

pulley bolt and the No. 2 timing belt idler pulley bolt.

WARNING

Having the crankshaft pulley at the wrong angle can cause the piston head and valve head to come into contact with each other during camshaft removal. Always set the crankshaft pulley at the described angle.

5. Using a special wrench, rotate the camshaft timing tube from left to right about 2-3 times, within only a 25°range of movement. Use a waste cloth to collect oil from the camshaft timing oil control valve installation hole.

6. Remove the LH camshafts first. With a hex wrench on the hex portion of the camshaft, rotate so that a 6mm service bolt can be inserted into the bolt hole in the rear face of the camshaft pulley into order to secure the camshaft in place. The bolt should be about 0.63-0.79 in. (16-20mm) long.

7. Align the timing mark (2 dots) of the camshaft drive gear by turning the camshaft with a hex wrench until the timing mark aligns.

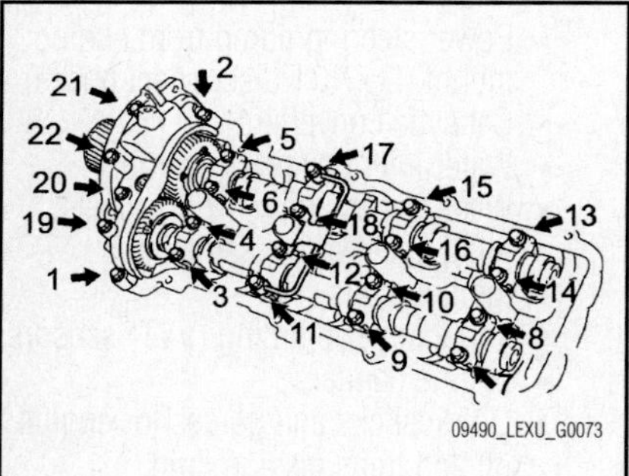

Fig. 47 LH camshaft bearing cap bolt loosening sequence—4.3L (3UZ-FE) Engine

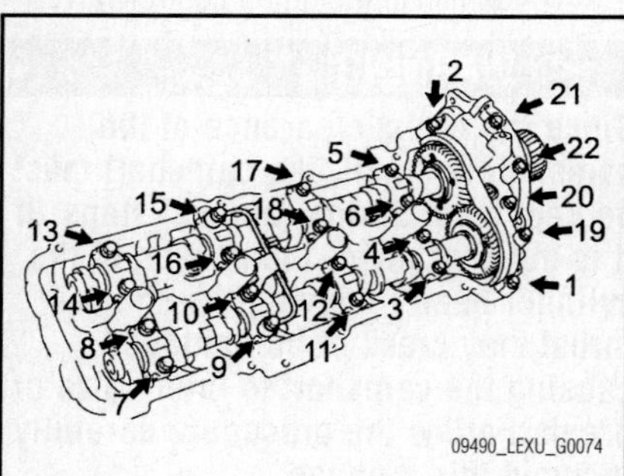

Fig. 48 RH camshaft bearing cap bolt loosening sequence—4.3L (3UZ-FE) Engine

8. Now, uniformly loosen the 22 camshaft bearing cap bolts, in several passes, following the sequence shown.

9. Remove the 22 bearing cap bolts, 4 seal washers, oil feed pipe, 9 bearing caps, the camshaft housing plug, the oil control valve filter, and both LH camshafts. Keep all parts in order for proper reinstallation.

10. Remove the RH camshafts. With a hex wrench on the hex portion of the camshaft, rotate so that a 6mm service bolt can be inserted into the bolt hole in the rear face of the camshaft pulley into order to secure the camshaft in place. The bolt should be about 0.63-0.79 in. (16-20mm) long.

11. Align the timing mark (1 dot) of the camshaft main gear about 10° angle by turning the camshaft with a hex wrench until the timing mark aligns.

12. Now, uniformly loosen the 22 camshaft bearing cap bolts, in several passes, following the sequence shown.

13. Remove the 22 bearing cap bolts, 4 seal washers, oil feed pipe, 9 bearing caps, the camshaft housing plug, the oil control valve filter, and both LH camshafts. Keep all parts in order for proper reinstallation.

To install:

WARNING

Since the thrust clearance of the camshaft is small, the camshaft must be kept level during removal steps. If it is not kept level, the portion of the cylinder head receiving the shaft thrust may crack or be damaged, causing the camshaft to later seize or break. Follow the procedure carefully to avoid this damage.

14. Ensure the crankshaft pulley is in position so that its timing mark is in line with the centers of the pulley bolt and idler pulley bolt.

WARNING

Having the crankshaft pulley at the wrong angle can cause the piston head and valve head to come into contact with each other during camshaft removal. Always set the crankshaft pulley at the described angle.

15. Apply grease to the thrust portion of the LH intake and exhaust camshafts. Align the timing marks (2 dots) of the camshaft drive and driven main gears. Place the camshafts into the LH cylinder head.

16. Apply new seal packing material around the opening of the camshaft housing plug. Install the camshaft housing plug and the oil control valve filter into the cylinder head, as shown.

17. Remove any old packing material, then install new packing material around the mounting edge (not in the grooves) of the front bearing cap.

18. Position the front bearing cap in

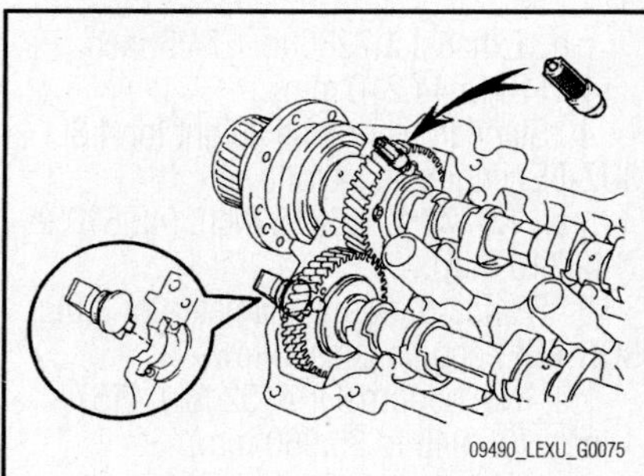

Fig. 49 Installing the camshaft housing plug and oil control valve filter for the LH camshaft—4.3L (3UZ-FE) Engine

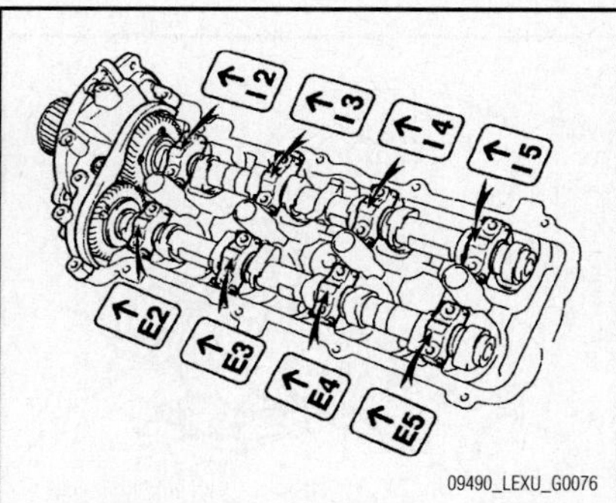

Fig. 50 Installing the camshaft bearing caps in sequence on the camshaft (left shown)—4.3L (3UZ-FE) Engine

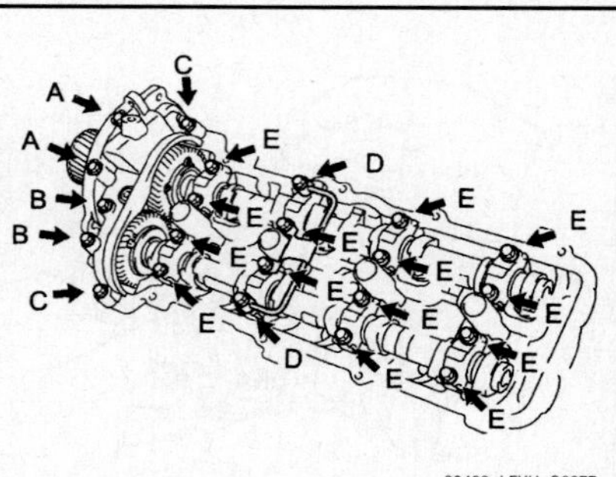

Fig. 51 Identifying the locations of camshaft bearing caps bolts on the LH camshaft—4.3L (3UZ-FE) Engine

place. This will determine the thrust portion of the camshaft.

a. Install the other bearing caps, in sequence shown, with the arrow marks facing forward.

b. Push the camshaft oil seal into place by pushing from the front of the engine. Install a new seal washer to the front bearing cap bolts.

c. Apply a light coat of oil to bearing cap bolt threads and under the heads of the bearing cap bolts "D" and "E", as shown. Do NOT apply engine oil under the heads of bearing cap bolts "A", "B" and "C".

d. Bolt lengths vary for each bearing cap. Refer to the illustration for each of the following bolts:

- Bolt "A" with seal washer is 3.70 in. (94mm)
- Bolt "B" with seal washer is 2.83 in. (72mm)
- Bolt "C" is 0.98 in. (25mm)
- Bolt "D" is 2.05 in. (52mm)
- Bolt "E" is 1.50 in. (38mm)

19. Install the oil feed pipe and the 22 bearing cap bolts in their respective locations. Uniformly tighten the 22 bearing cap bolts, in several passes, following the sequence as shown. Torque bolt "C" to a final torque of 66 inch lbs. (7.5 Nm). Torque all other bearing cap bolts to a final torque of 12 ft. lbs. (16 Nm).

20. Remove the service bolt installed in the rear face of the gear.

21. Repeat this entire procedure for the RH camshafts. Use the illustrations given for the LH camshafts, as the sequences are the same on the RH camshafts.

22. Check and adjust valve lash. See VALVE LASH.

23. Install or reconnect the following:

- Camshaft timing oil control valve.
- Semi-circular plugs in rear ends of each cylinder head
- Cylinder head covers, with new gaskets and packing material
- Engine hangers
- Variable valve timing sensors
- Oil dipsticks and tubes
- All remaining components in reverse of removal procedure.

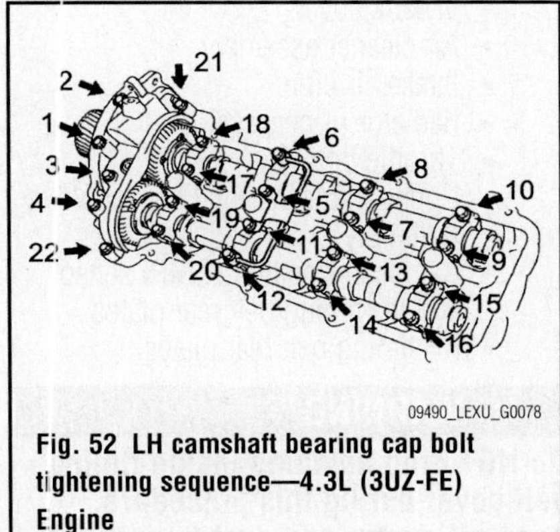

Fig. 52 LH camshaft bearing cap bolt tightening sequence—4.3L (3UZ-FE) Engine

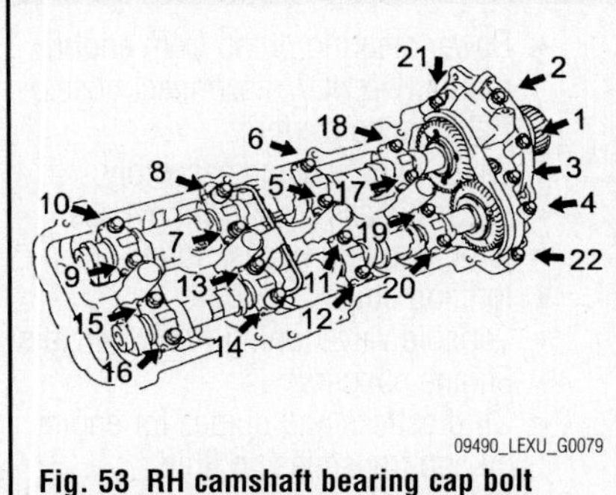

Fig. 53 RH camshaft bearing cap bolt tightening sequence—4.3L (3UZ-FE) Engine

CATALYTIC CONVERTER

REMOVAL & INSTALLATION

See Figure 54.

✱✱ CAUTION

- Wear protective gloves when removing the exhaust pipe.
- The exhaust pipe is extremely hot immediately after the engine has stopped.
- Confirm that the exhaust pipe has cooled down before removing it.

1. Remove the appropriate engine under covers and heat insulators.
2. Remove the 6 nuts, 2 converters and 2 gaskets.
3. Upon installation, install 2 new gaskets and the 2 converters with 6 new nuts. Tighten to 46 ft. lbs. (62 Nm).

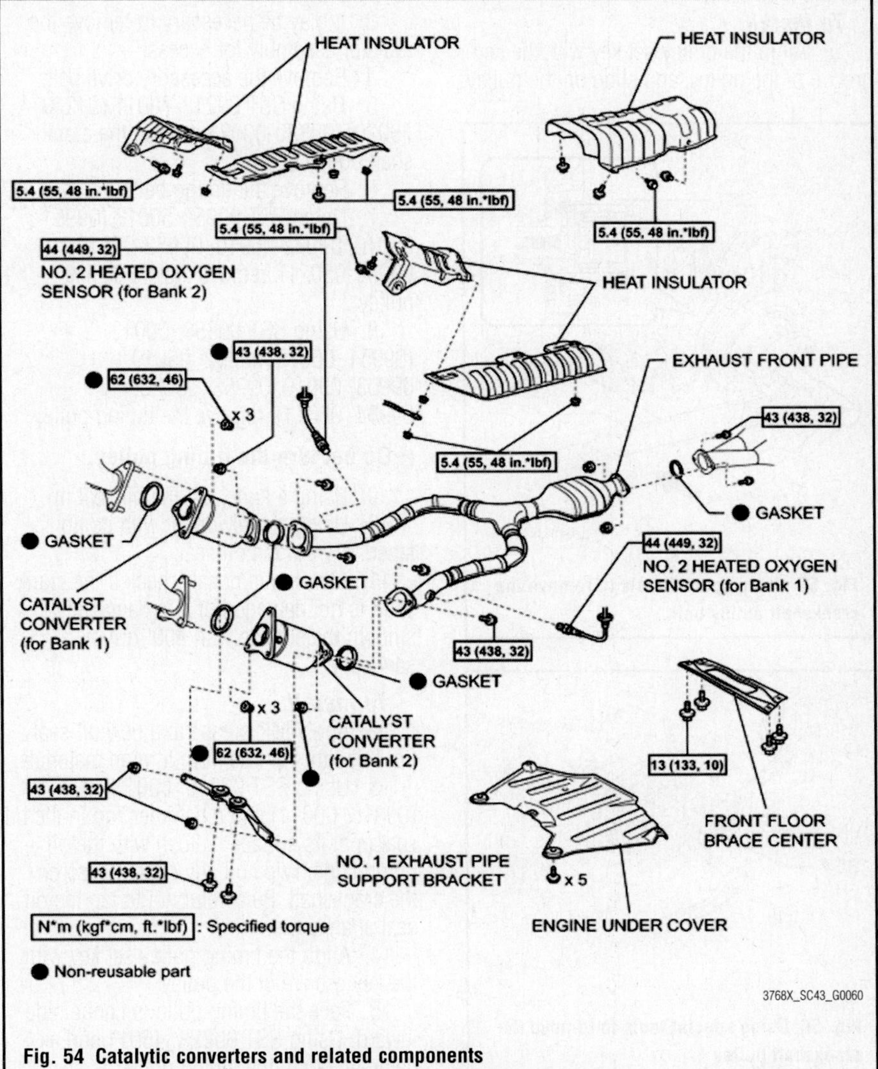

Fig. 54 Catalytic converters and related components

CRANKSHAFT DAMPER

REMOVAL & INSTALLATION

See Figures 55 and 56.

1. Before servicing the vehicle, refer to the Precautions section.
2. Disconnect the negative battery cable. Wait at least 90 seconds before performing any other work.
3. It may be necessary to remove the radiator assembly for access.
4. Remove the accessory drive belt(s).
5. Using SST 09213-70011 (91213-70020), 09330-00021, loosen the crankshaft pulley bolt.
6. Using SST 09950-50013 (09951-05010, 09952-05010, 09953-05010, 09954-05021), remove the crankshaft pulley.

➡Before using SST, apply lubricating oil on the threads and tip of the center bolt.

To install:

7. Align the pulley set key with the key groove of the pulley, and slide on the pulley.

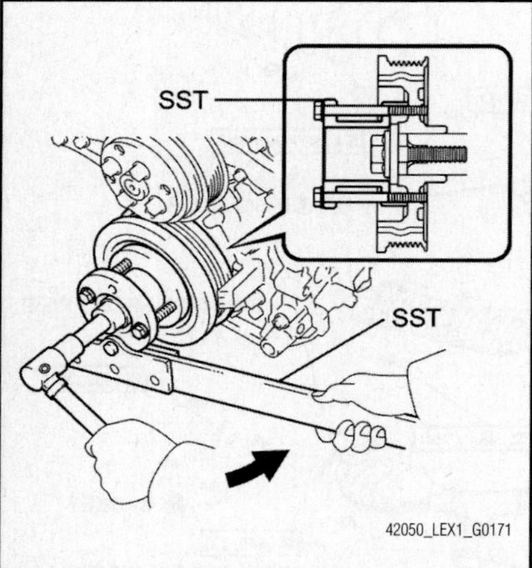

Fig. 55 Using special tools to remove the crankshaft pulley bolt

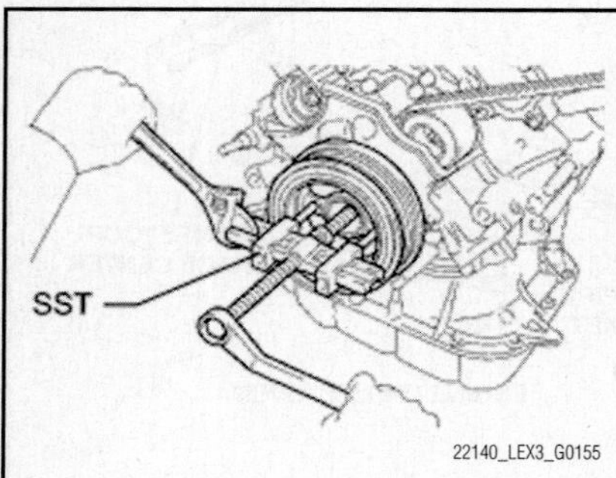

Fig. 56 Using special tools to remove the crankshaft pulley

8. Using SST 09213-70011 (91213-70020), 09330-00021, install the pulley bolt.
9. Tighten the crankshaft pulley bolt (depending on the engine) to the following values:
   - 3.0L (3GR-FSE) engine: 184 ft. lbs. (250 Nm).
   - 4.3L (3UZ-FE) engines: 181 ft. lbs. (245 Nm).
10. Install the accessory drive belt(s).
11. Install the radiator, if removed.
12. Connect the negative battery cable.

CRANKSHAFT FRONT SEAL

REMOVAL & INSTALLATION

See Figure 57.

1. Before servicing the vehicle, refer to the Precautions section.
2. Disconnect the negative battery cable. Wait at least 90 seconds before performing any other work.
3. It may be necessary to remove the radiator assembly for access.
4. Remove the accessory drive belt.
5. Using SST 09213-70011 (91213-70020), 09330-00021, loosen the crankshaft pulley bolt.
6. Remove the timing belt.
7. Using SST 09950-50013 (09951-05010, 09952-05010, 09953-05010, 09954-05021), remove the crankshaft pulley.
8. Using SST 09950_50013 (09951_05010, 09952_05010, 09953_05010, 09953_05020, 09954_05011), remove the timing pulley.

➡Do not turn the timing pulley.

9. Using a knife, cut the oil seal lip.
10. Using a screwdriver with its tip taped, pry out the oil seal.
11. After the removal, check if the crankshaft is not damaged. If it is damaged, smooth the surface with 400_grit sandpaper.

To install:

12. Apply MP grease to a new oil seal lip. Keep the lip free from foreign materials.
13. Using SST 09316_60011 (09316_00011) and a hammer, tap in the oil seal until its surface is flush with the oil pump edge. Wipe off any extra grease on the crankshaft. Be careful not to tap the oil seal at an angle.
14. Align the timing pulley set key with the key groove of the pulley.
15. Face the timing pulley's flange side inward. Using SST 09223_46011 and a hammer, tap in the timing pulley.

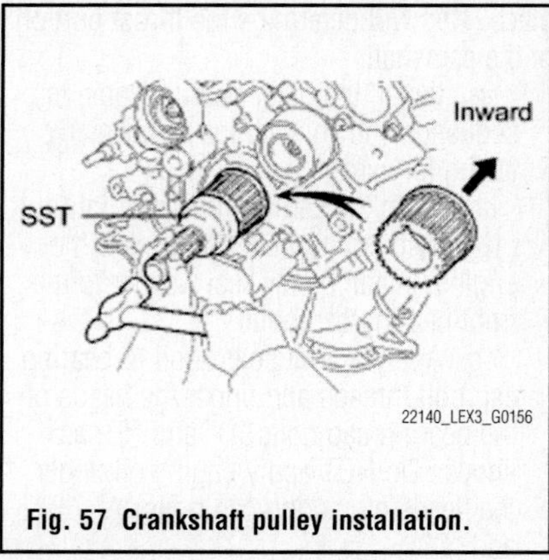

Fig. 57 Crankshaft pulley installation.

16. Install the timing belt.
17. Connect the negative battery cable.
18. Check for leaks.

CYLINDER HEAD

REMOVAL & INSTALLATION

See Figures 58 through 60.

1. Before servicing the vehicle, refer to the Precautions section.
2. Relieve the fuel system pressure.
3. Remove or disconnect the following, as applicable to each engine:
   - Engine undercover
   - Drain engine coolant
   - V-bank cover
   - Air cleaner assembly
   - Intake air pipe
   - Radiator (if necessary)
   - Throttle body
   - Upper and lower intake manifold assembly
   - Camshaft Position (CMP) sensor and LH timing belt rear plates
   - RH timing belt rear plates

WARNING

Do NOT drop anything inside timing belt cover during this procedure. Keep oil, water and dust from timing belt.

   - Power steering pump from engine mount (Do NOT disconnect hoses)
   - Catalytic converters
   - Water inlet housing assembly
   - Water bypass pipe, front bypass joint, and rear bypass joint
   - Ignition coils
   - Variable valve timing (VVT) sensors
   - Engine hangers
   - Oil dipsticks and guides for engine oil and transmission fluid
   - Cylinder head covers
   - Spark plugs
   - Camshafts

⁂ WARNING

Since the thrust clearance of the camshaft is small, the camshaft must be kept level during removal. If not, the portion of the cylinder head receiving the camshaft thrust may crack or be otherwise damaged, causing the camshaft to later seize or break. Follow the camshaft removal procedure carefully as given in this section.

- Both oxygen sensor connectors
- Ground wire from LH cylinder head
- Engine wire bracket for oxygen sensor on LH cylinder head

4. Uniformly loosen the 10 cylinder head bolts on each cylinder head, in several passes, following loosening sequence as shown.

⁂ WARNING

Use care so that no bolts or washers are dropped into the recesses or enclosed portions of the cylinder head or block.

5. Carefully lift the cylinder head from the locating dowels on the engine block. Place cylinder heads on wooden blocks on the workbench.

6. If necessary, exhaust manifolds may be removed from the cylinder heads at this time.

To install:

7. If removed, install exhaust manifolds to cylinder heads, with new gaskets. Ensure the white mark on the gasket is facing the manifold side.

8. Install and tighten the exhaust manifold bolts, in an alternating pattern, to a final torque of 32 ft. lbs. (44 Nm).

9. Install the heat shields.

10. With a new cylinder head gasket in place, carefully position the cylinder head onto the engine block locating dowels.

➡The cylinder head gaskets have a "3R" marks for the RIGHT cylinder head, and a "3L" mark for the LEFT cylinder head.

➡If any cylinder head bolt appears stretched or damaged, replace it. If a bolt will not reach final torque setting, replace it.

11. Apply a light coat of oil to the cylinder head bolt threads. Install the washers and insert the cylinder head bolts into position.

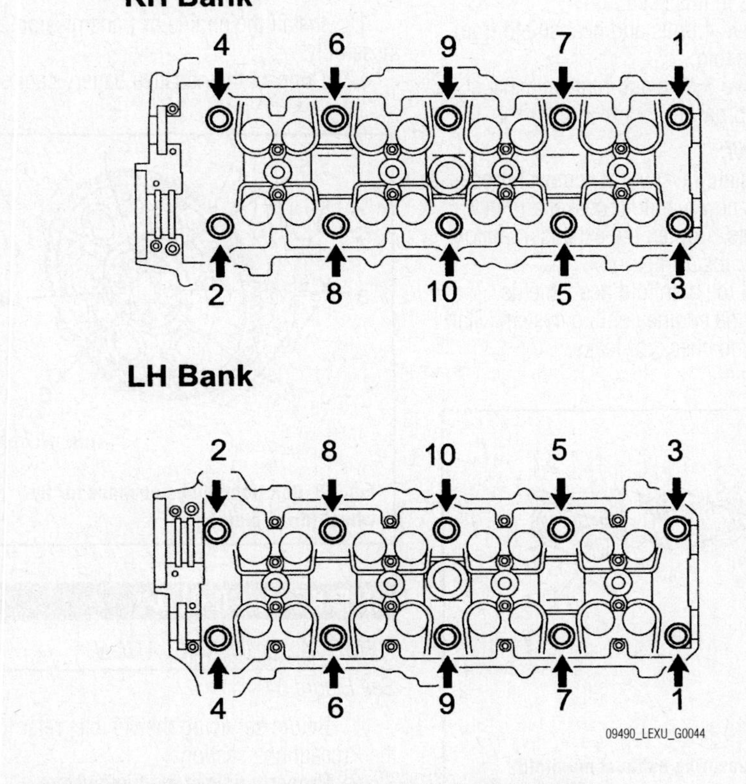

Fig. 58 Cylinder head bolt loosening sequence—4.3L (3UZ-FE) Engine

12. In several passes, following the tightening sequence shown, tighten the cylinder head bolts to a final torque of 44 ft. lbs. (59 Nm).

13. Once the bolts reach this setting, then place a white paint mark on the front of each bolt head. Using the torque wrench, turn each bolt, in the sequence shown, an additional 90 degrees, using the paint mark as a reference.

14. Install or reconnect the following:
- Engine wiring and ground straps
- Oxygen sensor wire bracket on LH cylinder head
- Spark plugs
- Camshafts

15. Inspect and adjust the valve lash.

16. Install or reconnect the following:
- Cylinder head covers, with new gaskets
- Engine hangers
- VVT sensors
- Engine and transmission dipsticks and tubes
- Ignition coils
- Water bypass joints and pipe; torque nuts and bolt to 13 ft. lbs. (18 Nm)
- Water inlet housing assembly
- Catalytic converters, with new gaskets and new nuts; torque nuts to 46 ft. lbs. (62 Nm)
- Power steering pump to mounting; torque bolts to 29 ft. lbs. (39 Nm) and nut to 32 ft. lbs. (43 Nm)
- LH timing belt rear plates and Camshaft Position (CMP) sensor; torque bolts to 66 inch lbs. (7.5 Nm)
- RH right belt rear plates; torque bolts to 66 inch lbs. (7.5 Nm)
- Camshaft pulleys
- Timing belt to camshaft pulleys
- Upper and lower intake manifold assembly, with new gaskets; torque bolts, in alternating pattern, to 13 ft. lbs. (18 Nm)
- Throttle body
- Radiator
- Intake air connector and air cleaner assembly

17. Refill the engine cooling system. Start the engine and check for leaks and proper operation.

18. Recheck the engine oil level.

19. Install the V-bank cover and the engine undercover.

20. Start the engine and check for leaks or abnormal conditions. Perform and road test. Then, recheck for leaks and recheck fluid levels.

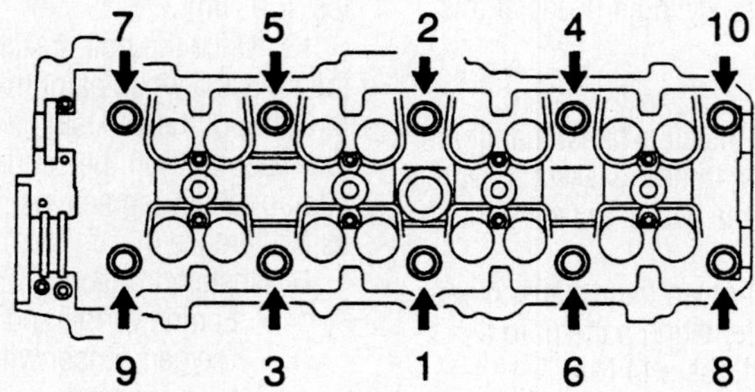

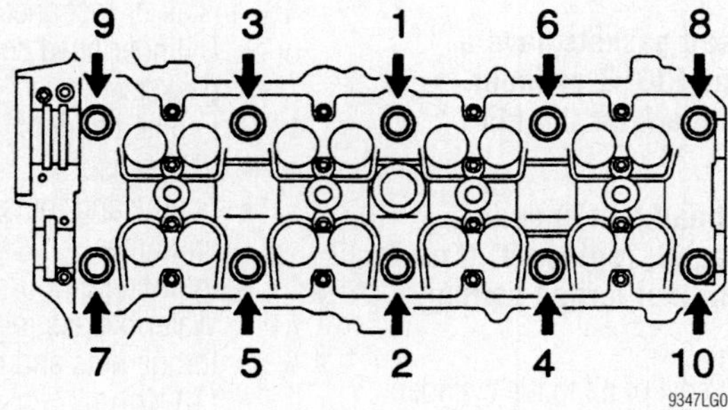

Fig. 60 Cylinder head torque sequence—4.3L (3UZ-FE) engine

EXHAUST MANIFOLD

REMOVAL & INSTALLATION

See Figures 61 and 62.

1. Before servicing the vehicle, refer to the Precautions section.
2. Remove or disconnect the following:
   - Negative battery cable
   - Engine undercover
   - Coolant
   - V-bank cover
   - Air cleaner assembly (if needed for access to exhaust manifold)
3. Remove cylinder heads. See CYLINDER HEADS in this section.
4. Remove 4 bolts and heat shield from exhaust manifold.
5. Remove 8 nuts and remove exhaust manifold and gasket.

To install:

6. To install, reverse the removal procedure. Install new manifold gasket and new retaining nuts. Tighten the exhaust manifold nuts to 32 ft. lbs. (44 Nm).
7. Install the manifold heat shields.
8. Refill the engine cooling system. Start the engine and check for leaks.

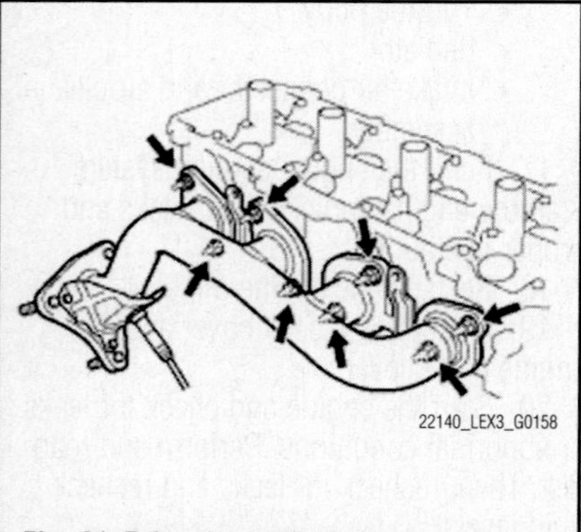

Fig. 61 Exhaust manifold view

67162-LEXU-G18

Fig. 62 Measuring exhaust manifold warpage

FLEXPLATE

REMOVAL & INSTALLATION

See Figure 63.

1. Before servicing the vehicle, refer to the Precautions section.
2. Disconnect the negative battery cable. Wait at least 90 seconds before performing any other work
3. Remove the engine or transmission assembly.
4. Using SST 09213-54015 (91651-60855), 09330-00021, hold the crankshaft.
5. Remove the 8 bolts and remove the rear spacer, drive plate and front spacer.

To install:

6. Using SST 09213-54015 (91651-60855), 09330-00021, hold the crankshaft.
7. Clean the bolts and bolt holes.
8. Apply adhesive (Part No. 08833-00070, THREE BOND 1324 or equivalent) to 2 or 3 threads of the bolts.
9. Install the rear spacer, drive plate and front spacer.
10. Install and uniformly tighten the 8 bolts in several passes, in the sequence shown, to the following torque values: 36 ft. lbs. (49 Nm) plus an additional 90°

➡Do not start the engine within an hour after installing.

11. Install the engine or transmission assembly.
12. Connect the negative battery cable.

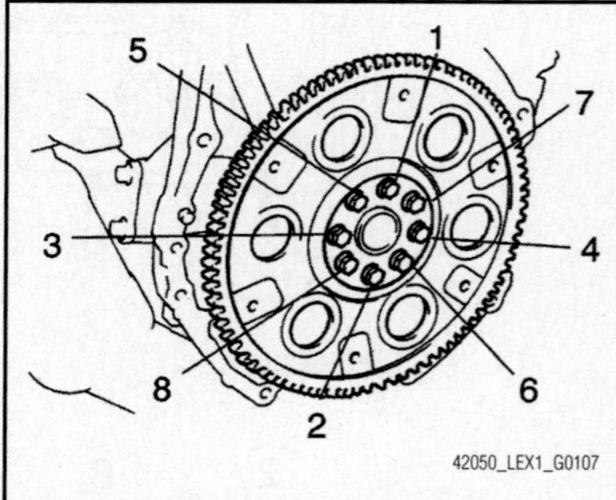

Fig. 63 Bolt tightening sequence for flywheel (drive plate).

INTAKE MANIFOLD

REMOVAL & INSTALLATION

See Figure 64.

1. Before servicing the vehicle, refer to the Precautions section.
2. Properly relieve the fuel system pressure.

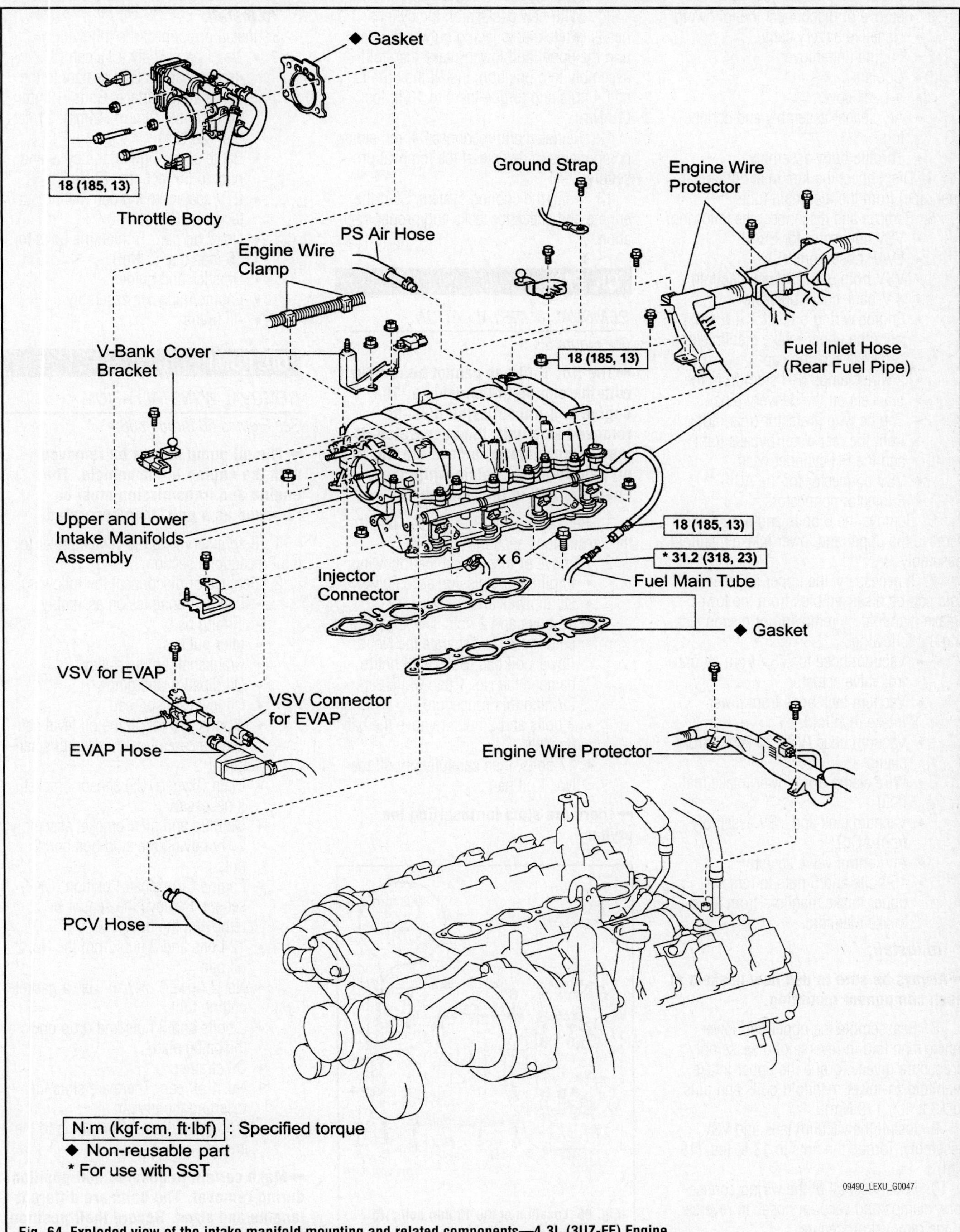

Fig. 64 Exploded view of the intake manifold mounting and related components—4.3L (3UZ-FE) Engine

3. Remove or disconnect the following:
   - Negative battery cable
   - Engine undercover
   - Coolant
   - V-bank cover
   - Air cleaner assembly and connectors
   - Throttle body assembly

4. Disconnect the fuel inlet hose (rear fuel pipe) from the fuel main tube.

5. Remove and disconnect the following:
   - VSV connector for EVAP
   - EVAP hose from VSV
   - VSV from upper intake manifold
   - 4 V-bank cover brackets
   - Engine wiring protector (LH side) from the upper intake manifold and camshaft bearing cap
   - 2 wire clamps (RH side) from the brackets on the delivery pipe
   - Engine wire protector (rear side) from the rear water bypass joint and the RH cylinder head
   - VSV connector for the ACIS
   - 8 injector connectors

6. Remove the 6 bolts and 4 nuts and remove the upper and lower intake manifold assembly.

7. If necessary, the upper intake manifold can be disassembled from the lower intake manifold by removing or disconnecting the following:
   - Vacuum hose for VSV from air control valve actuator
   - Vacuum tank hose from lower intake manifold
   - Vacuum hose (VSV for ACIS) from clamp
   - Wire clamp from lower intake manifold
   - Vacuum tank and VSV assembly from ACIS
   - Air control valve actuator
   - 15 bolts and 5 nuts to remove upper intake manifold from lower intake manifold

To install:

➡Always be sure to use new gaskets at each component mounting.

8. Reassemble the upper and lower intake manifold in reverse of disassembly procedure given. Torque the upper intake manifold-to-lower manifold bolts and nuts to 13 ft. lbs. (18 Nm).

9. Install the vacuum tank and VSV assembly. Torque the nuts to 13 ft. lbs. (18 Nm).

10. Reconnect all of the wiring connectors, clamps and vacuum hoses in reverse of the removal procedure.

11. With new gaskets on the cylinder heads (white marks facing outward), position the upper and lower intake manifold assembly into position. Install the 6 bolts and 4 nuts and torque them to 13 ft. lbs. (18 Nm).

12. Reinstall and reconnect all remaining components in reverse of the removal procedure.

13. Refill the cooling system. Start the engine and check for leaks and proper operation.

OIL PAN

REMOVAL & INSTALLATION

See Figure 65.

➡The No. 1 oil pan cannot be removed with the engine in the vehicle. The engine and transmission must be removed as a unit, then separated. It may be possible to remove the No. 2 oil pan from the vehicle while the engine is still in the vehicle.

1. Before servicing the vehicle, refer to the Precautions section.

2. Remove or disconnect the following:
   - Engine/transmission assembly
   - Oil dipstick and guide
   - 12 bolts and 2 nuts. Use a gasket-cutting tool to separate the No. 2 (lower) oil pan. Be careful not to damage the No. 1 pan while performing this procedure.
   - 3 bolts and 2 nuts; remove the baffle plate
   - 17 bolts, then carefully pry off the No. 1 oil pan

➡There are slots for inserting the prybar.

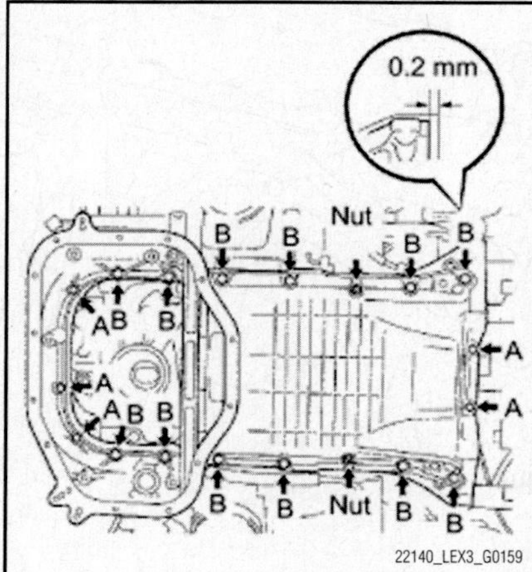

Fig. 65 Location of the 10 mm bolts (A) and 12 mm (B)

To install:

3. Install or connect the following:
   - No. 1 pan. Apply a ⅛inch (3–4mm) bead on RTV sealant to the pan mating surface. Bolts: 12mm: 66 inch lbs. (8mm); 14mm: 21 ft. lbs. (28 Nm)
   - Baffle plate. Tighten the bolts and nuts to 66 inch lbs. (8 Nm).
   - RTV sealant to the pan mating surface
   - No. 2 oil pan. Tighten the bolts to 66 inch lbs. (8 Nm)
   - Dipstick and guide
   - Engine/transaxle assembly
   - All fluids

OIL PUMP

REMOVAL & INSTALLATION

See Figures 66 through 68.

➡The oil pump cannot be removed with the engine in the vehicle. The engine and transmission must be removed as a unit, then separated.

1. Before servicing the vehicle, refer to the Precautions section.

2. Remove or disconnect the following:
   - Engine/transmission assembly
   - Timing belt
   - Idler pulleys
   - Crankshaft timing pulley
   - Oil dipstick and guide
   - Oil level sensor lead
   - 4 bolts and lift off the oil level sensor. Be careful not to drop this sensor.
   - Main Oxygen (O_2) sensor bracket, if necessary
   - Oil filter and filter bracket assembly by removing the stud bolt and 2 nuts
   - Engine Crankshaft Position (CKP) sensor. Remove the sensor by removing the bolt.
   - 12 bolts and 2 nuts from the No. 2 oil pan
   - No. 2 (lower) oil pan. Use a gasket-cutting tool
   - 2 bolts and 3 nuts and drop down the baffle plate
   - Oil strainer
   - No. 1 oil pan. There are slots for inserting the prybar.
   - 8 bolts holding the oil pump to the engine

➡Make certain to observe bolt position during removal. The bolts are different lengths and sizes. Record their position for proper reassembly.

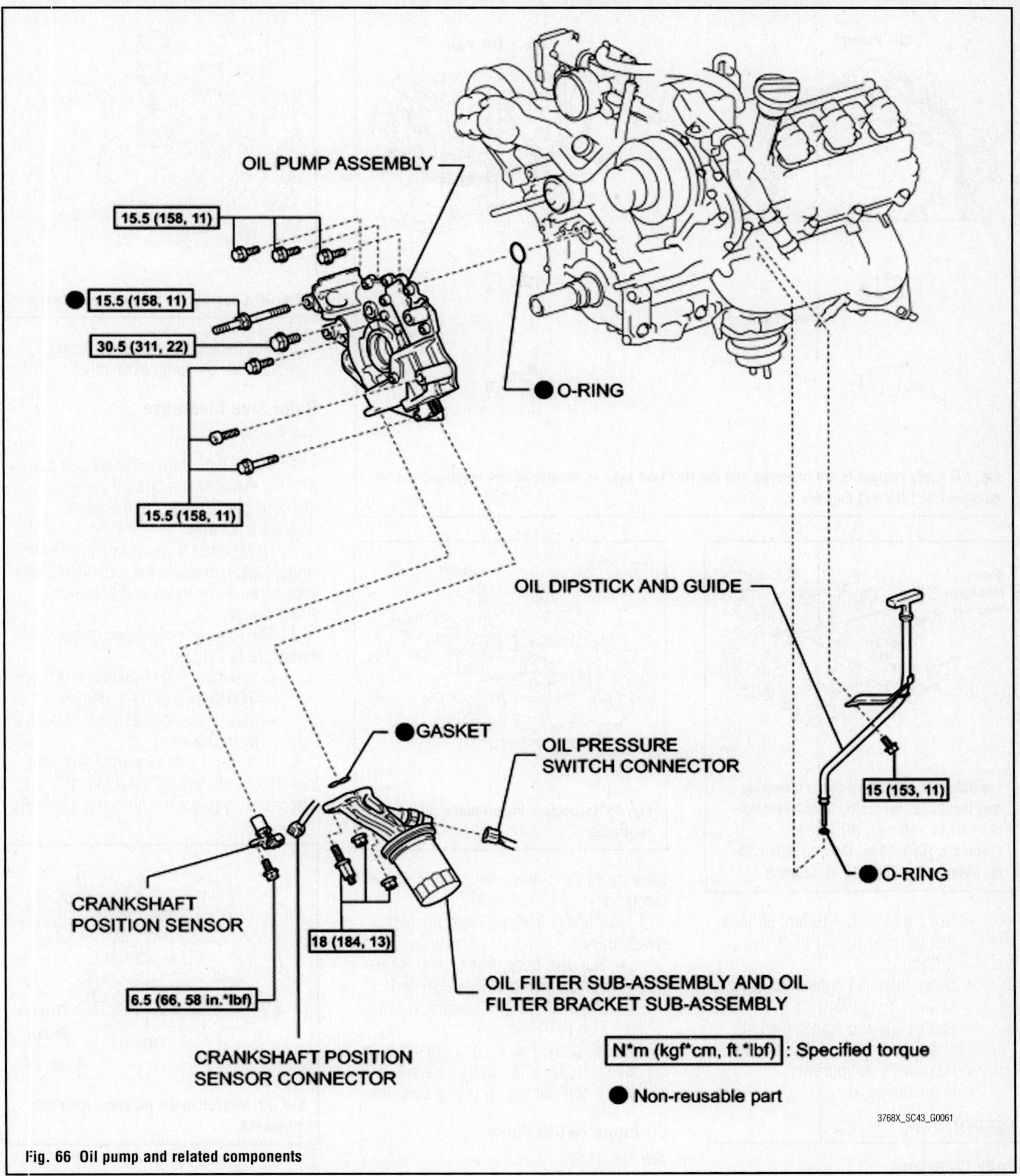

Fig. 66 Oil pump and related components

- Oil pump from the engine block
- O-ring from the block

To install:

➡Prior to installing the oil pump, lubricate the gears with clean engine oil.

3. Install or connect the following:
   - A 2–3mm wide (0.08–0.12 in.) bead of RTV sealant to the oil pump
   - New O-ring in position on the block
   - Oil pump on the engine
   - The 8 bolts in their correct locations. Tighten the bolts with 12mm or 6mm heads to 12 ft. lbs. (16 Nm) and the bolts with 14mm heads to 22 ft. lbs. (30 Nm).
   - A ⅛inch (3–4mm) bead of RTV sealant to the pan mating surface.

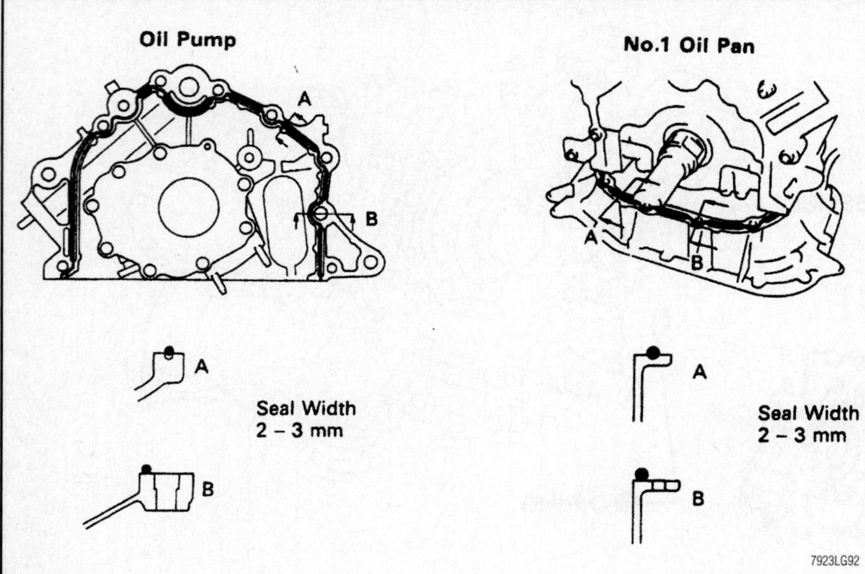

Fig. 67 Apply sealant to the oil pump and the No. 1 oil pan, as shown, before installing the oil pump—4.3L (3UZ-FE) Engines

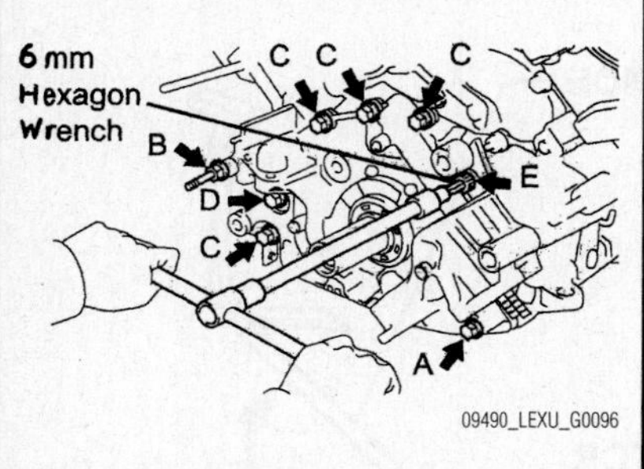

Fig. 68 4.3L engine oil pump mounting bolt locations, according to bolt lengths—(A) 1.97 in. (50mm), (B) 4.17 in. (106mm), (C) 1.18 in. (30mm), (D) 1.73 in. (44mm) and (E) 1.10 in. (28mm)

- No. 1 pan. Bolts–10mm: 66 inch lbs. (8 Nm); 12mm: 21 ft. lbs. (28 Nm)
- Oil strainer and tighten the bolts to 66 inch lbs. (8 Nm)
- Baffle plate and tighten the bolts and nuts to 66 inch lbs. (8 Nm)
- Remaining components
- Engine/transaxle

INSPECTION

Body Clearance

See Figure 69.

1. Install the rotors to the oil pump body with the marks facing upward and match-marks aligned. Check that the rotors revolves smoothly.
2. Using a feeler gauge, measure the body clearance between the driven rotor and pump body.

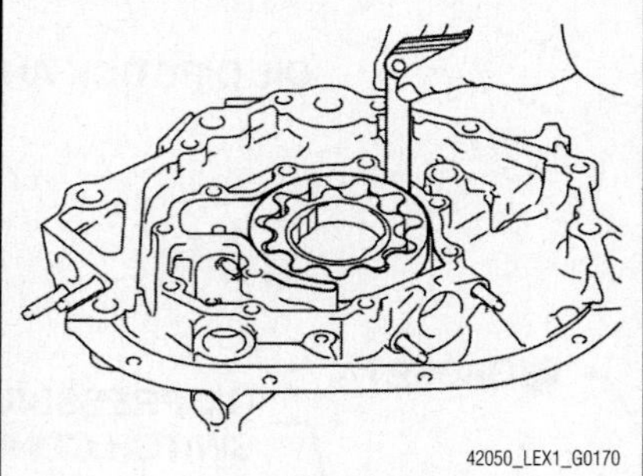

Fig. 69 Inspecting the oil pump rotor body clearance.

3. The body clearance measurements are as follows:
   - Standard body clearance: 0.0098–0.0128 inch (0.250–0.325mm)
   - Maximum body clearance: 0.325 inch (0.0128mm)
4. If the body clearance is greater than maximum, replace the rotors as a set. If necessary, replace the oil pump assembly.

Oil Pump Relief Valve

See Figure 70.

1. Inspect the oil pump relief valve as follows:
   a. Remove the plug, spring and relief valve.
   b. Apply a light coat of engine oil.
   c. Check that it falls smoothly into the valve hole by its own weight. If it doesn't, replace the relief valve. If necessary, replace the oil pump assembly.

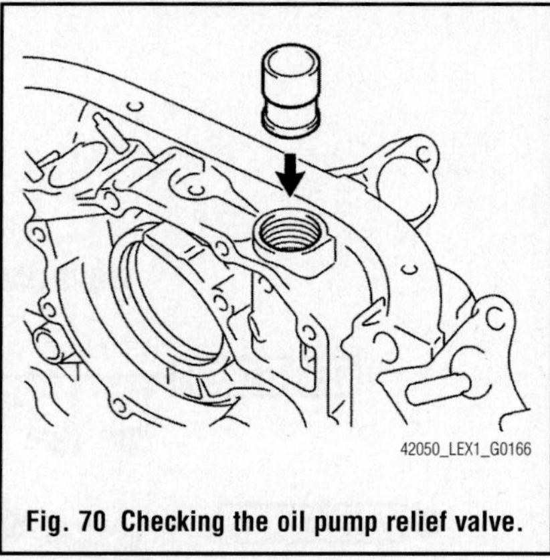

Fig. 70 Checking the oil pump relief valve.

Rotor Side Clearance

See Figure 71.

1. Install the rotors to the oil pump body with the marks facing upward and match-marks aligned. Check that the rotors revolves smoothly.
2. Using a feeler gauge and precision straight edge, measure the rotor side clearance between the rotors and precision straight edge.
3. The rotor side clearance measurements are as follows:
   - Standard side clearance: 0.0012–0.0035 inch (0.03–0.09mm)
   - Maximum side clearance: 0.0035 inch (0.09mm)
4. If the rotor side clearance is greater than maximum, replace the rotors as a set. If necessary, replace the oil pump assembly.

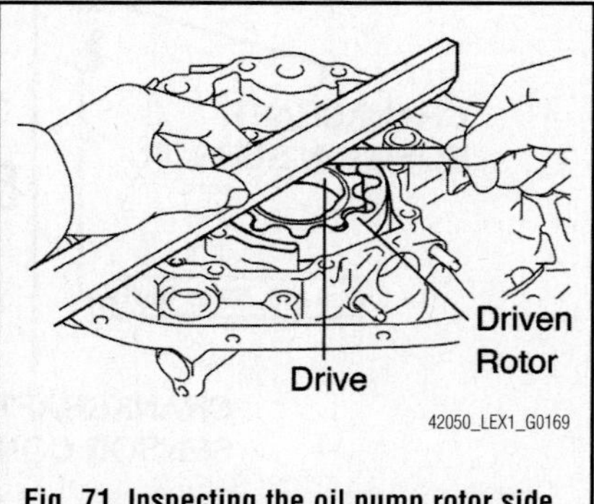

Fig. 71 Inspecting the oil pump rotor side clearance.

Rotor Tip Clearance

See Figures 72 and 73.

1. Install the rotors to the oil pump body with the marks facing upward and match-marks aligned. Check that the rotors revolves smoothly.

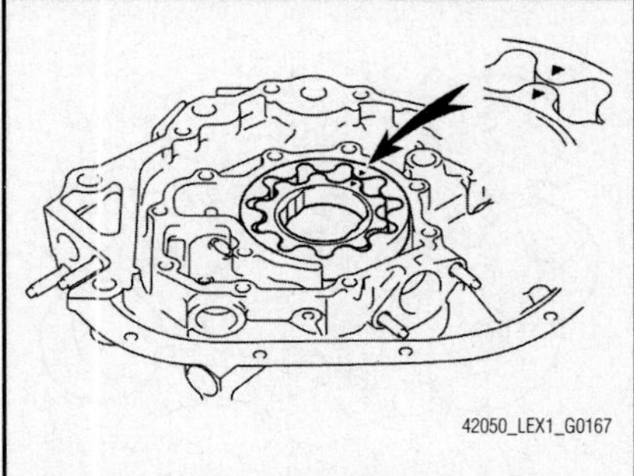

Fig. 72 Install the rotors to the oil pump body with the marks facing upward and matchmarks aligned.

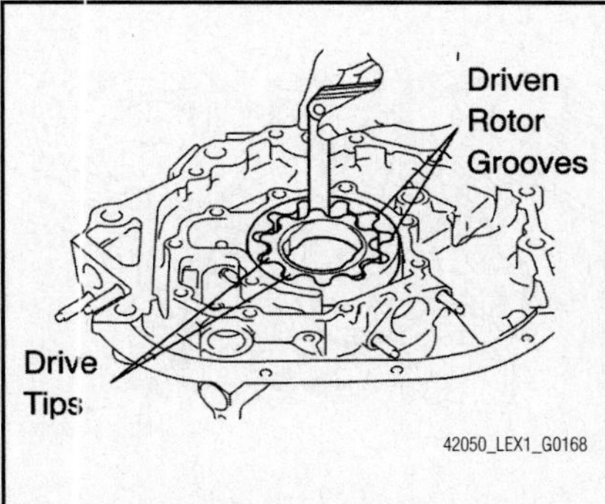

Fig. 73 Inspecting the oil pump rotor tip clearance.

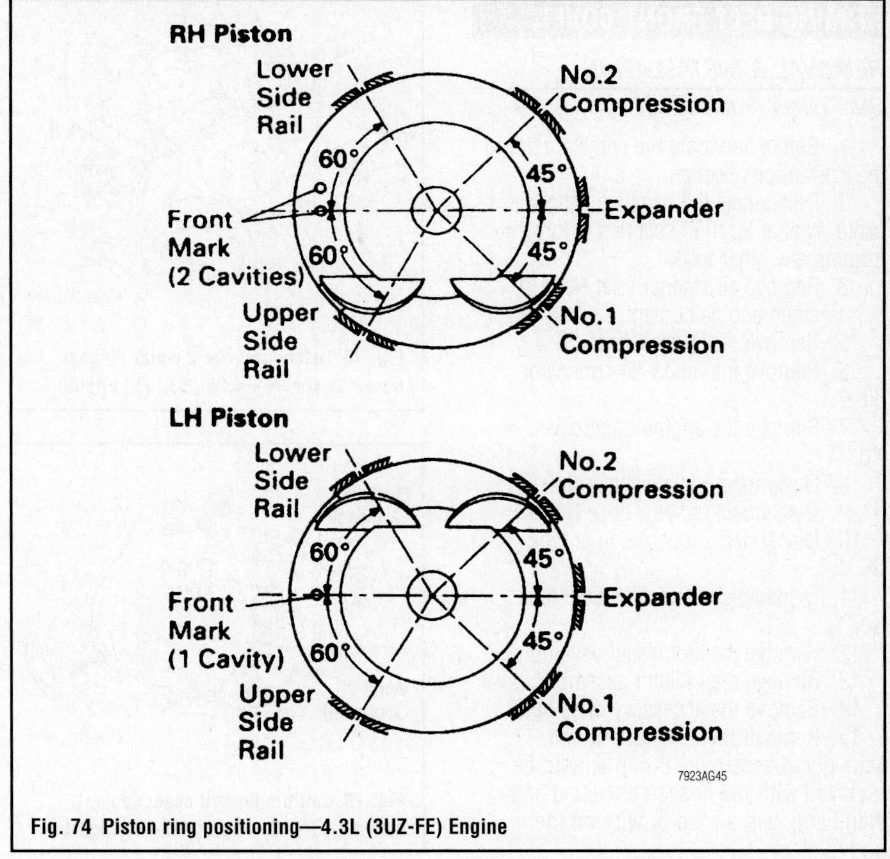

Fig. 74 Piston ring positioning—4.3L (3UZ-FE) Engine

2. Using a feeler gauge, measure the rotor tip clearance between the drive and driven rotor tips.

3. The rotor tip clearance measurements are as follows:

- Standard tip clearance: 0.0024–0.0071 inch (0.06–0.18mm)
- Maximum tip clearance: 0.0071 inch (0.18mm)

4. If the tip clearance is greater than maximum, replace the rotors as a set.

PISTON AND RING

POSITIONING

See Figure 74.

REAR MAIN SEAL

REMOVAL & INSTALLATION

See Figures 75 and 76.

There are 2 methods (1) and (2) to replace the oil seal.

Method (1)

1. If the rear oil seal retainer is removed from the cylinder block, perform the following:

- Using a screwdriver and hammer, tap out the oil seal.

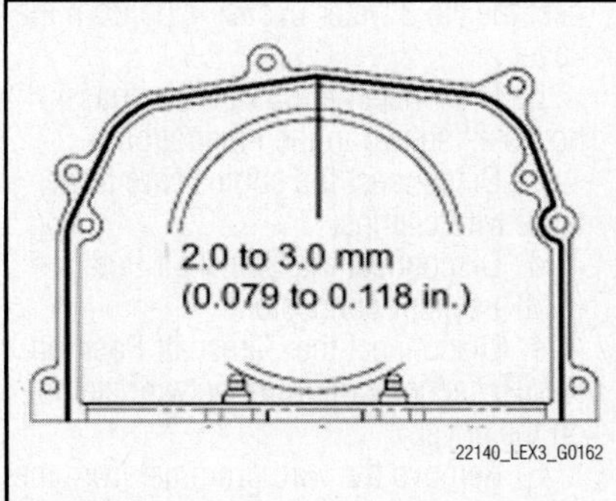

Fig. 75 Apply seal packing in a continuous bead as shown

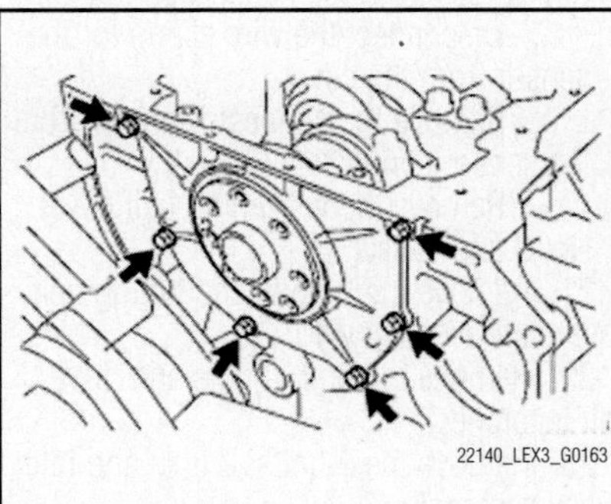

Fig. 76 Rear main oil seal retainer and mounting bolts

- Using SST 09223-15030, 09950-70010 (09951-07100) and a hammer, tap in a new oil seal until its surface is flush with the rear oil seal retainer edge.
- Apply MP grease to the oil seal lip.
- Apply seal packing in a continuous bead to the retainer.
- Install the oil seal retainer and tighten the bolts to 7 ft. lbs. (10 Nm).

Method (2)

2. If the rear seal retainer is installed on the cylinder block, perform the following:

- Using a knife, cut off the oil seal lip.
- Using a screwdriver, pry out the oil seal.

CAUTION

Be careful not to damage the crankshaft. Tape the screwdriver tip.

- Apply MP grease to a new oil seal lip.
- Using SST 09223-15030, 09950-70010 (09951-07100) and a hammer, tap in the oil seal until its surface is flush with the rear oil seal retainer edge.

TIMING BELT FRONT COVER

REMOVAL & INSTALLATION

See Figures 77 through 81.

1. Before servicing the vehicle, refer to the Precautions section.
2. Disconnect the negative battery cable. Wait at least 90 seconds before performing any other work.
3. Remove air cleaner inlet No. 1.
4. Drain engine coolant.
5. Remove the V-bank cover.
6. Remove the intake air connector pipe.
7. Remove the engine undercover No. 1.
8. Disconnect radiator hose No. 1.
9. Disconnect radiator hose No. 2.
10. Disconnect oil cooler inlet tube No. 1.
11. Disconnect oil cooler outlet tube No. 1.
12. Remove the air cleaner assembly.
13. Remove the radiator assembly.
14. Remove the accessory drive belt
15. Remove the 2 bolts, nut and vane pump assembly. Pump should be removed with the hoses connected and then hang with a rope or wire on the body's side.
16. Remove the alternator.
17. Remove the A/C compressor assembly. The cooler compressor with the magnetic clutch should be removed with the low-pressure and high-pressure hoses connected and then hang with a rope or wire on the body's side.
18. Disconnect the 2 PS air hoses from the clamp on the timing belt cover No. 2.
19. Remove the cap nut, 3 bolts, timing belt cover No. 2 and gasket.
20. Remove the left timing belt cover sub-assembly No. 3 as follows:
    a. Remove the cap nut, and disconnect the No.3 water bypass pipe from the cover.
    b. Disconnect the 2 water by-pass hoses as shown in the illustration.
    c. Disconnect the engine wire from the 2 wire clamps.
    d. Disconnect the Camshaft Position (CMP) sensor connector.
    e. Disconnect the Camshaft Position (CMP) sensor wire from the wire clamp on the cover.
    f. Remove the wire grommet from the cover.
    g. Remove the 4 bolts.
    h. Disconnect the cover from the timing plate and camshaft bearing cap.
    i. Disconnect the wire clamp for the sensor from the cover.
    j. Remove the connector holder from the sensor connector.
    k. Remove the left timing belt cover No. 3 and gasket.
21. Remove the 2 bolts and timing belt cover sub-assembly No. 2.
22. Remove the bolt, 2 nuts and drive belt tensioner.
23. Remove the 2 bolts, 2 nuts and idler pulley assembly.
24. Remove the crankshaft pulley (damper).
25. Remove the 4 bolts and timing belt No. 1 cover.

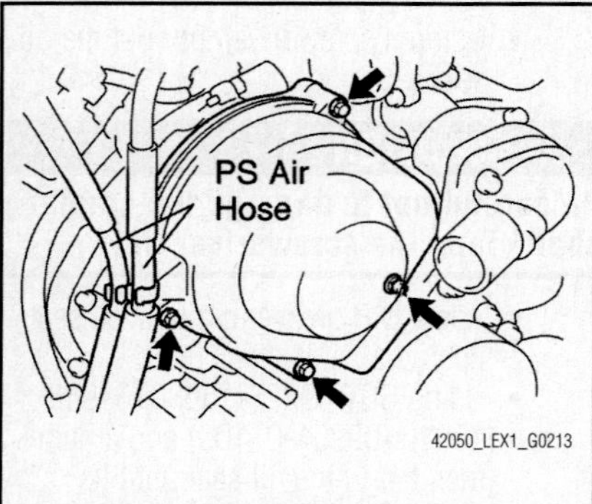

Fig. 77 Removal of timing belt cover No. 2—4.3L (3UZ-FE) engine

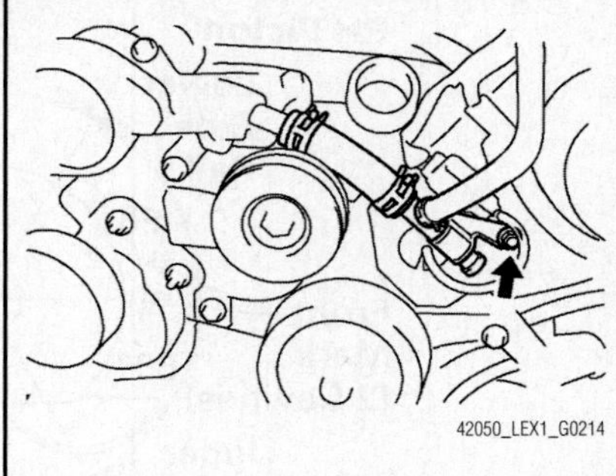

Fig. 78 Disconnect the 2 water by-pass hoses as shown—4.3L (3UZ-FE) engine

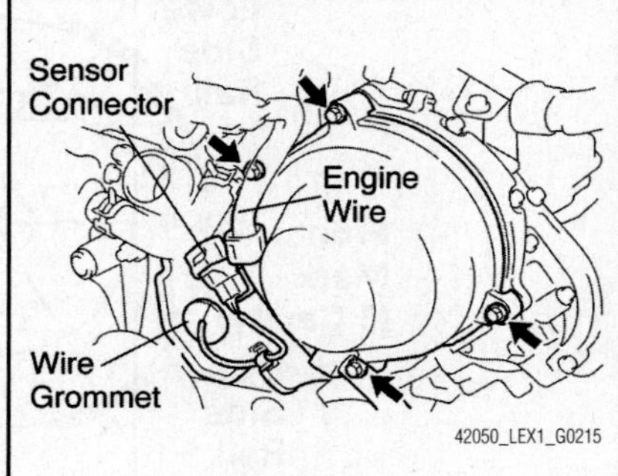

Fig. 79 Left timing belt cover sub-assembly No. 3—4.3L (3UZ-FE) engine

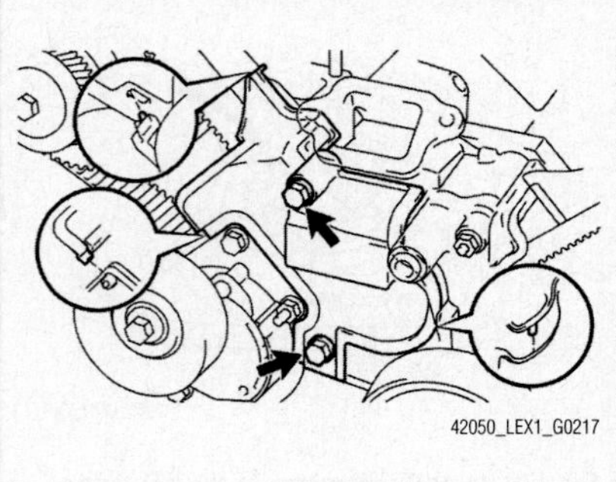

Fig. 80 Timing belt cover sub-assembly No. 2—4.3L (3UZ-FE) engine

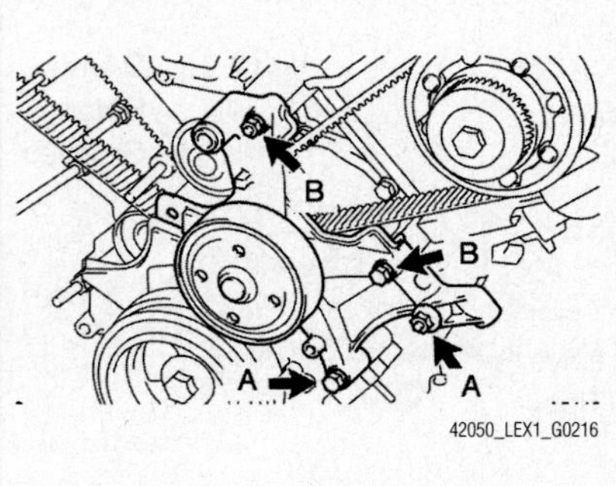

Fig. 81 Idler pulley assembly with 2 bolts and 2 nuts—4.3L (3UZ-FE) engine

To install:

26. Install the timing belt No. 1 cover with the 4 bolts and tighten to 66 inch lbs. (7.5 Nm).
27. Install the crankshaft pulley (damper).
28. Install the idler pulley assembly with the 2 bolts and 2 nuts. Tighten the bolts and nuts as follows:
    - 14mm head bolt A: 24 ft. lbs. (32 Nm)
    - 12mm head bolt B: 12 ft. lbs. (16 Nm)

    a. Each bolt length is indicated below.

    Bolt Length:
    - 4.49 inches (114mm) for 14mm head (A)
    - 4.17 inches (106mm) for 12mm head (B)
29. Install the drive belt tensioner with the bolt and 2 nuts. Tighten the bolt and nuts to 12 ft. lbs. (16 Nm). Use a bolt that is 4.18 inches (106mm) in length.
30. Fit the timing belt cover sub-assembly No. 2, matching the claws and pin with each part. Install the timing belt cover with the 2 bolts and tighten to 12 ft. lbs. (16 Nm). Use bolts that are 4.17 inches (106mm) in length.

31. Install the left timing belt cover sub-assembly No. 3 as follows:

a. Install the gasket to the cover.

b. Run the Camshaft Position (CMP) sensor wire through the cover hole.

c. Install the cover with the 4 bolts and tighten to 66 inch lbs. (7.5 Nm).

d. Install the wire grommet to the cover.

e. Install the sensor connector to the sensor holder.

f. Connect the sensor connector.

g. Install the sensor wire to the wire clamp on the cover.

h. Install the engine wire to the 2 wire clamps on the cover.

i. Connect the 2 water by-pass hoses, as shown in the illustration.

j. Install the No. 3 water by-pass pipe to the cover with the cap nut and tighten to 66 inch lbs. (7.5 Nm).

32. Install gasket to timing belt cover No. 2. Install the cover with the cap nut and 3 bolts and tighten to 66 inch lbs. (7.5 Nm).

33. Install the 2 PS air hoses to the clamp on the timing belt cover No. 2.

34. Install the A/C compressor, stay and wire bracket with the 3 bolts and nut. Tighten the bolts to 36 ft. lbs. (49 Nm). Tighten the nut to 21 ft. lbs. (29 Nm).

35. Remove the alternator.

36. Install the vane pump assembly with the 2 bolts and nuts. Alternately tighten the bolts and nut. Tighten the bolts to 29 ft. lbs. (39.2 Nm). Tighten the nut to 32 ft. lbs. (43 Nm).

37. Install the accessory drive belt

38. Install the radiator assembly.

39. Install the air cleaner assembly and tighten the fasteners to 44 inch lbs. (5 Nm).

40. Connect oil cooler outlet tube No. 1.

41. Connect oil cooler inlet tube No. 1.

42. Connect radiator hose No. 2.

43. Connect radiator hose No. 1.

44. Install air cleaner inlet No. 1 and tighten the fasteners to 44 inch lbs. (5 Nm).

45. Install the intake air connector pipe and tighten the fasteners to 44 inch lbs. (5 Nm).

46. Install the V-bank cover and tighten the fasteners to 44 inch lbs. (5 Nm).

47. Refill the engine cooling system and check for leaks.

48. Install the engine undercover No. 1.

49. Connect the negative battery cable.

TIMING BELT & SPROCKETS

REMOVAL & INSTALLATION

See Figures 82 and 83.

1. Remove all necessary components for access to the right-hand side No. 3 and No. 2, and left-hand side No. 2 timing belt covers, then remove the covers.

2. Turn the crankshaft pulley and align its groove with the timing mark **0** of the No. 1 timing cover. Check that the timing marks of the camshaft timing pulleys and timing belt rear plates are aligned. If not, turn the crankshaft 1 full revolution (360°).

3. Loosen crankshaft pulley bolt, then set the No. 1 cylinder to about 50°ATDC on the compression stroke. With the crankshaft pulley notch aligned with the **0** mark on the timing belt cover, turn the crankshaft pulley about 50°clockwise and put timing mark of the crank pulley in line with the centers of the pulley bolt and the No. 2 timing belt idler pulley bolt.

4. Remove or disconnect the following, as applicable for the vehicle:

- Timing belt tensioner. Using the proper tool, loosen the tension between the left side and right side timing pulleys by slightly turning the left side camshaft clockwise.
- Timing belt from the camshaft timing pulleys
- Power steering pump pulley
- Alternator
- Drive belt tensioner
- Bolt and timing pulleys, using the proper tool
- Bolt and the crankshaft pulley with the proper tool.
- Fan bracket
- Mounting bolts and the No. 1 timing belt cover
- 2 upper and lower timing belt covers
- Timing belt guide (No. 1 crank position sensor plate)
- Timing belt
- No. 1 and No. 2 timing belt idler pulleys, if necessary.

➡If the timing belt is to be reused, draw a directional arrow on the timing belt in the direction of engine rotation (clockwise) and place matchmarks on the timing belt and crankshaft gear to match the drilled mark on the pulley.

To install:

5. Align the installation mark on the timing belt with the drilled mark of the crankshaft timing pulley. Install the timing belt on the crankshaft timing pulley, No. 1 idler pulley and the No. 2 idler pulley.

➡If the old timing belt is being reinstalled, be sure the directional arrow is facing in the original direction and that the belt and crankshaft gear matchmarks are properly aligned.

6. Install the timing belt guide (No. 1 crank angle sensor plate) with the cup side facing forward. Replace the timing belt cover spacer.

7. Install the No. 1 timing belt cover and tighten the mounting bolts.

8. Align the pulley set key on the crankshaft with the key groove of the pulley. Install the pulley, using the proper tool to tap in the pulley. Tighten the pulley bolt to 181 ft. lbs. (245 Nm).

9. Align the knock pin on the right side camshaft with the knock pin of the timing pulley. Slide on the timing pulley with the right side mark facing forward. Tighten the bolt to 80 ft. lbs. (108 Nm).

10. Align the knock pin on the left side camshaft with the knock pin of the timing pulley. Slide on the timing pulley with the left side mark facing forward. Tighten the bolt to 80 ft. lbs. (108 Nm).

11. Turn the crankshaft pulley and align its groove with the **0** timing mark on the No. 1 timing belt cover. Using the proper tool, turn the crankshaft timing pulley and align the timing marks of the camshaft timing pulley and the timing belt rear plate.

12. Install the timing belt to the left side camshaft timing pulley by:

a. Using the proper tool, slightly turn the left side timing pulley clockwise. Align the installation mark of the timing belt with the timing mark of the camshaft timing pulley and hang the timing belt on the left side camshaft pulley.

b. Using the proper tool, align the timing marks of the left side camshaft pulley and the timing belt rear plate.

c. Check that the timing belt has tension between crankshaft timing pulley and the left side camshaft pulley.

13. Install the timing belt to the right side camshaft timing pulley by:

a. Using the proper tool, slightly turn the right side timing pulley clockwise. Align the installation mark of the timing belt with the timing mark of the camshaft timing pulley and hang the timing belt on the right side camshaft pulley.

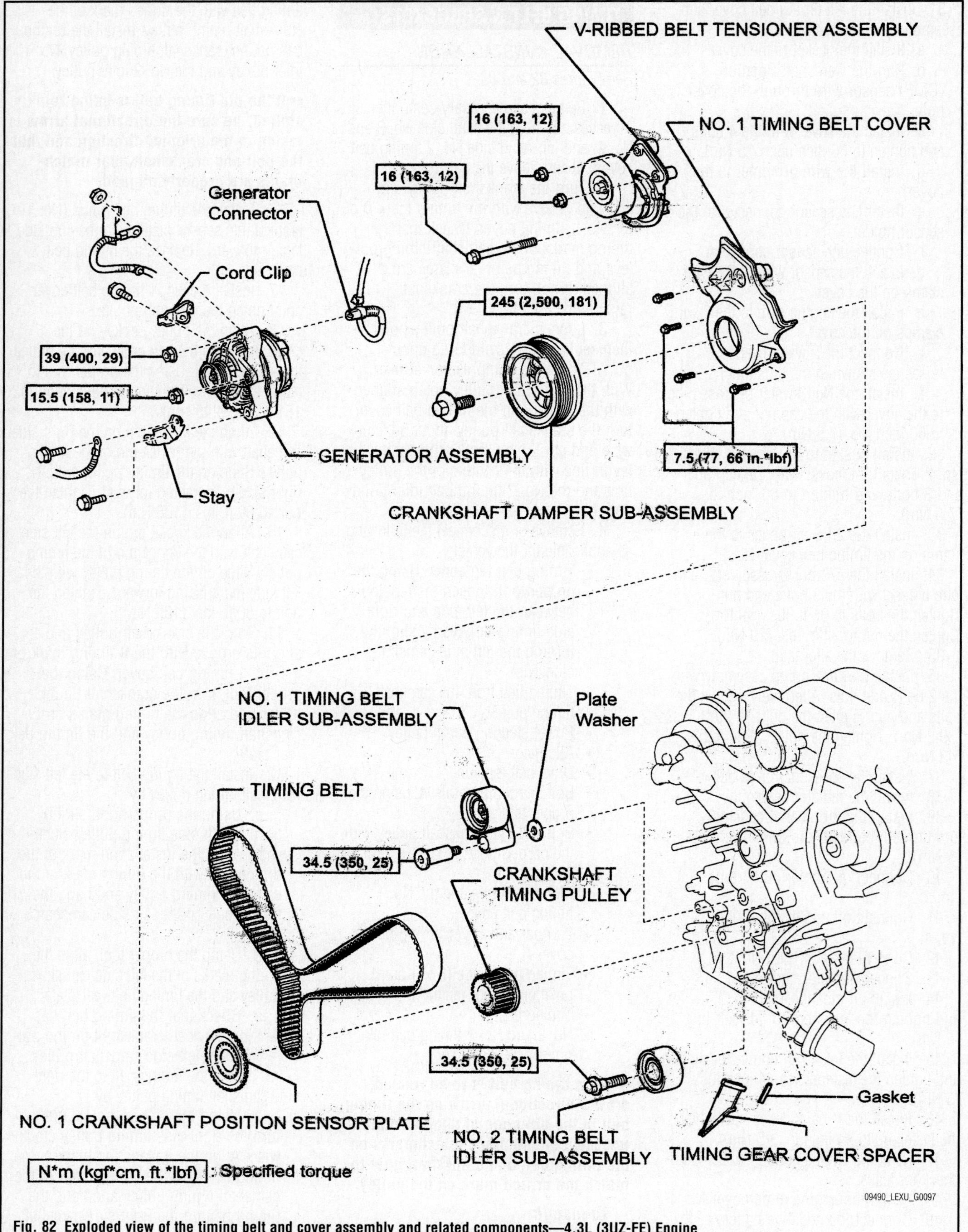

Fig. 82 Exploded view of the timing belt and cover assembly and related components—4.3L (3UZ-FE) Engine

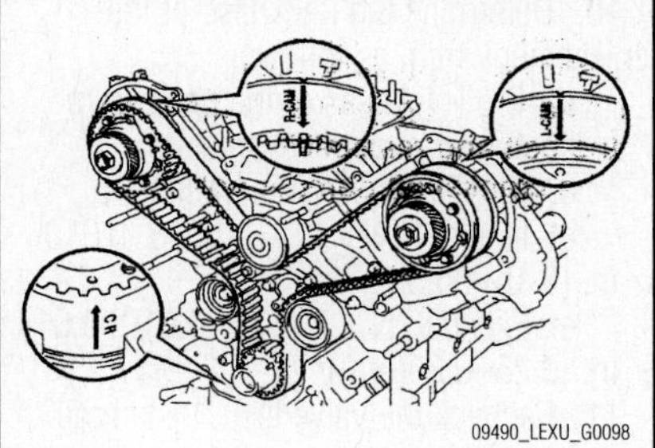

Fig. 83 Timing belt sprocket mark alignment for belt installation—4.3L (3UZ-FE) Engine

b. Using the proper tool, align the timing marks of the right side camshaft pulley and the timing belt rear plate.

c. Check that the timing belt has tension between the crankshaft timing pulley and the right side camshaft pulley.

14. The timing belt tensioner must be set prior to installation. The tensioner can be set as follows:

a. Place a plate washer between the tensioner and a block. Using a suitable press, press in the pushrod using 220–2205 lbs. (100–1000kg) of pressure.

b. Align the holes of the pushrod and housing, pass the proper tool (0.05 in. Allen wrench) through the holes to keep the setting position of the pushrod.

c. Release the press and install the dust boot on the tensioner.

15. Install the tensioner and tighten the bolts to 20 ft. lbs. (26 Nm). Remove the tool from the tensioner.

16. Turn the crankshaft pulley two complete revolutions from TDC-to-TDC. Always turn the crankshaft clockwise. Check that each pulley aligns with the timing marks.

17. Install all remaining components in the reverse order of removal.

VALVE LASH

ADJUSTMENT

See Figures 84 and 85.

1. Before servicing the vehicle, refer to the Precautions section.

2. Remove or disconnect the following, as applicable:

- Negative battery cable
- V-bank cover
- Intake air connector pipe
- Ignition coils

RH Bank

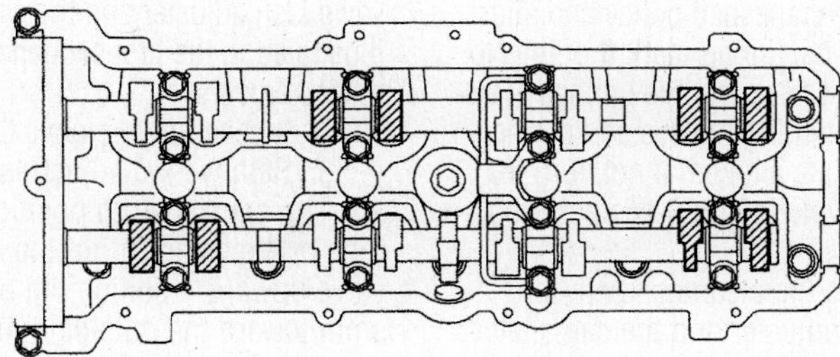

LH Bank

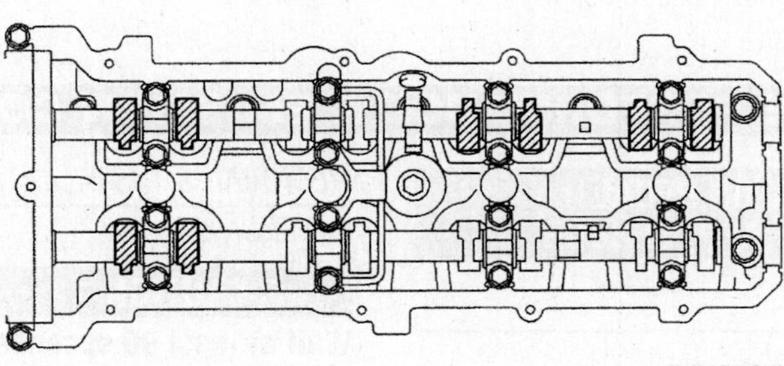

67162-LEXU-G113

Fig. 84 Adjust these valves FIRST—4.3L (3UZ-FE) Engine

RH Bank

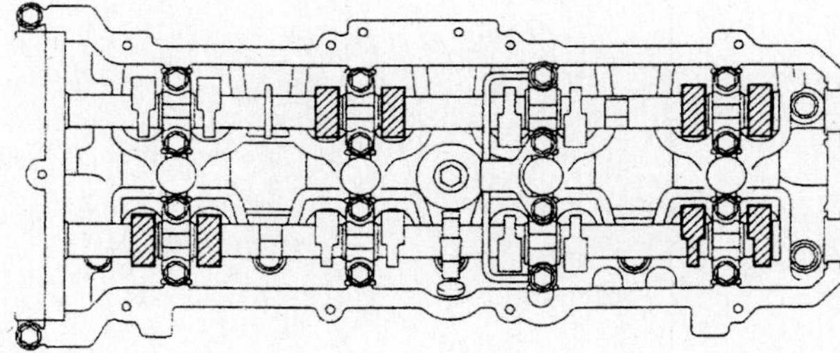

LH Bank

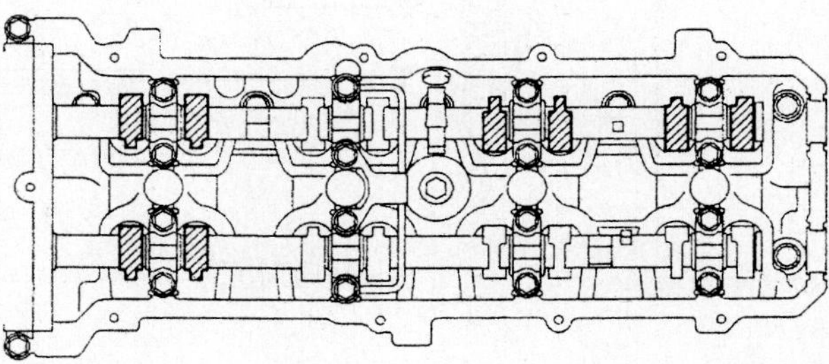

67162-LEXU-G114

Fig. 85 Adjust these valves SECOND—4.3L (3UZ-FE) Engine

- No. 3 timing belt covers
- Spark plug wires
- Cylinder head covers

3. Turn the crankshaft pulley and align its groove with the timing mark **0** of the No. 1 timing cover. Check that the timing marks of the camshaft timing pulleys and timing belt rear plates are aligned. If not, turn the crankshaft 1 revolution (360 degrees) and align the mark.

4. Measure the clearance between the valve lash adjuster and the camshaft on the valves, as illustrated, in the first sequence. Record the measurements.

    a. The intake valve lash cold is 0.006–0.010 in. (0.15–0.25mm).

    b. The exhaust valve lash cold is 0.010–0.014 in. (0.25–0.35mm).

5. Turn the crankshaft 1 full revolution (360 degrees) and align the mark.

6. Measure the clearance between the valve lash adjuster and the camshaft , as illustrated in the first sequence. Record the measurements.

7. If necessary, remove the camshafts.

8. Remove the adjusting shim and turn the crankshaft to position the cam lobe of the camshaft on the adjusting valve upward. Position the hole in the shim toward the outside of the cylinder head. Press down the valve lash adjuster with the proper tool and place the proper tool between the camshaft and the valve lash adjuster. Remove the tool.

9. Remove the adjusting shim with the proper tool.

10. Determine the thickness of the replacement shim as follows:

    a. T = Thickness of the used shim

    b. A = Measured valve lash

    c. N = Thickness of new shim

    d. Intake: N = T + (A - 0.006–0.010 in. (0.15–0.25mm))

    e. Exhaust: N = T + (A - 0.010–0.014 in. (0.25–0.35mm))

11. Recheck the valve lash. Install the cylinder head covers.

12. Connect the spark plug wires and install the No. 3 timing belt covers.

13. Install or reconnect all other components in reverse of removal procedure.

14. Connect the negative battery cable.

ENGINE PERFORMANCE & EMISSION CONTROLS

ACCELERATOR PEDAL POSITION (APP) SENSOR

LOCATION

See Figure 86.

Refer to the accompanying illustration.

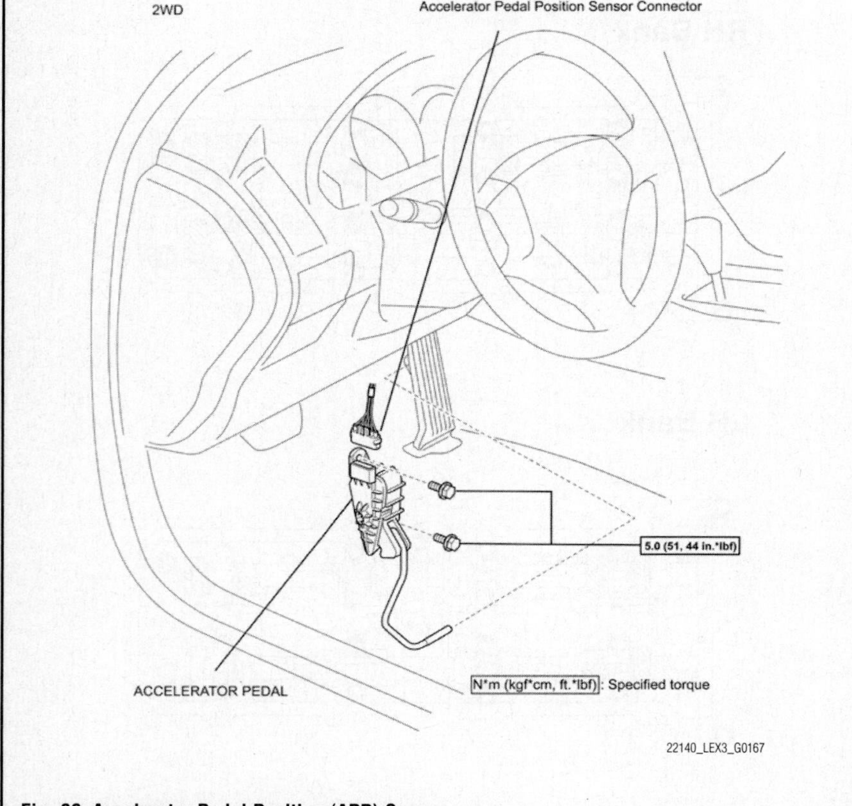

Fig. 86 Accelerator Pedal Position (APP) Sensor

REMOVAL & INSTALLATION

See Figure 87.

CAUTION

Wait at least 90 seconds after disconnecting the cable from the negative (-) battery terminal to prevent airbag and seat belt pre-tensioner activation.

1. Disconnect the accelerator pedal position sensor connector.

2. Remove the 2 bolts and accelerator pedal.

➡Do not disassemble the accelerator pedal.

To install:

3. Install the accelerator pedal with the 2 bolts.

4. Connect the accelerator pedal position sensor connector.

5. Perform initialization procedure. Certain systems need to be initialized after disconnecting and reconnecting the cable from the negative (-) battery terminal.

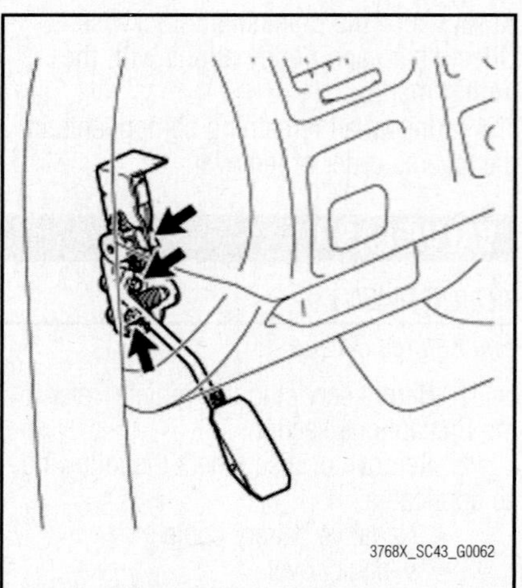

Fig. 87 Remove the 2 bolts and accelerator pedal

CAMSHAFT POSITION (CMP) SENSOR

LOCATION

See Figure 88.

Refer to the accompanying illustration.

REMOVAL & INSTALLATION

See Figure 89.

⁂ CAUTION

Wait at least 90 seconds after disconnecting the cable from the negative (-) battery terminal to prevent airbag and seat belt pre-tensioner activation.

1. Disconnect the negative battery cable.
2. Drain the engine coolant.
3. Remove the 7 clips and intake duct seal.
4. Remove the 3 clips and side cover.
5. Remove the 2 nuts and V-bank cover.
6. Remove the radiator hose inlet.
7. Remove the cap nut and disconnect the No. 3 water by-pass pipe from the timing belt cover.
8. Disconnect the 2 water by-pass hoses from the No. 3 water by-pass pipe.

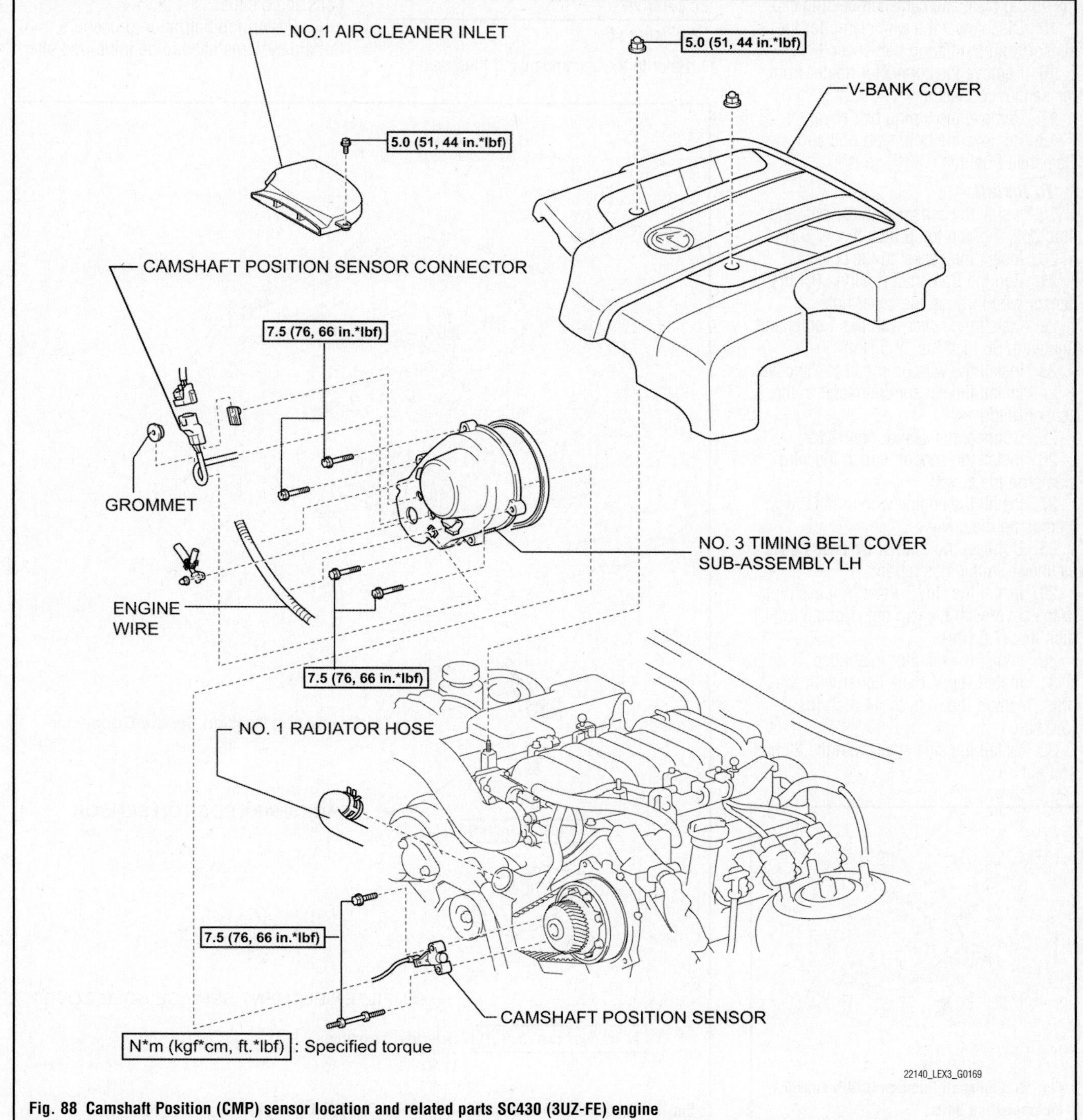

Fig. 88 Camshaft Position (CMP) sensor location and related parts SC430 (3UZ-FE) engine

9. Disconnect the engine wire from the 2 wire clamps.
10. Disconnect the Camshaft Position (CMP) sensor connector.
11. Disconnect the Camshaft Position (CMP) sensor wire from the wire clamp on the timing belt cover.
12. Remove the wire grommet from the timing belt cover.
13. Remove the 4 bolts.
14. Disconnect the timing belt cover from the timing plate and camshaft bearing cap.
15. Disconnect the wire clamp for the sensor from the timing belt cover.
16. Remove the connector holder from the sensor connector.
17. Remove the timing belt cover.
18. Remove the bolt, stud bolt and Camshaft Position (CMP) sensor.

To install:

19. Install the sensor with the bolt and stud bolt. Tighten to 66 inch lbs. (7.5 Nm).
20. Install the gasket to the cover.
21. Run the Camshaft Position (CMP) sensor wire through the cover hole.
22. Install the cover with the 4 bolts and tighten to 66 inch lbs. (7.5 Nm).
23. Install the wire grommet to the cover.
24. Install the sensor connector to the sensor holder.
25. Connect the sensor connector.
26. Install the sensor wire to the wire clamp on the cover.
27. Install the engine wire to the 2 wire clamps on the cover.
28. Connect the 2 water by-pass hoses as shown in the illustration.
29. Install the No. 3 water by-pass pipe to the cover with the cap nut. Tighten to 66 inch lbs. (7.5 Nm).
30. Install the radiator inlet hose.
31. Install the V-bank cover with the 2 nuts. Tighten the nuts to 44 inch lbs. (5.0 Nm).
32. Install the side cover with the 2 clips and nut.
33. Install the intake duct seal with the 7 clips.
34. Connect the negative battery cable.
35. Perform initialization procedure. Certain systems need to be initialized after disconnecting and reconnecting the cable from the negative (-) battery terminal.

CRANKSHAFT POSITION (CKP) SENSOR

LOCATION

See Figure 90.

Refer to the accompanying illustration.

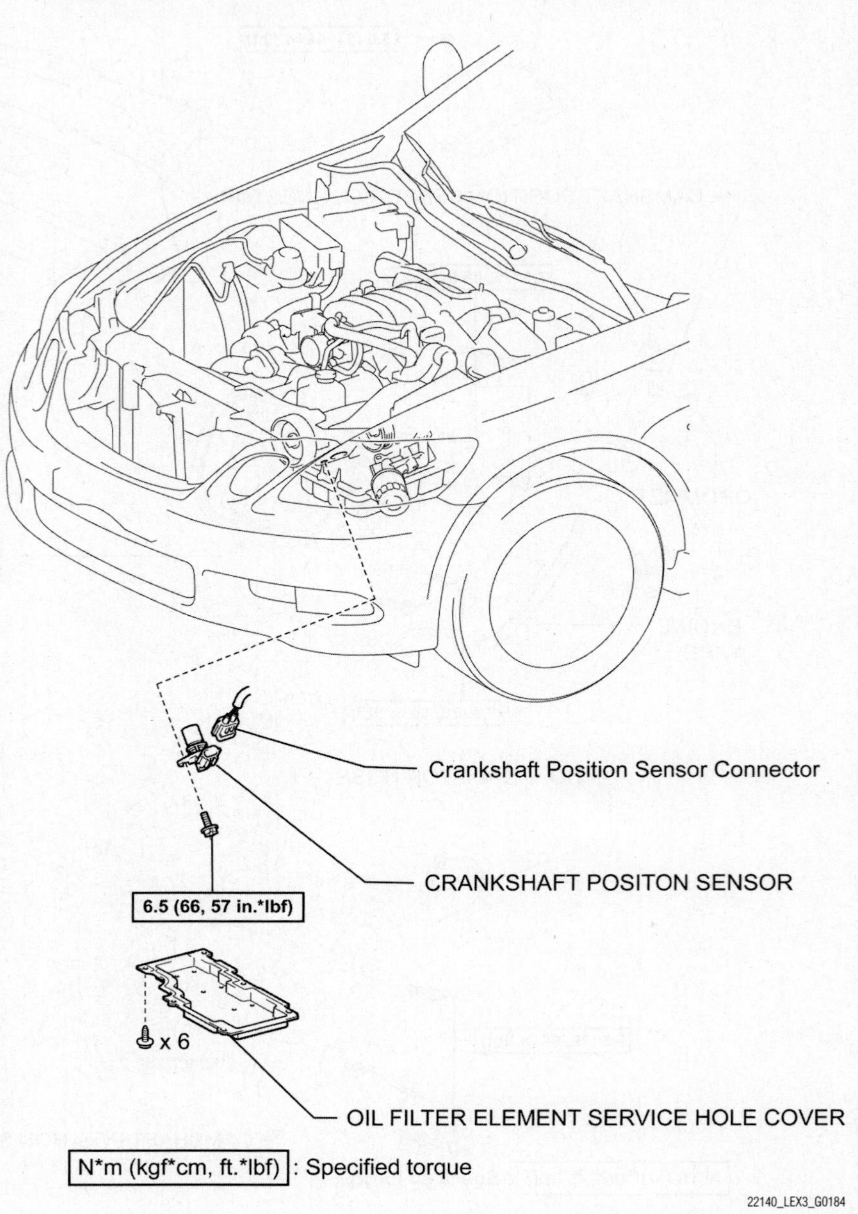

REMOVAL & INSTALLATION

See Figure 91.

1. Remove the engine undercover.
2. Remove the bolt and Crankshaft Position (CKP) sensor.

To install:

3. Install the crankshaft position sensor.
4. Tighten the mounting bolt to 57 inch lbs. (6.5 Nm).
5. Install the undercover with the 18 bolts and 5 clips.
6. Perform initialization procedure. Certain systems need to be initialized after

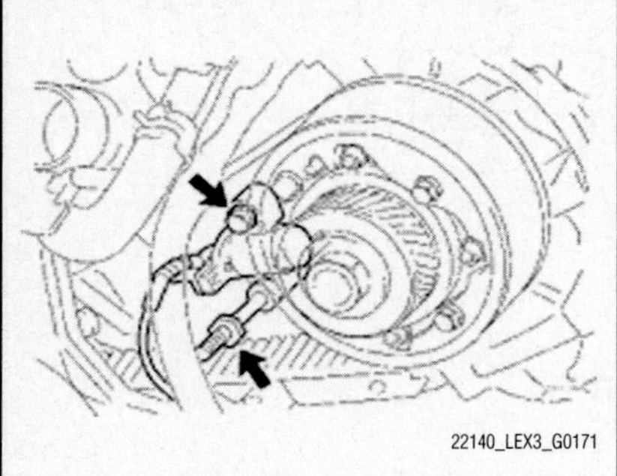

Fig. 89 Camshaft Position (CMP) sensor, and mounting bolts

Crankshaft Position Sensor Connector
CRANKSHAFT POSITON SENSOR
6.5 (66, 57 in.*lbf)
x 6
OIL FILTER ELEMENT SERVICE HOLE COVER
N*m (kgf*cm, ft.*lbf) : Specified torque
22140_LEX3_G0184

Fig. 90 Crankshaft Position (CKP) Sensor—4.3L (3UZ-FE) engine

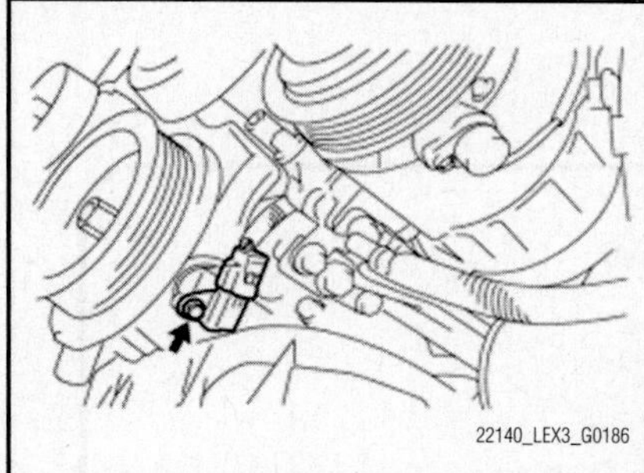

Fig. 91 Crankshaft Position (CKP) sensor and mounting bolt

disconnecting and reconnecting the cable from the negative (-) battery terminal.

ELECTRONIC CONTROL MODULE (ECM)

LOCATION

See Figure 92.

Refer to the accompanying illustration.

REMOVAL & INSTALLATION

See Figure 93.

1. Disconnect the negative battery cable.

CAUTION

Wait at least 90 seconds after disconnecting the cable from the negative (-) battery terminal to prevent airbag and seat belt pre-tensioner activation.

2. Using a clip remover, remove the 7 clips and cool air intake duct seal.
3. Remove the 3 clips and the left engine side cover.
4. Remove the 3 bolts and ECM cover.

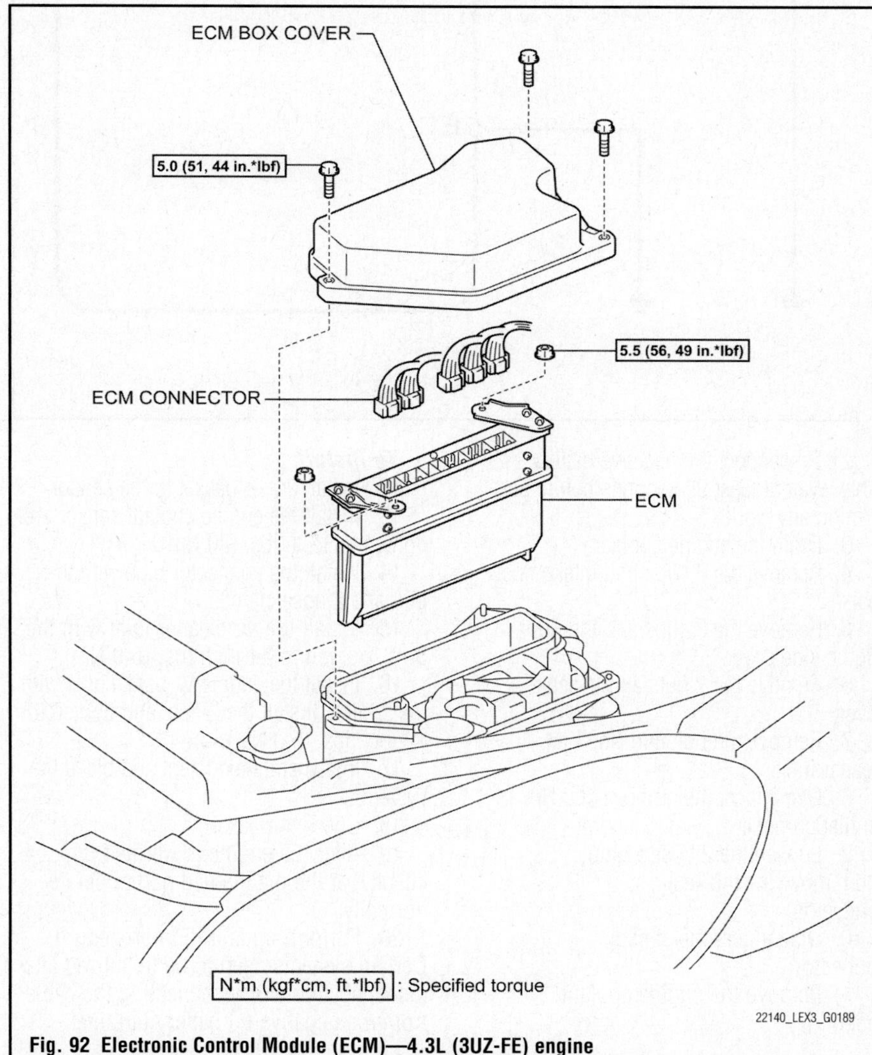

Fig. 92 Electronic Control Module (ECM)—4.3L (3UZ-FE) engine

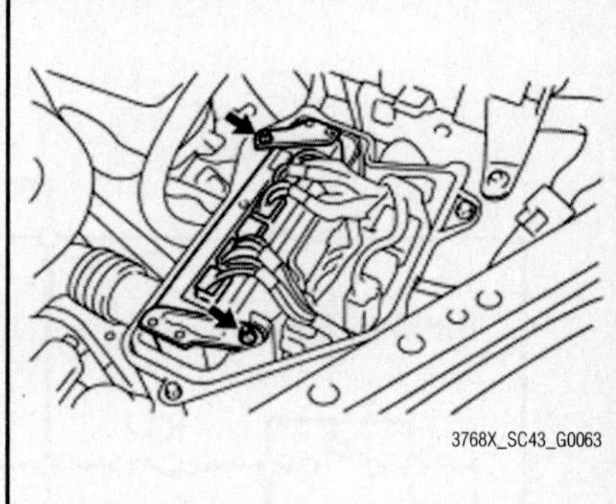

Fig. 93 Remove the 2 nuts and ECM from the ECM box

➡Be sure to prevent water intrusion to the ECM (connectors and screw parts).

5. Disconnect the 5 ECM connectors.
6. Remove the 2 nuts and ECM from the ECM box.

To install:

7. Insert the ECM into the ECM box.
8. Install the ECM to the ECM box with the 2 nuts.
9. Tighten the retaining nuts to nuts 49 inch lbs. (5.5 Nm).
10. Connect the 5 ECM connectors.
11. Install the ECM box cover with the 3 bolts.
12. Connect the negative battery cable.
13. Certain systems need to be initialized after disconnecting and reconnecting the cable from the negative (-) battery terminal.
14. After the engine is warmed up, check that the maintained parts operate normally.

TESTING

See Figure 94.

When the ignition switch is turned ON, the battery voltage is applied to terminal IGSW of the ECM. The ECM MREL output signal causes a current to flow to the coil, closing the contacts of the EFI (Marking: EFI MAIN) relay and supplying power to terminal +B of the ECM.

If the ignition switch is turned OFF, the ECM holds the EFI relay ON for a maximum of 2 seconds to allow for the initial setting of the throttle valve.

When the ignition switch is turned ON, voltage from the ECM's MREL terminal flows to the EFI relay. This causes the

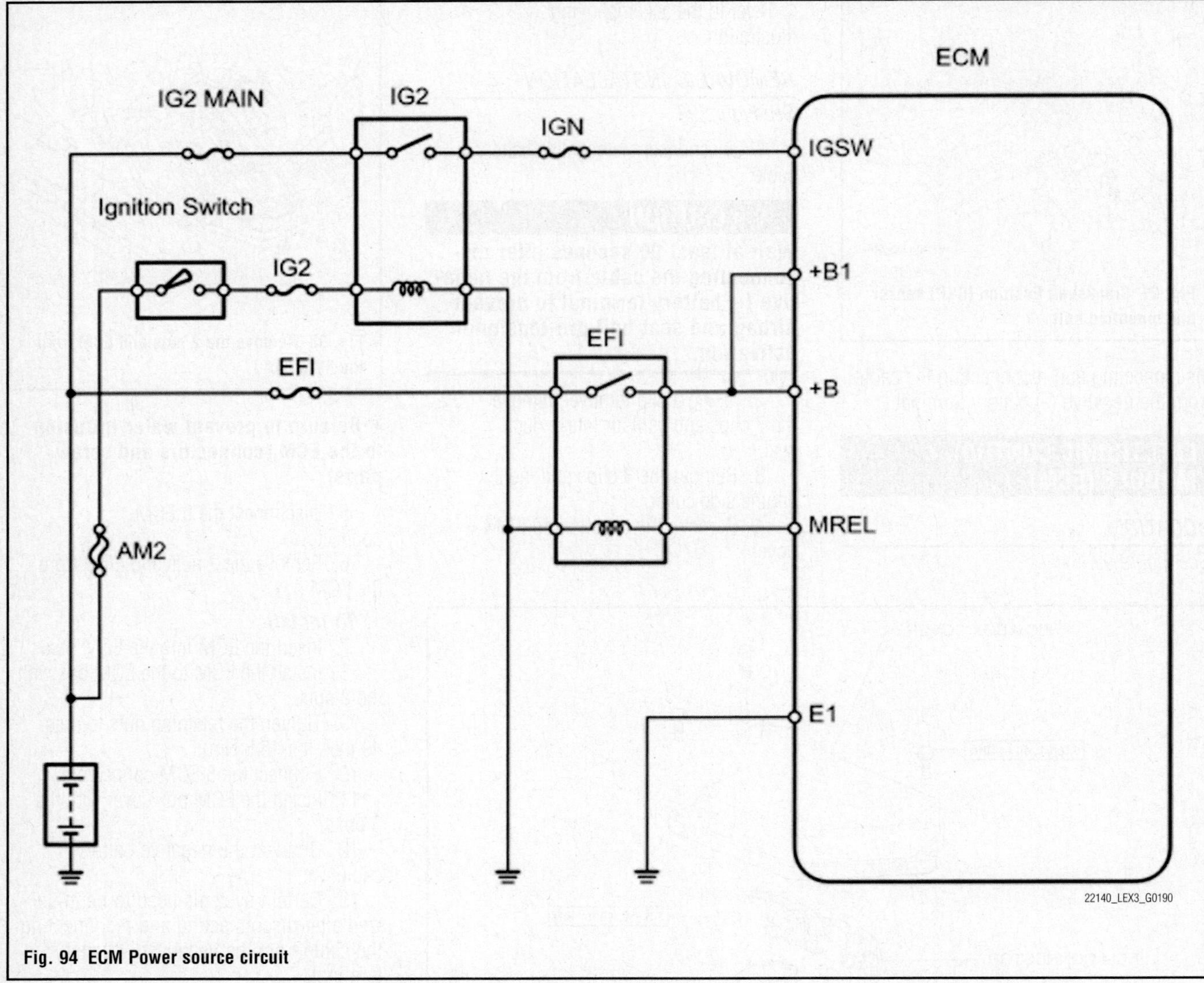

Fig. 94 ECM Power source circuit

contacts of the EFI relay to close, which supplies power to terminal +B or +B1 of the ECM.

1. Verify communication with the ECM.
2. Check the fuses, and EFI relay.
3. Check power supply and grounds to the ECM.

ENGINE COOLANT TEMPERATURE (ECT) SENSOR

LOCATION

See Figure 95.

Refer to the accompanying illustration.

REMOVAL & INSTALLATION

See Figure 96.

1. Before servicing the vehicle, refer to the Precautions section.
2. Disconnect the negative battery cable. Wait at least 90 seconds before performing any work.
3. Drain the engine coolant.
4. Remove the 7 clips and intake duct seal.
5. Remove the 3 clips and the right engine side cover.
6. Remove the 2 nuts and V-bank cover.
7. Remove the bolt and No. 1 air cleaner inlet.
8. Disconnect the air hose and No. 1 ventilation hose.
9. Loosen the 2 hose clamps and remove the intake air connector.
10. Disconnect the sensor connector.
11. Remove the engine coolant sensor.

To install:

12. Install a new gasket to the sensor.
13. Install the engine coolant sensor and tighten to 15 ft. lbs. (20 Nm).
14. Install the connector pipe with the bolt and 2 hose clamps.
15. Install the air cleaner inlet with the bolt. Tighten to 44 inch lbs. (5.0 Nm).
16. Install the engine V- bank cover with the 2 clips. Install the 2 nuts and tighten to 44 inch lbs. (5.0 Nm).
17. Install engine coolant and bleed the system.
18. Check for leaks.
19. After the engine is warmed up, check that the maintained parts operate normally.
20. Perform initialization procedure. Certain systems need to be initialized after disconnecting and reconnecting the cable from the negative (-) battery terminal.

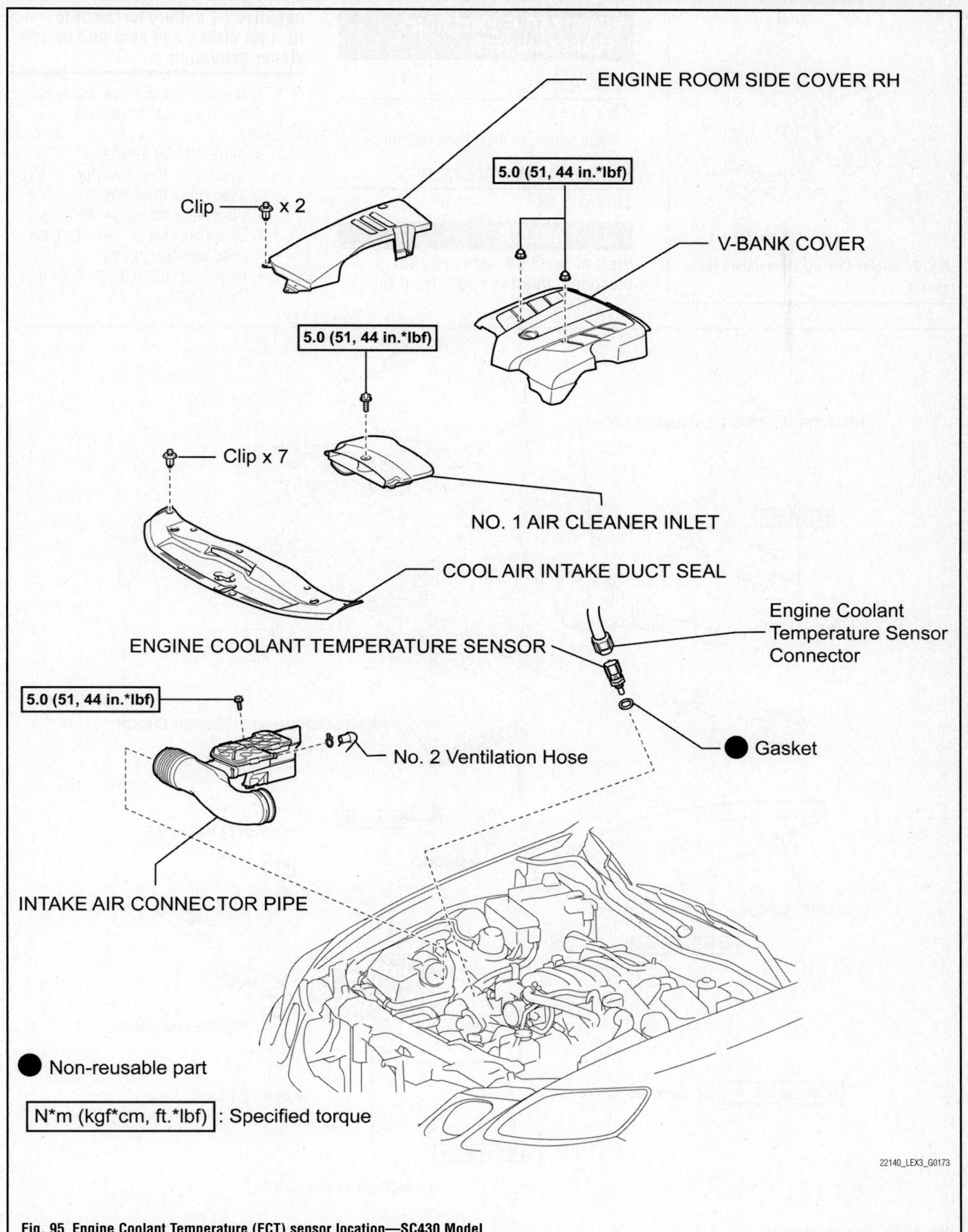

Fig. 95 Engine Coolant Temperature (ECT) sensor location—SC430 Model

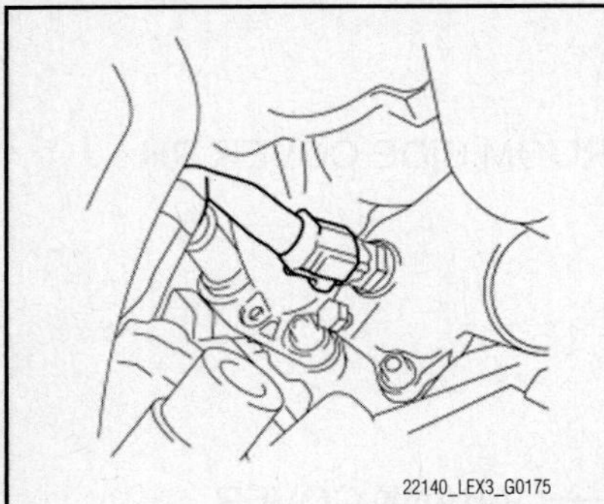

Fig. 96 Engine Coolant Temperature (ECT) sensor

EVAPORATIVE EMISSIONS (EVAP) CANISTER

LOCATION

See Figure 97.

Refer to the accompanying illustration.

REMOVAL & INSTALLATION

See Figure 98.

✲✲ CAUTION

Wait at least 90 seconds after disconnecting the cable from the negative (-) battery terminal to prevent airbag and seat belt pretensioner activation.

1. Disconnect the negative battery cable.
2. Remove the rear drive shaft assembly.
3. Remove canister assembly:
   a. Disconnect the connector.
   b. Disconnect the 3 hoses:
      - Disconnect the purge line hose.
      - Disconnect the air inlet line hose (leak detection pump).
      - Disconnect the vent line hose from the canister.

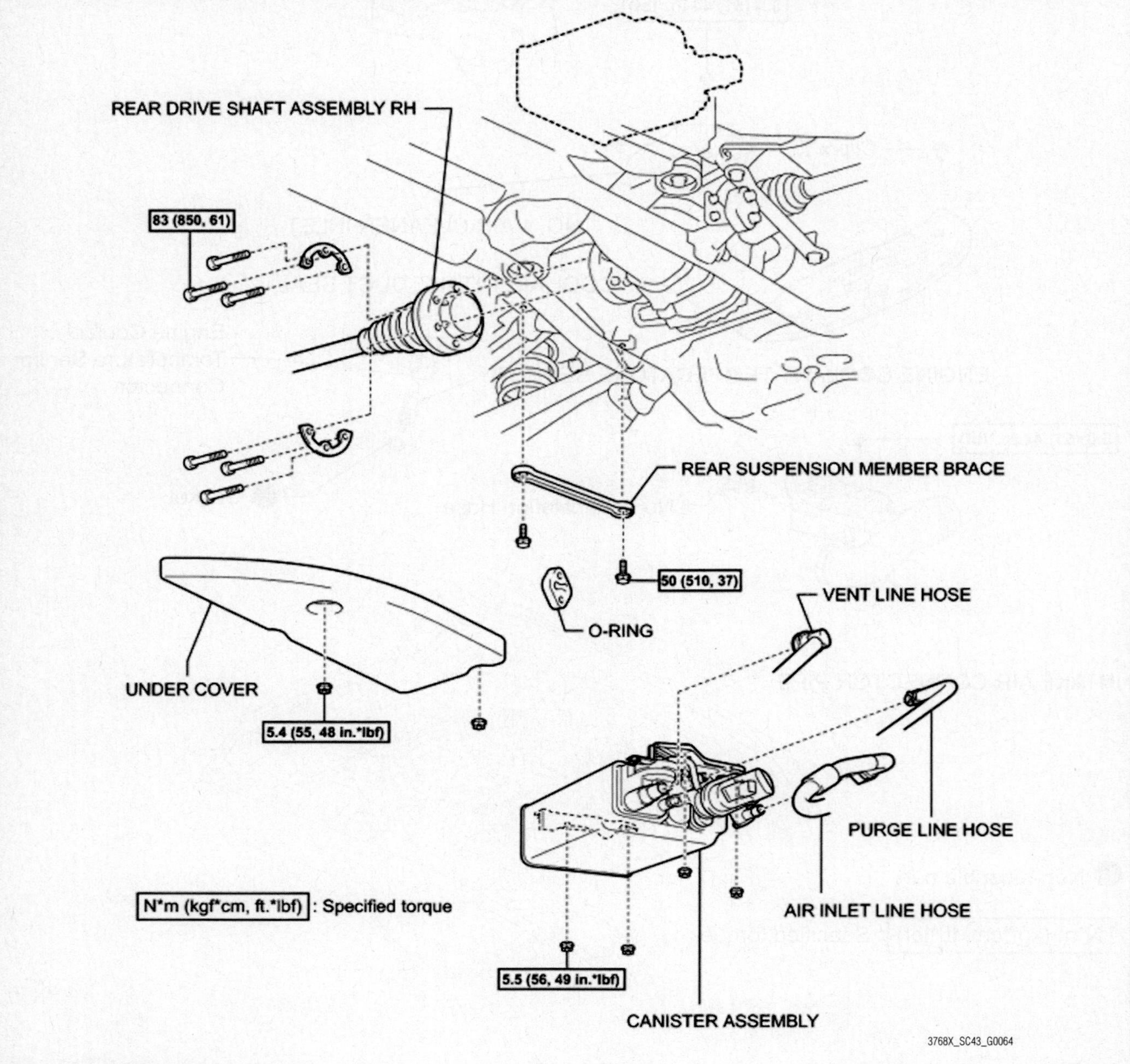

REAR DRIVE SHAFT ASSEMBLY RH
83 (850, 61)
REAR SUSPENSION MEMBER BRACE
50 (510, 37)
VENT LINE HOSE
O-RING
UNDER COVER
5.4 (55, 48 in.*lbf)
PURGE LINE HOSE
N*m (kgf*cm, ft.*lbf) : Specified torque
AIR INLET LINE HOSE
5.5 (56, 49 in.*lbf)
CANISTER ASSEMBLY
3768X_SC43_G0064

Fig. 97 EVAP canister location

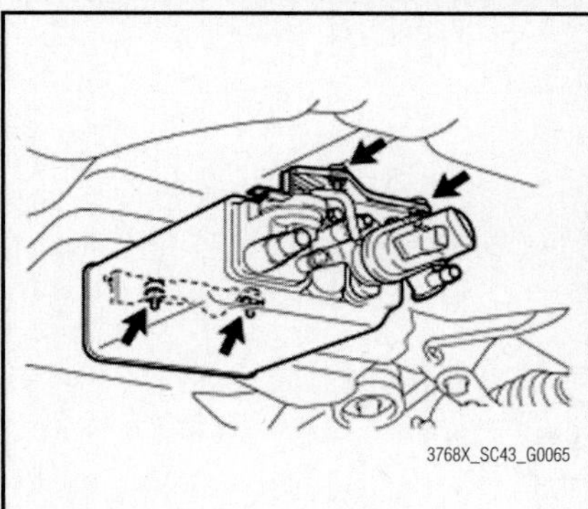

Fig. 98 Remove the 4 nuts and canister

4. Remove the 4 nuts and canister.
5. To install, reverse the removal procedure.
6. Perform initialization.

HEATED OXYGEN (HO2S) SENSOR

LOCATION

See Figures 99 and 100.

Refer to the accompanying illustrations.

REMOVAL & INSTALLATION

See Figure 101.

1. Disconnect the negative battery cable.
2. Wait at least 90 seconds after disconnecting the cable from the negative (-) battery terminal to prevent airbag and seat belt pre-tensioner activation.
3. Remove the console box.
4. Remove both engine undercovers.
5. Disconnect the sensor connector.
6. Disconnect the grommet and pull the sensor connector out of the cabin through the floor panel.
7. With a O2 socket remove the Heated Oxygen Sensor (HO2S).

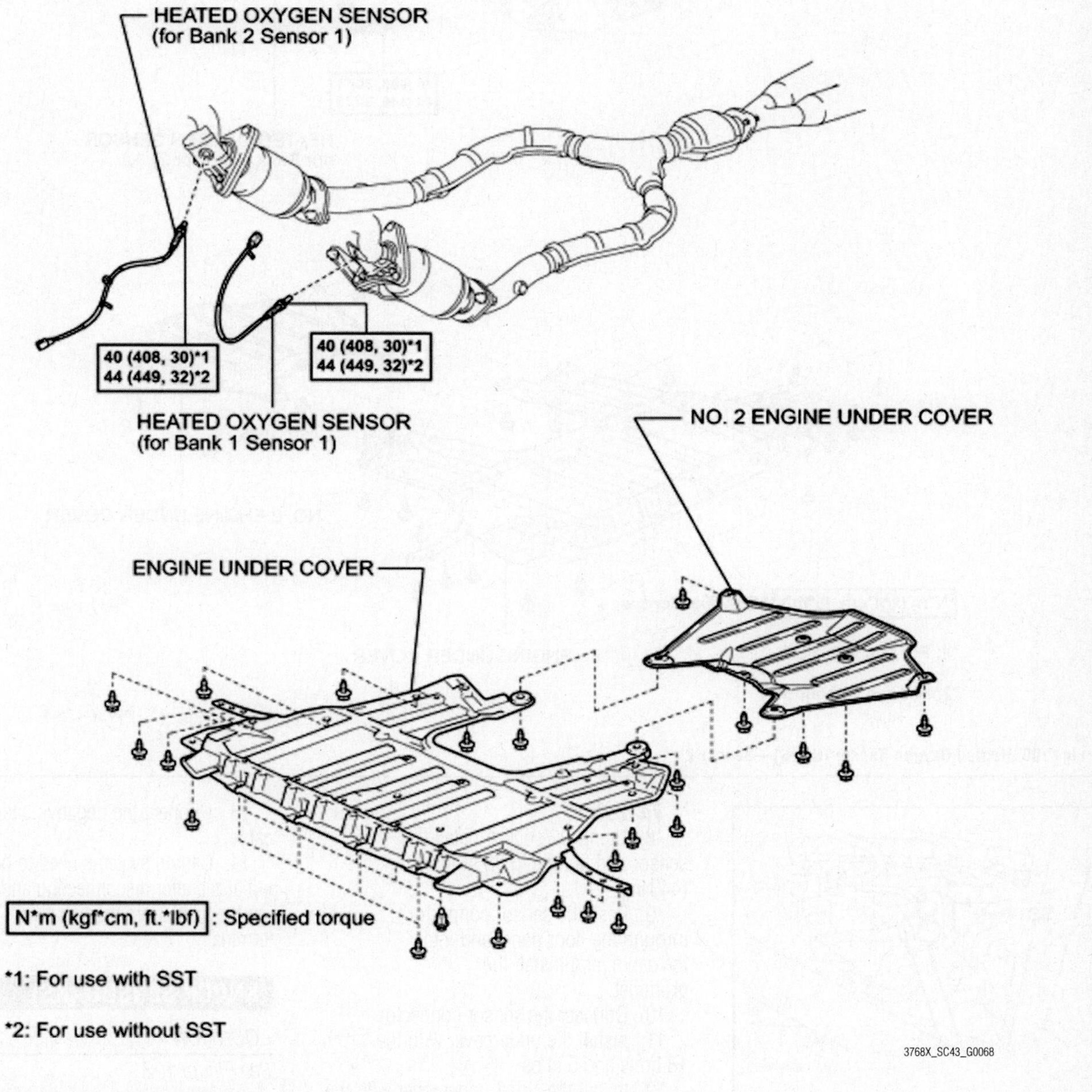

Fig. 99 Heated Oxygen Sensor (HO2S)—Sensor 1

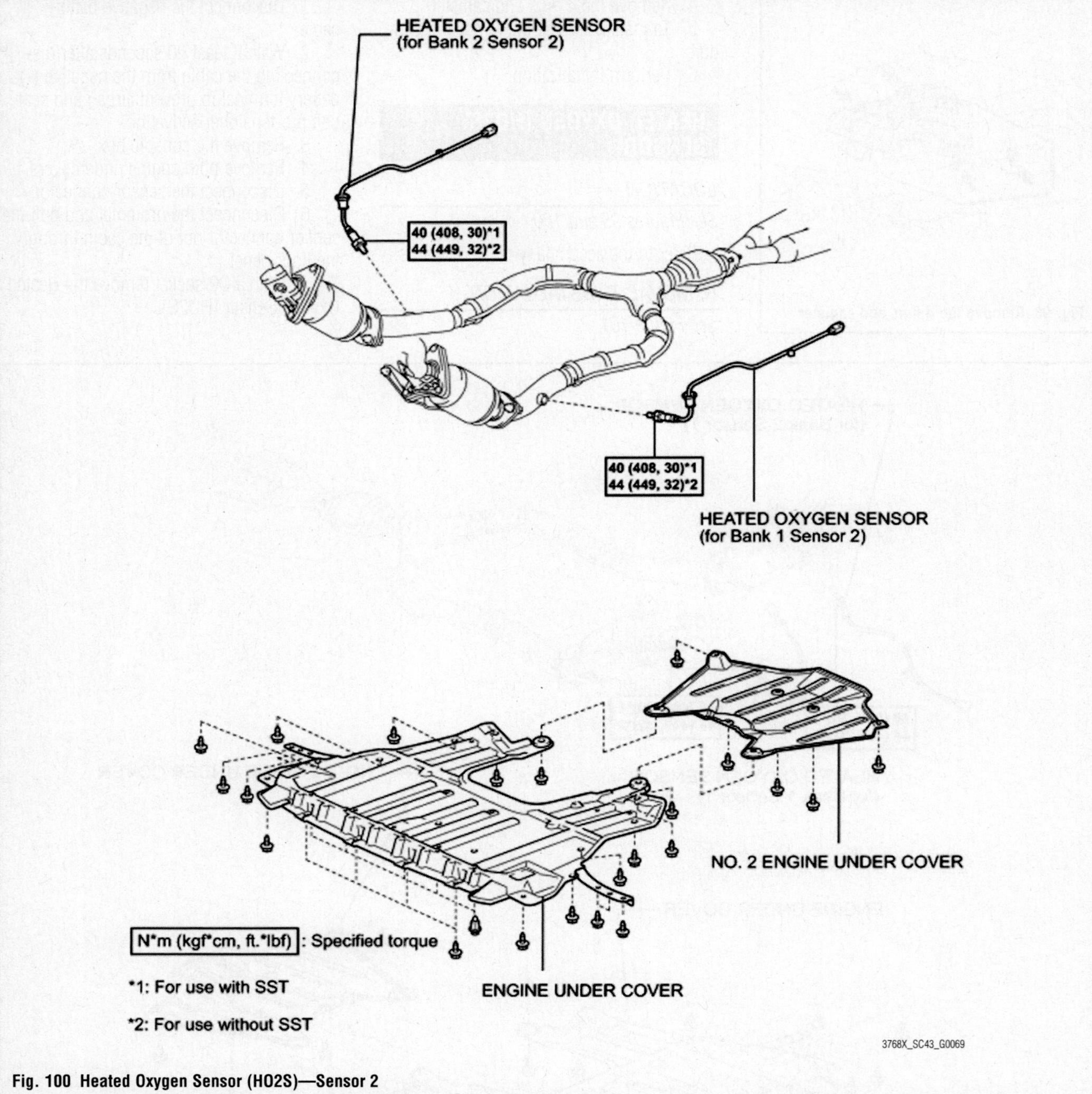

Fig. 100 Heated Oxygen Sensor (HO2S)—Sensor 2

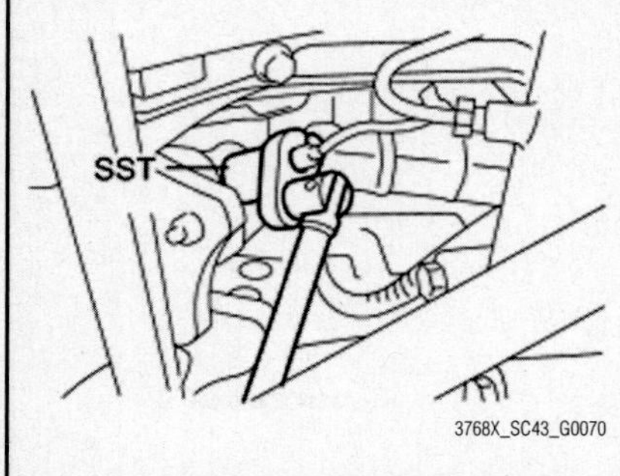

Fig. 101 With a O2 socket remove the HO2S

To install:

8. Using O2 socket install the sensor and tighten to 32 ft. lbs. (44 Nm).
9. Pass the sensor connector through the floor panel and into the cabin, and install the grommet.
10. Connect the sensor connector.
11. Install the undercover with the 18 bolts and 5 clips.
12. Install the No. 2 undercover with the 5 bolts.
13. Connect the negative battery cable.
14. Certain systems need to be initialized after disconnecting and reconnecting the cable from the negative (-) battery terminal.

KNOCK SENSOR (KS)

LOCATION

See Figure 102.

Refer to the accompanying illustration.

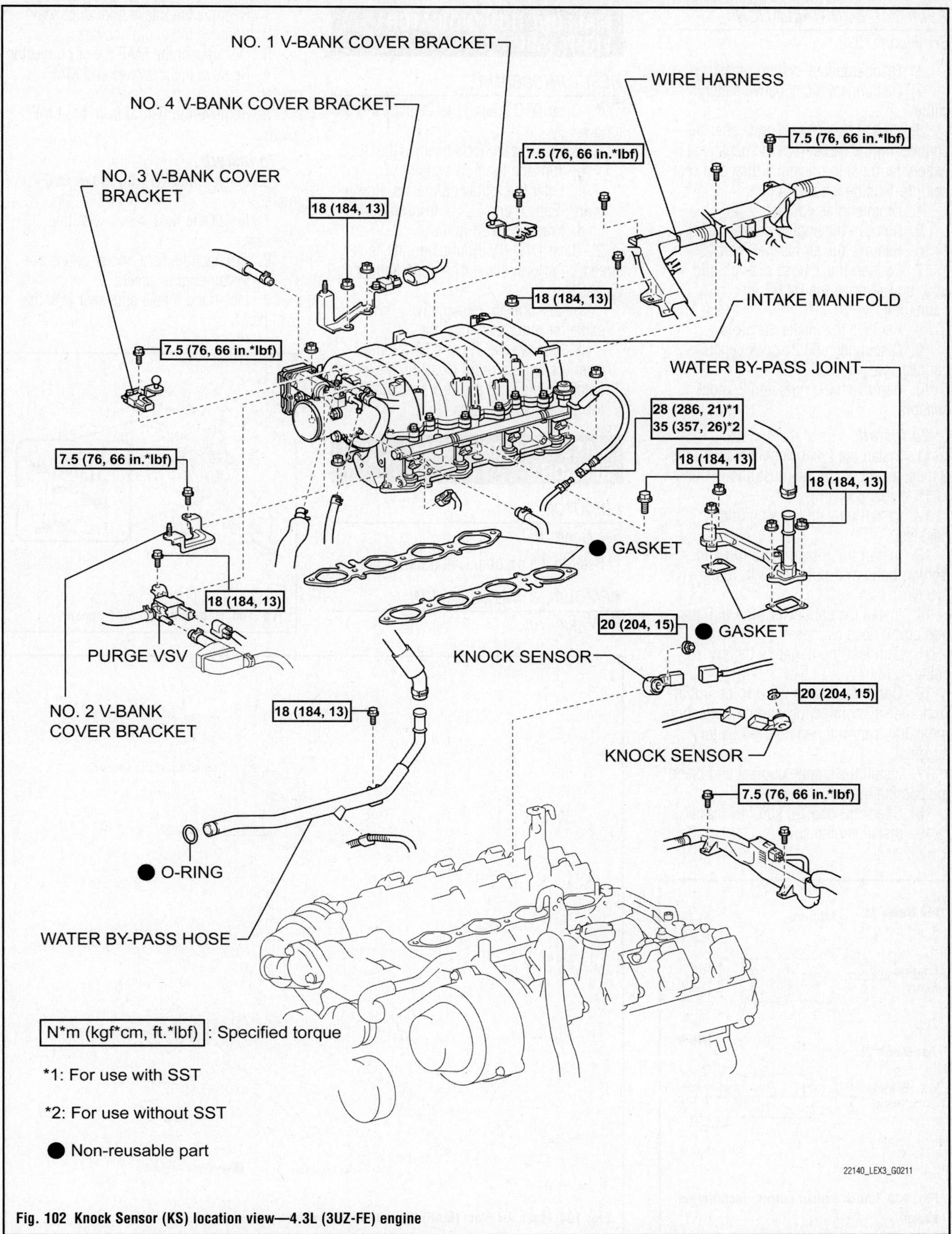

Fig. 102 Knock Sensor (KS) location view—4.3L (3UZ-FE) engine

REMOVAL & INSTALLATION

See Figure 103.

1. Discharge fuel system pressure.
2. Disconnect the negative battery cable.
3. Wait at least 90 seconds after disconnecting the cable from the negative (-) battery terminal to prevent airbag and seat belt pre-tensioner activation.
4. Drain engine coolant.
5. Remove the engine cover.
6. Remove the air cleaner inlet.
7. Loosen the 2 hose clamps and bolt, and remove the intake air connector.
8. Remove the intake manifold.
9. Disconnect the 2 knock sensor connectors.
10. Remove the 2 nuts and 2 knock sensors.

To install:

11. Install the 2 knock sensors with the 2 nuts, as shown in the illustration. Tighten to 15 ft. lbs. (20 Nm).
12. Install new intake manifold gaskets
13. Install the intake manifold and tighten bolts and nuts to 13 ft. lbs. (18 Nm).
14. Install the connector pipe with the bolt and 2 hose clamps.
15. Connect the negative battery cable.
16. Certain systems need to be initialized after disconnecting and reconnecting the cable from the negative (-) battery terminal.
17. Install the engine coolant and bleed the cooling system.
18. Check for coolant and fuel leaks.
19. Install engine cover.

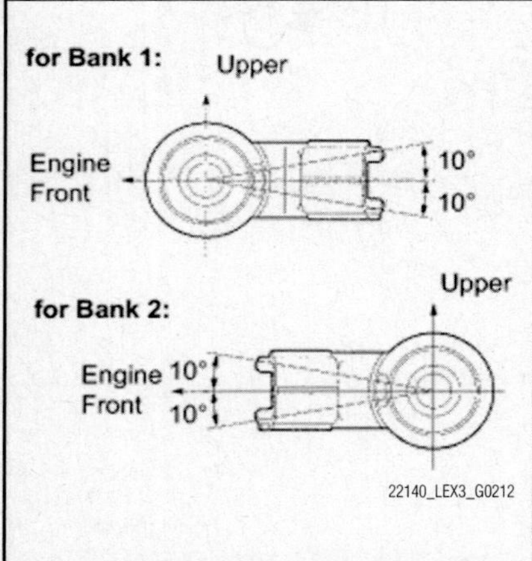

Fig. 103 Knock sensor correct installation shown

MALFUNCTION INDICATOR LIGHT (MIL)

RESET PROCEDURE

1. Clear DTC (Using the Techstream) as follows:
   a. Connect the Techstream to the DLC3.
   b. Turn the ignition switch ON.
   c. Enter the following menus: Powertrain / Engine and ECT / Trouble Codes.
   d. Press the YES button.
2. Clear DTC (Without using the Techstream). Perform either one of the following operations.
   a. Disconnect the negative (-) battery cable for more than 1 minute.
   b. Remove the EFI and ETCS fuses from the engine room No. 2 relay block located inside the engine compartment for more than 1 minute.

MASS AIR FLOW (MAF) SENSOR

LOCATION

See Figure 104.

Refer to the accompanying illustration.

REMOVAL & INSTALLATION

See Figure 105.

1. Remove the intake cool duct seal.
2. Remove the engine covers.
3. Disconnect the MAF meter connector.
4. Remove the 2 screws and MAF meter.
5. Remove the O-ring from the MAF meter.

To install:

6. Install a new O-ring to the MAF meter.
7. Install the MAF meter with the 2 screws.
8. Connect the MAF meter connector.
9. Install engine covers
10. Install the intake duct seal with the 7 clips.

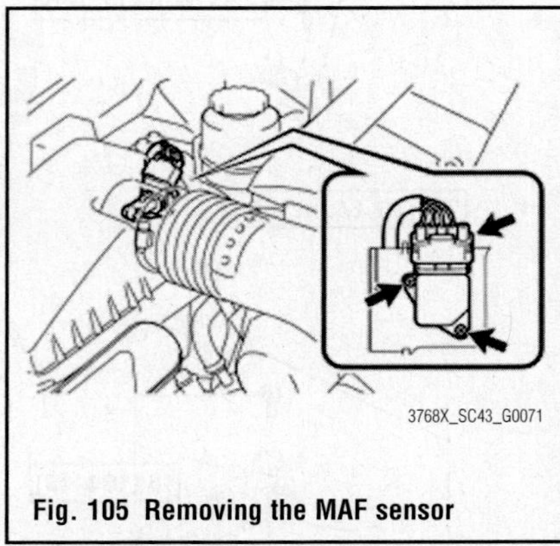

Fig. 105 Removing the MAF sensor

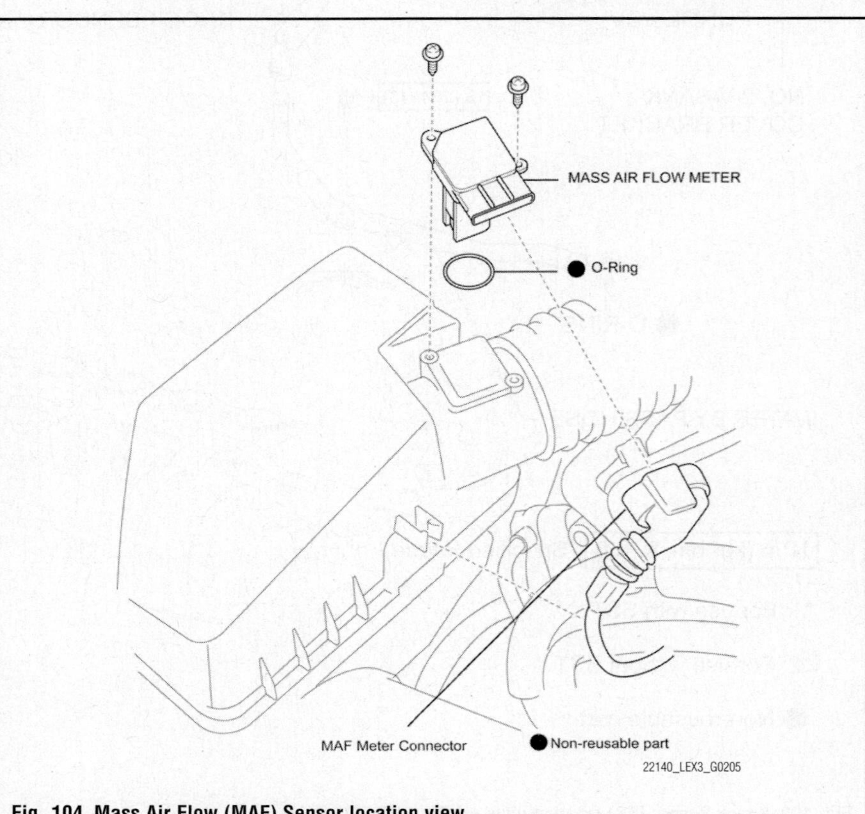

Fig. 104 Mass Air Flow (MAF) Sensor location view

POSITIVE CRANKCASE VENTILATION (PCV) VALVE

LOCATION

See Figure 106.

Refer to the accompanying illustration.

REMOVAL & INSTALLATION

See Figures 107 and 108.

1. Remove the 2 nuts and v bank cover.
2. Remove the bolt and no. 1 inlet.
3. Disconnect the air hose and ventilation hose.
4. Loosen the 2 hose clamps.
5. Remove the bolt and intake air connector.
6. Remove ventilation valve sub-assembly:

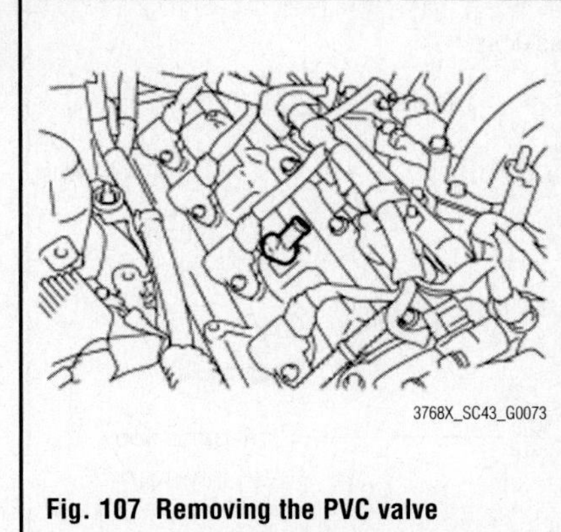

Fig. 107 Removing the PVC valve

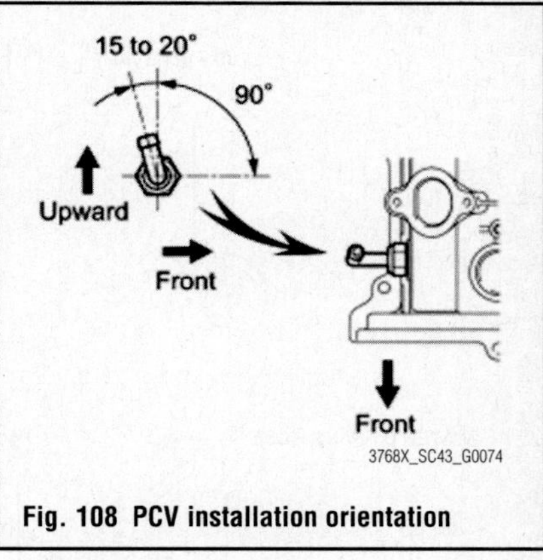

Fig. 108 PCV installation orientation

a. Disconnect the ventilation hose from the valve.
b. Remove ventilation valve cover.
c. Remove the valve.

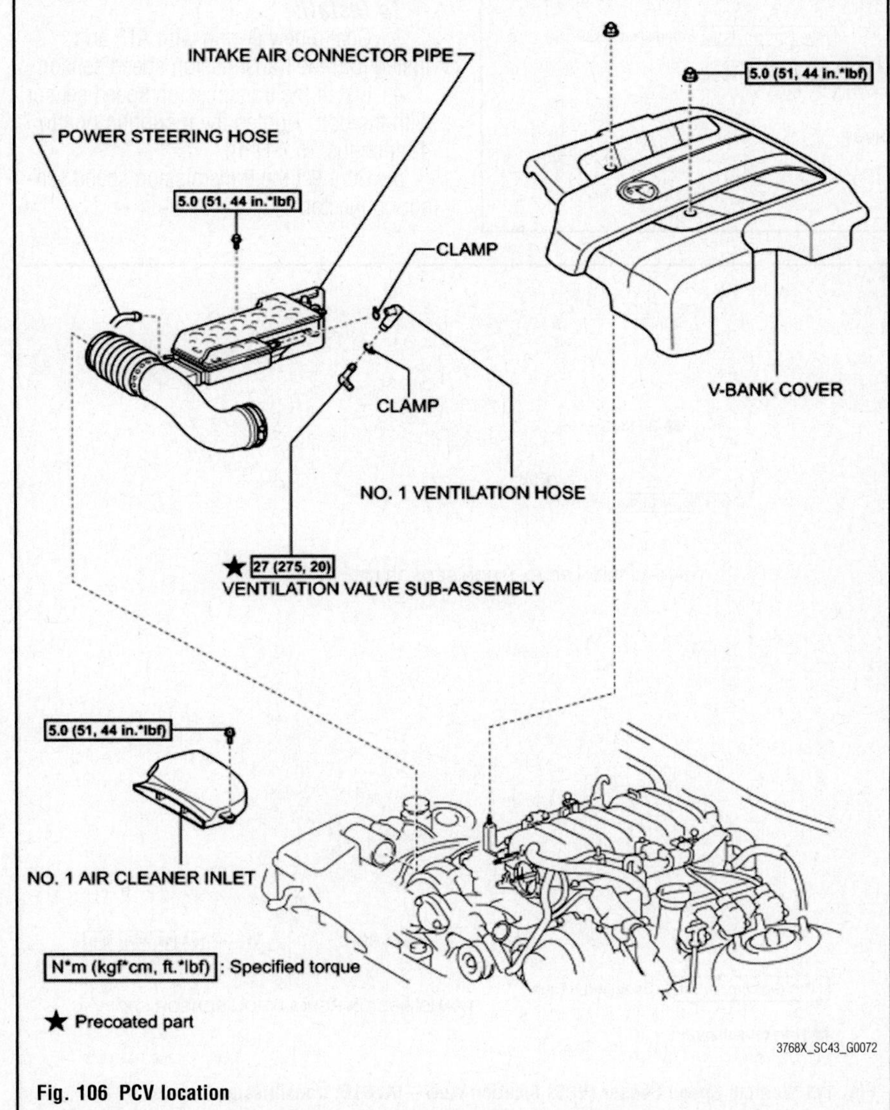

Fig. 106 PCV location

To install

7. Apply adhesive to 2 or 3 threads.

Adhesive: Toyota genuine adhesive 1324, three bond 1324 or equivalent

8. Install the valve. Tighten to 20 ft. lbs. (27 Nm).

➡After applying the specified torque, rotate the valve clockwise with the port facing the direction shown in the illustration.

9. Install the ventilation valve cover.
10. Connect the hose to the valve.
11. To complete installation, reverse remaining removal procedure.

THROTTLE POSITION SENSOR (TPS)

LOCATION

See Figure 109.

Refer to the accompanying illustration.

REMOVAL & INSTALLATION

See Figure 110.

1. Disconnect the negative battery cable.
2. Wait at least 90 seconds after disconnecting the cable from the negative (-) battery terminal to prevent airbag and seat belt pre-tensioner activation.
3. Using a clip remover, remove the 7 clips and duct seal.
4. Using a clip remover, remove the 3 clips.
5. Remove the 10 screws and under-cover.
6. Drain the engine coolant.
7. Remove the engine covers.
8. Disconnect the air hose and No. 1 ventilation hose.
9. Loosen the 2 hose clamps and bolt, and remove the intake air connector.
10. Disconnect the throttle body connector.

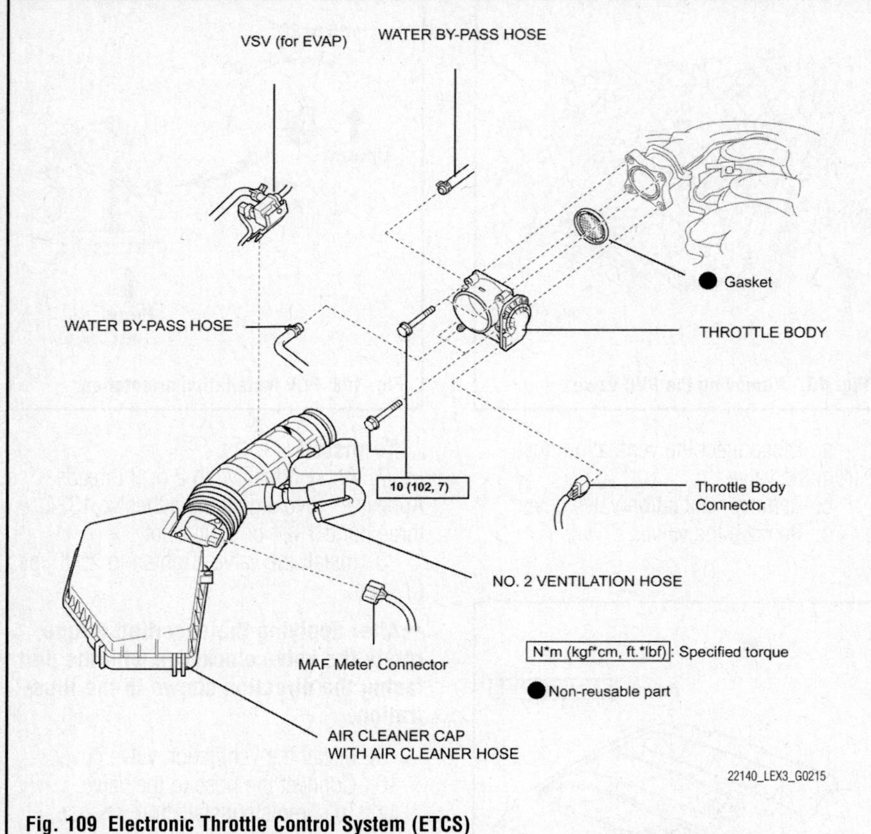

Fig. 109 Electronic Throttle Control System (ETCS)

11. Disconnect the ventilation hose.
12. Disconnect the 2 water by-pass hoses.
13. Remove the 2 bolts, 2 nuts, throttle body and gasket.

To install:

14. Install a new gasket and the throttle body with the 2 bolts and 2 nuts. Tighten to 13 ft. lbs. (18 Nm).
15. Connect the 2 water by-pass hoses.
16. Connect the ventilation hose.
17. Connect the connector.
18. Install the connector pipe with the bolt and 2 hose clamps.
19. Connect the air hose and No. 1 ventilation hose.

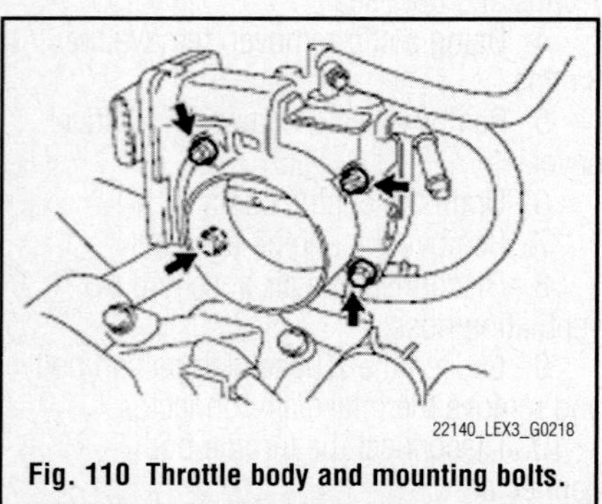

Fig. 110 Throttle body and mounting bolts.

20. Install the air cleaner inlet with the bolt.
21. Install the engine covers.
22. Connect the negative battery cable.
23. Install the engine coolant and bleed system.
24. After the engine is warmed up, check that the maintained parts operate normally.
25. Certain systems need to be initialized after disconnecting and reconnecting the cable from the negative (-) battery terminal

VEHICLE SPEED SENSOR (VSS)

REMOVAL & INSTALLATION

See Figure 111.

1. Disconnect the transmission speed sensor connector.
2. Remove the bolt and transmission speed sensor.

To install:

3. Coat a new O-ring with ATF and install it to the transmission speed sensor.
4. Install the transmission speed sensor with the bolt. Tighten the mounting bolt to 48 inch lbs. (5.4 Nm).
5. Connect the transmission speed sensor connector.

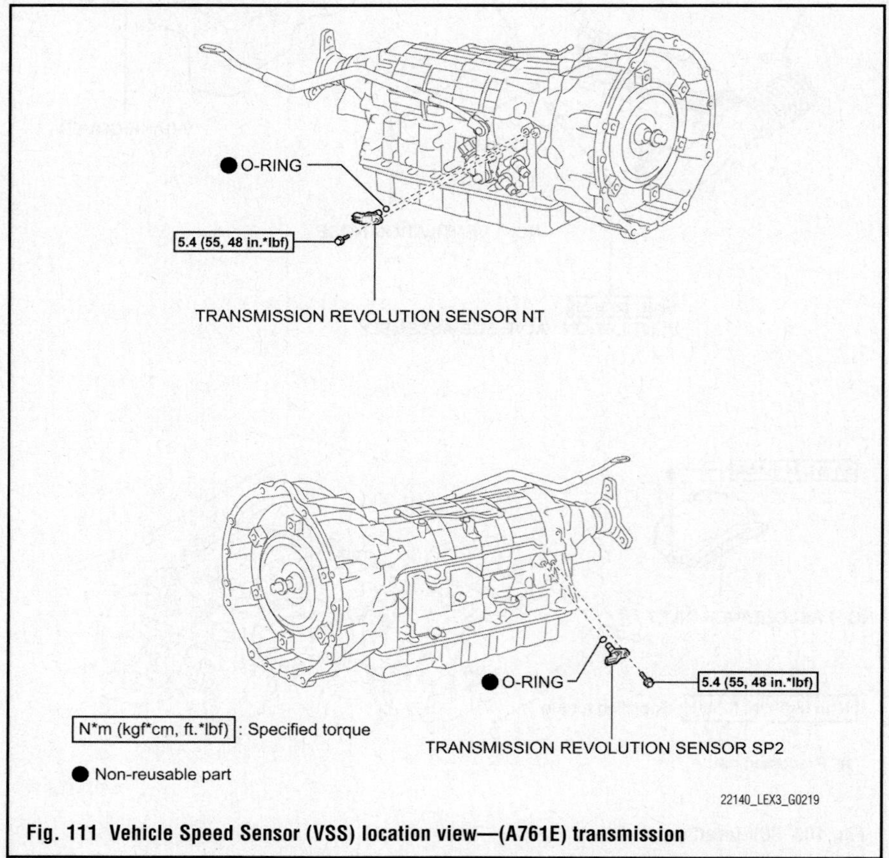

Fig. 111 Vehicle Speed Sensor (VSS) location view—(A761E) transmission

FUEL SYSTEM SERVICE PRECAUTIONS

Safety is the most important factor when performing not only fuel system maintenance, but any type of maintenance. Failure to conduct maintenance and repairs in a safe manner may result in serious personal injury or death. Work on a vehicle's fuel system components can be accomplished safely and effectively by adhering to the following rules and guidelines.

- To avoid the possibility of fire and personal injury, always disconnect the negative battery cable unless the repair or test procedure requires that battery voltage be applied.
- Always relieve the fuel system pressure prior to disconnecting any fuel system component (injector, fuel rail, pressure regulator, etc.) fitting or fuel line connection. Exercise extreme caution whenever relieving fuel system pressure to avoid exposing skin, face and eyes to fuel spray. Please be advised that fuel under pressure may penetrate the skin or any part of the body that it contacts.
- Always place a shop towel or cloth around the fitting or connection prior to loosening to absorb any excess fuel due to spillage. Ensure that all fuel spillage is quickly removed from engine surfaces. Ensure that all fuel-soaked cloths or towels are deposited into a flame-proof waste container with a lid.
- Always keep a dry chemical (Class B) fire extinguisher near the work area.
- Do not allow fuel spray or fuel vapors to come into contact with a spark or open flame.
- Always use a second wrench when loosening or tightening fuel line connection fittings. This will prevent unnecessary stress and torsion on fuel piping. Always follow the proper torque specifications.
- Always replace worn fuel fitting O-rings with new ones. Do not substitute fuel hose where rigid pipe is installed.

FUEL SYSTEM PRESSURE

RELIEVING

1. Before servicing the vehicle, refer to the Precautions section.
2. Remove the fuse for the electronic fuel pump.
3. Start the engine until the engine stalls.
4. Disconnect the negative battery terminal.
5. Place a catch-pan under the joint to be disconnected. A large quantity of fuel may be released when the joint is opened.
6. Wear eye or full-face protection.
7. Place a shop towel over the area and slowly release the joint using a wrench of the correct size.
8. Allow any fuel left in the line to bleed off slowly before fully disconnecting the joint.
9. Plug the opened lines immediately to prevent fuel spillage or the entry of dirt.
10. Dispose of the released fuel properly.
11. After connecting fuel lines, install the fuse for the fuel pump and start the engine.
12. Check for leaks and repair as needed.

FUEL FILTER

REMOVAL & INSTALLATION

The fuel filter is mounted in the fuel tank and is part of fuel pump unit housing.

FUEL PUMP

REMOVAL & INSTALLATION

See Figure 112.

1. Before servicing the vehicle, refer to the Precautions section.

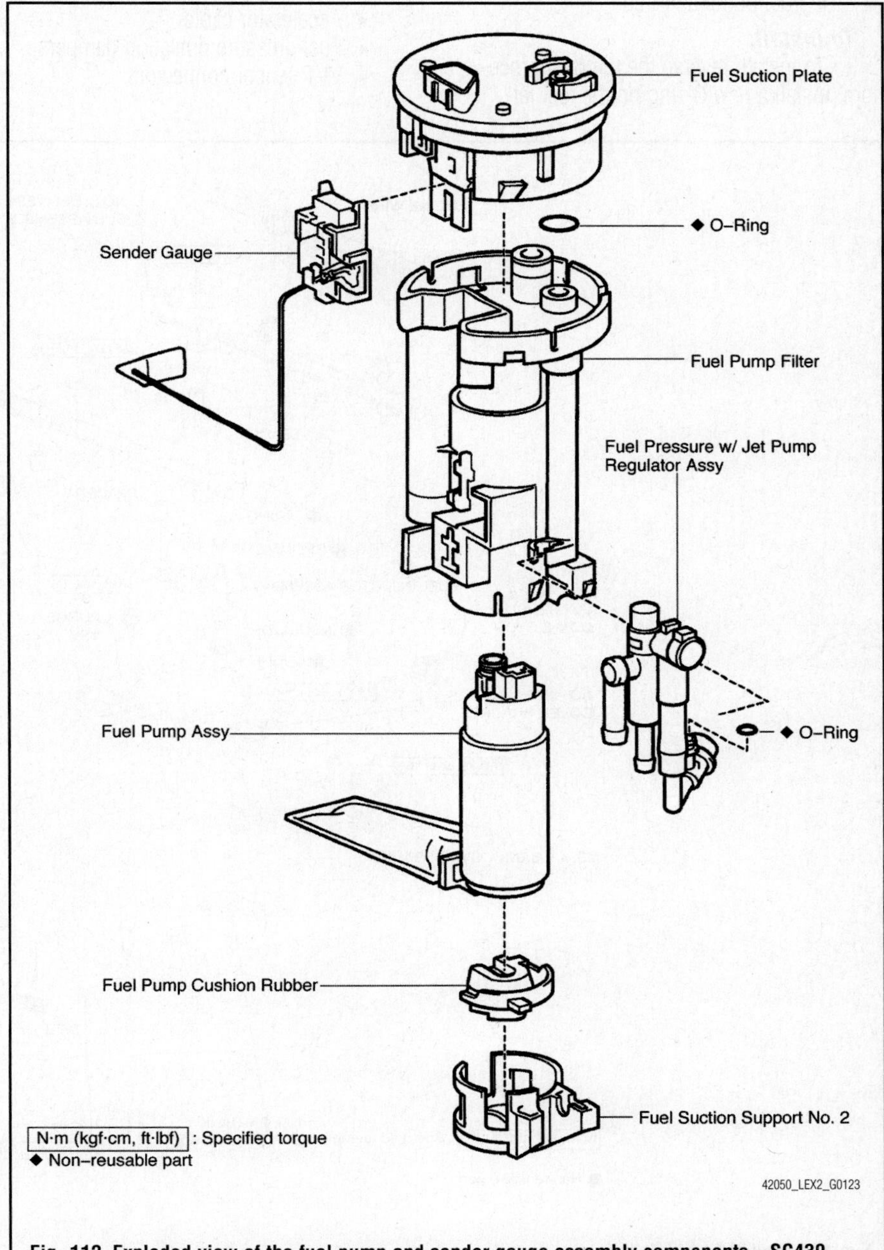

Fig. 112 Exploded view of the fuel pump and sender gauge assembly components—SC430

2. Remove or disconnect the following:
   - Negative battery cable. Wait at least 90 seconds before performing any other work.
   - Rear seat bottom and seat back
   - Fuel pump tube
   - 8 bolts and fuel tank vent tube set plate
   - Fuel pump and sensor gauge assembly
   - Fuel suction hose and support
   - Fuel pump cushion rubber
   - Fuel pressure w/jet pump regulator assembly
   - Fuel suction plate w/sender gauge
   - Fuel pump and filter

To install:

3. To install, reverse the removal procedure. Install a new O-ring on the fuel jet pump regulator assembly and a new gasket on the fuel pump/sender gauge assembly. Torque the 8 bolts securing the fuel tank set plate to 52 in. lbs (6.0 Nm).
4. Inspect fuel pump operation and check for fuel leaks.

FUEL RAIL AND INJECTOR

REMOVAL & INSTALLATION

See Figure 113.

1. Before servicing the vehicle, refer to the Precautions section.
2. Remove or disconnect the following:
   - V-bank cover
   - Intake air connector
   - Accelerator cable
   - Fuel pressure pulsation dampers.
   - VVT sensor connectors
   - Vacuum Switching Valve (VSV) for Evaporative Emissions (EVAP)
   - 2 nuts and accelerator cable bracket
   - 3 V-bank cover brackets
   - VSV connector for Acoustic Control Induction System (ACIS) from the No. 1 V-bank cover bracket
   - 4 bolts and 3 V-bank cover brackets
   - Engine wire from the delivery pipe
   - 2 wire clamps from the wire clamp bracket on the right-hand delivery pipe
   - 8 injector connectors
   - 4 nuts holding the delivery pipe to the intake manifold
   - 2 delivery pipes and 8 injector assemblies and 4 spacers
   - 2 O-rings, grommet and insulator from each injector

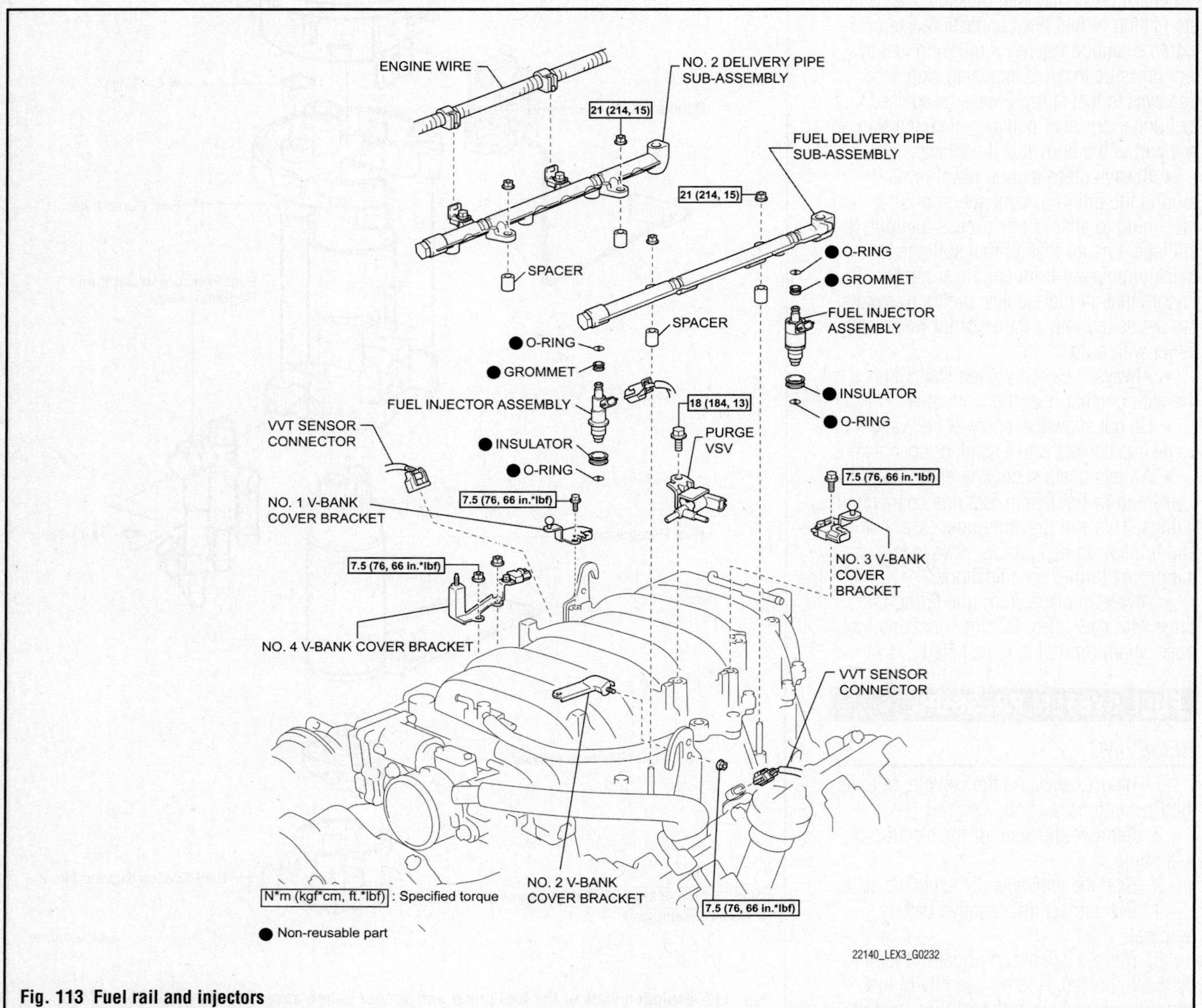

Fig. 113 Fuel rail and injectors

To install:

3. Install or connect the following:
   - A new insulator and grommet to each injector
   - A light coat of gasoline to new O-rings and install them to each injector
   - A light coat of gasoline on the place where a delivery pipe touches an O-ring of the injector
   - Injector, while turning the clockwise and counterclockwise, into the delivery pipe

➡Position the injector connector outward.

   - The 4 spacers in position on the intake manifold
   - A light coat of gasoline on the place where an intake manifold touches an O-ring
   - The delivery pipes in position on the intake manifold
   - Temporarily, the 3 bolts holding the delivery pipe to the intake manifold

➡Check that the injectors rotate smoothly. If the injectors do not rotate smoothly, the probable cause is incorrect installation of the O-rings. Replace the O-rings.

4. Tighten the 3 bolts holding the delivery pipe to the intake manifold. Tighten the bolts to 15 ft. lbs. (21 Nm).
5. Install or connect the following:
   - Engine wire protector with the 3 nuts
   - Injector connectors
   - Remaining components

FUEL TANK

REMOVAL & INSTALLATION

See Figure 114.

1. Before servicing the vehicle, refer to the Precautions section.
2. Detach the seat cushion's 2 front hooks from the vehicle body. Remove the seat cushion.
3. Discharge the fuel system pressure.
4. Disconnect the negative battery cable.
5. Wait at least 90 seconds after disconnecting the cable from the negative (-) battery terminal to prevent airbag and seat belt pre-tensioner activation.
6. Remove the rear floor service hole covers.
7. Remove the fuel lines and electrical connectors.
8. Remove the front center floor brace.
9. Remove the undercover.
10. Remove the center exhaust and muffler.
11. Remove the propeller shaft heat insulator.
12. Remove both differential support protectors.
13. Remove the propeller shaft assembly.
14. Disconnect and remove both rear parking brake cable assembly.
15. Remove the fuel tank filler tube.
16. Disconnect the vent hose from the canister.
17. With a helper and the use of transmission jack carefully lower the fuel tank.

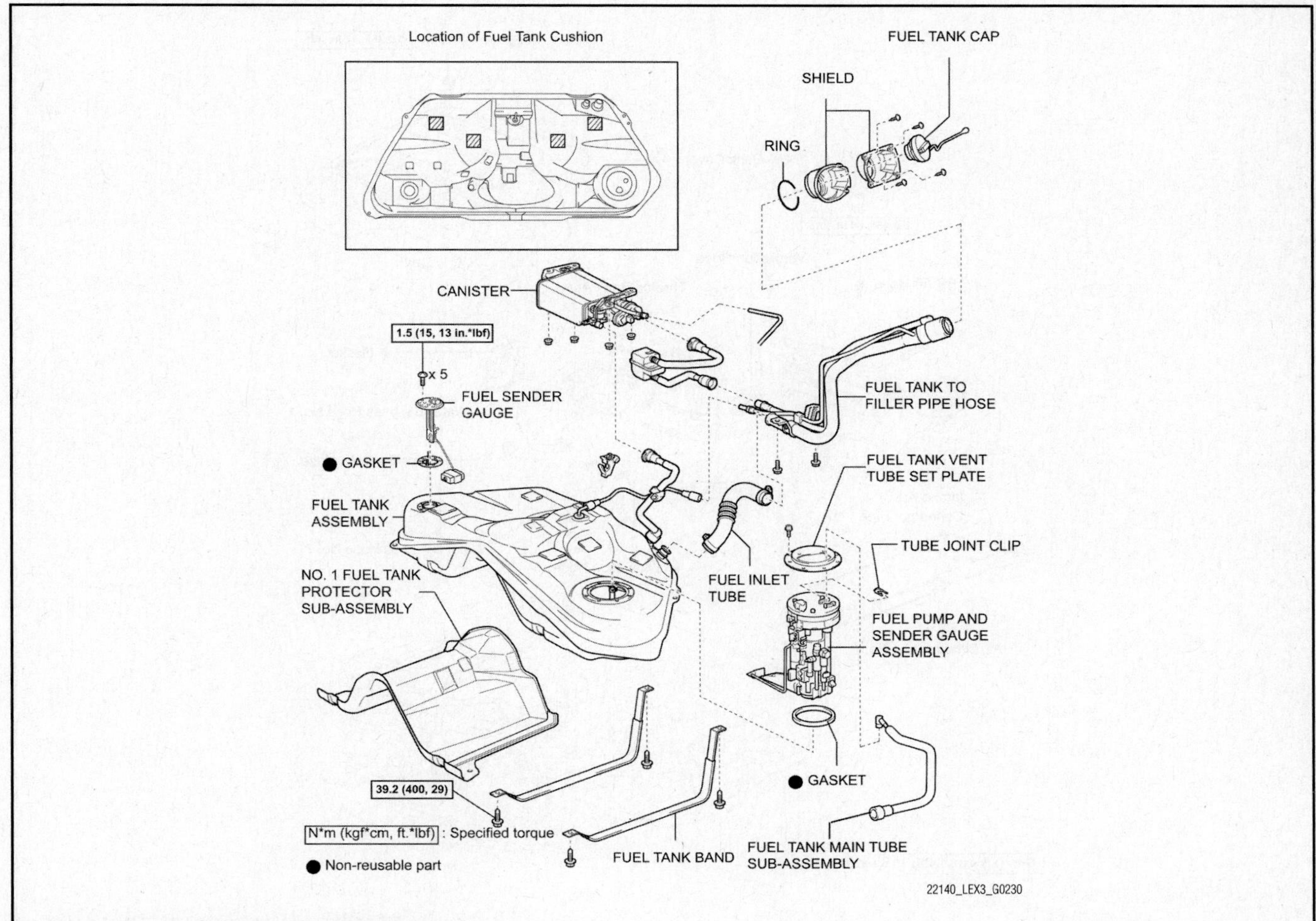

Fig. 114 Fuel tank assembly and related parts

To install:

18. With a helper and the use of transmission jack carefully raise the fuel tank
19. Install the fuel tank and fuel tank band with the 4 bolts. Tighten to 29 ft. lbs. (39 Nm).
20. Install fuel tank filler tube.
21. Connect the vent line hose to the canister.
22. Install both parking brake cable assembly.
23. Install rear propeller shaft assembly. Tighten the front mounting bolts to 58 ft. lbs. (79 Nm).
24. Temporarily install the 2 center support bearing set bolts with the adjusting washers.
25. Check that the center line of the bracket is at right angles to the shaft axial direction.
26. Tighten the center bearing mounting bolts to 36 ft. lbs. (49 Nm).
27. Tighten the front mounting bolts to 58 ft. lbs. (79 Nm).
28. Install both differential support protectors.
29. Install the front floor heat insulator with the 4 bolts. Tighten to 47 inch lbs. (5.4 Nm).
30. Install the propeller shaft heat insulator.
31. Install a new gasket and the exhaust pipe with the 2 nuts, 2 bolts and 2 rings. Tighten to 42 ft. lbs. (43 Nm).
32. Install the floor brace and tighten the 4 bolts to 10 ft. lbs. (13 Nm).
33. Install the muffler assembly.
34. Install the engine undercover.
35. Install fuel lines and electrical connectors.
36. Install both service hole covers.
37. Check for fuel pump operation and fuel leaks.
38. Install rear seat cushion assembly.
39. Check for exhaust leaks.

THROTTLE BODY

REMOVAL & INSTALLATION

See Figure 115.

1. Before servicing the vehicle, refer to the Precautions section.
2. Disconnect the negative battery cable. Wait at least 90 seconds before performing any other work.
3. Drain the engine coolant.
4. Remove the 2 nuts and V-bank cover.
5. Remove the air cleaner inlet No. 1.
6. Remove the intake air connector pipe as follows:
   - Disconnect the air hose and ventilation hose.
   - Remove the bolt.
   - Loosen the 2 hose clamps and remove the intake air connector pipe.
7. Remove the throttle body as follows:
   - Disconnect the water by-pass hose and water by-pass hose No. 7.
   - Disconnect the connector.
   - Remove the 2 bolts, 2 nuts, throttle body and gasket.

To install:

8. Install the throttle body as follows:

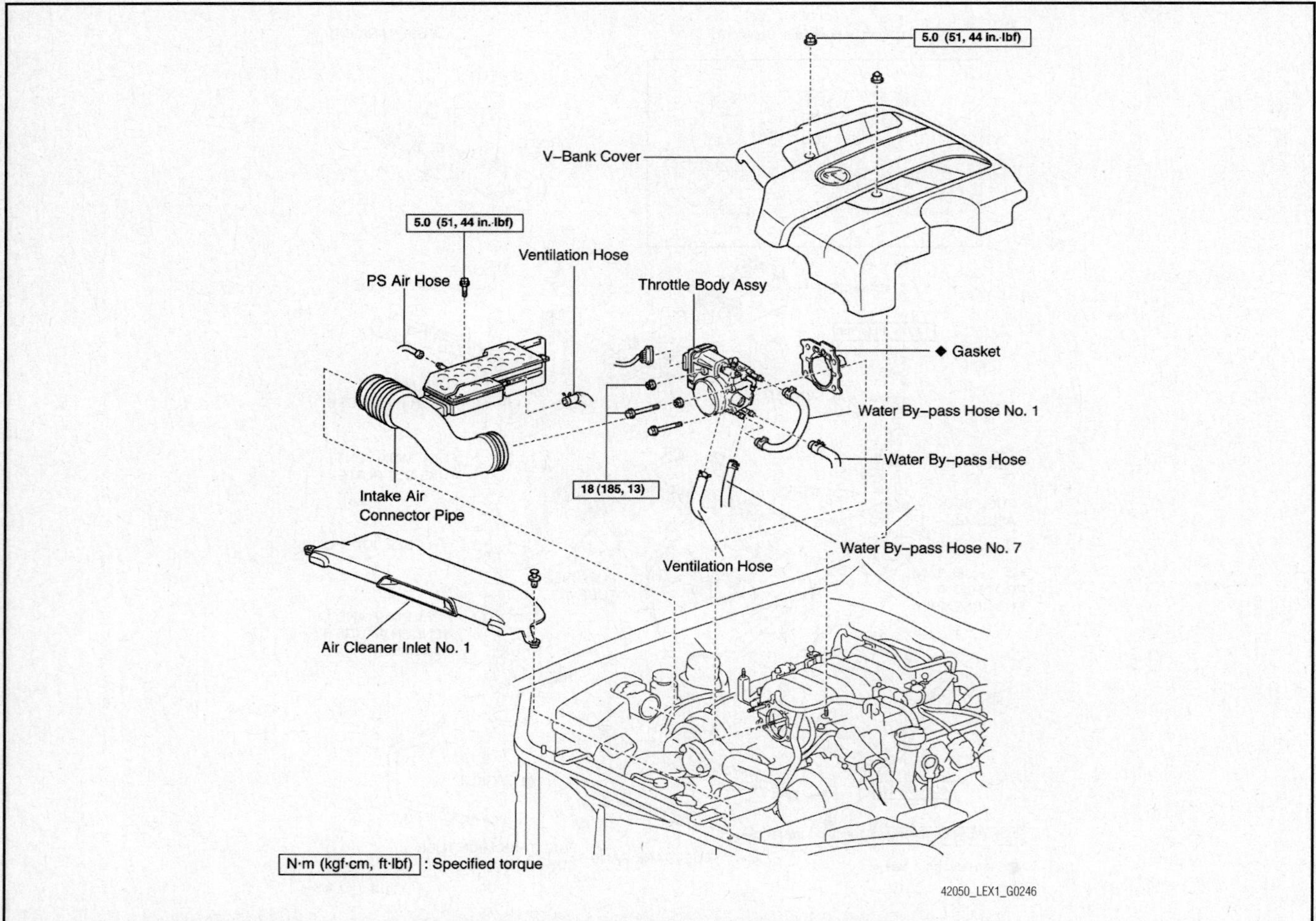

Fig. 115 Exploded view of the throttle body assembly and related components—4.3L (3UZ-FE) engine

- Install a new gasket and the throttle body with the 2 bolts and 2 nuts. Tighten the nuts and bolts to 13 ft. lbs. (18 Nm).
- Connect the water by-pass hose and water by-pass hose No. 7.
- Connect the connector.
- Install intake air connector pipe as follows:
- Install the intake air connector pipe with the bolt and 2 hose clamps. Tighten the hose clamps to 35 inch lbs. (4 Nm), and the bolt to 44 inch lbs. (5 Nm).
- Connect the air hose and ventilation hose.
- Install air cleaner inlet No. 1.
- Install the V-bank cover and tighten the nuts to 44 inch lbs. (5 Nm).
- Refill the engine cooling system.
- Connect the negative battery cable.
- Check the cooling system for leaks.
- Inspect the throttle control motor for operating sound as follows:
- Turn the ignition switch ON.
- When turning the accelerator pedal position sensor lever, check the running sound of the motor. The motor should be running smoothly without friction sounds. If operation is not as specified, check the throttle control motor, wiring and ECM.

9. Inspect the throttle position sensor as follows:
   - Inspect the throttle position sensor.
   - Connect the hand-held tester (with CAN VIM) to the DLC3.
   - Turn the ignition switch ON.
   - Check that the MIL does not light up.
   - Check that, under the CURRENT DATA, THROTTLE POS (throttle valve opening percentage)is within the standard value below.
   - Standard throttle valve opening percentage: 60% or more
10. If operation is not as specified, check the throttle position sensor, wiring and ECM.

HEATING & AIR CONDITIONING SYSTEM

BLOWER MOTOR

REMOVAL & INSTALLATION

See Figures 116 and 117.

1. Before servicing the vehicle, refer to the Precautions section.
2. Disconnect the negative battery cable. Wait 90 seconds before doing any further work while the airbag system de-energizes.
3. Remove the instrument panel assembly.
4. Remove the 2 screws and air duct No. 2.
5. Remove the blower assembly as follows:
   - Disconnect the connector and clamp from the blower.
   - Using a screwdriver, detach the claw and disconnect the connector.
   - Remove the 3 screws, bolt and nut.
   - Detach the claw and remove the blower.
   - Remove the 3 screws and recirculation damper servo.
   - Remove the 3 screws and blower w/ fan motor.

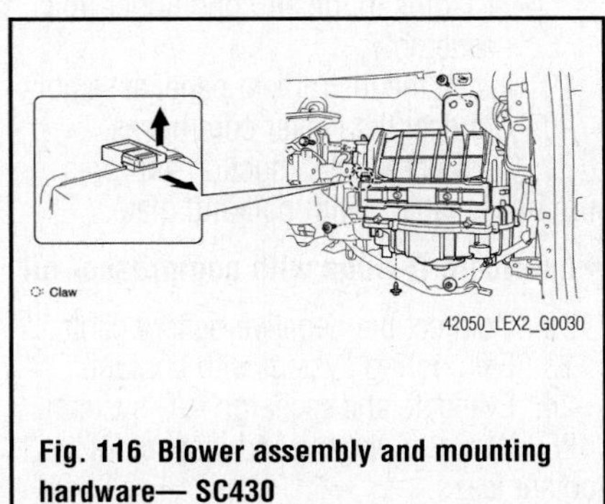

Fig. 116 Blower assembly and mounting hardware— SC430

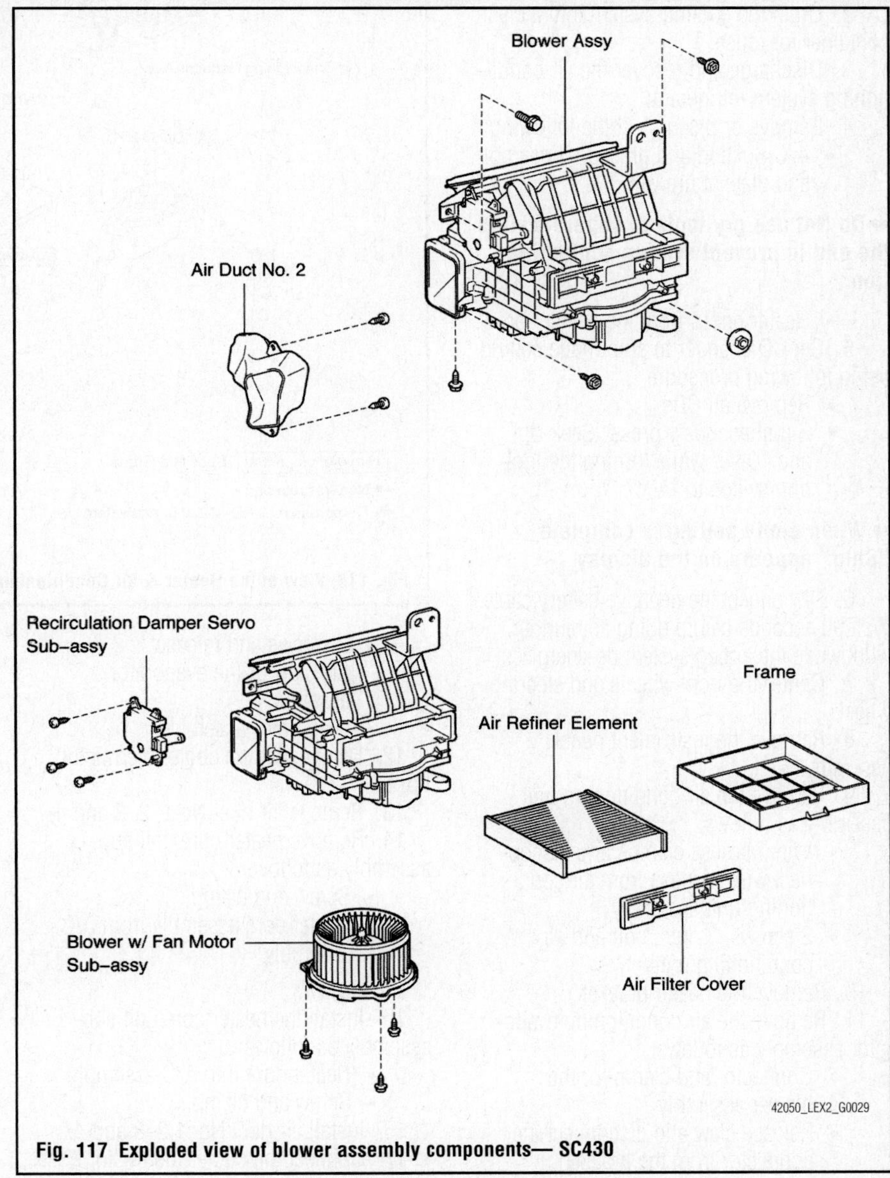

Fig. 117 Exploded view of blower assembly components— SC430

To install:

6. Install blower w/ fan motor.
7. Install the recirculation damper servo.
8. Install the blower assembly.
9. Install air duct No. 2.
10. Install the instrument panel assembly.
11. Connect the negative battery cable.

HEATER CORE

REMOVAL & INSTALLATION

See Figure 118.

➡Removal of the heater core requires removal of the entire heater air conditioning assembly.

1. Before servicing the vehicle, refer to the Precautions section.
2. Drain the cooling system into a clean container for reuse.
3. Discharge and recover the air conditioning system refrigerant.
4. Remove or disconnect the following:
   - A/C Suction and pressure hose bolt and plate at firewall

➡Do Not use pry tools to separate. Cap the end to prevent system contamination

   - Heater hoses from the heater core
5. Set CD changer to ship mode setting using following procedure
   - Remove all CDs
   - Simultaneously press "Seek Up" and "Disc" while turning the ignition switch to "Acc"

➡When mode setting is complete "Ship" appears on the display

6. Disconnect the negative battery cable. Wait 90 seconds before doing any further work while the airbag system de-energizes.
7. Center the front wheels and steering wheel.
8. Remove the instrument panel assembly.
9. Remove the air conditioning unit assembly as follows:
   - Wire harness clamps and disconnect wire harness from air conditioning unit assembly
   - 2 screws, 3 nuts, bolt and air conditioning unit
10. Remove the heater bracket.
11. Remove the air conditioning evaporator assembly as follows:
   - Connector and clamp for the blower assembly
   - Release claw and disconnect the connector from the bracket

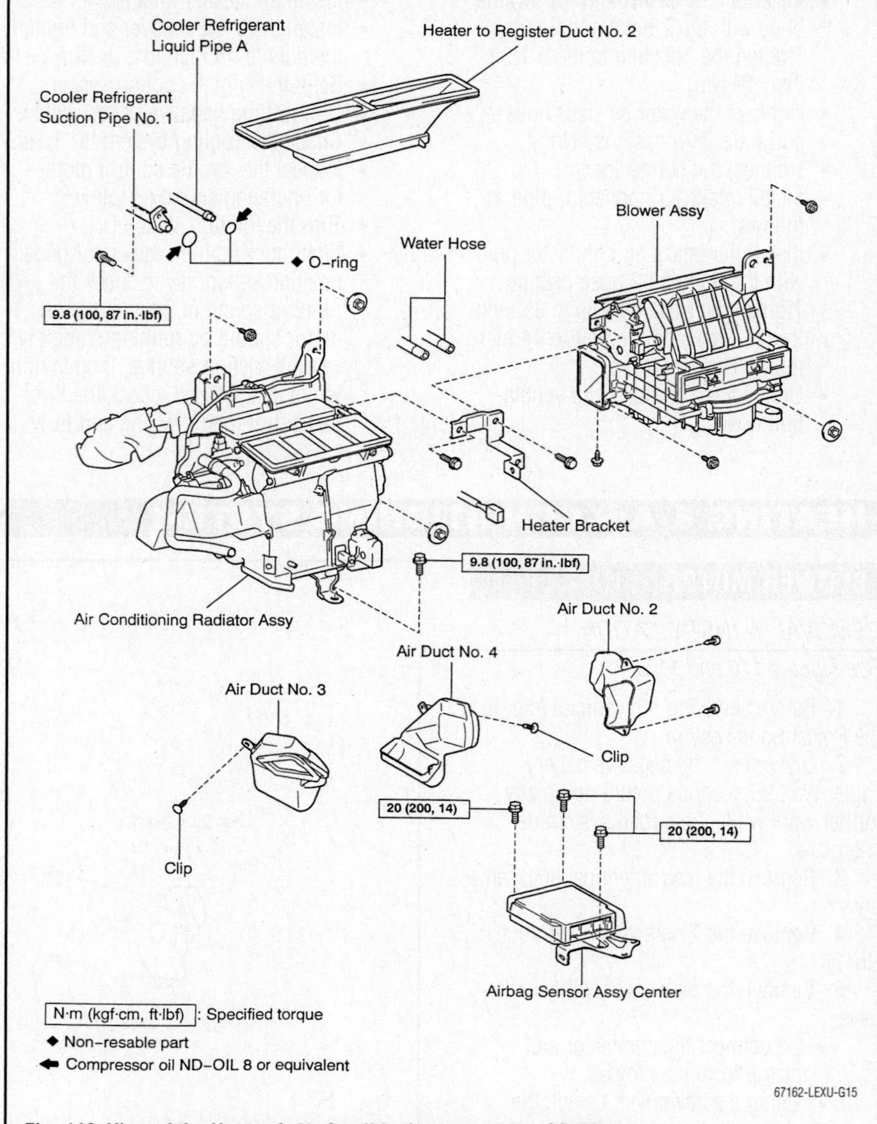

Fig. 118 View of the Heater & Air Conditioning assembly—SC430

   - 2 screws and release the claw or the evaporator assembly
   - evaporator assembly
12. Disconnect the cooler thermistor hose.
13. Remove air duct No.1, 2, 3 and 4.
14. Remove heater core unit sub-assembly as follows:
   - Screw and clamp
   - Heater core assembly from A/C assembly

To install:

15. Install the heater core unit sub-assembly as follows:
   - Heater core into A/C assembly
   - Screw and clamp
16. Install air duct No. 1,2, 3 and 4.
17. Connect the cooler thermistor hose.
18. Install the air conditioning unit assembly as follows:
   - 2 screws and 3 nuts and air conditioning unit
   - Bolt to air conditioning unit assembly. Torque: 87 inch lbs. (9.8 Nm)
   - Clamps to the air conditioner unit assembly
19. Install the instrument panel assembly.
20. Connect the heater core hoses.
21. Connect the A/C suction and pressure hoses, attach with bolt and plate.

➡Lubricate O-rings with compressor oil

22. Connect the negative battery cable.
23. Fill cooling system with coolant.
24. Evacuate and recharge A/C system.
25. Warm up engine and inspect for coolant leaks.

STEERING

POWER RACK & PINION STEERING GEAR

REMOVAL & INSTALLATION

See Figure 120.

1. Before servicing the vehicle, refer to the Precautions section.
2. Place the front wheels facing straight ahead.
3. Remove or disconnect the following:
   - Negative battery cable. Wait at least 90 seconds before performing any work.
   - Front wheels
   - Steering wheel switch & volume case
   - Steering wheel pad
   - Steering wheel column lower cover
   - Horn button
   - Steering wheel
   - Brake caliper
   - Tie rods from lower ball joint
   - Engine undercover
   - Front suspension member braces
   - Steering slide yoke. Match-mark with control valve shaft before removal
   - Oil feed tubes
4. Remove the following:
   - Tube clamps on rack assembly
   - Connector assembly
   - 4 gear assembly set bolts
   - Steering gear
   - Steering rack housing bracket
5. Match-mark and remove tie rod assemblies

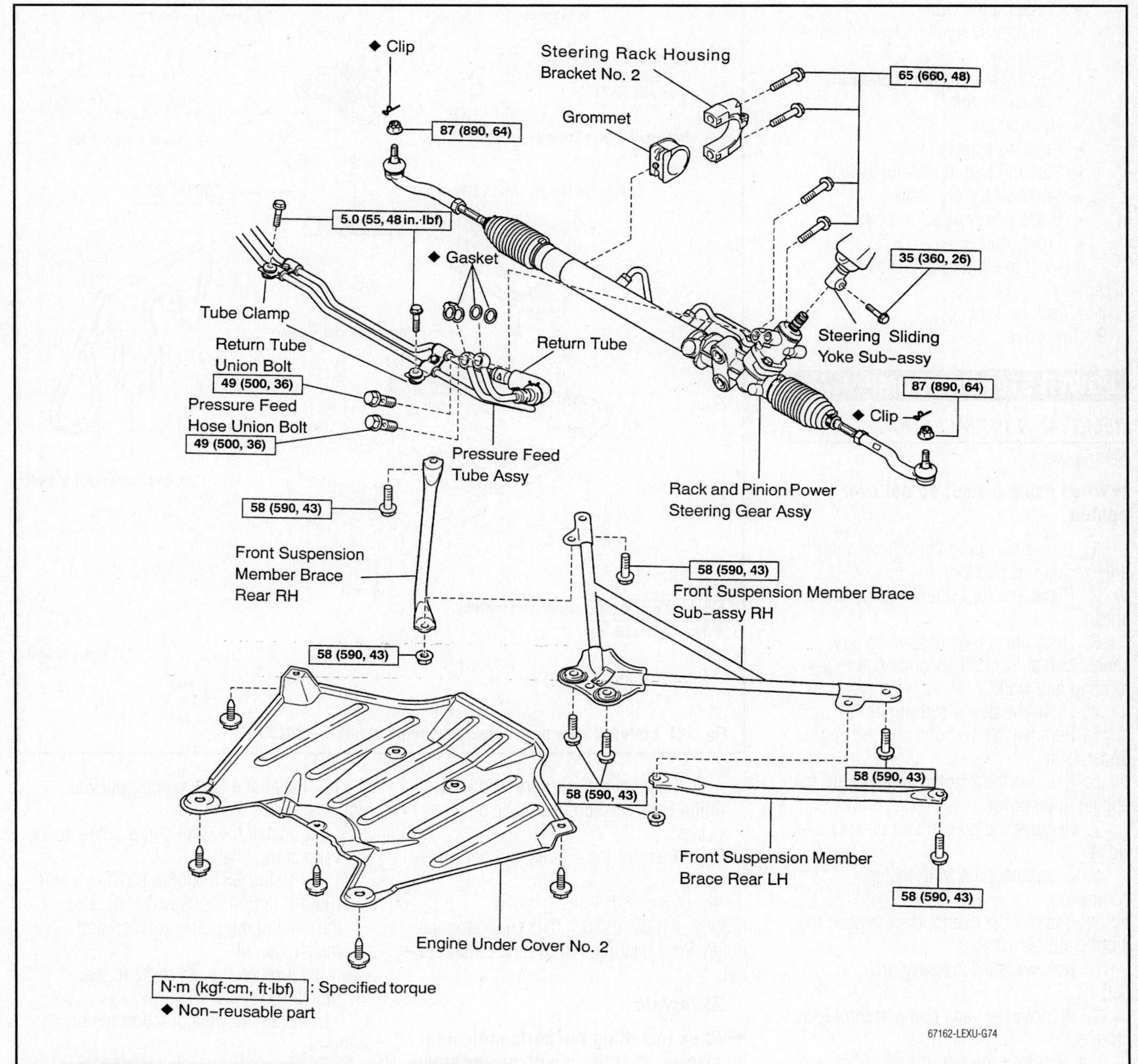

Fig. 120 Exploded view power steering gear assembly—SC430

To install:

6. Install or connect the following:
   - Tie rod assemblies. Align match marks
   - Oil feed tubes
   - Rack & Pinion gear assembly with 4 set bolts; torque to 48 ft. lbs (65 Nm)
   - Steering sliding yoke assembly
   - Pressure and return tubes
   - Tie rod assemblies to lower ball joint; torque to 64 ft. lbs. (87 Nm)
   - Brake calipers
   - Front suspension member braces
   - Engine undercover
   - Center spiral cable
   - Temporarily tighten steering wheel assembly
   - Center steering wheel and fully tighten set nut
   - Horn button
   - Steering wheel covers
   - Steering pad modulator switch
   - Negative battery cable
   - Bleed power steering system
   - Front wheels
7. Inspect toe in and adjust as necessary
8. Check for leaks.
9. Test drive

POWER STEERING PUMP

REMOVAL & INSTALLATION

See Figure 121.

➡When using a vise, do not over tighten.

1. Before servicing the vehicle, refer to the Precautions section.
2. Place the front wheels facing straight ahead.
3. Disconnect the negative battery cable. Wait at least 90 seconds before performing any work.
4. Drain the power steering fluid.
5. Remove the 16 bolts and the engine undercover.
6. Remove the 2 bolts and the right rear engine undercover.
7. Remove the bolt and air cleaner inlet No. 1.
8. Disconnect the MAF meter connector.
9. Remove the clamp, then remove the 3 bolts and air cleaner.
10. Remove the accessory drive belt.
11. Remove the vane pump assembly as follows:
    a. Remove the return hose, 2 vacuum hoses and 3 clips.

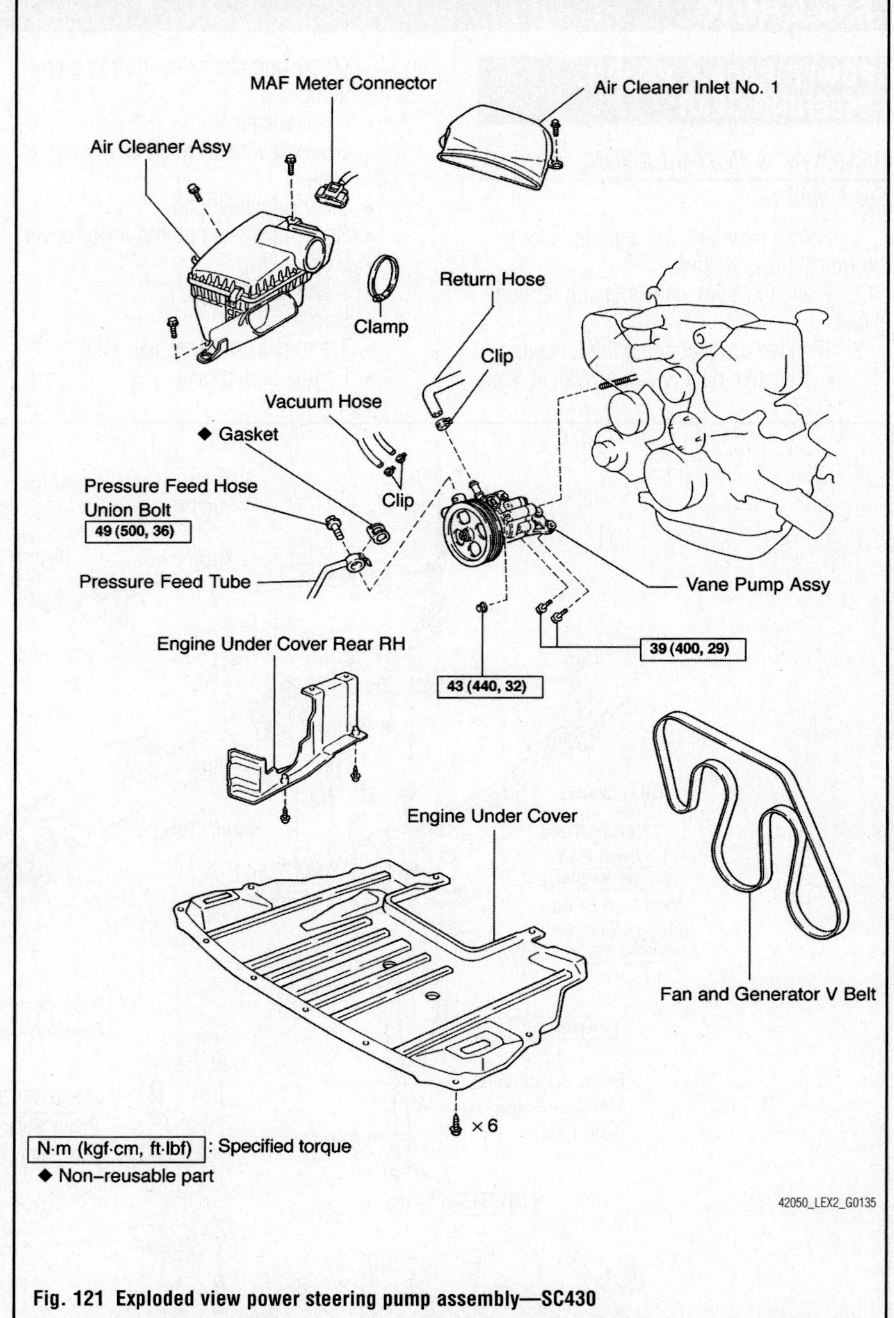

Fig. 121 Exploded view power steering pump assembly—SC430

b. Remove the pressure feed hose union bolt, pressure feed tube and gasket.

c. Remove the 2 bolts, nut and vane pump.

12. Using SST 09960_10010 (09962_01000, 09963_01000) to stop the pulley from rotating, remove the pulley set nut.

To install:

➡When installing the parts indicated by arrows, coat them with power steering fluid.

13. Install the vane pump pulley as follows:
    a. Install the vane pump pulley to the vane pump shaft.
    b. Using SST 09960-10010 (09962-01000, 09963-01000), stop the vane pump pulley rotation and install the nut.
    c. Tighten the nut to 32 ft. lbs. (43 Nm).
14. Install the vane pump assembly as follows:
    a. Install the power steering vane

pump with the nut. Tighten the nut to 32 ft. lbs. (43 Nm).

b. Install the 2 bolts and tighten them to 29 ft. lbs. (39 Nm).

c. Install the pressure feed tube and a new gasket to the pressure feed hose union bolt. Tighten the union bolt to 36 ft. lbs. (49 Nm).

d. Connect the return hose with the clip.

e. Connect the 2 vacuum hoses with the 2 clips.

15. Install the accessory drive belt.

16. Install the air cleaner assembly with the 3 bolts and install the clamp.

17. Connect the MAF meter connector.

18. Install the air cleaner inlet with the bolt.

19. Install the undercover with the 2 bolts.

20. Install the undercover with the 16 bolts.

21. Connect the negative battery cable.

22. Bleed the power steering system and inspect for leaks.

BLEEDING

1. Check the fluid level.
2. Jack up the front of the vehicle and support it with the stands.
3. Turn the steering wheel. With the engine stopped, turn the wheel slowly from lock to lock several times.
4. Lower the vehicle.
5. Start the engine. Run the engine at idle for a few minutes.
6. Turn the steering wheel. With the engine idling, turn the wheel to left or right full lock position and keep it there for 2 - 3 seconds, then turn the wheel to the opposite full lock position and keep it there for 2 - 3 seconds.
7. Repeat last step several times.
8. Stop the engine.
9. Check for foaming or emulsification. If the system has to be bled twice specifically because of foaming or emulsification, check for fluid leaks in the system.
10. Check the fluid level.

SUSPENSION

FRONT SUSPENSION

CONTROL LINKS

REMOVAL & INSTALLATION

See Figure 122.

1. Remove the front wheels.
2. Remove the bolt and nut, and disconnect the stabilizer LH bar from the stabilizer link.
3. Remove the nut and the LH stabilizer link from the lower arm.
4. Use the same procedures for the RH.

To install:

5. Install the LH stabilizer link to the lower arm with the nut. Tighten to 83 ft. lbs. (113 Nm).
6. Install the LH stabilizer link with the bolt and nut. Tighten to 38 ft. lbs (51 Nm).
7. Use the same procedures for the RH.
8. Install the front wheels. Tighten to 76 ft. lbs. (103 Nm).

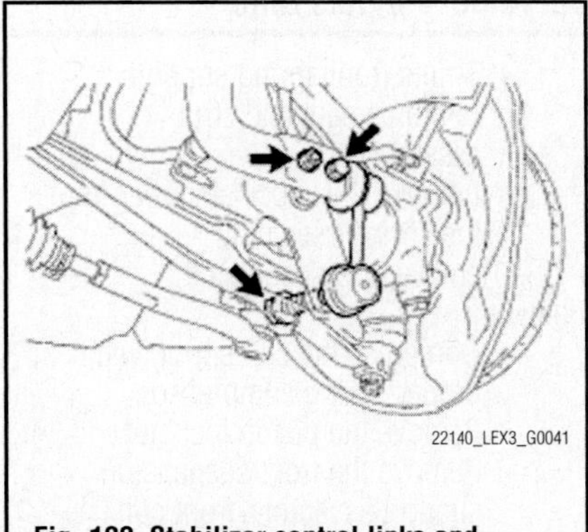

Fig. 122 Stabilizer control links and mounting nuts

LOWER BALL JOINT

REMOVAL & INSTALLATION

See Figure 123.

1. Before servicing the vehicle, refer to the Precautions section.
2. Remove the front wheel.
3. Remove the cotter pin and nut, remove the 2 bolts, then using SST 09628_62011, remove the lower ball joint from the lower control arm.
4. Remove the clip and nut, then using SST 09628_62011, disconnect the tie rod end from the lower ball joint.

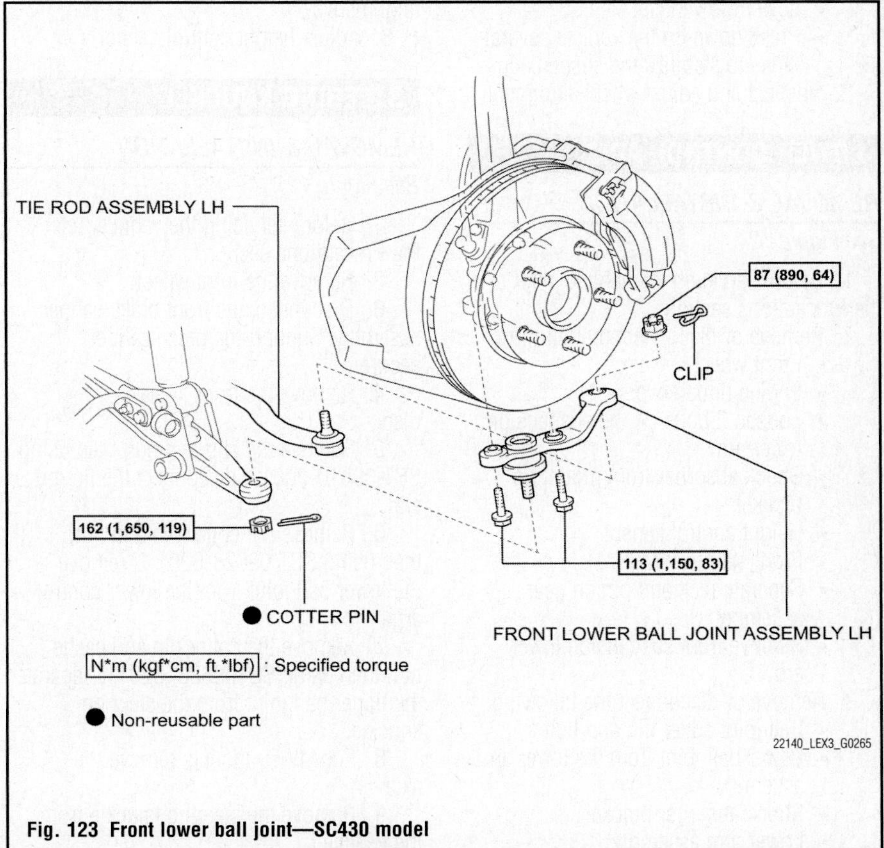

Fig. 123 Front lower ball joint—SC430 model

5. Inspect the lower control arm ball joint as follows:
- Flip the ball joint stud back and forth 5 times.
- Temporarily install the nut, and use a torque wrench to turn the nut continuously at a rate of 2 to 4 seconds per turn. Take the torque reading on the 5th turn.
- Turning torque: 0.9–26 inch lbs (0.1–3 Nm)
- Check the dust boots for cracks or grease leakage. If the value is not within the specified range, replace the front lower ball joint with a new one.

To install:

6. Install the lower ball joint to the lower suspension arm.
7. Install the nut and a new cotter pin. Tighten the nut to 119 ft. lbs. (162 Nm).
8. Install the lower ball joint with the 2 bolts and tighten to 83 ft. lbs. (113 Nm).
9. Install the tie rod to the lower ball joint. Install the nut and clip and tighten to 64 ft. lbs. (87 Nm).
10. Install the front wheels. Tighten to 76 ft. lbs. (103 Nm).
11. Stabilize the suspension as follows:
- Lower the vehicle.
- Press down on the vehicle several times to stabilize the suspension.
12. Inspect and adjust wheel alignment.

LOWER CONTROL ARM

REMOVAL & INSTALLATION

See Figure 124.

1. Before servicing the vehicle, refer to the Precautions section.
2. Remove or disconnect the following:
- Front wheel
- Engine undercovers
- Loosen 2 bolts on the suspension lower arm
- Shock absorber from mounting bracket
- Height control sensor
- Front stabilizer link
- Separate rack and pinion gear assembly
- Remove front suspension lower arm
3. Remove or disconnect the following:
- Ball joint cotter pin and bolt
- Lower ball joint from the lower arm assembly
- Shock absorber bracket
- Lower arm assembly
4. Matchmark the front and rear adjustment cams to the body and then remove the nuts and adjusting cams.
5. Lift out the lower control arm.

To install:

6. Install or connect the following:
- Shock absorber bracket. Torque: 44 ft. lbs. (59 Nm)
- Lower control arm to the body and temporarily install the adjusting cams and nuts. Do NOT tighten the nuts at this time.
- Lower control arm to the knuckle and tighten the ball joint nut to 119 ft. lbs. (162 Nm). Install a new cotter pin.
- 2 bolts on lower suspension arm to 121 ft. lbs. (164 Nm)
- Rack and pinion steering gear assembly nuts to 48 ft. lbs. (65 Nm)
- Front stabilizer link bolts and nuts (to stabilizer bar) to 38 ft. lbs. (51 Nm) and (to stabilizer link) to 116 ft. lbs. (157 Nm)
- Shock absorber assembly bolt to 116 ft. lbs. (157 Nm)
- Height control sensor
- Front wheel
7. Inspect and adjust front wheel alignment
8. Adjust height control sensor

STEERING KNUCKLE

REMOVAL & INSTALLATION

See Figure 125.

1. Before servicing the vehicle, refer to the Precautions section.
2. Remove the front wheel.
3. Disconnect the front brake caliper assembly. Support the brake caliper securely.
4. Remove the front brake disc.
5. Remove the clip and nut, then using SST 09610-20012, disconnect the tie rod end.
6. Remove the cotter pin and nut, then using SST 09628-6201 1, remove the lower ball joint from the lower control arm.
7. Remove the cotter pin and castle nut, then using SST09628_62011, separate the upper ball joint from the steering knuckle.
8. For (4WD) models remove the axle.
9. Remove the steering knuckle from the vehicle.
10. If necessary, remove the lower ball joint and hub and bearing assembly from the steering knuckle.

To install:

11. If necessary, install the hub and bearing assembly and lower ball joint to the steering knuckle.
12. Install the steering knuckle into the vehicle.
13. For (4WD) models install the axle.
14. Connect the upper ball joint to the steering knuckle. Install castle nut and new cotter pin.
15. Connect the lower ball joint to lower control arm. Install castle nut and new cotter pin.
16. Connect the tie rod end to the steering knuckle. Install castle nut and new cotter pin.
17. Install the front brake disc.
18. Install the front brake caliper assembly.
19. Install the front wheel.
20. Inspect and adjust wheel alignment.

STRUT

REMOVAL & INSTALLATION

See Figures 126 and 127.

1. Before servicing the vehicle, refer to the Precautions section.
2. Remove or disconnect the following:
- Tire and wheel assembly
- Speed sensor assembly
- Upper suspension arm from the steering knuckle
- Shock absorber from the mounting bracket

⁂ CAUTION

Loosen the piston rod lock nut. Do Not remove at this time

- 3 nuts front spring support reinforcement and strut assembly
- Remove the shock absorber from the spring assembly
3. Remove or disconnect the following:
- Compress the coil spring with proper spring compressor.
- Remove the piston lock nut.
- Remove the front suspension support assembly, front coil spring insulator and coil spring.

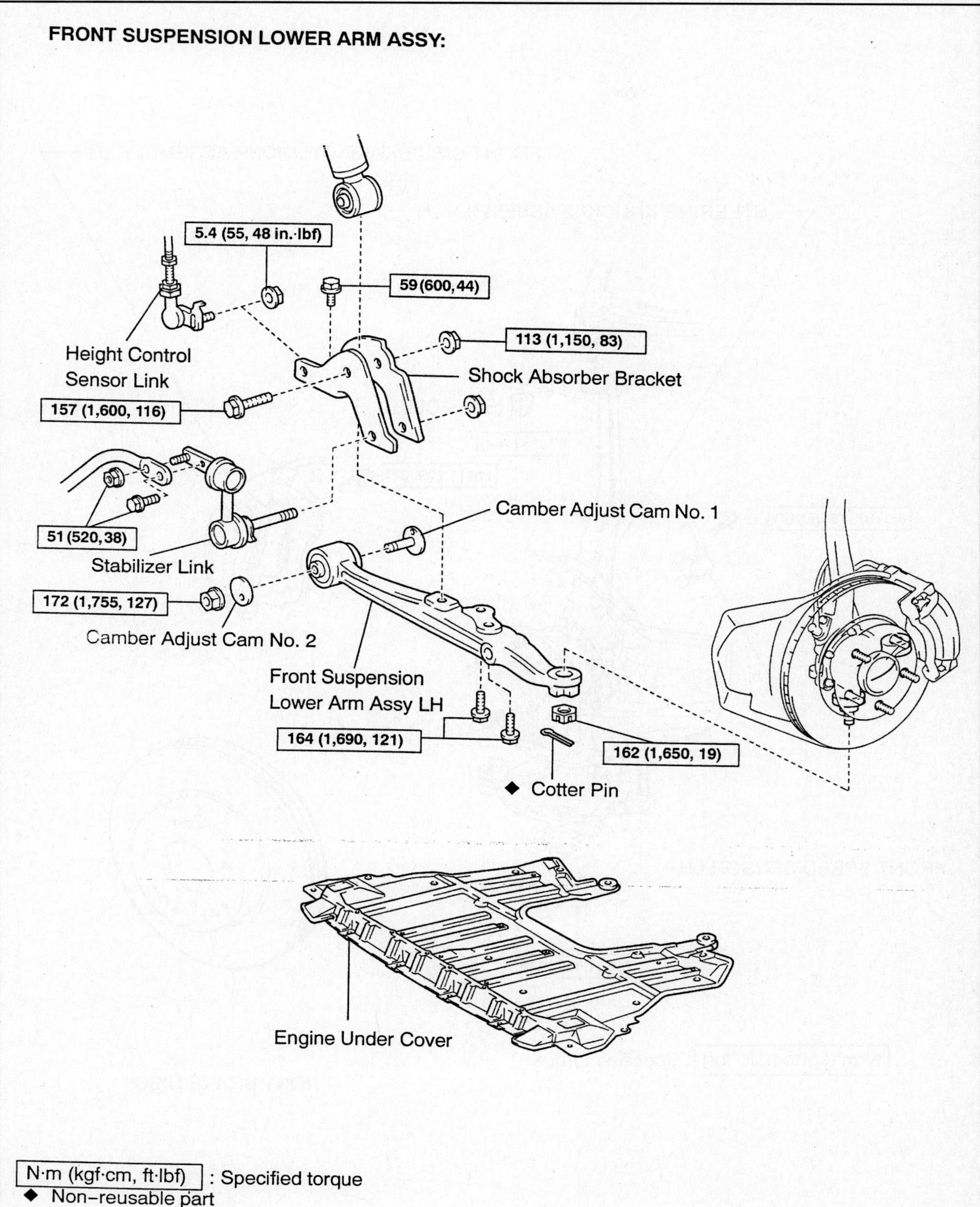

Fig. 124 Expanded view, lower front suspension arm—SC430

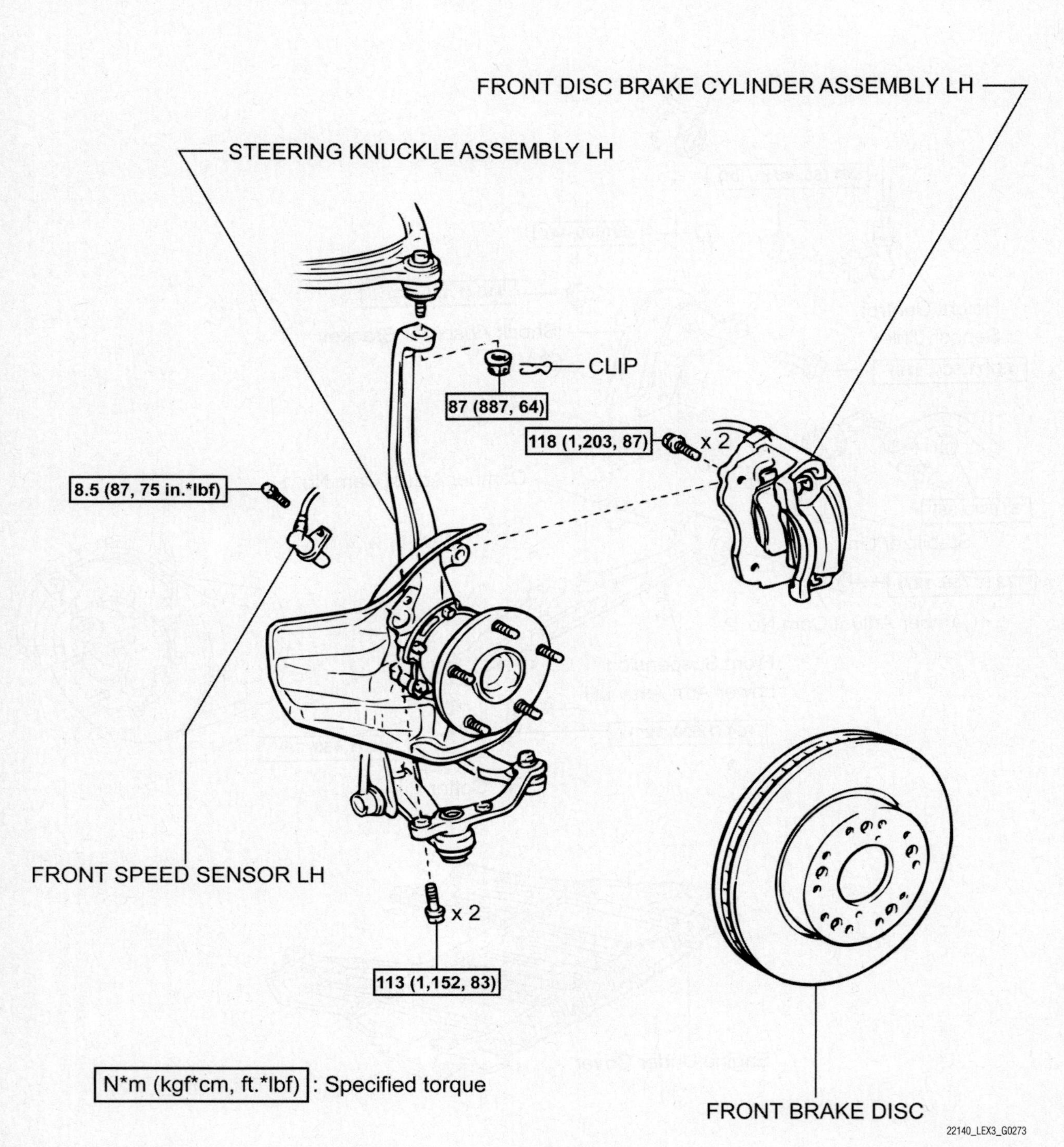

Fig. 125 Steering knuckle—SC430 model shown

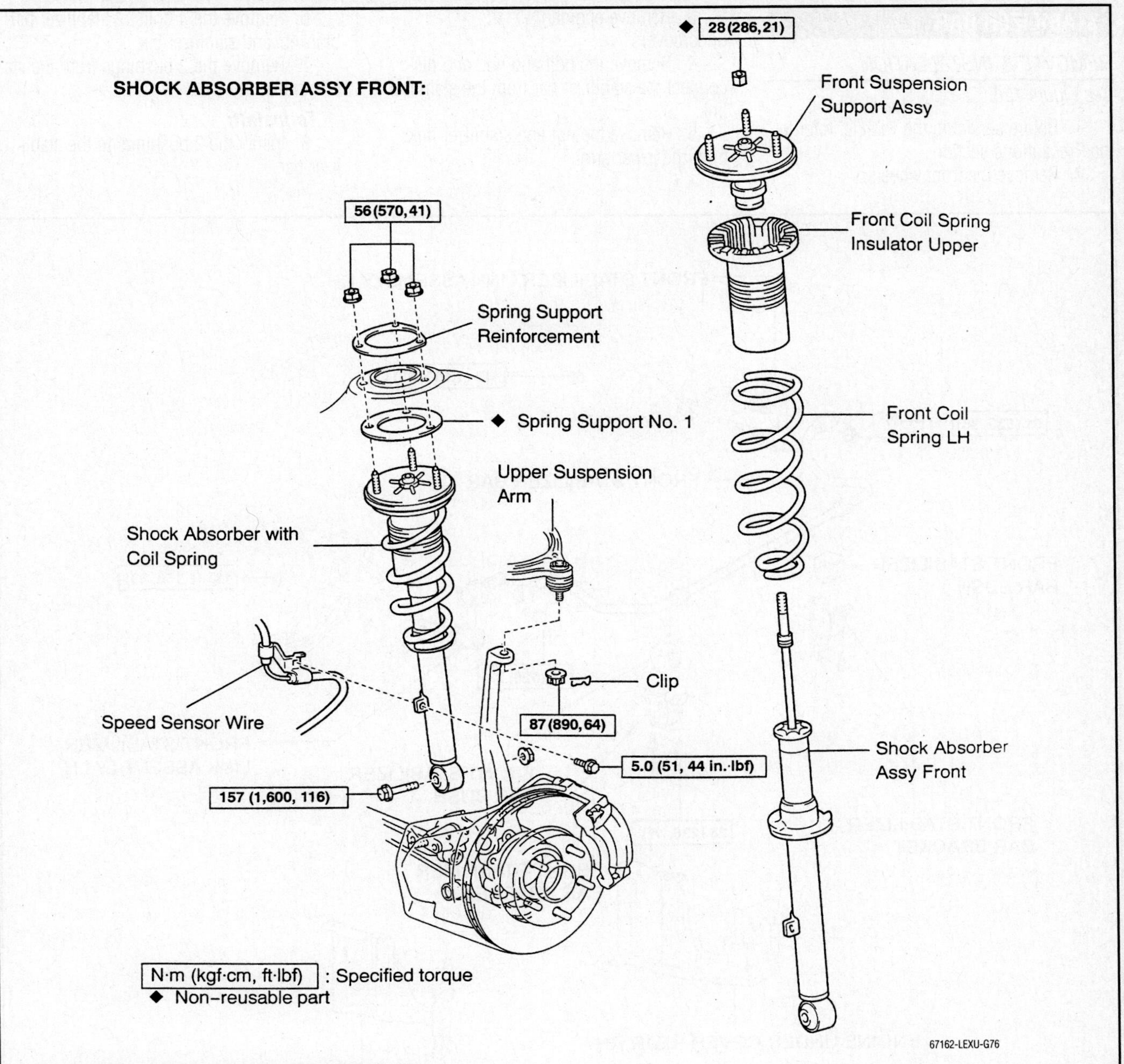

SHOCK ABSORBER ASSY FRONT:

28 (286, 21)
Front Suspension Support Assy
56 (570, 41)
Front Coil Spring Insulator Upper
Spring Support Reinforcement
Spring Support No. 1
Front Coil Spring LH
Upper Suspension Arm
Shock Absorber with Coil Spring
Clip
Speed Sensor Wire
87 (890, 64)
5.0 (51, 44 in.·lbf)
Shock Absorber Assy Front
157 (1,600, 116)
N·m (kgf·cm, ft·lbf) : Specified torque
◆ Non-reusable part
67162-LEXU-G76

Fig. 126 Exploded view of front strut assembly—SC430

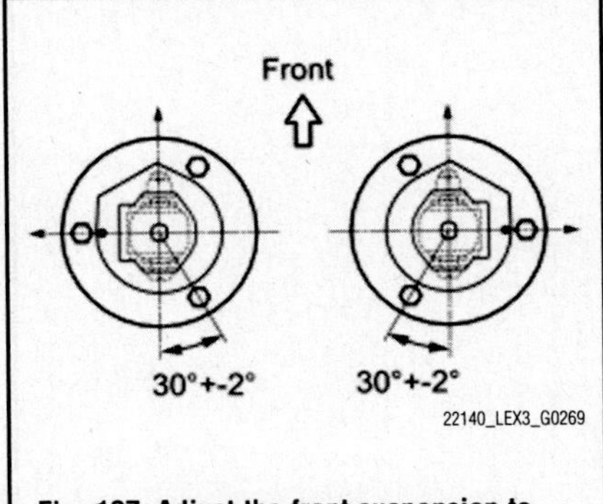

Fig. 127 Adjust the front suspension to position shown

To install:

4. Install the shock into the spring assembly.
5. Install the spring insulator.
6. Compress the coil spring using the proper spring compressor.
7. Install spring making sure that the spring seats properly.
8. Temporarily install a new lock nut.
9. Align the suspension support with the shock absorber lower bolt.
10. Install or connect the following:
   - 3 bolts attaching the strut assembly to the support assembly to 41 ft. lbs (56 Nm)
   - Fully tighten piston lock nut to 21 ft. lbs. (28 Nm)
   - Strut assembly to mounting bracket to 116 ft lbs. (157 Nm)
   - Upper suspension arm to steering knuckle to 64 ft. lbs. (87 Nm)
   - Speed sensor
   - Tire & wheel
11. Inspect and adjust front wheel alignment.

STABILIZER BAR

REMOVAL & INSTALLATION

See Figure 128.

1. Before servicing the vehicle, refer to the Precautions section.
2. Remove the front wheels.
3. Remove engine undercovers.
4. Remove the bolt and nut, and disconnect the stabilizer bar from the stabilizer link.
5. Remove the nut and stabilizer link from the lower arm.
6. Remove the 4 bolts, 2 stabilizer bar brackets and stabilizer bar.
7. Remove the 2 bushings from the stabilizer bar.

To install:

8. Install the 2 bushings to the stabilizer bar.

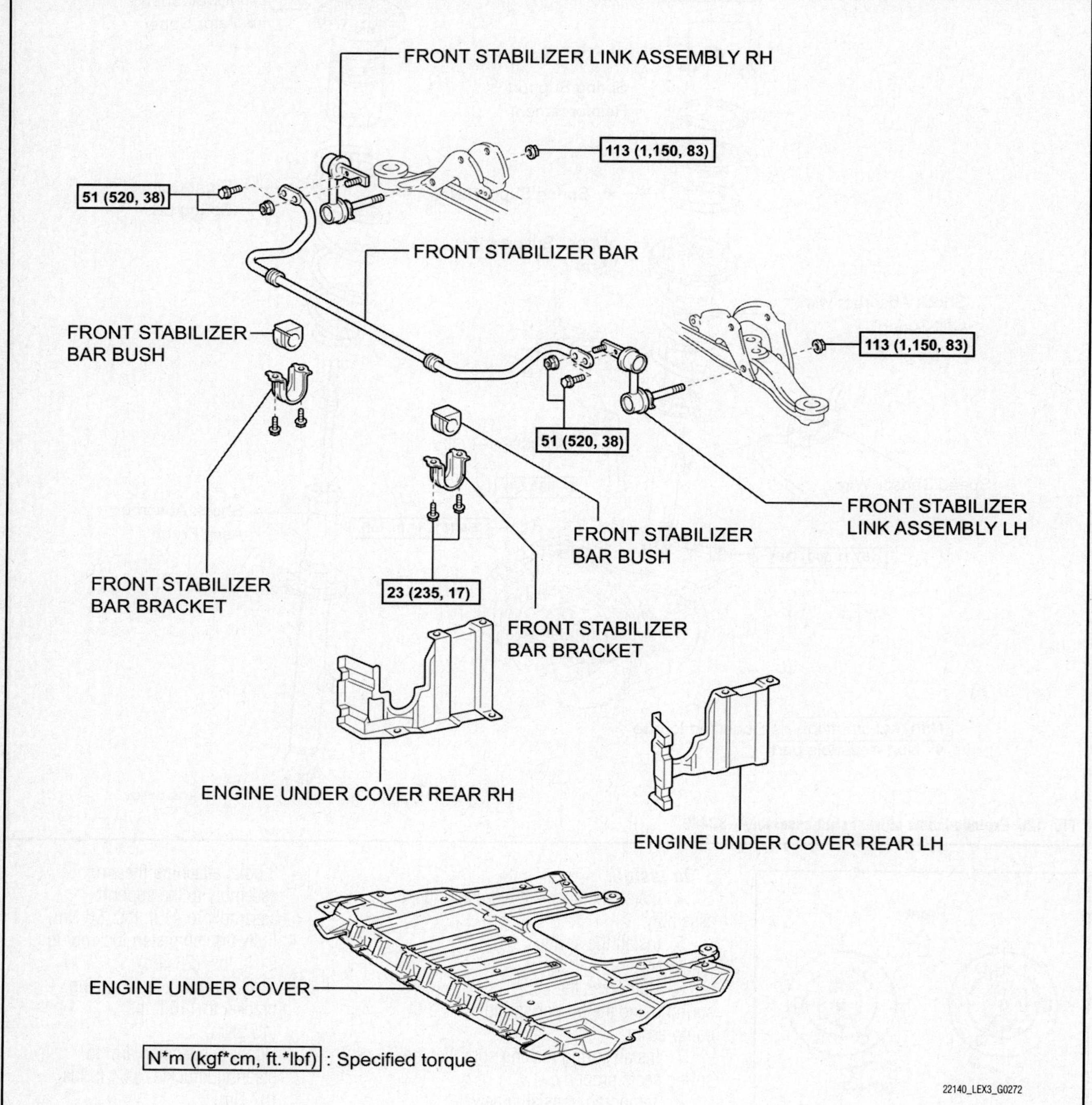

Fig. 128 Front stabilizer bar

9. Install the stabilizer bar and 2 stabilizer bar brackets with the 4 bolts.

10. Install the stabilizer link to the lower arm with the nut and tighten to 83 ft. lbs. (113 Nm).

11. Install the stabilizer link with the bolt and nut and tighten to 38 ft. lbs. (51 Nm).

12. Install engine undercover.

13. Install the front wheels.

14. Lower the vehicle.

15. Press down on the vehicle several times to stabilize the suspension.

UPPER BALL JOINT

REMOVAL & INSTALLATION

The upper ball joint is an integral part of the upper arm and is not replaced separately. The upper ball joint replacement is accomplished by replacing the upper arm.

UPPER CONTROL ARM

REMOVAL & INSTALLATION

See Figure 129.

1. Before servicing the vehicle, refer to the Precautions section.
2. Raise and safely support the vehicle.
3. Remove or disconnect the following:
   - Wheel
   - Strut
   - Anti-lock Brake System (ABS) speed sensor wire harness from the upper control arm by removing the bolt.
   - Mounting bolts holding the upper control arm to the vehicle
   - Upper control arm

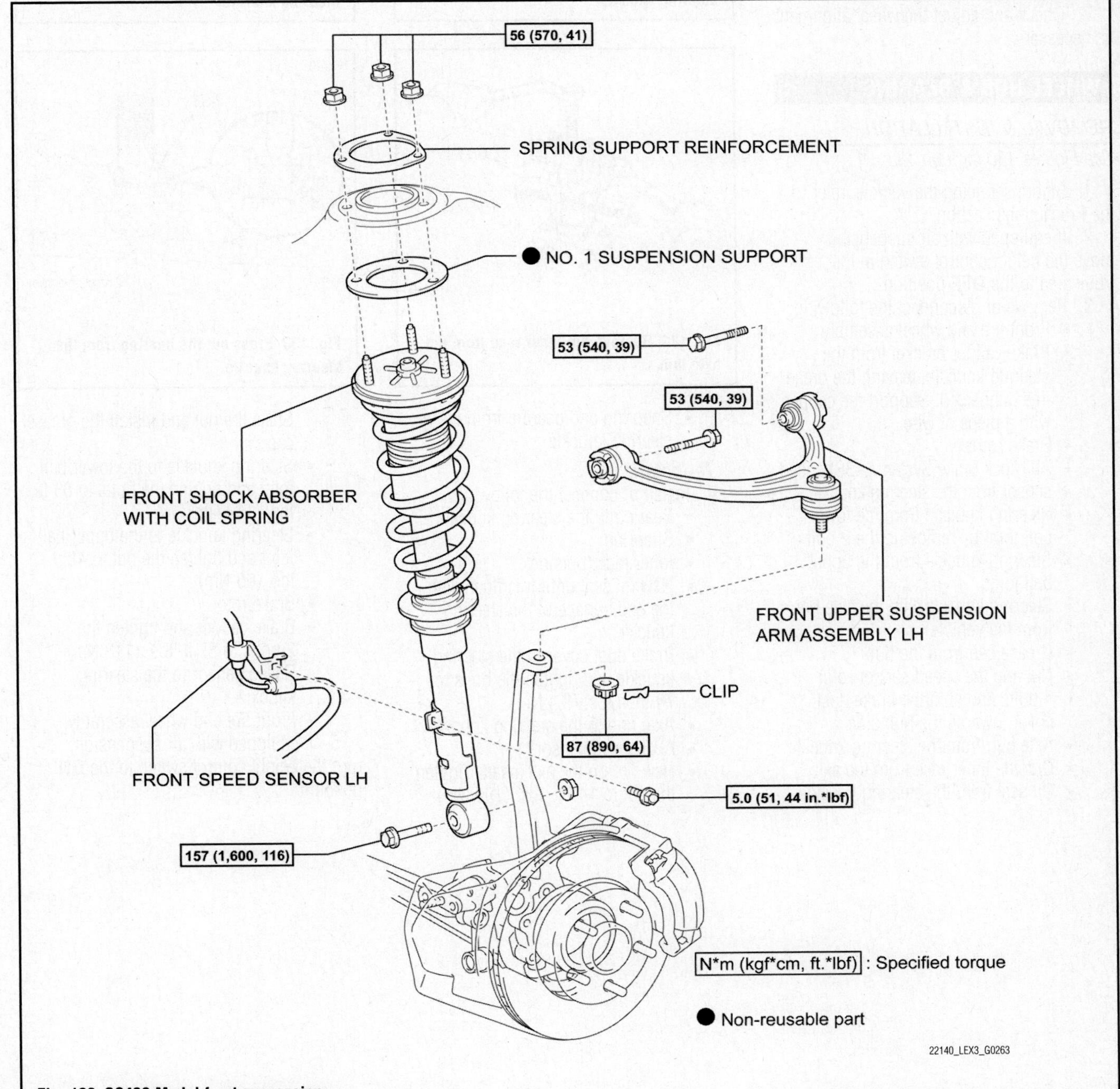

Fig. 129 SC430 Model front suspension

To install:

4. Install or connect the following:
   - Upper control arm and tighten the 2 mounting bolts to 39 ft. lbs. (53 Nm)
   - ABS speed sensor wire harness to the upper control arm with the attaching bolt. Tighten to 48 inch lbs. (5.0 Nm).
   - Strut
   - Wheel and tighten to 76 ft. lbs. (103 Nm).
5. Lower the vehicle.
6. Stabilize the suspension.
7. Check and adjust the wheel alignment as necessary.

WHEEL HUB & BEARING

REMOVAL & INSTALLATION

See Figures 130 through 133.

1. Before servicing the vehicle, refer to the Precautions section.
2. If equipped with air suspension, move the height control switch in the trunk area to the **OFF** position.
3. Remove or disconnect the following:
   - Front tire and wheel assembly
   - Brake caliper bracket from the steering knuckle, leaving the brake line connected. Support the caliper with a piece of wire.
   - Brake rotor
   - Anti-lock Brake System (ABS) speed sensor from the steering knuckle
   - Steering knuckle from the lower ball joint by removing the 2 bolts
   - Steering knuckle from the upper ball joint
   - Steering knuckle with the axle hub from the vehicle
   - Grease cap from the hub
   - Nut and the speed sensor rotor
   - 4 bolts and shift the brake dust cover towards the hub side
   - Axle hub from the steering knuckle
   - Outside inner race from the axle
   - Oil seal from the steering knuckle

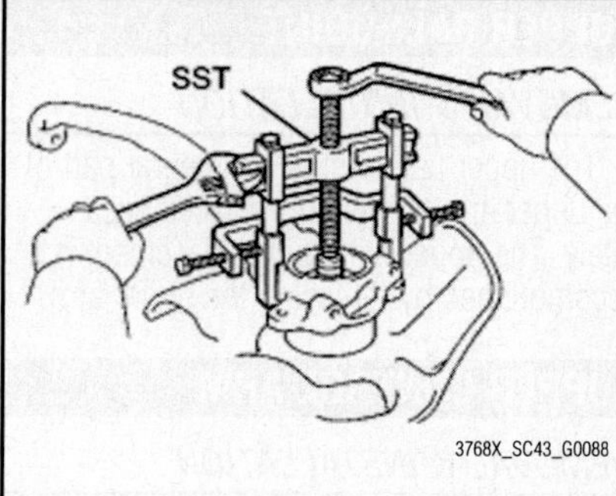

Fig. 130 Remove the axle hub from the steering knuckle

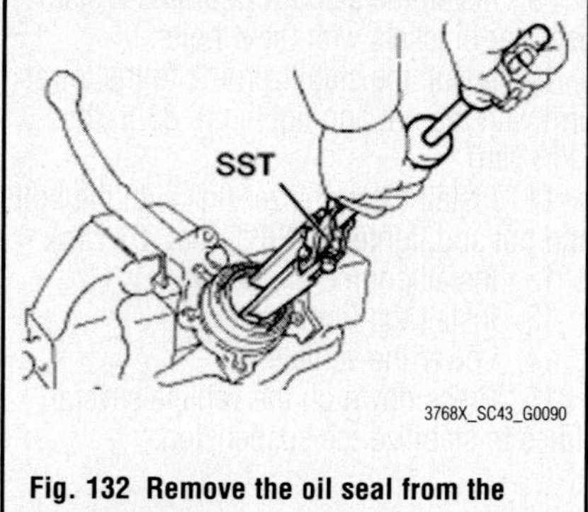

Fig. 132 Remove the oil seal from the steering knuckle

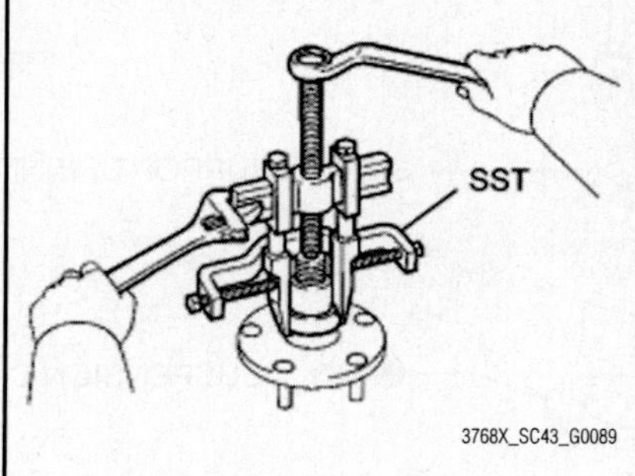

Fig. 131 Remove the inner race from the axle hub

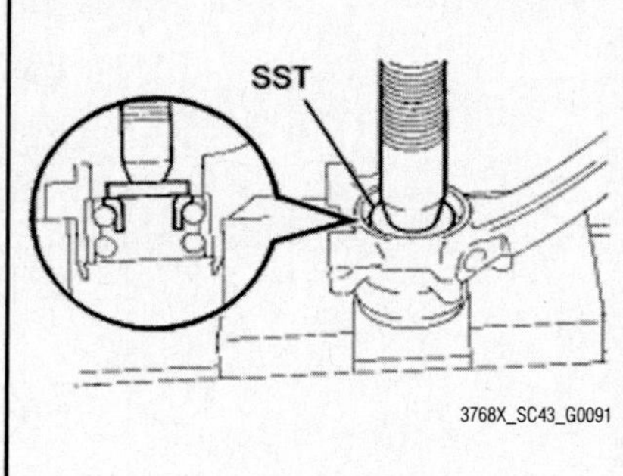

Fig. 133 Press out the bearing from the steering knuckle

   - Snapring and bearing from the steering knuckle

To install:

4. Install or connect the following:
   - Bearing in the steering knuckle
   - Snapring
   - Inner race (outside)
   - New oil seal until it is flush with the end surface of the steering knuckle
   - Brake dust cover to the steering knuckle and tighten the bolts to 74 inch lbs. (8.4 Nm)
   - Axle hub to the steering knuckle
   - ABS speed sensor
   - New nut on the axle shaft. Tighten the nut to 147 ft. lbs. (199 Nm). Stake the nut and install the grease cap.
   - Steering knuckle to the lower ball joint and tighten the bolts to 83 ft. lbs. (113 Nm)
   - Steering knuckle to the upper ball joint and tighten the nut to 48 ft. lbs. (65 Nm)
   - Brake rotor
   - Brake caliper and tighten the 2 bolts to 87 ft. lbs. (118 Nm)
   - Speed sensor to the steering knuckle
   - Front tire and wheel assembly
5. If equipped with air suspension, turn the height control switch to the **ON** position.

CONTROL ARMS/LINKS

REMOVAL & INSTALLATION

Toe Link

See Figures 134 and 135.

1. Remove the 10 bolts and ornament.
2. Remove the 5 nuts and wheel.
3. Remove the bolt and disconnect the rear speed sensor wire from the toe control link.
4. Remove the nut.
5. Using SST 09610-20012, disconnect the toe control link from the axle carrier.

➡**Be careful not to damage the dust cover.**

6. Place matchmarks on the adjusting cam and suspension member.
7. Remove the nut, camber adjusting cam No. 2, camber adjusting cam No. 1 and toe control link.

To install:

8. Align the matchmarks on the adjusting cam and suspension member.
9. Temporarily install the nut, camber adjusting cam No. 2, camber adjusting cam No. 1 and toe control link.

➡**After stabilizing the suspension, tighten the nuts.**

10. Connect the toe control link to the axle carrier with the nut. Tighten to 44 ft. lbs. (59 Nm).
11. Connect the rear speed sensor wire to the toe control link with the bolt.
12. To complete the installation, reverse remaining removal procedure.

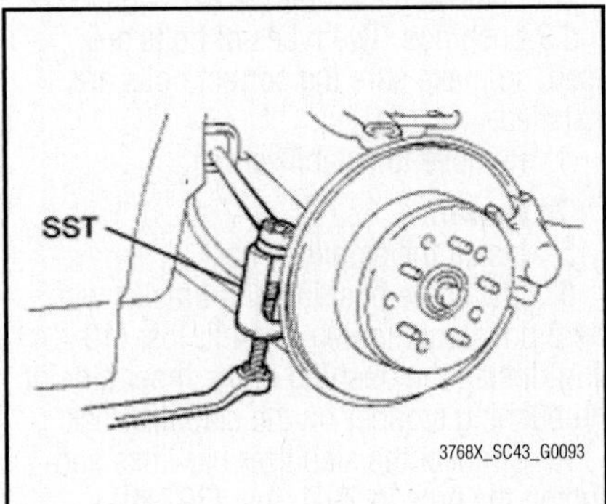

Fig. 134 Using SST 09610-20012, disconnect the toe control link from the axle carrier

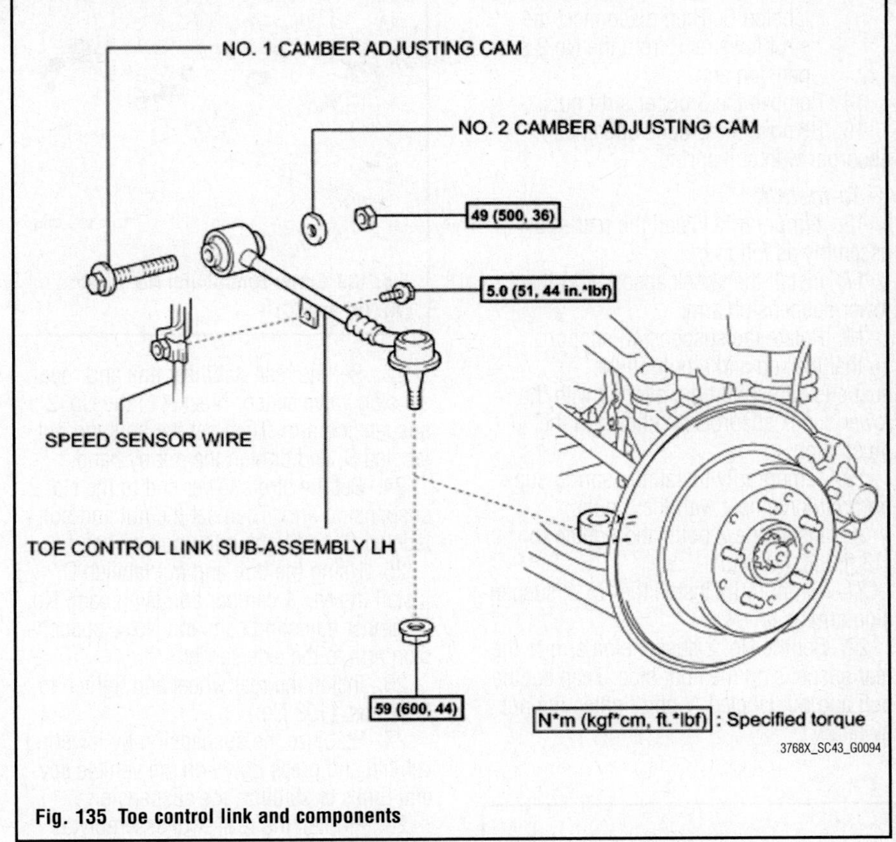

Fig. 135 Toe control link and components

13. Stabilize suspension. Refer to Shock Absorber.

SHOCK ABSORBER

REMOVAL & INSTALLATION

Strut

See Figures 136 through 139.

1. Before servicing the vehicle, refer to the Precautions section.
2. Remove the right and left tonneau cover sub-assembly.
3. Remove the rear floor finish plate.
4. Remove the left and right trim covers.
5. Remove the spare tire assembly.
6. Remove the tool box cover.
7. Remove the luggage compartment trim cover rear sub-assembly.
8. Remove the luggage compartment front trim cover.
9. Remove the rear tonneau cover assembly.
10. Remove the inner lower trim cover.
11. Remove the front and both side trim covers.
12. Remove the rear wheel.
13. Disconnect the rear No. 2 Suspension arm as follows:
    - Place matchmarks on the camber adjusting cam and No. 2 suspension arm.
    - Fix the bolt on the rear suspension member side, and loosen the nut labeled (A). Do not remove the nut.
    - Fix the bolt and remove the nut labeled (B). Then disconnect the rear stabilizer link and road sensing

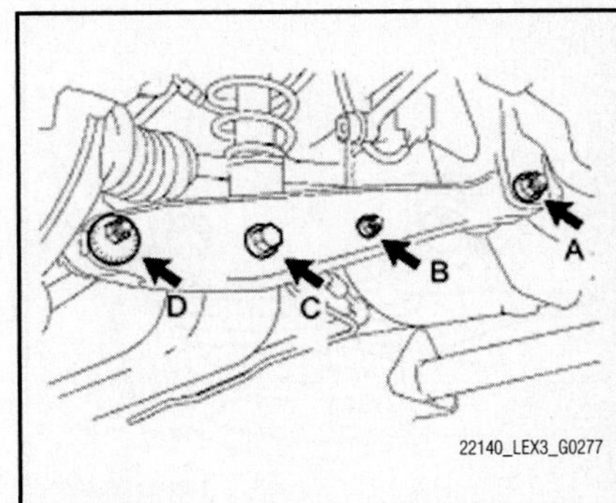

Fig. 136 No. 2 Suspension arm, nut (A) and (B) view

valve sensor bracket from the No. 2 suspension arm.

- Fix the nut and remove the bolt labeled C. Then disconnect the strut lower end from the No.2 suspension arm.

14. Remove the 3 upper strut nuts.
15. Remove the 2 bolts and shock absorber with coil spring.

To install:

16. Temporarily install the rear strut assembly as follows:
17. Install the shock absorber to the lower suspension arm.
18. Rotate the suspension support so that the rod and a bolt on the suspension support are aligned with the lower shock absorber as shown in the illustration.
19. Temporarily install the spring support reinforcement with the 3 nuts.
20. Install the 2 bolts and tighten to (13 ft. lbs. (18 Nm).
21. Temporarily install the No. 2 suspension arm as follows:
22. Set the No. 2 suspension arm to the rear suspension member side. Then set the bolt and nut labeled A, and tighten the nut by hand.

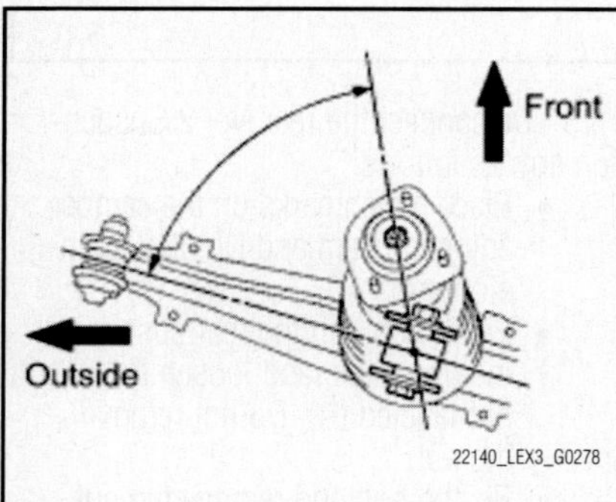

Fig. 137 Suspension support alignment view

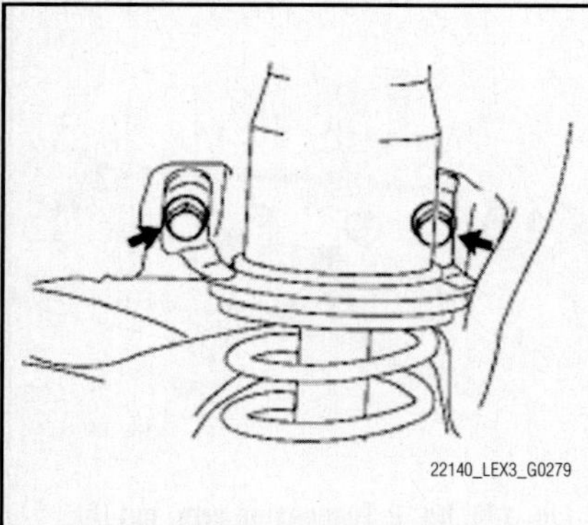

Fig. 138 Install the 2 bolts and tighten

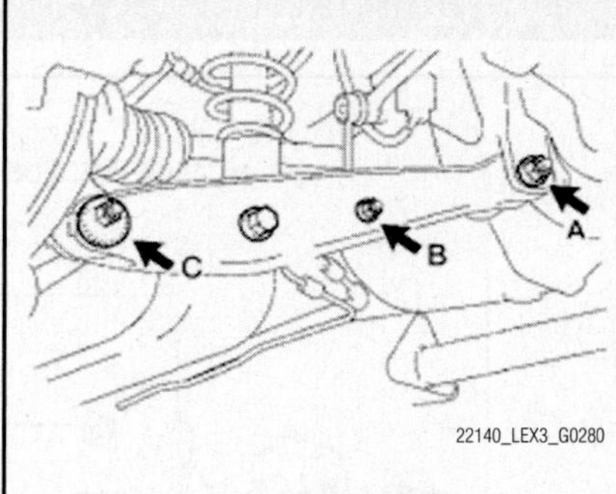

Fig. 139 Lower control arm No. 2 nuts (A), (B) and (C)

23. Set the rear stabilizer link and road sensing valve sensor bracket to the No. 2 suspension arm. Then set the bolt and nut labeled B, and tighten the nut by hand.
24. Set the struts lower end to the No. 2 suspension arm. Then set the nut and bolt labeled C, and tighten the bolt by hand.
25. Using the bolt and nut labeled D, install the No. 1 camber adjusting cam, No. 2 camber adjusting cam and No. 2 suspension arm to the axle carrier.
26. Install the rear wheel and tighten to 76 ft. lbs. (103 Nm).
27. Stabilize the suspension by lowering vehicle and press down on the vehicle several times to stabilize the suspension.
28. Tighten the rear strut assembly as follows:
    - Tighten the 3 upper nuts to 47 ft. lbs. (64 Nm).
    - Tighten the strut lock nut to 21 ft. lbs. (28 Nm).
    - Fix the lower strut assembly bolt and tighten the nut to 81 ft. lbs. (110 Nm).
29. Tighten the No. 2 suspension arm assembly. Fix the bolt and tighten the nuts labeled A, B and C.
30. Install the rear wheel and tighten to 76 ft. lbs. (103 Nm).
31. Inspect and adjust the rear wheel alignment

Coil Spring

See Figure 140.

1. Before servicing the vehicle, refer to the Precautions section.
2. Remove the strut assembly.
3. Remove the coil spring from strut as follows:
    - Install the coil spring compressor to the coil spring so that the 2 hooks on the compressor are fully extended.
    - Fully compress the coil spring.
    - Remove the lock nut.

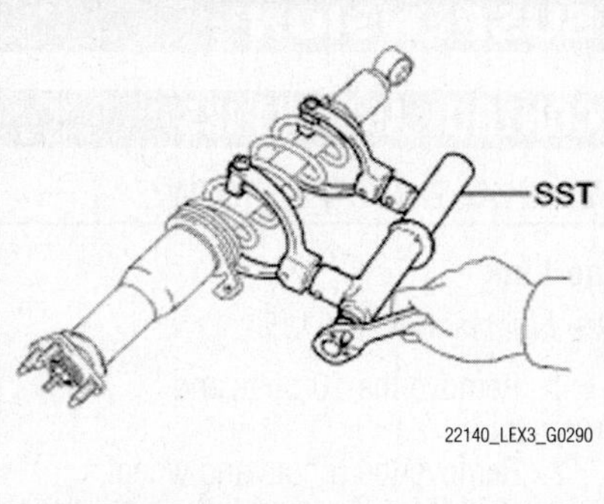

Fig. 140 Compress the coil spring with spring compressor tool

- Remove the washer, cushion, collar, suspension support, upper insulator, lower cup, cushion, spring bumper and coil spring.

To install:

4. Compress the coil spring.
5. SST Install the coil spring to the shock absorber. Fit the lower end of the coil spring into the recess of the spring seat of the shock absorber.
6. Install the spring bumper, lower cup, cushion, collar, upper insulator, suspension support, cushion and washer to the shock absorber. Temporarily install a new lock nut.
7. Install assembly and then tighten the lock nut to 21 ft. lbs. (28 Nm).

STABILIZER BAR

REMOVAL & INSTALLATION

See Figure 141.

1. Remove the rear wheels.
2. Remove the 2 nuts and left stabilizer bar link. If the ball joint turns together with the nut, use a hexagon wrench (5 mm) to hold the stud. Repeat for the right side.
3. Remove the 8 bolts, 2 No. 1 brackets and 2 bushings. Two types of bolts are used, so make sure the correct bolts are installed.
4. Remove the stabilizer bar.

To install:

5. Install the stabilizer bar.
6. Install the bushing and bracket with the 2 bolts and tighten to 14 ft. lbs. (19 Nm). Install the bushing to the inner side of the bushing stopper on the stabilizer bar.
7. Connect the stabilizer bar links and tighten the nuts to 29 ft. lbs. (39 Nm). Repeat for the right side.
8. Install the rear wheels.
9. Inspect and adjust rear wheel alignment.

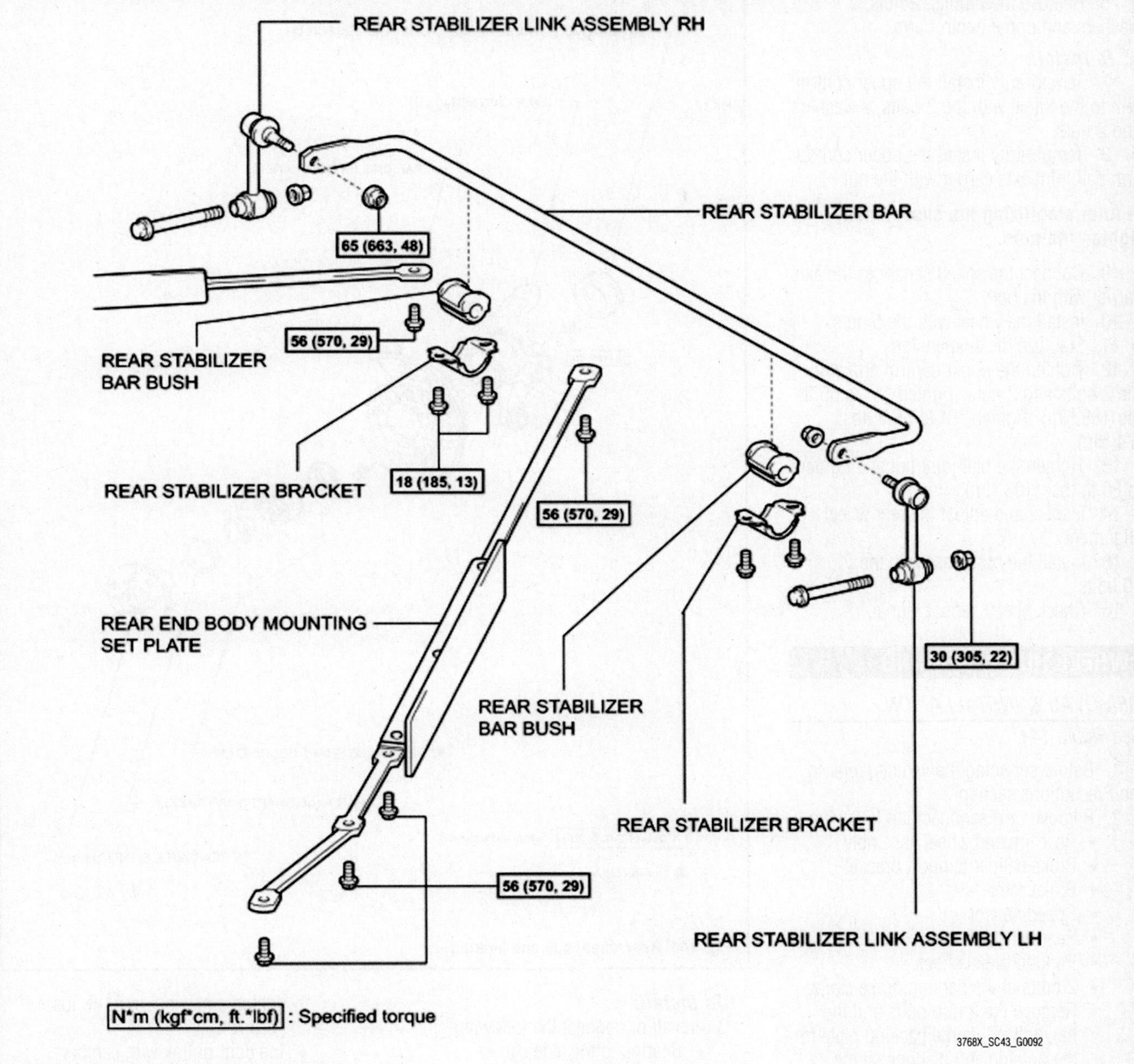

REAR STABILIZER LINK ASSEMBLY RH
REAR STABILIZER BAR
65 (663, 48)
56 (570, 29)
REAR STABILIZER BAR BUSH
18 (185, 13)
REAR STABILIZER BRACKET
56 (570, 29)
REAR END BODY MOUNTING SET PLATE
30 (305, 22)
REAR STABILIZER BAR BUSH
REAR STABILIZER BRACKET
56 (570, 29)
REAR STABILIZER LINK ASSEMBLY LH
N*m (kgf*cm, ft.*lbf) : Specified torque
3768X_SC43_G0092

Fig. 141 Rear stabilizer bar and components

UPPER CONTROL ARM

REMOVAL & INSTALLATION

See Figure 142 and *43*.

1. Remove the 10 bolts and ornament.
2. Remove the 5 nuts and wheel.
3. Remove the bolt and disconnect the speed sensor from the rear axle carrier.
4. Remove the nut.
5. Using SST 09628-00011, disconnect the upper control arm from the axle carrier.

➡Be careful not to damage the dust cover. Support the axle carrier.

SST
3768X_SC43_G0095

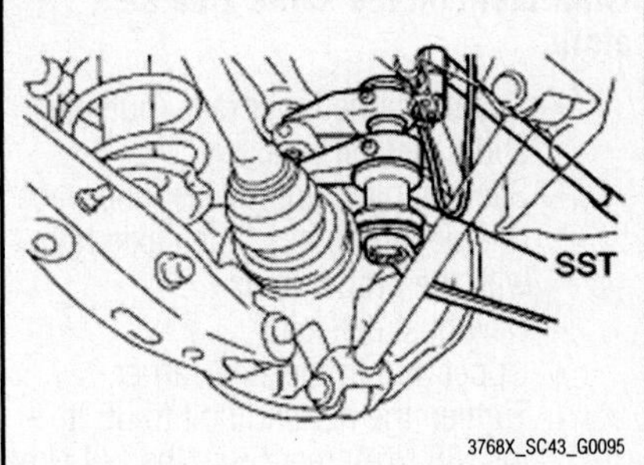

Fig. 142 Using SST 09628-00011, disconnect the upper control arm from the axle carrier

A
A
B
B
3768X_SC43_G0096

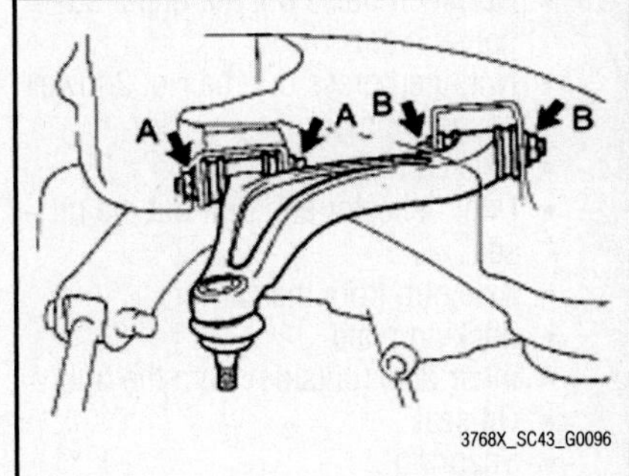

Fig. 143 Remove the 2 bolts, 2 nuts, 2 washers and upper control arm

6. Remove the 2 bolts, 2 nuts, 2 washers and upper control arm.

To install:

7. Temporarily install the upper control arm to the frame with the 2 bolts, 2 washers and 2 nuts.
8. Temporarily install the upper control arm ball joint axle carrier with the nut.

➡After stabilizing the suspension, tighten the nuts.

9. Connect the speed sensor to the axle carrier with the bolt.
10. Install the wheel with the 5 nuts.
11. Stabilize the suspension.
12. Tighten the upper control arm with the 2 bolts and 2 nuts. Tighten "A" to 65 ft. lbs. (88 Nm). Tighten "B" to 55 ft. lbs. (74 Nm).
13. Tighten the ball joint nut and tighten to 80 ft. lbs. (108 Nm).
14. Inspect and adjust the rear wheel alignment.
15. Install the ornament with the 10 bolts.
16. Check speed sensor signal.

WHEEL HUB & BEARING

REMOVAL & INSTALLATION

See Figure 144.

1. Before servicing the vehicle, refer to the Precautions section.
2. Remove or disconnect the following:
   - Rear tire and wheel assembly
   - Brake caliper support bracket
   - Brake rotor
   - Speed sensor
   - Rear halfshaft
   - Parking brake shoes
   - 2 bolts at the parking brake cable. Remove the 2 hub bolts and the hex bolt. Slide the backing plate to the outside and disconnect the parking brake cable.
   - Strut rod at the axle carrier
   - Nut, then press out the upper suspension arm.
   - Nut, then press out the No. 2 lower suspension arm
   - Axle carrier
   - Dust deflector and pull out the oil seal
   - Axle hub from the carrier
   - Backing plate
   - Inner race (outside) from the hub
   - Oil seal
   - Snapring
   - Bearing and inner race (inside)

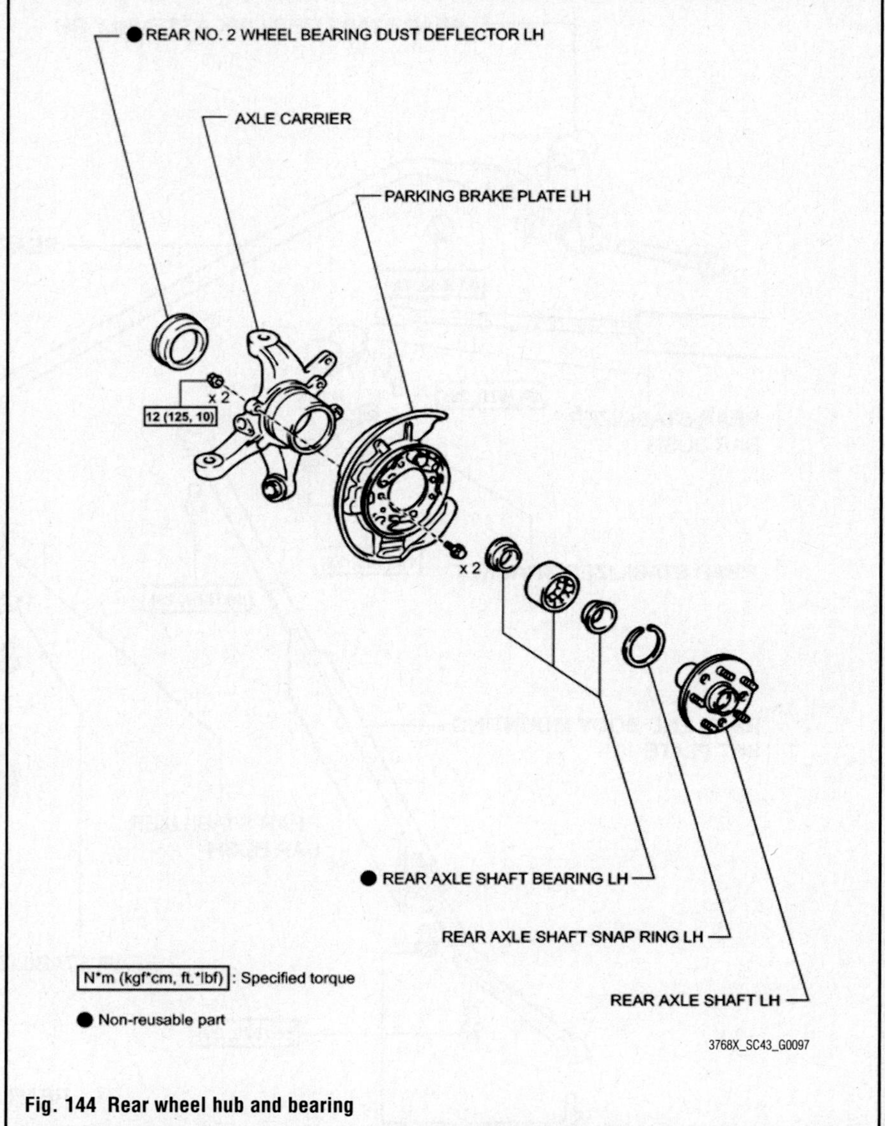

Fig. 144 Rear wheel hub and bearing

To install:

3. Install or connect the following:
   - Bearing to the axle carrier

➡If the inner races come loose from the bearing outer race, be sure to install them on the same side as before.

   - Snapring, the inner race (outside) and a new oil seal
   - Backing plate. Install the inner race (inside) and press in the axle hub with the proper tools.
   - New dust deflector
   - Upper arm to the axle carrier. Tighten the nut and bolt to 65 ft. lbs (88 Nm), rear 55 ft. lbs (74 Nm)
   - No. 2 lower arm to the carrier and tighten a new nut to 81 ft. lbs. (110 Nm).
   - Toe control link with camber adjusting cams. Tighten to 36 ft. lbs. (49 Nm). Stabilize and retighten to 44 ft. lbs. (59 Nm).
   - Rear drive shaft assembly
   - Parking brake cable and brake assembly
   - Install the parking brake shoes and the ABS sensor.
   - Brake rotor
   - Brake caliper to the rear axle carrier by installing the 2 bolts.
   - Tighten rear suspension arm assembly to 81 ft. lbs. (110 Nm)
4. Inspect and adjust to rear alignment
5. Perform speed sensor signal check.

LEXUS

Diagnostic Trouble Codes

33

DIAGNOSTIC TROUBLE CODES

OBD II VEHICLE APPLICATIONS

LEXUS

ES350
2009–2010
- 3.5L V6. . . . . 2GR-FE

GS350
2009–2010
- 3.5L V6. . . . . 2GR-FSE

GS460
2009–2010
- 4.6L V8. . . . . 1UR-FSE

GX470
2009
- 4.7L V8. . . . . 2UZ-FE

IS250
2009–2010
- 2.5L V6. . . . . 4GR-FSE

IS350
2009–2010
- 3.5L V6. . . . . 2GR-FSE

LS460
2009–2010
- 4.6L V8. . . . . 1UR-FSE

RX350
2009–2010
- 3.3L V6. . . . . 2GR-FE

SC430
2009–2010
- 4.3L V8. . . . . 3UZ-FE

OBD II Trouble Code List (P0XXX Codes)

DTC	Trouble Code Title, Conditions & Possible Causes
DTC: P0010 **1T ECM, MIL: Yes** **Year:** 2009, 2010 **Model:** All **Engine:** All V6, V8 **Transmission:** All	**Camshaft Position "A" Actuator Circuit (Bank 1):** All of the following conditions are met: - Starter: OFF Power switch: On (IG) Time after turning power switch off to on (IG): 0.5 seconds or more **Possible Causes:** • Open or short in camshaft timing oil control valve assembly (bank 1) circuit • Camshaft timing oil control valve assembly (bank 1) • ECM
DTC: P0011 **1T ECM, MIL: Yes** **Year:** 2009, 2010 **Model:** All **Engine:** All V6, V8 **Transmission:** All	**Camshaft Position "A" - Timing Over-Advanced or System Performance (Bank 1):** Valve timing is not adjusted in valve timing advance. Battery Voltage: 11V or more Engine: 500-4000 rpm ECT: 167-212 degrees F (75-100 degrees C) **Possible Causes:** • Valve timing • Oil Control Valve (OCV) for intake camshaft (Bank 1) • OCV filter (Bank 1) • Intake camshaft timing gear assembly (Bank 1) • ECM
DTC: P0012 **2T ECM, MIL: Yes** **Year:** 2009, 2010 **Model:** All **Engine:** All V6,V8 **Transmission:** All	**Camshaft Position "A" - Timing Over-Retarded (Bank 1):** Monitor runs whenever following DTCs not present: P0010, P0020 (VVT Oil Control Valve) P0016, P0018 (VVT System - Misalignment) P0102, P0103 (Mass Air Flow Sensor) P0115, P0117, P0118 (Engine Coolant Temperature Sensor) P0125 (Insufficient Coolant Temperature for Closed Loop Fuel Control) P0335 (Crankshaft Position Sensor) P0340 (Camshaft Position Sensor) Battery voltage: 11 V or higher Engine speed: 400 to 4000 rpm Engine coolant temperature sensor: 75°C (167°F) to 100°C (212°F) **Possible Causes:** • Valve timing • Camshaft timing oil control valve for intake camshaft • Oil control valve filter • Camshaft timing gear assembly • ECM
DTC: P0013 **1T ECM, MIL: Yes** **Year:** 2009, 2010 **Model:** All **Engine:** All V6, V8 **Transmission:** All	**Camshaft Position "B" Actuator Circuit / Open (Bank 1):** Monitor runs whenever following DTCs not present: None All of the following conditions are met: Starter: OFF Engine switch: On (IG) Time after engine switch off to on (IG): 0.5 seconds or more **Possible Causes:** • Open or short in OCV for exhaust camshaft (bank 1) circuit • OCV for exhaust camshaft (bank 1) • ECM
DTC: P0014 **2T ECM, MIL: Yes** **Year:** 2009, 2010 **Model:** All **Engine:** All V6, V8 **Transmission:** All	**Camshaft Position "B" - Timing Over-Advanced or System Performance (Bank 1):** Battery Voltage: 11V or more Engine: 500-4000 rpm Engine Coolant Temperature: 167-212 degrees F (75-100 degrees C) **Possible Causes:** • Valve timing • OCV for exhaust camshaft (bank 1, 2) • OCV filter (bank 1, 2) • Exhaust camshaft timing gear assembly (bank 1, 2) • ECM

DTC	Trouble Code Title, Conditions & Possible Causes
DTC: P0015 **1T ECM, MIL: Yes** **Year:** 2009, 2010 **Model:** All **Engine:** All V6, V8 **Transmission:** All	**Camshaft Position "B" - Timing Over-Retarded (Bank 1):** P0335 (CKP sensor) Battery voltage: 11 V or more Engine rpm: 500 to 4000 rpm ECT: 167 to 212°F (75 to 100°C) **Possible Causes:** • Valve timing • OCV for exhaust camshaft (bank 1, 2) • OCV filter (bank 1, 2) • Exhaust camshaft timing gear assembly • ECM
DTC: P0016 **2T ECM** **Year:** 2009,2010 **Model:** All **Engine:** All V6, V8 **Transmission:** All	**Crankshaft Position - Camshaft Position Correlation (Bank 1 Sensor A):** Engine speed: 800 to 1000 rpm **Possible Causes:** • Valve timing • Camshaft timing oil control valve assembly (bank 1, 2) • Oil control valve filter (bank 1, 2) • Camshaft timing gear assembly • ECM
DTC: P0017 **2T ECM, MIL: Yes** **Year:** 2009, 2010 **Model:** All **Engine:** All V6,V8 **Transmission:** All	**Crankshaft Position - Camshaft Position Correlation (Bank 1 Sensor B):** VVT feedback mode: Executing VVT: Maximum advance position Engine rpm: 500 to 1000 rpm **Possible Causes:** • Valve timing • Oil control valve for exhaust side (bank 1) • Camshaft timing exhaust gear assembly (bank 1)
DTC: P0018 **2T ECM** **Year:** 2010 **Model:** All **Engine:** All V6, V8 **Transmission:** All	**Crankshaft Position - Camshaft Position Correlation (Bank 2 Sensor A):** Engine speed: 800 to 1000 rpm **Possible Causes:** • Valve timing • Camshaft timing oil control valve assembly (bank 1, 2) • Oil control valve filter (bank 1, 2) • Camshaft timing gear assembly • ECM
DTC: P0019 **2T ECM, MIL: Yes** **Year:** 2009, 2010 **Model:** All **Engine:** All V6, V8 **Transmission:** All	**Crankshaft Position - Camshaft Position Correlation (Bank 2 Sensor B):** VVT feedback mode: Executing VVT: Maximum advanced position Engine rpm: 500 to 1000 rpm **Possible Causes:** • Valve timing • Camshaft timing oil control valve for exhaust camshaft • Camshaft timing oil control valve filter • Exhaust camshaft timing gear assembly • ECM
DTC: P0020 **1T ECM, MIL: Yes** **Year:** 2009, 2010 **Model:** All **Engine:** All V6, V8 **Transmission:** All	**Camshaft Position "A" Actuator Circuit (Bank 2):** All of the following conditions are met:- Starter: OFF Power switch: On (IG) Time after turning power switch off to on (IG): 0.5 seconds or more **Possible Causes:** • Open or short in camshaft timing oil control valve assembly (bank 2) circuit • Camshaft timing oil control valve assembly (bank 2) • ECM

DTC	Trouble Code Title, Conditions & Possible Causes
DTC: P0021 **1T ECM, MIL: Yes** **Year:** 2009, 2010 **Model:** All **Engine:** All V6, V8 **Transmission:** All	**Camshaft Position "A" - Timing Over-Advanced or System Performance (Bank 2):** Battery voltage: 11 V or more Engine rpm: 500 to 4000 rpm ECT: 167 to 212°F (75 to 100°C) **Possible Causes:** • Valve timing • Oil Control Valve (OCV) for intake camshaft (bank 1, 2) • OCV filter (bank 1, 2) • Intake camshaft (bank 1, 2) timing gear assembly • ECM
DTC: P0022 **2T ECT, MIL: Yes** **Year:** 2009, 2010 **Model:** All **Engine:** All V6, V8 **Transmission:** All	**Camshaft Position "A" - Timing Over-Retarded (Bank 2):** Battery Voltage: 11V or more Engine: 500-4000 rpm ECT: 167-212 degrees F (75-100 degrees C) **Possible Causes:** • Valve timing • OCV for intake camshaft (bank 1, 2) • OCV filter (bank 1, 2) • Intake camshaft timing gear assembly (bank 1, 2) • ECM
DTC: P0023 **1T ECM, MIL: Yes** **Year:** 2009, 2010 **Model:** All **Engine:** All V6, V8 **Transmission:** All	**Camshaft Position "B" Actuator Circuit / Open (Bank 2):** Starter: OFF Engine Switch: On (IG) Time after turning engine switch off to on (IG): 0.5 seconds or more **Possible Causes:** • Open or short in OCV for exhaust camshaft (bank 2) circuit • OCV for exhaust camshaft (bank 2) • ECM
DTC: P0024 **2T ECM, MIL: Yes** **Year:** 2009, 2010 **Model:** All **Engine:** All V6 **Transmission:** All	**Camshaft Position "B" - Timing Over-Advanced or System Performance (Bank 2):** Battery Voltage: 11V or more Engine: 500-4000 rpm Engine Coolant Temperature: 167-212 degrees F (75-100 degrees C) **Possible Causes:** • Valve timing • OCV for exhaust camshaft (bank 1, 2) • OCV filter (bank 1, 2) • Exhaust camshaft timing gear assembly (bank 1, 2) • ECM
DTC: P0025 **1T , MIL: Yes** **Year:** 2009, 2010 **Model:** All **Engine:** All V6, V8 **Transmission:** All	**Camshaft Position "B" - Timing Over-Retarded (Bank 2):** Battery voltage: 11 V or more Engine rpm: 500 to 4000 rpm ECT: 167 to 212°F (75 to 100°C) **Possible Causes:** • Valve timing • OCV for exhaust camshaft (bank 1, 2) • OCV filter (bank 1, 2) • Exhaust camshaft timing gear assembly • ECM
DTC: P0031 **1T ECM, MIL: Yes** **Year:** 2009, 2010 **Model:** All **Engine:** All V6, V8 **Transmission:** All	**Oxygen (A/F) Sensor Heater Control Circuit Low (Bank 1 Sensor 1):** Battery voltage: 10.5 V or more A/F sensor heater duty-cycle ratio: 50% or more Time after engine start: 10 seconds or more **Possible Causes:** • Open in A/F sensor heater (bank 1, 2 sensor 1) circuit • A/F sensor heater (bank 1, 2 sensor 1) • A/F fuse • Engine room No. 2 junction block • ECM

DTC	Trouble Code Title, Conditions & Possible Causes
DTC: P0032 **1T ECM** **Year:** 2009, 2010 **Model:** All **Engine:** All V6, V8 **Transmission:** All	**Oxygen (A/F) Sensor Heater Control Circuit High (Bank 1 Sensor 1):** Battery voltage: 10.5 V or more Heater output duty: 0% or more Time after engine start: 10 seconds or more Active heater off control: Not operating Active heater on control: Not operating **Possible Causes:** • Short in air fuel ratio sensor heater (bank 1, 2 sensor 1) circuit • Air fuel ratio sensor heater (bank 1, 2 sensor 1) • A/F fuse • Engine room junction block assembly (A/F relay) • ECM
DTC: P0037 **1T ECM, MIL: Yes** **Year:** 2009, 2010 **Model:** All **Engine:** All V6, V8 **Transmission:** All	Oxygen Sensor Heater Control Circuit Low (Bank 1 Sensor 2): Battery voltage: 10.5 to 20 V **Possible Causes:** • Open in HO2 sensor heater circuit • HO2 sensor heater • ECM
DTC: P0038 **1T ECM, MIL: Yes** **Year:** 2009, 2010 **Model:** All **Engine:** All V6, V8 **Transmission:** All	**Oxygen Sensor Heater Control Circuit High (Bank 1 Sensor 2):** Case 1: Battery voltage: 10.5 V or more, Engine: Running, Starter: OFF Case 2: Battery voltage: 10.5 V or more, nad less than 20 V **Possible Causes:** • Short in HO2 sensor heater circuit • HO2 sensor heater • Integration relay • ECM
DTC: P0051 **1T , MIL: Yes** **Year:** 2009, 2010 **Model:** All **Engine:** All V6, V8 **Transmission:** All	**Oxygen (A/F) Sensor Heater Control Circuit Low (Bank 2 Sensor 1):** Battery voltage: 10.5 V or more Heater output duty: 50% or more Time after engine start: 10 seconds or more Active heater off control: Not operating Active heater on control: Not operating **Possible Causes:** • Open in A/F sensor heater circuit (bank 1, 2, sensor 1) • A/F sensor heater (bank 1, 2, sensor 1) • A/F sensor heater relay • ECM
DTC: P0052 **1T ECM, MIL: Yes** **Year:** 2009, 2010 **Model:** All **Engine:** All V6, V8 **Transmission:** All	Oxygen (A/F) Sensor Heater Control Circuit High (Bank 2 Sensor 1): Time after engine start: 10 seconds or more **Possible Causes:** • Short in A/F sensor heater (bank 1, 2 sensor 1) circuit • A/F sensor heater (bank 1, 2 sensor 1) • A/F relay • ECM
DTC: P0057 **1T ECM, MIL: Yes** **Year:** 2009, 2010 **Model:** All **Engine:** All **Transmission:** All	Oxygen Sensor Heater Control Circuit Low (Bank 2 Sensor 2): Battery voltage: 10.5 V or more, and less than 20 VP0057 **Possible Causes:** • Open in Heated Oxygen (HO2) sensor heater circuit • HO2 sensor heater • Integration relay • ECM

DTC	Trouble Code Title, Conditions & Possible Causes
DTC: P0058 **1T ECM** **Year:** 2010 **Model:** All **Engine:** All V6, V8 **Transmission:** All	**Oxygen Sensor Heater Control Circuit High (Bank 2 Sensor 2):** Battery voltage: 10.5 V or more Engine: Running Starter: OFF Catalyst active air fuel ratio control: Not operating Time after heater on: 10 seconds or more Learned heater off current operation: Complete **Possible Causes:** • Short in heated oxygen sensor heater (bank 1, 2 sensor 2) circuit • Heated oxygen sensor heater (bank 1, 2 sensor 2) • ECM
DTC: P0085 **1T ECM, MIL: Yes** **Year:** 2009, 2010 **Model:** All **Engine:** All V8 **Transmission:** All	**Intake Air Temperature Sensor 2 Circuit High:** Monitor runs whenever following DTCs not present: None **Possible Causes:** • Open in IAT sensor (bank 2) circuit • IAT sensor (bank 2) (built into MAF meter [Bank 2]) • ECM
DTC: P0087 **1T ECM, MIL: Yes** **Year:** 2009, 2010 **Model:** All **Engine:** All V6, V8 **Transmission:** All	**Fuel Rail / System Pressure - Too Low:** Time after engine start: 0.2 seconds or more Time after fuel cut finished: 3 seconds or more Target fuel pressure: Small change **Possible Causes:** • Leak of fuel • Fuel pipe (Fuel tank - Fuel pump for high pressure [bank 1 or bank 2]) • Fuel pipe (Fuel pump for high pressure - Fuel injector [for direct injection]) • Fuel injector (for direct injection) • Fuel relief valve • Fuel pressure sensor • Fuel pump for low pressure • Fuel pump for high pressure (bank 1) • Fuel pump for high pressure (bank 2) • Injector driver (No. 1) • Injector driver (No. 2) • ECM
DTC: P0088 **T ECM, MIL: Yes** **Year:** 2009, 2010 **Model:** All **Engine:** All **Transmission:** All	**Fuel Rail / System Pressure - Too High:** Time after engine start: 0.2 seconds or more Time after fuel cut finished: 3 seconds or more Target fuel pressure: Small change **Possible Causes:** • Fuel pump for high pressure • Fuel pressure sensor • ECM
DTC: P0092 **1T ECM, MIL: Yes** **Year:** 2009, 2010 **Model:** All **Engine:** All V6 **Transmission:** All	**Camshaft Position Sensor "B" Circuit Low Input (Bank 2):** Starter: OFF Engine switch: On (IG) Time after engine switch off to on (IG): 2 seconds or more Exhaust VVT sensor verify pulse input fail (P0365, P0390): Not detected Battery voltage: 8 V or more **Possible Causes:** • Open or short in VVT sensor for exhaust camshaft circuit • VVT sensor for exhaust camshaft • Exhaust camshaft • ECM

DTC	Trouble Code Title, Conditions & Possible Causes
DTC: P0096 **2T ECM, MIL: Yes** **Year:** 2009, 2010 **Model:** All **Engine:** All V8 **Transmission:** All	**Intake Air Temperature Sensor 2 Circuit Range / Performance:** Monitor runs whenever following DTCs are not present: P0115, P0117, P0118 (Engine Coolant Temperature Sensor) P0102, P0103, P010C, P010D (Mass Air Flow Sensor) Battery voltage: 10.5 V or more After engine stop: Time after engine start: 10 seconds or more ECT sensor circuit: OK ECT in previous driving cycle: 70°C (158°F) or more MAF sensor circuit: OK Accumulated MAF amount in previous driving cycle: 2806 g or more ECT when 30 minutes elapsed after engine stop: -40°C (68°F) or more After cold engine start: Key-off duration: 5 hours Time after engine start: 10 seconds or more ECT sensor circuit: OK ECT: 70°C (158°F) or more MAF sensor circuit: OK Accumulated MAF amount: 2806 g or more One of the following conditions 1 or 2 is met: 1. Duration while engine load is low: 120 seconds or more 2. Duration while engine load is high: 10 seconds or more **Possible Causes:** • Mass air flow (MAF) meter (bank 1) • Mass air flow (MAF) meter (bank 2)
DTC: P0097 **1T ECM, MIL: Yes** **Year:** 2009, 2010 **Model:** All **Engine:** All V8 **Transmission:** All	**Intake Air Temperature Sensor 2 Circuit Low:** Monitor runs whenever following DTCs not present: None **Possible Causes:** • Short in IAT sensor (bank 2) circuit • IAT sensor (bank 2) (built into MAF meter [Bank 2]) • ECM
DTC: P0100 **T ECM, MIL: Yes** **Year:** 2009, 2010 **Model:** All **Engine:** All **Transmission:** All	**Mass or Volume Air Flow Circuit:** Monitor runs whenever following DTCs are not present:None **Possible Causes:** • Open or short Mass Air Flow (MAF) meter circuit • MAF meter • ECM
DTC: P0101 **2T ECM, MIL: Yes** **Year:** 2009, 2010 **Model:** All **Engine:** All **Transmission:** All	**Mass Air Flow Circuit Range / Performance Problem:** TP (Throttle position) sensor voltage: 0.24 to 2 V Engine: Running Battery voltage: 10.5 V or more ECT: 158°F (70°C) or more Estimated load: 30 to 70% **Possible Causes:** • Mass Air Flow (MAF) meter • Air induction system • PCV hose connections • EGR valve assembly
DTC: P0102 **T ECM, MIL: Yes** **Year:** 2009, 2010 **Model:** All **Engine:** All V6, V8 **Transmission:** All	Mass or Volume Air Flow Circuit Low Input: Monitor runs whenever following DTCs not present: None **Possible Causes:** • Open in MAF meter circuit • Short in MAF meter circuit • MAF meter • ECM

DTC	Trouble Code Title, Conditions & Possible Causes
DTC: P0103 **T ECM** **Year:** 2009, 2010 **Model:** All **Engine:** All V6 **Transmission:** All	**Mass or Volume Air Flow Circuit High Input:** Monitor runs whenever following DTCs not present: None **Possible Causes:** • Open or short in mass air flow meter sub-assembly circuit • Mass air flow meter sub-assembly • ECM
DTC: P0103 **T ECM, MIL: Yes** **Year:** 2009, 2010 **Model:** All **Engine:** All V8 **Transmission:** All	**Mass Air Flow Circuit High:** Monitor runs whenever following DTCs not present: None **Possible Causes:** • Open in MAF meter (bank 1, 2) circuit • Short in ground circuit • MAF meter (bank 1, 2) • ECM
DTC: P0107 **1T ECM** **Year:** 2009, 2010 **Model:** All **Engine:** All V6, V8 **Transmission:** All	**Manifold Absolute Pressure / Barometric Pressure Circuit Low Input:** Monitor runs whenever following DTCs not present: None Starter: Off Time after starter on to off: 2 seconds or more **Possible Causes:** • Open or short in manifold absolute pressure sensor circuit • Manifold absolute pressure sensor • ECM
DTC: P010C **T ECM, MIL: Yes** **Year:** 2009, 2010 **Model:** LS460 **Engine:** 4.6L V8 **Transmission:** All	**Mass or Volume Air Flow "A" Circuit Low Input:** Monitor runs whenever following DTCs not present: None **Possible Causes:** • Open in MAF meter (bank 1, 2) circuit • Short in ground circuit • MAF meter (bank 1, 2) • ECM
DTC: P010D **T ECM, MIL: Yes** **Year:** 2009, 2010 **Model:** LS460 **Engine:** 4.6L V8 **Transmission:** All	**Mass or Volume Air Flow "A" Circuit High Input:** Monitor runs whenever following DTCs not present: None **Possible Causes:** • Short in MAF meter circuit (+B circuit) • MAF meter (bank 1, 2) • ECM
DTC: P010F **2T ECM, MIL: Yes** **Year:** 2009, 2010 **Model:** LS460 **Engine:** 4.6L V8 **Transmission:** All	**Mass or Volume Air Flow Sensor:** Throttle position (TP sensor voltage): 0.24 to 3.6 V Time after engine start: 5 seconds or more Battery voltage: 10.5 V or more Engine coolant temperature: 70°C (158°F) or more Estimated load: 30 to 100% IAT sensor circuit: OK ECT sensor circuit: OK CKP sensor circuit: OK TP sensor circuit: OK Canister pressure sensor circuit: OK EVAP leak detection pump: OK EVAP vent valve: OK **Possible Causes:** • Mass Air Flow (MAF) meter (bank 1, 2) • Air induction system • Ventilation hose connections
DTC: P0110 **T ECM, MIL: Yes** **Year:** 2009 **Model:** All **Engine:** All **Transmission:** All	**Intake Air Temperature Circuit:** Monitor will run whenever these DTCs are not present: None The typical enabling condition is not available: - **Possible Causes:** • Open or short in Intake Air Temperature (IAT) sensor circuit • Intake Air Temperature (IAT) sensor (built into Mass Air Flow [MAF] meter) • ECM

DTC	Trouble Code Title, Conditions & Possible Causes
DTC: P0111 **2T ECM, MIL: Yes** **Year:** 2009, 2010 **Model:** IS250, IS350 **Engine:** 2.5L V6, 3.5L V6 **Transmission:** All	**Intake Air Temperature Sensor Gradient Too High :** Battery voltage: 10.5 V or more After engine stop: Time after engine start: 10 seconds or more ECT sensor circuit (P0115, P0117, P0118): OK ECT change since engine stop: Less than 180°C (324°F) ECT before engine stop: 70°C (158°F) or more Time that MAF is low before engine stop: Less than 70 minutes Accumulated MAF amount before engine stop: 3800 g or more Key-off duration: 30 minutes After cold enigne start: Key-off duration: 5 hours or more Time after engine start: 10 seconds or more ECT sensor circuit (P0115, P0117, P0118): OK ECT: 70°C (158°F) or more Accumulated MAF amount: 3800 g or more Either of the following conditions is met: Condition a or b a. Duration while engine load is low: 120 second or more b. Duration while engine load is high: 10 seconds or more **Possible Causes:** • MAF meter
DTC: P0111 **2T ECM, MIL: Yes** **Year:** 2009, 2010 **Model:** LS460 **Engine:** 4.6L V8 **Transmission:** All	**Intake Air Temperature Sensor 1 Circuit Range / Performance:** Monitor runs whenever following DTCs are not present: P0115, P0117, P0118 (Engine Coolant Temperature Sensor) P0102, P0103, P010C, P010D (Mass Air Flow Sensor) Battery voltage: 10.5 V or more After engine stop: Time after engine start: 10 seconds or more ECT sensor circuit: OK ECT in previous driving cycle: 70°C (158°F) or more MAF sensor circuit: OK Accumulated MAF amount in previous driving cycle: 2806 g or more ECT when 30 minutes elapsed after engine stop: -40°C (68°F) or more After cold engine start: Key-off duration: 5 hours Time after engine start: 10 seconds or more ECT sensor circuit: OK ECT: 70°C (158°F) or more MAF sensor circuit: OK Accumulated MAF amount: 2806 g or more One of the following conditions 1 or 2 is met: 1. Duration while engine load is low: 120 seconds or more 2. Duration while engine load is high: 10 seconds or more **Possible Causes:** • Mass air flow (MAF) meter (bank 1) • Mass air flow (MAF) meter (bank 2)

DTC	Trouble Code Title, Conditions & Possible Causes
DTC: P0111 **2T ECM, MIL: Yes** **Year:** 2009, 2010 **Model:** ES350, GS350 **Engine:** 3.5L V6 **Transmission:** All	**Intake Air Temperature Sensor Gradient Too High:** After Engine Stop: Time after engine start: 10 seconds or more Battery voltage: 10.5 V or more ECT change since engine stop: -40°F (-40°C) or more Accumulated MAF amount before engine stop: 2033 g or more Key-off duration: 30 minutes After Cold Engine Start: Key-off duration:5 hours Time after engine start: 10 seconds or more ECT: 158° F (70° C) or more Accumulated MAF amount: 2033 g or more One of the following conditions 1 or 2 is met: 1. Duration while engine load is low: 120 seconds or more 2. Duration while engine load is high: 10 seconds or more **Possible Causes:** • MAF Meter
DTC: P0112 **1T ECM** **Year:** 2010 **Model:** All **Engine:** All **Transmission:** All	**Intake Air Temperature Circuit Low Input:** Battery voltage: 8 V or more Power switch: On (IG) Time after power switch off to on (IG): More than 0.5 seconds **Possible Causes:** • Short in intake air temperature sensor circuit • Intake air temperature sensor (built into mass air flow meter sub-assembly) • ECM
DTC: P0113 **1T ECM, MIL: Yes** **Year:** 2009, 2010 **Model:** All **Engine:** All V6, V8 **Transmission:** All	**Intake Air Temperature Circuit High Input:** Battery voltage: 8 V or more Power switch: On (IG) Time after power switch off to on (IG): More than 0.5 seconds **Possible Causes:** • Open in intake air temperature sensor circuit • Intake air temperature sensor (built into mass air flow meter sub-assembly) • ECM
DTC: P0115 **1T ECM, MIL: Yes** **Year:** 2009, 2010 **Model:** All **Engine:** All V6, V8 **Transmission:** All	**Engine Coolant Temperature Circuit Malfunction:** Engine coolant temperature sensor voltage: Less than 0.14 V, or more than 4.91 V **Possible Causes:** • Open or short in ECT sensor circuit • ECT sensor • ECM

DTC	Trouble Code Title, Conditions & Possible Causes
DTC: P0116 **T , MIL: Yes** **Year:** 2009, 2010 **Model:** All **Engine:** All **Transmission:** All	**Engine Coolant Temperature Circuit Range / Performance Problem:** ECT Sensor cold start monitor: Monitor runs whenever following DTCs not present: P0097, P0098, P0112, P0113 (Intake Air Temperature Sensor) Battery voltage: 10.5 V or more Time after engine start: 1 second or more ECT at engine start: Less than 60°C (140°F) IAT sensor circuit: OK Soak time: 0 or more Accumulated MAF: 1534 g or more Fuel cut: OFF Difference between ECT at engine start and IAT: Less than 40°C (72°F) ECT Sensor soak monitor: Monitor runs whenever following DTCs not present: P0097, P0098, P0112, P0113 (Intake Air Temperature Sensor) Battery voltage: 10.5 V or more Engine: Running Soak time: 5 hours or more Either (a) or (b) condition met: (a) ECT: 60°C (140°F) or more (b) Accumulated MAF: 3420 g or more **Possible Causes:** • Thermostat • ECT sensor
DTC: P0117 **1T ECM, MIL: Yes** **Year:** 2009, 2010 **Model:** All **Engine:** All V6, V8 **Transmission:** All	**Engine Coolant Temperature Circuit Low Input:** Engine coolant temperature sensor voltage: Less than 0.14 V **Possible Causes:** • Short in ECT sensor circuit • ECT sensor • ECM
DTC: P0118 **1T ECM, MIL: Yes** **Year:** 2009, 2010 **Model:** All **Engine:** All V6, V8 **Transmission:** All	**Engine Coolant Temperature Circuit High Input:** Engine coolant temperature sensor voltage: More than 4.91 V **Possible Causes:** • Open in ECT sensor circuit • ECT sensor • ECM
DTC: P011B **2T ECM, MIL: Yes** **Year:** 2009, 2010 **Model:** LS460 **Engine:** 4.6L V8 **Transmission:** All	**Engine Coolant Temperature / Intake Air Temperature Correlation:** The monitor will run whenever these DTCs are not present: P2610 (ECM internal engine off timer) P0102, P0103, P010C, P010D (Mass Air Flow Sensor) Both of the following conditions are met: Conditions 1 and 2 1. All of the following conditions are met: Conditions (a), (b), (c) and (d) (a) After engine switch on (IG) and engine not running time: Less than 20 seconds (b) Soak Time: 7 hours or more (c) Battery voltage: 10.5 V or more (d) Time after engine start: 15 seconds or more 2. Either of the following conditions is met: Condition (a) and (b) (a) Minimum intake air temperature after engine start: -10°C (14°F) or more (b) Engine coolant temperature before engine start: -10°C (14°F) or more **Possible Causes:** • Intake air temperature sensor • Engine coolant temperature sensor • ECM

DTC	Trouble Code Title, Conditions & Possible Causes
DTC: P011B **2T ECM, MIL: Yes** **Year:** 2009, 2010 **Model:** All **Engine:** All V6, V8 **Transmission:** All	**Engine Coolant Temperature / Intake Air Temperature Correlation:** All of following conditions met: Conditions 1 and 2 1. All of the following conditions are met: Conditions (a), (b), (c) and (d) (a) After engine switch on (IG) and engine not running time: Less than 20 seconds (b) Soak Time: 7 hours or more (c) Battery voltage: 10.5 V or more (d) Time after engine start: 15 seconds or more 2. Either of the following conditions are met: Condition (a) and (b) (a) Minium IAT after engine start: 14° F (-10° C) or more (b) ECT before engine start: 14° F (-10° C) or more **Possible Causes:** • IAT sensor • ECT sensor • ECM
DTC: P0120 **1T ECM, MIL: Yes** **Year:** 2009 **Model:** GX470 **Engine:** 4.7L V8 **Transmission:** All	**Throttle / Pedal Position Sensor / Switch "A" Circuit Malfunction:** Monitor runs whenever following DTCs not present: None Either of following conditions A or B met: A. Ignition switch ON: 0.012 seconds or more B. Electronic throttle actuator power: ON **Possible Causes:** • Throttle position sensor (built into throttle body with motor assembly) • ECM
DTC: P0120 **1T ECM, MIL: Yes** **Year:** 2009, 2010 **Model:** LS460 **Engine:** 4.6L V8 **Transmission:** All	**Throttle / Pedal Position Sensor / Switch "A" Circuit Malfunction:** Monitor runs whenever following DTCs not present: None **Possible Causes:** • Throttle Position (TP) sensor (built into throttle body) • ECM
DTC: P0120 **1T ECM, MIL: Yes** **Year:** 2009, 2010 **Model:** ES350, GS350 **Engine:** 3.5L V6 **Transmission:** All	**Throttle / Pedal Position Sensor / Switch "A" Circuit Malfunction:** Either of the following conditions A or B is met: A. Engine switch on (IG): 0.012 seconds or more B. Electronic throttle actuator power: ON **Possible Causes:** • Throttle position sensor (built into throttle body) • ECM
DTC: P0120 **1T ECM, MIL: Yes** **Year:** 2009, 2010 **Model:** GS350, IS250, IS350 **Engine:** 2.5L V6, 3.5L V6 **Transmission:** All	**Throttle / Pedal Position Sensor / Switch "A" Circuit Malfunction:** Monitor runs whenever following DTCs not present: None **Possible Causes:** • Throttle Position (TP) sensor (built into throttle body) • ECM
DTC: P0121 **1T ECM, MIL: Yes** **Year:** 2009, 2010 **Model:** All **Engine:** All V6, V8 **Transmission:** All	**Throttle / Pedal Position Sensor / Switch "A" Circuit Range / Performance Problem:** Either of the following conditions A or B is met: A. Engine switch: On (IG) B. Electric throttle motor power: ON Throttle position sensor malfunction (P0120, P0122, P0123, P0220, P0222, P0223, P2135): not detected **Possible Causes:** • Throttle position sensor (built into throttle body) • Throttle position sensor circuit • ECM
DTC: P0122 **1T ECM, MIL: Yes** **Year:** 2009, 2010 **Model:** All **Engine:** All V6, V8 **Transmission:** All	**Throttle / Pedal Position Sensor / Switch "A" Circuit Low Input:** Either of the following conditions A or B is met: A. Engine switch on (IG): 0.012 seconds or more B. Electronic throttle actuator power: ON **Possible Causes:** • Throttle position sensor (built into throttle body) • Short in VTA1 circuit • Open in VC circuit • ECM

DTC	Trouble Code Title, Conditions & Possible Causes
DTC: P0123 **1T ECM, MIL: Yes** **Year:** 2009, 2010 **Model:** All **Engine:** All V6 **Transmission:** All	**Throttle / Pedal Position Sensor / Switch "A" Circuit High Input:** Either of the following conditions A or B is met: A. Engine switch on (IG): 0.012 seconds or more B. Electronic throttle actuator power: ON **Possible Causes:** • Throttle position sensor (built into throttle body) • Open in VTA1 circuit • Open in E2 circuit • Short between VC and VTA1 circuits • ECM
DTC: P0125 **T ECM, MIL: Yes** **Year:** 2009, 2010 **Model:** All **Engine:** All V6, V8 **Transmission:** All	**Insufficient Coolant Temperature for Closed Loop Fuel Control:** Monitor runs whenever following DTCs are not present: MAF sensor circuit fail (P0102, P0103) IAT sensor circuit fail (P0112, P0113) ECT sensor circuit fail (P0115, P0117, P0118) Thermostat fail (P0128) **Possible Causes:** • Engine coolant temperature sensor • Cooling system • Thermostat
DTC: P0128 **2T ECM, MIL: Yes** **Year:** 2009, 2010 **Model:** All **Engine:** All V6, V8 **Transmission:** All	**Coolant Thermostat (Coolant Temperature Below Thermostat Regulating Temperature):** Battery voltage: 11 V or more Either of the following conditions 1 or 2 is met: 1. All of the following conditions are met: * ECT at engine start - IAT at engine start: -15 to 7°C (-27 to 12.6°F) * ECT at engine start: -10 to 56°C (14 to 133°F) * IAT at engine start: -10 to 56°C (14 to 133°F) 2. All of the following conditions are met: * ECT at engine start - IAT at engine start: More than 7°C (12.6°F) * ECT at engine start: 56°C (133°F) or less * IAT at engine start: -10°C (14°F) or more Accumulated time that vehicle speed is 80 mph (128 km/h) or more: Less than 20 seconds **Possible Causes:** • Thermostat • Cooling system • ECT sensor • ECM

DTC	Trouble Code Title, Conditions & Possible Causes
DTC: P0136 **2T ECM, MIL: Yes** **Year:** 2009, 2010 **Model:** All **Engine:** All **Transmission:** All	**Oxygen Sensor Circuit Malfunction (Bank 1 Sensor 2):** Heated Oxygen Sensor Output Voltage (Output Voltage, High Voltage and Low Voltage): Active air-fuel ratio control: Executing Active air-fuel ratio control begins when all of following conditions met: Battery voltage: 11 V or more Engine coolant temperature: 75°C (167°F) or more Idling: OFF Engine RPM: Less than 3200 rpm A/F sensor status: Activated Fuel system status: Closed loop Fuel cut: OFF Engine load: 10 to 70% Shift position: 3rd or more Heated Oxygen Sensor Impedance (Low): Battery voltage: 11 V or more Estimated sensor temperature: Less than 700°C (1292°F) ECM monitor: Completed DTC P0606: Not set Heated Oxygen Sensor Impedance (High): Battery voltage: 11 V or more Estimated sensor temperature: 450 to 750°C (842 to 1382°F) or higher DTC P0606: Not set Heated Oxygen Sensor Output Voltage (Extremely High): Battery voltage: 11 V or more Time after engine start: 2 seconds or more **Possible Causes:** • Open or short in HO2 sensor (bank 1 sensor 2) circuit • HO2 sensor (bank 1 sensor 2) • HO2 sensor heater (bank 1 sensor 2) • Air-Fuel Ratio (A/F) sensor (bank 1 sensor 1) • Gas leakage from exhaust system
DTC: P0137 **2T ECM, MIL: Yes** **Year:** 2009, 2010 **Model:** All **Engine:** All **Transmission:** All	**Oxygen Sensor Circuit Low Voltage (Bank 1 Sensor 2):** Active A/F control: Performing Battery voltage: 11 V or more ECT: 167°F (75°C) or more Idle: OFF Engine rpm: Less than 3200 rpm A/F sensor status: Activatted Fuel system status: Closed loop Fuel cut: OFF Engineload: 10 to 70% Shift position: 4rd or more Battery voltage: 11 V or more Estimated rear HO2S temperature: 842-1382°F (450-750°C) P0607: Not preset **Possible Causes:** • Open in HO2 sensor (sensor 2) circuit • HO2 sensor (sensor 2) • HO2 sensor heater (sensor 2) • Air-Fuel Ratio (A/F) sensor (sensor 1) • Gas leakage from exhaust system

DTC	Trouble Code Title, Conditions & Possible Causes
DTC: P0138 **2T ECM, MIL: Yes** **Year:** 2009, 2010 **Model:** All **Engine:** All V6, V8 **Transmission:** All	**Oxygen Sensor Circuit High Voltage (Bank 1 Sensor 2):** Active A/F control: Performing Battery voltage: 11 V or more ECT: 167°F (75°C) or more Idle: OFF Engine rpm: Less than 3200 rpm A/F sensor status: Activatted Fuel system status: Closed loop Fuel cut: OFF Engineload: 10 to 70% Shift position: 4rd or more Battery voltage: 11 V or more Time after engine start: 2 seconds or more **Possible Causes:** • Short in HO2 sensor (sensor 2) circuit • HO2 sensor (sensor 2) • ECM internal circuit malfunction • A/F sensor (sensor 1)
DTC: P0139 **2T ECM** **Year:** 2009, 2010 **Model:** All **Engine:** All V6, V8 **Transmission:** All	**Oxygen Sensor Circuit Slow Response (Bank 1 Sensor 2):** Heated Oxygen Sensor Output Voltage (Output Voltage, High Voltage and Low Voltage): Active air fuel ratio control: Performing Active air fuel ratio control is performed when all of the following conditions met: - Battery voltage: 11 V or more Engine coolant temperature: 75°C (167°F) or more Idling: OFF Engine speed: Less than 3200 rpm Air fuel ratio sensor status: Activated Fuel system status: Closed loop Fuel cut: OFF Engine load: 10% or more, and less than 70% Heated Oxygen Sensor Impedance (Low): Battery voltage: 11 V or more Estimated heated oxygen sensor temperature: Less than 700°C (1292°F) ECM monitor: Completed DTC P0607: Not set Heated Oxygen Sensor Impedance (High): Battery voltage: 11 V or more Estimated heated oxygen sensor temperature: 450°C (842°F) or more and less than 750°C (1382°F) DTC P0607: Not set Heated Oxygen Sensor Output Voltage (Extremely high): Battery voltage: 11 V or more Time after engine start: 2 seconds or more Heated Oxygen Sensor Voltage During Fuel Cut: Engine coolant temperature: 75°C (167°F) or more Estimated catalyst temperature: 400°C (752°F) or more Fuel cut: ON **Possible Causes:** • Short in heated oxygen sensor (bank 1, 2 sensor 2) circuit • Heated oxygen sensor (bank 1, 2 sensor 2) • ECM

DTC	Trouble Code Title, Conditions & Possible Causes
DTC: P0141 **2T ECM, MIL: Yes** **Year:** 2009, 2010 **Model:** All **Engine:** All V6, V8 **Transmission:** All	**Oxygen Sensor Heater Circuit Malfunction (Bank 1 Sensor 2):** Either of the following conditions is met: Condition A or B A. All of the following conditions are met: Conditions a, b, c, d and e a. Battery voltage: 10.5 V or more b. Fuel cut: OFF c. Time after fuel cut ON to OFF: 30 seconds or more d. Accumulated heater ON time: 100 seconds or more e. Learned heater OFF current operation: Completed B. Duration that rear heated oxygen sensor impedance is less than 15 kW: 2 seconds or more **Possible Causes:** • HO2 sensor heater • ECM
DTC: P014C **2T ECM, MIL: Yes** **Year:** 2009, 2010 **Model:** All **Engine:** All **Transmission:** All	**A/F Sensor Slow Response - Rich to Lean Bank 1 Sensor 1:** Monitor runs whenever following DTCs not stored: None Active air fuel ratio control: Performing Active air fuel ratio control is performed when the following conditions are met: Battery voltage: 11 V or higher Engine coolant temperature: 70°C (158°F) or higher Idle: OFF Engine speed: 1000 to 4000 rpm Air fuel ratio sensor status: Activated Fuel-cut: OFF Engine load: 25 to 75% Shift position: 2nd or higher Catalyst monitor: Not executing Mass air flow: 6.25 to 18 g/sec. **Possible Causes:** • Air fuel ratio sensor (bank 1, 2 sensor 1) • Air fuel ratio sensor (bank 1, 2 sensor 1) heater • ECM
DTC: P014D **2T ECM, MIL: Yes** **Year:** 2009, 2010 **Model:** All **Engine:** All **Transmission:** All	**A/F Sensor Slow Response - Lean to Rich Bank 1 Sensor 1:** Monitor runs whenever following DTCs not stored: None Active air fuel ratio control: Performing Active air fuel ratio control is performed when the following conditions are met: Battery voltage: 11 V or higher Engine coolant temperature: 70°C (158°F) or higher Idle: OFF Engine speed: 1000 to 4000 rpm Air fuel ratio sensor status: Activated Fuel-cut: OFF Engine load: 25 to 75% Shift position: 2nd or higher Catalyst monitor: Not executing Mass air flow: 6.25 to 18 g/sec. **Possible Causes:** • Air fuel ratio sensor (bank 1, 2 sensor 1) • Air fuel ratio sensor (bank 1, 2 sensor 1) heater • ECM

DTC	Trouble Code Title, Conditions & Possible Causes
DTC: P014E **2T ECM, MIL: Yes** **Year:** 2009, 2010 **Model:** All **Engine:** All **Transmission:** All	**A/F Sensor Slow Response - Rich to Lean Bank 2 Sensor 1:** Monitor runs whenever following DTCs are not stored: None Active air fuel ratio control: Performing Active air fuel ratio control is performed when the following conditions are met: Battery voltage: 11 V or higher Engine coolant temperature: 167°F (75°C) or more Idle: OFF Engine speed: 1000-4000 rpm Air fuel ratio sensor status: Activated Fuel cut: OFF Engine load: 10-70% Shift position: 2nd or higher Catalyst monitor: Not yet Mass air flow: 5-15 g/sec. Rich to Lean Response rate deterioration level: 0.035 V or less **Possible Causes:** • Air fuel ratio sensor • Air fuel ratio sensor heater • ECM
DTC: P014F **2T ECM, MIL: Yes** **Year:** 2009, 2010 **Model:** All **Engine:** All **Transmission:** All	**A/F Sensor Slow Response - Lean to Rich Bank 2 Sensor 1:** Monitor runs whenever following DTCs are not stored: None Active air fuel ratio control: Performing Active air fuel ratio control is performed when the following conditions are met: Battery voltage: 11 V or higher Engine coolant temperature: 167°F (75°C) or more Idle: OFF Engine speed: 1000-4000 rpm Air fuel ratio sensor status: Activated Fuel cut: OFF Engine load: 10-70% Shift position: 2nd or higher Catalyst monitor: Not yet Mass air flow: 5-15 g/sec. Lean to Rich Response rate deterioration level: -0.038 V or higher **Possible Causes:** • Air fuel ratio sensor • Air fuel ratio sensor heater • ECM
DTC: P0156 **1T ECM, MIL: Yes** **Year:** 2009, 2010 **Model:** All **Engine:** All **Transmission:** All	**Oxygen Sensor Circuit Malfunction (Bank 2 Sensor 2):** Active A/F control is performed when the following conditions are met: Battery voltage: 11 V or more ECT: 75°C (167°F) or more Idle: OFF Engine rpm: Less than 3200 rpm A/F sensor status: Activated Fuel system status: Closed loop Fuel-cut: OFF Engine load: 10 to 75% Shift position: 4th or more Battery voltage: 11 V or more Estimated rear oxygen sensor temperature: Less than 700°C (1292°F) ECM monitor: Completed P0606: Not set **Possible Causes:** • Open or short in HO2 sensor (bank 1, 2 sensor 2) circuit • HO2 sensor (bank 1, 2 sensor 2) • HO2 sensor heater (bank 1, 2 sensor 2) • Air Fuel Ratio (A/F) sensor (bank 1, 2 sensor 1) • EFI relay • Gas leakage from exhaust system

DTC	Trouble Code Title, Conditions & Possible Causes
DTC: P0157 **2T ECM, MIL: Yes** **Year:** 2009, 2010 **Model:** All **Engine:** All **Transmission:** All	**Oxygen Sensor Circuit Low Voltage (Bank 2 Sensor 2):** Active A/F control: Performing Battery voltage: 11 V or more ECT: 167°F (75°C) or more Idle: OFF Engine rpm: Less than 3200 rpm A/F sensor status: Activatted Fuel system status: Closed loop Fuel cut: OFF Engineload: 10 to 70% Shift position: 4rd or more Battery voltage: 11 V or more Estimated rear HO2S temperature: 842-1382°F (450-750°C) P0607: Not preset **Possible Causes:** • Open in HO2 sensor (bank 1, 2 sensor 2) circuit • HO2 sensor (bank 1, 2 sensor 2) • HO2 sensor heater (bank 1, 2 sensor 2) • Engine room junction block (EFI relay) • Gas leakage from exhaust system
DTC: P0158 **1T ECM, MIL: Yes** **Year:** 2009, 2010 **Model:** All **Engine:** All **Transmission:** All	**Oxygen Sensor Circuit High Voltage (Bank 2 Sensor 2):** Active A/F control is performed when the following conditions are met: Battery voltage: 11 V or more ECT: 75°C (167°F) or more Idle: OFF Engine rpm: Less than 3200 rpm A/F sensor status: Activated Fuel system status: Closed loop Fuel-cut: OFF Engine load: 10 to 75% Shift position: 4th or more Battery voltage:11 V or more Time after engine start: 2 seconds or more **Possible Causes:** • Short in HO2 sensor (bank 1, 2 sensor 2) circuit • HO2 sensor (bank 1, 2 sensor 2) • ECM internal circuit malfunction • Air Fuel ratio (A/F) sensor (bank 1, 2 sensor 1)
DTC: P0159 **2T ECM, MIL: Yes** **Year:** 2009, 2010 **Model:** All **Engine:** All **Transmission:** All	**Oxygen Sensor Circuit Slow Response (Bank 2 Sensor 2):** ECT: 167°F (75°C) or more Estimated catalyst temperature: 752°F (400°C) or more Fuel cut: ON **Possible Causes:** • Short in HO2 sensor (bank 1, 2 sensor 2) • HO2 sensor (bank 1, 2 sensor 2) • ECM internal circuit malfunction

DTC	Trouble Code Title, Conditions & Possible Causes
DTC: P015A **2T ECM, MIL: Yes** **Year:** 2009, 2010 **Model:** All **Engine:** All **Transmission:** All	**A/F Sensor Delayed Response - Rich to Lean Bank 1 Sensor 1:** Monitor runs whenever following DTCs not stored: None Active air fuel ratio control: Performing Active air fuel ratio control is performed when the following conditions are met: Battery voltage: 11 V or higher Engine coolant temperature: 70°C (158°F) or higher Idle: OFF Engine speed: 1000 to 4000 rpm Air fuel ratio sensor status: Activated Fuel-cut: OFF Engine load: 25 to 75% Shift position: 2nd or higher Catalyst monitor: Not executing Mass air flow: 6.25 to 18 g/sec. **Possible Causes:** • Air fuel ratio sensor (bank 1, 2 sensor 1) • Air fuel ratio sensor (bank 1, 2 sensor 1) heater • ECM
DTC: P015B **2T ECM, MIL: Yes** **Year:** 2009, 2010 **Model:** All **Engine:** All **Transmission:** All	**A/F Sensor Delayed Response - Lean to Rich Bank 1 Sensor 1:** Monitor runs whenever following DTCs not stored: None Active air fuel ratio control: Performing Active air fuel ratio control is performed when the following conditions are met: Battery voltage: 11 V or higher Engine coolant temperature: 70°C (158°F) or higher Idle: OFF Engine speed: 1000 to 4000 rpm Air fuel ratio sensor status: Activated Fuel-cut: OFF Engine load: 25 to 75% Shift position: 2nd or higher Catalyst monitor: Not executing Mass air flow: 6.25 to 18 g/sec. **Possible Causes:** • Air fuel ratio sensor (bank 1, 2 sensor 1) • Air fuel ratio sensor (bank 1, 2 sensor 1) heater • ECM
DTC: P015C **2T ECM, MIL: Yes** **Year:** 2009, 2010 **Model:** All **Engine:** All **Transmission:** All	**A/F Sensor Delayed Response - Rich to Lean Bank 2 Sensor 1:** Monitor runs whenever following DTCs not stored: None Active air fuel ratio control: Performing Active air fuel ratio control is performed when the following conditions are met: Battery voltage: 11 V or higher Engine coolant temperature: 70°C (158°F) or higher Idle: OFF Engine speed: 1000 to 4000 rpm Air fuel ratio sensor status: Activated Fuel-cut: OFF Engine load: 25 to 75% Shift position: 2nd or higher Catalyst monitor: Not executing Mass air flow: 6.25 to 18 g/sec. **Possible Causes:** • Air fuel ratio sensor (bank 1, 2 sensor 1) • Air fuel ratio sensor (bank 1, 2 sensor 1) heater • ECM

DTC	Trouble Code Title, Conditions & Possible Causes
DTC: P015D **2T ECM, MIL: Yes** **Year:** 2009, 2010 **Model:** All **Engine:** All **Transmission:** All	**A/F Sensor Delayed Response - Lean to Rich Bank 2 Sensor 1:** Monitor runs whenever following DTCs are not stored: None Active air fuel ratio control: Performing Active air fuel ratio control is performed when the following conditions are met: Battery voltage: 11 V or higher Engine coolant temperature: 167°F (75°C) or more Idle: OFF Engine speed: 1000-4000 rpm Air fuel ratio sensor status: Activated Fuel cut: OFF Engine load: 10-70% Shift position: 2nd or higher Catalyst monitor: Not yet Mass air flow: 5-15 g/sec. Lean to Rich delay level: 230 msec or more **Possible Causes:** • Air fuel ratio sensor • Air fuel ratio sensor heater • ECM
DTC: P0161 **2T ECM, MIL: Yes** **Year:** 2009, 2010 **Model:** All **Engine:** All **Transmission:** All	**Oxygen Sensor Heater Circuit Malfunction (Bank 2 Sensor 2):** Case 1: Monitor runs whenever following DTCs not stored: P0031, P0032, P0051, P0052 (Air fuel ratio sensor heater) P0037, P0038, P0057, P0058 (Rear oxygen sensor heater) Battery voltage: 10.5 V or higher Fuel cut: OFF Time after fuel cut ON to OFF: 30 seconds or more Accumulated heater ON time: 100 seconds or more Learned heater OFF current operation: Complete Case 2: Monitor runs whenever following DTCs not stored: None Duration that rear heated oxygen sensor impedance is less than 15 kW: 2 seconds or more **Possible Causes:** • Open or short in heated oxygen sensor heater circuit • Heated oxygen sensor heater (sensor 2) • Integration relay (EFI MAIN) • ECM
DTC: P0171 **2T ECM, MIL: Yes** **Year:** 2009, 2010 **Model:** All **Engine:** V6, V8 **Transmission:** All	**System Too Lean (Bank 1):** Fuel system status: Closed loop Battery voltage: 11 V or more Either of the following conditions 1 or 2 is met: 1. Engine RPM: Less than 1100 rpm 2. Engine load: 10% or more Catalyst monitor: Not executed **Possible Causes:** • Intake system • Fuel injector assembly for port injection • Fuel injector assembly for direct injection • Mass air flow meter sub-assembly • Engine coolant temperature sensor • Fuel pressure • Gas leaks from exhaust system • Open or short in air fuel ratio sensor (bank 1 sensor 1) circuit • Air fuel ratio sensor (bank 1 sensor 1) • Air fuel ratio sensor heater (bank 1 sensor 1) • Air fuel ratio sensor (bank 1 sensor 1) heater circuits • PCV valve and hose • PCV hose connections • ECM

DTC	Trouble Code Title, Conditions & Possible Causes
DTC: P0172 **2T ECM, MIL: Yes** **Year:** 2009, 2010 **Model:** All **Engine:** All V6, V8 **Transmission:** All	**System Too Rich (Bank 1):** Fuel system status: Closed loop Battery voltage: 11 V or more Either of the following conditions 1 or 2 is met: 1. Engine RPM: Less than 1100 rpm 2. Engine load: 10% or more Catalyst monitor: Not executed **Possible Causes:** • Mass air flow meter sub-assembly • Engine coolant temperature sensor • Fuel injector assembly for port injection • Fuel injector assembly for direct injection • Ignition system • Fuel pressure • Gas leaks from exhaust system • Open or short in air fuel ratio sensor (bank 1 sensor 1) circuit • Air fuel ratio sensor (bank 1 sensor 1) • Air fuel ratio sensor heater (bank 1 sensor 1) • Air fuel ratio sensor heater (bank 1 sensor 1) circuits • ECM
DTC: P0174 **2T ECM, MIL: Yes** **Year:** 2009, 2010 **Model:** All **Engine:** All **Transmission:** All	**System Too Lean (Bank 2):** Fuel system status: Closed loop Battery voltage: 11 V or more Either following condition is met: 1. Engine RPM: less than 1100 rpm 2. Intake air amount per revolution: 0.22 g/rev or more Catalyst monitor: Not executed **Possible Causes:** • Air induction system • Injector blockage • MAF meter • ECT sensor • Fuel pressure • Gas leakage from exhaust system • Open or short in A/F sensor (bank 1, 2 sensor 1) circuit • A/F sensor (bank 1, 2 sensor 1) • A/F sensor heater (bank 1, 2 sensor 1) • A/F sensor heater relay • A/F sensor heater and A/F sensor heater relay circuits • PCV valve and hose • PCV hose connections • ECM
DTC: P0175 **2T ECM, MIL: Yes** **Year:** 2009, 2010 **Model:** All **Engine:** All **Transmission:** All	**System Too Rich (Bank 2):** Fuel system status: Closed loop Battery voltage: 11 V or more Either of the following conditions 1 or 2 is met: 1. Engine RPM: Less than 1100 rpm 2. Engine load: 10% or more Catalyst monitor: Not executed **Possible Causes:** • Mass air flow meter sub-assembly • Engine coolant temperature sensor • Fuel injector assembly for port injection • Fuel injector assembly for direct injection • Ignition system • Fuel pressure • Gas leaks from exhaust system • Open or short in air fuel ratio sensor (bank 2 sensor 1) circuit • Air fuel ratio sensor (bank 2 sensor 1) • Air fuel ratio sensor heater (bank 2 sensor 1) • Air fuel ratio sensor heater (bank 2 sensor 1) circuits • ECM

DTC	Trouble Code Title, Conditions & Possible Causes
DTC: P0190 **1T ECM, MIL: Yes** **Year:** 2009, 2010 **Model:** All **Engine:** All **Transmission:** All	**Fuel Rail Pressure Sensor Circuit:** Time after engine start: 5 seconds or more **Possible Causes:** • Open or short in fuel pressure sensor • Fuel pressure sensor • ECM
DTC: P0192 **1T ECM, MIL: Yes** **Year:** 2009, 2010 **Model:** All **Engine:** 2All **Transmission:** All	**Fuel Rail Pressure Sensor Circuit Low Input:** Time after engine start: 5 seconds or more **Possible Causes:** • Short in fuel pressure sensor • Fuel pressure sensor • ECM
DTC: P0193 **1T ECM, MIL: Yes** **Year:** 2009, 2010 **Model:** All **Engine:** All **Transmission:** All	**Fuel Rail Pressure Sensor Circuit High Input:** Time after engine start: 5 seconds or more **Possible Causes:** • Open or short in fuel pressure sensor • Fuel pressure sensor • ECM
DTC: P0200 **1T ECM, MIL: Yes** **Year:** 2009, 2010 **Model:** All **Engine:** All **Transmission:** All	**Injector Circuit / Open:** Time after engine switch off to on (IG): 1 second or more Engine speed: 4000 rpm or less Battery voltage: 10.5 V or more **Possible Causes:** • Open or short in injector driver (EDU) circuit • Injector driver (EDU) • Fuel injector assembly • ECM
DTC: P0201 **1T ECM, MIL: Yes** **Year:** 2009, 2010 **Model:** All **Engine:** All **Transmission:** All	**Injector Circuit / Open - (Cylinder 1):** Time after engine switch off to on (IG): 1 second or more Battery: 10.5 V or more Fuel injection start timing: - Injector driver fail signal: Not overlap Engine switch: on (IG) Injector driver relay: ON **Possible Causes:** • Open or short in Injector driver (No. 1, 2) circuit • Injector driver (No. 1, 2) • Fuel injector for direct injection assembly • Engine room No. 2 junction block • ECM
DTC: P0202 **1T ECM, MIL: Yes** **Year:** 2009, 2010 **Model:** All **Engine:** All **Transmission:** All	**Injector Circuit / Open - (Cylinder 2):** Time after engine switch off to on (IG): 1 second or more Battery: 10.5 V or more Fuel injection start timing: - Injector driver fail signal: Not overlap Engine switch: on (IG) Injector driver relay: ON **Possible Causes:** • Open or short in Injector driver (No. 1, 2) circuit • Injector driver (No. 1, 2) • Fuel injector for direct injection assembly • Engine room No. 2 junction block • ECM

DTC	Trouble Code Title, Conditions & Possible Causes
DTC: P0203 **1T ECM, MIL: Yes** **Year:** 2009, 2010 **Model:** All **Engine:** All **Transmission:** All	**Injector Circuit / Open - (Cylinder 3):** Time after engine switch off to on (IG): 1 second or more Battery: 10.5 V or more Fuel injection start timing: - Injector driver fail signal: Not overlap Engine switch: on (IG) Injector driver relay: ON **Possible Causes:** • Open or short in Injector driver (No. 1, 2) circuit • Injector driver (No. 1, 2) • Fuel injector for direct injection assembly • Engine room No. 2 junction block • ECM
DTC: P0204 **1T ECM, MIL: Yes** **Year:** 2009, 2010 **Model:** All **Engine:** All **Transmission:** All	**Injector Circuit / Open - (Cylinder 4):** Time after engine switch off to on (IG): 1 second or more Battery: 10.5 V or more Fuel injection start timing: - Injector driver fail signal: Not overlap Engine switch: on (IG) Injector driver relay: ON **Possible Causes:** • Open or short in Injector driver (No. 1, 2) circuit • Injector driver (No. 1, 2) • Fuel injector for direct injection assembly • Engine room No. 2 junction block • ECM
DTC: P0205 **1T ECM, MIL: Yes** **Year:** 2009, 2010 **Model:** All **Engine:** V6, V8 **Transmission:** All	**Injector Circuit / Open - (Cylinder 5):** Time after engine switch off to on (IG): 1 second or more Battery: 10.5 V or more Fuel injection start timing: - Injector driver fail signal: Not overlap Engine switch: on (IG) Injector driver relay: ON **Possible Causes:** • Open or short in Injector driver (No. 1, 2) circuit • Injector driver (No. 1, 2) • Fuel injector for direct injection assembly • Engine room No. 2 junction block • ECM
DTC: P0206 **1T ECM, MIL: Yes** **Year:** 2009, 2010 **Model:** All **Engine:** V6, V8 **Transmission:** All	**Injector Circuit / Open - (Cylinder 6):** Time after engine switch off to on (IG): 1 second or more Battery: 10.5 V or more Fuel injection start timing: - Injector driver fail signal: Not overlap Engine switch: on (IG) Injector driver relay: ON **Possible Causes:** • Open or short in Injector driver (No. 1, 2) circuit • Injector driver (No. 1, 2) • Fuel injector for direct injection assembly • Engine room No. 2 junction block • ECM

DTC	Trouble Code Title, Conditions & Possible Causes
DTC: P0207 **1T ECM, MIL: Yes** **Year:** 2009, 2010 **Model:** All **Engine:** V8 **Transmission:** All	**Injector Circuit / Open - (Cylinder 7):** Time after engine switch off to on (IG): 1 second or more Battery: 10.5 V or more Fuel injection start timing: - Injector driver fail signal: Not overlap Engine switch: on (IG) Injector driver relay: ON **Possible Causes:** • Open or short in Injector driver (No. 1, 2) circuit • Injector driver (No. 1, 2) • Fuel injector for direct injection assembly • Engine room No. 2 junction block • ECM
DTC: P0208 **1T ECM, MIL: Yes** **Year:** 2009, 2010 **Model:** All **Engine:** All V8 **Transmission:** All	**Injector Circuit / Open - (Cylinder 8):** Time after engine switch off to on (IG): 1 second or more Battery: 10.5 V or more Fuel injection start timing: - Injector driver fail signal: Not overlap Engine switch: on (IG) Injector driver relay: ON **Possible Causes:** • Open or short in Injector driver (No. 1, 2) circuit • Injector driver (No. 1, 2) • Fuel injector for direct injection assembly • Engine room No. 2 junction block • ECM
DTC: P0220 **1T ECM, MIL: Yes** **Year:** 2009, 2010 **Model:** All **Engine:** All V6, V8 **Transmission:** All	**Throttle / Pedal Position Sensor / Switch "B" Circuit:** Either of the following conditions A or B is met: A. Engine switch on (IG): 0.012 seconds or more B. Electronic throttle actuator power: ON **Possible Causes:** • Throttle position sensor (built into throttle body) • ECM
DTC: P0222 **1T ECM, MIL: Yes** **Year:** 2009, 2010 **Model:** All **Engine:** V6, V8 **Transmission:** All	**Throttle / Pedal Position Sensor / Switch "B" Circuit Low Input:** Monitor runs whenever following DTCs not present: None **Possible Causes:** • TP sensor (built into throttle body) • Short in VTA2 circuit • Open in VC circuit • ECM
DTC: P0223 **1T ECM, MIL: Yes** **Year:** 2009, 2010 **Model:** All **Engine:** All V6, V8 **Transmission:** All	**Throttle / Pedal Position Sensor / Switch "B" Circuit High Input:** Monitor runs whenever following DTCs not present: None **Possible Causes:** • TP sensor (built into throttle body) • Open in VTA2 circuit • Open in E2 circuit • Short between VC and VTA2 circuits • ECM
DTC: P0230 **1T ECM, MIL: Yes** **Year:** 2009, 2010 **Model:** All **Engine:** All **Transmission:** All	**Fuel Pump Primary Circuit:** Open or short in F/PMP relay circuit **Possible Causes:** • Open or short in F/PMP relay circuit • FUEL PUMP relay (F/PMP) • ECM

DTC	Trouble Code Title, Conditions & Possible Causes
DTC: P0300 **2T ECM, MIL: Yes** **Year:** 2009, 2010 **Model:** All **Engine:** All V6, V8 **Transmission:** All	**Random / Multiple Cylinder Misfire Detected:** Battery voltage: 8 V or more VVT system: Not operated by scan tool Engine RPM: 335 to 6600 rpm Either of the following conditions 1 and 2 is met: 1. ECT at engine start: More than -7°C (19°F) 2. ECT: More than 20°C (68°F) Fuel cut: OFF Monitor period of emission-related-misfire: First 1000 revolutions after engine start, or Check Mode: Crankshaft 1000 revolutions Except above: Crankshaft 1000 revolutions x 4 Monitor period of catalyst-damaged-misfire (MIL blinks): All of following conditions 1, 2 and 3 are met: Crankshaft 200 revolutions x 3 1. Driving cycles: 1st 2. Check mode: OFF 3. Engine RPM: Less than 2200 rpm Except above (MIL blinks immediately): Crankshaft 200 revolutions **Possible Causes:** • Open or short in engine wire harness • Connector connection • Vacuum hose connections • Ignition system • Injector for direct injector • Injector for port injector • Fuel pressure • Mass Air Flow (MAF) meter • Engine Coolant Temperature (ECT) sensor • Compression pressure • Valve clearance • Valve timing • Ventilation valve and hose • Ventilation hose connections • Air induction system • ECM

DTC	Trouble Code Title, Conditions & Possible Causes
DTC: P0301 **2T ECM, MIL: Yes** **Year:** 2009, 2010 **Model:** All **Engine:** All V6, V8 **Transmission:** All	**Cylinder 1 Misfire Detected:** Battery voltage: 8 V or more VVT system: Not operated by scan tool Engine RPM: 335 to 6600 rpm Either of the following conditions 1 and 2 is met: 1. ECT at engine start: More than -7°C (19°F) 2. ECT: More than 20°C (68°F) Fuel cut: OFF Monitor period of emission-related-misfire: First 1000 revolutions after engine start, or Check Mode: Crankshaft 1000 revolutions Except above: Crankshaft 1000 revolutions x 4 Monitor period of catalyst-damaged-misfire (MIL blinks): All of following conditions 1, 2 and 3 are met: Crankshaft 200 revolutions x 3 1. Driving cycles: 1st 2. Check mode: OFF 3. Engine RPM: Less than 2200 rpm Except above (MIL blinks immediately): Crankshaft 200 revolutions **Possible Causes:** • Open or short in engine wire harness • Connector connection • Vacuum hose connections • Ignition system • Injector for direct injector • Injector for port injector • Fuel pressure • Mass Air Flow (MAF) meter • Engine Coolant Temperature (ECT) sensor • Compression pressure • Valve clearance • Valve timing • Ventilation valve and hose • Ventilation hose connections • Air induction system • ECM

DTC	Trouble Code Title, Conditions & Possible Causes
DTC: P0302 **2T ECM, MIL: Yes** **Year:** 2009, 2010 **Model:** All **Engine:** All V6, V8 **Transmission:** All	**Cylinder 2 Misfire Detected:** Battery voltage: 8 V or more VVT system: Not operated by scan tool Engine RPM: 335 to 6600 rpm Either of the following conditions 1 and 2 is met: 1. ECT at engine start: More than -7°C (19°F) 2. ECT: More than 20°C (68°F) Fuel cut: OFF Monitor period of emission-related-misfire: First 1000 revolutions after engine start, or Check Mode: Crankshaft 1000 revolutions Except above: Crankshaft 1000 revolutions x 4 Monitor period of catalyst-damaged-misfire (MIL blinks): All of following conditions 1, 2 and 3 are met: Crankshaft 200 revolutions x 3 1. Driving cycles: 1st 2. Check mode: OFF 3. Engine RPM: Less than 2200 rpm Except above (MIL blinks immediately): Crankshaft 200 revolutions **Possible Causes:** • Open or short in engine wire harness • Connector connection • Vacuum hose connections • Ignition system • Injector for direct injector • Injector for port injector • Fuel pressure • Mass Air Flow (MAF) meter • Engine Coolant Temperature (ECT) sensor • Compression pressure • Valve clearance • Valve timing • Ventilation valve and hose • Ventilation hose connections • Air induction system • ECM

DTC	Trouble Code Title, Conditions & Possible Causes
DTC: P0303 **2T ECM, MIL: Yes** **Year:** 2009, 2010 **Model:** All **Engine:** All V6, V8 **Transmission:** All	**Cylinder 3 Misfire Detected:** Either of the following conditions 1 and 2 is met: 1. ECT at engine start: More than -7°C (19°F) 2. ECT: More than 20°C (68°F) Fuel cut: OFF Monitor period of emission-related-misfire: First 1000 revolutions after engine start, or Check Mode: Crankshaft 1000 revolutions Except above: Crankshaft 1000 revolutions x 4 Monitor period of catalyst-damaged-misfire (MIL blinks): All of following conditions 1, 2 and 3 are met: Crankshaft 200 revolutions x 3 1. Driving cycles: 1st 2. Check mode: OFF 3. Engine RPM: Less than 2200 rpm Except above (MIL blinks immediately): Crankshaft 200 revolutions **Possible Causes:** • Open or short in engine wire harness • Connector connection • Vacuum hose connections • Ignition system • Injector for direct injector • Injector for port injector • Fuel pressure • Mass Air Flow (MAF) meter • Engine Coolant Temperature (ECT) sensor • Compression pressure • Valve clearance • Valve timing • Ventilation valve and hose • Ventilation hose connections • Air induction system • ECM

DTC	Trouble Code Title, Conditions & Possible Causes
DTC: P0303 **2T ECM, MIL: Yes** **Year:** 2009, 2010 **Model:** All **Engine:** All V6, V8 **Transmission:** All	**Cylinder 3 Misfire Detected:** Battery voltage: 8 V or more VVT system: Not operated by scan tool Engine RPM: 450-6500 rpm Either of the following conditions (a) or (b) is met: (a) Engine coolant temperature at engine start: More than 19.4°F (-7°C) (b) Engine coolant temperature: More than 68°F (20°C) Fuel cut: OFF Monitor period of emission-related misfire: First 1000 revolutions after engine start, or Check Mode: Crankshaft 1000 revolutions Except above: Crankshaft 1000 revolutions X 4 Monitor period of catalyst-damaged misfire (MIL blinks): All of the following conditions 1, 2 and 3 are met: Crankshaft 200 revolutions 1. Driving cycles: 1st 2. Check mode: OFF 3. Enine RPM: Less than 2300 rpm Except above: Crankshaft 200 revolutions X 3 **Possible Causes:** • Open or short in engine wire harness • Connector connections • Vacuum hose connections • Ignition system • Injector • Fuel pressure • MAF meter • ECT sensor • Compression pressure • Valve timing • PCV valve and hose • PCV hose connections • Air induction system • ECM

DTC	Trouble Code Title, Conditions & Possible Causes
DTC: P0304 **2T ECM, MIL: Yes** **Year:** 2009, 2010 **Model:** All **Engine:** All V6, V8 **Transmission:** All	**Cylinder 4 Misfire Detected:** Battery voltage: 8 V or more VVT system: Not operated by scan tool Engine RPM: 335 to 6600 rpm Either of the following conditions 1 and 2 is met: 1. ECT at engine start: More than -7°C (19°F) 2. ECT: More than 20°C (68°F) Fuel cut: OFF Monitor period of emission-related-misfire: First 1000 revolutions after engine start, or Check Mode: Crankshaft 1000 revolutions Except above: Crankshaft 1000 revolutions x 4 Monitor period of catalyst-damaged-misfire (MIL blinks): All of following conditions 1, 2 and 3 are met: Crankshaft 200 revolutions x 3 1. Driving cycles: 1st 2. Check mode: OFF 3. Engine RPM: Less than 2200 rpm Except above (MIL blinks immediately): Crankshaft 200 revolutions **Possible Causes:** • Open or short in engine wire harness • Connector connection • Vacuum hose connections • Ignition system • Injector for direct injector • Injector for port injector • Fuel pressure • Mass Air Flow (MAF) meter • Engine Coolant Temperature (ECT) sensor • Compression pressure • Valve clearance • Valve timing • Ventilation valve and hose • Ventilation hose connections • Air induction system • ECM

DTC	Trouble Code Title, Conditions & Possible Causes
DTC: P0305 **2T ECM, MIL: Yes** **Year:** 2009, 2010 **Model:** All **Engine:** All V6, V8 **Transmission:** All	**Cylinder 5 Misfire Detected:** Battery voltage: 8 V or more VVT system: Not operated by scan tool Engine RPM: 450-6500 rpm Either of the following conditions (a) or (b) is met: (a) Engine coolant temperature at engine start: More than 19.4°F (-7°C) (b) Engine coolant temperature: More than 68°F (20°C) Fuel cut: OFF Monitor period of emission-related misfire: First 1000 revolutions after engine start, or Check Mode: Crankshaft 1000 revolutions Except above: Crankshaft 1000 revolutions X 4 Monitor period of catalyst-damaged misfire (MIL blinks): All of the following conditions 1, 2 and 3 are met: Crankshaft 200 revolutions 1. Driving cycles: 1st 2. Check mode: OFF 3. Enine RPM: Less than 2300 rpm Except above: Crankshaft 200 revolutions X 3 **Possible Causes:** • Open or short in engine wire harness • Connector connections • Vacuum hose connections • Ignition system • Injector • Fuel pressure • MAF meter • ECT sensor • Compression pressure • Valve timing • PCV valve and hose • PCV hose connections • Air induction system • ECM

DTC	Trouble Code Title, Conditions & Possible Causes
DTC: P0306 **2T ECM, MIL: Yes** **Year:** 2009, 2010 **Model:** All **Engine:** All V6, V8 **Transmission:** All	**Cylinder 6 Misfire Detected:** Battery voltage: 8 V or more VVT system: Not operated by scan tool Engine RPM: 335 to 6600 rpm Either of the following conditions 1 and 2 is met: 1. ECT at engine start: More than -7°C (19°F) 2. ECT: More than 20°C (68°F) Fuel cut: OFF Monitor period of emission-related-misfire: First 1000 revolutions after engine start, or Check Mode: Crankshaft 1000 revolutions Except above: Crankshaft 1000 revolutions x 4 Monitor period of catalyst-damaged-misfire (MIL blinks): All of following conditions 1, 2 and 3 are met: Crankshaft 200 revolutions x 3 1. Driving cycles: 1st 2. Check mode: OFF 3. Engine RPM: Less than 2200 rpm Except above (MIL blinks immediately): Crankshaft 200 revolutions **Possible Causes:** • Open or short in engine wire harness • Connector connection • Vacuum hose connections • Ignition system • Injector for direct injector • Injector for port injector • Fuel pressure • Mass Air Flow (MAF) meter • Engine Coolant Temperature (ECT) sensor • Compression pressure • Valve clearance • Valve timing • Ventilation valve and hose • Ventilation hose connections • Air induction system • ECM

DTC	Trouble Code Title, Conditions & Possible Causes
DTC: P0307 **2T ECM, MIL: Yes** **Year:** 2009, 2010 **Model:** All **Engine:** All V8 **Transmission:** All	**Cylinder 7 Misfire Detected:** Battery voltage: 8 V or more VVT system: Not operated by scan tool Engine RPM: 335 to 6600 rpm Either of the following conditions 1 and 2 is met: 1. ECT at engine start: More than -7°C (19°F) 2. ECT: More than 20°C (68°F) Fuel cut: OFF Monitor period of emission-related-misfire: First 1000 revolutions after engine start, or Check Mode: Crankshaft 1000 revolutions Except above: Crankshaft 1000 revolutions x 4 Monitor period of catalyst-damaged-misfire (MIL blinks): All of following conditions 1, 2 and 3 are met: Crankshaft 200 revolutions x 3 1. Driving cycles: 1st 2. Check mode: OFF 3. Engine RPM: Less than 2200 rpm Except above (MIL blinks immediately): Crankshaft 200 revolutions **Possible Causes:** • Open or short in engine wire harness • Connector connection • Vacuum hose connections • Ignition system • Injector for direct injector • Injector for port injector • Fuel pressure • Mass Air Flow (MAF) meter • Engine Coolant Temperature (ECT) sensor • Compression pressure • Valve clearance • Valve timing • Ventilation valve and hose • Ventilation hose connections • Air induction system • ECM

DTC	Trouble Code Title, Conditions & Possible Causes
DTC: P0308 **2T ECM, MIL: Yes** **Year:** 2009, 2010 **Model:** All **Engine:** All V8 **Transmission:** All	**Cylinder 8 Misfire Detected:** Battery voltage: 8 V or more VVT system: Not operated by scan tool Engine RPM: 335 to 6600 rpm Either of the following conditions 1 and 2 is met: 1. ECT at engine start: More than -7°C (19°F) 2. ECT: More than 20°C (68°F) Fuel cut: OFF Monitor period of emission-related-misfire: First 1000 revolutions after engine start, or Check Mode: Crankshaft 1000 revolutions Except above: Crankshaft 1000 revolutions x 4 Monitor period of catalyst-damaged-misfire (MIL blinks): All of following conditions 1, 2 and 3 are met: Crankshaft 200 revolutions x 3 1. Driving cycles: 1st 2. Check mode: OFF 3. Engine RPM: Less than 2200 rpm Except above (MIL blinks immediately): Crankshaft 200 revolutions **Possible Causes:** • Open or short in engine wire harness • Connector connection • Vacuum hose connections • Ignition system • Injector for direct injector • Injector for port injector • Fuel pressure • Mass Air Flow (MAF) meter • Engine Coolant Temperature (ECT) sensor • Compression pressure • Valve clearance • Valve timing • Ventilation valve and hose • Ventilation hose connections • Air induction system • ECM
DTC: P0327 **1T ECM, MIL: Yes** **Year:** 2009, 2010 **Model:** All **Engine:** All V6, V8 **Transmission:** All	**Knock Sensor 1 Circuit Low Input (Bank 1 or Single Sensor):** Battery voltage: 10.5 V or more Time after engine start: 5 seconds or more Engine switch: On (IG) Starter: OFF **Possible Causes:** • Short in knock sensor circuit • Knock sensor • ECM
DTC: P0328 **1T ECM, MIL: Yes** **Year:** 2009, 2010 **Model:** All **Engine:** All V6, V8 **Transmission:** All	**Knock Sensor 1 Circuit High Input (Bank 1 or Single Sensor):** Battery voltage: 10.5 V or more Time after engine start: 5 seconds or more Engine switch: On (IG) Starter: OFF **Possible Causes:** • Open in knock sensor circuit • Knock sensor • ECM
DTC: P032C **1T ECM, MIL: Yes** **Year:** 2009, 2010 **Model:** LS460 **Engine:** 4.6L V8 **Transmission:** All	**Knock Sensor 3 Circuit Low:** Battery voltage: 10.5 V or more Time after engine start: 5 seconds or more **Possible Causes:** • Short in knock sensor (bank 1 sensor 2) circuit • Knock sensor (bank 1 sensor 2) • ECM

DTC	Trouble Code Title, Conditions & Possible Causes
DTC: P032D **1T ECM, MIL: Yes** **Year:** 2009, 2010 **Model:** All **Engine:** All **Transmission:** All	**Knock Sensor 3 Circuit High:** Battery voltage: 10.5 V or more Time after engine start: 5 seconds or more **Possible Causes:** • Open in knock sensor (bank 1 sensor 2) circuit • Knock sensor (bank 1 sensor 2) • ECM
DTC: P0332 **1T ECM, MIL: Yes** **Year:** 2009, 2010 **Model:** All **Engine:** All **Transmission:** All	**Knock Sensor 2 Circuit Low Input (Bank 2):** Monitor runs whenever following DTCs not present: None Battery voltage: 10.5 V or higher Time after engine start: 5 seconds or more **Possible Causes:** • Short in knock sensor circuit • Knock sensor • ECM
DTC: P0333 **1T ECM, MIL: Yes** **Year:** 2009, 2010 **Model:** All **Engine:** All **Transmission:** All	**Knock Sensor 2 Circuit High Input (Bank 2):** Battery voltage: 10.5 V or more Time after engine start: 5 seconds or more Engine switch: On (IG) Starter: OFF **Possible Causes:** • Open in knock sensor circuit • Knock sensor 1 or 2 • ECM
DTC: P0335 **1T ECM, MIL: Yes** **Year:** 2009, 2010 **Model:** All **Engine:** All **Transmission:** All	**Crankshaft Position Sensor "A" Circuit:** CKP Sensor Range Check/Rationality: Time after starter OFF to ON: 3 seconds or more Battery voltage: 7 V or more Minimum battery voltage while starter ON: Less than 11 V Number of VVT sensor signal pulse: 6 times Camshaft position sensor circuit fail (P0340,P0342,P0343): Not detected CKP Sensor Verify Pulse Input (Case 1): Engine speed: 600 rpm or less Starter: OFF Time after starter ON to OFF: 3 seconds or more CKP Sensor Verify Pulse Input (Case 2): Starter: ON Minimum battery voltage starter ON: Below 11 V **Possible Causes:** • Open or short in CKP sensor circuit • CKP sensor • CKP sensor plate • ECM
DTC: P0337 **1T ECM, MIL: Yes** **Year:** 2009, 2010 **Model:** All **Engine:** All **Transmission:** All	**Crankshaft Position Sensor "A" Circuit Low Input:** Monitor runs whenever following DTCs not present: P0335 (Crankshaft Position Sensor) Battery voltage: 8 V or more Engine switch: on (IG) Starter: OFF **Possible Causes:** • Open or short in CKP sensor circuit • CKP sensor • CKP sensor plate • ECM

DTC	Trouble Code Title, Conditions & Possible Causes
DTC: P0339 **1T ECM, MIL: Yes** **Year:** 2009, 2010 **Model:** All **Engine:** All **Transmission:** All	**Crankshaft Position Sensor "A" Circuit Intermittent:** Monitor runs whenever following conditions are met: Engine speed 1,000 rpm or more. No Crankshaft Position (CKP) sensor signal for 0.05 seconds or more. 3 seconds or more have elapsed since starter signal switched from ON to OFF. **Possible Causes:** • Open or short in CKP sensor circuit • CKP sensor • CKP sensor plate • ECM
DTC: P033C **1T ECM, MIL: Yes** **Year:** 2009, 2010 **Model:** All **Engine:** All V8 **Transmission:** All	**Knock Sensor 4 Circuit Low Input:** Battery voltage: 10.5 V or more Time after engine start: 5 seconds or more **Possible Causes:** • Short in knock sensor (bank 2 sensor 2) circuit • Knock sensor (bank 2 sensor 2) • ECM
DTC: P033D **1T ECM, MIL: Yes** **Year:** 2009, 2010 **Model:** All **Engine:** All V8 **Transmission:** All	**Knock Sensor 4 Circuit High Input:** Battery voltage: 10.5 V or more Time after engine start: 5 seconds or more **Possible Causes:** • Open in knock sensor (bank 2 sensor 2) circuit • Knock sensor (bank 2 sensor 2) • ECM
DTC: P0340 **1T ECM, MIL: Yes** **Year:** 2009, 2010 **Model:** All **Engine:** All V6. V8 **Transmission:** All	**Camshaft Position Sensor Circuit Malfunction:** Camshaft Position Sensor / Crankshaft Position Sensor Range Check / Rationality: Monitor runs whenever following DTCs are not present : None Engine speed: 600 rpm or more Starter: OFF Crankshaft Position Sensor Range Check / Rationality: Starter: ON Minimum battery voltage: Less than 11 V **Possible Causes:** • Open or short in VVT sensor for intake camshaft circuit • VVT sensor for intake camshaft • Camshaft timing gear for intake camshaft • Jumped tooth of timing chain for intake camshaft • ECM
DTC: P0342 **1T ECM, MIL: Yes** **Year:** 2009, 2010 **Model:** All **Engine:** All **Transmission:** All	**Camshaft Position Sensor "A" Circuit Low Input (Bank 1 or Single Sensor):** Starter: OFF Engine switch: On (IG) Time after engine switch off to on (IG): 2 seconds or more VVT sensor verify pulse input fail (P0340): Not detected Battery voltage: 8 V or more **Possible Causes:** • Open or short in VVT sensor for intake side circuit • VVT sensor for intake side • Camshaft timing gear assembly for intake camshaft • ECM
DTC: P0343 **1T ECM, MIL: Yes** **Year:** 2009, 2010 **Model:** All **Engine:** All **Transmission:** All	**Camshaft Position Sensor "A" Circuit High Input (Bank 1 or Single Sensor):** Camshaft Position Sensor Range Check (Chattering, Low voltage, High voltage): Starter: OFF Time after engine switch off to on (IG): 2 seconds or more **Possible Causes:** • Open or short in VVT sensor for intake camshaft circuit • VVT sensor for intake camshaft • Camshaft timing gear for intake camshaft • Jumped tooth of timing chain for intake camshaft • ECM

DTC	Trouble Code Title, Conditions & Possible Causes
DTC: P0345 **1T ECM, MIL: Yes** **Year:** 2009, 2010 **Model:** All **Engine:** All **Transmission:** All	**Camshaft Position Sensor "A" Circuit (Bank 2):** Camshaft Position Sensor / Crankshaft Position Sensor Range Check / Rationality: Engine speed: 600 rpm or more Battery voltage: 8 V or more Starter: OFF Engine switch: On (IG) **Possible Causes:** • Open or short in VVT sensor for intake camshaft circuit • VVT sensor for intake camshaft • Camshaft timing gear for intake camshaft • Jumped tooth of timing chain for intake camshaft • ECM
DTC: P0347 **1T ECM, MIL: Yes** **Year:** 2009, 2010 **Model:** All **Engine:** All **Transmission:** All	**Camshaft Position Sensor "A" Circuit Low Input (Bank 2):** Starter: OFF Engine switch: On (IG) Time after engine switch off to on (IG): 2 seconds or more VVT sensor verify pulse input fail (P030, P0345): Not detected Battery voltage: 8 V or more **Possible Causes:** • Open or short in VVT sensor for intake side circuit • VVT sensor for intake side • Jumped tooth of timing chain for intake camshaft • ECM
DTC: P0348 **1T ECM, MIL: Yes** **Year:** 2009, 2010 **Model:** All **Engine:** All **Transmission:** All	**Camshaft Position Sensor "A" Circuit High Input (Bank 2):** Camshaft Position Sensor Range Check (Chattering, Low voltage, High voltage): Starter: OFF Time after engine switch off to on (IG): 2 seconds or more Battery Voltage: 8 V or more **Possible Causes:** • Open or short in VVT sensor for intake camshaft circuit • VVT sensor for intake camshaft • Camshaft timing gear for intake camshaft • Jumped tooth of timing chain for intake camshaft • ECM
DTC: P0351 **1T ECM, MIL: Yes** **Year:** 2009, 2010 **Model:** All **Engine:** All V6, V8 **Transmission:** All	**Ignition Coil "A" Primary / Secondary Circuit:** Monitor runs whenever following DTCs are not present: - Engine speed: 1500 rpm or less Either A or B condition is met: - A. Following conditions are met: (a) and (b) (a) Engine speed: 500 rpm or less (b) Battery voltage: 6 V or more B. Following conditions are met: (a), (b) and (c) (a) Engine speed: More than 500 rpm (b) Battery voltage: 10 V or more (c) Number of sparks after CPU is reset: 5 sparks or more **Possible Causes:** • Ignition system • Open or short in IGF1, IGF2 or IGT circuit (1 to 8) between ignition coil and ECM • No. 1 to No. 8 ignition coils • ECM

DTC	Trouble Code Title, Conditions & Possible Causes
DTC: P0352 **1T ECM, MIL: Yes** **Year:** 2009, 2010 **Model:** All **Engine:** All V6, V8 **Transmission:** All	**Ignition Coil "B" Primary / Secondary Circuit:** Either of the following condition A or B is met: A. Engine RPM: 1500 rpm or less B. Starter: OFF Either of the following condition C or D is met: C. Both of the following conditions are met: (a) Engine speed: 500 rpm or less (b) Battery voltage: 6 V or more D. All of the following conditions are met: (a) Engine speed: More than 500 rpm (b) Battery voltage: 10 V or more (c) Number of sparks after CPU reset: 5 sparks or more **Possible Causes:** • Ignition system • Open or short in IGF1 or IGT2 circuit (1 to 4) or (1 to 6) between ignition coil and ECM • No. 1 to No. 4 ignition coils or No. 1 to No. 6 ignition coils • ECM
DTC: P0353 **1T ECM, MIL: Yes** **Year:** 2009, 2010 **Model:** All **Engine:** All V6, V8 **Transmission:** All	**Ignition Coil "C" Primary / Secondary Circuit:** Monitor runs whenever following DTCs are not present: - Engine speed: 1500 rpm or less Either A or B condition is met: - A. Following conditions are met: (a) and (b) (a) Engine speed: 500 rpm or less (b) Battery voltage: 6 V or more B. Following conditions are met: (a), (b) and (c) (a) Engine speed: More than 500 rpm (b) Battery voltage: 10 V or more (c) Number of sparks after CPU is reset: 5 sparks or more **Possible Causes:** • Ignition system • Open or short in IGF1, IGF2 or IGT circuit (1 to 8) between ignition coil and ECM • No. 1 to No. 8 ignition coils • ECM
DTC: P0354 **1T ECM, MIL: Yes** **Year:** 2009, 2010 **Model:** All **Engine:** All V6, V8 **Transmission:** All	**Ignition Coil "D" Primary / Secondary Circuit:** Either of the following condition A or B is met: A. Engine RPM: 1500 rpm or less B. Starter: OFF Either of the following condition C or D is met: C. Both of the following conditions are met: (a) Engine speed: 500 rpm or less (b) Battery voltage: 6 V or more D. All of the following conditions are met: (a) Engine speed: More than 500 rpm (b) Battery voltage: 10 V or more (c) Number of sparks after CPU reset: 5 sparks or more **Possible Causes:** • Ignition system • Open or short in IGF1 or IGT4 circuit (1 to 4) or (1 to 6) between ignition coil and ECM • No. 1 to No. 4 ignition coils or No. 1 to No. 6 ignition coils • ECM

DTC	Trouble Code Title, Conditions & Possible Causes
DTC: P0355 **1T ECM, MIL: Yes** **Year:** 2009, 2010 **Model:** All **Engine:** All V6, V8 **Transmission:** All	**Ignition Coil "E" Primary / Secondary Circuit:** Either of the following condition A or B is met: A. Engine RPM: 1500 rpm or less B. Starter: OFF Either of the following condition C or D is met: C. Both of the following conditions are met: (a) Engine speed: 500 rpm or less (b) Battery voltage: 6 V or more D. All of the following conditions are met: (a) Engine speed: More than 500 rpm (b) Battery voltage: 10 V or more (c) Number of sparks after CPU reset: 5 sparks or more **Possible Causes:** • Ignition system • Open or short in IGF1 or IGT circuit (1 to 6) between ignition coil and ECM • No. 1 to No. 6 ignition coils • ECM
DTC: P0356 **1T ECM, MIL: Yes** **Year:** 2009, 2010 **Model:** All **Engine:** All V6, V8 **Transmission:** All	**Ignition Coil "F" Primary / Secondary Circuit:** Either of the following conditions is met: Condition A or B A. Both of the following conditions are met: Conditions a and b a. Engine speed: 500 rpm or less b. Battery voltage: 6 V or more B. All of the following conditions are met: Conditions d, e and f d. Engine speed: More than 500 rpm e. Battery voltage: 10 V or more f. Number of sparks after CPU is reset: 5 sparks or more **Possible Causes:** • Ignition system • Open or short in IGF1, IGF2 or IGT circuit (1 to 6) between ignition coil and ECM • No. 1 to No. 6 ignition coils • ECM
DTC: P0357 **1T ECM, MIL: Yes** **Year:** 2009, 2010 **Model:** All **Engine:** All V8 **Transmission:** All	**Ignition Coil "G" Primary / Secondary Circuit:** Monitor runs whenever following DTCs are not present: - Engine speed: 1500 rpm or less Either A or B condition is met: - A. Following conditions are met: (a) and (b) (a) Engine speed: 500 rpm or less (b) Battery voltage: 6 V or more B. Following conditions are met: (a), (b) and (c) (a) Engine speed: More than 500 rpm (b) Battery voltage: 10 V or more (c) Number of sparks after CPU is reset: 5 sparks or more **Possible Causes:** • Ignition system • Open or short in IGF1, IGF2 or IGT circuit (1 to 8) between ignition coil and ECM • No. 1 to No. 8 ignition coils • ECM

DTC	Trouble Code Title, Conditions & Possible Causes
DTC: P0358 **1T ECM, MIL: Yes** **Year:** 2009, 2010 **Model:** All **Engine:** All V8 **Transmission:** All	**Ignition Coil "H" Primary / Secondary Circuit:** Monitor runs whenever following DTCs are not present: - Engine speed: 1500 rpm or less Either A or B condition is met: - A. Following conditions are met: (a) and (b) (a) Engine speed: 500 rpm or less (b) Battery voltage: 6 V or more B. Following conditions are met: (a), (b) and (c) (a) Engine speed: More than 500 rpm (b) Battery voltage: 10 V or more (c) Number of sparks after CPU is reset: 5 sparks or more **Possible Causes:** • Ignition system • Open or short in IGF1, IGF2 or IGT circuit (1 to 8) between ignition coil and ECM • No. 1 to No. 8 ignition coils • ECM
DTC: P0365 **1T ECM, MIL: Yes** **Year:** 2009, 2010 **Model:** All **Engine:** All **Transmission:** All	**Camshaft Position Sensor "B" Circuit (Bank 1):** Camshaft Position Sensor Range Check (Chattering, Low voltage, High voltage): Starter: OFF Engine switch: On (IG) Time after engine switch off to on (IG): 2 seconds or more Battery voltage: 8 V or more Camshaft Position Sensor Range Check / Rationality: Engine speed: 600 rpm or more Battery voltage: 8 V or more Starter: OFF Engine switch: On (IG) **Possible Causes:** • Open or short in VVT sensor for exhaust camshaft circuit • VVT sensor for exhaust camshaft • Exhaust camshaft • Jumped tooth of timing chain • ECM
DTC: P0367 **1T ECM, MIL: Yes** **Year:** 2009, 2010 **Model:** All **Engine:** All **Transmission:** All	**Camshaft Position Sensor "B" Circuit Low Input (Bank 1):** Camshaft Position Sensor Range Check (Chattering, Low voltage, High voltage): Starter: OFF Engine switch: On (IG) Time after engine switch off to on (IG): 2 seconds or more Battery voltage: 8 V or more **Possible Causes:** • Open or short in VVT sensor for exhaust camshaft circuit • VVT sensor for exhaust camshaft • Exhaust camshaft • Jumped tooth of timing chain • ECM
DTC: P0368 **1T ECM, MIL: Yes** **Year:** 2009, 2010 **Model:** All **Engine:** All **Transmission:** All	**Camshaft Position Sensor "B" Circuit High Input (Bank 1):** Monitor runs whenever following DTCs not present: None VVT Sensor Range Check (Fluctuating, Low voltage, High voltage): Starter: OFF Engine switch on (IG) and time after engine switch changed from off to on (IG): 2 seconds or more Battery voltage: 8 V or more **Possible Causes:** • Open or short in VVT sensor for exhaust side circuit • VVT sensor for exhaust side • Exhaust camshaft • ECM

DTC	Trouble Code Title, Conditions & Possible Causes
DTC: P0390 **1T ECM, MIL: Yes** **Year:** 2009, 2010 **Model:** All **Engine:** All **Transmission:** All	**Camshaft Position Sensor "B" Circuit (Bank 2):** Engine speed: 600 rpm or more Starter: OFF Exhaust VVT sensor range check fail (P0367, P0368, P0392, P0393): Not detected Exhaust VVT sensor voltage: 0.3 V or more, and 4.7 V or less Battery voltage: 8 V or more Engine switch: On (IG) Time after engine switch off to on (IG): 0.5 seconds or more **Possible Causes:** • Open or short in VVT sensor for exhaust camshaft circuit • VVT sensor for exhaust camshaft • Exhaust camshaft • ECM
DTC: P0392 **1T ECM, MIL: Yes** **Year:** 2009, 2010 **Model:** All **Engine:** All **Transmission:** All	**Camshaft Position Sensor "B" Circuit Low Input (Bank 2):** Starter: OFF Ignition switch: ON Time after ignition switch off to on: 2 seconds or more Battery voltage: 8 V or more **Possible Causes:** • Open or short in VVT sensor for exhaust camshaft circuit • VVT sensor for exhaust camshaft • Exhaust camshaft • Jumped tooth of timing chain • ECM
DTC: P0393 **1T ECM, MIL: Yes** **Year:** 2009, 2010 **Model:** All **Engine:** All **Transmission:** All	**Camshaft Position Sensor "B" Circuit High Input (Bank 2):** Monitor runs whenever following DTCs not present: None VVT Sensor Range Check (Fluctuating, Low voltage, High voltage): Starter: OFF Engine switch on (IG) and time after engine switch changed from off to on (IG): 2 seconds or more Battery voltage: 8 V or more **Possible Causes:** • Open or short in VVT sensor for exhaust side circuit • VVT sensor for exhaust side • Exhaust camshaft • ECM
DTC: P0412 **1T ECM, MIL: Yes** **Year:** 2009 **Model:** GX470 **Engine:** 4.7L V8 **Transmission:** All	**Secondary Air Injection System Switching Valve "A" Circuit:** The monitor will run whenever this DTC is not present: None Case 1: Air pump: Operating Air switching valve: Operating Battery voltage: 8 V or more Ignition switch: ON Starter: OFF Case 2: Air pump: Not operating Air switching valve: Not operating Battery voltage: 8 V or more Ignition switch: ON Starter: OFF **Possible Causes:** • Open in air switching valve drive circuit • Short between air switching valve circuit and +B circuit • Air injection control driver (AID) • Air switching valve (ASV) • ECM

DTC	Trouble Code Title, Conditions & Possible Causes
DTC: P0418 **1T ECM, MIL: Yes** **Year:** 2009 **Model:** GX470 **Engine:** 4.7L V8 **Transmission:** All	**Secondary Air Injection System Control "A" Circuit:** The monitor will run whenever this DTC is not present: None Case 1: Air pump: Operating Air switching valve: Operating Battery voltage: 8 V or more Ignition switch: ON Starter: OFF Case 2: Air pump: Not operating Air switching valve: Not operating Battery voltage: 8 V or more Ignition switch: ON Starter: OFF **Possible Causes:** • Open in air pump drive circuit • Short between air pump circuit and +B circuit • Air injection control driver (AID) • ECM
DTC: P0420 **2T ECM, MIL: Yes** **Year:** 2009, 2010 **Model:** All **Engine:** All V6, V8 **Transmission:** All	**Catalyst System Efficiency Below Threshold (Bank 1):** Battery voltage: 11 V or more IAT: -10°C (14°F) or more Engine coolant temperature sensor: 75°C (167°F) or more Atmospheric pressure coefficient: 75 kPa (570 mmHg) or more Idling: OFF Engine RPM: Less than 3200 rpm A/F sensor status: Activated Fuel system status: Closed loop Engine load: 10 to 70% All of the following conditions are met: Condition 1, 2 and 3 1. Mass air flow rate: 2.9 to 75 g/sec. 2. Front catalyst temperature (estimated): 600 to 820°C (1112 to 1508°F) 3. Rear catalyst temperature (estimated): 300 to 830°C (572 to 1526°F) Shift position: 3rd or higher **Possible Causes:** • Gas leakage from exhaust system • A/F sensor (bank 1 sensor 1) • HO2 sensor (bank 1 sensor 2) • Exhaust manifold sub-assembly LH • Front exhaust pipe assembly

DTC	Trouble Code Title, Conditions & Possible Causes
DTC: P0430 **2T ECM, MIL: Yes** **Year:** 2009, 2010 **Model:** All **Engine:** All **Transmission:** All	**Catalyst System Efficiency Below Threshold (Bank 2):** Battery voltage: 11 V or more IAT: -10°C (14°F) or more ECT: 75°C (167°F) or more Atmospheric pressure: 100 kPa (750 mmHg) or more Idle: OFF Engine rpm: Less than 3200 rpm A/F sensor: Activated Fuel system status: Closed loop Engine load: 10 to 75% All of the following conditions are met: Condition 1, 2 and 3 1. MAF: 5 to 45 g/sec. 2. Front catalyst temperature (estimated): 620 to 800°C (1148 to 1472°F) 3. Rear catalyst temperature (estimated): 620 to 800°C (1148 to 1472°F) Rear HO2S monitor: Completed Shift position: 4th or more **Possible Causes:** • Gas leakage from exhaust system • A/F sensor (bank 1 sensor 1) • HO2 sensor (bank 1 sensor 2) • Exhaust manifold (TWC) • Front exhaust pipe assembly
DTC: P043E **2T ECM, MIL: Yes** **Year:** 2009, 2010 **Model:** All **Engine:** All V6, V8 **Transmission:** All	**Evaporative Emission System Reference Orifice Clog Up:** Monitor runs whenever following DTCs not present: None EVAP key-off monitor runs when all of following conditions met: Atmospheric pressure: 70 to 110 kPa-a (525 to 825 mmHg-a) Battery voltage: 10.5 V or more Vehicle speed: Below 2.5 mph (4 km/h) Engine switch: off Time after key off: 5 or 7 or 9.5 hours Canister pressure sensor malfunction (P0451, P0452 and P0453): Not detected Purge VSV: Not operated by scan tool Vent valve: Not operated by scan tool Leak detection pump: Not operated by scan tool Both of following conditions met before key off: Conditions 1 and 2 1. Duration that vehicle driven: 5 minutes or more 2. EVAP purge operation: Performed ECT: 4.4 to 35°C (40 to 95°F) IAT: 4.4 to 35°C (40 to 95°F) **Possible Causes:** • Canister pump module (reference orifice, leak detection pump, vent valve) • Connector/wire harness (canister pump module - ECM) • EVAP system hose (pipe from air inlet port to canister pump module, canister filter, fuel tank vent hose) • ECM

DTC	Trouble Code Title, Conditions & Possible Causes
DTC: P043F **2T ECM, MIL: Yes** **Year:** 2009, 2010 **Model:** All **Engine:** All V6, V8 **Transmission:** All	**Evaporative Emission System Reference Orifice High Flow:** Monitor runs whenever following DTCs not present: None EVAP key-off monitor runs when all of following conditions met: Atmospheric pressure: 70 to 110 kPa-a (525 to 825 mmHg-a) Battery voltage: 10.5 V or more Vehicle speed: Below 2.5 mph (4 km/h) Engine switch: off Time after key off: 5 or 7 or 9.5 hours Canister pressure sensor malfunction (P0451, P0452 and P0453): Not detected Purge VSV: Not operated by scan tool Vent valve: Not operated by scan tool Leak detection pump: Not operated by scan tool Both of following conditions met before key off: Conditions 1 and 2 1. Duration that vehicle driven: 5 minutes or more 2. EVAP purge operation: Performed ECT: 4.4 to 35°C (40 to 95°F) IAT: 4.4 to 35°C (40 to 95°F) **Possible Causes:** • Canister pump module (reference orifice, leak detection pump, vent valve) • Connector/wire harness (canister pump module - ECM) • EVAP system hose (pipe from air inlet port to canister pump module, canister filter, fuel tank vent hose) • ECM
DTC: P0441 **2T ECM, MIL: Yes** **Year:** 2009, 2010 **Model:** All **Engine:** All V6, V8 **Transmission:** All	**Evaporative Emission Control System Incorrect Purge Flow:** Monitor runs whenever following DTCs not present: None EVAP key-off monitor runs when all of the following conditions are met: Atmospheric pressure: 70 to 110 kPa-a (525 to 825 mmHg-a) Battery voltage: 10.5 V or more Vehicle speed: Below 2.5 mph (4 km/h) Engine switch: Off Time after key off: 5 or 7 or 9.5 hours EVAP pressure sensor malfunction (P0450, P0451, P0452 and P0453): Not detected EVAP canister purge valve: Not operated by scan tool EVAP canister vent valve: Not operated by scan tool EVAP leak detection pump: Not operated by scan tool Both of the following conditions are met before key off: Conditions 1 and 2 1. Duration that vehicle has been driven: 5 minutes or more 2. EVAP purge operation: Performed ECT: 4.4 to 35°C (40 to 95°F) IAT: 4.4 to 35°C (40 to 95°F) **Possible Causes:** • Purge VSV • Connector/wire harness (Purge VSV - ECM) • ECM • Canister pump module • Leakage from EVAP system • Leakage from EVAP line (Purge VSV - Intake manifold)
DTC: P0443 **1T ECM, MIL: Yes** **Year:** 2009, 2010 **Model:** All **Engine:** All **Transmission:** All	**Evaporative Emission Control System Purge Control Valve Circuit:** Monitor runs whenever following DTCs not present: None **Possible Causes:** • Open or short in purge VSV circuit • Purge VSV • ECM
DTC: P0450 **1T ECM, MIL: Yes** **Year:** 2009, 2010 **Model:** All **Engine:** All **Transmission:** All	**Evaporative Emission Control System Pressure Sensor / Switch:** Either of following conditions is met: Condition A or B A. Engine switch: On (IG) B. Soak timer: ON **Possible Causes:** • Canister pump module • ECM

DTC	Trouble Code Title, Conditions & Possible Causes
DTC: P0451 **2T ECM, MIL: Yes** **Year:** 2009, 2010 **Model:** All **Engine:** All V6, V8 **Transmission:** All	**Evaporative Emission Control System Pressure Sensor Range / Performance:** Noise monitor: Monitor runs whenever following DTCs not present: None Atmospheric pressure (absolute pressure): 70 to 110 kPa-a (525 to 825 mmHg-a) Battery voltage: 10.5 V or more Intake air temperature: 4.4 to 35°C (40 to 95°F) Canister pressure sensor malfunction (P0452, 0453): Not detected Either of following conditions met: A or B A. Engine condition: Running B. Time after key off: 5 or 7 or 9.5 hours Fixed/flat monitor: Monitor runs whenever following DTCs not present: None Battery voltage: 10.5 V or more Intake air temperature: 4.4 to 35°C (40 to 95°F) Canister pressure sensor malfunction (P0452, 0453): Not detected Atmospheric pressure (absolute pressure): 70 to 110 kPa-a (525 to 825 mmHg-a) Time after key off: 5 or 7 or 9.5 hours **Possible Causes:** • Canister pump module • Connector/wire harness (Canister pump module - ECM) • EVAP system hose (pipe from air inlet port to canister pump module, canister filter, fuel tank vent hose) • ECM
DTC: P0452 **2T ECM, MIL: Yes** **Year:** 2009, 2010 **Model:** All **Engine:** All V6, V8 **Transmission:** All	**Evaporative Emission Control System Pressure Sensor / Switch Low Input:** Monitor runs whenever following DTCs not present: None Either of following conditions met: (a) or (b) (a) Engine switch: on (IG) (b) Soak timer: ON **Possible Causes:** • Canister pump module • Connector/wire harness (Canister pump module - ECM) • EVAP system hose (pipe from air inlet port to canister pump module, canister filter, fuel tank vent hose) • ECM
DTC: P0453 **1T ECM, MIL: Yes** **Year:** 2009, 2010 **Model:** All **Engine:** All V6, V8 **Transmission:** All	**Evaporative Emission Control System Pressure Sensor / Switch High Input:** Either of following conditions is met: Condition A or B A. Engine switch: On (IG) B. Soak timer: ON **Possible Causes:** • Canister pump module (including canister pressure sensor) • Connector/wire harness (Canister pump module - ECM) • ECM

DTC	Trouble Code Title, Conditions & Possible Causes
DTC: P0455 **2T ECM, MIL: Yes** **Year:** 2009, 2010 **Model:** All **Engine:** All V6, V8 **Transmission:** All	**Evaporative Emission Control System Leak Detected (Gross Leak):** Monitor runs whenever following DTCs not present: None EVAP key-off monitor runs when all of following conditions met: Atmospheric pressure: 70 to 110 kPa-a (525 to 825 mmHg-a) Battery voltage: 10.5 V or more Vehicle speed: Below 2.5 mph (4 km/h) Engine switch: off Time after key off: 5 or 7 or 9.5 hours Canister pressure sensor malfunction (P0451, P0452 and P0453): Not detected Purge VSV: Not operated by scan tool Vent valve: Not operated by scan tool Leak detection pump: Not operated by scan tool Both of following conditions met before key off: Conditions 1 and 2 1. Duration that vehicle driven: 5 minutes or more 2. EVAP purge operation: Performed ECT: 4.4 to 35°C (40 to 95°F) IAT: 4.4 to 35°C (40 to 95°F) **Possible Causes:** • Fuel tank cap (loose) • Leakage from EVAP line (Canister - Fuel tank) • Leakage from EVAP line (Purge VSV - Canister) • Canister pump module • Leakage from fuel tank • Leakage from canister
DTC: P0456 **2T ECM, MIL: Yes** **Year:** 2009, 2010 **Model:** All **Engine:** All V6, V8 **Transmission:** All	**Evaporative Emission Control System Leak Detected (Very Small Leak):** EVAP key-off monitor runs when all of following conditions are met: Atmospheric pressure: 70 to 110 kPa (525 to 825 mmHg) Battery voltage: 10.5 V or more Vehicle speed: Below 2.5 mph (4 km/h) Engine switch: OFF Time after key off: 5, 7 or 9.5 hours EVAP pressure sensor malfunction (P0451, P0452 and P0453): Not detected Purge VSV: Not operated by scan tool Vent valve: Not operated by scan tool Leak detection pump: Not operated by scan tool Both of following conditions are met before key OFF: Conditions 1 and 2 1. Duration that vehicle is being driven: 5 minutes or more 2. EVAP purge operation: Performed ECT: 4.4 to 35°C (40 to 95°F): P0456 IAT:4.4 to 35°C (40 to 95°F): P0456 **Possible Causes:** • Fuel tank cap (loose) • Leakage from EVAP line (Canister - Fuel tank) • Leakage from EVAP line (Purge VSV - Canister) • Canister pump module • Leakage from fuel tank • Leakage from canister

DTC	Trouble Code Title, Conditions & Possible Causes
DTC: P0500 **1T TCM, MIL: Yes** **Year:** 2009, 2010 **Model:** All **Engine:** All V6, V8 **Transmission:** All	**Vehicle Speed Sensor "A":** The monitor will run whenever the following DTCs are not present: None Battery voltage: 8 V or more Engine switch: On (IG) Starter: OFF Engine speed: 2005 rpm or more (varies with throttle position) Throttle position sensor circuit: Not malfunction Fuel cut at high engine speed: OFF Transmission range: Not R position When either condition below is met: Condition (1) or (2): Condition (1): * ECT: 20°C (68°F) or more * ECT sensor circuit: Not malfunction * Time after park/neutral position switch ON to OFF: 10 sec. or more Condition (2): * ECT sensor: ECT is less than 20°C (68°F) or ECT sensor malfunction * Time after park/neutral position switch ON to OFF: 30 sec. or more When all conditions below are met: Condition (1), (2), (3) and (4) Condition (1): Transmission range switch: Not R position Condition (2): TCM indicated pressure value of SL1: Less than 1600 kPa (16.3 kgf/cm2, 232 psi) Condition (3): Oil pressure switch: ON Condition (4): When either condition below is met: Condition (a) or (b) * Condition (a): TCM selected gear: Not 1st gear * Condition (b): Input speed sensor: 0 rpm or more **Possible Causes:** • Speed sensor (SP2) • Transmission wire • Harness and connector • Combination meter assembly • TCM
DTC: P0502 **1T TCM, MIL: Yes** **Year:** 2009, 2010 **Model:** GS350 **Engine:** 3.5L V6 **Transmission:** All	**Vehicle Speed Sensor "A" Circuit Low:** The monitor will run whenever the following DTCs are not present: None Battery voltage: 8 V or more Engine switch: On (IG) Starter: OFF **Possible Causes:** • Speed sensor (SP2) • Transmission wire • Harness and connector • TCM
DTC: P0503 **1T TCM, MIL: Yes** **Year:** 2009, 2010 **Model:** All **Engine:** All **Transmission:** All	**Vehicle Speed Sensor "A" Intermittent / Erratic / High:** The monitor will run whenever the following DTCs are not present: None Battery voltage: 8 V or more Engine switch: On (IG) Starter: OFF **Possible Causes:** • Speed sensor (SP2) • Transmission wire • Harness and connector • TCM

DTC	Trouble Code Title, Conditions & Possible Causes
DTC: P0504 **1T ECM, MIL: Yes** **Year:** 2009, 2010 **Model:** All **Engine:** All **Transmission:** All	**Brake Switch "A" / "B" Correlation:** Conditions (a), (b) and (c) continue for 0.5 seconds or more: # (a) Engine switch on (IG) # (b) Brake pedal released # (c) STP signal OFF when ST1- signal OFF **Possible Causes:** • Short in stop light switch signal circuit • STOP fuse • Stop light switch • ECM
DTC: P0505 **2T ECM, MIL: Yes** **Year:** 2009, 2010 **Model:** All **Engine:** All V6, V8 **Transmission:** All	**Idle Control System Malfunction:** Monitor will run whenever these DTCs are not present: P0011, P0021 (VVT System - Advance) P0012, P0022 (VVT System - Retard) P0016, P0018 (VVT System - Misalignment) P0013, P0023 (Exhaust VVT Oil Control Valve) P0014, P0024 (Exhaust VVT System - Advance) P0015, P0025 (Exhaust VVT System - Retard) P0017, P0019 (Exhaust VVT System - Misalignment) P0010, P0020, P2614, P1023, P1360, P1361, P1362, P1363, P1364, P1365, P1366, P1367 (Motor Drive VVT System Control Module) P0031, P0032, P0051, P0052 (Air Fuel Ratio Sensor Heater) P2195, P2196, P2197, P2198, P2237, P2238, P2239, P2240, P2241, P2242, P2252, P2253, P2255, P2256, P2A00, P2A03 (Air Fuel Ratio Sensor) P0102, P0103, P010C, P010D, P010F (Mass Air Flow Sensor) P0115, P0117, P0118 (Engine Coolant Temperature Sensor) P0125 (Insufficient Coolant Temperature for Closed Loop Fuel Control) P0120, P0121, P0122, P0123, P0220, P0222, P0223, P2135 (Throttle Position Sensor) P0171, P0172, P0174, P0175 (Fuel System) P0300 - P0308 (Misfire) P0335 (Crankshaft Position Sensor) P0340, P1340 (Camshaft Position Sensor) P0340, P0342, P0343, P0345, P0347, P0348 (VVT Sensor) P0365, P0367, P0368, P0390, P0392, P0393 (Exhaust VVT Sensor) P0351 - P0358 (Igniter) P0500 (Vehicle Speed Sensor) P0705 (Shift lever position switch) Engine: Running **Possible Causes:** • ETCS • Air induction system • Ventilation hose connection • ECM

DTC	Trouble Code Title, Conditions & Possible Causes
DTC: P050A **2T ECM, MIL: Yes** **Year:** 2009, 2010 **Model:** All **Engine:** All V6, V8 **Transmission:** All	**Cold Start Idle Air Control System Performance:** Battery voltage: 8 V or more Time after engine start: 3 seconds or more Starter: OFF ECT at engine start: -10°C (14°F) or more ECT: -10°C to 50°C (14°F to 122°F) Engine idling time: 3 seconds or more Fuel-cut: OFF Vehicle speed: Less than 1.875 mph (3 km/h) Time after shift position changed: 1 second or more Atmospheric pressure: 76 kPa (570 mmHg) or more **Possible Causes:** • Throttle body assembly • MAF meter • Air induction system • PCV hose connections • VVT system • Air cleaner filter element • ECM
DTC: P050B **2T ECM, MIL: Yes** **Year:** 2009, 2010 **Model:** All **Engine:** All V6, V8 **Transmission:** All	**Cold Start Ignition Timing Performance:** Battery voltage: 8 V or more Time after engine start: 3 seconds or more Starter: OFF ECT at engine start: -10°C (14°F) or more ECT: -10°C to 60°C (14°F to 140°F) Engine idling time: 3 seconds or more Vehicle speed: Less than 1.875 mph (3 km/h) **Possible Causes:** • Throttle body assembly • Mass air flow meter • Air induction system • Ventilation hose connections • VVT-iE system • ECM
DTC: P0560 **1T ECM, MIL: Yes** **Year:** 2009, 2010 **Model:** All **Engine:** All V6, V8 **Transmission:** All	**System Voltage:** Stand by RAM: Initialize **Possible Causes:** • Faulty battery • Open in back-up power source circuit • EFI fuse • ECM
DTC: P0571 **T ECM, MIL: Yes** **Year:** 2009, 2010 **Model:** All **Engine:** All V6, V8 **Transmission:** All	**Brake Switch "A" Circuit:** When voltage of STP terminal and that of ST1- terminal of ECM are less than 1 V for 0.5 sec. or more. **Possible Causes:** • Stop light switch • Stop light switch circuit • ECM
DTC: P0604 **1T ECM, MIL: Yes** **Year:** 2009, 2010 **Model:** All **Engine:** All V6, V8 **Transmission:** All	**Internal Control Module Random Access Memory (RAM) Error:** The ECM continuously monitors its internal memory status. This self-check ensures that the ECM is functioning properly. It is diagnosed by internal "mirroring" of the main CPU and sub CPU to detect the Random Access Memory (RAM) errors. If outputs from these CPUs are different and deviate from the standards, the ECM will illuminate the MIL and set a DTC immediately. Monitor will run whenever this DTC is not present: None **Possible Causes:** • ECM

DTC	Trouble Code Title, Conditions & Possible Causes
DTC: P0606 **1T ECM, MIL: Yes** **Year:** 2009, 2010 **Model:** All **Engine:** All V6, V8 **Transmission:** All	**ECM/PCM Processor:** Monitor will run whenever this DTC is not present: None With the engine running. Estimated A/F sensor temperature 450 to 800°C (842 to 1,472°F). Estimated HO2S temperature 450 to 80. ECM CPUs malfunction. A/F sensor transistors malfunction. HO2S transistors malfunction. **Possible Causes:** • Exhaust gas leak • HO2 sensor • ECM
DTC: P0607 **T ECM, MIL: Yes** **Year:** 2009, 2010 **Model:** All **Engine:** All V6, V8 **Transmission:** All	**Control Module Performance:** Monitor runs whenever the following DTCs are not present: None Engine: Running **Possible Causes:** • Exhaust gas leak • HO2 sensor • ECM
DTC: P060A **T ECM, MIL: Yes** **Year:** 2009, 2010 **Model:** All **Engine:** All V6, V8 **Transmission:** All	**Internal Control Module Monitoring Processor Performance:** Monitor runs whenever following DTCs not present: None **Possible Causes:** • ECM
DTC: P060B **T ECM** **Year:** 2010 **Model:** All **Engine:** All **Transmission:** All	**Internal Control Module A/D Processing Performance:** Monitor runs whenever following DTCs not present: None **Possible Causes:** • ECM
DTC: P060D **T ECM, MIL: Yes** **Year:** 2009, 2010 **Model:** All **Engine:** All **Transmission:** All	**Internal Control Module Accelerator Pedal Position Performance:** None provided **Possible Causes:** • ECM
DTC: P060E **T ECM, MIL: Yes** **Year:** 2009, 2010 **Model:** All **Engine:** All V6, V8 **Transmission:** All	**Internal Control Module Throttle Position Performance:** DMA communication error: Not detected **Possible Causes:** • ECM
DTC: P0617 **1T TCM, MIL: Yes** **Year:** 2009, 2010 **Model:** All **Engine:** All **Transmission:** All	**Starter Relay Circuit High:** Monitor runs whenever these DTCs are not present: None Battery voltage: 10.5 V or more Vehicle speed: 12.43 mph (20 km/h) or more Engine speed: 1000 rpm or more **Possible Causes:** • Park/Neutral Position switch • Starter relay circuit • Cranking holding function circuit • ECM/TCM

DTC	Trouble Code Title, Conditions & Possible Causes
DTC: P062D **1T ECM, MIL: Yes** **Year:** 2009, 2010 **Model:** LS460 **Engine:** 4.6L V8 **Transmission:** All	**No. 1 Fuel Injector Driver Circuit Performance:** Time after engine switch off to on (IG): 1 second or more Battery: 10.5 V or more Fuel injection start timing: - Injector driver fail signal: Not overlap Engine switch: on (IG) Injector driver relay: ON **Possible Causes:** • Open or short in Injector driver (No. 1, 2) circuit • Injector driver (No. 1, 2) • Fuel injector for direct injection assembly • Engine room No. 2 junction block • ECM
DTC: P062D **1T ECM, MIL: Yes** **Year:** 2009, 2010 **Model:** GS350, IS350 **Engine:** 3.5L V6 **Transmission:** All	**No. 1 Fuel Injector Driver Circuit Performance:** Time after engine switch off to on (IG): 1 second or more Engine speed: 4000 rpm or less Battery voltage: 10.5 V or more **Possible Causes:** • Open or short in injector driver (EDU) circuit • Injector driver (EDU) • Fuel injector assembly • ECM
DTC: P062E **1T ECM, MIL: Yes** **Year:** 2009, 2010 **Model:** LS460 **Engine:** 4.6L V8 **Transmission:** All	**No. 2 Fuel Injector Driver Circuit Performance:** Time after engine switch off to on (IG): 1 second or more Battery: 10.5 V or more Fuel injection start timing: - Injector driver fail signal: Not overlap Engine switch: on (IG) Injector driver relay: ON **Possible Causes:** • Open or short in Injector driver (No. 1, 2) circuit • Injector driver (No. 1, 2) • Fuel injector for direct injection assembly • Engine room No. 2 junction block • ECM
DTC: P062F **1T ECM, MIL: Yes** **Year:** 2009, 2010 **Model:** All **Engine:** All **Transmission:** All	**Internal Control Module EEPROM Error:** Monitor runs whenever following DTCs not stored: None Time after engine start: 10 seconds or more Battery voltage: 8 V or higher Engine switch: On (IG) Starter: OFF **Possible Causes:** • ECM
DTC: P0630 **T ECM, MIL: Yes** **Year:** 2009, 2010 **Model:** All **Engine:** All **Transmission:** All	**VIN not Programmed or Mismatch - ECM / PCM:** Battery voltage: 8 V or more Engine switch: On (IG) Starter: OFF **Possible Causes:** • ECM
DTC: P0657 **T ECM, MIL: Yes** **Year:** 2009, 2010 **Model:** All **Engine:** All V6, V8 **Transmission:** All	**Actuator Supply Voltage Circuit / Open:** Monitor will run whenever this DTC is not present: None Engine switch: Front ON to OFF **Possible Causes:** • ECM

DTC	Trouble Code Title, Conditions & Possible Causes
DTC: P0662 **2T ECM, MIL: Yes** **Year:** 2009, 2010 **Model:** LS460 **Engine:** 4.6L V8 **Transmission:** All	**Intake Manifold Tuning Valve Control Circuit High:** Motor operation circuit (ECM interior) detects overcurrent or overheat **Possible Causes:** • Intake manifold • Intake air control valve actuator • Intake air control valve actuator circuit • ECM
DTC: P0664 **2T ECM** **Year:** 2009, 2010 **Model:** LS460 **Engine:** 4.6L V8 **Transmission:** All	**Intake Manifold Tuning Valve Control Circuit Low:** Open in intake air control valve actuator circuit **Possible Causes:** • Intake manifold • Intake air control valve actuator • Intake air control valve actuator circuit • ECM
DTC: P0705 **2T ECM** **Year:** 2009, 2010 **Model:** All **Engine:** All V6, V8 **Transmission:** All	**Transmission Range Sensor Circuit Malfunction (PRNDL Input):** The monitor will run whenever this DTC is not present: None Engine switch: on (IG) Battery voltage: 10.5 V or more **Possible Causes:** • Open or short in park/neutral position switch circuit • Park/neutral position switch • Open or short in transmission control switch circuit • Transmission control switch • TCM/ECM
DTC: P0710 **1T TCM, MIL: Yes** **Year:** 2009, 2010 **Model:** ES350, GS350 **Engine:** 3.5L V6 **Transmission:** All	**Transmission Fluid Temperature Sensor "A" Circuit:** The monitor will run whenever this DTC is not present: None The typical enablling condition is not available. **Possible Causes:** • Open or short in ATF temperature sensor circuit • Transmission wire • ATF temperature sensor • TCM
DTC: P0710 **1T TCM, MIL: Yes** **Year:** 2009, 2010 **Model:** RX350 **Engine:** 3.5L V6 **Transmission:** All	**Transmission Fluid Temperature Sensor "A" Circuit:** (a) and (b) are detected momentarily within 0.5 sec. when neither P0712 nor P0713 is detected (1-trip detection logic) (a) ATF temperature sensor resistance is less than 79 W. (b) ATF temperature sensor resistance is more than 156 kW. HINT: Within 0.5 sec., the malfunction switches from (a) to (b) or from (b) to (a) **Possible Causes:** • Open or short in ATF temperature sensor circuit • Transmission wire (ATF temperature sensor) • ECM

DTC	Trouble Code Title, Conditions & Possible Causes
DTC: P0711 **2T ECM, MIL: Yes** **Year:** 2009, 2010 **Model:** All **Engine:** All **Transmission:** All	**Transmission Fluid Temperature Sensor "A" Performance:** TFT (Transmission fluid temperature) sensor circuit: No circuit malfunction ECT (Engine coolant temperature) sensor circuit: No circuit malfunction IAT (Intake air temperature) sensor circuit: No circuit malfunction Turbine speed sensor circuit: No circuit malfunction Intermediate shaft speed sensor: No circuit malfunction Intermediate shaft speed sensor: No circuit malfunction Shift solenoid valve SL1 circuit: No circuit malfunction Shift solenoid valve SL2 circuit: No circuit malfunction Shift solennoid valve SL3 circuit: No circuit malfunction Shift solenoid valve SL4 circuit: No circuit malfunction (KCS) Knock control sensor circuit: No circuit malfunction (ETCS) Electronic throttle control system: No system down CAN communication system: Not system down Time after engine start: 16 min. and 40 sec. or more ECT (Engine Coolant Temperature): 5°F (-15°C) or more **Possible Causes:** • Transmission wire • ATF temperature sensor • TCM/ECM
DTC: P0712 **1T ECM, MIL: Yes** **Year:** 2009, 2010 **Model:** All **Engine:** All **Transmission:** All	**Transmission Fluid Temperature Sensor "A" Circuit Low Input:** The monitor will run whenever this DTC is not present: None The typical enabling condition is not available. **Possible Causes:** • Short in ATF temperature sensor circuit • Transmission wire • ATF temperature sensor • TCM/ECM
DTC: P0713 **1T ECM, MIL: Yes** **Year:** 2009, 2010 **Model:** All **Engine:** All **Transmission:** All	**Transmission Fluid Temperature Sensor "A" Circuit High Input:** The monitor will run whenever this DTC is not present: None The typical enabling condition is not available. **Possible Causes:** • Open in ATF temperature sensor circuit • Transmission wire • ATF temperature sensor • TCM/ECM
DTC: P0715 **T TCM, MIL: Yes** **Year:** 2009, 2010 **Model:** ES350, GS350 **Engine:** 3.5L V6 **Transmission:** All	**Input / Turbine Speed Sensor Circuit Malfunction:** Battery voltage: 8 V or more Engine switch: ON Starter: OFF **Possible Causes:** • Transmission revolution sensor (speed sensor NT) • TCM
DTC: P0717 **1T TCM, MIL: Yes** **Year:** 2009, 2010 **Model:** All **Engine:** All **Transmission:** All	**Turbine Speed Sensor Circuit No Signal:** Shift change: Shift change is completed and before starting next shift change operation ECM selected gear: 4th, 5th or 6th Output shaft rpm: 1,000 rpm or more NSW (STAR) switch: OFF R switch: OFF Engine: Running **Possible Causes:** • Open or short in speed sensor (NT) circuit • Speed sensor (NT) • ECM • Automatic transmission (clutch, brake or gear, etc.)

DTC	Trouble Code Title, Conditions & Possible Causes
DTC: P0722 **2T TCM, MIL: Yes** **Year:** 2009 **Model:** GX470 **Engine:** 4.7L V8 **Transmission:** All	**Output Speed Sensor Circuit No Signal:** Vehicle speed at vehicle speed sensor: 9 km/h (5.6 mph) or more **Possible Causes:** • Open or short in speed sensor (SP2) circuit • Speed sensor (SP2) • ECM • Automatic transmission assembly (clutch, brake or gear, etc.)
DTC: P0724 **2T TCM, MIL: Yes** **Year:** 2009, 2010 **Model:** All **Engine:** All **Transmission:** All	**Brake Switch "B" Circuit High:** The stop light switch remains on during GO and STOP 5 times. GO and STOP is defined as follows; The monitor will run whenever this DTC is not present: None GO: (Vehicle speed is 18.65 mph (30 km/h) or more): Once STOP: (Vehicle speed is less than 1.86 mph (3 km/h)): Once Starter: OFF Battery voltage: 8 V or more Engine switch: ON **Possible Causes:** • Short in stop light switch signal circuit • Stop light switch • ECM/TCM
DTC: P0729 **2T TCM, MIL: Yes** **Year:** 2009, 2010 **Model:** GS350, IS250, IS350 **Engine:** 2.5L V6, 3.5L V6 **Transmission:** All	**Gear 6 Incorrect Ratio:** The monitor will run whenever this DTC is not present: P0500 (VSS) P0748 (Trans solenoid (range)) Turbine speed sensor circuit: Not circuit malfunction Output speed sensor circuit: Not circuit malfunction Shift solenoid valve S1 circuit: Not circuit malfunction Shift solenoid valve S2 circuit: Not circuit malfunction Shift solenoid valve S3 circuit: Not circuit malfunction Shift solenoid valve S4 circuit: Not circuit malfunction Shift solenoid valve SR circuit: Not circuit malfunction Shift solenoid valve SL1 circuit: Not circuit malfunction Shift solenoid valve SL2 circuit: Not circuit malfunction ECT (Engine coolant temperature) sensor circuit: Not circuit malfunction KCS sensor circuit: Not circuit malfunction ETCS (Electric throttle control system): Not system down Transmission range: "D" ECT: 40°C (104°F) or more Spark advance from Max. retard timing by KCS control: 0°CA or more Engine: Starting **Possible Causes:** • Valve body is blocked up or stuck (sequence valve) • Shift solenoid valve SLT remains open or closed • Automatic transmission (clutch, brake or gear, etc.)

DTC	Trouble Code Title, Conditions & Possible Causes
DTC: P0741 **2T TCM, MIL: Yes** **Year:** 2009, 2010 **Model:** All **Engine:** All **Transmission:** All	**Torque Converter Clutch Solenoid Performance (Shift Solenoid Valve SL):** ECT (Engine coolant temperature): 104°F (40°C) or more Spark advance from Max. retard timing by KCS control: 0°CA or more Transmission range: "D" TFT (Transmission fluid temperature): 14°F (-10°C) or more TFT (Transmission fluid temperature): -10°C (14°F) or more TFT (Transmission fluid temperature) sensor circuit: No circuit malfunction ECT (Engine coolant temperature) sensor circuit:No circuit malfunction Turbine speed sensor circuit:No circuit malfunction Intermediate shaft speed sensor circuit: No circuit malfunction Shift solenoid valve SL1 circuit: No circuit malfunction Shift solenoid valve SL2 circuit: No circuit malfunction Shift solenoid valve SL3 circuit:No circuit malfunction Shift solenoid valve SL4 circuit: No circuit malfunction Shift solenoid valve SLU circuit: No circuit malfunction Shift solenoid valve SL circuit: No circuit malfunction (KCS) Knock control sensor circuit: No circuit malfunction (ETCS) Electronic throttle control system: Not system down CAN communication system: Not system down TCM selected gear:Not 1st Vehicle speed: 15.5 mph (25 km/h) or more Turbine speed/Output speed (NT/NO) with 1st: 3.304 to 7.724 Turbine speed/Output speed (NT/NO) with 2nd:1.901 to 2.340 Turbine speed/Output speed (NT/NO) with 3rd: 1.399 to 1.649 Turbine speed/Output speed (NT/NO) with 4th: 0.998 to 1.138 Turbine speed/Output speed (NT/NO) with 5th: 0.705 to 0.836 Turbine speed/Output speed (NT/NO) with 6th: 0.568 to 0.695 **Possible Causes:** • Shift solenoid valve SL (closed) • Valve body (blocked) • Torque converter clutch • Automatic transaxle (clutch, brake or gear etc.) • Line pressure is too low
DTC: P0746 **2T TCM, MIL: Yes** **Year:** 2009 **Model:** All **Engine:** All **Transmission:** All	**Pressure Control Solenoid "A" Performance (Shift Solenoid Valve SL1):** ETCS (Electric throttle control system): Not system down Transmission range: "D" Duration time from shifting "N" to "D": 4 sec. or more ECT: 40°C (104°F) or more Spark advance from Max. retard timing by KCS control: 0° CA or more Engine: Starting Malfunction A: ECM selected gear: 5th Vehicle speed: 2 km/h (1.2 mph) or more Throttle valve opening angle: 2.0% or more at 1,000 rpm (Varies with engine speed) Malfunction B: Current ECM selected gear: 5th Last ECM selected gear: 4th Continuous time of ECM selecting 4th gear: 2 sec. or more Malfunction C: Current ECM selected gear: 5th Last ECM selected gear: 4th **Possible Causes:** • Shift solenoid valve SL1 remains closed • Shift solenoid valve SR remains open or closed • Valve body is blocked • Automatic transmission (clutch, brake or gear, etc.)

DTC	Trouble Code Title, Conditions & Possible Causes
DTC: P0748 **1T TCM, MIL: Yes** **Year:** 2009, 2010 **Model:** All **Engine:** All **Transmission:** All	**Pressure Control Solenoid "A" Electrical (Shift Solenoid Valve SL1):** The monitor will run whenever this DTC is not present: P0115 - P0118 (ECT sensor) P0125 (Insufficient ECT for Closed Loop) P0500 (VSS) P0748 - P0798 (Trans solenoid (range)) Solenoid current cut status: Not cut CPU commanded duty: 19% or more Battery voltage: 11 V or more Engine Switch: ON Starter: OFF **Possible Causes:** • Open or short in shift solenoid valve SL1 circuit • Shift solenoid valve SL1 • ECM

DTC	Trouble Code Title, Conditions & Possible Causes
DTC: P0751 **2T TCM, MIL: Yes** **Year:** 2009 **Model:** All **Engine:** All **Transmission:** All	**Shift Solenoid "A" Performance (Shift Solenoid Valve S1):** ETCS (Electric throttle control system): Not system down Transmission range: "D": Duration time from shifting "N" to "D" 4 sec. or more ECT: 40°C (104°F) or more Spark advance from Max. retard timing by KCS control: 0° CA or more Engine: Starting OFF malfunction (A): ECM selected gear: 1st Vehicle speed: 2 to 40 km/h (1.2 to 24.9 mph) Throttle valve opening angle: 8.0% or more (When actual gear is 4th) 2.0% or more at 1,000 rpm (Varies with engine speed) OFF malfunction (B): Current ECM selected gear: 5th Last ECM selected gear: 4th Continuous time for ECM selecting 4th gear: 2 sec. or more Actual gear when ECM selected 4th gear: 4th OFF malfunction (C): Current ECM selected gear: 5th Last ECM selected gear: 4th ON malfunction (A): ECM selected gear: 1st Vehicle speed: 2 to 40 km/h (1.2 to 24.9 mph) Throttle valve opening angle: 8.0% or more (When actual gear is 4th) 2.0% or more at 1,000 rpm (Varies with engine speed) ON malfunction (B): ECM selected gear: 4th Vehicle speed: 2 km/h (1.2 mph) or more Throttle valve opening angle: 2.0% or more at 1,000 rpm (Varies with engine speed) ON malfunction (C): ECM selected gear: 3rd Vehicle speed: 2 km/h (1.2 mph) or more Throttle valve opening angle: 2.0% or more at 1,000 rpm (Varies with engine speed) ON malfunction (D): Current ECM selected gear: 5th Last ECM selected gear: 4th Vehicle speed (During transition from 4th to 5th gear): Less than 100 km/h (62.2 mph) ON malfunction (E): ECM selected gear: 5th Engine speed - Turbine speed (NE - NT): After transition from 4th to 5th gear) 150 rpm or less Vehicle speed (After transition from 4th to 5th gear): Less than 100 km/h (62.2 mph) **Possible Causes:** • Shift solenoid valve S1 remains open or closed • Valve body is blocked • Automatic transmission (clutch, brake or gear, etc.)

DTC	Trouble Code Title, Conditions & Possible Causes
DTC: P0756 **2T , MIL: Yes** **Year:** 2009, 2010 **Model:** GS350, IS250, IS350 **Engine:** 2.5L V6, 3.5L V6 **Transmission:** All	**Shift Solenoid "B" Performance (Shift Solenoid Valve S2):** The monitor will run whenever this DTC is not present: P0115 - P0118 (ECT sensor) P0125 (Insufficient ECT for Closed Loop) P0500 (VSS) P0748 - P0798 (Trans solenoid (range)) Turbine speed sensor circuit: Not circuit malfunction Output speed sensor circuit: Not circuit malfunction Shift solenoid valve S1 circuit: Not circuit malfunction Shift solenoid valve S2 circuit: Not circuit malfunction Shift solenoid valve S3 circuit: Not circuit malfunction Shift solenoid valve S4 circuit: Not circuit malfunction Shift solenoid valve SR circuit: Not circuit malfunction Shift solenoid valve SL1 circuit: Not circuit malfunction Shift solenoid valve SL2 circuit: Not circuit malfunction ECT (Engine coolant temperature) sensor circuit: Not circuit malfunction KCS sensor circuit: Not circuit malfunction ETCS (Electric throttle control system): Not system down Transmission range: "D" ECT: 40°C (104°F) or more Spark advance from Max. retard timing by KCS control: 0°CA or more Engine: Starting OFF Malfunction (A): ECM selected gear: 1st Vehicle speed: 1.2 mph (2 km/h) or more and less than 24.9 mph (40 km/h) Engine speed - Turbine speed (NE - NT): 150 rpm or more OFF Malfunction (B): ECM selected gear: 6th Vehicle speed: 1.2 mph (2 km/h) or more Throttle valve opening angle: 6.5% or more at 2,000 rpm (Conditions vary with engine speed) ON Malfunction (A): ECM selected gear: 5th Vehicle speed: 1.2 mph (2 km/h) or more Throttle valve opening angle: 6.5% or more at 2,000 rpm (Conditions vary with engine speed) ON Malfunction (B): ECM selected gear: 6th Vehicle speed: 1.2 mph (2 km/h) or more Throttle valve opening angle: 6.5% or more at 2,000 rpm (Conditions vary with engine speed) **Possible Causes:** • Shift solenoid valve S2 remains open • Valve body is blocked • Automatic transmission (clutch, brake or gear, etc.)

DTC	Trouble Code Title, Conditions & Possible Causes
DTC: P0761 **2T TCM, MIL: Yes** **Year:** 2009, 2010 **Model:** GS350, IS250, IS350 **Engine:** 2.5L V6, 3.5L V6 **Transmission:** All	**Shift Solenoid "C" Performance (Shift Solenoid Valve S3):** The monitor will run whenever this DTC is not present: P0115 - P0118 (ECT sensor) P0125 (Insufficient ECT for Closed Loop) P0500 (VSS) P0748 - P0798 (Trans solenoid (range)) Turbine speed sensor circuit: Not circuit malfunction Output speed sensor circuit: Not circuit malfunction Shift solenoid valve S1 circuit: Not circuit malfunction Shift solenoid valve S2 circuit: Not circuit malfunction Shift solenoid valve S3 circuit: Not circuit malfunction Shift solenoid valve S4 circuit: Not circuit malfunction Shift solenoid valve SR circuit: Not circuit malfunction Shift solenoid valve SL1 circuit: Not circuit malfunction Shift solenoid valve SL2 circuit: Not circuit malfunction ECT (Engine coolant temperature) sensor circuit: Not circuit malfunction KCS sensor circuit: Not circuit malfunction ETCS (Electric throttle control system): Not system down Transmission range: "D" ECT: 40°C (104°F) or more Spark advance from Max. retard timing by KCS control: 0°CA or more Engine: Starting OFF Malfunction (A): ECM selected gear: 2nd Vehicle speed: 1.2 mph (2 km/h) or more Output speed: 2nd → 1st down shift point or more Throttle valve opening angle: 6.5% or more at 2,000 rpm (Condition varies with engine speed) OFF Malfunction (B): ECM selected gear: 6th Vehicle speed: 1.2 mph (2 km/h) or more Throttle valve opening angle: 6.5% or more at 2,000 rpm (Conditions vary with engine speed) ON Malfunction (A): ECM selected gear: 4th Vehicle speed: 1.2 mph (2 km/h) or more Throttle valve opening angle: 6.5% or more at 2,000 rpm (Conditions vary with engine speed) ON Malfunction (B): Current ECM selected gear: 5th Last ECM selected gear: 4th Vehicle speed (During transition from 4th to 5th gear): Less than 62.2 mph (100 km/h) ON Malfunction (C): ECM selected gear: 5th Engine speed - Turbine speed (NE - NT) (After transition to 5th gear): 150 rpm or less Vehicle speed (After transition to 5th gear): Less than 62.2 mph (100 km/h) **Possible Causes:** • Shift solenoid valve S3 remains open • Shift solenoid valve SLT remains open or closed • Valve body is blocked • Automatic transmission (clutch, brake or gear, etc.)

DTC	Trouble Code Title, Conditions & Possible Causes
DTC: P0766 **2T TCM, MIL: Yes** **Year:** 2009, 2010 **Model:** All **Engine:** All **Transmission:** All	**Shift Solenoid "D" Performance (Shift Solenoid Valve S4):** The monitor will run whenever this DTC is not present: P0115 - P0118 (ECT sensor) P0125 (Insufficient ECT for Closed Loop) P0500 (VSS) P0748 - P0798 (Trans solenoid (range)) Turbine speed sensor circuit: Not circuit malfunction Output speed sensor circuit: Not circuit malfunction Shift solenoid valve S1 circuit: Not circuit malfunction Shift solenoid valve S2 circuit: Not circuit malfunction Shift solenoid valve S3 circuit: Not circuit malfunction Shift solenoid valve S4 circuit: Not circuit malfunction Shift solenoid valve SR circuit: Not circuit malfunction Shift solenoid valve SL1 circuit: Not circuit malfunction Shift solenoid valve SL2 circuit: Not circuit malfunction ECT (Engine coolant temperature) sensor circuit: Not circuit malfunction KCS sensor circuit: Not circuit malfunction ETCS (Electric throttle control system): Not system down Transmission range: "D" ECT: 40°C (104°F) or more Spark advance from Max. retard timing by KCS control: 0°CA or more Engine: Starting OFF malfunction (A) and P0776 ON malfunction (A): ECM selected gear: 5th Vehicle speed: 1.2 mph (2 km/h) or more Throttle valve opening angle: 6.5% or more at 2,000 rpm (Conditions vary with engine speed) **Possible Causes:** • Shift solenoid valve S4 remains closed • Shift solenoid valve SLT remains open or closed • Valve body is blocked (Brake control valve) • Automatic transmission (clutch, brake or gear, etc.)

DTC	Trouble Code Title, Conditions & Possible Causes
DTC: P0771 **2T TCM, MIL: Yes** **Year:** 2009, 2010 **Model:** RX350 **Engine:** 3.5L V6 **Transmission:** All	**Shift Solenoid "E" Performance (Shift Solenoid Valve SR):** The monitor will run whenever this DTC is not present: (Not circuit malfunction) P0712, P0713 (ATF temperature sensor circuit) P0115, P0117, P0118 (ECT sensor circuit) P0717 (Turbine speed sensor circuit) P0793 (Intermediate shaft speed sensor circuit) P0500 (Vehicle speed sensor circuit) P0748 (Shift solenoid valve SL1 circuit) P0778 (Shift solenoid valve SL2 circuit) P0798 (Shift solenoid valve SL3 circuit) P0982, P0983 (Shift solenoid valve S4 circuit) P0985, P0986 (Shift solenoid valve SR circuit) P2769, P2770 (Shift solenoid valve DSL circuit) P0120, P0121, P0122, P0123, P0220, P0222, P0223, P0505, P2102, P2103, P2111, P2112, P2118, P2119, P2135 (Electronic throttle system (if applicable)) ECT (Engine coolant temperature): 10°C (50°F) or more Transmission range: "D" TFT (Transmission fluid temperature): -20°C (-4°F) or more OFF Malfunction (A): ECM selected gear: 5th Throttle valve opening angle: 5% or more Vehicle speed: 10 km/h (6 mph) or more OFF Malfunction (B): ECM lock-up command: ON ECM selected gear: 3rd, 4th or 5th Vehicle speed: 25 km/h (16 mph) or more ON Malfunction (A): ECM lock-up command: OFF ON Malfunction (B): ECM selected gear: 1st Vehicle speed: Less than 40 km/h (25 mph) Throttle valve opening angle: 4.5% or more at engine speed 1900 rpm (Varies with engine speed) ON Malfunction (C): ECM selected gear: 3rd Throttle valve opening angle: 4.5% or more at engine speed 1900 rpm (Varies with engine speed) ON Malfunction (D): Duration time from shift command of ECM: 15 sec. or more ECM selected gear: 4th or 5th **Possible Causes:** • Shift solenoid valve SR remains open or closed • Valve body is blocked • Automatic transaxle (clutch, brake or gear, etc.)

DTC	Trouble Code Title, Conditions & Possible Causes
DTC: P0776 **2T TCM, MIL: Yes** **Year:** 2009, 2010 **Model:** All **Engine:** All **Transmission:** All	**Pressure Control Solenoid "B" Performance (Shift Solenoid Valve SL2):** ECT (Engine coolant temperature): 10°C (50°F) or more Transmission range: "D" TFT (Transmission fluid temperature): -20°C (-4°F) or more OFF Malfunction (A): ECM lock-up command: OFF Vehicle speed: 3rd: Less than 60 km/h (37 mph)" 4th, 5th: Less than 120 km/h (75 mph) Throttle valve opening angle: 7% or more OFF Malfunction (B): ECM selected gear: 1st Vehicle speed: Less than 40 km/h (25 mph) Throttle valve opening angle: 4.5% or more at engine speed 1900 rpm (Varies with engine speed) OFF Malfunction (C): ECM selected gear: 3rd Throttle valve opening angle: 4.5% or more at engine speed 1900 rpm (Varies with engine speed) OFF Malfunction (D): Duration time from shift command of ECM: 15 sec. or more ECM selected gear: 4th or 5th ON Malfunction (A): ECM selected gear: 1st Vehicle speed: Less than 40 km/h (25 mph) Throttle valve opening angle: 4.5% or more at engine speed 1900 rpm (Varies with engine speed) ON Malfunction (B): ECM selected gear: 3rd Throttle valve opening angle: 5% or more at output speed 1400 rpm (Varies with engine speed) Malfunction of pressure control solenoid "B" (SL2) and "C" (SL3): Not detected ON Malfunction (C): Throttle valve opening angle: 7.0% or more at output speed 1050 rpm (Varies with engine speed) Malfunction of pressure control solenoid "B" (SL2): Not detected **Possible Causes:** • Shift solenoid valve SL2 remains open or closed • Valve body is blocked • Automatic transaxle (clutch, brake or gear, etc.)
DTC: P0778 **1T TCM, MIL: Yes** **Year:** 2009, 2010 **Model:** All **Engine:** All **Transmission:** All	**Pressure Control Solenoid "B" Electrical (Shift Solenoid Valve SL2):** The monitor will run whenever this DTC is not present: None Engine switch: ON Starter: OFF Condition (A): Battery voltage: 12 V or more Condition (B): Battery voltage: 10 V or more and less than 12 V Target current: Less than 0.75 A Condition (C): Battery voltage: 8 V or more Target current: 0.25 A or more **Possible Causes:** • Open or short in shift solenoid valve SL2 circuit • Shift solenoid valve SL2 • ECM/TCM

DTC	Trouble Code Title, Conditions & Possible Causes
DTC: P0781 **2T TCM, MIL: Yes** **Year:** 2009, 2010 **Model:** All **Engine:** All **Transmission:** All	**1-2 Shift (1-2 Shift Valve):** The monitor will run whenever this DTC is not present: P0500 (VSS) P0748 - P0798 (Trans solenoid (range)) Turbine speed sensor circuit: Not circuit malfunction Output speed sensor circuit: Not circuit malfunction Shift solenoid valve S1 circuit: Not circuit malfunction Shift solenoid valve S2 circuit: Not circuit malfunction Shift solenoid valve S3 circuit: Not circuit malfunction Shift solenoid valve S4 circuit: Not circuit malfunction Shift solenoid valve SR circuit: Not circuit malfunction Shift solenoid valve SL1 circuit: Not circuit malfunction Shift solenoid valve SL2 circuit: Not circuit malfunction ECT (Engine coolant temperature) sensor circuit: Not circuit malfunction KCS sensor circuit: Not circuit malfunction ETCS (Electric throttle control system): Not system down Transmission range: "D" ECT: 40°C (104°F) or more Spark advance from Max. retard timing by KCS control: 0°CA or more Engine: Starting Condition (A): ECM selected gear: 2nd Vehicle speed: 1.2 mph (2 km/h) or more Output speed: 2nd → 1st down shift point or more Throttle valve opening angle: 6.5% or more at 2,000 rpm (Conditions vary with engine speed) Condition (B): ECM selected gear: 4th Vehicle speed: 1.2 mph (2 km/h) or more Throttle valve opening angle: 6.5% or more at 2,000 rpm (Conditions vary with engine speed) Condition (C): Current ECM selected gear: 5th Last ECM selected gear: 4th Vehicle speed (During transition from 4th to 5th gear): Less than 62.2 mph (100 km/h) Condition (D): ECM selected gear: 5th Engine speed -Turbine speed (NE - NT) After transition to 5th gear): 150 rpm or less Vehicle speed (After transition to 5th gear): Less than 62.2 mph (100 km/h) **Possible Causes:** • Shift solenoid valve SLT remains open or closed • Valve body is blocked up or stuck (1-2 shift valve) • Automatic transmission (clutch, brake or gear, etc.)
DTC: P0791 **T TCM, MIL: Yes** **Year:** 2009, 2010 **Model:** ES350, GS350 **Engine:** 3.5L V6 **Transmission:** All	**Intermediate Shaft Speed Sensor "A" Circuit:** The monitor will run whenever this DTC is not present: P0500 (VSS), P0748, P0778, P0798 (Shift solenoid valve (range)) Vehicle speed: 15.5 mph (25 km/h) or more Battery voltage: 8 V or more Engine switch: ON Starter: OFF **Possible Causes:** • Transmission revolution sensor (speed sensor NC) • TCM
DTC: P0793 **T TCM, MIL: Yes** **Year:** 2009, 2010 **Model:** All **Engine:** All **Transmission:** All	**Intermediate Shaft Speed Sensor "A":** The monitor will run whenever this DTC is not present: P0500 (VSS), P0748, P0778, P0798 (Shift solenoid valve (range)) Vehicle speed: 15.5 mph (25 km/h) or more Battery voltage: 8 V or more Engine switch: ON Starter: OFF **Possible Causes:** • Transmission revolution sensor (speed sensor NC) • TCM

DTC	Trouble Code Title, Conditions & Possible Causes
DTC: P0794 **1T TCM, MIL: Yes** **Year:** 2009 **Model:** GX470 **Engine:** 4.7L V8 **Transmission:** All	**Shift Solenoid "A" Control Circuit High (Shift Solenoid Valve S1):** The monitor will run whenever this DTC is not present: None Shift solenoid valve S1: OFF **Possible Causes:** • Open in shift solenoid valve S1 circuit • Shift solenoid valve S1 • ECM
DTC: P0796 **2T TCM, MIL: Yes** **Year:** 2009, 2010 **Model:** All **Engine:** All **Transmission:** All	**Pressure Control Solenoid "C" Performance (Shift Solenoid Valve SL3):** ECT (Engine coolant temperature): 10°C (50°F) or more Transmission range: "D" TFT (Transmission fluid temperature):-20°C (-4°F) or more OFF Malfunction (A): ECM selected gear: 4th or 5th Throttle valve opening angle: 4.5% or more at engine speed 1900 rpm (Varies with engine speed) OFF Malfunction (B): ECM selected gear: 4th Throttle valve opening angle: 5% or more Vehicle speed: 10 km/h (6 mph) or more ON Malfunction (A): ECM selected gear: 1st Vehicle speed: Less than 40 km/h (25 mph) Throttle valve opening angle: 4.5% or more at engine speed 1900 rpm (Varies with engine speed) ON Malfunction (B): ECM selected gear: 3rd Throttle valve opening angle: 5% or more at output speed 1400 rpm (Varies with engine speed) Malfunction of pressure control solenoid "B" (SL2) and "C" (SL3): Not detected ON Malfunction (C): Throttle valve opening angle: 7.0% or more at output speed 1050 rpm (Varies with engine speed) Malfunction of pressure control solenoid "B" (SL2): Not detected **Possible Causes:** • Shift solenoid valve SL3 remains open or closed • Valve body is blocked • Automatic transaxle (clutch, brake or gear, etc.)
DTC: P0798 **1T TCM, MIL: Yes** **Year:** 2009, 2010 **Model:** All **Engine:** All **Transmission:** All	**Pressure Control Solenoid "C" Electrical (Shift Solenoid Valve SL3):** The monitor will run whenever this DTC is not present: None Battery voltage: 10 V or more Ignition switch: ON Starter: OFF **Possible Causes:** • Open or short in shift solenoid valve SL3 circuit • Shift solenoid valve SL3 • ECM
DTC: P0872 **T TCM, MIL: Yes** **Year:** 2009, 2010 **Model:** ES350, GS350 **Engine:** 3.5L V6 **Transmission:** All	**Transmission Fluid Pressure Sensor / Switch "C" Circuit Low:** None available. **Possible Causes:** • ATF temperature sensor assembly (ATF pressure switch No. 1) • Transmission wire • TCM
DTC: P0873 **T TCM, MIL: Yes** **Year:** 2009, 2010 **Model:** ES350, GS350 **Engine:** 3.5L V6 **Transmission:** All	**Transmission Fluid Pressure Sensor / Switch "C" Circuit High:** None provided. **Possible Causes:** • ATF temperature sensor assembly (ATF pressure switch No. 1) • Transmission wire • TCM

DTC	Trouble Code Title, Conditions & Possible Causes
DTC: P0877 **T TCM, MIL: Yes** **Year:** 2009, 2010 **Model:** ES350, GS350 **Engine:** 3.5L V6 **Transmission:** All	**Transmission Fluid Pressure Sensor / Switch "D" Circuit Low:** None provided. **Possible Causes:** • ATF temperature sensor assembly (ATF pressure switch No. 2) • Transmission wire • TCM
DTC: P0878 **T TCM, MIL: Yes** **Year:** 2009, 2010 **Model:** ES350, GS350 **Engine:** 3.5L V6 **Transmission:** All	**Transmission Fluid Pressure Sensor / Switch "D" Circuit High:** None provided. **Possible Causes:** • ATF temperature sensor assembly (ATF pressure switch No. 2) • Transmission wire • TCM
DTC: P0894 **2T TCM, MIL: Yes** **Year:** 2009, 2010 **Model:** GS350, IS250, IS350 **Engine:** 2.5L V6, 3.5L V6 **Transmission:** All	**Transmission Component Slipping:** The monitor will run whenever this DTC is not present: None Turbine speed sensor circuit: Not circuit malfunction Output speed sensor circuit: Not circuit malfunction ATF temperature sensor circuit: Not circuit malfunction Shift solenoid valve S1 circuit: Not circuit malfunction Shift solenoid valve S2 circuit: Not circuit malfunction Shift solenoid valve S3 circuit: Not circuit malfunction Shift solenoid valve S4 circuit: Not circuit malfunction Shift solenoid valve SR circuit: Not circuit malfunction Shift solenoid valve SL1 circuit: Not circuit malfunction Shift solenoid valve SL2 circuit: Not circuit malfunction Shift solenoid valve SLT circuit: Not circuit malfunction ECT (Engine coolant temperature) sensor circuit: Not circuit malfunction KCS sensor circuit: Not circuit malfunction ETCS (Electric throttle control system): Not system down Transmission range: "D" ECT: 40°C (104°F) or more Spark advance from Max. retard timing by KCS control: 0°crankshaft angle or more Engine, Starting ATF temperature, 10°C (50°F) or more **Possible Causes:** • Shift solenoid valve SLT remains open or closed • Shift solenoid valve S1, S2, S3, S4 or SL2 remains open or closed • Gear 6 incorrect ratio (sequence valve) or 1-2 shift valve is stuck • Valve body is blocked • Automatic transmission (clutch, brake, gear, etc.)
DTC: P0973 **1T ECM, MIL: Yes** **Year:** 2009, 2010 **Model:** GS350, IS250, IS350 **Engine:** 2.5L V6, 3.5L V6 **Transmission:** All	**Shift Solenoid "A" Control Circuit Low (Shift Solenoid Valve S1):** The monitor will run whenever this DTC is not present: None Shift solenoid valve S1: ON **Possible Causes:** • Short in shift solenoid valve S1 circuit • Shift solenoid valve S1 • ECM
DTC: P0974 **1T ECM, MIL: Yes** **Year:** 2009, 2010 **Model:** All **Engine:** All **Transmission:** All	**Shift Solenoid "A" Control Circuit High (Shift Solenoid Valve S1):** The monitor will run whenever this DTC is not present: None Shift solenoid valve S1: OFF **Possible Causes:** • Open in shift solenoid valve S1 circuit • Shift solenoid valve S1 • ECM

DTC	Trouble Code Title, Conditions & Possible Causes
DTC: P0976 **1T TCM, MIL: Yes** **Year:** 2009, 2010 **Model:** All **Engine:** All **Transmission:** All	**Shift Solenoid "B" Control Circuit Low (Shift Solenoid Valve S2):** The monitor will run whenever this DTC is not present: None Shift solenoid valve S2: ON **Possible Causes:** • Short in shift solenoid valve S2 circuit • Shift solenoid valve S2 • ECM
DTC: P0977 **1T ECM, MIL: Yes** **Year:** 2009, 2010 **Model:** All **Engine:** All **Transmission:** All	**Shift Solenoid "B" Control Circuit High (Shift Solenoid Valve S2):** The monitor will run whenever this DTC is not present: None Shift solenoid valve S2: OFF **Possible Causes:** • Open in shift solenoid valve S2 circuit • Shift solenoid valve S2 • ECM
DTC: P0979 **1T ECM, MIL: Yes** **Year:** 2009, 2010 **Model:** All **Engine:** All **Transmission:** All	**Shift Solenoid "C" Control Circuit Low (Shift Solenoid Valve S3):** The monitor will run whenever this DTC is not present: None Shift solenoid valve S3: ON Battery voltage: 8 V or more Engine switch: ON Starter: OFF **Possible Causes:** • Short in shift solenoid valve S3 circuit • Shift solenoid valve S3 • ECM
DTC: P0980 **1T ECM, MIL: Yes** **Year:** 2009, 2010 **Model:** All **Engine:** All **Transmission:** All	**Shift Solenoid "C" Control Circuit High (Shift Solenoid Valve S3):** * Open in shift solenoid valve S3 circuit * Shift solenoid valve S3 * ECM **Possible Causes:** • The monitor will run whenever this DTC is not present: None • Shift solenoid valve S3: OFF • Battery voltage: 8 V or more • Engine switch: ON • Starter: OFF
DTC: P0982 **1T ECM, MIL: Yes** **Year:** 2009, 2010 **Model:** All **Engine:** All **Transmission:** All	**Shift Solenoid "D" Control Circuit Low (Shift Solenoid Valve S4):** The monitor will run whenever this DTC is not present: None Shift solenoid valve S4: ON Battery voltage: 8 V or more Engine switch: ON Starter: OFF **Possible Causes:** • Short in shift solenoid valve S4 circuit • Shift solenoid valve S4 • ECM
DTC: P0983 **1T ECM, MIL: Yes** **Year:** 2009, 2010 **Model:** All **Engine:** All **Transmission:** All	**Shift Solenoid "D" Control Circuit High (Shift Solenoid Valve S4):** The monitor will run whenever this DTC is not present: None Shift solenoid valve S4: OFF Battery voltage: 8 V or more Engine switch: ON Starter: OFF **Possible Causes:** • Open in shift solenoid valve S4 circuit • Shift solenoid valve S4 • ECM

DTC	Trouble Code Title, Conditions & Possible Causes
DTC: P0985 **1T ECM, MIL: Yes** **Year:** 2009, 2010 **Model:** All **Engine:** All **Transmission:** All	**Shift Solenoid "E" Control Circuit Low (Shift Solenoid Valve SR):** The monitor will run whenever this DTC is not present: None Shift solenoid valve SR: ON Battery voltage: 8 V or more Engine switch: ON Starter: OFF **Possible Causes:** • Short in shift solenoid valve SR circuit • Shift solenoid valve SR • ECM
DTC: P0986 **1T TCM, MIL: Yes** **Year:** 2009, 2010 **Model:** All **Engine:** All **Transmission:** All	**Shift Solenoid "E" Control Circuit High (Shift Solenoid Valve SR):** The monitor will run whenever this DTC is not present: None Shift solenoid valve SR: OFF Battery voltage: 8 V or more Ignition switch: ON Starter: OFF **Possible Causes:** • Open in shift solenoid valve SR circuit • Shift solenoid valve SR • ECM
DTC: P0989 **2T TCM, MIL: Yes** **Year:** 2009, 2010 **Model:** All **Engine:** All **Transmission:** All	**Transmission Fluid Pressure Sensor / Switch "E" Circuit Low:** ECT (Engine coolant temperature): 40°C (104°F) or more Spark advance from Max. retard timing by KCS control: 0°CA or more Transmission range: "D" TFT (Transmission fluid temperature): -10°C (14°F) or more TFT (Transmission fluid temperature) sensor circuit: No circuit malfunction ECT (Engine coolant temperature) sensor circuit: No circuit malfunction Turbine speed sensor circuit: No circuit malfunction Intermediate shaft speed sensor circuit: No circuit malfunction Shift solenoid valve SL1 circuit: No circuit malfunction Shift solenoid valve SL2 circuit: No circuit malfunction Shift solenoid valve SL3 circuit: No circuit malfunction Shift solenoid valve SL4 circuit: No circuit malfunction Shift solenoid valve SLU circuit: No circuit malfunction Shift solenoid valve SL circuit: No circuit malfunction (KCS) Knock control sensor circuit: No circuit malfunction (ETCS) Electronic throttle control system: Not system down CAN communication system: Not system down TCM selected gear: Not 1st Vehicle speed: 15.5 mph (25 km/h) or more Turbine speed/Output speed (NT/NO) with 1st: 3.304 to 7.724 Turbine speed/Output speed (NT/NO) with 2nd: 1.901 to 2.340 Turbine speed/Output speed (NT/NO) with 3rd: 1.399 to 1.649 Turbine speed/Output speed (NT/NO) with 4th: 0.998 to 1.138 Turbine speed/Output speed (NT/NO) with 5th: 0.705 to 0.836 Turbine speed/Output speed (NT/NO) with 6th: 0.568 to 0.695 TCM lock-up command: ON Engine speed - Turbine speed: Less than 35 rpm Throttle valve opening angle: 7% or more Vehicle speed: Less than 62 mph (100 km/h) Shift solenoid valve SLU: Not ON malfunction **Possible Causes:** • ATF temperature sensor assembly (ATF pressure switch No. 3) • Transmission wire • TCM

DTC	Trouble Code Title, Conditions & Possible Causes
DTC: P0990 **2T TCM, MIL: Yes** **Year:** 2009, 2010 **Model:** All **Engine:** All **Transmission:** All	**Transmission Fluid Pressure Sensor / Switch "E" Circuit High:** ((ETCS) Electronic throttle control system) U0100 (CAN communication system) ECT (Engine coolant temperature): 40°C (104°F) or more Spark advance from Max. retard timing by KCS control: 0°CA or more Transmission range: "D" TFT (Transmission fluid temperature): -10°C (14°F) or more TFT (Transmission fluid temperature) sensor circuit: No circuit malfunction ECT (Engine coolant temperature) sensor circuit: No circuit malfunction Turbine speed sensor circuit: No circuit malfunction Intermediate shaft speed sensor circuit: No circuit malfunction Shift solenoid valve SL1 circuit: No circuit malfunction Shift solenoid valve SL2 circuit: No circuit malfunction Shift solenoid valve SL3 circuit: No circuit malfunction Shift solenoid valve SL4 circuit: No circuit malfunction Shift solenoid valve SLU circuit: No circuit malfunction Shift solenoid valve SL circuit: No circuit malfunction (KCS) Knock control sensor circuit: No circuit malfunction (ETCS) Electronic throttle control system: Not system down CAN communication system: Not system down TCM selected gear: Not 1st Vehicle speed: 15.5 mph (25 km/h) or more Turbine speed/Output speed (NT/NO) with 1st: 3.304 to 7.724 Turbine speed/Output speed (NT/NO) with 2nd: 1.901 to 2.340 Turbine speed/Output speed (NT/NO) with 3rd: 1.399 to 1.649 Turbine speed/Output speed (NT/NO) with 4th: 0.998 to 1.138 Turbine speed/Output speed (NT/NO) with 5th: 0.705 to 0.836 Turbine speed/Output speed (NT/NO) with 6th: 0.568 to 0.695 TCM indicated pressure valve of SLU: Less than 4 kPa TCM lock-up command: OFF Shift solenoid valve SLU: Not malfunction Shift solenoid valve SL: Not OFF malfunction **Possible Causes:** • ATF temperature sensor assembly (ATF pressure switch No. 3) • Transmission wire • TCM

OBD II Trouble Code List (P1XXX Codes)

DTC	Trouble Code Title, Conditions & Possible Causes
DTC: P101D **1T ECM, MIL: Yes** **Year:** 2009, 2010 **Model:** All **Engine:** All V6, V8 **Transmission:** All	**A/F Sensor Heater Circuit Performance Bank 1 Sensor 1 Stuck ON:** Monitor runs whenever following DTCs not stored: P0031, P0051 (Air fuel ratio sensor heater) Battery voltage: 10.5 V or higher Time after engine start: 10 seconds or more Active heater OFF control: Not operating Active heater ON control: Not operating Air fuel ratio sensor heater duty-cycle: 10 to 60% Air fuel ratio sensor heater ON current: 0.8 A or higher **Possible Causes:** • ECM
DTC: P1023 **1T ECM, MIL: Yes** **Year:** 2009, 2010 **Model:** LS460 **Engine:** 4.6L V8 **Transmission:** All	**Camshaft Position Signal Output "B" Circuit:** Monitor runs whenever following conditions are met: None Battery voltage: 11 V or more Engine switch: on (IG) **Possible Causes:** • Camshaft timing control motor (bank 2)

DTC	Trouble Code Title, Conditions & Possible Causes
DTC: P1170 **2T ECM, MIL: Yes** **Year:** 2009, 2010 **Model:** All **Engine:** All **Transmission:** All	**Port Injector Fuel Performance:** Battery voltage: 11 V or more Either of the following conditions 1 or 2 is met: 1. Engine RPM: Less than 1100 rpm 2. Engine load: 10% or more Catalyst monitor: Not executed **Possible Causes:** • Fuel injector assembly for port injection • Fuel injector assembly for direct injection • Fuel pressure • ECM
DTC: P117B **2T ECM, MIL: Yes** **Year:** 2009, 2010 **Model:** All **Engine:** All **Transmission:** All	**Direct Injector Fuel Performance:** Fuel system status: Closed loop Battery voltage: 11 V or more Either of the following conditions 1 or 2 is met: 1. Engine RPM: Less than 1100 rpm 2. Engine load: 10% or more Catalyst monitor: Not executed **Possible Causes:** • Fuel injector assembly for port injection • Fuel injector assembly for direct injection • Fuel pressure • ECM
DTC: P1235 **1T ECM** **Year:** 2009, 2010 **Model:** All **Engine:** All **Transmission:** All	**High Pressure Fuel Pump Circuit:** Time after engine start: 5 seconds or more Output duty ratio: More than Output duty ratio change map value, and less than 95% Battery voltage: 10.5 V or more Engine switch: on (IG) Starter: OFF Output duty ratio change map value: - Engine speed is 500 rpm: 5% Engine speed is 2000 rpm: 20% Engine speed is 4000 rpm: 40% Engine speed is 6000 rpm: 60% Engine speed is 8000 rpm: 80% **Possible Causes:** • Open in fuel pump for high pressure (bank 1) circuit • Fuel pump for high pressure • Injector driver (No. 1) • ECM
DTC: P1236 **1T ECM** **Year:** 2009, 2010 **Model:** All **Engine:** All **Transmission:** All	**High Pressure Fuel Pump No. 2 Circuit:** Time after engine start: 5 seconds or more Output duty ratio: More than Output duty ratio change map value, and less than 95% Battery voltage: 10.5 V or more Engine switch: on (IG) Starter: OFF Output duty ratio change map value: - Engine speed is 500 rpm: 5% Engine speed is 2000 rpm: 20% Engine speed is 4000 rpm: 40% Engine speed is 6000 rpm: 60% Engine speed is 8000 rpm: 80% **Possible Causes:** • Open in fuel pump for high pressure (bank 2) circuit • Fuel pump for high pressure • Injector driver (No. 2) • ECM

DTC	Trouble Code Title, Conditions & Possible Causes
DTC: P1276 **T ECM, MIL: Yes** **Year:** 2009, 2010 **Model:** GS350, IS350 **Engine:** 3.5L V6 **Transmission:** All	**Port Injector Circuit No. 1:** Battery voltage: 8 V or more Fuel cut: OFF Port injector: Energized Engine switch: On (IG) **Possible Causes:** • Open or short in injector for port injection circuit • IG2 fuse • Injector for port injection
DTC: P1276 **T ECM, MIL: Yes** **Year:** 2009, 2010 **Model:** LS460 **Engine:** 4.6L V8 **Transmission:** All	**Port Injector Circuit No. 1:** Battery voltage: 8 V or more Fuel cut: OFF Port injector: Energized Engine switch: on (IG) **Possible Causes:** • Open or short in injector for port injection circuit • Injector for port injection • ECM
DTC: P1277 **T , MIL: Yes** **Year:** 2009, 2010 **Model:** GS350, IS350 **Engine:** 3.5L V6 **Transmission:** All	**Port Injector Circuit No. 2:** Battery voltage: 8 V or more Fuel cut: OFF Port injector: Energized Engine switch: On (IG) **Possible Causes:** • Open or short in injector for port injection circuit • IG2 fuse • Injector for port injection
DTC: P1277 **T ECM, MIL: Yes** **Year:** 2009, 2010 **Model:** LS460 **Engine:** 4.6L V8 **Transmission:** All	**Port Injector Circuit No. 2:** Battery voltage: 8 V or more Fuel cut: OFF Port injector: Energized Engine switch: on (IG) **Possible Causes:** • Open or short in injector for port injection circuit • Injector for port injection • ECM
DTC: P1278 **T ECM, MIL: Yes** **Year:** 2009, 2010 **Model:** GS350, IS350 **Engine:** 3.5L V6 **Transmission:** All	**Port Injector Circuit No. 3:** Battery voltage: 8 V or more Fuel cut: OFF Port injector: Energized Engine switch: On (IG) **Possible Causes:** • Open or short in injector for port injection circuit • IG2 fuse • Injector for port injection
DTC: P1278 **T ECM, MIL: Yes** **Year:** 2009, 2010 **Model:** LS460 **Engine:** 4.6L V8 **Transmission:** All	**Port Injector Circuit No. 3:** Battery voltage: 8 V or more Fuel cut: OFF Port injector: Energized Engine switch: on (IG) **Possible Causes:** • Open or short in injector for port injection circuit • Injector for port injection • ECM

DTC	Trouble Code Title, Conditions & Possible Causes
DTC: P1279 **T ECM, MIL: Yes** **Year:** 2009, 2010 **Model:** GS350, IS350 **Engine:** 3.5L V6 **Transmission:** All	**Port Injector Circuit No. 4:** Battery voltage: 8 V or more Fuel cut: OFF Port injector: Energized Engine switch: On (IG) **Possible Causes:** • Open or short in injector for port injection circuit • IG2 fuse • Injector for port injection
DTC: P1279 **T ECM, MIL: Yes** **Year:** 2009, 2010 **Model:** LS460 **Engine:** 4.6L V8 **Transmission:** All	**Port Injector Circuit No. 4:** Battery voltage: 8 V or more Fuel cut: OFF Port injector: Energized Engine switch: on (IG) **Possible Causes:** • Open or short in injector for port injection circuit • Injector for port injection • ECM
DTC: P127A **T ECM, MIL: Yes** **Year:** 2009, 2010 **Model:** GS350, IS350 **Engine:** 3.5L V6 **Transmission:** All	**Port Injector Circuit No. 5:** Battery voltage: 8 V or more Fuel cut: OFF Port injector: Energized Engine switch: On (IG) **Possible Causes:** • Open or short in injector for port injection circuit • IG2 fuse • Injector for port injection
DTC: P127A **T ECM, MIL: Yes** **Year:** 2009, 2010 **Model:** LS460 **Engine:** 4.6L V8 **Transmission:** All	**Port Injector Circuit No. 5:** Battery voltage: 8 V or more Fuel cut: OFF Port injector: Energized Engine switch: on (IG) **Possible Causes:** • Open or short in injector for port injection circuit • Injector for port injection • ECM
DTC: P127B **T ECM, MIL: Yes** **Year:** 2009, 2010 **Model:** GS350, IS350 **Engine:** 3.5L V6 **Transmission:** All	**Port Injector Circuit No. 6:** Battery voltage: 8 V or more Fuel cut: OFF Port injector: Energized Engine switch: On (IG) **Possible Causes:** • Open or short in injector for port injection circuit • IG2 fuse • Injector for port injection
DTC: P127B **T ECM, MIL: Yes** **Year:** 2009, 2010 **Model:** LS460 **Engine:** 4.6L V8 **Transmission:** All	**Port Injector Circuit No. 6:** Battery voltage: 8 V or more Fuel cut: OFF Port injector: Energized Engine switch: on (IG) **Possible Causes:** • Open or short in injector for port injection circuit • Injector for port injection • ECM

DTC	Trouble Code Title, Conditions & Possible Causes
DTC: P127C **T ECM, MIL: Yes** **Year:** 2009, 2010 **Model:** LS460 **Engine:** 4.6L V8 **Transmission:** All	**Port Injector Circuit #7:** Battery voltage: 8 V or more Fuel cut: OFF Port injector: Energized Engine switch: on (IG) **Possible Causes:** • Open or short in injector for port injection circuit • Injector for port injection • ECM
DTC: P127D **T ECM, MIL: Yes** **Year:** 2009, 2010 **Model:** LS460 **Engine:** 4.6L V8 **Transmission:** All	**Port Injector Circuit #8:** Battery voltage: 8 V or more Fuel cut: OFF Port injector: Energized Engine switch: on (IG) **Possible Causes:** • Open or short in injector for port injection circuit • Injector for port injection • ECM
DTC: P12FF **1T ECM, MIL: Yes** **Year:** 2009, 2010 **Model:** IS250 **Engine:** 2.5L V6 **Transmission:** All	**Electric Driver Unit:** Time after engine switch off to on (IG): 1 second or more Engine speed: 4000 rpm or less Battery voltage: 10.5 V or more **Possible Causes:** • Open or short in injector driver (EDU) circuit • Injector driver (EDU) • Fuel injector assembly • ECM
DTC: P1340 **1T ECM, MIL: Yes** **Year:** 2009, 2010 **Model:** LS460 **Engine:** 4.6L V8 **Transmission:** All	**Camshaft Position Sensor "A" (Bank 1 Sensor 2):** Camshaft position sensor range check: Starter: ON and not starter ON again Minimal battery voltage while starter ON: Less than 11 V Camshaft position/Crankshaft position misalignment: Monitor runs whenever following DTCs are not present: P0342, P0343, P1342, P1343 (VVT Sensor) Engine RPM: 600 rpm or more Starter: OFF Camshaft position sensor range: 0.3 to 4.7 V **Possible Causes:** • Open or short in camshaft position sensor circuit • Camshaft position sensor • Camshaft timing gear assembly (bank 2) • Jumped tooth of timing chain • ECM
DTC: P1340 **2T ECM, MIL: Yes** **Year:** 2009 **Model:** GX470 **Engine:** 4.7L V8 **Transmission:** All	**Camshaft Position Sensor "A" (Bank 1 Sensor 2):** The monitor will run whenever these DTCs are not present: None Camshaft position sensor range check: Starter: ON Minimal battery voltage while starter ON: Less than 11 V Camshaft position / Crankshaft position misalignment: Engine RPM: 600 rpm or more Starter: OFF **Possible Causes:** • Open or short in camshaft position sensor circuit • Camshaft position sensor • Camshaft timing pulley LH • ECM

DTC	Trouble Code Title, Conditions & Possible Causes
DTC: P1342 **1T ECM, MIL: Yes** **Year:** 2009, 2010 **Model:** LS460 **Engine:** 4.6L V8 **Transmission:** All	**Camshaft Position Sensor "A" Low Input (MRE):** Monitor runs whenever following DTCs are not present: P0340, P1340 (Camshaft Position Sensor) Time after engine switch off to on (IG): 2 seconds or more Battery voltage: 8 V or more Starter: OFF **Possible Causes:** • Open or short in camshaft position sensor circuit • Camshaft position sensor • Camshaft timing gear assembly (bank 2) • Jumped tooth of timing chain • ECM
DTC: P1343 **1T ECM, MIL: Yes** **Year:** 2009, 2010 **Model:** LS460 **Engine:** 4.6L V8 **Transmission:** All	**Camshaft Position Sensor "A" High Input (MRE):** Monitor runs whenever following DTCs are not present: P0340, P1340 (Camshaft Position Sensor) Time after engine switch off to on (IG): 2 seconds or more Battery voltage: 8 V or more Starter: OFF **Possible Causes:** • Open or short in camshaft position sensor circuit • Camshaft position sensor • Camshaft timing gear assembly (bank 2) • Jumped tooth of timing chain • ECM
DTC: p1360 **1T ECM, MIL: Yes** **Year:** 2009, 2010 **Model:** LS460 **Engine:** 4.6L V8 **Transmission:** All	**"A" Camshaft Position Actuator Circuit Open, Low, High Bank1:** Monitor runs whenever following DTCs not present: - Engine speed: 100 rpm or more Motor direction (calculation from CMP sensor): Forward Battery voltage: 11 V or more Engine switch: on (IG) **Possible Causes:** • Camshaft timing control motor (bank 1) • Wire harness and connector • Camshaft timing control motor power source
DTC: P1361 **1T ECM, MIL: Yes** **Year:** 2009, 2010 **Model:** LS460 **Engine:** 4.6L V8 **Transmission:** All	**"A" Camshaft Position Actuator Circuit Open, Low, High Bank2:** Monitor runs whenever following DTCs not present: - Engine speed: 100 rpm or more Motor direction (calculation from CMP sensor): Forward Battery voltage: 11 V or more Engine switch: on (IG) **Possible Causes:** • Camshaft timing control motor (bank 2) • Wire harness and connector • Camshaft timing control motor power source
DTC: P1362 **1T ECM, MIL: Yes** **Year:** 2009, 2010 **Model:** LS460 **Engine:** 4.6L V8 **Transmission:** All	**"B" Camshaft Position Actuator Circuit Open, Low, High Bank1:** Monitor runs whenever following DTCs not present: - CPU commanded motor direction: Forward Output motor speed: 100 rpm or more Motor direction (Calculation from CMP sensor): Forward Battery voltage: 11 V or more Engine switch: on (IG) **Possible Causes:** • Camshaft timing control motor (bank 1) • Wire harness and connector • Camshaft timing control motor power source

DTC	Trouble Code Title, Conditions & Possible Causes
DTC: P1363 **1T ECM** **Year:** 2009, 2010 **Model:** LS460 **Engine:** 4.6L V8 **Transmission:** All	**"B" Camshaft Position Actuator Circuit Open, Low, High Bank2:** Monitor runs whenever following DTCs not present: - CPU commanded motor direction: Forward Output motor speed: 100 rpm or more Motor direction (Calculation from CMP sensor): Forward Battery voltage: 11 V or more Engine switch: on (IG) **Possible Causes:** • Camshaft timing control motor (bank 2) • Wire harness and connector • Camshaft timing control motor power source
DTC: P1364 **1T ECM** **Year:** 2009, 2010 **Model:** LS460 **Engine:** 4.6L V8 **Transmission:** All	**"C" Camshaft Position Actuator Circuit Open, Low, High Bank1:** Monitor runs whenever following DTCs not present: None Battery voltage: 11 V or more Engine switch: on (IG) **Possible Causes:** • Camshaft timing control motor (bank 1) • Power supply of Camshaft timing control motor (bank 1) • Wire harness and connector
DTC: P1366 **1T , MIL: Yes** **Year:** 2009, 2010 **Model:** LS460 **Engine:** 4.6L V8 **Transmission:** All	**"E" Camshaft Position Actuator Circuit Open, Low, High Bank1:** Monitor runs whenever following DTCs not present: None Battery voltage: 11 V or more Engine switch: on (IG) **Possible Causes:** • Camshaft timing control motor (bank 1) • Wire harness and connector
DTC: P1367 **1T ECM, MIL: Yes** **Year:** 2009, 2010 **Model:** LS460 **Engine:** 4.6L V8 **Transmission:** All	**"E" Camshaft Position Actuator Circuit Open, Low, High Bank2:** Monitor runs whenever following DTCs not present: None Battery voltage: 11 V or more Engine switch: on (IG) **Possible Causes:** • Camshaft timing control motor (bank 2) • Wire harness and connector
DTC: P1440 **1T ECM, MIL: Yes** **Year:** 2009 **Model:** GX470 **Engine:** 4.7L V8 **Transmission:** All	**Secondary Air Injection Vacuum Switching Valve Circuit Malfunction Bank 1:** The monitor will run whenever these DTCs are not present: None Engine: Running No. 2 air switching valve (bank 1): Not operating **Possible Causes:** • Open or short in VSV for air injection system circuit (bank 1) • VSV power source • VSV for air injection system (bank 1) • ECM
DTC: P1441 **2T ECM, MIL: Yes** **Year:** 2009 **Model:** GX470 **Engine:** 4.7L V8 **Transmission:** All	**Stuck Open in Secondary Air Injection Vacuum Switching Valve Bank 1:** This monitor runs whenever these DTCs are not present: None **Possible Causes:** • VSV for air injection system circuit (bank 1) • No. 2 air switching valve (bank 1) • VSV for air injection system (bank 1) • ECM

DTC	Trouble Code Title, Conditions & Possible Causes
DTC: P1442 **2T ECM, MIL: Yes** **Year:** 2009 **Model:** GX470 **Engine:** 4.7L V8 **Transmission:** All	**Stuck Close in Secondary Air Injection Vacuum Switching Valve Bank 1:** This monitor runs whenever these DTCs are not present: None Atmospheric pressure: 45 kPa (420 mmHg) or more Battery voltage: 11.5 V or higher **Possible Causes:** • VSV for air injection system circuit (bank 1) • Vacuum hose (VSV for air injection system - No. 2 air switching valve) • Air injector pipe (No. 2 air switching valve - exhaust manifold) • No. 2 air switching valve (bank 1) • VSV for air injection system (bank 1) • ECM
DTC: P1443 **1T ECM, MIL: Yes** **Year:** 2009 **Model:** GX470 **Engine:** 4.7L V8 **Transmission:** All	**Secondary Air Injection Vacuum Switching Valve Circuit Malfunction Bank 2:** The monitor will run whenever these DTCs are not present: None Engine: Running No. 2 air switching valve (bank 2): Not operating **Possible Causes:** • Open or short in VSV for air injection system circuit (bank 2) • VSV power source • VSV for air injection system (bank 2) • ECM
DTC: P1444 **2T ECM, MIL: Yes** **Year:** 2009 **Model:** GX470 **Engine:** 4.7L V8 **Transmission:** All	**Stuck Open in Secondary Air Injection Vacuum Switching Valve Bank 2:** This monitor runs whenever these DTCs are not present: None **Possible Causes:** • VSV for air injection system circuit (bank 2) • No. 2 air switching valve (bank 2) • VSV for air injection system (bank 2) • ECM
DTC: P1445 **2T ECM, MIL: Yes** **Year:** 2009 **Model:** GX470 **Engine:** 4.7L V8 **Transmission:** All	**Stuck Close in Secondary Air Injection Vacuum Switching Valve Bank 2:** This monitor runs whenever these DTCs are not present: None Atmospheric pressure: 45 kPa (420 mmHg) or more Battery voltage: 11.5 V or higher **Possible Causes:** • VSV for air injection system circuit (Bank 2) • Vacuum hose (VSV for air injection system - No. 2 air switching valve) • Air injector pipe (No. 2 air switching valve - exhaust manifold) • No. 2 air switching valve (bank 2) • VSV for air injection system (bank 2) • ECM
DTC: P1570 **T ECM, MIL: Yes** **Year:** 2009, 2010 **Model:** RX350 **Engine:** 3.5L V6 **Transmission:** All	**Radar Sensor Malfunction:** The ECM detects a laser sensor malfunction signal for 0.15 sec. or more while the dynamic laser cruise control is in operation. **Possible Causes:** • Laser sensor
DTC: P1570 **T ECM, MIL: Yes** **Year:** 2009, 2010 **Model:** GS350, IS250, IS350 **Engine:** 2.5L V6, 3.5L V6 **Transmission:** All	**Radar Sensor Malfunction:** The ECM detects a radar sensor malfunction signal for 0.15 sec. or more while the dynamic radar cruise control is in operation. **Possible Causes:** • Millimeter wave radar sensor
DTC: P1572 **T ECM, MIL: Yes** **Year:** 2009, 2010 **Model:** RX350 **Engine:** 3.5L V6 **Transmission:** All	**Improper Aiming of Radar Sensor Beam Axis:** The ECU detects that the laser sensor beam axis is in an incorrect position (0.15 sec. or more) while the dynamic laser cruise control is in operation. **Possible Causes:** • Laser sensor

DTC	Trouble Code Title, Conditions & Possible Causes
DTC: P1572 **T ECM, MIL: Yes** **Year:** 2009, 2010 **Model:** GS350, IS250, IS350 **Engine:** 2.5L V6, 3.5L V6 **Transmission:** All	**Improper Aiming of Radar Sensor Beam Axis:** The ECU detects that the millimeter wave radar sensor beam axis is in an incorrect position (0.15 sec. or more) while the dynamic radar cruise control is in operation. **Possible Causes:** • Millimeter wave radar sensor
DTC: P1575 **T ECM, MIL: Yes** **Year:** 2009, 2010 **Model:** IS250, IS350 **Engine:** 2.5L V6, 3.5L V6 **Transmission:** All	**Warning Buzzer Malfunction:** The ECM receives a buzzer abnormal signal for 0.2 sec. or more while the dynamic radar cruise control is in operation. **Possible Causes:** • Skid control buzzer • Skid control buzzer circuit • ABS & traction actuator (Skid control ECU) • ECM
DTC: P1575 **T ECM, MIL: Yes** **Year:** 2009, 2010 **Model:** RX350 **Engine:** 3.5L V6 **Transmission:** All	**Warning Buzzer Malfunction:** The ECM receives a buzzer abnormal signal for 0.2 sec. or more while the dynamic laser cruise control is in operation. **Possible Causes:** • Skid control buzzer • Skid control buzzer circuit • ABS & traction actuator (Skid control ECU)
DTC: P1578 **T ECM, MIL: Yes** **Year:** 2009, 2010 **Model:** IS250, IS350 **Engine:** 2.5L V6, 3.5L V6 **Transmission:** All	**Brake System Malfunction:** The ECM receives a brake system error signal for 0.2 sec. or more while the dynamic radar cruise control is in operation. **Possible Causes:** • VSC system
DTC: P1578 **T ECM, MIL: Yes** **Year:** 2009, 2010 **Model:** RX350 **Engine:** 3.5L V6 **Transmission:** All	**Brake System Malfunction:** The ECM receives a brake system error signal for 0.2 sec. or more while the dynamic laser cruise control is in operation. **Possible Causes:** • VSC system
DTC: P1602 **1T ECM** **Year:** 2009 **Model:** LS460 **Engine:** 4.6L V8 **Transmission:** All	**Deterioration of Battery:** Battery power is 0% **Possible Causes:** • Battery • ECM back-up power source circuit
DTC: P1603 **1T ECM, MIL: Yes** **Year:** 2010 **Model:** LS460 **Engine:** 4.6L V8 **Transmission:** All	**Engine Stall History:** After monitoring for startability problems (P1604) finishes and 5 seconds or more elapse after starting the engine, with the engine running, the engine stops (the engine speed drops to 200 rpm or less) without the engine switch being operated for 0.5 seconds or more. **Possible Causes:** • Air leak in intake system • Purge VSV • Mass air flow meter • Engine coolant temperature sensor • Wire harness or connector • Air fuel ratio sensor • Power supply circuit (purge VSV, fuel injector for port injection, ignition coil assembly) • Fuel pump (for low pressure) • Fuel pump (for low pressure) control system • Fuel line • Throttle body • Camshaft timing oil control valve • VVT-iE system • Air conditioning system • Electrical load signal system • A/T system • Park/neutral position switch • ECM

DTC	Trouble Code Title, Conditions & Possible Causes
DTC: P1604 **1T ECM, MIL: Yes** **Year:** 2009, 2010 **Model:** LS460 **Engine:** 4.6L V8 **Transmission:** All	**Startability Malfunction:** The engine speed is below 500 rpm with the STA signal on for a certain amount of time. After the engine starts (engine speed is 500 rpm or more), the engine speed drops to 200 rpm or less within approximately 2 seconds. **Possible Causes:** • Immobiliser system • Engine assembly (excess friction, compression loss) • Starter assembly • Crankshaft position sensor • Camshaft position sensor • VVT sensor • Engine coolant temperature sensor • Fuel pump (for low pressure) • Fuel pump (for low pressure) control system • Fuel pipes • Fuel injector for port injection • Throttle body • Pressure regulator • Battery • Drive plate • Spark plug • Ignition coil circuit • Intake system • Camshaft timing oil control valve • Mass air flow meter • Air fuel ratio sensor • Valve timing • Fuel • Purge VSV • Intake valve • Exhaust valve • ECM
DTC: P1605 **1T ECM, MIL: Yes** **Year:** 2010 **Model:** LS460 **Engine:** 4.6L V8 **Transmission:** All	**Rough Idling:** After 5 seconds or more elapse after starting the engine, with the engine running, the engine speed drops to 400 rpm or less. **Possible Causes:** • Air leak in intake system • Purge VSV • Mass air flow meter • Engine coolant temperature sensor • Wire harness or connector • Air fuel ratio sensor • Power supply circuit (purge VSV, fuel injector for port injection, ignition coil) • Fuel pump (for low pressure) • Fuel pump (for low pressure) control system • Fuel line • Throttle body • Camshaft timing oil control valve • VVT-iE system • Knock sensor • Ignition coil • Fuel injector for port injection • Spark plug(s) • Air conditioning system • Electrical load signal system • A/T system • Park/neutral position switch • ECM
DTC: P1607 **1T ECM, MIL: Yes** **Year:** 2009, 2010 **Model:** All **Engine:** All **Transmission:** All	**Cruise Control Input Processor:** Monitor runs whenever the following DTCs are not present: None **Possible Causes:** • ECM

DTC	Trouble Code Title, Conditions & Possible Causes
DTC: P1613 **T ECM, MIL: Yes** **Year:** 2009 **Model:** GX470 **Engine:** 4.7L V8 **Transmission:** All	**Secondary Air Injection Driver Malfunction:** The monitor will run whenever these DTCs are not present: None Case 1: Battery voltage: 8 V or more Ignition switch: ON Starter: OFF Case 2: Either of following conditions is met: Condition 1 or 2 1. Air pump: Not operating 2. Air switching valve: Not operating Battery voltage: 8 V or more Ignition switch: ON Starter: OFF Case 3: Air pump: Operating Air switching valve: Operating Battery voltage: 8 V or more Ignition switch: ON Starter: OFF Case 4: Battery voltage: 8 V or more Ignition switch: ON Starter: OFF **Possible Causes:** • Air injection control driver (AID) • Open in air injection control driver ground circuit • Opeen or short in diagnostic information signal circuit (AID - ECM) • Open or short in air pump and air switching valve command
DTC: P1615 **T , MIL: Yes** **Year:** 2009, 2010 **Model:** All **Engine:** All V6, V8 **Transmission:** All	**Communication Error from Distance Control ECU to ECM:** While the dynamic radar cruise control is either preparing for operation or operating, if communication data from the distance control ECU is logically inconsistent for a certain amount of time, the ECM records this logical error code. **Possible Causes:** • Communication circuit • Distance control ECU • ECM
DTC: P1616 **T ECM, MIL: Yes** **Year:** 2009, 2010 **Model:** All **Engine:** All V6, V8 **Transmission:** All	**Communication Error from ECM to Distance Control ECU:** While the dynamic laser cruise control is either preparing for operation or operating, if the ECM continuously receives a logical error signal from the distance control ECU for more than a specific amount of time, the ECM records this logical error code. **Possible Causes:** • Communication circuit • ECM • Distance control ECU (Cruise control ECU)
DTC: P1617 **T ECM, MIL: Yes** **Year:** 2009, 2010 **Model:** All **Engine:** All V6, V8 **Transmission:** All	**Distance Control ECU Malfunction:** While the dynamic radar cruise control is either preparing for operation or operating, if a designation signal from the ECM and a designation return signal from the distance control ECU do not match or the distance control ECU is malfunctioning for more than a specific amount of time, the ECM records this trouble code. **Possible Causes:** • Distance control ECU
DTC: P1630 **T ECM, MIL: Yes** **Year:** 2009, 2010 **Model:** All **Engine:** All V6, V8 **Transmission:** All	**Communication Error from VSC to ECM:** While the dynamic radar cruise control is either preparing for operation or operating, if communication data from the skid control ECU is logically inconsistent for a certain amount of time, the ECM records this logical error code. **Possible Causes:** • Communication circuit • ABS & traction actuator (Skid control ECU) • ECM

DTC	Trouble Code Title, Conditions & Possible Causes
DTC: P1631 **T ECM, MIL: Yes** **Year:** 2009, 2010 **Model:** All **Engine:** All V6, V8 **Transmission:** All	**Communication Error from ECM to VSC:** While the dynamic laser cruise control is either preparing for operation or operating, if the ECM continuously receives a logical error signal from the skid control ECU for a certain amount of time, the ECM records this logical error code. **Possible Causes:** • Communication circuit • ECM • ABS & traction actuator (Skid control ECU)
DTC: P16A0 **T ECM, MIL: Yes** **Year:** 2009, 2010 **Model:** ES350 **Engine:** 3.5L V6 **Transmission:** All	**G Sensor Circuit:** While the power source voltage of the G sensor (VGS) is 4.8 V or more and less than 5.2 V, the GL1 voltage is less than 0.03 x VGS, or 0.94 x VGS or more for 10 seconds or more. While the power source voltage of the G sensor (VGS) is less than 1 V, the GL1 voltage is less than 0.03 x VGS for 10 seconds or more. Power source voltage of the G sensor is less than 4.8 V or 5.2 V or more for 10 seconds or more. Difference between the current 11-fold G sensor value and the last 11-fold G sensor value exceeds the specified value. The G sensor standard voltage in the ECU remains less than 2 V or 3 V or more for 10 seconds or more. **Possible Causes:** • Acceleration sensor • Acceleration sensor power source circuit • Open or short in acceleration sensor circuit • Active control engine mount ECU
DTC: P16A1 **T , MIL: Yes** **Year:** 2009, 2010 **Model:** ES350 **Engine:** 3.5L V6 **Transmission:** All	**Linear Solenoid Circuit or G Sensor Circuit:** While the active control engine mount ECU is controlling the actuators, problems (open circuit, short circuit, overcurrent, etc.) occur in the ECU, actuators, or the circuits. The vibration signal level from the acceleration sensor is lower than specified (problems with the sensor or the actuator may be stuck). **Possible Causes:** • Acceleration sensor • Front engine mounting insulator • Open or short in active control engine mount actuator circuit • Active control engine mount ECU
DTC: P16A2 **T ECM, MIL: Yes** **Year:** 2009, 2010 **Model:** ES350 **Engine:** 3.5L V6 **Transmission:** All	**Battery High Voltage:** IG voltage is more than 16 V for 0.5 seconds or more. **Possible Causes:** • ECU power source circuit • Active control engine mount ECU
DTC: P16A3 **T ECM, MIL: Yes** **Year:** 2009, 2010 **Model:** ES350 **Engine:** 3.5L V6 **Transmission:** All	**TACH Signal Communication:** TACH signals from the ECM remain interrupted for 1.3 seconds or more. **Possible Causes:** • Open or short in TACH signal circuit • Active control engine mount ECU • ECM
DTC: P16A4 **T ECM, MIL: Yes** **Year:** 2009, 2010 **Model:** ES350 **Engine:** 3.5L V6 **Transmission:** All	**High Speed CAN Communication Bus:** * Communication between the ECM and E-ACM ECU is interrupted. * Signals indicating ambient temperature of less than -40°C (-40°F) or more than 50°C (122°F) are received for 10.6 seconds or more. * Signals indicating coolant temperature of less than -40°C (-40°F) or more than 140°C (284°F) are received for 1.25 seconds or more. * Signals indicating that more than one shift position are ON or all shift positions are OFF for 30 seconds or more. **Possible Causes:** • ECM • Active control engine mount ECU • Skid control ECU • A/C amplifier • CAN communication system

OBD II Trouble Code List (P2XXX Codes)

DTC	Trouble Code Title, Conditions & Possible Causes
DTC: P2004 **T ECM, MIL: Yes** **Year:** 2009, 2010 **Model:** IS250 **Engine:** 2.5L V6 **Transmission:** All	**Intake Manifold Runner Control Stuck Open (Bank 1):** ECT: -10°C (14°F) or more IAT: -10°C (14°F) or more Engine switch: On (IG) Battery voltage: 8 V or more **Possible Causes:** • DC Motor for SCV circuit • DC Motor for SCV • SCV position sensor • SCV • ECM
DTC: P2006 **T ECM, MIL: Yes** **Year:** 2009, 2010 **Model:** IS250 **Engine:** 2.5L V6 **Transmission:** All	**Intake Manifold Runner Control Stuck Closed (Bank 1):** ECT: -10°C (14°F) or more IAT: -10°C (14°F) or more Engine switch: On (IG) Battery voltage: 8 V or more **Possible Causes:** • DC Motor for SCV circuit • DC Motor for SCV • SCV position sensor • SCV • ECM
DTC: P2009 **1T , MIL: Yes** **Year:** 2009, 2010 **Model:** IS250 **Engine:** 2.5L V6 **Transmission:** All	**Intake Manifold Runner Control Circuit Low (Bank 1):** Output signal duty: 100% or more Motor current - Last motor current: Less than 0.2 A Intake manifold runner control circuit range check high current fail (P2010): Not detected **Possible Causes:** • Open or short in DC motor for SCV circuit • Intake manifold (DC motor for SCV) • ECM
DTC: P2010 **1T ECM, MIL: Yes** **Year:** 2009, 2010 **Model:** IS250 **Engine:** 2.5L V6 **Transmission:** All	**Intake Manifold Runner Control Circuit High (Bank 1):** Output signal duty: 100% or more **Possible Causes:** • Open or short in DC motor for SCV circuit • Intake manifold (DC motor for SCV) • ECM
DTC: P2014 **T ECM, MIL: Yes** **Year:** 2009, 2010 **Model:** IS250 **Engine:** 2.5L V6 **Transmission:** All	**Intake Manifold Runner Position Sensor / Switch Circuit (Bank 1):** Monitor runs whenever following DTCs not present: None **Possible Causes:** • Open or short in SCV position sensor circuit • SCV position sensor • ECM
DTC: P2016 **T ECM, MIL: Yes** **Year:** 2009, 2010 **Model:** IS250 **Engine:** 2.5L V6 **Transmission:** All	**Intake Manifold Runner Position Sensor / Switch Circuit Low (Bank 1):** Monitor runs whenever following DTCs not present: None **Possible Causes:** • Open or short in SCV position sensor circuit • SCV position sensor • ECM
DTC: P2017 **T ECM, MIL: Yes** **Year:** 2009, 2010 **Model:** IS250 **Engine:** 2.5L V6 **Transmission:** All	**Intake Manifold Runner Position Sensor / Switch Circuit High (Bank 1):** Monitor runs whenever following DTCs not present: None **Possible Causes:** • Open or short in SCV position sensor circuit • SCV position sensor • ECM

<table>
<tr><th>DTC</th><th>Trouble Code Title, Conditions & Possible Causes</th></tr>
<tr><td>DTC: P2102
1T , MIL: Yes
Year: 2009, 2010
Model: All
Engine: All V6, V8
Transmission: All</td><td>Throttle Actuator Control Motor Circuit Low:
Throttle actuator: ON
Duty-cycle ratio to open throttle actuator: 80% or more
Throttle actuator power supply: 8 V or higher
Motor current change during latest 0.016 seconds: Less than 0.2 A
Possible Causes:
• Open in throttle actuator circuit
• Throttle actuator
• ECM</td></tr>
<tr><td>DTC: P2103
1T , MIL: Yes
Year: 2009, 2010
Model: All
Engine: All V6, V8
Transmission: All</td><td>Throttle Actuator Control Motor Circuit High:
Throttle actuator: ON
Either of the following conditions 1 or 2 is met:
1. Throttle actuator power supply: 8 V or higher
2. Throttle actuator power: ON
Possible Causes:
• Short in throttle actuator circuit
• Throttle actuator
• Throttle valve
• Throttle body assembly
• ECM</td></tr>
<tr><td>DTC: P2109
5T ECM, MIL: Yes
Year: 2009, 2010
Model: LS460
Engine: 4.6L V8
Transmission: All</td><td>Throttle / Pedal Position Sensor "A" Minimum Stop Performance:
Either condition is met:
* With atmospheric pressure 85 kPa (638 mmHg) or more (elevation 1400 m (4592 ft.) or less), when the engine coolant temperature is 45°C (113°F) or less at engine start, the engine is warmed up and conditions for ISC learning are met, the ISC learned value is approximately 3 times larger than normal even though the mass air flow during idling is normal.
* With atmospheric pressure 85 kPa (638 mmHg) or more (elevation 1400 m (4592 ft.) or less), when the engine switch has been turned to on (IG) for 1 hour or more, the engine is warmed up, conditions for ISC learning are met and the vehicle has been driven at a speed of 18.6 mph (30 km/h) or more at least once, the ISC learned value is approximately 3 times larger than normal even though the mass air flow during idling is normal.
Possible Causes:
• Throttle body</td></tr>
<tr><td>DTC: P2111
1T ECM, MIL: Yes
Year: 2009, 2010
Model: All
Engine: All V6, V8
Transmission: All</td><td>Throttle Actuator Control System - Stuck Open:
All of following conditions are met: -
System guard* judge condition: ON
Throttle actuator current: 2 A or more
Duty cycle to close throttle: 80% or more
Possible Causes:
• Throttle actuator
• Throttle body
• Throttle valve
• ECM</td></tr>
<tr><td>DTC: P2112
1T ECM, MIL: Yes
Year: 2009, 2010
Model: All
Engine: All V6, V8
Transmission: All</td><td>Throttle Actuator Control System - Stuck Closed:
All of the following conditions are met:
System guard* judge condition: ON
Throttle actuator current: 2 A or more
Duty-cycle to open throttle: 80% or more
*: System guard is ON when the following conditions set:
Throttle actuator: ON
Throttle actuator duty calculation: Executing
Throttle position sensor fail: Not detected
Throttle actuator current-cut operation: Not executing
Throttle actuator power supply: 4 V or more
Throttle actuator fail: Not detected
Possible Causes:
• Throttle actuator
• Throttle body assembly
• Throttle valve
• ECM</td></tr>
</table>

DTC	Trouble Code Title, Conditions & Possible Causes
DTC: P2118 **1T ECM, MIL: Yes** **Year:** 2009, 2010 **Model:** All **Engine:** All V6, V8 **Transmission:** All	Throttle Actuator Control Motor Current Range / Performance: Battery voltage: 8 V or more Throttle actuator power: ON **Possible Causes:** • Open in ETCS power source circuit • ETCS fuse • ECM
DTC: P2119 **1T ECM, MIL: Yes** **Year:** 2009, 2010 **Model:** All **Engine:** All V6, V8 **Transmission:** All	**Throttle Actuator Control Throttle Body Range / Performance:** System guard* judge condition: ON *System guard is ON when the following conditions are met: Throttle actuator: ON Throttle actuator duty calculation: Executing Throttle position sensor fail: Not detected Throttle actuator current-cut operation: Not executing Throttle actuator power supply: 4 V or more Throttle actuator fail: Not detected **Possible Causes:** • ETCS • ECM
DTC: P2120 **1T ECM, MIL: Yes** **Year:** 2009, 2010 **Model:** All **Engine:** All **Transmission:** All	**Throttle / Pedal Position Sensor / Switch "D" Circuit:** Engine switch: On (IG) Electronic throttle actuator power: ON **Possible Causes:** • Accelerator Pedal Position (APP) sensor • ECM
DTC: P2121 **1T ECM, MIL: Yes** **Year:** 2009, 2010 **Model:** ES350, GS350 **Engine:** 3.5L V6 **Transmission:** All	**Throttle / Pedal Position Sensor / Switch "D" Circuit Range / Performance:** Either of the following conditions met: Condition (a) or (b) (a) Engine switch: ON (b) Throttle actuator power: ON **Possible Causes:** • Accelerator Pedal Position (APP) sensor • ECM
DTC: P2122 **1T ECM, MIL: Yes** **Year:** 2009, 2010 **Model:** All **Engine:** All **Transmission:** All	**Throttle / Pedal Position Sensor / Switch "D" Circuit Low Input:** Engine switch: On (IG) Electronic throttle actuator power: ON **Possible Causes:** • APP sensor • Open in VCP1 circuit • Open or ground short in VPA circuit • ECM
DTC: P2123 **1T ECM, MIL: Yes** **Year:** 2009, 2010 **Model:** All **Engine:** All **Transmission:** All	**Throttle / Pedal Position Sensor / Switch "D" Circuit High Input:** Engine switch: On (IG) Electronic throttle actuator power: ON **Possible Causes:** • APP sensor • Open in EPA circuit • ECM
DTC: P2125 **1T ECM, MIL: Yes** **Year:** 2009, 2010 **Model:** All **Engine:** All **Transmission:** All	**Throttle / Pedal Position Sensor / Switch "E" Circuit:** Engine switch: On (IG) Electronic throttle actuator power: ON **Possible Causes:** • APP sensor • ECM

DTC	Trouble Code Title, Conditions & Possible Causes
DTC: P2127 **1T ECM, MIL: Yes** **Year:** 2009, 2010 **Model:** All **Engine:** All **Transmission:** All	**Throttle / Pedal Position Sensor / Switch "E" Circuit Low Input:** Engine switch: On (IG) Electronic throttle actuator power: ON **Possible Causes:** • APP sensor • Open in VCP2 circuit • Open or ground short in VPA2 circuit • ECM
DTC: P2128 **1T ECM, MIL: Yes** **Year:** 2009, 2010 **Model:** All **Engine:** All **Transmission:** All	**Throttle / Pedal Position Sensor / Switch "E" Circuit High Input:** Engine switch: On (IG) Electronic throttle actuator power: ON **Possible Causes:** • APP sensor • Open in EPA2 circuit • ECM
DTC: P2129 **1T ECM, MIL: Yes** **Year:** 2009, 2010 **Model:** All **Engine:** All **Transmission:** All	**Throttle / Pedal Position Sensor / Switch "E" Circuit High Input:** Engine switch: On (IG) Electronic throttle actuator power: ON **Possible Causes:** • APP sensor • Open in EPA2 circuit • ECM
DTC: P2135 **1T ECM, MIL: Yes** **Year:** 2009, 2010 **Model:** All **Engine:** All V6, V8 **Transmission:** All	**Throttle / Pedal Position Sensor / Switch "A" / "B" Voltage Correlation:** **Possible Causes:** • Short between VTA1 and VTA2 circuits • TP sensor (built into throttle body) • ECM
DTC: P2138 **1T ECM, MIL: Yes** **Year:** 2009, 2010 **Model:** All **Engine:** All **Transmission:** All	**Throttle / Pedal Position Sensor / Switch "D" / "E" Voltage Correlation:** Engine switch: On (IG) Electronic throttle actuator power: ON **Possible Causes:** • Short between VPA and VPA2 circuits • APP sensor • ECM
DTC: P2195 **2T ECM, MIL: Yes** **Year:** 2009, 2010 **Model:** All **Engine:** All V6, V8 **Transmission:** All	**Oxygen (A/F) Sensor Signal Stuck Lean (Bank 1 Sensor 1):** Sensor voltage detection monitor: Time after engine start: 30 seconds or more A/F sensor status: Activated Fuel system status: Closed-loop Sensor current detection monitor: Battery voltage: 11 V or more ECT: 75°C (167°F) or more Atmospheric pressure: 76 kPa (570 mmHg) or more Air-fuel ratio sensor status: Activated Continuous time of fuel cut: 4 to 10 seconds **Possible Causes:** • Open or short in A/F sensor (bank 1, 2 sensor 1) circuit • A/F sensor (bank 1, 2 sensor 1) • A/F sensor (bank 1, 2 sensor 1) heater • A/F sensor heater and relay circuits • Air induction system • Fuel pressure (low pressure side) • Fuel pressure (high pressure side) • Injector • Fuel pressure sensor • ECM

DTC	Trouble Code Title, Conditions & Possible Causes
DTC: P2196 **2T EGR1, MIL: Yes** **Year:** 2009, 2010 **Model:** All **Engine:** All V6, V8 **Transmission:** All	**Oxygen (A/F) Sensor Signal Stuck Rich (Bank 1 Sensor 1):** Sensor current detection monitor: Battery voltage: 11 V or more Atmospheric pressure: 76 kPa (570 mmHg) or higher Air-fuel ratio sensor status: Activated Engine coolant temperature: 75°C (167°F) or more Continuous time of fuel cut: 3 to 10 seconds **Possible Causes:** • Open or short in A/F sensor (bank 1 sensor 1) circuit • A/F sensor (bank 1 sensor 1) • A/F sensor (bank 1 sensor 1) heater • A/F sensor heater circuits • Air induction system • Fuel pressure (low pressure side) • Fuel pressure (high pressure side) • Fuel injector for port injection • Fuel injector for direct injection • ECM
DTC: P2197 **2T ECM, MIL: Yes** **Year:** 2009, 2010 **Model:** All **Engine:** All V6, V8 **Transmission:** All	**Oxygen (A/F) Sensor Signal Stuck Lean (Bank 2 Sensor 1):** Lean side malfunction: Time after engine start: 30 seconds or more A/F sensor status: Activated Fuel system status: Closed-loop Sensor current detection monitor: Battery voltage: 11 V or more Atmospheric pressure: 76 kPa (570 mmHg) or higher Air-fuel ratio sensor status: Activated Engine coolant temperature: 75°C (167°F) or more Continuous time of fuel cut: 3 to 10 seconds **Possible Causes:** • Open or short in A/F sensor (bank 2 sensor 1) circuit • A/F sensor (bank 2 sensor 1) • A/F sensor (bank 2 sensor 1) heater • A/F sensor heater circuits • Air induction system • Fuel pressure (low pressure side) • Fuel pressure (high pressure side) • Fuel injector for port injection • Fuel injector for direct injection • ECM
DTC: P2198 **2T ECM, MIL: Yes** **Year:** 2009, 2010 **Model:** All **Engine:** All V6, V8 **Transmission:** All	**Oxygen (A/F) Sensor Signal Stuck Rich (Bank 2 Sensor 1):** Time after engine start: 30 seconds or more A/F sensor status: Activated Fuel system status: Closed-loop Sensor current detection monitor: Battery voltage: 11 V or more Atmospheric pressure: 76 kPa (570 mmHg) or higher Air-fuel ratio sensor status: Activated Engine coolant temperature: 75°C (167°F) or more Continuous time of fuel cut: 3 to 10 seconds **Possible Causes:** • Open or short in A/F sensor (bank 2 sensor 1) circuit • A/F sensor (bank 2 sensor 1) • A/F sensor (bank 2 sensor 1) heater • A/F sensor heater circuits • Air induction system • Fuel pressure (low pressure side) • Fuel pressure (high pressure side) • Fuel injector for port injection • Fuel injector for direct injection • ECM

DTC	Trouble Code Title, Conditions & Possible Causes
DTC: P2237 **2T ECM, MIL: Yes** **Year:** 2009, 2010 **Model:** All **Engine:** All V6, V8 **Transmission:** All	**Oxygen (A/F) Sensor Pumping Current Circuit / Open (Bank 1 Sensor 1):** A/F sensor open circuit between A1A+ and A1A-/A2A+ and A2A-: Estimated sensor temperature: 450 to 550°C (842 to 1022°F) Engine: Running Battery voltage: 11 V or more A/F sensor low impedance: Estimated sensor temperature: 700 to 800°C (1292 to 1472°F) Engine coolant temperature: 0°C (32°F) or more Fuel cut: Not executed Other: Battery voltage: 11 V or more Engine switch: On (IG) Timing after engine switch is Off to On (IG): 5 seconds or more **Possible Causes:** • Open or short in A/F sensor (bank 1, 2 sensor 1) circuit • A/F sensor (bank 1, 2 sensor 1) • ECM
DTC: P2238 **2T ECM, MIL: Yes** **Year:** 2009, 2010 **Model:** All **Engine:** All V6, V8 **Transmission:** All	**Oxygen (A/F) Sensor Pumping Current Circuit Low (Bank 1 Sensor 1):** Open circuit between AF+ and AF-: Estimated sensor temperature: 1292-1472°F (700-800°C) ECT: 41°F (5°C) or more (varies with ECT at engine start) Fuel cut: Not executed **Possible Causes:** • Open or short in A/F sensor (bank 1, 2 sensor 1) circuit • A/F sensor (bank 1, 2 sensor 1) • A/F sensor heater • A/F sensor heater relay • A/F sensor heater and relay circuits • ECM
DTC: P2239 **T , MIL: Yes** **Year:** 2009, 2010 **Model:** All **Engine:** All V6, V8 **Transmission:** All	**Oxygen (A/F) Sensor Pumping Current Circuit High (Bank 1 Sensor 1):** Battery voltage: 11 V or more Engine switch: on (IG) Time after engine switch is off to on (IG): 5 seconds or more **Possible Causes:** • Open or short in A/F sensor (bank 1 sensor 1) circuit • A/F sensor (bank 1 sensor 1) • ECM
DTC: P2240 **2T ECM, MIL: Yes** **Year:** 2009, 2010 **Model:** All **Engine:** All V6, V8 **Transmission:** All	**Oxygen (A/F) Sensor Pumping Current Circuit / Open (Bank 2 Sensor 1):** Open circuit between AF+ and AF-: Estimated sensor temperature: 842-1022°F (450-550°C) Engine: Running Battery voltage: 11 V or more **Possible Causes:** • Open or short in A/F sensor (bank 1, 2 sensor 1) circuit • Air fuel ratio sensor (bank 1, 2 sensor 1) • ECM
DTC: P2241 **2T ECM, MIL: Yes** **Year:** 2009, 2010 **Model:** All **Engine:** All V6, V8 **Transmission:** All	**Oxygen (A/F) Sensor Pumping Current Circuit Low (Bank 2 Sensor 1):** Estimated sensor temperature: 700 to 800°C (1292 to 1472°F) Engine coolant temperature: -6°C (21.2°F) or higher Fuel cut: No executed **Possible Causes:** • Open or short in A/F sensor (bank 2 sensor 1) circuit • A/F sensor (bank 2 sensor 1) • A/F sensor (bank 2 sensor 1) heater • ECM

DTC	Trouble Code Title, Conditions & Possible Causes
DTC: P2242 **2T ECM, MIL: Yes** **Year:** 2009, 2010 **Model:** All **Engine:** All V6, V8 **Transmission:** All	**Oxygen (A/F) Sensor Pumping Current Circuit High (Bank 2 Sensor 1):** Battery voltage: 11 V or more Engine switch: ON Time after engine switch OFF to ON: 5 seconds or more **Possible Causes:** • Open or short in A/F sensor (bank 1, 2 sensor 1) circuit • A/F sensor (bank 1, 2 sensor 1) • A/F sensor heater • A/F sensor heater relay • A/F sensor heater and relay circuits • ECM
DTC: P2252 **2T ECM, MIL: Yes** **Year:** 2009, 2010 **Model:** All **Engine:** All V6, V8 **Transmission:** All	**Oxygen (A/F) Sensor Reference Ground Circuit Low (Bank 1 Sensor 1):** A/F sensor open circuit between A1A+ and A1A-/A2A+ and A2A-: Estimated sensor temperature: 450 to 550°C (842 to 1022°F) Engine: Running Battery voltage: 11 V or more A/F sensor low impedance: Estimated sensor temperature: 700 to 800°C (1292 to 1472°F) Engine coolant temperature: 0°C (32°F) or more Fuel cut: Not executed Other: Battery voltage: 11 V or more Engine switch: On (IG) Timing after engine switch is Off to On (IG): 5 seconds or more **Possible Causes:** • Open or short in A/F sensor (bank 1, 2 sensor 1) circuit • A/F sensor (bank 1, 2 sensor 1) • ECM
DTC: P2253 **2T ECM, MIL: Yes** **Year:** 2009, 2010 **Model:** All **Engine:** All V6, V8 **Transmission:** All	**Oxygen (A/F) Sensor Reference Ground Circuit High (Bank 1 Sensor 1):** A/F sensor open circuit between A1A+ and A1A-/A2A+ and A2A-: Estimated sensor temperature: 450 to 550°C (842 to 1022°F) Engine: Running Battery voltage: 11 V or more A/F sensor low impedance: Estimated sensor temperature: 700 to 800°C (1292 to 1472°F) Engine coolant temperature: 0°C (32°F) or more Fuel cut: Not executed Other: Battery voltage: 11 V or more Engine switch: On (IG) Timing after engine switch is Off to On (IG): 5 seconds or more **Possible Causes:** • Open or short in A/F sensor (bank 1, 2 sensor 1) circuit • A/F sensor (bank 1, 2 sensor 1) • ECM

DTC	Trouble Code Title, Conditions & Possible Causes
DTC: P2255 **2T ECM, MIL: Yes** **Year:** 2009, 2010 **Model:** All **Engine:** All V6, V8 **Transmission:** All	**Oxygen (A/F) Sensor Reference Ground Circuit Low (Bank 2 Sensor 1):** A/F sensor open circuit between A1A+ and A1A-/A2A+ and A2A-: Estimated sensor temperature: 450 to 550°C (842 to 1022°F) Engine: Running Battery voltage: 11 V or more A/F sensor low impedance: Estimated sensor temperature: 700 to 800°C (1292 to 1472°F) Engine coolant temperature: 0°C (32°F) or more Fuel cut: Not executed Other: Battery voltage: 11 V or more Engine switch: On (IG) Timing after engine switch is Off to On (IG): 5 seconds or more **Possible Causes:** • Open or short in A/F sensor (bank 1, 2 sensor 1) circuit • A/F sensor (bank 1, 2 sensor 1) • ECM
DTC: P2256 **2T ECM, MIL: Yes** **Year:** 2009, 2010 **Model:** All **Engine:** All V6, V8 **Transmission:** All	**Oxygen (A/F) Sensor Reference Ground Circuit High (Bank 2 Sensor 1):** Battery voltage: 11 V or more Engine switch: ON Time after engine switch OFF to ON: 5 seconds or more **Possible Causes:** • Open or short in A/F sensor (bank 1, 2 sensor 1) circuit • A/F sensor (bank 1, 2 sensor 1) • A/F sensor heater • A/F sensor heater relay • A/F sensor heater and relay circuits • ECM
DTC: P2401 **2T ECM, MIL: Yes** **Year:** 2009, 2010 **Model:** All **Engine:** All V6, V8 **Transmission:** All	**Evaporative Emission Leak Detection Pump Stuck OFF:** Monitor runs whenever following DTCs not present: None EVAP key-off monitor runs when all of following conditions met: Atmospheric pressure: 70 to 110 kPa-a (525 to 825 mmHg-a) Battery voltage: 10.5 V or more Vehicle speed: Below 2.5 mph (4 km/h) Engine switch: off Time after key off: 5 or 7 or 9.5 hours Canister pressure sensor malfunction (P0451, P0452 and P0453): Not detected Purge VSV: Not operated by scan tool Vent valve: Not operated by scan tool Leak detection pump: Not operated by scan tool Both of following conditions met before key off: Conditions 1 and 2 1. Duration that vehicle driven: 5 minutes or more 2. EVAP purge operation: Performed ECT: 4.4 to 35°C (40 to 95°F) IAT: 4.4 to 35°C (40 to 95°F) **Possible Causes:** • Canister pump module (reference orifice, leak detection pump, vent valve) • Connector/wire harness (canister pump module - ECM) • EVAP system hose (pipe from air inlet port to canister pump module, canister filter, fuel tank vent hose) • ECM

DTC	Trouble Code Title, Conditions & Possible Causes
DTC: P2402 **2T ECM, MIL: Yes** **Year:** 2009, 2010 **Model:** All **Engine:** All V6, V8 **Transmission:** All	**Evaporative Emission System Leak Detection Pump Control Circuit High:** Monitor runs whenever following DTCs not present: None EVAP key-off monitor runs when all of the following conditions are met: Atmospheric pressure: 70 to 110 kPa-a (525 to 825 mmHg-a) Battery voltage: 10.5 V or more Vehicle speed: Below 2.5 mph (4 km/h) Engine switch: Off Time after key off: 5 or 7 or 9.5 hours EVAP pressure sensor malfunction (P0450, P0451, P0452 and P0453): Not detected EVAP canister purge valve: Not operated by scan tool EVAP canister vent valve: Not operated by scan tool EVAP leak detection pump: Not operated by scan tool Both of the following conditions are met before key off: Conditions 1 and 2 1. Duration that vehicle has been driven: 5 minutes or more 2. EVAP purge operation: Performed ECT: 4.4 to 35°C (40 to 95°F) IAT: 4.4 to 35°C (40 to 95°F) **Possible Causes:** • Canister pump module (0.02 inch orifice, leak detection pump, vent valve) • Connector / wire harness (Canister pump module - ECM) • ECM
DTC: P2419 **2T ECM, MIL: Yes** **Year:** 2009, 2010 **Model:** All **Engine:** All V6, V8 **Transmission:** All	**Evaporative Emission System Switching Valve Control Circuit Low:** Monitor runs whenever following DTCs not present: None EVAP key-off monitor runs when all of following conditions met: Atmospheric pressure: 70 to 110 kPa-a (525 to 825 mmHg-a) Battery voltage: 10.5 V or more Vehicle speed: Below 2.5 mph (4 km/h) Engine switch: off Time after key off: 5 or 7 or 9.5 hours Canister pressure sensor malfunction (P0451, P0452 and P0453): Not detected Purge VSV: Not operated by scan tool Vent valve: Not operated by scan tool Leak detection pump: Not operated by scan tool Both of following conditions met before key off: Conditions 1 and 2 1. Duration that vehicle driven: 5 minutes or more 2. EVAP purge operation: Performed ECT: 4.4 to 35°C (40 to 95°F) IAT: 4.4 to 35°C (40 to 95°F) **Possible Causes:** • Canister pump module (reference orifice, leak detection pump, vent valve) • Connector/wire harness (canister pump module - ECM) • EVAP system hose (pipe from air inlet port to canister pump module, canister filter, fuel tank vent hose) • ECM

DTC	Trouble Code Title, Conditions & Possible Causes
DTC: P2420 **2T ECM, MIL: Yes** **Year:** 2009, 2010 **Model:** All **Engine:** All V6, V8 **Transmission:** All	**Evaporative Emission System Switching Valve Control Circuit High:** EVAP key-off monitor runs when all of following conditions met: Atmospheric pressure: 70 to 110 kPa-a (525 to 825 mmHg-a) Battery voltage: 10.5 V or more Vehicle speed: Below 2.5 mph (4 km/h) Engine switch: off Time after key off: 5 or 7 or 9.5 hours Canister pressure sensor malfunction (P0451, P0452 and P0453): Not detected Purge VSV: Not operated by scan tool Vent valve: Not operated by scan tool Leak detection pump: Not operated by scan tool Both of following conditions met before key off: Conditions 1 and 2 1. Duration that vehicle driven: 5 minutes or more 2. EVAP purge operation: Performed ECT: 4.4 to 35°C (40 to 95°F) IAT: 4.4 to 35°C (40 to 95°F) **Possible Causes:** • Canister pump module (Reference orifice, leak detection pump, vent valve) • Connector/wire harness (Canister pump module - ECM) • ECM
DTC: P2431 **1T ECM, MIL: Yes** **Year:** 2009 **Model:** GX470 **Engine:** 4.7L V8 **Transmission:** All	**Secondary Air Injection System Air Flow / Pressure Sensor Circuit Range / Performance Bank1:** The monitor will run whenever this DTC is not present: P2430, P2432, P2433 (Secondary Air Injection System Pressure Sensor) Starter: OFF Time after starter turned from ON to OFF: 2 seconds or more Battery voltage: 8 V or more Ignition switch: ON **Possible Causes:** • Pressure sensor • Open or short in pressure sensor circuit • ECM
DTC: P2432 **1T ECM, MIL: Yes** **Year:** 2009 **Model:** GX470 **Engine:** 4.7L V8 **Transmission:** All	**Secondary Air Injection System Air Flow / Pressure Sensor Circuit Low Bank1:** Starter: OFF Time after starter turned from ON to OFF: 2 seconds or more Battery voltage: 8 V or more Ignition switch: ON **Possible Causes:** • Pressure sensor • Open or short in pressure sensor circuit • ECM
DTC: P2433 **1T ECM, MIL: Yes** **Year:** 2009 **Model:** GX470 **Engine:** 4.7L V8 **Transmission:** All	**Secondary Air Injection System Air Flow / Pressure Sensor Circuit High Bank1:** Starter: OFF Time after starter turned from ON to OFF: 2 seconds or more Battery voltage: 8 V or more Ignition switch: ON **Possible Causes:** • Air pressure sensor • Open or short in air pressure sensor circuit • ECM

DTC	Trouble Code Title, Conditions & Possible Causes
DTC: P2440 **1T ECM, MIL: Yes** **Year:** 2009 **Model:** GX470 **Engine:** 4.7L V8 **Transmission:** All	**Secondary Air Injection System Switching Valve Stuck Open Bank1:** Case 1: Atmospheric pressure: 45 kPa (420 mmHg) or higher Battery voltage: 11.5 V or higher Case 2: Atmospheric pressure: 45 kPa (420 mmHg) or higher Battery voltage: 11.5 V or higher AIR pump: OFF Time after engine start: 10 seconds or more AIR valve bank 1: OFF AIR valve bank 2: OFF AIR status: OFF Engine load: 0% or more Intake air amount: 40 g/sec. or more IAT at engine start: -15°C (5°F) or higher ECT at engine start: Lower than 5°C (41°F) AIR valve: ON Engine RPM: Lower than 3750 rpm Case 3: Cumulative intake air amount: 109 g/sec. or more AIR pump: OFF AIR valve: OFF AIR valve bank 1: OFF AIR valve bank 2: OFF Engine RPM: Lower than 3750 rpm AIR status: OFF **Possible Causes:** • Air switching valve (ASV) • No. 2 air switching valve (Bank 1 and/or 2) • VSV for air injection system (bank 1 and/or 2) • Air injection control driver (AID) • Air injection control driver circuit • ECM
DTC: P2441 **2T ECM, MIL: Yes** **Year:** 2009 **Model:** GX470 **Engine:** 4.7L V8 **Transmission:** All	**Secondary Air Injection System Switching Valve Stuck Close Bank1:** Atmospheric pressure: 45 kPa (420 mmHg) or more Battery voltage: 11.5 V or higher **Possible Causes:** • Vacuum hoses (Throttle body - VSVs for air injection system) • Air switching valve • Air injector pipe (No. 2 air switching valve - exhaust manifold) • Air injection hose • No. 2 air switching valve (bank 1 and/or 2) • VSV for air injection system (bank 1 and/or 2) • Air injection control driver • Air injection control driver circuit • ECM

DTC	Trouble Code Title, Conditions & Possible Causes
DTC: P2444 **2T ECM, MIL: Yes** **Year:** 2009 **Model:** GX470 **Engine:** 4.7L V8 **Transmission:** All	**Secondary Air Injection System Pump Stuck On Bank1:** This monitor runs whenever these DTCs are not present P0010, P0020 (VVT Oil Control Valve) P0011, P0021 (VVT System - Advance) P0012, P0022 (VVT System - Retard) P0016, P0018 (VVT System - Misalignment) P0031, P0032, P0051, P0052 (Air Fuel Ratio Sensor Heater) P2195, P2196, P2197, P2198, P2237, P2238, P2239, P2240, P2241, P2242, P2252, P2253, P2255, P2256, P2A00, P2A03 (Air Fuel Ratio Sensor) P0100, P0102, P0103 (Mass Air Flow Sensor) P0110, P0112, P0113 (Intake Air Temperature Sensor) P0115, P0117, P0118 (Engine Coolant Temperature Sensor) P0125 (Insufficient Coolant Temperature for Closed Loop Fuel Control) P0120, P0121, P0122, P0123, P0220, P0222, P0223, P2135 (Throttle Position Sensor) P0171, P0172, P0174, P0175 (Fuel System) P0300 - P0308 (Misfire) P0327, P0328, P0332, P0333 (Knock Sensor) P0335 (Crankshaft Position Sensor) P1340 (Camshaft Position Sensor) P0340, P0345 (VVT Sensor) P0351 - P0358 (Igniter) P0500 (Vehicle Speed Sensor) P0722 (Output Speed Sensor) P2430, P2431, P2432, P2433 (Secondary Air Injection System Pressure Sensor) **Possible Causes:** • Short in air pump circuit • Air injection control driver (AID) • Pressure sensor • Open or short in pressure sensor circuit • ECM

DTC	Trouble Code Title, Conditions & Possible Causes
DTC: P2445 **2T ECM, MIL: Yes** **Year:** 2009 **Model:** GX470 **Engine:** 4.7L V8 **Transmission:** All	**Secondary Air Injection System Pump Stuck Off Bank1:** This monitor runs whenever these DTCs are not present P0010, P0020 (VVT Oil Control Valve) P0011, P0021 (VVT System - Advance) P0012, P0022 (VVT System - Retard) P0016, P0018 (VVT System - Misalignment) P0031, P0032, P0051, P0052 (Air Fuel Ratio Sensor Heater) P2195, P2196, P2197, P2198, P2237, P2238, P2239, P2240, P2241, P2242, P2252, P2253, P2255, P2256, P2A00, P2A03 (Air Fuel Ratio Sensor) P0100, P0102, P0103 (Mass Air Flow Sensor) P0110, P0112, P0113 (Intake Air Temperature Sensor) P0115, P0117, P0118 (Engine Coolant Temperature Sensor) P0125 (Insufficient Coolant Temperature for Closed Loop Fuel Control) P0120, P0121, P0122, P0123, P0220, P0222, P0223, P2135 (Throttle Position Sensor) P0171, P0172, P0174, P0175 (Fuel System) P0300 - P0308 (Misfire) P0327, P0328, P0332, P0333 (Knock Sensor) P0335 (Crankshaft Position Sensor) P1340 (Camshaft Position Sensor) P0340, P0345 (VVT Sensor) P0351 - P0358 (Igniter) P0500 (Vehicle Speed Sensor) P0722 (Output Speed Sensor) P2430, P2431, P2432, P2433 (Secondary Air Injection System Pressure Sensor) **Possible Causes:** • Air pump • Open in air pump circuit • Air injection system piping • Vacuum hose (pressure sensor - air switching valve) • Pressure sensor • Open or short in pressure sensor circuit • Air injection control driver • ECM
DTC: P2610 **2T ECM, MIL: Yes** **Year:** 2009, 2010 **Model:** All **Engine:** All V6, V8 **Transmission:** All	**ECM / PCM Internal Engine Off Timer Performance:** Case 1: Engine switch: ON Engine: Running Battery voltage: 8 V or more Starter: OFF Case 2: Internal engine OFF timer (elapsed time from engine stop): 10-30 minutes Battery voltage: 8 V or more Engine switch: ON Starter: OFF Case 3: Internal engine OFF timer (elapsed time from engine stop): 40 minutes Battery voltage: 8 V or more Engine switch: ON Starter: OFF **Possible Causes:** • ECM

DTC	Trouble Code Title, Conditions & Possible Causes
DTC: P2714 **2T , MIL: Yes** **Year:** 2009, 2010 **Model:** All **Engine:** All **Transmission:** All	**Pressure Control Solenoid "D" Performance (Shift Solenoid Valve SLT):** Transmission range: "D" Duration time from shifting "N" to "D": 4 sec. or more Engine: Starting Input turbine torque: 100 N*m or more Turbine speed: 250 rpm or more Output speed: 250 rpm or more **Possible Causes:** • Shift solenoid valve SLT (open) • Shift solenoid valve SL1, SL2, SL3 or SL4 (open or closed) • Valve body (blocked) • Torque converter clutch • Automatic transaxle (clutch, brake or gear, etc.)
DTC: P2716 **1T ECM, MIL: Yes** **Year:** 2009, 2010 **Model:** All **Engine:** All **Transmission:** All	**Pressure Control Solenoid "D" Electrical (Shift Solenoid Valve SLT):** Solenoid current cut status: Not cut CPU command duty: 19% or more Battery voltage: 11 V or more Engine switch: ON Starter: OFF **Possible Causes:** • Open or short in shift solenoid valve SLT circuit • Shift solenoid valve SLT • ECM
DTC: P2741 00 **1T TCM, MIL: Yes** **Year:** 2009 **Model:** GX470 **Engine:** 4.7L V8 **Transmission:** All	**Transmission Fluid Temperature Sensor "B" Circuit:** The monitor will run whenever the following DTCs are not present: None The typical enabling condition is not available: - **Possible Causes:** • Open or short in No. 2 ATF temperature sensor circuit • No. 2 ATF temperature sensor • ECM
DTC: P2742 **1T TCM, MIL: Yes** **Year:** 2009 **Model:** GX470 **Engine:** 4.7L V8 **Transmission:** All	**Transmission Fluid Temperature Sensor "B" Circuit Low Input:** The typical enabling condition is not available: - **Possible Causes:** • Short in No. 2 ATF temperature sensor circuit • No. 2 ATF temperature sensor • ECM
DTC: P2743 **1T TCM, MIL: Yes** **Year:** 2009 **Model:** GX470 **Engine:** 4.7L V8 **Transmission:** All	**Transmission Fluid Temperature Sensor "B" Circuit High Input:** Time after engine start: 15 minutes or more **Possible Causes:** • Open in No. 2 ATF temperature sensor circuit • No. 2 ATF temperature sensor • ECM

DTC	Trouble Code Title, Conditions & Possible Causes
DTC: P2757 **2T TCM, MIL: Yes** **Year:** 2009, 2010 **Model:** All **Engine:** All **Transmission:** All	**Torque Converter Clutch Pressure Control Solenoid Performance (Shift Solenoid Valve SLU):** Transmission shift position: "D" ECT (Engine coolant temperature): 40°C (104°F) or more Spark advance from Max. retard timing by KCS control: 0°CA or more Engine: Starting ECM selected gear: 2nd, 3rd, 4th, 5th or 6th Vehicle speed: 15.5 mph (25 km/h) or more Shift solenoid valve S1 circuit: Not circuit malfunction Shift solenoid valve S3 circuit: Not circuit malfunction Shift solenoid valve S4 circuit: Not circuit malfunction Shift solenoid valve SL2 circuit: Not circuit malfunction 1 - 2 shift valve: Not circuit malfunction Sequence valve: Not circuit malfunction OFF Malfunction (A): ECM lock-up command: ON (SLU pressure: 580 kPa (5.9 kgfcm2, 84 psi) or more) Duration time from lock-up on command: 3 sec. or more Vehicle speed: Less than 74.5 mph (120 km/h) OFF Malfunction (B): ECM selected gear: 2nd Vehicle speed: 1.2 mph (2 km/h) or more Output speed: 2nd → 1st down shift point or more Throttle valve opening angle: 6.5% or more at 2,000 rpm (Conditions vary with engine speed) ON Malfunction: ECM lock-up command: OFF (SLU pressure: less than 4 kPa (0.04 kgf cm2, 0.6 psi) Duration time from lock-up on command: 3 sec. or more Throttle valve opening angle: 6% or more Vehicle speed: Less than 74.5 mph (120 km/h) at 2nd gear (Varies with ECM selected gear) **Possible Causes:** • Shift solenoid valve SLU remains open or closed • Valve body is blocked • Torque converter clutch • Automatic transmission (clutch, brake or gear, etc.) • Line pressure is too low
DTC: P2759 **1T TCM, MIL: Yes** **Year:** 2009, 2010 **Model:** All **Engine:** All **Transmission:** All	**Torque Converter Clutch Pressure Control Solenoid Control Circuit Electrical (Shift Solenoid Valve SLU):** Engine switch: ON, Starter: OFF Condition (A): Battery voltage: 12 V or more Condition (B): Battery voltage: 10 V or more and less than 12 V Target current: Less than 0.75 A Condition (C): Battery voltage: 8 V or more Target current: 0.25 A or more **Possible Causes:** • Open or short in shift solenoid valve SLU circuit • Shift solenoid valve SLU • TCM
DTC: P2769 **2T TCM, MIL: Yes** **Year:** 2009, 2010 **Model:** All **Engine:** All **Transmission:** All	**Torque Converter Clutch Solenoid Circuit Low (Shift Solenoid Valve DSL):** Shift solenoid valve DSL: ON Solenoid current cut status: Not cut Battery voltage: 8 V or more Ignition switch: ON Starter: OFF **Possible Causes:** • Short in shift solenoid valve DSL circuit • Shift solenoid valve DSL • ECM/TCM

DTC	Trouble Code Title, Conditions & Possible Causes
DTC: P2770 **2T TCM, MIL: Yes** **Year:** 2009, 2010 **Model:** ES350, GS350 **Engine:** 3.5L V6 **Transmission:** All	**Open in Torque Converter Clutch Solenoid Circuit (Shift Solenoid Valve SL):** The monitor will run whenever the following DTCs are not present: None Solenoid: ON Time after solenoid ON to OFF: More than 0.008 sec. **Possible Causes:** • Open in shift solenoid valve SL circuit • Shift solenoid valve SL • TCM
DTC: P2772 **1T TCM, MIL: Yes** **Year:** 2009 **Model:** GX470 **Engine:** 4.7L V8 **Transmission:** All	**Four Wheel Drive (4WD) Low Switch Circuit Range / Performance:** Output speed sensor circuit: Not circuit malfunction Vehicle speed sensor circuit: Not circuit malfunction Transfer neutral position switch: OFF ON malfunction (A): Output speed (Transfer output speed): 1,000 to 3,000 rpm ON malfunction (B): Output speed (Transfer output speed): 143 rpm or more **Possible Causes:** • Short in transfer L4 position switch circuit • Transfer L4 position switch • ECM
DTC: P2808 **2T TCM, MIL: Yes** **Year:** 2009, 2010 **Model:** ES350, GS350 **Engine:** 3.5L V6 **Transmission:** All	**Pressure Control Solenoid "G" Performance (Shift Solenoid Valve SL4):** Transmission range: "D" TFT (Transmission fluid temperature): -10°C (14°F) or more TFT (Transmission fluid temperature) sensor circuit: No circuit malfunction ECT (Engine coolant temperature) sensor circuit: No circuit malfunction Turbine speed sensor circuit: No circuit malfunction Intermediate shaft speed sensor circuit: No circuit malfunction Shift solenoid valve SL1 circuit: No circuit malfunction Shift solenoid valve SL2 circuit: No circuit malfunction Shift solenoid valve SL3 circuit: No circuit malfunction Shift solenoid valve SL4 circuit: No circuit malfunction (KCS) Knock control sensor circuit: No circuit malfunction (ETCS) Electronic throttle control system: Not system down CAN communication system: Not system down **Possible Causes:** • Shift solenoid valve SL4 (open or closed) • Valve body (blocked) • Automatic transaxle (clutch, brake or gear etc.)
DTC: P2810 **1T TCM, MIL: Yes** **Year:** 2009, 2010 **Model:** ES350, GS350 **Engine:** 3.5L V6 **Transmission:** All	**Pressure Control Solenoid "G" Electrical (Shift Solenoid Valve SL4):** Engine switch: ON, Starter: OFF Condition (A): Battery voltage: 12 V or more Condition (B): Battery voltage: 10 V or more and less than 12 V Target condition: Less than 0.75 A Condition (C): Battery voltage: 8 V or more Target condition: 0.25 A or more **Possible Causes:** • Open or short in shift solenoid valve SL4 circuit • Shift solenoid valve SL4 • TCM

DTC	Trouble Code Title, Conditions & Possible Causes
DTC: P2A00 **T ECM, MIL: Yes** **Year:** 2009, 2010 **Model:** All **Engine:** All V6, V8 **Transmission:** All	**A/F Sensor Circuit Slow Response (Bank 1 Sensor 1):** Active A/F control: Performing Active A/F control is performed when the following conditions are met: Battery voltage: 11 V or more ECT: 75°C (167°F) or more Idle: OFF, Engine rpm: Less than 4000 rpm, A/F sensor status: Activated Fuel-cut: OFF, Engine load: 10 to 70%, Shift position: 2nd or more Catalyst monitor: Not yet, MAF: 5 to 14 g/sec. **Possible Causes:** • Open or short in A/F sensor circuit • A/F sensor • ECM
DTC: P2A03 **2T ECM, MIL: Yes** **Year:** 2009, 2010 **Model:** All **Engine:** All **Transmission:** All	**A/F Sensor Circuit Slow Response (Bank 2 Sensor 1):** Active A/F control: Performing Battery voltage: 11 V or more ECT: 167°F (75°C) or more Idle: OFF Engine rpm: Less than 4000 rpm A/F sensor status: Activated Fuel cut: OFF Engine load: 10 to 70 % Shift position: 2 or more Catalyst monitor: Not yet MAF: 2.5-15 g/sec **Possible Causes:** • Open or short in A/F sensor (bank 1, 2 sensor 1) circuit • A/F sensor (bank 1, 2 sensor 1) • ECM

OBD II Trouble Code List (U0XXX Codes)

DTC	Trouble Code Title, Conditions & Possible Causes
DTC: U0073/86 **T ECM, MIL: Yes** **Year:** 2009, 2010 **Model:** IS250, IS350 **Engine:** 2.5L V6, 3.5L V6 **Transmission:** All	**Control Module Communication Bus OFF:** 1. When the following continues for 5 seconds or more: * Signals from the 4WD control ECU are not received. 2. When the following occurs 10 times consecutively: * A communication malfunction (bus off) occurs one or more times within 0.1 second. **Possible Causes:** • Wire harness (CANL, CANH circuit) • 4WD control ECU
DTC: U0100 **1T TCM, MIL: Yes** **Year:** 2009, 2010 **Model:** ES350, GS350 **Engine:** 3.5L V6 **Transmission:** All	**Lost Communication with ECM / PCM "A":** Battery Voltage: 10.5 V or more Engine Switch: ON Starter: OFF **Possible Causes:** • TCM
DTC: U0100 **T ECM, MIL: Yes** **Year:** 2009, 2010 **Model:** IS250, IS350 **Engine:** 2.5L V6, 3.5L V6 **Transmission:** All	**Lost Communication with ECM / PCM "A":** While the dynamic radar cruise control is either preparing for operation or operating, if the ECM continuously receives a communication cut off signal from the ABS & traction actuator (skid control ECU) for a certain amount of time, the ECM records this communication cut off code. **Possible Causes:** • Communication circuit • ECM • ABS & traction actuator (Skid control ECU)

DTC	Trouble Code Title, Conditions & Possible Causes
DTC: U0100 **T ECM, MIL: Yes** **Year:** 2009, 2010 **Model:** RX350 **Engine:** 3.5L V6 **Transmission:** All	**Lost Communication with ECM / PCM "A":** While the dynamic laser cruise control is either preparing for operation or operating, if the ECM continuously receives a communication cut off signal from the distance control ECU for more than a specific amount of time, the ECM records this communication cut off code. **Possible Causes:** • Communication circuit • ECM • Distance control ECU (Cruise control ECU)
DTC: U0100 **1T , MIL: Yes** **Year:** 2009, 2010 **Model:** GS350, IS250, IS350 **Engine:** 2.5L V6, 3.5L V6 **Transmission:** All	**Lost Communication with ECM / PCM "A":** Monitor runs whenever following DTCs not present: None Battery Voltage: 10.5 V or more Engine Switch: ON Starter: OFF **Possible Causes:** • ECM
DTC: U0100/85 **T ECM, MIL: Yes** **Year:** 2009, 2010 **Model:** IS250, IS350 **Engine:** 2.5L V6, 3.5L V6 **Transmission:** All	**Lost Communication with ECM / PCM "A":** When both of the following continue for 2 seconds or more: * The voltage of the IG1 terminal is 10 V or more. * At a vehicle speed of 38 mph (60 km/h) or more, communication with the ECM cannot be performed continue for 2 seconds or more. **Possible Causes:** • Wire harness (CANL, CANH circuit) • 4WD control ECU • ECM
DTC: U0101 **T ECM, MIL: Yes** **Year:** 2009, 2010 **Model:** All **Engine:** All **Transmission:** All	**Lost Communication with TCM:** Battery voltage: 10.5 V or more Engine switch: ON Starter: OFF **Possible Causes:** • Open or short in TCM and ECM circuit • TCM • ECM
DTC: U0122 **T ECM, MIL: Yes** **Year:** 2009, 2010 **Model:** All **Engine:** All V6, V8 **Transmission:** All	**Lost Communication with Vehicle Dynamics Control Module:** While dynamic radar cruise control is either preparing for operation or operating, if communication data from the skid control ECU is invalid for a certain amount of time, the ECM records the communication cut off code. **Possible Causes:** • Communication circuit • ABS & traction actuator (Skid control ECU) • ECM
DTC: U0123 **T ECM, MIL: Yes** **Year:** 2009, 2010 **Model:** All **Engine:** All V6, V8 **Transmission:** All	**Communication Error of Yaw Rate Sensor:** While the dynamic laser cruise control is either preparing for operation or operating, the ECM continuously receives a yaw rate sensor malfunction signal for a certain amount of time. **Possible Causes:** • Communication circuit • Yaw rate sensor • Distance control ECU (Cruise control ECU) • ECM
DTC: U0126 **T ECM, MIL: Yes** **Year:** 2009, 2010 **Model:** All **Engine:** All V6, V8 **Transmission:** All	**Lost Communication with Steering Angle Sensor Module:** While the dynamic radar cruise control is either preparing for operation or operating, if the ECM continuously receives a steering sensor communication error signal for a certain amount of time, the distance control ECU sends a signal to the ECM to record this trouble code. **Possible Causes:** • Communication circuit • Steering angle sensor • Distance control ECU • ECM

DTC	Trouble Code Title, Conditions & Possible Causes
DTC: U0126/84 **T , MIL: Yes** **Year:** 2009, 2010 **Model:** IS250, IS350 **Engine:** 2.5L V6, 3.5L V6 **Transmission:** All	**Lost Communication with Steering Angle Sensor Module:** 1. When both of the following continue for 1 second or more: * The voltage of the IG1 terminal is 10 V or more. * Communication with the steering angle sensor cannot be performed. 2. When both of the following occur 10 times consecutively: * The voltage of the IG1 terminal is 10 V or more. * The condition that communication with the steering angle sensor cannot be performed occurs once within 5 seconds. **Possible Causes:** • Wire harness (CANL, CANH circuit) • 4WD control ECU • Steering angle sensor
DTC: U0129/83 **T ECM, MIL: Yes** **Year:** 2009, 2010 **Model:** IS250, IS350 **Engine:** 2.5L V6, 3.5L V6 **Transmission:** All	**Lost Communication with Brake System Control Module:** When both of the following continue for 3 seconds or more: * The voltage of the IG1 terminal is 10 V or more. * Communication with the skid control ECU cannot be performed. **Possible Causes:** • Wire harness (CANL, CANH circuit) • 4WD control ECU • Skid control ECU
DTC: U0235 **T ECM, MIL: Yes** **Year:** 2009, 2010 **Model:** All **Engine:** All **Transmission:** All	**Lost Communication with Cruise Control Front Distance Range Sensor:** The ECM detects a communication error signal (from the millimeter wave radar sensor to the distance control ECU) for 0.15 sec. or more while the dynamic radar cruise control is in operation. **Possible Causes:** • Communication circuit • Millimeter wave radar sensor • Distance control ECU
DTC: U0319 **1T ECM, MIL: Yes** **Year:** 2010 **Model:** LS460 **Engine:** 4.6L V8 **Transmission:** All	**Software Incompatible with Steering Effort Control Module or Internal Control Module EEPROM Error:** One of the following conditions is met for 3 seconds or more with the engine switch on (IG) (1 trip detection logic): 1. There is a problem with the communication between the ECM and steering control ECU when the ECM is storing information from the steering control ECU. 2. There is an ECM internal error. 3. The engine switch is turned off when the ECM is storing the sport package information.*1 **Possible Causes:** • CAN communication system • ECM • Steering control ECU

OBD II Trouble Code List (U1XXX Codes)

DTC	Trouble Code Title, Conditions & Possible Causes
DTC: U1101 **T ECM, MIL: Yes** **Year:** 2009, 2010 **Model:** All **Engine:** All **Transmission:** All	**Lost Communication with Distance Control ECU:** While the dynamic radar cruise control is either preparing for operation or operating, if communication data from the distance control ECU is invalid for a certain amount of time, The ECM records this communication cut off code. **Possible Causes:** • Communication circuit • Distance control ECU • ECM
DTC: U1102 **T ECM, MIL: Yes** **Year:** 2009, 2010 **Model:** All **Engine:** All **Transmission:** All	**Lost Communication with Radar Sensor:** The ECM detects a communication error signal (from the distance control ECU (cruise control ECU) to the millimeter wave radar sensor) for 0.15 sec. or more while the dynamic radar cruise control is in operation. **Possible Causes:** • Communication circuit • Millimeter wave radar sensor • Distance control ECU

ENGLISH TO METRIC CONVERSION: TORQUE

To convert foot-pounds (ft. lbs.) to Newton-meters (Nm), multiply the number of ft. lbs. by 1.36
To convert Newton-meters (Nm) to foot-pounds (ft. lbs.), multiply the number of Nm by 0.7376

ft. lbs.	Nm	ft. lbs.	Nm	ft. lbs.	Nm	ft. lbs.	Nm
0.1	0.1	34	46.2	76	103.4	118	160.5
0.2	0.3	35	47.6	77	104.7	119	161.8
0.3	0.4	36	49.0	78	106.1	120	163.2
0.4	0.5	37	50.3	79	107.4	121	164.6
0.5	0.7	38	51.7	80	108.8	122	165.9
0.6	0.8	39	53.0	81	110.2	123	167.3
0.7	1.0	40	54.4	82	111.5	124	168.6
0.8	1.1	41	55.8	83	112.9	125	170.0
0.9	1.2	42	57.1	84	114.2	126	171.4
1	1.4	43	58.5	85	115.6	127	172.7
2	2.7	44	59.8	86	117.0	128	174.1
3	4.1	45	61.2	87	118.3	129	175.4
4	5.4	46	62.6	88	119.7	130	176.8
5	6.8	47	63.9	89	121.0	131	178.2
6	8.2	48	65.3	90	122.4	132	179.5
7	9.5	49	66.6	91	123.8	133	180.9
8	10.9	50	68.0	92	125.1	134	182.2
9	12.2	51	69.4	93	126.5	135	183.6
10	13.6	52	70.7	94	127.8	136	185.0
11	15.0	53	72.1	95	129.2	137	186.3
12	16.3	54	73.4	96	130.6	138	187.7
13	17.7	55	74.8	97	131.9	139	189.0
14	19.0	56	76.2	98	133.3	140	190.4
15	20.4	57	77.5	99	134.6	141	191.8
16	21.8	58	78.9	100	136.0	142	193.1
17	23.1	59	80.2	101	137.4	143	194.5
18	24.5	60	81.6	102	138.7	144	195.8
19	25.8	61	83.0	103	140.1	145	197.2
20	27.2	62	84.3	104	141.4	146	198.6
21	28.6	63	85.7	105	142.8	147	199.9
22	29.9	64	87.0	106	144.2	148	201.3
23	31.3	65	88.4	107	145.5	149	202.6
24	32.6	66	89.8	108	146.9	150	204.0
25	34.0	67	91.1	109	148.2	151	205.4
26	35.4	68	92.5	110	149.6	152	206.7
27	36.7	69	93.8	111	151.0	153	208.1
28	38.1	70	95.2	112	152.3	154	209.4
29	39.4	71	96.6	113	153.7	155	210.8
30	40.8	72	97.9	114	155.0	156	212.2
31	42.2	73	99.3	115	156.4	157	213.5
32	43.5	74	100.6	116	157.8	158	214.9
33	44.9	75	102.0	117	159.1	159	216.2

METRIC TO ENGLISH CONVERSION: TORQUE

To convert foot-pounds (ft. lbs.) to Newton-meters (Nm), multiply the number of ft. lbs. by 1.36
To convert Newton-meters (Nm) to foot-pounds (ft. lbs.), multiply the number of Nm by 0.7376

Nm	ft. lbs.	Nm	ft. lbs.	Nm	ft. lbs.	Nm	ft. lbs.	Nm	ft. lbs.
0.1	0.1	34	25.0	76	55.9	118	86.8	160	117.6
0.2	0.1	35	25.7	77	56.6	119	87.5	161	118.4
0.3	0.2	36	26.5	78	57.4	120	88.2	162	119.1
0.4	0.3	37	27.2	79	58.1	121	89.0	163	119.9
0.5	0.4	38	27.9	80	58.8	122	89.7	164	120.6
0.6	0.4	39	28.7	81	59.6	123	90.4	165	121.3
0.7	0.5	40	29.4	82	60.3	124	91.2	166	122.1
0.8	0.6	41	30.1	83	61.0	125	91.9	167	122.8
0.9	0.7	42	30.9	84	61.8	126	92.6	168	123.5
1	0.7	43	31.6	85	62.5	127	93.4	169	124.3
2	1.5	44	32.4	86	63.2	128	94.1	170	125.0
3	2.2	45	33.1	87	64.0	129	94.9	171	125.7
4	2.9	46	33.8	88	64.7	130	95.6	172	126.5
5	3.7	47	34.6	89	65.4	131	96.3	173	127.2
6	4.4	48	35.3	90	66.2	132	97.1	174	127.9
7	5.1	49	36.0	91	66.9	133	97.8	175	128.7
8	5.9	50	36.8	92	67.6	134	98.5	176	129.4
9	6.6	51	37.5	93	68.4	135	99.3	177	130.1
10	7.4	52	38.2	94	69.1	136	100.0	178	130.9
11	8.1	53	39.0	95	69.9	137	100.7	179	131.6
12	8.8	54	39.7	96	70.6	138	101.5	180	132.4
13	9.6	55	40.4	97	71.3	139	102.2	181	133.1
14	10.3	56	41.2	98	72.1	140	102.9	182	133.8
15	11.0	57	41.9	99	72.8	141	103.7	183	134.6
16	11.8	58	42.6	100	73.5	142	104.4	184	135.3
17	12.5	59	43.4	101	74.3	143	105.1	185	136.0
18	13.2	60	44.1	102	75.0	144	105.9	186	136.8
19	14.0	61	44.9	103	75.7	145	106.6	187	137.5
20	14.7	62	45.6	104	76.5	146	107.4	188	138.2
21	15.4	63	46.3	105	77.2	147	108.1	189	139.0
22	16.2	64	47.1	106	77.9	148	108.8	190	139.7
23	16.9	65	47.8	107	78.7	149	109.6	191	140.4
24	17.6	66	48.5	108	79.4	150	110.3	192	141.2
25	18.4	67	49.3	109	80.1	151	111.0	193	141.9
26	19.1	68	50.0	110	80.9	152	111.8	194	142.6
27	19.9	69	50.7	111	81.6	153	112.5	195	143.4
28	20.6	70	51.5	112	82.4	154	113.2	196	144.1
29	21.3	71	52.2	113	83.1	155	114.0	197	144.9
30	22.1	72	52.9	114	83.8	156	114.7	198	145.6
31	22.8	73	53.7	115	84.6	157	115.4	199	146.3
32	23.5	74	54.4	116	85.3	158	116.2	200	147.1
33	24.3	75	55.1	117	86.0	159	116.9	201	147.8

ENGLISH/METRIC CONVERSION: TEMPERATURE

To convert Fahrenheit (F°) to Celsius (C°), take F° temperature and subtract 32, multiply the result by 5 and divide the result by 9
To convert Celsius (C°) to Fahrenheit (F°), take C° temperature and multiply it by 9, divide the result by 5 and add 32

F°	C°	F°	C°	C°	F°	C°	F°
-40	-40.0	150	65.6	-38	-36.4	46	114.8
-35	-37.2	155	68.3	-36	-32.8	48	118.4
-30	-34.4	160	71.1	-34	-29.2	50	122
-25	-31.7	165	73.9	-32	-25.6	52	125.6
-20	-28.9	170	76.7	-30	-22	54	129.2
-15	-26.1	175	79.4	-28	-18.4	56	132.8
-10	-23.3	180	82.2	-26	-14.8	58	136.4
-5	-20.6	185	85.0	-24	-11.2	60	140
0	-17.8	190	87.8	-22	-7.6	62	143.6
1	-17.2	195	90.6	-20	-4	64	147.2
2	-16.7	200	93.3	-18	-0.4	66	150.8
3	-16.1	205	96.1	-16	3.2	68	154.4
4	-15.6	210	98.9	-14	6.8	70	158
5	-15.0	212	100.0	-12	10.4	72	161.6
10	-12.2	215	101.7	-10	14	74	165.2
15	-9.4	220	104.4	-8	17.6	76	168.8
20	-6.7	225	107.2	-6	21.2	78	172.4
25	-3.9	230	110.0	-4	24.8	80	176
30	-1.1	235	112.8	-2	28.4	82	179.6
35	1.7	240	115.6	0	32	84	183.2
40	4.4	245	118.3	2	35.6	86	186.8
45	7.2	250	121.1	4	39.2	88	190.4
50	10.0	255	123.9	6	42.8	90	194
55	12.8	260	126.7	8	46.4	92	197.6
60	15.6	265	129.4	10	50	94	201.2
65	18.3	270	132.2	12	53.6	96	204.8
70	21.1	275	135.0	14	57.2	98	208.4
75	23.9	280	137.8	16	60.8	100	212
80	26.7	285	140.6	18	64.4	102	215.6
85	29.4	290	143.3	20	68	104	219.2
90	32.2	295	146.1	22	71.6	106	222.8
95	35.0	300	148.9	24	75.2	108	226.4
100	37.8	305	151.7	26	78.8	110	230
105	40.6	310	154.4	28	82.4	112	233.6
110	43.3	315	157.2	30	86	114	237.2
115	46.1	320	160.0	32	89.6	116	240.8
120	48.9	325	162.8	34	93.2	118	244.4
125	51.7	330	165.6	36	96.8	120	248
130	54.4	335	168.3	38	100.4	122	251.6
135	57.2	340	171.1	40	104	124	255.2
140	60.0	345	173.9	42	107.6	126	258.8
145	62.8	350	176.7	44	111.2	128	262.4

LENGTH CONVERSION

To convert inches (in.) to millimeters (mm), multiply the number of inches by 25.4

To convert millimeters (mm) to inches (in.), multiply the number of millimeters by 0.04

Inches	Millimeters	Inches	Millimeters	Inches	Millimeters	Inches	Millimeters
0.0001	0.00254	0.005	0.1270	0.09	2.286	4	101.6
0.0002	0.00508	0.006	0.1524	0.1	2.54	5	127.0
0.0003	0.00762	0.007	0.1778	0.2	5.08	6	152.4
0.0004	0.01016	0.008	0.2032	0.3	7.62	7	177.8
0.0005	0.01270	0.009	0.2286	0.4	10.16	8	203.2
0.0006	0.01524	0.01	0.254	0.5	12.70	9	228.6
0.0007	0.01778	0.02	0.508	0.6	15.24	10	254.0
0.0008	0.02032	0.03	0.762	0.7	17.78	11	279.4
0.0009	0.02286	0.04	1.016	0.8	20.32	12	304.8
0.001	0.0254	0.05	1.270	0.9	22.86	13	330.2
0.002	0.0508	0.06	1.524	1	25.4	14	355.6
0.003	0.0762	0.07	1.778	2	50.8	15	381.0
0.004	0.1016	0.08	2.032	3	76.2	16	406.4

ENGLISH/METRIC CONVERSION: LENGTH

To convert inches (in.) to millimeters (mm), multiply the number of inches by 25.4
To convert millimeters (mm) to inches (in.), multiply the number of millimeters by 0.04

Inches		Millimeters	Inches		Millimeters	Inches		Millimeters
Fraction	Decimal	Decimal	Fraction	Decimal	Decimal	Fraction	Decimal	Decimal
1/64	0.016	0.397	11/32	0.344	8.731	11/16	0.688	17.463
1/32	0.031	0.794	23/64	0.359	9.128	45/64	0.703	17.859
3/64	0.047	1.191	3/8	0.375	9.525	23/32	0.719	18.256
1/16	0.063	1.588	25/64	0.391	9.922	47/64	0.734	18.653
5/64	0.078	1.984	13/32	0.406	10.319	3/4	0.750	19.050
3/32	0.094	2.381	27/64	0.422	10.716	49/64	0.766	19.447
7/64	0.109	2.778	7/16	0.438	11.113	25/32	0.781	19.844
1/8	0.125	3.175	29/64	0.453	11.509	51/64	0.797	20.241
9/64	0.141	3.572	15/32	0.469	11.906	13/16	0.813	20.638
5/32	0.156	3.969	31/64	0.484	12.303	53/64	0.828	21.034
11/64	0.172	4.366	1/2	0.500	12.700	27/32	0.844	21.431
3/16	0.188	4.763	33/64	0.516	13.097	55/64	0.859	21.828
13/64	0.203	5.159	17/32	0.531	13.494	7/8	0.875	22.225
7/32	0.219	5.556	35/64	0.547	13.891	57/64	0.891	22.622
15/64	0.234	5.953	9/16	0.563	14.288	29/32	0.906	23.019
1/4	0.250	6.350	37/64	0.578	14.684	59/64	0.922	23.416
17/64	0.266	6.747	19/32	0.594	15.081	15/16	0.938	23.813
9/32	0.281	7.144	39/64	0.609	15.478	61/64	0.953	24.209
19/64	0.297	7.541	5/8	0.625	15.875	31/32	0.969	24.606
5/16	0.313	7.938	41/64	0.641	16.272	63/64	0.984	25.003
21/64	0.328	8.334	21/32	0.656	16.669	1/1	1.000	25.400
			43/64	0.672	17.066			

CHILTON® LABOR GUIDE

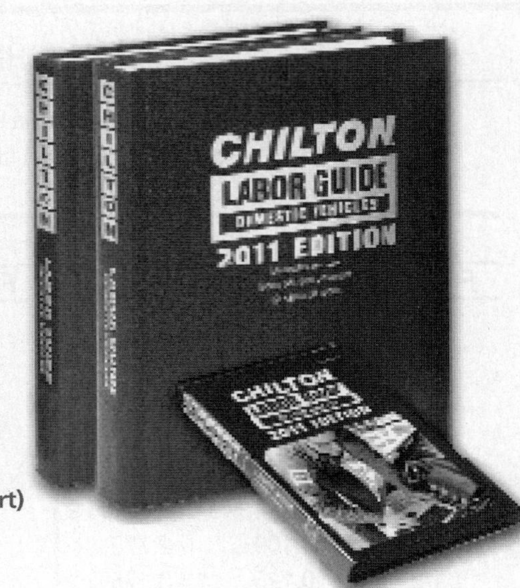

Whether you are looking for labor times in print, or on CD-ROM, Chilton is your source! Chilton's editors have carefully crafted the latest edition of the famous Chilton Labor Guide to bring you the most accurate repair information available. Chilton's editors consider warranty times, component locations, component type, the environment in which technicians work, the training they receive, and the tools they use when calculating a labor time. To allow for vehicle age, operating conditions, and type of service, the Chilton Labor Guide provides standard and severe service times, plus OEM warranty times. Vehicle makes and models conform to current Automotive Aftermarket Industry Association (AAIA) standards.

978-1-1115-4291-7 Chilton 2011 Labor Guide Manual Set (Domestic & Import)
978-1-1115-4294-8 Chilton 2011 Labor Guide CD-ROM (Domestic & Import)

CD-ROM FEATURES

- access labor times for 1981-2011 import and domestic vehicle models
- save time with automatically calculated labor charges, taxes, & parts as total job is estimated
- create professional estimates for your customer and worksheets for your technicians, printing them whenever needed
- keep track of customers, prior estimates, and your own parts or package jobs with less paper
- choose part names for estimates from an industry standard database to reduce typing
- estimate and track your work status with improved forms
- communicate easily with customers using re-designed printouts which show all labor and parts in an easy-to-read format.
- simplify adding parts to your estimate or work order with a helpful parts list
- locate information quick with a keyword search engine
- quickly locate work requests by day, week and month using the calendar feature

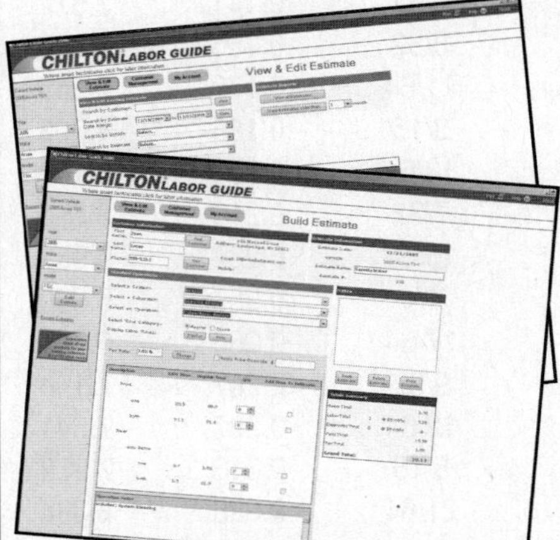

Manual FEATURES

- more than 2,500 pages of updated Chilton labor times split into two volumes includes vehicle information from 1981 to 2011
- trusted by more service professionals than any other labor guide
- less flipping though pages with separate domestic and imported vehicle manuals
- convenient tabs display contents by manufacturer and model
- easy-to-find manufacturers are arranged alphabetically within each volume
- search using two-indexes - labor operations and systems - in each model group
- page numbers include manufacturer code so you know where you are in the book

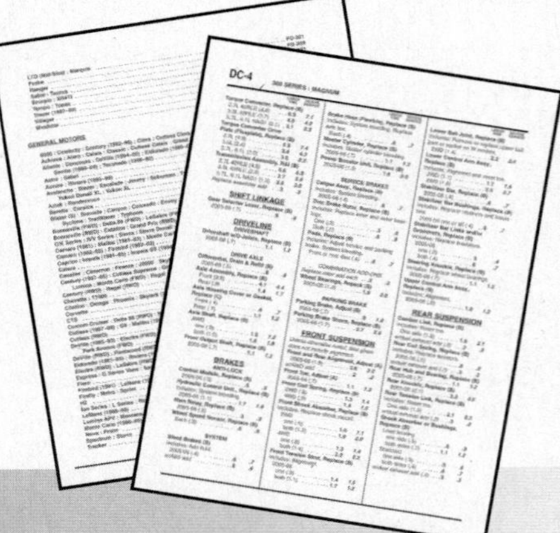

Chilton's labor times are so trusted, even a competing publisher uses them!

CHILTONPRO.COM

Where smart technicians click for service information.

ChiltonPRO is the alternative for professional technicians who want a cost-effective electronic automotive repair system. It combines Chilton's famous automotive repair information into one solution covering more than 20 years of domestic and imported vehicles. The information is delivered online and is updated regularly throughout the year.

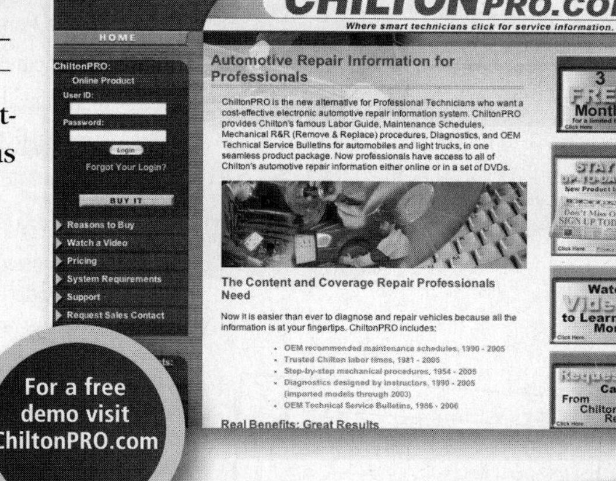

Online Monthly Payment
ISBN: 978-14180-3002-5

Online Annual Payment
ISBN: 978-14180-2876-3

For a free demo visit ChiltonPRO.com

ChiltonPRO FEATURES

- make repairs even easier with videos & animations which explain system operations & contribute to technician knowledge
- create better estimates using labor times developed with real-world factors
- save money by accurately identifying and solving engine performance problems
- save time with expert guidance through OBDII diagnostics
- increase efficiency by understanding system operation through detailed explanations and theory
- increase profits using Technical Service Bulletins (TSBs) to ensure that work is not going unperformed
- execute effective repairs by viewing cutaway diagrams and actual photos
- make better use of your time with information that can be found quicker using AAIA standards for year, make, and model
- increase confidence levels by always being able to print what you need
- eliminate guesswork with quick reference to critical specifications in helpful tables
- spend less on repair information

Coverage Includes:

- OEM recommended maintenance schedules, 1990–current
- Trusted Chilton labor times, 1981–current
- Step-by-step mechanical procedures, 1950s–current
- Diagnostics designed by instructors, 1990–current
- More than 75,000 OEM Technical Service Bulletins issued during the past 20 years

System Requirements:
Web browser

- Internet Explorer 6.0 or above (recommended)
- Firefox 2 or 3, or Safari
- High-speed internet connection
- Adobe Flash Player
- Adobe Shockwave Player
- Windows XP or Vista

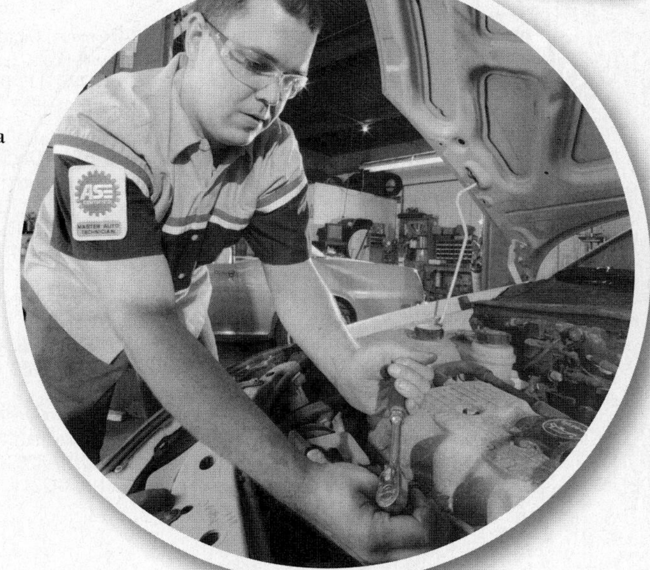

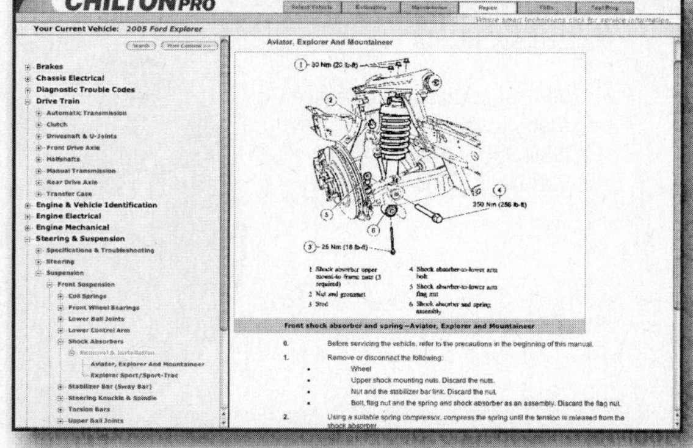

FOR CUSTOMER SUPPORT CALL **1-800-477-3692**

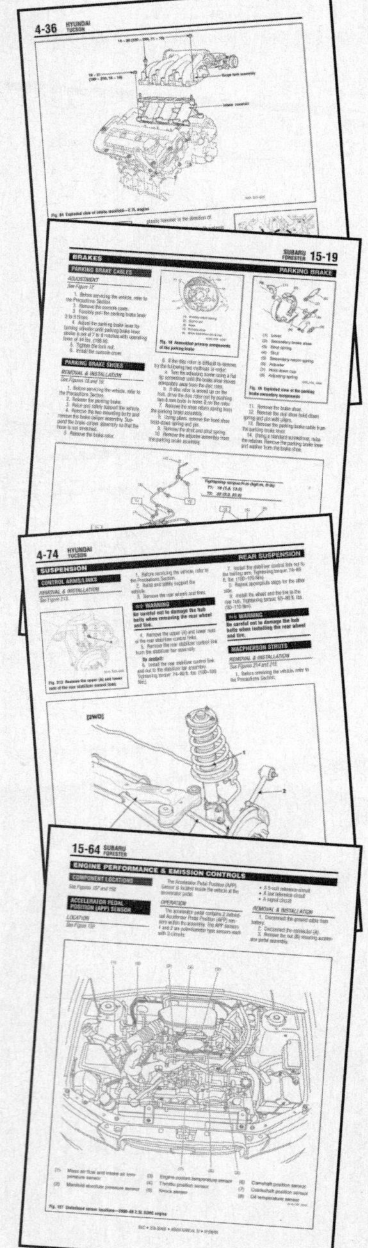

Chilton® 2010 Service Manuals

The Chilton 2010 Service Manuals now include even better graphics and expanded procedures! Chilton's editors have put together the most current automotive repair information available to assist users during daily repairs. These new manuals allow users to accurately and efficiently diagnose and repair late-model cars and trucks. Trust the step-by-step procedures and helpful illustrations that only Chilton can provide. The 2010 Service Manuals cover 2008 and 2009 models plus available 2010 models.

KEY FEATURES

- organized by vehicle manufacturer
- provides thousands of pages of expertly written content
- access new year, make, and model information without repeating previous edition's content
- comprehensive, technically detailed content, including exploded view illustrations, diagnostics and specification charts, arranged alphabetically by model group for quick, easy access

2010 EDITIONS

2010 Asian Service Manual Vol. 1*
ISBN 978-1-1110-3764-2
Part No. 163764

2010 Asian Service Manual Vol. 2*
ISBN 978-1-1110-3765-9
Part No. 163765

2010 Asian Service Manual Vol. 3*
ISBN 978-1-1110-3766-6
Part No. 163766

2010 Asian Service Manual Vol. 4*
ISBN 978-1-1110-3767-3
Part No. 163767

2010 Asian Service Manual Vol. 5*
ISBN 978-1-1110-3768-0
Part No. 163768

2010 European Service Manual*
ISBN 978-1-1110-3769-7
Part No. 163769

2010 Chrysler Service Manual, Volumes 1 & 2
ISBN 978-1-1110-3654-6
Part No. 163654

2010 Ford Service Manual, Vols. 1 & 2
ISBN 978-1-1110-3657-7
Part No. 163657

2010 General Motors Service Manuals, Vols. 1, 2, & 3
ISBN 978-1-111-03661-4
Part No. 163661

2008 EDITIONS

2008 Chrysler Service Manual, Vols. 1 & 2
ISBN 978-1-4283-2204-2
Part No. 142204

2008 Ford Service Manuals, Vols. 1 & 2
ISBN 978-1-4283-2208-0
Part No. 142208

2008 Edition General Motors Service Manuals, Vols. 1 & 2
ISBN 978-1-4283-2211-0
Part No. 142211

2008 Asian Service Manuals, Vols. 1-4
ISBN 978-1-4283-2214-1
Part No. 142214

2008 Asian Service Manual, Vol. 1
ISBN 978-1-4283-2215-8
Part No. 142215

2008 Asian Service Manual, Vol. 2
ISBN 978-1-4283-2216-5
Part No. 142216

2008 Asian Service Manual, Vol. 3
ISBN 978-1-4283-2217-2
Part No. 142217

2008 Asian Service Manual, Vol. 4
ISBN 978-1-4283-2218-9
Part No. 142218

2008 European Service Manual
ISBN 978-1-4283-2220-2
Part No. 142220

2006 EDITIONS

2006 DaimlerChrysler Diagnostic Service Manual
ISBN 978-1-4180-2118-4
Part No. 132118

2006 General Motors Diagnostic Service Manual
ISBN 978-1-4180-2120-7
Part No. 132120

2006 Asian Diagnostic Service Manual, Vol. 1
ISBN 978-1-4180-2913-5
Part No. 132913

2006 Asian Diagnostic Service Manual, Vol. 2
ISBN 978-1-4180-2914-2
Part No. 132914

2006 Asian Diagnostic Service Manual, Vol. 3
ISBN 978-1-4180-2915-9
Part No. 132915

2006 European Diagnostic Service Manual
ISBN 978-1-4180-2924-1
Part No. 132924

2006 DaimlerChrysler Mechanical Service Manual
ISBN 978-1-4180-0600-6
Part No. 130600

2006 Asian Mechanical Service Manual, Vol. 1
ISBN 978-1-4180-0947-2
Part No. 130947

2006 Asian Mechanical Service Manual, Vol. 2
ISBN 978-1-4180-0948-9
Part No. 130948

2006 Asian Mechanical Service Manual, Vol. 3
ISBN 978-1-4180-0949-6
Part No. 130949

2006 Asian Mechanical Service Manual, 3 Vol. Set
ISBN 978-1-4180-0603-7
Part No. 130603

2006 European Mechanical Service Manual
ISBN 978- 1-4180-0604-4
Part No. 130604

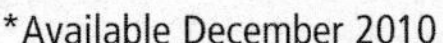

*Available December 2010

Chilton® Mechanical Service Manuals–Perennial Editions

These manuals contain repair and maintenance information for all major systems. Included are repair and overhaul procedures using thousands of illustrations.

CHILTON AUTO REPAIR MANUALS

1998-2002
ISBN 978-0-8019-9362-6/Part No. 9362
Covers all popular American and Canadian cars. An added feature includes scheduled maintenance interval charts.

1993-97
ISBN 978-0-8019-7919-4/Part No. 7919
Covers all popular American and Canadian cars.

1980-87
ISBN 978-0-8019-7670-4/Part No. 7670
Covers all popular American and Canadian cars.

CHILTON IMPORT AUTO REPAIR MANUALS

1998-2002
ISBN 978-0-8019-9363-3/Part No. 9363
Covers all popular Import cars. An added feature includes scheduled maintenance intervals charts.

1993-97
ISBN 978-0-8019-7920-0/Part No. 7920
Covers all popular Import cars.

1988-92
ISBN 978-0-8019-7907-1/Part No. 7907
Covers all popular Import cars.

1980-87
ISBN 978-0-8019-7672-8/Part No. 7672
Covers all popular Import cars.

CHILTON TRUCK AND VAN REPAIR MANUALS

1998-2002
ISBN 978-0-8019-9364-0/Part No. 9364
Covers popular U.S., Canadian, and Import Pick-Ups, Vans, and 4WDs. An added feature includes scheduled maintenance interval charts.

1993-97
ISBN 978-0-8019-7921-7/Part No. 7921
Covers popular U.S., Canadian, and Import Pick-Ups, Sport-Utilities, Vans, RVs and 4 wheel drives.

1991-95
ISBN 978-0-8019-7911-8/Part No. 7911
Covers popular U.S., Canadian, and Import Pick-Ups, Vans, RVs and 4 wheel drives.

1986-90
ISBN 978-08019-7902-6/Part No. 7902
Covers popular U.S., Canadian, and Import Pick-Us, Vans, RVs and 4 wheel drives.

1979-86
ISBN 978-08019-7655-1/Part No. 7655
Covers popular U.S., Canadian, and Import Pick-Ups, Vans, RVs and 4 wheel drives.

CHILTON SUV REPAIR MANUAL

1998-2002
ISBN 978-08019-9365-7/Part No. 9365
Covers popular U.S., Canadian, and import SUVs. An added feature includes scheduled maintenance intervals charts.

COLLECTOR'S SERIES

CHILTON AUTO REPAIR MANUAL 1964-1971
ISBN 978-08019-5974-5/Part No. 5974
1971-1978
ISBN 978-08019-7012-2/Part No. 7012

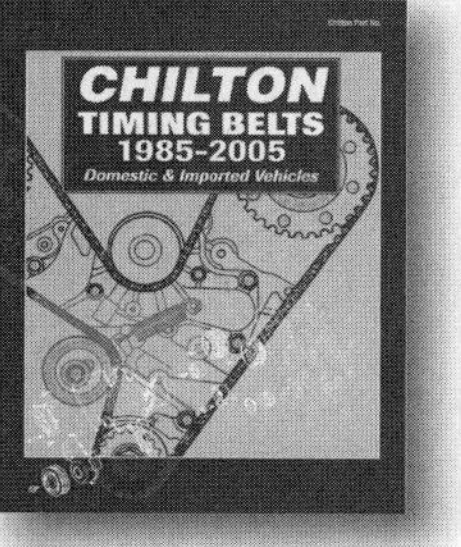

ISBN 978-1-4018-9880-9
Part No. 129880
544 pp, 8" x 11", SC, ©2006

Chilton Timing Belts, 1985-2005

Timing belt procedures can represent increased profits for automotive repair shops and service stations, and this manual contains all the information automotive technicians need to properly service timing belts on domestic and imported cars, vans, and light trucks through 2005 models. Clear, straightforward procedures, illustrations, and specifications help to communicate 20 years of vehicle applications for fast, accurate inspection, replacement, and tensioning of timing belts. Users will learn how to perform key procedures quickly and safely, while learning the correct labor time to charge for the service.

ALSO AVAILABLE:

Quick-Reference Manuals

The Chilton Professional Series offers *Quick-Reference Manuals* for the automotive professional, providing complete coverage on repair and maintenance, adjustments, and diagnostic procedures for specific systems and components.

KEY FEATURES

- step-by-step procedures
- detailed illustrations and exploded views
- easy-to-use manufacturer and model indexing
- handy specifications or data charts

Heater Core Service 1990-2000,
ISBN 978-0-8019-9311-4
Part No. 9311

Brake Specifications and Service 1990-2000
ISBN 978-0-8019-9312-1
Part No. 9312

Electric Cooling Fans, Accessory Drive Belts & Water Pumps, 1995-1999,
ISBN 978-0-8019-9126-4
Part No. 9126

Powertrain Codes & Oxygen Sensors, 1990-1999,
ISBN 978-0-8019-9127-1
Part No. 9127

ASE Test Preparation for Transit Bus

53939	(H1) Compressed Natural Gas Engines	978-1-4354-3939-9
36570	(H2) Diesel Engines	978-1-4180-6570-6
55376	(H3) Drive Train	978-1-4354-5376-0
34998	(H4) Brakes	978-1-4180-4998-0
44011	(H5) Suspension & Steering	978-1-4283-4011-4
34999	(H6) Electrical/Electronic Systems	978-1-4180-4999-7
36571	(H7) Heating, Ventilation, & Air Conditioning	978-1-4180-6571-3
53938	(H8) Preventive Maintenance	978-1-4354-3938-2

ASE Test Preparation for Truck Equipment

53935	(E1) Truck Equipment Installation & Repair	978-1-4354-3935-1
53936	(E2) Electronic Systems Installation & Repair	978-1-4354-3936-8
53937	(E3) Auxilary Power Systems Installation & Repair	978-1-4354-3937-5

Online ASE Test Preparation

Covers the A1-A8, X1, P2, L1, C1, X1, and T1-T8 & B2-B6 Certification Exams

Visit www.techniciantestprep.com for a free demo.

31305	*Online (A1) Engine Repair	978-1-4180-1305-9
31306	*Online (A2) Automatic Transmissions & Transaxles	978-1-4180-1306-6
31307	*Online (A3) Manual Drive Trains & Axles	978-1-4180-1307-3
31308	*Online (A4) Suspension & Steering	978-1-4180-1308-0
31309	*Online (A5) Brakes	978-1-4180-1309-7
31310	*Online (A6) Electrical/Electronic Systems	978-1-4180-1310-3
31311	*Online (A7) Heating & Air Conditioning	978-1-4180-1311-0
31312	*Online (A8) Engine Performance	978-1-4180-1312-7
31313	*Online (X1) Exhaust Systems	978-1-4180-1313-4
31314	*Online (P2) Automobile Parts Specialist	978-1-4180-1314-1
31315	*Online (L1) Advanced Engine Performance	978-1-4180-1315-8
31316	*Online (C1) Service Consultant	978-1-4180-1316-5
27897	Online (T1) Gasoline Engines	978-1-4018-7897-9
27898	Online (T2) Diesel Engines	978-1-4018-7898-6
27900	Online (T3) Drive Train	978-1-4018-7900-6
27901	Online (T4) Brakes	978-1-4018-7901-3
27903	Online (T5) Suspension & Steering	978-1-4018-7903-7
31879	Online (T6) Electrical/Electronic Systems	978-1-4180-1879-5
31880	Online (T7) Heating, Ventilation, & Air Conditioning	978-1-4180-1880-1
27906	Online (T8) Preventive Maintenance	978-1-4018-7906-8
54748	Online (B2) Painting & Refinishing	978-1-4354-4748-6
54749	Online (B3) Non-Structural Analysis & Damage Repair	978-1-4354-4749-3
54750	Online (B4) Structural Analysis and Repair	978-1-4354-4750-9
54751	Online (B5) Mechanical & Electrical Components	978-1-4354-4751-6
54752	Online (B6) Damage Analysis & Estimating	978-1-4354-4752-3

Switch between English & Spanish at the click of a button!

ASE Test Preparation in Spanish

Manuals

- Covers the A1-A8, L1, X1, P2, & B2-B6 Certification Exams in Spanish

Switch between English & Spanish at the click of a button!

- Covers the A1-A8, L1, X1, P2, & C1 exams online. Visit techniciantestprep.com

Online

21014	Spanish (A1) Engine Repair	978-1-4018-1014-6
21015	Spanish (A2) Transmissions and Transaxles	978-1-4018-1015-3
21016	Spanish (A3) Manual Drive Train and Axles	978-1-4018-1016-0
21017	Spanish (A4) Suspension and Steering	978-1-4018-1017-7
21018	Spanish (A5) Brakes	978-1-4018-1018-4
21019	Spanish (A6) Electrical/Electronic Systems	978-1-4018-1019-1
21020	Spanish (A7) Heating and Air Conditioning	978-1-4018-1020-7
21021	Spanish (A8) Engine Performance	978-1-4018-1021-4
21022	Spanish (L1) Advanced Engine Performance	978-1-4018-1022-1
21024	Spanish (X1) Exhaust Systems	978-1-4018-1024-5
21023	Spanish (P2) Parts Specialist	978-1-4018-1023-8
29255	Spanish (B2) Painting and Refinishin	978-1-4018-9255-5
22544	Spanish (B3) Non-Structural Analysis and Damage Repair	978-1-4018-2544-7
29131	Spanish (B4) Structural Analysis and Damage Repair	978-1-4018-9131-2
27759	Spanish (B5) Mechanical and Electrical Components	978-1-4018-7759-0
26573	Spanish (B6) Damage Analysis and Estimation	978-1-4018-6573-3

Complete Series

133954	ASE Manuals for Automotive (A1-A8, X1, P2, L1, C1)	978-1-4180-3954-7
136139	ASE Manuals for Automotive (A1-A8 & L1)	978-1-4180-6139-5
134197	ASE Manuals for Automotive (A1-A8, L1, & P2)	978-1-4180-4197-7
136237	ASE Manuals for Automotive (A1-A8)	978-1-4180-6237-8
136335	ASE Manuals for Automotive (A1-A8, L1, P2, & X1)	978-1-4180-6335-1
133447	Online ASE Manuals for Automotive (A1-A8, X1, P2, L1, C1)	978-1-4180-1344-8
134934	ASE Manuals for Medium/Heavy Duty Truck (T1-T8)	978-1-4180-4934-8
130611	Online ASE for Medium/Heavy Duty Truck (T1-T8)	978-1-4180-0611-2
125120	ASE Manuals for Collision (B2-B6)	978-1-4018-5120-0
24155	ASE Manuals for Collision in Spanish (B2-B6)	978-1-4018-4155-3
16283	ASE Manuals for Engine Machinist (M1-M3)	978-0-7668-6283-8

CSAT-Automotive Series

The online *Comprehensive Skill Assessment Tool-Automotive Series* helps instructors and trainers implement the necessary training programs for individual areas needing improvement over various key automotive topics. As a true skill gap analysis tool, within each key topic, strategic learning areas are measured for knowledge of theory, hands-on application, and diagnostic skill. Areas of strength and areas needing improvement are identified. The combined phases of education and training, and post-assessment allow instructors to track skill level growth and target specific areas needing development.

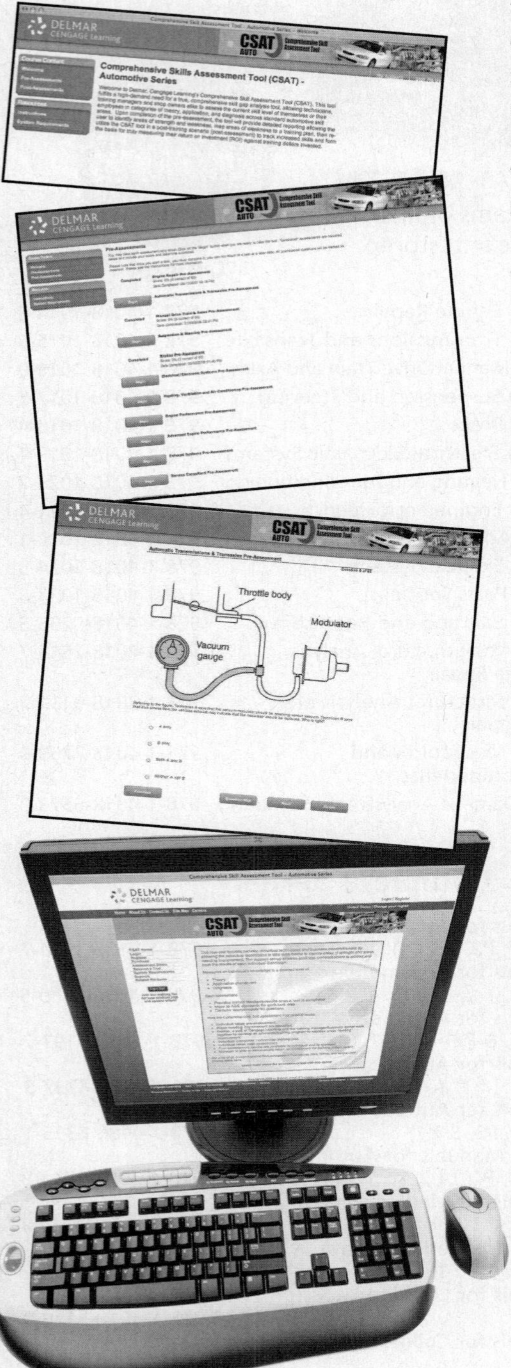

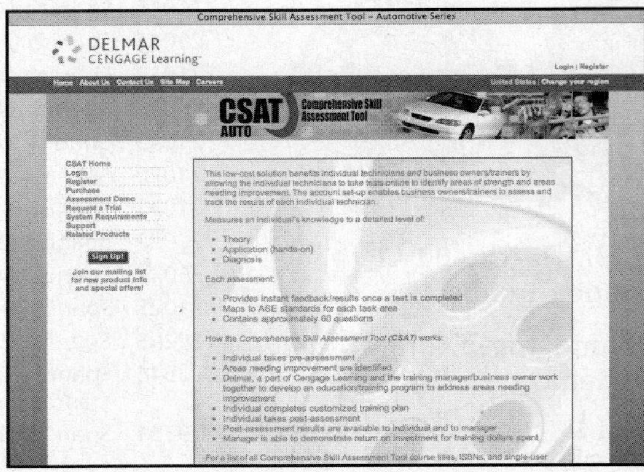

Courses Available in the CSAT Automotive Series

Parts Specialist
ISBN 978-1-4180-3225-8

Service Consultant
ISBN 978-1-4180-3223-4

Advanced Engine Performance
ISBN 978-1-4180-0073-8

Brakes
ISBN 978-1-4180-0069-1

Electrical/Electronic Systems
ISBN 978-1-4180-0070-7

Engine Performance
ISBN 978-1-4180-0072-1

Engine Repair
ISBN 978-1-4180-0065-3

Exhaust Systems
ISBN 978-1-4180-0074-5

Heating and Air Conditioning
ISBN 978-1-4180-0071-4

Manual Drive Train & Axles
ISBN 978-1-4180-0067-7

Suspension & Steering
ISBN 978-1-4180-0068-4

Transmissions & Transaxles
ISBN 978-1-4180-0066-0

All-in-One (contains questions from all eight core automotive areas in one product)
ISBN 978-1-4354-2825-6

FEATURES

- available tests include Engine Repair, Transmissions and Transaxles, Manual Drive Train and Axles, Suspension and Steering, Brakes, Electrical/Electronic Systems, Heating and Air Conditioning, Engine Performance, Advanced Engine Performance, and Exhaust Systems
- can be utilized by companies to measure the technical skill level of individuals against an "ideal" to identify areas of strength and creates a skill gap analysis to help users address areas needing improvement
- questions are written and reviewed by experts in the industry and offer users the opportunity to receive instant feedback
- account set-up that enables instructors and trainers to assess and track the results of individual students
- acts as a true return on investment (ROI) tool for companies to ensure they invest their training dollars in the most appropriate areas

Visit www.skillanalysis.com
for a free demo!

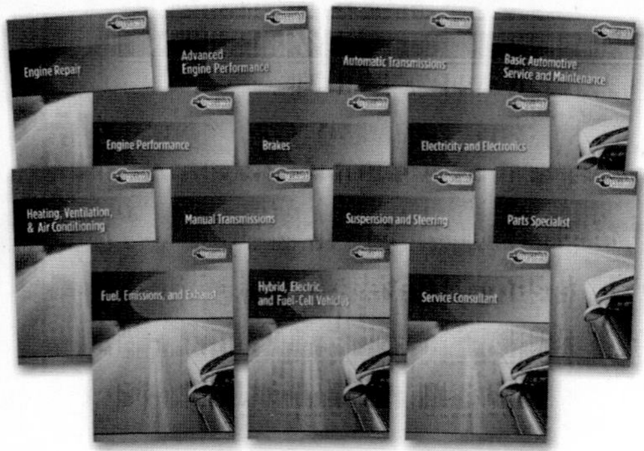

Professional Automotive Technician Training Series: PATTS

Delmar

Delmar, the leader in providing first-rate educational materials for automotive technicians, now offers this exciting self-paced learning series. Choose the delivery method that best suits your needs– CD-ROM or Web-based product – and receive more than 8.5 hours worth of quality instruction. Combining theory, diagnosis, and repair information into one easy-to-use training tool, this highly interactive product helps technicians receive the most applicable delivery method for their needs, regardless of technical infrastructure.

KEY FEATURES

- attention-grabbing animations and learner interactions keep users interested and engaged throughout the course of the program
- bookmarking technology enables users to track their progress from beginning to end
- periodic progress checks and end-of-section reviews are integrated throughout to ensure the highest level of retention
- a certificate of completion can be printed by users achieving a score of 80% or higher on the final review of the course
- all material is completely AICC and SCORM compliant
- all material follows the latest ASE and NATEF standards

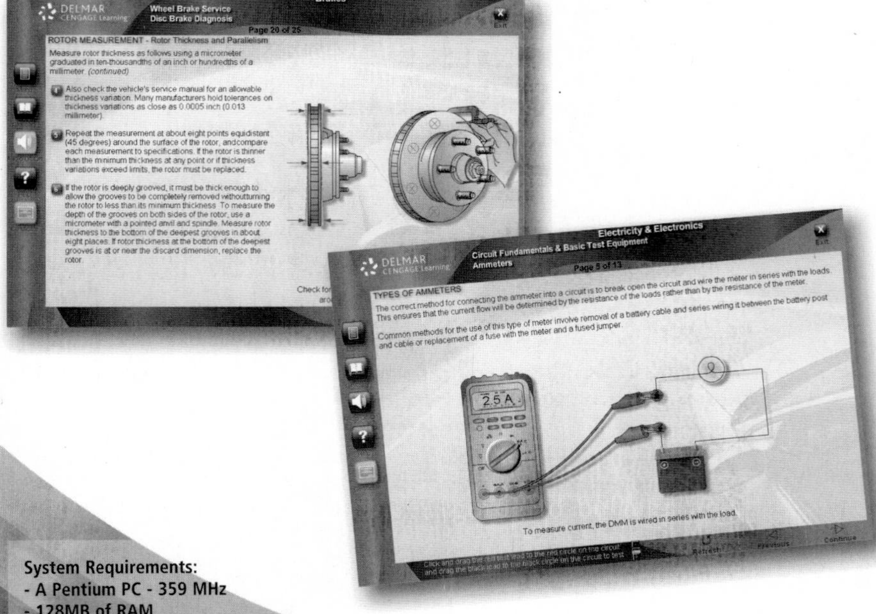

System Requirements:
- A Pentium PC - 359 MHz
- 128MB of RAM
- Windows 2000, Windows XP, Windows Vista
- Graphics adapter with Minimum 1024 x 768 display resolution, 32 bit depth
- Minimum Display Resolution 1024 x 768
- High Speed Internet Connection
- Internet Explorer 6, 7, or Firefox 2
- Not Mac Compatible

Basic Automotive Service and Maintenance Web Based Training
ISBN 978-1-4180-4101-4

Basic Automotive Service and Maintenance Computer Based Training
ISBN 978-1-4180-4100-7

Electricity and Electronics Web Based Training
ISBN 978-1-4180-4242-4

Electricity and Electronics Computer Based Training
ISBN 978-1-4180-4241-7

Brakes Web Based Training
ISBN 978-1-4180-4236-3

Brakes Computer Based Training
ISBN 978-1-4180-4235-6

Engine Performance Web Based Training
ISBN 978-1-4180-4240-0

Engine Performance Computer Based Training
ISBN 978-1-4180-4239-4

Suspension and Steering Web Based Training
ISBN 978-1-4180-4238-7

Suspension and Steering Computer Based Training
ISBN 978-1-4180-4237-0

Automatic Transmissions Web Based Training
ISBN 978-1-4180-4244-8

Automatic Transmissions Computer Based Training
ISBN 978-1-4180-4243-1

Service Consultant Web Based Training
ISBN 978-1-4180-4249-3

Service Consultant Computer Based Training
ISBN 978-1-4180-4247-9

Engine Repair Web Based Training
ISBN 978-1-4180-4254-7

Engine Repair Computer Based Training
ISBN 978-1-4180-4253-0

Parts Specialist Web Based Training
ISBN 978-1-4180-4252-3

Parts Specialist Computer Based Training
ISBN 978-1-4180-4250-9

Heating and Air Conditioning Web Based Training
ISBN 978-1-4180-4246-2

Heating and Air Conditioning Computer Based Training
ISBN 978-1-4180-4245-5

Manual Transmissions Web Based Training
ISBN 978-1-4180-4256-1

Manual Transmissions Computer Based Training
ISBN 978-1-4180-4255-4

Advanced Engine Performance Web Based Training
ISBN 978-1-4283-2098-7

Advanced Engine Performance Computer Based Training
ISBN 978-1-4283-2097-0

New Courses!

Fuels, Emissions, and Exhaust Computer Based Training
ISBN 978-1-4354-4148-4

Fuels, Emissions, and Exhaust Web Based Training
ISBN 978-1-4354-4147-7

Hybrid, Electric, and Fuel-Cell Vehicles Web Based Training
ISBN 978-1-4354-4144-6

Hybrid, Electric, and Fuel-Cell Vehicles Computer Based Training
ISBN 978-1-4354-4143-9

Visit www.techniciantraining.com for a free demo!